CRC Handbook of Chemistry and Physics

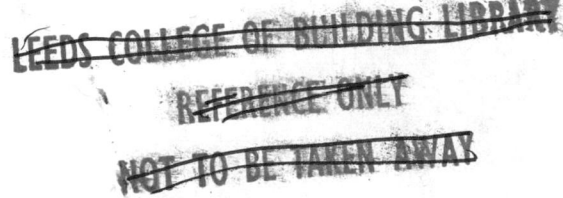

1ST
Student
Edition

Editor-in-Chief

Robert C. Weast, Ph.D

CRC Press, Inc.
Boca Raton, Florida

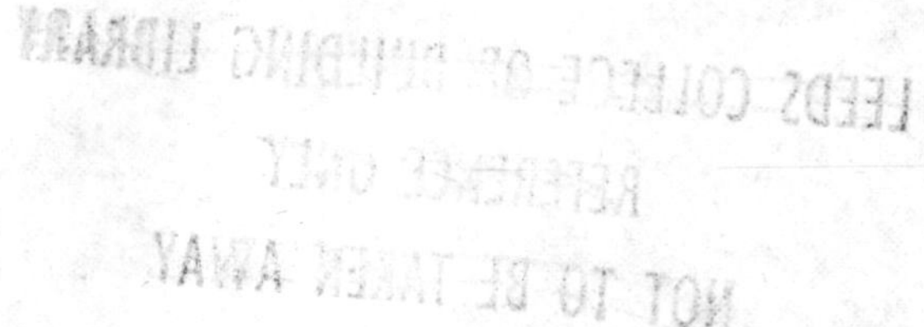

Library of Congress Cataloging-in-Publication Data

Handbook of chemistry and physics.

Bibliography: p.
Includes index.
1. Chemistry—Tables. 2. Physics—Tables.
I. Weast, Robert C., 1916-
QD65.H283 1987 540′.212 87-26820
ISBN 0-8493-0740-6

Direct all inquiries to CRC Press, Inc., 2000 Corporate Blvd., N.W., Boca Raton, Florida, 33431.

International Standard Book Number 0-8493-0740-6

Library of Congress Card Number 87-26820
Printed in the United States

PREFACE

Seventy-four years ago the *CRC Handbook of Chemistry and Physics* was first published. Its editor was Professor William R. Veazey of the Case School of Applied Science. Later he was an outstanding scientist, administrator, and officer of the Dow Chemical Company. Arthur Friedman, a 1907 graduate of Case, was the publisher and originator of the concept of a single-volume reference book to contain data for chemists and physicists as well as containing certain frequently used mathematical tables for these scientific disciplines. His company, the Chemical Rubber Company, sold rubber tubing, stoppers, aprons, and laboratory supplies primarily to colleges and high schools. The company introduced as a complement to the products being sold the *CRC Handbook of Chemistry and Physics*.

Over the past 74 years the book has been published continuously and has been revised 68 times. For several reasons during World Wars I and II the book was not revised annually. From 1915 through 1951 the book was edited by Professor Charles D. Hodgman of the Physics Department of the Case School of Applied Science, a school which was later called Case Institute of Technology, and still later, after the affiliation with Western Reserve University, Case Western Reserve University. Subsequent to Professor Hodgman's retirement, the *CRC Handbook of Chemistry and Physics* has been edited by the present editor who was a professor at Case Institute of Technology.

Chemistry and physics, always two closely related sciences, have been brought into more intimate relations by recent developments in research and our increasing understanding of matter and energy. To a growing extent people in either science are learning more of the other. One of the goals of the editor and the publisher is to provide a reference book which will assist in providing certain information to further this understanding.

The editor attempts to include material which has a high probability to find extended use in many branches of chemistry and physics and the closely allied sciences. This 1st Student Edition provides certain core data and information that are constant or which change only slightly over an extended period of time.

There continues to be progress in education. New information in all types of books, including reference books, is necessary to aid students in keeping pace with this progress. Therefore, the Student Edition will be revised from time to time.

The editor and publisher will greatly appreciate being notified of any corrections needed to help with the continuing improvement of the *CRC Handbook of Chemistry and Physics, Student Edition*.

Robert C. Weast
October, 1987

TABLE OF CONTENTS

MISCELLANEOUS MATHEMATICAL CONSTANTS

π CONSTANTS

π = 3.14159 26535 89793 23846 26433 83279 50288 41971 69399 37511
$1/\pi$ = 0.31830 98861 83790 67153 77675 26745 02872 40689 19291 48091
π^2 = 9.86960 44010 89358 61883 44909 99876 15113 53136 99407 24079
$\log_e \pi$ = 1.14472 98858 49400 17414 34273 51353 05871 16472 94812 91531
$\log_{10} \pi$ = 0.49714 98726 94133 85435 12682 88290 89887 36516 78324 38044
$\log_{10} \sqrt{2\pi}$ = 0.39908 99341 79057 52478 25035 91507 69595 02099 34102 92128

CONSTANTS INVOLVING e

e = 2.71828 18284 59045 23536 02874 71352 66249 77572 47093 69996
$1/e$ = 0.36787 94411 71442 32159 55237 70161 46086 74458 11131 03177
e^2 = 7.38905 60989 30650 22723 04274 60575 00781 31803 15570 55185
$M = \log_{10} e$ = 0.43429 44819 03251 82765 11289 18916 60508 22943 97005 80367
$1/M = \log_e 10$ = 2.30258 50929 94045 68401 79914 54684 36420 76011 01488 62877
$\log_{10} M$ = 9.63778 43113 00536 78912 29674 98645 −10

π^e AND e^π CONSTANTS

π^e = 22.45915 77183 61045 47342 71522
e^π = 23.14069 26327 79269 00572 90864
$e^{-\pi}$ = 0.04321 39182 63772 24977 44177
$e^{\frac{1}{2}\pi}$ = 4.81047 73809 65351 65547 30357
$i^i = e^{-\frac{1}{2}\pi}$ = 0.20787 95763 50761 90854 69556

NUMERICAL CONSTANTS

$\sqrt{2}$ = 1.41421 35623 73095 04880 16887 24209 69807 85696 71875 37695
$\sqrt[3]{2}$ = 1.25992 10498 94873 16476 72106 07278 22835 05702 51464 70151
$\log_e 2$ = 0.69314 71805 59945 30941 72321 21458 17656 80755 00134 36026
$\log_{10} 2$ = 0.30102 99956 63981 19521 37388 94724 49302 67881 89881 46211
$\sqrt{3}$ = 1.73205 08075 68877 29352 74463 41505 87236 69428 05253 81039
$\sqrt[3]{3}$ = 1.44224 95703 07408 38232 16383 10780 10958 83918 69253 49935
$\log_e 3$ = 1.09861 22886 68109 69139 52452 36922 52570 46474 90557 82275
$\log_{10} 3$ = 0.47712 12547 19662 43729 50279 03255 11530 92001 28864 19070

OTHER CONSTANTS

Euler's Constant γ = 0.57721 56649 01532 86061
$\log_e \gamma$ = −0.54953 93129 81644 82234
Golden Ratio ϕ = 1.61803 39887 49894 84820 45868 34365 63811 77203 09180

RELATION OF ANGULAR FUNCTIONS IN TERMS OF ONE ANOTHER

TRIGONOMETRIC FUNCTIONS

Function	sin α	cos α	tan α	cot α	sec α	csc α
sin α	$\sin\alpha$	$\pm\sqrt{1-\cos^2\alpha}$	$\frac{\tan\alpha}{\pm\sqrt{1+\tan^2\alpha}}$	$\frac{1}{\pm\sqrt{1+\cot^2\alpha}}$	$\frac{\pm\sqrt{\sec^2\alpha-1}}{\sec\alpha}$	$\frac{1}{\csc\alpha}$
cos α	$\pm\sqrt{1-\sin^2\alpha}$	$\cos\alpha$	$\frac{1}{\pm\sqrt{1+\tan^2\alpha}}$	$\frac{\cot\alpha}{\pm\sqrt{1+\cot^2\alpha}}$	$\frac{1}{\sec\alpha}$	$\frac{\pm\sqrt{\csc^2\alpha-1}}{\csc\alpha}$
tan α	$\frac{\sin\alpha}{\pm\sqrt{1-\sin^2\alpha}}$	$\frac{\pm\sqrt{1-\cos^2\alpha}}{\cos\alpha}$	$\tan\alpha$	$\frac{1}{\cot\alpha}$	$\pm\sqrt{\sec^2\alpha-1}$	$\frac{1}{\pm\sqrt{\csc^2\alpha-1}}$
cot α	$\frac{\pm\sqrt{1-\sin^2\alpha}}{\sin\alpha}$	$\frac{\cos\alpha}{\pm\sqrt{1-\cos^2\alpha}}$	$\frac{1}{\tan\alpha}$	$\cot\alpha$	$\frac{1}{\pm\sqrt{\sec^2\alpha-1}}$	$\pm\sqrt{\csc^2\alpha-1}$
sec α	$\frac{1}{\pm\sqrt{1-\sin^2\alpha}}$	$\frac{1}{\cos\alpha}$	$\pm\sqrt{1+\tan^2\alpha}$	$\frac{\pm\sqrt{1+\cot^2\alpha}}{\cot\alpha}$	$\sec\alpha$	$\frac{\csc\alpha}{\pm\sqrt{\csc^2\alpha-1}}$
csc α	$\frac{1}{\sin\alpha}$	$\frac{1}{\pm\sqrt{1-\cos^2\alpha}}$	$\frac{\pm\sqrt{1+\tan^2\alpha}}{\tan\alpha}$	$\pm\sqrt{1+\cot^2\alpha}$	$\frac{\sec\alpha}{\pm\sqrt{\sec^2\alpha-1}}$	$\csc\alpha$

Note: The choice of sign depends upon the quadrant in which the angle terminates.

HYPERBOLIC FUNCTIONS

Function	sinh x	cosh x	tanh x
sinh x =	$\sinh x$	$+\sqrt{\cosh^2 x-1}$	$\frac{\tanh x}{\sqrt{1-\tanh^2 x}}$
cosh x =	$\sqrt{1+\sinh^2 x}$	$\cosh x$	$\frac{1}{\sqrt{1-\tanh^2 x}}$
tanh x =	$\frac{\sinh x}{\sqrt{1+\sinh^2 x}}$	$\pm\frac{\sqrt{\cosh^2 x-1}}{\cosh x}$	$\tanh x$
cosech x =	$\frac{1}{\sinh x}$	$\pm\frac{1}{\sqrt{\cosh^2 x-1}}$	$\frac{\sqrt{1-\tanh^2 x}}{\tanh x}$
sech x =	$\frac{1}{\sqrt{1+\sinh^2 x}}$	$\frac{1}{\cosh x}$	$\sqrt{1-\tanh^2 x}$
coth x =	$\frac{\sqrt{1+\sinh^2 x}}{\sinh x}$	$\frac{\pm\cosh x}{\sqrt{\cosh^2 x-1}}$	$\frac{1}{\tanh x}$

Function	cosech x	sech x	coth x
sinh x =	$\frac{1}{\text{cosech}\, x}$	$\pm\frac{\sqrt{1-\text{sech}^2 x}}{\text{sech}\, x}$	$\frac{\pm 1}{\sqrt{\coth^2 x-1}}$
cosh x =	$\pm\frac{\sqrt{\text{cosech}^2 x+1}}{\text{cosech}\, x}$	$\frac{1}{\text{sech}\, x}$	$\pm\frac{\coth x}{\sqrt{\coth^2 x-1}}$
tanh x =	$\frac{1}{\sqrt{\text{cosech}^1 x+1}}$	$\pm\sqrt{1-\text{sech}^2 x}$	$\frac{1}{\coth x}$
cosech x =	$\text{cosech}\, x$	$\pm\frac{\text{sech}\, x}{\sqrt{1-\text{sech}^1 zsx}}$	$\pm\frac{\sqrt{\coth^2 x-1}}{1}$
sech x =	$\pm\frac{\text{cosec}\, x}{\sqrt{\text{cosech}^2 x+1}}$	$\text{sech}\, x$	$\pm\frac{\sqrt{\coth^2 x-1}}{\coth x}$
coth x =	$\sqrt{\text{cosech}^2 x+1}$	$\pm\frac{1}{\sqrt{1-\text{sech}^2 x}}$	$\coth x$

Whenever two signs are shown, choose + sign if x is positive, − sign if x is negative.

*Derivatives

In the following formulas u, v, w represent functions of x, while a, c, n represent fixed real numbers. All arguments in the trigonometric functions are measured in radians, and all inverse trigonometric and hyperbolic functions represent principal values.

1. $\frac{d}{dx}(a) = 0$

2. $\frac{d}{dx}(x) = 1$

3. $\frac{d}{dx}(au) = a\frac{du}{dx}$

4. $\frac{d}{dx}(u + v - w) = \frac{du}{dx} + \frac{dv}{dx} - \frac{dw}{dx}$

5. $\frac{d}{dx}(uv) = u\frac{dv}{dx} + v\frac{du}{dx}$

6. $\frac{d}{dx}(uvw) = uv\frac{dw}{dx} + vw\frac{du}{dx} + uw\frac{dv}{dx}$

7. $\frac{d}{dx}\left(\frac{u}{v}\right) = \frac{v\frac{du}{dx} - u\frac{dv}{dx}}{v^2} = \frac{1}{v}\frac{du}{dx} - \frac{u}{v^2}\frac{dv}{dx}$

8. $\frac{d}{dx}(u^n) = nu^{n-1}\frac{du}{dx}$

9. $\frac{d}{dx}(\sqrt{u}) = \frac{1}{2\sqrt{u}}\frac{du}{dx}$

10. $\frac{d}{dx}\left(\frac{1}{u}\right) = -\frac{1}{u^2}\frac{du}{dx}$

11. $\frac{d}{dx}\left(\frac{1}{u^n}\right) = -\frac{n}{u^{n+1}}\frac{du}{dx}$

12. $\frac{d}{dx}\left(\frac{u^n}{v^m}\right) = \frac{u^{n-1}}{v^{m+1}}\left(nv\frac{du}{dx} - mu\frac{dv}{dx}\right)$

13. $\frac{d}{dx}(u^n v^m) = u^{n-1}v^{m-1}\left(nv\frac{du}{dx} + mu\frac{dv}{dx}\right)$

14. $\frac{d}{dx}[f(u)] = \frac{d}{du}[f(u)] \cdot \frac{du}{dx}$

* Let $y = f(x)$ and $\frac{dy}{dx} = \frac{d[f(x)]}{dx} = f'(x)$ define respectively a function and its derivative for any value x in their common domain. The differential for the function at such a value x is accordingly defined as

$$dy = d[f(x)] = \frac{dy}{dx}dx = \frac{d[f(x)]}{dx}dx = f'(x)\,dx$$

Each derivative formula has an associated differential formula. For example, formula 6 above has the differential formula

$$d(uvw) = uv\,dw + vw\,du + uw\,dv$$

15. $\frac{d^2}{dx^2}[f(u)] = \frac{df(u)}{du} \cdot \frac{d^2u}{dx^2} + \frac{d^2f(u)}{du^2} \cdot \left(\frac{du}{dx}\right)^2$

16. $\frac{d^n}{dx^n}[uv] = \binom{n}{0} v \frac{d^n u}{dx^n} + \binom{n}{1} \frac{dv}{dx} \frac{d^{n-1}u}{dx^{n-1}} + \binom{n}{2} \frac{d^2 v}{dx^2} \frac{d^{n-2}u}{dx^{n-2}}$

$$+ \cdots + \binom{n}{k} \frac{d^k v}{dx^k} \frac{d^{n-k}u}{dx^{n-k}} + \cdots + \binom{n}{n} u \frac{d^n v}{dx^n}$$

where $\binom{n}{r} = \frac{n!}{r!(n-r)!}$ the binomial coefficient, n non-negative integer and $\binom{n}{0} = 1$.

17. $\frac{du}{dx} = \frac{1}{\frac{dx}{du}} \qquad \text{if } \frac{dx}{du} \neq 0$

18. $\frac{d}{dx}(\log_a u) = (\log_a e) \frac{1}{u} \frac{du}{dx}$

19. $\frac{d}{dx}(\log_e u) = \frac{1}{u} \frac{du}{dx}$

20. $\frac{d}{dx}(a^u) = a^u (\log_e a) \frac{du}{dx}$

21. $\frac{d}{dx}(e^u) = e^u \frac{du}{dx}$

22. $\frac{d}{dx}(u^v) = vu^{v-1} \frac{du}{dx} + (\log_e u) u^v \frac{dv}{dx}$

23. $\frac{d}{dx}(\sin u) = \frac{du}{dx}(\cos u)$

24. $\frac{d}{dx}(\cos u) = -\frac{du}{dx}(\sin u)$

25. $\frac{d}{dx}(\tan u) = \frac{du}{dx}(\sec^2 u)$

26. $\frac{d}{dx}(\cot u) = -\frac{du}{dx}(\csc^2 u)$

27. $\frac{d}{dx}(\sec u) = \frac{du}{dx} \sec u \cdot \tan u$

28. $\frac{d}{dx}(\csc u) = -\frac{du}{dx} \csc u \cdot \cot u$

29. $\frac{d}{dx}(\text{vers } u) = \frac{du}{dx} \sin u$

30. $\frac{d}{dx}(\text{arc sin } u) = \frac{1}{\sqrt{1 - u^2}} \frac{du}{dx}, \qquad \left(-\frac{\pi}{2} \leq \text{arc sin } u \leq \frac{\pi}{2}\right)$

31. $\dfrac{d}{dx}(\operatorname{arc}\cos u) = -\dfrac{1}{\sqrt{1-u^2}}\dfrac{du}{dx}, \qquad (0 \le \operatorname{arc}\cos u \le \pi)$

32. $\dfrac{d}{dx}(\operatorname{arc}\tan u) = \dfrac{1}{1+u^2}\dfrac{du}{dx}, \qquad \left(-\dfrac{\pi}{2} < \operatorname{arc}\tan u < \dfrac{\pi}{2}\right)$

33. $\dfrac{d}{dx}(\operatorname{arc}\cot u) = -\dfrac{1}{1+u^2}\dfrac{du}{dx}, \qquad (0 \le \operatorname{arc}\cot u \le \pi)$

34. $\dfrac{d}{dx}(\operatorname{arc}\sec u) = \dfrac{1}{u\sqrt{u^2-1}}\dfrac{du}{dx}, \qquad \left(0 \le \operatorname{arc}\sec u < \dfrac{\pi}{2}, -\pi \le \operatorname{arc}\sec u < -\dfrac{\pi}{2}\right)$

35. $\dfrac{d}{dx}(\operatorname{arc}\csc u) = -\dfrac{1}{u\sqrt{u^2-1}}\dfrac{du}{dx}, \qquad \left(0 < \operatorname{arc}\csc u \le \dfrac{\pi}{2}, -\pi < \operatorname{arc}\csc u \le -\dfrac{\pi}{2}\right)$

36. $\dfrac{d}{dx}(\operatorname{arc}\operatorname{vers} u) = \dfrac{1}{\sqrt{2u-u^2}}\dfrac{du}{dx}, \qquad (0 \le \operatorname{arc}\operatorname{vers} u \le \pi)$

37. $\dfrac{d}{dx}(\sinh u) = \dfrac{du}{dx}(\cosh u)$

38. $\dfrac{d}{dx}(\cosh u) = \dfrac{du}{dx}(\sinh u)$

39. $\dfrac{d}{dx}(\tanh u) = \dfrac{du}{dx}(\operatorname{sech}^2 u)$

40. $\dfrac{d}{dx}(\coth u) = -\dfrac{du}{dx}(\operatorname{csch}^2 u)$

41. $\dfrac{d}{dx}(\operatorname{sech} u) = -\dfrac{du}{dx}(\operatorname{sech} u \cdot \tanh u)$

42. $\dfrac{d}{dx}(\operatorname{csch} u) = -\dfrac{du}{dx}(\operatorname{csch} u \cdot \coth u)$

43. $\dfrac{d}{dx}(\sinh^{-1} u) = \dfrac{d}{dx}[\log(u + \sqrt{u^2+1})] = \dfrac{1}{\sqrt{u^2+1}}\dfrac{du}{dx}$

44. $\dfrac{d}{dx}(\cosh^{-1} u) = \dfrac{d}{dx}[\log(u + \sqrt{u^2-1})] = \dfrac{1}{\sqrt{u^2-1}}\dfrac{du}{dx}, \qquad (u > 1, \cosh^{-1} u > 0)$

45. $\dfrac{d}{dx}(\tanh^{-1} u) = \dfrac{d}{dx}\left[\dfrac{1}{2}\log\dfrac{1+u}{1-u}\right] = \dfrac{1}{1-u^2}\dfrac{du}{dx}, \qquad (u^2 < 1)$

46. $\dfrac{d}{dx}(\coth^{-1} u) = \dfrac{d}{dx}\left[\dfrac{1}{2}\log\dfrac{u+1}{u-1}\right] = \dfrac{1}{1-u^2}\dfrac{du}{dx}, \qquad (u^2 > 1)$

47. $\dfrac{d}{dx}(\operatorname{sech}^{-1} u) = \dfrac{d}{dx}\left[\log\dfrac{1+\sqrt{1-u^2}}{u}\right] = -\dfrac{1}{u\sqrt{1-u^2}}\dfrac{du}{dx}, \qquad (0 < u < 1, \operatorname{sech}^{-1} u > 0)$

48. $\dfrac{d}{dx}(\operatorname{csch}^{-1} u) = \dfrac{d}{dx}\left[\log\dfrac{1+\sqrt{1+u^2}}{u}\right] = -\dfrac{1}{|u|\sqrt{1+u^2}}\dfrac{du}{dx}$

49. $\frac{d}{dq}\int_p^q f(x)\,dx = f(q), \qquad [p \text{ constant}]$

50. $\frac{d}{dp}\int_p^q f(x)\,dx = -f(p), \qquad [q \text{ constant}]$

51. $\frac{d}{da}\int_p^q f(x,a)\,dx = \int_p^q \frac{\partial}{\partial a}[f(x,a)]\,dx + f(q,a)\frac{dq}{da} - f(p,a)\frac{dp}{da}$

INTEGRALS

ELEMENTARY FORMS

1. $\int a\,dx = ax$

2. $\int a \cdot f(x)\,dx = a\int f(x)\,dx$

3. $\int \phi(y)\,dx = \int \frac{\phi(y)}{y'}\,dy, \qquad \text{where } y' = \frac{dy}{dx}$

4. $\int (u+v)\,dx = \int u\,dx + \int v\,dx,$ where u and v are any functions of x

5. $\int u\,dv = u\int dv - \int v\,du = uv - \int v\,du$

6. $\int u\frac{dv}{dx}dx = uv - \int v\frac{du}{dx}dx$

7. $\int x^n\,dx = \frac{x^{n+1}}{n+1}, \qquad \text{except } n = -1$

8. $\int \frac{f'(x)\,dx}{f(x)} = \log f(x), \qquad (df(x) = f'(x)\,dx)$

9. $\int \frac{dx}{x} = \log x$

10. $\int \frac{f'(x)\,dx}{2\sqrt{f(x)}} = \sqrt{f(x)}, \qquad (df(x) = f'(x)\,dx)$

11. $\int e^x\,dx = e^x$

12. $\int e^{ax}\,dx = e^{ax}/a$

13. $\int b^{ax}\,dx = \frac{b^{ax}}{a\log b}, \qquad (b > 0)$

14. $\int \log x\,dx = x\log x - x$

15. $\int a^x \log a\,dx = a^x, \qquad (a > 0)$

16. $\int \frac{dx}{a^2+x^2} = \frac{1}{a}\tan^{-1}\frac{x}{a}$

17. $$\int \frac{dx}{a^2 - x^2} = \begin{cases} \frac{1}{a}\tanh^{-1}\frac{x}{a} \\ \text{or} \\ \frac{1}{2a}\log\frac{a+x}{a-x}, & (a^2 > x^2) \end{cases}$$

18. $$\int \frac{dx}{x^2 - a^2} = \begin{cases} -\frac{1}{a}\coth^{-1}\frac{x}{a} \\ \text{or} \\ \frac{1}{2a}\log\frac{x-a}{x+a}, & (x^2 > a^2) \end{cases}$$

19. $$\int \frac{dx}{\sqrt{a^2 - x^2}} = \begin{cases} \sin^{-1}\frac{x}{|a|} \\ \text{or} \\ -\cos^{-1}\frac{x}{|a|}, & (a^2 > x^2) \end{cases}$$

20. $$\int \frac{dx}{\sqrt{x^2 \pm a^2}} = \log\left(x + \sqrt{x^2 \pm a^2}\right)$$

21. $$\int \frac{dx}{x\sqrt{x^2 - a^2}} = \frac{1}{|a|}\sec^{-1}\frac{x}{a}$$

22. $$\int \frac{dx}{x\sqrt{a^2 \pm x^2}} = -\frac{1}{a}\log\left(\frac{a + \sqrt{a^2 \pm x^2}}{x}\right)$$

ABBREVIATIONS USED IN TABLE OF PHYSICAL CONSTANTS OF INORGANIC COMPOUNDS

a	acid
abs	absolute
ac. a	acetic acid
acet	acetone
act	active
al	alcohol
alk	alkali
amm	ammonium
amor	amorphous
anh	anhydrous
appr	approximately
aq	aqua, water
aq. reg	aqua regia
asym	asymmetrical
atm	atmospheres
bipyr	bipyramidal
bl	blue
blk	black
boil	boiling
br., brn	brown
brnsh	brownish
bz	benzene
c	cold
calc	calculated
carb	carbon
caust	caustic
chl	chloroform
choc	chocolate
cit. a	citric acid
col	colorless
coll	colloidal
com'l	commercial
comp	compounds
compl	completely
conc	concentrated
const	constant
cont	contains
corros	corrosive
cr	crystalline
cub	cubic
d., dec	decomposes
deliq	deliquescent
deriv	derivative
dibas	dibasic
di-H	dihydrogen
dil	dilute
dimorph	dimorphous
disg	disagreeable
dk	dark
doubt	doubtful
duct	ductile
effl	efflorescent
em	emerald
eth	ether
ev	evolves
evln	evolution
ex	excess
exist	existence
exp	explodes
extr	extreme(ly)
f., fr	from
feath	feathery
fl	flakes
floc	flocculent
fluo, fluores	fluorescent
form	formic
fum	fuming
fus	fused
fxd	fixed
gel., gelat	gelatinous
gl	glass
glac	glacial
glit	glittering
glob	globular
glyc	glycerin
gran	granular
greas	greasy
grn	green
h	hot
hex	hexagonal
ht	heat
hyd	hydrolyzed
hydx	hydroxides
hyg	hygroscopic
i	insoluble
ign	ignites
ind	indigo
indef	indefinite
infl., inflam	inflammable
infus	infusible
irid	iridescent
leaf	leaflets
lem	lemon
lgr	ligroin
lng	long
lq., liq	liquid
lt	light
lum	luminous
lust	lustrous
me., meth	methyl
met	metal or metallic
micr	microscopic
min	mineral
misc	miscible
mixt	mixture
mod	modifications
monbas	monobasic
mon-H	monohydrogen
monocl	monoclinic
near	nearly
need	needles
nit	nitrate
oct	octahedral
odorl	odorless
offen	offensive
olv	olive
opt	optical or optically
or	orange
ord	ordinary
org	organic
oxal	oxalate or oxalic
pa	pale
pet	petroleum
pl	plates
pois	poisonous
polymorph	polymorphous
powd	powder
ppt	precipitate
pr	prisms
press	pressure
prob	probably
prop	properties
purp	purple
pyr	pyridine
quad	quadrilateral
quest	questioned
rect	rectangular
redsh	reddish
reg	regular
rhbdr	rhombohedral
rhomb	rhombic, orthorhombic
s	soluble
satd	saturated
sld	solid
sensit	sensitive
sc	scales
sec	secondary
silv	silver
sl	slightly
sly	slowly
sm	small
sod	sodium
soln	solution
solv	solvents
spont	spontaneous
st	steel
stab	stable
subl	sublimes
suffoc	suffocating
sulfd	sulfides
sulf	sulfur
sym	symmetrical
tabl	tablets
tart. a	tartaric acid
tetr	tetragonal
tetrah	tetrahedral
tol	toluene
trac	trace, traces
trans	transparent
translu	translucent
tri., trig	trigonal
tribas	tribasic
tricl	triclinic
trim	trimetric
tr	transition point
turp	turpentine
unpleas	unpleasant
unst	unstable
v	very
vac	vacuum
var	various
viol	violent, violence
visc	viscous
vitr	vitreous
vlt	violet
volt., volat	volatizes
wh	white
wh. lt	white light
yel	yellow
yelsh	yellowish
∞	soluble in all proportions
$>$	above
$<$	below

PHYSICAL CONSTANTS OF INORGANIC COMPOUNDS

No.	Name	Synonyms and Formulae	Mol. wt.	Crystalline form, properties and index of refraction	Density or spec. gravity	Melting point, °C	Boiling point, °C	Solubility, in grams per 100 cc		
								Cold water	Hot water	Other solvents
a1	**Actinium**	Ac	227.0278	silv wh met, cub	...	1050	3200 ± 300	d to $Ac(OH)_3$	...	...
a2	bromide	$AcBr_3$	466.74	wh, hex	5.85	subl 800	...	s	...	...
a3	chloride, tri-	$AcCl_3$	333.39	wh cr, hex	4.81	subl 960	...	...	...	...
a4	fluoride, tri-	AcF_3	284.02	wh cr, hex	7.88	...	...	i	i	...
a5	hydroxide	$Ac(OH)_3$	278.05	wh	...	...	...	i	...	...
a6	iodide	AcI_3	607.74	wh	...	subl 700—800	...	s	...	...
a7	oxalate	$Ac_2(C_2O_4)_3$	718.11	...	...	...	...	i	...	...
a8	oxide, sesqui-	Ac_2O_3	502.05	wh cr, hex	9.19	...	...	i	...	...
a9	sulfide, sesqui-	Ac_2S_3	550.24	dk, cub	6.75	...	...	...	...	...
a10	**Aluminum**	Al	26.98154	silv wh duct met, cub	2.702	660.37	2467	i	i	s alk, HCl, H_2SO_4; i conc HNO_3, h ac a
a11	acetate, tri-	$Al(C_2H_3O_2)_3$	204.12	wh solid	...	d	...	v sl s	d	...
a12	acetylacetonate	$Al(C_5H_7O_2)_3$	324.31	col, monocl	1.27	193, subl	314	i	i	v s al; s eth, bz
a13	*ortho*arsenate	$AlAsO_4 \cdot 8H_2O$	310.02	wh powd	3.001	vac $-H_2O$	...	i	i	sl s a
a14	benzoate	$Al(C_7H_5O_2)_3$	390.33	wh cr powd	...	...	...	v sl s	...	...
a15	benzyloxide	Aluminum benzylate· $Al(C_7H_7O)_3$	348.38	...	...	59—60	$283—284^{0.6}$	...	...	...
a16	boride	AlB_{12}	156.70	dk red-blk, monocl	2.55_4^{18}	...	...	i	...	s hot HNO_3; i a, alk
a17	boride, di-	AlB_2	48.60	copper red, hex.	3.19	...	...	...	...	...
a18	bromate	$Al(BrO_3)_3 \cdot 9H_2O$	572.83	wh cr, hygr	...	62.3	d 100	s	s	sl s a
a19	bromide	$AlBr_3$ (or Al_2Br_6)	266.69	col, rhomb pl, deliq	2.64^{10} (fused)	97.5	263.3^{747}	s with viol	d	s al, acet, CS_2
a20	bromide, hexahydrate	$AlBr_3 \cdot 6H_2O$	374.79	col-yelsh need, deliq	2.54	93	d 135	s	d	s al, amyl al; sl s CS_2
a21	bromide, pentadecylhydrate	$AlBr_3 \cdot 15H_2O$	536.92	col need	...	−7.5	d 7	s	s	s al
a22	butoxide, tert-	$Al(C_4H_9O)_3$	246.33	wh cr.	1.0251_0^{20}	subl 180, m.p >300, sealed tube.	...	...	...	v s org solv
a23	carbide	Al_4C_3	143.96	yel-grn, hex.	2.36	stab to 1400	d 2200^{400}	d to CH_4	...	d dil a; i acet
a24	chlorate	$Al(ClO_3)_3 \cdot 6H_2O$	385.43	col, rhbdr, deliq	...	d	...	vs	vs	s dil HCl
a25	*per*chlorate	$Al(ClO_4)_3 \cdot 6H_2O$	433.43	col, hygr	2.020	82	$-6H_2O$, 178	s	s	...
a26	chloride	$AlCl_3$ (or Al_2Cl_6)	133.34	wh to col, hex, odor HCl, v deliq	fus 2.44^{25} liq 1.31^{200}	$190^{2.5}$ atm	d 262 182.7^{752} subl. 177.8	69.9^{15} with viol	s d	$100^{12.5}$ abs al; 0.072^{25} chl; $sCCl_4$, eth sl s bz
a27	chloride, hexahydrate	$AlCl_3 \cdot 6H_2O$	241.43	col, rhomb, deliq, 1.6	2.398	d 100	...	s	v s ev HCl	50 abs al; s eth; sl s HCl
a28	chloride, hexammine	$AlCl_3 \cdot 6NH_3$	235.52	col cr, hygr	1.412_4^{25}	d	...	s	...	...
a29	diethylmalonate deriv	$Al(C_7H_{11}O_4)_3$	504.46	...	1.084^{100}	98	...	i	...	s org solv
a30	ethoxide	$Al(C_2H_5O)_3$	162.16	wh cr.	1.142_0^{20}	134	205^{14}	d	s	v sl s al, eth
a31	α-ethylacetoacetate deriv	$Al(C_6H_9O_3)_3$	414.39	wh cr.	1.101^{80}	78—79	$190—200^{11}$	s	...	s lgr
a32	ferrocyanide	$Al_4[Fe(CN)_6]_3 \cdot 17H_2O$	1050.05	br powd	...	...	...	sl s	sl s	s dil a
a33	fluoride	AlF_3	83.98	col, tricl	2.882_4^{25}	1291 $subl^{760}$	...	0.559^{25}	s	i a, alk, al, acet
a34	fluoride	$AlF_3 \cdot 3\ 1/2\ H_2O$	147.03	wh cr powd	1.914_4^{25}	$-2H_2O$ 100	anhydr 250	i	sl s	d ac a
a35	fluoride, monohydrate	Nat. fluellite. $AlF_3 \cdot H_2O$	101.99	col, rhomb, 1.473, 1.490, 1.511	2.17	...	...	sl s	sl s	...
a36	fluosilicate	Nat topaz. $2AlFO \cdot SiO_2$	184.04	rhomb, 1.619, 1.620, 1.627	3.58	...	...	i	i	...
a37	hydroxide	Nat. boehmite. AlO(OH)	59.99	wh, orthorhomb microcr.	3.01	$-H_2O$, trans to γ-Al_2O_3	...	i	i	s h a, h alk
a38	hydroxide	Nat. diaspore. AlO(OH)	59.99	col. rhomb cr.	3.3—3.5	$-H_2O$, trans to Al_2O_3	...	i	i	s h a, h alk
a39	hydroxide	$Al(OH)_3$	78.00	wh, monocl	2.42	$-H_2O$, 300	...	i	i	s a, alk; i al
a40	iodide	AlI_3 (or Al_2I_6)	407.70	br pl, cont free I_2, deliq	3.98^{25}	191	360	s d	s	s al, eth, CS_2, liq NH_3
a41	iodide, hexahydrate	$AlI_3 \cdot 6H_2O$	515.79	wh-yel cr, hygr	2.63	d 185	3	v s	v s	s al, CS_2
a42	isopropoxide	$Al(C_3H_7O)_3$	204.25	wh cr.	1.0346_0^{20}	118.5	140.5^{8}	d	...	s al, bz, chl
a43	lactate	$Al(C_3H_5O_3)_3$	294.19	wh-yelsh powd	...	...	...	v s	...	...
a44	nitrate	$Al(NO_3)_3 \cdot 9H_2O$	375.13	col, rhomb, deliq, 1.54	...	73.5	d 150	63.7^{25}	v s d	100 al; s alk, acet, HNO_3. a
a45	nitride	AlN	40.99	wh cr, hex	3.26	>2200 (in N_2)	subl 2000	d (NH_3)	d	d a, alk
a46	oleate (com'l)	$Al(C_{18}H_{33}O_2)_3$(?)	871.36	wh powd, existence doubted except as basic salt	...	...	...	d	s	i al; v sl s bz
a47	oxalate	$Al_2(C_2O_4)_3 \cdot 4H_2O$	390.08	wh powd	...	...	...	i	i	i al; s a
a48	oxide	Al_2O_3	101.96	col, hex, 1.768, 1.760	3.965^{25}	2072	2980	i	...	v sl s a, alk
a49	oxide	α-Alumina, nat. corundum. Al_2O_3	101.96	col, rhomb cr, 1.765	3.97	2015 ± 15	2980 ± 60	0.000098^{29}	i	v sl s a, alk
a50	oxide	γ-Alumina. Al_2O_3	101.96	wh micr cr, 1.7	3.5—3.9	tr to α	...	i	i	sl s a, alk
a51	oxide, monohydrate	$Al_2O_3 \cdot H_2O$	119.98	col, rhomb, 1.624 ± 0.003	3.014	...	...	i	i	...
a52	oxide, trihydrate	Nat. gibbsite, hydraargillite. $Al_2O_3 \cdot 3H_2O$	156.01	wh monocl cr, 1.577, 1.577, 1.595	2.42	tr to $Al_2O_3 \cdot H_2O$ (Boehmite)	...	i	i	s h a, alk

PHYSICAL CONSTANTS OF INORGANIC COMPOUNDS (continued)

No.	Name	Synonyms and Formulae	Mol. wt.	Crystalline form, properties and index of refraction	Density or spec. gravity	Melting point, °C	Boiling point, °C	Solubility, in grams per 100 cc		
								Cold water	Hot water	Other solvents
a54	*meta*phosphate	$Al(PO_3)_3$	263.90	col, tetr	2.779			i	i	i a
a55	palmitate, mono-(com'l)	$Al(OH)_2C_{16}H_{31}O_2$	316.42	wh	1.095	200		i		s alk, hydrocarb
a56	1-phenol-4-sulfonate	$Al(C_6H_5O_4S)_3$	546.47	redsh-wh powd				s		s al, glyc
a57	phenoxide	$Al(C_6H_5O)_3$	306.30	grayish-wh cr mass	1.23	d 265		d		s al, eth, chl
a58	*ortho*phosphate	$AlPO_4$	121.95	wh rhomb pl, 1.546, 1.556, 1.578	2.566	>1500		i	i	s a, alk, al
a59	propoxide	$Al(C_3H_7O)_3$	204.25	wh cr.	1.0578^{20}_{0}	106	248^{14}	d	d	s al
a60	salicylate	$Al(C_7H_5O_3)_3$	438.33	redsh-wh powd				i		i al; s alk
a61	selenide	Al_2Se_3	290.84	lt brn powd, unstable in air	3.437^{15}_{4}			d	d	d a
a62	silicate	Nat. sillimanite, andalusite, cyanite. $Al_2O_3 \cdot SiO_2$	162.05	wh, rhomb, 1.66	3.247	1545 tr to $Al_2O_3 \cdot 2SiO_2$	>1545	i	i	d HF; i HCl; s fus alk
a63	silicate	Nat. mullite. $3Al_2O_3 \cdot 2SiO_2$	426.05	col, rhomb, 1.638, 1.642, 1.653	3.156	1920		i	i	i a, HF
a64	stearate, tri-	$Al(C_{18}H_{35}O_2)_3$	877.41	wh powd	1.010	103		i		s al, bz. turp, alk
a65	sulphate	$Al_2(SO_4)_3$	342.14	wh powd, 1.47	2.71	d 770		31.3^{0}	98.1^{100}	s dil a; sl s al
a66	sulfate, hydrate	Nat. alunogenite. $Al_2(SO_4)_3 \cdot 18H_2O$	666.41	col, monocl, 1.474, 1.467, 1.483	1.69^{17}	d 86.5		86.9^{0}	1104^{100}	i al
a67	sulfide	Al_2S_3	150.14	yel, hex, odor H_2S, d moist air	2.02^{13}	1100	subl 1500 (N_2)	d		s a; i acet
a68	thallium sulfate	Aluminum thallium alum. $AlTl(SO_4)_2 \cdot 12H_2O$	639.66	col, oct, 1.50112	2.325^{20}_{4}	91		4.84^{0}	65.19^{60}	
a69	**Americium**	Am	(243)	silvery, hex		994 ± 4	2607 (extrap)			s dil a
a70	bromide	$AmBr_3$	482.71	wh, orthorhomb		subl		s		
a71	chloride	$AmCl_3$	349.49	pink, hex	5.78	subl 850		s		
a72	fluoride	AmF_3	300.00	pink, hex	9.53			i		
a73	iodide	AmI_3	623.71	yel, orthoromb	6.9			s		
a74	oxide	Am_2O_3	534.00	redsh-brn, cub or tan, or hex.						s min a
a75	oxide, di-	AmO_2	275.00	blk, cub	11.68					s min a
a76	**Ammonia**	NH_3	17.03	col gas; liq, 0.817^{-79}, $1.325^{16.5}$	0.7710 g/ℓ 760 mm	−77.7	−33.35	89.9	7.4^{100}	13.20^{20} al; s eth, org solv
a77	Ammonia-d_3	Trideuterio ammonia. ND_3	20.05			−74	−30.9			
	Ammonium									
a78	acetate	$NH_4C_2H_3O_2$	77.08	wh cr, hygr	1.17^{20}_{4}	114	d	148^{4}	d	7.89^{15} MeOH; s al; s s acet
a79	acetate, hydrogen	$(NH_4)H(C_2H_3O_2)_2$	137.14	col need, deliq.		66		s		s al
a80	aluminum chloride	$NH_4Cl \cdot AlCl_3$	186.83	wh cr.		304		s		
a81	aluminum sulfate	$NH_4Al(SO_4)_2$	237.14	col, hex	2.45^{20}			s		s glyc; i al
a82	aluminum sulfate, hydrate	Nat. tschernigite. $NH_4Al(SO_4)_2 \cdot 12H_2O$	453.32	col, cub, 1.459	1.64	93.5	$-10H_2O$. 120	15^{20}	v s	s dil a; i al
a83	*ortho*arsenate	$(NH_4)_3AsO_4 \cdot 3H_2O$	247.08	rhomb cr		d, $-NH_3$		sl s		
a84	*ortho*arsenate, di-H	$NH_4H_2AsO_4$	158.97	col, tetr, 1.577, 1.522	2.311^{9}	d, $-NH_3^{300}$		33.74^{0}	122.4^{90}	
a85	*ortho*arsenate mono-H	$(NH_4)_2HAsO_4$	176.00	col, monocl, odor NH_3	1.989	d		s	d	i al
a86	*meta*arsenite	NH_4AsO_2	124.96	col, rhomb pr, hygr.				v s	d	i al, acet; sl s NH_4OH
a87	azide	NH_4N_3	60.06	col pl.	1.346	160	subl 134 expl	20.16^{30}	27.04^{40}	1.06^{20} 80% al; i eth, bz
a88	benzene sulfonate	$NH_4C_6H_5SO_3$	175.20	rhomb	1.342	d 271—275		98	320	19 c al; i eth, bz
a89	benzoate	$NH_4C_7H_5O_2$	139.15	col, rhomb	1.260	d 198	subl 160	$19.6^{14.5}$	83.3^{100}	1.63^{25}al; s glyc; i eth
a90	*penta*borate	Aluminum decaborate. $NH_4B_5O_8 \cdot 4H_2O$	272.14					7.03^{18}		
a91	*peroxy*borate	$NH_4BO_3 \cdot 1/2H_2O$	85.85	wh cr.		d		$1.55^{17.5}$		i al
a92	*tetra*borate	Ammonium biborate. $(NH_4)_2B_4O_7 \cdot 4H_2O$	263.37	col, tetr		d		7.27^{18}	52.68^{90}	sl s acet; i al
a93	bromate	NH_4BrO_3	145.94	col, hex		expl		v s	vs	
a94	bromide	NH_4Br	97.94	cub, coll hygr, 1.712^{25}	2.429	subl 452	235 vac	97^{25}	145.6^{100}	10^{78} al; s acet, eth, NH_3
a95	*di*bromoiodide	NH_4IBr_2	304.75	met-grn pr, hygr		198		v s	s	s eth
a96	bromoplatinate	$(NH_4)_2PtBr_6$	710.58	red-brn cub	4.265^{24}	d 145		0.59^{20}	0.36^{100}	
a97	bromoselenate	$(NH_4)_2SeBr_6$	594.46	red, oct cr.	3.326			d	d	sl s eth
a98	bromostannate	$(NH_4)_2SnBr_6$	634.19	col, cub	3.50	d		v s		
a99	cadmium chloride	$4NH_4Cl \cdot CdCl_2$	397.28	col, rhomb, 1.6038	2.01			s		
a100	calcium arsenate	$NH_4CaAsO_4 \cdot 6H_2O$	305.13	col, monocl.	1.905^{15}	d 140		0.02	s	s NH_4Cl; i NH_4OH
a101	calcium phosphate	$NH_4CaPO_4 \cdot 7H_2O$	279.20	1.561^{15}		d		i	d	s a
a102	carbamate	$NH_4NH_2CO_2$	78.07	col, rhomb		subl 60		v s	d	v s NH_4OH; sl s al; i acet
a103	carbamate acid carbonate	Sal volatile. $NH_4NH_2CO_2 \cdot NH_4HCO_3$	157.13	wh cr.		subl		25^{15}	67^{65}	d al; s glyc; i acet
a104	carbonate	$(NH_4)_2CO_3 \cdot H_2O$	114.10	col, cub		d 58		100^{15}	d	i al, NH_3, CS_2; s dil MeOH
a105	carbonate, hydrogen	Ammonium bicarbonate. NH_4HCO_3	79.06	col, rhomb or monocl, 1.423, 1.536, 1.555	1.58	107.5 (d 36—60)	subl	11.9^{0}	d	i al, acet

No.	Name	Synonyms and Formulae	Mol. wt.	Crystalline form, properties and index of refraction	Density or spec. gravity	Melting point, °C	Boiling point, °C	Solubility, in grams per 100 cc		
								Cold water	Hot water	Other solve
a106	cerium nitrate(ic)	$(NH_4)_2Ce(NO_3)_6$	548.23	or, monocl				141^{25}	227^{80}	s HNO_3 al
a107	cerium nitrate(ous)	$2NH_4NO_3·Ce(NO_3)_3·4H_2O$	558.28	monocl.		74		318.20	817.4^{65}	
a108	cerium sulfate(ous)	$(NH_4)_2SO_4·Ce_2(SO_4)_3·8H_2O$	844.67	monocl.	2.523	$-6H_2O$, 100	$-8H_2O$, 150	3.29^{49} (anhydr)		
a109	chlorate	NH_4ClO_3	101.49	col, monocl need.	1.80	102 expl		28.7^{0}	115^{75}	sl s al
a110	*per*chlorate	NH_4ClO_4	117.49	col, rhomb, 1.482	1.95	d		10.74^{0}	42.45^{85}	s acet; sl s al
a111	chloride	Sal ammoniac. NH_4Cl	53.49	col, cub, 1.642	1.527	subl 340	520	29.7^{0}	75.8^{100}	0.6^{19}al; s liq NI acet, eth
a112	chloroaurate	NH_4AuCl_4	356.82	yel, monocl or rhomb			520	s		sl s al
a113	chloroaurate, hydrate	$(NH_4AuCl_4)_4·5H_2O$	1517.34	yel, monocl		$-5H_2O$, 100		s		s al
a114	chlorogallate	NH_4GaCl_4	229.57	wh cr.		275		v s	v s	s al; i pet eth
a115	chloroiridate	$(NH_4)_2IrCl_6$	441.01	red-blk, cub.	2.856	d		0.69^{14}	4.38^{80}	i al; s HCl
a116	chloroiridite	$(NH_4)_3IrCl_6·1\ 1/2H_2O$	486.08					s		
a117	chloroosmate	$(NH_4)_2OsCl_6$	438.99	cub	2.93					
a118	chloropalladate	$(NH_4)_2PdCl_6$	355.21	red-brn, cub	2.418	d		sl s		
a119	chloropalladite	$(NH_4)_2PdCl_4$	284.31	olive grn, tetr.	2.17	d		s		i al
a120	*hexa*chloroplatinate	$(NH_4)_2PtCl_6$	443.87	yel, cub, 1.8	3.065	d		0.7^{15}	1.25^{100}	0.005 al; i eth, HCl
a121	chloroplatinite	$(NH_4)_2PtCl_4$	372.97	red, rhomb (tetr)	2.936	d 140—150		s	s	i al
a122	chloroplumbate	$(NH_4)_2PbCl_6$	455.99	yel, cub	2.925	d 120		sl s	d	s a
a123	chlorostannate	$(NH_4)_2SnCl_6$	367.48	wh, cub	2.4	d		$33.^{14.5}$	v s	
a124	*tetra*chlorozincate	$ZnCl_2·2NH_4Cl$	243.27	wh pl, rhomb, hygr.	1.879	d 150		v s		
a125	chromate	$(NH_4)_2CrO_4$	152.07	yel, monocl.	1.91^{12}	d 180		40.5^{30}	d	i al; sl s NH_3, a
a126	*di*chromate	$(NH_4)_2Cr_2O_7$	252.06	or, monocl	2.15^{25}	d 170		30.8^{15}	89^{30}	s al; i acet
a127	*peroxy*chromate	$(NH_4)_3CrO_8$	234.11	red-brn, cub		d 40	expl 50	sl s	d	i al, eth, sl s NI expl H_2SO_4
a129	chromium sulfate(ic)	$(NH_4)Cr(SO_4)_2·12H_2O$	478.33	grn or vlt, cub; 1.4842	1.72	94, $-9H_2O$, 100		21.2^{25}	32.8^{40}	s al, dil a
a130	citrate, di(*sec.*)	$(NH_4)_2HC_6H_5O_7$	226.19	wh gran or powd	1.48			100		sl s al
a130	citrate, di(*sec.*)	$(NH_4)_2HC_6H_5O_7$	226.19	wh gran or powd	1.48			100		sl s al
a131	citrate, tri-(*tert.*	$(NH_4)_3C_6H_5O_7$	243.22	wh cr, deliq.		d		v s		i al, eth, acet
a132	cobalt *ortho*phosphate(ous)	$(NH_4)CoPO_4·H_2O$	189.96	vlt cr powd				i	d	s a
a133	cobalt (II)-selenate	$(NH_4)_2SO_4·CoSeSO_4·6H_2O$	482.02	ruby-red, monocl, 1.526, 1.532, 1.541	2.228^{20}_{4}					
a134	cobalt sulfate(ous)	$(NH_4)_2SO_4·CoSO_4·6H_2O$	395.22	ruby-red, monocl, 1.490, 1.495, 1.503	1.902			20.5^{20}	45.4^{80}	i al
a135	copper chloride(ic)	$2NH_4Cl·CuCl_2·2H_2O$	277.47	blue, tetrag, 1.744, 1.724	1.993	d 110		33.8^{0}	99.3^{80}	s a, al; sl s NH_3
a136	copper iodide(ous)	$NH_4I·CuI·H_2O$	353.41	rhomb pl				d	d	s NH_4I
a137	cyanate	NH_4OCN	60.06	wh cr.	1.342^{20}_{4}	d 60		v s	d	sl s al; i eth
a138	cyanide	NH_4CN	44.06	col, cub	1.02^{100}	d 36	subl 40	v s	d	v s al
a139	cyanaurate	$NH_4Au(CN)_4·H_2O$	337.09	col pl.		d 200		v s		v s al; i eth
a140	cyanaurite	$NH_4Au(CN)_2$	267.04	col, cub		d 100		v s	v s	s al: i eth
a141	cyanoplatinite	$(NH_4)_2Pt(CN)_4·H_2O$	353.24	yel cr.				s		
a142	ethylsulfate	$NH_4C_2H_5SO_4$	143.16			99		s		
a143	ferricyanide	$(NH_4)_3Fe(CN)_6$	266.07	red cr, rhomb.		d		v s		
a144	ferrocyanide	$(NH_4)_4Fe(CN)_6·3H_2O$	338.15	yel, monocl, turns bl in air		d		s	d	i al
a145	fluoantimonite	$(NH_4)_2SbF_5$	252.82	col, rhomb		d, subl		108		
a146	fluoborate	NH_4BF_4	104.84	wh, rhomb	1.871^{15}	subl		25^{16}	97^{100}	s NH_4OH
a147	fluogallate	$(NH_4)_3GaF_6$	237.83	wh oct cr		d >250 − GaF_3			sl s	
a148	fluogermanate	$(NH_4)_2GeF_6$	222.66	col hex pr and bi-pyram, 1.428, 1.425	2.564^{25}_{25}			s		i al, MeOH
a149	fluophosphate, di-	$NH_4PO_2F_2$	119.01	col, rhomb		213		s	s	s al, acet
a150	fluophosphate, hexa	NH_4PF_6	163.00	col, pl	2.180^{18}	d		s	s	s al, acet
a151	fluoride	NH_4F	37.04	col, hex, deliq	1.009^{25}	subl		100^{0}	d	s al; i NH_3
a152	fluoride, hydrogen	NH_4HF_2	57.04	rhomb or tetr, deliq, 1.390	1.50	125.6		v s	v s	sl s al
a153	fluorosulfonate	NH_4FSO_3	117.10	wh need		d 245		v s	v s	sl s al; s MeOH
a154	fluosilicate	α-Nat. cryptohalite. $(NH_4)_2SiF_6$	178.15	α oct, β hex, col α 1.3696	α 2.011 β 2.152	d		18.16^{17}	55.5^{100}	sl s al; i acet
a155	fluosulfonate	NH_4SO_3F	117.10	col, need		244.7		s	s	sl s al; s MeOH; NH_4OH
a156	fluotitanate	$(NH_4)_2TiF_6$	197.95	hex pr		d		s	s	i al, eth
a157	fluozirconate	$(NH_4)_2ZrF_6$	241.29	rhomb, hex	1.154					
a158	fluozirconate	$(NH_4)_3ZrF_7$	278.32	col, cub	1.433			sl s		
a159	formate	NH_4CHO_2	63.06	wh, monocl, deliq	1.280	116	d 180	102^{0}	531^{80}	s al, eth, NH_3
a160	gallium sulfate	Ammonium gallium alum. $Ga(NH_4)(SO_4)_2·12H_2O$	496.06	cub, 1.468	1.777			30.84^{25}		0.00875^{25} 70% I
a161	*tri*hydrogen para *per*iodate	$(NH_4)_2H_3IO_6$	262.00	col, rhomb	2.85					
a162	hydroxide	NH_4OH	35.05	at ord temp in sol only		−77		s		
a163	iodate	NH_4IO_3	192.94	col, rhomb or monocl	3.309^{21}	d 150		2.06^{15}	14.5^{101}	

No.	Name	Synonyms and Formulae	Mol. wt.	Crystalline form, properties and index of refraction	Density or spec. gravity	Melting point, °C	Boiling point, °C	Solubility, in grams per 100 cc		
								Cold water	Hot water	Other solvents
a164	iodide	NH_4I	144.94	col, cub, hygr, 1.7031	2.514^{25}	subl 551	220 vac	154.2^{0}	250.3^{100}	v s al, acet, NH_3; sl s eth
a165	*tri*iodide	NH_4I_3	398.75	dk br, rhomb	3.749	d 175		s d	d	
a166	iodoplatinate	$(NH_4)_2PtI_6$	992.58	blk, cub	4.61					i al
a167	iridium chloride (III)	$(NH_4)_3IrCl_6 \cdot H_2O$	477.07	grn-br-gl, cub		d 350		10.5^{19}		
a168	iridium sulfate	$NH_4Ir(SO_4)_2 \cdot 12H_2O$	618.56	yel-red cr			106	s		
a169	iron (III) chloride	$2NH_4Cl \cdot FeCl_3 \cdot H_2O$	287.20	ruby-red, rhomb, hygr, 1.78	1.99	234		v s	v s	
a170	iron (III) fluoride	$3NH_4F \cdot FeF_3$	223.95	col to lt yel, oct	1.96			sl s	sl s	
a171	iron (II) selenate	$(NH_4)_2SeO_4 \cdot FeSeO_4 \cdot 6H_2O$	485.93	lt grn, monocl, 1.5226, 1.5260, 1.5334	2.191_4^{20}					
a172	iron (III) sulfate	$(NH_4)_2SO_4 \cdot Fe_2(SO_4)_3$	532.00	wh, hex	2.49^{22}	d 420		44.15^{25}		
a173	iron sulfate(ic)	$NH_4Fe(SO_4)_2 \cdot 12H_2O$	482.18	vlt, cub oct, 1.4854	1.71	39—41	$-12H_2O$, 230	124.0^{25}	400^{100}	i al; s dil a
a174	iron sulfate(ous)	$(NH_4)_2SO_4 \cdot FeSO_4 \cdot 6H_2O$	392.13	grn, monocl, 1.487, 1.492, 1.499	1.864_4^{20}	d 100—110		26.9^{20}	73.0^{80}	i al
a175	lactate	$NH_4C_3H_5O_3$	107.11	col-yelsh liq	$1.19—1.21^{15}$			∞		∞ al
a176	laurate, acid (mixt.)	$NH_4C_{12}H_{23}O_2 \cdot C_{12}H_{24}O_2$	417.67	wh		75	d	s	s	4.8^{7} al; sl s eth, acet
a177	magnesium arsenate	$NH_4MgAsO_4 \cdot 6H_2O$	289.35	col, tetrg, 1.608	1.932^{15}	d		0.038^{20}	0.024^{80}	s a; i al
a178	magnesium carbonate	$(NH_4)_2CO_3 \cdot MgCO_3 \cdot 4H_2O$	252.46	wh				s	v s	s a; i al
a179	magnesium chloride	$NH_4Cl \cdot MgCl_2 \cdot 6H_2O$	256.80	col, rhomb, deliq	1.456	$-2H_2O$, 100	d	16.7		d al
a180	magnesium chromate	$(NH_4)_2CrO_4 \cdot MgCrO_4 \cdot 6H_2O$	400.46	yel, monocl, 1.636, 1.637, 1.653	1.84	d		v s	v s	
a181	magnesium phosphate	Guanite, struvite $NH_4MgPO_4 \cdot 6H_2O$	245.41	col, rhomb, 1.495, 1.496, 1.504	1.711	d		0.0231^{0}	0.0195^{80}	v s dil a; i al
a182	magnesium selenate	$(NH_4)_2SeO_4 \cdot MgSeO_4 \cdot 6H_2O$	454.39	col monocl pr, 1.507, 1.509, 1.517	2.058_4^{20}			sl s		
a183	magnesium sulfate	Nat. boussingaulite. $(NH_4)_2SO_4 \cdot MgSO_4 \cdot 6H_2O$	360.59	col, monocl, 1.472, 1.473, 1.479	1.723	>120	d 250	17.92^{0}	64.7^{100} (anhydr)	
a184	*l*-malate, hydrogen	$NH_4HC_4H_4O_5$	151.12	col, rhomb	1.5	161	d	$32.2^{15.7}$		
a185	*per*manganate	NH_4MnO_4	136.97	rhomb	2.208^{10}	d 110		7.9^{15}	d	
a186	manganese phosphate(ic)	$NH_4MnPO_4 \cdot H_2O$	185.96	wh cr.				0.0031	0.05	i al, NH_4 salts
a187	manganese sulfate(ous)	$(NH_4)_2SO_4 \cdot MnSO_4 \cdot 6H_2O$	391.22	pa red, monocl, 1.480, 1.484, 1.491	1.83			51.3^{25}	v s	
a188	molybdate	$(NH_4)_2MoO_4$	196.01	col, monocl pr	2.276_4^{25}	d		s, d	d	s a; i al, NH_3 acet
a189	*para*molybdate	"Molybdic acid" com'l. $(NH_4)_6Mo_7O_{24} \cdot 4H_2O$	1235.86	col-yelsh, monocl	2.498	$-H_2O$, 90	d 190	43	d	i al; s a, alk
a190	*per*molybdate	$3(NH_4)_2 \cdot 1/2MoO_3 \cdot 2MoO_4 \cdot 6H_2O$	608.17	lt yel monocl pr	2.975	d 170		v s	v s	sl s al
a191	molybdotellurate	$(NH_4)_6TeMo_6O_{24} \cdot 7H_2O$	1321.56	col,, rhomb	2.78	d 550	d	s	s	
a192	myristate, acid (mixt.)	$NH_4C_{14}H_{27}O_2 \cdot C_{14}H_{28}O_2$	473.78	wh solid		75—90	d	sl s	s	s al; i eth
a193	nickel chloride	$NH_4Cl \cdot NiCl_2 \cdot 6H_2O$	291.18	grn, monocl, deliq	1.654			v s	v s	
a194	nickel sulfate	Double nickel salt. $(NH_4)_2SO_4 \cdot NiSO_4 \cdot 6H_2O$	394.97	dk bl-grn, monocl, 1.495, 1.501, 1.508	1.923			10.4^{20}	30^{80}	i al; s $(NH_4)_2SO_4$
a195	nitrate	NH_4NO_3	80.04	col, rhomb, (monocl >32.1^{0})	1.725^{25}	169.6	210^{11}	118.3^{0}	871^{100}	3.8^{20} al; 17.1^{20} MeOH s acet, NH_3; i eth
a196	nitratocerate	$(NH_4)_2Ce(NO_3)_6$	548.23	yel-red, monocl				142.6^{25}	232^{90}	s al; sl s HNO_3
a197	nitrite	NH_4NO_2	64.04	wh-yelsh cr	1.69	60—70 expl	30 subl vac	v s	d	s dil al; i eth
a198	oleate, acid (mixt.)	$NH_4C_{18}H_{33}O_2 \cdot C_{18}H_{34}O_2$	581.96	wh powd		d 78		s	s	80^{50} al; 13.3^{15} eth
a199	osmium chloride	$(NH_4)_2OsCl_6$	438.99	blk, oct	2.93^{25}	subl 170		d	d	s HCl
a200	oxalate	$(NH_4)_2C_2O_4 \cdot H_2O$	142.11	col, rhomb, 1.439, 1.546, 1.594	1.50	d		2.54^{0}	11.8^{50}	i NH_3
a201	oxalate, acid	Ammonium binoxalate. $NH_4HC_2O_4 \cdot H_2O$	125.08	col, rhomb	1.556	$-H_2O$, 170		s		i eth, bz
a202	oxalatoferrate (III)	$(NH_4)_3Fe(C_2O_4)_3 \cdot 3H_2O$	428.07	grn, monocl.	1.78	d 165		42.7^{0}	345^{100}	
a203	palladium (II)-chloride	$(NH_4)_2PdCl_4$	284.31	grnsh-yel, tetr	2.17	d		v s	v s	s dil al; i abs al
a204	palladium (IV)-chloride	$(NH_4)_2PdCl_6$	355.21	red oct cr	2.418	d $-Cl$		sl s	sl s	
a205	palmitate, (acid)- (mixt.)	$NH_4C_{16}H_{31}O_2 \cdot C_{16}H_{32}O_2$	529.89	yelsh soapy mass or yel powd		>100	d	sl s	s	8.8^{50} al; 0.23^{13} eth
a206	*metaper*iodate	NH_4IO_4	208.94	col, tetr	3.056^{18}	expl		2.7^{16}		
a207	*hypo*phosphate	$(NH_4)_2H_2P_2O_6$	196.04	col cr.		170		7^{25}	25^{100}	
a208	*ortho*phosphate	$(NH_4)_3PO_4 \cdot 3H_2O$	203.13	wh pr				26.1^{25}		sl s dil NH_4OH; i NH_3, acet
a209	*ortho*phosphate, di-H	$NH_4H_2PO_4$	115.03	col, tetr, 1.525, 1.479	1.803^{19}	190		22.7^{0}	173.2^{100}	i acet
a210	*ortho*phosphate-mono-H	$(NH_4)_2HPO_4$	132.06	col, monocl, 1.52	1.619	d 155	d	57.5^{10}	106.70^{70}	i al, acet, NH_3
a211	*hypo*phosphite	$NH_4H_2PO_2$	83.03	rhomb tabl	1.634	200	d 240	s	s	s al, NH_3; i acet
a212	*ortho*phosphite, di-H	$NH_4H_2PO_3$	99.03	col, monocl pr		123	d 145	171^{0}	260^{31}	i al
a213	phosphofluoride, hexa-	NH_4PF_6	163.00	col, cub	2.180_4^{18}	d		74.8^{20}		s al, acet; d h a
a214	phosphomolybdate	Ammonium molybdo phosphate. $(NH_4)_3P(Mo_3O_{10})_4$	1876.35	yel powd		d		sl s	sl s	s alk; i al, HNO_3

No.	Name	Synonyms and Formulae	Mol. wt.	Crystalline form, properties and index of refraction	Density or spec. gravity	Melting point, °C	Boiling point, °C	Solubility, in grams per 100 cc		
								Cold water	Hot water	Other solvents
a215	phosphotungstate	$(NH_4)_3P(W_3O_{10})_4$	2931.27	wh				sl s	sl s	
a216	picramate	$NH_4C_6H_4N_3O_5$	216.15	redsh-brn cr powd				s		
a217	picrate	$NH_4C_6H_2N_3O_7$	246.14	red or yel, rhomb	1.719	d	expl 423	1.1^{20}	s	sl s al
a219	praseodymium sulfate	$(NH_4)_2SO_4 \cdot Pr_2(SO_4)_3 \cdot 8H_2O$	846.24	cr	$2.531^{16.5}$	$-8H_2O$, 170		sl s		
a220	propionate	$NH_4C_3H_5O_2$	91.11	pr, deliq	1.108^{25}	45		v s		s al, ac a
a221	*per*rhenate	NH_4ReO_4	268.24	wh, hex pl	3.97	d	d	6.1^{20}	32.34^{80}	
a222	rhodanid	NH_4NCS	76.12	monocl cr, to rhomb at 92, 1.685	1.305	149.6	d 170	128^{0}	347^{60}	v s al; s MeOH, acet; i $CHCl_3$
a223	rhodium chloride	$(NH_4)_3RhCl_6 \cdot H_2O$	387.75	dk red rhomb need		$-H_2O$, 140		s	s	sl s al; s dil NH_4Cl
a224	rhodium sulfate	Ammonium rhodium alum. $Rh(NH_4)(SO_4)_2 \cdot 12H_2O$	529.24	orange, 1.5150		102				
a225	*d*-saccharate, hydrogen	$NH_4HC_6H_8O_8$	227.17	wh need or monocl pr				1.22^{15}	24.35^{100}	i c al; s h al
a226	salicylate	$NH_4C_7H_5O_3$	155.15	col, monocl			subl	111^{25}	v s	28.8^{25} al
a227	selenate	$(NH_4)_2SeO_4$	179.03	col, monocl, 1.561, 1.563, 1.585	2.194^{20}_{4}	d		117^{7}	197^{100}	i al, acet, NH_3
a228	selenate, hydrogen	NH_4HSeO_4	162.00	rhomb	2.162	d				
a229	selenide	$(NH_4)_2Se$	115.04	col, or wh cr		d		s		
a230	sodium phosphate, hydrate	Microsmic salt. $NH_4NaHPO_4 \cdot 4H_2O$	209.07	col, monocl, 1.439, 1.442, 1.469	1.574	d 97				
a231	stearate, acid (mixt.)	$NH_4(C_{18}H_{35}O_2 \cdot C_{18}H_{36}O_2$	586.00	wh cr.		d 110		v s		0.3^{25} al; 0.19^{25}eth; 0.08^{25} acet
a232	succinate	$(NH_4)_2 \cdot C_4H_4O_4$	152.15	col cr.	1.37			s		sl s al
a233	sulfamate	$NH_4NH_2SO_3$	114.12	large, pl, deliq		125	d 160	166.6^{10}	357^{50}	i al
a234	sulfate	Nat. mascagnite. $(NH_4)_2SO_4$	132.13	col, rhomb, 1.521, 1.523, 1.533	1.769^{50}	d 235		70.6^{0}	103.8^{100}	i al, acet, NH_3
a235	sulfate, hydrogen	Ammonium bisulfate. NH_4HSO_4	115.10	col, rhomb, 1.473	1.78	146.9	d	100	v s	sl s al; i acet
a236	*peroxy*disulfate	$(NH_4)_2S_2O_8$	228.19	col, monocl, 1.498, 1.502, 1.587	1.982	d 120		58.2^{0}	v s	
a237	sulfide, hydro-	NH_4HS	51.11	wh rhomb, 1.74	1.17	118^{19atm}	88.4^{19atm}	128.1^{0}	d	s al; NH_3
a238	sulfide, mono-	$(NH_4)_2S$	68.14	col yel cr (> − 18), hygr.		d		v s	d	s al; v s NH_3
a239	sulfide, penta-	$(NH_4)_2S_5$	196.39	yel pr		d 115		v s		s al, i eth, CS_2
a240	sulfite	$(NH_4)_2SO_3 \cdot H_2O$	134.15	col, monocl, 1.515	1.41^{25}	d 60—70	subl 150	32.4^{0}	60.4^{100}	sl s al; i acet
a241	sulfite, hydrogen	Ammonium bisulfite. NH_4HSO_3	99.10	rhomb pr, deliq	2.03	subl 150 (in N_2)		71.8^{0}	84.7^{60}	
a242	*dl*-tartrate	$(NH_4)_2C_4H_4O_6$	184.15	col, monocl, d, α 1.55, β 1.581	1.601	d		58.01^{15}	81.17^{60}	sl s al
a243	*dl*-tartrate, hydrogen	$NH_4HC_4H_4O_6$	167.12	col, monocl pr, 1.561, 1.591	1.636	d		2.35^{15}	3.24^{25}	i al, s a, alk
a244	tellurate	$(NH_4)_2TeO_4$	227.67	wh powd	$3.024^{24.5}$	d		s	s	i al; s dil a
a245	thallium chloride	$3NH_4Cl \cdot TlCl_3 \cdot 2H_2O$	507.25	col	2.39			s		
a246	thioantimonate	$(NH_4)_3SbS_4 \cdot 4H_2O$	376.17	yel pr		d		71.2^{0}	d	i al, eth
a247	thiocarbamate	$NH_4CS_2NH_2$	110.19	yel cr.		d 50		v s		s al; sl s eth
a248	thiocarbonate, tri-	$(NH_4)_2CS_3$	144.27	yel cr, hygr		subl		v s	d	sl s al, eth
a249	thiocyanate	NH_4SCN	76.12	col, monocl, deliq	1.305	149.6	d 170	128^{0}	v s	s al, acet, NH_3
a250	*di*thionate	$(NH_4)_2S_2O_6 \cdot 1/2H_2O$	205.20	col, monocl	1.704	d 130		135^{0}	v s	i al
a251	thiosulfate	$(NH_4)_2S_2O_3$	148.20	col, monocl, hygr	1.679	d 150		v s	103.3^{100}	i al; sl s acet
a252	titanium oxalate, basic	$(NH_4)_2TiO(C_2O_4)_2 \cdot H_2O$	294.01					v s		
a253	uranylcarbonate	$2(NH_4)_2CO_3 \cdot UO_2CO_3 \cdot 2H_2O$	558.24	yel, monocl	2.773	d 100		5.8^{19}	d	s$(NH_4)_2CO_3$, aq·SO_2
a254	uranylfluoride, penta-	$(NH_4)_3UO_2F_5$	419.14	tetr cr, 1.495	3.186	subl		s		
a255	valerate	$NH_4C_5H_9O_2$	119.16	col or wh cr		d		s		s al, eth
a256	*meta*vanadate	NH_4VO_3	116.98	wh-yelsh or col cr	2.326	d 200		0.52^{15}	6.95^{96}, d	
a257	vanadium sulfate	$NH_4V(SO_4)_2 \cdot 12H_2O$	477.28	red to bl	1.687	49		28.45^{20}		
a258	zinc sulfate	$(NH_4)_2SO_4 \cdot ZnSO_4 \cdot 6H_2O$	401.66	wh monocl, 1.489, 1.493, 1.499	1.931	d		7^{0} (anhydr)	42^{80} (anhydr)	
a259	**Antimony**	Sb	121.75 ± 3	silv wh met, hex	6.684^{25}	630.5	1750	i	i	s hot conc H_2SO_4, aq reg
a260	bromide, tri-	$SbBr_3$	361.46	col rhomb, 1.74	4.148^{23}	96.6	280	d	d	s HCl, HBr, CS_2, NH_3, al, acet
a261	chloride, penta-	$SbCl_5$	299.02	wh liq or monocl, 1.601^{14}	liq 2.336^{20}_{4}	2.8	79^{22}	d	d	s HCl, tarta, $CHCl_3$
a262	chloride, tri-	Butter of antimony. $SBCl_3$	228.11	col, rhomb, deliq	3.140^{25}	73.4	283	601.6^{0}	∞^{80}	s abs al, HCl, tart a, $CHCl_3$, bz, acet
a263	fluoride, penta-	SbF_5	216.74	col oily liq	2.99^{23} liq	7	149.5	s		s KF
a264	fluoride, tri-	SbF_3	178.75	col, rhomb	$4.379^{20.9}$	292	subl 319	384.7^{0}	563.6^{20}	i NH_3
a265	hydride	SbH_3	124.77	inflamm gas	gas 4.36^{15}; liq 2.26^{-25}	−88	-17.1^{751}	0.41^{0}		1500 mℓ A; 2500 mℓ CS_2
a266	iodide, penta-	SbI_5	756.27	br		79	400.6			
a267	iodide, tri-	SbI_3	502.46	ruby-red, hex, 2.78_{Li}, 2.36_{Li}	4.917^{17}	170	401	d	d	i al; s CS_2, bz, HI, HCl
a268	iodosulfide	SbSI	280.71	dk red		392	d	i	i	d conc HCl; i CS_2

No.	Name	Synonyms and Formulae	Mol. wt.	Crystalline form, properties and index of refraction	Density or spec. gravity	Melting point, °C	Boiling point, °C	Solubility, in grams per 100 cc Cold water	Hot water	Other solvents
a269	mercaptoacetamide	Antimony thioglycolamide. $Sb(C_2H_4NOS)_3$	392.11	wh cr.		139		200		
a270	nitrate, basic	$2Sb_2O_3 \cdot N_2O_5$(?)	691.01	wh glossy cr		d		d		d a; v sl s conc HNO_3
a271	nitride	SbN	135.76	or powd		d		d		
a272	oxide, penta-	Sb_2O_5 (or Sb_4O_{10})	323.50	yel powd	3.80 (dep on temp)	−O, 380 −O_2, 930		v sl s	v sl s	v sl s KOH, HCl, HI
a273	oxide, tetra-	Nat. cervantite. Sb_2O_4 (or $Sb_2O_3 \cdot Sb_2O_5$)	307.50	wh powd, 2.00	5.82	−O, 930		v sl s	v sl s	v sl s KOH, HCl, HI
a274	oxide, tri-	Nat. senarmontite. Sb_2O_3 (or Sb_4O_6)	291.50	wh, cub, 2.087	5.2	656	1550 subl	v sl s	sl s	s KOH, HCl, tart a, ac a
a275	oxide, tri-	Nat. valentinite. Sb_2O_3 (or Sb_4O_6)	291.50	col, rhomb, 2.18, 2.35, 2.35	5.67	656	1550	v sl s	sl s	s KOH, HCl, tart, a ac a
a276	III oxychloride(ous)	SbOCl	173.20	wh monocl		d 170		i	d	s acet, HCl, CS_2; i al, eth, $CHCl_3$
a277	III oxychloride(ous)	$Sb_4O_5Cl_2$	637.90	monocl.	5.01	d 320		i		s HCl; i al, eth
a278	oxyhydrate	$H_4Sb_2O_7$	359.53	wh, amorph.		−H_2O, 200		sl s	sl s	s alk
a279	III oxysulfate, di-(ous).	$Sb_2O_2SO_4$	371.56	wh	4.89			d	d	s, H_2SO_4
a280	potassium tartrate	Tartar emetic. $K(SbO)C_4H_4O_6 \cdot 1/2H_2O$	333.93	col cr.	2.6	−1/2H_2O, 100		8.3	33.3	i al; 6.7 glyc
a281	selenide	Sb_2Se_3	480.38	gray cr.		611		v sl s		s conc HCl
a282	III sulfate	$Sb_2(SO_4)_3$	531.67	wh powd, deliq	3.625^4	d		i	d	s a
a283	sulfide, penta-	Sb_2S_5	403.80	yel powd, prism	4.120	d 75		i	i	i al; s HCl, alk, NH_4HS
a284	sulfide, tri-	Nat. stibnite. Sb_2S_3	339.68	blk, rhomb, 3.194, 4.064, 4.303	4.64	550	*ca* 1150	0.000175^{18}		s al, NH_4HS. K_2S, HCl; i ac a
a285	sulfide, tri-	Sb_2S_3	339.68	yel-red, amorph	4.12	550	*ca* 1150	0.000175^{18}		s al, NH_4SH, K_2S, HCl; i ac a
a286	*d*-tartrate	$Sb_2(C_4H_4O_6)_3 \cdot 6H_2O$	795.81	wh cr powd				s		
a287	telluride, tri-	Sb_2Te_3	626.30	gray	6.50^{13}	629				s HNO_3, aq reg
a288	**Argon**	Ar	39.948	col inert gas,	1.784^0 g/ℓ cr: 1.65^{233}, liq: 1.40^{-186}	−189.2	−185.7	5.6^0 cm³	3.01^{50} cm³	
a289	**Arsenic**	As	74.9216	gray met, hex-rhomb	5.727^{14}	817 (28 atm.)	613 subl	i	i	s HNO_3
a290	**Arsenic**	As_4	299.69	yel, cub	2.026^{18}	d 358		i	i	s CS_2, bz
a291	**Arsenic acid,** meta-	$(HO)AsO_2$	123.93	wh, hygr		d	forms ortho-arsenic acid	d		
a292	ortho-	$H_3AsO_4 \cdot 1/2H_2O$	150.95	wh translu hygr cr	2.0—2.5	35.5	−H_2O 160	$302^{12.5}$	50	s al, alk, glyc
a293	pyro-	$H_4As_2O_7$	265.87	col pr		forms orthoarsenic acid 206 d				
a294	bromide, tri-	$AsBr_3$	314.63	col-yel hygr pr	3.54^{25}	32.8	221	d	d	s HCl, HBr, CS_2
a295	chloride, tri-	$AsCl_3$	181.28	oily liq or need, 1.621_F^{14}	liq 2.163^{20}	−8.5	130.2	d	d	s HBr, HCl, PCl_3, al, eth
a296	fluoride, penta-	AsF_5	169.91	col gas.	7.71 g/ℓ	−80	−53	s		s alk, al, eth, bz
a297	fluoride, tri-	AsF_3	131.92	oily liq.	liq 2.666	−8.5	-63^{752}	d	d	s al, eth, bz, NH_4OH
a298	hydride	Arsine. AsH_3	77.95	col gas.	gas 2.695 liq $1.689^{84.9}$	−116.3	−55 (d 300)	20 mℓ		s $CHCl_3$, bz
a299	iodide, di-	AsI_2	328.73	red pr		d 136		d		s al, eth, $CHCl_3$, bz, CS_2
a300	iodide, penta-	AsI_5	709.44		3.93	76				
a301	iodide, tri-	AsI_3	455.64	red hex, *ca.* 2.59, *ca.* 2.23	4.39^{13}	146	403	6^{25}	30 d	5.2 CS_2; s al, eth, bz, $CHCl_3$
a302	oxide, pent-	As_2O_5	229.84	wh amor, deliq.	4.32	d 315		150^{16}	76.7^{100}	s al, a, alk
a303	oxide, tri-	As_2O_3	197.84	amor or vitreous	3.738	312.3		3.7^{20}	10.14^{100}	s alk, alk carb, HCl
a304	oxide, tri-	Nat. arsenolite. As_2O_3 (or As_4O_6)	197.84	col, cub or fibr, 1.755	3.865^{25}	subl 193		1.2^2	11.46^{100}	s al, alk, HCl
a305	oxide, tri-	Nat. claudetite. As_2O_3 (or As_4O_6)	197.84	col, monocl, 1.871, 1.92, 2.01	4.15	193, subl 312.3	457.2	1.2^2	11.46^{100}	s al, alk, HCl
a306	(III)oxychloride(ous)	AsOCl	126.37	brnsh.		d		d	d	
a307	phosphide, mono-	AsP	105.90	br-red powd.		subl, d		d	d	sl s CS_2; s H_2SO_4, HCl; i al, eth, $CHCl_3$
a308	selenide	As_2Se_3	386.72	br cr	4.75	*ca.* 360		i	d	s alk
a309	sulfide, di-	Nat. realgar. As_2S_2	213.96	red-br monocl, 2.46, 2.59, 2.61	α 3.506^{19} β 3.254^{19}	α tr 267 β 307	565	i	i	s K_2S, $NaHCO_3$
a310	sulfide, penta-	As_2S_5	310.14	yel		subl, d 500		0.000136^0	i	s alk, alk sulf, HNO_3
a311	sulfide, tri-	Nat. orpiment. As_2S_3	246.03	yel or red, monocl, 2.4, 2.81, 3.02 (Li)	3.43	300	707	0.00005^{18}	sl s	s al, alk, alk carb
	Auric or **Aurous**	*See* **Gold**								
b1	**Barium**	Ba	137.33	yel-silv met	3.51^{20}	725	1640	d, ev H_2	d	s al; i bz
b2	acetate	$Ba(C_2H_3O_2)_2$	255.42	col cr.	2.468			58.8^0	75^{100}	
b3	acetate, hydrate	$Ba(C_2H_3O_2)_2 \cdot H_2O$	273.43	col tricl, 1.500, 1.517, 1.525	2.19	−H_2O, 150		76.4^{26}	75^{100}	sl s al
b4	amide	$Ba(NH_2)_2$	169.38	gray-wh cr		280		d	d	i liq NH_3
b5	*ortho*arsenate	$Ba_3(AsO_4)_2$	689.83		5.10	1605		0.055		s a, NH_4Cl
b6	*ortho*arsenate, mono-H	$BaHAsO_4 \cdot H_2O$	295.27	col, rhomb or monocl, 1.635	1.93^{15}	−H_2O, 150		sl s	d	s HCl

PHYSICAL CONSTANTS OF INORGANIC COMPOUNDS (continued)

No.	Name	Synonyms and Formulae	Mol. wt.	Crystalline form, properties and index of refraction	Density or spec. gravity	Melting point, °C	Boiling point, °C	Solubility, in grams per 100 cc		
								Cold water	Hot water	Other solvents
b7	arsenide	Ba_3As_2	561.83	br	4.1^{15}	...	...	d	...	d Cl_2, F_2, Br_2
b8	azide	$Ba(N_3)_2$	221.37	monocl pr	2.936	$-N_2$, 120	expl	17.3^{17}	...	abs al 0.017^{16}; i eth
b9	azide, hydrate	$Ba(N_3)_2 \cdot H_2O$	239.39	tricl, 1.7	...	expl	...	v s	v s	sl s al; i eth
b10	benzoate	$Ba(C_7H_5O_2)_2 \cdot 2H_2O$	415.59	col, nacreous leaf	...	$-2H_2O$, 100	...	s	...	sl s al
b11	boride, hexa-	BaB_6	202.19	met-blk, cub	4.36^{16}	2270	...	i	i	s HNO_3; i HCl
b12	bromate	$Ba(BrO_3)_2 \cdot H_2O$	411.15	col, monocl	3.99^{18}	d 260	...	0.3^{0}	5.67^{100}	i al; s acet
b13	bromide	$BaBr_2$	297.14	col cr, 1.75	4.781^{24}	847	d	104.1^{20}	149^{100}	v s al, MeOH
b14	bromide, dihydrate	$BaBr_2 \cdot 2H_2O$	333.17	col, monocl, 1.713, 1.727, 1.744	3.58^{24}	880, $(-H_2O$, 75)	$-2H_2O$, 120	151^{20}	204^{100}	v s MeOH, s al
b15	bromofluoride	$BaBr_2 \cdot BaF_2$	472.46	col pl.	4.96^{18}	...	...	d	d	i al; s conc HCl, conc HNO_3
b16	bromoplatinate	$BaPtBr_6 \cdot 10H_2O$	991.99	monocl.	3.71	...	...	...	...	...
b17	butyrate	$Ba(C_4H_7O_2)_2 \cdot 2H_2O$	347.56	col	...	...	...	37.42^{0}	42.12^{80}	...
b18	carbide	BaC_2	161.35	gray, tetr	3.75	...	...	d to C_2H_2		d a
b19	carbonate(α)	$BaCO_3$	197.34	wh, hex	4.43	$1740^{90}{}_{atm}$	d	0.002^{20}	0.006^{100}	s a, NH_4Cl; i al
b20	carbonate(β)	$BaCO_3$	197.34	...	...	tr to α, 982	...	0.0022^{18}	0.0065^{100}	s a, NH_4Cl; i al
b21	carbonate(γ)	Nat. witherite. $BaCO_3$	197.34	wh rhomb, 1.529, 1.676, 1.677	4.43	to β, 811	d 1450	0.0022^{18}	0.0065^{100}	s a, NH_4Cl; i al
b22	chlorate	$Ba(ClO_3)_2 \cdot H_2O$	322.25	col monocl, 1.5622, 1.577, 1.635	3.18	414 $(-H_2O$, 120)	$-O$, 250	27.4^{15}	111.2^{100}	sl s al, acet, HCl
b23	*per*chlorate	$Ba(ClO_4)_2$	336.23	col, hex	3.2	505	...	198.5^{25}	562.3^{100}	v s al
b24	*per*chlorate, hydrate	$Ba(ClO_4)_2 \cdot 3H_2O$	390.28	col, hex, 1.533	2.74	d 400	...	198^{33}	s	al 124^{62}
b25	chloride α	$BaCl_2$	208.24	col, monocl; 1.7303, 1.7367, 1.7420	3.856^{24}	tr to cub 962	1560	37.5^{26}	59^{100}	sl s HCl, HNO_3; v sl s al
b26	chloride β	$BaCl_2$	208.24	col, cub	3.917	963	1560	...	...	v sl s al
b27	chloride, hydrate	$BaCl_2 \cdot 2H_2O$	244.27	col, monocl, 1.629, 1.642, 1.658 n_D^{25}	3.097_4^{24}	$-2H_2O$, 113	37.5^{20}		58.7^{100}	sl s HCl, HNO_3; v sl s al
b28	*hypo*chlorite	$Ba(ClO)_2 \cdot 2H_2O$	276.27	col cr.	...	d	...	...	...	...
b29	chlorofluoride	$BaCl_2 \cdot BaF_2$	383.56	col, tetr, 16.40	4.51^{18}	...	...	d	d	i al; s conc HCl, HNO_3
b30	chlorofluoride	$BaPtCl_6H_2O$	653.22	or-yel, monocl	2.868	$-5H_2O$, 70	d	s	...	d a, al; i MeOH, eth
b31	chloroplatinite	$BaPtCl_4 \cdot 3H_2O$	528.27	dk red pr	2.868	$-3H_2O$, 150	...	v s	...	s al
b32	chromate	$BaCrO_4$	253.32	yel, rhomb	4.498^{15}	...	...	0.00034^{16}	0.00044^{28}	s min a
b33	*di*chromate	$BaCr_2O_7$	353.32	red, monocl.	...	...	...	sl s	...	s h conc H_2SO_4
b34	*di*chromate, hydrate	$BaCr_2O_7 \cdot 2H_2O$	389.35	bright red-yel need	...	$-2H_2O$, 120	...	d	...	s conc CrO_3 soln
b35	chromite	$BaO \cdot 4Cr_2O_3$	761.29	grn-blk, hex	5.4^{15}	...	...	i	i	s a, fus carb
b36	citrate	$Ba_3(C_6H_5O_7)_2 \cdot 7H_2O$	916.30	wh powd	...	$-7H_2O$, 150	...	0.0406^{18}	0.0572^{25}	sl s al; s HCl
b37	cyanide	$Ba(CN)_2$	189.37	wh cr powd	...	...	...	80^{14}	...	18^{14} 70% al
b38	cyanoplatinite	$BaPt(CN)_4 \cdot 4H_2O$	508.54	(a) monocl, yel, α 1.6704	a) 2.076	$-2H_2O$, 100	...	3^{16}	25^{100}	i al
...	...	...		(b) grn, rhomb	b) 2.085					
b39	dithionate	$Ba(SO_3)_2 \cdot 2H_2O$	333.48	col,, rhomb or monocl, 1.586, 1.595, 1.607	$4.536^{13.5}$	d 120	...	24.75^{18}	90.9^{100}	sl s al
b40	ethylsulfate	$Ba(C_2H_5SO_4)_2 \cdot 2H_2O$	423.60	wh lust leaf	...	...	...	s	...	sl s al
b41	ferrocyanide	$Ba_2Fe(CN)_6 \cdot 6H_2O$	594.70	yel, monocl	2.666	$-H_2O$, 40	...	0.17^{15}	0.9^{100}	i al
b42	fluogallate	$Ba_3(GaF_6)_2 \cdot H_2O$	797.43	wh cr.	4.06	$-1/2$ H_2O, 110; $-1/2$ H_2O, 230	...	i	...	s HF
b43	fluoride	BaF_2	175.33	col, cub, 1.4741 n_D^{25}	4.89	1355	2137	0.12^{25}	sl s	s a, NH_4Cl
b44	fluoroiodide	$BaF_2 \cdot BaI_2$	566.47	pl	5.21^{18}	...	...	d	d	i al; s conc HCl, conc NHO_3
b45	fluosilicate	$BaSiF_6$	279.41	rhomb need	4.29_4^{21}	d 300	...	0.026^{100}	0.09^{100}	i al; sl s a, NH_4Cl
b46	formate	$Ba(CHO_2)_2$	227.37	col, rhomb, 1.573, 1.597, 1.636	3.21	d	...	27.76^{0}	39.71^{90}	i al, eth
b47	gluconate	$Ba(C_6H_{11}O_7)_2 \cdot 3H_2O$	581.67	pr or rhomb leaf	...	-3 H_2O, 100; d 120	...	$3.3^{15.5}$	...	i al
b48	hydride	BaH_2	139.35	gray cr.	4.21^{0}	d 675	1400(?)	d to $Ba(OH)_2$ $+H_2$	...	d a
b49	hydroxide	$Ba(OH)_2 \cdot 8H_2O$	315.47	col, monocl, 1.471, 1.502, 1.50	2.18^{16}	78	$-8H_2O$, 78	5.6^{15}	94.7^{78}	sl s al; i acet
b50	hyponitrite	$BaN_2O_2 \cdot 4H_2O$	269.40	wh cr powd	2.742^{25}	...	...	...	...	...
b51	iodate	$Ba(IO_3)_2$	487.14	monocl.	4.998	d	...	0.008^{0}	197^{100}	s HNO_3, HCl
b52	iodate, hydrate	$Ba(IO_3)_2 \cdot H_2O$	505.15	col, monocl	4.657^{18}	$-H_2O$, 200	...	v sl s	sl s	s NHO_3, HCl; i al, acet, H_2SO_4
b53	iodide	BaI_2	391.14	col cr.	5.15_4^{25}	740	...	170^{0}	...	al 77^{20}
b54	iodide, hydrate	$BaI_2 \cdot 2H_2O$	427.17	col rhomb, deliq	5.15	$-H_2O$, 98.9; $-2H_2O$, 539; d 740	200^{15}	269^{100}	...	1.07^{15}al; s acet
b55	iodide, hydrate	$BaI_2 \cdot 6H_2O$	499.23	col, hex	...	25.7	...	410^{0}	v s	v s al
b56	laurate	$Ba(C_{12}H_{23}O_2)_2$	535.96	wh leaf cr	...	260	...	$0.008^{15.3}$	0.011^{50}	0.008^{25} al; 0.006^{25} eth

PHYSICAL CONSTANTS OF INORGANIC COMPOUNDS (continued)

No.	Name	Synonyms and Formulae	Mol. wt.	Crystalline form, properties and index of refraction	Density or spec. gravity	Melting point, °C	Boiling point, °C	Solubility, in grams per 100 cc		
								Cold water	Hot water	Other solvents
b57	*l*-malate	$BaC_4H_4O_5$	269.40					0.883^{20}	1.044^{80}	
b58	malonate	$BaC_3H_2O_4{\cdot}H_2O$	257.39	col				0.143^{0}	0.326^{80}	
b59	manganate	$BaMnO_4$	256.27	gray-grn, hex	4.85			v sl s		s a
b60	*per*-manganate	$Ba(MnO_4)_2$	375.20	br-vlt cr	3.77	d 200		62.5^{11}	75.4^{25}	d al
b61	methylsulfate	$Ba(CH_3SO_4)_2{\cdot}2H_2O$	395.55	col effl cr				s		s al
b62	molybdate	$BaMoO_4$	297.27	wh powd	4.65	1480		0.0058^{23}		sl s a
b63	myristate	$Ba(C_{14}H_{27}O_2)_2$	592.06					0.007^{25}	0.010^{50}	0.009^{25} al; 0.003^{25} eth 0.046^{15} MeOH
b64	nitrate	Nitrobarite. $Ba(NO_3)_2$	261.34	col cub, 1.572	3.24^{23}	592	d	8.7^{20}	34.2^{100}	i al; sl s a
b65	nitride	Ba_3N_2	440.00	yel-br	4.783^{25}_{4}		1000 vac	d	d	
b66	nitrite	$Ba(NO_2)_2$	229.34	col, hex	3.23^{23}	d 217		67.5^{20}	300^{100}	sl s al
b67	nitrite, hydrate	$Ba(NO_2)_2{\cdot}H_2O$	247.36	col-yelsh, hex	3.173^{20}	d 115		63^{20}	109.6^{80}	1.6 al; v s HCl; i acet
b68	oxalate	$BaCr_2O_4$	225.35	cr	2.658	d 400		0.0093^{18}	0.0228^{100}	i al; s NH_4Cl, a
b69	oxide	BaO	153.33	col, cub, wh-yelsh powd, 1.98	5.72	1918	*ca* 2000	3.48^{20}	90.8^{100}	s dil a, al; i acet, NH_3
b70	oxide, per-	BaO_2	169.33	wh-gray powd	4.96	450	−O, 800	v sl s	d	s dil a; i acet
b71	oxide, per-, hydrate	BaO_28H_2O	313.45	col, hex	2.292	$-8H_2O$, 100		0.168	d	s dil a; i al, eth, acet
b72	palmitate	$Ba(C_{16}H_{31}O_2)_2$	648.17	wh cr powd		d		0.004^{15}	0.007^{50}	$0.008^{16.5}$ al; 0.001^{15} eth
b73	*hypo*phosphate	$BaPO_3$	216.30	need				sl s		s al; v sl s ac a
b74	*ortho*phosphate di-	$BaHPO_4$	233.31	wh, rhomb, 1.635, 1.617	4.165^{15}	d 410^{710}		0.01—0.02		s a, NH_4Cl
b75	*ortho*phosphate, mono-	$Ba(H_2PO_4)_2$	331.30	tricl.	2.9^{4}			d	d	s a
b76	*ortho*phosphate, tri-	$Ba_3(PO_4)_2$	601.93	wh, cub	4.1^{16}			i	i	s a
b77	*pyro*phosphate	$Ba_2P_2O_7$	448.60	wh, rhomb	3.9^{20}			0.01	sl s	s a, NH_4 salts
b78	*hypo*phosphite	$Ba(H_2PO_2)_2{\cdot}H_2O$	285.32	wh, monocl.	2.90^{17}_{4}	d 100—150		30^{15}	33^{100}	i al
b79	propionate	$Ba(C_3H_5O_2)_2{\cdot}H_2O$	301.49	rhomb, β 1.518		d 300		48^{0}	67.9^{90}	0.08 al
b80	salycylate	$Ba(C_7H_9O_3)_2{\cdot}H_2O$	437.64	wh need				s		
b81	selenate	$BaSeO_4$	280.29	wh, rhomb	4.75	d		0.0118	0.138^{100}	s HCl; i HNO_3
b82	selenide	BaSe	216.29	wh cub disc, n_D, 2.268	5.02			d	d	d HCl
b83	*meta*silicate	$BaSiO_3$	213.41	col, rhomb, 1.673, 1.674, 1.678	4.399	1604		i	d	s HCl
b84	*meta*silicate, hydrate	$BaSiO_3{\cdot}6H_2O$	321.51	rhomb, 1.542, 1.548, 1.548	2.59			0.17	d	
b85	stearate	$Ba(C_{16}H_{35}O_2)_2$	704.28	wh powd				0.004^{15}	0.006^{50}	$0.005^{16.5}$ al, 0.008^{25} al, 0.001^{25} eth
b86	succinate	$BaC_4H_4O_4$	253.40	wh powd				0.421^{0}	0.237^{80}	sl s al
b87	sulfate	Nat. barite, prec. blanc fixe. $BaSO_4$	233.39	wh, rhomb (monocl), 1.637, 1.638, 1.649	4.50^{15}	1580	tr 1149 monocl	0.000222^{50} 0.000246^{25}	0.000336^{50} 0.000413^{100}	0.006 s 3% HCl; sl s H_2SO_4
b88	*peroxydi*sulfate	$BaS_2O_3{\cdot}4H_2O$	401.51	wh, monocl.		d		52.2^{0}	d	d al
b89	sulfide, hydro-	$Ba(HS)_2{\cdot}4H_2O$	275.53	yel, rhomb		d 50		s		i al
b90	sulfide, mono-	BaS	169.39	col, cub, n_D 2.155	4.25^{15}	1200		d	d	i al
b91	sulfide, tetra-	$BaS_4{\cdot}H_2O$	283.59	red or yel, rhomb	2.988	d 300		41^{15}	v s	i al, CS_2
b92	sulfide, tri-	BaS_3	233.51	yel-grn cr		d 554		s	s	
b93	sulfite	$BaSO_3$	217.39	col, cub hex		d		0.02^{20}	0.002^{80}	v s HCl
b94	tartrate	$BaC_4H_4O_6{\cdot}H_2O$	303.42	wh cr.	$2.980^{20.8}$			0.026^{18}	0.056^{30}	0.032^{18} al
b95	tellurate	$BaTeO_4{\cdot}3H_2O$	382.97	volum wh	4.2^{200}	d >200		sl s	sl s	s HCl, NHO_3
b96	*pyro*tellurate, hydrogen	$Ba(HTe_2O_7)_2{\cdot}H_2)$	891.75	volum ppt; hot:yel; cold: wh				s	s	s a
b97	telluride	BaTe	264.93	yel-wh, cub, disc, n_D 2.440	5.13					d a
b98	thiocarbonate	$BaCS_2$	245.42	yel, hex		d		1.08^{0}	d	i al
b99	thiocyanate	$Ba(SCN)_2{\cdot}2H_2$	289.52	need, deliq	2.286^{18}	$-H_2O$, 160		4.3^{20}	s	35^{20} al
b100	thiosulfate	BaS_2O_3	249,45	wh, rhomb		d220		0.2		
b101	thiosulfate, hydrate	$BaS_2O_3{\cdot}H_2O$	267.46	wh, rhomb	3.5^{18}	d 100		0.208^{20}		i al
b102	titanate	$BaTiO_3$	233.21	tetr and hex, 2.40	tetr 6.017, hex 5.806					
b103	tungstate	$BaWO_4$	385.18	col, tetr	5.04			sl s	sl s	d a
b104	*pyro*vanadate	$Ba_2V_2O_7$	488.54	wh cr.		863				
b105	**Beryllium**	Glucinum, Be	9.01218	grey met, hex	1.85^{20}	1278 ± 5	$2970^{(5\ mm)}$	i	sl s, d	s dil a, alk; i Hg
b106	acetate	$Be(C_2H_3O_2)_2$	127.10	col pl.		d 300		i		i al, eth, CCl_4
b107	acetate, basic	$Be_4O(C_2H_3O_2)_6$	406.32	oct	1.36^{4}	284	331	sl d	d	s chl, ac a; sl s al, eth
b108	acetate propionate, basic	$Be_3O(C_2H_3O_2)_3{\cdot}(C_3H_5O_2)_3$	448.40			127	330			
b109	aluminate	Nat. chrysoberyl. $BeAl_2O_4$	126.97	rhomb, 1.747, 1.748, 1.757	3.76	1870				i a
b110	aluminum silicate	Nat. beryl. $Be_3Al_2(SiO_3)_6$	537.50	transp, hex, col, 1.580, 1.547	2.66	1410 ± 100				i a
b111	aluminum silicate	Nat. euclase. $Be_2Al_2(SiO_4)_2{\cdot}(OH)_2$	290.17	monocl, 1.652, 1.655, 1.671	3.1					
b112	benzenesulfonate	$Be(C_6H_5O_3S)_2$	323.34	monocl.				v s	v s	v s al, acet, ac a; i eth, bz, CS_2, CCl_4
b113	*ortho*borate, basic	Nat. hamgergite, $Be_2(OH)BO_3$	93.84	rhomb, 1.560, 1,591, 1.631	2.35					
b114	bromide	$BeBr_2$	168.83	wh need, deliq	3.465^{25}	490 ± 10 subl	520	s	v s	s al, eth; i bz; 18.56 pyr

PHYSICAL CONSTANTS OF INORGANIC COMPOUNDS (continued)

No.	Name	Synonyms and Formulae	Mol. wt.	Crystalline form, properties and index of refraction	Density or spec. gravity	Melting point, °C	Boiling point, °C	Solubility, in grams per 100 cc		
								Cold water	Hot water	Other solvents
b115	butyrate, basic	$Be_4O(C_4H_7O_2)_6$	574.64				239^{19}			
b116	carbide	Be_2C	33.03	yel, hex	1.90^{15}	>2100 d		d	d	s a; d alk
b117	carbonate, basic	$BeCO_3 + Be(OH)_2$	112.05	wh powd				i	d	s a, alk
b118	chloride	$BeCl_2$	79.92	col need, deliq	1.899^{25}	405	520 (488)	v s	v s, d	v s al, eth, pyrl sl s bx, chl, CS_2, i NH_3, acet
b119	fluoride	BeF_2	47.01	col, amorph, <1.33	1.986^{25}	subl 800		α	α	sl s al; s H_2SO_4
b120	hydride	BeH_2	11.03	wh cr.		d 125		d	d	i eth, tol
b121	iodide	BeI_2	262.82	col need	4.325^{26}	510±10	590	d	d	s al, eth, CS_2
b122	nitrate	$Be(NO_3)_2{\cdot}3H_2O$	187.97	wh-yel cr, deliq	1.557	60	142	v s	v s	v s al
b123	nitride	Be_2N_2	55.05	col, cub		2200± 100	d 2240	d	d	d a, conc alk; i al
b124	oxalate	$BeC_2O_4{\cdot}3H_2O$	151.08	rhomb, β 1.487		$-H_2O$, 100; $-3H_2O$, 220	d 350	38.22^{26}		
b125	oxide	Nat. bromillite, BeO	25.01	wh hex, 1.719, 1.733	3.01	2530 ± 30	*ca*3900	0.00002^{30}		a conc H_2SO_4, fus KOH
b126	oxide	$BeO{\cdot}xH_2O$		wh amorph powd or gel		d		i	i	s a, alk, $(NH_4)_2CO_3$
b127	2,4-epntanedione deriv	Beryllium acetylacetonate $Be(C_5H_7O_3)_2$	207.23	wh, monocl	1.168^{4}	108	270	sl s	d	s al, eth, a
b128	*ortho*phosphate	$Be_3(PO_4)_2{\cdot}3H_2O$	271.03			$-H_2O$, 100		s	s	s ac a
b129	propionate basic	$Be_4O(C_3H_6O_2)_6$	490.48			120				
b130	selenate	$BeSeO_4{\cdot}4H_2O$	224.03	col, rhomb, 1.466, 1.501, 1,503	2.03	$-2H_2O$, 100; $-4H_2O$, 300		56.7^{25}	s	
b131	*di*silicate	Nat. bertrandite. $Be_4Si_2O_7(OH)_2$	238.23	rhomb, 1.591, 1.605, 1.604	2.6					
b132	*ortho*silicate	Nat. phenazite, Be_2SiO_4	110.11	tricl, col, 1.654, 1.670	3.0					i a
b133	stearate (com'l)	$Be(C_{18}H_{35}O_2)_2$	575.96	wh, waxy		45		i	i	i al; s eth, CCl_4
b134	sulfate	$BeSO_4$	105.07		2.443	d 550—600		i	d to $BeSO_4$, $4H_2O$	
b135	sulfate, hydrate	$BeSO_4{\cdot}H_2O$	177.13	col, tetr, 1,472, 1.440	$1.713^{10.5}$	$-2H_2O$, 100	$-4H_2O$, 400	42.4^{26}	100^{100}	sl a conc H_2SO_4; i al
b136	sulfide	BeS	41.07	reg	2.36			d	d	
b137	**Bismuth**	Bi	208.9804	rhomb silver-wh or redsh met	9.80	271.3	1560 ± 5^{760}	i	i	s h H_2SO_4, HNO, aq, reg; sl s h HCl
b138	acetate	$Bi(C_2H_3O_2)_3$	386.11	wh cr.		d		i	i	s ac a
b139	*ortho*arsenate	$BiAsO_4$	347.90	wh, monocl, 2.14,2.15, 2.18	7.14			i	i	sl s h conc HNO_3
b140	benzoate	$Bi(C_7H_5O_2)_3$	572.33	wh powd						s a; i eth
b141	bromide, tri-	$BiBr_3$	448,69	yel cr powd, deliq	5.72^{25}_{4}	218	453	d to BiOBr	d	i al; s HCl, HBr, eth; v s liq NH_3
a142	carbonate, basic	Bismuth oxycarbonate, $Bi_2O_2CO_3$	509.97	wh powd	6.86		d	i	i	s a
b143	chloride, tetra-	$BiCl_4$	350.79	col cr.		226				
b144	chloride, tri-	$BiCl_3$	315.34	wh cr, deliq.	4.75^{25}_{4}	230—232	447	d to BiOCl	d	s a, al, eth, acet
b145	*di*chromate, basic	$(BiO)_2{\cdot}Cr_2O_7$	665.95	yel-or-red cr				i	i	s a; i alk
b146	citrate	$BiC_6H_5O_7$	398.08	wh cr.	3.458	d		sl s	sl s	sl s al; a NH_4OH
b147	fluoride, tri-	BiF_3	265.98	gray cr, cub, 1.74	5.32^{20}	727		i		s inorg a; i liq NH_3, al
b148	gallate, basic	Com'l dermatol. $Bi(OH)_2{\cdot}C_7H_5O_5$	412.11	yel, amorph.		d		i		i al, eth
b149	hydroxide	$Bi(OH)_3$	260.00	wh amorph powd	4.36	$-H_2O$, 100; d 415	$-1^1/_2\ H_2O$, 400	0.00014	d	s a; i or sl s conc alk
b150	iodate	$Bi(IO_3)_3$	733.69	wh				i		i HNO_3
b151	iodide, di-	BiI_2	762.79	red need, rhomb		d 400	subl vac	s		s al, MeOH
b152	iodide, tri-	BiI_3	589.69	redsh, hex.	5.778^{15}	408	*ca* 500	i	d	3.5^{20}al; i a; sl s NH_3; s HCl, HI
b153	lactate, *dl*	$Bi(C_6H_9O_6){\cdot}7H_2O$	512.22	pr, need				v sl s		i al
b154	molybdate	$Bi_2(MoO_4)_3$	485.07	col tricl, sl hygr	2.83	d 30	$-5H_2O$, 80	d	d	v s HNO_3; s a, gluc; i al, 42^{19} acet
b156	nitrate, basic	$BiONO_3{\cdot}H_2O$	305.00	hex leaf	4.928^{18}	$-H_2O$, 105	$-HNO_3$, 260	i	i	s a; i al
b157	oxalate	$Bi_2(C_2O_4)_3{\cdot}7H_2O$	808.13	wh powd		$-6H_2O$, *ca*130		d		s inorg a; i al, eth
b158	oxide, mono-	BiO	224.98	dk-gray powd.	7.15^{19}	d *ca*180 (Bi_2O_3)		sl d	d	d dil a; s dil KOH
b159	oxide pent-	Bi_2O_5	497.96	dk red or br.	5.10	$-O$, 150	$-2O$, 357	i	i	s KOH
b160	oxide, tetr-	$Bi_2O_4{\cdot}2H_2O$	517.99	br powd	5.6	$-H_2O$, 110	$-2H_2O$, 180	i	i	s a
b161	oxide, tri-	Bi_2O_3	465.96	yel, rhomb	8.9	825 ± 3	1890 (?)	i	i	s a
b162	oxide tri-	Bi_2O_3	465.96	gray-blk, cub	8.20	tr 704		i	i	s a
b163	oxide, tri-	Bi_2O_3	465.96	wh-lt yel, rhomb, 1.91	8.55	860		i	i	sl s a
b164	oxybromide	$BiOBr$	304.88	col cr or wh powd	8.082^{15}	d red ht		i	i	i al; s a
b165	oxychloride	$BiOCl$	260.43	wh cr or powd, 2.15	7.72^{15}	red ht.		i	i	s a; i NH_3, tart a, acet
b166	oxyfluoride	$BiOF$	243.98	wh cr or powd	7.5^{20}_{20}	d red ht		i		s a
b167	oxyiodide	$BiOI$	351.88	red cr, tetr	7.922	d red ht		i	i	s a; i al, $CHCl_3$
b168	*ortho*phosphate	$BiPO_4$	303.95	wh, monocl	6.323^{15}	d		i	i	s HCl; i al, dil HNO_3

No.	Name	Synonyms and Formulae	Mol. wt.	Crystalline form, properties and index of refraction	Density or spec. gravity	Melting point, °C	Boiling point, °C	Solubility, in grams per 100 cc		
								Cold water	Hot water	Other solvents
b169	propionate, basic	$BiOC_3H_5O_2$	298.05	wh powd	...	...	...	i	...	v s dil HCl; i al
b170	salicylate, basic	Bismuth subsalicylate, $Bi(C_7H_5O_3)_3 \cdot Bi_2O_3$ (appr)	1086.28	wh micr cr (variable comp)	...	...	...	i	...	s a, alk; i al, eth
b171	selenide, tri-	Nat. guanajuatite, Bi_2Se_3	654.84	blk, rhomb	6.82	710	d	i	...	i alk
b172	silicate	Nat. eulytite, $2Bi_2O_3 \cdot 3SiO_2$	1112.17	yel, cub, 2.05	6.11	...	...	i	i	s d HCl, HNO_3
b173	sulfate	$Bi_2(SO_4)_3$	706.13	wh need	5.08^{15}	d 405	...	d	d	s a
b174	mono-sulfide	BiS	241.04	dk gray powd.	7.6—7.8	680 (in CO_2)	d	v sl s	...	...
b175	sulfide, tri-	Nat. bismuthinite, bismuthglance, Bi_2S_4	514.14	br-blk, rhomb, 1.340, 1.456, 1.459	7.39	d 685	...	0.000018^{18}	...	s HNO_3; i dil sl
b176	tartrate	$Bi_2(C_4H_4O_6)_3 \cdot 6H_2O$	970.27	wh powd	2.595^{25}_{5}	$-3H_2O$, 105	...	i	i	s a, alk, i al
b177	tellurate	Montanite. $Bi_2TeO_6 \cdot 2H_2O$	677.59	biaxial, β: 2.09	3.79	...	...	...	...	...
b178	telluride, tri-	Nat. tetradymite. Bi_2Te_3	800.76	gray, rhbdr	7.7^{20}_{4}	573	...	...	...	d HNO_3
b179	vanadate	Nat. pucherite, $Bi_2O_3 \cdot V_2O_5$	647.84	red-grn, rhomb, 2.41, 2.50, 2,51 (Li)	6.25^{25}	...	...	...	...	...
b180	**Bismuthic acid**	$HBiO_3$	257.99	red	5.75	$-H_2O$, 120	−2O, 357	i	i	s a, KOH
b181	**Borazole**	$B_3N_3H_6$	80.50	col liq	1^{-65}, 0.8614^{06}	−58	55	sl d	d	...
b182	**Boric Acid**, meta-	HBO_2	43.82	wh cr, cub, 1.691	2.486	236 ± 1	...	v sl s	sl s	...
b183	ortho-	Boracic acid, H_3BO_3	61.83	col, tricl, 1.337, 1.461, 1.462	1.435^{15}	169 ± 1 tr to HBO_2	$-1^1/_2H_2O$, 300	6.35^{30}	27.6^{100}	28^{20} glyc; 0.0078 eth; 5.56 al; 20. 20^{25} MeOH; 1.92^{25} liq NH_3; sl s acet
b184	tetra- (pyro-)	$H_2B_4O_7$	157.25	vitr or wh powd	...	...	...	s	s	s al
b185	fluo-	HBF_4	87.81	col liq	...	...	d 130	∞	∞	s al
b186	**Borinoaminoborine**	B_2H_7N	42.68	col liq	...	−66.5	76.2	...	...	s triborine triamine
b187	**Boron**	B	10.81	yel monocl or br amorph powd	2.34, 2.37 amorph	2300	2550	i	i	v sl s HNO_3
b188	arsenate	$BAsO_4$	149.73	wh cr tetrag, 1.681, 1.690	3.64	subl *ca* 700	...	v sl s	1.4^{100}	i al; s inorg a
b189	bromide, tri-	BBr_2	250.52	col fum liq, $n_D^{16.3}$ 1.5312	$2.6431^{18.4}_{4}$	−46	91.3 ± 0.25	d	...	s al, CCl_4
b190	bromide, di-, iodide,	BBr_2I	297.52	col liq	...	...	125	d	d	...
b191	bromide, mono-, diiodide	$BBrI_2$	344.52	col liq	...	...	180	d	d	...
b192	(di-) bromide, mono-, pentahydride	B_2H_4Br	106.56	col gas	...	−104	*ca* 10	hyd to HBO_2 HBr + H_2	...	...
b193	(tetra-) carbide	B_4C	55.26	blk rhbdr	2.52	2350	>3500	i	i	i a; s fus alk
b194	chloride, tri-	BCl_3	117.17	col fum liq, $1.419^{5.7}$ α line H_2	1.349^{11}_{4}	−107.3	12.5	...	d to HCl + H_3BO_3	d al
b196	fluoride dihydrate	$BF_3 \cdot 2H_2O$	103.84	col liq; n_{HE}^{20}	1.6316^{20}_{4}	6	...	d	d	s eth, dioxan
b197	hydride	Diborane, borethane, B_2H_6	27.67	col gas	liq; 0.447^{-11}	−1.65.5	−92.5	sl s d to H_3BO_3 + H_2	...	d 1.6^0 al; s NH_4OH, conc H_2SO_4
b198	hydride	Dihydrotetraborane, borobutane. B_4H_{10}	53.32	col gas, pois	0.56^{-35}	−120.8	16	sl s, d	...	s bz; d al
b199	hydride	Pentaborane. B_5H_9	63.12	col liq	0.66^0	−46.82	58.4	d	...	...
b200	hydride	Hexaborane. B_6H_{10}	74.94	col liq	0.69^0	−65	0^{72}	d	...	...
b201	hydride	Decaborane. $B_{10}H_{14}$	122.21	wh cr.	0.94^{25}	99.5	213	sl s	d	v s CS_2; s al, eth bz
b202	iodide, tri-	BI_3	391.52	col pl, hygr	3.35^{50}	49.9	210	d	d	d al; v s CS_2, bz, CCl_4
b203	nitride	BN	24.82	wh, hex	2.25	subl *ca* 3000	...	i	sl d	sl s h a
b204	oxide	B_2O_3	69.92	rhomb cr, 1.64, 1.61	2.46 ± 0.01	45 ± 2	*ca* 1860	sl s	s	...
b205	oxide glass	B_2O_3	69.62	col, vitr 1.485	1.812^{25}_{4}	*ca* 450	...	1.1^0	15.7^{100}	s al, a
b206	phosphide	BP	41.78	maroon powd.	...	ign 200	i	i	i all solv	
b207	*tri*selenide	B_2Se_3	258.50	yel-gray powd	...	...	...	d	d	...
b208	(hexa-) silicide	B_6Si	92.95	blk cr	2.47	...	...	i	...	s HNO_3; d H_2SO_4; i KOH
b209	(tri-) silicide	B_2Si	60.52	blk rhomb	2.52	...	...	i	...	sl s HNO_3; d H_2SO_4, KOH
b210	sulfide, penta-	B_2S_5	181.92	col, tetrag	1.85	390	...	d	d	d al
b211	sulfide, tri-	B_2S_3	117.80	wh cr or vitr	1.55	310	...	d	...	sl s PCl_3, SCl_2; d al
b212	**Borotungstic acid**	$H_5BW_{12}O_{40} \cdot 30H_{20}$	3402.48	tetr, cr	3	45—51	...	s	...	s al, eth
b213	**Bromic acid**	$HBrO_3$	128.91	known in sol only, col or yelsh	...	d 100	...	v s	d	...
b214	**Bromine**	Br_2	159.808	dk red liq, 1.661	2.928^{59}, 3.119^{20}	−7.2	58.78	4.17^0. 3.58^{20}	3.52^{50}	v s al, eth, chl, CS_2
b215	azide	Bromoazide, BrN_3	121.92	cr, red liq	...	*ca* 45	exp	...	...	s eth, KI; sl s bz, ligr
b216	chloride	BrCl	115.36	red-col liq or gas	...	*ca* −66 d 10	*ca* 5	s d	...	s eth, CS_2
b217	fluoride, mono-	BrF	98.90	red-br gas	...	d −33	−20	...	...	...
b218	fluoride, penta-	BrF_5	174.90	col liq	2.466^{25}	−61.3	40.5	d	d	...
b219	fluoride, tri-	BrF_3	136.90	col-gray-yel liq.	2.49^{135}	(−2) 8.8	135	d viol. to O_2, HOBr, HF, $HBrO_3$	...	d alk

PHYSICAL CONSTANTS OF INORGANIC COMPOUNDS (continued)

No.	Name	Synonyms and Formulae	Mol. wt.	Crystalline form, properties and index of refraction	Density or spec. gravity	Melting point, °C	Boiling point, °C	Solubility, in grams per 100 cc: Cold water	Hot water	Other solvents
b220	hydrate	$Br_2 \cdot 10H_2O$	339.96	red oct	1.49	d 6.8		s		
b221	oxide, di-	BrO_2	111.91	lt yel		d 0				
b222	oxide, mon-	Br_2O	175.81	dk br		−17 to −18				s, d CCl_4
b223	(tri-) oxide, oct-	Br_3O_8(or $Br_3O_8)n$	367.71	wh		stable at −40				
	Bromauric acid									
b224	**Bromauric acid**	$HAuBr_4 \cdot 5H_2O$	607.67	red-br cr		27.		v s		s al
b225	**Bromous acid, hypo-**	HBrO	96.91	known only in soln, col-yel		40 (vac)		s	s d	s al, eth, chl
b226	**Bromoplatinic acid**	$H_2PtBr_6 \cdot 9H_2O$	838.66	red monocl, deliq		d −100		v s	v s	
c1	**Cadmium**	Cd	112.41	hex silv-wh malleable met	8.642	320.9	765	i	i	s a, NH_4NO_3, h H_2SO_4
c2	acetate	$Cd(C_2H_3O_2)_2$	230.50	col	2.341	256	d	v s		s MeOH
c3	acetate, hydrate	$Cd(C_2H_3O_2)_2 \cdot 2H_2O$	266.53	col monocl, odor ac a	2.01	$-H_2O$, 130		v s	v s	v s al
c4	amide	$Cd(NH_2)_2$	144.46		3.05^{25}	d 120				
c5	ammonium chloride	$CdCl_2 \cdot NH_4Cl$	236.81	col need, rhomb	2.93	289		33.45^{16}	$43.99^{63.8}$	s al, MeOH
c6	ammonium sulfate	$Cd(NH_4)_2(SO_4)_2 \cdot 6H_2O$	448.69	col monocl pr	2.061_4^{20}	100 d $-H_2O$		s	s	
c7	arsenate, hydrogen	$CdHAsO_4 \cdot H_2O$	270.35		4.164_4^{15}	>120				
c8	arsenide	Cd_3As_2	487.07	dk gray cub	6.21_4^{15}	721		i	i	sl s HCl; s HNO_3; i aq reg
c9	benzoate	$Cd(C_7H_5O_2)_2 \cdot 2H_2O$	390.67					3.34^{20}		sl s al
c10	borate	$Cd(BO_3)_2 \cdot H_2O$	248.04	wh rhomb	3.758	d		125^{17}		i al
c11	borotungstate	$Cd_5(BW_{12}O_{40})_2 \cdot 18H_2O$	6600.30	yel tricl		75		1250^{19}	v s	
c12	bromide	$CdBr_2$	227.22	yel cr.	5.192^{25}	567	863	57^{10}	162^{104}	26.6^{15} sl; 0.4^{15} eth; s HCl; 1.6^{18} acet
c13	bromide, tetrahydrate	$CdBr_2 \cdot 4H_2O$	344.28	sm wh need, effl		tr 36		121^{10}	v s	25 al; s acet; sl s eth
c14	carbonate	$CdCO_3$	172.42	wh, trig	4.258^4	d <500		i	i	s a, KCN, NH_4 salts; i NH_3
c15	chlorate	$Cd(ClO_3)_2 \cdot 2H_2O$	315.34	col pr, deliq	2.28^{18}	80		298^0	487^{65}	s al, a, acet
c16	chloride	$CdCl_2$	183.32	col, hex	4.047^{25}	568	960	140^{20}	150^{100}	1.52^{15} al; $1.7^{15.5}$ MeOH; i acet, eth
c17	chloride	$CdCl_2 \cdot 2\ ^1/_2\ H_2O$	228.35	col monocl, 1.6513	3.327	tr 34		168^{20}	180^{100}	2.05^{15} MeOH; sl s al
c18	chloroacetate, di-	$Cd(C_2HCl_2O_2)_2 \cdot 2H_2O$	386.29	need	2.132^{15}					
c19	chloroacetate, mono-	$Cd(C_2H_2ClO_2)_2 \cdot 6H_2O$	407.48		1.942^{25}					
c20	chloroacetate, tri-	$Cd(C_2Cl_3O_2)_2 \cdot 1^1/_2H_2O$	464.19	rhomb	2.093^{25}					
c21	chloroplatinate	$CdPtCl_6 \cdot 3H_2O$	574.25	yel trig need	2.882	$-H_2O$, 170, d		s	s	
c23	chromite	$CdCr_2O_4$	280.40	grn to blk, cub	5.79^{17}			i	i	i a
c24	cyanide	$Cd(CN)_2$	167.45	cr		dec >200		1.7^{15}	s	s a, KCN, NH_4OH; i al
c25	ferrocyanide	$Cd_2Fe(CN)_6 \cdot xH_2O$						i	i	s HCl
c26	fluogallate	$[Cd(H_2O)_6] \cdot [GaG_5H_2O]$	403.23	col cr 1.45	2.79	$-5H_2O$, 110		v s		
c27	fluoride	CdF_2	150.41	wh cub, 1.56	6.64	1100	1758	4.35^{25}		s a, HF; i al, NH_3
c28	fluosilicate	$CdSeF_6 \cdot 6H_2O$	362.58	col hex.				s	s	s 50% al
c29	formate	$Cd(CHO_2)_2 \cdot 2H_2O$	238.47	monocl.	2.44	d		v s		
c30	fumarate	$CdC_4H_2O_4$	226.47					0.9^{30}		
c31	hydroxide	$Cd(OH)_2$	146.42	wh, trig or amorph	4.79^{15}	d 300		0.00026^{26}		s a, NH_4 salts; i alk
c32	iodate	$Cd(IO_3)_2$	462.22	wh cr.	6.43	d		s		s NHO_3, $NH_4 \cdot OH$
c33	idodie	CdI_2	366.22	grn-yel powd	5.670_4^{30}	387	796	86.2^{25}	125^{100}	110.5^{20} al; 41^{25} acet; sl s NH_3; 206.7^{25} MeOH
c34	lactate	$Cd(C_3H_5O_3)_2$	290.55	need				10	12.5	i al
c35	maleate	$Cd(C_4H_2O_4) \cdot 2H_2O$	262.50					0.66^{30}		
c36	*per*manganate	$Cd(MnO_4)_2 \cdot 6H_2O$	458.37		2.81	d 95		v s	v s	
c37	molybdate	$CdMoO_4$	272.35	yel pl.	5.347			sl s		s a, NH_4OH, KCN
c38	nitrate	$Cd(NO_3)_2$	236.42	col		350		109^0	326^{50} 682^{100}	v s a; s et ac
c39	nitrate, tetra-hydrate	$Cd(NO_3)_2 \cdot 4H_2O$	308.48	wh pr, need, hygr	2.455_4^{17}	59.4	132	215		s al, NH_3; i HNO_3
c40	nitrocovaltate (III)	Cadmium cobaltinitrite. $Cd_3[Co(NO_2)_6]_2$	1007.16	yel		d 175		sl s	v s	d a, alk, org solv
c41	oxalate	CdC_2O_4	200.43	col cr.	3.32_4^{18}	d 340				s a; i al
c42	oxalate, trihydrate	$CdC_2O_4 \cdot 3H_2O$	254.48	col cr.		d		0.005^{18}	0.009	
c43	oxide	CdO	128.41	br, amorph	6.95	>1500	d 900-1000	i	i	s a, NH_4 salts; i alk
c44	oxide	CdO	128.41	br cub, 1.49 (Li)	8.15	>1500	subl 1559	i	i	s a, NH_4 salts; i alk
c45	*ortho*phosphate	$Cd_3(PO_4)_2$	527.17	col amorph		1500		i		s a, NH_4 salts
c46	*pyro*posophate	$Cd_2P_2O_7$	398.76	wh cr leaf	4.965^{15}	above red heat		sl s	s	s a, NH_3
c47	phosphate, dihydrogen	$Cd(H_2PO_4)_2 \cdot 2H_2O$	342.42	col, tricl	2.74_4^{15}	d 100				i al, eth; s HCl
c48	phosphide	Cd_3P_2	399.18	grn, tetr need	5.60	700				s d HCl; s exp conc HNO_3
c49	potassium cyanide	$Cd(CN)_2 \cdot 2KCN$	294.68	col, glossy, oct.	1.847			33.3	100	i al
c50	potassium sulfate	$CdK_2(SO_4) \cdot 2H_2O$	322.69	tricl col tab	2.922^{16}			42.89^{16}	47.40^{40}	
c51	salicylate	$Cd(C_7H_6O_3)_2 \cdot H_2O$	404.66	wh need				sl s	s	s al, eth, glycerol a, NH_4OH
c52	selenate	$CdSeO_4 \cdot 2H_2O$	291.40	rhomb	3.63	$-1H_2O$, 100; $-2H_2O$, 170		v s		
c53	selenide	CdSe	191.37	grn-br or red powd, hex	5.81_4^{15}	>1350		i		d a
c54	*meta*silicate	$CdSiO_3$	188.49	col. rhomb, 1.739	4.93	1242		v sl s		

No.	Name	Synonyms and Formulae	Mol. wt.	Crystalline form, properties and index of refraction	Density or spec. gravity	Melting point, °C	Boiling point, °C	Solubility, in grams per 100 cc: Cold water	Hot water	Other solvents
c55	sulfate	$CdSO_4$	208.47	wh, rhomb	4.691_4^{20}	1000		75.5^0	60.8^{100}	i al, acet, NH_3
c56	sulfate, hydrate	$CdSO_4 \cdot H_2O$	226.48	monocl	3.79^{20}	tr 108		s	s	i al
c57	sulfate, hydrate	$CdSO_4 \cdot 7H_2O$	334.57	col. monocl	2.48	tr 4		s		i al
c58	sulfate hydrate	$3CdSO_4 \cdot 8H_2O$	769.53	col. monocl, 1.565	3.09	tr 41.5		113^0	s	
c59	sulfide	Nat. greenockite, CdS	144.47	yel-or hex, 2.506, 2.529	4.82	1750^{100atm}	subl in N_2, 980	0.00013^{18}	colloid	s a; v sl s NH_4OH
c60	sulfite	$CdSO_3$	192.47	cr		d		sl s		i al; s a, NH_4OH
c61	tartrate	$CdC_4H_4O_4$	260.48	wh cr powd				sl s		s a, NH_4OH
c62	telluride	CdTe	240.01	blk, cub	5.850^{15}	1121	1091	i		i a; d HNO_3
c63	tungstate	$CdWO_4$	360.26	yel cr				0.05		s NH_4OH
	Cadmium complexes									
c64	tetramminecadmium *per*rhenate	$[Cd)NH_3)_4](ReO_4)_2$	680.94		3.714_4^{25}					0.037 conc NH_4OH
c65	tetrapyridine cadmiumfluosilicate	$[Cd(C_5H_5N)_4]SiF_6$	570.89	wh, tricl	2.282					
c66	**Calcium**	Ca	40.08	silv wh soft met, cub	1.54	839 ± 2	1484	d to H_2 + $Ca(OH)_2$	d	s a, liq NH_3; sl s al; i bz
c67	acetate	$Ca(C_2H_3O_2)_2$	158.17	col cr; 1.55, 1.56, 1.57		d		37.4^0	29.7^{100}	sl s al
c68	acetate, dihydrate	$Ca(C_2H_3O_2)_2 \cdot 2H_2O$	194.20	col need		$-1H_2O$, 84		34.7^{20}	33.5^{50}	
c69	acetate, monohydrate	$Ca(C_2H_3O_2)_2 \cdot H_2O$	176.18	col need		d		43.6^0	34.3^{100}	sl s al
c70	aluminate	$CaAl_2O_4$ (or $CaO \cdot Al_2O_2$)	158.04	wh monocl, tricl or rhomb; 1.643, 1.665, 1.663	2.981^{25}	1600		d		s HCl; i HNO_3, H_2SO_4
c71	(tri-)aluminate	$Ca_3Al_2O_6$ (or $3CaO \cdot Al_2O_2$)	270.20	wh, cub, 1.710	3.038^{25}	d 1535		i		
c72	(tri-)aluminate hexahydrate	$3CaO \cdot Al_2O_3 \cdot 6H_2O$	378.29	col, oct, 1.603	2.52^{20}	d 700—800		d		
c73	aluminosilicate	$2CaO \cdot Al_2O_3 \cdot SiO_3$	274.20	col, tetr, 1.669, 1.658	3.048	1590 ± 2				d a
c74	aluminosilicate	Nat. anorthite. $CaAl_3 \cdot Si_2O_6$ (or $CaO \cdot Al_2O_3 \cdot 2SiO_2$)	278.21	wh, tricl, 1.5832	2.765	1551				
c75	*ortho*arsenate	$Ca_3(AsO_4)_2$	398.08	col amorph powd	3.620	1.455		0.013^{25}		
c76	arsenate, trihydrate	Nat. haidingerite, $2CaO \cdot As_2O_5 \cdot 3H_2O$	396.04	col, rhomb, 1.590, 1.602, 1.638	2.967					
c77	arsenide	Ca_3As_2	270.08	red cr	3.031^{25}	d		d	d	d a; s h HNO_3
c78	azide	$Ca(N_3)^2$	124.12	col, rhomb, hyg		$-3H_2O$, 110; exp 144-156		38.1^0	45^{15}	d a; s h HNO_3 0.211^{16} al; i eth
c79	benzoate	$Ca(C_7H_5O_2)_2 \cdot 3H_2O$	336.36	col, rhomb	1.436	$-3H_2O$, 110		2.7^0	8.3^{80}	
c80	*meta*borate	$Ca(BO_2)_2$	125.70	cil, flat rhomb pr, 1.550, 1.660, 1.680		1154		sl s		s a, NH_4 salts; sl s ac a
c81	*meta*borate, hexahydrate	$Ca(BO_2)_3 \cdot 6H_2O$	233.79	col, tetr, 1.520, 1.502	1.88			0.25^{20}		
c82	*tetra*borate	CaB_4O_7	195.32	readily vitrified		986				
c83	boride	CaB_6	104.94	blk, cub	2.3^{30}	2235		i	i	s HNO_3; sl s conc H_2SO_4
c84	bromide	$CaBr_2$	199.89	col, rhomb need, deliq	3.353^{25}	sl d 730	806—812	142^{30}	312^{106}	s al, acet, a; sl a NH_3, MeOH
c85	bromate	$Ca(BrO_3)_2 \cdot H_2O$	313.90	monocl cr	3.329	$-H_2O$, 180		v s	v s	
c86	bromide, hexahydrate	$CaBr_2 \cdot 6H_2O$	307.98	col, hex cr	2.295	38.2	149	594^0	1360^{25}	s al, acet a
c87	butyrate	$Ca(C_4H_7O_2)_2 \cdot 3H_2O$	268.32	col cr				s	sl s	
c88	carbide	CaC_2	64.10	col, tetr, 1.75	2.22	stab 25—447	2300	d	d	
c89	carbonate	Nat. aragonite, $CaCO_3$	100.09	col, rhomb, 1.530, 1.681, 1.685	2.930	tr to calcite 520	d 825	0.00153^{25}	0.00190^{75}	s a, NH_4Cl
c90	carbonate	Nat. calcite. $CaCO_3$	100.09	col, rhomb or hex. 1.6583, 1.4864	2.710^{18}	1339^{1025}	d 898.6	0.0014^{25}	0.0018^{75}	s a, NH_4Cl
c91	carbonate, hexahydrate	$CaCO_3 \cdot 6H_2O$	208.16	col, monocl, 1.460, 1.535, 1.545	1.771^0					
c92	chlorate	$Ca(ClO_3)_2$	206.98	wh cr, hyg		340 ± 10 (− some O)		s	s	s al, acet
c93	chlorate, dihydrate	$Ca(ClO_3)_2 \cdot 2H_2O$	243.01	wh-yelsh, rhomb. or monoc, deliq	2.711	$-H_2O$, 100		177.8^8	v s	s al, acet
c94	*per*chlorate	$(CaClO_4)_2$	238.98	col cr	2.651	d 270		188.6^{25}	v s	166.2_{26}^{25} al; 237.4 MeOH
c95	chloride	$CaCl_2$	110.99	col, cub, deliq 1.52	2.15_4^{25}	782	>1600	74.5^{20}	159^{100}	s al, acet, ace a
c96	chloride aluminate	$3CaO \cdot Al_2O_3 \cdot CaCl_2 \cdot 10H_2O$	561.34	col, monocl or hex, hex, 1.550, 1.535	1.892^{14}	$-H_2O$, 105	$-8H_2O$, 350	sl s	d	s a
c97	chloride, dihydrate	$CaCl_2 \cdot 2H_2O$	147.02	col cr	0.835			97.7^0	326^{60}	50^{80} al
c98	chloride, hexahydrate	$CaCl_2 \cdot 6H_2O$	219.08	col, trig, deliq, 1.417, 1.393	1.71^{25}	26.92	$-4H_2O$, 30, $-6H_2O$, 200	279^0	536^{20}	s al
c99	chloride, monohydrate	$CaCl_2 \cdot H_2O$	129.00	col cr, deliq		260		76.8^0	249^{100}	s al; i acet
c100	chloride fluoride *ortho*phosphate	$3Ca_3(PO_4)_2 \cdot CaClF$	1025.08	col cr, 1.634, 1.631	3.14	1270		v sl s		
c101	chlorite	$Ca(ClO_2)_2$	174.98	wh, cub	2.71			d	d	i al
c102	*hypo*chlorite	$Ca(ClO)_2$	142.98	wh powd or flat pl, 1.545, 1.69	2.35	d 100		s		i al

No.	Name	Synonyms and Formulae	Mol. wt.	Crystalline form, properties and index of refraction	Density or spec. gravity	Melting point, °C	Boiling point, °C	Solubility, in grams per 100 cc		
								Cold water	Hot water	Other solvents
c103	chlorite, basic	$Ca(ClO)_2 \cdot 2Ca(OH)_2$	291.17	wh, hex, 1.51, 1.585	2.10			sl s solns with 5—6 % avail Cl	d	d a
c104	*hypo*chlorite, basic	Bleaching powder, chlorinated lime, $Ca(ClO)_2 \cdot CaCl_2 \cdot xCa(OH)_2 \cdot xH_2O$	comp varies	wh powd strong Cl odor		d		d evin Cl		d a
c105	*hypo*chlorite, trihydrate	$Ca(ClO)_2 \cdot 3H_2O$	197.03	tetr pl, 1.535, 1.63	2.1	$-3H_2O$, 60				
c106	chromate	$CaCrO_4 \cdot 2H_2O$	192.10	yel, monocl pr		$-2H_2O$, 200		16.3^{20}	18.2^{45}	s a, al
c107	chromite	$CaCr_2O_4$	208.07	ol grn, cub need	4.8^{18}	2090		i	i	i a; s fus K_2CO_3
c108	cinnamate	$Ca(C_9H_7O_2)_2 \cdot 3H_2O$	388.43	col cr.				0.22^{2}	1.34^{100}	
c109	citrate	$Ca_3(C_6H_5O_7)_2 4H_2)$	570.50	wh need		$-4H_2O$, 120		0.085^{18}	0.096^{23}	0.0065^{18} al
c110	cyanamide	$CaCN_2$	80.10	col, hex, rhbdr		1300 subl >1150		d evl NH_3	d	
c111	cyanide	$Ca(CN)_2$	92.12	wh powd		d>350		d	d	
c112	cyanoplatinite	$CaPt(CN)_4 \cdot 5H_2O$	429.31	yel-grn fluoresc, rhomb, 1.6226		$-5H_2O$, 100		s		
c113	ferricyanide	$Ca_3[Fe(CN)_6]_2 12H_2O$	760.33	red need, deliq				v s	v s	
c114	ferrite, mono-	$CaO \cdot Fe_2O_3$	215.77	dk redsh r, rhomb, 2.58, 2.43 (Na)	5.08	1250		i	i	v sl s a
c115	ferrocyanide	$Ca_2Fe(CN)_6 11$ or $12H_2O$	490.28 or 508.30	yel tricl, 1.570, 1.582, 1.596	1.68	d		86.8^{25}	115^{65}	i al
c116	fluosilicate	$CaSiF_6$	182.16	col, tetr	2.66^{18}			sl s		s al, HF, HCl
c117	fluoride	Nat. fluorite. CaF_2	78.08	cool, cub luminisc w heat, 1.434	3.180	1423	*ca*2500	0.0016^{18}	0.0017^{26}	s HN_4 salts; sl s a; i acet
c118	fluosilicate, dihydrate	$CaSiF_6 \cdot 2H_2O$	218.19	col, tetrag	2.254			sl s d		s HCl, HF; i al
c119	formate	$Ca(CHO_2)_2$	130.12	col, rhomb, 1.510, 1.514, 1.578	2.015	d		16.2^{0}	18.4^{100}	i al
c120	fumarate	$CaC_4H_2O_4 \cdot 3H_2O$	208.18	col, rhomb				2.11^{30}		
c121	*d*gluconate	$Ca(C_6H_{11}O_7)_2 \cdot H_2O$	448.39	wh cr powd, need		$-H_2O$, 120		3.3^{15}		v sl s al
c122	glycerophosphate	$CaC_3H_5(OH)_2PO_4$	210.14	wh cr powd, hyg		d 170		2^{25}	sl s	i al
c123	hydride	CaH_2	42.10	wh, rhomb cr.	1.9	816 (in H_2) d *ca* 600		d H_2 + $Ca(OH)_2$		d a
c124	hydroxide	$Ca(OH)_2$	74.09	col, hex, 1.574, 1.545	2.24	$-H_2O$, 580	d	0.185^{0}	0.077^{100}	s NH_4 salts, a; i al
c125	hyponitrite	$CaN_2O_2 \cdot 4H_2O$	172.15	wh cr.	1.834	d 320				d dil a
c126	iodate	Nat. lautarite, $Ca(IO_3)_2$	389.89	col, monocl.	4.519^{15}	d 540		0.20^{15}	0.67^{90}	s HNO_3; i al
c127	iodate, hexahydrate	$Ca(IO_3)_2 \cdot 6H_2O$	497.98	col, rhomb		d 35		0.13^{0}	1.22^{100}	s HNO_3
c128	iodide	CaI_2	293.89	yelsh-wh, hex, deliq	3.956_4^{25}	784	*ca*1100	209^{20}	426^{100}	126^{20} MeOH; s al, acet, a
c129	iodide, hexahydrate	$CaI_2 \cdot 6H_2O$	401.98	yel, hex need	2.55	d 42	160	757^{0}	1680^{30}	s a, al, acet
c130	iron (III) aluminate	Calcium (tetra-) aluminoferite, nat. celite. $4CaO \cdot Fe_2O_3 \cdot Al_2O_3$	485.97	brn, rhomb, 1.98, 2.05, 2.08 all for λ	3.77	1418				
c131	isobutyrate	$Ca(C_4H_7O_2)_2 \cdot 5H_2O$	304.35	col powd				20	sl s	
c132	lactate	$Ca(C_3H_5O_3)_2 \cdot 5H_2O$	308.30	wh need, effl		$-3H_2O$, 100		3.1°	7.9^{30}	sl s a; i al, eth
c133	laurate	$Ca(C_{12}H_{23}O_2)_2 \cdot H_2O$	456.73	wh need, effl		182—183		0.004^{15}	0.055^{100}	0.059^{15}, 1.72^{78} al
c134	linoleate	$Ca(C_{18}H_{31}O_2)_2$	598.97	wh amorph powd				i		s al, eth
c135	magnesium carbonate	Nat. dolomite. $CaCO_3 \cdot MgCO_3$	184.40	col, trig, 1.6817, 1.5026	2.872	d 730—760		0.0078^{18}		
c136	magnesium *meta*silicate	Nat. diopside. $CaO \cdot MgO \cdot 2SiO_2$	216.55	col, monocl, 1.665, 1.672, 1.695	3.275	1391		i	i	i HCl
c137	magnesium *ortho*silicate	Nat. mervinite. $Ca_3Mg(SiO_4)_2$	328.71	col to pa grn, monocl, 1.708, 1.711, 1.718	3.150					
c138	dl-malate	$CaC_4H_4O_5 \cdot 3H_2O$	226.20	col, rhomb, 1.545, 1.555, 1.575				0.321°	$0.451^{37.5}$	i al
c139	*l*-malate	$CaC_4H_4O_5 \cdot 2H_2O$	208.18	col				0.812°	$1.224^{37.5}$	s al
c140	malate, dihydrogen	$Ca(HC_4H_4O_5)_2 \cdot 6H_2O$	414.33	rhomb, or wh cr powd, 1.493, 1.507, 1.545				sl s		
c141	maleate	$CaC_4H_2O_4 \cdot H_2O$	172.15	col rhomb, 1.495, 1.575, 1.640				2.89^{25}	3.21^{40}	
c142	malonate	$CaC_3H_2O_4 \cdot 4H_2O$	214.19					0.44°	0.72^{100}	
c143	*per*manganate	$Ca(MnO_4)_2 \cdot 5H_2O$	368.03	purp cr.	2.4	d		331^{14}	338^{25}	s NH_4OH
c144	α-methylbutyrate	Calcium ethylmethylacetate. $Ca(C_5H_9O_2)_2$	242.33					24.24°	25.65^{70}	
c145	molybdate	Nat. powellite $CaMoO_4$	200.02	col, tetr, 1.967, 1.978	4.38—4.53			i	d	s a i al, eth
c146	nitrate	$Ca(NO_3)_2$	164.09	col, cub, hyg	2.504^{18}	561		121.2^{18}	376^{100}	14^{15} al; s MeOH, liq NH_3, acet; i eth
c147	nitrate, tetrahydrate	$Ca(NO_3)_2 \cdot 4H_2O$	236.15	col, monocl, deliq, 1.465, 1.498, 1.504	α 1.896, β 1.82	α 42.7, β 39.7	d 132	266°	660^{30}	s al, acet
c148	nitrate, trihydrate	$Ca(NO_3)_2 \cdot 3H_2O$	218.14	col, tricl		51.1				
c149	nitride	Ca_3N_2	148.25	brn cr, hex	2.63^{17}	1195		d	d	s dil a; d abs al
c150	nitrite	$Ca(NO_2)_2 \cdot H_2O$	150.11	col-yelsh, hex, deliq	2.23^{34}	$-H_2O$, 100		45.9^{0}	89.6^{91}	sl s al
c151	nitrite, tetrahydrate	$Ca(NO)_2)_2 \cdot 4H_2O$	204.15	col cr, tetr.	1.674_0^0	$-2H_2O$, 44		74.9°	106^{42}	s al
c152	oleate	$Ca(C_{18}H_{33}O_2)_2$	603.00	wh wax-like cr		83—84		0.04^{25}	0.03^{50}	sl s eth

PHYSICAL CONSTANTS OF INORGANIC COMPOUNDS (continued)

No.	Name	Synonyms and Formulae	Mol. wt.	Crystalline form, properties and index of refraction	Density or spec. gravity	Melting point, °C	Boiling point, °C	Solubility, in grams per 100 cc		
								Cold water	Hot water	Other solvents
c153	oxalate	CaC_2O_4	128.10	col, cub	2.2^{4}	d		0.00067^{13}	0.0014^{95}	s a; i ac a
c154	oxalate, hydrate	$CaC_2O_4 \cdot H_2O$	146.11	col	2.2	$-H_2O$, 200		i	i	s a; i ac a
c155	oxide	Lime, calcia. CaO	56.08	col, cub, 1.838	3.25—3.38	2614	2850	0.131^{10} d	0.07^{80} d	s a
c156	oxide, per-	CaO_2	72.08	wh, tetr, 1.895	2.92^{25}_{4}	d 275		sl s		s a
c157	oxide, peroctahydrate	$CaO_2 \cdot 8H_2O$	216.20	wh, tetr, pearly	1.70	$-8H_2O$, 200	d 275 expl	sl s	d	s a, NH_4 salts; i al, eth
c158	palmitate	$Ca(C_{16}H_{31}O_2)_2$	550.92	wh or yelsh, wh fatty powd				0.003^{25}		v sl s al; 0.008^{25} eth
c159	1-phenol-4 sulfonate(p-)	$Ca[C_6H_4(OH)SO_3]_2 \cdot H_2O$	404.42	wh to pinkish powd				s		s al
c160	phenoxide	$Ca(OC_6H_5)_2$	226.29	redsh powd				sl s		sl s al
c161	*hypo*phosphate	$Ca_2P_2O_6 \cdot 2H_2O$	274.13	gel				i		s HCl
c162	*meta*phosphate	$Ca(PO_3)_2$	198.02	col, 1.588, 1.595	2.82	975		i	i	i a
c163	*ortho*phosphate, di-(sec)	Nat. brushite. $CaHPO_4 \cdot 2H_2O$	172.09	wh, tricl, 1.5576, 1.5457, 1.5392	2.306^{16}	$-H_2O$, 109		0.0316^{38}	0.075^{100}	i al, s a
c164	*ortho*phosphate, mono-(prim.)	$Ca(H_2PO_4)_2 \cdot H_2O$	252.07	col, tricl, deliq, 1.5292, 1.5176, 1.4392	2.220^{16}	$-H_2O$, 109	d 203	1.8^{30}	d	s a
c165	*ortho*phospate, tri-(tert.)	Nat. whitlockite. $Ca_3(PO)_4)_2$	310.18	wh amorph powd, 1.629, 1.626	3.14	1670		0.002	d	i al; s a
c166	*pyro*phosphate	$Ca_2P_2O_7$	254.10	col, biax, 1.585, 1.604	3.09	1230		i		s a
c167	*pyro*phosphate, pentahydrate	$Ca_2P_2O_7 \cdot 5H_2O$	344.18	col, monocl, 1.539, 1.545, 1.551	2.25			sl s		s a; i HN_4Cl
c168	phosphide	Ca_3P_2	182.19	gray lumps	2.51	*ca* 1600		d ev PH_3		s a; i al, eth, bz
c169	*hypo*phosphite	$Ca(H_2PO_2)_2$	170.06	wh-gray, monocl		d		15.4^{25}	12.5^{100}	i al
c170	*ortho*phosphite, di-	$2CaHPO_3 \cdot 3H_2O$	294.17					sl s	d	s NH_4Cl
c171	*ortho*plumbate	Ca_2PbO_4	351.36	red-br cr	5.71	d		i	d	s a
c172	propionate	$Ca(C_3H_5O_2)_2 \cdot H_2O$	204.24	col, monocl tabl				49^{0}	55.8^{100}	i al
c173	*l*-quinate	$Ca(C_7H_{11}O_6)_2 \cdot 10H_2O$	602.55	rhomb leaf		50, $-10H_2O$ 120		16^{18}		i al
c174	salicylate	$Ca(C_7H_5O_3)_2 \cdot 2H_2O$	350.34	wh, oct		$-2H_2O$, 120		4^{25}	s	s al
c175	selenate	$CaSeO_4$	183.04	col	2.88			7.9^{6}	5.4^{67}	
c176	selenate, dihydrate	$CaSeO_4 \cdot 2H_2O$	219.07	col, monocl	2.68					
c177	selenide	CaSe	119.04	cub, 2.274	3.57					
c178	*meta*silicate(α)	Nat. pseudowollastonite. $CaSiO_3$	116.16	col, monocl, 1.610, 1.611, 1.664	2.905	1540		0.0095^{17}		s HCl
c179	*meta*silicate(β)	Nat. wollastonite. $CaSiO_3$	116.16	col, monocl, 1.616, 1.629, 1.631	2.5	tr 1200				
c180	di-*ortho*silicate (I)	Ca_2SiO_4	172.24	col, monocl, 1.717, 1.735	3.27	2130				
c181	di-*ortho*silicate (II)	Ca_2SiO_4	172.24	col, rhomb, 1.717, 1.735	3.28	tr to (I) 1420				
c182	di-*ortho*silicate (III)	Ca_2SiO_4	172.24	col, monocl, 1.642, 1.645, 1.654	2.97	tr to 675				
c183	(tri-)silicate	Nat. alite. Ca_3SiO_5 or $(3CaO \cdot SiO_2)$	228.32	col, monocl, α 1.718, β 1.724		1900 (incogr)				
c184	silicide	$CaSi_2$	96.25		2.5			i	d	s a, alk
c185	stearate	$Ca(C_{18}H_{35}O_2)_2$	607.03	cr powd		179—180		0.004^{15}		i al, eth
c186	succinate	$CaC_4H_6O_4 \cdot 3H_2O$	212.22	col, 1.460, 1.540, 1.610				0.193^{10}	0.89^{80}	
c187	sulfate	Nat. anhydrite. $CaSO_4$	136.14	col, rhomb, or monocl, 1.569, 1.575, 1.613	2.960	monocl 1450	rhomb tr to monocl 1193	0.209^{30}	0.1619^{100}	s a, NH_4 salts, $Na_2S_2O_3$, glyc
c188	sulfate	Soluble anhydrite. $CaSO_4$	136.14	col, hex or tricl, 1.505, 1.548	2.61	tr to rhomb >200				
c189	sulfate half-hydrate	Plaster of Paris. $CaSO_4 \cdot {}^1/_2H_2O$	145.15	wh powd		$-{}^1/_2H_2O$, 163		0.3^{20}	sl s	s a NH_4 salts, $Na_2S_2O_3$, glyc
c190	sulfate dihydrate	Nat. gypsum. $CaSO_4 \cdot 2H_2O$	172.17	col, monocl, 1.521, 1.523, 1.530	2.32	$-1{}^1/_2H_2O$, 128	$-2H_2O$, 163	0.241	0.222^{100}	s a, NH_4 salts, $Na_2S_2O_3$, glyc
c191	sulfide	Nat. oldhamite. CaS	72.14	col, cub, 2.137	2.5	d		0.021^{15} d	0.048^{60} d	d a
c192	sulfide, hydro-	$Ca(HS)_2 \cdot 6H_2O$	214.31	col pr		d 15—18		v s		s al
c193	sulfite	$CaSO_3 \cdot {}^1/_2H_2O$	129.15	col, hex		$-{}^1/_2H_2O$, >250		0.0043^{18}	0.0011^{100}	s H_2SO_4
c194	sulfite, dihydrogen	$Ca(HSO_3)_2$	202.21	yelsh liq, strong SO_2, odor				s		s a
c195	*d*-tartrate	$CaC_4H_4O_6 \cdot 4H_2O$	260.21	col, rhomb, 1.525, 1.535, 1.550		d		0.0266^{0}	$0.0689^{37.5}$	sl s al
c196	*dl*-tartrate	$CaC_4H_4O_6 \cdot 4H_2O$	260.21	tricl, powd or need		$-4H_2O$, 200		0.0032^{0}	$0.0078^{37.5}$	s HCl; i ac, a
c197	*meso*tartrate	$CaC_4H_4O_6 \cdot 3H_2O$	242.20	wh, monocl or tricl pr		$-3H_2O < 170$		i	0.16^{100}	0.28^{18}, 0.85^{100} ac a
c198	telluride	CaTe	167.68	cub, 251, 2.58	4.873					
c199	tellurite	$CaTeO_3$	215.68	wh fl		>960		sls	s	s a
c200	thiocarbonate, tri-	$CaCS_3$	148.27	yel cr				s		i al
c201	thiocyanate	$Ca(SCN)_2 \cdot 3H_2O$	210.28	wh cr, deliq				v s	v s	v s al
c202	*di*-thionate	$Ca(SO_3)_2 \cdot 4H_2O$	272.26	col, trig, 1.5496	2.176			16^{0}	30^{30}	

No.	Name	Synonyms and Formulae	Mol. wt.	Crystalline form, properties and index of refraction	Density or spec. gravity	Melting point, °C	Boiling point, °C	Solubility, in grams per 100 cc: Cold water	Hot water	Other solvents
c203	thiosulfate	$CaS_2O_3 \cdot 6H_2O$	260.29	tricl.	1.872	d		100^{3}	d	s al
c204	*meta*titanate	Nat. perovskite. $CaTiO_3$	135.96	col, cub, rhomb, β 2.34	4.10	1975				
c205	tungstate	$CaWO_4$	287.93	wh, tetr, 1.9263, 1.9107	6.062^{20}			0.00064^{15}	0.00012^{100}	
c206	tungstate	Nat. scheelite. $CaWO_4$	287.93	col or w sc, tetr. 1.918, 1.934	6.06			0.2		i al, a; s NH_4Cl
c207	*meta*tungstate	$Ca_3H_4[H_2(W_2O_7)_6] \cdot 27H_2O$	3490.88	col, tric		$-7H_2O$, 105	$-10H_2O$, d			d a
c208	valerate	$Ca(C_5H_9O_2)_2$	242.33					8.28^{0}	7.39^{100}	
c209	*meta*zirconate	$CaZrO_3$	179.30	col, monocl	4.78	2550				
c210	**Carbon**	Diamond. C	12.011	col, cub, 2.4173	3.51	Diamond trans. to graphite		i	i	i a, alk
c211	carbon	Graphite C	12.011	blk, hex	2.25^{20}	subl 3652 – in vac. at 1500°C 97			i	s liq Fe; i a, alk
c212	carbon, amorphous	C	12.011	amorph, blk.	1.8—2.1	subl 3652—97		i	i	i a, alk
c213	(di)-bromide, hexa-	Hexabromomethane. C_2Br_6	503.45	rhomb pr, 1.740, 1.847, 1.863	3.823	148—149 d	210	i		s CS_2; v sl s al, eth
c214	bromide, tetra-	Tetrabromomethane. CBr_4	331.63	col, monocl or oct	3.42	tr to oct 48.4; m.p. 90.1	189.5	0.024^{30}		s al, eth, chl
c215	(di)-bromide, tetra-	Tetrabromethylene. C_2Br_4	343.64			57.5	227			
c216	(di)-chloride, hexa-	Hexachloro ethane. C_2Cl_6	236.74	col, rhmb, tricl or cub	2.091	subl 187		i		s al, eth, oils
c217	chloride, tetra-	Tetrachloromethane. CCl_4	153.82	col liq, 1.4601	1.5867^{20}_{20}	−23	76.8		v sl s	s al, bz, chl, eth
c218	(di)-chloride, tetra,-	Tetrachloroethylene. C_2Cl_4	165.83	col liq, eth odor, 1.5055	1.6311^{15}_{4}	−22.4	120.8			s al, eth
c219	fluoride, tetra-	Tetrafluoromethane. CF_4	88.00	col gas	1.96^{-184}	−184	−128	sl s		
c220	iodide, tetra-	Tetraiodomethane. CI_4	519.63	dk red, cub	4.34^{20}	d 171		i	d	s al, CS_2, eth, MeOH, bz
c221	oxide, di-	CO_2	44.01	col gas or col liq	1.977^{0} g/ℓ, liq 1.101^{-37}, solid 1.56^{-79}	$-56.6^{5.2atm}$	−78.5 subl	171.3^{0} cm³ 0.348^{0}g 0.145^{25}g	90.1^{20} cm³ 0.097^{40} g 0.058^{60}g	31^{15} cm³ al, s acet
c222	oxide, mon-	CO	28.01	col odorl pois gas	1.250^{0} g/ℓ liq 0.793	−199	−191.5	3.5^{0} cm	2.32^{20} cm³	s al, bz, ac a, Cu_2Cl_2
c223	oxide, sub-	C_3O_2	68.03	col gas or liq, 1.4538	liq 1.114^{0}	−111.3	7	d		
c224	oxysulfide	COS	60.07	col gas, pois	gas 1.073 g/ℓ° liq 1.24^{-87}	−138.2	−50.2	54^{20} mℓ		s al; v s CS_2
c225	selenide, di-	CSe_2	169.93	golden yel liq, 1.845^{20}	2.6626^{25}_{4}	−45.5	125—126	i		d al; s CS_2, tol
c226	selenide, sulfide	CSeS	123.03	yel oily liq	1.9874	−85	84.5	i	i	sl s al; s CS_2
c227	sulfide, di-	CS_2	76.13	col liq, inflamm, 1.62950^{18}	1.261^{22}_{20}	−110.8	46.3	0.22^{22}	0.14^{50}	s al, eth
c228	sulfide, mono-	CS	44.07	red powd	1.66	d −160		i		ial; s eth, CS_2
c229	sulfide, sub-	C_3S_2	100.15	red liq	1.274	−0.5	d 90			
c230	sulfide telluride	CSTe	171.67	yel-red liq	2.9^{-50}	−54	d >−54			s CS_2, bz
c231	sulfochloride	Thiophosgene. $CSCl_2$	114.98	yel-red liq	1.509^{15}		73.5			
c232	**Carbonic acid**	H_2CO_3	62.03	exists in solution only				s		
c233	**Carbonyl bromide**	Carbon oxybromide. $COBr_2$	187.82	col liq			64.5			
c234	**Carbonyl chloride**	Phosgene, carbon oxychloride, $COCl_2$	98.92	col gas, pois	1.392	−104	8.3	d		d al, a; v s bz, tol s ac a
c235	**Carbonyl fluoride, di-**	COF_2	66.01	col gas, hyg	sol 1.388^{-190} liq 1.139^{-114}	−114	−83.1	d		
c236	**Carbonyl selenide**	COSe	106.97	col gas, very pois	liq $1.812^{4.1}$	−124.4	−21.7	d		s $COCl_2$
c237	**Cerium**	Ce	140.12	gray met, cub or hex	hex 6.657 cub 6.757	799	3426	sl d	d	s dil min a, i alk
c238	(III) acetate	$Ce(C_2H_3O_2)_3$	317.25	col		d 308		20^{15}	12^{75}	
c239	(III) acetate hydrate	$Ce(C_2H_3O_2)_3 \cdot 1^1/_2H_2O$	344.28	wh-redsh cr powd		$-1^1/_2H_2O$, 115	d	26.5^{15}	16.2^{75}	
c240	boride, hexa-	CeB_6	204.98	blue met, cub.		2190	d	i	i	i HCl
c241	boride, tetra-	CeB_4	183.36	tetr	5.74					
c242	III bromate	$Ce(BrO_3)_3 \cdot 9H_2O$	685.96	redsh-wh, hex		49		s		
c243	bromide	$CeBr_3 \cdot H_2O$	397.85	col need, deliq		d		v s	v s	v s al
c244	carbide	CeC_2	164.14	red, hex	5.23			d	d	s a
c245	carbonate	$Ce_2(CO_3)_3 \cdot 5H_2O$	550.34	wh cr.				i		s a; sl s $(NH_4)_2CO_3$
c246	carbonate fluoride	Nat. bastnaesite. $CeFCO_3$	219.13	hex, 1.717, 1.817	5					
c247	chloride	$CeCl_3$	246.48	col cr, deliq.	3.92^{0}	848	1727	100	d	30 al; s acet
c248	citrate	$Ce(C_6H_5O_7) \cdot 3^1/_2H_2O$	392.27	wh powd				i		s dil min a
c249	(III) cyanoplatinite	$Ce_2[CN)_4]_3 \cdot 18H_2O$	1501.97	yel-bl lust, monocl	2.657	$-13^1/_2H_2O$ 100—110	d	s		
c250	(III) fluoride	CeF_3	197.12	wh, hex	6.16	1460	2300	i		
c251	(IV) fluoride	$CeF_4 \cdot H_2O$	234.13	col microcr, 1.614	4.77	*ca* 650	d	i		s a
c252	hydride	CeH_3	143.14	dk bl amorph powd		ign		d		

No.	Name	Synonyms and Formulae	Mol. wt.	Crystalline form, properties and index of refraction	Density or spec. gravity	Melting point, °C	Boiling point, °C	Solubility, in grams per 100 cc		
								Cold water	Hot water	Other solvents
c253	(III) hydroxide	$Ce(OH)_3$	191.14	wh gelat ppt						s a, $(NH_4)_2CO_3$; i alk
c254	(IV) iodate	$Ce(IO_3)_4$	839.73	yel cr.				0.015^{20}		
c255	(III) iodate	$Ce(IO_3)_3{\cdot}2H_2O$	700.86	cr				0.16^{25}		s HNO_3
c256	(III) iodide	$CeI_3{\cdot}9H_2O$	682.97	wh or redsh-wh cr		752	1397	v s		v s al
c257	(III) molybdate	$Ce_2(MoO_4)_3$	760.05	yel, tetr, 2.019, 2.007	4.83	973				
c258	(III) nitrate	$Ce(NO_3)_3{\cdot}6H_2O$	434.23	col or redsh cr (trac La, Di), deliq		$-3H_2O$, 150	d 200	v s	v s	50 al; s acet
c259	(IV) nitrate, basic	$Ce(OH)(NO_3)_3{\cdot}3H_2O$	397.19	long red need.				s		
c260	(III) oxalate	$Ce_2(C_2O_4)_3{\cdot}9H_2O$	706.44	yel-wh cr		d		v sl s		s H_2SO_4, HCl; i $H_2C_2O_4$, alk, eth, al
c261	(IV) oxide	$CeO_2{\cdot}xH_2O$		yelsh gelat ppt						s a; sl s alk carb, i alk
c262	(III) oxide	Ce_2O_3	328.24	gray-grn, trig	6.86	1692, ign 200		i	i	s H_2SO_4, i HCl
c263	(IV) oxide(di-)	Ceria. CeO_2	172.12	brn-wh, cub.	7.132^{23}	*ca* 2600		i	i	s H_2SO_4, HNO_3; i dil a
c264	oxychloride	CeOCl	191.57	purp leaf.				i		s dil a
c265	(III) 2,4- pentanedione.	Cerium acetylacetonate. $Ce(C_5H_7O_2)_3{\cdot}3H_2O$	491.49	lt yel cr ppt		131—132		d		v s al
c266	(III) *meto*phosphate	$Ce(PO_3)_3$	377.04	micr need	3.272					i a
c267	(III) *ortho*phosphate	Nat. monazite. $CePO_4$	235.09(?)	red, monocl or yel, rhomb, 1.795	5.22			i	i	s a; i al
c268	(III) salicylate	$Ce(C_7H_5O_3)_3$	551.46	wh-redsh wh powd				i		i al
c269	(III) selenate	$Ce_2(SeO_4)_3$.	709.11	rhomb	4.456			39.55^{0}	2.513^{00}	
c270	silicide	$CeSi_2$	196.29		5.67^{17}			i		
c271	(IV) sulfate	$Ce(SO_4)_2$	332.24	deep yel cr	3.91^{18}	d 195		sl d, forms basic salts		
c272	(III) sulfate.	$Ce_2(SO_4)_3$	568.41	col to grn, monocl or rhomb.	3.912	d 920^{746}		10.1^{0}	2.25^{100}	
c273	(IV) sulfate, dihydrate	$Ce(SO_4)_2{\cdot}4H_2O$	404.30	yel, rhomb				v s d		s dil H_2SO_4
c274	(III) sulfate, monohydrate	$Ce_2(SO_4)_3{\cdot}9H_2O$	730.55	hex need.	2.831			11.87^{15}	0.42^{30}	
c275	(III) sulfate, octahydrate	$Ce_2(SO_4)_3{\cdot}8H_2O$	712.54	pink cr, tricl	2.886^{17}	$-8H_2O$, 630		12^{20}	6^{50}	
c276	(III) sulfate pentahydarte	$Ce_2(SO_4)_3{\cdot}5H_2O$	658.49	monocl.	3.17			3.90^{50}	0.514^{100}	
c277	(III) sulfide	Ce_2S_3	376.42	red cr, br-dk powd, purp	5.020^{11}	d 2100 (vac)		i	d	s dil a
c278	(III) tungstate	$Ce_2(WO_4)_3$	1023.78	yel, tetr	6.77^{17}	1089				
	Cerium complexes									
c280	hexaantipyrinecerium perchlorate	$[Ce(C_{11}H_{12}N_2O)_6]{\cdot}(ClO_4)_3$	1567.85	col, hex cr		d 295—300		1.08^{20}		
c281	hexaantipyrinecerium iodide (III)	$Ce(C_{11}H_{12}N_2O)_6{\cdot}I_3$	1650.21	large yel cr		268—270		15.10^{20}		
c282	**Cesium**	Cs	132.9054	silv met cr hex	1.8785^{15}	28.40 ± 0.01	669.3	d		s liq NH_3
c283	acetate	$CsC_2H_3O_2$	191.95	deliq		194		$945.1^{-2.5}$	$1345.5^{88.5}$	
c284	aluminum sulfate	$CsAl(SO_4)_2{\cdot}12H_2O$	568.19	col, cub, 1.4587	1.97	117		0.34^{0}	42.54^{100}	s dil al
c285	amide	$CsNH_2$	148.93	wh need	3.44^{25}_{4}	262 ± 1		d		s liq NH_3
c286	azide	CsN_3	174.93	col need, deliq		310		224.2^{0}		1.037^{16} al; i eth
c287	benzoate	$CsC_7H_5O_2$	254.02					294.5^{0}	398.5^{100}	
c288	borofluoride	$CsBF_4$.	219.71	rhomb, 1.350.	3.20	550 d		1.6^{17}	*ca* 30^{100}	s dil NH_3
c289	borohydride	$CsBN_4$	147.75	wh, cub, 1.498.	2.404			v s		sl s al; i eth, bz
c290	bromate	$CsBrO_3$	260.81	hex, *ca* 2.15	4.109^{16}	*ca* 420 d		3.66^{25}	5.32^{35}	
c291	bormide, mono-	CsBr	212.81	col cub, 1.6984	4.44, liq 3.04^{700}	636	1300	124.3^{25}	v s	s a
c292	bromide, tri-	$CsBr_3$	372.62	rhomb		180				
c293	*di*bromochloride	$CsBr_2Cl$	328.17	yel-red, rhomb		191	150, $-Br_2$	s		d al, acet
c294	bromochloride iodide	CsIBrCl.	375.17	yel-red, rhomb		235	d 290	s		s al
c295	bromoiodide di-	$CsIBr_2$	419.62	rhomb	4.25	248	d 320	4.61^{20}		s al
c296	carbonate	Cs_2CO_3	325.82	col cr, deliq.		d 610		260.5^{15}	v s	11^{19} al; s eth
c297	carbonate, hydrogen	$CsHCO_3$	193.92	rhomb		$175 - {}^{1}/_{2}H_2O$		209.3^{15}	v s	s al
c298	chlorate	$CsClO_3$	216.36	sm cr.	3.57			$6.28^{19.8}$	76.5^{90}	s al
c299	*per*chlorate	$CsClO_4$	232.36	rhomb, at 219 cub, 1.4752, 1.4788, 1.4804	3.327^{4}	d 250		2.00^{25}	28.57^{99}	0.093^{25} al; 0.7878^{25} al; 0.150^{25} acet
c300	chlorobromide	$CsBrCl_2$	283.72	glossy-yel, rhomb.		205		s		d al, eth
c301	chloride	CsCl.	168.36	col, cub, deliq, 1.6418	3.988	645	1290	$162.22^{0.7}$	$259.56^{89.5}$	33.7^{25} MeOH; v s al; i acetone
c302	chloroiodide	$CsICl_2$	330.72	or, trig	3.86	230	d 290	s		s al
c303	chloroaurate	$CsAuCl_4$	471.68	yel, monocl				0.5^{10}	27.5^{100}	s al
c304	chlorobromide, di-	$CsBrCl_2$	283.72	glossy-yel, rhomb		205				
c305	chloro*di*bromide	$CsBr_2Cl$	328.17	yel		191				
c306	chloroiodide, di-	$CsICl_2$	330.72	or, trig	3.86	230	d 290	s		s al
c307	chloroplatinate	Cs_2PtCl_6	673.61	yel, cub	4.197 ± 0.004	d 570		0.024^{0}	0.377^{100}	i al
c309	chlorostannate	Cs_2SnCl_6	597.22	wh, cub	3.33					
c310	chromate	Cs_2CrO_4	381.80	yel pr, rhomb.	4.237			71.4^{13}	95.5^{30}	
c311	chromium sulfate	Cesium chromium alum. $Cs[Cr(H_2O)_6](SO_4)_2{\cdot}6H_2O$	593.20	vlt cr	2.064	116		9.4^{25}		
c312	cyanide	CsCN	158.92	very sm wh cr	2.93			v s	v s	
c313	fluoride	CsF	151.90	cub, deliq, 1.478 ± 0.005^{18}	4.115	682	1251	367^{18}		191^{15} MeOH; i Diox, Pyr

No.	Name	Synonyms and Formulae	Mol. wt.	Crystalline form, properties and index of refraction	Density or spec. gravity	Melting point, °C	Boiling point, °C	Solubility, in grams per 100 cc		
								Cold water	Hot water	Other solvents
c314	fluoride	$CsF \cdot 1^1/_2H_2O$	178.93			703		366.5^{18}		
c315	fluorogermanate	Cs_2GeF_6	452.39	isotrop cr, reg oct	4.10			sl s	v s	sl s a
c316	fluosilicate	Cs_2SiF_6	407.89	wh, cub	3.372^{17}			60^{17}	sl s	i al
c317	fluotellurite	$CsTeF_5$	355.50	col need				d	d	s HF soln
c318	formate	$CsCHO_2$	177.92		1.0169^{21}_{4}				$2012^{96.4}$	
c319	formate	$CsCHO_2 \cdot H_2O$	195.94			41, $-H_2O$				
c320	gallium selenate	$CaGa(SeO_4)_2 \cdot 2H_2O$	704.72	col cr.				4.14^{25}		
c321	gallium sulfate	$CsGa(SeO_4)_2 \cdot 12H_2O$	610.92	col cub, 1.46495		2.113		1.21^{25}		0.0035^{25} 75% al
c322	hydride	CsH	133.91	wh cr, cub	3.41	d		d	d	d a; i org solv
c323	hydrofluoride	CsF·HF	171.91	need, deliq		160		v s		v s a; i al
c324	hydrogencarbide	$CsHC_2$	157.94	trsp cr		300		d		
c325	hydroxide	CsOH	149.90	lt yel, deliq	3.675	272.3		395.5^{15}		s al
c326	iodate	$CsIO_3$	307.81	wh, monocl	4.85			2.6^{24}		
c327	*meta*periodate	$CsIO_4$	323.81	wh rhomb pl	4.259^{15}_{4}			2.15^{15}	s	
c328	iodide	CsI	259.81	rhomb, deliq, 1.7876	4.510^{25}_{4}	626	1280	44^{0}	160^{61}	s al
c329	iodide, penta-	CsI_5	767.43	bl, tricl		73				
c330	iodide, tri-	CsI_3	513.62	blk, rhomb	4.47	207.5		sl s	sl s	s al
c331	iodotetrachloride	$CsICl_4$	401.62	pale or needles	3.374^{-10}	228	d	sl s	sl s	
c332	iron (II) sulfate	$Cs_2SO_4 \cdot FeSO_4 \cdot 6H_2O$	621.86	lt grn, monocl, 1.500, 1.504, 1.509	2.791^{20}_{4}	*ca* 70		101.1^{25} (anhyd)		
c333	iron (III) sulfate	$CsFe(SO_4)_2 \cdot 12H_2O$	597.05	pa-vlt cr, 1.484	2.061^{20}	*ca* 90		s	s	
c334	magnesium sulfate	$Cs_2SO_4 \cdot MgSO_4 \cdot 6H_2O$	590.32	col, monocl, 1.486, 1.452	2.676^{20}_{4}					
c335	*per*manganate	$CsMnO_4$	251.84		3.597	d 320		0.097^{1}	1.27^{50}	
c336	mercury bromide(ic)	$CsBr \cdot 2HgBr_2$	933.61	rhomb				0.807^{17}		sl s al
c337	mercury chloride(ic)	$CsCl \cdot HgCl_2$	439.85	col, cub or rhomb, 1.792				1.44^{17}		i abs al
c338	nitrate	$CsNO^3$	194.91	col, hex or cub, 1.55, 1.56	3.685 liq 2.71^{500}	414	d	9.16^{0}	196.8^{100}	s acet, sl s al
c339	nitrate, hydrogen	$CsNO_3 \cdot NHO_3$	257.92	oct		100				
c340	nitrate, dihydrogen	$CsNO_3 \cdot 2HNO_3$	320.94	col pl.		32—36				
c341	nitrite	$CsNO_2$	178.91	yel cr.				v s	v s	
c342	oxalate	$Cs_2C_2O_4$	353.83		3.230^{15}			282.9^{25}		
c343	oxide	Cs_2O	281.81	or need	4.25	d 400; m.p. 490 (in N_2)		v s	d	s a
c344	oxide, per	Cs_2O_2	297.81	pa yel need	4.25	400	650. $-O_2$	s	d	s a
c345	oxide, tri-	Cesium oxide, sesqui- Cs_2O_3	313.81	choc br cr, cub.	4.25	400		d		s a
c346	phthalate, hydrogen	$CsHC_8H_4O_4$	298.03	rhomb	2.178					
c347	polonium chloride	Cs_2PoCl_6	687.53	cub, 1.86	3.82					
c348	rhodium sulfate	$CsRh(SO_4)_2 \cdot 12H_2O$	644.12	yel, oct	2.238	110—111		sl s		
				or cr	2.22^{20}_{4}	111		sl s	s	
c350	salicylate	$CsC_7H_5O_3$	270.02					196.2^{0}	1522^{100}	
c351	selenate	Cs_2SeO_4	408.77	rhomb, deliq, 1.5950, 1.5060, 1.5964	4.4528^{20}_{4}			244.8^{12}		
c352	sulfate	Cs_2SO_4	361.87	col rhomb, or hex, 1.560, 1.564, 1.566	4.243	1010	tr hex 600	167^{0}	220^{100}	i al, acet
c353	sulfate, hydrogen	$CsHSO_4$	229.97	col rhomb pr	3.352^{16}	d		s		
c354	sulfide	$Cs_2S \cdot 4H_2O$	369.93	wh cr, deliq.				v s	v s	
c355	sulfide, di-	Cs_2S_2	329.93	dk red, amorph.		460	>800	hgr		
c356	sulfide, di-	$Cs_2S_2 \cdot H_2O$	347.95	tetr				s		
c357	sulfide, hexa-	Cs_2S_6	458.17	br red		186				
c358	sulfide, penta-	Cs_2S_5	426.11		2.806^{15}	210				
c359	sulfide, tetra-	Cs_2S_4	394.05	yel		d 160				
c360	sulfide, tri-	Cs_2S_3	361.99	yel leaf		217	780			
c361	tartrate, hydrogen	$CsHC_4H_4O_6$	281.99	wh, rhomb cr.				9.7^{25}	98^{100}	
c362	*l*-tartrate	$Cs_2C_4H_4O_6$	413.88	col, trig	3.03^{14}			v s d	v s	
c363	vanadium sulfate	Cesium vanadium alum·$VCs(SO_4)_2 \cdot 12H_2O$	592.15	red, cub, 1.4780	2.033^{20}_{4}	82	$-12H_2O$, 230; d 300	0.464^{10}	sl s	
c364	**Chloramine, mono-**	NH_2Cl	51.48	yel liq		−66		s		s al, eth; v sl s CCl_4, bz
c365	**Chloric acid**	$HClO_3 \cdot 7H_2O$	210.57	known only as col sol.	$1.282^{14.2}$	<−20	d 40	v s		
c366	**Chloric acid, per**	$HClO_4$	100.46	col liq unstable	1.764^{22}	−112	39^{56}	∞		
c367	**Chloric acid, per**	Hydronium perchlorate·$HClO_4 \cdot H_2O$ or $(H_3O)^+(ClO_4)^-$	118.47	need, fairly stab	1.88, liq 1.776^{50}	50	exp 100	v s	v s	
c368	per, dihydrate	$HClO_4 \cdot 2HO$	136.49	stab liq.	1.65	−17.8	200	v s	v s	s al
c369	**Chlorine**	Cl_2	70.906	grnsh-yel gas, or liq, or rhomb cr; gas 1.000768, liq 1.367	3.214° g/ℓ	−100.98	−34.6	310^{10} cm³ 1.46° g	177^{30} cm³ 0.57^{30} g	s alk
c370	azide	chlor(o)azide Cln_3	77.47	gas, expl				sl s		d alk
c371	fluoride, mono-	ClF	54.45	col gas	1.62^{-100}	−154 ± 5	−100.8	d	d	
c372	fluoride, tri-	ClF_3	92.45	col gas	1.77^{13}	−83	11.3	d	d	
c373	hydrate	$Cl_2 \cdot 8H_2O$	215.03	lt yel, rhomb	1.23	d 9.6		i		s alk

No.	Name	Synonyms and Formulae	Mol. wt.	Crystalline form, properties and index of refraction	Density or spec. gravity	Melting point, °C	Boiling point, °C	Solubility, in grams per 100 cc		
								Cold water	Hot water	Other solvents
c374	oxide, di-	ClO_2	67.45	yel red gas, or red cr, expl.	3.09[11] g/ℓ	−59.5	9.9[731] exp	2000[4] cm³	d to $HClO_3$, Cl_2, O_2	s alk, H_2SO_4
c375	oxide, hept-	Cl_2O_7	182.90	col oil	...	−91.5	82	s d	...	s bz
c376	oxide, mono-	Cl_2O	86.91	yel-red gas, or red-br liq	3.89° g/ℓ	−20	3.8[766] exp	200 cm³	d to HOCl	s alk, H_2SO_4
c377	oxide, tetr-	ClO_4 or Cl_2O_8	99.45	...	...	...	d	s d	...	s bz
c378	chloroauric acid	$HAuCl_4{\cdot}4H_2O$	411.85	brt yel need, deliq	...	d	...	s	v s	s al, eth
c379	chloroplatinic acid	$H_2PtCl_6{\cdot}6H_2O$	517.91	red br pr, deliq	2.431	60	...	v s	v s	s al, eth
c380	**chlorostannic acid**	$H_2SnCl_6{\cdot}6H_2O$	441.52	col leaf	1.93	9	...	s	...	...
c381	**Chlorosulfonic acid**	$ClSO_3H$	116.52	col fum liq, 1.437[14]	1.766[18]	−80	158	d to H_2SO_4+ HCl	...	d al, a; i CS_2
c382	**Chlorotetroxy fluoride**	ClO_4F	118.45	col gas, v exp	...	−167.3	−15.9	...	...	...
c383	**Chloryl(per-)fluoride**	ClO_3F	102.45	gas	1.392[25]	−146	−46.8	...	...	...
c384	**Chromium**	Cr	51.996	steel gray, cub v hard	7.20[28]	1857 ± 20	2672	i	i	s dil H_2SO_4, HCl; i HNO_3, aq reg
c385	(II) acetate	$Cr(C_2H_3O_2)_2$	170.09	red cr	...	...	...	sl s	s	sl s al
c386	(III) acetate	$Cr(C_2H_3O_2)_3{\cdot}H_2O$	247.15	gray-grn powd or blsh-grn pasty mass	...	...	...	s	...	i al
c387	arsenide, mon-	CrAs	126.92	gray, hex	6.35[16]	...	...	i	i	i a
c388	boride, mono-	CrB	62.81	silv cr, orthorhomb	6.17	2760(?)	...	i	i	s fus Na_2O_2
c389	(II) bromide	$CrBr_2$	211.80	wh cr.	4.356	842	...	s	s	s al
c390	(III) bromide	$CrBr_3$	291.71	olv gr, hex	4.250^{25}_{4}	subl	...	i	s	v s al; d alk
c391	bromide, hexahydrate	$[CrBr_2(H_2O)_4]Br{\cdot}2H_2O$	399.80	grn cr, deliq	...	...	...	s	s tr to vlt	s al; i eth
c392	bromide, hexahydrate	$[Cr(H_2O)_6]Br_3$	399.80	blsh gray to vlt.	5.4[17]	...	...	v s	v s	i al
c393	(tri-)carbide, di-	Cr_3C_2	180.01	gray, rhomb	6.68	1980	3800	i	i	...
c394	carbonyl	$Cr(CO)_6$	220.06	col, orthorhomb	1.77	d 110	210 exp	i	i	i al, eth, ac a; sl s CHI_3, CCl_4
c395	(II) chloride	$CrCl_2$	122.90	wh need, deliq	2.878[25]	824	...	v s	v s	i al, eth
c396	(III) chloride	$CrCl_3$	158.36	vlt, trig	2.76[15]	*ca* 1150	subl 1300	i	sl s	i al, acet, MeOH, eth
c397	chloride, hexahydrate	$[Cr(H_2O)_4Cl_2]Cl{\cdot}2H_2O$	266.45	vlt, monocl	1.76	83	...	58.5[25]	s	s al; i eth; sl s acet
c398	(II) fluoride	CrF_2	89.99	grn, cr, monocl	4.11	1100	>1300	sl s	...	i al; s h HCl
c399	(III) fluoride	CrF_3	108.99	grn, rhomb	3.8	>1000	subl 1100–1200	i	...	i al, NH_3; sl s a; s HF
c400	(II) hydroxide	$Cr(OH)_2$	86.01	yel-br	...	v	...	d	...	s a
c401	iodate, hydrate	$[Cr(H_2O)_6]I_3{\cdot}3H_2O$	594.85	dk vlt cr, hygro	4.915^{25}_{4}	41 −HI	...	v s	v s	s al, acet; i CHI
c402	(II) iodide	CrI_2	305.80	graysh powd	5.196	856	subl vac 800	s	...	...
c403	(III) iodide	CrI_3	432.71	shiny blk cr.	4.915^{25}_{5}	>600	$-I_2$, vac 350	...	...	...
c404	(III) nitrate	$Cr(NO_3)_3{\cdot}7\frac{1}{2}H_2O$	373.13	br, monocl	...	100	d	s	s	...
c405	(III) nitrate	$Cr(NO_3)_3{\cdot}9H_2O$	400.15	purple, monocl.	...	60	d 100	s	s	s a, slk, al, acet
c406	nitride, mono-	CrN	66.00	cub or amorph	5.9	d 1700	...	i	i	sl s aq reg
c407	(II) oxlate	$CrC_2O_4{\cdot}H_2O$	158.03	yel cr powd.	2.468	...	...	sl s	s	s dil a
c408	(III) oxalate	$Cr_2(C_2O_4)_3{\cdot}6H_2O$	476.14	red, amorph, hyg	...	120, $-H_2O$ tr to grn	...	s	s	v s (red) al, eth; i (grn) al
c409	oxide, di-	CrO_2	83.99	br-blk powd.	...	300, −O	...	i	...	s HNO_3
c410	(II) oxide, mon-	CrO	68.00	blk powd	...	...	...	i	i	i dil HNO_3
c411	(III) oxide, sesqui-	Cr_2O_3	151.99	grn, hex, 2.551	5.21	2266 ± 25	4000	i	i	i a, alk, al
c412	(III) oxide, sesqui-	$Cr_2O_3{\cdot}xH_2O$	varies	vlt, amorph or bl-gray grn gel	...	...	...	i	i	s a, alk; sl s NH_4OH
c413	oxide, tri-	Chromic anhydride, "chromic acid", CrO_3	99.99	red, rhomb, deliq	2.70	196	d	61.7°	67.45[100]	s al, eth, H_2SO_4 HNO_3
c414	oxychloride	CrO_2Cl_2	154.90	dk red liq	1.911	−96.5	117	d	d	d al; s eth, ac a
c415	2,4-pentandione	Chrominium acetylacetonate. $Cr(C_5H_7O_2)_3$	349.32	...	...	216	340	i	...	s org solv; i lgr
c416	(III) *ortho*phosphate	$CrPO_4{\cdot}2H_2O$	183.00	vlt cr	2.42[32.5]	...	...	sl s	...	s a, alk; i ac a
c417	(III) *ortho*phosphate	$CrPO_4{\cdot}6H_2O$	255.06	vlt, tricl, 1.568, 1.591, 1.699	2.121[14]	100	...	...	...	...
c418	*pyro*phosphate	$Cr_4(P_2O_7)_3$	729.81	pa grn, monocl.	3.2	...	...	i	i	s alk
c419	phosphide, mono-	CrP	82.97	gray-blk cr	5.7[15]	...	...	i	...	s HNO_3, HF
c420	silicide	Cr_3Si_2	212.16	gray, tetr pr.	5.5°	...	...	i	i	s HCl, HF; i H_2SO_4, HNO_3
c421	(II) sulfate	$CrSO_4{\cdot}7H_2O$	274.16	bl cr	...	...	...	12.35°	d	s ls al; s NH_4OH
c422	(III) sulfate	$Cr_2(S)_4)_3$	392.16	vlt or red powd	3.012	...	...	i, s*	...	sl s al; i a
c423	(III) sulfate	$Cr_2(SO_4)_3{\cdot}15H_2O$	622.39	vlt, amorph sc	1.867[17]	100	$-10H_2O$, 100	s	d 67	i al
c424	(III) sulfate	$Cr_2(SO_4)_3{\cdot}18H_2O$	716.44	bl vlt, cub oct, 1.564	1.7[22]	$-12H_2O$, 100	...	120[20]	s	s al
c425	(II) sulfide, mono-	CrS	84.06	blk powd, hex	4.85	1550	...	i	...	v s a
c426	(III) sulfide, sesqui-	Cr_2S_3	200.17	brn-blk powd.	3.77[19]	−S, 1350	...	i, d	i	s HNO_3; d al
c427	(III) sulfite	$Cr_2(SO_3)_3$	344.17	grnsh-wh	2.2	d	...	...	...	...
c428	(II) tartrate	$CrC_4H_4O_6$	200.07	bl powd	2.33	...	...	i	i	sl s a; i ac a
c429	**Chromium complexes** *hex*ammine chromium-(III) chloride	$[Cr(NH_3)_6]Cl_3{\cdot}H_2O$	278.55	yel cr.	1.585	...	...	s	...	...
c430	*hex*aureachromium- (III) fluosilicate	$[Cr(CON_2H_4)_6]_2 \cdot [SiF_6]_3{\cdot}3H_2O$	1304.93	lt grn leaf	...	...	...	0.522[20]	...	i al

* Several chromic salts exist in two forms, a soluble and insoluble modification.

No.	Name	Synonyms and Formulae	Mol. wt.	Crystalline form, properties and index of refraction	Density or spec. gravity	Melting point, °C	Boiling point, °C	Solubility, in grams per 100 cc Cold water	Hot water	Other solvents
c431	*hex*aureachromium- (III) *per*rhenate	$[Cr(CON_2H_4)_6]\cdot(ReO_4)_3$	1162.44	grn need	2.652_4^{25}			1.786		0.667 al
c432	chloropentammine chromium chloride	$[Cr(NH_3)_5Cl]Cl_2$	243.51	red, oct	1.696			0.65^{16}		i HCl
c433	**Cobalt**	Co	58.9332	silv gray met, cub	8.9	1495	2870	i	i	s a
c434	(III) acetate	$Co(C_2H_3O_2)_3$	236.07	grn, oct		d 100		hydr		s a, glac ac a
c435	(II) acetate	$Co(C_2H_3O_2)_2\cdot4H_2O$	249.08	red-vlt, monocl, deliq, 1.542	1.705^{19}	$-4H_2O$, 140		s	s	s a, al
c436	aluminate	(approx) Thenard's blue. $CoAl_2O_4$	176.89	bl, cub				i	i	
c437	(II) *ortho*arsenate	$Co_3(AsO_4)_2\cdot8H_2O$	598.76	vlt-red, monocl, 1.626, 1.661, 1.669	3.178^{15}	d		i	i	s dil a, NH_4OH
c438	arsenic sulfide	Nat. cobaltite. CoAsS	165.91	gray-redsh	6.2—6.3	d				
c439	arsenide	Co_2As	192.79	cr powd	8.28	950		i	i	i CHl, H_2SO_4; s HNO_3, aq reg
c440	(II) benzoate	$Co(C_7H_5O_2)_2\cdot4H_2O$	373.23	gray red leaf		$-4H_2O$, 115		v s		
c441	boride, mono-	CoB	69.74	pr	7.25^{18}			d	d	s HNO_3, aq reg
c442	(II) bromate	$Co(BrO_3)_2\cdot6H_2O$	422.83	red, oct				45.5^{17}		s NH_4OH
c443	(II) bromide	$CoBr_2$	218.74	grn, hex. deliq	4.909_4^{25}	678 (in N_2)		66.7^{59}	68.1^{97}	77.1^{20} al; 58.6^{30} MeOH; s eth, acet
c444	(II) bromide hexahydrate	$CoBr_2\cdot6H_2O$	626.84	red-vlt pr, deliq	2.46	47—48, $-4H_2O$ 100	$-6H_2O$, 130	s red color	153.2^{97}	s blk color, al, a, eth
c445	bromoplatinate	$CoPtBr_6\cdot12H_2O$	949.62	trig	2.762					
c446	carbonate	Nat. spherocobaltite. $CoCO_3$	118.94	red, trig, 1.855, 1.60	4.13	d		i	i	s a; i NH_3
c447	(II) carbonate, basic	$2CoCO_3\cdot3Co(OH)_2\cdot H_2O$	534.74	vlt-red pr				i	d	s a, $(NH_4)_2CO_3$
c448	carbonyl tetra-	Dicobalt octacarbonyl. $[Co(CO)_4]_2$ or $Co_2(CO)_8$	341.95	or cr or dk br, microcr	1.73^{18}	51	d 52	i	i	sl s al; s CS_2, eth
c449	carbonyl, tri-	Tetracobalt dodecacarbonyl. $[Co(CO)_3]_4$ or $Co_4(CO)_{12}$	571.86	blk cr				sl s		s bz; d Br
c450	(II) chlorate	$Co(ClO_3)_2\cdot6H_2O$	333.93	red, cub, deliq, 1.55	1.92	50	d 100	558.3^0	v s	s al
c451	(II) *per*chlorate	$Co(ClO_4)_2$	257.83	red need 1.510, 1.490	3.327			100^0	115^{45}	s al, acet
c452	(II) *per*chlorate	$Co(ClO_4)_2\cdot5H_2O$	347.91	red, hex		143		100.13^0	115.10^{65}	v s al, acet i $CHCl_3$
c453	*per*chlorate	$Co(ClO_4)_2\cdot6H_2O$	365.93	red pr		d 1534	d	259^{18}		s al, acet
c454	(II) *per*chlorate	$Co(ClO_4)_2\cdot6H_2O$	365.93	red, oct, deliq, 1.55		d 182		255^{18}		s al, acet
c455	(II) chloride	$CoCl_2$	129.84	bl, hex, hygr	3.356_4^{36}	724 (in HCl gas)	1049	45^7	105^{96}	54.4 al; 8.6 acet; 38.5 MeOH; sl s eth
c456	(III) chloride	$CoCl_3$	165.29	red cr or yel cr.	2.94	subl		s		
c457	(II) chloride, dihydrate	$CoCl_2\cdot2H_2O$	165.87	red-vlt, monocl or tricl, 1.625, 1.671, 1.67	2.477_{26}^{25}			s	s	v sl s eth
c458	(II) chloride, hexahydrate	$CoCl_2\cdot6H_2O$	237.93	red, monocl.	1.924_{25}^{25}	86	$-6H_2O$, 110	76.7^0	190.7^{100}	v s (bl col) al; s acet; 0.29 eth
c459	chloroplatinate	$CoPtCl_6\cdot6H_2O$	574.82	trig	2.699	d				
c460	chlorostannate	$CoSnCl_6\cdot6H_2O$	498.43	rhomb or trig		d 100				
c461	(II) chromate	$CoCrO_4$	174.93	gray blk cr		d		i	d	s a, NH_4OH
c462	(II) citrate	$Co_3(C_6H_5O_7)_2\cdot2H_2O$	591.03	rose-red		$-2H_2O$, 150		0.8		
c463	(II) cyanide dihydrate	$Co(CN)_2\cdot2H_2O$	147.00	buff anhydr bl-vlt powd	anhydr 1.872^{25}	$-2H_2O$, 280	d 300	0.00418^{18}		s KCN, HCl, NH_4OH
c464	(II) cyanide, trihydrate	$Co(CN)_2\cdot3H_2O$	165.01	red-gray powd. amorph		$-3H_2O$, 250		i		s KCN
c465	(II) ferricyanide	$Co_3[Fe(CN)_6]_2$	600.71	red need				i		s NH_4OH; i HCl
c466	(II) ferrocyanide	$Co_2Fe(CN)_6\cdot xH_2O$		gray-grn				i		s KCN; i HCl
c467	(II) fluogallate	$[Co(H_2O)_6][GaF_5\cdot H_2O]$	349.75	pink cr, monocl (?), 1.45	2.35	$-5H_2O$, 110		sl s		d a
c468	(II) fluoride	CoF_2	96.93	pink monocl	4.46_4^{25}	*ca* 1200	1400	1.5^{25}	s	sl s a; i al, eth, bz
c469	(III) fluoride	CoF_3	115.93	br, hex	3.88			d to $Co(OH)_3$		i, al,eth, bz
c470	fluoride	$Co_2F_6\cdot7H_2O$	357.96	grn powd	2.314^{25}			d		s H_2SO_4
c471	(II) fluoride, tetrahydrate	$CoF_2\cdot4H_2O$	168.99	α: red, rhomb oct, β: rose cr powd	2.192_4^{25}	d 200		s	s	i al
c472	fluosilicate	$CoSiF_6\cdot6H_2O$	309.10	pink trig, 1.382, 1.387	2.113^{19}			$118.1^{21.5}$		
c473	(II) formate	$Co(CHO_2)_2\cdot2H_2O$	185.00	red cr	2.129^{22}	$-2H_2O$, 140	d 175	5.03^{20}		
c474	(III) hydroxide	$Co(OH)_2$	92.95	rose-red, rhomb	3.597^{15}	d		0.00032		s a, NH_4 salts; i alk
c475	(III) hydroxide	$Co_2O_3\cdot3H_2O$	219.91	blk-brn powd	4.46	d	$-H_2O$, 100	0.00032		s a; i al
c476	(II) iodate	$Co(IO_3)_2$	408.74	bl-vlt need	5.008^{18}	d 200		0.45^{18}	1.33^{100}	s HCl, HNO_3. h H_2SO_4
c477	(II) iodate, hexahydrate	$Co(IO_3)_2\cdot6H_2O$	516.83	red, oct	3.689^{21}	d 61	$-4H_2O$, 135	s		
c478[1]	(II) iodide (α) stable	CoI_2	312.74	blk hex, hyg	5.68	515(vac)	570 (vac)	159^0	420^{100}	v s al, acet
c478[2]	(II) iodide (β)	CoI_2	312.74	yel need, unstab	5.45^{25}	d 400		s		
c479	(II) iodide, dihydrate	$CoI_2\cdot2H_2O$	348.77	grn, deliq		d 100		376.2^{45}	s	
c480	(II) iodide, hexahydrate	$CoI_2\cdot6H_2O$	420.83	br-red hex, hygr	2.90	d 27, $-6H_2O$,		s	s	s al, eth, chl
c481	iodoplatinate	$CoPtI_6\cdot9H_2O$	1177.59	trig	3.618					

No.	Name	Synonyms and Formulae	Mol. wt.	Crystalline form, properties and index of refraction	Density or spec. gravity	Melting point, °C	Boiling point, °C	Solubility, in grams per 100 cc		
								Cold water	Hot water	Other solvents
c482	linoleate	$Co(C_{18}H_{31}O_2)_2$	617.82	br, amorph				i		s al,eth,acet
c483	(II) nitrate	$Co(NO_3)_2 \cdot 6H_2O$	291.03	red, monocl, 1.52	1.87^{25}_{4}	55—56	$-3H_2O$, 55	133.8^{0}	217^{80}	$100.0^{12.5}$ al; s acet; sl s NH_3
c484	nitrosylcarbonyl	$Co(NO)(CO)_3$	172.97	cherryred liq		−1.05	48.6; d 55	i		s al, eth, acet, bz
c485	(II) oleate	$Co(C_{18}H_{35}O_2)_2$	621.85	br, amorph						s al, eth, oils, bz
c486	(II) oxalate	CoC_2O_4	146.95	wh or redsh	3.021^{25}	d 250		i		s a, NH_4OH
c487	oxalate, dihydrate	$CoC_2O_4 \cdot 2H_2O$	182.98	pink cr		$-H_2O$, *ca* 190		v sl s	sl s	v sl s a; s NH_4OH
c488	(II) oxide	CoO	74.93	pink cub	6.45	1795 ± 20		i	i	s a; i al, NH_4OH
c489	(III) oxide	Co_2O_3	165.86	blk-gray, hex, or rhomb	5.18	d 895		i	i	s a; i al
c490	(II, III) oxide	Co_3O_4	240.80	blk, cub	6.07	tr to CoO 900—950		i	i	v sl s a; i sq reg
c491	palmitate	$Co(C_{16}H_{31}O_2)_2$	569.78			70.5				s pyr, hot CS_2, CCl_4; sl s eth; i MeOH, acet
c492	(II) *ortho*phosphate	$Co_3(PO_4)_2$	366.74	redsh cr	2.587^{25}			i	i	s H_3PO_4, NH_4OH
c493	(II) *ortho*phosphate, dihydrate	$Co_3(PO_4)_2 \cdot 2H_2O$	402.77	pink powd.				i		s H_3PO_4
c494	(II) *ortho*phosphate, octahydrate	$Co_3(PO_4)_2 \cdot 8H_2O$	510.86	redsh powd	2.769^{25}	$-8H_2O$, 200		sl s		s min a, H_3PO_4; i al
c495	phosphide	Co_2P	148.84	gray need	6.4^{15}	1386		i	i	s HNO_3, aq reg
c496	(II) propionate	$Co(C_3H_5O_2)_2 \cdot 3H_2O$	259.12	dk-red cr		*ca*250		anh 33.5^{11}		v s al
c497	(II) *per*rhenate	$Co(ReO_4)_2 \cdot 5H_2O$	649.42	dk pink		d		d		
c498	(II) selenate, heptahydrate	$CoSeO_4 \cdot 7H_2O$	328.00	monocl.	2.135					
c499	selenate, hexahydrate	$CoSeO_4 \cdot 6H_2O$	309.98	red, monocl, 1.5225	2.25^{17}			s	s	
c500	(II) selenate, pentahydrate	$CoSeO_4 \cdot 5H_2O$	291.97	ruby red, tricl-	2.512	d		v s		
c501	selenide, mono-	CoSe	137.89	yel, hex	7.65	red heat				sHNO_3,aq reg; i alk
c502	(II) *ortho*silicate	CO_2SiO_4	209.95	vlt cr, rhomb	4.63	1345		i	i	s dil HCl
c503	silicide	CoSi	87.02	rhomb		1395				s HCl; i HNO_3, H_2SO_4
c504	silicide,di-	$CoSi_2$	115.10	rhomb	5.3	1277				
c505	(di-)silicide	Co_2Si	145.95	gray cr	7.28°	1327				d a
c506	(II) *ortho*stannate	Co_2SnO_4	300.55	grnsh-bl, cub	6.30^{18}					i H_2SO_4; s h HCl
c507	(II) sulfate	$CoSo_4$	154.99	dk blsh, cub	3.71^{25}_{25}	d 735		36.2^{20}	83^{100}	1.04^{18} MeOH; i NH_3
c508	(II) sulfate, heptahydrate	Nat. bieberite. $CoSO_4 \cdot 7H_2O$	281.10	red-pink, monocl, 1.477, 1.483, 1.489	1.948^{25}_{25}	96.8	$-7H_2O$, 420	60.4^{3}	67^{70}	2.5^{3} al; 54.5^{18} MeOH
c509	(II) sulfate, hexahydrate	$CoSO_4 \cdot 6H_2O$	263.08	red, monocl, 1.531, 1.549, 1.552	2.019^{15}_{15}	$-2H_2O$, 95				
c510	(II) sulfate, monohydrate	$CoSO_4 \cdot H_2O$	173.01	red cr, 1.603, 1.639, 1.683	3.075^{25}	d		s	s	
c511	(III) sulfate	$Co_2(SO_4)_3 \cdot 18H_2O$	730.31	bl-grn		d 35		s d		s H_2SO_4; i pyr
c512	sulfide, di-	CoS_2	123.05	blk, cub	4.269			i		s HNO_3, aq reg
c513	sulfide, mono-	Nat. sycoporite. CoS.	90.99	redsh, silv-wh, oct	5.45^{18}	>1116		0.00038^{18}		sl s a
c514	(III) sulfide, sesqui	Co_2S_3	214.05	blk cr	4.8					d a, aq reg
c515	(tri-) sulfide	Cobalt sulfide, tetra-(ous, ic) Nat. linneite. Co_3S_4	305.04	dk gray, cub	4.86	d 480				
c516	(II) sulfite	$CoSO_3 \cdot 5H_2O$	229.07	red				i		s H_2SO_3
c517	tartrate	$CoC_4H_4O_6$	207.01	redsh, monocl				sl s		s dil a
c518	thiocyanate	$Co(SCN)_2 \cdot 3H_2O$	229.13	vlt, rhomb.		$-3H_2O$, 105		s		s al, MeOH, eth
c519	*ortho*titanate	Co_2TiO_4	229.74	grnsh-blk, cub.	5.07—5.12					s conc HCl; sl s dil HCl
c520	(II) tungstate	$CoWO_4$	306.78	bl-grn, monocl	8.42			i		s h conc a; sl s c dil a
	Cobalt complexes									
c521	hexammine cobalt (II) bromide	$CoBr_2 \cdot 6NH_3$	320.92	dk pink cr	1.871^{25}_{4}	d 258		d		
c522	diamminecobalt (II) chloride [α]	$CoCl_2 \cdot 2NH_3$	163.90	rose cr	2.097	273				
c523	diamminecobalt (II) chloride (β)	$CoCl_2 \cdot 2NH_3$	163.90	bl-vlt	2.073	tr to α, 210 (ín NH_3)				
c524	hexamminecobalt (II) chloride	$[Co(NH_3)_6]Cl_2$	232.02	rose red, oct	1.497	d		d		s NH_4OH; i abs al
c525	hexamminecobalt (III) chloride	$Co(NH_3)_6Cl_3$	267.48	wine-red, monocl	1.710^{25}_{4}	$-1NH_3$, 215		5.9^{10}	$12.74^{46.6}$	s conc HCl; i al,, NH_4OH
c526	hexamminecobalt (II) iodide	$CoI_2 \cdot 6NH_3$	414.93	dk pink, cub	2.096^{25}_{4}	141^{100mm}				
c527	hexamminecobalt (III) nitrate	$Co(NH_3)_6 \cdot (NO_3)_3$	347.13	yel, tetr	1.804^{25}_{4}			1.7^{25}	v s	v sl s dil a
c528	hexamminecobalt (III) *per*rhenate	$[Co(NH_3)_6](ReO_4)_3 \cdot 2H_2O$	947.76	or-yel pr	3.329^{25}			0.0469		
c529	hexamminecobalt (II) sulfate	$CoSO_4 \cdot 6NH_3$	257.17	pink powd.	1.654^{25}_{4}	d 116^{760}		d		v s dil NH_3
c530	hexamminecobalt (III) sulfate	$[Co(NH_3)_6]_2(SO_4)_3 \cdot 5H_2O$	700.48	dk yel, monocl	1.797^{25} anhydr	$-4H_2O$, 100	$-5H_2O$, 150	$1.4^{17.4}$		
c531	ammonium tetranitrodiammine (III) cobaltate	Erdmann's salt. $NH_4[Co(NH_3)_2(NO_2)_4]$	295.05	redsh-pa brn, rhomb, 1.78, 1.78. 1.74	1.876^{25}					

No.	Name	Synonyms and Formulae	Mol. wt.	Crystalline form, properties and index of refraction	Density or spec. gravity	Melting point, °C	Boiling point, °C	Solubility, in grams per 100 cc		
								Cold water	Hot water	Other solvents
c532	aquapentamminecobalt (III) chloride (roseo)	$[Co(NH_3)_5{\cdot}H_2O]Cl_3$	268.46	brick red cr	1.7^{25}	d 100	...	24.87^{25}	...	sl s HCl; i al
c533	aquapentamminecobalt (III)-sulfate (roseo)	$[Co(NH_3)_5H_2O]_2(SO_4)_3{\cdot}2H_2O$	638.33	red, tetr	1.854^{20}	$-3H_2O$, 99	d 110	$1^{17.2}$	1.72^{27}	s H_2SO_4
c534	cis-chloroaquotetrammine-cobalt (III) chloride	$[Co(NH_3)_4H_2O)Cl]Cl_2$	251.43	vlt, rhomb.	1.847	d	...	1.4^{0}	...	s a; i al
c535	chloropentamminecobalt-(III) chloride (purpureo)	$[Co(NH_3)_5Cl]Cl_2$	250.44	dk red-vlt, rhomb	1.819^{25}_{25}	d	...	0.4^{25}	$1.031^{46.6}$	s conc H_2SO_4; i al
c536	triethylenediammine-cobalt-(III) chloride	$Co[C_2H_4(NH_2)_2]_3Cl_3{\cdot}3H_2O$	399.64	br pr	1.542^{17}	256; $-3H_2O$, 100	...	v s	...	...
c537	trinitrotriamminecobalt	$CO(NH_3)_3(NO_2)_3$	248.04	yel, rhomb pl or leaf	1.992^{25}_{4}	d 158	exp 164	$0.177^{16.5}$	0.28^{25}	...
c538	trinitrotetramminecobalt-(III)nitrate	$[Co(NH_3)_4(NO_2)_2]NO_2$	265.07	hel, rhomb	1.922^{17}	...	...	3^{20}	...	...
c539	potassium tetranitrodiam-minecobaltate (III)	$K[Co(NH_3)_2(NO_2)_4]$	316.11	yel, rhomb	2.076^{15}	...	...	$1.758^{16.5}$	...	...
	Columbium	see Niobium.		...						
c540	**Copper**	Cu	63.546 ± 3	redsh met, cub	8.92	1083.4±0.2	2567	i	i	s HNO_3, h H_2SO_4; v sl s HCl, NH_4OH
c541	acetate, basic	Blue verdigris. $Cu(C_2H_3O_2)_2{\cdot}CuO.6H_2O$	369.27	grnsh-bl powd	...	...	...	sl s	...	s dil a, NH_4OH; sl s al
c542	(II) acetate	Neutral verdigris. $Cu(C_2H_3O_2)_2{\cdot}H_2O$	199.65	dk grn powd, 1.545, 1.550	1.882, anhydr-1.93	115	d 240	7.2	20	7.14 al; s eth
c543	(II) acetate metaarsenate	Paris green. $Cu(C_2H_3O_2)_2{\cdot}3Cu(AsO_2)_2$ (approx)	1013.80	em grn powd	...	...	...	i	...	s a, NH_4OH; i al
c544	(III) acetylide	Cu_2C_2	151.11	red, amorph, expl.	...	exp	...	v sl s	...	s a, KCN
c545	amine azide	$Cu(NH_3)_2(N_3)_2$	181.65	dk grn cr, exp	...	d 100—105	exp 202	i	d	d a; i MeOH
c546	(II) *di*amminechloride, di-	$Cu(NH_3)_2Cl_2$	168.51	grn cr	2.32^{25}_{4}	260—270	d 300	i	...	s NH_4OH; i abs a
c547	(II) *hex*amminechloride, di-	$Cu(NH_3)_6Cl_2$	236.64	bl, cub	1.48^{25}_{4}	...	...	v s	...	...
c548	*tetr*ammine dithionate	$[Cu(NH_3)_4]S_2O_6$	291.78	vlt-bl cr	...	d 160	...	s	d	...
c549	(II) *tetr*ammine nitrate	$[Cu(NH_3)_4](NO_3)_2$	225.68	dk-bl, oct	1.91^{25}_{4}	d 210 exp	...	s	...	...
c550	(II) amine nitrate	$[Cu(NH_3)_4](NO_2)_2$	223.68	vlt-bl, tetr	...	$-2NH_3$ 97	...	v s	...	...
c551	*tetr*ammine sulfate	Cuprum ammoniacale. $[Cu(NH_3)_4]SO_4{\cdot}H_2O$	245.74	dk-bl, rhomb, unstab	1.79^{25}_{4}	$-NH_3.H_2O$, 30	...	$18.05^{21.5}$	...	...
c552	(tri-)antimonide	Cu_3Sb	312.39	gray	8.51	687	...	...	...	...
c553	(II) *ortho*arsenate	$Cu_3(AsO_4)_2{\cdot}4H_2O$	540.54	blsh-grn	...	...	...	i	i	s a, NH_4OH
c554	(II) *ortho*arsenate, di-H	$Cu_5H_2(AsO_4)_4{\cdot}2H_2O$	911.45	bl	...	...	...	i	...	s a, NH_4OH
c555	arsenide	Cu_5As_2	467.57	bl, oct	7.56	...	...	i	i	s a, NH_4OH
c556	tri-arsenide	Nat. domeykite, Cu_3As.	265.56	hex	8.0	830	...	...	...	...
c557	(II) *ortho*arsenite, hydrogen(?)	Scheele's green. $CuHAsO_3$(?)	187.47	grn powd	...	d	...	i	i	s a, NH_4OH; i al
c558	(I) azide	CuN_3	105.57	col cr, v exp	3.26	...	...	0.00075^{20}	...	d conc H_2SO_4; s NH_4Cl
c559	(II) azide	$Cu(N_3)_2$	147.59	brn-red or brn-yel cr, exp	2.604	exp 215	...	0.008^{20}	...	v s dil a
c560	(II) benzoate	$Cu(C_7H_5O_2)_2{\cdot}2H_2O$	341.81	lt bl cr powd	...	$-H_2O$, 110	...	sl s	...	s dil a; sl s al
c561	(II) *meta*borate	$Cu(BO_2)_2$	149.16	blsh grn cr powd	3.859	...	...	s	...	...
c562	boride	Cu_3B_2	212.26	yel	8.116	...	...	...	...	...
c563	(II) bromate	$Cu(BrO_3)_2{\cdot}6H_2O$	427.44	bl-grn, cub	2.583	d 180	$-6H_2O$, 200	v s	...	s NH_4 OH
c564	(I) bromide	CuBr (or Cu_2Br_2)	143.45	wh, cub, 2.116.	4.98	492	1345	v sl s	d	s HBr, HCl, HNO_3, NH_4OH; i acet
c565	(II) bromide	$CuBr_2$	223.35	blk, monocl, deliq	4.77^{25}_{4}	498	...	v s	...	s al, acet, NH_3, pyr; i bz
c566	*trioxy*bromide	$CuBr_2{\cdot}3Cu(OH)_2$	516.04	em grn, rhomb	4.00	$-H_2O$, 210—215	d 240—250	i	d	s dil min a, NH_4OH; v s ac a
c567	(II) butyrate	$Cu(C_4H_7O_2)_2{\cdot}2H_2O$	273.77	dk grn cr	...	...	...	v sl s	...	s al, eth, NH_4OH, dil a
c568	(I) carbonate	Cu_2CO_3	187.10	yel	4.40	d	...	i	i	s a, NH_4OH
c569	(II) carbonate, basic	Nat. malachite. $CuCO_3{\cdot}Cu(OH)_2$	221.12	dk grn, monocl, 1.655, 1.875, 1.909	4.0	d 200	...	i	d	0.026 aq CO_2; s a, NH_4OH, KCN; i al
c570	(II) carbonate, basic	Nat. azurite, chessylite. $2CuCO_3{\cdot}Cu(OH)_2$	344.67	bl, monocl, 1.730, 1.758, 1.838	3.88	d 220	...	i	d	s NH_4OH, h $NaHCO_3$
c571	(II) chlorate	$Cu(ClO_3)_2{\cdot}6H_2O$	338.54	grn, cub, deliq	...	65	d 100	207^{0}	v s	s al, acet
c572	(II) chlorate, basic	$Cu(ClO_3)_2{\cdot}3Cu(OH)_2$	523.13	grn cr or amorph	3.55	d	...	i	i	s dil a
c573	*per*chlorate	$Cu(ClO_4)_2$	262.45	monocl, 1.495, 1.505, 1.522	2.225^{23}	82.3	...	s	s	...
c574	*per*chlorate, hexahydrate	$Cu(ClO_4)_2{\cdot}6H_2O$	370.54	lt bl, tricl, deliq, 2.505	2.225^{25}_{4}	82	d 120	ws	...	s al, eth
c575	(I) chloride(ous)	Nat. nantokite. CuCl (or Cu_2Cl_2)	99.00	wh, cub, 1.93	4.14	430	1490	0.0062	...	s HCl, NH_4OH, eth; i al
c576	(II) chloride	$CuCl_2$	134.45	br, yel powd, hygr	3.386^{25}_{4}	620	993 d to CuCl	70.6^{0}	107.9^{100}	53^{15} al; 68^{15} MeOH; s h H_2SO_4, acet
c577	(II) chloride, basic	$CuCl_2{\cdot}Cu(OH)_2$	232.01	yel-grn, hex.	3.78	$-H_2O$, 250	d red heat	d	d	...
c578	(II) chloride, dihydrate	Nat. eriochalcite. $CuCl_2{\cdot}2H_2O$	170.43	bl-grn, rhomb, deliq. 1.644, 1.683, 1.731	2.54	$-2H_2O$, 100	d	110.4^{0}	192.4^{100}	s al, NH_4OH
c579	chloride, thioureate	$CuCl{\cdot}3[CS(NH_2)_2]$	327.35	col pr, 1.758, 1.17719	1.73	168	...	v s	...	...

No.	Name	Synonyms and Formulae	Mol. wt.	Crystalline form, properties and index of refraction	Density or spec. gravity	Melting point, °C	Boiling point, °C	Solubility, in grams per 100 cc		
								Cold water	Hot water	Other solvents
c580	(II) chromate, basic	$CuCrO_4·2CuO·2H_2O$	374.66	yel br		$-2H_2O$, 260		i		s dil a, NH_4OH; i al
c581	(II) *di*chromate	$CuCr_2O_7·2H_2O$	315.56	blk cr, deliq	2.283	$-2H_2O$, 100		v s	d	s a, NH_4OH, al
c582	(I) chromite	$Cu_2Cr_2O_4$	295.08	gray blk cub pl.	5.24^{20}			i	i	s HNO_3
c583	(II) citrate	$2Cu_2C_6H_4O_7·5H_2O$	720.45	blsh grn powd		$-H_2O$, 100		i		s a, NH_4OH
c584	(I) cyanide	CuCN	89.56	wh, monocl pr	2.92	473 (in N_2)	d	i	i	s HCl, KCN, NH_4OH; sl s
c585	(II) cyanide	$Cu(CN)_2$	115.58	yel-grn powd		d		i		s a, alk, KCN, pyr,
c586	ethylacetoacetate	$Cu(C_6H_9O_3)_2$	321.82	grn need		192—193	subl	i		v s al, eth; 10^{80} bz
c587	(I) ferricyanide	$Cu_3Fe(CN)_6$	402.59	br red				i		i HCl; s NH_4OH
c588	(II) ferricyanide	$Cu_3[Fe(CN)_6]_2·14H_2O$	866.76	yel-grn				i		i HCl; s NH_4OH
c589	(II) ferrocyanide	Hatchett's brown $Cu_2Fe(CN)_6·xH_2O$		red brn				i	i	i a, NH_3; s NH_4OH
c590	(I) fluogallate	$[Cu(H_2O)_6][GaF_5·H_2O]$	354.36	pa bl, monocl(?), 1.45	2.20	$-5H_2O$, 110		sl s		s HF
c591	(I) fluoride	CuF (or Cu_2F_2)	82.54	red cr, (exist?)		908	subl, 1100	i		s HCl, HF; d HNO_3; i al
c592	(II) fluoride	CuF_2	101.54	wh, tricl	4.23	d 950		4.7^{20}	s	s dil min a; i al
c593	(II) fluoride dihydrate	$CuF_2·2H_2O$	137.57	bl, monocl	2.93^{25}_{4}	d		4.7^{20}	d	s HCl, HF, HNO_3, al; i acet, NH_3
c594	(I) fluosilicate	Cu_2SiF_6	269.17	red powd		d to SiF_4			d 100	
c595	(II) fluosilicate	$CuSiF_6·4H_2O$	277.68	monocl pr	2.158			42.8		
c596	(II) fluosilicate hexahydrate	$CuSiF_6·6H_2O$	313.71	bl, rhomb, deliq, 1.409, 1.408	2.207			232^{17}		0.16^{20} 92% al
c597	(II) formate	$Cu(CHO_2)_2$	153.58	bl, monocl	1.831			12.5	d	0.25 al
c598	(II) formate tetrahydrate	$Cu(CHO_2)_2·4H_2O$	225.64	bl cr	1.81	$-H_2O$, 130		6.2		s alk; sl s al
c599	(II) glycerine deriv	$Cu(C_2H_4NO_2)_2·H_2O$	229.68	bl need		$-H_2O$, 130		0.57^{15}		s alk
c600	hydride	CuH (or Cu_2H_2)	64.55	red-brn, (exist?)	6.38	d sl 55—60		i	d 65	d HCl
c601	(II) hydroxide	$Cu(OH)_2$	97.56	bl gel cr powd	3.368	$-H_2O$, d		i	d	s a, NH_4OH, KCN
c602	(II) *tri*hydroxychloride	γ: Paratacamite δ: atacamite $CuCl_2·3Cu(OH)_2$	427.13	γ: grn, hex, δ: grn rhomb; γ: 1.743, 1.849, δ: 1.861, 1.861, 1.880, grn lt	(γ) 3.75	$-H_2O$, 250		i	i	v s a
c603	(II) *tri*hydroxynitrate	$Cu(NO_3)_2·3Cu(OH)_2$	480.24	dk grn, rhomb or moncl.	rhomb, 3.41 monocl, 3.378	$-H_2O$ ~400		i	d	v s a
c604	(II) iodate	$Cu(IO_3)_2$	413.35	grn, moncl	5.241^{15}	d		0.1364^{15}	i	s dil HNO_3, dil H_2SO_4
c605	(II) iodate, basic	$Cu(OH)IO_3$	255.46	grn, rhomb	4.873	d 290		i	i	s dil H_2SO_4
c606	(II) iodate, monohydrate	Nat. bellingerite. $Cu_3(IO_3)_6·2H_2O$	431.37	bl, tricl	4.872	$-H_2O$, 248	d 290	0.33^{15}	0.65^{100}	s dil H_2SO_4, NH_4OH; i al, dil HNO_3
c607	*para*periodate	Cu_2HIO_6	351.00	grn cr powd.		d 110		i	i	s HNO_3, NH_4OH
c608	(I) iodide	Nat. marshite. CuI (or Cu_2I_2)	190.45	wh or brnsh-wh, cub, 2.346	5.62	605	1290	0.008^{18}		s dil HCl, KI, KCN, con c H_2SO_4, liq NH_3
c609	(II) lactate	$Cu(C_3H_5O_3)_2·2H_2O$	277.72	dk bl, monocl				16.7	45^{100}	s NH_4OH; sl s al
c610	(II) laurate	$Cu(C_{12}H_{23}O_2)_2$	462.17	lt bl powd		111—113		sl s	sl s	
c611	mercury iodide (α)	Cu_2HgI_4	835.30	red, tetr	6.116	*tr ca* 67		i		
c612	mercury iodide (β)	Cu_2HgI_4	835.30	choc, cub	6.102			i		
c613	(II) nitrate, hexahydrate	$Cu(NO_3)_2·H_2O$	295.65	bl cr, deliq	2.074	$-3H_2O$, 26.4		243.7^{0}	∞	s al
c614	(II) nitrate, trihydrate	$Cu(NO_3)_2·3H_2O$	241.60	bl cr, deliq	2.32^{25}_{4}	114.5	$-HNO_3$, 170	137.8^{0}	1270^{100}	$100^{12.5}$ al; v s liq NH_3
c615	nitride	Cu_3N	204.64	dk grn powd	5.84^{25}_{4}	d 300		d		d a
c616	(II) nitrite, basic	$Cu(NO_2)_2·3Cu(OH)_2$	448.24	grn powd		d 120		i	d	v s dil a; sl s al; s NH_4OH
c617	(II) *hypo*nitrite, basic	$Cu(NO)_2·Cu(OH)_2$	221.12	pea grn amorph, hygr.		d<100		i		s dil a; v s NH_4OH; d NaOH
c618	(II) nitroprusside	$CuFe(CN)_5NO·2H_2O$	315.52	wh-grnsh powd.				i		s alk; i al
c619	(II) oleate	$Cu(C_{18}H_{33}O_2)_2$	626.46	br powd or grn-bl mass, pois				i	i	s eth
c620	(II) oxalate	$CuC_2O_4·{}^1/_2H_2O$	160.57	bl wh.				0.00253^{25}		s NH_4OH; i ac a
c621	(I) oxide	Nat. cuprite. Cu_2O	143.09	red, oct cub, 2.705	6.0	1235	−O, 1800	i	i	s HCl, NH_4Cl, NH_4OH; sl s HNO_3; i al
c622	(II) oxide	Nat. tenorite. CuO	79.55	blk, monocl, β 2.63	6.3—6.49	1326		i	i	s a, NH_4Cl, KCN
c623	oxide, per-	$CuO_2·H_2O$	113.56	br or brnsh-blk cr		d 60		i		i al, s d a
c624	oxide, sub-	Cu_4O	270.18	olv grn, (exist?)		d		i		d a
c625	(II) oxychloride	Nat. atacamite. $Cu_2(OH)_3Cl$ (or $CuCl_2·3Cu(OH)_2$)	213.57	grn, orthorhomb.	3.76—3.78					
c626	(II) oxychloride	Brunswick green. $CuCl_2·3CuO·4H_2O$(?)	445.15	grn powd, or em grn to grnsh-blk, rhomb		$-H_2O$, 140			d 100	s a, NH_4OH
c627	(II) palmitate	$Cu(C_{16}H_{31}O_2)_2$	574.39	grn-bl powd.		120		i		s h bz, CS_2, CCl_4; sl s al, eth; i MeOH, acet
c628	2,4-pentanedione	Copper acetylacetonate $Cu(C_5H_7O_2)_2$	261.76	bl cr		>230	subl	i		sl s al; s chl
c629	(I) phenyl	C_6H_5Cu	140.65	col powd		d 80		d	d	i al, CS_2; s pyr

No.	Name	Synonyms and Formulae	Mol. wt.	Crystalline form, properties and index of refraction	Density or spec. gravity	Melting point, °C	Boiling point, °C	Solubility, in grams per 100 cc		
								Cold water	Hot water	Other solvents
c630	(II) *ortho*phosphate	$Cu_3(PO_4)_2 \cdot 3H_2O$	434.63	bl, rhomb		d		i	sl s	s a, NH_4OH, H_3PO_4; i NH_3
c631	(tri-) phosphide	Cu_3P	221.61	gray-blk	6.4—6.8	d		i		s HNO_3; i HCl
c632	(di-) pyridine chloride(di)	$Cu(C_5H_5N)_2Cl_2$	292.65	grn-bluish, monocl, 1.60, 1.75	1.76	d 263		s	d	sl s c al, chl
c633	(II) salicylate	$Cu(C_7H_5O_3)_2 \cdot 4H_2O$	409.84	bl-grn need				v s		v s al, NH_4OH
c634	(II) selenate	$CuSeO_4 \cdot 5H_2O$	296.58	bl, tricl, 1.56	2.559	$-4H_2O$, 50—100	$-5H_2O$, 150	25.7^{15}	d	s a, NH_4OH; v sl s acet; i a
c635	(I) selenide	Cu_2Se	206.05	blk, cub	6.749^{30}_{4}	1113				d HCl
c636	(II) selenide	CuSe	142.51	grn-blk hex pl, unstab	5.99	d red heat		i	i	sl s HCl, NH_4OH; s h HNO_3
c637	selenite	$CuSeO_3 \cdot 2H_2O$	226.53	bl-grn, rhomb	3.31^{25}_{4}	$-H_2O$, 100		i	i	
c638	silicide	Cu_4Si	282.27	wh met	7.53	850				i HCl; d HNO_3
c639	(II) stearate	$Cu(C_{18}H_{35}O_2)_2$	630.50	lt grn-bl amorph powd		125		i		s eth, h bz, chl, turp; sl s pyr; i MeOH, acet
c640	(I) sulfate	Cu_2SO_4	223.15	gray powd, 1.724, 1.733, 1.739	3.605	+O, 200		d		s conc HCl, NH_3 glac ac a
c641	(II) sulfate	Nat. hydrocyanite. $CuSO_4$	159.60	grn, wh, rhomb, 1.733	3.603	sl d above 200	d 650 to CuO	14.3^{0}	75.4^{100}	1.04^{18} MeOH; i al
c642	(II) sulfate, basic	Nat. brochantite. $CuSO_4 \cdot 3Cu(OH)_2$	452.29	grn, monocl, 1.728, 1.771, 1.800	3.78	d 300		i	i	s a, NH_4OH
c643	(II) sulfate, pentahydrate	Bluevitriol, nat. chalcanthite. $CuSO_4 \cdot 5H_2O$	249.68	bl, tricl, 1.514, 1.537, 1.543	2.284	$-4H_2O$, 110	$-5H_2O$, 150	31.6^{0}	203.3^{100}	15.6^{18} MeOH; i al
c644	(I) sulfide	Nat. chalcocite. Cu_2S	159.15	blk, rhomb	5.6	1100		$X10^{-14}$		s HNO_3, NH_4OH; i acet
c645	(II) sulfide	Nat. covellite. CuS	95.61	blk, monocl or hex, 1.45	4.6	tr 103	d 220	0.000033^{18}		s HNO_3, KCN, h HCl, H_2SO_4; i al, alk
c646	(I) sulfite, monohydrate	$Cu_2SO_3 \cdot H_2O$	225.17	red pr	4.46^{15}			i		d dil a; s NH_4OH
c647	(I) sulfite, monohydrate	$Cu_2SO_3 \cdot H_2O$	225.17	wh, hex	3.83^{15}	d		sl s		s HCl, NH_4OH; i al, eth
c648	(I, II) sulfite, dihydrate	Chevreul's salt. $Cu_2SO_3 \cdot CuSO_3 \cdot 2H_2O$	386.78	red cr	3.57	d 200		i	i	s HCl, NH_4OH; sl s HNO_3
c649	(II) tartrate	$CuC_4H_4O_6$	211.62	lt bl powd				v sl s		s a, alk
c650	(II) tartrate, trihydrate	$CuC_4H_4O_6 \cdot 3H_2O$	265.66	lt gray-bl powd.		d		0.02^{15}	0.14^{35}	s a, alk
c651	telluride	Cu_2Te	254.69	bl-blk, oct.	7.27					s Br + H_2O; i HCl, H_2SO_4
c652	telluride	Nat. rickardite. Cu_4Te_3	636.98	purp, tetr	7.54					
c653	tellurite	$CuTeO_3$	239.14	blk glass				i	i	s conc HCl
c654	(I) thiocyanate	CuSCN	121.62	wh	2.843	1084		0.0005^{18}		s NH_4OH; sl s ac a; i al; d min a
c655	(II) thiocyanate	$Cu(SCN)_2$	179.70	blk		d 100		d	d	s a, NH_4OH
c656	(II) tungstate	$CuWO_4 \cdot 2H_2O$	3437.42	lt-grn, oct		red heat		0.1^{15}		s NH_4OH; sl s ac a; i al; d min a
c657	xanthate	Copper ethylxanthogenate. $Cu(C_3H_5OS_2)_2$	305.93	yel ppt		d		i		s NH_4OH; v sl s al; i CS_2
	Copper complexes									
c658	diamminecopper (II) acetate	$Cu(C_2H_2O_2)_2 \cdot 2NH_3$	215.70	vlt bl cr		d *ca* 175		s d		s ac a, NH_4OH; i al
c659	tetrammine copper (II) sulfate	$[Cu(NH_3)_4]So_4 \cdot H_2O$	245.74	bl, rhomb	1.81	d 150		$18.5^{21.5}$	d	i al
c660	tetrapyridine copper- (II) fluosilicate	$[Cu(C_5H_5N)_4]SiF_6$	522.03	purp-bl, rhomb	2.108					
c661	tetrapyridine copper- (II) *per*rhenate	$[Cu(C_5H_5N)_4](ReO_4)_2$	880.36	bl cr, monocl	2.338			0.5555		
c662	**Cyanic acid** isocyanic acid	HOCN	43.03	liq	1.14^{20}_{4}			s d		s eth, bz, tol
c663	Cyanoauric acid	$HAu(CN)_4 \cdot 3H_2O$	356.09	tab		50	d	s		s al, eth
c664	**Cobalticyanic acid**	$[H_3Co(CN)_6]2 \cdot H_2O$	454.14	col need, deliq		d 100		s		s al, HCl, dil HNO_3, dil H_2SO_4
c665	**Cyanogen**	$(CN)_2$	52.04	col gas, pungent odor v pois	2.335 g/ℓ, liq: $0.9577^{-21.17}$	−27.9	−20.7	450^{20} cm^3		230 cm^3 al, 500 cm^3 eth
c666	**Cyanogen compounds**	See organic tables								
d1	**Deuterium**	Heavy hydrogen. D_2	4.032	col gas	liq $0.169^{-250.9}$	−254.6	−249.7	sl s		
d2	deuterium chloride	DCl	37.47	gas		−114.8	−81.6	24.1 cm^2 11.9^{25} cm^3	8.4^{50} cm^3 7.12^{50} cm^3	
d3	deuterium oxide	Heavy water. D_2O	20.03	col liq or hex cr, 1.33844^{20}	1.105^{20}_{4}	3.82	101.42^{760}			
d4	**Dysprosium**	Dy	162.50±3	met, hex.	8.5500	1412	2562	i	i	s a
d5	acetate	$Dy(C_2H_3O_2)_3 \cdot 4H_2O$	411.69	yel need		d 120		s		v sl s al
d6	bromate	$Dy(BrO_3)_3 \cdot 9H_2O$	708.34	yel hex need		78	$-6H_2O$, 110	v s		sl s al
d7	bromide	$DyBr_3$	402.21	col cr.		881	1480	s	s	
d8	carbonate	$Dy_2(CO_3)_3 \cdot 4H_2O$	577.09			$-3H_2O$, 15		i		
d9	chloride	$DyCl_3$	268.86	shining yel pl.	3.67^{0}_{4}	718	1500	s	s	
d10	chromate	$Dy_2(CrO_4)_3 \cdot 10H_2O$	853.13	yel cr.		$-3\frac{1}{2}H_2O$, 150	d		1.002^{25}	
d11	fluoride	DyF_3	219.50	col cr.		1360	>2200	i	i	i dil a

PHYSICAL CONSTANTS OF INORGANIC COMPOUNDS (continued)

No.	Name	Synonyms and Formulae	Mol. wt.	Crystalline form, properties and index of refraction	Density or spec. gravity	Melting point, °C	Boiling point, °C	Solubility, in grams per 100 cc: Cold water	Hot water	Other solvents
d12	iodide	DyI_3	543.21	yelsh grn cr		955	1320	s	s	
d13	nitrate	$Dy(NO_3)_3 \cdot 5H_2O$	438.59	yel cr		88.6		s		
d14	oxalate	$Dy_2(C_2O_4)_3 \cdot 10H_2O$	769.21	wh pr		$-H_2O$, 40		i	i	s dil a
d15	oxide	Dysprosia. Dy_2O_3	373.00	wh powd	7.81^{27}	2340 ± 10				grn soln a
d16	*ortho*phosphate	$DyPO_4 \cdot 5H_2O$	347.55	yelsh-wh powd		$-5H_2O$> 200		i		s dil a, ac a
d17	selenate	$Dy_2(SeO_4)_3 \cdot 8H_2O$	898.00	yel need		$-8H_2O$, 200		v s		i al
d18	sulfate	$Dy_2(SO_4)_3 \cdot 8H_2O$	757.30	bril yel cr		stab 110	$-8H_2O$, 360	5.072^{20}	3.34^{40}	
e1	**Erbium**	Er	167.26 ± 3	dk gray powd	9.006	1529	2863	i	i	s a
e2	acetate	$Er(C_2H_3O_2)_3 \cdot 4H_2O$	416.45	wh cr, tricl	2.114					
e3	bromide	$ErBr_3 \cdot 9H_2O$	569.11	vlt-rose cr		950	1460	s	s	
e4	chloride	$ErCl_3 \cdot 6H_2O$	381.71	pink cr, deliq		774	1500	s	s	s al
e5	fluoride	ErF_3	224.26	rose cr		1350	2200	i	i	i dil a
e6	iodide	ErI_3	547.97	vlt-red cr		1020	1280	s	s	
e7	nitrate	$Er(NO_3)_3 \cdot 5H_2O$	443.35	redsh cr		$-4H_2O$, 130		s		s al, eth, acet
e8	oxalate	$Er_2(C_2O_4)_3 \cdot 10H_2O$	778.73	redsh micr powd	2.64(?)	d 575		i	i	i dil a
e9	oxide	Erbia. Er_2O_3	382.52	rose red powd, tr to cub at 1300	8.640	infus		0.00049^{24}		sl s min a
e10	sulfate	$Er_2(SO_4)_3$	622.69	wh powd, hygr	3.678	d 630		43^{0}		
e11	sulfate, octahydrate	$Er_2(SO_4)_3 \cdot 8H_2O$	766.82	rose red, monocl	3.217	$-8H_2O$, 400		16^{20}	6.53^{40}	
e12	**Europium**	Eu	151.96	steel gray met, cub	5.2434	822	1597	i	i	
e13	(II) bromide	$EuBr_2$	311.77			677	1880	s	s	
e14	(III) bromide	$EuBr_3$	391.67			702	d	s	s	
e15	(II) chloride	$EuCl_2$	222.87	wh amorph		727	>2000	s	s	s a
e16	(III) chloride	$EuCl_3$	258.32	yel need	4.89^{20}	850				
e17	(II) fluoride	EuF_2	189.96	brt yel	6.495	1380	>2400	i	i	
e18	(III) fluoride	EuF_3	208.96	col		1390	2280	i	i	i dil a
e19	(II) iodide	EuI_2	405.77	br to olv grn cr	5.50^{26}_{4}	527	1580	s	s	
e20	(III) iodide	EuI_3	532.67			877	d	s	s	
e21	(III) nitrate	$Eu(NO_3)_3 \cdot 6H_2O$	446.07	col		85 (sealed tube)		v s	v s	
e22	oxide	Eu_2O_3	351.92	pa rose powd	7.42					
e23	(II) sulfate	$EuSO_4$	248.02	col, orthorhomb	4.989^{20}			i	i	i dil a
e24	(III) sulfate	$Eu_2(SO_4)_3 \cdot 8H_2O$	736.22	pa rose cr	4.95 (anh)	$-8H_2O$, 375		2.563^{20}	1.93^{40}	
	Ferric or ferrous	See *Iron*								
f1	**Ferricyanic acid**	$H_3Fe(CN)_6$	214.98	grn-brn need, deliq		d		s	s	s al
f2	**Ferrocyanic acid**	$H_4Fe(CN)_6$	215.98	wh need, bl in moist air		d		s	s	s al; i eth
f3	**Fluoboric acid**	HBF_4	87.81	col liq		d 130		∞	s	∞ al
f4	**Fluorophosphoric acid, di-**	HPO_2F_2	101.98	col fum liq	1.583^{26}_{4}	−75	116	s		
f5	**Fluorophosphoric acid, hexa-**	HPF_6	145.97	col fum liq	*ca* 1.65 (65%)	31 ($6H_2O$)				
f6	**Fluorophosphonic acid, mono-**	H_2PO_3F	99.99	col visc liq	1.818^{26}_{4}			∞		
f7	**Fluorine**	F	18.998403	grn yel gas, pois, 1.000195	1.69^{15}g/ℓ 1.51^{-188}	−219.62	−188.14	HF + O_3	d	
f8	(di-)oxide	F_2O	54.00	col gas or yel brn liq	liq $1.90^{-233.8}$	−223.8	−144.8	sl s d	i	sl s alk, a
f9	oxide, di-	Dioxygen fluoride, F_2O_2	70.00	brn gas, red liq, orange solid	sol 1.912^{-165} lil 1.45^{-57}	−163.5	−57			
f10	**Fluosilicic acid**	$H_2SiF_6 \cdot xH_2O$	hydr (not known)	col fum coros liq, 1.3465^{25} (60.97% soln)	1.4634^{25} (60.97% soln)		d	s	s	sl s alk
f11	dihydrate	$H_2SiF_6 \cdot 2H_2O$	180.12	wh cr, fum, deliq		d		s	s	s alk
f12	**Fluosulfonic acid**	HSO_3F	100.06	col liq	1.743^{15}	−87.3	165.5	s		
g1	**Gadolinium**	Gd	157.25 ± 3	col or lt yel met, hex	7.9004	1313	3266	i	i	s a
g2	acetate tetrahydrate	$Gd(C_2H_3O_2)_3 \cdot 4H_2O$	406.44	col. tricl	1.611			11.6^{25}		
g3	acetylacetonate, trihydrate	$Gd[CH(COCH_3)_2]_3 \cdot 3H_2O$	508.62			143.5—145		i	i	
g4	bromide, hexahydrate	$GdBr_3 \cdot 6H_2O$	505.05	rhomb pl	2.844^{16}			s	s	s HBr
g5	chloride	$GdCl_3$	263.61	col monocl, pr	4.52^{0}	609		s	s	
g6	chloride, hexahydrate	$GdCl_3 \cdot 6H_2O$	371.70	wh pr, deliq	2.424^{0}			s	s	
g7	iodide	GdI_3	537.96	citr yel		926	1340	s	s	
g8	fluoride	GdF_3	214.25					i		sl s h HF
g9	*di*methylphosphate	$Gd[(CH_3)_2PO_4]_3$	532.37					23.0^{25}	6.7^{100}	
g10	nitrate, hexahydrate	$Gd(NO_3)_3 \cdot 6H_2O$	451.36	tricl, deliq	2.332	91		v s	v s	s al
g11	nitrate, pentahydrate	$Gd(NO_3)_3 \cdot 5H_2O$	433.34	pr	2.406^{15}	92		i	i	v sl s conc HNO_3
g12	oxalate	$Gd_2(C_2O_4)_3 \cdot 10H_2O$	758.71	monocl		$-6H_2O$, 110		i		s HNO_3; v sl s H_2SO_4
g13	oxide	Gadolinia. Gd_2O_3	362.50	wh amorph powd, hygr	7.407^{15}	2330 ± 20		v sl s		s a
g14	selenate	$Gd_2(SeO_4)_3 \cdot 8H_2O$	887.50	monocl, pearly	3.309	$-8H_2O$, 130		s	s	
g15	sulfate	$Gd_2(SO_4)_3$	602.67	col	$4.139^{14.5}$	d 500		3.98^{0}	$2.26^{34.4}$	
g16	sulfate, octahydrate	$Gd_2(SO_4)_3 \cdot 8H_2O$	746.80	col, monocl	$3.010^{14.6}$			3.28^{20}	2.54^{40}	
g17	sulfide	Gd_2S_3	418.68	yel mass, hyg	3.8			d		d a
g18	**Gallium**	Ga	69.72	gray-blk orthorhomb, tendency to undercool	sol $5.904^{29.6}$ liq $6.095^{29.8}$	29.78	2403	i	i	s a, i alk

No.	Name	Synonyms and Formulae	Mol. wt.	Crystalline form, properties and index of refraction	Density or spec. gravity	Melting point, °C	Boiling point, °C	Solubility, in grams per 100 cc		
								Cold water	Hot water	Other solvents
g19	acetate, basic	$4Ga(C_2H_3O_2)_3 \cdot 2Ga_2O_3 \cdot 6H_2O$	1452.37	wh micro cr.		d 160		s d	d	i ac a
g20	acetylacetonate	2,4-Pentanedione deriv. $Ga(C_5H_7)_2)_3$	367.04	monocl or pl, rhomb or pyram, rhomb pyram	1.42, 1.41	194—95	subl 140^{10}	s	s	s acet
g21	arsenide	GaAs	144.64	dk-gray cub cr		1238				
g22	bromide, tri-	$GaBr_3$	309.43	wh cr.	3.69^{25}_{4}	121.5 ± 0.6	278.8	s	s	sl s NH_3
g23	bromide, tri-, hexammine	$GaBr_3 \cdot 6NH_3$	411.62	wh powd				d	d	sl s NH_2
g24	bromide, tri-, monammine	$GaBr_3 \cdot NH_3$	326.46	wh powd	3.11^{25}	124		d	d	sl s NH_3
g25	*per*chlorate	$Ga(ClO_4)_3 \cdot 6H_2O$	476.16			d 175		v s		v s al
g26	chloride, di-	$GaCl_2$	140.63	wh cr, deliq.		164	535	d	d	s bz
g27	chloride, tri-	$GaCl_3$	176.08	wh need, deliq	2.47^{26}_{4}; liq 2.36^{80}_{80}	77.9 ± 02	201.3	v s	v s	s bz, CCl_4, CS_2
g28	chloride, tri-, hexammine	$GaCl_3 \cdot 6NH_3$	278.26					d	d	s NH_3
g29	chloride, tri-, monoammine	$GaCl_3 \cdot NH_3$	193.11	wh powd	2.189^{25}	124		d	d	s NH_3
g30	ferrocyanide	$Ga_4[Fe(CN)_6]_3$	914.74			d		d	d	i conc HCl
g31	fluoride, tri-	GaF_3	126.72	wh powd	4.47 ± 0.01	subl 800 (in N_2)	*ca* 1000	0.002	i	v sl s dil a; s HF
g32	fluoride, tri-, trihydrate	$GaF_3 \cdot 3H_2O$	180.76	wh powd		$-H_2O$ (vac) 140		i	sl s	sl s dil H_2F_3; v s dil HCl
g33	fluoride, tri-, triammine	$GaF_3 \cdot 3HN_3$	177.81	wh powd		$-NH_3$, 100		d	d	
g34	hydride	Digallane, galloethane, Ga_2H_6	145.49	col liq		−21.4	139, d > 130	d	d	d a, alk
g35	hydroxide	$Ga(OH)_3$	120.74	wh		d 440		i	i	a dil a
g36	hydroxyquinoline deriv.	$Ga(C_9H_6NO)_3$	502.18	grn-yel cr		>150	subl vac	0.0001	0.0012	s a, alk; sl s al
g37	iodide, tri-	GaI_3	450.43	lt yel cr	4.15^{25}_{4}	212 ± 1	subl 345		d	
g38	iodide, tri-, hexammine	$GaI_3 \cdot 6NH_3$	552.62	wh powd				d	d	
g39	iodide, tri-, monoammine	$GaI_3 \cdot NH_3$	467.46	wh powd	3.635^{25}_{4}			d	d	
g40	nitrate	$Ga(NO_3)_3 \cdot \times H_2O$		wh cr, deliq.		d 110	d to Ga_2O_3 200	v s	v s	s abs al; i eth
g41	nitride	GaN	83.73	dk gray powd.	6.1	subl 800		i	i	i dil a; sl s h conc H_2So_4, h conc NaOH
g42	oxide, sesqui-(α)	Ga_2O_3	187.44	wh, hex, rhomb, 1.92, 1.95	6.44	1900; tr to β 600		i	i	s alk; v sl s h a
g43	oxide, sesqui-(β)	Ga_2O_3	187.44	monocl, rhomb.	5.88	1795 ± 15		i	i	s alk; v sl s ha a
g44	oxide, sesqui-, monohydrate	$Ga_2O_3 \cdot H_2O$	205.45	wh micr cr, orthorhom, 1.84	5.2	$-H_2O$, 400 tr to Ga_2O_3		i	i	sl s a; s alk
g45	oxide, sub-	Ga_2O	155.44	blk brn powd	4.77^{26}_{4}	>660	subl > 500	i	i	s a, alk
g46	oxalate	$Ga_2(C_2O_4)_3 \cdot 4H_2O$	475.56	wh micro cr, hygr		$-4H_2O$, 180	d 200	0.4		
g47	oxychloride	$6GaOCl \cdot 14H_2O$	979.25	oct				i		v a KOH; i dil HNO_3; s acet
g48	selenate	$Ga_2(SeO_4)_3 \cdot 16H_2O$	856.56	col, monocl or tricl cr				57.5^{25}	v s	
g49	selenide, mono-	GaSe	148.68	dk red-br greasy leaf	5.03^{25}_{4}	960 ± 10				
g50	selenide, sesqui-	Ga_2Se_3	376.32	rdsh-bl brittle, hard	4.92^{26}_{4}	> 1020 ± 10				
g51	selenide, sub-	Ga_2Se	218.40	bl	5.02^{25}_{4}					
g52	sulfate	$Ga_2(SO_4)_3$	427.61	wh powd		diss 690^{760}		v s	v s	s al; i eth
g53	sulfate, hydrate	$Ga_3(SO_4)_3 \cdot 18H_2O$	751.89	oct cr.				v s	v s	s 60% al; i eth
g54	sulfide, mono-	GaS	101.78	yel cr.	3.86^{25}_{4}	965 ± 10		i	d	s a, alk
g55	sulfide, sesqui-	Ga_2S_3	235.62	yel cr, or wh amorph	3.65^{25}	1255 ± 10		d	d	a s, alk
g56	sulfide, sub-	Ga_2S	171.50	dk gray	4.18^{25}_{4}	d vac 800		d	d	s a, alk
g57	telluride, mono-	GaTe	197.32	blk soft cr.	5.44^{25}_{4}	824 ± 2				
g58	telluride, sesqui-	Ga_2Te_3	522.24	blk brittle cr	5.57^{25}_{4}	790 ± 2				
	Germane									
g59	bromo-	GeH_3Br	155.52	col liq	$2.34^{29.5}$	−32	52	d	d	i al; d alk
g60	chloro-	GeH_3Cl	111.07	col liq	1.75^{-25}	−52	28.0	d	d	i al; d alk
g61	chloro trifluoro-	GeF_3Cl	165.04	col gas		−66.2	−20.63	d	d	s abs al
g62	dibromo-	GeH_2Br_2	234.41		2.80^{0}	−15.0	89.0	d	d	i al; d alk
g63	dichloro-	GeH_2Cl_2	145.51	col liq	1.90^{-68}	−68.0	69.5	d	d	i al; d alk
g64	dichlorodifluoro-	$GeCl_2F_2$	181.49	col gas		−51.8	−2.8	d		s abs al
g65	tribromo-	Germanium bromoform $GeHBr_3$	313.31	col liq		−24	d	d	d	d alk
g66	trichloro-	Germanium chloroform. $GeHCl_3$	179.96	col liq	1.93^{0}	−71	75.2 d	d	d	d alk
g67	trichlorofluoro-	$GeCl_3F$	197.95	col liq		−49.8	37.5	d		s abs al
g68	**Germanium**	Ge	72.59 ± 3	gray-wh met-cub	5.35^{20}_{20}	937.4	2830	i	i	s h H_2SO_4, aq reg; i alk
g69	bromide, di-	$GeBr_2$	232.40	col need or pl		122	d	d	d	s a, $GeBr_4$, al; i bz
g70	bromide, tetra-	$GeBr_4$	392.21	gray-wh oct, 1.6269	3.132^{29}_{29}	26.1	186.5	d	d	s abs al, eth, bz; i conc H_2SO_4
g71	chloride, di-	$GeCl_2$	143.50	wh powd		d to Ge + $GeCl_4$		d	d	s $GeCl_4$; i al, chl
g72	chloride, tetra-	$GeCl_4$	214.40	col liq, 1.464	1.8443^{30}	−49.5	84	d	d	s al, eth; v s dil HCl; i conc HCl conc H_2SO_4

No.	Name	Synonyms and Formulae	Mol. wt.	Crystalline form, properties and index of refraction	Density or spec. gravity	Melting point, °C	Boiling point, °C	Solubility, in grams per 100 cc Cold water	Hot water	Other solvents
g73	fluoride, di-	GeF_2	110.59	wh cr, hygr	...	d > 350	subl	s	v s	...
g74	fluoride, tetra-	GeF_4	148.58	col gas or liq, not liq at atm press	$2.46^{36.5}$	subl −37	...	d to GeO + H_2GeF_6	...	...
g75	fluoride, tetra-	$GeF_4 \cdot 3H_2O$	202.63	wh cr, deliq	...	d	...	s	s	...
g76	hydride	Digermane. Ge_2H_6	151.23	liq	1.98^{-100}	−109	29; d 215	d	...	s liq NH_3
g77	hydride	Trigermane. Ge_3H_3	225.84	col liq	2.2^{30}	−105.6	110.5; d 195	i	i	s CCl_4
g78	hydride, tetra-	Germane. GeH_4	76.62	col gas	1.523^{-142}	−165	−88.5; d 350	i	i	s liq NH_3, NaOCl; sl s h HCl
g79	imide	$Ge(NH)_2$	102.62	wh amorph powd	...	d 150	...	...	d to NH_3 + GeO_2	...
g80	iodide, di-	GeI_2	326.40	or hex pl	5.37	d	subl vac 240	s	s d	s conc HI, dil a; sl s CCl_4, chl; i CS_2
g81	iodide, tetra-	GeI_4	580.21	red-or, cub	4.322^{26}_{26}	144	d 440	s d	...	d al, acet; s CS_2, CCl_4, bz, MeOH
g82	(tri-) nitride, di-	Ge_3N_2	245.78	blk cr	...	...	subl 650	...	...	...
g83	(tri-) nitride, tetra-	Ge_3N_4	273.80	wh-lt brn powd.	5.25^{25}_{4}	d 450	...	i	i	i a, alk
g84	oxide, d-(insoluble)	GeO_2	104.59	tetr	6.239	1086 ± 5	...	i	...	sl s NaOH; i HCl
g85	oxide, di-(soluble)	GeO_2	104.59	col, hex, 1.650.	4.228^{25}	1115.0 ± 4	...	0.447^{25}	1.07^{100}	s a, alk; i HCl, HF; one form NaOH, NH_4OH; one form
g86	oxide, mono-	GeO	88.59	blk cr powd, 1.607	...	subl 710	...	i	i	s Cl_2 water, H_2O_2 + NH_4OH; i a, alk
g87	oxychloride	$GeOCl_2$	159.50	col liq	...	−56	d > 20	d	d	i all solv
g88	selenide	$GeSe_2$	230.51	orange, rhomb(?)	4.56^{25}	707 ± 3	d	i	i	v sl s a; sl s alk
g89	sulfide, di-	GeS_2	136.71	wh powd, or wh, orthorhomb	2.94^{14}	*ca* 800	subl > 600	0.45 d	d to GeO_2 + H_2S	s alk, alk sulf; i al, eth, a; 3.112 liq NH_3
g90	sulfide, mono-	GeS	104.65	yel-red amorph, or rhomb bipyram, blk	amorph: 3.31 rhomb: 4.01	530	subl 430	0.24	i	s HCl, alk or alk sulf; sl s NH_4OH; 0.0473 liq NH_3
	Glucinum	See **Beryllium**		...						
g91	**Gold**	Au	196.9665	yel duct met, cub, coll blue-viol	19.31	1064.43	3080	i	i	s aq reg, KCN, h H_2SO_4; i a
g92	(I) bromide	AuBr	276.87	yel-gray mass, or cr powd	7.9	d 115	...	i	i	d a; s NaCN
g93	(III) bromide	$AuBr_3$	436.68	gray powd, or br cr	...	97.5–Br, 160	...	sl s	...	s eth, al
g94	(I) chloride	AuCl	232.42	yel cr.	7.4	170 d to $AuCl_3$	d 289.5	v sl s d	d	s HCl, HBr
g95	(III) chloride	$AuCl_2$ or Au_2Cl_6	303.33	claret-red cr pr	3.9	d 254	subl 265	68	v s	s al, eth; sl s NH_3; i CS_2
g96	(I) cyanide	AuCN	222.98	lt yel cr powd	7.12^{25}	d	...	v sl s	v sl s	s KCN, NH_4OH; i eth, alk
g97	(III) cyanide	Cyanoauric acid. $Au(CN)_3 \cdot 3H_2O$ or $HAu(CN)_4 \cdot 3H_2O$	329.07 or 356.09	col pl, hygr	...	d 50	...	v s	d, v s	s al, eth
g98	(I) iodide	AuI	323.87	grnsh-yel powd.	8.25	d 120	...	v sl s	sl s d	s KI
g99	(III) iodide	AuI_3	577.68	dk grn	...	...	...	i	d	s iodides
g100	(III) nitrate, hydrogen	Nitratoauric acid. $AuH(NO_3)_4 \cdot 3H_2O$ or $H[Au(NO_3)_4] \cdot 3H_2O$	500.04	yel, tricl, oct	2.84	d 72	...	s, d	...	s HNO_3
g101	(III) oxide	Au_2O_3	441.93	...	...	–O, 160	–3O, 250	i	i	s HCl, conc HNO;3, NaCN
g102	(III) oxide	$Au_2O_3 \cdot xH_2O$	...	...	...	$-1^1/_2H_2O$, 250	...	5.7 × 10^{-11} 25	...	s HCl, NaCN, conc HNO_3
g103	phosphide	Au_2P_3	486.85	gray	6.67	...	d	...	...	i HCl, dil HNO_3
g104	selenide	Au_2Se_2	630.81	...	4.65^{22}	...	...	...	...	...
g105	(I) sulfide	Au_2S	425.99	br blk powd.	...	d 240	...	i fresh sol	ppt coll	s sq reg, KCN; i a
g106	(III) sulfide	Au_2S_3	490.11	br blk powd.	8.754	d 197	...	i	...	i al, eth, s Na_2S
g107	telluride, di-	Nat. krennerite. $AuTe_2$	452.17	1) rhomb, 2) monocl, 3) tricl yel earthy to massive	8.2—9.3	d 472	...	i	i	...
h1	**Hafnium**	Hf	178.49 ± 3	hex	13.31^{20}	2227 ± 20	4602	i	i	s HF
h2	bromide	$HfBr_4$	498.11	wh	...	subl 420	...	...	...	...
h3	carbide	HfC	190.50	...	12.20	*ca* 3890	...	i	...	...
h4	chloride	$HfCl_4$	320.30	wh	...	subl 319	...	d	...	s MeOH, acet
h5	fluoride	HfF_4	254.48	monocl, 1.56	...	...	...	...	...	...
h6	iodide	HfI_4	686.11	...	...	...	400 subl vac	...	...	...
h7	nitride	HfN	192.50	yel-brn, cub.	...	3305	...	...	...	...
h8	oxide	Hafnia. HfO_2	210.49	wh, cub	9.68^{20}	2758 ± 25	~5400(?)	i	i	...
h9	oxychloride	$HfOCl_2 \cdot 8H_2O$	409.52	col	...	...	...	s	...	...
h10	**Helium**	He	4.00260	col gas, inert odorless	0.1785^0 g/ℓ liq $0.147^{-270.3}$	-272.2^{26atm}	−268.9	0.94^0 cm³ 0.94^{25} cm³	1.05^{50} cm³ 1.21^{75} cm³	i al; absorbed by Pt
h11	**Holmium**	Ho	164.9304	met, hex.	8.7947	1474	2695	i	i	...
h12	bromide	$HoBr_3$	404.64	lt yel	...	914	1470	s	s	...
h13	chloride	$HoCl_3$	271.29	lt yel	...	718	1500	s	s	...
h14	iodide	HoI_3	545.64	lt yel	...	989	1300	s	s	...
h15	fluoride	HoF_3	221.93	lt yel	...	1143	>2200	i	i	i dil a

No.	Name	Synonyms and Formulae	Mol. wt.	Crystalline form, properties and index of refraction	Density or spec. gravity	Melting point, °C	Boiling point, °C	Solubility, in grams per 100 cc Cold water	Hot water	Other solvents
h16	oxalate	$Ho_2(C_2O_4)_3 \cdot 10H_2O$	774.07	pa tan		$-H_2O$, 40	d	i	i	i dil a
h17	oxide	Holmia. Ho_2O_3	377.86	tan				i	i	s a
h18	**Hydrazine**	HN_2NH_2	32.05	col liq or wh cr, 1.470^{22}	1.004^{25}	2.0	113.5	v s		s al
h19	azide	$N_2H_4 \cdot HN_3$	75.07	wh pr, deliq, 1.53, 1.76.		75.4		v s	v s	1.2^{23} al; 6.1^{23} MeOH; i CS_2 bz
h20	fluogermanate	$2N_2H_4 \cdot H_2GeF_6$	252.69	monocl pr, 1.452, 1.460, 1.464	2.406^{25}_{25}			s		
h21	fluosilicate	$N_2H_4 \cdot H_2SiF_6$	176.14	cr		d 186		v s		sl s al
h22	formate	$N_2H_4 \cdot 2CH_2O_2$	124.10			128		s		
h23	hydrate	$N_2H_4 \cdot H_2O$	50.06	col fum liq or cub cr, 1.42842	1.03^{21}	−40	118.5^{740}	∞	∞	s al; i eth, chl
h24	hydrochloride, di-	$N_2H_4 \cdot 2HCl$	104.97	col vitr, oct	1.42	198, −HCl	d 200	27.2^{32}	v s	sl s al
h25	hydrochloride, mono-	$N_2H_4 \cdot HCl$	68.51	wh need		89	d 240	v s		v sl s al; v s liq NH_3
h26	hydroiodide	Hydrazine monoiodide, $N_2H_4 \cdot HI$	159.96	col pr		124—126		s		
h27	nitrate, di-	$N_2H_4 \cdot 2HNO_3$	158.07	col cr.		104 (rapid heat); d 80 (slow heat)		v s	d	
h28	nitrate, mono-	$N_2H_4 \cdot HNO_3$	95.06	col dimorph need (α, β)		α70.71; β62.09	subl 140	174.9^{10}	2127^{60}	sl s al
h29	oxalate	$2N_2H_4 \cdot H_2C_2O_4$	154.13	wh need		148		200^{35}		0.0003^{22} al; i eth
h30	*per*chlorate	$N_2H_4 \cdot HClO_4 \cdot {}^1/_2H_2O$	141.51	exp	1.939	137	d 145	d		s al; i eth, bz, chl, CS_2
h31	*hypo*phosphate	$N_2H_4 \cdot 2H_2PO_3$	194.02			152				
h32	*ortho*phosphate	$N_2H_4 \cdot H_3PO_4$	130.04	cr, hygr		82		v s		
h33	*ortho*phosphite	$N_2H_4 \cdot 2H_3PO_3$	196.04			82				
h34	*ortho*phosphite	$N_2H_4 \cdot H_3PO_3$	114.04			36		v s		
h35	picrate	$N_2H_4 \cdot HC_6H_2N_3O_7 \cdot {}^1/_2H_2O$	270.16			201.3		s	s	
h36	selenate	$N_2H_4 \cdot H_2SeO_4$	177.02	col cr powd, unstab		exp		v sl s	v s	
h37	sulfate	$N_2H_4H_2SO_4$	130.12	col, rhomb	1.37	254	d	3.415^{25}	14.39^{80}	i al
h38	sulfate	$(N_2H_4)_2 \cdot H_2SO_4$	162.16	col cr, hygr		85		202.2^{25}	554.4^{60}	i al
h39	tartrate	$(N_2H_4)C_4H_6O_6$	182.13	col cr; $[\alpha]^{20}_4$ + 22.5.		182—183		6.0^0		
h40	**Hydroazoic acid**	Azoimide. HN_3	43.03	col liq	1.09^{25}_4	−80	37	∞	∞	s al, alk, eth
h41	**Hydrogen**	H_2	2.01588 ± 14	col gas, cub sol	gas 0.0899 g/ℓ li1 0.070	−259.14	−252.8	2.14^0 cm^3 1.91^{25} cm^3	0.85^{30} cm^3 1.89^{50} cm^3	6.925^0 cm^3 al
h42	antimonide	Stibine H_3Sb	124.77	col gas, pois	gas 5.30^0 g/ℓ liq 2.26^{-25}	−88.5	−17	20 cm^3	4 cm^3	1500 cm^3 al; 2500 cm^3 CS_2
h43	arsenide	Arsine H_3As	77.95	col gas, pois	3.484 g/ℓ	−113.5	−55, d 230	20 cm^3	sl s	sl s al, alk
h44	arsenide (solid)	H_2As_2	151.86	br powd		d 200		i	d	i al, eth, CS_2, alk; s HNO_3
h45	bismuthide	Bismuthine. H_3Bi	212.00	liq, v unstab			22			
h46	bromide	Hydrobromic acid. HBr	80.91	col gas or pa yel liq, 1.325	gas 3.5^0 g/ℓ liq 2.77^{-67}	−88.5	−67.0	221^0	130^{100}	s al
h47	bromide (const. boiling)	HBr(47%) + H_2O		col liq	1.49	−11	126			
h48	bromide, dihydrate	$HBr \cdot 2H_2O$	116.94	wh cr, col liq	2.11^{-15}	−11		s	s	
h49	bromide, monohydrate	$HBr \cdot H_2O$	98.93	col liq	1.78	stab	−3.6 to −15.5 between 1—2.5 atm			
h50	chloride	Hydrochloric acid. HCl	36.46	col gas or col liq, pois	$1.187^{-84.9}$ gas 1.00045 g/ℓ	−114.8	−84.9	82.3°	56.1^{60}	327 cm^3 al; s eth, bz
h51	chloride (const. boiling)	HCl(20.24%) + H_2O		col liq	1.097		110			
h52	chloride, dihydrate	$HCl \cdot 2H_2O$	72.49	col liq	$1.46^{18.3}$	−17.7	d	d	∞	s al
h53	chloride, monohydrate	$HCl \cdot H_2O$	54.48	col liq	1.48	−15.35		∞	∞	s al
h54	chloride, trihydrate	$HCl \cdot 3H_2O$	90.51	col liq		−24.4	d	∞		s al
h55	cyanide	Hydrocyanic acid. HCN	27.03	col liq or gas, pois, liq 1.2675^{10}	gas 0.901 g/ℓ liq 0.699^{22}	−14	26	∞	∞	∞ al; s eth
h56	fluoride	Hydrofluoric acid. HF	20.01	col fum cor liq, or gas; gas 1.90	$0.991^{19.54}$ liq 0.987	−83.1	19.54	∞	v s	
h57	fluoride (const. boiling)	HF(35.35%) + H_2O		col liq			120			
h58	iodide	Hydroiodic acid. HI	127.91	col gas, or pa yel liq, $n_D^{16.5}$ 1.466	gas 5.66^0 g/ℓ liq $2.85^{-4.7}$	−50.8	-35.38^4 atm	42.5^0 cm^3	v s	s al
h59	iodide (const. boiling)	HI(57%) + H_2O		col or pa yel fum liq	1.70^{15}		127^{774}			
h60	iodide, dihydrate	$HI \cdot 2H_2O$	163.94	col liq		−43		∞		
h61	iodide, tetrahydrate	$HI \cdot 4H_2O$	199.97	col liq		−36.5		∞		
h62	iodide, trihydrate	$HI \cdot 3H_2O$	181.96	col liq		−48		∞		
h63	oxide	Water. H_2O	18.01528	col liq or hex cr, liq 1.333 sol 1.309, 1.313	1.000^4_4	0.000	100.000			∞ al
h64	oxide, per-	H_2O_2	34.01	col liq; 1.414^{22}	1.4067^{25}	−0.41	150.2^{760}	∞		s al, eth; i pet eth
h65	phosphide	Phosphine, H_3P	34.00	col pois inflam gas or col liq, 1.317 liq	gas 1.529 g/ℓ liq 0.746^{-90}	−133.5	−87.4	26^{17} cm^3	i	s al, eth, Cu_2Cl_2

PHYSICAL CONSTANTS OF INORGANIC COMPOUNDS (continued)

No.	Name	Synonyms and Formulae	Mol. wt.	Crystalline form, properties and index of refraction	Density or spec. gravity	Melting point, °C	Boiling point, °C	Solubility, in grams per 100 cc		
								Cold water	Hot water	Other solvents
h66	phosphide	H_4P_2	65.98	col liq	1.012	−90	57.5^{735}	i	i	s al, turp
h67	phosphide	$(H_2P_4)_3$	377.73	yel solid	1.83^{19}	ign 160	d	i	i	i al; s P, P_2H_4
h68	sulfide	H_2S	34.08	col gas, infl, liq 1.374	1.539^{0} g/ℓ	−85.5	−60.7	437^{0} cm^3	186^{40} cm^3	9.54^{20} cm^3 al; s CS_2
h69	sulfide, di-	H_2S_2	66.14	yel oil, 1.885	1.334^{20}	−89.6	70.7^{760}	d	...	s bz, eth, CS_2; i al
h70	sulfide, penta-	H_2S_5	162.32	clear yel oil	1.67^{16}	−50	50^{4}	...	...	s bz, eth, CS_2; i al
h71	sulfide, tetra-	H_2S_4	130.26	lt yel liq	1.588^{15}	−85	...	...	...	...
h72	sulfide, tri-	H_2S_3	98.20	brt yel liq, 1.705^{15}	1.496^{15}	−52	d 90	...	...	s bz, eth, CS_2; i al
h73	telluride	H_2Te	129.62	col gas or yel need	gas 5.81 g/ℓ liq 2.57^{-20}	−49	−2	s unstab	...	s al, alk
h74	**Hydroxylamine**	NH_2OH	33.03	wh need or col liq, deliq	1.204	33.05	56.5	s	d	s a, al, MeOH; v sl s eth, chl, bz, CS_2
h75	acetate	$NH_2OH{\cdot}CH_3CO_2$	92.07	col cr.	...	87	subl 90	v s	...	...
h76	bromide	$NH_2OH{\cdot}HBr$	113.94	wh, monocl	2.35^{22}_{4}	...	...	v s	v s	i eth
h77	fluogermanate	$(NH_2OH)_2{\cdot}H_2GeF_6{\cdot}2H_2O$	290.69	monocl pr, 1.418, 1.438, 1.443	2.229^{25}	...	...	s	...	s abs al
h78	fluosilicate	$(NH_2OH)_2H_2SiF_6{\cdot}2H_2O$	246.18	scales	...	...	...	v s	...	i al
h79	formate	$NH_2OH{\cdot}HCO_2$	78.05	col need	...	76	d 80	v s	d	s h al; i eth
h80	hydrochloride	$NH_2OH{\cdot}HCl$	69.49	col, monocl	1.67^{17}	151	d	83^{17}	v s	4.43^{20} al; 16.4^{20} MeO H; s gluc; i eth
h81	iodide	$NH_2OH{\cdot}HI$	160.94	col need, hygr	...	83—84 exp	...	v s	d	v s MeO H; sl s eth
h82	nitrate	$NH_2OH{\cdot}HNO_3$	96.04	wh	...	48	d <100	v s	d	v s al
h83	*ortho*phosphate	$(NH_2OH)_3{\cdot}H_3PO_4$	197.08	...	...	148 exp	...	1.9^{20}	16.8^{90}	...
h84	sulfate	$(NH_2OH)_2{\cdot}H_2SO_4$	164.13	col, monocl	...	d 170	d	32.9^{0}	68.5^{20}	sl s al; s eth
i1	**Indium**	In	114.82	soft silv wh met, tetr	7.30^{20}	156.61	2080	i	i	s a; v sl s NaOH
i2	antimonide	InSb	236.57	cr	...	535	...	...	...	...
i3	arsenide	InAs	189.74	met cr	...	943	...	...	...	i a
i4	bromide, di-	$InBr_2$	274.63	pa yel solid	4.22^{25}	235	632 subl	d	...	s a
i5	bromide, mono-	InBr	194.72	red br solid	4.96^{25}	220	662 subl	d	...	s a
i6	bromide, tri-	$InBr_3$	354.53	wh to yel need, deliq	4.74^{25}	436 ± 2	subl	v s	...	...
i7	*per*chlorate	$In(ClO_4)_3{\cdot}8H_2O$	557.29	col cr, deliq.	...	ca 800	d 200	v s	d	s abs al; sl s eth
i8	chloride, di-	$InCl_2$	185.73	wh rhomb, deliq	3.655^{25}	235	550—570	d	d	...
i9	chloride, mono-	InCl	150.27	1) yel or 2) dk red deliq	4.19^{25} yel 4.18^{25} red	225 ± 1	608	d	d	s a
i10	chloride, tri-	$InCl_3$	221.18	wh pl, deliq.	3.46^{25}_{4}	586 subl 300	volat 600	v s	v s	sl s al, eth
i11	cyanide	$In(CN)_3$	192.87	wh ppt	...	...	...	unstab	...	i dil a; s HCN; v sl s NaOH
i12	fluoride	InF_3	171.82	col	4.39^{25} ± 1	1170 ± 10	>1200	0.040^{25}	...	...
i13	fluoride, trihydrate	$InF_3{\cdot}3H_2O$	225.86	cr	...	$-3H_2O$, 100	...	8.49^{22}	...	s a; i al, eth
i14	fluoride, nonahydrate	$InF_3{\cdot}9H_2O$	333.95	wh need	...	d	...	sl s	d	s HCl, HNO_3; i al, eth
i15	hydroxide	$In(OH)_3$	165.84	wh ppt	...	$-H_2O$<150	...	i	...	s a; v sl s NaOH; i NH_4OH
i16	iodate	$In(IO_3)_3$	639.53	wh cr.	...	...	d	0.067^{20}	...	s dil HNO_3, dil H_2SO_4; s d HCl
i17	iodide, di-	InI_2	368.63	...	4.71^{25}	212	...	...	...	s a
i18	iodide, mono-	InI	241.72	br red solid	5.31	351	711—715	...	sl d	i al, eth, chl; s dil a
i19	iodide, tri-	InI_3	495.53	carmine red, or yel cr	4.69	210	...	v unstable	s	s a, chl, bz yxl
i20	methylate	$In(CH_3)_3$	159.92	col cr.	1.568^{19}_{9}	...	...	d	...	d al, MeOH; s liq NH_3, eth; v s acet, bz
i21	nitrate	$In(NO_3)_3{\cdot}3H_2O$	354.88	pl, deliq	...	$-2H_2O$, 100	d	v s	...	s al
i22	nitrate	$In(NO_3)_3{\cdot}4^1/_2H_2O$	381.90	need, deliq	...	$-4^1/_2H_2O$	d	v s	...	s al
i23	oxide, mon-	InO	130.82	wh gray	...	...	...	i	...	s a
i24	oxide, sesqui-	In_2O_3	277.64	red brn, (h) pa yel (c) amorph and trig	7.179	...	volat 850	i	...	amorph s a; cr i a
i25	oxide, sub-	In_2O	245.64	blk cr	6.99^{25}_{4}	subl vac 565—700	...	...	...	s HCl
i26	phosphide	InP	145.79	brittle mass met	...	1070	...	...	...	v sl s min a
i27	selenate	$In_2(SeO_4)_3{\cdot}10H_2O$	838.67	cr, deliq	...	...	...	v s	...	...
i28	selenide, sesqui-	In_2Se_3	466.52	blk cr, or soft dk scales	5.67^{25}_{4}	890 ± 10	...	...	...	s, d conc a
i29	sulfate	$In_2(SO_4)_3$	517.81	wh gray powd, monocl pr, hygr	3.438	...	...	s	v s	...
i30	sulfate	$In_2(SO_4)_3{\cdot}9H_2O$	679.95	wh powd, hygr.	3.44	d 250	...	v s	...	...
i31	sulfate, dihydrate	$In_2(SO_4)_3{\cdot}H_2SO_4{\cdot}7H_2O$	741.99	rhomb cr	...	$-7H_2O$, $-H_2SO_4$, *ca* 250	...	...	...	...
i32	sulfide, mono-	InS	146.88	dk.	5.18^{25}	692 ± 5	subl vac 850	...	...	s HCl, HNO_3
i33	sulfide, sesqui-	In_2S_3	325.82	red cr or yel ppt	4.90	1050	subl *ca* 850 in high vac	i	...	s a; sl s Na_2S
i34	sulfide, sub-	In_2S	261.70	yel or blk need	5.87^{25}	653 ± 5	...	...	...	...
i35	sulfite, basic	$2In_2O_3{\cdot}3SO_2{\cdot}8H_2O$	891.58	wh cr.	...	$-3H_2O$, 100	$-8H_2O$, 260	i	...	...
i36	telluride, sesqui-	InTe	242.42	dk met, shiny	6.29^{25}_{4}	696 ± 2	...	...	...	i HCl, s HNO_3
i37	telluride	In_2Te_3	612.44	bl brittle cr	5.78	667	...	...	...	...

PHYSICAL CONSTANTS OF INORGANIC COMPOUNDS (continued)

No.	Name	Synonyms and Formulae	Mol. wt.	Crystalline form, properties and index of refraction	Density or spec. gravity	Melting point, °C	Boiling point, °C	Solubility, in grams per 100 cc Cold water	Hot water	Other solvents
i38	**Iodic acid**	HIO_3	175.91	col or pa yel cr powd, rhomb	4.629^0	d 110		286^0 310^{16}	473^{80} 576^{101}	v s 87% al; sl s HNO_3; i abs al, eth, chl
i39	metaper-	HIO_4	191.91	col		subl 110	d 138	v s, d		
i40	*ortho*paraper-	H_5IO_6 or $HIO_4 \cdot 2H_2O$	227.94	wh monocl, deliq			d 140	113	v s	s al, eth
i41	**Iodine**	I_2	253.809	vlt blk met lust, rhomb, 3.34	4.93	113.5	184.35^{atm}	0.029^{20} 0.030^{25}	0.078^{50}	20.5^{15} al; 16.46^{25} bz 20.6^{17} eth; s chl glyc, KI; 24^{25} eth 23^{25} MeOH 20.15^{25} CS_2; $2.91^{25}CCl_4$
i42	azide	Iod(o)azide. IN_3	168.92	yel, exp				s d		s $Na_2S_2O_3$
i43	bromide, mono-	IBr	206.81	dk gray cr	4.4157^0	(42) subl 50	d 116	s d		s al, eth, chl, CS_2
i44	bromide, tri-	IBr_3	366.62	br liq				s		s al
i45	chloride, mono-(α)	ICl	162.36	dk-red need, cub, red br oily liq	3.1822^0	27.2	97.4	d to HIO_3 + Cl		s al, eth, CS_2, HCl
i46	chloride, mono-(β)	ICl	162.36	brn red, rhombic 6 sided pl	liq 3.24^{34}	13.92	97.4; d 100	d		s al, eth, HCl
i47	chloride, tri-	ICl_3	233.26	yel brn, rhomb, red liq	3.117^{15}	101^{16atm}	d 77	s d		s al, eth, CCl_4, ac a, bz
i48	cyanide	ICN	152.92	wh cr.				sl s	sl s	s al, eth
i49	fluoride, hepta-	IF_7	259.89	col cr or liq	liq 2.8^6	5.5	4.5 subl	v s, d	d	d a, alk
i50	fluoride, penta-	IF_6	221.90	col liq	3.75	9.6	98	d	d	d a, ak
i51	oxide, di- (or tetra-)	IO_2 or I_2O_4	158.90	lem-yel cr	4.2^{18}_{10}	d slow 75 rap 130		d to HIO_3 + I_2		s H_2SO_4; sl s acet; i al, eth
i52	oxide, pent-	I_2O_5	333.81	wh trim	4.799^{25}_{4}	d 300—350		187.4^{13}	v s	i abs al, eth, chl, CS_2; sl s dil a
i53	(tetra-) oxide non-	I_4O_9	651.61	yel powd, hygr.			d 75			
i54	**Iodo platinic acid**	$H_2PtI_6 \cdot 9H_2O$	1120.66	blk, monocl, deliq		>100		v s, d	d	
i55	**Iodous acid, hypo-**	Iodine hydroxide. HOI	143.91	only in sol, yel to grayish				d	d	
i56	**Iridium**	Ir	192.22 ± 3	silv wh met, cub	22.421	2410	4130	i	i	sl s aq reg; i a, alk
i57	bromide, tetra-	$IrBr_4$	511.84	blk, deliq		d		s d		s al
i58	bromide, tri-	$IrBr_3 \cdot 4H_2O$	503.99	olv-grn cr		$-3H_2O$,100		v s		i al
i59	carbonyl	$Ir_2(CO)_8$	608.52	yel cr.		subl 160 (in CO_2)				s CCl_4
i60	carbonyl	$Ir_4(CO)_{12}$	1105.00	yel cr.		subl 250 (in CO_2)				i CCl_4
i61	carbonyl chloride	$Ir(CO)_2Cl_2$	319.15	col need		d 140		d		d HCl, KOH
i62	chloride, di-	$IrCl_2$	263.13	blk gray cr (exist?)		d 733		s		i a, alk
i63	chloride, tetra-	$IrCl_4$	334.00	dk-brn, amorph, hygr.		d		s	d	s al, dil HCl
i64	chloride, tri-	$IrCl_3$	298.58	olv-grn hex, or trig	5.30	d 763		i	i	i a, alk
i65	fluoride, hexa-	IrF_6	306.21	yel glass or tetr	6.0	44.4^{30}	53	d	d	
i66	iodide, tetra-	IrI_4	699.84	blk		d 100		i	i	i al, s KI
i67	iodide, tri-	IrI_3	572.93	grn		d		sl s	s	sl s al
i68	oxalic acid	$H_3[Ir(C_2O_4)_3 \cdot xH_2O$		pa yel cr.				v s	v s	sl s al; i eth
i69	oxide, di-	IrO_2	224.22	blk tetr or bl cr.	11.665	d 1100		0.0002^{20}		i a, alk
i70	oxide, di- hydrate	$IrO_2 \cdot 2H_2O$ or $Ir(OH)_4$	260.25	indigo bl cr		$-2H_2O$, 350		i	i	s HCl
i71	oxide, sesqui-	Ir_2O_3	432.44	bl-blk (exist?)		$-O$, 400		i		s H_2SO_4, h HCl; i alk
i72	oxide, sesqui-	$Ir_2O_2 \cdot xH_2O$		olive green		d		i		s a, alk
i73	phosphor chloride	IrP_3Cl_{12} or $IrCl_3 \cdot 3PCl_3$	710.58	yel pr		d 250		sl s	d 100	sl s, d al
i74	selenide	$IrSe_2$	350.14	dk gray cr powd		d 600—700 (in CO_2)				sl s aq reg; i a
i75	sulfate	$Ir_2(SO_4)_3 \cdot xH_2O$		yel pr		d		s		
i76	sulfide, di	IrS_2	256.34	br-blk	8.43^{25}_{4}	d 300		i		s aq reg; i a
i77	sulfide, hydro-	$Ir(HS)_3 \cdot 2H_2O$	327.45	choc br		d		i		s HNO_3
i78	sulfide, mono-	IrS	224.28	bl-blk			d	i		s k_2S, i a
i79	sulfide, sesqui-	Ir_2S_3	480.62	br-blk	9.64^{25}_{4}	d		sl s		s K_2S, HNO_3
i80	telluride	$IrTe_3$	575.02	dk gray cr.	9.5^{25}_{4}			i	i	i a; s h aq reg
	Iridium complexes									
i81	aquopentammine iridium-chloride	$[Ir(NH_3)_5(H_2O)]Cl_3$	366.29	wh micro cr.	2.474^{15}_{4}	$-H_2O$, 100		ca 75^{25}	d	i al, eth
i82	chloropentammine iridium chloride	$[Ir(NH_3)_5Cl]Cl_2$	383.73	pa yel, rhomb	$2.681^{15.5}_{4}$	d		sl s	s	i HCl
i83	hexammine iridium chloride	$[Ir(NH_3)_6]Cl_3$	400.76	col, rhomb	$2.434^{15.5}_{4}$			20		
i84	hexammine iridium nitrate	$[Ir(NH_3)_6](NO_3)_3$	480.42	col, tetr micro cr	2.395^{15}_{4}			1.7^{14}		
i85	nitratopentammine iridium nitrate	$[Ir(NH_3)_5(NO)_3)](NO_3)_2$	463.39	wh micro cr.	2.515^{18}_{4}	heat exp		0.28	2.5^{100}	
i86	**Iron**	Fe	55.847 ± 3	silv met, cub	7.86	1535	2750	i	i	s a; i alk, al, eth
i87	(II) acetate	$Fe(C_2H_3O_2)_2 \cdot 4H_2O$	246.00	lt grn need, monocl		d		v s		
i88	(III) acetate, basic	$FeOH(C_2H_3O_2)_2$	190.94	br red powd.				i		s a, al
i89	(III) acetylacetonate	$Fe(C_5O_2H_7)_3$	353.18	rubyred, rhomb	1.33	184		sl s	sl s	s al, acet, bz, chl
i90	(III) *ortho*arsenate	$Fe_3(AsO_4)_2 \cdot 6H_2O$	553.47	grn amorph powd		d		i	i	s dil HCl; sl s NH_4OH
i91	(III) *ortho*arsenate	Nat. scorodite. $FeAsO_4 \cdot 2H_2O$	230.80	grn, rhomb, 1.765, 1.774, 1.797	3.18	d		i	i	s HCl; i HNO_3

PHYSICAL CONSTANTS OF INORGANIC COMPOUNDS (continued)

No.	Name	Synonyms and Formulae	Mol. wt.	Crystalline form, properties and index of refraction	Density or spec. gravity	Melting point, °C	Boiling point, °C	Solubility, in grams per 100 cc: Cold water	Hot water	Other solvents
i92	(III) *ortho*arsenite, basic	$2FeAsO_3·Fe_2O_3·5H_2O$	607.30	br-yel powd		d		sl s		s a, alk
i93	(II) *pyro*arsenite	$Fe_2As_2O_5$	341.53	grn-wh				i		s NH_4OH
i94	arsenide	FeAs	130.77	wh	7.83	1020		v sl s		
i95	arsenide, di-	Arsenoferrite. $FeAs_2$	205.69	silv gray, cub.	7.4	990		i		sl s HNO_3; i HCl
i96	(III) benzoate	$Fe(C_7H_5O_2)_3$	419.19	br powd				i		s h al, h eth
i97	boride	FeB	66.66	gray cr	7.15^{18}			i		s NHO_3, h conc H_2SO_4
i98	(II) bromide	$FeBr_2$	215.66	grn-yel, hex	4.636^{25}	d 684 (?)		109^{10}	170^{96}	s al; sl s bz
i99	(III) bromide	$FeBr_3$ or Fe_2Br_6	295.56	dk red-brn, rhomb (?) deliq		subl d		s	s	s al, eth, sl s NH_3
i100	(III) bromide-, hexahydrate	$FeBr_3·6H_2O$	403.65	dk grn		27		v s	v s	s al, eth
i101	(III) cacodylate	$Fe[(CH_3)_2AsO_2]_3$	466.82	yelsh amorph powd				6.67		sl s al
i102	carbide	Fe_3C	179.55	gray, cub	7.694	1837		i	i	s a
i103	(II) carbonate	Nat. siderite. $FeCO_3$	115.86	gray, trig, 1.875, 1.633	3.8	d		0.0067^{25}		s CO_2 sol
i104	(II) carbonate, hydrate	$FeCO_3·H_2O$	133.87	amorph		d		sl s		s a, CO_2 sol
i105	carbonyl, ennea-	$Fe_2(CO)_9$	363.79	yel met cr, hex	2.085^{18}	d 80		i	i	v sl s al, MeOH; d HNO_3; i a
i106	carbonyl, penta-	$Fe(CO)_5$	195.90	visc yel liq	liq 1.457^{21}	−21	102.8^{749}	i		s al, eth, bz, alk, conc H_2SO_4
i107	carbonyl, tetra-	$Fe(CO)_4$	167.89	dk grn lust cr, tetr	1.996^{18}	d 140—150		i		s org solv, conc HNO_3 h H_2SO_4
i108	(II) *per*chlorate	$Fe(ClO_4)_2$	254.75	wh or grnsh-wh need, hygr		d>100		v s		
i109	(II) *per*chlorate hexahydrate	$Fe(ClO_4)_2·6H_2O$	362.84	grn, 1.493, 1.478		d>100		97.8^{0}	116.1^{60}	86.5^{20} al; s $HClO_4$
i110	*oxy*chloride	FeOCl	107.30	brn, rhomb	3.1	d 200				
i111	(II) chloride	Nat. lawrencite. $FeCl_2$	126.75	grn to yel, hex deliq, 1.567	3.16^{25}_{4}	670—674	subl	64.4^{10}	105.7^{100}	100 al; s acet; i eth
i112	(II) chloride, dihydrate	$FeCl_2·2H_2O$	162.78	grn, monocl	2.358					
i113	(II) chloride, tetrahydrate	$FeCl_2·4H_2O$	198.81	bl-grn, monocl, deliq	1.93			160.1^{10}	415.5^{100}	s al; sl s acet
i114	(III) chloride	Nat. molysite. $FeCl_3$ or Fe_2Cl_6	162.21	blk-brn hex	2.898^{25}_{4}	306	d 315	74.4^{0}	535.7^{100}	v s al, MeOH, eth; 63^{18} acet
i115	(III) chloride, hydrate	$FeCl_3·2^1/_2H_2O$	207.24	dk yel-red, rhomb, deliq		56		v s	v s	v s al, eth
i116	(III) chloride, hexahydrate	$FeCl_3·6H_2O$	270.30	br yel cr mass, v deliq		37	280—285	91.9^{20}	∞	s al, eth
i117	(II) chloride, hexammine	$FeCl_2·6NH_3$	228.94	wh powd	1.928^{25}_{4}					
i118	(II) chloroplatinate	$FePtCl_6·6H_2O$	571.74	yel, hex	2.714	d		v s	v s	
i119	(III) *di*chromate	$Fe_2(Cr_2O_7)_3$	759.66	red-brn gran				s		s a
i120	(II) chromite	$FeCr_2O_4$	223.84	brn-blk, cub	4.97^{20}			i	i	sl s a
i121	(II) citrate	$FeC_6H_6O_7·H_2O$	263.97	wh micr, rhomb		d 350 (in H_2)		sl s		s NH_4OH
i122	(III) citrate	$FeC_6H_5O_7·5H_2O$	335.03	red-brn scales				sl s	s	i al
i123	hydroxide	Nat. goethite. FeO(OH)	88.85	brn, blksh, rhomb, 2.260, 2.394, 2.400	4.28	$-^1/_2H_2O$, 136				s HCl
i124	(II) ferricyanide	$Fe_3[Fe(CN)_6]_2$(?)	591.45	deep bl.		d		i		i al, dil a
i125	(III) ferricyanide	Berlin green. $Fe[Fe(CN)_6]$	267.80	cub						
i126	(II) ferrocyanide	$Fe_2[Fe(CN)_6]$	323.65	bl-wh, amorph	1.601^{25}_{4}	d 100	d 430 (vac)	i		
i127	(III) ferrocyanide	$Fe_4[Fe(CN)_6]_3$	859.25	dk bl cr		d		i	i	s HCl, H_2SO_4; i al, eth
i128	(II) fluoride	FeF_2	93.84	wh, rhomb	4.09^{25}_{4}	>1000 (?)		sl s		s a; i al, eth
i129	(II) fluoride, tetrahydrate	$FeF_2·8H_2O$	237.97	grn-bl	4.20 (anh)	$-8H_2O$, 100		sl s	s	s HF, a; i al, eth
i130	(II) fluoride, tetrahydrate	$FeF_2·4H_2O$	165.90	wh, rhomb	2.095	d		v sl s		s a; sl s al, eth
i131	(III) fluoride	FeF_3 or Fe_2F_6	112.84	grn, rhomb	3.52	>1000		sl s	s	s a; i al, eth
i132	(III) fluoride, tetrahydrate	$FeF_3·4^1/_2H_2O$	193.91	yel cr.		$-3H_2O$, 100	d	sl s	s	i al
i133	(II) fluosilicate	$FeSiF_6·H_2O$	306.01	col, trig, 1.361, 1.385	1.961			128.2		
i134	(III) fluosilicate	$Fe_2(SiF_6)_3$ (exist ?)	537.92	flesh col, gel				s	s, d	
i135	(III) formate	$Fe(CHO_2)_3·H_2O$	208.92	red cr or powd				s		v sl s al
i136	(III) glycerophosphate	$Fe_3[C_3H_5(OH)_2·PO_4]_3$	677.72	yelsh-grn sc or powd				50^{25}		i al
i137	(II) hydrogen cyanide	$H_4[Fe(CN)_6]$	215.98	wh, rhomb	1.536^{25}_{4}	d 190		s	s	v s al; s a; i acet
i138	(III) hydrogen cyanide	$H_3[Fe(CN)_6]$	214.98	brn-yel need		d 50—60		s	s	v s al; i eth
i139	(II) hydroxide	$Fe(OH)_2$	89.86	pa grn hex, or wh amorph	3.4	d		0.00015^{18}		s a, NH_4Cl; i alk
i140	(III) hydrosulfate	Iron(ic) tetrasulfate. Nat. rhomboclase $Fe_2O_3·4SO_3·9H_2O$	642.06	wh to pink powd, 1.533, 1.550, 1.635	2.172	$-6H_2O$, 80		s		sl s abs al
i141	(III) iodate	$Fe(IO_3)_3$	580.56	grn yel powd	4.80^{20}_{4}	d 130		sl s		i dil HNO_3
i142	(II) iodide	FeI_2	309.66	gray hex, hygr	5.315	red heat		s	s	s al, acet
i143	(II) iodide, tetrahydrate	$FeI_2·4H_2O$	381.72	gray blk cr, deliq	2.873	d 90—98		v s	d	s al, eth
i144	(II) lactate	$Fe(C_3H_5O_3)_2·3H_2O$	288.03	grn-wh cr or powd		d		2.1^{10}	8.5^{100}	s alk citrate; v sl s al; i eith
i145	(III) lactate	$Fe(C_3H_5O_3)_3$	323.06	br, amorph, deliq				s	v s	i eth
i146	(III) malate	$Fe_2(C_4H_4O_5)_3$	507.91	br scales, hygr				s		s al

No.	Name	Synonyms and Formulae	Mol. wt.	Crystalline form, properties and index of refraction	Density or spec. gravity	Melting point, °C	Boiling point, °C	Solubility, in grams per 100 cc Cold water	Hot water	Other solvents
i147	methanoarsenate	$Fe_2(CH_3AsO_3)_3$	525.56	redsh br lustr scales				50		i al, eth
i148	(II) nitrate	$Fe(NO_3)_2{\cdot}6H_2O$	287.95	grn, rhomb		60.5		83.5^{20}	166.7^{61}	
i149	(III) nitrate	$Fe(NO_3)_3{\cdot}6H_2O$	349.95	cub		35	d	150^{0}	∞	
i150	(III) nitrate	$Fe(NO_3)_3{\cdot}9H_2O$	404.00	col-pa vlt, monocl, deliq	1.684	47.2	d 125	s	s	s al, acet; sl s HNO_3
i151	nitride	Fe_4N	237.39		6.57(?)					
i152	nitride	Fe_2N or Fe_4N_2	125.70	gray	6.35	d 200		i		s HCl, H_2SO_4
i153	nitrosyl carbonyl	$Fe(NO)_2(CO)_2$	171.88	dk red cr	1.56	18.5	d 50, 110	i		s org solv
i154	(III) oleate	$Fe(C_{18}H_{33}O_2)_3$	900.22	br-red fatty lumps				i		s a, al, eth
i155	(II) oxalate	$FeC_2O_4{\cdot}2H_2O$	179.90	pa yel, rhomb	2.28	d 190		0.022	0.026	s a
i156	(III) oxalate	$Fe_2(C_2O_4)_3{\cdot}5H_2O$	465.83	yel micro cr powd		d 100		v s	v s	s a; i al
i157	(II) oxide	Nat. wuestite. FeO	71.85	blk, cub, 2.32	5.7	1369 ± 1		i	i	s a; i al, alk
i158	(III) oxide	Nat. hematitie. Fe_2O_3	159.69	red-brn to blk, trig, 3.01, 2.94(Li)	5.24	1565		i	i	s HCl, H_2SO_4; sl s HNO_3
i159	oxide	Iron ferrosoferric, nat. magnetite. Fe_3O_4	231.54	blk, cub or red-blk powd, 2.42	5.18	1594 ± 5		i	i	s conc a; i al, eth
i160	(III) oxide, hydrate	$Fe_2O_3{\cdot}xH_2O$		red-brn amorph powd or gelat	2.44—3.60	all H_2O, 350—400		i	i	s a; i al, eth
i161	(II) *ortho*phosphate	Nat. vivianite. $Fe_3(PO_4)_2{\cdot}8H_2O$	501.61	wh-bl, monocl, 1.579, 1.603, 1.633	2.58			i	i	s a; i ac a
i162	(III) *ortho*phosphate	$FePO_4{\cdot}2H_2O$	186.85	pink, monocl	2.74	d		v sl s	0.67^{100}	s HCl, H_2SO_4; i HNO_3
i163	(III) *pyro*phosphate	$Fe_4(P_2O_7)_3{\cdot}9H_2O$	907.36	yel-wh powd				i		s a, alk citr
i164	phosphide, mono-	FeP	86.82	rhomb	6.07 (5.2^{20})					
i165	(di-)phosphide	Fe_2P	142.67	bl-gray cr or powd	6.56	1290		i	i	s aq reg, HNO_3 + HF; i dil a
i166	(tri-)phosphide	Fe_3P	198.51	gray	6.74	1100		i		
i167	(III) *hypo*phosphite	$Fe(H_2PO_2)_3$	250.81	wh or gray-wh powd		d		0.043^{25}	0.083^{100}	s alk citr
i168	(II) *meta*silicate	Nat. gruenerite. $FeSiO_3$	131.93	gray-grn, rhomb, 1.672, 1.697, 1.717	3.5	1146				
i169	*ortho*silicate	Nat. fayalite; Fe_2SiO_4	203.78	col, rhomb	4.34	1503 (?)		i	i	d HCl
i170	silicide	FeSi	83.93	yel-gray, oct	6.1			i	i	i aq reg
i171	(II) sulfate	Nat. szomolnokite. $FeSO_4{\cdot}H_2O$	169.92	off-wh, monocl	2.970^{25}			sl s	s	
i172	(III) sulfate	$Fe_2(SO_4)_3$	399.87	yel rhomb, hygr, 1.814	3.097^{18}	d 480		sl s	d	i H_2SO_4, NH_3
i173	(II) sulfate, heptahydrate	Nat. melanterite. $FeSO_4{\cdot}7H_2O$	278.01	bl-grn, monocl, 1.471, 1.478, 1.486	1.898	64 $-6H_2O$, 90	$-7H_2O$, 300	15.65	48.6^{50}	sl s al; s abs MeOH
i174	(II) sulfate, pentahydrate	Nat. siderotil. $FeSO_4{\cdot}5H_2O$	241.98	wh, tricl, 1.526, 1.536, 1.542	2.2	$-5H_2O$, 300		s	s	i al
i175	(II) sulfate, tetrahydrate	$FeSO_4{\cdot}4H_2O$	223.97	grn monocl pr, 1.533, 1.535	2.23—2.29					
i176	(III) sulfate, enneahydrate	Nat. coquimbite. $Fe_2(SO_4)_3{\cdot}9H_2O$	562.00	rhomb, deliq, 1.552, 1.558	2.1	$-7H_2O$, 175		440	d	s abs al
i177	sulfide, di-	Nat. pyrite. FeS_2	119.97	yel, cub	5.0	1171		0.00049		d HNO_3, dil a
i178	sulfide, di-	Nat. marcasite. FeS_2	119.97	yel, rhomb	4.87	tr 450	d	0.00049		d HNO_3, i dil a
i179	(II) sulfide	Nat. troilite. FeS	87.91	blk-brn, hex	4.74	1193—1199	d	0.00062^{18}	d	s d a; i NH_3
i180	(III) sulfide	Fe_2S_3	207.87	yel-grn	4.3	d		sl d	d FeS + S	d a
i181	(II) sulfite	$FeSO_3{\cdot}3H_2O$	189.95	grnsh or wh cr		d 250		s sl s		s SO_2 sol; i al
i182	tantalate	Nat. tapiolite. $Fe(TaO_3)_2$	513.74	lt brn, tetr, 2.27, 2.42 (Li)	7.33					
i183	*d*-tartrate	$FeC_4H_4O_6$	203.92	wh cr.				0.877^{16}	v sl s	v s min a; s NH_4OH
i184	(II) thiocyanate	$Fe(SCN)_2{\cdot}3H_2O$	226.05	grn, rhomb		d		v s		s al, eth, acet
i185	(III) thiocyanate	$Fe(SCN)_3$ or $Fe_2(SCN)_6$	230.08	blk-red, cub, deliq				v s	d	s al, eth, acet
i186	(II) thiosulfate	$FeS_2O_3{\cdot}5H_2O$	258.04	grn cr, deliq				v s	d	v s al
i187	tungstate	Nat. ferberite. $FeWO_4$	303.69	tetr, 2.40 (Li)	6.64					
i188	*meta*vanadate	$Fe(VO_3)_3$	352.67	grayish-brn powd.				i		s a; i al
k1	**Krypton**	Kr	83.80	inert gas	gas 3.736 g/ℓ liq 2.155 *at* −152.9	−156.6	−152.30 ± 0.10	11.0^{0} cm^3 6.0^{25} cm^3	4.67^{50} cm^3	
l1	**Lanthanum**	La	138.9055 ± 3	wh met, tarnish in air, α: hex, β: cut above 350	α 6.1453, β 6.17	921	3457	d	d	s min a; i conc H_2SO_4
l2	acetate	$La(C_2H_3O_2)_3{\cdot}1^1/_2H_2O$	343.06					16.88^{25}		
l3	boride, hexa-	LaB_6	203.77	purp met cub	2.61	2210	d	i	i	i HCl
l4	bromate	$La(BrO_3)_3{\cdot}9\ H_2O$	684.75	hex pr		37.5	−7 H_2O, 100	28.5^{16}		i al
l5	bromide	$LaBr_3{\cdot}7H_2O$	504.72	col cr.	5.057^{26} anh	783 ± 3 anh	1577	v s		v s al; i eth
l6	carbide	LaC_2	162.93	yel cr.	5.02			d	d	s H_2SO_4; i conc HNO_3
l7	carbonate	$La_2(CO_3)_3{\cdot}8\ H_2O$	601.96	wh	2.6—2.7			i	i	s dil a; sl s aq CO_2; i acet
l8	chloride	$LaCl_3$	245.26	wh cr, deliq.	3.842^{25}	860	>1000	v s	d	v s al, pyr; i eth, bz
l9	chloride, heptahydrate	$LaCl_3{\cdot}7H_2O$	371.37	wh, tricl, hygr		d 91		v s	v s	v s al
l10	hexaantipyrin *per*chlorate	$[La(C_{11}H_{12}N_2O)]_6{\cdot}(ClO_4)_3$	1566.63	col, hex cr		d 290—295		1.50^{20}		
l11	hydroxide	$La(OH)_3$	189.93	wh powd		d		i		s a

No.	Name	Synonyms and Formulae	Mol. wt.	Crystalline form, properties and index of refraction	Density or spec. gravity	Melting point, °C	Boiling point, °C	Solubility, in grams per 100 cc		
								Cold water	Hot water	Other solvents
112	iodate	$La(IO_3)_3$	663.62	col, cr				1.7^{25}		
113	iodide	LaI_3	519.62	gray-wh, rhomb, hygr.	5.63	772		v s		s acet
114	molybdate	$La_2(MoO_4)_3$	757.62	tetr	4.77^{16}	1181		0.00179^{25}	0.0033^{85}	
115	nitrate	$La(NO_3)_3{\cdot}6H_2O$	433.01	col cr, deliq, tricl		40	d 126	151.1^{25}	v s	v s al; s acet
116	oxalate	$La_2(C_2O_4)_3{\cdot}9H_2O$	704.01	wh cr.		d		0.00008^{25}		s min a
117	oxide	Lanthana. La_2O_3	325.81	wh rhomb, or amorph	6.51^{15}	2307	4200	0.0004^{29}	d	s a, NH_4Cl; i acet
118	sulfate	$La_2(SO_4)_3$	565.98	wh powd, hygr.	3.60^{15}	d 1150		3.0	0.69^{100}	sl s al; i acet
119	sulfate, hydrate	$La_2(SO_4)_3{\cdot}9H_2O$	728.12	col hex, 1.564	2.821	d white heat		3.8^{0}	0.87^{100}	sl s HCl; i al
120	sulfide	La_2S_3	373.99	red-yel cr, hex	4.911^{11}	2100—2150 vac		d	d	s a
121	**Lead**	Pb	207.20	silv-blsh wh soft met, cub, 2.01	11.3437^{16} Ra-Pb 11.288_{20}^{20} UPb 11.2960^{16}	327.502	1740	i	i	s HNO_3, h conc H_2SO_4
122	abietate	$Pb(C_{20}H_{29}O_2)_2$	810.10	brn lumps or yelsh-wh powd				i		
123	acetate	$Pb(C_2H_3O_2)_2$	325.29	wh cr.	3.25_4^{20}	280		44.30^{20}	221^{50}	s glyc; v sl s al
124	acetate, basic	$Pb_2(OH)(C_2H_3O_2)_3$	608.54	wh				v s		sl s al
125	acetate, basic	$Pb(C_2H_3O_2)_2{\cdot}3PbO{\cdot}H_2O$	1012.90	wh powd						
126	acetate, basic	$Pb(C_2H_3O_2)_2{\cdot}Pb(OH)_2{\cdot}H_2O$	584.52	wh, monocl				v s		v s al
127	acetate, decahydrate	$Pb(C_2H_3O_2)_2{\cdot}10H_2O$	505.44	wh, rhomb cr.	1.69	22		s	s	i al
128	acetate, trihydrate	Sugar of lead. $Pb(C_2H_3O_2)_2{\cdot}3H_2O$	379.34	wh, monocl, β 1.567	2.55	– H_2O, 75	d 200	45.61^{15}	200^{100}	i al
129	acetate, tetra-	$Pb(C_2H_3O_2)_4$	443.38	col, monocl	2.228^{17}	175		d		d al; s chl, h ac a
130	*di*antimonate	$Pb_2Sb_2O_7$	769.90	dk yel powd	6.72			i	i	v sl s HCl
131	*ortho*antimonate	$Pb_3(SbO_4)_2$	993.10	or-yel powd	6.58_4^{20}			i	i	v sl s HCl
132	*ortho*antimonate	Nat. monimolite. $Pb_3(SbO_4)_2$	993.10	orange powd	6.58_4^{20}			i		i dil a
133	*meta*arsenate	$Pb(AsO_3)_2$	453.04	hex tabl	6.42^{15}			d	d	s HNO_3
134	*ortho*arsenate	$Pb_3(AsO_4)_2$	899.44	wh cr, v pois	7.80	1042, sl d 1000		v sl s		s HNO_3
135	*ortho*arsenate, di-	Nat. schultenite. $PbHAsO_4$	347.13	monocl leaf, α 1.90, γ 1.97	5.79	d 720	– H_2O, 220	i	sl s	s HNO_3, caust alk
136	*ortho*arsenate, mono-	$Pb(H_2AsO_4)_2$	489.07	tricl, 1.74, 1.82	4.46^{15}	d 140		d		s HNO_3
137	*pyro*arsenate	$Pb_2As_2O_7$	676.24	rhomb, β 2.03	6.85_{15}^{15}	802		i	d	s HCl, HNO_3; i ac a
138	*meta*arsenite	$Pb(AsO_2)_2$	421.04	wh powd	5.85			i		s HNO_3
139	*ortho*arsenite	$Pb_3(AsO_3)_2{\cdot}xH_2O$		wh powd	5.85			i		s alk, HNO_3
140	azide	$Pb(N_3)_2$	291.24	col need, or powd.			expl 350	0.023^{18}	0.09^{70}	v s ac a; i NH_4OH
141	*meta*borate	$Pb(BO_2)_2{\cdot}H_2O$	310.83	wh cr powd	5.598, anhydr	– H_2O, 160		i	i	s a; i alk
142	borofluoride	$Pb(BF_4)_2$	380.81	cr pr				d		d al
143	bromate	$Pb(BrO_3)_2{\cdot}H_2O$	481.02	col. monocl	5.53	d 180		1.38^{20}	sl s	
144	bromide	$PbBr_2$	367.01	wh, rhomb	6.66	373	916	0.4554^{0} 0.8441^{20}	4.71^{100}	s a, KBr; sl s NH_3; i al
145	buryrate	$Pb(C_4H_7O_2)_2$	381.40	col scales, pois.		90		i	i	s dil HNO_3
146	caprate	$Pb(C_{10}H_{19}O_2)_2$	549.72			103—104		i	i	0.0029^{20} eth
147	caproate	$Pb(C_6H_{11}O_2)_2$	437.50			73—74				1.09^{25} eth
148	caprylate	Lead octoate. $Pb(C_3H_{15}O_2)_2$	493.61	wh leaf		83.5—84.5		i	i	s al; 0.0938 eth
149	carbonate	Nat. cerussite. $PbCO_3$	267.21	col, rhomb, 1.804, 2.076, 2.078	6.6	d 315		0.00011^{20}	d	s a, alk; i NH_3, al
150	carbonate, basic	White lead, hydrocerussite. $2PbCO_3{\cdot}Pb(OH)_2$	775.63	wh powd, or hex.	6.14	d 400		i	i	sl s aq CO_2; s HNO_2; i al
151	cerotate	$Pb(C_{26}H_{51}O_2)_2$	998.58	wh need		113		i		i al, eth; s bz
152	chlorate	$Pb(ClO_3)_2$	374.10	wh monocl, deliq	3.89	d 230		v s		s al
153	chlorate, hydrate	$Pb(ClO_3)_2{\cdot}H_2O$	392.12	wh, monocl, deliq.	4.037	d 110		151.3^{18}	171^{80}	s al
154	*per*chlorate	$Pb(ClO_4)_2{\cdot}3H_2O$	460.15	wh, rhomb	2.6	d 100		499.7^{26}		s al
155	chloride	Nat. cotunnite. $PbCl_2$	278.11	wh, rhomb, 2.199, 2.217, 2.260	5.85	501	950^{760}	0.99^{20}	3.34^{100}	sl s dil HCl,NH_3; i al; s NH_4 salts
156	chloride, tetra-	$PbCl_4$	349.01	yel oily liq	3.18^{0}	–15	expl 105	d (Cl_2)	d	s conc HCl
157	chloride, sulfide	$PbCl_2{\cdot}3PbS$	995.89	red				i	d	d a, alk; i dil a
158	chlorite	$Pb(ClO_2)_2$	342.10	yel, monocl		espl 126		0.095^{20}	0.42^{100}	s KOH
159	chromate	Nat. crocoite, chrome yellow. $PbCrO_4$	323.19	yel, monocl, 2.31, 2.37(Li), 2.66	6.12^{15}	844	d	0.000058^{25}	i	s a, alk; i ac a, NH_3
160	chromate, basic	Chrome red. $PbCrO_4{\cdot}PbO$	546.39	red cr powd	6.63			i	i	s a, alk
161	chromate, basic	$Pb_2(OH)_2CrO_4$	564.41	red amorph or cr	6.63	920		i	i	s KOH
162	*di*chromate	$PbCr_2O_7$	423.19	red cr				d		s a, alk
163	citrate	$Pb_3(C_6H_5O_7)_2{\cdot}3H_2O$	1053.85	wh cr powd				s		v sl s al
164	cyanate	$Pb(OCN)_2$	291.23	wh need		d		i	sl s	
165	cyanide	$Pb(CN)_2$	259.24	yelsh-wh powd, pois				sl s	s	s KCN
166	enanthate	$Pb(C_7H_{13}O_2)_2$	465.56	wh leaf		91.5		sl s		i al
167	ethylsulfate	$Pb(C_2H_5SO_4)_2{\cdot}2H_2O$	493.47	col liq, pois.				s		
168	ferricyanide	$Pb_3[Fe(CN)_6]_2{\cdot}5$ (or 6) H_2O	1135.58	blk-brn to red, monocl pr		–H_2O, 110—120 d		sl s	s, d 100	s alk, HNO_3
169	ferrite	$PbFe_2O_4$	382.89	hex		1530 d, 725				
170	ferrocyanide	$Pb_2Fe(CN)_6$ $3H_2O$	680.40	yelsh-wh powd.		–H_2O, 100		i		sl s H_2SO_4
171	fluoride	PbF_2	245.20	col, rhomb, pois	8.24	855	1290	0.064^{20}		s HNO_3; i acet, NH_3

No.	Name	Synonyms and Formulae	Mol. wt.	Crystalline form, properties and index of refraction	Density or spec. gravity	Melting point, °C	Boiling point, °C	Solubility, in grams per 100 cc: Cold water	Hot water	Other solvents
l72	fluorochloride	Nat. matlockite. PbFCl	261.65	wh, tetr, 2.145, 2.006	7.05	601		0.037^{25}	0.1081^{100}	
l73	fluosilicate	$PbSiF_6 \cdot 2H_2O$	385.31	col, monocl		d		s	v s	
l74	fluosilicate, tetrahydrate	$PbSiF_6 \cdot 4H_2O$	421.34	col, monocl		d<100				
l75	formate	$Pb(CHO_2)_2$	297.24	wh, rhomb, lust, 1.789, 1.852, 1.877	4.63	d 190		1.6^{15}	20^{100}	i al
l76	hydride, di-	PbH_2	209.22	gray powd		d				
l77	hydroxide	$Pb(OH)_2$	241.21	wh, amorph		d 145		0.0155^{20}	sl s	s a, alk; i ac a
l78	hydroxide	$Pb_2O(OH)_2$ or $2PbO \cdot H_2O$	464.41	wh cub, or amorph powd, pois	7.592	d 145		0.014	sl s	s alk, ac a, HNO_3
l79	iodate	$Pb(IO_3)_2$	557.01	wh	6.155^{20}	d 300		0.0012^{2}	0.003^{25}	sl s HNO_3; i NH_3
l80	*para*periodate	$PbHIO_5$	415.11	wh cr.		d 130		i	i	s dil HNO_3
l81	*para*periodate, hydrate	$PbHIO_5 \cdot H_2O$	433.12	amorph		$-H_2O$, 110		i	i	sl s dil HNO_3
l82	iodide, basic	$PbI_2 \cdot PbO \cdot H_2O$	702.22	rhomb cr	6.83^{20}	d 100				
l83	iodide, di-	PbI_2	461.01	yel hex powd, pois	6.16	402	954	0.044^{0} 0.063^{20}	0.41^{100}	s alk, KI; i al
l84	iodide, mono-	PbI	334.10	pa yel		d 300		0.1		
l85	isobutyrate	$Pb(C_4H_7O_2)_2$	381.40	wh pr		<100		9.1^{15}		
l86	lactate	$Pb(C_3H_5O_3)_2$	385.34	wh cr powd				s		s h al
l87	laurate	$Pb(C_{12}H_{23}O_2)_2$	605.83	chalky wh powd		104.7		0.009^{35}		0.008^{25} al; $0.007^{14.5}$ eth
l88	lignocenate	$Pb(C_{24}H_{47}O_2)_2$	942.47	wh powd		117		i		v s h bz; sl s al; i eth
l89	malate	$Pb(C_4H_4O_5) \cdot 3H_2O$	393.32	wh powd				sl s		v sl s al
l90	melissate	$Pb(C_{31}H_{61}O_2)_2$	1138.85	wh powd		115—116		i	i	s biol tol, ac a; sl s h bz, chl; i al, eth
l91	molybdate	Nat. wulfenite. $PbMoO_4$	367.14	col-lt yel, tetr pl	6.92_4^{25}	1060—1070			i	d conc H_2SO_4; s a, KOH; i al
l92	myristate	$Pb(C_{14}H_{27}O_2)_2$	661.93	wh powd		107		0.005^{35}	0.006^{50}	0.004^{25} al; $0.010^{14.5}$
l93	2-naphthalenesulfonate	$Pb(C_{10}H_7SO_3)_2$	621.65	wh cr powd, pois				i		s al
l94	nitrate	$Pb(NO_3)_2$	331.21	col, cub or monocl, pois, 1.782	4.53^{20}	d 470		37.65^{0} 56.5^{0}	127^{100}	8.77^{22} 43% al; s alk, NH_3
l95	nitrate, basic	$Pb(OH)NO_3$	286.21	wh rhomb cr	5.93	d 180		19.4^{19}	s	s a
l96	nitrite	$3PbO \cdot N_2O_3 \cdot H_2O$	763.63	lt yel powd				v s		s dil HNO_3
l97	oleate	$Pb(C_{18}H_{32}O_2)_2$	770.12					i		6.46^{20} eth; s pet eth; sl s al
l98	oxalate	PbC_2O_4	295.22	wh powd	5.28	d 300		0.00016^{18}		s HNO_3; i al
l99	oxide-, di.	Plattnerite. PbO_2	239.20	bz, tetr, ω 2.3(Li)	9.375	d 290		i	i	s dil HCl; sl s ac a; i al
l100	oxide, mono-	Litharge. PbO	223.20	yel, tetr	9.53	886		0.0017^{20}		s HNO_3. alk, Pb acet, NH_4Cl, $CaCl_2$, $SrCl_2$
l101	oxide, mono-	Massicot. PbO	223.20	yel, rhomb, 2.51, 2.61(Li), 2.71	8.0			0.0023^{22}	i	s alk
l102	oxide, red	Minium. Pb_3O_4	685.60	red cr sc, or amorph powd	9.1	d 500		i	i	s HCl, acet a; i al
l103	oxide, sesqui-	Pb_2O_3	462.40	or-yel powd, amorph		d 370		i	d	d a
l104	oxide, sub	Pb_2O	430.40	blk, amorph	8.342	d		i	i	s a, alk
l105	oxychloride	$PbCl_2 \cdot 3PbO$	947.70	yel				0.0056^{18}	0.07^{74}	
l106	oxychloride	Cassel yellow. $PbCl_2 \cdot 7PBO$	1840.50	yel cr, or powd				i		
l107	oxychloride	Fiedlerite. $2PbCl_2 \cdot PbO \cdot H_2O$	797.43	monocl, 1.816 2.1023, 2.026	5.88^{20}	d 150				s HNO_3
l108	oxychloride	Nat. laurionite. $PbCl_2 \cdot Pb(OH)_2$	519.32	rhomb	6.24	d 142				
l109	oxychloride	Matlockite. $PbCl_2 \cdot PbO$	501.31						i	s alkalies, hot conc HCl
l110	oxychloride	Nat. matlockite. $PbCl_2 \cdot Pb(OH)_2$	519.32	wh, tetrag, 2.04 2.15, 2.15	7.21	d 524		0.0095^{18}		s alk
l111	oxychloride	Nat. mendipite. $PbCl_2 \cdot 2PbO$	724.50	yel, rhomb, 2.24, 2.27, 2.31	7.08	693		i	i	s alk
l112	oxychloride	Nat. paralaurionite. $PbCl_2 \cdot PbO \cdot H_2O$	519.32	col to wh, monocl pr, 2.146	6.05^{15}	d 150				
l113	palmitate	$Pb(C_{16}H_{31}O_2)_2$	718.04	wh powd		112.3		0.005^{35}	0.007^{50}	s al; 0.148^{20} eth
l114	phenolate	Lead phenate, lead carbolate. $Pb(OH)OC_6H_5$	317.31	yelsh-wh powd				i		
l115	phenolsulfonate	Lead sulfocarbolate. $Pb[C_6H_4(OH)SO_3]_2 \cdot 5H_2O$	643.60	wh lustr need				s		s al
l116	*meta*phosphate	$Pb(PO_3)_2$	365.14	col cr.		800(?)			v sl s	
l117	*ortho*phosphate	$Pb_2(PO_4)_2$	811.54	col or wh powd, hex, 1.970, 1.936	6.9—7.3	1014		0.000014^{20}	i	s HNO_3, alk; i ac a, al
l118	*ortho*phosphate, di-	$PbHPO_4$	303.18	rhomb	5.661^{15}	d				s HNO_3, alk, NH_4Cl
l119	*ortho*phosphate, mono-	$Pb(H_2PO_4)_2$	401.17	need						s alk, dil HNO_3, h conc HCl
l120	phosphide	PbP_5	362.07	blk unstable, inflam		d 400 (vac)		d	d	d dil a
l122	*ortho*phosphite	$PbHPO_3$	287.18	wh, powd		d		i		s HNO_3

No.	Name	Synonyms and Formulae	Mol. wt.	Crystalline form, properties and index of refraction	Density or spec. gravity	Melting point, °C	Boiling point, °C	Solubility, in grams per 100 cc		
								Cold water	Hot water	Other solvents
1123	picrate	$Pb(C_6H_2N_3O_7)_2 \cdot H_2O$	681.41	yel, need	2.831^{20}	$-H_2O$, 130	expl	0.88^{15}		
1124	proprionate, tetra-	$Pb(C_3H_5O_2)_4$	499.49	solid		132				
1125	*pyro*phosphate	$Pb_2P_2O_7$	588.34	wh, rhomb	5.8^{20}	824		i	i	s HNO_3, KOH
1126	*pyro*phosphate, hydrate	$Pb_2P_2O_7 \cdot H_2O$	606.36	wh, rhomb		806 anhydr		i	d	s HNO_3, KOH, $Na_4P_2O_7$
1127	selenate	$PbSeO_4$	350.16	wh, rhomb	6.37_4^{20}	d		i	i	s conc a
1128	selenide	Nat. clausthalite. PbSe	286.16	gray, cub	8.10^{15}	1065		i		s HNO_3
1129	metasilicate	Nat. alamosite. $PbSiO_3$	283.28	col or wh, monocl	6.49	766		i		d a
1130	*ortho*silicate, di-	Nat. barysilite. $Pb_2Si_2O_7$	582.57	wh, trig, 2.070, 2.050	6.707			i	i	
1131	stearate	$Pb(C_{18}H_{35}O_2)_2$	774.15	wh powd		115.7		0.05^{35}	0.06^{50}	$0.005^{14.5}$ eth; i al
1132	sulfate	Nat. anglesite. $PbSO_4$	303.26	wh, monocl, or rhomb, 1.877, 1.822, 1.894	6.2	1170		0.00425^{25}	0.0056^{40}	s NH_4 salts; sl s conc H_2SO_4; i a
1133	sulfate, basic	Nat. lanarkite. $PbSO_4 \cdot PbO$	526.46	wh, monocl, 1.93, 1.99, 2.02	6.92	977		0.0044^{0}	v sl s	sl s H_2SO_4
1134	sulfate, hydrogen	$Pb(HSO_4)_2 \cdot H_2O$	419.35	wh cr.		d		0.0001^{18} d		sl s H_2SO_4
1135	*peroxydi*sulfate	$PbS_2O_3 \cdot 3H_2O$	453.36	deliq				v s		
1136	sulfide	Nat. galena. PbS	239.26	bl met cub, 3.921	7.5	1114		0.000086^{13}		s a; i al, KOH
1137	sulfite	$PbSO_3$	287.26	wh powd		d		i	i	s HNO_3
1138	tartrate, *dl*-	$PbC_4H_4O_6$	355.27	wh cr powd	2.53^{19}			0.0025^{20}	0.0074^{100}	s HNO_3, KOH; i al, ac a, NH_4 ac
1139	*di*thionate	$PbS_2O_6 \cdot 4H_2O$	439.38	trig, 1.635, 1.653	3.22	d		115.0^{20}		
1140	thiosulfate	PbS_2O_3	319.32	wr cr.	5.18	d		0.03		s a, $Na_2S_2O_3$
1141	*meta*titanate	$PbTiO_3$	303.08	yel. rhomb-pyr.	7.52			i	i	
1142	telluride	Nat. altaite. PbTe	334.80	Wh, cub	8.164_4^{20}	917				i a
1143	thiocyanate	$Pb(SCN)_2$	323.36	wh, monocl	3.82	d 190		0.05^{20}	0.2^{100}	s KCNS, HNO_3
1144	tungstate	Nat. stolzite. $PbWO_4$	455.05	tetr, 2.269, 2.182	8.23			i		i HNO_3, s KOH
1145	tungstate	Nat. raspite. $PbwO_4$	455.05	col, monocl, 2.27, 2.27, 2.30	1123			0.03		d a; i al
1146	*meta*vanadate	$Pb(VO_3)_2$	405.08	yel powd				sl s		d HCl; s dil HNO_3
1147	**Lithium**	Li	6.941 ± 3	silver white, soft	0.534^{20}	180.54	1342	d		
1148	acetate	$LiC_2H_3O_2 \cdot 2H_2O$	102.02	wh, rhomb, α 1.40, β 1.50		70	d	300^{15}	v s	21.5 al
1149	acetylsalicylate	$LiC_9H_7O_4$	186.09	wh powd hygr, d in moist air				100		25 al
1150	*meta*aluminate	$LiAlO_2$ (or $Li_2Al_2O_4$)	65.92	wh, rhomb, 1.604, 1.614	2.55_4^{25}	1900—2000		i		
1151	aluminum hydride	$LiAlH_4$	37.95	wh cr powd	0.917	d 125		d		ca 30 eth
1152	amide	$LiNH_2$	22.96	col need, cub	$1.178^{17.5}$	380—400	d 750—200 subl	s	d	sl s liq NH_3, al; i eth, bz
1153	antimonide	Li_3Sb	142.57		3.2^{17}	>950		d	d	d a
1154	*ortho*arsenate	Li_3AsO_4	159.74	wh powd, rhomb	3.07^{15}			v sl s		s dil ac a; i pyr
1155	azide	LiN_3	48.96	col cr, hygr		d 115—298		66.41^{16}		20.26^{16} abs al; i eth
1156	benzoate	$LiC_7H_5O_2$	128.06	wh cr or powd				33^{25}	40^{100}	7.7^{26} al, 10^{78} al
1157	*meta*borate	$LiBO_2$	49.75	wh, tricl	$1.397^{41.7}$	845		2.57^{20}	11.83^{80}	
1158	*meta*borate	$LiBO_2 \cdot 8H_2O$	193.87	col, trig	$1.38^{14.9}$	47				
1159	*penta*borate	$Li_2B_{10}O_{16} \cdot 8H_2O$	522.09	wh	1.72	300—350 $-8H_2O$		36.3^{45}	194^{100}	3.9^{30} al; 22^{33} glycerine; i bz
1160	*tetra*borate	$Li_2B_4O_7$	169.12	2h cr		930		2.89^{20}	5.45^{100}	i org solv
1161	borohydrate	$LiBH_4$	21.78	rhomb cr	0.66	d 279		s d		s eth
1162	borohydrate	$LiBH_4$	21.78	wh, orthorhomb	0.666	275 d		v sl s		d al; 2.5 eth
1163	bromide	LiBr	86.85	wh, cub, deliq, 1.784	3.464^{25}	550	1265	145^{4}	254^{90}	73^{40} al; 8 MeOH; s al, eth; sl s pyrid
1164	bromide, dihydrate	$LiBr \cdot 2H_2)$	122.88	wh cr.		$-1H_2O$; 44		$246.^{20}$	v s	s al
1165	carbide	Li_2C_2	37.90	wh cr or powd	1.65^{18}			d	d	s a
1166	carbonate	Li_2CO_2	73.89	wh, monocl, 1.428, 1.567, 1.572	2.11	723	d 1310^{760}	1.54^{0}	0.72^{100}	i al; acet
1167	carbonate, acid	Lithium bicarbonate. $LiHCO_3$	67.96	wh				5.5^{13}		
1168	chlorate	$LiClO_3$	90.39	col, rhomb need, deliq, α 1.63, γ 1.64	1.1190_4^{18} (18%Soln)	127.6	300 d	500^{27}		v s al; 0.142^{25} acetone
1169	chlorate	$LiClO_3 \cdot {}^1/_2H_2O$ (or ${}^1/_3H_2O$)	99.40	wh, tetr, deliq		65(?)	$-{}^1/_2H_2O$, 90 d 290	v s	v s	v s al
1170	perchlorate	$LiClO_4$	106.39	wh	2.428	236	430 d	60.0^{25}	150^{89}	152^{25} al; 182^{25} MeOH; 114^{25} eth; 137^{25} acetone
1171	perchlorate, trihydrate	$LiClO_4 \cdot 3H_2O$	160.44	wh, hex	1.841	95 deliq 236 (anhydr)	d 100 $-2H_2O$	130^{25}		72.9^{25} al; 156^{25} MeOH; 96.2^{25} acetone; 0.096^{25} eth
1172	chloride	LiCl	42.39	wh, cub, 1.662.	2.068^{25}	605	1325—1360	63.7^{0}	130^{96}	25.10^{30} al; 42.36^{25} MeOH; 4.11^{25} acetone; $0.538^{33.9}$ NH_4OH
1173	chloride, monohydrate	$LiCl \cdot H_2O$	60.41	wh cr, hydr	1.78	$-H_2O$ >98		86.2^{30}	2	s HCl
1174	chloroplatinate	$Li_2PtCl_6 \cdot 6H_2O$	529.77	or prism		$-6H_2O$, 180		v s	v s	v s al; i eth
1175	dichromate, dihydrate	$Li_2Cr_2O_7 \cdot 2H_2O$	265.90	orange-red cr, deliq	2.34^{30}	187 d	110 $-2H_2O$	187^{30}	278^{100}	s reacts al

No.	Name	Synonyms and Formulae	Mol. wt.	Crystalline form, properties and index of refraction	Density or spec. gravity	Melting point, °C	Boiling point, °C	Solubility, in grams per 100 cc		
								Cold water	Hot water	Other solvents
1177	citrate	$Li_3C_6H_5O_7{\cdot}4H_2O$	281.99	col cr or powd, deliq		$-4H_2O$, 105		74.5^{25}	66.7^{100}	sl s al, eth
1178	fluoride	LiF	25.94	wh, cub, 1.3915	2.635^{20}	845	1676	0.27^{18}		i al; s HF
1179	fluosilicate	$Li_2SiF_6{\cdot}2H_2O$	191.99	wh, monocl, 1.300, 1.296	2.33^{12}	$-2H_2O$, 100	d	73^{17}		s al; i eth, acet
1180	fluosulfonate	$LiSO_3F$	106.00	wh powd		360		v s	s	v s al, eth, acet; i ligorin
1181	formate, monohydrate	$H{\cdot}COOLi{\cdot}H_2O$	69.97	wh, rhomb	1.46	$-H_2O$, 94	d 230	27.85^{18}	57.05^{98}	sl s al, acet; i bz
1182	gallium hydride	$LiGaH_4$	80.69	wh cr.				d	d	s eth
1183	gallium nitride	Li_3GaN_2	118.56	lt gr powd.	3.35	d 800		d	d	s a, alk
1184	*meta*germanate	Li_2GeO_3	134.47	monocl, 1.7	3.53^{21}	1239		0.85^{25}		s a
1185	hydride	LiH	7.95	wh cr.	0.82	680		d		v sl s a
1186	hydroxide	LiOH	23.95	wh tetr, 1.464, 1.452	1.46	450	d 924	12.8^{20}	17.5^{100}	sl s al
1187	hydroxide, monohydrate	$LiOH{\cdot}H_2O$	41.96	wh monocl, 1.460, 1.524	1.51			22.3^{10}	26.8^{80}	sl s al; i eth
1188	iodate	$LiIO_3$	181.84	wh, hex, hygr	4.502^{32}_{4}			80.3^{18}		i al
1189	iodide	LiI	133.85	wh, cub, 1.955 ± 0.003	4.076	449	1180 ± 10	165^{20}	433^{80}	250.8^{20} al; 42.6^{18} acet 343.4^{20} MeOH; v s NH_4OH
1190	iodide, trihydrate	$LiI{\cdot}3H_2O$	187.89	col-yelsh, hex, hygr.	3.48	73 $-H_2O$	$-2H_2O$, 80 $-H_2O$, 300	151^{0}	201.2^{60}	s abs al, acet
1191	laurate	$LiC_{12}H_{23}O_2$	206.25	wh powd		229.2—229.8		$0.154^{16.3}$	0.178^{25}	0.322^{25} al; $0.008^{15.3}$
1192	*per*manganate	$LiMnO_4{\cdot}3H_2O$	179.92	cub	2.06	d 190		71.43^{16}		d alk
1193	molybdate	Li_2MoO_4	173.82	wh trig, hygr	2.66	705		v s		
1194	myristate	$LiC_{14}H_{27}O_2$	234.31			223.6—224.2		$0.027^{16.3}$ 0.036^{25}	0.062^{50}	$0.010^{15.3}$ eth; 0.331^{15} acet; 0.155^{20} al
1195	nitrate	$LiNO_3$	68.95	wh, trig, 1.735, 1.735	2.38	264	d 600	$89.8^{27.55}$	234^{100}	s NH_4OH, al; 37.15 pyridine
1196	nitrate, trihydrate	$LiNO_3{\cdot}3H_2O$	122.99	col need		$-2^1/_2H_2O$, 29.9	$-3H_2O$, 61.1	34.8^{0}	$57.48^{29.6}$	s al, MeOH, acet
1197	nitride	Li_3N	34.83	red-brn amorph, or blk-gray cr, cub		tr 840—850 (in N_2)				
1198	nitrite	$LiNO_2{\cdot}H_2O$	70.96	col flat need	1.615^{0}	>100	d	125^{0}	459^{50}	v s abs al
1199	oxalate	$Li_2C_2O_4$	101.90	col, rhomb, 1.465, 1.53, 1.696	$2.121^{17.5}$	d		$8^{19.5}$		i al, eth
1200	oxalate, acid	$LiHC_2O_4{\cdot}H_2O$	113.98			d		8^{17}		
1201	oxide	Li_2O	29.88	wh cr, cub, n_D 1.644	$2.013^{25.2}$	>1700	1200^{600}	6.67^{0} d	10.02^{100}	
1202	palmitate	$LiC_{16}H_{31}O_2$	262.36	wh powd		224.5		0.01^{18}	0.015^{25}	0.347^{15} acet; 0.077^{20} al; $0.005^{15.8}$ eth
1203	*meta*phosphate	$LiPO_3$	85.91	col pl.	2.461	red heat		i	i	s a
1204	*ortho*phosphate	Li_3PO_4	115.79	col, rhomb	$2.537^{17.5}$	837		0.039^{18}		s a, NH_4OH; i acet
1205	*ortho*phosphate	$Li_3PO_4{\cdot}{}^1/_2H_2O$	124.80	wh cr powd	2.41	$-{}^1/_2H_2O$, 100		0.04^{25}		s a
1206	phosphate, di- H	LiH_2PO_4	103.93	col cr, hygr	2.461	>100				
1207	potassium sulfate	$LiKSO_4$	142.10	col, hex; n_D 1.4723, 1.4717	2.393^{20}			s	s	
1208	potassium *dl*-tartrate	$LiKC_4H_4O_6{\cdot}H_2O$	212.13	col, monocl, β 1.523 (red)	1.610			s		
1209	salicylate	$LiC_7H_5O_3$	144.06	wh, powd, deliq		d		133.3		50 al
1210	selenide	$Li_2Se{\cdot}9H_2O$	254.98	col, rhomb, deliq				d		
1211	*meta*silicate	Li_2SiO_3	89.97	col, rhomb; α 1.584, γ 1.604	2.52^{25}_{4}	1204		i	s d	s dil HCl
1212	*ortho*silicate	Li_4SiO_4	119.85	col, rhomb; α 1.594, γ 1.614	2.392^{25}_{4}	1256		i	d	d a
1213	silicide	Li_6Si_2	97.82	bl cr, hygr	*ca.* 1.12	d 600 (vac)		d	d	d a; i NH_3 turp
1214	sodium fluoaluminate	$Li_3Na_3(AlF_6)_2$	371.74	cub cr, 1.3395	2.774	710		0.074^{18}		
1215	stearate	$LiC_{18}H_{35}O_2$	290.42	wh cr.	220.5—221.5			0.010^{18}		0.010^{25} al; 0.040^{18} eth; 0.457^{15} acet
1216	sulfate	Li_2SO_4	109.94	α monocl; β hex or rhomb, γ cub 500°C; β 1.465	2.221		845	26.1^{0}	23^{108}	i abs al, acet
1217	sulfate, hydrogen	$LiHSO_4$	104.01	col pr	2.123^{13}	120		d		
1218	sulfate, monohydrate	$Li_2SO_4{\cdot}H_2O$	127.95	col cr, monocl, 1.465, 1.477, 1.488	880			34.9^{25}	29.2^{100}	11.5^{30} al + H_2O (23.9% alco); i acet, pryidine
1219	sulfide	Li_2S	45.94	wh-yel, cub, deliq	1.66	900—975		v s	v s	v s al
1220	sulfide, hydro-	LiHS	40.01	wh powd, hygr.				s	s	s al
1221	sulfite, monohydrate	$Li_2SO_3{\cdot}H_2O$	111.96	wh need, α 1.53, γ 1.59		455 d	140 $-H_2O$	24.9^{30}	22^{80}	i org solv
1222	tartrate	$Li_2C_4H_4O_6{\cdot}H_2O$	179.97	wh cr powd				s		
1223	thallium *dl*-tartrate	$LiTlC_4H_4O_6{\cdot}2H_2O$	395.43	tricl.	3.144					
1224	thiocyanate	LiSCN	65.02	wh cr, deliq, n_D 1.333				v s		s methylacet
1225	*di*thionate	$Li_2S_2O_6{\cdot}2H_2O$	210.03	col, rhomb, 1.5602	2.158	d		v s		
1226	tungstate	Li_2WO_4	261.73	col, trig	3.71	742		v s	v s	d a; i al
1227	**Lutetium**	Cassiopeium. Lu	174.967	met, hex.	9.8404	1663	3395			

No.	Name	Synonyms and Formulae	Mol. wt.	Crystalline form, properties and index of refraction	Density or spec. gravity	Melting point, °C	Boiling point, °C	Solubility, in grams per 100 cc		
								Cold water	Hot water	Other solvents
1228	bromide	$LuBr_3$	414.68		1025	1400		s	s	
1229	chloride	$LuCl_3$	281.33	col cr.	3.98	905	subl 750	s		
1230	fluoride	LuF_3	231.96			1182	220	i	i	
1231	iodide	LuI_3	555.68			1050	1200	s	s	
1232	oxalate	$Lu_2(C_3O_4)_3{\cdot}6H_2O$	722.08	wh cr.		50 ($-H_2O$)		i	i	i dil a
1233	oxide	Lu_2O_3	397.93	cub cr	9.42					
1234	sulfate	$Lu_2(SO_4)_3{\cdot}8H_2O$	782.23	col cr.				42.27^{20}	16.93^{40}	
m1	**Magnesium**	Mg	24.305	silv wh met, hex	1.74^{5}	648.8	1107	i	d to $Mg(OH)_2$	s min a, conc. HF, NH_4 salts; i CrO_3, alk.
m2	acetate	$Mg(C_2H_3O_2)_2$	142.39	wh cr.	1.42	323 d		v s	v s	5.25^{15} MeOH
m3	acetate, tetrahydrate	$Mg(C_2H_3O_2)_2{\cdot}4H_2O$	214.46	col, monocl deliq, β 1.491	1.454	80		120^{15}	∞	v s al
m4	aluminate	Nat. spinel. $MgAl_2O_4$	142.27	col, cub, 1.723.	3.6	2135				sl s H_2SO_4; v sl s dil HCl; i HNO_3
m5	amide	$Mg(NH_2)_2$	56.35	gray powd.		d 350—400	d	d		v sl s liq NH_3; d al
m6	antimonate	$MgO{\cdot}Sb_2O_5{\cdot}12H_2O$	579.98	hex or tricl cr.	2.60 (hex)	$-12H_2O$, 200		v sl s		d HCl
m7	antimonide	Mg_3Sb_2	316.42	met hex pl	4.088^{25}_{4}	961		i		
m8	*ortho*arsenate	Nat. hoernesite. $Mg_3(AsO_4)_2{\cdot}8H_2O$	494.88	wh monocl	2.60—2.61					
m9	*ortho*arsenate	$Mg_3(AsO_4)_2{\cdot}22H_2O$	747.09	wh cr.	1.788	$-17H_2O$, 100	$-21H_2O$, 220	i	i	s a, NH_4Cl
m10	*ortho*arsenate, mono-H	Nat. roesslerite. $MgHAsO_4{\cdot}7H_2O$	290.34	monocl.	1.943^{15}	$-5H_2O$, 100		d		
m11	arsenide	Mg_3As_2	222.76	brn-red, cub	3.148^{25}_{4}	800		d	d	d dil a, al
m12	*ortho*arsenite	$Mg_3(AsO_3)_2$	318.75	wh				s	v s	s a, NH_4Cl; i NH_4OH
m13	benzoate	$Mg(C_7H_5O_2)_2{\cdot}3H_2O$	320.58	wh powd		$-3H_2O$, 110	d 200	6.16^{15}	19.6^{100}	s al
m14	bismuthide	Mg_3Bi_2	490.88	met, hex.	5.945^{25}_{4}	823				
m15	bismuth nitrate	$3\ Mg(NO_3)_2{\cdot}2Bi(NO_3)_2{\cdot}24H_2O$	1543.29	col cr, deliq.	2.32^{16}_{16}	71		d	d	s HNO_3
m16	diborate	Nat. ascharite. $Mg_2B_2O_5{\cdot}H_2O$	168.24	orthorhmb, 1.54	2.60—2.70					
m17	*meta*borate	Nat. pinnoite. $Mg(BO_2)_2{\cdot}3H_2O$	163.97	yel, tetr, pyram, 1.565, 1.575	2.27—2.30					
m18	metaborate, octahydrate	$Mg(BO_2)_2{\cdot}8H_2O$	254.04	col, tetr, 1.565, 1.575	2.30			i	v sl s	s a
m19	*ortho*borate	$Mg_3(BO_3)_2$	190.53	col, rhomb, 1.6527, 1.6537, 1.6548	2.99^{21}			i	i	s min a; i ac a
m20	boride	MgB_4	89.17	bl		d 1200 (vac)		d		sl s a
m21	bromate	$Mg(BrO_3)_2{\cdot}6H_2O$	388.20	col, cub, 1.514.	2.29	$-6H_2O$, 200	d	42^{18}	v s	i al
m22	bromide	$MgBr_2$	184.11	wh hex cr, deliq	3.72^{25}_{4}	700		101.50^{20}	125.6^{100}	6.9 al; 21.8^{20} MeOH
m23	bromide, hexahydrate	$MgBr_2{\cdot}6H_2O$	292.20	col, hex pr or need, hygr, fluo in x-rays	2.00	172.4		316^{0}	v s	s al, acet; sl s NH_3
m24	bromoplatinate	$MgPtBr_6{\cdot}12H_2O$	914.99	trig	2.802					
m25	carbonate	Nat. magnesite. $MgCO_3$	84.31	wh, trig, 1.717, 1.515	2.958	d 350	$-CO_2$, 900	0.0106		s a, aq + CO_2; i acet, NH_3
m26	carbonate, basic artinite	Nat. Artinite. $MgCO_3{\cdot}Mg(OH)_2{\cdot}3H_2O$	196.68	wh, rhomb, 1.489, 1.534, 1.557	2.02^{20}					
m27	carbonate, basic	Nat. hydromagnesite. $3MgCO_3{\cdot}Mg(OH)_2{\cdot}3H_2O$	365.31	wh, rhomb, 1.527, 1.530, 1.540	2.16	d		0.04	0.011	s a, NH_4 salts
m28	carbonate, pentahydrate	Nat. lansfordite. $MgCO_35H_2O$	174.39	wh, monocl, 1.456, 1.476, 1.502	1.73	d in air		0.176^{7}	0.375^{20}	s HCl, $MgSO_4$ soln
m29	carbonate, trihydrate	Nat. nesquehonite. $MgCO_3{\cdot}3H_2O$	138.36	col, rhomb need, 1.495, 1.501, 1.526	1.850	165		0.179^{16}	d	s a; 1.4 aq + CO_2
m30	chlorate	$Mg(ClO_3)_2{\cdot}6H_2O$	299.30	wh, rhomb need, deliq	1.80^{25}	35	d 120	128.6^{18}	v s	s al
m31	*per*chlorate	$Mg(ClO_4)_2$	223.21	deliq	2.21^{18}	d 251		99.3^{25}	v s	23.96^{25} al
m32	*per*chlorate, hexahydrate	$Mg(ClO_4)_2{\cdot}6H_2O$	331.30	wh, rhomb cr, 1.482, 1.458	1.98	185—190		v s	v s	
m33	*per*chlorate, hexammine	$Mg(ClO_4)_2{\cdot}6NH_3$	325.39	wh, cub	1.41^{20}_{4}					s liq NH_3; s d al
m34	chloride	$MgCl_2$	95.21	wh, lustr hex cr, 1.675, 1.59	2.316—2.33	714	1412	54.25^{20}	72.7^{100}	7.40^{30} al
m35	chloride, hexahydrate	Nat. bischofite. $MgCl_2{\cdot}6H_2O$	203.30	col, monocl, deliq, 1.495, 1.507, 1.528	1.569	d 116—118	d	167	367	s al
m36	chloropalladate	$MgPdCl_6{\cdot}6H_2O$	451.53	hex	2.12	d				
m37	chloroplatinate	$MgPtCl_6{\cdot}6H_2O$	540.19	yel, trig	3.692	$-H_2O$, 180		s	s	
m38	chlorostannate	$MgSnCl_6{\cdot}6H_2O$	463.80	tricl.	2.08	d 100				
m39	chromate	$MgCrO_4{\cdot}7H_2O$	266.41	yel, rhomb, 1.521, 1.550, 1.568	1.695		211.5^{18}	v s		
m40	chromite	$MgCr_2O_4$	192.29	dk-grn or red, cub	4.6^{20}			i	i	s conc H_2SO_4; i dl a, dil alk
m41	citrate, nono-H	$MgHC_6H_5O_7{\cdot}5H_2O$	304.49	wh gran powd				20^{25}	s	s a; i al
m42	cyanide	$Mg(CN)_2$	76.34			d 300 to $MgCN_2$	d 600	s	d	
m43	cyanoplatinite	$MgPt(CN)_4{\cdot}7H_2O$	449.56	red, pr	2.185^{16}	$-H_2O$, 45		v s	v s	i al, eth

PHYSICAL CONSTANTS OF INORGANIC COMPOUNDS (continued)

No.	Name	Synonyms and Formulae	Mol. wt.	Crystalline form, properties and index of refraction	Density or spec. gravity	Melting point, °C	Boiling point, °C	Solubility, in grams per 100 cc: Cold water	Hot water	Other solvents
m44	ferrite	$MgFe_2O_4$	200.00	blk, oct, 2.35	4.44—4.60	1750 ± 25	...	...	...	s conc HCl; i dil a, h HNO_3, al
m45	ferrocyanide	$Mg_2Fe(CN)_6 \cdot 12H_2O$	476.75	pa yel cr	...	d *ca* 200	...	33	...	i al
m46	fluoride	Nat. sellaite. MgF_2	62.30	col, tetr, faint vlt, lumin, 1.378, 1.390; 3.14	...	1261	2239	0076^{18}	i	s HNO_3; sl s a; i al
m47	fluosilicate	$MgSiF_6$	166.38	wh, cr or powd	...	...	...	65	...	...
m48	fluosilicate, hexahydrate	$MgSiF_6 \cdot 6H_2O$	274.47	wh, trig	1.788	d 120	...	$64.8^{17.5}$	...	i al
m49	formate	$Mg(CHO_2)_2 \cdot 2H_2O$	150.37	col, rhomb	...	$-2H_2O$, 100	...	14^{0} (anh)	24^{100} (anh)	i al, eth
m50	orthogermanate	Mg_2GeO_4	185.20	wh ppt	...	...	...	0.0016^{25}	...	s a; i alk
m51	germanide	Mg_2Ge	121.20	...	...	1115	...	...	...	...
m52	hydride	MgH_2	26.32	wh tetr cr or mass	...	d 280 (vac)	...	d viol	...	i eth
m53	hydroxide	Nat. brucite. $Mg(OH)_2$	58.32	col, hex pl, 1.559, 1.580	2.36	$-H_2O$, 350	...	0.0009^{18}	0.004^{100}	s a, NH_4 salts
m54	iodate	$Mg(IO_3)_2 \cdot 4H_2O$	446.17	wh, monocl	$3.3^{13.5}$	$-4H_2O$, 210	d	10.2^{20}	19.3^{100}	...
m55	iodide	MgI_2	278.11	wh, hex, deliq	4.43^{25}_{4}	d<637	...	148^{18}	164.9^{110}	s al, eth, NH_3
m56	iodide, octahydrate	$MgI_2 \cdot 8H_2O$	422.24	wh powd. deliq	...	d 41	...	81^{20}	90.3^{80}	s al, eth
m57	lactate	$Mg(C_3H_5O_3)_2 \cdot 3H_2O$	256.49	wh cr powd, v bitter taste	...	...	...	3.3	16.7^{100}	i al, eth
m58	laurate	$Mg(C_{12}H_{23}O_2)_2 \cdot 2H_2O$	458.96	wh lumps	...	150.4	...	0.007^{25}	0.041^{100}	0.415^{15} al; 0.012^{25} eth
m59	*per*manganate	$Mg(MnO_4)_2 \cdot 6H_2O$	370.27	dk purp need, deliq	2.18(?)	d	...	v s	d	s MeOH, ac a
m60	molybdate	$MgMoO_4$	184.24	rhomb, tricl	2.208	...	...	13.7^{25}	...	...
m61	myristate	$Mg(C_{14}H_{27}O_2)_2$	479.04	wh powd	...	131.6	...	0.006^{15}	0.014^{50}	0.189^{15} al; 0.007^{25} eth
m62	nitrate, dihydrate	$Mg(NO_3)_2 \cdot 2H_2O$	184.35	col pr	2.0256^{25}	129	...	s	s	s al, liq NH_3; sl s conc HNO_3
m63	nitrate, hexahydrate	$Mg(NO_3)_2 \cdot 6H_2O$	256.41	col, monocl, deliq.	1.6363^{25}_{4}	89	d 330	125	v s	s al, liq NH_3
m64	nitride	Mg_3N_2	100.93	grn-yel, powd or mass	2.712^{25}_{4}	d 800	subl 700 (vac)	d	d	s a; i al
m65	nitrite, trihydrate	$Mg(NO_2)_2 \cdot 3H_2O$	170.36	wh pr, hygr	...	d 100	...	s	...	s al
m66	oleate	$Mg(C_{18}H_{33}O_2)_2$	587.22	yel powd or mass	...	...	...	0.024^{25}	...	6.64^{20} al; s linseed oil; sl s eth
m67	oxalate	$MgC_2O_4 \cdot 2H_2O$	148.36	wh powd	2.45	d 150	...	0.07^{16}	0.08^{100}	s alk, a, oxalate
m68	oxide	Nat. periclase. MgO	40.30	col, cub, 1.736	3.58^{25}	2852	3600	0.00062	0.0086^{30}	s a, NH_4 salts, i al
m69	oxide, per-	MgO_2	56.30	wh powd	...	...	...	i	i	s a
m70	palmitate	$Mg(C_{16}H_{31}O_2)_2$	535.15	wh cr need or lumps	...	121.5	...	0.008^{25}	0.009^{50}	0.047^{25} al; 0.003^{25} eth
m71	*ortho*phosphate	$Mg_3(PO_4)_2$	262.86	rhomb pl, iridisc	...	1184	...	i	i	s NH_4 salts, i liq NH_3
m72	*ortho*phosphate	$Mg_3(PO_4)_2 \cdot 22H_2O$	659.19	col, monocl pr	1.640^{15}	$-18H_2O$, 100	d 200	v sl s	...	d a
m73	*ortho*phosphate, mono-H	Nat. newberyite. $MgHPO_4 \cdot 3H_2O$	174.33	wh, rhomb, 1.514, 1.518, 1.533	2.123^{15}	$-H_2O$, 205	d 550—650	sl s	...	s a
m74	*ortho*phosphate, mono-H, heptahydrate	$MgHPO_4 \cdot 7H_2O$	246.39	wh, monocl need	1.728^{15}	$-4H_2O$, 100	d 550—650	0.3	0.2	s a; i al
m75	*ortho*phosphate, octahydrate	Nat. bobierite. $Mg_3(PO_4)_2 \cdot 8H_2O$	406.98	wh, monocl pl, 1.510, 1.520, 1.543	2.195^{15}	$-5H_2O$, 150	$-8H_2O$, 400	v sl s	...	s NH_4 citrate
m76	*ortho*phosphate, tetrahydrate	$Mg_3(PO_4)_2 \cdot 4H_2O$	334.92	monocl.	1.64^{15}	...	...	0.0205	...	s a; i NH_4 salts
m77	phosphide	Mg_3P_2	134.86	yel-grn cub cr	2.055	...	...	d	d	d dil min a; sl d conc H_2SO_4
m78	*hypo*phosphite	$Mg(H_2PO_2)_2 \cdot 6H_2O$	262.37	wh, ditetrag	$1.59^{12.5}_{4}$	$-5H_2O$, 100	$-6H_2O$, 180	20^{25}	...	i al, eth
m79	*ortho*phosphite	$MgHPO_3 \cdot 3H_2O$	158.33	...	...	...	...	...	0.25	s a
m80	*pyro*phosphate	$Mg_2P_2O_7$	222.55	col, tab monocl, 1.602, 1.604, 1.615	2.559, (3.06)	1383	...	i	i	s a; i al
m81	platinocyanide	$MgPt(CN)_4 \cdot 7H_2O$	449.56	red cr, 1.561	2.185^{16}	$-2H_2O$, 45	...	s	s	s al; i eth
m82	salicylate	$Mg(C_7H_5O_3)_2 \cdot 4H_2O$	370.60	col or sl redsh cr powd, effl	...	...	...	s	...	s al
m83	selenate	$MgSeO_4 \cdot 6H_2O$	275.35	col, monocl, 1.468, 1.489, 1.491	1.928	...	...	v s	v s	...
m84	selenide	MgSe	103.27	lght gray powd or cr, 2.44	4.21	...	...	d	d	d a
m85	*meta*silicate	Nat. clinoenstatite. $MgSiO_3$	100.39	wh, monocl, α 1.651, γ 1.660	3.192^{25}_{4}	d 1557	...	i	i	v sl s HF
m86	*ortho*silicate	Nat. forsterite. Mg_2SiO_4	140.69	wh, orthorhmb, 1.65, 1.66, 1.67	3.21	1910	...	i	...	d h HCl
m87	(di-) silicide	Mg_2Si	76.70	blue cub	1.94	1102	...	i	d	s a, NH_4Cl, HCl
m88	silicofluoride	$MgSiF_6 \cdot 6H_2O$	274.47	wh hex-rhomb, 1.3439, 1.3602	1.788	d 100	...	60^{25}	5	s dil a, v sl s HF; i al
m89	stannide	Mg_2Sn	167.30	blsh-wh meet	...	778	...	s	...	s dil HCl
m90	stearate	$Mg(C_{18}H_{35}O_2)_2$	591.25	wh powd or lumps	...	86—88	...	0.003^{15} 0.004^{25}	0.008^{50}	0.020^{25} al; 0.003^{25} eth
m91	sulfate	$MgSO_4$	120.36	col, rhomb cr, 1.56	2.66	d 1124	...	26^{0}	73.8^{100}	s al, glyc; 1.16^{18} eth; i acet
m92	sulfate, heptahydrate	Epsom salt, nat. epsomite. $MgSO_4 \cdot 7H_2O$	246.47	col, rhomb or monocl, 1.433, 1.455, 1.461	1.68	$-6H_2O$, 150	$-7H_2O$, 200	71^{20}	91^{40}	sl s al, glyc

No.	Name	Synonyms and Formulae	Mol. wt.	Crystalline form, properties and index of refraction	Density or spec. gravity	Melting point, °C	Boiling point, °C	Solubility, in grams per 100 cc		
								Cold water	Hot water	Other solvents
m93	sulfate, monohydrate	Nat. kieserite, $MgSO_4 \cdot H_2O$	138.38	col, monocl pr, 1.523, 1.535, 1.586	2.445				68.4^{100}	
m94	sulfide	MgS	56.37	pa red-brn, cub, phosph, 2.271	2.84	d >2000		d	d	s a, PCl_3
m95	sulfite	$MgSO_3 \cdot 6H_2O$	212.45	wh, rhomb or hex, 1.511, 1.464 (hex)	1.725	$-6H_2O$, 200	d	1.25	s	i al, NH_3
m96	*d*-tartrate	$MgC_4H_4O_6 \cdot 5H_2O$	262.45	wh, rhomb	1.67	$-4H_2O$, 100	$-5H_2O$, 200	0.8^{18}	1.44^{90}	s min a; i al, NH_3
m97	*d*-tartrate, hydrogen	$Mg(HC_4H_4O_6)_2 \cdot 4H_2O$	394.53	wh, rhomb	1.72				1.893^{100}	
m98	telluride	$MgTe$	151.91	wh, hex cr	3.86			d		d a
m99	thiosulfate	$MgS_2O_3 \cdot 6H_2O$	244.51	col, rhomb pr.	1.818^{24}	$-3H_2O$, 170	d	v s	v s	i al
m100	thiotellurite	Mg_3TeS_5	360.82	pa yel cr mass				s	s	s al
m101	tungstate	$MgWO_4$	272.15	col, monocl	5.66			i		d a; i al
m102	**Manganese**	Mn	54.9380	gray-pink met, cub or tetr.	7.20	1244 ± 3	1962	d	d	s dil a
m103	(II) acetate	$Mn(C_3H_3O_2)_2$	1173.03	brn cr	1.74			s, d		s al
m104	acetate, tetrahydrate	$Mn(C_2H_3O_2)_2 \cdot 4H_2O$	245.09	pa red, monocl.	1.589			s		s al
m105	arsenide, mono-	$MnAs$	129.86	blk, hex	6.17—6.20 (5.55)	d 400		i	i	s HCl, aq reg
m106	arsenide, di-	Mn_2As	184.80			1400		i	i	s aq reg
m107	arsenide, tri-	Mn_2As_2	314.66	magnetic, (exist?)				i	i	s aq reg
m108	(II) benzoate	$Mn(C_7H_5O_2)_2 \cdot 4H_2O$	369.23	pa red pr				6.55^{15}		
m109	boride, di-	MnB_2	76.56	gray-vlt cr.	6.9			d	d	s a
m110	boride, mono-	MnB	65.75	cr powd	6.2^{15}					
m111	bromide, di-	$MnBr_2$	214.75	rose cr	4.385^{25}_{4}	d		127.3^{0}	228^{100}	i NH_3
m112	bromide, di-, tetrahydrate	$MnBr_2 \cdot 4H_2O$	286.81	α stable, rose monocl, deliq β labile, col, rhomb		d 64.3		296.7^{0}		
m113	carbide	Mn_3C	176.83	tetr	6.89^{17}			d	d	s a
m114	(II) carbonate	Nat rhodochrosite $MnCO_3$	114.95	rose, rhomb, lt brn in air	3.125	d		0.0065^{25}		s dil a, aq CO_2; i al, NH_3
m115	chloride, di-	Scacchite. $MnCl_2$	125.84	pink, cub cr, deliq	2.977^{25}_{4}	650	1190	72.3^{25}	123.8^{100}	s al; i eth, NH_3
m116	chloride, di-, tetrahydrate	$MnCl_2 \cdot 4H_2O$	197.91	rose, monocl, deliq	2.01	58	$-H_2O$, 106; $-4H_2O$, 198	151^{8}	656^{100}	s al; i eth
m117	chloride, tri-	$MnCl_3$	161.30	brn cr or grnsh-blk		d sl				s abs al
m118	chloroplatinate	$MnPtCl_6 \cdot 6H_2O$	570.83	trig	2.692	d				
m119	chromite	$MnCr_2O_4$	222.93	gray-blk, cub	4.97^{20}			i	i	i a
m120	(II) citrate	$Mn_3(C_6H_5O_7)_2$	543.02	wh-redsh powd.				v sl s		s Na-citr sol
m121	(II) ferrocyanide	$Mn_2Fe(CN)_6 \cdot 7H_2O$	447.94	grnsh-wh powd.				i		s HCl; i NH_4 salts
m122	fluogallate	$[Mn(H_2O)_6][GaF_5 \cdot H_2O]$	345.76	pink, orthorhomb, 1.45	2.22	d 230		v s		s HF
m123	fluosilicate	$MnSiF_6 \cdot 6H_2O$	305.11	rose, hex pr, 1.357, 1.374	1.903	d		140^{18}	v s	s al
m124	fluoride, di-	MnF_2	92.93	red, tetr, or redsh powd	3.98	856		0.66^{40}	0.48^{100}	s a; i al, eth
m125	fluoride, tri-	MnF_3	111.93	red cr	3.54	d		d	d	s a
m126	formate	$Mn(CHO_2)_2 \cdot 2H_2O$	181.00	rhomb	1.953	d		s	s	
m127	(II) glycerophosphate	$MnC_3H_7O_6P$	225.00	wh or sl red powd				sl s		s a, citr a; i al
m128	hydroxide	$MnO(OH)_2$	104.95	blk-brn, amorph (exist?)	2.58			v sl s		
m129	(II) hydroxide	Nat. pyrochroite. $Mn(OH)_2$	88.95	wh-pink, trig 1.723, 1.681	3.258^{13}	d		0.0002^{18}		s a, NH_4 salts; i alk
m130	(III) hydroxide	Magnanite. $MnO(OH)$	87.94	br-blk, rhomb. 2.24, 2.24, 2.53 (Li)	4.2—4.4	d		i	i	s HCl, h H_2SO_4
m131	iodide, di-	MnI_2	308.75	pink, hex cr, deliq, br. in air	5.0^{1}	638 (vac) d *ca*80	subl vac 500	s	s	0.02^{25} NH_3
m132	iodide, di-, tetrahydrate	$MnI_2 \cdot 4H_2O$	380.81	rose, monocl, deliq		d		s	v s	
m133	*hexa*iodoplatinate	$MnPtI_6 \cdot 9H_2O$	1173.58	trig	3.604^{20}_{4}	d				
m134	(II) nitrate	$Mn(NO_3)_2 \cdot 4H_2O$	251.01	col, or rose, monocl	1.82	25.8	129.4	426.4^{0}	∞	v s al
m135	(II) lactate	$Mn(C_3H_5O_3)_2 \cdot 3H_2O$	287.13	pa red, monocl.		d		10	v s	s al
m136	(II) oxalate	MnC_2O_4	142.96	wh cr powd	$2.43^{21.7}$	d 150		i	i	s a, NH_4Cl
m137	(II) oxalate, dihydrate	$MnC_2O_4 \cdot 2H_2O$	178.99	redsh-wh oct cr powd		$-2H_2O$, 100	d	0.0312^{25}	0.037^{36}	
m138	(II) oxalate, trihydrate	$MnC_2O_4 \cdot 3H_2O$	197.00	pink, tricl		$-H_2O$, 25				
m139	(II, III) oxide	Nat. hausmannite. Mn_3O_4	228.81	blk, tetr (rhomb), 2.46 (Li) 2.15 (Li)	4.856	1564		i	i	s HCl
m140	oxide, di-	Nat. pyrolusite. MnO_2	86.94	blk, rhomb, or brn-blk powd	5.026	−O, 535		i	i	s HCl; i HNO_3, acet
m141	oxide, hept-	Mn_2O_7	221.87	dk red oil, hyg, exp	2.396^{20}_{4}	5.9	d 55, exp 95	v s	d	s H_2SO_4
m142	(II) oxide, mono-	Nat. manganosite, MnO	70.94	grn, cub, 2.16	5.43—5.46 (3.7—3.9)			i	i	s a, NH_4Cl
m143	(III) oxide, sesqui-	Nat. braunite. Mn_2O_3	157.87	blk, cub (tetr)	4.50	−O, 1080		i	i	s a; i ac a
m144	oxide, tri-	MnO_3	102.94	redsh, deliq (exist?)		d		s	d	s alk, H_2SO_4
m145	(III) *meta*phosphate	$Mn_2(PO_3)_6 \cdot 2H_2O$	619.74					sl s	s	

PHYSICAL CONSTANTS OF INORGANIC COMPOUNDS (continued)

No.	Name	Synonyms and Formulae	Mol. wt.	Crystalline form, properties and index of refraction	Density or spec. gravity	Melting point, °C	Boiling point, °C	Solubility, in grams per 100 cc Cold water	Hot water	Other solvents
m146	(II) *ortho*phosphate	Nat. reddingite. $Mn_3(PO_4)_2·3H_2O$	408.80	rose or yelsh-wh rhomb, 1.651, 1.656, 1.683	3.102					
m147	(III) *ortho*phosphate	$MnP_4·H_2O$	167.92	gray cr powd		$-H_2O$, 300	d	i		s h conc H_2SO_4, conc HCl, molten H_3PO_4
m148	(II) *ortho*phosphate, di-H	$Mn(H_2PO_4)_2·2H_2O$	284.94			$-H_2O$, >100		s		i al
m149	(II) *ortho*phosphate mono-H	$MnHPO_4·3H_2O$	204.96	red, rhomb or pink powd, 1.656				sl s	d	s a; i al
m150	(II) *pyro*phosphate	$Mn_2P_2O_7$	283.82	br-pink, monocl, 1.695, 1.704, 1.710	3.707^{25}	1196		i		s a
m151	(II) *pyro*phosphate, trihydrate	$Mn_2P_2O_7·3H_2O$	337.87	wh, amorph powd				i		s $K_2P_2O_7$ sol, H_2SO_3 i acet
m152	phosphide, mono-	MnP	85.91	dk gray	5.39^{21}	1190		i	i	sl s HNO_3
m153	(tri-)phosphide, di-	Mn_3P_2	226.76	dk gray	5.12^{18}	1095		i	i	sl s dil HNO_3
m154	(II) *hypo*phosphite	$Mn(H_2PO_2)_2·H_2O$	202.93	rose cr or powd		$-H_2O$, >150		12.5	16.7	i al
m155	(II) *ortho*phosphite	$MnHPO_3·H_2O$	152.93	redsh		$-H_2O$, 200		sl s		s $MnSO_4$, $MnCl_2$
m156	selenate, dihydrate	$MnSeO_4·2H_2O$	233.93	rhomb	2.95—3.01			s	s	
m157	selenate, penta-hydrate	$MnSeO_4·5H_2O$	287.97	pa red, trig	2.33—2.39					
m158	selenide	MnSe	133.90	gray, cub	5.55^{15}			i		d dil a
m159	selenite	$MnSeO_3·2H_2O$	217.93	monocl cr				v sl s	v sl s	
m160	(II) *meta*slicate	Nat. rhodonite. $MnSiO_3$	131.02	red, tricl, 1.733, 1.740, 1.744	3.72^{25}	1323		i		i HCl
m161	silicide, di-	$MnSi_2$	111.11	gray, oct.	5.24^{13}			i	i	s HF, alk; i HNO_3, H_2SO_3
m162	silicide, mono-	MnSi	83.02	tetrah.	5.90^{15}	1280		i	i	s HF; v sl s a
m163	(di-)silicide,	Mn_3Si	137.96	quadr pr	6.20^{15}	1316		i	i	s HCl, NaOH: i HNO_3
m164	(II) sulfate	$MnSO_4$	151.00	redsh	3.25	700	d 850	52^{5}	70^{70}	s al; i eth
m165	(III) sulfate	$Mn_2(SO_4)_3$	398.05	grn cr, deliq, hex	3.24	d 160		d	d	s HCl, dil H_2SO_4; i conc. H_2SO_4, HNO_3
m166	(II) sulfate, dihydrate	$MnSO_4·2H_2O$	187.03	(exist?).	2.526^{15}	stab 57—117		85.25^{35}	106.8^{55}	
m167	(II) sulfate, heptahydrate	$MnSO_47H_2O$	277.10	red monocl or rhomb	2.09	$-7H_2O$, 280; stab + 9		172	118^{13}	i al
m168	(II) sulfate, hexahydrate	$MnSO_4·6H_2O$	259.09	(exist?).		stab + 5 to + 8		147.4	1.345^{38}	
m169	(II) sulfate, monohydrate	Nat. szmikite. $MnSO_4·H_2O$	169.01	pa pink monocl, 1.562, 1.595, 1.632	2.95	stab 57—117		98.47^{48}	79.8^{100}	
m170	(II) sulfate, pentahydrate	$MnSO_4·5H_2O$	241.07	rose, tricl, 1.495, 1.508, 1.514	2.103^{15}	stab 9—26		124^{0}	142^{54}	
m171	(II) sulfate, tetrahydrate	Common form. $MnSO_4·4H_2O$	223.06	pink, monocl or rhomb effl, 1.508, 1.522	2.107	stab 26—27		105.3^{0}	111.2^{54}	i al
m172	(II) sulfate, trihydrate	$MnSO_4·3H_2O$	205.04	(exist?).	2.356^{15}	stab 30—40		74.22^{5}	99.31^{57}	
m173	(II) sulfide	Nat. alabandite. MnS	87.00	grn cub or pink amorph, 2.70 (Li)	3.99	d		0.00047^{18}		s dil a, al; i $(NH_4)_2S$
m174	(II) sulfide	$3MnS·H_2O$	279.01	gray-pink		d		0.0006	i	s dil a; i $(NH_4)_2$ S
m175	(IV) sulfide	Nat. hauerite. MnS_2	119.06	blk cub, 2.69 (Li)	3.463	d		i	i	d HCl
m176	(II) tantalate	$Mn(TaO_3)_2$	512.83	blk, rhomb, 2.22, 2.25, 2.29	7.03					
m177	(II) tartrate	$MnC_4H_4O_6$	203.01	wh powd				v sl s		s dil a
m178	(II) thiocyanate	$Mn(SCN)_2·3H_2O$	225.14	deliq		$-3H_2O$, 160—170		s	v s	v s al
m179	(II) dithionate	$Mn(SO_3)_2$	215.05	tricl cr	1.757			s	v s	
m180	(II) titanate	Nat. pyrophanite. $MnTiO_3$	150.82	yel, trig, 2.481, 2.210	4.54	1360				
m181	valerate	$Mn(C_5H_9O_2)_2·2H_2O$	293.22	br powd				s		
m182	**Manganic acid, per-**	$HMnO_4$	118.84					v s	d	
m183	**Manganocyanic acid**	$H_4Mn(CN)_6$	215.08			d		i		v s al; i eth
m184	**Mercury**	Quicksilver. Hg	200.59 ± 3	silv wh met, liq.	13.5939^{20}_{4}	-38.87	356.58	i	i	s HNO_3; i dil HCl, HBr, HI cold H_2SO_4
m185	(I) acetate	$Hg_2(C_2H_3O_2)_2$	519.27	micaceous plates		d		0.75^{12}		s HNO_3, H_2SO_4; i al eth
m186	(II) acetate	$Hg(C_2H_3O_2)_2$	318.68	wh, sc or powd	3.270	d		25^{10}	100^{100}	s al, ac a
m187	(II) acetylide	$3HgC_2·H_2O$	691.85	wh powd	5.3	expl		i	i	i al
m188	(II) *ortho*arsenate	$Hg_3(AsO_4)_2$	879.61	yel				v sl s		s HCl, HNO_3
m189	(I) *ortho*arsenate mono-H	Hg_2HAsO_4	541.11					i		s HNO_3; i ac a, NH_4OH
m190	(I) azide	$Hg_2(N_3)_2$	485.22	wh cr.		expl, d by light		0.025		
m191	(II) benzoate	$Hg(C_7H_5O_2)_2·H_2O$	460.84	wh cr powd		165		1.2^{15}	2.5^{100}	s al, NaCl, NH_4Cl, bz
m192	(I) bromate	$Hg_2(BrO_3)_2$	656.98	cr		d		d		sl s HNO_3
m193	(II) bromate	$Hg(BrO_3)_2·2H_2O$	492.42	cr		d 130—140		0.15	1.6	s HCl, HNO_3, $Hg(NO_3)_2$
m194	(I) bromide	Hg_2Br_2	560.99	wh, yel, tetr	7.307	subl 345		0.000004^{26}		s a; i al, acet

No.	Name	Synonyms and Formulae	Mol. wt.	Crystalline form, properties and index of refraction	Density or spec. gravity	Melting point, °C	Boiling point, °C	Solubility, in grams per 100 cc		
								Cold water	Hot water	Other solvents
m195	(II) bromide	$HgBr_2$	360.40	col rhomb	6.109^{25} 5.12^{240}	236	322	0.61^{25}	4.0^{100}	15^0 al; s MeOH; v sl s eth
m196	bromide iodide	HgBrI	407.40	yel, rhomb		229	360			s al, eth
m197	(I) carbonate	Hg_2CO_3	461.19	yel br cr		d 130		0.0000045	d	s NH_4Cl; i al
m198	(II) carbonate, basic	$HgCO_3 \cdot 2HgO$	693.78	br red				i		s NH_4Cl, aq CO_2
m199	(I) chlorate	$Hg_2(ClO_3)_2$	568.08	wh, rhomb	6.409	d 250		s	d	s al, ac a
m200	(II) chlorate	$Hg(ClO_3)_2$	367.49	wh need	4.998	d		25		
m201	(I) chloride	Calomel. Hg_2Cl_2	472.09	wh, tetr, 1.973, 2.656	7.150	subl 400		0.00020^{25}	0.001^{43}	s aq reg, $Hg(NO_3)_2$; sl s HCl, h HNO_3; i al, eth
m202	(II) chloride	Corrosive sublimate. $HgCl_2$	271.50	col, rhomb or wh powd pois, 1.859	5.44^{25}, liq 4.44^{280}	276	302	6.9^{20}	48^{100}	33^{25} al; 4 eth; s ac a, pyr
m203	(I) chromate	Hg_2CrO_4	517.17	red need, or powd		d		v sl s	sl s	s HCN, HNO_3; i al, ac
m204	(II) chromate	$HgCrO_4$	316.58	red, rhomb		d		sl s, d	d	s HN_4Cl; d a; i acet
m205	(II) cyanide	$Hg(CN)_2$	252.63	col, tetr, or wh powd pois, 1.645	3.996	d		9.3^{14}	33^{100}	$25^{19.5}$ MeOH; 8 al; s NH_3, glyc; i bz
m206	(I) fluoride	Hg_2F_2	439.18	yel, cub	8.73_4^{15}	570	d	d to Hg_2O		
m207	(II) fluoride	HgF_2	238.59	col, cub	8.95^{15}	d 645	650	d		s HF, dil HNO_3
m208	(I) fluosilicate	$Hg_2SiF_6 \cdot 2H_2O$	579.29	col pr	2.134			sl s		i HCl
m209	(II) fluosilicate	$HgSiF_6 \cdot 6H_2O$	450.76	col, rhomb, deliq		d easily				
m210	(I) formate	$Hg_2(CHO_2)_2$	491.22	glist scales		d		0.4^{17}	d	i al
m211	(II) fulminate	$Hg(CNO)_2$	284.62	wh, cub	4.42	expl		sl s	s	s al, NH_4OH
m212	(I) iodate	$Hg_2(IO_3)_2$	750.99	yelsh		d 250		i	i	s dil HCl, conc HNO_3
m213	(II) iodate	$Hg(IO_3)_2$	550.40	wh, amorph powd				i		s HCl, NH_4Cl, NaCl, KI; i HNO_3
m214	(I) iodide	Hg_2I_2	654.99	yel, tetr or amorph powd	7.70	subl 40	d 290	v sl s		s KI, NH_4OH; i al, eth
m215	(II) iodide (α)	HgI_2	454.40	red, tetr	6.36_4^{25}	tr 127		0.01^{25}		3.18^{25} acet; 2.23^{25} al; s chl; d NH_4OH
m216	(II) iodide (β)	HgI_2	454.40	yel, rhomb cr or powd	6.094_4^{27}	259	354	v sl s	sl s	s eth, KI, $Na_2S_2O_3$; v sl s al
m217	(I) nitrate	$Hg_2(NO_3)_2 \cdot 2H_2O$	561.22	col, monocl, effl.	4.79^4	70		d	s, d	s dil HNO_3; i NH_4OH
m218	(II) nitrate	$Hg(NO_3)_2 \cdot {}^1/_2H_2O$	333.61	wh-yelsh cr or powd, deliq	4.39	79	d	v s	d	s acet, NH_3, NH_3; i al
m219	(II) nitrate	$Hg(NO_3)_2 \cdot H_2O$	342.62	col cr or wh powd, deliq	4.3			s		s HNO_3; i al
m220	(I) nitrite	$Hg_2(NO_2)_2$	493.19	yel	7.33	d 100		d		
m221	nitride	Hg_3N_2	629.78	br powd		expl		d		s NH_4OH, NH_4 salts; d a
m222	(I) oxalate	$Hg_2C_2O_4$	489.20					i	i	sl s HNO_3
m223	(II) oxalate	HgC_2O_4	288.61			d		0.0107^{20}		s HCl; sl s HNO_3
m224	(I) oxide	Hg_2O	417.18	blk or brnish-blk powd	9.8	d 100		i	i	s HNO_3
m225	(II) oxide	Nat. montroydite. HgO	216.59	yel or red, rhomb, 2.37, 2.5, 2.65	1.1^4	d 500		0.0053^{25}	0.0395^{100}	s a; i al, eth, acet, alk, NH_3
m226	(II) oxybromide	$HgBr_2 \cdot 3HgO$	1010.17	yel cr.				i	sl s	v s al
m227	(II) oxychloride	$HgCl_2 \cdot 2HgO$	704.67	red hex, or blk monocl	red 8.16—8.43 blk 8.53					
m228	(II) oxychloride	$HgCl_2 \cdot 3HgO$	921.26	yel, hex	7.93	d 260		i	d	
m229	(II) oxycyanide	$Hg(CN)_2 \cdot HgO$	469.21	wh need or cr powd	4.437^{19}	expl		1.25	s	
m230	(II) oxyfluoride	$HgF_2 \cdot HgO \cdot H_2O$	473.19	yel cr.		d 100		d		s dil HNO_3
m231	(II) oxyiodide	$HgI_2 \cdot 3HgO$	1104.17	yel br					d	s HI
m232	(II) selenide	Nat. tiemannite. HgSe	279.55	gray plates	8.266	vac subl		i		s aq reg
m233	(I) sulfate	Hg_2SO_4	497.24	col monocl, wh-yelsh powd	7.56	d	d	0.06^{25}	0.09^{100}	s HNO_3, H_2SO_4
m234	(II) sulfate	$HgSO_4$	296.65	col rhomb or wh powd	6.47	d		d		s a, NaCl; i al, acet, NH_3
m235	(II) sulfate, basic	$HgSO_4 \cdot 2HgO$	729.83	lem yel powd.	6.44		volat	0.003^{16}	sl s	s a; i al
m236	(I) sulfide	Hg_2S	433.24	blk		d		i		i al, $(NH_4)_2$ S
m237	(II) sulfide (α)	Cinnabar, vermillion, HgS	232.65	red cr hex, or powd, 2.854, 3.201	8.10	subl 583.5		0.000001^{18}		s aq reg, Na_2S; i al, HNO_3
m238	(II) sulfide (β)	Metacinnabar. HgS	232.65	blk, cub or amorph powd	7.73	583.5		i		s aq reg, Na_2S, alk; i al, HNO_3
m239	(I) tartrate	$Hg_2C_4H_4O_6$	549.25	yelsh-wh cr powd				i		i a
m240	(I) *ortho*tellurate	HgH_4TeO_6	428.22	trans, orthorhomb		d 20		slow d	rapid d	
m241	(II) *ortho*tellurate	Hg_3TeO_6	825.37	amber, cub		unalt at 140		i	i	s HCl, HNO_3
m242	(I) thiocyanate	$Hg_2(SCN)_2$	517.34			d		i		s HCl, KCNS
m243	(II) thiocyanate	$Hg(SCN)_2$	316.75	wh powd, pois.		d		0.07^{25}	s	s NH_4 salts, HCl, NH_3, KCN; sl s al, eth
m244	(I) tungstate	Hg_2WO_4	649.03	yel, amorph.		d		i	i	d a; i al
m245	(II) tungstate	$HgWO_4$	448.44	yel		d		i	d	d a; i al

No.	Name	Synonyms and Formulae	Mol. wt.	Crystalline form, properties and index of refraction	Density or spec. gravity	Melting point, °C	Boiling point, °C	Solubility, in grams per 100 cc: Cold water	Hot water	Other solvents
	Mercury nitrogen compounds									
m246	mercury (II) bromide, ammonobasic	$Hg(NH_2)Br$	296.52	wh powd		d		d		s NH_4OH; i al
m247	mercury (II) bromide, diammine	$Hg(NH_3)_2Br_2$	394.46	wh powd		180		d		s NH_4Cl, NH_4Br, NH_4I
m248	mercury (II) chloride ammonobasic	Infusible ppt. $Hg(NH_2)Cl$	252.07	wh powd or sm pr	5.70	infus		0.14	d 100	d a; i al
m249	mercury (II) chloride, aquobasic ammonobasic	Chloride of Millon's base. OHg_2NH_2Cl	468.65	pa yel or wh powd		d >120		sl s		s HCl, HNO_3
m250	mercury (II) chloride, diammine	Fusible white ppt. $Hg(NH_3)_2Cl_2$	305.56	rhombd		300		i	d	s a, KI
m251	mercury (II) iodide, ammonobasic	$Hg(NH_2)I$	343.52							i eth
m252	mercury (II) iodide aquobasic ammonobasic	Iodide of Millon's base. OHg_2NH_2I	560.11	yel to brn		>128	expl	i		s d, HCl, KI soln
m253	mercury (II) iodide, diammine	$Hg(NH_3)_2I_2$	488.46	col or pa yel powd or need				d		s NH_4OH
m254	**Millons's base**	$(HO)_2Hg_2NH_2OH$	468.22		4.083^{18}					
m255	**Molybdenum**	Mo	95.94	silv-wh met, or gray-blk powd, cub	10.2	2610	5560	i	i	s h conc HNO_3, h conc H_2SO_4, aq reg; sl s HCl; i HF, NH_3
m256	boride, (di-)	MoB_2	117.56	rhomb	7.12					
m257	boride, (mono-)	MoB	106.75	tetr	8.65					
m258	(di-) boride	Mo_2B	202.69	tetr	9.26					
m259	bromide, di-	$MoBr_2$ (or Mo_2Br_4)	255.75	yel red, amorph	$4.88^{17.5}$			i	i	s alk; i a, aq reg
m260	bromide, tetra-	$MoBr_4$	415.56	blk need, deliq		d	volat	v s		d alk
m261	bromide, tri-	$MoBr_3$	335.65	dk-grn need		d		i	i	d alk, NH_3; i a
m262	carbide, mono-	MoC	107.95	gray, hex	8.20^{20}	2692		i	i	sl s HNO_3, HF, h H_2SO_4, HCl; i alk hydr
m263	carbide(di-)	Mo_2C	203.89	wh, hex pr	8.9	2687		i		sl s HNO_3, HF h H_2SO_4 aq reg, HCl; i alk
m264	carbonyl	$Mo(CO)_6$	264.00	wh cr, rhomb diamagnet	1.96	d 150, without meltg	156.4^{766}	i	i	s bz; sl s eth
m265	carbonyl, tripyridine, tri-	$Mo(CO)_3(C_5H_5N)_3$	417.28	yel-brn cr		d				
m266	chloride, di-	$MoCl_2$ (or Mo_3Cl_6)	166.85	yel, amorph	3.714^{25}	d		i	i	s HCl, H_2SO_4, alk, NH_4OH, al, acet
m267	chloride, penta-	$MoCl_5$	273.21	grn-blk cr, trig, deliq	2.928	194	268	d	d	s conc min a, liq NH_3, CCl_4, chl; s d al, eth
m268	chloride, tetra-	$MoCl_4$	237.75	brn powd or cr, deliq		d	vol	d	d	s conc min a; d al, eth
m269	chloride, tri-	$MoCl_3$	202.30	dk red need or powd	3.578^{25}_{4}	d		i	sl d	s conc H_2SO_4, conc HNO_3; v sl s al, eth; i HCl; d alk
m270	fluoride, hexa-	MoF_6	209.93	col cr.	liq $2.55^{17.5}$	17.5^{406}	37^{760}	s d	d	s NH_4OH, alk
m271	hydro*tetra*chlorohydroxide, di-	$[Mo_3Cl_4(H_2O_2](OH)_2 \cdot 6H_2O$	607.77	lt yel cr		$-H_2O$ 35—300		i	i	i a, al
m272	hydroxide	$Mo(OH)_3$ (or $Mo_2O_3 \cdot 3H_2O$)	146.96	blk powd		d		0.2		sl s H_2SO_4, HCl; s 30% H_2O_2
m273	hydroxide	$MoO(OH)_3$ (or $Mo_2O_5 \cdot 3H_2O$)	162.96	br to blk powd				0.2 (coll)		s a, alk carb; i alk hydr
m274	(VI) hydroxide	$MoO_3 \cdot 2H_2O$	179.97	lt yel, monocl pr.	3.124^{15}			0.05^{15}		s dil H_2O_2, alk hydr; sl s a
m275	hydroxy*tetra*bromide, di-	$Mo_3Br_4(OH)_2$	641.45	red powd						s alk
m276	hydroxy*tetra*bromide, dioctahydrate	$Mo_3Br_4(OH)_2 \cdot 8H_2O$	785.57	golden yel cr		d	d	4lf		s HCl; d alk, HNO_3
m277	hydroxy*tetra*chloride, di-	$Mo_3Cl_4(OH)_2 \cdot 2H_2O$	499.68	pa yel, amorph					i	s conc a; i al
m278	iodide, di-	MoI_2	349.75	brn powd	$5.278^{25.4}$			i	d	v sl s a
m279	iodide, tetra-	MoI_4	603.56	blk cr		d 100				
m280	oxide, di-	MoO_2	127.94	lead gray, tetr or monocl	6.47			i	i	sl s h conc H_2SO_4; i alk, HCl, HF
m281	oxide, pent-	Mo_2O_5	271.88	vlt-bl powd (exist?)						s h H_2SO_4, h CHl
m282	oxide, pent-	"Molybd. blue" $Mo_2O_5 \cdot xH_2O$ (variations in Mo and O)		dk blue coll or powd	3.6^{18}	d				s a, MeOH; i acet, bz, chl
m283	oxide, sesqui-	Mo_2O_3	239.88	blk, opaque (exist?)				i	i	i a, alk, NH_4OH
m284	oxide, tri-	Molybdic anhydride	143.94	col or wh-yel, rhomb	4.692^{21}	795	subl 1155^{760}	0.1066^{18}	2.055^{70}	s a, alk sulf, NH_4OH
m285	oxydibromide, di-	MoO_2Br_2	287.75	yel-red tabl, deliq				s		
m286	oxy*tetra*chloride	$MoOCl_4$	253.75	grn cr, deliq		subl		s		
m287	oxy*tri*chloride	$MoOCl_3$	218.30	grn cr		subl 100		d		
m288	oxy*di*chloride, di-	MoO_2Cl_2	198.84	yelsh wh scaly cr	3.31^{17}	subl		s	s	s al, eth
m289	oxy*di*chloride, dihydrate	$MoO_2Cl_2 \cdot H_2O$	216.86	pa yel cr.		subl		s	s	sl s al, acet, eth
m290	oxy*hexa*chloride, tri-	$Mo_2O_3Cl_6$	452.60	rubyred or dk vlt cr		d		d		s eth

No.	Name	Synonyms and Formulae	Mol. wt.	Crystalline form, properties and index of refraction	Density or spec. gravity	Melting point, °C	Boiling point, °C	Solubility, in grams per 100 cc: Cold water	Hot water	Other solvents
m291	oxy*penta*chloride, tri-	$Mo_2O_3Cl_5$	417.14	dk brn-blk cr, deliq		melts easily	subl	s	s	
m292	oxychloride acid	$MoO(OH)_2Cl_2$	216.86	wh need, deliq		d 160		v s		s al, eth, acet
m293	oxy*di*fluoride, di-	MoO_2F_2	165.94	wh cr, hygr	3.49^{25}	subl 270		v s	v s	s al, MeOH; i eth, chl, tol
m294	oxy*tetra*fluoride	$MoOF_4$	187.93	col-wh, deliq	3.001^{25}	98	180	s		s al, eth, CCl_4, s d H_2So_4; v sl s bz
m295	*meta*phosphate	$Mo(PO_3)_6$	569.77	yel powd	3.28^{0}			i	i	sl s h aq reg; i HCL, HNO_3, H_2SO_4
m296	phosphide	MoP (or Mo_2P_2)	126.91	gray-grn cr powd	6.167					s h HNO_3
m297	phosphide	MoP_2	157.89	blk powd	5.35^{25}					s HNO_3, h conc H_2SO_4, aq reg; i conc HCl
m298	silicide	$MoSi_2$	152.11	gray, met, tetr	$6.31^{20.5}$					i a, aq reg; v s $HF + HNO_3$
m299	sulfide, di-	Nat. molybdenite. MoS_2	160.06	blk luster, hex.	4.80^{14}	1185	subl 450, d in air	i	i	s h H_2SO_4, aq reg, HNO_3; i dil a, conc H_2SO_4
m300	sulfide, penta	$Mo_2S_5 \cdot 3H_2O$	406.23	dk br powd		$-H_2O$, 135	d	i	i	s NH_4OH, alk, sulfides
m301	sulfide, sesqui-	Mo_2S_3	288.06	steel gray need	5.91^{15}	d 1100	vol 1200			i conc HCl; d h HNO_3
m302	sulfide, tetra-	MoS_4	224.18	brn powd		d		i	i	i a; s h H_2SO_4, alk sulfide
m303	sulfide, tri-	MoS_3	192.12	blk pl		d	d	sl s	s	s alk sulf, conc KOH
m304	**Molybdic acid**	$H_2MoO_4 \cdot H_2O$ (or $MoO_3 \cdot 2H_2O$)	179.97	yel, monocl	3.124^{15}	$-H_2O$, 70	d	0.133^{18}	2.568^{70}	s alk hydr, alk carb; sl s a
m305	anhydrous	H_2MoO_4 (or $MoO_3 \cdot H_2O$)	161.95	wh or sl yelsh, hex	3.112	$-H_2O$, 70		sl s	sl s	s alk, NH_4OH, H_2SO_4; i NH_3
m306	**Molybdic arsenic acid**	$As_2O_5 \cdot 6MoO_3 \cdot 18H_2O$	1417.74	col, trig	$2.493^{19.8}$	$-15H_2O$, 150		v s		s abs al; i chl, CS_2
m307	**Molybdic phosphoric acid**	$H_7[P(Mo_2O_7)_6] \cdot 28H_2O$	2365.71	yel, oct	2.53	78		d		
m308	**Molybdic silicic acid**	$H_8[Si(Mo_2O_7)_6] \cdot 28H_2O$	2363.83	yel, tetr		45	d 100	600^{14}		s dil a; i bz, chl, CS_2
n1	**Neodymium**	Nd	144.24 ± 3	silv-wh to yelsh met, hex to m.p. 868, cub from 868	hex 7.004 cub 6.80	1024	3027	d		
n2	acetate	$Nd(C_2H_3O_2)_3 \cdot H_2O$	339.39					26.2		
n3	acetylacetonate	$Nd[CH(COCH_3)_2]_3$	441.57	pink cr	1.618	150—152				
n4	*hexa*antipyrin *per*-chlorate	$[Nd(C_{11}H_{12}N_2O)_6] \cdot (ClO_4)_3$	1571.97	rose, hex cr		d 285—289		0.99^{20}		
n5	bromate	$Nd(BrO_3)_3 \cdot 9H_2O$	690.08	red, hex		66.7	$-9H_2O$, 150	151^{25}		
n6	bromide	$NdBr_3$	383.95	grn cr		684	540	sl s		
n7	carbide	NdC_2	168.26	yel, hex leaf	5.15	d		d	d	s dil a; i conc HNO_3
n8	chloride	$NdCl_3$	250.60	rose-vlt pr	4.134^{25}	784	1600	96.7^{13}	140^{100}	44.5 al; i eth, chl
n9	chloride, hexahydrate	$NdCl_3 \cdot 6H_2O$	358.69	red, rhomb		124	$-6H_2O$, 160	246^{13}	511^{100}	v s al
n10	chromate	$Nd_2(CrO_4)_3 \cdot 8H_2O$	780.58	yel cr.				0.027		
n11	fluoride	NdF_3	201.24	pa lilac		1410	2300	i	i	
n12	iodide	NdI_3	524.95	blk cr powd		775 ± 3	370	s	s	
n13	kojate	$Nd(C_6H_5O_4)_3$	567.55	lt choc		d 275		i		
n14	manganous nitrate	$2Nd(NO_3)_3 \cdot 3Mn(NO_3)_2 \cdot 24H_2O$	1629.72	vlt-red	2.114	77		77.4^{30}		
n15	magnesium nitrate	$2Nd(NO_3)_3 \cdot 3Mg(NO_3)_2 \cdot 24H_2O$	1537.82	vlt-red	2.020	109		69.5^{30}		
n16	dimethylphosphate	$Nd[(CH_3)_2PO_4]_3$	519.36	pa lilac, hex pl				56.1^{25}	22.3^{100}	
n17	molybdate	$Nd_2(MoO_4)_3$	768.29	tetr, 2.005	5.14^{18}	176				
n18	nickel nitrate	$2Nd(NO_3)_3 \cdot 3Ni(NO_3)_2 \cdot 24H_2O$	1640.98	blsh-grn	2.202	105.6		71.5^{30}		
n19	nitrate	$Nd(NO_3)_3 \cdot 6H_2O$	438.35	tricl				152.9^{25}		s al, acet
n20	nitride	NdN	158.25	blk powd				d		
n21	oxalate	$Nd_2(C_2O_4)_3 \cdot 10H_2O$	732.69	rose cr				0.000074^{25}		
n22	oxide	Neodymia. Nd_2O_3	336.48	lt bl powd, red fluores	7.24	~1900		0.00019^{29}	0.003^{75}	s a
n23	sulfate	$Nd_2(SO_4)_3 \cdot 8H_2O$	720.89	red, monocl, 1.41, 1.551, 1.562	2.85	1176		8^{20}	5.4^{40}	
n24	sulfide	Nd_2S_3	384.66	oliv grn powd	5.179^{11}	d		i	d	s dil a
n25	**Neon**	Ne	20.179	inert gas col sol, cub	gas: 0.9002^{0} g/ℓ; liq: $1.204^{-245.9}$	−248.67	−245.9	1.47^{20} cm^3		s liq O_2
n26	**Neptunium**	Np	237.0482	α: orthorhomb silvery β: tetr (above 278) υ: cub (above 500)	α; 20.45 β: 19.36^{313} γ: 18.0^{600}	630 ± 1 278 ± 5 stab 278—570				s HCl
n27	bromide, tri-	$NpBr_3$	476.76	α: hex β: grn orthorhomb	α 6.92	subl *ca* 800		s		
n28	chloride, tetra-	$NpCl_4$	378.86	red-brn tetr	4.92	538		s		
n29	chloride, tri-	$NpCl_3$	343.41	wh, hex	5.38	*ca*800		s		
n30	fluoride, hexa-	NpF_6	351.04	brn, orthorhomb		53	d	d		
n31	fluoride, tetra-	NpF_4	313.04	grn, monocl.	6.8			i		i conc HNO_3
n32	fluoride, tri-	NpF_3	294.04	purple, hex	9.12			i		
n33	iodide, tri-	NpI_3	617.76	brn, orthorhomb	6.82			s		
n34	oxide, di-	NpO_2	269.05	apple grn, cub	11.11			i		s conc a

No.	Name	Synonyms and Formulae	Mol. wt.	Crystalline form, properties and index of refraction	Density or spec. gravity	Melting point, °C	Boiling point, °C	Solubility, in grams per 100 cc		
								Cold water	Hot water	Other solvents
n35	(tri-) oxide, octa-	Np_3O_3	839.14	brn, cub		d 500				s HNO_3
n36	**Nickel**	Ni	58.69	silv met, cub	8.90	1455	2730	i	i	s dil HNO_3; sl s HCl, H_2SO_4; i NH_3
n37	acetate	$Ni(C_2H_3O_2)_2$	176.78	grn pr	1.798	d	16.6			i al
n38	acetate, tetrahydrate	$Ni(C_2H_3O_2)_2 \cdot 4H_2O$	248.84	grn pr	1.744	d	16			s dil al
n39	antimonide	Nat. breithauptite. NiSb	180.44	lt copper red, hex	7.54	1158	d 1400			
n40	*ortho*arsenate, octahydrate	$Ni_3(AsO_4)2 \cdot 8H_2O$	598.03	helsh-grn powd.	4.98			i		s a
n41	arsenide	Nat. niccolite. NiAs,	133.61	hex	7.57^0	968		i	i	s aq reg
n42	*ortho*arsenite, acid	$Ni_3H_6(AsO_3)_4 \cdot H_2O$	691.81	grn-wh.		d		i		s a, alk
n43	benzenesulfonate	$Ni(C_6H_5SO_3)_2 \cdot 6H_2O$	481.11	grn, monocl.	1.628^{25}	$-H_2O$	d	14.3^{18}	51.5^{22}	5.9 al; 4.5 eth
n44	boride	NiB	69.50	pr	7.39^{18}			d	d	s aq reg, HNO_3
n45	bromate	$Ni(BrO_3)_2 \cdot 6H_2O$	422.59	monocl.	2.575	d		28		
n46	bromide	$NiBr_2$	218.50	yel brn, deliq.	5.098^{27}	963		112.8°	155.1^{100}	s al, eth, NH_4OH
n47	bromide, trihydrate	$NiBr_2 \cdot 3H_2O$	272.54	yelsh-grn need, deliq		$-3H_2O$, 300		199°	315.7^{100}	s al, eth, NH_4OH
n48	bromoplatinate	$NiPtBr_6 \cdot 6H_2O$	841.29	trig	3.715					
n49	di-*N*-butyldithiocarbamate	(NBC) $Ni[(C_4H_9)_2NCSS]_2$	467.43	dk oliv grn powd	1.29	89—90		i	i	sl s bz, pet comp; i al
n50	carbide	Ni_3C	188.08	dk gray powd.	7.957^{25}					
n51	carbonate	$NiCO_3$	118.70	lt grn, rhomb		d		0.0093^{25}	i	s a
n52	carbonate, basic	$2NiCO_3 \cdot 3Ni(OH)_2 \cdot 4H_2O$	587.57	lt grn cr or brn powd		d		i	d	s a, NH_4 salts
n53	carbonate, basic	Zaratite. $NiCO_3$ $2Ni(OH)_2 \cdot 4H_2O$(?)	376.17	emerald grn, cub, 1.56—1.61	2.6			i	i	s h dil HCl, NH_4OH
n54	carbonyl	$Ni(CO)_4$	170.73	col, volat, inflamm, liq, or need	1.32^{17}	−25	43	$0.018^{9.8}$		s aq reg, al, eth, bz, HNO_3; i dil a, dil alk
n55	chlorate	$Ni(ClO_3)_2 \cdot 6H_2O$	333.68	dk red	2.07	d 80		0.9^{27}		
n56	*per*chlorate	$Ni(Clo_2)_2 \cdot 6H_2O$	365.68	grn, hex need, 1.518, 1.498		140		222.5°	273.7^{45}	s al, acet, chl
n57	chloride	$NiCl_2$	129.60	yel sc, deliq	3.55	1001	subl 973	64.2^{20}	87.6^{100}	s al, NH_4OH; i NH_3
n58	chloride, hexahydrate	$NiCl_2 \cdot 6H_2O$	237.69	grn, monocl, deliq ~1.57				254^{20}	599^{100}	v s al
n59	chloropalladate	$NiPdCl_6 \cdot 6H_2O$	485.92	hex	2.353					
n60	chloroplatinate	$NiPtCl_6 \cdot 6H_2O$	574.58	trig	2.798					
n61	cyanide	$Ni(CN)_2$	110.73	yel-brn				i	i	s KCN
n62	cyanide, tetrahydrate	$Ni(CN)_2 \cdot 4H_2O$	182.79	lt grn pl or powd pois		$-4H_2O$, 200	d	i	i	s KCN, NH_4OH, alk; sl s dil a
n63	ferrocyanide	$Ni_2Fe(CN)_6 \cdot \chi H_2O$		grn-wh.	1.982(?)			i		s KCN, NH_4OH; i HCl
n64	fluogallate	$[Ni(H_2O)_6][GaF_5 \cdot H_2O]$	349.51	pa grn, monocl(?) 1.45	2.45	$-5H_2O$, 110		sl s		s HF
n65	fluoride	NiF_2	96.69	grn, tetr	4.63	subl 1000 (in HF)		4^{25}		s a, alk, eth, NH_3
n66	fluosilicate	$NiSiF_6 \cdot 6H_2O$	308.86	grn, trig, 1.391, 1.407	2.134	d				
n67	formate, dihydrate	$Ni(CHO_2)_2 \cdot 2H_2O$	184.76	grn cr	2.154	d		s		
n68	(II) hydroxide	$Ni(OH)_2$ (or $NiO \cdot \chi H_2O$)	92.70	grn cr, or amorph	4.15(3.65)	d 230		0.013		s a, NH_4OH
n69	iodate	$Ni(IO_3)_2$	408.50	yel need	5.07			1.1^{30}	1.0^{30}	
n70	iodate, tetrahydrate	$Ni(IO_3)_2 \cdot 4H_2O$	480.56	hex		d 100		1.4^{30}	1.1^{90}	
n71	iodide	NiI_2	312.50	blk cr, deliq	5.834	797		124.2°	188.2^{100}	s al
n72	dimethylglyoxime	$Ni(HC_4H_6N_2O_2)_2$	288.91	scarlet red cr		subl 250		i	i	s a, abs al; i a ac, NH_4OH
n73	nitrate, hexahydrate	$Ni(NO_3)_2 \cdot 6H_2O$	290.79	grn, monocl, deliq	2.05	56.7	136.7	238.5°	v s	s al, NH_4OH
n74	oleate	$Ni(C_{13}H_{33}O_2)_2$	621.61	grn oil		18—20				
n75	oxalate	$NiC_2O_4 \cdot 2H_2O$	182.74	lt grn powd				i		s a, NH_4 salts; v sl s oxal a
n76	oxide, mono-	Nat. bunsenite. NiO	74.69	grn-blk, cub, 2.1818(red)	6.67	1984		i	i	s a, NH_4OH
n77	*ortho*phosphate	$Ni_3(PO_4)_2 \cdot 8H_2O$	510.13	apple grn pl or emerald cr gran		d		i	i	s a, NH_4 salts; i me acet. et acet
n78	*pyro*phosphate	$Ni_2P_2O_7 \cdot \chi H_2O$		grn	3.93 (anhydr)					s a, NH_4OH
n79	(di-)phosphide	Ni_2P	148.35	gray cr.	6.31^{15}	1112		i		s HNO_3 + HF; i a
n80	(penta)phosphide, (di)-	Ni_5P_2	355.40	need or tabl cr		1185				
n81	(tri-)phosphide, (di-)	Ni_3P_2	238.01	dk grn-blk	5.99			i	i	s HNO_3; i HCl
n82	*hypo*phosphite	$Ni(H_2PO_2)_2 \cdot 6H_2O$	296.76	grn	$1.82^{19.8}$	d 100		s		
n83	selenate	$NiSeO_4 \cdot 6H_2O$	309.74	grn, tetr, 1.5393	2.314			s		
n84	selenide	NiSe	137.65	wh or gray, cub	8.46	red heat		i		s aq reg, HNO_3; i a, HCl
n85	silicide	Ni_2Si	145.47		7.2^{17}	1309		i		i a
n86	stearate	$Ni(C_{18}H_{35}O_2)_2$	625.64	grn powd		100		i		s CCl_4, pyr; sl s acet; i MeOH, eth
n87	sulfate	$NiSO_4$	154.75	yel, cub	3.68	d 848^{760}		29.3°	87.3^{100}	i al, eth, acet
n88	sulfate, hepatahydrate	Morenosite. $NiSO_4 \cdot 7H_2O$	280.85	grn, rhomb, 1.467, 1.489, 1.492	1.948	99; $-H_2O$ 31.5	$-6H_2O$, 103	$75.6^{15.5}$	475.8^{100}	s al
n89	sulfate, hexahydrate	Single nickel salt. $NiSO_4 \cdot 6H_2O$	262.84	α: bl, tetr β: grn, monocl, 1.511, 1.487	2.07	tr: 53.3	$-6H_2O$, 103	65.52°	340.7^{100}	12.5 MeOH; v s al, NH_4OH
n90	sulfide, mono-	Nat. millerite. NiS	90.75	blk; trig or amorph	5.3—5.65	797		0.00036^{18}		s aq reg, HNO_3, KHS; sl s a
n91	sulfide, sub-	Heazlewoodite. Ni_3S_2	240.19	pa yelsh bronze met, lust	5.82	790		i		s HNO_3

No.	Name	Synonyms and Formulae	Mol. wt.	Crystalline form, properties and index of refraction	Density or spec. gravity	Melting point, °C	Boiling point, °C	Solubility, in grams per 100 cc		
								Cold water	Hot water	Other solvents
n92	(II, III) sulfide	Poydomite. Ni_3S_4	304.31	gray-blk, cub	4.7			i		s HNO_3
n93	sulfite	$NiSO_3{\cdot}6H_2O$	246.84	grn, tetrah				i		s HCl, H_2SO_4
	dithionate	$NiS_2O_6{\cdot}6H_2O$	326.90	grn, tricl	1.908	d				
n94	**Nickel Complexes**									
n95	diaquotetrammine nickel (II) nitrate	$[Ni(NH_3)_4(H_2O)_2]{\cdot}(NO_3)_2$	286.85	grn cr				s		i al
n96	hexamminenickel (II) bromide	$[Ni(NH_3)_6]Br_2$	320.68	vlt powd	1.837			v s	d	
n97	hexamminenickel (II) chlorate	$[Ni(NH_3)_6](ClO_3)_2$	327.78		1.52	180		d to $Ni(NH_3)_4$		
n98	hexamminenickel (II) chloride	$[Ni(NH_3)_6]Cl_2$	231.78	blsh, cub	1.468^{25}			s	d	s NH_4OH; i al
n99	hexamminenickel (II) iodide	$[Ni(NH_3)_6]I_2$	414.68	pa bl, cub	2.101	d		d		s NH_4OH
n100	hexamminenickel(II) nitrate	$[Ni(NH_3)_6](NO_3)_2$	284.88	bl, oct or cub				4.46		
n101	tetrapyridinnickel (II) fluosilic	$[Ni(C_5H_5N)_4]SiF_6$	517.17	bl grn, rhomb	2.307					
n102	**Niobium**	Columbium. Nb	92.9064	steel gray, lustr met cub, 1.80	8.57^{20}	2468 ± 10	5127	i	i	s fus alk; i HCl, HNO_3, aq reg
n103	boride	NbB_2	114.53	hex	6.97	2900(?)				
n104	bromide, penta-	$NbBr_5$	492.43	purp red		265.2	361.6	d		s al, ethyl bromide
n105	carbide	NbC	104.92	blk, cub or laven-der-gray powd	7.6	3500		i		s HNO_3, HF
n106	chloride, penta-	$NbCl_5$	270.17	yel-wh, deliq	2.75	204.7	254	d		s al, HCl, CCl_4
n107	fluoride, penta-	NbF_5	187.90	col, monocl pr, hygr.	3.293	72—73	236	d		s al; sl s chl, CS_2, H_2SO_4
n108	hydride	NbH	93.91	gray powd.	6.6	infus				s HF, conc H_2SO_4; i HCl, alk, HNO_3
n109	nitride	NbN	106.91	blk, cub	8.4	2573		i		s HF + HNO_3; i HNO_3
n110	oxalate, hydrogen	$Nb(HC_2O_4)_5$	538.04	col, monocl				d	d	s $H_2C_2O_4$; d al
n111	oxide, di-	NbO_2	124.91	blk	5.9			i	i	sl s alk; i a
n112	oxide, mon- (or di-)	NbO (or Nb_2O_2)	108.91	blk, cub	7.30			i	i	s a, alk; i al, NHO_3
n113	oxide, pent-	Nb_2O_5	265.81	wh, rhomb	4.47	1485 ± 5		i	i	s HF, alk; i a
n114	oxide, pent-, hydrate	$Nb_2O_5{\cdot}\chi H_2O$				d		i		s conc H_2SO_4, conc HCl, HF, alk; i NH_3
n115	oxide, tri-(seequi)	Nb_2O_3	233.81	bl-blk		1780				
n116	oxybromide	$NbOBr_3$	348.62	yel cr.		subl		d		s a
n117	oxychloride	$NbOCl_3$	215.26	col need		subl 400		s, d	d	s al, H_2SO_4; i HCl
n118	potassium fluoride	$NbOF_3{\cdot}2KF{\cdot}H_2O$	300.11	monocl leaf, lustr fatty				7.8	v s	
n119	**Nitric acid**	HNO_3	63.01	col liq, corr, pois, $1.397^{16.4}$	1.5027^{25}_{4}	−42	83	∞	∞	d al, viol; s eth
n120	const boil.	68% HNO_3 + 32% H_2O		col liq	1.41		120.5	∞	∞	
n121	**Nitrogen**	N_2	28.0134	col gas, col liq, sol cub cr	gas 1.2506 g/ℓ; liq $0.8081^{-195.8}$ sol $1.026^{-252.5}$	−209.86	−195.8	2.33° cm^3	1.42^{40} cm^3	sl s al
n122	chloride, tri-	NCl_3	120.37	yel oil or rhomb cr	1.653	<−40	<71, expl 95	i	d	s chl, bz, CCl_4, CS_2, PCl_3
n123	fluoride, tri-	NF_3	71.00	col gas	liq: 1.537^{-129}	−206.60	−128.8	v sl s		
n124	iodide, tri-	NI_3	394.72	blk, expl.		expl	subl vac	i	d	s KCNS, $Na_2S_2O_3$
n125	iodide, tri-, monoammine	$NI_3{\cdot}NH_3$	411.75	dk red, bhomb	3.5	d>20	expl	i	d	s HCl, KCNS, $Na_2S_2O_3$; i abs al
n126	oxide(ic)	NO	30.01	col gas, bl liq, sol liq 1.330^{-90}	gas 1.3402 g/ℓ liq; $1.269^{-150 2}$	−163.6	−151.8	7.34° cm^3	2.37^{60} cm^3	3.4 cm^3 H_2SO_4; 26.6 cm^3 al; s $FeSO_4$, CS_2
n127	oxide(ous)	N_2O	44.01	col gas or liq or cub cr, 1.005^{20}_{760}	1.977^{0}_{760} g/ℓ	−90.8	−88.5	130° cm^3	56.7^{25} cm^3	s al, eth, H_2SO_4
n128	oxide, pent-	Nitric anhydride. N_2O_5	108.01	wh, rhomb or hex	1.642^{18}	30	d 47	s	d to HNO_3	d chl
n129	oxide, tri-	NO_3	62.00	blsh gas		d at ord temp				s eth
n130	peroxide	Nitrogen oxide, di-. NO_2	46.01	col sol, yel liq or brn gas, 1.40^{20}	1.4494^{20}_{20}	−11.20	21.2	s, d		s alk, CS_2, chl
n131	(di-) oxide(tri-)	Nitrous anhydride. N_2O_3	76.01	red-bron gas, bl sol or liq	1.447^{2}	−102	d 3.5	s	d	s eth, a alk
n132	oxi (tri-) fluoride	NO_3F	81.00	col gas expl.	liq; $1.507^{-45.9}$ sol: $1.951^{-193.2}$	−175	−45.9	d		s acet; expl al, eth
n133	sulfide, penta-	N_2S_5	188.31	gray cr.		10—11	d	d	d	s CS_2, eth; i bz, al
n134	sulfide, tetra-	Tetranitrogen tetrasulfide sulfurnitride. N_4S_4	184.27	hel cr, 2.046	2.24^{18}	d 178				s al, bz, CS_2
n135	**Nitrosyl bromide**	NOBr	109.91	br gas or dk br liq	>1.0	−55.5	−2	d	d	s alk
n136	*per*chlorate	$NOClO_4H_2O$	147.47	rhomb, deliq	2.169	d 100		d		expl al, eth
n137	chloride	NOCl	65.46	yel gas or yel-red liq or cr	gas: 2.99 g/ℓ liq: 1.417^{-12}	−64.5	−5.5	d	d	s fum H_2SO_4
n138	fluoborate	$NOBF_4$	116.81	col, rhomb, cr, hygr.	2.185^{25}_{4}	subl $250^{0.01}$		d		

No.	Name	Synonyms and Formulae	Mol. wt.	Crystalline form, properties and index of refraction	Density or spec. gravity	Melting point, °C	Boiling point, °C	Solubility, in grams per 100 cc		
								Cold water	Hot water	Other solvents
n139	fluoride	NOF	49.00	col gas	2.176 g/ℓ	−134	−56			d to HNO_2 + HF
n140	**Nitrosylsulfuric acid**	Chamber crystals. $NOHSO_4$	127.07	col, rhomb		d 73.5		d		s H_2SO_4
n141	**Nitrosylsulfuric anhydride**	$(NOSO_3)_2O$	236.13	tetr		217	360	d		s H_2SO_4
n142	**Nitrous acid**	HNO_2	47.01	only in sol (pa bl)				d		
n143	hypo-	$H_2N_2O_2$	62.03	wh, sol		expl		s		
n144	**Nitryl** chloride	NO_2Cl	81.46	pa yel-br gas	gas: 2.57 g/ℓ liq: 1.32^{14}	<−31	5	d		
n145	fluoride	NO_2F	65.00	col gas, col sol.	2.90 g/ℓ	−139	−63.5	d		d al, eth, chl
o1	**Osmium**	Os	190.20	gray-blsh met, hex	22.48^{20}	2700	>5300	i	i	sl s aq reg, HNO_3; i NH_3
o2	carbonyl chloride	$Os(CO)_3Cl_2$	345.14	col pr	269—273	d 280		i	i	s NaOH; i a
o3	chloride, di-	$OsCl_2$	261.11	dk brn, deliq	d			i	sl d	s al, eth, HNO_3; sl s alk
o4	chloride, tetra-	$OsCl_4$	332.01	red br need		subl		sl s, d		i al
o5	chloride, tri-	$OsCl_3$	296.56	br, cub		d 500—600		v s		s alk, al, a; sl s eth
o6	chloride, tri-, trihydrate	$OsCl_3{\cdot}3H_2O$	250.60	dk gr cr		d		v s		s al
o7	fluoride, hexa-	OsF_6	304.19	brn cr		32.1	45.9	d	d	
o8	fluoride, tetra-	OsF_4	266.19	br powd				d	d	
o9	iodide	OsI_4	697.82	vlt-blk, hygr met, lust				v s	d	s al
o10	oxide, di-, brown	OsO_2	222.20	brn cr	$11.37^{21.4}$	30% tr to OsO_4, 500		i	i	i a
o11	oxide, di- black	OsO_2	222.20	blk powd	7.71^{21}	tr to br 350—400		i	i	s dil HCL
	oxide, mon-	OsO	206.20	blk				i	i	
o13	oxide, sesqui-	Os_2O_3	428.40	dkg brn		d		i	i	i a
o14	oxide, tetra-	OsO_4	254.20	col, monocl.	4.906^{22}	40.6	130	5.70^{10}	6.23^{25}	250 ± 10^{20} CCl_4; s al, eth, NH_4OH, $POCl_3$
o15	sulfide, di-	OsS_2	254.32	blk, cub	9.47	d		i	i	s HNO_3; i alk
o16	sulfide, tetra-	OsS_4	318.44	br blk (exist?)		d		i		s dil HNO_3; i $(NH_4)_2S$
o17	sulfite	$OsSO_3$	270.26	bl blk		d		i		s dil HCL, alk
o18	telluride	$OsTe_2$	445.40	gray-blk cr		*ca*600				i a; d dil HNO_3
o19	**Oxygen**	O_2	31.9988	col gas, sol hex cr	gas: 1.429° g/ℓ, liq: 1.149^{-183} sol: $1.426^{-252.5}$	−218.4	−182.962	4.89° cm^3 3.16^{25} cm^3	2.46^{50} cm^3 2.30^{100} cm^3	2.78^{25} al
o20	Fluoride	OF_2	54.00	col gas, unst	liq: $1.90^{-223.8}$	−223.8	−144.8	sl s, d	i	sl s a, alk
o21	**Ozone**	O_3	47.9982	col gas, or dk bl liq, or bl-blk cr, liq: 1.2226.	gas: 2.144° g/ℓ liq: $1.614^{-195.4}$ g/ℓ	−192.7 ± 2 1.0	−111.9	49° cm^3		s alk sol, oils
p1	**Palladium**	Pd	106.42	silv wh, met, cub	12.02^{20} $11.40^{22.5}$	1554	2970	i	i	s aq reg, h HNO_3, H_2SO_4; sl s HCl
p2	bromide	$PdBr_2$	266.23	red br	5.173^{16}	d		i	i	s H Br
p3^1	chloride	$PdCl_2$	177.33	dk red, cub need, deliq	4.0^{18}	d 500		s	s	s HBr; acet
p3^2	chloride, dihydrate	$PdCl_2{\cdot}2H_2O$	213.36	br pr, deliq		d		v s	v s	s HCl, acet
p4	cyanide	$Pd(CN)_2$	158.46	yelsh-wh.		d		i	i	s KCN, NH_4OH; i dil a
p5	fluoride, di-	PdF_2	144.42	br, etr	5.80	volat	d red heat	sl s, d		s HF
p6	fluoride, tri-	PdF_3	163.42	blk, rhomb	5.06	d	d	d	d	s HF
p7	hydride	Pd_2H (or Pd_4H_2)	213.85	silv met (exist?)	10.76	d				
p8	iodide	PdI_2	360.23	blk powd	6.003^{18}	d 350		i	i	s KI; i al, eth, dil HCl
p9	nitrate	$Pd(NO_3)_2$	230.43	br yel, rhomb, deliq		d		s, d		s HNO_3
p10	oxide, di-	$PdO_2{\cdot}\chi H_2O$		dull red		d $-H_2O$, $-O$		i	i,d	s a, alk
p11	oxide, mon-	PdO	122.42	grnsh-bl or amber mass, or blk powd	9.70^{20}_{4}	870		i	i	i aq reg
p12	oxide, mon-, hydrate	$PdO{\cdot}\chi H_2O$		yel to brn		d		i	i	s a, NH_3,NH_4Cl
p13	selenate	$PdSeO_4$	249.38	dk brn-red, rhomb, deliq	6.5	d red heat		v s	v s	i al, eth, alk; s NH_3
p14	selenide	PdSe	185.38	dk gray		<960		i		s aq reg
p15	selenide, di-	$PdSe_2$	264.34	olive gray, hex		<1000		i	i	v s aq reg; v sl s HNO_3; i alk
p16	silicide	PdSi	134.51	cr	7.31^{15}			i		
p17	sulfate	$PdSO_4{\cdot}2H_2O$	238.51	red-br cr, deliq		d		v s	d	
p18	sulfide, di-	PdS_2	170.54	dk br cr	4.7—4.8^{25}_{4}	d		i	i	s aq reg, $(NH_4)_2S$
p19	sulfide, mono-	PdS	138.48	brn-blk tetr	6.6^{25}_{4}	d 950		i	i	sl s HNO_3, aq reg; i HCl, $(NH_4)_2S$
p20	sulfide, sub-	Pd_2S	244.90	grn gray (exist?)	7.303^{15}	d 800		i	i	sl s a, aq reg
p21	telluride, di-	$PdTe_2$	361.62	silvery cr, hex				i	i	v s aq reg; s HNO_3; i alk

No.	Name	Synonyms and Formulae	Mol. wt.	Crystalline form, properties and index of refraction	Density or spec. gravity	Melting point, °C	Boiling point, °C	Solubility, in grams per 100 cc: Cold water	Hot water	Other solvents
	Palladium complexes . . .									
p22	diamminepalladium (II) hydroxide	$Pd(NH_3)_2(OH)_2$	174.50	yel micr cr		>105		v s	d	
p23	dichlorodiammine-palladium (II) *trans* (or α)	$Pd(NH_3)_2Cl_2$	211.39	yel, tetr	2.5	d		0.304^{10}	s, d	s, d a; s NH_4OH; i chl, acet
p24	tetramminepalladium (II) chloride	$Pd(NH_3)_4Cl_2{\cdot}H_2O$	263.46	col, tetr	1.91^{18}	d 120		v s		
p25	tetramminepalladium tetrachloropalladate	Vauquelin's salt. $Pd(NH_3)_4{\cdot}PdCl_4$	422.77	pink powd or need	2.489^{21}	tr yel 184 d above 192		i	sl s	sl s dil HCl; s KOH
p26	**Phospham**	PN_2H	60.00	wh, amorph		infus		i	d	s conc H_2SO_4, alk; i a
p27	**Phosphomolybdic acid**	Molybdophosphoric acid $P_2O_5{\cdot}20MoO_3{\cdot}51H_2O$	3939.49	yel, tetr		78—90		s	s	
	Phosphonium									
p28	bromide	PH_4Br	114.91	col, cub	gas: 2.464 g/ℓ	subl *ca* 30	38.8^{794}	d	d	
p29	chloride	PH_4Cl	70.46	col, cub		$28^{46\ atm}$	subl	d		
p30	iodide	PH_4I	161.91	col, tetr, deliq	2.86	18.5, subl 61.8	80	d		d, s a, alk
p31	sulfate	$(PH_4)_2SO_4$	166.07					d		
p32	**Phosphoramide**	Phosphorylamide. $PO(NH_2)_3$	94.04	wh, cr		d		40.66^{15}	i	sl, me
	Phosphoramide									
p33	difluoro-	$H_2PO_2F_2$	102.99	col, fum liq	1.583^{25}_{4}	−96.5 ± 0.1	115.9 sl d			
p34	hypo-	$H_4P_2O_4{\cdot}2H_2O$	198.01	col, rhomb deliq		70	d 100	d to H_3PO_3 + HPO_3		
p35	meta-	HPO_3	79.98	col, vitrous deliq	2.2—2.5	subl		d to H_3PO_4d		s al; i liq CO_2
p36	monofluo-	H_2PO_3F	99.99	oily, col liq	1.818	−80				
p37	ortho-	H_3PO_4	98.00	col, liq, or rhomb cr, deliq	1.834^{18}	42.35	−1/2 H_2O, 213	548	v s	s al
p38	ortho-	$2H_3PO_4{\cdot}H_2O$	214.01	col, hex pr deliq		29.32	d	v s		
p39	pyro-	$H_4P_2O_7$	177.98	col, need or liq, hygr.		61		709^{23}	d to H_3PO_4	v s al, eth
p40	**Phosphorus,** black	P_4	123.89504	blk, incombust	2.70					i CS_2, conc H_2SO_4
p41	red	P_4	123.89504	redsh-brn, cub, or amorph powd, (mix of col and vlt?)	2.34	590 $^{43\ atm}$	ign 200 280	v sl s	i	s abs al; i CS_2, eth, NH_3
p42	violet	P_4	123.89504	vlt, monocl	2.36	590				i org solv
p43	yellow	Phosphorus, white. P_4	123.89504	yel (or wh) cub or wax like solid 2.144	1.82^{20}	44.1	280	0.0003^{15}	sl s	0.3 al; $880^{10}CS_2$; s bz, NH_3, alk, eth, chl, tol
p44	bromide, penta-	PBr_5	430.49	yel, rhomb		d < 100	d 106	d		s CS_2, CCl_4, bz
p45	bromide, tri-	PBr_3	270.69	col, fum liq, $1.697^{26.6}$	2.852^{15}	−40	172.9	d		d al; s eth, chl, CS_2, CCl_4
p46	bromide(di-) chloride, tri-	PBr_2Cl_3	297.14	or cr		d 35		d		
p47	bromide(hepta-) chloride, di-	PBr_7Cl_2	661.21	pr				d		s PCl_3, PCl_5
p48	bromide(mono-)chloride, tetra-	$PBrCl_4$	252.69	yel cr				d		
p49	bromide(octa-) chloride, tri-	PBr_8Cl_3	776.56	brn need		25		d		
p50	bromide(di-) fluoride, tri-	PBr_2F_3	247.78	pa yel liq		−20	d 15	d		d glass
p51	bromide nitride	$(PNBr_2)_3$	614.37	col, rhomb		190	subl v 150	i		s eth; sl s chl, CS_2
p52	chloride, di-	PCl_2 (or P_2Cl_4?)	101.88	col liq		−28	180	hydr		
p53	chloride, penta-	PCl_5	208.24	yelsh-wh, tetr. fum	gas: 4.65^{296} g/ℓ	d 166.8 (press)	subl 162	d		d a; s CS_2, CCl_4
p54	chloride, tri-	PCl_3	137.33	col, fum liq, 1.516^{14}	1.574^{21}	−112	75.5^{749}	d	d	s eth, bz, chl, CS_2, CCl_4
p55	chloride(di-) fluoride, tri-	PCl_2F_3	158.87	col liq	5.4 g/ℓ	−8	10			
p56	chloride(tri-)iodide, di-	PCl_3I_2	391.14	red, hex		d 259		d		s CS_2
p57	chloride(di-)nitride	$(PNCl_2)_3$	347.66	rhomb	1.98	114	256.5	i	d	s al, eth, bz, chl, a ac, CS_2
p58	chloride(di-)nitride	$(PNCl_2)_4$	463.55		2.18^{24}_{24}	123.5	328.5			
p59	chloride(di-)nitride	$(PNCl_2)_5$	579.43			41	224^{13}, polym >250			
p60	chloride(di-)nitride	$(PNCl_2)_6$	695.32			90	262^{13}, polym >250			
p61	cyanide	$P(CN)_3$	109.03	wh need		subl 130		d		v s eth; sl s h bz
p62	fluoride, penta-	PF_5	125.97	col gas	5.805 g/ℓ	−83	−75	d		
p63	fluoride, tri-	PF_3	87.97	col gas	3.907 g/ℓ	−151.5	−101.5	d		s al; d alk
p64	hydride, tri-	Phosphine. PH_3	34.00	col gas, pois		−133	−87.7	o.26 vol 20		
p65	iodide, di.	P_2I_4	569.57	or, tricl		110		d	d	s CS_2
p66	iodide, tri-	PI_3	411.69	red, hex deliq	4.18	61	d	d	d	v s CS_2
p67	oxide, pent-	Phosphoric anhydride. P_2O_5 (or P_4O_{10})	141.94	wh, monocl or powd, v deliq	2.39	580—585	subl 300	d to H_3PO_4	d	s H_2SO_4; i acet, NH_3
p68	oxide, sesqui-	Phosphorus trioxide. P_4O_6 (or P_2O_3)	219.89	col, or wh powd, monocl cr, deliq	2.135^{2}s;l	23.8	175.4	d to H_3PO_3	d	s chl, bz, eth, CS_2
p69	oxide, tetra-	P_2O_4	125.95	col, rhomb, deliq	2.54^{23}	>100	180	v s to H_3PO_3	d	
p70	oxide, tri-	P_2O_3(or P_4O_6)	109.95	col, or wh powd, or monocl, deliq	2.135^{21}	23.8	173.8 (in N_2)	d to H_3PO_3	d	s CS_2, eth, chl, bs.

PHYSICAL CONSTANTS OF INORGANIC COMPOUNDS (continued)

No.	Name	Synonyms and Formulae	Mol. wt.	Crystalline form, properties and index of refraction	Density or spec. gravity	Melting point, °C	Boiling point, °C	Solubility, in grams per 100 cc: Cold water	Hot water	Other solvents
p71	oxybromide	$POBr_3$	286.69	col pl.	2.822	56	189.5	d		s H_2SO_4 CS_2, eth, bz, chl
p72	oxy*di*bromide chloride	$POBr_2Cl$	242.23	sol or liq	liq: 2.45^{50}	30	165	d		
p73	oxybromide chloride, di	$POBrCl_2$	197.78	tabl or liq	liq:2.104^{14}	13	137.6	d		
p74	oxychloride	$POCl_3$	153.33	col, fum liq, $1.460^{25.1}$	1.675	2	105.3	d	d	d al, a
p75	oxychloride	$P_2O_3Cl_4$	251.76	col fum liq	liq: 1.58^{7}	<-50	212	d		
p76	oxyfluoride	POF_3	103.97	col gas	4.69 g/ℓ	−68	−39.8	d		d al
p77	oxynitride	PON	60.98	wh, amorph		red heat		i	i	i a, alk
p78	oxysulfide	$P_4O_6S_4$	348.13	wh, tetr, deliq		102	295	d		50 CS_2
p79	selenide, penta-	P_2Se_5	456.75	dk red-blk need		d		d		s CCl_4; i CS_2
p80	(tetra-)selenide, tri-	P_4Se_3	360.78	or red cr	1.31	242	360—400			
p81	(tetra-)sulfide, hepta-	P_4S_7	348.32	lt yel cr	2.19^{17}	310	523			sl s CS_2
p82	sulfide, penta-	P_2S_5 (or P_4S_{10})	222.25	gray-yel cr, deliq	2.03	286—90	514	i	d	0.22 CS_2; s alk
p83	sesquisulfide	Tetraphosphorus trisulfide. P_4S_3	220.08	yel, rhomb	2.03^{17}	174	408	i	d	100^{17} CS_2; 11.1^{80} bz
p84	thiobromide	$PSBr_3$	302.75	yel, oct	2.85^{17}	38	d 212	s		s eth, CS_2, PCl_3
p85	thiochloride	$PSCl_3$	169.39	col, fum liq	1.668	−35	125	sl d	d	s CS_2
p86	thiocyanate	$P(SCN)_3$	205.21	liq	1.625^{18}	*ca* −4	265	d		s al, eth, bz, CS_2
	Phosphorous acid									
p87	hypo-	$H(H_2PO_2)$	66.00	col oily liq or deliq cr	1.493^{19}	26.5	d 130	s	v s	v s al, eth
p88	meta-	HPO_2	63.98	feather like cr				d		
p89	ortho-	$H_2(HPO_3)$	82.00	col-yel cr, deliq	$1.651^{21.2}$	73.6	d 200	309°	694^{40}	s al
p90	pyro-	$H_4P_2O_5$	145.98	need		38	d 120	d		
p91	**Phosphotungstic acid**	Tungstophosphoric acid $H_3[P(W_3O_{10})_4]\cdot 14H_2O$	3132.39	yel-grn cr, tricl.		d		s		s al, eth
p92	phosphotungstic acid	Dodecatungtophosphoric acid. $H_3[P(W_3O_{10})_4]\cdot 24H_2O$	3312.54	trig		89		s		
p93	**Platinic acid,** *hexa*chloro	$H_2PtCl_6\cdot 6H_2O$	517.91	red, brn, deliq	2.431	60		v s	v s	s al, eth
p94	*tetra*cyano	$H_2Pt(CN)_4$	301.17			d 100		v s	v s	v s al, eth, chl
p95	*hexa*hydroxy-	$H_2Pt(OH)_6$	299.14	yel need		$-2H_2O$, 100	$-3H_2O$, 120	i	v sl s	s H_2SiF_6, dil a, alk
p96	*hexa*iodo-	$H_2PtI_6\cdot 9H_2O$	1120.66	blk-red, deliq				d		
p97	**Platinum**	Pt	195.08 ± 3	silv met, cub	21.45^{20}	1772	3827 ± 100			s aq reg, fus alk
p98	arsenide	Nat. sperrylite. $PtAs_2$	344.92	gray, cub	11.8	d >800		sl d	sl d	v sl s a
p99	bromic acid	$H_2PtBr_6\cdot 9H_2O$	838.66	br-red, monocl, hygr.				v s	v s	v s al, eth
p100	(II) bromide, di-	$PtBr_2$	345.89	br	6.65^{25}_{4}	d 250		i	i	i al; s HBr, KBr, aq Br
p101	(IV) bromide, tetra-	$PtBr_4$	514.70	dk cr	5.69^{25}_{4}	d 180		0.41^{20}	sl s	v s al, eth, HBr
p102	carbonyl bromide	$[Pt_2(CO)_2]Br_4$	765.80	lt red, need, hygr	5.115^{25}_{4}	d 180		s d		s abs al, CCl_4, bz
p103	carbonyl chloride, di-	$Pt(CO)Cl_2$	294.00	yel need	4.2326^{25}_{4}	195, subl 240 in CO_2	d 300	d	d	s conc HCl H_2SO_4, al
p104	*di*carbonyl chloride, di-	$Pt(CO)_2Cl_2$	322.01	lt yel need.	3.4882^{25}_{4}	142	−CO, 210	d	d	d HCl; s CCl_4
p105	*di*platinum dicarbonyl tetrachloride	$Pt_2(CO)_2Cl_4$	587.99	or-yel need	4.235^{25}_{4}	195	subl 240 (in CO_2)	d		d HCl
p106	*di*platinum tricarbonyl tetrachloride	$Pt_2(CO)_3Cl_4$	616.00	or-yel need		130	d 250	d		s h CCl_4, d al
p107	carbonyl diiodide	PtI_2CO	476.90	red cr	5.157	d 140—150		d		s d, al; s bz
p108	carbonyl sulfide	Pt(CO)S	255.15	br-blk		d 300—400		d		d alk, al
p109	chloric acid	$H_2PtCl_6\cdot 6H_2o$	517.91	brn-red cr, hygr	2.431	60	d >115	s	s	s abs al, acet; v s eth
p110	(II) chloride, di-	$PtCl_2$	265.99	olive grn, hex	6.05	d 581 (in Cl_2)		v sl s		i al, eth; s HCl, NH_4OH
p111	(IV) chloride, tetra-	$PtCl_4$	336.89	br-red cr.	4.303^{25}_{4}	d 370 (in Cl_2)		58.7^{25}	v s	sl s al, NH_3; s acet; i eth
p112	(IV) chloride, tetra-, hydrage	$PtCl_4\cdot 5H_2O$	426.97	red, monocl.	2.43	$-H_2O$, 100		v s	v s	s al, eth
p113	chloride, tri-	$PtCl_3$	301.44	grnsh-blk	5.256^{25}	435		sl s	s	s h HCl; v sl s conc HCl
p114	*di*chlorocarbonyl, dichloride	$Pt(COCl_2)Cl_2$	364.90	yel cr.		d		v s		sl s al; v sl s CCl_4
p115	(II) cyanide, di-	$Pt(CN)_2$	247.12	yel-br cr				i	i	i al, a alk; s KCN
p116	(II) fluoride, di-	PtF_2	233.98	yelsh-grn				i	i	
p117	fluoride, hex-	PtF_6	309.07	dk red solid, very unstable		57.6				
p118	(IV) fluoride, tetra-	PtF_4	271.07	deep red, fused, or yel-lt brn cr, deliq		d red heat		s d	v s	s a, alk
p119	(II) hydroxide	$Pt(OH)_2$	229.09	blk		d		i	i	s HCl, HBr, alk; i H_2SO_4, dil HNO_3
p120	(II) hydroxide, hydrage	$Pt(OH)_2\cdot 2H_2O$	265.13			$-2H_2O$, 100		i	i	s conc a
p121	*mono*hydroxy chloric acid	$H_2[PtCl_5(OH)]\cdot H_2O$	409.38	red-brn cr, hygr						
p122	(II) iodide, di-	PtI_2	448.89	blk powd	6.403^{25}_{4}	d 360		i	i	i eth, a; s HI; sl s Na_2SO_3
p123	(IV) iodide, tetra-	PtI_4	702.70	brn, amorph, or blk cr.	6.064^{25}_{4}	d 130		s d		s al, acet, alk, HI, KI, liq NH_3
p124	iodide, tri-	PtI_3	575.79	blk, like graphite	7.414^{25}_{4}	d 270		i	i	i al, eth; s KI
p125	(II) oxide, mon-	PtO	211.08	vlt-blk	14.9^{15}	d 550		i	i	s HCl; i a, aq reg

No.	Name	Synonyms and Formulae	Mol. wt.	Crystalline form, properties and index of refraction	Density or spec. gravity	Melting point, °C	Boiling point, °C	Solubility, in grams per 100 cc		
								Cold water	Hot water	Other solvents
p126	(II) oxide, mon-, dihydrate	$PtO·2H_2O$	247.11			$-2H_2O$, 140—150				s conc HCl, conc H_2SO_4, conc HNO_3
p127	(IV) oxide, di-	PtO_2	227.08	blk	10.2	450		i	i	i a, aq reg
p128	(IV) oxide, di-dihydrate	Platinic hydroxide· $PtO_2·2H_2O$ or $Pt(OH)_4$	263.11			$-2H_2O$, 100		i	i	s HCl, aq reg, KOH
p129	(IV) oxide, di-. hydrate	$PtO_2·H_2O$	245.09					i	i	i aq reg, ac a, HCl; sl s NaOH
p130	(IV) oxide, di- trihydrate	$PtO_2·3H_2O$	281.12	ochre		d 300		i	i	i aq reg, HCl
p131	(IV) oxide, di-, tetrahydrate	Hydroxoplatinic acid. $PtO_2·4H_2O$ or $H_2Pt(OH)_6$	299.14	yel need		$-2H_2O$, 100	$-3H_2O$, 120	i	i	s a, dil alk
p132	(II, IV) oxide,	Pt_3O_4	649.24			d		i		i a, aq reg
p133	oxide, sesqui,	$Pt_2O_3·3H_2O$	492.20			$-H_2O$, 100		i	i	s conc H_2SO_4, caust alk
p134	oxide, tri-	PtO_3	243.09	redsh-brn powd						s HCl, H_2SO_3 sl s HNO_3, H_2SO_4
p135	*pyro*phosphate	PtP_2O_7	369.02	grn-yel	4.85	d 600		v sl s		
p136	phosphide	PtP_2	257.03	met shine	9.01^{25}_{4}	*ca* 1500		i	i	i a; v sl s aq reg
p137	selenide, di-	$PtSe_2$	353.00	blk or gray cr or amorph	7.65	d when dry				s aq reg; sl s HNO_3, H_2SO_4
p138	selenide, tri-	$PtSe_3$	431.96	bl flakes	7.15	d 140		i	i	i conc a, CS_2; s aq reg
p139	sulfate	$Pt(SO_4)_2·4H_2O$	459.26	yel pl				s	d	s al, eth, a
p140	(IV) sulfide, di-	PtS_2	259.20	blk-brn powd	7.66^{25}_{4}	d 225—250		i	i	s HCl, HNO_3; i $(NH_4)_2S$
p141	(II) sulfide, mono-	PtS	227.14	blk, tetr	10.04^{25}_{4}	d		i	i	s $(NH_4)_2$ S; i a, alk
p142	sulfide, sesqui-	Pt_2S_3	486.34	gray (exist?)	5.52	d		i	i	sl s aq reg; i a
p143	(III) sulfuric acid	$H_2[Pt_2(SO_4)_4(H_2O)_2]·9^1/_2H_2O$	983.58	or, red, tricl		d 150		s	s	d alk
p144	telluride	$PtTe_2$	450.28	gray, hex		1200—1300				sl s Na_2S, $(NH_4)_2S$
	Platinum Complexes									
p145	tetramine platinum (II) chloride	$[Pt(NH_3)_4]Cl_2·H_2O$	352.12	col, tetr, 1.672, 1.667	2.737	250, $-H_2O$, 100				
p146	tetrammineplatinum (II) chloroplatinite	Magnus, salt. $[Pt(NH_3)_4]PtCl_4$	600.09	grn or red, tetr	<4.1	d		sl s	sl s	
p147	tetrachlorodiammine platinum (IV) *trans-*	$[Pt(NH_3)_2]Cl_4$	370.95		3.3	200—216				
p148	tetrachlorodiammine platinum (IV), *cis-*	$[Pt(NH_3)_2]Cl_4$	370.95	or-yel, rhomb or hexag pl or need		240				
p149	**Plumbous, plumbic**	see Lead								
p150	**Plutonium**α	Pu	242.059	sil wh, monocl	19.84	641	3232			s HCl; i HNO_3, conc H_2SO_4
p151	β	Pu	239.0522	monocl	17.70	stab 122 ± 2 to 206 ± 3				
p152	γ	Pu	239.0522	orthorhomb	17.14	stab 206 ± 3 to 319 ± 5				
p153	δ	Pu	239.0522	cub	15.92	stab 319 ± 5 to 451 ± 4				
p154	δ′	Pu	239.0522	tetrag	16.00	stab 451 ± 4 to 476 ± 5				
p155	ε	Pu	239.0522	cub	16.51	stab 476 ± 5 to 639.5 ± 2				
p156	bromide, tri-	$PuBr_3$	478.76	grn, orthorhomb	6.69	681		s		
p157	chloride, tri-	$PuCl_3$	345.41	emerald grn, hex	5.70	760		s		s dil a
p158	fluoride, hexa-	PuF_6	353.04	redsh-brn, orthorhomb		50.75	62.3	d		
p159	fluoride, tetra-	PuF_4	315.05	pa brn, monocl	7.0 ± 0.2	1037				
p160	fluoride, tri-	PuF_3	296.05	purple, hex	9.32	1425(±3)		i		
p161	iodide, tri-	PuI_3	619.77	bright grn, orthorhomb	6.92	777		s		
p162	nitride	PuN	253.06	blk, cub	14.25			hydrol		s HCl, H_2SO_2
p163	oxalate	$Pu(C_2O_4)_2·6H_2O$	523.18	yel-grn				i		
p164	oxide, di-	PuO_2	271.05	yelsh-grn, cub	11.46					sl s h conc H_2SO_4, HNO_3, HF
p165	sulfate	$Pu(SO_4)_2$	431.17	light pink						s dil min a
p166	sulfate, tetrahydrate	$Pu(SO_4)_2·4H_2O$	503.23	coral pink		d 280				s dil min a
p167	**Polonium**	Po	(209)	α-Po: simple cub; β-Po: rhbr	9.4 (for β-Po)	254	962	sl s		s dil min a; v sl s dil KOH
p168	ammonium chloride	$(NH_4)_2PoCl_6$	457.78		2.76					
p169	*tetra*bromide	$PoBr_4$	528.60	bright red, cub		330 (in Br atm)	360^{200}			s al, acet; i bz, CCl_4
p170	*di*chloride	$PoCl_2$	279.89	ruby red, orthorhomb	6.50	subl 190				s dil HNO_3
p171	*tetra*chloride	$PoCl_4$	350.79	yel, monocl or tric		300 (in Cl atm)	390	s, d		s HCL; sl s al, acet
p172	*tetra*iodide	PoI_4	716.60	blk cr		200 (in N atm subl)				sl s al, acet; i bz, CCl_4
p173	*di*oxide	PoO_2	240.98	red, tetr		d 500				
p174	selenate	$2PoO_2·SeO_3$	608.92	wh powd		>400				s dil HCl
p175	sulfate, basic	$2PoO_2·SO_3$	562.02	wh powd		>400, d 550				s dil HCl

No.	Name	Synonyms and Formulae	Mol. wt.	Crystalline form, properties and index of refraction	Density or spec. gravity	Melting point, °C	Boiling point, °C	Solubility, in grams per 100 cc		
								Cold water	Hot water	Other solvents
p176	*di*sulfate	$Po(SO_4)_2$	401.10	purp		d 550				i al; v s dil HCl
p177	monosulfide	PoS	241.04	blk		d 275				i al; sl s dil HCl
p178	**Potassium**	Kalium. K	39.0983	cub silv met	0.86^{20}	63.25	760	d to KOH	d	d al; s a, Hg, NH_3
p179	acetate	$KC_2H_3O_2$	98.14	wh, lust powd, deliq	1.57^{25}	292		253^{20}	492^{62}	33 al; 24.24^{15} MeOH; s liq NH_3; i eth, acet
p180	acetate, acid	$K_2C_2H_3O_2 \cdot HC_2H_3O_2$	158.20	col, need or pl, hygr.		148	d 200	d	d	s al, acet
p181	acetyl salicylate	$KC_9H_7O_4 \cdot 2H_2O$	254.28			65				
p182	*meta*asluminate	$K_2Al_2O_4 \cdot 3H_2O$	250.20	col cr.				v s, d	v s, d	s alk; i al
p183	aluminosilicate	Nat. orthoclase. $KAlSi_3O_8$ (or $K_2O \cdot Al_2O_3 \cdot 6SiO_2$)	278.33	wh, monocl, 1.518, 1.524, 1.526	2.56	*ca*1200				
p184	aluminosilicate	Nat. microcline. $KAlSi_3O_8$ (or $K_2O \cdot Al_2O_3 \cdot 6SiO_2$).	278.33	wh, tricl, 1.522, 1.526, 1530	2.54—2.57	1140—1300				
p185	aluminosilicate	Nat. muscovite, white mica. $KAl_3Si_3O_{10} \cdot (OH)_2$ (or $K_2O \cdot 3Al_2O_3 \cdot 6SiO_2 \cdot 2H_2O$	398.31	col, monocl, 1.551, 1.587, 1.581	2.76—2.80	d		i		
p186	aluminum *meta*silicate	Nat. leucite. $KAlSi_2O_6$	218.25	col cr, 1.508	2.47	1686 ± 5		i	i	d a
p187	aluminum *ortho*silicate	Nat. kaliophilite. $KAlSiO_4$	158.16	col, hex or rhomb (hex→rhomb 2540°) hex: 1.532, 1.572; rhomb: 1.528, 1.536	2.5	*ca*1800 (rhomb)				
p188	aluminum sulfate	Nat. kalinite. $KAl(SO_4)_2 \cdot 12H_2O$	474.38	col, cub, oct or monocl, cub: 1.454, 1.4564; hex: 1.456, 1.429	1.757^{20}_{14}	92.5 $-9H_4O$, 64.5	$-12H_2O$, 200	11.4^{20}	v s	i al, acet; s dil a
p189	amide	Potassamide. KNH_2	55.12	col-wh, or yel-grn, hygr.		335	subl 400	d	d	d al; s liq NH_3
p190	*peroxy*almmine sulfonate	$(KSO_3)_2NO$	268.32	yel cr, expl				$o.62^3$, d	6.6^{29}, d	i al
p191	ammonium tartrate	$KNH_4C_4H_4O_6$	205.21	wh, cr powd				v s		
p192	antimonate, hydroxo-	"*Pyro*"-antimonate. $KSb(OH)_6 \cdot {}^1/_2H_2O$	271.90	wh gran or cr powed				2.82^{26}	s	
p193	antimonide	K_3Sb	239.04	yel-grn		812		d		d air
p194	antimony tartrate	$KSbC_4H_4O_7 \cdot {}^1/_2H_2O$	333.93	col, rhomb, 1.620, 1.636, 1.638	2.607	$-H_2O$, 100		$5.26^{8.7}$	35.7^{100}	i al; 6.67^{25} glyc
p195	*ortho*arsenate	K_3AsO_4	256.21	col need, deliq		1310		18.87	v s	4 al
p196	*ortho*arsenate, di-H	KH_2AsO_4	180.03	col, tetr, 1.567, 1.518	2.867	288		19^6	v s	i al; s NH_3, a; 52.5 glyc
p197	*ortho*arsenate, mono H	K_2HAsO_4	218.12	col monocl pr		d 300		18.86^6	s	i al
p198	*meta*arsenite	$KAsO_2$	146.02	wh powd, hygr.				s	s	sl s al
p199	*ortho*arsenite	K_3AsO_3	240.21	col need				v s		s al
p200	*meta*arsenite, acid	$KH(AsO_2)_2 \cdot H_2O$	271.96					s		sl s al
p201	aurate	$KAuO_2 \cdot 3H_2O$ (or $2H_2O$)	322.11	lt yel need.		d		s	d	s al
p202	azide	KN_3	81.12	col, tetr	2.04	350 (vac)		49.6^{17}	105.7^{100}	s al; i eth
p203	benzoate	$KC_7H_5O_2 \cdot 3H_2O$	214.26	wh cr powd.		$-3H_2O$, 110	d	52^{25}	112^{100}	s al
p204	*di*borane	Diboranidex. $K_2B_2H_6$	105.86	wh, cub cr, 1.493	1.18	subl 400, vac		d		
p205	*penta*borate	Pentaboranidex. $K_2B_5H_9$	141.32	wh powd		d<180		s, d	d	
p206	*di*borane, dihydroxy	$K_2B_2H_6O_2$	137.86	col, cub cr	1.39	d→K, 400—500		s, d		s al; d a
p207	*meta*borate	KBO_2 (or $K_2B_2O_4$)	81.91	col, hex 1.526, 1.450		950		71^{30}	v s	i al, eth
p208	*penta*borate	$KB_5O_8 \cdot 4H_2O$	293.20	col, rhomb		780			0.007°	
p209	*peroxy*borate	$KBO_3 \cdot {}^1/_2H_2O$	106.91	wh cr.		$-O_2$, 100	d 150	1.22°		i al, eth
p210	*tetra*borate	$K_2B_4O_7 \cdot 8H_2O$	377.55	col, monocl.	1.74 (anhydr)	d		26.7^{30}	v s	
p211	borohydride	KBH_4	53.94	wh, cub, 1.494.	1.178	d 500		19.3^{20}	v s	o.25 al; 0.56 MeOH; i eth
p213	boroxalate	$KHC_2O_4 \cdot HBO_2 \cdot 2H_2O$	207.97			$-H_2O$, 110		v s	v s	i, d al
p214	borotartrate	Solution: cream of tartar. $KC_4H_4BO_7$	213.98	wh cr powd.	1.832			v s		i al, eth, chl
p215	bromate	$KBrO_3$	167.00	col, trig	$3.27^{17.5}$	434 d 370		13.3^{40}	49.75^{100}	sl s al; i acet
p216	bromide	KBr	119.00	col, cub, sl hygr, 1.559	2.75^{25}	734	1435	53.48°	102^{100}	0.142^{25} al; sl s eth; s glyc
p217	bromoaurate	$K[AuBr_4]$	555.68	red-brn, rhomb		d 120		sl s		s al
p218	bromoaurate, dihydrate	$K[AuBr_4] \cdot 2H_2O$	591.71	vlt, monocl cr	4.08			19.5^{15}	204^{67}	s al, KBr; d eth
p219	bromoiodide, di-	$KIBr_2$	325.81	red, rhomb		60	d 180	v s		
p220	*hexa*bromoplatinate	$K_2[PtBr_6]$	752.70	dk red-brn, cub	4.66^{24}	d>400		2.02^{20}	10^{100}	i al
p221	*tetra*bromoplatinite	$K_2[PtBr_4]$	592.89	br, rhomb				v s	v s	
p222	bromoplatinite, dihydrate	$K_2[PtBr_4] \cdot 2H_2O$	628.92	blk, rhomb	3.747^{25}_{4}	$-H_2O$, vac		v s	v s	
p223	bromostannate	$K_2[SnBr_6]$	671.31	wh cr.	3.783					
p224	cacodylate	$K[(CH_3)_2AsO_2] \cdot H_2O$	194.10	wh cr.				s		sl s al; i eth
p225	cadmium cyanide	$K_2[Cd(CN)_4]$	294.68	col, oct	1.846	450		33.3	100^{100}	s al
p226	cadmium iodide	$2KI \cdot CdI_2 \cdot 2H_2O$	734.26	wh-yelsh cr powd, deliq	3.359			137^{15}		s a, al, eth
p227	calcium chloride	Chlorocalcite $KCl \cdot CaCl_2$	184.54	col cub, β 1.52	754			s		

No.	Name	Synonyms and Formulae	Mol. wt.	Crystalline form, properties and index of refraction	Density or spec. gravity	Melting point, °C	Boiling point, °C	Solubility, in grams per 100 cc: Cold water	Hot water	Other solvents
p228	calcium magnesium sulfate	Polyhalite. $K_2Ca_2Mg(SO_4)_4\cdot 2H_2O$	602.92	wh, trig, 1.548, 1.562, 1.567	2.775					
p229	calcium sulfate	Kaluszite, syngenite. $K_2Ca(SO_4)_2\cdot H_2O$	328.41	col, monocl, 1.500, 1.517, 1.518	2.60	1004		0.25	d	i al; s a
p230	*d*-camphorate	$K_2C_{10}H_{14}O_4\cdot 5H_2O$	366.49	col, need cluster, hygr.		$-5H_2O$, 110		260^{14}		s al
p231	carbide	KHC_2	64.13	col, rhomb cr.	1.37					
p232	carbonate	K_2CO_3	138.21	col, monocl, hygr, 1.531	2.428^{19}	891	d	112^{20}	156^{100}	i al, acet
p233	*peroxy*carbonate	$K_2C_2O_6\cdot H_2O$	216.23			200—300		s		
p234	carbonate, dihydrate	$K_2CO_3\cdot 2H_2O$	174.24	col, monocl, hygr, 1.380, 1.432, 1.573	2.043	$-H_2O$, 130		146.9	331^{100}	
p235	carbonate, hydrogen	$KHCO_3$	100.12	col, monocl, 1.482	2.17	d 100—200		22.4	60^{60}	i al
p236	carbonate, trihydrate	$2K_2CO_3\cdot 3H_2O$	330.46	col, monocl, 1.380, 1.482, 1.573	2.043			129.4	268.3^{100}	i al, conc NH_4OH
p237	carbonyl	$(KCO)_6$	402.65	gray-red		expl		expl		d al
p238	chlorate	$KClO_3$	122.55	col, monocl, 1.409, 1.517, 1.524	2.32	356	d 400	7.1^{20}	57^{100}	14.1^{100} 50% al; sl s glyc, liq NH_3; i acet; s alk
p239	*per*chlorate	$KClO_4$	138.55	col, rhomb, 1.4717, 1.4724, 1.476	2.52^{10}	610 ± 10	d 400	0.75°	21.8^{100}	v sl s al; i eth
p240	chloride	Nat. sylvite. KCl	74.55	cub, col 1.490	1.984	770	subl 1500	34.4	56.7^{100}	sl s al; s eth, glyc, alk
p241	*hypo*chlorite	KClO	90.55	in sol only		d		v s	v s	
p242	chloroaquoruthenate (III) penta-	$K_2[Ru(H_2O)Cl_5]$	374.55	rose pr		$-H_2O$, 200		s	s	sl s al
p243[1]	chloroaurate	$KAuCl_4$	377.88	yel, monocl	3.75	d 357		61.8^{20}	80.2^{60}	25 al; s a
p243[2]	chloroaurate, dihydrate	$K[AuCl_4]\cdot 2H_2O$	413.91	yel, rhomb pl				s	s	s al, eth
p244	chlorochromate	Peligot's salt. $KCrO_3Cl$.	174.55	red, monocl.	2.497	d		s, d		s a
p245	chlorohydroxoruthenate	$K_2[Ru(OH)Cl_5]$	373.54	brn-red cr		d		s, d	d	i al
p246	chloroiodate (III)	$KICl_4$	307.81	yel, rhomb	1.76^{45}	d		d		d eth
p247	chloriodide, di-	$KICl_2$	236.91	col, monocl		60	d 215	d		
p248	chloroiridate	K_2IrCl_6	483.13	blk, cub	3.546	d		125^{19}	6.67	i al, KCl, HN_4OH
p249	chloronitrosylruthenate (III) penta-	$K_2Ru(NO)Cl_5$	386.54	dk red, rhomb		d		12^{25}	80^{60}	i al
p250	chloroösmate (III)	$K_3OsCl_6\cdot 3H_2O$	574.26	dk red cr		$-3H_2O$, 150		v s		s a; i eth
p251	chloroösmate (IV)	K_2OsCl_6	481.11	red, cub		d		sl s	s	i al; s dil HCl
p252	chloropalladate	K_2PdCl_6	397.33	red, cub	2.738	d		sl s, d	d	i al; sl s HCl
p253	chloropalladite	K_2PdCl_4	326.43	red-brn, tetr (yel cub)	2.67	d 105		s	v s	s KCl, HN_4OH; i al
p254	*hexa*chloroplatinate	K_2PtCl_6	485.99	yel, cub, ~ 1.825	3.499^{24}	d 250		0.481^{2}	5.22^{100}	i al, eth
p255	*tetra*chloroplatinite	K_2PtCl_4	415.09	red-brn tetr, 1.64, 1.67	3.38	d		0.93^{16}	5.3^{100}	i al
p256	chloroplumbate	K_2PbCl_6	498.11	lt yel, cub		d 190		d		s h HCl
p257	chlororhenate (IV)	K_2ReCl_6	477.12	yel-grn, oct	3.34			0.8	d	d alk; sl s HCl
p258	chlororhenate (V)	K_2ReOCl_5	457.67	grn hex pl				d	d	s a; i al, eth
p259	chlororhodate, hexa-	$K_3RhCl_6\cdot 3H_2O$	486.96	red, tricl	3.291	d		d		sl s al, KCl
p260	chlororhodite, penta-	K_2RhCl_5	358.37	red, rhomb		d		sl s	d	i al
p261	chlororuthenate (IV)	K_2RuCl_6	391.98	blk, cub		d		s, d		i al
p262	chlorostannate	K_2SnCl_6	409.60	col, cub, 1.657	2.71			s	s	
p263	chlorotellurate	K_2TeCl_6	418.51	pale yel, octahedral				d	d	s HCl
p264	chromate	Nat. tarapacaite. K_2CrO_4	194.19	hel, rhomb, β 1.74	2.732^{18}	968.3		62.9^{20}	79.2^{100}	i al
p265	*di*chromate	$K_2Cr_2O_7$	294.18	red, monocl or tricl, 1.738	2.676^{25}_{4}	tricl→monocl 241.6 m.p. 398	d 500	4.9°	102^{100}	i al
p266	*peroxy*chromate	K_3CrO_8	297.29	brn-red, cub		d 170		sl s		i a, al, eth
p267	chromium sulfate	Potassium chromium alum. $K[Cr(SO_4)_2]\cdot 12H_2O$	499.39	vlt-ruby red, but, oct, 1.4814	1.826^{25}	89	$-10H_2O$, 100 $-12H_2O$, 400	24.39^{25}	50	i al; s dil a
p268	chromium chromate, basic	$K_2CrO_4\cdot 2[Cr(OH)\cdot CrO_4]$	564.18	vlt brn amorph powd	2.28^{14}	300		i		i al, acet a
p269	citrate	$K_3C_6H_5O_7\cdot H_2O$	324.41	wh cr.	1.98	d 230		167^{15}	199.7^{31}	sl s al; s glyc
p270	citrate, monobasic	$KH_2C_6H_5O_7$	230.22	wh cr powd				s		
p271	cobalt carbonate, hydrogen(ous)	$KHCO_3\cdot CoCO_3\cdot 4H_2O$	291.12	rose need				d		
p272	cobalt (II) cyanide	$K_4Co(CN)_6$	371.43	redsh-brn cr, deliq	2.039^{25}_{4}			v s	v s	d a; i al, eth, $CHCl_3$
p273	cobalt (III)cyanide	$K_3Co(CN)_6$	332.33	wh-yel, monocl pr	1.878^{25}_{4}			sl s	sl s	s dil HCL, dil HNO_3; sl s al
p274	cobaltinitrite	Fischer's salt. $K_3[Co(NO_2)_6]$	452.56	yel pr, cub				0.9^{17}	d	i al
p275	cobaltinitrite, hydrate	$K_3[Co(NO_2)_6]\cdot H_2O$	470.28	yel cr powd				i	s, d	s min a; sl s ac a; i al, eth
p276	cobaltinitrite, hydrate	$K_3[Co(NO_2)_6]\cdot 1\frac{1}{2}H_2O$	479.28	yel, tetr		d 200		0.089^{17}	sl s	i al, meth
p277	cobaltmalonate (II)	$K_2[Co(C_3H_2O_4)_2]$	341.22		2.234					
p278	cobalt sulfate (II)	$K_2SO_4\cdot CoSO_4\cdot 6H_2O$	437.34	red pr, monocl, 1.481, 1.487, 1.500	2.218			25.5°	108.4^{49}	

No.	Name	Synonyms and Formulae	Mol. wt.	Crystalline form, properties and index of refraction	Density or spec. gravity	Melting point, °C	Boiling point, °C	Solubility, in grams per 100 cc		
								Cold water	Hot water	Other solvents
p279	copperchloride	$KCl\cdot CuCl_2$	209.00	red need	2.86					
p280	cyanate	KOCN	81.12	col, tetrag	2.056^{20}	d 700—900		75^{25}	s	i al
p281	cyanide	KCN	65.12	col, cub, wh gran, deliq, very pois, 1.410	1.52^{16}	634.5		50	100	$0.88^{19.5}$ al; $4.91^{19.5}$ MeOH; s glyc
p282	cyanoargentate (I)	Potassium argentocyanide. $K[Ag(CN)_2]$	199.00	cub, 1.625	2.36			25^{20}	100	4.85% al; i a
p283	cyanoaurate	$K[Au(CN_2]$	288.10	col, rhomb	3.45			14.3	200	sl s al; i eth
p284	cyanoaurate (III)	$K[Au(CN)_4\cdot 1^1/_2H_2O$	367.16	col tabl		d 200		s	v s	s al
p285	cyanocadmate	$K_2[Cd(CN_4]$	294.68	col, cub	1.85			33	100^{100}	sl s al
p286	cyanochromate (III)	$K_3[Cr(CN)_6]$	325.40	yel, rhomb	1.71			30.9^{20}		i al
p287	cyanocobaltate (II)	$K_4[Co(CN)_6]$	371.43					s	s	i al, eth
p288	cyanocobaltate (III)	$K_3[CO(CN)_6]$	332.33	yel, monocl	1.906	d		s	s	i al; sl s NH_3
p289	cyanocuprate (I)	$K_3[Cu(CN)_4]$	284.91	col, rhbdr		d		v s		
p290	cyanomanganate (II)	$K_4[Mn(CN)_6]\cdot 3H_2O$	421.48	deep bl, tetr.				s	d	
p291	cyanomanganate (III)	$K_3[Mn(CN)_6]$	328.34	red, rhomb, 1.553, 1.555, 1.571 (Li)				s		
p292	cyanomercurate	$K_2[Hg(CN)_4]$	382.86	col, cr pois				s		s al
p293	cyanomolybdate	$K_4[Mo(CN)_8]\cdot 2H_2O$	496.51	yel, rhomb	2.337^{25}_{4} (anhyd)	$-H_2O$, 105-119		v s	v s	i eth; 0.0017^{20} al
p294	cyanonickelate (II)	$K_2[Ni(CN)_4]H_2O$	258.97	red-yel, monocl cr or powd	1.875^{11}	$-H_2O$, 100		s		d a
p295	cyanoosmite	$K_4[Os(CN)_6]\cdot 3H_2O$	556.75	col, hel, monocl, β 1.607		d		sl s	s	i al, eth
p296	cyanoplatinite	$K_2[Pt(CN)_4]\cdot 3H_2O$	431.39	col, yel, rhomb, blue fluor, deliq	2.455^{16}	$-3H_2O$, 100	d 400-600	sl s	s	sl s al, eth, H_2SO_4
p297	cyanotungstate (IV)	$K_4[W(CN)_8]\cdot 2H_2O$	584.42	lt yel-grn cr powd	1.989^{25}_{4} (anhydr)	$-2H_2O$, 115		130^{18}	s	i al, eth
p298	ethylsulfate	$KC_2H_5SO_4$	164.22	wh, monocl	1.843			s		s al
p299	ferricyanide	$K_3Fe(CN)_6$	329.25	red, monocl, 1.566, 1.569, 1,583	1.85^{25}	d		33^{4}	77.5^{100}	i al; s acet
p300	ferrocyanide	Yellow prussate of potash, $K_4Fe(CN)_6\cdot 3H_2O$	422.39	lem yel, monocl, β 1.577	1.85^{17}	$-3H_2O$, 70	d	27.8^{12} anhydr 14.5^{0}	$90.6^{96.3}$ anhydr 74^{98}	s acet; i al, eth, NH_2
p301	fluoberyllate	K_2BeF_4	163.20	col, rhomb		red heat		2^{20}	5.26^{100}	
p302	fluoborate	Nat. avogadrite, KBF_4	125.90	col, rhomb or cub, 1.324, 1.325, 1,325	2.498^{20}	d 350	d	0.44^{20}	6.27^{100}	sl s al, eth; i alk
p303	fluogermanate	K_2GeF_6	264.78	wh, hex		730	*ca* 835	0.542^{18}	2.58^{130}	
p304	fluomanganate (IV)	K_2MnF_6	247.13	yel, hex, tabl		d		d	d	s conc HCl
p305	fluoniobate, penta-	Potassium oxyniobate, $K_2NbOF_8\cdot H_2O$	300.11	col, monocl pl or leaf				7.69	s	s d conc HF
p306	fluorescein deriv	$K_2C_{20}H_{10}O_5$	408.49	yelsh-red powd.				s		
p307	fluoride	KF	58.10	col, cub deliq, 1.363	2.48	858	1505	92.3^{18}	v s	s HF, NH_2, i al
p308	fluoride, acid	KHF_2	78.10	col, cub, deliq	2.37	d *ca* 225	d	41^{21}	v s	S $KC_2H_3O_2$; i al
p309	fluoride, dihydrate,	$KF\cdot 2H_2)$	94.13	col, monco pr, deliq, 1.352	2.454	41	156	349.3^{18}	v s	s HF; i al
p310	*hex*fluorophosphate	KPF_6	184.06			*ca*575	d	9.3^{25}	20.6^{50}	
p311	fluorotungstate	$2KF\cdot WO_2F_2\cdot H_2O$	388.05	monocl.		$-H_2O$, red heat		6^{17}	s	
p312	fluosilicate	Nat. hieratite, K_2SiF_6	220.27	col, bub or hex 1.3991	hex 3.08 cub 2.665^{17}	d		$0.12^{17.5}$ 6.9^{19}	0.954^{100}	s HCl; i NH_3; v sl s al
p313	fluostannate	$K_2SnF_6\cdot H_2O$	328.89	monocl pr	3.053			3.7^{18}	33.3^{100}	i al, NH_3
p314	fluosulfonate	$KFSO_3$	138.15	short, thick pr		311	6.9^{19}			
p315	fluotantalate	K_2TaF_7	392.13	col, rhomb	4.56; 5.24			sl s, d		sl s HF
p316	fluotellurate, di-	$K_2TeO_3F_2\cdot 3H_2O$	345.84	micros oct, monocl		d		sl s	sl s	s HF
p317	fluothorate	$K_2ThF_6\cdot 4H_2O$	496.29	col		d		2.15	6.6	d a; i al
p318	fluotitanate	$K_2TiF_6\cdot H_2O$	258.08	col, monocl lust leaf		$-H_2O$, 32 m.p. 780	d	0.556^{0}	1.27^{21}	sl s min a; i NH_3
p319	fluozirconate	K_2ZrF_6	283.41	col, monocl, 1.466, 1.455	3.48			0.781^{2}	25^{100}	i NH_3
p320	formate	$KCHO_2$	84.12	col, rhomb deliq	1.91	167.5	d	331^{18}	657^{80}	s al; i eth
p321	gadolinium sulfate	$D_2SO_4\cdot Gd_2(SO_4)_3\cdot 2H_2O$	812.96	cr	3.503^{16}			s	s	s K_2SO_4
p322	gallium sulfate	$KGa(SO_4)_2\cdot 12H_2O$	517.12	col cr.	1.895			s		s a
p323	*di*germanate	$K_2Ge_2O_4$	303.37	wh cr.	$4.31^{21.6}$	>83		s		s a
p324	*meta*germanate	K_2GeO_2	198.78	wh cr.	$3.40^{21.5}$	823		s		s a
p325	*tetra*germanate	$K_2Ge_4O_9$	512.55	wh cr.	$4.12^{21.5}$	1083		s		s a
p326	glycerophosphate	$K_2C_3H_7PO_6$	248.26	colsl yelsh mass, hygr.				v s	v s	s al
p327	hydride	KH	40.11	wh need 1.453	1.47	d		d	d	i CS_2, eth, bz
p328	hydroxide	KOH	56.11	wh, rhomb deliq.	2.044	360.4 ±0.7	1320-1324	107^{15}	178^{100}	v s al; i eth, NH_2
p329	(tri-)hydroxylammine trisulfonate	$(KSO_3)_3\cdot NO\cdot 1\frac{1}{2}H_2O$	414	50col, nonocl pr		$-H_2O$, 100-200		4^{18}	sl d	
p330	*hexa*hydroxyplatinate	$K_2[Pt(OH)_6]$	375.32	yel, rhomb	5.18	d 160		s	s	i al
p331	imidolsulfonate	$(KSO_3)_2NH$	253.33	col, monocl	2.515	d 170-180	d 360-440, vac	1.3^{23}	d	i HNO_3
p332	iodate	KIO_3	214.00	col, monocl	3.93^{32}_{4}	560	d>100	4.74^{0}	32.3^{100}	s KI; i al, NH_3
p333	iodate, acid	$KIO_3\cdot HIO_3$	389.91	col monocl				1.33^{15}		i al

No.	Name	Synonyms and Formulae	Mol. wt.	Crystalline form, properties and index of refraction	Density or spec. gravity	Melting point, °C	Boiling point, °C	Solubility, in grams per 100 cc		
								Cold water	Hot water	Other solvents
p334	iodate, acid	$KIO_3 \cdot 2HIO_3$	565.82	col, tircl				4.15		
p335	*meta*periodate	KIO_4	230.00	col, tetr, 1.6205	3.618^{15}_{4}	582	−0, 300	0.66^{12}	s	v sl s KOH
p336	iodide	KI	166.00	col or wh, cub or gran, 1.677	3.13	681	1330	127.5^{6}	208^{100}	1.88^{25} al; 1.31^{25} acet; sl s eth; NH
p337	iodide, tri-	$KI_3 \cdot \frac{1}{2}H_2O$	428.82	dk bl, monocl,deliq	3.498	31	d 225	v s		s al, KI
p338	iodoaurate	$KAuI_4$	743.68	blk lust cr		d 150		s, d		s dil KI sol
p339	iodoridite	K_3IrI_6	1070.94	gr cr		d		i	i	i al
p340	iodomercurate (II) tetra-	K_2HgI_4 (or $KI \cdot HgI_2$)	786.40	yel cr, deliq				v s		i al
p341	iodomercurate (II) tri-	Potassium mercury iodide, $KHgI_3$ (or $KI \cdot HgI_2$)	620.40	yel pr, deliq		105		v s		341^{34} al; s KI sol, a eth
p342	iodoplatinate	K_2PtI_6	1034.70	blk, rect	4.96^{25}_{4}			s	s, d	i al
p343	iridium chloride	Potassium hexachloro iridate, K_2IrCl_4	483.13	redsh-blk, cub	3.549	d		1.12^{20}	s	i al
p344	iridium oxalate	$K_3[Ir(C_2O_4)_3] \cdot 4H_2O$	645.63	orange, tricl cr	2.510^{19}	$-H_2O$, 120	d 160	s	v s	i al, eth
p345	iron chloride (III)	Nat. erythrosiderite, $2KCl \cdot FeCl_3 \cdot H_2O$	329.32	red, orthorhomb	2.372					
p346	iron (III) oxalate	$K_2Fe(C_2O_4)_3 \cdot 3H_2O$	491,25	emerald grn, monocl, 1.5019, 1.5558, 1.5960	2.133^{20}_{4}	$-3H_2O$, 100	d 230	4.7^{0}	118^{100}	i al
p348	iron sulfate (III)	$KFe(SO_4)_3 \cdot 12H_2O$	599.30	vlt, cub oct, 1.452	1.83	33		$20^{12.6}$	v s	i al
p349	iron sulfate (III)	Nat. krausite. $K_2SO_4Fe_2(SO_4)_3 \cdot 24H_2O$	1006.49	pa yel-grn, monocl, 1.482	1.806	28	d 33			
p350	iron sulfate (II)	$K_2SO_4FeSO_4 \cdot 6H_2O$	434.25	grn pr, monocl, 1.476, 1,482, 1.497	2.169	d		s	s	
p351	iron sulfide	$KFeS_2$	159.07	purp. hex	2.563			d		
p352	lactate	$KC_3H_5O_3 \cdot xH_2O$						s		s al; i eth
p353	laurate	$KC_{12}H_{23}O_2$	238.41	amorph						4.5^{15} al
p354	laurate, acid (mixt)	$KC_{12}H_{23}O_2 \cdot C_{12}H_{24}O_2$	438.73	wh, wax like sol		160				$0.904^{13.5}$al
p355	lead chloride	Nat. pseudocotunnite, $2HCl \cdot PbCl_2$	427.21	yel		490		s		
p356	magnesium carbonate, hydrogen	$KHCO_3 \cdot MgCO_3 \cdot 4H_2O$	256.49	col, tricl or rhomb	2.98			d		
p357	magnesium chloride	Nat. carnalite $KCL \cdot MgCl_2 \cdot 6H_2O$	277.85	col, rhomb, deliq 1.466, 1.475, 1,494	1.61	265		64.5^{19}	d	d al
p358	magnesium chloride sulfate	Nat. kainite $[KMgCl(SO_4)]_4 \cdot 11H_2O$	977.82	col, monocl	2.131			79.56^{18}		i al, eth
p359	magnesium chromate	$K_2CrO_4 \cdot MgCrO_4 \cdot 2H_2O$	370.52	tricl	2.59					
p360	magnesium phosphate, hexahydrate	$KMgPO_4 \cdot 6H_2O$	266.47	wh, rhomb cr		$-5H_2O$, 110		d		
p361	magnesium selenate	$K_2SeO_4 \cdot MgSeO_4 \cdot 6H_2O$	496.51	col, monocl, 1.497, 1.499, 1,514	2.3645^{20}_{4}	$-2H_2O$, 33		s	s	
p362	magnesium sulfate	Nat. langbeinite. $K_2SO_4 \cdot 2MgSO_4$	414.98	tetrah	2.829	927				
p363	magnesium sulfate	Nat. leonite, $K_2SO_4 \cdot MgSO_4 \cdot 4H_2O$	366.68	col, monocl, 1.483, 1.487, 1,490	2.201^{20}			v s		
p364	magnesium sulfate	$K_2SO_4 \cdot MgSO_4 \cdot 6H_2O$	402.71	col, monocl 1.461, 1.463, 1.476	2.15	d 72		19.26^{0} 25^{20}	59.8^{75}	
p365	malate	$K_2C_4H_4O_4$	210.27	col, viscid mass				s		
p366	manganate	K_2MnO_4	197.13	grn, rhomb		d 190		d	d	s KOH
p367	*per*manganate	$KMnO_4$	158.03	purple rhomb, 1.59	2.703	d<240		6.38^{20}	25^{65}	d al; v s MeOH, acet; s H_2SO_4
p368	manganese chloride(ous)	Chloromanganokalite, $4KCl \cdot MnCl_2$	424.05	trig. 1.50	2.31			s	s	
p369	manganese sulfate(ic)	$KMn(SO_4)_2 \cdot 12H_2O$	502.33	vlt, cub, (oct)				d		
p370	manganese sulfate(ous)	Manganolongbeinite. $K_2SO_4 \cdot 2MnSO_4$	476.25	rose, tetrah, 1.572	3.02	850				
p371	mercury tartrate(ous)	$KHgC_4H_4O_6$	387.76	wh cr powd				i		i al
p372	methionate	Potassium methane disulfonate. $K_2CH_2(SO_3)_2$	252.34	monocl, β 1.539	2.376			s		
p373	methylsulfate	$2KCH_3SO_4 \cdot H_2O$	318.40	wh cr.				s		s al
p374	molybdate	K_2MoO_4	238.13	wh powd or 4-sid pr, deliq	2.91^{18}	919	d 1400	184.6^{25}	v s	i al
p375	*per*molybdate	$K_2O \cdot 3MoO_4 \cdot 3H_2O$	628.05	lt yel cr, monocl		d 180		sl s	s	v sl s al
p376	*tri*molybdate	$K_2O \cdot 3MoO_3 \cdot 3H_2O$	580.06	wh need		571 (anhydr)		0.22^{15}	s	
p377	molybdenum cyanate	$K_4[Mo(CN)_8] \cdot 2H_2O$	496.51	yel, rhomb	2.337^{25}_{4} (anhydr)	$-2H_2O$ 105-110		v s	v s	0.0017^{20} abs al; i eth; d HCl, H_2SO_4
p378	myristate, acid (mixt)	$KC_{14}H_{27}O_2 \cdot C_{14}H_{28}O_2$	494.84	wh, waxlike sol		153				$0.453^{13.5}$ al
p379	naphthalene -1,5-disulfonate	$K_2C_{10}H_6(SO_3)_2 \cdot 2H_2O$	400.50	monocl, 1.485, 1.669, 1.697	1.797			s		
p380	nickelsulfate	$K_2SO_4 \cdot NiSO_4 \cdot 6H_2O$	437.09	bl, monocl, 1.484, 1.492, 1.505	2.124	d < 100		7^{0}	60.8^{75}	
p381	nitrate	Saltpeter, KNO_3	101.10	col, rhomb or trig 1.335, 1.5056, 1.5064	2.109^{16}	tr-trig 129 m.p. 334	d 400	13.3^{0}	247^{100}	i dil al, eth; s liq NH_3, glyc
p382	nitride	K_3N	131.30	grnsh-blk		d		d		

No.	Name	Synonyms and Formulae	Mol. wt.	Crystalline form, properties and index of refraction	Density or spec. gravity	Melting point, °C	Boiling point, °C	Solubility, in grams per 100 cc		
								Cold water	Hot water	Other solvents
p383	nitrite	KNO_3	85.10	wh-yelsh pr, deliq	1.915	440	d	281^{0}	413^{100}	s hot al; v s liq NH_3
p384	*m*-nitrophenoxide	$KOC_4H_4NO_2 \cdot 2H_2O$	213.23	flat or need	1.691^{20}	$-H_2O$, 130	d	16.3^{15}		s al
p385	*p*-nitrophenoxide	$KOC_6H_4NO_2 \cdot 2H_2O$	213.23	yel leaf	1.652^{20}	$-2H_2O$, 130	d	7.5^{15}		sl s al
p386	nitroplatinite	$K_2Pt(NO_2)_4$	457.30	col, monocl			d	3.8^{15}	s	
p387	nitroprusside	$K_2[Fe(NO)(CN)_5]2H_2O$	330.17	red, monocl, hydr				100^{16}		s al
p388	nitrososulfate	$K_2SO_3(NO)^2$	218.27	col need		d 127, espl		$12^{14.5}$, d		i al
p389	oleate	$KC_{18}H_{33}O_2$	320.56	yelsh or brnsh soft mass or cr, 1.452				25	s	$4.315^{13.5}$ al; 100^{50} al; 3.5^{35} eth
p390	oleate, acid (mixt)	$KC_{18}H_{32}O_2 \cdot C_{18}H_{34}O_2$	603.02	wh, wax-like solid		95		s	s	$5.2^{13.5}$ al
p391	osmate	$K_2OsO_4 \cdot 2H_2O$	367.42	vlt, cub, hybr.		$-H_2O$, > 100		sl s	s, d	i al, eth
p392	osmiumchloride	K_2OsCl_6	481.11	blk, oct	3.42^{16}	d 600		s	s	i al; s HCl
p393	osmylchloride	$K_2OsO_2Cl_4$	442.21	red, tetr	3.42	d 200 (in H atm)		s; s; i al; s HCl		
p394	osmyloxalate	$K_2[OsO_2(C_2O_4)_2] \cdot 2H_2O$	512.47	brn need, tricl		$-H_2O$, 80	d 180	0.75^{15}	3.0^{15}	
p395	oxalate	$K_2C_2O_4 \cdot H_2O$	184.23	wh, monocl, 1.440, 1.485, 1.550	$2.127^{3.9}$	$-H_2O$, 100			33^{16}	
p396	oxalate, hydrogen	KHC_2O_4	128.13	col, monocl, 1.382, 1.553, 1.573	2.044	d		2.5	16.7^{100}	i al, eth
p396	oxalate, hydrogen, monohydrate	$KHC_2O_4 \cdot H_2O$	146.14	rhomb	$2.044^{3.9}$					
p398	oxalate, tetra-	$KHC_2O_4 \cdot H_2C_2O_4 \cdot 2H_2O$	254.19	col, tricl, 1.415, 1.536, 1.560	1.836	d		1.8^{13}		d al
p399	oxaloferrate (II)	$K_2 \cdot Fe(C_2O_4)_3] \cdot 2H_2O$	346.11	gold need		d		s	s	
p400	oxaloferrate (II)	$K[Fe(C_2O_4)_2] \cdot 2\frac{1}{2}H_2O$	316.02	br cr		d		92^{21}	d	i al
p401	oxaloferrate (III)	$K_2[Fe(C_2O_4)_3] \cdot 3H_2O$	491.25	grn, monocl.		$-3H_2O$, 100	d 230	4.7^{0}	117.7^{100}	i al, NH_3; s acet
p402	oxalatoplatinate	$K_2[Pt(C_2O_4)_2] \cdot 2H_2O$	485.35	col, monocl pr	3.04^{12}	$-H_2O$, 100		sl s	s	
p403	oxalatouranate (IV)	$K_4[U(C_2O_4)_4] \cdot 5H_2O$	836.58	yel, monocl	2.563					
p404	oxide, mon-	K_2O	94.20	col, cub, hygr	2.32^{0}	d 350		v s	v s	s al eth
p405	*per*oxide	K_2O_2	110.20	wh, amorph, deliq		490	d			
p406	oxide, super-	KO_2	71.10	yel, cub leaf	2.14	380	d	v s, d		d al
p407	oxide, tri- (sesqui-)	K_2O_3	126.20	red		430		ev O_2		d dil H_2SO_4
p408	palladium chloride	$K_2(PdCl_4)$	326.43	yel-grnsh-br cr tetr, 1.710, 1.523	2.67	524		s	s	sl s^{80} al
p409	palladium chloride	$K_2(PdCl_4)$	397.33	lt red, oct	2.738	d 170		v sl s	sl s	s HCl; i al
p410	palladium oxalate	$K_2[Pd(C_2O_4)_3] \cdot 4H_2O$	432.72	yel need		dec in air, $-4H_2O$, 80		0.833^{27}	$9.98^{32.9}$	
p411	palmitate, acid (mixt)	$KC_{16}H_{31}O_2 \cdot C_{16}H_{32}O_2$	550.95	wh, fatty sol		138				0.198^{13} al
p412	1-phenol-2-sulfonate (*o*-)	$KC_6H_4(OH)SO_3 \cdot H_2O$	230.28	rhomb, 1.527, 1.568, 1.647	1.87	400		s		s al
p413	1-phenol-4-sulfonate(*p*-)	$KC_6H_4(OH)SO_3$	212.26	rhomb, 1.571, 1.608, 1,694	1.87	>260				
p414	phenylsulfate	$KC_6H_5SO_4$	212.26	rhomb leaf		d 150-160	d	14^{15}		v sl s al
p415	*meta*phosphate, hexa-	$(KPO_3)_6$	708.42	col mass hygr	1.207	810	1320	s	s	i al
p416	*meta*phosphate, tetra-	$(KPO_3)_4 \cdot 2H_2O$	508.31	col cr.		$-2H_2O$, 100		100^{15}	s	
p417	*ortho*phosphate	K_3PO_4	212.27	col, rhomb, deliq	2.564^{17}	1340		90^{20}	s	i al
p418	*ortho*phosphate, di-H	KH_2PO_4	136.09	col, tetr, deliq, 1.510, 1.4864	2.338	252.6		33^{25}	83.5^{20}	i al
p419	*ortho*phosphate, mono-H	K_2HPO_4	174.18	wh, amorph, deliq		d		167^{20}	v s	v s al
p420	*pyro*phosphate	$K_4P_2O_7 \cdot 3H_20$	384.38	col, deliq	2.33	$-2H_2O$, 180	$-3H_2O$, 300	s	v s	i al
p421	*sub*phosphate	$K_2PO_3 \cdot 4H_2O$	229.23	col, rhomb		40	$1-4H_2O$, 150	v s	v s	
p422	hypophosphite	KH_2PO_2	104.09	wh, hex, deliq		d		200^{25}	330	v sl s abs al, NH_2; i eth; 11.1^{25} chl
p423	orthophosphite, di-H	KH_2PO_3	120.09	wh cr, deliq.		d		220^{20}	v s	i al
p424	phthalate, hydrogen	$KHC_8H_4O_4$	204.44	col, rhomb	1.636			10^{25}	33^{100}	
p425	picrate	$KC_6H_2N_3O_7$	267.20	yel-rdsh or grnsh, rhomb, 1.527, 1.903, 1.952	1.852	expl 310		0.5^{15}	25^{100}	0.184^{25} al
p426	Piperate	$KC_{12}H_9O_4$	256.30	lt yel cr powd				sl s	v s	
p427	platinate, hydroxo-	$K_2Pt(OH)_6$	375.32	yel, rhomb		d 160		s		i al
p428	platinorhodanide	$K_2[Pt(CNS)_6] \cdot 2H_2O$	657.77	red, rhomb				s	8^{60}	s h al
p429	platinum iodide, hexa-	K_2PtI_6	1034.70	blk, cub	4.963_{4}^{25}			s		sl s al
p430	plumbate, hydroxo-	$K_2Pb(OH)_6$	387.44	col, rhomb				d	d	s KOH
p431	praseodymium sulfate	$3K_2SO_4 \cdot Pr_2(SO_4)_3 \cdot H_2O$	1110.77	cr	3.275^{16}			sl s		s HCl, HNO_3
p432	propionate	$KC_3H_5O_2 \cdot H_2O$	130.19	wh cr, hygr, or wh leaf, deliq		$-H_2O$, 120		207^{16}	359	22.2^{13} 95% al
p433	propyl sulfate	$KC_3H_7SO_4$	178.24	wh cr powd				v s		
p434	*per*rhenate	$KReO_4$	289.30	wh, tetr, 1.643	4.887	550	1360-1370	1.21^{20}	14.0^{100}	v sl s al
p435	rhenium (IV) chloride	$K_2[Recl_6]$	477.12	yel-grn cr, oct	3.34	d		s	s	s a; i conc H_2SO_4; d h H_2SO_4
p437	rhenium (V) oxychloride	$K_2[ReOCl_4]$	457.67	yel-grn cr, rhomb or monocl, 1.52				s		sl s HCl; s H_2SO_4; i al, eth
p438	rhenium oxycyanide	$K_3[ReO_2(CN)_4]$	439.57	red cr, monocl	2.70_{4}^{25}	d 300-400, vac		s	s	v sl s al; i alk
p439	rhodium cyanide	$K_3[Rh(CH)_6]$	376.31	pa yel, monocl, 1.5498, 1.5513, 1.5634				v s	v s	

No.	Name	Synonyms and Formulae	Mol. wt.	Crystalline form, properties and index of refraction	Density or spec. gravity	Melting point, °C	Boiling point, °C	Solubility, in grams per 100 cc		
								Cold water	Hot water	Other solvents
p440	rhodium oxalate	$K_3[Rh(C_2O_4)_3]\cdot4\frac{1}{2}H_2O$	565.33	col, tricl	2.171_4^{20}	$-4\frac{1}{2}H_2O$, 190		v s	v s	
p441	rhodium sulfate	$KRh(SO_4)_3\cdot12H_2O$	646.36	yel, cub	2.23			s		
p442	ruthenate	$K_2RuO_6\cdot H_2O$	261.28	blk, tetr		$-H_2O$, 200	d 400 vac	v s	d	d a, al
p443	perruthenate	$KRuO_4$	204.17	blk, tetr		d 44		sl s	s, d	
p444	d-saccharate, acid	$KHC_6H_8O_8$	248.23	rhomb need				1.1^{6}	s	
p445	salicylate	$KC_7H_5O_3$	176.21	wh powd				s		s al
p446	santoninate	$KC_{15}H_{19}O_4$	302.41	wh cr powd, deliq				s		s al
p447	selenate	K_2SeO_4	221.15	col, rhomb, hygr, 1.535, 1.539, 1.545	3.066			110.5^{0}	122.2^{100}	
p448	selenide	K_2Se	157.16	wh cub, reddens in air, hygr	2.851^{15}			s d	s	
p449	selenite	K_2SeO_3	205.15	wh, deliq		d 875		s		sl s al
p450	selenocyanate	$KSeCN$	144.08	need, deliq	2.347	d 100		s	s	d a; s al
p451	selenocyanoplatinate	$K_2Pt(SeCN)_6$	903.14	rhomb	$3.378^{12.5}$	d 80				
p452	selenothionate	$K_2SeS_2O_4$	317.27	col, monocl pr		d 250		s, d		
p453	*di*silicate	$K_2Si_2O_5$	214.36	col, rhomb, 1.502, 1.513	2.456_4^{25}	1015 ± 10				
p454	*meta*silicate	K_2SiO_3	154.28	col, rhomb (?), 1.520, 1.528		976		s	s	i al
p455	*tetra*silicate	$K_2Si_4O_9\cdot H_2O$	352.55	wh, rhomb, α 1.495, β 1.535	2.417	d 400		s	s	i al
p456	*di*silicate, hydrogen	$KHSi_2O_5$	176.27	wh, rhomb, 1.480, 1.530	2.417_4^{15}	515				d HCl
p457	silicotungstate	$K_4SiW_{12}O_{40}\cdot18H_2O$	3354.93	col, hex		$-17H_2O$, 100		33.3^{20}	v s	v s acet; a MeOH; sl s al; eth, bz
p458	silver carbonate	$KAgCO_3$	206.98	rect pl	3.769	d		d	d	
p459	silver nitrate	$KNO_3\cdot AgNO_3$	270.98	monocl.	3.219	125		v s	v s	
p460	sodium antimony tartrate	$KNaSbC_4H_3O_7$	346.90	wh, scales or powd				s		
p461	sodium carbonate	$KNaCO_3\cdot6H_2O$	230.19	monocl, hygr. eff	$1.61\text{-}1.63^{14}$	$-6H_2O$, 100		185.2^{15}		
p462	sodium ferricyanide	$K_2Na[Fe(CN)_6]$	313.14	or-red, monocl		d		50^{25}	80^{80}	
p463	sodium nitrocobaltate (III)	$K_2Na[Co(NO_2)_6]\cdot H_2O$	454.17	yel cr.	1.633	135		0.07^{25}		i al
p464	sodium sulfate	$3K_2SO_4\cdot Na_2SO_4$	664.80	col, rhbdr	2.7			s	s	
p465	sodium tartrate	Rochelle salt, seignette salt, $KNaC_4H_4O_6\cdot4H_2O$	282.22	col, rhomb, 1.492, 1.493, 1.496	1.790	70-80	$-4H_2O$, 215	26^{0}	66^{26}	v sl s al
p466	sorbate	$KC_6H_7O_2$	150.22	col cr.	1.363_{20}^{25}	d 270		58.5^{25}		sl s MeOH
p467	stannate, hydroxo-	$K_2Sn(OH)_6$	298.93	col, trig	3.197			85^{10}	110.5^{20}	sl s KOH; i al, acet
p468	stearate	$KC_{18}H_{35}O_2$	322.57	wh cr powd					s	$0.145^{13.5}$ al; i eth, chl, CS_2
p469	stearate, acid (mixt)	$KC_{18}H_{35}O_2\cdot C_{18}H_{36}O_2$	607.06	wh powd		153		s	s	$0.091^{13.5}$ al
p470	strontium chromium oxalate(ic)	$KSrCr(C_2O_4)_3\cdot6H_2O$	550.86	grnsh blk cr.	2.155^{13}					
p471	styphnate	$KC_6H_2N_3O_8\cdot H_2O$	301.21	yel, monocl pr		$-H_2O$, 120	expl	1.54^{30}		v sl s al
p472	succinate	$K_2C_4H_4O_4\cdot3H_2O$	248.32	rhomb, hygr	1.564			s		
p473	succinate, hydrogen	$KHC_4H_4O_4$	156.18	monocl.	1.767	d 242				
p474	succinate, hydrogen, dihydrate	$KHC_4H_4O_4\cdot2H_2O$	192.21	rhomb, 1.417, 1.530, 1.533	1.616			s		s al
p475	succinate, hydrogen	$KHC_4H_4O_4\cdot C_4H_6O_4$	274.27	monocl.	1.56	162				
p476	sulfate	Nat. arcanite, K_2SO_4	174.25	col, rhomb or hex 1.494, 1.495, 1.497	2.662	tr 558 m. p. 1069	1689	12^{25}	24.1^{100}	i al, acet, CS_2
p477	*peroxydi*sulfate	$K_2S_2O_8$	270,31	col, tricl, 1.461, 1.467, 1.566	2.477	d<100		1.75^{0}	5.2^{20}	i al
p478	*pyro*sulfate	$K_2S_2O_7$	254.31	col need	2.512_4^{25}	>300	d	s	d	
p479	sulfate, hydrogen	Nat. mercallite, misenite, $KHSO_4$	136.16	col, rhomb, deliq 1.480	2.322	214	d	36.3^{0}	121.6^{100}	i al, acet
p480	sulfide, di-	K_2S_2	142.32	red-yel cr		470		s	d	s al
p481	sulfide, hydro-	KHS	72.17	yel, rhomb deliq	1.68-1.70	455		d	d	s al
p482	sulfide, mono-	K_2S	110.26	yel-br, cub, deliq	1.805^{14}	840		s	v s	s al, glyc; i eth
p483	sulfide, mono-, pentahydrate	$K_2S\cdot5H_2O$	200.33	col, rhomb		60	$-3H_2O$, 150	s		s al, glycl i eth
p484	sulfide, penta-	K_2S_5	238.50	or cr, hygr		206	d 300	v s	v s	sl s al
p485	sulfide, tetra-	K_2S_4	206.44	red-brn cr		145	d 850	s		s al
p486	sulfide, tetra-dihydrate	$K_2S_4\cdot2H_2O$	242.47	yel				s	s	sl s al
p487	sulfide, tri-	K_2S_3	174.38	br-yel cr		252		s	d	s al
p488	sulfide, di-, trihydrate	$K_2S_2\cdot3H_2O$	196.36	yel				v s	v s.	s al
p489	sulfite	$K_2SO_3\cdot2H_2O$	194.29	wh-yelsh, hex		d		100	<100	sl s al; i NH_3; d dil a
p490	*pyro*sulfite	Potassium *meta*sulphite $K_2S_2O_4$	222.31	col, monocl pl	2.34	d 190		sl s		sl s al; i eth
p491	sulfite, hydrogen	$KHSO_2$	120.16	col cr.		d 190		s	s	i al
p492	*d*-tartrate	$K_2C_4H_4O_6\cdot\frac{1}{2}H_2O$	235.28	col, monocl, β 1.526	1.98_4^{20}	$-H_2O$, 155	d 200-220	150^{14}	278^{100}	sl s al
p493	*dl*-tartrate	$K_2C_4H_4O_6\cdot2H_2O$	262.30	col, monocl.	1.984	$-2H_2O$, 100		100^{25}		
p494	*d*-tartrate, hydrogen	$KHC_4H_4O_6$	188.18	col, rhomb, 1.511, 1.550, 1,590	1.984^{18}			0.37	6.1^{100}	s a, alk; i al, ac a
p495	*dl*-tartrate, hydrogen;.	$KHC_4H_4O_6$	188.10	col, monocl.	1.954			0.42^{15}	7.0^{100}	i al; s min a
p496	*meta*tellurate	K_2TeO_4	269.79	soft glutinous mass		d 200		d		
p497	*ortho*tellurate	$K_2H_4TeO_6\cdot3H_2O$	359.87	col, rhomb, deliq		$-H_2O$	$-O$, 300	sl s	s	i al; sl s KOH
p498	telluride	K_2Te	205.80	col, cub, hygr	2.51			s, d	s	

PHYSICAL CONSTANTS OF INORGANIC COMPOUNDS (continued)

No.	Name	Synonyms and Formulae	Mol. wt.	Crystalline form, properties and index of refraction	Density or spec. gravity	Melting point, °C	Boiling point, °C	Solubility, in grams per 100 cc		
								Cold water	Hot water	Other solvents
p499	tellurite	K_2TeO_3	253.79	wh cr, deliq.		d 460-470		v s	v s	s hK_2CO_3, KOH
p500	tellurium chloride	K_2TeCl_6	418.51	yel, oct, hygr.	2.645			s d		s, d al; s dil HCl
p501	thioantimonate	$2K_3SbS_4{\cdot}4\frac{1}{2}H_2O$	815.64	yel cr, deliq.				300^{0}	400^{80}	i al
p502	thioarsenate	K_2AsS_4	320.46	wh cr, deliq.			d	v s		i al
p503	thioarsenite	K_3AsS_3	288.40			d		s		i al
p504	thiocarbonate, tri-	K_2CS_3	186.39	yel-red-brn cr, deliq		d		v s	s	s NH_3; sl s al; i eth
p505	thiocyanate	KNCS	97.18	col, rhomb pr, deliq	1.886^{14}	173.2	d 500	177.2^{0}	217^{20}	s al; 20.75^{22} acet; 0.18^{13} amyl al; v s liq NH_3
p506	*di*thionate	$K_2S_2O_6$	238.31	col, trig, 1.455, 1.515	2.278	d		6	66^{100}	i al
p507	*penta*thionate	$K_2S_4O_6{\cdot}1^1/_2H_2O$	361.52	col, rhomb. -1.63	2.112	d		s	d	i al
p508	*tetra*thionate	$K_2S_4O_6$	302.43	col, monocl, 1.6057	2.296			v s		i al
p509	*tri*thionate	$K_2S_3O_6$	207.37	col, rhomb, 1.475, 1.480, 1.487	2.304	d 30-40		s	d	i al
p510	thioplatinate	$K_2Pt_4S_4$	1050.88	bl gray cr	6.44^{15}	d, ign		i		d HCl
p511	thiosulfate, penta-hydrate	$3K_2S_2O_3{\cdot}5H_2O$	661.02	col, rhomb		d		$150.2^{17.2}$		
p512	*meta*thiostannate	$K_2SnS_3{\cdot}3H_2O$	347.11	dk brn oil	1.847^{18}	$-3H_2O$, 100		s		i al
p513	thiosulfate	$K_2S_2O_3{\cdot}{}^1/_3H_2O$	196.32	col, monocl, deliq	2.590 anhydr 2.23	$-H_2O$, 200	d	96.1^{0}	312^{90}	i al
p514	tungstate	$K_2WO_4{\cdot}2H_2O$	362.07	col, monocl, deliq	3.113	tr 388, 921		51.5	151.5	d a; i al
p515	*meta*tungstate	$K_6[H_2W_{12}O_{40}]{\cdot}18H_2O$	3407.06	hex		*ca*930		s	v s	d a
p516	*meta*tungstate	K_2UO_4	380.22	or-yel, rhomb.					i	v s a
p517	uranyl acetate	$KUO_2(C_2H_3O_2)_3{\cdot}H_2O$	504.28	tetr	3.296^{15} ($^1/_2H_2O$)	$-H_2O$), 275		s		
p518	uranyl carbonate	$2K_2CO_3{\cdot}UO_2CO_3$	606.45	yel, hex		$-CO_2$		7.4^{15}	d	s K_2CO_3 sol; i a
p519	uranyl sulfate	$K_2SO_4{\cdot}UO_2SO_4{\cdot}2H_2O$	576.37	yel, monocl	$3.363^{19.1}$	$-2H_2O$, 120		s		
p520	urate, acid	$KHC_5H_2N_4O_3$	206.20	wh powd				sl s		
p521	*meta*vanadate	KVO_3	138.04	col cr.				sl s	s	sl s KOH; i al
p522	vanadium sulfate	Potassium vanadium alum $KV(SO_4)_2{\cdot}12H_2O$	498.34	violet cubic	1.783^{20}_{4}	20	$-12H_2O$, 230	1984^{10}		
p523	*ethyl*xanthate	KC_2H_5OCSS	160.29	wh to pa yel cr or cr powd	$1.558^{21.5}$	d>200		v s	d	s al; i eth
p524	**Praseodymium** (α form)	Pr	140.9077	pa yel, met, hex up to 798	6.773	931	3512	d		s a
p525	(β form)	Pr	140.9077	cub	6.64	935	3127	d		s a
p526	acetate	$Pr(C_2H_3O_2)_3{\cdot}3H_2)$	372.09	grn need				v s		
p527	acetylacetonate	$Pr(C_5H_7O_2)_3$	438.24	cr ppt		146				s CS_2
p528	bromate	$Pr(BrO_3)_3{\cdot}9H_2O$	686.75	grn, hex		56.5	$-7H_2O$, 170	196^{25}		
p529	bromide	$PrBr_3$	380.62	grn cr powd.		691	1547	d, sl s		
p530	carbide	PrC_2	164.93	yel cr.	5.10	d		d	d	s dil a
p531	carbonate	$Pr_2(CO_3)_3{\cdot}8H_2O$	605.97	grn silky pl		$-6H_2O$, 100		i		s a
p532	chloride	$PrCl_2$	247.27	br grn need	4.02^{25}	786	1700	103.9^{12}	∞^{100}	v s al; 2.4 pyr; i chl eth
p533	chloride, heptahydrate	$PrCl_3{\cdot}7H_2O$	373.37	grn, tricl.	2.25^{17}	115		334^{13}	∞^{100}	s al, HCl
p534	hexantipyrine *per*chlorate	$[Pr(C_{11}H_{12}N_2O)_6]{\cdot}(ClO_4)_3$	1568.63	grn hex leaf.		d 286-291				
p535	iodide	PrI_3	521.62	gr cr, hygr		737		v s		
p536	molybdate	$Pr_2(MoO_4)_3$	761.63	grass-green tetr.	4.84	1030				
p537	oxalate	$Pr_2(C_2O_4)_3{\cdot}10H_2O$	726.03	lt grn cr				i		s a
p538	oxide, di-	PrO_2	172.91	br-bl powd	6.82	>350 tr to Pr_6O_{10}				
p539	oxide, sesqui-	Preseodymia, Pr_2O_3	329.81	yel-grn, amorph	7.07	d		0.000020^{29}		s a
p540	selenate	$Pr_2(Se)_4)_3$	710.69		4.30^{15}			36^{0}	3^{92}	
p541	sulfate	$Pr_2(SO_4)_3$	569.99	lt grn powd	3.72^{16}			23.7^{0} 17.7^{20}	1.02^{96}	
p542	sulfate, octahydrate	$Pr_2(SO_4)_3{\cdot}8H_2O$	714.11	grn, monocl, 1.540, 1.549, 1,561	$2.827^{13.3}$			17.4^{20}	sl s	
p543	sulfate, pentahydrate	$Pr_2(SO_4)_3{\cdot}5H_2O$	660.06	monocl pr	3.176^{16}				1.85^{3}	
p544	sulfide	Pr_2S_3	378.00	br powd	5.042^{11}	d		o	d	s dil a
p545	**Protactinium**	Pa	231.0359	gray met, tetrag	15.37	<1600				
p546	chloride	$PaCl_4$	372.85	yel-grn, tetrag		subl 400 in vac		s		s dil HCl
p547	fluoride	PaF_4	307.03	monocl.				i		
p548	oxide, di-	PaO_2	263.03	blk, cub						
p549	oxide, pent-	Pa_2O_5	542.07	wh, cub						s dil HF
r1	**Radium**	Ra	226.0254	silver wh met.	5(?)	700	<1140	d, ev H_2		d a
r2	bromide	$RaBr_2$	385.83	col-yelsh, moncl	5.79	728	subl 900	s	s	s al
r3	bromide dihydrate	$RaBr_2{\cdot}2H_2O$	421.86	wh, monocl		$-2H_2O$, 100		s	s	
r4	carbonate	$RaCO_3$	286.03	wh, or sl brnsh, monocl				i		d a
r5	chloride	$RaCl_2$	296.93	col-yelsh, monocl	4.91	1000		s	s	s al
r6	chloride, dihydrate	$RaCl_2{\cdot}2H_2O$	332.96	wh, monocl, discol		$-2H_2O$, 100		s	s	s HCl
r7	iodate	$Ra(IO_3)_2$	575.83					0.0175^{0}	0.170^{100}	
r8	nitrate	$Ra(NO_3)_2$	350.04	cr				13.9^{20}		
r9	sulfate	$RaSO_4$	322.08	col, rhomb				0.000002^{25}	0.000005^{45}	i a
r10	**Radon**	Niton, Radium emanation, Rn	(222)	col gas, opaque cr	gas 9.73 g/ℓ liq 4.4^{-62} sol 4.0	−71	−61.8	51.0^{0} cm^3 22.4^{25} cm^3	13.0^{60} cm^3	sl s al, org liqu

No.	Name	Synonyms and Formulae	Mol. wt.	Crystalline form, properties and index of refraction	Density or spec. gravity	Melting point, °C	Boiling point, °C	Solubility, in grams per 100 cc: Cold water	Hot water	Other solvents
r11	**Rhenium**	Re	186.207	met lust, hex	20.53	3180	5627 (est)	i	i	s dil HNO_3, H_2O_2; sl s H_2SO_4; i HCl
r12	bromide	$ReBr_3$	425.92	grn-blk cr		subl 500, vac				s dil H_2SO_4, HBr, liq NH_3
r13	carbonyl-penta-	$[Re(CO)_5]_2$	652.52	col. cub cr		d 250				v sl s org solv
r14	chloride, penta-	$ReCl_5$	363.47	dr grn to blk	4.9	d	d	d	d	S HCl, alk
r15	chloride, tetra-	$ReCl_4$	328.02	bok (exist. ?)			500	s d	s d	s HCl
r16	chloride, tri-	ReC_3	292.57	dk red, hex			>500	s	s	s a, alk, liq NH_3, al; sl s eth
r17	fluoride	ReF_4	262.20	de grn	5.383^{26}	124.5	d 500	d		s a
r18	fluoride, hexa-	ReF_6	300.20	pa yel, v hygr	liq 6.1573, sld $3.616^{18.8}$	18.8	47.6	2, d	s, d	d HNO_3, H_2SO_4
r19	iodide pentacarbonyl	ReI.5CO	453.16	yel, rhombd.		200	d 400, subl vac 90	i		s bz
r20	oxide,di-	ReO_2	218.21	blk	11.4_4^{25}	d 1000		i	i	s conc HCl, H_2O_2
r21	oxide, hept-	Re_2O_7	484.41	yel pl or hex or powd, hygr	6.103	ca 297	subl 250	v s	v s	v s al; s alk, a
r22	oxide, per-	Re_2O_3(?)	500.41	wh	8.4	145		v s	v s	s alk; sl s eth
r23	oxide, sesqui-	Dirhenium trioxide $Re_2O_3{\cdot}xH_2O$		unstable			d, ev H_2			
r24	oxide, tri-	ReO_3	234.21	red, blue, cub	6.9-7.4		d 400	i	i	s H_2O_2, HNO_3
r25	oxybromide, tri-	ReO_3Br	314.11	wh		39.5	163			
r26	oxychloride, tri-	ReO_3Cl	296.66	col liq	3.867_4^{20}	4.5	131^{760}	d	d	s CCl_4
r27	oxy*tetra*chloride	$ReOCl_4$	344.02	or need		29.3	223.00	d	d	
r28	oxy*tetra*fluoride	$ReOF_4$	278.20	wh	lq 3.717 sol 4.032	39.7	62.7			
r29	oxy*di*fluoride, di-	ReO_2F_2	256.20	col		156				
r30	sulfide, di-	ReS_2	250.33	blk tr leaf	7.506_4^{20}		d	i	i	s HNO_2; i al, alk, HCl
r31	sulfide, hepta-	Re_2S_7	596.83	blk powd	4.866		d	i	i	s HNO_3, H_2O_2, alk; i CHl
r32	**Rhodium**	Rh	102.9055	gray-wh, cub	12.4	1966 ± 3	3727 ± 100	i	i	s H_2SO_4 + HCl, conc H_2SO_4; sl s a, aq reg
r33	amminechloride, hexa-	$[Rh(NH_3)_4]Cl_3$	311.45	rhomb pl	2.008_5^{25}	$-NH_3$ 210, d		12.5^8	s	
r34	carbonylchloride, basic	$RhCl_2{\cdot}RhO{\cdot}3CO$	376.75	ruby red need.		subl 125.5		sl s	d	s CCl_4, ac a, bz
r35	chloride, tri-	$RhCl_3$	209.26	br-red powd, deliq		d 450-500	subl 800	i	i	a a, aq reg
r36	chloride, tri-	$RhCl_3{\cdot}xH_2O$		dk red, deliq		d 100		v s		s al, HCl; i eth
r37	fluoride, tri-	RhF_3	159.90	red, rhomb	5.38	>600 subl		i	i	i a, alk
r38	iodide, tri-	RhI_3	483.62	blk				i		
r39	nitrate	$Rh(NO_3)_3{\cdot}2H_2O$	324.95	red, deliq				s	s	i al
r40	oxide, di-	RhO_2	134.90	br				i	i	i a, alk
r41	oxide, di-, dihydrate	$RhO_2{\cdot}2H_2O$	170.93	olive grn.		d		i		s HCl, acet a, alk
r42	oxide, sesqui-	Rh_2O_3	253.81	gray cr or amorph	8.20	d 1100-1150		i	i	i a, aq reg, KOH
r43	oxide, sesqui-, pentahydrate	$Rh_2O_3{\cdot}5H_2O$	343.89	yel ppt		d		i	s	s a
r44	sulfate	$Rh_2(SO_4)_3{\cdot}4H_2O$	566.04	red		d		s	s	
r45	sulfate	$Rh_2(S)_4)_3{\cdot}12H_2O$	710.17	lt yel cr				v s	d	i al
r46	sulfate	$Rh_2(SO_4)_3{\cdot}15H_2O$	764.21	pa yel cr.		d		v s	d	i al, eth
r47	sulfide, hydro-	$Rh(HS)_3$	202.11	blk		d		i	d	s aq reg, aq Br; i NA_2S
r48	sulfide, mono-	RhS	134.97	gray-blk cr		d		i	i	i a, aq reg
r49	sulfide, sesqui-	Rh_2S_3	302.00	blk	6.40_4^{25}	d		i	i	i a, aq reg, aq Br
r50	sulfite	$Rh_2(SO_3)_3{\cdot}6H_2O$	554.08	yel cr.		d		s		i al
r51	**Rubidium**	Rb	85.4678 ± 3	soft, silver wh met	1.532 liq: $1.475^{36.5}$	38.89	686	d	d	d al; s a
r52	acetate	$RbC_2H_2O_2$	144.51	col, nacreous leaf, hygr.		246		$86^{44.7}$	$89.3^{99.4}$	
r53	aluminum sulfate	$RbAl(SO_4)_2{\cdot}12H_2O$	520.75	col, cub, oct, 1.457, 1.45232, 1.46618	1.867^0	99		2.59^{20}	43.25^{80}	
r54	azide	RbN_2	127.49	col need or plates	2.7876	d *ca*310		107.1^{16}		0.182^{16} al; i eth
r55	borofluoride	$RbBF_4$	172.27	very sm rhomb cr, 1.333	2.829^{30}	590	d 500	0.6^{17}	10^{100}	
r56	borohydride	$RbBH_4$	100.31	white, cubic	1.920	1.487		v s		sl s al, i ether bz
r57	bromate	$RbBrO_3$	213.37	cub	3.68	430		$.293^{25}$	5.08^{40}	
r58	bromide	RbBr	165.37	col cub, 1.5530	3.35 liq:2.79^{730}	693	1340	98^5	$205.2^{113.5}$	i al, sl s acet
r59	bromide, tri-	$RbBr_3$	325.18	red, rhomb		d 140				
r60	bromochloroiodide	RbIBrCl	327.73	rhomb		d 200				s al, d eth
r61	bromoiodide, di-	$RbIBr_2$	372.18	red, rhomb	3.84	225	d 265	s		s al, d eth
r62	carbonate	Rb_2Co_3	230.94	col cr, deliq.		837	d 740	450^{20}	s	0.7abs sl
r63	carbonate acid	$RbHCO_3$	146.48	wh, rhomb		d 175		53.73^{20}		2.0 al
r64	chlorate	$RbClO_3$	168.92	trim.	3.19			5.0^{19}	62.8^{100}	
r65	perchlorate	$RbClO_4$	184.92	rhomb, 1.4701	2.80	fus	d	0.5^0	18^{100}	0.009^{25} al, 0.06^{25} MeOH
r66	chloride	RbCl	120.92	col, cub, 1.493^{35}	2.80: liq: 2.088^{750}	718	1390	77^0	138.9^{100}	0.08^{25} al, 1.41^{25} MeOH; v sl s NH_3
r67	chlorobromide, di-	$RbBrCl_2$	236.29	rhomb		d 110				
r68	chloro*di*bromide	$RbBr_2Cl$	280.73	rhomb		76				

No.	Name	Synonyms and Formulae	Mol. wt.	Crystalline form, properties and index of refraction	Density or spec. gravity	Melting point, °C	Boiling point, °C	Solubility, in grams per 100 cc		
								Cold water	Hot water	Other solvents
r69	chloroiodide, di-	$RbICl_2$	283.28	dk orange, rhomb	...	180-200	d 265	v s	v s	...
r70	chloroplatinate	Rb_2PtCl_4	578.73	yel, cub	$3.94^{17.5}$	d	...	0.184^{0}	0.634^{100}	i al
r71	chloroplatinate, hexa-	$Rb_2[PtCl_6]$	578.73	yel, cub	$3.94^{17.5}$	d	...	0.014^{0}	0.33^{100}	i al
r72	chromate	Rb_2CrO_4	286.93	yel, rhomb, ~1.71	3.518	...	...	62^{0}	95.7^{60}	...
r73	*di*chromate	$Rb_2Cr_2O_7$	386.92	tricl or monocl, >1.95, 1.70	3.02 monocl 3.125 tricl		tricl: monocl:	4.96^{16} 5.42^{18}	27.3 28.1^{60}	...
r74	chromiumsulfate	$RbCr(SO_4)_2 \cdot 12H_2O$	545.76	vlt cub, 1.482	1.946	107	...	43.4^{25}	...	...
r75	cobalt (II) sulfate	$RbSO_4 \cdot CoSO_4 \cdot 6H_2O$	444.61	rubyred, monocl, 1.486, 1.491, 1.501	2.56^{15}	...	...	9.3^{25}	s	...
r76	coppersulfate	$Rb_2SO_4 \cdot CuSO_4 \cdot 6H_2O$	534.69	monocl, 1.489, 1.491, 1.504	2.57	...	...	10.28^{25}	...	...
r77	cyanide	RbCN	111.49	col cr powder	2.32	...	...	s	s	i al, eth
r78	galliumsulfate	$RbGa(SO_4)_2 \cdot 12H_2O$	563.49	col cr, 1.46579	1.962	...	...	s	...	...
679	fluoride	RbF	104.47	col, cubic, 1.398	3.557	795	1410	130.6^{18}	s dil HF; i al, eth, NH_2	
r80	fluorogermanate	Rb_2GeF_6	357.52	wh cr.	...	...	...	sl s	v s	...
r81	rluosilicate	Rb_2SiF_6	313.01	cub, oct	3.332^{20}	...	...	0.16^{20}	1.35^{100}	s a, i al
r82	fluosulfonate	$RbFSO_3$	184.52	need	...	304	...	...	...	...
r83	iodate	$RbIO_2$	260.37	monocl or cub	$4.33^{19.5}$	d	...	2.1^{23}	...	v s HCl
r84	*meta*periodate	$RbIO_4$	276.37	tetr	3.918^{18}	...	...	0.65^{13}	...	...
r85	iodide	RbI	212.37	col, cub, 1.6474	3.55; liq: 2.87^{825}	647	1300	$152.^{17}$	163^{25}	s liq NH_3; 0.674^{25} acet
r86	iodide, tri-	RbI_3	466.18	blk, rhomb	4.03^{22}	190	...	s	...	...
r87	iodide, cmpd, with SO_2	$RbI \cdot 4SO_2$	468.61	lemon yel	...	13.5	...	...	...	...
r88	iron (II) selenate	$Rb_2SeO_4 \cdot FeSeO_4 \cdot 6H_2O$	620.79	blue-grn monocl, prism, 1.513, 1.520, 1.532	2.819	...	...	...	...	...
r89	iron (III) selenate	$RbFe(SeO_4)_2 \cdot 12H_2O$	643.41	cub, 1.507^{18}	2.31^{11}	45	$-12H_2O$, 100	...	...	...
r90	iron (II) sulfate	$Rb_2SO_4 \cdot FeSO_4 \cdot 6H_2O$	526.99	gr monocl, prism, 1.4815, 1.4874, 1.4977	2.516	d 60	...	24.2^{25} (anhydr)	...	...
r91	iron (III) sulfate	$RbFe(SO_4)_2 \cdot 12H_2O$	549.61	cub, 1.4823	1.91-1.95	48.53	$4.55^{6.6}$	52.6^{90}	...	...
r92	hydride	RbH	86.48	col need	2.60	d 300	...	d	d	d a
r93	hydroxide	RbOH	102.48	gray-wh, deliq	3.203^{11}	301 ± 0.9	...	180^{15}	v s	s al
r94	neodymium nitrate	$2(?)RbNO_3 \cdot Nd(NO_3)_3 \cdot 4H_2O$	697.26	redsh-vlt pl	2.56	47	$-4H_2O$, 60	...	...	...
r95	nitrate	$RbNO_3$	147.47	col, hex cub, rhomb or tricl, hygr, 1.51, 1.52, 1.524	3.11: liq: 2.395^{100}	tr cub 161.4 m.p. 310	tricl rhom 219	44.28^{16}	452^{100}	sl s acet: v s HNC
r96	nitrate, hydrogen	$RbNO_3 \cdot HNO_3$	210.49	tetr	...	62	...	...	...	...
r97	nitrate, hydrogen	$RbNO_3 \cdot 2HNO_3$	273.50	col need	...	45	...	...	...	...
r98	magnesium sulfate	$Rb_2SO_4 \cdot MgSO_4 \cdot 6H_2O$	495.45	col, monocl, 1.467, 1.469, 1.478	2.386_4^{20}	...	...	20.2^{25} (anhydr)	...	...
r99	permanganate	$RbMnO_4$	204.40	cr	$3.235^{10.4}$	d 295	...	0.5^{0}	4.7^{60}	...
r100	oxide, mon-	Rb_2O	186.94	col-yel, cub	3.72	d 400	...	s d	s d	s liq NH_3
r101	oxide, per-	Rb_2O_2	202.93	yel, cub	3.65^{0}	570	d 1011^{760} mm	dec to RbOH + H_2	...	...
r102	oxide, super	RbO_2, unstable	117.47	yel plates	3.80	432	d 1157^{1} atm	...	...	...
r103	oxide (tetr-)	Rb_2O_4	234.93	dk orange cr, deliq	...	dec 500 vac	...	...	...	...
r104	oxide, tri- (sesqui)	Rb_2O_3 (or Rb_4O_6)	218.93	blk cub	3.53^{0}	489	...	s d	...	...
r105	praseodymium nitrate	$2RbNO_3 \cdot Pr(NO_3)_3 \cdot 4H_2O$	693.93	grnsh, monocl, need hygr	2.50	63.5	$-4H_2O$, 60	...	...	...
r106	selenate	Rb_2SeO_4	313.89	col, rhomb, 1.5515, 1.5537, 1.5582	3.90	...	...	159^{12}	...	...
r107	silicofluoride	Rb_2SiF_6	313.01	col, oct or hex	3.3383_4^{20}	...	...	sl s	s	s a, i al
r108	sulfate	Rb_2SO_4	266.99	col, rhomb hex, 1.513, 1.513, 1.514	3.613_4^{20} liq2.53^{1100}	1060 trig 653	*ca* 1700	42.4^{10}	81.8^{100}	i acet; sl s NH_2
r109	sulfate, hydrogen	$RbHSO_4$	182.53	rhomb, 1.473	2.892^{16}	<red heat	...	...	...	...
r110	sulfide, di-	Rb_2S_2	235.06	dk red	...	420	volat > 850	...	...	...
r111	sulfide, hexa-	Rb_2S_6	363.30	brn-red	...	201	...	...	...	...
r112	sulfide, mono-	Rb_2S	203.00	wh-pale yel	2.912	530 d vac	...	v s	v s	...
r113	sulfide, mono-tetrahydrate	$Rb_2S \cdot 4H_2O$	275.06	cr, deliq	...	...	...	v s	v s	...
r114	sulfide, penta-	RbS_4	331.24	red, rhomb, deliq	2.628^{15}	225	...	d	...	s 70· al; i eth, chl
r115	sulfide, tri-	Rb_2S_3	267.12	redsh yel	...	213	...	...	...	...
r116	tartrate, d & l	$Rb_2C_4H_6$	319.01	trig	2.658_4^{20}	...	...	200^{25}	...	i toluol
r117	*dl*-tartrate, hydrogen,	$RbHC_4H_4O_6$	234.55	trim pr	2.282	201 d	...	1.18^{25}	11.7^{100}	...
r118	vanadium sulfate	Rubidium vanadium alum. $RbV(SO_4) \cdot 12H_2O$	544.71	yellow, cubic, 1.4689	1.915_4^{20}	64	230-$12H_2O$, 300 dec	2.56^{10}	...	...
r119	**Ruthenium**	Ru	101.07 ± 3	gray-wh or silv brittle met, hex	12.30	2310	3900	i	i	i sq reg,a, al; s fu alk
r120	carbonyl, penta-	$Ru(CO)_6$	241.12	col liq	...	−22	...	...	...	s al, bz
r121	chloride, tetra-	$RuCl_4 \cdot 5H_2O$	332.96	rdsh-br cr, hygr	...	d	...	s	...	s al

No.	Name	Synonyms and Formulae	Mol. wt.	Crystalline form, properties and index of refraction	Density or spec. gravity	Melting point, °C	Boiling point, °C	Solubility, in grams per 100 cc Cold water	Hot water	Other solvents
r122	chloride, tri-	$RuCl_3$	207.43	br cr, deliq	3.11	d > 500		i	d	sl s al; s HCl; i CS_2
r123	fluoride penta-	RuF_5	196.06	dk grn cr	$2.963^{16.5}$	101	250	d	d	
r124	hydroxide	$Ru(OH)_3$	152.09	blk powd				v sl s		s a; i alk
r125	oxide, di-	RuO_2	133.07	dk bl, tetr	6.97	d		i	i	i a; s fus alk
r126	oxide, tetr-	RuO_4	165.07	yel, rhomb need	3.29^{21}	25.5	d 108	2.033^{20}	2.249^{76}	s al, a alk, CCl_4
r127	oxychloride ammoniated	Ruthenium red. $Ru_2(OH)_2Cl_4{\cdot}7NH_3{\cdot}3H_2O$	551.23	brn-red powd				s		
r128	silicide	RuSi.	129.16	met pr	5.40^4			i	i	s HNO_3 + HF
r129	sulfide	Nat. laurite RuS_2	165.19	gray-blk, cub	6.99	d 1000		i	i	i a; s fus alk
s1	**Samarium**	Sm.	150.36 ± 3	wh-gray met, hex	7.520	1077	1791	i	i	s a
s2	acetate	$Sm(C_2H_3O_2)_3{\cdot}3H_2O$	381.54		1.94			15^{25}		
s3	acetylacetonate	$Sm(C_6H_7O_2)_3$	447.69	cr mass		146-147		i		
s4	benzylacetonate	$Sm(C_{10}H_9O_2)_3{\cdot}2H_2O$	669.93	straw color		103-105		i		s org solv
s5	bromate	$Sm(BrO_3)_3{\cdot}9H_2O$	696.20	yel, hex		75	$-9H_2O$, 150	114^{25}		v sl s al
s6	(II) bromide	$SmBr_2$	310.17	dk brn	5.1	508	1880	d		
s7	(III) bromide	$SmBr_3{\cdot}6H_2O$	498.16	yel cr, deliq.	2.971^{22}	640				
s8	carbide	SmC_2	174.38	yel, hex	5.86			d	d	s, d a
s9	(II) chloride	$SmCl_2$	221.27	red-brn cr	4.56^{25}	740		s, d		i al, CS_2
s10	(III) chloride	SmC_3	256.72	yelsh-wh cr, hygr	4.46^{18}	678 ± 1	d	92.4^{10}	99.9^{50}	v s al; 6.4^{25} pyr
s11	(III) chloride, hexahydrate	$SmCl_3{\cdot}6H_2O$	364.81	grn-yel, tricl, hygr	2.383	$-5H_2O$, 100				
s12	chromate	$Sm_2(CrO_4)_3{\cdot}8H_2O$	792.82	yel				0.043^{25}		
s13	(II) fluoride	SmF_2	188.36			1306	>2400			
s14	(III) fluoride	SmF_3	207.36			1306	2323	i	i	
s15	hydroxide	$Sm(OH)_3$	201.38	pa yel powd				i		s a; i alk
s17	(II) iodide	SmI_2	404.17	dk brn		527	1580	d		
s18	kojate	$Sm(C_6H_5O_4)_3$	573.67			d 275		i		
s19	(III) methyl-phosphate, di-	$Sm[(CH_3)_2PO_4]_3$	525.48	cream col, hex pr				35.2^{25}	10.8^{95}	
s20	(III) molybdate	$Sm_2(MoO_4)_3$	780.53	vlt, rhomb oct	5.36					
s21	(III) nitrate)	$Sm(NO_3)_3{\cdot}6H_2O$	444.47	pa yel, tricl	2.375	78-79		v s		
s22	(III) oxalate	$Sm_2(C_2O_4)_3{\cdot}10H_2O$	744.93	wh cr.				0.000054		s H_2SO_4
s23	oxide, sesqui-	Samaria. Sm_2O_2	348.72	wh-yelsh powd	8.347			i		v s a
s24	(III) sulfate	$Sm_2(SO_4)_3{\cdot}8H_2O$	733.02	lt yel, monocl, 1.543, 1.552, 1.563	2.930	$-8H_2O$, 450		2.67^{30} 4.4^{25}	1.99^{40}	
s25	(III) sulfide	Sm_2S_3	396.90	yelsh-pink	5.729	1900			d	d dil a
s26	sulfate, basic	$Sm_2O_2SO_4$	428.78	yel powd		d 1100		i		i dil H_2SO_4
s27	**Scandium**	Sc	44.9559	silv met, cubic or hex	2.9890	1541	2831	d, ev H_2		
s28	acetylacetonate	$Sc(C_5H_7O_2)_3$	342.28	col pl.		187.5	subl 210-215			s al, bz, chl
s29	bromide	$ScBr_3$	284.67		3.914	subl>1000				
s30	chloride	$ScCl_2$	151.31	col cr.	2.39_4^{25}	939	subl 800-850	v s	v s	i abs al
s31	hydroxide	$Sc(OH)_3$	95.98	col amorph				i		s dil a
s32	nitrate	'Sc$(NO_3)_3$	230.97	col, deliq		150		s		s al
s33	nitrate, tetrahydrate	$Sc(NO_3)_3{\cdot}4H_2O$	303.03	col, pr, deliq		$-4H_2O$, 100		v s		
s34	oxalate	$Sc_2(C_2O_4)_3{\cdot}5H_2O$	444.05	cr		$-4H_2O$, 140				
s35	oxide	Scandia. ScO_2	137.91	wh powd	3.864			i	i	s h a
s36	sulfate	$Sc_2(SO_4)_3$	378.08	col cr.	2.579	d		10.3^{25}	v s	
s37	sulfate, hexahydrate	$Sc_2(SO_4)_3{\cdot}6H_2O$	486.18			$-4H_2O$, 100 $-6H_2O$, 250		v s		
s38	sulfate, pentahydrate	$Sc_2(SO_4)_3{\cdot}5H_2O$	468.16		2.519	$-3H_2O$, 100		54.6^{25}		
s39	**Selenic acid**	H_2SeO_4	144.97	wh, hex prism, hygr.	3.004_4^{15}	58 eas undercools	d 260	1300^{30}	∞^{60}	s H_2SO_4; d al; i NH_3
s40	monohydrate	$H_2SeO_4{\cdot}H_2O$	162.99	wh, need	2.627_4^{15} liq 2.3564_4^{15}	26 eas undercools	205	v s	v s	
s41	tetrahydrate	$H_2SeO_4{\cdot}4H_2O$	217.03	col liq		51.7 eas undercools	$-H_2O$, 172^{35}	α		s H_2SO_4; d org solv
s42	**Selenium**	Se	78.96 ± 3	blsh-gray, met hex	4.81_4^{20}	217	684 ± 1.0	i	i	s H_2SO_4 $CHCl_3$; i al; v sl s CS_2
s43	selenium	Se	78.96 ± 3	red, monocl prism	4.50	170-180 traf to hex	684.8	i	i	$0.1^{46.6}$ CS_2; s H_2SO_4, HNO_3
s44	selenium	Se	78.96 ± 3	red amorph, blk vitr	red 4.26 blk 4.28	tr to hex, 60-80	684.8	i	i	s H_2SO_4, CS_2 bz
s45	bromide, "mono-"	Diselenium dibromide Se_2Br_2	317.73	dk red liq	3.604^{15}		227d	d	d	d al; s CS_2, $CHCl_3$, C_2H_5Br
s46	bromide, tetra-	$SeBr_4$	398.58	or-red-brn cr		d 75		d	d	s CS_2, chl, C_2H_5Br, HCl
s47	bromide (mono-) chloride, tri-	$SeBrCl_3$	265.22	yel br cr		190				i Cs_2
s48	bromide (tri-) chloride	$SeBe_2Cl$	354.13	or cr, hygr		d				v sl s CS_2
s49	carbide	SeC_2	102.98	yel liq, 1.845	2.682_4^{20}	45.5	$125—126^{760}$	i		s CS_2, eth, CCl_4, bz, al
s50	chloride "mono"-	Diselenium dichloride, Se_2Cl_2	228.83	br red liq, 1.596	2.77_4^{25}	−85	d 130	d	d	d al, eth; s CS_2, chl, CCl_4, bz
s51	chloride, tetra-	$SeCl_4$	220.77	wh-yel, cub, deliq, 1.807	$3.78\text{-}3.85^{360}$	205, subl 170—196	d 288	d	d	i al, eth, CS_2: s $POCl_3$; d a, alk
s52	fluoride, hexa-	SeF_6	192.95	col gas, 1.895	3.25^{-28} g/ℓ	−39, subl −46.6	−34.5	s d		
s53	fluoride, tetra-	SeF_4	154.95	col liq or wh cr		m.p. −13.8 frz −90	>100	d	d	

No.	Name	Synonyms and Formulae	Mol. wt.	Crystalline form, properties and index of refraction	Density or spec. gravity	Melting point, °C	Boiling point, °C	Solubility, in grams per 100 cc		
								Cold water	Hot water	Other solvents
s54	hydride	H_2Se	80.98	col gas, pois	gas 3.664^{760} air; liq: $2.004^{-41.5}$	60.4	−41.5	3.77^{4}	$270^{22.5}$	s CS_2, $COCl_2$
s55	iodide, "mono"-	Diselenium diiodide, Se_2I_2	411.73	steelgray cr (exist ?)	...	68-70	d 100	d	d	...
s56	nitride	Se_4N_4	371.87	amorph, or yel-brickred, hygr.	...	expl 160—200	d	u	sl d	i al, eth; v sl s acet ac, bz
s57	oxide, di-	SeO_2	110.96	wh, monocl, col, tetr, pois, >1.76	3.95^{15}_{15}	340—350 subl 315—317	...	38.4^{14}	82.5^{65}	6.67^{14} al; $4.35^{15.3}$ acet; $1.11^{13.9}$ac a; s bz
s58	oxide, tri-	SeO_3	126.96	pa yel cub or fiber, deliq	3.6	118	d 180	d, v s	d, v s	s al, conc H_2SO_4; i eth, bz, chl CCl_4
s59	oxybromide	Selenyl bromide. $SeOBr_2$	254.77	red yel cr	liq 3.38^{50}	41.6	217^{710} d	d	...	s CS_2, CCl_4, chl, H_2SO_4, bz
s60	oxychloride	$SeOCl_2$	165.87	col-yel, liq 1.651^{20}	2.42^{22}	8.5	176.4	d	...	s CS_2, CCl_4, chl, bz
s61	sulfur oxy*tetra*chloride	$SeSO_3Cl_4$	300.83	hex pr	...	165	183	d	...	...
s62	oxyfluoride	Selenyl fluoride, $SeOF_2$	132.96	col liq	2.67	4.6	124	d	...	s al, CCl_4
s63	sulfide, di-	SeS_2	143.08	br red-yel	...	<100	d	i	...	d aq reg, HNO_3; s $(NH_4)_2S$
s64	sulfide, "mono"-	SeS	111.02	or-yel tabl or powd	3.056^{0}	d 118—119	...	i	i	s CS_2; i ethl d al
s65	sulfur oxide	$SeSO_2$	159.02	grn pr or yel powd	...	$-SO_2$, 40	...	d, 118	...	s H_2SO_4; i SO_3
s67	**Selenious acid**	H_2SeO_3	128.97	col, hex, deliq	3.004^{15}_{4}	d 70	$-H_2O$	167^{20}	v s	v s al; i NH_3
s68	**Silane,** bromo-	SiH_3Br	111.01	col gas, expl in air	1.72^{-80} 1.533^{0}	−94	1.9	...	...	...
s69	bromotrichloro-	$SiBrCl_3$	214.35	col liq	1.826	−62	80.3	d	d	...
s70	chloro-	SiH_3Cl	66.56	col gas	gas: 3.033 g/ℓ liq: 1.145^{-113}	−118.1	−30.4	...	...	...
s71	dibromo-	SiH_2Br_2	189.91	col liq, inflam	2.17^{0}	−70.1	66	d	...	d alk
s72	dibromo-	Silicobromoform, $SiHBr_2$	268.81	col liq, inflam	2.7^{17}	−73	109	d	d	d NH_3
s73	dibromodichloro-	$SiBr_2Cl_2$	258.80	col liq	2.172^{25}_{4}	−45.5	104	d	d	...
s74	dichloro-	SiH_2Cl_2	101.01	gas	gas 4.599g/ℓ liq 1.42^{-122}	−122	8.3	d	d	...
s75	dichloro difluoro-	$SiCl_2F_2$	136.99	gas	6.2784 g/ℓ	−144 ± 2	−31.7 ± 0.2	d	d	...
s76	(hexa-) hexaoxocyclo-	Siloxane $Si_6O_3H_6$	222.56	wh, pl	1.32^{20}	d 140, inflam	...	sl d	d	...
s77	monochloro trifluoro-	$SiClF_3$	120.53	gas	5.455 g/ℓ	−138.0 ± 2	−70.0 ± 0.2d	d	d	...
s78	monoiodo-	SiH_2I	158.01	col liq	$2.035^{14.8}$	−57.0	45.5	d	...	...
s79	(tri-)nitrilo-	Silicylamine, tri-$(SiH_3)_3N$	107.33	col inflam liq	0.895^{-106}	−105.6	...	...	...	...
s81	tribromochloro-	$SiBr_3Cl$	303.25	col liq	2.497^{25}_{4}	−20.8 ± 1	126-128	d	d	...
s82	trichloro-	Silicochloroform, $SiHCl_3$	135.45	col liq	1.34	−126.5	33^{768} mm	d	d	s CS_2, CCl_4, chl, bz
s83	trichloroiodo-	$SiCl_3I$	261.35	col liq	...	>−60	113.5	d	...	...
s84	trifluoro-	Silicofluoroform, $SiHF_3$	86.09	col gas	3.86^{0} g/ℓ	−131.4	*ca*-95	d	d	d al, eth, alk; s tol
s85	triiodo-	Silicoiodoform. $SiHI_2$	409.81	red liq	3.314^{20}	8	220	d	d	s CS_2, bz
s86	**Silicane cyanate**	$Si(OCN)_4$	196.15	sol or liq	1.414^{20}_{4}	34.5 ± 0.5	247.2 ± 0.5 0.5^{760}	d	...	...
s87	diimide	$Si(NH)_2$	58.11	wh powd	...	d 900	...	...	...	...
s88	isocyanate	$Si(NCO)_4$	196.15	sol or liq	1.434^{25}_{4}	26.0 ± 0.5	185.6 ± 0.3^{760}	d	...	i acet; s bz, CCl_4, CS_2
s89	**Silicic acid, di-**	$H_2Si_2O_5$	138.18	col cr.	...	d 150	...	i	i	s NH_3, HF
s90	meta-	H_2SiO_3	78.10	col, amorph.	...	d room temp	...	i	i	s NH_3, HF, h alk
s91	**Silicon**	Si	28.0855 ± 3	steel gray, large to micr cr, cub.	2.32—2.34	1410	2355	i	i	s HF + HNO_3; i HF
s92	acetate, tetra-	$Si(C_2H_3O_2)_4$	264.26	col cr, hygr	...	subl 110 d 160—170	148^{6} mm	d	...	d al; sl s acet, bz
s93	bromide, tetra-	Tetrabromosilane. $SiBr_4$	347.70	col fum liq, sol cub	liq: 2.7715^{25}_{4} sol: 3.292^{-79}	5.4	154	d	d	d H_2SO_4
s94	(di-) bromide, hexa-	Si_2Br_6	535.60	wh, rhomb	...	95	240	d	d	s CS_2; d KOH
s95	bromide(di-) sulfide	$SiSBr_2$	219.95	col pl.	...	93	$150^{18.3}$ mm	d	d	s bz, CS_2
s96	carbide	SiC	40.10	col-blk, hex or cub, 2.654, 2.697	3.217	~2700, subl, d	...	i	i	i a; s fus KOH
s97	chloride, tetra-	Tetrachlorosilane. $SiCl_4$	169.90	col fum liq	liq: 1.483^{20}, sol: 1.90^{-97}gas: 7.59 g/ℓ	−70	57.57	d	d	d al
s98	(di-)chloride, hexa-	Hexachlorodisilane. Si_2Cl_6	268.89	col liq, 1.4748^{18}	1.58^{0}	−1	145^{769}	d	d	d al
s99	chloride(di-) sulfide	$SiSCl_2$	131.05	col pr	...	75	$92^{22.5}$	d	d	s CCl_4, CS_2 bz
s100	chloride(tri-) sulfide, hydro-	$SiCl_3HS$	167.51	col liq	1.45	...	96—100	d	d	d al
s101	fluoride, tetra-	Tetrafluorosilane. SiF_4	104.08	col gas	gas: 4.69 g/$ℓ^{760}$ liq: 1.66^{-95}	−90.2	−86	d	d	s abs al, HF; i eth
s102	(di-)fluoride, hexa	Hexa-fluorodisilane. Si_2F_6	170.16	gas	7.759 g/ℓ	−18.7	−18.5	d	d	...
s103	hydride	Silane, silicane. SiH_4	32.12	col gas	liq: 0.68^{-85} gas: 1.44 g/ℓ	−185	-111.8^{760} mm	i	...	d KOH
s104	hydride	Disilane, disilicane. Si_2H_6	62.22	col gas	gas: 2.865 g/ℓ liq: 0.686^{-20}	−132.5	−14.5	sl d	...	s al, bz, CS_2

PHYSICAL CONSTANTS OF INORGANIC COMPOUNDS (continued)

No.	Name	Synonyms and Formulae	Mol. wt.	Crystalline form, properties and index of refraction	Density or spec. gravity	Melting point, °C	Boiling point, °C	Solubility, in grams per 100 cc		
								Cold water	Hot water	Other solvents
s105	hydride	Trisilane, trisilanepropane. Si_3H_8	92.32	col liq	liq: 0.743^{0}; gas: 4.15 g/ℓ^{760}	−117.4	52.9	d	d	d CCl_4
s106	hydride	Tetrasilane, tetrasilane butane. Si_4H_{10}	122.42	col liq	liq: 0.79^{0} gas 5.48 g/ℓ$^{0.760}$	−108	84.3	d	...	...
s107	iodide, tetra-	Tetraiodosilane. SiI_4	535.70	col, cub	4.198	120.5	287.5	d	...	2.2^{27} CS_2
s108	(di-)iodide, hexa-	Hexaiodosilane. Si_2I_6	817.60	col, hex	...	d 250	d	d	d	19^{19} CS_2
s109	nitride	Si_3N_4	140.28	gray-wh amorph powd	3.44	1900 press	...	i	i	s HF
s110	oxide, di-	Nat. cristobalite. SiO_2	60.08	col, cub or tetr, 1.487, 1.484	2.32	1723 ± 5	2230 (2590)	i	i	s HF; v sl s alk
s111	oxide, di-	Nat. lechatelierite. SiO_2	60.08	col, amorph, vitr, 1.4588	2.19	...	2230 (2590)	i	i	s HF; v sl s alk
s112	oxide, di-	Nat. opal. $SiO_2 \cdot xH_2O$	...	col. amorph 1.41—1.46	2.17—2.20	>1600	...	i	i	s HF; v sl s alk
s113	oxide, di-	Nat. tridymite. SiO_2	60.08	col, rhomb, 1.469, 1.470, 1.471	2.26^{25}_{4}	1703	2230 (2590)	i	i	s HF; v sl s alk
s114	oxide, di-	Nat. quartz. SiO_2	60.08	col, hex, 1.544, 1.553	2.635—2.660	1610	2230 (2590)	i	i	s HF; v sl s alk
s115	oxide, mon-	SiO	44.08	wh, cub	2.13	>1702	1880	i	i	s dil HF + HNO_3
s116	oxychloride	Chlorosiloxane. Si_2OCl_6	284.89	col liq	...	28.1 ± 0.2	137	d	...	∞ CS_2, CCl_4
s117	oxyfluoride	Si_2OF_6	186.16	col gas	1.358 liq	−47.8 ± 0.5	−23.3	d	d	d alk
s118	sulfide, di-	SiS_2	92.21	wh need, rhomb	2.02	subl 1090	white heat	d	...	d al, liq NH_3; s dil alk; i bz
s119	sulfide, mono-	SiS	60.15	yel need	1.853^{15}	subl 940^{20}	...	d	d	d al, alk
s120	thiocyanate	$Si(CNS)_4$	260.40	wh, rhomb need	1.409^{20}_{4}	143.8	314.2	d	...	d al, a, alk; i eth, CS_2, $CHCl_3$
s121	**Silicotungstic acid**	$H_4[Si(W_3O_{10})_4] \cdot 26H_2O$	3346.69	wh-sl yel cr, deliq	...	...	...	v s	v s	v s al
s122	**Silicyl oxide**	Disiloxane $(SiH_3)_2O$	78.22	col gas	gas: 3.491 g/ℓ liq: 0.881^{-80}	−144	−15.2	v sl s	sl d	...
s123	**Siloxane, (di-), oxide**	$[H(O)Si]_2 \cdot O$	106.19	wh volum subst	...	expl *ca* 300	...	sl s	...	s, d HF; d al
s124	**Silver**	Ag	107.8682 ± 3	wh met, cub 0.54	10.5^{20}	961.93	2212	i	i	s HNO_3, h H_2SO_4, KCN; i alk
s125	acetate	$AgC_2H_3O_2$	166.91	wh pl.	3.259^{15}	d	...	1.02^{20}	2.52^{80}	s dil HNO_3
s126	acetylide	Ag_2C_2	239.76	wh ppt	...	expl	...	i	...	s a; sl s al
s127	*ortho*arsenate	Ag_3AsO_4	462.52	dk red, cub	6.657^{25}	d	...	0.00085^{20}	...	s NH_4OH, ac a
s128	*ortho*arsenite	Ag_3AsO_3	446.52	yel, powd	...	d 150	...	0.00115^{20}	i	s ac a, NH_4OH, HNO_3; i al
s129	azide	AgN_3	149.89	wh rhomb pr, expl	...	252	297	i	0.01^{100}	s KCN, dil HNO_3; sl s NH_4OH
s130	benzoate	$AgC_7H_5O_2$	228.98	wh powd	...	...	...	0.262^{25}	s	0.017 al
s131	*tetra*borate	$Ag_2B_4O_7 \cdot 2H_2O$	407.00	wh cr.	...	...	...	sl s	...	s a
s132	bromate	$AgBrO_3$	235.77	col, tetr, 1.874, 1.920	5.206	d	...	0.196^{25}	1.33^{90}	s NH_4OH; sl s HNO_3
s133	bromide	Bromyrite: AgBr	187.77	pa yel, 2.253	6.473^{25}	432	d>1300	8.4 × 10^{-6}	0.00037^{100}	s KCN, $Na_2Sr_2O_3$, NaCl sol; sl s NH_4OH; i al
s134	carbonate	Ag_2CO_3	275.75	yel powd	6.077	d 218	...	0.0032^{20}	0.05^{100}	s NH_4OH, $Na_2S_2O_3$; i al
s135	chlorate	$AgClO_3$	191.32	wh, tetr	4.430^{20}_{4}	230	d 270	10^{15}	50^{80}	sl s al
s136	*per*chlorate	$AgClO_4$	207.32	wh, cr, deliq	2.806^{25}	d 486	...	557^{25}	s	s al; 101 tol; 5.28 bz
s137	chloride	Nat. cerargyrite. AgCl	143.32	wh, cub, 2.071	5.56	455	1550	0.000089^{10}	0.0021^{100}	s NH_4OH, $Na_2S_2O_3$, KCN
s138	chlorite	$AgClO_2$	175.32	yel cr.	...	105 expl	...	0.45^{25}	2.13^{100}	...
s139	chromate	Ag_2CrO_4	331.73	red, monocl.	5.625	...	...	0.0014^{0}	0.008^{70}	s NH_4OH, KCN
s140	*di*chromate	$Ag_2Cr_2O_7$	431.72	red, tricl.	4.770	d	...	0.0083^{15}	d	s a, NH_4OH, KCN
s141	citrate	$Ag_3C_6H_5O_7$	512.71	wh need	...	d	...	0.028^{18}	sl s	s a, NH_4OH, KCN, $Na_2S_2O_3$
s142	cyanate	AgOCN	149.89	col	4.00	d	...	sl s	s	s KCN, HNO_3, NH_4OH
s143	cyanide	AgCN	133.89	wh, hex	3.95	d 320	...	0.000023^{20}	...	s HNO_3, NH_4OH, KCN, $Na_2S_2O_3$
s144	ferricyanide	$Ag_3Fe(CN)_6$	535.56	...	...	...	...	0.000066^{20}	...	i a; s NH_4OH, h $(NH_4)_2CO_3$
s145	ferrocyanide	$Ag_4Fe(CN)_6 \cdot H_2O$	661.44	wh	...	...	...	i	i	s KCN; i a, NH_4 salts, NH_4OH
s146	fluogallate	$Ag_3[GaF_6] \cdot 10H_2O$	687.47	col, orthorhomb cr, 1.493	2.90	...	...	v s	...	i al
s147	fluoride	AgF	126.87	yel, cub, deliq	$5.852^{15.5}$	435	*ca* 1159	$182^{15.5}$	205^{108}	sl s NH_4OH
s148	fluoride, di-	AgF_2	145.87	brn, rhomb	4.57—4.58	690	d 700	d	d	...
s149	(di-)fluoride	Ag_2F	234.73	yel, hex	8.57	d 90	...	d	...	...
s150	fluosilicate	$Ag_2SiF_6 \cdot 4H_2O$	429.87	wh powd or col cr, deliq	...	>100	d	v s	...	...
s151	fulminate	$Ag_2C_2N_2O_2$	299.77	need	...	expl	...	0.075^{13}	s	i NHO_3; s NH_4OH
s152	iodate	$AgIO_3$	282.77	col, rhomb	$5.525^{16.5}$	>200	d	0.03^{10}	0.019^{60}	s HNO_3, NH_4OH, KI
s153	*per*iodate	$AgIO_4$	298.77	or yel, tetrag	5.57	d 180	...	d	...	s HNO_3
s154	iodide(α)	miersite AgI	234.77	ye tetr 2.02 ± .02	5.683^{30}_{4}	tr 146 to β	...	2.8 × $10^{-7.25}$	2.5 × $10^{-6.60}$	s KCN, $Na_2S_2O_3$, KI; sl s NH_4OH
s155	iodide (β)	iodyrite AgI	234.77	pa ye hex	$6.010^{14.6}_{4}$	558	1506	...	...	...
s156	iodomercurate (α)	Ag_2HgI_4	923.94	yel, tetrag	6.02	tr to β 50.7	...	i	...	s KI, KCN; i dil a
s157	iodomercurate (β)	Ag_2HgI_4	923.94	red, cub	5.90	d 158	...	i	...	s KI, KCN; i dil a

No.	Name	Synonyms and Formulae	Mol. wt.	Crystalline form, properties and index of refraction	Density or spec. gravity	Melting point, °C	Boiling point, °C	Solubility, in grams per 100 cc		
								Cold water	Hot water	Other solvents
s158	hydrogen(tri-) *para*periodate	$Ag_2H_3IO_6$	441.66	yel, rhomb	5.68^{25}	60 d		1.68^{25}		s HNO_3
s159	hyponitrite	$Ag_2N_2O_2$	275.75	yel	5.75^{30}	d 110		v sl s		d HNO_3, H_2SO_4
s160	lactate	$AgC_3H_5O_3 \cdot H_2O$	214.95	wh or sl gray cr, powd				*ca* 7.7		
s161	laurate	$AgC_{12}H_{23}O_2$	307.18	wh, greasy powd		212.5				0.007^{25} al; 0.008^{15} eth
s162	levunilate	$AgC_5H_7O_8$	222.98	leaf				0.67^{17}	d	
s163	*per*manganate	$AgMnO_4$	226.80	dk vlt, monocl	4.27^{25}	d		0.55^{0}	$1.69^{28.5}$	d al
s164	mercury iodide (α)	Ag_2HgI_4	923.94	yel, tetrag	6.02	trst 50.7		i		
s165	mercury iodide (β)	Ag_2HgI_4	923.94	red, cub	5.90	158 d		i		
s166	myristate	$AgC_{14}H_{27}O_2$	335.24			211		0.007^{25}		0.006^{25} al; 0.007^{15} eth
s167	nitrate	$AgNO_3$	169.87	col, rhomb, 1.729, 1.744, 1.788	4.352^{19}	212	d 444	122^{0}	952^{190}	s eth, glyc; v sl s abs al
s168	nitrite	$AgNO_2$	153.87	wh, rhomb	4.453^{26}	d 140		0.155^{0}	1.363^{60}	s ac a, NH_4OH; i al
s169	nitroplatinite	$Ag_2[Pt(NOp_2)_4]$	594.84	yel-brn monocl pr		d 100		sl s	s	
s170	nitroprusside	$Ag_2[FeNO(CN)_5]$	431.68	lt pink				i		s NH_4OH; i al, HNO_3
s171	oxalate	$Ag_2C_2O_4$	303.76	col cr.	5.029^{4}	expl 140		0.00339^{18}		s KCN, NH_4OH, a
s172	oxide	Ag_2O	231.74	br-blk, cub	$7.143^{16.6}$	d 230		0.0013^{20}	0.0053^{80}	s a, KCN, NH_4OH, al
s173	oxide, per	Ag_2O_2 (or AgO)	247.74	gray-blk, cub	7.44	d>100		i		s H_2SO_4, HNO_3, NH_4OH
s174	palmitate	$AgC_{16}H_{31}O_2$	363.29	wh, greasy powd		209		0.0012^{20}	0.006^{25}	0.007^{15} eth; 0.006^{25} al
s175	*meta*phosphate	$AgPO_3$	186.84	wh, amorph	6.37	*ca*482		i		s HNO_3, NH_4OH
s176	*ortho*phosphate	Ag_3PO_4	418.58	yel, cub	6.370^{25}	849		$0.00065^{19.5}$		s a, KCN, NH_4OH, NH_3
s177	*ortho*phosphate, mono-H	Ag_2HPO_4	311.72	wh, trig	1.8036	d 110				
s178	*pyro*phosphate	$Ag_4P_2O_7$	605.42	wh	$5.306^{7.5}$	585		i	i	s a, NH_4OH, KCN, ac a
s179	propionate	$AgC_3H_5O_2$	180.94	wh leaf or need	2.687^{25}_{4}			0.842^{20}	2.04^{80}	
s180	*per*rhenate	$AgReO_4$	358.07	wh cr, tetrag or rhomb	7.05	430		0.32^{20}		
s181	salicylate	$AgC_7H_5O_3$	244.98	wh to redsh-wh cr				sl s		s al
s182	selenate	Ag_2SeO_4	358.69	wh, orthorhomb cr	5.72			0.118^{20}		
s183	selenide	Ag_2Se	294.70	thin gray pl, cub	8.0	880	d	i		s NH_4OH, h HNO_3
s184	stearate	$AgC_{18}H_{35}O_2$	391.34	wh powd amorph		205		0.006^{20}		0.006^{25} al; 0.006^{25} eth
s185	sulfate	Ag_2SO_4	311.79	wh, rhomb. 1.7583, 1.7748, 1.7852	$5.45^{29.2}$	652	d 1085	0.57^{0}	1.41^{100}	s a, NH_4OH; i al
s186	sulfide	Nat. acanthite. Ag_2S	247.80	gray-blk, rhomb	7.326	tr 175	d		v sl s	s KCN, conc H_2SO_4, HNO_3
s187	sulfide	Nat. argentite. Ag_2S	247.80	blk, cub	7.317	825	d	8.4×10^{-15}		s KCN, a
s188	sulfite	Ag_2SO_3	295.79	wh cr.		d 100		v sl s		s a, NH_4OH, KCN; i HNO_3
2189	*d*-tartrate	$Ag_2C_4H_4O_6$	363.81	wh, scales	3.423^{15}	d		0.2^{18}	0.203^{25}	s a, KCN, NH_4OH
s190	*ortho*tellurate, tetra-H	$Ag_2H_4TeO_6$	443.36	straw yel, rhomb bipyr		d>200		i	i	s KCN, NH_4OH
s191	telluride	Nat. hessite, Ag_2Te	343.34	gray, cub	8.5	955		i	i	s KCN, NH_4OH
s192	tellurite	Ag_2TeO_3	391.33	yel-wh ppt		250-bl 450-pa yel		i	i	s KCN, NH_3
s193[1]	thioantimonite	Nat. pyrargyrite. Ag_3SbS_3	541.53	red, trig, 3.084 2.881 (Li)	5.76	486		i	i	s HNO_3
s193[2]	thioarsenite	Nat. proustite. Ag_3AsS_3	494.71	scarlet red, trig, 3.088, 2.792	5.49	490		i	i	s HNO_3
s194	thiocyanate	AgSCN	165.95	col cr.		d		0.000021^{25}	0.00064^{100}	s NH_4OH; i a
s195	*di*-thionate	$Ag_2S_2O_6 \cdot 2H_2O$	411.88	rhomb cr, ~1.662	3.61					
s196	thiosulfate	$Ag_2S_2O_3$	327.85	wh cr.		d		sl s		s $Na_2S_2O_3$, NH_4OH
s197	tungstate	Ag_2WO_4	463.58	pa yel cr.				0.015^{15}		s KCN, NH_4OH, HNO_3
	Silver complex									
s198	diamminesilver *per*rhenate	$[Ag(NH_3)_2]ReO_4$	392.13	col monocl cr.	3.901					1.618 conc NH_4OH
s199	**Sodium**	Na	22.98977	silv, met cub, 4.22	0.97	97.81 ± 0.03	882.9	d to NaOH + H_2		d al; i eth, bz
s200	acetate	$NaC_2H_3O_2$	82.03	wh gr powd, monocl, 1.464	1.528	324		119^{0}	170.15^{100}	sl s al
s201	acetate trihydrate	$NaC_2H_3O_2 \cdot 3H_2O$	136.08	col, monocl pr, effl, β 1.464	1.45	58	123, $-3H_2O$, 120	76.2^{0}	138.8^{50}	2.1^{18} al; s eth
s202	alumina trisilicate	Nat. albite. $NaAlSi_3O_8 \cdot$ (or $Na_2O \cdot Al_2O_3 \cdot 6SiO_2$)	262.22	col, tricl, 1.525, 1.529, 1.536	2.61	1100			sl d	s HCl; d dil al
s203	*meta*aluminate	$NaAlO_2$	81.97	wh amorph powd, hygr, 1.566, 1.595, 1.580		1800		s	v s	i al
s204	aluminum chloride	$NaCl \cdot AlCl_3$	191.78	wh-yelsh cr powd, hygr.		185		s	s	
s205	aluminum *meta*-silicate	Nat. jadeite. $Na_2O \cdot Al_2O_3 \cdot 4SiO_2$	404.28	col, monocl	3.3	1000—1060		i	i	d HCl

PHYSICAL CONSTANTS OF INORGANIC COMPOUNDS (continued)

No.	Name	Synonyms and Formulae	Mol. wt.	Crystalline form, properties and index of refraction	Density or spec. gravity	Melting point, °C	Boiling point, °C	Solubility, in grams per 100 cc		
								Cold water	Hot water	Other solvents
s206	aluminum *ortho*-silicate	Nat. nephelite. $Na_2O{\cdot}Al_2O_3{\cdot}2SiO_2$	284.11	col, hex, 1.537 ± 0.002	2.619^{21}	1526		i	d	d a
s207	aluminum sulfate	$NaAl(SO_4)_2{\cdot}12H_2O$	458.27	col, cub oct, 1.4388	1.6754^{20}	61		110^{15} (anhydr)	146^{30} (anhydr)	
s208	amide	Sodamide. $NaNH_2$	39.01	wh, conchoid fract		210	400	d	d	d hot al; 0.1 liq NH_3
s209	ammonium phosphate	Microcosmic salt, stercorite. $NaNH_4HPO_4{\cdot}4H_2O$	209.07	col, monoc, 1.439, 1.441, 1.469	1.554	d 79		16.7	100	i al, acet
s210	ammonmium sulfate	$NaNH_4SO_4{\cdot}2H_2O$	173.12	wh, rhomb	1.63^{15}	d 80		s	s	
s211	ammonium tartrate	$NaNH_4C_4H_4O_6{\cdot}4H_2O$	261.16	wh, rhomb	1.590			21.09^{0}		
s212	*meta*antimonate	Leuconine. $NaSbO_3$	192.74	wh powd				i	s	s Na_2S sol
s213	antimonate, hydroxy	"Pyroantimonate". $NaSb(OH)_6$	246.78	pseudo cub				$0.03^{12.3}$	0.3^{100}	sl s al
s214	*pyro*antimonate, dihydro-	$Na_2H_2Sb_2O_7{\cdot}6H_2O$(?)	511.58	wh, tetrag		d 280		i	0.28^{100}, d	
s215	antimonide	Na_3Sb	190.72	blk powd or bl cr, inflamm		856		d		sl s NH_3
s216	*meta*antimonite	$NaSbO_2{\cdot}3H_2O$	230.78	col, rhomb	2.864	d		d		
s217	*meta*arsenate	$NaAsO_3$	145.91	rhomb, effl, 1.479, 1.502, 1.527	2.301	615		v s		
s218	*ortho*arsenate	$Na_3AsO_4{\cdot}12H_2O$	424.07	col, trig or hex prism, 1.457, 1.466	1.752—1.804	86.3		$38.9^{15.5}$		1.67 al; 50^{15} glyc
s219	*ortho*arsenate, di-H	$NaH_2AsO_4{\cdot}H_2O$	181.94	col, rhomb or monocl, 1.583, 1.553, 1.507	2.53	130, $-H_2O$, 100	d 200—280	s		
s220	*ortho*arsenate, mono-H	$Na_2HAsO_4{\cdot}7H_2O$	312.01	col, monocl, pois, 1.462, 1.466, 1.478	1.88	130, $-5H_2O$, 50	d 180	5.46^{0}	100^{100}	s glyc; sl s al
s221	*ortho*arsenate, mono-H	$Na_2HAsO_4{\cdot}12H_2O$	402.09	col, monocl, effl. 1.445, 1.466, 1.451	1.736	28	$-12H_2O$, 100	56^{14}	140.7^{30}	sl s al; i liq Cl
s222	*pyro*arsenate	$Na_4As_2O_7$	353.80	wh cr.	2.205	850	d 1000	v s		
s223	arsenate fluoride	$2Na_3AsO_4{\cdot}NaF{\cdot}19H_2O$	800.06	wh, cub, 1.4657, 1.4693, 1.4726	2.849^{25}			10^{75}		
s224	arsenite	Sodium metaarsenite (?) (com'l) $NaAsO_2$ (or mixt with Na_3AsO_3)	129.91	gray-wh powd, pois	1.87			v s	v s	sl s al
s225	arsenotartrate	$Na(AsO)C_4H_4O_6{\cdot}2^1/_2H_2O$	307.02	shiny cr, pois.		$-2^1/_2H_2O$, 275	d 275	6.5^{19}		i al
s226	azide	NaN_3	65.01	col, hex	1.846^{20}	d Na + N	d in vac	41.7^{17}		0.314^{16} al; s liq NH_3, i eth
s227	barbital	$NaC_8H_{11}N_2O_3$	206.18	wh powd				20^{25}	40^{100}	sl s al; i eth
s228	benzenesulfonate	$NaC_6H_5SO_3$	180.15	wh cr.				35.8^{30}	v s	
s229	benzoate	$NaC_7H_5O_2$	144.11	col cr, or wh amorph, or gran powd				66^{20}	74.2^{100}	1.64^{25} al
s230	*meta*bismuthate	$NaBiO_3$	297.97	yel-brn powd (com'l), yel (pure)				i	d	d a
s231	*meta*boratc	$NaBO_2$	65.80	col, hex pr	2.464	966	1434	26^{20}	36^{35}	
s232	*meta*borate, tetrahydrate	$NaBO_2{\cdot}4H_2O$	137.86	tricl, coll		57	$-H_2O$, 120	v s	v s	
s233	*meta*borate, peroxyhydrate	Sodium perborate (com'l). $NaBO_2{\cdot}H_2O_2{\cdot}3H_2O$	153.86	col, monocl		63.0	$-H_2O$, 130—150	2.55^{15}	3.75^{32}	s a, al, glyc
s234	*tetra*borate	$Na_2B_4O_7$	201.22	cr, 1.5010	2.367	741	d 1575	1.06^{0}	8.79^{40}	i al
s235	*tetra*borate, decahydrate	Borax. $Na_2B_4O_7{\cdot}10H_2O$	381.37	col, monocl, effl, 1.447, 1.469, 1.472	1.73	75, $-8H_2O$, 60	$-10H_2O$, 320	2.01^{0}	170^{100}	v sl s al; s glyc; i a
s236	*tetra*borate, pentahydrate	$Na_2B_4O_7{\cdot}5H_2O$	291.29	col, cub or hex, deliq, 1.461	1.815	$-H_2O$, 120		22.65^{65} (anhydr)	52.3^{100}	
s237	borohydride	$NaBH_4$	37.83	white cub, 1.542	1.074	400 dec		55^{25}	v s	4 al; 16.4 MeOH; s pyr; i eth
s238	bromate	$NaBrO_3$	150.89	col, cub, 1.594.	$3.339^{17.5}$	381		27.5^{0}	90.9^{100}	i al
s239	bromide	$NaBr$	102.89	col, cub, hygr, 1.6412	3.203^{25}_{4}	747	1390	116.0^{50}	121^{100}	sl s al
s240	bromide, dihydrate	$NaBr{\cdot}2H_2O$	138.92	col, monocl pr	2.176	$-2H_2O$, 51		79.5^{0}	$118.6^{80.5}$	2.31^{25} al; s liq NH_3; 17.42^{15} MeOH
s241	bromoaurate	$NaAuBr_4{\cdot}2H_2O$	575.60	br-blk cr.				s		
s242	bromoiridite	$Na_3IrBr_6{\cdot}12H_2O$	956.80	dk grn, rhomb, effl		100	$-H_2O$, 150			s NH_4OH
s243	bromoplatinate	$Na_2PtBr_6{\cdot}6H_2O$	828.58	dk red, tricl.	3.323	d 150		v s	v s	v s al
s244	cacodylate	$Na[(CH_3)_2AsO_2]{\cdot}3H_2O$	214.03	wh		ca 60	$-H_2O$, 120	$200^{15\text{-}20}$		40^{25} al; $100^{15\text{-}20}$ 90% al
s245	calcium sulfate	$Na_2Ca(SO_4)_2{\cdot}2H_2O$	314.21	col, monocl need.	2.64	$-2H_2O$, 80		d	d	
s246	*d*-camphorate	$Na_2C_{10}H_{14}O_4{\cdot}3H_2O$	298.24	wh need, hygr		$-3H_2O$, 100		122^{14}		s al
s247	carbide	Na_2C_2	70.00	wh powd	1.575^{15}	*ca* 700		d	d	s a; d al
s248	carbonate	Na_2CO_3	105.99	wh powd, hygr, 1.535	2.532	851	d	7.1^{0}	45.5^{100}	sl s abs al; i acet
s249	carbonate, decahydrate	Washing soda. $Na_2CO_3{\cdot}10H_2O$	286.14	wh, monocl, 1.405, 1.425, 1.440	1.44^{15}	32.5—34.5	$-H_2O$, 33.5	21.52^{0}	421^{104}	i al
s250	carbonate, heptahydrate	$Na_2CO_3{\cdot}7H_2O$	232.10	rhomb bipyr, effl	1.51	$-H_2O$, 32		16.90	33.9^{35}	
s251	carbonate, monohydrate	Crystal carbonate, thermonatrite, $Na_2CO_3{\cdot}H_2O$	124.00	col, rhomb, deliq, 1.506, 1.509	2.25	$-H_2O$, 100		33	52.08	14^{25} glyc; i al, eth

PHYSICAL CONSTANTS OF INORGANIC COMPOUNDS (continued)

No.	Name	Synonyms and Formulae	Mol. wt.	Crystalline form, properties and index of refraction	Density or spec. gravity	Melting point, °C	Boiling point, °C	Solubility, in grams per 100 cc: Cold water	Hot water	Other solvents
s252	carbonate, sesqui-	$Na_2CO_3 \cdot NaHCO_3 \cdot 2H_2O$	226.03	col, monocl, 1.5073	2.112	d	...	13^{0}	42^{100}	...
s253	carbonate hydrogen	$NaHCO_3$	84.01	wh, monocl pr, 1.500	2.159	$-CO_2$, 270	...	6.9^{0}	16.4^{60}	sl s al
s254	chlorate	$NaClO_3$	106.44	col, cub or trig, 1.513	2.490^{15}	248—261	d	79^{0}	230^{100}	s al, liq NH_3, glyc
s255	*per*chlorate	$NaClO_4$	122.44	wh, rhomb, deliq, 1.4606, 1.4617, 1.4731	...	d 482	d	s	v s	s al
s256	*per*chlorate, hydrate	$NaClO_4 \cdot H_2O$	140.46	col rhbdr, deliq	2.02	130	d 482	209^{15}	284^{50}	s al
s257	chloride	Common salt, nat. halite. NaCl	58.44	col, cub, 1.5442	2.165^{25}_{4}	801	1413	35.7^{0}	39.12^{100}	sl s al, liq, NH_3; s glyc; i HCl
s258	chlorite	$NaClO_2$	90.44	wh, cr, hygr	...	d 180—200	...	39^{17}	55^{60}	...
s259	hypochlorite, pentahydrate	$NaOCl \cdot 5H_2O$	164.52	col	...	18	...	29.3^{0}	94.2^{23}	...
s260	*hypo*chlorite	NaOCl	74.44	in solution only	...	...	...	...	...	...
s261	*hypo*chlorite, dihydrate	$NaOCl \cdot 2^1/_2H_2O$	119.48	col, hygr	...	57.5	...	v s	...	...
s262	chloroaurate	$NaAuCl_4 \cdot 2H_2O$	397.80	yel, rhomb, ω 1.545 ϵ> 1.75	...	d 100	...	150^{10}	990^{60}	v s al, eth
s263	chloroiridate	$Na_2IrCl_6 \cdot 6H_2O$	559.01	dull red-blk, tricl	...	d 600	...	v s	v s	sl s al
s264	chloroiridite	$Na_3IrCl_6 \cdot 12H_2O$	690.09	dk grn cr	...	$-H_2O$, 50	...	31.46^{15}	307.26^{85}	...
s265	chloroosmate	$Na_2OsCl_6 \cdot 2H_2O$	484.93	or-red, rhomb pr	...	...	...	v s	...	s al
s266	chloropalladite	$Na_2PdCl_4 \cdot 3H_2O$	348.26	br-red cr, deliq	...	...	...	v s	...	s al
s267	chloroplatinate	Na_2PtCl_6	453.78	or-yel powd, hygr	...	tr 150—160	...	s	v s	s al
s268	*hexa*chloroplatinate	$Na_2PtCl_6 \cdot 6H_2O$	561.87	or-red, tricl	2.500	$-6H_2O$, 100	...	66^{15}	v s	11.9 al, MeOH; i eth
s269	chloroplatinite	$Na_2PtCl_4 \cdot 4H_2O$	454.93	red pr	...	100	$-H_2O$, 150	s	...	s al
s270	chlororhodite, hexa-	Na_3RhCl_6	384.59	red, tricl	...	d>550	...	v s	...	...
s271	chlororhodite, hexa-, hydrate	$Na_3RhCl_6 \cdot 18H_2O$	708.87	garnet red, oct, effl	...	d 904, effl	...	v s	...	i al
s272	chromate	Na_2CrO_4	161.97	yel, rhomb bipyram	2.710—2.736	...	...	87.3^{30}	...	sl s al; s MeOH
s273	chromate decahydrate	$Na_2CrO_4 \cdot 10H_2O$	342.13	yel, monocl, deliq	1.483	19.92	...	50^{10}	126^{100}	sl s al; i ac a
s274	*di*chromate	$Na_2Cr_2O_7 \cdot 2H_2O$	298.00	red, monocl pr, deliq, 1.661, 1.699, 1.751	2.52^{13}	$-2H_2O$, 100 356.7 (anhydr)	400 (anhydr)	238^{0} (anhydr) 180^{20}	508^{80} (anhydr) 433^{98}	i al
s275	*peroxy*chromate	Na_3CrO_8	248.96	or pl	...	d 115	...	sl s	...	i al, eth
s276	cinnamate	$NaC_9H_7O_2$	170.14	wh cr powd	...	...	...	9.1	5^{100}	0.625 90% al; s glyc
s277	citrate, dihydrate	$Na_3C_6H_5O_7 \cdot 2H_2O$	294.10	wh cr, gran or powd	...	$-2H_2O$, 150	...	72^{25}	167^{100}	0.625 90% al; s glyc
s278	citrate, pentahydrate	$Na_2C_6H_5O_7 \cdot 5(or\ 5^1/_2)H_2O$	348.15	wh, rhomb	$1.857^{23.5}$	$-5H_2O$, 150	d	92.6^{25}	250^{100}	sl s al
s279	cobaltinitrite	$Na_3Co(NO_2)_6$	403.94	yelsh-brnsh cr powd	...	...	...	v s, d	...	sl s al; d min a; i dil ac a
s280	cyanamide, mono-	$NaHCN_2$	64.02	wh cr powd, hygr	...	...	...	v s	...	...
s281	cyanate	NaOCN	65.01	col need	1.937^{20}	d 700 vac	...	s	s	v sl s eth, bz
s282	cyanide	NaCN	49.01	col, cub, deliq, pois, 1.452	...	563.7	1496	48^{10}	82^{35}	sl s al; s NH_3
s283	cyanoaurite	Sodium aurocyanide. $NaAu(CN)_2$	271.99	...	...	...	...	s	...	...
s284	cyanocuprate (I)	$NaCu(CN)_2$	138.57	...	1.013^{20}	d 100	...	s	...	...
s285	cyanoplatinite	$Na_2[Pt(CN)_4] \cdot 3H_2O$	399.18	col, tricl	2.646	$-3H_2O$, 120—125	...	s	s	s al
s286	enanthate	Sodium heptanoate. $NaC_7H_{13}O_2$	152.17	wh cr powd or leaf	...	240—350	...	s	...	s al
s287	ethyl acetoacetate	$NaC_6H_9O_3$	152.13	need	...	d	...	d	...	s eth
s288	ethyl sulfate	$NaC_2H_5SO_4 \cdot H_2O$	166.12	wh, hex pl, deliq	...	d	...	164^{17}	...	d alk, H_2SO_4; 142al
s289	ferrate (III)	Ferrite. $Na_23Fe_2O_4$	221.67	br, hex pl or need	4.05	...	...	d	...	v s dil HCl
s290	ferricyanide	$Na_3Fe(CN)_6 \cdot H_2O$	298.94	red cr, deliq	...	...	...	18.9^{0}	67^{100}	i al
s291	ferrocyanide	Yellow prussiate of soda. $Na_4Fe(CN)_6 \cdot 10H_2O$	484.07	pa yel, monocl, 1.519, 1.530, 1.544	1.458	...	...	31.85^{20}	156.5^{98}	i al
	...	Sodium hexacvanoferrate (II).		...						
s292	fluoaluminate	Na_3AlF_6	209.94	col, monocl, β 1.364	2.90	1000	...	sl s	...	i HCl.; d alk
s293	fluoantimonate	$NaSbF_6$	258.73	rhomb	3.375^{18}	<1360	...	128.6^{20}	...	s al, acet
s294	fluoberyllate	Na_2BeF_4	130.99	wh, rhomb or monocl	...	d	...	1.47^{18}	2.94^{100}	...
s295	fluoborate	$NaBF_4$	109.79	wh, rhomb	2.47^{20}	sl d 384	d	108^{26}	210^{100}	sl s al; d H_2SO_4
s296	fluoride	Nat. villiaumite. NaF	41.99	col, cub or tetr, 1.336	2.558^{41}	993	1695	4.22^{18}	...	s HF; v sl s al
s297	fluoride, hydrogen	$NaF \cdot HF$	61.99	col, or wh cr powd, rhdr	2.08	...	...	s	s	...
s298	fluoride *ortho*phosphate	$NaF \cdot Na_3PO_4 \cdot 12H_2O$	422.11	...	2.2165	...	...	12^{25}	57.5^{50}	...
s299	fluoroacetate, mono-	$NaC_2H_2FO_2$	100.02	wh powd	...	200	...	111^{25}	...	1.4^{25} al; 5^{25} MeOH; 0.04^{25} acet; 0.0049^{25} CCl_4
s300	fluorophosphate, hexa-	$NaPF_6 \cdot H_2O$	185.97	...	2.369^{19}	...	...	103.2^{0}	...	...
s301	fluorophosphate, mono-	Na_2PO_3F	143.95	col	...	ca 625	...	25	...	...
s302	fluosilicate	Na_2SiF_6	188.06	col, hex, 1.312, 1.309	2.679	d	...	0.0652^{17}	2.46^{100}	i al
s303	fluosulfonate	$NaSO_3F$	122.05	shiny leaf, hygr	...	d red heat	...	s	...	s al, acet; i eth

PHYSICAL CONSTANTS OF INORGANIC COMPOUNDS (continued)

No.	Name	Synonyms and Formulae	Mol. wt.	Crystalline form, properties and index of refraction	Density or spec. gravity	Melting point, °C	Boiling point, °C	Solubility, in grams per 100 cc		
								Cold water	Hot water	Other solvents
s304	formaldehydesulfoxylate	$NaHSO_2 \cdot CH_2O \cdot 2H_2O$	154.11	rhomb pr, hygr.	...	64	d	v s	...	d a; s al, alk
s305	formate	$NaCHO_2$	68.01	col, monocl, deliq	1.92^{20}	253	d	97.2^{20}	160^{100}	sl s al; i eth
s306	2-furanacrylate	$NaC_7H_5O_3$	160.10	lt brn powd	1.919	d	...	s	s	sl s al; i eth
s307	*meta*germanate	Na_2GeO_3	166.57	wh, monocl, deliq, 1.59	3.31^{22}	1083	...	...	d	s a
s308	*meta*germanate, heptahydrate	$Na_2GeO_3 \cdot 7H_2O$	292.67	col, rhomb	...	83	...	24.6^{0}	...	s a
s309	(mono-) d-glutamate	$NaC_5H_8NO_4$	169.11	wh cr.	...	d	...	v s	...	sl s al
s310	glycerophosphate, monohydrate	$Na_2C_3H_7O_6P \cdot H_2O$	234.05	yelsh visc liq; wh cr or powd.	...	...	...	s	...	s al
s311	glycerophosphate, pentahydrate	$Na_2C_3H_7O_6P \cdot 5^1/_2H_2O$	315.12	wh pl, sc or powd	...	>130	...	v s	...	i al
s312	gold sulfide	$NaAuS \cdot 4H_2O$	324.08	col, monocl.	...	d	...	s	...	s al
s313	hydride	NaH	24.00	silver need, 1.470	0.92	d 800	...	d	d	s molten Na; i CS_2, CCl_4, NH_3, bz
s314	hydroxide	NaOH	40.00	wh, deliq, 1.3576	2.130	318.4	1390	42^{0}	347^{100}	v s al, glyc; i acet, eth
s315	iodate	$NaIO_3$	197.89	wh, rhomb	$4.277^{17.5}$	d	...	9^{20}	34^{100}	i al; s ac a
s316	*meta*periodate	$NaIO_4$	213.89	col, tetr	4.174	d 300	...	14.44^{25}	$38.9^{51.5}$	s H_2SO_4, HNO_3, ac a
s317	*meta*periodate, trihydrate	$NaIO_4 \cdot 3H_2O$	267.94	col, rhombdsh, effl	3.219_4^{18}	d 175	...	18.78^{25}	$36.4^{34.5}$	...
s318	*para*periodate	Na_5IO_6	337.85	wh	...	800 d	...	d	...	...
s319	(tri-)*para*periodate	$Na_3H_2IO_6$	293.89	col, hexag.	...	...	...	sl s	...	s con NaOH sol
s320	iodide	NaI	149.89	col, cub, 1.7745	3.667_4^{25}	661	1304	184^{25}	302^{100}	42.57^{25} al; 39.9^{25} acet; s glyc
s321	iodide, dihydrate	$NaI \cdot 2H_2O$	185.92	col, monocl.	$2.448^{20.8}$	752	...	317.9^{0}	1550^{100}	v s NH_3
s322	iodoplatinate	$Na_2PtI_6 \cdot 6H_2O$	1110.58	brn, monocl.	3.707	...	...	v s	...	s al
s323	iridium chloride	Sodium hexachloroiridate. $Na_3IrCl_6 \cdot 12H_2O$	690.09	olive cr, rhomb or trig-rhomb	...	50	...	s	s	i al
s324	iron (III) nitrosopentacyanide	$Na_2[Fe(CN)_5NO] \cdot 2H_2O$	297.95	ruby red, rhomb, 1.605, 1.575, 1.56.	1.687^{25}	$-H_2O$, 100	d 160	40^{16}	s	...
s325	iron (III) oxalate	$Na_3[Fe(C_2O_4)_3] \cdot 5^1/_2H_2O$	487.96	grn, monocl.	$1.973^{17.5}$	$-4H_2O$	d 300	32^{0}	182^{100}	...
s326	iron (III) sulfate	$3Na_2SO_4 \cdot Fe_2(SO_4)_3 \cdot 6H_2O$	934.07	wh, trig, 1.558, 1.613	2.5	$-6H_2O$, 100	...	d v sl	...	i al
s327	lactate	$NaC_3H_5O_3$	112.06	col or yelsh liq, very hygr	...	17	d 140	v s	...	s al; i eth
s328	lithium sulfate	$Na_3Li(SO_4)_2 \cdot 6H_2O$	376.12	col, ditrig	2.009	$-6H_2O$, 50	...	s	s	...
s329	magnesium carbonate	$Na_2CO_3 \cdot MgCO_3$	190.30	wh, rhomb	2.729^{15}	677 CO_2 1240/atm	...	d	d	...
s330	magnesium sulfate	Nat. bloedite. $Na_2SO_4 \cdot MgSO_4 \cdot 4H_2O$	334.46	col, monocl, 1.486, 1.488, 1.489	2.23	...	...	s	...	...
s331	magnesium tartrate	$Na_2Mg(C_4H_4O_6)_2 \cdot 10H_2O$	546.58	wh, monocl pr or powd	...	...	...	s	s	...
s332	manganate	$Na_2MnO_4 \cdot 10H_2O$	345.07	grn, monocl.	...	17	...	s	d	...
s333	*per*manganate	$NaMnO_4$	141.93	red cr. deliq	...	d	...	v s	v s	...
s334	*per*manganate, trihydrate	$NaMnO_4 \cdot 3H_2O$	195.97	purp cr, deliq.	2.47	d 170	...	v s	v s	s NH_3; d alk
s335	methanearsenate	$Na_2CH_3AsO_2 \cdot 6H_2O$	292.03	wh cr powd.	...	130—140	...	ca 100	...	sl s al; i bz, eth. oils
s336	methoxide	$CH_3ONa \cdot 2CH_3OH$	118.11	wh powd	...	d, $-CH_3OH$	...	s, d	...	s CH_3OH
s337	methylsulfate	$NaCH_3SO_4 \cdot H_2O$	152.10	col cr, hygr.	...	...	...	s	...	s al
s338	molybdate	Na_2MoO_4	205.92	opaque wh	3.28^{18}	687	...	s 44.3	84^{100}	...
s339	molybdate, dihydrate	$Na_2MoO_4 \cdot 2H_2O$	241.95	wh, rhbdr	3.28(?)	$-2H_2O$, 100	...	56.2^{0}	115.5^{100}	i meth acet
s340	*deca*molybdate	$Na_2Mo_{10}O_{31} \cdot 21H_2O$	1879.68	wh, monocl pr	...	...	...	sl s	0.842^{100}	...
s341	*di*molybdate	$Na_2Mo_2O_7$	349.86	wh need	...	612	...	sl s	sl s	...
s342	*octa*molybdate	$Na_2Mo_8O_{25}17H_2O$	1519.74	monocl cr	...	$-H_2O$, 20	...	v s	v s	...
s343	*para*molybdate	$Na_6Mo_7O_{24} \cdot 22H_2O$	1589.84	col, monocl, effl	...	700 $-H_2O$, 100—120	...	117.9^{30} (anhydr)	s	...
s344	*tetra*molybdate	$Na_2Mo_4O_{13} \cdot 6H_2O$	745.82	yel need	...	...	...	39.8^{21}	v s	...
s345	*tri*molybdate	$Na_2Mo_3O_{10} \cdot 7H_2O$	619.90	need acicular	...	528 $-6H_2O$, 100—120	...	3.878^{20}	13.7^{100}	...
346	*meta*niobate	$Na_2Nb_2O_6 \cdot 7H_2O$	453.90	pseudo-cub	4.512—4.559	$-H_2O$, 100	...	s	...	...
s347	nitrate	Soda niter. $NaNO_3$	84.99	col, trig or rhbdr, 1.587, 1.336	2.261	306.8	d 380	92.1^{25}	180^{100}	s al, MeOH; v s NH_3; v sl s acet; sls glyc
s348	nitride	Na_3N	82.98	dk gray	...	d 300	...	d	...	...
s349	nitrite	$NaNO_2$	69.00	col-yel, rhomb pr, hygr.	2.168^{0}	271	d 320	81.5^{15}	163^{100}	0.3^{20} eth; 4.4^{20} MeOH; 3 abs al; v s NH_3
s350	*hypo*nitrite	$Na_2N_2O_2$	105.99	...	1.728^{25}	d 300	...	d	...	i al
s351	*p*-nitrophenoxide	$NaOC_6H_4NO_2 \cdot 4H_2O$	233.15	yel, monocl pr	...	$-2H_2O$, 36 $-4H_2O$, 120	d	5.97^{25}	...	sl s al
s352	nitroplatinite	$Na_2Pt(NO_2)_4$	425.08	pa yel rhomb or monocl pr, effl	...	...	...	s	s	...
s353	nitroprusside	$Na_2[Fe(NO)(CN)_5] \cdot 2H_2O$	297.95	red, rhomb	1.72	...	...	40^{16}	...	s al
s354	oleate	$NaC_{18}H_{38}O_2$	304.45	wh cr, or yel amorph gran.	...	232—235	...	10^{12}	...	s al; sl s eth
s355	oxalate	$Na_2C_2O_4$	134.00	col cr, or wh powd	2.34	d 250—270	...	3.7^{20}	6.33^{100}	i al, eth
s356	oxalate, hydrogen	$NaHC_2O_4 \cdot H_2O$	130.03	wh, monocl.	...	$-H_2O$, 100	d 200	1.7^{15}	21^{100}	...
s357	oxalatoferrate (III)	$Na_3Fe(C_2O_4)_3 \cdot xH_2O$	388.88 $+xH_2O$	grn, monocl cr	$1.973^{17.5}$	$-H_2O$, 100—120	...	32.5	182^{100}	...

No.	Name	Synonyms and Formulae	Mol. wt.	Crystalline form, properties and index of refraction	Density or spec. gravity	Melting point, °C	Boiling point, °C	Solubility, in grams per 100 cc		
								Cold water	Hot water	Other solvents
s358	oxide, mon-	Na_2O	61.98	wh-gray, deliq	2.27	subl 1275		d	d	d al
s359	oxide, per-	Na_2O_2	77.98	yel-wh powd	2.805	d 460	d 657	s	d	i al
s360	oxide, per-, octahydrate	$Na_2O_2 \cdot 8H_2O$	222.10	wh, hex		d 30	d	s	d	i al
s361	palmitate	$NaC_{16}H_{31}O_2$	278.41	wh cr.		270				
s362	pentobarbital	$NaC_{11}H_{17}N_2O_3$	248.26					s	s, d	s al
s363	phenobarbital	$NaC_{12}H_{11}N_2O_3$	254.22	wh				v s		s al; i eth, chl
s364	1-phenol-4-sulfonate (*p*-)	$NaC_6H_4(OH)SO_3 \cdot 2H_2O$	232.18	col, monocl or gran, sl effl		d		23.8^{25}	125^{100}	0.75^{25} al; 20^{25}glyc
s365	phenoxide	$NaOC_6H_5$	116.09	wh cr need, deliq				v s		s al, acet; d a
s366	phenylcarbonate	$NaC_7H_5O_3$	160.10	col powd		d 120		d		d acet
s367	*hypo*phosphate	$Na_4P_2O_6 \cdot 10H_2O$	430.06	col, monocl, 1.477, 1.482, 1.504	1.823	d		1.49^{25}	5.46^{50}	
s368	*hypo*phosphate, di-H	$Na_2H_2P_2O_6 \cdot 6H_2O$	314.03	col, monocl, 1.468, 1.490, 1.504	1.849	250 (anhydr)	$-6H_2O$, 100	2.35	25	s dil H_2SO_4, NH_4OH; i al
s369	*meta*phosphate, hexa-	Graham's salt. $(NaPO_3)_6$	611.77	col glass, 1.482 ± 0.002				v s		
s370	*meta*phosphate, tri-, hexahydrate	Knorre's salt. $(NaPO_3)_3 \cdot 6H_2O$	413.98	col, tricl, effl, 1.433, 1.442, 1.446		53; $-6H_2O$, 50		s		
s371	*ortho*phosphate	$Na_3PO_4 \cdot 10H_2O$	344.09	col, oct	$2.536^{17.5}$ (anhydr)	100		8.8 (anhydr)		
s372	*ortho*phosphate	$Na_3PO_4 \cdot 12H_2O$	380.12	col, trig, 1.446, 1.452	1.62^{20}	d 73.3—76.7	$-12H_2O$, 100	1.5^{0}	157^{70}	i CS_2, al
s373	*ortho*phosphate, di-H	$NaH_2PO_4 \cdot H_2O$	137.99	col, rhomb, 1.456, 1.458, 1.487	2.040	$-H_2O$, 100	d 204	59.9^{0}	427^{100}	v sl s eth, chl, tol; i al
s374	*ortho*phosphate, di-H	$NaH_2PO_4 \cdot 2H_2O$	156.01	col, rhomb, 1.4629	1.91	60		v s	v s	
s375	*ortho*phosphate, mono-H	Sörensen's sodium phosphate. $Na_2HPO_4 \cdot 2H_2O$	177.99	rhomb bisphenoidal, 1.463	2.066^{16}	$-2H_2O$, 95		100^{50}	117^{80}	
s376	*ortho*phosphate, mono-H	$Na_2HPO_4 \cdot 7H_2O$	268.07	col, monocl pr, 1.442	1.679	$-5H_2O$, 48.1		104^{40}		i al
s377	*ortho*phosphate, mono-H	$Na_2HPO_4 \cdot 12H_2O$	358.14	col, rhomb or monocl, eff, wh powd, 1.432, 1.436, 1.437	1.52	$-5H_2O$, 35.1	$-12H_2O$, 100	4.15	87.4^{34}	i al
s378	*pyro*phosphate	$Na_4P_2O_7$	265.90	wh cr, 1.425	2,534	880		3.16^{0}	40.26^{100}	
s379	*pyro*phosphate	$Na_4P_2O_7 \cdot 10H_2O$	446.06	col, monocl, 1.450, 1.453, 1.460	1.815—1.836	$-H_2O$, 93.8 m.p. 880		5.41^{0}	93.11^{100}	i al, NH_3
s380	*pyro*phosphate, di-H	$Na_2H_2P_2O_7 \cdot H_2O$	330.03	monocl, 1.4599, 1.4646, 1.4649	1.85	$-H_2O$,220		6.9^{0}	35^{40}	
s381	phosphide	Na_3P	99.94	red		d		d, PH_3		
s382	*hypo*phosphite	$NaH_2PO_2 \cdot H_2O$	105.99	col, monocl, deliq		d viol		100^{25}	667^{100}	v s al; s glyc; sl s NH_3 NH_4OH
s383	*ortho*phosphite, di-H	$NaH_2PO_3 \cdot 2^1/_2H_2O$	149.02	col, monocl, 1.419, 1.431, 1.449		42	$-2^1/_2H_2O$, 100	56^{0}	193^{42}	
s384	*ortho*phosphite, mono-H	$Na_2HPO_2 \cdot 5H_2O$	216.04	wh, rhomb deliq, β, 1.443	53	d 200-250		s	v s	i al, NH_4OH
a385	*tri*phosphate	Sodium tripolyphosphate, $Na_5P_3O_{10}$	367.86	powd and gran				14.5^{25}	32.5^{100}	
s386	phthalate	$Na_2C_8H_4O_4$	210.10	wh powd or pearly pl		$-H_2O$, 150				
s387	platinate, hydroxo-	$Na_2Pt(OH)_6$	343.10	yel or red-brn. hex		$-3H_2O$, 150—170	d	s		i al; sl s HCl
s388	platinum cyanide	$Na_2[Pt(CN)_4] \cdot 3H_2O$	399.18	col, triel	2.646	$-H_2O$, 120—125		s		s al
s389	plumbate, hydroxo-	$Na_2Ph(OH)_6$	355.22	yel-wh lumps, hygr				d to PbO_2		d a; s alk
s390	potassium(*dl*)- tartrate	$NaKC_4H_4O_6$	210.16	col, triel		90—100	d 200	47.4_6 (anhydr)	v s	
s391	propionate	$NaC_2H_5O_2$	96.06	wh, gran powd				s		s al
s392	*per*rhenate	$NaReO_4$	273.19	col, hex pl, hygr	5.39	300(in O_2) d 440 (vac)		100^{20}		s al
s393	*pyrohypo*rhenate	$Na_4Re_2O_7 \cdot H_2O$	594.38	sandy yel cr.				0.004		
s394	rhodiumchloride	$Na_2RhCl_6 \cdot 12H_2O$	600.78	dk red cr, monocl pr		$-12H_2O$, 120		v s	v s	i al
s395	rhodiumnitrite	$Na_3[Rh(NO_2)_6]$	447.91	wh cr.		d 360		40^{17}	s	i al; d a
s396	*per*ruthenate	$NaRuO_4 \cdot H_2O$	206.07	blk cr, lamellar		d 440 vac		v s	d	
s397	salicylate	$NaC_7H_5O_3$	160.10	wh cr powd				111^{15}	125^{25}	17^{15} al; 25 glyc
s398	selenate	Na_2SeO_4	188.94	col, rhomb	$3.213^{17.4}$			84^{35}	72.8^{100}	
s399	selenate, decahydrate	$Na_2SeO_4 \cdot 10H_2O$	369.09	col, monocl	1.603—1.620	*ra*32 trans		43.5^{20}	340^{100}	
s400	selenide	Na_2Se	124.94	wh to red, cr, deliq	2.625^{10}	>875		d		i NH_3
s401	selenite	$Na_2SeO_3 \cdot 5H_2O$	263.01	wh cr, tetrag				s	s	i al
s402	silicate	Waterglass. $Na_2O \cdot xSiO_2 (r = 3 - 5)$		col, amorph, deliq				s	s	i al, K and Na salts
s403	*di*silicate	$Na_2Si_2O_5$	182.15	rhomb pearly luster 1.500, 1.510		874		s	s	
s404	*meta*silicate	Na_2SiO_3	122.06	col, monocl, α 1.518, γ 1.527	2.4	1088		s	s, d	i al, K and Na salts

PHYSICAL CONSTANTS OF INORGANIC COMPOUNDS (continued)

No.	Name	Synonyms and Formulae	Mol. wt.	Crystalline form, properties and index of refraction	Density or spec. gravity	Melting point, °C	Boiling point, °C	Solubility, in grams per 100 cc		
								Cold water	Hot water	Other solvents
s405	*meta*silicate	$Na_2SiO_3{\cdot}9H_2O$	284.20	col, rhomb bi-pyramid, effl		40—48	$-6H_2O$, 100	v s	v s	s dil NaOH; i al, a
s406	*ortho*silicate	Na_4SiO_4	184.04	col, hex, 1.530		1018		s	s	
s407	silicotungstate, dodeca-	$Na_4[Si(W_3O_{10})_4]{\cdot}20H_2O$	3326.53	col, tricl		$-7H_2O$, 100	d	v s	v s	s a, sl s al
s408	stannate, hydroxo-	$Na_2Sn(OH)_6$	266.71	col, hex or wh powd, or lumps		$-3H_2O$, 140		$61.3^{15.5}$	50^{100}	i al, acet
s409	stearate	$NaC_{18}H_{35}O_2$	306.46	wh fatty powd				s	s	s h al
s410	succinate	$Na_2C_4H_4O_4{\cdot}6H_2O$	270.14	wh, gran or powd		$-6H_2O$, 120		21.45^0	86.63^{75}	v sl s al
s411	succinate, tetrahydroxy-	Sodium dihydroxy tartrate. $Na_2C_4H_4O_8{\cdot}3H_2O$	280.10			d		0.032^0	d	d min a; i al, eth
s412	sulfanilate	$NaC_6H_4(NH_2)SO_3$	195.17	wh. lust cr leaf.				s		
s413	sulfate, anhydr	Na_2SO_4	142.04	monocl (between *ca*160—185), 1.480		884; tr to hex *ca* 241		s	42—5^{100}	s HI
s414	sulfate, anhydr	Nat. thenardite. Na_2SO_4	142.04	orthorhomb, 1.484, 1.477, 1,471	2.68			4.76^0	42.7^{100}	s glyc; i al
s415	sulfate, decahydrate	Glauber's salt, mirabilite. $Na_2SO_4{\cdot}10H_2O$	322.19	col, monocl, effl, 1.394, 1.396, 1.398	1.464	32.38	$-10H_2O$, 100	11^0	92.7^{30}	i al
s416	sulfate, heptahydrate	$Na_2SO_4{\cdot}7H_2O$	268.14	wh, rhomb or tetrag		tr to anhydr 24.4		19.5^0	44^{20}	i al
s417	*pyro*sulfite	Sodium metabisulfite, $Na_2S_2O_5$	190.10	wh powd or cr ($+7H_2O$)	1.4	>d 150		54^{20}	81.7^{100}	sl s al; s glyc
s418	*pyro*sulfate	$Na_2S_2O_7$	222.10	wh, transluc cr, deliq	2.658^{25}	400.9	d 460	s		s fum H_2SO_4
s419	sulfate hydrogen	$NaHSO_4$	120.06	col, tricl	2.435^{13}	>315	d	28.6^{25}	100^{100}	sl s al; i NH_3
s420	sulfate hydrogen, monohydrate	$NaHSO_4{\cdot}H_2O$	138.07	col, monocl, deliq ~ 1.46	$2.103_4^{13.5}$	58.54 ± 0.5		*ca* 67, d	d	d al
s421	sulfide, hydro-	NaHS	56.06	col, rhomb or wh gran cr, deliq		350		vs		s al
s422	sulfide, hydrodihydrate	$NaHS{\cdot}2H_2O$	92.09	col need, deliq		d		s	s	d a; s al
s423	sulfide, mono-	Na_2S	78.04	wh cr, deliq.	1.856^{14}	1180		15.4^{10}	57.2^{90}	d a; sl s al; i eth
s424	sulfide, monohydrate.	$Na_2S{\cdot}9H_2O$	240.18	col, tetr, deliq	1.427_4^{16}	d 920		47.5^{10}	96.7^{10}	d, sl s al
s425	sulfide, penta-	Na_2S_5	206.28	yel (exist ?)		251.8		s	s	s al
s426	sulfide, tetra-	Na_2S_4	174.22	yel, cub, hygr		275	d	s		s al
s427	sulfide, hydrotrihydrate	$NaHS{\cdot}3H_2O$	110.11	col, lust rhomb cr		22	d	s	s	s al
s428	sulfite	Na_2SO_3	126.04	wh powd or hex, prism, 1.565, 1.515	$2.633^{15.4}$	d red heat	d	12.54^0	28.3^{80}	sl s al; i liq Cl_2, NH_3
s429	sulfite hydrate	$Na_2SO_3{\cdot}7H_2O$	252.14	col, monocl, effl	1.539^{15}	$-7H_2O$, 150	d	32.8^0	196^{40}	sl s al
s430	*hydro*sulfite	Dithionite, hyposulfite $Na_2S_2O_4{\cdot}2H_2O$	210.13	col, monocl(?) cr, or yel-wh powd		d 52		25.4^{20}	d	d a; s alk; i al
s431	sulfite, hydrogen	$NaHSO_3$	104.06	wh, monocl, yel in sol, 1.526	1.48	d		v s	v s	sl s al
s432	*d*(& *l*)-tartrate	$Na_2C_4H_4O_6{\cdot}2H_2O$	230.08	col, rhomb. 1.545, 1.49	1.818	$-2H_2O$, 150		29^6	66^{43}	i al
s433	*d*-tartrate, hydrogen	$NaHC_4H_4O_6{\cdot}H_2O$	190.09	wh cr powd, rhomb, 1.53, 1.54, 1.60		$-H_2O$, 100	d 234	6.7^{18}	9.2^{30}	
s434	*di*-tartrate hydrogen	$NaHC_4H_4O_6{\cdot}H_2O$	190.09	col, monocl or tricl, 1.53, 1.54, 1.60		$-H_2O$, 100	d 219	8.9^{19}		
s435	*ortho*tellurate, tetra-H	$Na_2H_4TeO_6$	273.61	col, hex pl		d	d	0.77^{18}	2^{100}	s h dil HNO_3; i NaOH
s436	telluride	Na_2Te	173.58	wh cr powd very hygr, d in air	2.90	953		v s, d	v s, d	
s437	tellurite	Na_2TeO_3	221.58	wh, rhomb pr.				sl s	s	
s438	thioantimonate	Schlippe's salt, $Na_2SBS_4{\cdot}9H_2O$	481.10	pa yel, cub	1.806	87	d 234	20.15^0	100^{100}	i al, eth
s439	thioarsenate	$Na_3AsS_4{\cdot}8H_2O$	416.25	yel, monocl, β 1.6802		d		v s	d	i al
s440	thiocarbonate, tri	$Na_2CS_2{\cdot}H_2O$	172.19	yel need, deliq		d 75		s	d	s al; i eth, bz
s441	thiocyanate	NaSCN	81.07	col, rhomb deliq pois, ~1.625		287		$139.3^{21.3}$	225^{100}	v s al, acet
s442	*di*thionate	$Na_2S_2O_6{\cdot}2H_2O$	242.13	col, rhomb, 1.482, 1.495, 1.519	2.189	$-2H_2O$, 110	$-SO_2$, 267	47.6^{16}	90.9^{100}	s HCl; i al
s443	thiosulfate	$Na_2S_2O_3$	158.10	col, monocl	1,667			50	231^{100}	i al
s444	thiosulfate, pentahydrate	"Hypo", sodium hyposulfite. $Na_2S_2O_3{\cdot}5H_2O$	248.17	col, monocl, effle, 1.489, 1.508, 1.536	1.729^{17}	40—45 d 48	$-5H_2O$, 100	79.4^0	291.1^{45}	s NH_3; i al
s445	thiosulfosurate (I)	$Na_3[AuS_2O_3)_2]{\cdot}2H_2O$	526.20	wh cr, monocl	3.09	$-H_2O$, 150	d	50		i al
s446	*tri*titanate	$Na_2Ti_2O_7$	301.62	wh need, monocl	3.35—3.50	1128		i		s h HCl
s447	tungstate	Na_2WO_4	293.83	wh, rhomb	4.179	698		57.5^0 73.2^{21}	96.9^{100}	
s448	tungstate, dihydrate	$Na_2WO_4/2H_2O$	329.86	col pl, rhomb. 1.5533	3.23—3.25	$-2H_2O$, 100 anhydr 698		41^0	123.5^{100}	sl s NH_3; i al, a
s449	*meta*tungstate	$Na_2O{\cdot}4WO_2{\cdot}10H_2O$	1169.52	col, oct		706.6		s	v s	i a
s450	*para*tungstate	$Na_6W_7O_{34}{\cdot}16H_2O$	2097.12	col, tricl	3.987	$-12H_2O$, 100; $-16H_2O$, 300		8	d	

No.	Name	Synonyms and Formulae	Mol. wt.	Crystalline form, properties and index of refraction	Density or spec. gravity	Melting point, °C	Boiling point, °C	Solubility, in grams per 100 cc: Cold water	Hot water	Other solvents
s451	*meta*uranate	Na_2UO_4	348.01	gr yel or red pl, rhomb pr	...	...	...	i	i	s dil a, alk carb
s452	uranyl acetate	$NaUO_2(C_2H_2O_2)_3$	470.15	yel, tetr pr, 1.501	2.56	...	...	...	...	...
s453	uranyl carbonate	$2Na_2CO_2 \cdot UO_2CO_3$	542.01	yel cr.	...	d 400	...	sl s	...	i al
s454	urate	$Na_2C_5H_2N_4O_2 \cdot H_2O$	230.09	wh gran powd or hard cr nodules	...	...	...	...	1.3^{100}	v sl s 90% al
s455	urate, acid	$NaHC_5H_2N_4O_3$	190.09	wh gran powd	...	...	...	0.083	0.8^{100}	
s456	valerate	$NaC_5H_9O_2$	124.12	wh cr or mass, hygr.	...	140	...	s	...	s al
s457	*meta*vanadate	$NaVO_3$	121.93	col, monocl pr	...	630	...	21.1^{25}	38.8^{75}	...
s458	*ortho*vanadate	Na_3VO_4	183.91	col, hex pr	...	850—866	...	s	...	i al
s459	*ortho*vanadate, decahydrate	$Na_2VO_4 \cdot 10H_2O$	364.06	wh, cub or hex cr, 1.5305, 1.5398, 1.5475	...	...	...	s	s	...
s460	*ortho*vanadate, hexadecylhydrate	$Na_2VO_4 \cdot 16H_2O$	472.15	col need	...	866 (anhydr)	...	v s	d	i al
s461	*pyro*vanadate	$Na_4V_2O_7$	305.87	col, hex	...	632—654	...	s	s	i al
s462	ethylxanthate	NaC_2H_5OCSS	144.18	yelsh powd	...	...	...	s	...	s al
s463	zinc uranyl acetate	$NaZn(UO_2)_3(C_2H_3O_2)_9 \cdot 9H_2O$	1591.99	monocl cr. α 1.475, γ 1.480	...	...	...	i	...	s al
	Stannous	See under tin		...						
	Stannic	See under tin		...						
s464	**Strontium**	Sr	87.62	silv wh to pa yel met	2.6^{20}	769	1384	d	d	s a, al, liq NH_3
s465	acetate	$Sr(C_2H_3O_2)_2$	205.71	wh cr.	2.099	d	...	36.9	36.4^{97}	0.26^{15} MeOH
s466	acetate	$Sr(C_2H_3O_2)_2 \cdot H_2O$	214.72	wh cr powd	...	$-{}^1/_2H_2O$ 150	...	s	...	sl s al
s467	*ortho*arsenate, acid	$SrHAsO_4 \cdot H_2O$	245.56	rhomb, need	3.606^{15}; 4.035 (anhydr)	$-H_2O$, 125	...	$0.284^{15.5}$d	s a	
s468	*ortho*arsenite	$Sr_2(AsO_3)_2 \cdot 4H_2O$	580.76	cr, or wh powd	...	...	...	sl s	...	s a; sl s al
s469	borate, *tetra-*	$SrB_4O_7 \cdot 4H_2O$	314.92	...	...	...	...	...	77^{100}	s HNO_3; NH_4 salts
s470	boride, hexa-	SrB_6	152.48	blk, cub	3.39^{15}	2235	...	i	i	s HNO_2; i HCl
s471	bromate	$Sr(BrO_3)_2 \cdot H_2O$	361.44	col yelsh, monocl, hygr.	3.773	$-H_2O$, 120	d 240	33^{16}	...	...
s472	bromide	$SrBr_2$	247.43	wh, hex need, hygr, 1,575	4.216^{24}	643	d	100^{20}	222.5^{100}	s al, amyl al
s473	bromide, hexahydrate	$SrBr_2 \cdot 6H_2O$	355.52	col, hex, hygr	2.386^{25}_{4}	tr to $2H_2O$, 88.6	$-6H_2O$, >180	204.2^0	∞	63.9^{30} al; 113.4^{30} MeOH; 0.6^{30} acet; i eth
s474	carbide	SrC_2	111.64	blk, tetr	3.2	>1700	...	d	d	d a
s475	carbonate	Nat. strontianite. $SrCO_3$	147.63	col, rhomb, or wh powd trfrs to −hex at 926, 1.516, 1.664, 1.666	3.70	1497^{69atm}	$-CO_2$, 1340	0.0011^{18}	0.065^{100}0	.12 aq CO_2; s a, NH_4 salts
s476	chlorate	$Sr(ClO_3)_2$	254.52	col, rhomb. or wh powd, 1.516, 1.605, 1.626	3.152	d 120	...	174.9^{18}	v s	s dil al; i abs al
s477	*per*chlorate	$Sr(ClO_4)_2$	286.52	col cr, hygr	...	...	...	310^{25}	v s	212 MeOH; 181 al; i eth
s478	chloride	$SrCl_2$	158.53	col, cub 1.650^{25}	3.052	875	1250	53.8^{20}	100.8^{100}	v sl s abs al, acet; i NH_3
s479	chloride, dihydrate	$SrCl_2 \cdot 2H_2O$	194.56	transp leaf, 1.594, 1.595, 1.617	2.6715^{25}	...	...	...	...	...
s480	chloride, fluoride	$SrCl_2 \cdot SrF_2$	284.14	col, tetr. 1.651, 1.627	4.18	962	...	d	d	s conc HNO_3, conc HCl, i al
s481	chloride, hexahydrate	$SrCl_2 \cdot 6H_2O$	266.62	col, trig, 1.536, 1.487	1.93	115, $-4H_2O$, 60	$-6H_2O$, 100	106.2^0	205.8^{40}	3.8^6 al
s482	chromate	$SrCrO_4$	203.61	yel, monocl	3.895^{15}	...	...	0.12^{15}	3^{100}	s HCl, HNO_3, ac a, NH_4 salts
s483	cyanide	$Sr(CN)_2 \cdot 4H_2O$	211.72	wh. rhomb. deliq	...	d	...	v s	...	...
s484	cyanoplatinite	$Sr[Pt(CN_4] \cdot 5H_2O$	476.85	col, monocl pr, 1.696	...	$-5H_2O$, 150	...	...	...	s abs al
s485	glycerophosphate	$SrC_2H_7O_6P$	257.68	wh powd	...	...	...	sl s	...	i al
s486	ferrocyanide	$Sr_2Fe(CN)_6 \cdot 15H_2O$	657.42	yel, monocl	...	...	...	50	100	...
s487	fluoride	SrF_2	125.62	col, cub or wh powd, 1.442	4.24	1473	2489^{760}	0.011^0	0.012^{27}	a hot HCl; i HF, al acet
s488	fluosilicate	$SrSiF_6 \cdot 2H_2O$	265.73	monocl.	$2.99^{17.5}$d	...	3 $.2^{15}$	v s	s HCl; 0.065^{16} 50% al	
s489	formate	$Sr(CHO_2)_2$	177.66	col, rhomb, 1.559, 1.547, 1.598	2.693	71.9	...	9.1^0	34.4^{100}	...
490	formate, dihydrate	$Sr(CHO_2)_2 \cdot 2H_2O$	213.69	col, rhomb, 1.484, 1.521, 1.538	2.25	d, $-2H_2O$, 100	...	$11.62^{36.6}$	26.57^{100}	i al, eth
s491	hydride	SrH_2(?)	89.64	wh, rhomb. hygr	3.72	d 675	subl 1000 (in H_2)	d	d	d al
s492	hydroxide	$Sr(OH)_2$	121.63	wh, deliq	3.625	375 (in H_2)	$-H_2O$, 710	0.41^0	21.83^{100}	s a, NH_4Cl
s493	hydroxide, octahydrate	$Sr(OH)_2 \cdot 8H_2O$	265.76	col, tetr, deliq. 1.499, 1.476	1.90	$-8H_2O$, 100	...	0.90^0	47.71^{100}	s a, NH_4Cl; i acet
s494	iodate	$Sr(IO_2)_2$	437.43	tricl.	5.045^{16}	...	...	0.03^{15}	0.8^{100}	...

No.	Name	Synonyms and Formulae	Mol. wt.	Crystalline form, properties and index of refraction	Density or spec. gravity	Melting point, °C	Boiling point, °C	Solubility, in grams per 100 cc		
								Cold water	Hot water	Other solvents
s495	iodide	SrI_2	341.43	col pl.	4.549^{26}_{4}	515	d	165.3^{0}	383^{100}	4.5^{30} al; 0.31^{0} NH_4OH; a MeOH
s496	iodide, hexahydrate	$SrI_2 \cdot 6H_2O$	449.52	col-yelah, hex, deliq	2.672^{25}	d 90		448.9^{0}	∞	s al; i eth
s497	lactate	$Sr(C_3H_5O_3)2 \cdot 3H_2O$	319.81	wh cr or gran powd		$-3H_2O$, 120		25	200^{100}	sl s al
s498	*per*manganate	$Sr(MnO_4)_2 \cdot 3H_2O$	379.54	purpl, cub	2.75	d 175		270^{0}	291^{18}	
s499	molybdate	$SrMoO_4$	247.56	col, tetr ~1.91	4.54^{26}_{25}	d		0.0104^{17}		s a
s500	nitrate	$Sr(NO_3)_2$	211.63	col, cub	2.986	570		70.9^{18}	100^{00}	0.012 aba al; v a NH_3; al s acet
s501	nitrate, tetrahydrate	$Sr(NO_3)_2 \cdot 4H_2O$	283.69	col, monocl	2.2	$-4H_2O$, 100	1100 tr SrO	60.43^{0}	206.5^{100}	s liq NH_3; v sl s abs al, acet; i HNO_3
s502	nitride	Sr_3N_2	290.87					d	d	s HCl
s503	nitrite	$Sr(NO_2)_2 \cdot H_2O$	197.65	col, hex, 1.588	2.408^{0}_{0}	$-H_2O$>100	d 240	58.9^{0}	182^{100}	0.42^{30} 90% al
s504	*hypo*nitrite	$SrN_2O_2 \cdot 5H_2O$	237.71	wh need	2.173^{25}_{4}			v sl s	sl s	v sl s NH_3
s505	oxalate	$SrC_2O_4 \cdot H_2O$	193.65	col cr.		$-H_2O$, 150		0.0051^{18}	0.15^{100}	s HCl, HNO_3
s506	oxide	SrO	103.62	gray-wh, cub, 1.810	4.7	2430	~3000	0.69^{30}	22.85^{100}	30 fus KOH; sl s al; i eth, acet
s507	oxide, per-	SrO_2	119.62	wh powd	4.56	d 215^{760}	0.018^{20}	d	v s al, NH_4Cl; i acet	
s508	oxide, per-, octahydrate	$SrO_2 \cdot 8H_2O$	263.74	col cr.	1.951	$-8H_2O$, 100	d	0.018^{20}	d	s NH_4Cl; i al, acet, NH_4OH
s509	*ortho*phosphate, di-	$SrHPO_4$	183.60	col, rhomb	3.544^{15}	1.62		i	i	s a, NH_4 salts
s510	salicylate	$Sr(C_7H_5O_3)_2 \cdot 2H_2O$	397.88	col cr.		d		5.6^{25}	28.6^{100}	1.5^{25}, 9.5^{78} al
s511	selenate	$SrSeO_4$	230.58	col, rhomb	4.23			i	i	s hot HCl; i HNO_3
s512	selenide	SrSe	166.58	wh, cub, 2.220	4.38			d	d	s HCl
s513	*meto*silicate	$SrSiO_2$	163.70	col, pr monocl, 1.599, 1.637	3.65	1580		i	i	
s514	*ortho*silicate	$SrSiO_4$	179.70	monocl, 1.728, 1.732, 1.758	3.84	>1750				
s515	sulfate	Nat. celestite, $SrSO_4$	183.68	col, rhomb, 1.622, 1.624, 1.631	3.96	1605		0.0113^{0}	0.014^{30}	sl s a; i al, dil H_2SO_4
s516	sulfate, hydrogen	$Sr(HSO_4)_2$	281.75	col		d		d		$14^{70}H_2SO_4$
s517	sulfide, hydro	$Sr(HS)_2$	153.76	col, cub need, 2.107		d		s	d	
s518	sulfide, mono	SrS	119.68	col, lt gray, cub, 2.107	3.70^{15}	>2000		i	d	d a
s519	sulfide, tetra-	$SrS_4 \cdot 6H_2O$	323.95	redsh cr, hygr		25	$-4H_2O$, 100	s	s	s al
s520	sulfite	$SrSO_2$	167.68	col cr.		d		0.0033^{17}		v s H_2SO_4; s a, HCl
s521	tartrate	$SrC_4H_4O_6 \cdot 4H_2O$	307.75	wh, monocl	1.966			0.112^{0}	0.755^{85}	s dil HCl, dil HNO_3
s522	telluride	SrTe	215.22	wh, cub, 2.408	4.83					
s523	thiocyanate	$Sr(SCN)_2 \cdot 3H_2O$	257.82	deliq		$-3H_2O$, 100	d 160—170	v s		v s al
s524	thiosulfate	$SrS_2O_2 \cdot 5H_2O$	289.81	monocl need	2.17^{17}	$-4H_2O$, 100		2.5^{12}	57^{100}	i al
s525	*di*thionate	$SrS_2O_4 \cdot 4H_2O$	319.80	trig. 1.530, 1.525	2.373	$-4H_2O$, 78		22^{16}	67^{100}	i al
s526	tungstate	$SrWO_4$	335.47	col, tetr	6.187	d		0.14^{15}		id a; i al
s527	**Sulfamic acid**	Amidosulfuric, amino-sulfonic acid, NH_2SO_3H	97.09	col, rhomb	2.126^{25}	200 d	d	14.68	47.08^{60}	v sl s al, eth, acet; i CS_2, CCl_4
s528	**Sulfamide**	Sulfuryl amide, $SO_2(NH_2)_2$	96.10	rhomb pl	1.611	91.5	d 250	s		s al
s529	**Sulfur**(α)	S_8	256.48	yel, rhomb, 1.957	2.07^{20}	112.8 95.5 (revers.) 444.6	444.674	i	i	23^{0} CS_2; sl s tol, al, bz, eth, liq NH_3; s CCl_4
s530	(β)	S_8	256.48	pa yel, monocl	1.96	119.0	444.674	i	i	70 CS_2; s al, bz
s531	(γ)	S_8	256.48	pa yel, amorph	1.92	*ca* 120	444.6	i	i	i CS_2
s532	bromide, mono-	S_2Br_2	223.93	red liq, 1.730	2.63	−40	$54^{0.2}$	d	d	s CS_2
s533	chloride, di-	SCl_2	102.97	dk red liq. 1.557^{11}	1.621^{15}_{15}	−78	d 59			s CCl_4, bz; d al, eth
s534	chloride, mono-	S_2Cl_2	135.03	yel-red liq, 1.666^{14}	1.678	−80	135.6	d	d	s bz, eth, CS_2
s535	chloride, tetra-	SCl_4	173.87	yel-br liq		−30	d −15	d	d	
s536	fluoride, hexa-	SF_6	146.05	col gas	gas 6.602 g/ℓ liq 1.88^{-50}	−50.5	−63.8 (subl)	sl s	sl s	s al, KOH
s537	fluoride, mono-	S_2F_2	102.12	col gas	liq 1.5^{-100}	−120.5	−38.4	d	d	d KOH
s538	fluoride, tetra-	SF_4	108.05	gas (exist ?)		−124	−40	d	d	
s539	(di) fluoride, deca-	S_2F_{10}	254.10	col liq	2.08^{0}_{4}	−92	29			d fus caust
s540	(tetra-) nitride, di-	S_4N_2	156.25	red liq or gray solid	1.901^{18}	23	d 100 expl	i		s eth; sl s al, CS_2
s541	(tetra-) nitride, tetra-	S_4N_4	184.27	or red, monocl	2.22^{15}	subl 179	expl 160	d		s CS_2, chl, bz, NH_3; sl s al, eth
s542	(tri)-*di*nitrogen dioxide	$S_2N_2O_2$	156.19	pa yel cr.		100.7	d	i		s al, bz
s543	oxide, di-	SO_2	64.06	col gas or liq suffoc odor	gas 2.927 g/ℓ liq 1.434	−72.7	−10	22.8^{0}	0.58^{90}	s al, ac a, H_2SO_4
s544	oxide, hept-	Sulfur oxide, per-S_2O_7	176.12	visc liq, or need		0	subl 10	d	d	s H_2SO_4
s545	oxide, mono-	SO (or S_2O_2)	48.06	col gas		d	d	d		
s546	oxide, sesqui-	S_2O_2	112.12	bl-grn cr.		d 70—95		d	d	s al, eth, fum H_2SO_4
s547	oxide, tetra-	Sulfurperoxide, SO_4	96.06	wh		d 0—3		s, d		d dil H_2SO_4
s548	oxide, tri-(α)	SO_2	80.06	silky fibr need, stable modific.	1.97^{20}	16.83	44.8	d	d	forms fum H_2SO_4
s549	oxide, tri-(β)	$(SO_3)_2$	160.12	asbestos like fiber metastable		62.4	50 (subl)	d	d	forms fum H_2SO_4

No.	Name	Synonyms and Formulae	Mol. wt.	Crystalline form, properties and index of refraction	Density or spec. gravity	Melting point, °C	Boiling point, °C	Solubility, in grams per 100 cc		
								Cold water	Hot water	Other solvents
s550	oxide, tri-(γ)	SO_2	80.06	vitreous, orthorhomb, metastable	liq 1.920_4^{20} sld 2.29^{-10}	16.8	44.8	d	d	forms fum H_2SO_4
s551	oxy*tetra*chloride, mono-	S_2OCl_4	221.93	dk red liq	1.656^0	...	60	d	d	d al
s552	oxy*tetra*chloride, tri-	$S_2O_2Cl_4$	253.93	wh, rhomb need or pl	...	d 57	...	d	d	d al
s553	*tri*thiazyl chloride	S_4N_2Cl	205.71	pa yel cr	...	d 170 (vac)	...	d	d	...
s554	**Sulfuric acid**	H_2SO_4	98.07	col liq	1.841 (96—98%)	10.36 (100%) 3.0 (98%)	330±0.5 (100%)	∞ ev heat	∞	d al
s555	dihydrate	$H_2SO_4 \cdot 2H_2O$	134.10	col liq, 1.405	1.650^0	−38.9	167	∞	∞	d al, eth
s556	hexahydrate	$H_2SO_4 \cdot 6H_2o$	206.17	liq	...	−54	...	v s	v s	...
s557	monohydrate	$H_2SO_4 \cdot H_2O$	116.09	col liq or monocl cr, 1.438	1.788	8.62	290	∞	∞	d al
s558	octahydrate	$H_2SO_4 \cdot 8H_2O$	242.20	liq	...	−62	...	v s	v s	...
s559	peroxidi-	Per(di-)sulfuric acid, $H_2S_2O_8$	194.13	hygr cr	...	d 65	d	d	d	s al, eth, H_2SO_4
s560	peroximono-	Permonosulfuric acid, Caro's acid H_2SO_5	114.07	...	...	d 45	...	d	d	s H_2PO_4
s561	pyro-	$H_2S_2O_7$	178.13	col cr, hygr	1.9^{20}	35	d	d	d	d al
s562	tetrahydrate	$H_2SO_4 \cdot 4H_2O$	170.13	...	...	−27	...	∞	∞	d al,eth
s563	**Sulfurous acid**	H_2SO_2	82.07	in sol only	*ca* 1.03	...	...	s	...	s al, eth, ac a
s564	**Sulfuryl** chloride	SO_2Cl_2	134.96	col liq, 1.444	1.6674_4^{20}	−54.1	69.1	d	d	s bz, ac a
s565	chloride fluoride	SO_2ClF	118.51	col gas	1.623^0 g/ℓ	−124.7	7.1	d	...	...
s566	fluoride	SO_2F_2	102.06	col gas	gas 3.72 g/ℓ liq 1.7	−136.7	−55.4	10^9	...	s al, CC_4; sl s alk
s567	pyro-, chloride	$S_2O_5Cl_2$	215.02	col liq, 1.937^{20}	gas 9.6 g/ℓ .iq 1.818_4^{11}	−39 to −37	152.5	d	d	d a
t1	**Tantalum**	Ta	180.9479	gray black hard metal, cub or powd	met 16.6^{20} powd 14.401	2996	5425 ± 100	i	i	s HF, fus alk; i a
t2	boride, di-	TaB_2	202.57	...	11.15	3000(?)	...	...	...	...
t3	bromide	$TaBr_5$	580.47	yel cr	4.67	265	348.8	d	d	s abs al, eth
t4	carbide	TaC	192.96	blk, cub	13.9	3880	5500	i	i	sl s H_2SO_4. HF
t5	chloride, penta-	$TaCl_5$	358.21	lt yel, vitr cr powd	3.68^{27}	216	242	d	...	s abs al, H_2SO_4
t6	fluoride	TaF_8	275.94	col, tetrag, deliq	4.74	96.8	229.5	s	...	s HF, eth
t7	nitride	TaN	194.95	br bronze or blk, hex	16.30	3360 ± 50	...	i	i	sl s aq reg, HF, HNO_3
t8	oxide, pent-	Ta_2O_5	441.89	col, rhomb	8.2	1872 ± 10	...	i	i	s fus $KHSO_4$, HF; i a
t9	oxide, pent- hydrate	Tantalic acid $Ta_2O_5xH_2O$	...	col gel	...	...	...	s	...	s alk, exc conc HNO_3; i a
t10	oxide, tetr-	Ta_2O_4 (or TaO_2)	425.89	dk gray powd	...	oxidizes	...	i	i	i a
t11	sulfide	Ta_2S_4 (or TaS_2)	490.14	blk powd or cr	...	>1300	...	i	i	al s HF + HNO_3; i HCl
t12	**Telluric acid, ortho-**	$Te(OH)_4$ or $H_2TeO_4 \cdot 2H_2O$	229.64	wh, monocl pr	3.071	136	...	s	s	sl a dil a, HNO_3; i abs al, acet, eth
t13	**Telluric acid**	$Te(OH)_4$ or H_6TeO_6	229.64	wh cub	3.158	136	...	s	s	sl s dil a, HNO_3; i abs al, acet, eth
t14	**Tellurium**	Te	127.60 ± 3	br blk, amorph, 1.0025	6.00	449.5 ± 0.3	989.8 ± 3.8	i	i	s H_2SO_4; HNO_3, aq reg, KCN, KOH; i HCl, CS_2
t15	**Tellurium**	Te	127.60 ± 3	rhomb silv wh met, 1.0025	6.25	452	1390	i	i	s H_2SO_4; HNO_3, aq keg, KCN KOH; i HCl, CS_2
t16	bromide, di-	$TeBr_2$	287.41	brn to gray grn, need, unstable	...	210	339	d	...	s eth; al s a; d NaOH
t17	bromide, tetra-	$TeBr_4$	447.22	or cr	4.31_4^{15}	380 ± 6	d 421	sl s	d	s eth, a tart a, NaOH
t18	chloride, di-	$TeCl_2$	198.51	blk cr or amorph, unstable	7.05	209 ± 5	327	d	d	s min a, tart a; d NaOH
t19	chloride, tetra	$TeCl_4$	269.41	wh to yel cr, deliq	3.26^{18} 2.559^{232}	224	380^{760}	s d	s d	s HCl, bz, al, chl, CCl_4; i CS_2
t20	ethoxide	$Te(OC_2H_5)_4$	307.84	...	...	20	$107—107.5^{5.5}$	...	...	...
t21	fluoride, hexa-	TeF_4	241.59	col gas unpleas odor	sol 4.006^{-191} liq 2.56^{-36}	−36	+35.5	d	d	d a, alk
t22	fluoride, tetra-	TeF_4	203.59	wh cr hygr	...	subl	>97	d	d	...
t23	hydride	H_2Te	129.62	col gas pois	4.49	−48.9	-2.2^{760}	v s	s	d al
t24	iodide, di-	TeI_2	381.41	blk cr (exist ?)	...	subl	...	i	i	...
t25	iodide, tetra,-	TeI_4	635.22	blk cr	5.403_4^{15}	280	d	sl s	d	s alk, aq NH_3, HI
t26	methoxide	$Te(OCH_3)_4$	251.74	solid	...	...	123—124	...	...	...
t27	oxide, di-	Tellurite, TeO_2	159.60	wh, tetr or rhomb, 2.00, 2.18(Li), 2.35	tetr 5.67^{15} rhomb 5.91^0	733	1245	i	i	s HCl, hot HNO_3, alk; i NH_4OH
t28	oxide, mon-	TeO	143.60	blk, amorph (exist ?)	5.682	d 370 (in CO_2)	d	i	i	s dil a, H_2SO_4, KOH
t29	oxide, tri-	TeO_3	175.60	α yel amorph β gray cr	α 5.0752_4^{105} β 6.21	d 395	...	i	i	d conc HCl; s hot KOH; i a, al
t30	sulfide	TeS_2	191.72	red blk powd amorph (exist ?)	...	...	...	i	...	i a; s alk sulf
t31	sulfoxide	$TeSO_3$	207.66	deep red amorph	...	d 30	d	d	d	s H_2SO_4
t32	**Tellurous acid**	H_2TeO_3(?)	177.61	wh flocks, indef not isolated	3.05	d 40	...	0.00067	d	s a, NaOH; sl s NH_4OH; i al

PHYSICAL CONSTANTS OF INORGANIC COMPOUNDS (continued)

No.	Name	Synonyms and Formulae	Mol. wt.	Crystalline form, properties and index of refraction	Density or spec. gravity	Melting point, °C	Boiling point, °C	Solubility, in grams per 100 cc		
								Cold water	Hot water	Other solvents
t33	**Terbium**	Tb	158.9254	silv-gray met, hex	8.2294	1360 ± 4	3123	i	i	s a
t34	bromide	$TbBr_3$	398.64	...	...	827	1490	s	s	...
t35	chloride hexahydrate	$TbCl_3 \cdot 6H_2O$	373.38	col pr cr, deliq	4.35 (anhydr)	588 (anhydr)	$-H_2O$, 180-200 (in HCl gas)	v s	...	...
t36	fluoride	TbF_3	215.92	...	...	1172	2280(?)	i	i	i dil a
t37	iodide	TbI_3	539.64	...	...	946	>1300	s	s	...
t38	*di*methylphosphate	$Tb[(CH_3)_2PO_4]_3$	534.05	...	...	...	...	12.6^{25}	8.07^{40}	...
t39	nitrate	$Tb(NO_3)_3 \cdot 6H_2O$	453.03	col, monocl cr	...	893	...	s	...	...
t40	oxalate	$Tb_2(C_2O_4)_3 \cdot 10H_2O$	762.06	wh cr.	2.60	$-H_2O$, 40	...	i	i	i dil a
t41	oxide	Terbia. Tb_2O_3	365.85	wh solid	...	...	...	...	...	s dil a
t42	oxide, per-	Tb_4O_7	747.70	dk-brn or blk solid	...	$-O2$	...	i	i	s hot conc a
t43	sulfate	$Tb_2(SO_4)_3 \cdot 8H_2O$	750.15	wh cr.	...	$-8HI_2O$, 360	...	3.561^{20}	2.51^{40}	...
t44	**Thallium**	Tl	204.383	bl-wh met, tetr	11.85	303.5	1457 ± 10	i	i	s HNO_3. H_2SO_4; sl s HCl
t45	acetate	$TlC_2H_2O_2$	263.43	silk wh cr, deliq	3.765^{137}	131	...	v s	...	v s al, $CHCl_3$; i acet
t46	aluminum sulfate	$TlAl(SO_4)_2 \cdot 12H_2O$	639.66	oct, 1.488	2.306^{20}	91	...	11.78^{25}	...	...
t47	azide	TlN_3	246.40	yel, tetr	...	330 (vac)	...	0.1712^{0}	0.3^{16}	i al, eth
t48	bromate	$TlBrO_2$	332.29	col, need	...	...	...	0.35^{20}	s	s dil al
t49	bromide, di-	Bromothallate(ous). Tl_2Br_4 or $Tl^{I}_3[Tl^{III}Br_4]$	728.38	yel need	...	...	...	d	d	...
t50	bromide, mono-	TlBr	284.29	yel-wh, cub, 2.4—2.8	$7.557^{17.3}$	480	815	0.05^{25}	0.25^{68}	s al; i HB_2, acet
t51	bromide, tri-	$TlBr_2$	444.10	yel, deliq. unstable	...	d	...	s	v s	v s al
t52	carbonate	Tl_2CO_2	468.78	col, monocl	7.11	273	...	$4.03^{15.5}$	27.2^{100}	i abs al, eth, acet
t53	chlorate	$TlClO_2$	287.83	need (rhomb ?)	5.047^{9}	...	...	2^{0}	57.31^{100}	...
t54	*per*chlorate	$TlClO_4$	303.83	col, rhomb	4.89	501	d	20.5^{30}	167^{100}	sl s al
t55	chloride	TlCl	239.84	wh reg discol in air, 2.247	7.004^{30}_{4}	430	720	$0.29^{15.6}$	$2.41^{99.35}$	i al, acet; d a
t56	chloride, tri-	$TlCl_3$	310.74	hex pl, hygr	...	25	d	v s	...	s al, eth
t57	chloride, tri-	$TlCl_3 \cdot H_2O$	328.76	col, need	...	$-H_2O$, 60	d 100	v s	d	v s al, eth
t58	chloride, tri-	$TlCl_3 \cdot 4H_2O$	382.80	col, need	...	37	$-4H_2O$, 100	86.2^{17}	d	s al, eth
t59	chloroplatinate	Tl_2PtCl_6	816.56	pale or cr	5.76^{17}	...	...	0.0064^{15}	0.05^{100}	...
t60	chromate	Tl_2CrO_4	524.76	yel	...	...	...	0.03^{60}	0.2^{100}	sl s a, alk; i ac a
t61	*di*chromate	$Tl_2Cr_2O_7$	624.75	red	...	...	...	i	...	d a
t62	chromium sulfate	Thallium chromium alum. $Tl[Cr(H_2O)_6](SO_4)_2 \cdot 6H_2O$	664.68	vlt cr	2.394	92	...	163.8^{25}	...	...
t63	cyanate	TlCNO	246.40	col, need	5.487^{20}_{4}	...	...	s	v s	sl s al
t64	cyanide	TlCN	230.40	tabl	6.523	d	...	$16.8^{28.5}$	s a	
t65	ethoxide	$(TlOC_2H_5)_4$	997.78	col liq	3.522	−3	d 80	s d	...	9.11^{2}ls5 al; s bz; i liq NH_3
t66	ethylate	$TlOC_2H_5$	249.44	liq, 1.6714^{30}	3.493^{20}_{4}	−3	d 130	...	...	sl s al; s eth
t67	ferrocyanide	$Tl_4Fe(CN)_6 \cdot 2H_2O$	1065.52	yel, tricl	4.641	...	...	0.37^{18}	3.93^{101}	...
t68	fluogallate	$Tl_2(GaF_5H_2O)$	591.49	col, orthorhomb.	6.44	...	...	s	...	...
t69	fluoride, mono-	TlF	223.38	col, cub, oct	8.23^{4}	327	655	78.6^{15}	d	sl s al
t70	fluoride, tri-	TlF_3	261.38	olive grn.	8.36^{25}	d 550	...	d	...	i conc HCl
t71	fluosilicate	$Tl_2SiF_6 \cdot 2H_2O$	586.87	hex pl	5.72	...	...	v s	...	...
t72	formate	$TlHCO_2$	249.40	col, need, hygr.	4.967^{104}	101	...	500^{10}	...	v a MeOH, sl s al; i $ChCl_3$
t73	(I) hydroxide	TlOH	221.39	pa yel, need	...	d139	...	25.9^{0}	52^{40}	s al
t74	iodate	$TlIO_3$	379.29	wh need	...	...	...	0.058^{20}	sl s	sl s HNO_3
t75	iodide (α)	TlI	331.29	yel, rhomb	7.29	tr to (β) 170	...	0.0006^{20}	0.12^{100}	s liq NH_3
t76	iodide (β)	TlI	331.29	red, cub	$7.098^{14.7}$	440	823	i	i	i al
t77	iodide, tri-	TlI_3	585.10	blk, lust rhomb.	...	d	...	s	...	...
t78	iron (III) sulfate	$TlFe(SO_4)_2 \cdot 12H_2O$	668.53	pink, oct, n_D^{17} 1.524	2.351^{15}	$-H_2O$ *ca* 100	...	36.15^{25} (anhydr)	...	...
t79	magnesium sulfate	$Tl_2SO_4 \cdot MgSO_4 \cdot 6H_2O$	733.28	wh dull cr, 1.5660, 1.5836, 1.5900	3.573^{20}_{4}	$-6H_2O$, 40	...	d 0	...	...
t80	methoxide	$TlOCH_3$	235.42	wh cr powd	...	d>120	...	s d	...	1.70^{25} CH_3OH; 3.16^{24} bz
t81	molybdate	Tl_2MoO_4	568.70	wh powd or cr	...	vol red heat	...	i	v sl s	i al; s alk carb, conc NH_4OH, HF
t82	myristate	$TlC_{14}H_{27}O_2$	431.75	wh powd	...	120—3	...	...	...	0.52^{25} 50% al
t83	(I) nitrate (α)	$TlNO_3$	266.39	cubic	...	206	430	9.55^{20}	4.13^{100}	i al, s acet
t84	(I) nitrate (β)	$TlNO_3$	266.39	trig	...	tr 145 to (α)	...	...	...	...
t85	(I) nitrate (γ)	$TlNO_3$	266.39	rhomb, α 1.817	$5.556^{21.4}$	tr 75 to (β)	...	3.91^{0}	414^{100}	i al; s acet
t86	(III) nitrate	$Tl(NO_3)_3$	390.40	cr	...	...	...	s	...	...
t87	(III) nitrate	$Tl(NO_3)_3 \cdot 3H_2O$	444.44	col, rhomb, deliq	...	...	s 100	d	d	...
t88	nitrite	$TlNO_2$	250.39	yel micro cr.	...	182	...	32.10^{25}	95.78^{96}	i a
t89	oleate	$TlC_{18}H_{33}O_2$	485.84	wh cr clusters	...	131—2	...	0.05^{15}	0.3^{80}	3.0^{25} al
t90	oxalate	$Tl_2C_2O_4$	496.79	monocl pr	6.31	...	...	1.48^{15}	9.02^{100}	
t91	oxalate, tetra-	$TlH_3(C_2O_4)_2 \cdot 2H_2O$	419.48	tricl, leaf, 1.5097, 1.6319, 1.6538	2.992^{17}	d 100	...	76.9^{23}	v s	i cold al; s hot al
t92	(I) oxide	Tl_2O	424.77	blk, deliq	9.52^{16}	300	1080^{760} −O, 1865	v s d to TlOH	...	s a, al
t93	(III) oxide	Tl_2O_2	456.76	col, amorph pr, hex	hex 10.19^{22} am 9.65^{21}	717 ± 5	−2O, 875	i	i	s a; i alk
t94	palmitate	$TlC_{16}H_{31}O_2$	459.80	crn need	...	115—117	...	0.01^{15}	0.07^{60}	1.04^{45} al

No.	Name	Synonyms and Formulae	Mol. wt.	Crystalline form, properties and index of refraction	Density or spec. gravity	Melting point, °C	Boiling point, °C	Solubility, in grams per 100 cc Cold water	Hot water	Other solvents
t95	phenoxide	$TlOC_6H_5$	297.49	wh cr.		233—5		d		s hot bz; sl s lgr
t96	*ortho*phosphate	Tl_3PO_4	708.12	col, need	6.89^{10}			0.5^{15}	0.67^{100}	i al, s NH_4 salts
t97	*ortho*phosphate, (di)-β.	TlH_2PO_4	301.37	monocl.	4.726	*ca*190		sl s	sl s	i al
t98	*pyro*phosphate	Tl_4Pli2O_7	991.48	monocl pr	6.786^{20}	>120		40		
t99	picrate	$TlC_4H_2N_3O_7$	432.48	red, monocl or yel tricl	red 3.164^{17} yel 2.993^{17}	expl 723—725		0.135^{0}	2.43^{70}	0.40 CH_2OH
t100	rhodanide	TlCNS	262.46	glossy leaflets, rhomb, tetr cr	4.954^{30}_{4}	d low temp		0.393^{25}		0.024^{0} liq SO_2; i acet; a MeOH
t101	selenate	Tl_2SeO_4	551.72	rhomb need, 1.949, 1.959, 1.964		>400		2.13^{10}	8.5^{50}	i al, eth
t102	selenide	Tl_2Se	487.73	gray leaf.	9.05^{25}_{4}	340		i		s a; i acet a
t103	silver nitrate	$TlNO_2 \cdot AgNO_3$	436.26	wh cr powd		75		s		
t104	stearate	$TlC_{18}H_{35}O_2$	487.86	need		119		0.005^{15}	0.095^{75}	0.18^{15} al, 0.060^{50} al
t105	(I) sulfate	Tl_2SO_4	504.82	col, rhomb. 1.860, 1.867, 1.885	6.77	632	d	4.87^{30}	19.14^{100}	
t106	(I) sulfate hydrogen	$TlHSO_4$	301.45	pr need		120 d				v sl s dil H_2SO_4
t107	(III) sulfate	$Tl_2(SO_4)_3 \cdot 7H_2O$	823.05	col leaf		$-6H_2O$, 220		d	d	s dil H_2SO_4
t108	(I) sulfide	Tl_2S	440.83	bl-blk tetr	8.46	448.5	d	0.02^{20}	sl s	s a; i alk, acet
t109	(III) sulfide	Tl_2S_3	504.95	blk, amorph.		260 (in N_2)	d	i	i	s hot H_2SO_4
t110	sulfite	Tl_2SO_3	488.82	wh cr.	6.427			3.34^{15}	v s	i al
t111	tartrate(dl)	$Tl_2C_4H_4O_6$	556.84	monocl.	4.659	d 165		13.3^{15}		
t112	*meta*tellurate	Tl_2TeO_4	600.36	heavy wh ppt	$6.760^{17.6}$	red heat		sl s	sl s	
t113	thiocyanate	TlSCN	262.46	col, tetr	4.956^{30}			0.315^{30}	0.727^{40}	i al
t114	*di*thionate	$Tl_2S_2O_4$	568.88	monocl.	5.57^{20}_{4}	d		41.8^{19}		
t115	thiosulfate	$Tl_2S_2O_3$	520.88	wh rhomb cr		d 130		sl s	v s	
t116	*meta*vanadate	$TlVO_3$	303.32	gray cr.	6.09^{17}	424		0.87^{11}	0.21^{100}	
t117	*pyro*vanadate	$Tl_4V_2O_7$	1031.41	light yel	8.21^{19}	454		0.2^{14}	0.26^{100}	
t118	**Thiocarbonyl chloride**	Thiophoagene. $CSCl_2$	114.98	red yel liq, 1.5442	1.509^{15}		73.5	d		d al; s eth
t119	**Thiocarbonyl chloride, tetra-**	$CSCl_4$	185.88	yel	1.712^{13}		146—147		d	
t120	**Thiocyanic acid(iso)**	HSCN(HNCS)	59.09	col mass or gas		>−110	polym to solid −90	v s		v s al, eth, bz
t121	**Thiocyanogen**	$(SCN)_2$	116.16	liq. or yel solid		−2 to −3		d		s al, eth, CS_2, CCl_4
t122	**Thionyl** bromide	$SOBr_2$	207.87	or, yel liq	2.68^{18}	−52	138^{773}, 68^{40}	d	d	s bz, chl, CS_2, CCl_4
t123	chloride	$SOCl_2$	118.97	col, or yel liq; 1.527^{10}	1.655^{10}_{4}	−105	78.8^{746} d 140	d	d	d a, al, alk; s bz, chl
s124	chloride fluoride	SOClF	102.51	gas		−139.5	12.2			
t125	fluoride	SOF_2	86.06	col, gas	gas 2.93 g/ℓ liq 1.780^{-100}	−110.5	−43.8	d	d	s eth, bz, chl, acet, $AsCl_3$
t126	**Thiophosphoramide**	Thiophosphorylamide. $PS(NH_2)_3$	111.10	yel wh cr	1.7^{13}	d 200		14.3^{25}	d	sl, me
t127	**Thiophosphoryl** bromide	$PSBr_3$	302.75	yel, cub	2.85^{17}	37.8	125^{25}	d		s eth, CS_2, PCl_3
t128	thiophosphoryl bromide, hydrate	$PSBr_3 \cdot H_2O$	320.76	yel cr.	2.794^{18}	35				
t129	thiophosphoryl bromide (mono-) chloride, di-	$PSBrCl_2$	213.84	yel liq	2.12^{0}	−30	d 150	d		
t130	thiophosphoryl bromide (di-) chloride	$PSBr_2Cl$	258.29	pa grn fum liq	2.48^{0}	−60	95^{60}	d		
t131	thiophosphoryl chloride	$PSCl_3$	169.39	col liq, 1.563 (c)	1.635	−35	125	d		s bz, CS_2, CCl_4
t132	thiophosphoryl fluoride	PSF_3	120.03	gas		$3.8^{7.6atm}$	d	sl s d		s eth; i bz, CS_2
t133	**Thiosulfuric acid**	$H_2S_2O_3$	114.13	in sol only				s		
t134	**Thorium**	Th	232.0381	gray, cub radioactive	11.7			i	i	s HCl, H_2SO_4, aq reg; sl s HNO_3
t135	boride, hexa-	ThB_6	296.90	dk viol-blk met, cub	6.4^{15}	2195		i	i	s HNO_3; i H_2SO_4, HCl, HF, aq alk
t136	boride, tetra	ThB_4	275.28	tetr pr	7.5^{15}			i	i	s HNO_3 HCl, hot H_2SO_4
t137	bromide	$ThBr_4$	551.65	col cr, hygr	5.67	subl 610	725	s	s	
t138	carbide	ThC_2	256.06	yel, tetr	8.96^{18}	2655 ± 25	*ca* 5000 (?)	d		v sl s conc a
t139	carbonate	$Th(CO_3)_2$	352.06	exist?.				i	d	s conc Na_2CO_3
t140	chloride	$ThCl_4$	373.85	wh, rhomb, deliq	4.59	770 ± 2 subl 820	d 928	v s	v s	s al, a, KCl; sl s eth
t142	*tetra*cyanoplatinate	$Th[Pt(CN)_4]_2 \cdot 16H_2O$	1118.58	yel-grn, rhomb.	2.460			sl s	s	
t143	fluoride	ThF_4	308.03	wh cub powd	6.32^{24}	>900				sl d dil H_2SO_4, HCl; i conc H_2SO_4
t144	fluoride	$ThF_4 \cdot 4H_2O$	380.09	cr		$-H_2O$, 100	$-2H_2O$, 140 — 200	0.017^{25}		i HF
t145	hydroxide	$Th(OH)_4$	300.07	wh gelat		d		i	i	s a; i alk, HF
t146	iodate	$Th(IO_3)_4$	931.65					i	id	s dil H_2SO_4; i dil HNO_3
t147	iodide, tetra-	ThI_4	739.66	yel		566	839	s		
t148	nitrate	$Th(NO_3)_4$	480.06	plates, deliq.		d 500		v s		s al
t149	nitrate	$Th(NO_3)_4 \cdot 4H_2O$	552.12	col cr.		swells		v s		v s al; sl s acet; 36.9 eth
t150	nitrate	$Th(NO_3)_4 \cdot 12H_2O$	696.24	col leaf, deliq		d		v s		v s al, a
t151	nitride	Th_3N_4	752.14	dk brn powd, or blk cr.				sl d	d	s HCl
t152	oxalate	$Th(C_2O_4)_2$	408.08	wh cr.	4.637^{16}	d		0.0017^{17}	0.0017^{50}	s h aq $(NH_4)_2C_2O_4$; sl s a

No.	Name	Synonyms and Formulae	Mol. wt.	Crystalline form, properties and index of refraction	Density or spec. gravity	Melting point, °C	Boiling point, °C	Solubility, in grams per 100 cc		
								Cold water	Hot water	Other solvents
t153	oxalate	$Th(C_2O_4)_2 \cdot 6H_2O$	516.17	wh amorph powd				i		s Na_2CO_3, $(NH_4)_2C_2O_4$ sol; i HNO_3
t154	oxide, di-	Thorianite. ThO_2	264.04	wh cub, 2.20 (liq)	9.86	3220 ± 50	4400	i	i	s hot H_2SO_4; i dil a, alk
t155	oxysulfide	$ThOS$	280.10	yel cr.	6.44^{0}	d		i		s aq reg; sl s HNO_3
t156	2,4-pentanedione	Thorium acetylacetonate. $Th(C_5H_7O_2)_4$	628.48	col cr.		171 subl 160^{10}	260 — 270^{10}	sl s		v s al, chl; s eth
t157	*hypo*phosphate	$ThP_2O_6 \cdot 11H_2O$	588.15	wh amorph ppt		$-11H_2O$, 160		i	i	i a, alk
t158	*meta*phosphate	$Th(PO_3)_4$	547.93	col rhomb pr	$4.08^{16.4}$					
t159	*ortho*phosphate	$Th_3(PO_4)_4 \cdot 4H_2O$	1148.06	wh gelat				i	i	s 30^{0} HCl; i a
t160	picrate	$Th(C_6H_2N_3O_7)_4 \cdot 10H_2O$	1324.58					0.305^{25}		
t161	selenate	$Th(SeO_4)_2 \cdot 9H_2O$	680.09	col, monocl	3.026	$-8H_2O$, 200	d 1500	0.5^{0}	2.0^{100}	
t162	*ortho*silicate	Thorite $ThSiO_4$	324.12	col, tetr, 1.80, 1.81	6.82^{16}			v sl s		i a
t163	silicide	$ThSi_2$	288.21	blk, tetr	7.96^{16}					s hot HCl; sl s H_2SO_4
t164	sulfate	$Th(SO_4)_2$	424.15	wh cr, hygr	4.225^{17}			s	s	i a; v s $NH_4C_2H_3O_2$
t165	sulfate	$Th(SO_4)_2 \cdot 4H_2O$	496.21	wh need, or cr powd		$-4H_2O$, 400		9.41^{17} (anhydr)	2.54^{50} (anhydr)	i a
t166	sulfate	$Th(SO_4)_2 \cdot 6H_2O$	532.24					1.63^{15}	6.64^{60}	i a
t167	sulfate	$Th(SO_4)_2 \cdot 8H_2O$	568.28	monocl, prism, 1.5168		$-4H_2O$, 42		1.88^{25}	3.71^{44}	i a
t168	sulfate	$Th(SO_4)_2 \cdot 9H_2O$	586.29	wh monocl	2.77	$-9H_2O$, 400		1.57^{20}	6.67^{55}	i a
t169	sulfide	ThS_2	296.16	dk brn-blk cr	7.30_4^{25}	1925 ± 50 (vac)		i	d 200	s hot aq reg; sl s a
t170	*pyro*vanadate	$ThV_2O_7 \cdot 6H_2O$	554.01	yel				i	i	s conc a
t171	**Thulium**	Tm	168.9342	silv wh met, hex	9.3208	1545	1947	i	i	
t172	bromide	$TmBr_3$	408.65			952	1440	s	s	
t173	chloride	$TmCl_3 \cdot 7H_2O$	401.40	grn cr, deliq		824	1440	v s		v s al
t174	fluoride	TmF_3	225.93			1158	>2200	i	i	s dil a
t175	iodide	TmI_3	549.65	brt yel cr		1015	1260	s	s	
t176	oxalate	$Tm_2(C_2O_4)_3 \cdot 6H_2O$	710.02	grn-wh ppt		$-H_2O$, 50		i		s alk oxal sol; i dil a
t177	oxide	Thulia. Tm_2O_3	385.87	grn-wh powd						sl s min a
t178	**Tin** gray	Sn	118.69 ± 3	gray, cub	5.75	231.9681	2270	i	i	s HCl, H_2SO_4, aq reg, alk; sl s dil HNO_3
t179	**Tin** white	Sn	118.69 ± 3	wh met, tetr	7.28	231.88 stable 13.2 — 161	2260	i	i	s HCl, H_2SO_4, aq reg, alk; sl s dil HNO_3
t180	**Tin** brittle	Sn	118.69 ± 3	wh, rhomb	6.52 — 56	231.89 stable > 161	2260	i	i	s HCl, H_2SO_4, aq reg, alk; sl s dil HNO_3
t181	(II) acetate	$Sn(C_2H_3O_2)_2$	236.78	yelsh powd		182	240	d		s dil HCl
t182	*pyro*arsenate	$Sn_2As_2O_7$	499.22	flocculent ppt		d AS_2O_3 + SnO_2		i	i	i conc ac a
t183	(II) bromide	$SnBr_2$	278.50	pa yel, rhomb	5.117^{17}	215.5	620	85.2^{0}	222.5^{100}	s al, eth, acet
t184	(IV) bromide	$SnBr_4$	438.31	col, rhomb pyr, deliq	liq 3.343^{35}	31	202^{734}	s d	d	s acet, PCl_3, $AsBr_3$
t185	bromide chloride (tri-)	$SnBrCl_3$	304.95	col liq	2.51^{13}	−31	50^{30}			
t186	bromide (di-) chloride (di-)	$SnBr_2Cl_2$	349.40		2.82^{13}	−20	65^{30} d 191	d	d	
t187	bromide (tri-) chloride	$SnBr_2Cl$	393.86	liq	3.12^{13}	1	73^{30}			
t188	bromide (di-) iodide (di-)	$SnBr_2I_2$	532.31	or-red, hex pl.	3.631^{15}	50	225	s	d < 80	
t189	(II) chloride	$SnCl_2$	189.60	wh, rhomb	3.95_4^{25}	246	652	83.9^{0}	259.8^{15} d	s al, eth, acet, et acet, me acet, pyr
t190	(II) chloride dihydrate	$SnCl_2 \cdot 2H_2O$	225.63	wh, monocl	$2.710^{15.5}$	37.7	d	d	d	s al, eth, acet, glac ac a
t191	(IV) chloride	$SnCl_4$	260.50	col liq, solid cub, 1.512	liq 2.226	−33	114.1	s	d	s eth
t192	(IV) chloride pentahydrate	$SnCl_4 \cdot 5H_2O$	350.58	monocl cr		stable 19 — 56		s		
t193	(IV) chloride tetrahydrate	$SnCl_4 \cdot 4H_2O$	332.56	opaque			stable 56 — 83	s		
t194	(IV) chloride trihydrate	$SnCl_4 \cdot 3H_2O$	314.55	col, monocl cr		80	stable 64 — 83	s		
t195	(IV) chloride *di*ammine	$SnCl_4 \cdot 2NH_3$	294.56	cr				s		d HCl
t196	chloride (tri-) bromide	$SnCl_3Br$	304.95	col liq	2.51^{13}	−31	50^{30}			
t197	chloride (di-) iodide (di-)	$SnCl_2I_2$	443.40	red mobile liq	3.287^{15}		297	s	d	s chl, bz CS_2
t198	(IV) chloride nitrosyl-chloride	$SnCl_4 \cdot 2NOCl$	391.42	pa yel, oct cr	2.60	180		d	d	
t199	(IV) chromate	$Sn(CrO_4)_2$	350.68	br yel cr powd		d		s		
t200	(II) ferricyanide	$Sn_3[Fe(CN)_6]_2$	779.98	wh		d		i	i	s HCl
t201	(II) ferrocyanide	$Sn_2Fe(CN)_6$	449.33	wh gel				i	i	d HCl
t202	(IV) ferrocyanide	$SnFe(CN)_6$	330.64					i	i	d h HCl
t203	(II) fluoride	Fluoristan. SnF_2	156.69	wh, monocl cr				s		
t204	(IV) fluoride	SnF_4	194.68	wh, monocl cr, hygr.	4.780^{19}	705 subl		v s	d	
205	hydride	Stannane. SnH_4	122.72	gas		d −150	−52			s $AgNO_3$, $HgCl_2$, conc alk, conc H_2SO_4

No.	Name	Synonyms and Formulae	Mol. wt.	Crystalline form, properties and index of refraction	Density or spec. gravity	Melting point, °C	Boiling point, °C	Solubility, in grams per 100 cc		
								Cold water	Hot water	Other solvents
t206	iodide	SnI_2	372.50	yelsh-red to red monocl need	5.285	320	717	0.98^{30}	4.03^{100}	v s NH_4OH, HI soln
t207	(IV) iodide	SnI_4	626.31	or red cub, 2.106	4.473^{0}	144.5	364.5	s	d	141.1^{25} CS_2; 6.03^{15} CCl_4; 17.88^{25} bz
t208	(II) nitrate	$Sn(NO_3)_2 \cdot 20H_2O$	603.01	col leaf		−20		d	d	d HNO_3
t209	(II) nitrate, basic	$SnO \cdot Sn(NO_3)_2$	377.39	wh cr mass		d > 100 expl		d	d	
t210	(IV) nitrate	$Sn(NO_3)_4$	366.71	silky need		d 50		d		
t211	(II) oxide, mon-	SnO	134.69	blk, cub (tetr)	6.446^{0}	d 1080^{600}		i	i	s a, alk; sl s NH_4Cl
t212	oxide, mon-hydrate	$SnO \cdot xH_2O$		wh powd or yel-low-brn cr					d to SnO	d a; alk; s alk carb; i NH_4OH
t213	(IV) oxide, di-	Nat. cassiterite. SnO_2	150.69	wh, tetr, (also hex or rhomb), 1.997, 2.093	6.95	1630	subl 1800 — 1900	i	i	d KOH, NaOH; i aq reg
t214	oxide, di-hydrate	α-Stannic acid or "ordinary" stannic acid. $SnO_2 \cdot xH_2O$		amorph or gel				i	i	s a, alk, K_2CO_3
t215	oxide, di-hydrate	β-Stannic acid or "meta" stannic acid. $SnO_2 \cdot xH_2O$		wh, amorph or gel				i	i	i a, K_2Co_3; sol alk
t216	(II) *meta*phosphate	$Sn(PO_3)_2$	276.63	amorph mass	$3.380^{22.8}$					
t217	(II) *ortho*phosphate	$Sn_3(PO_4)_2$	546.01	wh, amorph.	3.823^{17}			i	i	d a, alk
t218	(II) *ortho*phosphate, di-H	$Sn(H_2PO_4)_2$	312.66	wh, rhomb cr.	$3.167^{22.8}$	d	d		d	
t219	(II) *ortho*phosphate, mono-H	$SnHPO_4$	214.67	cr	$3.476^{15.5}$	stabl > 100	d	i	i	s dil min a
t220	(II) *pyro*phosphate	$Sn_2P_2O_7$	411.32	amorph powd.	$4.009^{16.4}$					s conc a
t221	phosphide, mono-	SnP	149.66	silv wh.	6.56	d	d	i	i	s HCl; i HNO_3
t222	phosphide, tri-	SnP_3	211.61	cr	4.10^{0}	<415 d to Sn_4P_3		i	i	d HNO_3; i HCl
t223	*tetra*phosphide, tri-	Sn_4P_3	567.68	wh cr.	5.181	d < 480		i	i	d fixed alk hydr, HCl
t224	phosphorus chloride	$SnCl_4 \cdot PCl_5$	468.74	col cr.		subl 200		d	d	
t225	(II) selenide	SnSe	197.65	steelgray cr	6.179^{0}	861		i	i	d HCl, HNO_3, aq reg, alk sulf
t226	(II) sulfate	$SnSO_4$	214.75	wh-yelsh cr powd		>360 (SO_2)		33^{25}		s H_2SO_4
t227	(IV) sulfate	$Sn(SO_4)_2 \cdot 2H_2O$	346.84	wh, hex pr, deliq				v s	d	s eth, dil H_2SO_4, HCl
t228	(II) sulfide	SnS	150.75	gray-blk cub, monocl	5.22^{25}	882	1230	0.000002^{18}		d HCl, alk, $(NH_4)_2S$
t229	(IV) sulfide	Mosaic gold. SnS_2	182.81	gold yel, hex	4.5	d 600		0.0002^{18}		d alk sulf, aq reg, alk hydr, PCl_5, $SnCl_2$; i a
t230	(IV) sulfur chloride	$SnCl_4 \cdot 2SCl_4$	608.25	yel cr.		37	d < 40	d	d	s eth, bz, CS_2, ethyl acet; d HNO_3
t231	tartrate	$SnC_4H_4O_6$	266.76	heavy wh powd				s		v s dil HCl
t232	(II) telluride	SnTe	246.29	gray cr.	6.48	780	d	i	i	d alk sulf
t233	(IV) telluride	$SnTe_2$	373.89	blk, flocc ppt.				i	i	d dil a, alk
t234	**Titanic acid, ortho-**	α-Titanic acid. H_2TiO_4	113.89	wh		d		v sl s d		s dil HCl, dil H_2SO_4, conc alk
t235	**Titanium**	Ti	47.88 ± 3	α hex, tr β cub 838, silv gray	4.5^{20}	1660 ± 10	3287	i	i	s dil a
t236	boride, di-	TiB_2	69.50	hex	4.50	2900				
t237	bromide, di-	$TiBr_2$	207.69	blk powd	4.31	d > 500		s ev H_2		
t238	bromide, tetra-	$TiBr_4$	367.50	or yel, deliq	2.6	39	230	d		s abs al, abs eth
t239	bromide, tri-	$TiBr_3 \cdot 6H_2O$	395.68	redsh-viol or dk blue cr, deliq		115	d 400	v s		v s al, acet
t240	carbide	TiC	59.89	gr met, cub	4.93	3140 ± 90	4820	i	i	s aq reg, HNO_3
t241	chloride, di-	$TiCl_2$	118.79	lt br-blk, hex, deliq	3.13	subl H_2	d 475 vac	d		s al, i eth, chl, CS_2
t242	chloride, tetra-	$TiCl_4$	189.69	lt yel liq, $1.61^{10.5}$	liq 1.726 sol 2.06^{-79}	−25	136.4	s	d	s dil HCl, al
t243	chloride, tri-	$TiCl_3$	154.24	dk viol, deliq	2.64	d 440	660^{108}	s	s	v s al; s HCl; i eth
t244	fluoride, tetra	TiF_4	123.87	wh powd, hygr	$2.798^{20.5}$	>400 (pressure)	284 (subl)	s d		s H_2SO_4, al, C_6H_5N; i eth
t245	fluoride, tri-	TiF_3	104.88	purp-red or vlt	3.40	1200	1400	red s vlt i		
t246	hydride	TiH_2	49.90	gray powd.	3.9^{12}	d 400				
t247	iodide, di-	TiI_2	301.69	blk, hygr	4.99	600	1000	d		d alk; s conc HF, conc HCl
t248	iodide, tetra-	TiI_4	555.50	red, cub	4.3	150	377.1	v s	d	
t249	nitride	TiN	61.89	yel-bronze, cub	5.22	2930		i	i	sl s hot aq reg + HF
t250	oxalate	$Ti_2(C_2O_4)_3 \cdot 10H_2O$	539.97	yel pr				s	s	i al, eth
t251	oxide, di-	Nat. brookite. TiO_2	79.88	wh, rhomb, 2.583, 2.586, 2.741	4.17	1825		i	i	s H_2SO_4, alk; i a
t252	oxide, di-	Nat. octahedrite, anatase. TiO_2	79.88	br-blk, tetr, 2.554, 2.493	3.84			i	i	s H_2SO_4, alk; i a
t253	oxide, di-	Nat. rutile. TiO_2	79.88	col, tetr, 2.616, 2.903	4.26	1830 — 1850	2500 — 3000	i	i	s H_2SO_4, alk; i a
t254	oxide, mon-	TiO	63.85	yel blk, pr.	4.93	1750	>3000			s dil H_2SO_4; i HNO_3
t255	oxide, sesqui-	Ti_2O_3	143.76	vlt blk, trig	4.6	2130 d		i	i	s H_2SO_4; i HCl, HNO_3
t256	phosphide	TiP	78.85	gray, met	3.95^{25}			i	i	i a
t257	sulfate	$Ti_2(SO_4)_3$	383.93	green powd				i	i	s dil a; i al, eth, conc H_2SO_4

No.	Name	Synonyms and Formulae	Mol. wt.	Crystalline form, properties and index of refraction	Density or spec. gravity	Melting point, °C	Boiling point, °C	Solubility, in grams per 100 cc		
								Cold water	Hot water	Other solvents
t258	sulfate, basic	$TiOSO_4$	159.94	wh or sl yelsh powd, 1.80 — 1.89	...	...	...	d	...	...
t259	sulfide, di-	TiS_2	112.00	yel sc	3.22^{20}	...	...	hyd sl	d in steam	d HCl; s dil HNO_3, H_2SO_4
t260	sulfide, mono-	TiS	79.94	redsh solid	4.12	...	...	i	...	s conc H_2SO_4; i HCl, HF, dil H_2SO_4
t261	sulfide, sesqui	Ti_2S_3	191.94	grayish-blk cr.	3.584	...	...	i	i	s conc H_2SO_4, conc HNO_3; i dil H_2SO_4, dil HCl
t262	**Tungsten**	Wolfram. W	183.85 ± 3	gray-blk, cub	19.35_{4}^{20}	3410 ± 20	5660	i	i	v sl s HNO_3, H_2SO_4, aq reg; s HNO_3 + HF, fus NaOH + $NaNO_3$; i HF, KOH
t263	arsenide	WAs_2	333.69	blk cr	6.9^{18}	d red heat	...	i	...	d hot HNO_3, hot H_2SO_4
t264	boride, di-	WB_2	205.47	silvery, oct	10.77	*ca* 2900	...	i	i	s aq reg
t265	bromide, di-	WBr_2	343.66	bl-blk need	...	d 400	...	d	...	...
t266	bromide, penta-	WBr_5	583.37	vlt-brn need, hygr	...	276	333	d	...	s abs al, chl, eth, alk
t267	bromide, hexa-	WBr_6	663.27	bl-blk, need	6.9	232	...	i	d	s abs a, eth, CS_2, NH_4OH
t268	carbide	WC	195.86	blk, hex	15.63^{18}	2870 ± 50	6000	i	...	s HNO_3 + HF, aq reg
t269	(di-)carbide	W_2C	379.71	blk, hex	17.15	2860	6000	i	...	s HNO_3 + HCl
t270	carbonyl	$W(CO)_6$	351.91	col, rhomb cr.	2.65	d ~ 150	175^{766}	i	i	s fum HNO_3; v sl
t271	chloride, di-	WCl_2	254.76	gray, amorph	5.436	...	...	d	...	s al, eth, bz
t272	chloride, hexa-	WCl_6	396.57	dk bl, cub	3.52_{4}^{25}	275	346.7	...	d^{60}	s al, eth, bz, CCl_4; v s CS_2, POCl
t273	chloride, penta-	WCl_5	361.12	blk, deliq	3.875_{4}^{25}	248	275.6	...	d to W_2O_5	v sl s CS_2
t274	chloride, tetra-	WCl_4	325.66	gray, deliq	4.624_{4}^{25}	d	...	d	...	...
t275	fluoride, hexa-	WF_6	297.84	col gas, or lt yel liq.	liq 3.44 gas 12.9 g/ℓ	2.5^{420}	17.5	d	d	s alk
t276	iodide, di-	WI_2	437.66	br-gr, amorph	6.799_{4}^{25}	d	...	i	d	s alk; i al CS_2
t277	iodide, tetra-	WI_4	691.47	blk, cr	5.2^{18}	d	...	i	d	s abs al; i eth, chl, turp
t278	nitride, di-	WN_2	211.86	brn, cub	...	above 400 (vac)	...	d	d	...
t279	oxide, di-	WO_2	215.85	br, cub.	12.11	1500 — 1600 (in N_2)	*ca* 1430 subl 800	i	i	s a, KOH
t280	oxide, pent-	Mineral blue. W_2O_5 or W_4O_{11}	447.70 or 911.39	blue-vlt, tricl	...	subl 800 — 900	*ca* 1530 d 2000	i	i	i a
t281	oxide, tri-	WO_3	231.85	yel, rhomb, or yel- or powd	7.16	1473	...	i	i	s hot alk; sl s HF; i a
t282	oxy*di*bromide, di-	WO_2Br_2	375.66	red, prism	...	d	...	...	...	...
t283	oxy*tetra*bromide	$WOBr_4$	519.47	blk, deliq	...	277	327	d	d	...
t284	oxy*tetra*chloride	$WOCl_4$	341.66	red, need	...	211	227.5	d	d	s CS_2, S_2Cl_2, bz
t285	oxy*di*chloride, di-	WO_2Cl_2	286.75	lt yel tabl	...	266	...	s	d	i al; s NH_4OH, alk
t286	oxy*tetra*fluoride-	WOF_4	275.84	col pl, hygr	...	110	187.5	d	...	sl s CS_2; i CCl_4
t287	phosphide	WP	214.82	gray, prism	8.5	...	...	i	...	s HNO_3 + HF; i alk, HCl
t288	phosphide	WP_2	245.80	blk cr	5.8	d	...	i	i	s HNO_3 + HF, aq reg; i al, eth
t289	phosphide	W_2P	398.67	dk gray prism	5.21	d	...	...	...	s fus Na_2CO_3 + $NaNO_3$; i a, aq reg
t290	silicide	WSi_2	240.02	blue, gray, tetrag	9.4	above 900	...	i	i	s HNO_3 + HF; i aq reg
t291	sulfide, di-	Nat. tungstenite. WS_2	247.97	dk gray, hexag	7.5^{10}	d 1250	...	i	...	s HNO_3 + HF, fus alk; i al
t292	sulfide, tri-	WS_3	280.03	choc brn powd	...	...	...	sl s	s	s alk
t293	**Tungstic acid,** meta-	$H_2W_4O_{13}·9H_2O$	1107.55	col, tetrag	3.93	d 50	...	88.57^{22}	$111.87^{43.5}$	$110.76^{24.3}$ eth; s al
t294	**Tungstic acid,** ortho-	H_2WO_4	249.86	yel powd, 2.24	5.5	$-H_2O$, 100	1473	i	sl s	s alk, HF, NH_3; i most a
t295	**Tungstic acid,** ortho-	$H_2WO_4·H_2O$	267.88	wh	...	$H_2W_2O_7$ *at* 100	...	sl s	...	s alk
u1	**Uranic acid** meta-	Uranyl hydroxide. H_2UO_4 (or $UO_2(OH)_2$)	304.04	yel, rhomb, or powd	5.926	$-H_2O$ 250 — 300	...	i	i	s a, alk carb
u2	**Uranium**	U	238.0289	silvery, cubic, radioactive	19.05 ± 0.02^{25}	1132.3 ± 0.8	3818	i	i	s a; i alk, al
u3	boride, di-	UB_2	259.65	hex	12.70	2365	...	...	...	...
u4	bromide tetra-	UBr_4	557.64	br leaf, deliq	5.35	516	792^{760}	v s	v s	s d al, MeOH; i bz; s liq HN_3
u5	bromide tri-	UBr_3	477.74	dk brn need, hygr	6.53	730	volat	s	...	d al
u6	*di*carbide	UC_2	262.05	met cr	11.28^{16}	2350 — 2400	4370^{760}	d	d	i al; d dil inorg a
u7	chloride, penta-	UCl_5	415.29	dk green, gray need, red by trans light, hydr	3.81 (?)	d 300	...	d	...	s abs als, a acet, NH_4Cl; d ac a; i bz, eth
u8	chloride, tetra-	UCl_4	379.84	dk grn met, cub oct, hygr	4.87	590 ± 1	792^{760}	v s	s	s al, acet, ac a; i eth, $CHCl_3$
u9	chloride, tri-	UCl_3	344.39	dk red need, hygr	5.44_{4}^{25}	842 ± 5	...	s	s	s MeOH, acet, glac acet a; i eth

No.	Name	Synonyms and Formulae	Mol. wt.	Crystalline form, properties and index of refraction	Density or spec. gravity	Melting point, °C	Boiling point, °C	Solubility, in grams per 100 cc: Cold water	Hot water	Other solvents
u10	fluoride, hexa-	UF_6	352.02	col cr, deliq, monocl	4.68^{21}	64.5 — 64.8	56.2^{765}	d		d al, eth; s CCl_4, chl; i CS_2
u11	fluoride, tetra-	UF_4	314.02	green, tricl need	6.70 ± 0.10	960 ± 5		v sl s		i dil a, alk; s conc a, conc alk
u12	fluoride, tri-	UF_3	295.02	blk cr or fused		d above 1000		sl d		v sl s dil inorg a
u13	hydride	UH_3	241.05	blk-brn powd	10.95			i	i	i al, acet, liq NH_3; sl s dil HCl; d HNO_3
u14	hydride	UH_3	241.05	blk powd, cub	11.4					
u15	iodide, tetra-	UI_4	745.65	blk, need	5.6^{15}	506	759	s	s d	
u16	nitride, mono-	UN	252.04	br powd	14.31	*ca* 2630 ± 50				i HCl, H_2SO_4
u17	oxide, di-	UO_2	270.03	br-blk rhomb, or cub	10.96	2878 ± 20		i	i	s HNO_3, conc H_2SO_4
u18	oxide, per-	$UO_4{\cdot}2H_2O$	338.06	pa yel cr, hygr		d 115		0.0006^{20}	0.008^{90}	d HCl
u19	oxide, tri-	Uranyl oxide. UO_3	286.03	yel-red powd	7.29	d		i		s HNO_3, HCl
u20	*tri*-oxide, oct-	U_2O_3	842.08	olive green-blk	8.30	d 1300 to UO_2		i	i	s HNO_3, H_2SO_4
u21	(IV) sulfate	$U(SO_4)_2{\cdot}4H_2O$	502.21	grn, rhomb		$-4H_2O$, 300		23^{11}	9^{63} (anhydr)	s dil a
u22	(IV) sulfate	$U(SO_4)_2{\cdot}8H_2O$	574.27			d 90		11.3^{18}	58.2^{62}	i al; s dil a
u23	(IV) sulfate	$U(SO_4)_2{\cdot}9H_2O$	592.28	grnsh, monocl		$-7H_2O$, 230	$-9H_2O$ red heat			s dil H_2SO_4
u24	sulfide, di-	US_2	302.15	gray-blk, tetr	7.96^{25}	>1100	oxidizes	sl d		s conc HCl; d HNO_3 v s al
u25	sulfide, mono-	US	270.09	blk amorph powd	10.87	above 2000				i HCl, HNO_3
u26	sulfide, sesqui-	U_2S_3	572.24	gray blk, rhomb need		ign				s + O aq reg conc HNO_3; i dil a
u27	**Uranyl acetate**	$UO_2(C_2H_3O_2)_2{\cdot}2H_2O$	424.15	yel, rhomb	2.893^{15}	$-2H_2O$, 110	d 275	7.694^{15}	d	v s al
u28	benzoate	$UO_2(C_7H_5O_2)_2$	512.26	yel powd				sl s		sl s al
u29	bromide	UO_2Br_2	429.84	grn-yel need, hygr				s d		s al, eth
u30	*per*chlorate	$UO_2(ClO_4)_2{\cdot}6H_2O$	577.02	yel cr, deliq, rhomb		90 d 110				
u31	chloride	UO_2Cl_2	340.93	yel, deliq		578	d	320^{18}	v s	s al, amyl al, eth
u32	formate	$UO_2(CHO_2)_2{\cdot}H_2O$	378.08	yel, oct	3.695^{19}	$-H_2O$, 110		7.2^{15}		sl s form a; 0.74^{15} MeOH, 2.37^{15} acet
u33	iodate	$UO_2(IO_3)_2$	619.83	yel, rhomb	5.2	d 250		s	s	i HNO_3
u34	iodate	$UO_2(IO_3)_2{\cdot}H_2O$	637.85	α prismatic, stable, β pyramidal	α 5.220^{18} β 5.052^{18}			α 0.1049^{18} β 0.1214^{18}		. . .
u35	iodide	UO_2I_2	523.84	red, deliq		d in air				s al, eth, bz
u36	nitrate	$UO_2(NO_3)_2{\cdot}6H_2O$	502.13	yel, rhomb, deliq, 1.4967	2.807^{13}	60.2 d 100	118		∞ 60	v s al, eth, ac a, acet, MeOH
u37	oxalate	$UO_2C_2O_4{\cdot}3H_2O$	412.09	yel cr.		$-H_2O$, 110		0.8^{14}	3.3^{100}	s inorg a, alk, oxal a
u38	phosphate, mono-H	$UO_2HPO_4{\cdot}4H_2O$	438.07	yel pl, tetr.				i	i	s HNO_3, aq Na_2CO_3; i ac a
u39	potassium carbonate	$UO_2CO_3{\cdot}2K_2CO_3$	606.45	yel cr.		$-CO_2$, 300		7.4^{15}	d	i al
u40	sodium carbonate	$UO_2CO_3{\cdot}2Na_2CO_3$	542.01	yel cr.				sl s		i al
u41	sulfate	$UO_2SO_4{\cdot}3H_2O$	420.13	yel-grn cr	$3.28^{16.5}$	d 100		$20.5^{15.5}$	22.2^{100}	24.3^{13} conc H_2SO_4; 30^{13} conc HCl
u42	sulfate	$2(UO_2SO_4){\cdot}7H_2O$	858.28	yel		anh 300		sl s		s H_2SO_3
u43	sulfide	UO_2S	302.09	brn-blk, tetr.		d 40-50		sl s	s d	s dil a, dil al, $(NH_4)_2CO_3$; i abs al
u44	sulfite	$UO_2SO_3{\cdot}4H_2O$	422.15	pa-gr cr						s H_2SO_3
v1	**Vanadic acid,** meta	HVO_3	99.95	yel sc				i		s a, alk,; i NH_4OH
v2	tetra-	$H_2V_4O_{11}$	381.78	br amorph				i		s a, alk, NH_4OH
v3	**Vanadium**	V	50.9415	lt gray met, cub, 3.03	5.96	1890 ± 10	3380	i	i	s aq reg, HNO_3, H_2SO_4, HF; i HCl, alk
v4	boride, di-	VB_2	72.56	hex	5.10					
$v5^1$	bromide, tri-	VBr_3	290.65	grn-blk, deliq.	4.00^{18}	d		s		s al, eth; i HBr
$v5^2$	carbide	VC	62.95	blk. cub	5.77	2810	3900	i		s HNO_3, fus KNO_3; i HCl, H_2SO_4
v6	chloride, di-	VCl_2	121.85	grn, hex, deliq	3.23^{18}			s d	s d	s al, eth
v7	chloride, tetra-	VCl_4	192.75	red-br liq	1.816^{30}	− 28 ± 2	148.5^{755}	s d		s abs al, eth, chl, acet a
v8	chloride, tri-	VCl_3	157.30	pink cr, deliq.	3.00^{18}	d		s d	s d	s abs al, eth
v9	fluoride, penta.	VF_5	145.93		2.177^{19}		111.2^{758}			s al
v10	fluoride, tetra-	VF_4	126.94	br yel	2.975^{23}	d 325		s		s acet; sl s al, chl
v11	fluoride, tri-	VF_3	107.94	grn, rhomb	3.363^{19}	>800	subl	i		s al, chl, CS_2
v12	fluoride, tri-	$VF_3{\cdot}3H_2O$	161.98	dk gr, rhomb		$-3H_2O$, 100		s	v s d	i abs al
v13	iodide, di-	VI_2	304.75	vlt-rose, hex	5.44	750-800 subl vac		s		i al, CCl_4, CS_2, bz
v14	iodide, tri-	$VI_3{\cdot}6H_2O$	539.75	gr cr, deliq		d		v s		s al
v15	nitride	VN	64.95	blk, cub	6.13	2320		i		sl s aq reg
v16	oxide	VO (or V_2O_2)	66.94	lt gray cr	5.758^{14}	ign		i	i	s a
v17	oxide, di- (or tetr)-	VO_2 (or V_2O_4)	82.94	bl cr	4.339	1967		i	i	s a, alk
v18	oxide, pent-	V_2O_5	181.88	yel-red, rhomb, 1.46, 1.52, 1.76	3.357^{18}	690	d 1750	0.8^{20}		s a, alk, i abs al
v19	oxide, sesqui	Vanadium trioxide. V_2O_3	149.88	blk cr	4.87_4^{18}	1970		sl s	s	s HNO_3, HF, alk
v20	oxybromide	VOBr	146.84	vlt, oct.	4.00^{18}	d 480		v sl s		s acet, anhydr eth, acet
v21	oxy di-bromide	$VOBr_2$	226.75	br powd, deliq		d 180		s		

PHYSICAL CONSTANTS OF INORGANIC COMPOUNDS (continued)

No.	Name	Synonyms and Formulae	Mol. wt.	Crystalline form, properties and index of refraction	Density or spec. gravity	Melting point, °C	Boiling point, °C	Solubility, in grams per 100 cc		
								Cold water	Hot water	Other solvents
v22	oxy*tri*bromide	$VOBr_3$	306.65	red liq	$2.933^{14.5}$	d 180	130^{100}	s		
v23	oxychloride	$VOCl$	102.39	yel brn powd	$2.824, 3.64^{20}$		127	i		v s HNO_3
v24	oxy*di*chloride	$VOCl_2$	137.85	grn, deliq	2.88^{13}			d		s dil HNO_3
v25	oxy*tri*chloride	$VOCl_3$	173.30	yel liq	1.829	−77 ± 2	126.7	s d		s al, eth, ac a
v26	oxy*di*fluoride	VOF_2	104.94	yel	3.396^{19}	d				sl s acet
v27	oxy*tri*fluoride	VOF_3	123.94	yel-wh, hygr	2.459^{19}	300	480			
v28	silicide, di-	VSi_2	107.11	met pr	4.42			i	i	s HF; i al, eth, a
v29	(*dl*-)silicide	V_2Si	129.97	silv wh pr	5.48^{17}			i	i	s HF; i al, eth, a
v30	sulfate (hypovanadous)	$VSO_4 \cdot 7H_2O$	273.11	vit, monocl		d in air				
v31	sulfide, mono- or (di-)	VS (or V_2S_2)	83.00	blk pl (exist ?)	4.20	d				s hot H_2SO_4, HNO_3; sl s KSH; i HCl, alk
v32	sulfide, penta-	V_2S_5	262.18	blk-grn powd	3.0	d		i		s HNO_3, alk sulf, alk
v33	sulfide, sesqui- or (tri-)	V_2S_3	198.06	grn-blk pl, or powd	4.72^{21}	d>600		i		s alk sulf; sl s, alk, HCl, HNO_3. H_2SO_4
v34	**Vanadyl sulfate**	$VOSO_4$	163.00	bl				v s		
w1	**Water**	H_2O	18.01528	col liq, or col hex cr	liq 1.000_4 sld 0.9168^0	0.00	100.00			s al
w2	**Water heavy**	Deuterium oxide. D_2O.	20.0312	col liq or hex cr, 1.33844^{20}	1.105^{20}_4	3.82	101.42	∞	∞	∞ al; sl s eth
w3	**Wolfram**	See tungsten								
x1	**Xenon**	Xe	131.29 ± 3	col inert gas	gas 5.887 g/ℓ ± 0.009 liq 3.52^{-100} solid 2.7^{-140}	−111.9	−107.1 ± 3	24.1° cm^3, 8.4^{50}, 11.9^{25} cm^3 7.12^{00}		
y1	**Ytterbium**	Yb	173.04 ± 3	cub	6.9654 up to 789 6.54 above 789	819 ± 5	1194	i		s a
y2	(III) acetate	$Yb(C_2H_3O_2)_3 \cdot 4H_2O$	422.23	hex pl	2.09	$-4H_2O$, 100		v s	v s	
y3	(II) bromide	$YbBr_2$	332.85		5.91^{25}_4	677	1800	s	s	s dil a
y4	(III) bormide	$YbBr_3$	412.75	col cr.		956	d	s	s	
y5	(II) chloride	$YbCl_2$	243.95	grn-yel cr	5.08	702	1900	s	s	s dil a
y6	(III) chloride	$YbCl_3 \cdot 6H_2O$	387.49	grn, rhomb cr, deliq	2.575	865 $-6H_2O$, 180		v s	v s	s abs al
y7	(II) fluoride	YbF_2	211.04			1052	2380	i	i	
y8	fluoride	YbF_3	230.04			1157	2200	i	i	i dil a
y9	(II) iodide	YbI_2	426.85	lt yel, hex cr	5.40^{25}_4	780 ± 4	1300 d(700) vac	s	s	s dil a
y10	(III) iodide	YbI_3	553.75	gold yel cr		d 700	d	s	s	s dil a
y11	(III) oxalate	$Yb_2(C_2O_4)_3 \cdot 10H_2O$	790.29	col cr.	2.644			0.00033^{25}		sl s dil a
y12	(III) oxide	Ytterbia. Yb_2O_3	394.08	col	9.17			i	i	s h dil a
y13	(III) selenate	$Yb_2(SeO_4)_3 \cdot 8H_2O$	919.08	hex pl	3.30			s d	s	
y14	(III) selenite	$Yb_2(SeO_3)_3$	726.95					i		
y15	(III) sulfate	$Yb_2(SO_4)_3$	634.25	col cr.	3.793	d 900		44.2^0	4.7^{100}	
y16	(III) sulfate, octohydrate	$Yb_2(SO_4)_3 \cdot 8H_2O$	778.38	prism	3.286			35.9^{25}	21.1^{40}	
y17	**Yttrium**	Y	88.9059	gray-blk met, hex	4.4689	1522	3338	sl d	d	v s dil a; s h KOH
y18	acetate	$Y(C_2H_3O_2)_3 \cdot 4H_2O$	338.10	col, tricl					9.03^{25}	
y19	bromate	$Y(BrO_3)_3 \cdot 9H_2O$	634.75	hex pr		74	$-6H_2O$, 100	168^{25}		sl s al; i eth
y20	bromide	YBr_3	328.62	deliq		904		v s		s al; i eth
y21	bromide hydrate	$YBr_3 \cdot 9H_2O$	490.76	col tabl, deliq				v s		sl s al; i eth
y22	carbide	YC_2	112.93	yel., microcr	4.13^{18}			d		
y23	carbonate	$Y_2(CO_3)_3 \cdot 3H_2O$	411.89	wh-redsh powd.						s dil min a, $(NH_4)_2CO_3$; sl s aq CO_2; i al, eth
y24	chloride	YCl_3	195.26	shiny wh leaf	2.67	721	1507	78^{10}	82^{50}	60.1^{15} al; 60.6^{15} pyr
y25	chloride, hexahydrate	$YCl_3 \cdot 6H_2O$	303.36	redsh-wh, rhomb, deliq	2.18^{18}	$-5H_2O$, 100		217^{20}	235^{50}	s al; i eth
y26	chloride, monohydrate	$YCl_3 \cdot H_2O$	213.28	col cr.		$-H_2O$, 160		v s		
y27	fluoride	YF_3	145.90	gelat	4.01	1387		i		v sl s dil a
y28	hydroxide	$Y(OH)_3$	139.93	wh-yel gelat or powd		d		i	i	s a, NH_4Cl; i alk
y29	iodide	YI_3	469.62	wh, cr, deliq		1004	$650\text{-}700^{0.02}$	v s		s al, acet; sl s eth
y30	molybdate	$Y_2(MoO_4)_3 \cdot 4H_2O$	729.69	grayish or yelsh, tetr pl, 2.03	4.79^{16}_{16}	1347				
y31	nitrate, hexahydrate	$Y(NO_3)_3 \cdot 6H_2O$	383.01	col, redsh cr, deliq	2.68	$-3H_2O$, 100		$134.7^{22.5}$		v s al, eth, HNO_3
y32	nitrate, tetrahydrate	$Y(NO_3)_3 \cdot 4H_2O$	346.98	redsh-wh pr	2.682			s		s al, HNO_3
y33	oxalate	$Y_2(C_2O_4)_3 \cdot 9H_2O$	604.01	wh cr powd		d		0.0001		sl s HCl
y34	oxide	Yttria. Y_2O_3	225.81	col-yelsh, cub or powd	5.01	2410		0.00018^{29}		s a; i alk
y35	sulfate	$Y_2(SO_4)_3$	465.98	wh powd	2.52	d 1000		5.38^{25}	s	s sat K_2SO_4 sol
y36	sulfate, octahydrate	$Y_2(SO_4)_3 \cdot 8H_2O$	610.11	col-redsh, monocl, 1.543, 1.549, 1.576	2.558	$-8H_2O$, 120	d 700	7.47^{16} (anhydr)	1.99^{95} (anhydr)	i al, alk; s conc H_2SO_4
y37	sulfide	Y_2S_3	273.99	yel-gr powd						d a
y38	**Yttrium hexaantipyrine** *per*chlorate	$[Y(C_{11}H_{12}N_2O)_6](ClO_4)_3$	1516.63	col, nex cr		d 293-296		0.55^{20}		
y39	hexaantipyrine iodide	$[Y(C_{11}H_{12}N_2O)_6]I_3$	1598.99	col cr.		280-282		4.65^{20}	i	s a, alk, ac a
z1	**Zinc**	Zn	65.38	bluish-wh met, hex	7.14	419.58	907	i		s a, alk, ac a
z2	acetate	$Zn(C_2H_3O_2)_2$	183.47	col, monocl	1.84	d 200	subl vac	30^{20}	44.6^{100}	2.8^{25} al; 166.79^{79}
z3	acetate, dihydrate	$Zn(C_2H_3O_2)_2 \cdot 2H_2O$	219.50	col. monocl, β 1.494	1.735	237	$-2H_2O$,100	31.1^{20}	66.6^{100}	2 al

No.	Name	Synonyms and Formulae	Mol. wt.	Crystalline form, properties and index of refraction	Density or spec. gravity	Melting point, °C	Boiling point, °C	Solubility, in grams per 100 cc: Cold water	Hot water	Other solvents
z4	acetylacetonate	$Zn(C_5H_7O_2)_2$	263.60	need	...	138	subl	v s d	...	v s bz, acet; s al
z5	aluminate	Nat. gahnite. $ZnAl_2O_4$	183.34	cub, grn 1.78	4.58	...	...	i	i	i a; sl s alk
z6	amide	$Zn(NH_2)_2$	97.43	wh powd, amorph	2.13^{25}	d 200 vac	...	d	d	i al, eth
z7	antimonide	Zn_3Sb_2	439.64	silv wh, rhomb pr	6.33	570	...	d	...	...
z8	*ortho*arsenate	Nat. koettigite. $Zn_3(AsO_4)_2·8H_2O$	618.10	monocl, 1.662, 1.683, 1.717	3.309^{15}	$-1H_2O$, 100	...	i	i	s HNO_3, H_3PO_4, alk
z9	*ortho*arsenate, basic	Nat. adamite. $Zn_3(AsO_4)_2·Zn(OH)_2$	573.37	col, rhomb	4.475^{15}	d 250	...	...	...	...
z10	*ortho*arsenate, hydrogen	$ZnHAsO_4·H_2O$	277.37	wh, rhomb	...	$-H_2O$, 327	...	d	d	...
z11	arsenide	Zn_3As_2	345.98	met-gray, tetr	5.528	1015	...	i	...	d a
z12	benzoate	$Zn(C_7H_5O_2)_2$	307.61	wh powd	...	...	...	2.46^{20}	1.44^{20}	...
z13	borate	$3ZnO·2B_2O_3$	383.37	wh tricl cr, or amorph powd	cr 4.22 powd 3.64	980	...	s	...	cr i HCl; amorph; s HCl
z14	bromate	$Zn(BrO_3)_2·6H_2O$	429.28	wh, cub, 1.5452	2.566	100	$-6H_2O$, 200	v s	∞	...
z15	bromide	$ZnBr_2$	225.19	col, rhomb, hygr n_D^{18} 1.5452	4.201^{25}_4	394	650	447^{20}	675^{100}	v s al, eth, acet; s NH_4OH
z16	butyrate	$Zn(C_4H_7O_2)_2·2H_2O$	275.61	wh pr	...	...	...	10.7^{16}	d	...
z17	caproate	$Zn(C_6H_{11}O_2)_2$	295.68	...	...	...	...	$1.03^{24.5}$	...	...
z18	carbonate	Nat. smithsonite. $ZnCO_3$	125.39	col, trig, 1.818, 1.618	4.398	$-CO_2$, 300	...	0.001^{15}	...	s a, alk, NH_4 salts; i NH_3, acet, pyr
z19	chlorate	$Zn(ClO_3)_2·4H_2O$	304.34	col yelsh, cub, deliq	2.15	d 60	d	262^{20}	v s	167 al; s acet, eth, glyc
z20	chlorate, per-	$Zn(ClO_4)_2·6H_2O$	372.37	wh, rhomb, deliq, 1.508, 1.480	2.252 ± 0.01	105-107	d 200	s	...	s al
z21	chlordie	$ZnCl_2$	136.29	wh, hex, deliq, 1.681, 1.713	2.91^{25}	283	732	432^{25}	615^{100}	$100^{12.5}$ al; v s eth; i NH_3
z22	chloroplatinate	$ZnPtCl_6.6H_2O$	581.27	yel, trig, hygr	2.717^{12}	d 160	...	v s	v s	v s al; d H_2SO_4
z23	chromate	$ZnCrO_4$	181.37	lem-yel pr	3.40	...	...	i	d	s a, liq NH_3; i acet
z24	chromate	$ZnCr_2O_4$	233.37	dk grn to black, cub	5.30^{15}	...	...	...	...	...
z25	*di*chromate	$ZnCr_2O_7·3H_2O$	335.41	redsh-brn cr, or or-yel powd, hygr	...	...	...	v s	d	i al, eth; s a
z26	citrate	$Zn_3(C_6H_5O_7)_2·2H_2O$	610.37	...	...	...	...	sl s	...	...
z27	cyanide	$Zn(CN)_2$	117.42	col, rhomb	1.852	d 800	...	0.0005^{20}	...	s alk, KCN, NH_3, i al
z28	ferrate (III)	Ferrite. $ZnFe_2O_4$	241.07	blk, oct	5.33^{20}	1590	...	...	...	s conc HCl; i dil a, alk
z29	ferrocyanide	$Zn_2Fe(CN)_6$	342.71	wh powd	1.85^{25}_4	...	...	i	...	s excess alk; i dil a
z30	ferrocyanide, trihydrate	$Zn_2Fe(CN)_6·3H_2O$	396.76	wh powd	...	d	...	i	i	i al HCl; d NaOH; s NH_4OH; v sl s NH_3
z31	fluoride	ZnF_2	103.38	col, monocl or tricl	4.95^{25}_4	872	ca 1500	1.62^{20}	s	s hot a, NH_4OH; i al, NH_3
z32	fluoride, tetrahydrate	$ZnF_2·4H_2O$	175.44	col, rhomb	2.255	$-4H_2O$, 100	tr to ZNO, 3000	1.6^{18}	s	s a, alk, NH_4OH
z33	fluosilicate	$ZnSiF_6·6H_2O$	315.55	col, hex pr, 1.3824, 1.3956	2.104	d 100	...	v s	...	...
z34	formaldehydesulfoxylate	$Zn(HSO_2·CH_2O)_2$	255.57	rhomb pr	...	d	...	v s	v s	d a; i al
z35	formaldehydesulfoxylate, basic	$Zn(OH)HSO_2·CH_2O$	177.48	rhomb pr	...	d	...	i	i	d a; i al
z36	formate	$Zn(CHO_2)_2$	155.42	col, cr	2.368	d	...	3.80	62^{100}	...
z37	formate	$Zn(CHO_2)_2·2H_2O$	191.45	wh, monocl, 1.513, 1.526, 1.566	2.207^{20}	$-2H_2O$, 140	d	5.2^{20}	38^{100}	i al
z38	gallate	$ZnGa_2O_4$	268.82	wh fine cr, 1.74	6.15 calc	<800	...	i	i	i org solv; s dil a, NH_4OH
z39	glycerophosphate	$ZnC_3H_7O_6P$	235.44	wh amorph powd	...	...	...	s	...	i al, eth
z40	hydroxide(ε)	$Zn(OH)_2$	99.39	col, rhomb	3.053	d 125	...	v sl s	...	s a, alk
z41	iodate	$Zn(IO_3)_2$	415.19	wh, need	5.063^{25}	d	...	0.87	1.31	s alk, HNO_3
z42	iodate, dihydrate	$Zn(IO_3)_2·2H_2O$	451.22	wh, cr powd	4.223^{25}_4	$-H_2O$, 200	...	0.877	1.32	s HNO_3, NH_4OH
z43	iodide	ZnI_2	319.19	col, hexag	4.7364^{25}_4	446	d 624	432^{18}	511^{100}	s a, al, eth, NH_3, $(NH_4)_2CO_3$
z44	*d*-lactate	$Zn(C_3H_5O_3)_2·2H_2O$	279.55	...	...	...	...	5.7^{15}	9^{33}	0.104 h 98% al
z45	*di*-lactate	$Zn(C_3H_5O_3)_2·3H_2O$	297.57	wh, rhomb cr	...	...	...	1.67^{106}	16.7^{100}	v sl s al
z46	laurate	$Zn(C_{12}H_{23}O_2)_2$	464.01	wh powd	...	128	...	0.01^{15}	0.019^{100}	0.010^{15} al
z47	*per*manganate	$Zn(MnO_4)_2·6H_2O$	411.34	vlt-br or bl, deliq	2.47	$-5H_2O$, 100	...	33.3	v s	d al, a
z48	nitrate, trihydrate	$Zn(NO_3)_2·3H_2O$	243.44	col, need	...	45.5	...	327.3^{40}	...	...
z49	nitrate, hexahydrate	$Zn(NO_3)_2·6H_2O$	297.48	col, tetrag	2.065^{14}	36.4	$-6H_2O$, 105-131	184.3^{20}	∞	v s al
z50	nitride	Zn_3N_2	224.15	gray	6.22^{25}_4	...	...	d	...	s HCl
z51	oleate	$Zn(C_{13}H_{33})_2)_2$	628.30	wax-like solid	...	70	...	i	...	s al, eth, bz, CS_2; al s acet
z52	oxalate	$ZnC_2O_4·2H_2O$	189.43	wh powd	3.28^{25}_4	d 100	...	0.00079^{18}	...	s a, alk
z53	oxide	Nat. zincite. ZnO	81.38	wh, hex, 2.008, 2.029	5.606	1975	...	0.00016^{29}	...	s a, alk, NH_4Cl; i al, NH_3
z54	oxide, per-	$ZnO_2·{}^1/_2H_2O$	106.39	yelsh, powd.	3.00 ± 0.08	$-O_2$, vac	...	sl d	d	d al, eth, acet
z55	1-phenol-4-sulfonate(p)	$Zn(C_6H_5SO_4)_2·8H_2O$	555.83	col cr or fine wh powd, effl	...	$-8H_2O$, 125	...	62.5	250^{100}	55.6^{25} al
z56	*ortho*phosphate	$Zn_3(PO_4)_2$	386.08	col, rhomb	3.998^{15}	900	...	i	i	s a, NH_4OH; i al
z57	*ortho*phosphate, dihydrogen	$Zn(H_2PO_4)_2·2H_2O$	295.39	tricl.	...	d 100	...	d	...	...
z58	*ortho*phosphate, octahydrate	$Zn_3(PO_4)_2·8H_2O$	530.20	rhomb pl	3.109^{15}	...	...	i	...	s alk

No.	Name	Synonyms and Formulae	Mol. wt.	Crystalline form, properties and index of refraction	Density or spec. gravity	Melting point, °C	Boiling point, °C	Solubility, in grams per 100 cc		
								Cold water	Hot water	Other solvents
z59	*ortho*phosphate, tetrahydrate	α-Hopeite. $Zn_3(PO_4)_2{\cdot}4H_2O$	458.14	col, rhomb, 1.572, 1.591, 1.59	3.04	tr > 105		i	i	v s a, NH_4OH, HN_4 salts
z60	*ortho*phosphate tetrahydrate	β-Hopeite. $Zn_3(PO_4)_2{\cdot}4H_2O$	458.14	col, rhomb, 1.574, 1.582, 1.582	3.03	tr > 140		i	i	v s a, NH_4OH, NH_4 salts
z61	*ortho*phosphate tetrahydrate	Parahopeite. $Zn_3(PO_4)_2{\cdot}4H_2O$	458.14	col, tricl, 1.614, 1.625, 1.665	3.75	tr > 163		i	i	v s a, NH_4OH, NH_4 salts
z62	*pyro*phosphate	$Zn_2P_2O_7$	304.70	wh powd	3.75^{23}			i	i	s a, alk, NH_4OH
z63	phosphide	Zn_3P_2	258.09	dk gray, tetrag, pois	4.55^{13}	>420	1100; subl in H_2	d		d H_2SO_4 ev H_3P s HNO_3; s (viol) dil a; i al
z64	hypophosphite	$Zn(H_2PO_2)_2{\cdot}H_2O$	213.37	col, cr powd, hygr				s		s alk
z65	picrate	$Zn(C_6H_2N_3O_7)_2{\cdot}8H_2O$	665.70	yel cr powd, expl		expl		s		
z66	salicylate	$Zn(C_7H_5O_3)_2{\cdot}3H_2O$	393.66	need				5^{20}		s al
z67	selenate	$ZnSeO_4{\cdot}5H_2O$	298.41	wh, tricl	2.591^{20}_{4}	d > 50		s		
z68	selenide	ZnSe	144.34	yelsh to redsh, cub, 2.89	5.42^{15}_{4}	>1100		i		s a; d NHO_3
z69	silicate	Nat. hemimorphite. $2ZnO{\cdot}SiO_2{\cdot}H_2O$	240.86	rhomb, or trigon 1.614, 1.617, 1.636	3.45			i	i	
z70	*meta*silicate	$ZnSiO_3$	141.46	col, rhomb	3.42	1437		i		i a
z71	*ortho*silicate	Nat. willemite. Zn_2SiO_4	222.84	trig, 1.694, 1.723	4.103	1509		i	i	s acet a
z72	stearate	$Zn(C_{18}H_{35}O_2)_2$	632.33	light powd		130		i		i al, eth
z73	sulfate	Nat. zinkosite. $ZnSO_4$	161.44	sol, rhomb, 1.658, 1.669, 1.670	3.54^{25}_{4}	d 600		s	s	sl s al; s MeOH, glyc
z74	sulfate, heptahydrate	Nat. goslarite. $ZnSO_4{\cdot}7H_2O$	287.54	col, rhomb, effl, 1.457, 1.480, 1.484	1.957^{25}_{4}	100	$-7H_2O$, 280	96.5^{20}	663.6^{100}	sl s al, glyc
z75	sulfate, hexahydrate	$ZnSO_4{\cdot}6H_2O$	269.53	col, monocl or tetrag	2.072^{15}	$-5H_2O$, 70		s	117.5^{40}	
z76	sulfide,(α)	Nat. wurtzite. ZnS	97.44	col, hex, 2.356, 2.378	3.98	1700 ± 20^{50}	atm 1185	0.00069^{18}		v s a; i ac a
z77	sulfide(β)	Nat. sphalerite. ZnS	97.44	col, cub, 2.368	4.102^{25}	tr 1020		0.000065^{18}		v s a
z78	sulfide, monohydrate	$ZnS{\cdot}H_2O$	115.46	yelsh-wh powd	3.98	1049		i		s a
z79	sulfite	$ZnSO_3{\cdot}2H_2O$	181.47	wh, cr powd		$-2H_2O$, 100	d 200	0.16	d	i al; s H_2SO_3
z80	tartrate	$ZnC_4H_4O_6{\cdot}H_2O$ (or $2H_2O$)	231.47	wh powd				0.055^{30}		s KOH, NaOH
z81	tellurate	Zn_3TeO_6	419.74	wh, gran ppt				i	i	s a
z82	telluride	ZnTe	192.98	red, cub, 3.56	6.34^{15}	1238.5		d		s d a
z83	thiocyanate	$Zn(SCN)_2$	181.54	wh powd, deliq				s		s al, NH_4OH
z84	valerate	$Zn(C_5H_9O_2)_2{\cdot}2H_2O$	303.66	wh glist sc or powd				$2.6^{24\text{-}25}$	s	*ca*2.5 al; v sl s eth
	Zinc Complexes									
z85	diamminezinc chloride	$[Zn(NH_3)_2]Cl$	170.35	col, rhomb, 1.625, 1.590	2.10	210.8	d 271	d		
z86	tetrammine perrhenate	$[Zn(NH_3)_4](ReO_4)_2$	633.91	wh, cub cr	3.608^{25}_{4}					0.1852 conc NH_4OH
z87	tetrapyridine fluosilicate	$[Zn(C_5H_5N)_4]SiF_6$	523.86	wh, rhomb	2.197					
z88	**Zirconium**	Zr	91.22	silver gray, met	6.49	1852 ± 2	4377	i	i	s HF, aq reg; sl s a
z89	boride, di.	ZrB_2	112.84	hex	6.085	*ca*3200				
z90	bromide, di	$ZrBr_2$	251.03	blk powd, ign in air		d 350		d ev H_2		
z93	bromide, tetra-	$ZrBr_4$	410.84	wh cr powd, deliq		450 ± 1^{15atm}	357 subl	i d		s liq NH_3, acetone; i bz, CCl_4
z94	bromide, tri-	$ZrBr_3$	330.93	bl-blk powd		d 350		d ev H_2		
z95	carbide	ZrC	103.23	gray met, cub	6.73	3540	5100	i		sl s conc H_2SO_4
z96	carbonate, basic	$3ZrO_2{\cdot}CO_2{\cdot}H_2O$	431.68	wh, amorph powd				i		s a
z97	chloride, di-	$ZrCl_2$	162.13	blk	3.6^{18}	d 350		d ev H_2		
z98	chloride, tetra-	$ZrCl_4$	233.03	wh cr.	2.803^{15}	437^{25atm}	subl 331	s	d	s al, eth, conc HCl
z99	chloride, tri-	$ZrCl_3$	197.58	br cr	3.00^{18}	d 350		d ev H_2		s $-H_2$ conc al; i org cpd
z100	fluoride	ZrF_4	167.21	wh hex, 1.59	4.43	subl ~ 600		1.388^{25}	d	sl s HF
z101	hydride	ZrH_2	93.24	gray-blk powd						s di HF, conc a
z102	hydroxide	$Zr(OH)_4$	159.25	wh amorp powd	3.25	$-2H_2O$, 500		0.02	i	s min a
z103	iodide	ZrI_4	598.84	wh need, hygr		499 ± 2 6.3^{atm}	d ~ 600	s d	s	d al; s eth; v sl s CS_2, bz; i liq NH_3
z104	nitrate	$Zr(NO_3)_4{\cdot}5H_2O$	429.32	col cr, deliq, 1.60, 1.61				v s		s al
z105	nitride	ZrN	105.23	yel-brn cr	7.09	2980 ± 50		i	i	sl s inorg ac; s conc H_2SO_4 HF, aq reg
z106	oxide	Nat. baddeleyite. ZrO_2	123.22	col-yel-brn, monocl, 2.13, 2.19, 2.20	5.89	*ca* 2700	*ca* 5000	i	i	s H_2SO_4, HF
z107	oxide	Zirconia. ZrO_2 HfO_2 < 2%	123.22	wh, monocl below 1000°, cub, above	5.6	2715		i	i	s H_2SO_4, HF
z108	oxide	Zirconium hydroxide, zirconic acid. $ZrO_2{\cdot}xH_2O$		gel or wh amorph powd	3.25	$-2H_2O$, 550		0.02		s acids; i al, alk
z109	phosphide	ZrP_2	153.17	gray, brittle	4.77^{25}_{4}			i		v s conc hot H_2SO_4
z110	selenate	$Zr(SeO_4)_2{\cdot}4H_2O$	449.20	hex trsp cr				s		sl s al, conc a
z111	selenite	$Zr(SeO_3)_2$	345.14	wh sm cr	4.3	d ~ 400		i	i	sl s H_2SO_4
z112	*ortho*silicate	Zircon, hyacinth. $ZrSiO_4$	183.30	tetr, var colors, 1.92-96; 1.97-2.02	4.56	2550		i	i	i a, aq reg, alk

No.	Name	Synonyms and Formulae	Mol. wt.	Crystalline form, properties and index of refraction	Density or spec. gravity	Melting point, °C	Boiling point, °C	Solubility, in grams per 100 cc		
								Cold water	Hot water	Other solvents
z113	silicide	$ZrSi_2$	147.39	steel gray rhomb, lust met	4.88^{22}	....	....	i	i	s HF; i inorg a, aq reg
z114	sulfate	$Zr(SO_4)_2$	283.34	microcr powd, hygr	3.22^{16}	410 d	....	s	....	....
z115	sulfate	$Zr(SO_4)_2{\cdot}4H_2O$	355.40	wh cr powd, rhomb	3.22^{16}	$-3H_2O$, 135-150	....	v s	....	i al
z116	sulfide	ZrS_2	155.34	steelgray cr, hexag	3.87	~ 1550	....	i	i	i a
z117	**Zirconyl bromide**	$ZrOBr_2{\cdot}xH_2O$	....	brill need, deliq	....	$-H_2O$, 120	....	s	....	s hot conc HBr
z118	chloride	$ZrOCl_2{\cdot}8H_2O$	322.25	wh, need, tetr, effl, 1.552, 1.563	....	$-6H_2O$, 150	$-8H_2O$, 210	s	d	s al, eth; sl s HCl
z119	iodide	$ZrOI_2{\cdot}8H_2O$	505.15	col, need, hygr	....	d	....	v s	v s	s al; v s eth
z120	sulfide	Zirconium sulfoxide. ZrOS	139.28	yel powd	4.87	ign in air	....	i	i	....

PHYSICAL CONSTANTS OF MINERALS

Compiled by Ralph Kretz

The following table presents data for many of the more common minerals.

In order to avoid duplication and save space, very few cross references are given in the body of the table. If the name sought is not found in the table, consult the **synonym index** given below.

Specific gravities are given at normal atmospheric temperatures, a more precise statement being valueless considering the large variations in natural minerals.

Hardness is given in terms of Mohs' scale. (See under Hardness.)

Indices of refraction for the sodium line, $\lambda = 5893$ Å, unless otherwise indicated. Li, $\lambda = 6708$ Å. Indices will invariably be given in the order ω, ϵ or α, β, γ. Uniaxial crystals are considered positive if $\epsilon > \omega$, negative if $\omega > \epsilon$. Biaxial crystals are considered positive if β is nearer α in value than it is γ and negative if β is nearer γ than α.

ABBREVIATIONS

Abbreviation	Meaning of abbreviation	Abbreviation	Meaning of abbreviation	Abbreviation	Meaning of abbreviation
bl	blue	grn	green	rhbdr	rhombohedral
blk	black	grnsh	greenish	rhomb	rhombic
blksh	blackish	hex	hexagonal	somet	sometimes
blsh	bluish	iridesc	iridescent	tarn	tarnishes
br	brown	monocl.	monoclinic	tetr	tetragonal
brnsh	brownish	oft	often	tricl	triclinic
col	colorless	pa	pale	vlt	violet
cub	cubic	purp	purple	wh	white
dk	dark	(R)	radioactive	yel	yellow
Fe	Fe, ferrous iron	redsh	redish	yelsh	yellowish
Fe^{+3}	Fe, ferric iron				

SYNONYM INDEX

Compound sought	Listed
Acmite	Aegirine
Agate	Quartz (impure)
Aluminum hydroxide	Boehmite, Diaspore, Gibbsite
Amphibole	Actinolite, Anthophyllite, Cummingtonite, Glaucophane, Hornblende, Riebeckite, Tremolite
Antimony oxide	Senarmontite, Valentinite
Antimony sulfide	Stibnite
Arsenic oxide	Arsenolite, Claudetite
Arsenic sulfide	Orpiment, Realgar
Barium carbonate	Witherite
Barium sulfate	Barite
Barytes	Barite
Bauxite	Gibbsite, Boehmite, Diaspore
Brimstone	Sulfur
Bronzite	Orthopyroxene
Cadmium sulfide	Greenockite
Calamine	Hemimorphite
Calcium carbonate	Aragonite, Calcite, Vaterite
Calcium sulfate	Anhydrite, Gypsum
Calcium sulfide	Oldhamite
Carborundum	Moissanite
Chalcedony	Quartz (impure, fibrous)
Chinaclay	Kaolinite
Chloanthite	Skutterodite
Chromespinel	Chromite
Chrysolite	Serpentine
Clinoptolite	Heulandite
Clayminerals	Illite, Kaolinite, Montmorillonite
Clinochlore	Chlorite
Cobaltbloom	Erythrite
Copper chloride	Nantokite
Copper oxide	Cuprite
Copper sulfide	Chalcocite, Covellite, Digenite
Emerald	Beryl
Emery	Mixture of Corundum, Magnetite and other minerals
Epsom salt	Epsomite
Feldspar	Orthoclase, Microcline, Anorthoclase, Albite, Oligoclase, Andesine, Anorthite
Fibrolite	Sillimanite
Flint	Quartz (impure)
Fluorapatite	Apatite
Fluorspar	Fluorite
Garnet	Almandine, Pyrope, Spessartite, Andradite, Grossularite, Uvarovite, Hydrogrossularite
Garnierite	Serpentine (Ni-bearing)
Glauber salt	Mirabilite
Hyacinth	Zircon
Iceland spar	Calcite
Idocrase	Vesuvianite
Iron carbonate	Siderite
Iron hydroxide	Goethite, Lepidocrocite
Iron oxide	Hematite, Magnetite
Iron spinel	Hercynite
Iron sulfide	Marcasite, Pyrite, Pyrrhotite
Lapis lazuli	Lazurite
Lead carbonate	Cerussite
Lead chloride	Cotunnite
Lead chromate	Crocoite
Lead oxide	Litharge, Minium
Lead sulfate	Anglesite
Lead sulfide	Galena
Limonite	Goethite (impure)
Lithiophyllite	Triphylite
Lithium mica	Lepidolite
Lodestone	Magnetite
Magnesium carbonate	Magnesite
Magnesium hydroxide	Brucite
Magnesium oxide	Periclase
Magnesium sulfate	Kieserite
Manganese carbonate	Rhodochrosite
Manganese hydroxide	Pyrochroite
Manganese oxide	Hausmannite, Manganosite, Pyrolusite
Manganese sulfide	Alabandite
Meerschaum	Serpentine
Mica	Muscovite, Paragonite, Phlogopite, Biotite, Lepidolite
Native copper	Copper
Native gold	Gold
Nickel oxide	Bunsenite
Nickel sulfide	Millerite
Orthite	Allanite
Penninite	Chlorite
Peridote	Olivine
Pistacite	Epidote
Pitchblende	Uraninite
Plagioclase	Albite, Oligoclase, Andesine, Anorthite
Potassium chloride	Sylvite
Potassium sulfate	Arcanite
Pyroxene	Diopsite, Angite, Aegirine, Jadeite, Pigeonite, Eustatite, Orthopyroxene
Rocksalt	Halite
Ruby	Corundum
Sapphire	Corundum
Silica	Christobalite, Quartz, Tridymite
Silver chloride	Cerargyrite
Silver iodide	Jodyrite, Miersite
Silver sulfide	Acanthite, Argentite
Smalltite	Skutterotite
Soapstone	Mixture of Talc and other minerals
Sodium chloride	Halite
Sodium sulfate	Thenardite
Strontium carbonate	Strontianite
Strontium sulfate	Celestite
Thorium oxide	Thorianite
Tin oxide	Cassiterite
Titanite	Sphene
Titanium oxide	Anatase, Brookite, Rutile
Uranium oxide	Uraninite
Zeolite	Natrolite, Mesolite, Scolecite, Thomasonite, Harmatome, Eddingtonite, Heulandite, Stilbite, Phillipsite, Chabazite, Gmelinite, Levyn, Laumontite, Mordenite
Zincblende	Sphalerite
Zinc carbonate	Smithsonite
Zinc oxide	Zincite
Zinc spinel	Gahnite
Zinc sulfide	Sphalerite, Wurtzite
Zirconium oxide	Baddeleyite

PHYSICAL CONSTANTS OF MINERALS

Name	Formula	Sp. gr.	Hardness	Crystalline form and color	Index of refraction (Na) η; ω ϵ α β γ
Acanthite	AgS	7.2–7.3	2–2.5	rhomb.(?), iron-blk.	....
Actinolite	$Ca_2((Mg,Fe)_5Si_8O_{22}(OH,F)_2$	3.02–3.44	5–6	monocl., pa. to dk. grn.	1.599–1.688, 1.612–1.697, 1.622–1.705
Aegirine	$NaFe^{+3}Si_2O_6$	3.55–3.60	6	monocl., dk. grn. to grnsh. blk.	1.750–1.776, 1.780–1.820, 1.800–1.836
Åkermanite	$Ca_2MgSi_2O_7$	2.944	5–6	tetr., col., gray-grn., br.	1.632, 1.640
Alabandite	MnS	4.050	3.5–4	cub., iron-blk., tarn., br.	....
Albite	$NaAlSi_3O_8$	2.63	6–6.5	tricl., col., wh., somet. yel., pink, grn.	1.527, 1.531, 1.538
Allanite	$(Ca,Mn,Ce,La,Y,Th)_2(Fe,Fe^{+3},Ti)(Al,Fe^{+3})_2Si_3O_{12}(OH)$	3.4–4.2	5–6.5	monocl., pa. br. to blk.	1.690–1.791, 1.700–1.815, 1.706–1.828
Allemontite	$AsSb$	5.8–6.2	3–4	hex., tin-wh. to redsh., gray, tarn. gray–brnsh. blk.	....
Almandine	$Fe_3Al_2Si_3O_{12}$	4.318	6–7.5	cub., red, dk. red, blk.	1.830
Altaite	$PbTe$	8.15	3	cub., tin-wh., yelsh., tarn. bronze-yel.	....
Aluminite	$Al_2(SO_4)(OH)_4.7H_2O$	1.66–1.82	1–2	monocl.(?), wh.	1.459, 1.464, 1.470
Alunite	$(K,Na)Al_3(SO_4)_2(OH)_6$	2.6–2.9	3.5–4	rhbdr., wh., gray, yel., redsh., br.	1.572, 1.592
Alunogen	$Al_2(SO_4)_3.18H_2O$	1.77	1.5–2	tricl., col., wh., yelsh. wh., redsh. wh.	1.459–1.475, 1.461–1.478, 1.470–1.485
Amblygonite	$(Li,Na)Al(PO_4)(F,OH)$	3.0–3.1	5.5–6	tricl., wh., yelsh. wh., grnsh. wh., blsh. wh., gray	1.591, 1.604, 1.613
Analcite	$NaAlSi_2O_6.H_2O$	2.24–2.29	5.5	cub., wh., pink, gray	1.479–1.493
Anatase	TiO_2	3.90	5.5–6	tetr., br., yelsh. br., redsh. br., bl., blk., grn., gray	2.5612, 2.4880
Andalusite	Al_2OSiO_4	3.13–3.16	6.5–7.5	rhomb., pink, wh., red	1.629–1.640, 1.633–1.644, 1.638–1.650
Andesine	$([NaSi]_{0.7-0.5}[CaAl]_{0.3-0.5})AlSi_2O_8$	2.65–2.68	6–6.5	tricl., wh., gray, grn.	1.544–1.555, 1.548–1.558, 1.551–1.563
Andorite	$PbAgSb_3S_6$	5.33–5.37	3–3.5	rhomb., dk. steel gray, somet. tarn. yel. or iridesc.	....
Andradite	$Ca_3Fe_2^{+3}Si_3O_{12}$	3.859	6–7.5	cub., brnsh. red, blk., somet. yel., grn.	1.887
Anglesite	$PbSO_4$	6.37–6.39	2.5–3	rhomb., col., wh., somet. gray, yelsh., grn. tinge	1.8771, 1.8826, 1.8937
Anhydrite	$CaSO_4$	2.98	3.5	rhomb., col., blsh. wh., vlt.	1.5698, 1.5754, 1.6136
Ankerite	$Ca(Fe,Mg,Mn)(CO_3)_2$	2.8–3.1	3.5–4	rhbdr., br., yelsh. br., grnsh. br., pink	1.690–1.750, 1.510–1.548
Anorthite	$CaAl_2Si_2O_8$	2.76	6–6.5	tricl., wh., yel., grn., blk.	1.577, 1.585, 1.590
Anorthoclase	$(Na,K)AlSi_3O_8$	2.56–2.60	6	tricl., col., wh.	1.523, 1.528, 1.529
Anthophyllite	$(Mg,Fe)_7Si_8O_{22}(OH,F)_2$	2.85–3.57	5.5–6	rhomb., wh., gray, grn., br., yelsh. br., dk. br.	1.596–1.694, 1.605–1.710, 1.615–1.722
Antimony	Sb	6.61–6.72	3–3.5	hex., tin-wh.	....
Apatite	$Ca_5(PO_4)_3(OH,F,Cl)$	3.1–3.35	5	hex., grn., wh., yel., br. red, bl.	1.629–1.667, 1.624–1.666
Apophyllite	$KFCa_4Si_8O_{20}.8H_2O$	2.33–2.37	4.5–5	tetr., col., wh., pink, pa. yel., pa. grn.	1.534–1.535, 1.535–1.537
Aragonite	$CaCO_3$	2.94–2.95	3.5–4	rhomb., col., wh.	1.530–1.531, 1.680–1.681, 1.685–1.686
Arcanite	K_2SO_4	2.663	....	rhom., col., wh.	1.4935, 1.4947, 1.4973
Argentite	Ag_2S	7.2–7.4	2–2.5	cub., blksh. lead gray	....
Arsenic	As	5.63–5.78	3.5	hex., tin-wh., tarn. dk. gray	....
Arsenolite	As_2O_3	3.86–3.88	1.5	cub., wh., somet. blsh., yelsh., redsh. tinge	1.755
Arsenopyrite	$FeAsS$	5.9–6.2	5.5–6	monocl., silver-wh., to steel gray	....
Atacamite	$Cu_2(OH)_3Cl$	3.74–3.78	3–3.5	rhomb., grn., dk. grn., blksh. grn.	1.831, 1.861, 1.880
Augelite	$Al_2(PO_4)(OH)_3$	2.696	4.5–5	monocl., col., wh., yelsh. wh., rose	1.5736, 1.5759, 1.5877
Augite	$(Ca,Mg,Fe,Fe^{+3},Ti,Al)_2(Si,Al)_2O_6$	3.23–3.52	5.5–6	monocl., pa. br., br., purp. br., grn., blk.	1.671–1.735, 1.672–1.741, 1.703–1.761
Autunite	$Ca(UO_2)_2(PO_4)_2.10–12H_2O$	3.1–3.2	2–2.5	tetr., yel., somet. grnsh. yel. to pa. grn.	1.577, 1.553
Axinite	$(Ca,Mn,Fe)_3Al_2BO_3Si_4O_{12}(OH)$	3.26–3.36	6.5–7	tricl., br., yelsh.	1.674–1.693, 1.681–1.701, 1.684–1.704
Azurite	$Cu_3(OH)_2(CO_3)_2$	3.77	3.5–4	monocl., azure bl., dk. bl., pa. bl.	1.730, 1.758, 1.838
Baddeleyite	ZrO_2	5.4–6.02	6.5	monocl., col., yel., gr., redsh. br., br., blk.	2.13, 2.19, 2.20
Barite	$BaSO_4$	4.50	3–3.5	rhomb., col., wh., somet. br., dk. br., gray	1.6362, 1.6373, 1.6482
Benitoite	$BaTi(SiO_3)_3$	3.65	6–6.5	rhbdr., bl., purp., col.	1.757, 1.804
Bertrandite	$Be_4Si_2O_7(OH)_2$	2.6	6	rhomb., col.	1.589, 1.602, 1.613
Beryl	$Be_3Al_2Si_6O_{18}$	2.66–2.83	7.5–8	hex., col., wh., blsh. grn., grnsh. yel., yel., bl.	1.565–1.590, 1.567–1.598
Beryllonite	$NaBe(PO)_4$	2.81	5.5–6	monocl., col., wh., pa. yel.	1.5520, 1.5579, 1.561
Biotite	$K(Mg,Fe)_3AlSi_3O_{10}(OH,F)_2$	2.7–3.3	2.5–3	monocl., blk., dk. br., redsh. br.	1.565–1.625–1.605–1.696, 1.605–1.696
Bismuth	Bi	9.70–9.83	2–2.5	rhbdr., silver-wh. to redsh wh.	....
Bismuthinite	Bi_2S_3	6.75–6.81	2	rhomb., lead gray to tin-wh., tarn. yel. or iridesc.	....
Bixbyite	$(Mn,Fe)_2O_3$	4.945	6–6.5	cub., blk.	....
Bloedite	$Na_2Mg(SO_4)_2.4N_2O$	2.22–2.28	2.5–3	monocl., col., somet. blsh.-grn. or redsh.	1.483, 1.486, 1.487
Boehmite	$AlO(OH)$	3.01–3.06	3.5–4	rhomb., wh.	1.64–1.65, 1.65–1.66, 1.65–1.67
Boracite	$Mg_3B_7O_{13}Cl$	2.91–2.97	7–7.5	rhomb., col., wh., gray, yel., blsh.-grn., grn.	1.66, 1.66, 1.67
Borax	$Na_2B_4O_7.10H_2O$	1.715	2–2.5	monocl., col., wh., gray, blsh. or grnsh-wh.	1.4466, 1.4687, 1.4717
Bornite	Cu_5FeS_4	5.06–5.08	3	cub., copper red to pinchbeck br., tarn. purp., iridesc.	....
Boulangerite	$Pb_5Sb_4S_{11}$	6.0–6.2	2.5–3	monocl., blsh. lead gray, oft, with yel. spots	....
Bournonite	$PbCuSbS_3$	5.80–5.86	2.5–3	rhomb., steel gray to blk.	....
Braggite	PtS	10.0	....	tetr., steel gray	....
Braunite	$(Mn,Si)_2O_3$	4.72–4.83	6–6.5	tetr., brns. blk. to steel gray	....
Bravoite	$(Ni,Fe)S_2$	4.62	5.5–6	cub., steel gray	....
Breithauptite	$NiSb$	8.23	5.5	hex., pa. copper red to vlt., tarn.	....
Brochantite	$Cu_4(SO_4)(OH)_6$	3.79	3.5–4	monocl., emerald-grn. to blksh. grn., pa. grn.	1.728, 1.771, 1.800
Bromyrite	$AgBr$	6.47	2.5	cub., col., gray, yelsh., grnsh.-br.	2.253
Brookite	TiO_2	4.08–4.20	5.5–6	rhomb., br., yelsh. br., redsh. br., blk.	2.5831, 2.5843, 2.7004
Brucite	$Mg(OH)_2$	2.38–3.40	2.5	hex., wh., pa. grn., gray, bl., yel., br.	1.560–1.590, 1.580–1.600
Bunsenite	NiO	6.898	5.5	cub., dk. pistachio-grn.	(Li) 2.37
Cacoxenite	$Fe_4(PO_4)_3(OH)_3.12H_2O$	2.2–2.4	3–4	hex., yel. to brnsh.-yel., redsh. yel., somet. grnsh.	1.575–1.585, 1.635–1.656
Calcite	$CaCO_3$	2.715–2.94	3	rhbdr., col., wh., somet. gray, yel., pink, bl.	1.658–1.740, 1.486–1.550
Caledonite	$Cu_2Pb_5(SO_4)_3(CO_3)(OH)_6$	5.75–5.77	2.5–3	rhomb., dk. grn., blsh. grn.	1.815–1.821, 1.863–1.869, 1.906–1.912
Calomel	$HgCl$	7.15	1.5	tetr., col., wh., gray, yelsh. wh., br.	1.973, 2.656
Cancrinite	$(Na,Ca)_{7-8}Al_6Si_6O_{24}(CO_3SO_4Cl)_{1.5-2}.1–5H_2O$	2.51–2.42	5–6	hex., col., wh., pa. bl., pa. grn., yel., redsh.	1.528–1.507, 1.503–1.495
Carnalite	$KMgCl_3.6H_2O$	1.602	2.5	rhomb., col., wh., oft. redsh., somet. yel., bl.	1.466, 1.475, 1.494
Carnotite	$K_2(UO_2)_2(VO_4)_2.3H_2O$	....	1–2	rhomb. or monocl., bright yel., yel., grnsh. yel.	1.75, 1.92, 1.95
Cassiterite	SnO_2	6.99	6–7	tetr., yelsh. or redsh. br., brnsh.-blk.	2.006, 2.0972
Celestite	$SrSO_4$	3.96	3–3.5	rhomb., col., wh.. pa. bl., redsh., grnsh., brnsh.	1.621–1.622, 1.623–1.624, 1.630–1.631
Celsian	$BaAl_2Si_2O_8$	3.10–3.39	6–6.5	monocl., col., wh., yel.	1.579–1.587, 1.583–1.593, 1.588–1.600
Cervantite	$Sb_2O_4(?)$	6.64	4–5	rhomb.(?)., yel., wh., somet. redsh.-wh.	....
Cerargyrite	$AgCl$	5.55	2.5	cub., col., gray, grnsh.-br., tarn. purp., yelsh.	2.071
Cerussite	$PbCO_3$	6.53–6.57	3–3.5	rhomb., col., wh., gray, somet. bl., blk., grn.	1.8036, 2.0765, 2.0786
Chabazite	$(Ca,Na_2)Al_2Si_4O_{12}.6H_2O$	2.05–2.10	4.5	rhbdr., redsh.-wh., wh., yelsh., grnsh.	1.470–1.494
Chalcocite	Cu_2S	5.5–5.8	2.5–3	rhomb., blksh., lead gray	....
Chalcanthite	$CuSO_4.5H_2O$	2.28	2.5	tricl., dk. bl. to sky bl., somet. grnsh.	1.514, 1.537, 1.543
Chalcopyrite	$CuFeS_2$	4.1–4.3	3.5–4	tetr., brass-yel., tarn., iridisc.	....
Chiolite	$Na_5Al_3F_{14}$	3.00	3.5–4	tetr., wh. to col.	1.349, 1.342
Chlorite	$(Mg,Al,Fe)_{12}(Si,Al)_8O_{20}(OH)_{16}$	2.6–3.3	2–3	monocl., grn., wh., yel., pink, br., red	1.57–1.66, 1.57–1.67, 1.57–1.67
Chloritoid	$(Fe,Mg,Mn)_2(AlFe^{+3})Al_3O_2SiO_4(OH)_4$	3.51–3.80	6.5	monocl., tricl., dk. grn.	1.713–1.730, 1.719–1.734, 1.723–1.740

Name	Formula	Sp. gr.	Hardness	Crystalline form and color	Index of refraction (Na) η; ω ϵ α β γ
Chondrodite	$Mg(OH,F)_2.2Mg_2SiO_4$	3.16–3.26	6.5	monocl., yel., br., red	1.592–1.615, 1.602–1.627, 1.621–1.646
Chromite	$FeCr_2O_4$	4.5–5.1	5.5	cub., blk.	2.16
Chrysoberyl	$BeAl_2O_4$	3.65–3.85	8.5	rhomb., grn., yel.	1.746, 1.748, 1.756
Chrysocolla	$CuSiO_3.2H_2O$	~2.4	2	rhomb., (?)., grn., bl., br., blk.	1.575, 1.597, 1.598
Cinnabar	HgS	8.090	2–2.5	hex., red, brnsh. red., gray	(Li) 2.814, 3.143
Claudetite	As_2O_3	4.15	2.5	monocl., col. to wh.	1.87, 1.92, 2.01
Clinozoisite	$Ca_2Al_3Si_3O_{12}(OH)$	3.21–3.38	6.5	monocl., col., pa. yel., gray, grn.	1.670–1.715, 1.674–1.725, 1.690–1.734
Cobaltite	CoAsS	6.33	5.5	cub., silver wh., redsh., steel gray, blk.	...
Colemanite	$Ca_2B_6O_{11}.5H_2O$	2.42–2.43	4.5	monocl., col., wh., yelsh. wh., gray	1.586, 1.592, 1.614
Columbite	$(Fe,Mn)(Cb,Ta)_2O_6$	5.15–5.25	6	rhomb., iron blk. to br. blk.	...
Connellite	$Cu_{19}(SO_4)Cl_4(OH)_{32}.3H_2O(?)$	3.36	3	hex., azure bl.	1.724–1.738, 1.746–1.758
Copiapite	$(Fe,Mg)Fe_4^{+3}(SO_4)_6(OH)_2.20H_2O$	2.08–2.17	2.5–3	tricl., yel., grnsh. yel.	1.51–1.53, 1.53–1.55, 1.58–1.60
Copper	Cu	8.95	2.5–3	cub., red	...
Coquimbite	$Fe_2(SO_4)_3.9H_2O$	2.10–2.12	2.5	hex., pa. vlt. to dk. amethystine, yelsh., grnsh.	1.53–1.55, 1.55–1.57
Cordierite	$Al_3(Mg,Fe)_2Si_5AlO_{18}$	2.53–2.78	7	rhomb., gray-bl., bl., dk. bl.	1.522–1.558, 1.524–1.574, 1.527–1.578
Corundum	Al_2O_3	4.022	9	hex., col., bl., yel., purp., grn., pink, red	1.767–1.772, 1.759–1.763
Cotunnite	$PbCl_2$	5.80	2.5	rhomb., col. to wh., somet. yelsh., grnsh.	2.199, 2.217, 2.260
Covellite	CuS	4.6–4.76	1.5–2	hex., indigo bl., dk. bl., iridesc. brass yel. to red	...
Cristobalite	SiO_2	2.33	6–7	tetr.(?)., col., wh., yel.	1.487, 1.484
Crocoite	$PbCrO_4$	5.96–6.02	2.5–3	monocl., red, orange red, orange yel.	2.29, 2.36, 2.66
Cryolite	Na_3AlF_6	2.96–2.98	2.5	monocl., col. to wh., brnsh., redsh., blk.	1.338, 1.338, 1.339
Cryolithionite	$Na_3Li_3Al_2F_{12}$	2.77	2.5–3	cub., col. to wh.	1.3395
Cubanite	$CuFe_2S_3$	4.03–4.18	3.5	rhomb., brass to bronze yel.	...
Cummingtonite	$(Mg,Fe)_7Si_8O_{22}(OH)_2$	3.2–3.5	5–6	monocl., dk. grn., br.	1.635–1.665, 1.644–1.675, 1.655–1.698
Cuprite	Cu_2O	6.14	3.5–4	cub., red, somet. blk.	...
Danburite	$CaSi_2B_2O_8$	3.0	7	rhomb., pa. yel., col., dk. yel., yelsh. br.	1.63, 1.63–1.64, 1.63–1.64
Datolite	$CaBSiO_4(OH)$	2.96–3.00	5–5.5	monocl. col., wh., yelsh., grnsh., pinksh.	1.622–1.626, 1.649–1.654, 1.666–1.670
Daubreelite	Cr_2FeS_4	3.80–3.82	?	cub., blk.	...
Derbylite	$Fe_6Ti_6Sb_2O_{23}(?)$	4.53	5	rhomb., pitch blk.	2.45, 2.45, 2.51
Diamond	C	3.50–3.53	10	cub., col., pa. yel. to dk. yel., pa. br. to dk. br., wh., blsh. wh.	2.4175
Diaspore	AlO(OH)	3.3–3.5	6.5–7	rhomb., wh., graysh. wh., col.	1.682–1.706, 1.705–1.725, 1.730–1.752
Digenite	Cu_2xS	5.546	2.5–3	cub., bl. to blk.	...
Diopside	$CaMgSi_2O_6$	3.22–3.38	5.5–6.5	monocl., wh., pa. grn., dk. grn.	1.664–1.695, 1.672–1.701, 1.695–1.721
Dioptase	$Cu_6Si_6O_{18}.6H_2O$	3.5	5	rhbdr., emerald grn.	1.64–1.66, 1.70–1.71
Dolomite	$CaMg(CO_3)_2$	2.86	3.5–4	rhbdr., wh., oft. yel. or br. tinge, col.	1.679, 1.500
Douglasite	$K_2FeCl_4.2H_2O(?)$	2.16	...	pa. grn., tarn. brnsh. red	1.485–1.491, 1.497–1.503
Dyscrasite	Ag_3Sb	9.67–9.81	3.5–4	rhomb., silver wh., tarn. gray, yelsh. or blksh.	...
Eddingtonite	$BaAl_2Si_3O_{10}.4H_2O$	2.7–2.8	...	rhomb. or monocl., col., pink, br. wh.	1.541, 1.553, 1.557
Eglestonite	Hg_4OCl_2	8.4	2.5	cub., yel., orange-yel. to dk. brnsh., tarn. bl.	2.47–2.51
Emplectite	$CuBiS_2$	6.38	2	rhomb., gray to tin wh.	...
Empressite	AgTe	7.510	3–3.5	pa. bronze	...
Enargite	Cu_3AsS_4	4.4–4.5	3	rhomb., gray-blk. to iron-blk.	...
Enstatite	$MgSiO_3$	3.209	5–6	rhomb., col., gray, grn., yel., brn.	1.650–1.662, 1.653–1.671, 1.658–1.680
Epidote	$Ca_2Fe^{+3}Al_2Si_3O_{12}(OH)$	3.38–3.49	6	monocl., grn., yel., gray	1.715–1.751, 1.725–1.784, 1.734–1.797
Epsomite	$MgSO_4.7H_2O$	1.675–1.679	2–2.5	rhomb., col., wh. pink, grn.	1.4325, 1.4554, 1.4609
Erythrite	$(Co,Ni)_3(AsO_4)_2.8H_2O$	3.06	1.5–2.5	monocl., crimson-red, red, pa. pink	1.626, 1.661, 1.699
Eucairite	CuAgSe	7.6–7.8	2.5	silver wh. to lead gray	...
Euclasite	$BeAlSiO_4(OH)$	3.0–3.1	7.5	monocl., col., pa. grn., bl.	1.651, 1.655, 1.671
Eudialite	$(Na,Ca,Fe)_6ZrSi_6O_{18}(OH,Cl)(?)$	2.8–3.1	5–6	hex., pa. pink, red, br.	1.59–1.61, 1.59–1.61
Eulytite	$Bi_4Si_3O_{12}$	6.6	4.5	cub., br., yel., gray	2.05
Euxenite	$(Y,Ca,Ce,U,Th)(Cb,Ta,Ti)_2O_6$	5.0–5.9	5.5–6.5	rhomb., blk., grnsh. or brnsh. tint.	~2.2
Fayalite	Fe_2SiO_4	4.392	6.5	rhomb., grnsh., yelsh.	1.827, 1.869, 1.879
Ferberite	$FeWO_4$	7.51	4–4.5	monocl., br. to blk.	(Li)2.37–2.43
Fergussonite	$(Y,Er,Ce,Fe)(Cb,Ta,Ti)O_4$	5.6–5.8	5.5–6.5	tetr., gray, yel., br., dk. br.	2.1
Fluorite	CaF_2	3.18	4	cub., bl., purp., wh., col., yel., grn.	1.433–1.435
Forsterite	Mg_2SiO_4	3.222	7	rhomb., wh., grnsh., yelsh.	1.635, 1.651, 1.670
Franklinite	$ZnFe_2^{+3}O_4$	5.07–5.34	5.5–6.5	Cub., blk. to br.-blk.	(Li) ~2.36
Gahnite	$ZnAl_2O_4$	4.62	7.5–8	cub., dk. bl.-grn., somet. yelsh. or brnsh.	1.79–1.81
Galena	PbS	7.57–7.59	2.5–2.75	cub. lead gray	...
Galenabismuthite	$PbBi_2S_4$	7.04	2.5–3.5	rhomb., pa. gray to tin-wh., lead gray, somet. tarn., yel. or irid.	...
Ganomalite	$(Ca,Pb)_{10}(OH,Cl)_2(Si_2O_7)_3$	5.4–5.7	3–4	hex., col., gray	1.910, 1.945
Gaylussite	$Na_2Ca(CO_3)_2.5H_2O$	1.991	2.5–3	monocl., col. to yelsh. wh., graysh. wh., wh.	1.4435, 1.5156, 1.5233
Gehlenite	$Ca_2Al_2SiO_7$	3.038	5–6	tetr., col., gray-grn., br.	1.669, 1.658
Geikielite	$MgTiO_3$	4.05	5–6	rhbdr., brnsh blk., blsh.	2.31, 1.95
Gibbsite	$Al(OH)_3$	2.38–2.42	2.5–3.5	monocl., wh., graysh., grnsh. or redsh.-wh.	1.56–1.58, 1.56–1.58, 1.58–1.60
Glauberite	$Na_2Ca(SO_4)_2$	2.75–2.85	2.5–3	monocl., gray, yel.sh., somet. col., redsh.	1.515, 1.535, 1.536
Glauconite	$(K,Na,Ca)_{1.2-2}(Fe^{+3},Al,Fe,Mg)_4Si_{7-7.6}Al_{1-0.4}O_{20}(OH)_4.nH_2O$	2.4–2.95	2	monocl., col., ye.lsh. grn., grn., blsh. gray	1.592–1.610, 1.614–1.641, 1.614–1.641
Glaucophane	$Na_2Mg_3Al_2Si_8O_{22}(OH)_2$	3.08–3.30	6	monocl., gray, lavender bl.	1.606–1.661, 1.622–1.667, 1.627–1.670
Gmelinite	$(Ca,Na_2)Al_2Si_4O_{12}.6H_2O$	~2.1	4.5	rhbdr., wh., redsh.-wh., yelsh., grnsh.	1.476–1.494, 1.474–1.480
Goethite	FeO(OH)	3.3–4.3	5–5.5	rhomb., blksh.-br., yelsh. or redsh.-br., yel.	2.260–2.275, 2.393–2.409, 2.398–2.515
Gold	Au	19.3	2.5–3	cub., yel.	...
Goslarite	$ZnSO_4.7H_2O$	1.978	2–2.5	rhomb., col., wh., somet. br., grn., bl.	1.4568, 1.4801, 1.4844
Graphite	C	2.09–2.23	1–2	hex., iron-blk. to steel gray	...
Greenockite	CdS	4.9	3–3.5	hex., yel. to orange	2.506, 2.529
Grossularite	$Ca_3Al_2Si_3O_{12}$	3.594	6–7.5	cub., wh., yel., grn., br., red	1.734
Gummite (R)	$UO_3.H_2O$	3.9–6.4	2.5–5	yel., orange, redsh.-yel., red, br. blk.	...
Gypsum	$CaSO_4.2H_2O$	2.30–2.37	2	monocl., wh., col., somet. gray, red, yel., br.	1.519–1.521, 1.523–1.526, 1.529–1.531
Halite	NaCl	2.16–2.17	2.5	cub., col., wh., orange, red	1.544
Hambergite	$Be_2(OH)(BO_3)$	2.36	7.5	rhomb., col. to gray, wh., yel.	1.56, 1.59, 1.63
Hanksite	$Na_{22}K(SO_4)_9(CO_3)_2Cl$	2.562	3–3.5	hex., col., somet. pa.-yelsh. or gray	1.481, 1.461
Harmotome	$BaAl_2Si_6O_{16}.6H_2O$	2.41–2.47	4.5	monocl., or rhomb., col., wh., pink, gray, yel.	1.503–1.508, 1.505–1.509, 1.508–1.514
Hausmannite	Mn_3O_4	4.83–4.85	5.5	tetr., brnsh.-blk.	(Li) 2.46, 2.15
Haüyne	$(Na,Ca)_{4-8}Al_6Si_6O_{24}(SO_4,S)_{1-2}$	2.44–2.50	5.5–6	cub., wh., gray, grn., bl.	1.496–1.505
Hedenbergite	$CaFeSi_2O_6$	3.50–3.56	6	monocl., brnsh.-grn., dk. grn., blk.	1.716–1.726, 1.723–1.730, 1.741–1.751
Helvite	$Mn_4Be_3Si_3O_{12}S$	3.20–3.44	6	cub., yel., br., redsh.-brn.	1.728–1.749
Hematite	Fe_2O_3	5.26	5–6	rhbdr., steel gray, dull red to bright red	3.22, 2.94
Hemimorphite	$Zn_4Si_2O_7(OH)_2.H_2O$	3.45	5	rhomb., col., wh., pa. bl., pa. grn., br.	1.614, 1.617, 1.636
Hercynite	$FeAl_2O_4$	4.40	7.5–8	cub., blk.	1.835
Herderite	$CaBe(PO_4)(Fe,OH)$	2.95–3.01	5–5.5	monocl., col. to pa. yel. or grnsh.-wh.	1.592, 1.612, 1.621
Hessite	Ag_2Te	8.24–8.45	2–3	monocl., (<149.5°), cub. (>149.5°)., gray	
Heulandite	$(Ca,Na_2)Al_2Si_7O_{18}.6H_2O$	2.1–2.2	3.5–4	pseudo-monocl., col., wh., yel., pink, red, gray, br.	1.491–1.505, 1.493–1.503, 1.500–1.512

PHYSICAL CONSTANTS OF MINERALS (Continued)

Name	Formula	Sp. gr.	Hardness	Crystalline form and color	Index of refraction (Na) η; ω ε; ω β γ
Hopeite	$Zn_3(PO_4)_2.4H_2O$	3.0–3.1	3.25	rhomb., col. to grayish-wh., pa. yel.	1.57–1.59, 1.58–1.60, 1.58–1.60
Hornblende	$(Ca,Na,K)_{2-3}(Mg,Fe,Fe^{+3}Al)_5Si_6(Si,Al)_2O_{22}(OH,F)_2$	3.02–3.45	5–6	monocl., grn., dk. grn., blk.	1.615–1.705, 1.618–1.714, 1.632–1.730
Huebnerite	$MnWO_4$	7.12	4–4.5	monocl., yel.-br. to red br., somet. br., blk.	2.17, 2.22, 2.32
Humite	$Mg(OH,F)_2.3Mg_2SiO_4$	3.2–3.32	6	rhomb., yel., orange	1.607–1.643, 1.619–1.653, 1.639–1.675
Huntite	$Mg_3Ca(CO_3)_4$	2.696	....	rhomb.(?)., wh.	....
Hydrogrossularite	$Ca_3Al_2Si_2O_8(SiO_4)_{1-m}(OH)_{4m}$	3.594–3.13	6–7.5	cub., wh., buff, pa. grn., gray, pink	1.734–1.675
Hydromagnesite	$Mg_4(OH)_2(CO_3.3H_2O$	2.236	3.5	monocl., col. to wh.	1.520–1.526, 1.524–1.530, 1.544–1.546
Illite	$K_{1-1.5}Al_4Si_{7-6.5}Al_{1-1.5}O_{20}(OH)_4$	2.6–2.9	1–2	monocl., wh.	1.54–1.57, 1.57–1.61, 1.57–1.61
Ilmenite	$FeTiO_3$	4.68–4.76	5–6	rhbdr., iron-blk.	....
Iodyrite	AgI	5.69	1.5	hex., col. on exposure to light, yel., br.	2.21, 2.22
Jadeite	$NaAlSi_2O_6$	3.24–3.43	6	monocl., col., wh., grn., grnsh. bl.	1.640–1.658, 1.645–1.663, 1.652–1.673
Jamesonite	$Pb_4FeSb_6S_{14}$	5.63	2.5	monocl., gray-blk., somet. tarn. iridesc.	....
Jarosite	$KFe_3(SO_4)_2(OH)_6$	2.91–3.26	2.5–3.5	rhbdr., ocherous, amber yel. to dk. br.	1.820, 1.715
Kainite	$KMg(SO_4)Cl\cdot 3H_2O$	2.15	2.5–3	monocl., col., gray bl., vlt., yelsh., redsh.	1.494, 1.505, 1.516
Kaliophyllite	$KAlSiO_4$	2.61	6	hex., col.	1.532, 1.537
Kaolinite	$Al_4Si_4O_{10}(OH)_8$	2.61–2.68	2–2.5	tricl. or monocl., wh., redsh.-wh., grnsh.-wh.	1.533–1.565, 1.559–1.569, 1.560–1.570
Kernite	$Na_2B_4O_7.4H_2O$	1.908	2.5	monocl., col., wh.	1.454, 1.472, 1.488
Kieserite	$MgSO_4.H_2O$	2.571	3.5	monocl., col., gray, wh., yelsh.	1.520, 1.533, 1.584
Kyanite	Al_2OSiO_4	3.53–3.65	5.5–7	tricl., bl., wh., gray, grn., yel., pink	1.712–1.718, 1.721–1.723, 1.727–1.734
Lanarkite	$Pb_2(SO_4)O$	6.92	2–2.5	monocl., gray to grnsh. wh., pa. yel.	1.925–1.931, 2.004–2.010, 2.033–2.039
Lanthanite	$(La,Ce)_2(CO_3)_38H_2O$	2.69–2.74	2.5–3	rhomb., col. to wh., pink, yelsh.	1.51–1.53, 1.584–1.590, 1.610–1.616
Laumontite	$CaAl_2Si_4O_{12}.4-3.5H_2O$	2.2–2.3	3–3.5	monocl., col., wh., red, yel., brn.	1.502–1.514, 1.512–1.522, 1.514–1.525
Laurionite	$Pb(OH)Cl$	6.24	3–3.5	rhomb., col. to wh.	2.08, 2.12, 2.16
Lawsonite	$CaAl_2(OH)_2Si_2O_7.H_2O$	3.05–3.10	6	rhomb., col., wh.	1.655, 1.674–1.675, 1.684–1.686
Lazulite	$(Mg,Fe)Al_2(PO_4)_2(OH)_2$	3.08–3.38	5.5–6	monocl., bl., blsh. wh., dk. bl., blsh. grn.	1.604–1.626, 1.626–1.654, 1.637–1.663
Lazurite	$Na_4SSi_3Al_3O_{12}$	2.38–2.45	5–5.5	cub., berlin bl., azure bl., grnsh. bl., vlt.	1.500
Leadhillite	$Pb_4(SO_4)(CO_3)_2(OH)_2$	6.55	2.5–3	monocl., col. to wh., gray, pa. grn., pa. bl., yelsh.	1.87, 2.00, 2.01
Lepidocrocite	FeO(OH)	4.05–4.31	5	rhomb., ruby-red to red-br.	1.94, 2.20, 2.51
Lepidolite	$K_2(Li,Al)_{5-6}Si_{6-7}Al_{2-1}O_{20}(OH,F)_4$	2.80–2.90	2.5–4	monocl., col., pa. pink, pa. purp.	1.525–1.548, 1.551–1.585, 1.554–1.587
Leucite	$KAlSi_2O_6$	2.47–2.50	5.5–6	tetr., (pseudo-cub.) wh., gray	1.508–1.511
Levyne	$(Ca,Na_2)Al_2Si_4O_{12}.6H_2O$	~2.1	4.5	rhbdr., wh., redsh. wh., yelsh., grnsh.	1-496–1.505, 1.491–1.500
Litharge	PbO	9.14	2	tetr., red	(Li) 2.665, 2.535
Loellingite	$FeAs_2$	7.39–7.41	5–5.5	rhomb., silver wh. to steel-gray	....
Magnesite	$MgCO_3$	2.98–3.44	3.5–4.5	rhbdr., wh., col., somet. yel., br.	1.700–1.782, 1.509–1.563
Magnetite	Fe_3O_4	5.175	5.5–6.5	cub., blk. to br.-blk.	2.42
Malachite	$Cu_2(OH)_2(CO_3)$	4.03–4.07	3.5–4	monocl., bright grn. to dk. grn., blksh. grn.	1.652–1.658, 1.872–1.878, 1.906–1.912
Manganite	MnO(OH)	4.32–4.43	4	monocl., dk. steel-gray to iron-blk.	(Li) 2.25, 2.25, 2.53
Manganosite	MnO	5.364	5.5	cub., emerald grn., tarn. bl.	....
Marcasite	FeS_2	4.887	6–6.5	rhomb., pa. bronze-yel., tin-wh.	....
Marialite	$Na_4Al_3Si_9O_{24}Cl$	2.50–2.62	5–6	tetr., col., wh., pa. grnsh. yel., gray, br.	1.546–1.550, 1.540–1.541
Marshite	CuI	5.68	2.5	cub., col. to pa. yel., on exposure to light, red	2.346
Mascagnite	$(NH_4)_2SO_4$	1.768	2–2.5	rhomb., col., gray, yelsh.	1.5202, 1.5230, 1.5330
Matlockite	PbFCl	7.12	2.5–3	tetr., col. or yel. to pa. amber, grnsh.	2.145, 2.006
Meionite	$Ca_4Al_6Si_6O_{24}CO_3$	2.78	5–6	tetr., col., wh., pa. grnsh. yel., gray, br.	1.590–1.600, 1.556–1.562
Melanterite	$FeSO_4.7H_2O$	1.898	2	monocl., grn., grnsh. bl., grnsh. wh.	1.47, 1.48, 1.49
Melilite	$(Ca,Na,K)_2(Mg,Fe,Fe^{+3},Al,Si)_3O_7$	2.95–3.05	5–6	tetr., yelsh., br., grn.-br.	1.624–1.666, 1.616–1.661
Mellite	$Al_2C_{12}O_{12}.18H_2O$	1.64	2–2.5	tetr., yel., redsh., brnsh., somet. wh.	1.5393, 1.5110
Mendipite	$Pb_3O_2Cl_2$	7.24	2.5	rhomb., col. to wh., gray, oft. yel., red, bl. tinge	2.22–2.26, 2.25–2.29, 2.29–2.33
Mesolite	$Na_2Ca_2(Al_2Si_3O_{10}).8H_2O$	~2.26	5	monocl., col., wh., gray, yel., pink, red	β = 1.504–1.508
Metacinnabar	HgS	7.65	3	cub., graysh.-blk.	....
Microcline	$KAlSi_3O_8$	2.56–2.63	6–6.5	tricl., col., wh., pink, red, yel., grn.	1.514–1.529, 1.518–1.533, 1.521–1.539
Microlite	$(Na,Ca)_2Ta_2O_6(O,OH,F)$	4.2–6.4	5–5.5	cub., pa. yel. to br., somet. red, grn.	~2.0
Miersite	AgI	5.64–5.68	2.5	cub., canary-yel.	2.18–2.22
Millerite	NiS	5.3–5.7	3–3.5	hex., pa. brass-yel. to bronze-yel., gray, tarn. iridesc.	....
Mimetite	$Pb_5(AsO_4,PO_4)_3Cl$	7.24	3.5–4	hex. pa. yel. to yelsh. br., orange-yel., wh.	2.147, 2.128
Minium	Pb_3O_4	8.9–9.2	2.5	scarlet red, bl. red, somet. yel. tint.	(Li) 2.40–2.44
Mirabilite	$Na_2SO_4.10H_2O$	1.490	1.5–2	monocl., col. to wh.	1.391–1.397, 1.393–1.399, 1.395–1.401
Moissanite	SiC	3.218	9.5	hex., grn. to blk., somet. blsh., red	2.647–2.649, 2.689–2.693
Molybdenite	MoS_2	4.62–4.73	1–1.5	hex., lead-gray	....
Monazite	$(Ce,La,Th)PO_4$	5.0–5.3	5	monocl., yel., br., redsh. br.	1.774–1.800, 1.777–1-801, 1.828–1.851
Monetite	$CaH(PO_4)$	2.929	3.5	tricl., wh., pa. yelsh.-wh.	1.587, ~1.615, 1.640
Monticellite	$CaMgSiO_4$	3.08–3.27	5.5	rhomb., col.	1.639–1.654, 1.646–1.664, 1.653–1.674
Montmorillonite	$(0.5Ca,Na)_{0.7}(Al,Mh,Fe)_4(Si,Al)_8O_{20}(OH)_4.nH_2O$	2–3	1–2	monocl., wh., yel., grn.	1.48–1.61, 1.50–1.64, 1.50–1.64
Montroydite	HgO	11.23	2.5	rhomb., dk. red to brnsh. red, br.	(Li) 2.37, 2.5, 2.65
Mordenite	$(Na_2,K_2Ca)Al_2Si_{10}O_{24}.7H_2O$	2.12–2.15	3–4	rhomb., col., wh., red, yel., br.	1.472–1.483, 1.475–1.485, 1.477–1.487
Muscovite	$KAl_2Si_3AlO_{10}(OH,F)_2$	2.77–2.88	2.5–3	monocl., col., pa. grn., pa. red, pa. br.	1.552–1.574, 1.582–1.610, 1.587–1.616
Nantokite	CuCl	4.136	2.5	cub., col. to wh., grayish, grn.	1.925–1.935
Natrolite	$Na_2Al_2Si_3O_{10}.2H_2O$	2.20–2.26	5	rhomb., col., wh., gray, yel., pink, red	1.473–1.483, 1.476–1.486, 1.485–1.496
Nepheline	$Na_3KAl_4Si_4O_{16}$	2.56–2.665	5.5–6	hex., col., wh., gray	1.529–1.546, 1.526–1.542
Newberyite	$MgH(PO_4).3H_2O$	2.10	3.0–3.5	rhomb., col.	1.511–1.517, 1.514–1.520, 1.530–1.536
Niccolite	NiAs	7.784	5–5.5	hex., pa. copper-red, tarn. gray to blk.	....
Nosean	$Na_8Al_6Si_6O_{24}SO_4$	2.30–2.40	5.5	cub., gray, bl., br.	1.495
Oldhamite	CaS	2.58	4	cub., pa. chestnut-br.	2.137
Oligoclase	$([NaSi]_{0.9-0.7}[CaAl]_{0.1-0.3})AlSi_2O_8$	2.63–2.65	6–6.5	tricl., col., wh., gray, grnsh., pink	1.533–1.544, 1.537–1.548, 1.543–1.552
Olivenite	$Cu_2(AsO_4)(OH)$	3.9–4.5	3	rhomb., olive grn., grnsh.-br., br., gray	1.75–1.78, 1.79–1.82, 1.83–1.87
Olivine	$(Mg,Fe)SiO_4$	3.22–4.39	6.5–7	rhomb., olive grn., grayish grn. to yelsh. br.	1.63–1.83, 1.65–1.87, 1.67–1.88
Opal	$SiO_2.nH_2O$	1.73–2.16	~6	col., wh., yel., br., red, grn., bl., blk., amorp.	1.41–1.46
Orpiment	As_2S_3	3.49	1.5–2	monocl., yel., brnsh. yel.	(Li) 2.4, 2.81, 3.02
Orthoclase	$KAlSi_3O_8$	2.55–2.63	6–6.5	monocl., col., wh., pink, red, yel., grn.	1.518–1.529, 1.522–1.533, 1.522–1.539
Orthopyroxene	$(Mg,Fe)SiO_3$	3.209–3.96	5–6	rhomb., col., gray, grn., yel., dk. brn.	1.650–1.768, 1.653–1.770, 1.658–1.788
Paragonite	$NaAl_2Si_3AlO_{10}(OH)_2$	2.85	2.5	monocl., col., pa. yel.	1.564–1.580, 1.594–1.609, 1.600–1.609
Parisite	$(Ce,La,Na)FCO_3.CaCO_3$	4.42	4.5	hex., brnsh., yel.	1.672, 1.771
Pectolite	$Ca_2NaH(SiO_3)_3$	2.86–2.90	4.5–5	tricl., col., wh.	1.595–1.610, 1.605–1.615, 1.632–1.645
Penfieldite	$Pb_4Cl_6(OH)_2$	6.6		hex., wh.	2.13, 2.21
Pentlandite	$(Fe,Ni)_9S_8$	4.6–5.0	3.5–4	cub., pa. bronze-yel.	....
Percylite	$PbCuCl_2(OH)_2(?)$	?	2.5	cub(?), sky bl.	2.04–2.06
Periclase	MgO	3.55–3.68	5.5	cub., col. to gray-wh., yel., brnsh. yel., grn., bl.	1.7350
Pekovskite	$CaTiO_3$	3.97–4.26	5.5	pseudo cub., blk., gray-blk., brnsh. bl., redsh. br., br., yel.	2.30–2.38
Petalite	$LiAlSi_4O_{10}$	2.412–2.422	6.5	monocl., wh., gray, somet. pink, grn.	1.504–1.507, 1.510–1.513, 1.516–1.523

Name	Formula	Sp. gr.	Hard-ness	Crystalline form and color	Index of refraction (Na) η; ω ϵ α β γ
Pharmacosiderite	$Fe_3(AsO_4)_2(OH)_3.5H_2O$	2.797	2.5	cub., olive-grn. to yel., br., redsh.	1.676–1.704
Phenakite	Be_2SiO_4	2.98	7.5	rhbder., col., rose, yel., br.	1.654, 1.670
Phillipsite	$(0.5Ca,Na,K)_3Al_3Si_5O_{16}.6H_2O$	2.2	4–4.5	monocl. or rhomb., col., wh., pink, gray, yel.	1.483–1.504, 1.484–1.509, 1.496–1.514
Phlogopite	$KMg_3AlSi_3O_{10}(OH,F)_2$	2.76–2.90	2–2.5	monocl., col., yelsh.-br., grn., redsh.-br., br.	1.530–1.590, 1.557–1.637, 1.558–1.637
Phosgenite	$Pb_2(CO_3)Cl_2$	6.133	2–3	tetr., yelsh. wh. to yelsh. br., br., somet. wh., rose, gray	2.1181, 2.1446
Piemontite	$Ca_2(Mn,Fe^{+3},Al)_2AlSi_3O_{12}(OH)$	3.45–3.52	6	monocl., redsh. brn., blk.	1.732–1.794, 1.750–1.807, 1.762–1.829
Pigeonite	$(Mg,Fe,Ca)(Mg,Fe)Si_2O_6$	3.30–3.46	6	monocl., br., grnsh. br., blk.	1.682–1.722, 1.684–1.722, 1.705–1.751
Platinum	Pt	14–19	4–4.5	cub., whitish, steel gray to dk. gray	....
Pollucite	$CsAlSi_2O_6$	2.9	6.5	tetr., (pseudo-cub.) col.	1.507–1.527
Polybasite	$(Ag,Cu)_{16}Sb_2S_{11}$	6.0–6.2	2–3	monocl., iron-blk.	....
Powellite	$Ca(Mo,W)O_4$	4.21–4.25	3.5–4	tetr., straw-yel., br., grnsh., somet. gray, bl., blk.	1.959–1.982, 1.967–1.993
Prehnite	$Ca_2Al_2Si_3O_{10}(OH)_2$	2.90–2.95	6–6.5	rhomb., pa. grn., yel., gray, wh.	1.611–1.632, 1.615–1.642, 1.632–1.665
Proustite	Ag_3AsS_3	5.57	2–2.5	rhbdr., scarlet-vermillion	3.0877, 2.7924
Pseudobrookite	Fe_2TiO_5	4.33–4.39	6	rhomb., dk. red-br. to brnsh. blk. and blk.	(Li) 2.38, 2.39, 2.42
Psilomelane	$BaMn^{+2}Mn_8^{+4}O_{16}(OH)_4$	4.71	5–6	rhomb., iron-blk. to steel-gray	....
Pumpellyite	$Ca_4(Mg,Fe,Mn)(Al,Fe^{+3},Ti)_5(OH)_3Si_6O_{23}.2H_2O$	3.18–3.23	6	monocl., grn., blsh. grn., br.	1.674–1.702, 1.675–1.715, 1.688–1.722
Pyrargyrite	Ag_3SbS_3	5.85	2.5	rhbdr., deep red	(Li) 3.084, 2.881
Pyrite	FeS_2	5.018	6–6.5	cub., pa. brass-yel., tarn. iridesc.	....
Pyrochlore	$NaCaCb_2O_6F$	4.2–6.4	5–5.5	cub., br. to blk., yelsh., redsh. or blksh. br.	....
Pyrochroite	$Mn(OH)_2$	3.23–3.27	2.5	hex., col. to pa. grn. or bl., tarn. br. to blk.	1.72, 1.68
Pyrolusite	MnO_2	5.04–5.08	6–6.5	tetr., pa. steel-gray, iron-gray, blk., blsh.	....
Pyromorphite	$Pb_5(PO_4,AsO_4)_3Cl$	7.00–7.08	3.5–4	hex., grn., yel., br., orange, brnsh. red., gray	2.058, 2.048
Pyrope	$Mg_3Al_2Si_3O_{12}$	3.582	6–7.5	cub., red, pink	1.714
Pyrophyllite	$Al_2Si_4O_{10}(OH)_2$	2.65–2.90	1–2	monocl., wh., yel., pa. bl., gray-grn., brnsh.-grn.	1.534–1.556, 1.568–1.589, 1.596–1.601
Pyrrhotite	$Fe_{1-0.8}S$	4.58–4.65	3.5–4.5	hex., bronze-yel. to br., tarn., somet. iridesc.	....
Quartz	SiO_2	2.65	7	rhbdr., col., wh., blk., purp., grn., bl., rose	1.544, 1.553
Rammelsbergite	$NiAs_2$	7.0–7.2	5.5–6	tin. wh., redsh. tinge	....
Raspite	$PbWO_4$	8.46	2.5–3	monocl., yelsh. br., pa. yel., gray	1.25–1.29, 1.25–1.29, 1.28–1.32
Realgar	AsS	3.56	1.5–2	monocl., aurora-red to orange-yel.	2.538, 2.684, 2.704
Riebeckite	$Na_2Fe_3Fe_2^{+3}Si_8O_{22}(OH,F)_2$	3.02–3.42	5	monocl., dk. bl., bl.	1.654–1.701, 1.662–1.711, 1.668–1.717
Rhodochrosite	$MnCO_3$	3.70	3.5–4	rhbdr., pink, red, br., brnsh.-yel.	1.816, 1.597
Rhodonite	$(Mn,Fe,Ca)SiO_3$	3.57–3.76	5.5–6.5	tricl., pink to brnsh. red	1.711–1.738, 1.716–1.741, 1.724–1.751
Rutile	TiO_2	4.23–5.5	6–6.5	tetr., redsh. brn. to red, somet. yelsh., blsh.	2.605–2.613, 2.899–2.901
Safflorite	$(Co,Fe)As_2$	7.0–7.5	4.5–5	rhomb., tin-wh., tarn. dk. gray	....
Samarskite	$(Y,Er,Ce,U,Ca,Fe,Pb,Th)(Cb,Ta,Ti,Sn)_2O_6$	5.69	5–6	rhomb., velvet blk., somet. brnsh. tint	~2.20
Sapphirine	$(Mg,Fe)_2Al_4O_6SiO_4$	3.40–3.58	7.5	monocl., pa. bl., pa. grn.	1.701–1.717, 1.703–1.720, 1.705–1.724
Scapolite	$(Na,Ca)_4Al_3(Al,Si)_3Si_6O_{24}(Cl,F,OH,CO_3,SO_4)$	2.50–2.78	5–6	tetr., col., wh., pa. grnsh. yel., gray, bl.	1.546–1.600, 1.540–1.562
Scheelite	$CaWO_4$	6.08–6.12	4.5–5	tetr., yelsh. wh., pa. yel., brnsh., col., wh., gray	1.920, 1.936
Scolecite	$CaAl_2Si_3O_{10}.3H_2O$	2.25–2.29	5	monocl., col., wh., gray, yel., pink, red	1.507–1.513, 1.516–1.520, 1.517–1.521
Scorodite	$Fe^{+3}(AsO_4).2H_2O$	3.28	3.5–4	rhomb., pa. grn., gray grn., br. somet. col., blsh., yel.	1.784, 1.795, 1.814
Sellaite	MgF_2	3.15	5	tetr., col. to wh.	1.378, 1.390
Senarmontite	Sb_2O_3	5.50	2–2.5	pseudo-cub., col., gray-wh.	2.087
Serpentine	$Mg_3Si_2O_5(OH)_4$	~2.55	2.5–3.5	monocl., wh., yel., gray, grn., blsh. grn.	1.53–1.57, 1.56, 1.54–1.57
Siderite	$FeCO_3$	3.96	4–4.5	rhbdr., yelsh. br., br., dk. br.	1.875, 1.635
Sillimanite	Al_2OSiO_4	3.23–3.27	6.5–7.5	rhomb., col., wh., yelsh., br., grnsh.	1.654–1.661, 1.658–1.662, 1.637–1.683
Silver	Ag	10.1–11.1	2.5–3	cub., wh., tarn. gray or blk.	....
Skutterudite	$(Co,Ni)As_3$	6.1–6.9	5.5–6	cub., between tin-wh. and silver-gray, tarn. gray or iridesc.	....
Smithsonite	$ZnCO_3$	4.42–4.44	4–4.5	rhbdr., grayish wh. to dk. gray, grnsh., brnsh. wh.	1.848, 1.621
Sodalite	$Na_8Al_6Si_6O_{24}Cl_2$	2.27–2.33	5.5–6	cub., bl., grn., yel., gray, pink	1.483–1.487
Sperrylite	$PtAs_2$	10.58	6–7	cub., tin-wh.	....
Spessartite	$Mn_3Al_2Si_3O_{12}$	4.190	6–7.5	cub., blk., dk. red, brnsh. red., bl., yelsh. orange	1.800
Sphalerite	ZnS	3.9–4.1	3.5–4	cub., br., blk., yel., red, wh.	2.369
Sphene	$CaTiSiO_4(O,OH,F)$	3.45–3.55	5	monocl., col., yel., grn., br., blk.	1.843–1.950, 1.870–2.034, 1.943–2.110
Spinel	$MgAl_2O_4$	3.55	7.5–8	cub., grn., red, bl., br. to col.	1.719
Spodumene	$LiAlSi_2O_6$	3.03–3.22	6.5–7	monocl., col., gray-wh., pa. bl., pa. grn., yelsh.	1.648–1.663, 1.655–1.669, 1.662–1.679
Stannite	Cu_2FeSn_4	4.3–4.5	4	tetr., steel gray to iron blk.	....
Staurolite	$(Fe,Mg)_2(AlFe^{+3})_9O_6SiO_4(O,OH)_2$	3.74–3.83	7.5	monocl., brn., redsh., yelsh.	1.739–1.747, 1.745–1.753, 1.752–1.761
Stercorite	$Na(NH_4)H(PO_4).4H_2O$	1.615	2	tricl., wh., yelsh., brnsh.	1.439, 1.442, 1.469
Stibiotantalite	$Sb(Ta,Cb)O_4$	5.7–7.5	5.5	rhomb., dk. br. to pa. yel.-br., red-br., grnsh.-yel.	2.38, 2.41, 2.46
Stibnite	Sb_2S_3	4.61–4.65	2	rhomb., lead-gray to steel-gray	....
Stilbite	$(Ca,Na_2K_2)Al_2Si_7O_{18}.7H_2O$	2.1–2.2	3.5–4	monocl., col., wh., yel., pink, red, gray, br.	1.484–1.500, 1.492–1.507, 1.494–1.513
Stilpnomelane	$(K,Na,Ca)_{0-1.4}(Fe^{+3}Fe,Mg,Al)_{6-8}Si_8O_{20}(OH)_4(O,OH,H_2O)_{4-8}$	2.59–2.96	3–4	monocl., br., dk. br., redsh. br., blk., dk. grn.	1.543–1.634, 1.576–1.745, 1.576–1.745
Stolzite	$PbWO_4$	7.9–8.4	2.5–3	tetr., redsh. br., yelsh. gray, straw-yel., grnsh.	2.26–2.28, 2.18–2.20
Strengite	$Fe^{+3}(PO_4).2H_2O$	2.90	3.5–4.5	rhomb., red, carmine, vlt., near col.	1.707, 1.719, 1.741
Strontianite	$SrCO_3$	3.72	3.5	rhomb., col., wh., yel., grnsh., brnsh.	1.516–1.520, 1.664–1.667, 1.666–1.669
Struvite	$Mg(NH_4)(PO_4).6H_2O$	1.71	2	rhomb., col., somet. yelsh., brnsh.	1.495, 1.496, 1.504
Sulfur	S	2.07	1.5–2.5	rhomb., yel., brnsh., grnsh., redsh., gray	1.9579, 2.0377, 2.2452
Sylvanite	$(Ag,Au)Te_2$	8.161	1.5–2	monocl., steel-gray to silver-wh.	....
Sylvite	KCl	1.99	2	cub., col., wh., somet. grayish, blsh., yelsh., red	1.49031
Talc	$Mg_3Si_4O_{10}(OH)_2$	2.58–2.83	1	monocl., col., wh., pa. grn., dk. grn., br.	1.539–1.550, 1.589–1.594, 1.589–1.600
Tantalite	$(Fe,Mn)(Ta,Cb)_2O_6$	7.90–8.00	6.5	rhomb., iron-bl. to br.-blk.	2.26, 2.32, 2.43
Tapiolite	$FeTa_2O_6$	7.9	6–6.5	tetr., blk.	(Li) 2.27, 2.42
Tellurobismuthite	Bi_2Te_3	7.800–7.830	1.5–2	rhbdr., pa. lead-gray	....
Terlinguaite	Hg_2OCl	8.725	2.5	monocl., yel. to grnsh.-yel., somet. br.	(Li) 2.33–2.37, 2.62–2.66, 2.64–2.68
Tetrahedrite	$(Cu,Fe)_{12}Sb_4S_{13}$	4.6–5.1	3–4.5	cub., flint-gray to iron-blk. to dull-blk.	....
Thenardite	Na_2SO_4	2.664	2.5–3	rhomb., col., grayish-wh., yelsh., yelsh. br., redsh.	1.464–1.471, 1.473–1.477, 1.481–1.485
Thermonatrite	$Na_2CO_3.H_2O$	2.255	1–1.5	rhomb., col. to wh., grayish, yelsh.	1.420, 1.506, 1.524
Thomsenolite	$NaCaAlF_6.H_2O$	2.981	2	monocl., col. to wh., somet. brnsh., redsh.	1.4072, 1.4136, 1.4150
Thomsonite	$NaCa_2([Al,Si]_5O_{10})_2.6H_2O$	2.10–2.39	5–5.5	rhomb., col., wh., pink, br.	1.497–1.530, 1.513–1.533, 1.518–1.544
Thorianite (R)	ThO_2	9.7	6.5	cub., dk. gray to brnsh.-blk., blk.	~2.20
Thorite (R)	$ThSiO_4$	5.2–5.4	4.5–5	tetr., orange-yel., brnsh. to blk.	~1.8
Topaz	$Al_2SiO_3(OH,F)_2$	3.49–3.57	8	rhomb., col., wh., yel., gray, grn., red, bl.	1.606–1.629, 1.609–1.631, 1.616–1.638

Name	Formula	Sp. gr.	Hard-ness	Crystalline form and color	Index of refraction (Na) η; ω ϵ ω β γ
Torbernite (R)	$Cu(UO_2)_2(PO_4)_2.8\text{–}12H_2O$	3.22	2–2.5	tetr., various shades of grn.	1.592, 1.582
Tourmaline	$Na(Mg,Fe,Mn,Li,Al)_3Al_6Si_6O_{18}(BO_3)_3(OH,F)_4$	3.03–3.25	7	rhbdr., blk., bl., grn., yel., red, col., br.	1.635–1.675, 1.610–1.650
Tremolite	$Ca_2Mg_5Si_8O_{22}(OH,F)_2$	3.0	5–6	monocl., col., gray, wh.	1.599, 1.612, 1.622
Tridymite	SiO_2	2.27	7	rhomb., col., wh.	1.471–1.479, 1.472–1.480, 1.474–1.483
Triphyllite-Lithiophyllite	$Li(Fe,Mn)PO_4$	3.34–3.58	4–5	rhomb., blsh. or grnsh. gray to yelsh. br., br.	1.66–1.70, 1.67–1.70, 1.68–1.71
Troegerite (R)	$(UO_2)_3(AsO_4)_2.12H_2O$	....	2–3	tetr., lemon-yel.	1.58–1.59, 1.625–1.635
Trona	$Na_3H(CO_3)_2.2H_2O$	2.14	2.5–3	monocl., gray or yelsh. wh., col.	1.412, 1.492, 1.540
Turquois	$Cu(Al,Fe^{+3})_6(PO_4)_4(OH)_8.4H_2O$	2.6–3.2	4.5–6	tricl., bl., grn., grnsh.-gray	1.61–1.78, 1.62–1.84, 1.65–1.84
Ullmannite	$NiSbS$	6.61–6.69	5–5.5	cub., steel-gray to silver-wh.	....
Uraninite (R)	UO_2	8.0–11	5–6	cub., steel-blk., brnsh.-blk., grayish, grn.	....
Uvarovite	$Ca_3Cr_2Si_3O_{12}$	3.90	6–7.5	cub., emerald-grn.	1.86
Valentinite	Sb_2O_3	5.76	2.5–3	rhomb., col. to wh., somet. yelsh., redsh., gray, br.	2.18, 2.35, 2.35
Vanadinite	$Pb_5(VO_4)_3Cl$	6.5–7.1	2.75–3	hex., orange-red, red, brnsh.-red, br., brnsh.-yel., yel.	2.416, 2.350
Variscite-Strengite	$(AlFe^{+3})(PO_4).2H_2O$	2.57–2.87	3.5–4.5	rhomb., pa. grn., grn., blsh.-grn., red, vlt., col.	1.563–1.707, 1.588–1.719, 1.594–1.741
Vaterite	$CaCO_3$	2.645	....	hex., col.	1.550, 1.640–1.650
Vermiculite	$(Mg,Ca)_{0.7}(Mg,Fe^{+3}Al)_6(Al,Si)_8O_{20}(OH)_4.8H_2O$	~2.3	~1.5	monocl., col., yel., grn., br.	1.525–1.564, 1.545–1.583, 1.545–1.583
Vesuvianite	$Ca_{10}(Mg,Fe)_2Al_4(Si_2O_7)_2(SiO_4)_5(OH,F)_4$	3.33–3.43	6–7	tetr., yel., grn., br.	1.700–1.746, 1.703–1.752
Villiaumite	NaF	2.79	2–2.5	cub., carmine, (nat.), col. (artif.)	1.327
Vivianite	$Fe_3(PO_4)_2.8H_2O$	2.67–2.69	1.5–2	monocl., col., tarn. pa. bl., grnsh. bl., dk. bl., blsh. blk.	1.579–1.616, 1.602–1.656, 1.629–1.675
Wagnerite	$Mg_2(PO_4)F$	3.15	5–5.5	monocl., yel., gray, somet. red, grn.	1.568, 1.572, 1.582
Wavellite	$Al_3(OH)_3(PO_4)_2.5H_2O$	2.36	3.25–4	rhomb., grnsh. wh., grn. to yel., somet. br., bl., wh.	1.520–1.535, 1.526–1.543, 1.545–1.561
Whewellite	$Ca(C_2O_4).H_2O$	2.23	2.5–3	monocl., col., somet. yelsh., brnsh.	1.491, 1.554, 1.650
Willemite	Zn_2SiO_4	3.9–4.1	5.5	rhbdr., wh., yel., grn., red, gray, br.	1.691, 1.719
Witherite	$BaCO_3$	4.29–4.30	3.5	rhomb., col., wh., gray, yelsh. br.	1.529, 1.676, 1.677
Wolframite	$(Fe,Mn)WO_4$	7.12–7.51	4–4.5	monocl., dk. gray, brnsh. blk. to iron blk.	(Li) ~2.26, 2.32, 2.42
Wollastonite	$CaSiO_3$	2.87–3.09	4.5–5	tricl., wh., col., gray, pa. grn.	1.616–1.640, 1.628–1.650, 1.631–1.653
Wulfenite	$PbMoO_4$	6.5–7.0	2.75–3	tetr., orange-yel. to yel., gray, grn., br., red	2.403, 2.283
Wurtzite	ZnS	3.98	3.5–4	hex., brnsh. blk.	2.356, 2.378
Xenotime	$Y(PO_4)$	4.4–5.1	4–5	tetr., yelsh. br. to redsh. br., somet. gray, wh., pa. yel., grnsh.	1.721, 1.816
Zeunerite (R)	$Cu(UO_2)_2(AsO_4)_2.10\text{–}16H_2O$	....	....	tetr.	1.602–1.610
Zincite	ZnO	5.64–5.68	4	hex., orange-yel. to dk. red, somet. yel.	2.013, 2.029
Zircon	$ZrSiO_4$	4.6–4.7	7.5	tetr., redsh. br., yel., gray, grn., col.	1.923–1.960, 1.968–2.015
Zoisite	$Ca_2Al_3Si_3O_{12}(OH)$	3.15–3.365	6	rhomb., gray, grnsh., brnsh.	1.685–1.705, 1.688–1.710, 1.697–1.725

X-Ray Crystallographic Data, Molar Volumes, and Densities of Minerals and Related Substances

From U.S. Geological Survey Bulletin 1248 by
Richard A. Robie, Philip M. Bethke and Keith M. Beardsley

An extensive list of references and the bases for the calculations and the selection of data are given in the above referenced Bulletin. Bulletin 1248 may be obtained from the Superintendent of Documents, U.S. Government Printing Office, Washington, D.C., 20402.

Z; The number of gram formula weights per unit cell.

r; Indicates the data were obtained at an unspecified room temperature and may be taken as 25° +5°C.

*; Indicates the measurements were made on a natural specimen which may have deviated slightly from the listed formula. Densities for these minerals were calculated using the formula weight for the stoichiometric phase.

hex-R; Rhombohedral symmetry. To distinguish from true hexagonal symmetry.

X-Ray Crystallographic Data of Minerals

	Name and formula	Crystal system	Space group	Structure type	Z	a_o	b_o	c_o	α_o	β_o	γ_o	Cell volume 10^{-24} cm³	Molar volume cm³	Molar volume cal bar⁻¹	X-Ray density grams cm⁻³	Temp. °C	
	Elements																
1	Silver Ag	cubic	Fm3m(225)	face-centered cubic	4	4.0862 ±.0002						68.227 ± .010	10.272 ± .002	.24556 ±.00008	10.501 ±.002	25	1
2	Arsenic As	hex-R	R$\bar{3}$m(166)	arsenic	6	3.760 ±.001		10.555 ±.003				129.23 ± .08	12.972 ± .002	.31007 ±.00023	5.776 ±.004	26	2
3	Gold Au	cubic	Fm3m(225)	face-centered cubic	4	4.0786 ±.0002						67.847 ± .010	10.215 ± .002	.24420 ±.00008	19.282 ±.003	25	3
4	Bismuth Bi	hex-R	R$\bar{3}$m(166)	arsenic	6	4.5459 ±.0010		11.8622 ±.0030				212.29 ± .11	21.309 ± .011	.50934 ±.00030	9.8071 ±.0050	26	4
5	Diamond C*	cubic	Fd3m(227)	diamond	8	3.5670 ±.0001						45.385 ± .004	3.4166 ± .0003	.08170 ±.00005	3.5155 ±.0003	25	5
6	Graphite C*	hex.	C6/mmc(194)	graphite	4	2.4612 ±.0001		6.7079 ±.0010				35.189 ± .006	5.2982 ± .0009	.12668 ±.00007	2.2670 ±.0004	15	6
7	Copper Cu	cubic	Fm3m(225)	face-centered cubic	4	3.6150 ±.0005						47.242 ± .020	7.1128 ± .0030	.17005 ±.00012	8.9331 ±.0037	25	7
8	α-Iron Fe	cubic	Im3m(229)	body-centered cubic	2	2.8664 .0005						23.551 ± .012	7.0918 ± .0037	.16954 ±.00013	7.8748 ±.0041	25	8
9	Nickel Ni	cubic	Fm3m(225)	face-centered cubic	4	3.5238 ±.0005						43.756 ± .019	6.5880 ± .0028	.15750 ±.00011	8.9117 ±.0038	25	9
10	Lead Pb	cubic	Fm3m(225)	face-centered cubic	4	4.9505 ±.0005						121.32 ± .04	18.267 ± .006	.43663 ±.00018	11.342 ±.003	25	10
11	Platinum Pt	cubic	Fm3m(225)	face-centered cubic	4	3.9231 ±.0005						60.379 ± .023	9.0909 ± .0035	.21732 ±.00013	21.460 ±.008	25	11
12	orthorhombic Sulfur S	orth.	Fddd(70)	S_8 ring molecules	128	10.4646 ±.0020	12.8660 ±.0020	24.4860 ±.0040				3296.73 ± .97	15.511 ± .005	.37078 ±.00015	2.0671 ±.0006	25	12
13	monoclinic Sulfur S	mon.	$P2_1/c$(14)	S_8 ring molecules	48	11.04 ±.03	10.98 ±.03	10.92 ±.03		96.73 ±.50		1314.6 ± 6.4	16.49 ± .08	.3943 ±.0020	1.944 ±.009	103	13
14	rhombohedral Sulfur S	hex-R	R$\bar{3}$(148)	S_6 ring molecules	18	10.818 ±.002		4.280 ±.001				433.78 ± .19	14.514 ± .006	.34693 ±.00020	2.2092 ±.0010	r	14
15	Antimony Sb	hex-R	R$\bar{3}$m(166)	arsenic	6	4.310 ±.001		11.279 ±.003				181.45 ± .09	18.213 ± .010	.43535 ±.00028	6.685 ±.004	26	15
16	Selenium Se	hex.	$P3_121$(152) $P3_221$(154)		3	4.3642 ±.0008		4.9588 ±.0008				81.793 ± .033	16.420 ± .007	.39249 ±.00020	4.8088 ±.0019	26	16
17	Silicon Si	cubic	Fd3m(227)	diamond	8	5.4305 ±.0003						160.15 ± .03	12.056 ± .002	.28819 ±.00009	2.3296 ±.0004	25	17
18	β-Tin (white) Sn	tet.	$I4_1$/amd(141)		4	5.8315 ±.0008		3.1813 ±.0006				108.18 ± .04	16.289 ± .005	.38935 ±.00017	7.2867 ±.0024	26	18
19	Tellurium Te	hex.	$P3_121$(152) $P3_221$(154)		3	4.4570 ±.0008		5.9290 ±.0010				102.00 ± .04	20.476 ± .008	.48944 ±.00024	6.2316 ±.0025	25	19

X-Ray Crystallographic Data of Minerals

	Name and formula	Crystal system	Space group	Structure type	Z	a_o	b_o	c_o	α_o	β_o	γ_o	Cell volume 10^{-24} cm³	Molar volume cm³	Molar volume cal bar⁻¹	X-Ray density grams cm⁻³	Temp. °C	
20	Zinc Zn	hex.	$P6_3/mmc$(194)	hexagonal close packed	2	2.665 ± .001		4.947 ± .001				30.428 ± .024	9.162 ± .007	.2190 ± .0002	7.134 ± .006	25	20
	Sulfides, arsenides, tellurides, selenides, and sulfosalts																
21	Shandite β-$Ni_3Pb_2S_2$*	hex-R	$R\bar{3}m$(166)		3	5.576 ± .010		13.658 ± .010				367.76 ± 1.35	73.83 ± .27	1.765 ± .007	8.867 ± .033	r	21
22	High-Argentite Ag_2S I	cubic			4	6.269 ± .020						246.4 ± 2.4	37.09 ± .36	.8866 ± .0085	6.680 ± .064	600	22
23	Argentite Ag_2S II	cubic			2	4.870 ± .008						115.5 ± .6	34.78 ± .17	.8313 ± .0041	7.125 ± .035	189	23
24	Acanthite Ag_2S III	mon.	$P2_1/c$(14)		4	4.228 ± .002	6.928 ± .005	7.862 ± .003		99.58 ± .30		227.08 ± .29	34.19 ± .04	.8172 ± .0011	7.248 ± .009	25	24
25	High-Naumanite Ag_2Se	cubic			2	4.993 ± .016						124.48 ± 1.20	37.48 ± .36	.8959 ± .0087	7.862 ± .076	170	25
26	Ag_2Te I	cubic			2	5.29 ± .01						148.0 ± .8	44.58 ± .26	1.065 ± .006	7.702 ± .044	825	26
27	Ag_2Te II	cubic			4	6.585 ± .010						285.54 ± 1.30	42.99 ± .20	1.028 ± .005	7.986 ± .036	250	27
28	Hessite Ag_2Te III	mon.	$P2_1/c$(14)		4	8.09 ± .02	4.48 ± .01	8.96 ± .02		123.33 ± .30		271.33 ± 1.43	40.85 ± .22	.9764 ± .0052	8.405 ± .044	r	28
29	$Ag_{1.55}Cu_{.45}S$ I	cubic			4	6.110 ± .010						228.10 ± 1.12	34.34 ± .17	.8209 ± .0041	6.635 ± .033	300	29
30	$Ag_{1.55}Cu_{.45}S$ II	cubic			2	4.825 ± .005						112.33 ± .35	33.83 ± .11	.8085 ± .0026	6.736 ± .021	116	30
31	Jalpaite $Ag_{1.55}Cu_{.45}S$ III	tet.			16	8.673 ± .004		11.756 ± .006				884.30 ± .93	33.286 ± .035	.79559 ± .00088	6.8455 ± .0072	r	31
32	$Ag_{.93}Cu_{1.07}S$ I	cubic			4	5.961 ± .009						211.82 ± .96	31.89 ± .14	.7623 ± .0035	6.283 ± .029	196	32
33	$Ag_{.93}Cu_{1.07}S$ II	hex.			2	4.138 ± .004		7.105 ± .007				105.36 ± .23	31.73 ± .07	.7583 ± .0017	6.316 ± .014	100	33
34	Stromeyerite $Ag_{.93}Cu_{1.07}S$ III	orth.	Cmcm(63)		4	4.066 ± .002	6.628 ± .003	7.972 ± .004				214.84 ± .18	32.35 ± .03	.7732 ± .0007	6.194 ± .005	r	34
35	Eucairite $AgCuSe$	orth.	pseudo $P4/nmm$(129)		10	4.105 ± .010	20.35 ± .02	6.31 ± .01				527.12 ± 1.62	31.75 ± .10	.7588 ± .0024	7.887 ± .024	r	35
36	Petzite Ag_3AuTe_2*	cubic	$I4_132$(214)		8	10.38 ± .02						1118.4 ± 6.5	84.19 ± .49	2.012 ± .012	9.214 ± .053	r	36
37	Maldonite Au_2Bi	cubic	Fd3m(227)	Cu_2Mg	8	7.958 ± .002						503.98 ± .38	37.94 ± .03	.9068 ± .0007	15.891 ± .012	r	37
38	High-Digenite Cu_2S I	cubic			4	5.725 ± .010						187.64 ± .98	28.25 ± .15	.6753 ± .0036	5.633 ± .030	465	38
39	High-Chalcocite Cu_2S II	hex.			2	3.961 ± .004		6.722 ± .007				91.34 ± .21	27.50 ± .06	.6574 ± .0015	5.786 ± .013	152	39
40	Chalcocite Cu_2S III	orth.	Ab2m(39)		96	11.881 ± .004	27.323 ± .010	13.491 ± .004				4379.5 ± 2.5	27.475 ± .016	.65671 ± .00043	5.7924 ± .0034	r	40
41	Digenite $Cu_{1.79}S$ (Cu rich side)	cubic		deformed fluorite	4	5.5695 ± .0010						172.76 ± .09	26.012 ± .014	.6217 ± .0004	5.605 ± .003	25	41
42	Digenite $Cu_{1.77}S$ (S rich side)	cubic		deformed fluorite	4	5.5542 ± .0010						171.34 ± .09	25.798 ± .014	.6166 ± .0004	5.602 ± .003	25	42
43	Berzelianite Cu_2Se	cubic			4	5.85 ± .01						200.2 ± 1.0	30.14 ± .15	.7205 ± .0037	6.835 ± .035	170	43
44	High-Bornite Cu_5FeS_4*	cubic			1	5.50 ± .01						166.4 ± .9	100.2 ± .5	2.395 ± .013	5.008 ± .027	240	44

X-Ray Crystallographic Data of Minerals

	Name and formula	Crystal system	Space group	Structure type	Z	a_o	b_o	c_o	α_o	β_o	γ_o	Cell volume 10^{-24} cm³	Molar volume cm³	Molar volume cal bar⁻¹	X-Ray density grams cm⁻³	Temp. °C	
45	Metastable Bornite Cu_5FeS_4	cubic			8	10.94 ±.02						1309.34 ± 7.18	98.57 ± .54	2.356 ±.013	5.091 ±.028	r	45
46	Low-Bornite Cu_5FeS_4*	tet.	$P\bar{4}2_1c$(144)		16	10.94 ±.02		21.88 ±.04				2618.7 ±10.7	98.57 ± .40	2.356 ±.010	5.091 ±.021	r	46
47	Umangite Cu_3Se_2	tet.	P4/mmm(123)		2	6.402 ±.010		4.276 ±.010				175.25 ± .68	52.77 ± .21	1.261 ±.005	6.604 ±.026	r	47
48	Heazelwoodite Ni_3S_2	hex-R	R32(155)		3	5.746 ±.001		7.134 ±.002				203.98 ± .09	40.95 ± .02	.9788 ±.0005	5.867 ±.003	r	48
49	Maucherite $Ni_{11}As_8$	tet.	$P4_12_12$(92)		4	6.870 ±.001		21.81 ±.01				1029.36 ± .56	154.98 ± .08	3.7043 ±.0021	8.0343 ±.0044	r	49
50	Pentlandite $Fe_{5.25}Ni_{3.75}S_8$	cubic	Fm3m(225)		4	10.196 ±.010						1059.96 ± 3.12	159.59 ± .47	3.8144 ±.0113	4.823 ±.014	r	50
51	Pentlandite $Fe_{4.75}Ni_{5.25}S_8$	cubic	Fm3m(225)		4	10.095 ±.010						1028.77 ± 3.06	154.89 ± .46	3.702 ±.011	4.998 ±.015	r	51
52	Sternbergite $AgFe_2S_3$*	orth.	Ccmm(63)		8	11.60 ±.02	12.675 ±.020	6.63 ±.01				974.81 ± 2.71	73.39 ± .20	1.754 ±.005	4.303 ±.012	r	52
53	Argentopyrite $AgFe_2S_3$*	orth.	Pmmm(47)		4	6.64 ±.01	11.47 ±.02	6.45 ±.02				491.2 ± 1.9	73.96 ± .29	1.768 ±.007	4.269 ±.017	r	53
54	Realgar AsS*	mon.	$P2_1/m$(11)		16	9.29 ±.05	13.53 ±.05	6.57 ±.03		106.55 ±.30		791.6 ± 6.4	29.80 ± .24	.7122 ±.0058	3.591 ±.029	r	54
55	Oldhamite CaS	cubic	Fm3m(225)	rock salt	4	5.689 ±.006						184.12 ± .58	27.722 ± .088	.6626 ±.0021	2.602 ±.008	r	55
56	Greenockite CdS	hex.	$P6_3mc$(186)	zincite	2	4.1354 ±.0010		6.7120 ±.0010				99.407 ± .050	29.934 ± .015	.71549 ±.00041	4.8261 ±.0024	r	56
57	Hawleyite CdS	cubic	$F\bar{4}3m$(216)	sphalerite	4	5.833 ±.002						198.46 ± .20	29.88 ± .03	.7142 ±.0008	4.835 ±.005	r	57
58	(hypothetical) CdS	cubic	Fm3m(225)	rock salt	4	5.516 ±.002						167.83 ± .18	25.27 ± .03	.6040 ±.0007	5.717 ±.006	r	58
59	Cadmoselite CdSe	hex.	$P6_3mc$(186)	zincite	2	4.2977 ±.0010		7.0021 ±.0010				112.00 ± .05	33.727 ± .016	.80614 ±.00044	5.6738 ±.0028	r	59
60	CdTe	cubic	$F\bar{4}3m$(216)	sphalerite	4	6.4805 ±.0006						272.16 ± .08	40.977 ± .012	.97943 ±.00032	5.8569 ±.0016	25	60
61	(hypothetical) CoS	cubic	$F\bar{4}3m$(216)	sphalerite	4	5.339 ±.001						152.19 ± .09	22.91 ± .02	.5477 ±.0004	3.971 ±.002	r	61
62	Chalcopyrite ($CuFeS_2$) $CuFeS_{1.90}$	tet.	I42d(122)		4	5.2988 ±.0010		10.434 ±.005				292.96 ± .18	44.109 ± .027	1.0543 ±.0007	4.0878 ±.0025	r	62
63	Cubanite $CuFe_2S_3$*	orth.	Pcmn(62)		4	6.46 ±.01	11.12 ±.01	6.23 ±.01				447.53 ± 1.08	67.38 ± .16	1.611 ±.004	4.026 ±.010	r	63
64	Covellite CuS	hex.	$P6_3/mmc$(194)		6	3.792 ±.001		16.34 ±.01				203.48 ± .16	20.42 ± .02	.4882 ±.0005	4.682 ±.001	r	64
65	Klockmannite CuSe	hex.		deformed covellite	78	14.206 ±.010		17.25 ±.05				3014.8 ± 9.7	23.28 ± .08	.5564 ±.0018	6.122 ±.020	r	65
66	Troilite FeS	hex.	$P6_3/mmc$(194)	niccolite	2	3.446 ±.003		5.877 ±.001				60.439 ± .106	18.20 ± .03	.4350 ±.0008	4.830 ±.009	28	66
67	Pyrrhotite $Fe_{.980}S$	hex.	$P6_3/mmc$(194)	defect niccolite	2	3.446 ±.001		5.848 ±.002				60.14 ± .04	18.11 ± .01	.4329 ±.0003	4.793 ±.003	28	67
68	Pyrrhotite $Fe_{.885}S$	hex.	$P6_3/mmc$(194)	defect niccolite	2	3.440 ±.001		5.709 ±.003				58.507 ± .046	17.62 ± .02	.4211 ±.0004	4.625 ±.004	28	68
69	(hypothetical) FeS	cubic	$F\bar{4}3m$(216)	sphalerite	4	5.455 ±.001						162.32 ± .09	24.44 ± .01	.5842 ±.0004	3.597 ±.002	r	69
70	(hypothetical) FeS	hex.	$P6_3mc$(186)	zincite	2	3.872 ±.001		6.345 ±.002				82.38 ± .05	24.81 ± .02	.5930 ±.0004	3.544 ±.002	r	70

X-Ray Crystallographic Data of Minerals

	Name and formula	Crystal system	Space group	Structure type	Z	a_o	b_o	c_o	α_o	β_o	γ_o	Cell volume 10^{-24} cm^3	Molar volume cm^3	Molar volume cal bar^{-1}	X-Ray density grams cm^{-3}	Temp. °C	
71	Cinnabar HgS	hex.	P$3_1$21(152) P$3_2$1(154)	cinnabar	3	4.149 ±.001		9.495 ±.002				141.55 ± .07	28.416 ± .015	.6792 ±.0004	8.187 ±.004	r	71
72	Metacinnabar HgS	cubic	F$\bar{4}$3m(216)	sphalerite	4	5.8517 ±.0010						200.38 ± .10	30.169 ± .016	.7211 ±.0004	7.712 ±.004	r	72
73	Tiemannite HgSe	cubic	F$\bar{4}$3m(216)	sphalerite	4	6.0853 ±.0050						225.34 ± .56	33.928 ± .084	.8110 ±.0020	8.239 ±.020	r	73
74	Coloradoite HgTe	cubic	F$\bar{4}$3m(216)	sphalerite	4	6.4600 ±.0006						269.59 ± .08	40.590 ± .011	.97016 ±.00032	8.0855 ±.0023	r	74
75	Alabandite MnS	cubic	Fm3m(225)	rock salt	4	5.2234 ±.0005						142.51 ± .04	21.457 ± .006	.51289 ±.00019	4.0546 ±.0012	r	75
76	(hypothetical) MnS	cubic	F$\bar{4}$3m(216)	sphalerite	4	5.611 ±.002						176.65 ± .19	26.60 ± .03	.6357 ±.0007	3.271 ±.004	r	76
77	(hypothetical) MnS	hex.	P6_3mc(186)	zincite	2	3.986 ±.001		6.465 ±.002				88.96 ± .05	26.79 ± .02	.6403 ±.0004	3.248 ±.002	r	77
78	Niccolite NiAs	hex.	P6_3/mmc(194)	niccolite	2	3.618 ±.001		5.034 ±.001				57.07 ± .03	17.18 ± .01	.4108 ±.0003	7.776 ±.005	r	78
79	Millerite NiS	hex-R	R3m(160)		9	9.616 ±.001		3.152 ±.001				252.41 ± .10	16.891 ± .006	.40374 ±.00020	5.3743 ±.0020	r	79
80	Breithauptite NiSb	hex.	P6_3/mmc(194)	niccolite	2	3.942 ±.001		5.155 ±.001				69.37 ± .04	20.89 ± .01	.4994 ±.0004	8.639 ±.005	r	80
81	Galena PbS	cubic	Fm3m(225)	rock salt	4	5.9360 ±.0005						209.16 ± .05	31.492 ± .008	.75272 ±.00024	7.5973 ±.0019	26	81
82	Clausthalite PbSe	cubic	Fm3m(225)	rock salt	4	6.1255 ±.0005						229.84 ± .06	34.605 ± .009	.82713 ±.00025	8.2690 ±.0020	r	82
83	Teallite $PbSnS_2$	orth.	Pbnm(62)	GeS	2	4.266 ±.003	11.419 ±.007	4.090 ±.002				199.24 ± .21	59.996 ± .063	1.4340 ±.0016	6.501 ±.007	r	83
84	Altaite PbTe	cubic	Fm3m(225)	rock salt	4	6.4606 ±.0005						269.66 ± .06	40.601 ± .009	.97043 ±.00027	8.2459 ±.0019	r	84
85	Cooperite PtS	tet.	P4_2/mmc(131)		2	3 4699 ±.0006		6.1098 ±.0010				73.563 ± .028	22.152 ± .008	.5295 ±.0003	10.254 ±.004	r	85
86	Herzenbergite SnS	orth.	Pbnm(62)	GeS	4	4.328 ±.002	11.190 ±.004	3.978 ±.001				192.66 ± .12	29.01 ± .02	.6933 ±.0005	5.197 ±.003	r	86
87	Sphalerite ZnS	cubic	F$\bar{4}$3m(216)	sphalerite	4	5.4093 ±.0005						158.28 ± .04	23.831 ± .007	.56962 ±.00020	4.0885 ±.0011	r	87
88	Wurtzite ZnS	hex.	P6_3mc(186)	zincite	2	3.8230 ±.0010		6.2565 ±.0010				79.190 ± .043	23.846 ± .013	.56998 ±.00036	4.0859 ±.0022	r	88
89	Stilleite ZnSe	cubic	F$\bar{4}$3m(216)	sphalerite	4	5.6685 ±.0005						182.14 ± .05	27.424 ± .007	.65548 ±.00022	5.2630 ±.0014	r	89
90	ZnTe	cubic	F$\bar{4}$3m(216)	sphalerite	4	6.1020 ±.0006						227.20 ± .07	34.209 ± .010	.81765 ±.00029	5.6410 ±.0017	r	90
91	Orpiment As_2S_3*	mon.	P2_1/n(14)		4	11.49 ±.02	9.59 ±.02	4.25 ±.01		90.45 ±.30		468.3 ± 1.7	70.51 ± .25	1.685 ±.006	3.490 ±.013	r	91
92	Bismuthinite Bi_2S_3	orth.	Pbnm(62)	stibnite	4	11.150 ±.004	11.300 ±.004	3.981 ±.001				501.59 ± .28	75.520 ± .043	1.8050 ±.0011	6.8081 ±.0038	26	92
93	Tellurobismuthite Bi_2Te_3	hex-R	R$\bar{3}$m(166)	Bi_2Te_2S	3	4.3835 ±.0020		30.487 ±.003				507.33 ± .47	101.85 ± .09	2.4342 ±.0023	7.862 ±.007	25	93
94	Stibnite Sb_2S_3	orth.	Pbnm(62)	stibnite	4	11.229 ±.004	11.310 ±.004	3.8389 ±.0010				487.54 ± .28	73.406 ± .042	1.7545 ±.0010	4.6276 ±.0026	25	94
95	Linnaeite Co_3S_4	cubic	Fd3m(227)	spinel	8	9.401 ±.001						830.85 ± .27	62.548 ± .020	1.4950 ±.0005	4.8772 ±.0016	r	95
96	Greigite Fe_3S_4	cubic	Fd3m(227)	spinel	8	9.876 ±.002						963.26 ± .59	72.52 ± .04	1.733 ±.001	4.079 ±.003	r	96

X-Ray Crystallographic Data of Minerals

	Name and formula	Crystal system	Space group	Structure type	Z	a_o	b_o	c_o	α_o	β_o	γ_o	Cell volume 10^{-24} cm^3	Molar volume cm^3	Molar volume cal bar^{-1}	X-Ray density grams cm^{-3}	Temp. °C	
97	Daubreeite $FeCr_2S_4$	cubic	Fd3m(227)	spinel	8	9.966 ±.005						989.83 ± 1.49	74.52 ± .11	1.781 ±.003	3.866 ±.006	r	97
98	Violarite $FeNi_2S_4$	cubic	Fd3m(227)	spinel	8	9.464 ±.005						847.66 ± 1.34	63.81 ± .10	1.525 ±.002	4.725 ±.008	r	98
99	Polymidite Ni_3S_4	cubic	Fd3m(227)	spinel	8	9.480 ±.001						851.97 ± .27	64.138 ± .020	1.5330 ±.0005	4.7458 ±.0015	r	99
100	Co-Safflorite $CoAs_2$	mon.		deformed marcasite	2	5.049 ±.002	5.872 ±.002	3.127 ±.001		90.45 ±.20		92.706 ± .057	27.92 ± .02	.6672 ±.0005	7.479 ±.005	26	100
101	Safflorite $(CO_{.5}Fe_{.5})As_2$	orth.	Pnnm(58)	marcasite	2	5.231 ±.002	5.953 ±.002	2.962 ±.002				92.237 ± .078	27.775 ± .024	.6639 ±.0006	7.461 ±.006	26	101
102	Cobaltite $CoAsS$*	cubic	$P2_13$(198)	NiSbS	4	5.60 ±.05						175.62 ± 4.70	26.44 ± .71	.6320 ±.0170	6.275 ±.168	r	102
103	Glaucodot $(Co,Fe)AsS$*	orth.	Cmmm(65)		24	6.64 ±.05	28.39 ±.10	5.64 ±.05				1063.2 ±12.9	26.68 ± .32	.6377 ±.0078	6.161 ±.075	r	103
104	Cattierite CoS_2	cubic	Pa3(205)	pyrite	4	5.5345 ±.0005						169.53 ± .05	25.524 ± .007	.61009 ±.00021	4.8213 ±.0013	r	104
105	Trogtalite $CoSe_2$	cubic	Pa3(205)	pyrite	4	5.8588 ±.0010						201.11 ± .10	30.279 ± .016	.72374 ±.00042	7.1618 ±.0037	r	105
106	Loellingite $FeAs_2$	orth.	Pnnm(58)	marcasite	2	5.300 ±.002	5.981 ±.002	2.882 ±.001				91.357 ± .056	27.51 ± .02	.6576 ±.0005	7.477 ±.005	26	106
107	Arsenopyrite $FeAsS$*	tri.	$P\bar{1}$(2)		4	5.760 ±.010	5.690 ±.005	5.785 ±.005	90.00 ±.20	112.23 ±.20	90.00 ±.20	175.51 ± .44	26.42 ± .07	.6316 ±.0016	6.162 ±.015	r	107
108	Gudmundite $FeSbS$*	mon.	$B2_1/d$(14)		8	10.00 ±.05	5.93 ±.03	6.73 ±.03		90.00 ±.50		399.09 ± 3.35	30.04 ± .25	.7181 ±.0061	6.978 ±.059	r	108
109	Pyrite FeS_2	cubic	Pa3(205)	pyrite	4	5.4175 ±.0005						159.00 ± .04	23.940 ± .007	.57221 ±.00020	5.0116 ±.0014	r	109
110	Marcasite FeS_2*	orth.	Pnnm(58)	marcasite	2	4.443 ±.002	5.423 ±.002	3.3876 ±.0015				81.622 ± .060	24.579 ± .018	.58749 ±.00047	4.8813 ±.0036	25	110
111	Ferroselite $FeSe_2$	orth.	Pnnm(58)	marcasite	2	4.801 ±.005	5.778 ±.005	3.587 ±.004				99.50 ± .17	29.96 ± .05	.7162 ±.0013	7.134 ±.013	r	111
112	Frohbergite $FeTe_2$	orth.	Pnnm(58)	marcasite	2	5.265 ±.005	6.265 ±.005	3.869 ±.002				127.62 ± .17	38.43 ± .05	.9185 ±.0013	8.094 ±.011	r	112
113	Hauerite MnS_2	cubic	Pa3(205)	pyrite	4	6.1014 ±.0006						227.14 ± .07	34.198 ± .010	.81741 ±.00029	3.4816 ±.0010	28	113
114	Molybdenite MoS_2	hex.	$P6_3/mmc$(194)	molybdenite	2	3.1604 ±.0010		12.295 ±.002				106.35 ± .07	32.025 ± .021	.76547 ±.00055	4.9982 ±.0033	26	114
115	Rammelsbergite $NiAs_2$	orth.	Pnnm(58)	marcasite	2	4.757 ±.002	5.797 ±.004	3.542 ±.002				97.645 ± .096	29.41 ± .03	.7030 ±.0007	7.091 ±.007	26	115
116	Pararammelsbergite $NiAs_2$	orth.	Pbca(61)		8	5.75 ±.01	5.82 ±.01	11.428 ±.02				382.42 ± 1.15	28.79 ± .09	.6882 ±.0021	7.244 ±.022	r	116
117	Gersdorfite $NiAsS$	cubic	$P2_13$(198)		4	5.693 ±.001						184.51 ± .10	27.78 ± .01	.6640 ±.0004	5.964 ±.003	26	117
118	Vaesite NiS_2	cubic	Pa3(205)	pyrite	4	5.6873 ±.0005						183.96 ± .05	27.697 ± .007	.66203 ±.00022	4.4350 ±.0012	r	118
119	$NiSe_2$	cubic	Pa3(205)	pyrite	4	5.9604 ±.0010						211.75 ± .11	31.882 ± .016	.76204 ±.00043	6.7948 ±.0034	20	119
120	Melonite $NiTe_2$	hex.	$P\bar{3}ml$(164)	cadmium iodide	1	3.869 ±.010		5.308 ±.010				68.81 ± .38	41.44 ± .23	.9905 ±.0055	7.575 ±.042	84	120
121	Sperrylite $PtAs_2$	cubic	Pa3(205)	pyrite	4	5.968 ±.005						212.56 ± .53	32.00 ± .08	.7650 ±.0020	10.778 ±.027	r	121
122	Laurite RuS_2	cubic	Pa3(205)	pyrite	4	5.60 ±.02						175.6 ± 1.9	26.44 ± .28	.6320 ±.0068	6.248 ±.067	r	122

X-Ray Crystallographic Data of Minerals

	Name and formula	Crystal system	Space group	Structure type	Z	a_o	b_o	c_o	α_o	β_o	γ_o	Cell volume 10^{-24} cm³	Molar volume cm³	Molar volume cal bar⁻¹	X-Ray density grams cm⁻³	Temp. °C	
123	Tungstenite WS_2	hex.	$P6_3/mmc$(194)	molybdenite	2	3.154 ±.001		12.362 ±.004				106.50 ± .08	32.069 ± .023	.76652 ±.00059	7.7325 ±.0055	26	123
124	Co-Skutterudite $CoAs_{3-x}$ $CoAs_{2.95}$	cubic	Im3(204)		8	8.2060 ±.0010						552.58 ± .20	41.599 ± .015	.99428 ±.00041	6.7298 ±.0025	r	124
125	Fe-Skutterudite $FeAs_{3-x}$ $FeAs_{2.95}$	cubic	Im3(204)		8	8.1814 ±.0010						547.62 ± .20	41.226 ± .015	.98537 ±.00041	6.7158 ±.0025	r	125
126	Ni-Skutterudite $NiAs_{3-x}$ $NiAs_{2.95}$	cubic	Im3(204)		8	8.3300 ±.0010						578.01 ± .21	43.513 ± .016	1.0400 ±.0004	6.4286 ±.0023	r	126
127	Tennantite $Cu_{12}As_4S_{13}$	cubic	$I\bar{4}3m$(217)	tetrahedrite	2 2	10.190 ±.004						1058.09 ± 1.25	318.62 ± .38	7.1652 ±.0090	4.642 ±.006	r	127
128	Tetrahedrite $Cu_{12}Sb_4S_{13}$	cubic	$I\bar{4}3m$(217)	tetrahedrite	2 2	10.327 ±.004						1101.3 ± 1.3	331.64 ± .39	7.9266 ±.0094	5.024 ±.006	r	128
129	Enargite Cu_3AsS_4	orth.	Pnn2(34)		2	6.426 ±.005	7.422 ±.005	6.144 ±.005				293.03 ± .38	88.24 ± .12	2.109 ±.003	4.463 ±.006	26	129
130	Luzonite Cu_3AsS_4*	tet.	$I\bar{4}2m$(121)		2	5.289 ±.005		10.440 ±.008				292.04 ± .60	87.94 ± .18	2.1019 ±.0043	4.478 ±.009	26	130
131	Famatimite Cu_3SbS_4*	tet.	$I\bar{4}m$(121)		2	5.384 ±.005		10.770 ±.008				312.19 ± .62	94.01 ± .19	2.2469 ±.0045	4.687 ±.009	26	131
132	Proustite Ag_3AsS_3	hex-R	R3c(161)		6	10.816 ±.001		8.6948 ±.0013				880.89 ± .21	88.420 ± .021	2.1133 ±.0006	5.595 ±.001	26	132
133	Pyrargyrite Ag_3SbS_3	hex-R	R3c(161)		6	11.052 ±.002		8.7177 ±.0020				922.18 ± .40	92.564 ± .040	2.2124 ±.0010	5.8506 ±.0025	26	133
134	Miargyrite $AgSbS_2$*	mon.	Cc(9)		8	12.862 ±.013	4.111 ±.004	13.220 ±.010		98.63 ±.15		691.10 ± 1.14	52.027 ± .086	1.244 ±.002	5.646 ±.009	r	134
	Oxides and hydroxides																
135	Corundum Al_2O_3	hex-R	$R\bar{3}c$(167)	corundum	6	4.7591 ±.0004		12.9894 ±.0030				254.78 ± .07	25.575 ± .007	.61128 ±.00022	3.9869 ±.0011	25	135
136	Boehmite AlO(OH)*	orth.	Cmcm(63)	lepidocrocite	4	2.868 ±.003	12.227 ±.003	3.700 ±.003				129.75 ± .17	19.535 ± .026	.46695 ±.00067	3.071 ±.004	26	136
137	Diaspore AlO(OH)*	orth.	Pbnm(62)		4	4.401 ±.005	9.421 ±.005	2.845 ±.002				117.96 ± .17	17.760 ± .026	.4245 ±.0007	3.378 ±.005	r	137
138	Gibbsite $Al(OH)_3$	mon.	$P2_1/n$(14)		8	9.719 ±.002	5.0705 ±.0010	8.6412 ±.0010		94.57 ±.25		424.49 ± .20	31.956 ± .015	.7638 ±.0004	2.441 ±.001	r	138
139	Arsenolite As_2O_3	cubic	Fd3m(227)	diamond	16	11.074 ±.005						1358.0 ± 1.8	51.118 ± .069	1.2218 ±.0017	3.870 ±.005	25	139
140	Claudetite As_2O_3	mon.	$P2_1/n$(14)		4	5.339 ±.002	12.984 ±.005	4.5405 ±.0010		94.27 ±.10		313.88 ± .19	47.259 ± .028	1.1296 ±.0007	4.1863 ±.0025	25	140
141	Bromellite BeO	hex.	$P6_3mc$(186)	zincite	2	2.6979 ±.0005		4.3772 ±.0005				27.592 ± .011	8.3086 ± .0032	.19862 ±.00012	3.0104 ±.0012	26	141
142	Bismite α-Bi_2O_3	mon.	$P2_1/c$(14)	pseudo orthorhombic	8	8.166 ±.005	13.827 ±.010	5.850 ±.004		90.00 ±.20		660.53 ± .77	49.73 ± .06	1.1885 ±.0014	9.371 ±.011	25	142
143	Lime CaO	cubic	Fm3m(225)	rock salt	4	4.8108 ±.0005						111.34 ± .03	16.764 ± .005	.40071 ±.00017	3.3453 ±.0010	26	143
144	Portlandite $Ca(OH)_2$	hex.	$P\bar{3}ml$(164)	CdI_2	1	3.5933 ±.0005		4.9086 ±.0020				54.888 ± .027	33.056 ± .016	.79011 ±.00043	2.2415 ±.0011	26	144
145	Monteponite CdO	cubic	Fm3m(225)	rock salt	4	4.6953 ±.0010						103.51 ± .07	15.585 ± .010	.37254 ±.00028	8.2386 ±.0053	27	145
146	Cerianite CeO_2	cubic	Fm3m(225)	fluorite	4	5.4110 ±.0020						158.43 ± .18	23.853 ± .026	.57016 ±.00068	7.216 ±.008	26	146
147	CoO	cubic	Fm3m(225)	rock salt	4	4.260 ±.002						77.31 ± .11	11.64 ± .02	.2782 ±.0004	6.438 ±.009	26	147

X-Ray Crystallographic Data of Minerals

	Name and formula	Crystal system	Space group	Structure type	Z	a_o	b_o	c_o	α_o	β_o	γ_o	Cell volume 10^{-24} cm^3	Molar volume cm^3	Molar volume cal bar^{-1}	X-Ray density grams cm^{-3}	Temp. °C	
148	Eskolaite Cr_2O_3	hex-R	R$\bar{3}$c(167)	corundum	6	4.9607 ±.0020		13.599 ±.010				289.82 ± .32	29.090 ± .032	.6953 ±.0008	5.225 ±.006	r	148
149	Tenorite CuO	mon.	C2/c(15)		4	4.684 ±.005	3.425 ±.005	5.129 ±.005		99.47 ±.17		81.16 ± .17	12.22 ± .03	.2921 ±.0007	6.509 ±.014	26	149
150	Cuprite Cu_2O	cubic	Pn3m(224)		2	4.2696 ±.0010						77.833 ± .055	23.437 ± .016	.56021 ±.00044	6.1047 ±.0043	26	150
151	Wustite $Fe_{.953}O$	cubic	Fm3m(225)	defect rock salt	4	4.3088 ±.0003						79.996 ± .017	12.044 ± .003	.28791 ±.00011	5.7471 ±.0012	17	151
152	Hematite Fe_2O_3	hex-R	R$\bar{3}$c(167)	corundum	6	5.0329 ±.0010				13.7492 ±.0010		301.61 ± .12	30.274 ± .012	.72361 ±.00034	5.2749 ±.0021	25	152
153	Magnetite Fe_3O_4	cubic	Fd3m(227)	spinel	8	8.3940 ±.0005						591.43 ± .11	44.524 ± .008	1.0642 ±.0002	5.2003 ±.0009	22	153
154	Goethite α-FeO(OH)*	orth.	Pbnm(62)		4	4.596 ±.005	9.957 ±.010	3.021 ±.003				138.2 ± .2	20.82 ± .04	.4975 ±.0009	4.269 ±.008	r	154
155	Lepidocrocite γ-FeO(OH)*	orth.	Amam(63)		4	3.868 ±.010	12.525 ±.010	3.066 ±.003				148.54 ± .43	22.364 ± .064	.5346 ±.0016	3.973 ±.011	r	155
156	α-Ga_2O_3	hex-R	R$\bar{3}$c(167)	corundum	6	4.9793 ±.0010		13.429 ±.003				288.34 ± .13	28.943 ± .013	.69179 ±.00036	6.4762 ±.0030	24	156
157	Low-germania GeO_2	tet.	P4/mnm(136)	rutile	2	4.3963 ±.0010		2.8626 ±.0010				55.327 ± .032	16.660 ± .010	.39824 ±.00027	6.2777 ±.0036	25	157
158	High-germania GeO_2	hex.	$P3_121(152)$ $P3_221(154)$	α-quartz	3	4.987 ±.002		5.652 ±.002				121.73 ± .11	24.438 ± .021	.58413 ±.00056	4.2797 ±.0038	26	158
159	Ice H_2O	hex.	$P6_3/mmc(194)$		4	4.5212 ±.0010		7.3666 ±.0010				130.41 ± .06	19.635 ± .009	.46932 ±.00026	.9175 ±.0004	0	159
160	Hafnia HfO_2	mon.	$P2_1/c(14)$	baddeleyite	4	5.1156 ±.0010	5.1722 ±.0010	5.2948 ±.0010		99.18 ±.08		138.30 ± .06	20.823 ± .008	.49772 ±.00025	10.108 ±.004	r	160
161	Montroydite HgO	orth.	Pnma(62)		4	6.608 ±.003	5.518 ±.003	3.519 ±.003				128.3 ± .1	19.32 ± .02	.4618 ±.0006	11.21 ±.01	25	161
162	Periclase MgO	cubic	Fm3m(225)	rock salt	4	4.2117 ±.0005						74.709 ± .027	11.248 ± .004	.26889 ±.00014	3.5837 ±.0013	25	162
163	Brucite $Mg(OH)_2$	hex.	P$\bar{3}$ml(164)	CdI_2	1	3.147 ±.004		4.769 ±.004				40.90 ± .11	24.63 ± .07	.5888 ±.0016	2.368 ±.006	26	163
164	Manganosite MnO	cubic	Fm3m(225)	rock salt	4	4.4448 ±.0005						87.813 ± .030	13.221 ± .004	.31604 ±.00015	5.3653 ±.0018	26	164
165	Pyrolusite MnO_2	tet.	P4/mnm(136)	rutile	2	4.388 ±.003		2.865 ±.002				55.16 ± .08	16.61 ± .02	.3971 ±.0007	5.234 ±.008	r	165
166	Bixbyite Mn_2O_3	cubic	Ia3(206)	Tl_2O_3	16	9.411 ±.005						833.5 ± 1.3	31.37 ± .05	.7499 ±.0012	5.032 ±.008	25	166
167	Hausmanite Mn_3O_4	tet.	$I4_1/amd(141)$		8	8.136 ±.005		9.422 ±.005				623.68 ± .84	46.95 ± .06	1.1222 ±.0016	4.873 ±.007	20	167
168	Molybdite MoO_3	orth.	Pbnm(62)		4	3.962 ±.002	13.858 ±.005	3.697 ±.004				202.98 ± .25	30.56 ± .04	.7305 ±.0010	4.710 ±.006	26	168
169	Bunsenite NiO	cubic	Fm3m(225)	rock salt	4	4.177 ±.002						72.88 ± .10	10.97 ± .02	.2623 ±.0004	6.809 ±.010	26	169
170	Litharge PbO red	tet.	P4/nmm(129)		2	3.9759 ±.0040		5.023 ±.004				79.40 ± .17	23.91 ± .05	.5715 ±.0013	9.334 ±.020	27	170
171	Massicot PbO yellow	orth.	Pb2a(32)		4	5.489 ±.003	4.755 ±.004	5.891 ±.004				153.8 ± .2	23.15 ± .03	.5533 ±.0007	9.641 ±.012	27	171
172	Minium Pb_3O_4	tet.	$P4_2/mbc(135)$		4	8.815 ±.005		6.565 ±.003				510.13 ± .62	76.81 ± .09	1.836 ±.002	8.926 ±.009	25	172
173	Senarmontite Sb_2O_3	cubic	Fm3m(225)	arsenic trioxide	16	11.152 ±.003						1386.9 ± 1.1	52.206 ± .042	1.2478 ±.0011	5.5837 ±.0045	26	173

X-Ray Crystallographic Data of Minerals

	Name and formula	Crystal system	Space group	Structure type	Z	a_o	b_o	c_o	α_o	β_o	γ_o	Cell volume 10^{-24} cm³	Molar volume cm³	Molar volume cal bar⁻¹	X-Ray density grams cm⁻³	Temp. °C	
174	Valentinite Sb_2O_3	orth.	Pccn(56)	antimony trioxide	4	4.914 ±.002	12.468 ±.005	5.421 ±.004				332.13 ± .31	50.007 ± .047	1.1952 ±.0012	5.8292 ±.0054	25	174
175	Cervantite Sb_2O_4	cubic	Fd3m(227)		8	10.305 ±.005						1094.3 ± 1.6	82.38 ± .12	1.9690 ±.0029	3.733 ±.005	26	175
176	Selenolite SeO_2	tet.	$P4_2/mbc$(135) $P4_2bc$(106)		8	8.35 ±.01		5.08 ±.01				354.2 ± 1.1	26.66 ± .08	.6373 ±.0020	4.161 ±.013	26	176
177	α-Quartz SiO_2*	hex.	$P3_121$(152) $P3_221$(154)		3	4.9136 ±.0001		5.4051 ±.0001				113.01 ± .01	22.688 ± .001	.54229 ±.00007	2.6483 ±.0001	25	177
178	β-Quartz SiO_2*	hex.	$P6_422$(181) $P6_222$(180)		3	4.999 ±.001		5.4592 ±.0020				118.15 ± .06	23.718 ± .013	.5669 ±.0004	2.533 ±.002	575	178
179	α-Cristobalite SiO_2	tet.	$P4_12_12$(92) $P4_32_12$(96)		4	4.971 ±.003		6.918 ±.003				170.95 ± .22	25.739 ± .033	.61521 ±.00083	2.3344 ±.0030	25	179
180	β-Cristobalite SiO_2	cubic	Fd3m(227)		8	7.1382 ±.0010						363.72 ± .15	27.381 ± .012	.65447 ±.00032	2.1944 ±.0009	405	180
181	Keatite SiO_2	tet.	$P4_12_12$(92) $P4_32_12$(96)		12	7.456 ±.003		8.604 ±.005				478.3 ± .5	24.01 ± .02	.5738 ±.0006	2.503 ±.003	r	181
182	β-Tridymite SiO_2	hex.	$P\bar{6}2c$(172) $P6_3/mmc$(194)		4	5.0463 ±.0020		8.2563 ±.0030				182.08 ± .16	27.414 ± .024	.65527 ±.00062	2.1917 ±.0019	405	182
183	Coesite SiO_2*	mon.	B2/b(15)		16	7.152 ±.001	12.379 ±.002	7.152 ±.001		120.00 ±.17		548.37 ± .95	20.641 ± .036	.49338 ±.00090	2.9110 ±.0050	25	183
184	Stishovite SiO_2*	tet.	P4/mnm(136)	rutile	2	4.1790 ±.0010		2.6649 ±.0010				46.540 ± .028	14.014 ± .009	.33500 ±.00025	4.2874 ±.0026	r	184
185	Melanophlogite SiO_2*	cubic	Pm3n(223)	clathrate type	46	13.402 ±.004						2407.2 ± 2.2	31.516 ± .028	.75325 ±.00072	1.9065 ±.0017	r	185
186	Cassiterite SnO_2	tet.	P4/mnm(136)	rutile	2	4.738 ±.003		3.188 ±.003				71.57 ± .11	21.55 ± .03	.5151 ±.0009	6.992 ±.011	26	186
187	Tellurite TeO_2*	orth.	Pbca(61)	tellurite	8	5.607 ±.003	12.034 ±.005	5.463 ±.003				368.61 ± .32	27.750 ± .024	.66328 ±.00062	5.7514 ±.0050	25	187
188	Paratellurite TeO_2	tet.	$P4_12_12$(92) $P4_32_12$(96)		4	4.810 ±.002		7.613 ±.002				176.14 ± .15	26.52 ± .02	.6339 ±.0006	6.018 ±.005	25	188
189	Thorianite ThO_2	cubic	Fm3m(225)	fluorite	4	5.5952 ±.0005						175.16 ± .05	26.373 ± .007	.63038 ±.00021	10.012 ±.003	25	189
190	Rutile TiO_2	tet.	P4/mnm(136)		2	4.5937 ±.0005		2.9618 ±.0010				62.500 ± .025	18.820 ± .008	.44986 ±.00023	4.2453 ±.0017	25	190
191	Anatase TiO_2	tet.	$I4_1/amd$(141)		4	3.785 ±.002		9.514 ±.006				136.30 ± .17	20.522 ± .025	.4905 ±.0007	3.893 ±.005	r	191
192	Brookite TiO_2*	orth.	Pcab(61)		8	5.456 ±.002	9.182 ±.005	5.143 ±.003				257.6 ± .2	19.40 ± .02	.4636 ±.0005	4.119 ±.004	r	192
193	Titanium sesquioxide Ti_2O_3	hex-R	$R\bar{3}c$(167)	corundum	6	5.149 ±.002		13.642 ±.010				313.2 ± .3	31.44 ± .03	.7515 ±.0009	4.574 ±.005	r	193
194	Uraninite UO_2	cubic	Fm3m(225)	fluorite	4	5.4682 ±.0010						163.51 ± .09	24.618 ± .014	.58843 ±.00037	10.969 ±.006	26	194
195	Karelianite V_2O_3	hex-R	$R\bar{3}c$(167)	corundum	6	4.952 ±.002		14.002 ±.010				297.36 ± .32	29.848 ± .032	.71342 ±.00081	5.0216 ±.0054	r	195
196	Zincite ZnO	hex.	$P6_3mc$(186)	zincite	2	3.2495 ±.0005		5.2069 ±.0005				47.615 ± .015	14.338 ± .005	.34273 ±.00016	5.6750 ±.0018	25	196
197	Baddeleyite ZrO_2	mon.	$P2_1/c$(14)	baddeleyite	4	5.1454 ±.0010	5.2075 ±.0010	5.3107 ±.0010		99.23 ±.08		140.46 ± .06	21.148 ± .009	.50548 ±.00025	5.8267 ±.0023	r	197
	Multiple oxides																
198	Spinel $MgAl_2O_4$	cubic	Fd3m(227)	spinel	8	8.080 ±.002						527.5 ± .4	39.71 ± .03	.9492 ±.0008	3.583 ±.003	26	198

X-Ray Crystallographic Data of Minerals

No.	Name and formula	Crystal system	Space group	Structure type	Z	a_o	b_o	c_o	α_o	β_o	γ_o	Cell volume 10^{-24} cm³	Molar volume cm³	Molar volume cal bar⁻¹	X-Ray density grams cm⁻³	Temp. °C	
199	Hercynite $FeAl_2O_4$	cubic	Fd3m(227)	spinel	8	8.150 ±.004						541.3 ± .6	40.75 ± .05	.9740 ±.0011	4.265 ±.005	25	199
200	Galaxite $MnAl_2O_4$	cubic	Fd3m(227)	spinel	8	8.258 ±.002						563.2 ± .4	42.39 ± .03	1.013 ±.001	4.078 ±.003	25	200
201	Gahnite $ZnAl_2O_4$	cubic	Fd3m(227)	spinel	8	8.0848 ±.0020						528.45 ± .39	39.783 ± .030	.95088 ±.00075	4.6083 ±.0034	26	201
202	Magnetite $FeFe_2O_4$	cubic	Fd3m(227)	spinel	8	8.3940 ±.0005						591.43 ± .11	44.524 ± .008	1.0642 ±.0002	5.2003 ±.0009	22	202
203	Jacobsite $MnFe_2O_4$	cubic	Fd3m(227)	spinel	8	8.499 ±.002						613.9 ± .4	46.22 ± .03	1.105 ±.001	4.990 ±.004	25	203
204	Trevorite $NiFe_2O_4$	cubic	Fd3m(227)	spinel	8	8.339 ±.003						579.9 ± .6	43.65 ± .05	1.043 ±.001	5.370 ±.006	25	204
205	Picrochromite $MgCr_2O_4$	cubic	Fd3m(227)	spinel	8	8.333 ±.003						578.6 ± .6	43.56 ± .05	1.041 ±.001	4.415 ±.005	26	205
206	Ilmenite $FeTiO_3$	hex-R	$R\bar{3}$(148)	ilmenite	6	5.093 ±.005		14.055 ±.020				315.73 ± .75	31.69 ± .08	.7574 ±.0019	4.788 ±.012	r	206
207	Geikielite $MgTiO_3$	hex-R	$R\bar{3}$(148)	ilmenite	6	5.054 ±.005		13.898 ±.010				307.44 ± .65	30.86 ± .07	.7376 ±.0016	3.896 ±.008	26	207
208	Pyrophanite $MnTiO_3$	hex-R	$R\bar{3}$(148)	ilmenite	6	5.155 ±.005		14.18 ±.01				326.3 ± .7	32.76 ± .07	.7829 ±.0017	4.605 ±.010	r	208
209	Cobalt Titanate $CoTiO_3$	hex-R	$R\bar{3}$(148)	ilmenite	6	5.066 ±.001		13.918 ±.005				309.34 ± .17	31.05 ± .02	.7422 ±.0004	4.986 ±.003	r	209
210	Perovskite $CaTiO_3$	orth.	Pcmn(62)	perovskite	4	5.3670 ±.0010	7.6438 ±.0010	5.4439 ±.0010				223.33 ± .07	33.626 ± .010	.80371 ±.00028	4.0439 ±.0012	r	210
211	Chrysoberyl $BeAl_2O_4$	orth.	Pmnb(62)	olivine	4	5.4756 ±.0020	9.4041 ±.0030	4.4267 ±.0020				227.94 ± .15	34.320 ± .023	.82031 ±.00059	3.6997 ±.0025	25	211
	Halides																
212	Halite NaCl	cubic	Fm3m(225)	rock salt	4	5.6402 ±.0002						179.43 ± .02	27.015 ± .003	.64571 ±.00011	2.1634 ±.0002	26	212
213	Sylvite KCl	cubic	Fm3m(225)	rock salt	4	6.2931 ±.0002						249.23 ± .02	37.524 ± .004	.89690 ±.00013	1.9868 ±.0002	25	213
214	Villiaumite NaF	cubic	Fm3m(225)	rock salt	4	4.6342 ±.0005						99.523 ± .032	14.984 ± .005	.35818 ±.00016	2.8021 ±.0009	25	214
215	Chlorargyrite AgCl	cubic	Fm3m(225)	rock salt	4	5.5491 ±.0005						170.87 ± .05	25.727 ± .007	.61493 ±.00021	5.5710 ±.0015	26	215
216	Bromargyrite AgBr	cubic	Fm3m(225)	rock salt	4	5.7745 ±.0005						192.55 ± .05	28.991 ± .008	.69294 ±.00022	6.4772 ±.0017	26	216
217	Nantockite CuCl	cubic	$F\bar{4}3m$(216)	sphalerite	4	5.416 ±.003						158.87 ± .26	23.92 ± .04	.5717 ±.0010	4.139 ±.007	25	217
218	Marshite CuI	cubic	$F\bar{4}3m$(216)	sphalerite	4	6.0507 ±.0010						221.52 ± .11	33.353 ± .017	.7972 ±.0004	5.710 ±.003	26	218
219	Miersite AgI	cubic	$F\bar{4}3m$(216)	sphalerite	4	6.4963 ±.0010						274.16 ± .10	41.278 ± .020	.9866 ±.0004	5.688 ±.003	r	219
220	Iodargyrite AgI	hex.	$P6_3mc$(186)	zincite	2	4.5955 ±.0010		7.5005 ±.0033				137.18 ± .10	41.308 ± .030	.9873 ±.0009	5.683 ±.004	25	220
221	Calomel HgCl	tet.	14/mm(139)		4	4.478 ±.005		10.910 ±.005				218.77 ± .50	32.939 ± .075	.7873 ±.0018	7.166 ±.016	26	221
222	Fluorite CaF_2	cubic	Fm3m(225)	fluorite	4	5.4638 ±.0004						163.11 ± .04	24.558 ± .005	.58701 ±.00017	3.1792 ±.0007	25	222
223	Sellaite MgF_2	tet.	$P4_2/mnm$(136)	rutile	2	4.621 ±.001		3.050 ±.001				65.13 ± .04	19.61 ± .01	.4688 ±.0003	3.177 ±.002	18	223

X-Ray Crystallographic Data of Minerals

	Name and formula	Crystal system	Space group	Structure type	Z	a_o	b_o	c_o	α_o	β_o	γ_o	Cell volume 10^{-24} cm^3	Molar volume cm^3	Molar volume cal bar^{-1}	X-Ray density grams cm^{-3}	Temp. °C	
224	Chloromagnesite $MgCl_2$	hex-R	R3m(166)		3	3.632 ±.004		17.795 ±.016				203.29 ± .48	40.81 ± .10	.9754 ±.0024	2.333 ±.006	r	224
225	Lawrencite $FeCl_2$	hex-R	R$\bar{3}$m(166)		3	3.593 ±.003		17.58 ±.09				196.55 ± 1.06	39.46 ± .21	.9431 ±.0051	3.212 ±.017	r	225
226	Scacchite $MnCl_2$	hex-R	R$\bar{3}$m(166)		3	3.711 ±.002		17.59 ±.07				209.79 ± .86	42.11 ± .17	1.007 ±.004	2.988 ±.012	r	226
227	Cotunnite $PbCl_2$	orth.	Pnmb(62)		4	4.535 ±.005	7.62 ±.01	9.05 ±.01				312.74 ± .64	47.09 ± .10	1.1254 ±.0023	5.906 ±.012	26	227
228	Matlockite PbFCl	tet.	P4/nmm(129)		2	4.106 ±.005		7.23 ±.01				121.89 ± .34	36.70 ± .10	.8773 ±.0025	9.853 ±.028	26	228
229	Cryolite Na_3AlF_6*	mon.	P2$_1$/n(14)		2	5.40 ±.01	5.60 ±.01	7.776 ±.010		90.18 ±.25		235.1 ± .7	70.81 ± .20	1.692 ±.005	2.965 ±.009	r	229
230	Neighborite $NaMgF_3$	orth.	Pcmn(62)	perovskite	4	5.363 ±.001	7.676 ±.001	5.503 ±.001				226.54 ± .07	34.11 ± .01	.8152 ±.0003	3.058 ±.001	18	230
	Carbonates and nitrates																
231	Calcite $CaCO_3$	hex-R	R$\bar{3}$c(167)	calcite	6	4.9899 ±.0010		17.064 ±.002				367.96 ± .15	36.934 ± .015	.88278 ±.00041	2.7100 ±.0011	26	231
232	Otavite $CdCO_3$	hex-R	R$\bar{3}$c(167)	calcite	6	4.9204 ±.0010		16.298 ±.003				341.72 ± .15	34.300 ± .015	.81983 ±.00041	5.0265 ±.0022	26	232
233	Cobalticalcite $CoCO_3$	hex-R	R$\bar{3}$c(167)	calcite	6	4.6581 ±.0010		14.958 ±.003				281.07 ± .13	28.213 ± .013	.67435 ±.00036	4.2159 ±.0020	26	233
234	Siderite $FeCO_3$	hex-R	R$\bar{3}$c(167)	calcite	6	4.6887 ±.0010		15.373 ±.003				292.68 ± .14	29.378 ± .014	.70219 ±.00037	3.9436 ±.0018	26	234
235	Magnesite $MgCO_3$	hex-R	R$\bar{3}$c(167)	calcite	6	4.6330 ±.0010		15.016 ±.003				279.13 ± .13	28.018 ± .013	.66969 ±.00036	3.0095 ±.0014	26	235
236	Rhodochrosite $MnCO_3$	hex-R	R$\bar{3}$c(167)	calcite	6	4.7771 ±.0010		15.664 ±.003				309.57 ± .14	31.073 ± .014	.74272 ±.00039	3.6992 ±.0017	26	236
237	Nickelous Carbonate $NiCO_3$	hex-R	R$\bar{3}$c(167)	calcite	6	4.5975 ±.0010		14.723 ±.002				269.51 ± .12	27.052 ± .012	.64660 ±.00034	4.3886 ±.0020	26	237
238	Smithsonite $ZnCO_3$	hex-R	R$\bar{3}$(167)	calcite	6	4.6528 ±.0010		15.025 ±.003				281.69 ± .13	28.275 ± .013	.67583 ±.00037	4.4343 ±.0021	26	238
239	Dolomite $CaMg(CO_3)_2$*	hex-R	R$\bar{3}$(148)	calcite	3	4.8079 ±.0010		16.010 ±.003				320.50 ± .15	64.341 ± .029	1.5378 ±.0008	2.8661 ±.0013	26	239
240	Huntite $Mg_3Ca(CO_3)_4$*	hex-R	R32(155)	calcite	3	9.498 ±.003		7.816 ±.004				610.63 ± .50	122.58 ± .10	2.9299 ±.0024	2.880 ±.002	26	240
241	Norsethite $BaMg(CO_3)_2$*	hex-R	R32(155)	calcite	3	5.020 ±.005		16.75 ±.02				365.6 ± .8	73.39 ± .17	1.754 ±.004	3.838 ±.009	r	241
242	Vaterite $CaCO_3$	hex.			6	7.135 ±.005		8.524 ±.007				375.80 ± .61	37.72 ± .06	.9016 ±.0015	2.653 ±.004	r	242
243	Witherite $BaCO_3$	orth.	Pnam(62)	aragonite	4	6.430 ±.005	8.904 ±.005	5.314 ±.005				304.24 ± .41	45.81 ± .06	1.095 ±.002	4.308 ±.006	26	243
244	Aragonite $CaCO_3$	orth.	Pnam(62)	aragonite	4	5.741 ±.005	7.968 ±.005	4.959 ±.005				226.85 ± .33	34.15 ± .05	.8164 ±.0012	2.930 ±.004	26	244
245	Cerussite $PbCO_3$	orth.	Pnam(62)	aragonite	4	6.152 ±.005	8.436 ±.005	5.195 ±.005				269.61 ± .38	40.59 ± .06	.9702 ±.0014	6.582 ±.009	26	245
246	Strontianite $SrCO_3$	orth.	Pnam(62)	aragonite	4	6.029 ±.005	8.414 ±.005	5.107 ±.005				259.07 ± .37	39.01 ± .06	.9323 ±.0014	3.785 ±.005	26	246
247	Shortite $Na_2Ca_2(CO_3)_3$	orth.	Amm2(38)		2	4.961 ±.005	11.03 ±.02	7.12 ±.01				389.6 ± 1.0	117.3 ± .3	2.804 ±.007	2.610 ±.007	r	247
248	Malachite $Cu_2(OH)_2CO_3$	mon.	P2$_1$/a(14)		4	9.502 ±.007	11.974 ±.007	3.240 ±.003		98.75 ±.25		364.35 ± .54	54.86 ± .08	1.311 ±.002	4.030 ±.006	25	248

X-Ray Crystallographic Data of Minerals

	Name and formula	Crystal system	Space group	Structure type	Z	a_o	b_o	c_o	α_o	β_o	γ_o	Cell volume 10^{-24} cm³	Molar volume cm³	Molar volume cal bar⁻¹	X-Ray density grams cm⁻³	Temp. °C	
249	Azurite $Cu_3(OH)_2(CO_3)_2$	mon.	$P2_1/a(14)$		2	5.008 ±.005	5.844 ±.005	10.336 ±.005		92.45 ±.25		302.22 ± .43	91.01 ± .13	2.1752 ±.0031	3.787 ±.005	25	249
250	Niter KNO_3	orth.	Pnam(62)	aragonite	4	6.431 ±.005	9.164 ±.005	5.414 ±.005				319.07 ± .42	48.04 ± .06	1.148 ±.002	2.105 ±.003	26	250
251	Soda Niter $NaNO_3$	hex-R	$R\bar{3}c(167)$	calcite	6	5.0696 ±.0010		16.829 ±.005				374.57 ± .19	37.598 ± .019	.89866 ±.00049	2.2606 ±.0011	25	251
252	Gerhardite $Cu_2(NO_3)(OH)_3$	orth.	$P2_12_12_1(19)$		4	6.075 ±.004	13.812 ±.008	5.592 ±.004				469.21 ± .53	70.65 ± .08	1.689 ±.002	3.399 ±.004	r	252
	Sulfates and borates																
253	Barite $BaSO_4$	orth.	Pnma(62)	barite	4	8.878 ±.005	5.450 ±.005	7.152 ±.003				346.05 ± .40	52.10 ± .06	1.245 ±.002	4.480 ±.005	26	253
254	Anhydrite $CaSO_4$	orth.	Amma(63) Ccmm(63)	anhydrite	4	6.991 ±.005	6.996 ±.005	6.238 ±.005				305.09 ± .39	45.94 ± .06	1.098 ±.002	2.964 ±.004	26	254
255	Anglesite $PbSO_4$	orth.	Pnma(62)	barite	4	8.480 ±.005	5.398 ±.005	6.958 ±.003				318.50 ± .38	47.95 ± .06	1.146 ±.002	6.324 ±.008	25	255
256	Celestite $SrSO_4$	orth.	Pnma(62)	barite	4	8.359 ±.005	5.352 ±.005	6.866 ±.005				307.17 ± .41	46.25 ± .06	1.105 ±.002	3.972 ±.005	26	256
257	Zinkosite $ZnSO_4$	orth.	Pnma(62)	barite	4	8.588 ±.008	6.740 ±.006	4.770 ±.005				276.10 ± .46	41.57 ± .07	.9936 ±.0017	3.883 ±.006	25	257
258	Arcanite KS_2SO_4	orth.	Pnma(62)	arcanite	4	5.772 ±.005	10.072 ±.005	7.483 ±.004				435.03 ± .49	65.50 ± .07	1.566 ±.002	2.661 ±.003	25	258
259	Mascagnite $(NH_4)_2SO_4$	orth.	Pnma(62)	arcanite	4	7.782 ±.005	5.993 ±.005	10.636 ±.005				496.04 ± .57	74.68 ± .09	1.7851 ±.0021	1.7693 ±.0020	25	259
260	Thenardite Na_2SO_4	orth.	Fddd(70)	thenardite	8	5.863 ±.005	12.304 ±.005	9.821 ±.005				708.47 ± .76	53.33 ± .06	1.275 ±.002	2.663 ±.003	25	260
261	Gypsum $CaSO_4.2H_2O$*	mon.	C2/c(15)		4	5.68 ±.01	15.18 ±.01	6.29 ±.01		113.83 ±.22		496.1 ± 1.5	74.69 ± .22	1.785 ±.005	2.305 ±.007	r	261
262	Epsomite $MgSO_4.7H_2O$	orth.	$P2_12_12_1(19)$		4	11.86 ±.01	11.99 ±.01	6.858 ±.007				975.22 ± 1.53	146.83 ± .23	3.5094 ±.0055	1.679 ±.003	25	262
263	Goslarite $ZnSO_4.7H_2O$	orth.	$P2_12_12_1(19)$	epsomite	4	11.779 ±.005	12.050 ±.005	6.822 ±.003				968.29 ± .72	145.79 ± .11	3.4845 ±.0026	1.9723 ±.0015	25	263
264	Mirabilite $Na_2SO_4.10H_2O$	mon.	$P2_1/c(14)$		4	11.51 ±.01	10.38 ±.01	12.83 ±.01		107.75 ±.17		1459.9 ± 2.6	219.8 ± .4	5.253 ±.009	1.466 ±.003	24	264
265	Chalcanthite $CuSO_4.5H_2O$	tri.	$P\bar{1}(2)$		2	6.1045 ±.0050	10.72 ±.01	5.949 ±.007	97.57 ±.17	107.28 ±.17	77.43 ±.17	361.88 ± .72	108.97 ± .22	2.6045 ±.0052	2.2912 ±.0046	r	265
266	Brochantite $Cu_4SO_4(OH)_6$*	mon.	$P2_1/c(14)$		4	13.066 ±.010	9.85 ±.01	6.022 ±.010		103.27 ±.25		754.3 ± 1.8	113.6 ± .2	2.715 ±.006	3.982 ±.009	r	266
267	Syngenite $K_2Ca(SO_4)_2.H_2O$	mon.	$P2_1/m(11)$		2	9.775 ±.005	7.156 ±.005	6.251 ±.005		104.00 ±.25		424.27 ± .68	127.76 ± .20	3.0535 ±.0049	2.5707 ±.0041	r	267
268	Alunite $KAl_3(SO_4)_2(OH)_6$	hex-R	R3m(160)		3	6.982 ±.005		17.32 ±.01				731.2 ± 1.1	146.8 ± .2	3.508 ±.005	2.822 ±.004	r	268
269	Natroalunite $NaAl_3(SO_4)_2(OH)_6$	hex-R	R3m(160)		3	6.974 ±.005		16.69 ±.01				702.99 ± 1.09	141.1 ± .2	3.373 ±.005	2.821 ±.004	r	269
270	Hexahydrite $MgSO_4.6H_2O$	mon.	C2/c(15)		8	10.110 ±.005	7.212 ±.004	24.41 ±.01		98.30 ±.10		1761.2 ± 1.6	132.58 ± .12	3.1689 ±.0029	1.7232 ±.0015	r	270
271	Leonhardtite $MgSO_4.4H_2O$	mon.	$P2_1/n(14)$		4	5.922 ±.006	13.604 ±.004	7.905 ±.005		90.85 ±.20		636.78 ± .78	95.88 ± .12	2.2915 ±.0029	2.0071 ±.0025	r	271
272	Melanterite $FeSO_4.7H_2O$	mon.	$P2_1/c(14)$		4	14.072 ±.010	6.503 ±.007	11.041 ±.010		105.57 ±.15		973.29 ± 1.69	146.54 ± .25	3.5025 ±.0061	1.8972 ±.0033	r	272
273	Vanthoffite $MgSO_4.3Na_2SO_4$	mon.	$P2_1/c(14)$		2	9.797 ±.003	9.217 ±.003	8.199 ±.003		113.50 ±.10		678.96 ± .65	204.45 ± .20	4.8866 ±.0047	2.6730 ±.0025	r	273

X-Ray Crystallographic Data of Minerals

No.	Name and formula	Crystal system	Space group	Structure type	Z	a_o	b_o	c_o	α_o	β_o	γ_o	Cell volume 10^{-24} cm^3	Molar volume cm^3	Molar volume cal bar^{-1}	X-Ray density grams cm^{-3}	Temp. °C	No.
274	Dolerophanite $Cu_2O(SO_4)$	mon.	C2/m(15)		4	9.355 ±.010	6.312 ±.005	7.628 ±.005		122.29 ±.10		380.77 ± .70	57.33 ± .11	1.3703 ±.0026	4.171 ±.008	r	274
275	Retgersite $NiSO_4.4H_2O$	tet.	$P4_12_12$(92) $P4_32_1$(96)		4	6.782 ±.004		18.28 ±.01				840.80 ± 1.09	126.59 ± .16	3.0257 ±.0040	2.076 ±.003	25	275
276	Colemanite $CaB_3O_4(OH)_3.H_2O$*	mon.	$P2_1/a$(14)		4	8.743 ±.004	11.264 ±.002	6.102 ±.003		110.12 ±.08		564.26 ± .49	84.957 ± .073	2.0306 ±.0018	2.4194 ±.0021	r	276
277	Borax $Na_2B_4O_7.10H_2O$	mon.	C2/c(15)		4	11.858 ±.005	10.674 ±.005	12.197 ±.005		106.68 ±.03		1478.8 ± 1.1	222.66 ± .17	5.3217 ±.0041	1.7128 ±.0013	r	277
278	Kernite $Na_2B_4O_7.4H_2O$	mon.	$P2_1/c$(14)		4	7.022 ±.003	9.151 ±.004	15.676 ±.008		108.83 ±.25		953.40 ± 1.61	143.55 ± .24	3.4309 ±.0058	1.9038 ±.0032	r	278
279	Hambergite $Be_2BO_3.(OH,F)$*	orth.	Pbca(61)		8	9.755 ±.001	12.201 ±.001	4.426 ±.001				526.79 ± .14	39.658 ± .011	.9479 ±.0003	2.3663 ±.0006	r	279
	Phosphates, molybdates, and tungstates																
280	Berlinite $AlPO_4$	hex.	$P3_121$(152) $P3_221$(154)	α-quartz	3	4.942 ±.005		10.97 ±.007				232.03 ± .50	46.58 ± .10	1.113 ±.002	2.618 ±.006	25	280
281	Xenotime YPO_4	tet.	$I4_1$/amd(141)	zircon	4	6.885 ±.005		5.982 ±.005				283.57 ± .48	42.69 ± .07	1.020 ±.002	4.307 ±.008	26	281
282	Hydroxylapatite $Ca_5(PO_4)_3OH$	hex.	$P6_3/m$(176)	apatite	2	9.418 ±.003		6.883 ±.003				528.7 ± .5	159.2 ± .2	3.805 ±.004	3.155 ±.004	r	282
283	Fluorapatite $Ca_5(PO_4)_3F$	hex.	$P6_3/m$(176)	apatite	2	9.3684 ±.0030		6.8841 ±.0030				523.25 ± .41	157.56 ± .12	3.7659 ±.0030	3.2007 ±.0025	25	283
284	Chlorapatite $Ca_5(PO_4)_3Cl$	hex.	$P6/_3m$(176)	apatite	2	9.629 ±.005		6.777 ±.003				544.16 ± .61	163.86 ± .19	3.916 ±.004	3.178 ±.004	r	284
285	Carbonate-apatite $Ca_{10}(PO_4)_6CO_3H_2O$	hex.	$P6_3/m$(176)	apatite	1	9.436 ±.010		6.883 ±.010				530.74 ± 1.36	319.6 ± .8	7.640 ±.020	3.281 ±.008	r	285
286	Turquois $CuAl_6(PO_4)_4(OH)_8.4H_2O$*	tri.	$P\bar{1}$(2)		1	7.424 ±.008	7.629 ±.008	9.910 ±.010	68.61 ±.20	69.71 ±.20	65.08 ±.20	461.40 ± 1.12	277.9 ± .7	6.6416 ±.0162	2.927 ±.007	r	286
287	Powellite $CaMoO_4$	tet.	$I4_1/a$(100)	scheelite	4	5.226 ±.005		11.43 ±.007				312.17 ± .63	47.00 ± .09	1.1234 ±.0023	4.256 ±.009	25	287
288	Wulfenite $PbMoO_4$	tet.	$I4_1/a$(100)	scheelite	4	5.435 ±.005		12.110 ±.007				357.72 ± .69	53.859 ± .104	1.2873 ±.0025	6.816 ±.013	25	288
289	Scheelite $CaWO_4$	tet.	$I4_1/a$(100)	scheelite	4	5.242 ±.005		11.372 ±.005				312.49 ± .61	47.049 ± .092	1.1245 ±.0023	6.120 ±.012	25	289
290	Stolzite $PbWO_4$	tet.	$I4_1/a$(100)	scheelite	4	5.4616 ±.0030		12.046 ±.005				359.32 ± .42	54.100 ± .064	1.2931 ±.0016	8.4110 ±.0099	25	290
291	Ferberite $FeWO_4$	mon.	P2/c(13)	wolframite	2	4.732 ±.004	5.708 ±.003	4.965 ±.004		90.00 ±.05		134.11 ± .17	40.38 ± .05	.9652 ±.0013	7.520 ±.010	r	291
292	Huebnerite $MnWO_4$	mon.	P2/c(13)	wolframite	2	4.834 ±.004	5.758 ±.005	4.999 ±.004		91.18 ±.10		139.11 ± .20	41.89 ± .06	1.001 ±.002	7.228 ±.010	r	292
293	Wolframite $Fe_{.5}Mn_{.5}WO_4$	mon.	P2/c(13)	wolframite	2	4.782 ±.004	5.731 ±.004	4.982 ±.004		90.57 ±.10		136.53 ± .18	41.11 ± .06	.9826 ±.0014	7.376 ±.010	r	293
294	Sanmartinite $ZnWO_4$	mon.	P2/c(13)	wolframite	2	4.691 ±.003	5.720 ±.003	4.925 ±.003		89.36 ±.20		132.14 ± .14	39.79 ± .04	.9511 ±.0010	7.872 ±.008	25	294
	Ortho and ring structure silicates																
295	Forsterite Mg_2SiO_4	orth.	Pbnm(62)	olivine	4	4.758 ±.002	10.214 ±.003	5.984 ±.002				290.81 ± .18	43.786 ± .027	1.0465 ±.0007	3.2136 ±.0020	25	295
296	Fayalite Fe_2SiO_4	orth.	Pbnm(62)	olivine	4	4.817 ±.005	10.477 ±.005	6.105 ±.010				308.11 ± .62	46.389 ± .093	1.1088 ±.0023	4.3928 ±.0088	r	296
297	Tephroite Mn_2SiO_4*	orth.	Pbnm(62)	olivine	4	4.871 ±.005	10.636 ±.005	6.232 ±.005				322.87 ± .45	48.612 ± .067	1.1619 ±.0017	4.1545 ±.0058	r	297

X-Ray Crystallographic Data of Minerals

	Name and formula	Crystal system	Space group	Structure type	Z	a_o	b_o	c_o	α_o	β_o	γ_o	Cell volume 10^{-24} cm³	Molar volume cm³	Molar volume cal bar⁻¹	X-Ray density grams cm⁻³	Temp. °C	
298	Lime Olivine γCa_2SiO_4	orth.	Pbnm(62)	olivine	4	5.091 ±.010	11.371 ±.020	6.782 ±.010				392.61 ± 1.19	59.11 ± .18	1.4129 ±.0043	2.914 ±.009	r	298
299	Nickel Olivine Ni_2SiO_4	orth.	Pbnm(62)	olivine	4	4.727 ±.002	10.121 ±.005	5.915 ±.002				282.98 ± .21	42.61 ± .03	1.0184 ±.0008	4.917 ±.004	r	299
300	Cobalt Olivine Co_2SiO_4	orth.	Pbnm(62)	olivine	4	4.782 ±.002	10.301 ±.005	6.003 ±.002				295.70 ± .21	44.52 ± .03	1.0642 ±.0008	4.716 ±.003	r	300
301	Monticellite $CaMgSiO_4$	orth.	Pbnm(62)	olivine	4	4.827 ±.005	11.084 ±.005	6.376 ±.005				341.13 ± .47	51.362 ± .071	1.2276 ±.0017	3.046 ±.004	r	301
302	Kerschsteinite $CaFeSiO_4$	orth.	Pbnm(62)	olivine	4	4.886 ±.005	11.146 ±.005	6.434 ±.010				350.39 ± .67	52.756 ± .101	1.2609 ±.0025	3.564 ±.007	r	302
303	Knebelite $MnFeSiO_4$*	orth.	Pbnm(62)	olivine	4	4.854 ±.010	10.602 ±.010	6.162 ±.010				317.11 ± .88	47.74 ± .13	1.1412 ±.0032	4.249 ±.012	r	303
304	Glauchroite $CaMnSiO_4$	orth.	Pbnm(62)	olivine	4	4.944 ±.004	11.19 ±.01	6.529 ±.005				361.2 ± .9	54.38 ± .14	1.2997 ±.0032	3.441 ±.009	r	304
305	Fluor-Norbergite $Mg_2SiO_4.MgF_2$	orth.	Pnmb(62)		4	8.727 ±.005	10.271 ±.010	4.709 ±.002				422.09 ± .51	63.551 ± .077	1.5190 ±.0019	3.194 ±.004	25	305
306	Chondrodite $2Mg_2SiO_4.MgF_2$*	mon.	$P2_1/c$(14)		2	7.89 ±.03	4.743 ±.020	10.29 ±.03		109.03 ±.30		364.0 ± 2.4	109.6 ± .7	2.620 ±.017	3.136 ±.021	r	306
307	Fluor-Humite $3Mg_2SiO_4.MgF_2$	orth.	Pnma(62)		4	10.243 ±.005	20.72 ±.02	4.735 ±.002				1004.9 ± 1.2	151.31 ± .18	3.6163 ±.0042	3.2017 ±.0037	25	307
308	Clinohumite $4Mg_2SiO_4.MgF_2$*	mon.	$P2_1/c$(14)		2	13.68 ±.04	4.75 ±.02	10.27 ±.02		100.83 ±.50		655.5 ± 3.8	197.4 ± 1.1	4.717 ±.027	3.167 ±.018	r	308
309	Grossularite $Ca_3Al_2Si_3O_{12}$	cubic	Ia3d(230)	garnet	8	11.851 ±.001						1664.43 ± .42	125.30 ± .03	2.9948 ±.0008	3.595 ±.001	25	309
310	Uvarovite $Ca_3Cr_2Si_3O_{12}$	cubic	Ia3d(230)	garnet	8	11.999 ±.002						1727.57 ± .86	130.05 ± .07	3.1084 ±.0016	3.848 ±.002	26	310
311	Andradite $Ca_3Fe_2Si_3O_{12}$	cubic	Ia3d(230)	garnet	8	12.048 ±.001						1748.82 ± .44	131.65 ± .03	3.1466 ±.0008	3.860 ±.001	25	311
312	Goldmanite $Ca_3V_2Si_3O_{12}$	cubic	Ia3d(230)	garnet	8	12.070 ±.005						1758.42 2.19	132.38 ± .16	3.1639 ±.0040	3.765 ±.005	r	312
313	Almandite $Fe_3Al_2Si_3O_{12}$	cubic	Ia3d(230)	garnet	8	11.526 ±.001						1531.21 ± .40	115.27 ± .04	2.7551 ±.0008	4.318 ±.001	25	313
314	Pyrope $Mg_3Al_2Si_3O_{12}$	cubic	Ia3d(230)	garnet	8	11.459 ±.001						1504.67 ± .39	113.27 ± .03	2.7074 ±.0008	3.559 ±.001	25	314
315	Spessartite $Mn_3Al_2Si_3O_{12}$	cubic	Ia3d(230)	garnet	8	11.621 ±.001						1569.39 ± .41	118.15 ± .03	2.8238 ±.0008	4.190 ±.001	25	315
316	Zircon $ZrSiO_4$*	tet.	I4/amd(141)	zircon	4	6.604 ±.005		5.979 ±.005				260.76 ± .45	39.261 ± .068	.9384 ±.0017	4.669 ±.008	25	316
317	Thorite $ThSiO_4$	tet.	I4/amd(141)	zircon	4	7.143 ±.004		6.327 ±.003				322.82 ± .39	48.60 ± .06	1.1617 ±.0015	6.668 ±.008	r	317
318	Coffinite $USiO_4$	tet.	I4/amd(141)	zircon	4	6.995 ±.004		6.263 ±.005				306.45 ± .43	46.140 ± .064	1.103 ±.002	7.155 ±.010	r	318
319	Kyanite Al_2SiO_5*	tri.	$P\bar{1}$(2)		4	7.123 ±.001	7.848 ±.002	5.564 ±.008	89.92 ±.15	101.25 ±.08	105.97 ±.08	292.83 ± .45	44.09 ± .07	1.054 ±.002	3.675 ±.006	25	319
320	Andalusite Al_2SiO_5*	orth.	Pnnm(58)		4	7.7959 ±.0050	7.8983 ±.0020	5.5583 ±.0020				342.25 ± .27	51.530 ± .040	1.2316 ±.0010	3.145 ±.002	25	320
321	Sillimanite Al_2SiO_5*	orth.	Pbnm(62) Pnma(62)		4	7.4843 ±.0030	7.6730 ±.0030	5.7711 ±.0040				331.42 ± .30	49.899 ± .044	1.1927 ±.0011	3.248 ±.003	25	321
322	3.2 Mullite $3AL_2O_3.2SiO_2$	orth.			3/4	7.557 ±.002	7.6876 ±.0020	2.8842 ±.0010				167.56 ± .09	134.55 ± .07	3.2159 ±.0016	3.166 ±.002	r	322
323	2.1 Mullite $2Al_2O_3.SiO_2$	orth.	Pbam(55)		6/5	7.5788 ±.0020	7 6909 ±.0020	2.8883 ±.0010				168.35 ± .09	84.492 ± .043	2.0195 ±.0011	3.125 ±.002	r	323

X-Ray Crystallographic Data of Minerals

No.	Name and formula	Crystal system	Space group	Structure type	Z	a_o	b_o	c_o	α_o	β_o	γ_o	Cell volume 10^{-24} cm^3	Molar volume cm^3	Molar volume cal bar^{-1}	X-Ray density grams cm^{-3}	Temp. °C	No.
324	Staurolite $Fe_2Al_9Si_4O_{22}(OH)_2$*	mon.	C2/m(15)		2	7.90 ±.10	16.65 ±.15	5.63 ±.10		90.00 ±.25		740.5 ±17.5	223.0 ± 5.3	5.330 ±.126	3.825 ±.090	r	324
325	Topaz $Al_2(SiO_4)(OH)$*	orth.	Pmnb(62)		4	8.394 ±.005	8.792 ±.007	4.649 ±.003				343.10 ± .41	51.66 ± .06	1.2347 ±.0015	3.563 ±.005	26	325
326	Phenacite Be_2SiO_4*	hex-R	R$\bar{3}$(148)	phenacite	18	12.472 ±.005		8.252 ±.005				1111.6 ± 1.1	37.194 ± .037	.8890 ±.0009	2.960 ±.003	25	326
327	Willemite Zn_2SiO_4	hex-R	R$\bar{3}$(148)	phenacite	18	13.94 ±.01		9.309 ±.003				1566.6 ± 2.3	52.42 ± .08	1.253 ±.002	4.251 ±.006	25	327
328	Dioptase CuH_2SiO_4*	hex-R	R$\bar{3}$(148)	phenacite	18	14.61 ±.02		7.80 ±.01				1441.9 ± 4.4	48.24 ± .15	1.153 ±.004	3.247 ±.010	r	328
329	Larnite β-Ca_2SiO_4*	mon.	$P2_1/n(14)$		4	5.48 ±.02	6.76 ±.02	9.28 ±.02		94.55 ±.33		342.7 ± 1.8	51.60 ± .27	1.233 ±.006	3.338 ±.017	r	329
330	Akermanite $Ca_2MgSi_2O_7$	tet.	P$\bar{4}2_1$m(113)	melilite	2	7.8435 ±.0030		5.010 ±.003				308.22 ± .30	92.812 ± .090	2.2183 ±.0022	2.9375 ±.0029	r	330
331	Gehlenite $Ca_2Al_2SiO_7$	tet.	P$\bar{4}2_1$m(113)	melilite	2	7.690 ±.003		5.0675 ±.0030				299.67 ± .29	90.239 ± .088	2.1568 ±.0022	3.0387 ±.0030	r	331
332	Fe-Gehlenite $Ca_2Fe_2SiO_7$	tet.	P$\bar{4}2_1$m(113)	melilite	2	7.54 ±.01		4.855 ±.005				276.01 ± .79	83.12 ± .24	1.9865 ±.0057	3.994 ±.011	r	332
333	Hardystonite $Ca_2ZnSi_2O_7$*	tet.	P$\bar{4}2_1$m(113)	melilite	2	7.87 ±.03		5.01 ±.02				310.3 ± 2.7	93.44 ± .80	2.233 ±.019	3.357 ±.029	r	333
334	Sodium Melilite $NaCaAlSi_2O_7$	tet.	P$\bar{4}2_1$m(113)	melilite	2	8.511 ±.005		4.809 ±.003				348.35 ± .46	104.90 ± .14	2.507 ±.003	2.462 ±.003	r	334
335	Beryl $Be_3Al_2(Si_6O_{18})$*	hex.	P6/mmc(192)	beryl	2	9.215 ±.005		9.192 ±.005				675.98 ± .82	203.55 ± .25	4.8651 ±.0060	2.641 ±.003	25	335
336	Indialite high Cordierite $Mg_2Al_3(AlSi_5O_{18})$	hex.	P6/mmc(192)	beryl	2	9.7698 ±.0030		9.3517 ±.0030				773.02 ± .54	232.78 ± .16	5.5636 ±.0039	2.513 ±.002	25	336
337	Low Cordierite $Mg_2Al_3(AlSi_5O_{18})$	orth.	Cccm(66)		4	9.721 ±.003	17.062 ±.006	9.339 ±.003				1548.96 ± .88	233.22 ± .13	5.5741 ±.0032	2.508 ±.001	25	337
338	Fe-Indialite $Fe_2Al_3(AlSi_5O_{18})$	hex.	P6/mmc(192)	beryl	2	9.860 ±.010		9.285 ±.010				781.75 ± 1.80	235.40 ± .54	5.6264 ±.0130	2.753 ±.006	r	338
339	Fe-Cordierite $Fe_2Al_3(AlSi_5O_{18})$	orth.	Cccm(66)	cordierite	4	9.726 ±.010	17.065 ±.010	9.287 ±.010				1541.40 ± 2.47	232.08 ± .37	5.5468 ±.0089	2.792 ±.005	r	339
340	Mn-Indialite $Mn_2Al_3(AlSi_5O_{18})$	hex.	P6/mmc(192)	beryl	2	9.925 ±.010		9.297 ±.010				793.11 ± .81	238.8 ± .5	5.708 ±.013	2.706 ±.006	r	340
341	Sapphirine $Mg_2Al_4O_6SiO_4$*	mon.	$P2_1/c(14)$		8	11.26 ±.03	14.46 ±.03	9.95 ±.02		125.33 ±.50		1321.7 ± 9.7	99.50 ± .73	2.378 ±.017	3.464 ±.025	r	341
342	Elbaite $NaLiAl_{7.67}B_3Si_6O_{27}(OH)_4$*	hex-R	R3m(160)	tourmaline	3	15.842 ±.010		7.009 ±.010				1523.4 ± 2.9	305.82 ± .58	7.3093 ±.0140	3.271 ±.006	r	342
343	Schorl $NaFe_3Al_6B_3Si_6O_{27}(OH)_4$*	hex-R	R3m(160)	tourmaline	3	16.032 ±.010		7.149 ±.010				1591.3 ± 3.0	319.45 ± .60	7.635 ±.014	3.297 ±.006	r	343
344	Dravite $NaMg_3Al_6B_3Si_6O_{27}(OH)_4$	hex-R	R3m(160)	tourmaline	3	15.942 ±.010		7.224 ±.010				1589.99 ± 2.97	319.19 ± .60	7.629 ±.014	3.004 ±.006	r	344
345	Uvite $CaMg_4Al_5B_3Si_6O_{27}(OH)_4$	hex-R	R3m(160)	tourmaline	3	15.86 ±.01		7.19 ±.01				1566.3 ± 2.9	314.4 ± .6	7.515 ±.014	3.095 ±.006	r	345
346	Sphene $CaTiSiO_5$*	mon.	A2/a(15)		4	7.07 ±.01	8.72 ±.01	6.56 ±.01		113.95 ±.25		369.61 ± 1.13	55.65 ± .17	1.330 ±.004	3.523 ±.011	r	346
347	Datolite $CaBSiO_4(OH)$*	mon.	$P2_1/c(14)$		4	9.62 ±.03	7.60 ±.03	4.84 ±.02		90.15 ±.25		353.9 ± 2.3	53.28 ± .35	1.273 ±.008	3.003 ±.020	r	347
348	Euclase $AlBeSiO_4(OH)$*	mon.	$P2_1/a(14)$		4	4.763 ±.005	14.29 ±.02	4.618 ±.005		100.25 ±.10		309.30 ± .64	46.57 ± .10	1.113 ±.002	3.116 ±.007	r	348
349	Chloritoid $H_2FeAl_2SiO_7$*	mon.	C2/c(15)		8	9.48 ±.01	5.48 ±.01	18.18 ±.01		101.77 ±.25		924.6 ± 2.2	69.61 ± .16	1.664 ±.004	3.619 ±.008	r	349

X-Ray Crystallographic Data of Minerals

	Name and formula	Crystal system	Space group	Structure type	Z	a_o	b_o	c_o	α_o	β_o	γ_o	Cell volume 10^{-24} cm³	Molar volume cm³	Molar volume cal bar⁻¹	X-Ray density grams cm⁻³	Temp. °C	
350	Hemimorphite $Zn_4(OH)_2Si_2O_7.H_2O$*	orth.	Imm2(35)		2	8.370 ±.005	10.719 ±.005	5.120 ±.005				459.36 ± .57	138.32 ± .17	3.306 ±.004	3.482 ±.004	25	350
351	Zoisite $Ca_2Al_3(SiO_4)_3OH$	orth.	Pnma(62)		4	16.15 ±.01	5.581 ±.005	10.06 ±.01				906.74 ± 1.34	136.52 ± .20	3.263 ±.005	3.328 ±.005	r	351
352	Clinozoisite $Ca_2Al_3(SiO_4)_3OH$	mon.	$P2_1/m(11)$		2	8.887 ±.007	5.581 ±.005	10.14 ±.01		115.93 ±.33		452.30 ± 1.45	136.20 ± .44	3.255 ±.010	3.336 ±.011	r	352
353	Epidote $Ca_2Al_{1.5}Fe_{1.5}(SiO_4)_3OH$*	mon.	$P2_1/m(11)$		2	8.89 ±.02	5.63 ±.01	10.19 ±.02		115.40 ±.30		460.72 ± 1.97	138.7 ± .6	3.316 ±.014	3.587 ±.015	r	353
354	Piemontite $Ca_2Al_{1.5}Mn_{1.5}(SiO_4)_3OH$*	mon.	$P2_1/m(11)$		2	8.95 ±.02	5.70 ±.01	9.41 ±.02		115.70 ±.50		432.56 ± 2.38	130.3 ± .7	3.113 ±.017	3.810 ±.021	r	354
355	Lawsonite $CaAl_2Si_2O_7(OH)_2.H_2O$	orth.	Cccm(63)		4	8.787 ±.005	5.836 ±.005	13.123 ±.008				672.96 ± .80	101.32 ± .12	2.4217 ±.0029	3.101 ±.004	r	355
	Chain and band structure silicates																
356	Enstatite $MgSiO_3$*	orth.	Pcab(61)		16	8.829 ±.010	18.22 ±.01	5.192 ±.005				835.21 ± 1.32	31.44 ± .05	.7514 ±.0012	3.194 ±.005	r	356
357	Clinoenstatite $MgSiO_3$	mon.	$P2_1/c(15)$		8	9.620 ±.005	8.825 ±.005	5.188 ±.005		108.33 ±.17		418.10 ± .66	31.47 ± .05	.7523 ±.0012	3.190 ±.005	r	357
358	Protoenstatite $MgSiO_3$	orth.	Pbcn(60)		8	9.25 ±.01	8.74 ±.01	5.32 ±.01				430.10 ± 1.05	32.38 ± .08	.7739 ±.0019	3.101 ±.008	r	358
359	High Clinoenstatite $MgSiO_3$	tri.			8	10.000 ±.005	8.934 ±.004	5.170 ±.003	88.27 ±.05	70.03 ±.04	91.01 ±.04	433.72 ± .40	32.65 ± .03	.7804 ±.0008	3.075 ±.003		359
360	Clinoferrosilite $FeSiO_3$	mon.	$P2_1/c(14)$		8	9.7085 ±.0010	9.0872 ±.0011	5.2284 ±.004		108.43 ±.05		437.60 ± .15	32.943 ± .011	.7874 ±.0003	4.005 ±.002	r	360
361	Orthoferrosilite $FeSiO_3$	orth.	Pcab(61)	enstatite	16	9.080 ±.002	18.431 ±.004	5.238 ±.001				876.6 ± .54	33.00 ± .02	.7887 ±.0008	3.998 ±.004	r	361
362	Diopside $CaMg(SiO_3)_2$	mon.	C2/c(15)	diopside	4	9.743 ±.005	8.923 ±.005	5.251 ±.003		105.93 ±.25		438.97 ± .69	66.09 ± .10	1.580 ±.003	3.277 ±.005	r	362
363	Hedenbergite $CaFe(SiO_3)_2$*	mon.	C2/c(15)	diopside	4	9.854 ±.010	9.024 ±.010	5.263 ±.010		104.23 ±.33		453.64 ± 1.28	68.30 ± .19	1.632 ±.005	3.632 ±.010	r	363
364	Johannsenite $CaMn(SiO_3)_2$*	mon.	C2/c(15)	diopside	4	9.83 ±.03	9.04 ±.03	5.27 ±.02		105.00 ±.50		452.35 ± 2.87	68.11 ± .43	1.628 ±.010	3.629 ±.023	r	364
365	Ureyite $NaCr(SiO_3)_2$	mon.	C2/c(15)	diopside	4	9.550 ±.016	8.712 ±.007	5.273 ±.008		107.44 ±.16		418.6 ± 1.1	63.02 ± .16	1.506 ±.004	3.605 ±.009	r	365
366	Jadeite $NaAl(SiO_3)_2$*	mon.	C2/c(15)	diopside	4	9.409 ±.005	8.564 ±.005	5.220 ±.005		107.50 ±.20		401.15 ± .67	60.40 ± .10	1.444 ±.002	3.347 ±.006	r	366
367	Acmite (Aegirine) $NaFe(SiO_3)_2$	mon.	C2/c(15)	diopside	4	9.658 ±.005	8.795 ±.005	5.294 ±.005		107.42 ±.20		429.06 ± .70	64.60 ± .11	1.544 ±.003	4.411 ±.007	r	367
368	Ca Tschermak Molecule $CaAl_2SiO_6$	mon.	C2/c(15)	diopside	4	9.615 ±.005	8.661 ±.005	5.272 ±.003		106.12 ±.20		421.77 ± .59	63.50 ± .09	1.518 ±.002	3.435 ±.005	r	368
369	Spodumene $LiAl(SiO_3)_2$	mon.	C2/c(15)	diopside	4	9.451 ±.002	8.387 ±.002	5.208 ±.001		110.07 ±.03		387.7 ± .1	58.37 ± .02	1.395 ±.001	3.188 ±.001	r	369
370	β-Spodumene $LiAl(SiO_3)_2$	tet.	$P4_32_12(96)$ $P4_12_12(92)$		4	7.5332 ±.0008		9.1540 ±.0008				519.48 ± .12	78.215 ± .018	1.8694 ±.0005	2.379 ±.001	r	370
371	Pectolite $Ca_2NaH(SiO_3)_3$*	tri.	$P\bar{1}(2)$		2	7.99 ±.01	7.04 ±.01	7.02 ±.01	90.05 ±.25	95.27 ±.25	102.47 ±.25	383.84 ± .99	115.58 ± .30	2.763 ±.007	2.876 ±.007	r	371
372	Wollastonite $CaSiO_3$*	tri.	$P\bar{1}(2)$		6	7.94 ±.01	7.32 ±.01	7.07 ±.01	90.03 ±.25	95.37 ±.25	103.43 ±.25	397.82 ± 1.03	39.93 ± .10	.9544 ±.0025	2.909 ±.008	r	372
373	Parawollastonite $CaSiO_3$*	mon.	$P2_1(4)$		12	15.417 ±.004	7.321 ±.002	7.066 ±.003		95.40 ±.10		793.98 ± .47	39.85 ± .02	.9524 ±.0006	2.915 ±.002	r	373
374	Pseudowollastonite $CaSiO_3$*	tri.			24	6.90 ±.02	11.78 ±.02	19.65 ±.02	90.00 ±.30	90.80 ±.30	90.00 ±.30	1597.0 ± 5.6	40.08 ± .14	.9579 ±.0034	2.899 ±.010	r	374

X-Ray Crystallographic Data of Minerals

	Name and formula	Crystal system	Space group	Structure type	Z	a_o	b_o	c_o	α_o	β_o	γ_o	Cell volume 10^{-24} cm^3	Molar volume cm^3	Molar volume cal bar^{-1}	X-Ray density grams cm^{-3}	Temp. °C	
375	Rhodonite $MnSiO_3$*	tri.	$P\bar{1}(2)$		10	7.682 ±.002	11.818 ±.003	6.707 ±.002	92.36 ±.05	93.95 ±.05	105.66 ±.05	583.77 ± .31	35.158 ± .019	.8403 ±.0005	3.727 ±.002	r	375
376	Bustamite $CaMn(SiO_3)_2$*	tri.	$A\bar{1}(2)$		6	7.736 ±.003	7.157 ±.003	13.824 ±.010	90.52 ±.25	94.58 ±.25	103.87 ±.25	740.38 ± 1.08	74.32 ± .11	1.776 ±.003	3.326 ±.005	r	376
377	Pyroxmangite $MnFe(SiO_3)_2$*	tri.	$P\bar{1}(2)$		7	7.56 ±.02	17.45 ±.05	6.67 ±.02	84.00 ±.30	94.30 ±.30	113.70 ±.30	800.77 ± 4.29	68.90 ± .36	1.647 ±.009	3.817 ±.020	r	377
378	Tremolite $Ca_2Mg_5[Si_8O_{22}](OH)_2$*	mon.	C2/m(12)	tremolite	2	9.840 ±.010	18.052 ±.020	5.275 ±.010		104.70 ±.25		906.34 ± 2.43	272.92 ± .73	6.523 ±.018	2.977 ±.008	r	378
379	Fluor-tremolite $Ca_2Mg_5[Si_8O_{22}]F_2$	mon.	C2/m(12)	tremolite	2	9.781 ±.007	18.01 ±.01	5.267 ±.005		104.52 ±.25		898.18 ± 1.56	270.46 ± .47	6.464 ±.011	3.018 ±.005	20	379
380	Ferrotremolite $Ca_2Fe_5[Si_8O_{22}](OH)_2$	mon.	C2/m(12)	tremolite	2	9.97 ±.01	18.34 ±.02	5.30 ±.01		104.50 ±.10		938.24 ± 2.92	282.53 ± .69	6.753 ±.017	3.434 ±.008	r	380
381	Grunerite $Fe_7[Si_8O_{22}](OH)_2$	mon.	C2/m(12)	tremolite	2	9.572 ±.005	18.44 ±.01	5.342 ±.007		101.77 ±.25		923.08 ± 1.63	277.96 ± .49	6.644 ±.012	3.603 ±.006	r	381
382	Cummingtonite (hypo.) $Mg_7[Si_8O_{22}](OH)_2$	mon.	C2/m(12)	tremolite	2	9.476 ±.010	17.935 ±.010	5.292 ±.005		102.23 ±.25		878.97 ± 1.58	264.68 ± .47	6.326 ±.011	2.950 ±.005	r	382
383	Riebeckite $Na_2Fe_3Fe_2[Si_8O_{22}](OH)_2$	mon.	C2/m(12)	tremolite	2	9.729 ±.020	18.065 ±.020	5.334 ±.010		103.31 ±.25		912.29 ± 2.89	274.71 ± .87	6.566 ±.021	3.407 ±.011	r	383
384	Magnesioriebeckite $Na_2Mg_3Fe_2[Si_8O_{22}](OH)_2$	mon.	C2/m(12)	tremolite	2	9.733 ±.010	17.946 ±.020	5.299 ±.010		103.30 ±.25		900.74 ± 2.37	271.24 ± .71	6.483 ±.017	3.102 ±.008	r	384
385	Gaucophane I $Na_2Mg_3Al_2[Si_8O_{22}](OH)_2$	mon.	C2/m(12)	tremolite	2	9.748 ±.010	17.915 ±.020	5.273 ±.010		102.78 ±.25		898.04 ± 2.35	270.42 ± .71	6.463 ±.017	2.898 ±.008	r	385
386	Glaucophane II $Na_2Mg_3Al_2[Si_8O_{22}](OH)_2$	mon.	C2/m(12)	tremolite	2	9.663 ±.010	17.696 ±.020	5.277 ±.010		103.67 ±.10		876.79 ± 2.17	264.02 ± .65	6.310 ±.016	2.968 ±.007	r	386
387	Fluor-edenite $NaCa_2Mg_5[AlSi_7O_{22}]F_2$	mon.	C2/m(12)	tremolite	2	9.847 ±.005	18.00 ±.01	5.282 ±.005		104.83 ±.25		905.03 ± 1.51	272.53 ± .46	6.514 ±.011	3.076 ±.005	r	387
388	Fluor-richterite $Na_2CaMg_5[Si_8O_{22}]F_2$	mon.	C2/m(12)	tremolite	2	9.823 ±.005	17.96 ±.01	5.268 ±.005		104.33 ±.25		900.47 ± 1.48	271.15 ± .45	6.481 ±.011	3.033 ±.005	r	388
389	Anthophyllite $Mg_7[Si_8O_{22}](OH)_2$	orth.	Pnma(62)		4	18.61 ±.02	18.01 ±.06	5.24 ±.01				1756.3 ± 7.0	264.4 ± 1.1	6.320 ±.025	2.953 ±.012	r	389
	Framework structure silicates																
390	Microcline $KAlSi_3O_8$	tri.	$C\bar{1}(2)$		4	8.582 ±.002	12.964 ±.005	7.222 ±.002	90.62 ±.10	115.92 ±.10	87.68 ±.10	722.06 ± .67	108.72 ± .10	2.5984 ±.0025	2.560 ±.002	r	390
391	High Sanidine $KAlSi_3O_8$	mon.	C2/m(12)		4	8.615 ±.002	13.031 ±.003	7.177 ±.002		115.98 ±.10		724.28 ± .69	109.05 ± .10	2.6064 ±.0025	2.552 ±.002	r	391
392	Orthoclase $KAlSi_3O_8$*	mon.	C2/m(12)		4	8.562 ±.003	12.996 ±.004	7.193 ±.003		116.02 ±.15		719.25 ± 1.02	108.29 ± .15	2.5883 ±.0037	2.570 ±.004	r	392
393	Fe-Sanidine $KFeSi_3O_8$	mon.	C2/m(12)		4	8.689 ±.008	13.12 ±.01	7.319 ±.007		116.10 ±.30		749.28 ± 2.24	112.81 ± .34	2.6964 ±.0081	2.723 ±.008	r	393
394	Fe-Microcline $KFeSi_3O_8$	tri.	$C\bar{1}(2)$		4	8.68 ±.01	13.10 ±.01	7.340 ±.007	90.75 ±.25	116.05 ±.25	86.23 ±.25	748.09 ± 1.92	112.63 ± .29	2.692 ±.007	2.727 ±.007	r	394
395	Low Albite $NaAlSi_3O_8$	tri.	$C\bar{1}(2)$		4	8.139 ±.002	12.788 ±.003	7.160 ±.002	94.27 ±.10	116.57 ±.10	87.68 ±.10	664.65 ± .60	100.07 ± .09	2.3918 ±.0022	2.620 ±.002	26	395
396	High Albite (Analbite) $NaAlSi_3O_8$	tri.	$C\bar{1}(2)$		4	8.160 ±.002	12.870 ±.003	7.106 ±.002	93.54 ±.10	116.36 ±.10	90.19 ±.10	667.00 ± .60	100.43 ± .09	2.4003 ±.0022	2.611 ±.002	r	396
397	Anorthite $CaAl_2Si_2O_8$	tri.	$P\bar{1}(2)$	primitive cell	8	8.177 ±.002	12.877 ±.003	14.169 ±.003	93.17 ±.02	115.85 ±.02	91.22 ±.02	1338.9 ± .6	100.79 ± .04	2.4090 ±.0011	2.760 ±.001	r	397
398	Synthetic $CaAl_2Si_2O_8$	hex.	$P6_3/mcm(193)$		2	5.10 ±.02		14.72 ±.02				331.57 ± 2.64	99.85 ± .79	2.386 ±.019	2.786 ±.022	r	398
399	Synthetic $CaAl_2Si_2O_8$	orth.	$P2_12_12(18)$		2	8.22 ±.02	8.60 ±.02	4.83 ±.01				341.44 ± 1.35	102.82 ± .41	2.457 ±.010	2.706 ±.011	r	399

X-Ray Crystallographic Data of Minerals

	Name and formula	Crystal system	Space group	Structure type	Z	a_o	b_o	c_o	α_o	β_o	γ_o	Cell volume 10^{-24} cm^3	Molar volume cm^3	Molar volume cal bar^{-1}	X-Ray density grams cm^{-3}	Temp. °C	
400	Celsian $BaAl_2Si_2O_8$*	mon.	$I2_1/c(15)$		8	8.627 ±.010	13.045 ±.010	14.408 ±.020		115.20 ±.25		1467.1 ± 4.2	110.45 ± .31	2.640 ±.008	3.400 ±.010	r	400
401	Paracelsian $BaAl_2Si_2O_8$*	mon.	$P2_1/a(14)$		4	8.58 ±.02	9.583 ±.020	9.08 ±.02		90.00 ±.50		746.6 ± 2.9	112.4 ± .4	2.687 ±.010	3.340 ±.013	r	401
402	Banalsite $BaNa_2Al_4Si_4O_{16}$*	orth.			4	8.50 ±.02	9.97 ±.02	16.72 ±.03				1416.9 ± 5.1	213.3 ± .8	5.099 ±.018	3.092 ±.011	r	402
403	Danburite $CaB_2Si_2O_8$*	orth.	Pnam(62)		4	8.04 ±.02	8.77 ±.02	7.74 ±.02				545.8 ± 2.3	82.17 ± .35	1.964 ±.008	2.992 ±.013	r	403
404	Low Nepheline $NaAlSiO_4$	hex.	$C6_3(178)$		8	9.986 ±.005		8.330 ±.004				719.38 ± .80	54.16 ± .06	1.294 ±.002	2.623 ±.003	r	404
405	High Carnegeite $NaAlSiO_4$	cubic			4	7.325 ±.004						393.03 ± .64	59.18 ± .10	1.414 ±.002	2.401 ±.004	750	405
406	Kaliophilite natural $KAlSiO_4$*	hex.	$P6_322(182)$		54	26.930 ±.010		8.522 ±.004				5352.4 ± 4.7	59.69 ± .05	1.427 ±.001	2.650 ±.002	r	406
407	Kaliophilite synthetic $KAlSiO_4$	hex.	$P6_3(173)$ $P6_322(182)$		2	5.180 ±.002		8.559 ±.004				198.89 ± .18	59.89 ± .05	1.431 ±.001	2.641 ±.002	r	407
408	Kalsilite $KAlSiO_4$	hex.	$P6_3(173)$		2	5.1597 ±.0020		8.7032 ±.0030				200.66 ± .17	60.424 ± .051	1.4442 ±.0031	2.618 ±.002	r	408
409	Leucite $KAlSi_2O_6$	tet.	$I4_1/a(100)$		16	13.074 ±.003		13.738 ±.003				2348.23 ± 1.19	88.389 ± .045	2.1126 ±.0011	2.469 ±.001	25	409
410	High Leucite $KAlSi_2O_6$*	cubic	Ia3d(230)		16	13.43 ±.05						2422.3 ±27.1	91.18 ± 1.02	2.179 ±.024	2.394 ±.027	625	410
411	Fe-Leucite $KFeSi_2O_6$	tet.	$I4_1/a(100)$		16	13.205 ±.002		13.970 ±.003				2435.98 ± .91	91.692 ± .034	2.1915 ±.0009	2.695 ±.001	25	411
412	Petalite $LiAlSi_4O_{10}$*	mon.	$P2_1/n(14)$		2	11.32 ±.03	5.14 ±.01	7.62 ±.01		105.90 ±.20		426.41 ± 1.57	128.4 ± .5	3.069 ±.011	2.385 ±.009	r	412
413	Marialite $Na_4Al_3Si_9O_{24}Cl$	tet.	I4/m(87) P4/m(83)		2	12.064 ±.008		7.514 ±.004				1093.6 ± 1.6	329.3 ± .5	7.871 ±.011	2.566 ±.004	r	413
414	Meionite $Ca_4Al_6Si_6O_{24}CO_3$	tet.	I4/m(87) P4/m(83)		2	12.174 ±.008		7.652 ±.015				1134.07 ± 2.68	341.5 ± .8	8.162 ±.019	2.737 ±.007	r	414
	Sheet structure silicates																
415	Muscovite $KAl_2[AlSi_3O_{10}](OH)_2$*	mon.	C2/c(15)	$2M_2$ mica	4	5.203 ±.005	8.995 ±.005	20.030 ±.010		94.47 ±.33		934.57 ± 1.21	140.71 ± .18	3.363 ±.004	2.831 ±.004	r	415
416	Paragonite $NaAl_2[AlSi_3O_{10}](OH)_2$*	mon.	C2/c(15)	$2M_1$ mica	4	5.13 ±.03	8.89 ±.05	19.32 ±.10		95.17 ±.50		877.52 ± 8.47	132.1 ± 1.3	3.158 ±.031	2.893 ±.028	r	416
417	Lepidolite $K_2Al_3Li_2[AlSi_7O_{20}](OH)_4$*	mon.	C2/c(15)	$2M_2$ mica	2	9.2 ±.1	5.3 ±.1	20.0 ±.2		98.00 ±.50		965.7 ±23.2	290.8 ± 7.0	6.950 ±.167	2.698 ±.065	r	417
418	Phlogopite $KMg_3[AlSi_3O_{10}](OH)_2$	mon.	Cm(8)	1M mica	2	5.326 ±.010	9.210 ±.010	10.311 ±.010		100.17 ±.10		497.83 ± 1.19	149.91 ± .36	3.5830 ±.0086	2.784 ±.007	r	418
419	Fluor-phlogopite $KMg_3[AlSi_3O_{10}]F_2$	mon.	Cm(8)	1M mica	2	5.299 ±.005	9.188 ±.005	10.135 ±.005		99.92 ±.10		486.07 ± .60	146.37 ± .18	3.498 ±.004	2.878 ±.004	r	419
420	Annite $KFe_3[AlSi_3O_{10}](OH)_2$	mon.	Cm(8)	1M mica	2	5.391 ±.010	9.350 ±.005	10.313 ±.020		99.70 ±.25		512.40 ± 1.45	154.30 ± .44	3.688 ±.010	3.318 ±.009	r	420
421	Ferriannite $KFe_3[FeSi_3O_{10}](OH)_2$	mon.	C2/m(12)		2	5.430 ±.002	9.404 ±.003	10.341 ±.006		100.07 ±.20		519.92 ± .51	156.56 ± .15	3.7419 ±.0037	3.454 ±.003	r	421
422	Margarite $CaAl_2[Al_2Si_2O_{10}](OH)_2$*	mon.	C2/c(15)	2M mica	4	5.13 ±.02	8.92 ±.03	19.50 ±.05		95.00 ±.50		888.9 ± 5.2	133.8 ± .8	3.199 ±.019	2.975 ±.017	r	422
423	Talc $Mg_3Si_4O_{10}(OH)_2$*	mon.	C2/c(15)	$2M_1$	4	5.287 ±.007	9.158 ±.010	18.95 ±.01		99.50 ±.20		904.94 ± 1.71	136.25 ± .26	3.2565 ±.0062	2.784 ±.005	r	423
424	Pyrophyllite $Al_2Si_4O_{10}(OH)_2$*	mon.	C2/c(15)	$2M_1$	4	5.14 ±.02	8.90 ±.02	18.55 ±.03		99.92 ±.20		835.9 ± 4.0	125.9 ± .6	3.008 ±.015	2.863 ±.014	r	424

X-Ray Crystallographic Data of Minerals

	Name and formula	Crystal system	Space group	Structure type	Z	a_o	b_o	c_o	α_o	β_o	γ_o	Cell volume 10^{-24} cm³	Molar volume cm³	Molar volume cal bar⁻¹	X-Ray density grams cm⁻³	Temp. °C	
425	Minnesotaite $Fe_3Si_4O_{10}(OH)_2$*	mon.	C2/c(15)		4	5.4 ±.1	9.42 ±.04	19.4 ±.1		100.00 ±.50		971.8 ±19.2	146.3 ± 2.9	3.497 ±.069	3.239 ±.064	r	425
426	Dickite $Al_2Si_2O_5(OH)_4$*	mon.	Cc(9)		4	5.150 ±.002	8.940 ±.003	14.736 ±.005		103.58 ±.10		659.49 ± .49	99.30 ± .07	2.3733 ±.0018	2.600 ±.002	r	426
427	Kaolinite $Al_2Si_2O_5(OH)_4$*	tri.	P1(1)		2	5.155 ±.007	8.959 ±.010	7.407 ±.008	91.68 ±.35	104.87 ±.35	89.93 ±.35	330.48 ± .86	99.52 ± .26	2.3785 ±.0062	2.594 ±.007	r	427
428	Nacrite $Al_2Si_2O_5(OH)_4$*	mon.	Cc(9)		4	8.909 ±.010	5.146 ±.010	15.697 ±.020		113.70 ±.25		658.9 ± 2.1	99.21 ± .32	2.3713 ±.0076	2.602 ±.008	r	428
	Zeolites																
429	Analcite $NaAlSi_2O_6.H_2O$	cubic	Ia3d(230)		16	13.733 ±.005						2589.98 ± 2.83	97.49 ± .11	2.3301 ±.0026	2.258 ±.003	r	429
430	Natrolite $Na_2Al_2Si_3O_{10}.2H_2O$*	orth.	Fdd2(43)		8	18.30 ±.02	18.63 ±.02	6.60 ±.01				2250.1 ± 4.8	169.39 ± .37	4.049 ±.009	2.245 ±.005	r	430

HEAT CAPACITY OF ROCK FORMING MINERALS

The units of heat capacity at constant pressure, C_p, in this table are cal kg⁻¹deg⁻¹. Values of these units are given for several temperatures with the temperatures being in degrees C.

Heat capacity at other temperatures may be calculated by use of the equation $C_p = a + bT - cT^{-5}$. The units of these constants are cal kg⁻¹deg⁻¹.

Mineral	Heat capacity at various temperatures -200	0	200	400	800	1200	Constants for heat capacity equation a	10^3b	10^5c
Albite	–	0.1695	0.236	0.26	0.286	–	1.018	0.187	0.268
Amphibole	–	0.177	0.246	0.27	0.296	–	1.067	1.183	0.281
Apatite	–	0.24	–	–	–	–	–	–	–
Arsenopyrite	–	0.103 at 55°C	–	–	–	–	–	–	–
Asbestos	–	0.195	–	–	–	–	–	–	–
Barite	0.047	0.1076	0.12	0.1315	0.1555	–	0.383	0.253	0
Cassiterite	–	0.0814	0.103	0.115	0.132	–	0.387	0.157	0.007
Chalcopyrite	–	0.129 at 50°C	–	–	–	–	–	–	–
Diamond	–	0.104	0.253	0.328	0.445	–	0.754	1.067	0.454
Dolomite	–	0.222 at 60°C	–	–	–	–	–	–	–
Fluorite	0.0525	0.203	0.213	0.222	0.24	0.263	0.798	0.204	0
Galena	0.034	0.0496	0.0528	0.0562	–	–	0.188	0.007	0
Garnet	–	0.177 at 58°C	–	–	–	–	–	–	–
β-Graphite	–	0.152	0.282	0.348	0.45	–	0.932	0.913	0.4077
Hematite	–	0.146	0.189	0.215	0.258	–	0.640	0.420	0.111
Ice	0.156	0.492	–	–	–	–	–	–	–
Kaolinite	–	0.222	0.244	–	–	–	0.806	0.463	0.0
Labradorite	–	0.196 at 60°C	–	–	–	–	–	–	–
Magnesite	0.0385	0.207	–	–	–	–	–	–	–
Magnetite	–	0.1435	0.1985	0.222	–	–	0.744	0.340	0.177
Mica (mono-crystal)	–	0.208	–	–	–	–	–	–	–
Microcline	–	0.163	0.227	0.248	0.342	–	0.988	0.166	0.263
Oligoclase	–	0.2035 at 60°C	–	–	–	–	–	–	–
Olivine	–	0.189 at 36°C	–	–	–	–	–	–	–
Orthoclase	–	0.146	0.226	0.251	0.347	–	0.043	0.124	0.351
Pyrite	0.0179	0.1195	0.142	0.165	–	–	0.373	0.466	0
Pyroxene	–	0.18	0.246	0.275	–	–	0.973	0.336	0.233
α-Quartz	0.0414	0.167	0.232	0.2695	–	–	0.7574	0.607	0.168
β-Quartz	–	–	–	–	0.28	0.3165	0.763	0.383	0
Serpentine		0.227	–	–	–	–	–	–	–
Siderite	0.056	0.163	–	–	–	–	–	–	–
Talc		0.208 at 59°C	–	–	–	–	–	–	–
Zircon	–	0.146 at 60°C	–	–	–	–	–	–	–

RESISTIVITIES OF SEMICONDUCTING MINERALS (ZERO FREQUENCY)

Native elements	ϱ (ohm-m)
Diamond (C)	2.7
Sulfides	
Argentite, Ag_2S	1.5 to 2.0×10^{-3}
Bismuthinite, Bi_2S_3	3 to 570
Bornite, $Fe_2S_3 \cdot nCu_2S$	1.6 to 6000×10^{-6}
Chalcocite, Cu_2S	80 to 100×10^{-6}
Chalcopyrite, $Fe_2S_3 \cdot Cu_2S$	150 to 9000×10^{-6}
Covellite, CuS	0.30 to 83×10^{-6}
Galena, PbS	6.8×10^{-6} to 9.0×10^{-2}
Haverite, MnS_2	10 to 20
Marcasite, FeS_2	1 to 150×10^{-3}
Metacinnabarite, 4HgS	2×10^{-6} to 1×10^{-3}
Millerite, NiS	2 to 4×10^{-7}
Molybdenite, MoS_2	0.12 to 7.5
Pentlandite, $(Fe, Ni)_9S_8$	1 to 11×10^{-6}
Pyrrhotite, Fe_7S_8	2 to 160×10^{-6}
Pyrite, FeS_2	1.2 to 600×10^{-3}
Sphalerite, ZnS	2.7×10^{-3} to 1.2×10^{4}
Antimony-sulfur compounds	
Berthierite, $FeSb_2S_4$	0.0083 to 2.0
Boulangerite, $Pb_5Sb_4S_{11}$	2×10^{3} to 4×10^{4}
Cylindrite, $Pb_3Sn_4Sb_2S_{14}$	2.5 to 60
Franckeite, $Pb_5Sn_3Sb_2S_{14}$	1.2 to 4
Hauchecornite, $Ni_9(Bi, Sb)_2S_8$	1 to 83×10^{-6}
Jamesonite, $Pb_4FeSb_6S_{14}$	0.020 to 0.15
Tetrahedrite, Cu_3SbS_3	0.30 to 30,000
Arsenic-sulfur compounds	
Arsenopyrite, FeAsS	20 to 300×10^{-6}
Cobaltite, CoAsS	6.5 to 130×10^{-3}
Enargite, Cu_3AsS_4	0.2 to 40×10^{-3}
Gersdorffite, NiAsS	1 to 160×10^{-6}
Glaucodote, (Co, Fe)AsS	5 to 100×10^{-6}
Antimonide	
Dyscrasite, Ag_3Sb	0.12 to 1.2×10^{-6}
Arsenides	
Allemonite, $SbAs_3$	70 to 60,000
Lollingite, $FeAs_2$	2 to 270×10^{-6}
Nicollite, NiAs	0.1 to 2×10^{-6}
Skutterudite, $CoAs_3$	1 to 400×10^{-6}
Smaltite, $CoAs_2$	1 to 12×10^{-6}
Tellurides	
Altaite, PbTe	20 to 200×10^{-6}
Calavarite, $AuTe_2$	6 to 12×10^{-6}
Coloradoite, HgTe	4 to 100×10^{-6}
Hessite, Ag_2Te	4 to 100×10^{-6}
Nagyagite, $Pb_6Au(S, Te)_{14}$	20 to 80×10^{-6}
Sylvanite, $AgAuTe_4$	4 to 20×10^{-6}
Oxides	
Braunite, Mn_2O_3	0.16 to 1.0
Cassiterite, SnO_2	4.5×10^{-4} to 10,000
Cuprite, Cu_2O	10 to 50
Hollandite, $(Ba, Na, K)Mn_8O_{16}$	2 to 100×10^{-3}
Ilmenite, $FeTiO_3$	0.001 to 4
Magnetite, Fe_3O_4	52×10^{-6}
Manganite, $MnO \cdot OH$	0.018 to 0.5
Melaconite, CuO	6000
Psilomelane, $KMnO \cdot MnO_2 \cdot nH_2O$	0.04 to 6000
Pyrolusite, MnO_2	0.007 to 30
Rutile, TiO_2	29 to 910
Uraninite, UO	1.5 to 200

From Carmichael, R. S., ed., *Handbook of Physical Properties of Rocks*, Vol. I, CRC Press, 1982.

MINERALS ARRANGED IN ORDER OF INCREASING VICKERS HARDNESS NUMBERS

Mineral species	Mean	Range	Remarks
Graphite	12	12	
Molybdenite	17	16—19	⊥ to cleavage
	23	21—28	∥ to cleavage
Bismuth	18	16—19	
Tellurbismuth	21	20—21	
Argentite	24	20—30	
Hessite	33	28—41	
Orpiment	38	23—52	
Electrum	40	34—44	
Stromeyerite	41	38—44	
Altaite	51	48—57	
Gold	51	50—52	
Silver	53	48—63	
Realgar	56	53—60	
Digenite	61	56—67	
Arsenic	63	57—69	
Pyrargyrite	71	50—97	⊥ to cleavage
	106	98—126	∥ to cleavage
Covellite	72	69—78	
Galena	76	71—84	
Pyrolusite	76	76	Average hardness ⊥ to fibers
	252	252	Average hardness ∥ to fibers
	279	256—346	Isotropic sections
	292	225—405	Microcrystalline
Stibnite	77	42—109	
Chalcophanite	81	71—85	⊥ to cleavage
	124	103—165	∥ to cleavage
	133	110—178	Isotropic sections
Chalcocite	84	68—98	
Antimony	89	83—99	
Jamesonite	99	96—105	Granular allotriomorphic sections
	113	105—121	Prismatic sections
Bornite	103	97—105	
Bismuthinite	107	92—119	
Miargyrite	110	104—123	
Sylvanite	110	102—125	
Kobellite	116	69—173	
Proustite	123	109—135	
Platinum	126	125—127	
Copper	134	120—143	
Naumannite	148	115—185	
Zincite	154	150—157	⊥ to cleavage
	304	295—318	∥ to cleavage
Pearceite	160	153—164	
Enargite	160	133—185	⊥ to cleavage
	272	245—346	∥ to cleavage
Boulangerite	166	157—183	
Dyscrasite	167	162—178	
Berthierite	171	155—185	
Zinkenite	178	162—207	
Emplectite	191	168—213	∥ to elongation
	222	197—238	⊥ to elongation
Bournonite	192	185—199	
Chalcopyrite	194	186—219	
Blende	198	186—209	
Cuprite	199	192—218	
Stannite	210	197—221	
Cubanite	213	199—228	
Pentlandite	215	202—230	
Tenorite	236	209—254	
Millerite	236	225—256	Isotropic sections
	254	235—280	∥ to elongation
	348	318—376	⊥ to elongation
Pyrrhotite	248	230—259	Anistropic sections
	303	280—318	Isotropic sections
Alabandite	251	240—266	
Coffinite	258	236—333	
Chalcostibite	276	264—285	
Niccolite	336	328—348	Anistropic sections
	446	433—455	Isotropic sections
Tennantite	338	320—361	
Freibergite	345	317—375	
Scheelite	348	285—429	
Tetrahedrite	351	328—367	
Famatinite	363	333—397	
Wolframite	373	357—394	
Manganite	410	367—459	
Carrollite	463	351—566	
Lollingite	486	421—556	⊥ to elongation
	825	739—920	∥ to elongation
Siegenite	524	503—533	
Ullmannite	525	498—542	
Betafite	525	503—560	
Ilmenite	536	519—553	Possible differences in composition
	681	659—703	
Goethite	554	525—620	Microcrystalline
	803	772—824	Coarsely crystalline
Magnetite	560	530—599	
Breithauptite	563	542—584	
Psilomelane	572	503—627	
Hausmannite	587	541—613	
Braunite	595	584—605	

MINERALS ARRANGED IN ORDER OF INCREASING VICKERS HARDNESS NUMBERS (Continued)

Mineral species	Mean	Range	Remarks
Pyrochlore	613	572—665	
Hollandite	620	560—724	
Skutterudite	653	589—724	
Gersdorffite	698	665—743	
Maucherite	704	685—724	
Euxenite	707	599—782	
Rammelsbergite	712	687—778	
Brannerite	720	710—730	
Pitchblende	720	673—803	Fresh specimens, oxidation produces marked decrease in hardness
Lepidocrocite	724	690—782	
Jacobsite	734	724—745	
Davidite	745	707—803	
Hematite	755	739—822	Microcrystalline
	1,000	920—1,062	Coarsely crystalline
Pararammelsbergite	772	762—803	
Coronadite	784	767—813	
Columbite-tantalite	803	724—882	
Uraninite	808	782—839	
Maghemite	946	894—988	
Bixbyite	1,018	1,003—1,033	
Thorianite	1,918	988—1,115	
Cassiterite	1,053	1,027—1,075	
Arsenopyrite	1,094	1,048—1,127	
Bravoite	1,097	1,003—1,288	
Marcasite	1,113	941—1,288	
Glaucodot	1,124	1,071—1,166	
Rutile	1,139	1,074—1,210	
Pyrite	1,165	1,027—1,240	
Cobaltite	1,200	1,176—1,226	
Chromite	1,206	1,195—1,210	

From Carmichael, R. S., ed., *Handbook of Physical Properties of Rocks,* Vol. I, CRC Press, 1982.

SOLUBILITY PRODUCT CONSTANTS

James C. Chang

The following solubility product constants are calculated from the free energies of formation of the substances as solids and those of the aqueous ions at their standard states of m = 1. Thus, for the reaction

$$M_mX_n(s) \leftrightarrows mM^{n+}(aq) + nX^{m-}(aq),$$

$$\Delta G^\circ = m\Delta G_f^\circ(M^{n+}, aq) + n\Delta G_f^\circ(X^{m-}, aq) - \Delta G^\circ(M_mX_n, s)$$

where M_mX_n is the slightly soluble substance, M^{n+} and X^{m-} are the two ions produced in solution by the dissociation of $M_m X_n$. Then the solubility product constant, K_{sp}, is calculated by using the equation

$$\ln K_{sp} = -\frac{\Delta G^\circ}{RT}$$

The values in the following table are for K_{sp} at 25°C.

Substance	Formula	Solubility Product
Aluminum phosphate	$AlPO_4$	9.83×10^{-21}
Barium carbonate	$BaCO_3$	2.58×10^{-9}
Barium chromate	$BaCrO_4$	1.17×10^{-10}
Barium fluoride	BaF_2	1.84×10^{-7}
Barium hydroxide 8-hydrate	$Ba(OH)_2 \cdot 8H_2O$	2.55×10^{-4}
Barium iodate	$Ba(IO_3)_2$	4.01×10^{-9}
Barium iodate 1-hydrate	$Ba(IO_3)_2 \cdot H_2O$	1.67×10^{-9}
Barium sulfate	$BaSO_4$	1.07×10^{-10}
Bismuth arsenate	$BiAsO_4$	4.43×10^{-10}
Bismuth sulfide	Bi_2S_3	1.82×10^{-99}
Cadmium arsenate	$Cd_3(AsO_4)_2$	2.17×10^{-33}
Cadmium carbonate	$CdCO_3$	6.18×10^{-12}
Cadmium fluoride	CdF_2	6.44×10^{-3}
Cadmium hydroxide	$Cd(OH)_2$	5.27×10^{-15}
Cadmium iodate	$Cd(IO_3)_2$	2.49×10^{-8}
Cadmium oxalate 3-hydrate	$CdC_2O_4 \cdot 3H_2O$	1.42×10^{-8}
Cadmium phosphate	$Cd_3(PO_4)_2$	2.53×10^{-33}
Cadmium sulfide	CdS	1.40×10^{-29}
Calcium carbonate	$CaCO_3$	4.96×10^{-9}
Calcium fluoride	CaF_2	1.46×10^{-10}
Calcium hydroxide	$Ca(OH)_2$	4.68×10^{-6}
Calcium iodate	$Ca(IO_3)_2$	6.47×10^{-6}
Calcium iodate 6-hydrate	$Ca(IO_3)_2 \cdot 6H_2O$	7.54×10^{-7}
Calcium oxalate 1-hydrate	$CaC_2O_4 \cdot H_2O$	2.34×10^{-9}
Calcium phosphate	$Ca_3(PO_4)_2$	2.07×10^{-33}
Calcium sulfate	$CaSO_4$	7.10×10^{-5}
Cobalt(II) arsenate	$Co_3(AsO_4)_2$	6.79×10^{-29}
Cobalt(II) hydroxide (pink)	$Co(OH)_2$	1.09×10^{-15}
Cobalt(II) hydroxide (blue)	$Co(OH)_2$	5.92×10^{-15}
Cobalt(II) iodate 2-hydrate	$Co(IO_3)_2 \cdot 2H_2O$	1.21×10^{-2}
Cobalt(II) phosphate	$Co_3(PO_4)_2$	2.05×10^{-35}
Copper(I) bromide	$CuBr$	6.27×10^{-9}
Copper(I) chloride	$CuCl$	1.72×10^{-7}
Copper(I) iodide	CuI	1.27×10^{-12}
Copper(I) sulfide	Cu_2S	2.26×10^{-48}
Copper(I) thiocyanate	$CuSCN$	1.77×10^{-13}
Copper(II) arsenate	$Cu_3(AsO_4)_2$	7.93×10^{-36}
Copper(II) iodate 1-hydrate	$Cu(IO_3)_2 \cdot H_2O$	6.94×10^{-8}
Copper(II) oxalate	CuC_2O_4	4.43×10^{-10}
Copper(II) phosphate	$Cu_3(PO_4)_2$	1.39×10^{-37}
Copper(II) sulfide	CuS	1.27×10^{-36}
Iron(II) carbonate	$FeCO_3$	3.07×10^{-11}
Iron(II) fluoride	FeF_2	2.36×10^{-6}
Iron(II) hydroxide	$Fe(OH)_2$	4.87×10^{-17}
Iron(II) sulfide	FeS	1.59×10^{-19}
Iron(III) hydroxide	$Fe(OH)_3$	2.64×10^{-39}
Iron(III) phosphate 2-hydrate	$FePO_4 \cdot 2H_2O$	9.92×10^{-29}
Lead bromide	$PbBr_2$	6.60×10^{-6}
Lead carbonate	$PbCO_3$	1.46×10^{-13}
Lead chloride	$PbCl_2$	1.17×10^{-5}
Lead fluoride	PbF_2	7.12×10^{-7}
Lead hydroxide	$Pb(OH)_2$	1.42×10^{-20}
Lead iodate	$Pb(IO_3)_2$	3.68×10^{-13}
Lead iodide	PbI_2	8.49×10^{-9}
Lead oxalate	PbC_2O_4	8.51×10^{-10}
Lead sulfate	$PbSO_4$	1.82×10^{-8}
Lead sulfide	PbS	9.04×10^{-29}
Lead thiocyanate	$Pb(SCN)_2$	2.11×10^{-5}
Lithium carbonate	Li_2CO_3	8.15×10^{-4}
Magnesium carbonate	$MgCO_3$	6.82×10^{-6}
Magnesium carbonate 3-hydrate	$MgCO_3 \cdot 3H_2O$	2.38×10^{-6}
Magnesium carbonate 5-hydrate	$MgCO_3 \cdot 5H_2O$	3.79×10^{-6}
Magnesium fluoride	MgF_2	7.42×10^{-11}
Magnesium hydroxide	$Mg(OH)_2$	5.61×10^{-12}
Magnesium oxalate 2-hydrate	$MgC_2O_4 \cdot 2H_2O$	4.83×10^{-6}
Magnesium phosphate	$Mg_3(PO_4)_2$	9.86×10^{-25}
Manganese(II) carbonate	$MnCO_3$	2.24×10^{-11}
Manganese(II) hydroxide	$Mn(OH)_2$	2.06×10^{-13}
Manganese(II) iodate	$Mn(IO_3)_2$	4.37×10^{-7}
Manganese(II) oxalate 2-hydrate	$MnC_2O_4 \cdot 2H_2O$	1.70×10^{-7}
Manganese(II) sulfide	MnS	4.65×10^{-14}
Mercury(I) bromide	Hg_2Br_2	6.41×10^{-23}
Mercury(I) carbonate	Hg_2CO_3	3.67×10^{-17}
Mercury(I) chloride	Hg_2Cl_2	1.45×10^{-18}
Mercury(I) fluoride	Hg_2F_2	3.10×10^{-6}
Mercury(I) iodide	Hg_2I_2	5.33×10^{-29}
Mercury(I) oxalate	$Hg_2C_2O_4$	1.75×10^{-13}
Mercury(I) sulfate	Hg_2SO_4	7.99×10^{-7}
Mercury(I) thiocyanate	$Hg_2(SCN)_2$	3.12×10^{-20}
Mercury(II) hydroxide	$Hg(OH)_2$	3.13×10^{-26}
Mercury(II) iodide	HgI_2	2.82×10^{-29}
Mercury(II) sulfide (black)	HgS	6.44×10^{-53}
Mercury(II) sulfide (red)	HgS	2.00×10^{-53}
Nickel(II) carbonate	$NiCO_3$	1.42×10^{-7}

Substance	Formula	Solubility Product
Nickel(II) hydroxide	$Ni(OH)_2$	5.47×10^{-16}
Nickel(II) iodate	$Ni(IO_3)_2$	4.71×10^{-5}
Nickel(II) phosphate	$Ni_3(PO_4)_2$	4.73×10^{-32}
Nickel(II) sulfide	NiS	1.07×10^{-21}
Palladium(II) sulfide	PdS	2.03×10^{-58}
Palladium(II) thiocyanate	$Pd(SCN)_2$	4.38×10^{-23}
Platinum(II) sulfide	PtS	9.91×10^{-74}
Potassium hexachloroplatinate(IV)	$K_2[PtCl_6]$	7.48×10^{-6}
Potassium perchlorate	$KClO_4$	1.05×10^{-2}
Silver(I) acetate	$AgC_2H_3O_2$	1.94×10^{-3}
Silver(I) arsenate	Ag_3AsO_4	1.03×10^{-22}
Silver(I) bromate	$AgBrO_3$	5.34×10^{-5}
Silver(I) bromide	AgBr	5.35×10^{-13}
Silver(I) carbonate	Ag_2CO_3	8.45×10^{-12}
Silver(I) chloride	AgCl	1.77×10^{-10}
Silver(I) chromate	Ag_2CrO_4	1.12×10^{-12}
Silver(I) cyanide	AgCN	5.97×10^{-17}
Silver(I) iodate	$AgIO_3$	3.17×10^{-8}
Silver(I) iodide	AgI	8.51×10^{-17}
Silver(I) oxalate	$Ag_2C_2O_4$	5.40×10^{-12}
Silver(I) phosphate	Ag_3PO_4	8.88×10^{-17}
Silver(I) sulfate	Ag_2SO_4	1.20×10^{-5}
Silver(I) sulfide (α-form)	Ag_2S	6.69×10^{-50}
Silver(I) sulfide (β-form)	Ag_2S	1.09×10^{-49}
Silver(I) sulfite	Ag_2SO_3	1.49×10^{-14}
Silver(I) thiocyanate	AgSCN	1.03×10^{-12}
Strontium arsenate	$Sr_3(AsO_4)_2$	4.29×10^{-19}
Strontium carbonate	$SrCO_3$	5.60×10^{-10}
Strontium fluoride	SrF_2	4.33×10^{-9}
Strontium iodate	$Sr(IO_3)_2$	1.14×10^{-7}
Strontium iodate 1-hydrate	$Sr(IO_3)_2 \cdot H_2O$	3.58×10^{-7}
Strontium iodate 6-hydrate	$Sr(IO_3)_2 \cdot 6H_2O$	4.65×10^{-7}
Strontium sulfate	$SrSO_4$	3.44×10^{-7}
Tin(II) hydroxide	$Sn(OH)_2$	5.45×10^{-27}
Tin(II) sulfide	SnS	3.25×10^{-28}
Zinc arsenate	$Zn_3(AsO_4)_2$	3.12×10^{-28}
Zinc carbonate	$ZnCO_3$	1.19×10^{-10}
Zinc carbonate 1-hydrate	$ZnCO_3 \cdot H_2O$	5.41×10^{-11}
Zinc fluoride	ZnF_2	3.04×10^{-2}
Zinc hydroxide (γ-form)	$Zn(OH)_2$	6.86×10^{-17}
Zinc hydroxide (β-form)	$Zn(OH)_2$	7.71×10^{-17}
Zinc hydroxide (ϵ-form)	$Zn(OH)_2$	4.12×10^{-17}
Zinc iodate	$Zn(IO_3)_2$	4.29×10^{-6}
Zinc oxalate 2-hydrate	$ZnC_2O_4 \cdot 2H_2O$	1.37×10^{-9}
Zinc sulfide	ZnS	2.93×10^{-25}

PROPERTIES OF RARE EARTH METALS

F. H. Spedding

Symbol	M. P. °C[1]	B. P. °C[1]	Heat of Vaporization $\Delta H_{v,0}$ Kcal/g atm[2]	Density	Atomic Vol. (cm³/mole)	Metallic Radius Å	Electrical Resistivity Polycrystalline Wire 298°K (ohm-cm x 10^{-6})[3]	Residual Resistivity Wire 4.2°K (ohm-cm x 10^{-6})[3]	Compressibility cm²/kg x 10^{-6} **
Sc	1541	2831	89.9	2.989	15.041	1.640	52	3	2.26
Y	1522	3338	101.3	4.469	19.894	1.801	59	2	2.68
La	921	3457	103.1	6.145	22.603	1.879	61-80*	SC	4.04
Ce	799	3426	101.1	γ=6.767	20.400	1.820	70-80	10	4.10
				β=6.657	21.049				
Pr	931	3512	85.3	6.773	20.805	1.828	68	1	3.21
Nd	1021	3068	78.5	7.007	20.585	1.821	65	7	3.0
Pm	1168	2700 est.							
Sm	1077	1791	49.2	7.520	20.001	1.804	91	7	3.34
Eu	822	1597	41.9	5.243	28.981	1.984	91	1	8.29
			ΔH°298						
Gd	1313	3266	95.3	7.900	19.904	1.801	127	1	2.56
Tb	1356	3123	93.4	8.229	19.312	1.783	114	4	2.45
Dy	1412	2562	70.0	8.550	19.006	1.774	100	5	2.55
Ho	1474	2695	72.3	8.795	18.753	1.766	88	3	2.47
Er	1529	2863	76.1	9.066	18.450	1.757	71	3	2.39
Tm	1545	1947	55.8	9.321	18.124	1.746	74	3	2.47
Yb	819	1194	36.5	6.965	24.843	1.939	28	2	7.39
Lu	1663	3395	102.2	9.840	17.781	1.735	60	2	2.38

* Crystal usually mixture of α hcp and fcc lattice. SC - Superconductor

** Best values in author's opinion. Many numbers are taken from P. W. Bridgman's publication.

[1] Corrected for new temperature scale. Best values Ames Laboratory 2/1/72.

[2] Values from Thermodynamic Properties of Metals and Alloys (Review) R. Hultgren, R. Orr and K. Kelley. Supplements 1967.

[3] Weighted average of original papers 1/1/72.

Symbol	Crystal Structure at Room Temperature 25°C		Allotropic Forms Transition Temperatures Expressed in °C	
Sc	Hex. (to 1335°)	a=3.3088A, c=5.2680A	bcc(above 1335°)	
Y	Hex. (to 1478°)	a=3.6482A, c=5.7318A	bcc(above 1478°)	a=4.08A
La	Hex. (to 310°) (usually contains fcc also)	a=3.7740A, c=12.171A	fcc(310°-865°)	
			bcc(above 865°)	a=4.26A
Ce	fcc(~ 0° to 726°)	a=5.160A	fcc(below-157° on cooling) (up to -94° on heating)	a=4.85A
			Hex. (below-23° on cooling) (up to 168° on heating)	a=3.68A c=11.92A
Pr	Hex. (to 795°)	a=3.6721A, c=11.832A	bcc(above 795°)	a=4.12A
Nd	Hex. (to 863°)	a=3.6583A, c=11.7966A	bcc(above 863°)	a=4.13A
Sm	*Rhom (to 926°)	a=8.9834A, α=23° 49.5′	bcc(above 926°)	a=4.07A
Eu	bcc	a=4.5827A		
Gd	Hex. (to 1235°)	a=3.6336A, c=5.7810A	bcc(above 1235°)	a=4.05A†
Tb	Hex. (to 1289°)	a=3.6055A, c=5.6966A	bcc(above 1289°)	a=4.02A †
Dy	Hex. (to 1381°)	a=3.5915A, c=5.6501A	bcc(above 1381°)	a=3.98A †
Ho	Hex.	a=3.5778A, c=5.6178A	Not present pure metals	
Er	Hex.	a=3.5592A, c=5.5850A	Not present pure metals	
Tm	Hex.	a=3.5375A, c=5.5540A	Not present pure metals	
Yb	Hex. (to 795°)	a=5.4848A,	bcc(above 795°)	a=4.44A
Lu	Hex.	a=3.5052A, c=5.5494A	Not present pure metals	

† Extrapolated from magnesium alloy studies. Hex. refers to close packed hexagonal. La, Ce, Pr, and Nd have a stacking order ABAC, the other rare earths ABAB. fcc refers to close packed face centered cubic with a stacking order ABC, ABC, bcc refers to body centered cubic. These high temperature forms are very soft and deform very easily.

* The rhombic Sm cell can be expressed as hexagonal with a=3.6290A, c=26.207 and possess a stacking order ABABCBCAC

TABLE OF THE ISOTOPES (1985 UPDATE)

Compiled by
Russell L. Heath
Idaho National Engineering Laboratory
EG&G Idaho Inc.
Idaho Falls, Idaho

The following information is provided to assist in the use of the Table of Isotopes. This compilation of nuclear data presents a selected set of currently adopted values for experimental quantities which characterize the decay of radioactive nuclides. The approach used for the presentation of gamma rays and particles emitted in the decay of radionuclides was to include the major photons and beta groups which dominate the observed energy spectrum of photons and particles emitted in the decay of a given nuclide. In general, gamma rays are listed which span the first 2 decades on a relative intensity scale. To provide the applied spectroscopist with information on the relative intensity of X-rays emitted in electron-capture decay or associated with the internal conversion process, the intensity of the most intense K ($K_{\alpha 1}$) or L (L_β) X-ray line is listed when the X-ray component represents a significant contribution to the total photon spectrum. Values used in this case represent calculated intensities derived from level-scheme information. Values adopted and presented in the table have been compiled from original published data and evaluated data sets in major data compilations. A list of the major references used in this evaluation is given below. The effective literature cutoff date for data in this edition of the Table is 1984.

TABLE LAYOUT

Column no.	Column title	Description
1	Isotope	This column lists the isotopes with atomic number and mass number. Isomers are indicated by the addition of the letter m. The elements and their corresponding atomic number are also listed.
2	A	Mass number
3	Z	Atomic number
4	Isotopic abundance	Isotopic abundance in percent
5	Atomic mass	Lists values for atomic mass in the physical scale. Values of atomic mass are all relative to the mass of the Carbon 12 isotope which has been assigned a value of 12.000000 atomic mass units.
6	Half-life	Half-life in decimal notation. The notation used is: ns = nanoseconds, μs = microseconds, ms = milliseconds, s = seconds, m = minutes, d = days, and y = years.
7	Decay mode	Observed modes of decay for all radioactive species. Symbols used are: α = alpha particle emission; β− = negative beta emission; β+ = positron emission; E.C. = orbital electron emission; I.T. = isomeric transition from upper to lower isomeric state; n = neutron emission; and S.F. = spontaneous fission. Contained in ( ) are values for the % decay by a given decay mode.
8	Decay energy	The currently adopted values for total disintegration energy (Q) in MeV. Where more than one mode of decay exists, separate values are given.
9	Particle energy	The end-point energies of beta particle transitions or discrete energies of alpha particles are given in MeV. Estimated experimental error for alpha particle energies is indicated in ( ), representing variation in the least significant figure.
10	Particle intensity	The intensity of beta groups or alpha particle transitions are given in percent of total decay rate.

Column no.	Column title	Description
11	Thermal neutron cross section	Cross sections for thermal-neutron capture is given in barns (10^{-24} cm^2) or in millibarns (mb). Where neutron capture for a given nucleus may produce two isomers, the separate cross sections are listed. Fission cross sections are listed for the heavy elements and are designated by the symbol σ_f.
12	Spin	Values for the nuclear spin or angular momentum of the ground state of the isotope is given in units $h/2\pi$.
13	μ	Values of the magnetic dipole moments of the isotopes are given in nuclear magnetons, with diamagnetic correction.
14	Gamma-ray energy	Gamma-ray energies, together with estimated experimental error are given in MeV. The abbreviation ann.rad. refers to 511.006 keV photons associated with the annihilation of positrons in matter. Where positron emission is a significant component in the decay, annihilation radiation will be present.
15	Gamma-ray intensity	Values for gamma-ray intensity are given in percent of total decay rate for a given gamma-ray transition. When complete level-scheme information is not available relative photon intensities are indicated by the symbol +.

GENERAL NUCLEAR DATA REFERENCES

The following references represent the major sources of compiled nuclear data:

1. Nuclear Data Sheets, edited by The National Nuclear Data Center for the International Network for Nuclear Structure Data Evaluation, Academic Press, New York, Vol. 1—36.
2. **Lederer, C. M. and Shirley, V., Eds.,** *Table of Isotopes*, 7th ed., Wiley Interscience, New York, 1978.
3. **Browne, E. and Firestone, R.,** *Radioactivity Handbook*, to be published by Isotopes Project, LBL.
4. Chart of the Nuclides, 1983 Ed., prepared by Walker, F. W., Miller, D. G., and Feiner, F., Knolls Atomic Power Lab. operated by General Electric Co.
5. **Wapstra, A. H. and Audi, G.,** Atomic mass table, *Nuclear Physics A*, (preprint), July 1984.
6. **Mughabghab, S. F., Divadeenam, M., and Holden, N. E.,** *Neutron Cross Sections*, Part A, Academic Press, 1981; **Mughabghab, S. F.,** *Neutron Cross Section*, Part B, Academic Press, 1984.
7. Gamma-ray Spectrum Catalogue — Ge(Li) and Si(Li) Spectrometry, DOE Report ANCR-1000, 1974.
8. **Reus, U. and Westmeier, W.,** Catalogue of gamma rays from radioactive decay, *Atomic Data and Nuclear Data Tables*, 29, No. 2, September 1983.
9. ENSDF — a computer file of evaluated nuclear data maintained by the National Nuclear Data Center at Brookhaven National Laboratory.
10. ENDF-B Data File summary documentation, Ed. by Kinsey, R., Report BNL-NCS-17541(ENDF-201), National Nuclear Data Center, BNL, 1979.

Isotope	A	Z	% Natural abundance	Atomic mass	Half-life	Decay mode	Decay energy (MeV)	Particle energy (MeV)	Particle intensity	Thermal neutron cross section	Spin ($h/2\pi$)	μ Nucl. mag. moment	Gamma-ray energy (MeV)	Gamma-ray intensity
n	1			1.008665	12m.	β-	0.7825	0.7825	100%		1/2	-1.9131		
H		1		1.00797						0.332 ± 2 mb				
${}_1H^1$	1	1	99.985%	1.007825							1/2	+2.79284		
${}_1H^2$	2	1	0.015%	2.0140						0.51 ± 0.01mb	1	+0.85743		
${}_1H^3$	3	1		3.01605	12.26 yr	β-	0.01861	0.01861	100%	≤6μb	1/2	+2.97896		
He				4.002603										
${}_2He^3$	3	2	0.00014%	3.01603						<0.1 mb	1/2	-2.12762		
${}_2He^4$	4	2	99.99986%	4.00260						0	3/2-			
${}_2He^5$	5	2		5.01222										
${}_2He^6$	6	2		6.018886	0.808 s.	β-	3.5097	3.5097	100%		0+			
${}_2He^8$	8	2		8.03392	0.122 s.	β-	14	13	88%		0+		0.99	88%
						n			12%					
Li				6.9409										
${}_3Li^5$	5	3		5.01254							3/2-			
${}_3Li^6$	6	3	7.5%	6.015121						38 ± 3 mb	1	+.822056		
${}_3Li^7$	7	3	92.5%	7.016003						45 ± 5 mb	3/2	+3.25644		
${}_3Li^8$	8	3		8.022485	0.844 s.	β-,α	16.005	12.5	100%		2+	+1.6532		
								α(1.6)						
${}_3Li^9$	9	3		9.026789	0.178 s.	β-	13.6068	13.5	75%		3/2-			
						β-		11	25%					
						n2α(35%)		(n)0.3	96 †					
Be				9.012182						7.6 ± 0.8 mb				
${}_4Be^6$	6	4		6.019725							0+			
${}_4Be^7$	7	4		7.016928	53.29d	EC	0.862				3/2-		0.47759	10.35%
${}_4Be^8$	8	4		8.005305	0.067 fs	2α		0.046			0+			
${}_4Be^9$	9	4	100%	9.012182						7.6 ± 0.8 mb	3/2-	-1.1776		
${}_4Be^{10}$	10	4		10.013534	1.6 x 10^6y	β-	0.556	0.555	100%		0+			
${}_4Be^{11}$	11	4		11.021658	13.8 s.	β-,β-α	11.48	11.48	61%		1/2+		1.7722	0.28%
	11	4											2.1248	33%
	11	4											2.8931	0.09%
	11	4											4.6663	2%
	11	4											5.019	0.5%
	11	4											5.8518	2.1%
	11	4											6.7905	4.5%
	11	4											7.9747	1.7%
B				10.81003										
${}_5B^8$	8	5		8.024605	0.772 s.	β+,2α		13.7(β+)	93%		2+	1.0355	ann. rad.	
${}_5B^9$	9	5		9.013328	0.85 as	p2α								
${}_5B^{10}$	10	5	19.8%	10.012937	σ_α3837b						3+	+1.8007		
${}_5B^{11}$	11	5	80.2%	11.009305						5 ± 3 mb	3/2-	+2.6886		
${}_5B^{12}$	12	5		12.014352	0.0202 s.	β-	13.369				1+	+1.0031	4.439	1.3%
						β-α(1.6%)								
${}_5B^{13}$	13	5		13.01778	0.0173 s.	β-	13.436	13.4			3/2-	+3.17778	3.68	7.6%
	13	5				β-n(.25%)		2.43(n)	0.09%					
	13	5						3.55(n)	0.16%					

TABLE OF THE ISOTOPES (Continued)

Isotope	A	Z	% Natural abundance	Atomic mass	Half-life	Decay mode	Decay energy (MeV)	Particle energy (MeV)	Particle intensity	Thermal neutron cross section	Spin (h/2π)	μ Nucl. mag. moment	Gamma-ray energy (MeV)	Gamma-ray intensity
C				12.0111						3.5 mb.				
$_6C^9$	9	6		9.031039	127 ms.	β+,p,2α	16.497						ann.rad.	
$_6C^{10}$	10	6		10.01686	19.3 s.	β+	3.650	1.865			0+		ann. rad.	100%
	10	6											0.71829 ± 0.0001	98.5%
	10	6											1.02178 ± 0.0002	1.5%
$_6C^{11}$	11	6		11.01143	20.3 m.	β+,E.C.	1.982	0.9608	99%		3/2-	-0.964	ann.rad.	99+%
$_6C^{12}$	12	6	98.90%	12.000000						3.5 mb.	0+			
$_6C^{13}$	13	6	1.10%	13.003355						1.4 mb.	1/2-	+0.70241		
$_6C^{14}$	14	6		14.003241	5730 y.	β-	0.15648	0.1565	100%		0+			
$_6C^{15}$	15	6		15.010599	2.45 s.	β-	9.772	4.51	68%		1/2+		5.29887 ± 0.0001	68%
	15	6						9.82	32%				7.3011 ± 0.0005	0.008%
	15	6											8.3129 ± 0.001	0.032%
	15	6											9.0500 ± 0.001	0.031%
$_6C^{16}$	16	6		16.014701	0.75 s.	β-,n	8.012							
N				14.0067						1.91 b.				
$_7N^{12}$	12	7		12.018613	11.00 ms.	β+,β+α	17.338	16.38	95%		1+		ann.rad.	
	12	7											4.4389	2.1%
$_7N^{13}$	13	7		13.005738	9.97 m.	β+	2.2205	1.190	100%		1/2-	0.32224		
$_7N^{14}$	14	7	99.63%	14.003074						1.83 b.	1+	+0.40376		
$_7N^{15}$	15	7	0.37%	15.000108						0.02 mb.	1/2-	-0.28319		
$_7N^{16}$	16	7		16.006099	7.13 s.	β-,α	10.4187	4.27 β	68%		2-		6.129170 ± 0.0004	69%
	16	7				β-		10.44 β	26%				7.11515 ± 0.0001	5%
	16	7				α		1.7					2.75	1%
$_7N^{17}$	17	7		17.008450	4.17 s.	β-,β-n	8.680	3.7	100%		1/2-		0.871	3%
	17	7						0.4-1.7n	95%				2.1842 ± 0.	0.3%
$_7N^{18}$	18	7		18.014081	0.63 s.	β-	14.057	9.4	100%				0.82	60%
	18	7											1.65	60%
	18	7											1.982	100%
	18	7											2.47	40%
$_7N^{19}$	19	7		19.017040	0.42 s.	β-	12.53						2.47	
O				15.9994						0.28 mb.				
$_8O^{13}$	13	8		13.02810	8.9 ms.	β+,p	17.77	1.56 (p)					ann. rad.	
$_8O^{14}$	14	8		14.008595	70.6 s.	β+	5.1430	1.81	99%		0+		ann.rad.	
	14	8											2.31264 ± 0.0001	99%
$_8O^{15}$	15	8		15.003065	122 s.	β+	2.754	1.723	100%		1/2-	0.7189	ann.rad.	
$_8O^{16}$	16	8	99.762%	15.994915						0.19 mb.	0+			
$_8O^{17}$	17	8	0.038%	16.999131						0.4 mb.	5/2+	-1.89379		
$_8O^{18}$	18	8	0.200%	17.999160						0.16 mb.	0+			
$_8O^{19}$	19	8		19.003577	26.9 s.	β-	4.819	3.25	60%		5/2+		0.197	100%
	19	8						4.60	40%				1.3569 ± 0.001	55%
	19	8											1.4437 ± 0.001	3%
	19	8											1.5544 ± 0.001	1%
$_8O^{20}$	20	8		20.004075	13.5 s.	β-	3.82				0+		1.057	
$_8O^{21}$	21	8		21.008730	3.14 s.	β-	8.17						(0.21 - 4.2)	
F				18.998403						9.6 mb.				
$_9F^{17}$	17	9		17.002095	64.7 s.	β+	2.761	1.75			5/2+	+4.7223	ann.rad.	
$_9F^{18}$	18	9		18.000937	109.8 m.	β+,E.C.	1.655	0.635	97%				ann.rad.	
$_9F^{19}$	19	9	100%	18.998403						9.6 mb.	1+	+2.62887		
$_9F^{20}$	20	9		19.999981	11.0 s.	β-	7.029	5.398	100%		2+	+2.0935	1.6326 ± 0.0008	100%
	20	9											3.3343 ± 0.0007	0.02%
$_9F^{21}$	21	9		20.999948	4.33 s.	β-	5.686	3.7	8%		5/2+		0.3505 ± 0.0005	71%
	21	9						5.0	63%				1.3951 ± 0.0003	7%
	21	9						5.4	29%				1.746	
$_9F^{22}$	22	9		22.003030	4.23 s.	β-	10.85	3.48	15%		4+		1.2746 ± 0.0003	100%
	22	9						4.67	7%				1.9000 ± 0.0006	8.7%
	22	9						5.50	62%				2.0826 ± 0.0005	82%
	22	9											2.1661 ± 0.0005	62%
	22	9											2.2839 ± 0.0007	5.1%
	22	9											2.9877 ± 0.0009	7.0%
	22	9											3.9835 ± 0.001	1.2%
	22	9											4.2479 ± 0.001	1.0%
	22	9											4.3661 ± 0.001	11.3%
$_9F^{23}$	23	9		23.003600	2.2 s.	β-	8.51				5/2+		0.49289 ± 0.0007	6%
	23	9											0.81523 ± 0.0005	12%
	23	9											1.01672 ± 0.0005	10%
	23	9											1.70144 ± 0.0001	48%
	23	9											1.82225 ± 0.0002	24%
	23	9											1.91932 ± 0.0005	9%
	23	9											2.12877 ± 0.000	34%
	23	9											2.41434 ± 0.0004	7.5%
	23	9											2.73424 ± 0.0005	6%
	23	9											3.43139 ± 0.0004	12%
	23	9											3.83071 ± 0.0004	3.3%
Ne				20.179						40 mb.				
$_{10}Ne^{17}$	17	10		17.017690	109 ms.	β+,p	14.526	1.4-10.6			1/2-		ann.rad.	
$_{10}Ne^{18}$	18	10		18.005710	1.67 s.	β+	4.45	3.416	92%		0+		ann.rad.	
	18	10											0.659 ± 0.001	0.14%
	18	10											1.0413 ± 0.001	7.8%
$_{10}Ne^{19}$	19	10		19.001879	17.22 s.	β+	3.238	2.24	99%		1/2+	-1.887	ann.rad.	
	19	10											1.35692 ± 0.0001	0.0019%

TABLE OF THE ISOTOPES (Continued)

Isotope	A	Z	% Natural abundance	Atomic mass	Half-life	Decay mode	Decay energy (MeV)	Particle energy (MeV)	Particle intensity	Thermal neutron cross section	Spin (h/2π)	μ Nucl. mag. moment	Gamma-ray energy (MeV)	Gamma-ray intensity
$_{10}Ne^{20}$	20	10	90.51%	19.992435							0+			
$_{10}Ne^{21}$	21	10	0.21%	20.993843						<1.5 b.	3/2+	-0.66179		
$_{10}Ne^{22}$	22	10	9.22%	21.991383						48 mb.	0+			
$_{10}Ne^{23}$	23	10		22.994465	37.2 s.	β-	4.376	3.95	32%		5/2+	-1.08	0.440 ± 0.001	33%
	23	10						4.39	67%				1.639 ± 0.003	1%
$_{10}Ne^{24}$	24	10		23.993613	3.38 m.	β-	2.468	1.10	8%		0+		0.4722 ± 0.0001	100%
	24	10						1.98	92%				0.87435 ± 0.0001	9%
$_{10}Ne^{25}$	25	10		24.997690	0.61 s.	β-	7.2	6.3			1/2+		0.0885 ± 0.0001	96%
	25	10						7.3					0.97977 ± 0.0001	20%
Na				22.98997						0.53 b.				
$_{11}Na^{19}$	19	11		19.013879	0.03 s.	β+,p	11.18							
$_{11}Na^{20}$	20	11		20.007344	0.446 s.	β+	13.89				2+	+0.3694	ann.rad.	
	20	11				α		2.15					1.633	
$_{11}Na^{21}$	21	11		20.997650	22.5 s.	β+	3.547	2.50	95%		3/2+	+2.3863	ann.rad.	
	21	11											0.351	5.1%
$_{11}Na^{22}$	22	11		21.994434	2.605 y.	β+(90%)	2.842	0.545	90%		3+	+1.746	ann.rad.	
	22	11				E.C.(10%)							1.2745 ± 0.00005	99.9%
$_{11}Na^{23}$	23	11	100%	22.989767						0.53 b.	3/2+	+2.21752		
$_{11}Na^{24m}$	24	11			20.2 ms.	I.T.,β-					1+		0.4723	100%
$_{11}Na^{24}$	24	11		23.990961	14.97 h.	β-	5.514	1.389	>99%		4+	+1.6903	1.3686 ± 0.0003	100%
	24	11											2.7541 ± 0.0003	100%
	24	11											3.8672 ± 0.0003	0.06%
$_{11}Na^{25}$	25	11		24.989953	59.3 s.	β-	3.833	2.6	7%		5/2+	+3.683	0.38966 ± 0.0001	13.5%
	25	11						3.15	25%				0.58506 ± 0.0001	13%
	25	11						4.0	65%				0.9752 ± 0.0004	15%
	25	11											1.3797 ± 0.0005	0.003%
	25	11											1.6117	0.1%
$_{11}Na^{26}$	26	11		25.992586	1.07 s.	β-	9.31							
$_{11}Na^{27}$	27	11		26.993940	0.29 s.	β-	8.96	7.95					0.98477 ± 0.0002	86%
	27	11				β-,n							1.6985 ± 0.0005	14%
$_{11}Na^{28}$	28	11		27.978780	30 ms.	β-	13.9	12.3			1+		1.475 ± 0.01	30%
	28	11				β-,n							2.380 ± 0.02	16%
$_{11}Na^{29}$	29	11		29.002830	43 ms.	β-,n	13.4	11.5					1.510 ± 0.02	7%
	29	11											2.100 ± 0.02	4%
	29	11											2.570 ± 0.03	12%
	29	11											3.16 ± 0.04	3%
$_{11}Na^{30}$	30	11		30.008800	53 ms.	β-	18.1							
$_{11}Na^{31}$	31	11		31.012680	17 ms.	15.7								
	31	11				β-,n								
Mg				24.305						63 mb.				
$_{12}Mg^{20}$	20	12		20.018864	0.1 s.	β+,p	10.73							
$_{12}Mg^{21}$	21	12		21.011716	122 ms.	β+,p	13.10				5/2+			
$_{12}Mg^{22}$	22	12		21.999574	3.86 s.	β+	4.788	3.05			0+		0.0739 ± 0.001	60%
	22	12											0.5830 ± 0.007	100%
	22	12											1.2797 ± 0.001	6%
	22	12											1.9347 ± 0.002	9%
$_{12}Mg^{23}$	23	12		22.994124	11.32 s.	β+	4.058	3.09	92%		3/2+		0.438 ± 0.001	9%
$_{12}Mg^{24}$	24	12	78.99%	23.985042						53.mb.	0+			
$_{12}Mg^{25}$	25	12	10.00%	24.985837						18 mb.	5/2+	-0.85545		
$_{12}Mg^{26}$	26	12	11.01%	25.982593						36 mb.	0+			
$_{12}Mg^{27}$	27	12		26.984341	9.45 m.	β-	2.610	1.59	41%		1/2+		0.17068 ± 0.00001	1%
	27	12						1.75	58%				0.84376 ± 0.00003	73%
	27	12						2.65	0.3%				1.01443 ± 0.0004	30%
$_{12}Mg^{28}$	28	12		27.983876	21.0 h.	β-	1.832	0.459	95%		0+		0.0306 ± 0.0006	95%
	28	12											0.4006 ± 0.0002	38%
	28	12											0.9417 ± 0.0004	38%
	28	12											1.3422 ± 0.0002	57%
	28	12											1.3726 ± 0.0002	5%
	28	12											1.5894 ± 0.0004	5%
$_{12}Mg^{29}$	29	12		28.98848	1.3 s.	β-	7.49	5.4			3/2+		0.9603 ± 0.0005	52 +
	29	12											1.3981 ± 0.001	64
	29	12											1.4300 ± 0.001	34
	29	12											1.7539 ± 0.0007	22
	29	12											2.2237 ± 0.0004	100
	29	12											2.865 ± 0.001	1
$_{12}Mg^{30}$	30	12		29.990230	0.33s.	β-	6.10				0+			
$_{12}Mg^{31}$	31	12		30.995930	0.25 s.	β-	11.2							
Al				26.98154						233 mb.				
$_{13}Al^{22}$	22	13		22.079370	70 ms.	β+	18.5						ann.rad.	
	22	13				β+,p								
$_{13}Al^{23}$	23	13		23.007265	0.47 s.	β+	12.24						ann.rad.	
	23	13				β+,p								
$_{13}Al^{24}$	24	13		23.999941	2.07 s.	β+	13.87	3.40	48%		4+		1.078 ± 0.002	16%
	24	13						4.42	41%				1.368 ± 0.002	96%
	24	13						6.80	3%				2.753 ± 0.002	43%
	24	13						8.74	8%				3.205 ± 0.002	4%
	24	13											3.505 ± 0.002	2%
	24	13											3.886 ± 0.002	6%
	24	13											4.200 ± 0.002	4.4%

Isotope	A	Z	% Natural abundance	Atomic mass	Half-life	Decay mode	Decay energy (MeV)	Particle energy (MeV)	Particle intensity	Thermal neutron cross section	Spin (h/2π)	μ Nucl. mag. moment	Gamma-ray energy (MeV)	Gamma-ray intensity
	24	13											4.237 ± 0.002	3.6%
	24	13											4.315 ± 0.003	15%
	24	13											4.640 ± 0.003	3.6%
	24	13											5.177 ± 0.003	1%
	24	13											5.392 ± 0.003	20%
	24	13											7.0662 ± 0.002	41%
	24	13											7.928 ± 0.003	1.4%
$_{13}Al^{25}$	25	13		24.990429	7.17 s.	β+	4.277	3.27			5/2+	3.6455	ann.rad.	
	25	13											1.6115 ± 0.0002	100 +
	25	13											0.975 ± 0.002	5
$_{13}Al^{26m}$	26	13			6.34 s.	β+		3.2			0+		ann.rad.	
$_{13}Al^{26}$	26	13		25.986892	7.2 x 10^5y	β+ (82%)	4.005	1.16			5+		ann.rad.	
	26	13				E.C. (18%)							1.80865 ± 0.00007	99.8%
	26	13											1.12967 ± 0.0001	2.5%
	26	13											2.938 ± 0.002	.24%
$_{13}Al^{27}$	27	13	100%							233 mb.	5/2+	+3.64150		
$_{13}Al^{28}$	28	13		27.981910	2.25 m.	β-	4.642	2.865	100%		3+	2.791	1.7778 ± 0.0006	100%
$_{13}Al^{29}$	29	13		28.980446	6.5 m.	β-	3.68	1.4	30%		5/2+		1.2732 ± 0.0008	89%
	29	13						2.5	70%				2.0282 ± 0.0008	4%
	29	13											2.4262 ± 0.0008	7%
$_{13}Al^{30}$	30	13		29.982940	3.69 s.	β-	8.54	5.05					1.26313 ± 0.00003	35%
	30	13											1.3115 ± 0.0006	2.5%
	30	13											1.7330 ± 0.0005	2%
	30	13											2.23525 ± 0.00005	65%
	30	13											2.5951 ± 0.0005	5%
$_{13}Al^{31}$	31	13		30.983800	0.64 s.	β-	7.9	6.25					0.6281 ± 0.0003	10 +
	31	13											0.75223 ± 0.0003	18
	31	13											1.56449 ± 0.0003	17
	31	13											1.69473 ± 0.0003	59
	31	13											2.31664 ± 0.0004	73
	31	13											2.7876 ± 0.0005	4
Si				28.0855						171 mb.				
$_{14}Si^{24}$	24	14		24.011546	0.1 ms.	β+,p	10.82						ann.rad.	
$_{14}Si^{25}$	25	14		25.004109	220 ms.	β+,p	12.74						ann.rad.	
$_{14}Si^{26}$	26	14		25.992330	2.20 s.	β+	5.065	3.282			0+		ann.rad.	
	26	14											0.8294 ± 0.0008	22%
	26	14											1.6223 ± 0.001	2.6%
	26	14											1.8442 ± 0.002	0.3%
$_{14}Si^{27}$	27	14		26.986704	4.14 s.	β+	4.812	3.85	100%		5/2+		ann.rad.	
	27	14					1.45						2.211 ± 0.005	0.2%
$_{14}Si^{28}$	28	14	92.23%	27.976927						171 mb.	0+			
$_{14}Si^{29}$	29	14	4.67%	28.976495						0.1 b.	1/2+	-0.5553		
$_{14}Si^{30}$	30	14	3.10%	29.973770						0.107 b.	0+			
$_{14}Si^{31}$	31	14		30.975362	2.62 h.	β-	1.49	1.471	99.9%		3/2+		1.2662 ± 0.0005	0.07%
$_{14}Si^{32}$	32	14		31.974148	≈100 y.	β-	0.227	0.213	100%		0+			
$_{14}Si^{33}$	33	14		32.977920	6.2 s.	5.77	3.92						1.4313 ± 0.0005	13 +
	33	14											1.84769 ± 0.0005	100
	33	14											2.538 ± 0.002	10
$_{14}Si^{34}$	34	14		33.97636	2.8 s.	β-	4.7	3.09			0+		0.42907 ± 0.0005	60 +
	34	14											1.17852 ± 0.0002	64
	34	14											1.60756 ± 0.0005	36
P		15		30.97376						0.180 b.				
$_{15}P^{26}$	26	15		26.012080	≈20 ms.	β+,p								
$_{15}P^{28}$	28	15		27.992313	270 ms.	β+	14.33	3.94	13%		3+		ann.rad.	
	28	15						5.25	13%				1.779 ± 0.002	98%
	28	15						6.96	16%				2.839 ± 0.002	2.8%
	28	15						8.8	7%				3.040 ± 0.002	3.2%
	28	15						11.49	52%				4.498 ± 0.002	12%
	28	15											6.021 ± 0.002	1.9%
	28	15											6.810 ± 0.002	2%
	28	15											7.537 ± 0.002	9%
	28	15											7.933 ± 0.002	2%
$_{15}P^{29}$	29	15		28.981803	4.14 s.	β+	4.944	3.945	98%		1/2+	1.2349	ann.rad.	
	29	15											1.28	0.8%
	29	15											2.43	0.2%
$_{15}P^{30}$	30	15		29.978307	2.50 m.	β+	4.226	3.245	99.9%		1+		ann.rad.	
	30	15											2.230 ± 0.003	0.07%
$_{15}P^{31}$	31	15	100%	30.973762						0.233 b.	1/2+	+1.13160		
$_{15}P^{32}$	32	15		31.973907	14.28 d.	β-	1.710	1.710	100%		1+	-0.2524		
$_{15}P^{33}$	33	15		32.971725	25.3 d.	β-	0.249	0.249	100%		1/2+			
$_{15}P^{34}$	34	15		33.973636	12.4 s.	β-	5.37	3.2	15%		1+		1.78 -4.1 (weak)	
	34	15						5.1	85%				2.127 ± 0.005	15%
$_{15}P^{35}$	35	15		34.973232	47.s.	β-	3.9	2.34	100%				1.5722 ± 0.001	100%
$_{15}P^{36}$	36	15		35.977570	5.9 s.	β-	9.8						0.902	
	36	15											3.291	
S		16		32.066						0.52 b.				
$_{16}S^{29}$	29	16		28.996610	0.19 s.	β+	13.79				5/2+		ann.rad.	
	29	16				β+,p								

Isotope	A	Z	% Natural abundance	Atomic mass	Half-life	Decay mode	Decay energy (MeV)	Particle energy (MeV)	Particle intensity	Thermal neutron cross section	Spin (h/2π)	μ Nucl. mag. moment	Gamma-ray energy (MeV)	Gamma-ray intensity
$_{16}S^{30}$	30	16		29.984903	1.18 s.	β+	6.144	4.42	78%		0+		ann.rad.	
	30	16						5.08	20%				0.678	79%
$_{16}S^{31}$	31	16		30.979554	2.55 s.	β+	5.396	4.39	99%		1/2+	0.48793	ann.rad.	
	31	16											1.2662 ± 0.0005	1.2%
	31	16											3.135 ± 0.005	0.03%
	31	16											3.505 ± 0.005	0.01%
$_{16}S^{32}$		16	95.02%	31.972070						0.52 b.	0+			
$_{16}S^{33}$	33	16	0.75%	32.971456						1.4 mb.	3/2+	+0.64382		
$_{16}S^{34}$	34	16	4.21%	33.967866						0.2 b.	0+			
$_{16}S^{35}$	35	16		34.969031	87.2 d.	β-	0.167	0.1674	100%		3/2+	+1.00		
$_{16}S^{36}$	36	16	0.02%	35.967080						0.23 b.	0+			
$_{16}S^{37}$	37	16		36.971125	5.05 m.	β-	4.865	1.64	94%				0.9083 ± 0.0004	0.06%
	37	16						4.75	5.6%				3.1033 ± 0.00002	94.2%
	37	16											3.7416 ± 0.0005	0.2%
$_{16}S^{38}$	38	16		37.971162	2.84 h.	β-	2.94	1.00			0+		0.1962 ± 0.0004	0.2%
	38	16											1.8459 ± 0.0005	2.4%
	38	16											1.9421 ± 0.0003	84%
	38	16											2.7516 ± 0.0005	1.6%
$_{16}S^{39}$	39	16		38.975310	11.5 s.	β-	6.8						1.301	
	39	16											1.697	
Cl		17		35.453						33.5 b.				
$_{17}Cl^{31}$	31	17		30.992410	0.15 s.	β+	12.0						ann.rad.	
	31	17				β+,p								
$_{17}Cl^{32}$	32	17		31.985690	297 ms.	β+	12.69	4.75	25%		1+		ann.rad.	
	32	17						6.18	10%				1.548 ± 0.002	3.5%
	32	17						7.48	14%				2.2305 ± 0.001	92%
	32	17						9.47	50%				2.4638 ± 0.001	4%
	32	17						11.6	1%				2.885 ± 0.001	1%
	32	17											3.3175 ± 0.001	2.4%
	32	17											4.281 ± 0.001	2.5%
	32	17											4.433 ± 0.001	0.8%
	32	17											4.694 ± 0.001	2.7%
	32	17											4.770 ± 0.001	20%
	32	17											5.549 ± 0.002	1.6%
	32	17											7.194 ± 0.003	0.4%
$_{17}Cl^{33}$	33	17		32.977451	2.51 s.	β+	5.583	4.51	98%		3/2+		ann.rad.	
	33	17											0.8405 ± 0.001	0.56%
	33	17											1.9661 ± 0.0005	0.56%
	33	17											2.8665 ± 0.0005	0.56%
$_{17}Cl^{34m}$	34	17			32.2 m.	β+		1.35	24%		3+		ann.rad.	
	34	17						2.47	28%					
	34	17				I.T.							0.1457 ± 0.0008	42%
	34	17											1.1758 ± 0.0005	10%
	34	17											2.1276 ± 0.0005	42%
	34	17											3.3037 ± 0.001	10%
$_{17}Cl^{34}$	34	17		33.973763	1.53 s.	β+	5.492	4.50	100%		0+		ann.rad.	
$_{17}Cl^{35}$	35	17	75.77%	34.968852						43.6 b.	3/2+	+0.82187		
$_{17}Cl^{36}$	36	17		35.968306	3 x 10^5y.	β-	0.7093	0.7093	98%		2+	+1.28547		
	36	17				β+,E.C.	1.142	0.115	0.002%				ann.rad.	
$_{17}Cl^{37}$	37	17	24.23%	36.965903						0.4 b.	3/2+	+0.68412		
$_{17}Cl^{38m}$	38	17			0.70 s.	I.T.					5-		0.67138 ± 0.0002	100%
$_{17}Cl^{38}$	38	17		37.968010	37.2 m.	β-	4.917	1.11	31%		2-	2.05	1.64216 ± 0.0001	31%
	38	17						2.77	11%				2.16760 ± 0.0002	42%
	38	17						4.91	58%					
$_{17}Cl^{39}$	39	17		38.968005	55.7 m.	β-	3.44	1.91	85%		3/2+		0.25026 ± 0.0001	47%
	39	17						2.18	8%				0.98579 ± 0.0001	3.2%
	39	17						3.45	7%				1.09097 ± 0.0001	4.0%
	39	17											1.26720 ± 0.0005	54%
	39	17											1.51736 ± 0.0008	38%
$_{17}Cl^{40}$	40	17		39.970440	1.35 m.	β-	7.5				2-		0.6431 ± 0.0003	6%
	40	17											0.6591 ± 0.0002	2%
	40	17											0.8810 ± 0.0003	2.4%
	40	17											1.0629 ± 0.0002	2.4%
	40	17											1.4608 ± 0.0001	77%
	40	17											1.5889 ± 0.0002	7%
	40	17											1.7465 ± 0.0002	2.5%
	40	17											1.7978 ± 0.0002	2.4%
	40	17											2.0505 ± 0.0004	1%
	40	17											2.2201 ± 0.0002	7.5%
	40	17											2.4579 ± 0.0002	5%
	40	17											2.6219 ± 0.0002	14%
	40	17											2.8402 ± 0.0002	17%
	40	17											3.1010 ± 0.0002	11%
	40	17											3.9186 ± 0.0002	4%
	40	17											5.8799 ± 0.0008	4%
$_{17}Cl^{41}$	41	17		40.970590	34 s.	β-	5.67	3.8					(0.167 - 1.359)	
Ar		18		39.948						0.66 b.				
$_{18}Ar^{32}$	32	18		31.997660	≈0.1 s.	β+,p	11.2						ann. rad.	
$_{18}Ar^{33}$	33	18		32.989930	17 ms.	β+	11.62	3.12			1/2+		ann.rad.	
	33	18				β+,p							0.810 ± 0.002	48%
$_{18}Ar^{34}$	34	18		33.980269	0.844 s.	β+	6.061	5.0	95%		0+		ann.rad.	
	34	18											0.4608 ± 0.001	0.8%

TABLE OF THE ISOTOPES (Continued)

Isotope	A	Z	% Natural abundance	Atomic mass	Half-life	Decay mode	Decay energy (MeV)	Particle energy (MeV)	Particle intensity	Thermal neutron cross section	Spin (h/2π)	μ Nucl. mag. moment	Gamma-ray energy (MeV)	Gamma-ray intensity
	34	18											0.6658 ± 0.001	2.5%
	34	18											2.5795 ± 0.001	0.8%
	34	18											3.1290 ± 0.001	1.3%
$_{18}Ar^{35}$	35	18		34.975256	1.77 s.	β+	5.965	4.94	93%		3/2+	+0.633	ann.rad.	
	35	18											1.2185 ± 0.0005	1.22%
	35	18											1.763 ± 0.001	0.25%
	35	18											2.964 ± 0.001	0.2%
	35	18											3.003 ± 0.005	0.1%
$_{18}Ar^{36}$	36	18	0.337%	35.967545						5.5 mb.	0+			
$_{18}Ar^{37}$	37	18		36.966776	34.8 d.	E.C.					3/2+	+0.95		
$_{18}Ar^{38}$	38	18	0.63%	37.962732							0+			
$_{18}Ar^{39}$	39	18		38.964314	269 y.	β-	0.565	0.565	100%	600 b.	7/2-	-1.3		
$_{18}Ar^{40}$	40	18	99.60%	39.962384						0.65 b.	0+			
$_{18}Ar^{41}$	41	18		40.964501	1.83 h.	β-	2.492	1.198		0.5 b.	7/2-		1.29364 ± 0.00005	99%
	41	18											1.6770 ± 0.0003	0.05%
$_{18}Ar^{42}$	42	18		41.963050	33 y.	β-	0.60	0.60	100%		0+			
$_{18}Ar^{43}$	43	18		42.965670	5.4 m.	β-	4.6						0.4791 ± 0.002	10 +
	43	18											0.7380 ± 0.0001	43
	43	18											0.9752 ± 0.0001	100
	43	18											1.4400 ± 0.0003	39
	43	18											2.3455 ± 0.0005	28
$_{18}Ar^{44}$	44	18		43.96365	11.9 m.	β-	3.54				0+		0.182	
	44	18											1.703	
	44	18											1.866	
$_{18}Ar^{45}$	45	18		44.968090	21 s.	β-	6.9				7/2-		0.0610	
	45	18											1.020	
	45	18											3.707	
$_{18}Ar^{46}$	46	18		45.968090	8.3 s.	β-	5.70				0+		1.944	
K		19		39.0983						2.1 b.				
$_{19}K^{35}$	35	19		34.988011	0.19 s.	β+	11.88				3/2+		ann.rad.	
	35	19				β+,p							1.751	
	35	19											2.5698	
	35	19											2.9827	
$_{19}K^{36}$	36	19		35.981293	0.342 s.	β+	12.81	5.3	42%		2+	0.548	ann.rad.	
	36	19						9.9	44%				1.97044 ± 0.0005	82%
	36	19											2.17029 ± 0.0002	3%
	36	19											2.20783 ± 0.0005	30%
	36	19											2.43343 ± 0.0002	32%
	36	19											2.47046 ± 0.0004	5%
	36	19											4.44079 ± 0.0003	8%
	36	19											6.61213 ± 0.0004	7%
$_{19}K^{37}$	37	19		36.973377	1.23 s.	β+	6.149	5.13			3/2+	+0.20321	ann.rad.	
	37	19											2.7944 ± 0.0008	2%
	37	19											3.602 ± 0.002	0.05%
$_{19}K^{38m}$	38	19			0.926 s.	β+	6.742	5.02	100%		0+		ann.rad.	
$_{19}K^{38}$	38	19		37.969080	7.63 m.	β+	5.913	2.60	99.8%		3+	+1.374	ann.rad.	
	38	19											2.1675 ± 0.0003	99.8%
	38	19											3.9356 ± 0.0005	0.2%
$_{19}K^{39}$	39	19	93.2581%	38.963707						2.10 b.	3/2+	+0.39146		
$_{19}K^{40}$	40	19	0.0117%	39.963999	1.25x10^{9}y.	β-	1.32	1.312	89%		4-	-1.298	ann.rad.	
	40	19				β+,E.C.	1.50		10.7%				1.46081 ± 0.00005	10.5%
$_{19}K^{41}$	41	19	6.7302%	40.961825						1.46 b.	3/2+	+0.21487		
$_{19}K^{42}$	42	19		41.962402	12.36 h.	β-	3.523	1.97	19%		2-	-1.1425	0.31260 ± 0.0002	0.3%
	43	19						3.523	81%				1.5246 ± 0.0003	18.9%
$_{19}K^{43}$	43	19		42.960717	22.3 h.	β-	1.82	0.465	8%		3/2+	0.163	0.2211 ± 0.0002	4%
	43	19						0.825	87%				0.3729 ± 0.0002	88%
	43	19						1.24	3.5%				0.3971 ± 0.0002	11%
	43	19						1.814	1.3%				0.6178 ± 0.0002	81%
$_{19}K^{44}$	44	19		43.96156	22.1 m.	β-	5.66	5.66	34%		2-		0.368207 ± 0.0001	2.2%
	44	19											0.65135 ± 0.00001	2.0%
	44	19											0.72649 ± 0.00001	2.4%
	44	19											0.74763 ± 0.00001	1.2%
	44	19											0.87653 ± 0.00003	1.1%
	44	19											1.01955 ± 0.00007	5.5%
	44	19											1.02474 ± 0.00002	6%
	44	19											1.10799 ± 0.00001	5%
	44	19											1.12608 ± 0.00001	7%
	44	19											1.15700 ± 0.00001	58%
	44	19											1.24475 ± 0.00005	8%
	44	19											1.75263 ± 0.00001	4%

Isotope	A	Z	% Natural abundance	Atomic mass	Half-life	Decay mode	Decay energy (MeV)	Particle energy (MeV)	Particle intensity	Thermal neutron cross section	Spin (h/2π)	μ Nucl. mag. moment	Gamma-ray energy (MeV)	Gamma-ray intensity
	44	19											1.77797 ± 0.00002	2%
	44	19											2.14423 ± 0.00008	7%
	44	19											2.15079 ± 0.00002	22%
	44	19											2.51899 ± 0.00002	9%
	44	19											2.65641 ± 0.00003	4.6%
	44	19											3.39551 ± 0.00004	1.6%
	44	19											3.66136 ± 0.00002	6%
$_{19}K^{45}$	45	19		44.960696	17.3 m.	β-	4.20	1.1	23%		3/2 +	0.1734	0.1743 ± 0.0005	80%
	45	19						2.1	69%				1.2607 ± 0.0008	7%
	45	19						4.0	8%				1.7056 ± 0.0006	69%
	45	19											2.3542 ± 0.0005	14%
	45	19											2.5988 ± 0.001	3%
$_{19}K^{46}$	46	19		45.961976	107 s.	β-	7.72	6.3			2-		1.347 ± 0.001	91%
	45	19											1.439 ± 0.002	3%
	45	19											1.670 ± 0.002	3%
	45	19											1.780 ± 0.002	8%
	45	19											2.274 ± 0.002	8%
	45	19											3.015 ± 0.005	9%
	45	19											3.700 ± 0.005	28%
$_{19}K^{47}$	47	19		46.961677	17.5 s.	β-	6.65	4.1	99%		1/2 +		0.56474 ± 0.0003	15%
	47	19						6.0	1%				0.58575 ± 0.0003	85%
	47	19											2.01313 ± 0.0003	100%
$_{19}K^{48}$	48	19		47.965514	69 s.	β-	12.1	5.0			(2-)		0.67122 ± 0.0001	4%
	48	19											0.6723 ± 0.00005	20%
	48	19											0.78016 ± 0.0001	32%
	48	19											0.7931 ± 0.0001	10%
	48	19											0.86275 ± 0.0001	4.5%
	48	19											1.3009 ± 0.0002	9%
	48	19											1.5378 ± 0.0001	15%
	48	19											2.2830 ± 0.0003	3%
	48	19											2.3881 ± 0.0001	11%
	48	19											2.7889 ± 0.0001	18%
	48	19											3.06229 ± 0.0003	5%
	48	19											3.83153 ± 0.00007	80%
	48	19											4.5072 ± 0.0003	4%
	48	19											6.6137 ± 0.0005	14%
	48	19											7.3009 ± 0.0005	2%
$_{19}K^{49}$	49	19		48.966940	1.3 s.	β-	11						2.025	
	49	19											2.252	
$_{19}K^{50}$	50	19			≈0.7 s.	β-	16							
$_{19}K^{51}$					0.38 s.	β-								
Ca		20		40.078						0.43 b.				
$_{20}Ca^{36}$	36	20		35.993090	0.1 s.	β+	11.0						ann.rad.	
	36	20				β+,n								
$_{20}Ca^{37}$	37	20		36.985873	175 ms.	β+	11.6				3/2 +		ann.rad.	
	37	20				β+,n								
$_{20}Ca^{38}$	38	20		37.976318	0.45 s.	β+	6.74				0+		ann.rad.	
	38	20											1.5677 ± 0.0005	25%
	38	20											3.210 ± 0.002	1%
$_{20}Ca^{39}$	39	20		38.970718	0.86 s.	β+	6.531	5.49	100%		3/2 +	1.02168	ann.rad.	
$_{20}Ca^{40}$	40	20	96.941%	39.962591						0.41 b.	0+			
$_{20}Ca^{41}$	41	20		40.962278	1 x 10^5 y.	E.C.	0.421				7/2-	-1.595		
$_{20}Ca^{42}$	42	20	0.647%	41.958618						0.7 b.	0+			
$_{20}Ca^{43}$	43	20	0.135%	42.958766						6 b.	7/2-	-1.3173		
$_{20}Ca^{44}$	44	20	2.086%	43.955480						0.8 b.	0+			
$_{20}Ca^{45}$	45	20		44.956185	163.8 d.	β-	0.257	0.257	100%		7/2-			
$_{20}Ca^{46}$	46	20	0.004%	45.953689						0.7 b.	0+			
$_{20}Ca^{47}$	47	20		46.954543	4.536 d.	β-	1.988	0.684	84%		7/2-		0.4889 ± 0.0003	9%
	47	20						1.98	16%				0.8079 ± 0.0003	9%
	47	20											1.29680 ± 0.0002	77%
$_{20}Ca^{48}$	48	20	0.187%	47.952533						1.1 b.	0+			
$_{20}Ca^{49}$	49	20		48.955672	8.72 m.	β-	5.263	0.89	7%		3/2-		3.0844 ± 0.0001	92%
	49	20						1.95	92%				4.0719 ± 0.0001	7%
$_{20}Ca^{50}$	50	20		49.957519	14 s.	β-	4.97	3.12			0+		0.2569	
	50	20											(0.0715 - 1.591)	
$_{20}Ca^{51}$	51	20		50.961420	10 s.	β-	0.728							
Sc		21		44.95591						27.2 b.				
$_{21}Sc^{40}$	40	21		39.977963	0.182 s.	β+	14.32	5.73	50%		4-		ann.rad.	
	40	21						7.53	15%				0.7556 ± 0.0008	41%
	40	21						8.76	15%				1.126 ± 0.003	12%
	40	21						9.58	20%				1.8778 ± 0.0007	25%
	40	21											2.0458 ± 0.0007	25%
	40	21											3.1679 ± 0.0007	12%
	40	21											3.7356 ± 0.0008	100%
	40	21											3.920 ± 0.001	13%

Isotope	A	Z	% Natural abundance	Atomic mass	Half-life	Decay mode	Decay energy (MeV)	Particle energy (MeV)	Particle intensity	Thermal neutron cross section	Spin (h/2π)	μ Nucl. mag. moment	Gamma-ray energy (MeV)	Gamma-ray intensity
$_{21}Sc^{41}$	41	21		40.96250	0.596 s.	β+	6.495	5.61	100%		7/2-	5.43	ann.rad.	
$_{21}Sc^{42m}$	42	21			61.6 s.	β+		2.82			7+		ann.rad.	
	42	21											0.4375 ± 0.0005	100%
	42	21											1.2270 ± 0.0005	100%
	42	21											1.5245 ± 0.0005	100%
$_{21}Sc^{42}$	42	21		41.965514	0.68 s.	β+	6.424	5.32	100%		0+		ann.rad.	
$_{21}Sc^{43}$	43	21		42.961150	3.89 h.	β+,E.C.	2.221	0.82	22%		7/2-	+4.62	ann.rad.	
	43	21						1.22	78%				0.3729 ± 0.0001	22%
$_{21}Sc^{44m}$	44	21			58.6 h.	I.T.	0.27				6+	+3.88	0.27124 ± 0.0001	98.4%
	44	21				E.C.	3.926						1.0018 ± 0.00003	1.4%
	44	21											1.12606 ± 0.00003	1.4%
	44	21											1.15700 ± 0.0003	1.4%
$_{21}Sc^{44}$	44	21		43.959404	3.93 h.	β+, E.C.	3.655	1.47			0+	+2.56	ann.rad.	
	44	21											1.15700 ± 0.00001	100%
	44	21											1.49945 ± 0.00002	1%
$_{21}Sc^{45}$	45	21	100%	44.955910						(0.1 + 17)b.	7/2-	+4.756		
$_{21}Sc^{46m}$	46	21			18.7 s.	I.T.	0.14253				1-		0.14253 ± 0.00002	62%
$_{21}Sc^{46}$	46	21		45.995170	83.8 d.	β-	2.367	0.357	100%		4+	+3.03	0.88925 ± 0.00003	100%
	46	21											1.12051 ± 0.00001	100%
$_{21}Sc^{47}$	47	21		46.952408	3.42 d.	β-	0.601	0.439	69%		7/2-	+5.34	0.15938 ± 0.00001	68%
	47	21						0.601	31%					
$_{21}Sc^{48}$	48	21		47.952235	43.7 h.	β-	3.99	0.655			6+		0.98350 ± 0.0001	100%
	48	21											1.03750 ± 0.00001	97%
	48	21											1.21285 ± 0.00007	2%
	48	21											1.31209 ± 0.00003	100%
$_{21}Sc^{49}$	49	21		48.950022	57.3 m.	β-	2.005	2.00	99.9%		7/2-		1.7619 ± 0.0003	0.05%
$_{21}Sc^{50}$	50	21		49.952186	1.71 m.	β-	3.05	76%			(5+)		0.5235 ± 0.0001	88%
	50	21						3.60	24%				1.1210 ± 0.0001	100%
	50	21											1.5537 ± 0.0002	100%
$_{21}Sc^{51}$	51	21		50.953602	12.4 s.	β-	6.51	4.4			7/2-		0.7177 ± 0.0004	7%
	51	21						5.0					0.9072 ± 0.0004	9%
	51	21											1.2938 ± 0.0004	6%
	51	21											1.4373 ± 0.0004	52%
	51	21											1.5675 ± 0.0004	15%
	51	21											2.0511 ± 0.0004	8%
	51	21											2.1441 ± 0.0004	31%
Ti		22		47.88						6.1 b.				
$_{22}Ti^{41}$	41	22		40.983150	80 ms.	β+,p	12.94				3/2+		ann.rad.	
$_{22}Ti^{42}$	42	22		41.973031	0.20 s.	β+	7.001	6.0					ann.rad.	
	42	22											0.6107 ± 0.0005	56%
$_{22}Ti^{43}$	43	22		42.968523	0.49 s.	β+	6.867	5.80			7/2-		ann.rad.	
$_{22}Ti^{44}$	44	22		43.959689	47 y.	E.C.	0.265				0+		0.06785 ± 0.00004	88%
	44	22											0.07838 ± 0.00004	93%
$_{22}Ti^{45}$	45	22		44.958124	3.078 h.	β+ (86%)	2.063	1.04			7/2-	0.095	ann.rad.	
	45	22				E.C. (14%)							(0.36-1.66)weak	
$_{22}Ti^{46}$	46	22	8.0%	45.952629						0.6 b.	0+			
$_{22}Ti^{47}$	47	22	7.3%	46.951764						1.7 b.	5/2-	-0.7885		
$_{22}Ti^{48}$	48	22	73.8%	47.947947						7.9 b.	0+			
$_{22}Ti^{49}$	49	22	5.5%	48.947871						2.2 b.	7/2-	-1.10417		
$_{22}Ti^{50}$	50	22	5.4%	49.944792						0.177 b.	0+			
$_{22}Ti^{51}$	51	22		50.946616	5.76 m.	β-	2.472	1.50	92%		3/2-		0.3197 ± 0.0002	93%
	51	22						2.13					0.6094 ± 0.0003	1%
	51	22											0.9291 ± 0.0003	6%
$_{22}Ti^{52}$	52	22		51.946898	1.7 m.	β-	1.97	1.8	100%		0+		0.0170 ± 0.0005	100%
	52	22											0.12445 ± 0.00005	100%
$_{22}Ti^{53}$	53	22		52.949730	33 s.	β-	5.02	(2.2-3)			3/2-		0.1008 ± 0.0001	20%
	53	22											0.1276 ± 0.0001	45%
	53	22											0.2284 ± 0.0001	39%
	53	22											0.6796 ± 0.001	4%
	53	22											1.001 ± 0.001	4%
	53	22											1.3211 ± 0.001	6%
	53	22											1.4217 ± 0.001	10%
	53	22											1.6755 ± 0.0005	45%
	53	22											(1.72-2.8)	
V		23		50.9415						5.06 b.				
$_{23}V^{44}$	44	23		43.974450	0.09 s.	β+,α	13.7						ann.rad.	
$_{23}V^{46}$	46	23		45.960198	0.422 s.	β+	7.05	6.03	100%		0+		ann.rad.	
$_{23}V^{47}$	47	23		46.954906	31.3 m.	β+,E.C.	2.927	1.90	99+%		3/2-		ann.rad.	
	47	23											1.7949 ± 0.0008	0.19%
	47	23											(0.2-2.16)weak	

TABLE OF THE ISOTOPES (Continued)

Isotope	A	Z	% Natural abundance	Atomic mass	Half-life	Decay mode	Decay energy (MeV)	Particle energy (MeV)	Particle intensity	Thermal neutron cross section	Spin (h/2π)	μ Nucl. mag. moment	Gamma-ray energy (MeV)	Gamma-ray intensity
$_{23}V^{48}$	48	23		47.952257	15.98 d.	β+	4.015	0.698	50%		4+	1.63	ann.rad.	
	48	23											0.94410 ± 0.00002	8%
	48	23											0.98350 ± 0.00002	100%
	48	23											(1.3-2.4)weak	
$_{23}V^{49}$	49	23		48.948517	331 d.	E.C.	0.601				7/2-	4.47		
$_{23}V^{50}$	50	23	0.25%	49.947161	>3.9x10^{17}y	E.C., β-				0.1 b.	6+	+3.34745		
$_{23}V^{51}$	51	23	99.75%	50.943962						4.9 b.	7/2-	+5.1514		
$_{23}V^{52}$	52	23		51.944778	3.76 m.	β-	3.976	2.47			3/2-		1.4341 ± 0.0001	100%
$_{23}V^{53}$	53	23		52.944340	1.61 m.	β-	3.436	2.52			7/2-		1.0060 ± 0.0005	90%
	53	23											1.2891 ± 0.0003	10%
$_{23}V^{54}$	54	23		53.946442	49.8 s.	β-	7.04	1.00	5%		(5+)		0.564 ± 0.002	4%
	54	23						2.00	12%				0.8351 ± 0.0001	100%
	54	23						2.95	45%				0.986 ± 0.001	82%
	54	23						5.20	11%				1.462 ± 0.002	7%
	54	23											1.784 ± 0.003	7%
	54	23											2.255 ± 0.003	50%
	54	23											2.353 ± 0.003	12%
	54	23											3.170 ± 0.005	12%
$_{23}V^{55}$	55	23		54.947240	6.5 s.	β-	6.0	6.0					0.517	
	55	23											0.8806	
Cr		24								3.1 b.				
$_{24}Cr^{45}$	45	24		44.979110	0.05 s.	β+,p	12.4				7/2-		ann.rad.	
$_{24}Cr^{46}$	46	24		45.968360	≈0.26 s.	β+	7.61						ann.rad.	
$_{24}Cr^{47}$	47	24		46.962905	460 ms.	β+	7.45						ann.rad.	
$_{24}Cr^{48}$	48	24		47.954033	21.6 h.	E.C.	1.65						ann.rad.	
	48	24											0.116 ± 0.002	95%
	48	24											0.305 ± 0.010	100%
$_{24}Cr^{49}$	49	24		48.951338	42.1 m.	β+,E.C.	2.627	1.39			5/2-	0.476	ann.rad.	
	49	24						1/45					0.06229 ± 0.00001	0.04%
	49	24						1.54					0.09064 ± 0.00001	51%
	49	24											0.15293 ± 0.00001	27%
	49	24											(0.2-1.6)weak	
$_{24}Cr^{50}$	50	24	4.35%	49.946046						15.8 b.	0+			
$_{24}Cr^{51}$	51	24		50.944768	27.70 d.	E.C.	0.751				7/2-	-0.934	0.320076 ± 0.0001	10.2%
$_{24}Cr^{52}$	52	24	83.79%	51.940509						0.8 b.	0+			
$_{24}Cr^{53}$	53	24	9.50%	52.940651						18 b.	3/2-	-0.47454		
$_{24}Cr^{54}$	54	24	2.36%	53.938882						0.36 b.	0+			
$_{24}Cr^{55}$	55	24		54.940842	3.50 m.	β-	2.603	2.5			3/2-		1.5282 ± 0.0002	0.04%
	55	24											2.2518 ± 0.0003	0.004%
$_{24}Cr^{56}$	56	24		55.940643	5.9 m.	β-	1.62	1.50	100%		0+		0.026 ± 0.002	100%
	56	24											0.083 ± 0.003	100%
$_{24}Cr^{57}$	57	24		56.943440	21 s.	β-	≈4.7	3.3			3/2-	0.0834		
	57	24						3.5					0.850	
	57	24											1.752	
Mn		25		54.9380						13.3 b.				
$_{25}Mn^{49}$	49	25		48.951338	0.38 s.	β+	7.72	6.69			5/2-		ann.rad.	
$_{25}Mn^{50m}$	50	25			1.7 m.	β+	7.887	3.54			5+		ann.rad.	
	50	25											1.0980 ± 0.0002	100%
	50	25											1.2824 ± 0.0005	33%
	50	25											1.4433 ± 0.0002	70%
	50	25											1.9445 ± 0.0005	4%
	50	25											3.1152 ± 0.001	1%
$_{25}Mn^{50}$	50	25		49.954239	0.283 s.	β+	7.632	6.61			0+		ann.rad.	
$_{25}Mn^{51}$	51	25		50.948213	46.2 m.	β+,E.C.	3.209	2.2			5/2-	3.568	ann.rad.	
	51	25											0.7491 ± 0.0001	0.26%
	51	25											1.1480 ± 0.0001	0.1%
	51	25											1.1644 ± 0.0001	0.1%
	51	25											2.00135 ± 0.001	0.05%
$_{25}Mn^{52m}$	52	25			21.1 m.	β+(98%)	5.09	2.631			2+	0.0076	ann.rad.	
	52	25				I.T.(2%)	0.378						0.3778 (I.T.)	
	52	25											1.43406 ± 0.00001	98%
	52	25											(0.7-4.8)weak	
$_{25}Mn^{52}$	52	25		51.945568	5.59 d.	β+	4.712	0.575			6+	+3.0621	ann.rad.	
	52	25				E.C.							0.74421 ± 0.00001	90%
	52	25											0.84816 ± 0.00005	3%
	52	25											1.2462 ± 0.0003	4%
	52	25											1.3336 ± 0.0001	5%
	52	25											1.43406 ± 0.00001	100%
$_{25}Mn^{53}$	53	25		52.941291	3.7x10^{6} y.	E.C.	0.596			70 b.	7/2-	5.024		
$_{25}Mn^{54}$	54	25		53.940361	312 d.	E.C.	1.377				3+	+3.2818	0.83403 ± 0.00005	100%
$_{25}Mn^{55}$	55	25	100%	54.938047						13.3 b.	5/2-	+3.4687		
$_{25}Mn^{56}$	56	25		55.938906	2.579 h.	β-	3.696	0.718	18%		3+	+3.2266	0.84675 ± 0.0002	98.9%

TABLE OF THE ISOTOPES (Continued)

Isotope	A	Z	% Natural abundance	Atomic mass	Half-life	Decay mode	Decay energy (MeV)	Particle energy (MeV)	Particle intensity	Thermal neutron cross section	Spin (h/2π)	μ Nucl. mag. moment	Gamma-ray energy (MeV)	Gamma-ray intensity
								1.028	34%				1.81072 ± 0.00004	27%
	56	25											2.11305 ± 0.00005	14.5%
	56	25											2.5229 ± 0.0005	1%
$_{25}Mn^{57}$	57	25		56.938285	1.45 m.	β-	2.691				5/2-			
$_{25}Mn^{58}$	58	25		57.940060	65 s.	β-	6.32	3.8			3+		0.45916 ± 0.0002	20%
	58	25						5.1					0.81076 ± 0.00001	82%
	58	25											0.86394 ± 0.00003	14%
	58	25											1.26574 ± 0.00005	8%
	58	25											1.32309 ± 0.00005	53%
	58	25											1.67472 ± 0.00007	10%
	58	25											2.42245 ± 0.0001	1%
	58	25											2.63815 ± 0.0001	1.2%
$_{25}Mn^{59}$	59	25		58.940440	4.6 s.	β-	5.18	4.5					0.471	
	59	25											0.531	
	59	25											0.726	
$_{25}Mn^{60}$	60	25		59.943210	1.8 s.	β-	8.5	5.7			3+		0.824	
	60	25											1.969	
$_{25}Mn^{62}$	62	25			0.9 s.	β-					(3+)		0.877	
	62	25											0.942	
	62	25											1.299	
Fe		26		55.847						2.56 b.				
$_{26}Fe^{49}$	49	26			0.08 s.	β+	13.1						ann.rad.	
$_{26}Fe^{51}$	51	26		50.956825	0.25 s.	β+	8.02						ann.rad.	
$_{26}Fe^{52m}$	52	26			46 s.	β+	4.4						ann.rad.	
	52	26											(0.622-2.286)	
$_{26}Fe^{52}$	52	26		51.948114	8.28 h.	β+(57%)	2.37	0.804			0+		ann.rad.	
	52	26				E.C.(43%)							0.16868 ± 0.00001	99%
	52	26				I.T.							0.377 (I.T.)	
$_{26}Fe^{53m}$	53	26			2.54 s.	I.T.	3.0407				19/2-		0.7011 ± 0.0001	99%
	53	26											1.0115 ± 0.0001	87%
	53	26											1.3281 ± 0.0001	87%
	53	26											2.3396 ± 0.0001	13%
$_{26}Fe^{53}$	53	26		52.945310	8.51 m.	β+	3.774	2.40	42%		7/2-		ann.rad.	
	53	26						2.80	57%				0.3779 ± 0.0001	42%
	53	26											(1.2 - 3.2)weak	
$_{26}Fe^{54}$	54	26	5.8%	53.939612						2.3 b.	0+			
$_{26}Fe^{55}$	55	26		54.938296	2.7 y.	E.C.	0.2314				3/2-			
$_{26}Fe^{56}$	56	26	91.72%	55.934939						2.6 b.	0+			
$_{26}Fe^{57}$	57	26	2.2%	56.935396						2.5 b.	1/2-	+0.09044		
$_{26}Fe^{58}$	58	26	0.28%	57.933277						1.26 b.	0+			
$_{26}Fe^{59}$	59	26		58.934877	44.51 d.	β-	1.565	0.273	48%		3/2-	0.29	0.14265 ± 0.00002	1%
	59	26						0.475	51%				0.19234 ± 0.00006	3%
	59	26											1.09922 ± 0.00002	56%
	59	26											1.29156 ± 0.00003	43%
	59	26											1.48178 ± 0.00006	0.06%
$_{26}Fe^{60}$	60	26		59.93408	≈10^5 y.	β-	0.243	0.184	100%		0+		0.0586 ± 0.0005	100%(IT)
$_{26}Fe^{61}$	61	26		60.936748	6.0 m.	β-	3.97	2.5	13%		3/2-		0.12034 ± 0.0001	4.4%
	61	26						2.63	54%				0.29790 ± 0.00007	22%
	61	26						2.80	31%				1.02742 ± 0.0001	43%
	61	26											1.20507 ± 0.0001	44%
	61	26											1.64595 ± 0.0001	7%
	61	26											2.0116 ± 0.0002	4%
$_{26}Fe^{62}$	62	26		61.936773	68 s.	β-	2.50	2.5	100%		0+		0.5061 ± 0.0001	100%
$_{26}Fe^{63}$	63	26		62.94075	4.9 s.	β-							0.995	
	63	26											1.365	
	63	26											1.427	
Co		27		58.9332						37.2 b.				
$_{27}Co^{53m}$	53	27			0.25 s.	β+,p					19/2-		ann.rad.	
$_{27}Co^{53}$	53	27		52.954225	0.26 s.	β+	8.30				7/2-		ann.rad.	
$_{27}Co^{54m}$	54	27			1.46 m.	β+	8.44	4.25	100%		7+		ann.rad.	
	54	27											0.411 ± 0.001	99%
	54	27											1.130 ± 0.001	100%
	54	27											1.408 ± 0.001	100%
$_{27}Co^{54}$	54	27		53.948460	0.19 s.	β+	8.242	7.34	100%		0+		ann.rad.	
$_{27}Co^{55}$	55	27		54.942001	17.5 h.	β+	3.452	0.53			7/2-	+4.822	ann.rad.	
	55	27				E.C.		1.03					0.0918 ± 0.0003	2.7%
	55	27						1.50					0.4772 ± 0.0002	20%
	55	27											0.9315 ± 0.0003	75%
	55	27											1.3167 ± 0.0003	7%

Isotope	A	Z	% Natural abundance	Atomic mass	Half-life	Decay mode	Decay energy (MeV)	Particle energy (MeV)	Particle intensity	Thermal neutron cross section	Spin (h/2π)	μ Nucl. mag. moment	Gamma-ray energy (MeV)	Gamma-ray intensity
	55	27											1.3700 ± 0.0005	3%
	55	27											1.4087 ± 0.0003	16%
$_{27}Co^{56}$	56	27		55.939841	77.7 d.	β+	4.566	1.459	18%		0+	3.830	ann.rad.	
	56	27				E.C.							0.84678 ± 0.00006	99.9%
	56	27											1.03783 ± 0.00007	14%
	56	27											1.23828 ± 0.00004	68%
	56	27											1.36022 ± 0.00002	4%
	56	27											1.77149 ± 0.00005	16%
	56	27											2.01536 ± 0.00005	3%
	56	27											2.0349 ± 0.00005	8%
	56	27											2.59858 ± 0.00008	17%
	56	27											3.20230 ± 0.0001	3%
	56	27											3.25360 ± 0.0001	7.4%
	56	27											3.27325 ± 0.0001	1.7%
	56	27											3.54805 ± 0.0002	0.18%
	56	27											3.60060 ± 0.0004	0.16%
$_{27}Co^{57}$	57	27		56.936294	271 d.	E.C.	0.836				7/2-	+4.733	0.01441 ± 0.00005	11%
	57	27											0.12206 ± 0.00002	85.6%
	57	27											0.13647 ± 0.00003	10%
$_{27}Co^{58m}$	58	27			9.1 h.	I.T.					5+		0.024889 ± 0.00002	0.035%
$_{27}Co^{58}$	58	27		57.935755	70.91 d.	β+	2.30				2+	+4.044	ann.rad.	
	58	27				E.C.							0.810755 ± 0.00003	99%
	58	27											0.86347 ± 0.0001	0.7%
	58	27											1.67473 ± 0.00006	0.52%
$_{27}Co^{59}$	59	27	100%	58.933198						(20 + 17) b.	7/2-	+4.627		
$_{27}Co^{60m}$	60	27			10.48 m.	I.T.(99.8%	0.059				2+	+4.40	0.058603 ± 0.00001	2.0%
	60	27				β-(0.02%)	1.56							
$_{27}Co^{60}$	60	27		59.933819	5.272 y.	β-	2.824	0.315	99.7%	2 b.	5+	+3.799	1.173210 ± 0.00002	100%
	60	27											1.332470 ± 0.00002	100%
$_{27}Co^{61}$	61	27		60.932478	1.65 h.	β-	1.322	1.22	95%		7/2-		0.067415 ± 0.00001	86%
	61	27											0.8417 ± 0.0005	0.7%
	61	27											0.9092 ± 0.0005	3%
$_{27}Co^{62m}$	62	27			13.9 m.	β-		0.88	25%		5+		1.1635 ± 0.0003	70%
	62	27						2.88	75%				1.1730 ± 0.0003	98%
	62	27											1.7191 ± 0.0003	7%
	62	27											2.0039 ± 0.0003	19%
	62	27											2.1049 ± 0.0003	6%
$_{27}Co^{62}$	62	27		61.934060	1.5 m.	β-	5.32	1.03	10%		2+		1.1292 ± 0.0003	13%
	62	27						1.76	5%				1.1730 ± 0.0003	83%
	62	27						2.9	20%				1.9851 ± 0.001	3%
	62	27						4.05	60%				2.3020 ± 0.001	19%
	62	27											2.3458 ± 0.001	1%
	62	27											3.159 ± 0.002	1%
$_{27}Co^{63}$	63	27		62.933614	27.5 s.	β-	3.67	3.6			7/2-		0.08713 ± 0.0001	49%
	63	27											0.1556 ± 0.0001	1.8%
	63	27											0.9817 ± 0.0003	2.6%
	63	27											1.0691 ± 0.0001	1.6%
	63	27											2.1745 ± 0.0005	1.2%
$_{27}Co^{64}$	64	27		63.935812	0.30 s.	β-	8.12	7.0			1+			
Ni		28		58.67						37.2 b.				
$_{28}Ni^{53}$	53	28		52.968430	0.05 s.	β+,p	13.2				7/2-		ann.rad.	
$_{28}Ni^{55}$	55	28		54.951336	0.19 s.	β+	8.7	7.66			7/2-		ann.rad.	
$_{28}Ni^{56}$	56	28		55.943124	6.10 d.	E.C.	2.14				0+		0.15838 ± 0.00003	98.8%
	56	28											0.26950 ± 0.00002	36%
	56	28											0.48044 ± 0.00002	32%
	56	28											0.74995 ± 0.00003	49%
	56	28											0.81185 ± 0.00003	87%
	56	28											1.56180 ± 0.00005	14%
$_{28}Ni^{57}$	57	28		56.39799	36.1 h.	β+	3.265	0.712	10%		3/2-	0.88	ann.rad.	
	57	28				E.C.		0.849	76%				0.12719 ± 0.00002	13.6%

Isotope	A	Z	% Natural abundance	Atomic mass	Half-life	Decay mode	Decay energy (MeV)	Particle energy (MeV)	Particle intensity	Thermal neutron cross section	Spin (h/2π)	μ Nucl. mag. moment	Gamma-ray energy (MeV)	Gamma-ray intensity
	57	28											1.37759 ± 0.00004	78%
	57	28											1.75748 ± 0.00008	7%
	57	28											1.91943 ± 0.00008	15%
$_{28}Ni^{58}$	58	28	68.27%	57.935346						4.6 b.	0+			
$_{28}Ni^{59}$	59	28		58.934349	≈7.6x10^4y.	E.C.					3/2-			
$_{28}Ni^{60}$	60	28	26.10%	59.930788						2.9 b.	0+			
$_{28}Ni^{61}$	61	28	1.13%	60.931058						2.4 b.	3/2-	-0.75002		
$_{28}Ni^{62}$	62	28	3.59%	61.928346						14.5 b.	0+			
$_{28}Ni^{63}$	63	28		62.929669	100 y.	β-	0.065	0.065			$^1/_2$-			
$_{28}Ni^{64}$	64	28	0.91%	63.927968						1.55 b.	0+			
$_{28}Ni^{65}$	65	28		64.930086	2.52 h.	β-	2.134	0.65	30%		5/2-	0.69	0.36627 ± 0.00003	5%
	65	28						1.020	11%				1.11553 ± 0.00004	16%
	65	28						2.140	58%				1.48184 ± 0.00005	23%
$_{28}Ni^{66}$	66	28		65.929116	54.8 h.	β-	0.24				0+			
$_{28}Ni^{67}$	67	28		66.9315709	20 s.	β-	3.56	3.8					0.1406 ± 0.0008	39 +
	67	28											0.2080 ± 0.0006	68
	67	28											0.5531 ± 0.0004	43
	67	28											0.7085 ± 0.0005	92
	67	28											0.7515 ± 0.002	23
	67	28											0.7791 ± 0.0004	27
	67	28											0.8741 ± 0.0004	88
	67	28											1.0722 ± 0.0005	100
	67	28											1.1004 ± 0.0007	31
	67	28											1.6539 ± 0.0004	100
	67	28											1.7602 ± 0.0005	39
	67	28											1.809 ± 0.001	30
	67	28											1.938 ± 0.001	25
	67	28											1.975 ± 0.001	65
Cu		29		63.546						3.78 b.				
$_{29}Cu^{58}$	58	29		57.944538	3.21 s.	β+	8.563	4.5	15%		1+		ann.rad.	
	58	29				E.C.		7.439	83%				0.0403 ± 0.0004	5%
	58	29											1.4483 ± 0.0002	11%
	58	29											1.4546 ± 0.0002	16%
$_{29}Cu^{59}$	59	29		58.939503	82 s.	β+	4.801	1.9			3/2-		ann.rad.	
	59	29						3.75					0.3393 ± 0.0001	8%
	59	29											0.8780 ± 0.0001	12%
	59	29											1.3015 ± 0.0001	15%
	59	29											(0.4 - 2.6)weak	
$_{29}Cu^{60}$	60	29		59.937366	23.2 m.	β+	6.127	2.00	69+		2+	+1.219	ann.rad.	
	60	29				E.C.		3.00	18				0.4673 ± 0.0002	3.6%
	60	29						3,92	6				0.9524 ± 0.0002	2.8%
	60	29											1.0352 ± 0.0002	3.8%
	60	29											1.3325 ± 0.0002	88%
	60	29											1.8618 ± 0.0003	5%
	60	29											1.9369 ± 0.0003	2.2%
	60	29											2.1589 ± 0.0002	3.6%
	60	29											2.7461 ± 0.0003	1.1%
	60	29											3.1941 ± 0.0003	2.1%
	60	29											(0.4 - 5.0)weak	
$_{29}Cu^{61}$	61	29		60.933461	3.41 h.	β+	2.239	0.56	3%		3/2-	+2.14	ann.rad.	
	61	29						0.94	5%				0.06711 ± 0.0002	6%
	61	29					1.15	2%					0.28370 ± 0.0002	13%
	61	29						1.220	51%				0.3729 ± 0.0005	2.3%
	61	29											0.65604 ± 0.0002	11%
	61	29											0.90868 ± 0.0004	1.2 %
	61	29											1.8516 ± 0.0003	4.8%
	61	29											(0.5 - 2.1)weak	
$_{29}Cu^{62}$	62	29		61.932586	9.74 m.	β+(98%)	3.95	2.93	98%		1+	-0.380	ann.rad.	
	62	29				E.C.							1.17302 ± 0.0001	0.6%
	62	29											(0.87 - 3.37)weak	
$_{29}Cu^{63}$	63	29	69.17%	62.939598						4.47 b.	3/2-	+2.2233		
$_{29}Cu^{64}$	64	29		63.929765	12.701 h.	β-(39%)	0.578	0.578			1+	-0.217	ann.rad.	
	64	29				β+(19%)	1.675	0.65					1.3459 ± 0.0003	0.6%
	64	29				E.C.(41%)								
$_{29}Cu^{65}$	65	29	30.83%	64.927793						2.17 b.	3/2-	+2.3817		
$_{29}Cu^{66}$	66	29		65.928872	5.10 m.	β-	2.642	1.65	6%		1+	-0.282	0.8330 ± 0.001	0.15%
	66	29						2.7	94%				1.0392 ± 0.0002	8%
$_{29}Cu^{67}$	67	29		66.927747	61.9 h.	β-	0.58	0.395	56%		3/2-		0.09125 ± 0.0001	7%
	67	29						0.484	23%				0.09325 ± 0.0001	17%
	67	29						0.577	20%				0.18453 ± 0.0001	47%
	67	29											0.30022 ± 0.0001	1.5%
$_{29}Cu^{68m}$	68	29			3.8 m.	I.T.(86%)					6-		0.0843 ± 0.0005	70%
	68	29				β-(14%)	1.8						0.1112 ± 0.0005	18%
	68	29											0.5259 ± 0.0005	74%
	68	29											0.6369 ± 0.0005	8%
	68	29											1.0410 ± 0.0005	8%
	68	29											1.3403 ± 0.0005	12%

Isotope	A	Z	% Natural abundance	Atomic mass	Half-life	Decay mode	Decay energy (MeV)	Particle energy (MeV)	Particle intensity	Thermal neutron cross section	Spin (h/2π)	μ Nucl. mag. moment	Gamma-ray energy (MeV)	Gamma-ray intensity
$_{29}Cu^{68}$	68	29		67.929620	31 s.	β-	4.45	3.5	40%		1+		0.8059 ± 0.0005	0.8%
	68	29						4.6	31%				1.0774 ± 0.0005	58%
	68	29											1.2613 ± 0.0005	17%
	68	29											1.8832 ± 0.0005	1.2%
	68	29											(0.15 - 2.34)weak	
$_{29}Cu^{69}$	69	29		68.929425	3.0 m.	β-	2.67	2/48	80%		3/2-		0.5307 ± 0.0003	3%
	69	29											0.6490 ± 0.0005	1.5%
	69	29											0.8340 ± 0.0005	6%
	69	29											1.0065 ± 0.0008	10%
	69	29											1.1795 ± 0.0001	1%
$_{29}Cu^{70m}$	70	29			46 s.	β-		2.52	10%		5-		0.3865 ± 0.0004	8%
	70	29											0.8848 ± 0.0002	100%
	70	29											0.9017 ± 0.0002	90%
	70	29											1.1087 ± 0.0004	8%
	70	29											1.2517 ± 0.0005	60%
	70	29											1.6906 ± 0.0006	5%
	70	29											2.0614 ± 0.0005	4%
	70	29											3.062 ± 0.002	1.4%
$_{29}Cu^{70}$	70	29		69.932386	5 s.	β-	6.58	5.42	54%		1+		0.8848 ± 0.0002	54%
	70	29						6.09	46%					
$_{29}Cu^{71}$	71	29		70.932560	20 s.	β-					3/2-		0.490	
$_{29}Cu^{72}$	72	29			6.6 s.						(1+)		0.652	
$_{29}Cu^{73}$	73	29			3.9 s.	β-							0.450	
Zn		30		65.39						1.1 b.				
$_{30}Zn^{57}$	57	30		56.964990	0.04 s.	β+,p	15						ann.rad.	
$_{30}Zn^{59}$	59	30		58.949270	184 ms.	β+,p	9.1	8.1			3/2-		ann.rad.	
	59	30											0.491	
	59	30											0.914	
$_{30}Zn^{60}$	60	30		59.941830	2.4 m.	β+(97%)	4.16				0+		ann.rad.	
	60	30				E.C.(3%)							0.0614 ± 0.0005	24%
	60	30											0.2734 ± 0.0005	10%
	60	30											0.3344 ± 0.0005	9%
	60	30											0.3646 ± 0.0003	3%
	60	30											0.6703 ± 0.0004	68%
$_{30}Zn^{61}$	61	30		60.939514	89.1 s.	β+	5.64	4.38	68%		3/2-		ann.rad.	
	61	30											0.2664 ± 0.0004	16%
	61	30											0.4752 ± 0.0003	7.4%
	61	30											0.6904 ± 0.0004	1.5%
	61	30											0.9700 ± 0.0003	2.4%
	61	30											1.1854 ± 0.0003	1.5%
	61	30											1.6605 ± 0.0004	7.4%
	61	30											1.9971 ± 0.0005	1%
$_{30}Zn^{62}$	62	30		61.934332	9.26 h.	β+(3%)	1.63	0.66	7%		0+		ann.rad.	
	62	30				E.C.(93%)							0.04094 ± 0.0006	25%
	62	30											0.5075 ± 0.0004	15%
	62	30											0.5481 ± 0.0004	15%
	62	30											0.59665 ± 0.00001	24%
	62	30											(0.2 - 1.5)weak	
$_{30}Zn^{63}$	63	30		62.933214	38.1 m.	β+(93%)	3.367	1.02			3/2-	-0.28164	ann.rad.	
	63	30				E.C.(7%)		1.40					0.66962 ± 0.00005	8.4%
	63	30						1.71					0.96206 ± 0.00005	6.6%
	63	30						2.36	84%				1.13067 ± 0.0002	0.01%
	63	30											1.3270 ± 0.0003	0.07%
	63	30											1.3926 ± 0.0004	0.1%
	63	30											1.41208 ± 0.00005	0.76%
	63	30											1.54704 ± 0.0005	0.13%
	63	30											(0.24 - 3.1)weak	
$_{30}Zn^{64}$	64	30	48.6%	63.929145						0.76 b.	0+			
$_{30}Zn^{65}$	65	30		64.929243	243.8 d.	β+(98%)	1.352	0.325			5/2-	+0.7690	ann.rad.	
	65	30				E.C.(1.5%)							1.11552 ± 0.00002	50.8%
$_{30}Zn^{66}$	66	30	27.9%	65.926034						0.9 b.	0+			
$_{30}Zn^{67}$	67	30	4.1%	66.927129						7.25 b.	5/2-	+0.8755		
$_{30}Zn^{68}$	68	30	18.8%	67.924846						(0.072 + 0.9) b.	0+			
$_{30}Zn^{69m}$	69	30			13.8 h.	I.T. (99+%)	0.439				9/2+		0.4390 ± 0.0002	95%
$_{30}Zn^{69}$	69	30		68.926552	57 m	β-	0.905	0.905	99.9%		1/2-		0.318	0.001%
$_{30}Zn^{70}$	70	30	0.6%	69.925325						(0.008 + 0.083) b.	0+			
$_{30}Zn^{71m}$	71	30			3.97 h.	β-		1.45			9/2+		0.12148 ± 0.00005	2.9%
	71	30											0.14260 ± 0.00005	5.4%
	71	30											0.38628 ± 0.00005	92%
	71	30											0.48734 ± 0.00005	60%
	71	30											0.51155 ± 0.00005	28%

Isotope	A	Z	% Natural abundance	Atomic mass	Half-life	Decay mode	Decay energy (MeV)	Particle energy (MeV)	Particle intensity	Thermal neutron cross section	Spin (h/2π)	μ Nucl. mag. moment	Gamma-ray energy (MeV)	Gamma-ray intensity
	71	30											0.59607 ± 0.00005	27%
	71	30											0.62019 ± 0.00005	56%
	71	30											0.9647 ± 0.0001	4%
	71	30											1.1074 ± 0.0002	2.1%
	71	30											1.7596 ± 0.0002	1%
	71	30											2.3177 ± 0.0002	0.1%
$_{30}Zn^{71}$	71	30		70.927727	2.4 m.						1/2-		0.12152 ± 0.00005	2.7%
	71	30											0.3900 ± 0.0003	3.6%
	71	30											0.5116 ± 0.0001	30%
	71	30											0.9103 ± 0.0001	7.5%
	71	30											1.1200 ± 0.0001	2.1%
	71	30											(0.39 - 2.29)weak	
$_{30}Zn^{72}$	72	30		71.926856	β-	0.457	0.25	14%			0+		0.0164 ± 0.0003	8%
	72	30						0.30	86%				0.1887 ± 0.0001	2%
	72	30											0.1447 ± 0.0001	83%
	72	30											0.1915 ± 0.0002	9.4%
$_{30}Zn^{73}$	73	30		72.929780	24 s.	β-	4.70	4.7			3/2-		0.216 ± 0.001	100 +
	73	30											0.496 ± 0.001	26
	73	30											0.911 ± 0.001	26
$_{30}Zn^{74}$	74	30		73.929461	96 s.	β-	2.4	2.1					0.0503 ± 0.0005	18%
	74	30											0.0531 ± 0.0005	10%
	74	30											0.0573 ± 0.0005	80%
	74	30											0.0861 ± 0.0005	4%
	74	30											0.1167 ± 0.0001	4%
	74	30											0.1400 ± 0.0005	37%
	74	30											0.1904 ± 0.0005	27%
	74	30											0.3473 ± 0.0005	6.3%
$_{30}Zn^{75}$	75	30		74.932690	10.2 s.	β-	6.1							
$_{30}Zn^{76}$	76	30		75.932940	5.7 s.	β-	4.0	3.6					0.56	
	76	30											1.10	
$_{30}Zn^{77}$	77	30		76.936750	1.4 s.	β-	7.5	4.8			7/2+		0.189	
	77	30											0.473	
$_{30}Zn^{78}$	78	30		77.937780	1.5 s.	β-	6.0						0.1817	
	78	30											0.2248	
Ga		31		69.723						2.9 b.				
$_{31}Ga^{62}$	62	31		61.944178	0.116 s.	β+	9.17	8.3			0+		ann.rad.	
	63	31				E.C.								
$_{31}Ga^{63}$	63	31		62.939140	32 s.	β+	5.5	4.5					ann.rad.	
	63	31				E.C.							0.1930 ± 0.0002	5%
	63	31											0.2480 ± 0.0002	3.4%
	63	31											0.6271 ± 0.0002	10%
	63	31											0.6370 ± 0.0002	11%
	63	31											0.6501 ± 0.0002	4.5%
	63	31											0.7685 ± 0.0002	2%
	63	31											1.0652 ± 0.0004	45 %
	63	31											1.6917 ± 0.0005	3%
$_{31}Ga^{64}$	64	31		63.936836	2.63 m.	β+	7.16	2.79			0+		ann.rad.	195%
	64	31						6.05					0.80785 ± 0.0001	14%
	64	31											0.91878 ± 0.0001	8%
	64	31											0.99152 ± 0.0001	43%
	64	31											1.38727 ± 0.0001	12%
	64	31											1.6175 ± 0.0002	1.6&
	64	31											1.79943 ± 0.0001	3.5%
	64	31											1.9958 ± 0.0003	1.6%
	64	31											2.2704 ± 0.0001	2.1%
	64	31											3.3659 ± 0.0001	13%
	64	31											3.4251 ± 0.0001	4.0%
	64	31											3.7951 ± 0.0001	1%
	64	31											4.454 ± 0.0001	0.7%
$_{31}Ga^{65}$	65	31		64.932738	15.2 m.	β+(86%)	3.256	0.82	10+		3/2-		ann.rad.	
	65	31				E.C.		1.39	19				0.0538 ± 0.0002	5%
	65	31						2.113	56				0.0611 ± 0.0002	12%
	65	31						2.237	15				0.1151 ± 0.0002	55%
	65	31											0.1530 ± 0.0002	96%
	65	31											0.2069 ± 0.0002	39%
	65	31											0.7518 ± 0.0002	8.2%
	65	31											0.7689 ± 0.0002	1/3%
	65	31											0.7946 ± 0.0002	0.25%
	65	31											0.9097 ± 0.0002	0.5%
	65	31											0.9322 ± 0.0002	1.8%
	65	31											1.0474 ± 0.0003	0.9%
	65	31											1.2288 ± 0.0002	0.7%
	65	31											1.3547 ± 0.0002	0.8%
	65	31											2.2121 ± 0.0003	0.13%
	65	31											(0.06 - 2.4)weak	
$_{31}Ga^{66}$	66	31		65.931590	9.4 h.	β+(56%)	5.175	0.74	1%		0+		ann.rad.	
	66	31				E.C.(43%)		1.84	54%				0.8337 ± 0.0003	6.1%
	66	31						4.153	51%				1.03935 ± 0.00008	38%
	66	31											1.2322 ± 0.0002	0.5%
	66	31											1.3334 ± 0.0001	1.2%

Isotope	A	Z	% Natural abundance	Atomic mass	Half-life	Decay mode	Decay energy (MeV)	Particle energy (MeV)	Particle intensity	Thermal neutron cross section	Spin (h/2π)	μ Nucl. mag. moment	Gamma-ray energy (MeV)	Gamma-ray intensity
	66	31											1.8992 ± 0.0001	0.43%
	66	31											1.9187 ± 0.0001	2.4%
	66	31											2.1902 ± 0.0001	5.7%
	66	31											2.4225 ± 0.0001	1.8%
	66	31											2.7523 ± 0.0001	23%
	66	31											3.3813 ± 0.0002	1.4%
	66	31											3.4225 ± 0.0002	0.8%
	66	31											4.0865 ± 0.0001	1.1%
	66	31											4.2955 ± 0.0002	3.5%
	66	31											4.8066 ± 0.0002	1.5%
	66	31											(0.29 - 4.8)weak	
$_{31}Ga^{67}$	67	31		66.928204	78.25 h.	E.C.	1.001				3/2-	+1.8507	0.09128 ± 0.00002	4%
	67	31											0.09332 ± 0.0002	38%
	67	31											0.18459 ± 0.0004	23%
	67	31											0.20896 ± 0.0006	3%
	67	31											0.30024 ± 0.0006	19%
	67	31											0.3936 ± 0.0006	5.6%
	67	31											0.8880 ± 0.0002	0.2%
$_{31}Ga^{68}$	68	31		67.927981	68.1 m.	β+(90%)	2.921	1/83			1+	0.01175	ann.rad.	
	68	31				E.C.(10%)							1.0774 ± 0.0001	3%
	68	31											(0.57 - 2.33)weak	
$_{31}Ga^{69}$	69	31	60.1%	68.925580						1.7 b.	3/2-	+2.01659		
$_{31}Ga^{70}$	70	31		69.926028	21.1 m.	E.C.(0.2%)	0.655				1+		0.1755 ± 0.0005	0.15%
	70	31				β-(99.8%)	1.653	1.65	99%				1.042 ± 0.005	0.48%
$_{31}Ga^{71}$	71	31	39.9%	70.924700						4.7 b.	3/2-	+2.56227		
$_{31}Ga^{72}$	72	31		71.926365	13.95 h.	β-	3.99	0.64	40%		3-	-0.13224	0.60005 ± 0.00005	5.8%
	72	31						1.51	9%				0.62986 ± 0.00005	24%
	72	31						2.52 8%					0.89422 ± 0.0001	10%
	72	31						3.15	11%				1.0507 ± 0.0001	7%
	72	31											1.2309 ± 0.0002	1.4%
	72	31											1.2601 ± 0.0002	1.15%
	72	31											1/2768 ± 0.0002	1.6%
	72	31											1/2768 ± 0.0002	1.6%
	72	31											1.4640 ± 0.0001	3.6%
	72	31											1.5968 ± 0.0002	4.3%
	72	31											1.8611 ± 0.0001	5.3%
	72	31											2.1095 ± 0.0002	1%
	72	31											2.2016 ± 0.0002	26%
	72	31											2.4910 ± 0.0002	7.5%
	72	31											2.5077 ± 0.0002	12.8%
	72	31											(0.11 - 3.3)weak	
$_{31}Ga^{73}$	73	31		72.925169	4.87 h.	1.59					3/2-		0.05344 ± 0.00005	10%
	73	31											0.0687 ± 0.0002	2%
	73	31											0.29732 ± 0.00005	47%
	73	31											0.32570 ± 0.00007	7%
	73	31											0.73942 ± 0.00005	2%
	73	31											0.9936 ± 0.0005	0.1%
	73	31											(0.01 - 1.00)weak	
$_{31}Ga^{74m}$	74	31			9.5 s.	I.T.					1+		0.0565 ± 0.0001	75%
$_{31}Ga^{74}$	74	31		73.926940	8.1 m.	β-	5.4	2.6			3-		0.59588 ± 0.00004	91%
	74	31											0.6042 ± 0.00001	2.9%
	74	31											0.60840 ± 0.00005	14.5%
	74	31											0.8678 ± 0.0001	9%
	74	31											1.10134 ± 0.00006	5.5%
	74	31											1.2043 ± 0.0001	7.5%
	74	31											1.3322 ± 0.0003	1.8%
	74	31											1.7448 ± 0.0001	4.8%
	74	31											1.0406 ± 0.0001	5.5%
	74	31											2.2790 ± 0.0001	2.4%
	74	31											2.3535 ± 0.0001	44%
	74	31											2.58007 ± 0.0001	1.3%
	74	31											2.97090 ± 0.0001	1%
	74	31											(0.49 - 3.99)weak	
$_{31}Ga^{75}$	75	31		74.926499	2.10 m.	β-	3.39	3.3			3/2-		0.1770	10 +
	75	31											0.2528	100
	75	31											0.5747	32
	75	31											0.8854	11
	75	31											0.9272	7
	75	31											1.2485	5
	75	31											1.5011	5
	75	31											1.5430	1
$_{31}Ga^{76}$	76	31		75.928670	29.1 s.	β-	6.8				3-		0.4310 ± 0.0005	9%
	76	31											0.54551 ± 0.00003	26%

Isotope	A	Z	% Natural abundance	Atomic mass	Half-life	Decay mode	Decay energy (MeV)	Particle energy (MeV)	Particle intensity	Thermal neutron cross section	Spin (h/2π)	μ Nucl. mag. moment	Gamma-ray energy (MeV)	Gamma-ray intensity
	76	31											0.56293 ± 0.00003	65.8%
	76	31											0.8472 ± 0.0001	3.5%
	76	31											0.97650 ± 0.00005	4.5%
	76	31											1.10841 ± 0.00008	18%
	76	31											1.2080 ± 0.0001	1.3%
	76	31											2.1295 ± 0.0001	2.2%
	76	31											2.2144 ± 0.0001	1.1%
	76	31											2.3569 ± 0.0001	2.4%
	76	31											2.5786 ± 0.0001	2.2%
	76	31											2.6192 ± 0.0001	2.1%
	76	31											2.9199 ± 0.0001	9.1%
	76	31											3.1414 ± 0.0001	4.2%
	76	31											3.3888 ± 0.0002	3%
	76	31											3.9517 ± 0.0002	4.2%
	76	31											(1.0 - 4.2)weak	
$_{31}$Ga^{77}	77	31		76.928700	13 s.	β-	5/3	5/2					0.469	
	77	31											0.459	
$_{31}$Ga^{78}	78	31		77.931760	5.09 s.	β-	7/9				3+		0.619	
	78	31											1.025	
	78	31											1.186	
$_{31}$Ga^{79}	79	31		78.932530	3.0 s.	β-	6.8	4.6					2.18	
$_{31}$Ga^{80}	80	31		79.936250	1.66 s.	β-	10	10						
$_{31}$Ga^{81}	81	31		80.937750	1.23 s.	β-	8.3	5.1					0.217	
$_{31}$Ga^{82}	82	31			1.9 s.	β-	7.4							
$_{31}$Ga^{83}	83	31			1.2 s.	β-								
Ge		32								2.2 b.				
$_{32}$Ga^{64}	64	32		63.941570	63 s.	β+	4.4	3.0			0+		ann.rad.	
	64	32				E.C.							0.0651 ± 0.0002	2%
	64	32											0.1282 ± 0.0002	11%
	64	32											0.3841 ± 0.0003	4.7%
	64	32											0.4270 ± 0.0003	37%
	64	32											0.6671 ± 0.0003	17%
	64	32											0.7745 ± 0.0003	7%
$_{32}$Ge^{65}	65	32		64.939440	31 s.	β+	6.2	0.82	10+				ann.rad.	
	65	32				E.C.		1.39	19				0.0621 ± 0.0005	27%
	65	32						2.113	56				0.1908 ± 0.0002	10%
	65	32						2.237	15				0.4591 ± 0.0005	10%
	65	32											0.5877 ± 0.0002	3%
	65	32											0.6187 ± 0.0004	1.6%
	65	32											0.6497 ± 0.0002	33%
	65	32											0.8091 ± 0.0002	21%
	65	32											1.0759 ± 0.0003	0.8%
	65	32											1.2371 ± 0.0003	1.3%
	65	32											2.0996 ± 0.0004	1.5%
	65	32											(0.4 - 3.2)weak	
$_{32}$Ge^{66}	66	32		65.933847	2.27 h.	β+(27%)	2.10				0+		ann.rad.	
	66	32				E.C.(73%)							0.0224 ± 0.0002	1.6%
	66	32											0.0483 ± 0.0002	1.0%
	66	32											0.04389 ± 0.0001	29%
	66	32											0.06512 ± 0.0001	6%
	66	32											0.10885 ± 0.00003	10.5%
	66	32											0.18203 ± 0.00004	5.7%
	66	32											0.19020 ± 0.00003	5.7%
	66	32											0.27297 ± 0.00004	11%
	66	32											0.33805 ± 0.00004	10%
	66	32											0.38185 ± 0.00005	28%
	66	32											0.47062 ± 0.00006	7.4%
	66	32											0.4720 ± 0.0001	3%
	66	32											0.53674 ± 0.00007	6%
	66	32											0.70594 ± 0.00003	4.3%
$_{32}$Ge^{67}	67	32		66.932737	19.0 m.	β+(96%)	4.22	1.6			1/2-		ann.rad.	
	67	32				E.C.(4%)		2.3					0.16701 ± 0.00005	84%
	67	32						3.15					0.3595 ± 0.0002	1.5%
	67	32											0.7282 ± 0.0005	2.5%
	67	32											0.8283 ± 0.0003	3.0%
	67	32											0.9112 ± 0.0003	3.1%
	67	32											0.9148 ± 0.0003	3.0%
	67	32											1.0818 ± 0.0003	1%
	67	32											1.4728 ± 0.0003	4.9%
	67	32											1.8094 ± 0.0006	1.3%
	67	32											(0.4 - 3.7)weak	

Isotope	A	Z	% Natural abundance	Atomic mass	Half-life	Decay mode	Decay energy (MeV)	Particle energy (MeV)	Particle intensity	Thermal neutron cross section	Spin (h/2π)	μ Nucl. mag. moment	Gamma-ray energy (MeV)	Gamma-ray intensity
$_{32}Ge^{68}$	68	32		67.928096	288 d.	E.C.	0.11				0+		Ga k x-ray	39%
$_{32}Ge^{69}$	69	32		68.927969	39.1 h.	β+(36%)	2.225	0.70			5/2-	0.735	ann.rad.	72%
	69	32				E.C.(64%)		1.2					0.3184 ± 0.0002	1.25%
	69	32											0.5739 ± 0.0002	12%
	69	32											0.8717 ± 0.0002	11%
	69	32											1.1064 ± 0.0002	27%
	69	32											1.3362 ± 0.0002	3%
	69	32											(0.2 - 2.04)weak	
$_{32}Ge^{70}$	70	32	20.5%	69.924250						3.3 b.	0+			
$_{32}Ge^{71}$	71	32		70.924953	11.2 d.	E.C.	0.236				1/2-	+0.547		
$_{32}Ge^{72}$	72	32	27.4%	71.922079						1.0 b.	0+			
$_{32}Ge^{73}$	73	32	7.8%	72.923463						14 b.	9/2+	-0.87946		
$_{32}Ge^{74}$	74	32	36.5%	73.921177						(0.16 + 0.36)b.	0+			
$_{32}Ge^{75m}$	75	32			48 s.	I.T.					7/2+		0.13968 ± 0.00003	39%
$_{32}Ge^{75}$	75	32		74.922858	82.8 m.	β-	1.178	1.19			1/2-	+0.510	0.26461 ± 0.00005	11%
	75	32											0.41931 ± 0.00005	0.2%
$_{32}Ge^{76}$	76	32	7.8%	75.921401						(.09 + .06)b.	0+			
$_{32}Ge^{77m}$	77	32			53 s.	I.T.(20%)					1/2-		0.1597 ± 0.0001	10%
	77	32				β-(80%)	2.861	2.9					0.1948 ± 0.0001	0.2%
	77	32											0.21551 ± 0.0001	21%
$_{32}Ge^{77}$	77	32		76.923548	11.30 h.	β-	2.70	0.71	23%		7/2+		0.2108 ± 0.0005	29%
	77	32						1.38	35%				0.2156 ± 0.0005	27%
	77	32						2.19	42%				0.2645 ± 0.0005	53%
	77	32											0.3674 ± 0.0005	12%
	77	32											0.4163 ± 0.0005	20%
	77	32											0.5579 ± 0.0005	15%
	77	32											0.6316 ± 0.0005	7%
	77	32											0.7141 ± 0.0005	7.5%
	77	32											0.8101 ± 0.0005	2.2%
	77	32											1.0851 ± 0.0005	6.4%
	77	32											1.3684 ± 0.0005	3.1%
	77	32											1.9996 ± 0.0005	0.6%
	77	32											2.3415 ± 0.0005	0.5%
	77	32											(0.15 - 2.37)weak	
$_{32}Ge^{78}$	78	32		77.922853	1,45 h.	-	0.95	0.70			0+		0.2773 ± 0.0005	96%
	78	32											0.2939 ± 0.0005	4%
$_{32}Ge^{79m}$	79	32			19 s.	β-								
$_{32}Ge^{79}$	79	32		78.925360	42 s.	β-	4.1	4.0	20%		1/2-		0.2164 ± 0.0004	5%
	79	32						4.3	80%				0.2304 ± 0.0004	25%
	79	32											0.5427 ± 0.0004	15%
	79	32											0.6331 ± 0.0004	3%
	79	32											0.7450 ± 0.0004	5%
	79	32											0.7818 ± 0.0005	6%
$_{32}Ge^{80}$	80	32		79.925520	29 s.	β-	2/69	2.4			0+		0.1104 ± 0.0004	6%
	80	32											0.2656 ± 0.0004	25%
	80	32											0.9372 ± 0.0004	4%
	80	32											1.0140 ± 0.0004	2.5%
	80	32											1.1160 ± 0.0004	2.5%
	80	32											1.2561 ± 0.0005	3%
	80	32											1.5643 ± 0.0005	4.5%
$_{32}Ge^{81m}$	81	32			7.6 s.	β-		3.75			9/2+		0.3362 ± 0.0004	
	81	32											0.7935 ± 0.0004	
$_{32}Ge^{81}$	81	32		80.928820	7.6 s.	β-	6.2	3.44			1/2+		0.1976 ± 0.0004	21 +
	81	32											0.3362 ± 0.0004	100
$_{32}Ge^{82}$	82	32		81.929810	4.6 s.	β-	4.7				0+		1.093	
$_{32}Ge^{83}$	83	32		82.934250	1.9 s.	β-	7.4							
As		33		74.9216						4.5 b.				
$_{33}As^{67}$	67	33		66.939190	43 s.	β+	6.0	5.0			5/2-		0.121	
	67	33				E.C.							0.123	
	67	33											0.244	
$_{33}As^{68}$	68	33		67.936790	2.53 m.	β+	8.1				3+		ann.rad.	
	68	33											0.6135 ± 0.0005	6%
	68	33											0.6512 ± 0.0002	24%
	68	33											0.7626 ± 0.0002	23%
	68	33											1.0165 ± 0.0001	66%
	68	33											1.2534 ± 0.0004	1%
	68	33											1.2635 ± 0.0005	4%
	68	33											1.4125 ± 0.0003	12%
	68	33											1.622 ± 0.001	3%
	68	33											1.7787 ± 0.0002	23%
	68	33											2.008 ± 0.001	3%
	68	33											2.457 ± 0.001	3%
	68	33											3.058 ± 0.003	1%
	68	33											3.220 ± 0.003	0.6%
$_{33}As^{69}$	69	33		68.932280	15.1 m.	β+(98%)	4.02	2.95			5/2-		ann.rad.	
	69	33				E.C.(2%)							0.0868 ± 0.0005	1.5%
	69	33											0.1458 ± 0.0003	2.4%
	69	33											0.2327 ± 0.0003	5%
	69	33											0.2871 ± 0.0005	0.9%
	69	33											0.3741 ± 0.0005	0.7%
	69	33											0.3981 ± 0.0005	0.6%

TABLE OF THE ISOTOPES (Continued)

Isotope	A	Z	% Natural abundance	Atomic mass	Half-life	Decay mode	Decay energy (MeV)	Particle energy (MeV)	Particle intensity	Thermal neutron cross section	Spin (h/2π)	μ Nucl. mag. moment	Gamma-ray energy (MeV)	Gamma-ray intensity
$_{33}As^{70}$	70	33		69.930929	52.6 m.	β+(84%)	6.22	1.44			4+	2.1	ann.rad.	
	70	33				E.C.(16%)		2.14					0.1753 ± 0.0005	2.6%
	70	33						2.89					0.5952 ± 0.0005	16%
	70	33											0.6684 ± 0.0005	21%
	70	33											0.7448 ± 0.0005	21%
	70	33											0.9057 ± 0.0005	12%
	70	33											1.0395 ± 0.0007	82%
	70	33											1.0993 ± 0.0005	4.4%
	70	33											1.1143 ± 0.001	21%
	70	33											1.1181 ± 0.0007	3.2%
	70	33											1.3394 ± 0.0008	8.5%
	70	33											1.4125 ± 0.0005	9%
	70	33											1.4961 ± 0.0005	1.6%
	70	33											1.7079 ± 0.0007	18%
	70	33											1.7813 ± 0.0007	3.9%
	70	33											2.0077 ± 0.001	3%
	70	33											(0.17 - 4.4)weak	
$_{33}As^{71}$	71	33		70.927114	62 h.	β+(32%)	2.013				5/2-	+1.6735	ann.rad.	
	71	33				E.C.(68%)							0.1749 ± 0.0002	84%
	71	33											0.3274 ± 0.0002	2.7%
	71	33											0.5000 ± 0.0002	2.8%
	71	33											1.0957 ± 0.0002	4.2%
	71	33											1.2127 ± 0.0003	0.3%
	71	33											1.2988 ± 0.0003	0.2%
$_{33}As^{72}$	72	33		71.926755	26.0 h.	β+(77%)	4.355	0.669	5 +		2-	-2.1578	ann.rad.	
	72	33						1.884	12				0.6299 ± 0.0001	8%
	72	33						2.498	62				0.83395 ± 0.00005	80%
	72	33						3.339	19				1.0507 ± 0.0001	9.6%
	72	33											1.4640 ± 0.0001	1.1%
	72	33											1.4758 ± 0.0002	0.5%
	72	33											2.1059 ± 0.0002	0.6%
	72	33											2.1095 ± 0.0002	0.3%
	72	33											2.2016 ± 0.0002	0.5%
	72	33											2.5077 ± 0.0002	0.3%
	72	33											(0.1 - 4.0)weak	
$_{33}As^{73}$	73	33		72.923827	80.3 d.	E.C.	0.346				3/2-		0.013263 ± 0.00001	0.1%
	73	33											0.053437 ± 0.00001	10.5%
	73	33											Se k x-ray	90%
$_{33}As^{74}$	74	33		73.923827	17.78 d.	β+(31%)	2.562	0.94	26%		2-	-1.597	ann.rad.	
	74	33				E.C.(37%)	1.53	3%					0.59588 ± 0.0001	60%
	74	33				β-	1.354	0.71	16%				0.6084 ± 0.0001	0.6%
	74	33						1.35	16%				0.6348 ± 0.0001	15%
	74	33											1.2043 ± 0.0001	0.25%
$_{33}As^{75}$	75	33	100%	74.921594						4.5 b.	3/2-	+1.43947		
$_{33}As^{76}$	76	33		75.922393	26.3 h.	β-	0.54	3%			2-	-0.906	0.5591 ± 0.0001	45%
	76	33						1.184	2%				0.5632 ± 0.0001	1.2%
	76	33						1.785	8%				0.65703 ± 0.0005	5.7%
	76	33						2.410	36%				1.21272 ± 0.0001	1.3%
	76	33						2.97	51%				1.21602 ± 0.0001	3.4%
	76	33											1.2285 ± 0.0001	1.2%
	76	33											(0.3 - 2.6)weak	
$_{33}As^{77}$	77	33		76.920646	38.8 h.	β-	0.6904	0.70	98%		3/2-		0.0880 ± 0.0003	0.27%
	77	33											0.2391 ± 0.0002	1.6%
	77	33											0.2500 ± 0.0003	0.4%
	77	33											0.52078 ± 0.0001	0.43%
$_{33}As^{78}$	78	33		77.921830	1.515 h.	β-	4.21	3.00	12%		2-		0.3543 ± 0.0003	1.7%
	78	33						3.70	17%				0.5454 ± 0.0003	2.5%
	78	33						4.42	37%				0.6136 ± 0.0003	54%
	78	33											0.6862 ± 0.0003	1.8%
	78	33											0.6954 ± 0.0003	18%
	78	33											0.8276 ± 0.0003	7.2%
	78	33											1.0800 ± 0.0003	1.8%
	78	33											1.1445 ± 0.0003	2.1%
	78	33											1.2399 ± 0.0003	5.6%
	78	33											1.3088 ± 0.0003	10%
	78	33											1.3731 ± 0.0003	3.3%
	78	33											1.5300 ± 0.0003	2.5%
	78	33											1.7143 ± 0.0003	1.8%
	78	33											1.8360 ± 0.0003	1.4%
	78	33											1.9216 ± 0.0003	1.6%
	78	33											1.9955 ± 0.0003	1.1%
	78	33											2.6825 ± 0.0003	1.5%
$_{33}As^{79}$	79	33		78.920946	9.0 m.	β-	2.28	1.80	95%		3/2-		0.0955 ± 0.0005	16%
	79	33											0.3645 ± 0.0005	1.9%
	79	33											0.4320 ± 0.0005	1.5%
	79	33											0.3645 ± 0.0005	1.5%
	79	33											0.8785 ± 0.0008	1.4%
$_{33}As^{80}$	80	33		79.922528	16 s.	β-	5.58	3.38			1+		0.6662 ± 0.0002	42%
	80	33											1.2072 ± 0.0002	4%
	80	33											1.6454 ± 0.0002	7%
	80	33											2.3578 ± 0.0005	0.9%
	80	33											(2.5 - 3.0)weak	

Isotope	A	Z	% Natural abundance	Atomic mass	Half-life	Decay mode	Decay energy (MeV)	Particle energy (MeV)	Particle intensity	Thermal neutron cross section	Spin (h/2π)	μ Nucl. mag. moment	Gamma-ray energy (MeV)	Gamma-ray intensity
$_{33}As^{81}$	81	33		80.922131	33 s.	β-	3.87				3/2-		0.4676 ± 0.0002	20%
	81	33											0.4911 ± 0.0002	8%
	81	33											0.5211 ± 0.0002	1%
	81	33											1.4060 ± 0.0002	1%
	81	33											2.8324 ± 0.0002	0.3%
$_{33}As^{82m}$	83	33			14 s.	β-		3.6			5-		0.3435 ± 0.001	23%
	82	33											0.5605 ± 0.0001	14%
	82	33											0.6544 ± 0.0001	72%
	82	33											0.8151 ± 0.0004	7%
	82	33											0.8186 ± 0.0004	27%
	82	33											1.0799 ± 0.0004	23%
	82	33											1.7180 ± 0.0005	3%
	82	33											1.7313 ± 0.0002	27%
	82	33											1.8954 ± 0.0002	38%
$_{33}As^{82}$	82	33		81.924769	19 s.	β-	7.52	7.2	80%		1+		0.6544 ± 0.0001	15%
	82	33											0.7552 ± 0.0002	2%
	82	33											1.7313 ± 0.0002	3.8%
	82	33											1.9709 ± 0.0003	1.5%
	82	33											2.3462 ± 0.001	1.6%
	82	33											2.3534 ± 0.001	1.7%
	82	33											2.4412 ± 0.001	1.6%
	82	33											2.6038 ± 0.001	1.5%
	82	33											2.7227 ± 0.001	1.5%
	82	33											2.8348 ± 0.001	1.5%
	82	33											3.6688 ± 0.001	1.2%
	82	33											3.7730 ± 0.001	0.9%
$_{33}As^{83}$	83	33		82.924980	13 s.	β-	5.5						0.7345	100 +
	83	33											0.8338	19
	83	33											1.1131	34
	83	33											1.8954	18
	83	33											2.0767	28
	83	33											2.2029	22
	83	33											2.8579	16
$_{33}As^{84}$	84	33		83.929060	5.5 s.	β-,n	9.8				1-		0.6671 ± 0.0002	21%
	84	33											1.4439 ± 0.0005	49%
	84	33											1.8437 ± 0.0002	3%
	84	33											2.0866 ± 0.0003	5%
	84	33											2.4612 ± 0.0003	4%
	84	33											3.0379 ± 0.0005	1.2%
	84	33											4.9459 ± 0.001	1.1%
	84	33											5.0877 ± 0.001	0.7%
	84	33											5.1510 ± 0.001	0.8%
$_{33}As^{85}$	85	33		84.931820	2.03 s.	β-,n					3/2-		0.667 ± 0.001	42 +
	85	33											1.1115 ± 0.0005	12
	85	33											1.4551 ± 0.0002	100
	85	33											3.7494 ± 0.0007	3
$_{33}As^{86}$	86	33			0.9 s.	β-,n							0.704	
Se		34		78.96						11.7 b.				
$_{34}Se^{69}$	69	34		68.939570	27.4 s.	β+	6.79	5.006					ann.rad.	
	69	34				E.C.							0.0664 ± 0.0004	27%
	69	34											0.0982 ± 0.0004	63%
	69	34											0.69114 ± 0.0005	14%
$_{34}Se^{70}$	70	34		69.933880	41.1 m.	β+	2.8				0+		ann.rad.	
	70	34											0.03205 ± 0.00005	1.9%
	70	34											0.04951 ± 0.00005	35%
	70	34											0.13254 ± 0.00005	2.6%
	70	34											0.13563 ± 0.0005	2.6%
	70	34											0.2027 ± 0.0001	4.8%
	70	34											0.2441 ± 0.0001	2.8%
	70	34											0.3767 ± 0.0002	9.6%
	70	34											0.4262 ± 0.0002	29%
	70	34											0.49969 ± 0.0001	1.4%
$_{34}Se^{71}$	71	34		70.932270	4.7 m.	β+	4.8	3.4	36%		5/2-		ann.rad.	
	71	34				E.C.							0.1472 ± 0.0003	47%
	71	34											0.7241 ± 0.0003	6%
	71	34											0.8309 ± 0.0003	13%
	71	34											0.8711 ± 0.0003	7%
	71	34											0.9784 ± 0.0003	4.7%
	71	34											1.0960 ± 0.0003	10%
	71	34											1.2432 ± 0.0003	7%
	71	34											1.265 ± 0.0003	1.7%
$_{34}Se^{72}$	72	34		71.927110	8.4 d.	E.C.	0.33				0+		0.0460 ± 0.0002	57%
$_{34}Se^{73m}$	73	34				I.T.(73%)	0.0257	0.85			3/2-		ann.rad.	36%
	73	34				β+(27%)	2.77	1.45					0.0257 ± 0.0002	27%
	73	34						1.70					0.1807 ± 0.0001	0.5%
	73	34											0.2538 ± 0.0001	2.5%
	73	34											0.3204 ± 0.0001	0.8%
	73	34											0.3934 ± 0.0001	1.6%
	73	34											0.4016 ± 0.0001	1.2%
	73	34											0.5775 ± 0.0001	1.2%
	73	34											1.0778 ± 0.0002	0.6%

TABLE OF THE ISOTOPES (Continued)

Isotope	A	Z	% Natural abundance	Atomic mass	Half-life	Decay mode	Decay energy (MeV)	Particle energy (MeV)	Particle intensity	Thermal neutron cross section	Spin (h/2π)	μ Nucl. mag. moment	Gamma-ray energy (MeV)	Gamma-ray intensity
$_{34}Se^{73}$	73	34		72.926768	7.1 h.	β+(65%)	2.74	0.80			9/2+		ann.rad.	125%
	73	34				E.C.(35%)		1.32	95%				0.0670 ± 0.0001	72%
	73	34						1.68	1%				0.3609 ± 0.0001	97%
	73	34											(0.6 - 1.5)weak	
$_{34}Se^{74}$	74	34	0.9%	73.922475						52 b.	0+			
$_{34}Se^{75}$	75	34		74.922521	118.5 d.	E.C.	0.864				5/2+	0.67	0.09673 ± 0.00001	3%
	75	34											0.121115 ± 0.00001	15.8%
	75	34											0.136000 ± 0.00001	55%
	75	34											0.198596 ± 0.00001	1.4%
	75	34											0.264651 ± 0.00001	58%
	75	34											0.279528 ± 0.00001	15.9%
	75	34											0.303913 ± 0.00001	1.3%
	75	34											0.400646 ± 0.00001	11.6%
$_{34}Se^{76}$	76	34	9.0%	75.919212						(21 + 64) b.	0+			
$_{34}Se^{77m}$	77	34			17.4 s.	I.T.					7/2+		0.1619 ± 0.0002	52%
$_{34}Se^{77}$	77	34	7.6%	76.919912						42 b.	1/2-	+0.53506		
$_{34}Se^{78}$	78	34	23.5%							(0.43 + 0.2)b.	0+			
$_{34}Se^{79m}$	79	34			3.89 m.	I.T.					1/2-		0.09573 ± 0.00003	9.5%
$_{34}Se^{79}$	79	34		78.918498	6x10^{4} y.	β-	0.149				7/2+	-1.018		
$_{34}Se^{80}$	80	34	49.6%	79.916520						(0.07 + 0.54) b.	0+			
$_{34}Se^{81m}$	81	34			57.3 m.	I.T.(99%)	0.1031				7/2+		0.1031 ± 0.0003	9.7%
	81	34											0.2602 ± 0.0002	0.06%
	81	34											0.27599 ± 0.0001	0.06%
$_{34}Se^{81}$	81	34		803917990	18.5 m.	β-	1.59	1.6	98%		1/2-		0.27594 ± 0.00005	0.85%
	81	34											0.29008 ± 0.00005	0.75%
	81	34											0.5524 ± 0.0001	0.12%
	81	34											0.56604 ± 0.00005	0.25%
	81	34											0.6498 ± 0.0001	0.25%
	81	34											0.82827 ± 0.00005	0.32%
$_{34}Se^{82}$	82	34	9.4%	81.916698						(0.039 + 0.005) b.	0+			
$_{34}Se^{83m}$	83	34			70 s.	β-	1.78				1/2-		0.35666 ± 0.00006	17%
	83	34											0.7990 ± 0.0001	11%
	83	34						2.88					0.9879 ± 0.0001	15%
	83	34						3.92					0.9976 ± 0.0001	1.1%
	83	34											1.0206 ± 0.0001	1.9%
	83	34											1.0305 ± 0.0001	20.6%
	83	34											1.0634 ± 0.0002	3.5%
	83	34											1.6600 ± 0.0001	1.7%
	83	34											2.0514 ± 0.0002	10.7%
$_{34}Se^{83}$	83	34		82.919117	22.3 m.	β-	3.67	0.93			9/2+		0.22516 ± 0.00006	33%
	83	34						1.51					0.35666 ± 0.00006	69%
	83	34											0.51004 ± 0.00008	45%
	83	34											0.7180 ± 0.0001	16%
	83	34											0.7990 ± 0.0001	16%
	83	34											0.8666 ± 0.0001	9.1%
	83	34											0.8836 ± 0.0001	7.2%
	83	34											1.0641 ± 0.0001	6%
	83	34											1.0820 ± 0.0001	2.8%
	83	34											1.1917 ± 0.0001	4.2%
	83	34											1.299 ± 0.0001	6.0%
	83	34											1.8948 ± 0.0001	7.2%
	83	34											2.2902 ± 0.0003	9.5%
	83	34											2.3374 ± 0.0003	3.5%
$_{34}Se^{84}$	84	34		83.918463	3.3 m.	β-	1.83	1.41	100%		0+		0.4088 ± 0.0005	100%
$_{34}Se^{85}$	85	34		84.922260	32 s.	β-	6.2	5.9			5/2+		0.3450 ± 0.001	22%
	85	34											0.6094 ± 0.001	41%
	85	34											0.941 ± 0.001	2.2%
	85	34											0.954 ± 0.001	2.9%
	85	34											1.207 ± 0.001	2.9%
	85	34											1.373 ± 0.001	1%
	85	34											1.428 ± 0.001	2.2%
	85	34											2.237 ± 0.001	1.3%
	85	34											2.418 ± 0.001	1%
	85	34											2.456 ± 0.001	1%
	85	34											3.376 ± 0.001	3.2%
	85	34											3.657 ± 0.001	1.5%
	85	34											3.685 ± 0.001	1%

TABLE OF THE ISOTOPES (Continued)

Isotope	A	Z	% Natural abundance	Atomic mass	Half-life	Decay mode	Decay energy (MeV)	Particle energy (MeV)	Particle intensity	Thermal neutron cross section	Spin (h/2π)	μ Nucl. mag. moment	Gamma-ray energy (MeV)	Gamma-ray intensity
	85	34											3.775 ± 0.001	1%
$_{34}Se^{86}$	86	34		85.924270	15 s.	β-	5.1				5/2+		0.7881 ± 0.001	4 +
	86	34											1.1183 ± 0.001	5
	86	34											1.4003 ± 0.0006	14
	86	34											2.0124 ± 0.001	24
	86	34											2.2416 ± 0.001	17
	86	34											2.4433 ± 0.0008	100
	86	34											2.6619 ± 0.001	49
$_{34}Se^{87}$	87	34		86.928390	5.6 s.	β-	≈7						0.468 ± 0.001	100 +
	87	34				n							1.4979 ± 0.001	23
$_{34}Se^{88}$	88	34			1.5 s.	β-,n	≈7						0.5346	
$_{34}Se^{89}$	89	34			0.41 s.	β-,n								
Br		35		79.904						6.8 b.				
$_{35}Br^{72}$	72	35		71.936630	1.31 m.	β+	≈9.0				3		ann.rad.	
	72	35											0.3799 ± 0.0003	3.4%
	72	35											0.4547 ± 0.0003	14%
	72	35											0.7528 ± 0.0004	2.5%
	72	35											0.7748 ± 0.0003	7%
	72	35											0.8620 ± 0.0002	70%
	72	35											1.0547 ± 0.0003	3.5%
	72	35											1.0616 ± 0.0003	5%
	72	35											1.1364 ± 0.0004	17%
	72	35											1.3167 ± 0.0003	17%
	72	35											1.5098 ± 0.0004	2.7%
	72	35											1.5713 ± 0.0004	3.8%
	72	35											1.7240 ± 0.0005	3.4%
	72	35											2.3719 ± 0.0007	7%
$_{35}Br^{73}$	73	35		72.931680	3.4 m.	β+	4.6	3.7			3/2-		ann.rad.	
	73	35											0.0649 ± 0.0001	100 +
	73	35											0.1255 ± 0.0001	23
	73	35											0.2751 ± 0.0002	10
	73	35											0.3352 ± 0.0002	34
	73	35											0.4006 ± 0.0002	20
	73	35											0.6995 ± 0.0002	40
	73	35											0.8487 ± 0.0004	20
	73	35											0.9136 ± 0.0002	19
	73	35											0.9307 ± 0.0002	22
	73	35											0.9956 ± 0.0002	7
$_{35}Br^{74m}$	74	35			41.5 m.	β+		4.5			4-		ann.rad.	200%
	74	35						5.2					0.6152 ± 0.0001	8%
	74	35											0.6343 ± 0.0002	19%
	74	35											0.6348 ± 0.0001	98%
	74	35											0.7285 ± 0.0001	38%
	74	35											0.8389 ± 0.0002	6%
	74	35											1.2495 ± 0.0002	7.6%
	74	35											1.2691 ± 0.0002	8.8%
	74	35											1.7149 ± 0.0003	6.5%
	74	35											2.2838 ± 0.0003	3%
	74	35											2.3119 ± 0.0003	3.7%
	74	35											3.9576 ± 0.0006	3.8%
	74	35											(0.2 - 4.38)weak	
$_{35}Br^{74}$	74	35		73.929898	25.3 m.	β+	6.92						ann.rad.	
	74	35											0.6341 ± 0.0002	20%
	74	35											0.6348 ± 0.0001	68%
	74	35											1.0228 ± 0.0001	5.6%
	74	35											1.2689 ± 0.0001	7.6%
	74	35											1.8428 ± 0.0002	2.5%
	74	35											2.1306 ± 0.0002	2%
	74	35											2.3961 ± 0.0002	8%
	74	35											2.6152 ± 0.0002	8%
	74	35											2.6616 ± 0.0003	5.5%
	74	35											2.7708 ± 0.0005	2.4%
	74	35											3.2499 ± 0.0005	6.7%
	74	35											3.6246 ± 0.0003	6%
	74	35											3.6319 ± 0.0005	2.6%
	74	35											3.9727 ± 0.0002	2.5%
	74	35											4.3796 ± 0.0004	4.5%
	74	35											(0.2 - 4.7)weak	
$_{35}Br^{75}$	75	35		74.925753	98 m.	β+(76%)	3.0				3/2-		ann.rad.	150%
	75	35				E.C.(24%)							0.14119 ± 0.0001	7%
	75	35											0.28650 ± 0.0002	92%
	75	35											0.29285 ± 0.0002	2.8%
	75	35											0.37739 ± 0.0001	4.1%
	75	35											0.43175 ± 0.0001	4%
	75	35											0.57293 ± 0.0001	2%
	75	35											0.73394 ± 0.0001	1.6%
	75	35											0.91205 ± 0.0001	1.1%
	75	35											0.95210 ± 0.0001	1.7%
	75	35											(0.1 - 1.56)weak	
$_{35}Br^{76m}$	76	35			1.49 s.	I.T.	5.05				4+		0.104548 ± 0.00005	
	76	35											0.05711 ± 0.00005	
$_{35}Br^{76}$	76	35		75.924528	16.1 h.	β+(57%)	4.956	1.9			1-	0.5482	ann.rad.	130%

Isotope	A	Z	% Natural abundance	Atomic mass	Half-life	Decay mode	Decay energy (MeV)	Particle energy (MeV)	Particle intensity	Thermal neutron cross section	Spin (h/2π)	μ Nucl. mag. moment	Gamma-ray energy (MeV)	Gamma-ray intensity
	76	35				E.C.(43%)		3.15					0.47291 ± 0.00006	1.9%
	76	35						3.68					0.55911 ± 0.00005	74%
	76	35											0.56322 ± 0.00005	3%
	76	35											0.65700 ± 0.00005	5.3%
	76	35											1.12985 ± 0.00006	4.5%
	76	35											1.2160 ± 0.00005	9%
	76	35											1.22865 ± 0.00006	2.1%
	76	35											1.4752 ± 0.00006	2.4%
	76	35											1.85368 ± 0.00005	14.9%
	76	35											2.11127 ± 0.00008	2.4%
	76	35											2.79272 ± 0.00006	5.4%
	76	35											2.95055 ± 0.00005	7.6%
	76	35											(0.4 - 4.6)weak	
$_{35}Br^{77m}$	77	35			4.3 m.	I.T.	0.1059				9/2+		0.1059 ± 0.0002	13.7%
$_{35}Br^{77}$	77	35		76.921378	57.0 h.	E.C.(99%)	1.365				1+		ann.rad.	
	77	35				β+(0.74%)							0.08759 ± 0.0007	1.3%
	77	35											0.16183 ± 0.00008	1.0%
	77	35											0.20040 ± 0.00007	1.1%
	77	35											0.23898 ± 0.00007	23%
	77	35											0.24977 ± 0.00007	2/9%
	77	35											0.29723 ± 0.00008	4.0%
	77	35											0.30376 ± 0.00009	2.0%
	77	35											0.43947 ± 0.00006	1.5%
	77	35											0.52069 ± 0.00006	22%
	77	35											0.57464 ± 0.00008	1.1%
	77	35											0.57891 ± 0.00007	2.8%
	77	35											0.58548 ± 0.00007	1.5%
	77	35											0.75535 ± 0.00007	1.6%
	77	35											0.81779 ± 0.00006	2.0%
	77	35											1/00505 ± 0.00006	0.9%
	77	35											(0.08 - 1.2)weak	
$_{35}Br^{78}$	78	35		77.921144	6.46 m.	β+(92%)	3.3574	1.2			1+		ann.rad.	185%
	78	35				E.C.(8%)		2.5					0.61363 ± 0.00006	13.6%
	78	35											0.8848 ± 0.0001	0.07%
	78	35											1.3086 ± 0.0001	0.04%
	78	35											1.7210 ± 0.0002	0.04%
	78	35											1.9239 ± 0.0002	0.05%
	78	35											2.4767 ± 0.0004	0.02%
	78	35											(0.7 - 3.0)weak	
$_{35}Br^{79m}$	79	35			4.86 s.	I.T.	0.207				9/2+	+2.106	0.2072 ± 0.0004	76%
$_{35}Br^{79}$	79	35	50.69%	78.918336						(2.5 + 8.2)b.	3/2-	+2.1064		
$_{35}Br^{80m}$	80	35			4.42 h.	I.T.	0.04885				5-	+1.3177	Br k x-ray	93%
	80	35											0.03705 ± 0.00002	39%
	80	35											0.04885 ± 0.00003	0.5%
$_{35}Br^{80}$	80	35		79.918528	17.6 m.	β-(92%)	2.00	1.38 β-	7.6%		1+	0.5140	ann.rad.	
	80	35				E.C.(5.7%)	1.870	1.99 β-	82%				0.6162 ± 0.0005	6.7%
	80	35				β+(2.6%)		0.85 β+	2.8%				0.6394 ± 0.0002	0.2%
	80	35											0.6658 ± 0.0002	1.1%
	80	35											0.7038 ± 0.0002	0.13%
	80	35											1.2561 ± 0.0004	0.09%
$_{35}Br^{81}$	81	35	49.31%	80.916289						(2.4 + 0.26)b.	3/2-	+2.2706		
$_{35}Br^{82m}$	82	35				I.T.(98%)	0.046				2-		0.046	0.2%
	82	35				β-(2%)	3.139						0.6985 ± 0.0002	0.025%
	82	35											0.77645 ± 0.0002	0.2%
	82	35											1.4748 ± 0.0002	0.02%
$_{35}Br^{82}$	82	35		81.916802	35.30 h.	β-	3.093	0.444			5-	+1.6270	0.221411 ± 0.00002	2.3%

Isotope	A	Z	% Natural abundance	Atomic mass	Half-life	Decay mode	Decay energy (MeV)	Particle energy (MeV)	Particle intensity	Thermal neutron cross section	Spin (h/2π)	μ Nucl. mag. moment	Gamma-ray energy (MeV)	Gamma-ray intensity
	82	35											0.55432 ± 0.00001	71%
	82	35											0.61905 ± 0.00002	43%
	82	35											0.69832 ± 0.00002	29%
	82	35											0.77649 ± 0.00003	83%
	82	35											0.82781 ± 0.00002	24%
	82	35											1.04398 ± 0.00003	27%
	82	35											1.3747 ± 0.00005	27%
	82	35											1.47482 ± 0.00008	17%
$_{35}Br^{83}$	83	35		82.915179	2.39 h.	β-	0.98	0.395	1%		3/2-		0.52041 ± 0.00005	0.06%
	83	35						0.925	99%				0.52964 ± 0.00001	1.3%
	83	35											0.5526 ± 0.0001	0.02%
$_{35}Br^{84m}$	84	35			6.0 m.	β-		2.2	100%		(6-)		0.4240 ± 0.001	100%
	84	35											0.4472 ± 0.001	2%
	84	35											0.8816 ± 0.001	98%
	84	35											1.0160 ± 0.001	1%
	84	35											1.4628 ± 0.0005	97%
	84	35											1.8972 ± 0.0008	2%
$_{35}Br^{84}$	84	35		83.916503	31.8 m.	β-	4.65	2.70	11%		2-		0.6048 ± 0.0003	1.7%
	84	35						3.81	20%				0.7365 ± 0.0003	1.2%
	84	35						4.63	34%				0.8022 ± 0.0002	5.7%
	84	35											0.8816 ± 0.0001	42%
	84	35											1.0159 ± 0.0003	6.2%
	84	35											1.2133 ± 0.0002	2.6%
	84	35											1.4638 ± 0.0007	2.0%
	84	35											1.7412 ± 0.0004	1.6%
	84	35											1.8775 ± 0.0004	2.1%
	84	35											1.8976 ± 0.0002	14.9%
	84	35											2.0296 ± 0.0005	2.1%
	84	35											2.4841 ± 0.0003	6.7%
	84	35											2.8241 ± 0.0004	1.1%
	84	35											3.0454 ± 0.0004	2.5%
	84	35											3.2353 ± 0.0005	2.0%
	84	35											3.3658 ± 0.0004	2.9%
	84	35											3.9275 ± 0.0004	6.8%
$_{35}Br^{85}$	85	35		84.915612	2.87 m.	β-	2.5				3/2-		0.30486 ± 0.00002	14%
	85	35											0.80241 ± 0.0001	2.5%
	85	35											0.92463 ± 0.00005	1.6%
	85	35											1.7270 ± 0.0001	0.4%
	85	35											1.8325 ± 0.0001	0.1%
	85	35											(0.09 - 2.4)weak	
$_{35}Br^{86}$	86	35		85.918800	β-	7.6					(2-)		0.5012 ± 0.0005	1.6%
	86	35											0.6853 ± 0.0001	1.1%
	86	35											0.78496 ± 0.0001	3.6%
	86	35											1.21702 ± 0.0001	6.6%
	86	35											1.2861 ± 0.0001	7.7%
	86	35											1.38973 ± 0.0001	10.2%
	86	35											1.46509 ± 0.0001	7.4%
	86	35											1.53424 ± 0.0001	7.7%
	86	35											1.56460 ± 0.0001	62%
	86	35											1.9663 ± 0.0001	6%
	86	35											2.34937 ± 0.0001	10%
	86	35											2.75106 ± 0.0001	19%
	86	35											3.0090 ± 0.0003	1%
	86	35											3.7588 ± 0.0003	1%
	86	35											3.7831 ± 0.0003	1.1%
	86	35											4.4012 ± 0.0003	1%
	86	35											4.88512 ± 0.0002	1.2%
	86	35											5.40580 ± 0.0002	5.6%
	86	35											5.5176 ± 0.0003	3.6%
	86	35											6.2116 ± 0.0003	0.8%
	86	35											(0.5 - 6.8)weak	
$_{35}Br^{87}$	87	35		86.920690	56.1 s.	β-	6.8	6.1			3/2-		0.42182 ± 0.00006	5%
	87	35											0.5294 ± 0.0001	1.2%
	87	35											0.53190 ± 0.00007	8%
	87	35											0.6105 ± 0.0001	1.1%
	87	35											0.9528 ± 0.0001	1%
	87	35											1.0213 ± 0.0001	1.9%
	87	35											1.36089 ± 0.0001	5%
	87	35											1.41983 ± 0.00009	32%
	87	35											1.4494 ± 0.0001	1.8%
	87	35											1.4762 ± 0.0001	12%

Isotope	A	Z	% Natural abundance	Atomic mass	Half-life	Decay mode	Decay energy (MeV)	Particle energy (MeV)	Particle intensity	Thermal neutron cross section	Spin (h/2π)	μ Nucl. mag. moment	Gamma-ray energy (MeV)	Gamma-ray intensity
	87	35											1.5777 ± 0.0001	9%
	87	35											1.6073 ± 0.0001	1.8%
	87	35											1.9784 ± 0.0001	1%
	87	35											2.07156 ± 0.0001	3.5%
	87	35											2.4526 ± 0.0001	1%
	87	35											2.7051 ± 0.0001	2.6%
	87	35											2.8212 ± 0.0001	2.5%
	87	35											2.8366 ± 0.0001	2.4%
	87	35											3.0273 ± 0.0001	1.9%
	87	35											3.9173 ± 0.0001	2.9%
	87	35											4.1826 ± 0.0001	7%
	87	35											4.7848 ± 0.0001	27%
	87	35											4.9618 ± 0.0002	2.8%
	87	35											5.1205 ± 0.0002	0.8%
	87	35											5.1955 ± 0.0002	0.7%
	87	35											5.2013 ± 0.0003	1.0%
	87	35											(0.2 - 6.1)weak	
$_{35}Br^{88}$	88	35		87.924080	16.4 s.	β-	9.0				1-		0.7649 ± 0.0005	1.6%
	88	35				n							0.7753 ± 0.0001	77%
	88	35											0.7934 ± 0.0005	1.2%
	88	35											0.8021 ± 0.0001	16%
	88	35											1.0537 ± 0.0001	1.8%
	88	35											1.3690 ± 0.0001	1.1%
	88	35											1.4407 ± 0.0001	5%
	88	35											1.5670 ± 0.0001	2.3%
	88	35											1.6442 ± 0.0001	3.1%
	88	35											2.6245 ± 0.0001	1.8%
	88	35											2.9457 ± 0.0002	2.4%
	88	35											3.0194 ± 0.0002	1.2%
	88	35											3.9322 ± 0.0002	5%
	88	35											4.0217 ± 0.0002	2.6%
	88	35											4.1479 ± 0.0001	4%
	88	35											4.5628 ± 0.0002	3.2%
	88	35											4.7134 ± 0.0002	1.1%
	88	35											4.7212 ± 0.0005	1%
	88	35											4.985 ± 0.0005	1%
	88	35											5.0193 ± 0.0005	2.1%
	88	35											(0.1 - 6.99)weak	
$_{35}Br^{89}$	89	35		88.926550	4.4 s.	β-	8.5				3/2-		0.7753	
	89	35				n							1.0978	
$_{35}Br^{90}$	90	35		89.931010	1.9 s.	β-	10.7	8.3			2-		0.6555	18 +
	90	35				n		9.8					0.7071	100
	90	35											1.3626	27
	90	35											2.1282	17
	90	35											2.2528	16
	90	35											3.3454	10
$_{35}Br^{91}$	91	35			0.54 s.	β-(90%)							0.263	
	91	35				β-n(10%)							0.803	
$_{35}Br^{92}$	92	35			0.36 s.	β-							0.740	
	92	35				β-n								
Kr		36	83.80							25 b.				
$_{36}Kr^{72}$	72	36		71.942060	17 s.	β+	5.1				0+		ann.rad.	190%
	72	36				E.C.							0.1626 ± 0.0002	8%
	72	36											0.2522 ± 0.0002	3%
	72	36											0.3100 ± 0.0002	15%
	72	36											0.4150 ± 0.0002	19%
	72	36											0.5766 ± 0.0002	6%
$_{36}Kr^{73}$	73	36		72.938920	27 s.	β+	6.7						ann.rad.	
	73	36				E.C.							0.1511 ± 0.0004	13%
	73	36				β+,p							0.1781 ± 0.0003	66%
	73	36											0.2136 ± 0.0004	9%
	73	36											0.2413 ± 0.0004	7%
	73	36											0.3036 ± 0.0004	4%
	73	36											0.3292 ± 0.0004	4%
	73	36											0.3919 ± 0.0004	7%
	73	36											0.4736 ± 0.0004	11%
$_{36}Kr^{74}$	74	36		73.933290	11.5 m.	β+	3.3				0+		ann.rad.	74%
	74	36				E.C.							0.00985 ± 0.00002	5%
	74	36											0.0628 ± 0.0001	10%
	74	36											0.06740 ± 0.00001	1.2%
	74	36											0.08970 ± 0.0001	31%
	74	36											0.0938 ± 0.0001	3%
	74	36											0.1234 ± 0.0001	9.3%
	74	36											0.1403 ± 0.0001	3.3%
	74	36											0.14970 ± 0.0001	1%
	74	36											0.2030 ± 0.0001	19%
	74	36											0.2169 ± 0.0001	10%
	74	36											0.2339 ± 0.0001	5.3%
	74	36											0.2967 ± 0.0001	1%
	74	36											0.3065 ± 0.0001	10.5%
	74	36											0.6091 ± 0.0001	1%
	74	36											0.7013 ± 0.0001	1.7%

Isotope	A	Z	% Natural abundance	Atomic mass	Half-life	Decay mode	Decay energy (MeV)	Particle energy (MeV)	Particle intensity	Thermal neutron cross section	Spin (h/2π)	μ Nucl. mag. moment	Gamma-ray energy (MeV)	Gamma-ray intensity
	74	36											(0.02 - 0.9)weak	
$_{36}Kr^{75}$	75	36		74.931029	4.5 m.	β^+	5.0	3.2					ann.rad.	190%
	75	36				E.C.							0.0884 ± 0.0001	1.6%
	75	36											0.1325 ± 0.0001	31%
	75	36											0.1533 ± 0.0001	3.7%
	75	36											0.1547 ± 0.0002	9.6%
	75	36											0.7931 ± 0.0002	0.8%
	75	36											1.3463 ± 0.0004	0.65%
	75	36											1.6018 ± 0.0004	0.84%
$_{36}Kr^{76}$	76	36		75.925959	14.8 h.	E.C.	1.35				0+		Br k x-ray	51%
	76	36											0.0455 ± 0.0001	18%
	76	36											0.1032 ± 0.0001	3%
	76	36											0.2522 ± 0.0002	6.5%
	76	36											0.2718 ± 0.0002	4%
	76	36											0.2992 ± 0.0002	1%
	76	36											0.3099 ± 0.0002	2%
	76	36											0.3158 ± 0.0002	38%
	76	36											0.3353 ± 0.0002	5%
	76	36											0.4065 ± 0.0002	11%
	76	36											0.4521 ± 0.0002	8.8%
	76	36											0.5528 ± 0.0003	1.3%
	76	36											0.5815 ± 0.0004	0.9%
	76	36											0.5825 ± 0.0004	0.9%
	76	36											(0.03 - 1.07)	
$_{36}Kr^{77}$	77	36		76.924610	1.24 h.	β^+(80%)	0.90				5/2+		ann.rad.	
	77	36				E.C>(20%)		1.55					0.1062 ± 0.0002	1.2%
	77	36						1.70					0.1297 ± 0.0001	84%
	77	36						1.87					0.1465 ± 0.0001	3.9%
	77	36											0.2762 ± 0.0002	3.2%
	77	36											0.3122 ± 0.0002	3.6%
	77	36											0.6060 ± 0.0001	0.4%
	77	36											0.7346 ± 0.0002	0.4%
	77	36											(0.1 - 2.3)weak	
$_{36}Kr^{78}$	78	36	0.35%							(0.17 + 6) b.	0+			
$_{36}Kr^{79m}$	79	36			50 s.	I.T.	0.1300				7/2+		Kr x-ray	
	79	36											0.13001 ± 0.00001	27%
$_{36}Kr^{79}$	79	36		78.920084	35.0 h.	β^+(7%)	1.63				1/2-		ann.rad.	
	79	36				E.C.(93%)							0.21702 ± 0.0001	2.4%
	79	36											0.2613 ± 0.0001	12.7%
	79	36											0.2995 ± 0.0001	1.7%
	79	36											0.3063 ± 0.0001	2.6%
	79	36											0.3890 ± 0.0001	1.6%
	79	36											0.39756 ± 0.0001	9.5%
	79	36											0.6061 ± 0.0001	8.1%
	79	36											0.8320 ± 0.0001	1.3%
	79	36											1.11514 ± 0.0003	0.4%
	79	36											1.33213 ± 0.0001	0.44%
	79	36											(0.13 - 1.3)weak	
$_{36}Kr^{80}$	80	36	2.25%	79.916380						(4.6 + 8) b.	0+			
$_{36}Kr^{81m}$	81	36			13.3 s.	I.T.	0.1903				1/2-		0.19030 ± 0.0001	67%
$_{36}Kr^{81}$	81	36		80.916590	2.1x10^5 y.	E.C.	0.276				7/2+		Br k x-ray	53%
	81	36											0.2760 ± 0.0001	3.6%
$_{36}Kr^{82}$	82	36	11.6%	81.913482						(16 + 20) b.	0+			
$_{36}Kr^{83m}$	83	36			1.86 h.	I.T.	0.04155				1/2-		Kr k x-ray	13.2%
	83	36											0.00940 ± 0.00002	4.9%
	83	36											0.03216 ± 0.00002	0.04%
$_{36}Kr^{83}$	83	36	11.5%	82.914135						1800 b.	9/2+	-0.9707		
$_{36}Kr^{84}$	84	36	57.0%	83.911507						(0.09 + 0.042) b.	0+			
$_{36}Kr^{85m}$	85	36			4.48 h.	β^-(79%)		0.83	79%		1/2-		0.30487 ± 0.00002	14%
	85	36				I.T.(21%)	0.304						0.15118 ± 0.00001	79%
$_{36}Kr^{85}$	85	36		84.912531	10.72 y.	β^-	0.687	0.15	0.4%		9/2+	1.005	0.51399 ± 0.00002	0.43%
$_{36}Kr^{86}$	86	36	17.3%	85.910616						0.003 b.	0+			
$_{36}Kr^{87}$	87	36		86.913360	76.3 m.	β^-	3.886	0.93	4%		5/2+		0.40258 ± 0.00002	50%
	87	36						1.33	8%				0.6739 ± 0.0001	1.1%
	87	36					1.45	5%					0.84545 ± 0.00004	8%
	87	36						3.0	7%				1.1747 ± 0.0001	1.1%
	87	36					3.49	43%					1.7405 ± 0.0001	2%
	87	36						3.89	30%				2.01181 ± 0.0001	2.1%
	87	36											2.4084 ± 0.0002	0.5%
	87	36											2.5548 ± 0.0002	9%
	87	36											2.5583 ± 0.0002	4%
	87	36											3.3084 ± 0.0005	0.5%
$_{36}Kr^{88}$	88	36		87.914453	2.84 h.	β^-	2.91				0+		0.0275 ± 0.00001	2%
	88	36											0.16598 ± 0.00004	3.2%

Isotope	A	Z	% Natural abundance	Atomic mass	Half-life	Decay mode	Decay energy (MeV)	Particle energy (MeV)	Particle intensity	Thermal neutron cross section	Spin (h/2π)	μ Nucl. mag. moment	Gamma-ray energy (MeV)	Gamm ray intensi
	88	36											0.19632 ± 0.00002	26.6%
	88	36											0.36223 ± 0.00001	2.3%
	88	36											0.83482 ± 0.00001	13.3%
	88	36											0.98578 ± 0.00002	1.3%
	88	36											1.14133 ± 0.00006	1.3%
	88	36											1.17951 ± 0.00003	1.0%
	88	36											1.25067 ± 0.00004	1.1%
	88	36											1.3695 ± 0.0002	1.5%
	88	36											1.5184 ± 0.0001	2.2%
	88	36											1.5298 ± 0.0001	1.3%
	88	36											2.0298 ± 0.0003	4.6%
	88	36											2.0354 ± 0.0001	3.8%
	88	36											2.1958 ± 0.0001	1.3%
	88	36											2.2318 ± 0.0001	3.5%
	88	36											2.39202 ± 0.0001	3.5%
$_{36}Kr^{89}$	89	36		88.917640	3.16 m.	β-	4.99	3.8			5/2+		0.19746 ± 0.00003	1.9%
	89	36						4.6					0.2209 ± 0.0001	2%
	89	36						4.9					0.3450 ± 0.0001	1.2%
	89	36											0.3561 ± 0.0001	4.2%
	89	36											0.3693 ± 0.0001	1.4%
	89	36											0.4114 ± 0.0001	2.6%
	89	36											0.4975 ± 0.0001	6.8%
	89	36											0.4986 ± 0.0003	1.2%
	89	36											0.5769 ± 0.0001	5.8%
	89	36											0.5858 ± 0.0001	16.8%
	89	36											0.6962 ± 0.0001	1.8%
	89	36											0.7384 ± 0.0001	4.3%
	89	36											0.8671 ± 0.0001	6.0%
	89	36											0.9043 ± 0.0001	7.3%
	88	36											1.1078 ± 0.0001	3%
	89	36											1.1161 ± 0.0001	1.7%
	89	36											1.2737 ± 0.0001	1.4%
	89	36											1.3243 ± 0.0001	3.12%
	89	36											1.4728 ± 0.0001	7%
	89	36											1.5010 ± 0.0001	1.3%
	89	36											1.5300 ± 0.0001	1.3%
	89	36											1.5337 ± 0.0001	5.2%
	89	36											1.6937 ± 0.0001	4.5%
	89	36											1.9034 ± 0.0001	1%
	89	36											2.0122 ± 0.0001	1.6%
	89	36											2.8662 ± 0.0002	1.8%
	89	36											3.1403 ± 0.0002	1.1%
	89	36											3.3617 ± 0.0002	1.1%
	89	36											3.5329 ± 0.0002	1.4%
	89	36											(0.2 - 4.7)weak	
$_{36}Kr^{90}$	90	36		89.919520	32.3 s.	β-	4.39	2.6	77%		0+		0.12182 ± 0.00003	33.5%
	90	36						2.8	6%				0.49263 ± 0.00005	1.2%
	90	36											0.5395 ± 0.0001	30.7%
	90	36											0.5544 ± 0.0001	5.0%
	90	36											0.6191 ± 0.0001	1.1%
	90	36											0.7313 ± 0.0001	1.5%
	90	36											1.1187 ± 0.0001	38.9%
	90	36											1.4238 ± 0.0001	2.9%
	90	36											1.5378 ± 0.0001	9.6%
	90	36											1.5522 ± 0.0001	2.2%
	90	36											1.6582 ± 0.0001	1.3%
	90	36											1.7801 ± 0.0001	6.7%
	90	36											2.1275 ± 0.0001	1.4%
	90	36											2.7267 ± 0.0001	0.9%
	90	36											(0.1 - 4.2)weak	
$_{36}Kr^{91}$	91	36		90.923380	8.6 s.	β-	6.4	4.33			5/2+		0.10878 ± 0.00004	42%
	91	36						4.59					0.4120 ± 0.0001	2.3%
	91	36						4.98					0.50658 ± 0.0001	19%
	91	36						5.4					0.5556 ± 0.0001	2%
	91	36											0.6129 ± 0.0001	7.6%
	91	36											0.6301 ± 0.0001	2.2%
	91	36											0.6624 ± 0.0001	1.3%
	91	36											0.7610 ± 0.0001	1.0%
	91	36											0.8749 ± 0.0001	1.3%
	91	36											1.0249 ± 0.0002	2.9%
	91	36											1.1087 ± 0.0001	7.1%
	91	36											1.1368 ± 0.0001	1.0%
	91	36											1.1780 ± 0.0001	1.3%
	91	36											1.3043 ± 0.0001	1.24%

Isotope	A	Z	% Natural abundance	Atomic mass	Half-life	Decay mode	Decay energy (MeV)	Particle energy (MeV)	Particle intensity	Thermal neutron cross section	Spin (h/2π)	μ Nucl. mag. moment	Gamma-ray energy (MeV)	Gamma-ray intensity
	91	36											1.5016 ± 0.0001	4.8%
	91	36											1.6667 ± 0.0001	1.0%
	91	36											2.4843 ± 0.0001	2.8%
	91	36											2.7358 ± 0.0002	1.5%
	91	36											2.9818 ± 0.0002	1.3%
	91	36											3.0568 ± 0.0002	0.9%
	91	36											3.1135 ± 0.0002	2.1%
	91	36											(0.1 - 4.4)weak	
$_{36}Kr^{92}$	92	36		91.926270	1.84 s.	β-	6.2						0.1424 ± 0.0001	64%
	92	36				n							0.3168 ± 0.0001	5.8%
	92	36											0.3423 ± 0.0001	2.1%
	92	36											0.4847 ± 0.0001	3.2%
	92	36											0.54830 ± 0.0001	14%
	92	36											0.6237 ± 0.0001	1.3%
	92	36											0.8126 ± 0.0001	14.5%
	92	36											0.8763 ± 0.0001	4.2%
	92	36											1.0442 ± 0.0001	4.7%
	92	36											1.21860 ± 0.0001	59.8%
	92	36											1.3608 ± 0.0001	3.5%
	92	36											1.8968 ± 0.0002	0.8%
	92	36											2.0390 ± 0.0002	0.4%
	92	36											2.8328 ± 0.0002	0.3%
	92	36											3.1995 ± 0.0002	0.4%
	92	36											(0.15 - 3.7)weak	
$_{36}Kr^{93}$	93	36		92.931130	1.29 s.	β-	8.5	7.1					0.1820 ± 0.0001	5.4%
	93	36				n							0.2523 ± 0.0002	19.6%
	93	36											0.2536 ± 0.0002	41%
	93	36											0.26683 ± 0.00005	21%
	93	36											0.32309 ± 0.00002	24%
	93	36											0.4966 ± 0.0001	1.8%
	93	36											0.5702 ± 0.0001	1.2%
	93	36											0.8204 ± 0.001	3.7%
	93	36											1.0262 ± 0.0001	2.2%
	93	36											1.2150 ± 0.0001	1.8%
	93	36											1.2388 ± 0.0001	1.1%
	93	36											1.2961 ± 0.0001	1.9%
	93	36											1.3879 ± 0.0001	1.3%
	93	36											1.4353 ± 0.0001	1.0%
	93	36											1.5058 ± 0.0001	2.2%
	93	36											1.5962 ± 0.0001	1.4%
	93	36											1.6271 ± 0.0001	2.0%
	93	36											1.6411 ± 0.0001	1.4%
	93	36											1.6978 ± 0.0001	1.4%
	93	36											1.7425 ± 0.0001	1.3%
	93	36											1.9618 ± 0.0001	1.8%
	93	36											2.0189 ± 0.0001	1.4%
	93	36											2.0354 ± 0.0001	1.8%
	93	36											2.1815 ± 0.0001	1.2%
	93	36											2.3499 ± 0.0001	7.4%
	93	36											2.4960 ± 0.0001	2.3%
	93	36											2.5613 ± 0.0001	1.0%
	93	36											2.6026 ± 0.0001	4.2%
	93	36											2.8559 ± 0.0001	2.2%
	93	36											3.2267 ± 0.0002	1.0%
	93	36											3.4607 ± 0.0006	0.7%
$_{36}Kr^{94}$	94	36			0.21 s.	β-	7.5						0.1353 ± 0.001	13 +
	94	36				n							0.1874 ± 0.001	37
	94	36											0.2196 ± 0.001	67
	94	36											0.2881 ± 0.001	33
	94	36											0.3208 ± 0.001	26
	94	36											0.3546 ± 0.001	28
	94	36											0.3590 ± 0.001	39
	94	36											0.3948 ± 0.001	21
	94	36											0.4022 ± 0.001	20
	94	36											0.5933 ± 0.001	29
	94	36											0.6293 ± 0.0001	100
	94	36											0.6957 ± 0.001	25
Rb		37		85.4678						0.38 b.				
$_{37}Rb^{75}$	75	37		74.938510	17 s.	β+	6.6	2.31					ann. rad.	71%
	75	37											0.0628 ± 0.0001	10%
	75	37											0.08970 ± 0.0001	31%
	75	37											0.0938 ± 0.0001	3.3%
	75	37											0.1234 ± 0.0001	9.3%
	75	37											0.1403 ± 0.0001	9%
	75	37											0.1497 ± 0.0001	2.1%
	75	37											0.2030 ± 0.0001	19/4%
	75	37											0.2169 ± 0.0001	10%
	75	37											0.2339 ± 0.0001	5.3%
	75	37											0.2967 ± 0.0001	11%
	75	37											0.3065 ± 0.0001	10.5%
	75	37											0.6091 ± 0.0001	1%
	75	37											0.7013 ± 0.0001	1.7%

Isotope	A	Z	% Natural abundance	Atomic mass	Half-life	Decay mode	Decay energy (MeV)	Particle energy (MeV)	Particle intensity	Thermal neutron cross section	Spin (h/2π)	μ Nucl. mag. moment	Gamma-ray energy (MeV)	Gamma-ray intensity
	75	37											0.9699 ± 0.0001	0.25%
	75	37											(0.06 - 1.1)weak	
$_{37}Rb^{76}$	76	37		75.934960	17 s.	β^+	8.2	5.2					ann.rad.	425 +
	76	37											0.3452	16
	76	37											0.3549	25
	76	37											0.4235	100
	76	37											0.800	7
	76	37											0.885	9
	76	37											0.919	8
	76	37											0.974	4
	76	37											1/173	4
	76	37											1.219	5
$_{37}Rb^{77}$	77	37		76.930280	3.8 m.	β^+	5.10	3.86			3/2-		ann.rad.	
	77	37											0.0665 ± 0.0001	66%
	77	37											0.1785 ± 0.0003	26%
	77	37											0.3933 ± 0.0003	12%
	77	37											0.6085 ± 0.0008	3.6%
	77	37											0.6265 ± 0.0004	4.3%
	77	37											0.7798 ± 0.0008	1.2%
	77	37											0.9587 ± 0.0004	2.4%
	77	37											0.9708 ± 0.0004	4%
	77	37											0.9883 ± 0.0004	2.4%
	77	37											(0.04 - 1.8)weak	
$_{37}Rb^{78m}$	78	37			5.7 m.	I.T.	0.1034				4-		ann.rad.	247 +
	78	37				β^+							0.10336 ± 0.0001	8
	78	37				E.C.							0.4553 ± 0.0001	100
	78	37											0.6647 ± 0003	44
	78	37											0.6927 ± 0.0003	13
	78	37											0.7251 ± 0.0003	7
	78	37											0.7534 ± 0.0003	8
	78	37											1.1994 ± 0.0003	9
	78	37											1.2325 ± 0.0001	2
	78	37											1.5301 ± 0.0005	5
	78	37											1.6303 ± 0.0004	6.6
	78	37											1.6446 ± 0.0004	9
	78	37											1.8530 ± 0.0005	7
	78	37											1.9440 ± 0.0005	9
	78	37											2.0136 ± 0.0005	5
	78	37											2.1180 ± 0.0007	4
	78	37											2.6273 ± 0.0007	4
	78	37											3.0830 ± 0.0007	1.7
	78	37											3.3180 ± 0.001	1.5
$_{37}Rb^{78}$	78	37		77.928090	17.66 m.	β^+	7.02				0+		ann.rad.	200 +
	78	37				E.C.							0.4553 ± 0003	100
	78	37											0.5624 ± 0.0003	17
	78	37											0.6353 ± 0.0003	2.3
	78	37											0.6647 ± 0.0003	5
	78	37											0.6929 ± 0.0003	20
	78	37											0.7351 ± 0.0005	3
	78	37											0.8591 ± 0.0005	5
	78	37											1.1481 ± 0.0003	12
	78	37											1.3006 ± 0.0005	5
	78	37											1.7810 ± 0.001	11
	78	37											2.4202 ± 0.0007	7
	78	37											2.4953 ± 0.0007	4
	78	37											2.5152 ± 0.0007	6
	78	37											2.8930 ± 0.001	3
	78	37											2.9830 ± 0.001	7
	78	37											3.0830 ± 0.0007	5
	78	37											3.4380 ± 0.0007	20
	78	37											3.5410 ± 0.001	2
	78	37											3.5740 ± 0.0015	2
	78	37											3.8930 ± 0.001	4
$_{37}Rb^{79}$	79	37		78.923954	23 m.	β^+(84%)	3.53				5/2-		ann.rad.	168%
	79	37				E.C.(16%)							0.01788 ± 0.00006	1.6%
	79	37											0.13601 ± 0.00002	10%
	79	37											0.14349 ± 0.00005	11%
	79	37											0.14723 ± 0.00003	8%
	79	37											0.15484 ± 0.00002	6%
	79	37											0.16068 ± 0.00002	7%
	79	37											0.18282 ± 0.0001	16%
	79	37											0.35066 ± 0.00006	7%
	79	37											0.39765 ± 0.00005	5%
	79	37											0.50530 ± 0.00004	13%
	79	37											0.62208 ± 0.0001	7%

Isotope	A	Z	% Natural abundance	Atomic mass	Half-life	Decay mode	Decay energy (MeV)	Particle energy (MeV)	Particle intensity	Thermal neutron cross section	Spin (h/2π)	μ Nucl. mag. moment	Gamma-ray energy (MeV)	Gamma-ray intensity
	79	37											0.68812 ± 0.00004	24%
	79	37											(0.14 - 2.0)weak	
$_{37}\text{Rb}^{80}$	80	37		79.922519	34.s.	β+	5.71	3.86	22%		1+	-0.0834	ann.rad.	198%
	80	37						4/7	74%				0.6167 ± 0.0005	25%
	80	37											0.6396 ± 0.0005	1.5%
	80	37											0.7043 ± 0.0005	1.9%
	80	37											1.2571 ± 0.001	0.6%
$_{37}\text{Rb}^{81m}$	81	37			32 m.	I.T.	0.85	1.4			9/2+		ann.rad.	46%
	81	37				β+,E.C.							0.085	
$_{37}\text{Rb}^{81}$	81	37		80.918990	4.58 h.	β+(27%)	2.24	1.05			3/2-	+2.05	ann.rad.	46%
	81	37				E.C.(73%)							0.19030 ± 0.00005	66%(D)
	81	37											0.44614 ± 0.00002	19%
	81	37											0.45671 ± 0.00003	2.4%
	81	37											0.53760 ± 0.00004	1.5%
	81	37											(0.05 - 1.5)weak	
$_{37}\text{Rb}^{82m}$	82	37			6.47 h.	β+(26%)		0.80			5-	+1.5	ann.rad.	
	82	37				E.C.(74%)							0.5542 ± 0.0005	63%
	82	37											0.6189 ± 0.0004	37%
	82	37											0.6982 ± 0.0004	24%
	82	37											0.7768 ± 0.0004	82%
	82	37											0.8278 ± 0.0004	21%
	82	37											1/0079 ± 0.0006	6.9%
	82	37											1.0442 ± 0.0006	33%
	82	37											1.3172 ± 0.0006	26%
	82	37											1.4749 ± 0.0006	17%
	82	37											(0.1 - 2.3)weak	
$_{37}\text{Rb}^{82}$	82	37		81.918195	1.273 m.	β+(96%)	4.36	3.3			1+	+1.6434	ann.rad.	191%
	82	37				E.C.(4%)							0.7665 ± 0.0002	13.6%
	82	37											1.3952 ± 0.0003	0.5%
	82	37											(0.6 - 2.9)weak	
$_{37}\text{Rb}^{83}$	83	37		82.915144	86.2 d.	E.C.	0.96				5/2-	+1.43	Kr x-ray	60%
	83	37											0.009396 ± 0.00009	6%
	83	37											0.52039 ± 0.00003	46%
	83	37											0.52958 ± 0.00001	30%
	83	37											0.55254 ± 0.00002	16.6%
	83	37											0.6485 ± 0.0002	0.1%
	83	37											0.7892 ± 0.0002	0.8%
	83	37											0.7986 ± 0.0002	0.3%
$_{37}\text{Rb}^{84m}$	84	37			20.3 m.	I.T.	0.216				6-		0.2163 ± 0.0002	29%
	84	37											0.2482 ± 0.0002	64.5%
	84	37											0.4645 ± 0.0002	53%
$_{37}\text{Rb}^{84}$	84	37		83.91439	32.9 d.	β+(22%)	2.682	0.780	11%		2-	-1.297	ann.rad.	
	84	37				E.C.(75%)		1.658	11%				0.88160 ± 0.00002	74%
	84	37				β-(3%)	0.893	0.893					1.01614 ± 0.00002	0.35%
	84	37											1.89773 ± 0.00004	1%
$_{37}\text{Rb}^{85}$	85	37	72.17%	84.911794						(0.05 + 0.43)b.	5/2-	+1.35302		
$_{37}\text{Rb}^{86m}$	86	37			1.018 m.	I.T.	0.5560				6-		0.5558 ± 0.0002	98%
$_{37}\text{Rb}^{86}$	86	37		85.911172	18.63 d.	β-	1.774	1.774	8.8%		2-	-1.6920	1.0768 ± 0.0005	8.8%
$_{37}\text{Rb}^{87}$	87	37	27.83%	86.909187	4.9×10^{10}y.	β-	0.273	0.273	100%		3/2-	+2.7512		
$_{37}\text{Rb}^{88}$	88	37		87.911326	17.7 m.	β-	5.32	5.32		1.0 b.	2-	0.508	0.89803 ± 0.00004	14%
	88	37											1.83601 ± 0.00005	21.4%
	88	37											2.11889 ± 0.00007	0.42%
	88	37											2.57772 ± 0.00006	0.2%
	88	37											2.57772 ± 0.00006	0.2%
	88	37											2.67781 ± 0.00005	1.95 %
	88	37											3.00945 ± 0.00007	0.24%
	88	37											3.21848 ± 0.00008	0.21%
	88	37											3.21850 ± 0.0001	2.2%
	88	37											3.486 ± 0.0001	1.35%
	88	37											4.7424 ± 0.0002	1.5%
$_{37}\text{Rb}^{89}$	89	37		88.912278	β-	4.50	1.26	38%			3/2-		0.6571 ± 0.0001	1.0%
	89	37						1.9	5%				1.03188 ± 0.0002	58%
	89	37						2.2	34%				1.2481 ± 0.0007	43%
	89	37						4.49	18%				1.5381 ± 0.0001	2.6%
	89	37											2.0075 ± 0.0001	2.4%

Isotope	A	Z	% Natural abundance	Atomic mass	Half-life	Decay mode	Decay energy (MeV)	Particle energy (MeV)	Particle intensity	Thermal neutron cross section	Spin ($h/2\pi$)	μ Nucl. mag. moment	Gamma-ray energy (MeV)	Gamma-ray intensity
	89	37											2.1960 ± 0.0001	13%
	89	37											2.5701 ± 0.0001	10%
	89	37											2.7072 ± 0.0001	2%
	89	37											3.5088 ± 0.0002	1.1%
$_{37}Rb^{90m}$	90	37			4.28 m.	β-	4.50	1.7			4-		0.1069 ± 0.0001	0.22%
	90	37						6.5					0.3145 ± 0.0003	0.9%
	90	37											0.5512 ± 0.0002	0.9%
	90	37											0.7207 ± 0.0001	0.5%
	90	37											0.8242 ± 0.0001	8.9%
	90	37											0.83169 ± 0.00005	47%
	90	37											0.8720 ± 0.0001	0.5%
	90	37											0.9524 ± 0.0001	1.7%
	90	37											1.0607 ± 0.0001	7.8%
	90	37											1.1405 ± 0.0001	0.9%
	90	37											1.2428 ± 0.0001	3.1%
	90	37											1.2718 ± 0.0001	1.6%
	90	37											1.3754 ± 0.00004	17.1%
	90	37											1.3772 ± 0.0001	2.3%
	90	37											1.4890 ± 0.0004	0.4%
	90	37											1.6035 ± 0.0002	0.5%
	90	37											1.6656 ± 0.0001	4.9%
	90	37											1.6962 ± 0.0001	1.7%
	90	37											1.7389 ± 0.0001	1.9%
	90	37											1.8382 ± 0.0001	0.8%
	90	37											2.1283 ± 0.0001	5.3%
	90	37											2.2565 ± 0.0001	0.7%
	90	37											2.4973 ± 0.0001	0.7%
	90	37											2.5923 ± 0.0001	0.6%
	90	37											2.6178 ± 0.0001	0.6%
	90	37											2.7527 ± 0.0001	11.8%
	90	37											2.8344 ± 0.0001	1.9%
	90	37											3.0321 ± 0.0005	0.4%
	90	37											3.2051 ± 0.0002	1.1%
	90	37											3.3170 ± 0.0001	14.7%
	90	37											3.3708 ± 0.0004	0.4%
$_{37}Rb^{90}$	90	37		89.914811	2.6 m.	β-	6.57				1-		0.8317 ± 0001	27.8%
	90	37											0.9978 ± 0.0001	0.3%
	90	37											1.0386 ± 0.0001	0.2%
	90	37											1.0607 ± 0.0004	6.6%
	90	37											1.8041 ± 0.0001	0.4%
	90	37											1.8923 ± 0.0001	0.4%
	90	37											2.1393 ± 0.0002	0.3%
	90	37											2.2075 ± 0.0001	0.3%
	90	37											2.2163 ± 0.0002	0.34%
	90	37											2.4739 ± 0.0002	.42%
	90	37											3.2951 ± 0.0002	0.6%
	90	37											3.3170 ± 0.0001	0.6%
	90	37											3.3619 ± 0.0001	0.68%
	90	37											3.3832 ± 0.0001	4.7%
	90	37											3.5342 ± 0.0001	2.8%
	90	37											3.8144 ± 0.0001	0.4%
	90	37											4.1355 ± 0.0002	4.7%
	90	37											4.6464 ± 0.0002	1.6%
	90	37											5.1874 ± 0.0002	0.81%
	90	37											5.3330 ± 0.0002	0.30%
$_{37}Rb^{91}$	91	37		90.916485	58.4 s.	β-	5.86				3/2-		0.09363 ± 0.0002	32%
	91	37											0.3454 ± 0.0001	7.9%
	91	37											0.4391 ± 0.0001	2.0%
	91	37											0.5932 ± 0.0001	1.2%
	91	37											0.6028 ± 0.0001	2.7%
	91	37											0.9485 ± 0.0001	1.1%
	91	37											1.0420 ± 0.0001	2.1%
	91	37											1.1372 ± 0.0001	3.7%
	91	37											1.3678 ± 0.0001	0.7%
	91	37											1.4822 ± 0.0001	1.4%
	91	37											1.6158 ± 0.0001	2.3%
	91	37											1.7402 ± 0.0001	1.3%
	91	37											1.8492 ± 0.0001	3.1%
	91	37											1.9716 ± 0.0001	6.4%
	91	37											2.5059 ± 0.0002	1.3%
	91	37											2.5642 ± 0.0002	11.9%
	91	37											2.9257 ± 0.0002	1.5%
	91	37											3.4465 ± 0.0002	1.4%
	91	37											3.5997 ± 0.0002	9.9%
	91	37											3.6391 ± 0.0002	1.1%
	91	37											3.6437 ± 0.0002	0.75%
	91	37											3.8443 ± 0.0003	0.98%
	91	37											4.0782 ± 0.0002	3.9%
	91	37											4.2654 ± 0.0002	1.4%
$_{37}Rb^{92}$	92	37		91.919661		β-	8.12	8.1	94%		1-		0.5698 ± 0.0001	0.7%
	92	37											0.8147 ± 0.0001	4.0%
	92	37											1.3846 ± 0.0003	0.44%
	92	37											1.7123 ± 0.0002	0.53%
	92	37											2.8206 ± 0.0002	0.76%

Isotope	A	Z	% Natural abundance	Atomic mass	Half-life	Decay mode	Decay energy (MeV)	Particle energy (MeV)	Particle intensity	Thermal neutron cross section	Spin (h/2π)	μ Nucl. mag. moment	Gamma-ray energy (MeV)	Gamma-ray intensity
	92	37											3.1100 ± 0.0005	0.12%
	92	37											4.6377 ± 0.001	0.3%
	92	37											5.1881 ± 0.0008	0.3%
	92	37											5.55842 ± 0.001	0.2%
	92	37											5.6322 ± 0.001	0.24%
	92	37											(0.1 - 6.1)	
$_{37}Rb^{93}$	93	37			5.85 s.	β-	7.47	7.4					0.2134 ± 0.0005	4.5%
	93	37				n(1%)							0.2192 ± 0.0006	1.9%
	93	37											0.4326 ± 0.0001	11.8%
	93	37											0.7099 ± 0.0001	3.6%
	93	37											0.9861 ± 0.0001	4.6%
	93	37											1.1482 ± 0.0001	1.0%
	93	37											1.2383 ± 0.0001	1.0%
	93	37											1.3852 ± 0.0001	3.9%
	93	37											1.5629 ± 0.0001	0.7%
	93	37											1.6129 ± 0.0001	1.1%
	93	39											1.8085 ± 0.0001	1.9%
	93	37											1.8697 ± 0.0001	1.3%
	93	37											2.0541 ± 0.0001	0.9%
	93	37											2.2294 ± 0.0001	0.64%
	93	37											2.5052 ± 0.0002	0.55%
	93	37											2.7050 ± 0.0002	0.7%
	93	37											2.8613 ± 0.0002	0.75%
	93	37											3.4582 ± 0.0002	2.5%
	93	37											3.8040 ± 0.0002	1.1%
	93	37											3.8676 ± 0.0002	1.7%
	93	37											3.9343 ± 0.0002	0.7%
	93	37											4.2712 ± 0.0002	0.23%
	93	37											4.8751 ± 0.0003	0.12%
$_{37}Rb^{94}$	94	37		93.926432	2.73 s.	β-	10.2	9.3					0.6777 ± 0.0003	4 +
	93	37				n(10%)							0.8369 ± 0.0001	100
	94	37											1.0894 ± 0.0002	21
	94	37											1.3091 ± 0.0001	16
	94	37											1.5775 ± 0.0001	39
$_{37}Rb^{95}$	95	37		94.929352	0.38 s.	β-	9.3	8.6					0.2040	22 +
	95	37				n(8%)							0.3289	17
	95	37											0.3522	100
	95	37											0.6602	10
	95	37											0.6808	36
	95	37											0.7691	11
$_{37}Rb^{96}$	96	37		95.934370	0.20 s.	β-	10.3	10.8					0.412 ± 0.001	4 +
	96	37				n(13%)							0.593 ± 0.001	8
	96	37											0.606 ± 0.001	6
	96	37											0.691 ± 0.001	10
	96	37											0.813 ± 0.001	100
	96	37											0.978 ± 0.001	8
	96	37											1.036 ± 0.001	9
	96	37											1.179 ± 0.001	4
	96	37											1.335 ± 0.001	4
$_{37}Rb^{97}$	97	37		96.937440	0.17 s.	β-							0.167 ± 0.001	100 +
	97	37				n(27%)							0.418 ± 0.001	18
	97	37											0.519 ± 0.001	38
	97	37											0.585 ± 0.001	79
	97	37											0.599 ± 0.001	56
	97	37											0.652 ± 0.001	21
	97	37											0.697 ± 0.001	21
	97	37											1.258 ± 0.001	52
$_{37}Rb^{98}$	98	37		97.941960	0.13 s.	β-								
	98	37				n(13%)								
Sr		38		87.62						1.2 b.				
$_{38}Sr^{79}$	79	38		78.929860	2.1 m.	β+	5.2						ann.rad.	
	79	38											0.1889 ± 0.0003	100 +
	79	38											0.4442 ± 0.0003	37
	79	38											0.5599 ± 0.0008	63
	79	38											0.6112 ± 0.00008	26
$_{38}Sr^{80}$	80	38		79.924650	106 m.	β+	2.00				0+		ann.rad.	
	80	38											0.1750 ± 0.0005	26 +
	80	38											0.2359 ± 0.0008	11
	80	38											0.3788 ± 0.0005	11
	80	38											0.4141 ± 0.0005	8
	80	38											0.5534 ± 0.0005	18
	80	38											0.5890 ± 0.0005	100
$_{38}Sr^{81}$	81	38		80.923270	22.2 m.	β+(87%)	3.99	2.43			1/2+		ann.rad.	173%
	81	38				E.C.(13%)		2.68					0.1478 ± 0.0003	30%
	81	38											0.1423 ± 0.0003	4%
	81	38											0.1534 ± 0.0003	36%
	81	38											0.1883 ± 0.0003	21%
	81	38											0.3865 ± 0.0005	4%
	81	38											0.4435 ± 0.0003	17%
	81	38											0.5746 ± 0.0003	6%
	81	38											0.7015 ± 0.001	1.4%
	81	38											0.7213 ± 0.0003	2%
	81	38											0.9093 ± 0.0003	3.0%

Isotope	A	Z	% Natural abundance	Atomic mass	Half-life	Decay mode	Decay energy (MeV)	Particle energy (MeV)	Particle intensity	Thermal neutron cross section	Spin (h/2π)	μ Nucl. mag. moment	Gamma-ray energy (MeV)	Gamma-ray intensity
	84	39						3.15	7				0.6583 ± 0.0002	4.5%
	84	39											0.6606 ± 0.0002	11.5%
	84	39											0.7931 ± 0.0002	89%
	84	39											0.9744 ± 0.0002	79%
	84	39											0.9942 ± 0.0002	4.2%
	84	39											1.0398 ± 0.0002	58%
	84	39											1.2550 ± 0.0002	6.7%
	84	39											1.2626 ± 0.0002	2.5%
	84	39											1.3309 ± 0.0002	3.1%
	84	39											1.4534 ± 0.0002	1.8%
	84	39											1.5028 ± 0.0002	6.8%
	84	39											1.6145 ± 0.0002	1.8%
	84	39											1.6546 ± 0.0002	2.6%
	84	39											1.7444 ± 0.0002	2.2%
	84	39											1.7636 ± 0.0002	1.9%
	84	39											1.9180 ± 0.0004	2.3%
	84	39											2.2953 ± 0.0004	1.9%
	84	39											2.3095 ± 0.0004	1.1%
	84	39											(0.2 - 3.3)	
$_{39}Y^{85m}$	85	39			4.9 h.	β+(70%)					9/2+		ann.rad.	
	85	39				E.C.(30%)							0.2317 ± 0.0001	23%(D)
	85	39											0.5044 ± 0.0002	1.5%
	85	39											0.5356 ± 0.0002	3.5%
	85	39											0.5467 ± 0.0002	1.2%
	85	39											0.5684 ± 0.0002	1.7%
	85	39											0.6119 ± 0.0002	1%
	85	39											0.6980 ± 0.0002	1.2%
	85	39											0.7673 ± 0.0001	3.7%
	85	39											0.7686 ± 0.0002	1.2%
	85	39											0.7879 ± 0.0002	1.4%
	85	39											1.0301 ± 0.0002	2.1%
	85	39											1.1232 ± 0.0002	1.8%
	85	39											2.1238 ± 0.0002	5%
	85	39											2.1721 ± 0.0002	2.3%
	85	39											2.3517 ± 0.0002	0.5%
	85	39											2.7822 ± 0.0003	0.35%
	85	39											(0.1 - 3.1)	
$_{39}Y^{85}$	85	39		84.916437	2.6 h.	β+(55%)	3.26	1.54			1/2-		ann.rad.	
	85	39				E.C.(45%)							0.2317 ± 0.0001	80%(D)
	85	39											0.5045 ± 0.0002	64%
	85	39											0.9140 ± 0.0002	6.8%
	85	39											1.2780 ± 0.0005	0.3%
	85	39											1.3208 ± 0.0005	0.35%
	85	39											(0.07 - 1.4)	
$_{39}Y^{86m}$	86	39			48 m.	I.T.(99%)					8+		ann.rad.	
	86	39				β+							0.0102 ± 0.0001	
	86	39				E.C.							0.2080 ± 0.0003	94%
	86	39											(0.09 - 1.1)	
$_{39}Y^{86}$	86	39		85.914893	14.74 h.	β+	5.24				4		ann.rad.	
	86	39				E.C.							0.3070 ± 0.0001	3%
	86	39											0.3829 ± 0.0002	3.6%
	86	39											0.4431 ± 0.0001	17%
	86	39											0.5806 ± 0.0001	4.8%
	86	39											0.6088 ± 0.0001	2%
	86	39											0.6277 ± 0.0001	33%
	86	39											0.7033 ± 0.0001	15%
	86	39											0.7774 ± 0.0001	12%
	86	39											0.8260 ± 0.0001	3%
	86	39											1.0240 ± 0.0001	3.8%
	86	39											1.0766 ± 0.0001	82%
	86	39											1.1531 ± 0.0001	31%
	86	39											1.8017 ± 0.0001	1.6%
	86	39											1.8544 ± 0.0001	16%
	86	39											1.9207 ± 0.0001	21%
	86	39											2.5679 ± 0.0002	2.3%
	86	39											2.6101 ± 0.0002	1.2%
	86	39											(0.1 - 3.8)	
$_{39}Y^{87m}$	87	39			13 h.	I.T.(98%)					9/2+		0.3807 ± 0.0005	78%(D)
	87	39				β+(0.7%)		1.15	0.7%					
	87	39				E.C.								
$_{39}Y^{87}$	87	39		86.910882	80.3 h.	E.C. (99+%)	1.861	0.78					0.3880 ± 0.0001	90%(D)
	87	39											0.4870 ± 0.0001	92%
$_{39}Y^{88}$	88	39		87.909508	106.61 d.	E.C. (99+%)	3.623	0.76			4-		ann.rad.	0.4%
	88	39				β+(0.2%)							0.89802 ± 0.00002	92%
	88	39											1.83601 ± 0.00003	99.3%
	88	39											2.73404 ± 0.00005	0.5%
	88	39											3.2190 ± 0.0002	0.007%
$_{39}Y^{89m}$	89	39			15.7 s.	I.T.	0.909				9/2+		0.9092 ± 0.0002	99.1%
$_{39}Y^{89}$	89	39	100%	88.905849						(0.001 + 1.28)b.	1/2-	-0.1373		

TABLE OF THE ISOTOPES (Continued)

Isotope	A	Z	% Natural abundance	Atomic mass	Half-life	Decay mode	Decay energy (MeV)	Particle energy (MeV)	Particle intensity	Thermal neutron cross section	Spin (h/2π)	μ Nucl. mag. moment	Gamma-ray energy (MeV)	Gamma-ray intensity
$_{39}Y^{90m}$	90	39			3.19 h.	I.T. (99+%)	0.68204				7+		0.2025 ± 0.0001	97%
	90	39				β-(0.002%)							0.4794 ± 0.0001	91%
	90	39											0.6820 ± 0.0002	0.4%
$_{39}Y^{90}$	90	39		89.907152	64.0 h.	β-	2.283	2.283			2-	-1.630		
$_{39}Y^{91m}$	91	39			49.7 m.	I.T.	0.555				9/2+		0.55562 ± 0.00005	95%
$_{39}Y^{91}$	91	39		90.907303	58.5 d.	β-	1.546	1.546			1/2-	0.1641	1.208 ± 0.001	0.3%
$_{39}Y^{92}$	92	39		91.908917	3.54 h.	β-	3.62	3.64			2-		0.4485 ± 0.0001	2.4%
	92	39											0.4926 ± 0.0001	0.4%
	92	39											0.5611 ± 0.0001	2.4%
	92	39											0.8443 ± 0.0001	1.2%
	92	39											0.9128 ± 0.0001	0.6%
	92	39											0.9345 ± 0.0001	13.9%
	92	39											1.1324 ± 0.0001	0.2%
	92	39											1.4054 ± 0.0001	4.7%
	92	39											1.8473 ± 0.0001	0.35%
	92	39											(0.4 - 3.3)	
$_{39}Y^{93m}$	93	39			0.82 s.	I.T.	0.759				9/2+		0.1684 ± 0.0001	51%
	93	39											0.5902 ± 0.0001	100%
$_{39}Y^{93}$	93	39		92.909571	10.2 h.	β-	2.89	2.89	90%		1/2-		0.2669 ± 0.0002	6.8%
	93	39											0.6802 ± 0.0001	0.6%
	93	39											0.9471 ± 0.0001	2.0%
	93	39											1.4254 ± 0.0001	0.24%
	93	39											1.45050 ± 0.0001	0.34%
	93	39											1.9178 ± 0.0002	1.4%
	93	39											2.1846 ± 0.0002	0.15%
	93	39											2.1908 ± 0.0002	0.17%
$_{39}Y^{94}$	94	39		93.911597	18.7 m.	β-	4.92	4.92			2-		0.3816 ± 0.0001	2%
	94	39											0.5509 ± 0.0001	5%
	94	39											0.7526 ± 0.0001	1.4%
	94	39											0.9188 ± 0.0001	56%
	94	39											1.1389 ± 0.0001	65
	94	39											1.6714 ± 0.0001	2.5%
	94	39											2.1406 ± 0.0002	1%
	94	39											2.8463 ± 0.0003	0.4%
	94	39											2.8987 ± 0.0006	0.1%
	94	39											(0.3 - 4.1)	
$_{39}Y^{95}$	95	39		94.912814	10.3 m.	β-	4.45				1/2-		0.4324 ± 0.0002	2%
	95	39											0.9542 ± 0.0002	19%
	95	39											1.0485 ± 0.0004	1%
	95	39											1.1740 ± 0.0004	0.7%
	95	39											1.3243 ± 0.0003	5.3%
	95	39											1.6185 ± 0.0008	1.7%
	95	39											1.8050 ± 0.0005	1.5%
	95	39											1.8928 ± 0.0005	0.7%
	95	39											1.9406 ± 0.0005	2.8%
	95	39											2.1760 ± 0.0005	8.2%
	95	39											2.2529 ± 0.0006	1.4%
	95	39											2.3733 ± 0.0008	0.8%
	95	39											2.6330 ± 0.0008	5.1%
	95	39											3.1298 ± 0.001	0.7%
	95	39											3.2502 ± 0.001	1.3%
	95	39											3.5770 ± 0.001	7.6%
	95	39											3.6840 ± 0.001	0.4%
	95	39											3.8870 ± 0.002	0.3%
$_{39}Y^{96m}$	96	39			9.8 s.	β-							0.1467 ± 0.0002	36%
	96	39											0.1744 ± 0.0002	2%
	96	39											0.2268 ± 0.0002	2%
	96	39											0.2891 ± 0.0002	1%
	96	39											0.3636 ± 0.0002	22%
	96	39											0.4753 ± 0.0002	3.3%
	96	39											0.6174 ± 0.0002	55%
	96	39											0.6316 ± 0.0002	7.6%
	96	39											0.6526 ± 0.0002	1.5%
	96	39											0.9062 ± 0.0002	18%
	96	39											0.9150 ± 0.0002	59%
	96	39											0.9603 ± 0.0002	4.2%
	96	39											0.9793 ± 0.0002	3.6%
	96	39											1.1071 ± 0.0002	48%
	96	39											1.1850 ± 0.0002	3.4%
	96	39											1.2227 ± 0.0002	26%
	96	39											1.7507 ± 0.0002	89%
	96	39											1.8976 ± 0.0002	5.2%
	96	39											2.2255 ± 0.0002	5.8%
$_{39}Y^{96}$	96	39		95.915940	6.2 s.	β-	7.12	7.12			0-		1.594 ± 0.0005	25%
$_{39}Y^{97m}$	97	39			1.21 s.	β-	7.4	4.8			9/2+		0.1614	77%
	97	39						6.0					0.9700	43%
	97	39											1.1030	92%
	97	39											1.2441	9%
	97	39											1.2642	3%
	97	39											1.4000	5%
$_{39}Y^{97}$	97	39		96.918120	3.7 s.	β-					1/2-		0.2969	1.3%
	97	39											0.5448	1%
	97	39											0.7560	1.1%

Isotope	A	Z	% Natural abundance	Atomic mass	Half-life	Decay mode	Decay energy (MeV)	Particle energy (MeV)	Particle intensity	Thermal neutron cross section	Spin (h/2π)	μ Nucl. mag. moment	Gamma-ray energy (MeV)	Gamma-ray intensity
	97	39											1.1030	5%
	97	39											1.2910	5.7%
	97	39											1.4000	4.5%
	97	39											1.8870	1.8%
	97	39											1.9960	7.4%
	97	39											2.7431	6.5%
	97	39											3.2876	18%
	97	39											3.4013	14%
	97	39											3.5495	3.1%
$_{39}Y^{98m}$	98	39			2.0 s.	β-	9.8	8.7					0.2415	7%
	98	39											0.2531	4%
	98	39											0.5830	18%
	98	39											0.6205	75%
	98	39											0.6473	55%
	98	39											0.7526	7%
	98	39											1.2228	97%
	98	39											1.5907	3%
	98	39											1.7873	4.2%
	98	39											1.8016	45%
$_{39}Y^{98}$	98	39		97.922300	0.65 s.	β-	8.9	5.5			1+		0.2131	1.3%
	98	39											0.2686	2.3%
	98	39											0.5216	0.7%
	98	39											1.2228	11%
	98	39											1.5907	4.4%
	98	39											1.7441	1.3%
	98	39											2.4206	1.5%
	98	39											2.9413	5.3%
	98	39											3.2037	0.7%
	98	39											3.2283	1.3%
	98	39											3.3100	2.2%
	98	39											3.3757	0.6%
	98	39											3.4686	0.6%
	98	39											4.4501	3.1%
$_{39}Y^{99}$	99	39		98.924720	1.5 s.	β-	7.6				1/2-		0.1217	44%
	99	39											0.1307	6.2%
	99	39											0.1940	2.2%
	99	39											0.2764	2.6%
	99	39											0.4060	1.8%
	99	39											0.4538	4.8%
	99	39											0.5362	8.8%
	99	39											0.5754	11%
	99	39											0.6000	6.2%
	99	39											0.6026	4.4%
	99	39											0.6399	3.5%
	99	39											0.7242	19.8%
	99	39											0.7300	1.3%
	99	39											0.7822	4.8%
	99	39											0.9301	4.4%
	99	39											1.0130	7.9%
Zr		40		91.224						0.184 b.				
$_{40}Zr^{82}$	82	40		81.931100	2.5 m.	β+	4.0						ann.rad.	
$_{40}Zr^{83m}$	83	40			8 s.	β+							ann.rad.	
$_{40}Zr^{83}$	83	40		82.928760	44 s.	β+	6.0						ann.rad.	
	83	40											0.0556	
	83	40											0.1050	10%
	83	40											0.2560	8%
	83	40											0.303	7%
	83	40											0.474	9%
	83	40											0.791	4%
	83	40											1.525	9%
$_{40}Zr^{84}$	84	40		83.923320	28 m.	β+	2.7				0+		ann.rad.	
	84	40				E.C.							0.0449	
	84	40											0.1125	
	84	40											0.3729	
	84	40											0.667	
$_{40}Zr^{85m}$	85	40			10.9 s.	I.T.	0.292				1/2-		ann.rad.	
	85	40				β+,E.C.							0.2922 ± 0.0003	100 +
	85	40											0.4165 ± 0.0003	9
$_{40}Zr^{85}$	85	40		84.921470	7.9 m.	β+	4.4				7/2+		ann.rad.	
	85	40				E.C.							0.2663 ± 0.0002	23%
	85	40											0.4163 ± 0.0002	25%
	85	40											0.4543 ± 0.0002	41%
	85	40											1.1984 ± 0.0002	4.5%
	85	40											1.7682 ± 0.0003	1.8%
	85	40											1.8762 ± 0.0003	0.4%
	85	40											1.9557 ± 0.0003	0.4%
$_{40}Zr^{86}$	86	40		85.916290	16.5 h.	E.C.	1.3				0+		0.0280 ± 0.0005	20%
	86	40											0.243 ± 0.001	96%
	86	40											0.612 ± 0.001	5%
$_{40}Zr^{87m}$	87	40			14.0 s.	I.T.	0.3362				1/2-		0.1352 ± 0.0003	27%
	87	40											0.2010 ± 0.0003	97%
$_{40}Zr^{87}$	87	40		86.914817	1.73 h.	β+	3.67				9/2+		ann.rad.	
	87	40				E.C.							0.3811 ± 0.0002	93%(D)
	87	40											0.793 ± 0.001	0.6%

Isotope	A	Z	% Natural abundance	Atomic mass	Half-life	Decay mode	Decay energy (MeV)	Particle energy (MeV)	Particle intensity	Thermal neutron cross section	Spin (h/2π)	μ Nucl. mag. moment	Gamma-ray energy (MeV)	Gamma-ray intensity
	87	40											1.210 ± 0.001	1.3%
	87	40											1.228 ± 0.001	4%
	87	40											1.808 ± 0.001	0.2%
	87	40											2.220 ± 0.001	0.4%
	87	40											2.616 ± 0.002	0.2%
$_{40}Zr^{88}$	88	40		87.910225	83.4 d.	E.C.	0.67				0+		0.3929 ± 0.001	97%
$_{40}Zr^{89m}$	89	40			4.18 m.	I.T.(94%)	0.5878				1/2-		ann.rad.	
	89	40				β+(1.5%)							0.5878 ± 0.0002	94%
	89	40				E.C.(4.7%)							0.9092 ± 0.0001	6%(D)
$_{40}Zr^{89}$	89	40		88.908890	78.4 h.	β+(23%)	2.83	0.9			9/2+		ann.rad.	46%
	89	40				E.C.(77%)							1.7129 ± 0.0008	0.7%
$_{40}Zr^{90m}$	90	40			0.809 s.	I.T.					5-		0.1326 ± 0.0002	5%
	90	40											2.1862 ± 0.0001	16%
	90	40											2.3189 ± 0.0001	84%
$_{40}Zr^{90}$	90	40	51.45%	89.904703						0.05 b.	0+			
$_{40}Zr^{91}$	91	40	11.27%	90.905644						0.9 b.	5/2+	-1.3036		
$_{40}Zr^{92}$	92	40	17.17%	91.905039						0.2 b.	0+			
$_{40}Zr^{93}$	93	40		92.906474	$1.5x10^6$ y.	β-	0.08			≈1 b.	5/2+		0.0304	(D)
$_{40}Zr^{94}$	94	40	17.33%	93.906314						0.05 b.	0+			
$_{40}Zr^{95}$	95	40		94.908042	64.03 d.	β-	1.124	0.360	55%		5/2+		0.23569 ± 0.00005	0.9%
	95	40						0.396	44%				0.72418 ± 0.00001	44%
	95	40											0.75672 ± 0.00001	55%
$_{40}Zr^{96}$	96	40	2.78%	95.908275						0.022 b.	0+			
$_{40}Zr^{97}$	97	40		96.910950	16.8 h.	β-	2.658	1.91			1/2-		0.2541 ± 0.0002	1.2%
	97	40											0.3554 ± 0.0001	2.3%
	97	40											0.5076 ± 0.0001	5%
	97	40											0.6024 ± 0.0002	1.4%
	97	40											0.7434 ± 0.0003	93%(D)
	97	40											1.0213 ± 0.0003	1.3%
	97	40											1.1479 ± 0.0001	2.6%
	97	40											1.2761 ± 0.0001	1.0%
	97	40											1.3627 ± 0.0001	1.3%
	97	40											1.7505 ± 0.0001	1.3%
$_{40}Zr^{98}$	98	40		97.912735	30.7 s.	β-	2.24	2.2	100%		0+			
$_{40}Zr^{99}$	99	40		98.916540	2.1 s.	β-	4.6	3.5			1/2+		0.0284	1.7%
	99	40											0.0558	2.2%
	99	40											0.0818	2.8%
	99	40											0.1792	5.6%
	99	40											0.3872	7.8%
	99	40											0.4150	5.0%
	99	40											0.4617	11.8%
	99	40											0.4691	56%
	99	40											0.4899	0.8%
	99	40											0.5460	45%
	99	40											0.5940	27%
	99	40											0.6279	2.2%
	99	40											0.6500	2.2%
$_{40}Zr^{100}$	100	40		99.917750	7.1 s.	β-	3.3				0+		0.4005	
	100	40											0.5042	
$_{40}Zr^{101}$	101	40		100.921520	2.0 s.	β-	5.8	6.2					0.1194	
	101	40											0.2057	
	101	40											0.2089	
Nb		41								1.15 b.				
$_{41}Nb^{86}$	86	41		85.925310	1.45 m.	β+	8.0						ann.rad.	
	86	41											0.751	100 +
	86	41											1.003	57
$_{41}Nb^{87m}$	87	41			3.7 m.	β+					1/2-		ann.rad.	
	87	41				E.C.							0.1352 ± 0.0003	100%(D)
	87	41											0.2010 ± 0.0003	100%(D)
$_{41}Nb^{87}$	87	41		86.920370	2.6 m.	β+5.2							ann.rad.	
	87	41				E.C.							0.2010 ± 0.0003	26%(D)
	87	41											0.4706 ± 0.0002	19%
	87	41											0.6000 ± 0.0006	2%
	87	41											0.6165 ± 0.0002	9%
	87	41											0.9142 ± 0.0003	6%
	87	41											1.0665 ± 0.0004	10%
	87	41											1.6832 ± 0.0003	4%
	87	41											1.8842 ± 0.0003	9%
	87	41											2.1533 ± 0.0007	0.9%
$_{41}Nb^{88m}$	88	41			7.8 m.	β+					4-		ann.rad.	
	88	41				E.C.							0.2625 ± 0.0005	10%
	88	41											0.3996 ± 0.0003	42%
	88	41											0.4510 ± 0.0003	24%
	88	41											0.5341 ± 0.0003	13%
	88	41											0.6382 ± 0.0003	25%
	88	41											0.7607 ± 0.0003	16%
	88	41											0.9184 ± 0.0005	11%
	88	41											1.0569 ± 0.0003	90%
	88	41											1.0825 ± 0.0003	58%
	88	41											1.3992 ± 0.0005	5.5%
	88	41											1.8179 ± 0.0006	10%

Isotope	A	Z	% Natural abundance	Atomic mass	Half-life	Decay mode	Decay energy (MeV)	Particle energy (MeV)	Particle intensity	Thermal neutron cross section	Spin (h/2π)	μ Nucl. mag. moment	Gamma-ray energy (MeV)	Gamma-ray intensity
	88	41											1.9754 ± 0.0008	6%
$_{41}Nb^{88}$	88	41		87.917950	14.3 m.	β+	7.2	3.2			8+		ann.rad.	
	88	41				E.C.							0.0767 ± 0.0002	23%
	88	41											0.2714 ± 0.0002	53%
	88	41											0.3994 ± 0.0001	33%
	88	41											0.5029 ± 0.0002	64%
	88	41											0.6711 ± 0.0002	65%
	88	41											1.0570 ± 0.0001	100%
	88	41											1.0828 ± 0.0001	100%
	88	41											1.5430 ± 0.0006	1%
	88	41											2.3119 ± 0.0007	0.8%
	88	41											(0.07 - 2.5)	
$_{41}Nb^{89m}$	89	41			122 m.	β+		3.3			9/2+		0.5324 ± 0.0001	0.5%
	89	41				E.C.							0.5880 ± 0.0001	10%(D)
	89	41											1.1270 ± 0.0002	2%
	89	41											1.2590 ± 0.0002	1%
	89	41											1.3323 ± 0.0002	1%
	89	41											1.5114 ± 0.0003	1.8%
	89	41											1.6272 ± 0.0002	3.6%
	89	41											2.5723 ± 0.0004	3%
	89	41											3.0927 ± 0.0002	3.1%
	89	41											(0.17 - 4.0)	
$_{41}Nb^{89}$	89	41		88.913449	66 m.	β+(74%)	4.44	2.8			1/2-		ann.rad.	150%
	89	41				E.C.(26%)							0.5074 ± 0.0007	85%
	89	41											0.5880 ± 0.0002	100%
	89	41											0.7696 ± 0.0005	6%
	89	41											1.2775 ± 0.002	1.6%
$_{41}Nb^{90m}$	90	41			18.8 s.	I.T.	0.1246				4-		0.002	
	90	41											0.1225 ± 0.0001	64%
$_{41}Nb^{90}$	90	41		89.911263	14.6 h.	β+(53%)	6.111	0.86	5%		8+	4.941	ann.rad.	100%
	90	41				E.C.(47%)		1.5	92%				0.1412 ± 0.0001	64%
	90	41											0.3713 ± 0.0001	1.8%
	90	41											0.8277 ± 0.0001	1%
	90	41											0.8906 ± 0.0001	1.9%
	90	41											1.1292 ± 0.0001	91%
	90	41											1.6117 ± 0.0001	3%
	90	41											2.1862 ± 0.0001	18%
	90	41											2.3189 ± 0.0001	82%
	90	41											(0.1 - 3.3)	
$_{41}Nb^{91m}$	91	41			62 d.	I.T.(97%)					1/2-		0.1045 ± 0.0005	0.6%
	91	41				E.C.(3%)							1.2050 ± 0.0007	3.4%
$_{41}Nb^{91}$	91	41		90.906991	7 x 10^2 y.						5/2+		Mo k x-ray	54%
$_{41}Nb^{92m}$	92	41			10.13 d.	E.C. (99+%)					2+	6.114	0.9126 ± 0.0002	1.7%
	92	41											0.9345 ± 0.0001	99%
	92	41											1.8475 ± 0.0003	0.8%
$_{41}Nb^{92}$	92	41		91.907192	3x10^7 y.	E.C.	2.006				7+		0.5611 ± 0.0001	100%
	92	41											0.9345 ± 0.0001	100%
$_{41}Nb^{93m}$	93	41			13.6 y.	I.T.	0.0304				1/2-		Nb x-ray	
	93	41											0.0304	
$_{41}Nb^{93}$	93	41	100%	92.906377						1.15 b.	9/2+	+6.1705		
$_{41}Nb^{94m}$	94	41			6.26 m.	I.T. (99+%)	2.086				3+		Nb k x-ray	
	94	41				β-(0.5%)							0.0409 ± 0.0002	0.07%
	94	41											0.87109 ± 0.00002	0.5%
$_{41}Nb^{94}$	94	41		93.907280	2.4x10^4y.	β-	2.04	0.47			6+		0.70263 ± 0.00002	99%
	94	41											0.87109 ± 0.00002	100%
$_{41}Nb^{95m}$	95	41			3.61 d.	I.T.(97.5%	0.2357				1/2-		0.2040 ± 0.0001	2%
	95	41				β-(2.5%)							0.2356 ± 0.0005	26.1%
$_{41}Nb^{95}$	95	41		94.906835	34.98 d.	β-	0.160	0.926			9/2+	6.123	0.76578 ± 0.00002	99.8%
$_{41}Nb^{96}$	96	41		95.908100	23.4 h.	β-	3.187	0.5	10%		6+		0.2191 ± 0.0002	3.8%
	96	41						0.75	90%				0.2414 ± 0.0002	3.9%
	96	41											0.3501 ± 0.0002	1.1%
	96	41											0.3718 ± 0.0001	2.8%
	96	41											0.4600 ± 0.0006	28.2%
	96	41											0.4807 ± 0.0008	6.3%
	96	41											0.5689 ± 0.0006	55.7%
	96	41											0.7195 ± 0.0002	7.3%
	96	41											0.7782 ± 0.0001	96.8%
	96	41											0.8102 ± 0.0007	10%
	96	41											0.8124 ± 0.0003	3.4%
	96	41											0.8476 ± 0.0003	1.6%
	96	41											1.0913 ± 0.0006	49.3%
	96	41											1.2002 ± 0.0006	20%
	96	41											1.4977 ± 0.0008	3%
$_{41}Nb^{97m}$	97	41			54 s.	I.T.	0.7434	98%			6+		0.7434 ± 0.0001	98%
$_{41}Nb^{97}$	97	41		96.908096	73.6 m.	β-	1.934	1.2	98%		9/2+	7.3	0.4809 ± 0.0002	0.15%
	97	41											0.6579 ± 0.0001	98.3%
	97	41											1.0245 ± 0.0001	1.1%
	97	41											1.2686 ± 0.0001	0.2%
	97	41											1.5156 ± 0.0001	0.2%

Isotope	A	Z	% Natural abundance	Atomic mass	Half-life	Decay mode	Decay energy (MeV)	Particle energy (MeV)	Particle intensity	Thermal neutron cross section	Spin (h/2π)	μ Nucl. mag. moment	Gamma-ray energy (MeV)	Gamma-ray intensity
$_{41}Nb^{98m}$	98	41			51 m.	β-							0.1726 ± 0.0006	1.7%
	98	41											0.3354 ± 0.0006	9.3%
	98	41											0.4344 ± 0.0006	1.0%
	98	41											0.6448 ± 0.0006	5.3%
	98	41											0.7131 ± 0.0006	9.7%
	98	41											0.7227 ± 0.0003	69.7%
	98	41											0.7874 ± 0.0003	93%
	98	41											0.7916 ± 0.0006	6.4%
	98	41											0.8235 ± 0.0005	2.5%
	98	41											0.8336 ± 0.0005	10.9%
	98	41											0.9097 ± 0.0006	1.2%
	98	41											0.9937 ± 0.0006	1.3%
	98	41											0.9966 ± 0.0006	1.7%
	98	41											1.0243 ± 0.0006	1.0%
	98	41											1.1689 ± 0.0006	17.1%
	98	41											1.2302 ± 0.0006	1.5%
	98	41											1.3176 ± 0.0006	1.0%
	98	41											1.3355 ± 0.0006	1.4%
	98	41											1.4326 ± 0.0006	5.5%
	98	41											1.4367 ± 0.0006	1.9%
	98	41											1.4671 ± 0.0006	1.0%
	98	41											1.5121 ± 0.0006	5.1%
	98	41											1.5412 ± 0.0006	2.1%
	98	41											1.5412 ± 0.0006	2.1%
	98	41											1.5465 ± 0.0006	3.0%
	98	41											1.7019 ± 0.0006	9.1%
	98	41											1.8851 ± 0.0006	3.0%
	98	41											1.9454 ± 0.0006	1.5%
	98	41											1.8905 ± 0.0006	3.5%
$_{41}Nb^{98}$	98	41		97.910330	2.8 s.	β-	4.59	4.6			1+		0.6451 ± 0.0003	0.8%
	98	41											0.7874 ± 0.0003	3.2%
	98	41											0.9717 ± 0.0003	0.8%
	98	41											1.0243 ± 0.0003	1.6%
	98	41											1.4324 ± 0.0003	0.8%
$_{41}Nb^{99m}$	99	41			2.6 m.	β-		3.2			1/2-		0.0978 100 +	
	99	41				I.T.							0.1375	22
	99	41											0.254	57
	99	41											0.264	11
	99	41											0.352	42
	99	41											0.451	20
	99	41											0.525	13
	99	41											0.549	10
	99	41											0.554	8
	99	41											0.598	12
	99	41											0.631	19
	99	41											0.655	8
	99	41											0.673	11
	99	41											0.793	20
	99	41											0.889	5.5
	99	41											0.905	8
	99	41											0.945	9.5
	99	41											1.100	6.5
	99	41											1.259	7
	99	41											1.317	5
	99	41											1.475	11
	99	41											1.698	11
	99	41											1.735	8.5
	99	41											2.010	6.5
	99	41											2.241	8
	99	41											2.544	10
	99	41											2.642	50
	99	41											2.693	14
	99	41											2.734	15
	99	41											2.791	7.5
	99	41											2.854	48
	99	41											3.010	4
$_{41}Nb^{99}$	99	41		98.911619	15.0 s.	β-	3.62	3.5	100%		9/2+		0.0978 ± 0.0005	50%
	99	41											0.1375 ± 0.0005	90%
$_{41}Nb^{100m}$	100	41			1.5 s.	β-							Nb k x-ray	
	100	41											0.159	
	100	41											0.6364	
	100	41											1.0637	
$_{41}Nb^{100}$	100	41		99.914180	3.1 s.	β-	6.2	5.8					0.5354	100 +
	100	41											0.6001	46
	100	41											0.9661	
	100	41											1.0637	
	100	41											1.2803	
	100	41											1.5658	
$_{41}Nb^{101}$	101	41		100.915320	7.1 s.	β-	4.6	4.3					0.1105 ± 0.0005	
	101	41											0.1577 ± 0.0005	
	101	41											0.1806 ± 0.0005	
	101	41											0.2762 ± 0.0005	
	101	41											0.2897 ± 0.0005	
	101	41											0.4409 ± 0.0005	
	101	41											0.4659 ± 0.0005	

Isotope	A	Z	% Natural abundance	Atomic mass	Half-life	Decay mode	Decay energy (MeV)	Particle energy (MeV)	Particle intensity	Thermal neutron cross section	Spin (h/2π)	μ Nucl. mag. moment	Gamma-ray energy (MeV)	Gamma-ray intensity
	101	41											0.7969 ± 0.0005	
	101	41											0.8100 ± 0.0005	
$_{41}Nb^{102m}$	102	41			1.3 s.	β-								
$_{40}Nb^{102}$	102	41		101.918040	4.3 s.	β-	7.2						0.2960 ± 0.0002	
	102	41											0.3976 ± 0.0002	
	102	41											0.4006	
	102	41											0.5514	
	102	41											0.8474	
	102	41											0.9490	
	102	41											1.2354	
	102	41											1.6330	
	102	41											1.7375	
	102	41											2.1844	
$_{41}Nb^{103}$	103	41		102.919370	1.5 s.	β-	5.5							
Mo		42		95.94						2.60 b.				
$_{42}Mo^{88}$	88	42		87.921820	8.0 m.	β+	3.6				0+		ann.rad.	
	88	42				E.C.							0.0800 ± 0.0005	80 +
	88	42											0.1399 ± 0.0005	60
	88	42											0.1707 ± 0.0005	100
$_{42}Mo^{89m}$	89	42			0.19 s.	I.T.	0.119				1/2-		0.119	
	89	42											0.268	
$_{42}Mo^{89}$	89	42		88.919480	2.2 m.	β+					9/2+		ann.rad.	
	89	42				E.C.							0.659	
	89	42											0.803	
	89	42											1.155	
	89	42											1.272	
$_{42}Mo^{90}$	90	42		89.913933	5.67 h.	β+(25%)	2.49	1.085			0+		ann.rad.	50%
	90	42				E.C>(75%)							0.04274 ± 0.00001	2%
	90	42											0.12237 ± 0.00005	64%
	90	42											0.1629 ± 0.0001	6%
	90	42											0.2031 ± 0.0001	6%
	90	42											0.25734 ± 0.00005	78%
	90	42											0.3232 ± 0.0001	6%
	90	42											0.4454 ± 0.0002	6%
	90	42											0.9415 ± 0.0004	6%
	90	42											0.9902 ± 0.0006	1%
	90	42											1.2713 ± 0.0006	4%
	90	42											1.3874 ± 0.0005	2%
	90	42											1.4546 ± 0.0007	2%
$_{42}Mo^{91m}$	91	42			65 s.	I.T.(50%)	0.653				1/2-		ann.rad.	75%
	91	42				β+, EC (50%)		2.5					0.6529 ± 0.0001	48%
	91	42						2.8					1.2081 ± 0.0001	20%
	91	42						2.8					1.5080 ± 0.0001	24%
	91	42						4.0					2.2407 ± 0.0004	0.8%
$_{42}Mo^{91}$	91	42		90.911755	15.5 m.	β+(94%)	4.44	3.44	94%		9/2-		ann.rad.	190%
	91	42				E.C.(6%)							1.6373 ± 0.0001	0.3%
	91	42											2.6321 ± 0.0002	0.1%
	91	42											3.0286 ± 0.0002	0.08%
	91	42											(0.1 - 4.2)	
$_{42}Mo^{92}$	92	42	14.84%	91.906808							0+			
$_{42}Mo^{93m}$	93	42			6.9 h.	I.T. (99+%)	2.425				21/2+	+9.21	0.26306 ± 0.00001	58%
	93	42											0.68461 ± 0.00008	99.7%
	93	42											1.47711 ± 0.00002	99%
$_{42}Mo^{93}$	93	42		92.906813	3.5x10³ y.	E.C.	0.406				5/2+		0.0304	(D)
$_{42}Mo^{94}$	94	42	9.25%	93.905085							0+			
$_{42}Mo^{95}$	95	42	15.92%	94.905840						14.5 b.	5/2+	-0.9133		
$_{42}Mo^{96}$	96	42	16.68%	95.904678							0+			
$_{42}Mo^{97}$	97	42	9.55%	96.906020						2 b.	5/2+	-0.9335		
$_{42}Mo^{98}$	98	42	24.13%	97.905406						0.132 b.	0+			
$_{42}Mo^{99}$	99	42		98.907711	65.94 h.	β-	1.357	0.45	14%		1/2+		0.144048 ± 0.0002	88%
	99	42						0.84	2%				0.18109 ± 0.0003	6.0%
	99	42						1.21	84%				0.36644 ± 0.0005	1.2%
	99	42											0.73947 ± 0.0005	12.6%
	99	42											0.77787 ± 0.00002	0.1%
	99	42											0.82298 ± 0.0004	0.1%
	99	42											0.9607 ± 0.0001	0.07%
$_{42}Mo^{100}$	100	42	9.63%	99.907477						0.195 b.	0+			
$_{42}Mo^{101}$	101	42		100.910345	14.6 m.	β-	2.81	2.23			1/2+		0.0063 ± 0.0001	80%
	101	42											0.0093 ± 0.0001	
	101	42											0.19193 ± 0.00004	22%
	101	42											0.5058 ± 0.0001	11.4%
	101	42											0.5908 ± 0.0001	
	101	42											0.6955 ± 0.0001	6.6%
	101	42											0.7130 ± 0.0001	

Isotope	A	Z	% Natural abundance	Atomic mass	Half-life	Decay mode	Decay energy (MeV)	Particle energy (MeV)	Particle intensity	Thermal neutron cross section	Spin (h/2π)	μ Nucl. mag. moment	Gamma-ray energy (MeV)	Gamma-ray intensity
	101	42											0.8774 ± 0.0001	3.1%
	101	42											0.9342 ± 0.0001	3.8%
	101	42											1.0123 ± 0.0001	11.4%
	101	42											1.1609 ± 0.0001	
	101	42											1.2511 ± 0.0001	
	101	42											1.3821 ± 0.0001	
	101	42											1.5323 ± 0.0001	1.1%
	101	42											1.5990 ± 0.0001	1.6%
	101	42											1.6736 ± 0.0001	1.5%
	101	42											1.7599 ± 0.0001	1.0%
	101	42											1.8404 ± 0.0001	1.2%
	101	42											2.0319 ± 0.0001	6.1%
$_{42}Mo^{102}$	102	42		101.910297	11.2 m.	β-	1.04	1.2			0+		0.1493	89 +
	102	42											0.2116	100
	102	42											0.2243	32
$_{42}Mo^{103}$	103	42		102.913470	68 s.	β-	4.0						0.1028 ± 0.0002	
	103	42											0.1440 ± 0.0002	
	103	42											0.2511 ± 0.0002	
$_{42}Mo^{104}$	104	42		103.913600	60 s.	β-	2.0				0+		0.0686 ± 0.0001	100 +
	104	42											0.0931 ± 0.0001	14
	104	42											0.3760 ± 0.0003	12
	104	42											0.4239 ± 0.0004	21
$_{42}Mo^{105m}$	105	42			30 s.	β-							0.078	
	105	42											0.161	
	105	42											0.251	
$_{42}Mo^{105}$	105	42		104.917190	50 s.	β-	5.5						0.0642	
	105	42											0.0856	
	105	42											0.2495	
$_{42}Mo^{106}$	106	42		105.917950	8.4 s.	β-	3.2				0+		0.1894 ± 0.0002	22 +
	106	42											0.3644 ± 0.0002	6
	106	42											0.3723 ± 0.0002	12
Tc		43												
$_{43}Tc^{90}$	90	43			49.2 s.	β+							ann.rad.	
													0.9479	
	90	43											1.0542	
$_{43}Tc^{90}$	90	43		89.923810	8.3 s.	β+	8.8	7.0	15%		1+		ann.rad.	199%
	90	43						7.9	95%				0.9479	
$_{43}Tc^{91m}$	91	43			3.3 m.	β+					$^1/_2$+		ann.rad.	170%
	91	43				E.C.							0.3375 ± 0.0002	1.1%
	91	43											0.4832 ± 0.0006	1%
	91	43											0.5487 ± 0.0003	1.6%
	91	43											0.8110 ± 0.0005	5%
	91	43											1.1111 ± 0.0001	3.1%
	91	43											1.3620 ± 0.0001	4.0%
	91	43											1.6052 ± 0.0001	7.8%
	91	43											1.6339 ± 0.0001	9.1%
	91	43											1.9023 ± 0.0001	6%
	91	43											2.4509 ± 0.0001	13.5%
	91	43											2.7164 ± 0.0001	1.8%
	91	43											2.7813 ± 0.0002	3.1%
	91	43											2.8878 ± 0.0002	1.4%
$_{43}Tc^{92}$	92	43		91.915257	4.4 m.	β+	6.2				9/2+		ann.rad.	200%
	92	43				E.C.					8+		0.0850 ± 0.0005	7.6%
	92	43											0.1475 ± 0.0005	55%
	92	43											0.2437 ± 0.0005	15%
	92	43											0.3293 ± 0.0005	78%
	92	43											0.7731 ± 0.0005	97%
	92	43											1.5096 ± 0.0005	100%
	92	43											2.1588 ± 0.001	1.8%
	92	43											2.3083 ± 0.001	1.4%
	92	43											3.0261 ± 0.001	0.5%
	92	43											4.3684 ± 0.001	0.4%
	92	43											4.5723 ± 0.001	0.4%
$_{43}Tc^{93m}$	93	43			43 m.	I.T.(13%)					$^1/_2$-		0.3918 ± 0.0001	60%
	93	43				E.C.(20%)							0.9437 ± 0.0005	2.5%
	93	43											1.4922 ± 0.0005	1.6%
	93	43											2.6445 ± 0.0003	13%
	93	43											3.1290 ± 0.0005	1.9%
	93	43											3.2205 ± 0.0007	0.9%
	93	43											3.2982 ± 0.0008	0.35%
$_{43}Tc^{93}$	93	43		92.910246	2.83 h.	β+(13%)	3.193	0.80			9/2+	6.15	ann.rad.	26%
						E.C.(87%)							1.3629 ± 0.0001	66%
	93	43											1.4771 ± 0.0001	9.6%
	93	43											1.5203 ± 0.0001	23%
	93	43											1.5388 ± 0.0002	0.6%
	93	43											2.7306 ± 0.0002	0.3%
	93	43											(0.1 - 3.0)	
$_{43}Tc^{94m}$	94	43			52 m.	β+(72%)	4.26				2+		ann.rad.	140%
	94	43				E.C.(28%)							0.8710 ± 0.0001	94%
	94	43											0.9932 ± 0.0001	2.3%
	94	43											1.5221 ± 0.0002	4.8%
	94	43											1.8686 ± 0.0001	6.1%
	94	43											2.7401 ± 0.0003	3.7%
	94	43											3.1291 ± 0.0005	1.5%

Isotope	A	Z	% Natural abundance	Atomic mass	Half-life	Decay mode	Decay energy (MeV)	Particle energy (MeV)	Particle intensity	Thermal neutron cross section	Spin (h/2π)	μ Nucl. mag. moment	Gamma-ray energy (MeV)	Gamma-ray intensity
$_{43}Tc^{94}$	94	43		93.909654	4.88 h.	β+(11%) E.C.(89%)	4.26				7+	5.20	ann.rad.	22%
	94	43											0.4491 ± 0.0002	2.6%
	94	43											0.5321 ± 0.0001	2.6%
	94	43											0.7026 ± 0.0001	100%
	94	43											0.8496 ± 0.0001	98%
	94	43											0.8710 ± 0.0001	100%
	94	43											0.9161 ± 0.0001	7.4%
	94	43											1.5917 ± 0.0003	2.4%
$_{43}Tc^{95m}$	95	43			61 d.	I.T.(4%))					$^1/_2$-		ann.rad.	
	95	43				β+(0.3%)		0.5					0.0389 ± 0.0001	1%
	95	43				E.C.(96%)		0.7					0.2041 ± 0.0001	66%
	95	43											0.5821 ± 0.0001	32%
	95	43											0.7658 ± 0.0001	3.7%
	95	43											0.7862 ± 0 0001	9.1%
	95	43											0.8206 ± 0.0002	5%
	95	43											0.8351 ± 0.0002	28%
	95	43											1.0392 ± 0.0002	3%
$_{43}Tc^{95}$	95	43		94.907657	20.0 h.	E.C.(100%)	1.69				9/2+	9.058	0.7657 ± 0.0001	93%
	95	43											0.8699 ± 0.0001	0.3%
	95	43											0.9478 ± 0.0001	2.2%
	95	43											1.0738 ± 0.0001	4%
$_{43}Tc^{96m}$	96	43			52 m.	I.T.(90%)	1.69				4+		0.0342 ± 0.0002	
	96	43				β+, EC (2%)							0.7782 ± 0.0001	1.9%
	96	43											1.2002 ± 0.0001	1.1%
$_{43}Tc^{96}$	96	43		95.907870	4.3 d.	E.C.	2.97	2.0			7+	+5.37	Mo k x-ray	55%
	96	43											0.3143 ± 0.0001	2.4%
	96	43											0.3165 ± 0.0001	1.4%
	96	43											0.5689 ± 0.0001	1%
	96	43											0.7782 ± 0.0001	99%
	96	43											0.8125 ± 0.0001	82%
	96	43											0.8498 ± 0.0001	98%
	96	43											1.0913 ± 0.0002	1.1%
	96	43											1.12168 ± 0.0001	15%
	96	43											1.2002 ± 0.0002	0.4%
$_{43}Tc^{97m}$	97	43			90 d.	I.T.	0.0965				$^1/_2$-		Tc k x-ray	41%
	97	43											0.0965	0.32%
$_{43}Tc^{97}$	97	43		96.906364	2.6x10^6 y.	E.C.(100%)	0.320				9/2+		Mo k x-ray	54%
$_{43}Tc^{98}$	98	43		97.907215	4.2x10^6 y.	β-	1.79	0.597	100%		6+		0.65241 ± 0.00005	100%
	98	43											0.74535 ± 0.00005	100%
$_{43}Tc^{99m}$	99	43			6.01 h.	I.T.(100%)	0.142				$^1/_2$-		Tc k x-ray	6.7%
	99	43											0.14049 ± 0.00001	89%
	99	43											0.14261 ± 0.00001	0.04%
$_{43}Tc^{99}$	99	43		98.906254	2.13x10^{5}y.	β-	0.293	0.293	100%		9/2+	+5.6847		
$_{43}Tc^{100}$	100	43		99.907657	15.8 s.	β-	3.203	2.2			1+		0.5396 ± 0.0001	7%
	100	43						2.9					0.5908 ± 0.0001	5.7%
	100	43					3.3						1.5122 ± 0.0003	0.4%
	100	43											(0.3 - 2.6)weak	
$_{43}Tc^{101}$	101	43		100.907327	14.2 m.	β-	1.62	1.32			9/2+		0.1272 ± 0.0001	2.9%
	101	43											0.1841 ± 0.0001	
	101	43											0.3068 ± 0.0001	88%
	101	43											0.5314 ± 0.0001	1%
	101	43											0.5451 ± 0.0001	6%
	101	43											0.7156 ± 0.0001	0.7%
	101	43											(0.1 - 0.93)weak	
$_{43}Tc^{102m}$	102	43			4.4 m.	I.T.(2%)	4.8				5+		0.4184 ± 0.0001	4%
	102	43				β-(98%)							0.4752 ± 0.0001	85$
	102	43											0.4972 ± 0.0006	5.7%
	402	43											0.6281 ± 0.0006	25%
	402	43											0.6302 ± 0.0005	15$
	402	43											0.6969 ± 0.0009	6%
	102	43											1.0464 ± 0.0003	12%
	102	43											1.1033 ± 0.0004	12%
	102	43											1.1131 ± 0.0006	2%
	102	43											1.1976 ± 0.0005	7%
	102	43											1.2925 ± 0.0003	4%
	102	43											1.3386 ± 0.0003	4%
	102	43											1.5962 ± 0.0008	2.7%
	102	43											1.6163 ± 0.0007	15%
	102	43											1.7112 ± 0.0001	2.7%
	102	43											1.8107 ± 0.0001	5.6%
	102	43											2.2257 ± 0.001	5.5%
	102	43											2.2447 ± 0.001	11.4%
	102	43											2.4384 ± 0.001	4.4%
$_{43}Tc^{102}$	102	43		101.909208	5.3 s.	β-	4.5	3.4			1+		0.4686 ± 0.0001	0.8%
	102	43						4.2					0.4751 ± 0.0001	6%
	102	43											0.6281 ± 0.0001	0.7%
	102	43											0.6368 ± 0.0001	0.4%
	102	43											1.1032 ± 0.0001	0.35%
	102	43											1.1055 ± 0.0002	0.7%
	102	43											1.3620 ± 0.0002	0.3%

Isotope	A	Z	% Natural abundance	Atomic mass	Half-life	Decay mode	Decay energy (MeV)	Particle energy (MeV)	Particle intensity	Thermal neutron cross section	Spin (h/2π)	μ Nucl. mag. moment	Gamma-ray energy (MeV)	Gamma-ray intensity
$_{43}Tc^{103}$	103	43		102.909172	54 s.	β-	2.64	2.0					0.1361 ± 0.0001	15%
	103	43						2.2					0.1743 ± 0.0001	2.6%
	103	43											0.2104 ± 0.0001	9%
	103	43											0.3435 ± 0.0002	3.8%
	103	43											0.3464 ± 0.0001	16%
	103	43											0.3886 ± 0.0001	2%
	103	43											0.4032 ± 0.0001	2%
	103	43											0.5012 ± 0.0001	2%
	103	43											0.5629 ± 0.0001	6%
	103	43											0.6612 ± 0.0001	0.7%
	103	43											0.9024 ± 0.0003	0.6%
	103	43											(0.13 - 1.0)weak	
$_{43}Tc^{104}$	104	43		103.911460	18.3 m.	β-	5.6	2.4			(3)		0.3483 ± 0.0001	15%
	104	43						3.3					0.3580 ± 0.0001	89%
	104	43						4.4					0.5305 ± 0.0001	16%
	104	43											0.5351 ± 0.0001	15%
	104	43											0.8844 ± 0.0001	11%
	104	43											0.8931 ± 0.0001	10%
	104	43											1.1574 ± 0.0001	3%
	104	43											1.2818 ± 0.0001	2%
	104	43											1.3805 ± 0.0001	1.7%
	104	43											1.3966 ± 0.0001	2.4%
	104	43											1.5413 ± 0.0001	1%
	104	43											1.5967 ± 0.0001	4%
	104	43											1.6124 ± 0.0001	6%
	104	43											1.6768 ± 0.0001	8%
	104	43											1.7369 ± 0.0001	2%
	104	43											1.9110 ± 0.0001	2%
	104	43											1.9771 ± 0.0002	1.6%
	104	43											2.0157 ± 0.0001	1.8%
	104	43											2.1238 ± 0.0001	2.2%
	104	43											2.1905 ± 0.0001	1.8%
	104	43											2.4655 ± 0.0002	1.2%
	104	43											2.6085 ± 0.0002	1.6%
	104	43											3.1492 ± 0.0002	1.2%
	104	43											(0.3 - 3.7)	
$_{43}Tc^{105}$	105	43		104.911820	7.6 m.	β-	3.7	3.4			5/2 +		0.1079 ± 0.0001	9.6%
	105	43											0.1384 ± 0.0001	3%
	105	43											0.1432 ± 0.0001	11%
	105	43											0.1578 ± 0.0001	1.8%
	105	43											0.1593 ± 0.0001	7%
	105	43											0.2256 ± 0.0001	2%
	105	43											0.2520 ± 0.0001	4%
	105	43											0.2726 ± 0.0001	2.4%
	105	43											0.3215 ± 0.0001	7.6%
	105	43											0.3223 ± 0.0001	1%
	105	43											0.3583 ± 0.0003	1.7%
	105	43											0.4419 ± 0.0003	1.2%
	105	43											0.4459 ± 0.0002	1.1%
	105	43											0.4628 ± 0.0001	3%
	105	43											0.4663 ± 0.0003	1%
	105	43											0.4801 ± 0.0005	1%
	105	43											0.4906 ± 0.0003	1.6%
	105	43											0.5779 ± 0.0002	1.6%
	105	43											0.5402 ± 0.0002	1.8%
	105	43											0.7393 ± 0.0003	1.1%
	105	43											0.8960 ± 0.0005	1%
	105	43											1.0084 ± 0.0005	1.2%
	105	43											1.3663 ± 0.0003	2.2%
	105	43											1.5106 ± 0.0005	1.6%
	105	43											1.5601 ± 0.0005	1.4%
	105	43											2.0539 ± 0.0005	1.0%
	105	43											2.1554 ± 0.0005	1.5%
$_{43}Tc^{106}$	106	43		105.914510	36 s.	β-	6.7						0.2703 ± 0.0001	100%
	106	43											0.5222 ± 0.0001	7%
	106	43											0.7923 ± 0.0001	5%
	106	43											1.5043 ± 0.0001	1.2%
	106	43											1.6155 ± 0.0001	1.7%
	106	43											1.9694 ± 0.0001	9%
	106	43											2.2393 ± 0.0001	14%
	106	43											2.7014 ± 0.0001	2%
	106	43											2.7770 ± 0.0002	2%
	106	43											2.7893 ± 0.0002	8%
	106	43											2.9163 ± 0.0002	3%
	106	43											2.9459 ± 0.0002	3%
	106	43											3.0471 ± 0.0002	3%
	106	43											3.1864 ± 0.0002	5%
	106	43											3.2595 ± 0.0002	3%
	106	43											3.3642 ± 0.0003	1%
	106	43											3.5510 ± 0.0003	1%
$_{43}Tc^{107}$	107	43		106.915230	21.2 s.	β-	4.8						0.1027 ± 0.0001	21%
	107	43											0.1063 ± 0.0001	8%
	107	43											0.1421 ± 0.0001	3%
	107	43											0.1455 ± 0.0001	2%
	107	43											0.1770 ± 0.0001	9%

Isotope	A	Z	% Natural abundance	Atomic mass	Half-life	Decay mode	Decay energy (MeV)	Particle energy (MeV)	Particle intensity	Thermal neutron cross section	Spin (h/2π)	μ Nucl. mag. moment	Gamma-ray energy (MeV)	Gamma-ray intensity
	107	43											0.1997 ± 0.0001	2%
	107	43											0.2915 ± 0.0001	4%
	107	43											0.3225 ± 0.0001	1%
	107	43											0.3354 ± 0.0001	1%
	107	43											0.3545 ± 0.0001	2%
	107	43											0.3603 ± 0.0001	3%
	107	43											0.4587 ± 0.0001	6%
	107	43											0.4898 ± 0.0001	1%
	107	43											0.5954 ± 0.0001	5%
	107	43											0.8569 ± 0.0001	1%
	107	43											1.1181 ± 0.0001	1%
	107	43											1.2187 ± 0.0001	1%
	107	43											1.5732 ± 0.0001	1%
	107	43											1.7419 ± 0.0001	0.5%
	107	43											2.3789 ± 0.0003	0.5%
	107	43											2.5023 ± 0.0003	0.6%
	107	43											2.5374 ± 0.0003	0.6%
$_{43}Tc^{108}$	108	43		107.918420	5.0 s.	β-	7.8						0.2422 ± 0.0001	82%
	108	43											0.4656 ± 0.0001	14%
	108	43											0.7078 ± 0.0001	11%
	108	43											0.7326 ± 0.0001	10%
	108	43											1.1180 ± 0.0002	5%
	108	43											1.4170 ± 0.0001	4%
	108	43											1.5835 ± 0.0001	10%
	108	43											1.7604 ± 0.0001	2%
Ru		44		101.07						2.6 b.				
$_{44}Ru^{92}$	92	44		91.920120	3.7 m.	β+ (53%)	4.5				0+		ann.rad.	106%
	92	44				E.C.(47%)							0.1346 ± 0.0001	63%
	92	44											0.2138 ± 0.0001	92%
	92	44											0.2593 ± 0.0001	89%
	92	44											0.4507 ± 0.0001	6.6%
	92	44											0.8670 ± 0.0001	11%
	92	44											0.9102 ± 0.0001	3%
	92	44											0.9450 ± 0.0003	2.6%
	92	44											0.9472 ± 0.0003	2.6%
	92	44											1.2196 ± 0.0001	6%
	92	44											1.2291 ± 0.0001	3%
	92	44											1.4036 ± 0.0002	1.6%
	92	44											1.5176 ± 0.0003	1.8%
	92	44											1.6976 ± 0.00001	3.5%
	92	44											2.0597 ± 0.0002	3%
	92	44											2.3023 ± 0.001	1%
$_{44}Ru^{93m}$	93	44			10.8 s.	I.T.(21%)					1/2-		ann.rad.	
	93	44				β+, EC (79%)							0.7344 ± 0.0001	23%
	93	44											0.9283 ± 0.0002	1.7%
	93	44											1.1112 ± 0.0001	27%
	93	44											1.3962 ± 0.0001	40%
	93	44											2.0931 ± 0.0002	20%
$_{44}Ru^{93}$	93	44		92.917050	60 s.	β+	6.3				9/2+		ann.rad.	
	93	44				E.C.							0.6807 ± 0.0001	5.9%
	93	44											1.4349 ± 0.0001	0.7%
	93	44											1.8014 ± 0.0001	0.4%
	93	44											3.2343 ± 0.0002	0.3%
	93	44											3.9147 ± 0.0002	0.3%
	93	44											4.3895 ± 0.0003	0.1%
	93	44											(0.5- 4.2)weak	
$_{44}Ru^{94}$	94	44		93.911361	52 m.	E.C.(100%)	1.59				5/2+		0.3672 ± 0.0005	80%
	94	44											0.5247 ± 0.0005	2%
	94	44											0.8922 ± 0.0005	20%
$_{44}Ru^{95}$	95	44		94.910414	1.64 h.	E.C.(85%)	2.57				5/2+		ann.rad.	
	95	44				β+(15%)							0.2904 ± 0.0001	3.5%
	95	44											0.3010 ± 0.0001	2%
	95	44											0.3364 ± 0.0001	71%
	95	44											0.6268 ± 0.0001	18%
	95	44											0.7485 ± 0.0001	1.6%
	95	44											0.8063 ± 0.0001	4.1%
	95	44											0.8422 ± 0.0001	1.3%
	95	44											0.8890 ± 0.0001	1.9%
	95	44											1.0507 ± 0.0001	2.6%
	95	44											1.0968 ± 0.0001	21%
	95	44											1.1787 ± 0.0002	5%
	95	44											1.4106 ± 0.0001	2.5%
	95	44											1.4593 ± 0.0001	2.1%
	95	44											1.7854 ± 0.0002	0.6%
	95	44											1.9881 ± 0.0002	0.7%
	95	44											2.3445 ± 0.0002	1.4%
$_{44}Ru^{96}$	96	44	5.52%	95.907599						0.2 b.	0+			
$_{44}Ru^{97}$	97	44		96.907556	2.89 d.	E.C.	1.12				5/2+	0.687	Tc k x-ray	58%
	97	44											0.1088 ± 0.0001	0.1%
	97	44											0.2157 ± 0.0001	10%
	97	44											0.3245 ± 0.0001	86%
	97	44											0.4606 ± 0.0001	10%
	97	44											0.5693 ± 0.0001	0.9%

Isotope	A	Z	% Natural abundance	Atomic mass	Half-life	Decay mode	Decay energy (MeV)	Particle energy (MeV)	Particle intensity	Thermal neutron cross section	Spin (h/2π)	μ Nucl. mag. moment	Gamma-ray energy (MeV)	Gamma-ray intensity
$_{44}Ru^{98}$	98	44	1.88%	97.905287							0+			
$_{44}Ru^{99}$	99	44	12.7%	98.905939						5 b.	5/2+	-0.6413		
$_{44}Ru^{100}$	100	44	12.6%	99.904219						5.8 b.	0+			
$_{44}Ru^{101}$	101	44	17.0%	100.905582						5 b.	5/2+	-0.7188		
$_{44}Ru^{102}$	102	44	31.6%	101.904348						1.2 b.	0+			
$_{44}Ru^{103}$	103	44		102.906323	39.24 d.	β-	0.767	0.117	5%		5/2+	0.67	0.05329 ± 0.00001	0.36%
	103	44						0.225	91%				0.29498 ± 0.00001	0.36%
	103	44						0.725	3%				0.4438 ± 0.0001	0.31%
	103	44											0.49708 ± 0.00001	86%
	103	44											0.55704 ± 0.00004	0.8%
	103	44											0.61033 ± 0.00001	5.3%
	103	44											(0.04 - 1.6)weak	
$_{44}Ru^{104}$	104	44	18.7%	103.905424						0.35 b.	0+			
$_{44}Ru^{105}$	105	44		104.907744	4.44 h.	β-	1.917	1.109	22%	0.4 b.	3/2+		0.12968 ± 0.00007	5.7%
	105	44						1.134	13%				0.1491 ± 0.0001	1.8%
	105	44						1.187	49%				0.2629 ± 0.0001	6.6%
	105	44											0.31664 ± 0.00005	11%
	105	44											0.3502 ± 0.0001	1%
	105	44											0.3934 ± 0.0001	3.8%
	105	44											0.4135 ± 0.0001	2.2%
	105	44											0.46943 ± 0.00005	17.6%
	105	44											0.4993 ± 0.0003	2%
	105	44											0.6562 ± 0.0001	2%
	105	44											0.67634 ± 0.00006	15.7%
	105	44											0.72420 ± 0.00005	47%
	105	44											0.87585 ± 0.00005	2.5%
	105	44											0.90763 ± 0.00006	0.5%
	105	44											0.9694 ± 0.0001	2.1%
	105	44											1.0174 ± 0.0001	0.3%
	105	44											1.3214 ± 0.0001	0.2%
	105	44											1.6983 ± 0.0002	0.2%
	105	44											(0.1 - 1.8)	
$_{44}Ru^{106}$	106	44		105.907321	372.6 d.	β-	0.039	0.0394	100%	0.15 b.	0+			
$_{44}Ru^{107}$	107	44		106.910130	3.8 m.	β-	3.2	2.1					0.1939 ± 0.0002	10.9%
	107	44						3.2					0.3741 ± 0.0003	3.5%
	107	44											0.4055 ± 0.0002	2.6%
	107	44											0.4625 ± 0.0002	4.3%
	107	44											0.4891 ± 0.0002	1.4%
	107	44											0.5791 ± 0.0002	2.6%
	107	44											0.8488 ± 0.0002	5.3%
	107	44											1.0429 ± 0.0002	1.9%
	107	44											1.2724 ± 0.0002	2.2%
$_{44}Ru^{108}$	108	44		107.910140	4.6 m.	β-	1.2	1.2					0.0923 ± 0.0004	2%
	108	44											0.1651 ± 0.0004	32%
	108	44											0.4339 ± 0.0004	46%
	108	44											0.4975 ± 0.0004	7%
	108	44											0.5112 ± 0.0004	1.8%
	108	44											0.6189 ± 0.0004	15%
	108	44											0.9324 ± 0.0004	4%
$_{44}Ru^{109}$	109	44		108.913240	35 s.	β-	4.2						0.1164	
	109	44											0.3584	
$_{44}Ru^{110}$	110	44		109.913760	15 s.	β-	2.5						0.1121 ± 0.0003	100 +
	110	44											0.3737 ± 0.0003	100
	110	44											0.4397 ± 0.0003	10
	110	44											0.5729 ± 0.0003	4
	110	44											0.7967 ± 0.0005	10
	110	44											0.8134 ± 0.0005	3
	110	44											1.0962 ± 0.0005	2
Rh		45		102.9055						145 b.				
$_{45}Rh^{94m}$	94	45			71 s.	β+		6.4					ann.rad.	
	94	45											0.1264 ± 0.0002	4%
	94	45											0.3117 ± 0.0001	12%
	94	45											0.4381 ± 0.0002	7%
	94	45											0.4926 ± 0.0003	4%
	94	45											0.5529 ± 0.0003	2%
	94	45											0.7562 ± 0.0001	51%
	94	45											1.0681 ± 0.0003	5%
	94	45											1.0752 ± 0.0002	31%
	94	45											1.1107 ± 0.0002	3%
	94	45											1.4307 ± 0.0001	100%
	94	45											1.5397 ± 0.0003	4%
	94	45											1.8043 ± 0.001	2%

TABLE OF THE ISOTOPES (Continued)

Isotope	A	Z	% Natural abundance	Atomic mass	Half-life	Decay mode	Decay energy (MeV)	Particle energy (MeV)	Particle intensity	Thermal neutron cross section	Spin (h/2π)	μ Nucl. mag. moment	Gamma-ray energy (MeV)	Gamma-ray intensity
	94	45											1.9025 ± 0.001	2%
	94	45											2.1245 ± 0.001	2%
	94	45											2.7786 ± 0.001	1.1%
	94	45											3.0077 ± 0.001	1.1%
	94	45											3.2103 ± 0.001	1%
	94	45											3.2560 ± 0.001	2%
$_{45}Rh^{94}$	94	45		93.921670	25.8 s.	β+	9.6				8+		ann.rad.	
	94	45											0.1461 ± 0.0001	75%
	94	45											0.3117 ± 0.0001	97%
	94	45											0.4381 ± 0.0002	3%
	94	45											0.7562 ± 0.0001	100%
	94	45											1.4307 ± 0.0001	100%
	94	45											2.0995 ± 0.001	2%
	94	45											2.1245 ± 0.001	1.1%
	94	45											3.2103 ± 0.001	0.9%
	94	45											3.2560 ± 0.001	1.4%
$_{45}Rh^{95m}$	95	45			1.96 m.	I.T.(88%)		0.5433			1/2+		ann.rad.	
	95	45				β+, EC (12%)							0.5433 ± 0.0003	80%
	95	45											0.7837 ± 0.0004	8%
	95	45											2.8210 ± 0.0008	0.8%
	95	45											3.1862 ± 0.0008	0.9%
	95	45											3.8244 ± 0.0007	1.3%
$_{45}Rh^{95}$	95	45		94.915900	5.0 m.	β+	5.1	0.7	1.3 +		9/2+		ann.rad.	
	95	45						1.04	12				0.2293 ± 0.0003	2%
	95	45						1.33	3.5				0.4103 ± 0.0003	1%
	95	45											0.6225 ± 0.0005	2.6%
	95	45											0.6610 ± 0.0003	105%
	95	45											0.6776 ± 0.0003	6%
	95	45											0.7644 ± 0.0007	2%
	95	45											0.8950 ± 0.0003	2%
	95	45											0.9416 ± 0.0003	72%
	95	45											1.3170 ± 0.0003	3%
	95	45											1.3520 ± 0.0003	21%
	95	45											1.4893 ± 0.0003	3.4%
	95	45											1.4947 ± 0.0003	5%
	95	45											1.5245 ± 0.0005	2%
	95	45											2.1210 ± 0.0003	1.6%
	95	45											2.7918 ± 0.0003	2.4%
	95	45											3.0632 ± 0.0005	1%
	95	45											(0.2 - 3.8)	
$_{45}Rh^{96m}$	96	45			1.51 m.	I.T.(60%)	0.0520				2+		ann.rad.	
	96	45				β+, EC (40%)		4.70					Tc,Ru x-rays	
	96	45											0.0520 ± 0.0001	0.2%
	96	45											0.6853 ± 0.0001	4%
	96	45											0.8087 ± 0.0002	2.5%
	96	45											0.8326 ± 0.0001	39%
	96	45											1.0985 ± 0.0002	8%
	96	45											1.4512 ± 0.0002	1.3%
	96	45											1.6921 ± 0.0002	6.4%
	96	45											1.7436 ± 0.0002	1%
	96	45											1.9073 ± 0.0002	1%
	96	45											2.1636 ± 0.0002	2.5%
	96	45											2.2575 ± 0.0003	1.7%
	96	45											2.4591 ± 0.0003	0.7%
	96	45											(0.4 - 3.3)	
$_{45}Rh^{96}$	96	45		95.914515	9.9 m.	β+	6.44	3.3			5+		ann.rad.	
	96	45				E.C.							0.4299 ± 0.0002	2%
	96	45											0.6315 ± 0.0001	80%
	96	45											0.6441 ± 0.0001	5%
	96	45											0.6853 ± 0.0001	98%
	96	45											0.7418 ± 0.0001	32%
	96	45											0.8007 ± 0.0002	3%
	96	45											0.8326 ± 0.0001	100%
	96	45											0.9155 ± 0.0002	1%
	96	45											0.9441 ± 0.0002	2%
	96	45											1.0703 ± 0.0002	1.7%
	96	45											1.2279 ± 0.0003	8.7%
	96	45											1.2306 ± 0.0003	7.5%
	96	45											1.2421 ± 0.0002	1.3%
	96	45											1.2757 ± 0.0002	3.1%
	96	45											1.5667 ± 0.0005	2%
	96	45											1.5887 ± 0.0005	1%
	96	45											1.6055 ± 0.0002	2.7%
	96	45											1.6487 ± 0.0002	2%
	96	45											1.6921 ± 0.0002	1.9%
	96	45											1.7325 ± 0.0002	5.7%
	96	45											1.7886 ± 0.0002	1.9%
	96	45											1.8596 ± 0.0002	1.6%
	96	45											1.9630 ± 0.0002	1.2%
	96	45											(0.2 - 3.4)	
$_{45}Rh^{97m}$	97	45			46 m.	I.T.(5%)		1.8			1/2-		ann.rad.	
	97	45				β+, EC (95%)		2.1					0.1886 ±	51%

Isotope	A	Z	% Natural abundance	Atomic mass	Half-life	Decay mode	Decay energy (MeV)	Particle energy (MeV)	Particle intensity	Thermal neutron cross section	Spin (h/2π)	μ Nucl. mag. moment	Gamma-ray energy (MeV)	Gamma-ray intensity
	97	45						2.5					0.4215 ± 0.0003	13%
	97	45											0.5278 ± 0.0002	9%
	97	45											0.5824 ± 0.0003	2.6%
	97	45											0.7190 ± 0.0003	3%
	97	45											0.7711 ± 0.0003	4.8%
	97	45											0.9085 ± 0.0005	2%
	97	45											0.9955 ± 0.0003	3.0%
	97	45											1.0134 ± 0.0003	3.6%
	97	45											1.1838 ± 0.0006	4.5%
	97	45											1.1870 ± 0.0006	4.5%
	97	45											1.4264 ± 0.0004	2.3%
	97	45											1.5866 ± 0.0004	8.3%
	97	45											1.7184 ± 0.0006	2.4%
	97	45											2.0074 ± 0.0005	4%
	97	45											2.0361 ± 0.0005	4%
	97	45											2.1223 ± 0.0007	3%
	97	45											2.2452 ± 0.0005	13%
	97	45											2.6088 ± 0.0006	2%
	97	45											2.6478 ± 0.0006	3%
	97	45											3.3741 ± 0.0006	1%
$_{45}Rh^{97}$	97	45		96.911320	31.0 m.	β+	3.51	1.8			9/2+		ann.rad.	
	97	45						2.1					0.1886 ± 0.0002	1%
	97	45						2.52					0.3892 ± 0.0004	1%
	97	45											0.4515 ± 0.0003	75%
	97	45											0.7772 ± 0.0004	1.6%
	97	45											0.8073 ± 0.0004	1.4%
	97	45											0.8398 ± 0.0003	12%
	97	45											0.8788 ± 0.0003	9.3%
	97	45											1.0537 ± 0.0004	1.7%
	97	45											1.2280 ± 0.0005	1.1%
	97	45											1.3101 ± 0.0006	1.1%
	97	45											1.91310 ± 0.0009	0.7%
	97	45											(0.2 - 3.5)	
$_{45}Rh^{98m}$	98	45			3.5 m.	β+		3.4			2+		ann.rad.	
	98	45											0.6154 ± 0.0001	5%
	98	45											0.6524 ± 0.0001	96%
	98	45											0.7452 ± 0.0002	78%
	98	45											1.1440 ± 0.0005	8%
	98	45											1.4149 ± 0.0005	4%
$_{45}Rh^{98}$	98	45		97.910716	8.6 m.	β+(90%)	5.06	2.8			5+		ann.rad.	180%
	98	45						3.45					0.6524 ± 0.0001	94%
	98	45											0.7453 ± 0.0001	5%
	98	45											0.7623 ± 0.0002	78%
	98	45											1.1644 ± 0.0004	1.1%
	98	45											1.4149 ± 0.0004	1.1%
	98	45											1.8170 ± 0.0004	4.8%
$_{45}Rh^{99m}$	99	45			4.7 h.	β+(8%)		.74			9/2+		ann.rad.	16%
	99	45				E.C.(92%)							0.2766 ± 0.0004	1.6%
	99	45											0.3408 ± 0.0004	69%
	99	45											0.5282 ± 0.0004	1.4%
	99	45											0.6178 ± 0.0004	12%
	99	45											0.7193 ± 0.0004	1.2%
	99	45											0.9366 ± 0.0004	2.2%
	99	45											1.2612 ± 0.0004	11%
$_{45}Rh^{99}$	99	45		98.908192	16.1 d.	β+(4%)	2.09	0.54			1/2-		ann.rad.	
	99	45				E.C.(97%)		0.68					0.0894 ± 0.0001	31%
	99	45											0.3224 ± 0.0004	4%
	99	45											0.3530 ± 0.0004	32%
	99	45											0.5277 ± 0.0004	41%
	99	45											0.6180 ± 0.0004	5%
	99	45											0.8066 ± 0.0004	1.4%
	99	45											0.9145 ± 0.0004	1.4%
	99	45											(0.1 - 2.0)	
$_{45}Rh^{100m}$	100	45			4.7 m.	I.T.(99%)					5+		ann.rad.	
	100	45				β+(0.4%)					5+		0.0748 ± 0.0005	70%
	100	45											0.2647 ± 0.0005	30%
	100	45											0.5396 ± 0.001	0.4%
	100	45											0.6869 ± 0.0005	0.2%
$_{45}Rh^{100}$	100	45		99.908116	20.8 h.	β+	3.63	1.3			1-		0.4462 ± 0.0001	11%
	100	45				E.C.		2.1					0.5396 ± 0.0001	78%
	100	45						2.6					0.5882 ± 0.0002	4%
	100	45											0.5908 ± 0.0001	1.4%
	100	45											0.8225 ± 0.0002	20%
	100	45											1.1071 ± 0.0002	13%
	100	45											1.3476 ± 0.0002	4.8%
	100	45											1.3621 ± 0.0001	15%
	100	45											1.5534 ± 0.0002	20%
	100	45											1.6275 ± 0.0003	1.6%
	100	45											1.9297 ± 0.0002	12%
	100	45											2.3761 ± 0.0003	35%
	100	45											2.5302 ± 0.0002	2.7%
$_{45}Rh^{101m}$	101	45			4.34 d.	E.C.(92%)					9/2+	+5.51	Rh k x-ray	51%
	101	45				I.T.(8%)	0.1573						0.1272 ± 0.0001	0.6%
	101	45											0.1573 ± 0.0001	0.2%
	101	45											0.3069 ± 0.0001	86%
	101	45											0.5451 ± 0.0001	4%

Isotope	A	Z	% Natural abundance	Atomic mass	Half-life	Decay mode	Decay energy (MeV)	Particle energy (MeV)	Particle intensity	Thermal neutron cross section	Spin (h/2π)	μ Nucl. mag. moment	Gamma-ray energy (MeV)	Gamma-ray intensity
$_{45}Rh^{101}$	101	45		100.906159	3.3 y.	E.C.	0.54				$^1/_2$-		Ru k x-ray	66%
	101	45											0.1272 ± 0.0001	73%
	101	45											0.1980 ± 0.0002	71%
	101	45											0.2950 ± 0.0003	0.7%
	101	45											0.3252 ± 0.0002	13%
	101	45											0.4880 ± 0.0003	0.4%
$_{45}Rh^{102m}$	102	45			206 d.	I.T.(5%)							ann.rad.	28%
	102	45				β-(19%)							0.4686 ± 0.0001	3%
	102	45				β+(14%)							0.4751 ± 0.0001	44%
	102	45				E.C.(62%)							0.5566 ± 0.0001	2%
	102	45											0.6280 ± 0.0001	4%
	102	45											1.1032 ± 0.0001	2.8%
	102	45											(0.4 - 1.6)	
$_{45}Rh^{102}$	102	45		101.906814	2.9 y.	E.C.	2.33				(6+)	4.11	Ru k x-ray	55%
	102	45											0.4152 ± 0.0002	2%
	102	45											0.4185 ± 0.0002	9%
	102	45											0.4204 ± 0.0002	3%
	102	45											0.4751 ± 0.0001	94%
	102	45											0.6280 ± 0.0001	8%
	102	45											0.6313 ± 0.0001	55%
	102	45											0.6924 ± 0.0002	1.6%
	102	45											0.6956 ± 0.0003	2.8%
	102	45											0.6975 ± 0.0001	43%
	102	45											0.7668 ± 0.0001	34%
	102	45											1.0466 ± 0.0001	34%
	102	45											1.1032 ± 0.0001	45%
	102	45											1.1128 ± 0.0001	19%
$_{45}Rh^{103}$	103	45	100%	102.905500						(11 + 134)b.	$^1/_2$-	-0.0884		
$_{45}Rh^{103m}$	103	45			56.12 ± 0.01 m	I.T.(100%)					7/2+		Rh k x-ray	7%
													0.03975 ± 0.00001	0.07%
$_{45}Rh^{104m}$	104	45			4.36 m.	I.T. (99+%)				800 b.	5+		Rh k x-ray	55%
	104	45											0.0514 ± 0.0001	48%
	104	45											0.0775 ± 0.0001	2%
	104	45											0.0971 ± 0.0001	3%
	104	45											0.5558 ± 0.0001	2.4%(D)
	104	45											0.7678 ± 0.0001	0.1%
$_{45}Rh^{104}$	104	45		103.906651	41.8 s.	β-(99+%)	2.44	1.88	2%		1+		0.3581 ± 0.0002	0.02%
	104	45				E.C.(0.4%)	1.14	2.44	98%				0.5558 ± 0.0001	2%
	104	45											1.2370 ± 0.0001	0.07%
	104	45											(0.35 - 1.8)weak	
$_{45}Rh^{105m}$	105	45			45 s.	I.T.	1.298				$^1/_2$-		Rh k x-ray	35%
	105	45											0.1296 ± 0.0001	20%
$_{45}Rh^{105}$	105	45		104.905686	35.4 h.	β-	0.57	0.247	30%	5x10^3 b.	7/2+	+4.428	0.2801 ± 0.0002	0.2%
	105	45						0.567	70%				0.3061 ± 0.0002	5.1%
	105	45											0.3189 ± 0.0001	19.2%
$_{45}Rh^{106m}$	106	45			2.18 h.	β-		0.92			6+		0.2217 ± 0.0001	6.5%
	106	45											0.2286 ± 0.0001	2%
	106	45											0.3910 ± 0.0001	3.5%
	106	45											0.4062 ± 0.0001	12%
	106	45											0.4296 ± 0.0001	13%
	106	45											0.4510 ± 0.0001	24%
	106	45											0.5119 ± 0.0001	86%
	106	45											0.6012 ± 0.0001	3%
	106	45											0.6162 ± 0.0001	20%
	106	45											0.6460 ± 0.0001	2.8%
	106	45											0.6904 ± 0.0001	2%
	106	45											0.7031 ± 0.0001	4.5%
	106	45											0.7173 ± 0.0001	29%
	106	45											0.7484 ± 0.0001	19%
	106	45											0.7932 ± 0.0001	5.7%
	106	45											0.8043 ± 0.0001	13%
	106	45											0.8084 ± 0.0001	7.5%
	106	45											0.8247 ± 0.0001	14%
	106	45											0.8473 ± 0.0001	3.6%
	106	45											1.0197 ± 0.0002	2%
	106	45											1.0458 ± 0.0001	31%
	106	45											1.1280 ± 0.0001	14%
	106	45											1.1994 ± 0.0001	11%
	106	45											1.2229 ± 0.0001	8%
	106	45											1.3944 ± 0.0002	2.8%
	106	45											1.5277 ± 0.0002	18%
	106	45											1.5724 ± 0.0002	6.7%
	106	45											1.7228 ± 0.0002	2%
	106	45											1.8390 ± 0.0002	2%
$_{45}Rh^{106}$	106	45		105.907279	29.8 s.	β-	3.54	2.4	2%		1+		0.51186 ± 0.00001	21%
	106	45						3.0	12%				0.61612 ± 0.00005	0.8%
	106	45						3.54	79%				0.62187 ± 0.00005	9.8%
	106	45											1.0504 ± 0.0001	1.5%
	106	45											1.12807 ± 0.00005	0.4%
	106	45											1.5622 ± 0.0001	0.16%
	106	45											1.7664 ± 0.0001	0.03%
	106	45											1.9885 ± 0.0001	0.03%

TABLE OF THE ISOTOPES (Continued)

Isotope	A	Z	% Natural abundance	Atomic mass	Half-life	Decay mode	Decay energy (MeV)	Particle energy (MeV)	Particle intensity	Thermal neutron cross section	Spin (h/2π)	μ Nucl. mag. moment	Gamma-ray energy (MeV)	Gamma-ray intensity
	106	45											2.1126 ± 0.0001	0.035%
	106	45											2.3660 ± 0.0001	0.03%
	106	45											(0.05 - 3.04)	
$_{45}Rh^{107}$	107	45		106.906751	21.7 m.	β-	1.2	65%			5/2+		0.2776 ± 0.0002	2%
	107	45						1.5	17%				0.3028 ± 0.0002	66%
	107	45											0.3128 ± 0.0002	5%
	107	45											0.3218 ± 0.0002	2%
	107	45											0.3482 ± 0.0002	2%
	107	45											0.3673 ± 0.0002	2%
	107	45											0.3925 ± 0.0002	9%
	107	45											0.5677 ± 0.0002	1%
	107	45											0.6701 ± 0.0002	2%
$_{45}Rh^{108m}$	108	45			17 s.	β-	4.5				1+		0.4339 ± 0.0004	43%
	0.8	45											0.4973 ± 0.0004	6%
	108	45											0.5112 ± 0.0004	2%
	108	45											0.6189 ± .0004	14%
	108	45											0.9324 ± 0.0004	3%
$_{45}Rh^{108}$	108	45		107.908650	6.0 m.	β-	4.4	1.57					0.4046 ± 0.0002	27%
	108	45											0.4339 ± 0.0002	92%
	108	45											0.4973 ± 0.0002	23%
	108	45											0.5811 ± 0.0002	58%
	108	45											0.6146 ± 0.0002	28%
	108	45											0.7230 ± 0.0002	7%
	108	45											0.9014 ± 0.0002	30%
	108	45											0.9313 ± 0.0002	7%
	108	45											0.9471 ± 0.0002	50%
	108	45											1.0927 ± 0.0004	3%
	108	45											1.2343 ± 0.0004	8%
	108	45											1.5280 ± 0.0004	1%
	108	45											1.8156 ± 0.0004	6%
$_{45}Rh^{109}$	109	45		108.908734	81 s.	β-	2.58				5/2+		0.1134 ± 0.0001	6%
	109	45											0.1780 ± 0.0001	8%
	109	45											0.2153 ± 0.0001	2%
	109	45											0.2450 ± 0.0001	1.3%
	109	45											0.2492 ± 0.0001	6%
	109	45											0.2763 ± 0.0001	2.2%
	109	45											0.2914 ± 0.0001	7.8%
	109	45											0.3254 ± 0.0002	1.5%
	109	45											0.3268 ± 0.0001	56%
	109	45											0.3780 ± 0.0001	1.3%
	109	45											0.4261 ± 0.0001	8%
	109	45											(0.1 - 1.6)	
$_{45}Rh^{110m}$	110	45			3.1 s.	β-		≈5					0.3737 ± 0.0003	100%
	110	45											0.4397 ± 0.0003	10%
	110	45											0.5729 ± 0.0005	4%
	110	45											0.7967 ± 0.0005	9.5%
	110	45											0.8134 ± 0.0005	3%
	110	45											1.0962 ± 0.0005	2%
$_{45}Rh^{110}$	110	45		109.910960	29 s.	β-	5.4	2.6					0.3737 ± 0.0002	91%
	110	45											0.3985 ± 0.0002	15%
	110	45											0.4400 ± 0.0002	26%
	110	45											0.4788 ± 0.0005	4%
	110	45											0.5312 ± 0.0005	2%
	110	45											0.5463 ± 0.0002	36%
	110	45											0.5849 ± 0.0002	17%
	110	45											0.6534 ± 0.0002	17%
	110	45											0.6877 ± 0.0002	28%
	110	45											0.8137 ± 0.0002	9%
	110	45											0.8381 ± 0.0002	22%
	110	45											0.8905 ± 0.0002	13%
	110	45											0.9045 ± 0.0002	27%
	110	45											0.9796 ± 0.0005	5%
	110	45											1.0483 ± 0.0005	8%
	110	45											1.0865 ± 0.0005	3%
	110	45											1.2165 ± 0.0005	7%
	110	45											1.2309 ± 0.0005	13%
	110	45											1.3921 ± 0.0005	4.6%
	110	45											1.5258 ± 0.001	2%
	110	45											1.5792 ± 0.001	2%
	110	45											1.5936 ± 0.0005	6%
	110	45											1.8717 ± 0.001	1%
	110	45											1.8851 ± 0.0005	4%
$_{45}Rh^{111}$	111	45		110.911630	11 s.	β-							0.3753	
$_{45}Rh^{112}$	112	45		111.914410	0.8 s.	β-							0.3489 ± 0.0002	
Pd		46		106.42						6.9 b.				
$_{46}Pd^{96}$	96	46		95.918010	2.0 m.	E.C.	3.3						0.1248	100 +
	96	46											0.4995	42
$_{46}Pd^{97}$	97	46		96.916480	3.1 m.	β+,E.C.	4.8				5/2+		ann.rad.	
	97	46											0.2653 ± 0.0001	50%
	97	46											0.4752 ± 0.0001	24%
	97	46											0.7927 ± 0.0001	12%
	97	46											0.9337 ± 0.0001	2%
	97	46											0.9403 ± 0.0003	3%
	97	46											1.0536 ± 0.0005	3%
	97	46											1.0554 ± 0.0005	5%

Isotope	A	Z	% Natural abundance	Atomic mass	Half-life	Decay mode	Decay energy (MeV)	Particle energy (MeV)	Particle intensity	Thermal neutron cross section	Spin (h/2π)	μ Nucl. mag. moment	Gamma-ray energy (MeV)	Gamma-ray intensity
	97	46											1.0585 ± 0.0005	2.4%
	97	46											1.1718 ± 0.0003	3%
	97	46											1.2378 ± 0.0005	2%
	97	46											1.4942 ± 0.0002	5.5%
	97	46											1.5198 ± 0.0005	2%
	97	46											1.6387 ± 0.0003	3.4%
	97	46											1.6411 ± 0.0003	3%
	97	46											1.7596 ± 0.0001	6%
	97	46											1.7972 ± 0.0005	1%
	97	46											1.8468 ± 0.0003	2.6%
	97	46											1.9939 ± 0.0005	1%
	97	46											2.0295 ± 0.0005	2.5%
	97	46											2.4284 ± 0.001	1%
	97	46											2.6844 ± 0.001	1%
	97	46											2.9746 ± 0.001	1%
	97	46											3.3421 ± 0.001	0.7%
	97	46											(0.2 - 3.4)	
$_{46}Pd^{98}$	98	46		97.912722	18 m.	β+	1.87				0+		ann.rad.	
	98	46				E.C.							0.0677 ± 0.0002	18 +
	98	46											0.1068 ± 0.0002	26
	98	46											0.1125 ± 0.0002	100
	98	46											0.1745 ± 0.0002	21
	98	46											0.6630 ± 0.0004	53
	98	46											0.7257 ± 0.0004	10
	98	46											0.8379 ± 0.0004	30
$_{46}Pd^{99}$	99	46		98.911763	21.4 m.	β+(49%)	3.36	1.58			5/2+		ann.rad.	
	99	46				E.C.(51%)		1.93					0.1360 ± 0.0001	73%
	99	46						2.18					0.2636 ± 0.0001	15%
	99	46											0.3867 ± 0.0001	2.8%
	99	46											0.3998 ± 0.0001	3.6%
	99	46											0.4103 ± 0.0002	1.3%
	99	46											0.4273 ± 0.0001	2.0%
	99	46											0.6504 ± 0.0005	1.3%
	99	46											0.6528 ± 0.0005	1.4%
	99	46											0.6531 ± 0.0003	2.6%
	99	46											0.6734 ± 0.0002	6.9%
	99	46											0.7866 ± 0.0002	3.3%
	99	46											0.8098 ± 0.0002	2.0%
	99	46											1.0134 ± 0.0003	1.4%
	99	46											1.0996 ± 0.0004	1.0%
	99	46											1.2564 ± 0.0004	1.0%
	99	46											1.3356 ± 0.0004	4.6%
	99	46											1.6805 ± 0.001	0.6%
	99	46											1.7176 ± 0.0005	0.6%
	99	46											2.2462 ± 0.0007	0.8%
	99	46											2.5363 ± 0.001	0.6%
	99	46											(0.2 - 2.85)	
$_{46}Pd^{100}$	100	46		99.908527	3.6 d.	E.C.	0.38				0+		0.03271 ± 0.0001	1.6 +
	100	46											0.0421 ± 0.0001	1.5
	100	46											0.0748 ± 0.0001	98
	100	46											0.0840 ± 0.0001	100
	100	46											0.1261 ± 0.0001	11
	100	46											0.1588 ± 0.0003	2
$_{46}Pd^{101}$	101	46		100.908287	8.4 h.	β+(5%)	1.98				5/2+		ann.rad.	
	101	46				E.C.(95%)							0.0244 ± 0.0001	4%
	101	46											0.2697 ± 0.0001	6%
	101	46											0.2963 ± 0.0001	19%
	101	46											0.5660 ± 0.00014	3.5%
	101	46											0.5904 ± 0.0001	12%
	101	46											0.7238 ± 0.0001	2%
	101	46											1.2020 ± 0.0001	1.5%
	101	46											1.2890 ± 0.0001	2.3%
$_{46}Pd^{102}$	102	46	1.02%	101.905634						3 b.	0+			
$_{46}Pd^{103}$	103	46		102.906114	16.97 d.	E.C.	0.572				5/2+		Rh k x-ray	64%
	103	46											0.03975 ± 0.0002	0.07 (D)
	103	46											0.3575 ± 0.0001	0.02
	103	46											0.4971 ± 0.0001	0.004
$_{46}Pd^{104}$	104	46	11.14%	103.904029							0+			
$_{46}Pd^{105}$	105	46	22.33%	104.905079						22 b.	5/2+	-0.642		
$_{46}Pd^{106}$	106	46	+27.33%	105.903478						(0.015 + 28)b.	0+			
$_{46}Pd^{107m}$	107	46			20.9 s.	I.T.	0.2149				11/2-		Pd k x-ray	16%
	107	46											0.2149 ± 0.0005	69%
$_{46}Pd^{107}$	107	46			6.5x10^{6} y.	β-	0.033	0.033			5/2+			
$_{46}Pd^{108}$	108	46	26.46%	107.903895						(0.19 + 8) b.	0+			
$_{46}Pd^{109m}$	109	46			4.68 m.	I.T.	0.189				11/2-		Pd x-ray	
	109	46											0.18903 ± 0.0002	56%
$_{46}Pd^{109}$	109	46		108.905954	β-	1.116	1.028				5/2+		0.08803 ± 0.0002	3.6%
	109	46											0.31134 ± 0.0003	0.03%
	109	46											0.6024 ± 0.0001	0.08%
	109	46											0.6363 ± 0.0001	0.01%
	109	46											0.6472 ± 0.0001	0.02%
	109	46											0.7813 ± 0.0001	0.01%
	109	46											(0.08 - 1.0)weak	
$_{46}Pd^{110}$	110	46	11.72%	109.905167						0.02 b.	0+			
$_{46}Pd^{111m}$	111	46			5.5 h.	I.T.(73%)	0.172				11/2-		0.0704 ± 0.0001	8%

Isotope	A	Z	% Natural abundance	Atomic mass	Half-life	Decay mode	Decay energy (MeV)	Particle energy (MeV)	Particle intensity	Thermal neutron cross section	Spin (h/2π)	μ Nucl. mag. moment	Gamma-ray energy (MeV)	Gamma-ray intensity
	111	46				β-(27%)							0.1722 ± 0.0001	33%
	111	46											0.3913 ± 0.0003	5%
	111	46											0.4135 ± 0.0003	1.7%
	111	46											0.4155 ± 0.0003	1.5%
	111	46											0.5256 ± 0.0001	1.2%
	111	46											0.5750 ± 0.0001	3%
	111	46											0.6328 ± 0.0002	3%
	111	46											0.6942 ± 0.0001	2%
	111	46											0.7622 ± 0.0001	1.2%
	111	46											1.1159 ± 0.0002	1%
	111	46											1.2825 ± 0.0002	1%
	111	46											1.6911 ± 0.0002	1.2%
	111	46											(0.1 - 1.97)	
$_{46}Pd^{111}$	111	46		110.907660	22 m.	β-	2.20	2.12	95%		5/2+		0.0598 ± 0.0001	0.5%(D)
	111	46											0.2454 ± 0.0001	0.5%
	111	46											0.3761 ± 0.0001	0.4%
	111	46											0.5800 ± 0.0001	0.8%
	111	46											0.6504 ± 0.0001	0.55%
	111	46											1.3885 ± 0.0002	0.54%
	111	46											1.4590 ± 0.0003	0.56%
$_{46}Pd^{112}$	112	46		111.907323	21.03 h.	β-	0.29	0.28			0+		0.0185 ± 0.0005	27%
$_{46}Pd^{113m}$	113	46			89 s.	β-							0.0958 ± 0.0002	
$_{46}Pd^{113}$	113	46		112.910110	98 s.	β-	3.4				5/2+		0.0958 ± 0.0002	100 +
	113	46											0.2220 ± 0.0002	37
	113	46											0.4824 ± 0.0004	62
	113	46											0.5679 ± 0.0003	21
	113	46											0.6436 ± 0.0002	81
	113	46											0.7394 ± 0.0002	76
	113	46											0.8695 ± 0.0006	4.2
$_{46}Pd^{114}$	114	46		113.910310	2.48 m.	β-	1.5				0+		0.1266 ± 0.0002	3%
	114	46											0.2320 ± 0.0002	3.5%
	114	46											0.5582 ± 0.0002	12%(D)
	114	46											0.5760 ± 0.0002	9%(D)
$_{46}Pd^{115}$	115	46		114.913590	47 s.	β-	4.6						0.1255 ± 0.0003	64 +
	115	46											0.2554 ± 0.0003	59
	115	46											0.3040 ± 0.0002	32
	115	46											0.3428 ± 0.0002	100
	115	46											0.3606 ± 0.0003	28
	115	46											0.5944 ± 0.0003	34
$_{46}Pd^{116}$	116	46		115.914000	12.7 s.	β-	2.6						0.1015 ± 0.0002	8%
	116	46											0.1147 ± 0.0001	88%
	116	46											0.1778 ± 0.0002	12%
	116	46											0.2161 ± 0.0002	2%
	116	46											0.2795 ± 0.0003	6%
Ag		47		107.8682						63.6 b.				
$_{47}Ag^{96}$	96	47			5.1 s.	β+							ann.rad.	
	96	47				E.C.							0.1248	100 +
	96	47											0.4995	42
$_{47}Ag^{97}$	97	47		96.923890	19 s.	β+							ann.rad.	
	97	47				E.C.							0.6862 ± 0.0001	100 +
	97	47											1.2941 ± 0.0002	53
$_{47}Ag^{98}$	98	47		97.921560	46.7 s.	β+	8.6				5+		ann.rad.	
	98	47				E.C.							0.5711 ± 0.0002	59%
	98	47											0.6611 ± 0.0002	12%
	98	47											0.6786 ± 0.0002	88%
	98	47											0.8631 ± 0.0002	100%
$_{47}Ag^{99m}$	99	47			11 s.	I.T.(100%)					1/2-		Ag k x-ray	26%
	99	47											0.1636 ± 0.0003	37%
	99	47											0.3426 ± 0.0002	99%
$_{47}Ag^{99}$	99	47		98.917590	2.07 m.	β+(87%)	3.5				9/2+		ann.rad.	
	99	47				E.C.(13%)							0.2199 ± 0.0001	4%
	99	47											0.2645 ± 0.0001	63%
	99	47											0.4637 ± 0.0001	1%
	99	47											0.5682 ± 0.0001	3.8%
	99	47											0.5962 ± 0.0001	1.5%
	99	47											0.6360 ± 0.0001	1.4%
	99	47											0.6870 ± 0.0001	3.2%
	99	47											0.8056 ± 0.0001	12%
	99	47											0.8157 ± 0.0003	7%
	99	47											0.8323 ± 0.0001	13%
	99	47											0.8385 ± 0.0001	2%
	99	47											0.8640 ± 0.0001	4%
	99	47											0.9632 ± 0.0003	1%
	99	47											1.5319 ± 0.0001	4.5%
	99	47											1.5404 ± 0.0001	1.4%
	99	47											1.5853 ± 0.0005	1%
	99	47											1.8734 ± 0.0001	2.3%
	99	47											1.8810 ± 0.0002	1.3%
	99	47											1.9071 ± 0.0004	1.2%
	99	47											2.7085 ± 0.0004	0.9%
	99	47											3.1818 ± 0.0004	1.3%
	99	47											(0.2 - 3.5)	
$_{47}Ag^{100m}$	100	47			2.3 m.	β+					2+		ann.rad.	
	100	47				E.C.							0.6657 ± 0.0002	93%

Isotope	A	Z	% Natural abundance	Atomic mass	Half-life	Decay mode	Decay energy (MeV)	Particle energy (MeV)	Particle intensity	Thermal neutron cross section	Spin (h/2π)	μ Nucl. mag. moment	Gamma-ray energy (MeV)	Gamma-ray intensity
	100	47											0.9222 ± 0.0002	10%
	100	47											1.5877 ± 0.0003	7%
	100	47											1.6941 ± 0.0003	13%
	100	47											1.8208 ± 0.0008	6%
	100	47											1.9560 ± 0.0007	5%
	100	47											2.0130 ± 0.001	1%
	100	47											2.1190 ± 0.0005	11%
$_{47}Ag^{100}$	100	47		99.916140	2.0 m.	β+	7.1	5.4			5+		ann.rad.	
	100	47				E.C.							0.2807 ± 0.0002	10%
	100	47											0.4503 ± 0.0002	19%
	100	47											0.6657 ± 0.0002	100%
	100	47											0.7309 ± 0.0002	8%
	100	47											0.7508 ± 0.0002	84%
	100	47											0.7732 ± 0.0002	23%
	100	47											0.8625 ± 0.0002	4%
	100	47											0.8905 ± 0.0002	3.6%
	100	47											0.9607 ± 0.0002	1.9%
	100	47											1.0538 ± 0.0002	13%
	100	47											1.1157 ± 0.0002	4.3%
	100	47											1.2607 ± 0.0005	5.2%
	100	47											1.2780 ± 0.0002	3.4%
	100	47											1.4053 ± 0.0002	3.6%
	100	47											1.5039 ± 0.0002	13.6%
	100	47											1.6860 ± 0.0003	8%
	100	47											1.7679 ± 0.0004	1%
	100	47											2.1190 ± 0.0005	2.9%
	100	47											2.2148 ± 0.0005	1.9%
$_{47}Ag^{101m}$	101	47			3.1 s.	I.T.	0.23				$^{1}/_{2}$-		Ag k x-ray	47%
	101	47											0.0981 ± 0.0002	62%
	101	47											0.1762 ± 0.0002	47%
$_{47}Ag^{101}$	101	47		100.912810	11.1 m.	β+(69%)	4.2	1.08			9/2+		ann.rad.	135%
	101	47				E.C.(31%)		1.56					0.2610 ± 0.0001	52%
	101	47						2.18					0.2747 ± 0.0002	2%
	101	47						2.73					0.3269 ± 0.0002	2%
	101	47						3.38					0.4392 ± 0.0003	3%
	101	47											0.5076 ± 0.0004	2%
	101	47											0.5433 ± 0.0002	2%
	101	47											0.5880 ± 0.0002	10%
	101	47											0.6673 ± 0.0001	10%
	101	47											0.6778 ± 0.0002	4%
	101	47											0.7347 ± 0.0002	3%
	101	47											0.8932 ± 0.0002	1%
	101	47											0.9383 ± 0.0002	1%
	101	47											0.9443 ± 0.0002	1%
	101	47											1.0936 ± 0.0002	2.6%
	101	47											1.1739 ± 0.0002	9%
	101	47											1.2053 ± 0.0002	2.6%
	101	47											2.0531 ± 0.0003	0.8%
	101	47											2.6991 ± 0.0003	0.4%
	101	47											(0.2 - 3.1)	
$_{47}Ag^{102m}$	102	47			7.8 m.	β+(38%)					2+	+4.14	ann.rad.	76%
	102	47				E.C.(13%)							0.5567 ± 0.0002	42%
	102	47				I.T.(49%)							0.9777 ± 0.0003	2.6%
	102	47											1.3878 ± 0.0003	2.6%
	102	47											1.4611 ± 0.0004	4.5%
	102	47											1.5348 ± 0.0004	2.6%
	102	47											1.5888 ± 0.0004	1.2%
	102	47											1.6923 ± 0.0004	2.2%
	102	47											1.8347 ± 0.0003	9.8%
	102	47											2.0178 ± 0.0004	2.8%
	102	47											2.0545 ± 0.0004	6.6%
	102	47											2.1594 ± 0.0004	5%
	102	47											2.6821 ± 0.0004	1.7%
	102	47											2.7165 ± 0.0004	1.8%
	102	47											3.2386 ± 0.0004	5%
$_{47}Ag^{102}$	102	47		101.911950	13 m.	β+(78%)	5.9	2.26			5+		ann.rad.	156%
	102	47				E.C.(22%)							0.5567 ± 0.0002	98%
	102	47											0.7194 ± 0.0002	58%
	102	47											0.8354 ± 0.0003	14%
	102	47											0.8657 ± 0.0003	3.7%
	102	47											0.8915 ± 0.0003	4%
	102	47											0.9777 ± 0.0003	2%
	102	47											1.2571 ± 0.0003	13%
	102	47											1.3057 ± 0.0004	2%
	102	47											1.4733 ± 0.0004	2.7%
	102	47											1.5227 ± 0.0004	2.7%
	102	47											1.5348 ± 0.0004	2.3%
	102	47											1.5558 ± 0.0004	2.6%
	102	47											1.5816 ± 0.0003	14%
	102	47											1.7446 ± 0.0003	17%
	102	47											1.8007 ± 0.0004	2.8%
	102	47											2.2429 ± 0.0005	1%
	102	47											2.6130 ± 0.0004	3.5%
	102	47											2.7269 ± 0.0005	1.4%
	102	47											3.3980 ± 0.0006	1.4%

Isotope	A	Z	% Natural abundance	Atomic mass	Half-life	Decay mode	Decay energy (MeV)	Particle energy (MeV)	Particle intensity	Thermal neutron cross section	Spin (h/2π)	μ Nucl. mag. moment	Gamma-ray energy (MeV)	Gamma-ray intensity
	102	47											3.4065 ± 0.0006	1.7%
$_{47}Ag^{103m}$	103	47			5.7 s.	I.T.	0.134				1/2-		Ag k x-ray	33%
	103	47											0.1344 ± 0.0001	21%
$_{47}Ag^{103}$	103	47		102.908980	66 m.	β+(28%)	2.67				7/2+	+4.47	ann.rad.	56%
	103	47				E.C.(72%)							0.1187 ± 0.0001	31%
	103	47											0.1482 ± 0.0001	28%
$_{47}Ag^{104m}$	104	47			33 m.	β+(43%)		2.70			2+	+3.7	ann.rad.	86%
	104	47				E.C.(24%)							0.5558 ± 0.0001	60%
	104	47				I.T.(33%)							0.7657 ± 0.0002	1%
	104	47											1.2388 ± 0.0003	2.6%
	104	47											1.3418 ± 0.0003	1%
	104	47											1.7208 ± 0.0004	1.4%
	104	47											2.1392 ± 0.0005	1%
	104	47											2.2767 ± 0.0004	1.6%
	104	47											2.7295 ± 0.0005	0.8%
	104	47											3.2136 ± 0.0005	1%
	104	47											3.4078 ± 0.0005	1%
	104	47											(0.5 - 3.4)	
$_{47}Ag^{104}$	104	47		103.908623	69 m.	β+(16%)	4.28	0.99			5+	+4.0	ann.rad.	32%
	104	47				E.C.(84%)							0.5558 ± 0.0001	92%
	104	47											0.6232 ± 0.0002	2.5%
	104	47											0.7405 ± 0.0002	7%
	104	47											0.7587 ± 0.0002	10%
	104	47											0.8579 ± 0.0002	10%
	104	47											0.8630 ± 0.0003	7%
	104	47											0.9080 ± 0.0003	4.5%
	104	47											0.9233 ± 0.0005	7%
	104	47											0.9259 ± 0.0005	12%
	104	47											0.9416 ± 0.0003	25%
	104	47											1.0753 ± 0.0003	2%
	104	47											1.2652 ± 0.0003	4.3%
	104	47											1.3418 ± 0.0003	7.3%
	104	47											1.5266 ± 0.0003	7.1%
	104	47											1.6258 ± 0.0003	5%
	104	47											1.7818 ± 0.0004	3%
	104	47											(0.18 - 2.27)	
$_{47}Ag^{105m}$	105	47			7.2 m.	I.T.(98%)	0.025				7/2+		Ag x-ray	1%
	105	47				E.C.(2%)							0.3063 ± 0.0001	0.2%
	105	47											0.3192 ± 0.0001	0.9%
	105	47											(0.1 - 1.0)weak	
$_{47}Ag^{105}$	105	47		104.906520	41.3 d.	E.C.	1.34				1/2-	0.1014	0.0640 ± 0.0001	11%
	105	47											0.2804 ± 0.0001	31%
	105	47											0.3192 ± 0.0001	4%
	105	47											0.3315 ± 0.0001	4%
	105	47											0.3445 ± 0.0001	42%
	105	47											0.3726 ± 0.0001	2%
	105	47											0.4434 ± 0.0001	11%
	105	47											0.6179 ± 0.0001	1%
	105	47											0.6445 ± 0.0001	10%
	105	47											0.6507 ± 0.0001	2%
	105	47											0.8075 ± 0.0001	1%
	105	47											1.0879 ± 0.0001	3.6%
$_{47}Ag^{106m}$	106	47			8.5 d.	E.C.					6+	3.71	Pd k x-ray	58%
	106	47											0.2217 ± 0.0001	6.6%
	106	47											0.2286 ± 0.0001	2%
	106	47											0.3910 ± 0.0001	4%
	106	47											0.4062 ± 0.0001	13%
	106	47											0.4296 ± 0.0001	13%
	106	47											0.4510 ± 0.0001	28%
	106	47											0.5118 ± 0.0001	88%
	106	47											0.6162 ± 0.0001	22%
	106	47											0.6802 ± 0.0001	2%
	106	47											0.7031 ± 0.0001	4%
	106	47											0.7173 ± 0.0001	29%
	106	47											0.7484 ± 0.0001	21%
	106	47											0.7932 ± 0.0001	6%
	106	47											0.8043 ± 0.0001	12%
	106	47											0.8084 ± 0.0001	4%
	106	47											0.8247 ± 0.0001	15%
	106	47											0.8478 ± 0.0001	4%
	106	47											1.0458 ± 0.0001	30%
	106	47											1.1280 ± 0.0001	12%
	106	47											1.1994 ± 0.0001	11%
	106	47											1.2229 ± 0.0001	7%
	106	47											1.5277 ± 0.0001	16%
	106	47											1.5724 ± 0.0002	7%
	106	47											1.8390 ± 0.0001	1%
$_{47}Ag^{106}$	106	47		105.906662	24.0 m.	β+(59%)	2.98		1.96		1+	+2.85	ann.rad.	120%
	106	47				E.C.(41%)							0.5119 ± 0.0001	17%
	106	47											0.6219 ± 0.0001	0.3%
	106	47											0.8735 ± 0.0001	0.2%
	106	47											1.0503 ± 0.0001	0.16%
$_{47}Ag^{107m}$	107	47			44.2 s.	I.T.	0.093				7/2+		Ag x-ray	
	107	47											0.0931 ± 0.0001	4.7%
$_{47}Ag^{107}$	107	47	51.84%	106.905092						(0.35 + 38)b.	1/2-	-0.1135		

Isotope	A	Z	% Natural abundance	Atomic mass	Half-life	Decay mode	Decay energy (MeV)	Particle energy (MeV)	Particle intensity	Thermal neutron cross section	Spin (h/2π)	μ Nucl. mag. moment	Gamma-ray energy (MeV)	Gamma-ray intensity
$_{47}Ag^{108m}$	108	47			$1.3x10^2$ y.	E.C.(92%)					6+	3.580	Ag k x-ray	11%
	108	47				I.T.(8%)	0.079						Pd k x-ray	54%
	108	47											0.0791 ± 0.0001	6.5%
	108	47											0.43392 ± 0.00004	91%
	108	47											0.61427 ± 0.00005	91%
	108	47											0.72290 ± 0.00005	91%
$_{47}Ag^{108}$	108	47		107.905952	2.42 m.	β-(97%)		1.02	1.7%		1+	+2.6884	ann.rad.	
	108	47				E.C.(2%)		1.65	96%				0.43392 ± 0.00004	0.5%
	108	47				β+(1%)		0.88	0.3%				0.61885 ± 0.00005	0.25%
	108	47											0.63298 ± 0.00005	1.75%
$_{47}Ag^{109m}$	109	47			39.8 s.	I.T.	0.088				7/2+		Ag k x-ray	28%
	109	47											0.0880 ± 0.0001	3.6%
$_{47}Ag^{109}$	109	47	48.16%	108.904757						(4.6 + 87) b.	1/2-	-0.1305		
$_{47}Ag^{110m}$	110	47			249.8 d.	β-(99%)				80 b.	6+	+3.607	0.4468 ± 0.0001	3.7%
	110	47				I.T.(1%)	0.1164						0.6203 ± 0.0001	2.8%
	110	47											0.65774 ± 0.00002	95%
	110	47											0.6776 ± 0.0001	11%
	110	47											0.6870 ± 0.0001	6.5%
	110	47											0.70667 ± 0.00002	16.7%
	110	47											0.7443 ± 0.0001	4.7%
	110	47											0.76393 ± 0.00002	22%
	110	47											0.81802 ± 0.00002	7.3%
	110	47											0.88467 ± 0.00002	73%
	110	47											0.93748 ± 0.00002	34%
	110	47											1.38427 ± 0.00003	24%
	110	47											1.47575 ± 0.00003	4%
	110	47											1.50501 ± 0.00003	13%
	110	47											1.56226 ± 0.00003	1.2%
$_{47}Ag^{110}$	110	47		109.906111	24.6 s.	β-	2.89	2.22	5%		1+	2.7271	0.65774 ± 0.00002	4.5%
	110	47						2.89	95%				0.8154 ± 0.0002	0.04%
	110	47											1.1257 ± 0.0001	0.015%
$_{47}Ag^{111m}$	111	47			65 s.	I.T.(99%)	0.059				7/2+		Ag k x-ray	16%
	111	47				β-(1%)							0.0598 ± 0.0001	0.5%
	111	47											0.1713 ± 0.0001	0.1%
	111	47											0.2454 ± 0.0001	0.5%
	111	47											0.6201 ± 0.0003	0.1%
$_{47}Ag^{111}$	111	47		110.905295	7.47 d.	β-	1.037	1.035		3 b.	1/2-	-0.146	0.2454 ± 0.0001	1.2%
	111	47											0.3421 ± 0.0001	6.7%
$_{47}Ag^{112}$	112	47		111.907010	3.14 h.	β-	3.96	3.94			2-	0.0547	0.6067 ± 0.0002	3%
	112	47											0.6174 ± 0.0002	42%
	112	47											0.6948 ± 0.0002	3%
	112	47											0.8512 ± 0.0002	1%
	112	47											1.3123 ± 0.0002	1%
	112	47											1.3877 ± 0.0002	5%
	112	47											1.6136 ± 0.0002	2.8%
	112	47											2.1062 ± 0.0002	2.4%
	112	47											2.5068 ± 0.0002	1%
	112	47											(0.4 - 2.9)	
$_{47}Ag^{113m}$	113	47			68 s.	I.T.(80%)	0.043				7/2+		0.1422 ± 0.0002	1.6%
	113	47				β-(20%)							0.2983 ± 0.0001	6%
	113	47											0.3161 ± 0.0001	10%
	113	47											0.3923 ± 0.0002	6%
	113	47											0.5838 ± 0.0003	2%
	113	47											0.7083 ± 0.0004	2%
$_{47}Ag^{113}$	113	47		112.906558	5.3 h.	β-	2.01	2.01			1/2-	0.159	0.2588 ± 0.0001	1.6%
	113	47											0.2986 ± 0.0001	10%
	113	47											0.3163 ± 0.0001	1.3%
	113	47											0.6723 ± 0.0001	0.9%
	113	47											0.6906 ± 0.0001	0.7%
$_{47}Ag^{114}$	114	47		113.908760	4.5 s.	β-	5.0	4.9			1+		0.5582 ± 0.0002	12%
	114	47											0.5760 ± 0.0002	1%
	114	47											1.3041 ± 0.0005	0.7%
	114	47											1.9946 ± 0.0005	1%
$_{47}Ag^{115m}$	115	47			18.7 s.	β-					7/2+		0.1134 ± 0.0002	11 +
	115	47											0.1315 ± 0.0002	89
	115	47											0.2288 ± 0.0002	100
	115	47											0.2753 ± 0.0002	6
	115	47											0.3887 ± 0.0002	46
	115	47											0.451 ± 0.001	2

TABLE OF THE ISOTOPES (Continued)

Isotope	A	Z	% Natural abundance	Atomic mass	Half-life	Decay mode	Decay energy (MeV)	Particle energy (MeV)	Particle intensity	Thermal neutron cross section	Spin (h/2π)	μ Nucl. mag. moment	Gamma-ray energy (MeV)	Gamma-ray intensity
	115	47											0.4734 ± 0.0005	2
$_{47}Ag^{115}$	115	47		114.908800	20.0 m.	β-	3.14				1/2-		0.1316 ± 0.0002	3%
	115	47											0.2128 ± 0.0001	4.4%
	115	47											0.2291 ± 0.0001	18%
	115	47											0.3261 ± 0.0001	2%
	115	47											0.3722 ± 0.0001	2%
	115	47											0.4727 ± 0.0001	4%
	115	47											0.6491 ± 0.0001	3%
	115	47											0.6981 ± 0.0001	2%
	115	47											1.5069 ± 0.0003	1.2%
	115	47											1.8416 ± 0.0003	1.8%
	115	47											1.9269 ± 0.0003	1.3%
	115	47											2.1132 ± 0.0003	2.8%
	115	47											(0.13 - 2.49)	
$_{47}Ag^{116m}$	116	47			10 s.	I.T.(2%) β-(98%)		3.2			5+		0.1027 ± 0.0002	2%
	116	47											0.2549 ± 0.0003	7%
	116	47											0.2643 ± 0.0003	5%
	116	47											0.4577 ± 0.0005	2%
	116	47											0.5134 ± 0.0002	92%
	116	47											0.6670 ± 0.0003	8%
	116	47											0.6995 ± 0.0002	8%
	116	47											0.7055 ± 0.0002	61%
	116	47											0.7088 ± 0.0003	20%
	116	47											0.8068 ± 0.0001	16%
	116	47											0.9312 ± 0.0004	4%
	116	47											0.9744 ± 0.0006	2%
	116	47											1.0289 ± 0.0003	30%
	116	47											1.2130 ± 0.0003	6%
	116	47											1.4086 ± 0.0004	2%
$_{47}Ag^{116}$	116	47		115.911200	2.68 m.	β-	5.0				2-		0.5134 ± 0.0002	76%
	116	47											0.6399 ± 0.0003	3%
	116	47											0.6993 ± 0.0002	11%
	116	47											0.7058 ± 0.0002	2%
	116	47											0.8668 ± 0.0004	1.4%
	116	47											0.9935 ± 0.0005	1.7%
	116	47											1.1285 ± 0.0006	2.5%
	116	47											1.2126 ± 0.0006	7%
	116	47											1.3041 ± 0.0005	5.5%
	116	47											1.4018 ± 0.0005	1.2%
	116	47											1.4077 ± 0.0006	3%
	116	47											1.4371 ± 0.0006	1.6%
	116	47											1.5696 ± 0.0006	1%
	116	47											1.6046 ± 0.0006	2%
	116	47											1.6414 ± 0.0006	2%
	116	47											1.6910 ± 0.0006	1.7%
	116	47											2.0039 ± 0.0006	1.9%
	116	47											2.0908 ± 0.0006	1.1%
	116	47											2.1344 ± 0.0006	1.7%
	116	47											2.2463 ± 0.0006	2.4%
	116	47											2.2890 ± 0.0006	1.7%
	116	47											2.3487 ± 0.0006	1.1%
	116	47											2.4779 ± 0.0006	12%
	116	47											2.5009 ± 0.0006	1%
	116	47											2.6618 ± 0.0006	4.2%
	116	47											2.7032 ± 0.0006	1.8%
	116	47											2.8281 ± 0.0007	1.3%
	116	47											2.8341 ± 0.0007	2.5%
$_{47}Ag^{117m}$	117	47			5.3 s.	β-		3.3			7/2+		0.1354 ± 0.0001	46%
	117	47											0.2981 ± 0.0001	20%
	117	47											0.3221 ± 0.0001	7%
	117	47											0.3377 ± 0.0001	9%
	117	47											0.3868 ± 0.0001	38%
	117	47											0.5221 ± 0.0001	9%
	117	47											0.5578 ± 0.0001	2%
	117	47											0.6373 ± 0.0001	1%
	117	47											0.6846 ± 0.0001	7%
	117	47											0.7548 ± 0.0001	1%
	117	47											0.7863 ± 0.0001	2%
	117	47											0.8199 ± 0.0001	2%
	117	47											1.2204 ± 0.0002	1%
$_{47}Ag^{117}$	117	47		116.911700	73 s.	β-	4.17				1/2-		0.1354 ± 0.0001	23%
	117	47											0.1571 ± 0.0001	8%
	117	47											0.3072 ± 0.0001	2%
	117	47											0.3123 ± 0.0001	6%
	117	47											0.3377 ± 0.0001	10%
	117	47											0.4426 ± 0.0001	7%
	117	47											0.4677 ± 0.0001	2%
	117	47											0.5299 ± 0.0001	1%
	117	47											0.6651 ± 0.0001	1%
	117	47											0.7795 ± 0.0001	2%
	117	47											1.6090 ± 0.0001	4%
	117	47											1.6576 ± 0.0001	2%
	117	47											1.6962 ± 0.0001	1%
	117	47											1.7488 ± 0.0001	1%
	117	47											1.8544 ± 0.0001	2%

Isotope	A	Z	% Natural abundance	Atomic mass	Half-life	Decay mode	Decay energy (MeV)	Particle energy (MeV)	Particle intensity	Thermal neutron cross section	Spin (h/2π)	μ Nucl. mag. moment	Gamma-ray energy (MeV)	Gamma-ray intensity
	117	47											1.9955 ± 0.0001	4%
	117	47											2.0133 ± 0.0002	4%
	117	47											1.0354 ± 0.0002	1%
	117	47											2.0567 ± 0.0002	3%
	117	47											2.1921 ± 0.0002	2%
	117	47											2.2459 ± 0.0002	3%
	117	47											2.5141 ± 0.0002	1%
	117	47											2.8883 ± 0.0002	2.6%
$_{47}Ag^{118m}$	118	47			2.8 s.	β-(59%)							0.1277 ± 0.0001	7%
	118	47				I.T.(41%)	0.1277						0.4878 ± 0.0001	59%
	118	47											0.6771 ± 0.0001	58%
	118	47											0.7709 ± 0.0001	20%
	118	47											0.8083 ± 0.0001	6%
	118	47											1.0586 ± 0.0002	32%
$_{47}Ag^{118}$	118	47		117.914570	4.0 s.	β-	7.1						0.4878 ± 0.0001	81%
	118	47											0.6771 ± 0.0001	34%
	118	47											0.7815 ± 0.0001	6%
	118	47											0.7978 ± 0.0001	7%
	118	47											1.0586 ± 0.0002	2%
	118	47											1.2700 ± 0.0003	4%
	118	47											1.9390 ± 0.0003	4%
	118	47											2.1015 ± 0.0003	8%
	118	47											2.7792 ± 0.0003	6%
	118	47											2.7894 ± 0.0003	9%
	118	47											3.2259 ± 0.0003	11%
$_{47}Ag^{119}$	119	47		118.915630	2.1 s.	β-	5.4				7/2+		0.0674 ± 0.0001	6%
	119	47											0.1990 ± 0.0001	7%
	119	47											0.2134 ± 0.0001	7%
	119	47											0.3662 ± 0.0001	10%
	119	47											0.3706 ± 0.0001	3%
	119	47											0.3991 ± 0.0001	9%
	119	47											0.4071 ± 0.0001	2%
	119	47											0.4827 ± 0.0001	2%
	119	47											0.4979 ± 0.0001	3%
	119	47											0.5439 ± 0.0001	3%
	119	47											0.6264 ± 0.0002	11%
	119	47											0.6282 ± 0.0002	2%
	119	47											0.6544 ± 0.0003	3%
	119	47											0.6561 ± 0.0002	1%
	119	47											0.6604 ± 0.0001	6%
	119	47											0.7374 ± 0.0001	1%
	119	47											0.7792 ± 0.0001	4%
	119	47											0.8254 ± 0.0001	2%
	119	47											0.8514 ± 0.0001	2%
	119	47											1.0085 ± 0.0001	2%
	119	47											1.0265 ± 0.0001	6%
	119	47											1.1733 ± 0.0002	1%
	119	47											1.3748 ± 0.0002	1%
	119	47											1.5266 ± 0.0002	1%
	119	47											1.8983 ± 0.0002	2%
	119	47											2.0282 ± 0.0002	1%
	119	47											2.0607 ± 0.0005	4%
	119	47											2.4709 ± 0.0004	1%
	119	47											2.7864 ± 0.0002	2%
	119	47											2.9518 ± 0.0003	1.5%
$_{47}Ag^{120m}$	120	47			0.32 s.	β-							0.2030 ± 0.0002	
	120	47				I.T.							0.5059 ± 0.0002	
	120	47											0.6978 ± 0.0002	
	120	47											0.8300 ± 0.0002	
	120	47											0.9258 ± 0.0002	
$_{47}Ag^{120}$	120	47		119.918650	1.2 s.	β-	8.2						0.5059 ± 0.0002	
	120	47											0.6978 ± 0.0002	
	120	47											0.8171 ± 0.0002	
	120	47											1.3231 ± 0.0002	
$_{47}Ag^{121}$	121	47		120.919970	0.8 s.	β-	6.4						0.1150 ± 0.0001	21 +
	121	47											0.1464 ± 0.0004	2
	121	47											0.1785 ± 0.0001	10
	121	47											2.2030 ± 0.0003	2
	121	47											0.2736 ± 0.0001	8
	121	47											0.3148 ± 0.0001	100
	121	47											0.3537 ± 0.0001	57
	121	47											0.3622 ± 0.0001	11
	121	47											0.3696 ± 0.0001	17
	121	47											0.3720 ± 0.0001	9
	121	47											0.4306 ± 0.0001	12
	121	47											0.5007 ± 0.0001	24
	121	47											0.8020 ± 0.0002	5
	121	47											0.8172 ± 0.0002	11
	121	47											1.1570 ± 0.0002	8
	121	47											1.1705 ± 0.0003	6
	121	47											1.1959 ± 0.0002	15
	121	47											1.3711 ± 0.0002	4
	121	47											1.5105 ± 0.0002	17
	121	47											1.8120 ± 0.0003	5
	121	47											2.2052 ± 0.0005	7

Isotope	A	Z	% Natural abundance	Atomic mass	Half-life	Decay mode	Decay energy (MeV)	Particle energy (MeV)	Particle intensity	Thermal neutron cross section	Spin (h/2π)	μ Nucl. mag. moment	Gamma-ray energy (MeV)	Gamma-ray intensity
	121	47											2.5194 ± 0.0005	4
	121	47											(0.11 - 2.5)many	
$_{47}Ag^{122}$	122	47			1.5 s.	β-							0.5695 ± 0.0001	96%
	122	47				n							0.6502 ± 0.0001	20%
	122	47											0.7597 ± 0.0001	32%
	122	47											0.7884 ± 0.0003	12%
	122	47											1.3678 ± 0.0005	4%
	122	47											1.4231 ± 0.0009	3%
Cd		48		112.41						2.45 b.				
$_{48}Cd^{99}$	99	48		98.924860	16 s.	β+,E.C.							ann.rad.	
$_{48}Cd^{100}$	100	48		99.920230	1.1 m.	β+,E.C.	≈4.0						ann.rad.	
	100	48											0.0935 ± 0.0005	
	100	48											0.1238 ± 0.0005	
	100	48											0.1388 ± 0.0005	
	100	48											0.1781 ± 0.0005	
	100	48											0.2198 ± 0.0005	
	100	48											0.3676 ± 0.0005	
	100	48											0.4275 ± 0.0005	
	100	48											0.5670 ± 0.0005	
	100	48											0.9353 ± 0.0005	
$_{48}Cd^{101}$	101	48		100.918740	1.2 m.	β+(83%)	5.5				5/2+		In k x-ray	
	101	48				E.C.(17%)							0.0985 ± 0.0002	47%
	101	48											0.3089 ± 0.0002	2%
	101	48											0.5234 ± 0.0002	5%
	101	48											0.6379 ± 0.0002	2%
	101	48											0.6869 ± 0.0002	4%
	101	48											0.7058 ± 0.0002	2%
	101	48											0.9247 ± 0.0002	7%
	101	48											1.0226 ± 0.0002	2%
	101	48											1.1873 ± 0.0002	5%
	101	48											1.2030 ± 0.0002	5%
	101	48											1.2589 ± 0.0002	8%
	101	48											1.3318 ± 0.0002	2%
	101	48											1.4171 ± 0.0002	6%
	101	48											1.6314 ± 0.0002	2%
	101	48											1.6909 ± 0.0002	3%
	101	48											1.6967 ± 0.0002	4%
	101	48											1.7225 ± 0.0002	11%
	101	48											1.8597 ± 0.0002	4%
	101	48											1.9609 ± 0.0002	2%
	101	48											1.9902 ± 0.0002	1.4%
	101	48											2.1300 ± 0.0002	1%
	101	48											2.8419 ± 0.0002	2%
$_{48}Cd^{102}$	102	48		101.914440	5.5 m.	β+(27%)	2.4				0+		ann.rad.	
	102	48				E.C.(73%)							0.0974 ± 0.0002	3%
	102	48											0.1160 ± 0.0002	6%
	102	48											0.1204 ± 0.0002	2%
	102	48											0.2133 ± 0.0002	4%
	102	48											0.3603 ± 0.0002	4%
	102	48											0.4148 ± 0.0002	7%
	102	48											0.4810 ± 0.0002	61%
	102	48											0.5051 ± 0.0002	9%
	102	48											0.6757 ± 0.0002	4%
	102	48											1.0366 ± 0.0002	12%
	102	48											1.3598 ± 0.0002	5%
$_{48}Cd^{103}$	103	48		102.913451	7.7 m.	β+(33%)	4.17				5/2+		ann.rad.	
	103	48				E.C.(67%)							Ag k x-ray	
	103	48											0.1344 ± 0.0001	3%
	103	48											0.3870 ± 0.0001	3%
	103	48											0.5630 ± 0.0004	1.7%
	103	48											0.6262 ± 0.0004	1.8%
	103	48											0.9631 ± 0.0004	2%
	103	48											1.0799 ± 0.0001	5.7%
	103	48											1.0993 ± 0.0001	1.8%
	103	48											1.3117 ± 0.0001	1.9%
	103	48											1.4487 ± 0.0001	5.8%
	103	48											1.4618 ± 0.0001	12%
	103	48											1.4763 ± 0.0001	2%
	103	48											1.7485 ± 0.0001	1.5%
	103	48											1.8220 ± 0.0001	1.1%
	103	48											1.8342 ± 0.0001	1.0%
	103	48											1.8800 ± 0.0001	3.5%
	103	48											1.9302 ± 0.0001	2.0%
	103	48											2.0119 ± 0.0001	1.3%
	103	48											2.0225 ± 0.0001	1.1%
	103	48											2.1330 ± 0.0002	2%
	103	48											2.1995 ± 0.0002	1.5%
	103	48											2.2451 ± 0.0002	1.2%
	103	48											2.3737 ± 0.0002	1.6%
	103	48											2.4011 ± 0.0002	1.2%
	103	48											2.6814 ± 0.0003	1.5%
	103	48											(0.1 - 2.8)	
$_{48}Cd^{104}$	104	48		103.909851	58 m.	E.C.	1/40				0+		Ag k x-ray	
	104	48											0.0666 ± 0.0002	2.2%
	104	48											0.0835 ± 0.0002	47%

Isotope	A	Z	% Natural abundance	Atomic mass	Half-life	Decay mode	Decay energy (MeV)	Particle energy (MeV)	Particle intensity	Thermal neutron cross section	Spin (h/2π)	μ Nucl. mag. moment	Gamma-ray energy (MeV)	Gamma-ray intensity
	104	48											0.5590 ± 0.0002	6.7%
	104	48											0.6257 ± 0.0002	2.1%
	104	48											0.7093 ± 0.0002	20%
$_{48}Cd^{105}$	105	48		104.909459	55.3 m.	β+(26%) E.C.(74%)	2.74	1.69			5/2+		Ag k x-ray	
	105	48											0.3469 ± 0.0001	4%
	105	48											0.4332 ± 0.0001	3%
	105	48											0.6072 ± 0.0001	4%
	105	48											0.6485 ± 0.0001	1.6%
	105	48											0.9341 ± 0.0001	1.3%
	105	48											0.9618 ± 0.0001	4.7%
	105	48											1.0716 ± 0.0001	1.3%
	105	48											1.3025 ± 0.0001	4%
	105	48											1.3885 ± 0.0001	2.7%
	105	48											1.4161 ± 0.0005	1.6%
	105	48											1.5578 ± 0.0001	2%
	105	48											1.6358 ± 0.0001	1%
	105	48											1.6653 ± 0.0001	1.3%
	105	48											1.6933 ± 0.0001	3.5%
	105	48											1.8975 ± 0.0001	1.4%
	105	48											1.9331 ± 0.0001	1.6%
	105	48											2.2728 ± 0.0002	1.0%
	105	48											2.3333 ± 0.0001	2%
	105	48											(0.25 - 2.4)	
$_{48}Cd^{106}$	106	48	1.25%	105.906461						1 b.	0+			
$_{48}Cd^{107}$	107	48		106.906613	6.5 h.	E.C. (99+%)	1.42				5/2+	-0.6144	Ag k x-ray	
	107	48				β+							0.0931 ± 0.0001	5%
	107	48											0.7965 ± 0.0001	0.06%
	107	48											0.8289 ± 0.0001	0.16%
$_{48}Cd^{108}$	108	48	0.89%	107.904176						1.1 b.	0+			
$_{48}Cd^{109}$	109	48		108.904953	462.3 d.	E.C.	0.184			700 b.	5/2+	-0.8270	Ag k x-ray	
	109	48											0.08804 ± 0.00008	3.6%
$_{48}Cd^{110}$	110	48	122.49%	109.903005						(0.1 + 11) b.	0+			
$_{48}Cd^{111m}$	111	48			48.7 m.	I.T.					11/2-		Cd k x-ray	
	111	48											0.15082 ± 0.00001	29%
	111	48											0.24539 ± 0.00002	94%
$_{48}Cd^{111}$	111	48	12.80%	110.904182						24 b.	½+	-0.5943		
$_{48}Cd^{112}$	112	48	24.13%	111.902758						2.2 b.	0+			
$_{48}Cd^{113m}$	113	48			13.7y.	β-(99.9%)	0.59	0.59	99.9%		11/2-	-1.087	0.2637 ± 0.0003	0.02%
$_{48}Cd^{113}$	113	48	12.22%	112.904400						$2x10^4$b.	½+	-0.6217		
$_{48}Cd^{114}$	114	48	28.73%	113.903357						(0.04 + 0.3)b.	0+			
$_{48}Cd^{115m}$	115	48			44.6 d.	β-	1.614	0.68	1.6%		11/2-	-1.042	0.48450 ± 0.0005	0.3%
	115	48						1.62	97%				0.93381 ± 0.0003	2%
	115	48											1.13261 ± 0.00004	0.1%
	115	48											1.29064 ± 0.00004	0.9%
$_{48}Cd^{115}$	115	48		114.905430	53.5 h.	β-	1.441	0.58	42%		½+	-0.648	0.23141 ± 0.00003	0.7%
	115	48						1.11	58%				0.26085 ± 0.00003	1.9%
	115	48											0.33624 ± 0.00003	50%(D)
	115	48											0.49227 ± 0.00003	8.5%
	115	48											0.52780 ± 0.00003	29%
$_{48}Cd^{116}$	116	48	7.49%	115.904754						(0.2 + 0.8)b.	0+			
$_{48}Cd^{117m}$	117	48			3.4 h.	β-		0.67			11/2-		0.1586 ± 0.0002	90%(D)
	117	48											0.3669 ± 0.0001	3%
	117	48											0.4609 ± 0.0001	1.6%
	117	48											0.5529 ± 0.0002	100%
	117	48											0.5644 ± 0.0001	15%
	117	48											0.6318 ± 0.0001	2.8%
	117	48											0.7127 ± 0.0001	1%
	117	48											0.7481 ± 0.0001	4%
	117	48											0.8604 ± 0.0001	8%
	117	48											0.9314 ± 0.0001	3.6%
	117	48											1.0291 ± 0.0001	12%
	117	48											1.0660 ± 0.0001	23%
	117	48											1.2346 ± 0.0001	11%
	117	48											1.3393 ± 0.0005	2%
	117	48											1.3655 ± 0.0001	1.5%
	117	48											1.4329 ± 0.0001	13%
	117	48											1.9973 ± 0.0001	26%
	117	48											2.0964 ± 0.0001	7.4%
	117	48											2.3228 ± 0.0001	7.9%
	117	48											2.4174 ± 0.0001	1%
$_{48}Cd^{117}$	117	48		116.907228	2.49 h.	β-	2.53	0.67	51%		½+		0.2209 ± 0.0001	1.2%
	117	48						1.29	10%				0.2733 ± 0.0001	29%
	117	48											0.3445 ± 0.0001	18%
	117	48											0.4342 ± 0.0001	10%

Isotope	A	Z	% Natural abundance	Atomic mass	Half-life	Decay mode	Decay energy (MeV)	Particle energy (MeV)	Particle intensity	Thermal neutron cross section	Spin (h/2π)	μ Nucl. mag. moment	Gamma-ray energy (MeV)	Gamma-ray intensity
	117	48											0.8318 ± 0.0001	2.3%
	117	48											0.8807 ± 0.0001	4%
	117	48											1.0517 ± 0.0001	3.8%
	117	48											1.1166 ± 0.0001	1%
	117	48											1.1424 ± 0.0001	1.7%
	117	48											1.2479 ± 0.0001	1.2%
	117	48											1.2600 ± 0.0001	1.1%
	117	48											1.3033 ± 0.0001	18%
	117	48											1.3376 ± 0.0001	1.6%
	117	48											1.4087 ± 0.0001	1.3%
	117	48											1.5622 ± 0.0001	1.4%
	117	48											1.5766 ± 0.0001	11%
	117	48											1.7069 ± 0.0001	1%
	117	48											1.7231 ± 0.0001	2%
$_{48}Cd^{118}$	118	48		117.906914	50.3 m.	β-	0.75	0.75			0+			
$_{48}Cd^{119m}$	119	48			2.20 m.	β-					11/2-		0.1056 ± 0.0001	3.3%
	119	48											0.3603 ± 0.0002	1%
	119	48											0.4115 ± 0.0003	2%
	119	48											0.4224 ± 0.0001	10%
	119	48											0.5850 ± 0.0001	4.8%
	119	48											0.6330 ± 0.0003	1.5%
	119	48											0.7090 ± 0.0005	1.2%
	119	48											0.7208 ± 0.0002	18%
	119	48											0.8177 ± 0.0001	1.3%
	119	48											0.9025 ± 0.0005	1.1%
	119	48											0.9232 ± 0.0001	6.9%
	119	48											0.9963 ± 0.0002	2%
	119	48											1.0250 ± 0.0001	25%
	119	48											1.1019 ± 0.0001	9.8%
	119	48											1.1855 ± 0.0002	2.9%
	119	48											1.2037 ± 0.0001	13%
	119	48											1.3441 ± 0.0002	6.8%
	119	48											1.3608 ± 0.0002	1.5%
	119	48											1.3641 ± 0.0002	5.2%
	119	48											1.3771 ± 0.001	1.2%
	119	48											1.4363 ± 0.0003	5.2%
	119	48											1.4742 ± 0.0004	2%
	119	48											1.6685 ± 0.0002	3%
	119	48											1.7019 ± 0.0002	2.8%
	119	48											1.7729 ± 0.0005	1.3%
	119	48											1.9607 ± 0.0006	1.1%
	119	48											2.0213 ± 0.0002	22%
	119	48											2.1043 ± 0.0002	6%
	119	48											2.4226 ± 0.0003	4.5%
	119	48											2.52033 ± 0.0003	1.1%
$_{48}Cd^{119}$	119	48		118.909890	2.69 m.	β-	3.79	3.5			1/2+		0.1340 ± 0.0001	8%
	119	48											0.2929 ± 0.0001	40%
	119	48											0.3429 ± 0.0001	19%
	119	48											0.4462 ± 0.0003	2%
	119	48											0.7730 ± 0.0002	1.8%
	119	48											0.7847 ± 0.0003	1%
	119	48											0.9413 ± 0.0001	3%
	119	48											1.0184 ± 0.0002	1%
	119	48											1.0503 ± 0.0001	7%
	119	48											1.1326 ± 0.0002	1%
	119	48											1.2878 ± 0.0001	3%
	119	48											1.3169 ± 0.0002	10%
	119	48											1.6097 ± 0.0001	12%
	119	48											1.7140 ± 0.0002	2.5%
	119	48											1.7338 ± 0.0001	9%
	119	48											1.7637 ± 0.0001	10%
	119	48											2.0266 ± 0.0004	1.5%
	119	48											2.0565 ± 0.0003	2.5%
	119	48											2.3564 ± 0.0003	7%
	119	48											2.5483 ± 0.0001	2%
$_{48}Cd^{120}$	120	48		119.909852	50.8 s.	β-	1.8	1.5			0+			
$_{48}Cd^{121m}$	121	48			8 s.	β-					11/2-		0.1008 ± 0.0001	3%
	121	48											0.4201 ± 0.0001	4%
	121	48											0.4471 ± 0.0001	3%
	121	48											0.5722 ± 0.0001	2%
	121	48											0.9525 ± 0.0001	5%
	121	48											0.9878 ± 0.0001	14%
	121	48											1.0209 ± 0.0001	19%
	121	48											1.0695 ± 0.0001	2%
	121	48											1.1393 ± 0.0001	6%
	121	48											1.1815 ± 0.0001	12%
	121	48											1.2713 ± 0.0001	3%
	121	48											1.3367 ± 0.0001	2%
	121	48											1.3820 ± 0.0001	3%
	121	48											1.4569 ± 0.0001	1.8%
	121	48											1.4675 ± 0.0001	3%
	121	48											1.4873 ± 0.0002	2%
	121	48											1.5041 ± 0.0001	4%
	121	48											2.0594 ± 0.0001	21%
	121	48											2.1148 ± 0.0001	2%

Isotope	A	Z	% Natural abundance	Atomic mass	Half-life	Decay mode	Decay energy (MeV)	Particle energy (MeV)	Particle intensity	Thermal neutron cross section	Spin (h/2π)	μ Nucl. mag. moment	Gamma-ray energy (MeV)	Gamma-ray intensity
	121	48											2.2918 ± 0.0001	2%
	121	48											2.3319 ± 0.0001	2.8%
	121	48											2.3648 ± 0.0001	8%
	121	48											2.3698 ± 0.0002	1%
	121	48											2.4550 ± 0.0001	4.8%
	121	48											2.5108 ± 0.0001	2.8%
	121	48											2.5623 ± 0.0001	3.7%
$_{48}Cd^{121}$	121	48		120.913100	13.5 s.		β-	5.0			(3/2+		0.2102 ± 0.0001	3%
	121	48											0.3242 ± 0.0001	49%
	121	48											0.3492 ± 0.0001	13%
	121	48											0.4025 ± 0.0001	4%
	121	48											1.4411 ± 0.0002	2%
	121	48											0.5947 ± 0.0002	2%
	121	48											0.6506 ± 0.0002	4%
	121	48											0.6736 ± 0.0002	3%
	121	48											0.7653 ± 0.0001	6%
	121	48											0.9098 ± 0.0002	2%
	121	48											0.9786 ± 0.0003	2%
	121	48											0.9878 ± 0.0001	2%
	121	48											1.0403 ± 0.0001	18%
	121	48											1.0960 ± 0.0002	6%
	121	48											1.1499 ± 0.0002	2%
	121	48											1.2774 ± 0.0003	3%
	121	48											1.2969 ± 0.0001	4%
	121	48											1.3152 ± 0.0001	7%
	121	48											1.3236 ± 0.0003	1%
	121	48											1.3279 ± 0.0003	2%
	121	48											1.4512 ± 0.0002	1%
	121	48											1.5841 ± 0.0001	5%
	121	48											1.6271 ± 0.0002	2%
	121	48											1.6475 ± 0.0002	4%
	121	48											1.6618 ± 0.0002	1.6%
	121	48											1.6989 ± 0.0001	5.7%
	121	48											1.8226 ± 0.0002	1.6%
	121	48											1.8348 ± 0.0001	3%
	121	48											1.8540 ± 0.0001	3%
	121	48											1.8853 ± 0.0002	2%
$_{48}Cd^{122}$	122	48		121.913500	5.8 s.	β-	3.0				0+			
$_{48}Cd^{124}$	124	48			0.9 s.	β-					0+		0.0365 ± 0.0001	5%
	124	48											0.0628 ± 0.0001	23%
	124	48											0.1433 ± 0.0001	13%
	124	48											0.1799 ± 0.0001	50%
In		49		114.82						194 b.				
$_{49}In^{102}$	102	49		101.92440	24 s.	E.C.	9.2						0.1566	10%
	102	49											0.3965	12%
	102	49											0.5930	30%
	102	49											0.7768	100%
	102	49											0.8614	96%
	102	49											0.9237	10%
$_{49}In^{103}$	103	49		102.920110	1.1 m.	β+,E.C.	6.2				9/2+		ann.rad.	
	103	49											0.1879 ± 0.0001	100 +
	103	49											0.2020 ± 0.0001	19
	103	49											0.6995 ± 0.001	10
	103	49											0.7200 ± 0.001	32
	103	49											0.7399	19
$_{49}In^{104}$	104	49		103.918440	1.82 m.	β+,E.C.	≈8.0				(6+)		ann.rad.	
	104	49											0.3212 ± 0.0002	3.5%
	104	49											0.4739 ± 0.0002	5%
	104	49											0.5026 ± 0.0008	4%
	104	49											0.5330 ± 0.0003	3%
	104	49											0.6222 ± 0.0005	12%
	104	49											0.6580 ± 0.0002	100%
	104	49											0.8341 ± 0.0003	100%
	104	49											0.8781 ± 0.0002	28%
	104	49											0.9433 ± 0.0006	17%
	104	49											1.0002 ± 0.0006	10%
	104	49											1.1249 ± 0.0003	2%
	104	49											1.2816 ± 0.0003	2.5%
	104	49											1.7023 ± 0.0003	1.2%
	104	49											2.0062 ± 0.0003	2%
$_{49}In^{105m}$	105	49			43 s.	I.T.					1/2-		In k x-ray	
	105	49											0.6740 ± 0.0001	94%
$_{49}In^{105}$	105	49		104.914558	4.9 m.	β+,E.C.					9/2+		0.1310 ± 0.0001	43%
	105	49											0.1956 ± 0.0001	5%
	105	49											0.2282 ± 0.0003	1%
	105	49											0.2600 ± 0.0001	14%
	105	49											0.4735 ± 0.0003	1%
	105	49											0.5694 ± 0.0003	1.8%
	105	49											0.6038 ± 0.0001	10%
	105	49											0.6394 ± 0.0002	4.7%
	105	49											0.6680 ± 0.0001	7.6%
	105	49											0.7017 ± 0.0003	1%
	105	49											0.7702 ± 0.0003	1.6%
	105	49											0.8323 ± 0.0001	6.9%

Isotope	A	Z	% Natural abundance	Atomic mass	Half-life	Decay mode	Decay energy (MeV)	Particle energy (MeV)	Particle intensity	Thermal neutron cross section	Spin (h/2π)	μ Nucl. mag. moment	Gamma-ray energy (MeV)	Gamma-ray intensity
	108	49											1.2996 ± 0.0002	16%
	108	49											1.4862 ± 0.0003	4%
	108	49											1.6066 ± 0.0003	8%
$_{49}In^{109m}$	109	49			1.3 m.	I.T.	0.649				1/2-		In k x-ray	3.7%
	109	49											0.6498 ± 0.0002	94%
$_{49}In^{109}$	109	49		108.907133	4.2 h.	β+(8%)	2.03	0.79			9/2+	+5.53	ann.rad.	
	109	49				E.C.(92%)							Cd k x-ray	60%
	109	49											0.2035 ± 0.0002	73%
	109	49											0.3475 ± 0.0003	2%
	109	49											0.4262 ± 0.0003	4%
	109	49											0.6136 ± 0.0004	2.5%
	109	49											0.6235 ± 0.0004	6%
	109	49											0.6498 ± 0.0004	3%
	109	49											1.1491 ± 0.0006	4.3%
	109	49											1.6223 ± 0.0008	2.1%
$_{49}In^{110m}$	110	49			4.9 h.	E.C.					7+	10.5	Cd k x-ray	60%
	110	49											0.4611 ± 0.0001	2.2%
	110	49											0.4618 ± 0.0001	4.6%
	110	49											0.5819 ± 0.0001	8.4%
	110	49											0.5842 ± 0.0001	6.4%
	110	49											0.6417 ± 0.0001	26%
	110	49											0.6577 ± 0.0001	97%
	110	49											0.6776 ± 0.0004	4%
	110	49											0.7074 ± 0.0001	29%
	110	49											0.7599 ± 0.0001	3%
	110	49											0.8180 ± 0.0001	2%
	110	49											0.8447 ± 0.0001	3%
	110	49											0.8847 ± 0.0001	92%
	110	49											0.9375 ± 0.0001	67%
	110	49											0.9972 ± 0.0001	10%
	110	49											1.1174 ± 0.0001	4%
	110	49											1.4758 ± 0.0001	1.2%
	110	49											(0.1 - 1.98)	
$_{49}In^{110}$	110	49		109.907230	69 m.	β+(62%)	3.94	2.26			2+		ann.rad.	
	110	49				E.C.(38%)							Cd k x-ray	23%
	110	49											0.6577 ± 0.0001	98%
	110	49											1.1258 ± 0.0001	1%
	110	49											2.1295 ± 0.0001	2%
	110	49											2.2115 ± 0.0001	1.8%
	110	49											2.3175 ± 0.0001	1.3%
	110	49											(0.6 - 3.6)	
$_{49}In^{111m}$	111	49			7.7 m.	I.T.	0.357				1/2-	+5.53	In k x-ray	7.4%
	111	49											0.5372 ± 0.0001	87%
$_{49}In^{111}$	111	49		110.905109	2.806 d.	E.C.							Cd k x-ray	68%
	111	49											0.1712 ± 0.0001	94%
	111	49											0.2453 ± 0.0001	90%
$_{49}In^{112m}$	112	49			20.9 m.	I.T.	0.155				4+		In k x-ray	47%
	112	49											0.1555 ± 0.0002	13%
$_{49}In^{112}$	112	49		111.905536	14.4 m.	β+(22%)	2.588				1+		ann.rad.	
	112	49				E.C.(34%)							Cd k x-ray	20%
	112	49											0.6064 ± 0.0001	1.2%
	112	49											0.6171 ± 0.0001	5%
	112	49											0.8509 ± 0.0002	0.2%
	112	49											1.2531 ± 0.0002	0.2%
$_{49}In^{113m}$	113	49			1.657 h.	I.T.	0.3917				1/2-	-0.210	In k x-ray	20%
	113	49											0.3917 ± 0.0001	64%
$_{49}In^{113}$	113	49	4.3%	112.904061						(8 + 3.9)b.	9/2+	+5.523		
$_{49}In^{114m}$	114	49			49.51 d.	I.T.(97%)	0.190				5+	+4.7	In k x-ray	28%
	114	49				E.C.(3%)	1.6						0.19027 ± 0.00003	16%
	114	49											0.55843 ± 0.00003	3.4%
	114	49											0.72524 ± 0.00003	3.4%
	114	49											1.29983 ± 0.00007	0.16%
$_{49}In^{114}$	114	49		113.904916	71.9 s.	β-(97%)	1.986	1.986			1+	+1.7	Cd k x-ray	2%
	114	49				E.C.(3%)	1.452						0.5584 ± 0.0002	0.07%
	114	49											0.5727 ± 0.0001	0.004%
	114	49											1.2998 ± 0.0001	0.14%
$_{49}In^{115m}$	115	49			4.486 h.	I.T.(95%)	0.336	0.83			1/2-	-0.255	In k x-ray	28%
	115	49				β-(5%)							0.3362 ± 0.0001	46%
	115	49											0.4974 ± 0.0001	0.05%
$_{49}In^{115}$	115	49		114.903880	$4.4x10^{14}$y.	β-	0.496	1348		(87 + 75 + 41)b.	9/2+	+5.534		
$_{49}In^{116m2}$	116	49			2.16 s.	I.T.	0.162				8-		In k x-ray	28%
	116	49											0.1624 ± 0.0001	37%
$_{49}In^{116m1}$	116	49			54.1 m.	β-					5+	+4.3	0.13792 ± 0.0003	3%
	116	49											0.41688 ± 0.00002	29%
	116	49											0.81865 ± 0.00002	11%
	116	49											1.09723 ± 0.00002	56%
	116	49											1.29349 ± 0.00005	84%

Isotope	A	Z	% Natural abundance	Atomic mass	Half-life	Decay mode	Decay energy (MeV)	Particle energy (MeV)	Particle intensity	Thermal neutron cross section	Spin (h/2π)	μ Nucl. mag. moment	Gamma-ray energy (MeV)	Gamma-ray intensity
	105	49											0.8554 ± 0.0003	1%
	105	49											0.9633 ± 0.0003	1%
	105	49											1.1396 ± 0.0003	1.3%
	105	49											1.1903 ± 0.0003	1.2%
	105	49											1.2560 ± 0.0003	2.4%
	105	49											1.3870 ± 0.0002	5.1%
$_{49}In^{106m}$	106	49			6.2 m.	β+(85%)					3+		ann.rad.	
	106	49				E.C.(15%)							0.6326 ± 0.0001	92%
	106	49											0.8611 ± 0.0001	11%
	106	49											1.0838 ± 0.0002	3%
	106	49											1.1626 ± 0.0001	1.6%
	106	49											1.6209 ± 0.0001	4.9%
	106	49											1.7164 ± 0.0001	19%
	106	49											1.7379 ± 0.0003	1.5%
	106	49											1.9336 ± 0.0001	8.3%
	106	49											1.9975 ± 0.0003	2%
	106	49											2.0873 ± 0.0002	2%
	106	49											2.2569 ± 0.0002	5%
	106	49											2.8621 ± 0.0005	1.5%
	106	49											2.9182 ± 0.0003	4%
	106	49											3.4945 ± 0.0001	2%
$_{49}In^{106}$	106	49		105.913490	5.3 m.	β+(65%)					6+		ann.rad.	
	106	49				E.C.(35%)							0.2259 ± 0.0003	7%
	106	49											0.4331 ± 0.0003	2.3%
	106	49											0.5246 ± 0.0005	2%
	106	49											0.5412 ± 0.0002	13%
	106	49											0.5524 ± 0.0001	25%
	106	49											0.5586 ± 0.0005	2%
	106	49											0.5925 ± 0.0005	3%
	106	49											0.6107 ± 0.0002	4%
	106	49											0.6232 ± 0.0005	2%
	106	49											0.6327 ± 0.0001	100%
	106	49											0.7535 ± 0.0005	2%
	106	49											0.8368 ± 0.0005	2%
	106	49											0.8611 ± 0.0001	96%
	106	49											0.9978 ± 0.0001	48%
	106	49											1.0091 ± 0.0002	30%
	106	49											1.1390 ± 0.0005	2.4%
	106	49											1.4719 ± 0.0003	3%
	106	49											1.7801 ± 0.0005	1.5%
$_{49}In^{107m}$	107	49			51 s.	I.T.	0.678				1/2-		In k x-ray	3%
	107	49											0.6785 ± 0.0003	94%
$_{49}In^{107}$	107	49		106.910284	32.5 m.	β+(35%)	3.4	2.3			9/2+		ann.rad.	
	107	49				E.C(65%)							Cd k x-ray	41%
	107	49											0.2050 ± 0.0001	48%
	107	49											0.3209 ± 0.0001	10%
	107	49											0.3653 ± 0.0001	4%
	107	49											0.5055 ± 0.0001	12%
	107	49											0.7280 ± 0.0001	3.3%
	107	49											0.8090 ± 0.0001	3.3%
	107	49											1.2683 ± 0.0001	5.5%
	107	49											1.9222 ± 0.0001	1.8%
	107	49											2.0648 ± 0.0003	1.7%
	107	49											2.2848 ± 0.0002	1.1%
	107	49											2.3040 ± 0.0004	1.2%
	107	49											(0.2 - 2.99)	
$_{49}In^{108m}$	108	49			40 m.	β+(53%)					3+		ann.rad.	
	108	49				E.C.(47%)							Cd k x-ray	28%
	108	49											0.6329 ± 0.0002	76%
	108	49											0.8455 ± 0.0004	2.4%
	108	49											0.9685 ± 0.0005	4.3%
	108	49											1.5294 ± 0.0005	7.3%
	108	49											1.6012 ± 0.0003	4%
	108	49											1.7321 ± 0.0004	4%
	108	49											1.8519 ± 0.0005	3%
	108	49											1.9863 ± 0.0005	12%
	108	49											2.0483 ± 0.0004	3%
	108	49											3.0468 ± 0.0004	2.4%
	108	49											3.4522 ± 0.0005	9.1%
	108	49											3.8255 ± 0.002	2.3%
$_{49}In^{108}$	108	49		107.909678	57 m.	β+(33%)	5.13	1.3			6+		ann.rad.	
	108	49				E.C.(67%)							Cd k x-ray	41%
	108	49											0.2429 ± 0.0002	37%
	108	49											0.2667 ± 0.0003	3%
	108	49											0.3259 ± 0.0002	13%
	108	49											0.5689 ± 0.0002	5.3%
	108	49											0.6331 ± 0.0001	100%
	108	49											0.6489 ± 0.0002	5%
	108	49											0.7311 ± 0.0003	9%
	109	49											0.7547 ± 0.0005	3%
	108	49											0.8756 ± 0.0002	93%
	108	49											1.0331 ± 0.0002	24%
	108	49											1.0568 ± 0.0002	30%
	108	49											1.0935 ± 0.0003	5%
	108	49											1.1985 ± 0.0005	4%

Isotope	A	Z	% Natural abundance	Atomic mass	Half-life	Decay mode	Decay energy (MeV)	Particle energy (MeV)	Particle intensity	Thermal neutron cross section	Spin (h/2π)	μ Nucl. mag. moment	Gamma-ray energy (MeV)	Gamma-ray intensity
	116	49											1.50752 ± 0.00005	10%
	116	49											1.7524 ± 0.0001	2.5%
	116	49											2.11221 ± 0.00006	15%
$_{49}In^{116}$	116	49		115.905264	14.1 s.	β-	3.27	3.3	99%		1+		0.46313 ± 0.00003	0.25%
	116	49											1.2526 ± 0.0005	0.03%
	116	49											1.29349 ± 0.00005	1.3%
$_{49}In^{117m}$	117	49			1.933 h.	β-(53%)	1.769	1.77			$^1/_2$-	0.25	In k x-ray	13%
	117	49				I.T.(47%)							0.15855 ± 0.00001	81%
	117	49											0.31531 ± 0.00002	19%
	117	49											0.55294 ± 0.00002	75%(D)
$_{49}In^{117}$	117	49		116.904517	43.1 m.	β-	1.453	0.74			9/2+		0.15855 ± 0.00001	86%
	117	49											0.3966 ± 0.0004	0.14%
	117	49											0.55294 ± 0.00002	99%
$_{49}In^{118m2}$	118	49			8.5 s.	I.T.(98%)					(8-)		In k x-ray	31%
	118	49				β-(2%)							0.1382 ± 0.0005	22%
	118	49											0.2537 ± 0.0001	1.4%
	118	49											1.0507 ± 0.0001	1.5%
	118	49											1.2296 ± 0.0001	1.5%
$_{49}In^{118m1}$	118	49			4.40 m.	β-					5+		0.2086 ± 0.0003	2.3%
	118	49											0.4458 ± 0.0002	6%
	118	49											0.4744 ± 0.0004	3%
	118	49											0.5602 ± 0.001	1.3%
	118	49											0.6373 ± 0.0004	3.5%
	118	49											0.6833 ± 0.0002	55%
	118	49											1.0970 ± 0.0004	3%
	118	49											1.1732 ± 0.0006	1%
	118	49											1.2295 ± 0.0002	96%
	118	49											1.2591 ± 0.0005	4%
	118	49											2.0423 ± 0.001	3%
$_{49}In^{118}$	118	49		117.906120	5.0 s.	β-	4.2	4.2			1+		0.5282 ± 0.0004	0.7%
	118	49											1.1734 ± 0.0005	0.4%
	118	49											1.2295 ± 0.0002	5%
	118	49											2.0432 ± 0.0005	0.1%
$_{49}In^{119m}$	119	49			18.0 m.	β-(97%)		2.7			$^1/_2$-		0.3114 ± 0.0001	1%
	119	49				I.T.(3%)	0.311						0.7631 ± 0.0001	2.5%(D)
$_{49}In^{119}$	119	49		118.905819	2.4 m.	β-	2.34	1.6			9/2+		0.0239 ± 0.0001	16%
	119	49											0.6495 ± 0.0001	0.5%
	119	49											0.7631 ± 0.0001	99%
	119	49											1.2149 ± 0.0001	0.4%
$_{49}In^{120m}$	120	49			3 s.	β-		536			1+		0.7042 ± 0.0006	1.4%
	120	49											1.1725 ± 0.0003	19%
	120	49											2.0398 ± 0.001	2%
	120	49											2.3902 ± 0.001	1%
$_{49}In^{120}$	120	49		119.907890	44 s.	β-	5.3	2.2			(5+)		0.4146 ± 0.0003	2%
	120	49						3.1					0.5924 ± 0.0005	2%
	120	49											0.6371 ± 0.0004	3%
	120	49											0.7029 ± 0.0004	2%
	120	49											0.7134 ± 0.0002	7%
	120	49											0.8637 ± 0.0002	32%
	120	49											0.9849 ± 0.0004	2%
	120	49											1.0232 ± 0.0002	54%
	120	49											1.1714 ± 0.0002	96%
	120	49											1.1840 ± 0.0004	2.4%
	120	49											1.2945 ± 0.0003	12%
	120	49											1.4723 ± 0.00033	4%
	120	49											1.8871 ± 0.0005	5%
	120	49											2.0081 ± 0.0005	6%
	120	49											2.1790 ± 0.0005	3%
	120	49											2.6061 ± 0.0009	2%
	120	49											(0.4 - 2.7)	
$_{49}In^{121m}$	121	49			3.9 m.	β-(99%)		5.3			$^1/_2$-		0.0601 ± 0.0002	20%
	121	49				I.T.(1%)	0.313						0.3136 ± 0.0001	0.5%
	121	49											0.9256 ± 0.0001	1%
	121	49											1.0412 ± 0.0005	1%
	121	49											1.1022 ± 0.0005	1%
	121	49											1.1204 ± 0.0002	0.5%
	122	49											2.8038 ± 0.0007	0.1%
	121	49											2.8643 ± 0.0007	0.1%
$_{49}In^{121}$	121	49		120.907847	23 s.	β-	3.36				9/2+		0.2620 ± 0.0001	8%
	121	49											0.6573 ± 0.0001	7%
	121	49											0.8693 ± 0.0001	1%
	121	49											0.9193 ± 0.0001	4%
	121	49											0.9256 ± 0.0001	87%
	121	49											1.0928 ± 0.0004	0.3%
$_{49}In^{122m}$	122	49			1.5 s.	β-		5.3			(1+)		1.0131 ± 0.0001	3%
	122	49											1.1403 ± 0.0001	29%

TABLE OF THE ISOTOPES (Continued)

Isotope	A	Z	% Natural abundance	Atomic mass	Half-life	Decay mode	Decay energy (MeV)	Particle energy (MeV)	Particle intensity	Thermal neutron cross section	Spin (h/2π)	μ Nucl. mag. moment	Gamma-ray energy (MeV)	Gamma-ray intensity
	122	49											1.3897 ± 0.0001	2%
	122	49											2.0656 ± 0.0002	2%
	122	49											2.7591 ± 0.0002	3%
	122	49											2.9757 ± 0.0004	0.8%
	122	49											3.8197 ± 0.0003	0.3%
$_{49}In^{122}$	122	49		121.910280	10.1 s.	β-	6.37	4.4					0.2391 ± 0.0002	2%
	122	49											0.6435 ± 0.0002	3%
	122	49											0.8194 ± 0.0002	9%
	122	49											0.8313 ± 0.0002	8%
	122	49											0.9024 ± 0.0002	6%
	122	49											0.9745 ± 0.0001	14%
	122	49											1.0014 ± 0.0001	54%
	122	49											1.0131 ± 0.0001	12%
	122	49											1.0915 ± 0.0003	10%
	122	49											1.1403 ± 0.0001	100%
	122	49											1.1363 ± 0.0001	26%
	122	49											1.1903 ± 0.0001	28%
	122	49											1.3010 ± 0.0001	7%
	122	49											2.0931 ± 0.0002	4%
	122	49											2.4419 ± 0.0004	1%
$_{49}In^{123m}$	123	49			48 s.	β-		4.6			(1/2-)		0.1258 ± 0.0001	38%
	123	49											1.170 ± 0.001	0.1%
	123	49											3.234 ± 0.003	0.1%
$_{49}In^{123}$	123	49		122.910450	6.0 s.	β-	4.4	3.3			(9/2+)		0.6188 ± 0.0003	3%
	123	49											0.8455 ± 0.0002	1%
	123	49											1.0197 ± 0.0002	32%
	123	49											1.1305 ± 0.0002	63%
	123	49											1.3823 ± 0.0003	1%
	123	49											2.0012 ± 0.0005	0.3%
$_{49}In^{124m}$	124	49			2.4 s.		β-				8-		0.1029 ± 0.0001	45%
	124	49											0.1203 ± 0.0001	38%
	124	49											0.2431 ± 0.0001	11%
	124	49											0.2535 ± 0.0001	4%
	124	49											0.3635 ± 0.0001	17%
	124	49											0.8497 ± 0.0002	2%
	124	49											0.9154 ± 0.0002	3%
	124	49											0.9559 ± 0.0001	12%
	124	49											0.9699 ± 0.0001	52%
	124	49											1.0729 ± 0.0001	47%
	124	49											1.1168 ± 0.0001	15%
	124	49											1.1316 ± 0.0001	100%
	124	49											1.1990 ± 0.0001	9%
	124	49											1.3599 ± 0.0001	39%
	124	49											1.4401 ± 0.0001	9%
	124	49											1.1847 ± 0.0002	2%
	124	49											1.8560 ± 0.0004	1%
	124	49											2.6976 ± 0.0004	3%
$_{49}In^{124}$	124	49		123.912980	3.2 s.	β-	7.61	5			3+		0.7070 ± 0.0001	2%
	124	49											0.9699 ± 0.0001	3%
	124	49											0.9978 ± 0.0001	21%
	124	49											1.0899 ± 0.0001	3%
	124	49											1.1316 ± 0.0001	68%
	124	49											1.3147 ± 0.0001	4%
	124	49											1.4707 ± 0.0001	6%
	124	49											1.5813 ± 0.0001	2%
	124	49											1.7435 ± 0.0002	2%
	124	49											2.0825 ± 0.0002	3%
	124	49											2.4264 ± 0.0002	1%
	124	49											3.2142 ± 0.0002	21%
	124	49											3.2641 ± 0.0002	1%
	124	49											3.7615 ± 0.0003	1%
	124	49											3.9170 ± 0.0003	2%
	124	49											(0.3 - 4.6)	
$_{49}In^{125m}$	125	49			12.2 s.	β-		5.5			1/2-		0.1876 ± 0.0005	100 +
$_{49}In^{125}$	125	49		124.913670	2.33 s.	β-	5.5	4.1			9/2+		0.4260 ± 0.0001	2%
	125	49											0.6179 ± 0.0001	8%
	125	49											0.7446 ± 0.0001	6%
	125	49											0.8271 ± 0.0001	2%
	125	49											0.9365 ± 0.0001	3%
	125	49											1.0318 ± 0.0001	10%
	125	49											1.3350 ± 0.0001	76%
	125	49											1.5582 ± 0.0004	1%
$_{49}In^{126m}$	126	49			1.45 s.			4.9			(8-)		0.9086 ± 0.0001	4%
	126	49											0.9696 ± 0.0001	15%
	126	49											1.1357 ± 0.0001	2%
	126	49											1.1411 ± 0.0001	56%
	126	49											1.5710 ± 0.0001	3%
	126	49											1.6872 ± 0.0001	2%
	126	49											2.1053 ± 0.0002	2%
	126	49											2.11035 ± 0.0002	2%
	126	49											2.3704 ± 0.0002	2%
$_{49}In^{126}$	126	49		125.916470	1.53 s.	β-	8.1	4.2			3+		0.1118 ± 0.0001	88%
	126	49											0.2585 ± 0.0001	9%
	126	49											0.2693 ± 0.0001	6%
	126	49											0.3159 ± 0.0001	12%

Isotope	A	Z	% Natural abundance	Atomic mass	Half-life	Decay mode	Decay energy (MeV)	Particle energy (MeV)	Particle intensity	Thermal neutron cross section	Spin (h/2π)	μ Nucl. mag. moment	Gamma-ray energy (MeV)	Gamma-ray intensity
	126	49											0.4439 ± 0.0001	2%
	126	49											0.5014 ± 0.0001	6%
	126	49											0.5717 ± 0.0001	3%
	126	49											0.7883 ± 0.0001	8%
	126	49											0.9058 ± 0.0001	11%
	126	49											0.9086 ± 0.0001	100%
	126	49											0.9627 ± 0.0001	2%
	126	49											0.9774 ± 0.0002	3%
	126	49											1.0648 ± 0.0001	5%
	126	49											1.1411 ± 0.0001	100%
	126	49											1.1925 ± 0.0001	4%
	126	49											1.2359 ± 0.0001	2%
	126	49											1.3674 ± 0.0001	3%
	126	49											1.3780 ± 0.0001	23%
	126	49											1.4069 ± 0.0001	3%
	126	49											1.4954 ± 0.0003	2%
	126	49											1.5644 ± 0.0001	2%
	126	49											1.6117 ± 0.0001	6%
	126	49											1.6365 ± 0.0001	30%
	126	49											1.7583 ± 0.0002	5%
	126	49											2.5601 ± 0.0003	4%
	126	49											2.8286 ± 0.0003	5%
$_{49}In^{127m}$	127	49			3.76 s.	β-		6.4			(1/2-)		0.2523 ± 0.0003	77%
	127	49											0.8328 ± 0.0003	4%
	127	49											0.9484 ± 0.0003	2%
	127	49											1.0851 ± 0.0003	3%
	127	49											3.074 ± 0.001	6%
$_{49}In^{127}$	127	49		126.917320	1.12 s.	β-	6.5	4.9			(9/2+)		0.4680 ± 0.0003	1%
	127	49											0.6387 ± 0.0003	3%
	127	49											0.6461 ± 0.0003	8%
	127	49											0.7154 ± 0.0003	2%
	127	49											0.7926 ± 0.0003	2%
	127	49											0.8051 ± 0.0003	8%
	127	49											0.9563 ± 0.0003	6%
	127	49											0.9637 ± 0.0003	5%
	127	49											1.0486 ± 0.0003	7%
	127	49											1.0947 ± 0.0003	5%
	127	49											1.2141 ± 0.0003	1%
	127	49											1.5558 ± 0.0003	2%
	127	49											1.5977 ± 0.0003	67%
	127	49											20.190 ± 0.001	0.5%
$_{49}In^{128m}$	128	49			0.835 s.	β-		5.4			(8-)		0.1205 ± 0.0001	11%
	128	49											0.2572 ± 0.0001	4%
	128	49											0.3212 ± 0.0001	10%
	128	49											0.4577 ± 0.0001	2%
	128	49											0.8315 ± 0.0001	12%
	128	49											1.0549 ± 0.0001	6%
	128	49											1.1688 ± 0.0001	12%
	128	49											1.7800 ± 0.0001	3%
	128	49											1.8670 ± 0.0001	32%
	128	49											1.9739 ± 0.0001	19%
	128	49											2.1221 ± 0.0001	4%
$_{49}In^{128}$	128	49		127.920560	0.9 s.		β-				3+		0.9352 ± 0.0001	8%
	128	49											1.0895 ± 0.0001	7%
	128	49											1.1688 ± 0.0001	50%
	128	49											1.4643 ± 0.0001	2%
	128	49											1.5877 ± 0.0002	2%
	128	49											1.7393 ± 0.0001	2%
	128	49											1.8167 ± 0.0001	2%
	128	49											2.1041 ± 0.0001	6%
	128	49											2.2585 ± 0.0001	3%
	128	49											3.0511 ± 0.0002	2%
	128	49											3.5198 ± 0.0002	17%
	128	49											3.8862 ± 0.0002	4%
	128	49											3.9548 ± 0.0002	4%
	128	49											4.0380 ± 0.0002	2%
	128	49											4.2970 ± 0.0002	12%
$_{49}In^{129m}$	129	49			1.25 s.	β-(98%) n(2%)		≈7.5					0.3153 ± 0.0003	68%
	129	49											0.9067 ± 0.0003	4%
	129	49											0.9732 ± 0.0003	1%
	129	49											1.2220 ± 0.0003	6%
	129	49											1.2885 ± 0.0003	2%
$_{49}In^{129}$	129	49		128.921600	0.59 s.	β-	7.6	5.5					0.2853 ± 0.0003	1%
	129	49											0.7288 ± 0.0003	5%
	129	49											0.7693 ± 0.0003	9%
	129	49											1.0083 ± 0.0003	6%
	129	49											1.0546 ± 0.0003	4%
	129	49											1.0745 ± 0.0003	2%
	129	49											1.0957 ± 0.0003	3%
	129	49											1.1010 ± 0.0003	1%
	129	49											1.3487 ± 0.0003	2%
	129	49											1.7814 ± 0.0003	2%
	129	49											1.8650 ± 0.0003	32%
	129	49											2.1180 ± 0.0003	44%
	129	49											2.5460 ± 0.001	2%

Isotope	A	Z	% Natural abundance	Atomic mass	Half-life	Decay mode	Decay energy (MeV)	Particle energy (MeV)	Particle intensity	Thermal neutron cross section	Spin (h/2π)	μ Nucl. mag. moment	Gamma-ray energy (MeV)	Gamma-ray intensity
$_{49}In^{130m}$	130	49			0.53 s.	β-							0.0892 ± 0.0001	17%
	130	49											0.1298 ± 0.0001	7%
	130	49											0.1380 ± 0.0001	9%
	130	49											0.4082 ± 0.0001	7%
	130	49											0.7744 ± 0.0001	39%
	130	49											0.8070 ± 0.0001	5%
	130	49											1.2212 ± 0.0001	76%
	130	49											1.3402 ± 0.0001	3%
	130	49											1.4292 ± 0.0002	2%
	130	49											2.0283 ± 0.0001	11%
	130	49											2.3171 ± 0.0001	13%
	130	49											2.4099 ± 0.0002	2%
	130	49											3.1840 ± 0.0003	8%
	130	49											3.2417 ± 0.0003	4%
	130	49											4.0415 ± 0.0003	3%
$_{49}In^{130}$	130	49		129.924870	0.51 s.	β-	≈10.2				10-		0.0892 ± 0.0001	41%
	130	49											0.1298 ± 0.0001	65%
	130	49											0.1380 ± 0.0001	20%
	130	49											0.2191 ± 0.0001	2%
	130	49											0.7744 ± 0.0001	54%
	130	49											0.8070 ± 0.0001	2%
	130	49											0.9526 ± 0.0001	17%
	130	49											1.2212 ± 0.0001	65%
	130	49											1.9052 ± 0.0001	80%
	130	49											1.9458 ± 0.0001	7%
	130	49											2.0283 ± 0.0001	4%
	130	49											2.0915 ± 0.0002	5%
	130	49											2.8985 ± 0.0003	3%
$_{49}In^{131}$	131	49		130.926410	0.28 s.	β-	8.8	6.4			(9/2+)		0.3328 ± 0.0002	
	131	49											2.433	
$_{49}In^{132}$	132	49			0.22 s.	β-		≈5					0.1320	22%
	132	49											0.2992	63%
	132	49											0.3747	74%
	132	49											0.4791	24%
	132	49											0.5764	22%
	132	49											2.2863	19%
	132	49											2.3797	31%
	132	49											4.0406	65%
	132	49											4.3513	26%
	132	49											4.4158	8%
Sn		50		118.710						0.63 b.				
$_{50}Sn^{106}$	106	50		105.917030	2.1 m.	β+(20%)	3.3						ann.rad.	
	136	50				E.C.(80%)							In k x-ray	27%
	106	50											0.1223 ± 0.0001	15%
	106	50											0.2241 ± 0.0002	14%
	106	50											0.2532 ± 0.0001	29%
	106	50											0.3262 ± 0.0003	7%
	106	50											0.3865 ± 0.0002	51%
	106	50											0.4772 ± 0.0002	32%
	106	50											0.7123 ± 0.0003	17%
	106	50											0.8639 ± 0.0003	11%
	106	50											1.097 ± 0.001	1%
	106	50											1.1896 ± 0.0003	17%
$_{50}Sn^{107}$	107	50		106.915870	2.9 m.	E.C.	5.2						0.4218	5 +
	107	50				β+							0.6105	3
	107	50											0.6785 ± 0.0003	100(D)
	107	50											1.0013 ± 0.0003	29
	107	50											1.1290 ± 0.0002	100
	107	50											1.1860	12
	107	50											1.358	6
	107	50											1.396	21
	107	50											1.424	10
	107	50											1.542	30
	107	50											1.808	25
	107	50											2.116	10
	107	50											2.216	8
	107	50											2.316	6
	107	50											2.547	10
	107	50											2.825	13
	107	50											3.060	7
$_{50}Sn^{108}$	108	50		107.911880	10.3 m.	β+(1%)	2.05				0+		In k x-ray	66%
	108	50				E.C.(99%)							0.1046 ± 0.0004	12%
	108	50											0.1678 ± 0.0003	18%
	108	50											0.2357 ± 0.0002	6%
	108	50											0.2724 ± 0.0003	41%
	108	50											0.3965 ± 0.0002	58%
	108	50											0.6692 ± 0.0004	20%
	108	50											0.8293 ± 0.0005	3%
	108	50											0.8587 ± 0.0006	2%
	108	50											0.8891 ± 0.0005	3%
	108	50											1.6544 ± 0.0005	2%
	108	50											1.6848 ± 0.0006	2%
$_{50}Sn^{109}$	109	50		108.911294	18.0 m.	β+(9%)	3.88				7/2+		ann.rad.	
	109	50				E.C.(91%)							In k x-ray	56%

Isotope	A	Z	% Natural abundance	Atomic mass	Half-life	Decay mode	Decay energy (MeV)	Particle energy (MeV)	Particle intensity	Thermal neutron cross section	Spin (h/2π)	μ Nucl. mag. moment	Gamma-ray energy (MeV)	Gamma-ray intensity
	109	50											0.3312 ± 0.0002	10%
	109	50											0.3845 ± 0.0003	3%
	109	50											0.4374 ± 0.0003	3%
	109	50											0.5219 ± 0.0002	3%
	109	50											0.6234 ± 0.0004	2%
	109	50											0.6498 ± 0.0002	32%(D)
	109	50											0.7909 ± 0.0003	2%
	109	50											1.0264 ± 0.0002	5%
	109	50											1.0390 ± 0.0002	5%
	109	50											1.0992 ± 0.0002	31%
	109	50											1.1192 ± 0.0003	3%
	109	50											1.3213 ± 0.0002	12%
	109	50											1.4620 ± 0.0006	2%
	109	50											1.4636 ± 0.0004	3%
	109	50											1.4642 ± 0.0002	4%
	109	50											1.4887 ± 0.0002	4%
	109	50											1.5744 ± 0.0002	6%
	109	50											1.8898 ± 0.0003	2%
	109	50											1.9111 ± 0.0002	6%
	109	50											2.0552 ± 0.0003	2%
	109	50											2.1956 ± 0.0002	1.5%
	109	50											2.5418 ± 0.0003	2.7%
	109	50											2.7854 ± 0.0003	1.8%
	109	50											2.8586 ± 0.0002	1%
$_{50}Sn^{110}$	110	50		109.907858	4.0 h.	E.C.	0.58				0+		In k x-ray	61%
	110	50											0.283 ± 0.001	97%
$_{50}Sn^{111}$	111	50		110.907741	35.3 m.	β+(31%)	2.45	1.5			7/2+		In k x-ray	42%
	111	50				E.C.(69%)							0.7620 ± 0.0001	1.4%
	111	50											1.1530 ± 0.0001	2.5%
	111	50											1.6105 ± 0.0002	1.2%
	111	50											1.9147 ± 0.0002	1.9%
	111	50											2.1071 ± 0.0002	0.4%
	111	50											2.1795 ± 0.0003	0.3%
	111	50											2:2121 ± 0.0002	0.2%
	111	50											2.3233 ± 0.0002	0.3%
$_{50}Sn^{112}$	112	50	1.0%	111.904826						(0.3 + 0.7)b.	0+			
$_{50}Sn^{113m}$	113	50			21.4 m.	I.T.(92%)	0.077				7/2+		Sn k x-ray	36%
	113	50				E.C.(8%)							In x-ray	5%
	113	50											0.0774 ± 0.0001	0.5%
$_{50}Sn^{113}$	113	50		112.905176	115.1 d.	E.C.					1/2+		In k x-ray	80%
	113	50											0.25511 ± 0.00001	1.9%
	113	50											0.39169 ± 0.00001	64%
	113	50											0.6380 ± 0.0001	0.001%
$_{50}Sn^{114}$	114	50	0.7%	113.902784						0.1 b.	0+			
$_{50}Sn^{115}$	115	50	0.4%	114.903348						30 b.	1/2+	-0.918		
$_{50}Sn^{116}$	116	50	14.7%	115.901747						(0.006+0.1) b.	0+			
$_{50}Sn^{117m}$	117	50			13.6 d.	I.T.	0.3146				11/2-		Sn k x-ray	54%
	117	50											0.15602 ± 0.00002	2%
	117	50											0.15856 ± 0.00002	86%
$_{50}Sn^{117}$	117	50	7.7%	116.902956						2.3 b.	1/2+	-1.000		
$_{50}Sn^{118}$	118	50	24.3%	117.901609						(0.054 + 0.2) b.	0+			
$_{50}Sn^{119m}$	119	50			293 d.	I.T.	0.0895				11/2-	+0.67	Sn k x-ray	
	119	50											0.02387 ± 0.00002	16%
	119	50											0.0657 ± 0.0001	0.02%
$_{50}Sn^{119}$	119	50	8.6%	118.903310						2 b.	1/2+	-1.046		
$_{50}Sn^{120}$	120	50	32.4%	119.902200						(0.001 + 0.16) b.	0+			
$_{50}Sn^{121m}$	121	50			≈55y.	I.T.(78%)	0.006				11/2-		Sn k x-ray	
	121	50				β-(22%)	0.394	0.354					0.03715 ± 0.0004	2%
$_{50}Sn^{121}$	121	50		120.904238	27.0 h.	β-	0.388	0.383	100%		3/2+	0.70		
$_{50}Sn^{122}$	122	50	4.6%	121.903440						(0.16 + 0.001)b.	0+			
$_{50}Sn^{123m}$	123	50			40.1 m.	β-	1.42	1.26	99%		3/2+		0.1603 ± 0.0001	86%
	123	50											0.3814 ± 0.0004	0.04%
$_{50}Sn^{123}$	123	50		122.905722	129.2 d.	β-	1.398	1.42	99.4%		11/2-		0.1603 ± 0.0001	0.002%
	123	50											1.0302 ± 0.0001	0.03%
	123	50											1.0886 ± 0.0001	0.6%
$_{50}Sn^{124}$	124	50	5.66%	123.905274						(0.13 + 0.004) b.	0+			
$_{50}Sn^{125m}$	125	50			9.52 m.	β-		2.03	98%		3/2+		0.3321 ± 0.0001	97%
	125	50											0.5896 ± 0.0005	0.2%
	125	50											1.4040 ± 0.0005	0.7%
	125	50											1.4839 ± 0.0005	0.2%
$_{50}Sn^{125}$	125	50		124.907785	β-	2.352	2.35	82%			11/2-		0.3321 ± 0.0001	1.3%
	125	50											0.4698 ± 0.0001	1.4%
	125	50											0.8008 ± 0.0001	1%
	125	50											0.8225 ± 0.0001	4%
	125	50											0.9155 ± 0.0001	4%
	125	50											1.0671 ± 0.0001	9%
	125	50											1.0877 ± 0.0001	1.1%
	125	50											1.0892 ± 0.0001	4.3%

TABLE OF THE ISOTOPES (Continued)

Isotope	A	Z	% Natural abundance	Atomic mass	Half-life	Decay mode	Decay energy (MeV)	Particle energy (MeV)	Particle intensity	Thermal neutron cross section	Spin (h/2π)	μ Nucl. mag. moment	Gamma-ray energy (MeV)	Gamma-ray intensity
	125	50											2.0018 ± 0.0001	1.8%
$_{50}Sn^{126}$	126	50		125.907654	≈10^5y.	β-	0.38	0.25	100%		0+		0.0643 ± 0.0001	9.6%
	126	50											0.0869 ± 0.0001	9%
	126	50											0.0876 ± 0.0001	37%
	126	50											0.4148 ± 0.0002	98%(D)
	126	50											0.6663 ± 0.0002	100%(D)
	126	50											0.6950 ± 0.0002	100%(D)
$_{50}Sn^{127m}$	127	50			4.15 m.	β-		2.72			1/2+		0.4909 ± 0.0004	90%
	127	50											1.3480 ± 0.001	5%
	127	50											1.5640 ± 0.002	4%
$_{50}Sn^{127}$	127	50		126.910355	2.1 h.	β-	3.2	2.42			11/2-		0.1197 ± 0.0004	2%
	127	50						3.2					0.1692 ± 0.0004	2%
	127	50											0.2625 ± 0.0004	2%
	127	50											0.2662 ± 0.0004	2%
	127	50											0.2843 ± 0.0004	3%
	127	50											0.4382 ± 0.0004	6%
	127	50											0.4909 ± 0.0004	5%
	127	50											0.4932 ± 0.0004	5%
	127	50											0.5454 ± 0.0004	2%
	127	50											0.5833 ± 0.0004	3%
	127	50											0.5923 ± 0.0004	2%
	127	50											0.8059 ± 0.0004	8%
	127	50											0.8231 ± 0.0004	11%
	127	50											0.8247 ± 0.0004	6%
	127	50											0.8595 ± 0.0004	8%
	127	50											0.9792 ± 0.0004	7%
	127	50											1.0361 ± 0.0004	2%
	127	50											1.0933 ± 0.0007	4%
	127	50											1.0956 ± 0.0004	19%
	127	50											1.1143 ± 0.0004	38%
	127	50											1.1604 ± 0.0004	2%
	127	50											1.5843 ± 0.0004	2%
	127	50											2.0034 ± 0.0005	5%
	127	50											2.3174 ± 0.0005	1%
	127	50											2.5849 ± 0.0005	1.6%
	127	50											2.6959 ± 0.0005	1.6%
	127	50											2.8464 ± 0.0005	1%
$_{50}Sn^{128}$	128	50		127.910560	59.1 m.	β-	1.29	0.48			0+		0.0321 ± 0.0002	4%
	128	50						0.63					0.0457 ± 0.0002	13%
	128	50											0.0751 ± 0.0002	27%
	128	50											0.1527 ± 0.0002	6%
	128	50											0.1604 ± 0.0002	2%
	128	50											0.4044 ± 0.0002	6%
	128	50											0.4823 ± 0.0002	58%
	128	50											0.5573 ± 0.0002	16%
	128	50											0.6805 ± 0.0002	16%
$_{50}Sn^{129m}$	129	50			6.9 m.	β-					11/2-		1.1611 ± 0.0003	
$_{50}Sn^{129}$	129	50		128.913440	2.5 m.	β-	4.0				3/2+		0.6456 ± 0.0003	
$_{50}Sn^{130m}$	130	50		1.7 m.	β-		3.0				(7-)		0.0847 ± 0.0001	14%
	130	50											0.1449 ± 0.0001	34%
	130	50											0.3113 ± 0.0001	14%
	130	50											0.5436 ± 0.0002	10%
	130	50											0.8992 ± 0.0002	17%
	130	50											0.9624 ± 0.0005	3%
$_{50}Sn^{130}$	130	50		129.913920	3.7 m.	β-	2.1	1.10			0+		0.0700 ± 0.0001	36%
	130	50											0.1925 ± 0.0002	71%
	130	50											0.2292 ± 0.0002	24%
	130	50											0.3413 ± 0.0002	2%
	130	50											0.4347 ± 0.0002	14%
	130	50											0.5505 ± 0.0002	3%
	130	50											0.7431 ± 0.0001	19%
	130	50											0.7798 ± 0.0001	59%
	130	50											0.8723 ± 0.0003	1%
$_{50}Sn^{131m}$	131	50			39 s.	β-							0.0823 ± 0.0001	21 +
	131	50											0.3043 ± 0.0001	32
	131	50											0.4500 ± 0.0001	90
	131	50											0.7985 ± 0.0001	86
	131	50											0.8851 ± 0.0001	9
	131	50											1.0734 ± 0.0001	9
	131	50											1.1415 ± 0.0003	12
	131	50											1.2026 ± 0.0002	19
	131	50											1.2260 ± 0.0001	100
	131	50											1.2292 ± 0.0001	30
	131	50											1.4811 ± 0.0001	12
	131	50											1.9311 ± 0.0001	9
	131	50											2.0293 ± 0.0002	5
	131	50											2.0825 ± 0.0002	5
	131	50											2.1864 ± 0.0002	4
	131	50											(0.08 - 3.21)	
$_{50}Sn^{131}$	131	50		130.916940	61 s.	β-	4.6						see Sn^{131m}	
$_{50}Sn^{132}$	132	50		131.917760	40 s.	β-	3.1	1.76					0.0855 ± 0.0001	49%
	132	50											0.2467 ± 0.0001	42%
	132	50											0.3402 ± 0.0001	43%
	133	50											0.5287 ± 0.0002	2%
	132	50											0.5488 ± 0.0002	2%

Isotope	A	Z	% Natural abundance	Atomic mass	Half-life	Decay mode	Decay energy (MeV)	Particle energy (MeV)	Particle intensity	Thermal neutron cross section	Spin (h/2π)	μ Nucl. mag. moment	Gamma-ray energy (MeV)	Gamma-ray intensity
	132	50											0.6519 ± 0.0002	2%
	132	50											0.8985 ± 0.0001	42%
	132	50											0.9922 ± 0.0001	38%
	132	50											1.0778 ± 0.0003	2%
	132	50											1.2388 ± 0.0002	13%
Sb		51								5.4 b.				
$_{51}Sb^{109}$	109	51		108.918143	18.3 s.	β+	6.83						ann.rad.	
	109	51				E.C.							0.2467	2 +
	109	51											0.2610	2
	109	51											0.5448	11
	109	51											0.6645	63
	109	51											0.6876	19
	109	51											0.9254	100
	109	51											0.9506	2
	109	51											1.0617	75
	109	51											1.0780	4
	109	51											1.3435	2
	109	51											1.3435	2
	109	51											1.4958	30
$_{51}Sb^{110}$	110	51		109.916770	23.5 s.	β+	8.4	6.9			3+		ann.rad.	
	110	51				E.C.							0.6365 ± 0.0004	4%
	110	51											0.7515 ± 0.0004	4%
	110	51											0.8271 ± 0.0003	9%
	110	51											0.9089 ± 0.0003	8%
	110	51											0.9847 ± 0.0001	31%
	110	51											1.0258 ± 0.0004	2%
	110	51											1.2117 ± 0.0001	92%
	110	51											1.2433 ± 0.0003	13%
	110	51											1.4825 ± 0.0004	4%
	110	51											1.6095 ± 0.0005	2%
	110	51											1.7359 ± 0.0005	7%
	110	51											1.7653 ± 0.0005	4%
	110	51											1.9709 ± 0.0006	4.9%
	110	51											2.0291 ± 0.0006	3.7%
	110	51											2.1208 ± 0.0008	7.3%
	110	51											2.2349 ± 0.0008	2%
	110	51											2.6732 ± 0.001	2%
$_{51}Sb^{111}$	111	51		110.913220	75 s.	β+(87%)	5.2	3.3			5/2+		ann.rad.	
	111	51				E.C.(13%)							0.1002 ± 0.0001	3%
	111	51											0.1545 ± 0.0001	71%
	111	51											0.3888 ± 0.0001	4%
	111	51											0.4891 ± 0.0001	42%
	111	51											0.6436 ± 0.0002	2%
	111	51											0.7554 ± 0.0002	5%
	111	51											0.7778 ± 0.0002	3%
	111	51											0.8974 ± 0.0003	2%
	111	51											1.0326 ± 0.0001	10%
	111	51											1.1475 ± 0.0002	4%
	111	51											1.8413 ± 0.0003	0.5%
$_{51}Sb^{112}$	112	51		111.912411	51 s.	β+(90%)	7.06	4.75			3+		ann.rad.	
	112	51				E.C.(10%)							0.6700 ± 0.0004	4%
	112	51											0.8946 ± 0.0002	2.6%
	112	51											0.9909 ± 0.0001	14%
	112	51											1.0980 ± 0.0002	2%
	112	51											1.2571 ± 0.0001	96%
	112	51											1.5664 ± 0.0002	1.6%
	112	51											1.7102 ± 0.0002	1.3%
	112	51											2.1609 ± 0.0002	1%
	112	51											2.24 791 ± 0.0002	1%
	112	51											(0.3 - 3.6)	
$_{51}Sb^{113}$	113	51		112.909372	6.7 m.	β+(65%)	3.89	2.4			5/2+		ann.rad.	
	113	51				E.C.(35%)							Sn k x-ray	20%
	113	51											0.3324 ± 0.0001	14%
	113	51											0.4980 ± 0.0001	77%
	113	51											0.9406 ± 0.0001	2.5%
	113	51											1.0133 ± 0.0001	0.9%
	113	51											1.5563 ± 0.0002	1%
$_{51}Sb^{114}$	114	51		113.909110	3.5 m.	β+(78%)	6.09	4.0			3/2+		ann.rad.	
	114	51				E.C.(22%)							Sn k x-ray	9%
	114	51											0.3272 ± 0.0001	7%
	114	51											0.7173 ± 0.0001	5%
	114	51											0.8876 ± 0.0001	18%
	114	51											0.9748 ± 0.0001	3%
	114	51											1.2999 ± 0.0001	98%
	114	51											1.5600 ± 0.0002	1%
	114	51											1.6438 ± 0.0001	1.3%
	114	51											1.9079 ± 0.0001	1%
	114	51											1.9262 ± 0.0001	1.7%
	114	51											2.2398 ± 0.0002	1%
$_{51}Sb^{115}$	115	51		114.906601	32.1 m.	β+(67%)	3.03	1.50			5/2+		ann.rad.	
	115	51				E.C.(33%)							Sn k x-ray	27%
	115	51											0.4973 ± 0.0001	98%
	115	51											1.2366 ± 0.0002	0.6%

Isotope	A	Z	% Natural abundance	Atomic mass	Half-life	Decay mode	Decay energy (MeV)	Particle energy (MeV)	Particle intensity	Thermal neutron cross section	Spin (h/2π)	μ Nucl. mag. moment	Gamma-ray energy (MeV)	Gamma-ray intensity
	115	51											1.2799 ± 0.0002	0.3%
	115	51											1.6338 ± 0.0002	0.3%
$_{51}Sb^{116m}$	116	51			60 m.	β+(78%)		1.2			8-		ann.rad.	
	116	51				E.C.(22%)							Sn k x-ray	51%
	116	51											0.09982 ± 0.00002	32%
	116	51											0.13552 ± 0.00005	29%
	116	51											0.4073 ± 0.0001	42%
	116	51											0.5429 ± 0.0002	52%
	116	51											0.8440 ± 0.0001	12%
	116	51											0.9725 ± 0.0001	72%
	116	51											1.0724 ± 0.0001	28%
	116	51											1.2935 ± 0.0001	100%
$_{51}Sb^{116}$	116	51		115.906800	16 m.	β+(50%)	4.71	1.3			3+		ann.rad.	
	116	51				E.C.(50%)		2.3					Sn k x-ray	20%
	116	51											0.93180 ± 0.00005	26%
	116	51											1.29354 ± 0.00004	85%
	116	51											2.22533 ± 0.00007	17%
$_{51}Sb^{117}$	117	51		116.904841	2.80 h.	β+(2%)	1.76	0.57			5/2+	+2.67	Sn k x-ray	44%
	117	51				E.C.(98%)							0.1586 ± 0.00001	86%
	117	51											0.8614 ± 0.0001	0.3%
	117	51											1.0045 ± 0.0002	0.2%
	117	51											1.0206 ± 0.0005	0.1%
	117	51											1.0210 ± 0.0005	0.1%
$_{51}Sb^{118m}$	118	51			5.00 h.	E.C.(99%)					8-		Sn k x-ray	57%
	118	51											0.0410 ± 0.0001	18%
	118	51											0.25368 ± 0.00001	99%
	118	51											1.05069 ± 0.00003	97%
	118	51											1.09151 ± 0.00008	4%
	118	51											1.22964 ± 0.00004	100%
$_{51}Sb^{118}$	118	51		117.905534	3.6 m.	β+(74%)	3.66	2.65			1+		ann.rad.	150%
	118	51				E.C.(26%)							Sn k x-ray	10%
	118	51											0.5282 ± 0.00004	0.4%
	118	51											0.8269 ± 0.0006	0.4%
	118	51											1.22964 ± 0.00006	2.5%
	118	51											1.2670 ± 0.0005	0.6%
$_{51}Sb^{119}$	119	51		118.903948	36.1 h.	E.C.	0.59				5/2+	+3.45	Sn k x-ray	39%
	119	51											0.0239 ± 0.0001	16%
$_{51}Sb^{120m}$	120	51			5.76 d.	E.C.					8-		Sn k x-ray	50%
	120	51											0.0898 ± 0.0002	80%
	120	51											0.19730 ± 0.00003	88%
	120	51											1.02301 ± 0.00004	99%
	120	51											1.1130 ± 0.0006	1.3%
	120	51											1.17121 ± 0.00006	100%
$_{51}Sb^{120}$	120	51		119.903821	15.9 m.	β+(41%)	2.68	1.75			1+	+2.3	ann.rad.	
	120	51				E.C.(59%)							Sn k x-ray	23%
	120	51											0.7038 ± 0.0003	0.15%
	120	51											0.9886 ± 0.0007	0.06%
	120	51											1.17121 ± 0.00005	1.7%
$_{51}Sb^{121}$	121	51	57.3%	120.903821						(0.05 + 6.2)b.	5/2+	+3.359		
$_{51}Sb^{122m}$	122	51			4.21 m.	I.T.	0.162				8-		Sb x-ray	42%
	122	51											0.0614 ± 0.0001	57%
	122	51											0.0761 ± 0.0001	20%
$_{51}Sb^{122}$	122	51		121.905179	2.71 d.	β-(98%)	1.982	1.41	65%		2-	-1.90	0.56409 ± 0.00005	71%
	122	51				β+(2%)	1.622	1.98	26%				0.69277 ± 0.00008	3.7%
	122	51											1.14050 ± 0.0001	0.6%
	122	51											1.2569 ± 0.0001	0.8%
$_{51}Sb^{123}$	123	51	42.7%	122.904216						(0.02 + 0.04 + 4.1)b.	7/2+	+2.547		
$_{51}Sb^{124m2}$	124	51			20.3 m.	I.T.	0.035							
$_{51}Sb^{124m1}$	124	51			96 s.	I.T.(80%)		1.2			5+		0.4984 ± 0.0001	20%
	124	51				β-(20%)		1.7					0.6027 ± 0.0001	20%
	124	51											0.6458 ± 0.0001	20%
	124	51											1.1010 ± 0.0002	0.3%
$_{51}Sb^{124}$	124	51		123.905038	60.20 d.	β-	2.905	0.61	52%	17 b.	3-		0.60271 ± 0.0001	98%
	124	51						2.3	23%				0.64583 ± 0.00001	7%
	124	51											0.7093 ± 0.0001	1.4%
	124	51											0.71376 ± 0.0000'	2.4%

Isotope	A	Z	% Natural abundance	Atomic mass	Half-life	Decay mode	Decay energy (MeV)	Particle energy (MeV)	Particle intensity	Thermal neutron cross section	Spin (h/2π)	μ Nucl. mag. moment	Gamma-ray energy (MeV)	Gamma-ray intensity
	124	51											0.72277 ± 0.00002	11%
	124	51											0.96819 ± 0.00002	1.9%
	124	51											1.04511 ± 0.00002	1.9%
	124	51											1.3255 ± 0.0001	1.5%
	124	51											1.3681 ± 0.0001	2.5%
	124	51											1.4366 ± 0.0001	1.2%
	124	51											1.69094 ± 0.00004	49%
	124	51											2.09089 ± 0.00004	5.6%
$_{51}Sb^{125}$	125	51		124.905252	2.76 y.	β-	0.767	0.13	30%		$^1/_2$+		0.0355 ± 0.0001	6%
	125	51						0.30	45%				0.17632 ± 0.00002	6.8%
	125	51						0.62	13%				0.38044 ± 0.00002	1.5%
	125	51											0.42786 ± 0.00002	29%
	125	51											0.46336 ± 0.00001	10%
	125	51											0.60060 ± 0.00001	18%
	125	51											0.60672 ± 0.00003	5%
	125	51											0.63595 ± 0.00003	11%
	125	51											0.67144 ± 0.00004	2%
$_{51}Sb^{126m2}$	126	51			11 s.	I.T.					3-		L x-ray	
	126	51											0.0227 ± 0.0001	0.13%
$_{51}Sb^{126m1}$	126	51			19.0 m.	β-(86%)					5+		0.4148 ± 0.0002	86%
	126	51				I.T.(14%)							0.6663 ± 0.0002	86%
	126	51											0.6950 ± 0.0002	86%
	126	51											0.9282 ± 0.0003	1.3%
	126	51											1.0348 ± 0.0002	1.8%
$_{51}Sb^{126}$	126	51		125.907250	12.4 d.	β-	3.66				8-		0.2786 ± 0.0002	2%
	126	51											0.2965 ± 0.0003	4%
	126	51											0.4148 ± 0.0002	84%
	126	51											0.5738 ± 0.0002	7%
	126	51											0.5930 ± 0.0002	7%
	126	51											0.6563 ± 0.0002	2%
	126	51											0.6663 ± 0.0002	100%
	126	51											0.6950 ± 0.0002	100%
	126	51											0.6970 ± 0.0002	30%
	126	51											0.7205 ± 0.0002	54%
	126	51											0.8567 ± 0.0002	18%
	126	51											0.9893 ± 0.0002	7%
	126	51											1.0348 ± 0.0002	1%
	126	51											1.2130 ± 0.0002	2.4%
$_{51}Sb^{127}$	127	51		126.906919	3.84 d.	β-	1.58	0.89			7/2+		0.2524 ± 0.0003	8%
	127	51						1.1					0.2908 ± 0.0005	2%
	127	51						1.50					0.4121 ± 0.0005	4%
	127	51											0.4451 ± 0.0005	4%
	127	51											0.4370 ± 0.0004	25%
	127	51											0.5433 ± 0.0005	3%
	127	51											0.6035 ± 0.0005	4%
	127	51											0.6857 ± 0.0005	35%
	127	51											0.6985 ± 0.0005	3%
	127	51											0.7222 ± 0.0005	2%
	127	51											0.7837 ± 0.0005	14%
	127	51											0.9244 ± 0.0009	0.5%
$_{51}Sb^{128m}$	128	51			10 m.	β-(96%)					5+		0.3140 ± 0.0001	92%
	128	51				I.T.(4%)							0.5941 ± 0.0001	3%
	128	51											0.7432 ± 0.0001	96%
	128	51											0.7539 ± 0.0001	96%
	128	51											0.7876 ± 0.0001	7%
	128	51											0.8440 ± 0.0003	2%
	128	51											0.9083 ± 0.0002	2%
	128	51											1.0409 ± 0.0003	1%
	128	51											1.1417 ± 0.0003	0.8%
	128	51											1.1580 ± 0.0003	1.7%
$_{51}Sb^{128}$	128	51		127.909180	9.1 h.	β-	4.39	2.3			8-		0.2148 ± 0.0002	2%
	128	51											0.3141 ± 0.0001	61%
	128	51											0.3177 ± 0.0002	3%
	128	51											0.3224 ± 0.0002	3%
	128	51											0.5265 ± 0.0001	45%
	128	51											0.6287 ± 0.0001	31%
	128	51											0.6362 ± 0.0001	36%
	128	51											0.6542 ± 0.0002	17%
	128	51											0.6671 ± 0.0003	2%
	128	51											0.6839 ± 0.0003	3%
	128	51											0.6929 ± 0.0003	2%
	128	51											0.7276 ± 0.0003	4%

Isotope	A	Z	% Natural abundance	Atomic mass	Half-life	Decay mode	Decay energy (MeV)	Particle energy (MeV)	Particle intensity	Thermal neutron cross section	Spin (h/2π)	μ Nucl. mag. moment	Gamma-ray energy (MeV)	Gamma-ray intensity
	128	51											0.7433 ± 0.0001	100%
	128	51											0.7540 ± 0.0001	100%
	128	51											0.8136 ± 0.0002	13%
	128	51											0.8458 ± 0.0004	2%
	128	51											0.8780 ± 0.0004	3%
	128	51											1.0475 ± 0.0004	3%
	128	51											1.0786 ± 0.0004	2%
	128	51											1.1127 ± 0.0004	2%
	128	51											1.1816 ± 0.0004	4%
	128	51											1.3780 ± 0.0004	1.8%
$_{51}Sb^{129m}$	129	51			17.7 m.	β-							0.4338	
	129	51											0.6578	
	129	51											0.7598	
$_{51}Sb^{129}$	129	51		128.909146	4.4 h.	β-	2.38	1.4					0.0278 ± 0.0001	19%(D)
	129	51						1.5					0.1808 ± 0.0005	3%
	129	51						1.8					0.3594 ± 0.0005	3%
	129	51											0.4596 ± 0.0001	8%(D)
	129	51											0.5447 ± 0.0003	18%
	129	51											0.6337 ± 0.0005	3%
	129	51											0.6543 ± 0.0005	3%
	129	51											0.6836 ± 0.0003	6%
	129	51											0.7610 ± 0.0005	4%
	129	51											0.7734 ± 0.0006	3%
	129	51											0.8128 ± 0.0005	45%
	129	51											0.8762 ± 0.0007	3%
	129	51											0.9146 ± 0.0005	21%
	129	51											0.9664 ± 0.0006	8%
	129	51											1.0301 ± 0.0006	13%
	129	51											1.2085 ± 0.0007	1%
	129	51											1.5687 ± 0.0008	0.8%
	129	51											1.6546 ± 0.001	1%
	129	51											1.7365 ± 0.001	6%
$_{51}Sb^{130m}$	130	51			6.3 m.	β-	2.6	2.12					0.1023 ± 0.0001	41%
	130	51											0.3485 ± 0.0002	5%
	130	51											0.6271 ± 0.0003	5%
	130	51											0.6477 ± 0.0003	3%
	130	51											0.6974 ± 0.0003	4%
	130	51											0.7489 ± 0.0003	4%
	130	51											0.7934 ± 0.0001	86%
	130	51											0.8163 ± 0.0002	12%
	130	51											0.8394 ± 0.0001	100%
	130	51											0.9208 ± 0.0001	4%
	130	51											0.9422 ± 0.0004	3%
	130	51											1.0175 ± 0.0002	3%
	130	51											1.0465 ± 0.0004	3%
	130	51											1.0717 ± 0.0004	2%
	130	51											1.1028 ± 0.0004	4%
	130	51											1.1420 ± 0.0004	6%
	130	51											1.1773 ± 0.0004	2%
	130	51											1.2000 ± 0.0004	4%
	130	51											1.5980 ± 0.0005	3%
	130	51											2.1166 ± 0.0008	1%
$_{51}Sb^{130}$	130	51		129.911590	38.4 m.	β-	5.0	2.9			8-		0.1823 ± 0.0001	65%
	130	51											0.2580 ± 0.0002	4%
	130	51											0.2853 ± 0.0002	3%
	130	51											0.3033 ± 0.0002	2%
	130	51											0.3309 ± 0.0001	78%
	130	51											0.4554 ± 0.0002	5%
	130	51											0.4680 ± 0.0001	18%
	130	51											0.4836 ± 0.0003	2%
	130	51											0.5067 ± 0.0003	2%
	130	51											0.6267 ± 0.0003	3%
	130	51											0.6547 ± 0.0003	2%
	130	51											0.6809 ± 0.0003	6%
	130	51											0.7320 ± 0.0001	22%
	130	51											0.7394 ± 0.0001	100%
	130	51											0.8394 ± 0.0001	100%
	130	51											0.9349 ± 0.0002	19%
	130	51											1.0002 ± 0.0004	2%
	130	51											1.0895 ± 0.0004	4%
	130	51											1.1414 ± 0.0004	2%
	130	51											1.2923 ± 0.0004	4%
	130	51											1.4437 ± 0.0005	2%
	130	51											1.5819 ± 0.0008	2%
	130	51											1.7626 ± 0.0005	2%
	130	51											1.9974 ± 0.0005	2%
$_{51}Sb^{131}$	131	51		130.911950	23.0 m.	β-	3.2	1.3			7/2 +		0.6423 ± 0.0001	22%
	131	51						3.0					0.6579 ± 0.0004	7%
	131	51											0.7263 ± 0.0001	4%
	131	51											0.8546 ± 0.0005	3%
	131	51											0.9331 ± 0.0001	25%
	131	51											0.9434 ± 0.0001	44%
	131	51											1.1236 ± 0.0002	8%
	131	51											1.2074 ± 0.0001	4%
	131	51											1.2338 ± 0.0002	2%

TABLE OF THE ISOTOPES (Continued)

Isotope	A	Z	% Natural abundance	Atomic mass	Half-life	Decay mode	Decay energy (MeV)	Particle energy (MeV)	Particle intensity	Thermal neutron cross section	Spin (h/2π)	μ Nucl. mag. moment	Gamma-ray energy (MeV)	Gamma-ray intensity
	131	51											1.2675 ± 0.0002	3%
	131	51											1.7220 ± 0.0005	2.3%
	131	51											1.8544 ± 0.0003	4%
	131	51											2.1799 ± 0.0004	2%
	131	51											2.3356 ± 0.0004	2%
	131	51											2.3989 ± 0.0007	1%
	131	51											2.6623 ± 0.0002	1%
$_{51}Sb^{132m}$	132	51			3.07 m.	β-		3.0			4+		0.1034 ± 0.0001	14%
	132	51						4.0					0.3538 ± 0.0002	3%
	132	51											0.3823 ± 0.0001	8%
	132	51											0.4368 ± 0.0002	3%
	132	51											0.4473 ± 0.0002	2%
	132	51											0.6099 ± 0.0003	2%
	132	51											0.6356 ± 0.0002	10%
	132	51											0.6968 ± 0.0001	86%
	132	51											0.8141 ± 0.0003	5%
	132	51											0.8166 ± 0.0002	11%
	132	51											0.9739 ± 0.0001	98%
	132	52											0.9896 ± 0.0001	15%
	132	51											1.0932 ± 0.0003	5%
	132	51											1.1335 ± 0.0002	6%
	132	51											1.1522 ± 0.0004	3%
	132	51											1.1965 ± 0.0004	3%
	132	51											1.2133 ± 0.0004	2%
	132	51											1.4363 ± 0.0004	2%
	132	51											1.5135 ± 0.0005	2%
	132	51											1.6445 ± 0.0008	2%
	132	51											1.7880 ± 0.0008	3%
	132	51											2.2804 ± 0.0008	1%
	132	51											2.5883 ± 0.0008	1%
$_{51}Sb^{132}$	132	51		131.914410	4.12 m.	β-	5.5				8-		0.1034 ± 0.0001	35%
	132	51											0.1506 ± 0.0001	66%
	132	51											0.2760 ± 0.0002	4%
	132	51											0.2930 ± 0.0002	4%
	132	51											0.3686 ± 0.0002	7%
	132	51											0.3823 ± 0.0001	7%
	132	51											0.4965 ± 0.0002	13%
	132	51											0.6968 ± 0.0001	100%
	132	51											0.8819 ± 0.0003	6%
	132	51											0.9739 ± 0.0001	100%
	132	51											1.0415 ± 0.0003	18%
	132	51											1.1669 ± 0.0004	10%
	132	51											1.3783 ± 0.0004	4%
	132	51											1.7637 ± 0.0008	
	132	51											1.8546 ± 0.0008	2%
	132	51											2.664 ± 0.001	4%
$_{51}Sb^{133}$	133	51		132.915150	2.5 m.	β-	4.0	1.0			7/2+		0.4235 ± 0.0005	6%
	133	51											0.6318 ± 0.0005	19%
	133	51											0.8165 ± 0.0005	15%
	133	51											0.8385 ± 0.0005	8%
	133	51											1.0250 ± 0.0005	2%
	133	51											1.0764 ± 0.0005	30%
	133	51											1.3050 ± 0.0005	2%
	133	51											1.4900 ± 0.001	1%
	133	51											1.7282 ± 0.001	5%
	133	51											2.7517 ± 0.001	9%
$_{51}Sb^{134m}$	134	51			0.85 s.	β-		8.4						
$_{51}Sb^{134}$	134	51		133.920550	10.4 s.	β-	8.4	6.1					0.1152 ± 0.0001	49%
	134	51											0.2970 ± 0.0001	97%
	134	51											0.7063 ± 0.0001	57%
	134	51											1.2791 ± 0.0001	100%
$_{51}Sb^{135}$	135	51		134.924520	1.71 s.	β-					7/2+		1.127	
	135	51											1.279	
Te		52								5.4 b.				
$_{52}Te^{108}$	108	52		107.929550	2.1 s.	α(68%)					0+			
	108	52				β+, EC (32%)	6.9							
$_{52}Te^{109}$	109	52		108.927370	4.2 s.	β+, EC (96%)	8.6							
	109	52				α(4%)								
$_{52}Te^{110}$	110	52		109.922560	18.5 s.	β+,E.C.	5.3				0+		ann.rad.	
	110	52											0.2191 ± 0.0006	
	110	52											0.6059 ± 0.0006	
$_{52}Te^{111}$	111	52		110.921130	19.3 s.	β+,E.C.	≈7				(7/2+)		ann.rad.	
	111	52											0.267	
	111	52											0.322	
	111	52											0.341	
$_{52}Te^{112}$	112	52		111.917020	2.0 m.	β+,E.C.	4.3				0+		ann.rad.	
	112	52											0.0386 ± 0.0003	16 +
	112	52											0.1042 ± 0.0003	27
	112	52											0.1327 ± 0.0002	23
	112	52											0.1674 ± 0.0002	7
	112	52											0.2364 ± 0.0004	9
	112	52											0.2962 ± 0.0002	86
	112	52											0.3509 ± 0.0003	36

Isotope	A	Z	% Natural abundance	Atomic mass	Half-life	Decay mode	Decay energy (MeV)	Particle energy (MeV)	Particle intensity	Thermal neutron cross section	Spin (h/2π)	μ Nucl. mag. moment	Gamma-ray energy (MeV)	Gamma-ray intensity
	112	52											0.3727 ± 0.0002	100
	112	52											0.4187 ± 0.0002	57
	112	52											0.4769 ± 0.0003	14
	112	52											0.4940 ± 0.0003	30
	112	52											0.6904 ± 0.0004	10
	112	52											0.7430 ± 0.0002	11
	112	52											0.7973 ± 0.0002	24
	112	52											0.8074 ± 0.0004	9
	112	52											0.8201 ± 0.0002	17
	112	52											0.8819 ± 0.0003	10
	112	52											0.9248 ± 0.0006	11
	112	52											0.9284 ± 0.0004	11
	112	52											0.9713 ± 0.0002	23
	112	52											1.2824 ± 0.0009	17
	112	52											1.2872 ± 0.0008	10
	112	52											1.5026 ± 0.0006	15
	112	52											1.6576 ± 0.0003	14
	112	52											1.9637 ± 0.0004	17
$_{52}Te^{113}$	113	52		112.915920	1.7 s.	β+(85%)	≈ 6.1	4.5			(7/2+)		ann.rad.	
	113	51				E.C.(15%)							Sb k x-ray	6%
	113	52											0.6448 ± 0.0002	6%
	113	52											0.8144 ± 0.0003	22%
	113	52											1.0181 ± 0.0004	12%
	113	52											1.1812 ± 0.0004	12%
	113	52											1.2567 ± 0.0005	5.5%
	113	52											1.5503 ± 0.0007	2.2%
	113	52											1.8681 ± 0.0009	2.4%
	113	52											2.0937 ± 0.001	3%
	113	52											2.1155 ± 0.001	2%
	113	52											2.2212 ± 0.0009	2%
	113	52											2.5352 ± 0.0005	2.6%
	113	52											2.5524 ± 0.0009	1.5%
	113	52											2.6065 ± 0.0005	1.8%
$_{52}Te^{114}$	114	52		113.912230	15 m.	β+(40%)	2.7				0+		ann.rad.	
	114	52				E.C.(60%)							Sb k x-ray	55%
	114	52											0.0838 ± 0.0001	7%
	114	52											0.0903 ± 0.0001	10%
	114	52											0.2446 ± 0.0001	3%
	114	52											0.4972 ± 0.0001	3%
	114	52											0.7266 ± 0.0002	4%
	114	52											1.4176 ± 0.0002	3%
	114	52											1.8417 ± 0.0006	2%
	114	52											1.8973 ± 0.0005	4%
$_{52}Te^{115m}$	115	52			6.7 m.	β+(45%)					(1/2+)		ann.rad.	
	115	52				E.C.(55%)							Sb k x-ray	20%
	115	52											0.5487 ± 0.0002	4%
	115	52											0.6106 ± 0.0002	4%
	115	52											0.7236 ± 0.0001	18%
	115	52											0.7704 ± 0.0001	35%
	115	52											1.0319 ± 0.0002	8%
	115	52											1.0987 ± 0.0001	9%
	115	52											1.1557 ± 0.0004	3%
	115	52											1.2793 ± 0.0002	10%
	115	52											1.3508 ± 0.0002	8%
	115	52											1.4081 ± 0.0003	3%
	115	52											1.4917 ± 0.0003	3%
	115	52											1.5041 ± 0.0002	10%
	115	52											1.5617 ± 0.0004	4%
	115	52											1.6548 ± 0.0004	6%
	115	52											1.9360 ± 0.0003	3%
	115	52											2.1044 ± 0.0002	8%
	115	52											2.2153 ± 0.0004	6%
$_{52}Te^{115}$	115	52		114.911700	5.8 m.	β+(45%)	4.8	2.7			7/2+		ann.rad.	
	115	52				E.C.(55%)							Sb k x-ray	20%
	115	52											0.3745 ± 0.0002	3%
	115	52											0.5684 ± 0.0001	3%
	115	52											0.6033 ± 0.0001	4%
	115	52											0.6570 ± 0.0001	7%
	115	52											0.7236 ± 0.0001	32%
	115	52											1.0986 ± 0.0001	17%
	115	52											1.2905 ± 0.0001	6%
	115	52											1.3004 ± 0.0003	2%
	115	52											1.3268 ± 0.0001	22%
	115	52											1.3806 ± 0.0001	24%
	115	52											1.5997 ± 0.0001	3%
	115	52											(0.22 - 2.7)	
$_{52}Te^{116}$	116	52		115.908450	2.5 h.	E.C.	≈ 1.6				0+		Sb k x-ray	60%
	116	52											0.0937 ± 0.0001	30%
	116	52											0.1030 ± 0.0001	2%
	116	52											0.6287 ± 0.0001	3%
	116	52											0.6379 ± 0.0002	0.7%
	116	52											1.0553 ± 0.0002	0.6%
$_{52}Te^{117}$	117	52		116.908630	62 m.	E.C.(75%)	3.53	1.75			1/2+		ann.rad.	
	117	52				β+(25%)							Sb k x-ray	30%
	117	52											0.9197 ± 0.0006	65%

Isotope	A	Z	% Natural abundance	Atomic mass	Half-life	Decay mode	Decay energy (MeV)	Particle energy (MeV)	Particle intensity	Thermal neutron cross section	Spin (h/2π)	μ Nucl. mag. moment	Gamma-ray energy (MeV)	Gamma-ray intensity
	117	52											0.9239 ± 0.0007	6%
	117	52											0.9967 ± 0.0007	4%
	117	52											1.0907 ± 0.0007	7%
	117	52											1.5651 ± 002	1%
	117	52											1.7164 ± 0.0007	16%
	117	52											2.3000 ± 0.0007	11%
$_{52}Te^{118}$	118	52		117.905908	6.00 d.	E.C.					0+		Sb k x-ray	40%
$_{52}Te^{119m}$	119	52			4.7 d.	E.C.					11/2-		Sb k x-ray	40%
	119	52											0.15360 ± 0.00003	66%
	119	52											0.2705 ± 0.0001	28%
	119	52											0.9126 ± 0.0001	6%
	119	52											0.9422 ± 0.0001	5%
	119	52											0.9764 ± 0.0001	3%
	119	52											0.9793 ± 0.0001	3%
	119	52											1.0132 ± 0.0001	2.5%
	119	52											1.0484 ± 0.0001	3%
	119	52											1.0957 ± 0.0001	2%
	119	52											1.1368 ± 0.0001	8%
	119	52											1.21271 ± 00005	66%
	119	52											1.3664 ± 0.0002	1%
	119	52											2.0896 ± 0.0001	5%
$_{52}Te^{119}$	119	52		118.906411	16.05 h.	β+(2%)	2.29	0.63			1/2+		ann.rad.	
	119	52				E.C.(98%)							Sb k x-ray	40%
	119	52											0.6440 ± 0.0001	84%
	119	52											0.6998 ± 0.0001	10%
	119	52											1.4132 ± 0.0001	1%
	119	52											1.7497 ± 0.0001	4%
	119	52											1.1056 ± 0.0001	0.6%
	119	52											1.1770 ± 0.0001	0.7%
$_{52}Te^{120}$	120	52	0.096%	119.904048						0.3 b.	0+			
$_{52}Te^{121m}$	121	52			I.T.(89%)	0.29							Te k x-ray	19%
	121	52			E.C.(11%)								0.2122 ± 0.0001	83%
	121	52											1.1021 ± 0.0001	2.5%
$_{52}Te^{121}$	121	52		120.904947	16.8 d.	E.C.	1.05				1/2+		Sb k x-ray	40%
	121	52											0.4705 ± 0.0001	1%
	121	52											0.5076 ± 0.0001	18%
	121	52											0.5731 ± 0.0001	80%
$_{52}Te^{122}$	122	52	2.60%	121.903054						3 b.	0+			
$_{52}Te^{123m}$	123	52			119.7 d.	I.T.	0.247				11/2-		Te k x-ray	26%
	123	52											0.0885 ± 0.0001	0.09%
	123	52											0.1590 ± 0.0001	84%
$_{52}Te^{123}$	123	52	0.903%	122.904271	$1.3x10^{13}$y.	E.C.	0.052			420 b.	1/2+	-0.7359		
$_{52}Te^{124}$	124	52	4.816%	123.902823						(0.04 + 7) b.	0+			
$_{52}Te^{125m}$	125	52			58 d.	I.T.	0.145				11/2-	+0.7	Te k x-ray	63%
	125	52											0.0355 ± 0.0001	6.7%
	125	52											0.1093 ± 0.0001	0.3%
$_{52}Te^{125}$	125	52	7.14%	124.904433						1.6 b.	1/2+	-0.8871		
$_{52}Te^{126}$	126	52	18.95%	125.903314						(0.13 + 0.9) b.	0+			
$_{52}Te^{127m}$	127	52			109 d.	I.T.(98%)	0.088				11/2-		Te k x-ray	19%
	127	52				β-(2%)	0.77						0.0883 ± 0.0001	0.08%
$_{52}Te^{127}$	127	52		126.905227	9.5 h.	β-	0.697	0.697			3/2+		0.3603 ± 0.0001	0.1%
$_{52}Te^{128}$	128	52	31.69%	127.904463						(0.015 + 0.20) b.	0+			
$_{52}Te^{129m}$	129	52			33.4 d.	I.T.(63%)	0.105	0.91			11/2-		Te k x-ray	15%
	129	52				β-(37%)		1.60					0.45984 ± 0.00004	4.5%(D)
	129	52											0.6959 ± 0.0001	3%
	129	52											0.7296 ± 0.0001	0.7%
$_{52}Te^{129}$	129	52		128.906594	69.5 m.	β-	1.499	0.99	9%		3/2-		0.0278 ± 0.0001	16%
	129	52						1.45	89%				0.45984 ± 00004	7%
	129	52											0.48728 ± 0.00003	1.3%
	129	52											0.80198 ± 0.00004	0.2%
	129	52											1.08378 ± 0.00004	0.5%
	129	52											1.11157 ± 0.00004	0.2%
$_{52}Te^{130}$	130	52	33.80%	129.906229	$2.4x10^{21}$y.					(0.02 + 0.22) b.	0+			
$_{52}Te^{131m}$	131	52			32.4 h.	β-(78%)	2.4	0.42			11/2-		0.0811 ± 0.0001	4%
	131	52				I.T.(22%)	0.18						0.1021 ± 0.0001	8%
	131	52											0.14973 ± 0.00002	20%(D)
	131	52											0.20066 ± 0.00004	7%
	131	52											0.24095 ± 0.00004	7.6%
	131	52											0.33431 ± 0.00003	9.5%
	131	52											0.66506 ± 0.00005	4%
	131	52											0.77369 ± 0.00004	38%

Isotope	A	Z	% Natural abundance	Atomic mass	Half-life	Decay mode	Decay energy (MeV)	Particle energy (MeV)	Particle intensity	Thermal neutron cross section	Spin (h/2π)	μ Nucl. mag. moment	Gamma-ray energy (MeV)	Gamma-ray intensity
	131	52											0.78249 ± 0.00004	7.8%
	131	52											0.79375 ± 0.00003	14%
	131	52											0.82278 ± 0.00004	6%
	131	52											0.85225 ± 0.00005	21%
	131	52											0.9099 ± 0.0001	3%
	131	52											1.059 ± 0.001	1.5%
	131	52											1.12551 ± 0.00006	11%
	131	52											1.20657 ± 0.00006	9.7%
	131	52											1.6457 ± 0.0001	1.2%
	131	52											1.8876 ± 0.0001	1.3%
	131	52											2.0011 ± 0.0001	2%
$_{52}Te^{131}$	131	52		131.908528	25.0 m.	β-	2.249	1.35	12%		3/2+		0.14973 ± 0.00002	69%
	131	52						1.69	22%				0.45327 ± 0.00004	18%
	131	52						2.14	60%				0.49269 ± 0.00005	5%
	131	52											0.60205 ± 0.00005	4%
	131	52											0.65426 ± 0.00005	1.5%
	131	52											0.94857 ± 0.00004	2.3%
	131	52											0.99719 ± 0.00004	3.3%
	131	52											1.14698 ± 0.00006	5%
$_{52}Te^{132}$	132	52		131.908517	78.2 h.	β-	2.249	1.35	12%		3/2+		0.049725 ± 0.00005	14%
	132	52						1.69	22%				0.11198 ± 0.00005	1.8%
	132	52											0.11645 ± 0.00004	1.9%
	132	52											0.22830 ± 0.00002	88%
$_{52}Te^{133m}$	133	52			55.4 m.	β-(82%)		2.4	30%		11/2-		Te k x-ray	
	133	52				I.T.(18%)	0.334						0.0949 ± 0.0002	3%
	133	52											0.1689 ± 0.0001	5%
	133	52											0.2134 ± 0.0001	2%
	133	52											0.3121 ± 0.0003	17%(D)
	133	52											0.3341 ± 0.0001	7%
	133	52											0.4290 ± 0.0001	2%
$_{52}Te^{133}$	133	52		132.910910	12.5 m.	β-	3.0	2.3	25%				0.3121 ± 0.0003	73%
	133	52											0.4079 ± 0.0003	31%
	133	52											0.7201 ± 0.0004	7%
	133	52											0.7874 ± 0.0005	6%
	133	52											0.8445 ± 0.0006	3%
	133	52											0.9311 ± 0.0005	5%
	133	52											1.0003 ± 0.0008	6%
	133	52											1.0210 ± 0.0006	3%
	133	52											1.0618 ± 0.0008	1%
	133	52											1.2520 ± 0.0007	1%
	133	52											1.3334 ± 0.0005	10%
	133	52											1.7175 ± 0.0006	3%
	133	52											1.8818 ± 0.0007	1.5%
$_{52}Te^{134}$	134	52		133.911520	42 m.	β-	1.6	0.6			0+		0.0794 ± 0.0001	21%
	134	52						0.7					0.1809 ± 0.0001	18%
	134	52											0.2012 ± 0.0001	9%
	134	52											0.2105 ± 0.0001	22%
	134	52											0.2780 ± 0.0001	21%
	134	52											0.4351 ± 0.0001	19%
	134	52											0.4610 ± 0.0001	10%
	134	52											0.4646 ± 0.0001	4.7%
	134	52											0.5660 ± 0.0001	18%
	134	52											0.6363 ± 0.0002	2%
	134	52											0.6658 ± 0.0001	1%
	134	52											0.7130 ± 0.0001	4.7%
	134	52											0.7426 ± 0.0001	15%
	134	52											0.7672 ± 0.0001	29%
	134	52											0.8441 ± 0.0001	1%
	134	52											0.9255 ± 0.0001	1.5%
$_{52}Te^{135}$	135	52		134.916420	19.2 s.	β-	6.0	5.4					0.267	8 +
	135	52											0.603	100
	135	52											0.870	23
$_{52}Te^{136}$	136	52		135.920120	18 s.	β-	5.1	2.5			0+		0.0873 ± 0.0002	13%
	136	52											0.1350 ± 0.0003	3%
	136	52											0.3326 ± 0.0002	21%
	136	52											0.3561 ± 0.0004	2%
	136	52											0.4913 ± 0.0003	3%

Isotope	A	Z	% Natural abundance	Atomic mass	Half-life	Decay mode	Decay energy (MeV)	Particle energy (MeV)	Particle intensity	Thermal neutron cross section	Spin (h/2π)	μ Nucl. mag. moment	Gamma-ray energy (MeV)	Gamma-ray intensity
	136	52											0.5423 ± 0.0003	2%
	136	52											0.5786 ± 0.0002	20%
	136	52											0.6307 ± 0.0002	12%
	136	52											0.7382 ± 0.0002	6%
	136	52											1.3412 ± 0.0005	2%
	136	52											1.5669 ± 0.0005	1%
	136	52											2.0779 ± 0.0003	25%
	136	52											2.4969 ± 0.0005	5%
	136	52											2.5694 ± 0.0003	17%
	136	52											2.6048 ± 0.0006	1%
	136	52											2.6560 ± 0.0006	1%
	136	52											2.8040 ± 0.0006	2%
	136	52											3.0495 ± 0.0006	2%
	136	52											3.2351 ± 0.0004	17%
$_{52}Te^{137}$	137	52		136.925410	4 s.	β-(98%)							0.2436 ± 0.0003	
	137	52				n(2%)								
I		53								6.2 b.				
$_{53}I^{110}$	110	53		109.935060	0.65 s.	β+, EC (83%)	≈ 12						ann.rad.	
	110	53				α(17%)								
	110	53				p (11%)								
$_{53}I^{111}$	111	53		110.930250	7.5 s.	β+,E.C.	≈ 8.5						ann.rad.	
	111	53											0.2665 ± 0.0006	
	112	53											0.3215 ± 0.0006	
	112	53											0.3412 ± 0.0006	
$_{53}I^{112}$	112	53		111.927940	3.4 s.	β+,E.C.	≈ 10						ann.rad.	
	112	53											0.6889 ± 0.0006	
	112	53											0.7869 ± 0.0006	
$_{53}I^{113}$	113	53		112.923650	≈ 5.9 s.	β+,E.C.	≈ 7.2						ann.rad.	
	113	53											0.0550 ± 0.0002	32 +
	113	53											0.1600 ± 0.0002	14
	113	53											0.2165 ± 0.0002	7
	113	53											0.3204 ± 0.0002	33
	113	53											0.3515 ± 0.0002	43
	113	53											0.4061 ± 0.0002	8
	113	53											0.4625 ± 0.0002	100
	113	53											0.5230 ± 0.0005	7
	113	53											0.5674 ± 0.0002	36
	113	53											0.6086 ± 0.0005	6
	113	53											0.6224 ± 0.0002	74
	113	53											0.6280 ± 0.0002	13
	113	53											0.6514 ± 0.0005	3
	113	53											0.6902 ± 0.0005	8
	113	53											0.6962 ± 0.0005	3
	113	53											0.7740 ± 0.0005	8
	113	53											0.7982 ± 0.0002	12
	113	53											0.8021 ± 0.0005	8
	113	53											0.8960 ± 0.0005	10
	113	53											0.9291 ± 0.0003	8
	113	53											1.1610 ± 0.0005	9
	113	53											1.4224 ± 0.0003	11
$_{53}I^{114}$	114	53		113.921790	2.1 s.	β+,E.C.	≈ 8.9						ann.rad.	
	114	53											0.6826 ± 0.0006	
	114	53											0.7088 ± 0.0006	
$_{53}I^{115}$	115	53		114.918090	28 s.	β+,E.C.	≈ 6.0						ann.rad.	
	115	53											0.275	
	;15	53											0.284	
	115	53											0.460	
	115	53											0.709	
$_{53}I^{116}$	116	53		115.916780	2.9 s.	β+(97%)	7.8	6.7			1+		ann.rad.	
	116	53				E.C.(3%)							0.5402 ± 0.0004	1.2%
	116	53											0.6789 ± 0.0003	8.3%
$_{53}I^{117}$	117	53		116.913460	2.3 m.	β+,E.C.	4.3	3.3			(5/2+)		ann.rad.	
	117	53											0.2744	27 +
	117	53											0.3032	2
	117	53											0.3259	100
	117	53											0.4972	1
	117	53											0.6831	2
	117	53											0.8373	2
$_{53}I^{118m}$	118	53			≈ 8.5 m.	β+,E.C.		4.9					ann.rad.	
	118	53				I.T.							0.104	
	118	53											0.5998 ± 0.0003	100 +
	118	53											0.6052 ± 0.0004	100
	118	53											0.6138 ± 0.0007	54
$_{53}I^{118}$	118	53		117.912780	14.3 m.	β+,E.C.	≈ 6.4						ann.rad.	
	118	53											0.3524 ± 0.0005	3%
	118	53											0.5448 ± 0.0004	12%
	118	53											0.5518 ± 0.0006	2%
	118	53											0.6052 ± 0.0004	95%
	118	53											0.7407 ± 0.0005	2%
	118	53											1.1499 ± 0.0005	5%
	118	53											1.2570 ± 0.0006	4%
	118	53											1.3384 ± 0.0005	12%
$_{53}I^{119}$	119	53		118.910030	19.2 m.	β+(54%)	3.4	2.1			(5/2+)		ann.rad.	

Isotope	A	Z	% Natural abundance	Atomic mass	Half-life	Decay mode	Decay energy (MeV)	Particle energy (MeV)	Particle intensity	Thermal neutron cross section	Spin (h/2π)	μ Nucl. mag. moment	Gamma-ray energy (MeV)	Gamma-ray intensity
	119	53				E.C.(46%)							Te k x-ray	
	119	53											0.2575 ± 0.0001	90%
	119	53											0.3206 ± 0.0001	2%
	119	53											0.5570 ± 0002	2%
	119	53											0.6356 ± 0.0001	3%
	119	53											0.7062 ± 0.0002	1.3%
	119	53											1.0034 ± 0.0002	0.5%
$_{53}I^{120m}$	120	53			53 m.	β+(80%)							ann.rad.	
	120	53				E.C.(20%)							Te k x-ray	
	120	53											0.4257 ± 0.0005	3%
	120	53											0.5604 ± 0.0003	100%
	120	53											0.6011 ± 0.0003	8%
	120	53											0.6147 ± 0.0003	67%
	120	53											0.6545 ± 0.0005	2%
	120	53											0.7039 ± 0.0005	2%
	120	53											0.7632 ± 0.0004	3%
	120	53											0.8818 ± 0.0005	2%
	120	53											0.9213 ± 0.0004	4%
	120	53											1.0315 ± 0.0006	1%
	120	53											1.0399 ± 0.0005	6%
	120	53											1.0592 ± 0.0005	5%
	120	53											1.1586 ± 0.0006	3%
	120	53											1.1973 ± 0.0006	2%
	120	53											1.2613 ± 0.0007	2%
	120	53											1.3346 ± 0.0007	4%
	120	53											1.3459 ± 0.0004	19%
	120	53											1.3635 ± 0.0007	4%
	120	53											1.4021 ± 0.0007	4%
	120	53											1.4050 ± 0.0005	9%
	120	53											1.7614 ± 0.001	4%
	120	53											1.7758 ± 0.001	5%
	120	53											1.8083 ± 0.001	4%
	120	53											2.4032 ± 0.001	7%
	120	53											2.4628 ± 0.0015	4%
	120	53											2.6025 ± 0.002	3%
	120	53											2.8110 ± 0.0015	4%
	120	53											2.8643 ± 0.002	2%
	120	53											2.9329 ± 0.0015	4%
	120	53											3.1051 ± 0.0015	2%
$_{53}I^{120}$	120	53		119.909840	1.35 h.	β+(81%)	5.4				2-		ann.rad.	
	120	53				E.C.(19%)							Tek x-ray	8%
	120	53											0.5427 ± 0.0003	1%
	120	53											0.5604 ± 0.0003	73%
	120	53											0.6011 ± 0.0003	6%
	120	53											0.6411 ± 0.0003	9%
	120	53											1.2016 ± 0.0005	2%
	120	53											1.5230 ± 0.0004	11%
	120	53											1.5347 ± 0.0005	2%
	120	53											2.1880 ± 0.001	1.4%
	120	53											2.4548 ± 0.0005	2%
	120	53											2.4918 ± 0.001	1%
	120	53											2.5644 ± 0.001	2%
	120	53											2.9329 ± 0.0015	0.7%
	120	53											(0.43 - 3.1)	
$_{53}I^{121}$	121	53		120.907394	2.12 h.	β+(13%)	2.28	1.2			5/2+		ann.rad.	
	121	53				E.C.(87%)							Te k x-ray	37%
	121	53											0.2122 ± 0.0001	85%
	121	53											0.5321 ± 0.0001	5.4%
	121	53											0.5988 ± 0.0001	1.4%
	121	53											(0.14 - 1.1)weak	
$_{53}I^{122}$	122	53		121.907595	3.6 m.	β+	4.23	3.1			1+		ann.rad.	
	122	53				E.C.							Te k x-ray	10%
	122	53											0.5641 ± 0.0001	18%
	122	53											0.6928 ± 0.0001	1.3%
	122	53											0.7933 ± 0.0001	1.3%
	122	53											1.7469 ± 0.0001	0.33%
	122	53											2.1923 ± 0.0001	0.26%
$_{53}I^{123}$	123	53		122.905594	13.1 h.	E.C.	1.23				5/2+		Te k x-ray	46%
	123	53											0.1590 ± 0.0001	83%
	123	53											0.4400 ± 0.0001	0.4%
	123	53											0.5290 ± 0.0001	1.4%
	123	53											0.5385 ± 0.0001	0.4%
$_{53}I^{124}$	124	53		123.906207	4.17 d.	β+(23%)	3.16	1.5			2-		ann.rad.	
	124	53				E.C.(77%)		2.1					Te k x-ray	31%
	124	53											0.6027 ± 0.0001	61%
	124	53											0.7228 ± 0.0001	10%
	124	53											1.3255 ± 0.0001	1.4%
	124	53											1.3760 ± 0.0001	1.7%
	124	53											1.5095 ± 0.0001	3%
	124	53											1.6910 ± 0.0001	10%
	124	53											2.0910 ± 0.0001	0.6%
	124	53											2.2833 ± 0.0001	0.7%
	124	53											2.7469 ± 0.0001	0.5%
$_{53}I^{125}$	125	53		124.904620	59.9 d.	E.C.	0.178			900 b.	5/2+	+3.0	Te k x-ray	74%
	125	53											0.0355 ± 0.0001	6.7%

TABLE OF THE ISOTOPES (Continued)

Isotope	A	Z	% Natural abundance	Atomic mass	Half-life	Decay mode	Decay energy (MeV)	Particle energy (MeV)	Particle intensity	Thermal neutron cross section	Spin (h/2π)	μ Nucl. mag. moment	Gamma-ray energy (MeV)	Gamma-ray intensity
$_{53}I^{126}$	126	53		125.905624	13.0 d.	E.C.				9 x 10^3 b.	2-		ann.rad.	
	126	53				β+	2.16	1.13					Te k x-ray	22%
	126	53				β-	1.25	0.86					0.3887 ± 0.0002	32%
	126	53						1,25					0.6622 ± 0.0002	31%
	126	53											0.7538 ± 0.0002	3.9%
	126	53											0.8799 ± 0.0002	0.7%
	126	53											1.4201 ± 0.0004	0.3%
$_{53}I^{127}$	127	53		126.904473						6.2 b.	5/2+	+2.808		
$_{53}I^{128}$	128	53		127.905810	25.00 m.	β-	2.128	2.12					Te k x-ray	2%
	128	53				E.C.	1.257						0.44287 ± 0.00002	16%
	128	53											0.52658 ± 0.00003	1.5%
	128	53											0.74321 ± 0.00004	0.15%
	128	53											0.96943 ± 0.00007	0.38%
$_{53}I^{129}$	129	53		128.904986	1.6x10^7y.	β-	0.193	0.15		(20 + 10) b.	7/2+	+2.617	Xe k x-ray	
	129	53											0.0396 ± 0.0001	7.5%
$_{53}I^{130m}$	130	53		129.906713	9.0 m.	I.T.(83%)	0.048				2+		I k x-ray	14%
	130	53				β-(17%)							0.5361 ± 0.0001	17%
	130	53											0.5861 ± 0.0001	1%
	130	53											1.6141 ± 0.0001	0.5%
$_{53}I^{130}$	130	53		129.906713	12.36 h.	β-	2.98	1.04		18 b.	5+		0.4180 ± 0.0001	34%
	130	53											0.5361 ± 0.0001	99%
	130	53											0.5391 ± 0.0001	1.4%
	130	53											0.5861 ± 0.0001	1.7%
	130	53											0.6685 ± 0.0001	96%
	130	53											0.7395 ± 0.0001	82%
	130	53											1.1575 ± 0.0001	11%
	130	53											1.2721 ± 0.0001	0.8%
$_{53}I^{131}$	131	53		130.906114	8.040 d.	β-	0.971	0.606		80 b.	7/2+	+2.74	0.08017 ± 0.0005	2.6%
	131	53											0.17725 ± 0.0009	0.27%
	131	53											0.27248 ± 0.00008	0.9%
	131	53											0.28431 ± 0.00003	6%
	131	53											0.32574 ± 0.00005	0.25%
	131	53											0.36446 ± 0.00003	81%
	131	53											0.50300 ± 0.00005	0.4%
	131	53											0.63699 ± 0.00005	7.3%
	131	53											0.64266 ± 0.00005	0.2%
	131	53											0.72288 ± 0.00005	1.3%
$_{53}I^{132m}$	132	53			83 m.	β-(14%)	3.58	0.80			4+	3.08	I k x-ray	13%
	132	53				I.T.(86%)		1.03					0.0980 ± 0.001	4%
	132	53						1.2					0.5059 ± 0.0001	5%
	132	53						1.6					0.52264 ± 0.00003	16%
	132	53						2.16					0.63019 ± 0.00003	14%
	132	53											0.6506 ± 0.0002	2.7%
	132	53											0.66768 ± 0.00003	99%
	132	53											0.6698 ± 0.0003	5%
	132	53											0.6716 ± 0.0004	5.2%
	132	53											0.72695 ± 0.00003	6.5%
	132	53											0.77260 ± 0.00003	76%
	132	53											0.8098 ± 0.0001	3%
	132	53											0.81228 ± 0.00001	5.6%
	132	53											0.87697 ± 0.00001	1.1%
	132	53											0.95457 ± 0.00003	18%
	132	53											1.13602 ± 0.00004	3%
	132	53											1.1436 ± 0.0001	1.4%
	132	53											1.1732 ± 0.0001	1.1%
	132	53											1.2953 ± 0.00001	2%
	132	53											1.37201 ± 0.00005	2.5%
	132	53											1.39895 ± 0.00005	7.1%
	132	53											1.44254 ± 0.00005	1.4%
	132	53											1.92096 ± 0.00008	1.2%

Isotope	A	Z	% Natural abundance	Atomic mass	Half-life	Decay mode	Decay energy (MeV)	Particle energy (MeV)	Particle intensity	Thermal neutron cross section	Spin (h/2π)	μ Nucl. mag. moment	Gamma-ray energy (MeV)	Gamma-ray intensity
	132	53											2.00225 ± 0.00007	1.1%
	132	53											2.0866 ± 0.0001	0.2%
	132	53											2.1724 ± 0.0001	0.2%
$_{53}I^{133m}$	133	53			9 s.	I.T.	1.63				19/2-		I kx-ray	35%
	133	53											0.0730 ± 0.001	4%
	133	53											0.6474 ± 0.0001	100%
	133	53											0.9126 ± 0.0001	100%
$_{53}I^{133}$	133	53		132.907780	20.8 h.	β-	1.76	1.23	85%		7/2+	+2.84	0.51056 ± 0.00002	1.8%
	133	53											0.52989 ± 0.00002	86%
	133	53											0.61794 ± 0.00003	0.5%
	133	53											0.68031 ± 0.00004	0.65%
	133	53											0.70661 ± 0.00002	1.5%
	133	53											0.76836 ± 0.00004	0.46%
	133	53											0.82061 ± 0.00003	0.15%
	133	53											0.85636 ± 0.00003	1.2%
	133	53											0.87537 ± 0.00002	45%
	133	53											0.909	0.2%
	133	53											1.05231 ± 0.00005	0.56%
	133	53											1.06017 ± 0.00005	0.14%
	133	53											1.23653 ± 0.00003	1.5%
	133	53											1.29833 ± 0.00003	2.3%
	133	53											1.35054 ± 0.00008	0.15%
$_{53}I^{134m}$	134	53			3.5 m.	I.T.(98%)	0.316				8-		I k x-ray	39%
	134	53				β-(2%)							0.0444 ± 0.0002	10%
	134	53											0.2719 ± 0.0003	79%
$_{53}I^{134}$	134	53		133.909850	52.5 m.	β-	4.2	1.25			4+		0.1354 ± 0.0001	4%
	134	53											0.2355 ± 0.0001	2%
	134	53											0.40545 ± 0.00002	7%
	134	53											0.4333 ± 0.0001	4%
	134	53											0.5144 ± 0.0001	2%
	134	53											0.5408 ± 0.0001	7.6%
	134	53											0.5954 ± 0.0001	11%
	134	53											0.6218 ± 0.0001	11%
	134	53											0.6280 ± 0.0001	2%
	134	53											0.67734 ± 0.00003	7.8%
	134	53											0.7307 ± 0.0001	1.8%
	134	53											0.76667 ± 0.00003	4.1%
	134	53											0.84702 ± 0.00003	95%
	134	53											0.8573 ± 0.0001	7%
	134	53											0.88409 ± 0.00002	65%
	134	53											0.9479 ± 0.0001	4%
	134	53											0.9747 ± 0.0001	4.8%
	134	53											1.0403 ± 0.0001	1.9%
	134	53											1.07255 ± 0.00003	15%
	134	53											1.13616 ± 0.00004	9.2%
	134	53											1.4552 ± 0.0001	2.3%
	134	53											1.61380 ± 0.00004	4.3%
	134	53											1.71419 ± 0.00005	2.7%
	134	53											1.80684 ± 0.00004	5.5%
$_{53}I^{135}$	135	53		134.910023	6.585 h.	β-	2.71	0.9			7/2+		0.2884 ± 0.0001	3%
	135	53						1.3					0.41768 ± 0.00003	3.5%
	135	53											0.52658 ± 0.00002	14%(D)
	135	53											0.54658 ± 0.00002	7.1%
	135	53											0.83686 ± 0.00002	6.7%
	135	53											0.97233 ± 0.00004	1.2%

Isotope	A	Z	% Natural abundance	Atomic mass	Half-life	Decay mode	Decay energy (MeV)	Particle energy (MeV)	Particle intensity	Thermal neutron cross section	Spin (h/2π)	μ Nucl. mag. moment	Gamma-ray energy (MeV)	Gamma-ray intensity
	135	53											1.03877 ± 0.00004	7.9%
	135	53											1.10160 ± 0.00005	1.6%
	135	53											1.12402 ± 0.00003	3.6%
	135	53											1.13156 ± 0.00003	22.5%
	135	53											1.26046 ± 0.00003	28.6%
	135	53											1.45766 ± 0.00005	8.6%
	135	53											1.67817 ± 0.00004	9.5%
	135	53											1.70658 ± 0.00004	4.1%
	135	53											1.79133 ± 0.00004	7.7%
	135	53											1.83080 ± 0.00005	0.60%
	135	53											1.9274 ± 0.0001	0.30%
	135	53											2.04610 ± 0.00006	0.90%
	135	53											2.25518 ± 0.00008	0.60%
	135	53											2.40877 ± 0.00005	0.95%
$_{53}I^{136m}$	136	53			45 s.	β-		4.7			6-		0.1973 ± 0.0001	78%
	136	53						5.2					0.3468 ± 0.0001	3%
	136	53											0.3701 ± 0.0001	17%
	136	53											0.3814 ± 0.0001	100%
	136	53											0.7500 ± 0.0001	6%
	136	53											0.8126 ± 0.0001	3%
	136	53											0.9141 ± 0.0002	3%
	136	53											1.3130 ± 0.0001	100%
	136	53											(0.16 - 2.36)	
$_{53}I^{136}$	136	53		135.914650	83.6 s.	β-	6.9	4.7			2-		0.3447 ± 0.0001	2.4%
	136	53											0.9765 ± 0.0002	2.7%
	136	53											1.2468 ± 0.0001	2.3%
	136	53											1.3130 ± 0.0001	67%
	136	53											1.3211 ± 0.0001	25%
	136	53											1.5364 ± 0.0001	1%
	136	53											1.9622 ± 0.0003	2.3%
	136	53											2.2896 ± 0.0002	10%
	136	53											2.4146 ± 0.0002	6.9%
	136	53											2.6342 ± 0.0002	6.8%
	136	53											2.8689 ± 0.0002	4%
	136	53											2.9563 ± 0.0003	0.7%
	136	53											3.1411 ± 0.0003	0.7%
	136	53											3.2118 ± 0.0003	0.5%
	136	53											4.2695 ± 0.0002	0.4%
	136	53											(0.3 - 6.1)	
$_{53}I^{137}$	137	53		136.917870	24.5 s.	β-	5.9	5.0			(7/2+)		0.6010 ± 0.0001	5%
	137	53											1.2180 ± 0.0001	13%
	137	53											1.2201 ± 0.00022	3%
	137	53											1.3026 ± 0.0001	4.4%
	137	53											1.5122 ± 0.0001	1.2%
	137	53											1.5343 ± 0.0001	3.2%
	137	53											1.7661 ± 0.0001	1.2%
	137	53											1.8730 ± 0.0001	1.5%
	137	53											2.0298 ± 0.0001	1.7%
	137	53											2.6297 ± 0.0001	0.8%
	137	53											3.1944 ± 0.0002	0.75%
	137	53											3.7956 ± 0.0002	0.85%
	137	53											3.9962 ± 0.0003	0.5%
	137	53											(0.25 - 4.4)weak	
$_{53}I^{138}$	138	53		137.922370	6.4 s.	β-	7.8	6.9					0.4836 ± 0.0001	5%
	138	53						7.4					0.5888 ± 0.0001	77%
	138	53											0.8307 ± 0.0001	2.2%
	138	53											0.8702 ± 0.0002	4%
	138	53											0.8752 ± 0.0001	13%
	138	53											1.2775 ± 0.0001	3.3%
	138	53											1.3143 ± 0.0001	1.4%
	138	53											1.4269 ± 0.0003	1.6%
	138	53											1.6733 ± 0.0001	1.7%
	138	53											1.8093 ± 0.0002	2.8%
	138	53											2.2622 ± 0.0001	5.3%
	138	53											2.5724 ± 0.0002	1.6%
	138	53											2.8356 ± 0.0002	1.7%
	138	53											3.3103 ± 0.0002	1.5%
	138	53											3.4963 ± 0.0002	1.0%
	138	53											4.1820 ± 0.0002	0.9%
	138	53											4.3189 ± 0.0002	0.6%
	138	53											(0.4 - 5.3)	
$_{53}I^{139}$	139	53		138.926050	2.3 s.	β-	6.8						0.192	

Isotope	A	Z	% Natural abundance	Atomic mass	Half-life	Decay mode	Decay energy (MeV)	Particle energy (MeV)	Particle intensity	Thermal neutron cross section	Spin (h/2π)	μ Nucl. mag. moment	Gamma-ray energy (MeV)	Gamma-ray intensity
	139	53				n							0.198	
	139	53											0.273	
	139	53											0.382	
	139	53											0.386	
	139	53											0.468	
	139	53											0.683	
	139	53											1.313	
$_{53}I^{140}$	140	53			0.86 s.	β-							0.372	
	140	53				n							0.377	
	140	53											0.457	
Xe		54		131.29						25 b.				
$_{54}Xe^{114}$	114	54		113.928170	10.3 s.	β+,E.C.	≈ 6.0				0+		ann.rad.	
	114	54											0.1031 ± 0.0002	
	114	54											0.1616 ± 0.0002	
	114	54											0.3085 ± 0.0002	
	114	54											0.6826 ± 0.0006	(D)
	114	54											0.7088 ± 0.0006	(D)
$_{54}Xe^{115}$	115	54		114.926280	18 s.	β+,E.C.	≈ 7.6						ann.rad.	
$_{54}Xe^{116}$	116	54		115.921610	57 s.	β+,E.C.	≈ 4.5	3.3			0+		ann.rad.	
	116	54											0.1042 ± 0.0002	100 +
	116	54											0.1916 ± 0.0002	38
	116	54											0.2264 ± 0.0002	29
	116	54											0.2477 ± 0.0003	40
	116	54											0.3000 ± 0.0004	12
	116	54											0.3107 ± 0.0004	42
	116	54											0.4127 ± 0.0002	36
	116	54											0.9230 ± 0.001	25
$_{54}Xe^{117}$	117	54		116.920250	61 s.	β+,E.C.	≈ 6.4						ann.rad.	
	117	54											0.0737	15 +
	117	54											0.0949	12
	117	54											0.1121	5
	117	54											0.1171	41
	117	54											0.1554	17
	117	54											0.1607	34
	117	54											0.2033	4
	117	54											0.2214	100
	117	54											0.2570	17
	117	54											0.	2947
	117	54											0.3034	23
	117	54											0.3158	26
	117	54											0.3532	20
	117	54											0.4392	20
	117	54											0.5190	55
	117	54											0.6097	4.5
	117	54											0.6389	50
	117	54											0.6613	56
	117	54											1.5232	24
$_{54}Xe^{118}$	118	54		117.916210	4 m.	β+,E.C.	≈ 3.2	2.7			0+		ann.rad.	
	118	54											0.0535 ± 0.0005	100 +
	118	54											0.0600 ± 0.0005	90
	118	54											0.1199 ± 0.0005	76
	118	54											0.1505 ± 0.0005	44
	118	54											0.2740 ± 0.0005	30
$_{54}Xe^{119}$	119	54		118.915390	5.8 m.	β+,E.C.	5.0	3.5					0.0873 ± 0.0006	43 +
	119	54											0.0910 ± 0.0006	16
	119	54											0.0960 ± 0.0006	38
	119	54											0.1000 ± 0.0006	95
	119	54											0.1417 ± 0.0006	11
	119	54											0.1466 ± 0.0006	8
	119	54											0.2082 ± 0.0006	60
	119	54											0.2318 ± 0.0006	100
	119	54											0.2357 ± 0.0006	10
	119	54											0.2948 ± 0.0006	8
	119	54											0.3080 ± 0.0006	11
	119	54											0.3205 ± 0.0006	12
	119	54											0.4377 ± 0.0006	13
	119	54											0.4615 ± 0.0006	97
	119	54											0.5365 ± 0.0006	9
	119	54											0.6930 ± 0.0006	12
$_{54}Xe^{120}$	120	54		119.911940	40 m.	β+, EC (97%)	2.0				0+		I k x-ray	55%
	120	54				β+(3%)							0.0251 ± 0.0002	30%
	120	54											0.0726 ± 0.0002	9%
	120	54											0.0772 ± 0.0002	4%
	120	54											0.1760 ± 0.0003	5%
	120	54											0.1781 ± 0.0002	7%
	120	54											0.2956 ± 0.0002	1.1%
	120	54											0.3359 ± 0.0002	1%
	120	54											0.4243 ± 0.0003	1.2%
	120	54											0.4492 ± 0.0003	1.7%
	120	54											0.5294 ± 0.0003	1.4%
	120	54											0.5556 ± 0.0003	1.5%
	120	54											0.5904 ± 0.0003	1.6%

Isotope	A	Z	% Natural abundance	Atomic mass	Half-life	Decay mode	Decay energy (MeV)	Particle energy (MeV)	Particle intensity	Thermal neutron cross section	Spin (h/2π)	μ Nucl. mag. moment	Gamma-ray energy (MeV)	Gamma-ray intensity
	120	54											0.6311 ± 0.0003	1.0%
	120	54											0.6789 ± 0.0002	1.6%
	120	54											0.7484 ± 0.0004	1.1%
	120	54											0.7533 ± 0.0003	1.4%
	120	54											0.7625 ± 0.0003	4.5%
	120	54											0.9655 ± 0.0003	1.2%
	120	54											(0.1 - 1.03)many	
$_{54}Xe^{121}$	121	54		120.911450	39 m.	β+(44%)	3.8	2.8			5/2+		ann.rad.	
	121	54				E.C.(56%)							I k x-ray	33%
	121	54											0.0801 ± 0.0002	3%
	121	54											0.0957 ± 0.0002	6%
	121	54											0.1328 ± 0.0002	15%
	121	54											0.1758 ± 0.0003	6%
	121	54											0.2527 ± 0.0002	18%
	121	54											0.3007 ± 0.0002	4%
	121	54											0.3105 ± 0.0002	8%
	121	54											0.4452 ± 0.0002	11%
	121	54											0.5291 ± 0.0004	2%
	121	54											0.6497 ± 0.0003	2%
	121	54											0.8425 ± 0.0003	1%
	121	54											0.9310 ± 0.0003	2%
	121	54											0.9580 ± 0.0003	1.5%
	121	54											1.0356 ± 0.0003	1.5%
	121	54											1.1864 ± 0.0005	0.9%
	121	54											1.5408 ± 0.0004	0.8%
	121	54											1.6315 ± 0.0004	0.8%
	121	54											2.5447 ± 0.0005	1.5%
	121	54											2.6217 ± 0.0005	1.1%
	121	54											2.6434 ± 0.0005	3.1%
	121	54											2.7977 ± 0.0004	1.2%
	121	54											(0.1 - 3.1)many	
$_{54}Xe^{122}$	122	54		121.908170	20.0 h.	E.C.	0.7				0+		I k x-ray	41%
	122	54											0.1486 ± 0.0001	2.6%
	122	54											0.3501 ± 0.0001	7.7%
	122	54											0.4166 ± 01.0001	1.9%
$_{54}Xe^{123}$	123	54		122.908469	2.0 h.	β+(23%)	2.68	1.51			½+		ann.rad.	
	123	54				E.C.(77%)							I k x-ray	37%
	123	54											0.1489 ± 0.0002	49%
	123	54											0.1781 ± 0.0002	15%
	123	54											0.3302 ± 0.0002	8%
	123	54											0.8996 ± 0.0004	2.4%
	123	54											1.0934 ± 0.0003	2.8%
	123	54											1.1131 ± 0.0003	1.6%
	123	54											1.8073 ± 0.0003	1.2%
	123	54											(0.1 - 2.1)weak	
$_{54}Xe^{124}$	124	54	0.10%	123.905894						(28 + 140)b.				
$_{54}Xe^{125m}$	125	54			57 s.	I.T.	0.253				(9/2-)		Xe k x-ray	33%
	125	54											0.1111 ± 0.001	62%
	125	54											0.141 ± 0.001	20%
$_{54}Xe^{125}$	125	54		124.906397	17.1 h.	E.C.	1.68	0.47			½+		I k x-ray	54%
	125	54											0.0550 ± 0.0001	6%
	125	54											0.1884 ± 0.0001	55%
	125	54											0.2434 ± 0.0001	29%
	125	54											0.4538 ± 0.0001	4%
	125	54											0.8465 ± 0.0004	1%
	125	54											1.1810 ± 0.0004	0.6%
$_{54}Xe^{126}$	126	54	0.09%	125.904281						(0.4 + 3) b.	0+			
$_{54}Xe^{127m}$	127	54			69 s.	I.T.	0.297				(9/2-)		Xe k x-ray	29%
	127	54											0.1246 ± 0.0003	69%
	127	54											0.1725 ± 0.0003	38%
$_{54}Xe^{127}$	127	54		126.905182	36.341 d.	E.C.	0.66				½+		I k x-ray	46%
	127	54											0.0576 ± 0.0001	1.2%
	127	54											0.1453 ± 0.0001	4.3%
	127	54											0.1721 ± 0.0001	25%
	127	54											0.2029 ± 0.0001	68%
	127	54											0.3750 ± 0.0001	17%
$_{54}Xe^{128}$	128	54	1.91%	127.903531						(0.5 + 4) b.	0+			
$_{54}Xe^{129m}$	129	54			8.88 d.	I.T.	0.236				11/2-		Xe k x-ray	67%
	129	54											0.0396 ± 0.0001	7.5%
	129	54											0.1966 ± 0.0001	4.6%
$_{54}Xe^{129}$	129	54	26.4%	128.904780						21 b.	½+	-0.7768		
$_{54}Xe^{130}$	130	54	4.1%	129.903509						(0.4 + 5) b.	0+			
$_{54}Xe^{131m}$	131	54			11.92 d.	I.T.	0.164				11/2-	+0.6908	Xe k x-ray	28%
	131	54											0.16398 ± 0.00005	2%
$_{54}Xe^{131}$	131	54	21.2%	130.905072						90 b.	3/2+			
$_{54}Xe^{132}$	132	54	26.9%	131.904144						(0.05 + 0.4) b.	0+			
$_{54}Xe^{133m}$	133	54			2.19 d.	I.T.	0.233				11/2-		Xe k x-ray	30%
	133	54											0.23325 ± 0.00005	10%
$_{54}Xe^{133}$	133	54		132.905888	5.25 d.	β-	0.427	0.346	99%		3/2+		Cs k x-ray	24%
	133	54											0.080998 ± 0.0001	36%
	133	54											0.1606 ± 0.0001	0.06%

Isotope	A	Z	% Natural abundance	Atomic mass	Half-life	Decay mode	Decay energy (MeV)	Particle energy (MeV)	Particle intensity	Thermal neutron cross section	Spin (h/2π)	μ Nucl. mag. moment	Gamma-ray energy (MeV)	Gamma-ray intensity
$_{54}Xe^{134}$	134	54	10.4%	133.905395						(0.003 + 0.26) b.	0+			
$_{54}Xe^{135m}$	135	54			15.3 m.	I.T.					11/2-		Xe k x-ray	7%
	135	54											0.52658 ± 0.00002	80%
$_{54}Xe^{135}$	135	54		134.907130	9.10 h.	β-	1.16	0.91		2.6x10⁶ b.	3/2+		0.24975 ± 0.00002	90%
	135	54											0.3584 ± 0.0001	0.2%
	135	54											0.4080 ± 0.0001	0.4%
	135	54											0.60807 ± 0.00004	2.9%
$_{54}Xe^{136}$	136	54	8.9%	135.907214						0.26 b.	0+			
$_{54}Xe^{137}$	137	54		136.911557	3.84 m.	β-	4	18	3.7		7/2-		0.45549 ± 0.00005	31%
	137	54						4.2					0.8489 ± 0.0001	0.6%
	137	54											0.9822 ± 0.0001	0.2%
	137	54											1.2732 ± 0.0001	0.2%
	137	54											1.7834 ± 0.0001	0.4%
	137	54											2.8498 ± 0.0001	0.2%
$_{54}Xe^{138}$	138	54		137.913980	14.1 m.	β-	2.7	0.8			0+		0.1538 ± 0.0001	6%
	138	54						2.4					0.2426 ± 0.0001	3.5%
	138	54											0.2583 ± 0.0001	31%
	138	54											0.3964 ± 0.0001	6%
	138	54											0.4014 ± 0.0001	2.2%
	138	54											0.4345 ± 0.0001	20%
	138	54											0.9171 ± 0.0001	0.9%
	138	54											1.1143 ± 0.0001	1.5%
	138	54											1.76826 ± 0.0001	16.7%
	138	54											1.8509 ± 0.0001	1.5%
	138	54											2.0047 ± 0.0001	5.4%
	138	54											2.0158 ± 0.0001	12.3%
	138	54											2.0792 ± 0.0001	1.4%
	138	54											2.2523 ± 0.0001	2.3%
$_{54}Xe^{139}$	139	54		138.918740	40.4 s.	β-	5.0	4.5			7/2-		0.1750 ± 0.0001	18%
	139	54						5.0					0.2186 ± 0.0001	52%
	139	54											0.2254 ± 0.0001	3%
	139	54											0.2898 ± 0.0001	8%
	139	54											0.2965 ± 0.0001	20%
	139	54											0.3935 ± 0.0001	6%
	139	54											0.4915 ± 0.0001	1.3%
	139	54											0.6128 ± 0.0001	5.1%
	139	54											0.7238 ± 0.0001	1.7%
	139	54											0.7324 ± 0.0001	1.6%
	139	54											0.7880 ± 0.0001	3.1%
	139	54											1.2065 ± 0.0001	0.6%
	139	54											1.2429 ± 0.0001	0.5%
	139	54											1.3449 ± 0.0001	1.1%
	139	54											1.5202 ± 0.0001	0.8%
	139	54											1.6703 ± 0.0001	1.0%
	139	54											1.8960 ± 0.0001	0.6%
	139	54											2.0859 ± 0.0001	0.6%
	139	54											2.3288 ± 0.0001	1.6%
	139	54											(0.1 - 3.37)weak	
$_{54}Xe^{140}$	140	54		139.921620	13.6 s.	β-	4.1				0+		0.0801 ± 0.0001	5%
	140	54											0.1125 ± 0.0001	4%
	140	54											0.1184 ± 0.0001	5%
	140	54											0.2120 ± 0.0001	2%
	140	54											0.2810 ± 0.0001	1.5%
	140	54											0.3900 ± 0.0001	1.5%
	140	54											0.4387 ± 0.0001	2.7%
	140	54											0.4618 ± 0.0001	1.6%
	140	54											0.5149 ± 0.0002	1.1%
	140	54											0.5189 ± 0.0002	1.1%
	140	54											0.5478 ± 0.0002	1.2%
	140	54											0.5573 ± 0.0001	5.3%
	140	54											0.6080 ± 0.0001	2.4%
	140	54											0.6220 ± 0.0001	8%
	140	54											0.6273 ± 0.0002	1%
	140	54											0.6392 ± 0.0002	1.4%
	140	54											0.6534 ± 0.0001	5.1%
	140	54											0.7741 ± 0.0001	4%
	140	54											0.8055 ± 0.0001	21%
	140	54											0.8798 ± 0.0001	3%
	140	54											0.9250 ± 0.0002	1.5%
	140	54											0.9890 ± 0.0001	3%
	140	54											1.1371 ± 0.0001	2.2%
	140	54											1.1767 ± 0.0002	1.2%
	140	54											1.2091 ± 0.0001	1.4%
	140	54											1.3091 ± 0.0001	6.7%
	140	54											1.4137 ± 0.0001	13%
	140	54											1.4276 ± 0.0001	1.2%
	140	54											1.8859 ± 0.0001	0.4%
	140	54											(0.04 - 2.3)weak	
$_{54}Xe^{141}$	141	54		140.926610	1.72 s.	β-	6.2	4.9					0.1187 ± 0.0001	12%
	141	54											0.1876 ± 0.0001	2.6%

Isotope	A	Z	% Natural abundance	Atomic mass	Half-life	Decay mode	Decay energy (MeV)	Particle energy (MeV)	Particle intensity	Thermal neutron cross section	Spin (h/2π)	μ Nucl. mag. moment	Gamma-ray energy (MeV)	Gamma-ray intensity
	141	54											0.4591 ± 0.0001	4.5%
	141	54											0.4678 ± 0.0001	2.9%
	141	54											0.5399 ± 0.0001	5.2%
	141	54											0.5566 ± 0.0002	4.7%
	141	54											0.7553 ± 0.0001	1.2%
	141	54											0.9095 ± 0.0001	22%
	141	54											0.9800 ± 0.0001	1%
	141	54											1.0281 ± 0.0001	1.7%
	141	54											1.0519 ± 0.0001	1.0%
	141	54											1.5570 ± 0.0002	2.9%
	141	54											(0.05 - 2.55)weak	
$_{54}Xe^{142}$	142	54		141.929630	1.2 s.	β-	5.0	3.7			0+		0.0338 ± 0.0001	25 +
	142	54						4.2					0.0729 ± 0.0001	27
	142	54											0.1575 ± 0.0001	17
	142	54											0.1651 ± 0.0001	22
	142	54											0.1917 ± 0.0001	36
	142	54											0.1974 ± 0.0002	7
	142	54											0.2038 ± 0.0001	92
	142	54											0.2117 ± 0.0004	4
	142	54											0.2191 ± 0.0003	5
	142	54											0.2507 ± 0.0001	32
	142	54											0.2867 ± 0.0001	17
	142	54											0.2919 ± 0.0001	17
	142	54											0.3091 ± 0.0001	27
	142	54											0.3347 ± 0.0001	12
	142	54											0.3530 ± 0.0002	13
	142	54											0.3799 ± 0.0001	10
	142	54											0.3942 ± 0.0001	17
	142	54											0.4065 ± 0.0001	11
	142	54											0.4145 ± 0.0001	47
	142	54											0.4324 ± 0.0002	12
	142	54											0.4382 ± 0.0002	6
	142	54											0.4531 ± 0.0001	20
	142	54											0.4682 ± 0.0001	24
	142	54											0.5382 ± 0.0001	77
	142	54											0.5477 ± 0.0002	7
	142	54											0.5578 ± 0.0002	7
	142	54											0.5718 ± 0.0001	100
	142	54											0.6056 ± 0.0001	22
	142	54											0.6181 ± 0.0001	72
	142	54											0.6448 ± 0.0001	63
	142	54											0.6646 ± 0.0001	17
	142	54											0.6722 ± 0.0002	7
	142	54											0.7355 ± 0.0004	6
	142	54											0.7374 ± 0.0002	13
	142	54											0.8012 ± 0.0002	6
	142	54											0.8074 ± 0.0003	6
	142	54											0.8914 ± 0.0001	11
	142	54											0.9912 ± 0.0001	9
	142	54											0.1568 ± 0.0002	7
	142	54											1.1874 ± 0.0002	6
	142	54											1.2192 ± 0.0002	6
	142	54											1.2270 ± 0.0001	19
	142	54											1.2580 ± 0.0001	8
	142	54											1.3001 ± 0.0001	31
	142	54											1.3123 ± 0.0001	21
	142	54											1.4106 ± 0.0001	6
	142	54											1.9020 ± 0.0002	8
Cs		55		132.9054						29 b.				
$_{55}Cs^{114}$	114	55		113.941270	0.57 s.	β+,E.C.							ann.rad.	
	114	55											0.6826 ± 0.0006	
	114	55											0.7088 ± 0.0006	
$_{55}Cs^{115}$	115	55		114.936070	1.4 s.	β+,E.C.							ann.rad.	
$_{55}Cs^{116m}$	116	55			0.7 s.	β+,E.C.							ann.rad.	
	116	55											0.3935 ± 0.0002	
$_{55}Cs^{116}$	116	55		115.933110	3.8 s.	β+,E.C.	≈ 11.1				5		ann.rad.	
	116	55											0.3222 ± 0.0004	4%
	116	55											0.3935 ± 0.0002	93%
	116	55											0.4583 ± 0.0003	3%
	116	55											0.5243 ± 0.0002	75%
	116	55											0.5412 ± 0.0003	6%
	116	55											0.5602 ± 0.0003	7%
	116	55											0.6113 ± 0.0003	6%
	116	55											0.6151 ± 0.0003	30%
	116	55											0.6223 ± 0.0003	10%
	116	55											0.6393 ± 0.0003	7%
	116	55											0.6774 ± 0.0006	3%
	116	55											0.9037 ± 0.0008	3%
	116	55											0.9059 ± 0.0008	2%
	116	55											0.9112 ± 0.0004	3%
	116	55											0.9656 ± 0.0006	1%
	116	55											1.0158 ± 0.0004	3%
	116	55											1.0339 ± 0.0008	2%
	116	55											1.0449 ± 0.0006	1%

Isotope	A	Z	% Natural abundance	Atomic mass	Half-life	Decay mode	Decay energy (MeV)	Particle energy (MeV)	Particle intensity	Thermal neutron cross section	Spin (h/2π)	μ Nucl. mag. moment	Gamma-ray energy (MeV)	Gamma-ray intensity
	116	55											1.0615 ± 0.0004	7%
	116	55											1.0721 ± 0.0006	2%
	116	55											1.0807 ± 0.0004	7%
	116	55											1.1680 ± 0.0008	3%
	116	55											1.2470 ± 0.0008	2%
	116	55											1.3215 ± 0.0008	2%
	116	55											1.4460 ± 0.0008	1.5%
$_{55}Cs^{117}$	117	55		116.928900	6.7 s.	β+,E.C.	≈ 8.1						ann.rad.	
$_{55}Cs^{118}$	118	55		117.926740	15 s.	β+,E.C.	≈ 9.8						ann.rad.	
	118	55											0.3372 ± 0.0002	100 +
	118	55											0.4727 ± 0.0002	37
	118	55											0.4930 ± 0.0002	5
	118	55											0.5559 ± 0.0002	5
	118	55											0.5862 ± 0.0003	3
	118	55											0.5865 ± 0.0002	15
	118	55											0.5906 ± 0.0002	11
	118	55											0.6765 ± 0.0002	3
	118	55											0.9281 ± 0.0003	4
	118	55											1.0218 ± 0.0002	4
	118	55											1.0288 ± 0.0002	5
	118	55											1.1120 ± 0.0002	1.5
	118	55											1.1437 ± 0.0002	1.6
$_{55}Cs^{119}$	119	55		118.922490	38 s.	β+,E.C.	6.6				9/2+		ann.rad.	
	119	55											0.169	67 +
	119	55											0.176	81
	119	55											0.224	100
	119	55											0.257	57
	119	55											0.314	38
	119	55											0.390	16
	119	55											0.667	17
$_{55}Cs^{120}$	120	55		119.920800	64 s.	β+,E.C.	8.2						ann.rad.	
	120	55											0.3224 ± 0.0001	100 +
	120	55											0.3955 ± 0.0003	3
	120	55											0.4735 ± 0.0001	30
	120	55											0.5251 ± 0.0001	3
	120	55											0.5534 ± 0.0001	19
	120	55											0.5858 ± 0.0003	4
	120	55											0.6012 ± 0.0002	11
	120	55											0.6051 ± 0.0002	4
	120	55											0.7018 ± 0.0002	2
	120	55											0.8758 ± 0.0003	6
	120	55											0.9491 ± 0.0003	7
	120	55											1.2750 ± 0.0008	7
	120	55											1.3894 ± 0.0003	2
	120	55											1.4451 ± 0.0003	2
	120	55											1.6727 ± 0.0008	2
	120	55											1.9820 ± 0.0005	1.5
	120	55											2.0560 ± 0.0008	1
	120	55											2.3154 ± 0.0004	0.9
	120	55											2.4673 ± 0.0004	0.7
	120	55											(0.3 - 3.28)	
$_{55}Cs^{121m}$	121	55			121 s.	I.T.(60%)		4	45		(9/2+)		ann.rad.	
	121	55				β+(40%)							0.1794 ± 0.0001	10%
	121	55											0.1961 ± 0.0001	12%
	121	55											0.2345 ± 0.0001	2%
	121	55											0.4146 ± 0.0002	2%
	121	55											0.4273 ± 0.0001	4%
	121	55											0.4598 ± 0.0001	5%
	121	55											0.5540 ± 0.0002	0.8%
$_{55}Cs^{121}$	121	55		120.917250	136 s.	β+,E.C.	5.4	4.4			3/2+		ann.rad.	
	121	55											0.1537 ± 0.0001	1%
	121	55											(0.08 - 0.56)weak	
$_{55}Cs^{122m}$	122	55			4.4 m.						8		ann.rad.	
	122	55											0.3311 ± 0.0002	91%
	122	55											0.3710 ± 0.0002	3%
	122	55											0.4971 ± 0.0002	77%
	122	55											0.5120 ± 0.0004	8%
	122	55											0.5601 ± 0.0002	14%
	122	55											0.5740 ± 0.0002	5%
	122	55											0.6385 ± 0.0002	61%
	122	55											0.6541 ± 0.0003	7%
	122	55											0.6846 ± 0.0003	4%
	122	55											0.7506 ± 0.0002	11%
	122	55											0.8155 ± 0.0004	2.4%
	122	55											0.8430 ± 0.0002	3%
	122	55											0.8829 ± 0.0002	7%
	122	55											0.9459 ± 0.0003	4%
	122	55											0.9942 ± 0.0003	2%
	122	55											1.0888 ± 0.0003	4%
	122	55											1.0978 ± 0.0002	11%
	122	55											1.2981 ± 0.0004	4.5%
	122	55											1.3801 ± 0.0005	3.4%
	122	55											1.4057 ± 0.0004	8%
	122	55											1.4600 ± 0.0004	4%

TABLE OF THE ISOTOPES (Continued)

Isotope	A	Z	% Natural abundance	Atomic mass	Half-life	Decay mode	Decay energy (MeV)	Particle energy (MeV)	Particle intensity	Thermal neutron cross section	Spin (h/2π)	μ Nucl. mag. moment	Gamma-ray energy (MeV)	Gamma-ray intensity
	122	55											1.4955 ± 0.0004	3%
	122	55											1.6846 ± 0.0004	2%
	122	55											2.0521 ± 0.0004	2.5%
	122	55											2.1021 ± 0.0005	2.3%
	122	55											(0.27 - 2.22)weak	
$_{55}Cs^{122}$	122	55		121.916090	21 s.	β+,E.C.	7.4	5.8			(1+)		ann.rad.	
	122	55											0.3311 ± 0.0002	100 +
	122	55											0.4971 ± 0.0002	2
	122	55											0.5120 ± 0.0004	9
	122	55											0.8179 ± 0.0002	6
	122	55											0.8430 ± 0.0002	4
	122	55											0.8829 ± 0.0002	1.7
	122	55											1.0359 ± 0.0003	1
	122	55											1.3852 ± 0.0003	1
	122	55											1.4955 ± 0.0004	1
	122	55											1.7344 ± 0.0004	1
	122	55											2.1991 ± 0.0007	1
$_{55}Cs^{123m}$	123	55			1.7 s.	I.T.					11/2-		Cs k x-ray	25%
	123	55											0.0640 ± 0.0001	4%
	123	55											0.0946 ± 0.0001	25%
$_{55}Cs^{123}$	123	55		122.912990	5.9 m.	β+(75%)	4.12	2.4			1/2+		ann.rad.	
	123	55				E.C.(25%)		3.0					Xe k x-ray	17%
	123	55											0.0834 ± 0.0001	2.7%
	123	55											0.0974 ± 0.0001	13%
	123	55											0.2619 ± 0.0001	1.7%
	123	55											0.3071 ± 0.0001	2.7%
	123	55											0.5964 ± 0.0002	7.4%
	123	55											0.6109 ± 0.0002	2.3%
	123	55											0.6441 ± 0.0002	2%
	123	55											0.7415 ± 0.0001	2%
	123	55											1.1762 ± 0.0004	1%
	123	55											1.2732 ± 0.0002	1.8%
$_{55}Cs^{124}$	124	55		123.912270	30.8 s.	β+(92%)	5.8				1+		ann.rad.	
	124	55				E.C.(8%)							Xe k x-ray	4%
	124	55											0.3539 ± 0.0002	41%
	124	55											0.4925 ± 0.0004	3%
	124	55											0.8462 ± 0.0002	1%
	124	55											0.9418 ± 0.0001	4%
	124	55											1.3362 ± 0.0002	0.6%
	124	55											1.6890 ± 0.0003	0.5%
	124	55											2.0199 ± 0.0002	0.8%
$_{55}Cs^{125}$	125	55		124.909725	45 m.	β+(40%)	3.06	2.05			1/2+	+1.41	ann.rad.	
	125	55				E.C.(60%)							Xe k x-ray	26%
	125	55											0.112	9%
	125	55											0.412	5%
	125	55											0.526	24%
	125	55											0.540	3%
	125	55											0.600	3%
	125	55											0.712	3.5%
	125	55											0.922	0.8%
	125	55											0.995	0.5%
	125	55											1.158	0.4%
	125	55											1.579	0.3%
	125	55											1.698	0.3%
	125	55											2.116	0.8%
$_{55}Cs^{126}$	126	55		125.909465	1.	64 m.	β+ (81%)	4.83	3.4		1+		ann.rad.	
	126	55				E.C.(19%)		3.7					Xe k x-ray	8%
	126	55											0.3886 ± 0.0001	42%
	126	55											0.4912 ± 0.0002	5%
	126	55											0.8798 ± 0.0003	1.5%
	126	55											0.9252 ± 0.0002	5%
	126	55											1.6786 ± 0.0005	0.7%
	126	55											2.0673 ± 0.0005	0.3%
$_{55}Cs^{127}$	127	55		126.907428	6.2 h.	β+(96%)	2.11	0.65			1/2+	+1.46	Xe k x-ray	8%
	127	55				E.C.(4%)							0.1247 ± 0.0002	16%
	127	55											0.2871 ± 0.0002	3.4%
	127	55											0.4119 ± 0.0002	58%
	127	55											0.4623 ± 0.0002	4%
	127	55											0.5872 ± 0.0002	3.5%
	127	55											0.8066 ± 0.0004	0.3%
	127	55											0.9314 ± 0.0003	0.3%
$_{55}Cs^{128}$	128	55		127.907755	3.62 m.	β+(68%)	3.93	2.4			1+		ann.rad.	
	128	55				E.C.(32%)		2.9					Xe k x-ray	13%
	128	55											0.4429 ± 0.0001	26%
	128	55											0.5266 ± 0.0001	2%
	128	55											0.9695 ± 0.0001	0.6%
	128	55											1.1401 ± 0.0001	1%
$_{55}Cs^{129}$	129	55		128.906027	32.3 h.	E.C.	1.17				1/2+	+1.479	Xe k x-ray	55%
	129	55											0.0396 ± 0.0001	3%
	129	55											0.2786 ± 0.0001	1.3%
	129	55											0.3182 ± 0.0001	2.5%
	129	55											0.3719 ± 0.0001	31%
	129	55											0.4115 ± 0.0001	23%
	129	55											0.5489 ± 0.0001	3.5%

Isotope	A	Z	% Natural abundance	Atomic mass	Half-life	Decay mode	Decay energy (MeV)	Particle energy (MeV)	Particle intensity	Thermal neutron cross section	Spin (h/2π)	μ Nucl. mag. moment	Gamma-ray energy (MeV)	Gamma-ray intensity
	129	55											0.5885 ± 0.0001	0.6%
$_{55}Cs^{130}$	130	55		129.906753	29.2 m.	β+(55%)	3.02	1.98			1+	+1.4	ann.rad.	
	130	55				E.C.(43%)							Xe k x-ray	22%
	130	55				β-(1.6%)	0.44	0.44	1.6%				0.5361 ± 0.0001	4%
	130	55											0.5861 ± 0.0001	0.5%
	130	55											0.8945 ± 0.0002	0.4%
	130	55											1.6150 ± 0.0002	0.3%
	130	55											1.6874 ± 0.0002	0.2%
	130	55											1.7070 ± 0.0002	0.1%
	130	55											1.9973 ± 0.0003	0.2%
$_{55}Cs^{131}$	131	55		130.905444	9.69 d.	E.C.	0.35				5/2+	+3.54	Xe k x-ray	39%
$_{55}Cs^{132}$	132	55		131.906431	6.47 d.	E.C.(98%)					3+	+2.22	Xe k x-ray	39%
	132	55				β+(0.3%)							0.4646 ± 0.0002	1.9%
	132	55				β-(2%)							0.6302 ± 0.0002	1%
	132	55											0.66769 ± 0.00002	97%
	132	55											1.13605 ± 0.00002	0.5%
	132	55											1.31791 ± 0.00002	0.6%
$_{55}Cs^{133}$	133	55	100%	132.905429						(2.6 + 27) b.	7/2+	+2.579		
$_{55}Cs^{134m}$	134	55			2.91 h.	I.T.	0.139				8-	+1.096	Cs k x-ray	16%
	134	55											0.0112 ± 0.0001	1%
	134	55											0.12749 ± 0.00001	13%
$_{55}Cs^{134}$	134	55		133.906696	2.065 y.	β-	2.06	0.089	27%	140 b.	4+	+2.990	0.56327 ± 0.00002	8.4%
	134	55						0.658	70%				0.56935 ± 0.00002	15.4%
	134	55											0.60473 ± 0.00003	97.6%
	134	55											0.79584 ± 0.00003	85.4%
	134	55											0.80194 ± 0.00005	8.7%
	134	55											1.03864 ± 0.00006	1.0%
	134	55											1.16798 ± 0.00006	1.8%
	134	55											1.36519 ± 0.00005	3.0%
$_{55}Cs^{135m}$	135	55			53 m.	I.T.	1.627				19/2-		0.7869 ± 0.0001	99%
	135	55											0.8402 ± 0.0002	96%
$_{55}Cs^{135}$	135	55		134.905885	$3x10^6$ y.	β-	0.205	0.205	100%	8.7 b.	7/2+	+2.729		
$_{55}Cs^{136m}$	136	55			19 s.	I.T.					8			
$_{55}Cs^{136}$	136	55		135.907289	13.1 d.	β-	2.55	0.341			5+	+3.70	0.06691 ± 0.00001	12%
	136	55											0.08629 ± 0.00005	6%
	136	55											0.15322 ± 0.00005	7.5%
	136	55											0.17656 ± 0.00005	14%
	136	55											0.27365 ± 0.00004	13%
	136	55											0.34057 ± 0.00005	47%
	136	55											0.81850 ± 0.00004	86%
	136	55											1.04807 ± 0.00007	65%
	136	55											1.23534 ± 0.00005	20%
$_{55}Cs^{137}$	137	55		136.907073	30.17 y.	β-	1.17	0.514	95%		7/2+	+2.838	Ba k x-ray	4%
	137	55											0.66164 ± 0.00001	85.1%(D)
$_{55}Cs^{138m}$	138	55			2.9 m.	I.T.(75%)	0.080				6-		Cs k x-ray	20%
	138	55				β-(25%)							0.0799 ± 0.0003	0.1%
	138	55											0.1125 ± 0.0003	2%
	138	55											0.1917 ± 0.0002	20%
	138	55											0.3249 ± 0.0001	1.5%
	138	55											0.4628 ± 0.0001	25%
	138	55											1.43579 ± 0.00004	25%
$_{55}Cs^{138}$	138	55		137.911004	β-	5.38	3.2				3-	+0.5	0.1381 ± 0.0001	2%
	138	55											0.2278 ± 0.0001	1.5%
	138	55											0.40844 ± 0.00005	4.7%
	138	55											0.46269 ± 0.00003	31%
	138	55											0.54685 ± 0.00003	11%
	138	55											0.87170 ± 0.00003	5.1%

Isotope	A	Z	% Natural abundance	Atomic mass	Half-life	Decay mode	Decay energy (MeV)	Particle energy (MeV)	Particle intensity	Thermal neutron cross section	Spin (h/2π)	μ Nucl. mag. moment	Gamma-ray energy (MeV)	Gamma-ray intensity
	138	55											1.00969 ± 0.00003	30%
	138	55											1.43579 ± 0.00003	76.3%
	138	55											1.5553 ± 0.0001	0.4%
	138	55											2.21788 ± 0.00006	15.2%
	138	55											2.58285 ± 0.0001	0.2%
	138	55											2.63929 ± 0.00008	7.6%
	138	55											3.3390 ± 0.0003	0.15%
	138	55											3.3669 ± 0.0003	0.23%
$_{55}Cs^{139}$	139	55		138.913349	β-	4.21	4.2				7/2+		0.6272 ± 0.0001	0.6%
	139	55											1.2832 ± 0.0001	7.7%
	139	55											1.4207 ± 0.0001	0.8%
	139	55											1.6807 ± 0.0001	0.6%
	139	55											2.1109 ± 0.0001	0.7%
	139	55											2.3499 ± 0.0001	0.6%
	139	55											2.5318 ± 0.0001	0.45%
	139	55											2.6058 ± 0.0001	0.3%
	139	55											(0.4 - 3.66)weak	
$_{55}Cs^{140}$	140	55		139.917256	63.7 s.	β-	6.22	5.7			1-		0.5283 ± 0.0001	4%
	140	55						6.2					0.6023 ± 0.0001	70%
	140	55											0.6722 ± 0.0001	1.5%
	140	55											0.9084 ± 0.0001	11%
	140	55											1.0082 ± 0.0002	1%
	140	55											1.1300 ± 0.0001	3.1%
	140	55											1.2005 ± 0.0001	6%
	140	55											1.2216 ± 0.0001	3%
	140	55											1.3914 ± 0.0002	2.2%
	140	55											1.4220 ± 0.0002	1.1%
	140	55											1.6349 ± 0.0002	3.2%
	140	55											1.7073 ± 0.0002	1.6%
	140	55											1.8533 ± 0.0002	4%
	140	55											2.1016 ± 0.0002	3.9%
	140	55											2.2373 ± 0.0002	3.9%
	140	55											2.2684 ± 0.0002	1.7%
	140	55											2.3306 ± 0.0002	4.6%
	140	55											2.4297 ± 0.0002	1.7%
	140	55											2.5220 ± 0.0002	4.1%
	140	55											3.0540 ± 0.0002	1.5%
	140	55											3.4515 ± 0.0003	0.6%
	140	55											(0.41 - 3.94)weak	
$_{55}Cs^{141}$	141	55		140.920006	24.9 s.	β-	5.26	5.2			7/2+		Ba k x-ray	33%
	141	55											0.0485 ± 0.0001	10%
	141	55											0.5551 ± 0.00001	4.7%
	141	55											0.5616 ± 0.0001	5.8%
	141	55											0.5887 ± 0.0001	5.1%
	141	55											0.6919 ± 0.0001	3.7%
	141	55											1.0618 ± 0.0001	1.3%
	141	55											1.1406 ± 0.0002	1.3%
	141	55											1.1471 ± 0.0002	3.8%
	141	55											1.1536 ± 0.0002	1.2%
	141	55											1.1715 ± 0.0002	1.1%
	141	55											1.1775 ± 0.0002	1.3%
	141	55											1.1940 ± 0.0002	5.4%
	141	55											1.8940 ± 0.0004	0.5%
	141	55											1.9407 ± 0.0003	0.65%
	141	55											2.0596 ± 0.0006	0.5%
	141	55											2.0660 ± 0.0004	0.5%
	141	55											3.0722 ± 0.0003	0.7%
	141	55											(0.05 - 3.33)many	
$_{55}Cs^{142}$	142	55		141.924220	1.8 s.	β-	7.32	6.9					0.3596 ± 0.0001	100 +
	142	55											0.9668 ± 0.0001	28
	142	55											1.1759 ± 0.0001	10
	142	55											1.3265 ± 0.0001	34
	142	55											1.4222 ± 0.0002	3
	142	55											1.9821 ± 0.0002	4.4
	142	55											2.3978 ± 0.0002	2.6
	142	55											3.2834 ± 0.0003	2.1
	142	55											3.5733 ± 0.0003	2.3
$_{55}Cs^{143}$	143	55		142.927220	1.78 s.	β-	6.1				(3/2+)		0.1955 ± 0.0001	1%
	143	55											0.2324 ± 0.0001	14%
	143	55											0.2633 ± 0.0001	5.6%
	143	55											0.2727 ± 0.0001	5.6%
	143	55											0.3064 ± 0.0001	10%
	143	55											0.4666 ± 0.0002	7%
	143	55											0.5274 ± 0.0002	4%
	143	55											0.5348 ± 0.0002	2%
	143	55											0.5707 ± 0.0002	2%
	143	55											0.6052 ± 0.0002	2.4%
	143	55											0.6267 ± 0.0002	4%
	143	55											0.6599 ± 0.0002	7%
	143	55											0.6617 ± 0.0002	7%
	143	55											0.7293 ± 0.0002	2%

Isotope	A	Z	% Natural abundance	Atomic mass	Half-life	Decay mode	Decay energy (MeV)	Particle energy (MeV)	Particle intensity	Thermal neutron cross section	Spin (h/2π)	μ Nucl. mag. moment	Gamma-ray energy (MeV)	Gamma-ray intensity
	143	55											0.7927 ± 0.0002	1%
	143	55											0.8337 ± 0.0002	2%
	143	55											0.8679 ± 0.0002	1%
	143	55											1.0213 ± 0.0002	0.4%
	143	55											1.2081 ± 0.0002	0.4%
	143	55											1.9775 ± 0.0002	2.6%
	143	55											(0.17 - 1.98)many	
$_{55}Cs^{144}$	144	55		143.931930	1.01 s.	β-	8.5	8			1		0.1993 ± 0.0001	47%
	144	55											0.3083 ± 0.0002	2.2%
	144	55											0.3309 ± 0.0002	5%
	144	55											0.5598 ± 0.0002	9.2%
	144	55											0.6392 ± 0.0002	9.6%
	144	55											0.7587 ± 0.0002	9.6%
	144	55											0.8204 ± 0.0003	2%
	144	55											1.0095 ± 0.0004	1.4%
	144	55											1.0889 ± 0.0004	1.4%
	144	55											1.1160 ± 0.0004	1.7%
	144	55											1.3190 ± 0.0005	1.5%
	144	55											2.0130 ± 0.0006	2.6%
	144	55											2.1395 ± 0.0008	2.6%
	144	55											2.1765 ± 0.0008	2.2%
	144	55											2.4095 ± 0.001	3.2%
	144	55											2.7110 ± 0.0015	2.2%
	144	55											3.057 ± 0.003	1%
$_{55}Cs^{145}$	145	55		144.935320	0.58 s.	β-	7.8	7.4			3/2+		0.1126 ± 0.0001	10%
	145	55											0.1755 ± 0.0001	20%
	145	55											0.1990 ± 0.0001	13%
	145	55											0.2410 ± 0.0001	5.3%
	145	55											0.4357 ± 0.0001	8.5%
	145	55											0.4548 ± 0.0001	4.5%
	145	55											0.4921 ± 0.0001	2.2%
	145	55											0.5471 ± 0.0001	5.1%
	145	55											0.7532 ± 0.0001	3%
Ba		56		137.33						1.3 b.				
$_{56}Ba^{120}$	120	56		119.926490	≈ 32 s.	β+,E.C.	≈ 5.3				0+		ann.rad.	
	120	56											0.051	
	120	56											0.182	
$_{56}Ba^{121}$	121	56		120.924700	30 s.	β+,E.C.	≈ 6.9						ann.rad.	
$_{56}Ba^{122}$	122	56		121.920170	2.0 m.	β+,E.C.	≈ 3.8				0+		ann.rad.	
$_{56}Ba^{123}$	123	56		122.919210	2.7 m.	β+,E.C.	≈ 5.7						ann.rad.	
	123	56											0.0306 ± 0.0006	56 +
	123	56											0.0583 ± 0.0006	2
	123	56											0.0639 ± 0.0006	14
	123	56											0.0927 ± 0.0006	51
	123	56											0.1161 ± 0.0006	54
	123	56											0.1200 ± 0.0006	23
	123	56											0.1235 ± 0.0006	69
	123	56											0.1370 ± 0.0006	23
$_{56}Ba^{124}$	124	56		123.915380	11.4 m.	β+,E.C.	≈ 2.9						ann.rad.	
	124	56											0.1568 ± 0.0005	3%
	124	56											0.1695 ± 0.0003	21%
	124	56											0.1888 ± 0.0003	11%
	124	56											0.2116 ± 0.0004	4%
	124	56											0.2531 ± 0.0003	5%
	124	56											0.2716 ± 0.0005	8%
	124	56											0.7155 ± 0.001	4%
	124	56											0.9330 ± 0.001	3%
	124	56											1.0480 ± 0.001	3%
	124	56											1.2160 ± 0.001	13%
$_{56}Ba^{125}$	125	56		124.914640	3.5 m.	β+,E.C.							ann.rad.	
	125	56											0.0550 ± 0.0006	48 +
	125	56											0.0631 ± 0.0006	8
	125	56											0.0776 ± 0.0006	100
	125	56											0.0854 ± 0.0006	82
	125	56											0.1001 ± 0.0006	6
	125	56											0.1080 ± 0.0006	8
	125	56											0.1409 ± 0.0006	86
$_{56}Ba^{126}$	126	56		125.911260	99 m.	β+(2%) E.C.(98%)	≈ 1.8				0+		Cs k x-ray	42%
	126	56											0.2179 ± 0.0001	4%
	126	56											0.2336 ± 0.0001	20%
	126	56											0.2410 ± 0.0001	6%
	126	56											0.2576 ± 0.0001	8%
	126	56											0.2812 ± 0.0002	3%
	126	56											0.3283 ± 0.0002	2%
	126	56											0.4893 ± 0.0002	3%
	126	56											0.5389 ± 0.0002	2%
	126	56											0.6818 ± 0.0002	4.6%
	126	56											0.7098 ± 0.0003	1.6%
	126	56											0.8416 ± 0.0005	1%
	126	56											0.8639 ± 0.0002	1%
	126	56											0.9768 ± 0.0002	2%
	126	56											0.9934 ± 0.0003	2%
	126	56											1.0354 ± 0.0003	2%
	126	56											1.0520 ± 0.0003	1%

Isotope	A	Z	% Natural abundance	Atomic mass	Half-life	Decay mode	Decay energy (MeV)	Particle energy (MeV)	Particle intensity	Thermal neutron cross section	Spin (h/2π)	μ Nucl. mag. moment	Gamma-ray energy (MeV)	Gamma-ray intensity
	126	56											1.2108 ± 0.0003	2%
	126	56											1.2418 ± 0.0003	1%
	126	56											1.2930 ± 0.0003	4%
$_{56}Ba^{127}$	127	56		126.911130	12 m.	β+(54%)	3.5				1/2+		ann.rad.	
	127	56				E.C.(46%)							Cs k x-ray	26%
	127	56											0.1148 ± 0.0003	9%
	127	56											0.1808 ± 0.0003	12%
	127	56											1.2010 ± 0.0003	1.6%
	127	56											1.5001 ± 0.0003	0.4%
	127	56											1.5660 ± 0.0003	0.4%
	127	56											(0.07 - 2.5)weak	
$_{56}Ba^{128}$	128	56		127.908237	2.43 d.	E.C.	0.45				0+		Cs k x-ray	41%
	128	56											0.27344 ± 0.00005	14%
$_{56}Ba^{129m}$	129	56			2.1 h.	E.C.(98%)					7/2+		Cs k x-ray	45%
	129	56				β+(2%)							0.1769 ± 0.0003	6%
	129	56											0.1823 ± 0.0001	47%
	129	56											0.2023 ± 0.0001	16%
	129	56											0.2143 ± 0.0001	4%
	129	56											0.3924 ± 0.0004	6%
	129	56											0.4202 ± 0.0002	12%
	129	56											0.4596 ± 0.0002	3%
	129	56											0.4816 ± 0.0002	3%
	129	56											0.5432 ± 0.0002	2%
	129	56											0.5468 ± 0.0003	5%
	129	56											0.5661 ± 0.0003	3%
	129	56											0.5969 ± 0.0001	7%
	129	56											0.6789 ± 0.0001	7%
	129	56											0.7486 ± 0.0002	3%
	129	56											0.7806 ± 0.0002	3%
	129	56											0.8040 ± 0.0003	2%
	129	56											0.8203 ± 0.0003	2%
	129	56											0.8725 ± 0.0001	3%
	129	56											0.8927 ± 0.0001	10%
	129	56											0.9573 ± 0.0003	2%
	129	56											0.9997 ± 0.0001	4%
	129	56											1.0351 ± 0.0003	4%
	129	56											1.0447 ± 0.0003	7.5%
	129	56											1.0476 ± 0.0006	3%
	129	56											1.1224 ± 0.0002	3%
	129	56											1.2092 ± 0.0002	4%
	129	56											1.2218 ± 0.0001	4%
	129	56											1.4593 ± 0.0001	26%
	129	56											1.6238 ± 0.0002	5%
$_{56}Ba^{129}$	129	56		128.908642	2.5 h.	β+(20%)	2.43				1/2+		ann.rad.	
	129	56				E.C.(80%)							Cs k x-ray	35%
	129	56											0.1291 ± 0.0001	6%
	129	56											0.2143 ± 0.0001	10%
	129	56											0.2208 ± 0.0001	6%
	129	56											0.5541 ± 0.0002	1.5%
	129	56											1.1646 ± 0.0002	1.1%
	129	56											1.8304 ± 0.0002	0.5%
	129	56											1.9540 ± 0.0002	0.5%
$_{56}Ba^{130}$	130	56	0.106%	129.906282						(2.5 + 9) b.	0+			
$_{56}Ba^{131m}$	131	56			14.6 m.	I.T.	0.187				9/2-		Ba k x-ray	25%
	131	56											0.0790 ± 0.0002	1.2%
	131	56											0.1085 ± 0.0002	55%
$_{56}Ba^{131}$	131	56		130.906902	11.8 d.	E.C.	1.36				1/2+		Cs k x-ray	52%
	131	56											0.12381 ± 0.00001	29%
	131	56											0.13360 ± 0.00001	2.2%
	131	56											0.21608 ± 0.00001	20%
	131	56											0.37324 ± 0.00001	14%
	131	56											0.49636 ± 0.00001	47%
	131	56											0.58499 ± 0.00002	1.2%
	131	56											0.62016 ± 0.00002	1.4%
	131	56											0.9239 ± 0.0001	0.7%
	131	56											1.0476 ± 0.0001	1.3%
$_{56}Ba^{132}$	132	56	0.101%	131.905042						(0.6 +7) b.	0+			
$_{56}Ba^{133m}$	133	56			38.9 h.	I.T.	0.288				11/2-		Ba k x-ray	28%
	133	56											0.2761 ± 0.0002	17%
$_{56}Ba^{133}$	133	56		132.905988	10.53 y.	E.C.	0.52				1/2+		Cs k x-ray	63%
	133	56											0.0796 ± 0.0001	2%
	133	56											0.08099 ± 0.00001	33%
	133	56											0.27639 ± 0.00001	7.3%
	133	56											0.30285 ± 0.00001	19%

Isotope	A	Z	% Natural abundance	Atomic mass	Half-life	Decay mode	Decay energy (MeV)	Particle energy (MeV)	Particle intensity	Thermal neutron cross section	Spin (h/2π)	μ Nucl. mag. moment	Gamma-ray energy (MeV)	Gamma-ray intensity
	133	56											0.35600 ± 0.00002	62%
	133	56											0.38385 ± 0.00002	8.8%
$_{56}Ba^{134}$	134	56	2.417%	133.904486						(0.16 + 2) b.	0+			
$_{56}Ba^{135m}$	135	56			28.7 h.	I.T.	0.268				11/2-		Ba k x-ray	28%
	135	56											0.2682 ± 0.0001	16%
$_{56}Ba^{135}$	135	56	6.592%	134.905665						(0.014 + 6) b.	3/2+	+0.8365		
$_{56}Ba^{136m}$	136	56			0.306 s.	I.T.	2.0305				7-		Ba k x-ray	16%
	136	56											0.1639 ± 0.0001	31%
	136	56											0.8185 ± 00.0001	100%
	136	56											1.0481 ± 0.0001	100%
$_{56}Ba^{136}$	136	56	7.854%	135.904553						(0.010 + 0.4) b.	0+			
$_{56}Ba^{137m}$	137	56			2.552 m.	I.T.	0.66165				11/2-		Ba k x-ray	4%
	137	56											0.66164 ± 0.00002	90%
$_{56}Ba^{137}$	137	56	11.23%	136.905812						5.1 b.	3/2+	+0.9357		
$_{56}Ba^{138}$	138	56	71.70%	137.905232						0.4 b.	0+			
$_{56}Ba^{139}$	139	56		138.908826	1.41 ± 0.01 h.	β-	2.3	2.2	27%	6 b.	7/2-		0.16585 ± 0.00001	24%
	139	56						2.3	72%				1.2544 ± 0.0001	0.03%
	139	56											1.42033 ± 0.00005	0.26%
$_{56}Ba^{140}$	140	56		139.910581	12.76 d.	β-	1.03	0.48		1.6 b.	0+		0.16268 ± 0.00001	7.1%
	140	56						1.0					0.30485 ± 0.00001	4.7%
	140	56											0.42372 ± 0.00001	3.3%
	140	56											0.43757 ± 0.00001	2%
	140	56											0.53727 ± 0.00002	24.5%
$_{56}Ba^{141}$	141	56		140.914363	18.3 m.	β- 2.73	3.23	2.59			3/2-		0.1903 ± 0.0001	46%
	141	56											0.2770 ± 0.0001	24%
	141	56											0.3042 ± 0.0001	25%
	141	56											0.3437 ± 0.0001	14%
	141	56											0.4576 ± 0.0001	4%
	141	56											0.4621 ± 0.0001	4%
	141	56											0.4673 ± 0.0001	5.6%
	141	56											0.6252 ± 0.0001	3.2%
	141	56											0.6479 ± 0.0001	6%
	141	56											0.7390 ± 0.0001	4.2%
	141	56											0.8761 ± 0.0001	3.3%
	141	56											1.1975 ± 0.0002	4.3%
	141	56											1.3240 ± 0.0004	0.8%
	141	56											1.4370 ± 0.0003	0.8%
	141	56											1.6823 ± 0.0002	1.4%
	141	56											(0.1 - 2.5)many	
$_{56}Ba^{142}$	142	56		141.916360	10.7 m.	β-	2.13	1.0			0+		0.23152 ± 0.00004	11.5%
	142	56					1.7						0.25512 ± 0.00004	20%
	142	56											0.3090 ± 0.0001	25%
	142	56											0.3638 ± 0.0001	4.5%
	142	56											0.4250 ± 0.0001	5.5%
	142	56											0.5998 ± 0.0001	1.8%
	142	56											0.8402 ± 0.0001	3.5%
	142	56											0.8949 ± 0.0001	12%
	142	56											0.9488 ± 0.0001	10%
	142	56											1.0009 ± 0.0001	8.9%
	142	56											1.0785 ± 0.0001	10.5%
	142	56											1.0936 ± 0.0001	2.6%
	142	56											1.1265 ± 0.0001	1.8%
	142	56											1.2022 ± 0.0001	6.0%
	142	56											1.2040 ± 0.0001	15.5%
	142	56											1.3799 ± 0.0001	3.9%
$_{56}Ba^{143}$	143	56		142.920480	15 s.	β-	4.2				3/2-		0.1786 ± 0.0001	1.5%
	143	56											0.21148 ± 0.00003	10%
	143	56											0.2912 ± 0.0001	3.5%
	143	56											0.4315 ± 0.0001	1.2%
	143	56											0.7190 ± 0.0001	1.6%
	143	56											0.7988 ± 0.0001	5.6%
	143	56											0.8952 ± 0.0001	1.5%
	143	56											0.9250 ± 0.0001	1.8%
	143	56											0.9805 ± 0.0001	3.8%
	143	56											1.0103 ± 0.0001	3.2%
	143	56											1.1964 ± 0.0001	2.4%
	143	56											1.6492 ± 0.0002	0.35%
	143	56											(0.17 - 2.4)weak	
$_{56}Ba^{144}$	144	56		143.922840	11.5 s.	β-	3.0				0+		La k x-ray	40%
	144	56											0.0690 ± 0.0001	4%
	144	56											0.0818 ± 0.0001	7%

Isotope	A	Z	% Natural abundance	Atomic mass	Half-life	Decay mode	Decay energy (MeV)	Particle energy (MeV)	Particle intensity	Thermal neutron cross section	Spin (h/2π)	μ Nucl. mag. moment	Gamma-ray energy (MeV)	Gamma-ray intensity
	144	56											0.10386 ± 0.00005	25%
	144	56											0.1113 ± 0.0001	6%
	144	56											0.1150 ± 0.0001	3%
	144	56											0.1566 ± 0.0001	16%
	144	56											0.1728 ± 0.0001	16%
	144	56											0.2076 ± 0.0002	2.4%
	144	56											0.2282 ± 0.0002	2%
	144	56											0.2594 ± 0.0002	4%
	144	56											0.2893 ± 0.0002	2%
	144	56											0.2917 ± 0.0003	3%
	144	56											0.3734 ± 0.0002	2%
	144	56											0.3882 ± 0.0001	17%
	144	56											0.43048 ± 0.00005	21%
	144	56											0.5158 ± 0.0002	8%
	144	56											0.5412 ± 0.0002	5%
	144	56											0.5704 ± 0.0002	3.8%
	144	56											0.5834 ± 0.0002	3%
	144	56											0.7032 ± 0.0003	1.4%
	144	56											0.7851 ± 0.0003	1.4%
$_{56}Ba^{145}$	145	56		144.926960	4.0 s.	β-	4.9	4.9			(5/2-)		La k x-ray	25%
	145	56											0.0656 ± 0.0002	6%
	145	56											0.0918 ± 0.0002	8%
	145	56											0.09709 ± 0.00005	20%
	145	56											0.1618 ± 0.0002	3.6%
	145	56											0.1892 ± 0.0002	1.8%
	145	56											0.2542 ± 0.0002	1.8%
	145	56											0.2860 ± 0.0002	1.6%
	145	56											0.3032 ± 0.0002	3.6%
	145	56											0.3255 ± 0.0002	2.0%
	145	56											0.3344 ± 0.0002	1.0%
	145	56											0.3437 ± 0.0002	1.2%
	145	56											0.3521 ± 0.0002	1.0%
	145	56											0.3788 ± 0.0002	6.5%
	145	56											0.4175 ± 0.0002	6%
	145	56											0.4775 ± 0.0002	2%
	145	56											0.5330 ± 0.0002	2.6%
	145	56											0.5714 ± 0.0002	1.6%
	145	56											0.5785 ± 0.0002	2.0%
	145	56											0.5985 ± 0.0002	3.6%
	145	56											0.6838 ± 0.0002	1.4%
	145	56											0.7306 ± 0.0002	1.6%
	145	56											0.8435 ± 0.0002	1.2%
	145	56											1.1104 ± 0.0003	1.4%
$_{56}Ba^{146}$	146	56		145.930120	2.2 s.	β-	4.3	3.9			0+		0.0644 ± 0.0001	16 +
	146	56											0.2513 ± 0.0002	30
	146	56											0.3270 ± 0.0003	38
	146	56											0.3329 ± 0.0002	100
	146	56											0.3622 ± 0.0003	
$_{56}Ba^{147}$	147	56		146.934230	0.70 s.	β-	≈ 6.1							
$_{56}Ba^{148}$	148	56		147.937190	0.47 s.	β-,n	≈ 4.9							
La		57		138.9055						8.98 b.				
$_{57}La^{125}$	125	57			≈ 76 s.	β+,E.C.					11/2-		ann.rad.	
	125	57											0.0436	
	125	57											0.0676	
$_{57}La^{126}$	126	57			1.0 m.	β+,E.C.							ann.rad.	
	126	57											0.2561 ± 0.001	
	126	57											0.340 ± 0.005	
	126	57											0.4555 ± 0.001	
	126	57											0.6214 ± 0.001	
$_{57}La^{127}$	127	57		126.916280	3.8 m.	β+,E.C.	≈ 4.8						ann.rad.	
	127	57											0.025	
	127	57											0.0562	
$_{57}La^{128}$	128	57		127.915320	4.6 m.	β+(80%) E.C.(20%)	≈ 6.9				(5-)		ann.rad.	
	128	57											Ba k x-ray	10%
	128	57											0.2841 ± 0.0001	87%
	128	57											0.4399 ± 0.0003	2.1%
	128	57											0.4757 ± 0.0005	2%
	128	57											0.4793 ± 0.0001	54%
	128	57											0.4879 ± 0.0002	10%
	128	57											0.5670 ± 0.0002	3.9%
	128	57											0.6005 ± 0.0002	10%
	128	57											0.6090 ± 0.0003	8.2%
	128	57											0.6266 ± 0.0002	3.8%
	128	57											0.6325 ± 1.0002	5.6%
	128	57											0.6436 ± 0.0002	15%
	128	57											0.8845 ± 0.0002	8.1%
	128	57											0.9150 ± 0.0003	3.5%
	128	57											0.9389 ± 0.0003	2.6%
	128	57											1.0363 ± 0.0003	2%
	128	57											1.0404 ± 0.0002	10%
	128	57											1.0532 ± 0.0002	10%

Isotope	A	Z	% Natural abundance	Atomic mass	Half-life	Decay mode	Decay energy (MeV)	Particle energy (MeV)	Particle intensity	Thermal neutron cross section	Spin (h/2π)	μ Nucl. mag. moment	Gamma-ray energy (MeV)	Gamma-ray intensity
	128	57											1.0704 ± 0.0002	4.5%
	128	57											1.0882 ± 0.0002	9%
	128	57											1.31009 ± 0.0003	4.6%
	128	57											1.2761 ± 0.0005	5%
	128	57											1.4123 ± 0.0003	3.5%
	128	57											1.5059 ± 0.0004	3.6%
	128	57											1.7555 ± 0.0004	1.2%
	128	57											1.9196 ± 0.0004	1.2%
	128	57											2.2120 ± 0.0006	0.9%
$_{57}La^{129}$	129	57		128.912640	11.60 m.	β^+(58%)	3.72	2.4			3/2 +		ann.rad.	
	129	57				E.C.(42%)							Ba k x-ray	23%
	129	57											0.1105 ± 0.0001	17%
	129	57											0.2538 ± 0.0001	8%
	129	57											0.2786 ± 0.0001	24%
	129	57											0.3184 ± 0.0001	2%
	129	57											0.3465 ± 0.0001	5%
	129	57											0.4486 ± 0.0001	5%
	129	57											0.4570 ± 0.0001	8%
	129	57											0.4582 ± 0.0001	2%
	129	57											0.6013 ± 0.0002	1%
	129	57											0.6178 ± 0.0002	1%
	129	57											(0.1 - 1.8)	
$_{57}La^{130}$	130	57		129.912400	8.7 m.	β^+(78%)	$\approx$ 5.7				(3-)		ann.rad.	
	130	57				E.C.(22%)							Ba k x-ray	10%
	130	57											0.3573 ± 0.0001	81%
	130	57											0.4529 ± 0.0001	3.7%
	130	57											0.5444 ± 0.0001	18%
	130	57											0.5506 ± 0.0001	27%
	130	57											0.5694 ± 0.0001	3%
	130	57											0.5758 ± 0.0003	3%
	130	57											0.6494 ± 0.0001	1.7%
	130	57											0.7180 ± 0.0001	2.9%
	130	57											0.9079 ± 0.0001	17%
	130	57											0.9748 ± 0.0001	3%
	130	57											1.0036 ± 0.0001	8%
	130	57											1.1200 ± 0.0001	2%
	130	57											1.1708 ± 0.0001	3.9%
	130	57											1.1772 ± 0.0001	2.2%
	130	57											1.2006 ± 0.0001	3%
	130	57											1.4387 ± 0.0001	2.4%
	130	57											1.5254 ± 0.0001	7%
	130	57											1.7216 ± 0.0001	2.1%
	130	57											2.7521 ± 0.0003	1.0%
	130	57											2.7967 ± 0.0004	1.3%
	130	57											2.8101 ± 0.0003	1.4%
$_{57}La^{131}$	131	57		130.910080	59 m.	β^+(76%)	3.0	1.4			3/2 +		ann.rad.	
	131	57				E.C.(24%)		1.9					Ba k x-ray	40%
	131	57											0.1085 ± 0.0002	23%
	131	57											0.1609 ± 0.0002	2%
	131	57											0.2097 ± 0.0004	3%
	131	57											0.2575 ± 0.0004	3%
	131	57											0.3658 ± 0.0006	16%
	131	57											0.4184 ± 0.0007	6%
	131	57											0.4542 ± 0.0007	6%
	131	57											0.5263 ± 0.0008	10%
	131	57											0.5617 ± 0.0005	1.3%
	131	57											0.5941 ± 0.0005	1.5%
	131	57											0.6111 ± 0.0005	1.0%
	131	57											0.8660 ± 0.001	1.2%
	131	57											0.9742 ± 0.001	0.7%
$_{57}La^{132m}$	132	57			24 m.	I.T.(76%)					6-		La k x-ray	19%
	132	57				β^+, EC (24%)							0.1352 ± 0.0002	44%
	132	57											0.2376 ± 0.0005	4%
	132	57											0.2856 ± 0.0005	7%
	132	57											0.4645 ± 0.0001	22%
	132	57											0.5671 ± 0.0001	4%
	132	57											0.6631 ± 0.0001	5%
	132	57											0.6977 ± 0.0001	4%
	132	57											0.8993 ± 0.0001	7%
	132	57											1.0317 ± 0.0001	3%
	132	57											1.0466 ± 0.00014	6%
$_{57}La^{132}$	132	57		131.910100	4.8 h.	β^+(40%)	4.71	2.6			2-		ann.rad.	
	132	57				E.C.(60%)		3.2					Ba k x-ray	24%
	132	57											0.4645 ± 0.0001	77%
	132	57											0.5158 ± 0.0001	5%
	132	57											0.5404 ± 0.0001	8%
	132	57											0.5671 ± 0.0001	16%
	132	57											0.6631 ± 0.0001	9%
	132	57											0.8993 ± 0.0001	5%
	132	57											1.0317 ± 0.0001	8%
	132	57											1.0466 ± 0.0001	3%
	132	57											1.2212 ± 0.0001	3%
	132	57											1.5337 ± 01.0001	1.5%
	132	57											1.6046 ± 0.0001	4%

Isotope	A	Z	% Natural abundance	Atomic mass	Half-life	Decay mode	Decay energy (MeV)	Particle energy (MeV)	Particle intensity	Thermal neutron cross section	Spin (h/2π)	μ Nucl. mag. moment	Gamma-ray energy (MeV)	Gamma-ray intensity
	132	57											1.9099 ± 0.0001	9%
	132	57											2.1028 ± 0.0001	6%
	132	57											2.3914 ± 0.0001	1.0%
	132	57											2.7547 ± 0.0001	1.6%
	132	57											3.1990 ± 0.0001	0.7%
$_{57}La^{133}$	133	57		132.908140	3.91 h.	β+(4%)	2.0	1.2			5/2+		Ba k x-ray	38%
	133	57				E.C.(96%)							0.2788 ± 0.0001	1.9%
	133	57											0.2901 ± 0.0001	1.1%
	133	57											0.3024 ± 0.0001	1.2%
	133	57											0.5653 ± 0.0001	0.5%
	133	57											0.6183 ± 0.0001	0.8%
	133	57											0.6328 ± 0.0004	0.9%
$_{57}La^{134}$	134	57		133.908460	6.5 m.	β+(63%)	3.70	2.67			1+		ann.rad.	
	134	57				E.C.(37%)							Ba k x-ray	15%
	134	57											0.6047 ± 0.0001	5%
	134	57											1.5549 ± 0.0001	0.4%
	134	57											(0.5 - 1.9)weak	
$_{57}La^{135}$	135	57		134.906953	19.5 h.	E.C.	1.20				5/2+		Ba k x-ray	45%
	135	57											0.4805 ± 0.0001	1.5%
	135	57											0.5878 ± 0.0001	0.11%
	135	57											0.8745 ± 0.0001	0.16%
$_{57}La^{136}$	136	57		135.907630	9.87 m.	β+(36%)	2.9	1.9			1+		ann.rad.	
	136	57				E.C.(64%)							Ba k x-ray	26%
	136	57											0.8185 ± 0.0001	2.3%
	136	57											1.3230 ± 0.0001	0.3%
$_{57}La^{137}$	147	57		136.906460	6x10^{4}y.	E.C>	0.61				7/2+			
$_{57}$La138	138	57	0.09%	137.907105	1.06x10^{11}y	β-(34%)	1.04	0.26			5+	+3.707	Ba k x-ray	13%
	138	57				E.C.(66%)	1.75						0.7887 ± 0.0001	34%
	138	57											1.4359 ± 0.0001	66%
$_{57}La^{139}$	139	57	99.91%	138.906346						8.94 b.	7/2+			
$_{57}La^{140}$	140	57		139.909471	40.28 h.	β-	3.761	1.24	19%	2.7 b.	3-		0.32876 ± 0.00005	20.5%
	140	57						1.36	41%				0.43252 ± 0.00002	2.9%
	140	57						1.68	20%				0.48701 ± 0.00003	45%
	140	57						2.16	10%				0.75165 ± 0.00003	4.5%
	140	57											0.81577 ± 0.00003	24%
	140	57											0.86784 ± 0.00003	5.6%
	140	57											0.91954 ± 0.00004	2.7%
	140	57											0.92519 ± 0.00004	2.7%
	140	57											1.59617 ± 0.00006	95%
	140	57											2.34780 ± 0.00006	0.85%
	140	57											2.52132 ± 0.00006	3.5%
	140	57											2.54714 ± 0.00006	0.1%
	140	57											2.8995 ± 0.0002	0.07%
	140	57											3.1185 ± 0.0002	0.03%
$_{57}La^{141}$	141	57		140.910896	3.93 h.	β-	2.45	2.43			(7/2+)		1.3545 ± 0.0001	1.6%
	141	57											1.6933 ± 0.0001	0.07%
	141	57											2.2670 ± 0.0002	0.04%
$_{57}La^{142}$	142	57		141.914090	92 m.	β-	4.52	1.98			2-		0.6412 ± 0.0001	47%
	142	57						2.11					0.8948 ± 0.0001	8%
	142	57											1.0114 ± 0.0001	4%
	142	57											1.0437 ± 0.0001	2.7%
	142	57											1.1602 ± 0.0001	1.7%
	142	57											1.2331 ± 0.0001	1.9%
	142	57											1.3629 ± 0.0001	2.1%
	142	57											1.5458 ± 0.0001	3%
	142	57											1.7229 ± 0.0002	1.5%
	142	57											1.7564 ± 0.0001	2.7%
	142	57											1.9013 ± 0.0001	7.2%
	142	57											2.0042 ± 0.0002	0.9%
	142	57											2.0255 ± 0.0002	1.0%
	142	57											2.0552 ± 0.0001	2.2%
	142	57											2.1004 ± 0.0002	1.0%
	142	57											2.1872 ± 0.0001	3.7%
	142	57											2.3977 ± 0.0001	13%
	142	57											2.5426 ± 0.0001	10%
	142	57											2.6668 ± 0.0002	1.8%
	142	57											2.9718 ± 0.0003	3.1%
	142	57											3.6121 ± 0.0002	0.9%
	142	57											3.6327 ± 0.0002	1.0%
	142	57											(0.17 - 3.8)many	
$_{57}La^{143}$	143	57		142.915920	14.1 m.	β-	3.30	3.3					0.6203 ± 0.0001	1.0%
	143	57											0.6214 ± 0.0001	0.65%
	143	57											0.6437 ± 0.0001	0.66%

Isotope	A	Z	% Natural abundance	Atomic mass	Half-life	Decay mode	Decay energy (MeV)	Particle energy (MeV)	Particle intensity	Thermal neutron cross section	Spin (h/2π)	μ Nucl. mag. moment	Gamma-ray energy (MeV)	Gamma-ray intensity
	143	57											0.7981 ± 0.0001	0.5%
	143	57											1.1461 ± 0.0002	0.36%
	143	57											1.1485 ± 0.0002	0.5%
	143	57											1.5564 ± 0.0001	0.43%*
	143	57											1.9614 ± 0.0001	0.43%
	143	57											2.5001 ± 0.0001	0.31%
	143	57											2.6247 ± 0.0001	0.13%
$_{57}La^{144}$	144	57		143.919650	40 s.	β-	5.6	4.3					0.3973 ± 0.0002	94%
	144	57						4.6					0.4314 ± 0.0002	3.6%
	144	57											0.5411 ± 0.0002	39%
	144	57											0.5849 ± 0.0003	7.9%
	144	57											0.7054 ± 0.0004	4.1%
	144	57											0.7352 ± 0.0003	7.0%
	144	57											0.8448 ± 0.0002	22%
	144	57											0.9522 ± 0.0003	2.9%
	144	57											0.9688 ± 0.0005	3.3%
	144	57											0.9785 ± 0.0005	1.9%
	144	57											1.0527 ± 0.0003	2%
	144	57											1.2763 ± 0.0005	1.6%
	144	57											1.2943 ± 0.0005	6.5%
	144	57											1.4314 ± 0.0004	4.2%
	144	57											1.4896 ± 0.0006	1.4%
	144	57											1.5235 ± 0.00074	3.5%
	144	57											1.6737 ± 0.0006	1.4%
	144	57											1.7555 ± 0.0008	1.0%
	144	57											1.9423 ± 0.0009	1.7%
	144	57											1.9964 ± 0.0007	2.8%
	144	57											2.0078 ± 0.0009	1.2%
	144	57											2.0505 ± 0.001	1.3%
	144	57											2.3530 ± 0.001	1.9%
	144	57											2.6627 ± 0.001	1.9%
	144	57											2.8652 ± 0.0012	1.0%
$_{57}La^{145}$	145	57		144.921650	25 s.	β-	4.1	4.1					Pr k x-ray	20%
	145	57											0.0700 ± 0.0002	12%
	145	57											0.1182 ± 0.0002	4%
	145	57											0.1641 ± 0.0001	3%
	145	57											0.1698 ± 0.0002	3.5%
	145	57											0.3558 ± 0.0002	4.2%
	145	57											0.4474 ± 0.0002	3.5%
	145	57											0.5052 ± 0.0002	1.9%
	145	57											0.6718 ± 0.0002	2.0%
	145	57											0.7435 ± 0.0002	1.6%
	145	57											0.7865 ± 0.0002	1.9%
	145	57											0.8835 ± 0.0002	0.9%
	145	57											0.8896 ± 0.0002	1.1%
	145	57											0.9320 ± 0.0002	3.1%
	145	57											1.0307 ± 0.0003	1.9%
	145	57											1.0508 ± 0.0003	1.6%
	145	57											1.2380 ± 0.0003	1.0%
	145	57											1.5965 ± 0.0003	1.3%
	145	57											1.8195 ± 0.0003	3.4%
	145	57											1.9461 ± 0.0003	1.0%
	145	57											2.0878 ± 0.0003	0.9%
	145	57											2.1552 ± 0.0003	1.0%
	145	57											2.2047 ± 0.0003	0.9%
	145	57											2.3594 ± 0.0003	1.5%
	145	57											2.3771 ± 0.0005	0.67%
	145	57											2.4792 ± 0.0003	0.8%
	145	57											2.5426 ± 0.0003	0.8%
$_{57}La^{146}$	146	57		145.925530	10 s.	β-	6.2						0.2585 ± 0.0001	100 +
	146	57											0.4099 ± 0.0001	7
	146	57											0.6661 ± 0.0001	10
	146	57											0.7023 ± 0.0001	10
	146	57											0.7847 ± 0.0001	5
	146	57											0.9246 ± 0.0001	12
	146	57											1.0159 ± 0.0001	5
	146	57											1.2744 ± 0.0002	2
	146	57											1.3821 ± 0.0002	3
	146	57											1.4982 ± 0.0002	2.2
	146	57											1.7567 ± 0.0003	1.2
$_{57}La^{147}$	147	57		146.928100	4.1 s.	β-	4.8	4.3-4.6					0.1176 ± 0.0003	15%
	147	57											0.1869 ± 0.0003	7.1%
	147	57											0.2150 ± 0.0003	8%
	147	57											0.2357 ± 0.0003	3.1%
	147	57											0.2735 ± 0.0003	2.5%
	147	57											0.2834 ± 0.0003	2.9%
	147	57											0.3532 ± 0.0003	2%
	147	57											0.3996 ± 0.0005	2%
	147	57											0.4384 ± 0.0003	6%
	147	57											0.4953 ± 0.0005	2%
	147	57											0.5074 ± 0.0005	1%
	147	57											0.5168 ± 0.0005	3%
	147	57											0.5709 ± 0.0005	1%
	147	57											0.5984 ± 0.0005	1%
	147	57											0.7098 ± 0.0005	0.6%

Isotope	A	Z	% Natural abundance	Atomic mass	Half-life	Decay mode	Decay energy (MeV)	Particle energy (MeV)	Particle intensity	Thermal neutron cross section	Spin (h/2π)	μ Nucl. mag. moment	Gamma-ray energy (MeV)	Gamma-ray intensity
$_{57}La^{148}$	148	57		147.931390	≈ 2.6 s.	β-	≈ 6.5						0.1584 ± 0.0001	56%
	148	57											0.2524 ± 0.0001	1.7%
	148	57											0.2949 ± 0.0001	6.7%
	148	57											0.3790 ± 0.0001	4%
	148	57											0.6018 ± 0.0001	7.7%
	148	57											0.6829 ± 0.0001	6.5%
	148	57											0.7603 ± 0.0001	8.6%
	148	57											0.7771 ± 0.0001	7.2%
	148	57											0.8313 ± 0.0001	5.2%
	148	57											0.9583 ± 0.0001	4.0%
	148	57											0.9899 ± 0.0001	9.4%
	148	57											1.3386 ± 0.0001	1.8%
	148	57											1.4315 ± 0.0001	1.3%
	148	57											1.8905 ± 0.0001	1.2%
	148	57											1.9854 ± 0.0001	2.5%
	148	57											1.9947 ± 0.0001	3.3%
	148	57											2.0307 ± 0.0001	1.2%
	148	57											2.0931 ± 0.0001	7.1%
	148	57											2.2191 ± 0.0002	1.5%
	148	57											2.3914 ± 0.0001	3.9%
$_{57}La^{149}$	149	57		148.934310	1.2 s.	β-	≈ 5.1							
Ce		58		140.12						0.6 b.				
$_{58}Ce^{129}$	129	58			≈ 3.5 m.	β+,E.C.							ann.rad.	
	129	58											0.0675 ± 0.0001	
$_{58}Ce^{130}$	130	58			25 m.	β+,E.C.	≈ 2.3				0+		ann.rad.	
	130	58											La k x-ray	15%
$_{58}Ce^{131m}$	131	58			5 m.	β+,E.C.							ann.rad.	
	131	58											0.2304	36 +
	131	58											0.3955	
	131	58											0.4213	54
$_{58}Ce^{131}$	131	58		130.914270	9.5 m.	β+,E.C.							ann.rad.	
	131	58											0.119	
	131	58											0.169	
	131	58											0.414	
$_{58}Ce^{132}$	132	58		131.911490	3.5 h.	E.C.	≈ 1.3				0+		La k x-ray	40%
	132	58											0.1554	11%
	132	58											0.1821	79
	132	58											0.1901	3
	132	58											0.2167	5
	132	58											0.2514	2
	132	58											0.2799	2
	132	58											0.3027	2
	132	58											0.3296	3
	132	58											0.3681	1
	132	58											0.4244	1
	132	58											0.4315	1
	132	58											0.4515	1.5
	132	58											0.5762	1
$_{58}Ce^{133m}$	133	58			97 m.	β+,E.C.					1/2+		ann.rad.	
	133	58											0.0769 ± 0.0005	35 +
	133	58											0.0973 ± 0.0001	100
	133	58											0.1740 ± 0.0005	1
	133	58											0.3767 ± 0.0003	2
	133	58											0.5577 ± 0.0003	25
$_{58}Ce^{133}$	133	58		132.911360	5.4 h.	β+(8%) E.C.(92%)	≈ 3.0	1.3			9/2-		ann.rad.	
	133	58											La k x-ray	55%
	133	58											0.0584 ± 0.0001	19%
	133	58											0.0879 ± 0.0001	5%
	133	58											0.1308 ± 0.0001	18%
	133	58											0.3464 ± 0.0001	4%
	133	58											0.4326 ± 0.0001	3.5%
	133	58											0.4442 ± 0.0001	2.3%
	133	58											0.4755 ± 0.0001	3.2%
	133	58											0.4722 ± 0.0001	39%
	133	58											0.5104 ± 0.0001	21%
	133	58											0.5238 ± 0.0001	3.1%
	133	58											0.5419 ± 0.0001	2.9%
	133	58											0.6118 ± 0.0001	2.6%
	133	58											0.6447 ± 0.0001	2.0%
	133	58											0.6895 ± 0.0001	4.1%
	133	58											0.7845 ± 0.0001	9.6%
	133	58											0.9510 ± 0.0001	1.3%
	133	58											0.9901 ± 0.0001	2.9%
	133	58											1.3772 ± 0.0001	1.7%
	133	58											1.4322 ± 0.0001	1.2%
	133	58											1.4948 ± 0.0001	3.2%
	133	58											1.5004 ± 0.0001	4.8%
	133	58											1.5266 ± 0.0001	2.5%
	133	58											1.5846 ± 0.0001	2.4%
	133	58											1.7202 ± 0.0002	1.3%
	133	58											1.7694 ± 0.0001	1.2%
	133	58											1.8873 ± 0.0003	1.0%
	133	58											2.0182 ± 0.0001	1.4%
	133	58											2.1192 ± 0.0002	1.2%
$_{58}Ce^{134}$	134	58		133.908890	76 h.	E.C.	≈ 0.4				0+		La k x-ray	40%

Isotope	A	Z	% Natural abundance	Atomic mass	Half-life	Decay mode	Decay energy (MeV)	Particle energy (MeV)	Particle intensity	Thermal neutron cross section	Spin (h/2π)	μ Nucl. mag. moment	Gamma-ray energy (MeV)	Gamma-ray intensity
	134	58											0.1304 ± 0.0001	0.21%
	134	58											0.1623 ± 0.0001	0.23%
	134	58											0.6047 ± 0.0001	5%(D)
$_{58}Ce^{135m}$	135	58			20 s.	I.T.	0.446				11/2-		Ce k x-ray	
	135	58											0.0826 ± 0.0001	23%
	135	58											0.1497 ± 0.0001	23%
	135	58											0.2134 ± 0.0001	78%
	135	58											0.2961 ± 0.0001	17%
$_{58}Ce^{135}$	135	58		134.909117	17.8 h.	β+(1%)	2.02	0.80			1/2+		La k x-ray	43%
	135	58				E.C.(99%)							0.0345 ± 0.0001	1.9%
	135	58											0.2065 ± 0.0001	8%
	135	58											0.2656 ± 0.0001	42%
	135	58											0.3001 ± 0.0001	23%
	135	58											0.4836 ± 0.0001	1.9%
	135	58											0.5181 ± 0.0001	13.6%
	135	58											0.5723 ± 0.0001	11%
	135	58											0.5771 ± 0.0001	5.1%
	135	58											0.6046 ± 0.0001	2.9%
	135	58											0.6068 ± 0.0001	19%
	135	58											0.6658 ± 0.0001	3.3%
	135	58											0.7836 ± 0.0001	11%
	135	58											0.8284 ± 0.0001	5.2%
	135	58											0.8714 ± 0.0001	3.2%
	135	58											0.9059 ± 0.0001	1.6%
	135	58											1.1841 ± 0.0001	1.1%
$_{58}Ce^{136}$	136	58	0.19%	135.907140						(1 + 6) b.	0+			
$_{58}Ce^{137m}$	137	58			34.4 h.	I.T.(99%)	0.254				11/2-		Ce k x-ray	29%
	137	58				E.C.(0.8%)							0.1693 ± 0.0001	0.4%
	137	58											0.2543 ± 0.0001	11%
	137	58											0.8248 ± 0.0001	0.4%
$_{58}Ce^{137}$	137	58		136.907780	9.0 h.	β+	1.22				3/2+		La k x-ray	40%
	137	58											0.0106 ± 0.0001	0.6%
	137	58											0.4332 ± 0.0001	0.06%
	137	58											0.4366 ± 0.0001	0.33%
	137	58											0.4472 ± 0.0001	2.2%
$_{58}Ce^{138}$	138	58	0..25%	137.905985						(0.015 + 1.1) b.	0+			
$_{58}Ce^{139m}$	139	58			56 s.	I.T.	0.7542				11/2-		Ce k x-ray	3%
	139	58											0.7542 ± 0.0001	92.5%
$_{58}Ce^{139}$	139	58		138.906631							3/2+		La k x-ray	42%
	139	58											0.16585 ± 0.00003	80%
$_{58}Ce^{140}$	140	58	88.48%	139.905433						0.58 b.	0+			
$_{58}Ce^{141}$	141	58		140.908271	32.5 d.	β-	0.581	0.444	69%		7/2-		Pr k x-ray	9%
	141	58						0.582	31%				0.14544 ± 0.00003	48.4%
$_{58}Ce^{142}$	142	58	11.08%	141.909241						0.95 b.	0+			
$_{58}Ce^{143}$	143	58		142.912383	33.0 h.	β-	1.462	0.74	12%		3/2-		Pr k x-ray	34%
	143	58						1.110	47%				0.0574 ± 0.0001	12%
	143	58											0.2316 ± 0.0001	2%
	143	58											0.2933 ± 0.0001	43%
	143	58											0.3506 ± 0.0001	3.3%
	143	58											0.4904 ± 0.0001	2.1%
	143	58											0.6645 ± 0.0001	5.6%
	143	58											0.7220 ± 0.0001	5.3%
	143	58											0.8804 ± 0.0001	1.0%
	143	58											1.1030 ± 0.0002	0.4%
$_{58}Ce^{144}$	144	58		143.913643	284.4 d.	β-	0.318	0.185	20%		0+		Pr k x-ray	4%
	144	58						0.238	5%				0.0801 ± 0.0001	1.1%
	144	58											0.1335 ± 0.0001	11%
$_{58}Ce^{145}$	145	58		144.917230	2.9 m.	β-	2.5	1.7	24%		5/2-		Pr k x-ray	31%
	145	58						2.1	76%				0.0627 ± 0.0001	15%
	145	58											0.0320 ± 0.0001	2.6%
	145	58											0.2845 ± 0.0001	9.7%
	145	58											0.3514 ± 0.0001	6.3%
	145	58											0.4236 ± 0.0001	4.6%
	145	58											0.4397 ± 0.0001	7.3%
	145	58											0.4924 ± 0.0001	1.5%
	145	58											0.5123 ± 0.0001	1.3%
	145	58											0.6562 ± 0.0001	1.3%
	145	58											0.7245 ± 0.0001	64%
	145	58											0.7832 ± 0.0001	2.5%
	145	58											0.8599 ± 0.0002	2.1%
	145	58											1.1107 ± 0.0001	2.7%
	145	58											1.1481 ± 0.0001	10%
	145	58											1.2105 ± 0.0002	1%
$_{58}Ce^{146}$	146	58		145.918670	13.6 m.	β-	1.0	0.75	90%		0+		Pr k x-ray	7%
	146	58											0.0986 ± 0.0001	3%
	146	58											0.1009 ± 0.0001	2.6%
	146	58											0.1335 ± 0.0001	8.3%
	146	58											0.1413 ± 0.0001	3.3%
	146	58											0.2105 ± 0.0001	5.0%
	146	58											0.2182 ± 0.0001	21%
	146	58											0.2509 ± 0.0001	2.6%
	146	58											0.2646 ± 0.0001	9.1%
	146	58											0.3167 ± 0.0001	57%
	146	58											0.3515 ± 0.0001	2.3%

Isotope	A	Z	% Natural abundance	Atomic mass	Half-life	Decay mode	Decay energy (MeV)	Particle energy (MeV)	Particle intensity	Thermal neutron cross section	Spin (h/2π)	μ Nucl. mag. moment	Gamma-ray energy (MeV)	Gamma-ray intensity
	146	58											0.4157 ± 0.0001	1.3%
	146	58											0.5030 ± 0.0001	1.0%
$_{58}Ce^{147}$	147	58		146.922530	56 s.	β-	3.3	3.3					0.0930 ± 0.0003	4%
	147	58											0.1987 ± 0.0003	1.7%
	147	58											0.2183 ± 0.0003	1.9%
	147	58											0.2687 ± 0.0003	5.5%
	147	58											0.2891 ± 0.0003	1.0%
	147	58											0.3591 ± 0.0003	1.3%
	147	58											0.3741 ± 0.0003	3.1%
	147	58											0.4520 ± 0.0003	2.3%
	147	58											0.4671 ± 0.0003	2.3%
	147	58											0.5779 ± 0.0003	1.0%
	147	58											0.5804 ± 0.0003	1.7%
	147	58											0.7011 ± 0.0003	1.3%
	147	58											0.8223 ± 0.0003	1.3%
$_{58}Ce^{148}$	148	58		147.924410	48 s.	β-	2.0	1.66			0+		0.0904 ± 0.0005	15 +
	148	58											0.0985 ± 0.0005	76
	148	58											0.1052 ± 0.0005	30
	148	58											0.1168 ± 0.0005	18
	148	58											0.1212 ± 0.0005	72
	148	58											0.1301 ± 0.0005	3
	148	58											0.1683 ± 0.0005	4
	148	58											0.1917 ± 0.0005	8
	148	58											0.1957 ± 0.0005	40
	148	58											0.2337 ± 0.0005	6
	148	58											0.2697 ± 0.0005	29
	148	58											0.2738 ± 0.0005	33
	148	58											0.2918 ± 0.0005	100
	148	58											0.3250 ± 0.0005	45
	148	58											0.3327 ± 0.0005	5
	148	58											0.3472 ± 0.0005	9
	148	58											0.3693 ± 0.0005	14
	148	58											0.3744 ± 0.0005	4
	148	58											0.3906 ± 0.0005	3
	148	58											0.3997 ± 0.0005	6
	148	58											0.4220 ± 0.0005	22
	148	58											0.5207 ± 0.0005	2
$_{58}Ce^{149}$	149	58		148.927760	5.2 s.	β-	≈ 3.5						0.0577 ± 0.0003	100 +
	149	58											0.0864 ± 0.0003	20
	149	58											0.3800 ± 0.0003	34
	149	58											0.3900 ± 0.0003	2
	149	58											0.4600 ± 0.0003	2
	149	58											0.7028 ± 0.0003	2
	149	58											0.8645 ± 0.0003	8
	149	58											0.8927 ± 0.0003	8
$_{58}Ce^{150}$	150	58		149.929670	4.4 s.	β-	≈ 3.1						0.1099 ± 0.0003	100 +
$_{58}Ce^{151}$	151	58		150.933170	1.0 s.	β-							0.0526 ± 0.0002	
	151	58											0.0848 ± 0.0001	
	151	58											0.0968 ± 0.0002	
	151	58											0.1186 ± 0.0001	
Pr		59		140.9077						11.4 b.				
$_{59}Pr^{132}$	132	59		131.919120	1.6 m.	β+,E.C.	≈ 7.2						ann.rad.	
	132	59											0.325 ± 0.001	
	132	59											0.496 ± 0.001	
	132	59											0.533 ± 0.001	
$_{59}Pr^{133}$	133	59		132.916190	6.7 m.	β+,E.C.	≈ 4.6				5/2+		ann.rad.	
	133	59											0.074	74 +
	133	59											0.1343	100
	133	59											0.2419	40
	133	59											0.2767	6
	133	59											0.3156	85
	133	59											0.3308	43
	133	59											0.3626	12
	133	59											0.4605	17
	133	59											0.4650	50
	133	59											0.6449	10
	133	59											0.8536	3
	133	59											1.4610	3
	133	59											1.4948	8
	133	59											1.8033	3
	133	59											1.8312	3.5
	133	59											1.8641	5
	133	59											1.8755	4
$_{59}Pr^{134m}$	134	59			11 m.	β+,E.C.							ann.rad.	
	134	59											0.294	
	134	59											0.460	
	134	59											0.495	
	134	59											0.632	
$_{59}Pr^{134}$	134	59		133.915440	17 m.	β+,E.C.	≈ 6.1				2+		ann.rad.	
	134	59											0.294	100 +
	134	59											0.460	15
	134	59											0.495	60
	134	59											0.495	60
	134	59											0.632	10
$_{59}Pr^{135}$	135	59		134.913140	25 m.	β+,E.C.	3.6				3/2+		ann.rad.	

Isotope	A	Z	% Natural abundance	Atomic mass	Half-life	Decay mode	Decay energy (MeV)	Particle energy (MeV)	Particle intensity	Thermal neutron cross section	Spin (h/2π)	μ Nucl. mag. moment	Gamma-ray energy (MeV)	Gamma-ray intensity
	135	59											0.0826 ± 0.0001	50 +
	135	59											0.2135 ± 0.0001	48
	135	59											0.2961 ± 0.0001	100
	135	59											0.4843 ± 0.0003	6
	135	59											0.5832 ± 0.0002	30
	135	59											0.6138 ± 0.0002	6
	135	58											0.6209 ± 0.0002	6
	135	58											0.6975 ± 0.0002	5
	135	59											0.8069 ± 0.0002	4
	135	59											0.9341 ± 0.0002	3
	135	59											1.0169 ± 0.0002	2
	135	59											1.5389 ± 0.0004	4
	135	59											1.7519 ± 0.0006	2
$_{59}Pr^{136}$	136	59		135.912640	13.1 m.	β+(57%)	5.10	2.98			2+		ann.rad.	
	136	59				E.C.(43%)							Ce k x-ray	18%
	136	59											0.4608 ± 0.0002	7.7%
	136	59											0.5398 ± 0.0002	52%
	136	59											0.5522 ± 0.0002	76%
	136	59											1.0008 ± 0.0003	5%
	136	59											1.0925 ± 0.0003	18%
	136	59											1.3597 ± 0.0004	1%
	136	59											1.4250 ± 0.0003	1%
	136	59											1.5148 ± 0.0004	1.9%
	136	59											1.6028 ± 0.0003	3.9%
	136	59											1.8990 ± 0.0005	1%
	136	59											2.0668 ± 0.0003	3%
	136	59											2.2407 ± 0.0004	0.7%
	136	59											2.3136 ± 0.0004	0.6%
	136	59											2.4508 ± 0.0003	0.7%
$_{59}Pr^{137}$	137	59		136.910680	77 m.	β+(26%)	2.70	1.68			5/2+		ann.rad.	
	137	59				E.C.(74%)							Ce k x-ray	30%
	137	59											0.4339 ± 0.0002	1.3%
	137	59											0.5140 ± 0.0002	1.1%
	137	59											0.8367 ± 0.0001	1.8%
	137	59											1.0886 ± 0.0002	0.4%
	137	59											(0.16 - 1.8)	
$_{59}Pr^{138m}$	138	59			2.1 h.	β+(24%)		1.65			7-		ann.rad.	
	138	59				E.C.(76%)							Ce k x-ray	36%
	138	59											0.3027 ± 0.0001	80%
	138	59											0.3909 ± 0.0001	6.1%
	138	59											0.5475 ± 0.0001	5.2%
	138	59											0.6359 ± 0.0001	1.8%
	138	59											0.7887 ± 0.0001	99%
	138	59											1.0378 ± 0.0001	100%
	138	59											(0.07 - 2.0)	
$_{59}Pr^{138}$	138	59		137.910748	1.5 m.	β+(75%)	4.44	3.44			1+		ann.rad.	150%
	138	59				E.C.(25%)							Ce k x-ray	10%
	138	59											0.6882 ± 0.0001	0.8%
	138	59											0.7887 ± 0.0001	2.4%
	138	59											1.4478 ± 0.0002	0.1%
	138	59											1.5511 ± 0.0001	0.4%
$_{59}Pr^{139}$	139	59		138.908917	4.41 h.	β+(8%)	2.13	1.09			5/2+		ann.rad.	16%
	139	59				E.C.(92%)							Ce k x-ray	40%
	139	59											0.2551 ± 0.0001	0.2%
	139	59											1.3473 ± 0.0001	0.4%
	139	59											1.6307 ± 0.0001	0.3%
$_{59}Pr^{140}$	140	59		139.909071	3.39 m.	β+(51%)	3.39	2.37			1+		ann.rad.	100%
	140	59				E.C.(49%)							Ce k x-ray	20%
	140	59											0.3069 ± 0.0002	0.2%
	140	59											1.5965 ± 0.0001	0.5%
$_{59}Pr^{141}$	141	59	100%	140.907647						(3.9 + 7.5) b.	5/2+	+4.3		
$_{59}Pr^{142m}$	142	59			14.6 m.	I.T.	0.004	c.e.			5/2-			
$_{59}Pr^{142}$	142	59		141.910039	19.13 h.	β-	2.160	0.58	4%	20 b.	2-		0.5088 ± 0.0005	0.02%
	142	59						2.16	96%				1.57580 ± 0.00005	3.7%
$_{59}Pr^{143}$	143	59		142.910814	13.58 d.	β-	0.934	0.935		90 b.	7/2+		0.7420 ± 0.0001	0.00001%
$_{59}Pr^{144m}$	144	59			7.2 m.	I.T. (99+%)	0.059				3-		Pr k x-ray	16%
	144	59				β-							0.0590 ± 0.0001	0.08%
	144	59											0.6965 ± 0.0001	0.04%
	144	59											0.8142 ± 0.0002	0.04%
$_{59}Pr^{144}$	144	59		143.913301	β-	2.997	0.807	1%			0-		0.69649 ± 0.00002	1.3%
	144	59					2.30	1.1%					1.48912 ± 0.00004	0.27%
	144	59					2.997	98%					2.18562 ± 0.00005	0.7%
$_{59}Pr^{145}$	145	59		144.914501	5.98 h.	β-	1.81	1.81	97%		7/2+		0.0725 ± 0.00011	0.3%
	145	59											0.6758 ± 0.0001	0.5%
	145	59											0.7483 ± 0.0001	0.52%
	145	59											0.9790 ± 0.0001	0.2%
	145	59											1.1503 ± 0.0001	0.2%
$_{59}Pr^{146}$	146	59		145.917570	24.1 m.	β-	4.1	2.6	30%				0.4539 ± 0.0001	48%
	146	59						3.7	10%				0.6017 ± 0.0001	3%
	146	59						4.1	40%				0.7357 ± 0.0001	7.5%

TABLE OF THE ISOTOPES (Continued)

Isotope	A	Z	% Natural abundance	Atomic mass	Half-life	Decay mode	Decay energy (MeV)	Particle energy (MeV)	Particle intensity	Thermal neutron cross section	Spin (h/2π)	μ Nucl. mag. moment	Gamma-ray energy (MeV)	Gamma-ray intensity
	146	59											0.7889 ± 0.0001	6.3%
	146	59											0.9229 ± 0.0001	2.3%
	146	59											1.0168 ± 0.0001	1.2%
	146	59											1.3767 ± 0.0001	4.4%
	146	59											1.5247 ± 0.0001	45.6%
	146	59											1.6904 ± 0.0002	0.6%
	146	59											2.2275 ± 0.0003	0.5%
	146	59											2.2252 ± 0.0001	1%
	146	59											2.3566 ± 0.0002	0.8%
	146	59											2.6816 ± 0.0003	0.4%
$_{59}$Pr^{147}	147	59		146.918980	13.4 m.	β-	2.68	1.5			5/2+		0.0780 ± 0.0001	10%
	147	59						2.1					0.1279 ± 0.0001	9%
	147	59											0.3146 ± 0.0001	24%
	147	59											0.3357 ± 0.0001	6%
	147	59											0.4778 ± 0.0001	5%
	147	59											0.5548 ± 0.0001	8%
	147	59											0.5779 ± 0.0001	16%
	147	59											0.6413 ± 0.0001	19%
	147	59											0.9960 ± 0.0001	1.6%
	147	59											1.1365 ± 0.0001	1.6%.
	147	59											1.2611 ± 0.0002	5.3%
	147	59											1.3004 ± 0.0001	3%
	147	59											1.3245 ± 0.0003	1%
	147	59											1.4168 ± 0.0002	0.3%
	147	59											1.5433 ± 0.0003	0.3%
	147	59											1.5433 ± 0.0003	0.3%
	147	59											1.5465 ± 0.0008	0.3%
	147	59											1.5935 ± 0.0003	0.3%
	147	59											1.6237 ± 0.0003	0.3%
	147	59											1.6734 ± 0.0003	0.3%
	147	59											1.7932 ± 0.0002	0.3%
$_{59}$Pr^{148m}	148	59			2.0 m.	β-		4.0			(4)		0.5503 ± 0.0001	22%
	148	59											0.6113 ± 0.0001	1%
	148	59											0.8964 ± 0.0001	1%
	148	59											0.9149 ± 0.0001	11%
	148	59											1.4651 ± 0.0001	22%
$_{59}$Pr^{148}	148	59		147.922210	2.28 m.	β-	5.0	4.8			(1)		0.3017 ± 0.0001	58%
	148	59											0.6150 ± 0.0003	2%
	148	59											0.6975 ± 0.0001	4%
	148	59											0.7212 ± 0.0005	4%
	148	59											0.8693 ± 0.0002	4%
	148	59											1.0230 ± 0.0002	5%
	148	59											1.2486 ± 0.0002	3%
	148	59											1.3578 ± 0.0002	5%
	148	59											1.3817 ± 0.0003	2%
	148	59											1.9080 ± 0.0003	1%
	148	59											2.1304 ± 0.0002	1.7%
	148	59											2.6297 ± 0.0004	1%
$_{59}$Pr^{149}	149	59		148.923792	2.3 m.	β-	3.3	3.3	55%		(5/2+)		0.1085 ± 0.0001	9.5%
	149	59											0.1385 ± 0.0001	11%
	149	59											0.1623 ± 0.0001	3%
	149	59											0.1651 ± 0.0001	10%
	149	59											0.2077 ± 0.0002	3%
	149	59											0.2583 ± 0.0001	5.7%
	149	59											0.3164 ± 0.0001	3%
	149	59											0.3213 ± 0.0001	2.5%
	149	59											0.3330 ± 0.0001	6%
	149	59											0.3660 ± 0.0001	3.1%
	149	59											0.4063 ± 0.0001	2.4%
	149	59											0.4330 ± 0.0001	2.4%
	149	59											0.4746 ± 0.0001	2.8%
	149	59											0.5174 ± 0.0001	4.8%
	149	59											0.6230 ± 0.0001	1.8%
	149	59											0.6625 ± 0.0001	1.8%
	149	59											0.7491 ± 0.0001	1.4%
	149	59											0.7820 ± 0.0002	1.3%
$_{59}$Pr^{150}	150	59		149.926360	6.2 s.	β-	≈ 5.0				1+		0.1302 ± 0.0003	100 +
	150	59											0.2512 ± 0.0003	13
	150	59											0.5459 ± 0.0003	16
	150	59											0.8044 ± 0.0003	63
	150	59											0.8527 ± 0.0003	25
	150	59											0.9315 ± 0.0003	18
	150	59											1.0616 ± 0.0003	12
$_{59}$Pr^{151}	151	59		150.927910	4 s.	β-	≈ 3.7						0.1640 ± 0.0001	
$_{59}$Pr^{152}	152	59		151.930700	3.2 s.	β-	≈ 5.4						0.0726	
	151	59											0.164	
	152	59											0.285	
Nd		60		144.24						49 b.				
$_{60}$Nd^{133}	133	60			1.2 m.	β+,E.C.							ann.rad.	
	133	60											0.061	
	133	60											0.106	
	133	60											0.166	
	133	60											0.227	
	133	60											0.251	

Isotope	A	Z	% Natural abundance	Atomic mass	Half-life	Decay mode	Decay energy (MeV)	Particle energy (MeV)	Particle intensity	Thermal neutron cross section	Spin (h/2π)	μ Nucl. mag. moment	Gamma-ray energy (MeV)	Gamma-ray intensity
	133	60											0.369	
$_{60}Nd^{134}$	134	60		133.918870	≈ 8.5 m.	β+(17%)	≈ 2.8				0+		ann.rad.	34%
	134	60				E.C.(83%)							Pr k x-ray	40%
	134	60											0.0901	2%
	134	60											0.1012	2%
	134	60											0.1631	58%
	134	60											0.2168	12%
	134	60											0.2889	13%
	134	60											0.4679	3%
	134	60											0.4835	2%
	134	60											0.9920	2%
	134	60											1.000	4%
$_{60}Nd^{135}$	135	60		134.918190	12 m.	β+(65%)	≈ 4.8				9/2-		ann.rad.	130%
	135	60				E.C.(35%)							Pr k x-ray	35%
	135	60											0.0415 ± 0.0001	23%
	135	60											0.1126 ± 0.0001	5%
	135	60											0.1647 ± 0.0001	4%
	135	60											0.1851 ± 0.0001	3%
	135	60											0.2041 ± 0.0001	51%
	135	60											0.2060 ± 0.0003	3%
	135	60											0.2454 ± 0.0002	3%
	135	60											0.2561 ± 0.0002	3%
	135	60											0.2719 ± 0.0002	2%
	135	60											0.3728 ± 0.0002	2%
	135	60											0.4411 ± 0.0002	15%
	135	60											0.4519 ± 0.0002	4%
	135	60											0.4758 ± 0.0002	8%
	135	60											0.5016 ± 0.0002	10%
	135	60											0.5937 ± 0.0004	4%
	135	60											0.6165 ± 0.0003	2%
	135	60											0.9666 ± 0.0007	3%
	135	60											1.1721 ± 0.0007	1%
	135	60											1.4807 ± 0.0007	1%
	135	60											1.7520 ± 0.001	2%
$_{60}Nd^{136}$	136	60		135.915010	50.7 m.	E.C.(94%)	2.21	1.04			0+		Pr kx-ray	56%
	136	60				β+(6%)							0.0401 ± 0.0002	20%
	136	60											0.1091 ± 0.0001	33%
	136	60											0.1446 ± 0.0002	2%
	136	60											0.1492 ± 0.0001	9%
	136	60											0.4766 ± 0.0001	1.6%
	136	60											0.5749 ± 0.0001	12%
	136	60											0.9724 ± 0.0002	1.1%
$_{60}Nd^{137m}$	137	60			1.6 s.	I.T.	0.5196				11/2-		Nd k x-ray	30%
	137	60											0.1084 ± 0.0005	34%
	137	60											0.1775 ± 0.0005	57%
	137	60											0.2337 ± 0.0005	64%
	137	60											0.2861 ± 0.0005	21%
$_{60}Nd^{137}$	137	60		136.914760	38 m.	β+(40%)	3.80	1.7	20%		1/2+		ann.rad.	80%
	137	60				E.C.(60%)		2.40		20%			Pr k x-ray	45%
	137	60											0.0755 ± 0.0001	17%
	137	60											0.2382 ± 0.0001	4%
	137	60											0.3066 ± 0.0002	10%
	137	60											0.5051 ± 0.0003	9%
	137	60											0.5806 ± 0.0001	13%
	137	60											0.7616 ± 0.0002	9%
	137	60											0.7816 ± 0.0001	9%
	137	60											0.9257 ± 0.0002	7%
	137	60											0.9272 ± 0.0002	3%
	137	60											1.2431 ± 0.0002	1.4%
	137	60											1.6264 ± 0.0002	0.9%
	137	60											1.8131 ± 0.0002	0.8%
	137	60											2.0573 ± 0.0002	1%
$_{60}Nd^{138}$	138	60		137.911820	5.1 h.	E.C.	≈ 1.1				0+		Pr k x-ray	40%
	138	60											0.1995 ± 0.0001	0.6%
	138	60											0.3258 ± 0.0001	3%
$_{60}Nd^{139m}$	139	60			5.5 h.	I.T.(12%)	0.231	1.7			11/2-		Nd k x-ray	3%
	139	60				β+(88%)							Pr k x-ray	53%
	139	60											0.1139 ± 0.0001	34%
	139	60											0.3624 ± 0.0001	2%
	139	60											0.4038 ± 0.0001	2%
	139	60											0.5476 ± 0.0001	2%
	139	60											0.7012 ± 0.0001	3%
	139	60											0.7081 ± 0.0001	22%
	139	60											0.7382 ± 0.0002	30%
	139	60											0.7965 ± 0.0003	4%
	139	60											0.8020 ± 0.0003	6%
	139	60											0.8096 ± 0.0003	5%
	139	60											0.8278 ± 0.0003	9%
	139	60											0.9101 ± 0.0003	6.5%
	139	60											0.9822 ± 0.0002	22%
	139	60											1.0062 ± 0.0004	2.7%
	139	60											1.0123 ± 0.0003	2.3%
	139	60											1.0752 ± 0.0003	3%
	139	60											1.1053 ± 0.0003	2.3%
	139	60											1.3223 ± 0.0004	1.6%

TABLE OF THE ISOTOPES (Continued)

Isotope	A	Z	% Natural abundance	Atomic mass	Half-life	Decay mode	Decay energy (MeV)	Particle energy (MeV)	Particle intensity	Thermal neutron cross section	Spin (h/2π)	μ Nucl. mag. moment	Gamma-ray energy (MeV)	Gamma-ray intensity
	139	60											2.0609 ± 0.0003	4%
$_{60}Nd^{139}$	139	60		138.911920	30 m.	β+(25%)	2.80	1.77			3/2+		ann.rad.	50%
	139	60				E.C.(75%)							Pr k x-ray	31%
	139	60											0.4050 ± 0.0001	6%
	139	60											0.4755 ± 0.0003	1%
	139	60											0.6217 ± 0.0003	1%
	139	60											0.6690 ± 0.0003	1.3%
	139	60											0.9169 ± 0.0003	1.3%
	139	60											0.9234 ± 0.0003	1.1%
	139	60											1.0742 ± 0.0004	2.1%
	139	60											1.4055 ± 0.0005	0.5%
$_{60}Nd^{140}$	140	60		139.909306	3.37 d.	E.C.	0.22				0+		Pr k x-ray	40%
$_{60}Nd^{141m}$	141	60			61 s.	I.T. (99+%)	0.757				11/2-		Nd k x-ray	3%
	141	60											0.7565 ± 0.0003	91.5%
$_{60}Nd^{141}$	141	60		140.909594	2.5 h.	E.C.(98%)	1.81	0.79			3/2+		Pr k x-ray	39%
	141	60				β+(2%)							0.1454 ± 0.0001	0.2%
	141	60											1.1269 ± 0.0002	0.8%
	141	60											1.1473 ± 0.0002	0.3%
	141	60											1.2926 ± 0.0002	0.5%
	141	60											1.2986 ± 0.0002	0.13%
$_{60}Nd^{142}$	142	60	27.13%	141.907719						19 b.	0+			
$_{60}Nd^{143}$	143	60	12.18%	142.909810						330 b.	7/2-	-1.08		
$_{60}Nd^{144}$	144	60	23.80%	143.910083	2.1x10^{15}y.					3.6 b.	0+			
$_{60}Nd^{145}$	145	60	8.30%	144.912570						45 b.	7/2-	-0.66		
$_{60}Nd^{146}$	146	60	17.19%	145.913113						1.4 b.	0+			
$_{60}Nd^{147}$	147	60		146.916097	10.99 d.	β-	0.895	0.805		400 b.	5/2-	0.59	Pr k x-ray	23%
	147	60											0.09111 ± 0.00002	28%
	147	60											0.12048 ± 0.00005	0.4%
	147	60											0.19644 ± 0.00005	0.2%
	147	60											0.27537 ± 0.00002	0.8%
	147	60											0.31941 ± 0.00002	1.95%
	147	60											0.39816 ± 0.00002	0.9%
	147	60											0.43989 ± 0.00002	1.2%
	147	60											0.48924 ± 0.00003	0.15%
	147	60											0.53102 ± 0.00002	13%
	147	60											0.58934 ± 0.00004	0.04%
	147	60											0.59480 ± 0.00003	0.27%
	147	60											0.68052 ± 0.00002	0.03%
	147	60											0.68590 ± 0.00004	0.81%
$_{60}Nd^{148}$	148	60	5.76%	147.916889						2.5 b.	0+			
$_{60}Nd^{149}$	149	60		148.920145	1.73 h.	β-	1.688	1.03	25%		5/2-		Pr k x-ray	16%
	149	60						1.13	26%				0.0970 ± 0.0001	1.4%
	149	60											0.11432 ± 0.00002	19%
	149	60											0.15588 ± 0.00001	5.9%
	149	60											0.1886 ± 0.0001	1.8%
	149	60											0.1989 ± 0.0001	1.4%
	149	60											0.2081 ± 0.0001	2.5%
	149	60											0.21131 ± 0.00001	26%
	149	60											0.2402 ± 0.0001	3.9%
	149	60											0.2677 ± 0.0001	6.0%
	149	60											0.2702 ± 0.0001	11%
	149	60											0.32656 ± 0.00001	4.6%
	149	60											0.3492 ± 0.0001	1.4%
	149	60											0.42355 ± 0.00001	7.5%
	149	60											0.4436 ± 0.0001	1.1%
	149	60											1.54051 ± 0.00001	6.6%
	149	60											1.65483 ± 0.00002	7.9%
	149	60											1.2342 ± 0.0001	0.32%
	149	60											(0.06 - 1.6)	
$_{60}Nd^{150}$	150	60	5.64%	149.920887						2.5 b.	0+			
$_{60}Nd^{151}$	151	60		150.923825	12.4 m.	β-	2.443	1.2			(3/2+)		Pm k x-ray	14%
	151	60											0.1168 ± 0.0001	46%
	151	60											0.1389 ± 0.0001	8.7%
	151	60											0.1707 ± 0.0001	3.8%
	151	60											0.1751 ± 0.0001	7.2%

Isotope	A	Z	% Natural abundance	Atomic mass	Half-life	Decay mode	Decay energy (MeV)	Particle energy (MeV)	Particle intensity	Thermal neutron cross section	Spin (h/2π)	μ Nucl. mag. moment	Gamma-ray energy (MeV)	Gamma-ray intensity
	151	60											0.2557 ± 0.0001	15%
	151	60											0.3006 ± 0.0001	1.9%
	151	60											0.4023 ± 0.0001	2%
	151	60											0.4235 ± 0.0001	6.4%
	151	60											0.4606 ± 0.0001	1.1%
	151	60											0.5852 ± 0.0001	1.6%
	151	60											0.6778 ± 0.0001	7.7%
	151	60											0.7364 ± 0.0002	7.2%
	151	60											0.7394 ± 0.0007	1.5%
	151	60											0.7555 ± 0.0005	1.4%
	151	60											0.7975 ± 0.0002	5.5%
	151	60											0.8411 ± 0.0005	1.1%
	151	60											0.9141 ± 0.0008	1.2%
	151	60											1.0164 ± 0.0002	2.9%
	151	60											1.1221 ± 0.0003	4.6%
	151	60											1.1806 ± 0.0002	15%
	151	60											(0.10 - 1.9)many	
$_{60}\mathrm{Nd}^{152}$	152	60		151.924680	11.4 m.	β-	1.1				0+		0.0160 ± 0.0005	8%
	152	60											0.0746 ± 0.0002	1%
	152	60											0.2501 ± 0.0002	22%
	152	60											0.2785 ± 0.0002	32%
	152	60											0.2946 ± 0.0002	4%
$_{60}\mathrm{Nd}^{154}$	154	60		153.929400	≈ 40 s.	β-	≈ 2.5						0.40	
	154	60											0.70	
Pm		61												
$_{61}\mathrm{Pm}^{134}$	134	61			24 s.	β+,E.C.							ann.rad.	
	134	61											0.294	100 +
	134	61											0.460	15
	134	61											0.495	60
	134	61											0.632	10
$_{61}\mathrm{Pm}^{135}$	135	61			0.8 m.	β+,E.C.					11/2-		0.129	
	135	61											0.1987	100 +
	135	61											0.271	
	135	61											0.3622	20
	135	61											0.465	
$_{61}\mathrm{Pm}^{136}$	136	61		135.923980	1.8 m.	β+(89%)	≈ 7.9				(5/2+)		ann.rad.	180%
	136	61				E.C.(11%)							Nd k x-ray	7%
	136	61											0.3028 ± 0.0005	14%
	136	61											0.3700 ± 0.0005	10%
	136	61											0.3735 ± 0.0003	89%
	136	61											0.4880 ± 0.0005	9.2%
	136	61											0.6027 ± 0.0003	50%
	136	61											0.6780 ± 0.0008	7%
	136	61											0.6930 ± 0.001	3%
	136	61											0.6960 ± 0.001	10%
	136	61											0.7704 ± 0.0003	18%
	136	61											0.8150 ± 0.0003	18%
	136	61											0.8580 ± 0.0003	31%
	136	61											0.8621 ± 0.0005	7.7%
	136	61											1.0597 ± 0.0005	14%
	136	61											1.0700 ± 0.0008	2.6%
$_{61}\mathrm{Pm}^{137}$	137	61		136.920450	2.4 m.	β+,E.C.	≈ 5.1				(11/2-		ann.rad.	
	137	61											0.0870 ± 0.0002	12%
	137	61											0.1086 ± 0.0002	73%
	137	61											0.1775 ± 0.0002	84%
	137	61											0.2687 ± 0.0003	18%
	137	61											0.3251 ± 0.0005	6%
	137	61											0.3288 ± 0.0005	7%
	137	61											0.3523 ± 0.0003	2.7%
	137	61											0.3706 ± 0.0003	6%
	137	61											0.3892 ± 0.0003	6%
	137	61											0.4106 ± 0.0005	15%
	137	61											0.4140 ± 0.0005	4%
	137	61											0.4573 ± 0.0005	9%
	137	61											0.4592 ± 0.0005	4%
	137	61											0.4707 ± 0.0003	8%
	137	61											0.5060 ± 0.001	15%
	137	61											0.5251 ± 0.0004	3%
	137	61											0.5296 ± 0.0004	5%
	137	61											0.5338 ± 0.0004	11%
	137	61											0.5488 ± 0.0003	8%
	137	61											0.5656 ± 0.0003	5%
	137	61											0.5810 ± 0.001	28%
	137	61											0.6908 ± 0.0003	6%
	137	61											0.7595 ± 0.0005	2%
	137	61											0.8368 ± 0.0006	2%
	137	61											0.9230 ± 0.0005	5%
	137	61											1.0647 ± 0.0005	2%
	137	61											1.0922 ± 0.0005	9%
	137	61											1.1899 ± 0.0005	2.4%
	137	61											1.2847 ± 0.0005	14%
$_{61}\mathrm{Pm}^{138}$	138	61		137.919340	3.2 m.	β+(50%)	≈ 7.0	3.9			3+		ann.rad.	100%
	138	61				E.C.(50%)							Nd k x-ray	21%
	138	61											0.4372 ± 0.0002	9.5%

Isotope	A	Z	% Natural abundance	Atomic mass	Half-life	Decay mode	Decay energy (MeV)	Particle energy (MeV)	Particle intensity	Thermal neutron cross section	Spin (h/2π)	μ Nucl. mag. moment	Gamma-ray energy (MeV)	Gamma-ray intensity
	138	61											0.4931 ± 0.0002	20%
	138	61											0.5209 ± 0.0002	91%
	138	61											0.7290 ± 0.0002	35%
	138	61											0.7406 ± 0.0003	6%
	138	61											0.8103 ± 0.0003	2.8%
	138	61											0.8290 ± 0.0003	6%
	138	61											0.9306 ± 0.0002	4.7%
	138	61											0.9721 ± 0.0003	4%
	138	61											1.0116 ± 0.0003	3.5%
	138	61											1.0140 ± 0.0003	7%
	138	61											1.1346 ± 0.0003	2.3%
	138	61											1.2791 ± 0.0003	10%
	138	61											1.4828 ± 0.0003	2%
	138	61											1.6753 ± 0.0003	3%
	138	61											1.8029 ± 0.0005	1.6%
	138	61											2.6050 ± 0.0004	2.7%
	138	61											3.4605 ± 0.0004	2.7%
	138	61											3.4799 ± 0.0004	1%
$_{61}Pm^{139}$	139	61		138.916780	4.1 m.	β+(68%)	4.4	3.0			(5/2+)		ann.rad.	
	139	61				E.C.(32%)							Nd k x-ray	14%
	139	61											0.3678 ± 0.0002	3%
	139	61											0.4028 ± 0.0002	12%
	139	61											0.4631 ± 0.0002	3.5%
	139	61											0.7565 ± 0.0003	1.7%
	139	61											1.7046 ± 0.0003	0.6%
	139	61											(0.27 - 2.4)	
$_{61}Pm^{140m}$	140	61			5.9 m.	β+(70%)					7/2-		ann.rad.	140%
	140	61				E.C.(30%)							Nd k x-ray	15%
	140	61											0.4199 ± 0.0002	92%
	140	61											0.7738 ± 0.0002	100%
	140	61											1.0283 ± 0.0002	100%
	140	61											1.1975 ± 0.0002	3.8%
	140	61											2.1457 ± 0.0005	0.75%
$_{61}Pm^{140}$	140	61		139.915820	9.2 s.	β+(89%)	6.0	5.0	74%		1+		ann.rad.	180%
	140	61				E.C.(11%)							Nd k x-ray	5%
	140	61											0.7738 ± 0.0002	5.3%
	140	61											1.4898 ± 0.0005	1.0%
$_{61}Pm^{141}$	141	61		140.913600	20.9 m.	β+(52%)	3.72				5/2+		ann.rad.	100%
	141	61				E.C.(48%)							Nd k x-ray	21%
	141	61											0.1937 ± 0.0001	1.6%
	141	61											0.8862 ± 0.0001	2.4%
	141	61											1.2233 ± 0.0001	4.6%
	141	61											1.3455 ± 0.0001	1.3%
	141	61											1.4031 ± 0.0001	0.7%
	141	61											1.5647 ± 0.0001	0.8%
	141	61											2.0738 ± 0.0001	0.6%
$_{61}Pm^{142}$	142	61		141.912970	40.5 s.	β+(86%)	4.9	3.8			1+		ann.rad.	170%
	142	61				E.C.(20%)							Nd k x-ray	9%
	142	61											0.6414 ± 0.0005	0.6%
	142	61											1.5758 ± 0.0004	3.3%
	142	61											2.3843 ± 0.0006	0.1%
$_{61}Pm^{143}$	143	61		142.910930	265 d.	E.C.	1.04				5/2+		Nd k x-ray	40%
	143	61											0.7420 ± 0.0001	38%
$_{61}Pm^{144}$	144	61		143.912588	363 d.	E.C.	2.333				5-		Nd k x-ray	41%
	144	61											0.4768 ± 0.0001	42%
	144	61											0.6180 ± 0.0001	99%
	144	61											0.6965 ± 0.0001	100%
	144	61											0.7786 ± 0.0001	1.5%
	144	61											0.8141 ± 0.0001	0.6%
$_{61}Pm^{145}$	145	61		144.912743	17.7 y.	E.C.	0.161				5/2+		Nd k x-ray	37%
	145	61											0.0672 ± 0.0001	0.55%
	145	61											0.0723 ± 0.0001	1.8%
$_{61}Pm^{146}$	146	61		145.914708	5.53 y.	E.C.(63%)	1.48			8 x 10^3 b.	3-		Nd k x-ray	25%
	146	61				β-(37%)	1.542	0.795					0.4538 ± 0.0002	62%
	146	61											0.6333 ± 0.0003	2.5%
	146	61											0.7362 ± 0.0004	22%
	146	61											0.7474 ± 0.0003	36%
$_{61}Pm^{147}$	147	61		146.915135	2.6234 y.	β-	0.225	0.224		(85 + 97) b.	7/2+	± 2.7	0.1213 ± 0.0001	0.003%
	147	61											0.1974 ± 0.0001	0.0001%
$_{61}Pm^{148m}$	148	61			41.3 d.	β-(95%)	2.6	0.4	60%	1.1 x 10^4 b.	6-		0.0985 ± 0.0001	2.5%
	148	61				I.T.(5%)	0.137	0.5	17%				0.2881 ± 0.0001	13%
	148	61						0.7	21%				0.3116 ± 0.0001	3.9%
	148	61											0.4141 ± 0.0001	19%
	148	61											0.4328 ± 0.0001	5.4%
	148	61											0.5013 ± 0.0001	6.7%
	148	61											0.5503 ± 0.0001	96%
	148	61											0.5997 ± 0.0001	13%
	148	61											0.6113 ± 0.0001	5.5%
	148	61											0.6300 ± 0.0001	89%
	148	61											0.7257 ± 0.0001	33%
	148	61											0.9153 ± 0.0001	17%
	148	61											1.0138 ± 0.0001	20%
$_{61}Pm^{148}$	148	61		147.917473	5.37 d.	β-	2.47	1.02		2 x 10^3 b.	1-	+2.0	0.5503 ± 0.0001	22%
	148	61											0.6113 ± 0.0001	1%
	148	61											0.8964 ± 0.0001	1%

TABLE OF THE ISOTOPES (Continued)

Isotope	A	Z	% Natural abundance	Atomic mass	Half-life	Decay mode	Decay energy (MeV)	Particle energy (MeV)	Particle intensity	Thermal neutron cross section	Spin (h/2π)	μ Nucl. mag. moment	Gamma-ray energy (MeV)	Gamma-ray intensity
	148	61											0.9149 ± 0.0001	11%
	148	61											1.4651 ± 0.0001	22%
$_{61}Pm^{149}$	149	61		148.918332	53.1 h.	β-	1.073	0.78	9%	1400 b.	7/2+		0.2859 ± 0.0001	3.1%
	149	61						1.062	90%				0.5909 ± 0.0003	0.07%
	149	61											0.8305 ± 0.0002	0.03%
	149	61											0.8332 ± 0.0002	0.03%
	149	61											0.8594 ± 0.0003	0.1%
	149	61											0.8819 ± 0.0005	0.02%
$_{61}Pm^{150}$	150	61		149.920981	2.69 h.	β-	3.45	1.6			(1-)		0.3339 ± 0.0001	69%
	150	61						2.1					0.4065 ± 0.0001	5.6%
	150	61											0.7122 ± 0.0001	4.4%
	150	61											0.7375 ± 0.0001	2.3%
	150	61											0.8318 ± 0.0001	12%
	150	61											0.8600 ± 0.0001	3.4%
	150	61											0.8764 ± 0.0001	7.4%
	150	61											1.1658 ± 0.0001	16%
	150	61											1.1939 ± 0.0001	5%
	150	61											1.2233 ± 0.0001	2.9%
	150	61											1.3245 ± 0.0001	18%
	150	61											1.3793 ± 0.0001	3.2%
	150	61											1.7364 ± 0.0001	7%
	150	61											1.9367 ± 0.0001	1.5%
	150	61											(0.25 - 2.9)	
$_{61}Pm^{151}$	151	61		150.921203	28.4 h.	β-	1.187	0.84		700 b.	5/2+	± 1.8	0.1000 ± 0.0001	2.5%
	151	61											0.1048 ± 0.0001	3.5%
	151	61											0.1636 ± 0.0001	1.5%
	151	61											0.1677 ± 0.0001	7.8%
	151	61											0.1772 ± 0.0001	3.6%
	151	61											0.2090 ± 0.0001	1.6%
	151	61											0.2324 ± 0.0001	1.0%
	151	61											0.2401 ± 0.0001	3.6%
	151	61											0.2751 ± 0.0001	6.6%
	151	61											0.3239 ± 0.0001	1.2%
	151	61											0.3401 ± 0.0001	22%
	151	61											0.3449 ± 0.0001	2.1%
	151	61											0.4408 ± 0.0001	1.5%
	151	61											0.4457 ± 0.0001	4.0%
	151	61											0.6361 ± 0.0001	1.4%
	151	61											0.7176 ± 0.0001	4.0%
	151	61											0.7527 ± 0.0001	1.2%
	151	61											0.7726 ± 0.0001	0.8%
	151	61											0.8078 ± 0.0001	0.5%
$_{61}Pm^{152m2}$	152	61			15 m.	β-,I.T.					(>6)		0.1218 ± 0.0001	
	152	61											0.1374 ± 0.0005	
	152	61											0.2003 ± 0.0005	
	152	61											0.2299 ± 0.0003	
	152	61											0.2447 ± 0.0001	
	152	61											0.3404 ± 0.0001	
	152	61											0.3604 ± 0.0005	
	152	61											1.2140 ± 0.001	
	152	61											1.2338 ± 0.0001	
	152	61											1.4375 ± 0.0003	
$_{61}Pm^{152m1}$	152	61			7.52 m.	β-					(4-)		0.1218 ± 0.0001	45%
	152	61											0.2447 ± 0.0001	78%
	152	61											0.3404 ± 0.0001	31%
	152	61											0.6562 ± 0.0003	3.5%
	152	61											0.6883 ± 0.0002	2.3%
	152	61											0.7808 ± 0.0001	4.2%
	152	61											0.8102 ± 0.0002	5.2%
	152	61											1.0051 ± 0.0001	2.9%
	152	61											1.0971 ± 0.0001	29%
	152	61											1.1121 ± 0.0001	4%
	152	61											1.4375 ± 0.0001	23%
	152	61											2.2007 ± 0.0002	0.7%
$_{61}Pm^{152}$	152	61		151.923490	4.1 m.	β-	3.5	3.45	20%		1+		0.1218 ± 0.0001	16%
	152	61						3.50	60%				0.6959 ± 0.0001	1.3%
	152	61											0.8414 ± 0.0001	2.2%
	152	61											0.9609 ± 0.0004	1.9%
	152	61											0.9633 ± 0.0004	1.8%
	152	61											1.3212 ± 0.0002	0.6%
	152	61											(0.12 - 2.1)	
$_{61}Pm^{153}$	153	61		152.924134	5.4 m.	β-	1.90	1.65			(5/2-)		0.0910 ± 0.0003	3.5%
	153	61											0.1198 ± 0.0001	6%
	153	61											0.1273 ± 0.0001	14%
	153	61											0.1294 ± 0.0001	1.8%
	153	61											0.1754 ± 0.0001	2%
	153	61											0.1829 ± 0.0001	2.7%
$_{61}Pm^{154m}$	154	61			2.7 m.	β-		2.0					0.0820 ± 0.0001	15%
	154	61											0.1848 ± 0.0001	30%
	154	61											0.2311 ± 0.0003	10%
	154	61											0.2802 ± 0.0002	11%
	154	61											0.3589 ± 0.0003	3%
	154	61											0.5467 ± 0.0002	10%
	154	61											0.7428 ± 0.0004	2.8%
	154	61											0.7453 ± 0.0004	3.2%

Isotope	A	Z	% Natural abundance	Atomic mass	Half-life	Decay mode	Decay energy (MeV)	Particle energy (MeV)	Particle intensity	Thermal neutron cross section	Spin (h/2π)	μ Nucl. mag. moment	Gamma-ray energy (MeV)	Gamma-ray intensity
	154	61											0.8342 ± 0.0003	3.5%
	154	61											0.8396 ± 0.0002	2.1%
	154	61											0.9305 ± 0.0002	4.9%
	154	61											1.2047 ± 0.0006	1.6%
	154	61											1.2737 ± 0.0005	1.7%
	154	61											1.3586 ± 0.0002	8.8%
	154	61											1.3940 ± 0.0002	3.4%
	154	61											1.4403 ± 0.0002	20%
	154	61											1.4577 ± 0.0004	3.3%
	154	61											1.5494 ± 0.0006	2.6%
	154	61											1.5513 ± 0.0006	1.6%
	154	61											1.6256 ± 0.0002	3.7%
	154	61											0.6561 ± 0.0002	3.7%
	154	61											1.7339 ± 0.0003	2.0%
	154	61											1.7976 ± 0.0004	2.0%
	154	61											1.8409 ± 0.0002	2.9%
	154	61											2.0589 ± 0.0002	5.5%
	154	61											2.1409 ± 0.0002	3.1%
$_{61}Pm^{154}$	154	61		153.926500	1.7 m.	β-	4.0	1.9					0.0820 ± 0.0001	12%
	154	61											0.1848 ± 0.0001	4.7%
	154	61											0.7543 ± 0.0006	4.5%
	154	61											0.8396 ± 0.0002	12%
	154	61											0.8915 ± 0.0002	6.6%
	154	61											0.9111 ± 0.0003	4.5%
	154	61											0.9216 ± 0.0002	8.3%
	154	61											0.9700 ± 0.0002	5.0%
	154	61											1.0176 ± 0.0002	9.9%
	154	61											1.0962 ± 0.0003	5.6%
	154	61											1.1481 ± 0.0002	9.0%
	154	61											1.1778 ± 0.0008	3.6%
	154	61											1.3940 ± 0.0002	12%
	154	61											2.0589 ± 0.0002	19%
	154	61											2.1409 ± 0.0002	11%
	154	61											2.3477 ± 0.0003	1.7%
	154	61											2.5106 ± 0.0002	1.3%
	154	61											(0.08 - 2.8)	
$_{61}Pm^{155}$	155	61		154.927960	48 s.	β-	≈ 3.1				(5/2)		0.0531 ± 0.0005	0.94%
	155	61											0.4098 ± 0.00002	2.2%
	155	61											0.7254 ± 0.0002	5.3%
	155	61											0.7620 ± 0.0003	1.5%
	155	61											0.7786 ± 0.0002	7.8%
Sm		62		150.36						59 x 10^2 b.				
$_{62}Sm^{138}$	138	62		137.923420	3.0 m.	β+,E.C.	≈ 3.5				0+		ann.rad.	
	138	62											0.0536	
	138	62											0.0747	
$_{62}Sm^{139m}$	139	62			≈ 9.5 s.	I.T.(94%)	0.457				(11/2-		Sm k x-ray	29%
	139	62				β+(6%)							0.1118 ± 0.0003	23%
	139	62											0.1553 ± 0.0003	33%
	139	62											0.1901 ± 0.0003	37%
	139	62											0.2673 ± 0.0003	36%
$_{62}Sm^{139}$	139	62		138.922600	2.6 m.	β+(75%)	5.4	4.1			(1/2+)		Pm k x-ray	14%
	139	62				E.C.(25%)							0.3678 ± 0.0002	3%
	139	62											0.4028 ± 0.0002	12%
	139	62											0.4631 ± 0.0002	3.5%
	139	62											0.7565 ± 0.0003	1.7%
	139	62											0.8158 ± 0.0002	1.0%
	139	62											0.9816 ± 0.0004	0.9%
	139	62											1.7046 ± 0.0003	0.6%
	139	62											(0.27 - 2.4)	
$_{62}Sm^{140}$	140	62		139.919040	14.8 m.	β+,E.C.	≈ 3.0	1.9			0+		ann.rad.	
	140	62											Pm k x-ray	35%
	140	62											0.1141 ± 0.0005	1.5%
	140	62											0.1201 ± 0.0005	3%
	140	62											0.1396 ± 0.0005	8%
	140	62											0.2207 ± 0.0005	2%
	140	62											0.2255 ± 0.0005	13%
	140	62											0.3398 ± 0.0005	2%
	140	62											0.3448 ± 0.0005	1.1%
	140	62											1.13800 ± 0.0007	1.5%
	140	62											1.2745 ± 0.0007	1.3%
	140	62											1.5298 ± 0.0007	1%
	140	62											(0.07 - 1.7)	
$_{62}Sm^{141m}$	141	62			22.6 m.	β+(32%)		1.6			11/2-		ann.rad.	64%
	141	62				E.C.(68%)		2.19					Pm k x-ray	36%
	141	62				I.T.(0.3%)	0.1758						0.1966 ± 0.0003	75%
	141	62											0.4318 ± 0.0001	41%
	141	62											0.5385 ± 0.0003	8.5%
	141	62											0.6387 ± 0.0001	2.7%
	141	62											0.6768 ± 0.0003	1.4%
	141	62											0.6846 ± 0.0002	8%
	141	62											0.7257 ± 0.0005	1.5%
	141	62											0.7503 ± 0.0003	1.6%
	141	62											0.7774 ± 0.0003	21%
	141	62											0.7859 ± 0.0001	7%

Isotope	A	Z	% Natural abundance	Atomic mass	Half-life	Decay mode	Decay energy (MeV)	Particle energy (MeV)	Particle intensity	Thermal neutron cross section	Spin (h/2π)	μ Nucl. mag. moment	Gamma-ray energy (MeV)	Gamma-ray intensity
	141	62											0.8059 ± 0.0001	3.6%
	141	62											0.8371 ± 0.0002	3.6%
	141	62											0.7850 ± 0.0001	1.3%
	141	62											0.8965 ± 0.0001	1.5%
	141	62											0.9113 ± 0.0003	9.3%
	141	62											0.9247 ± 0.0001	2.3%
	141	62											0.9833 ± 0.0003	7.4%
	141	62											1.0091 ± 0.0004	3%
	141	62											1.1084 ± 0.0002	1.3%
	141	62											1.1176 ± 0.0002	3.3%
	141	62											1.1451 ± 0.0002	8.9%
	141	62											1.4634 ± 0.0006	1.8%
	141	62											1.4903 ± 0.0001	9.4%
	141	62											1.7864 ± 0.0004	11.1%
	141	62											2.0737 ± 0.0002	1.4%
$_{62}Sm^{141}$	141	62		140.918473	10.2 m.	β+(52%)	4.55	3.2			1/2+		ann.rad.	104%
	141	62				E.C.(48%)							Pm k x-ray	23%
	141	62											0.3244 ± 0.0002	2.5%
	141	62											0.4039 ± 0.0001	2.5%
	141	62											0.4382 ± 0.0001	38%
	141	62											1.0571 ± 0.0002	3.3%
	141	62											1.0919 ± 0.0002	2.6%
	141	62											1.2926 ± 0.0002	6.8%
	141	62											1.4639 ± 0.0004	1.9%
	141	62											1.4957 ± 0.0002	1.8%
	141	62											1.6007 ± 0.0003	4.0%
	141	62											2.0378 ± 0.0003	2.8%
$_{62}Sm^{142}$	142	62		141.915206	72.5 m.	β+(6%)	2.1	1.0			0+		ann.rad.	12%
	142	62				E.C.(94%)							Pm k x-ray	38%
$_{62}Sm^{143m}$	143	62			66 s.	I.T.(99%)	0.7540				11/2-		Sm k x-ray	4%
	143	62				E.C.(0.2%)							0.2718 ± 0.0004	0.2%
	143	62											0.6886 ± 0.0004	0.2%
	143	62											0.7540 ± 0.002	90%
$_{62}Sm^{143}$	143	62		142.914626	8.83 m.	β+(46%)	3.44	2.47			3/2+		ann.rad.	92%
	143	62				E.C.(54%)							Pm k x-ray	22%
	143	62											0.2718 ± 0.0003	0.3%
	143	62											1.0565 ± 0.0002	1.8%
	143	62											1.1734 ± 0.0004	0.4%
	143	62											1.5149 ± 0.0002	0.6%
$_{62}Sm^{144}$	144	62	3.1%	143.911998						0.7 b.	0+			
$_{62}Sm^{145}$	145	62		144.913409	340 d.	E.C.	0.621				7/2-		Pm k x-ray	72%
	145	62											0.0613 ± 0.0001	12%
	145	62											0.4924 ± 0.0002	0.003%
$_{62}Sm^{146}$	146	62		145.913053	$1.03x10^{8}$y.	α		2.50			0+			
$_{62}Sm^{147}$	147	62	15.0%	146.914895	$1.08x10^{11}$y.	α		2.23		57 b.	7/2-	-0.813		
$_{62}Sm^{148}$	148	62	11.3%	147.914820	$7x10^{15}$ y.	α		1.96		3 b.	0+			
$_{62}Sm^{149}$	149	62	13.8%	148.917181	10^{16} y.	α				4 10^{4} b.	7/2-	-0.66		
$_{62}Sm^{150}$	150	62	7.4%	149.917273						103 b.	0+			
$_{62}Sm^{151}$	151	62		150.919929	90 y.	β-	0.076	0.076		1.5 x 10^{4} b.	5/2-		0.02154 ± 0.00001	0.03%
$_{62}Sm^{152}$	152	62	26.7%	151.919729						208 b.	0+			
$_{62}Sm^{153}$	153	62		152.922094	46.7 h.	β-	0.810	0.64		400 b.	3/2+		Eu k x-ray	31%
	153	62						0.71					0.069676 ± 0.00004	5.3%
	153	62											0.10318 ± 0.00001	28%
	153	62											0.17286 ± 0.00001	0.7%
$_{62}Sm^{154}$	154	62	22.7%	153.922206						8 b.	0+			
$_{62}Sm^{155}$	155	62		154.924636	22.2 m.	β1.622	1.52				3/2-		Eu k x-ray	9%
	155	62											0.10432 ± 0.00005	75%
	155	62											0.1414 ± 0.0001	2%
	155	62											0.2457 ± 0.0001	3.8%
	155	62											0.5225 ± 0.0002	0.15%
$_{62}Sm^{156}$	156	62		155.925518	9.4 h.	β-	0.71	0.43			0+		0.0381 ± 0.0001	3%
	156	62						0.71					0.0872 ± 0.0002	24%
	156	62											0.1657 ± 0.0004	15%
	156	62											0.2038 ± 0.0001	23%
	156	62											0.2464 ± 0.0003	1.3%
	156	62											0.2687 ± 0.0006	2.5%
	156	62											0.2907 ± 0.0005	3%
$_{62}Sm^{157}$	157	62		156.928210	8.1 m.	β-	2.6	2.4			3/2-		Eu k x-ray	9%
	157	62											0.0767 ± 0.0003	2%
	157	62											0.1210 ± 0.0002	6%
	157	62											0.1964 ± 0.0002	22%
	157	62											0.1978 ± 0.0002	62%
	157	62											0.2631 ± 0.0002	1.6%
	157	62											0.3175 ± 0.0003	1.6%
	157	62											0.3942 ± 0.0002	13%
	157	62											0.8440 ± 0.0002	6%
	157	62											1.3861 ± 0.0003	1.2%
	157	62											1.4630 ± 0.0003	3.4%
$_{62}Sm^{158}$	158	62		157.930000	5.5 m.	β-	≈ 1.9				0+		0.1002 ± 0.0003	4.6%
	158	62											0.1490 ± 0.0003	4.7%
	158	62											0.1777 ± 0.0003	4%

Isotope	A	Z	% Natural abundance	Atomic mass	Half-life	Decay mode	Decay energy (MeV)	Particle energy (MeV)	Particle intensity	Thermal neutron cross section	Spin (h/2π)	μ Nucl. mag. moment	Gamma-ray energy (MeV)	Gamma-ray intensity
	158	62											0.1894 ± 0.0003	15%
	158	62											0.1907 ± 0.0003	4%
	158	62											0.2241 ± 0.0003	8.5%
	158	62											0.2266 ± 0.0003	5.2%
	158	62											0.2297 ± 0.0003	6.7%
	158	62											0.2854 ± 0.0003	1.7%
	158	62											0.2997 ± 0.0003	2.1%
	158	62											0.3213 ± 0.0003	8.3%
	158	62											0.3245 ± 0.0003	11%
	158	62											0.3268 ± 0.0003	2%
	158	62											0.3386 ± 0.0003	3.7%
	158	62											0.3617 ± 0.0003	6.6%
	158	62											0.3636 ± 0.0003	12%
	158	62											0.5512 ± 0.0003	3%
	158	62											0.7914 ± 0.0003	1.6%
	158	62											1.1269 ± 0.0003	1.2%
Eu		63								4600 b.				
$_{63}Eu^{141m}$	141	63			3.3 s.	β+(58%)					11/2-		ann.rad.	118%
	141	63				E.C.(9%)							Eu k x-ray	4%
	141	63				I.T.(33%)	0.0964						0.0964 ± 0.0002	0.7%
	141	63											0.3940 ± 0.0002	0.6%
	141	63											0.8829 ± 0.0002	0.5%
	141	63											1.5953 ± 0.0003	0.4%
	141	63											(0.09 - 1.6)	
$_{63}Eu^{141}$	141	63		140.924870	40 s.	β+(81%)	6.0				5/2+		ann.rad.	170%
	141	63				E.C.(15%)							Sm k x-ray	7%
	141	63											0.3695 ± 0.0002	2.5%
	141	63											0.3829 ± 0.0002	4.5%
	141	63											0.3845 ± 0.0002	8.5%
	141	63											0.3940 ± 0.0002	14%
	141	63											0.3956 ± 0.0002	2.5%
	141	63											0.5931 ± 0.0002	4.5%
	141	63											0.5979 ± 0.0002	1.9%
	141	63											0.6059 ± 0.0002	1.5%
	141	63											0.8829 ± 0.0002	1.1%
	141	63											0.9961 ± 0.0003	0.6%
	141	63											0.9998 ± 0.0003	0.4%
	141	63											1.2454 ± 0.0003	0.4%
	141	63											1.7662 ± 0.0005	0.6%
$_{63}Eu^{142m}$	142	63			1.22 m.	β+(83%)		4.8					ann.rad.	160%
	142	63				E.C.(17%)							Sm k x-ray	7%
	142	63											0.5400 ± 0.0002	5%
	142	63											0.5566 ± 0.0002	86%
	142	63											0.5637 ± 0.0002	8%
	142	63											0.6287 ± 0.0002	4%
	142	63											0.7680 ± 0.0002	99%
	142	63											1.0161 ± 0.0002	11%
	142	63											1.0233 ± 0.0002	91%
	142	63											1.3419 ± 0.0002	3%
	142	63											1.7001 ± 0.0003	0.8%
	142	63											2.2584 ± 0.0002	0.6%
$_{63}Eu^{142}$	142	63		141.923150	β+(94%)	≈ 7.5	7.0				1+		ann.rad.	190%
	142	63				E.C.(6%)							0.7680 ± 0.0002	11%
	142	63											0.8896 ± 0.0003	1.5%
	142	63											1.2874 ± 0.0003	1.5%
	142	63											1.6581 ± 0.0005	1.5%
	142	63											1.7541 ± 0.0004	1.5%
	142	63											2.0555 ± 0.001	0.5%
$_{63}Eu^{143}$	143	63		142.920150	2.62 m.	β+(72%)	5.1	4.1			5/2+		ann.rad.	140%
	143	63				E.C.(28%)		5.1					Sm k x-ray	12%
	143	63											0.1077 ± 0.0001	2%
	143	63											0.8053 ± 0.0002	1%
	143	63											1.4684 ± 0.0002	1.1%
	143	63											1.5368 ± 0.0003	3.3%
	143	63											1.6073 ± 0.0002	1%
	143	63											1.8049 ± 0.0002	1.6%
	143	63											1.9127 ± 0.0002	2.1%
	143	63											2.1045 ± 0.0002	0.9%
$_{63}Eu^{144}$	144	63		143.918792	10.2 s.	β+(86%)	6.32	5.2			1+		ann.rad.	4%
	144	63				E.C.(13%)							Sm k x-ray	5%
	144	63											0.8177 ± 0.0002	1.5%
	144	63											1.6601 ± 0.0002	9.6%
	144	63											2.4233 ± 0.0002	1%
$_{63}Eu^{145}$	145	63		144.916267	5.93 d.	β+(2%)	2.72	0.80			5/2+		ann.rad.	4%
	145	63				E.C.(98%)	1.70						Sm k x-ray	41%
	145	63											0.5426 ± 0.0002	4.2%
	145	63											0.6535 ± 0.0001	15%
	145	63											0.7648 ± 0.0002	1.6%
	145	63											0.8937 ± 0.0002	66%
	145	63											1.6587 ± 0.0002	16%
	145	63											1.8044 ± 0.00022	1.1%
	145	63											1.8768 ± 0.0002	1.4%
	145	63											1.9970 ± 0.0002	7%
$_{63}Eu^{146}$	146	63		145.917215	4.58 d.	β+(5%)	3.87	1.47			4-		ann.rad.	10%

Isotope	A	Z	% Natural abundance	Atomic mass	Half-life	Decay mode	Decay energy (MeV)	Particle energy (MeV)	Particle intensity	Thermal neutron cross section	Spin (h/2π)	μ Nucl. mag. moment	Gamma-ray energy (MeV)	Gamma-ray intensity
	146	63				E.C.(95%)							Sm k x-ray	38%
	146	63											0.4305 ± 0.0001	4.8%
	146	63											0.5223 ± 0.0002	4.9%
	146	63											0.6336 ± 0.0001	43%
	146	63											0.6341 ± 0.0001	37%
	146	63											0.6655 ± 0.0001	6.9%
	146	63											0.7025 ± 0.0001	6.5%
	146	63											0.7032 ± 0.0002	8.5%
	146	63											0.7047 ± 0.0002	2.7%
	146	63											0.7470 ± 0.0001	98%
	146	63											0.9002 ± 0.0003	3.7%
	146	63											1.0583 ± 0.0002	6.7%
	146	63											1.1760 ± 0.0002	2.1%
	146	63											1.2969 ± 0.0003	5.6%
	146	63											1.4088 ± 0.0004	3.2%
	146	63											1.5396 ± 0.0002	6.1%
	146	63											1.9310 ± 0.0002	1.2%
	146	63											2.0808 ± 0.0003	2.2%
	146	63											2.0808 ± 0.0003	2.2%
	146	63											2.4365 ± 0.0003	1.0%
	146	63											(0.27 - 2.64)	
$_{63}Eu^{147}$	147	63		146.916742	24.3 d.	E.C.(99.%)					5/2+		Sm k x-ray	52%
	147	63				β+(0.4%)							0.12113 ± 0.0008	23%
	147	63											0.19725 ± 0.0009	26%
	147	63											0.6014 ± 0.0001	6.8%
	147	63											0.6776 ± 0.0001	11%
	147	63											0.7988 ± 0.0001	5.5%
	147	63											0.8571 ± 0.0001	3.1%
	147	63											0.9331 ± 0.0001	3.6%
	147	63											0.9559 ± 0.0001	3.9%
	147	63											1.0772 ± 0.0001	6.4%
	147	63											1.2559 ± 0.0002	1.0%
$_{63}Eu^{148}$	148	63		147.918125	54.5 d.	E.C.	3.12				5-		Sm k x-ray	41%
	148	63											0.2415 ± 0.0002	1%
	148	63											0.4139 ± 0.0002	19%
	148	63											0.4327 ± 0.0002	2.8%
	148	63											0.5503 ± 0.0001	99%
	148	63											0.5532 ± 0.0001	17%
	148	63											0.5719 ± 0.0001	9.1%
	148	63											0.6113 ± 0.0001	19%
	148	63											0.6299 ± 0.0001	71%
	148	63											0.6543 ± 0.0001	2%
	148	63											0.7257 ± 0.0001	13%
	148	63											0.8700 ± 0.0001	5.5%
	148	63											0.9135 ± 0.0001	2.4%
	148	63											0.9305 ± 0.0001	2.9%
	148	63											0.9673 ± 0.0001	2.9%
	148	63											1.0341 ± 0.0001	7.9%
	148	63											1.1469 ± 0.0001	1.9%
	148	63											1.1833 ± 0.0001	1.7%
	148	63											1.3285 ± 0.0002	1.2%
	148	63											1.3446 ± 0.0001	3.6%
	148	63											1.6215 ± 0.0001	4.6%
	148	63											1.6504 ± 0.0001	3.7%
$_{63}Eu^{149}$	149	63		148.917926	93.1 d.	E.C.	0.69				5/2+		Sm k x-ray	39%
	149	63											0.2545 ± 0.0002	0.6%
	149	63											0.2770 ± 0.0002	3.3%
	149	63											0.3275 ± 0.0002	3.9%
	149	63											0.3500 ± 0.0003	0.3%
	149	63											0.5059 ± 0.0002	0.55%
	149	63											0.5285 ± 0.0002	0.53%
	149	63											0.5359 ± 0.0003	0.05%
	149	63											0.5584 ± 0.0003	0.05%
$_{63}Eu^{150m}$	150	63			36 y.	E.C.					4,5-		Sm k x-ray	43%
	150	63											0.3340 ± 0.0001	94%
	150	63											0.4394 ± 0.0001	79%
	150	63											0.5055 ± 0.0001	4.7%
	150	63											0.5843 ± 0.0001	52%
	150	63											0.7122 ± 0.0001	1.1%
	150	63											0.7374 ± 0.0001	9.4%
	150	63											0.7480 ± 0.0001	5.0%
	150	63											0.8692 ± 0.0001	2.1%
	150	63											1.0490 ± 0.0001	5.2%
	150	63											1.1706 ± 0.0001	1.3%
	150	63											1.1971 ± 0.0001	1.1%
	150	63											1.2469 ± 0.0001	1.9%
	150	63											1.3437 ± 0.0001	2.5%
	150	63											1.4855 ± 0.0001	1.8%
	150	63											(0.25 - 1.8)	
$_{63}Eu^{150}$	150	63		149.919702	12.6 h.	β-(92%)	1.009	1.010			0-		Sm k x-ray	3%
	150	63				β+(0.4%)	2.29	1.24					0.3339 ± 0.0001	3.7%
	150	63				E.C.(8%)							0.4065 ± 0.0001	2.4%
	150	63											0.7122 ± 0.0001	0.1%
	150	63											0.8319 ± 0.0001	0.2%
	150	63											0.9214 ± 0.0001	0.2%

TABLE OF THE ISOTOPES (Continued)

Isotope	A	Z	% Natural abundance	Atomic mass	Half-life	Decay mode	Decay energy (MeV)	Particle energy (MeV)	Particle intensity	Thermal neutron cross section	Spin (h/2π)	μ Nucl. mag. moment	Gamma-ray energy (MeV)	Gamma-ray intensity
	150	63											1.1657 ± 0.0001	0.2%
	150	63											1.2233 ± 0.0001	0.2%
	150	63											1.9637 ± 0.0001	0.1%
$_{63}Eu^{151}$	151	63	47.8%	150.919847						(4 + 3300 + 5900) b.	5/2 +	+ 3.464		
$_{63}Eu^{152m2}$	152	63			96 m.	I.T.	0.1478				8-		Eu k x-ray	12%
	151	63											0.0898 ± 0.0001	70%
$_{63}Eu^{152m1}$	152	63			9.3 h.	β-(72%)		1.85			0-		Sm k x-ray	13%
	152	63				E.C.(28%)		0.89					0.12178 ± 0.00001	7.2%
	152	63											0.34427 ± 0.00001	2.4%
	152	63											0.84153 ± 0.00002	14.5%
	152	63											0.96334 ± 0.00002	12%
	152	63											1.31461 ± 0.00002	0.9%
	152	63											1.38900 ± 0.00002	0.8%
$_{63}Eu^{152}$	152	63		151.921742	13.4 y.	E.C.(72%)	1.876	0.69			3-	± 1.924	Sm k x-ray	38%
	152	63				β-(28%)	1.822	1.47					Gd k x-ray	11%
	152	63											0.12178 ± 0.00001	28%
	152	63											0.24470 ± 0.00001	7.5%
	152	63											0.34427 ± 0.00001	27%
	152	63											0.44396 ± 0.00001	3.1%
	152	63											0.77887 ± 0.00001	13%
	152	63											0.86737 ± 0.00002	4.2%
	152	63											0.96404 ± 0.00001	14.6%
	152	63											1.08583 ± 0.00002	9.9%
	152	63											1.11209 ± 0.00002	13.6%
	152	63											1.2129 ± 0.0001	1.4%
	152	63											1.2991 ± 0.0001	1.6%
	152	63											1.40802 ± 0.00003	21%
$_{63}Eu^{153}$	153	63	52.2%	152.921225						350 b.	5/2 +	+ 1.530		
$_{63}Eu^{154m}$	154	63			46.1 m.	I.T.	≈ 0.16				8-		Eu k x-ray	6%
	154	63											0.0358 ± 0.0001	10%
	154	63											0.0682 ± 0.0001	36%
	154	63											0.1009 ± 0.0001	26%
$_{63}Eu^{154}$	154	63		153.922975	8.5 y.	β-(99.9%)	1.978	0.27	29%		3-	± 2.000	Gd k x-ray	13%
	154	63				EC.(0.02%)	0.728	0.58	38%				0.12299 ± 0.00001	40%
	154	63						0.84	17%				0.2477 ± 0.0001	6.6%
	154	63						0.98	4%				0.59178 ± 0.00002	4.8%
	154	63						1.87	11%				0.6924 ± 0.0001	1.7%
	154	63											0.72331 ± 0.00001	19.7%
	154	63											0.7568 ± 0.0001	4.3%
	154	63											0.8732 ± 0.0001	11.5%
	154	63											0.9963 ± 0.10.3%	
	154	63											1.2745 ± 0.0001	35.5%
	154	63											1.5965 ± 0.0002	1.8%
$_{63}Eu^{155}$	155	63		154.922889	4.73 y.	β-	0.25	0.25	13%		5/2 +		Gd k x-ray	12%
	155	63											0.0600 ± 0.0001	1.2%
	155	63											0.0865 ± 0.0001	33%
	155	63											0.1053 ± 0.0001	22%
$_{63}Eu^{156}$	156	63		155.924752	15.2 d.	β-	2.45	0.30	11%		1 +		0.08397 ± 0.0005	9%
	156	63						0.49	30%				0.5995 ± 0.0001	2%
	156	63						1.2	12%				0.64623 ± 0.00008	6.7%
	156	63						2.45	31%				0.723441 ± 0.0001	58%
	156	63											0.8118 ± 0.0001	10%
	156	63											0.8670 ± 0.0001	1.4%
	156	63											0.9444 ± 0.0001	1.4%
	156	63											0.9605 ± 0.0001	1.6%
	156	63											1.0651 ± 0.0001	5.2%
	156	63											1.0792 ± 0.0001	4.9%
	156	63											1.1535 ± 0.0001	7.2%
	156	63											1.1542 ± 0.0001	5.0%
	156	63											1.2307 ± 0.0001	8.5%
	156	63											1.2424 ± 0.0001	7%
	156	63											1.2774 ± 0.0001	3.1%
	156	63											1.3664 ± 0.0001	1.7%

Isotope	A	Z	% Natural abundance	Atomic mass	Half-life	Decay mode	Decay energy (MeV)	Particle energy (MeV)	Particle intensity	Thermal neutron cross section	Spin (h/2π)	μ Nucl. mag. moment	Gamma-ray energy (MeV)	Gamma-ray intensity
	156	63											1.8770 ± 0.0002	1.6%
	156	63											1.9377 ± 0.0001	2.1%
	156	63											1.9660 ± 0.0001	4.1%
	156	63											2.0266 ± 0.0001	3.5%
	156	63											2.0977 ± 0.0001	4.0%
	156	63											2.1810 ± 0.0001	2.3%
	156	63											2.1868 ± 0.0001	3.7%
	156	63											2.2056 ± 0.0001	1%
	156	63											2.2699 ± 0.0001	1.1%
$_{63}Eu^{157}$	157	63		156.925418	15.15 h.	β-	1.36	0.91			(5/2+)		Gd k x-ray	32%
	157	63						1.30	41%				0.0545 ± 0.0001	4%
	157	63											0.064 ± 0.001	22%
	157	63											0.320 ± 0.001	3%
	157	63											0.373 ± 0.001	10%
	157	63											0.401 ± 0.001	1%
	157	63											0.413 ± 0.001	17%
$_{63}Eu^{158}$	158	63		157.927800	β-	3.5	2.5				(1-)		0.0795 ± 0.0001	11%
	158	63											0.1820 ± 0.0001	1.9%
	158	63											0.5280 ± 0.0001	1.3%
	158	63											0.6064 ± 0.0001	3.3%
	158	63											0.7430 ± 0.0001	3.0%
	158	63											0.8241 ± 0.0001	1.1%
	158	63											0.8976 ± 0.0001	10%
	158	63											0.9065 ± 0.0001	1.5%
	158	63											0.9225 ± 0.0002	1.6%
	158	63											0.9442 ± 0.0001	25%
	158	63											0.9530 ± 0.0001	1.6%
	158	63											0.9621 ± 0.0001	1.6%
	158	63											0.9771 ± 0.0001	13.6%
	158	63											0.9870 ± 0.0001	1.1%
	158	63											1.0054 ± 0.0003	1.0%
	158	63											1.1076 ± 0.0001	4.3%
	158	63											1.1165 ± 0.0001	1.0%
	158	63											1.2636 ± 0.0002	2.0%
	158	63											1.3479 ± 0.0001	1.4%
	158	63											1.8846 ± 0.0002	1.0%
	158	63											1.9445 ± 0.0002	1.3%
	158	63											2.3677 ± 0.0003	0.66%
	158	63											2.4474 ± 0.0004	0.63%
$_{63}Eu^{159}$	159	63		158.929084	18 m.	β-	2.51	2.4			(5/2+)		0.0678 ± 0.0001	33%
	159	63						2.6					0.0786 ± 0.0001	15%
	159	63											0.0804 ± 0.0004	2%
	159	63											0.0957 ± 0.0001	12%
	159	63											0.1464 ± 0.0001	6%
	159	63											0.1598 ± 0.0002	2%
	159	63											0.1769 ± 0.0001	2%
	159	63											0.2275 ± 0.0003	3%
	159	63											0.6022 ± 0.00022	1.5%
	159	63											0.6134 ± 0.0002	2%
	159	63											0.6595 ± 0.0001	2%
	159	63											0.6649 ± 0.0001	5%
	159	63											0.6766 ± 0.0001	3%
	159	63											0.6819 ± 0.0001	4%
	159	63											0.7265 ± 0.0003	1%
	159	63											0.7443 ± 0.0002	1.6%
	159	63											0.7539 ± 0.0002	1.6%
	159	63											0.8047 ± 0.0002	4%
	159	63											1.0948 ± 0.0002	2%
	159	63											1.1284 ± 0.0003	0.9%
	159	63											1.59202 ± 0.0002	1.1%
$_{63}Eu^{160}$	160	63		159.931880	53 s.	β-	≈ 4.4	2.7			(0-)		0.0753 ± 0.0001	17 +
	160	63						4.1					0.1735 ± 0.0002	32
	160	63											0.2666 ± 0.0003	8
	160	63											0.3020 ± 0.0003	31
	160	63											0.3980 ± 0.0003	39
	160	63											0.4131 ± 0.0003	76
	160	63											0.5155 ± 0.0003	100
	160	63											0.6578 ± 0.0003	6
	160	63											0.7370 ± 0.0003	9
	160	63											0.8217 ± 0.0003	68
	160	63											0.9110 ± 0.0003	97
	160	63											0.9246 ± 0.0003	60
	160	63											0.9953 ± 0.0005	32
	160	63											1.1480 ± 0.0005	33
Gd		64		157.25						49 x 10^3 b.				
$_{64}Gd^{143m}$	143	64			1.83 m.	β+(67%)					11/2-		ann.rad.	145%
	143	64				E.C.(33%)							Eu k x-ray	42%
	143	64				I.T.							0.1176 ± 0.0001	6.4%
	143	64											0.2719 ± 0.0001	83%
	143	64											0.3895 ± 0.0001	3.4%
	143	64											0.5880 ± 0.0001	15%
	143	64											0.6681 ± 0.0001	9.5%
	143	64											0.7586 ± 0.0001	5.4%
	143	64											0.7999 ± 0.0001	10%

Isotope	A	Z	% Natural abundance	Atomic mass	Half-life	Decay mode	Decay energy (MeV)	Particle energy (MeV)	Particle intensity	Thermal neutron cross section	Spin (h/2π)	μ Nucl. mag. moment	Gamma-ray energy (MeV)	Gamma-ray intensity
	143	64											0.8244 ± 0.0001	4.9%
	143	64											0.8905 ± 0.0001	1.7%
	143	64											0.9070 ± 0.0001	2.1%
	143	64											0.9165 ± 0.0001	4.2%
	143	64											0.9849 ± 0.0001	2%
	143	64											1.0083 ± 0.0001	1.3%
	143	64											1.0414 ± 0.0001	3%
	143	64											1.0873 ± 0.0001	1.6%
	143	64											1.2192 ± 0.0001	4.1%
	143	64											1.3736 ± 0.0001	1.1%
	143	64											1.3867 ± 0.0001	1.2%
	143	64											1.4046 ± 0.0001	2.8%
	143	64											1.6293 ± 0.0001	1.9%
	143	64											1.7932 ± 0.0001	2.6%
	143	64											1.8071 ± 0.0001	7.5%
	143	64											1.8203 ± 0.0001	3.0%
	143	64											1.8860 ± 0.0002	0.7%
$_{64}Gd^{143}$	143	64		142.926490	39 s.	β+(82%)	≈ 5.9				1/2+		ann.rad.	160%
	143	64				E.C.(18%)							Eu k x-ray	13%
	143	64											0.2048 ± 0.0001	19%
	143	64											0.2588 ± 0.0001	75%
	143	64											0.4637 ± 0.0001	9.9%
	143	64											0.8129 ± 0.0001	5.4%
	143	64											1.2842 ± 0.0004	1%
	143	64											1.4648 ± 0.0004	0.9%
$_{64}Gd^{144}$	144	64		143.922760	4.5 m.	β+(45%)	≈ 3.7	3.3			0+		ann.rad.	90%
	144	64				E.C.(55%)							Eu k x-ray	22%
	144	64											0.3332 ± 0.0005	12%
	144	64											0.3470 ± 0.0005	4%
	144	64											0.6220 ± 0.0005	2%
	144	64											0.6298 ± 0.0005	4%
	144	64											0.6419 ± 0.0005	2%
	144	64											0.8677 ± 0.0005	2%
$_{64}Gd^{145m}$	145	64			85 s.	I.T.(95%)	0.749				11/2-		0.0273 ± 0.0001	4.5%
	145	64				β+(4%)	5.7						0.3295 ± 0.0003	4.4%
	145	64											0.3866 ± 0.0003	4.1%
	145	64											0.7214 ± 0.0004	83%
$_{64}Gd^{145}$	145	64		144.921690	23 m.	β+(33%)	5.07	2.5			1/2+		ann.rad.	66%
	145	64				E.C.(67%)							Eu k x-ray	27%
	145	64											0.3299 ± 0.0001	2.7%
	145	64											0.8044 ± 0.0001	8.6%
	145	64											0.9526 ± 0.0001	1.5%
	145	64											1.0408 ± 0.0001	9.9%
	145	64											1.0723 ± 0.0001	2.8%
	145	64											1.6001 ± 0.0001	1.8%
	145	64											1.7579 ± 0.0001	34%
	145	64											1.8806 ± 0.0001	33%
	145	64											2.4948 ± 0.0001	1.3%
	145	64											2.6422 ± 0.0001	2.1%
	145	64											(0.32 - 3.69)	
$_{64}Gd^{146}$	146	64		145.918304	48.3 d.	E.C. (99.9%)	1.02	0.35			0+		Eu k x-ray	96%
	146	64				β+(0.2%)							0.1147 ± 0.0001	44%
	146	64											0.1155 ± 0.0001	44%
	146	64											0.1546 ± 0.0001	46%
$_{64}Gd^{147}$	147	64		146.918943	38.1 h.	E.C. (99.8%)	2.08	0.93			7/2-		Eu k x-ray	48%
	147	64				E.C.(0.2%)							0.2293 ± 0.0001	64%
	147	64											0.3099 ± 0.0001	3.7%
	147	64											0.3185 ± 0.0001	2%
	147	64											0.3463 ± 0.0001	1.9%
	147	64											0.3699 ± 0.0001	17%
	147	64											0.3960 ± 0.0001	34%
	147	64											0.4849 ± 0.0001	2.8%
	147	64											0.5591 ± 0.0001	6.2%
	147	64											0.6191 ± 0.0001	3.5%
	147	64											0.6252 ± 0.0001	4.5%
	147	64											0.7550 ± 0.0001	1.9%
	147	64											0.7658 ± 0.0001	10%
	147	64											0.7764 ± 0.0001	5%
	147	64											0.7780 ± 0.0001	4.6%
	147	64											0.8616 ± 0.0001	1.7%
	147	64											0.8934 ± 0.0001	737%
	147	64											0.9289 ± 0.0001	19%
	147	64											1.0692 ± 0.0001	6.5%
	147	64											1.1307 ± 0.0001	5.7%
	147	64											1.2357 ± 0.0001	1.0%
	147	64											(0.1 - 1.8)	
$_{64}Gd^{148}$	148	64		147.918113	75 y.	α		3.1828			0+			
$_{64}Gd^{149}$	149	64		148.919344	9.3 d.	E.C.	1.32				7/2-		Eu k x-ray	52%
	149	64											0.1496 ± 0.0002	42%
	149	64											0.2605 ± 0.0003	1%
	149	64											0.2720 ± 0.0001	2.6%
	149	64											0.2985 ± 0.0003	23%
	149	64											0.3465 ± 0.0003	18%

Isotope	A	Z	% Natural abundance	Atomic mass	Half-life	Decay mode	Decay energy (MeV)	Particle energy (MeV)	Particle intensity	Thermal neutron cross section	Spin (h/2π)	μ Nucl. mag. moment	Gamma-ray energy (MeV)	Gamma-ray intensity
	149	64											0.4964 ± 0.0003	1.3%
	149	64											0.5164 ± 0.0003	2%
	149	64											0.5342 ± 0.0003	2.4%
	149	64											0.6452 ± 0.0003	1.1%
	149	64											0.7482 ± 0.0003	6.2%
	149	64											0.7886 ± 0.0003	5.3%
	149	64											0.9391 ± 0.0004	1.6%
$_{64}Gd^{150}$	150	64		149.918662	$1.8x10^6$ y.	α		2.73			0+			
$_{64}Gd^{151}$	151	64		150.920346	≈ 120 d.	E.C.	0.48				7/2-		Eu k x-ray	43%
	151	64											0.1536 ± 0.0001	5.1%
	151	64											0.1747 ± 0.0001	2.4%
	151	64											0.2432 ± 0.0001	4.6%
	151	64											0.3074 ± 0.0001	0.8%
$_{64}Gd^{152}$	152	64	0.20%	151.919786						10 b.	0+			
$_{64}Gd^{153}$	153	64		152.921745	241.6 d.	E.C.				$3 x 10^4$ b.	3/2-		Eu k x-ray	61%
	153	64											0.06968 ± 0.00002	2.3%
	153	64											0.09743 ± 0.00001	30%
	153	64											0.10318 ± 0.00001	22%
$_{64}Gd^{154}$	154	64	2.18%	153.920861						80 b.	0+			
$_{64}Gd^{155}$	155	64	14.80%	154.922618						$61 x 10^3$ b.	3/2-	-0.27		
$_{64}Gd^{156}$	156	64	20.47%	155.922118						2 b.	0+			
$_{64}Gd^{157}$	157	64	15.65%	156.923956						$2.55 x 10^5$ b.	3/2-	-0.36		
$_{64}Gd^{158}$	158	64	24.84%	157.924099						2.4 b.	0+			
$_{64}Gd^{159}$	159	64		158.926384	18.6 h.	β-	0.60	11%			3/2-		Tb k x-ray	10%
	159	64						0.89	26%				0.05845 ± 0.00005	2.3%
	159	64						0.95	63%				0.36351 ± 0.00001	10.8%
$_{64}Gd^{160}$	160	64	21.86%	159.927049						0.8 b.	0+			
$_{64}Gd^{161}$	161	64		160.929664	3.7 m.	β-	1.958	1.56	85%	$3 x 10^4$ b.	5/2-		Tb k x-ray	25%
	161	64											0.0563 ± 0.0001	3.8%
	161	64											0.0774 ± 0.0001	1.1%
	161	64											0.1023 ± 0.0001	14%
	161	64											0.1652 ± 0.0001	2.6%
	161	64											0.2836 ± 0.0001	6%
	161	64											0.3149 ± 0.0001	23%
	161	64											0.3381 ± 0.0001	1.7%
	161	64											0.3609 ± 0.0001	61%
	161	64											0.4801 ± 0.0001	2.74%
	161	64											0.5295 ± 0.0001	1.3%
$_{64}Gd^{162}$	162	64		161.931010	8.4 m.	β-	1.4	1.0			0+		0.4030	
	162	64											0.4421	
$_{64}Gd^{163}$	163	64			68 s.	β-							0.2868	
	163	64											0.214	
	163	64											1.685	
Tb		65		158.9254						23 b.				
$_{65}Tb^{145}$	145	65		144.928940	30 s.	β+,E.C.	≈ 6.6						ann.rad.	
	145	65											0.2003 ± 0.0003	7%
	145	65											0.2466 ± 0.0003	4%
	145	65											0.2577 ± 0.0003	39%
	145	65											0.2685 ± 0.0003	3%
	145	65											0.5240 ± 0.0003	10%
	145	65											0.5370 ± 0.0003	23%
	145	65											0.5721 ± 0.0003	14%
	145	65											0.6980 ± 0.0003	5%
	145	65											0.9085 ± 0.0003	7%
	145	65											0.9351 ± 0.0003	5%
	145	65											0.9876 ± 0.0003	37%
	145	65											1.0149 ± 0.0003	5%
	145	65											1.1093 ± 0.0003	14%
	145	65											1.3880 ± 0.0003	6%
	145	65											1.4325 ± 0.0003	10%
	145	65											1.4467 ± 0.0003	15%
$_{65}Tb^{146}$	146	65		145.927150	23 s.	β+(76%) E.C.(24%)	≈ 8.3				(4-)		ann.rad.	150%
	146	65											Gd k x-ray	10%
	146	65											0.4410 ± 0.0001	13%
	146	65											0.6550 ± 0.0002	2%
	146	65											0.9876 ± 0.0004	1%
	146	65											1.0319 ± 0.0004	3%
	146	65											1.0789 ± 0.0002	50%
	146	65											1.5795 ± 0.0002	97%
	146	65											1.9718 ± 0.0003	3%
	146	65											3.1396 ± 0.0004	11%
$_{65}Tb^{147m}$	147	65			1.8 m.	β+(35%) E.C.(65%)					11/2-		ann.rad.	70%
	147	65											Gd k x-ray	26%
	147	65											1.1789 ± 0.0004	2%
	147	65											1.3977 ± 0.0002	83%
	147	65											1.7978 ± 0.0003	14%

Isotope	A	Z	% Natural abundance	Atomic mass	Half-life	Decay mode	Decay energy (MeV)	Particle energy (MeV)	Particle intensity	Thermal neutron cross section	Spin (h/2π)	μ Nucl. mag. moment	Gamma-ray energy (MeV)	Gamma-ray intensity
$_{65}Tb^{147}$	147	65		146.923820	1.6 h.	β+(42%)	≈ 4.6				5/2+		ann.rad.	85%
	147	65				E.C.(58%)							Gd k x-ray	26%
	147	65											0.1197 ± 0.0004	4%
	147	65											0.1398 ± 0.0004	20%
	147	65											0.3474 ± 0.0006	1.7%
	147	65											0.4070 ± 0.0004	1.4%
	147	65											0.5472 ± 0.0004	2.0%
	147	65											0.5547 ± 0.0004	3.7%
	147	65											0.6944 ± 0.0004	31%
	147	65											1.1522 ± 0.0004	72%
	147	65											1.6281 ± 0.0004	2.7%
	147	65											1.9483 ± 0.0004	1.4%
	147	65											2.5619 ± 0.0004	1.7%
	147	65											2.6814 ± 0.0004	2.6%
$_{65}Tb^{148m}$	148	65			2.2 m.	β+(25%)					9+		ann.rad.	50%
	148	65				E.C.(75%)							Gd k x-ray	33%
	148	65											0.1295 ± 0.0002	2%
	148	65											0.1429 ± 0.0002	2%
	148	65											0.3945 ± 0.0001	86%
	148	65											0.4817 ± 0.0001	3%
	148	65											0.4888 ± 0.0001	5%
	148	65											0.6319 ± 0.0001	94%
	148	65											0.7530 ± 0.0001	2%
	148	65											0.7845 ± 0.0001	99%
	148	65											0.8081 ± 0.0006	3%
	148	65											0.8824 ± 0.0001	91%
$_{65}Tb^{148}$	148	65		147.924140	60 m.	β+,EC					2-		ann.rad.	
	148	65											Gd k x-ray	
	148	65											0.4888 ± 0.0001	22 +
	148	65											0.6319 ± 0.0001	15
	148	65											0.7845 ± 0.0001	100
	148	65											1.0781 ± 0.0002	12
	148	65											1.4025 ± 0.001	2
	148	65											1.491 ± 0.001	5
	148	65											1.8626 ± 0.0005	8
	148	65											(0.14 - 3.8)weak	
$_{65}Tb^{149m}$	149	65			4.2 m.	E.C.(88%)					11/2-		ann.rad.	24%
	149	65				β+(12%)							Gd k x-ray	36%
	149	65											0.1650 ± 0.0001	7%
	149	65											0.6307 ± 0.0003	2.6%
	149	65											0.7960 ± 0.0001	92%
$_{65}Tb^{149}$	149	65		148.923248	4.15 h.	β+(4%)	3.7	1.8			(1/2+)		Gd k x-ray	38%
	149	65				α(16%)		3.97					0.1650 ± 0.0001	27%
	149	65											0.1872 ± 0.0002	4.3%
	149	65											0.3522 ± 0.0001	30%
	149	65											0.3886 ± 0.0001	19%
	149	65											0.4648 ± 0.0001	5.7%
	149	65											0.6521 ± 0.0001	16%
	149	65											0.8171 ± 0.0001	12%
	149	65											0.8534 ± 0.0001	16%
	149	65											0.8619 ± 0.0001	8.4%
	149	65											1.1755 ± 0.0001	3.5%
	149	65											1.3412 ± 0.0001	2.3%
	149	65											1.6403 ± 0.0001	3.2%
	149	65											1.8274 ± 0.0001	1.2%
	149	65											2.0079 ± 0.0001	0.8%
	149	65											2.9613 ± 0.0001	0.8%
	149	65											(0.1 - 3.2)weak	
$_{65}Tb^{150m}$	150	65			6.0 m.	β+(17%)							ann.rad.	35%
	150	65				E.C.(83%)							Gd k x-ray	37%
	150	65											0.1620 ± 0.0002	7%
	150	65											0.3431 ± 0.0001	25%
	150	65											0.4124 ± 0.0002	10%
	150	65											0.4153 ± 0.0002	4%
	150	65											0.4384 ± 0.0001	42%
	150	65											0.4557 ± 0.0002	12%
	150	65											0.4963 ± 0.0001	23%
	150	65											0.510 ± 0.0.001	26%
	150	65											0.5665 ± 0.0001	22%
	150	65											0.6380 ± 0.0001	99%
	150	65											0.6484 ± 0.0003	18%
	150	65											0.6504 ± 0.0003	69%
	150	65											0.7899 ± 0.0004	2%
	150	65											0.8275 ± 0.0001	41%
$_{65}Tb^{150}$	150	65		149.923669	3.3 h.	β+,E.C.	≈ 4.7				2-		ann.rad.	
	150	65											0.4963 ± 0.0001	
	150	65											0.5691 ± 0.0001	2.5%
	150	65											0.6380 ± 0.0001	72%
	150	65											0.6504 ± 0.0002	4%
	150	65											0.7925 ± 0.0003	4.4%
	150	65											0.8803 ± 0.0001	3%
	150	65											1.2917 ± 0.0001	1.6%
	150	65											1.4305 ± 0.0001	2.4%
	150	65											1.4536 ± 0.0001	2.4%
	150	65											1.5185 ± 0.0002	2.3%

TABLE OF THE ISOTOPES (Continued)

Isotope	A	Z	% Natural abundance	Atomic mass	Half-life	Decay mode	Decay energy (MeV)	Particle energy (MeV)	Particle intensity	Thermal neutron cross section	Spin (h/2π)	μ Nucl. mag. moment	Gamma-ray energy (MeV)	Gamma-ray intensity
	150	65											1.5927 ± 0.0001	1.6%
	150	65											1.7881 ± 0.0001	1.6%
	150	65											2.0917 ± 0.0003	1.4%
	150	65											2.1487 ± 0.0003	1.0%
	150	65											2.2078 ± 0.0003	1.0%
	150	65											2.4263 ± 0.0003	0.9%
	150	65											(0.3 - 4.29)weak	
$_{65}Tb^{151m}$	151	65			≈ 50 s.	I.T.(95%)					11/2-		0.0229 ± 0.0001	3%
	151	65				β+, E.C. (7%)							0.0495 ± 0.0001	24%
	151	65											0.3797 ± 0.0001	7%
	151	65											0.5224 ± 0.0001	1.7%
	151	65											0.8305 ± 0.0001	3.7%
$_{65}Tb^{151}$	151	65		150.923100	17.6 h.	β+(1%)	≈ 4.7	3.7			1/2+		Gd k x-ray	60%
	151	65				E.C.(99%)							0.1083 ± 0.0001	25%
	151	65											0.1804 ± 0.0001	11%
	151	65											0.1921 ± 0.0002	4%
	151	65											0.2517 ± 0.0001	26%
	151	65											0.2870 ± 0.0001	25%
	151	65											0.3804 ± 0.0002	4%
	151	65											0.3953 ± 0.0002	10%
	151	65											0.4265 ± 0.0002	4%
	151	65											0.4437 ± 0.0002	10%
	151	65											0.4790 ± 0.0003	16%
	151	65											0.5873 ± 0.0001	17%
	151	65											0.6048 ± 0.0001	3%
	151	65											0.6166 ± 0.0001	10%
	151	65											0.7038 ± 0.0002	4%
	151	65											0.7311 ± 0.0002	9%
	151	65											(0.1 - 1.8)weak	
$_{65}Tb^{152m}$	152	65			4.1 m.	I.T.(79%)	0.5018				(6+)		Tb k x-ray	40%
	152	65				E.C.(21%)	4.35						Gd k x-ray	9%
	152	65											0.0480 ± 0.0002	1%
	152	65											0.0589 ± 0.0002	7%
	152	65											0.1596 ± 0.0001	16%
	152	65											0.2354 ± 0.0001	4%
	152	65											0.2772 ± 0.0001	8.5%
	152	65											0.2833 ± 0.0001	60%
	152	65											0.3443 ± 0.0001	20%
	152	65											0.3859 ± 0.0001	3%
	152	65											0.4111 ± 0.0001	18%
	152	65											0.4719 ± 0.0001	12%
	152	65											0.5194 ± 0.0001	5%
	152	65											0.5326 ± 0.0001	4%
	152	65											0.5862 ± 0.0002	1%
	152	65											0.6474 ± 0.0002	4%
	152	65											0.7260 ± 0.0002	3%
	152	65											1.1062 ± 0.0002	3%
	152	65											1.1669 ± 0.0002	4%
$_{65}Tb^{152}$	152	65		151.923919	17.6 h.	β+(20%)	3.85	2.45			2-		ann.rad.	40%
	152	65				E.C.(80%)		2.82					Gd k x-ray	33%
	152	65											0.2711 ± 0.0001	8%
	152	65											0.3443 ± 0.0001	57%
	152	65											0.4111 ± 0.0001	3.6%
	152	65											0.5863 ± 0.0001	8%
	152	65											0.7033 ± 0.0002	2.2%
	152	65											0.7649 ± 0.0001	2.6%
	152	65											0.7789 ± 0.0001	5%
	152	65											0.9741 ± 0.0002	2.8%
	152	65											1.1092 ± 0.0002	2.3%
	152	65											1.2991 ± 0.0001	1.9%
	152	65											1.3147 ± 0.0002	1.2%
	152	65											1.9042 ± 0.0002	1.7%
	152	65											2.4050 ± 0.0003	1.3%
	152	65											(0.2 - 2.88)	
$_{65}Tb^{153}$	153	65		152.923440	2.34 d.	E.C.	1.58				5/2+		Gd k x-ray	55%
	153	65											0.0829 ± 0.0001	6%
	153	65											0.0876 ± 0.0001	2%
	153	65											0.1022 ± 0.0001	6%
	153	65											0.1097 ± 0.0001	7%
	153	65											0.1704 ± 0.0001	7%
	153	65											0.2119 ± 0.0001	32%
	153	65											0.2496 ± 0.0001	2.4%
	153	65											0.3036 ± 0.0001	1.0%
	153	65											0.8354 ± 0.0002	1.2%
	153	65											0.9451 ± 0.0002	1.1%
	153	65											0.9917 ± 0.0002	1.2%
	153	65											(0.05 - 1.1)weak	
$_{65}Tb^{154m2}$	154	65			23 h.	E.C.(98%)					(7-)		Gd k x-ray	61%
	154	65				I.T.(2%)							0.1231 ± 0.0001	44%
	154	65											0.1413 ± 0.0001	7%
	154	65											0.1720 ± 0.0001	5%
	154	65											0.2259 ± 0.0001	27%
	154	65											0.2479 ± 0.0001	81%
	154	65											0.2658 ± 0.0001	4%

Isotope	A	Z	% Natural abundance	Atomic mass	Half-life	Decay mode	Decay energy (MeV)	Particle energy (MeV)	Particle intensity	Thermal neutron cross section	Spin ($h/2\pi$)	μ Nucl. mag. moment	Gamma-ray energy (MeV)	Gamma-ray intensity
	154	65											0.2675 ± 0.0003	4%
	154	65											0.3467 ± 0.0001	71%
	154	65											0.4268 ± 0.0001	18%
	154	65											0.4792 ± 0.0002	4%
	154	65											0.5064 ± 0.0002	4%
	154	65											0.5180 ± 0.0001	3.9%
	154	65											0.6423 ± 0.0002	4.2%
	154	65											0.6495 ± 0.0001	7%
	154	65											0.8732 ± 0.0001	3.4%
	154	65											0.8927 ± 0.0001	4.8%
	154	65											0.9930 ± 0.0002	17%
	154	65											0.9963 ± 0.0001	3%
	154	65											1.0047 ± 0.0001	4%
	154	65											1.0612 ± 0.0003	4%
	154	65											1.1407 ± 0.0002	2.3%
	154	65											1.1934 ± 0.0009	3%
	154	65											1.4199 ± 0.0001	47%
$_{65}Tb^{154m1}$	154	65			9 h.	β+(78%)					(3-)		Gd k x-ray	41%
	154	65				I.T.(22%)							0.1231 ± 0.0001	31%
	154	65											0.2479 ± 0.0001	22%
	154	65											0.5180 ± 0.0001	6%
	154	65											0.5401 ± 0.0001	20%
	154	65											0.6495 ± 0.0001	11%
	154	65											0.6765 ± 0.0001	3%
	154	65											0.6924 ± 0.0001	3%
	154	65											0.7567 ± 0.0001	2.7%
	154	65											0.8732 ± 0.0001	9.3%
	154	65											0.8927 ± 0.0001	3%
	154	65											0.9963 ± 0.0001	8.7%
	154	65											1.0047 ± 0.0001	11%
	154	65											1.1287 ± 0.0002	1.6%
	154	65											1.1407 ± 0.0002	1.4%
	154	65											1.1521 ± 0.0006	2.2%
	154	65											1.2581 ± 0.0002	1.6%
	154	65											1.2884 ± 0.0002	1.4%
	154	65											1.4906 ± 0.0002	1.0%
	154	65											1.9650 ± 0.0001	2%
	154	65											2.1538 ± 0.0002	1%
	154	65											(0.12 - 2.57)many	
$_{65}Tb^{154}$	153	65		153.924690	22 h.	E.C.(99%)	3.56	1.86			(0-)		Gd k x-ray	49%
	154	65				β+(1%)		2.45					0.1231 ± 0.0001	28%
	154	65											0.5576 ± 0.0001	5.8%
	154	65											0.6924 ± 0.0001	3.4%
	154	65											0.7051 ± 0.0001	5%
	154	65											0.7221 ± 0.0002	8.2%
	154	65											0.8732 ± 0.0001	5.6%
	154	65											0.8783 ± 0.0002	3%
	154	65											0.9963 ± 0.0001	5.2%
	154	65											1.1181 ± 0.0003	2.5%
	154	65											1.1232 ± 0.0002	6.1%
	154	65											1.2744 ± 0.0001	11%
	154	65											1.2913 ± 0.0002	7.4%
	154	65											1.4146 ± 0.0002	2.0%
	154	65											1.9966 ± 0.0001	8.0%
	154	65											2.0419 ± 0.0001	2.1%
	154	65											2.0641 ± 0.0001	7.6%
	154	65											2.1197 ± 0.0002	4.5%
	154	65											2.1872 ± 0.0001	11%
	154	65											2.3075 ± 0.0002	1.6%
	154	65											2.3425 ± 0.0003	1.6%
	154	65											2.3453 ± 0.0003	1.6%
	154	65											2.4305 ± 0.0001	2.3%
	154	65											2.4862 ± 0.0002	1.4%
	154	65											2.9000 ± 0.0004	1.0%
	154	65											3.0232 ± 0.0003	0.8%
	154	65											(0.12 - 3.14)many	
$_{65}Tb^{155}$	155	65		154.923499	5.3 d.	E.C.	0.82				3/2+		Gd k x-ray	55%
	155	65											0.0453 ± 0.0001	1.5%
	155	65											0.08654 ± 0.00001	29%
	155	65											0.10530 ± 0.00001	23%
	155	65											0.1486 ± 0.0001	2.4%
	155	65											0.1613 ± 0.0001	2.5%
	155	65											0.16330 ± 0.00002	4.1%
	155	65											0.18008 ± 0.00002	6.8%
	155	65											0.3407 ± 0.0001	1.1%
	155	65											0.36738 ± 0.00002	2.1%
$_{65}Tb^{156m2}$	156	65			24 h.	I.T.					(4+)		Tb k x-ray	
	156	65											0.0496 ± 0.0001	73%
$_{65}Tb^{156m1}$	156	65			5.0 h.	I.T.	0.0884				(0+)		Tb k x-ray	
	156	65											0.0884 ± 0.0001	1%

Isotope	A	Z	% Natural abundance	Atomic mass	Half-life	Decay mode	Decay energy (MeV)	Particle energy (MeV)	Particle intensity	Thermal neutron cross section	Spin (h/2π)	μ Nucl. mag. moment	Gamma-ray energy (MeV)	Gamma-ray intensity
$_{65}Tb^{156}$	156	65		155.924742	5.3 d.	E.C.	2.438				3-	± 1.4	Gd k x-ray	54%
	156	65											0.08896 ± 0.00001	19%
	156	65											0.19921 ± 0.00002	40%
	156	65											0.2625 ± 0.0001	5.7%
	156	65											0.2965 ± 0.0001	4.5%
	156	65											0.35645 ± 0.00002	14%
	156	65											0.42244 ± 0.00003	8%
	156	65											0.53435 ± 0.00002	67%
	156	65											0.7801 ± 0.0001	2.4%
	156	65											0.9259 ± 0.0001	3.4%
	156	65											1.1541 ± 0.0001	10%
	156	65											1.1589 ± 0.0001	7.3%
	156	65											1.22245 ± 0.00007	31%
	156	65											1.3345 ± 0.0001	2.5%
	156	65											1.4217 ± 0.0001	12%
	156	65											1.6461 ± 0.0001	3.8%
	156	65											1.8454 ± 0.0001	4.1%
	156	65											2.0143 ± 0.0002	1.1%
$_{65}Tb^{157}$	157	65		156.924023	≈ 150 y.	E.C.	0.058				3/2+		Gd k x-ray	14%
	157	65											0.0545 ± 0.0001	0.02%
$_{65}Tb^{158m}$	158	65			10.5 s.	I.T.	0.11				0-		Gd k x-ray	25%
	158	65											0.0110 ± 0.0001	0.9%
$_{65}Tb^{158}$	158	65		157.925411	≈ 150 y.	E.C.(80%)	1.216				3-	± 1.74	Gd k x-ray	39%
	158	65				β-(20%)	0.936						0.0795 ± 0.0001	11%
	158	65											0.0989 ± 0.0001	4.6%
	158	65											0.1820 ± 0.0001	9.2%
	158	65											0.7801 ± 0.0002	9.3%
	158	65											0.9442 ± 0.0001	43%
	158	65											0.9621 ± 0.0001	20%
	158	65											1.1076 ± 0.0001	2.1%
	158	65											1.1871 ± 0.0002	1.6%
$_{65}$Tb159	159	65	100%	158.925342						23.0 b.	3/2+	+1.95		
$_{65}Tb^{160}$	160	65		159.927163	72.4 d.	β-	1.834	0.57	47%	600 b.	3-		Dy k x-ray	11%
	160	65						0.87	27%				0.08678 ± 0.00001	13%
	160	65											0.19703 ± 0.00001	5.2%
	160	65											0.21564 ± 0.00001	4.0%
	160	65											0.29857 ± 0.00001	27.5%
	160	65											0.30956 ± 0.00004	0.9%
	160	65											0.87936 ± 0.00002	30%
	160	65											0.9623 ± 0.0001	10%
	160	65											0.96615 ± 0.00002	25.5%
	160	65											1.17793 ± 0.00002	15.5%
	160	65											1.1999 ± 0.00005	2.4%
	160	65											1.27185 ± 0.00003	7.6%
	160	65											1.31216 ± 0.00005	3.0%
$_{65}Tb^{161}$	161	65		160.927566	6.91 d.	β-	0.591	0.46	23%		3/2+		Dy k x-ray	11%
	161	65						0.52	66%				0.02565 ± 0.00005	21%
	161	65						0.6	10%				0.04892 ± 0.00005	15%
	161	65											0.05720 ± 0.00005	1.6%
	161	65											0.07458 ± 0.00004	9.8%
	161	65											0.08793 ± 0.00004	0.2%
$_{65}Tb^{162}$	162	65		161.929510	7.6 m.	β-2.5	1.3				(1/2-)		Dy k x-ray	10%
	162	65											0.0807 ± 0.0001	8.8%
	162	65											0.1850 ± 0.0001	2.7%
	162	65											0.1853 ± 0.0001	15%
	162	65											0.2600 ± 0.0001	81%
	162	65											0.6974 ± 0.0001	2.6%
	162	65											0.8075 ± 0.0001	43%
	162	65											0.8823 ± 0.0001	14%
	162	65											0.8882 ± 0.0001	39%
$_{65}Tb^{163}$	163	65		162.930550	19.5 m.	β-	1.70	1.3			3/2+		Dy k x-ray	4%
	163	65											0.2509 ± 0.0002	6.7%
	163	65											0.3163 ± 0.0002	8%
	163	65											0.3385 ± 0.0002	4.5%

Isotope	A	Z	% Natural abundance	Atomic mass	Half-life	Decay mode	Decay energy (MeV)	Particle energy (MeV)	Particle intensity	Thermal neutron cross section	Spin (h/2π)	μ Nucl. mag. moment	Gamma-ray energy (MeV)	Gamma ray intensity
	163	65											0.3479 ± 0.0001	6%
	163	65											0.3511 ± 0.0001	26%
	163	65											0.3542 ± 0.0002	4.6%
	163	65											0.3865 ± 0.0002	4.5%
	163	65											0.3897 ± 0.0002	24%
	163	65											0.4019 ± 0.0002	2.5%
	163	65											0.4151 ± 0.0002	5.4%
	163	65											0.4219 ± 0.0002	11%
	163	65											0.4277 ± 0.0002	3.5%
	163	65											0.4624 ± 0.0002	2.2%
	163	65											0.4754 ± 0.0002	2.9%
	163	65											0.4945 ± 0.0002	72%
	163	65											0.5075 ± 0.0002	4.6%
	163	65											0.5331 ± 0.0002	9.5%
	163	65											0.5596 ± 0.0002	2%
	163	65											0.5840 ± 0.0002	7%
	163	65											0.6084 ± 0.0002	3.7%
	163	65											0.6301 ± 0.0002	1.1%
	163	65											0.8334 ± 0.0001	1.0%
$_{65}Tb^{164}$	164	65		163.933320	3.0 m.	β-	3.86	1.7			(5+)		Dy k x-ray	13%
	164	65											0.0734 ± 0.0001	8%
	164	65											0.1488 ± 0.0001	4%
	164	65											0.1689 ± 0.0005	24%
	164	65											0.2111 ± 0.0001	6%
	164	65											0.2157 ± 0.0001	20%
	164	65											0.2591 ± 0.0002	4%
	164	65											0.2775 ± 0.0001	8%
	164	65											0.2947 ± 0.0001	6.37%
	164	65											0.3448 ± 0.0005	5%
	164	65											0.4103 ± 0.0002	6%
	164	65											0.5485 ± 0.0002	8%
	164	65											0.5859 ± 0.0002	3.9%
	164	65											0.6110 ± 0.0002	19%
	164	65											0.6473 ± 0.0005	5.8%
	164	65											0.6737 ± 0.0002	9%
	164	65											0.6885 ± 0.0002	20%
	164	65											0.7548 ± 0.0002	22%
	164	65											0.7617 ± 0.0002	16%
	164	65											0.7826 ± 0.0002	5%
	164	65											0.8430 ± 0.001	3%
	164	65											0.845 ± 0.001	5%
	164	65											0.9660 ± 0.0005	1.5%
	164	65											1.1043 ± 0.001	1.1%
	164	65											1.1662 ± 0.001	1.9%
	164	65											1.1694 ± 0.001	2.3%
	164	65											1.2898 ± 0.0005	5.6%
	164	65											1.3775 ± 0.0005	5%
	164	65											1.4339 ± 0.0005	8.1%
	164	65											1.6567 ± 0.001	1%
	164	65											2.5110 ± 0.0015	1.3%
$_{65}Tb^{165}$	165	65			2.1 m.	β-	≈ 3.0				3/2+		0.5389	
	165	65											1.1785	
	165	65											1.2920	
	165	65											1.6648	
Dy		66		162.50						920 b.				
$_{66}Dy^{147m}$	147	66			58 s.	I.T.(40%)					(11/2-		Dy k x-ray	
	147	66				β+, EC(60%)							0.072 ± 0.001	5%
	147	66											0.6787 ± 0.0002	33%
$_{66}Dy^{147}$	147	66		146.930680	≈ 80 s.	E.C.,β+	≈ 6.4				1/2+		ann.rad.	
	147	66											0.1007	
	147	66											0.2534	
	147	66											0.3653	
$_{66}Dy^{148}$	148	66		147.927020	3.1 m.	β+(4%)	2.9	1.2			0+		ann.rad.	
	148	66				E.C.(96%)							Tb k x-ray	38%
	148	66											0.6202 ± 0.0001	100%
$_{66}Dy^{149}$	149	66		148.927110	4.2 m.	β+,E.C.	≈ 3.6				(7/2-)		ann.rad.	
	149	66											0.1008 ± 0.0001	100 +
	149	66											0.1063 ± 0.0001	51
	149	66											0.2534 ± 0.0001	50
	149	66											0.6536 ± 0.0001	60
	149	66											0.7365 ± 0.0001	19
	149	66											0.7417 ± 0.0001	17
	149	66											0.7753 ± 0.0001	35
	149	66											0.7894 ± 0.0001	65
	149	66											1.2742 ± 0.0003	18
	149	66											1.7765 ± 0.0003	79
	149	66											1.8062 ± 0.0003	64
$_{66}Dy^{150}$	150	66		149.925577	7.17 m.	β+, EC(67%)	≈ 1.8				0+		Tb k x-ray	26%
	150	66				α (33%)		4.233					0.3967 ± 0.0002	66%
$_{66}Dy^{151}$	151	66		150.926032	17 m.	β+(5%)	2.76				7/2-		Tb k x-ray	38%
	151	66				E.C.(89%)							0.1764 ± 0.0001	11%
	151	66				α (6%)		4.067					0.3030 ± 0.0001	1.4%
	151	66											0.3861 ± 0.0001	20%

Isotope	A	Z	% Natural abundance	Atomic mass	Half-life	Decay mode	Decay energy (MeV)	Particle energy (MeV)	Particle intensity	Thermal neutron cross section	Spin (h/2π)	μ Nucl. mag. moment	Gamma-ray energy (MeV)	Gamma-ray intensity
	151	66											0.4322 ± 0.0001	4.1%
	151	66											0.4632 ± 0.0001	2.5%
	151	66											0.4766 ± 0.0001	8.2%
	151	66											0.5463 ± 0.0001	15%
	151	66											0.6892 ± 0.0001	2.6%
	151	66											0.7003 ± 0.0001	1.9%
	151	66											0.7556 ± 0.0001	2.1%
	151	66											0.8339 ± 0.0001	2.3%
	151	66											0.8455 ± 0.0001	2.1%
	151	66											0.9837 ± 0.0001	2.1%
	151	66											1.0104 ± 0.0001	3.1%
	151	66											1.1143 ± 0.0001	2.7%
	151	66											1.1298 ± 0.0001	2.3%
	151	66											1.1418 ± 0.0001	2.1%
	151	66											1.4757 ± 0.0001	2.2%
	151	66											1.5381 ± 0.0001	2.0%
	151	66											1.5931 ± 0.0001	2.7%
	151	66											1.7016 ± 0.0001	4.7%
	151	66											(0.16 - 2.09)weak	
$_{66}Dy^{152}$	152	66		151.924716	2.3 h.	E.C.	0.74				0+		Tb k x-ray	40%
	152	66				α		3.63					0.2569 ± 0.0001	97.5%
$_{66}Dy^{153}$	153	66		152.925769	6.3 h.	β+(1%)	2.170	0.89			(7/2-)		Tb k x-ray	90%
	153	66				E.C.(99%)							0.0807 ± 0.0001	11%
	153	66				α (0.01%)		3.46					0.0997 ± 0.0001	10%
	153	66											0.1476 ± 0.0001	3.7%
	153	66											0.2137 ± 0.0001	11%
	153	66											0.2442 ± 0.0001	4.2%
	153	66											0.2543 ± 0.0001	8.3%
	153	66											0.2747 ± 0.0004	6.9%
	153	66											0.2967 ± 0.0001	1.0%
	153	66											0.3237 ± 0.0001	1.1%
	153	66											0.3895 ± 0.0001	1.5%
	153	66											0.4156 ± 0.0001	1.1%
	153	66											0.4341 ± 0.0001	1.2%
	153	66											0.4487 ± 0.0001	1.0%
	153	66											0.4714 ± 0.0001	1.3%
	153	66											0.5105 ± 0.0002	1.1%
	153	66											0.5372 ± 0.0001	1.3%
	153	66											0.5937 ± 0.0001	1.1%
	153	66											0.6598 ± 0.0001	1.1%
	153	66											1.0240 ± 0.0001	1.1%
	153	66											1.0499 ± 0.0001	1.1%
	153	66											1.1043 ± 0.0001	1.0%
	153	66											(0.08 - 1.66)weak	
$_{66}Dy^{154}$	154	66		153.924429	≈ 3 x 10^6y.	α		2.87			0+			
$_{66}Dy^{155}$	155	66		154.925747	10 h.	β+(2%)	2.09	0.845			3/2-		Tb k x-ray	48%
	155	66				E.C.(98%)							0.0655 ± 0.0001	1.8%
	155	66											0.0903 ± 0.0001	1.1%
	155	66											0.1614 ± 0.0001	1.1%
	155	66											0.1846 ± 0.0001	3.4%
	155	66											0.2269 ± 0.0001	69%
	155	66											0.2711 ± 0.0001	1.2%
	155	66											0.4842 ± 0.0001	1.1%
	155	66											0.4986 ± 0.0001	1.8%
	155	66											0.5084 ± 0.0001	1.2%
	155	66											0.6411 ± 0.0001	1.3%
	155	66											0.6642 ± 0.0001	2.3%
	155	66											0.9055 ± 0.0001	2.5%
	155	66											0.9997 ± 0.0001	2.4%
	155	66											1.0899 ± 0.0001	2.8%
	155	66											1.1555 ± 0.0001	2.1%
	155	66											1.1662 ± 0.0001	1.7%
	155	66											1.2512 ± 0.0001	0.9%
	155	66											1.3678 ± 0.0001	0.8%
	155	66											1.6650 ± 0.0001	0.9%
$_{66}Dy^{156}$	156	66	0.06%	155.925277						33 b.	0+			
$_{66}Dy^{157}$	157	66		156.925460	8.1 h.	E.C.	1.34				3/2-	-0.30	Tb k x-ray	43%
	157	66											0.1822 ± 0.0001	2%
	157	66											0.3262 ± 0.0001	93%
$_{66}Dy^{158}$	158	66	0.10%	157.924403						40 b.	0+			
$_{66}Dy^{159}$	159	66		158.925735	144 d.	E.C.	0.366				3/2-		Tb k x-ray	48%
	159	66											0.0582 ± 0.0001	2%
	159	66											0.3262 ± 0.0001	93%
$_{66}Dy^{160}$	160	66	2.34%	159.925193						60 b.	0+			
$_{66}Dy^{161}$	161	66	18.9%	160.926930						580 b.	5/2+	-0.48		
$_{66}Dy^{162}$	162	66	25.5%9	161.926795						180 b.	0+			
$_{66}Dy^{163}$	163	66	24.9%	162.928728						130 b.	5/2-	+0.673		
$_{66}Dy^{164}$	164	66	28.2%	163.929171						(1.7 + 10^3) b.	0+			
$_{66}Dy^{165m}$	165	66			1.26 m.	I.T.(98%)	0.108			2 x 10^3 b.	1/2-		Dy k x-ray	4.8%
	165	66				β-(2%)							0.1082 ± 0.0001	3%
	165	66											0.1538 ± 0.0001	0.2%
	165	66											0.3617 ± 0.0001	0.5%
	165	66											0.5155 ± 0.0001	1.5%
$_{66}Dy^{165}$	165	66		164.931700	2.33 h.	β-	1.286	1.29		3.6 x 10^3 b.	7/2+	0.51	Ho k x-ray	5%
	165	66											0.09468 ± 0.00001	3.6%

Isotope	A	Z	% Natural abundance	Atomic mass	Half-life	Decay mode	Decay energy (MeV)	Particle energy (MeV)	Particle intensity	Thermal neutron cross section	Spin (h/2π)	μ Nucl. mag. moment	Gamma-ray energy (MeV)	Gamma-ray intensity
	165	66											0.27974 ± 0.00001	0.5%
	165	66											0.36166 ± 0.00003	0.8%
	165	66											0.63340 ± 0.00003	0.6%
	165	66											0.71534 ± 0.00004	0.5%
	165	66											1.07964 ± 0.00001	0.07%
$_{66}Dy^{166}$	166	66		165.932803	81.6 h.	β-	0.48	0.40			0+		Ho k x-ray	25%
	166	66											0.0282 ± 0.0001	1%
	166	66											0.0825 ± 0.0001	13%
	166	66											0.3717 ± 0.0001	0.5%
	166	66											0.4260 ± 0.0001	0.5%
$_{66}Dy^{167}$	167	66		166.935650	6.2 m.	β-	2.35	1.78			(1/2-)		Ho k x-ray	20%
	167	66											0.1332 ± 0.0001	3%
	167	66											0.2500 ± 0.0001	10%
	167	66											0.2593 ± 0.0001	28%
	167	66											0.3103 ± 0.0001	25%
	167	66											0.5697 ± 0.0002	48%
	167	66											0.7071 ± 0.0002	1%
	167	66											0.9970 ± 0.0002	0.5%
	167	66											(0.06 - 1.4)	
$_{66}Dy^{168}$	168	66			≈ 8.5 m.	β-					0+		Ho k x-ray	14%
	168	66											0.1435 ± 0.0002	8%
	168	66											0.1925 ± 0.0002	31%
	168	66											0.4430 ± 0.0005	16%
	168	66											0.4867 ± 0.0002	22%
	168	66											0.6302 ± 0.0003	15%
Ho		67		164.9304						65 b.				
$_{67}Ho^{148}$	148	67		147.937340	9 s.	β+,E.C.	9.61						ann.rad.	
	148	67											0.5043 ± 0.0003	17 +
	148	67											0.6615 ± 0.0002	69
	148	67											1.6883 ± 0.0002	100
$_{67}Ho^{149}$	149	67		148.933600	21 s.	β+,E.C.	6.04				(9+)		ann.rad.	
	149	67											1.0733 ± 0.0001	13 +
	149	67											1.0911 ± 0.0001	100
	149	67											1.5836 ± 0.0002	9
$_{67}Ho^{150m}$	150	67			26 s.	β+,E.C.					(9+)		ann.rad.	
	150	67											0.3939 ± 0.0001	93%
	150	67											0.4112 ± 0.0002	7%
	150	67											0.5511 ± 0.0001	88%
	150	67											0.6243 ± 0.0002	3%
	150	67											0.6534 ± 0.0001	100
	150	67											0.8034 ± 0.0001	100
$_{67}Ho^{150}$	150	67		149.933200	≈ 88 s.	β+,E.C.							ann.rad.	
	150	67											0.5913 ± 0.0002	31%
	150	67											0.6534 ± 0.0003	30%
	150	67											0.8034 ± 0.0001	100%
$_{67}Ho^{151m}$	151	67			48 s.	β+, EC(87%)							ann.rad.	
	151	67				α(13%)		4.605					0.2102 ± 0.0002	
	151	67											0.4889 ± 0.0004	
	151	67											0.6948 ± 0.0002	
	151	67											0.7762 ± 0.0001	
$_{67}Ho^{151}$	151	67		150.931510	β+, EC(80%)	≈ 5.10							ann.rad.	
	151	67				α(20%)		4.519					0.3522 ± 0.0004	
	151	67											0.5274 ± 0.0001	100 +
	151	67											0.9676 ± 0.0003	3
	151	67											1.0471 ± 0.0001	
$_{67}Ho^{152m}$	152	67			51 s.	β+, EC(90%)					(9+)		ann.rad.	
	152	67				α(10%)		4.453					0.4929 ± 0.0004	53%
	152	67											0.6138 ± 0.0001	90%
	152	67											0.6474 ± 0.0001	90%
	152	67											0.6835 ± 0.0001	77%
	152	67											0.75850 ± 0.0002	10%
$_{67}Ho^{152}$	152	67		151.931580	2.4 m.	β+, EC(88%)	6.4				(3+)		ann.rad.	
	152	67				α(12%)		4.387					0.6140 ± 0.0001	88%
	152	67											0.6476 ± 0.0001	14%
$_{67}Ho^{153m}$	153	67			2.0 m.	β+, EC (99+%)							ann.rad.	
	153	67				α		3.91					0.2958 ± 0.0001	100 +
	153	67											0.3346 ± 0.0001	45
	153	67											0.3661 ± 0.0001	4
	153	67											0.4381 ± 0.0001	16
	153	67											0.6383 ± 0.0001	29
	153	67											1.0872 ± 0.0002	5
	153	67											1.2770 ± 0.001	10
$_{67}Ho^{153}$	153	67		152.930195	9.3 m.	β+, EC (99+%)	4.12						ann.rad.	

Isotope	A	Z	% Natural abundance	Atomic mass	Half-life	Decay mode	Decay energy (MeV)	Particle energy (MeV)	Particle intensity	Thermal neutron cross section	Spin (h/2π)	μ Nucl. mag. moment	Gamma-ray energy (MeV)	Gamma-ray intensity
	153	67				α		4.01					0.0905 ± 0.0001	6%
	153	67											0.1089 ± 0.0001	99%
	153	67											0.1215 ± 0.0001	2%
	153	67											0.1618 ± 0.0001	95%
	153	67											0.1990 ± 0.0002	5%
	153	67											0.2302 ± 0.0001	58%
	153	67											0.2590 ± 0.0002	12%
	153	67											0.2707 ± 0.0001	78%
	153	67											0.3659 ± 0.0001	100%
	153	67											0.3917 ± 0.0002	10%
	153	67											0.4054 ± 0.0003	4%
	153	67											0.4202 ± 0.0002	17%
	153	67											0.4565 ± 0.0002	46%
	153	67											0.5510 ± 0.0002	9%
	153	67											0.5537 ± 0.0002	23%
	153	67											0.5656 ± 0.0002	22%
$_{67}Ho^{154m}$	154	67			3.2 m.	β+,E.C.					(8+)		ann.rad.	
	154	67											0.2894 ± 0.0002	5%
	154	67											0.3095 ± 0.0002	4%
	154	67											0.3346 ± 0.0001	94%
	154	67											0.3466 ± 0.0002	10%
	154	67											0.4058 ± 0.0004	3%
	154	67											0.4069 ± 0.0001	19%
	154	67											0.4124 ± 0.0001	79%
	154	67											0.4347 ± 0.0002	2%
	154	67											0.4433 ± 0.0002	5%
	154	67											0.4771 ± 0.0001	55%
	154	67											0.5047 ± 0.0002	16%
	154	67											0.5238 ± 0.0001	18%
	154	67											0.5706 ± 0.0001	10%
	154	67											0.7251 ± 0.0001	12%
	154	67											0.7328 ± 0.0002	3%
	154	67											0.7406 ± 0.0002	2%
	154	67											0.8141 ± 0.0001	14%
	154	67											0.9591 ± 0.0003	2%
	154	67											0.9683 ± 0.0003	2.5%
	154	67											0.9929 ± 0.0003	5%
	154	67											0.9997 ± 0.0003	2.3%
	154	67											1.1385 ± 0.0003	1.0%
	154	67											1.2488 ± 0.0002	18%
$_{67}Ho^{154}$	154	67		153.930610	12 m.	β+,E.C.	5.76				(3+)		ann.rad.	135%
	154	67											Dy k x-ray	15%
	154	67											0.3262 ± 0.0003	4.7%
	154	67											0.3346 ± 0.0001	83%
	154	67											0.4125 ± 0.0001	15%
	154	67											0.5049 ± 0.0003	1.5%
	154	67											0.5700 ± 0.0008	11%
	154	67											0.6925 ± 0.0002	5.1%
	154	67											0.6955 ± 0.0003	1%
	154	67											0.7297 ± 0.0003	1.2%
	154	67											0.8734 ± 0.0001	12%
	154	67											0.9052 ± 0.0002	2.2%
	154	67											0.9997 ± 0.0002	3.2%
	154	67											1.0271 ± 0.0002	5.6%
	154	67											1.0859 ± 0.0003	1.6%
	154	67											1.1732 ± 0.0002	1.8%
	154	67											1.3006 ± 0.0002	3.7%
	154	67											1.4204 ± 0.0003	2.0%
	154	67											1.4983 ± 0.0003	1.4%
	154	67											1.5103 ± 0.0003	1.8%
	154	67											1.6565 ± 0.0003	1.4%
	154	67											1.8341 ± 0.0004	1.0%
	154	67											1.8493 ± 0.0004	0.9%
	154	67											1.9378 ± 0.0005	0.6%
	154	67											2.0103 ± 0.0006	0.5%
$_{67}Ho^{155}$	155	67		154.929078	48 m.	β+(6%)	3.10				(5/2+)		ann.rad.	12%
	155	67				E.C.(94%)							Dy k x-ray	47%
	155	67											0.0474 ± 0.0001	3%
	155	67											0.1039 ± 0.0001	2.7%
	155	67											0.1363 ± 0.0001	4.1%
	155	67											0.1385 ± 0.0001	1.1%
	155	67											0.1608 ± 0.0001	1.4%
	155	67											0.1630 ± 0.0001	1.2%
	155	67											0.1851 ± 0.0001	2.8%
	155	67											0.2009 ± 0.0001	1.9%
	155	67											0.2024 ± 0.0001	2.1%
	155	67											0.2084 ± 0.0001	2.0%
	155	67											0.2189 ± 0.0001	1.8%
	155	67											0.2478 ± 0.0001	1.9%
	155	67											0.2622 ± 0.0001	1.4%
	155	67											0.3097 ± 0.0001	1.2%
	155	67											0.3254 ± 0.0001	3.4%
	155	67											0.3692 ± 0.0001	1.0%
	155	67											0.3829 ± 0.0002	1.2%
	155	67											0.4086 ± 0.0001	1.5%

Isotope	A	Z	% Natural abundance	Atomic mass	Half-life	Decay mode	Decay energy (MeV)	Particle energy (MeV)	Particle intensity	Thermal neutron cross section	Spin (h/2π)	μ Nucl. mag. moment	Gamma-ray energy (MeV)	Gamma-ray intensity
	155	67											0.8971 ± 0.0001	1.2%
	155	67											1.0235 ± 0.0001	1.3%
	155	67											(0.06 - 2.24)weak	
$_{67}Ho^{156m}$	156	67			56 m.	I.T.	0.0352				1+		ann.rad.	50%
	156	67				β+(25%)		1.8			1+		Dy k x-ray	47%
	156	67				E.C.(75%)		2.9					0.1378 ± 0.0001	52%
	156	67											0.2666 ± 0.0001	54%
	156	67											0.3664 ± 0.0001	11%
	156	67											0.6844 ± 0.0002	5%
	156	67											0.6911 ± 0.0002	4.3%
	156	67											0.7644 ± 0.0002	3.6%
	156	67											0.8846 ± 0.0002	7.1%
	156	67											0.8909 ± 0.0002	2.7%
	156	67											0.9317 ± 0.0002	3.0%
	156	67											1.0310 ± 0.0002	3.2%
	156	67											1.1221 ± 0.0001	3.4%
	156	67											1.2236 ± 0.0002	2.4%
	156	67											1.4169 ± 0.0002	1.1%
	156	67											1.4540 ± 0.0002	1.1%
	156	67											1.4723 ± 0.0002	1.1%
	156	67											2.0310 ± 0.0004	1.0%
	156	67											2.0362 ± 0.0004	0.7%
	156	67											2.0534 ± 0.0005	0.7%
	156	67											2.4178 ± 0.0005	1.5%
	156	67											(0.28 - 2.9)weak	
$_{67}Ho^{156}$	156	67		155.929640	≈ 2 m.	β+,E.C.	5.00				(5+)		ann.rad.	
	156	67											0.1378	
	156	67											0.2665	
$_{67}HO^{157}$	157	67		156.928190	12.6 m.	β+(5%)	2.54	1.18			7/2-		ann.rad.	10%
	157	67				E.C.(95%)							Dy k x-ray	71%
	157	67											0.0611 ± 0.0001	5%
	157	67											0.0865 ± 0.0001	5%
	157	67											0.1477 ± 0.0001	2%
	157	67											0.1531 ± 0.0001	3%
	157	67											0.1624 ± 0.0001	1%
	157	67											0.1881 ± 0.0001	4%
	157	67											0.1934 ± 0.0001	7%
	157	67											0.2722 ± 0.0002	4%
	157	67											0.2800 ± 0.0001	21%
	157	67											0.3202 ± 0.0001	2%
	157	67											0.3411 ± 0.0001	16%
	157	67											0.5083 ± 0.0001	3%
	157	67											0.5556 ± 0.0001	3%
	157	67											0.7086 ± 0.0002	1.4%
	157	67											0.8354 ± 0.0002	1.1%
	157	67											0.8700 ± 0.0002	1.0%
	157	67											0.8966 ± 0.0002	4.2%
	157	67											1.2111 ± 0.0002	2.3%
$_{67}Ho^{158m2}$	158	67			21 m.	β+,E.C.					(9+)		ann.rad.	
	158	67											0.0981 ± 0.0001	
	158	67											0.1664 ± 0.0002	
	158	67											0.2182 ± 0.0001	
	158	67											0.3205 ± 0.0001	
	158	67											0.4062 ± 0.0001	
	158	67											0.9774 ± 0.0001	
	158	67											1.0532	
	158	67											0.4846	
$_{67}Ho^{158m1}$	158	67			27 m.	I.T.(44%)					2-		ann.rad.	
	158	67				E.C.(56%)							Dy k x-ray	67%
	158	67											0.0989 ± 0.0001	31%
	158	67											0.2182 ± 0.0001	46%
	158	67											0.9945 ± 0.0001	3%
	158	67											1.1615 ± 0.0001	1%
	158	67											1.2982 ± 0.0001	1%
	158	67											1.6237 ± 0.0001	2%
	158	67											1.7905 ± 0.0001	8%+
	158	67											2.2213 ± 0.0002	1.6%
	158	67											2.5456 ± 0.0002	1.2%
	158	67											2.6054 ± 0.0002	2.4%
$_{67}Ho^{158}$	158	67		157.928930	11.3 m.	β+(8%)	4.22	1.30			5+		ann.rad.	
	158	67				E.C.(92%)							Dy k x-ray	53%
	158	67											0.0989 ± 0.0001	53%
	158	67											0.2182 ± 0.0001	43%
	158	67											0.3205 ± 0.0001	8%
	158	67											0.7274 ± 0.0001	3%
	158	67											0.7314 ± 0.0001	4%
	158	67											0.8466 ± 0.0001	7.6%
	158	67											0.8474 ± 0.0001	18%
	158	67											0.8505 ± 0.0001	15%
	158	67											0.9457 ± 0.0001	16%
	158	67											0.9464 ± 0.0001	9%
	158	67											0.9488 ± 0.0001	24%
	158	67											0.9976 ± 0.0001	4%
	158	67											1.0648 ± 0.0001	3%
	158	67											1.1842 ± 0.0001	1%

Isotope	A	Z	% Natural abundance	Atomic mass	Half-life	Decay mode	Decay energy (MeV)	Particle energy (MeV)	Particle intensity	Thermal neutron cross section	Spin (h/2π)	μ Nucl. mag. moment	Gamma-ray energy (MeV)	Gamma-ray intensity
	158	67											1.2110 ± 0.0001	1.5%
	158	67											1.4634 ± 0.0001	2.3%
	158	67											1.5781 ± 0.0001	5.8%
	158	67											2.0651 ± 0.0001	2.3%
	158	67											2.1194 ± 0.0001	1.7%
	158	67											2.2016 ± 0.0002	3.4%
$_{67}$Ho^{159m}	159	67			8.3 s.	I.T.	0.206				$^1/_2$+		Ho k x-ray	10%
	159	67											0.1660 ± 0.0001	5%
	159	67											0.2059 ± 0.0001	40%
$_{67}$Ho^{159}	159	67		158.927706	33 m.	E.C.	1.836						Dy k x-ray	73%
	159	67											0.0567 ± 0.0002	5%
	159	67											0.1006 ± 0.0002	3.7%
	159	67											0.1210 ± 0.0002	33%
	159	67											0.1320 ± 0.0002	22%
	159	67											0.1558 ± 0.0002	2%
	159	67											0.1731 ± 0.0002	2%
	159	67											0.1776 ± 0.0002	6%
	159	67											0.1863 ± 0.0002	3%
	159	67											0.2177 ± 0.0002	3%
	159	67											0.2529 ± 0.0002	13%
	159	67											0.3096 ± 0.0002	14%
	159	67											0.8387 ± 0.0002	3%
	159	67											(0.06 - 1.2)weak	
$_{67}$Ho^{160m}	160	67			4.9 h.	I.T.(67%)	0.060				2-		0.0868 ± 0.0001	14%
	160	67				E.C.(33%)	3.35						0.1970 ± 0.0001	12%
	160	67											0.5385 ± 0.0003	4%
	160	67											0.6455 ± 0.0003	14%
	160	67											0.6464 ± 0.0003	39%
	160	67											0.7281 ± 0.0003	30%
	160	67											0.7529 ± 0.0003	2.6%
	160	67											0.7652 ± 0.0003	3.7%
	160	67											0.8715 ± 0.0004	6.6%
	160	67											0.8791 ± 0.0004	19%
	160	67											0.9619 ± 0.0004	17%
	160	67											0.9658 ± 0.0004	17%
	160	67											1.0684 ± 0.0005	2.7%
	160	67											1.1985 ± 0.0006	2.2%
	160	67											1.2715 ± 0.0006	2.8%
	160	67											1.2854 ± 0.0006	1.7%
	160	67											1.370 ± 0.001	1%
	160	67											1.4310 ± 0.0007	1%
	160	67											2.5430 ± 0.001	1.2%
	160	67											2.6127 ± 0.001	1.1%
	160	67											2.6305 ± 0.0015	1.3%
	160	67											2.6732 ± 0.0014	1.6%
$_{67}$Ho^{160}	160	67		159.928720	25.6 m.	β+,E.C.	3.29	0.57			5+		See Ho^{166m}	
	160	67											0.7282	
	160	67											0.8794	
$_{67}$Ho^{161m}	161	67			6.7 s.	I.T.	0.211						Ho k x-ray	10%
	161	67											0.2112 ± 0.0001	44%
$_{67}$Ho^{161}	161	67		160.927849	2.5 h.	E.C.	0.856				7/2-		Dy k x-ray	44%
	161	67											0.0256 ± 0.0001	27%
	161	67											0.0592 ± 0.0001	1.2%
	161	67											0.0774 ± 0.0001	2.5%
	161	67											0.1031 ± 0.0001	3.3%
$_{67}$Ho^{162m}	162	67			68 m.	I.T.(61%)					6-		Dy k x-ray	61%
	162	67				E.C.(39%)							Ho k x-ray	23%
	162	67											0.0578 ± 0.0001	4%
	162	67											0.0807 ± 0.0001	11%
	162	67											0.1850 ± 0.0001	29%
	162	67											0.2828 ± 0.0001	11%
	162	67											0.9372 ± 0.0002	11%
	162	67											1.1249 ± 0.0002	1.2%
	162	67											1.2200 ± 0.0002	22%
$_{67}$Ho^{162}	162	67		161.929092	15 m.	E.C.(96%)	0.295				1+		Dy k x-ray	48%
	162	67				β+(4%)							0.0807 ± 0.0001	8%
	162	67											1.3196 ± 0.0002	3.7%
	162	67											1.3728 ± 0.0002	0.8%
$_{67}$Ho^{163m}	163	67			1.09 s.	I.T.	0.298				($^1/_2$+)		Ho k x-ray	5.7%
	163	67											0.2798 ± 0.0001	77.5%
$_{67}$Ho^{163}	163	67		162.928731	33 ± .2 y.	E.C.	0.004				7/2-		Dy M x-rays	
$_{67}$Ho^{164m}	164	67			37.5 ± 1 m.	I.T.	0.140				(6-)		Ho k x-ray	37%
	164	67											0.0373 ± 0.0001	11%
	164	67											0.0566 ± 0.0001	6.7%
	164	67											0.0940 ± 0.0001	0.15%
$_{67}$Ho^{164}	164	67		163.930285	29 ± 0.5 m.	E.C.(58%)	1.029				1+		Dy k x-ray	25%
	164	67				β-(42%)	1.013						0.0734 ± 0.0001	1.8%
	164	67											0.0914 ± 0.0001	2.5%
$_{67}$Ho^{165}	165	67	100%	164.930319						(3.5 + 62) b.	7/2-	+4.173		
$_{67}$Ho^{166m}	166	67			1.2 ± 0.2 x 10^3y	β-					7-	4.1	Er k x-ray	20%
	166	67											0.08057 ± 0.00001	13%
	166	67											0.18407 ± 0.00002	74%

Isotope	A	Z	% Natural abundance	Atomic mass	Half-life	Decay mode	Decay energy (MeV)	Particle energy (MeV)	Particle intensity	Thermal neutron cross section	Spin (h/2π)	μ Nucl. mag. moment	Gamma-ray energy (MeV)	Gamma-ray intensity
	166	67											0.2159 ± 0.0001	2.6%
	166	67											0.28046 ± 0.00002	30%
	166	67											0.3007 ± 0.0001	3.8%
	166	67											0.3657 ± 0.0001	2.5%
	166	67											0.4109 ± 0.0001	12%
	166	67											0.4515 ± 0.0001	3.1%
	166	67											0.4648 ± 0.0001	1.2%
	166	67											0.5298 ± 0.0001	10%
	166	67											0.5710 ± 0.0001	6%
	166	67											0.6115 ± 0.0001	1.4%
	166	67											0.6705 ± 0.0001	5.8%
	166	67											0.6912 ± 0.0001	1.6%
	166	67											0.71169 ± 0.00004	59%
	166	67											0.7523 ± 0.0001	13%
	166	67											0.7788 ± 0.0001	3%
	166	67											0.81031 ± 0.00004	63%
	166	67											0.8306 ± 0.0001	11%
	166	67											0.9509 ± 0.0001	3.1%
	166	67											1.2414 ± 0.0001	1%
	166	67											1.4270 ± 0.0001	0.6%
$_{67}Ho^{166}$	166	67		165.932281	1.117 ± 0.002 d.	β-	1.854	1.776	48%		0-		Er k x-ray	5%
	166	67						1.855	51%				0.08057 ± 0.00002	6%
	166	67											1.37943 ± 0.00005	0.9%
	166	67											1.5819 ± 0.0001	0.2%
	166	67											1.6624 ± 0.0001	0.1%
$_{67}Ho^{167}$	167	67		166.933127	3.1 ± 0.1 h.	β-	0.97	0.31	43%		(7/2-)		Er k x-ray	12%
	167	67						0.61	21%				0.0793 ± 0.0002	2%
	167	67						0.96	15%				0.0835 ± 0.0002	2%
	167	67						0.97	15%				0.2379 ± 0.0002	5%
	167	67											0.3213 ± 0.0002	24%
	167	67											0.3465 ± 0.0002	57%
	167	67											0.3862 ± 0.0002	3%
	167	67											0.4030 ± 0.0002	3%
	167	67											0.4600 ± 0.0002	2%
$_{67}Ho^{168}$	168	67		167.935290	3.0 ± 0.1 m.	β-	2.7	2.0			3+		Er k x-ray	9%
	168	67											0.0798 ± 0.0001	10%
	168	67											0.1843 ± 0.0001	6%
	168	67											0.1982 ± 0.0001	2%
	168	67											0.4475 ± 0.0001	1.4%
	168	67											0.6317 ± 0.0001	3%
	168	67											0.7306 ± 0.0001	1.5%
	168	67											0.7413 ± 0.0001	36%
	168	67											0.8159 ± 0.0001	18%
	168	67											0.8211 ± 0.0001	34%
	168	67											1.3718 ± 0.0001	1.3%
	168	67											(0.08 - 2.34)weak	
$_{67}Ho^{169}$	169	67		168.936869	4.7 ± 0.1 m.	β-	2.12	1.2			(7/2-)			
	169	67						2.0					0.1496 ± 0.0003	7 +
	169	67											0.1519 ± 0.0003	26
	169	67											0.6289 ± 0.0003	13
	169	67											0.6765 ± 0.0002	20
	169	67											0.7170 ± 0.0002	15
	169	67											0.7610 ± 0.0002	48
	169	67											0.7784 ± 0.0002	47
	169	67											0.7884 ± 0.0001	100
	169	67											0.8494 ± 0.0006	5
	169	67											0.8529 ± 0.0002	536
	169	67											0.8664 ± 0.0002	21
	169	67											0.8764 ± 0.0003	10
$_{67}Ho^{170m}$	170	67			43 ± 2 s.	β-							0.0787 ± 0.0002	100 +
	170	67											0.1816 ± 0.0002	8
	170	67											0.4820 ± 0.0003	13
	170	67											0.5409 ± 0.0002	13
	170	67											0.6998 ± 0.0003	7.6
	170	67											0.8123 ± 0.0002	59
	170	67											0.8812 ± 0.0002	12
	170	67											0.9594 ± 0.0005	7
	170	67											1.0227 ± 0.0004	9
	170	67											1.1875 ± 0.0003	15
	170	67											1.2663 ± 0.0007	7.9
	170	67											1.8940 ± 0.0003	27
	170	67											1.9401 ± 0.0003	6.2
	170	67											1.9726 ± 0.0003	21
	170	67											1.9925 ± 0.0005	3
	170	67											2.6061 ± 0.0004	2
	170	67											2.6465 ± 0.0004	2
	170	67											2.6848 ± 0.0004	3

Isotope	A	Z	% Natural abundance	Atomic mass	Half-life	Decay mode	Decay energy (MeV)	Particle energy (MeV)	Particle intensity	Thermal neutron cross section	Spin (h/2π)	μ Nucl. mag. moment	Gamma-ray energy (MeV)	Gamma-ray intensity
	170	67											2.7151 ± 0.0008	1.5
	170	67											2.7892 ± 0.0015	0.7
$_{67}Ho^{170}$	170	67		169.939620	2.8 ± 0.2 m.	β^-							Er k x-ray	29%
	170	67											0.0787 ± 0.0001	13%
	170	67											0.0947 ± 0.0001	3%
	170	67											0.1035 ± 0.0001	5%
	170	67											0.1239 ± 0.0001	4%
	170	67											0.14125 ± 0.0001	2%
	170	67											0.1654 ± 0.0001	4%
	170	67											0.1816 ± 0.0001	27%
	170	67											0.2274 ± 0.0001	4%
	170	67											0.2582 ± 0.0001	43%
	170	67											0.2804 ± 0.0001	3%
	170	67											0.2834 ± 0.0001	3%
	170	67											0.4132 ± 0.0002	4%
	170	67											0.4774 ± 0.0002	4%
	170	67											0.7504 ± 0.0002	6%
	170	67											0.7863 ± 0.0005	6%
	170	67											0.8435 ± 0.0002	3%
	170	67											0.8547 ± 0.0005	14%
	170	67											0.8670 ± 0.0002	2.5%
	170	67											0.8902 ± 0.0002	25%
	170	67											0.9321 ± 0.0002	42%
	170	67											0.9346 ± 0.0005	4%
	170	67											0.9414 ± 0.0005	24%
	170	67											0.9574 ± 0.0003	4%
	170	67											0.9765 ± 0.0003	3%
	170	67											1.0247 ± 0.0004	1.7%
	170	67											1.1118 ± 0.0003	2.4%
	170	67											1.1387 ± 0.0002	24%
	170	67											1.1530 ± 0.0003	2%
	170	67											1.2260 ± 0.0003	4%
	170	67											1.3069 ± 0.0003	0.5%
Er		68		167.26						160 b.				
$_{68}Er^{150}$	150	68		149.997710	20 s.	β^+(36%)	4.2				0+		ann.rad.	125%
	150	68				E.C.(64%))							Ho k x-ray	
	150	68											0.4758 ± 0.0003	99%
$_{68}Er^{151}$	151	68		150.937200	23 s.	β^+,E.C.	5.3						ann.rad.	
$_{68}Er^{152}$	152	68		151.934920	10.3 s.	β^+, EC (10%)	3.12				0+		ann.rad.	
	152	68				α(90%)		4.804						
$_{68}Er^{153}$	153	68		152.934870	37.1 s.	α(53%)		4.674					ann.rad.	
	153	68				β^+, EC (47%)		4.35						
$_{68}Er^{154}$	154	68		153.932772	3.7 m.	β^+, EC (99+%)	2.014				0+		ann.rad.	
	154	68				α(0.5%)		4.166						
$_{68}Er^{155}$	155	68		154.933060	5.3 m.	β^+, EC (47%)					(7/2-)		ann.rad.	94%
	155	68				E.C.(53%)							Ho k x-ray	32%
	155	68											0.1101 ± 0.0001	7%
	155	68											0.1238 ± 0.0001	2%
	155	68											0.1851 ± 0.0001	1%
	155	68											0.2011 ± 0.0001	2%
	155	68											0.2340 ± 0.0001	3%
	155	68											0.2415 ± 0.0002	5%
	155	68											0.3287 ± 0.0002	1%
	155	68											0.3586 ± 0.0003	1%
	155	68											0.4227 ± 0.0001	1.5%
	155	68											0.4526 ± 0.0002	1.8%
	155	68											0.5122 ± 0.0002	2.8%
$_{68}Er^{156}$	156	68		155.931290	20 m.	β^+,E.C.	1.53				0+		ann.rad.	
	156	68											0.0298 ± 0.0001	17 +
	156	68											0.0352 ± 0.0001	100
	156	68											0.0522 ± 0.0001	
	156	68											0.1336 ± 0.0004	44
$_{68}Er^{157}$	157	68		156.931910	24 m.	β^+,E.C.	3.47				3/2-		ann.rad.	
	157	68											0.117	
	157	68											0.385	
	157	68											1.320	
	157	68											1.660	
	157	67											1.820	
	157	68											2.000	
$_{68}Er^{158}$	158	68		157.930010	2.3 h.	E.C. (99.5%)	1.00	0.74			0+		Ho k x-ray	71%
	158	68				β^+(0.5%)							0.0719 ± 0.0001	10%
	158	68											0.2486 ± 0.0001	3%
	158	68											0.3108 ± 0.0001	1.6%
	158	68											0.3868 ± 0.0001	7%
	158	68											0.5161 ± 0.0003	1%
$_{68}Er^{159}$	159	68		158.930678	36 m.	β^+(7%)	2.77				3/2-		ann.rad.	14%
	159	68				E.C.(93%)							Ho k x-ray	42%
	159	68											0.1660 ± 0.0001	4%

Isotope	A	Z	% Natural abundance	Atomic mass	Half-life	Decay mode	Decay energy (MeV)	Particle energy (MeV)	Particle intensity	Thermal neutron cross section	Spin (h/2π)	μ Nucl. mag. moment	Gamma-ray energy (MeV)	Gamma-ray intensity
	159	68											0.2523 ± 0.0003	4%
	159	68											0.3146 ± 0.0001	1%
	159	68											0.5054 ± 0.0001	2%
	159	68											0.5517 ± 0.0002	2.5%
	159	68											0.5810 ± 0.0001	4.6%
	159	68											0.6245 ± 0.0001	35%
	159	68											0.6493 ± 0.0001	25%
	159	68											0.9425 ± 0.0004	1%
	159	68											1.8383 ± 0.0007	1%
	159	68											2.0016 ± 0.0007	0.9%
	159	68											(0.07 - 2.5)many	
$_{68}Er^{160}$	160	68		159.929080	28.6 h.	E.C.	0.33				0+		Ho k x-ray	38%
	160	68											(0.05 - 0.96)weak	
$_{68}Er^{161}$	161	68		160.929996	3.24 h.	E.C.	2.00				3/2-	-0.370	Ho k x-ray	46%
	161	68											0.2015 ± 0.0001	1%
	161	68											0.3148 ± 0.0001	2.6%
	161	68											0.5926 ± 0.0001	3%
	161	68											0.8265 ± 0.0001	61%
	161	68											0.8650 ± 0.0002	1.2%
	161	68											0.9317 ± 0.0001	1.9%
	161	68											1.1745 ± 0.0003	1.2%
	161	68											(0.07 - 1.74)weak	
$_{68}Er^{162}$	162	68	0.14%	161.928775						19 b.	0+			
$_{68}Er^{163}$	163	68		162.930030	75.1 m.	E.C.	1.21				5/2-	+0.57	Ho k x-ray	40%
	163	68											0.4361 ± 0.0001	0.03%
	163	68											0.4399 ± 0.0001	0.03%
	163	68											1.1135 ± 0.0003	0.05%
$_{68}Er^{164}$	164	68	1.61%	163.929198						13 b.	0+			
$_{68}Er^{165}$	165	68		164.930723	10.36 h.	E.C.	0.377				3/2-		Ho k x-ray	39%
$_{68}Er^{166}$	166	68	33.6%	165.930290						(15 + 5) b.	0+			
$_{68}Er^{167m}$	167	68			2.28 s.	I.T.	0.208				1/2-		Er k x-ray	10%
	167	68											0.2078 ± 0.0001	42%
$_{68}Er^{167}$	167	68	22.95%	166.932046						670 b.	7/2+	-0.5665		
$_{68}Er^{168}$	168	68	26.8%	167.932368						2.7 b.	0+			
$_{68}Er^{169}$	169	68		168.934588	9.4 ± .02 d.	β-	0.351	0.35	≈ 100%		1/2-	+0.515	Tm k x-ray	0.006%
	169	68											0.1098 ± 0.0001	0.0013%
	169	68											0.1182 ± 0.0001	0.0001%
$_{68}Er^{170}$	170	68	14.9%	169.935461						5.7 b.	0+			
$_{68}Er^{171}$	171	68		170.938027	7.52 ± 0.03 h.	β-					5/2-	0.70	Tm k x-ray	24%
	171	68											0.11160 ± 0.00002	20%
	171	68											0.11669 ± 0.00001	2.3%
	171	68											0.12409 ± 0.00001	9%
	171	68											0.29591 ± 0.00003	29%
	171	68											0.30832 ± 0.00001	64%
	171	68											0.79634 ± 0.00008	0.6%
	171	68											0.90795 ± 0.00008	0.6%
	171	68											(0.08 - 1.4)weak	
$_{68}Er^{172}$	172	68		171.939353	2.05 ± 0.02 d.	β-	0.889	0.28	48%				Tm k x-ray	17%
	172	68						0.36	46%				0.0597 ± 0.0001	2.9%
	172	68											0.0681 ± 0.0001	3.5%
	172	68											0.1278 ± 0.0001	2.3%
	172	68											0.2027 ± 0.0001	1.1%
	172	68											0.3835 ± 0.0001	2.5%
	172	68											0.4073 ± 0.0001	45%
	172	68											0.4460 ± 0.0001	3.1%
	172	68											0.4754 ± 0.0001	1.1%
	172	68											0.6101 ± 0.0001	47%
$_{68}Er^{173}$	173	68		172.942280	1.40 ± 0.1 m.	β-	2.50				(7/2-)		Tm k x-ray	34%
	173	68											0.0942 ± 0.0002	5%
	173	68											0.1161 ± 0.0002	19%
	173	68											0.1186 ± 0.0002	2%
	173	68											0.1224 ± 0.0001	21%
	173	68											0.1928 ± 0.0002	46%
	173	68											0.1992 ± 0.0002	48%
	173	68											0.8008 ± 0.0006	10%
	173	68											0.8952 ± 0.0004	54%
Tm		69		168.9342						105 b.				
$_{69}Tm^{152}$	152	69		151.944460	5.2 s.	β+,E.C.	8.88						ann.rad.	
$_{69}Tm^{153}$	153	69		152.941830	1.6 s.	β+, EC (10%)	6.49						ann.rad.	
	153	69				α(90%)		5.11						
$_{69}Tm^{154m}$	154	69			3.4 s.	β+, EC (15%)							ann.rad.	
	154	69				α		5.03						

Isotope	A	Z	% Natural abundance	Atomic mass	Half-life	Decay mode	Decay energy (MeV)	Particle energy (MeV)	Particle intensity	Thermal neutron cross section	Spin (h/2π)	μ Nucl. mag. moment	Gamma-ray energy (MeV)	Gamma-ray intensity
$_{69}Tm^{154}$	154	69		153.941360	8.3 s.	β+, EC (56%)	7.99						ann.rad.	
	154	69				α (44%)		4.96						
$_{69}Tm^{155}$	155	69		154.939010	25 s.	β+,E.C.	5.55						0.0315	5 +
	155	69				α		4.45					0.0638 ± 0.0001	3
	155	69											0.0881 ± 0.0002	17
	155	69											0.1520 ± 0.0001	7
	155	69											0.1716 ± 0.0001	2
	155	69											0.2268 ± 0.0002	100
	155	69											0.2476 ± 0.0002	6
	155	69											0.3153 ± 0.0003	2
	155	69											0.3235 ± 0.0003	8
	155	69											0.3790 ± 0.0004	4
	155	69											0.4334 ± 0.0003	3
	155	69											0.5187 ± 0.0004	3
	155	69											0.5320 ± 0.0005	20
	155	69											0.5333 ± 0.0005	5
	155	69											0.5757 ± 0.0003	2
	155	69											0.6067 ± 0.0002	11
$_{69}Tm^{156m}$	156	69			19 s.	α		4.46						
$_{69}Tm^{156}$	156	69		155.938840	80 s.	β+,E.C.	7.03						ann.rad.	
	156	69				α		4.23					0.3446 ± 0.0001	86%
	156	69											0.4208 ± 0.0001	1.6%
	156	69											0.4529 ± 0.0001	17%
	156	69											0.5860 ± 0.0002	14%
	156	69											0.6089 ± 0.0002	1.4%
	156	69											0.7000 ± 0.0002	1.2%
	156	69											0.8762 ± 0.0002	2.3%
	156	69											0.8985 ± 0.0002	1.3%
	156	69											0.9304 ± 0.0001	5.0%
	156	69											0.9590 ± 0.0001	8.8%
	156	69											1.0068 ± 0.0002	3.1%
	156	69											1.0171 ± 0.0003	1.1%
	156	69											1.2208 ± 0.0002	2.9%
	156	69											0.2261 ± 0.0003	1.2%
	156	69											1.2861 ± 0.0002	2.7%
	156	69											1.3661 ± 0.0003	1.5%
	156	69											1.5163 ± 0.0003	1.2%
	156	69											1.5180 ± 0.0004	1.9%
	156	69											1.5651 ± 0.0003	1.6%
	156	69											1.6640 ± 0.0004	1.2%
	156	69											0.6700 ± 0.0003	1.4%
	156	69											1.8253 ± 0.0003	0.8%
$_{69}Tm^{157}$	157	69		156.936880	3.6 m.	β+,E.C.	4.63						ann.rad.	
	157	69				α		3.97					0.1104 ± 0.0001	9%
	157	69											0.1312 ± 0.0002	4%
	157	69											0.1754 ± 0.0002	3%
	157	69											0.1960 ± 0.0001	3%
	157	69											0.2416 ± 0.0001	7%
	157	69											0.2475 ± 0.0001	3%
	157	69											0.3080 ± 0.0002	2%
	157	69											0.3484 ± 0.0002	9%
	157	69											0.3570 ± 0.0002	7%
	157	69											0.3578 ± 0.0002	5%
	157	69											0.3606 ± 0.0002	4%
	157	69											0.3674 ± 0.0002	4%
	157	69											0.3707 ± 0.0001	5%
	157	69											0.3810 ± 0.0001	2%
	157	69											0.3855 ± 0.0001	10%
	157	69											0.3873 ± 0.0002	2%
	157	69											0.4550 ± 0.0002	10%
	157	69											0.4846 ± 0.0002	3%
	157	69											0.5250 ± 0.0002	4%
	157	69											0.5353 ± 0.0002	4%
	157	69											0.5491 ± 0.0003	5%
	157	69											0.5556 ± 0.0003	3%
	157	69											0.5750 ± 0.0001	3%
	157	69											0.6855 ± 0.0002	2%
	157	69											0.9234 ± 0.0003	1%
	157	69											0.9564 ± 0.0003	1%
	157	69											1.2624 ± 0.0005	2%
	157	69											(0.1 - 1.58)weak	
$_{69}Tm^{158}$	158	69		157.936990	4.0 m.	β+, EC (74%)	6.50				(2-)		ann.rad.	150%
	158	69				E.C.(26%)							Er k x-ray	18%
	158	69											0.1921 ± 0.0001	68%
	158	69											0.3351 ± 0.0001	19%
	158	69											0.6143 ± 0.0001	1.9%
	158	69											0.6280 ± 0.0001	7.4%
	158	69											0.6566 ± 0.0001	1.9%
	158	69											0.7969 ± 0.0002	1.2%
	158	69											0.8148 ± 0.0001	1.3%
	158	69											0.8201 ± 0.0001	3.6%
	158	69											0.8512 ± 0.0001	5.2%
	158	69											0.9891 ± 0.0001	4.0%
	158	69											1.0651 ± 0.0001	1.7%

TABLE OF THE ISOTOPES (Continued)

Isotope	A	Z	% Natural abundance	Atomic mass	Half-life	Decay mode	Decay energy (MeV)	Particle energy (MeV)	Particle intensity	Thermal neutron cross section	Spin (h/2π)	μ Nucl. mag. moment	Gamma-ray energy (MeV)	Gamma-ray intensity
	158	69											1.1498 ± 0.0001	8.4%
	158	69											1.2259 ± 0.0001	1.5%
	158	69											1.3340 ± 0.0001	3.6%
	158	69											1.4186 ± 0.0001	1.5%
	158	69											1.5505 ± 0.0001	1.8%
	158	69											1.5772 ± 0.0001	1.0%
	158	69											(0.18 - 2.81)weak	
$_{69}Tm^{159}$	159	69		158.934970	9.0 m.	β+(23%)	4.00				5/2+		ann.rad.	45%
	159	69				E.C.(77%)							Er k x-ray	79%
	159	69											0.0591 ± 0.0001	5%
	159	69											0.0848 ± 0.0001	7%
	159	69											0.1197 ± 0.0001	3%
	159	69											0.1277 ± 0.0001	5%
	159	69											0.1441 ± 0.0001	2%
	159	69											0.1609 ± 0.0001	4%
	159	69											0.1629 ± 0.0001	2%
	159	69											0.1965 ± 0.0001	2%
	159	69											0.2202 ± 0.0001	5%
	159	69											0.2477 ± 0.0001	2%
	159	69											0.2527 ± 0.0001	2%
	159	69											0.2713 ± 0.0001	6%
	159	69											0.2890 ± 0.0001	5%
	159	69											0.3483 ± 0.0001	4%
	159	68											0.3749 ± 0.0002	2%
	159	69											0.4085 ± 0.0002	2%
	159	69											0.4503 ± 0.0001	1%
	159	69											0.4617 ± 0.0001	1%
	159	69											0.5417 ± 0.0002	1%
	159	69											1.2701 ± 0.0003	0.8%
	159	69											(0.05 - 1.27)weak	
$_{69}Tm^{160}$	160	69		159.935090	9.2 m.	β+(15%)	5.60				1-		ann.rad.	30%
	160	69				E.C.(85%)							Er k x-ray	
	160	69											0.1264	100 +
	160	69											0.2642	26
	160	69											0.5971	5
	160	69											0.6175	5
	160	69											0.6401	5
	160	69											0.7285	36
	160	69											0.7678	8
	160	69											0.7977	7
	160	69											0.8544	23
	160	69											0.8614	20
	160	69											0.8820	6
	160	69											1.0077	7
	160	69											1.2491	8
	160	69											1.2641	4
	160	69											1.2697	8
	160	69											1.3685	24
	160	69											1.3947	10
	160	69											1.4606	12
	160	69											1.5264	11
	160	69											1.5366	3
	160	69											1.5869	3
	160	69											1.7685	2
	160	69											1.8944	2
	160	69											2.0685	3
	160	69											2.1334	3
	160	69											2.2022	4
$_{69}Tm^{161}$	161	69		160.933430	38 m.	β+,E.C.	3.20				7/2+		ann.rad.	98%
	161	69											Er k x-ray	25%
	161	69											0.0595 ± 0.0001	5%
	161	69											0.0844 ± 0.0001	9%
	161	69											0.0998 ± 0.0001	2.4%
	161	69											0.1059 ± 0.0001	3.4%
	161	69											0.1126 ± 0.0001	3%
	161	69											0.1256 ± 0.0001	1.6%
	161	69											0.1289 ± 0.0001	3%
	161	69											0.1439 ± 0.0001	4%
	161	69											0.1534 ± 0.0001	3%
	161	69											0.1578 ± 0.0001	2%
	161	69											0.1720 ± 0.0001	5%
	161	69											0.1902 ± 0.0001	3.4%
	161	69											0.2071 ± 0.0001	2%
	161	69											0.2129 ± 0.0001	3.2%
	161	69											0.2157 ± 0.0001	1.6%
	161	69											0.2181 ± 0.0001	1%
	161	69											0.2446 ± 0.0001	1%
	161	69											0.2525 ± 0.0001	1.6%
	161	69											0.2655 ± 0.0001	1%
	161	69											0.3538 ± 0.0001	1%
	161	69											0.3695 ± 0.0001	1.4%
	161	69											0.3726 ± 0.0001	1%
	161	69											0.5236 ± 0.0004	0.8%
	161	69											1.0032 ± 0.0004	0.7%
	161	69											1.6481 ± 0.0003	19%

Isotope	A	Z	% Natural abundance	Atomic mass	Half-life	Decay mode	Decay energy (MeV)	Particle energy (MeV)	Particle intensity	Thermal neutron cross section	Spin (h/2π)	μ Nucl. mag. moment	Gamma-ray energy (MeV)	Gamma-ray intensity
	161	69											1.7880 ± 0.0003	1.7%
	161	69											1.8500 ± 0.0003	1.6%
	161	69											1.8547 ± 0.0003	0.8%
	161	69											1.8941 ± 0.0004	0.7%
	161	69											(0.04 - 2.15)many	
$_{69}$Tm^{162m}	162	69			24 s.	I.T.(90%)					5+		Tm k x-ray	27%
	162	69				β+, EC (10%)							Er k x-ray	5%
	162	69											0.0669 ± 0.0001	7%
	162	69											0.1020 ± 0.0001	3%
	162	69											0.2275 ± 0.0001	5%
	162	69											0.3775 ± 0.0002	1.6%
	162	69											0.7100 ± 0.0001	3.6%
	162	69											0.7987 ± 0.0002	5.6%
	162	69											0.8115 ± 0.0001	6.6%
	162	69											0.9003 ± 0.0004	6.8%
$_{69}$Tm^{162}	162	69		161.933920	21.7 m.	β+(8%)	4.99				1-		ann.rad.	16%
	162	69				E.C.(92%)							Er k x-ray	48%
	162	69											0.1020 ± 0.0001	17%
	162	69											0.2275 ± 0.0001	7%
	162	69											0.5707 ± 0.0001	2.1%
	162	69											0.6723 ± 0.0002	6.9%
	162	69											0.7987 ± 0.0001	9.1%
	162	69											0.8998 ± 0.0004	5.6%
	162	69											0.9007 ± 0.0004	6.5%
	162	69											0.9851 ± 0.0001	1.1%
	162	69											1.0690 ± 0.0001	1.1%
	162	69											1.1000 ± 0.0001	1.4%
	162	69											1.2500 ± 0.0001	4.8%
	162	69											1.2547 ± 0.0001	1.5%
	162	69											1.3184 ± 0.0001	5.6%
	162	69											1.3522 ± 0.0001	3.4%
	162	69											1.4042 ± 0.0001	2.8%
	162	69											1.5064 ± 0.0001	1.4%
	162	69											1.9747 ± 0.0001	1.2%
	162	69											2.0158 ± 0.0001	1.1%
	162	69											2.1402 ± 0.0001	1.3%
	162	69											2.2317 ± 0.0001	0.8%
	162	69											3.2979 ± 0.0002	0.65%
	162	69											3.5746 ± 0.0002	0.4%
	162	69											(0.1 - 3.75)many	
$_{69}$Tm^{163}	163	69		162.932648	1.8 h.	E.C.(98%)	2.44				1/2+	0.081	Er k x-ray	76%
	163	69				β+(1%)							0.0692 ± 0.0001	11%
	163	69											0.1043 ± 0.0001	19%
	163	69											0.1901 ± 0.0001	1.3%
	163	69											0.2396 ± 0.0001	4.1%
	163	69											0.2414 ± 0.0001	9.3%
	163	69											0.2414 ± 0.0001	9.3%
	163	69											0.2752 ± 0.0001	2.4%
	163	69											0.2997 ± 0.0001	4.1%
	163	69											0.3457 ± 0.0002	1.1%
	163	69											0.3934 ± 0.0002	1.2%
	163	69											0.4041 ± 0.0002	1.0%
	163	69											0.4712 ± 0.0002	3.8%
	163	69											0.5502 ± 0.0002	1.5%
	163	69											0.5796 ± 0.0002	1.7%
	163	69											0.6659 ± 0.0002	1.7%
	163	69											0.7798 ± 0.0002	0.7%
	163	69											0.9452 ± 0.0002	0.7%
	163	69											1.1300 ± 0.0001	1.9%
	163	69											1.2048 ± 0.0001	2.4%
	163	69											1.2240 ± 0.0001	2.0%
	163	69											1.2649 ± 0.0001	4.9%
	163	69											1.3182 ± 0.0001	1.4%
	163	69											1.3743 ± 0.0001	4.2%
	163	69											1.3869 ± 0.0001	1.0%
	163	69											1.3974 ± 0.0001	7.1%
	163	69											1.4343 ± 0.0001	7.6%
	163	69											1.4558 ± 0.0001	3.4%
	163	69											1.4656 ± 0.0002	1.8%
	163	69											1.4694 ± 0.0002	2.7%
	163	69											1.7492 ± 0.0001	0.9%
	163	69											1.8037 ± 0.0001	1.2%
$_{69}$Tm^{164m}	164	69			5.1 m.	I.T.(80%)					6-		0.0914 ± 0.0001	12%
	164	69				β+, EC (20%)							0.1394 ± 0.0001	2.9%
	164	69											0.2081 ± 0.0001	18%
	164	69											0.2405 ± 0.0001	8.6%
	164	69											0.3149 ± 0.0001	11%
	164	69											0.4102 ± 0.0001	1.6%
	164	69											0.5470 ± 0.0001	5.1%
	164	69											0.7689 ± 0.0001	1.9%
	164	69											0.8207 ± 0.0001	1.5%
	164	69											0.8549 ± 0.0001	1.2%
$_{69}$Tm^{164}	164	69		163.933451	2.0 m.	β+(36%)	3.96	2.94			1+		ann.rad.	70%

Isotope	A	Z	% Natural abundance	Atomic mass	Half-life	Decay mode	Decay energy (MeV)	Particle energy (MeV)	Particle intensity	Thermal neutron cross section	Spin (h/2π)	μ Nucl. mag. moment	Gamma-ray energy (MeV)	Gamma-ray intensity
	164	69				E.C.(64%)							Er k x-ray	31%
	164	69											0.0914 ± 0.0001	6.7%
	164	69											0.2081 ± 0.0001	1.2%
	164	69											0.7689 ± 0.0001	1.4%
	164	69											0.8603 ± 0.0001	1.1%
	164	69											1.1546 ± 0.0002	1.7%
	164	69											1.6107 ± 0.0001	1.1%
	164	69											1.6744 ± 0.0001	1.0%
	164	69											1.8625 ± 0.0001	0.5%
	164	69											2.0816 ± 0.0001	1.6%
	164	69											2.3836 ± 0.0001	0.4%
$_{69}Tm^{165}$	165	69		164.932432	30.06 h.	E.C.	1.59				1/2+	0.139	Er k x-ray	52%
	165	69											0.0472 ± 0.0001	17%
	165	69											0.0544 ± 0.0001	7.2%
	165	69											0.1136 ± 0.0001	1.6%
	165	69											0.2189 ± 0.0001	3.3%
	165	69											0.24296 ± 0.00005	3.5%
	165	69											0.2924 ± 0.0001	1.3%
	165	69											0.29728 ± 0.00005	14%
	165	69											0.3469 ± 0.0001	3.1%
	165	69											0.3565 ± 0.0001	2.7%
	165	69											0.3894 ± 0.0001	2.8%
	165	69											0.4483 ± 0.0002	2.6%
	165	69											0.46024 ± 0.00006	4.0%
	165	69											0.4873 ± 0.0002	1.2%
	165	69											0.5424 ± 0.0002	1.7%
	165	69											0.5639 ± 0.0002	2.4%
	165	69											0.5897 ± 0.0002	2.3%
	165	69											0.80636 ± 0.00008	8.4%
	165	69											1.1313 ± 0.0002	1.4%
	165	69											1.18456 ± 0.00008	2.6%
	165	69											1.4272 ± 0.0003	0.9%
$_{69}Tm^{166}$	166	69		165.933561	7.70 h.	E.C.(98%)	3.047				2+	0.092	Er k x-ray	51%
	166	69				β+(2%)							0.0806 ± 0.0001	11%
	166	69											0.1844 ± 0.0001	18%
	166	69											0.2152 ± 0.0001	6.1%
	166	69											0.4596 ± 0.0001	2.7%
	166	69											0.5499 ± 0.0001	3.7%
	166	69											0.5988 ± 0.0001	2.2%
	166	69											0.6743 ± 0.0001	6.7%
	166	69											0.6748 ± 0.0001	2.8%
	166	69											0.6912 ± 0.0001	8.0%
	166	69											0.7043 ± 0.0002	1.1%
	166	69											0.7053 ± 0.0001	12%
	166	69											0.7578 ± 0.0001	2.7%
	166	69											0.7789 ± 0.0001	21%
	166	69											0.7859 ± 0.0001	11%
	166	69											0.8103 ± 0.0001	1.2%
	166	69											0.8756 ± 0.0001	4.7%
	166	69											1.1523 ± 0.0002	1.8%
	166	69											1.1766 ± 0.0002	11%
	166	69											1.2353 ± 0.0002	2.1%
	166	69											1.2734 ± 0.0001	17%
	166	69											1.3006 ± 0.0001	1.6%
	166	69											1.3469 ± 0.0002	1.3%
	166	69											1.3741 ± 0.0001	6.7%
	166	69											1.5050 ± 0.0002	1.0%
	166	69											1.6529 ± 0.0002	1.2%
	166	69											1.8680 ± 0.0002	4.8%
	166	69											1.8953 ± 0.0002	1.5%
	166	69											2.0524 ± 0.0002	20%
	166	69											2.0796 ± 0.0002	7.5%
	166	69											2.0923 ± 0.0002	1.9%
$_{69}Tm^{167}$	167	69		166.932848	9.25 d.	E.C.	0.747				1/2+	-0.197	Er k x-ray	48%
	167	69											0.0571 ± 0.0001	3.5%
	167	69											0.20778 ± 0.00008	41%(D)
	167	69											0.5315 ± 0.0001	1.6%
$_{69}Tm^{168}$	168	69		167.934170	93.1 d.	E.C.	1.680				3+		Er k x-ray	47%
	168	69											0.0798 ± 0.0001	11%
	168	69											0.0992 ± 0.0001	4.4%
	168	69											0.1843 ± 0.0001	16.4%
	168	69											0.19825 ± 0.00002	50%
	168	69											0.4475 ± 0.0001	22%
	168	69											0.5468 ± 0.0001	2.4%
	168	69											0.6317 ± 0.0001	7.7%
	168	69											0.6457 ± 0.0001	1.4%
	168	69											0.7203 ± 0.0001	11%
	168	69											0.7306 ± 0.0001	5%

Isotope	A	Z	% Natural abundance	Atomic mass	Half-life	Decay mode	Decay energy (MeV)	Particle energy (MeV)	Particle intensity	Thermal neutron cross section	Spin (h/2π)	μ Nucl. mag. moment	Gamma-ray energy (MeV)	Gamma-ray intensity
	168	69											0.74132 ± 0.00003	11%
	168	69											0.81595 ± 0.00003	46%
	168	69											0.8211 ± 0.0001	11%
	168	69											0.8299 ± 0.0001	6.2%
	168	69											0.91490 ± 0.00003	2.9%
	168	69											1.27741 ± 0.00005	1.6%
$_{69}Tm^{169}$	169	69	100%	168.934212						105 b.	1/2+	-0.2316		
$_{69}Tm^{170}$	170	69		169.935198	128.6 ± 0.3 d.	β-(99.8%)	0.968	0.883	24%	92 b.	1-	0.2476	Yb k x-ray	2%
	170	69				E.C.(0.2%)	0.314	0.968	76%				0.08425 ± 0.00003	3.3%
$_{69}Tm^{171}$	171	69		170.936427	1.92 ± 0.1 y.	β-	0.096	0.03	2%		1/2+	0.2303	0.06674 ± 0.00001	0.14%
	171	69						0.096	98%					
$_{69}Tm^{172}$	172	69		171.938397	2.65 ± 0.01 d.	β-	1.88	1.79	36%		2-		Yb k x-ray	5%
	172	69						1.88	29%				0.07879 ± 0.00001	6.5%
	172	69											0.18156 ± 0.00001	2.7%
	172	69											0.91211 ± 0.00001	1.4%
	172	69											1.38722 ± 0.00002	5.5%
	172	69											1.46601 ± 0.00002	4.5%
	172	69											1.4705 ± 0.0001	1.9%
	172	69											1.52982 ± 0.00002	5.1%
	172	69											1.60861 ± 0.00003	4.0%
$_{69}Tm^{173}$	173	69		172.939596	8.24 ± 0.08 h.	β-	1.29	0.80	21%		1/2+		Yb k x-ray	3%
	173	69						0.86	71%				0.3988 ± 0.0005	88%
	173	69											0.4613 ± 0.0005	6.9%
$_{69}Tm^{174}$	174	69		173.942180	5.4 ± 0.1 m.	β-	3.09	0.70	14%		(4-)		Yb k x-ray	18%
	174	69						1.20	83%				0.07664 ± 0.00004	9.1%
	174	69											0.17669 ± 0.00004	66%
	174	69											0.27332 ± 0.00008	86%
	174	69											0.3666 ± 0.0001	92%
	174	69											0.49433 ± 0.00009	11.4%
	174	69											0.62845 ± 0.00009	2.7%
	174	69											0.99205 ± 0.00008	87%
	174	69											1.2419 ± 0.0001	1.7%
	174	69											1.2654 ± 0.0001	2.2%
	174	69											(0.08 - 1.6)	
$_{69}Tm^{175}$	175	69		174.942180	15.2 ± 0.5 m.	β-	2.40	0.9	36%		(1/2+)		Yb k x-ray	10%
	175	69						1.9	23%				0.36396 ± 0.00001	13%
	175	69											0.3946 ± 0.0002	3.3%
	175	69											0.51487 ± 0.00001	65%(D)
	175	69											0.63926 ± 0.00001	6.1%
	175	69											0.81143 ± 0.00001	4.3%
	175	69											0.85808 ± 0.00005	5.7%
	175	69											0.94125 ± 0.00004	14.2%
	175	69											0.98247 ± 0.00004	9.9%
	175	69											1.3770 ± 0.0002	3.0%
	175	69											1.5251 ± 0.0005	1.3%
$_{69}Tm^{176}$	176	69		175.946750	1.9 ± 0.1 m.	β-	3.90				(4+)		Yb k x-ray	18%
	176	69											0.0817 ± 0.001	12%
	176	69											00.1898 ± 0.0001	44%
	176	69											0.2344 ± 0.0001	3.2%
	176	69											0.2383 ± 0.0001	2.5%
	176	69											0.2398 ± 0.0001	7.9%
	176	69											0.2929 ± 0.0002	3.5%
	176	69											0.2996 ± 0.0001	3.2%

TABLE OF THE ISOTOPES (Continued)

Isotope	A	Z	% Natural abundance	Atomic mass	Half-life	Decay mode	Decay energy (MeV)	Particle energy (MeV)	Particle intensity	Thermal neutron cross section	Spin (h/2π)	μ Nucl. mag. moment	Gamma-ray energy (MeV)	Gamma-ray intensity
	176	69											0.3303 ± 0.0002	8.6%
	176	69											0.3435 ± 0.0001	6.9%
	176	69											0.3819 ± 0.0001	23%
	176	69											0.4106 ± 0.0002	4.6%
	176	69											0.4570 ± 0.0002	2.8%
	176	69											0.6216 ± 0.0002	3.4%
	176	69											0.9005 ± 0.0003	2.6%
	176	69											1.0499 ± 0.0002	7.0%
	176	69											1.0691 ± 0.0002	33%
	176	69											1.0881 ± 0.0002	5.7%
	176	69											1.1787 ± 0.0002	2.9%
	176	69											1.2541 ± 0.0003	2.0%
	176	69											1.2609 ± 0.0002	2.3%
	176	69											1.5893 ± 0.0002	2.8%
	176	69											1.9711 ± 0.0002	2.4%
	176	69											2.6214 ± 0.0005	2.9%
	176	69											2.8716 ± 0.0003	2.1%
	176	69											2.9142 ± 0.0005	4.3%
Yb		70		173.04						35 b.				
$_{70}Yb^{154}$	154	70		153.946190	0.40 s.	β,EC(7%)							ann.rad.	
	154	70				α(93%)		5.32						
$_{70}Yb^{155}$	155	70		154.945530	1.7 s.	β+, EC (16%)							ann.rad.	
	155	70				α(84%)		5.19						
$_{70}Yb^{156}$	156	70		155.942690	24 s.	β+, EC (21%)	3.7				0+		ann.rad.	
	156	70				α(79%)		4.69						
$_{70}Yb^{157}$	157	70		156.942430	39 s.	β+, EC (99+%)	5.2						ann.rad.	
	157	70				α(0.5%)		4.69						
$_{70}Yb^{158}$	158	70		157.939858	1.5 m.	β+,E.C.	≈ 2.7				0+		ann.rad.	
	158	70											0.0741 ± 0.0001	100 +
	158	70											0.1477 ± 0.0001	1.7
	158	70											0.1603 ± 0.0001	2.1
	158	70											0.2526 ± 0.0002	3.3
$_{70}Yb^{159}$	159	70		158.939950	12 s.	E.C.,β+	≈ 4.4						Tm k x-ray	61%
	159	70											0.0777 ± 0.0001	7
	159	70											0.1131 ± 0.0001	12
	159	70											0.1661 ± 0.0001	100
	159	70											0.1761 ± 0.0001	14
	159	70											0.1772 ± 0.0001	20
	159	70											0.1919 ± 0.0001	4
	159	70											0.1937 ± 0.0001	5
	159	70											0.1976 ± 0.0001	5
	159	70											0.2391 ± 0.0001	10
	159	70											0.3297 ± 0.0001	18
	159	70											0.3903 ± 0.0001	18
	159	70											0.4972 ± 0.0003	9
$_{70}Yb^{160}$	160	70		159.937670	4.8 m.	β+,E.C.	≈ 2.1				0+		ann.rad.	
	160	70											0.0342 ± 0.0001	3 +
	160	70											0.0420 ± 0.0001	7
	160	70											0.0982 ± 0.0001	3
	160	70											0.1322 ± 0.0001	14
	160	70											0.1404 ± 0.0001	22
	160	70											0.1737 ± 0.0001	100
	160	70											0.1744 ± 0.0001	13
	160	70											0.2158 ± 0.0001	48
	160	70											0.3200 ± 0.0002	3
	160	70											0.3276 ± 0.0002	6
	160	70											0.3730 ± 0.0001	10
	160	70											0.3863 ± 0.0002	3
	160	70											0.3894 ± 0.0002	5
	160	70											0.5821 ± 0.0002	3
$_{70}Yb^{161}$	161	70		160.937940	4.2 m.	β+,E.C.	≈ 4.1				3/2-		ann.rad.	
	161	70											Tm k x-ray	44%
	161	70											0.0782 ± 0.0001	41%
	161	70											0.1403 ± 0.0001	3%
	161	70											0.1444 ± 0.0001	5.5%
	161	70											0.1883 ± 0.0001	4.2%
	161	70											0.2985 ± 0.0001	1.6%
	161	70											0.3147 ± 0.0001	3.2%
	161	70											0.3301 ± 0.0001	3.3%
	161	70											0.3447 ± 0.0003	1.7%
	161	70											0.3810 ± 0.0003	2%
	161	70											0.4582 ± 0.0002	3.4%
	161	70											0.5555 ± 0.0002	1.7%
	161	70											0.5605 ± 0.0002	2.5%
	161	70											0.5697 ± 0.0002	6.7%
	161	70											0.5999 ± 0.0001	31%
	161	70											0.6315 ± 0.0001	16%
	161	70											0.6591 ± 0.0001	3.8%
	161	70											1.0072 ± 0.0004	1.1%
	161	70											1.0427 ± 0.0002	1.2%
	161	70											1.1456 ± 0.0005	1.1%
	161	70											1.1825 ± 0.0005	1%

TABLE OF THE ISOTOPES (Continued)

Isotope	A	Z	% Natural abundance	Atomic mass	Half-life	Decay mode	Decay energy (MeV)	Particle energy (MeV)	Particle intensity	Thermal neutron cross section	Spin (h/2π)	μ Nucl. mag. moment	Gamma-ray energy (MeV)	Gamma-ray intensity
	161	70											1.3649 ± 0.0005	1.1%
	161	70											1.5178 ± 0.0005	1.1%
$_{70}Yb^{162}$	162	70		161.935860	18.9 ± 0.2 m.	β+,E.C.	1.81				0+		ann.rad.	
	162	70											Tm k x-ray	47%
	162	70											0.1188 ± 0.0001	25%
	162	70											0.1635 ± 0.0001	36%
$_{70}Yb^{163}$	163	70		162.936270	11.05 ± 0.25m.	β+(26%)	3.37	1.4			3/2-		ann.rad.	52%
	163	70											Tm k x-ray	36%
	163	70											0.0636 ± 0.0001	7%
	163	70											0.1232 ± 0.0001	2.1%
	163	70											0.1615 ± 0.0001	1.1%
	163	70											0.3262 ± 0.0001	1.7%
	163	70											0.6872 ± 0.0001	1.7%
	163	70											0.8603 ± 0.0001	10.8%
	163	70											1.3318 ± 0.0001	0.7%
	163	70											1.6891 ± 0.0001	0.8%
	163	70											1.7467 ± 0.0002	1.8%
	163	70											1.9078 ± 0.0001	1.6%
	163	70											(0.06 - 1.9)weak	
$_{70}Yb^{164}$	164	70		163.934530	1.26 ± 0.03 h.	E.C.	1.00				0+		Tm k x-ray	38%
	164	70											0.0914 ± 0.0001	6.9%(D)
	164	70											0.6752 ± 0.0001	0.3%
$_{70}Yb^{165}$	165	70		164.935398	9.9 ± 0.3 m.	β+(10%)	2.76	1.58			(5/2-)		ann.rad.	10%
	165	70				E.C.(90%)							Tm k x-ray	60%
	165	70											0.0801 ± 0.0001	33%
	165	70											0.1181 ± 0.0001	1.6%
	165	70											0.1473 ± 0.0001	0.7%
	165	70											0.9567 ± 0.0001	0.7%
	165	70											1.0903 ± 0.0001	3%
	165	70											1.5013 ± 0.0001	0.4%
$_{70}Yb^{166}$	166	70		165.933875	2.36 ± .004 d.	E.C.	0.292				0+		Tm k x-ray	67%
	166	70											0.0828 ± 0.0001	15%
	166	70											0.1844 ± 0.0001	21%
	166	70											0.7789 ± 0.0001	25%
	166	70											1.2734 ± 0.0001	20%(D)
	166	70											2.0524 ± 0.0002	24%(D)
$_{70}Yb^{167}$	167	70		166.934946	17.5 ± 0.2 m.	β+(0.5%)	1.954	0.639			5/2-		Tm k x-ray	91%
	167	70				E.C. (99.5%)							0.06296 ± 0.00008	5%
	167	70											0.10616 ± 0.00004	22%
	167	70											0.11337 ± 0.00002	55%
	167	70											0.1166 ± 0.0001	2.8%
	167	70											0.1435 ± 0.0001	2.1%
	167	70											0.17633 ± 0.00006	20%
	167	70											0.1772 ± 0.0001	2.7%
	167	70											1.0371 ± 0.0001	0.6%
$_{70}Yb^{168}$	168	70	0.13%	167.933894						2.3 x 10^3 b.	0+			
$_{70}Yb^{169m}$	169	70			46 ± 2 s.	I.T.	0.0242				$^1/_2$-		Yb L x-ray	16%
	169	70											0.0242 ± 0.0001	0.0004%
$_{70}Yb^{169}$	169	70		168.935186	32.02 ± 0.01d.	E.C.	0.908			3.6 x 10^3 b.	7/2+		Tm k x-ray	95%
	169	70											0.06306 ± 0.00003	45%
	169	70											0.09365 ± 0.00001	2.7%
	169	70											0.10977 ± 0.00001	18%
	169	70											0.1182 ± 0.0001	1.9%
	169	70											0.13051 ± 0.00001	11.5%
	169	70											0.17718 ± 0.00002	22%
	169	70											0.19795 ± 0.00003	36%
	169	70											0.26106 ± 0.00003	1.8%
	169	70											0.30772 ± 0.00003	11.1%
$_{70}Yb^{170}$	170	70	3.05%	169.934759						10 b.	0+			
$_{70}Yb^{171}$	171	70	14.3%	170.936323						50 b.	$^1/_2$-			
$_{70}Yb^{172}$	172	70	21.9%	171.936378						1 b.	0+			
$_{70}Yb^{173}$	173	70	16.12%	172.938208						17 b.	5/2-			
$_{70}Yb^{174}$	174	70	31.8%	173.938859							65 b.	0+		
$_{70}Yb^{175}$	175	70		174.941273	4.19 ± 0.01 d.	β-	0.468	0.467			7/2-		Lu k x-ray	2%
	175	70											0.11378 ± 0.00001	1.9%

Isotope	A	Z	% Natural abundance	Atomic mass	Half-life	Decay mode	Decay energy (MeV)	Particle energy (MeV)	Particle intensity	Thermal neutron cross section	Spin (h/2π)	μ Nucl. mag. moment	Gamma-ray energy (MeV)	Gamma-ray intensity
	175	70											0.28248 ± 0.00001	3.1%
	175	70											0.39629 ± 0.00002	6.5%
$_{70}Yb^{176m}$	176	70			11.4 ± 0.5 s.	I.T.	1.051				(8-)		Yb k x-ray	31%
	176	70											0.0821 ± 0.0002	12%
	176	70											0.0961 ± 0.0003	72%
	176	70											0.1901 ± 0.0002	76%
	176	70											0.2929 ± 0.0003	93%
	176	70											0.3897 ± 0.0004	97%
$_{70}Yb^{176}$	176	70	12.7%	175.942564						3 b.	0+			
$_{70}Yb^{177m}$	177	70			6.41 ± 0.02 s.	I.T.	0.3315				1/2-		Yb k x-ray	38%
	177	70											0.1131 ± 0.0001	6.6%
	177	70											0.2084 ± 0.0001	11%
	177	70											0.2497 ± 0.0001	0.2%
	177	70											0.3213 ± 0.0001	0.2%
$_{70}Yb^{177}$	177	70		176.945253	1.9 ± 0.1 h.	β-	1.398	1.40			9/2+		Lu k x-ray	7%
	177	70											0.1216 ± 0.0001	3%
	177	70											0.1504 ± 0.0001	20%
	177	70											0.9417 ± 0.0003	1%
	177	70											1.0801 ± 0.0003	5.5%
	177	70											1.2414 ± 0.0003	3.4%
$_{70}Yb^{178}$	178	70		177.946639	1.23 ± 0.05 h.	β-	0.630	0.25			0+			
$_{70}Yb^{179}$	179	70			8 m.								0.1415 ± 0.0004	6 +
	179	70											0.1473 ± 0.0003	14
	179	70											0.3246 ± 0.0004	20
	179	70											0.3516 ± 0.0003	43
	179	70											0.3815 ± 0.0003	26
	179	70											0.4111 ± 0.0003	17
	179	70											0.4265 ± 0.0006	6
	179	70											0.4312 ± 0.0004	8
	179	70											0.4711 ± 0.0005	6
	179	70											0.5001 ± 0.0004	11
	179	70											0.6125 ± 0.0003	100
	179	70											0.6430 ± 0.0004	11
	179	70											0.9942 ± 0.001	4
	179	70											1.0244 ± 0.001	7
Lu		71		174.967						84 b.				
$_{71}Lu^{154}$	154	71		153.957460	1.0 s.	β+,E.C.	10.560							
$_{71}Lu^{155}$	155	71		154.954080	0.07 s.	E.C.	7.97							
	155	71				α		5.66						
$_{71}Lu^{156m}$	156	71			0.21 s.	β+,E.C.							ann.rad.	
	156	71				α		5.57						
$_{71}Lu^{156}$	156	71		155.953070	0.5 s.	β+,E.C.	9.67						ann.rad.	
	156	71				α		5.45						
$_{71}Lu^{157}$	157	71		156.949940	5.5 s.	β+, EC (94%)	6.99						ann.rad.	
	157	71				α		5.00						
$_{71}Lu^{158}$	158	71		157.949290	10 s.	β+, EC (99%)	8.78						ann.rad.	
	158	71				α		4.67					0.3682 ± 0.0001	100 +
	158	71											0.4770 ± 0.0002	21
$_{71}Lu^{159}$	159	71		158.946480	12 s.	β+,E.C.	6.08						ann.rad.	
	159	71											0.1505 ± 0.0001	100 +
	159	71											0.1875 ± 0.0001	25
	159	71											0.3693 ± 0.0001	19
$_{71}Lu^{160}$	160	71		159.946040	35 s.	β+,E.C.	7.80						ann.rad.	
	160	71											0.2434 ± 0.0001	100 +
	160	71											0.3756 ± 0.0002	8
	160	71											0.3957 ± 0.0002	30
	160	71											0.5773 ± 0.0002	13
	160	71											0.7044 ± 0.0002	6
	160	71											0.7382 ± 0.0002	7
	160	71											0.8201 ± 0.0003	9
	160	71											0.8707 ± 0.0004	9
$_{71}Lu^{161}$	161	71		160.943630	1.2 m.	β+,E.C.	5.30						ann.rad.	
	161	71											0.0437 ± 0.0003	70 +
	161	71											0.0671 ± 0.0002	48
	161	71											0.0868 ± 0.0002	17
	161	71											0.1003 ± 0.0001	95
	161	71											0.1052 ± 0.0001	28
	161	71											0.1108 ± 0.0001	100
	161	71											0.1562 ± 0.0001	49
	161	71											0.1701 ± 0.00002	14
	161	71											0.1771 ± 0.0002	14
	161	71											0.2046 ± 0.0002	30
	161	71											0.2111 ± 0.0002	20
	161	71											0.2218 ± 0.0002	20
	161	71											0.2562 ± 0.0003	49
$_{71}Lu^{162}$	162	71		161.943470	1.4 m.	β+,E.C.	7.09						ann.rad.	
	162	71											0.1666 ± 0.0001	100 +

Isotope	A	Z	% Natural abundance	Atomic mass	Half-life	Decay mode	Decay energy (MeV)	Particle energy (MeV)	Particle intensity	Thermal neutron cross section	Spin (h/2π)	μ Nucl. mag. moment	Gamma-ray energy (MeV)	Gamma-ray intensity
	162	71											0.3209 ± 0.0001	20
	162	71											0.6314 ± 0.0001	28
	162	71											0.6564 ± 0.0002	7
	162	71											0.8253 ± 0.0002	18
	162	71											0.8398 ± 0.0003	8
$_{71}Lu^{163}$	163	71		162.941240	4.1 ± 0.2 m.	β+,E.C.	4.63						ann.rad.	
	163	71											0.0539	82
	163	71											0.0581	43
	163	71											0.0750	2
	163	71											0.0792	5
	163	71											0.0935	5
	163	71											0.1023	9
	163	71											0.1504	44
	163	71											0.1631	100
	163	71											0.1674	9
	163	71											0.2066	6
	163	71											0.2213	20
	163	71											0.2527	8
	163	71											0.3026	24
	163	71											0.3135	25
	163	71											0.3169	13
	163	71											0.3717	49
	163	71											0.3818	10
	163	71											0.3912	6
	163	71											0.3954	8
	163	71											0.4002	4
	163	71											0.4530	11
	163	71											0.4568	6
	163	71											0.4611	7
	163	71											0.4827	6
	163	71											0.4843	19
	163	71											0.5382	12
	163	71											0.5623	16
	163	71											0.5665	7
	163	71											0.6331	8
	163	71											0.6951	15
$_{71}Lu^{164}$	164	71		163.941290	3.17 ± 0.03 m.	β+,E.C.	6.30	1.6					0.1238 ± 0.0002	100 +
	164	71						3.8					0.2621 ± 0.0002	32
	164	71											0.5520 ± 0.0002	12
	164	71											0.6082 ± 0.0002	6
	164	71											0.6880 ± 0.0003	6
	164	71											0.7404 ± 0.0002	38
	164	71											0.7479 ± 0.0002	16
	164	71											0.8521 ± 0.0003	9
	164	71											0.8639 ± 0.0002	29
	164	71											0.8804 ± 0.0002	21
	164	71											0.9494 ± 0.0003	7
	164	71											0.9796 ± 0.0004	3
	164	71											1.0738 ± 0.0005	10
	164	71											1.1148 ± 0.0004	4
	164	71											1.1994 ± 0.0004	8
	164	71											1.2124 ± 0.0003	14
	164	71											1.2925 ± 0.0004	6
	164	71											1.3356 ± 0.0006	11
	164	71											1.3760 ± 0.0004	6
	164	71											1.3895 ± 0.0004	6
	164	71											1.5134 ± 0.0005	6
$_{71}Lu^{165}$	165	71		164.939480	11.8 ± 0.5 m.	β+,E.C.	3.80	2.06			1/2+		ann.rad.	
	165	71											0.0393 ± 0.0001	8%
	165	71											0.1206 ± 0.0001	25%
	165	71											0.1324 ± 0.0001	23%
	165	71											0.1742 ± 0.0001	12%
	165	71											0.2036 ± 0.0001	10%
	165	71											0.2174 ± 0.0001	5%
	165	71											0.2534 ± 0.0001	4%
	165	71											0.2710 ± 0.0001	5%
	165	71											0.3565 ± 0.0001	5%
	165	71											0.3605 ± 0.0001	8%
	165	71											0.3605 ± 0.0001	8%
	165	71											0.3725 ± 0.3%	3%
	165	71											0.5523 ± 0.0002	2%
	165	71											0.6091 ± 0.0002	2%
	165	71											0.6866 ± 0.0002	2.5%
	165	71											0.7535 ± 0.0002	2.2%
	165	71											1.0734 ± 0.0003	1.9%
	165	71											1.5600 ± 0.0003	1.9%
	165	71											1.6016 ± 0.0002	4.0%
	165	71											1.6135 ± 0.0002	4.0%
	165	71											1.7344 ± 0.0003	2.2%
	165	71											1.8019 ± 0.0004	1.9%
	165	71											(0.04 - 2.0)weak	
$_{71}Lu^{166m2}$	166	71			2.1 ± 0.1 m	β+(35%)					(0-)		ann.rad.	70%

Isotope	A	Z	% Natural abundance	Atomic mass	Half-life	Decay mode	Decay energy (MeV)	Particle energy (MeV)	Particle intensity	Thermal neutron cross section	Spin (h/2π)	μ Nucl. mag. moment	Gamma-ray energy (MeV)	Gamma-ray intensity
	166	71				E.C.(65%)							Yb k x-ray	26%
	166	71											0.1024 ± 0.0001	11%
	166	71											0.2281 ± 0.0001	4%
	166	71											1.0673 ± 0.0002	15%
	166	71											1.2494 ± 0.0008	13%
	166	71											1.2566 ± 0.0001	23%
	166	71											1.4775 ± 0.0003	2.7%
	166	71											1.5297 ± 0.0001	11%
	166	71											1.9232 ± 0.0004	2.4%
	166	71											1.9963 ± 0.0002	3.3%
	166	71											2.0986 ± 0.0002	16%
	166	71											2.3246 ± 0.0003	9%
$_{71}Lu^{166m1}$	166	71			1.4 ± 0.1 m.	β+, EC (58%)					(3-)		ann.rad.	
	166	71				I.T.(42%)	0.0344						0.1024 ± 0.0001	21%
	166	71											0.2281 ± 0.0001	26%
	166	71											0.2861 ± 0.0001	19%
	166	71											0.4213 ± 0.0001	4%
	166	71											0.5260 ± 0.0001	5%
	166	71											0.5709 ± 0.0001	5.5%
	166	71											0.5810 ± 0.0006	2%
	166	71											0.6432 ± 0.0001	6%
	166	71											0.7051 ± 0.0001	7.5%
	166	71											0.7088 ± 0.0001	2.4%
	166	71											0.8119 ± 0.0001	17%
	166	71											0.8301 ± 0.0001	18%
	166	71											0.8322 ± 0.0001	4.5%
	166	71											0.8664 ± 0.0004	2.1%
	166	71											0.9324 ± 0.0001	14%
	166	71											0.9368 ± 0.0001	14%
	166	71											0.9846 ± 0.0006	4%
	166	71											1.2769 ± 0.0002	2%
	166	71											1.2835 ± 0.0002	6.6%
	166	71											1.3544 ± 0.0002	2%
	166	71											1.5049 ± 0.0006	2%
	166	71											1.6787 ± 0.0004	2.2%
	166	71											1.8013 ± 0.0006	1.7%
	166	71											1.9740 ± 0.0006	1.0%
$_{71}Lu^{166}$	166	71		165.939760	2.8 ± 0.2 m.	β+(25%)	2.2				(6-)		ann.rad.	50%
	166	71				E.C.(75%)							Yb k x-ray	51%
	166	71											0.0676 ± 0.0001	4%
	166	71											0.1024 ± 0.0001	25%
	166	71											0.2087 ± 0.0001	4%
	166	71											0.2281 ± 0.0001	77%
	166	71											0.2485 ± 0.0001	4.8%
	166	71											0.27440 ± 0.0001	9.9%
	166	71											0.2763 ± 0.0001	14%
	166	71											0.3375 ± 0.0001	41%
	166	71											0.3601 ± 0.0001	3.6%
	166	71											0.3679 ± 0.0001	31%
	166	71											0.3830 ± 0.0001	3.0%
	166	71											0.4303 ± 0.0001	5%
	166	71											0.4747 ± 0.0001	2.7%
	166	71											0.5376 ± 0.0001	8.1%
	166	71											0.5777 ± 0.0001	4.0%
	166	71											0.6293 ± 0.0001	7%
	166	71											0.6599 ± 0.0001	3.7%
	166	71											0.7944 ± 0.0001	3%
	166	71											0.8145 ± 0.0001	6.7%
	166	71											0.8322 ± 0.0001	6%
	166	71											0.8376 ± 0.0001	2.7%
	166	71											0.8606 ± 0.0001	3.3%
	166	71											0.9368 ± 0.0001	5.7%
	166	71											0.9974 ± 0.0001	18%
	166	71											1.0563 ± 0.0006	2.1%
	166	71											1.0673 ± 0.0002	2.5%
	166	71											1.1224 ± 0.0001	4.0%
	166	71											1.1748 ± 0.0002	4.4%
	166	71											1.2907 ± 0.0002	9.7%
	166	71											1.3544 ± 0.0002	1.7%
	166	71											1.4596 ± 0.0001	7.8%
	166	71											1.4873 ± 0.0004	1.1%
	166	71											1.6266 ± 0.0003	0.9%
	166	71											1.6858 ± 0.0003	0.5%
$_{71}Lu^{167}$	167	71		166.938310	51.5 ± 0.1 m.	β+(2%)	3.13	2.1			7/2+		Yb k x-ray	50%
	167	71				E.C.(98%)							0.0297 ± 0.0001	15%
	167	71											0.0339 ± 0.0001	3%
	167	71											0.0787 ± 0.0001	1.5%
	167	71											0.1450 ± 0.0001	2.2%
	167	71											0.1789 ± 0.0001	2.6%
	167	71											0.1887 ± 0.0001	2%
	167	71											0.2132 ± 0.0001	3.5%
	167	71											0.2392 ± 0.0001	8.2%

Isotope	A	Z	% Natural abundance	Atomic mass	Half-life	Decay mode	Decay energy (MeV)	Particle energy (MeV)	Particle intensity	Thermal neutron cross section	Spin (h/2π)	μ Nucl. mag. moment	Gamma-ray energy (MeV)	Gamma-ray intensity
	167	71											0.2585 ± 0.0001	1.4%
	167	71											0.2618 ± 0.0001	1.3%
	167	71											0.2787 ± 0.0001	1.9%
	167	71											0.3177 ± 0.0001	1.5%
	166	71											0.4011 ± 0.0001	2.5%
	167	71											0.4454 ± 0.0001	1.0%
	167	71											0.5733 ± 0.0008	1.1%
	166	71											0.7848 ± 0.0001	0.6%
	167	71											0.9884 ± 0.0001	0.8%
	167	71											1.1885 ± 0.0001	1.2%
	167	71											1.2272 ± 0.0002	1.2%
	167	71											1.2672 ± 0.0001	3.3%
	167	71											1.5068 ± 0.0001	2.4%
	167	71											1.6445 ± 0.0001	1.2%
	167	71											1.9414 ± 0.0001	1.3%
	167	71											1.9740 ± 0.0001	1.2%
	167	71											2.0131 ± 0.0002	1.2%
	167	71											(0.03 - 2.0)weak	
$_{71}Lu^{168m}$	168	71			6.7 ± 0.4 m.	β+(12%)					3+		ann.rad.	24%
	168	71				E.C.(88%)							Yb k x-ray	48%
	168	71											0.0877 ± 0.0001	13%
	168	71											0.1988 ± 0.0001	28%
	168	71											0.2987 ± 0.0001	2.6%
	168	71											0.7303 ± 0.0003	1.9%
	168	71											0.7525 ± 0.0008	2.0%
	168	71											0.7805 ± 0.0003	3.7%
	168	71											0.8535 ± 0.0002	4.8%
	168	71											0.8846 ± 0.0002	14%
	168	71											0.8960 ± 0.0002	16%
	168	71											0.9792 ± 0.0002	20%
	168	71											0.9838 ± 0.0002	12%
	168	71											1.0326 ± 0.0002	11%
	168	71											1.0717 ± 0.0003	3.7%
	168	71											1.0838 ± 0.0003	5.5%
	168	71											1.1368 ± 0.0002	12%
	168	71											1.2199 ± 0.0002	11%
	168	71											1.2335 ± 0.0002	3.3%
	168	71											1.2645 ± 0.00003	3.0%
	168	71											1.3377 ± 0.0002	4.2%
	168	71											1.3639 ± 0.0002	3.9%
	168	71											1.4208 ± 0.0002	10%
	168	71											1.4635 ± 0.0003	2.5%
	168	71											2.1414 ± 0.0005	2.8%
	168	71											2.3401 ± 0.0002	1.1%
$_{71}Lu^{168}$	168	71		167.938690	5.5 ± 0.1 m.	β+(6%)	4.47	1.2			(6-)		ann.rad.	12%
	168	71				E.C.(94%)							Yb k x-ray	
	168	71											0.1114 ± 0.0002	16%
	168	71											0.1124 ± 0.0002	16%
	168	71											0.1566 ± 0.0002	7.3%
	168	71											0.1796 ± 0.0002	6.0%
	168	71											0.2236 ± 0.0001	8.2%
	168	71											0.2286 ± 0.0002	17%
	168	71											0.3247 ± 0.0002	7.4%
	168	71											0.3483 ± 0.0002	18%
	168	71											0.3874 ± 0.0002	3.4%
	168	71											0.4011 ± 0.0004	6.7%
	168	71											0.4794 ± 0.0004	2.4%
	168	71											0.5398 ± 0.0002	11%
	168	71											0.8600 ± 0.0003	3.1%
	168	71											1.1850 ± 0.0002	11%
	168	71											1.2335 ± 0.0005	2.6%
	168	71											1.3875 ± 0.0002	6%
	168	71											1.4135 ± 0.0003	3.9%
	168	71											1.4836 ± 0.0002	17%
	168	71											1.6860 ± 0.0005	4.8%
$_{71}Lu^{169m}$	169	71			2.7 ± 0.2 m.	I.T.	0.0290				3+		Lu L x-ray	17%
	169	71											0.0290 ± 0.0002	0.001%
$_{71}Lu^{169}$	169	71		168.937648	1.419 ± 0.002 d.	E.C.	2.293	1.271			7/2+		Yb k x-ray	53%
	169	71											0.0874 ± 0.0001	2.1%
	169	71											0.19121 ± 0.0001	18%
	169	71											0.3786 ± 0.0001	1.8%
	169	71											0.8898 ± 0.0001	4.6%
	169	71											0.9606 ± 0.0001	20%
	169	71											1.0075 ± 0.0001	1.6%
	169	71											1.0603 ± 0.0001	1.6%
	169	71											1.1849 ± 0.0001	1.9%
	169	71											1.2833 ± 0.0001	1.8%
	169	71											1.3388 ± 0.0001	1.4%
	169	71											1.3790 ± 0.0001	2.8%
	169	71											1.4497 ± 0.0001	8.6%
	169	71											1.4634 ± 0.0001	1.3%
	169	71											1.4668 ± 0.0001	2.9%

Isotope	A	Z	% Natural abundance	Atomic mass	Half-life	Decay mode	Decay energy (MeV)	Particle energy (MeV)	Particle intensity	Thermal neutron cross section	Spin (h/2π)	μ Nucl. mag. moment	Gamma-ray energy (MeV)	Gamma-ray intensity
	169	71											(0.08 - 2.1)weak	
$_{71}Lu^{170m}$	170	71			0.7 ± 0.1 s.	I.T.	0.0929						Lu L x-ray	
	170	71											0.04449 ± 0.00006	0.85%
	170	71											0.0484 ± 0.0001	0.4%
$_{71}Lu^{170}$	170	71		169.938452	2.01 ± 0.03 d.	E.C.	3.44	2.44			0+		Yb k x-ray	46%
	170	71											0.19319 ± 0.00004	2.1%
	170	71											1.57227 ± 0.00001	1.2%
	170	71											0.58711 ± 0.00001	12.7%
	170	71											0.5908 ± 0.0001	15%
	170	71											0.93886 ± 0.00005	1.6%
	170	71											0.98512 ± 0.00005	5.4%
	170	71											0.98721 ± 0.00005	1.7%
	170	71											0.9996 ± 0.0001	1.5%
	170	71											1.00317 ± 0.00005	3.4%
	170	71											1.0543 ± 0.0001	4.6%
	170	71											1.0615 ± 0.0001	2.1%
	170	71											1.13368 ± 0.00006	1.0%
	170	71											1.13862 ± 0.00003	3.5%
	170	71											1.21841 ± 0.00006	1.4%
	170	71											1.22556 ± 0.00005	4.8%
	170	71											1.2571 ± 0.0001	1.4%
	170	71											1.28029 ± 0.00004	7.9%
	170	71											1.29476 ± 0.00006	2.8%
	170	71											1.30746 ± 0.00005	1.1%
	170	71											1.34101 ± 0.00004	3.2%
	170	71											1.36460 ± 0.00004	4.5%
	170	71											1.39565 ± 0.00006	2.2%
	170	71											1.40521 ± 0.00006	2.5%
	170	71											1.42816 ± 0.00004	3.4%
	170	71											1.45032 ± 0.00004	1.6%
	170	71											1.45532 ± 0.00005	1.1%
	170	71											1.45988 ± 0.00007	1.0%
	170	71											1.51246 ± 0.00004	2.5%
	170	71											1.9557 ± 0.0001	1.3%
	170	71											2.0400 ± 0.0001	2.5%
	170	71											2.0419 ± 0.0001	5.9%
	170	71											2.12621 ± 0.00005	5.0%
	170	71											2.1912 ± 0.0001	1.6%
	170	71											2.36417 ± 0.00004	1.4%
	170	71											2.66390 ± 0.00005	1.2%
	170	71											2.69145 ± 0.00008	2.2%
	170	71											2.74821 ± 0.00005	2.1%
	170	71											2.7832 ± 0.0001	1.0%
	170	71											2.8544 ± 0.0001	1.7%
	170	71											2.93982 ± 0.00005	1.5%
	170	71											2.9657 ± 0.0001	1.2%
	170	71											3.0309 ± 0.0001	1.3%
	170	71											(0.1 - 3.38)many	
$_{71}Lu^{171m}$	171	71			1.3 ± 0.3 m.	I.T.	0.0711				1/2-		Lu k x-ray	61%
	171	71											0.07119 ± 0.0001	0.02%
$_{71}Lu^{171}$	171	71		170.937911	8.24 ± 0.03 d.	E.C.	1.481	0.362			7/2+	2.03	Yb k x-ray	64%

Isotope	A	Z	% Natural abundance	Atomic mass	Half-life	Decay mode	Decay energy (MeV)	Particle energy (MeV)	Particle intensity	Thermal neutron cross section	Spin (h/2π)	μ Nucl. mag. moment	Gamma-ray energy (MeV)	Gamma-ray intensity
	171	71											0.01939 ± 0.00001	14%
	171	71											0.06674 ± 0.00002	2.5%
	171	71											0.072387 ± 0.00002	2.0%
	171	71											0.075899 ± 0.00002	6.1%
	171	71											0.085611 ± 0.00002	1.0%
	171	71											0.66744 ± 0.00001	11%
	171	71											0.68931 ± 0.00001	2.4%
	171	71											0.71268 ± 0.00001	1.1%
	171	71											0.73983 ± 0.00001	48%
	171	71											0.78072 ± 0.00001	4.3%
	171	71											0.84001 ± 0.00001	3.0%
	171	71											0.85311 ± 0.00001	2.5%
	171	71											1.2822 ± 0.0001	0.3%
	171	71											(0.02 - 1.3)weak	
$_{71}Lu^{172m}$	172	71			3.7 ± 0.5 m.	I.T.	0.0419				1-		Lu L x-rays	
	172	71											0.04186 ± 0.00004	0.004%
$_{71}Lu^{172}$	172	71		171.939085	6.70 ± 0.03 d.	E.C.	2.524				4-	2.25	Yb k x-ray	57%
	172	71											0.07879 ± 0.00001	11%
	172	71											0.0966 ± 0.0002	5.1%
	172	71											0.11276 ± 0.00002	1.5%
	172	71											0.18156 ± 0.00001	20%
	172	71											0.20342 ± 0.00002	4.8%
	172	71											0.26993 ± 0.00004	1.8%
	172	71											0.27974 ± 0.00003	1.1%
	172	71											0.32392 ± 0.00004	1.4%
	172	71											0.37251 ± 0.00002	2.6%
	172	71											0.37756 ± 0.00002	3.2%
	172	71											0.41033 ± 0.00002	2.0%
	172	71											0.43256 ± 0.00002	1.5%
	172	71											0.49046 ± 0.00001	1.9%
	172	71											0.52828 ± 0.00002	3.9%
	172	71											0.54020 ± 0.00004	1.3%
	172	71											0.69737 ± 0.00002	5.8%
	172	71											0.81012 ± 0.00002	16%
	172	71											0.90079 ± 0.00002	29%
	172	71											0.91211 ± 0.00001	15%
	172	71											0.92909 ± 0.00005	3.1%
	172	71											1.00278 ± 0.00002	5.3%
	172	71											1.02241 ± 0.00005	1.5%
	172	71											1.0808 ± 0.0004	1.1%
	172	71											1.09367 ± 0.00001	63%
	172	71											1.11307 ± 0.00005	1.9%
	172	71											1.48898 ± 0.00004	1.1%
	172	71											1.58416 ± 0.00002	2.5%
	172	71											1.62195 ± 0.00002	2.1%

Isotope	A	Z	% Natural abundance	Atomic mass	Half-life	Decay mode	Decay energy (MeV)	Particle energy (MeV)	Particle intensity	Thermal neutron cross section	Spin (h/2π)	μ Nucl. mag. moment	Gamma-ray energy (MeV)	Gamma-ray intensity
	172	71											(0.07 - 2.2)weak	
$_{71}Lu^{173}$	173	71		172.938929	1.37 ± 0.01 y.	E.C.	0.675				7/2+		Yb k x-ray	45%
	173	71										2.34	0.07860 ± 0.00002	7.8%
	173	71											0.10066 ± 0.00001	3.1%
	173	71											0.17132 ± 0.00003	1.8%
	173	71											0.27198 ± 0.00004	13%
	173	71											0.63586 ± 0.00002	0.9%
$_{71}Lu^{174m}$	174	71			142 ± 2 d.	I.T.(99.3%)	0.17086				6-	2.34	Lu k x-ray	33%
	174	71				E.C.(0.7%)							0.067055 ± 0.00008	6.8%
	174	71											0.1767 ± 0.0001	0.5%
	174	71											0.2733 ± 0.0001	0.6%
	174	71											0.99205 ± 0.00008	0.6%
$_{71}Lu^{174}$	174	71		173.940336	3.31 ± 0.05 y.	E.C.	1.378				1-	1.94	Yb k x-ray	42%
	174	71											0.07664 ± 0.00004	5.8%
	174	71											1.2419 ± 0.0001	6.5%
$_{71}Lu^{175}$	175	71	97.40%	174.940770						(16 + 9) b.	7/2+	+2.2327		
$_{71}Lu^{176m}$	176	71			3.63 ± 0.01 h.	β-	1.313	1.229			1-	+0.318	Hf k x-ray	5%
	176	71						1.317					0.088372 ± 0.00009	8.9%
$_{71}Lu^{176}$	176	71	2.59%	175.942679	3.6 x 10^{10}y.					(5 + 2300) b.	7-	+3.19	Hf k x-ray	16%
	176	71											0.08837 ± 0.00001	13%
	176	71											0.20187 ± 0.00003	84%
	176	71											0.30691 ± 0.00005	93%
$_{71}Lu^{177m}$	177	71			160 ± 0.3 d.	I.T.(22%)	0.9702				23/2-	2.75	Lu k x-ray	9.7%
	177	71				β-(78%)							Hf k x-ray	58%
	177	71											0.10534 ± 0.00001	11%
	177	71											0.11295 ± 0.00001	21%
	177	71											0.12164 ± 0.00002	6.3%
	177	71											0.12850 ± 0.00004	15%
	177	71											0.13670 ± 0.00001	1.4%
	177	71											0.14717 ± 0.00003	3.8%
	177	71											0.15329 ± 0.00003	17.8%
	177	71											0.17186 ± 0.00005	5.2%
	177	71											0.17440 ± 0.00001	12.7%
	177	71											0.17700 ± 0.00002	3.4%
	177	71											0.20410 ± 0.00005	14.4%
	177	71											0.20836 ± 0.00001	61%
	177	71											0.21443 ± 0.00001	6.6%
	177	71											0.21809 ± 0.00001	3.2%
	177	71											0.22847 ± 0.00005	37%
	177	71											0.23384 ± 0.00001	5.6%
	177	71											0.24965 ± 0.00002	6.1%
	177	71											0.26879 ± 0.00001	3.6%
	177	71											0.28179 ± 0.00005	14%
	177	71											0.2915 ± 0.0001	1.0%
	177	71											0.29645 ± 0.00003	5.4%
	177	71											0.29905 ± 0.00003	1.6%
	177	71											0.30550 ± 0.00002	1.7%

Isotope	A	Z	% Natural abundance	Atomic mass	Half-life	Decay mode	Decay energy (MeV)	Particle energy (MeV)	Particle intensity	Thermal neutron cross section	Spin (h/2π)	μ Nucl. mag. moment	Gamma-ray energy (MeV)	Gamma-ray intensity
	177	71											0.31371 ± 0.00001	1.2%
	177	71											0.31903 ± 0.00001	11%
	177	71											0.32769 ± 0.00001	17.4%
	177	71											0.3417 ± 0.0001	1.8%
	177	71											0.36743 ± 0.00001	3.2%
	177	71											0.37850 ± 0.00001	28%
	177	71											0.38504 ± 0.00004	2.9%
	177	71											0.41366 ± 0.00001	17%
	177	71											0.41853 ± 0.00001	20%
	177	71											0.46583 ± 0.00005	2.3%
$_{71}Lu^{177}$	177	71		176.943752	6.71 d.	β-	0.497	0.497			7/2+	+2.239	0.11295 ± 0.00001	6.4%
	177	71											0.20836 ± 0.00001	11%
$_{71}Lu^{178m}$	178	71			23 m.	β-					(9-)		0.2166 ± 0.0001	2.5%
	178	71											0.3317 ± 0.0001	11.6%
$_{71}Lu^{178}$	178	71		177.945963	28.5 m.	β-	2.11	2.03			1+		Hf k x-ray	4%
	178	71											0.0932 ± 0.0001	6.6%
	178	71											1.2692 ± 0.0001	1.0%
	178	71											1.3099 ± 0.0001	1.5%
	178	71											1.3408 ± 0.0001	4.7%
	178	71											(0.09 - 1.7)weak	
$_{71}Lu^{179}$	179	71		178.947260	4.6 h.	β-	1.35	1.35			7/2+		0.2143 ± 0.0001	12%
	179	71											0.3377 ± 0.0001	0.2%
$_{71}Lu^{180}$	180	71		179.949870	5.7 ± 0.1 m.	β-	3.1	1.49					0.09331 ± 0.00006	13%
	180	71											0.21525 ± 0.00001	21%
	180	71											0.31651 ± 0.00005	14.9%
	180	71											0.40795 ± 0.00005	50%
	180	71											0.9830 ± 0.0001	2.2%
	180	71											1.1068 ± 0.0001	23%
	180	71											1.1982 ± 0.0001	15%
	180	71											1.2001 ± 0.0001	26%
	180	71											1.2995 ± 0.0001	14%
	180	71											1.4349 ± 0.0003	2.0%
	180	71											1.5147 ± 0.0001	8.0%
	180	71											1.8885 ± 0.0009	1.2%
$_{71}Lu^{181}$	181	71			3.5 ± 0.3 m.	β-					(7/2+)		0.0458 ± 0.0002	6.5%
	181	71											0.0530 ± 0.0002	4%
	181	71											0.0988 ± 0.0002	3.5%
	181	71											0.1056 ± 0.0003	4%
	181	71											0.1250 ± 0.0004	3.2%
	181	71											0.1530 ± 0.0003	2.6%
	181	71											0.2059 ± 0.0003	16%
	181	71											0.2404 ± 0.0004	4.5%
	181	71											0.3293 ± 0.0003	5.0%
	181	71											0.3344 ± 0.0004	3.7%
	181	71											0.3418 ± 0.0004	3.2%
	181	71											0.4637 ± 0.0005	4.5%
	181	71											0.5749 ± 0.0003	15%
	181	71											0.5899 ± 0.0001	3.2%
	181	71											0.6525 ± 0.0004	2.5%
	181	71											0.6999 ± 0.0004	4.1%
	181	71											0.8054 ± 0.0003	8.6%
	181	71											0.8584 ± 0.0003	7.6%
$_{71}Lu^{182}$	182	71			2.0 m.	β-	≈ 4.1						0.0978 ± 0.0002	14%
	182	71											0.2240 ± 0.0005	4%
	182	71											0.7208 ± 0.0005	29%
	182	71											0.8081 ± 0.0005	14%
	182	71											0.8182 ± 0.0005	29%
Hf		72		178.49						104 b.				
$_{72}Hf^{158}$	158	72		157.954590	2.9 s.	E.C.(54%)	≈ 4.9				0+			
	158	72				α(46%)		5.27						
$_{72}Hf^{159}$	159	72		158.953740	5.6 s.	β+, EC (88%)	6.76						ann.rad.	
	159	72				α(12%)		5.09						
$_{72}Hf^{160}$	160	72		159.950550	12 s.	β+, EC (97%)	4.21				0+		ann.rad.	
	160	72				α	4.78							
$_{72}Hf^{161}$	161	72		160.950110	17 s.	α		4.60						
$_{72}Hf^{162}$	162	72		161.947204	37.6 s.	β+,E.C.	3.48				0+		ann.rad.	

TABLE OF THE ISOTOPES (Continued)

Isotope	A	Z	% Natural abundance	Atomic mass	Half-life	Decay mode	Decay energy (MeV)	Particle energy (MeV)	Particle intensity	Thermal neutron cross section	Spin (h/2π)	μ Nucl. mag. moment	Gamma-ray energy (MeV)	Gamma-ray intensity
	162	72											0.1739 ± 0.0001	100 +
	162	72											0.1963 ± 0.0001	25
	162	72											0.4101 ± 0.0001	17
$_{72}Hf^{163}$	163	72		162.946980	40 s.	β+,E.C.	5.35						ann.rad.	
	163	72											0.0454 ± 0.0001	48 +
	163	72											0.0621 ± 0.0001	64
	163	72											0.0710 ± 0.0001	100
	163	72											0.0849 ± 0.0001	1
	163	72											0.1331 ± 0.0001	24
	163	72											0.1622 ± 0.0002	16
	163	72											0.2333 ± 0.0001	17
	163	72											0.4961 ± 0.0001	13
	163	72											0.5203 ± 0.0001	19
	163	72											0.5352 ± 0.0002	4
	163	72											0.6882 ± 0.0001	33
$_{72}Hf^{166}$	166	72		165.942250	6.8 m.	E.C.(93%) β+(7%)	2.32						ann.rad.	
	166	72											Lu k x-ray	48%
	166	72											0.0788 ± 0.0001	42%
	166	72											0.0930 ± 0.0002	3%
	166	72											0.2446 ± 0.0004	1.6%
	166	72											0.2839 ± 0.0002	1.6%
	166	72											0.3068 ± 0.0004	1.8%
	166	72											0.3418 ± 0.0001	4.8%
	166	72											0.3776 ± 0.0005	4.1%
	166	72											0.4079 ± 0.0001	4.6%
	166	72											0.4830 ± 0.0001	4.2%
$_{72}Hf^{167}$	167	72		166.942600	2.05 m.	β+(40%) E.C.(60%)	4.00				(5/2-)		ann.rad.	80%
	167	72											Lu k x-ray	24%
	167	72											0.1399 ± 0.0002	3.1%
	167	72											0.1754 ± 0.0002	4.9%
	167	72											0.3152 ± 0.0001	81%
$_{72}Hf^{168}$	168	72		167.940730	25.9 m.	β+,E.C.	1.90				0+		ann.rad.	
	168	72											0.1572	70 +
	168	72											0.1838	100
	168	72											0.1988 ± 0.0001	38
$_{72}Hf^{169}$	169	72		168.941240	3.25 m.	E.C.(85%) β+(15%)	3.35				(5/2-)		ann.rad.	30%
	169	72											Lu k x-ray	38%
	169	72											0.1236 ± 0.0002	4.1%
	169	72											0.3695 ± 0.0002	10.2%
	169	72											0.4929 ± 0.0001	89%
$_{72}Hf^{170}$	170	72		169.939740	16.0 h.	E.C.	1.20				0+		Lu k x-ray	58%
	170	72											0.0985 ± 0.0001	4%
	170	72											0.0999 ± 0.0001	2.5%
	170	72											0.1202 ± 0.0001	19%
	170	72											0.1647 ± 0.0001	33%
	170	72											0.2081 ± 0.0002	3.4%
	170	72											0.2255 ± 0.0002	1.1%
	170	72											0.2914 ± 0.0002	1.3%
	170	72											0.3089 ± 0.0003	2.6%
	170	72											0.4813 ± 0.0002	4.7%
	170	72											0.5016 ± 0.0002	4.7%
	170	72											0.5402 ± 0.0002	3.1%
	170	72											0.5729 ± 0.0002	18.5%
	170	72											0.6207 ± 0.0002	22.9%
$_{72}Hf^{171}$	171	72		170.940490	12.1 ± 0.4 h.	E.C.,β+	2.40				7/2+		ann.rad.	
	171	72											Lu k x-ray	58%
	171	72											0.1221 ± 0.0001	13%
	171	72											0.1370 ± 0.0001	7%
	171	72											0.1471 ± 0.0001	2.4%
	171	72											0.2691 ± 0.0001	2.2%
	171	72											0.2958 ± 0.0001	7.9%
	171	72											0.3475 ± 0.0001	9.7%
	171	72											0.4695 ± 0.0001	5.5%
	171	72											0.5401 ± 0.0002	2.1%
	171	72											0.6620 ± 0.0001	15%
	171	72											0.6660 ± 0.0001	4.0%
	171	72											0.7883 ± 0.0002	1.8%
	171	72											0.8525 ± 0.0001	5.0%
	171	72											1.0714 ± 0.0002	12%
	171	72											1.1616 ± 0.0003	2.2%
	171	72											1.2926 ± 0.0003	1.1%
	171	72											1.3008 ± 0.0003	1.4%
	171	72											1.3085 ± 0.0004	1.1%
	171	72											1.3402 ± 0.0003	1.7%
	171	72											1.5050 ± 0.0003	1.4%
	171	72											1.5580 ± 0.0003	1.4%
	171	72											1.7473 ± 0.0003	2.1%
	171	72											1.8350 ± 0.0003	1.4%
	171	72											2.0195 ± 0.0004	2.3%
$_{72}Hf^{172}$	172	72		171.939460	1.87 ± 0.03 y.	E.C.	0.350				0+		Lu k x-ray	57%
	172	72											0.02399 ± 0.00005	20%
	172	72											0.06735 ± 0.0001	5.3%

Isotope	A	Z	% Natural abundance	Atomic mass	Half-life	Decay mode	Decay energy (MeV)	Particle energy (MeV)	Particle intensity	Thermal neutron cross section	Spin (h/2π)	μ Nucl. mag. moment	Gamma-ray energy (MeV)	Gamma-ray intensity
	172	72											0.08175 ± 0.00005	4.5%
	172	72											0.1141 ± 0.0001	2.6%
	172	72											0.1229 ± 0.0001	1.1%
	172	72											0.12582 ± 0.00005	11.3%
	172	72											0.1279 ± 0.0001	1.5%
$_{72}Hf^{173}$	173	72		172.940650	23.6 ± 0.1 h.	E.C.	1.60				1/2-		Lu k x-ray	55%
	173	72											0.12367 ± 0.00002	83%
	173	72											0.13495 ± 0.00002	4.8%
	173	72											0.13963 ± 0.00003	12%
	173	72											0.1620 ± 0.0001	6.5%
	173	72											0.29697 ± 0.00002	34%
	173	72											0.30656 ± 0.00002	6.3%
	173	72											0.31124 ± 0.00002	11%
	173	72											0.89910 ± 0.00006	1.0%
	173	72											1.0340 ± 0.0001	0.4%
	173	72											1.0387 ± 0.0001	0.3%
	173	72											1.2056 ± 0.0001	0.3%
	173	72											(0.1 - 2.1)weak	
$_{72}Hf^{174}$	174	72	0.16%	173.940044						500 b.	0+			
$_{72}Hf^{175}$	175	72		174.941507	70 ± 2 d.	E.C.	0.686				5/2-	0.70	Lu k x-ray	47%
	175	72											0.08936 ± 0.0001	2.3%
	175	72											0.34340 ± 0.00008	87%
	175	72											0.43275 ± 0.00008	1.6%
$_{72}Hf^{176}$	176	72	5.2%	175.941406						26 b.	0+			
$_{72}Hf^{177m2}$	177	72			51.4 ± 0.5 m.	I.T.	2.740				37/2-		Hf k x-ray	29%
	177	72											0.2140 ± 0.0001	40%
	177	72											0.2951 ± 0.0001	68%
	177	72											0.2951 ± 0.0001	6.8%
	177	72											0.3115 ± 0.0001	58%
	177	72											0.3267 ± 0.0001	65%
	177	72											0.5724 ± 0.0001	7%
	177	72											0.6065 ± 0.0001	11%
	177	72											0.6382 ± 0.0001	20%
$_{72}Hf^{177m1}$	177	72			1.08 ± 0.06 s.	1.315					23/2+		Hf k x-ray	75%
	177	72											0.10534 ± 0.00001	15%
	177	72											0.11295 ± 0.00001	27%
	177	72											0.12849 ± 0.00001	20%
	177	72											0.15329 ± 0.00001	23%
	177	72											0.17440 ± 0.00002	16%
	177	72											0.17700 ± 0.00002	4.3%
	177	72											0.20410 ± 0.00001	19%
	177	72											0.20836 ± 0.00006	79%
	177	72											0.21443 ± 0.00001	8%
	177	72											0.22847 ± 0.00005	48%
	177	72											0.23384 ± 0.00001	7%
	177	72											0.24965 ± 0.00002	8%
	177	72											0.28179 ± 0.00001	18%
	177	72											0.29645 ± 0.00003	7%
	177	72											0.29905 ± 0.00003	2.0%
	177	72											0.30550 ± 0.00003	2.2%
	177	72											0.31371 ± 0.00001	1.6%
	177	72											0.32769 ± 0.00001	22%

Isotope	A	Z	% Natural abundance	Atomic mass	Half-life	Decay mode	Decay energy (MeV)	Particle energy (MeV)	Particle intensity	Thermal neutron cross section	Spin (h/2π)	μ Nucl. mag. moment	Gamma-ray energy (MeV)	Gamma-ray intensity
	177	72											0.37851 ± 0.00001	37%
	177	72											0.38504 ± 0.00004	4%
	177	72											0.41853 ± 0.00001	25%
	177	72											0.46583 ± 0.00005	2.8%
$_{72}Hf^{177}$	177	72	18.6%	176.943217						(1 + 370) b.	7/2-	+0.7935		
$_{72}Hf^{178m2}$	178	72			31 ± 1 y.	I.T.					16+		Hf k x-ray	47%
	178	72											0.08886 ± 0.00002	62%
	178	72											0.09316 ± 0.00001	17%
	178	72											0.21342 ± 0.00001	81%
	178	72											0.21665 ± 0.00001	64%
	178	72											0.23738 ± 0.00002	9%
	178	72											0.25761 ± 0.00002	174%
	178	72											0.29680 ± 0.00003	10%
	178	72											0.32555 ± 0.00002	94%
	178	72											0.42635 ± 0.00002	97%
	178	72											0.45403 ± 0.00002	16%
	178	72											0.49499 ± 0.00002	69%
	178	72											0.53499 ± 0.00003	9%
	178	72											0.57418 ± 0.00003	84%
$_{72}Hf^{178m1}$	178	72			4.0 ± 0.2 s.	I.T.					8-		Hf k x-ray	35%
	178	72											0.08886 ± 0.00002	62%
	178	72											0.09316 ± 0.00001	17%
	178	72											0.21342 ± 0.00001	81%
	178	72											0.32555 ± 0.00002	94%
	178	72											0.42635 ± 0.00002	97%
$_{72}Hf^{178}$	178	72	27.1%	177.943696						(50 + 30) b.	0+			
$_{72}Hf^{179m2}$	179	72			25.1 d.	I.T.	1.1057				25/2-		Hf k x-ray	56%
	179	72											0.1227 ± 0.0001	27%
	179	72											0.1461 ± 0.0001	26%
	179	72											0.1698 ± 0.0001	19%
	179	72											0.1928 ± 0.0002	21%
	179	73											0.2170 ± 0.0002	8.8%
	179	72											0.2366 ± 0.0002	18%
	179	72											0.2575 ± 0.0003	3.2%
	179	72											0.2689 ± 0.0002	11%
	179	72											0.3160 ± 0.0002	20%
	179	72											0.3626 ± 0.0002	38%
	179	72											0.4098 ± 0.0003	21%
	179	72											0.4537 ± 0.0003	66%
$_{72}Hf^{179m1}$	179	72			18.7 s.	I.T.	0.375				1/2-		Hf k x-ray	28%
	179	72											0.1607 ± 0.0001	2.8%
	179	72											0.2141 ± 0.0001	95%
	179	72											0.3748 ± 0.0001	0.005%
$_{72}Hf^{179}$	179	72	13.74%	178.945812						(45 + 41) b.	9/2+	-0.6409		
$_{72}Hf^{180m}$	180	72			5.519 ± 0.004 h.	I.T.	1.1416				8-	+8.7	Hf k x-ray	18%
	180	72											0.0575 ± 0.0001	48%
	180	72											0.0933 ± 0.0001	17%
	180	72											0.2152 ± 0.0001	82%
	180	72											0.3323 ± 0.0001	94%
	180	72											0.4432 ± 0.0001	85%
	180	72											0.5007 ± 0.0001	13%
$_{72}Hf^{180}$	180	72	35.2%	179.946545						13 b.	0+			
$_{72}Hf^{181}$	181	72		180.949096	42.4 ± 0.06 d.	β-	1.027	0.408		30 b.	1/2-		Ta k x-ray	13%
	181	72											0.13294 ± 0.00007	36%
	181	72											0.13617 ± 0.00007	6%
	181	72											0.34583 ± 0.00007	15%
	181	72											0.48200 ± 0.00005	81%

Isotope	A	Z	% Natural abundance	Atomic mass	Half-life	Decay mode	Decay energy (MeV)	Particle energy (MeV)	Particle intensity	Thermal neutron cross section	Spin (h/2π)	μ Nucl. mag. moment	Gamma-ray energy (MeV)	Gamma-ray intensity
$_{72}Hf^{182m}$	182	72			62 m.	β-(54%)	1.60	0.49	43%		8-		Hf k x-ray	8%
	182	72				I.T.(46%)	1.1729	0.95	10%				0.0509 ± 0.0002	13%
	182	72											0.0978 ± 0.0002	14%
	182	72											0.1143 ± 0.0002	7%
	182	72											0.1328 ± 0.0002	3%
	182	72											0.1432 ± 0.0002	4.7%
	182	72											0.1468 ± 0.0002	4.0%
	182	72											0.1734 ± 0.0002	3.0%
	182	72											0.1787 ± 0.0002	2%
	182	72											0.2244 ± 0.0002	38%
	182	72											0.3396 ± 0.0002	6.2%
	182	72											0.3441 ± 0.0002	46%
	182	72											0.4558 ± 0.0002	20%
	182	72											0.5066 ± 0.0002	24%
	182	72											0.6032 ± 0.0002	6%
	182	72											0.6133 ± 0.0002	1.2%
	182	72											0.6276 ± 0.0002	1.1%
	182	72											0.7997 ± 0.0002	10%
	182	72											0.8231 ± 0.0002	3%
	182	72											0.9428 ± 0.0002	21%
$_{72}Hf^{182}$	182	72		181.950550	9 x 10^6 y.	β-	0.431				0+		Ta k x-ray	7%
	182	72											0.1143 ± 0.0001	3%
	182	72											0.1561 ± 0.0001	7%
	182	72											0.2704 ± 0.0001	80%
$_{72}Hf^{183}$	183	72		182.953530	64 m.	β-	2.01	1.18	68%		3/2-		Ta k x-ray	13%
	183	72						1.54	25%				0.0732 ± 0.0001	38%
	183	72											0.3159 ± 0.0001	1%
	183	72											0.3979 ± 0.0001	3%
	183	72											0.4591 ± 0.0001	27%
	183	72											0.7837 ± 0.0001	65%
	183	72											1.4702 ± 0.0001	2.7%
$_{72}Hf^{184}$	184	72		183.955440	4.1 h.	β-	1.3	0.74	38%		0+		Ta k x-ray	15%
	184	72						0.85	16%				0.0414 ± 0.0002	10%
	184	72						1.10	46%				0.0439 ± 0.0002	6%
	184	72											0.0479 ± 0.0002	1%
	184	72											0.1391 ± 0.0002	48%
	184	72											0.1810 ± 0.0002	15%
	184	72											0.3449 ± 0.0002	38%
Ta		73		180.9479						20.5 b.				
$_{73}Ta^{159}$	159	73		158.962860	0.6 s.	β+, EC (20%)	8.49						ann.rad.	
	159	73				α(80%)		5.60						
$_{73}Ta^{160}$	160	73		159.961630		β+,E.C.	10.3						ann.rad.	
	160	73				α		5.41						
$_{73}Ta^{161}$	161	73		160.958210		β+,E.C.	7.55						ann.rad.	
	161	73				α		5.15						
$_{73}Ta^{164}$	164	73		163.953370	13.6 s.	β+	8.35						ann.rad.	
	164	73				α		4.62					0.2110	
	164	73											0.3768	
$_{73}Ta^{166}$	166	73		165.950280	32 s.	β+(82%)	7.48						ann.rad.	160%
	166	73				E.C.(18%)							Hf k x-ray	16%
	166	73											0.1587 ± 0.0002	53%
	166	73											0.3117 ± 0.0003	28%
	166	73											0.5360 ± 0.0004	4.0%
	166	73											0.5524 ± 0.0004	3.0%
	166	73											0.5945 ± 0.0003	3.5%
	166	73											0.6514 ± 0.0004	8.5%
	166	73											0.7428 ± 0.0004	7.0%
	166	73											0.7500 ± 0.0005	5.5%
	166	73											0.8101 ± 0.0004	9.8%
	166	73											0.8474 ± 0.0004	7.2%
	166	73											0.8622 ± 0.0006	3.7%
	166	73											0.8641 ± 0.0005	4.9%
	166	73											0.9062 ± 0.0006	6.1%
	166	73											0.9770 ± 0.0008	2.5%
	166	73											1.0549 ± 0.001	4.4%
	166	73											1.1738 ± 0.001	5%
	166	73											1.2883 ± 0.0012	3.1%
	166	73											1.4470 ± 0.002	3%
$_{73}Ta^{167}$	167	73		166.948080	3 m.	β+,E.C.	5.10						ann.rad.	
$_{73}Ta^{168}$	168	73		167.947820	2.4 m.	β+(77%)	6.60						ann.rad.	150%
	168	73				E.C.(23%)							Hf k x-ray	21%
	168	73											0.1239 ± 0.0002	37%
	168	73											0.2615 ± 0.0002	28%
	168	73											0.3711 ± 0.0004	4%
	168	73											0.5270 ± 0.0006	2.7%
	168	73											0.6464 ± 0.0008	3%
	168	73											0.7502 ± 0.0006	10%
	168	73											0.7730 ± 0.0008	6%
	168	73											0.8156 ± 0.0008	2%
	168	73											0.8340 ± 0.0008	2%
	168	73											0.8741 ± 0.0008	6%
	168	73											0.8975 ± 0.001	2%
	168	73											0.9072 ± 0.001	6%

Isotope	A	Z	% Natural abundance	Atomic mass	Half-life	Decay mode	Decay energy (MeV)	Particle energy (MeV)	Particle intensity	Thermal neutron cross section	Spin (h/2π)	μ Nucl. mag. moment	Gamma-ray energy (MeV)	Gamma-ray intensity
	168	73											0.9342 ± 0.0012	3%
	168	73											0.9866 ± 0.001	5%
	168	73											1.0581 ± 0.001	2%
	168	73											1.2481 ± 0.002	2%
	168	73											1.2824 ± 0.002	2%
	168	73											1.4063 ± 0.002	1.6%
	168	73											1.4414 ± 0.002	1.1%
	168	73											1.6682 ± 0.003	1.1%
$_{73}Ta^{169}$	169	73		168.946020	5 m.	β+,E.C.	4.45						ann.rad.	
	169	73											0.0288 ± 0.0001	100 +
	169	73											0.0382 ± 0.0001	25
	169	73											0.0777 ± 0.0001	7
	169	73											0.1328 ± 0.0001	9
	169	73											0.1535 ± 0.0001	35
	169	73											0.1770 ± 0.0001	10
	169	73											0.1878 ± 0.0002	5
	169	73											0.1924 ± 0.0001	43
	169	73											0.2300 ± 0.0001	12
	169	73											0.3945 ± 0.0001	15
	169	73											0.4040 ± 0.0002	9
	169	73											0.4408 ± 0.0001	17
	169	73											0.5204 ± 0.0002	9
	169	73											0.5290 ± 0.0002	11
	169	73											0.5474 ± 0.0003	9
	169	73											0.5950 ± 0.0002	26
$_{73}Ta^{170}$	170	73		169.945970	6.76 ± 0.06 m.	β+(70%)	5.80				(3+)		ann.rad.	140%
	170	73				E.C.(35%)							Hf k x-ray	22%
	170	73											0.1008 ± 0.0002	21%
	170	73											0.2212 ± 0.0002	16%
	170	73											0.6650 ± 0.0003	1%
	170	73											0.7655 ± 0.0002	1%
	170	73											0.8348 ± 0.0004	1.5%
	170	73											0.8604 ± 0.0002	7.4%
	170	73											0.9870 ± 0.0003	5.8%
	170	73											1.1190 ± 0.0006	1.1%
	170	73											1.3442 ± 0.0006	1.4%
$_{73}Ta^{171}$	171	73		170.944680	23.4 m.	β+,E.C.	3.90				(5/2-)		0.0496 ± 0.0001	100 +
	171	73											0.0619 ± 0.0001	9
	171	73											0.0667 ± 0.0001	4
	171	73											0.0807 ± 0.0001	4
	171	73											0.0920 ± 0.0001	11
	171	73											0.1171 ± 0.0001	5
	171	73											0.1524 ± 0.0001	6
	171	73											0.1663 ± 0.0001	19
	171	73											0.1755 ± 0.0001	16
	171	73											0.3524 ± 0.0001	3
	171	73											0.4067 ± 0.0001	5
	171	73											0.4444 ± 0.0001	16
	171	73											0.4547 ± 0.0001	4
	171	73											0.4713 ± 0.0002	9
	171	73											0.4927 ± 0.0002	15
	171	73											0.5018 ± 0.0002	23
	171	73											0.5064 ± 0.0002	54
	171	73											0.5223 ± 0.0002	11
	171	73											0.5380 ± 0.0002	15
	171	73											0.5545 ± 0.0002	7
	171	73											0.5709 ± 0.0002	3
	171	73											0.6068 ± 0.0002	4
	171	73											0.6217 ± 0.0002	4
	171	73											0.7676 ± 0.0002	9
	171	73											0.7889 ± 0.0002	4
	171	73											0.9871 ± 0.0002	9
	171	73											1.0078 ± 0.0002	3
	171	73											(0.05 - 1.02)many	
$_{73}Ta^{172}$	172	73		171.944740	36.8 ± 0.3 m.	β+(25%)	4.92				(3-)		ann.rad.	50%
	172	73				E.C.(75%)							Hf k x-ray	40%
	172	73											0.21396 ± 0.00005	52%
	172	73											0.3187 ± 0.0002	5.0%
	172	73											0.5035 ± 0.0001	1.3%
	172	73											0.6431 ± 0.0001	2.2%
	172	73											0.7760 ± 0.0001	2.4%
	172	73											0.8203 ± 0.0001	3.0%
	172	73											0.8571 ± 0.0001	4.1%
	172	73											0.9523 ± 0.0001	1.8%
	172	73											0.9800 ± 0.0001	3.7%
	172	73											0.9954 ± 0.0001	2.1%
	172	73											1.0500 ± 0.0001	2.2%
	172	73											1.0752 ± 0.0001	3.5%
	172	73											1.0856 ± 0.0001	7.6%
	172	73											1.10923 ± 0.00006	14%
	172	73											1.18646 ± 0.00005	2.5%

Isotope	A	Z	% Natural abundance	Atomic mass	Half-life	Decay mode	Decay energy (MeV)	Particle energy (MeV)	Particle intensity	Thermal neutron cross section	Spin (h/2π)	μ Nucl. mag. moment	Gamma-ray energy (MeV)	Gamma-ray intensity
	172	73											1.2404 ± 0.0001	2.0%
	172	73											1.2656 ± 0.0002	2.5%
	172	73											1.2775 ± 0.0001	2.7%
	172	73											1.3303 ± 0.0001	7.6%
	172	73											1.3869 ± 0.0001	2.5%
	172	73											1.4796 ± 0.0002	2.2%
	172	73											1.5443 ± 0.0001	6.2%
	172	73											(0.09 - 3.8)many	
$_{73}Ta^{173}$	173	73		172.943650	3.65 ± 0.05 h.	β+(24%)	2.80				(5/2-)		ann.rad.	48%
	173	73				E.C.(76%)							Hf k x-ray	56%
	173	73											0.06972 ± 0.00007	6%
	173	73											0.17219 ± 0.00006	17%
	173	73											0.18058 ± 0.00007	2.1%
	173	73											0.7011 ± 0.0001	1.2%
	173	73											1.0299 ± 0.0001	1.6%
	173	73											1.2082 ± 0.0001	2.7%
	173	73											1.4322 ± 0.0003	0.6%
	173	73											(0.06 - 2.7)weak	
$_{73}Ta^{174}$	174	73		173.944340	1.18 ± 0.05 h.	β+(27%)	4.00				(3+)		ann.rad.	54%
	174	73				E.C.(73%)							Hf k x-ray	44%
	174	73											0.09089 ± 0.00002	16%
	174	73											0.20638 ± 0.00003	58%
	174	73											0.31080 ± 0.00004	1.0%
	174	73											0.76472 ± 0.00004	1.3%
	174	73											0.97110 ± 0.00005	1.2%
	174	73											1.15135 ± 0.00005	1.1%
	174	73											1.20582 ± 0.00004	4.8%
	174	73											1.22831 ± 0.00004	1.4%
	174	73											1.35773 ± 0.00006	0.8%
	174	73											(0.09 - 3.64)many	
$_{73}Ta^{175}$	175	73		174.943650	10.5 ± 0.2 h.	E.C.	2.00				7/2+		Hf k x-ray	64%
	175	73											0.0816 ± 0.0001	5.7%
	175	73											0.1046 ± 0.0001	3.0%
	175	73											0.1261 ± 0.0001	5.5%
	175	73											0.1410 ± 0.0001	2.2%
	175	73											0.2077 ± 0.0001	13.3%
	175	73											0.2671 ± 0.0001	10%
	175	73											0.3487 ± 0.0001	11%
	175	73											0.4368 ± 0.0002	3.8%
	175	73											0.4754 ± 0.0002	1.9%
	175	73											0.8578 ± 0.0002	3.0%
	175	73											0.9987 ± 0.0002	2.4%
	175	73											1.1439 ± 0.0002	1.1%
	175	73											1.2255 ± 0.0002	2.4%
	175	73											1.7121 ± 0.0002	1.1%
	175	73											1.7218 ± 0.0003	1.1%
	175	73											1.7447 ± 0.0002	1.3%
	175	73											1.7936 ± 0.0002	4.4%
	175	73											1.8263 ± 0.0002	1.2%
$_{73}Ta^{176}$	176	73		175.944730	8.08 ± 0.07 h.	E.C.	3.10				1-		Hf k x-ray	45%
	176	73											0.08837 ± 0.00001	11%
	176	73											0.20187 ± 0.00003	5.5%
	176	73											0.46623 ± 0.00005	1.1%
	176	73											0.50775 ± 0.00008	1.4%
	176	73											0.52152 ± 0.00005	2%
	176	73											0.61121 ± 0.00005	1.2%
	176	73											0.61690 ± 0.00004	1.0%
	176	73											0.71053 ± 0.00004	5.2%
	176	73											1.02317 ± 0.00004	2.6%
	176	73											.15735 ±).00003	24.6%

TABLE OF THE ISOTOPES (Continued)

Isotope	A	Z	% Natural abundance	Atomic mass	Half-life	Decay mode	Decay energy (MeV)	Particle energy (MeV)	Particle intensity	Thermal neutron cross section	Spin (h/2π)	μ Nucl. mag. moment	Gamma-ray energy (MeV)	Gamma-ray intensity
	176	73											1.19023 ± 0.00006	4.4%
	176	73											1.22503 ± 0.00003	5.5%
	176	73											1.25298 ± 0.00003	3.0%
	176	73											1.26887 ± 0.00006	1.3%
	176	73											1.29106 ± 0.00004	1.28%
	176	73											1.34135 ± 0.00004	3.2%
	176	73											1.35748 ± 0.00004	1.9%
	176	73											1.55507 ± 0.00004	3.9%
	176	73											1.58402 ± 0.00004	5.1%
	176	73											1.61627 ± 0.00005	1.2%
	176	73											1.63084 ± 0.00005	1.7%
	176	73											1.63371 ± 0.00005	2.8%
	176	73											1.64344 ± 0.00004	2.3%
	176	73											1.69653 ± 0.00005	4.5%
	176	73											1.72208 ± 0.00005	3.1%
	176	73											1.82370 ± 0.00004	4.3%
	176	73											1.86287 ± 0.00004	3.8%
	176	73											2.04485 ± 0.00006	1.3%
	176	73											2.83193 ± 0.00007	4.2%
	176	73											2.92031 ± 0.00007	2.1%
$_{73}Ta^{177}$	177	73		176.944460	2.36 ± 0.01 d.	E.C.	1.158				7/2+		Hf k x-ray	42%
	177	73											0.11295 ± 0.00001	7.2%
	177	73											0.20836 ± 0.00001	1.0%
	177	73											0.42460 ± 0.00005	0.1%
	177	73											0.74591 ± 0.00005	0.2%
	177	73											1.0577 ± 0.0001	0.3%
	177	73											(0.07 - 1.06)weak	
$_{73}Ta^{178m}$	178	73			2.45 ± 0.05 h.	E.C.					(7-)		Hf k x-ray	75%
	178	73											0.08886 ± 0.00002	62%
	178	73											0.09316 ± 0.00003	17.4%
	178	73											0.21342 ± 0.00002	81%
	178	73											0.32555 ± 0.00002	94%
	178	73											0.33166 ± 0.00006	32%
	178	73											0.42635 ± 0.00002	97%
$_{73}Ta^{178}$	178	73		177.945750	9.3 ± 0.03 m.	E.C.(99%)	1.910				1+		ann.rad.	2%
	178	73				β+(1%)							Hf k x-ray	42%
	178	73											0.09316 ± 0.00003	6.7%
	178	73											1.10614 ± 0.00007	1.53%
	178	73											1.18345 ± 0.00004	0.17%
	178	73											1.3409 ± 0.0001	1.0%
	178	73											1.3506 ± 0.0001	1.2%
	178	73											1.4961 ± 0.0001	0.27%
$_{73}Ta^{179}$	179	73		178.945930	1.82 ± 0.05 y.	E.C.	0.110				7/2+		Hf k x-ray	29%
$_{73}Ta^{180m}$	180	73			8.15 ± 0.01 h.	E.C.(87%)	0.865				1+		Hf k x-ray	36%
	180	73				β-(13%)	0.710	0.61	3%				W k x-ray	0.3%
	180	73						0.71	10%				0.09333 ± 0.00006	5%

TABLE OF THE ISOTOPES (Continued)

Isotope	A	Z	% Natural abundance	Atomic mass	Half-life	Decay mode	Decay energy (MeV)	Particle energy (MeV)	Particle intensity	Thermal neutron cross section	Spin (h/2π)	μ Nucl. mag. moment	Gamma-ray energy (MeV)	Gamma-ray intensity
	180	73											0.10340 ± 0.00001	0.7%
$_{73}Ta^{180}$	180	73	0.012%	179.947462						600 b.	(9-)			
$_{73}Ta^{181}$	181	73	99.998%	180.947992						(0.011 + 20) b.	7/2+	+2.370		
$_{73}Ta^{182m}$	182	73			15.9 m.	I.T.	0.5198				10-		Ta k x-ray	46%
	182	73											0.14678 ± 0.00002	36%
	182	73											0.17157 ± 0.00002	47%
	182	73											0.18495 ± 0.00002	23%
	182	73											0.31837 ± 0.00005	6.5%
$_{73}Ta^{182}$	182	73		181.950149	114.5 d.	β-	1.814	0.25	30%		3-	2.6	W k x-ray	17%
	182	73						0.44	20%				0.06775 ± 0.00001	41%
	182	73						0.52	40%				0.10010 ± 0.00001	14%
	182	73											0.11367 ± 0.00002	1.9%
	182	73											0.15243 ± 0.00001	7.1%
	182	73											0.15639 ± 0.00001	2.7%
	182	73											0.17939 ± 0.00001	3.1%
	182	73											0.19836 ± 0.00001	1.5%
	182	73											0.22211 ± 0.00001	7.6%
	182	73											0.22932 ± 0.00001	3.6%
	182	73											0.26407 ± 0.00001	3.6%
	182	73											1.12127 ± 0.00003	35.0%
	182	73											1.18902 ± 0.00003	16.4%
	182	73											1.22138 ± 0.00003	27.4%
	182	73											1.23099 ± 0.00003	11.6%
	182	73											1.25739 ± 0.00003	1.5%
	182	73											1.27370 ± 0.00003	0.67%
	182	73											1.28913 ± 0.00003	1.4%
	182	73											1.3427 ± 0.0001	0.26%
	182	73											1.37381 ± 0.00003	0.23%
$_{73}Ta^{183}$	183	73		182.951369	5.1 d.	β-	1.07	0.45	5%		7/2+		W k x-ray	44%
	183	73						0.62	91%				0.0847 ± 0.0001	1.3%
	183	73											0.0991 ± 0.0001	6.6%(D)
	183	73											0.1079 ± 0.0001	11%(D)
	183	73											0.1441 ± 0.0001	2.5%
	183	73											0.1613 ± 0.0001	8.9%
	183	73											0.1623 ± 0.0001	4.9%
	183	73											0.2099 ± 0.0001	4.5%
	183	73											0.2443 ± 0.0001	8.6%
	183	73											0.2461 ± 0.0001	27%
	183	73											0.2917 ± 0.0001	3.8%
	183	73											0.3131 ± 0.0002	7.3%
	183	73											0.3540 ± 0.0001	11.4%
$_{73}Ta^{184}$	184	73		183.954005	8.7 h.	β-	2.86	1.11	15%		(5-)		W k x-ray	15%
	184	73						1.17	81%				0.1112 ± 0.0001	24%
	184	73											0.1613 ± 0.0001	3.3%
	184	73											0.2153 ± 0.0001	12%
	184	73											0.2267 ± 0.0001	6.8%
	184	73											0.2444 ± 0.0001	3.6%
	184	73											0.2528 ± 0.0001	49%
	184	73											0.3180 ± 0.0001	23%
	184	73											0.3843 ± 0.0001	12.8%
	184	73											0.4140 ± 0.0001	74%
	184	73											0.4611 ± 0.0001	11%
	184	73											0.5367 ± 0.0001	13%
	184	73											0.6420 ± 0.0001	1.4%
	184	73											0.7921 ± 0.0001	15%
	184	73											0.8948 ± 0.0001	11%
	184	73											0.9033 ± 0.0001	15%
	184	73											0.9209 ± 0.0001	33%
	184	73											1.1101 ± 0.0001	2.3%
	184	73											1.1738 ± 0.0001	4.9%
$_{73}Ta^{185}$	185	73		184.955553	49 m.	β-	1.994	1.21	5%		(7/2+)		W k x-ray	23%
	185	73						1.77	81%				0.0697 ± 0.0002	2%

Isotope	A	Z	% Natural abundance	Atomic mass	Half-life	Decay mode	Decay energy (MeV)	Particle energy (MeV)	Particle intensity	Thermal neutron cross section	Spin (h/2π)	μ Nucl. mag. moment	Gamma-ray energy (MeV)	Gamma-ray intensity
	185	73											0.1078 ± 0.0001	2.7%
	185	73											0.1473 ± 0.0001	1.1%
	185	73											0.1739 ± 0.0001	22%
	185	73											0.1776 ± 0.0001	26%
	185	73											0.2435 ± 0.0001	3.7%
	185	73											0.3944 ± 0.0005	0.8%
	185	73											0.5417 ± 0.0005	0.8%
	185	73											0.5887 ± 0.001	0.8%
$_{73}Ta^{186}$	186	73		185.958540	10.5 m.	β-	3.9	2.2			(3-)		W k x-ray	15%
	186	73											0.1223 ± 0.0001	23%
	186	73											0.1979 ± 0.0001	59%
	186	73											0.2149 ± 0.0001	50%
	186	73											0.2925 ± 0.0005	4%
	186	73											0.3075 ± 0.0001	11%
	186	73											0.3092 ± 0.0001	3%
	186	73											0.4177 ± 0.0002	15%
	186	73											0.4570 ± 0.001	2.5%
	186	73											0.5106 ± 0.0005	44%
	186	73											0.5672 ± 0.0003	4.0%
	186	73											0.6153 ± 0.0002	33%
	186	73											0.7375 ± 0.0003	34%
	186	73											0.7392 ± 0.0003	11.8%
	186	73											0.7594 ± 0.0005	2%
	186	73											0.7998 ± 0.0005	3%
	186	73											0.8300 ± 0.0005	2%
	186	73											0.8841 ± 0.001	2%
	186	73											0.9230 ± 0.001	1.4%
	186	73											(0.09 - 1.5)	1.4%
W		74								18.4 b.				
$_{74}W^{160}$	160	74		159.968480	0.08 s.	α		5.92			0+			
$_{74}W^{161}$	161	74		160.967140	0.41 s.	β+, EC (18%)	8.34							
	164	74				α(82%)		5.78						
$_{74}W^{162}$	162	74		161.963290	1.39 s.	β+, EC (54%)	5.72				0+			
	162	74				α(46%)		5.54						
$_{74}W^{163}$	163	74		162.962270	2.8 s.	β+, EC (59%)	7.54							
	163	74				α(41%)		5.38						
$_{74}W^{164}$	164	74		163.958820	6 s.	β+, EC (97%)	5.08				0+		ann.rad.	
	164	74				α(3%)		5.15						
$_{74}W^{165}$	165	74		164.958110	≈ 5.1 s.	β+, EC (99%)	6.86						ann.rad.	
	165	74				α(1%)		4.91						
$_{74}W^{166}$	166	74		165.955020	16 s.	β+, EC (99%)	4.42				0+		ann.rad.	
	166	74				α(1%)		4.74						
$_{74}W^{172}$	172	74		171.947430	≈ 6.7 m.	β+,E.C.	2.50						0.0359 ± 0.0003	39 +
	172	74											0.0396 ± 0.0003	10
	172	74											0.1093 ± 0.0002	6
	172	74											0.1302 ± 0.0002	27
	172	74											0.1452 ± 0.0005	5
	172	74											0.1538 ± 0.0003	10
	172	74											0.1749 ± 0.0003	21
	172	74											0.3244 ± 0.0002	7
	172	74											0.4234 ± 0.0002	6
	172	74											0.4576 ± 0.0002	100
	172	74											0.6236 ± 0.0002	28
	172	74											0.6360 ± 0.0003	5
	172	74											0.7708 ± 0.0006	4
$_{74}W^{173}$	173	74		172.947710	16.1 ± 0.5 m.	E.C.	3.78						0.0499	
	173	74											0.1057	
	173	74											0.3652	
	173	74											0.707	
$_{74}W^{174}$	174	74		173.946160	29 ± 1 m.	E.C.	1.70				0+		0.0354 ± 0.0001	148 +
	174	74											0.0619 ± 0.0004	20
	174	74											0.0964 ± 0.0001	11
	174	74											0.1252 ± 0.0001	81
	174	74											0.1365 ± 0.0001	78
	174	74											0.1437 ± 0.0001	24
	174	74											0.1627 ± 0.0001	14
	174	74											0.1740 ± 0.0001	5
	174	74											0.1930 ± 0.0001	56
	174	74											0.2020 ± 0.0001	41
	174	74											0.2164 ± 0.0002	7
	174	74											0.2334 ± 0.0001	32
	174	74											0.2395 ± 0.0001	13
	174	74											0.2898 ± 0.0002	10
	174	74											0.3287 ± 0.0001	100
	174	74											0.3398 ± 0.0001	36
	174	74											0.3549 ± 0.0001	21
	174	74											0.3645 ± 0.0001	37

Isotope	A	Z	% Natural abundance	Atomic mass	Half-life	Decay mode	Decay energy (MeV)	Particle energy (MeV)	Particle intensity	Thermal neutron cross section	Spin (h/2π)	μ Nucl. mag. moment	Gamma-ray energy (MeV)	Gamma-ray intensity
	174	74											0.3770 ± 0.0001	57
	174	74											0.3785 ± 0.0001	84
	174	74											0.4288 ± 0.0001	123
	174	74											0.4722 ± 0.0001	4
	174	74											0.5475 ± 0.0001	4
	174	74											0.5676 ± 0.0001	4
	174	74											0.8350 ± 0.0001	6
$_{74}W^{175}$	175	74		174.946770	34 ± 1 m.	E.C.	2.90				$^{1}/_{2}$-		0.01498 ± 0.00002	
	175	74											0.03641 ± 0.00002	
	175	74											0.05138 ± 0.00002	
	175	74											0.1211 ± 0.0001	
	175	74											0.1491 ± 0.0001	
	175	74											0.1667 ± 0.0001	
	175	74											0.2703 ± 0.0001	
$_{74}W^{176}$	176	74		175.945590	2.5 ± 0.2 h.	β+,E.C.	0.800				0+		0.03358 ± 0.00004	0.08 +
	176	74											0.06129 ± 0.00004	9
	176	74											0.08414 ± 0.0004	5
	176	74											0.09487 ± 0.00004	9
	176	74											0.10020 ± 0.00005	100
$_{74}W^{177}$	177	74		176.946610	2.21 ± 0.04 h.	E.C.	2.00				($^{1}/_{2}$-)		Ta k x-ray	78%
	177	74											0.15505 ± 0.00004	59%
	177	74											0.15594 ± 0.00004	4%
	177	74											0.18569 ± 0.00007	16%
	177	74											0.42694 ± 0.00004	13%
	177	74											0.5284 ± 0.0001	2.4%
	177	74											0.6116 ± 0.0001	6%
	177	74											0.6473 ± 0.0001	2.5%
	177	74											1.0149 ± 0.0001	4.8%
	177	74											1.0364 ± 0.0001	10%
	177	74											1.0668 ± 0.0001	3%
	177	74											1.1825 ± 0.0001	3.7%
$_{74}W^{178}$	178	74		177.945840	21.5 ± 0.1 d.	E.C.	0.089				0+		Ta k x-ray	13%
$_{74}W^{179m}$	179	74			6.4 m.	I.T.(99.7%)	0.222				($^{1}/_{2}$-)		W k x-ray	27%
	179	74				E.C.(0.3%)							0.2220 ± 0.0001	8.6%
	179	74											0.2387 ± 0.0003	0.2%
	179	74											0.2817 ± 0.0003	0.2%
$_{74}W^{179}$	179	74		178.947067	37.5 m.	E.C.	1.060				(7/2-)		Ta k x-ray	39%
	179	74											0.0307 ± 0.0001	28%
	179	74											0.0339 ± 0.0002	0.2%
$_{74}W^{180}$	180	74	0.13%	179.946701						30 b.	0+			
$_{74}W^{181}$	181	74		180.948192	121 ± 0.2 d.	E.C.	0.187				9/2+		Ta k x-ray	33%
	181	74											0.13617 ± 0.00007	0.032%
	181	74											0.15221 ± 0.00002	0.08%
$_{74}W^{182}$	182	74	26.3%	181.948202						21 b.	0+			
$_{74}W^{183m}$	183	74			5.15 s.	I.T.					(11/2+		W k x-ray	63%
	183	74											0.0465 ± 0.0001	6%
	183	74											0.0526 ± 0.0001	7%+
	183	74											0.0991 ± 0.0001	9%
	183	74											0.1025 ± 0.0001	2.3%
	183	74											0.1605 ± 0.0001	4.9%
$_{74}W^{183}$	183	74	14.3%	182.950220						10 b.	$^{1}/_{2}$-	+0.11778		
$_{74}W^{184}$	184	74	30.67%	183.950928						(0.002 + 1.8) b.	0+			
$_{74}W^{185m}$	185	74			1.65 m.	I.T.	0.1974				11/2+		W k x-ray	4%
	185	74											0.0659 ± 0.0001	5.8%
	185	74											0.1315 ± 0.0001	4.3%
	185	74											0.1737 ± 0.0001	3.3%
$_{74}W^{185}$	185	74		184.953416	74.8 d.	β-	0.433	0.433	99.9%		3/2-		0.12536 ± 0.00003	0.019%
$_{74}W^{186}$	186	74	28.6%	185.954357										
$_{74}W^{187}$	187	74		186.957153	23.9 h.	β-	1.312	0.624		60 b.	3/2-	0.688	Re k x-ray	14%
	187	74						1.315					0.0725 ± 0.0001	13%
	178	74											0.13424 ± 0.00003	10%
	187	74											0.47951 ± 0.00003	25%
	187	74											0.55151 ± 0.00001	5.9%
	187	74											0.61824 ± 0.00004	7.3%

Isotope	A	Z	% Natural abundance	Atomic mass	Half-life	Decay mode	Decay energy (MeV)	Particle energy (MeV)	Particle intensity	Thermal neutron cross section	Spin (h/2π)	μ Nucl. mag. moment	Gamma-ray energy (MeV)	Gamma-ray intensity
	187	74											0.68572 ± 0.00004	32%
	187	74											0.77295 ± 0.00007	4.8%
$_{74}W^{188}$	188	74		187.958480	69.4 d.	β-	0.349	0.349	99%		0+		0.0636 ± 0.0001	0.1%
	188	74											0.2271 ± 0.0001	0.2%
	188	74											0.2907 ± 0.0001	0.4%
$_{74}W^{189}$	189	74		188.961900	11.5 m.	β-	2.50	1.4			(3/2-(		0.258 ± 0.0003	100 +
	189	74						2.5					0.417 ± 0.004	96
	189	74											0.550 ± 0.001	28
	189	74											0.855 ± 0.015	20
	189	74											0.955 ± 0.020	17
$_{74}W^{190}$	190	74		189.963210	30 m.	β-	1.3	0.95			0+		Re k x-ray	54%
	190	74											0.1576 ± 0.0001	39%
	190	74											0.1621 ± 0.0001	11%
Re		75		186.207						90 b.				
$_{75}Re^{162}$	162	75		161.976060	0.10 s.	α		6.12						
$_{75}Re^{163}$	163	75		162.971970	0.26 s.	β+,E.C.	9.04							
	163	75				α		5.92						
$_{75}Re^{164}$	164	75		163.970590	0.9 s.	β+,E.C.	10.9							
	164	75				α		5.78						
$_{75}Re^{165}$	165	75		164.961890	2.4 s.	β+, EC (87%)	8.18							
	165	75				α		5.51						
$_{75}Re^{166}$	166	75		165.965740	2.2 s.	β+,E.C.	9.98							
	166	75				α		5.50						
$_{75}Re^{167}$	167	75		166.962590	2.0 s.	β+,E.C.	7.52							
	167	75				α		5.35						
$_{75}Re^{168}$	168	75		167.961540	2.9 s.	β+,E.C.	9.08							
	168	75				α		5.14						
$_{75}Re^{170}$	170	75		169.958040	8 s.	β+,E.C.	8.15						0.1560 ± 0.0004	57%
	170	75											0.3055 ± 0.0004	85%
	170	75											0.4125 ± 0.0004	50%
$_{75}Re^{172m}$	172	75			15 s.	β+,E.C.							ann.rad.	
	172	75											0.1234 ± 0.0001	45 +
	172	75											0.2537 ± 0.0002	100
	172	75											0.3504 ± 0.0005	55
	172	75											0.4194 ± 0.0003	10
$_{75}Re^{172}$	172	75		171.953180	2.3 m.	β+,E.C.	7.29						ann.rad.	
	172	75											0.1234 ± 0.0007	100 +
	172	75											0.2537 ± 0.0002	74
	172	75											0.7430 ± 0.0002	19
$_{75}Re^{174}$	174	75		173.953180	2.3 m.	β+,E.C.	6.54						ann.rad.	
	174	75											0.1119 ± 0.0004	29%
	174	75											0.2430 ± 0.0004	23%
	174	75											0.3490 ± 0.0004	11%
$_{75}Re^{175}$	175	75		174.951430	4.6 m.	β+,E.C.	4.35						ann.rad.	
$_{75}Re^{176}$	176	75		175.951500	5.3 m.	β+,E.C.	5.50				(3+)		ann.rad.	
	176	75											0.1089 ± 0.0003	26%
	176	75											0.2406 ± 0.0003	48%
$_{75}Re^{177}$	177	75		176.950370	14 m.	E.C.(78%)	3.50				(5/2-)		ann.rad.	44%
	177	75				β+(22%)							W k x-ray	62%
	177	75											0.0797 ± 0.0001	7%
	177	75											0.0843 ± 0.0002	6%
	177	75											0.0949 ± 0.0001	4%
	177	75											0.1014 ± 0.0002	3%
	177	75											0.1968 ± 0.0002	8%
	177	75											0.2098 ± 0.0003	3%
	177	75											0.7081 ± 0.0006	2%
	177	75											0.7234 ± 0.0006	2%
	177	75											1.7705 ± 0.0008	2%
	177	75											1.9112 ± 0.0008	1%
	177	75											1.9646 ± 0.0008	3%
	177	75											1.9861 ± 0.0008	1%
$_{75}Re^{178}$	178	75		177.950850	13.2 m.	β+(11%)	4.66	3.3			(3)		ann.rad.	22%
	178	75				E.C.(89%)							W k x-ray	47%
	178	75											0.1059 ± 0.0003	23%
	178	75											0.2373 ± 0.0003	45%
	178	75											0.7779 ± 0.0004	4%
	178	75											0.9391 ± 0.0005	9%
	178	75											0.9766 ± 0.0005	3%
	178	75											1.1108 ± 0.0004	2.7%
	178	75											1.1306 ± 0.0004	3.3%
	178	75											1.2553 ± 0.0004	1.4%
	178	75											1.2756 ± 0.0004	1.7%
	178	75											1.3115 ± 0.0002	1.2%
	178	75											1.4500 ± 0.0005	1.1%
	178	75											1.5984 ± 0.0004	1.3%
	178	75											1.5984 ± 0.0004	1.3%
	178	75											2.9576 ± 0.0005	0.9%
	178	75											3.1686 ± 0.0005	0.9%
$_{75}Re^{179}$	179	75		178.949960	19.7 m.	E.C.(99%)	2.69	0.95			(5/2+)		W k x-ray	56%
	179	75				β+(1%)							0.1199 ± 0.0001	4.7%
	179	75											0.1891 ± 0.0001	7.3%
	179	75											0.2900 ± 0.0001	26%

Isotope	A	Z	% Natural abundance	Atomic mass	Half-life	Decay mode	Decay energy (MeV)	Particle energy (MeV)	Particle intensity	Thermal neutron cross section	Spin (h/2π)	μ Nucl. mag. moment	Gamma-ray energy (MeV)	Gamma-ray intensity
	179	75											0.2963 ± 0.0001	8.7%
	179	75											0.3089 ± 0.0002	3.2%
	179	75											0.4018 ± 0.0001	7.0%
	179	75											0.4154 ± 0.0001	10%
	179	75											0.4302 ± 0.0001	27%
	179	75											0.4648 ± 0.0001	3.7%
	179	75											0.4773 ± 0.0001	9.0%
	179	75											0.4983 ± 0.0001	5.6%
	179	75											0.8326 ± 0.0001	2.9%
	179	75											1.3713 ± 0.0001	1.0%
	179	75											1.5604 ± 0.0001	3.1%
	179	75											1.6803 ± 0.0001	13%
	179	75											1.8087 ± 0.0001	2.2%
$_{75}Re^{180}$	180	75		179.950780	2.45 m.	E.C.(92%) β+(8%)	3.79	1.76			(1/2-)		ann.rad.	16%
	180	75											W k x-ray	45%
	180	75											0.1036 ± 0.0001	22%
	180	75											0.8254 ± 0.0001	9.8%
	180	75											0.9028 ± 0.0001	89%
	180	75											(0.07 - 2.2)weak	
$_{75}Re^{181}$	181	75		180.950020	20 h.	E.C.	1.70				5/2+	3.242	W k x-ray	61%
	181	75											0.1775 ± 0.0002	1.6%
	181	75											0.3186 ± 0.0003	1.1%
	181	75											0.3319 ± 0.0003	1.3%
	181	75											0.3561 ± 0.0003	1.7%
	181	75											0.3607 ± 0.0003	12%
	181	75											0.3655 ± 0.0003	56%
	181	75											0.5578 ± 0.0004	2.1%
	181	75											0.6390 ± 0.0004	6.4%
	181	75											0.6512 ± 0.0004	1.0%
	181	75											0.6618 ± 0.0004	3.0%
	181	75											0.8052 ± 0.0004	3.1%
	181	75											0.9074 ± 0.0005	1.0%
	181	75											0.9536 ± 0.0005	3.5%
	181	75											1.0002 ± 0.0005	3.3%
	181	75											1.0094 ± 0.0005	2.4%
	181	75											1.0756 ± 0.0005	1.0%
	181	75											1.4407 ± 0.0005	1.9%
$_{75}Re^{182m}$	182	75			12.7 h.	E.C.		0.55			2+		W k x-ray	52%
	182	75						1.74					0.0677 ± 0.0001	38%
	182	75											0.1004 ± 0.0001	14%
	182	75											0.1524 ± 0.0001	6.7%
	182	75											0.2293 ± 0.0001	2.1%
	182	75											0.4703 ± 0.0002	2.0%
	182	75											0.8949 ± 0.0002	2.1%
	182	75											1.1214 ± 0.0002	32%
	182	75											1.1892 ± 0.0002	15%
	182	75											1.2215 ± 0.0002	25%
	182	75											1.2311 ± 0.0002	1.3%
	182	75											1.2573 ± 0.0002	1.4%
	182	75											1.2893 ± 0.0002	1.2%
	182	75											(0.06 - 2.2)weak	
$_{75}Re^{182}$	182	75		181.951210	64 h.	E.C.	2.80				(7+)	0.399	W k x-ray	91%
	182	75											0.0678 ± 0.0001	25%
	182	75											0.1001 ± 0.0001	16%
	182	75											0.1137 ± 0.0001	5.3%
	182	75											0.1308 ± 0.0001	8.1%
	182	75											0.1338 ± 0.0001	2.6%
	182	75											0.1489 ± 0.0001	1.9%
	182	75											0.1524 ± 0.0001	9.2%
	182	75											0.1564 ± 0.0001	7.8%
	182	75											0.1692 ± 0.0001	12%
	182	75											0.1729 ± 0.0001	3.9%
	182	75											0.1785 ± 0.0001	2.5%
	182	75											0.1794 ± 0.0001	3.3%
	182	75											0.1914 ± 0.0001	7.3%
	182	75											0.1983 ± 0.0001	4.4%
	182	75											0.2143 ± 0.0001	1.2%
	182	75											0.2175 ± 0.0001	3.5%
	182	75											0.2216 ± 0.0001	7.0%
	182	75											0.2221 ± 0.0001	9.2%
	182	75											0.2262 ± 0.0001	3.3%
	182	75											0.2293 ± 0.0001	30%
	182	75											0.2475 ± 0.0001	5.5%
	182	75											0.2564 ± 0.0001	10%
	182	75											0.2641 ± 0.0001	3.9%
	182	75											0.2763 ± 0.0001	9.5%
	182	75											0.2814 ± 0.0001	6.2%
	182	75											0.2866 ± 0.0001	7.6%
	182	75											0.3391 ± 0.0001	6.0%
	182	75											0.3511 ± 0.0001	11%
	182	75											1.0017 ± 0.0001	2.7%
	182	75											1.0762 ± 0.0002	11%
	182	75											1.1133 ± 0.0001	5.1%
	182	75											1.1213 ± 0.0001	24%
	182	75											1.1890 ± 0.0001	9.8%

TABLE OF THE ISOTOPES (Continued)

Isotope	A	Z	% Natural abundance	Atomic mass	Half-life	Decay mode	Decay energy (MeV)	Particle energy (MeV)	Particle intensity	Thermal neutron cross section	Spin (h/2π)	μ Nucl. mag. moment	Gamma-ray energy (MeV)	Gamma-ray intensity
	182	75											1.2214 ± 0.0001	19%
	182	75											1.2310 ± 0.0001	16%
	182	75											1.3427 ± 0.0002	2.8%
	182	75											1.4273 ± 0.0002	11%
$_{75}Re^{183}$	183	75		182.950817	70 d.	E.C.	0.556				(5/2+)		W k x-ray	62%
	183	75											0.09908 ± 0.00001	2.7%
	183	75											0.10793 ± 0.00001	2.2%
	183	75											0.10972 ± 0.00001	2.9%
	183	75											0.16232 ± 0.00001	23%
	183	75											0.20880 ± 0.00001	3%
	183	75											0.2461 ± 0.0001	1.3%
	183	75											0.29172 ± 0.00001	3.2%
$_{75}Re^{184m}$	184	75			165 d.	I.T.(75%)	0.188				8+		Re k x-ray	24%
	184	75				E.C.(25%)							0.1047 ± 0.0001	13%
	184	75											0.16127 ± 0.00001	6.6%
	184	75											0.2165 ± 0.0001	9.6%
	184	75											0.31800 ± 0.00001	5.9%
	184	75											0.38425 ± 0.00001	3.2%
	184	75											0.53667 ± 0.00001	3.4%
	184	75											0.92093 ± 0.00002	8.3%
	184	75											(0.10 - 1.1)weak	
$_{75}Re^{184}$	184	75		183.952530	38 d.	E.C.	1.492			9 x 10^3 b.	3-	2.499	W k x-ray	45%
	184	75											0.1112 ± 0.0001	17%
	184	75											0.25284 ± 0.00001	3.0%
	184	75											0.6419 ± 0.0001	1.9%
	184	75											0.79207 ± 0.00002	37%
	184	75											0.8948 ± 0.0001	16%
	184	75											0.90328 ± 0.00003	38%
	184	75											(0.1 - 1.4)weak	
$_{75}Re^{185}$	185	75	37.40%	184.952951						111 b.	5/2+			
$_{75}Re^{186m}$	186	75			2.0 x 10^5 y.	I.T>	0.150						Re k x-ray	
	186	75											0.0590 ± 0.0001	18.6%
	186	75											0.0993 ± 0.0001	1.1%
$_{75}Re^{186}$	186	75		185.954984	β-(92%)	1.074	0.973	21%			1-	+1.739	W k x-ray	3%
	186	75				E.C.(8%)	0.585	1.07	71%				0.1227 ± 0.0001	0.7%
	186	75											0.1372 ± 0.0001	9%
	186	75											0.7675 ± 0.0001	0.03%
$_{75}Re^{187}$	187	75	62.6%	186.955744	4.5 x 10^{10}y.	β-	0.0025	0.0025		(2.8 + 75) b.	5/2+	+3.2197		
$_{75}Re^{188m}$	188	75			18.6 m.	I.T.	0.172				(6-)		Re k x-ray	31%
	188	75											0.0925 ± 0.0001	5.1%
	188	75											0.1059 ± 0.0001	11%
	188	75											0.1560 ± 0.0001	0.6%
	186	75											0.1695 ± 0.0001	0.1%
$_{75}Re^{188}$	188	75		187.958106	16.98 h.	β-	2.210	1.962	20%		1-	+1.788	Os k x-ray	2%
	188	75						2.118	79%				0.15502 ± 0.00002	15%
	188	75											0.47798 ± 0.00005	1.0%
	188	75											0.63312 ± 0.00004	1.2%
	188	75											0.82952 ± 0.00005	0.4%
	188	75											0.93141 ± 0.00006	0.6%
$_{75}Re^{189}$	189	75		188.959219	24 h.	β-	1.01	1.01			(5/2+)		0.1471 ± 0.0001	1.4%
	189	75											0.1854 ± 0.0001	2.1%
	189	75											0.2167 ± 0.0001	6.0%
	189	75											0.2194 ± 0.0001	5.0%
	189	75											0.2451 ± 0.0001	3.7%
	189	75											0.2759 ± 0.0001	0.34%
	189	75											0.5634 ± 0.0001	0.6%
$_{75}Re^{190m}$	190	75			3.0 h.	β-(51%)					(6-)		Re k x-ray	14%
	190	75				I.T.(49%)							0.1191 ± 0.0001	11%
	190	75											0.2238 ± 0.0001	14%
	190	75											0.2238 ± 0.0001	14%
	190	75											0.2829 ± 0.0001	2.4%
	190	75											0.2948 ± 0.0001	3.0%
	190	75											0.3902 ± 0.0001	5.3%
	190	75											0.4316 ± 0.0001	8.85%
	190	75											0.4908 ± 0.0001	3.8%
	190	75											0.5026 ± 0.0001	3.7%
	190	75											0.5186 ± 0.0001	7.3%

Isotope	A	Z	% Natural abundance	Atomic mass	Half-life	Decay mode	Decay energy (MeV)	Particle energy (MeV)	Particle intensity	Thermal neutron cross section	Spin (h/2π)	μ Nucl. mag. moment	Gamma-ray energy (MeV)	Gamma-ray intensity
	190	75											0.5587 ± 0.0005	5.9%
	190	75											0.6309 ± 0.0002	9.2%
	190	75											0.6731 ± 0.0001	10%
	190	75											0.7686 ± 0.0001	3.7%
	190	75											0.9582 ± 0.0001	2.2%
	190	75											(0.1 - 1.79)weak	
$_{75}Re^{190}$	190	75		189.961850	3.0 m.	β-	3.2	1.8			(2-)		Os k x-ray	6.8%
	190	75											0.1867 ± 0.0001	49%
	190	75											0.2238 ± 0.0001	25%
	190	75											0.3611 ± 0.0001	15%
	190	75											0.3712 ± 0.0001	21%
	190	75											0.3974 ± 0.0001	8.2%
	190	75											0.4072 ± 0.0001	15%
	190	75											0.4316 ± 0.0001	18%
	190	75											0.5580 ± 0.0001	29%
	190	75											0.5693 ± 0.0001	26%
	190	75											0.6051 ± 0.0001	66%
	190	75											0.6309 ± 0.0002	19%
	190	75											0.7686 ± 0.0001	2.9%
	190	75											0.8290 ± 0.0001	23%
	190	75											0.8391 ± 0.0001	7.8%
	190	75											1.2002 ± 0.0001	3.1%
	190	75											1.3870 ± 0.0002	1.3%
	190	75											1.4375 ± 0.0003	0.7%
	190	75											1.7945 ± 0.0003	0.5%
$_{75}Re^{191}$	191	75		190.963112	≈ 9.8 m.	β-	2.042	1.8						
$_{75}Re^{192}$	192	75		191.965870	16 s.	β-	4.10	≈ 2.5					0.2058 ± 0.0001	
	192	75											0.2832	
	192	75											0.4673	
	192	75											0.4890	
	192	75											0.7505	
Os		76								15 b.				
$_{76}Os^{166}$	166	76		165.972470	0.18 s.	β+, EC (28%)		6.27			0+		ann. rad.	
	166	76				α(72%)		5.98						
$_{76}Os^{167}$	167	76		166.971290	0.7 s.	β+, EC (76%)	8.11						ann.rad.	
	167	76				α(24%)		5.84						
$_{76}Os^{168}$	168	76		167.967670	2.2 s.	β+, EC (51%)	7.57				0+		ann. rad.	
	168	76				α(49%)								
$_{76}Os^{169}$	169	76		168.966850	3.3 s.	β+, EC (89%)	5.16						ann.rad.	
	169	76				α(17%)		5.57						
$_{76}Os^{170}$	170	76		169.963571	7.1 ± 0.2 s.	β+,E.C.	6.78				0+		ann.rad.	
	170	76				α		5.40						
$_{76}Os^{171}$	171	76		170.962890	7.9 ± 0.6 s.	β+, EC (98%)	6.78						ann.rad.	
	171	76				α(2%)		5.24						
$_{76}Os^{172}$	172	76		171.960000	19 ± 2 s.	β+, EC (99%)	4.43				0+		ann.rad.	
	172	76				α(1%)		5.10					0.177	100 +
	172	76											0.187	50
	172	76											0.276	25
	172	76											0.285	30
$_{76}Os^{173}$	173	76		172.957120	≈ 16 ± 5 s.	β+,E.C.	6.01						ann.rad.	
	173	76				α(0.02%)		4.94					ann.rad.	
$_{76}Os^{174}$	174	76		173.957120	44 ± 4 s.	β+,E.C.	3.67				0+		0.118 ± 0.001	100 +
	174	76				α(0.02%)		4.76					0.138 ± 0.001	25
	174	76											0.158 ± 0.001	
	174	76											0.302	26
	174	76											0.325	43
	174	76											0.372	20
	174	76											0.387	10
$_{76}Os^{175}$	175	76		174.956980	1.4 ± 0.1 m.	β+,,E.C.	5.17						0.125	100 +
	175	76											0.170	6
	175	76											0.181	11
	175	76											0.226	4
	175	76											0.248	9
	175	76											0.3.8	3
	175	76											0.410	5
$_{76}Os^{176}$	176	76		175.954880	3.6 ± 0.5 m.	β+,E.C.	3.15				0+		0.8155 ± 0.001	36 +
	176	76											0.7758 ± 0.0001	98
	176	76											0.8573 ± 0.00001	69
	176	76											1.2093 ± 0.0001	71
	176	76											1.2909 ± 0.0001	100
$_{76}Os^{177}$	177	76		176.954980	2.8 ± 0.3 m.	β+,E.C.	4.30				(1/2-)		0.0848 ± 0.0002	100 +
	177	76											0.1572 ± 0.0002	20
	177	76											0.1958 ± 0.0002	61
	177	76											0.3002 ± 0.0002	29
	177	76											0.4110 ± 0.0002	17

Isotope	A	Z	% Natural abundance	Atomic mass	Half-life	Decay mode	Decay energy (MeV)	Particle energy (MeV)	Particle intensity	Thermal neutron cross section	Spin (h/2π)	μ Nucl. mag. moment	Gamma-ray energy (MeV)	Gamma-ray intensity
	177	76											0.4570 ± 0.0002	21
	177	76											0.5394 ± 0.0002	22
	177	76											0.5762 ± 0.0003	13
	177	76											0.64492 ± 0.0004	10
	177	76											0.6861 ± 0.0001	12
	177	76											0.7333 ± 0.0003	26
	177	76											0.7914 ± 0.0004	13
	177	76											0.9524 ± 0.0005	18
	177	76											1.2686 ± 0.0006	33
	177	76											1.3368 ± 0.0006	11
	177	76											1.7439 ± 0.0006	12
$_{76}$Os^{178}	178	76		177.953250	5.0 ± 0.4 m.	β+,E.C.	2.24				0+		ann.rad.	
	178	76											0.3200 ± 0.001	4 +
	178	76											0.3508 ± 0.001	24
	178	76											0.5331 ± 0.001	52
	178	76											0.5518 ± 0.001	29
	178	76											0.5946 ± 0.0008	72
	178	76											0.6006 ± 0.001	41
	178	76											0.613 ± 0.001	22
	178	76											0.6325 ± 0.001	40
	178	76											0.6850 ± 0.0012	65
	178	76											0.9687 ± 0.0008	100
	178	76											1.3311 ± 0.0012	94
$_{76}$Os^{179}	179	76		178.953830	7 m.	β+,E.C.	3.61						ann.rad.	
	179	76											0.0654 ± 0.0001	100 +
	179	76											0.1657 ± 0.0002	7
	179	76											0.2186 ± 0.0002	17
	179	76											0.5328 ± 0.00041	10
	179	76											0.5938 ± 0.0003	16
	179	76											0.6334 ± 0.0005	5
	179	76											0.68947 ± 0.0005	11
	179	76											0.6975 ± 0.0005	5
	179	76											0.7453 ± 0.0005	6
	179	76											0.7508 ± 0.0005	5
	179	76											0.7594 ± 0.0003	8
	179	76											0.8177 ± 0.0003	6
	179	76											0.9684 ± 0.0003	14
	179	76											1.3110 ± 0.0004	10
	179	76											1.3303 ± 0.0004	13
	179	76											1.3642 ± 0.0005	3
	179	76											1.3835 ± 0.0005	4
	179	76											1.4295 ± 0.0005	3
	179	76											1.4488 ± 0.0005	3
$_{76}$Os^{180}	180	76		179.952390	21.7 ± 0.6 m.	β+,E.C.	1.510				0+		Re k x-ray	42%
	180	76											0.0202 ± 0.0001	17 +
	180	76											0.0316 ± 0.0002	
	180	76											0.0482 ± 0.0002	
	180	76											0.0499 ± 0.0002	
	180	76											0.0544 ± 0.0002	
	180	76											0.0746 ± 0.0002	
	180	76											0.1040 ± 0.0002	
	180	76											0.1070 ± 0.0002	
	180	76											0.1137 ± 0.0002	
	180	76											0.1838 ± 0.0002	
	180	76											0.2001 ± 0.0002	
	180	76											0.2182 ± 0.0002	
	180	76											0.2500 ± 0.0002	
	180	76											0.3194 ± 0.0002	
	180	76											0.3290 ± 0.0002	
	180	76											0.3491 ± 0.0002	
	180	76											0.4013 ± 0.0002	
	180	76											0.4857 ± 0.0002	
	180	76											0.6670 ± 0.0002	
	180	76											0.7174 ± 0.0002	
$_{76}$Os^{181m}	181	76			2.7 ± 0.1 m.	β+,E.C.	1.8				(7/2-)		0.11794 ± 0.00004	28 +
	181	76											0.14493 ± 0.00006	100
	181	76											1.1187 ± 0.0001	4.2
	181	76											1.4679 ± 0.001	1.3
$_{76}$Os^{181}	181	76		180.953270	1.75 ± 0.5 h.	E.C.	3.03				(1/2-)		ann.rad.	
	181	76											0.11794 ± 0.00004	55%
	181	76											0.16712 ± 0.00005	3.0%
	181	76											0.23868 ± 0.00001	44%
	181	76											0.24277 ± 0.00006	6.1%
	181	76											0.75120 ± 0.00002	3.2%
	181	76											0.7512 ± 0.0002	3.2%
	181	76											0.7590 ± 0.0002	2.4%

TABLE OF THE ISOTOPES (Continued)

Isotope	A	Z	% Natural abundance	Atomic mass	Half-life	Decay mode	Decay energy (MeV)	Particle energy (MeV)	Particle intensity	Thermal neutron cross section	Spin (h/2π)	μ Nucl. mag. moment	Gamma-ray energy (MeV)	Gamma-ray intensity
	181	76											0.7875 ± 0.0004	5.3%
	181	76											0.8267 ± 0.0002	20%
	181	76											0.9549 ± 0.0005	5.1%
	181	76											1.0603 ± 0.0002	5.7%
	181	76											1.1107 ± 0.0005	2.1%
	181	76											1.1814 ± 0.0008	1.0%
	181	76											1.3051 ± 0.0003	1.8%
	181	76											1.3465 ± 0.0005	1.1%
	181	76											1.3852 ± 0.0009	1.2%
	181	76											1.4923 ± 0.0004	1.0%
	181	76											1.5679 ± 0.0004	1.0%
	181	76											1.5726 ± 0.0003	1.1%
	181	76											1.7053 ± 0.0003	1.4%
	181	76											1.7397 ± 0.0003	1.2%
	181	76											1.9816 ± 0.0002	1.2%
	181	76											(0.07 - 2.64)many	
$_{76}Os^{182}$	182	76		181.952120	21.5 h.	E.C.	0.850				0+		Re k x-ray	43%
	182	76											0.1308 ± 0.0001	3.5%
	182	76											0.1802 ± 0.0001	37%
	182	76											0.2633 ± 0.0001	7.0%
	182	76											0.2743 ± 0.0001	1.8%
	182	76											0.5100 ± 0.0001	55%
$_{76}Os^{183m}$	183	76			9.9 h.	E.C.(84%)					1/2-		Os k x-ray	2.5%
	183	76				I.T.(16%)							Re k x-ray	34%
	183	76											0.4845 ± 0.0001	1.6%
	183	76											0.8784 ± 0.0005	1.6%
	183	76											0.9548 ± 0.0003	1.1%
	183	76											1.0347 ± 0.0003	6.5%
	183	76											1.1020 ± 0.0003	50%
	183	76											1.1080 ± 0.0003	23%
$_{76}Os^{183}$	183	76		182.953290	13 h.	E.C.	2.30				9/2+		Re k x-ray	72%
	183	76											0.1144 ± 0.0001	21%
	173	76											0.1679 ± 0.0001	7.7%
	183	76											0.2363 ± 0.0001	2.2%
	183	76											0.3818 ± 0.0001	77%
	183	76											0.8510 ± 0.0002	3.9%
	183	76											0.8876 ± 0.0003	1.1%
	183	76											0.8875 ± 0.0003	1.1%
	183	76											1.1633 ± 0.0004	1.2%
	183	76											1.4389 ± 0.0006	0.5%
$_{76}Os^{184}$	184	76	0.02%	183.952488						30 x 10^2 b.	0+			
$_{76}Os^{185}$	185	76		184.954041	93.6 d.	E.C.	1.015				1/2-		Re k x-ray	35%
	185	76											0.5921 ± 0.0001	1.3%
	185	76											0.6461 ± 0.0001	81%
	185	76											0.7174 ± 0.0001	4.1%
	185	76											0.8748 ± 0.0001	6.6%
	185	76											0.8805 ± 0.0001	5.0%
$_{76}Os^{186}$	186	76	1.58%	185.953830	2 x 10^{15} y.	α		≈ 2.75		80 b.	0+			
$_{76}Os^{187}$	187	76	1.6%	186.955741						320 b.	1/2-	+0.0646		
$_{76}Os^{188}$	188	76	13.3%	187.955860						5 b.	0+			
$_{76}Os^{189m}$	189	76			5.8 h.	I.T.	0.0308				9/2-		Os L x-ray	13%
	189	76											0.0308 ± 0.0001	0.0003%
$_{76}Os^{189}$	189	76	16.1%	188.958137						(0.0002 + 20) b.	3/2+	0.6599		
$_{76}Os^{190m}$	190	76			9.9 m.	I.T.	1.705				10-		Os k x-ray	9.8%
	190	76											0.1867 ± 0.0001	70%
	190	76											0.3611 ± 0.0001	95%
	190	76											0.5026 ± 0.0001	98%
	190	76											0.6161 ± 0.0002	98.5%
$_{76}Os^{190}$	190	76	26.4%	189.958436						(9 + 4) b.	0+			
$_{76}Os^{191m}$	191	76			13.1 h.	I.T.	0.0744				3/2-		Os k x-ray	4.1%
	191	76											0.0744 ± 0.0001	0.07%
$_{76}Os^{191}$	191	76		190.960920	15.4 d.	β-	0.313	0.140	100%		9/2-		Ir k x-ray	28%(D)
	191	76											0.1294 ± 0.0001	26%(D)
$_{76}Os^{192m}$	192	76			6.1 s.	I.T.	2.0154				(10-)		Os k x-ray	22%
	192	76											0.2058	69%
	192	76											0.2832	6.5%
	192	76											0.2924	4.2%
	192	76											0.3024	54%
	192	76											0.3068	5.8%
	192	76											0.3745	24%
	192	76											0.4204	6.1%
	192	76											0.4522	4.4%
	192	76											0.4531	59%
	192	76											0.4845	5.9%
	192	76											0.4890	15%
	192	76											0.5083	12%
	192	76											0.5632	12%
	192	76											0.5692	72%
	192	76											0.5883	4.1%
	192	76											0.6057	11%
	192	76											0.6195	11%
	192	76											0.6240	1.7%
$_{76}Os^{192}$	192	76	41.0%	191.961467						2.0 b.	0+			
$_{76}Os^{193}$	193	76		192.964138	β-	1.14	1.04	20%			3/2-	1.30	Ir k x-ray	69.5%
	193	76											0.1389 ± 0.0001	4.3%

Isotope	A	Z	% Natural abundance	Atomic mass	Half-life	Decay mode	Decay energy (MeV)	Particle energy (MeV)	Particle intensity	Thermal neutron cross section	Spin (h/2π)	μ Nucl. mag. moment	Gamma-ray energy (MeV)	Gamma-ray intensity
	193	76											0.2804 ± 0.0001	1.2%
	193	76											0.3216 ± 0.0001	1.3%
	193	76											0.3875 ± 0.0001	1.3%
	193	76											0.4605 ± 0.0001	3.9%
	193	76											0.5574 ± 0.0001	1%
$_{76}Os^{194}$	194	76		193.965173	6.0 y.	β-	0.097	0.054	33%		0+		Ir L x-ray	11%
	194	76						0.096	67%				0.0429 ± 0.0002	5.4%
$_{76}Os^{195}$	195	76		194.968110	6.5 m.	β-	≈ 2.0	≈ 2.0						
$_{76}Os^{196}$	196	76		195.969620	34.9 m.	β-	≈ 0.84	0.84			0+	0.1262	5%	
	196	76											0.2071 ± 0.0002	2.4%
	196	76											0.2578 ± 0.0002	2.3%
	196	76											0.3154 ± 0.0002	2.5%
	196	76											0.4079 ± 0.0002	5.9%
	196	76											0.6291 ± 0.0004	1.6%
$_{77}Ir^{170}$	170	77		169.974970	1.05 ± 0.1 s.	α		6.03						
$_{77}Ir^{171}$	171	77		170.971700	1.6 ± 0.1 s.	α		5.91						
$_{77}Ir^{172}$	172	77		171.970550	2.1 ± 0.1 s.	α		5.811						
$_{77}Ir^{173}$	173	77		172.967560	3.0 ± 0.1 s.	α		5.665						
$_{77}Ir^{174}$	174	77		173.966660	4 ± 1 s.	α		5.478						
$_{77}Ir^{175}$	175	77		174.964150	4.5 ± 1.0 s.	α		5.393						
$_{77}Ir^{176}$	176	77		175.963480	8 ± 1 s.	α		5.118						
$_{77}Ir^{177}$	177	77		176.961350	21 ± 2 s.	α		5.011						
$_{77}Ir^{178}$	178	77		177.961050	12 ± 2 s.	β+,E.C.	7.26						0.1320 ± 0.0005	71 +
	178	77											0.2667 ± 0.0003	100
	178	77											0.2700 ± 0.0001	5
	178	77											0.3633 ± 0.0004	35
	178	77											0.3987 ± 0.0004	13
	178	77											0.4329 ± 0.0005	5
	178	77											0.5329 ± 0.0005	5
	178	77											0.5469 ± 0.0005	6
	178	77											0.6250 ± 0.0005	15
	178	77											0.6395 ± 0.0004	13
	178	77											0.7002 ± 0.0004	8
	178	77											0.7329 ± 0.0004	5
	178	77											0.8649 ± 0.0004	9
	178	77											0.9000 ± 0.0004	13
	178	77											1.0176 ± 0.0005	5
	178	77											1.2015 ± 0.0004	7
$_{77}Ir^{179}$	179	77		178.959190	4 m.	E.C.	4.99							
$_{77}Ir^{180}$	180	77		179.959260	1.5 ± 0.1 m.	E.C.	6.40						0.1321 ± 0.0003	40%
	180	77											0.2765 ± 0.0003	42%
	180	77											0.4928 ± 0.0003	2.9%
	180	77											0.6141 ± 0.0004	2.1%
	180	77											0.6445 ± 0.0005	7.0%
	180	77											0.6990 ± 0.0005	9.4%
	180	77											0.7883 ± 0.0004	3.8%
	180	77											0.8463 ± 0.0005	2.4%
	180	77											0.8703 ± 0.0005	8.6%
	180	77											0.8905 ± 0.0004	9.1%
	180	77											0.9689 ± 0.0005	3.1%
	180	77											1.0143 ± 0.0005	1.1%
	180	77											1.0648 ± 0.0004	6.0%
	180	77											1.3306 ± 0.0005	4.4%
$_{77}Ir^{181}$	181	77		180.957640	4.9 ± 0.1 m.	β+,E.C.	4.07				(7/2+)		ann.rad.	
	181	77											0.0196 ± 0.0002	6 +
	181	77											0.0653 ± 0.0002	20
	181	77											0.0938 ± 0.0002	29
	181	77											0.1025 ± 0.0002	25
	181	77											0.1076 ± 0.0002	100
	181	77											0.1235 ± 0.0002	28
	181	77											0.1846 ± 0.0002	28
	181	77											0.2185 ± 0.0005	14
	181	77											0.2270 ± 0.0002	58
	181	77											0.2316 ± 0.0002	30
	181	77											0.3090 ± 0.0002	14
	181	77											0.3189 ± 0.0002	46
	181	77											0.3505 ± 0.0002	7
	181	77											0.3752 ± 0.0002	16
	181	77											0.5755 ± 0.0002	9
	181	77											0.7001 ± 0.0002	9
	181	77											1.1823 ± 0.0003	9
	181	77											1.1926 ± 0.0003	11
	181	77											1.3471 ± 0.0003	13
	181	77											1.3810 ± 0.0003	13
	181	77											1.5288 ± 0.0003	29
	181	77											1.5450 ± 0.0003	6
	181	77											1.5656 ± 0.0003	13
	181	77											1.5934 ± 0.0003	9
	181	77											1.6396 ± 0.0003	52
	181	77											1.6464 ± 0.0003	27
	181	77											1.6525 ± 0.0003	17
	181	77											1.7149 ± 0.0003	6
$_{77}Ir^{182}$	182	77		181.957970	15 m.	β+(44%)	5.45						ann.rad.	

Isotope	A	Z	% Natural abundance	Atomic mass	Half-life	Decay mode	Decay energy (MeV)	Particle energy (MeV)	Particle intensity	Thermal neutron cross section	Spin (h/2π)	μ Nucl. mag. moment	Gamma-ray energy (MeV)	Gamma-ray intensity
	182	77				E.C.(56%)							Os k x-ray	33%
	182	77											0.1273 ± 0.0003	35%
	182	77											0.2363 ± 0.0003	9.1%
	182	77											0.2370 ± 0.0002	43%
	182	77											0.3931 ± 0.0002	3%
	182	77											0.4000 ± 0.0003	3.1%
	182	77											0.7643 ± 0.0002	5.6%
	182	77											0.7901 ± 0.0003	3.2%
	182	77											0.8909 ± 0.0002	5.7%
	182	77											0.9123 ± 0.0002	8.8%
	182	77											1.0633 ± 0.0003	2.2%
	182	77											1.1180 ± 0.0006	2.6%
	182	77											1.2516 ± 0.0005	1.9%
	182	77											1.6520 ± 0.0006	2.5%
$_{77}Ir^{183}$	183	77		182.956710	56 m.	β+,E.C.	3.19						ann.rad.	
	183	77											0.0877 ± 0.0002	63 +
	183	77											0.1022 ± 0.0002	18
	183	77											0.1368 ± 0.0002	17
	183	77											0.1657 ± 0.0002	14
	183	77											0.1945 ± 0.0002	23
	183	77											0.2285 ± 0.0002	100
	183	77											0.2367 ± 0.0002	29
	183	77											0.2397 ± 0.0002	26
	183	77											0.2506 ± 0.0002	10
	183	77											0.2544 ± 0.0002	23
	183	77											0.2824 ± 0.0002	70
	183	77											0.3144 ± 0.0002	11
	183	77											0.3422 ± 0.0002	29
	183	77											0.3477 ± 0.0002	29
	183	77											0.4122 ± 0.0002	19
	183	77											0.4577 ± 0.0002	6
	183	77											0.4619 ± 0.0002	5
	183	77											0.4984 ± 0.0002	15
	183	77											0.6174 ± 0.0002	8
	183	77											0.6551 ± 0.0002	20
	183	77											0.6708 ± 0.0002	10
	183	77											0.6922 ± 0.0002	33
	183	77											0.7061 ± 0.0002	5
	183	77											0.7248 ± 0.0002	5
	183	77											0.8001 ± 0.0002	33
	183	77											0.8966 ± 0.0002	18
$_{77}Ir^{184}$	184	77		183.957560	3.0 h.	β+(12%)	4.72	2.3			5		ann.rad.	24%
	184	77				E.C.(88%)		2.9					Os k x-ray	48%
	184	77						3.3					0.11968 ± 0.0001	30%
	184	77											0.2640 ± 0.0001	67%
	184	77											0.3904 ± 0.0001	26%
	184	77											0.4931 ± 0.0001	5.8%
	184	77											0.5029 ± 0.0002	2.9%
	184	77											0.5397 ± 0.0001	6.7%
	184	77											0.6012 ± 0.0001	3.2%
	184	77											0.6266 ± 0.0001	2.4%
	184	77											0.8239 ± 0.0001	3.8%
	184	77											0.8413 ± 0.0002	7.9%
	184	77											0.9429 ± 0.0002	3.6%
	184	77											0.9441 ± 0.0002	2.7%
	184	77											0.9613 ± 0.0002	12%
	184	77											1.0445 ± 0.0002	5.3%
	184	77											1.0622 ± 0.0002	2.9%
	184	77											1.1053 ± 0.0002	5.3%
	184	77											1.0622 ± 0.0003	2.9%
	184	77											1.1053 ± 0.0002	5.3%
	184	77											1.2369 ± 0.0001	2.1%
	184	77											1.2478 ± 0.0001	2.6%
	184	77											1.3343 ± 0.0003	2.3%
	184	77											1.4579 ± 0.0002	1.4%
	184	77											1.6725 ± 0.0003	3.7%
	184	77											2.0630 ± 0.0004	4.5%
	184	77											2.2430 ± 0.0006	0.9%
$_{77}Ir^{185}$	185	77		184.956730	14 h.	β+(3%)					(5/2-)		ann.rad.	6%
	185	77				E.C.(97%)							Os k x-ray	55%
	185	77											0.0974 ± 0.0002	4.1%
	185	77											0.1007 ± 0.0002	2.4%
	185	77											0.1536 ± 0.0002	2.0%
	185	77											0.1582 ± 0.0002	2.4%
	185	77											0.2238 ± 0.0002	2.1%
	185	77											0.2543 ± 0.0002	13%
	185	77											0.5392 ± 0.0002	1.3%
	185	77											0.6462 ± 0.0002	1.2%
	185	77											1.6418 ± 0.0005	1.1%
	185	77											1.6683 ± 0.0005	3.6%
	185	77											1.7322 ± 0.0005	2.7%
	185	77											1.7384 ± 0.0005	2.4%
	185	77											1.8288 ± 0.0005	9.8%
	185	77											1.8700 ± 0.0005	1.2%
$_{77}Ir^{186m}$	186	77			1.7 h.	E.C.					(2-)		Os k x-ray	45%
	186	77											0.1371 ± 0.0001	30%

Isotope	A	Z	% Natural abundance	Atomic mass	Half-life	Decay mode	Decay energy (MeV)	Particle energy (MeV)	Particle intensity	Thermal neutron cross section	Spin (h/2π)	μ Nucl. mag. moment	Gamma-ray energy (MeV)	Gamma-ray intensity
	186	77											0.2969 ± 0.0001	11%
	186	77											0.6303 ± 0.0001	21%
	186	77											0.6363 ± 0.0001	2.1%
	186	77											0.7126 ± 0.0001	4.4%
	186	77											0.7675 ± 0.0001	24%
	186	77											0.7732 ± 0.0001	16%
	186	77											0.7832 ± 0.0001	2.3%
	186	77											0.8441 ± 0.0001	2.7%
	186	77											0.9334 ± 0.0001	2.3%
	186	77											0.9380 ± 0.0001	2.6%
	186	77											0.9870 ± 0.0001	12.6%
	186	77											1.0463 ± 0.0001	1.2%
	186	77											1.6172 ± 0.0002	4.8%
	186	77											1.7111 ± 0.0002	2.3%
	186	77											1.7544 ± 0.0003	5.3%
	186	77											2.1870 ± 0.0002	3.4%
	186	77											2.2241 ± 0.0004	1.2%
$_{77}Ir^{186}$	186	77		185.957943	15.7 h.	E.C.(98%)	3.83				(5+)		Os k x-ray	48%
	186	77				β+(2%)							0.1372 ± 0.0001	41%
	186	77											0.2968 ± 0.0001	62%
	186	77											0.3577 ± 0.0001	1.9%
	186	77											0.4348 ± 0.0001	34%
	186	77											0.5844 ± 0.0001	5.4%
	186	77											0.6303 ± 0.0001	4.9%
	186	77											0.6362 ± 0.0001	6.9%
	186	77											0.7673 ± 0.0001	5.3%
	186	77											0.7731 ± 0.0001	8.8%
	186	77											0.8413 ± 0.0001	5.1%
	186	77											0.8466 ± 0.0002	6.3%
	186	77											0.9332 ± 0.0001	5.3%
	186	77											0.9436 ± 0.0001	8.6%
	186	77											1.0571 ± 0.0001	3.1%
	186	77											1.1879 ± 0.0001	2.0%
	186	77											1.3144 ± 0.0001	2.0%
	186	77											1.6474 ± 0.0001	4.7%
	186	77											1.7010 ± 0.0001	2.1%
	186	77											2.2420 ± 0.0002	1.4%
	186	77											2.8352 ± 0.0003	0.8%
	186	77											(0.13 - 3.0)weak	
$_{77}Ir^{187}$	187	77		186.957350	10.5 h.	E.C.	1.500				3/2+		Os k x-ray	65%
	187	77											0.0743 ± 0.0001	4.6%
	187	77											0.1777 ± 0.0001	2.8%
	187	77											0.1874 ± 0.0001	1.8%
	187	77											0.4009 ± 0.0003	4.1%
	187	77											0.4271 ± 0.0002	4.4%
	187	77											0.4917 ± 0.0001	1.4%
	187	77											0.5015 ± 0.0001	1.6%
	187	77											0.6109 ± 0.0001	4.0%
	187	77											0.7997 ± 0.0002	0.9%
	187	77											0.9128 ± 0.0001	5.0%
	187	77											0.9774 ± 0.0002	3.1%
	187	77											0.9873 ± 0.0002	2.8%
$_{77}Ir^{188}$	188	77		187.958830	41.4 h.	β+	2.79		1.13		(2-)		Os k x-ray	44%
	188	77				E.C. (99+%)		1.64					0.1550 ± 0.0001	30%
	188	77											0.4780 ± 0.0001	15%
	188	77											0.6330 ± 0.0001	18%
	188	77											0.6349 ± 0.0002	5.0%
	188	77											0.8294 ± 0.0001	5.2%
	188	77											1.2098 ± 0.0001	7.0%
	188	77											1.4354 ± 0.0002	1.5%
	188	77											1.4572 ± 0.0002	1.8%
	188	77											1.4652 ± 0.0002	1.3%
	188	77											1.5745 ± 0.0002	2.6%
	188	77											1.7157 ± 0.0002	6.2%
	188	77											1.9441 ± 0.0002	3.9%
	188	77											2.0498 ± 0.0002	5.0%
	188	77											2.0596 ± 0.0004	7.0%
	188	77											2.0969 ± 0.0004	5.7%
	188	77											2.0991 ± 0.0004	4.8%
	188	77											2.1937 ± 0.0004	2.4%
	188	77											2.2146 ± 0.0002	19%
$_{77}Ir^{189}$	189	77		188.958712	13.2 d.	E.C.	0.535				3/2+		Os k x-ray	38%
	189	77											0.0952 ± 0.0001	0.4%
	189	77											0.1859 ± 0.0001	0.2%
	189	77											0.1974 ± 0.0001	0.3%
	189	77											0.2167 ± 0.0001	0.5%
	189	77											0.2194 ± 0.0001	0.5%
	189	77											0.2335 ± 0.0001	0.3%
	189	77											0.2449 ± 0.0001	6.0%
	189	77											0.2758 ± 0.0001	0.5%
$_{77}Ir^{190m2}$	190	77			3.2 h.	β+, EC (95%)					(11-)			
	190	77				I.T.(5%)								
$_{77}Ir^{190m1}$	190	77			1.2 h.	I.T.	0.0263						Ir L x-ray	14%

Isotope	A	Z	% Natural abundance	Atomic mass	Half-life	Decay mode	Decay energy (MeV)	Particle energy (MeV)	Particle intensity	Thermal neutron cross section	Spin (h/2π)	μ Nucl. mag. moment	Gamma-ray energy (MeV)	Gamma-ray intensity
$_{77}Ir^{190}$	190	77		11.8 d.	E.C.	2.0					(4+)		Os k x-ray	45%
	190	77											0.1867 ± 0.0001	48%
	190	77											0.1969 ± 0.0002	2.5%
	190	77											0.2338 ± 0.0001	3.6%
	190	77											0.2948 ± 0.0001	6.2%
	190	77											0.3611 ± 0.0001	13%
	190	77											0.3712 ± 0.0001	22%
	190	77											0.3800 ± 0.0001	2.0%
	190	77											0.3974 ± 0.0001	6.3%
	190	77											0.4072 ± 0.0001	27%
	190	77											0.4206 ± 0.0001	1.6%
	190	77											0.4316 ± 0.0001	2.6%
	190	77											0.4478 ± 0.0001	2.5%
	190	77											0.5186 ± 0.0001	33%
	190	77											0.5580 ± 0.0001	29%
	190	77											0.6051 ± 0.0001	38%
	190	77											0.6309 ± 0.0001	2.8%
	190	77											0.7262 ± 0.0001	3.6%
	190	77											0.7686 ± 0.0001	2.1%
	190	77											0.8290 ± 0.0001	3.3%
	190	77											1.0360 ± 0.0002	2.3%
	190	77											(0.2 - 1.4)weak	
$_{77}Ir^{191m}$	191	77		4.93 s.	I.T.	0.1714					11/2-		Ir k x-ray	28%
	191	77											0.0824 ± 0.0001	0.02%
	191	77											0.1294 ± 0.0001	25.7%
$_{77}Ir^{191}$	191	77	37.3%	190.960584						(0.2 + 310) b.	3/2+	+0.1461		
$_{77}Ir^{192m2}$	192	77			241 y.	I.T.	0.161				(9+)		Ir k x-ray	
$_{77}Ir^{192m1}$	192	77			1.44 m.	I.T.	0.0580				(1+)		Ir L x-ray	11%
	192	77											0.0580 ± 0.0004	0.04%
	192	77											0.2959 ± 0.0001	0.002%
	192	77											0.3165 ± 0.0001	0.01%
$_{77}Ir^{192}$	192	77		191.962580	73.83 d.	β-	1.454			14 x 10^2 b.	(4-)	+1.880	Pt k x-ray	5%
	192	77											0.20577 ± 0.00001	3.2%
	192	77											0.29595 ± 0.00001	29%
	192	77											0.30844 ± 0.00001	30%
	192	77											0.31649 ± 0.00001	83%
	192	77											0.46806 ± 0.00001	48%
	192	77											0.48457 ± 0.00001	3.2%
	192	77											0.58857 ± 0.00001	4.6%
	192	77											0.60440 ± 0.00001	8.4%
	192	77											0.88452 ± 0.00002	0.3%
	192	77											1.06148 ± 0.00004	0.05%
$_{77}Ir^{193m}$	193	77			10.6 d.	I.T.	0.0802				11/2-		Ir L x-ray	12%
	193	77											0.0803 ± 0.0001	0.005%
$_{77}Ir^{193}$	193	77	62.7%	192.962917						111 b.	3/2+	+0.1591		
$_{77}Ir^{194m}$	194	77			171 d.	β-					11		Pt k x-ray	8%
	194	77											0.1117 ± 0.0005	8.9%
	194	77											0.3284 ± 0.0001	93%
	194	77											0.3388 ± 0.0005	55%
	194	77											0.3908 ± 0.0005	35%
	194	77											0.4829 ± 0.0001	97%
	194	77											0.5624 ± 0.0005	70%
	194	77											0.6005 ± 0.0005	62%
	194	77											0.6878 ± 0.0005	59%
	194	77											1.0118 ± 0.0005	3.6%
$_{77}Ir^{194}$	194	77	193.965069	19.2 h.	β-	2.248	1.92	9%			1-	0.37	0.2935 ± 0.0001	2.5%
	194	77						2.25	86%				0.3284 ± 0.0001	13%
	194	77											0.6451 ± 0.0001	1.2%
	194	77											0.9387 ± 0.0001	0.6%
	194	77											1.1508 ± 0.0001	0.6%
	194	77											1.1835 ± 0.0001	0.3%
	194	77											1.4689 ± 0.0001	0.2%
	194	77											(0.1 - 2.2)weak	
$_{77}Ir^{195m}$	195	77			3.9 h.	β-		0.41			(11/2-		Pt k x-ray	36%
	195	77						0.97					0.0989 ± 0.0001	10%
	195	77											0.1297 ± 0.0001	1.7%
	195	77											0.1728 ± 0.0001	5.0%
	195	77											0.2018 ± 0.0002	1.4%
	195	77											0.2113 ± 0.0001	2.2%
	195	77											0.2392 ± 0.0001	1.8%
	195	77											0.2516 ± 0.0001	1.8%
	195	77											0.2878 ± 0.0002	1.0%
	195	77											0.2903 ± 0.0002	1.9%
	195	77											0.3065 ± 0.0001	2.2%
	195	77											0.3199 ± 0.0001	9.6%
	195	77											0.3564 ± 0.0002	1.8%

TABLE OF THE ISOTOPES (Continued)

Isotope	A	Z	% Natural abundance	Atomic mass	Half-life	Decay mode	Decay energy (MeV)	Particle energy (MeV)	Particle intensity	Thermal neutron cross section	Spin (h/2π)	μ Nucl. mag. moment	Gamma-ray energy (MeV)	Gamma-ray intensity
	195	77											0.3593 ± 0.0002	4.6%
	195	77											0.3649 ± 0.0001	9.5%
	195	77											0.4090 ± 0.0001	1.4%
	195	77											0.4329 ± 0.0001	9.6%
	195	77											0.4812 ± 0.0001	2.7%
	195	77											0.5754 ± 0.0001	1.5%
	195	77											0.6849 ± 0.0001	9.6%
	195	77											0.8009 ± 0.0001	1.0%
${}_{77}Ir^{195}$	195	77		194.965966	2.8 h.	β-	1.118	1.0	80%		(3/2+)		Pt k x-ray	28%
	195	77						1.11	13%				0.0989 ± 0.0001	9.7%
	195	77											0.1297 ± 0.0001	1.4%
	195	77											0.2113 ± 0.0001	1.5%
${}_{77}Ir^{196m}$	196	77			1.40 h.	β-		1.16					Pt k x-ray	10%
	196	77											0.1033 ± 0.0002	16%
	196	77											0.3557 ± 0.0002	94%
	196	77											0.3935 ± 0.0002	97%
	196	77											0.4209 ± 0.0003	2.5%
	196	77											0.4471 ± 0.0002	94%
	196	77											0.5214 ± 0.0002	96%
	196	77											0.6335 ± 0.0003	1.1%
	196	77											0.6473 ± 0.0002	91%
	196	77											0.6939 ± 0.0002	4.2%
	196	77											0.7273 ± 0.0002	2.6%
	196	77											0.8356 ± 0.0002	6.3%
	196	77											1.4825 ± 0.0004	2.3%
${}_{77}Ir^{196}$	196	77		195.968370	52 s.	β-	3.210	2.1	15%		0-		0.3329 ± 0.0002	4.3%
	196	77						3.2	80%				0.3557 ± 0.0002	19%
	196	77											0.4468 ± 0.0002	4.5%
	196	77											0.7796 ± 0.0002	10%
	196	77											1.0470 ± 0.0002	1%
	196	77											1.4684 ± 0.0002	0.8%
	196	77											1.5642 ± 0.0002	0.9%
${}_{77}Ir^{197m}$	197	77			8.9 m.	β-					(11/2-		0.3465	(D)
	197	77				I.T.							See Ir^{197}	
${}_{77}Ir^{197}$	197	77		196.969629	5.85 m.	β-	2.156	1.5			(3/2+)		0.0531 ± 0.0001	9 +
	197	77						2.0					0.1351 ± 0.0001	27
	197	77											0.2689 ± 0.0001	13
	197	77											0.2996 ± 0.0001	24
	197	77											0.3783 ± 0.0001	38
	197	77											0.4306 ± 0.0001	61
	197	77											0.4568 ± 0.0002	37
	197	77											0.4697 ± 0.0001	100
	197	77											0.4964 ± 0.0004	36
	197	77											0.5091 ± 0.0003	21
	197	77											0.5272 ± 0.0001	24
	197	77											0.5392 ± 0.0001	14
	197	77											0.5420 ± 0.0001	11
	197	77											0.7153 ± 0.0001	9
	197	77											0.8091 ± 0.0001	32
	197	77											0.8159 ± 0.0001	45
	197	77											0.8664 ± 0.0001	13
	197	77											0.9394 ± 0.0001	21
	197	77											0.9871 ± 0.0001	15
	197	77											1.3432 ± 0.0001	21
${}_{77}Ir^{198}$	198	77		197.972160	8 s.	β-	4.00						0.4074 ± 0.0003	100 +
	198	77											0.5070 ± 0.0003	76
Pt		78		195.08						10 b.				
${}_{78}Pt^{172}$	172	78		171.977220	0.10 s.	α		6.31			0+			
${}_{78}Pt^{173}$	173	78		172.976280	0.34 s.	β+,E.C.	8.12							
	173	78				α		6.20						
${}_{78}Pt^{174}$	174	78		173.972811	0.90 ± 0.01 s.	β+, EC (17%)	5.73				0+			
	174	78				α(83%)		6.040						
${}_{78}Pt^{175}$	175	78		174.972130	2.52 ± 0.08 s.	β+, EC (65%)	7.43						0.0774 ± 0.0008	
	175	78				α(35%)		5.831	5%				0.1354 ± 0.0008	
	175	78						5.96	54%				0.2128 ± 0.0008	
	175	78						6.038	5%					
${}_{78}Pt^{176}$	176	78		175.968930	6.3 ± 0.1 s.	β+, EC (60%)	5.08				0+		ann.rad.	
	176	78				α(40%)		5.528	0.6%				0.2277	
	176	78						5.750	41%					
${}_{78}Pt^{177}$	177	78		176.968360	11 ± 2 s.	E.C.(91%)	6.	53					0.0908	
	177	78				α(9%)		5.485	3%					
	177	78						5.525	6%					
${}_{78}Pt^{178}$	178	78		177.965700	21.0 ± 0.7 s.	E.C.(93%)	4.34				0+			
	178	78				α(7%)		5.286	0.2%					
	178	78						5.442	7%					
${}_{78}Pt^{179}$	179	78		178.965270	≈ 43 s.	β+,E.C.	5.66							
	179	78				α		5.16						
${}_{78}Pt^{180}$	180	78		179.963130	52 ± 3 s.	β+, EC (99.7%	3.61				0+			
	180	78				α(0.3%)		5.140						
${}_{78}Pt^{181}$	181	78		180.963100	51 ± 5 s.	β+,E.C.	5.08							

TABLE OF THE ISOTOPES (Continued)

Isotope	A	Z	% Natural abundance	Atomic mass	Half-life	Decay mode	Decay energy (MeV)	Particle energy (MeV)	Particle intensity	Thermal neutron cross section	Spin (h/2π)	μ Nucl. mag. moment	Gamma-ray energy (MeV)	Gamma-ray intensity
$_{78}Pt^{182}$	182	78		181.961160	2.7 m.	β+,E.C.	2.97				0+		ann.rad.	
	182	78											0.1360 ± 0.0002	100 +
	182	78											0.1460 ± 0.0015	15
	182	78											0.1860 ± 0.0015	7
	182	78											0.2100 ± 0.0015	12
$_{78}Pt^{183m}$	183	78		182.961630	43 s.	β+,E.C.					(7/2-)		ann.rad.	
	183	78				I.T.							0.3132 ± 0.0003	28 +
	183	78											0.3164 ± 0.0003	53
	183	78											0.3290 ± 0.0003	36
	183	78											0.6296 ± 0.0003	100
	183	78											0.6453 ± 0.0003	23
$_{78}Pt^{183}$	183	78		182.961630	7 m.	β+,E.C.	4.58						ann.rad.	
$_{78}Pt^{184}$	184	78		183.959920	17.3 m.	β+,E.C.	2.20						ann.rad.	
	184	78											0.0926 ± 0.0003	16 +
	184	78											0.1170 ± 0.0004	7
	184	78											0.1395 ± 0.0003	6
	184	78											0.1445 ± 0.0004	4
	184	78											0.1495 ± 0.0006	4
	184	78											0.1549 ± 0.0003	100
	184	78											0.1616 ± 0.0004	6
	184	78											0.1829 ± 0.0004	8
	184	78											0.1919 ± 0.0003	94
	184	78											0.2093 ± 0.0004	6
	184	78											0.2165 ± 0.0003	17
	184	78											0.3946 ± 0.0003	16
	184	78											0.5484 ± 0.0003	77
	184	78											0.6107 ± 0.0009	12
	184	78											0.7312 ± 0.0004	43
$_{78}Pt^{185m}$	185	78			33 m.	β+,E.C.					1/2-			
$_{78}Pt^{185}$	185	78		184.960700	71 m.	β+,E.C.	3.700				(9/2+)		ann.rad.	
	185	78											0.0857 ± 0.0001	4 +
	185	78											0.1056 ± 0.0001	6
	185	78											0.1198 ± 0.0001	15
	185	78											0.1353 ± 0.0001	80
	185	78											0.1974 ± 0.0001	74
	185	78											0.2068 ± 0.0002	6
	185	78											0.2126 ± 0.0001	12
	185	78											0.2296 ± 0.0001	100
	185	78											0.2430 ± 0.0002	6
	185	78											0.2512 ± 0.0003	8
	185	78											0.2551 ± 0.0002	51
	185	78											0.2644 ± 0.0002	8
	185	78											0.2943 ± 0.0001	7
	185	78											0.3001 ± 0.0002	8
	185	78											0.3354 ± 0.0002	13
	185	78											0.3845 ± 0.0002	15
	185	78											0.4188 ± 0.0002	6
	185	78											0.4598 ± 0.0002	7
	185	78											0.4650 ± 0.0002	25
	185	78											0.5849 ± 0.0002	17
	185	78											0.6408 ± 0.0002	7
	185	78											0.7205 ± 0.0002	20
	185	78											0.7353 ± 0.0002	8
	185	78											0.8376 ± 0.0003	7
	185	78											0.8952 ± 0.0003	7
	185	78											0.9625 ± 0.0004	5
	185	78											1.2928 ± 0.0004	4
	185	78											1.3958 ± 0.0004	4
$_{78}Pt^{186}$	186	78		185.959360	2.0 h.	β+,E.C.	2.320				0+		ann.rad.	
	186	78											0.1805 ± 0.0004	1.7 +
	186	78											0.2808 ± 0.0004	2.4
	186	78											0.3667 ± 0.0004	3.3
	186	78											0.6115 ± 0.0004	8.6
	186	78											0.6892 ± 0.0003	100
$_{78}Pt^{187}$	187	78		186.960470	2.35 h.	β+,E.C.		2.90			3/2		ann.rad.	
	187	78											Ir k x-ray	61%
	187	78											0.1064 ± 0.0001	8%
	187	78											0.1100 ± 0.0001	5%
	187	78											0.1220 ± 0.0001	2.8%
	187	78											0.1869 ± 0.0001	3.5%
	187	78											0.2015 ± 0.0003	6.8%
	187	78											0.2476 ± 0.0001	3.5%
	187	78											0.2849 ± 0.0001	5.2%
	187	78											0.3048 ± 0.0001	4.1%
	187	78											0.4272 ± 0.0001	2.1%
	187	78											0.6296 ± 0.0001	2.6%
	187	78											0.7092 ± 0.0001	4.9%
	187	78											0.8193 ± 0.0002	3.5%
	187	78											0.9127 ± 0.0003	1.7%
	187	78											1.1455 ± 0.0003	1.6%
	187	78											1.1570 ± 0.0007	1.1%
	187	78											1.2554 ± 0.0003	2.2%
$_{78}Pt^{188}$	188	78		187.959386	10.2 d.	E.C.	0.518				0+		Ir k x-ray	49%
	188	78											0.1876 ± 0.0001	19%
	188	78											0.1951 ± 0.0001	19%

Isotope	A	Z	% Natural abundance	Atomic mass	Half-life	Decay mode	Decay energy (MeV)	Particle energy (MeV)	Particle intensity	Thermal neutron cross section	Spin (h/2π)	μ Nucl. mag. moment	Gamma-ray energy (MeV)	Gamma-ray intensity
	188	78											0.3814 ± 0.0001	7.5%
	188	78											0.4233 ± 0.0001	4.4%
$_{78}Pt^{189}$	189	78		188.960817	10.9 h.	β+,E.C.	1.961						Ir k x-ray	55%
	189	78											0.0943 ± 0.0001	4.7%
	189	78											0.1411 ± 0.0001	2.6%
	189	78											0.1867 ± 0.0001	1.4%
	189	78											0.2435 ± 0.0001	4.4%
	189	78											0.3005 ± 0.0001	2.3%
	189	78											0.3177 ± 0.0001	2.0%
	189	78											0.5449 ± 0.0001	3.6%
	189	78											0.5688 ± 0.0001	4.4%
	189	78											0.6076 ± 0.0001	5.1%
	189	78											0.6271 ± 0.0001	1.5%
	189	78											0.7214 ± 0.0001	5.8%
	189	78											(0.09 - 1.47)weak	
$_{78}Pt^{190}$	190	78	0.01%	189.959917						800 b.	0+			
$_{78}Pt^{191}$	191	78		190.961665	2.96 d.	E.C.	1.021				(3/2-)		Ir k x-ray	66%
	191	78											0.0824 ± 0.0001	4.9%
	191	78											0.0965 ± 0.0001	3.3%
	191	78											0.1294 ± 0.0001	3.2%
	191	78											0.1722 ± 0.0001	3.5%
	191	78											0.3687 ± 0.0001	1.6%
	191	78											0.3512 ± 0.0001	3.4%
	191	78											0.3599 ± 0.0001	6.0%
	191	78											0.4094 ± 0.0001	8.0%
	191	78											0.4565 ± 0.0001	3.4%
	191	78											0.5389 ± 0.0001	13.7%
	191	78											0.6241 ± 0.0001	1.4%
$_{78}Pt^{192}$	192	78	0.79%	191.961019						(1 + 9) b.	0+			
$_{78}Pt^{193m}$	193	78			4.33 d.	I.T.	0.1498				13/2+		Pt k x-ray	7%
	193	78											0.1355 ± 0.0001	0.11%
$_{78}Pt^{193}$	193	78		192.962977	50 y.	E.C.	0.057				(1/2-)		Ir k x-rays	
$_{78}Pt^{194}$	194	78	32.9%	193.962655						(0.1 + 1.1) b.	0+			
$_{78}Pt^{195m}$	195	78			4.02 d.	I.T.	0.2952				13/2+	0.597	Pt k x-ray	39%
	195	78											0.0989 ± 0.0001	11%
	195	78											0.1297 ± 0.0001	2.8%
$_{78}Pt^{195}$	195	78	33.8%	194.964766						29 b.	1/2-	+0.6095		
$_{78}Pt^{196}$	196	78	25.3%	195.964926						(0.05 + 0.7) b.	0+			
$_{78}Pt^{197m}$	197	78			95.4 m.	I.T.(97%)					13/2+		Pt k x-ray	24%
	197	78				β-(3%)							0.0530 ± 0.0001	1.1%
	197	78											0.3465 ± 0.0002	11%
$_{78}Pt^{197}$	197	78		196.967315	18.3 h.	β-	0.719				1/2-	0.51	Au k x-ray	17%
	197	78											0.1914 ± 0.0001	3.7%
	197	78											0.2688 ± 0.0001	0.2%
$_{78}Pt^{198}$	198	78	7.2%	197.967869						(0.3 + 3.5) b.	0+			
$_{78}Pt^{199m}$	199	78			13.5 s.	I.T.	0.424				13/2+		Pt k x-ray	3.4%
	199	78											0.3919 ± 0.0001	85%
$_{78}Pt^{199}$	199	78		198.970552	30.8 m.	β-	1.688	0.90	18%		(5/2-)		0.0772 ± 0.0001	1.5%
	199	78						1.14	14%				0.18579 ± 0.00002	3.3%
	199	78											0.19169 ± 0.00003	2.4%
	199	78											0.24646 ± 0.00003	2.2%
	199	78											0.31703 ± 0.00003	4.9%
	199	78											0.4681 ± 0.0001	1.0%
	199	78											0.4747 ± 0.0001	1.1%
	199	78											0.49375 ± 0.00003	5.7%
	199	78											0.54298 ± 0.00004	15%
	199	78											0.71455 ± 0.00003	1.9%
	199	78											0.96831 ± 0.00004	1.1%
$_{78}Pt^{200}$	200	78		199.971417	12.5 ± 0.3 h.	β-	0.690				0+		Au k x-ray	10%
	200	78											0.13590 ± 0.00009	3.1%
	200	78											0.20004 ± 0.00004	0.6%
	200	78											0.22747 ± 0.00004	2.0%
	200	78											0.24371 ± 0.00003	2.4%
	200	78											0.33024 ± 0.00003	1.1%
$_{78}Pt^{201}$	201	78		200.974500	2.5 ± 0.1 m.	β-	2.660				(5/2-)		0.070	
	201	78											0.152	
	201	78											0.222	
	201	78											1.760	
Au		79		196.9665						98.7 b.				
$_{79}Au^{176}$	176	79		175.980060	1.3 ± 0.3 s.	β+,E.C.	10.37							

Isotope	A	Z	% Natural abundance	Atomic mass	Half-life	Decay mode	Decay energy (MeV)	Particle energy (MeV)	Particle intensity	Thermal neutron cross section	Spin (h/2π)	μ Nucl. mag. moment	Gamma-ray energy (MeV)	Gamma-ray intensity
	176	79				α		6.260	80 +					
	176	79						6.290	20					
$_{79}Au^{177}$	177	79		176.976920	1.3 ± 0.4 s.	α		6.115						
	176	79						6.150						
$_{79}Au^{178}$	178	79		177.975760	2.6 ± 0.5 s.	α		5.920						
$_{79}Au^{179}$	179	79		178.973170	7.5 s.	α		5.85						
$_{79}Au^{180}$	180	79		179.972310	8.1 ± 0.3 s.	E.C.	8.55						0.1522 ± 0.0003	100 +
	180	79											0.2564 ± 0.0003	30
	180	79											0.3240 ± 0.0003	18
	180	79											0.3434 ± 0.0003	14
	180	79											0.4505 ± 0.0005	7
	180	79											0.5242 ± 0.0003	44
	180	79											0.5524 ± 0.0004	7
	180	79											0.6765 ± 0.0004	20
	180	79											0.7077 ± 0.0005	4
	180	79											0.8084 ± 0.0004	30
	180	79											0.8597 ± 0.0006	35
	180	79											1.0321 ± 0.0007	23
$_{79}Au^{181}$	181	79		180.970130	11.4 ± 0.5 s.	E.C.(99%)	6.55	5.482						
$_{79}Au^{182}$	182	79		181.969580	21 s.	β+,E.C.	7.85						ann.rad.	
	182	79											0.1549 ± 0.0002	46%
	182	79											0.2649 ± 0.0003	18%
	182	79											0.3449 ± 0.0003	2.7%
	182	79											0.3561 ± 0.0003	1.2%
	182	79											0.5126 ± 0.0004	4.1%
	182	79											0.6138 ± 0.0004	2.2%
	182	79											0.6388 ± 0.0004	1.2%
	182	79											0.6672 ± 0.0003	3.1%
	182	79											0.7870 ± 0.0002	6.2%
	182	79											0.8553 ± 0.0002	6.6%
	182	79											0.8997 ± 0.0004	1.2%
	182	79											1.0264 ± 0.0003	3.4%
	182	79											1.0843 ± 0.0004	1.4%
	182	79											(0.13 - 1.4)weak	
$_{79}Au^{183}$	183	79		182.967660	42 s.	E.C.	7.09						0.1630 ± 0.0001	52%
	184	79											0.2730 ± 0.0001	42%
	184	79											0.3625 ± 0.0001	18%
	184	79											0.4327 ± 0.0004	2.0%
	184	79											0.4353 ± 0.0004	2.0%
	184	79											0.4860 ± 0.0001	6.2%
	184	79											0.5921 ± 0.0002	3.4%
	184	79											0.6487 ± 0.0002	3.2%
	184	79											0.6645 ± 0.0002	2.0%
	184	79											0.7770 ± 0.0002	6.9%
	184	79											0.8312 ± 0.0003	2.2%
	184	79											0.8440 ± 0.0002	5.6%
	184	79											0.8710 ± 0.0003	3.9%
	184	79											1.0096 ± 0.0003	2.7%
	184	79											1.0713 ± 0.0003	2.5%
	184	79											1.0903 ± 0.0003	1.9%
	184	79											1.2457 ± 0.0003	2.4%
	184	79											1.3087 ± 0.0003	1.5%
	184	79											1.3976 ± 0.0003	1.8%
	184	79											1.5256 ± 0.0003	1.2%
	184	79											1.7138 ± 0.0004	1.5%
	184	79											1.7138 ± 0.0004	1.5%
	184	79											1.7546 ± 0.0003	3.3%
	184	79											1.8142 ± 0.0003	2.8%
	184	79											2.1963 ± 0.0003	1.5%
	184	79											2.1963 ± 0.0003	1.5%
	184	79											2.4752 ± 0.0004	2.0%
	184	79											2.4909 ± 0.0003	1.0%
$_{79}Au^{185m}$	185	79			6.8 m.	β+,E.C.								
	185	79				I.T.	0.145							
$_{79}Au^{185}$	185	79		184.965800	4.3 m.	β+,E.C.	4.76				(5/2-)		ann.rad.	
$_{79}Au^{186m}$	186	79			2 m.	β+,E.C.							0.1915 ± 0.0001	
$_{79}Au^{186}$	186	79		185.966100	10.7 m.	β+,E.C.	6.28				3		ann.rad.	
	186	79											0.1915 ± 0.0001	56%
	186	79											0.2988 ± 0.0001	23%
	186	79											0.4156 ± 0.0002	7.7%
	186	79											0.6070 ± 0.0002	4.8%
	186	79											0.6765 ± 0.0003	3.0%
	186	79											0.7654 ± 0.0003	9.6%
	186	79											0.7987 ± 0.0004	4.8%
	186	79											0.8816 ± 0.0003	1.9%
	186	79											1.2162 ± 0.0003	3.4%
	186	79											1.2892 ± 0.0005	1.9%
	186	79											1.7259 ± 0.0004	1.1%
	186	79											1.7376 ± 0.0004	1.7%
	186	79											2.0246 ± 0.0005	1.8%
	186	79											2.0356 ± 0.0005	2.6%
$_{79}Au^{187}$	187	79		186.964460	8.2 m.	β+,E.C.	3.72				1/2		ann.rad.	
	187	79											0.0512 ± 0.0001	1.9%
	187	79											0.0653 ± 0.0001	1.9%

TABLE OF THE ISOTOPES (Continued)

Isotope	A	Z	% Natural abundance	Atomic mass	Half-life	Decay mode	Decay energy (MeV)	Particle energy (MeV)	Particle intensity	Thermal neutron cross section	Spin (h/2π)	μ Nucl. mag. moment	Gamma-ray energy (MeV)	Gamma-ray intensity
	187	79											0.1811 ± 0.0001	1.9%
	187	79											0.1853 ± 0.0001	1.6%
	187	79											0.1903 ± 0.0001	1.8%
	187	79											0.2474 ± 0.0001	1.8%
	187	79											0.2474 ± 0.0001	1.8%
	187	79											0.2510 ± 0.0001	1.9%
	187	79											0.3516 ± 0.0001	1.7%
	187	79											0.3903 ± 0.0001	2.6%
	187	79											0.4262 ± 0.0001	3.2%
	187	79											0.5601 ± 0.0003	2.7%
	187	79											0.6210 ± 0.0003	3.1%
	187	79											0.7069 ± 0.0003	3.7%
	187	79											0.7213 ± 0.0003	2.1%
	187	79											0.8344 ± 0.0003	2.0%
	187	79											0.9152 ± 0.0003	5.2%
	187	79											1.1897 ± 0.0003	1.9%
	187	79											1.2668 ± 0.0003	4.2%
	187	79											1.3191 ± 0.0003	3.4%
	187	79											1.3321 ± 0.0003	12%
	187	79											1.4081 ± 0.0003	4.7%
	187	79											1.4521 ± 0.0003	2.8%
	187	79											1.9605 ± 0.0003	1.7%
	187	79											1.9881 ± 0.0003	2.1%
	187	79											2.0270 ± 0.0003	1.9%
	187	79											2.0528 ± 0.0003	1.3%
	187	79											2.0813 ± 0.0003	1.3%
$_{79}Au^{188}$	188	79		187.965080	8.8 m.	β+,E.C.	5.30				(1-)		ann.rad.	
	188	79											0.2660 ± 0.0001	100 +
	188	79											0.3308 ± 0.0001	5
	188	79											0.3404 ± 0.0001	24
	188	79											0.4055 ± 0.0001	9.1%
	188	79											0.5334 ± 0.0003	5.9
	188	79											0.6061 ± 0.0001	16
	188	79											0.6708 ± 0.0001	8
	188	79											0.6791 ± 0.0001	2
	188	79											0.9491 ± 0.0001	2.4
	188	79											1.0470 ± 0.0001	1.9
	188	79											1.0843 ± 0.0001	6.6
	188	79											1.1153 ± 0.0001	5
	188	79											1.1705 ± 0.0001	2.6
	188	79											1.3126 ± 0.0001	3.0
	188	79											1.3601 ± 0.0001	4.1
	188	79											1.5104 ± 0.0001	2.7
	188	79											1.5450 ± 0.0001	2.3
	188	79											1.8825 ± 0.0002	1.5
	188	79											2.0300 ± 0.0001	2.5
	188	79											2.2319 ± 0.0001	2.6
	188	79											2.4469 ± 0.0002	1.4
	188	79											2.6268 ± 0.0003	1.8
	188	79											2.7810 ± 0.0002	2.2
$_{79}Au^{189m}$	189	79			4.6 m.	β+,E.C.					11/2-		0.1667 ± 0.0002	100 +
	189	79											0.3211 ± 0.0005	19
$_{79}Au^{189}$	189	79		188.963720	28.7 m.	E.C.(96%)	2.700				1/2+		ann.rad.	
	189	79				β+(4%)							Pt k x-ray	53%
	189	79											0.2157 ± 0.0001	3.2%
	189	79											0.2220 ± 0.0001	5.0%
	189	79											0.2257 ± 0.0001	2.1%
	189	79											0.2537 ± 0.0002	2.0%
	189	79											0.2975 ± 0.0003	3.0%
	189	79											0.3482 ± 0.0001	8.4%
	189	79											0.44121 ± 0.0001	704%
	189	79											0.4478 ± 0.0001	11%
	189	79											0.5295 ± 0.0001	6.5%
	189	79											0.5295 ± 0.0001	6.5%
	189	79											0.6311 ± 0.0005	2.4%
	189	79											0.7133 ± 0.0001	21%
	189	79											0.8128 ± 0.0003	13%
	189	79											1.0715 ± 0.0003	5.5%
	189	79											1.1605 ± 0.0003	7.0%
	189	79											1.1769 ± 0.0007	3.3%
$_{79}Au^{190}$	190	79		189.964685	43 m.	β+(2%)	4.442				1-	0.066	ann.rad.	4%
	190	79				E.C.(98%)							Pt k x-ray	41%
	190	79											0.2958 ± 0.0001	72%
	190	79											0.3018 ± 0.0001	24%
	190	79											0.3189 ± 0.0001	4.7%
	190	79											0.4412 ± 0.0001	3.8%
	190	79											0.5977 ± 0.0001	9.0%
	190	79											0.6208 ± 0.0001	2.4%
	190	79											0.6250 ± 0.0002	3.0%
	190	79											1.0575 ± 0.0004	3.2%
	190	79											1.3953 ± 0.0002	2.1%
	190	79											1.44131 ± 0.0004	2.9%
	190	79											1.7849 ± 0.0004	2.1%
	190	79											2.3824 ± 0.0002	3.8%
	190	79											2.4163 ± 0.0002	3.3%
	190	79											2.7527 ± 0.0002	2.7%

Isotope	A	Z	% Natural abundance	Atomic mass	Half-life	Decay mode	Decay energy (MeV)	Particle energy (MeV)	Particle intensity	Thermal neutron cross section	Spin (h/2π)	μ Nucl. mag. moment	Gamma-ray energy (MeV)	Gamma-ray intensity
	190	79											2.9598 ± 0.0009	1.0%
$_{79}$Au^{191m}	191	79			0.9 s.	I.T.	0.2663				(11/2-		Au k x-ray	9%
	191	79											0.2414 ± 0.0005	13%
	191	79											0.2526 ± 0.0004	61%
$_{79}$Au^{191}	191	79		190.963630	3.2 h.	E.C.	1.830				3/2+	0.138	Pt k x-ray	52%
	191	79											0.1665 ± 0.0001	3.1%
	191	79											0.1941 ± 0.0001	2.6%
	191	79											0.2064 ± 0.0001	2.1%
	191	79											0.2540 ± 0.0001	2.4%
	191	79											0.2716 ± 0.0001	2.4%
	191	79											0.2779 ± 0.0001	6.8%
	191	79											0.2804 ± 0.0001	2.8%
	191	79											0.2389 ± 0.0001	6.3%
	191	79											0.2935 ± 0.0001	2.7%
	191	79											0.3539 ± 0.0001	2.9%
	191	79											0.3869 ± 0.0001	3.4%
	191	79											0.3903 ± 0.0001	2.6%
	191	79											0.3998 ± 0.0001	4.5%
	191	79											0.4137 ± 0.0001	3.5%
	191	79											0.4214 ± 0.0001	3.2%
	191	79											0.4780 ± 0.0001	3.7%
	191	79											0.4876 ± 0.0001	2.6%
	191	79											0.5864 ± 0.0001	16%
	191	79											0.6742 ± 0.0001	6.4%
	191	79											(0.08 - 1.3)weak	
$_{79}$Au^{192}	192	79		191.964793	5.0 h.	β+(5%)	3.515	2.19			1/2-	0.0079	ann.rad.	10%
	192	79				E.C.(95%)		2.49					Pt k x-ray	41%
	192	79											0.2959 ± 0.0001	30%
	192	79											0.3084 ± 0.0001	4.6%
	192	79											0.3165 ± 0.0001	78%
	192	79											0.4681 ± 0.0001	2.3%
	192	79											0.4772 ± 0.0002	1.5%
	192	79											0.5826 ± 0.0001	3.6%
	192	79											0.6043 ± 0.0001	1.4%
	192	79											0.6124 ± 0.0001	5.9%
	192	79											0.7591 ± 0.0002	2.2%
	192	79											0.8787 ± 0.0002	1.1%
	192	79											1.0615 ± 0.0003	1.2%
	192	79											1.1269 ± 0.00023	2.0%
	192	79											1.1402 ± 0.0002	3.5%
	192	79											1.4229 ± 0.0002	4.0%
	192	79											1.5766 ± 0.0003	3.0%
	192	79											1.6244 ± 0.0003	2.3%
	192	79											1.7066 ± 0.0003	2.6%
	192	79											1.7231 ± 0.0002	4.2%
$_{79}$Au^{193m}	193	79			3.9 s.	I.T.	0.2901				11/2-		Au k x-ray	11%
	193	79											0.2197 ± 0.0001	3.8%
	193	79											0.2580 ± 0.0001	66%
$_{79}$Au^{193}	193	79		192.964050	17.6 h.	E.C.	1.000				3/2+	0.140	Pt k x-ray	55%
	193	79											0.1125 ± 0.0001	2.1%
	193	79											0.1735 ± 0.0001	2.9%
	193	79											0.1862 ± 0.0001	10%
	193	79											0.2556 ± 0.0001	6.7%
	193	79											0.2682 ± 0.0001	3.9%
	193	79											0.4390 ± 0.0001	1.9%
	193	79											0.4913 ± 0.0001	0.7%
$_{79}$Au^{194}	194	79		193.965348	39.5 h.	β+(3%)	2.609	1.49			1-	0.074	ann.rad.	6%
	194	79				E.C.(97%)							Pt k x-ray	39%
	194	79											0.2935 ± 0.0001	11%
	194	79											0.3284 ± 0.0001	63%
	194	79											0.3649 ± 0.0001	1.5%
	194	79											0.4828 ± 0.0001	1.2%
	194	79											0.5288 ± 0.0001	1.7%
	194	79											0.6220 ± 0.0001	1.8%
	194	79											0.6451 ± 0.0001	2.3%
	194	79											0.9387 ± 0.0001	1.2%
	194	79											0.9483 ± 0.0001	2.3%
	194	79											1.1041 ± 0.0001	2.1%
	194	79											1.1508 ± 0.0001	1.4%
	194	79											1.1753 ± 0.0001	2.1%
	194	79											1.2188 ± 0.0001	1.2%
	194	79											1.3422 ± 0.0001	1.2%
	194	79											1.4689 ± 0.0001	6.7%
	194	79											1.5924 ± 0.0002	1.1%
	194	79											1.5958 ± 0.0001	1.8%
	194	79											1.8559 ± 0.0002	1.8%
	194	79											1.8570 ± 0.0002	1.6%
	194	79											1.9242 ± 0.0001	2.1%
	194	79											2.0437 ± 0.0001	3.8%
$_{79}$Au^{195m}	195	79			30.5 s.	I.T.	0.3186				11/2-		Au k x-ray	11%
	195	79											0.2004 ± 0.0001	1.7%
	195	79											0.2617 ± 0.0001	68%
$_{79}$Au^{195}	195	79		194.965013	186.1 d.	E.C.	0.230				3/2+	0.148	Pt k x-ray	48%
$_{79}$Au^{196m2}	196	79			9.7 h.	I.T.	0.5954				12-	5.35	Au k x-ray	42%
	196	79											0.1478 ± 0.0001	42%

TABLE OF THE ISOTOPES (Continued)

Isotope	A	Z	% Natural abundance	Atomic mass	Half-life	Decay mode	Decay energy (MeV)	Particle energy (MeV)	Particle intensity	Thermal neutron cross section	Spin (h/2π)	μ Nucl. mag. moment	Gamma-ray energy (MeV)	Gamma-ray intensity
	196	79											0.1684 ± 0.0001	7.6%
	196	79											0.1883 ± 0.0001	37%
	196	79											0.2855 ± 0.0001	4.3%
	196	79											0.3162 ± 0.0001	2.9%
$_{79}Au^{196m1}$	196	79			8.1 s.	I.T.	0.0846				8+		0.0847 ± 0.0001	0.3%
$_{79}Au^{196}$	196	79		195.966544	6.18 d.	E.C.(92%)	0.507				2-	+0.514	Pt k x-ray	37%
$_{79}Au^{197m}$					7.8 s.	I.T.	0.4094				11/2-		Au k x-ray	12%
	197	79				β-(8%)	0.686						0.1302 ± 0.0001	3.1%
	197	79											0.2018 ± 0.0001	1.1%
	197	79											0.2790 ± 0.0001	71%
	197	79											0.4091 ± 0.0001	0.1%
$_{79}Au^{197}$	197	79	100%	196.966543						(0 + 98.7) b.	3/2+	+0.1457		
$_{79}Au^{198m}$	198	79			2.30 d.	I.T.	0.812				(12-)		Au k x-ray	43%
	198	79											0.0972 ± 0.0001	69%
	198	79											0.1803 ± 0.0001	51%
	198	79											0.2041 ± 0.0001	41%
	198	79											0.2419 ± 0.0001	77%
	198	79											0.3338 ± 0.0002	15%
$_{79}Au^{198}$	198	79		197.968217	2.693 d.	β-	1.372	0.290	1%	26 x 10^3 b.	2-	+0.5934	Hg k x-ray	1.4%
	198	79						0.961	99%				0.411794 ± 0.00001	95.5%
	198	79											0.67587 ± 0.00002	1.1%
	198	79											1.08766 ± 0.00002	0.2%
$_{79}Au^{199}$	199	79		198.968740	3.14 d.	β-	0.453	0.25	22%	30 b.	3/2+	+0.2715	Hg k x-ray	7.6%
	199	79					0.296	72%					0.15837 ± 0.00001	37%
	199	79						0.462	6%				0.20820 ± 0.00001	8.4%
$_{79}Au^{200m}$	200	79			18.7 h.	β-(84%)	1.0	0.56			12-	6.10	Au k x-ray	
	200	79				I.T.(16%)							0.1111 ± 0.0001	1.8%
	200	79											0.1203 ± 0.0001	1.0%
	200	79											0.1332 ± 0.0002	2.8%
	200	79											0.1373 ± 0.0003	1.2%
	200	79											0.1446 ± 0.0003	1.0%
	200	79											0.1461 ± 0.0002	3.5%
	200	79											0.1812 ± 0.0001	55%
	200	79											0.2185 ± 0.0001	1.6%
	200	79											0.2559 ± 0.0001	71%
	200	79											0.3328 ± 0.0004	12%
	200	79											0.36797 ± 0.0001	77%
	200	79											0.4978 ± 0.0001	73%
	200	79											0.5793 ± 0.0001	72%
	200	79											0.7595 ± 0.0001	66%
	200	79											0.9042 ± 0.0001	7.7%
$_{79}Au^{200}$	200	79		199.970670	48.4 m.	β-	2.210	0.7	15%		1-		0.3679 ± 0.0001	19%
	200	79						2.2	77%				1.2254 ± 0.0001	11%
	200	79											1.2629 ± 0.0001	3.1%
	200	79											(0.3 - 1.6)weak	
$_{79}Au^{201}$	201	79		200.971645	26 m.	β-	1.275	1.27	82%		3/2+		0.1674 ± 0.0001	1.0%
	201	79											0.3851 ± 0.0002	0.6%
	201	79											0.5170 ± 0.0003	1.3%
	201	79											0.5426 ± 0.0002	1.9%
	201	79											0.6132 ± 0.0003	1.2%
	201	79											0.6450 ± 0.0004	0.7%
$_{79}Au^{202}$	202	79		201.973840	28 s.	β-	3.00				(1-)		0.4396 ± 0.0001	10%
	202	79											0.9086 ± 0.0004	2%
	202	79											1.1254 ± 0.0004	2.5%
	202	79											1.2037 ± 0.0004	2.1%
	202	79											1.3065 ± 0.0005	2.3%
$_{79}Au^{203}$	203	79		202.975145	53 s.	β-	2.14	≈ 1.9			3/2+		0.690	10%
$_{79}Au^{204}$	204	79		203.978300	40 s.	β-	4.5						0.4366 ± 0.0002	91%
	204	79											0.6919 ± 0.0002	23%
	204	79											0.7230 ± 0.0003	22%
	204	79											1.3921 ± 0.0004	24%
	204	79											1.4048 ± 0.001	4%
	204	79											1.4148 ± 0.0005	8%
	204	79											1.5113 ± 0.0004	28%
	204	79											1.5531 ± 0.0004	8%
	204	79											1.7042 ± 0.0006	5%
	204	79											1.8283 ± 0.001	3%
	204	79											1.8416 ± 0.001	3%
Hg		80		200.59						374 b.				
$_{80}Hg^{178}$	178	80		177.982476	0.26 s.	E.C.(50%)	6.25				0+			
	178	80				α(50%)		6.43						
$_{80}Hg^{179}$	179	80		178.981630	1.09 s.	E.C.	7.88							
	179	80				α		6.29						
$_{80}Hg^{180}$	180	80		179.978250	2.9 s.	E.C.	5.54				0+		0.1250 ± 0.0004	10 +
	180	80				α		6.12					0.3005 ± 0.0003	100
	180	80											0.3812 ± 0.0004	69
	180	80											0.4050 ± 0.0005	17
	180	80											0.4505 ± 0.0005	16

TABLE OF THE ISOTOPES (Continued)

Isotope	A	Z	% Natural abundance	Atomic mass	Half-life	Decay mode	Decay energy (MeV)	Particle energy (MeV)	Particle intensity	Thermal neutron cross section	Spin (h/2π)	μ Nucl. mag. moment	Gamma-ray energy (MeV)	Gamma-ray intensity
	180	80											0.4799 ± 0.0004	23
$_{80}Hg^{181}$	181	80		180.977720	3.6 s.	β+, EC (74%)	7.07				($^1/_2$-)	+0.5071	0.0663 ± 0.001	
	181	80				α(26%)							0.0811 ± 0.001	
	181	80											0.0924 ± 0.001	
	181	80											0.1474 ± 0.001	
	181	80											0.1587 ± 0.001	
	181	80											0.2142 ± 0.001	
	181	80											0.2398 ± 0.001	
$_{80}Hg^{182}$	182	80		181.974750	11 s.	β+, EC (85%)	4.81				0+		0.1289 ± 0.0001	100 +
	182	80				α(15%)		5.87					0.2168 ± 0.001	75
	182	80											0.4126 ± 0.001	53
$_{80}Hg^{183}$	183	80		182.974350	8.8 s.	β+, EC (77%)	6.24				$^1/_2$	+0.524	0.0714 ± 0.001	
	183	80				α		5.83					0.0874 ± 0.001	
	183	80						5.91					0.1538 ± 0.001	
$_{80}Hg^{184}$	184	80		183.971810	30.9 s.	β+, EC (99%)	3.98				0+		0.0915 ± 0.0003	4.7 +
	184	80				α(1%)		5.54					0.1265 ± 0.0003	5
	184	80											0.1460 ± 0.0003	5
	184	80											0.1560 ± 0.0003	91
	184	80											0.1591 ± 0.0003	7
	184	80											0.1701 ± 0.0003	2
	184	80											0.2362 ± 0.0002	100
	184	80											0.2590 ± 0.0001	8
	184	80											0.2623 ± 0.0001	7
	184	80											0.2951 ± 0.0001	16
	184	80											0.2951 ± 0.0001	16
	184	80											0.3924 ± 0.0002	11
	184	80											0.4219 ± 0.0002	6
$_{80}Hg^{185m}$	185	80			21 s.	β+, EC, IT, α		5.37			13/2+		0.211	
	185	80											0.292	
$_{80}Hg^{185}$	185	80		184.971900	50 s.	β+, EC (95%)	5.68				$^1/_2$-		0.0236	
	185	80											0.0358	
	185	80											0.0958	
	185	80											0.1074	
	185	80											0.1078	
	185	80											0.1291	
	185	80											0.1810	
	185	80											0.1937	
	185	80											0.2052	
	185	80											0.2112	
	185	80											0.2125	
	185	80											0.2229	
	185	80											0.2442	
	185	80											0.2587	
	185	80											0.2701	
	185	80											0.2701	
	185	80											0.2887	
	185	80											0.2924	
	185	80											0.2924	
	185	80											(0.02 - 0.55)weak	
$_{80}Hg^{186}$	186	80		185.969350	1.4 m.	β+,E.C.	3.03				0+		0.1119 ± 0.0004	87 +
	186	80											0.2278 ± 0.0004	5
	186	80											0.2518 ± 0.0004	100
	186	80											0.3496 ± 0.0005	2.3
$_{80}Hg^{187m}$	187	80			1.7 m.	β+,E.C.					13/2+		See Hg^{187}	
$_{80}Hg^{187}$	187	80		186.969760	2.4 m.	β+,E.C.	4.94				3/2-	-0.593	0.1034 ± 0.0002	32 +
	187	80											0.2034 ± 0.0002	19
	187	80											0.2055 ± 0.0002	10
	187	80											0.2208 ± 0.0002	24
	187	80											0.2334 ± 0.0002	100
	187	80											0.2403 ± 0.0002	33
	187	80											0.27151 ± 0.0002	31
	187	80											0.2985 ± 0.0002	11
	187	80											0.3229 ± 0.0002	12
	187	80											0.3347 ± 0.0002	16
	187	80											0.3763 ± 0.0002	38
	187	80											0.4387 ± 0.0002	11
	187	80											0.4495 ± 0.0002	29
	187	80											0.4620 ± 0.0002	10
	187	80											0.4703 ± 0.0002	29
	187	80											0.4727 ± 0.0002	11
	187	80											0.4758 ± 0.0002	22
	187	80											0.4766 ± 0.0002	11
	187	80											0.4996 ± 0.0002	19
	187	80											0.5254 ± 0.0002	30
	187	80											0.6250 ± 0.0002	14
	187	80											1.9981 ± 0.0008	11
	187	80											2.0126 ± 0.0008	10
	187	80											2.0754 ± 0.0009	13

Isotope	A	Z	% Natural abundance	Atomic mass	Half-life	Decay mode	Decay energy (MeV)	Particle energy (MeV)	Particle intensity	Thermal neutron cross section	Spin (h/2π)	μ Nucl. mag. moment	Gamma-ray energy (MeV)	Gamma-ray intensity
	187	80											2.1765 ± 0.001	20
$_{80}Hg^{188}$	188	80		187.967580	3.2 m.	β+,E.C.	2.340				0+		0.0988 ± 0.00074	12 +
	188	80											0.1148 ± 0.0007	37
	188	80											0.1346 ± 0.0007	11
	188	80											0.1424 ± 0.0007	20
	188	80											0.1824 ± 0.0007	18
	188	80											0.1858 ± 0.0007	13
	188	80											0.1900 ± 0.0007	100
	188	80											0.2540 ± 0.0007	10
	188	80											0.3360 ± 0.0007	10
	188	80											0.3453 ± 0.0007	10
	188	80											0.5238 ± 0.0007	19
$_{80}Hg^{189m}$	189	80			8.6 m.	E.C.					13/2+		0.0780 ± 0.0001	63 +
	189	80											0.1665 ± 0.0001	36
	189	80											0.2366 ± 0.0002	29
	189	80											0.2976 ± 0.0002	34
	189	80											0.3210 ± 0.0002	100
	189	80											0.3876 ± 0.0002	36
	189	80											0.3987 ± 0.0002	17
	189	80											0.4345 ± 0.0002	47
	189	80											0.4591 ± 0.0002	10
	189	80											0.4839 ± 0.0004	11
	189	80											0.4996 ± 0.0002	10
	189	80											0.5026 ± 0.0002	20
	189	80											0.5124 ± 0.0004	38
	189	80											0.5399 ± 0.0003	16
	189	80											0.5655 ± 0.0002	48
	189	80											0.6000 ± 0.0002	22
	189	80											0.7370 ± 0.0002	17
	189	80											1.2793 ± 0.0008	14
	189	80											2.0214 ± 0.0004	11
	189	80											2.0252 ± 0.0005	16
	189	80											2.0339 ± 0.0003	19
	189	80											(0.08 - 2.10)many	
$_{80}Hg^{189}$	189	80		188.968230	7.6 m.	E.C>	4.20				3/2-	-0.6086	0.2005 ± 0.0002	7 +
	189	80											0.2038 ± 0.0002	71
	189	80											0.2291 ± 0.0002	13
	189	80											0.2386 ± 0.0002	100
	189	80											0.2485 ± 0.0002	95
	189	80											0.2790 ± 0.0002	15
$_{80}Hg^{190}$	190	80		189.966400	20 m.	E.C.	1.60				0+		0.1296	1%
	190	80											0.1426	52%
	190	80											0.1547	2%
	190	80											0.1715	3.6%
$_{80}Hg^{191m}$	191	80			51 m.	β+(6%)					13/2+		ann.rad.	12%
	191	80				E.C.(94%)							Au k x-ray	50%
	191	80											0.2741 ± 0.001	13%
	191	80											0.3316 ± 0.0009	4%
	191	80											0.3570 ± 0.0004	5%
	191	80											0.3710 ± 0.0004	6%
	191	80											0.4097 ± 0.0009	3%
	191	80											0.4203 ± 0.0006	18%
	191	80											0.5215 ± 0.0007	4%
	191	80											0.5361 ± 0.0006	8%
	191	80											0.5787 ± 0.0004	17%
	191	80											0.6106 ± 0.0006	5%
	191	80											0.6710 ± 0.0007	3%
	191	80											0.7180 ± 0.0007	3%
	191	80											0.8870 ± 0.0008	2%
	191	80											0.9964 ± 0.0008	2%
	191	80											1.1093 ± 0.001	2%
	191	80											1.2844 ± 0.001	2%
	191	80											1.3298 ± 0.001	3%
	191	80											1.4478 ± 0.001	1.5%
	191	80											1.5036 ± 0.001	2%
	191	80											1.5488 ± 0.001	2%
	191	80											1.7394 ± 0.001	1.4%
	191	80											1.9081 ± 0.001	1.5%
	191	80											(0.07 - 1.9)many	
$_{80}Hg^{191}$	191	80			49 m.	β+,E.C.					(3/2-)		0.1963 ± 0.0002	67 +
	191	80											0.2247 ± 0.0002	60
	191	80											0.2408 ± 0.0002	44
	191	80											0.2524 ± 0.0002	100
	191	80											0.3314 ± 0.0005	39
	191	80											0.5214 ± 0.0006	9
	191	80											0.778 ± 0.001	6
$_{80}Hg^{192}$	192	80		191.965650	4.9 h.	E.C.	0.800				0+		Au k x-ray	53%
	192	80											0.1019	1%
	192	80											0.1572	6%
	192	80											0.1864	2.9%
	192	80											0.2454	1.5%
	192	80											0.2748	45%
	192	80											0.3065	4.8%

Isotope	A	Z	% Natural abundance	Atomic mass	Half-life	Decay mode	Decay energy (MeV)	Particle energy (MeV)	Particle intensity	Thermal neutron cross section	Spin (h/2π)	μ Nucl. mag. moment	Gamma-ray energy (MeV)	Gamma-ray intensity
$_{80}Hg^{193m}$	193	80			11.8 h.	β+, EC (91%)					13/2+	-1.0584	Hg k x-ray	53%
	193	80				I.T.(9%)	0.2901						0.1866 ± 0.0001	2.2%
	193	80											0.2181 ± 0.0001	6%
	193	80											0.2198 ± 0.0001	3.3%
	193	80											0.2580 ± 0.0001	58%
	193	80											0.2908 ± 0.0001	1.9%
	193	80											0.3419 ± 0.0001	3.0%
	193	80											0.3453 ± 0.0002	2.0%
	193	80											0.4076 ± 0.0001	37%
	193	80											0.4996 ± 0.0001	5.5%
	193	80											0.5351 ± 0.0001	4.5%
	193	80											0.5371 ± 0.0001	4.7%
	193	80											0.5733 ± 0.0001	31%
	193	80											0.6006 ± 0.0001	4.7%
	193	80											0.8701 ± 0.0002	2.9%
	193	80											0.8778 ± 0.0002	4.8%
	193	80											0.9324 ± 0.0002	15%
	193	80											0.9946 ± 0.0002	3.5%
	193	80											1.2413 ± 0.0002	5.6%
	193	80											1.3255 ± 0.0002	4.9%
	193	80											1.3395 ± 0.0002	4.7%
	193	80											1.3651 ± 0.0002	3.1%
	193	80											1.4861 ± 0.0003	3.4%
	193	80											1.6394 ± 0.0003	3.4%
	193	80											1.6485 ± 0.0003	2.6%
	193	80											(0.1 - 1.96)many	
$_{80}Hg^{193}$	193	80		192.966560	3.8 h.	2.339					3/2-	-0.6276	0.1866 ± 0.0001	16%
	193	80											0.2181 ± 0.0001	3.7%
	193	80											0.2580 ± 0.0001	14%
	193	80											0.3816 ± 0.0001	11%
	193	80											0.4295 ± 0.0001	11%
	193	80											0.5810 ± 0.0001	4%
	193	80											0.7461 ± 0.0002	2.3%
	193	80											0.7892 ± 0.0002	4.7%
	193	80											0.8278 ± 0.0002	4.0%
	193	80											0.8611 ± 0.0002	13%
	193	80											1.0405 ± 0.0003	2.3%
	193	80											1.0807 ± 0.0003	3.8%
	193	80											1.1188 ± 0.0002	8.3%
	193	80											1.2764 ± 0.0003	2.2%
	193	80											1.6034 ± 0.0003	2.1%
	193	80											1.8156 ± 0.0004	2.6%
	193	80											1.8622 ± 0.0004	1.5%
	193	80											1.9766 ± 0.0004	1.8%
$_{80}Hg^{194}$	194	80		193.965391	520 y.	E.C.	0.040				0+		Au L x-rays	
$_{80}Hg^{195m}$	195	80			40.0 h.	I.T.(54%)	0.3186				13/2+	-1.044	Hg k x-ray	2.3%
	195	80				E.C.(46%)							Au k x-ray	51%
	195	80											0.2617 ± 0.0001	33%
	195	80											0.3879 ± 0.0001	2.3%
	195	80											0.5603 ± 0.0001	7.5%
	195	80											0.7798 ± 0.0001	5.0%(D)
$_{80}Hg^{195}$	195	80		194.966640	9.5 h.	E.C.	1.520				1/2-	+0.542	Au k x-ray	40%
	195	80											0.0614 ± 0.0001	6.4%
	195	80											0.1801 ± 0.0001	2%
	195	80											0.2071 ± 0.0001	1.6%
	195	80											0.2617 ± 0.0001	1.6%
	195	80											0.5851 ± 0.0001	2.0%
	195	80											0.5997 ± 0.0001	1.8%
	195	80											0.7798 ± 0.0001	7.0%
	195	80											1.1110 ± 0.0001	1.5%
	195	80											1.1724 ± 0.0001	1.3%
$_{80}Hg^{196}$	196	80	0.15%	195.965807						(120 + 3100) b.	0+			
$_{80}Hg^{197m}$	197	80			23.8 h.	I.T.(93%)	0.2989				13/2+	-1.0277	Hg k x-ray	17%
	197	80											Au k x-ray	13%
	197	80											0.13398 ± 0.00005	34%
	197	80											0.1650 ± 0.0001	0.3%
$_{80}Hg^{197}$	197	80		196.967187	64.1 h.	E.C.	0.600				1/2-	+0.5274	Au k x-ray	36%
	197	80											0.07735 ± 0.00002	18%
	197	80											0.19136 ± 0.00004	0.5%
	197	80											0.26871 ± 0.00003	0.04%
$_{80}Hg^{198}$	198	80	10.1%	197.966743						(0.01 + 1.9) b.	0+			
$_{80}Hg^{199m}$	199	80			42.6 m.	I.T.	0.532				13/2+	-1.015	Hg k x-ray	32%
	199	80											0.15841 ± 0.00002	52%
	199	80											0.37386 ± 0.00003	14%
	199	80											0.4134 ± 0.0002	0.03%
$_{80}Hg^{199}$	199	80	17%	198.968254						2.2 x 10^3 b.	1/2-	+0.5059		
$_{80}Hg^{200}$	200	80	23.1%	199.968300						< 60 b.	0+			
$_{80}Hg^{201}$	201	80	13.2%	200.970277						8 b.	3/2-	-0.5602		
$_{80}Hg^{202}$	202	80	29.65%	201.970617						4.9 b.	0+			

Isotope	A	Z	% Natural abundance	Atomic mass	Half-life	Decay mode	Decay energy (MeV)	Particle energy (MeV)	Particle intensity	Thermal neutron cross section	Spin (h/2π)	μ Nucl. mag. moment	Gamma-ray energy (MeV)	Gamma-ray intensity
$_{80}Hg^{203}$	203	80		202.972848	46.6 ± 0.02 d.	β-	0.492	0.213	100%		5/2-	+0.8489	Tl k x-ray	6.3%
	203	80											0.279188 ± 0.00001	81.5%
$_{80}Hg^{204}$	204	80	6.8%	203.973467						0.4 b.	0+			
$_{80}Hg^{205}$	205	80		204.976047	5.2 ± 0.1 m.	β-	1.534	1.33	4%		$^{1}/_{2}$-	+0.601	0.20378 ± 0.00003	2.2%
	205	80											(0.2 - 1.4)weak	
$_{80}Hg^{206}$	206	80		205.977489	8.5 ± 0.1 m.	β-	1.309	0.935	34%		0+		Tl k x-ray	4%
	206	80						1.3	63%				0.3052 ± 0.0002	27%
	206	80											0.6502 ± 0.0002	2.5%
Tl		81		204.383						3.4 b.				
$_{81}Tl^{184}$	184	81		183.981670	11 s.	β+, EC (98%)	9.19						0.2868 ± 0.0003	39 +
	184	81				α(2%)		6.16					0.3399 ± 0.0003	25
	184	81						6.16					0.3667 ± 0.0003	100
	184	81											0.4188 ± 0.0003	9
	184	81											0.5342 ± 0.0003	17
	184	81											0.5541 ± 0.0003	5
	184	81											0.6083 ± 0.0003	11
	184	81											0.6168 ± 0.0003	8
	184	81											0.7222 ± 0.0003	3.5
$_{81}Tl^{185m}$	185	81			1.8 s.	I.T.	0.453				(9/2-)		0.1688 ± 0.0005	13 +
	185	81				α	5.97						0.2840 ± 0.0005	100
	185	81						6.01						
$_{81}Tl^{186m}$	186	81			4 s.	I.T.	0.374						0.3738 ± 0.0003	79%
$_{81}Tl^{186}$	186	81		185.978510	28 s.	β+,E.C.	8.53						0.3567 ± 0.0003	29%
	186	81											0.4026 ± 0.0003	45%
	186	81											0.4053 ± 0.0002	91%
	186	81											0.4241 ± 0.0002	13%
	186	81											0.4592 ± 0.0003	2.5%
	186	81											0.5975 ± 0.0002	4.2%
	186	81											0.6075 ± 0.0003	6.1%
	186	81											0.6755 ± 0.0003	14%
	186	81											0.7702 ± 0.0003	4.6%
	186	81											0.7884 ± 0.0004	2.6%
	186	81											0.8111 ± 0.0004	2.1%
	186	81											0.8264 ± 0.0004	2.2%
	186	81											1.2097 ± 0.0005	1.0%
	186	81											1.2476 ± 0.0005	1.3%
	186	81											1.2726 ± 0.0005	1.3%
$_{81}Tl^{187m}$	187	81			16 s.	I.T.	≈ 0.33				(9/2+)		0.2995 ± 0.0003	
$_{81}Tl^{187}$	187	81		186.976240	45 s.	β+,E.C.	6.04							
$_{81}Tl^{188m}$	188	81			71 s.	β+,E.C.					(7+)		Hg k x-ray	37%
	188	81											0.2917 ± 0.0001	3.5%
	188	81											0.3012 ± 0.0001	4.8%
	188	81											0.3269 ± 0.0001	9.4%
	188	81											0.3858 ± 0.0001	3.2%
	188	81											0.4129 ± 0.0001	88%
	188	81											0.4241 ± 0.0001	3.4%
	188	81											0.4527 ± 0.0001	2.5%
	188	81											0.4607 ± 0.0001	7.2%
	188	81											0.4682 ± 0.0001	5.0%
	188	81											0.5043 ± 0.0001	23%
	188	81											0.5693 ± 0.0001	3.4%
	188	81											0.5740 ± 0.0001	3.9%
	188	81											0.5921 ± 0.0001	61%
	188	81											0.7724 ± 0.0001	12%
	188	81											0.7952 ± 0.0001	10%
	188	81											0.8811 ± 0.0001	7.5%
	188	81											0.9048 ± 0.0001	11%
	188	81											1.0420 ± 0.0001	3.0%
	188	81											1.1705 ± 0.0004	2.1%
	188	81											1.4456 ± 0.00001	1.0%
$_{81}Tl^{188}$	188	81		187.975880	≈ 70 s.	β+,E.C.	7.73				(2-)		See Tal^{188m}	
	188	81											0.4129 ± 0.0001	
$_{81}Tl^{189m}$	189	81			1.4 m.	β+,E.C.					(9/2-)		0.2156	90 +
	189	81											0.2284	50
	189	81											0.3175	100
	189	81											0.4452	14
$_{81}Tl^{189}$	189	81		188.980780	2.3 m.	β+,E.C.	5.20				($^{1}/_{2}$+)		0.3337	100 +
	189	81											0.4510	49
	189	81											0.5223	27
	189	81											0.9422	69
$_{81}Tl^{190m}$	190	81			3.7 m.	β+,E.C.		4.2			(7+)		0.1968	4.5 +
	190	81											0.2401	3.4
	190	81											0.3053	15
	190	81											0.4164	91
	190	81											0.5439	5.6
	190	81											0.5570	6.3
	190	81											0.6154	4.2
	190	81											0.6254	82
	190	81											0.6838	7.0

Isotope	A	Z	% Natural abundance	Atomic mass	Half-life	Decay mode	Decay energy (MeV)	Particle energy (MeV)	Particle intensity	Thermal neutron cross section	Spin (h/2π)	μ Nucl. mag. moment	Gamma-ray energy (MeV)	Gamma-ray intensity
	190	81											0.6921	4.7
	190	81											0.7311	37
	190	81											0.8397	24
	190	81											1.0999	4.3
$_{81}Tl^{190}$	190	81		189.973490	2.6 m.	β+,E.C.	6.60	5.7			(2-)		0.4164	87 +
	190	81											0.6254	12
	190	81											0.6838	9.3
	190	81											1.0999	8.1
$_{81}Tl^{191m}$	191	81			5.2 m.	β+, EC (98%)					(9/2+)		0.2157 ± 0.0001	100 +
	191	81											0.2647 ± 0.0001	51
	191	81											0.3256 ± 0.0003	67
	191	81											0.3359 ± 0.0003	45
	191	81											0.3743 ± 0.0005	16
	191	81											0.3781 ± 0.0003	27
	191	81											0.4775 ± 0.0003	11
	191	81											0.5351 ± 0.0004	10
	191	81											0.5631 ± 0.0003	20
	191	81											0.5797 ± 0.0004	35
	191	81											0.6153 ± 0.0004	12
	191	81											0.6390 ± 0.0004	16
$_{81}Tl^{192m}$	192	81			10.8 m.	β+,E.C.					(7+)		0.1740 ± 0.0001	12%
	192	81											0.3839 ± 0.0002	3.0%
	192	81											0.4228 ± 0.0001	96%
	192	81											0.4517 ± 0.0003	2.8%
	192	81											0.5841 ± 0.0001	2.1%
	192	81											0.6348 ± 0.0001	88%
	192	81											0.7455 ± 0.0001	32%
	192	81											0.7863 ± 0.0001	38%
	192	81											1.1130 ± 0.0002	4.2%
	192	81											1.2505 ± 0.0003	1.3%
	192	81											1.3451 ± 0.0003	1.0%
	192	81											1.3655 ± 0.0003	1.3%
	192	81											1.4218 ± 0.0002	1.9%
$_{81}Tl^{192}$	192	81		191.972120	9.4 m.	β+,E.C.	6.02				(2-)		0.3975 ± 0.0003	6.5%
	192	81											0.4228 ± 0.0001	81%
	192	81											0.6908 ± 0.0001	12%
	192	81											0.7967 ± 0.0003	2.4%
	192	81											1.1711 ± 0.0004	2.0%
	192	81											1.4218 ± 0.0002	2.6%
	192	81											1.6335 ± 0.0002	2.0%
	192	81											1.6589 ± 0.0002	2.2%
	192	81											1.9084 ± 0.0003	1.6%
	192	81											2.0545 ± 0.001	1.2%
	192	81											2.1163 ± 0.0003	2.6%
	192	81											2.1674 ± 0.0004	1.2%
	192	81											2.2627 ± 0.0004	1.4%
	192	81											2.3000 ± 0.0004	1.9%
$_{81}Tl^{193m}$	193	81			2.1 m.	I.T.(75%)					(9/2-)		0.3650 ± 0.0001	67%
$_{81}Tl^{193}$	193	81		192.970520	22 m.	β+,E.C.	3.68				(1/2+)		0.2077 ± 0.0002	20 +
	193	81											0.2744 ± 0.0002	13
	193	81											0.2849 ± 0.0002	22
	193	81											0.3244 ± 0.0001	100
	193	81											0.3351 ± 0.0001	26
	193	81											0.3440 ± 0.0001	42
	193	81											0.4935 ± 0.0002	12
	193	81											0.6364 ± 0.0003	18
	193	81											0.6529 ± 0.0003	10
	193	81											0.6761 ± 0.0002	48
	193	81											0.6923 ± 0.0003	10
	193	81											0.7525 ± 0.0004	12
	193	81											0.7704 ± 0.0004	13
	193	81											0.9947 ± 0.0003	11
	193	81											1.0447 ± 0.0003	59
	193	81											1.1303 ± 0.0003	12
	193	81											1.2054 ± 0.0003	10
	193	81											1.2560 ± 0.0003	10
	193	81											1.5793 ± 0.001	45
$_{81}Tl^{194m}$	194	81			32.8 m.	β+(20%)	≈ 0.30				(7+)		ann.rad.	40%
	194	81				E.C.(80%)							Hg k x-ray	49%
	194	81											0.1110 ± 0.0001	6.4%
	194	81											0.2554 ± 0.0001	9.2%
	194	81											0.3198 ± 0.0001	3.9%
	194	81											0.4282 ± 0.0003	96%
	194	81											0.6363 ± 0.0003	98%
	194	81											0.6503 ± 0.0003	7%
	194	81											0.7350 ± 0.0003	22%
	194	81											0.7490 ± 0.0003	77%
$_{81}Tl^{194}$	194	81		193.970920	33 m.	β+,E.C.	5.15.				2-		0.3955 ± 0.0005	1.9%
	194	81											0.4039 ± 0.0007	2.3%
	194	81											0.4282 ± 0.0003	92%
	194	81											0.6363 ± 0.0003	21%
	194	81											0.6452 ± 0.0002	12%
	194	81											1.0403 ± 0.0005	5%
	194	81											1.0733 ± 0.0005	4%
$_{81}Tl^{195m}$	195	81			3.6 s.	I.T.	0.483				9/2-		Tl k x-ray	3.2%

TABLE OF THE ISOTOPES (Continued)

Isotope	A	Z	% Natural abundance	Atomic mass	Half-life	Decay mode	Decay energy (MeV)	Particle energy (MeV)	Particle intensity	Thermal neutron cross section	Spin (h/2π)	μ Nucl. mag. moment	Gamma-ray energy (MeV)	Gamma-ray intensity
	195	81											0.0990 ± 0.0001	0.6%
	195	81											0.3836 ± 0.0001	91%
$_{81}Tl^{195}$	195	81		194.969630	1.13 h.	E.C.(97%)	7.78	≈ 1.8			½+		ann.rad.	6%
	195	81				β+(3%)							Hg k x-ray	41%
	195	81											0.2422 ± 0.0001	4.5%
	195	81											0.2792 ± 0.0001	3.9%
	195	81											0.3006 ± 0.0001	2.5%
	195	81											0.5584 ± 0.0001	2.7%
	195	81											0.5635 ± 0.0001	11%
	195	81											0.8147 ± 0.0001	2%
	195	81											0.8845 ± 0.0001	10.5%
	195	81											0.9216 ± 0.0001	2.3%
	195	81											0.9675 ± 0.0001	2.2%
	195	81											1.1003 ± 0.0001	2.4%
	195	81											1.1217 ± 0.0001	2.6%
	195	81											1.2695 ± 0.0001	2.5%
	195	81											1.3639 ± 0.0001	9.1%
	195	81											1.5116 ± 0.0001	1.5%
	195	81											1.7059 ± 0.0002	1.7%
	195	81											1.7782 ± 0.0003	1.1%
	195	81											1.9778 ± 0.0002	1.8%
	195	81											2.0148 ± 0.0002	1.0%
	195	81											2.285 ± 0.001	0.64%
	195	81											2.5133 ± 0.0002	0.5%
	195	81											(0.13 - 2.5)many	
$_{81}Tl^{196m}$	196	81			1.4 h.	β+, EC	4.9				(7+)		0.0840 ± 0.0001	7%
						(95%)								
	196	81											0.3015 ± 0.0002	4%
	196	81											0.4261 ± 0.0001	91%
	196	81											0.6353 ± 0.0001	51%
	196	81											0.6954 ± 0.0005	41%
	196	81											1.0364 ± 0.001	2%
	196	81											(0.08 - 1.0)	
$_{81}Tl^{196}$	196	81		195.970460	1.8 h.	β+(15%)	4.34				2-		ann.rad.	30%
	196	81				E.C.(85%)							Hg k x-ray	34%
	196	81											0.4257 ± 0.0002	84%
	196	81											0.6105 ± 0.0005	12%
	196	81											0.6352 ± 0.0005	9.8%
	196	81											0.9646 ± 0.001	3.6%
	196	81											1.0362 ± 0.001	2.6%
	196	81											1.3890 ± 0.0005	2.5%
	196	81											1.4958 ± 0.0005	8.2%
	196	81											1.5530 ± 0.0007	4.8%
	196	81											1.6214 ± 0.002	4.9%
	196	81											1.6967 ± 0.002	3.0%
	196	81											1.7755 ± 0.001	2.8%
	196	81											2.0113 ± 0.0015	2.8%
	196	81											2.1278 ± 0.0025	2.8%
	196	81											2.2120 ± 0.002	3.4%
	196	81											(0.03 - 2.4)weak	
$_{81}Tl^{197m}$	197	81			0.54 s.	I.T.(53%)	0.608						Tl k x-ray	33%
	197	81				β+, EC							0.2262 ± 0.0003	5.2%
						(47%)								
	197	81											0.2596 ± 0.0003	2.9%
	197	81											0.2828 ± 0.0002	28%
	197	81											0.4118 ± 0.0001	55%
	197	81											0.4418 ± 0.0003	2.1%
	197	81											0.4896 ± 0.0003	4.4%
	197	81											0.5192 ± 0.0003	3.5%
	197	81											0.5872 ± 0.0002	51%
	197	81											0.6367 ± 0.0002	55%
	197	81											0.7673 ± 0.0003	1.1%
$_{81}Tl^{197}$	197	81		196.969498	2.83 h.	β+(1%)	2.15				½+	+1.58	Hg k x-ray	43%
	197	81				E.C.(99%)							0.1522 ± 0.0001	7.2%
	197	81											0.3086 ± 0.0002	2.2%
	197	81											0.4258 ± 0.0001	13%
	197	81											0.4331 ± 0.0001	2.5%
	197	81											0.5780 ± 0.0001	4.4%
	197	81											1.6743 ± 0.0002	1.4%
	197	81											0.7015 ± 0.0001	1.0%
	197	81											0.7921 ± 0.0001	1.7%
	197	81											0.8572 ± 0.0001	2.0%
	197	81											0.9827 ± 0.0001	1.2%
	197	81											1.3853 ± 0.0001	1.2%
	197	81											1.4113 ± 0.0001	4.5%
$_{81}Tl^{198m}$	198	81			1.87 h.	β+, EC					7+	0.64	Hg k x-ray	33%
						(53%)								
	198	81				I.T.(47%)	0.5347						Tl k x-ray	14%
	198	81											0.2262 ± 0.0003	5.2%
	198	81											0.2596 ± 0.0003	2.9%
	198	81											0.2828 ± 0.0002	28%
	198	81											0.4118 ± 0.0001	55%
	198	81											0.4896 ± 0.0003	4.4%
	198	81											0.5192 ± 0.0003	3.5%
	198	81											0.5872 ± 0.0002	57%
	198	81											0.6367 ± 0.0002	55%

Isotope	A	Z	% Natural abundance	Atomic mass	Half-life	Decay mode	Decay energy (MeV)	Particle energy (MeV)	Particle intensity	Thermal neutron cross section	Spin (h/2π)	μ Nucl. mag. moment	Gamma-ray energy (MeV)	Gamma-ray intensity
	198	81											0.7673 ± 0.0003	1.1%
$_{81}Tl^{198}$	198	81		197.940460	5.3 h.	β+(1%)	3.46	1.4			2-	0.00	Hg k x-ray	40%
	198	81						2.1					0.4118 ± 0.0001	82%
	198	81						2.4					0.6367 ± 0.0002	10%
	198	81											0.6759 ± 0.0001	11%
	198	81											1.0076 ± 0.0003	2.7%
	198	81											1.0876 ± 0.0003	2.4%
	198	81											1.2006 ± 0.0002	9.7%
	198	81											1.3122 ± 0.0002	4.7%
	198	81											1.4206 ± 0.0003	8.0%
	198	81											1.4354 ± 0.0003	3.5%
	198	81											1.4470 ± 0.0003	4.3%
	198	81											1.4896 ± 0.0003	2.6%
	198	81											1.5936 ± 0.0002	2.1%
	198	81											1.7208 ± 0.0003	2.8%
	198	81											1.8326 ± 0.0003	4.2%
	198	81											1.8993 ± 0.0003	2.2%
	198	81											2.0402 ± 0.0002	8.4%
	198	81											2.1905 ± 0.0003	2.7%
	198	81											2.4682 ± 0.0003	1.1%
	198	81											(0.23 - 2.8)weak	
$_{81}Tl^{199}$	199	81		198.969870	7.4 h.	E.C.	1.500				1/2-	+1.60	Hg k x-ray	46%
	199	81											0.1584 ± 0.0001	4.9%
	199	81											0.2082 ± 0.0001	12%
	199	81											0.2473 ± 0.0001	9.2%
	199	81											0.2841 ± 0.0001	2.2%
	199	81											0.3339 ± 0.0001	1.7%
	199	81											0.4034 ± 0.0001	1.5%
	199	81											0.4555 ± 0.0001	12%
	199	81											0.4923 ± 0.0001	1.5%
	199	81											0.7504 ± 0.0001	1.0%
	199	81											1.0129 ± 0.0001	1.7%
$_{81}Tl^{200}$	200	81		199.970934	1.09 ± 0.01 d.	E.C.	2.454	1.07			2-	0.04	Hg k x-ray	39%
	200	81						1.44					0.36799 ± 0.00001	87%
	200	81											0.57932 ± 0.00005	14%
	200	81											0.66145 ± 0.00004	2.3%
	200	81											0.8284 ± 0.0001	11%
	200	81											0.88618 ± 0.00005	2.0%
	200	81											1.2057 ± 0.0001	30%
	200	81											1.2255 ± 0.0001	3.4%
	200	81											1.3631 ± 0.0001	3.4%
	200	81											1.5150 ± 0.0001	4.0%
	200	81											1.6045 ± 0.0001	1.2%
	200	81											(0.11 - 2.3)many	
$_{81}Tl^{201}$	201	81		200.970794	3.05 ± 0.01 d.	E.C.	0.482				1/2+	+1.61	Hg k x-ray	38%
	201	81											0.13528 ± 0.00003	2.7%
	201	81											0.16582 ± 0.00004	0.2%
	201	81											0.16740 ± 0.00004	9.4%
$_{81}Tl^{202}$	202	81		201.972085	12.23±0.02 d.	E.C.	1.367				2-	0.06	Hg k x-ray	38%
	202	81											0.43957 ± 0.00001	91%
	202	81											0.52014 ± 0.00007	0.9%
	202	81											0.95971 ± 0.00007	0.1%
$_{81}Tl^{203}$	203	81	29.52%	202.972320						11.4 b.	1/2+	+1.6222		
$_{81}Tl^{204}$	204	81			3.78 ± 0.02 y.	β-(97%)	0.763	0.763	97%		2-	0.0908	Hg k x-ray	0.7%
	204	81				E.C.(3%)	0.345							
$_{81}Tl^{205}$	205	81	70.476%	204.974401						0.10 b.	1/2+	+1.6382		
$_{81}Tl^{206m}$	206	81			3.76 ± 0.02 m.	I.T.	2.644				12-		Tl k x-ray	20%
	206	81											0.2166 ± 0.001	89%
	206	81											0.2477 ± 0.001	9.7%
	206	81											0.2661 ± 0.0002	86%
	206	81											0.4534 ± 0.0008	94%
	206	81											0.4576 ± 0.0008	22%
	206	81											0.5644 ± 0.0008	13%
	206	81											0.6866 ± 0.0006	100%
	206	81											1.0219 ± 0.0008	76%
	206	81											1.1400 ± 0.0008	7.5%
$_{81}Tl^{206}$	206	81		205.976084	4.20 ± 0.02 m.	β-	1.531	1.53	99.9%		0-		Pb k x-ray	0.03%
	206	81											0.80313 ± 0.00005	0.005%

Isotope	A	Z	% Natural abundance	Atomic mass	Half-life	Decay mode	Decay energy (MeV)	Particle energy (MeV)	Particle intensity	Thermal neutron cross section	Spin ($h/2\pi$)	μ Nucl. mag. moment	Gamma-ray energy (MeV)	Gamma-ray intensity
$_{81}Tl^{207m}$	207	81			1.33 ± 0.1 s.	I.T.	1.350				11/2-		Tl k x-ray	12.5%
	207	81											0.3501 ± 0.002	79%
	207	81											1.0000 ± 0.002	87%
$_{81}Tl^{207}$	207	81		206.977404	4.77 ± 0.02 m.	β-	1.427	1.43	99.8%		1/2+		0.89723 ± 0.00007	0.24%
$_{81}Tl^{208}$	208	81		207.981988	3.052 ± 0.003 m.	β-	4.994	1.28	23%		(5+)		Pb k x-ray	3.6%
	208	81						1.52	22%				0.27728 ± 0.00006	6.8%
	208	81						1.796	51%				0.51061 ± 0.00002	22%
	208	81											0.58302 ± 0.00002	86%
	208	81											0.86030 ± 0.00006	12%
	208	81											2.61448 ± 0.00005	99.8%
$_{81}Tl^{209}$	209	81		208.985334	2.20 ± 0.07 m.	β-	3.976	1.8	100%		(1/2+)		Pb k x-ray	10%
	209	81											0.1172 ± 0.0001	81%
	209	81											0.4651 ± 0.0002	81%
	209	81											1.5669 ± 0.0001	98%
$_{81}Tl^{210}$	210	81		209.990056	1.30 ± 0.03 m.	β-	5.490	1.3	25%		(5+)		Pb k x-ray	4.6%
	210	81						1.9	56%				0.081 ± 0.003	2.0%
	210	81											0.2981 ± 0.001	79%
	210	81											0.79788 ± 0.0001	99%
	210	81											0.860 ± 0.002	7%
	210	81											1.068 ± 0.001	12%
	210	81											1.110 ± 0.001	7%
	210	81											1.208 ± 0.001	17%
	210	81											1.315 ± 0.001	21%
	210	81											1.408 ± 0.001	5%
	210	81											2.008 ± 0.001	7%
	210	81											2.268 ± 0.001	3%
	210	81											2.358 ± 0.002	8%
	210	81											2.428 ± 0.002	9%
Pb		82		207.2						0.171 b.				
$_{82}Pb^{184}$	184	82		183.988120	0.6 s.	α		6.63			0+			
$_{82}Pb^{185}$	185	82		184.987490	4.1 s.	α		6.34						
	185	82						6.40						
	185	82						6.40						
	185	82						6.48						
$_{82}Pb^{186}$	186	82		185.984300	8 s.	β+, EC (95%)	5.39				0+			
	186	82				α(5%)		6.32						
$_{82}Pb^{187m}$	187	82			18.3 s.	E.C.					13/2+		0.1930 ± 0.0003	15 +
	187	82				α		6.08					0.3314 ± 0.0003	60
	187	82											0.3435 ± 0.0003	75
	187	82											0.3934 ± 0.0003	100
$_{82}Pb^{187}$	187	82		186.983830	15.2 s.	β+,E.C.	7.07	5.99			(1/2-)		0.0674 ± 0.0003	
	187	82						6.19					0.2080 ± 0.0003	
	187	82											0.2755 ± 0.0003	
	187	82											0.2995 ± 0.0003	
	187	82											0.4487 ± 0.0003	
	187	82											0.7477 ± 0.0003	
$_{82}Pb^{188}$	188	82		187.980970	24 s.	E.C.(78%)	4.740				0+		0.1850	49%
	188	82				α(22%)		5.98					0.7582	29%
$_{82}Pb^{189}$	189	82		188.980780	51 s.	E.C.	6.50							
	189	82				α		5.58						
$_{82}Pb^{190}$	190	82		189.978070	β+(13%)	4.27					0+		ann.rad.	26%
	190	82				E.C.(86%)							Tl k x-ray	47%
	190	82				α(0.9%)		5.58					0.1415 ± 0.0005	11%
	190	82											0.1512 ± 0.0001	9.0%
	190	82											0.1582 ± 0.0002	1.7%
	190	82											0.1932 ± 0.0002	1.3%
	190	82											0.2105 ± 0.0002	3.6%
	190	82											0.2742 ± 0.0001	3.1%
	190	82											0.3627 ± 0.0002	1.9%
	190	82											0.3764 ± 0.0001	7.1%
	190	82											0.3817 ± 0.0002	1.8%
	190	82											0.5660 ± 0.0002	4.7%
	190	82											0.5983 ± 0.0002	8.1%
	190	82											0.7394 ± 0.0002	4.1%
	190	82											0.7909 ± 0.0002	3.0%
	190	82											0.9422 ± 0.0001	34%
	190	82											1.2355 ± 0.0002	4.6%
	190	82											1.8545 ± 0.0003	0.7%
$_{82}Pb^{191m}$	191	82			2.2 m.	β+,E.C.					13/2+		ann.rad.	
	191	82											0.3250 ± 0.0002	27 +
	191	82											0.34112 ± 0.0002	20

Isotope	A	Z	% Natural abundance	Atomic mass	Half-life	Decay mode	Decay energy (MeV)	Particle energy (MeV)	Particle intensity	Thermal neutron cross section	Spin (h/2π)	μ Nucl. mag. moment	Gamma-ray energy (MeV)	Gamma-ray intensity
	191	82											0.3871 ± 0.0002	100
	191	82											0.4040 ± 0.0002	11
	191	82											0.5606 ± 0.0002	27
	191	82											0.6135 ± 0.0002	40
	191	82											0.7057 ± 0.0002	16
	191	82											0.7122 ± 0.0002	46
	191	82											0.8739 ± 0.0002	23
	191	82											1.0933 ± 0.0002	15
$_{82}Pb^{191}$	191	82		190.978160	1.3 m.	β+,E.C.	5.90						ann.rad.	
	191	82											0.9368 ± 0.0002	
$_{82}Pb^{192}$	192	82		191.975790	≈ 3.5 m.	β+,E.C.	3.42				0+		ann.rad.	
	192	82											0.1675 ± 0.0001	14 +
	192	82											0.2131 ± 0.0003	4
	192	82											0.2149 ± 0.0003	5
	192	82											0.2507 ± 0.0002	5
	192	82											0.3710 ± 0.0002	8
	192	82											0.4141 ± 0.0003	6
	192	82											0.6082 ± 0.0001	18
	192	82											0.7816 ± 0.0003	9
	192	82											1.1954 ± 0.0002	48
$_{82}Pb^{193}$	193	82		192.976120	5.8 m.	β+,E.C.	5.22				13/2+		ann.rad.	
	193	82											0.3650	100 +
	193	82											0.3922	21
	193	82											0.7165	7
	193	82											0.7361	5
	193	82											0.7558	3
$_{82}Pb^{194}$	194	82		193.973980	10 m.	β+,E.C.	2.85				0+		ann.rad.	
	194	82											0.2036 ± 0.0005	
$_{82}Pb^{195m}$	195	82			15 m.	β+(8%)					13/2+		ann.rad.	16%
	195	82				E.C.(92%)							Tl k x-ray	44%
	195	82											0.3132 ± 0.0001	7%
	195	82											0.3836 ± 0.0001	92%(D)
	195	82											0.3942 ± 0.0001	44%
	195	82											0.4284 ± 0.0001	4%
	195	82											0.6076 ± 0.0002	8%
	195	82											0.7077 ± 0.0002	14%
	195	82											0.7422 ± 0.0002	4.2%
	195	82											0.8784 ± 0.0002	24%
	195	82											1.0679 ± 0.0002	6.3%
$_{82}Pb^{195}$	195	82		194.974480	37 m.	β+,E.C.	4.52						ann.rad.	
	195	82											0.0.3836 ± 0.0001	100 +
	195	82											0.3937 ± 0.0003	7
	195	82											0.7776 ± 0.0002	6
	195	82											0.8354 ± 0.0002	3
	195	82											0.8712 ± 0.0002	2
	195	82											0.8834 ± 0.0003	4
$_{82}Pb^{196}$	196	82		195.972680	37 m.	β+,E.C.	2.060				0+		Tl k x-ray	51%
	196	82											0.1977 ± 0.0005	11%
	196	82											0.2400 ± 0.0005	8%
	196	82											0.2531 ± 0.0005	27%
	196	82											0.3022 ± 0.0005	4%
	196	82											0.3665 ± 0.0005	11%
	196	82											0.4939 ± 0.0005	6%
	196	82											0.5021 ± 0.0005	26%
	196	82											0.9541 ± 0.0008	4%
$_{82}Pb^{197m}$	197	82			43 m.	E.C.(79%)					13/2+		Tl k x-ray	43%
	197	82				β+(2%)							0.3079 ± 0.0002	3%
	197	82				I.T.(19%)	0.3193						0.3877 ± 0.0001	25%
	197	82											0.4162 ± 0.0001	2%
	197	82											0.5578 ± 0.0001	3.5%
	197	82											0.6956 ± 0.0001	9.5%
	197	82											0.7241 ± 0.0001	3.7%
	197	82											0.7743 ± 0.0001	14%
	197	82											0.8933 ± 0.0001	2.1%
	197	82											0.9577 ± 0.0001	5.8%
	197	82											1.1177 ± 0.0001	3.2%
	197	82											1.4972 ± 0.0001	1.0%
	197	82											(0.2 - 2.2)weak	
$_{82}Pb^{197}$	197	82		196.973360	≈ 8 m.	E.C.(97%)	3.60				(3/2-)		Tl k x-ray	42%
	197	82				β+(3%)							0.3755 ± 0.0001	14%
	197	82											0.3858 ± 0.0001	56%
	197	82											0.7611 ± 0.0001	15%
	197	82											0.8716 ± 0.0001	6.8%
	197	82											0.8961 ± 0.0001	5.6%
	197	82											0.9017 ± 0.0001	3.5%
	197	82											1.0928 ± 0.0001	3.7%
	197	82											1.1561 ± 0.0001	4.2%
	197	82											1.2612 ± 0.0001	9.2%
	197	82											1.2889 ± 0.0001	5.4%
	197	82											1.6746 ± 0.0001	3.3%
	197	82											1.8540 ± 0.0001	6.8%
	197	82											2.3455 ± 0.0001	4.9%
$_{82}Pb^{198}$	198	82		197.971960	2.4 h.	E.C.	1.44				0+		Tl k x-ray	49%
	198	82											0.1734 ± 0.0001	18%
	198	82											0.2595 ± 0.0001	5.8%

Isotope	A	Z	% Natural abundance	Atomic mass	Half-life	Decay mode	Decay energy (MeV)	Particle energy (MeV)	Particle intensity	Thermal neutron cross section	Spin (h/2π)	μ Nucl. mag. moment	Gamma-ray energy (MeV)	Gamma-ray intensity
	198	82											0.2903 ± 0.0001	36%
	198	82											0.3654 ± 0.0001	19%
	198	82											0.3820 ± 0.0001	5.6%
	198	82											0.3977 ± 0.0001	2.9%
	198	82											0.5750 ± 0.0001	3.1%
	198	82											0.6490 ± 0.0001	1.8%
	198	82											0.7430 ± 0.0003	1.5%
	198	82											0.8653 ± 0.0001	5.9%
$_{82}Pb^{199m}$	199	82			12.2 m.	I.T.(93%)	0.4248				13/2+		Pb k x-ray	21%
	199	82				β+, EC (7%)							0.4255 ± 0.0005	18.5%
$_{82}Pb^{199}$	199	82		198.972870	1.5 h.	E.C.(99%)	2.800				5/2-		Tl k x-ray	42%
	199	82				β+(1%)							0.3534 ± 0.0001	14%
	199	82											0.4005 ± 0.0001	1.9%
	199	82											0.7202 ± 0.0001	9.5%
	199	82											0.7539 ± 0.0001	2.3%
	199	82											0.7620 ± 0.0001	3.3%
	199	82											0.7815 ± 0.0001	2.7%
	199	82											0.8748 ± 0.0001	2.4%
	199	82											0.9379 ± 0.0001	3.1%
	199	82											1.0292 ± 0.0001	2.4%
	199	82											1.1210 ± 0.0001	2.2%
	199	82											1.1350 ± 0.0001	11.5%
	199	82											1.2391 ± 0.0001	3.1%
	199	82											1.3827 ± 0.0001	4.2%
	199	82											1.5020 ± 0.0001	3.1%
	199	82											1.6584 ± 0.0001	8.2%
	199	82											1.7497 ± 0.0001	3.4%
	199	82											(0.22 - 2.4)weak	
$_{82}Pb^{200}$	200	82		199.971790	21.5 ± 0.4 h.	E.C.	0.800				0+		Tl k x-ray	51%
	200	82											0.14763 ± 0.00002	38%
	200	82											0.23562 ± 0.00002	4.3%
	200	82											0.25717 ± 0.00002	4.5%
	200	82											0.26837 ± 0.00002	4.0%
	200	82											0.28916 ± 0.00005	1.1%
	200	82											0.28992 ± 0.00002	1.7%
	200	82											0.45052 ± 0.00003	3.3%
$_{82}Pb^{201m}$	201	82			1.02 ± 0.03 m.	I.T.	0.6291				13/2+		Pb k x-ray	15%
	201	82											0.6288 ± 0.0005	54%
$_{82}Pb^{201}$	201	82		200.972830	9.33 ± 0.03 h.	E.C.	1.90				5/2-		Tl k x-ray	44%
	201	82											0.33120 ± 0.00003	79%
	201	82											0.36131 ± 0.00003	9.9%
	201	82											0.40607 ± 0.00004	2.0%
	201	82											0.58462 ± 0.00004	3.6%
	201	82											0.69252 ± 0.00003	4.3%
	201	82											0.76738 ± 0.00004	3.2%
	201	82											0.90764 ± 0.00005	5.7%
	201	82											0.94594 ± 0.00004	7.4%
	201	82											1.07009 ± 0.00005	1.1%
	201	82											1.09858 ± 0.00004	1.8%
	201	82											1.23884 ± 0.00005	1.2%
	201	82											1.27714 ± 0.00004	1.6%
	201	82											(0.11 - 1.8)weak	
$_{82}Pb^{202m}$	202	82			3.62 ± 0.03 h.	I.T.(90%)	2.170				9-		Pb k x-ray	2.3%
	202	82				β+(10%)							Tl k x-ray	3%
	202	82											0.42219 ± 0.00003	86%
	202	82											0.45979 ± 0.00007	8.6%
	202	82											0.49055 ± 0.00007	9.1%

Isotope	A	Z	% Natural abundance	Atomic mass	Half-life	Decay mode	Decay energy (MeV)	Particle energy (MeV)	Particle intensity	Thermal neutron cross section	Spin ($h/2\pi$)	μ Nucl. mag. moment	Gamma-ray energy (MeV)	Gamma-ray intensity
	202	82											0.65753 ± 0.00003	32%
	202	82											0.78700 ± 0.00006	50%
	202	82											0.96271 ± 0.00005	92%
$_{82}Pb^{202}$	202	82		201.972134	5.3 x 10^4 y.	E.C.	0.046				0+		Tl L x-ray	16%
$_{82}Pb^{203m}$	203	82			6.3 ± 0.2 s.	I.T.	0.8252				13/2+		Pb k x-ray	7.6%
	203	82											0.8203 ± 0.0002	6.4%
	203	82											0.8252 ± 0.0001	71%
$_{82}Pb^{203}$	203	82		202.973365	2.169 ± 0.004d	E.C.	0.974				5/2-		Tl k x-ray	42%
	203	82											0.279188 ± 0.00003	80.1%
	203	82											0.40131 ± 0.00001	3.4%
	203	82											0.68050 ± 0.00001	0.7%
$_{82}Pb^{204m}$	204	82			1.12 ± 0.01 h.	I.T.	2.185				9-		Pb k x-ray	4.5%
	204	82											0.37481 ± 0.00006	89%
	204	82											0.89922 ± 0.00007	99%
	204	82											0.91175 ± 0.00007	94%
$_{82}Pb^{204}$	204	82	1.4%	203.973020						0.66 b.	0+			
$_{82}Pb^{205}$	205	82		204.974458	1.51 x 10^7y.	E.C.	0.053						Tl L x-ray	16%
$_{82}Pb^{206}$	206	82	24.1%	205.974440						(0.006 + 0.025) b.	0+			
$_{82}Pb^{207m}$	207	82			0.796 s.	I.T.	1.632				13/2+		Pb k x-ray	4.8%
	207	82											0.56915 ± 0.00002	98%
	207	82											1.06310 ± 0.00002	89%
$_{82}Pb^{207}$	207	82	22.1%	206.975872						0.70 b.	1/2-	+0.5926		
$_{82}Pb^{208}$	208	82	52.4%	207.976627						0.5 mb.	0+			
$_{82}Pb^{209}$	209	82		208.981065	3.25 ± 0.02 h.	β-	0.644	0.645	100%		9/2+			
$_{82}Pb^{210}$	210	82		209.984163	22.3 ± 0.2 y.	β-	0.063	0.017	81%		0+			
	210	82						0.061	19%					
$_{82}Pb^{211}$	211	82		210.988735	36.1 ± 0.2 m.	β-	1.379	0.57	5%		(9/2+)		0.40486 ± 0.00003	3.8%
	211	82						1.36	92%				0.42700 ± 0.00003	1.7%
	211	82											0.83186 ± 0.00003	3.8%
	211	82											(0.09 - 1.27)weak	
$_{82}Pb^{212}$	212	82		211.991871	10.64±0.01 h.	β-	0.574	0.28	83%		0+		Bi k x-ray	18%
	212	82						0.57	12%				0.23858 ± 0.00001	43.6%
	212	82											0.30003 ± 0.00001	3.3%
$_{82}Pb^{213}$	213	82		212.996510	10.2 ± 0.3 m.	β-	2.00							
$_{82}Pb^{214}$	214	82		213.999798	26.8 ± 0.9 m.	β-	1.032	0.67	48%		0+		Bi k x-ray	11%
	214	82						0.73	42%				0.24192 ± 0.00003	7.5%
	214	82											0.29509 ± 0.00002	19.2%
	214	82											0.35187 ± 0.00004	37%
	214	82											0.78583 ± 0.00002	1.1%
Bi		83		208.9804						0.034 b.				
$_{83}Bi^{190}$	190	83		189.988480	5.4 s.	β+, EC (10%)	9.70							
	190	83				α(90%)		6.45						
$_{83}Bi^{191}$	191	83		190.986110	13 s.	β+, EC (60%)	0.370							
	191	83				α(40%)		6.32						
$_{83}Bi^{192}$	192	83		191.985400	42 s.	β+, EC (80%)	8.95							
	192	83				α(20%)		6.06						
$_{83}Bi^{193m}$	193	83			3.5 s.	β+,E.C.								
	193	83				α		6.48						
$_{83}Bi^{193}$	193	83		192.983180	64 s.	β+, EC (40%)	6.58							
	193	83				α(60%)		5.91						
$_{83}Bi^{194}$	194	83		193.982540	1.8 m.	β+, EC (99.9%)	7.98				(10-)		0.1661	46 +
	194	83				α(0.1%)							0.1740	28

Isotope	A	Z	% Natural abundance	Atomic mass	Half-life	Decay mode	Decay energy (MeV)	Particle energy (MeV)	Particle intensity	Thermal neutron cross section	Spin (h/2π)	μ Nucl. mag. moment	Gamma-ray energy (MeV)	Gamma-ray intensity
	194	83											0.2802	70
	194	83											0.421	55
	194	83											0.5754	87
	194	83											0.9650	100
$_{83}Bi^{195m}$	195	83			1.5 m.	β+, EC (94%)								
	195	83				α(6%)		6.11						
$_{83}Bi^{195}$	195	83		194.980700	2.8 m.	β+, EC (99.8%)	5.80							
	195	83				α(0.2%)		5.45						
$_{83}Bi^{196}$	196	83		195.980690	4.5 m.	E.C.	7.46						0.1376 ± 0.0003	10 +
	196	83											0.3368 ± 0.0003	16
	196	83											0.3720 ± 0.0006	46
	196	83											0.6880 ± 0.0005	62
	196	83											1.0486 ± 0.0005	100
$_{83}Bi^{197}$	197	83		196.978880	≈ 10 m.	β+,E.C.					1/2+			
$_{83}Bi^{198m}$	198	83			7.7 s.	I.T.	0.2485				(10-)		0.2485 ± 0.0005	38%
$_{83}Bi^{198}$	198	83		197.979000	β+,E.C.	6.55					(7+)		0.0900	8 +
	198	83											0.1381	2
	198	83											0.1976	80
	198	83											0.3179	37
	198	83											0.4343	7
	198	83											0.5465	3
	198	83											0.5624	79
	198	83											0.9173	5
	198	83											1.0635	100
$_{83}Bi^{199m}$	199	83			24.7 m.	β+,E.C.							ann.rad.	
$_{83}Bi^{199}$	199	83		198.977520	27 m.	β+,E.C.	4.32				9/2-		0.7203 ± 0.0005	1.9%
	199	83											0.7794 ± 0.0005	1.9%
	199	83											0.8374 ± 0.0005	13%
	199	83											0.8417 ± 0.0005	16%
	199	83											0.9141 ± 0.0005	2.1%
	199	83											0.9264 ± 0.0005	7.6%
	199	83											0.9460 ± 0.0005	15%
	199	83											0.9661 ± 0.0005	2.8%
	199	83											0.9775 ± 0.0005	2.3%
	199	83											1.0228 ± 0.0005	6.3%
	199	83											1.0340 ± 0.0005	8.4%
	199	83											1.0528 ± 0.0005	10%
	199	83											1.1370 ± 0.0005	7.9%
	199	83											1.1464 ± 0.0005	6.4%
	199	83											1.2122 ± 0.0005	6.2%
	199	83											1.3056 ± 0.0005	10%
	199	83											1.4486 ± 0.0005	2.5%
	199	83											1.5059 ± 0.0005	7.8%
	199	83											1.5173 ± 0.0005	2.1%
	199	83											1.5405 ± 0.0005	1.3%
	199	83											1.7808 ± 0.0005	1.4%
	199	83											1.9216 ± 0.0005	1.3%
	199	83											2.0215 ± 0.0005	1.5%
	199	83											2.0587 ± 0.0005	1.9%
	199	83											2.6669 ± 0.0005.	1.0%
	199	83											(0.12 - 3.2)many	
$_{83}Bi^{200m}$	200	83			31 m.	β+,E.C.					(2+)		0.2453 ± 0.0001	4.3%
	200	83											0.4198 ± 0.0001	20%
	200	83											0.4624 ± 0.0001	36%
	200	83											0.7127 ± 0.0001	1.5%
	200	83											1.0265 ± 0.0002	85%
	200	83											1.7395 ± 0.0002	3.7%
$_{83}Bi^{200}$	200	83		199.978090	36 m.	E.C.(90%)	5.86				7+		ann.rad.	20%
	200	83				β+(10%)							Pb k x-ray	49%
	200	83											0.2452 ± 0.0001	46%
	200	83											0.4198 ± 0.0001	91%
	200	83											0.4623 ± 0.0001	98%
	200	83											0.5455 ± 0.0002	4.5%
	200	83											0.6478 ± 0.0004	2.6%
	200	83											0.7810 ± 0.0005	2.0%
	200	83											0.9316 ± 0.0005	2.6%
	200	83											1.0265 ± 0.0002	100%
$_{83}Bi^{201m}$	201	83			59.1 ± 0.6 m.	I.T.	0.846				(1/2+)		Bi k x-ray	9 +
	201	83				β+,E.C.							0.8464	100
$_{83}Bi^{201}$	201	83		200.976930	1.80 ± 0.05 h.	E.C.	3.810				9/2-		Pb k x-ray	51%
	201	83											0.6288 ± 0.0005	24%
	201	83											0.7859 ± 0.0004	9.7%
	201	83											0.9015 ± 0.0005	8.5%
	201	83											0.9357 ± 0.0004	11%
	201	83											1.0138 ± 0.0007	11%
	201	83											1.3255 ± 0.001	6.1%
	201	83											1.5030 ± 0.0007	1.0%
	201	83											1.6509 ± 0.0005	5.9%
	201	83											(0.13 - 2.4)weak	
$_{83}Bi^{202}$	202	83		201.977660	1.72 h.	β+(3%)	5.15				5+		ann.rad.	6%
	202	83				E.C.(97%)							Pb k x-ray	43%

Isotope	A	Z	% Natural abundance	Atomic mass	Half-life	Decay mode	Decay energy (MeV)	Particle energy (MeV)	Particle intensity	Thermal neutron cross section	Spin (h/2π)	μ Nucl. mag. moment	Gamma-ray energy (MeV)	Gamma-ray intensity
	202	83											0.16815 ± 0.00004	4.8%
	202	83											0.24896 ± 0.00004	3.1%
	202	83											0.32018 ± 0.00005	3.1%
	201	83											0.34651 ± 0.00003	4.6%
	202	83											0.56931 ± 0.00003	4.8%
	202	83											0.57860 ± 0.00004	7.3%
	202	83											0.67622 ± 0.00003	1.9%
	202	83											0.85261 ± 0.00007	2.3%
	202	83											0.85848 ± 0.00004	1.6%
	202	83											0.92734 ± 0.00004	7.1%
	202	83											1.24553 ± 0.00003	2.8%
	202	83											1.55667 ± 0.00005	1.9%
	202	83											1.78056 ± 0.00006	0.7%
	202	83											2.3226 ± 0.000106	0.2%
	202	83											2.3408 ± 0.0007	0.2%
	202	83											(0.08 - 3.5)weak	
$_{83}Bi^{203}$	203	83		202.976830	11.76 ± 0.05h.	E.C. (99.8%)	3.22				9/2-	+4.62	Pb k x-ray	42%
	203	83				β+(0.2%)		1.35					0.1865 ± 0.0002	3.1%
	203	83											0.2642 ± 0.0003	5.2%
	203	83											0.3818 ± 0.0004	1.3%
	203	83											0.5694 ± 0.0004	1.2%
	203	83											0.6337 ± 0.0004	1.3%
	203	83											0.7224 ± 0.0003	4.8%
	203	83											0.8162 ± 0.0005	4.0%
	203	83											0.8203 ± 0.0002	30%
	203	83											0.8473 ± 0.0003	8.5%
	203	83											0.8666 ± 0.0005	1.5%
	203	83											0.8969 ± 0.0004	13%
	203	83											0.9334 ± 0.0004	1.4%
	203	83											1.0339 ± 0.0003	8.8%
	203	83											1.1986 ± 0.0004	2.0%
	203	83											1..2031 ± 0.0004	1.5%
	203	83											1.2537 ± 0.0005	1.3%
	203	83											1.5067 ± 0.0003	3.7%
	203	83											1.5365 ± 0.0004	7.5%
	203	83											1.5523 ± 0.0004	1.5%
	203	83											1.5929 ± 0.0005	1.09%
	203	83											1.6796 ± 0.0003	8.85%
	203	83											1.7198 ± 0.0003	3.4%
	203	83											1.7485 ± 0.0004	1.9%
	203	83											1.8475 ± 0.0003	11.4%
	203	83											1.8882 ± 0.0003	1.9%
	203	83											1.8931 ± 0.0003	8.2%
	203	83											1.9282 ± 0.001	1.1%
	203	83											2.0114 ± 0.0005	1.8%
	203	83											(0.1 - 2.9)many	
$_{83}Bi^{204}$	204	83		203.977740	11.2 ± 0.1 h.	E.C.	4.39				6+	+4.28	Pb k x-ray	44%
	204	83											0.17617 ± 0.00006	1.1%
	204	83											0.21608 ± 0.00007	1.4%
	204	83											0.21951 ± 0.00008	2.3%c
	204	83											0.24909 ± 0.00006	2.1%
	204	83											0.28926 ± 0.00007	2.8%
	204	83											0.37481 ± 0.00006	81%
	204	83											0.44056 ± 0.00005	2.5%
	204	83											0.53284 ± 0.00007	1.3%
	204	83											0.66161 ± 0.00007	2.6%
	204	83											0.67085 ± 0.00006	10.6%
	204	83											0.79130 ± 0.00007	3.2%

Isotope	A	Z	% Natural abundance	Atomic mass	Half-life	Decay mode	Decay energy (MeV)	Particle energy (MeV)	Particle intensity	Thermal neutron cross section	Spin (h/2π)	μ Nucl. mag. moment	Gamma-ray energy (MeV)	Gamma-ray intensity
	204	83											0.89922 ± 0.00007	98%
	204	83											0.91175 ± 0.00007	14%
	204	83											0.91231 ± 0.00008	11.1%
	204	83											0.91834 ± 0.00008	10.8%
	204	83											0.98409 ± 0.00005	58%
	204	83											1.11140 ± 0.00006	1.4%
	204	83											1.20392 ± 0.00008	2.1%
	204	83											1.21181 ± 0.00006	3.1%
	204	83											1.27481 ± 0.00007	2.2%
	204	83											1.70355 ± 0.00009	2.0%
	204	83											1.75534 ± 0.00006	1.2%
	204	83											1.89640 ± 0.00009	1.3%
	204	83											2.6808 ± 0.0002	0.4%
	204	83											2.8374 ± 0.0001	0.2%
$_{83}Bi^{205}$	205	83		204.977365	15.31 ± 0.04d.	E.C.	2.708				9/2-	4.16	Pb k x-ray	35%
	205	83											0.54986 ± 0.00003	0.3%
	205	83											0.57060 ± 0.00003	0.43%
	205	83											0.57974 ± 0.00003	0.54%
	205	83											0.70347 ± 0.00003	31%
	205	83											0.98764 ± 0.00003	1.6%
	205	83											1.04375 ± 0.00003	0.75%
	205	83											1.19004 ± 0.00004	0.23%
	205	83											1.61435 ± 0.00004	0.23%
	205	83											1.76435 ± 0.00004	32.5%
	205	83											1.86171 ± 0.00003	0.61%
	205	83											1.90345 ± 0.00004	0.25%
$_{83}Bi^{206}$	206	83		205.978478	6.243 ± 0.003d	E.C.	3.761				6+		Pb k x-ray	54%
	206	83											0.18403 ± 0.00002	15.8%
	206	83											0.34353 ± 0.00002	23.4%
	206	83											0.39803 ± 0.00002	10.7%
	206	83											0.49700 ± 0.00003	15.3%
	206	83											0.51619 ± 0.00003	40.7%
	206	83											0.53748 ± 0.00003	30.4%
	206	83											0.62053 ± 0.00003	5.8%
	206	83											0.63228 ± 0.00003	4.5%
	206	83											0.65721 ± 0.00003	1.9%
	206	83											0.80313 ± 0.00005	98.9%
	206	83											0.88100 ± 0.00003	66.2%
	206	83											0.89503 ± 0.00003	15.7%
	206	83											1.01856 ± 0.00003	7.6%
	206	83											1.09825 ± 0.00003	13.5%
	206	83											1.40510 ± 0.00004	1.4%
	206	83											1.59525 ± 0.00003	5.0%
	206	83											1.71878 ± 0.00003	31.8%

Isotope	A	Z	% Natural abundance	Atomic mass	Half-life	Decay mode	Decay energy (MeV)	Particle energy (MeV)	Particle intensity	Thermal neutron cross section	Spin (h/2π)	μ Nucl. mag. moment	Gamma-ray energy (MeV)	Gamma-ray intensity
	206	83											1.87898 ± 0.00005	2.0%
$_{83}Bi^{207}$	207	83		206.978446	32.2 ± 0.1 y.	E.C.	2.398				9/2-	4.10	Pb k x-ray	36%
	207	83											0.56915 ± 0.00002	97.8%
	207	83											1.06310 ± 0.00002	74.9%
	207	83											1.76971 ± 0.00004	6.9%
$_{83}Bi^{208}$	208	83		207.979717	3.68 x 10^5y.	E.C.	2.878				5+		Pb k x-ray	32%
	208	83											2.61435 ± 0.0001	99.8%
$_{83}Bi^{209}$	209	83	100%	208.980374						(10 mb + 24 mb)	9/2-	+4.110	6	
$_{83}Bi^{210m}$	210	83			3.0 x 10^6 y.	α		4.420(3)	0.29%		9-		Tl k x-ray	6.6%
	210	83						4.569(3)	3.9%				0.2661 ± 0.0002	50%
	210	83						4.584(3)	1.4%				0.3052 ± 0.0002	28%
	210	83						4.908(4)	39%				0.6502 ± 0.0002	3.6%
	210	83						4.946(3)	55%					
$_{83}Bi^{210}$	210	83		209.984095	5.01 ± 0.01 d.	β-	1.16	1.16	99%		1-	-0.044	0.2661 ± 0.0002	4×10^{-5}
	210	83											0.352 ± 0.0002	6×10 −
$_{83}Bi^{211}$	211	83		210.987255	2.14 ± 0.02 m.	α(99.7%)		6.279	16%		9/2-		Tl k x-ray	1.3%
	211	83				β-(0.3%)	0.584	6.623	84%				0.3501 ± 0.0002	12.8%
$_{83}Bi^{212m2}$	212	83			9 ± 1 m.	β-					(15-)			
$_{83}Bi^{212m1}$	212	83			25 ± 1 m.	α(93%)		6.300	40%		(9-)		0.120 ± 0.001	
	212	83				β-(7%)		6.340	53%				0.233 ± 0.001	
	212	83											0.275 ± 0.001	
	212	83											0.404 ± 0.001	
	212	83											0.727 ± 0.001	
$_{83}Bi^{212}$	212	83		211.991255	1.009 ± 0.001 h.	β-(64%)	2.248				(1-)		Tl k x-ray	0.13%
	212	83				α(36%)		6.051	25%				Po k x-ray	0.1%
	212	83						6.090	9.6%				0.2881 ± 0.0001	0.34%
	212	83											0.4528 ± 0.0001	0.36%
	212	83											0.72725 ± 0.00005	6.6%
	212	83											0.78551 ± 0.00005	1.1%
	212	83											0.87342 ± 0.00006	0.37%
	212	83											0.89342 ± 0.00006	0.37%
	212	83											0.9522 ± 0.0001	0.18%
	212	83											1.0787 ± 0.0001	0.53%
	212	83											1.51275 ± 0.00006	0.31%
	212	83											1.62066 ± 0.00006	1.5%
	212	83											1.8059 ± 0.0001	0.1%
$_{83}Bi^{211}$	211	83		212.994359	45.6 ± 0.1 m.	β-(98%)	1.422	1.02	31%		9/2-		Po k x-ray	1.3%
	213	83				α(2%)		1.42	66%				0.3288 ± 0.0001	0.2%
	213	83						5.549	0.16%				0.44034 ± 0.00002	16%
	213	833						5.869	2.0%				0.80727 ± 0.00004	0.26%
	213	83											1.10006 ± 0.00005	0.28%
$_{83}Bi^{214}$	214	83		213.998691	19.9 ± 0.4 m.	β-	3.27						0.60931 ± 0.00001	46.1%
	214	83											0.66544 ± 0.00002	1.6%
	214	83											0.76835 ± 0.00001	4.9%
	214	83											0.80615 ± 0.00002	1.2%
	214	83											0.93404 ± 0.00002	3.2%
	214	83											1.12027 ± 0.00002	15%
	214	83											1.15518 ± 0.00002	1.7%
	214	83											1.23810 ± 0.00002	5.96%
	214	83											1.28095 ± 0.00002	1.5%
	214	83											1.37766 ± 0.00002	4.0%
	214	83											1.40148 ± 0.00004	1.4%
	214	83											1.40797 ± 0.00004	2.5%
	214	83											1.50922 ± 0.00003	2.2%

Isotope	A	Z	% Natural abundance	Atomic mass	Half-life	Decay mode	Decay energy (MeV)	Particle energy (MeV)	Particle intensity	Thermal neutron cross section	Spin (h/2π)	μ Nucl. mag. moment	Gamma-ray energy (MeV)	Gamma-ray intensity
	214	83											1.66126 ± 0.00002	1.1%
	214	83											1.72958 ± 0.00002	3.0%
	214	83											1.76449 ± 0.00002	15.9%
	214	83											1.84741 ± 0.00003	2.1%
	214	83											2.20409 ± 0.00011	4.99%
	214	83											2.44768 ± 0.00004	1.55%
	214	83											(0.19 - 3.27)many	
$_{83}Bi^{215}$	215	83		215.001930	7.4 ± 0.6 m.	β-	2.250							
Po		84												
$_{84}Po^{194}$	194	84		193.988180	0.7 s.	α		6.85			0+			
$_{84}Po^{195m}$	195	84			2.0 s.	α		6.70						
$_{84}Po^{195}$	195	84		194.988010	≈ 4.5 s.	α		6.61						
$_{84}Po^{196}$	196	84		195.985540	≈ 5.5 s.	α(95%)		6.520			0+			
	196	84				β+, EC (5%)								
$_{84}Po^{197m}$	197	84			26 s.	α(84%)		6.385			13/2+			
	197	84				β+, EC (16%)								
$_{84}Po^{197}$	197	84		196.985600	56 s.	α(44%)		6.282			(3/2-)			
	197	84				β+, EC (56%)								
$_{84}Po^{198}$	198	84		197.983360	1.76 m.	α(70%)		6.182			0+			
	198	84				β+, EC (30%)								
$_{84}Po^{199m}$	199	84			4.2 m.	β+, EC (51%)					13/2+		ann.rad.	
	199	84				α(39%)		6.059					0.2745	7.4%
	199	84											0.4998 ± 0.0005	25%
	199	84											1.0020 ± 0.0005	60%
$_{84}Po^{199}$	199	84		198.983610	5.2 m.	β+, EC (88%)	5.67				(3/2-)		Bi k x-ray	38%
	199	84				α(12%)		5.952					0.1877 ± 0.0005	7.5%
	199	84											0.2291 ± 0.0005	4.8%
	199	84											0.2335 ± 0.0005	5.5%
	199	84											0.2460 ± 0.0006	4.2%
	199	84											0.2607 ± 0.0005	4.0%
	199	84											0.3616 ± 0.0005	22%
	199	84											0.3978 ± 0.0004	4%
	199	84											0.4749 ± 0.0005	7%
	199	84											0.5068 ± 0.0005	3.7%
	199	84											0.9984 ± 0.0005	15%
	199	84											1.0214 ± 0.0006	24%
	199	84											1.0344 ± 0.0005	47%
$_{84}Po^{200}$	200	84		199.981700	11.5 ± 0.1 m.	β+, EC (85%)	3.370				0+		0.14748 ± 0.00001	4.4%
	200	84				α(15%)		5.863					0.32792 ± 0.00009	2.6%
	200	84											0.43007 ± 0.00012	4.8%
	200	84											0.4343 ± 0.00001	9.3%
	200	84											0.6176 ± 0.0001	19.7%
	200	84											0.6709 ± 0.0001	34%
	200	84											0.6956 ± 0.0002	5.5%
	200	84											0.7966 ± 0.0001	7.9%
	200	84											0.8499 ± 0.0001	4.9%
	200	84											0.8758 ± 0.0001	1.8%
	200	84											0.8958 ± 0.0002	1.5%
	200	84											0.9146 ± 0.0002	1.2%
	200	84											0.9455 ± 0.0001	1..1%
	200	84											1.0845 ± 0.0002	3.8%
	200	84											1.1729 ± 0.0002	1.1%
	200	84											1.2856 ± 0.0001	1.2%
	200	84											1.3876 ± 0.0002	1.0%
	200	84											1.8019 ± 0.0002	1.3%
$_{84}Po^{201m}$	201	84			8.9 ± 0.2 m.	β+, EC (57%)					13/2+		Bi k x-ray	19%
	201	84				I.T.(40%)	0.418						Po k x-ray	44%
	201	84				α(3%)		5.786					0.2726 ± 0.0004	2.8%
	201	84											0.4123 ± 0.0005	15.1%
	201	84											0.4179 ± 0.0003	33%
	201	84											0.9670 ± 0.0005	34%
$_{84}Po^{201}$	201	84		200.982190	15.3 ± 0.2 m.	β+, EC (98%)	4.90				3/2-		Bi k x-ray	38%
	201	84				α(2%)		5.683(3)					0.2056 ± 0.0003	8.0%
	201	84											0.2229 ± 0.0004	5.5%
	201	84											0.2250 ± 0.0004	12%
	201	84											0.2390 ± 0.0005	8.1%
	201	84											0.4285 ± 0.0004	8.9%

Isotope	A	Z	% Natural abundance	Atomic mass	Half-life	Decay mode	Decay energy (MeV)	Particle energy (MeV)	Particle intensity	Thermal neutron cross section	Spin (h/2π)	μ Nucl. mag. moment	Gamma-ray energy (MeV)	Gamma-ray intensity
	201	84											0.5375 ± 0.0004	5.5%
	201	84											0.5520 ± 0.0004	6.3%
	201	84											0.6390 ± 0.0004	5.4%
	201	84											0.8483 ± 0.0005	13%
	201	84											0.8904 ± 0.0004	5.4%
	201	84											0.9048 ± 0.0005	29%
	201	84											1.1639 ± 0.0005	3.7%
	201	84											1.2060 ± 0.0005	3.3%
$_{84}Po^{202}$	202	84		201.980680	44.7 ± 0.5 m.	β+, EC (98%)	2.820				0+		0.0410 ± 0.0001	3.0%
	201	84				α(2%)		5.588					0.1656 ± 0.0001	8.7%
	202	84											0.2135 ± 0.0002	3.4%
	202	84											0.3158 ± 0.0002	14%
	202	84											0.3365 ± 0.0002	2%
	202	84											0.4275 ± 0.0002	1.6%
	202	84											0.4581 ± 0.0002	3.8%
	202	84											0.5061 ± 0.0002	4.4%
	202	84											0.5977 ± 0.0002	2.6%
	202	84											0.6433 ± 0.0002	3.6%
	202	84											0.6884 ± 0.0003	51%
	202	84											0.7126 ± 0.0002	4.6%
	202	84											0.7168 ± 0.0002	6.1%
	202	84											0.7903 ± 0.0005	7.2%
	202	84											0.9736 ± 0.0002	4.9%
	202	84											1.1684 ± 0.0005	2.0%
	202	84											1.2148 ± 0.0005	1.7%
$_{84}Po^{203m}$	203	84			1.2 ± 0.2 m.	I.T.(96%)	0.6414				13/2+		Bi k x-ray	2.0%
	203	84				β,EC(4%)							Po k x-ray	14%
	203	84											0.5770 ± 0.0005	2.4%
	203	84											0.6414 ± 0.0001	50%
	203	84											0.9049 ± 0.0005	4.4%
$_{84}Po^{203}$	203	84		202.981370	34.8 ± 0.1 m.	β+,E.C.	4.24				5/2-		0.17516 ± 0.00006	3.0%
	203	84											0.18951 ± 0.00008	3.9%
	203	84											0.21477 ± 0.00006	14.5%
	203	84											0.41942 ± 0.00006	2.5%
	203	84											0.4861 ± 0.0001	2.1%
	203	84											0.64776 ± 0.00007	2.1%
	203	84											0.82291 ± 0.00007	2.4%
	203	84											0.8835 ± 0.001	2.0%
	203	84											0.89350 ± 0.00008	19.0%
	203	84											0.90863 ± 0.00007	56%
	203	84											1.09095 ± 0.00007	19.6%
	203	84											1.20233 ± 0.00007	4.7%
	203	84											1.33758 ± 0.00009	3.0%
	203	84											1.35282 ± 0.00008	1.4%
	203	84											1.8175 ± 0.0001	1.1%
	203	84											1.9308 ± 0.0005	0.9%
	203	84											2.0295 ± 0.0003	0.6%
	203	84											2.2369 ± 0.0001	0.6%
$_{84}Po^{204}$	204	84		203.980280	3.53 ± 0.02 h.	E.C.	2.37				0+		Bi k x-ray	65%
	204	84											0.1370 ± 0.0003	12.0%
	204	84											0.20368 ± 0.00007	3.4%
	204	84											1.2300 ± 0.0001	3.0%
	204	84											0.2702 ± 0.0001	31%
	204	84											0.3049 ± 0.0003	3.4%
	204	84											0.3167 ± 0.0001	4.9%
	204	84											0.4270 ± 0.0008	2.2%
	204	84											0.45194 ± 0.00007	2.7%
	204	84											0.4601 ± 0.0001	1.5%
	204	84											0.5349 ± 0.0003	12.9%
	204	84											0.5401 ± 0.0004	1.8%
	204	84											0.6807 ± 0.0001	9.2%
	204	84											0.6951 ± 0.0004	2.5%
	204	84											0.76265 ± 0.00007	11.4%
	204	84											0.8844 ± 0.0001	34%
	204	84											1.0162 ± 0.0001	24.6%
	204	84											1.0400 ± 0.0003	10.8%
	204	84											(0.11 - 1.9)weak	

Isotope	A	Z	% Natural abundance	Atomic mass	Half-life	Decay mode	Decay energy (MeV)	Particle energy (MeV)	Particle intensity	Thermal neutron cross section	Spin (h/2π)	μ Nucl. mag. moment	Gamma-ray energy (MeV)	Gamma-ray intensity
$_{84}Po^{205}$	205	84		204.981150	1.80 ± 0.04 h.	β+,E.C.	3.53				5/2-	+0.26	Bi k x-ray	45%
	205	84											0.21202 ± 0.00007	3.6%
	205	84											0.26108 ± 0.00007	4.0%
	205	84											0.59983 ± 0.00009	2.6%
	205	84											0.61426 ± 0.00007	1.6%
	205	84											0.62478 ± 0.0006	1.0%
	205	84											0.83681 ± 0.00006	19%
	205	84											0.84983 ± 0.00007	25%
	205	84											0.87241 ± 0.00007	37%
	205	84											1.00124 ± 0.00007	29%
	205	84											1.2391 ± 0.0001	4.6%
	205	84											1.5137 ± 0.0002	2.1%
	205	84											1.5519 ± 0.0001	2.9%
	205	84											1.7297 ± 0.0001	1.6%
	205	84											1.8112 ± 0.0001	1.2%
	205	84											2.1689 ± 0.0002	0.4%
	205	84											(0.12 - 2.77)weak	
$_{84}Po^{206}$	206	84		205.980456	8.8 ± 0.1 d.	E.C.(95%)	1.843				0+		Bi k x-ray	46%
	206	84				α(5%)		5.223					0.28644 ± 0.00002	24%
	206	84											0.31156 ± 0.00002	4.2%
	206	84											0.33844 ± 0.00002	19.2%
	206	84											0.46334 ± 0.00002	1.8%
	206	84											0.51134 ± 0.00002	24%
	206	84											0.52252 ± 0.00003	15.7%
	206	84											0.80737 ± 0.00002	22.7%
	206	84											0.86096 ± 0.00002	3.5%
	206	84											0.98027 ± 0.00003	7.1%
	206	84											1.00716 ± 0.00002	3.1%
	206	84											1.03228 ± 0.00002	32.9%
	206	84											1.19102 ± 0.00003	0.5%
	206	84											0.31871 ± 0.00002	0.65%
	206	84											(0.11 - 1.5)weak	
$_{84}Po^{207m}$	207	84			2.8 ± 0.2 s.	I.T.	1.383				19/2		Po k x-ray	28%
	207	84											0.2682 ± 0.001	45%
	207	84											0.30074 ± 0.00006	33%
	207	84											0.81448 ± 0.00006	99%
$_{84}Po^{207}$	207	84		206.981570	5.83 ± 0.07 h.	2.910					5/2-	+0.27	Bi k x-ray	41%
	207	84											0.24962 ± 0.00007	1.6%
	207	84											0.34521 ± 0.00002	2.0%
	207	84											0.36953 ± 0.00008	1.9%
	207	84											0.40570 ± 0.00006	10.1%
	207	84											0.68764 ± 0.00007	2.0%
	207	84											0.74263 ± 0.00006	29.2%
	207	84											0.91176 ± 0.00007	18.0%
	207	84											0.99225 ± 0.00007	60%
	207	84											1.14833 ± 0.00006	6.1%
	207	84											1.37244 ± 0.00008	1.4%
	207	84											2.06008 ± 0.00008	1.44%
$_{84}Po^{208}$	208	84		207.981222	2.898 y.	α	5.213	4.233	0.00	02	0+			

Isotope	A	Z	% Natural abundance	Atomic mass	Half-life	Decay mode	Decay energy (MeV)	Particle energy (MeV)	Particle intensity	Thermal neutron cross section	Spin (h/2π)	μ Nucl. mag. moment	Gamma-ray energy (MeV)	Gamma-ray intensity
	208	84						5.1158	100%					
$_{84}Po^{209}$	209	84		208.982404	105 ± 5 y.	α	4.976	4.624	0.56	%	1/2-	+0.77	0.26049 ± 0.00003	0.17%
	208	84						4.879	99.2	%			0.8964 ± 0.0002	0.25%
$_{84}Po^{210}$	210	84		209.982848	138.4 d.	α	5.407	4.516	0.00	1%	0+		0.80313 ± 0.00005	0.001%
	210	84						5.304	100%					
$_{84}Po^{211m}$	211	84			25.5 ± 0.3 s.	α		7.273	91%		25/2	+	Pb k x-ray	4.5%
	211	84						7.994	1.7%				0.32808 ± 0.00007	0.01%
	211	84						8.316	0.25	%			0.56915 ± 0.00002	92%
	211	84						8.875	7.0%				0.89723 ± 0.00007	1.6%
	211	84											1.06310 ± 0.00002	83.2%
$_{84}Po^{211}$	211	84		210.986627	0.52 s.	α	7.594	6.570	0.54	%	9/2+		0.56915 ± 0.00002	0.53%
	211	84						6.892	0.55	%			0.89723 ± 0.00007	0.52%
	211	84						7.450	98.9	%				
$_{84}Po^{212m}$	212	84			45.1 ± 0.6 s.	α		8.514	2.0%		16+			
	212	84						9.086	1.0%					
	212	84						11.650	97%					
$_{84}Po^{212}$	212	84		211.988842	0.3 μs.	α	8.953	8.784			0+			
$_{84}Po^{213}$	213	84		212.992833	≈ 4.2 μs.	α	8.537	7.614	0.00	3%	9/2+			
	213	84						8.375	100%					
$_{84}Po^{214}$	214	84		213.995176	163 μs.	α	7.833	6.904	0.01	%	0+			
	214	84						7.686	99.9	9%				
$_{84}Po^{215}$	215	84		214.999419	1.78 ms.	α	7.526	6.950	0.02	%	(9/2	+)		
	215	84						6.957	0.03	%				
	215	84						7.386	100%					
$_{84}Po^{216}$	216	84		216.001889	0.15 s.	α	6.906	5.895	0.00	2%	0+			
	216	84						6.778	99.9	9%				
$_{84}Po^{217}$	217	84		217.006260	<10 s.	α	6.662	6.539						
$_{84}Po^{218}$	218	84			3.11 ± 0.02 m.	α	6.114	5.181	1.00	%	0+			
	218	84						6.002	100%					
At		85												
$_{85}At^{196}$	196	85		195.995730	0.3 s.	α		7.06						
$_{85}At^{197}$	197	85		196.993410	0.4 s.	β+,E.C.	7.28				(9/2-)			
	197	85				α		6.96						
$_{85}At^{198m}$	198	85			1.5 s.	β+, EC (75%)								
	198	85				α(25%)		6.85						
$_{85}At^{198}$	198	85		197.992550	≈ 4.9 s.	α		6.75						
$_{85}At^{199}$	199	85		198.990580	7.0 s.	β+, EC (8%)	6.500				9/2-			
	199	85				α(92%)		6.64						
$_{85}At^{200m}$	200	85			4.3 ± 0.3 s.	β+, EC (80%)					10-			
	200	85				α(20%)		6.536						
$_{85}At^{200}$	200	85		199.990370	43 ± 5 s.	β+, EC (65%)	8.08				5+			
	200	85				α(35%)		6.412	21%					
	200	85						6.465	14%					
$_{85}At^{201}$	201	85		200.988440	1.48 ± 0.05 s.	β+, EC (29%)	5.82				9/2-			
	201	85				α(71%)	6.474	6.344						
$_{85}At^{202m}$	202	85			1.1 s.	I.T.	0.391							
$_{85}At^{202}$	202	85		201.988420	3.02 ± 0.05 m.	β+, EC (88%)	7.21				5+		ann.rad.	
	202	85				α(12%)		6.135	7.7%				0.4413 ± 0.0003	41%
	202	85						6.225	4.3%				0.5697 ± 0.0004	81%
	202	85											0.6753 ± 0.0005	87%
$_{85}At^{203}$	203	85		202.986790	7.4 ± 0.2 m.	β+, EC (69%)	5.040				9/2-		0.1458 ± 0.0001	14 +
	203	85				α(31%)	6.210	6.088					0.2459 ± 0.0002	48
	203	85											0.3616 ± 0.0003	23
	203	85											0.4169 ± 0.0001	14
	203	85											0.5319 ± 0.0001	18
	203	85											0.6088 ± 0.0001	20
	203	85											0.6414 ± 0.0001	53
	203	85											0.6562 ± 0.0001	30
	203	85											0.7379 + 0.0001	42
	203	85											0.8458 ± 0.0001	30
	203	85											0.8804 ± 0.0001	41
	203	85											1.0020 ± 0.0001	86
	203	85											1.0340 ± 0.0001	100
$_{85}At^{204}$	204	85		203.987210	9.2 ± 0.2 m.	β+, EC (95%)	6.45				(5+)		Po k x-ray	27%
	204	85				α(5%)		5.951					0.3271 ± 0.0007	4.7%
	204	85											0.3367 ± 0.0007	5.6%

Isotope	A	Z	% Natural abundance	Atomic mass	Half-life	Decay mode	Decay energy (MeV)	Particle energy (MeV)	Particle intensity	Thermal neutron cross section	Spin (h/2π)	μ Nucl. mag. moment	Gamma-ray energy (MeV)	Gamma-ray intensity
	204	85											0.4254 ± 0.0003	66%
	204	85											0.4904 ± 0.001	4.7%
	204	85											0.5156 ± 0.0003	90%
	204	85											0.5888 ± 0.001	8.5%
	204	85											0.6084 ± 0.0006	19.8%
	204	85											0.6837 ± 0.0005	94%
	204	85											0.7621 ± 0.0007	4.7%
	204	85											0.8427 ± 0.0007	8.5%
$_{85}At^{205}$	205	85		204.986000	26.2 ± 0.5 m.	β+, EC (90%)	4.51				(9/2	-)	Po k x-ray	47%
	205	85				α(10%)	6.020	5.902					0.1543 ± 0.0002	2.4%
	205	85											0.1610 ± 0.0001	1.2%
	205	85											0.3114 ± 0.0001	3.4%
	205	85											0.4488 ± 0.0001	1.4%
	205	85											0.5207 ± 0.0001	3.5%
	205	85											0.6179 ± 0.0001	1.9%
	205	85											0.6209 ± 0.0001	4.6%
	205	85											0.6596 ± 0.0001	2.0%
	205	85											0.6696 ± 0.0001	8.1%
	205	85											0.6729 ± 0.0001	3.0%
	205	85											0.7194 ± 0.0001	27%
	205	85											0.7832 ± 0.0001	1.6%
	205	85											0.8724 ± 0.0005	2.0%
	205	85											1.3254 ± 0.0003	1.1%
	205	85											1.4758 ± 0.0003	0.6%
	205	85											1.4792 ± 0.0003	0.6%
$_{85}At^{206}$	206	85		205.986580	29.4 ± 0.3 m.	β+, EC (99%)	5.70				5+		Po k x-ray	35%
	206	85				α(1%)	5.881	5.703					0.20186 ± 0.0006	5.4%
	206	85											0.23354 ± 0.00006	3.1%
	206	85											0.25658 ± 0.00004	4.4%
	206	85											0.2756 ± 0.0001	2.0%
	206	85											0.27899 ± 0.00004	2.6%
	206	85											0.38678 ± 0.00006	2.6%
	206	85											0.39561 ± 0.00004	48%
	206	85											0.47716 ± 0.00003	86%
	206	85											0.52742 ± 0.00006	2.9%
	206	85											0.56562 ± 0.00006	3.2%
	206	85											0.70071 ± 0.00003	98%
	206	85											0.70464 ± 0.00006	6.0%
	206	85											0.73375 ± 0.00004	10.1%
	206	85											0.86831 ± 0.00004	7.6%
	206	85											0.92303 ± 0.00006	5.6%
	206	85											1.01394 ± 0.00007	2.9%
	206	85											1.04811 ± 0.00007	2.2%
	206	85											1.05938 ± 0.00004	3.4%
	206	85											1.12489 ± 0.00004	1.8%
	206	85											0.1969 ± 0.0001	1.5%
	206	85											0.2576 ± 0.0001	1.2%
	206	85											1.44616 ± 0.00007	1.3%
	206	85											1.63746 ± 0.00009	1.2%
	206	85											1.93813 ± 0.00007	1.3%
$_{85}At^{207}$	207	85		206.985730	1.81 h.	β+, EC (90%)	3.88				9/2-		Po k x-ray	46%
	207	85				α(10%)	5.873	5.758					0.16801 ± 0.00007	1.1%
	207	85											0.30074 ± 0.00006	9.7%
	207	85											0.35733 ± 0.00007	1.8%
	207	85											0.4220 ± 0.0001	1.4%
	207	85											0.45681 ± 0.00007	1.3%
	207	85											0.45966 ± 0.00008	1.1%

TABLE OF THE ISOTOPES (Continued)

Isotope	A	Z	% Natural abundance	Atomic mass	Half-life	Decay mode	Decay energy (MeV)	Particle energy (MeV)	Particle intensity	Thermal neutron cross section	Spin (h/2π)	μ Nucl. mag. moment	Gamma-ray energy (MeV)	Gamma-ray intensity
	207	85											0.46720 ± 0.00005	5.3%
	207	85											0.52991 ± 0.00006	2.5%
	207	85											0.58842 ± 0.00006	14.7%
	207	85											0.61720 ± 0.00001	1.1%
	207	85											0.62680 ± 0.00007	1.7%
	207	85											0.63740 ± 0.00007	1.7%
	207	85											0.64809 ± 0.00006	3.3%
	207	85											0.65848 ± 0.00006	5.0%
	207	85											0.67065 ± 0.00007	2.8%
	207	85											0.67521 ± 0.00007	4.9%
	207	85											0.6936 ± 0.0001	1.6%
	207	85											0.72119 ± 0.00005	4.9%
	207	85											0.81448 ± 0.00006	33%
	207	85											0.90721 ± 0.00007	4.0%
	207	85											0.96059 ± 0.00009	1.7%
	207	85											0.99401 ± 0.00006	1.7%
	207	85											1.07780 ± 0.00007	1.4%
	207	85											1.11529 ± 0.00006	3.3%
	207	85											1.22582 ± 0.00007	1.1%
	207	85											1.39640 ± 0.00006	1.0%
	207	85											1.67682 ± 0.00007	2.0%
	207	85											1.71270 ± 0.00009	1.0%
	207	85											1.7310 ± 0.0001	2.8%
	207	85											2.7122 ± 0.0002	0.9%
$_{85}At^{208}$	208	85		207.986510	1.63 ± 0.03 h.	β+, EC (99%)	4.93				(6+)		Po k x-ray	38%
	208	85				α(1%)	5.752	5.626	0.01	%			0.1770 ± 0.0006	46%
	208	85						5.641	0.53	%			0.2060 ± 0.0008	5.3%
	208	85											0.5170 ± 0.0007	7.0%
	208	85											0.6311 ± 0.0008	4.3%
	208	85											0.6601 ± 0.0001	90%
	208	85											0.6852 ± 0.001	98%
	208	85											0.8081 ± 0.0008	8.4%
	208	85											0.8450 ± 0.0007	21%
	208	85											0.8961 ± 0.001	6.0%
	208	85											0.9861 ± 0.0002	9%
	208	85											0.9931 ± 0.0007	14%
	208	85											1.0281 ± 0.001	27%
	208	85											1.2311 ± 0.002	3.3%
	208	85											1.2801 ± 0.0008	3.8%
	208	85											1.5391 ± 0.0007	1.6%
	208	85											2.0281 ± 0.0002	1.5%
	208	85											2.6361 ± 0.002	2.3%
$_{85}At^{209}$	209	85		208.986149	5.41 ± 0.05 h.	β+, EC (96%)	4.93				(6+)		Po k x-ray	38%
	209	85				α(4%)	5.757	5.647	4.1%				0.10422 ± 0.00007	2.4%
	209	85											0.19505 ± 0.00006	22.6%
	209	85											0.23916 ± 0.00006	12.4%
	209	85											0.54503 ± 0.00007	91%
	209	85											0.55103 ± 0.00005	4.9%
	209	85											0.78189 ± 0.00006	83%
	209	85											0.79020 ± 0.00006	63%
	209	85											0.86399 ± 0.00006	2.1%
	209	85											0.90315 ± 0.00006	3.6%

Isotope	A	Z	% Natural abundance	Atomic mass	Half-life	Decay mode	Decay energy (MeV)	Particle energy (MeV)	Particle intensity	Thermal neutron cross section	Spin (h/2π)	μ Nucl. mag. moment	Gamma-ray energy (MeV)	Gamma-ray intensity
	209	85											1.10351 ± 0.00007	5.4%
	209	85											1.17076 ± 0.00006	3.1%
	209	85											1.26261 ± 0.0006	1.9%
	209	85											1.58168 ± 0.00006	1.8%
	209	85											1.76713 ± 0.00005	0.5%
	209	85											(0.1 - 2.6)many	
$_{85}At^{210}$	210	85		209.987126	8.1 ± 0.4 h.	E.C. (99.8%)	3.98				5+		Po k x-ray	40%
	210	85				α(0.2%)	5.632	5.361	0.05%				0.24535 ± 0.00008	79%
	210	85						5.442	0.05%				0.52758 ± 0.00007	1.1%
	210	85											0.81723 ± 0.00009	1.7%
	210	85											0.8527 ± 0.0001	1.4%
	210	85											0.95577 ± 0.00007	1.8%
	210	85											1.18143 ± 0.00009	99%
	210	85											1.43678 ± 0.00006	29%
	210	85											1.48335 ± 0.00005	46%
	210	85											1.59956 ± 0.00006	13%
	210	85											2.25401 ± 0.00009	1.5%
	210	85											(0.04 - 2.4)weak	
$_{85}At^{211}$	211	85		210.987469	7.21 ± 0.01 h.	E.C.(58%)	0.784				9/2-		Po k x-ray	19%
	211	85				α(42%)	5.980	5.211	0.00	4%			0.66956 ± 0.00007	0.003%
	211	85						5.868	42%				0.6870 ± 0.0001	0.25%
	211	85											0.74263 ± 0.00006	0.0009
$_{85}At^{212m}$	212	85			0.12 s.	α		7.837	65%		(9-)			
	212	85						7.897	33%					
$_{85}At^{212}$	212	85		211.990725	0.3 s.	α	7.828	7.058	0.4%		(1-)			
	212	85						7.088	0.6%					
	212	85						7.618	15%					
	212	85						7.681	84%					
$_{85}At^{213}$	213	85		212.992911	0.11 μs.	α	9.254	9.080			9/2-			
$_{85}At^{214m}$	214	85			0.7 μs.	α		8.762			(9-)			
$_{85}At^{214}$	214	85		213.976347	0.56 μs.	α	8.987	8.819	100%		(1-)			
$_{85}At^{215}$	215	85		214.998638	100 μs.	α	8.178	7.626	0.04	5%	(9/2-)		0.40486 ± 0.00003	0.045%
	215	85						8.023	99.9%					
$_{85}At^{216}$	216	85		216.002390	300 μs.	α	7.947	7.595	0.2%		(1-)			
	216	85						7.697	2.1%					
	216	85						7.800	97%					
$_{85}At^{217}$	217	85		217.004694	32.3 ± 0.4 μs.	α	7.202	6.812	0.06%	(9/2-)		0.2595 ± 0.0008		
	217	85						7.067	99.9%				0.3345 ± 0.0008	
	217	85											0.5940 ± 0.0008	
$_{85}At^{218}$	218	85		218.008684	1.6 ± 0.4 s.	α	6.883	6.654	6%					
	218	85						6.695	90%					
	218	85						6.748	4%					
$_{85}At^{219}$	219	85		219.011300	54 s.	α	6.390	6.275						
Rn		86												
$_{86}Rn^{200}$	200	86		199.995700	1.0 ± 0.2 s.	α(98%)		6.909			0+			
	200	86				E.C.(2%)	8.080							
$_{86}Rn^{201m}$	201	86			3.8 ± 0.4 s.	E.C.(10%)					13/2+			
	201	86				α(90%)		6.770						
$_{86}Rn^{201}$	201	86		200.995570	7.0 ± 0.4 s.	α(80%)	6.860	6.721			(3/2-)			
	201	86				E.C.(20%)	6.65							
$_{86}Rn^{202}$	202	86		201.993230	9.9 ± 0.2 s.	α(12%)	6.771	6.636(3)			0+			
	202	86				E.C.(88%)	4.48							
$_{86}Rn^{203m}$	203	86			28 s.	α		6.548(3)			13/2+			
$_{86}Rn^{203}$	203	86		202.993330	45 ± 3 s.	α(66%)	6.629	6.498			0			
	203	86				E.C.(34%)	6.09							
$_{86}Rn^{204}$	204	86		203.991330	1.24 ± 0.03 m.	α(68%)	6.546	6.417(3)			0+			
	204	86				E.C.(32%)	3.84							
$_{86}Rn^{205}$	205	86		204.991650	2.83 ± 0.1 m.	α(23%)	6.390	6.123(3)	0.02%		(5/2-)		0.2652 ± 0.0007	100 +
	205	86				E.C.(77%)	5.27	6.262(3)	23%				0.3553 ± 0.0008	4
	205	86											0.4648 ± 0.0008	25
	205	86											0.6205 ± 0.0008	25
	205	86											0.6753 ± 0.001	20
	205	86											0.7300 ± 0.0008	20

Isotope	A	Z	% Natural abundance	Atomic mass	Half-life	Decay mode	Decay energy (MeV)	Particle energy (MeV)	Particle intensity	Thermal neutron cross section	Spin (h/2π)	μ Nucl. mag. moment	Gamma-ray energy (MeV)	Gamma-ray intensity
$_{86}Rn^{206}$	206	86		205.990140	5.67 ± 0.2 m.	α(68%)	6.384	6.258(3)			0+		0.06170 ± 0.0009	14 +
	206	86				E.C.(32%)	3.32						0.0968 ± 0.0001	5
	206	86											0.1009 ± 0.0002	4
	206	86											0.1337 ± 0.0001	5
	206	86											0.1862 ± 0.0003	7
	206	86											0.1954 ± 0.0001	12
	206	86											0.2080 ± 0.0001	26
	206	86											0.2131 ± 0.0004	11
	206	86											0.2906 ± 0.0003	7
	206	86											0.3019 ± 0.0002	53
	206	86											0.3245 ± 0.0001	100
	206	86											0.3504 ± 0.0003	13
	206	86											0.3711 ± 0.0002	52
	206	86											0.3862 ± 0.0001	63
	206	86											0.4356 ± 0.0001	5
	206	86											0.4439 ± 0.0003	28
	206	86											0.4582 ± 0.0006	5
	206	86											0.4654 ± 0.0002	4
	206	86											0.4822 ± 0.0002	59
	206	86											0.4853 ± 0.0003	31
	206	86											0.4973 ± 0.0001	104
	206	86											0.5271 ± 0.0002	25
	206	86											0.5363 ± 0.0003	17
	206	86											0.6318 ± 0.0003	15
	206	86											0.6429 ± 0.0007	6.7
	206	86											0.7166 ± 0.0007	6.8
	206	86											0.7382 ± 0.0006	15
	206	86											0.7568 ± 0.0006	11
	206	86											0.7728 ± 0.0003	60
	206	86											0.7948 ± 0.0004	10
$_{86}Rn^{207}$	207	86		206.990690	9.3 ± 0.2 m.	β+, EC (77%)	4.62				5/2-		At k x-ray	21%
	207	86				α(23%)	6.252	5.995(4)	0.02%				0.32947 ± 0.00004	3.0%
	207	86						6.068(3)	0.15%				0.34455 ± 0.00004	45%
	207	86						6.126(3)	22.8%				0.36767 ± 0.00008	2.5%
	207	86											0.40267 ± 0.00004	11.8%
	207	86											0.55323 ± 0.0001	1.2%
	207	86											0.62873 ± 0.00007	1.1%
	207	86											0.63159 ± 0.00009	2.9%
	207	86											0.6433 ± 0.0002	1.2%
	207	86											0.6472 ± 0.0001	1.8%
	207	86											0.6858 ± 0.0001	1.2%
	207	86											0.6971 ± 0.0001	2.4%
	207	86											0.74723 ± 0.00005	14.1%
	207	86											0.7754 ± 0.0001	2.0%
	207	86											0.85343 ± 0.00009	2.3%
	207	86											0.8927 ± 0.0007	1.0%
	207	86											0.9086 ± 0.0001	1%
	207	86											0.97328 ± 0.00008	2.5%
	207	86											0.9992 ± 0.0002	1.2%
	207	86											(0.18 - 1.47)weak	
$_{86}Rn^{208}$	208	86		207.989610	24.4 ± 0.1 m.	α(60%)	6.260	5.469(2)	0.00	3%	0+			
	208	86				E.C.(40%)	2.88	6.140(2)	60%					
$_{86}Rn^{209}$	209	86		208.990370	28.5 ± 0.1 m.	β+(83%)	3.930	2.16	2.3%		5/2-		At k x-ray	36%
	209	86				α(17%)		5.887(3)	0.04%				0.27933 ± 0.00007	1.1%
	209	86						5.898(3)	0.02%				0.33753 ± 0.00003	14.7%
	209	86						6.039(2)	16.9%				0.40841 ± 0.00003	51%
	209	86											0.46154 ± 0.00006	1.5%
	209	86											0.67293 ± 0.00004	3.3%
	209	86											0.68942 ± 0.00004	9.8%
	209	86											0.74594 ± 0.00003	23.1%
	209	86											0.79481 ± 0.00005	3.4%
	209	86											0.85590 ± 0.00004	4.9%
	209	86											1.03811 ± 0.00005	4.2%

Isotope	A	Z	% Natural abundance	Atomic mass	Half-life	Decay mode	Decay energy (MeV)	Particle energy (MeV)	Particle intensity	Thermal neutron cross section	Spin (h/2π)	μ Nucl. mag. moment	Gamma-ray energy (MeV)	Gamma-ray intensity
	209	86											1.05460 ± 0.00005	1.7%
	209	86											1.06567 ± 0.00007	1.7%
	209	86											1.15893 ± 0.00006	0.85%
	209	86											1.5432 ± 0.0001	0.8%
	209	86											1.9258 ± 0.0003	0.3%
	209	86											2.1142 ± 0.0002	0.33%
	209	86											2.6428 ± 0.0002	0.32%
	209	86											(0.18 - 3.2)many	
$_{86}Rn^{210}$	210	86		209.989669	2.4 ± 0.1 h.	α(96%)	6.157	5.351(2)	0.00	5%	0+		At k x-ray	2.3%
	210	86				E.C.(4%)	2.368	6.039(2)	96%				0.19625 ± 0.00007	0.32%
	210	86											0.23324 ± 0.00006	0.5%
	210	86											0.45824 ± 0.00006	1.6%
	20	86											0.57104 ± 0.00006	0.8%
	210	86											0.64868 ± 0.00006	0.82%
	210	86											0.76148 ± 0.00006	0.52%
	210	86											0.95773 ± 0.00007	0.3%
	210	86											(0.14 - 1.7)weak	
$_{86}Rn^{211}$	211	86		210.990576	14.6 ± 0.2 h.	β+, EC (74%)	2.894				1/2-		At k x-ray	30%
	211	86				α(26%)	5.964	5.619(1)	0.7%				0.16877 ± 0.00006	6.8%
	211	86						5.784(1)	16.4%				0.25022 ± 0.00006	6.1%
	211	86						5.851(1)	8.8%				0.37049 ± 0.00008	1.4%
	211	86											0.41632 ± 0.00006	3.5%
	211	86											0.44209 ± 0.00006	23%
	211	86											0.67412 ± 0.00007	46%
	211	86											0.67839 ± 0.00006	29%
	211	86											0.85377 ± 0.00006	4.7%
	211	86											0.86600 ± 0.00007	8.0%
	211	86											0.93481 ± 0.00006	3.7%
	211	86											0.94666 ± 0.00006	3.7%
	211	86											0.94744 ± 0.00007	5.1%
	211	86											1.12668 ± 0.00006	22.5%
	211	86											1.36298 ± 0.00005	33.1%
	211	86											1.53985 ± 0.00008	4.8%
	211	86											1.9926 ± 0.0001	0.5%
	211	86											(0.11 - 2.7)weak	
$_{86}Rn^{212}$	212	86		211.990697	24 ± 2 m.	α	6.385	5.587(4)	0.05%		0+			
	212	86						6.260(4)	99.95%					
$_{86}Rn^{213}$	213	86		212.993856	25.0 ± 0.2 ms.	α	8.243	7.552(8)	1.0%		9/2+			
	213	86						8.087(8)	99%					
$_{86}Rn^{214m}$	214	86			7.3 ns.	α(4%)		10.63(3)			(8+)			
	214	86				I.T.	1.626							
$_{86}Rn^{214}$	214	86		213.996347	0.27 μs.	α	9.209	9.037(9)			0+			
$_{86}Rn^{215}$	215	86		214.998720	2.3 μs.	α	8.840	8.674(8)			(9/2+)			
$_{86}Rn^{217}$	217	86		217.003902		α	7.885	7.500	0.1%		9/2+			
	215	86						7.742(4)	100%					
$_{86}Rn^{218}$	218	86	218.005580	35 ± 6 ms.		α	7.267	6.534(1)	0.16%		0+			
	218	86						7.133(1)	99.8%					
$_{86}Rn^{219}$	219	86		219.009479	3.96 s.	α	6.946(1)	6.3130(5)	0.05%		(5/2+)		Po k x-ray	0.9%
	219	86						6.425(3)	7.5%				0.13057 ± 0.00006	0.13%
	219	86						6.5309(4)	0.12%				0.27113 ± 0.00005	9.9%
	219	86						6.5531(3	) 12.2%				0.40170 ± 0.00006	6.6%
	219	86						6.8193(3)	81%				(0.1 - 1.05)weak	
$_{86}Rn^{220}$	220	86		220.011368	55.6 ± 0.1 s.	α	6.404	5.7486(5)	0.07%		0+			
	220	86						6.2883(1)	99.9%					
$_{86}Rn^{221}$	221	86		221.015470	25 m.	α(22%)	6.148	5.778(3)	1.8%				Fr L x-ray	1.6%

Isotope	A	Z	% Natural abundance	Atomic mass	Half-life	Decay mode	Decay energy (MeV)	Particle energy (MeV)	Particle intensity	Thermal neutron cross section	Spin (h/2π)	μ Nucl. mag. moment	Gamma-ray energy (MeV)	Gamma-ray intensity
	221	86				β-(78%)	1.150	5.788(3)	2.2%				0.07384 ± 0.00004	0.53%
	221	86						6.037(3)	18%				0.08323	8.2%
	221	86											0.0610	13.6%
	221	86											0.09727	4.9%
	221	86											0.09982	2.8%
	221	86											0.10060	1.6%
	221	86											0.10836 ± 0.00003	2.2%
	211	86											0.11156 ± 0.00003	2.1%
	211	86											0.15008 ± 0.00003	4.4%
	211	86											0.18639 ± 0.00004	20.4%
	211	86											0.21686 ± 0.00004	2.3%
	211	86											0.254 ± 0.0003	2%
	211	86											0.26468 ± 0.00004	1.1%
	211	86											0.27927 ± 0.00003	1.8%
$_{86}Rn^{222}$	222	86		222.017570	3.82 d.	α	5.590	4.987(1)	0.08%	0.7 b.	0+		0.510 ± 0.002	0.07%
	222	86						5.4897(3)	99.9%					
$_{86}Rn^{223}$	223	86			43 m.	β-								
$_{86}Rn^{224}$	224	86			1.78 ± 0.005h.	β-					0+		0.1085 ± 0.0005	3.3 +
	224	86											0.1132 ± 0.0003	3.3
	224	86											0.2026 ± 0.0003	4.4
	224	86											0.2562 ± 0.0003	3.0
	224	86											0.2601 ± 0.0001	23
	224	86											0.2655 ± 0.0001	21
	224	86											0.3719 ± 0.0003	3.5
	224	86											0.398 ± 0.001	5.6
	224	86											0.402 ± 0.001	5.7
$_{86}Rn^{225}$	225	86			4.5 m.	β-								
$_{86}Rn^{226}$	226	86			6.0 m.	β-								
Fr		87												
$_{87}Fr^{201}$	201	87		201.004110	48 ms.	α	7.54	7.388(15)						
$_{87}Fr^{202}$	202	87		202.003300	0.34 s.	α	7.590	7.250(20)						
$_{87}Fr^{203}$	203	87		203.001000	0.55 s.	α	7.280	7.132(5)						
$_{87}Fr^{204}$	204	87		204.000670	2.1 s.	α	7.170	6.967(5)	30 +					
	204	87						7.027(5)	70					
$_{87}Fr^{205}$	205	87		204.998610	3.96 s.	α	7.050	6.914(5)						
$_{87}Fr^{206}$	206	87		205.99846s	16.0 ± 0.1 s.	α	7.416	6.789(5)						
$_{87}Fr^{207}$	207	87		206.996800	14.8 ± 0.1 s.	α	6.900	6.766(5)			9/2-			
$_{87}Fr^{208}$	208	87		207.997080	59 s.	α(77%)	6.770	6.636(5)			7			
	208	87				E.C.(23%)	6.960							
$_{87}Fr^{209}$	209	87		208.995878	50.0 ± 0.3 s.	α(89%)	5.130	6.646(3)			9/2-			
	209	87				E.C.(11%)	6.778							
$_{87}Fr^{210}$	210	87		209.996340	3.2 m.	α	6.670	6.543(5)			6+		0.2030 ± 0.0008	35 +
	210	87											0.2562 ± 0.0001	11
	210	87											0.4252 ± 0.0001	10
	210	87											0.461 ± 0.001	11
	210	87											0.6438 ± 0.0008	100
	210	87											0.733 ± 0.001	10
	210	87											0.8175 ± 0.0008	60
	210	87											0.9008 ± 0.0007	30
$_{87}Fr^{211}$	211	87		210.995490	3.1 m.	α	6.660	6.534(5)			9/2-		0.220 ± 0.0008	9 +
	211	87				E.C.	4.570						0.2799 ± 0.001	34
	211	87											0.4389 ± 0.001	20
	211	87											0.5389 ± 0.001	100
	211	87											0.9169 ± 0.001	55
	211	87											0.9819 ± 0.0008	20
$_{87}Fr^{212}$	212	87		211.996130	20.0 ± 0.6 m.	E.C.(57%)	5.070	6.076(3)	0.17%		(5+)		Rn x-ray	14%
	212	87				α(43%)	6.529	6.127(3)	0.43%				0.0789	2.4%
	212	87						6.173(4)	0.5%				0.08107	14%
	212	87						6.183(3)	0.6%				0.08152	4%
	212	87						6.261(1)	16%				0.08378	24%
	212	87						6.335(1)	4%				0.09468	8%
	212	87						6.343(1)	1.3%				0.1383 ± 0.0001	7.7%
	212	87						6.383(1)	10%				0.3091 ± 0.0002	1.0%
	212	87						6.406(1)	9.5%				0.3115 ± 0.0002	1.3%
	212	87											0.5320 ± 0.0005	2.8%
	212	87											0.8019 ± 0.0015	3.5%
	212	87											1.0473 ± 0.0014	7.2%
	212	87											1.1784 ± 0.0020	1.3%
	212	87											1.1856 ± 0.0014	14%
	212	87											1.2748 ± 0.0020	46%
$_{87}Fr^{213}$	213	87		212.996165	34.6 ± 0.3 s.	α	6.905	6.775(2)			9/2-			

Isotope	A	Z	% Natural abundance	Atomic mass	Half-life	Decay mode	Decay energy (MeV)	Particle energy (MeV)	Particle intensity	Thermal neutron cross section	Spin (h/2π)	μ Nucl. mag. moment	Gamma-ray energy (MeV)	Gamma-ray intensity
$_{87}Fr^{214m}$	214	87			3.4 ms.	α		7.594(5)	0.5%		9-			
	214	87						7.708(5)	1.1%					
	214	87						7.963(5)	0.7%					
	214	87						8.046(5)	0.9%					
	214	87						8.476(4)	51%					
	214	87						8.547(4)	46%					
$_{87}Fr^{214}$	214	87		213.998948	5.1 ms.	α	8.587	7.409(3)	0.3%		(1-)			
	214	87						7.605(8)	1.0%		214	87		
	214	87						7.940(3)	1.0%					
	214	87						8.355(3)	4.7%					
	214	87						8.427(3)	93%					
$_{87}Fr^{215}$	215	87		215.000310	0.12 μs.	α	9.537	9.360(8)			(9/2-)			
$_{87}Fr^{216}$	217	87		216.003178	0.7 μs.	α	9.175	9.005(10)						
$_{87}Fr^{217}$	217	87		217.004609	22 μs.	α	8.471	8.315(8)			(9/2-)			
$_{87}Fr^{218}$	218	87		218.007553	0.7 ms.	α	8.014	7.384(10)	0.5%					
	218	87						7.542(15)	1.0%					
	218	87						7.572(10)	5%					
	218	87						7.732(10)	0.5%					
	218	87						7.867(2)	93%					
$_{87}Fr^{219}$	219	87		219.009242	21 ± 1 s.	α	8.132	6.802(2)	0.25%		(9/2-)			
	219	87						6.967(2)	0.6%					
	219	87						7.146(2)	0.25%					
	219	87						7.313(2)	99%					
$_{87}Fr^{220}$	220	87		220.012293	27.4 ± 0.3 s.	α	6.800	6.389(1)	0.3%		1		0.0450 ± 0.0003	2.3 +
	220	87						6.413(1)	1.2%				0.061 ± 0.004	0.4%
	220	87						6.438(2)	0.24%				0.1060 ± 0.0004	1.7%
	220	87						6.483(1)	1.3%				0.1539 ± 0.0004	1.0%
	220	87						6.490(2)	0.6%				0.1617 ± 0.0004	1.5%
	220	87						6.519(1)	0.6%					
	220	87						6.527(1)	3%					
	220	87						6.535(1)	2.5%					
	220	87						6.582(1)	10%					
	220	87						6.630(2)	6%					
	220	87						6.641(1)	12%					
	220	87						6.686(1)	61%					
$_{87}Fr^{221}$	221	87		221.014230	4.9 ± 0.2 m.	α	6.457	5.9393(7)	0.17%		(5/2-)		At k x-ray	1.4%
	221	87						5.9797(7)	0.49%				0.0995 ± 0.0001	0.10%
	221	87						6.0751(7)	0.15%				0.21798 ± 0.00004	10.9%
	221	87						6.1270(7)					0.4091 ± 0.0002	0.13%
	221	87						6.2433(3)	1.3%					
	221	87						6.3410(7)	83.4%					
$_{87}Fr^{222}$	222	87		14.4 m.	β-	2.060	1.78				2			
	222	87				α	5.850							
$_{87}Fr^{223}$	223	87		223.019733	21.8 ± 0.4 m.	β-	1.147	1.17	65%		(3/2+)		0.05014 ± 0.00004	33%
	223	87											0.07972 ± 0.00003	8.9%
	223	87											0.08543	2.4%
	223	87											0.08543	2.4%
	223	87											0.08847	4.0%
	223	87											0.09991	1.5%
	223	87											0.20495 ± 0.00005	1.1%
	223	87											0.23482 ± 0.00005	3.7%
	223	87											0.31918 ± 0.00007	0.54%
	223	87											0.3693 ± 0.0001	0.11%
	223	87											0.7758 ± 0.0001	0.40%
	223	87											(0.13 - 0.93)weak	
$_{87}Fr^{224}$	224	87		224.023220	2.7 ± 0.2 m.	β-	2.830				1		0.13150 ± 0.00006	83 +
	224	87											0.2057 ± 0.0002	14
	224	87											0.21575 ± 0.00006	180
	224	87											0.7625 ± 0.0002	12
	224	87											0.8018 ± 0.0003	6
	224	87											0.8367 ± 0.0002	58
	224	87											0.8810 ± 0.0002	5
	224	87											0.9683 ± 0.0002	5
	224	87											1.1619 ± 0.0002	5
	224	87											1.2983 ± 0.0002	5
	224	87											1.3402 ± 0.0003	25
	224	87											1.3777 ± 0.0002	17
	224	87											1.4356 ± 0.0002	10
	224	87											1.6521 ± 0.0005	6
	224	87											(0.1 - 2.21)weak	
$_{87}Fr^{225}$	225	87		225.025590	3.9 m.	β-	1.850							
$_{87}Fr^{226}$	226	87		226.029200	48 s.	β-	3.540						0.18606 ± 0.00004	66 +
	226	87											0.25373 ± 0.00004	83
	226	87											1.0069 ± 0.0002	15

Isotope	A	Z	% Natural abundance	Atomic mass	Half-life	Decay mode	Decay energy (MeV)	Particle energy (MeV)	Particle intensity	Thermal neutron cross section	Spin (h/2π)	μ Nucl. mag. moment	Gamma-ray energy (MeV)	Gamma-ray intensity
	226	87											1.0489 ± 0.0002	15
	226	87											1.3219 ± 0.0002	8.5
	226	87											1.3889 ± 0.002	4.4
$_{87}Fr^{227}$	227	87		227.031770	2.4 m.	β-	2.420							
$_{87}Fr^{228}$	228	87		228.035570	39 s.	β-	4.200							
$_{87}Fr^{229}$	229	87			50 s.	β-								
Ra		88												
$_{88}Ra^{206}$	206	88		206.003800	0.4 s.	α	7.416	7.272(5)			0+			
$_{88}Ra^{207}$	207	88		207.003740	1.3 ± 0.2 s.	α	7.270	7.133(5)						
$_{88}Ra^{208}$	208	88		208.001750	1.4 ± 0.4 s.	α	7.273	7.133(5)			0+			
$_{88}Ra^{209}$	209	88		209.001930	4.6 ± 0.2 s.	α	7.150	7.008(5)						
$_{88}Ra^{210}$	210	88		210.000430	3.7 ± 0.2 s.	α	7.610	7.020(5)			0+			
$_{88}Ra^{211}$	211	88		211.000860	13 ± 2 s.	α	7.046	6.912(5)			(5/2-)			
	211	88				E.C.								
$_{88}Ra^{212}$	212	88		211.999760	13.0 ± 0.5 s.	α	7.033	6.901(2)			0+			
$_{88}Ra^{213}$	213	88		213.000330	2.7 m.	E.C.(20%)	3.880				(1/2-)		0.1024 ± 0.0001	0.3%
	213	88				α(80%)	6.860	6.521(3)	4.8%				0.11010 ± 0.00009	6.4%
	213	88						6.622(3)	39%				0.2125 ± 0.0001	1.1%
	213	88						6.730(3)	36%					
$_{88}Ra^{214}$	214	88		214.000079	2.46 ± 0.03 s.	α	7.272	7.136(4)			0+			
$_{88}Ra^{215}$	215	88		215.002695	1.59 ± 0.09 ms.	α	8.864	7.883(6)	2.8%		(9/2+)			
	215	88						8.171(3)	1.4%					
	215	88						8.700(3)	95.9%					
$_{88}Ra^{216}$	216	88		216.003509	0.18 μs.	α	9.526	9.349(8)			0+			
$_{88}Ra^{217}$	217	88		217.006294	1.6 ± 0.2 μs.	α	9.161	8.992(8)			9/2-			
$_{88}Ra^{218}$	218	88		218.007117	14 ± s μs.	α	8.547	8.390(8)			0+			
$_{88}Ra^{219}$	219	88		219.010053		α	8.132	7.680(10)	65%					
	219	88						7.982(9)	35%					
$_{88}Ra^{220}$	220	88		220.011004	23 ± 5 ms.	α	7.593	6.998(7)	1.0%		0+		0.465 ± 0.004	1.0%
	220	88						7.455(7)	99%					
$_{88}Ra^{221}$	221	88		221.013889	28 ± 2 s.	α	6.879	6.254(10)	0.7%					
	221	88						6.578(5)	3%					
	221	88						6.585(3)	8%					
	221	88						6.608(3)	35%					
	221	88						6.669(3)	21%					
	221	88						6.758(3)	31%					
$_{88}Ra^{222}$	22	88		222.015353	38.0 ± 0.5 s.	α	6.590	6.237(2)	3.0%		0+			
	222	88						6.556(2)	97%					
$_{88}Ra^{223}$	223	88		223.018501	11.43 ± 0.02d.	α	5.979	5.287(1)	0.15%		(1/2+)		Rn k x-ray	25%
	223	88						5.338(1)	0.13%				0.12231 ± 0.00006	1.2%
	223	88						5.365(1)	0.13%				0.14418 ± 0.00003	3.3%
	223	88						5.433(5)	2.3%				0.15418 ± 0.00003	5.6%
	223	88						5.502(1)	1.0%				0.15859 ± 0.00003	0.67%
	223	88						5.540(1)	9.2%				0.26939 ± 0.00003	14%
	233	88						5.607(3)	24%				0.32388 ± 0.00003	3.9%
	233	88						5.716(3)	52%				0.33328 ± 0.00004	2.8%
	233	88						5.747(1)	9%				0.44494 ± 0.00005	1.3%
	223	88						5.857(1)	0.32%				(0.10 - 0.71)weak	
	223	88						5.872(1)	0.85%					
$_{88}Ra^{224}$	224	88		224.020186	3.66 ± 0.04 d.	α	5.789	5.034(10)	0.00	3%	0+		Rn k x-ray	0.2%
	224	88						5.047(1)	0.007%				0.2407 ± 0.0001	3.9%
	224	88						5.164(5)	0.007%				0.4093 ± 0.0007	0.004%
	224	88						5.449(2)	4.9%				0.6501 ± 0.0007	0.007%
	224	88						5.685(2)	95%					
$_{88}Ra^{225}$	225	88		225.023604	14.8 ± 0.2 d.	β-	0.371	0.32	100%		(3/2+)		Ac k x-ray	6%
	225	88											0.0434 ± 0.0017	29%
$_{88}Ra^{266}$	226	88		226.025402	1600 ± 7 y.	α	4.870	4.194(1)	0.001%		0+		Rn k x-ray	0.3%
	226	88						4.343(1)	0.006%				0.1861 ± 0.0001	3.3%
	226	88						4.601(1)	5.5%				0.2624 ± 0.0002	0.005%
	226	88						4.784(1)	94%					
$_{88}Ra^{227}$	227	88		227.029170	42.2 ± 0.5 m.	β-	1.324	1.03			(3/2+)		Ac L x-ray	25%
	227	88						1.30					Ac k x-ray	43%
	227	88											0.02739 ± 0.00001	17%
	227	88											0.08767	2.6
	227	88											0.0988	4.3%

Isotope	A	Z	% Natural abundance	Atomic mass	Half-life	Decay mode	Decay energy (MeV)	Particle energy (MeV)	Particle intensity	Thermal neutron cross section	Spin (h/2π)	μ Nucl. mag. moment	Gamma-ray energy (MeV)	Gamma-ray intensity
	227	88											0.10261	1.5%
	227	88											0.23062 ± 0.00007	1.4%
	227	88											0.25843 ± 0.00006	2.0%
	227	88											0.27743 ± 0.00006	2.9%
	227	88											0.28367 ± 0.00002	3.5%
	227	88											0.30007 ± 0.00002	5.3%
	227	88											0.30267 ± 0.00002	3.8%
	277	88											0.33007 ± 0.00002	2.9%
	227	88											0.40789 ± 0.00004	2.5%
	227	88											0.48703 ± 0.00009	2.5%
	227	88											0.5013 ± 0.0001	1.0%
	227	88											0.5164 ± 0.0001	1.5%
	227	88											0.6117 ± 0.0002	1.3%
$_{88}Ra^{228}$	228	88		228.031064	5.75 ± 0.03 y.	β-	0.045				0+		0.0135 ± 0.001	100 +
	228	88											0.016	
$_{88}Ra^{229}$	229	88		229.034870	4.0 ± 0.2 m.	β-	1.760	1.76			(3/2+)			
$_{88}Ra^{230}$	230	88		230.036990	1.55 ± 0.05 h.	β-	0.700	0.7			0+		0.0631 ± 0.0001	40 +
	230	88											0.0720 ± 0.0001	113
	230	88											0.0921 ± 0.0001	21
	230	88											0.1011 ± 0.0001	16
	230	88											0.1107 ± 0.0001	3
	230	88											0.1343 ± 0.0001	4.5
	230	88											0.1479 ± 0.0001	5.6
	230	88											0.1841 ± 0.0001	11
	230	88											0.1892 ± 0.0001	11
	230	88											0.2028 ± 0.0001	31
	230	88											0.2118 ± 0.0001	11
	230	88											0.2516 ± 0.0001	10
	230	88											0.2852 ± 0.0001	18
	230	88											0.44898 ± 0.00007	15
	230	88											0.4580 ± 0.0001	18
	230	88											0.4698 ± 0.0001	29
	230	88											0.4787 ± 0.0001	24
	230	88											0.5092 ± 0.0001	6
Ac		89												
$_{89}Ac^{210}$	210	89		210.009230	0.35 s.	α	7.610	7.462(8)						
$_{89}Ac^{211}$	221	89		211.007590	≈ 0.25 s.	α	7.620	7.480(8)						
$_{89}Ac^{212}$	212	89		212.007760	0.93 s.	α	7.520	7.379(8)						
$_{89}Ac^{213}$	213	89		213.006530	0.8 s.	α	7.500	7.364(8)			(9/2-)			
$_{89}Ac^{214}$	214	89		214.006840	8.2 ± 0.2 s.	α(86%)	7.350	7.007(8)	3 +		(5+)			
	214	89				E.C.(14%)		7.082(5)	38					
	214	89						7.214(5)	45					
$_{89}Ac^{215}$	215	89		215.006410	0.17 ± 0.01 s.	α	7.750	7.604(5)			(9/2-)			
$_{89}Ac^{216m}$	216	89			0.33 ms.	α		8.198(8)	1.7%		(9-)			
	216	89						8.283(8)	2.5%					
	216	89						9.028(5)	49%					
	216	89						9.106(5)	46%					
$_{89}Ac^{216}$	216	89		216.008650	≈ 0.33 ms.	α	9.241	8.990(2)	10%		(1)			
	216	89						9.070(8)	90%					
$_{89}Ac^{217m}$	217	89			0.4 ± 0.1 μs.	α		10.540	100%					
$_{89}Ac^{217}$	217	89		217.009322	0.11 μs.	α	9.832	9.650(10)	100%		9/2-			
$_{89}Ac^{218}$	218	89		218.001620	0.27 ± 0.04 μs	α	9.380	9.205(15)						
$_{89}Ac^{219}$	219	89		219.012390	7 ± 2 μs.	α	8.830	8.664(10)			(9/2-)			
$_{89}Ac^{220}$	220	89		220.014740	26.1 ± 0.5 ms.	α	8.350	7.610(20)	23%					
	220	89						4.680(20)	21%					
	220	89						7.790(10)	13%					
	220	89						7.850(10)	24%					
	220	89						7.985(10)	4%					
	220	89						8.005(10)	5%					
	220	89						8.060(10)	6%					
	220	89						8.195(10)	3%					
$_{89}Ac^{221}$	221	89		221.015570	52 ± 2 ms.	α	7.790	7.170(10)	2%					
	221	89						7.375(10)	10%					
	221	89						7.440(15)	20%					
	220	89						7.645(10)	70%					
$_{89}Ac^{222m}$	222	89			1.10 ± 0.05 m.	α(>89%)		6.710(20)	7%					
	222	89				E.C.(1%)		6.750(20)	13%					

Isotope	A	Z	% Natural abundance	Atomic mass	Half-life	Decay mode	Decay energy (MeV)	Particle energy (MeV)	Particle intensity	Thermal neutron cross section	Spin (h/2π)	μ Nucl. mag. moment	Gamma-ray energy (MeV)	Gamma-ray intensity
	222	89				I.T.(<10%)		6.810(20)	24%					
	222	89						6.840(20)	9%					
	222	89						6.890(20)	13%					
	222	89						6.970(20)	7%					
	222	89						7.000(20)	13%					
$_{89}Ac^{222}$	222	89		222.017824	4.2 s.	α	7.141	6.967(10)	6%					
	222	89						7.013(2)	94%					
$_{89}Ac^{223}$	223	89		223.019128	2.2 ± 0.1 m.	α(99%)	6.783	6.131(2)	0.12%		(5/2-)		0.0725 ± 0.0009	0.2%
	223	89				E.C.(1%)	0.584	6.177(2)	0.94%				0.0839 ± 0.0007	0.2%
	223	89						6.293(1)	0.47%				0.0927 ± 0.0007	0.2%
	223	89						6.326(1)	0.3%				0.0990 ± 0.0006	0.2%
	223	89						6.332(2)	0.14%				0.1917 ± 0.0007	0.25%
	223	89						6.360(1)	0.22%				0.2158 ± 0.0009	0.15%
	223	89						6.397(1)	0.13%				0.3588 ± 0.0011	0.10%
	233	89						6.448(1)	0.2%				0.4768 ± 0.0015	0.14%
	223	89						6.473(1)	3.1%					
	223	89						6.523(2)	0.6%					
	223	89						6.528(1)	3.1%					
	223	89						6.563(1)	13.6%					
	223	89						6.582(3)	0.3%					
	223	89						6.646(1)	44%					
	223	89						6.661(1)	31%					
$_{89}Ac^{224}$	224	89		224.021685	2.9 ± 0.2 h.	E.C.(90%)	1.397	5.841(1)	0.5	+			Ra L kx-ray	18%
	224	89				α(10%)	6.323	5.860(1)	0.75				Ra k x-ray	35%
	224	89						5.875(1)	1.7				0.08426 ± 0.00005	1.1%
	224	89						5.941(1)	4.4				0.13150 ± 0.00006	20%
	224	89						6.000(1)	6.7				0.1571 ± 0.0003	0.5%
	224	89						6.013(1)	1.4				0.21575 ± 0.00006	44%
	224	89						6.056(1)	22				0.2619 ± 0.0003	0.2%
	224	89						6.138(1)	26				(0.03 - 0.37)weak	
	224	89						6.154(1)	1.0					
	224	89						6.204(1)	12					
	224	89						6.210(1)	20					
$_{89}Ac^{225}$	225	89		225.023205	10.0 ± 0.1 d.	α	5.935	5.286(1)	0.2%		3/2		Fr k x-ray	2.1%
	225	89						5.444(3)	0.1%				0.9958 ± 0.0004	0.6%
	225	89						5.554(1)	0.1%				0.9982 ± 0.0006	1.7%
	225	89						5.608(1)	1.1%				0.1084 ± 0.0001	0.3%
	225	89						5.636(1)	4.5%				0.1116 ± 0.0001	0.33%
	225	89						5.681(1)	1.4%				0.1451 ± 0.0001	0.13%
	225	89						5.722(1)	2.9%				0.1539 ± 0.0001	0.15%
	225	89						5.731(1)	10%				0.15724 ± 0.00003	0.31%
	225	89						5.791(1)	9%				0.18799 ± 0.00005	0.46%
	225	89						5.793(1)	18%				0.19575 ± 0.00003	0.14%
	225	89											0.2162 ± 0.001	0.34%
	225	89											0.21686 ± 0.00004	0.42%
	225	89											0.25351 ± 0.00004	0.10%
	225	89											0.4524 ± 0.0001	0.11%
	225	89											(0.025 - 0.52)weak	
$_{89}Ac^{226}$	226	89		226.026084	1.2 d.	E.C.(17%)	0.635				(1-)		Ra k x-ray	6.0%
	226	89				β-(83%)	1.117						Th k x-ray	1.8%
	226	89				α(0.006%)	5.510	5.399(5)	0.00	6%			0.07218 ± 0.0000	3 0.6%
	226	89											0.15816 ± 0.00003	17.3%
	226	89											0.18606 ± 0.0001	4.6%
	226	89											0.23034 ± 0.00003	29.6%
	226	89											0.25373 ± 0.00001	5.8%
$_{89}Ac^{227}$	227	89		227.027750	21.77 ± 0.03y.	β-(98.6%)	0.041	β0.0455	54%		(3/2-)	+1.1	0.01520 ± 0.00009	0.035%
	227	89				α(1.4%)	5.043	α4.869(1)	0.09%				0.0698 ± 0.0001	0.02%
	227	89						4.938(1)	0.52%				0.0997 ± 0.0005	0.03%
	227	89						4.951(1)	0.65%				0.1600 ± 0.0004	0.02%
	227	89											(0.009 - 0.17)weak	
$_{89}Ac^{228}$	228	89		228.031015	6.13 h.	β-	2.142	1.11	32%		(3+)		Th L x-ray	20%
	228	89						1.85	12%				Th k x-ray	5.6%
	228	89						2.18	11%				0.12903 ± 0.00007	2.9%
	228	89											0.20939 ± 0.00007	4.1%
	228	89											0.27026 ± 0.00008	3.8%
	228	89											0.32807 ± 0.00009	3.5%
	228	89											0.33842 ± 0.00006	12.4%

Isotope	A	Z	% Natural abundance	Atomic mass	Half-life	Decay mode	Decay energy (MeV)	Particle energy (MeV)	Particle intensity	Thermal neutron cross section	Spin (h/2π)	μ Nucl. mag. moment	Gamma-ray energy (MeV)	Gamma-ray intensity
	228	89											0.40962 ± 0.00008	2.2%
	228	89											0.46310 ± 0.00007	4.6%
	228	89											0.5815 ± 0.0002	3%
	228	89											0.75528 ± 0.00008	1.3%
	228	89											0.77228 ± 0.00007	1.1%
	228	89											0.7948 ± 0.0001	4.6%
	228	89											0.83560 ± 0.00009	1.7%
	228	89											0.91116 ± 0.00003	29%
	228	89											0.96897 ± 0.00005	17%
	228	89											1.4592 ± 0.0001	1.1%
	228	89											1.4960 ± 0.0003	1.0%
	228	89											1.5882 ± 0.0001	3.6%
	228	89											1.6304 ± 0.0002	1.9%
	228	89											(0.2 - 1.96)many	
$_{89}Ac^{229}$	229	89		229.032980	1.05 ± 0.01 h.	β-	1.140	1.1			(3/2+)		0.07450 ± 0.00002	8 +
	229	89											0.11715 ± 0.00002	15
	229	89											0.13533 ± 0.00001	34
	229	89											0.14635 ± 0.00002	35
	229	89											0.16451 ± 0.00001	100
	229	89											0.2392 ± 0.0002	4
	229	89											0.24529 ± 0.00002	9
	229	89											0.24866 ± 0.00001	9.2
	229	89											0.25201 ± 0.00008	24
	229	89											0.26188 ± 0.00008	39
	229	89											0.2747 ± 0.0002	1.2
	229	89											0.27806 ± 0.00001	2.5
	229	89											0.2849 ± 0.0002	4.3
	229	89											0.2878 ± 0.0001	6
	229	89											0.28795 ± 0.00001	2.7
	229	89											0.29132 ± 0.00001	11
	229	89											0.31713 ± 0.00001	22
	229	89											0.32051 ± 0.00001	6.4
	229	89											0.3228 ± 0.0002	3
	229	89											0.3320 ± 0.0001	2.0
	229	89											0.3656 ± 0.0002	2.7
	229	89											0.4046 ± 0.0001	9
	229	89											0.4065 ± 0.0001	6
	229	89											0.4228 ± 0.0001	7
	229	89											0.4359 ± 0.0001	6.3
	229	89											0.4492 ± 0.0001	16
	229	89											0.4784 ± 0.0001	17
	229	89											0.5085 ± 0.0002	37
	229	89											0.5267 ± 0.0001	6
	229	89											0.5399 ± 0.0001	20
	229	89											0.5635 ± 0.0001	6
	229	89											0.56916 ± 0.00008	91%
	229	89											0.5758 ± 0.0001	5
	229	89											0.60497 ± 0.0012	23
$_{89}Ac^{230}$	230	89		230.936240	2.03 ± 0.05 m.	β-	2.900	1.4			1+		Th k x-ray	0.8%
	230	89											0.12091 ± 0.00002	0.3%
	230	89											0.39769 ± 0.00005	0.4%
	230	89											0.45497 ± 0.00003	8.9%
	230	89											0.50820 ± 0.00003	5.1%
	230	89											0.58178 ± 0.00008	0.5%
	230	89											0.62885 ± 0.00009	0.24%
	230	89											0.72820 ± 0.00004	0.5%

TABLE OF THE ISOTOPES (Continued)

Isotope	A	Z	% Natural abundance	Atomic mass	Half-life	Decay mode	Decay energy (MeV)	Particle energy (MeV)	Particle intensity	Thermal neutron cross section	Spin (h/2π)	μ Nucl. mag. moment	Gamma-ray energy (MeV)	Gamma-ray intensity
	230	89											0.78143 ± 0.00004	0.4%
	230	89											0.78899 ± 0.00007	0.53%
	230	89											0.8167 ± 0.0001	0.32%
	230	89											0.8671 ± 0.0001	0.47%
	230	89											0.89275 ± 0.00005	0.7%
	230	89											0.95199 ± 0.00004	0.83%
	230	89											1.22681 ± 0.00005	0.96%
	230	89											1.24396 ± 0.00007	3.5%
	230	89											1.30258 ± 0.00006	0.54%
	230	89											1.34772 ± 0.00005	1.6%
	230	89											1.37535 ± 0.00006	1.2%
	230	89											1.69170 ± 0.00008	0.6%
	230	89											1.77525 ± 0.00007	1.1%
	230	89											1.89666 ± 0.00007	0.52%
	230	89											1.90273 ± 0.00009	0.75%
	230	89											1.9138 ± 0.0001	0.56%
	230	89											1.94989 ± 0.00007	1.25%
	230	89											2.00094 ± 0.00008	0.4%
	230	89											2.0986 ± 0.0001	0.52%
	230	89											2.12280 ± 0.00009	0.58%
	230	89											(0.12 - 2.5)many	
$_{89}Ac^{231}$	231	89		231.038550	7.5 ± 0.1 m.	β-	2.100	2.1	100%		($^1/_2$+)		0.14379 ± 0.00001	9 +
	231	89											0.18574 ± 0.00001	45
	231	89											0.19893 ± 0.00002	6
	231	89											0.22140 ± 0.00002	52
	231	89											0.22088 ± 0.00002	11
	231	89											0.2721 ± 0.0001	7
	231	89											0.28250 ± 0.00001	100
	231	89											0.3070 ± 0.0001	80
	231	89											0.3688 ± 0.0001	38
	231	89											0.3722 ± 0.0001	4.4
	231	89											0.3759 ± 0.0002	4.1
	231	89											0.4003 ± 0.0002	2.6
	231	89											0.4079 ± 0.0001	8.5
	231	89											0.5282 ± 0.0002	2.3
	231	89											0.5546 ± 0.0001	3.9
$_{89}Ac^{232}$	232	89		232.042130	35 ± 5 s.	β-	3.800				(2-)			
Th		90		232.0381						7.4 b.				
$_{90}Th^{212}$	212	90		212.012890	≈ 30 ms.	α		7.80			0+			
$_{90}Th^{213}$	213	90		213.012940	0.14 ± 0.04 s.	α	7.840	7.692(10)						
$_{90}Th^{214}$	214	90		214.011430	0.86 ± 0.1 s.	α	7.825	7.677(10)			0+			
$_{90}Th^{215}$	215	90		215.011690	1.2 ± 0.2 s.	α	7.660	7.33(10)	8%		($^1/_2$-)			
	215	90						7.395(8)	52%					
	215	90						7.524(8)	40%					
$_{90}Th^{216}$	216	90		216.011030	28 ± 2 ms.	α	8.071	7.921(8)			0+			
$_{90}Th^{217}$	217	90		217.013050	252 ± 7 μs.	α	9.424	9.250(10)						
$_{90}Th^{218}$	218	90		218.013252	0.11 μs.	α	9.847	9.665(10)			0+			
$_{90}Th^{219}$	219	90		219.015510	1.05 ± 0.03 μs	α	9.510	9.340(20)						
$_{90}Th^{220}$	220	90		220.015724	9.7 ± 0.6 μs.	α	8.953	8.790(20)			0+			
$_{90}Th^{221}$	221	90		221.018160	1.68 ± 0.06 ms	α	8.628	7.743(8)	6%					
	221	90						8.146(5)	56%					
	221	90						8.4272(5)	39%					
$_{90}Th^{222}$	222	90		222.018447	2.8 ± 0.3 ms.	α	8.129	7.982(8)			0+			
$_{90}Th^{223}$	223	90		223.020659	0.66 ± 0.02 s.	α	7.454	7.287(10)	60%					
	223	90						7.317(10)	40%					

Isotope	A	Z	% Natural abundance	Atomic mass	Half-life	Decay mode	Decay energy (MeV)	Particle energy (MeV)	Particle intensity	Thermal neutron cross section	Spin (h/2π)	μ Nucl. mag. moment	Gamma-ray energy (MeV)	Gamma-ray intensity
$_{90}Th^{224}$	224	90		224.021449	1.04 ± 0.5 s.	α	7.305	6.768(5)	1.2%					
	224	90						6.997(5)	19%					
	224	90						7.170(5)	79%					
$_{90}Th^{225}$	225	90		225.023922	8.0 m.	E.C.(10%)	0.668				(3/2+)			
	225	90				α(90%)	6.920	6.441(2)	15 +					
	225	90						6.479(2)	43					
	225	90						6.501(3)	14					
	225	90						6.627(3)	3					
	225	90						6.650(5)	3					
	225	90						6.700(5)	2					
	225	90						6.743(3)	7					
	225	90						6.796(2)	9					
$_{90}Th^{226}$	226	90		226.024885	31 m.	α	6.454	6.026(1)	0.2%		0+		Ra k x-ray	0.5%
	226	90						6.041(1)	0.19	%			0.11110 ± 0.00003	3.3%%
	226	90						6.098(1)	1.3%				0.13100 ± 0.00004	0.28%
	226	90						6.2283(4)	23%				0.19028 ± 0.00005	0.11%
	216	90						6.3375(4)	75%				0.20621 ± 0.00005	0.19%
	216	90											0.24210 ± 0.00004	0.87%
	216	90											(0.1 - 0.8)weak	
$_{90}Th^{227}$	227	90		227.027703	18.72 d.	α	6.146				(3/2+)		Ra L x-ray	21%
	227	90											Ra k x-ray	3.1%
	227	90											0.02987 ± 0.00002	0.1%
	227	90											0.04373 ± 0.00004	0.23%
	227	90											0.04985 ± 0.00003	0.2%
	227	90											0.05014 ± 0.00004	8.5%
	227	90											0.06236 ± 0.00004	0.24%
	227	90											0.7972 ± 0.00003	2.1%
	227	90											0.09393 ± 0.00004	1.4%
	227	90											0.11312 ± 0.00005	0.15%
	227	90											0.11319 ± 0.00006	0.56%
	227	90											0.11717 ± 0.00006	0.17%
	227	90											0.14144 ± 0.00007	0.13%
	227	90											0.20420 ± 0.00005	0.23%
	227	90											0.20604 ± 0.00005	0.23%
	227	90											0.21058 ± 0.00005	1.1%
	227	90											0.23597 ± 0.00004	11.2%
	227	90											0.25012 ± 0.00005	0.37%
	225	90											0.25246 ± 0.00006	0.11%
	227	90											0.25466 ± 0.00005	0.8%
	227	90											0.25624 ± 0.00003	6.7%
	227	90											0.26272 ± 0.00005	0.10%
	227	90											0.27295 ± 0.00004	0.49%
	227	90											0.28131 ± 0.00005	0.16%
	227	90											0.28611 ± 0.00003	1.59%
	227	90											0.29654 ± 0.00005	0.43%
	227	90											0.29997 ± 0.00004	2.1%
	227	90											0.30034 ± 0.00009	0.20%
	227	90											0.30451 ± 0.00004	1.09%
	227	90											0.31257 ± 0.00005	0.47%
	227	90											0.31482 ± 0.00005	0.46%
	277	90											0.32984 ± 0.00004	2.73%

Isotope	A	Z	% Natural abundance	Atomic mass	Half-life	Decay mode	Decay energy (MeV)	Particle energy (MeV)	Particle intensity	Thermal neutron cross section	Spin (h/2π)	μ Nucl. mag. moment	Gamma-ray energy (MeV)	Gamma-ray intensity
	227	90											0.34244 ± 0.00005	0.38%
	227	90											0.35048 ± 0.00008	0.11%
	227	90											(0.02 - 1.02)weak	
$_{90}Th^{228}$	228	90		228.028715	1.913 y.	α	5.520	5.1770(2)	0.18%		0+			
	228	90						5.2114(1)	0.4%					
	228	90						5.3405(1)	26.7%					
	228	90						5.4233(1)	73%					
$_{90}Th^{229}$	229	90		229.031755	7.3 x 10^3 y.	α	5.168	4.689(1)	0.15%		5/2+	+0.46		
	229	90						4.7618(5)	0.63%					
	229	90						4.7979(5)	1.27%					
	229	90						4.809(20)	0.22%					
	229	90						4.8140(5)	9.3%					
	229	90						4.833(25)	0.29%					
	229	90						4.838(5)	4.8%					
	229	90						4.845(5)	56%					
	229	90						4.861(25)	0.2%					
	229	90						4.9008(5)	10.2%					
	229	90						4.9301(5)	0.11%					
	229	90						4.9678(5)	5.97					
	229	90						4.9786(5)	3.2%					
	229	90						5.0354(5)	0.24%					
	229	90						5.050(25)	5.2%					
	229	90						5.0525(5)	1.6%					
	229	90						5.0774(5)	0.01					
$_{90}Th^{230}$	230	90		230.033127	7.54 x 10^4y.	α	4.771	4.4383(6)	0.03%		0+			
	230	90						4.4798(6)	0.12%					
	230	90						4.6211(6)	23.4%					
	230	90						4.6876(6)	76.3%					
$_{90}Th^{231}$	231	90		231.036298	25.2 h.	β-	0.389	0.138	22%		5/2+		Pa L x-ray	37%
	231	90						0.218	20%				Pa k x-ray	0.6%
	231	90						0.305	52%				0.02564 ± 0.00001	15%
	231	90											0.084203 ± 0.00009	6.6%
	231	90											0.08995 ± 0.00001	0.9%
	231	90											0.10225 ± 0.00001	0.4%
	231	90											0.10816 ± 0.	0.23%
	231	90											0.16311 ± 0.00001	0.15%
	231	90											(0.02 - 0.35)weak	
$_{90}Th^{232}$	232	90	100%	232.038054	1.4 x 10^{10}y.	α	4.081	3.830(10)	0.2%	7	4 b.	0+	0.0590 ± 0.0001	0.19%
	232	90						3.952(5)	23%				0.124 ± 0.001	0.04%
	232	90						4.010(5)	77%					
$_{90}Th^{233}$	233	90		233.041577	22.3 m.	β-	1.243	1.245		1500 b.	1/2+		Pa L x-ray	4.6%
	233	90								σf 15 b.			Pa k x-ray	0.8%
	233	90											0.02938 ± 0.00001	2.6%
	233	90											0.08653 ± 0.00001	2.6%
	233	90											0.08805 ± 0.00004	0.21%
	233	90											0.09288 ± 0.00004	0.51%
	233	90											0.09472 ± 0.00002	0.9%
	233	90											0.09586 ± 0.00004	0.8%
	233	90											0.10816 ± 0.00004	0.30%
	233	90											0.11189 ± 0.00004	0.10%
	233	90											0.16251 ± 0.00004	0.17%
	233	90											0.16258 ± 0.00008	0.15%
	233	90											0.16918 ± 0.00004	0.15%
	233	90											0.17078 ± 0.00008	0.13%
	233	90											0.19054 ± 0.00007	0.13%
	233	90											0.3598 ± 0.0001	0.12%
	233	90											0.44117 ± 0.00009	0.23%
	233	90											0.44784 ± 0.00009	0.15%
	233	90											0.45930 ± 0.00001	1.4%
	233	90											0.4907 ± 0.0001	0.17%
	233	90											0.4989 ± 0.0001	0.21%
	233	90											0.5953 ± 0.0009	0.16%
	233	90											0.66978 ± 0.00008	0.12%

Isotope	A	Z	% Natural abundance	Atomic mass	Half-life	Decay mode	Decay energy (MeV)	Particle energy (MeV)	Particle intensity	Thermal neutron cross section	Spin (h/2π)	μ Nucl. mag. moment	Gamma-ray energy (MeV)	Gamma-ray intensity
	233	90											0.7645 ± 0.0008	0.12%
	233	90											0.8901 ± 0.0001	0.14%
	233	90											(0.02 - 1.2)many	
$_{90}Th^{234}$	234	90		234.043593	24.10 d.	β-	0.270	0.102	20%		0+		Pa L x-ray	4%
	234	90						0.198	72%				0.06329 ± 0.00002	3.8%
	234	90											0.09235 ± 0.00003	2.7%
	234	90											0.09278 ± 0.00003	2.7%
	234	90											0.11280 ± 0.00003	0.24%
$_{90}Th^{235}$	235	90		235.047510	6.9 ± 0.2 m.	β-	1.940						0.4162 ± 0.0010	
	235	90											0.6594 ± 0.001	
	235	90											0.7272 ± 0.001	
	235	90											0.747 ± 0.001	
	235	90											0.9318 ± 0.001	
$_{90}Th^{236}$	236	90			37.1 ± 0.2 m.	β-	≈ 1						Pa k x-ray	3.3%
	236	90											0.1107 ± 0.0005	2.4%
	236	90											0.11189	0.40%
	236	90											0.1127 ± 0.0005	0.6%
	236	90											0.1316 ± 0.001	0.5%
	236	90											0.2296 ± 0.001	0.4%
Pa		91												
$_{91}Pa^{216}$	216	91		216.018960	0.20 s.	α	8.010	7.720						
	216	91						7.820						
	216	91						7.920						
$_{91}Pa^{217m}$	217	91			1.6 ms.	α		10.160(20)						
$_{91}Pa^{217}$	217	91		217.018250	4.9 ms.	α	8.490	8.340(10)						
$_{91}Pa^{218}$	218	91		218.019960	0.12 ms.	α		9.54						
	218	91						9.61						
$_{91}Pa^{222}$	222	91		222.023560	≈ 4.3 ms.	α	8.700	8.180	50%					
	222	91							8.33	0 20%				
	222	91						8.540	30%					
$_{91}Pa^{223}$	223	91		223.023950	6 ms.	α	8.340	8.006(10)	55%					
	223	91						8.196(10)	45%					
$_{91}Pa^{224}$	224	91		224.025530	0.95 s.	α	7.630	7.490(10)	100%					
$_{91}Pa^{225}$	225	91		225.026090	1.8 ± 0.3 s.	α	7.380	7.195(10)	30%					
	225	91						7.245(10)	70%					
$_{91}Pa^{226}$	226	91		226.027928	1.8 ± 0.2 s.	α(74%)	6.987	6.728(10)	0.7%					
	226	91				E.C.(26%)	2.834	6.823(10)	35%					
	226	91						6.863(10)	39%					
$_{91}Pa^{227}$	227	91		227.028797	38.3 m.	α(85%)	6.582	6.357(4)	7%		(5/2-)		0.0649 ± 0.001	5.3%
	227	91				E.C.(15%)	1.020	6.376(10)	2.2%				0.0669 ± 0.001	1.0%
	227	91						6.401(4)	8%				0.1100 ± 0.001	1.7%
	227	91						6.416(4)	13%					
	227	91						6.423(10)	10%					
	227	91						6.465(4)	43%					
$_{91}Pa^{228}$	228	91		228.030773	22 ± 1 h.	E.C.(98%)					(3+)		Th k x-ray	35%
	228	91				α(2%)		5.779	0.23%				0.20939 ± 0.00007	1.7%
	228	91						5.805	0.15%				0.27026 ± 0.00008	2.1%
	228	91						6.078	0.4%				0.28202 ± 0.00008	1.2%
	228	91						6.105	0.25%				0.32767 ± 0.00009	2%
	288	91						6.118	0.22%				0.32807 ± 0.00009	1.9%
	228	91											0.33248 ± 0.00009	1.6%
	228	91											0.33842 ± 0.00006	5.1%
	228	91											0.40962 ± 0.00008	6.4%
	228	91											0.46310 ± 0.00007	13.2%
	228	91											0.5815 ± 0.0002	1.0%
	228	91											0.7553 ± 0.0001	1.2%
	228	91											0.77228 ± 0.00007	1.2%
	228	91											0.7948 ± 0.0001	2.0%
	228	91											0.8306 ± 0.0001	1.9%
	228	91											0.8356 ± 0.0001	2.8%
	228	91											0.8404 ± 0.0001	1.0%
	228	91											0.8944 ± 0.0001	2.6%
	228	91											0.9043 ± 0.0001	2.8%
	228	91											0.91116 ± 0.00003	16.0%
	228	91											0.9457 ± 0.0008	1.8%
	228	91											0.96464 ± 0.00008	9.4%

Isotope	A	Z	% Natural abundance	Atomic mass	Half-life	Decay mode	Decay energy (MeV)	Particle energy (MeV)	Particle intensity	Thermal neutron cross section	Spin (h/2π)	μ Nucl. mag. moment	Gamma-ray energy (MeV)	Gamma-ray intensity
	228	91											0.96897 ± 0.00005	9.7%
	228	91											0.9757 ± 0.0001	1.6%
	228	91											1.2466 ± 0.0002	0.9%
	228	91											1.4592 ± 0.0001	0.72%
	228	91											1.5882 ± 0.0001	2.4%
	228	91											1.7385 ± 0.0002	0.64%
	228	91											1.7579 ± 0.0001	0.5%
	228	91											1.8351 ± 0.0001	0.64%
	228	91											1.8869 ± 0.0001	1.5%
	228	91											(0.1 - 1.96)many	
$_{91}Pa^{229}$	229	91		229.032073	1.4 ± 0.4 d.	E.C. (99.8%)	0.296				(5/2)		0.04244 ± 0.00001	
	229	91				α(0.2%)	5.836	5.536(2)	0.02%				(0.024 - 0.18)weak	
	229	91						5.579(2)	0.09%					
	229	91						5.668(2)	0.05%					
$_{91}Pa^{230}$	230	91		230.034527	17.4 ± 0.5 d.	E.C.(90%)		0.51		σf 1500 b.	(2-)		Th L x-ray	25%
	230	91				β-(10%)							Th k x-ray	30%
	230	91											0.39769 ± 0.00005	1.8%
	230	91											0.39996 ± 0.00006	0.61%
	230	91											0.44379 ± 0.00003	5.4%
	230	91											0.45477 ± 0.00003	6.1%
	230	91											0.46359 ± 0.00005	0.8%
	230	91											0.50820 ± 0.00003	3.5%
	230	91											0.51860 ± 0.00005	1.9%
	230	91											0.57116 ± 0.00006	1.0%
	230	91											0.72820 ± 0.00004	1.8%
	230	91											0.78143 ± 0.00004	1.4%
	230	91											0.89876 ± 0.00003	6.0%
	230	91											0.91856 ± 0.00005	8.0%
	230	91											0.95199 ± 0.00004	28%
	230	91											0.9591 ± 0.0002	0.5%
	230	91											0.95651 ± 0.00005	1.7%
	230	91											1.00974 ± 0.00005	1.7%
	230	91											1.02613 ± 0.00006	1.4%
	230	91											1.07467 ± 0.00006	0.73%
$_{91}Pa^{231}$	231	91		231.035880	3.27 x 10^4y.	α	5.148	4.6781(5)	1.5%		3/2-	2.01	Ac L x-ray	22%
	231	91						4.7102(5)	1.0%				Ac k x-ray	0.8%
	231	91						4.7343(5)	8.4%				0.01899 ± 0.00002	0.33%
	231	91						4.8513(5)	1.4%				0.027396 ± 0.00009	9.3%
	231	91						4.9339(5)	3%				0.03823 ± 0.00001	0.15%
	231	91						4.9505(5)	22.8%				0.04639 ± 0.00001	0.21%
	231	91						4.9858(5)	1.4%				0.25586 ± 0.00002	0.1%
	231	91						5.0131(5)	25.4%				0.26029 ± 0.00003	0.18%
	231	91						5.0292(5)	20%				0.28367 ± 0.00002	1.6%
	231	91						5.0318(5)	2.5%				0.30007 ± 0.00002	2.4%
	231	91						5.0587(5)	11%				0.30264 ± 0.00002	0.6%
	231	91											0.31306 ± 0.00002	0.13%
	231	91											0.33007 ± 0.00002	1.3%
	231	91											0.34087 ± 0.00002	0.17%
	231	91											0.35727 ± 0.00002	0.15%
	231	91											0.4271 ± 0.0001	0.10%
	231	91											(0.02 - 0.61)many	
$_{91}Pa^{232}$	232	91		232.038565	1.31 d.	β-	1.34			500 b.	(2-)		U k x-ray	1.8%

Isotope	A	Z	% Natural abundance	Atomic mass	Half-life	Decay mode	Decay energy (MeV)	Particle energy (MeV)	Particle intensity	Thermal neutron cross section	Spin (h/2π)	μ Nucl. mag. moment	Gamma-ray energy (MeV)	Gamma-ray intensity
	232	91								σf 700 b.			0.10900 ± 0.00001	2.8%
	232	91											0.15009 ± 0.00001	11%
	232	91											0.18414 ± 0.00001	1.3%
	232	91											0.38792 ± 0.00001	7.0%
	232	91											0.42196 ± 0.00001	2.5%
	232	91											0.45369 ± 0.00001	8.6%
	232	91											0.47243 ± 0.00001	4.2%
	232	91											0.51565 ± 0.00001	5.5%
	232	91											0.56323 ± 0.00001	3.7%
	232	91											0.58142 ± 0.00001	6.0%
	232	91											0.81925 ± 0.00001	7.5%
	232	91											0.86386 ± 0.00003	2.2%
	232	91											0.86683 ± 0.00001	5.8%
	232	91											0.89439 ± 0.00001	20%
	232	91											0.96934 ± 0.00001	42%
	232	91											(0.10 - 1.17)weak	
$_{91}Pa^{233}$	233	91		233.040242	27.0 ± 0.1 d.	β-	0.572	0.15	40%		3/2-	+3.5	U L x-ray	19%
	233	91						0.256	60%				U k x-rau	16%
	233	91											0.07534 ± 0.00001	1.2%
	233	91											0.08665 ± 0.00002	1.8%
	233	91											0.30017 ± 0.00002	6.2%
	233	91											0.31201 ± 0.00002	36%
	233	91											0.34059 ± 0.00002	4.2%
	233	91											0.39866 ± 0.00002	1.2%
	233	91											0.41593 ± 0.00002	1.5%
$_{91}Pa^{234m}$	234	91			1.17 m.	β-(99.9%)	2.29				(0-)		U k x-ray	0.2%
	234	91				I.T.(0.13%)							0.25818 ± 0.00003	0.06%
	234	91											0.74282 ± 0.00002	0.056%
	234	91											0.76641 ± 0.00001	0.21%
	234	91											0.78629 ± 0.00002	0.03%
	234	91											1.00100 ± 0.00003	0.65%
	234	91											1.7378 ± 0.0003	0.014%
	234	91											1.8317 ± 0.0004	0.011%
	234	91											(0.06 - 1.96)many	
$_{91}Pa^{234}$	234	91		234.043303	6.70 ± 0.05 h.	β-	2.199	0.51			(4+)		U L x-ray	52%
	234	91											U k x-ray	25%
	234	91											0.06278 ± 0.00004	3.2%
	234	91											0.09985 ± 0.00001	4.8%+
	234	91											0.1255 ± 0.0002	1.0%
	234	91											0.1312 ± 0.0002	20%
	234	91											0.15269 ± 0.00003	6.7%
	234	91											0.1859 ± 0.0002	2.0%
	234	91											0.20096 ± 0.00005	1.1%
	234	91											0.20320 ± 0.00003	1.1%
	234	91											0.2266 ± 0.0002	5.9%
	234	91											0.2272 ± 0.0002	5.5%
	234	91											0.2489 ± 0.0002	2.8%
	234	91											0.2721 ± 0.0002	1.0%
	234	91											0.2938 ± 0.0002	3.9%
	234	91											0.3700 ± 0.0002	2.9%
	234	91											0.4586 ± 0.0002	1.5%
	234	91											0.5067 ± 0.0002	1.6%

TABLE OF THE ISOTOPES (Continued)

Isotope	A	Z	% Natural abundance	Atomic mass	Half-life	Decay mode	Decay energy (MeV)	Particle energy (MeV)	Particle intensity	Thermal neutron cross section	Spin (h/2π)	μ Nucl. mag. moment	Gamma-ray energy (MeV)	Gamma-ray intensity
	234	91											0.5136 ± 0.0003	1.3%
	234	91											0.5652 ± 0.0004	1.4%
	234	91											0.5683 ± 0.0002	3.0%
	234	91											0.5695 ± 0.0002	11%
	234	91											0.6648 ± 0.001	1.3%
	234	91											0.6666 ± 0.0002	1.6%
	234	91											0.6697 ± 0.0003	1.4%
	234	91											0.6927 ± 0.0004	1.5%
	234	91											0.6988 ± 0.0002	4.6%
	234	91											0.70602 ± 0.00009	3.0%
	234	91											0.7332 ± 0.0002	8.6%
	234	91											0.7380 ± 0.0004	1.0%
	234	91											0.7428 ± 0.0002	2.4%
	234	91											0.7804 ± 0.0004	1.1%
	234	91											0.78629 ± 0.00002	1.4%
	234	91											0.7936 ± 0.001	1.5%
	234	91											0.7953 ± 0.0003	3.8%
	234	91											0.80587 ± 0.00009	3.4%
	234	91											0.8258 ± 0.0002	4.0%
	234	91											0.8314 ± 0.0002	5.5%
	234	91											0.8805 ± 0.0002	4.0%
	234	91											0.88053 ± 0.00004	9.0%
	234	91											0.88324 ± 0.00004	12.0%
	234	91											0.8986 ± 0.0002	4.0%
	234	91											0.9256 ± 0.0002	11%
	234	91											0.9258 ± 0.0002	2.9%
	234	91											0.92671 ± 0.00004	9.0%
	234	91											0.94602 ± 0.00003	8.1%
	234	91											0.9841 ± 0.0002	1.9%
	234	91											1.3528 ± 0.0002	1.7%
	234	91											1.3941 ± 0.0002	3.0%
	234	91											1.4527 ± 0.0002	1.0%
	234	91											1.6682 ± 0.0007	1.2%
	234	91											1.6942 ± 0.0004	1.2%
	234	91											1.9260 ± 0.0006	1.5%
	234	91											(0.02 - 1.99)many	
$_{91}Pa^{235}$	235	91		235.045430	24.1 m.	β-	1.4	1.4	97%		(3/2-)		0.0308 ± 0.0003	
	235	91											0.0367 ± 0.0002	
	235	91											0.05162 ± 0.00002	
	235	91											0.12928 ± 0.00002	
	235	91											0.34494 ± 0.00003	
	235	91											0.37502 ± 0.00002	
	235	91											0.38017 ± 0.00003	
	235	91											0.39312 ± 0.00004	
	235	91											0.41369 ± 0.00002	
	235	91											0.63796 ± 0.00008	
	235	91											0.64598 ± 0.00007	
	235	91											0.65218 ± 0.00007	
	235	91											0.65893 ± 0.00007	
$_{91}Pa^{236}$	236	91		236.048890	9.1 m.	β-	3.100	1.1	40%		(1-)		U k x-ray	2.6%
	236	91						2.0	50%				0.64235 ± 0.00005	29%
	236	91						3.1	10%				0.68759 ± 0.00005	7.8%
	236	91											1.5601 ± 0.001	1.9%
	236	91											1.7630 ± 0.0001	5.4%
	236	91											1.8082 ± 0.0007	2.0%
	236	91											2.0416 ± 0.0007	1.6%
	236	91											(0.04 - 2.18)weak	
$_{91}Pa^{237}$	237	91		237.051140	8.7 ± 0.2 m.	β-	2.25	1.1	60%		(1/2+)		0.4986 + 0.0001	2.4%
	237	91						1.6	30%				0.5293 ± 0.0001	14.8%
	237	91						2.3	10%				0.5407 ± 0.0001	9.3%
	237	91											0.5549 ± 0.0001	1.5%
	237	91											0.8536 ± 0.0001	34%
	237	91											0.8650 ± 0.0001	15%
	237	91											(0.04 - 1.4)weak	

Isotope	A	Z	% Natural abundance	Atomic mass	Half-life	Decay mode	Decay energy (MeV)	Particle energy (MeV)	Particle intensity	Thermal neutron cross section	Spin (h/2π)	μ Nucl. mag. moment	Gamma-ray energy (MeV)	Gamma-ray intensity
$_{91}Pa^{238}$	238	91		238.055040	2.3 m.	β-	3.960	1.2			(3-)		0.10350 ± 0.00004	12 +
	238	91						1.7					0.1785 ± 0.0005	11
	238	91											0.2179 ± 0.0005	14
	238	91											0.3698 ± 0.0005	12
	238	91											0.2930 ± 0.001	12
	238	91											0.3964 ± 0.0004	18
	238	91											0.4369 ± 0.0004	16
	238	91											0.4484 ± 0.0004	76
	238	91											0.4961 ± 0.0005	19
	238	91											0.4889 ± 0.0004	20
	238	91											0.5019 ± 0.0005	26
	238	91											0.5471 ± 0.0004	40
	238	91											0.6057 ± 0.0005	10
	238	91											0.6236 ± 0.001	19
	238	91											0.6350 ± 0.0004	88
	238	91											0.6800 ± 0.0004	73
	238	91											0.6870 ± 0.0004	54
	238	91											0.8058 ± 0.0004	44
	238	91											0.8491 ± 0.0005	14
	238	91											0.8637 ± 0.0005	54
	238	91											0.8857 ± 0.0004	45
	238	91											0.9049 ± 0.0005	23
	238	91											0.9111 ± 0.0004	19
	238	91											0.9526 ± 0.0005	21
	238	91											0.9571 ± 0.0005	18
	238	91											1.01446 ± 0.0004	100
	238	91											1.0602 ± 0.0005	45
	238	91											1.0834 ± 0.0003	50
	238	91											1.8892 ± 0.0004	17
	238	91											(0.04 - 2.5)many	
U		92		238.0289						7.57 b. σf 4.20 b.				
$_{92}U^{226}$	226	92		226.029170	0.50 ± 0.2 s.	α	7.560	7.430			0+			
$_{92}U^{227}$	227	92		227.030990	1.1 ± 0.3 m.	α	7.200	6.870						
$_{92}U^{228}$	228	92		228.031356	9.1 ± 0.2 m.	α	6.803	6.404(6)	0.6+		0+		0.095	1.6%
	228	92						6.440(5)	0.7				0.152	0.2%
	228	92						6.589(5)	29				0.187	0.3%
	228	92						6.681(6)	70				0.246	0.4%
$_{92}U^{229}$	229	92		229.033474	58 ± 3 m.	E.C.(80%)	1.305	6.223	3 +		(3/2+)			
	229	92				α(20%)	6.473	6.297(3)	11					
	229	92						6.332(3)	20					
	229	92						6.360(3)	64					
$_{92}U^{230}$	230	92		230.033921	20.8 d.	α	5.992	5.5866(3)	0.01%		0+		Th L x-ray	6.0%
	230	92						5.6624(3)	0.26%				0.07218 ± 0.00003	0.60%
	230	92						5.6663(3)	0.38%				0.15421 ± 0.00003	0.12%
	230	92						5.8178(3)	32%				0.23034 ± 0.00004	0.12%
	230	92						5.8887(3)	67%					
$_{92}U^{231}$	231	92		231.036270	4.2 ± 0.1 d.	E.C>					(5/2-)		Pa L x-ray	41%
	231	92											Pa k x-ray	27%
	231	92											0.02564 ± 0.00001	13%
	231	92											0.08420 ± 0.00001	6.0%
	231	92											0.21793 ± 0.00002	0.8%
	231	92											0.23598 ± 0.00002	0.2%
	231	92											0.31100 ± 0.00003	0.06%
$_{92}U^{232}$	231	92		232.037130	68.9 ± 0.1 y.	α	5.414	4.9979(1)	0.003%	73 b.	0+			
	232	92						5.1367(1)	0.3%					
	232	92						5.2635(1)	31%					
	232	92						5.3203(1)	69%					
$_{92}U^{233}$	233	92		233.039628	1.59 x 10^5y.	α	4.909	4.5097(8)	0.01%	466 b.	5/2+	+0.55	Th L x-ray	3.3%
	233	92						4.5130(8)	0.02%	σf 529 b.			0.04244 ± 0.00001	0.06%
	233	92						4.6323(8)	0.01%				0.09714 ± 0.00002	0.02%
	233	92						4.6642(8)	0.04%					
	233	92						4.6809(8)	0.01%					
	233	92						4.7014(8)	0.06%					
	233	92						4.7292(8)	1.6%					
	233	92						4.7541(8)	0.16%					
	233	92						4.758(10)	0.02%					
	233	92						4.7830(8)	13.2%					
	233	92						4.7960(8)	0.3%					
	233	92						4.804(8)	0.05%					

Isotope	A	Z	% Natural abundance	Atomic mass	Half-life	Decay mode	Decay energy (MeV)	Particle energy (MeV)	Particle intensity	Thermal neutron cross section	Spin (h/2π)	μ Nucl. mag. moment	Gamma-ray energy (MeV)	Gamma-ray intensity
	233	92						4.8247(8)	84.4%					
$_{92}U^{234}$	234	92		234.040946	2.45 x 10^5y.	α	4.856	4.604(1)	0.24%	100 b.	0+		0.05323 ± 0.00003	0.12%
	234	92						4.7231(1)	27.5%				0.12091 ± 0.00002	0.04%
	234	92						4.776(1)	72.5%					
$_{92}U^{235m}$	235	92			26 ± 2 m.	I.T.	0.0007				$^1/_2$+			
$_{92}U^{235}$	235	92	0.720%	235.043924	7.04 x 10^8y.	α	4.6793	4.1525(9)	0.9%	98 b.	7/2-	-0.35	Th L x-ray	15%
	235	92						4.2157(9)	5.7%	σf 583 b.			Th k x-ray	5.5%
	235	92						4.3237(9)	4.6%				0.10917 ± 0.00001	1.5%
	235	92						4.3641(9)	11%				0.14378 ± 0.00001	10.5%
	235	92						4.370(4)	6%				0.16338 ± 0.00001	4.7%
	235	92						4.3952(9)	55%				0.18574 ± 0.00001	53%
	235	92						4.4144(9)	2.1%				0.20213 ± 0.00001	1.0%
	235	92						4.5025(9)	1.7%				0.20533 ± 0.00001	4.7%
	235	92						4.5558(9)	4.2%				0.22140 ± 0.00002	0.1%
	235	92						4.5970(9)	5.0%				(0.03 - 0.79)weak	
$_{92}U^{236}$	236	92		236.045562	2.34 x 10^7y.	α	4.569	4.332(8)	0.26%	5.1 b.	0+		Th L x-ray	3.4%
	236	92						4.445(5)	26%				0.04937 ± 0.00001	0.08%
	236	92						4.494(3)	74%				0.11275 ± 0.00001	0.02%
$_{92}U^{237}$	237	92		237.048724	6.75 ± 0.01 d.	β-	0.519	0.24		400 b.	$^1/_2$+		Np L x-ray	30%
	237	92						0.25					Np k x-ray	26%
	237	92											0.02634 ± 0.00001	2.3%
	237	92											0.05953 ± 0.00001	33%
	237	92											0.06482 ± 0.00001	1.2%
	237	92											0.16459 ± 0.00001	1.8%
	237	92											0.20801 ± 0.00001	22%
	237	92											0.33236 ± 0.00001	1.2%
	237	92											0.37092 ± 0.00002	0.11%
$_{92}U^{238}$	238	92		238.050784	4.46 x 10^9y.	α		4.039(5)	0.23	% 2.68 b.	0+		Th L x-ray	4.1%
	238	92						4.147(5)	23%				0.04955 ± 0.00006	0.07%
	238	92						4.196(5)	77%					
$_{92}U^{239}$	239	92		239.054289	23.54 ± 0.05m.	β-	1.264	1.2		22 b.	5/2+		0.04354 ± 0.00001	4.4%
	239	92						1.3		σf 15 b.			0.07467 ± 0.00001	52%
	239	92											0.11770 ± 0.00003	0.12%
	239	92											0.66225 ± 0.00002	0.20%
	239	92											0.74805 ± 0.00003	0.10%
	239	92											0.81927 ± 0.00003	0.15%
	239	92											0.84410 ± 0.00003	0.17%
$_{92}U^{240}$	240	92		240.056587	14.1 ± 0.2 h.	β-	0.50	0.36			0+		Np L x-ray	18%
	240	92											0.04410 ± 0.00007	1.7%
$_{92}U^{242}$	242	92			16.8 m.	β-							0.05558 ± 0.0005	3.7%
	242	92											0.06760 ± 0.00005	9.2%
	242	92											0.1604 ± 0.0001	0.75%
	242	92											0.1820 ± 0.0001	0.70%
	242	92											0.32972 ± 0.00009	0.75%
	242	92											0.5729 ± 0.0001	1.8%
	242	92											0.5849 ± 0.0001	1.8%
Np		93												
$_{93}Np^{228}$	228	93			1.0 m.	S.F.								
$_{93}Np^{229}$	229	93		229.036230	4.0 ± 0.2 m.	α	7.010	6.890(20)						
$_{93}Np^{230}$	230	93		230.937810	4.6 m.	E.C.(97%)	3.620							
	230	93				α(3%)		6.660(20)						
$_{93}Np^{231}$	231	93		231.038240	48.8 ± 0.2 m.	E.C.(98%)	1.84				5/2		0.2629 ± 0.0003	2.8 +
	231	93				α(2%)	6.368	6.280	2%				0.3475 ± 0.0003	3.6

Isotope	A	Z	% Natural abundance	Atomic mass	Half-life	Decay mode	Decay energy (MeV)	Particle energy (MeV)	Particle intensity	Thermal neutron cross section	Spin (h/2π)	μ Nucl. mag. moment	Gamma-ray energy (MeV)	Gamma-ray intensity
	231	93											0.3703 ± 0.0002	9.8
	231	93											0.4201 ± 0.0003	1.0
	231	93											0.4838 ± 0.0005	1.6
	231	93											0.7369 ± 0.0002	1.2
	231	93											0.8364 ± 0.0004	0.4
	231	93											0.8511 ± 0.0003	0.7
	231	93											1.1072 ± 0.0002	0.5
$_{93}Np^{232}$	232	93		232.040020	14.7 m.	E.C.(99%)	2.70				(4-)		U L x-ray	35%
	232	93											U k x-ray	38%
	232	93											0.2229 ± 0.0002	2.2%
	232	93											0.2822 ± 0.0002	19.8%
	232	93											0.3268 ± 0.0002	52%
	232	93											0.75486 ± 0.00003	4.5%
	232	93											0.81415 ± 0.00007	4.1%
	232	93											0.81925 ± 0.00001	32.6%
	232	93											0.86386 ± 0.00003	19.7%
	232	93											0.86683 ± 0.00001	25.1%
	232	93											1.0371 ± 0.0002	3.3%
	232	93											1.1255 ± 0.0002	1.5%
$_{93}Np^{233}$	233	93		233.040800	36.2 ± 0.1 m.	E.C.	1.090				(5/2+)		U L x-ray	18%
	233	93											U k x-ray	35%
	233	93											0.2344 ± 0.0002	0.15%
	233	93											0.25846 ± 0.00003	0.10%
	233	93											0.2804 ± 0.0001	0.13%
	233	93											0.29887 ± 0.00003	0.48%
	233	93											0.31201 ± 0.00002	0.70%
	233	93											0.5061 ± 0.0002	0.15%
	233	93											0.5465 ± 0.0002	0.28%
$_{93}Np^{234}$	234	93		234.042888	4.4 ± 0.1 d.	β+,E.C.	1.808	0.79		σf 900 b.	(0+)		U L x-ray	17%
	234	93											U k x-ray	29%
	234	93											0.45092 ± 0.00004	1.3%
	234	93											0.6255 ± 0.0003	1.1%
	234	93											0.74282 ± 0.00002	5.0%
	234	93											0.78629 ± 0.00002	3.0%
	234	93											1.00100 ± 0.00003	1.4%
	234	93											1.19374 ± 0.00003	5.5%
	234	93											1.23721 ± 0.00003	2.2%
	234	93											1.3920 ± 0.0003	2.0%
	234	93											1.4354 ± 0.0003	6.3%
	234	93											1.5272 ± 0.0002	11.5%
	234	93											1.5587 ± 0.0003	18.4%
	234	93											1.5707 ± 0.0002	5.6%
	234	93											1.6022 ± 0.0003	9.7%
$_{93}Np^{235}$	235	93		235.044056	1.08 y.	E.C. (99.9%)	0.123			150 b.	5/2+		U k x-ray	16%
	235	93				α(0.001%)	5.191							
$_{93}Np^{236m}$	236	93			22.5 h.	E.C.(52%)					(1-)		U L x-ray	10%
	236	93				β-(48%)							Pu L x-ray	1.5%
	236	93											U k x-ray	18%
	236	93											0.64235 ± 0.00005	0.9%
	236	93											0.68759 ± 0.00005	0.25%
$_{93}Np^{236}$	236	93		236.046550	1.2 x 10^5 y.	E.C.(91%)	0.99			σf 2500 b.	(6-)		U L x-ray	57%
	236	93				β-(9%)	0.54						U k x-ray	33%
	236	93											0.10423 ± 0.00001	7.5%
	236	93											0.16031 ± 0.00001	28%
$_{93}Np^{237}$	237	93		237.048167	2.14 x 10^6y.	α	4.957	4.5779(5)	0.4%	180 b.	5/2+	+3.14	Pa L x-ray	25%
	237	93						4.6395(5)	6.2%	σf 0.02 b.			Pa k x-ray	2.6%
	237	93						4.659(2)	0.6%				0.029378 ± 0.00009	13%
	237	93						4.6645(5)	3.3%				0.08653 ± 0.00001	13%
	237	93						4.6971(7)	0.5%				0.09472 ± 0.00002	0.8%
	237	93						4.7071(5)	1.0%				0.11758 ± 0.00002	0.16%
	237	93						4.766(5)	8%				0.14323 ± 0.00002	0.40%

Isotope	A	Z	% Natural abundance	Atomic mass	Half-life	Decay mode	Decay energy (MeV)	Particle energy (MeV)	Particle intensity	Thermal neutron cross section	Spin (h/2π)	μ Nucl. mag. moment	Gamma-ray energy (MeV)	Gamma-ray intensity
	237	93						4.7715(5)	25%				0.15142 ± 0.00002	0.24%
	237	93						4.7884(5)	47%				0.19504 ± 0.00003	0.20%
	237	93						4.8040(5)	1.6%				0.21241 ± 0.00002	0.15%
	237	93						4.8173(5)	2.5%				(0.03 - 0.28)weak	
	237	93						4.8734(5)	2.6%					
$_{93}Np^{238}$	238	93		238.050941	2.117 d.	β-	1.291	1.2		σf 2100 b.	2+		Pu L x-ray	16%
	238	93											Pu k x-ray	0.3%
	238	93											0.1019 ± 0.0001	0.27%
	238	93											0.11990 ± 0.00007	0.11%
	238	93											0.56103 ± 0.00003	0.11%
	238	93											0.88258 ± 0.00002	0.9%
	238	93											0.91870 ± 0.00002	0.6%
	238	93											0.92339 ± 0.00002	2.9%
	238	93											0.98447 ± 0.00002	28%
	238	93											1.02588 ± 0.00002	9.6%
	238	93											1.02855 ± 0.00002	20%
$_{93}Np^{239}$	239	93		239.052933	2.35 d.	β-	0.721	0.341	30%	(30 + 30)b.	5/2+		Pu L x-ray	25%
	239	93						0.438	48%				Pu k x-ray	24%
	239	93											0.10613 ± 0.00001	23%
	239	93											0.20975 ± 0.00001	3.3%
	239	93											0.228186 ± 0.00002	10.7%
	239	93											0.27760 ± 0.00002	14.2%
	239	93											0.31588 ± 0.00001	1.6%
	239	93											0.33431 ± 0.00001	2.0%
	239	93											(0.04 - 0.50)weak	
$_{93}Np^{240m}$	240	93			7.22 m.	β-(99.9%)		2.18			(1-)		0.25143 ± 0.00005	0.9%
	240	93				I.T.(0.1%)							0.26333 ± 0.00006	1.1%
	240	93											0.30296 ± 0.00004	1.2%
	240	93											0.55454 ± 0.00003	22%
	240	93											0.59735 ± 0.00003	12.6%
	240	93											0.75864 ± 0.00004	1.2%
	240	93											0.81787 ± 0.00006	1.3%
	240	93											0.85750 ± 0.00004	0.5%
	240	93											0.91604 ± 0.00005	1.0%
	240	93											0.93805 ± 0.00007	1.3%
	240	93											0.96158 ± 0.00007	0.14%
	240	93											1.44533 ± 0.00007	0.36%
	240	93											1.49685 ± 0.00006	1.3%
	240	93											1.53967 ± 0.00006	0.79%
	240	93											1.63326 ± 0.00009	0.14%
$_{93}Np^{240}$	240	93		240.056050	1.03 h.	β-	2.090	0.89			5+		0.1471 ± 0.0003	1.5%
	240	93											0.15262 ± 0.00002	9.0%
	240	93											0.175 ± 0.001	6.5%
	240	93											0.1930 ± 0.0002	7.3%
	240	93											0.2708 ± 0.0003	9.0%
	240	93											0.4482 ± 0.0002	18%
	240	93											0.4669 ± 0.0002	2.2%
	240	93											0.5664 ± 0.0002	29%
	240	93											0.6008 ± 0.0002	22%
	240	93											0.847 ± 0.001	5%
	240	93											0.8674 ± 0.0002	9.0%
	240	93											0.8449 ± 0.001	4.0%

Isotope	A	Z	% Natural abundance	Atomic mass	Half-life	Decay mode	Decay energy (MeV)	Particle energy (MeV)	Particle intensity	Thermal neutron cross section	Spin (h/2π)	μ Nucl. mag. moment	Gamma-ray energy (MeV)	Gamma-ray intensity
	240	93											0.8963 ± 0.0003	14%
	240	93											0.9591 ± 0.0002	2.5%
	240	93											0.9741 ± 0.0002	23%
	240	93											0.98764 ± 0.00004	4.0%
	240	93											1.1672 ± 0.0002	5%
	240	93											1.1802 ± 0.0002	0.7%
$_{93}Np^{241}$	241	93		241.058250	13.9 ± 0.2 m.	β-	1.5	1.3			5/2+		0.1330 ± 0.0009	98 +
	241	93											0.1740 ± 0.0009	100
	241	93											0.280	
$_{93}Np^{242m}$	242	93			5.5 ± 0.1 m.	β-					6+		0.15910 ± 0.00008	32 +
	242	93											0.2651 ± 0.0001	24
	242	93											0.78570 ± 0.00008	100
	242	93											0.9448 ± 0.0001	63
	242	93											1.104 ± 0.0001	0.6
$_{93}Np^{242}$	242	93		242.061640	2.2 ± 0.2 m.	β-	2.7	2.7			(1+)		0.6209 ± 0.0001	0.9%
	242	93											0.6477 ± 0.0002	0.27%
	242	93											0.6853 ± 0.0001	0.35%
	242	93											0.73620 ± 0.00004	5.0%
	242	93											0.78074 ± 0.00004	2.6%
	242	93											0.81395 ± 0.00008	1.2%
	242	93											1.0076 ± 0.0002	0.15%
	242	93											1.0346 ± 0.0002	0.3%
	242	93											1.0938 ± 0.0001	1.1%
	242	93											1.1103 ± 0.0002	0.35%
	242	93											1.1374 ± 0.0001	1.2%
	242	93											1.47340 ± 0.00006	2.2%
	242	93											1.51794 ± 0.00006	1.2%
	242	93											1.55114 ± 0.00008	0.35%
	242	93											1.8596 ± 0.0002	0.55%
	242	93											1.9500 ± 0.0001	1.75%
	242	93											1.9702 ± 0.0001	0.52%
	242	93											1.9924 ± 0.0003	0.2%
	242	93											(0.04 - 2.37)weak	
Pu		94												
$_{94}Pu^{232}$	232	94		232.041169	34 m.	E.C.>80%	1.070				0+			
	232	94				α <20%	6.716	6.542(10)	38 +					
	232	94						6.600(10)	62					
$_{94}Pu^{233}$	233	94		233.042970	E.C.(99.9%)	2.020								
	233	94				α(0.1%)	6.416	6.300(20)	0.1%				0.1503 ± 0.0003	15 +
	233	94											0.1804 ± 0.0003	12
	233	94											0.1911 ± 0.0002	13
	233	94											0.2076 ± 0.0002	24
	233	94											0.2218 ± 0.0003	12
	233	94											0.2353 ± 0.0002	100
	233	94											0.4571 ± 0.0002	10
	233	94											0.4781 ± 0.0002	14
	233	94											0.5002 ± 0.0002	39
	233	94											0.5038 ± 0.0002	21
	233	94											0.5125 ± 0.0002	13
	233	94											0.5243 ± 0.0002	13
	233	94											0.5346 ± 0.0002	90
	233	94											0.5587 ± 0.0002	27
	233	94											0.6880 ± 0.0003	33
	233	94											0.8308 ± 0.0003	11
	233	94											0.9779 ± 0.0002	13
	233	94											0.9917 ± 0.0002	23
	233	94											1.0008 ± 0.0002	18
	233	94											1.0039 ± 0.0003	31
	233	94											1.0123 ± 0.0002	28
	233	94											1.0281 ± 0.0003	6.6
	233	94											1.0352 ± 0.0003	5.7
$_{94}Pu^{234}$	234	94		234.043299	8.8 h.	E.C.(94%)	0.383				0+			
	234	94				α(6%)	6.310	6.035(3)	0.024%					
	234	94						6.149(3)	1.9%					
	234	94						6.200(3)	4.0%					
$_{94}Pu^{235}$	235	94		235.045260	25.6 m.	E.C. (99+%)	1.13				(5/2+)			
	235	94				α(0.003%)	5.957	5.850(20)	0.003					
$_{94}Pu^{236}$	236	94		236.046032	2.85 ± 0.01 y.	α	5.867	5.4519(8)	0.002%	σf 160 b.	0+			
	236	94						5.6138(7)	0.18%					
	236	94						5.7210(7)	32%					
	236	94						5.7677(7)	68%					
$_{94}Pu^{237}$	237	94		237.048401	45.1 d.	E.C. (99.9%)	0.218				7/2-		Np L x-ray	21%

Isotope	A	Z	% Natural abundance	Atomic mass	Half-life	Decay mode	Decay energy (MeV)	Particle energy (MeV)	Particle intensity	Thermal neutron cross section	Spin (h/2π)	μ Nucl. mag. moment	Gamma-ray energy (MeV)	Gamma-ray intensity
	237	94				α(0.003%)	5.747	5.334(4)	0.0015				Np k x-ray	20%
	237	94						5.356(4)	0.0006				0.026344 ± 0.00001	0.23%
	237	94						5.650(4)	0.0007				0.03319 ± 0.00001	0.08%
	237	94											0.05954 ± 0.00001	3.3%
	237	94											(0.03 - 0.5)weak	
$_{94}Pu^{238}$	238	94		238.049554	87.74 y.	α	5.593	5.3583(1)	0.10%	540 b.	0+		U k x-ray	5.2%
	238	94						5.465(1)	28.3%	σf 18 b.			0.04347 ± 0.00001	0.04%
	238	94						5.4992(1)	71.6%				(0.04 - 1.1)weak	
$_{94}Pu^{239}$	239	94		239.052157	2.411 x 10^4y	α	5.244	5.0542(4)	0.023%	269 b.	1/2+	+0.203	U k x-ray	0.006%
	239	94						5.0752(4)	0.056%	σf 742 b.			0.05162 ± 0.00002	0.03%
	239	94						5.1047(4)	10.6%				0.05682 ± 0.00003	0.001%
	239	94						5.1428(4)	15.1%				0.12928 ± 0.00002	0.006%
	239	94						5.1555(4)	73.2%				0.37502 ± 0.00002	0.002%
	239	94											0.41369 ± 0.00002	0.0015%
$_{94}Pu^{240}$	240	94		240.053808	6537 ± 10 y.	α	5.255	5.0212(1)	0.07%	290 b.	0+		U L x-ray	5%
	240	39						5.1237(1)	26.4%				0.04524 ± 0.00001	0.045%
	240	94						5.1681(1)	73.5%				0.10423 ± 0.00001	0.007%
	240	94											(0.04 - 0.97)weak	
$_{94}Pu^{241}$	241	94		241.056845	14.4 ± 0.2 y.	β-(99+%)	0.021	4.8532(7)	3x10^{-4}%	360 b.	5/2+	-0.683	0.14854 ± 0.00001	0.00018%
	241	94				α(0.002%)	5.139	4.8966(7)	0.002%	σf 1010 b.				
$_{94}Pu^{242}$	242	94		242.058737	3.76 x 10^5y.	α	4.983	4.7546(7)	0.098%	19 b.	0+		U L x-ray	4.1%
	242	94						4.8564(7)	22.4%				0.04491 ± 0.00001	0.04%
	242	94						4.9006(7)	78%				0.10350 ± 0.00004	0.008%
	242	94											0.15880 ± 0.00008	0.0004%
$_{94}Pu^{243}$	243	94		243.061998	4.95 h.	β-	0.580	0.49	21%	90 b.	7/2+		Am L x-ray	6.2%
	243	94						0.58	60%	σf 200 b.			0.0417 ± 0.0002	0.8%
	243	94											0.06710 ± 0.0002	0.2%
	243	94											0.0839 ± 0.0002	23%
	243	94											0.10547 ± 0.00005	0.18%
	243	94											0.3564 ± 0.0002	0.13%
	243	94											0.3817 ± 0.0002	0.55%
	243	94											0.4232 ± 0.0002	0.01%
$_{94}Pu^{244}$	244	94		244.064199	8.2 x 10^7y.	α(99.9%)	4.665	4.546(1)	19.4%	1.7 b.	0+		U L x-ray	0.35%
	244	94				S.F.(0.1%)		4.589(1)	80.5%				0.0439 ± 0.0008	0.003%
$_{94}Pu^{245}$	245	94		245.067820	10.5 ± 0.1 h.	β-	1.28	0.93	57%		(9/2-)		Am L x-ray	6.6%
	245	94						1.21	11%				Am k x-ray	12%
	245	94											0.2804 ± 0.0001	1.3%
	245	94											0.30832 ± 0.00008	4.9%
	245	94											0.32752 ± 0.00001	25%
	245	94											0.37677 ± 0.00001	3.2%
	245	94											0.49169 ± 0.00001	2.7%
	245	94											0.56014 ± 0.00001	5.4%
	245	94											0.6302 ± 0.0001	2.7%
	245	94											0.8000 ± 0.0001	1.6%
	245	94											0.91063 ± 0.00002	1.4%
	245	94											0.93852 ± 0.00002	1.0%
	245	94											0.98770 ± 0.00004	1.3%
	245	94											1.0183 ± 0.0001	1.0%
	245	94											(0.03 - 1.2)weak	
$_{94}Pu^{246}$	246	94		246.070171	10.85 ± 0.02d.	β-	0.374	0.150	85%		0+		Am L x-ray	23%
	246	94						0.35	10%				Am k x-ray	22%
	246	94											0.02756 ± 0.00002	3.5%
	246	94											0.04379 ± 0.00002	25%
	246	94											0.17992 ± 0.00002	9.7%
	246	94											0.22371 ± 0.00002	23%

Isotope	A	Z	% Natural abundance	Atomic mass	Half-life	Decay mode	Decay energy (MeV)	Particle energy (MeV)	Particle intensity	Thermal neutron cross section	Spin (h/2π)	μ Nucl. mag. moment	Gamma-ray energy (MeV)	Gamma-ray intensity
	246	94											0.25553 ± 0.00002	0.2%
Am		95												
$_{95}Am^{237}$	237	95		237.050050	1.22 h.	E.C.(99.98)	1.540				(5/2-)		Pu k x-ray	40%
	237	95				α(0.02%)	6.20	6.042(5)	0.02%				0.14559 ± 0.00001	0.5%
	237	95											0.28026 ± 0.00001	47%
	237	95											0.32101 ± 0.00002	1.4%
	237	95											0.42585 ± 0.00007	1.9%
	237	95											0.43845 ± 0.00007	8.3%
	237	95											0.47355 ± 0.00007	4.3%
	237	95											0.6553 ± 0.0002	1.3%
	237	95											0.9089 ± 0.0001	2.6%
$_{95}Am^{238}$	238	95		238.051980	1.63 h.	E.C.	2.26				1+		Pu L x-ray	27%
	238	95				α(0.0001%)	6.04	5.940	0.0001%				Pu k x-ray	36%
	238	95											0.35767 ± 0.00003	2.1%
	238	95											0.56103 ± 0.00003	10.9%
	238	95											0.60511 ± 0.00003	7.6%
	238	95											0.91870 ± 0.00002	23%
	238	95											0.94137 ± 0.00004	2.2%
	238	95											0.96278 ± 0.00002	28%
	238	95											1.2662 ± 0.0003	1.7%
	238	95											1.5772 ± 0.0001	2.9%
	238	95											1.6364 ± 0.0001	1.3%
$_{95}Am^{239}$	239	95		239.053016	11.901 h.	E.C.(99.99)	0.800				5/2-		Pu L x-ray	45%
	239	95				α(0.01%)	5.924	5.734(2)	0.001%				Pu k x-ray	61%
	239	95						5.776(2)	0.008%				0.18172 ± 0.00001	1.1%
	239	95											0.20975 ± 0.00001	3.5%
	239	95											0.22638 ± 0.00001	3.3%
	239	95											0.22818 ± 0.00001	11.4%
	239	95											0.27760 ± 0.00001	15%
$_{95}Am^{240}$	240	95		240.055278	50.9 h.	E.C.	1.369				(3-)		Pu L x-ray	38%
	240	95				α	5.592	5.378(1)	1.6×10^{-5}				Pu k x-ray	28%
	240	95											0.09558 ± 0.00001	1.5%
	240	95											0.88878 ± 0.00004	25%
	240	95											0.98764 ± 0.00004	73%
	240	95											(0.1 - 1.3)weak	
$_{95}Am^{241}$	241	95		241.056823	432.2 y.	α	5.637	5.2443(1)	0.002%	(50 + 550) b.	5/2-	+1.61	Np L x-ray	20%
	241	95						5.3221(1)	0.015%	σf 3.2 b.			0.02634 ± 0.00001	2.4%
	241	95						5.3884(1)	1.4%				0.033192 ± 0.00001	0.12%
	241	95						5.4431(1)	12.8%				0.059536 ± 0.00001	35.7%
	241	95						5.4857(1)	85.2%				(0.03 - 0.95)weak	
	241	95						5.5116(1)	0.20%					
	241	95						5.5442(1)	0.34%					
$_{95}Am^{242m}$	242	95			141 ± 2 y.	I.T.(99.5%)	0.048				5-		Am L x-ray	12%
	242	95				α(0.5%)	5.62	5.1413(4)	0.026%				0.04863 ± 0.00005	0.00013%
	242	95						5.2070(2)	0.4%				0.08648 ± 0.00003	0.04%
	242	95											0.10944 ± 0.00009	0.024%
	242	95											0.13497 ± 0.00006	0.01%
	242	95											0.16304 ± 0.00004	0.023%
$_{95}Am^{242}$	242	95		242.059541	16.01 ± 0.02h.	β-(83%)	0.661	0.63	46%		1-	+0.3878	Pu L x-ray	4.9%
	242	95				E.C.(17%)		0.67	37%				Cm L x-ray	8.6%
	242	95											Pu k x-ray	5.8%
	242	95											0.0422 ± 0.0001	0.04%
	242	95											0.04453 ± 0.00001	0.014%
$_{95}Am^{243}$	243	95		243.061375	7.37 x 10^3y.	α	5.438	5.1798(5)	1.1%	(74 + 4) b.	5/2-	+1.61	0.04354 ± 0.00001	5.1%

Isotope	A	Z	% Natural abundance	Atomic mass	Half-life	Decay mode	Decay energy (MeV)	Particle energy (MeV)	Particle intensity	Thermal neutron cross section	Spin (h/2π)	μ Nucl. mag. moment	Gamma-ray energy (MeV)	Gamma-ray intensity
	243	95						5.2343(5)	11%	σf 0.2 b.			0.07467 ± 0.00001	60%
	243	95						5.2766(5)	88%				0.08657 ± 0.00003	0.30%
	243	95						5.394(5)	0.12%				0.11770 ± 0.00003	0.6%
	243	95						5.3500(5)	0.16%				0.14197 ± 0.00004	0.11%
$_{95}Am^{244m}$	244	95			26 m.	β-	1.498			σf 1600 b.	(1-)		0.0429 ± 0.0001	0.02%
$_{95}Am^{244}$	244	95		244.064279	10.1 h.	β-	1.427			σf 2300 b.			Am L x-ray	50%
	244	95											Cm k x-ray	3.6%
	244	95											0.0994 ± 0.0001	5%
	244	95											0.1540 ± 0.0008	18%
	244	95											0.7460 ± 0.0008	67%
	244	95											0.9000 ± 0.0008	28%
$_{95}Am^{245}$	245	95		245.066444	2.05 ± 0.01 h.	β-	0.894	0.65	19%		(5/2+)		Cm L x-ray	3.4%
	245	95						0.90	77%				Cm k x-ray	5.7%
	245	95											0.04287 ± 0.00002	0.06%
	245	95											0.24106 ± 0.00003	0.33%
	245	95											0.25299 ± 0.00002	6.1%
	245	95											0.29587 ± 0.00002	0.23%
$_{95}Am^{246m}$	246	95			25.0 m.	β-	1.31	79%			2-		Cm L x-ray	13%
	246	95						1.60	14%				Cm k x-ray	1.9%
	246	95						2.1	7%				0.27002 ± 0.00003	1.0%
	246	95											0.73442 ± 0.00002	1.2%
	246	95											0.79881 ± 0.00002	25%
	246	95											0.83358 ± 0.00002	1.8%
	246	95											1.03600 ± 0.00002	13%
	246	95											1.06201 ± 0.00002	17%
	246	95											1.07885 ± 0.00002	28%
	246	95											1.08517 ± 0.00002	1.5%
	246	95											1.27472 ± 0.00004	0.27%
	246	95											1.59068 ± 0.00003	0.5%
	246	95											1.66165 ± 0.00003	0.22%
	246	95											(0.04 - 2.29)weak	
$_{95}Am^{246}$	246	95		246.069770	39 m.	β-	2.38	1.2			(7-)		Cm L x-ray	63.%
	246	95											Cm k x-ray	4.6%
	246	95											0.1529 ± 0.0001	25%
	246	95											0.2046 ± 0.0008	36%
	246	95											0.6289 ± 0.001	2.7%
	246	95											0.6786 ± 0.0008	53%
	246	95											0.7558 ± 0.0004	13%
	246	95											0.78131 ± 0.00003	4.0%
	246	95											0.8389 ± 0.0014	2%
$_{95}Am^{247}$	247	95		247.072170	22 ± 3 m.	β-	1.700						Cm L x-ray	13%
	247	95											Cm k x-ray	21%
	247	95											0.2267 ± 0.0007	5.8%
	247	95											0.2853 ± 0.0002	23%
Cm		96												
$_{96}Cm^{238}$	238	96		238.053020	2.4 ± 0.1 h.	E.C. (>90%)	0.970				0+			
	238	96				α(<10%)	6.632	6.520(50)	<10%					
$_{96}Cm^{239}$	239	96		239.054840	≈ 3 h.	E.C.	1.700						0.0407 ± 0.0005	
	239	96											0.1466 ± 0.0005	
	239	96											0.1874 ± 0.0004	
$_{96}Cm^{240}$	240	96		240.055503	27 ± 1 d.	α	6.397	5.989	0.014%		0+			
	240	96						6.147	0.05%					
	240	96						6.2478(6)	28.8%					
	240	96						6.2906(6)	70.6%					
$_{96}Cm^{241}$	241	96		241.057645	32.8 ± 0.2 d.	E.C.(99%)	0.765				1/2+		Am k x-ray	35%
	241	96				α(1%)	6.184	5.8842(4)	0.12%				0.13241 ± 0.00001	3.9%
	241	96						5.9291(4)	0.18%				0.16505 ± 0.00001	3.0%
	241	96						5.9389(4)	0.69%				0.18028 ± 0.00001	0.5%

Isotope	A	Z	% Natural abundance	Atomic mass	Half-life	Decay mode	Decay energy (MeV)	Particle energy (MeV)	Particle intensity	Thermal neutron cross section	Spin (h/2π)	μ Nucl. mag. moment	Gamma-ray energy (MeV)	Gamma-ray intensity
	241	96											0.20588 ± 0.00001	2.8%
	241	96											0.43063 ± 0.00001	4.1%
	241	96											0.46327 ± 0.00001	1.2%
	241	96											0.47181 ± 0.00001	71%
	241	96											0.63686 ± 0.00001	1.5%
$_{96}Cm^{242}$	242	96		242.058830	162.9±0.1 d.	α.	6.216	5.9694(1)	0.035%	20 b.	0+		Pu L x-ray	4.8%
	242	96						6.069(1)	25%				0.04408 ± 0.00002	0.03%
	242	96						6.1129(1)	74%				0.10189 ± 0.00002	0.002%
	242	96											0.15742 ± 0.00006	0.001%
	242	96											(0.04 - 1.2)weak	
$_{96}Cm^{243}$	243	96		243.061381	28.5 ± 0.2 y.	α	6.167	5.6815(5)	0.2%		5/2+	0.41	Pu L x-ray	19%
	243	96						5.6856(5)	1.6%				Pu k x-ray	23%
	243	96						5.7420(5)	10.6%				0.10612 ± 0.00001	0.26%
	243	96						5.7859(5)	73.3%				0.20975 ± 0.00001	3.3%
	243	96						5.9922(5)	6.5%				0.22819 ± 0.00001	10.6%
	243	96						6.0103(5)	1.0%				0.27760 ± 0.00001	14%
	243	96						6.0589(5)	5%				0.28546 ± 0.00001	0.73%
	243	96						6.0666(5)	1.5%				0.33431 ± 0.00001	0.023%
	243	96											(0.04 - 0.7)wcak	
$_{96}Cm^{244}$	244	96		244.062747	18.11 y.	α	5.902	5.6656(1)	0.02%	15 b.	0+		Pu L x-ray	4%
	244	96						5.7528(1)	24%	σf 1.0 b.			0.04282 ± 0.00001	0.02%
	244	96						5.8050(1)	76%				0.09885 ± 0.00001	0.001%
	244	96											0.15262 ± 0.00002	0.001%
$_{96}Cm^{245}$	245	96		245.065483	8.5 x 10^3 y.	α	5.623	5.235(10)	0.3%	360 b.	7/2+	0.5	Pu L x-ray	55%
	245	96						5.3038(10)	5.0%	σf 2100 b.			Pu k x-ray	34%
	245	96						5.3620(7)	93%				0.04195 ± 0.00003	0.35%
	245	96						5.4927(11)	0.8%				0.13299 ± 0.00003	2.8%
	245	96						5.5331(11)	0.6%				0.13606 ± 0.00006	0.11%
	245	96											0.17494 ± 0.00004	9.5%
	245	96											0.18982 ± 0.00006	0.2%
$_{96}Cm^{246}$	246	96		246.067218	4.78 x 10^3y.	α	5.476	5.343(3)	21%	1.2 b.	0+		Pu L x-ray	3.9%
	246	96						5.386(3)	79%	σf 0.2 b.			0.04453 ± 0.00001	0.027%
$_{96}Cm^{247}$	247	96		247.070347	1.56 x 10^7y.	α	5.352	4.818(4)	4.7%	60 b.	9/2-	0.37	Pu k x-ray	2.1%
	247	96						4.8690(20)	71%	σf 80 b.	9/2-		0.2792 ± 0.0008	3.4%
	247	96						4.941(4)	1.6%				0.2886 ± 0.0007	2.0%
	247	96						4.9820(20)	2.0%				0.3471 ± 0.0008	1%
	247	96						5.1436(20)	1.2%				0.4035 ± 0.0005	72%
	247	96						5.2104(20)	5.7%					
	247	96						5.2659(20)	13.8%					
$_{96}Cm^{248}$	248	96		248.072343	3.4 x 10^5 y.	α(92%)	5.162	4.931(5)	0.07%	2.6 b.	0+			
	248	96				S.F.(8%)		5.0349(2)	16.5%	σf 0.4 b.				
	248	96						5.0784(2)	75	1%				
$_{96}Cm^{249}$	249	96		249.075948	64.15 m.	β-	0.902	0.9		2 b.	1/2+		Bk k x-ray	0.2%
	249	96											0.36897 ± 0.00004	0.35%
	249	96											0.51846 ± 0.00004	0.09%
	249	96											0.56039 ± 0.00005	0.84%
	249	96											0.62191 ± 0.00005	0.18%
	249	96											0.63431 ± 0.00005	1.5%
	249	96											0.65277 ± 0.00005	0.14%
$_{96}Cm^{250}$	250	96		250.078352	≈ 7.4x10^3 y.	S.F.				80 b.	0+			
	250	96				α	5.27							
$_{96}Cm^{251}$	251	96		251.082290	16.8 m.	β-	1.420	0.90	16%		(1/2+)		0.3896 ± 0.0002	1.3%
	251	96											0.4381 ± 0.0003	1.2%
	251	96											0.5299 ± 0.0002	1.6%
	251	96											0.5425 ± 0.0002	10.9%

Isotope	A	Z	% Natural abundance	Atomic mass	Half-life	Decay mode	Decay energy (MeV)	Particle energy (MeV)	Particle intensity	Thermal neutron cross section	Spin (h/2π)	μ Nucl. mag. moment	Gamma-ray energy (MeV)	Gamma-ray intensity
	251	96											0.5624 ± 0.0002	1.0%
	251	96											0.9782 ± 0.0002	1.0%
Bk		97												
$_{97}Bk^{242}$	242	97		242.061940	7.0 ± 0.1 m.	E.C.	2.900							
$_{97}Bk^{243}$	243	97		243.062997	4.5 h.	E.C. (99.8%)	1.505	6.542(4)	0.03%		(3/2-)		0.1466 ± 0.0005	0.01%
	243	97				α(0.15%)	6.871	6.5738(2)	0.04%				0.1874 ± 0.0004	0.06%
	243	97						6.7180(22)	0.02%				0.755 ± 0.002	10%
	243	97						6.7581(20)	0.02%				0.840 ± 0.004	3.0%
	243	97											0.946 ± 0.002	8%
$_{97}Bk^{244}$	244	97		244.065160	4.4 h.	E.C. (99.99%)	2.250				(4-)		0.1445 ± 0.001	7 +
	244	97				α(0.01%)	6.778	6.625(4)	0.003%				0.1876 ± 0.0003	16
	244	97						6.667(4)	0.003%				0.2176 ± 0.0003	100
	244	97											0.3335 ± 0.0005	10
	244	97											0.4905 ± 0.0005	18
	244	97											0.7461 ± 0.0008	8
	244	97											0.9815 ± 0.001	114
	244	97											0.9215 ± 0.001	22
	244	97											0.988 ± 0.001	5
	244	97											1.178 ± 0.001	5
	244	97											1.233 ± 0.001	4
	244	97											1.252 ± 0.001	3
	244	97											1.505 ± 0.005	3
$_{97}Bk^{245}$	245	97		245.066357	4.94 d.	E.C. (99.9%)	0.814				3/2-		Cm L x-ray	32%
	245	97				α(0.1%)	6.453	5.8851(5)	0.03%				Cm k x-ray	56%
	245	97						6.1176(9)	0.01%				0.25299 ± 0.00002	29%
	245	97						6.1467(5)	0.02%				0.3809 ± 0.0001	2.4%
	245	97						6.3087(5)	0.014%				0.3851 ± 0.0001	0.6%
	245	97						6.3492(5)	0.018%					
$_{97}Bk^{246}$	246	97		246.068720	1.80 d.	E.C>	1.400				(2-)		Cm L x-ray	38%
	246	97											Cm k x-ray	30%
	246	97											0.73442 ± 0.00002	3.2%
	246	97											0.79881 ± 0.00002	61%
	246	97											0.83358 ± 0.00002	4.9%
	246	97											1.06201 ± 0.00002	2.9%
	246	97											1.07885 ± 0.00002	3.8%
	246	97											1.08142 ± 0.00002	5.8%
	246	97											1.12427 ± 0.00002	4.3%
$_{97}Bk^{247}$	247	97		247.070300	1.4 x 10^3 y.	α	5.889	5.465(5)	1.5%		(3/2-)		0.04175 ± 0.0002	1%
	247	97						5.501(5)	7%				0.0839 ± 0.0002	40%
	247	97						5.532(5)	45%				0.268 ± 0.005	30%
	247	97						5.6535(20)	5.5%					
	247	97						5.678(2)	13%					
	247	97						5.712(2)	17%					
	247	97						5.753(2)	4.3%					
	247	97						5.794(2)	5.5%					
$_{97}Bk^{248}$	248	97		248.073106	23.7 ± 0.2 h.	β-(70%)	0.710	0.86			(1-)		Cm L x-ray	5.7%
	248	97				E.C.(30%)	0.860						Cf L x-ray	5.0%
	248	97											Cm k x-ray	9.8%
	248	97											Cf k x-ray	0.02%
	248	97											0.5507 ± 0.0001	5.0%
$_{97}Bk^{249}$	249	97		249.074980	320 ± 6 d.	β-	0.125	100%		710 b.	7/2+	2.0		
	249	97				α(0.001%)	5.525	5.3899(6)	0.0002%					
	249	97						5.4174(6)	0.001%					
$_{97}Bk^{250}$	250	97		250.078312	3.22 h.	β-	1.781	0.74		σf 1000 b.	2-		Cf L x-ray	10%
	250	97											Cf k x-ray	0.5%
	250	97											0.88996 ± 0.00001	1.5%
	250	97											0.92947 ± 0.00002	1.2%
	250	97											0.98912 ± 0.0001	45%
	250	97											1.02863 ± 0.00002	4.9%
	250	97											1.03184 ± 0.00001	36%
	250	97											(0.04 - 1.6) weak	
$_{97}Bk^{251}$	251	97		251.080760	57 m.	β-	1.100				(3/2-)		0.02481 ± 0.00001	+
	251	97											0.1528 ± 0.0001	39
	251	97											0.1776 ± 0.0001	100
Cf		98												

Isotope	A	Z	% Natural abundance	Atomic mass	Half-life	Decay mode	Decay energy (MeV)	Particle energy (MeV)	Particle intensity	Thermal neutron cross section	Spin (h/2π)	μ Nucl. mag. moment	Gamma-ray energy (MeV)	Gamma-ray intensity
$_{98}Cf^{240}$	240	98		240.062280	1.06 ± 0.1 m.	α	7.719	7.590(10)			0+			
$_{98}Cf^{241}$	241	98		241.063520	≈ 3.8 ± 0.7 m.	E.C.	3.080							
	241	98				α	7.60	7.335(5)						
$_{98}Cf^{242}$	242	98		242.063690	3.5 ± 0.2 m.	α	7.509	7.351(6)	20%		0+			
	242	98						7.385(4)	80%					
$_{98}Cf^{243}$	243	98		243.065390	10.7 ± 0.5 m.	E.C.(86%)	2.230	7.060(6)	20%		0+			
	243	97				α(14%)	7.40	7.170	4%					
$_{98}Cf^{244}$	244	98		244.065979	19.4 ± 0.6 m.	α	7.328	7.168(5)	25%		0+			
	244	98						7.210(5)	75%					
$_{98}Cf^{245}$	245	98		245.068037	43.6 ± 0.8 m.	α(30%)	7.255	6.886						
	245	98				E.C.(70%)	1.565	6.983						
	245	98						7.036						
	245	98						7.084						
	245	98						7.137(2)						
$_{98}Cf^{246}$	246	98		246.068800	36 h.	α	6.869	6.6156(10)	0.18%		0+		Cm L x-ray	4.2%
	246	98						6.7086(7)	21.8%				0.04221 ± 0.00001	0.02%
	246	98						6.7501(7)	78.0%				0.0945 ± 0.001	0.008%
	246	98											0.147 ± 0.004	0.004%
$_{98}Cf^{247}$	247	98		247.071020	3.11 ± 0.03 h.	E.C. (99.96%)	0.670				7/2+		Bk k x-ray	32%
	247	98				α(0.04%)	6.55	6.301(5)					0.2941 ± 0.0001	1.0%
	247	98											0.4070 ± 0.0001	0.2%
	247	98											0.4179 ± 0.0001	0.34%
	247	98											0.4778 ± 0.0001	0.55%
$_{98}Cf^{248}$	248	98		248.072183	334 ± 3 d.	α	6.369	6.220(5)	17%		0+			
	248	98						6.262(5)	83%					
$_{98}Cf^{249}$	249	98		249.074844	351 y.	α	6.295	5.7582(2)	3.7%	500 b.	9/2-		Cm L x-ray	10%
	249	98						5.8119(2)	84%	σf 1600 b.			Cm k x-ray	3.3%
	249	98						5.8488(2)	1.0%				0.25299 ± 0.00002	2.5%
	249	98						5.9029(2)	2.8%				0.33351 ± 0.00003	14.4%
	249	98						5.9451(2)	4.0%				0.38832 ± 0.00003	66%
	249	98						6.1401(2)	1.1%					
	249	98						6.1940(2)	2.2%					
$_{98}Cf^{250}$	250	98		250.076400	13.1 ± 0.1 y.	α	6.129	5.8913(4)	0.3%	2 x 10^3 b.	0+		Cm L x-ray	2.9%
	250	98						5.9889(4)	15%				0.04285 ± 0.00001	0.014%
	250	98						6.0310(4)	84.5%					
$_{98}Cf^{251}$	251	98		251.079580	8.9 x 10^2 y.	α	6.172	5.56448(7)	1.5%	2.9 x 10^3 b.	½+			
	251	98						5.632(1)	4.5%	σf 4.8 x 10^3 b.				
	251	98						5.648(1)	3.5%					
	251	98						5.6773(6)	35%					
	251	98						5.762(3)	3.8%					
	251	98						5.7937(7)	2.0%					
	251	98						5.8124(8)	4.2%					
	251	98						5.8514(6)	27%					
	251	98						6.0140(7)	11.6%					
	251	98						6.0744(7)	2.7%					
$_{98}Cf^{252}$	252	98		252.081621	2.64 y.	α(96.9%)	6.217	5.7977(1)	0.23%	20 b.	0+		Cm L x-ray	2.9%
	251	98				S.F.(3.1%)		6.0756(4)	15.2%	σf 32 b.			0.04339 ± 0.00002	0.015%
	252	98						6.1184(4)	81.6%				0.1002 ± 0.0001	0.01%
$_{98}Cf^{253}$	253	98		253.085127	17.8 d.	β-(99.7%)	0.29	0.27	100%	18 b.	(7/2+)			
	253	98				α(0.3%)	6.126	5.921(5)	0.02%	f 1300 b.				
	253	98						5.979(5)	0.29%					
$_{98}Cf^{254}$	254	98		254.087318	60.5 d.	S.F. (99.7%)				5 b.	0+			
	254	98				α(0.3%)	5.930	5.792(5)	0.05%					
	254	98						5.834(5)	0.26%					
$_{98}Cf^{255}$	255	98			1.4 h.	β-								
Es		99												
$_{99}Es^{243}$	243	99		243.069470	21 s.	α(>30%)	8.10	7.890(20)	>30%					
	243	99				E.C. (<70%)	3.810							
$_{99}Es^{244}$	244	99		244.070810	37 ± 4 s.	E.C.(76%)	4.500							
	244	99				α(4%)	7.84	7.570(20)	4%					
$_{99}Es^{245}$	245	99		245.071260	1.33 ± 0.1 m.	α(40%)	7.858	7.730(20)						
	245	99				E.C.(60%)	3.000							
$_{99}Es^{246}$	246	99		246.072920	≈ 7.7 ± 0.5 m.	E.C.(90%)	3.840							
	246	99				α(10%)	7.70	7.350(20)						
$_{99}Es^{247}$	247	99		247.073590	4.7 ± 0.3 m.	E.C.(93%)	2.400							
	247	99				α(7%)	7.441	7.320(20)						

TABLE OF THE ISOTOPES (Continued)

Isotope	A	Z	% Natural abundance	Atomic mass	Half-life	Decay mode	Decay energy (MeV)	Particle energy (MeV)	Particle intensity	Thermal neutron cross section	Spin (h/2π)	μ Nucl. mag. moment	Gamma-ray energy (MeV)	Gamma-ray intensity
$_{99}Es^{248}$	248	99		248.075440	27 ± 3 m.	E.C. (99.7%)	3.030							
	248	99				α(0.3%)	7.15	6.870(10)						
$_{99}Es^{249}$	249	99		249.076340	1.70 ± 0.1 h.	E.C. (99.4%)	1.395				(7/2+)			
	249	99				α(0.6%)	6.881	6.770(5)						
$_{99}Es^{250m}$	250	99			2.2 h.	E.C.							Cf L x-ray	26%
	250	99											Cf k x-ray	34%
	250	99											0.30395 ± 0.00003	0.1%
	250	99											0.62614 ± 0.00002	1.0%
	250	99											0.82883 ± 0.00002	5.5%
	250	99											0.98912 ± 0.00001	13.5%
	250	99											1.03184 ± 0.00001	10.6%
	250	99											1.16725 ± 0.00003	3.0%
	250	99											1.17550 ± 0.00002	1.5%
	250	99											1.20177 ± 0.00003	1.2%
	250	99											1.61525 ± 0.00002	1.8%
	250	99											1.65797 ± 0.00002	1.1%
$_{99}Es^{250}$	250	99		250.078660	8.6 ± 0.1 h.	E.C.	2.100						Cf L x-ray	120%
	250	99											Cf k x-ray	71%
	250	99											0.082283 ± 0.00006	2.6%
	250	99											0.08509 ± 0.00006	1.1%
	250	99											0.14069 ± 0.00006	4.7%
	250	99											0.22297 ± 0.00001	1.8%
	250	99											0.24686 ± 0.00005	3.8%
	250	99											0.30339 ± 0.00005	22.3%
	250	99											0.34948 ± 0.00005	20.4%
	250	99											0.38381 ± 0.00005	14%
	250	99											0.71228 ± 0.00005	1.3%
	250	99											0.76399 ± 0.00002	4.0%
	250	99											0.81009 ± 0.00002	9.1%
	250	99											0.82883 ± 0.00002	74%
	250	99											0.86315 ± 0.00002	5.1%
	250	99											0.88662 ± 0.00002	1.3%
$_{99}Es^{251}$	251	99		251.079986	1.38 d.	E.C. (99.5%)	0.379				(3/2-)			
	251	99				α(0.5%)	6.597	6.4625(13)	0.05%					
	251	99						6.492(1)	0.4%					
$_{99}Es^{252}$	252	99		252.082944	1.29 y.	α(76%)	6.739	6.051(1)	0.8%		(5-)			
	252	99				E.C.(24%)	1.12	6.238(1)	0.43%					
	252	99						6.266(3)	0.57%					
	252	99						6.4224(12)	0.34%					
	252	99						6.4827(12)	1.66%					
	252	99						6.5621(12)	10.3%					
	252	99						6.6316(12)	61.0%					
$_{99}Es^{253}$	253	99		253.084818	20.47 d.	α	6.739	6.2497(1)	0.04%	160 b.	7/2+	4.10	0.04180 ± 0.00004	0.05%
	253	99						6.4314(1)	0.06%				0.3871 ± 0.0001	0.018%
	253	99						6.4793(1)	0.08%				0.3892 ± 0.0001	0.026%
	253	99						6.4972(1)	0.26%				(0.03 - 1.1)weak	
	253	99						6.5405(1)	0.85%					
	253	99						6.5514(1)	0.7%					
	253	99						6.5916(1)	6.6%					
	253	99						6.6241(1)	0.8%					
	253	99						6.6327(5)	89.8%					
$_{99}Es^{254m}$	254	99			1.64 d.	β-(99.6%)	1.16	1.127		1 b.	2+		Fm L x-ray	13%
	254	99				α(0.3%)	6.67	6.3821(9)	0.25%	σf 1800 b.	2+		Fm k x-ray	0.8%
	254	99						6.5558(9)	0.02%				0.58432 ± 0.00002	2.9%
	254	99						6.5907(9)	0.01%				0.64879 ± 0.00002	28.9%

Isotope	A	Z	% Natural abundance	Atomic mass	Half-life	Decay mode	Decay energy (MeV)	Particle energy (MeV)	Particle intensity	Thermal neutron cross section	Spin (h/2π)	μ Nucl. mag. moment	Gamma-ray energy (MeV)	Gamma-ray intensity
	254	99											0.68867 ± 0.00002	12.4%
	254	99											0.69377 ± 0.00002	24.7%
$_{99}Es^{254}$	254	99		254.088019	275 d.	α	6.617	6.3475(10)	0.75%	< 40 b.	(7+)		0.0426 ± 0.0001	0.15%
	254	99						6.3573(10)	2.6%	σf 2900 b.			0.0650 ± 0.0012	2.0%
	254	99						6.4161(10)	1.8%				0.3167 ± 0.0013	0.15%
	254	99						6.4266(12)	93.1%					
	254	99						6.4801(13)	0.23%					
$_{99}Es^{255}$	255	99		255.090270	39.8 ± 0.1 d.	β-(92%)	0.300			60 b.	(7/2+)			
	255	99				α(8%)	6.436	6.213(10)	0.20%					
	255	99				S.F. (0.004%)		6.260(10)	0.8%					
	255	99						6.2995(15)	7.0%					
$_{99}Es^{256m}$	256	99			7.6 h.	β-					(8-)		0.1114	5.7%
	256	99											0.1726	30%
	256	99											(0.05 - 1.1)weak	
$_{99}Es^{256}$	256	99		256.093560	25 m.	β-	1.676				(1+)			
Fm		100												
$_{100}Fm^{243}$	243	100		243.074460	0.18 s.	α		8.546(25)						
$_{100}Fm^{244}$	244	100		244.074120	3.7 ms.	S.F.					0+			
$_{100}Fm^{245}$	245	100		245.076250	4 s.	α	8.40	8.150(20)						
$_{100}Fm^{246}$	246	100		246.075290	1.1 ± 0.2 s.	α(92%)	8.373	8.240(20)			0+			
	246	100				S.F.(8%)								
$_{100}Fm^{247m}$	247	100			9.2 ± 1.0 s.	α		8.180(30)						
$_{100}Fm^{247}$	247	100		247.076800	35 ± 4 s.	α	8.20	7.870(50)	70 +					
	247	100						7.930(50)	30					
$_{100}Fm^{248}$	248	100		248.077171	36 ± 3 s.	α(99.9%)	8.001	7.830(20)	20%		0+			
	248	100				S.F.(0.1%		7.870(20)	80%					
$_{100}Fm^{249}$	249	100		249.078910	≈ 2.6 m.	E.C.					(7/2+)			
	249	100				α	7.70	7.530(20)						
$_{100}Fm^{250m}$	250	100			1.8 ± 0.1 s.	I.T.								
$_{100}Fm^{250}$	250	100		250.079509	30 m.	α	7.548	7.430(30)			0+			
$_{100}Fm^{251}$	251	100		251.081590	5.3 h.	E.C.(98%)	1.49				(9/2-)			
	251	100				α(2%)	7.424	6.7817(8)	0.09%					
	251	100						6.8324(8)	1.57%					
	251	100						7.3051(8)	0.02%					
$_{100}Fm^{252}$	252	100		252.082466	25.4 h.	α	7.154	6.999(8)	15%		0+			
	252	100						7.040(9)	85%					
$_{100}Fm^{253}$	253	100		253.085173	E.C.(88%)	0.334					½+		Es k x-ray	20%
	253	100				α(12%)	7.200	6.544(2)	0.18%				0.14497 ± 0.00005	0.2%
	253	100						6.633(4)	0.31%				0.2719 ± 0.0004	2.6%
	253	100						6.653(4)	0.29%					
	253	100						6.6757(14)	2.78%					
	253	100						6.8467(13)	1.0%					
	253	100						6.9010(13)	1.18%					
	253	100						6.9433(13)	5.12%					
	253	100						7.0245(13)	0.8%					
$_{100}Fm^{254}$	254	100		254.086846	3.24 h.	α	7.303	7.050(4)	0.9%	80 b.	0+			
	254	100				S.F. (0.06%)		7.147(4)	14%					
	254	100						7.189(4)	84.9%					
$_{100}Fm^{255}$	255	100		255.089948	20.1 h.	α	7.240	6.8916(5)	0.62%	26 b.	7/2+			
	255	100						6.9635(5)	5.0%	σf 3400 b.				
	255	100						7.0225(5)	93.4%					
	255	100						7.0800(5)	0.40%					
$_{100}Fm^{256}$	256	100		256.091767	2.63 h.	S.F.(92%)				50 b.	0+			
	256	100				α(18%)	7.025	6.92						
$_{100}Fm^{257}$	257	100		257.075099	100.5 d.	α(99.8%)				5800 b.	(9/2+)		0.0616 ± 0.0001	1.4%
	257	100				S.F.(0.2%)				σf 3000 b.			0.1794 ± 0.0001	8.7%
	257	100											0.2410 ± 0.0001	11%
$_{100}Fm^{258}$	258	100			0.38 ms.	S.F.								
Md		101												
$_{101}Md^{248}$	248	101		248.082750	7 ± 3 s.	E.C.(80%)	5.210							
	248	101				α(20%)	8.60	8.320(20)	15%					
	248	101						8.360(30)	5%					
$_{101}Md^{249}$	249	101		249.082950	24 ± 4 s.	E.C > (<80%)	3.760							
	249	101				α(>20%)	8.46	8.030(20)						
$_{101}Md^{250}$	250	101		250.084380	≈ 52 ± 6 s.	E.C.(94%)	4.54							
	250	101				α(6%)	8.25	7.750(20)	4 +					
	250	101						7.820(30)	2					
$_{101}Md^{251}$	251	101		251.084830	4.0 m.	E.C. (>94%)	3.020							
	251	101				α(<6%)	8.05	7.550(20)						
$_{101}Md^{252}$	252	101		252.086470	≈ 2.3 m.	E.C. (>50%)	3.73							
	252	101				α(<50%)	7.85	7.73						
$_{101}Md^{254m}$	254	101			≈ 28 m.	E.C.								
$_{101}Md^{254}$	254	101		254.089630	10 ± 3 m.	E.C.	2.600							
$_{101}Md^{255}$	255	101		255.091081	27 ± 2 m.	E.C.(92%)	1.055				(7/2-)			
	255	101				α(8%)	7.911	7.326(5)						

Isotope	A	Z	% Natural abundance	Atomic mass	Half-life	Decay mode	Decay energy (MeV)	Particle energy (MeV)	Particle intensity	Thermal neutron cross section	Spin (h/2π)	μ Nucl. mag. moment	Gamma-ray energy (MeV)	Gamma-ray intensity
$_{101}Md^{256}$	256	101		256.093960	76 m.	E.C.(90%)	2.041							
	256	101				α(10%)	7.483	7.14	16 +					
	256	101						7.22	63					
$_{101}Md^{257}$	257	101		257.095580	≈ 5.2 h.	E.C.(90%)	0.450				(7/2-)			
	257	101				α(10%)	7.60	7.068(5)						
$_{101}Md^{258m}$	258	101			43 m.	E.C.					(1-)			
$_{101}Md^{258}$	258	101		258.098570	56 d.	α	7.40	6.716(5)	72%		(8-)			
	258	101						6.79(1)	28%					
$_{101}Md^{259}$	259	101			1.6 h.	S.F.								
No		102												
$_{102}No^{250}$	250	102			250 μs.	S.F.					0+			
$_{102}No^{251}$	251	102		251.088870	0.8 s.	α		8.600(20)	80%					
	251	102						8.680(20)	20%					
$_{102}No^{252}$	252	102		252.088949	2.3 ± 0.2 s.	α(73%)	8.551	8.372(8)	18%		0+			
	252	102				S.F.(27%)		8.415(6%)	55%					
$_{102}No^{253}$	253	102		253.090530	1.7 ± 0.3 m.	α	8	40	8.010(20)		(9/2-)			
$_{102}No^{254m}$	254	102			0.28 s.	I.T.								
$_{102}No^{254}$	254	102		254.090953	≈ 55 s.	α	8.235	8.100(20)			0+			
$_{102}No^{255}$	255	102		255.093260	3.1 m.	α(62%)	8.445	7.620(10)	1.7%		1/2+			
	255	102				E.C.(38%)		7.717(11)	1.5%					
	255	102						7.771(7)	5.5%					
	255	102						7.879(11)	2.6%					
	255	102						7.927(7)	7.3%					
	255	102							8.007(11)	3.9%				
	255	102						8.077(9)	7.3%					
	255	102						8.121(6)	27.9%					
	255	102						8.266(8)	3.1%					
	255	102						8.312(9)	1.2%					
$_{102}No^{256}$	256	102		256.094252	3.2 s.	α	8.554	8.430			0+			
$_{102}No^{257}$	257	102		257.096850	25 s.	α	8.452	8322(2)	55%		(7/2+)			
	257	102						8.27(2)	26%					
	257	102						8.32(2)	19%					
$_{102}No^{258}$	258	102		258.098150	≈ 1.2 ms.	S.F.					0+			
$_{102}No^{259}$	259	102		259.100931	≈ 58 m.	α(78%)	7.794	7.443(10)	13 +		(9/2+)			
	259	102				E.C.(22%)		7.488(10)	39					
	259	102						7.521(10)	23					
	259	102						7.593(10)	14					
	259	102						7.673(10)	11					
Lr		103												
$_{103}Lr^{253}$	253	103		253.095150	≈ 1.4 s.	α		8.721(20)						
	253	103						8.805(20)						
$_{103}Lr^{254}$	254	103		254.096320	≈ 20 ± 10 s.	α		8.455(20)						
$_{103}Lr^{255}$	255	103		255.096670	22 ± 4 s.	α	8.80	8.370(13)	60%					
	255	103						8.429(20)	40%					
$_{103}Lr^{256}$	256	103		256.098490	28 s.	α(99.7%)	8.554	8.43						
	256	103				S.F.(0.3%)								
$_{103}Lr^{257}$	257	103		257.099480	0.65 s.	α	9.30	8.796(13)	15%		7/2+			
	257	103						8.861(12)	85%					
$_{103}Lr^{258}$	258	103		258.101710	4.3 s.	α	9.00	8.54(2)	10%					
	258	103						8.589(10)	45%					
	258	103						8.614(10)	35%					
	258	103						8.648(10)	10%					
$_{103}Lr^{259}$	259	103		259.102900	≈ 5.4 s.	α	8.70	8.46(2)	100%					
$_{103}Lr^{260}$	260	103		260.105320	3 m.	α	8.30	8.04(2)	100%					
Rf		104												
$_{104}Rf^{257}$	257	104		257.102950	4.8 s.	α	9.20	8.663						
	257	104						8.720						
	257	104						8.778						
	257	104						8.824						
	257	104						8.870						
	257	104	8.951											
	257	104						9.016						
$_{104}Rf^{258}$	258	104		258.103430	11 ms.	S.F.								
$_{104}Rf^{259}$	259	104		259.105530	≈ 3.1 s.	α	9.20	8.77(2)						
	259	104						8.86						
$_{104}Rf^{260}$	260	104		260.106300	≈ 20 ms.	S.F.								
$_{104}Rf^{261}$	261	104		261.108690	≈ 65 s.	α	8.60	8.29						
$_{104}Rf^{262}$	262	104			≈ 63 ms.									
Ha		105												
$_{105}Ha^{257}$	257	105		257.107770	≈ 1 s.	α		8.96						
	257	105						9.08						
	257	105						9.16						
$_{105}Ha^{258}$	258	105		258.109020	4 s.	α		9.02						
	258	105						9.09						
	258	105						9.18						
$_{105}Ha^{259}$	259	105		259.109580	≈ 1.2 s.	S.F.								
$_{105}Ha^{260}$	260	105		260.111040	1.5 s.	α		9.05						
	260	105				S.F.		9.08						
	260	105						9.13						
$_{105}Ha^{261}$	261	105		261.111820	≈ 1.8 s.	α		8.93						

Isotope	A	Z	% Natural abundance	Atomic mass	Half-life	Decay mode	Decay energy (MeV)	Particle energy (MeV)	Particle intensity	Thermal neutron cross section	Spin (h/2π)	μ Nucl. mag. moment	Gamma-ray energy (MeV)	Gamma-ray intensity
	261	105				S.F.								
$_{105}Ha^{262}$	262	105		262.113760	34 s.	S.F.								
	262	105				α		8.45						
	262	105						8.53						
	262	105						8.67						
106		106												
	259	106			≈ 7 ms.	S.F.								
	263	106			0.8 s.	S.F.								
	261	106			≈ 1 ms.	α								
	262	106			≈ 115 ms.	α		9.70						
109		109												
	266	109			5 ms.			11.10						

CRYOGENIC PROPERTIES OF GASES

Property	Conditions	He	Ne	Ar	Kr	Xe	H_2	CH_4	NH_3	N_2	O_2	F_2
		Gas										
Density	32°F, 1 atm, lb/ft³	0.01114	0.0562	0.1113	0.234	0.368	0.00561	0.0448	0.0481	0.0781	0.0892	0.106
	0°C, 1 atm, kg/m³	0.1784	0.9002	1.783	3.748	5.895	0.0899	0.718	0.770	1.251	1.429	1.698
Boiling point	°F, 1 atm	-452.08	-410.89	-302.3	-242.1	-160.8	-423.2	-263.2	-28.03	-320.4	-297.35	-306.7
	°C, 1 atm	-268.934	-246.048	-185.7	-152.90	-107.1	-252.87	-164.0	-33.35	-195.8	-182.97	-188.14
	°K, 1 atm	4.216	27.10	87.45	120.25	166.05	20.28	109.15	239.80	77.35	90.18	85.01
Melting point	°F, 1 atm	-458.0[a]	-415.6	-308.6	-249.9	-169.4	-434.5	-296.46	-107.9	-345.87	-361.1	-363.3
	°C, 1 atm	-272.2[a]	-248.67	-189.2	-156.6	-111.9	-259.14	-182.48	-77.7	-209.86	-218.4	-219.62
	°K, 1 atm	0.95[a]	24.48	83.95	116.55	161.25	14.01	90.67	195.45	63.29	54.75	53.53
Vapor density at boiling point	lb/ft³	0.999	0.593	0.368	0.518	0.606	0.0830	0.1124	0.0556	0.288	0.279	
	kg/m³	16.002	9.499	5.895	8.298	9.707	1.329	1.8004	0.8906	4.613	4.4692	
Liquid density at boiling point	lb/ft³	7.803	74.91	86.77	149.8	193.5	4.37	26.47	42.58	50.19	71.23	94.4
	kg/m³	125.	1200.	1390.	2400.	3100.	70.0	424.	682.1	804.	1142	1512
Vapor pressure of solid at melting point	lb/in.²		6.25	9.98	10.6	11.8	1.04	1.35	0.87	1.86	0.038	0.002
	kg/m²		323.	516	549.	612.	54.	70.	45.2	96.4	2.0	0.12
	(N/m²) × 10⁴		4.34	6.93	7.36	8.20	0.723	0.938	0.604	1.29	0.0026	0.00014
Heat of vaporization at boiling point	Btu/lb	10.3	37.4	70.0	46.4	41.4	194.4	248.4	588.6	85.7	91.588	73.7
	kcal/kg	5.72	20.8	38.9	25.8	23.0	108	138	327	47.6	50.88	40.9
	(J/kg) × 10³	23.932	87.027	162.76	107.95	96.23	451.9	577.4	1368.2	199.2	212.9	171.1
Heat of fusion at melting point	Btu/lb	1.8	7.2	12.1	7.0	5.9	25.2	26.1	152.1	11.0	5.9	5.8
	kcal/kg	1.0	4.0	6.7	3.9	3.3	14.0	14.5	84.0	6.1	3.27	3.2
	(J/kg) × 10³	4.184	16.74	28.03	16.3	13.8	58.6	60.7	351.5	25.5	13.7	13.4
C_p	59°F, 1 atm, Btu/lb-°F or 15°C, 1 atm, kcal/kg-°C	1.25[b]	0.25[c]	0.125	0.06[c]	0.04[c]	3.39	0.528	0.523	0.248	0.220	0.180
	288.15°K, 1 atm, (J/kg) × 10³	5.23[b]	1.05[c]	0.523	0.251[c]	0.167[c]	14.2	2.21	2.188	1.038	0.9205	0.753
C_p/C_v	15–20°C, 1 atm; 288–293°K, 1 atm	1.66[b]	1.64[c]	1.67	1.68[c]	1.66[c]	1.41	1.31	1.31	1.40	1.40	
Critical temperature	°F	-450.2	-397.7	-188.5	-82.7	61.9	-399.8	-116.5	270.3	-232.8	-181.3	-200.2
	°C	-267.9	-228.7	-122.5	-63.7	16.6	-239.9	-82.5	132.4	-147.1	-118.57	-129.0
	°K	5.25	44.45	150.65	209.45	289.75	33.25	190.65	405.55	126.05	154.58	144.15
Critical pressure	lb/in.² (absolute)	33.2	394.6	705.2	798	855	188.1	672	1639	492.3	731.4	808.3
	kg/cm²	2.33	27.7	49.6	46.1	60.1	13.2	47.2	115.5	34.6	51.4	56.8
	(kg/m²) × 10³	23.3	277	496	561	601	132	472	1155	346	514	568
	(N/m²) × 10⁴	23.1	274.1	489.9	554.4	594	130.7	466.9	1139	342.	508.1	561.6

Note: For conversion factors see p. B-439.

[a] At 26 atmospheres.
[b] At -292°F or -180°C.
[c] Approximate.

CONVERSION FACTORS FOR TABLE OF CRYOGENIC PROPERTIES OF GASES

To convert from	To	Multiply by
lbs ft^{-3}	kg m^{-3}	16.018
lbs in.$^{-2}$	N m^{-2}	6894.8
lbs ft^{-2}	N m^{-2}	47.880
BTU lb^{-1}	J kg^{-1}	2324.4
cal g^{-1}	J kg^{-1}	4184
cal g^{-1} °F	J kg^{-1} °C	4184

VISCOSITY AND THERMAL CONDUCTIVITY OF NITROGEN AT CRYOGENIC TEMPERATURES

The viscosity and thermal conductivity of nitrogen gas for the temperature range 5 K–135 K have been computed from the second Chapman-Enskog approximation. Quantum effects, which become appreciable at the lower temperatures, are included by utilizing collision integrals based on quantum theory. A Lennard-Jones (12-6) potential was assumed. The computations yield viscosities about 20% lower than those predicted for the high end of this temperature range by the method of corresponding states, but the agreement is excellent when the computed values are compared with existing experimental data.

T, °K	η, micropoise	λ, $\frac{\mu\text{cal}}{\text{cm sec °K}}$
4.575	4.16639	1.10990
5.49	4.85525	1.29307
6.405	5.54773	1.47715
7.320	6.24722	1.63156
8.235	6.95025	1.85020
9.15	7.65067	2.03666
13.725	10.95231	2.91729
18.3	13.85738	3.69313
22.875	16.55204	4.41068
27.45	19.21670	5.11847
32.025	21.9423	5.84208
36.60	24.7606	6.59077
41.175	27.67445	7.36545
45.75	30.67422	8.16348
54.9	36.8747	9.81383
64.05	43.2469	11.51005
82.35	56.14634	14.94298
91.5	62.54767	16.64610
137.25	92.96598	24.74580

From Pearson, W. E., NASA Technical Note D-7565, National Aeronautics and Space Administration, 1974 (available from Superintendent of Documents, U.S. Government Printing Office, Washington, D.C.).

SOLUBILITY OF NITROGEN AND AIR IN WATER

Reproduced from *Journal of Physical and Chemical Reference Data,* 13, 565, 1984 with permission of the copywrite owners, the American Chemical Society, the American Institute of Physics and the authors, Dr. Rubin Battino and Dr. Timothy R. Rettich.

SOLUBILITY OF NITROGEN IN WATER FOR PARTIAL PRESSURE OF GAS OF 0.101325 MPa

T/K	V°/mℓ mol^{-1}	10^5 x_1	10^2 ℓ	T/K	V°/mℓ mol^{-1}	10^5 x_1	10^2 ℓ
273.15	18.018	1.908	2.373	313.15	18.156	0.9981	1.413
278.15	18.016	1.695	2.147	318.15	18.193	0.9585	1.376
283.15	18.020	1.524	1.965	327.15	18.233	0.9273	1.349
288.15	18.031	1.386	1.818	328.15	18.276	0.9033	1.331
293.15	18.047	1.274	1.698	333.15	18.323	0.8855	1.321
298.15	18.068	1.183	1.601	338.15	18.372	0.8735	1.319
303.15	18.094	1.108	1.523	343.15	18.425	0.8666	1.324
308.15	18.123	1.047	1.461	348.15	18.480	0.8644	1.336

The solubility of nitrogen at temperatures above 350 K and pressures up to 100 MPa has been examined by several researchers. The following equation was used to fit their data in terms of both temperatures and pressure:

$$\begin{aligned}\ln \chi_1 = & -43.0160 + 48.5244/\tau + 13.9321\ln \tau \\ & +0.970040\ln (P/\mathrm{MPa}) \\ & -0.00048296(P/\mathrm{MPa}) \\ \tau = & \; T/100\ \mathrm{K}\end{aligned}$$

AIR SOLUBILITIES IN WATER

T/K	$10^5\chi_1$	Ostwald Coefficient (ℓ)	T/K	$10^5\chi_1$	Ostwald Coefficient (ℓ)
273.15	2.316	0.02882	333.15	0.9920	0.01480
278.15	2.042	0.02587	338.15	0.9694	0.01464
283.15	1.824	0.02352	343.15	0.9521	0.01455
288.15	1.647	0.02151	348.15	0.9398	0.01453
293.15	1.504	0.01999	353.15	0.9314	0.01456
298.15	1.388	0.01880	358.15	0.9368	0.01465
303.15	1.293	0.01778	363.15	0.9249	0.01477
308.15	1.215	0.01696	368.15	0.9258	0.01493
313.15	1.151	0.01629	373.15	0.9305	0.01516
318.15	1.099	0.01576			
323.15	1.055	0.01535	293.15	1.503	0.02003
328.15	1.020	0.01503			

AIR SOLUBILITIES IN WATER AT ELEVATED PRESSURES AS S (cm^3(STP) PER GRAM WATER)

T/K	1.0 MPa	5.0 MPa	10.0 MPa	15.0 MPa	20.0 MPa	25.0 MPa
273.15	0.207	1.165	2.29	3.26	4.06	4.71
278.15	0.187	1.053	2.07	2.95	3.67	4.26
283.15	0.171	0.962	1.89	2.69	3.36	3.89
288.15	0.158	0.887	1.74	2.48	3.09	3.59
293.15	0.147	0.825	1.62	2.31	2.88	3.33
298.15	0.137	0.773	1.52	2.16	2.70	3.12
303.15	0.130	0.729	1.43	2.04	2.54	2.95
308.15	0.123	0.693	1.36	1.94	2.42	2.80
313.15	0.118	0.664	1.31	1.86	2.32	2.68
318.15	0.114	0.639	1.26	1.79	2.23	2.58
323.15	0.110	0.619	1.22	1.73	2.16	2.50
328.15	0.107	0.603	1.19	1.69	2.10	2.44
333.15	0.105	0.591	1.16	1.65	2.06	2.39
338.15	0.103	0.581	1.14	1.63	2.03	2.35
343.15	0.102	0.574	1.13	1.61	2.00	2.32

DEFINITIVE RULES FOR NOMENCLATURE OF ORGANIC CHEMISTRY

IUPAC Rules

These rules were taken from the IUPAC's *Nomenclature of Organic Chemistry*, 1979 Edition, published by Pergamon Press, Inc., Maxwell House, Fairview Park, Elmford, New York 10523, U.S.A. Permission to reproduce this information was granted by IUPAC. Such permission is gratefully acknowledged.

The IUPAC rules cover a wide variety of organic compounds. Those types of compounds for which rules of nomenclature are presented in the IUPAC publication are:

Acyclic hydrocarbons
Monocyclic hydrocarbons
Fused polycyclic hydrocarbons
Bridged hydrocarbons
Spiro hydrocarbons
Hydrocarbon ring assemblies
Cyclic hydrocarbons with side chains
Terpene hydrocarbons
Heterocyclic spiro compounds
Heterocyclic ring assemblies
Bridged heterocyclic systems
Halogen derivatives
Alcohols, phenols and their derivatives
Aldehydes, ketones and their derivatives
Carboxylic acids and their derivatives
Compounds containing bivalent sulfur
Sulfur halides, sulfoxides, sulfones, and sulfur acids and their derivatives
Compounds containing selenium or tellurium linked to an organic radical
Groups containing one nitrogen atom
Groups containing more than one nitrogen atom
Radical names
Coordination compounds
Organometallic compounds
Chains and rings with regular patterns of hetero atoms
Organic compounds containing phosphorus, arsenic, antimony, or bismuth
Organosilicon compounds
Organoboron compounds
Sterochemistry:
  Types of isomerism
  cis-trans isomerism
  Fused rings
  Chirality
  Conformations
  Stereoformulae
General rules for naming natural products and related compounds
Isotopically modified compounds, symbols, definitions and formulae for
Names for isotopically modified compounds
Numbering of isotopically modified compounds
Locants for nuclides in isotopically modified compounds

The information which follows and which was taken from the IUPAC publication deals nearly exclusively with the nomenclature of hydrocarbons.

A. HYDROCARBONS

ACYCLIC HYDROCARBONS

Rule A-1. Saturated Unbranched-chain Compounds and Univalent Radicals

1.1—The first four saturated unbranched acyclic hydrocarbons are called methane, ethane, propane, and butane. Names of the higher members of these series consist of a numerical term, followed by "-ane" with elision of terminal "a" from the numerical term. Examples of these names are shown in the table below. The generic name of saturated acyclic hydrocarbons (branched or unbranched) is "alkane".

Examples of names:

(n = total number of carbon atoms)

n		n		n	
1	Methane	15	Pentadecane	29	Nonacosane
2	Ethane	16	Hexadecane	30	Triacontane
3	Propane	17	Heptadecane	31	Hentriacontane
4	Butane	18	Octadecane	32	Dotriacontane
5	Pentane	19	Nonadecane	33	Tritriacontane
6	Hexane	20	Icosane*	40	Tetracontane
7	Heptane	21	Henicosane	50	Pentacontane
8	Octane	22	Docosane	60	Hexacontane
9	Nonane	23	Tricosane	70	Heptacontane
10	Decane	24	Tetracosane	80	Octacontane
11	Undecane	25	Pentacosane	90	Nonacontane
12	Dodecane	26	Hexacosane	100	Hectane
13	Tridecane	27	Heptacosane	132	Dotriacontahectane
14	Tetradecane	28	Octacosane		

1.2—Univalent radicals derived from saturated unbranched acyclic hydrocarbons by removal of hydrogen from a terminal carbon atom are named by replacing the ending "-ane" of the name of the hydrocarbon by "-yl". The carbon atom with the free valence is numbered as 1. As a class, these radicals are called normal, or unbranched chain, alkyls.

Examples:

Pentyl $\overset{5}{C}H_3—\overset{4}{C}H_2—\overset{3}{C}H_2—\overset{2}{C}H_2—\overset{1}{C}H_2—$

Undecyl $\overset{11}{C}H_3—[\overset{10-2}{C}H_2]_9—\overset{1}{C}H_2—$

Rule A-2. Saturated Branched-chain Compounds and Univalent Radicals

2.1—A saturated branched acyclic hydrocarbon is named by prefixing the designations of the side chains to the name of the longest chain present in the formula.

Example:

$$\overset{5}{C}H_3—\overset{4}{C}H_2—\underset{\underset{CH_3}{|}}{\overset{3}{C}H}—\overset{2}{C}H_2—\overset{1}{C}H_3$$

3-Methylpentane

The following names are retained for unsubstituted hydrocarbons only:

Isobutane	$(CH_3)_2CH—CH_3$
Isopentane	$(CH_3)_2CH—CH_2—CH_3$
Neopentane	$(CH_3)_4C$
Isohexane	$(CH_3)_2CH—CH_2—CH_2—CH_3$

2.2—The longest chain is numbered from one end to the other by Arabic numerals, the direction being so chosen as to give the lowest numbers possible to the side chains. When series of locants containing the same number of terms are compared term by term, that series is "lowest" which contains the lowest number on the occasion of the first difference. This principle is applied irrespective of the nature of the substituents.

Examples:

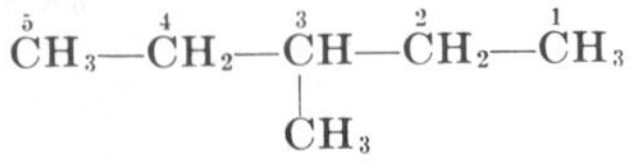

$$\overset{5}{C}H_3-\overset{4}{C}H_2-\underset{\underset{CH_3}{|}}{\overset{3}{C}H}-\overset{2}{C}H_2-\overset{1}{C}H_3$$

3-Methylpentane

$$\overset{6}{C}H_3-\underset{\underset{CH_3}{|}}{\overset{5}{C}H}-\overset{4}{C}H_2-\underset{\underset{CH_3}{|}}{\overset{3}{C}H}-\underset{\underset{CH_3}{|}}{\overset{2}{C}H}-\overset{1}{C}H_3$$

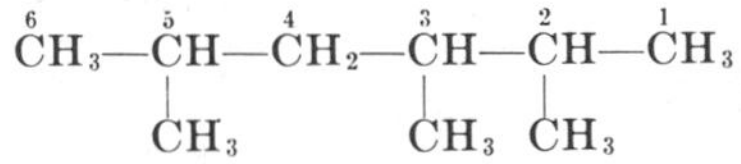

2,3,5-Trimethylhexane (not 2,4,5-Trimethylhexane)

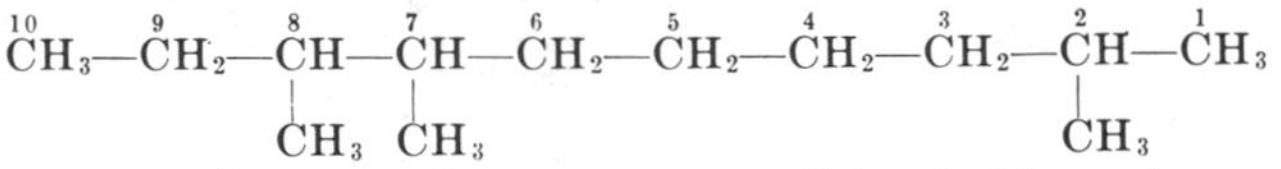

$$\overset{10}{C}H_3-\overset{9}{C}H_2-\underset{\underset{CH_3}{|}}{\overset{8}{C}H}-\underset{\underset{CH_3}{|}}{\overset{7}{C}H}-\overset{6}{C}H_2-\overset{5}{C}H_2-\overset{4}{C}H_2-\overset{3}{C}H_2-\underset{\underset{CH_3}{|}}{\overset{2}{C}H}-\overset{1}{C}H_3$$

2,7,8-Trimethyldecane (not 3,4,9-Trimethyldecane)

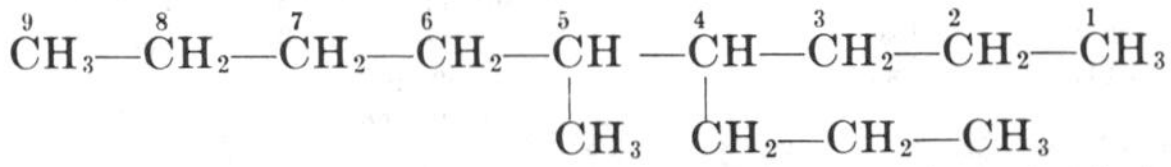

$$\overset{9}{C}H_3-\overset{8}{C}H_2-\overset{7}{C}H_2-\overset{6}{C}H_2-\underset{\underset{CH_3}{|}}{\overset{5}{C}H}-\underset{\underset{CH_2-CH_2-CH_3}{|}}{\overset{4}{C}H}-\overset{3}{C}H_2-\overset{2}{C}H_2-\overset{1}{C}H_3$$

5-Methyl-4-propylnonane (not 5-Methyl-6-propylnonane since 4,5 is lower than 5,6)

2.25—Univalent branched radicals derived from alkanes are named by prefixing the designation of the side chains to the name of the unbranched alkyl radical possessing the longest possible chain starting from the carbon atom with the free valence, the said atom being numbered as 1.

Examples:

1-Methylpentyl	$\overset{5}{C}H_3\overset{4}{C}H_2\overset{3}{C}H_2\overset{2}{C}H_2\overset{1}{C}H(CH_3)-$
2-Methylpentyl	$CH_3CH_2CH_2CH(CH_3)CH_2-$
5-Methylhexyl	$(CH_3)_2CHCH_2CH_2CH_2CH_2-$

The following names may be used for the unsubstituted radicals only:

Isopropyl	$(CH_3)_2CH-$
Isobutyl	$(CH_3)_2CHCH_2-$
sec-Butyl	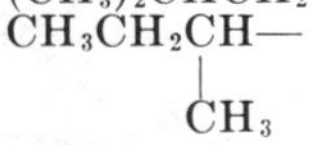 $CH_3CH_2\underset{\underset{CH_3}{\vert}}{C}H-$
tert-Butyl	$(CH_3)_3C-$
Isopentyl	$(CH_3)_2CHCH_2CH_2-$
Neopentyl	$(CH_3)_3CCH_2-$
tert-Pentyl	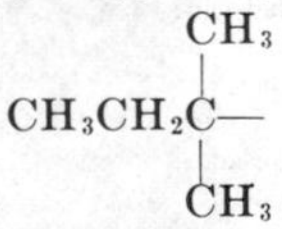 $CH_3CH_2\overset{\overset{CH_3}{\vert}}{\underset{\underset{CH_3}{\vert}}{C}}-$
Isohexyl	$(CH_3)_2CHCH_2CH_2CH_2-$

2.3—If two or more side chains of different nature are present, they are cited in alphabetical order.* The alphabetical order is decided as follows:
(i) The names of simple radicals are first alphabetized and the multiplying prefixes are then inserted.

Example:

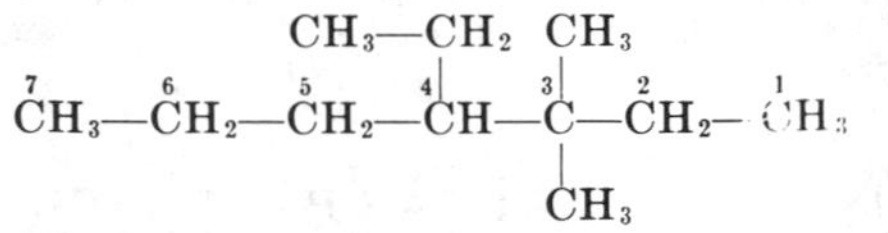

$$\overset{7}{C}H_3-\overset{6}{C}H_2-\overset{5}{C}H_2-\overset{\overset{CH_3-CH_2}{|}}{\overset{4}{C}H}-\overset{\overset{CH_3}{|}}{\underset{\underset{CH_3}{|}}{\overset{3}{C}}}-\overset{2}{C}H_2-\overset{1}{C}H_3$$

ethyl is cited before methyl, thus 4-Ethyl-3.3-dimethylheptane

(ii) The name of a complex radical is considered to begin with the first letter of its complete name.

Example:

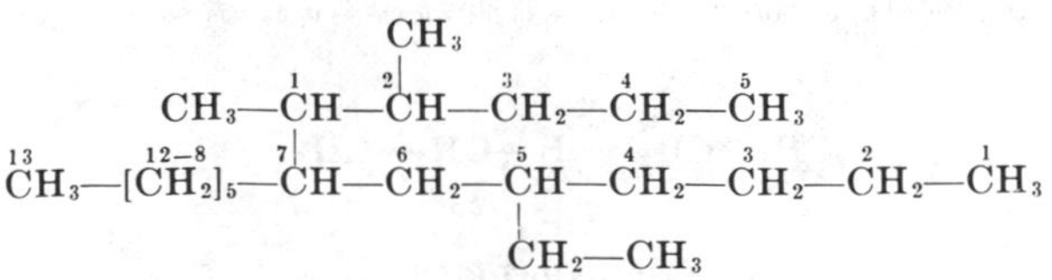

$$CH_3-\overset{1}{C}H-\overset{\overset{CH_3}{|}}{\overset{2}{C}H}-\overset{3}{C}H_2-\overset{4}{C}H_2-\overset{5}{C}H_3$$

$$\overset{13}{C}H_3-\overset{12-8}{[CH_2]_5}-\overset{7|}{C}H-\overset{6}{C}H_2-\underset{\underset{CH_2-CH_3}{|}}{\overset{5}{C}H}-\overset{4}{C}H_2-\overset{3}{C}H_2-\overset{2}{C}H_2-\overset{1}{C}H_3$$

dimethylpentyl (as a complete single substituent) is alphabetized under "d", thus 7-(1.2-Dimethylpentyl)-5-ethyltridecane

(iii) In cases where names of complex radicals are composed of identical words, priority for citation is given to that radical which contains the lowest locant at the first cited point of difference in the radical.

* Use of an order of complexity given as alternative in the First and Second Editions is abandoned.

Example:

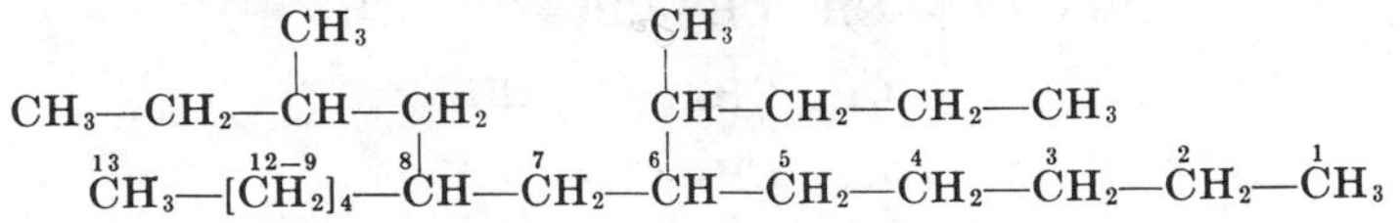

```
                  CH3                 CH3
                  |                   |
CH3—CH2—CH—CH2                        CH—CH2—CH2—CH3
                  |                   |
     CH3—[CH2]4—CH—CH2—CH—CH2—CH2—CH2—CH2—CH3
```

6-(1-Methylbutyl)-8-(2-Methylbutyl)tridecane

2.3—If two or more side chains are in equivalent positions, the one to be assigned the lower number is that cited first in the name.

Examples:

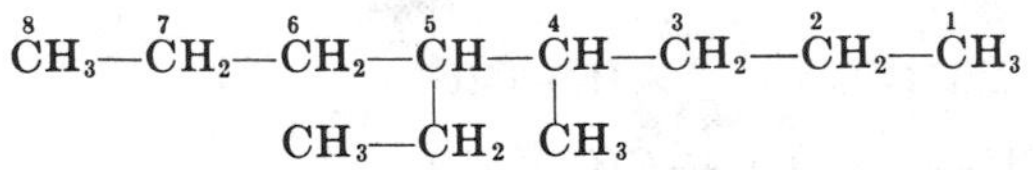

```
CH3—CH2—CH2—CH—CH—CH2—CH2—CH3
             |   |
        CH3—CH2 CH3
```

4-Ethyl-5-methyloctane

```
CH3—CH2—CH2—CH—CH—CH2—CH2—CH3
             |   |
            CH2 CH—CH3
             |   |
        CH3—CH2 CH3
```

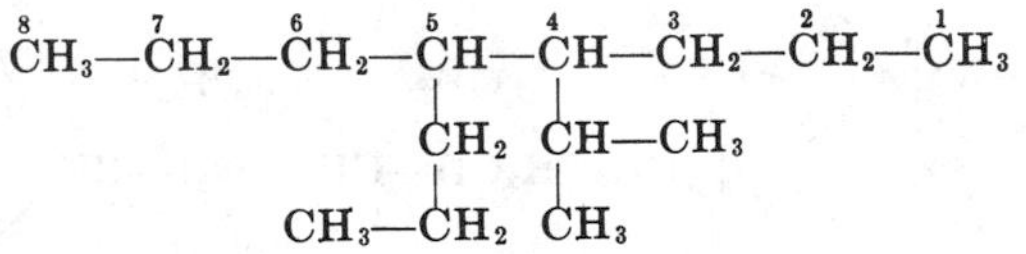

4-Isopropyl-5-propyloctane

2.5—The presence of identical unsubstituted radicals is indicated by the appropriate multiplying prefix di-, tri-, tetra-, penta-, hexa-, hepta-, octa-, nona-, deca-, undeca-, etc.

Example:

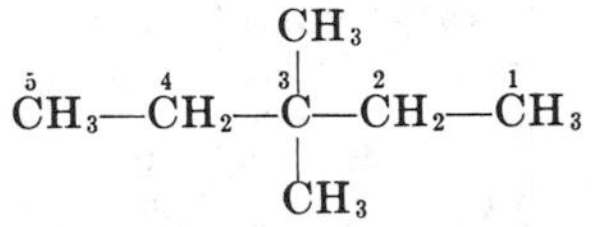

```
          CH3
          |
CH3—CH2—C—CH2—CH3
          |
          CH3
```

3,3-Dimethylpentane

The presence of identical radicals each substituted in the same way may be indicated by the appropriate multiplying prefix bis-, tris-, tetrakis-, pentakis-, etc. The complete expression denoting such a side chain may be enclosed in parentheses or the carbon atoms in side chains may be indicated by primed numbers.

Examples:

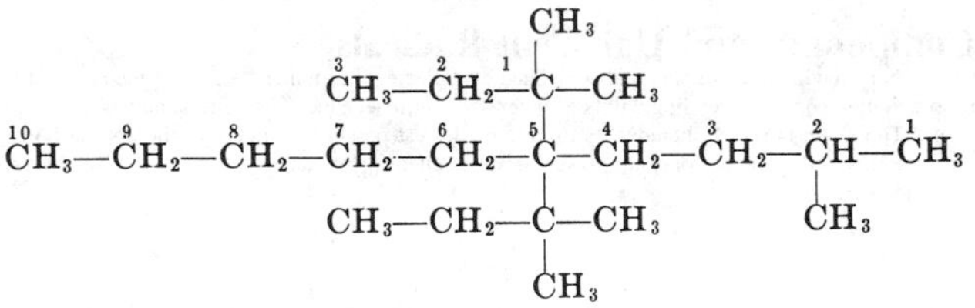

```
                               CH3
                               |
                    CH3—CH2—C—CH3
                               |
CH3—CH2—CH2—CH2—CH2—C—CH2—CH2—CH—CH3
                               |              |
                    CH3—CH2—C—CH3       CH3
                               |
                               CH3
```

(a) Use of parentheses and unprimed numbers: 5,5-Bis(1,1-dimethylproply)-2-methyldecane
(b) Use of primes: 5,5-Bis-1′,1′-dimethylpropyl-2-methyldecane

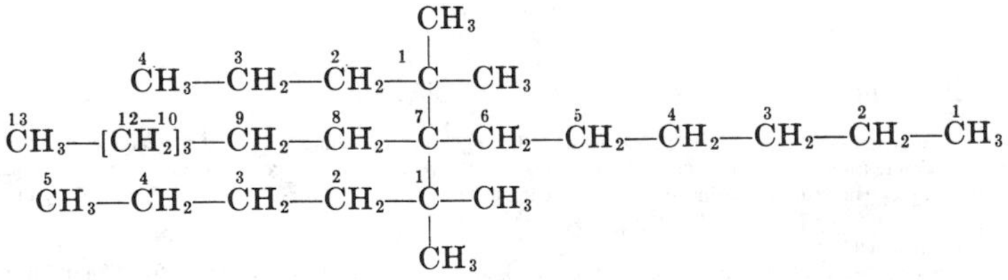

```
                                 CH3
                                 |
          CH3—CH2—CH2—C—CH3
                                 |
CH3—[CH2]3—CH2—CH2—C—CH2—CH2—CH2—CH2—CH2—CH3
                                 |
          CH3—CH2—CH2—CH2—C—CH3
                                 |
                                 CH3
```

(a) Use of parentheses and unprimed numbers: 7-(1,1-Dimethylbutyl)-7-(1,1-dimethylpentyl)tridecane
(b) Use of primes: 7-1′,1′-Dimethylbutyl-7-1″,1″-dimethylpentyltridecane

2.6—If chains of equal length are competing for selection as main chain in a saturated branched acyclic hydrocarbon, then the choice goes in series to:

(a) The chain which has the greatest number of side chains.

Example:

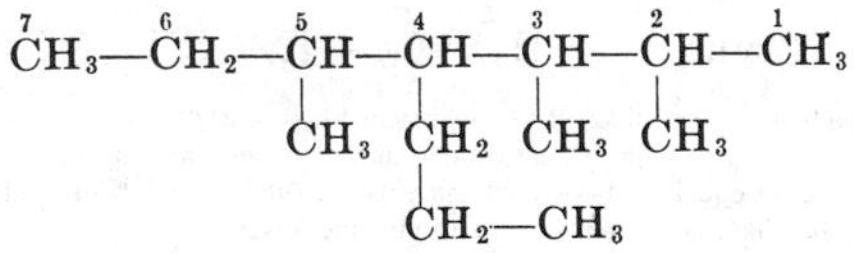

```
CH3—CH2—CH—CH—CH—CH—CH3
         |   |   |   |
        CH3 CH2 CH3 CH3
             |
             CH2—CH3
```

2,3,5-Trimethyl-4-propylheptane

(b) The chain whose side chains have the lowest-numbered locants.

Example:

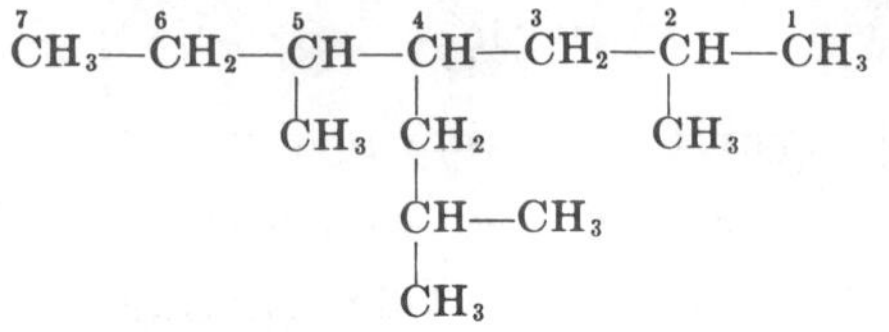

```
  7     6     5    4     3     2     1
CH3—CH2—CH—CH—CH2—CH—CH3
            |    |           |
           CH3  CH2         CH3
                 |
                CH—CH3
                 |
                CH3
```

4-Isobutyl-2,5-dimethylheptane

(c) The chain having the greatest number of carbon atoms in the smaller side chains.

Example*:

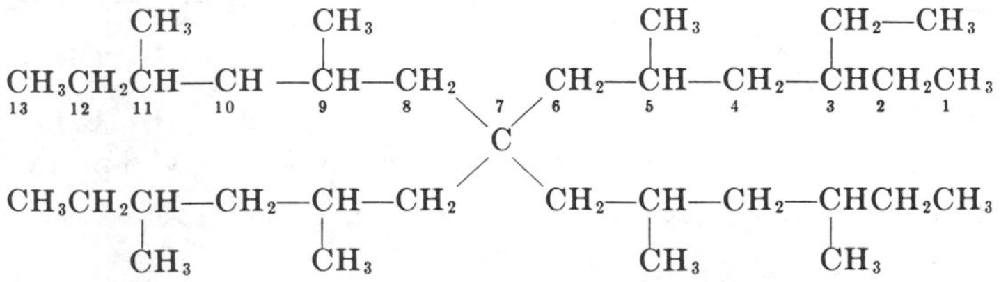

```
     CH3          CH3                    CH3          CH2—CH3
      |            |                      |            |
CH3CH2CH—CH  —CH—CH2       CH2—CH—CH2—CHCH2CH3
13  12   11   10     9    8  \  7 /  6   5    4   3  2   1
                              C
                            /   \
CH3CH2CH—CH2—CH—CH2       CH2—CH—CH2—CHCH2CH3
      |           |                 |          |
     CH3         CH3               CH3        CH3
```

7,7-Bis(2,4-dimethylhexyl)-3-ethyl-5,9,11-trimethyltridecane

(d) The chain having the least branched side chains.

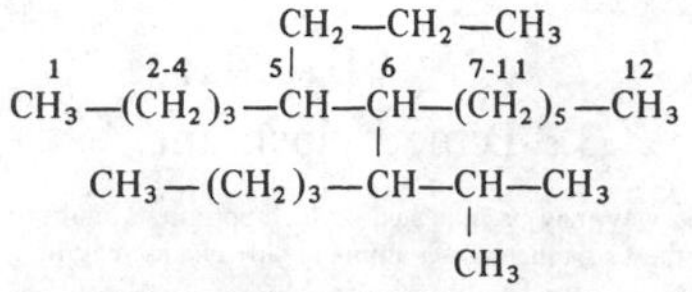

```
                    CH2—CH2—CH3
  1       2-4       5|    6     7-11      12
CH3—(CH2)3—CH—CH—(CH2)5—CH3
                          |
      CH3—(CH2)3—CH—CH—CH3
                          |
                         CH3
```

6-(1-Isopropylpentyl)-5-propyldodecane

Rule A-3. Unsaturated Compounds and Univalent Radicals

3.1—Unsaturated unbranched acyclic hydrocarbons having one double bond are named by replacing the ending "-ane" of the name of the corresponding saturated hydrocarbon with the ending "-ene". If there are two or more double bonds, the ending will be "-adiene", "-atriene", etc. The generic names of these hydrocarbons (branched or unbranched) are "alkene", "alkadiene", "alkatriene", etc. The chain is so numbered** as to give the lowest possible numbers to the double bonds. When, in cyclic compounds or their substitution products, the locants of a double bond differ by unity, only the lower locant is cited in the name; when they differ by more than unity, one locant is placed in parentheses after the other (see Rules **A-31.3** and **A-31.4**).

Examples:

2-Hexene	$\overset{6}{C}H_3—\overset{5}{C}H_2—\overset{4}{C}H_2—\overset{3}{C}H=\overset{2}{C}H—\overset{1}{C}H_3$
1,4-Hexadiene	$\overset{6}{C}H_3—\overset{5}{C}H=\overset{4}{C}H—\overset{3}{C}H_2—\overset{2}{C}H=\overset{1}{C}H_2$

The following nonsystematic names are retained:

Ethylene $CH_2=CH_2$ Allene $CH_2=C=CH_2$

3.2—Unsaturated unbranched acyclic hydrocarbons having one triple bond are named by replacing the ending "-ane" of the name of the corresponding saturated hydrocarbon with the ending "-yne". If there are two or more triple bonds, the ending will be "-adiyne", "-atriyne", etc. The generic names of these hydrocarbons (branched or unbranched) are "alkyne", "alkadiyne", "alkatriyne", etc. The chain is so numbered as to give the lowest possible numbers to the triple bonds. Only the lower locant for a triple bond is cited in the name of a compound.

The name "acetylene" for $HC \equiv CH$ is retained.

3.3—Unsaturated unbranched acyclic hydrocarbons having both double and triple bonds are named by replacing the ending "-ane" of the name of the corresponding saturated hydrocarbon with the ending "-enyne", "-adienyne", "-atrienyne", "-enediyne", etc. Numbers as low as possible are given to double and triple bonds even though this may at times give "-yne" to lower number than "-ene". When there is a choice in numbering, the double bonds are given the lowest numbers.

Examples:

1,3-Hexadien-5-yne	$H\overset{6}{C}\equiv\overset{5}{C}—\overset{4}{C}H=\overset{3}{C}H—\overset{2}{C}H=\overset{1}{C}H_2$
3-Penten-1-yne	$\overset{5}{C}H_3—\overset{4}{C}H=\overset{3}{C}H—\overset{2}{C}\equiv\overset{1}{C}H$
1-Penten-4-yne	$H\overset{5}{C}\equiv\overset{4}{C}—\overset{3}{C}H_2—\overset{2}{C}H=\overset{1}{C}H_2$

3.4—Unsaturated branched acyclic hydrocarbons are named as derivatives of the unbranched hydrocarbons which contain the maximum number of double and triple bonds. If there are two or more chains competing for selection as the chain with the maximum number of unsaturated bonds, then the choice goes to (1) that one with the greatest number of carbon atoms; (2) the number of carbon atoms being equal, that one containing the maximum number of double bonds. In other respects, the same principles apply as for naming saturated branched acyclic hydrocarbons. The chain is so numbered as to give the lowest possible numbers to double and triple bonds in accordance with Rule **A-3.3.**

* Here the choice lies between two possible main chains of equal length, each containing six side chains in the same positions. Listing in increasing order, the number of carbon atoms in the several side chains of the first choice as shown and of the alternate second choice results as follows:

first choice	1, 1, 1, 2, 8, 8
second choice	1, 1, 1, 1, 8, 9

** The expression, "the greatest number of carbon atoms in the smaller side chains", is taken to mean the largest side chain at the first point of difference when the size of the side chains is examined step by step. Thus, the selection in this case is made at the fourth step where 2 is greater than 1.
Only the lower locant for a double bond is cited in the name of an acyclic compound.

Examples:

3,4-Dipropyl-1,3-hexadien-5-yne

$$\begin{array}{l} \qquad\quad CH_2{-}CH_2{-}CH_3 \\ \overset{6}{CH}{\equiv}\overset{5}{C}{-}\overset{4}{C}{=}\overset{3}{C}{-}\overset{2}{CH}{=}\overset{1}{CH_2} \\ \qquad\qquad CH_2{-}CH_2{-}CH_3 \end{array}$$

5-Ethynyl-1,3,6-heptatriene

$$\begin{array}{l} \overset{7}{CH_2}{=}\overset{6}{CH}{-}\overset{5}{CH}{-}\overset{4}{CH}{=}\overset{3}{CH}{-}\overset{2}{CH}{=}CH_2 \\ \qquad\qquad C{\equiv}CH \end{array}$$

5,5-Dimethyl-1-hexene

$$\begin{array}{l} \qquad\quad CH_3 \\ \overset{6}{CH_3}{-}\overset{5}{C}{-}\overset{4}{CH_2}{-}\overset{3}{CH_2}{-}\overset{2}{CH}{=}\overset{1}{CH_2} \\ \qquad\quad CH_3 \end{array}$$

4-Vinyl-1-hepten-5-yne

$$\begin{array}{l} \overset{7}{CH_3}{-}\overset{6}{C}{\equiv}\overset{5}{C}{-}\overset{4}{CH}{-}\overset{3}{CH_2}{-}\overset{2}{CH}{=}\overset{1}{CH_2} \\ \qquad\qquad\quad CH{=}CH_2 \end{array}$$

The name "isoprene" is retained for the unsubstituted compound only:

$$\begin{array}{l} \qquad\qquad\quad CH_3 \\ CH_2{=}CH{-}C{=}CH_2 \end{array}$$

3.5—The names of univalent radicals derived from unsaturatd acyclic hydrocarbons have the endings "-enyl", "-ynyl", "-dienyl", etc., the positions of the double and triple bonds being indicated where necessary. The carbon atom with the free valence is numbered as 1.

Examples:

Ethynyl	$CH{\equiv}C{-}$
2-Propynyl	$CH{\equiv}C{-}CH_2{-}$
1-Propenyl	$CH_3{-}CH{=}CH{-}$
2-Butenyl	$CH_3{-}CH{=}CH{-}CH_2{-}$
1,3-Butadienyl	$CH_2{=}CH{-}CH{=}CH{-}$
2-Pentenyl	$CH_3{-}CH_2{-}CH{=}CH{-}CH_2{-}$
2-Penten-4-ynyl	$CH{\equiv}C{-}CH{=}CH{-}CH_2{-}$

Exceptions: The following names are retained:

Vinyl (for ethenyl)	$CH_2{=}CH{-}$
Allyl (for 2-propenyl)	$CH_2{=}CH{-}CH_2{-}$
Isopropenyl (for 1-methylvinyl)	$CH_2{=}C(CH_3){-}$ (for unsubstituted radical only)

3.6—When there is a choice for the fundamental chain of a radical, that chain is selected which contains (1) the maximum number of double and triple bonds; (2) the largest number of carbon atoms; and (3) the largest number of double bonds.

Examples:

$$\begin{array}{l} \overset{10}{CH_3}{-}\overset{9}{CH}{=}\overset{8}{CH}{-}\overset{7}{CH}{=}\overset{6}{CH}{-}\overset{5}{CH}{-}\overset{4}{CH}{=}\overset{3}{CH}{-}\overset{2}{C}{\equiv}\overset{1}{C}{-} \\ \qquad\qquad\qquad\qquad CH_2{-}CH_2{-}CH{=}CH{-}CH_3 \end{array}$$

5-(3-Pentenyl)-3,6,8-decatrien-1-ynyl

$$\begin{array}{l} \overset{12}{CH_3}{-}\overset{11}{CH_2}{-}\overset{10}{C}{\equiv}\overset{9}{C}{-}\overset{8}{CH}{=}\overset{7}{CH}{-}\overset{6}{CH}{-}\overset{5}{CH}{=}\overset{4}{CH}{-}\overset{3}{CH}{=}\overset{2}{CH}{-}\overset{1}{CH_2}{-} \\ \qquad\qquad\qquad\qquad\qquad CH{=}CH{-}CH{=}CH{-}CH_3 \end{array}$$

6-(1,3-Pentadienyl)-2,4,7-dodecatrien-9-ynyl

$$\begin{array}{l} \overset{11}{CH_3}{-}\overset{10}{CH}{=}\overset{9}{CH}{-}\overset{8}{CH}{=}\overset{7}{CH}{-}\overset{6}{CH}{-}\overset{5}{CH}{=}\overset{4}{CH}{-}\overset{3}{CH}{=}\overset{2}{CH}{-}\overset{1}{CH_2} \\ \qquad\qquad\qquad\qquad CH{=}CH{-}C{\equiv}C{-}CH_3 \end{array}$$

6-(1-Penten-3-ynyl)-2,4,7,9-undecatetraenyl

$$\begin{array}{l} \overset{4}{CH_3}{-}\overset{3}{CH}{=}\overset{2}{C}{-}\overset{1}{CH_2}{-} \\ \qquad\qquad CH_2{-}CH_2{-}CH_2{-}CH_2{-}CH_2{-}CH_2{-}CH_2{-}CH_2{-}CH_3 \end{array}$$

2-Nonyl-2-butenyl

Rule A-4. Bivalent and Multivalent Radicals*

4.1—Bivalent and trivalent radicals derived from univalent acyclic hydrocarbon radicals whose authorized names end in "-yl" by removal of one or two hydrogen atoms from the carbon atom with the free valences are named by adding "-idene" or "-idyne", respectively, to the name of the corresponding univalent radical. The carbon atom with the free valence is numbered as 1.

The name "methylene" is retained for the radical $CH_2{=}$.

* Rule **D-4.14** introduces an alternate method of naming radicals derived from any position of unbranched chains or ring systems by adding "-yl", "-diyl", "-triyl", etc. to the name of the chain or ring system with elision of "e" before "-yl". Examples: 2-pentanyl $CH_3{-}CH_2{-}CH_2{-}CH{-}CH_3$; 1,6-hexanediyl $-CH_2{-}(CH_2)_4{-}CH_2{-}$.

Examples:

Methylidyne[1]	$CH\equiv$
Ethylidene	$CH_3—CH=$
Ethylidyne	$CH_3—C\equiv$
Vinylidene	$CH_2=C=$
Isopropylidene[2]	$(CH_3)_2C=$

4.2—The names of bivalent radicals derived from normal alkanes by removal of a hydrogen atom from each of the two terminal carbon atoms of the chain are ethylene, trimethylene, tetramethylene, etc.

Examples:

Pentamethylene	$—CH_2—CH_2—CH_2—CH_2—CH_2—$
Hexamethylene	$—CH_2—CH_2—CH_2—CH_2—CH_2—CH_2—$

Names of the substituted bivalent radicals are derived in accordance with Rules **A-2.2** and **A-2.25**.

Example:

Ethylethylene $—\overset{2}{C}H_2—\overset{1}{C}H—$ (with $CH_2—CH_3$ on C-1)

The name "propylene" is retained:

$$CH_3—CH(—)—CH_2—$$

4.3—Bivalent radicals similarly derived from unbranched alkenes, alkadienes, alkynes, etc., by removing a hydrogen atom from each of the terminal carbon atoms are named by replacing the endings "-ene", "-diene", "-yne", etc., of the hydrocarbon name by "-enylene", "-dienylene", "-ynylene", etc., the positions of the double and triple bonds being indicated where necessary.

Example:

Propylene $—\overset{3}{C}H_2—\overset{2}{C}H=\overset{1}{C}H—$

The name "vinylene" is retained (for ethenylene):

$$—CH=CH—$$

Names of the substituted bivalent radicals are derived in accordance with Rule **A-3.4**.

Example:

4 Propyl-2-pentenylene $—\overset{5}{C}H_2—\overset{4}{C}H—\overset{3}{C}H=\overset{2}{C}H—\overset{1}{C}H_2—$ (with $CH_2—CH_2—CH_3$ on C-4)

4.4—Trivalent, quadrivalent, and higher-valent acyclic hydrocarbon radicals of two or more carbon atoms with the free valences at each end of a chain are named by adding to the hydrocarbon name the terminations "-yl" for single free valence, "-ylidene" for a double, and "-ylidyne" for a triple free valence on the same atom (the final "e" in the name of the hydrocarbon is elided when followed by a suffix beginning with "-yl"). If different types are present in the same radical, they are cited and numbered in the order "-yl", "-ylidene", "-ylidyne".

Examples:

Butanediylidene	$=\overset{4}{C}H—\overset{3}{C}H_2—\overset{2}{C}H_2—\overset{1}{C}H=$
Butanediylidyne	$\equiv\overset{4}{C}—\overset{3}{C}H_2—\overset{2}{C}H_2—\overset{1}{C}\equiv$
1-Propanyl-3-ylidene	$=\overset{3}{C}H—\overset{2}{C}H_2—\overset{1}{C}H_2—$
Propadienediylidene	$=\overset{3}{C}=\overset{2}{C}=\overset{1}{C}=$
2-Pentenediylidyne	$\equiv\overset{5}{C}—\overset{4}{C}H_2—\overset{3}{C}H=\overset{2}{C}H—\overset{1}{C}\equiv$
1-Butanyliden-4-ylidyne	$\equiv\overset{4}{C}—\overset{3}{C}H_2—\overset{2}{C}H_2—\overset{1}{C}H=$

4.5—Multivalent radicals containing three or more carbon atoms with free valences at each end of a chain and additional free valences at intermediate carbon atoms are named by adding the endings "-triyl", "-tetrayl", "-diylidene", "diyl-ylidene", etc., to the hydrocarbon name.

Examples:

$—\overset{3}{C}H_2—\overset{2}{C}H(—)—\overset{1}{C}H_2—$

1,2,3-Propanetriyl

$—\overset{3}{C}H_2—\overset{2}{C}(=)—\overset{1}{C}H_2—$

1,3-Propanediyl-2-ylidene

MONOCYCLIC HYDROCARBONS

Rule A-11. Unsubstituted Compounds and Radicals*

11.1—The names of saturated monocyclic hydrocarbons (with no side chains) are formed by attaching the prefix "cyclo" to the name of the acyclic saturated unbranched hydrocarbon with the same number of carbon atoms. The generic name of saturated monocyclic hydrocarbons (with or without side chains) is "cycloalkane".

[1] The group =CH– may be referred to as the "methine" group.
[2] For unsubstituted radical only.
* See footnote to Rule A-4.

Examples:

Cyclopropane

Cyclohexane

11.2—Univalent radicals derived from cycloalkanes (with no side chains) are named by replacing the ending "-ane" of the hydrocarbon name by "-yl", the carbon atom with the free valence being numbered as 1. The generic name of these radicals is "cycloalkyl".

Examples:

Cyclopropyl

Cyclohexyl

11.3—The names of unsaturated monocyclic hydrocarbons (with no side chains) are formed by substituting "-ene", "-adiene", "-atriene", "-yne", "-adiyne", etc., for "-ane" in the name of the corresponding cycloalkane. The double and triple bonds are given numbers as low as possible as in Rule **A-3.3**.

Examples:

Cyclohexene

1,3-Cyclohexadiene

1-Cyclodecen-4-yne

The name "benzene" is retained.

11.4—The names of univalent radicals derived from unsaturated monocyclic hydrocarbons have the endings "-enyl", "-ynyl", "-dienyl", etc., the positions of the double and triple bonds being indicated according to the principles of Rule **A-3.3**. The carbon atom with the free valence is numbered as 1, except as stated in the rules for terpenes (see Rules **A-72** to **A-75**).

Examples:

2-Cyclopenten-1-yl

2,4-Cyclopentadien-1-yl

The radical name "phenyl" is retained.

11.5—Names of bivalent radicals derived from saturated or unsaturated monocyclic hydrocarbons by removal of two atoms of hydrogen from the same carbon atom of the ring are obtained by replacing the endings "-ane", "-ene", "-yne", by "-ylidene", "-enylidene" and "-ynylidene", respectively. The carbon atom with the free valences is numbered as 1, except as stated in the rules for terpenes.

Examples:

Cyclopentylidene

2,4-Cyclohexadien-1-ylidene

11.6—Bivalent radicals derived from saturated or unsaturated monocyclic hydrocarbons by removing a hydrogen atom from each of two different carbon atoms of the ring are named by replacing the endings "-ane", "-ene", "-diene", "-yne", etc., of the hydrocarbon name by "-ylene", "-enylene", "-dienylene", "-ynylene", etc., the positions of the double and triple bonds and of the points of attachment being indicated. Preference in lowest numbers is given to the carbon atoms having the free valences.

Examples:

1,3-Cyclopentylene

3-Cyclohexen-1,2-ylene

2,5-Cyclohexadien-1,4-ylene

The name "phenylene" is retained:

Phenylene (*p*-shown)

Rule A-12. Substituted Aromatic Compounds

12.1—The following names for monocyclic substituted aromatic hydrocarbons are retained:

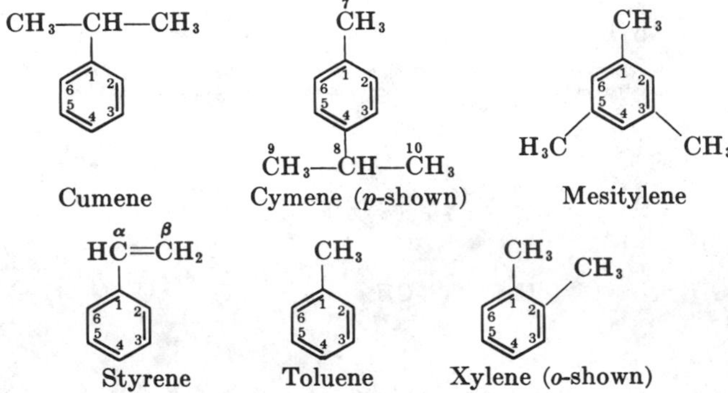

Cumene

Cymene (*p*-shown)

Mesitylene

Styrene

Toluene

Xylene (*o*-shown)

12.2—Other monocyclic substituted aromatic hydrocarbons are named as derivatives of benzene or of one of the compounds listed in Part **.1** of this rule. However, if the substituent introduced into such a compound is identical with one already present in that compound, then the substituted compound is named as a derivative of benzene (see Rule **61.4**).

12.3—The position of substituents is indicated by numbers except that *o-(ortho),m-(meta)* and *p-(para)* may be used in place of 1,2-, 1,3-, and 1,4-, respectively, when only two substituents are present. The lowest numbers possible are given to substituents, choice between alternatives being governed by Rule **A-2** so far as applicable, except that when names are based on those of compounds listed in Part **.1** of this rule the first priority for lowest numbers is given to the substituent(s) already present in those compounds.

Examples:

1-Ethyl-4-pentylbenzene or *p*-Ethylpentylbenzene

1,4-Diethylbenzene or *p*-Diethylbenzene

4-Ethylstyrene or *p*-Ethylstyrene

1,4-Divinylbenzene or *p*-Divinylbenzene, not *p*-Vinylstyrene

1,2,3-Trimethylbenzene, not Methylxylene nor Dimethyltoluene

1,2-Dimethyl-3-propylbenzene or 3-Propyl-*o*-xylene

1-Ethyl-2-propyl-3-butylbenzene (Order of complexity) or 1-Butyl-3-ethyl-2-propylbenzene (Alphabetical order)

12.4—The generic name of monocyclic and polycyclic aromatic hydrocarbons is "arene".

Rule A-13. Substituted Aromatic Radicals

13.1—Univalent radicals derived from monocyclic substituted aromatic hydrocarbons and having the free valence at a ring atom are given the names listed below. Such radicals not listed below are named as substituted phenyl radicals. The carbon atom having the free valence is numbered as 1.

Phenyl C_6H_5—

Cumenyl (*m*-shown) Mesityl

Tolyl (*o*-shown) Xylyl (2,3-shown)

13.2—Since the name phenylene (*o*-, *m*- or *p*-) is retained for the radical $-C_6H_4-$ (exception to Rule **A-11.6**), bivalent radicals formed from substituted benzene derivatives and having the free valences at ring atoms are named as substituted phenylene radicals. The carbon atoms having the free valences are numbered 1,2-, 1,3-, or 1,4- as appropriate.

13.3—The following trivial names for radicals having a single free valence in the side chain are retained:

Benzyl	$C_6H_5-\overset{\alpha}{C}H_2-$
Benzhydryl (alternative to Diphenylmethyl)	$(C_6H_5)_2\overset{\alpha}{C}H-$
Cinnamyl	$C_6H_5-\overset{\gamma}{C}H=\overset{\beta}{C}H-\overset{\alpha}{C}H_2-$
Phenethyl	$C_6H_5-\overset{\beta}{C}H_2-\overset{\alpha}{C}H_2-$
Styryl	$C_6H_5-\overset{\beta}{C}H=\overset{\alpha}{C}H-$
Trityl	$(C_6H_5)_3C-$

13.4—Multivalent radicals of aromatic hydrocarbons with the free valences in the side chain are named in accordance with Rule **A-4**.

Examples:

Benzylidyne	$C_6H_5-C\equiv$
Cinnamylidene	$C_6H_5-\overset{\gamma}{C}H=\overset{\beta}{C}H-\overset{\alpha}{C}H=$

13.5—The generic names of univalent and bivalent aromatic hydrocarbon radicals are "aryl" and "arylene", respectively.

FUSED POLYCYCLIC HYDROCARBONS

Rule A-21. Trivial and Semi-trivial names

21.1—The names of polycyclic hydrocarbons with maximum number of noncumulative* double bonds end in "-ene". The names listed on pp. C-9 and C-10 are retained.

21.2—The names of hydrocarbons containing five or more fused benzene rings in a straight linear arrangement are formed from a numerical prefix as specified in Rule **A-1.1** followed by "-acene". (Examples on pp. C-9 and C-10).

Examples:

$CH_2=C=C=C=CH_2$
Cumulative

$CH_3-CH=CH-CH=CH-CH=CH_2$
or

Non-cumulative

Examples (to Rule **A-21.2**):

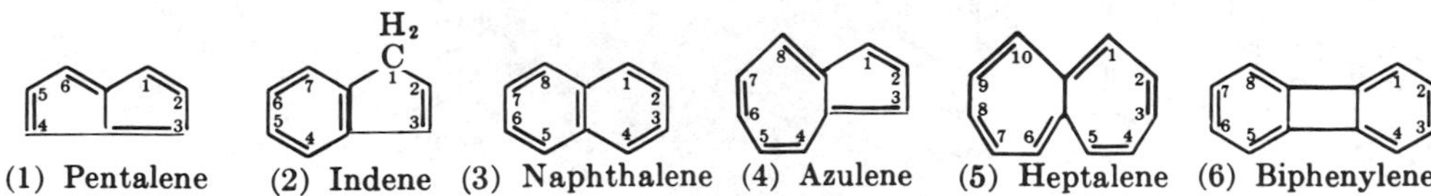

Pentacene **Hexacene**

The following list contains the names of polycyclic hydrocarbons which are retained (see Rule **A-21.1**). This list is not limiting.

(1) **Pentalene** (2) **Indene** (3) **Naphthalene** (4) **Azulene** (5) **Heptalene** (6) **Biphenylene**

* Cumulative double bonds are those present in a chain in which at least three contiguous carbon atoms are joined by double bonds; non-cumulative double bonds comprise every other arrangement of two or more double bonds in a single structure. The generic name "cumulene" is given to compounds containing three or more cumulative double bonds.

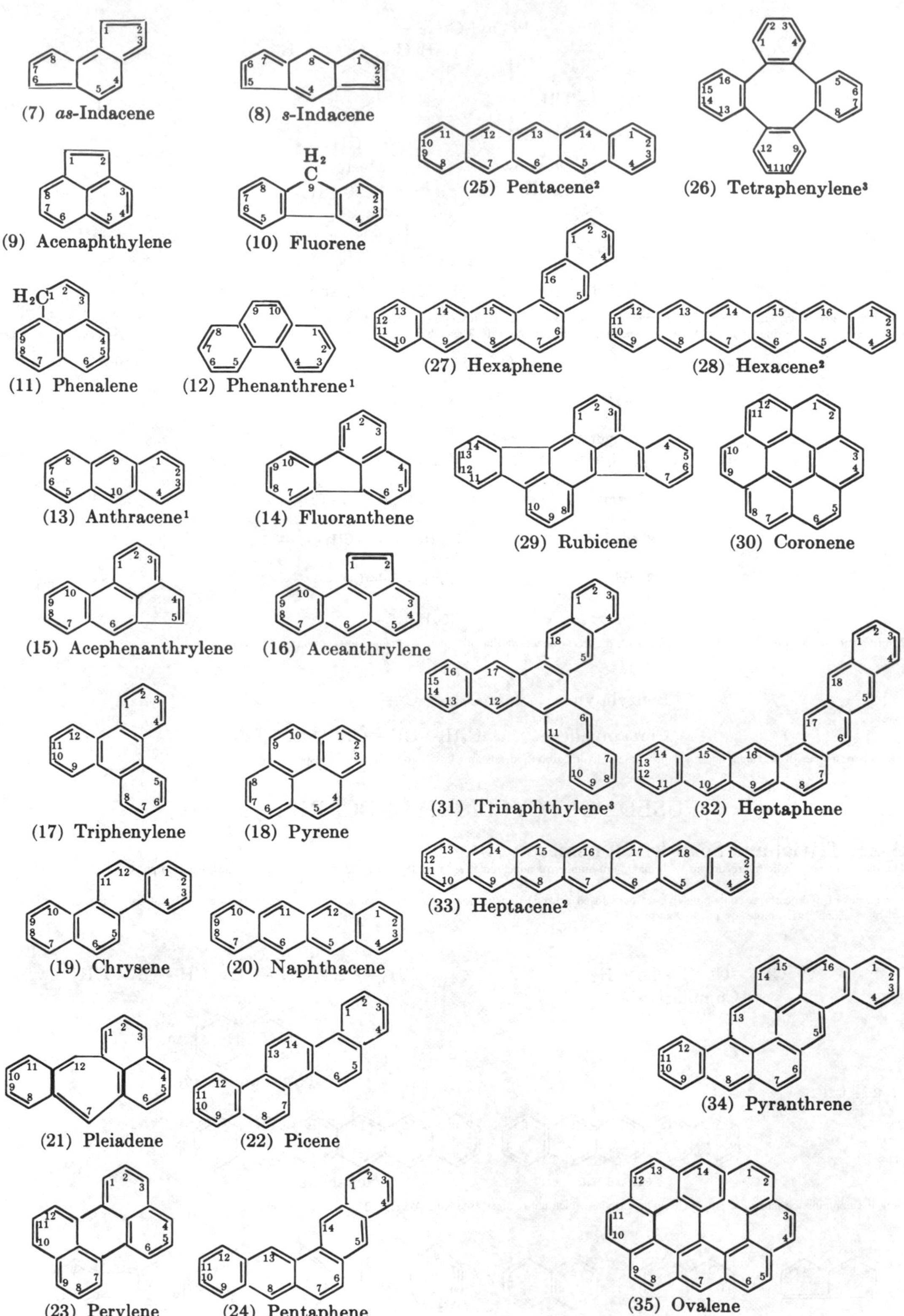

[1] Denotes exception to systematic numbering.
[2] See Rule **A-21.2**
[3] For isomer shown only.

21.3—"*Ortho*-fused"* or "*ortho-* and *peri*-fused"** polycyclic hydrocarbons with maximum number of noncumulative double bonds which contain at least two rings of five or more members and which have no accepted trivial name such as those of Part **.1** of this rule, are named by prefixing to the name of a component ring or ring system (the base component) designations of the other components. The base component should contain as many rings as possible (provided it has a trivial name), and should occur as far as possible from the beginning of the list of Rule **A-21.1.** The attached components should be as simple as possible.

Example:

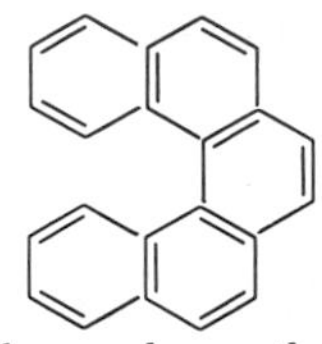

Dibenzophenanthrene

(not Naphthophenanthrene; benzo is "simpler" than naphtho, even though there are two benzo rings and only one naphtho)

21.4—The prefixes designating attached components are formed by changing the ending "-ene" of the name of the component hydrocarbon into "-eno"; e.g., "pyreno' (from pyrene). When more than one prefix is presented, they are arranged in alphabetical order. The following common abbreviated prefixes are recognized (see list in Part **.1** of this rule):

Acenaphtho	from	Acenaphthylene	Naphtho	from	Naphthalene
Anthra	from	Anthracene	Perylo	from	Perylene
Benzo	from	Benzene	Phenanthro	from	Phenanthrene

For monocyclic prefixes other than "benzo", the following names are recognized, each to represent the form with the maximum number of noncumulative double bonds: cyclopenta, cyclohepta, cycloocta, cyclonona, etc. When the base component is a monocyclic system, the ending "-ene" signifies the maximum number of noncumulative double bonds, and thus does not denote one double bond only.***

Examples:

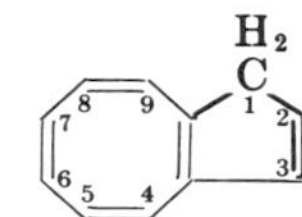

1*H*-Cyclopentacycloöctene

Benzocycloöctene

21.5—Isomers are distinguished by lettering the peripheral sides of the base component *a, b, c*, etc., beginning with "*a*" for the side "1,2", "*b*" for "2,3" (or in certain cases "2,2*a*") and lettering every side around the periphery. To the letter as early in the alphabet as possible, denoting the side where fusion occurs, are prefixed, if necessary, the numbers of the positions of attachment of the other component. These numbers are chosen to be as low as is consistent with the numbering of the component, and their order conforms to the direction of lettering of the base component (see Examples II and IV). When two or more prefixes refer to equivalent positions so that there is a choice of letters, the prefixes are cited in alphabetical order according to Rule A-**21.4** and the location of the first cited prefix is indicated by a letter as early as possible in the alphabet (see Example V). The numbers and letters are enclosed in square brackets and placed immediately after the designation of the attached component. This expression merely defines the manner of fusion of the components.

Examples:

I Benz[*a*]anthracene

II Anthra[2,1-*a*]naphthacene

III Dibenz[*a,j*]anthracene (not Naphtho[2,1-*b*]phenanthrene)

IV Indeno[1,2-*a*]indene

* Polycyclic compounds in which two rings have two, and only two, atoms in common are said to be "*ortho*-fused". Such compounds have *n* common faces and 2*n* common atoms (Example I).

** Polycyclic compounds in which one ring contains two, and only two, atoms in common with each of two or more rings of a contiguous series of rings are said to be "*ortho-* and *peri*-fused". Such compounds have *n* common faces and less than 2*n* common atoms (Examples II and III).

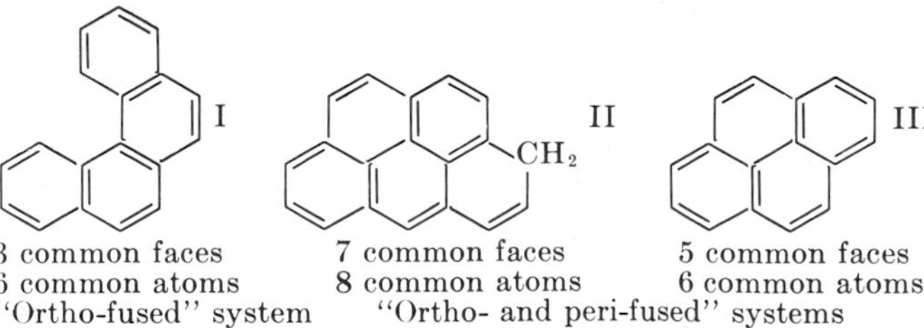

I: 3 common faces, 6 common atoms, "Ortho-fused" system

II: 7 common faces, 8 common atoms; III: 5 common faces, 6 common atoms — "Ortho- and peri-fused" systems

*** The final "o" of acenaphtho, benzo, naphtho and perylo and the "a" of the monocyclic prefixes cyclopropa, cyclopenta, cyclohepta, etc. are elided before another vowel, as benz(o)[*a*]anthracene. In all other cases the final "o" or "a" is retained.

1*H*-Benzo[*a*]cyclopent[*j*]anthracene

The completed system consisting of the base component and the other components is then renumbered according to Rule **A-22**, the enumeration of the component parts being ignored.

Example:

Benzene Pentaphene Benzene → 9-*H*-Dibenzo [*de,rst*] pentaphene

21.6—When a name applies equally to two or more isomeric condensed parent ring systems with the maximum number of noncumulative double bonds and when the name can be made specific by indicating the position of one or more hydrogen atoms in the structure, this is accomplished by modifying the name with a locant, followed by italic capital *H* for each of these hydrogen atoms. Such symbols ordinarily precede the name. The said atom or atoms are called "indicated hydrogen". The same principle is applied to radicals and compounds derived from these systems.*

Examples:

3*H*-Fluorene 2*H*-Indene

Rule A-22. Numbering

22.1—For the purposes of numbering, the individual rings of a polycyclic "*ortho*-fused" or "*ortho*- and *peri*-fused" hydrocarbon system are normally drawn as follows:

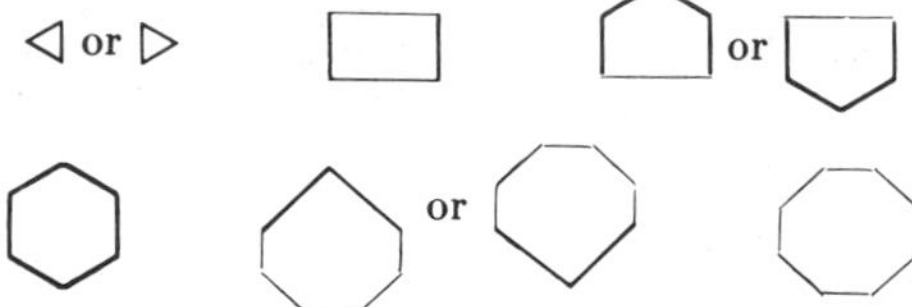

and the polycyclic system is oriented so that (a) the greatest number of rings are in a horizontal row and (b) a maximum number of rings are above and to the right of the horizontal row (upper right quadrant). If two or more orientations meet these requirements, the one is chosen which has as few rings as possible in the lower left quadrant.

Example:

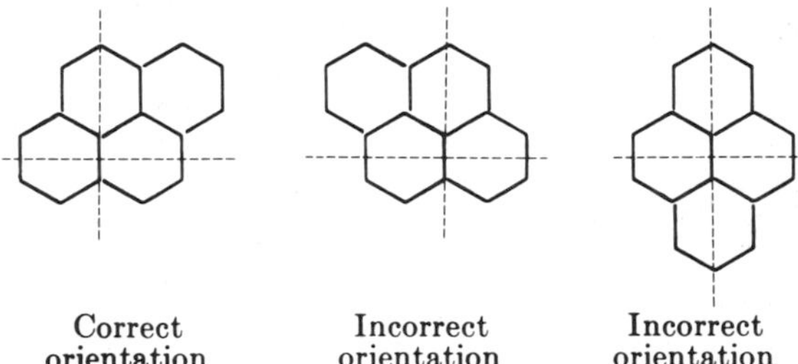

Correct orientation   Incorrect orientation   Incorrect orientation

The system thus oriented is numbered in a clockwise direction commencing with the carbon atom not engaged in ring-fusion in the most counter-clockwise position of the uppermost ring, or if there is a choice, of the uppermost ring farthest to the right, and omitting atoms common to two or more rings.

Example:

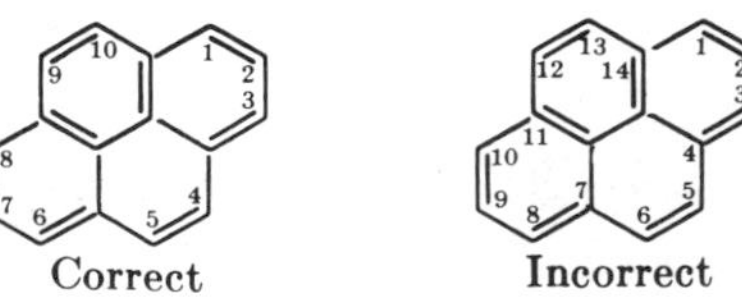

Correct   Incorrect

22.2—Atoms common to two or more rings are designated by adding roman letters "a", "b", "c", etc., to the number of the position immediately preceding. Interior atoms follow the highest number, taking a clockwise sequence wherever there is a choice.

Example:

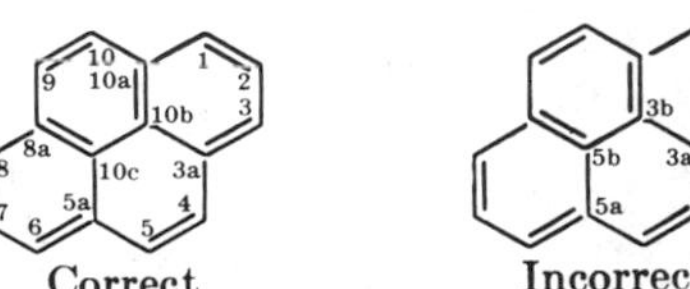

Correct   Incorrect

* See Rule B-5.12 for examples of "indicated hydrogen" in monocyclic rings.

22.3—When there is a choice,* carbon atoms common to two or more rings follow the lowest possible numbers.

Examples

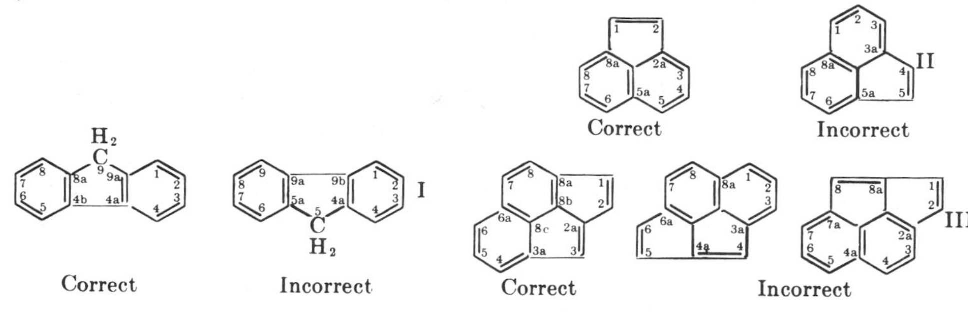

Note: I. 4, 4, 8, 9 is lower than 4, 5, 9, 9.
II. 2, 5, 8 is lower than 3, 5, 8.
III. 2, 3, 6, 8 is lower than 3, 4, 6, 8 or 2, 4, 7, 8.

22.4—When there is a choice, the carbon atoms which carry an indicated hydrogen atom are numbered as low as possible.

Example:

Correct Incorrect

22.5—The following are recommended exceptions to the above rules on numbering:

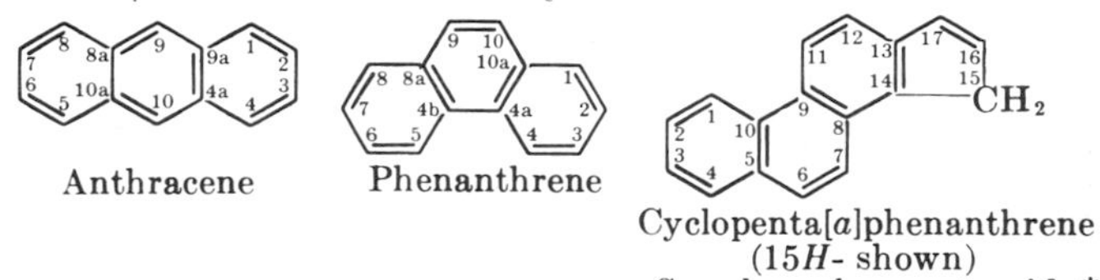

Anthracene Phenanthrene Cyclopenta[*a*]phenanthrene (15*H*- shown) See also rules on steroids**

Rule A-23. Hydrogenated Compounds

23.1—The names of "*ortho*-fused" or "*ortho*- and *peri*-fused" polycyclic hydrocarbons with less than maximum number of noncumulative double bonds are formed from a prefix "dihydro-", "tetrahydro-", etc., followed by the name of the corresponding unreduced hydrocarbon. The prefix "perhydro-" signifies full hydrogenation. When there is a choice for *H* used for indicated hydrogen it is assigned the lowest available number.

Examples:

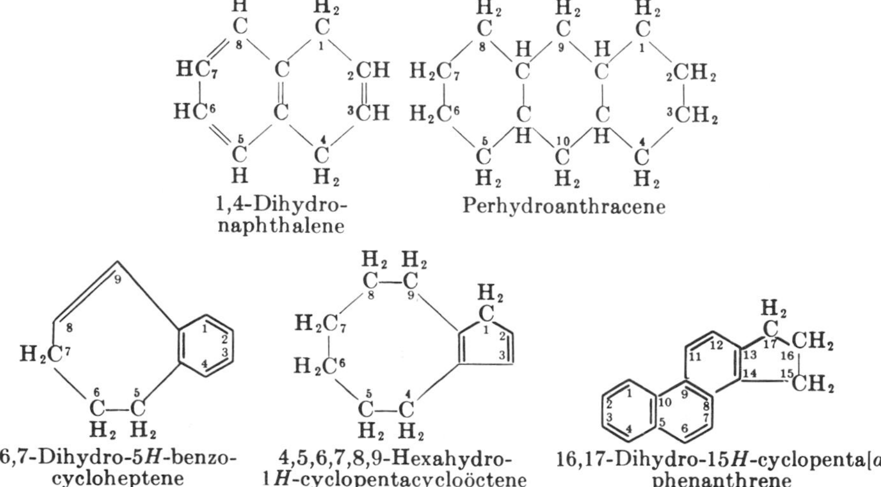

1,4-Dihydronaphthalene; Perhydroanthracene; 6,7-Dihydro-5*H*-benzocycloheptene; 4,5,6,7,8,9-Hexahydro-1*H*-cyclopentacyclooctene; 16,17-Dihydro-15*H*-cyclopenta[*a*]phenanthrene

* If after the requirements of the rules on orientation (cf Rule **A-22.1**) are met a choice remains, Rule **A-22.3** is applied.

** Definitive Rules for Nomenclature of Steroids, *Pure and Applied Chemistry*, Vol. 31, Nos. 1—2, 1972, pp. 285—322.

Exceptions: The following names are retained:

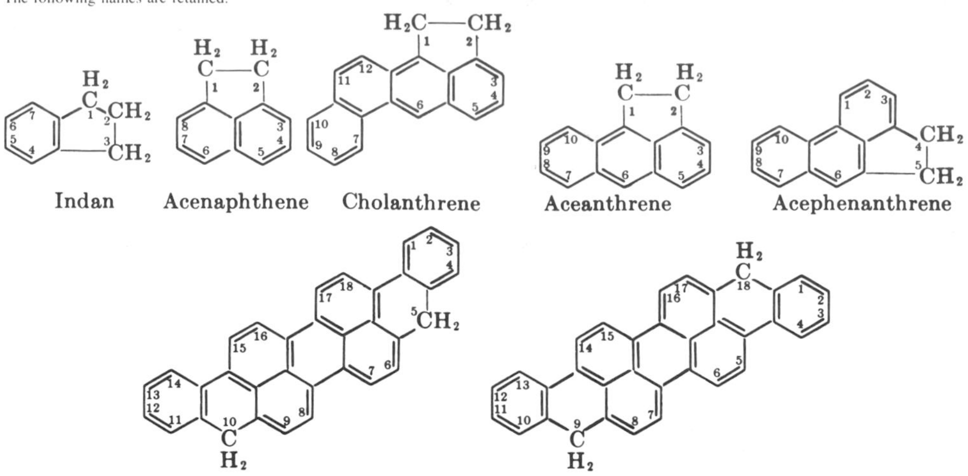

Indan Acenaphthene Cholanthrene Aceanthrene Acephenanthrene

Violanthrene Isoviolanthrene

23.2—When there is a choice, the carbon atoms to which hydrogen atoms are added are numbered as low as possible.

Example:

Correct **Incorrect**

23.3—Substituted polycyclic hydrocarbons are named according to the same principles as substituted monocyclic hydrocarbons (see Rules **A-12** and **A-61**).

23.5 (Alternate to part of Rule **A-23.1**)—The names of "*ortho*-fused" polycyclic hydrocarbons which have (a) less than the maximum number of noncumulative double bonds, (b) at least one terminal unit which is most conveniently named as an unsaturated cycloalkane derivative, and (c) a double bond at the positions where rings are fused together, may be derived by joining the name of the terminal unit to that of the other component by means of a letter "o" with elisions of a terminal "e". The abbreviations for fused aromatic systems laid down in Rule **A-21.4** are used, and the exceptions of Rule **A-23.1** apply.

Examples:

1,2-Benzo-1,3-cycloheptadiene **1,2-Cyclopenta-1′,3′-dienocyclooöctene** **1,2-Cyclopentenophenanthrene**

Rule A-24. Radical Names* from Trivial and Semitrivial Names

24.1—For radicals derived from polycyclic hydrocarbons, the numbering of the hydrocarbon is retained. The point or points of attachment are given numbers as low as is consistent with the fixed numbering of the hydrocarbon.

24.2—Univalent radicals derived from "*ortho*-fused" or "*ortho*- and *peri*-fused" polycyclic hydrocarbons with names ending in "-ene" by removal of a hydrogen atom from an aromatic or alicyclic ring are named in principle by changing the ending "-ene" of the names of the hydrocarbons to "-enyl".

Examples:

2-Indenyl **1-Pyrenyl** **1-Acenaphthenyl**

Exceptions:

Naphthyl (2-shown) **Anthryl (2-shown)** **Phenanthryl (2-shown)** **5,6,7,8-Tetrahydro-2-naphthyl**

* See footnote to Rule A-4

24.3—Bivalent radicals derived from univalent polycyclic hydrocarbon radicals whose names end in "-yl" by removal of one hydrogen atom from the carbon atom with the free valence are named by adding "-idene" to the name of the corresponding univalent radical.

Examples:

1-Acenaphthenylidene

1(4*H*)-Naphthylidene
(for 4*H* see Rule **A-21.6**)
or 1,4-Dihydro-1-naphthylidene

24.4—Bivalent radicals derived from "*ortho*-fused" or "*ortho*- and *peri*-fused" polycyclic hydrocarbons by removal of a hydrogen atom from each of two different carbon atoms of the ring are named by changing the ending "yl" of the univalent radical name to "-ylene" or by adding "-diyl" to the name of the ring system Multivalent radicals, similarly derived, are named by adding "-triyl", "-tetrayl", etc., to the name of the ring system.

Examples:

2,7-Phenanthrylene
or 2,7-Phenanthrenediyl

1,4,5,8-Anthracenetetrayl

Rule A-28. Radical Names for Fused Cyclic Systems with Side Chains

28.1—Radicals formed from hydrocarbons consisting of polycyclic systems and side chains are named according to the principles of the preceding rules.

BRIDGED HYDROCARBONS

EXTENSION OF THE VON BAEYER SYSTEM

Rule A-31. Bicyclic Systems

31.1—Saturated alicyclic hydrocarbon systems consisting of two rings only, having two or more atoms in common, take the name of an open chain hydrocarbon containing the same total number of carbon atoms preceded by the prefix "bicyclo-". The number of carbon atoms in each of the three bridges* connecting the two tertiary carbon atoms is indicated in brackets in descending order.

Examples:

Bicyclo [1.1.0]-butane

Bicyclo [3.2.1]-octane

Bicyclo [5.2.0]nonane

31.2—The system is numbered commencing with one of the bridgeheads, numbering proceeding by the longest possible path to the second bridgehead; numbering is then continued from this atom by the longer unnumbered path back to the first bridgehead and is completed by the shortest path from the atom next to the first bridgehead.

Examples:

Bicyclo [3.2.1]octane

Bicyclo [4.3.2]undecane

Note: Longest path 1, 2, 3, 4, 5
Next longest path 5, 6, 7, 1
Shortest path 1, 8, 5

31.3—Unsaturated hydrocarbons are named in accordance with the principles set forth in Rule **A-11.3**. When after applying Rule **A-31.2** a choice in numbering remains unsaturation is given the lowest numbers.

Examples:

Bicyclo [2.2.1]hept-2-ene

Bicyclo[12.2.2]octadeca-1(16),14,17-triene
or Bicyclo[12.2.2]octadeca-14,16(1),17-triene
(See Rule **A-3.1** for double locants)

31.4—Radicals derived from bridged hydrocarbons are named in accordance with the principles set forth in Rule **A-11**. The numbering of the hydrocarbon is retained and the point or points of attachment are given numbers as low as is consistent with the fixed numbering of the saturated hydrocarbon.

* A bridge is a valence bond or an atom or an unbranched chain of atoms connecting two different parts of a molecule. The two tertiary carbon atoms connected through the bridge are termed "bridgeheads".

Examples:

Bicyclo[3.2.1]oct-2-yl

Bicyclo[2.2.2]oct-5-en-2-yl

Bicyclo[5.5.1]tridec-1(12)-en-3-yl
or Bicyclo[5.5.1]tridec-12(1)-en-3-yl
(See Rule **A-3.1** for double locants)

Rule A-32. Polycyclic Systems

32.11—Cyclic hydrocarbon systems consisting of three or more rings may be named in accordance with the principles stated in Rule **A-31.** The appropriate prefix "tricyclo-", "tetracyclo-", etc., is substituted for "bicyclo-" before the name of the open-chain hydrocarbon containing the same total number of carbon atoms. Radicals derived from these hydrocarbons are named according to the principles set forth in Rule **A-31.4**.

32.12—A polycyclic system is regarded as containing a number of rings equal to the number of scissions required to convert the system into an open-chain compound.

32.13—The word "cyclo" is followed by brackets containing, in decreasing order, numbers indicating the number of carbon atoms in: the two branches of the main ring, the main bridge, and the secondary bridges.

Examples:

Tricyclo[2.2.1.0^{1}]heptane

Tricyclo[5.3.1.1^{1}]dodecane

32.21—The main ring and the main bridge form a bicyclic system whose numbering is made in compliance with Rule **A-31.**

32.22—The location of the other or so-called secondary bridges is shown by superscripts following the number indicating the number of carbon atoms in the said bridges.

32.23—For the purpose of numbering, the secondary bridges are considered in decreasing order. The numbering of any bridge follows from the part already numbered, proceeding from the highest-numbered bridgehead. If equal bridges are present, the numbering begins at the highest-numbered bridgehead.

32.31—When there is a choice, the following criteria are considered in turn until a decision is made:

(a) the main ring shall contain as many carbon atoms as possible, two of which must serve as bridgeheads for the main bridge.

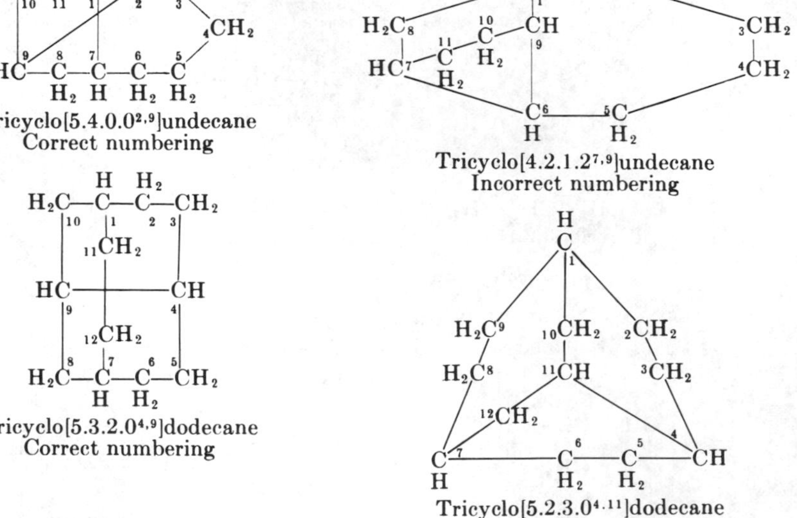

Tricyclo[5.4.0.$0^{2,9}$]undecane
Correct numbering

Tricyclo[4.2.1.$2^{7,9}$]undecane
Incorrect numbering

Tricyclo[5.3.2.$0^{4,9}$]dodecane
Correct numbering

Tricyclo[5.2.3.$0^{4,11}$]dodecane
Incorrect numbering

For location and numbering of the secondary bridge see Rules **A-32.22, A-32.23, A-32.31.**

(b) The main bridge shall be as large as possible

Tricyclo[7.3.2.0^{5,13}]tetradecane
Correct numbering

Tricyclo[7.3.1.1^{5,13}]tetradecane
Incorrect numbering

(c) The main ring shall be divided as symmetrically as possible by the main bridge.

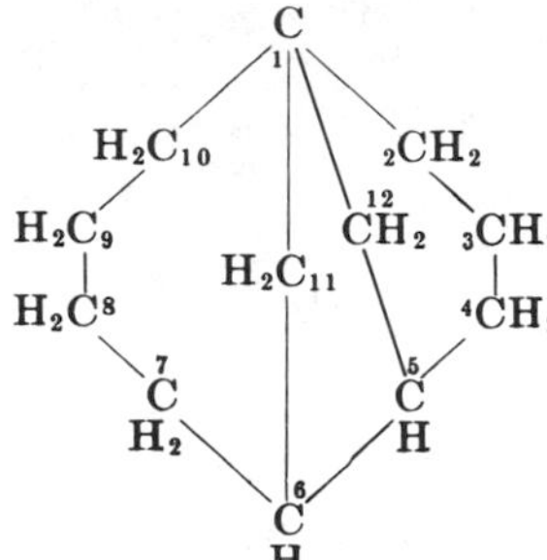

Tricyclo[4.4.1.1^{1,5}]dodecane:
Correct numbering

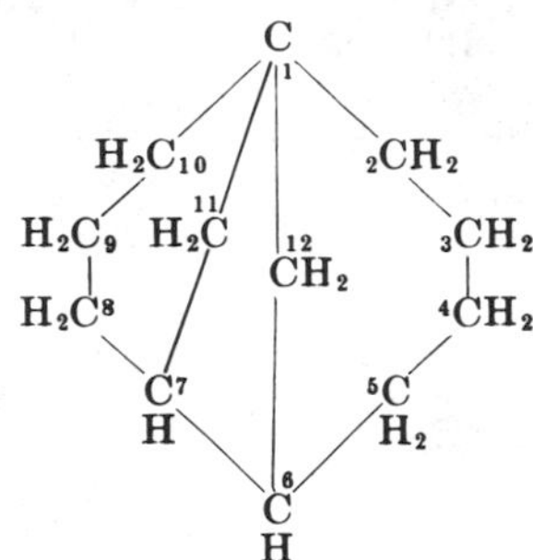

Tricyclo[5.3.1.1^{1,6}]dodecane:
Incorrect numbering

(d) The superscripts locating the other bridges shall be as small as possible (in the sense indicated in Rule **A-2.2**).

Tricyclo[5.5.1.0^{3,11}]tridecane
Correct numbering

Tricyclo[5.5.1.0^{5,9}]tridecane
Incorrect numbering

Rule A-34. Hydrocarbon Bridges

34.1—Polycyclic hydrocarbon systems which can be regarded as "*ortho*-fused" or "*ortho*- and *peri*-fused" systems according to Rule **A-21** and which, at the same time, have other bridges*, are first named as "*ortho*-fused" or "*ortho*- and *peri*-fused" systems. The other bridges are then indicated by prefixes derived from the name of the corresponding hydrocarbon by replacing the final "-ane", "-ene", etc., by "-ano", "-eno", etc., and their positions are indicated by the points of attachment in the parent compound. If bridges of different types are present, they are cited in alphabetical order.

Examples of bridge names:

Butano	$-CH_2-CH_2-CH_2-CH_2-$	Etheno	$-CH=CH-$
Benzeno (*o*-, *m*-, *p*-)	$-C_6H_4-$	Methano	$-CH_2-$
Ethano	$-CH_2-CH_2-$	Propano	$-CH_2-CH_2-CH_2-$

Examples

1,4-Dihydro-1,4-methanopentalene

9,10-Dihydro-9,10-(2-buteno)-anthracene

7,14-Dihydro-7,14-ethano-dibenz[*a*,*h*]anthracene

* The term "bridge", when used in connection with an "*ortho*-fused" or "*ortho*- and *peri*-fused" polycyclic system as defined in the note to Rule **A-31.1** also includes "bivalent cyclic systems".

34.2—The parent "*ortho*-fused" or "*ortho*- and *peri*-fused" system is numbered as prescribed in Rule A-22. Where there is a choice, the position numbers of the bridgeheads should be as low as possible. The remaining bridges are then numbered in turn starting each time with the bridge atom next to the bridgehead possessing the highest number.

Example:

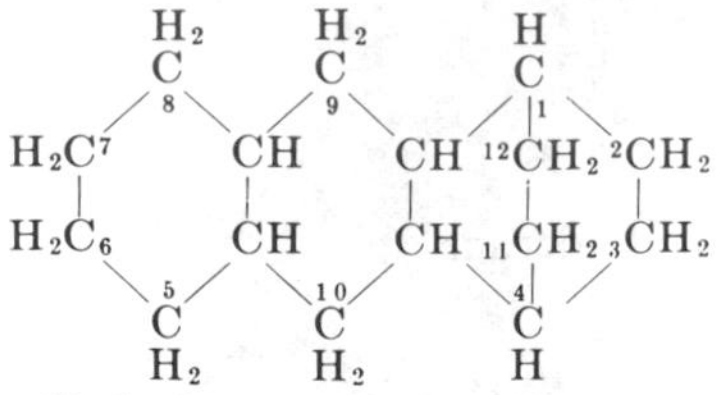

not

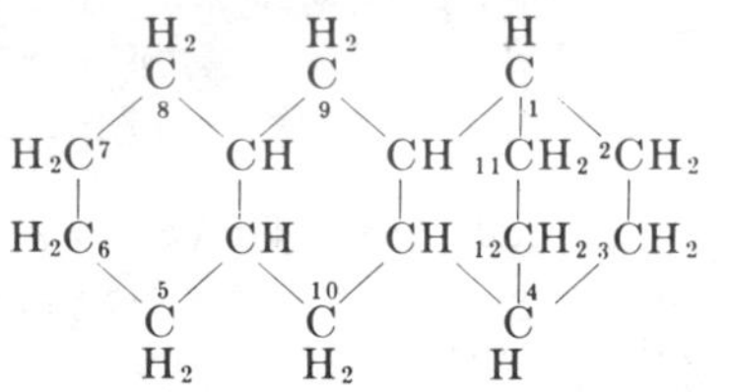

or

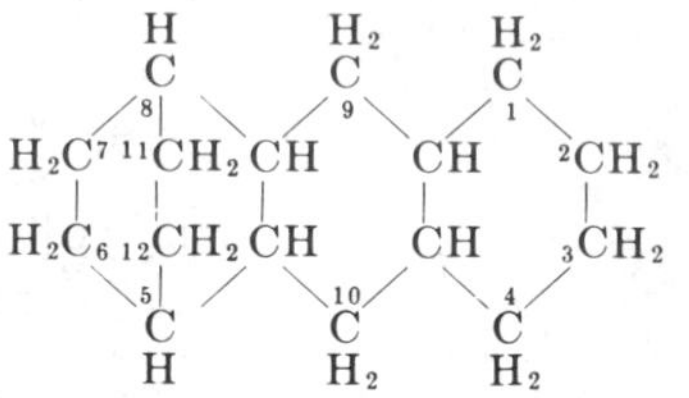

Perhydro-1,4-ethanoanthracene

34.3—When there is a choice of position numbers for the points of attachment for several individual bridges, the lowest numbers are assigned to the bridgeheads in the order of citation of the bridges and the bridge atoms are numbered according to the preceding rule.

Example:

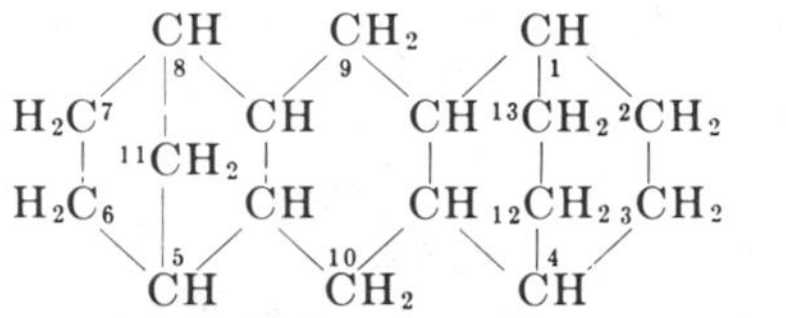

Perhydro-1,4-ethano-5,8-methanoanthracene

34.4—When the bridge is formed from a bivalent cyclic hydrocarbon radical, low numbers are given to the carbon atoms constituting the shorter bridge and numbering proceeds around the ring.

Example:

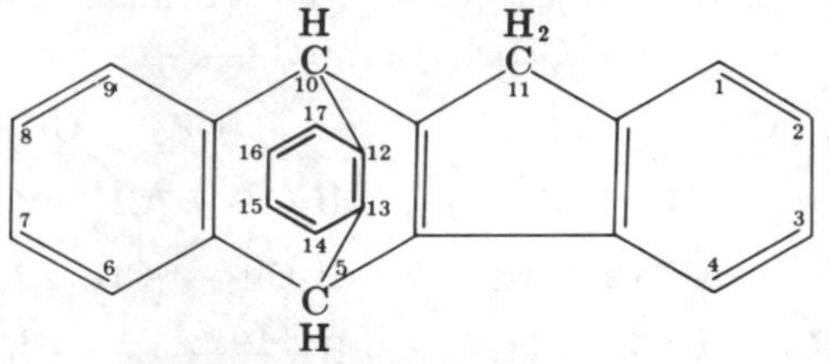

10,11-Dihydro-5,10-*o*-benzeno-5*H*-benzo[*b*]fluorene

34.5—Names for radicals derived from the bridged hydrocarbons considered in Rule A-34.1 are constructed in accordance with the principles set forth in Rule A-24. The abbreviated radical names naphthyl, anthryl, phenanthryl, naphthylene, etc., permitted as exceptions to Rules A-24.2 and A-24.4 are replaced in such cases by the regularly formed names naphthalenyl, anthracenyl, phenanthrenyl, naphthalenediyl, etc.

Examples:

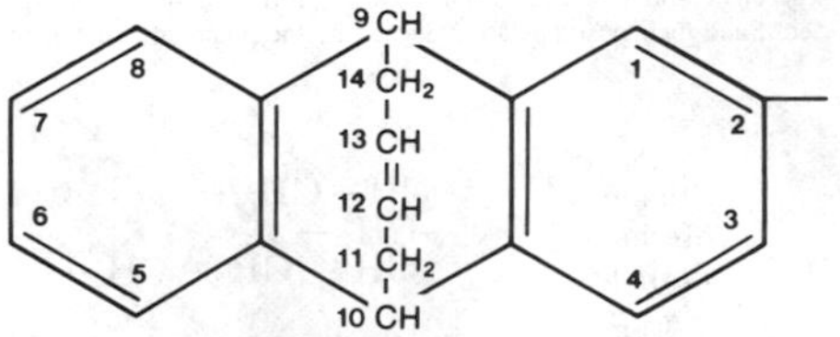

9,10-Dihydro-9,10-[2]butenoanthracen-2-yl

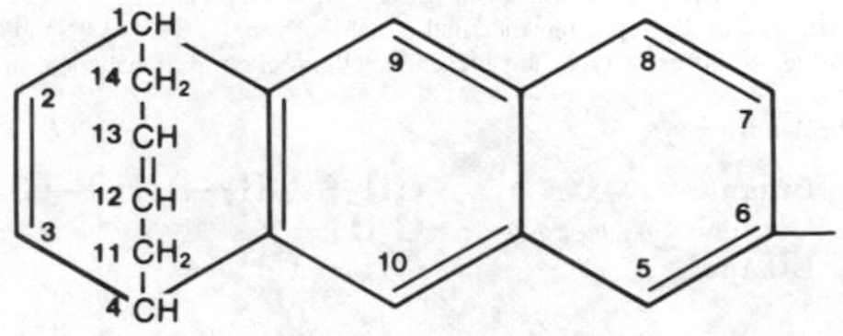

1,4-Dihydro-1,4-[2]butenoanthracen-6-yl

SPIRO HYDROCARBONS

A "spiro union" is one formed by a single atom which is the only common member of two rings. A "free spiro union" is one constituting the only union direct or indirect between two rings.* The common atom is designated as the "spiro atom". According to the number of spiro atoms present, the compounds are distinguished as monospiro-, dispiro-, trispirocompounds, etc. The following rules apply to the naming of compounds containing free spiro unions.

Rule A-41. Compounds: Method 1

41.1 — Monospiro compounds consisting of only two alicyclic rings as components are named by placing "spiro" before the name of the normal acyclic hydrocarbon of the same total number of carbon atoms. The number of carbon atoms linked to the spiro atom in each ring is indicated in ascending order in brackets placed between the spiro prefix and the hydrocarbon name.

* An example of a compound where the spiro union is *not* free is:

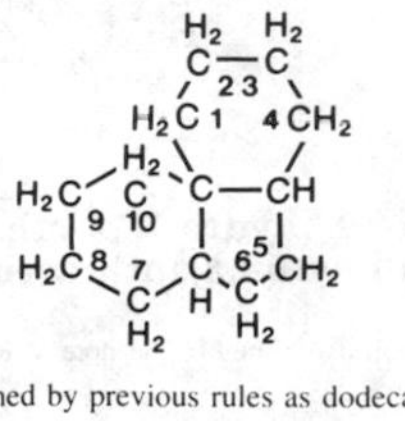

This compound is named by previous rules as dodecahydrobenz[*c*]indene.

Examples:

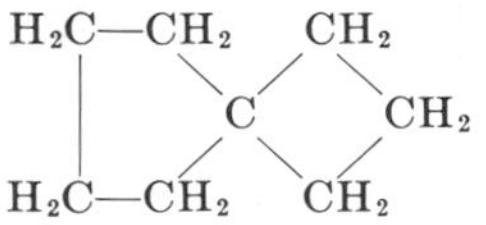

Spiro[3.4]octane

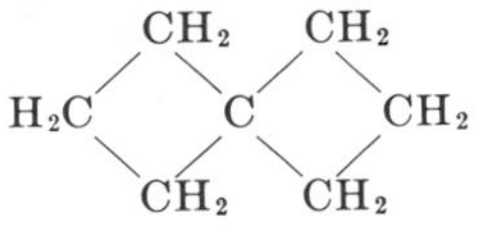

Spiro[3.3]heptane

41.2 — The carbon atoms in monospiro hydrocarbons are numbered consecutively starting with a ring atom next to the spiro atom, first through the smaller ring (if such be present) and then through the spiro atom and around the second ring.

Example:

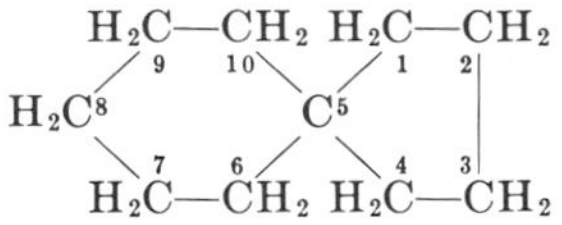

Spiro[4.5]decane

41.3 — When unsaturation is present, the same enumeration pattern is maintained, but in such a direction around the rings that the double and triple bonds receive numbers as low as possible in accordance with Rule **A-11.**

Example:

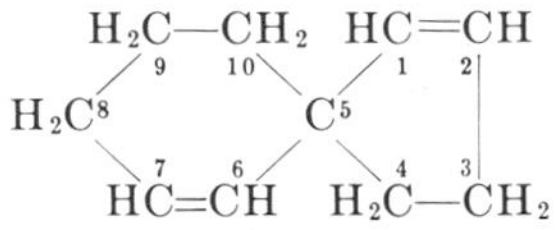

Spiro[4.5]deca-1,6-diene

41.4 — If one or both components of the monospiro compound are fused polycyclic systems, "spiro" is placed before the names of the components arranged in alphabetical order and enclosed in brackets. Established numbering of the individual components is retained. The lowest possible number is given to the spiro atom, and the numbers of the second component are marked with primes. The position of the spiro atom is indicated by placing the appropriate numbers between the names of the two components.

Example:

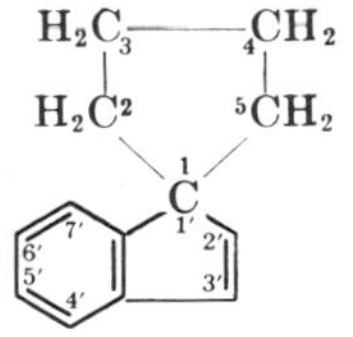

Spiro[cyclopentane-1,1′-indene]

41.5 — Monospiro compounds containing two similar polycyclic components are named by placing the prefix "spirobi" before the name of the component ring system. Established enumeration of the polycyclic system is maintained and the numbers of one component are distinguished by primes. The position of the spiro atom is indicated in the name of the spiro compound by placing the appropriate locants before the name.

Example:

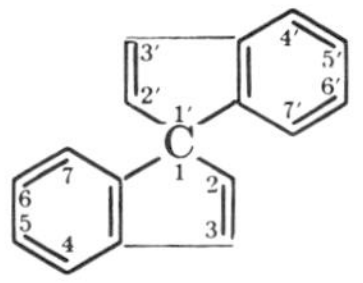

1,1′-Spirobiindene

41.6 — Polyspiro compounds consisting of a linear assembly of three or more alicyclic systems are named by placing "dispiro-", "trispiro-", "tetraspiro-", etc., before the name of the unbranched-chain acyclic hydrocarbon of the same total number of carbon atoms. The numbers of carbon atoms linked to the spiro atoms in each ring are indicated in brackets in the same order as the numbering proceeds about the ring. Numbering starts with a ring atom next to a terminal spiro atom and proceeds in such a way as to give the spiro atoms as low numbers as possible after numbering all the carbon atoms of the first ring linked to the terminal spiro atom.

Example:

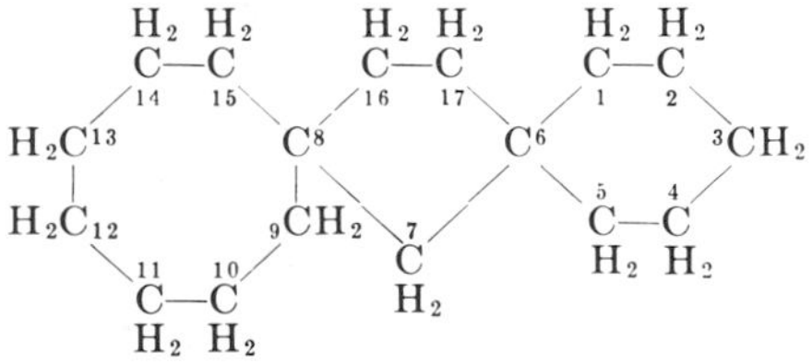

Dispiro[5.1.7.2]heptadecane

41.7 — Polycyclic compounds containing more than one spiro atom and at least one fused polycyclic component are named in accordance with Part **.4** of this rule by replacing "spiro" with "dispiro", "trispiro", etc., and choosing the end components by alphabetical order.

Example:

Dispiro[fluorene-9,1′-cyclohexane-4′,1″-indene]

Rule A-42. Compounds: Method 2

42.1 (Alternate to Rules **A-41.1** and **A-41.2**) — When two dissimilar cyclic components are united by a spiro union, the name of the larger component is followed by the affix "spiro" which, in turn, is followed by the name of the smaller component. Between the affix "spiro" and the name of each component system is inserted the number denoting the spiro position in the appropriate ring system, these numbers being as low as permitted by any fixed enumeration of the component. The components retain their respective enumerations but numerals for the component mentioned second are primed. Numerals 1 may be omitted when a free choice is available for a component.

Examples:

Cyclopentanespiro-cyclobutane

Cyclohexanespirocyclo-pentane

2*H*-Indene-2-spiro-1′-cyclopentane

42.2 (Alternate to **A-41.3**) — Rule **A-41.3** applies also with appropriate different enumeration, where nomenclature is according to Rule **A-42.1** but the spiro junction has priority for lowest numbers over unsaturation.

Example:

2-Cyclohexenespiro-(2′-cyclopentene)

42.3 (Alternate to **A-41.5**) — The nomenclature of Rule **A-41.5** is applied also to monocyclic components with identical saturation, the spiro union being numbered 1.

Example:

Spirobicyclohexane but 2-Cyclohexenespiro-(3′-cyclohexene)

42.4 (Alternate to **A-41.6** and **A-41.7**) — Polycyclic compounds containing more than one spiro atom are named in accordance with Rule **A-42.1** starting from the senior end-component irrespective of whether the components are simple or fused rings.

Examples:

Cycloöctanespirocyclopentane-3′-spirocyclohexane

Fluorene-9-spiro-1′-cyclohexane-4′-spiro-1″-indene

Rule A-43. Radicals

43.1 — Radicals derived from spiro hydrocarbons are named according to the principles set forth in Rules **A-11** and **A-24.**

Examples:

* "Seniority" in respect to spiro compounds is based on the principles: (i) an aggregate is senior to a monocycle; (ii) of aggregates, the senior is that containing the largest number of individual rings; (iii) of aggregates containing the same number of individual rings, the senior is that containing the largest ring; and (iv) if aggregates consist of equal numbers of equal rings the senior is the first occurring in the alphabetical list of names.

Spiro[4.5]deca-1,6-dien-2-yl
(cf. Rules **A-41.3** and **A-11**)
or 2-Cyclohexenespiro-2′-cyclopenten-3′-yl (cf. Rule **A-42.2**)

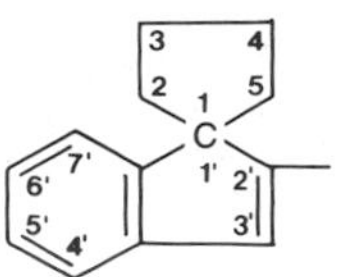

Spiro[cyclopentane-1,1′-inden]-2′-yl
(cf. Rules **A-41.4** and **A-24**)

HYDROCARBON RING ASSEMBLIES

Rule A-51. Definition

51.1—Two or more cyclic systems (single rings or fused systems) which are directly joined to each other by double or single bonds are named "ring assemblies" when the number of such direct ring junctions is one less than the number of cyclic systems involved.

Examples:

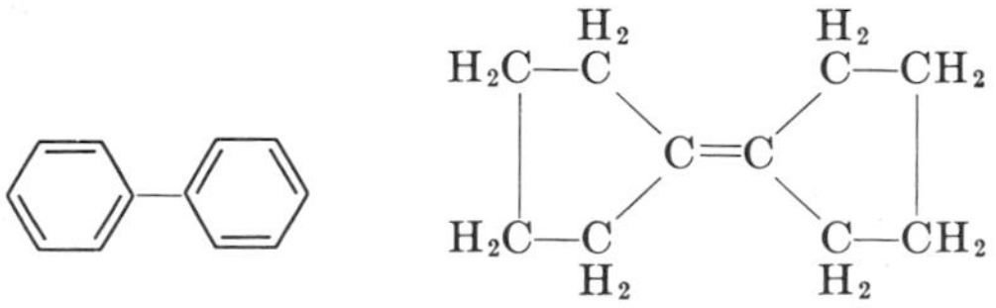

$H_2C—C(H_2)$, $C(H_2)—CH_2$, C=C, $H_2C—C(H_2)$, $C(H_2)—CH_2$

Ring assemblies

Fused polycyclic system

Rule A-52. Two Identical Ring Systems

52.1—Assemblies of two identical cyclic hydrocarbon systems are named in either of two ways: (a) by placing the prefix "bi-" before the name of the corresponding radical, or (b) for systems joined by a single bond by placing the prefix "bi-" before the name of the corresponding hydrocarbon. In each case, the numbering of the assembly is that of the corresponding radical or hydrocarbon, one system being assigned unprimed numbers and the other primed numbers. The points of attachment are indicated by placing the appropriate locants before the name.

Examples:

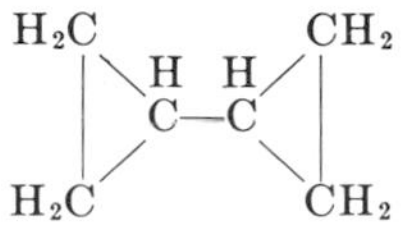

H_2C CH_2 H H C—C H_2C CH_2

1,1′-Bicyclopropyl
or 1,1′-Bicyclopropane

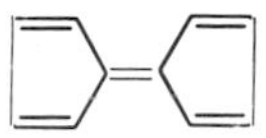

1,1′-Bicyclopentadienylidene
or $\Delta^{1,1'}$-Bicyclopentadienylidene*
(cf. footnote to Rule **B-1.2**)

52.2—If there is a choice in numbering, unprimed numbers are assigned to the system which has the lower-numbered point of attachment.

Example:

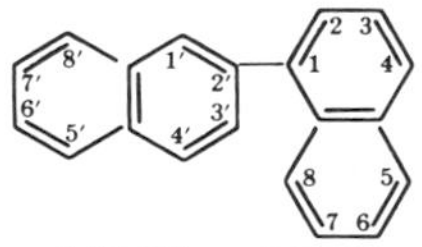

1,2′-Binaphthyl
or 1,2′-Binaphthalene

52.3—If two identical hydrocarbon systems have the same point of attachment and contain substituents at different positions, the locants of these substituents are assigned according to Rule **A-2.2**; for this purpose an unprimed number is considered lower than the same number when primed. Assemblies of primed and unprimed numbers are arranged in ascending numerical order.

Examples:

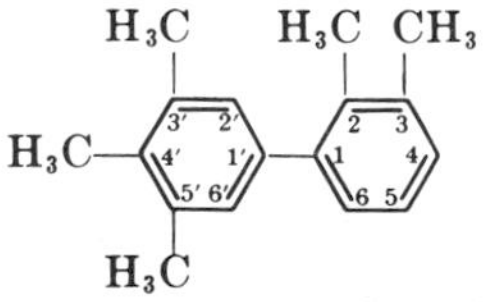

H_3C H_3C CH_3 H_3C H_3C

2,3,3′,4′,5′-
Pentamethylbiphenyl
(not 2′,3,3′,4,5-
Pentamethylbiphenyl)

$CH_3—CH_2—CH_2$ $CH_2—CH_3$

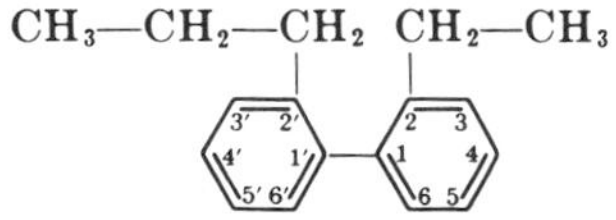

2-Ethyl-2′-
propylbiphenyl

52.4—The name "biphenyl" is used for the assembly consisting of two benzene rings.

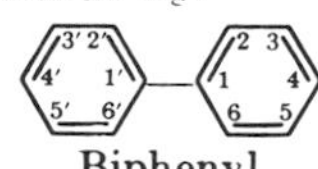

Biphenyl

CYCLIC HYDROCARBONS WITH SIDE CHAINS**

Rule A-61. General Principles

61.1—Hydrocarbons more complex than those envisioned in Rule **A-12**, composed of cyclic nuclei and aliphatic chains, are named according to one of the methods given below. Choice is made so as to provide the name which is the simplest permissible or the most appropriate for the chemical intent.

* A Greek capital delta (Δ) followed by superscript locants is used to denote the double bond.
** Note: cf. Rules **A-12** and **A-13**.

61.2—When there is no generally recognized trivial name for the hydrocarbon, then (1) the radical name denoting the aliphatic chain is prefixed to the name of the cyclic hydrocarbon, or (2) the radical name for the cyclic hydrocarbon is prefixed to the name of the aliphatic compound. Choice between these methods is made according to the more appropriate of the following principles: (a) the maximum number of substitutions into a single unit of structure; (b) treatment of a smaller unit of structure as a substituent into a larger. Numbering of double and triple bonds in chains or nonaromatic rings is assigned according to the principles of Rule **A-3**; numbering and citation of substituents are effected as described in Rule **A-2.**

61.3—In accordance with the principle (a) of Part **.2** of this rule, hydrocarbons containing several chains attached to one cyclic nucleus are generally named as derivatives of the cyclic compound; and compounds containing several side chains and/or cyclic radicals attached to one chain are named as derivatives of the acyclic compound.

Examples:

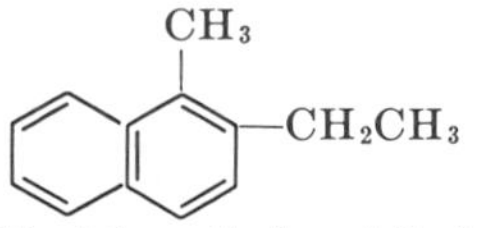

2-Ethyl-1-methylnaphthalene

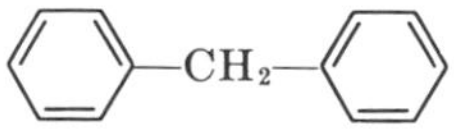

Diphenylmethane

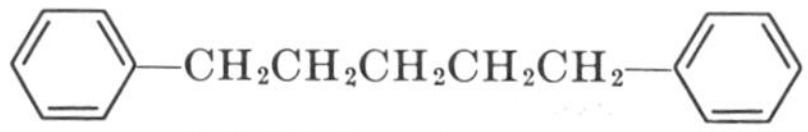

1,5-Diphenylpentane

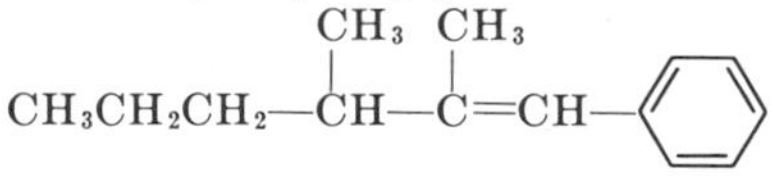

2,3-Dimethyl-1-phenyl-1-hexene

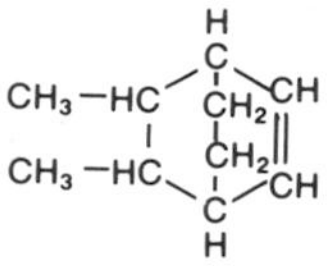

5,6-Dimethylbicyclo[2.2.2] oct-2-ene

TERPENE HYDROCARBONS

Owing to long-established custom, terpenes are given exceptional treatment in these rules.

Rule A-71. Acyclic Terpenes

71.1—The acyclic terpene hydrocarbons are named in a manner similar to that used for other unsaturated acyclic hydrocarbons when compounds with known structures are involved.

Example:

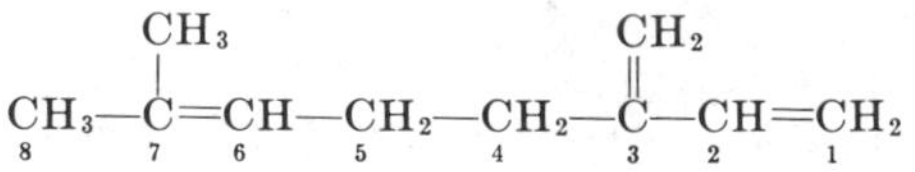

$$CH_3{-}C(CH_3){=}CH{-}CH_2{-}CH_2{-}C({=}CH_2){-}CH{=}CH_2$$

7-Methyl-3-methylene-1,6-octadiene

Rule A-72. Cyclic Terpenes

72.1—The following structural types with their special names and special systems of numbering are used as the basis for the specialized nomenclature of monocyclic and bicyclic terpene hydrocarbons. The name "bornane" replaces camphane and bornylane; "norbornane" replaces norcamphane and norborynlane.*

Fundamental terpene types:

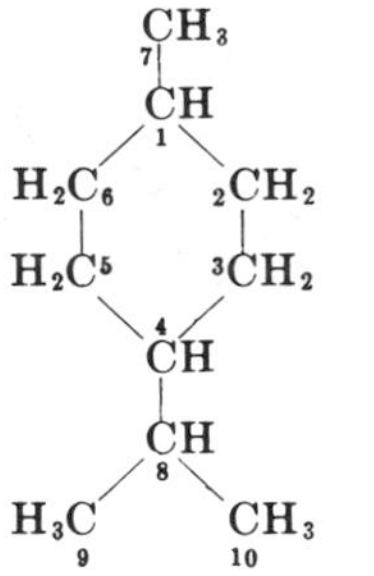

I
Menthane (*p*-form)

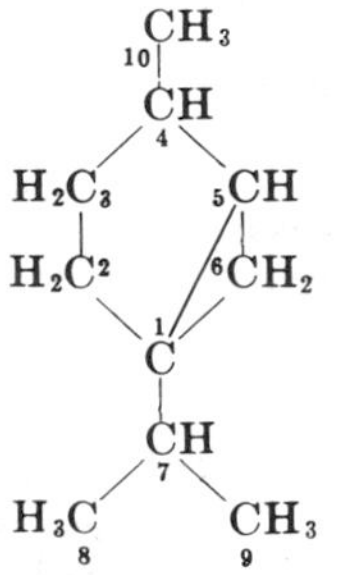

II
Thujane

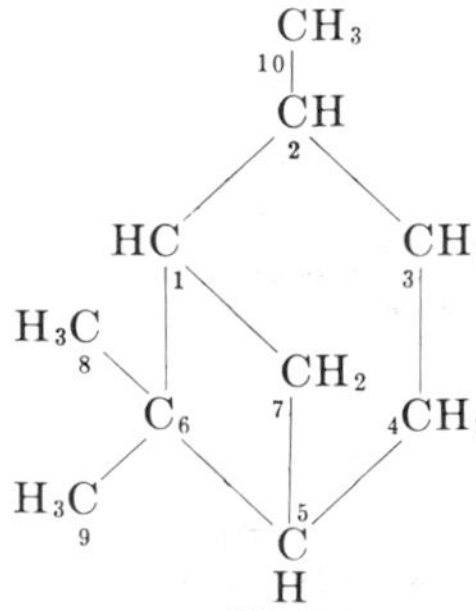

IV
Pinane

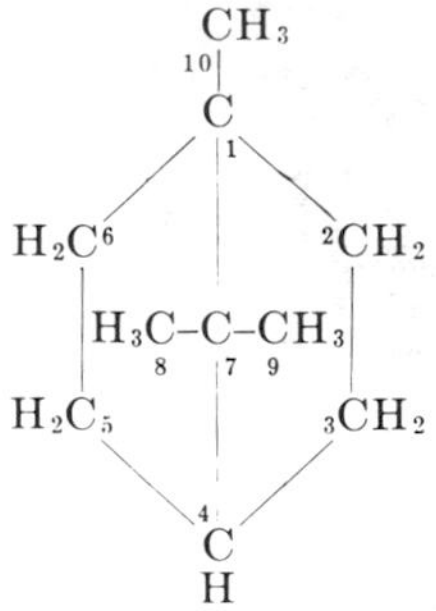

V
Bornane

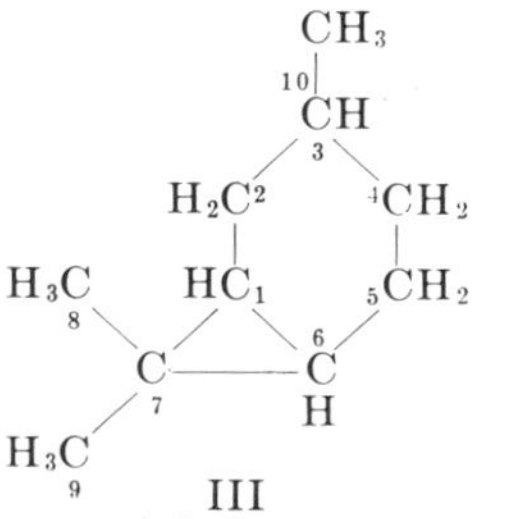

III
Carane

Nor-structures:

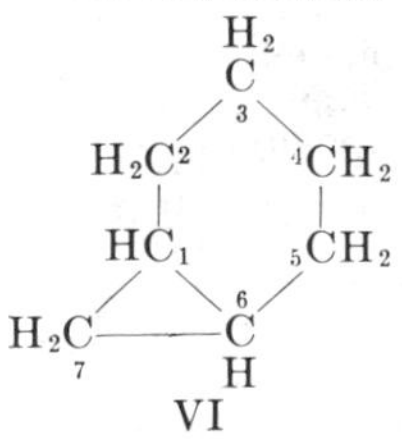

VI
Norcarane

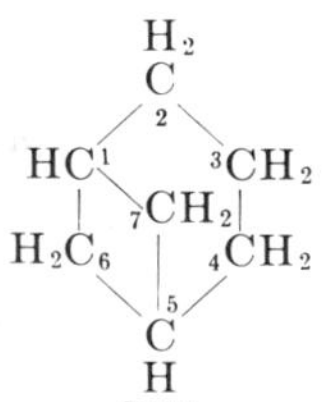

VII
Norpinane

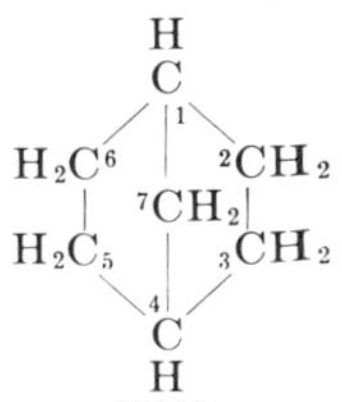

VIII
Norbornane

Rule A-73. Monocyclic Terpenes

73.1—Menthane Type: Monocyclic terpene hydrocarbons of this type (*ortho-*, *meta-*, and *para*-isomers) are named menthane, menthene, menthadiene, etc., and are given the fixed numbering of menthane (Formula I). Such compounds substituted by additional alkyl groups are named in accordance with Rules **A-11** and **A-61**.

* These names have been superseded (cf. Rule **F-4.2**).

Examples:

m-Menthane 1-p-Menthene 1,4(8)-p-Menthadiene

73.2—Tetramethylcyclohexane Type: Monocyclic terpene hydrocarbons of this type are named systematically as derivatives of cyclohexane, cyclohexene, and cyclohexadiene (see Rule **A-11**).

Examples:

1,1,2,3-Tetramethylcyclohexane 1,2,3,3-Tetramethylcyclohexene 1,5,5,6-Tetramethyl-1,3-cyclohexadiene

Rule A-74. Bicyclic Terpenes

74.1—Bicyclic terpene hydrocarbons having the skeleton of Formula II or this skeleton and additional side chains except methyl or isopropyl (or methylene if one methylene group is already present) are named as thujane, thujene, thujadiene, etc., and are given the fixed numbering shown for thujane (Formula II). Other hydrocarbons containing the thujane ring-skeleton are named from bicyclo[3.1.0]hexane and are given systematic bicyclo numbering (cf. Rule **A-31**).

Examples:

4(10)-Thujene **1-Isopropyl-2,4-dimethylenebicyclo-[3.1.0]hexane** **5-Isopropylbicyclo[3.1.0]hex-2-ene**

74.2—Bicyclic terpene hydrocarbons having the skeleton of Formula III, IV, or V and additional side chains except methyl (or methylene if one methylene group is already present) are named, respectively, as carane, carene, caradiene, etc.; pinane, pinene, pinadiene, etc.; bornane, bornene, bornadiene, etc. They are given, respectively, the fixed numbering shown for carane (Formula III), pinane (Formula IV), and bornane (Formula V). Other hydrocarbons containing the ring-skeleton of carane, pinane, orbornane are named, respectively, from norcarane (Formula VI), norpinane (Formula VII), or norbornane (Formula VIII). These names are preferred to those from bicyclo[4.1.0]heptane, bicyclo[3.1.1]heptane, or bicyclo[2.2.1]heptane. The nor-names* are given systematic bicyclo numbering (cf. Rule **A-31**).

Examples:

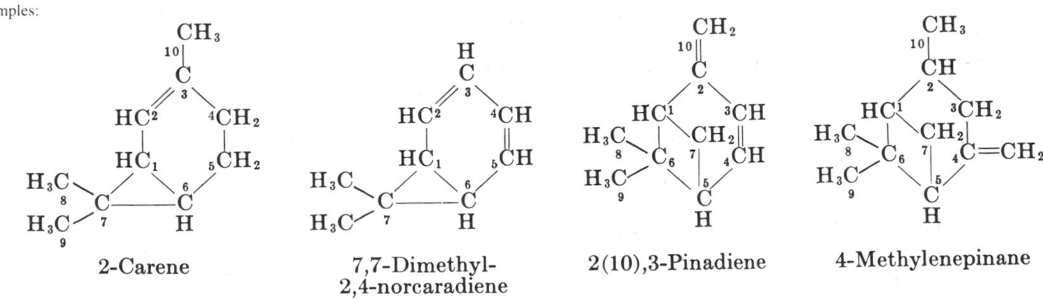

2-Carene 7,7-Dimethyl-2,4-norcaradiene 2(10),3-Pinadiene 4-Methylenepinane

* These names have been superseded (cf. Rule **F-4.1**).

ILLUSTRATIVE PREFIXES

acetamido (acetylamino)	CH_3CONH-
acetimido (acetylimino)	$CH_3C(=NH)-$
acetoacetamido	CH_3COCH_2CONH-
acetoacetyl	CH_3COCH_2CO-
acetonyl	CH_3COCH_2
acetonylidene	$CH_3COCH=$
acetyl	CH_3CO-
acrylyl	$CH_2=CHCO-$
adipyl (from adipic acid)	$-OC(CH_2)_4CO-$
alanyl (from alanine)	$CH_3CH(NH_2)CO-$
β-alanyl	$H\ N(CH_2)_2CO-$
allophanoyl	$H_2NCONHCO-$
allyl (2-propenyl)	$CH_2=CHCH_2-$
allylidene (2-propenylidene)	$CH_2=CHCH=$
amidino (aminoiminomethyl)	$H_2NC(=NH)-$
amino	H_2N-
amyl (pentyl)	$CH_3(CH_2)_4-$
anilino (phenylamino)	C_6H_5NH-
anisidino	$CH_3OC_6H_4NH-$
anisyl (from anisic acid)	$CH_3OC_6H_4CO-$
anthranoyl (2-aminobenzoyl)	$2-H_2NC_6H_4CO-$
arsino	AsH_2-
azelaoyl (from azelaic acid)	$-OC(CH_2)_7CO-$
azido	N_3-
azino	$=NN=$
azo	$-N=N-$
azoxy	$-N(O)N-$
benzal	$C_6H_5CH=$
benzamido (benzylamino)	C_6H_5CONH-
benzhydryl (diphenylmethyl)	$(C_6H_5)_2CH-$
benzimido (benzylimino)	$C_6H_5C(=NH)-$
benzoxy (benzoyloxy)	C_6H_5COO-
benzoyl	C_6H_5CO-
benzyl	$C_6H_5CH_2-$
benzylidine	$C_6H_5CH=$
benzyldyne	$C_6H_5C\equiv$
biphenylyl	$C_6H_5C_6H_5-$
biphenylene	$-C_6H_4C_6H_4-$
butoxy	C_4H_9O-
sec-butoxy	$C_2H_5CH(CH_3)O-$
tert-butoxy	$(CH_3)_3CO-$
butyl	$CH_3(CH_2)_3-$
iso-butyl (3-methylpropyl)	$(CH_3)_2(CH_2)_2-$
sec-butyl (1-methylpropyl)	$C_2H_5CH(CH_3)-$
tert-butyl (1,1, dimethylethyl)	$(CH_3)_3C-$
butyryl	C_3H_7CO-
caproyl (from caproic acid)	$CH_3(CH_2)_4CO-$
capryl (from capric acid)	$CH_3(CH_2)_8CO-$
caprylyl (from caprylic acid)	$CH_3(CH_2)_6CO-$
carbamido	$H_2NCONH-$
carbamoyl (aminocarbonyl)	H_2NCO-
carbamyl (aminocarbonyl)	H_2NCO-
carbazoyl (hydrazinocarbonyl)	$H_2NNHCO-$
carbethoxy	$C_2H_5O_2C-$
carbobenzoxy	$C_6H_5CH_2O_2C-$
carbonyl	$-C=O-$
carboxy	$HOOC-$

cetyl	$CH_3(CH_2)_{15}-$
chloroformyl (chlorocarbonyl)	$ClCO-$
cinnamyl (3-phenyl-2-propenyl)	$C_6H_5CH=CHCH_2-$
cinnamoyl	$C_6H_5CH=CHCO-$
cinnamylidene	$C_6H_5CH=CHCH=$
cresyl (hydroxymethylphenyl)	$HO(CH_3)C_6H_4-$
crotoxyl	$CH_3CH=CHCO-$
crotyl (2-butenyl)	$CH_3CH=CHCH_2$
cyanamido (cyanoamino)	$NCNH-$
cyanato	$NCO-$
cyano	$NC-$
decanedioyl	$-OC(CH_2)_8CO-$
decanoly	$CH_3(CH_2)_8CO-$
diazo	$N_2=$
diazoamino	$-NHN=N-$
disilanyl	H_3SiSiH_2-
disiloxanoxy	$H_3SiOSiH_2O-$
disulfinyl	$-S(O)S(O)-$
dithio	$-SS-$
enanthyl	$CH_3(CH_2)_5CO-$
epoxy	$-O-$
ethenyl (vinyl)	$CH_2=CH-$
ethinyl	$HC\equiv C-$
ethoxy	C_2H_5O-
ethyl	CH_3CH_2-
ethylthio	C_2H_5S-
formamido (formylamino)	$HCONH-$
formyl	$HCO-$
fumaroyl (from fumaric acid)	$-OCCH=CHCO-$
furfuryl (2-furanylmethyl)	$OC_4H_3CH_2-$
furfurylidene (2-furanylmethylene)	$OC_4H_3CH=$
furyl (furanyl)	OC_4H_3-
glutamyl (from glutamic acid)	$-OC(CH_2)_2CH(NH_2)CO-$
glutaryl (from glutaric acid)	$-OC(CH_2)CO-$
glycidyl (oxiranylmethyl)	CH_2-CHCH_2-
glycinamido	H_2NCH_2CONH-
glycolyl (hydroxyacetyl)	$HOCH_2CO-$
glycyl (aminoacetyl)	H_2NCH_2CO-
glyoxylyl (oxoacetyl)	$HCOCO-$
guanidino	$H_2NC(=NH)NH-$
guanyl	$H_2NC(=NH)-$
heptadecanoyl	$CH_3(CH_2)_{15}CO-$
heptanamido	$CH_3(CH_2)_{15}CONH-$
heptanedioyl	$-OC(CH_2)_5CO-$
heptanoyl	$CH_3(CH_2)_5CO-$
hexadecanoyl	$CH_3(CH_2)_4CO-$
hexamethylene	$-(CH_2)_6-$
hexanedioyl	$-OC(CH_2)_4CO-$
hippuryl (N-benzoylglycyl)	$C_6H_5CONHCH_2CO-$
hydantoyl	$H_2NCONHCH_2CO-$
hydrazino	N_2NNH-
hydrazo	$-HNNH-$
hydrocinnamoyl	$C_6H_5(CH_2)_2CO-$

hydroperoxy	$HOO-$
hydroxamino	$HONH-$
hydroxy	$HO-$
imino	$HN=$
iodoso	$OI-$
isoamyl (isopentyl)	$(CH_3)_2CH(CH_2)_2-$
isobutenyl (2-methyl-1-propenyl)	$(CH_3)_2C=CH-$
isobutoxy	$(CH_3)_2CHCH\ O-$
isobutyl	$(CH_3)_2CHCH_2-$
isobutylidene	$(CH_3)_2CHCH=$
isobutyryl	$(CH_3)_2CHCO-$
isocyanato	$OCN-$
isocyano	$CN-$
isohexyl	$(CH_3)_2CH(CH_2)_3-$
isoleucyl (from isoleucine)	$C_2H_3CH(CH_3)CH(NH_2)CO-$
isonitroso	$HON=$
isopentyl	$(CH_3)_2CH(CH_2)_2-$
isopentylidene	$(CH_3)_2CHCH_2CH=$
isopropenyl	$H_2C=C(CH_3)-$
isopropoxy	$(CH_3)_2CHO-$
isopropyl	$(CH_3)_2CH-$
isopropylidene	$(CH_3)_2C=$
isothiocyanato (isothiocyano)	$SCN-$
isovaleryl (from isovaleric acid)	$(CH_3)_2CHCH_2CO-$
keto (oxo)	$O=$
lactyl (from lactic acid)	$CH_3CH(OH)CO-$
lauroyl (from lauric acid)	$CH_3(CH_2)_{10}CO-$
leucyl (from leucine)	$(CH_3)_2CHCH_2CH(NH_2)CO-$
levulinyl (From levulinic acid)	$CH_3CO(CH_2)_2CO-$
malonyl (from malonic acid)	$-OCCH_2CO-$
mandelyl (from mandelic acid)	$C_6H_5CH(OH)CO-$
mercapto	$HS-$
methacrylyl (from methacrylic acid)	$CH_2=C(CH_3)CO-$
methallyl	$CH_2=C(CH_3)CH_2-$
methionyl (from methionine)	$CH_3SCH_2CH_2CH(NH_2)CO-$
methoxy	CH_3O-
methyl	H_3C-
methylene	$H_2C=$
methylenedioxy	$-OCH_2O-$
methylenedisulfonyl	$-O_2SCH_2SO_2-$
methylol	$HOCH_2-$
methylthio	CH_3S-
myristyl (from myristic acid)	$CH_3(CH_2)_{12}CO-$
naphthal	$(C_{10}H_7)CH=$
naphthobenzyl	$(C_{10}H_7)CH_2-$
naphthoxy	$(C_{10}H_7)O-$
naphthyl	$(C_{10}H_7)-$
naphthylidene	$(C_{10}H_6)=$
neopentyl	$(CH_3)_3CCH_2-$
nitramino	O_2NNH-
nitro	O_2N-
nitrosamino	$ONNH-$
nitrosimino	$ONN=$
nitroso	$ON-$
nonanoyl (from nonanoic acid)	$CH_3(CH_2)_7CO-$
oleyl (from oleic acid)	$CH_3(CH_2)_7CH=CH(CH_2)_7CO-$
oxalyl (from oxalic acid)	$-OCCO-$
oxamido	$H_2NCOCONH-$
oxo (keto)	$O=$
palmityl (from palmitic acid)	$CH_3CH_2)_{14}CO-$
pelargonyl (from pelargonic acid)	$CH_3(CH_2)_7CO-$
pentamethylene	$-(CH_2)_5-$
pentyl	$CH_3(CH_2)_4-$
phenacyl	$C_6H_5COCH_2-$
phenacylidene	$C_6H_5COCH=$
phenanthryl	$(C_{14}H_9)-$
phenethyl	$C_6H_5CH_2CH_2-$
phenoxy	C_6H_5O-
phenyl	C_6H_5-
phenylene	$-C_6H_4-$
phenylenedioxy	$-OC_6H_4O-$
phosphino	H_2P-
phosphinyl	$H_2P(O)-$
phospho	O_2P-
phosphono	$(HO)_2P(O)-$
phthalyl (from phthalic acid)	$1,2-C_6H_4(CO-)_2$
picryl (2,4,6-trinitrophenyl)	$2,4,6-(NO_2)_3C_6H_2-$
pimelyl (from pimelic acid)	$-OC(CH_2)_5CO-$
piperidino	$C_5H_{10}N-$
piperidyl (piperidinyl)	$(C_5H_{10}N)-$
piperonyl	$3,4-(CH_2O_2)C_6H_3CH_2-$
pivalyl (from pivalic acid)	$(CH_3)_3CCO-$
prenyl (3-methyl-2-butenyl)	$(CH_3)_2C=CHCH_2-$
propargyl (2-propynyl)	$HC\equiv CCH_2-$
propenyl	$CH_2=CHCH_2-$
iso-propenyl	$(CH_3)_2C=$
propionyl	CH_3CH_2CO-
propoxy	$CH_3CH_2CH_2O-$
propyl	$CH_3CH_2CH_2-$
iso-propyl	$(CH_3)_2CH-$
propylidene	$CH_3CH_2CH=$
pyridino	C_5H_5N-
pyridyl (pyridinyl)	$(C_5H_4N)-$
pyrryl (pyrrolyl)	$(C_3H_4N)-$
salicyl (2-hydroxybenzoyl)	$2-HOC_6H_4CO-$
selenyl	$HSe-$
seryl (from serine)	$HOCH_2CH(NH_2)CO-$
siloxy	H_3SiO-
silyl	H_3Si-
silylene	$H_2Si=$
sorbyl (from sorbic acid)	$CH_3CH=CHCH=CHCO-$
stearyl (from stearic acid)	$CH_3(CH_2)_{16}CO-$
styryl	$C_6H_5CH=CH-$
suberyl (from suberic acid)	$-OC(CH_2)_6CO-$
succinamyl	$H_2NCOCH_2CH_2CO-$
succinyl (from succinic acid)	$-OCCH_2CH_2CO-$
sulfamino	$HOSO_2NH-$
sulfamyl	H_2NSO-
sulfanilyl	$4-H_2NC_6H_4SO_2-$
sulfeno	$HOS-$
sulfhydryl (mercapto)	$HS-$
sulfinyl	$OS=$
sulfo	HO_3S-
sulfonyl	$-SO_2-$
terephthalyl	$1,4-C_6H_4(CO-)_2$
tetramethylene	$-(CH_2)_4-$
thenyl	$(C_4H_3S)CH-$
thienyl	$(C_4H_3S)-$
thiobenzoyl	C_6H_5CS-
thiocarbamyl	H_2NCS-
thiocarbonyl	$-CS-$
thiocarboxy	$HOSC-$
thiocyanato	$NCS-$
thionyl (sulfinyl)	$-SO-$
thiophenacyl	$C_6H_5CSCH_2-$
thiuram (aminothioxomethyl)	H_2NCS-
threonyl (from threonine)	$CH_3CH(OH)CH(NH_2)CO-$
toluidino	$CH_3C_6H_4NH-$
toluyl	$CH_3C_6H_4CO-$
tolyl (methylphenyl)	$CH_3C_6H_4-$
α-tolyl	$C_6H_5CH_2-$
tolylene (methylphenylene)	$(CH_3C_6H_3)=$
α-tolylene	$C_6H_5CH=$
tosyl [(4-methylphenyl) sulfonyl)]	$4-CH_3C_6H_4SO_2-$
triazano	$H_2NNHNH-$
trimethylene	$-(CH_2)_3-$
triphenylmethyl (trityl)	$(C_6H_5)_3C-$
tyrosyl (from tyrosine)	$4-HOC_6H_4CH_2CH(NH_2)CO-$
ureido	$H_2NCONH-$
valeryl (from valeric acid)	C_4H_9CO
valyl (from valine)	$(CH_3)_2CHCH(NH_2)CO-$
vinyl	$CH_2=CH-$
vinylidene	$CH_2=C=$
xenyl (biphenylyl)	$C_6H_5C_6H_4-$
xylidino	$(CH_3)_2C_6H_3NH-$
xylyl (dimethylphenyl)	$(CH_3)_2C_6H_3-$
xylylene	$-CH_2C_6H_4CH_2-$

ORGANIC RING COMPOUNDS

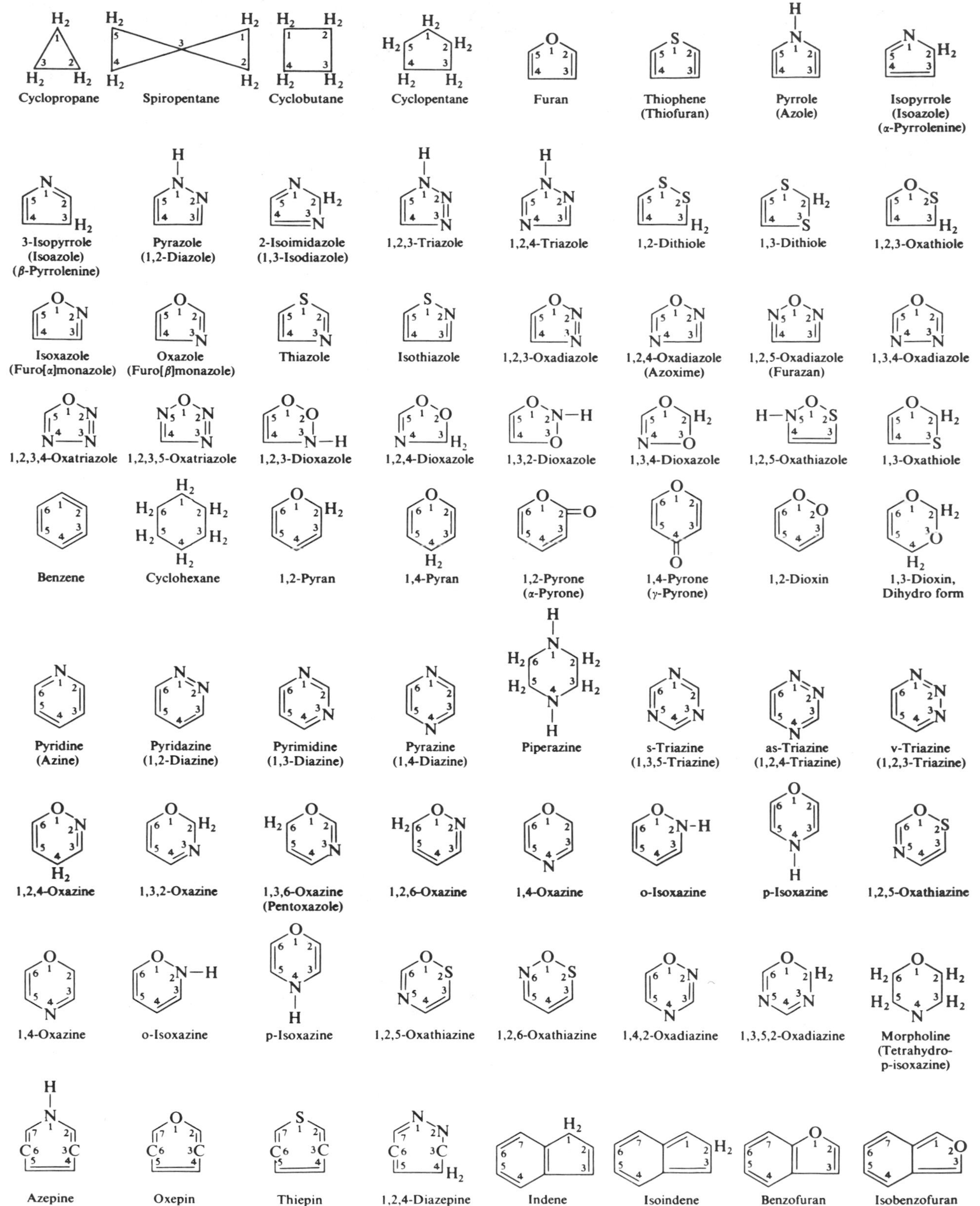
Cyclopropane
Spiropentane
Cyclobutane
Cyclopentane
Furan
Thiophene (Thiofuran)
Pyrrole (Azole)
Isopyrrole (Isoazole) (α-Pyrrolenine)
3-Isopyrrole (Isoazole) (β-Pyrrolenine)
Pyrazole (1,2-Diazole)
2-Isoimidazole (1,3-Isodiazole)
1,2,3-Triazole
1,2,4-Triazole
1,2-Dithiole
1,3-Dithiole
1,2,3-Oxathiole
Isoxazole (Furo[α]monazole)
Oxazole (Furo[β]monazole)
Thiazole
Isothiazole
1,2,3-Oxadiazole
1,2,4-Oxadiazole (Azoxime)
1,2,5-Oxadiazole (Furazan)
1,3,4-Oxadiazole
1,2,3,4-Oxatriazole
1,2,3,5-Oxatriazole
1,2,3-Dioxazole
1,2,4-Dioxazole
1,3,2-Dioxazole
1,3,4-Dioxazole
1,2,5-Oxathiazole
1,3-Oxathiole
Benzene
Cyclohexane
1,2-Pyran
1,4-Pyran
1,2-Pyrone (α-Pyrone)
1,4-Pyrone (γ-Pyrone)
1,2-Dioxin
1,3-Dioxin, Dihydro form
Pyridine (Azine)
Pyridazine (1,2-Diazine)
Pyrimidine (1,3-Diazine)
Pyrazine (1,4-Diazine)
Piperazine
s-Triazine (1,3,5-Triazine)
as-Triazine (1,2,4-Triazine)
v-Triazine (1,2,3-Triazine)
1,2,4-Oxazine
1,3,2-Oxazine
1,3,6-Oxazine (Pentoxazole)
1,2,6-Oxazine
1,4-Oxazine
o-Isoxazine
p-Isoxazine
1,2,5-Oxathiazine
1,4-Oxazine
o-Isoxazine
p-Isoxazine
1,2,5-Oxathiazine
1,2,6-Oxathiazine
1,4,2-Oxadiazine
1,3,5,2-Oxadiazine
Morpholine (Tetrahydro-p-isoxazine)
Azepine
Oxepin
Thiepin
1,2,4-Diazepine
Indene
Isoindene
Benzofuran (Coumarone)
Isobenzofuran

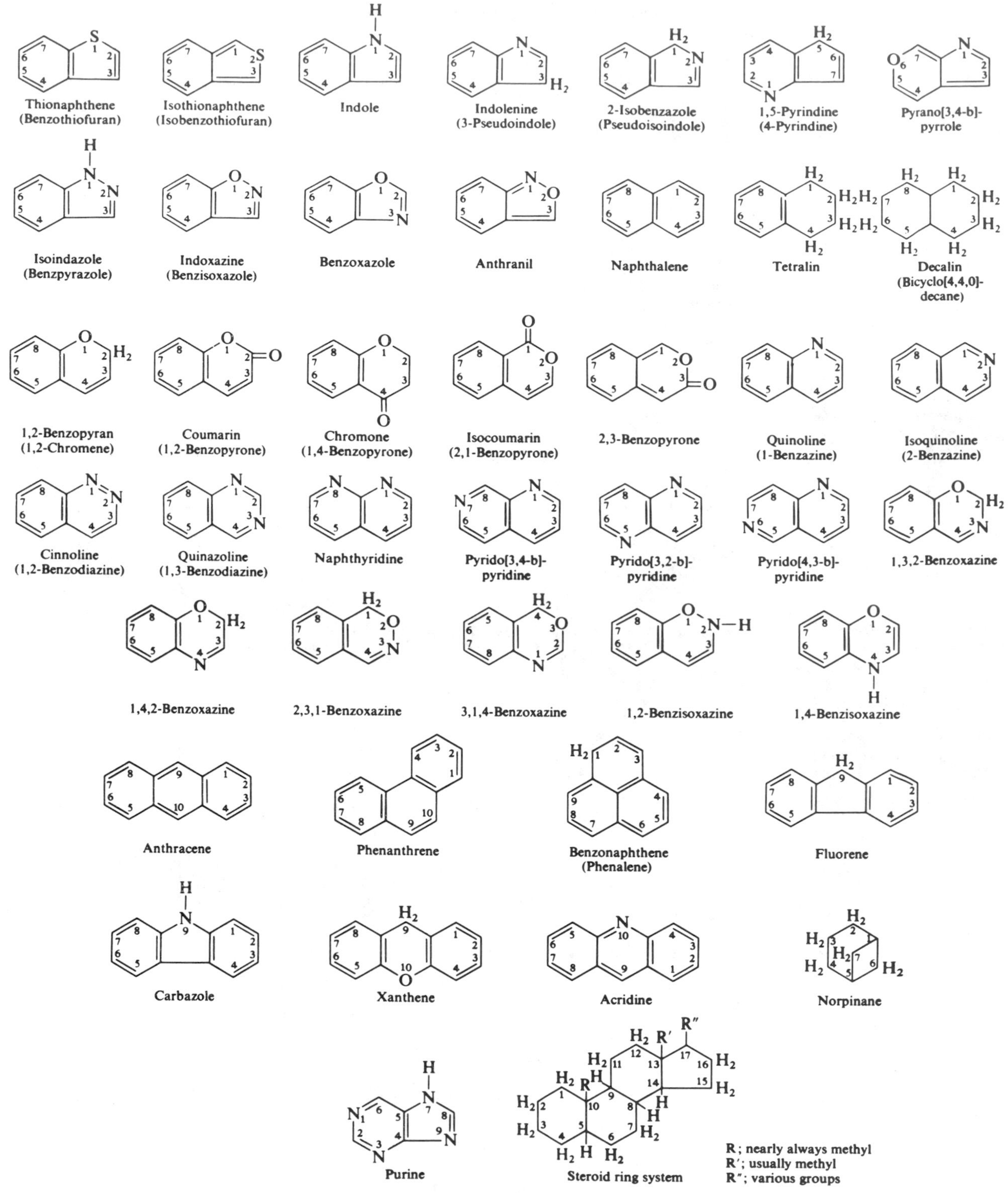

A more extensive listing of ring compounds and systems may be found in the following:

Chemical Abstracts, annual subject index.

The Ring Index, Patterson and Capell, Reinhold Publishing Company, 1940.

Lexikon der Kohlenstoffverbindungen, Richter, Leopold Voss, 1910.

The numbering system for the compounds listed above is that used in The Ring Index.

EXPLANATION OF TABLE PHYSICAL CONSTANTS OF ORGANIC COMPOUNDS

This table is a compilation of data on some 15,000 compounds of interest to chemists working in industrial or academic research as well as those in other areas having an occasional need for physical constant data.

An effort has been made to use names which are in most common usage with cross-reference to I.U.C. derived names. As far as possible, compounds are listed alphabetically, but derivatives are listed under the parent compound heading. Thus esters are listed under the name of the parent acid.

Frequent use is made of synonyms. As an example $C_3H_7CO_2H$ is listed under both butyric acid and butanoic acid but the derivatives, for the most part, are listed under butyric acid. As a general rule, if a derivative of a compound is not listed under the name in the first column, look for it to be listed as a derivative of the synonym.

Each compound has been given a number which is used in the formula, melting point, and boiling point indexes found at the end of the table.

References to original literature or other Compendia are given whenever possible (see Table of Abbreviations). The most important of these are, B (Berichte der Deutsche Chemische Gesselschaft), and A (Atlas for Spectral Data and Physical Constants).

Structural Formulas are given whenever possible. More complex formulas are given at the end of the Table.

Molecular Weights are obtained from recently published values of Atomic Weights.

The color of crystalline compounds is listed by abbreviations (see Table of Abbreviations) associated with the crystalline form. Solid compounds are considered as crystalline if not otherwise stated. Data on solvents of crystallization are given in brackets. (see Table of Abbreviations)

The following are specific examples:

pl	**Plates**
pl (al)	Plates obtained from alcohol as solvent
pl (al + 1½)	Plates obtained from alcohol with 1½ mol of alcohol of crystallization
pl (aq ace + 2w)	Solvent of crystallization is one of components of solvent, e.g. 2 mol of water
(al)	Crystalized from alcohol; crystalline form not reported.

Specific Rotation follows common usage e.g.

$[\alpha]^{25}_{D} = -25.8$ (w,c = 4) w = water; c = 4%

Melting Points are rounded to the nearest 0.1°C. A second, but probably less reliable, melting point is listed in brackets for a few compounds.

Boiling Points are at atmospheric pressure (760 mm) unless otherwise indicated by a superscript.

Density (specific Gravity). In most cases specific gravities are given. Thus $0.86^{20}{}_{4}$ indicates the density of the liquid at 20°C relative to the density of water at 4°C. Where only one temperature is given as a superscript, the value is in grams per milliliter at the indicated temperature.

The **Refractive Index** is reported for the D line of the sodium spectrum (η_D) at the temperature indicated by the superscript. In a few cases other spectral lines are indicated.

Common Solvents for each compound are given as a guide to solubility. No distinction is made between soluble and very soluble primarily because definitions for these terms differ widely among different investigators. Solvents in which the compounds are soluble in all proportions are underlined.

Three indexes are located at the end of the Table. These are:

1. **Empirical Formula Index** which is arranged according to increasing C, H, and remaining elements in alphabetical order. Hydrates are entered under the formula of the anhydrous compound. Salts, complexes, etc. are found under their total formula.
2. **Melting Point Index** which is arranged according to increasing melting point or freezing point. Only the lower temperature of the melting point range is given. Melting points are rounded off to the nearest unit.
3. **Boiling Point Index** which is arranged according to increasing boiling point of compounds of which the boiling point is given at a pressure near to 1 atmosphere, (i.e., 700 to 780 mm Hg). Only the lower temperature of the boiling range is given. No special entries are made for the remarks on the boiling point contained in the main table. Boiling points listed in the Index are rounded off to the nearest unit.

RULES FOR THE NAMING OF COMPOUNDS

1. Compounds are given common names when these are widely used. These compounds are usually cross-indexed to I.U.C. derived names. e.g., $HO_2CCH_2CH_2CO_2H$ is listed as succinic acid and also as butandioic acid. Physical constants are listed for both entries.
2. Derivatives of a parent compound having one principal function are listed under this compound; e.g. Ethyl butyrate is listed under butyric acid. Chloroacetic acid has a separate entry with many derivatives listed under it. However, 2-chlorobutyric acid is listed as a derivative of butyric acid because there are a limited number of chlorobutyric acid derivatives.
3. Whenever a derivative is not found under the name of the parent compound, look for it under the name of the synonym, or as a principal entry. Derivatives of derivatives are not in the table.

Aliphatic Hydrocarbons

4. I.U.C. nomenclature is used.

a) Saturated hydrocarbons are named as alkanes. Branched chain compounds are named as derivatives of the longest chain.

b) Hydrocarbons having carbon-carbon double bonds are named as alkenes, alkadienes, alkatrienes, etc., with the position of the double bonds indicated by suitable numbers. Branched chain compounds are named as derivatives of the longest chain that contains the maximum number of double bonds, e.g.

$CH_3CH{=}CHCH_3$	2-Butene
$CH_2{=}CHCH{=}CH_2$	1.3-Butadiene
$CH_2{=}CHCH(C_3H_7)CH{=}CH_2$	3-Propyl+1.4-pentadiene

c) Hydrocarbons having carbon-carbon triple bonds are named as alkynes. Branched chain alkynes are named as derivatives of the longest chain containing triple bonds. The position of the triple bond is indicated by suitable numbers, e.g.

$CH_3CH_2C{\equiv}CH$ 1-Butyne $CH_3CH_2CH(C{\equiv}CH)CH_2CH_3$ 3-Ethyl-1-pentyne

d) When both double and triple bonds are found in the same molecule, the double bond takes precedence in naming.

$CH_3CH_2CH = CHC \equiv CH$ 3 - Pentene - 1 - yne

5. **Cyclic Hydrocarbons** — Naphthenes — are given I.U.C. derived names.

Cyclohexane

Cyclopentadiene

Cyclohexene

Cyclobutane

6. **Bridged Hydrocarbon** ring systems appear under the headings bicyclo-, tricyclo-. Most such compounds however, are listed under common names.

Bicyclo [3.1.0] hexane

7. **Spiro-, dispiro-,** etc. hydrocarbons and derivatives appear under the spiro-headings.

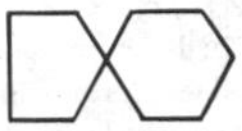

Spiro-[4,5-decane]

8. **Aromatic Hydrocarbons** are listed under their common names, e.g. Benzene (C_6H_6), Toluene ($C_6H_5CH_3$), Xylene [$C_6H_4(CH_3)_2$,] Naphthalene ($C_{10}H_8$) etc. Aromatic hydrocarbons having aliphatic side chains are listed as derivatives, e.g.

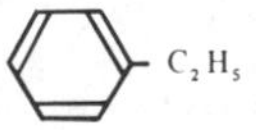

C_2H_5 Benzene, ethyl

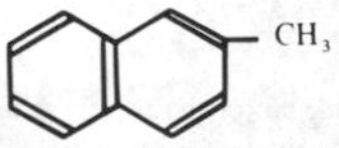

CH_3 Naphthalene, β-methyl

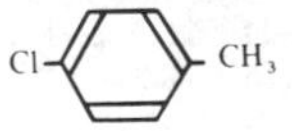

Cl CH_3 Toluene, *p*-chloro

When two or more benzene rings, phenyl groups, are attached to a single carbon atom or to more than one carbon atom of a hydrocarbon molecule, they are named as phenyl derivatives.

$(C_6H_5)_2CH_2$ Diphenyl methane $(C_6H_5)_3CH$ Triphenyl methane

$C_6H_5CH_2CH_2C_6H_5$ 1 2,-Diphenyl ethane

9. **Heterocyclic Systems.** Common names are used or when these are lacking, systematic names using *oxo, azo, thio,* etc. to indicate an oxygen, nitrogen or sulfur atom in the cyclic structure.

The following common names are used; furan, pyran, pyrrole, pyrazole, imidazole, piperidine, pyridine, pyrazine, pyrimidine, pyrazidine, thiophene, etc.

10. **Radicals.** Radical names such as methyl, ethyl, isopropyl, phenyl, tolyl, benzyl, xylyl, allyl, styryl, etc. are commonly used. (See List of Substituent Prefixes)

Compounds Containing Functional Groups

Order of Precedence of Functions. When a compound contains one or more functional groups, the parent compound is designated according to the following order of precedence: acid, aldehyde, ketone, alcohol, amine, ether, sulfide, sulfone, and sulfoxide.

ACIDS AND DERIVATIVES

1. **Common names** have long been used in naming acids and well established names are used in this table, e.g. Acetic acid, Caprioc acid, Benzoic acid, Palmitic acid, Oleic acid, Stearic acid, Succinic acid, Tartaric acid, etc.
2. **Acids having no well-established common names** are names as alkanoic acids, alkenoic acids, alkane dioic acids, alkene dioic acids, etc., e.g.

$CH_3(CH_2)_7CO_2H$ Nonanoic acid $CH_2{=}CHCH{=}CHCO_2H$ 2.4-Pentadienoic acid

$CH_3CH{=}CHCH_2CO_2H$ 3-Pentenoic acid $HO_2CCH_2C(C_4H_9){=}CHCO_2H$ 2-Pentenedioic acid, -3-butyl

$(CH_3)_2CH{=}CHCO_2H$ 2-Butenoic acid,-3-methyl

Use the synonym name to find derivatives not listed under the name of the acid in the first column.

The term *-iso* is used to indicate branching at the end of the hydrocarbon chain. Thus $(CH_3)_2CHCO_2H$ is named Isobutyric acid. Listing of this acid and its derivative is found following Butyric acid — alphabetically under "B".

Polycarboxylic acids use the carboxylic acid suffix added to the longest carbon chain that does not contain these groups, e.g.

$HO_2CCH_2CH(CO_2H)CH_2CO_2H$	1 2,3-Propane tricarboxylic acid

3. **Cyclic Acids.** After the name of the ring, the suffix carboxylic acid is attached.

COOH — 2-Cyclohexene carboxylic acid

COOH, COOH, CH_3 — 1,2-Cyclohexane dicarboxylic acid, -3-methyl

4. **Thioacids** are named as thio-, dithio-, thiolo-, or thiono-acids. The latter two prefixes are used only when it is certain that the hydroxyl or ketonic oxygen is replaced by the sulfur atom, e.g.,

CH_3CSOH	Thioacetic acid	CH_3CH_2COSH	Thiolopropionic acid
$(CH_3)_2CHCS_2H$	Dithioisobutyric acid	$CH_3CH_2C(S)OH$	Thionopropionic acid

5. **Imidic, Hydroxamic Acids.** Only simple compounds of these types appear in the Table. The following are examples

$CH_3CH_2CH(=NH)OH$	Butanimidic acid	$CH_3CH_2C(=NOH)OH$	Propanehydroxamic acid
$C_6H_5C(=NH)OH$	Benzimidic acid	$C_6H_5C(=NOH)OH$	Benzohydroxamic acid
C(=NH)OH (N–H)	2-Pyrrolecarboxyimidic acid	C(=NOH)OH (O)	2-Furancarbohydroxamic acid

6. **Sulfonic and Sulfinic Acids.** Compounds containing the $-SO_3H$ group are named as sulfonic acids. When a carboxyl group is also present this group is designated by the prefix sulfo-.

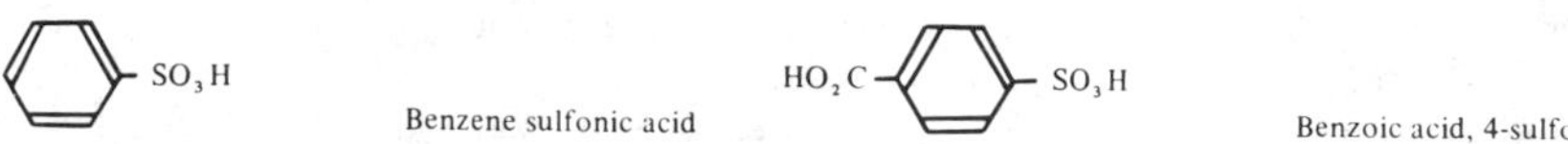

Compounds containing the $-SO_2H$ group are named as sulfinic acids.

7. **Keto Acids.** When the carbonyl function is present in the principal chain, the compounds are designated as oxo- acids.

$CH_3COCH_2CH_2CO_2H$	Valeric acid, 4-oxo or, Pentanoic acid, 4-oxo

Common names are listed when practical, e.g., $CH_3COCH_2CO_2H$ — Acetoacetic acid.

8. **Amino acids.** In most cases common names are used. Otherwise they are named as aminoalkanoic acids, aminoalkenoic acids, etc.

9. **Ortho acids (or esters)** are listed under ortho-, e.g.,

$HC(OCH_3)_3$	Orthoformic acid, trimethyl ester

10. **Acid halides and amides** are listed independently from the acids.

CH_3COCl	Acetyl chloride	$C_6H_5CON(C_2H_5)_2$	Benzamide, -*N,N*-diethyl
C_3H_7COBr	Butyryl bromide or Butanoyl bromide	$C_6H_5SO_2NH_2$	Benzenesulfonamide
C_6H_5COCl	Benzoyl chloride	$C_6H_5SO_2NHCH_3$	Benzenesulfonamide, -*N*-Methyl
$C_2H_5CONH_2$	Propionamide	$CH_3CONHCOCH_3$	Acetamide, -*N*-acetyl
$C_6H_5CONHCH_3$	Benzamide, -*N*-Methyl		

N-substituted amides in which the substituent is phenyl are named as amides. e.g.,

$CH_3CONHC_6H_5$	Acetanilide	$C_2H_5CONHC_6H_5$	Butyranalide

11. **Esters.** Esters are listed under the name of the acid, usually the common name.

$CH_3CO_2C_2H_5$	Acetic acid, ethyl ester	$C_2H_5O_2CCH_2CH_2CO_2C_2H_5$	Succinic acid, diethyl ester
$C_6H_5CO_2C_4H_9$	Benzoic acid, butyl ester	$(CH_3)_2C{=}CHCO_2CH$	2-Butenoic acid, 4-Methyl-methyl ester

Esters of complicated or polyhydric alcohols may appear under the name of the alcohol.

$$\begin{array}{l} CH_2O_2CC_{17}H_{35} \\ | \\ CHO_2CC_{17}H_{35} \\ | \\ CH_2O_2CC_{17}H_{35} \end{array}$$

Glyceryl tristearate or Stearin

12. **Lactams and Lactones** are listed under the names of the corresponding amino- or hydroxy acids.
13. **Nitriles** are listed independently from the corresponding acids and are designated as nitriles, e.g.,

C_3H_7CN	Butyronitrile	$NCCH_2CH_2CN$	succinonitrile
$CH_3CH(NH_2)CN$	Propionitrile,-2-amino	$2\text{-}ClC_6H_4CN$	Benzonitrile,-o-chloro

14. **Imides** are listed independently from the corresponding acids, e.g.,

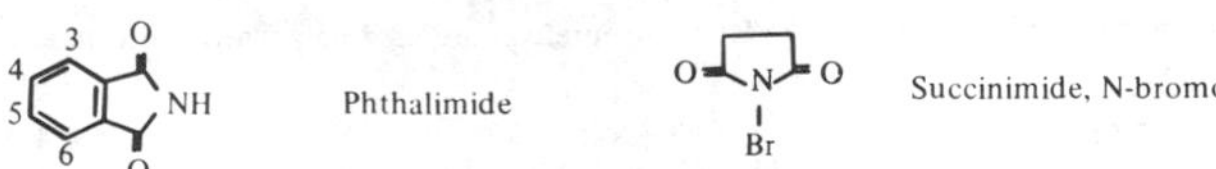

ALDEHYDES

1. Common names are listed for aldehydes. These are derived by dropping the -ic ending of the related acid and adding aldehyde, e.g.,

CH_3CHO	Acetaldehyde	$2\text{-}CH_3C_6H_4CHO$	*o*-Tolualdehyde
$(CH_3)_2CHCHO$	Isobutyraldehyde	$CH_3CH{=}CHCHO$	Crotonaldehyde
C_6H_5CHO	Benzaldehyde	$C_6H_4(CHO)_2$	Phthaldehyde

2. Where common names are not widely used, aldehydes are named as alkanals, alkenals, etc. e.g.,

$CH_3CH_2CH_2CHO$	Butanal	$CH_2{=}CHCH{=}CHCHO$	2,4-Butadienal
$(CH_3)_2C{=}CHCHO$	2-Butenal,-3-methyl	$OHCCH(CH_3)CHCHO$	Butanedial, 2-methyl

3. Compounds having a -CHO group attached to a cyclic structure are frequently named as carboxaldehydes, e.g.,

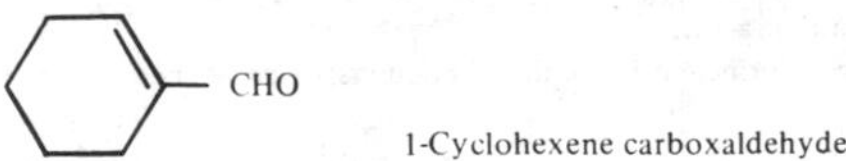

1-Cyclohexene carboxaldehyde

4. The -CHO group is frequently referred to as formyl when a carboxyl group is also present. The −C=O group may also be designated as oxo, e.g.,

OHC–C₆H₄–COOH — Benzoic acid,-4-formyl

$OHCCH_2CH_2COOH$	Butanoic acid,-4-oxo

5. Acetals and hemi-acetals are listed under the corresponding aldehyde or ketone, e.g.,

$CH_3CH(OC_2H_5)_2$	Acetaldehyde, diethyl acetal	$CH_3CH_2CH(OH)OCH_3$	Propionaldehyde, methyl-hemi-acetal
$CH_3CH_2CH(CH_3)CH(OCH_3)_2$	Butanal,-2-methyl, dimethyl acetal		

KETONES

1. Ketones having only aliphatic groups are listed as dialkyl ketones, or as alkanones, alkenones, etc., e.g.,

CH_3COCH_3	Dimethyl ketone or acetone	$CH_2{=}CHCOCH_3$	Methyl vinyl ketone or 3-butene-2-one
$(CH_3)_2CHCOCH_3$	Methyl isopropyl ketone or 2-butanone,-3-methyl		

When the derivative being sought is not found under the name appearing in the first column, look for it under the synonym name or as a principal entry.

2. When one of the groups attached to the carbonyl group (−C=O) is phenyl the compound is named as phenyl ketone or as a phenone, e.g.,

$C_6H_5COCH_3$	Acetophenone or methyl phenyl ketone	$C_6H_5COC_3H_7$	Butyrophenone or propyl phenyl ketone
$C_6H_5COC_6H_5$	Benzophenone or diphenyl ketone		

3. When cyclic structures are attached to the carbonyl group, the ketones may be named as acetyl-, propionyl-, butyryl-, etc. derivatives of the cyclic compound.

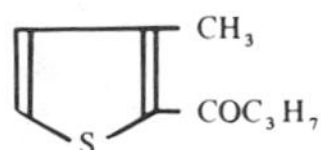

Thiophene, -3-methyl-2-butyryl

4. Thioketones are named as thiones.

$C_6H_5CSCH_3$	Acetothiophenone	$C_6H_5CSC_6H_5$	Dibenzothiophenone

5. Cyclic ketones and thiones are given the name of the parent compound followed by the ending -one or -thione.

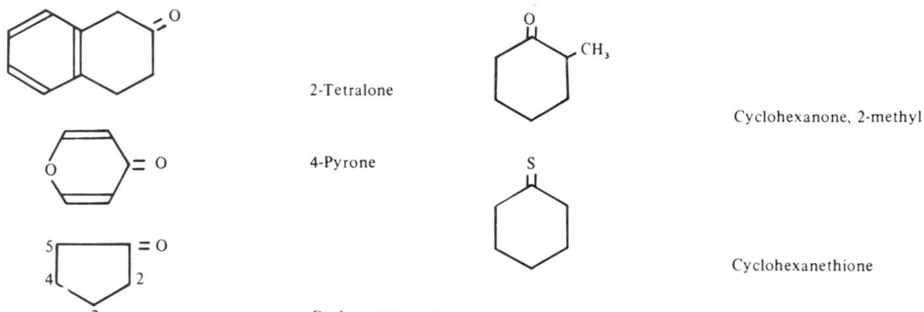

2-Tetralone

Cyclohexanone, 2-methyl

4-Pyrone

Cyclohexanethione

Cyclopentanone

6. Quinones, the name is reserved for structures of the type

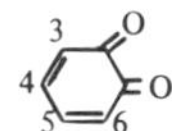

Orthoquinones or 1,2-quinones

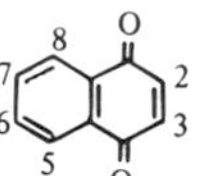

or

5 6 4 1 O O 3 2

Paraquinones or 1,4-quinones

Examples are:

1,4-Benzoquinone

CH_3

1,4-Benzoquinone, -2-methyl

1,2-Naphthoquinone

COOH

1,4-Benzoquinone, -2-carboxylic acid

ALCOHOLS

1. Alcohols are named as alkyl alcohols, or as alkanols, alkenols, alkanediols, etc.

CH_3CH_2OH	Ethyl alcohol ethanol	$(CH_3)_3COH$	Tert-butyl alcohol 2-Propanol, 2-methyl
$(CH_3)_2CHCH_2OH$	Isobutyl alcohol 1-propanol,-2-methyl	OH	Cyclohexanol
$CH_3CH_2CH(OH)CH_3$	Sec-butyl alcohol 2-butanol		
		$CH_3CH(OH)CH(OH)CH_3$	2,3-Butanediol

Phenyl substituted alcohols are named as phenyl-, diphenyl-, alkanols, etc.

$(C_6H_5)_2CHOH$	Methanol,–diphenyl

2. Iso-, sec-, and tert-alcohols and derivatives are listed sequentially after the name of the normal alcohol, e.g. isobutyl, sec-butyl and tert-butyl alcohols are listed as butyl alcohols.

3. Common names are frequently used in naming alcohols, e.g.

CH_2=$CHCH_2OH$	Allyl alcohol	$C_{11}H_{23}CH_2OH$	Lauryl alcohol
$HOCH_2CH_2OH$	Ethylene glycol	$C_{17}H_{35}CH_2OH$	Stearyl alcohol
$HOCH_2CH(OH)CH_2OH$	Glycerol		

4. Thioalcohols are names as thiols or mercaptans

C_2H_5SH	Ethanethiol ethyl mercaptan	$(CH_3)_2CHCH_2SH$	1-Propanethiol,-3-methyl isobutyl mercaptan
$HOCH_2CH_2SH$	Ethanedithiol		

PHENOLS

1. Hydroxyaromatic compounds are named as phenols. They are much more acidic than alcohols and deserve separate listing, e.g.,

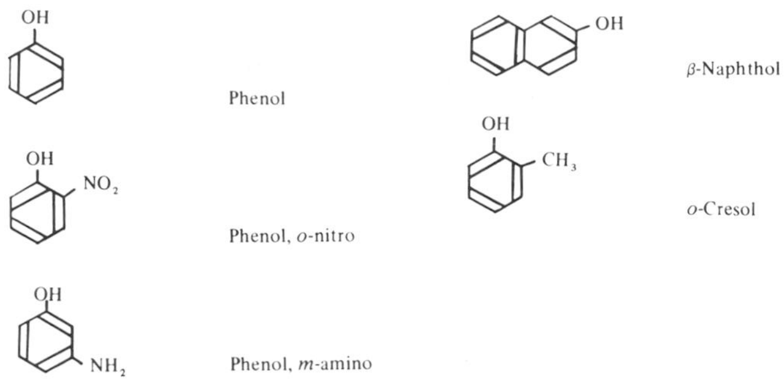

2. Common names are given to aromatic diphenols or triphenols and some mono phenols, e.g.,

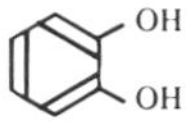

OH OH Catechol

OH OH Hydroquinone

OH OH Resorcinol

OH O_2N NO_2 NO_2 Picric acid

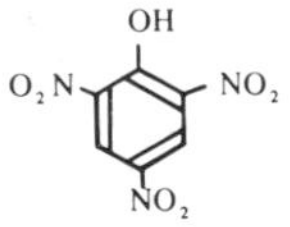

AMINES

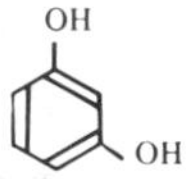

1. Alkyl amines, dialkyl amines and trialkyl amines are listed alphabetically according to the alkyl groups attached. The smallest group is named first and thus determines the listing. Refer to the synonym if necessary, e.g.,

CH_3NH_2	Methyl amine	$(CH_3)_2NH$	Dimethyl amine
$C_4H_9NH_2$	Butyl amine	$C_3H_7NHC_2H_5$	Ethyl propyl amine
		$(C_2H_5)_3N$	Triethyl amine
NH_2	Cyclohexyl amine	$(C_2H_5)_2NCH_3$	Methyl diethyl amine
		$H_2NCH_2CH_2CH_2NH_2$	1.3-Propane diamine

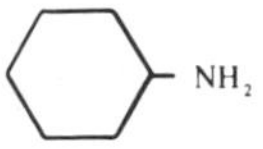

2. When more complex radical names are found the amines are names as amino alkanes, amino alkenes, etc.

$CH_3CH_2CH(NH_2)CH_3$	Butane,-2-amino	$(CH_3)_2CH_2CH(NH_2)CH_3$	Butane,-4-methyl-2-amino

3. Amino substituted heterocyclic compounds are names as amino- derivatives, e.g.,

O NH_2 Furan,-2-amino

S $NHCH_3$ Thiophene,-3-methylamino

4. Amino substituted toluene and xylene, respectively are named as toluidines and xylidines.

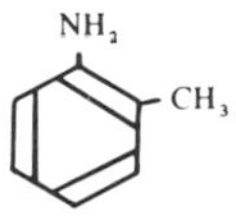

NH_2 CH_3 *o*-Toluidine

NH_2 CH_3 NO_2 *o*-Toluidine,-4-nitro

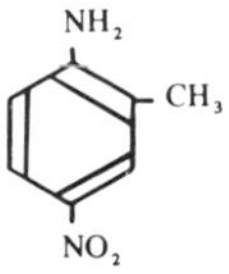

$NHCH_3$ / CH_3 — *p*-Toluidine,-N-Methyl

NHC_6H_5 / CH_3 — *m*-Toluidine, -N-phenyl

ETHERS

1. Dialkyl ethers are listed alphabetically with the smaller group being named first, e.g.,

$C_2H_5-O-C_2H_5$	Diethyl ether	$CH_3-O-C_6H_5$	Methyl phenyl ether
$C_2H_5-O-i-C_3H_7$	Ethyl isopropyl ether	$CH_3-O-CH{=}CH_2$	Methyl vinyl ether
$(CH_3)_3C-O-C_5H_{11}$	Tert-butyl pentyl ether	$C_6H_5CH_2-O-CH_2C_6H_5$	Dibenzyl ether

2. Alkyl ethers of aromatic or heterocyclic compounds are named as alkoxides, e.g.,

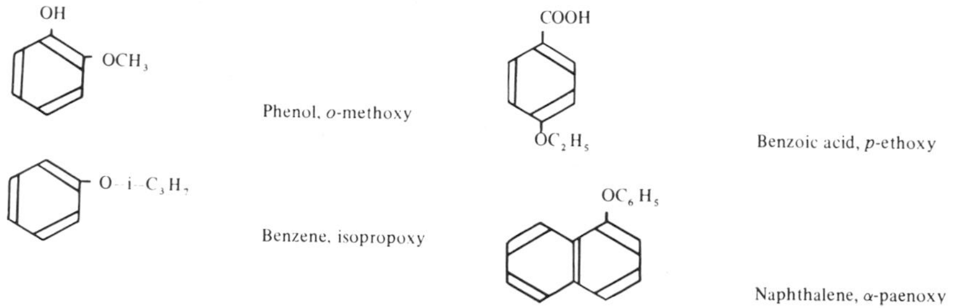

Phenol, *o*-methoxy

Benzoic acid, *p*-ethoxy

Benzene, isopropoxy

Naphthalene, α-paenoxy

SULFIDES, DISULFIDES, SULFOXIDES AND SULFONES

1. These compounds are named in the same way as ethers. The alkyl, or aryl groups are named alphabetically, beginning with the smaller group, e.g.,

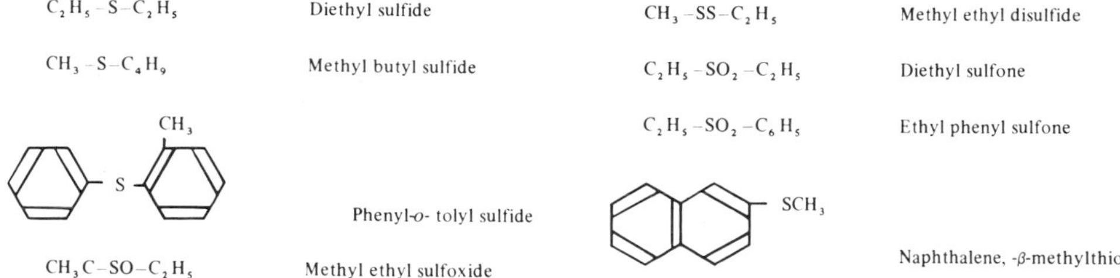

$CH_3C-SO-C_2H_5$ Methyl ethyl sulfoxide

There may be times when the compound can be located more readily by the name of the synonym.

SYMBOLS AND ABBREVIATIONS

[α]	specific rotation
>	above, more than
<	below, less than
?	unkown
aa	acetic acid
abs	absolute
ac	acid
Ac	acetyl
ace	acetone
AFCL	Aliphatic Fluorine Compounds
al	alcohol[1]
ALD	Aldrich Handbook of Organic Chemicals and Biochemicals
alk	alkali
Am	J. Am. Chem. Soc.
Am	amyl (pentyl)
amor	amorphous
anh	anhydrous
aqu	aqueous
as	asymmetric
Atlas	Atlas of Spectral Data and Physical Constants for Organic Compounds
atm	atmospheres
b	boiling
B	Beilstein
Ber	Chem. Ber.
bipym	bipyramidal
bk	black[2]
bl	blue[2]
BOSC	Bayant, et al., Organosilicon Compounds
br	brown[2]
bt	bright
Bu	butyl
bz	Benzene
CAS	Chemical Abstracts
c	percentage concentration
ca	about (circa)
chl	chloroform
co	columns
col	colorless
con	concentrated
cor	corrected
cr	crystals
cy	cyclohexane
d	decomposes
D	line in the spectrum of sodium (subscript)
D, d	dextro[3]
dd	slight decomposition
dil	diluted
diox	dioxane
distb	distillable
dk	dark
Dl, dl	racemic[3]
dlq	deliquescent
DMF	dimethyl formamide
E	Elsevier's
eff	efforescent
Et	ethyl
eth	ether[4]
exp	explodes
extrap	extrapolated
fl	flakes
flam	flammable
flr	fluorescent
fr	freezes
fr. p.	freezing point
fum	fuming
gel	gelatinous
gl	glacial
gold	golden
gr	green[3]
gran	granular
gy	gray[3]
h	hot
H	Helv. Chim. Acta
hex	hexagonal
HDOC	Helibron Dictionary of Organic Compounds
hp	heptane
htng	heating
hx	hexane
hyd	hydrate
hyg	hygroscopic
i	insoluble
i-	iso-
ign	ignites
in	inactive
inflam	inflammable
infus	infusible
irid	iridescent
iso	isooctaine
J	J. Chem. Soc.
JOC	J. Org. Chem.
KHOC	Kaufman Handbook of Organometallic Compounds
L, l	levo[3]
la	large
lf	leaf
liq	ligroin
liq	liquid
lo	long
lt	light
m	melting
m-	meta-
M	molar (concentration)
M	Merck Index, 7th Edition
mcl	monoclinic
Me	methyl
met	metallic
micr	microsopic
min	mineral
mod	modification
mut	mutarotatory
n	normal chain, refractive index
N	normal (concentration)
N	nitrogen[5]
nd	needles
o-	ortho-
oct	octahedral
og	orange[2]
ord	ordinary
org	organic
orh	orthorhombic
p-	para-
pa	pale
par	partial
PCHE	Egloff Physical Constants of Hydrocarbons
peth	petroleum ether
pk	pink[2]
Ph	phenyl
pl	plates
pr	prisms
Pr	propyl
Prak	J. Prak. Chem.
purp	purple[2]
pw	powder
Py	pyrimidine
pym	pyramids
rac	racemic
rect	rectangular
red	red
res	resinous
rh	rhombic
rhd	rhombodohedral
s	soluble
sc	scales
sec	secondary
sf	softens
sh	shoulder
silv	silvery
sl	slightly
so	solid
sol	solution
solv	solvent
Solvents	Entries in this column medium type means soluble; entries in bold means very soluble
sph	sphenoidal
st	stable
sub	sublimes
suc	supercooled
sulf	sulfuric acid
sym	symmetrical
syr	syrup
ta	tablets
tcl	triclinic
tert	tertiary
Tet	Tetrahedron
tetr	tetragonal
THF	tetrahydrofuran
to	toluene
tr	transparent
trg	trigonal
undil	undiluted
uns	unsymmetrical
unst	unstable
v	very
vac	vacuum
var	variable
vap	vapor
vic	vicinal
visc	viscous
volat	volatile or volatilises
vt	violet[2]
w	water
wh	white[1]
wr	warm
wx	waxy
ye	yellow[2]
xyl	xylene

[1] Generally means ethyl alcohol.

[2] The abbreviation of a color ending in "sh" is to be read as ending with the suffix "-ish," e.g., grsh means greenish.

[3] *D, L* generally means configuration and *d, l* generally mean optical rotation, but there are many examples in the chemical literature for which the meaning of these symbols is ambiguous and/or interchangeable.

[4] Generally means diethyl ether.

[5] *N* indicates a position in the molecule.

BEILSTEIN REFERENCES

"Beilstein is the short name for "Beilsteins Handbuch der Organischen Chemie". This Handbuch contains the most complete set of data on organic compounds in the world. The 1st edition was published as two volumes during the years 1881 to 1882. That edition contained data for about 15,000 compounds described on 2200 pages. The 2nd edition of Beilstein consisted of 3 volumes and 4080 pages. The 2nd edition appeared in the years 1885 to 1889. The 3rd edition was published over the years 1892 to 1906. It consisted of 8 volumes published on approximately 11,000 pages. The 4th edition of Beilstein began to be published in 1909. This 4th edition may be described as follows:

The Series of the Beilstein Handbook (4th edition)

Series	Abbreviation	Period of literature completely covered	Colour of label on spine
Basic Series	H	up to 1909	green
Supplementary Series I	E I	1910—1919	red
Supplementary Series II	E II	1920—1929	white
Supplementary Series III	E III	1930—1949	blue
Supplementary Series III/IV	E III/IV*	1930—1959	blue/black
Supplementary Series IV	E IV	1950—1959	black

* Volumes 17 to 27 of Supplementary Series III and IV, covering the heterocyclic compounds are combined in a joint issue.

Preparations are currently in progress for the Fifth Supplementary Series (E V), which will cover the literature from 1960 to 1979.

Each of these series of the 4th edition comprises 27 volumes (or groups of volumes) in which the individual compounds are arranged according to the "Beilstein System". Which compounds are dealt with in which Beilstein volumes is shown in the following two tables:

The Main Divisions of the Beilstein Handbook

Main Division	Volume No.	System No.
A. *Acyclic* Compounds	1—4	1—449
B. *Isocyclic* Compounds	5—16	450—2358
C. *Heterocyclic* Compounds	17—27	2359—4720

Contents of the 27 Volumes of the Beilstein Handbook

Type of registry compound	Feature of the functional group		A (Acyclics)	B (Isocyclics)	C (Heterocyclics): 1O*	2O*, 3O*,...	1N	2N	3N, 4N,...	1N, 1O*, 1N, 2O*, 2N, 1O*, 2N, 2O*, further heteroatoms**
1 Compounds without functional groups		–		5			20			
2 Hydroxy-compounds		–OH	1	6	17			23		
3 Oxo-compounds	=O	=O		7			21	24		
		=O + –OH		8						
4 Carboxylic acids	$\langle^{O}_{OH}$	$\langle^{O}_{OH}$	2	9						
		$\langle^{O}_{OH}$ + –OH; $\langle^{O}_{OH}$ + =O; $\langle^{O}_{OH}$ + =O + –OH	3	10						
5 Sulfinic acids		$-SO_2H$								
6 Sulfonic acids		$-SO_3H$								
7 Seleninic acids, Selenonic acids, Tellurinic acids		$-SeO_2H$ and $-SeO_3H$, $-TeO_2H$	4	11	18	19	22	25	26	27
8 Amines	$-NH_2$	$-NH_2$		12						
		$[-NH_2]_n$; $-NH_2$ + –OH		13						
		$-NH_2$ + =O; $-NH_2$ + $\langle^{O}_{OH}$; $-NH_2$ + ...		14						
9 Hydroxylamines and Dihydroxyamines		–NH–OH, $-N\langle^{OH}_{OH}$								
10 Hydrazines		$-NH-NH_2$		15						
11 Azo-compounds		–N=NH								
12 Diazonium compounds		$-N{\equiv}N]^{\oplus}$								
13 Compounds with groups of 3 or more N-atoms		$-NH-NH-NH_2$, $-N(NH_2)_2$, $-N{=}N-NH_2$, etc.								
14 Compounds containing carbon directly bonded to P, As, Sb, and Bi	e.g.	$-PH_2$, PH–OH, $-P(OH)_2$, $-PH_4$, ..., $-PO(OH)_2$		16						
15 Compounds containing carbon directly bonded to Si, Ge, and Sn	e.g.	$-SiH_3$, $-SiH_2(OH)$, ...								
16 Compounds containing carbon directly bonded to elements of the 3rd–1st A-groups of the periodic table	e.g.	$-BH_2$, –BH(OH), ..., $-Mg^{\oplus}$								
17 Compounds containing carbon directly bonded to elements of the 1st–8th B-groups of the periodic table	e.g.	–HgH, $-Hg^{\oplus}$, ...								

* Instead of O also S, Se, Te.
** e.g. B, Si, P, but not S, Se, Te.

Additional informational material on Beilstein is available upon request free of charge from

Springer-Verlag KG
Heidelberger Platz 3
D-1000 Berlin 33

Springer Verlag New York Inc.
175 Fifth Avenue
New York, New York 10010

Included in such information are:

Beilstein Reference Chart ("Contents of Beilstein Handbook of Organic Chemistry")
Brochure, "What is Beilstein"
How to Use Beilstein
Beilstein Wordbook (German/English)
Beilstein Outline (List of all available volumes; revised annually)

The Beilstein references listed in this edition of the *CRC Handbook of Chemistry and Physics* were revised and updated early in 1980 by personnel of Beilstein.

PHYSICAL CONSTANTS OF ORGANIC COMPOUNDS

No.	Name, Synonyms, and Formula	Mol. wt.	Color, crystalline form, specific rotation and λ_{max} (log ε)	b.p. °C	m.p. °C	Density	n_D	Solubility	Ref.
1	Abietic acid or sylvic acid $C_{20}H_{30}O_2$	302.46	mcl pl (al-w)	250	173-4			al, et, ace, bz	$B9^{3}$, 2904
2	Abietic acid, methyl ester $C_{21}H_{32}O_2$	316.48		$225\text{-}6^{16}$		$1.0491^{20/4}$	1.5344	al	$B9^{3}$, 2907
3	Acenaphthanthracene or Naphtho-2′,3′,4,5-acenaphthene $C_{20}H_{14}$	254.33	pa ye lf (lig)	192.5-3.5				bz, lig	$B5^{3}$, 2468
4	Acenaphthene $C_{12}H_{10}$	154.21		279	96.2	$1.0242^{90/4}$	1.6048^{95}	al, bz	$B5^{4}$, 1834
5	Acenaphthene, 1-amino $C_{12}H_{11}N$	169.23	cr (peth)	135	sub			al, bz, CS_2	$B12^{2}$, 764
6	Acenaphthene, 3-amino $C_{12}H_{11}N$	169.23	pl (al), nd (peth)		81.5			al, chl	$B12^{3}$, 3210
7	Acenaphthene, 4-amino $C_{12}H_{11}N$	169.23	nd (al,w)		87			bz, lig, w	$B12^{3}$, 3212
8	Acenaphthene, 5-amino $C_{12}H_{11}N$	169.23	nd (lig) red in air		108			al	$B12^{3}$, 3212
9	Acenaphthene, 5-bromo $C_{12}H_9Br$	233.11	pl (al)	335	52	$1.4392^{52/4}$	1.6565^{54}	al	$B5^{4}$, 1839
10	Acenaphthene, 5-chloro $C_{12}H_9Cl$	188.66	pl or nd (al)	$319.2^{770}, 163^{13}$	70.5	$1.1954^{20/4}$	1.6288^{7}	al	$B5^{4}$, 1837
11	Acenaphthene, 5-iodo $C_{12}H_9I$	280.11	nd (al)		65	$1.6738^{62/4}$	1.6909^{65}	al, bz	$B5^{1}$, 276
12	Acenaphthene, 5,6-dinitro $C_{12}H_8N_2O_4$	244.21			220-4				$B5^{3}$, 1784
13	Acenaphthene, 3-nitro $C_{12}H_9NO_2$	199.21	gr-ye nd (aa)		151.5			aa	$B5^{3}$, 1783
14	Acenaphthene, 5-nitro $C_{12}H_9NO_2$	199.21			103-4			w, al, eth, lig	$B5^{4}$, 1840
15	Acenaphthene, 2a,3,4,5-tetrahydro $C_{12}H_{14}$	158.24		246	12			chl	$B5^{3}$, 1385
16	Acenaphthene, 5-carboxylic acid or 5-Acenaphthoic acid $C_{13}H_{10}O_2$	198.22	nd (bz or lig)		220-1			bz, lig	$B9^{3}$, 3289
17	Acenaphthenequinone $C_{12}H_{16}O_2$	182.18	ye nd (aa)		261			al, bz	$B7^{3}$, 3796
18	3-Acenaphthene sulfonic acid $C_{12}H_9SO_3H$	234.27	hyg nd (bz)		87-9				$B11^{3}$, 435
19	1-Acenaphthenone or 1-oxoacenaphthene $C_{12}H_8O$	168.19	nd (al)		121			w, al, bz, chl	$B7^{3}$, 2046
20	Acenaphthylene $C_{12}H_8$	152.20	pr (eth), pl (al)	265-75	92-3	$0.8988^{16/2}$		al, eth, bz	$B5^{4}$, 2138
21	Acetaldehyde CH_3CHO	44.05		20.8	-121	$0.7834^{18/4}$	1.3316^{20}	**w, al, eth, ace, bz**	$B1^{4}$, 3094
22	Acetaldehyde, 2,4-dinitrophenylhydrazone $CH_3CH{=}NNHC_6H_3(NO_2)_2\text{-}2,4$	224.18	ye sc (al)		168.5			eth, ace, bz, ch	$B15^{3}$, 426
23	Acetaldehyde, amino, diethyl acetal $H_2NCH_2CH(OC_2H_5)_2$	133.19		163		0.9159^{25}	1.4170^{20}	w, al, eth, chl	$B4^{4}$, 1918
24	Acetaldehyde, ammonia or 1-aminoethanol $CH_3CH(OH)NH_2$	61.08	rh (eth-al)	110d	97			w	$C5^{2}$
25	Acetaldehyde, bromo, diethyl acetal $BrCH_2CH(OC_2H_5)_2$	197.07		$180, 66^{18}$		$1.280^{20/4}$	1.4376^{20}	al, eth	$B1^{4}$, 3151
26	Acetaldehyde, bromo, dimethyl acetal $BrCH_2CH(OCH_3)_2$	169.02		$148.50\text{-}50^{15}$		1.5049^{184}	1.4450^{20}	eth, ace, chl	$B1^{3}$, 2672
27	Acetaldehyde, chloro $ClCH_2CHO$	78.50		$85\text{-}5.5^{748}$				eth	$B1^{4}$, 3134
28	Acetaldehyde, chloro, diethyl acetal $ClCH_2CH(OC_2H_5)_2$	152.62		$157.4, 71\text{-}2^{35}$		$1.068^{20/4}$	1.4150^{20}	**al, eth,** bz	$B1^{4}$, 3134
29	Acetaldehyde, chloro, dimethyl acetal $ClCH_2CH(OCH_3)_2$	124.57		127-8		$1.068^{20/4}$	1.4150^{20}	al, bz	$B1^{4}$, 3134
30	Acetaldehyde, *bis*(2-chloroethyl) acetal $CH_3CH(OCH_2CH_2Cl)_2$	187.07		$194\text{-}6, 106\text{-}7^{14}$		$1.1737^{20/4}$	1.45266^{20}	al, eth	$B1^{4}$, 3104
31	Acetaldehyde, diacetate or Ethylidene diacetate $CH_3CH(O_2CCH_3)_2$	146.14		$169, 65\text{-}7^{10}$	18.9	1.3985^{25}	1.070^{25}	al, eth	$B2^{4}$, 282
32	Acetaldehyde, dichloro Cl_2CHCHO	112.94		90-1				al	$B1^{4}$, 3140
33	Acetaldehyde, dichloro, diethyl acetal $Cl_2CHCH(OC_2H_5)_2$	187.07		$183\text{-}4, 67\text{-}71^{12}$		1.1381^{14}		al, eth	$B1^{4}$, 3140

PHYSICAL CONSTANTS OF ORGANIC COMPOUNDS (Continued)

No.	Name, Synonyms, and Formula	Mol. wt.	Color, crystalline form, specific rotation and λ_{max} (log ε)	b.p. °C	m.p. °C	Density	n_D	Solubility	Ref.
34	Acetaldehyde, dichloro, hydrate $Cl_2CHCH(OH)_2$	130.96	cr (bz), ta	97	56-7			w, al, eth	B1, 614
35	Acetaldehyde, diethyl acetal $CH_3CH(OC_2H_5)_2$	118.18		$103.2, 21^{22}$		$0.8314^{20/4}$	1.3834^{20}	w, al, eth, ace	$B1^4$, 3103
36	Acetaldehyde, diethyl mercaptal $CH_3CH(SC_2H_5)_2$	150.30		$183\text{-}5, 73^{19}$		0.9706^{20}	1.5025^{20}	eth, ace, bz	$B1^4$, 3162
37	Acetaldehyde, dimethyl acetal $CH_3CH(OCH_3)_2$	90.12		64.5	−113.2	$0.85015^{20/4}$	1.3668^{20}	w, al, eth, ace, bz	$B1^4$, 3103
38	Acetaldehyde, dimethylamino, diethyl acetal $(CH_3)_2NCH_2CH(OC_2H_5)_2$	161.24	cr (bz), ye	170-1		0.885^{7}	1.4129^{20}	w, al, eth, ace	$B1^3$, 870
39	Acetaldehyde, 2,4-dinitrophenylhydrazone $CH_3CH{=}NNHC_6H_3(NO_2)_2$-2,4	224.18	ye sc (al)		168.5			eth, ace, bz	$B15^3$, 426
40	Acetaldehyde, diphenyl $(C_6H_5)_2CHCHO$	196.25		157.5^{5}		$1.1061^{21/4}$	1.5920^{21}	al, eth, ace	$B7^3$, 2117
41	Acetaldehyde, ethoxy $C_2H_5OCH_2CHO$	88.11		71-3		$0.942^{20/4}$	1.3956^{20}	w, al, ace	$B1^3$, 3181
42	Acetaldehyde, hydroxy or Glycolaldehyde $HOCH_2CHO$	60.05	pl		97	1.366^{100}	1.4772^{19}	w, al	$B1^4$, 3955
43	Acetaldehyde, hydroxy, diethyl acetal $HOCH_2CH(OC_2H_5)_2$	134.18		$167, 57\text{-}8^{8}$		$0.888^{24/4}$	1.4073^{20}	al, eth	$B1^4$, 3958
44	Acetaldehyde, methoxy CH_3OCH_2CHO	74.08		92.3^{770}		$1.005^{25/4}$	1.3950^{20}	w, al, eth, ace	$B1^4$, 3955
45	Acetaldehyde, methoxy, diethyl acetal $CH_3OCH_2CH(OC_2H_5)_2$	148.20		144-8			$1.3990^{20/D}$	al, eth, ace	$B1^4$, 3958
46	Acetaldehyde, oxime or Acetaldoxime $CH_3CH{=}NOH$	59.07	nd	115	47	$0.9656^{20/4}$	1.42567^{20}	al, eth	$B1^4$, 3121
47	Acetaldehyde, phenyl or α-Tolualdehyde $C_6H_5CH_2CHO$	120.15		$195, 88^{18}$	33-4	$1.0272^{20/4}$	1.5255^{20}	al, eth, bz	$B7^3$, 1003
48	Acetaldehyde, semicarbazone $CH_3CH{=}NNHCONH_2$	101.11			163	$1.300^{0/4}$		al	$B3^4$, 178
49	Acetaldehyde, phenyl, dimethyl acetal $C_6H_5CH_2CH(OCH_3)_2$	166.22		193-4					$B7^3$, 1006
50	Acetaldehyde, phenylhydrazone $CH_3CH{=}NNHC_6H_5$	134.18		$133\text{-}6^{21}$	98-101			al	$B15^3$, 79
51	Acetaldehyde, tribromo or bromal Br_3CCHO	280.74		$174, 61^{9}$		$2.6650^{25/4}$	1.5939^{20}	al, eth, ace	$B1^4$, 3155
52	Acetaldehyde, tribromo, hydrate or Bromal hydrate $Br_3CCH(OH)_2$	298.76	mcl pr (w + 1)	d	53.5	$2.5662^{40/4}$		al, eth	$B1^4$, 3155
53	Acetaldehyde, trichloro or chloral Cl_3CCHO	147.39		97.8	-57.5	$1.5121^{20/4}$	1.45572^{20}	w, al, eth	$B1^4$, 3142
54	Acetaldehyde, trichloro, diethyl acetal $Cl_3CCH(OC_2H_5)_2$	221.51		$205, 84\text{-}5^{10}$		$1.266^{25/4}$	1.4586^{25}	al, eth	$B1^4$, 3144
55	Acetaldehyde, trichloro, ethylhemiacetal $Cl_3CCH(OH)OC_2H_5$	193.46		$115\text{-}6^{760}$	56-7	1.143^{40}		w, al, eth	$B1^4$, 3144
56	Acetaldehyde, trichloro, hydrate or chloral hydrate $Cl_3CCH(OH)_2$	165.40	mcl pl (w)	$96.3^{764/d}$	57	$1.9081^{20/4}$		w, al, eth, ace, bz	$B1^4$, 3143
57	Acetaldehyde, trimethyl or Pivaldehyde $(CH_3)_3CCHO$	86.13		77-8	6	0.7923^{17}	1.3791^{20}	al, eth	$B1^4$, 3295
58	Acetamide CH_3CONH_2	59.07	trg mcl (al-eth)	$221.2, 120^{20}$	82.3	$0.9986^{85/4}$	1.4278^{78}	w, al	$B2^4$, 399
59	Acetamide, *N*-acetyl or diacetamide $CH_3CONHCOCH_3$	101.11	nd (eth)	$223.5, 113^{12}$	79			w, al, eth	$B2^4$, 416
60	Acetamide, *N*-acetyl-*N*-ethyl or Diacetylethylamine $CH_3CON(C_2H_5)COCH_3$	129.16		195-9		1.0092^{20}	1.4513^{20}	al	$B4^4$, 352
61	Acetamide, *N*-acetyl-*N*-methyl or Diacetylmethylamine $CH_3CON(CH_3)COCH_3$	115.31		$194.5, 114.5^{61}$	-25	$1.0663^{25/4}$	1.4502^{25}	w	$B4^4$, 183
62	Acetamide, N-acetyl-N-phenyl or Diacetanilide $(CH_3CO)_2NC_6H_5$	177.20	ta (lig)	$200^{100}\ 142^{41}$	37-8			al, bz	$B12^3$, 472
63	Acetamide, *N*-(2-aminoethyl) $CH_3CONHCH_2CH_2NH_2$	102.14		128^{3}	51			w, al, bz	$B4^4$, 1193
64	Acetamide, *N*-benzyl $CH_3CONHCH_2C_6H_5$	149.19		157^{2}	61			al, eth	$B12^5$, 2255
65	Acetamide, *N*-bromo, hydrate $CH_3CONHBr \cdot H_2O$	155.98	bt ye pl (w + 1)		108-9				$B2^1$, 82
66	Acetamide, *N*-butyl $CH_3CONHC_4H_9$	115.18		229			1.4388^{25}		$B4^4$, 565

No.	Name, Synonyms, and Formula	Mol. wt.	Color, crystalline form, specific rotation and λ_{max} (log ε)	b.p. °C	m.p. °C	Density	n_D	Solubility	Ref.
67	Acetamide, *N*-butyl-*N*-phenyl $CH_3CON(C_4H_9)C_6H_5$	191.27		$281, 141^{10}$	24.5	$0.9912^{20/4}$	1.5146^{20}	chl	$B12^3$, 467
68	Acetamide, chloro or chloroacetamide $ClCH_2CONH_2$	93.51		$224\text{-}5^{747}$	121			w, al	$B2^4$, 490
69	Acetamide, *N*-cyanomethyl $CH_3CONHCH_2CN$	98.10			79-81			w	$B4^2$, 790
70	Acetamide, *N,N*-diacetyl or Triacetamide $CH_3CON(COCH_3)_2$	143.14	nd (eth)		79			eth	$B2^4$, 416
71	Acetamide, dibenzyl $(C_6H_5CH_2)_2CHCONH_2$	239.32	nd (al, w)	259^{18}	129			al, eth	$B9^3$, 3356
72	Acetamide, dichloro $Cl_2CHCONH_2$	127.96		$233\text{-}4^{745}$	99.4			w, al, eth	$B2^4$, 505
73	Acetamide, *N,N*-diethyl $CH_3CON(C_2H_5)_2$	115.18		185-6		$0.9130^{17.4/D}$	$1.4374^{17.4/D}$	w, al, eth, ace, bz	$B4^4$, 349
74	Acetamide, *N,N*-dimethyl $CH_3CON(CH_3)_2$	87.12		$165^{758}, 84^{22}$	-20	$0.9366^{25/4}$	1.4380^{20}	w, al, ace, bz, eth	$B4^4$, 180
75	Acetamide, *N,N*-dipropyl $CH_3CON(C_3H_7)_2$	143.23		$209\text{-}10, 101^{16}$		$0.8992^{17.4/4}$	$1.4419^{17.4}$	al	$B4^4$, 476
76	Acetamide, *N*-ethyl $CH_3CONHC_2H_5$	87.12		$205, 104\text{-}5^{18}$		$0.942^{4.5/4}$		w, al, chl	$B4^4$, 347
77	Acetamide, hydroxy $HOCH_2CONH_2$	75.07		120				w	$B3^4$, 597
78	Acetamide, *N*-(2-hydroxyethyl) or *N*-acetylethanolamine $CH_3CONHCH_2CH_2OH$	103.12		$166\text{-}7^{8}$	63.5	$1.1079^{25/4}$	1.4674^{20}	w, ace	$B4^4$, 1535
79	Acetamide, *N*-methyl $CH_3CONHCH_3$	73.09	nd	$204\text{-}6, 95^{14}$	28	$0.9571^{25/4}$	1.4301^{20}	w, al, eth, ace, bz	$B4^4$, 176
80	Acetamide, *N*-methyl-*N*-α-naphthyl $CH_3CON(CH_3)\alpha C_{10}H_7$	199.25	nd (lig)		95-7			al, eth	$B12^3$, 2867
81	Acetamide, *N*-methyl-*N*-(4-nitrophenyl) $CH_3CON(CH_3)(C_6H_4NO_2\text{-}4)$	194.19	pl (w)		95-7			al, eth	$B12^3$, 1595
82	Acetamide, phenoxy $C_6H_5OCH_2CONH_2$	151.16			101.5			al	$B6^4$, 638
83	Acetamide, phenyl $C_6H_5CH_2CONH_2$	135.17			157			al, eth	$B9^3$, 2193
84	Acetamide, *N*-phenyl or Acetanilide $CH_3CONHC_6H_5$	135.17		304	114.3	1.2190^{15}		al, eth, ace, bz, chl	$B12^3$, 459
85	Acetamide, thiono CH_3CSNH_2	75.13			115-6			w, al	$B2^4$, 565
86	Acetamide, trichloro Cl_3CCONH_2	162.40		$238\text{-}9^{740}$	142			al, eth	$B2^4$, 520
87	Acetamidine $CH_3C(NH_2){=}NH$	58.08		80^{10}	63-65			al	$B2^4$, 428
88	Acetamidine, *N,N*'-diphenyl $CH_3C(NC_6H_5)NHC_6H_5$	210.28	nd (al)		131-2			eth	$B12^3$, 471
89	Acetamidine, hydrochloride $CH_3C(NH_2)NH\cdot HCl$	94.54	nd or pr (al)		177-8			w	$B2^4$, 429
90	Acetanilide or N-phenylacetamide $CH_3CONHC_6H_5$	135.17	rh or pl (w)	304	114.3	1.2190^{15}		al, eth, ace, bz	$B12^3$, 459
91	Acetoacetanilide $CH_3COCH_2CONHC_6H_5$	177.20			86			al, eth	$B12^3$, 993
92	Acetanilide, *o*-amino $2\text{-}H_2NC_6H_4NHCOCH_3$	150.18			132			w, al, eth	$B13^3$, 41
93	Acetanilide, *m*-amino $3\text{-}H_2NC_6H_4NHCOCH_3$	150.18	nd or pl (bz)		87-9			w, al, bz, eth, ace	$B13^3$, 75
94	Acetanilide, *p*-amino $4\text{-}H_2NC_6H_4NHCOCH_3$	150.18	nd (w)	267	165-8			al, eth, w	$B13^3$, 166
95	Acetanilide, *N*-bromo $C_6H_5N(Br)COCH_3$	214.06	ye pl (peth)		88			chl, lig	$B12^3$, 1078
96	Acetanilide, *o*-bromo $2\text{-}Br\text{-}C_6H_4NHCOCH_3$	214.06	nd (al)		99			al, eth	$B12^3$, 1417
97	Acetanilide, *m*-bromo $3\text{-}BrC_6H_4\text{-}NHCOCH_3$	214.06	nd (aq,al)		87.5			al, eth	$B12^3$, 1424
98	Acetanilide, *p*-bromo $4\text{-}BrC_6H_4NHCOCH_3$	214.06	nd (60% al)		168	1.717		al	$B12^3$, 1437
99	Acetanilide, *N*-butyl $C_6H_5N(C_4H_9)COCH_3$	191.27		$281, 140^{10}$	24.5	$0.9912^{20/4}$	1.5146^{20}		$B12^3$, 467
100	Acetanilide, *p*-butyl $4\text{-}C_4H_9C_6H_4NHCOCH_3$	191.27	wh, pl (al)		105			al	$B12^3$, 2715

PHYSICAL CONSTANTS OF ORGANIC COMPOUNDS (Continued)

No.	Name, Synonyms, and Formula	Mol. wt.	Color, crystalline form, specific rotation and λ_{max} (log ε)	b.p. °C	m.p. °C	Density	n_D	Solubility	Ref.
101	**Acetanilide, N-chloro** $C_6H_5N(Cl)COCH_3$	169.61	nd (dil aa) pl (peth-chl)		91				B12[3], 1077
102	Acetanilide, *o*-chloro $2\text{-}ClC_6H_4NHCOCH_3$	169.61	nd (dil aa)		87-8 (sub)			al, eth, bz	B12[3], 1288
103	Acetanilide, 2-chloro-3-methyl or 3-acetamino-2-chloro-toluene $2\text{-}Cl\text{-}3\text{-}CH_3\text{-}C_6H_3NHCOCH_3$	183.64	nd (al)		133-4			al, bz	B12[3], 1991
104	Acetanilide, 2-chloro-4-nitro $2\text{-}Cl\text{-}4\text{-}NO_2C_6H_3NHCOCH_3$	214.61	pr (al)		139-40				B12[3], 1663
105	Acetanilide, 2-chloro-5-nitro $2\text{-}Cl\text{-}5\text{-}NO_2C_6H_3NHCOCH_3$	214.61	nd (al)		156				B12[2], 398
106	**Acetanilide, *m*-chloro** $3\text{-}ClC_6H_4NHCOCH_3$	169.61	nd		79			al, eth, bz	B12[3], 1309
107	Acetanilide, 3-chloro-2-methyl or 2-acetamino-6-chloro-toluene $3\text{-}Cl\text{-}6\text{-}CH_3C_6H_3NHCOCH_3$	183.64	nd (dil al)		157-9			al, bz	B12[2], 1919
108	Acetanilide, 3-chloro-4-methyl or 4-acetamino-2-chloro-toluene $3Cl\text{-}4\text{-}CH_3\text{-}C_6H_3\text{-}NHCOCH_3$	183.64	tcl cr		105			al	B12[2], 530
109	Acetanilide, 3-chloro-5-methyl $3\text{-}Cl\text{-}5\text{-}CH_3C_6H_3NHCOCH_3$	183.64	nd (al)		151			al	B12, 871
110	Acetanilide, 3-chloro-6-methyl or 2-acetamino-4-chloro-toluene $3\text{-}Cl\text{-}6\text{-}CH_3C_6H_3NHCOCH_3$	183.64	nd (w)		139-40			al, eth, w	B12[3], 1912
111	Acetanilide, 3-chloro-4-nitro $3\text{-}Cl\text{-}4\text{-}NO_2C_6H_3NHCOCH_3$	214.61	pa ye nd (al)		145				B12[3], 1666
112	Acetanilide, *p*-chloro $4\text{-}ClC_6H_4NHCOCH_3$	169.61	nd (aq aa) ta (al ace) cr (w)		179	1.385^{224}		al, eth	B12[3], 1336
113	Acetanilide, 4-chloro-2-methyl or 2-acetamino-5-chloro-toluene $4\text{-}Cl\text{-}2\text{-}CH_3C_6H_3NHCOCH_3$	183.64	lf (al)		140			al, bz	B12[3], 1914
114	Acetanilide, 4-chloro-3-methyl or 5-acetamino-2-chloro-toluene $4\text{-}Cl\text{-}3\text{-}CH_3C_6H_3NHCOCH_3$	183.64	lf (al)		91.2-.7			al, bz	B12[2], 473
115	Acetanilide, 4-chloro-2-nitro $4\text{-}Cl\text{-}2\text{-}NO_2C_6H_3NHCOCH_3$	214.61	ye nd (al)		104				B12[3], 1652
116	Acetanilide, 4-chloro-3-nitro $4\text{-}Cl\text{-}3\text{-}NO_2C_6H_3NHCOCH_3$	214.61	ye nd (al)		150				B12[2], 357
117	Acetanilide, *a*-cyano $C_6H_5NHCOCH_2CN$	160.18	nd (al)		199-200				B12[3], 558
118	Acetanilide, 4-cyclohexyl $4\text{-}C_6H_{11}\text{-}C_6H_4NHCOCH_3$	217.31	nd (peth)		129			al, eth, chl	B12[3], 2819
119	Acetanilide, 2,4-dimethyl or 2,4-Acetoxylide $2,4\text{-}(CH_3)_2C_6H_3NHCOCH_3$	163.22	nd (al)	170^{10}	128-9			al	B12[3], 2474
120	Acetanilide, 4,6-dimethyl-2-nitro $2\text{-}NO_2\text{-}4,6\text{-}(CH_3)_2C_6H_2NHCOCH_3$	208.22	yesh nd (w)		172-3			al, bz, w	B12[3], 2491
121	Acetanilide, 2,3-dinitro $2,3\text{-}(NO_2)_2\text{-}C_6H_3\text{-}NHCOCH_3$	225.16	nd (al)		187			al, eth, bz	B12[3], 1680
122	Acetanilide, 2,4-dinitro $2,4\text{-}(NO_2)_2C_6H_3NHCOCH_3$	225.16	ye nd (al or bz)		125-6			al, eth, bz	B12[3], 1686
123	Acetanilide, 2,5-dinitro $2,5\text{-}(NO_2)_2C_6H_3NHCOCH_3$	225.16	nd (al)		121			al	B12[3], 1704
124	Acetanilide, 2,6-dinitro $2,6\text{-}(NO_2)_2C_6H_3NHCOCH_3$	225.16	nd (aa)		197			al, aa	B12[3], 1705
125	Acetanilide, 3,4-dinitro $3,4\text{-}(NO_2)_2C_6H_3NHCOCH_3$	225.16	ye cr (al) nd (w)		144-5			al, w	B12[3], 1705
126	Acetanilide, 3,5-dinitro $3,5\text{-}(NO_2)_2C_6H_3NHCOCH_3$	225.16	ye nd (dil aa, w)		191			al, w, aa	B12[3], 1706
127	Acetamide, *N,N*-dipropyl $CH_3CON(C_3H_7)_2$	143.23		209-10, 101^{18}		$0.8992^{17.4/4}$	$1.4419^{17.4}$	al	B4[4], 476
128	Acetanilide, 2-ethoxy or *o*-Acetophenetidide $2\text{-}(C_2H_5O)\text{-}C_6H_4NHCOCH_3$	179.22	lf (dil al)	>250	79			al, eth	B13[3], 779
129	Acetanilide, 2-ethoxy-4-nitro $2\text{-}(C_2H_5O)4\text{-}(NO_2)C_6H_3NHCOCH_3$	224.22	nd (al)		202			al	B13[2], 194
130	Acetanilide, 2-ethoxy-5-nitro $2(C_2H_5O)\text{-}5(NO_2)C_6H_3NHCOCH_3$	224.22	ye nd (al)		199			bz, al	B13[3], 881

PHYSICAL CONSTANTS OF ORGANIC COMPOUNDS (Continued)

No.	Name, Synonyms, and Formula	Mol. wt.	Color, crystalline form, specific rotation and λ_{max} (log ε)	b.p. °C	m.p. °C	Density	n_D	Solubility	Ref.
131	Acetanilide, 3-ethoxy or *m*-accetophenetidide $3-(C_2H_5O)-C_6H_4NHCOCH_3$	179.22	gy pl (w)	...	97-9	...	...	al, eth, w	B13^3, 950
132	Acetanilide, 4-ethoxy or Phenacitin $4-(C_2H_5O)C_6H_4NHCOCH_3$	179.22	mcl pr	dic	137-8	...	1.571	al, ace	B13^3, 1057
133	Acetanilide, 4-ethoxy-2-nitro $4-C_2H_5O-2NO_2C_6H_3NHCOCH_3$	224.22	ye nd (w)	...	104	...	...	al, eth, ace, chl	B13^3, 1208
134	Acetanilide, 4-ethoxy-3-nitro $4-(C_2H_5O)-3-NO_2C_6H_3NHCOCH_3$	224.22	nd (dil al)	...	123	...	...	al, ace, bz	B13^3, 1202
135	Acetanilide, 5-ethoxy-2-nitro $5-(C_2H_5O)-2-NO_2C_6H_3NHCOCH_3$	224.22	nd (al)	...	95	...	...	al, eth, ace, bz, chl	B13^3, 976
136	Acetanilide, *N*-ethyl $C_6H_5N(C_2H_5)COCH_3$	163.22	(w) rh (eth)	258^{731}	55	...	...	w, eth	B12^3, 466
137	Acetanilide, *N*-ethyl-3-nitro $3-NO_2-C_6H_4N(C_2H_5)COCH_3$	208.22	pa ye nd (dil al)	...	88-9	...	...	al, bz	B12, 704
138	Acetanilide, *N*-ethyl-4-nitro $4-NO_2-C_6H_4-N(C_2H_5)COCH_3$	208.22	lf or pr (dil al)	...	118-9	...	...	al, bz, eth	B12^3, 1596
139	Acetanilide, *o*-hydroxy or *o*-Acetamidophenol $2-HOC_6H_4NHCOCH_3$	151.16	pl (dil al)	...	208	...	...	al, eth, bz	B13^3, 778
140	**Acetanilide, *m*-hydroxy or *m*-Acetamidophenal** $3-HO-C_6H_4-NHCOCH_3$	151.16	nd (w)	...	148-9	...	...	w, al, eth, bz	B13^3, 950
141	Acetanilide, *p*-hydroxy or p-acetamidophenol $4-HOC_6H_4NHCOCH_3$	151.17	nd (bz-peth)	...	151	...	...	al, eth, bz	B13^3, 1056
142	Acetanilide, *p*-iodo $4-I-C_6H_4NHCOCH_3$	261.06	ta (w) pr (w, al)	...	184.5	...	...	al, eth, bz	B12^3, 1496
143	Acetanilide, *o*-methyl or *o*-acetotoluidide $2-CH_3-C_6H_4NHCOCH_3$	149.19	nd (al)	296	110	1.168^{15}	...	al, eth, ace, bz	B12^3, 1853
144	Acetanilide, *m*-methyl or *m*-acetotoluidide $3-CH_3-C_6H_4NHCOCH_3$	149.19	nd (w)	303,182-3^{14}	65.5	1.141^{15}	...	al, eth	B12^3, 1962
145	Acetanilide, *p*-methyl or *p*-acetotoluidide $4-CH_3-C_6H_4NHCOCH_3$	149.19	mcl cr or nd (dil al)	307 sub	148.5	1.212^{15}	...	al, eth, bz	B12^3, 2051
146	Acetanilide, *N*-methyl-*o*-hydroxy $2-HOC_6H_4N(CH_3)COCH_3$	165.19	nd (bz-peth)	...	151	...	...	al, eth, bz	B13^3, 783
147	Acetanilide, *N*-methyl-*p*-hydroxy $4-HOC_6H_4N(CH_3)COCH_3$	165.19	sc (w)	...	245	...	...	al, eth	B13^1, 162
148	Acetanilide, *o*-methoxy or *o*-Acetanisidine $2-CH_3OC_6H_4-NHCOCH_3$	163.19	nd (w)	303-5	87-8	...	...	al, eth, ace, w	B13^3, 778
149	Acetanilide, 2-methoxy-3-nitro $2-CH_3O-3-NO_2C_6H_3NHCOCH_3$	210.19	pa ye pr (dil aa MeOH)	...	103.4	...	...	MeOH, al	B13^2, 195
150	Acetanilide, 2-methoxy-4-nitro $2-CH_3O-4-NO_2C_6H_3NHCOCH_3$	210.19	pa ye cr (AcOEt)	...	153-4	...	...	al, bz	B13^3, 890
151	Acetanilide, 2-methoxy-5-nitro $2-CH_3O-5-NO_2C_6H_3NHCOCH_3$	210.19	nd (w)	...	178	...	...	...	B13^3, 881
152	Acetanilide, 2-methoxy-6-nitro $2-CH_3O-6-NO_2C_6H_3NHCOCH_3$	210.19	pa ye nd (dil al)	...	158-9	...	...	bz, al	B13^3, 876
153	Acetanilide, *m*-methoxy or *m*-acetanisidine $3-CH_3OC_6H_4NHCOCH_3$	165.19	nd or pl (w)	...	81	...	...	w, al, eth, ace	B13^3, 950
154	Acetanilide, 3-methoxy-2-nitro $3-CH_3O-2NO_2C_6H_3NHCOCH_3$	210.19	br (al) amor	...	265 sub	...	...	al, bz	B13^1, 136
155	Acetanilide, 3-methoxy-4-nitro $3-CH_3O-4NO_2C_6H_3NHCOCH_3$	210.19	ye nd (w)	...	165	...	...	al	B13^3, 978
156	Acetanilide, 3-methoxy-5-nitro $3-CH_3O-5-NO_2C_6H_3NHCOCH_3$	210.19	nd (aa)	...	201	...	...	al, eth, bz, aa	B13^2, 216
157	Acetanilide, *p*-methoxy or *p*-acetanisidine $4-CH_3OC_6H_4NHCOCH_3$	165.19	pl (w)	...	130-2	...	...	al, ace, eth, w	B13^3, 1056
158	Acetanilide, 4-methoxy-2-nitro $4-CH_3O-2-NO_2C_6H_3NHCOCH_3$	210.19	ye nd (al)	...	116.5-7	...	...	al, eth, bz	B13^3, 1208
159	Acetanilide, 4-methoxy-3-nitro $4-CH_3O-3-NO_2C_6H_3NHCOCH_3$	210.19	og-ye nd (w or dil al)	...	153	...	...	al, lig	B13^1, 186
160	Acetanilide, 5-methoxy-2-nitro $5-CH_3O-2-NO_2C_6H_3NHCOCH_3$	210.19	wh nd (al)	...	125	...	...	al, lig, w	B13^3, 976
161	Acetanilide, *N*-methyl or Exalgin $C_6H_5N(CH_3)COCH_3$	149.19	nd (eth) pr (al) lf (lig)	253^{712}	102-4	1.0036$^{105/4}$	1.576	w, al, eth, lig	B12^3, 465
162	Acetanilide, *N*-methyl-4-hydroxy-3-nitro $3-NO_2-4-HOC_6H_3N(CH_3)NOCH_3$	210.19	(al)	...	161-2	...	...	al	B13^1, 186
163	Acetanilide, *N*-methyl-4-nitro $4-NO_2-C_6H_4N(CH_3)COCH_3$	194.19	pl (w)	...	153	...	...	al, eth	B12^1, 352

No.	Name, Synonyms, and Formula	Mol. wt.	Color, crystalline form, specific rotation and λ_{max} (log ε)	b.p. °C	m.p. °C	Density	n_D	Solubility	Ref.
164	Acetanilide, *N*-methyl-2-methyl or *N*-methyl-*o*-acetotoluidide $2\text{-}CH_3C_6H_4N(CH_3)COCH_3$	163.22	....	260	55-6	....	....	al	$B12^3$, 1854
165	Acetanilide, *N*-methyl-3-methyl or *N*-methyl-*m*-acetotoluidide $3\text{-}CH_3\text{-}C_6H_4N(CH_3)COCH_3$	163.22	cr	....	64	....	....	al	$B12^3$, 1963
166	Acetanilide, *N*-methyl-4-methyl or *N*-methyl-*p*-acetotoluidide $4\text{-}CH_3\text{-}C_6H_4N(CH_3)COCH_3$	163.22	lf (eth-al)	263	83	....	....	al, eth, lig	$B12^3$, 2052
167	Acetanilide, 4-methyl-2-nitro $4\text{-}CH_3\text{-}2\text{-}NO_2C_6H_3NHCOCH_3$	194.19	dk ye nd (peth)	....	96	....	....	....	$B12^3$, 2178
168	Acetanilide, *o*-nitro $2\text{-}NO_2C_6H_4NHCOCH_3$	180.16	ye pr (lig) lf (dil al)	$100^{0.1}$	94	1.419^{15}	....	w, al, eth, lig	$B12^3$, 1523
169	Acetanilide, *m*-nitro $3\text{-}NO_2C_6H_4NHCOCH_3$	180.16	wh lf (al)	$100^{0.008}$	154-6	....	....	w, al, chl	$B12^3$, 1551
170	Acetanilide, *p*-nitro $4\text{-}NO_2C_6H_4NHCOCH_3$	180.16	ye pr (w)	$100^{0.008}$	216 (217)	....	....	al, eth, aa, lig	$B12^3$, 1594
171	Acetanilide, *p*-octyl $4\text{-}(C_8H_{17})C_6H_4NHCOCH_3$	247.38	lf or pr (al)	....	94	....	....	al, eth	B12, 1185
172	Acetanilide, *N*-phenyl or *N*-acetyldiphenylamine $(C_6H_5)_2NCOCH_3$	211.26	rh or nd (w or lig)	sub	103	....	....	al	$B12^3$, 468
173	Acetanilide, *N*-propyl $C_6H_5N(C_3H_7)COCH_3$	177.25	mcl lf (eth, lig)	266^{712}	49 (56)	....	....	al, eth	$B12^3$, 466
174	Acetanilide, *N*-isopropyl $C_6H_5N(i\text{-}C_3H_7)COCH_3$	177.22	lf (lig)	$262\text{-}3^{712}$	39	....	....	lig	$B12^3$, 466
175	Acetic acid or Ethanoic acid CH_3CO_2H	60.05	rh (hyg)	117.9, 17^{10}	16.6	$1.0492^{20/4}$	1.3716^{20}	**w, al, ace, bz**	$B2^4$, 94
176	Acetic acid, allyl ester or Allyl acetate $CH_3CO_2CH_2CH{=}CH_2$	100.12	....	103.5	......	$0.9276^{20/4}$	1.4049^{20}	**al, eth,** ace, w	$B2^4$, 180
177	Acetic acid, anhydride or Acetic anhydride $(CH_3CO)_2O$	102.09	....	139.55, 44^{15}	-73.1	$1.0820^{20/4}$	1.39006^{20}	al, **eth**, bz, w	$B2^4$, 386
178	Acetic acid, anhydride, trifluoro or Trifluoro acetic anhydride $(F_3CO)_2O$	210.03	....	39.5-40.1	-65	$1.490^{25/4}$	1.269^{25}	eth, aa	$B2^4$, 469
179	Acetic acid, benzyl ester or Benzyl acetate $CH_3CO_2CH_2C_6H_5$	150.18	....	215.5, $93\text{-}4^{10}$	-51.3	$1.0550^{20/4}$	1.5232^{20}	**al**, eth, ace	$B6^4$, 2262
180	Acetic acid, dibenzyl or Dibenzylacetic acid $(C_6H_5CH_2)_2CHCO_2H$	240.3	pl (peth), dil (aa), nd (w)	235^{18}	89	....	....	al, eth, bz	$B9^3$, 3356
181	Acetic acid, dibenzyl, methyl ester or Methyl dibenzylacetate $(C_6H_5CH_2)_2CHCO_2CH_3$	254.33	nd (al)	....	42-3	....	....	al, peth	$B9^2$, 475
182	Acetic acid, 2-bromo ethyl ester or (2-Bromoethyl) acetate $CH_3CO_2CH_2CH_2Br$	167.00	....	162-3	-13.8	$1.514^{20/4}$	1.457^{25}	w, al, **eth**, chl	$B2^4$, 136
183	Acetic acid, bromomethyl ester $CH_3CO_2CH_2Br$	152.98	....	$130\text{-}3^{750}$	......	$1.6350^{20/4}$	1.4520^{20}	al, eth, chl	$B2^4$, 280
184	Acetic acid, 2-butenyl ester or Crotyl acetate $CH_3CO_2CH_2CH{=}CHCH_3$	114.14	....	132	......	$0.9192^{20/4}$	1.4181^{20}	al, eth, bz	$B2^4$, 183
185	Acetic acid, butyl ester or Butyl acetate $CH_3CO_2C_4H_9$	116.16	....	126.5	-77.9	$0.8825^{20/4}$	1.3941^{20}	**al, eth**, bz	$B2^4$, 143
186	Acetic acid, *iso*-butyl ester or *iso*-Butyl acetate $CH_3CO_2\text{-}i\text{-}C_4H_9$	116.16	....	117.2	-98.58	$0.8712^{20/4}$	1.3902^{20}	**al, eth**, ace	$B2^4$, 149
187	Acetic acid, *sec*-butyl ester *d* or *sec*-butyl acetate $CH_3CO_2CH(CH_3)CH_2CH_3$	116.16	$[\alpha]^{20/D}$ + 25.43	112	......	$0.8758^{16/4}$	1.3877^{20}	al, eth, ace	$B2^3$, 241
188	Acetic acid, *sec*-butyl ester *dl* or *sec*-butyl acetate *dl* $CH_3CO_2CH(CH_3)CH_2CH_3$	116.16	....	112.2	......	$0.8716^{20/4}$	1.3888^{20}	al, eth, ace	$B2^4$, 148
189	Acetic acid, *sec*-butyl ester *l* or *sec*-butyl acetate *l* $CH_3CO_2CH(CH_3)CH_2CH_3$	116.16	$[\alpha]^{19/546}$ – 20.2	116-7	......	$0.87301^{19/4}$	1.3899^{18}	al, eth, ace	$B2^3$, 241
190	Acetic acid, *tert*-butyl ester or *tert*-butyl acetate $CH_3CO_2C(CH_3)_3$	116.16	....	97-8	......	$0.8665^{20/4}$	1.3855^{20}	al, eth, aa	$B2^4$, 151
191	Acetic acid, (1-chloroethyl) ester $CH_3CO_2CHCl\text{-}CH_3$	122.55	....	121.3^{746}	......	$1.110^{20/4}$	1.409^{20}	eth	$B2^4$, 282
192	Acetic acid, (2 chloroethyl) ester $CH_3CO_2CH_2CH_2Cl$	122.55	....	145, 50^{18}	......	$1.178^{20/4}$	1.4234^{20}	**al, eth**	$B2^4$, 136
193	Acetic acid, (2-chloro-*iso*-propyl) ester $CH_3CO_2CCl(CH_3)_2$	136.58	....	149-50	......	1.0788^{20}	1.4223^{20}	al, eth	$B2^4$, 286
194	Acetic acid, chloromethyl ester $CH_3CO_2CH_2Cl$	108.52	....	$115\text{-}6^{751}$	......	$1.194^{20/4}$	1.409^{20}	al, eth	$B2^4$, 280

PHYSICAL CONSTANTS OF ORGANIC COMPOUNDS (Continued)

No.	Name, Synonyms, and Formula	Mol. wt.	Color, crystalline form, specific rotation and λ_{max} (log ε)	b.p. °C	m.p. °C	Density	n_D	Solubility	Ref.
195	Acetic acid, (3-chloropropyl) ester $CH_3CO_2CH_2CH_2CH_2Cl$	136.58		163-5^{747}, 62-3^{10}		1.250^{19}	1.431^{20}	al, eth, ace	B2^4, 140
196	Acetic acid, cyclohexyl ester or Cyclohexyl acetate $CH_3CO_2C_6H_{11}$	142.20		173, 62-3^{12}		0.9698$^{20/4}$	1.4401^{20}	**al, eth**	B6^4, 36
197	Acetic acid, cyclopentyl ester or Cyclopentyl acetate $CH_3CO_2C_5H_9$	128.17		51.5-2.5^{12}		0.9522^{16}			B6^4, 7
198	Acetic acid, decyl ester or Decyl acetate $CH_3CO_2C_{10}H_{21}$	220.32		244, 125.8^{155}	-15.05	0.8671$^{20/4}$	1.4273^{20}	al, eth, bz, aa	B2^4, 168
199	Acetic acid, 1,3-dichloro-*iso*-propyl ester $CH_3CO_2CH(CH_2Cl)_2$	171.02		202-8, 81^{15}		1.281^{20}	1.4542^{20}	al, eth, chl	B2^3, 233
200	Acetic acid, 1,1-dimethyl butyl ester $CH_3CO_2C(CH_3)_2C_3H_7$	144.21		152-3^{755}, 34.5^{10}		0.8798$^{18/4}$	1.4068^{10}	al, eth	B2^3, 257
201	Acetic acid, (2,4-dinitrophenyl) ester $CH_3CO_2[C_6H_3(NO_2)_2$-2.4]	226.15	cr (MeOH)		72-3				B6^4, 1380
202	Acetic acid, 3,5,-dinitro phenyl ester $CH_3CO_2[C_6H_3(NO_2)_2$-3,5]	226.15	cr (bz-peth)		126-7			bz, aa	B6, 258
203	Acetic acid, dodecyl ester $CH_3CO_2C_{12}H_{25}$	228.38		260-70			**1.4439^{20}**		B2^4, 170
204	Acetic acid, ethyl ester or Ethyl acetate $CH_3CO_2C_2H_5$	88.11		77.06	-83.6	0.9003$^{20/4}$	1.3723^{20}	w, **al, eth,** ace, bz	B2^4, 127
205	Acetic acid, 2-ethylbutyl ester $CH_3CO_2CH_2CH(C_2H_5)_2$	144.21		162-3,63^{20}		0.8790$^{20/4}$	1.4109^{20}	al, eth	B2^4, 161
206	Acetic acid, 2-ethyl hexyl ester $CH_3CO_2CH_2CH(C_2H_5)C_4H_9$	172.27		199,95^{25}	-93	0.8734$^{20/20}$	1.4204^{20}	al, eth	B2^4, 166
207	Acetic acid, 2-ethoxyethyl ester or Cellosolve acetate $CH_3CO_2CH_2CH_2OH$	132.16		156.4,49^{12}	-61.7	0.9740$^{20/4}$	1.4054^{20}	w, **al, eth**, ace	B2^4, 214
208	Acetic acid, 3-ethyl-3-pentyl ester $CH_3CO_2C(C_2H_5)_3$	158.24		160-3				al	B2, 134
209	**Acetic acid, 9-fluorenyl** **$C_{13}H_9CH_2CO_2H$**	224.26	mcl nd (al)	218-20$^{\prime\prime}$	138-9				B9^3, 3448
210	Acetic acid, furfuryl ester or Furfuryl acetate $CH_3CO_2CH_2(C_4H_3O)$	140.14		177		1.1175$^{20/4}$	1.4327^{20}	al, eth	B17^4, 1246
211	Acetic acid, 2-furylester $CH_3CO_2(C_4H_3$-O)	126.11		102-4$^{0.4}$	68-9			w, bz, peth, MeOH	B17^4, 1220
212	Acetic acid, 1-heptylester $CH_3CO_2C_7H_{15}$	158.24		192.4,96^{28}	-50.2	0.8750$^{15/4}$	1.4150^{20}	al, eth	B2^4, 162
213	Acetic acid, 4-heptyl ester $CH_3CO_2CH(C_3H_7)_2$	158.24		170-2,60-70^{17}		0.8742$^{0/0}$	1.4105^{19}	eth	B2^4, 163
214	Acetic acid, hexadecylester or Cetyl acetate $CH_3CO_2C_{16}H_{33}$	284.48		220-5^{205}	α,-18.5, β, 24.2	0.8574$^{25/4}$	1.4438^{20}		B2^4, 171
215	Acetic acid, hexylester or Hexyl acetate $CH_3CO_2C_6H_{13}$	144.21		171.5,61.5^{12}	-80.9	0.8779$^{15/4}$	1.4092^{20}	al, eth	B2^4, 159
216	Acetic acid, 2-hexyl ester *d* $CH_3CO_2CH(CH_3)C_4H_9$	144.21		158,57^{20}		0.8658$^{18/4}$	1.4014^{25}	al, eth	B2^2, 145
217	Acetic acid, 2-hexyl ester *dl* $CH_3CO_2CH(CH_3)C_4H_9$	144.21		157-8		0.8651$^{15/15}$	1.4014^{25}	al, eth	B2^3, 256
218	Acetic acid, 3-hexyl ester *d* $CH_3CO_2CH(C_2H_5)C_3H_7$	144.21	$[\alpha]^{20/}_{D}$ + 0.55	149-51		0.8672$^{20/4}$	1.4037^{20}	al, eth	B2^4, 160
219	Acetic acid, hydrazide $CH_3CONHNH_2$	74.08		129^{18}	67			w, al, eth	B2^4, 435
220	Acetic acid, (2-hydroxyethyl) ester or Glycol mono acetate $CH_3CO_2CH_2CH_2OH$	104.11		187-9		1.108^{15}		**w, al, eth**	B2^4, 214
221	**Acetic acid, menthyl ester *l* or Menthyl acetate *l*** **$CH_3CO_2C_{10}H_{19}$**	198.31	$[\alpha]^{20/D}$- 79.42	109^{10}		0.9185$^{20/4}$	1.4469^{20}		B6^4, 153
222	Acetic acid, (4-methoxybenzyl) ester $CH_3CO_2CH_2C_6H_4OCH_3$-4	180.20		110-7		1.104-7$^{25/25}$			B6^3, 4549
223	Acetic acid, (2-methoxyethyl) ester $CH_3CO_2CH_2CH_2OCH_3$	118.13		144-5,40-1^{12}		1.0090$^{19/19}$	1.4002^{20}	w, al, eth	B2^4, 214
224	Acetic acid, (2-methoxy phenyl) ester $CH_3CO_2C_6H_4OCH_3$-2	166.18		134-3^{18}	31-2			al, eth	B6^3, 4227
225	Acetic acid, methyl ester or Methyl acetate $CH_3CO_2CH_3$	74.08		57	-98.1	0.9330$^{20/4}$	1.3595^{20}	al, eth, w, ace, bz, chl	B2^4, 122
226	Acetic acid, 2-methyl-2-butyl ester or *tert*-amyl acetate $CH_3CO_2C(CH_3)_2CH_2CH_3$	130.19		124-4.5		0.8740^{20}	1.4010^{20}	al, eth, ace	B2^4, 156
227	Acetic acid, 3-methyl butyl ester or *iso*Pentyl acetate $CH_3CO_2CH_2CH_2CH(CH_3)_2$	130.19		142	-78.5	0.8670$^{20/4}$	1.4003^{20}	**al, eth,** ace	B2^4, 157

No.	Name, Synonyms, and Formula	Mol. wt.	Color, crystalline form, specific rotation and λ_{max} (log ε)	b.p. °C	m.p. °C	Density	n_D	Solubility	Ref.
228	Acetic acid, methylene diester or Methanediol diacetate $CH_2(O_2CCH_3)_2$	132.13	...	164-5	-23	$1.136^{20/4}$	1.4025^{24}	al, eth	B2[4], 280
229	Acetic acid, (2-methyl-3-heptyl) ester CH_3CO_2-$CH(C_4H_9)CH(CH_3)_2$	172.27	...	172	...	6.875^{20}	1.4166^{20}	al	B2, 134
230	Acetic acid, 3-methyl-2-heptyl ester $CH_3CO_2CH(CH_3)CH(CH_3)C_4H_9$	172.27	...	185	...	$0.8545^{21/4}$	1.418^{21}	al	B2[2], 146
231	Acetic acid, 6-methyl-2-heptyl ester CH_3CO_2-$CH(CH_3)(CH_2)_3CH(CH_3)_2$	172.27	...	187-8^{768}	...	0.8474^{20}	1.4137^{20}	al	B2, 135
232	Acetic acid, 6-methyl-3-heptyl ester $CH_3CO_2CH(C_2H_5)CH_2CH_2CH(CH_3)_2$	172.27	...	184-5	...	0.8554^{20}	1.41602^{20}	al	B2, 135
233	Acetic acid, (2-methyl-1-naphthyl) ester 1-CH_3CO_2-$C_{10}H_6$-CH_3-2	200.24	nd (eth-peth)	...	81-2	...	...	eth	B6[3], 3027
234	Acetic acid, (4-methyl-l-naphthyl) ester 1-$CH_3CO_2C_{10}H_6CH_3$-4	200.24	nd (eth-peth)	...	86.8	...	...	...	C64, 19519
235	Acetic acid, (6-methyl-1-naphthyl) ester or 5-acetoxy-2-methyl naphthalene 5-$CH_3CO_2C_{10}H_6CH_3$-(2)	200.24	...	124^{2}	...	...	...	...	B6[3], 3028
236	Acetic acid, (7-methyl-l-naphthyl) ester or l-Acetoxy-7-methyl naphthalene 1-$CH_3CO_2C_{10}H_6$-CH_3-7	200.24	...	188^{15}	39.41	...	...	...	B6[3], 3030
237	Acetic acid, (2-methyl-3-pentyl) ester $CH_3CO_2CH(C_2H_5)CH(CH_3)_2$	144.21	...	148.5^{747}	...	0.8688^{20}	...	al, eth	B2[4], 160
238	Acetic acid, (3-methyl-3-pentyl) ester $CH_3CO_2C(C_2H_5)_2CH_3$	144.21	...	148	...	$0.8834^{20/20}$	1.4109^{18}	al, eth	B2[4], 161
239	Acetic acid, (4-methyl-2-pentyl) ester $CH_3CO_2CH(CH_3)CH_2CH(CH_3)_2$	144.21	...	147-8	...	0.8805^{0}	1.3980^{20}	al, eth	B2[4], 161
240	Acetic acid, 2-methylpentyl ester $CH_3CO_2CH_2CH(CH_3)C_3H_7$	144.21	...	162.2^{746}	...	$0.8717^{25/25}$	...	al, eth	B2[4], 160
241	Acetic acid, α-naphthyl ester or α-naphthyl acetate CH_3CO_2-α-$C_{10}H_7$	186.21	nd or pl (al)	...	49	...	...	al, eth	B6[3], 2928
242	Acetic acid, β-naphthyl ester or β-naphthyl acetate CH_3CO_2-β-$C_{10}H_7$	186.21	nd (al)	...	70	...	...	al, eth	B6[3], 2982
243	Acetic acid, (2-nitrophenyl) ester $CH_3CO_2C_6H_4NO_2$-2	181.15	nd or pr (lig)	253d	40-1	...	...	al, eth, ace, bz, w	B6[4], 1256
244	Acetic acid, 3-nitrophenyl ester CH_3CO_2-$C_6H_4NO_2$-3	181.15	nd (peth)	...	55-6	...	...	al, w, lig	B6[4], 1273
245	Acetic acid, (4-nitro phenyl) ester $CH_3CO_2C_6H_4NO_2$-4	181.15	lf (dil al)	...	81-2	...	...	al, bz, w, lig	B6[4], 1298
246	Acetic acid, octadecyl ester or Stearyl acetate $CH_3CO_2C_{18}H_{36}$	313.51	...	222-3^{15}	34.5	...	...	al	B2[4], 171
247	Acetic acid, octyl ester or Octyl acetate $CH_3CO_2C_8H_{17}$	172.27	...	210, 112-3^{30}	-38.5	$0.8705^{20/4}$	1.4150^{20}	al, eth	B2[4], 165
248	Acetic acid, 2-octylester *d* $CH_3CO_2CH(CH_3)C_6H_{13}$	172.27	$[\alpha]^{20/}{}_{D}$ + 7.00	196	...	$0.8606^{15/4}$	1.4141^{20}	al, eth, bz	B2[3], 260
249	Acetic acid, 2-octylester *dl* $CH_3CO_2CH(CH_3)C_6H_{13}$	172.27	...	194.5^{749}	...	$0.8626^{14/4}$	1.4146^{20}	al, eth	B2[4], 165
250	Acetic acid, 2-octyl ester *l* $CH_3CO_2CH(CH_3)C_6H_{13}$	172.27	$[\alpha]^{14.5/D}$ - 6.0	196, 89-90^{17}	...	$0.8570^{20/4}$	1.4140^{20}	al, eth	B2[4], 165
251	Acetic acid, 3-octyl ester *l* $CH_3CO_2CH(C_2H_5)C_5H_{11}$	172.27	$[\alpha]^{20/D}$ − 4.3	...	191-1.5, 56.5^{2}	$0.8641^{20/4}$	1.4152^{20}	al, eth, CS_2	B2[4], 166
252	Acetic acid, pentyl ester or Pentyl acetate $CH_3CO_2C_5H_{11}$	130.19	...	149.25	-70.8	$0.8756^{20/4}$	1.4023^{20}	al, eth	B2[4], 152
253	Acetic acid, 2-pentyl ester *d* or See-pentyl acetate *d* $CH_3CO_2CH(CH_3)C_3H_7$	130.19	$[\alpha]^{20/D}$ + 17.16 (undil)	130-1	...	$0.8692^{18/4}$	1.3960^{20}	al, eth, ace	B2[1], 60
254	Acetic acid, 2-pentyl ester *dl* $CH_3CO_2CH(CH_3)C_3H_7$	130.19	...	134	...	$0.8692^{18/4}$	1.3960^{20}	al, eth	B2[4], 155
255	Acetic acid, 2-pentyl ester *l* $CH_3CO_2CH(CH_3)C_3H_7$	130.19	$[\alpha]^{20/D}$ + 3.30	142	...	0.8803^{15}	1.4012^{20}	al, eth	B2, 132
256	Acetic acid, 3-pentyl ester $CH_3CO_2CH(C_2H_5)_2$	130.19	...	132^{741}	...	$0.8712^{20/4}$	1.4005^{20}	al, eth	B2[4], 155
257	Acetic acid, phenacyl ester or ω-aceto aceto phenone $CH_3CO_2CH_2COC_6H_5$	178.19	rh pl	270	49-9.5	$1.1169^{65/4}$	1.5036^{65}	al, eth	B8[3], 301
258	Acetic acid, (phenoxyethyl) ester $CH_3CO_2CH_2CH_2OC_6H_5$	180.20	...	109-12^{3}	...	1.1082^{20}	1.5080^{20}	...	B6[4], 575
259	Acetic acid, phenyl ester $CH_3CO_2C_6H_5$	136.15	...	195.7, 75-6^{8}	...	$1.0780^{20/4}$	1.5035^{20}	al, eth	B6[4], 611

PHYSICAL CONSTANTS OF ORGANIC COMPOUNDS (Continued)

No.	Name, Synonyms, and Formula	Mol. wt.	Color, crystalline form, specific rotation and λ_{max} (log ε)	b.p. °C	m.p. °C	Density	n_D	Solubility	Ref.
260	Acetic acid, 2-phenyl ethyl ester or β-phenyl ethyl acetate $CH_3CO_2CH_2CH_2C_6H_5$	164.20	...	$232.6, 153^{76}$	-31.1	$1.0883^{20/4}$	1.5171^{20}	al, eth	$B6^3$, 1709
261	Acetic acid, *iso*-propenyl ester $CH_3CO_2\text{-}C(CH_3)=CH_2$	100.12	...	$92\text{-}4^{732}$	-92.9	0.9090^{20}	1.4033^{20}	al, eth, ace	$B2^4$, 179
262	Acetic acid, propyl ester or Propyl acetate $CH_3CO_2C_3H_5$	102.12	...	101.6	-95	$0.8878^{20/4}$	1.3842^{20}	**al, eth**	$B2^4$, 138
263	Acetic acid, *iso*-propyl-ester or *iso*-propyl acetate $CH_3CO_2CH(CH_3)_2$	102.13	...	90	-73.4	$0.8718^{20/4}$	1.3773^{20}	al, **eth**, ace, w	$B2^4$, 141
264	Acetic acid, tetrahydro furfuryl ester $CH_3CO_2CH_2(C_4H_7O)$	143.17	...	204-7	...	$1.0624^{20/4}$	1.4350^{25}	**w, al, eth,** chl	$B17^4$, 1103
265	Acetic acid, 2-thionyl or 2-Thiophene acetic acid $(2\text{-}C_4H_3S)CH_2CO_2H$	142.18	cr (w)	...	76	...	...	w, al, eth	$B18^4$, 4062
266	Acetic acid, 2-toly estor or *o*-Cresyl acetate $CH_3CO_2\text{-}C_6H_4CH_3\text{-}2$	150.18	...	$208, 89^{10}$	...	1.0533^{15}	1.5002^{20}	al, eth	$B6^4$, 1960
267	Acetic acid, 3-tolyl ester or *m*-Cresyl acetate $CH_3CO_2C_6H_4CH_3\text{-}3$	150.18	...	212	...	$1.043^{20/4}$	1.4978^{20}	**al, eth, bz**	$B6^4$, 2047
268	Acetic acid, 4-tolyl ester or *p*-Cresyl acetate $CH_3CO_2C_6H_4CH_3\text{-}4$	150.18	...	$212.5, 107\text{-}9^5$	...	$1.0512^{17/4}$	1.5163^{22}	al, eth, chl	$B6^4$, 2112
269	Acetic acid, 1,2,2-trimethyl propyl ester *d* $CH_3CO_2CH(CH_3)C(CH_3)_3$	144.21	$[\alpha]^{25/D}$ + 16.2	141^{756}	...	$0.856^{20/4}$	1.4001^{25}	eth	$B2^3$, 258
270	Acetic acid, vinyl ester or Vinyl acetate $CH_3CO_2CH=CH_2$	86.09	...	72.3	-93.2	$0.9317^{20/4}$	1.3959^{20}	**al**, eth, ace, bz, chl	$B2^4$, 176
271	Aceto acetamide, *n*-phenyl or Acetoacetanilide $CH_3COCH_2CONHC_6H_5$	177.20	pl or nd (bz or lig)	...	86	...	...	al, eth, bz, chl, liq	$B12^3$, 993
272	**Aceto acetamide, *N*-2-tolyl or Acetoacet-2-toluide $CH_3COCH_2CONH(C_6H_4CH_3\text{-}2)$**	191.23	pr (AcOEt)	...	107-8	...	...	al, bz	$B12^3$, 1895
273	Acetoacetamide, *N*-3-tolyl or Acetoacet-3-toluide $CH_3COCH_2CONH(C_6H_4CH_3\text{-}3)$	191.23	pl (bz-peth)	...	57-8	...	...	al, bz	$B12^3$, 1983
274	**Aceto acetamide, *N*-4-tolyl $CH_3COCH_2CONH(C_6H_4CH_3\text{-}4)$**	191.23	pr (AcOEt)	...	95	...	...	al, eth	$B12^3$, 2129
275	Aceto acetanilide, α-bromo $CH_3COCHBrCONHC_6H_5$	256.10	lf (al)	...	138d	...	...	al, eth, chl	$B12^4$, 276
276	Aceto acetanilide, α-chloro $CH_3COCHClCONHC_6H_5$	211.65	nd (al)	...	137.5	...	...	al, bz, chl, MeOH	$B12^3$, 995
277	Acetoacetic acid or 3-Oxobutanoic acid CH_3COCH_2COOH	102.09	syr	<100d	...	...	...	**w, al,** eth	$B3^4$, 1527
278	Acetoacetic acid-α-acetyl, ethyl ester or Ethyl α-acetyl acetoacetate $(CH_3CO)_2CHCO_2C_2H_5$	172.18	...	$209\text{-}11, 104^{16}$	...	1.1045^{204}	1.4690^{20}	al, eth, bz	$B3^4$, 1781
279	**Acetoacetic acid, allyl ester or Allyl acetoacetate $CH_3COCH_2CO_2CH=CH_2$**	128.13	...	$194\text{-}5^{737}, 66.5^{14}$	-85	$1.0385^{20/20}$	1.4398^{20}	**al, bz,** w, lig	$B3^4$, 1538
280	Aceto acetic acid, α-benzylidene, ethyl ester or Ethyl α-benzylidine acetoacetate $CH:COC(=CHC_6H_5)CO_2C_2H_5$	216.24	rh pl or pyr (dil al)	295-7	60-1	...	...	chl, al, eth, bz	$B10^3$, 3158
281	Aceto-acetic acid, α-bromo, ethyl ester or Ethyl α-bromoaceto acetate $CH_3COCHBrCO_2C_2H_5$	209.04	...	210-5d, $104\text{-}10^{15}$	...	$1.4294^{16/4}$	1.463^{14}	al, eth	$B3^4$, 1551
282	Aceto acetic acid, γ-bromo, ethyl ester $BrCH_2COCH_2CO_2H_5$	209.04	...	$114\text{-}7^{14}, 56^{0.005}$	...	$1.4840^{20/4}$	1.5281^{20}	al, eth	$B3^4$, 1551
283	Acetoacetic acid, butyl ester or Butyl aceto acetate $CH_3COCH_2CO_2C_4H_9$	158.20	...	$127^{50}, 85^8$	-35.6	$0.9671^{25/4}$	1.4137^{20}	al, bz, lig	$B3^4$, 1536
284	Acetoacetic acid, *iso*-butyl ester or *iso*-Butyl acetoacetate $CH_3COCH_2CO_2CH_2CH(CH_3)_2$	158.20	...	198-202	...	0.932^{23}	...	...	$B3^4$, 1536
285	Acetoacetic acid, α-chloro, ethyl ester $CH_3COCHClCO_2C_2H_5$	164.59	...	$197^{748}, 86\text{-}9^{12}$	...	$1.191^{14/17}$	1.4414^{20}	al, eth	$B3^4$, 1549
286	Acetoacetic acid, 2-chloro,ethyl estor $ClCH_2COCH_2CO_2C_2H_5$	164.59	...	220d, 115^{14}	-8	$1.2157^{20/4}$	1.4546^{17}	al, eth, ace, bz, chl	$B3^4$, 1550
287	Acetoacetic acid, α,α-dibromo, ethylester or Ethyl α, α-dibromoaceto acetate $CH_3COCBr_2CO_2C_2H_5$	287.94	...	$120\text{-}4^{13}$	...	...	...	al, eth	$B3^2$, 427
288	Acetoacetic acid, α,α-dichloro, ethyl ester $CH_3COCCl_2CO_2C_2H_5$	199.03	...	$205\text{-}7^{756}, 91^{11}$	...	$1.293^{16/17}$	1.4492^{17}	al	$B3^2$, 427
289	Aceto acetic acid, α,α-dimethyl, ethyl ester $CH_3COC(CH_3)_2CO_2C_2H_5$	158.20	...	184^{760}	...	$0.9777^{20/20}$	1.4180^{20}	al, eth	$B3^4$, 1594
290	Acetoacetic acid, ethyl ester or Ethylacetoacetate $CH_3COCH_2COOC_2H_5$	130.14	...	$180.4, 74^{14}$	<-80	$1.0282^{20/4}$	1.4194^{20}	w, **al, eth**. bz, chl	$B3^4$, 1528

PHYSICAL CONSTANTS OF ORGANIC COMPOUNDS (Continued)

No.	Name, Synonyms, and Formula	Mol. wt.	Color, crystalline form, specific rotation and λ_{max} (log ε)	b.p. °C	m.p. °C	Density	n_D	Solubility	Ref.
291	Aceto acetic acid, ethylester (enol form) or Ethyl aceteo-acetate $CH_3C(OH){=}CHCOOC_2H_5$	130.14				1.0119^{10}	1.4432^{20}		$B3^4$, 1528
292	Acetoacetic acid, ethyl ester (keto form) or Ethylaceto acetate $CH_3COCH_2COOC_2H_5$	130.14			-39	$1.0368^{10/4}$	1.4171^{20}		$B3^4$, 1528
293	Acetoacetic acid, *α*-ethyl, ethyl ester or Ethyl ethyl aceto acetate $CH_3COCH(C_2H_5)CO_2C_2H_5$	158.20		$190, 58^{3}$		$0.9847^{16/4}$	1.4214^{25}	al, eth	$B3^4$, 1592
294	Acetoacetic acid, *α*-ethyl, methyl ester $CH_3COCH(C_2H_5)CO_2CH_3$	144.17		$182, 79\text{-}80^{14}$		0.995^{14}		al, eth, ace	$B3^4$, 1592
295	Aceto acetic acid, methyl ester or Methyl aceto acetate $CH_3COCH_2CO_2CH_3$	116.12		$171.7, 60^{8}$	27-8	$1.0762^{20/4}$	1.4184^{20}	w, al, eth	$B3^4$, 1527
296	Aceto acetic acid, *α*-methyl, ethyl ester or Methyl ethyl aceto acetate $CH_3COCH(CH_3)CO_2C_2H_5$	144.17		$187, 44^{2}$		$0.9941^{20/4}$	1.4185^{20}	al, eth, ace	$B3^4$, 1573
297	Aceto acetic acid, *α*-methyl, methyl ester $CH_3COCH(CH_3)CO_2CH_3$	130.14		$177.4, 80^{20}$		$1.0247^{25/25}$	1.416^{24}	al, eth	$B3^3$, 1225
298	Aceto acetic acid, *α*-phenyl, ethyl ester $CH_3COCH(C_6H_5)CO_2C_2H_5$	206.24		156^{22}		$1.0855^{20/4}$	1.5176^{20}	al, eth	$B10^3$, 3047
299	Acetoacetic acid, *iso*-propyl ester or *iso*-Propyl aceto acetate $CH_3COCH_2CO_2CH(CH_3)_2$	144.17		$185\text{-}7, 75\text{-}6^{15}$	-27.3	$0.9835^{20/4}$	1.4173^{20}	**al, eth**, lig, w	$B3^4$, 1535
300	Acetoacetic acid, *α*-*iso*-propyl, ethyl ester $CH_3COCH(iC_3H_7)CO_2C_2H_5$	172.22		$201^{758}, 97\text{-}8^{20}$		$0.9648^{18/4}$	$1.4256^{18/4}$	**al, eth**	$B3^4$, 1608
301	Acetoacetic acid, 4,4,4-trifluoro, ethyl ester or ethyl-trifluoro aceto acetate $F_3CCOCH_2CO_2C_2H_5$	184.11		131.5^{757}		$1.2586^{15.5}$	1.37830	al, eth	$B3^4$, 1548
302	Acetoacetonitrile CH_3COCH_2CN	83.09		120-5				al, ace	$B3^4$, 1545
303	Acetoacetonitrile, *α*-phenyl $CH_3COCH(C_6H_5)CN$	159.19	pr (bz), cr (dil al or ace-peth)		90-1			al, eth, bz, chl	$B10^3$, 3048
304	Aceto acetyl chloride CH_3COCH_2COCl	120.54			-50			eth	$B3^4$, 1545
305	Acetoin or 3-hydroxy-2-butanone $CH_3CHOHCOCH_3$	88.11		143,37.1	-72	$1.0062^{20/20}$	1.4171^{20}	**w,** ace	$B1^4$, 3991
306	Acetone or 2-Propanone CH_3COCH_3	58.08		56.2	-95.35	$0.7899^{20/4}$	1.3588^{20}	**w, al, eth, bz, chl**	$B1^4$, 3180
307	Acetone azine $(CH_3)_2C{=}NN{=}C(CH_3)_2$	112.17		133	-12.5	0.83899^{20}	1.4535^{20}	**w, al, eth,** ace	$B1^4$, 3207
308	Acetone, amino, hydrochloride or Acetonylamine hydrochloride $CH_3COCH_2NH_2{:}HCl$	109.56			75			w, al, eth	$B4^3$, 877
309	Acetone, bromo or Bromoacetone CH_3COCH_2Br	136.98		$136.5^{725}, 31.5^{8}$	-36.5	1.634^{23}	1.4697^{15}	al, eth, ace	$B1^4$, 3223
310	Acetone, chloro or Chloroacetone CH_3COCH_2Cl	92.53		119^{763}	-44.5	1.15^{20}		w, al, eth, chl	$B1^4$, 3215
311	Acetone, 1-chloro-3-phenyl $C_6H_5CH_2COCH_2Cl$	168.62	nd (chl)	$159\text{-}61^{17}$	72-3				$B7^3$, 1041
312	Acetone, 1,3-diamino,dihydrochloride $(H_2NCH_2)_2CO\cdot 2HCl$	161.03	(dil al, dil aa), pr (w + l)		180d			w	$B4^2$, 763
313	Acetone, 1,1-dichloro $CH_3COCHCl_2$	126.97		$120, 47^{76}$		$1.3051^{18/15}$		al, **eth**	$B1^4$, 3218
314	Acetone, 1,3-dichloro $(ClCH_2)_2CO$	126.97	pr or nd	$173.4, 86\text{-}8^{12}$	45	$1.3826^{46/4}$	1.4716^{46}	w, al, eth	$B1^4$, 3219
315	Acetone, diethyl acetal or 2,2-Diothoxypropane $(CH_3)_2C(OC_2H_5)_2$	132.20		$114, 48^{60}$		$0.8200^{21/4}$	1.3891^{20}	al, eth, ace, bz	$B1^4$, 3200
316	Acetone, diethyl amino $CH_3COCH_2N(C_2H_5)_2$	129.20		$155\text{-}6d, 64^{16}$		$0.8620^{20/4}$	1.4249^{20}	**w, al, eth**	$B4^3$, 877
317	Acetone, 1,3-dihydroxy $(HOCH_2)_2CO$	90.08			89-91			**w, al,** ace	$B1^4$, 4119
318	Acetone, 2,4-dinitrophenyl hydrazone $(CH_3)_2C{=}NNH[C_6H_3(NO_2)_2\text{-}2,4]$	238.20	ye nd or pl (al)		128			bz, al, eth	$B15^3$, 427
319	Acetone, 1,1-diphenyl $CH_3COCH(C_6H_5)_2$	210.28		$306\text{-}7^{750}, 174\text{-}6^{10}$	46		1.5361^{16}		$B7^3$, 2171

No.	Name, Synonyms, and Formula	Mol. wt.	Color, crystalline form, specific rotation and λ_{max} (log ε)	b.p. °C	m.p. °C	Density	n_D	Solubility	Ref.
320	Acetone, 1,3-diphenyl or Dibenzyl ketone $(C_6H_5CH_2)_2CO$	210.28	cr (al, peth)	331, 112-25^{01}	35	1.195$^{0/4}$	...	al, eth, peth	B7^{3}, 2160
321	Acetone, dipropylamino $CH_3COCH_2N(C_3H_7)_2$	157.26	...	70-2^{6}	...	...	1.4267$^{25/D}$	...	B4^{3}, 877
322	Acetone, fluoro CH_3COCH_2F	76.07	...	75	...	...	1.3700$^{20/D}$	...	B1^{4}, 3213
323	Acetone, hexachloro or Perchloroacetone $(Cl_3C)_2CO$	264.75	...	202-4, 110^{40}	-2	1.444$^{12/12}$	1.5112^{20}	bz	B1^{4}, 3223
324	Acetone, hexafluoro $(CF_3)_2CO$	166.02	...	-28	-129	...	...	...	B1^{4}, 3215
325	Acetone, hydroxy or Acetol CH_3COCH_2OH	74.08	...	145-6d, 5 4^{18}	-7	1.0824$^{20/20}$	1.4295^{20}	w, al, eth	B1^{4}, 3977
326	Acetone, iodo or Iodoacetone CH_3COCH_2I	183.98	yesh (lig)	62^{12}	...	2.17^{15}	...	al	B1^{4}, 3226
327	Acetone, iodo, oxime or Iodoacetoxime $CH_3(C{=}NOH)CH_2I$	198.99	pr (peth)	...	64.5	...	...	...	B1, 660
328	Acetone, 3-methoxyphenyl $(3\text{-}CH_3OC_6H_4)CH_2COCH_3$	164.20	...	258-60, 95-7$^{0.7}$	...	1.0812^{0}	1.5230^{25}	...	B8^{3}, 397
329	Acetone, 4-methoxyphenyl $(4\text{-}CH_3OC_6H_4)CH_2COCH_3$	164.20	...	267-9, 142^{14}	<-15	1.0670$^{18/4}$	1.5233^{20}	al, eth	B8^{3}, 398
330	Acetone, 4-nitrophenylhydrazone $(CH_3)_2C{=}NNHC_6H_4NO_2\text{-}4$	193.21	...	...	152 (149)	...	...	al, eth	B15^{3}, 427
331	Acetone, oxime or Acetoxime $(CH_3)_2C{=}NOH$	73.09	pr (al)	134.8^{728}, 61^{20}	61	0.9113$^{62/4}$	1.4156^{20}	al, ace, w, liq	B1^{4}, 3202
332	Acetone, pentabromo $CHBr_2COCBr_3$	372.40	...	...	79-80	...	...	...	B1^{4}, 3226
333	Acetone, pentachloro $Cl_3CCOCHCl_2$	230.31	cr (w + 4)	192^{253}, 98^{40}	2.1 (an-kyd)	1.69$^{15/15}$	...	...	B1^{4}, 3222
334	Acetone, phenacyl or Phenacylacetone $C_6H_5COCH_2CH_2COCH_3$	176.22	ye oil	162^{12}	...	...	1.5250^{30}	ace	B7^{3}, 3509
335	Acetone, phenoxy $C_6H_5OCH_2COCH_3$	150.18	...	229-30, 117-9^{20}	...	1.0903$^{20/4}$	1.5228^{2}	eth, ace	B6^{4}, 604
336	Acetone, phenyl or methyl benzyl ketone $CH_3COCH_2C_6H_5$	134.18	...	216.5, 101^{14}	-15	1.0157$^{20/4}$	1.5168^{20}	al, eth, bz	B7^{3}, 1036
337	Acetone, phenyl hydrazone $(CH_3)_2C{=}NNHC_6H_5$	148.21	rh	163^{50}	42	...	...	al, eth	B15^{3}, 88
338	Acetone, semi carbazone $(CH_3)_2C{=}NNHCONH_2$	115.13	nd (w, ace)	...	190-1d	...	...	al, eth	B3^{4}, 179
339	Acetone, 1,1,1,3-tetrachloro $CH_2ClCOCCl_3$	195.86	liq (an hyd) pr (w + 4)	183, 71-2^{11}	46(+ 4 w) 65	1.624$^{15/4}$	1.497^{18}	eth, ace	B1^{4}, 3222
340	Acetone, 1,1,3,3-tetrachloro $(Cl_2CH)_2CO$	195.86	...	180-2^{718}	...		...	al, eth, ace, bz	B1^{4}, 3222
341	Acetone, 1,1,1-trichloro CH_3COCCl_3	161.42	...	149^{764}, 28^{10}	...	1.435$^{20/4}$	1.4635^{17}	al, eth	B1^{4}, 3222
342	Acetonitrile CH_3CN	41.05	...	81.6	-45.7	0.7857^{20}	1.34423^{20}	**w, al, eth, ace, bz**	B2^{4}, 419
343	Acetonitrile, cyclohexylidone $C_6H_{10}{=}CHCN$	121.18	...	107-8^{22}	...	0.9483$^{15/4}$	1.4832^{25}	al, eth	B9^{3}, 163
344	Acetonitrile, dibenzyl $(C_6H_5CH_2)_2CHCN$	221.30	lf or pr (al)	...	89-91	...	...	al, eth	B9^{3}, 3359
345	Acetonitrile, dichloro Cl_2CHCN	109.94	...	112-3	...	1.369^{20}	1.4391^{25}	al	B2^{4}, 506
346	Acetonitrile, diphenyl $(C_6H_5)_2CHCN$	193.25	pr (eth), lf (di al)	181-4^{12}	72-3	...	...	al, eth	B9^{3}, 3304
347	Acetonitrile, methoxy CH_3OCH_2CN	71.08	...	118.1	...	0.9492$^{20/4}$	1.3831^{20}	al, eth, ace	B3^{3}, 399
348	Acetonitrolic acid $CH_3C({=}NOH)NO_2$	104.07	ye rh (w, al, eth)	...	87-8d	...	...	w, al, eth, ace	B2^{4}, 434
349	Acetonyl acetate or 2-oxopropyl acetate $CH_3CO_2CH_2COCH_3$	116.12	...	170-1^{755}, 63^{11}	...	1.0757$^{20/4}$	1.4141^{20}	w, al eth	B2^{4}, 297
350	Acetonyl acetone or 2,5-Hexanedione $CH_3COCH_2CH_2COCH_3$	114.14	...	194^{754}, 89^{25}	-5.5	0.9737$^{20/4}$	1.4421^{20}	w, **al, eth**, ace, bz	B1^{4}, 3688
351	Acetophenone or Methyl phenyl ketone $C_6H_5COCH_3$	120.15	mcl pr or pl	202.6, 79^{10}	20.5	1.0281$^{20/4}$	1.53718^{20}	al, eth, ace, bz, chl	B7^{3}, 936
352	Acetophenone, 2-amino or o-acetyl aniline $2\text{-}H_2NC_6H_4COCH_3$	135.17	ye cr	250-2d, 135^{17}	20	...	1.6160^{20}	eth	B14^{3}, 80

PHYSICAL CONSTANTS OF ORGANIC COMPOUNDS (Continued)

No.	Name, Synonyms, and Formula	Mol. wt.	Color, crystalline form, specific rotation and λ_{max} (log ε)	b.p. °C	m.p. °C	Density	n_D	Solubility	Ref.
353	Acetophenone, 3-amino or *m*-acetyl aniline $3\text{-}H_2NC_6H_4COCH_3$	135.17	pa ye pl (al), lf (eth)	289-90	98-9	...	...	...	B14[1], 88
354	Acetophenone, 4-amino or *pp*-acetyl aniline $4\text{-}H_2NC_6H_4COCH_3$	135.17	ye mcl pr (al)	293-5, 195-200^{15}	106	...	...	al, eth	B14[1], 93
355	Acetophenone, *p*-amino, *α*-chloro or *p*-Amino phenacyl chloride $4\text{-}H_2NC_6H_4COCH_2Cl$	169.61	ye pl	...	148	...	...	...	B14[1], 99
356	Acetophenone, 4-amino-3-chloro- $4\text{-}H_2N\text{-}3\text{-}Cl\text{-}C_6H_3COCH_3$	169.61	pr (chl-peth)	...	92	...	...	...	B14, 49
357	Acetophenone, 3-amino-4-methoxy $3(H_2N)\text{-}4(CH_3O)C_6H_3COCH_3$	165.19	pr (al)	...	102	...	...	al, eth, bz	B14[1], 548
358	Acetophenone, *α*-bromo or Phenacyl bromide $C_6H_5COCH_2Br$	199.05	nd (al), rh pr (dil al), pl (peth)	135^{18}	50-1	$1.647^{20/4}$	...	al, eth, bz, chl	B7[3], 979
359	Acetophenone, 2-bromo or 2-Bromophenyl methyl ketone $3\text{-}BrC_6H_4COCH_3$	199.05	ye	$131\text{-}5^{20}$	...	...	1.5678^{20}	...	B7[3], 976
360	Acetophenone, 3-bromo or -3-bromophenyl methyl ketone $3\text{-}BrC_6H_4OCH_3$	199.05	...	131^{16}	7-8	...	1.5755^{20}	ace, bz	B7[3], 977
361	Acetophenone, 4-bromo or 4-Bromo phenyl methyl ketone $4\text{-}BrC_6H_4COCH_3$	199.05	lf (al)	255.5^{736}, 130^{11}	50-1	1.647	...	al, eth, bz, aa	B7[3], 977
362	Acetophenone, *α*-bromo-3-chloro or 3-Chlorophenacyl bromide $3\text{-}ClC_6H_4COCH_2Br$	233.49	nd	...	395-40	...	...	al	B7[3], 982
363	Acetophenone, *α*-bromo-4-chloro or 4-Chlorophenacyl bromide $4\text{-}ClC_6H_4COCH_2Br$	233.49	nd	...	96-7	...	...	...	B7[3], 983
364	Acetophenone, 4-bromo-*α*-chloro or 4-Bromophenacyl chloride $4\text{-}BrC_6H_4COCH_2Cl$	233.49	nd (al)	...	116-7	...	...	al	B7[3], 982
365	Acetophenone, *α* bromo-4-methyl or 4-Methyl phenacyl bromide $4\text{-}CH_3C_6H_4COCH_2Br$	213.07	nd or lf (al)	$155\text{-}9^{14}$	51	...	...	al, eth	B7[3], 1060
366	Acetophenone, 4-*tert*-butyl $4\text{-}(CH_3)_3CC_6H_4COCH_3$	176.26	...	$136\text{-}8^{20}$	...	0.9705^{0}	$1.518^{15/D}$	...	B7[3], 1171
367	Acetophenone, 4-*tert* butyl-2,6 dimethyl, 3.5-dinitro or Musk ketone $C_{14}H_{18}N_2O_5$	294.31	ye	...	134.5-6.5	...	...	...	B7[3], 1228
368	Acetophenone, *α*-chloro or Phenacyl chloride $C_6H_5COCH_2Cl$	154.60	pl (dil al), rh, lf (peth)	247, $139\text{-}41^{14}$	56.5	$1.324^{15/4}$	...	al, eth, ace, bz	B7[3], 967
369	Acetophenone, 2-chloro $2\text{-}ClC_6H_4COCH_3$	154.60	...	$227\text{-}8^{738}$, 113^{18}	...	$1.2016^{17/4}$	1.685^{25}	eth	B7[3], 1008
370	Acetophenone, 3-chloro $3\text{-}ClC_6H_4COCH_3$	154.60	...	$241\text{-}5^{744}$, 127-31^{30}	...	1.2130^{04}	1.5494^{20}	al, eth, ace	B7[2], 218
371	Acetophenone, 4-chloro $4\text{-}ClC_6H_4COCH_3$	154.60	...	273, 106^{10}	20	$1.1922^{20/4}$	1.5550^{20}	al, eth	B7[3], 1008
372	Acetophenone, *α*-chloro-4-methyl or 4-Methyl phenacyl chloride $4\text{-}CH_3C_6H_4COCH_2Cl$	168.62	nd (al)	260-3, 113^{4}	57-8	...	...	al, eth	B7[3], 1065
373	Acetophenone, *α*-chloro-2,4-dimethyl $2,4(CH_3)_2C_6H_3COCH_2Cl$	182.65	nd	...	62	...	...	al, eth, ace	B7[3], 1106
374	Acetophenone, *αα*-dibromo or Phenacylidene bromide $C_6H_5COCHBr_2$	277.94	...	$159\text{-}60^{13}$	36-7	...	...	al, eth, chl	B7[3], 985
375	Acetophenone, *α*, 4-dibromo or 4-Bromophenacyl bromide $4\text{-}BrC_6H_4COCH_2Br$	277.94	nd (al)	...	110-2	...	...	eth, al	B7[3], 985
376	Acetophenone, *αα*-dichloro or Phenacylidine chloride $C_6H_5COCHCl_2$	189.04	amor	249, 143^{25}	20-1	1.340^{16}	1.5686^{20}	al, bz	B7[3], 972
377	Acetophenone, *α*,4-dichloro or *p*-Chorophenacyl chloride $4\text{-}ClC_6H_4COCH_2Cl$	189.04	nd (al)	270	101-2	...	...	al, bz	B7[3], 972
378	Acetophenone, 2,4-dichloro $2,4\text{-}Cl_2\text{-}C_6H_3COCH_3$	189.04	...	245-7, $140\text{-}50^{15}$	33-4	...	1.5640^{20}	...	B7[3], 969
379	Acetophenone, 3,4-dichloro $3,4\text{-}Cl_2\text{-}C_6H_3COCH_3$	189.04	nd (peth)	135^{12}	76	...	...	...	B7[3], 971

No.	Name, Synonyms, and Formula	Mol. wt.	Color, crystalline form, specific rotation and λ_{max} (log ε)	b.p. °C	m.p. °C	Density	n_D	Solubility	Ref.
380	Acetophenone, 2,3-dihydroxy or 3-Acetylcatechol $2,3(HO)_2C_6H_3COCH_3$	152.15	ye pr (bz-lig)	.....	97-8	.....	.....	al	B8^3, 2080
381	Acetophenone, 2,4-dihydroxy or Resacetophenone $2,4\text{-}(HO)_2C_6H_3COCH_3$	152.15	nd or lf	.....	147	.....	1.1800^{141}	al, bz	B8^3, 2082
382	Acetophenone, 2,5-dihydroxy or 2- Acetyl hydroquinone $2,5(HO)_2C_6H_3COCH_3$	152.15	ye-gr nd (dil al or w)	.....	204-5	.....	.....	al	B8^3, 2100
383	Acetophenone, 3,4-dihydroxy or 4-Acetyl catechol $3,4(HO)_2C_6H_3COCH_3$	152.15	nd (w or chl)	.....	115-6	.....	.....	.....	B8^3, 2108
384	Acetophenone, 3,5-dihydroxy or 5-Acetyl resorcinol $3,5(HO)_2C_6H_3COCH_3$	152.15	cr (w)	.....	147-8	.....	.....	al, eth, ace, w	B8^2, 301
385	Acetophenone, 3,4-dimethoxy or Aceto veratrone $3,4(CH_3O)_2C_6H_3COCH_3$	180.20	pr (dil al)	286-8, 160-2^{15}	51	.....	.....	w, al, chl, bz	B8^3, 2110
386	Acetophenone, 3,5-dimethoxy-4-hydroxy or Aceto syringone $3,5(CH_3O)_2\text{-}4\text{-}(HO)C_6H_2COCH_3$	196.20	nd (w) pr (peth)	.....	122-3	.....	.....	al, eth, ace	B8^3, 2118
387	Acetophenone, 2,4-dimethyl $2,4(CH_3)_2C_6H_3COCH_3$	148.20	.....	228, 110^{13}	.....	1.0121^{15}	1.5340^{20}	al	B7^3, 1105
388	Acetophenone, 2,5-dimethyl $2,5,(CH_3)_2C_6H_3COCH_3$	148.20	.....	232-3, 107^{13}	.....	0.9963$^{19/4}$	1.5291^{20}	al, eth, bz	B7^3, 1104
389	Acetophenone, 3,4-dimethyl $3,4\text{-}(CH_3)_2C_6H_3COCH_3$	148.20	.....	246-7, 213^{310}	.....	1.0090$^{14/4}$	1.5413^{15}	al, eth, bz	B7^3, 1104
390	Acetophenone, 3-dimethylamino $3\text{-}(CH_3)_2NC_6H_4COCH_3$	163.22	.....	148^{13}	43	.....	.....	ace	B14, 45
391	Acetophenone, 4-dimethylamino $4(CH_3)_2NC_6H_4COCH_3$	163.22	nd (w, Peth)	172-5^{11}	105.5	.....	.....	eth, w, lig	B14^3, 94
392	Acetophenone, 2-ethoxy or 2-Acetylphenetole $2\text{-}C_2H_5OC_6H_4COCH_3$	164.20	pr (dil al), pl (lig)	243-4	43	1.0036^{78}	.....	al, eth, lig	B885
393	Acetophenone, 4-ethoxy or 4-Acetylphenetole $4\text{-}C_2H_5OC_6H_4COCH_3$	164.20	pl (eth)	.....	39	.....	.....	al, eth	B8^3, 280
394	Acetophenone, 4 fluoro $4\text{-}F\text{-}C_6H_4COCH_3$	138.14	.....	196^{760}, 79^{10}	-45	1.1382$^{25/4}$	1.5081^{25}	.....	B7^3, 961
395	Acetophenone, furfurylidene $C_6H_5COCH{=}CH(C_4H_3O)$	198.22	.....	317, 181^{29}	.....	1.1140^{20}	.....	al, eth	B17^2, 377
396	Acetophenone, α-hydroxy or Phenacylalcohol $C_6H_5COCH_2OH$	136.15	hex pl (al or eth), pl (w or dil al)	124-6^{12} (sub), 56^1	90	1.0963$^{99/4}$	.....	al, eth, lig	B8^3, 298
397	Acetophenone, α-hydroxy, acetate or Phenacyl acetate $C_6H_5COCH_2O_2CCH_3$	178.19	rh pl (eth, lig or peth)	270, 150-2^{19}	49.6	1.1169$^{65/4}$	.....	al, eth, bz, chl	B8^3, 301
398	Acetophenone, 2-hydroxy or 2-Acetylphenol $2\text{-}HOC_6H_4COCH_3$	136.15	.....	218, 106^{17}	4-6	1.1307$^{20/4}$	1.5584^{20}	al, eth, aa	B8^3, 261
399	Acetophenone, 3-hydroxy or 3-Acetylphenol $3\text{-}HOC_6H_4COCH_3$	136.15	nd or lf	296^{756}, 153^5	96	1.0992^{109}	1.5348^{109}	al, eth, bz	B8^3, 272
400	Acetophenone, 4-hydroxy or 4 Acetylphenol $4\text{-}HOC_6H_4COCH_3$	136.15	nd (eth, dil al)	147-8^3	109-10	1.1090^{109}	1.5577^{109}	al, eth	B8^3, 276
401	Acetophenone, α hydroxy-4-methoxy or 4-methoxy phenacyl alcohol $4\text{-}CH_3OC_6H_4COCH_2OH$	166.18	pl (dil al)	.....	104	.....	.....	al, bz	B8^3, 2120
402	Acetophenone, 2-hydroxy-3-methoxy or o Aceto vanillon $2(HO)\text{-}3(CH_3O)C_6H_3COCH_3$	166.18	pa ye nd (peth or eth-peth)	.....	53-4	.....	.....	eth, bz	B8^{3}2, 081
403	Acetophenone, 2-hydroxy-4-methoxy or Peonol $2\text{-}HO\text{-}4\text{-}CH_3OC_6H_3COCH_3$	166.18	nd (al)	158^{20}	52-3	1.3102^{81}	1.5452^{81}	al, eth, bz, chl	B8^3, 2084
404	Acetophenone, 2-hydroxy- 5-methoxy $2\text{-}HO\text{-}5\text{-}CH_3OC_6H_3COCH_3$	166.18	pa ye pr (dil al)	.....	52	.....	.....	bz, al	B14, 477
405	Acetophenone, 3-hydroxy-4-methoxy or Iso aceto-vanillone $3\text{-}HO\text{-}4\text{-}CH_3OC_6H_3COCH_3$	166.18	cr (eth lig) or cr (w + 1)	.....	67-8 (+ 1w) 91 anhyd	.....	.....	eth, w	B8^3, 2109
406	Acetophenone, 4-hydroxy-α-methoxy $4\text{-}HO\text{-}C_6H_4COCH_2OCH_3$	166.18	nd (bz), pr (w + 1)	.....	130-1	.....	.....	al, ace, eth, bz	B8^2, 302
407	Acetophenone, 4-hydroxy-3-methoxy or Resacetophenone-2-methylether iso peonol $4\text{-}HO\text{-}3\text{-}CH_3O\text{-}C_6H_3COCH_3$	166.18	nd (w)	.....	138	.....	.....	bz, w	B8^1, 617
408	Acetophenone, 4-hydroxy-3-methoxy or Acetovanillon $4\text{-}HO\text{-}3\text{-}CH_3OC_6H_3COCH_3$	166.18	pr (w)	295-300, 233-5$^{15\text{-}20}$	115	.....	.....	al, eth, ace, bz, chl	B8^3, 2108
409	Acetophenone, α-iodo or Phenacyl iodide $C_6H_5COCH_2I$	246.05	.....	158-60^{15}	34.4	.....	.....	ace	B7^3, 989

No.	Name, Synonyms, and Formula	Mol. wt.	Color, crystalline form, specific rotation and λ_{max} (log ε)	b.p. °C	m.p. °C	Density	n_D	Solubility	Ref.
410	Acetophenone, 2-iodo $2I\text{-}C_6H_4COCH_3$	246.05	...	139-140^{12}	...	1.746$^{20/4}$	1.6180^{20}	bz	B7^3, 988
411	Acetophenone, 3-iodo $3\text{-}I\text{-}C_6H_4COCH_3$	246.05	...	129^{8}	...	...	1.622^{20}	bz	B7^1, 988
412	Acetophenone, 4-iodo $4\text{-}I\text{-}C_6H_4COCH_3$	246.05	...	153^{18}	85	...	...	al, bz, eth, lig	B7^3, 989
413	Acetophenone, 2-Methoxy or 2,Acetylanisole $2\text{-}CH_3OC_6H_4COCH_3$	150.18	ye	245	...	1.0897$^{20/4}$	1.5393^{20}	al, ace	B8^3, 263
414	Acetophenone, 3-methoxy $3\text{-}CH_3OC_6H_4COCH_3$	150.18	...	240, 125-6^{12}	95-6	1.0343^{19}	1.5410^{20}	al, ace, w, eth	B8^3, 273
415	Acetophenone, 4-methoxy $4\text{-}CH_3O_6H_4COCH_3$	150.18	...	258, 138-9^{15}	38-9	1.0818$^{41/4}$	1.547^{41}	al, eth, ace	B8^3, 277
416	Acetophenone, 2-methyl or 2-acetyl toluene $2CH_3C_6H_4COCH_3$	134.18	...	214, 89-92^{10}	...	1.026$^{20/4}$	1.5276^{20}	...	B7^3, 1052
417	Acetophenone, 3-methyl or 3 Acetyltoluene $3\text{-}CH_3C_6H_4COCH_3$	134.18	...	220^{766}, 109^{12}	...	1.0070$^{20/4}$	1.5270^{20}	al, eth, ace	B7^3, 1057
418	Acetophenone, 4-methyl or 4-acetyl toluene $4\text{-}CH_3C_6H_4COCH_3$	134.18	nd	226, 113^{11}	28	1.0051$^{20/4}$	1.5335^{20}	al, eth, bz	B7^3, 1060
419	Acetophenone, 2-nitro $2\text{-}NO_2C_6H_4COCH_3$	165.15	...	158^{16}	28-9	...	1.5468^{20}	al, eth	B7^3, 990
420	Acetophenone, 3-nitro $3\text{-}NO_2C_6H_4COCH_3$	165.15	nd (al)	202, 167^{18}	81	...	...	eth, w	B7^3, 991
421	Acetophenone, 4-nitro $4\text{-}NO_2C_6H_4COCH_3$	165.15	yesh pr (al)	...	80-2	...	...	al, eth	B7^3, 993
422	Acetophenone oxime $C_6H_5C(:NOH)CH_3$	135.17	nd (w)	245, 119^{20}	60	...	...	al, eth, ace, bz, lig	B7^3, 954
423	Acetophenone, 5-*iso*-propyl-2-methyl or Carvacryl methyl ketone $5\text{-}i\text{-}C_3H_7\text{-}2\text{-}CH_3C_6H_3COCH_3$	176.26	...	249.50	<-20	0.956$^{20/4}$	1.5181^{20}	...	B7^3, 1176
424	Acetophenone, $\alpha\alpha\alpha$-trichloro $C_6H_5COCCl_3$	223.49	...	256-7, 145^{25}	...	1.425^{16}	...	al, eth	B7^3, 974
425	Acetophenone, $\alpha\alpha\alpha$ trifluoro $C_6H_5COCF_3$	174.12	...	152^{730}	-40	1.279^{20}	1.4583^{20}	...	B7^3, 962
426	Acetophenone, 2,4,5-trimethyl $2,4,5\text{-}(CH_3)_3C_6H_2COCH_3$	162.23	...	246-7, 137-8^{20}	10-1	1.0039$^{15/4}$	1.541^{15}	al, eth, bz, aa	B7^3, 1145
427	Acetophenone, 2,4,6-trimethyl $2,4,6\text{-}(CH_3)_3C_6H_2COCH_3$	162.23	...	240.5^{735}, 120^{12}	...	0.9754$^{20/4}$	1.5175^{20}	al, eth, ace, bz	B7^3, 1137
428	Acetoxyacetic acid $CH_3CO_2CH_2CO_2H$	118.09	nd (bz)	144-5^{12}	67-8	...	...	w, al, eth, ace, chl	B3^4, 576
429	Acetoxy acetic acid, ethylester or Ethyl acetoxyacetate $CH_3CO_2CH_2CO_2H_5$	146.14	...	179	...	1.0880$^{20/4}$	1.4112^{20}	al, eth, aa	B3^4, 580
430	(2-Acetoxypropyl) trimethyl ammonium chloride *D* or *o*-acetyl-B-methyl choline chloride $CH_3CH(O_2CCH_3)CH_2N+(CH_3)_3Cl\text{-}$	195.69	hyg. $[\alpha]_D+41.9$	d	172-3	...	...	w, al	B4^4, 1671
431	(2-Acetoxypropyl) trimethyl ammonium chloride *L* or *o*-Acetyl-β-methyl choline chloride $CH_3CH(O_2CCH_3)CH_2N+(CH_3)_3Cl\text{-}$	195.69	$[\alpha]_D\text{-}41.9$	d	172-3	...	...	...	B4^4, 1671
432	Acetyl acetone or 2-4-Pentanedione $CH_3COCH_2COCH_3$	100.12	...	139^{746}	-23	0.9721$^{25/4}$	1.4494^{20}	w, **al, eth, ace, chl**	B1^4, 3662
433	Acetyl benzoyl peroxide $CH_3CO\text{-}OO\text{-}COC_6H_5$	180.16	wb nd (lig)	85-100 (exp)	40-1	...	...	eth	B9^3, 1051
434	Acetyl bromide CH_3COBr	122.95	ye in air	76	-98	1.6625$^{16/4}$	1.45376^{16}	**eth**, ace, **bz, chl**	B2^4, 398
435	Acetyl chloride CH_3COCl	78.50	...	50.9	-112	1.1051$^{20/4}$	1.38976^{20}	**eth, ace, bz, chl**	B2^4, 395
436	Acetyl chloride, dichloro $Cl_2CHCOCl$	147.39	...	108-10	...	1.5315$^{16/4}$	1.4591^{20}	eth	B2^4, 504
437	Acetyl chloride, ethoxy $C_2H_5OCH_2COCl$	122.55	...	123-4, 49-50^{37}	...	1.1170	1.4204^{20}	eth, ace	B3^3, 396
438	Acetyl chloride, methoxy CH_3OCH_2COCl	108.52	...	99, 46-9^{62}	...	1.1871$^{20/4}$	1.4196^{20}	eth, ace, chl	B3^3, 396
439	Acetyl fluoride CH_3COF	62.04	...	20.8	...	1.002$^{15/4}$	...	**al, eth**, ace, bz	B2^4, 393
440	Acetyl iodide CH_3COI	169.95	...	108, 36^{50}	...	2.0674$^{20/4}$	1.5491^{20}	eth	B2^4, 399
441	Acetyl isothiocyanate CH_3CONCS	101.12	...	132-3, 30^{9}	...	1.1523$^{13/4}$	1.5231^{18}	eth	B3^3, 278

PHYSICAL CONSTANTS OF ORGANIC COMPOUNDS (Continued)

No.	Name, Synonyms, and Formula	Mol. wt.	Color, crystalline form, specific rotation and λ_{max} (log ε)	b.p. °C	m.p. °C	Density	n_D	Solubility	Ref.
442	Acetyl peroxide $CH_3CO\text{-}OO\text{-}COCH_3$	118.09	nd (eth), lf	63^{21}	30 (26)	...	...	al, eth	$B2^4$, 392
443	Acetylene $HC{\equiv}CH$	26.04	...	-84.0	-80.8	$0.6208^{-82/4}$	1.00051^0	ace, bz, chl	$B1^4$, 939
444	Acetylene-bromo $BrC{\equiv}CH$	104.93	...	4.7	...	...	...	eth	$B1^3$, 919
445	Acetylene-chloro $ClC{\equiv}CH$	60.48	...	-30	-126	...	...	...	$B1^4$, 957
446	Acetylene-dibromo $BrC{\equiv}CBr$	183.83	nd	76 (exp)	-25	...	...	al, eth, ace, bz	$B1^3$, 919
447	Acetylene-dichloro $ClC{\equiv}CCl$	94.93	...	exp	-66	...	...	al, eth, ace	$B1^4$, 957
448	Acetylene-diido $IC{\equiv}CI$	277.83	rh nd (lig)	32-3 (exp)	81-2	...	...	al, eth, ace, bz	$B1^4$, 958
449	Acetylene-diphenyl $C_6H_5C{\equiv}CC_6H_5$	178.23	mcl pr or pl (al)	300, 170^{19}	62.5	$0.9657^{100/4}$	...	eth	$B5^3$, 2119
450	Acetylene, *bis*(l-hydroxycyclohexyl) $(C_6H_{10}OHC)_2$	222.33	nd (ccl_4)	182^{13}	112-3	...	...	al, eth, ace	$B6^3$, 4741
451	Acetylene, l-phenyl-2-methyl $C_6H_5C{\equiv}CCH_3$	116.16	...	181	...	0.942^{15}	$1.563^{15/}{}_D$	...	$B5^2$, 408
452	Aconic acid $C_5H_3O_4$	127.08	lf (eth), rh (w, al)	d	164	...	...	w, al	$B18^4$, 5333
453	Aconine $C_{25}H_{41}NO_9$	499.60	amor, $[\alpha]^{20/D}$ + 23	...	132	...	...	w, al	$B21^4$, 2899
454	Aconitomide $H_2NCOCH_2C(CONH_2){=}CHCONH_2$	171.16	ye nd (w)	...	sinters 260	...	...	w	B2, 853
455	Aconitic acid, *cis* or 1, 2, 3-Propene tricarboxylic acid $HO_2CCH_2C(CO_2H){=}CHCO_2H$	174.11	nd (w)	...	130	...	...	w	$B2^4$, 2405
456	Aconitic acid, *trans* or 1, 2, 3-Propene tricarboxylic acid $HO_2CCH_2C(CO_2H){=}CHCO_2H$	174.11	lf (w), nd (w, eth)	...	198-9	...	...	w, al	$B2^4$, 2405
457	Aconitic acid, triethyl ester *trans* $C_2H_5O_2CH_2C(CO_2C_2H_5) = CHCO_2C_2H_5$	258.27	...	275d, 159^9	...	$1.1064^{20/4}$	1.4556^{20}	al, eth	$B2^4$, 2405
458	Aconitic acid, trimethylester $CH_3O_2CH_2C(CO_2CH_3) = CHCO_2CH_3$	216.19	...	270-1, 161^{14}	...	...	...	al, eth	$B2^4$, 2405
459	Aconitic acid, tripropylester $C_3H_7O_2CH_2C(CO_2C_3H_7) = CHCO_2C_3H_7$	300.35	...	195^{13}	...	$1.050^{25/4}$	1.4521^{20}	al, eth	$B2^4$, 2405
460	Aconitine $C_{34}H_{47}NO_{11}$	645.75	rh lf $[\alpha]^{20/D}$ + 19 (chl)	...	204	...	...	al, bz, chl	$B21^4$, 2901
461	Aconitine, hydrobromide $C_{34}H_{47}NO_{11} \cdot HBr \cdot 1\frac{1}{2} H_2O$	753.68	yesh pr (w), $[\alpha]_D$-31 (w,c=2.3)	...	209.10	...	...	w, al, eth	$B21^4$, 2903
462	Aconitine, hydrochloride $C_{34}H_{47}NO_{11} \cdot HCl \cdot 3\frac{1}{2} H_2O$	745.26	(w + 3 1/2), $[\alpha]_D$-30.5 (w)	...	170-2	...	...	w, al, eth	$B21^4$, 2902
463	Aconitine, nitrate $C_{34}H_{47}NO_{11}HNO_3$	708.76	$[\alpha]^{20/D}$-35 (2% aq soln)	...	ca. 200d	...	...	w	$B21^4$, 2903
464	Acridine or 2,3,5,6-Dibenzopyridine $C_{13}H_9N$	179.22	rh nd or pr (al)	345-6	111	$1.005^{20/4}$	...	al, eth, bz	$B20^4$, 3987
465	Acridine, 2-amino $C_{13}H_{10}N_2$,2-$H_2NC_{13}H_8N$	194.24	ye nd (w or al)	...	213-4	...	...	al, eth	$B22^4$, 4984
466	Acridine, 2-amino-5(4-aminophenyl) or Chrysaniline $C_{19}H_{15}N_3.2H_2O$	321.38	ye nd (95% al)	...	260-7	...	...	...	$B22^4$, 5513
467	Acridine, 3-amino $C_{13}H_{10}N_2$,3-$NH_2C_{13}H_8N$	194.24	ye nd (w or al + 1)	...	213-4	...	...	al, eth	$B22^4$, 4987
468	Acridine, 4-amino $C_{13}H_{10}N_2$,4-$NH_2C_{13}H_8N$	194.24	red-br pr (peth), og nd (MeOH)	183-4	108	...	...	al, eth, ace, bz	$B22^4$, 4989
469	Acridine, 4 amino,-hydrochloride $C_{13}H_{10}N_2.HCl$	230.70	ye nd (dil al)	...	234d	...	...	w, al	$B22^2$, 376
470	Acridine, 9-amino $C_{13}H_{10}N_2$	194.24	ye nd (ace or al)	...	241 (cor)	...	...	...	$B21^4$, 4174
471	Acridine, 9-chloro $C_{13}H_8ClN$	213.67	nd (al)	sub	122	...	...	al, w	$B20^4$, 3995
472	Acridine, 3,6-diamino $C_{13}H_{11}N_3$	209.25	ye nd (al or w)	...	284-6	...	...	al, eth, w	$B22^4$, 5487

PHYSICAL CONSTANTS OF ORGANIC COMPOUNDS (Continued)

No.	Name, Synonyms, and Formula	Mol. wt.	Color, crystalline form, specific rotation and λ_{max} (log ε)	b.p. °C	m.p. °C	Density	n_D	Solubility	Ref.
473	Acridine, 6,9-diamino-2-ethoxy or Rivanol $C_{15}H_{15}N_3O$	255.30	ye nd		123-44d				B22^4, 6679
474	Acridine, 6,9-dichloro-2-methoxy $C_{14}H_9NOCl_2$	278.14			163-5				B21^4, 1553
475	Acridine, 9,10-dihydro or Acridan, carbazine $C_{13}H_{11}N$	181.24	pl or pr (al)	sub 300d	169-71			al, eth, ace	B20^4, 3885
476	Acridine, 9,10-dihydro-9-oxo or Acridone $C_{13}H_9NO$	195.22	ye lf (al)	>354	154			ace	B21^4, 4171
477	Acridine, 4-hydroxy $C_{13}H_9NO$	195.22			117			al, eth	B21^4, 1562
478	Acridine, 2-methyl $C_{13}H_{11}N$	193.25	ye nd (dil al)		134			al, eth, bz	B20^4, 4037
479	Acridine, 9 phenyl $C_{19}H_{13}N$	255.32	lf ye nd (al)	401-4 sub	184				B20^4, 4355
480	Acrolein or Propenal $CH_2=CHCHO$	56.06		52.5-3.5	-86.9	$0.8410^{20/4}$	1.4017^{20}	w, al, eth, ace	B1^4, 3435
481	Acrolein, 2-chloro $CH_2=C(Cl)CHO$	90.51		40^{30}		1.199^{20}	1.463^{20}	**al**, eth	B1^4, 3440
482	Acrolein, diethylacetal or 3.3-diethoxy-1-propene $CH_2=CHCH(OC_2H_5)_2$	130.19		123.5		0.8543^{15}	1.4000^{20}	eth, CCl_4	B1^4, 3437
483	Acrolein 3(2-furyl) or Furacrolein $(2\text{-}C_4H_3O)\text{-}CH=CHCHO$	122.12	ye or wh nd	> 200d	54			**al**, eth	B17^4, 4695
484	Acrolein, 2-methyl or methacrolein $CH_2=C(CH_3)CHO$	70.09		68.4		$0.837^{20/4}$	1.4144^{20}	**w, al, eth**	B1^4, 3455
485	Acrylamide $CH_2=CHCONH_2$	71.08	lf (bz)		84-5			w, al, eth, chl	B2^4, 1471
486	Acrylic acid or Propenoic acid $CH_2=CHCO_2H$	72.06		141.6, 48.5^{15}	13	$1.0511^{20/4}$	1.4224^{20}	**w, al, eth**, ace, bz	B2^4, 1455
487	Acrylic acid, allyl ester or Allyl acrylate $CH_2=CHCO_2CH_2CH=CH_2$	112.13		172-4		$0.9441^{20/4}$	1.4320^{20}	al, eth	B2^4, 1468
488	Acrylic acid, β-benzoyl $C_6H_5COCH=CHCO_2H$	176.17	nd or pr (to)		99			al, eth, to	B10^1, 3144
489	Acrylic acid, benzyl ester or Benzyl acrylate $CH_2=CHCO_2CH_2C_6H_5$	162.19		228		$1.0573^{20/4}$	1.5143^{20}	al, eth, ace	B6^1, 1481
490	Acrylic acid, 2-bromo, ethyl ester or Ethyl 2-bromo acrylate $CH_2=C(Br)CO_2C_2H_5$	179.01		155-8		1.4581^{25}	1.4660^{25}		B2^4, 1487
491	Acrylic acid, butyl ester or Butylacrylate $CH_2=CHCO_2C_4H_9$	128.17		146-8,39^{10}	-64.6	$0.8898^{20/4}$	1.4185^{20}	al, eth, ace	B2^4, 1463
492	Acrylic acid, *iso*butylester or *iso*-Butylacrylate $CH_2=CHCO_2CH_2CH(CH_3)_2$	128.17		132		$0.8896^{20/4}$	1.4150^{20}	al, eth	B2^4, 1465
493	Acrylic acid, 2 chloro, ethyl ester or Ethyl-2-chloroacrylate $CH_2=C(Cl)CO_2C_2H_5$	134.56		$51\text{-}3^{18}$		$1.1404^{20/4}$	1.4384^{20}	al, eth	B2^4, 1482
494	Acrylic acid, 2-chloro, methyl ester or Methyl-2-chloroacrylate $CH_2=C(Cl)CO_2CH_3$	120.54		52^{51}		$1.189^{20/4}$	1.4420^{20}	eth	B2^4, 1482
495	Acrylic acid, 3-chloro-*cis* or *cis*-3-chloropropenoic acid $ClCH=CHCO_2H$	106.51	lf or nd (HCl)	$107^{17.5}$	63-4			al, eth	B2^4, 1481
496	Acrylic acid, 3-chloro-*trans* or *trans*-3-chloropropenoic acid $ClCH=CHCO_2H$	106.51	lf	94^{18}	86			al, eth	**B2^4, 1481**
497	Acrylic acid, cyclohexyl ester or Cyclohexyl acrylate $CH_2=CHCOOC_6H_{11}$	154.21		$182\text{-}4^{750}$, 88^{20}		$1.0275^{20/4}$	1.4673^{20}	**al, eth**, chl	**B6^4, 38**
498	Acrylic acid, 2,3-dichloro $ClCH=C(Cl)CO_2H$	140.95	mcl pr (chl)		87-8			w, al, eth, ace, chl	B2^4, 1484
499	Acrylic acid, 3,3-dichloro $Cl_2C=CHCO_2H$	140.95	nd (peth) pr (chl)	sub	76-7			eth, chl	B2^4, 1484
500	Acrylic acid, 2,3-diphenyl or α-phenyl cinnamic acid $C_6H_5CH=C(C_6H_5)CO_2H$	224.26	nd (lig, dil al)	sub	172.5-3			al, eth	B9^1, 3414
501	Acrylic acid, ethyl ester or Ethyl acrylate $CH_2=CH_1CO_2C_2H_5$	100.12		99.8	-71.2	$0.9234^{20/4}$	1.4068^{20}	**al, eth**, chl	**B2^4, 1460**
502	Acrylic acid, 3-(2-furyl)-*cis* or 2- Furanacrylic acid *cis* $(2\text{-}C_4H_3O)CH=CHCO_2H$	138.12	wh pr or pl		103-4			eth	**B18^4, 4143**
503	Acrylic acid, 3-(2-furyl) *trans* or 2-Furanacrylic acid *trans* $(2\text{-}C_4H_3O)CH=CHCO_2H$	138.12	nd (w)	286 (sub)	141		1.5286^{20}	al, eth, bz	**B18^4, 4173**

No.	Name, Synonyms, and Formula	Mol. wt.	Color, crystalline form, specific rotation and λ_{max} (log ε)	b.p. °C	m.p. °C	Density	n_D	Solubility	Ref.
504	Acrylic acid, 3-(2-furyl) *cis*, benzylester or Benzyl-2-furanacrylate $(2\text{-}C_4H_3O)CH=CHCH_2C_6H_5$	138.12	pa ye	$201\text{-}3^{12}$	42-5		1.5872^{25}	eth, ace, bz, al	$B18^4$, 4147
505	Acrylic acid, 3-(2-furul)*cis*,butyl ester or Butyl-2-furanacrylate $(2\text{-}C_4H_3O)CH=CHCO_2C_4H_9$	194.23		$147\text{-}50^{15}$, $117\text{-}8^{3}$		1.045^{20}	1.5129^{20}	al, ace	$B18^4$, 4146
506	Acrylic acid, 3-(2-furyl) *cis*, pentylester or Pentyl-2-furanacrylate $(2C_4H_3O)CH=CHCO_2C_5H_{11}$	208.26		$116\text{-}8^{2}$		$1.0322^{20/4}$	1.5289^{24}	al, eth	$B18^4$, 4146
507	Acrylic acid, 3-(2-furyl) *cis*, propyl ester or Propyl-2-furan-acrylate $(2\text{-}C_4H_3O)CH=CHCO_2C_3H_7$	180.20		$91\text{-}4^{3}$		$1.07\,44^{20/4}$	1.5392^{24}	al, eth, bz	$B18^4$, 4146
508	Acrylic acid, 3-(2-furyl) *cis*, methyl ester or Methyl-2-furanacrylate $(2\text{-}C_4H_3O)CH=CHCO_2CH_3$	152.15		227.5^{774}, 112^{15}	35-7		1.4447^{20}	al, eth, bz	$B18^4$, 4144
509	Acrylic acid, methyl ester or Methyl acrylate $CH_2=CHCO_2CH_3$	86.09		80.5	< -75	$0.9535^{20/4}$	1.4040^{20}	al, eth, ace, bz	$B2^4$, 1457
510	Acrylic acid, 2 methylbutyl ester or (2-methylbutyl) acrylate $CH_2=CHCO_2CH_2CH(CH_3)CH_2CH_3$	142.20		160, 52^{11}		$0.8936^{20/4}$	1.4240^{20}	al, eth	$B2^4$, 1525
511	Acrylic acid, 3-(α-naphthyl) *cis* $\alpha\text{-}C_{10}H_7CH=CHCO_2H$	198.22	cal pl (al)		156			al	$B9^3$, 3284
512	Acrylic acid, 3-(α-naphthyl) *trans* $\alpha\text{-}C_{10}H_7CH=CHCO_2H$	198.22	nd (al,w,aa)	sub	211-2			eth, chl	$B9^3$, 3284
513	Acrylic acid, 3 (β-naphthyl) *trans* $\beta\text{-}C_{10}H_7CH=CHCO_2H$	198.22	nd (al, w)		210			al	$B9^3$, 3288
514	Acrylic acid, 2-phenyl or 2-Phenylpropenoic acid, Atropic acid $CH_2=C(C_6H_5)CO_2H$	148.17	lf (al), nd (w)	267d	106-7			al, eth, bz, chl	$B9^3$, 2751
515	Acrylic acid, trichloro $CCl_2=CClCO_2H$	175.40		158			$1.5271^{18.5}$	bz	$B2^4$, 1486
516	Acrylonitrile $CH_2=CHCN$	53.06		77.5-9		$0.8060^{20/4}$	1.3911^{20}	al, eth, ace, bz	$B2^4$, 1473
517	Acrylonitrile, 3,3 diphenyl $(C_6H_5)_2C=CHCN$	205.26		$125\text{-}8^{0.15}$			1.6352^{20}		$B9^3$, 3439
518	Acrylonitrile, triphenyl $(C_6H_5)_2C=C(C_6H_5)CN$	281.36	nd (al), pr		166-7			eth, al	$B9^3$, 3630
519	Acrylylchloride $CH_2=CHCOCl$	90.51		75-6		$1.1136^{20/4}$	1.4343^{20}	chl	$B2^4$, 1471
520	Acrylyl chloride, trichloro $CCl_2=CClCOCl$	193.84		158			$1.5271^{18.5}$	bz	$B2^4$, 1486
521	Actidione *l* or Cycloheximide $C_{15}H_{23}NO_4$	281.33	pl (al) $[\alpha]^{25/D}$ -33 (chl C=l)		144-5			al	$B20^4$, 1406
522	Adamantane or Tricyclo *3,3,1,1*^{3,7}decane $C_{10}H_{16}$	136.24	nd (sub)	sub	268 (sealed)	1.07	1.568	bz	$B5^4$, 469
523	Adenosine or Adenine, 9-β-D-ribofuranosyladenine $C_{10}H_{13}N_5O_4$	267.24	nd (w + 1 1/2), $[\alpha]^{20/D}$ -60.0 (w,c=1)		235-6 (anh)			w	B31, 27
524	5-Adenylic acid or Adenosine-5-phosphate $C_{10}H_{14}N_5O_7P$	347.22	pw, nd, (w, dil al) cr $[\alpha]^{50/D}$ -41.78		195-208d (sealed tube)			w, al	B31, 27
525	Adipaldehyde or 1,6-Hexanedial $OHC(CH_2)_4CHO$	114.14		$92\text{-}4^{9}$	-8	$1.003^{19/4}$	1.4350^{20}	al, eth, bz, aa	$B1^4$, 3686
526	Adipamic acid or Adipic acid monoamide $H_2NCO(CH_2)_4CO_2H$	145.16	nd (w)		161-2				$B2^4$, 1972
527	Adipamide $H_2NCO(CH_2)_4CONH_2$	144.17	pl		220			al	$B2^4$, 1972
528	Adipic acid or Hexanedioc acid $HO_2C(CH_2)_4CO_2H$	146.14	mcl pr (w, ace, lig)	265^{100}	153	$1.360^{25/4}$		al, eth	$B2^4$, 1956
529	Adipic acid, 2-amino *-dl* $HO_2C(CH_2)_3CH(NH_2)CO_2H$	161.16	pl (w)		206 (anh)				$B4^4$, 1555
530	Adipic acid, dibutylester or di-butyl adipate $C_4H_9O_2C(CH_2)_4CO_2C_4H_9$	258.36		165^{10}	-32.4	$0.9615^{20/4}$	1.4369^{20}	al, eth	$B2^4$, 1961
531	Adipic acid, diisobutylester or Di-*iso*butyl adipate $i\text{-}C_4H_9O_2C(CH_2)_4CO_2\text{-}i\text{-}C_4H_9$	258.36		$186\text{-}8^{15.7}$		0.9543^{19}	1.4301^{20}		$B2^4$, 1962

No.	Name, Synonyms, and Formula	Mol. wt.	Color, crystalline form, specific rotation and λ_{max} (log ε)	b.p. °C	m.p. °C	Density	n_D	Solubility	Ref.
532	Adipic acid, diethyl ester or diethyl adipate $C_2H_5O_2C(CH_2)_4CO_2C_2H_5$	202.25	...	245	-19.8	$1.0076^{20/4}$	1.4272^{20}	al, eth	B2^{4}, 1960
533	Adipic acid, di-2 ethyl butyl ester or di-2-ethyl butyl adipate $(C_2H_5)_2CHCH_2O_2C(CH_2)_4CO_2CH_2CH(C_2H_5)_2$	314.47	...	200^{10}	-15	$0934^{25/4}$	1.4434^{20}	al, ace, aa	B2^{4}, 1964
534	Adipic acid, di-(2 ethylhexyl) ester or di-(2-ethylhexyl) adipate $C_8H_{17}O_2C(CH_2)_4CO_2C_8H_{17}$	370.57	...	214^{5}	-67.8	$0.922^{25/4}$	1.4474^{20}	al, eth, ac, aa	B2^{4}, 1964
535	Adipic acid, dimethyl ester or dimethyl adipate $CH_3O_2C(CH_2)_4CO_2CH_3$	174.20	cr	115^{13}	10.3	$1.0600^{20/4}$	1.4283^{20}	al, eth, aa	B2^{4}, 1959
536	Adipic acid, dipropyl ester or dipropyl adipate $C_3H_7O_2C(CH_2)_4CO_2C_3H_7$	230.30	...	151^{11}	-15.7	$0.9790^{20/4}$	1.4314^{20}	al, eth, chl	B2^{4}, 1961
537	Adipic acid, di *iso*-propyl ester or di *iso*-propyladipate $i\text{-}C_3H_7O_2C(CH_2)_4CO_2\text{-}i\text{-}C_3H_7$	230.30	...	$120^{6.5}$	-1.1	0.9569^{20}	1.4247^{20}	al, eth, ace, aa	B2^{4}, 1961
538	Adipic acid, monoethyl ester or Ethyladipate $HO_2C(CH_2)_4CO_2C_2H_5$	174.20	hyg (eth-peth)	285, 170^{17}	29	$0.9796^{20/4}$	1.4311^{20}	al, eth	B2^{4}, 1960
539	Adipic acid, mono methyl ester or methyl adipate $HO_2C(CH_2)_4CO_2CH_3$	160.17	lf(Me_3N-MeOH)	158^{10}	9	1.0623^{20}	1.4283^{20}	al	B2^{3}, 1717
540	Adipic acid, 2-methyl $HO_2C(CH_2)_3CH(CH_3)CO_2H$	160.17	cr (peth-bz)	209^{13}	93	...	...	w, al, eth, chl	B2^{4}, 2010
541	Adipic acid, 3-methyl *d* $HO_2C(CH_2CH(CH_3)CH_2CO_2H$	160.17	cr (chl-bz)	230^{30}	94-4.5	...	...	w, al, eth, ace, bz	B2^{4}, 2014
542	Adipic acid, 3-methyl *dl* $HO_2CCH_2CH_2CH(CH_3)CH_2CO_2H$	160.17	nd (bz), cr (ace-bz)	190-200^{12}	97	...	...	w, al, eth, ace, bz	B2^{4}, 2014
543	Adipic acid, 2-oxo $HO_2C(CH_2)_3COCO_2H$	160.13	cr (al or eth)	...	127	...	...	w, al, ace	B3^{4}, 1821
544	Adipic acid, 3-oxo $HO_2CCH_2CH_2COCH_2CO_2H$	160.13	pl (ace - chl)	...	123	...	...	ace	B3^{4}, 1822
545	Adiponitrile $NC(CH_2)_4CN$	108.14	nd (eth)	295, 180^{20}	1	0.9676^{20}	1.4380^{20}	al, chl	B2^{4}, 1975
546	Adipoyl chloride or Adipilacid, dichloride $ClCO(CH_2)_4COCl$	183.03	...	126^{12}	...	...	...	...	B2^{4}, 1972
547	Adonitol or Adonite, Ribitol $HOCH_2(CHOH)_3CH_2OH$	152.15	pr (w), nd (al)	...	104	...	...	w, al	B1^{4}, 2832
548	Adrenaline, *d* or *d*-epinephrine $C_9H_{13}NO_3$	183.21	$[\alpha]^{20/D}$ + 50.5 (HCl)	...	215d	...	...	aa	B13^{3}, 2384
549	Adrenaline, *l* or *l*-epinephrine $C_9H_{13}NO_3$	183.21	br (in air) pw $[\alpha]^{20/D}$-53 (aq. Hcl)	...	211-2	...	...	aa	B13^{3}, 2384
550	Adrenalone $C_9H_{11}NO_3$ 3,4$(HO)_2C_6H_3COCH_2NHCH_3$	181.19	nd	...	235-6d	...	...	...	B14^{3}, 614
551	Adrenochrome *dl* $C_9H_9NO_3$	179.18	red-br rods (MeOH-HCO_2H)	...	125 (anh)	...	...	w, al	B21^{4}, 6436
552	Ajmalicene or Py-tetrahydro serpentine $C_{21}H_{24}N_2O_3$	352.43	pr (MeOH) $[\alpha]^{24/D}$ -60 (chl, C=0.5)	...	258-d 261-3(vac)	...	...	MeOH	J76, 1332
553	Ajmaline $C_{20}H_{26}N_2O_2$	326.44	pl (+ 3.5 w, aq AcOEt) $[\alpha]^{20/D}$ + 144(Chl)	...	158-60 (hyd) 250-7 (anh)	...	...	al, chl, eth	J1954, 1242
554	α-Alanine-*D* or l-α-aminopropionic acid $CH_3CH(NH_2)CO_2H$	89.09	nd (w, al) $[\alpha]^{25/D}$-13.6 (6N HCl,C=1)	sub	314d	...	...	w, al	B4^{3}, 1219
555	α Alanine-*DL* or dl-α-aminopropionic acid $CH_3CH(NH_2)CO_2H$	89.09	orh pr or nd (w)	sub 258	295-6d	1.424	...	w, al	B4^{3}, 1222
556	α-Alanine-*L* or d-α amino propioniuc acid $CH_3CH(NH_2)CO_2H$	89.09	rh (w), $[\alpha]^{25/D}$ + 2.8 (w, C=6)	sub 160-5	314d	1.432^{22}	...	w, al	B4^{3}, 1208
557	α-Alanine. *N*-alanvl-*l* $H_2NCH(CH_3)CONHCH(CH_3)CO_2H$	160.17	lf $[\alpha]^{20/D}$-21.6 (w)	...	298	...	...	w	B4^{3}, 1218

No.	Name, Synonyms, and Formula	Mol. wt.	Color, crystalline form, specific rotation and λ_{max} (log ε)	b.p. °C	m.p. °C	Density	n_D	Solubility	Ref.
558	α-Alanine, *N*-benzoyl-*d* $CH_3CH(NHCOC_6H_5)CO_2H$	193.20	pl (w), $[\alpha]^{20/D}$ -2.4 (w,c=1)	152-4					$B9^3$, 1141
559	α-Alanine, *N*-benzoyl-*dl* $CH_3CH(NHCOC_6H_5)CO_2H$	193.20	pl or pr, lf (eth)	d	165-6			w, al	$B9^3$, 1142
560	α-Alanine, *N*-benzoyl-*l* $CH_3CH(NHCOC_6H_5)CO_2H$	193.20	$[\alpha]^{20/D}$ + 24 (w, c=1)		151				$B9^3$, 1141
561	α-Alanine, *N*-(Carboxymethyl)-*dl* $HO_2CCH_2NHCH(CH_3)CO_2H$	147.13	cr (dil al or w)		222-3			w	$B4^3$, 1250
562	α-Alanine, *N*-(4-Chlorophenyl), nitrile $CH_3CH[NH(C_6H_4Cl\text{-}4)]CN$	180.64	lf (eth-peth)		114.5			al, eth, bz, chl	B12, 617
563	**α-Alanine, *N,N*-diethyl, nitrile** $CH_3CH[N(C_2H_5)_2]CN$	126.20		81^{27}		$0.857^{10/4}$		w, al, eth	$B4^3$, 1238
564	α-Alanine, 3(3,4 dihydroxyphenyl)-*L* or *l*-Dopa $3,4\text{-}(HO)_2C_6H_3CH_2CH(NH_2)COOH$	197.19	pl (dil al), pr or nd (w + SO_2) $[\alpha]^{15/D}$ −39.5 (w, p = 1.3)		285.5d			w	$B14^3$, 1629
565	α-Alanine, ethyl ester, hydrochloride $CH_3CH(NH_2)CO_2C_2H_5.HCl$	153.61	pr or hyg nd (al)	d	87-8			w, al, eth	$B4^3$, 1231
566	α-Alanine, *N*-fumaryl-*DL* or Fumaro alanide $HO_2CCH{=}CHCONHCH(CH_3)CO_2H$	187.15	nd		229d			al	$B4^3$, 1248
567	α-Alanine, *N*- methyl-*dl* $CH_3CH(NHCH_3)CO_2H$	103.12	rh pr (aks al)	sub 292	280 (d)				$B4^3$, 1235
568	β Alanine $H_2NCH_2CH_2CO_2H$	89.09	nd, rh pr (al)		207d	1.437^{19}		w	$B4^3$, 1259
569	β Alanine, ethyl ester, hydrochloride $N_2NCH_2CH_2CO_2C_2H_5HCl$	153.61			65.3			w	$B4^1$, 499
570	β Alanine, *N*-methyl $CH_3NHCH_2CH_2CO_2H$	103.12	pl (al, + 1w)		146 (anh)			w, al	$B4^3$, 1264
571	β Alanine, *N*-methyl, ethyl ester $CH_3NHCH_2CH_2CO_2C_2H_5$	131.17		80^{21}		$1.0082^{20/20}$	1.4443^{20}	al, eth	$B4^3$, 1264
572	β Alanine, *N*-methyl, nitrile or β Methylamino propionitrile $CH_3NHCH_2CH_2CN$	84.13		$101\text{-}4^{49}$		$0.8992^{20/4}$	1.4320^{20}	w, ace, bz, chl	$B4^3$, 1264
573	β Alanine, *N*-phenyl $C_6H_5NHCH_2CH_2CO_2H$	165.19	nd (bz or dil al), lf (dil, al, bz, eth)		92			al, eth, bz	$B12^3$, 933
574	**Aldrin of Octalene** $C_{12}H_8Cl_6$	364.91			104			al, eth, ace, bz	$B5^3$, 1385
575	Alizarin or 1,2-dihydroxy anthraquinone $1,2\text{-}(HO)_2C_{14}H_6O_2$	240.22	og or red tcl nd or pr (al, sub)	430 sub	289-90 (cor)			al, eth, ace, bz	$B8^3$, 3767
576	Alkanin-*l* $C_{16}H_{16}O_5$	288.30	red,br,pr, (bz,sub), nd(eth-al) $[\alpha]^{20/cd}$ -157 (bz)	sub $140^{0.001}$	149			ace, al, eth	$B8^3$, 4089
577	Allantoic acid or Dicarbamoacetic acid $(H_2NCONH)_2CHCO_2H$	176.13	nd, lf (MeOH)		173d			al	$B3^4$, 1492
578	Allantoin or Glyoxyldiureide $C_4H_6N_4O_3$	158.12	mcl pl or pr (w)		238.40			al, w	$B25^2$, 379
579	Allanturic acid or Glyoxalurea $C_3H_3N_2O_2$	101.08	amor, hyg pw		180 turns br				$B25^2$, 388
580	Allene or Propadiene $CH_2{=}C{=}CH_2$	40.06		-34.5	-136		1.4168	bz, peth	$B1^4$, 966
581	Allene, tetra fluoro $F_2C{=}C{=}CF_2$	112.03		-38					$B1^4$, 968
582	Allicin or S-oxodiallyl disulfide $CH_2{=}CHCH_2SS(O)CH_2CH{=}CH_2$	162.26		d		1.112^{20}	1.561^{20}	w	$B4^4$, 7
583	**D-Allitol or D-Allodulcitol** $HOCH_2(CHOH)_4CH_2OH$	182.17			150-1			w	**$B1^4$, 2839**

No.	Name, Synonyms, and Formula	Mol. wt.	Color, crystalline form, specific rotation and λ_{max} (log ε)	b.p. °C	m.p. °C	Density	n_D	Solubility	Ref.
584	Alloluecine-*l* ... $CH_3CH_2CH(CH_3)CH(NH_2)COOH$	131.18	lf (w) [α] $^{20/D}$-14.2 (w,c=2)	...	280-1d	...	...	w	$B4^{3}$, 1462
585	Allomucic acid or Tetrahydroxy adipic acid ... $HOOC(CHOH)_4COOH$	210.14	nd or pr (w)	...	198-200(d)	...	...	w	$B3^{3}$, 1116
586	α-Allonic acid-γ-lactone ... $C_6O_6H_{10}$	178.14	pr (al) [α] $^{20/D}$ -6.8 (w,c=11)	...	...	...	...	w, al	$B18^{4}$, 3024
587	Alloocimene-A or 2,6-Dimethyl-2,3,6-Octateiene (4-trans, 6-trans) ... **$CH_3CH = C(CH_3)CH = CHCH = C(CH_3)_2$**	136.24	...	188^{750},91^{20}	-35.4	$0.8118^{20/4}$	1.5446^{20}	...	$B1^{4}$, 1106
588	Alloocimene-B or 2,6-Dimethyl-2,4,6-Octatriene (4-*trans*-6-*cis*) ... $CH_3CH=C(CH_3)CH=CHCH=C(CH_3)_2$	136.24	...	89^{20}	-20.6	0.8060^{20}	1.5446^{20}	...	$B1^{4}$, 1106
589	Allophanic acid, ethyl ester ... $H_2NCONHCO_2C_2H_5$	132.12	nd (w), (bz)	d	195	...	...	...	$B3^{4}$, 127
590	Allophanonitrile, 3-phenyl or 1-Cyano-3-phenyl urea ... $C_6H_5NHCONHCN$	161.16	nd	...	125	...	...	al	$B12^{3}$, 781
591	D-Allose (β-anomer) ... $C_6H_{12}O_6$,$HOCH_2(CHOH)_4CHO$	180.16	cr (w) [α] $_D$ 0.58→14.4 (w,c=5) (mut)	...	128	...	...	w	$B1^{4}$, 4299
592	L-Allose ... $C_6H_{12}O_6$	180.16	pr (dil al) [α] $^{20/D}$ -0.58 (al)	...	128-9	...	...	w	$B1^{4}$, 4300
593	Alloxan or Mesoxalurea ... $C_4H_2N_2O_4$	142.07	...	sub (vac)	256 d (anh)	...	...	w, al, ace, bz, aa	$B24^{2}$, 301
594	Alloxanic acid ... $C_4H_4N_2O_5$	160.09	tcl pr (eth)	...	162-3d	...	...	w, al	$B25^{2}$, 266
595	Alloxantin ... $C_8H_6N_4O_8$	286.16	rh pr (w + 2)	...	253-5d	...	...	...	$B26^{2}$, 335
596	Allyl alcohol or 1-Propene-3-ol ... $CH_2=CHCH_2OH$	58.08	...	97.1	-129	$0.8540^{20/4}$	$1.4135^{20/}{}_D$	**w, al, eth,** chl	$B1^{4}$, 2079
597	Allyl alcohol, 2-bromo ... $CH_2=CBr\text{-}CH_2OH$	136.98	...	153-4^{755}, 62^{11}	...	1.621^{18}	1.500^{18}	eth, chl	$B1^{4}$, 2094
598	Allyl alcohol, 2-chloro ... $CH_2=C(Cl)CH_2OH$	92.53	...	136-40, 47^{10}	...	$1.1618^{20/4}$	1.4588^{20}	...	$B1^{4}$, 2091
599	Allyl alcohol, 3-chloro ... $ClCH=CHCH_2OH$	92.53	...	(i) 146^{746}, (ii) 153^{756}	...	(i) $1.1769^{20/4}$, (ii) $1729^{20/4}$	(i) 1.4738^{20}, (ii) 1.4664^{20}	...	$B1^{4}$, 2090
600	Allyl alcohol, 3-(3,5-dimethoxy-4-hydroxyphenyl) or Sinaphyl alcohol. Syringenin ... *[$3,5(CH_3O)_2\text{-}4\text{-}HOC_6H_2$]* $CH=CHCH_2OH$	210.23	nd (eth-peth)	...	66-7	...	...	eth	$B6^{3}$, 6690
601	Allyl alcohol, 3-(4-hydroxy phenyl) or *P*-coumaryl alcohol $4\text{-}HOC_6H_4CCH=CHCH_2OH$	150.18	pr (dil al)	...	124	...	...	al, eth, ace, bz	C45, 3359
602	Allyl alcohol, 1-phenyl ... $CH_2=CH\text{-}CH(C_6H_5)OH$	134.18	...	215-6, 111^{18}	...	$1.0251^{21/0}$	1.5406^{20}	al, eth, bz, chl	$B6^{3}$, 2417
603	Allyl alcohol, 3-phenyl *trans* or Cinnamyl alcohol ... $C_6H_5CH=CHCH_2OH$	134.18	wh nd (eth-peth)	257.5, 127-8^{10}	34	$1.0440^{20/4}$	1.5819^{20}	al, eth	$B6^{3}$, 2401
604	Allylamine or 3-Aminopropene ... $H_2NCH_2CH=CH_2$	57.10	...	58	...	$0.7621^{20/4}$	1.4205^{20}	**w, al, eth**, chl	$B4^{4}$, 1057
605	Allylamine, *N*-methyl or Allylmethyl amine ... $CH_2CH_2NHCH_3$	58.10	...	65	...	...	1.4065^{20}	w, al, eth, ace	$B4^{4}$, 1058
606	Allylamine, *N*-phenyl or *N*-allylaniline ... $CH_2CH_2NHC_6H_5$	120.17	...	217-9^{736}, 105-8^{12}	...	$0.982^{25/4}$	1.563^{20}	al, eth	$B12^{3}$, 277
607	Allylamine, *N*-iso-Propyl or Allyl-iso-propyl amine ... $CH_2CH_2NHCH(CH_3)_2$	86.16	...	96-7	...	...	$1.4140^{25/}$	...	$B4^{4}$, 1059
608	Allyl benzene, α-chloro ... $C_6H_5CHClCH=CH_2$	152.62	...	212-4, 97^{18}	...	$1.073^{14/4}$	1.545^{14}	eth, ace, bz	$B5^{4}$, 1363
609	Allylbromide or 3-Bromopropene ... $BrCH_2CH=CH_2$	120.98	...	70^{752}	119.4	$1.398^{20/4}$	1.4697^{20}	**al, eth**	$B1^{4}$, 754
610	Allyl chloride or 3-chloropropene ... $ClCH_2CH=CH_2$	76.53	...	45	134.5	$0.9376^{20/4}$	1.4157^{20}	**al, eth, ace, bz, lig**	$B1^{4}$, 738
611	Allyl ether or Diallyl ether ... $(CH_2=CHCH_2)_2O$	98.14	...	94	...	$0.8260^{20/4}$	1.4163^{20}	**al, eth**, ace	$B1^{4}$, 2086

No.	Name, Synonyms, and Formula	Mol. wt.	Color, crystalline form, specific rotation and λ_{max} (log ε)	b.p. °C	m.p. °C	Density	n_D	Solubility	Ref.
612	Allyl 4-chlorophenyl ether $4\text{-}ClC_6H_4OCH_2CH{=}CH_2$	168.62		106-7^{12}		1.131^{15}	1.5348^{25}	al, eth, bz	B6^4, 825
613	Allyl ethyl ether $CH_2{=}CHCH_2OC_2H_5$	86.13		66		0.7651$^{20/4}$	1.3881^{20}	**al, eth**, ace	B1^4, 2083
614	Allyl isocyanide $CH_2{=}CHCH_2NC$	67.09		98		0.794$^{17/4}$		**al, eth**	B4; 208
615	Allyl mercaptan $CH_2{=}CHCH_2SH$	74.14		67-8		0.925$^{23/4}$	1.4832^{20}	**al, eth**, Chl	B1^4, 2095
616	Allyl methyl ether $CH_2{=}CHCH_2OCH_3$	72.11		46		0.77$^{11/11}$	1.3778^{20}	**al, eth**, ace	B1^4, 2083
617	Allyl *iso*pentyl ether or Allyl-isoamyl-ether $CH_2{=}CHCH_2OCH_2CH_2CH(CH_3)_2$	128.21		120-2				**al, eth**	B1^2, 477
618	Allyl phenyl ether $CH_2{=}CHCH_2OC_6H_5$	134.17		191.7, 74^5		0.9811$^{20/4}$	1.5223^{20}	al, **eth**	B6^4, 562
619	Allyl propyl ether $CH_2{=}CHCH_2OC_3H_7$	100.16		90-2		0.7764^{20}	1.3919^{20}	al, **eth**, ace	B1^4, 2083
620	Allyl *iso*propyl ether $CH_2{=}CHCH_2OCH(CH_3)_2$	100.16		83-4		0.7764^{20}	1.3946^{20}	al, **eth**, ace	B1^4, 2083
621	Allyl 2-tolyl ether $(2\text{-}CH_3C_6H_4)OCH_2CH{=}CH_2$	148.20		205-8, 85^{12}		0.9698$^{15/4}$	1.5188^{15}		B6^4, 1946
622	Allyl 3-tolyl ether $(3\text{-}CH_3C_6H_4)OCH_2CH{=}CH_2$	148.20		211-4, 93.5^{13}		0.9564^{20}	1.5179^{20}	bz	B6^4, 2041
623	Allyl 4-tolyl ether $(4\text{-}CH_3C_6H_4)OCH_2CH{=}CH_2$	148.20		214.5, 97-8^{16}		0.9728$^{15/15}$	1.5157^{24}	bz	B6^4, 2101
624	Allyl vinyl ether $CH_2{=}CHCH_2\text{-}O\text{-}CH{=}CH_2$	84.12		65^{740}		0.8050$^{20/4}$	1.4062^{20}	eth, ace	B1^4, 2085
625	Allyl isothiocyanate or Allyl mustard oil $CH_2{=}CHCH_2NCS$	99.15		152, 44^{12}	−80	1.0126$^{20/4}$	1.5306^{20}	**al, eth**, bz	B4^4, 1081
626	Allyl sulfide $(CH_2{=}CHCH_2)_2S$	114.21		139^{768}, 35^6	−85	0.8877$^{27/4}$	1.4870^{25}	**al, eth**	B1^4, 2097
627	Allyl sulfoxide $(CH_2{=}CHCH_2)_2SO$	130.20		112-5^{12}	23.5	1.0261$^{20/4}$	1.5115^{20}	al, eth, ace	B1^4, 2097
628	Allyl thio cyanate $CH_2{=}CHCH_2SCN$	99.15		161		1.056^{15}		al, eth	B3^4, 332
629	Allyl (2,4,6-tribromophenyl) ether $(2,4,6,Br_3C_6H_2)OCH_2CH{=}CH_2$	370.87			33-4				B6^2, 194
630	Allyl trisulfide $(CH_2{=}CHCH_2)_2S_3$	178.33		112-22^{16}		1.0845^{15}		**eth**	B1, 441
631	Aloetic Acid $C_{15}H_4N_4O_{11}$	416.22	og-ye nd (aa), ye cr (w + l)		285d				B8, 525
632	Aloin or Barbaloin $C_{21}H_{22}O_9$	418.40	ye nd (al), $[\alpha]_D$ −8.3 (dil al)		148.5			w, al, ace	B18^4, 3630
633	Alstonine $C_{21}H_{20}N_2O_3$	348.40	ye nd (ace)		205-10d				B27^2, 824
634	D-Altrose (β-anomer) $C_6H_{12}O_6$	180.16	pr (MeOH-al), $[\alpha]^{20/}{}_D$ + 11.7 →33.1 (w, c = 7) (mut)		103.5			w	B1^4, 4300
635	L-Altrose (β-anomer) $C_6H_{12}O_6$	180.16	pr (al, aa) −28.55→ −32.30 (w) (mut)		107-9.5				B1^4, 4301
636	Amalic acid or Tetramethyl alloxantin $C_{12}H_{14}N_4O_8$	324.27	cr (w)		245d				B26^2, 336
637	Amarine $C_{21}H_{18}N_2$	298.39	pr (eth, bz-lig)	198d	136			al, eth, bz	B23^2, 274
638	Amarine hydrate $C_{21}H_{18}N_2 \cdot \frac{1}{2}H_2O$	307.39	pr (aq al + ½ w)		106			w	B23^2, 274
639	Amaron or Benzoin-imide $C_{28}H_{20}N_2$	384.48	tcl nd or pr (ace, al), nd (aa)	sub	2-46			ace, bz, chl	B23^2, 304
640	Aminoacetamide or Glycinamide $H_2NCH_2CONH_2$	74.08	hyg nd (chl)		67-8			w, al, ace	B4^3, 1118

No.	Name, Synonyms, and Formula	Mol. wt.	Color, crystalline form, specific rotation and λ_{max} (log ε)	b.p. °C	m.p. °C	Density	n_D	Solubility	Ref.
641	Aminoacetamide, *N'*(4-ethoxyphenyl) or Glycine-*p*-phenetidide $H_2NCH_2CONH\text{-}C_6H_4\text{-}OC_2H_5\text{-}4)$	194.23	nd (+lw)		100.5 (anh)			al, eth	B13[1], 179
642	Aminoacetamide Hydrochloride $H_2NCH_2CONH_2\text{-}HCl$	110.54	nd (al)		203-4			w, al	B4[3], 1119
643	Aminoacetamide, *N*-phenyl $C_6H_5NHCH_2CONH_2$	151.19			127-8			w, al	B12[3], 915
644	Aminoacetic acid or Glycine $H_2NCH_2CO_2H$	75.07	mcl or trg pr (dil al)		262d	1.607		w	B4[3], 1097
645	Aminoacetic acid, *N*-acetyl or Aceturic acid $CH_3CONHCH_2CO_2H$	117.10	lo nd (w, MeOH)		206			al, ace, w	B4[3], 1150
646	Aminoacetic acid, *N*-acetyl-*N*-phenyl or *N*-phenyl aceturic acid $CH_3CON(C_6H_5)CH_2CO_2H$	193.25	lf (w)		194			w	B12[3], 919
647	Aminoacetic acid, *N*-(4 aminophenyl) hydrate $(4\text{-}H_2N\text{-}C_6H_4)NHCH_2CO_2H\text{-}H_2O$	184.20	pl (dil aa)		222-3d			w	B13[1], 34
648	Aminoacetic acid, *N*-benzoyl or Hippuric acid $C_6H_5CONHCH_2COOH$	179.18	pr (w or al)		190-3	$1.371^{20/4}$		w, al	B9[3], 1123
649	Aminoacetic acid, *N*-benzyl or *N*-benzyl glycine $C_6H_5CH_2NHCH_2CO_2H$	165.19	nd (w)		198-9				B12[3], 2283
650	Aminoacetic acid, *N*-benzyl, ethyl ester or Ethyl-*N*-benzyl glycinate $C_6H_5CH_2NHCH_2CO_2C_2H_5$	193.25		$175\text{-}9^{50}$			1.5041^{20}		B12[3], 2283
651	Aminoacetic acid, *N*-bromoacetyl-*N*-phenyl $BrCH_2CON(C_6H_5)CH_2COOH$	272.10	pl (w)		153d			al, bz	B12, 477
652	Aminoacetic acid, *N*-(2-carboxyphenyl) or Anthranilidoacetic acid $(2\text{-}HO_2CC_6H_4)NHCH_2CO_2H$	179.18	nd (MeOH)		218-20			al, eth	B14[3], 938
653	Aminoacetic acid, *N*-chloroacetyl-*N*-phenyl $ClCH_2CON(C_6H_5)CH_2CO_2H$	227.65	pl or pr (bz)		132-3			al, bz	B12, 476
654	Aminoacetic acid, *N*-chloroacetyl-*N*-phenyl,methyl ester $ClCH_2CON(C_6H_5)CH_2CO_2CH_3$	241.67	pr (lig)		59-60			al, eth, bz	B12, 477
655	**Aminoacetic acid, *N*-ethyl of Ethyl aminoacetic acid** $C_2H_5NHCH_2CO_2H$	103.12	pl (al)		181-2d			w, al	B4[3], 1134
656	Aminoacetic acid, ethyl ester or Ethyl glycinate $H_2NCH_2CO_2C_2H_5$	103.12		$148\text{-}9^{750}$, $57\text{-}8^{18}$		$1.0275^{10/4}$	1.4242^{10}	**w, al, eth, ace bz, lig**	B4[3], 1116
657	Aminoacetic acid, ethyl ester,hydrochloride or Ethyl glycinate hydrochloride $H_2NCH_2CO_2C_2H_5\text{-}HCl$	139.58	nd (al)	sub	144			w, al	B4[3], 1117
658	Aminoacetic acid, *N*,*N*-(dicarbethoxy),ethyl ester $(C_2H_5O_2C)_2NCH_2CO_2C_2H_5$	247.25	pr (peth)	$152\text{-}3^{10}$	36.5			al, eth, bz	B4, 365
659	Aminoacetic acid, *N*,*N*-dimethyl $(CH_3)_2NCH_2CO_2H$	103.12	hyg nd (PrOH)		185-6			w, al, eth	B4[3], 1124
660	Aminoacetic acid hydrazide $H_2NCH_2CONHNH_2$	89.10		dec 150	80.5			chl	B4[3], 1121
661	Aminoacetic acetic acid, hydrochloride or Glycine hydrochloride $H_2NCH_2CO_2H\text{-}HCl$	111.53	hyg rh nd (w)		200-1			w	B4[3], 1111
662	Aminoacetic acid, *N*,*N*-bis(2-hydroxyethyl) $(HOCH_2CH_2)_2NCH_2CO_2H$	163.17	nd (al)		193-5d			w	B4[3], 1145
663	Aminoacetic acid, *N*-(4-hydroxyphenyl) $(4\text{-}HOC_6H_4)NHCH_2CO_2H$	167.16	pl (w)		245-7d				B13[3], 1132
665	Aminoacetic acid, *N*-methyl,hydrochloride $CH_3NHCH_2CO_2H \cdot HCl$	125.56			166-8d			w	B4, 345
666	Aminoacetic acid, methyl ester or Methyl glycinate, Methylamine acetate $NH_2CH_2CO_2CH_3$	89.09		130d, 45^{20}				al	B4[3], 1115
667	Aminoacetic acid, *N*,*N*-methylene di or *N*,*N*-methylene diamino acetic acid $CH_2(NHCH_2CO_2H)_2$	162.15	pl		199d			al	B4[3], 1148
668	Aminoacetic acid, *N*-(2-naphthyl) $2\text{-}C_{10}H_7\text{-}NHCH_2COOH$	201.22	(w)		134-5			**al, ace, w, aa eth**	B12, 1298
669	Aminoacetic acid, *N*-(2-nitro phenyl) $(2\text{-}O_2N\text{-}C_6H_4)\text{-}NH\text{-}CH_2CO_2H$	196.16	dk red pr (al)		192-3d			al	B12, 695
670	Aminoacetic acid, *N*-phenyl or Anilino-acetic acid $C_6H_5NHCH_2CO_2H$	151.16			127-8			w, al	B12[3], 914

No.	Name, Synonyms, and Formula	Mol. wt.	Color, crystalline form, specific rotation and λ_{max} (log ε)	b.p. °C	m.p. °C	Density	n_D	Solubility	Ref.
671	Aminoacetic acid, *N*-phenyl,ethyl ester or Ethyl-*N*-phenyl amino acetate $C_6H_5NHCH_2CO_2C_2H_5$	179.22	lf (dil al)	273-4, 163^{18}	58	...	...	al, eth	$B12^3$, 915
672	Aminoacetic acid, *N*-phenyl,methyl ester or Methyl-*N*-phenyl amino acetate $C_6H_5NHCH_2CO_2CH_3$	165.19	nd (al)	...	48	...	...	al, eth	$B12^1$, 263
673	Amino acetic acid, *N*-phthaloyl or Phthalimide-*N*-acetic acid $C_{10}H_7NO_4$	205.17	nd or pr (w or al)	...	193	...	...	al, eth	$B21^4$, 5176
674	Aminoacetic acid, *N*-phthaloyl,ethyl ester or Ethyl-*N*-phthaloyl amino acetate o-$C_6H_4(CO)_2N$-$CH_2CO_2C_2H_5$	233.23	nd (w, al or eth)	300	112-3	...	...	al, eth, bz, chl	$B21^4$, 5177
675	Aminoacetic acid, *N*-succinyl,ethyl ester or Ethyl succinimide-*N*-acetate $[C_2H_4(CO)_2N]$-$CH_2CO_2C_2H_5$	185.18	nd (eth)	198^{12}	67	...	...	w, al	$B21^2$, 305
676	Amino acetonitrile H_2NCH_2CN	56.07	...	58^{15}	...	...	...	al	$B4^3$, 1120
677	Amino acetonitrite-*N,N*-diethyl $(C_2H_5)_2NCH_2CN$	112.17	...	170, 70^{24}	...	...	1.4260^{20}	w, al	$B4^3$, 1136
678	Amino acetonitrile-*N,N*-dimethyl $(CH_3)_2NCH_2CN$	84.12	...	137-8, 42^{22}	...	$0.8650^{20/0}$	1.4095^{20}	w, al	$B4^3$, 1126
679	Amino acetonitrile-*N*-ethyl $C_2H_5NHCH_2CN$	84.12	...	166-7, 81-3^{29}	...	...	...	al, eth	$B4^2$, 787
680	Aminoacetonitrile-*N*-phenyl $C_6H_5NHCH_2CN$	132.16	pl (lig-eth)	...	48	...	...	al, bz	$B12^3$, 916
681	Aminoacetyl chloride-*N*-phthaloyl or *O-phthalinido-N*-acetyl chloride $[o$-$C_6H_4(CO)_2N]CH_2COCl$	223.62	nd (lig)	...	84-5	...	...	al, bz	$B21^4$, 5180
682	2-Aminobenzoic acid or Anthanilic acid 2-$H_2NC_6H_4CO_2H$	137.14	lf (al)	sub	146-7	1.412^{20}	...	w, al, eth	$B14^3$, 879
683	3-Aminobenzamide 3-$H_2NC_6H_4CONH_2$	136.15	ye mcl nd (+lw), nd (bz)	...	79—80 (hyd), 113—4 (anh)	...	...	w, al, eth	$B14^3$, 998
684	3-Aminobenzoic acid-*N*-acetyl 3-$CH_3CONHC_6H_4CO_2H$	179.18	nd (al)	...	248-50	...	...	al	$B14^3$, 1006
685	3-Amino benzoic acid-*N*-acetyl-6-ethoxy 3-(CH_3CONH)-6-$(C_2H_5O)C_6H_3CO_2H$	223.23	nd (w)	...	190	...	...	al	B14, 583
686	3-Aminobenzoic acid-6-chloro 3-H_2N-6-Cl-$C_6H_3CO_2H$	171.58	...	...	188	1.519^{15}	...	al	$B14^3$, 1017
687	3-Aminobenzoic acid-*N,N*-dimethyl 3-$(CH_3)_2NC_6H_4CO_2H$	165.19	nd (w)	...	151	...	...	w, al, eth	B14, 392
688	3-Aminobenzoic acid-*N*-ethyl 3-$(C_2H_5NH)C_6H_4CO_2H$	165.19	nd or pr (dil al)	...	154	...	...	al, eth, ace	B14, 393
689	3-Amino benzoic acid, ethyl ester or Ethyl 3-Amino benzoate 3-$H_2NC_6H_4CO_2C_2H_5$	165.19	...	294, 160-1^5	...	$1.1248^{22/4}$	1.5600^{22}	al, eth	$B14^3$, 993
690	3-Amino benzoic acid, 2-hydroxy or 3-Amino salicylic acid 3-H_2N-2-$HOC_6H_3CO_2H$	153.14	...	...	235d	...	...	...	$B14^3$, 1434
691	3-Aminobenzoic acid, 4-hydroxy 3-H_2N-4-$HOC_6H_3CO_2H$	153.14	pr (w + l)	...	210 (anh)	...	...	w	$B14^2$, 360
692	3-Aminobenzoic acid, 4-hydroxy,methyl ester 3-H_2N-4-$HOC_6H_3CO_2CH_3$	167.16	(i) nd (bz or aa) (st) (ii) nd (chl) (unst)	...	(i)143, (ii)111	...	...	al, eth	$B14^3$, 1477
693	3-Aminobenzoic acid, methyl ester or Methyl 3-amino benzoate 3-$H_2NC_6H_4CO_2CH_3$	151.16	...	152-3^{11}	39	1.232^{20}	...	eth, al, bz, lig	$B14^3$, 993
694	3-Amino benzoic acid, *N*-methyl 3-$CH_3NHC_6H_4CO_2H$	151.16	pl (peth)	...	127	...	...	al, ace, bz, chl	$B14^1$, 559
695	3-Amino benzoic acid, *N*-methyl,methyl ester 3-$CH_3NHC_6H_4CO_2CH_3$	165.19	cr (al)	...	72	...	...	al, eth	B14, 392
696	3-Amino benzoic acid, 4-methyl 3-H_2N-4-CH_3-$C_6H_3CO_2H$	151.16	nd (al)	...	164-6	...	...	w, al	$B14^2$, 291
697	3-Aminobenzoic acid, 6-methyl 3-H_2N-6-$CH_3C_6H_3CO_2H$	151.16	pr (w)	...	196	...	...	w, al	$B14^2$, 290

PHYSICAL CONSTANTS OF ORGANIC COMPOUNDS (Continued)

No.	Name, Synonyms, and Formula	Mol. wt.	Color, crystalline form, specific rotation and λ_{max} (log ε)	b.p. °C	m.p. °C	Density	n_D	Solubility	Ref.
698	3-Aminobenzoic acid, 2-nitro $3-H_2N-2-NO_2C_6H_3CO_2H$	182.14	ye nd (w, dil al)		156-7			al, eth, ace	B14, 414
699	3-Aminobenzoic acid, 4-nitro $3-H_2N-4-NO_2C_6H_3CO_2H$	182.14	red pl or nd (al)		298d			al, eth, ace	B14, 415
700	3-Amino benzoic acid, 5-nitro $3-H_2N-5-NO_2C_6H_3CO_2H$	182.14	ye pr (w)		209-10			al	$B14^3$, 565
701	3-Aminobenzoic acid, 6-nitro $3-H_2N-6-NO_2C_6H_3CO_2H$	182.14	ye nd or pr (w)		235d			al, **ace**	$B14^3$, 1021
702	3-Aminobenzoic acid, 2,4,6-tribromo $3-H_2N-2,4,6-Br_3C_6HCO_2H$	373.83	nd (w)		171.5-3			**al**	$B14^3$, 1019
703	3-Aminobenzomitrile or m-Cyanoaniline $3-H_2NC_6H_4CN$	118.14	nd (dil al or CCl_4)	286-96	53-4			**al, eth, ace**	$B14^3$, 1001
704	3-Aminobenzonitrile, 2-methyl $3-NH_2-2-CH_3C_6H_3CN$	132.16	rh cr (al)		90			w	$B1^4$, 477
705	3-Amino benzonitrile, 4-methyl $3NH_2-4-CH_3C_6H_3N$	132.16	pr (al)		81-2			al, eth, **ace, bz**	B14, 487
706	3-Aminobenzo nitrile, 5-methyl $3-H_2N-5-CH_3C_6H_3CN$	132.16	nd (lig)		75			al	$B14^1$, 600
707	3-Aminobenzonitrile, 6-methyl $3-H_2N-6-CH_3C_6H_3CN$	132.16	nd (peth)	$100-10^{22}$	88			al	$B14^2$, 290
708	4-Aminobenzamide $4-H_2NC_6H_4CONH_2$	136.15	ye cr (+ 1/4 w)		183			al, eth	$B14^3$, 1061
709	**4-Aminobenzamide, N(2-diethyl aminoehtyl) hydrochloride or Procainamide, hydrochloride** $4-H_2NC_6H_4CONH[CH_2CH_2N(C_2H_5)]_2 \cdot HCl$	313.85	cr		165-9			**w**, al	C54, 428
710	4-Amino benzamide, N-phenyl $4-H_2NC_6H_4CONHC_6H_5$	212.25			135-6			al	$B14^3$, 1062
711	4-Aminobenzoic acid $4-H_2NC_6H_4CO_2H$	137.14	mcl pr (w)		188-9	$1.374^{20/4}$		al, eth, w	$B14^3$, 1023
712	4-Aminobenzoic acid, N-acetyl $4-(CH_3CONH)C_6H_4CO_2H$	179.18	nd (aa)		256.5			al	$B14^3$, 1112
713	4-Aminobenzoic acid, butyl ester or Butesin $4-H_2NC_6H_4CO_2C_4H_9$	193.25	(al or bz)	$173-4^8$	58			al, eth, bz, chl	$B14^3$, 1027
714	4-Aminobenzoic acid, *iso*-butyl ester $4-H_2NC_6H_4CO_2-i-C_4H_9$	193.25			64.5				$B14^3$, 1027
715	4-Aminobenzoic acid, 2-chloro or 4-Amino-2-chlorobenzoic acid $4-H_2N-2-ClC_6H_3CO_2H$	171.58			dec 213			al	$B14^3$, 1153
716	4-Aminobenzoic acid, 3,5-dichloro $4-H_2N-3,5-Cl_2C_6H_2CO_2H$	206.03	(al)		291			al, eth, bz	$B14^3$, 1156
717	4-Aminobenzoic acid, (2-diethylaminoethyl) ester or Novocaine $4-H_2NC_6H_4CO_2[CH_2CH_2N(C_2H_5)_2]$	236.31	nd (w + 2), pl (lig or eth)		51, 61 (anh)			al, eth, bz, chl	$B14^3$, 1037
718	**4-Aminobenzoic acid, 2-(diethylaminoethyl) ester, hydrochloride or Novocaine hydrochloride** $4-H_2N-C_6H_4CO_2[CH_2CH_2N(C_2H_5)_2] \cdot HCl$	272.77	nd (al), mcl or tcl pl (w)		156	0.707^{17}		**w**, al	$B14^3$, 1037
719	4-Aminobenzoic acid, 3,5-diiodo $4-H_2N-3,5-I_2C_6H_2CO_2H$	388.93	nd (aa-NH_3)		> 350				$B14^3$, 1161
720	4-Aminobenzoic acid, 3,5-diido, ethyl ester $4-H_2N-3,5-I_2C_6H_2CO_2C_2H_5$	416.98	nd (al)		148				B14, 439
721	4-Aminobenzoic acid, N,N-dimethyl or 4-(Dimethyl amino) benzoic acid $4-(CH_3)_2NC_6H_4CO_2H$	165.19	nd (al)		242.5—3.5			al	$B14^3$, 1082
722	4-Aminobenzoic acid, ethyl ester or Benzocaine $4-H_2NC_6H_4CO_2C_2H_5$	165.19	nd (w), rh (eth)	310	92			**al, eth**, chl	$B14^3$, 1025
723	4-Aminobenzoic acid, N-ethyl $4-(C_2H_5NH)C_6H_4CO_2H$	165.19	cr (bz)		177-8			al, eth, ace, bz	$B14^3$, 1086
724	4-Aminobenzoic acid, 2-hydroxy or 4-Aminosalicylic acid $4-H_2N-2-HOC_6H_3CO_2H$	153.14	nd pl (al-eth)		dec 150-1			w, al, eth, ace	$B14^3$, 1436
725	4-Aminobenzoic acid, 2-hydroxy, methyl ester $2-HO-4-H_2NC_6H_3CO_2CH_3$	167.16			120-1				$B14^3$, 1437
726	4-Aminobenzoic acid, methyl ester or Methyl-4-aminobenzoate $4-H_2NC_6H_4CO_2CH_3$	151.16	lf or nd (aq MeOH)		114			Chl	$B14^3$, 1025
727	4-Aminobenzoic acid, N-methyl $4-CH_3NHC_6H_4CO_2H$	151.16	nd (bz w or dil al)		168 (159)			**al, eth**, w, bz	$B14^3$, 1081
728	4-Aminobenzoic acid, N-methyl,methyl ester $4-CH_3NHC_6H_4CO_2CH_3$	165.19	pl (dil al or lig)		95.5			al, eth	$B14^3$, 1081

PHYSICAL CONSTANTS OF ORGANIC COMPOUNDS (Continued)

No.	Name, Synonyms, and Formula	Mol. wt.	Color, crystalline form, specific rotation and λ_{max} (log ε)	b.p. °C	m.p. °C	Density	n_D	Solubility	Ref.
729	4-Aminobenzoic acid, 2-methyl $4\text{-}H_2N\text{-}2\text{-}CH_3C_6H_3CO_2H$	151.16	nd (al)	....	dec 165	....	....	al	$B14^{3}$,1200
730	**4-Aminobenzoic acid, 3-methyl** **$4\text{-}H_2N\text{-}3\text{-}CH_3C_6H_3CO_2H$**	151.16	nd (w)	....	170	....	....	w	$B14^{2}$, 290
731	4-Aminobenzoic acid, 2-nitro $4\text{-}H_2N\text{-}2\text{-}NO_2C_6H_3CO_2H$	182.14	rd nd (w), pr (dil aa)	....	239d	....	....	al, aa	$B14^{3}$, 1161
732	4-Aminobenzoic acid, 3-nitro $4\text{-}H_2N\text{-}3\text{-}NO_2C_6H_3CO_2H$	182.14	rd-ye nd (al)	....	284d	....	....	ace, aa	$B14^{3}$, 1161
733	4-Aminobenzoic acid, phenyl ester $4\text{-}H_2NC_6H_4CO_2C_6H_5$	213.24	nd (al)	....	173	....	....	al, eth	$B14^{1}$, 568
734	4-Aminobenzoic acid, propyl ester or Propaesin $4\text{-}H_2NC_6H_4CO_2C_3H_7$	179.22	pr	....	75	....	....	al, eth, bz, chl	$B14^{3}$, 1026
735	4-Aminobenzonitrile or *p*-Cyanoaniline $4\text{-}H_2NC_6H_4CN$	118.14	pr or pl (w)	....	86	....	....	al, eth, ace, bz, aa	$B14^{3}$, 1079
736	4-Aminobenzonitrile, 2-methyl $2\text{-}CH_3\text{-}4\text{-}H_2NC_6H_3CN$	132.16	rh cr (al)	....	90	....	....	al	$B14^{1}$, 598
737	4-Aminobenzonitrile, 3-methyl $4\text{-}H_2N\text{-}3\text{-}CH_3C_6H_3CN$	132.16	nd (w)	....	95	....	....	....	$B14^{1}$, 598
738	Ammelide or Cyanuromonoamide $C_3H_4N_4O_2$	128.09	mcl pr (w)	....	d	....	....	....	$B26^{2}$, 132
739	Ammeline or Cyanurodiamide $C_3H_5N_5O$	127.11	nd (aq Na_2CO_3)	....	d	....	....	....	$B26^{2}$, 132
740	Amphetamine (d) or Dexedrine, β-phenyl-*iso*-propyl amine $C_6H_5CH_2CH(NH_2)CH_3$	135.21	$[\alpha]^{12/}{}_D$ + 37.6	203-4, 80^{10}	....	$0.949^{15/4}$	1.4704^{20}	al, eth	$B12^{3}$, 2664
741	Amphetamine (dl) or Benzedrine $C_6H_5CH_2CH(NH_2)CH_3$	135.21	....	203, 97^{20}	....	$0.9306^{25/4}$	1.518^{26}	al, chl	$B12^{3}$, 2664
742	**Amphetamine, sulfate (d) or Dexedrine sulfate** **$[C_6H_5CH_2CH(NH_2)CH_3]_2\cdot H_2SO_4$**	368.49	$[\alpha]^{20/}{}_D$ + 22 (w, c=8)	....	> 300	$1.15^{25/4}$	....	w	$B12^{3}$, 2665
743	**Amphetamine, sulfate (dl) or Benzedrine sulfate** **$[C_6H_5CH_2CH(NH_2)CH_3]_2\cdot H_2SO_4$**	368.49	....	....	280—1	$1.15^{25/4}$	....	w	$B12^{3}$, 2665
744	Amygdalin or Mandelonitrile-β-gentiobioside $C_{20}H_{27}NO_{11}$	457.43	....	....	223—6	....	....	w	$B17^{4}$, 3614
745	**Amygdalin trihydrate** **$C_{20}H_{27}NO_{11}\cdot 3H_2O$**	511.48	rh (w + 3) $[\alpha]^{20/}{}_D$ −40 (w, c=1)	....	200 (remelts 125—30)	....	....	w, al	B31, 400
746	α-Amyrin or α-Amyrenol $C_{30}H_{50}O$	426.73	nd (al), $[\alpha]^{17/}{}_D$ + 91.6 (bz, c = 1.3) + 83.5 (chl)	$243^{0.5}$	186	....	....	al, eth, bz, aa	$B6^{3}$, 2889
747	β-Amyrin or β-Amyrenol $C_{30}H_{50}O$	426.73	nd (lig or al), $[\alpha]^{19/}{}_D$ + 99.3 (bz, c = 1.3) + 88.4 (chl)	$260^{0.5}$	197	....	....	....	$B6^{3}$, 2894
748	Anabasine (l) or l-2-(3-Pyridyl)piperidine $C_{10}H_{14}N_2$	162.23	$[\alpha]^{20/}{}_D$ −83.1	276, 105^{2}	9	$1.0455^{20/20}$	1.5430^{20}	w, al, eth, bz	$B23^{2}$, 113
749	Anagyrine or Monolupine $C_{15}H_{20}N_2O$	244.34	pa ye glass, $[\alpha]^{25/}{}_D$ −168 (al, c = 4.8)	260—70^{12}	....	....	....	al, w, eth, bz	$B24^{2}$, 84
750	**Analgen or 5-benxamide-8-ethoxyquinoline** **$C_{18}O_2N_2H_{16}$**	292.34	ye nd (al)	....	206	....	....	....	B22, 503
751	Anatabine *l* or 2-(3-Pyridyl)-1,2,3,6-tetrahydropyridine $C_{10}H_{12}N_2$	160.22	$[\alpha]^{17}{}_D$− 177.8	145-6^{10}	....	$1.091^{19/4}$	1.5676^{20}	w, al, eth, bz	C31, 3055
752	Androstane or Etiocholane $C_{19}H_{32}$	260.46	lf (ace-MeOH), or (hl-ace)	75-$80^{0.01}$(sub)	50-2	....	....	al, eth, ace, chl, lig	$B5^{4}$, 1211
753	Androsterone or 3-α-hydroxy-17-keto androstane $C_{19}H_{30}O_2$	290.45	lf or nd (al, ace), $[\alpha]^{20}{}_D$ + 94.6 (abs al, c=0.71)	....	185 (cor)	....	....	al,eth, ace, bz	$B7^{3}$, 586

PHYSICAL CONSTANTS OF ORGANIC COMPOUNDS (Continued)

No.	Name, Synonyms, and Formula	Mol. wt.	Color, crystalline form, specific rotation and λ_{max} (log ε)	b.p. °C	m.p. °C	Density	n_D	Solubility	Ref.
754	Anemonin or Anemone camphor $C_{10}H_8O_4$	192.17	rh pl (chl), nd (al or bz)		158			chl	$B19^4$, 1986
755	Anethole or 43Propenyl anisole $4\text{-}CH_3OC_6H_4CH{=}CHCH_3$	148.20	lf (al)	234.5^{763}, 115^{12}	21.35	$0.9882^{20/4}$	1.5615^{20}	**al, eth**, ace, bz	$B6^3$, 2395
756	Angelic acid or *cis*-2-Methyl-2-butenoic acid $CH_3CH{=}C(CH_3)CO_2H$	100.12	mcl pr or nd	185, $88\text{-}9^{10}$	45-6	$0.9834^{49/4}$	1.4434^{47}	eth, al	$B2^4$, 1551
757	Anhalamine $C_{11}H_{15}NO_3$	209.25	nd (al)		187-8			al, ace	$B21^4$, 2521
758	Anhalonidine $C_{12}H_{17}NO_3$	223.27	oct (bz, eth)		160-1			w, al	$B21^4$, 2524
759	Anhalonine *d* $C_{12}H_{15}NO_3$	221.26	$[\alpha]^{25/}{}_D+$ 56.7 (chl)	$140^{0.02}$	84.5-5			w, al, eth, peth	$B27^2$, 542
760	Anhalonine *dl* $C_{12}H_{15}NO_3$	221.26	rh nd (peth)		85-5.5			al, w, eth, peth	$B27^2$, 542
761	Anhalonine *l* $C_{12}H_{15}NO_3$	221.26	nd (peth), $[\alpha]^{25}{}_D-$ 56.3 (chl, c=4)	$140^{0.02}$				w, al, eth, peth	$B27^2$, 542
762	Anhalonine-hydrochloride *l* $C_{12}H_{15}NO_3HCl$	257.72	rh pr, $[\alpha]^{25}{}_D-$ 40.5 (al)		254-5d			al, w	$B27^2$, 542
763	Anhydroecgonine *dl* or Ecgonidine $C_9H_{13}NO_2$	167.21	cr (MeOH, MeOH-eth)		226-30d			w	$B22^4$, 284
764	Anhydroecgonine hydrochloride or Ecgonidine hydrochloride $C_9H_{13}NO_2 \cdot HCl$	203.67	rh nd (al)		240-1			al, eth, w	$B22^4$, 284
765	Aniline or Phenylamine $C_6H_5NH_2$	93.13		184, 68.3^{10}	−6.3	$1.02173^{20/4}$	1.5863^{20}	**al, eth, ace, bz** lig	$B12^3$, 217
766	Aniline, benzene sulfonate $C_6H_5NH_2\text{-}C_6H_5SO_3H$	251.30	nd		240d			w, al	$B12^2$, 75
767	Aniline, *m*-benzyl or (3-aminophenyl)phenyl methane $3\text{-}H_2NC_6H_4CH_2C_6H_5$	183.25	cr (lig)		46			lig	$B12^3$, 3215
768	Aniline, *p*-benzyl or (4-aminophenyl)phenyl methane $4\text{-}(C_6H_5CH_2)C_6H_4NH_2$	183.25	mcl (lig)	300	34-5	1.038^{25}		al, eth, lig	$B12^3$, 3215
769	Aniline, *N*-benzylidene or Benzalaniline $C_6H_5N{=}CHC_6H_5$	181.24	pa ye nd (CS_2), pl (dil al)	310	54	$1.038^{55/4}$	1.600^{100}	al, eth	$B12^3$, 319
770	Aniline, 2-bromo $2\text{-}BrC_6H_4NH_2$	172.02		229, 110.5^{19}	32	$1.578^{20/4}$	1.6113^{20}	al, eth	$B12^3$, 1415
771	Aniline, 2-bromo-4,6-dichloro $2\text{-}Br\text{-}4,6\text{-}Cl_2C_6H_2NH_2$	240.91	nd (al)	273	83.5			al, bz, chl	$B12^2$, 335
772	Aniline, 2-bromo-3,5-dinitro $2\text{-}Br\text{-}3,5\text{-}(NO_2)_2C_6H_2NH_2$	262.02	ye lf (al)		181			al	$B12^3$, 1728
773	Aniline, 2-bromo-4,5-dinitro $2\text{-}Br\text{-}4,5\text{-}(NO_2)_2C_6H_2NH_2$	262.02	pa ye (al)		186			al	B12, 762
774	Amiline, 2-bromo-4,6-dinitro $2\text{-}Br\text{-}4,6\text{-}(NO_2)_2C_6H_2NH_2$	262.02	ye nd (aa or al)	sub	153-4			w, ace	$B12^3$, 1724
775	Aniline, 2-bromo-4-nitro $2\text{-}Br\text{-}4\text{-}(NO_2)C_6H_3NH_2$	217.02	ye nd (al)		104.5			al, aa	$B12^2$, 403
776	Aniline, 2-bromo-5-nitro $2\text{-}Br\text{-}5\text{-}(NO_2)C_6H_2NH_2$	217.02	pa ye nd (al)		141			w, al, eth	$B12^3$, 1674
777	Aniline, 3-bromo $3\text{-}BrC_6H_4NH_2$	172.02		251, 130^{12}	18.5	$1.5793^{20.4/4}$	$1.6260^{20.4}$	al, eth	$B12^3$, 1421
778	Aniline, 3-bromo-4-nitro $3\text{-}Br\text{-}4\text{-}(NO_2)_2C_6H_3NH_2$	217.02	ye nd (al)		175-6			al	$B12^3$, 1676
779	Aniline, 4-bromo $4\text{-}BrC_6H_4NH_2$	172.02	rh bip yrm nd (60 % al)	d	66.4	$1.4970^{100/4}$		al, eth	$B12^3$, 1429
780	Aniline, 4-bromo-2,3-dichloro $4\text{-}Br\text{-}2,3\text{-}Cl_2C_6H_2NH_2$	240.91	nd		77.5			al, eth, bz	B12, 653
781	Aniline, 4-bromo-2,5-dichloro $4\text{-}Br\text{-}2,5\text{-}Cl_2C_6H_2NH_2$	240.91	nd		91			al, eth	$B12^3$, 1471
782	Aniline, 4-bromo-3,5-dichloro $4\text{-}Br\text{-}3,5\text{-}Cl_2C_6H_2NH_2$	240.91	nd (w + al)		129			al, bz, chl	B12, 654

PHYSICAL CONSTANTS OF ORGANIC COMPOUNDS (Continued)

No.	Name, Synonyms, and Formula	Mol. wt.	Color, crystalline form, specific rotation and λ_{max} (log ε)	b.p. °C	m.p. °C	Density	n_D	Solubility	Ref.
783	Aniline, 4-bromo-2,5-dinitro $4\text{-Br-2,5-(NO}_2)_2C_6H_2NH_2$	262.02	ye cr (al)	...	186	...	...	al	B12, 761
784	Aniline, 4-bromo-2,6-dinitro $4\text{-Br-2,6-(NO}_2)_2C_6H_2NH_2$	262.02	og-red pl (abs.al)	...	163 (cor)	...	...	al	$B12^3$, 1726
785	Aniline, 4-bromo-2-nitro $4\text{-Br-2-(NO}_2)C_6H_3NH_2$	217.02	og ye nd (w)	sub	111.5	...	...	al	$B12^3$, 1670
786	Aniline, 4-bromo-3-nitro $4\text{-Br-3-(NO}_2)C_6H_3NH_2$	217.02	nd (al)	...	132	...	...	al, eth, chl, aa	$B12^3$, 1674
787	Aniline, 5-bromo-2,4-dinitro $5\text{-Br-2,4-(NO}_2)_2C_6H_2NH_2$	262.02	pa ye nd (dil al)	...	178.4	...	...	al, eth, bz	$B12^3$, 1723
788	Aniline, 5-bromo-2-nitro $5\text{-Br-2-(NO}_2)C_6H_3NH_2$	217.02	rd-ye nd (dil al)	...	151-2	...	...	al	$B12^3$, 1671
789	Aniline, 6-bromo-2,3-dinitro $6\text{-Br-2,3-(NO}_2)_2C_6H_2NH_2$	262.02	dk rd cr (al)	...	158	...	...	...	B12, 760
790	Aniline, 6-bromo-2-nitro $6\text{-Br-2-(NO}_2)C_6H_3NH_2$	217.02	og or ye nd (dil al)	...	74.5	1.988	...	al	$B12^2$, 402
791	Aniline, *N*-butyl or Butyl phenyl amine $C_6H_5NHC_4H_9$	149.24	...	241.6, $118\text{-}20^{15}$	-14.4	$0.93226^{20/4}$	1.53412^{20}	al, eth	$B12^3$, 267
792	Aniline, *N*-isobutyl or Isobutyl phenyl amine $C_6H_5NH\text{-i-}C_4H_9$	149.24	...	231-3, 109^{13}	...	$0.940^{15/4}$	1.5328^{20}	bz, eth	$B12^3$, 269
793	Aniline, *N-sec*-butyl or *Sec*-butyl phenyl amine $C_6H_5NHCH(CH_3)C_2H_5$	149.24	...	225,$112\text{-}4^{22}$	...	...	1.5333^{20}	...	$B12^3$, 269
794	Aniline, *N-tert*-butyl or *tert*-Butyl phenyl amine $C_6H_5NHC(CH_3)_3$	149.24	...	$214\text{-}6^{753}$, $93\text{-}8^{14}$	...	1.5270^{20}	...	al, ace, bz	$B12^3$, 270
795	Aniline, 2-butyl or *o*-Butyl aniline $2\text{-}C_4H_9C_6H_4NH_2$	149.24	ye oil	$122\text{-}5^{12}$	...	$0.953^{20/4}$	...	al, ace, bz	$B12^3$, 2714
796	Aniline, 2-Sec-butyl $2\text{-}CH_3CH_2CH(CH_3)C_6H_4NH_2$	149.24	...	$120\text{-}2^{16}$	...	0.9574^{20}	...	al, ace, bz	$B12^3$, 2721
797	Aniline, 2-*tert*-butyl $2\text{-}(CH_3)_3CC_6H_4NH_2$	149.24	...	233-5	...	0.977^{15}	1.5453^{20}	al, eth, bz	$B12^3$, 2726
798	Aniline, 4-butyl or *p*-Butyl aniline $4\text{-}C_4H_9C_6H_4NH_2$	149.24	pa ye	261, 133^{14}	...	$0.945^{20/4}$	...	al, aa, bz	$B12^3$, 2715
799	Aniline, 4-*iso*-butyl $4\text{-}tert\text{-}C_4H_9C_6H_4NH_2$	149.24	pa ye	238^{762},112^{11}	...	$0.949^{15/4}$	...	**al, ace** bz	$B12^3$, 2715
800	Aniline, 4-*sec*-butyl $4\text{-}Sec\text{-}C_4H_9C_6H_4NH_2$	149.24	...	238^{762},118^{15}	...	$0.949^{15/4}$	1.5360^{20}	eth, bz	$B12^2$, 635
801	Aniline, 4-*tert*-butyl $4\text{-}(CH_3)_3CC_6H_4NH_2$	149.24	ye rd (peth)	240^{740}	17	$0.9525^{15/4}$	1.5380^{20}	**al, eth, bz**	$B12^3$, 2728
802	Aniline, 3-*tert*-butyl $3\text{-}(CH_3)CC_6H_4NH_2$	149.24	...	229^{708}	...	...	...	al, eth, bz	$B12^2$, 637
803	Aniline, 2-chloro-(α) $2\text{-}ClC_6H_4NH_2$	127.57	...	208.8	-14	$1.21253^{20/4}$	1.58951^{2}ls0	**al,** eth, ace	$B12^3$, 1281
804	Aniline, 2-chloro (β) $2\text{-}ClC_6H_4NH_2$	127.57	...	208.8, 84.6^{10}	-1.9	$1.21266^{20/4}$	1.5889^{20}	**al** eth, ace, **bz**	$B12^3$, 1281
805	Aniline, 2-chloro,hydrochloride $2\text{-}ClC_6H_4NH_2\text{-}HCl$	164.03	pl(w,aq.al)	...	235	1.505^{18}	...	w	$B12^2$, 315
806	Aniline, 2-chloro-4-nitro $2\text{-Cl-4-(NO}_2)C_6H_3NH_2$	172.57	ye nd (lig-CS_2)	...	108	...	...	al, eth	$B12^3$, 1662
807	Aniline, 3-chloro $3\text{-}ClC_6H_4NH_2$	127.57	...	229.9, 101.2^{10}	-10.3	$1.21606^{20/4}$	1.59414^{20}	**al, eth, ace, bz**	$B12^3$, 1303
808	Aniline, 3-chloro-2,6-dinitro $3\text{-Cl-2,6-(NO}_2)_2C_6H_2NH_2$	217.57	og-ye nd (al)	...	112.	...	...	al	$B12^3$, 1719
809	**Aniline, 3-chloro, hydrochloride** $3\text{-}ClC_6H_4NH_2\cdot HCl$	164.03	pl		222	...	...	w, al	$B12^2$, 320
810	Aniline, 3-chloro-4-nitro $3\text{-Cl-4-(NO}_2)C_6H_3NH_2$	172.57	ye lf (bz)	...	156-7	...	...	al, eth	$B12^3$, 1665
811	Aniline, 4-chloro $4\text{-}ClC_6H_4NH_2$	127.57	rh pr	232	72.5	$1.429^{19/4}$	1.5546^{87}	al, eth, w	$B12^3$, 1325
812	Aniline, 4-chloro-2,6-dinitro $4\text{-Cl-2,6-(NO}_2)_2C_6H_2NH_2$	217.57	og-ye nd (al)	...	147 (cor)	...	...	al	$B12^3$, 1720
813	Aniline, 4-chloro-2-methoxy. or 4-Chloro-*o*-anisidine $4\text{-Cl 2-(CH}_3O)C_6H_3NH_2$	157.60	nd or pr (dil.al)	260	52	...	...	al, eth, bz	$B13^2$, 184
814	Aniline, 4-chloro-2-nitro $4\text{-Cl-2-(NO}_2)C_6H_3NH_2$	172.57	dk og-ye pr (dil al)	...	116-7	...	...	al, eth, aa	$B12^3$, 1649
815	Aniline, 4-chloro-3-nitro $4\text{-Cl-3-NO}_2C_6H_3NH_2$	172.57	ye nd or pr (w) nd (peth)	...	103	...	...	al, eth, ace	$B12^3$, 1660

PHYSICAL CONSTANTS OF ORGANIC COMPOUNDS (Continued)

No.	Name, Synonyms, and Formula	Mol. wt.	Color, crystalline form, specific rotation and λ_{max} (log ε)	b.p. °C	m.p. °C	Density	n_D	Solubility	Ref.
816	Aniline, 5-chloro-2-ethoxy or 5-Chloro-*o*-phenetidine 5-Cl-2-$(C_2H_5O)C_6H_3NH_2$	171.63	nd (dil.al)	...	42	...	...	al, bz, aa	B13, 383
817	Aniline, 5-chloro-2-methoxy or 5-Chloro-*o*-anisidine 5-Cl-2-$(CH_3O)C_6H_3NH_2$	157.60	nd (dil al)	...	84	...	...	al	B13^{2}, 183
818	Aniline, 5-chloro-2-nitro 5-Cl-2-$(NO_2)C_6H_3NH_2$	172.57	gold ye nd (CS_2) ye lf (al, bz)	sub	126.5	...	...	al, eth	B12^{3}, 1655
819	Aniline, 5-chloro-3-nitro 5-Cl-3-$(NO_2)C_6H_3NH_2$	172.57	og.ye nd (al)	...	133-4	...	...	al, eth	B12, 732
820	Aniline, 6-chloro-2-nitro 6-Cl-2-$(NO_2)C_6H_3NH_2$	172.57	ye nd (dil al)	...	76	...	...	al	B12^{3}, 1658
821	Aniline, 6-chloro-3-nitro 6-Cl-3-$(NO_2)C_6H_3NH_2$	172.57	ye nd (lig)	...	121	...	...	al, eth, ace	B12^{3}, 1661
822	Aniline, *N*-cyclohexyl or Phenylcyclohexylamine *N*-cyclohexyl aniline $C_6H_5NHC_6H_{11}$	175.27	mcl pr	279^{764}	16	$1.0155^{20/4}$	1.5610^{20}	al, eth, bz	B12^{2}, 98
823	Aniline, 4-cyclohexyl 4-$C_6H_{11}C_6H_4NH_2$	175.27	pl (lig)	...	55	...	...	al,bz	B12^{3}, 2819
824	Aniline, 2,3-dibromo 2,3-$Br_2C_6H_3NH_2$	250.92	pl (dil al)	...	43	...	...	al, eth, aa	B12, 655
825	Aniline, 2,4-dibromo 2,4-$Br_2C_6H_3NH_2$	250.92	rh bi pym (chl)nd or lf (dil al)	156^{24}	79.5-80.5	2.260^{20}	...	al, eth	B12^{3}, 1471
826	**Aniline, 2,4-dibromo-6-nitro** $Br_2NO_2C_6H_2NH_2$	295.92	ye cr	...	128	...	...	...	B12^{2}, 403
827	Aniline, 2,5-dibromo 2,5-$Br_2C_6H_3NH_2$	250.92	pr (al)	...	53-5	...	...	al, eth	B12^{3}, 1474
828	Aniline, 2,6-dibromo 2,6-$Br_2C_6H_3NH_2$	250.92	nd (al)	262-4	87-8	...	...	al, eth, bz, chl	B12^{3}, 1474
829	Aniline, 2,6-dibromo-4-nitro 2,6-Br_2-4-$(NO_2)C_6H_2NH_2$	295.92	ye nd (al or aa)	207	...	...	...	aa	B12^{3}, 1678
830	Aniline, 3,4-dibromo 3,4-$Br_2C_6H_3NH_2$	250.92	lf (dil al)	100 sub	81	...	...	al	B12^{1}, 329
831	Aniline, 3,5-dibromo 3,5-$Br_2C_6H_3NH_2$	250.92	nd (dil al)	...	57	...	...	al, eth, bz	B12^{3}, 1475
832	Aniline, 3,5-dibromo-4-methoxy 3,5-Br_2-4-$CH_3OC_6H_2NH_2$	280.95	pl (lig)	...	66	...	...	al, eth, ace, bz	B13^{3}, 1195
833	Aniline, 4,6-dibromo-2-ethoxy 4,6-Br_2-2$(C_2H_5O)C_6H_2NH_2$	294.97	pr (dil al)	...	52	...	...	al, lig	B13^{3}, 867
834	Aniline, *N,N*-dibutyl or Dibutylphenylamine $C_6H_5N(C_4H_9)_2$	205.34	...	274.8, 138.8^{10}	−32.2	$0.9037^{20/4}$	1.5186^{20}	al, eth, ace, bz	B12^{3}, 269
835	Aniline, 2,3-dichloro 2,3-$Cl_2C_6H_3NH_2$	162.02	nd (lig)	252	24	...	...	al, eth, ace	B12, 621
836	Aniline, 2,4-dichloro 2,4-$Cl_2C_6H_3NH_2$	162.02	pr (ace) nd (dil al) (lig)	245	63-4	...	...	al, eth	B12^{3}, 1389
837	Aniline, 2,5-dichloro 2,5-$Cl_2C_6H_3NH_2$	162.02	nd (lig)	251	50	...	...	al, eth, bz	B12^{3}, 1397
838	Aniline, 2,6-dichloro 2,6-$Cl_2C_6H_3NH_2$	162.02	...	...	39	...	...	al, eth	B12^{2}, 337
839	Aniline, 2,6-dichloro-4-ethoxy or 4-Amino-3,5-dichloro-phenetole 2,6-Cl_2-4-$(C_2H_5O)C_6H_2NH_2$	206.07	nd (dil al)	275	46	...	...	al, eth, bz	B13^{2}, 276
840	Aniline, 2,6-dichloro-4-nitro or Dichloran 2,6-Cl_2-4-$(NO_2)C_6H_2NH_2$	207.02	ye nd (al, aa)	...	191	...	...	al	B12^{3}, 1669
841	Aniline, 3,4-dichloro 3,4-$Cl_2C_6H_3NH_2$	162.02	nd (lig)	272, 145^{15}	72	...	...	al, eth	B12^{3}, 1402
842	Aniline, 3,5-dichloro 3,5-$Cl_2C_6H_3NH_2$	162.02	nd (lig, dil al)	260^{741}	51-3	...	...	al, eth, bz, chl	B12^{2}, 337
843	Aniline, 3,5-dichloro-4-ethoxy or 4-Amino-2,6-dichloro-phenetole 3,5-Cl_2-4$(C_2H_5O)C_6H_2NH_2$	206.07	nd (peth)	...	105-7	...	...	al, eth, bz	B13^{2}, 275
844	Aniline, *N,N*-diethyl or Diethylphenylamine $C_6H_5N(C_2H_5)_2$	149.24	ye oil	216.27, 92^{10}	−38.8	$0.93507^{20/4}$	1.5409^{20}	al, eth, ace	B12^{3}, 260
845	Aniline, 2-diethylamino 2-$[(C_2H_5)_2N]C_6H_4NH_2$	164.25	oil	312^{244}, 127^{25}	...	...	...	al, ace, bz	B13^{3}, 34

No.	Name, Synonyms, and Formula	Mol. wt.	Color, crystalline form, specific rotation and λ_{max} (log ε)	b.p. °C	m.p. °C	Density	n_D	Solubility	Ref.
846	Aniline, 3-diethylamino $3-[(C_2H_5)_2N]C_6H_4NH_2$	164.25		276-8, 117[4]				al	B13[2], 26
847	Aniline, 4-diethylamino $4-[(C_2H_5)_2N]C_6H_4NH_2$	164.25	ye oil	260-2, 139-40[10]				bz	B13[3], 113
848	Aniline, *N,N*-diethyl-3-bromo $3-BrC_6H_4N(C_2H_5)_2$	228.13		142[10]				al, ace	B12[1], 315
849	Aniline, *N,N*-diethyl-4-bromo $4-BrC_6H_4N(C_2H_5)_2$	228.13	nd or pr	270	33			al, eth	B12[2], 347
850	Aniline, *N,N*-diethyl-4-chloro $4-ClC_6H_4N(C_2H_5)_2$	183.68	nd (al)	251-3, 95-6[1.5]	45.5-6.5			al	B12[3], 1329
851	Aniline, *N,N*-diethyl-2-ethoxy or *N,N*-diethyl-*o*-phentidine $2-(C_2H_5O)C_6H_4N(C_2H_5)_2$	193.29		231-3				al, eth, bz	B13, 365
852	Aniline, *N,N*-diethyl-3-ethoxy or *N,N*-diethyl-*m*-phenetidine $3-(C_2H_5O)C_6H_4N(C_2H_5)_2$	193.29		268-70, 145[14]			1.5325[25]	al, bz, aa	B13[3], 942
853	Aniline, *N,N*-diethyl-3-nitro $3-NO_2C_6H_4N(C_2H_5)_2$	194.23	ye	288-90					B12[3], 1545
854	Aniline, *N,N*-diethyl-4-nitro $4-NO_2C_6H_4N(C_2H_5)_2$	194.23	ye nd (lig) pl (al)		77-8	1.225		al	B12[3], 1585
855	Aniline, *N,N*-diethyl-4-nitroso $4-ONC_6H_4N(C_2H_5)_2$	178.23	gr mcl pr (eth) gr lf (ace)		87-8	1.24[15/4]		al, eth, ace	B12[3], 1585
856	Aniline, 3,4-dihydroxy $3,4-(HO)_2C_6H_3NH_2$	125.13	br-vt nd (al-bz)		124-5d			w, al, eth	B13[2], 464
857	Aniline, 3,5-dihydroxy or Phloramin $3,5-(HO)_2C_6H_3NH_2$	125.13	nd		146-52			al	B13, 787
858	Aniline, 2,4-diiodo $2,4-I_2C_6H_3NH_2$	344.92	br nd or rh cr (al)		95-6	2.748		al, eth, ace, bz, chl	B12[3], 1508
859	Aniline, 2,4-diiodo-5-nitro $2,4-I_2-5(NO_2)C_6H_2NH_2$	389.92	pa ye mcl pl		125				B12, 747
860	Aniline, 2,5-diiodo $2,5-I_2C_6H_3NH_2$	344.92	nd (al)		88-9			al, eth, ace	B12, 675
861	Aniline, 2,6-diiodo $2,6-I_2C_6H_3NH_2$	344.92	nd (al)		122			al, eth, ace, bz	B12, 675
862	Aniline, 2,6-diiodo-3-nitro $2,6-I_2-3-(NO_2)C_6H_2NH_2$	389.92	ye nd (dil al)		145.5			al	B12, 747
863	Aniline, 2,6-diiodo-4-nitro $2,6-I_2-4-(NO_2)C_6H_2NH_2$	389.92	pa ye lf or nd (bz)		245			bz	B12[3], 1680
864	Aniline, 3,4-diiodo $3,4-I_2C_6H_3NH_2$	344.92	pa ye lf or pr (bz-peth)		74.5			al, eth, bz	B12, 675
865	Aniline, 3,5-diiodo $3,5-I_2C_6H_3NH_2$	344.92	nd (al)		110			al, eth, chl	B12[1], 337
866	Aniline, 4,6-diiodo-2-nitro $4,6-I_2-2-(NO_2)C_6H_2NH_2$	389.92	ye nd (ace)		154			eth, ace, bz, chl	B12[1], 361
867	Aniline, 4,6-diiodo-3-nitro $4,6-I_2-3-(NO_2)C_6H_2NH_2$	389.92	pa ye nd (eth-al)		149			eth	B12, 747
868	Aniline, 2,3-dimethoxy or 3-Amino veratrole $2,3-(CH_3O)_2C_6H_3NH_2$	153.18		137[15]				w	B13[3], 2104
869	Aniline, 2,4-dimethoxy $2,4-(CH_3O)_2C_6H_3NH_2$	153.18	pl (lig)		33.5			al, eth, lig	B13[3], 2131
870	Aniline, 2,6-dimethoxy $2,6-(CH_3O)_2C_6H_3NH_2$	153.18	pl (al) lf (peth)	146[23]	75			al, eth, bz, lig	B13[3], 2128
871	Aniline, 3,4-dimethoxy or 4-Amino veratroile $3,4-(CH_3O)_2C_6H_3NH_2$	153.18	lf (eth)		87-8			eth	B13[3], 2105
872	Aniline, *N,N*-dimethyl $C_6H_5N(CH_3)_2$	121.18	pa ye	194, 77[13]	2.45	0.9557[20/4]	1.5582[20]	al, eth, ace, bz, chl	B12[3], 245
873	Aniline, 2-dimethylamino $2-[(CH_3)_2N]C_6H_4NH_2$	136.20	oil	218[751], 117[25]		0.995[22]		al, eth, ace, bz	B13[3], 33
874	Aniline, 3-dimethylamino $3-[(CH_3)_2N]C_6H_4NH_2$	136.20		268-70[740], 138[10]	<-20	0.995[25]		al, eth	B13[3], 68
875	Aniline, 4-dimethylamino $4-[(CH_3)_2N]C_6H_4NH_2$	136.20	nd (bz)	263, 158[11]	53	1.036[20/4]		al, eth, bz, w, chl	B13[3], 109
876	Aniline, *N,N*-dimethyl-2-bromo $2-BrC_6H_4N(CH_3)_2$	200.08		107—8[14]		1.3880[25/25]	1.5768[25]	al	B12[3], 1416

PHYSICAL CONSTANTS OF ORGANIC COMPOUNDS (Continued)

No.	Name, Synonyms, and Formula	Mol. wt.	Color, crystalline form, specific rotation and λ_{max} (log ε)	b.p. °C	m.p. °C	Density	n_D	Solubility	Ref.
877	Aniline, *N,N*-dimethyl-3-bromo $3BrC_6H_4N(CH_3)_2$	200.08	....	239, 126^{14}	11	....	....	al, aa	B12^3, 1422
878	Aniline, *N,N*-dimethyl-4-bromo $4\text{-}BrC_6H_4N(CH_3)_2$	200.08	lf (al)	264	55	....	....	al, eth	B12^3, 1432
879	Aniline, *N,N*-dimethyl-2-chloro $2\text{-}ClC_6H_4N(CH_3)_2$	155.63	....	203, 98-9^{18}	......	1.1067$^{20/4}$	1.5578^{20}	al, bz	B12^3, 1283
880	Aniline, *N,N*-dimethyl-4-chloro $4\text{-}ClC_6H_4N(CH_3)_2$	155.63	nd (al)	231	35.5	....	....	al	B12^3, 1329
881	Aniline, *N,N*-dimethjyl-4-chloro-3-nitro $4\text{-}Cl\text{-}3(NO_2)C_6H_3N(CH_3)_2$	200.62	ye nd (dil al)	....	81.5-2.5	....	....	al, lig	B12^3, 1660
882	Aniline, *N,N*-dimethyl, hydrochloride $C_6H_5N(CH_3)_2\text{-}HCl$	157.66	hyg pl (w) (bz)	....	85-95	1.1156$^{19/4}$	....	w, al, chl	B12^3, 251
883	Aniline, *N,N*-dimethyl-2-nitro $2\text{-}(NO_2)C_6H_4N(CH_3)_2$	166.18	ye-og	146^{20}	fp-20	1.1794$^{20/4}$	1.6102^{20}	w, al, eth, chl	B12^3, 1516
884	Aniline, *N,N*-dimethyl-3-nitro $3\text{-}NO_2C_6H_4N(CH_3)_2$	166.18	og-ye or red nd pr (eth or eth-al)	280-5	60-1	....	....	al, eth	B12^3, 1544
885	Aniline, *N,N*-dimethyl-4-nitro $4\text{-}NO_2C_6H_4N(CH_3)_2$	166.18	ye nd (al)	....	164.5	....	....	al, eth, aa	B12^3, 1583
886	Aniline, *N,N*-dimethyl-4-nitroso $4\text{-}ON\text{-}C_6H_4N(CH_3)_2$	150.18	gr-pl (eth)	....	92.5-3.5	1.145^{20}	....	al, eth	B12^3, 1509
887	Aniline, 2,3-dimethyl or 2,3-xylidine $2,3\text{-}(CH_3)_2C_6H_3NH_2$	121.18	....	221-2, 106^{15}	<−15	0.9931^{20}	1.5684^{20}	al, eth	B12^3, 2438
888	Aniline, 2,3-dimethyl-4-nitro or 4-Nitro-2,3-xylidine $2,3\text{-}(CH_3)_2\text{-}4\text{-}(NO_2)C_6H_2NH_2$	166.18	ye pr (al)	....	114	....	....	al	B12^3, 2442
889	Aniline, 2,3-dimethyl-5-nitro or 5-Nitro-2,3-xylidine $2,3\text{-}(CH_3)_2\text{-}5NO_2C_6H_3NH_2$	166.18	pa ye nd (al)	....	111-2	....	....	al	B12^1, 479
890	Aniline, 2,3-dimethyl-6-nitro or 6-Nitro-2,3-xylidine $2,3\text{-}(CH_3)_2\text{-}6\text{-}NO_2C_6H_2NH_2$	168.18	red pl (al)	....	118-9	....	....	al	B12^3, 2442
891	Aniline, 2,4-dimethyl or 2,4-xylidine $2,4\text{-}(CH_3)_2C_6H_3NH_2$	121.18	....	214, 91^{10}	−14.3	0.9723$^{20/4}$	1.5569^{20}	al, eth, bz	B12^3, 2469
892	Aniline, 2,4-dimethyl-3-nitro or 3-Nitro-2,4-xylidine $2,4\text{-}(CH_3)_2\text{-}3\text{-}NO_2C_6H_2NH_2$	166.18	ye nd	....	81-2	....	....	al, lig	B12^2, 612
893	Aniline, 2,4-dimethyl-5-nitro or 5-Nitro-2,4-xylidine $2,4\text{-}(CH_3)_2\text{-}5\text{-}NO_2C_6H_2NH_2$	166.18	og ye nd (al)	....	123	....	....	al	B12^2, 612
894	Aniline, 2,4-dimethyl-6-nitro or 6-Nitro-2,4-xylidine $2,4\text{-}(CH_3)_2\text{-}6\text{-}NO_2C_6H_2NH_2$	166.18	og red nd or pl (lig)	....	76	....	....	al	B12^2, 612
895	Aniline, 2,5 dimcthyl oı 2,5-Xylidine $2,5(CH_3)_2C_6H_3NH_2$	121.18	ye lf (lig)	214, 97-100^{10}	15.5	0.9790$^{21/4}$	1.5591^{21}	eth	B12^3, 2503
896	Aniline, 2,6-dimethyl or 2,6-Xylidine $2,6(CH_3)_2C_6H_3NH_2$	121.18	....	214^{739}	11.2	0.9842^{20}	1.5610^{20}	al, eth	B12^3, 2462
897	Aniline, 2,6-dimethyl-3-nitro or 3-Nitro-2,6-xylidine $2,6\text{-}(CH_3)_2\text{-}3\text{-}NO_2C_6H_2NH_2$	166.18	ye nd (dil al)	....	81-2	....	....	al	B12^2, 605
898	Aniline, 2,6-dimethyl-4-nitro or 4-Nitro-2,6-xylidine $2,6\text{-}(CH_3)_2\text{-}4\text{-}NO_2C_6H_2NH_2$	166.18	og-red nd or pl (lig)	....	76	....	....	al	B12^3, 2469
899	Aniline, 3,4-dimethyl or 3,4-Xylidine $3,4\text{-}(CH_3)_2C_6H_3NH_2$	121.18	pl or pr (lig)	228	51	1.076^{18}	....	eth, lig	B12^3, 2443
900	Aniline, 3,4-dimethyl-2-nitro or 2-Nitro-3,4-xylidine $3,5\text{-}(CH_3)_2\text{-}2\text{-}NO_2C_6H_2NH_2$	166.18	red pr (al)	....	65-6	....	....	al	B12^3, 2454
901	Aniline, 3,4-dimethyl-5-nitro or 5-Nitro-3,4-xylidine $3,4\text{-}(CH_3)_2\text{-}5\text{-}NO_2C_6H_2NH_2$	166.18	og lf (al)	....	74-5	....	....	al	B12, 1106
902	Aniline, 3,5-dimethyl or 3,5-Xylidine $3,5\text{-}(CH_3)_2C_6H_3NH_2$	121.18	....	220-1, 99-100^{20}	9.8	0.9706$^{20/4}$	1.5581^{20}	eth	B12^3, 2495
903	Aniline, 3,5-dimethyl-2-nitro or 2-Nitro-3,5-xylidine $3,5\text{-}(CH_3)_2\text{-}2\text{-}NO_2C_6H_2NH_2$	166.18	ye nd (lig)	....	56	....	....	....	B12^2, 613
904	Aniline, 3,5-dimethyl-4-nitro or 4-Nitro-3,5-xylidine $3,5\text{-}(CH_3)_2\text{-}4\text{-}NO_2C_6H_2NH_2$	166.18	og pr (bz), lf (lig)	....	133	....	....	al, bz	B12^3, 2500
905	Aniline, 4,5-dimethyl-2-nitro or 2-Nitro-4,5-xylidine $4,5\text{-}(CH_3)_2\text{-}2\text{-}NO_2C_6H_2NH_2$	166.18	red-br pr (al)	....	140	....	....	ace, bz	B12^3, 2455
906	Aniline, 2,3-dinitro $2,3\text{-}(NO_2)_2C_6H_3NH_2$	183.12	....	....	178	1.646^{50}	....	al, eth	B12^3, 1680
907	Aniline, 2,4-dinitro $2,4\text{-}(NO_2)_2C_6H_3NH_2$	183.12	ye nd (dil ace), gr-ye ta (al)	....	180 (188)	1.615^{14}	....	....	B12^3, 1681
908	Aniline, 2,6-dinitro $2,6\text{-}(NO_2)_2C_6H_3NH_2$	183.12	gold lf (50% aa), ye nd (al)	....	141-2	....	....	**eth, bz**	B12^3, 1704

No.	Name, Synonyms, and Formula	Mol. wt.	Color, crystalline form, specific rotation and λ_{max} (log ε)	b.p. °C	m.p. °C	Density	n_D	Solubility	Ref.
909	Aniline, 3,5-dinitro 3,5-$(NO_2)_2C_6H_3NH_2$	183.12			16.3	1.601^{50}		**al, eth**	$B12^3$, 1705
910	Aniline, *N,N*-dipropyl $C_6H_5N(C_3H_7)_2$	177.29	lf ye	243, 127^{10}		0.9104^{20}	1.5271^{20}	al, eth, ace, bz	$B12^3$, 266
911	Aniline, 2-ethoxy or *o*-Phenetidine 2-$(C_2H_5O)C_6H_4NH_2$	137.18		232.5, $127\text{-}8^{14}$	<-21		1.5560^{20}	al, eth	$B13^3$, 756
912	Aniline, 3-ethoxy or *m*-Phenetidine 3-$(C_2H_5O)C_6H_4NH_2$	137.18		248, $127\text{-}8^{11}$				al, eth	$B13^3$, 932
913	Aniline, 4-ethoxy or *p*-Phenetidine 4-$(C_2H_5O)C_6H_4NH_2$	137.18		254, 125^{11}	2.4	$1.06521^{16/4}$	1.5528^{20}	al, eth	$B13^3$, 996
914	Aniline, *N*-Ethyl or Ethylphenylamine $C_6H_5NHC_2H_5$	121.18		204.7, 97.5^{18}	-63.5	$0.9625^{20/4}$	1.5559^{20}	**al, eth**, ace, bz	$B12^3$, 255
915	Aniline, *N*-ethyl-2-chloro 2-$ClC_6H_4NHC_2H_5$	155.63		219^{726}		$1.104^{20/4}$			$B12^3$, 1284
916	Aniline, *N*-ethyl, hydrochloride $C_6H_5NHC_2H_5$-HCl	157.64	nd		178.5	1.0085^{182}		w, al, chl	$B12^3$, 257
917	Aniline, *N*-ethyl-*N*-methyl $C_6H_5N(CH_3)(C_2H_5)$	135.21		203-5, $93\text{-}5^{13}$		$0.9193^{55/4}$		al, eth	$B12^3$, 258
918	Aniline, *N*-ethyl-*N-nitroso or N*-ethylphenylnitrosamine $C_6H_5N(NO)C_2H_5$	150.18	yesh	$119\text{-}20^{15}$		1.0874^{20}		aa	$B12^3$, 1119
919	Aniline, *o*-ethyl 2-$C_2H_4C_6H_4NH_2$	121.18		209-10	-43	$0.983^{22/4}$	1.5584^{22}	al, eth	$B12^3$, 2374
920	Aniline, *m*-ethyl 3-$C_2H_5C_6H_4NH_2$	121.18		$214\text{-}5^{764}$, $93\text{-}5^{6}$	-64	0.9896		al, eth	$B12^3$, 2379
921	Aniline, *p*-ethyl 4-$C_2H_5C_6H_4NH_2$	121.18		217-8, 92.3^{10}	-4.87	$0.9679^{20/4}$	1.5554^{20}	al, eth	$B12^3$, 2380
922	Aniline, *o*-fluoro 2-$FC_6H_4NH_2$	111.12	pa ye	$174\text{-}6^{757}$	-28.5	1.1513^{21}	1.5421^{20}	al, eth	$B12^3$, 1273
923	Aniline, *m*-fluoro 3-$FC_6H_4NH_2$	111.12		187-9, $82\text{-}3^{18}$		1.1561^{19}	1.5436^{20}	al, eth	$B12^3$, 1274
924	Aniline, *p*-fluoro 4-$FC_6H_4NH_2$	111.12	pa ye	$180.5\text{-}2.5^{757}$, 85^{19}	-0.8	$1.1725^{20/4}$	$1.5195^{20/4}$	al, eth	$B12^3$, 1276
925	Anilinehydrochloride $C_6H_5NH_2$-HCl	129.59	lf or nd	245	198	1.2215^{4}		w, al	$B12^3$, 232
926	Aniline, *o*-iodo 2-$IC_6H_4NH_2$	219.02	nd (dil al)		60-1			al, eth, ace	$B12^3$, 1487
927	Aniline, *m*-iodo 3-$IC_6H_4NH_2$	219.02	lf or nd	280	33		1.6811^{20}	al, chl	$B12^3$, 1489
928	Aniline, *p*-iodo 4-$IC_6H_4NH_2$	219.02	nd (w)		67-8			al, eth	$B12^3$, 1493
929	Aniline, *o*-methoxy or *o*-Anisidine 2-$(CH_3O)C_6H_4NH_2$	123.15		224, 90^{4}	6.2	$1.0923^{20/4}$	1.5715^{10}	al, eth, ace, bz	$B13^3$, 754
930	Aniline, 2-methoxy-3-nitro 2-(CH_3O)-3-NO_2-$C_6H_3NH_2$	168.15	pa ye nd (lig)		67			al, lig	$B13^2$, 195
931	Aniline, *m*-methoxy or *m*-Anisidine 3-$(CH_3O)C_6H_4NH_2$	123.15		251 (cor)	-1	$1.096^{20/4}$	1.5794^{20}	al, eth, ace, bz	$B13^3$, 932
932	Aniline, *p*-methoxy or *p*-Anisidine 4-$(CH_3O)C_6H_4NH_2$	123.15	ta (w), rh pl	243, 115^{13}	57.2	$1.071^{57/4}$	1.5559^{60}	w, al, eth, ace, bz	$B13^3$, 994
933	Aniline, *N*-methyl or Methylphenyl amine $C_6H_5NHCH_3$	107.16		196.25, 86^{15}	-57	$0.9891^{20/4}$	1.5684^{20}	al, eth	$B12^3$, 240
934	Aniline, *N*-methyl, hydrochloride or Methylphenyl amine hydrochloride $C_6H_5NHCH_3$-HCl	143.62	nd (chl-eth)	122-3		$1.0660^{131/4}$		w, al	$B12^3$, 244
935	Aniline, *N*-(3-methylbutyl) or *N*-isopentylaniline C_6H_5NH-i-C_5H_{11}	163.26		254-5, $126\text{-}7^{14}$		$0.8912^{55/4}$	1.5305^{20}	**al, eth**	$B12^3$, 272
936	Aniline, *N*-methyl-2-chloro 2-$ClC_6H_4NHCH_3$	141.60		218, 106^{20}		1.1735^{12}	1.5780^{25}	al, ace, bz	$B12^3$, 1283
937	Aniline, *N*-methyl-4-chloro 4-$ClC_6H_4NHCH_3$	141.60		239^{764}, 120^{20}		1.614^{20}	1.5835^{20}	al, bz	$B12^3$, 1329
938	Aniline, *N*-methyl,-2-nitro or Methyl-2-nitrophenylamine 2-$O_2NC_6H_4NHCH_3$	152.15	red or og nd (peth)	d	38			al, eth, ace, bz, lig	$B12^3$, 1516
939	Aniline, *N*-methyl,3-nitro 3-$NO_2C_6H_4NHCH_3$	152.15	red-ye nd or pr (al), cr (lig)		68			al, eth, bz	$B12^3$, 1544
940	Aniline, *N*-methyl-4-nitro 4-$NO_2C_6H_4NHCH_3$	152.15	br-ye pr (al), cr (eth)	d	152	$1.201^{155/4}$		al, bz	$B12^3$, 1584

No.	Name, Synonyms, and Formula	Mol. wt.	Color, crystalline form, specific rotation and λ_{max} (log ε)	b.p. °C	m.p. °C	Density	n_D	Solubility	Ref.
941	Aniline, *N*-methyl-*N*-nitroso or *N*-Nitroso methylphenyl-amine $C_6H_5N(NO)CH_3$	136.15	ye	225d, 121^{13}	14.7	$1.1240^{20/4}$	1.57688^{20}	al, eth	$B12^3$, 1119
942	Aniline, *N*-methyl-4-nitroso $4\text{-}ONC_6H_4NHCH_3$	136.15	bl pl (bz)		118			al, eth, chl	$B7^3$, 3370
943	Aniline, *N*-methyl-4-*iso*-Propyl or Cuminyl amine $4\text{-}(CH_3)_2CHC_6H_4NHCH_3$	149.24	oil	$225\text{-}7^{734}$, $111\text{-}2^{11}$				al, eth	$B12^3$, 2741
944	Aniline, *N*-methyl-2,4,5-trimethyl $2,4,5\text{-}(CH_3)_3C_6H_2NHCH_3$	149.24	nd (dil al)		85			al, chl	B12, 1152
945	Aniline, *N*-methyl-3,4,5-trimethyl $3,4,5\text{-}(CH_3)_3C_6H_2NHCH_3$	149.24	lf (w)		123				B12, 1176
946	Aniline, *N*-methyl, *N*,2,4,6-tetranitro or Tetryl $2,4,6(NO_2)_3C_6H_2N(NO_2)CH_3$	287.15	ye pr (al)	exp 187	131-2	1.57^{10}		ace, bz	$B12^3$, 1738
947	Aniline, 2-methyl-5-*iso*-propyl or Carvacryl amine $2\text{-}CH_3\text{-}5\text{-}i\text{-}C_3H_7C_6H_3NH_2$	149.24		241, 118^{12}	-16	$0.9942^{20/4}$	1.5387^{20}	al, eth	$B12^3$, 2733
948	Aniline, 4-methylamino $4\text{-}(CH_3NH)C_6H_4NH_2$	122.17	lf (eth-peth)	257-9, 162^{20}	36			w, al, eth, ace, bz	$B13^3$, 108
949	Aniline, 4-methylthio $4\text{-}(CH_3S)C_6H_4NH_2$	139.22		272-3, 140^{15}		$1.1379^{20/4}$	1.6395^{20}	al, eth, ace, bz	$B13^3$, 1221
950	Aniline nitrate $C_6H_5NH_2\text{-}HNO_3$	156.15	rh		d190	1.356^4		w, al, eth	$B12^3$, 232
951	Aniline, *N*-nitro $C_6H_5NHNO_2$	138.13	lf (peth)	exp	43			w, al, eth, ace, bz	$B16^2$, 343
952	Aniline, 2-nitro $2\text{-}(NO_2)C_6H_4NH_2$	138.13	gold-ye pl or nd	284, $165\text{—}6^{28}$	71.5	1.442^{15}		al, eth, ace, bz	$B12^3$, 1513
953	Aniline, 2-nitro-5-methoxy $2\text{-}NO_2\text{-}5\text{-}(CH_3O)C_6H_3NH_2$	168.16	br nd	sub	131			al	$B13^3$, 975
954	Aniline, 3-nitro $3\text{-}(NO_2)C_6H_4NH_2$	138.13	ye nd, rh bipym (w)	305-7d, $100^{0.16}$	114	$1.1747^{160/4}$		al, eth, ace	$B12^3$, 1541
955	Aniline, 4-nitro $4\text{-}(NO_2)C_6H_4NH_2$	138.13	pa ye mcl nd (w)	331.7, $106^{0.03}$	148-9	$1.424^{20/4}$		al, eth, ace, chl	$B12^3$, 1580
956	Aniline, 5-nitro-2-propoxy $5\text{-}(NO_2)\text{-}2\text{-}(C_3H_7O)C_6H_3NH_2$	196.21	og (PrOH-peth)		49			al	$B13^3$, 879
957	Aniline, 4-nitroso $4\text{-}(ON)C_6H_4NH_2$	122.12	bl nd (bz)	dec	173-4			w, al	$B7^3$, 3370
958	Aniline, 4-octyl $4\text{-}C_8H_{17}C_6H_4NH_2$	205.35		310-1, $170\text{-}2^{17}$	20			eth	$B12^3$, 2770
959	Aniline oxalate(mono) $C_6H_5NH_2\text{-}H_2C_2O_4$	183.16	ta (aq al)		150-1			w, al, bz	$B12^3$, 236
960	Aniline oxalate(di) $(C_6H_5NH_2)_2\text{-}H_2C_2O_4$	276.29	tcl pr (w)		174-5d			w	$B12^3$, 236
961	Aniline, pentabromo $C_6Br_5NH_2$	487.61	nd (al-to)		265-6			al	$B12^3$, 1487
962	Aniline, pentachloro $C_6Cl_5NH_2$	265.35	nd (al)		232			al, eth, lig	$B12^3$, 1415
963	Aniline, pentafluoro $C_6F_5NH_2$	183.08		153-4	34				CAS53, 3112
964	Aniline, pentamethyl $(CH_3)_5C_6NH_2$	163.26	mcl pr (al)	277-8	152-3			al, eth	$B12^3$, 2758
965	Aniline, *N*-pentyl or Pentylphenyl amine $C_6H_5NHC_5H_{11}$	163.26		260-2					$B12^3$, 271
966	Aniline, 2-phenoxy $2\text{-}(C_6H_5O)C_6H_4NH_2$	185.23	cr (lig)	$307\text{-}8^{728}$, $172\text{-}3^{14}$	44-5			al, eth, ace, bz	$B13^3$, 758
967	Aniline, 3-phenoxy $3\text{-}(C_6H_5O)C_6H_4NH_2$	185.23	pr (lig)	315, $190\text{-}1^{14}$	37			al, eth, ace, bz	$B13^3$, 933
968	Aniline, 4-phenoxy or 4-Aminodiphenylether $4\text{-}(C_6H_5O)C_6H_4NH_2$	185.23	nd (w), cr (dil al)	$187\text{-}9^{14}$	85-6			al, eth, w	$B13^3$, 999
969	Aniline phosphate $C_6H_5NH_2\text{-}H_3PO_4$	191.12	nd (al), lf (w-al)		180			w, al	B12, 117
970	Aniline-picrate $C_{12}H_{10}N_4O_7$	322.23	ye or red mcl pr (w), dk gr		181	1.558		al	$B12^3$, 236
971	Aniline, *N*-propyl $C_6H_5NHC_3H_7$	135.21		222 (cor), 100^{11}		$0.9443^{20/4}$	1.5428^{20}	al, eth	$B12^3$, 264
972	Aniline, *o*-propyl $2\text{-}C_3H_7C_6H_4NH_2$	135.21		226^{758}, 116^{15}		$0.9602^{20/4}$	1.5427^{20}	al, eth	$B12^3$, 2657
973	Aniline, *p*-propyl $4\text{-}C_3H_7C_6H_4NH_2$	135.21		224-6					$B12^3$, 2658

PHYSICAL CONSTANTS OF ORGANIC COMPOUNDS (Continued)

No.	Name, Synonyms, and Formula	Mol. wt.	Color, crystalline form, specific rotation and λ_{max} (log ε)	b.p. °C	m.p. °C	Density	n_D	Solubility	Ref.
974	Aniline, *N*-isopropyl or Phenylisopropyl amine $C_6H_5NH\text{-}i\text{-}C_3H_7$	135.21		$203, 72\text{-}5^{14}$			1.5380^{20}	al, eth	$B12^3$, 266
975	Aniline, 2-isopropyl $2\text{-}i\text{-}C_3H_7C_6H_4NH_2$	135.21		$270\text{-}1^{745}, 95^{13}$		$0.9760^{12/4}$		eth, bz	$B12^3$, 2683
976	Aniline, 2-isopropyl-5-methyl or Thymylamine $2\text{-}i\text{-}C_3H_7\text{-}5\text{-}CH_3C_6H_3NH_2$	149.24	oil	238-42				al, eth	$B12^3$, 2741
977	Aniline, 4-isopropyl or Cumidine $4\text{-}(CH_3)_2CHC_6H_4NH_2$	135.21		225	-63	$0.953^{20/4}$	1.3415^{20}	al, eth, bz	$B12^3$, 2684
978	Aniline sulfate(mono) $C_6H_5NH_2\text{-}H_2SO_4$	191.20	lf (w + 1/2)		162				$B12^3$, 232
979	Aniline sulfate *(di)* $(C_6H_5NH_2)_2\text{-}H_2SO_4$	284.33	lf (al)		d	1.377^4		w	$B12^3$, 232
980	Aniline, 2,3,4,5-tetrachloro $2,3,4,5\text{-}Cl_4\text{-}C_6HNH_2$	230.91	nd (al)		118-20			al, eth, bz	$B12^2$, 340
981	Aniline, 2,3,5,6-tetrachloro $2,3,5,6\text{-}Cl_4C_6HNH_2$	230.91	nd (lig, al)		108			al, eth	$B12^3$, 1414
982	Aniline, 2,3,4,5-tetramethyl or Prehnidine $2,3,4,5\text{-}(CH_3)_4C_6HNH_2$	149.24	lf (w)	259-60	70			al, eth, lig	$B12^3$, 2743
983	Aniline, 2,3,4,6-tetramethyl or isoduridine $2,3,4,6\text{-}(CH_3)_4C_6HNH_2$	149.24		255	23-4	0.978^{24}		al	$B12^3$, 2744
984	Aniline, 2,4,5-tribromo $2,4,5\text{-}Br_3C_6H_2NH_2$	329.83	nd (al)		80-1 (85)			al, eth, bz	B12, 662
985	Aniline, 2,4,6-tribromo $2,4,6\text{-}Br_3\text{-}C_6H_2NH_2$	329.83	nd (al, bz)	300	122	$2.35^{20/20}$		ace, chl	$B12^3$, 1477
986	Aniline, 2,4,6-tribromo,hydrobromide $2,4,6\text{-}Br_3C_6H_2NH_2\text{-}HBr$	410.75	nd	sub	195-6				$B12^1$, 330
987	Aniline, 3,3,5-tribromo $3,4,5\text{-}Br_3C_6H_2NH_2$	329.83	nd (al)		123			al, eth	$B12^3$, 1487
988	Aniline, 2,3,4-trichloro $2,3,4\text{-}Cl_3C_6H_2NH_2$	196.46	nd (lig)	292^{774}	73			al	B12, 626
989	Aniline, 2,4,5-trichloro $2,4,5\text{-}Cl_3C_6H_2NH_2$	196.46	nd(lig or 50% al)	ca 270	96.5			al, eth	$B12^3$, 1409
990	Aniline, 2,4,6-trichloro or *sym*-Trichloroaniline $2,4,6\text{-}Cl_3C_6H_2NH_2$	196.46	cr (al), nd (lig or peth)	262^{746}	78.5			al, eth	$B12^3$, 1409
991	Aniline, 2,3,5-triido $2,3,5\text{-}I_3C_6H_2NH_2$	470.82	nd		116			al, bz	B12, 676
992	Aniline, 2,3,6-triido $2,3,6\text{-}I_3C_6H_2NH_2$	470.82	nd (al or al-eth)		116.8			al	B12, 676
993	Aniline, 2,4,6-triido $2,4,6\text{-}I_3C_6H_2NH_2$	470.82	ye nd or pl (al), pr (aa)		185.5			aa	$B12^3$, 1508
994	Aniline, 3,4,5-triido $3,4,5\text{-}I_3C_6H_2NH_2$	470.82	nd (al-ace)		174.5d			al, eth, ace,bz	B12, 676
995	Aniline, 2,4,5-trimethyl or Pseudocumidine $2,4,5,\text{-}(CH_3)_3C_6H_2NH_2$	135.21	nd (w)	234-5	68	0.957		al, w	$B12^3$, 2700
996	Aniline, 2,4,6-trimethyl or Mesidine $2,4,6\text{-}(CH_3)_3C_6H_2NH_2$	135.21		232-3	-5	1.5495^{20}			$B12^3$, 2708
997	Aniline, 2,4,6-trinitro or Pieramide $2,4,6\text{-}(NO_2)_3C_6H_2NH_2$	228.12	dk ye pr (aa)	exp	192-5	1.762^{12}		ace, bz	$B12^3$, 1737
998	*o*-Anisaldehyde or 2-Methoxybenzaldehyde $2\text{-}(CH_3O)C_6H_4CHO$	136.16	pr	$243\text{-}4, 124\text{-}5^{18}$	37-8	$1.1326^{20/4}$	1.5600^{20}	al, eth, ace, bz, chl	$B8^3$, 143
999	*m*-Anisaldehyde or 3-Methoxybenzaldehyde $3\text{-}(CH_3O)C_6H_4CHO$	136.16		$230, 62^1$		$1.1187^{20/4}$	1.5530^{20}	al, eth, ace, bz, chl	$B8^3$, 199
1000	*p*-Anisaldehyde or 4-Methoxtbenzaldehyde $4\text{-}(CH_3O)C_6H_4CHO$	136.16		$249.5, 83^2$	0	$1.1191^{15/4}$	1.5730^{20}	w, eth, ace, bz chl	$B8^3$, 218
1001	*o*-Anisamide $2\text{-}(CH_3O)C_6H_4CONH_2$	151.16	nd (bz), pr (eth), pl (w)		129			eth, bz	$B10^3$, 155
1002	*p*-Anisamide $4\text{-}(CH_3O)C_6H_4CONH_2$	151.16	nd or ta (w)		$295, 183^{40}$	166-7	16.5	w, al	$B10^3$, 341
1003	*o*-Anisic acid or 2-Methoxybenzoic acid $2\text{-}(CH_3O)C_6H_4CO_2H$	152.15	pl (w)	200	101			al, eth, bz, chl	$B10^3$ 97
1004	*o*-Anisic acid, ethyl ester $2\text{-}(CH_3O)C_6H_4CO_2C_2H_5$	180.20		$261, 134\text{-}6^{12}$		$1.1124^{20/4}$	1.5224^{20}	al, eth	$B10^3$, 116

PHYSICAL CONSTANTS OF ORGANIC COMPOUNDS (Continued)

No.	Name, Synonyms, and Formula	Mol. wt.	Color, crystalline form, specific rotation and λ_{max} (log ε)	b.p. °C	m.p. °C	Density	n_D	Solubility	Ref.
1005	*o*-Anisic acid, methyl ester $2\text{-}(CH_3O)C_6H_4CO_2CH_3$	166.18		245, 127^{11}		$1.1511^{19/4}$	$1.534^{19.5}$	al	B10³, 109
1006	*m*-Anisic acid or 3-Methoxybenzoic acid $3\text{-}(CH_3O)C_6H_4CO_2H$	152.15							B10³, 244
1007	*m*-Anisic acid, ethyl ester $3\text{-}(CH_3O)C_6H_4CO_2C_2H_5$	180.20		260-1, 110^{5}		$1.0993^{20/4}$	1.5161^{20}	al, eth	B10³, 250
1008	*m*-Anisic acid, methyl ester $3\text{-}(CH_3O)C_6H_4CO_2CH_3$	166.18		252, $121\text{-}4^{10}$		$1.13101^{20/4}$	1.5224^{20}	al	B10³, 250
1009	*p*-Anisic acid or 4-Methoxybenzoic acid $4\text{-}(CH_3O)C_6H_4CO_2H$	152.15	nd (w)	$170\text{-}2^{10}$	184			al, eth, bz	B10³, 280
1010	*p*-Anisic acid, butyl ester $4\text{-}(CH_3O)C_6H_4CO_2C_4H_9$	208.26		183^{40}		$1.054^{16.5}$	$1.5141^{16.5}$		B10³, 307
1011	*p*-Anisic acid, ethyl ester $4\text{-}(CH_3O)C_6H_4CO_2C_2H_5$	180.20		269-70, $136\text{-}7^{13}$	7-8	$1.1038^{20/4}$	1.5254^{20}	al, eth	B10³, 300
1012	*p*-Anisic acid, methyl ester $4\text{-}(CH_3O)C_6H_4CO_2CH_3$	166.18	fl (al or eth)	256, 160^{20}	49			al, eth	B10³, 297
1013	*o*-Anisidine or 2-Methoxy aniline $2\text{-}(CH_3O)C_6H_4NH_2$	123.15		224, 90^{4}	6.2	$1.0923^{20/4}$	1.5713^{20}	al, eth, ace, bz	B13³, 754
1014	*o*-Anisidine, 4-nitro or 2-Methoxy-4-nitro aniline $2\text{-}(CH_3O)\text{-}4(NO_2)C_6H_3NH_2$	168.15	pa ye nd (dil al)		139-40	1.2112^{150}		al, ace	B13³, 887
1015	*o*-Anisidine, 5-nitro or 2-Methoxy-5-nitro aniline $2\text{-}(CH_3O)\text{-}5\text{-}NO_2\text{-}C_6H_3NH_2$	168.15	og-red nd (al, eth, w)		118	1.2068^{156}		al, eth, ace, bz, aa	B13³, 878
1016	*o*-Anisidine, 6-nitro or 2-Methoxy-6-nitro aniline $2\text{-}(CH_3O)\text{-}6\text{-}NO_2\text{-}C_6H_3NH_2$	168.15	ye pa red nd (al)		76			al, bz	B13³, 875
1017	*m*-Anisidine or Methoxy aniline $3\text{-}(CH_3O)C_6H_4NH_2$	123.15		251		$1.096^{20/4}$	1.5794^{20}	al, ace, eth, bz	B13³, 932
1018	*m*-Anisidine, 2-nitro or 3-Methoxy-2-nitro aniline $3\text{-}(CH_3O)\text{-}2\text{-}(NO_2)C_6H_3NH_2$	168.15	ye nd (bz)		124 (143)			al, ace	B13³, 974
1019	*m*=Anisidine, 4-nitro or 3-Methoxy-4-nitro aniline $3(CH_3O)\text{-}4(NO_2)C_6H_3NH_2$	168.15	ye nd (al)	sub	169			al, ace	B13³, 975
1020	*m*-Anisidine, 5-nitro or 3-Methoxy-5-nitro aniline $3\text{-}(CH_3O)\text{-}5(NO_2)C_6H_3NH_2$	168.15	og cr (w)		120	1.2034^{150}		al, ace, bz	B13³, 977
1021	*p*-Anisidine or 4-Methoxy aniline $4\text{-}(CH_3O)C_6H_4NH_2$	123.15	ta (w), rh pl	243 (cor), 115^{13}	57.2	$1.071^{57/4}$	1.5559^{67}	al, eth, w, ace, bz	B13³, 994
1022	*p*-Anisidine hydrochloride or 4-Methoxy aniline,hydrochloride $4\text{-}(CH_3O)C_6H_4NH_2\text{-}HCl$	159.62	lf, nd		236			w, al	B13², 223
1023	*p*-Anisidine, 2-nitro or 4-Methoxy-2-nitro aniline $4(CH_3O)\text{-}2(NO_2)C_6H_3NH_2$	168.15	dk red pr (w or al)		129			w, al, eth, ace	B13³, 1203
1024	*p*-Anisidine, 3-nitro or 4-Methoxy-3-nitro aniline $4(CH_3O)\text{-}3(NO_2)C_6H_3NH_2$	168.15	red (eth), og pr or pl (eth-lig)		57			al, eth, ace, bz	B13², 284
1025	Anisole or Methoxybenzene, Methyl phenyl ether $C_6H_5OCH_3$	108.14		155	-37.5	$0.9961^{20/4}$	1.5179^{20}	al, eth, ace, bz	B6⁴, 548
1026	Anisole, 2acetyl or 2-Methoxy acetophenone $2\text{-}(CH_3CO)C_6H_4OCH_3$	150.18	ye	245		$1.0897^{20/4}$	1.5393^{20}	al, ace	B8³, 263
1027	Anisole, 3-acetyl or 3-Methoxy acetophenone $3\text{-}(CH_3CO)C_6H_4OCH_3$	150.18		240, $125\text{-}6^{12}$	95-6	1.0343^{19}	1.5410^{20}	w, al, ace	B8³, 273
1028	Anisole, 4-acetyl or 4-Methoxy acetophenone $4\text{-}(CH_3CO)C_6H_4OCH_3$	150.18	pl (eth)	258,$138\text{-}9^{15}$	38-9	$1.0818^{41/4}$	1.547^{41}	al, eth, ace	B8³ 277
1029	Anisole, 4-acetyl-2-nitro or 4-Methoxy-3-nitro acetophenone $4\text{-}(CH_3CO)\text{-}2(NO_2)C_6H_3OCH_3$	195.17	nd (al)		99.5			al, eth, ace, bz	B8³, 295
1030	Anisole, 2-bromo $2\text{-}BrC_6H_4OCH_3$	187.04		216 (223), 94^{10}	2.5	$1.5018^{20/4}$	1.5727^{20}	al, eth	B6³, 736
1031	Anisole, 3-bromo $3\text{-}Br\text{-}C_6H_4OCH_3$	187.04		$210\text{-}1^{752}$, 105^{16}		1.5635^{20}		al, eth, bz	B6⁴, 1043
1032	Anisole, 4-bromo $4\text{-}BrC_6H_4OCH_3$	187.04		215, 100^{16}	13-4	$1.4564^{20/4}$	1.5642^{20}	al, eth, chl	B6⁴, 1044
1033	Anisole, 2-chloro $2\text{-}ClC_6H_4OCH_3$	142.58		198.5, $90\text{-}1^{16}$	-26.8	$1.1911^{20/4}$	1.5480^{20}		B6⁴, 785
1034	Anisole, 3-chloro-2-nitro $3\text{-}Cl\text{-}2\text{-}(NO_2)C_6H_3OCH_3$	187.58			56			eth, bz, chl	B6⁴, 1347
1035	Anisole, 3-chloro $3\text{-}ClC_6H_4OCH_3$	142.58		193-4, 70^{9}		$1.1759^{12/4}$	1.5365^{20}	al, eth	B6⁴, 811
1036	Anisole, 4-chloro $4\text{-}ClC_6H_4OCH_3$	142.58		197.5, 75^{10}	−18	$1.201^{20/4}$	1.5390^{20}	al, eth, chl	B6⁴, 822

No.	Name, Synonyms, and Formula	Mol. wt.	Color, crystalline form, specific rotation and λ_{max} (log ε)	b.p. °C	m.p. °C	Density	n_D	Solubility	Ref.
1037	Anisole, 4-chloro-2-nitro 4-Cl-2-$(NO_2)C_6H_3OCH_3$	187.58	ye nd or pr (al)	......	98	......	......	al, MeOH	B6^{4}, 1348
1038	Anisole, 2-cyclohexyl 2-$C_6H_{11}C_6H_4OCH_3$	190.29	......	267-8.5	......	1.007^{18}	1.5365^{18}	......	B6^{3}, 2493
1039	Anisole, 4-cyclohexyl 4-$C_6H_{11}C_6H_4OCH_3$	190.29	......	275-6^{748}	57-8	......	......	......	B6^{3}, 2502
1040	Anisole, 2,4-dichloro 2,4-$Cl_2C_6H_3OCH_3$	177.03	pr	235, 125^{10}	28-9	......	......	al, eth	B6^{4}, 855
1041	Anisole, 2,3-dimethyl or 3-Methoxy-*o*-xylene 2,3-$(CH_3)_2C_6H_3OCH_3$	136.19	......	199, 85^{18}	29	0.9596^{40}	1.5120^{40}	al, eth, ace, bz	B6^{3}, 1723
1042	Anisole, 2,4-dimethyl 2,4-$(CH_3)_2C_6H_3OCH_3$	136.19	......	192, 83-4^{15}	......	0.9740$^{16/4}$	1.5190^{16}	al, eth, bz	B6^{3}, 1744
1043	Anisole, 2,5-dimethyl 2,5-$(CH_3)_2C_6H_3OCH_3$	136.19	......	194^{773}	......	0.9693$^{13/4}$	1.5182^{15}	al, eth, bz, peth	B6^{3}, 1772
1044	Anisole, 2,6-dimethyl 2,6-$(CH_3)_2C_6H_3OCH_3$	136.19	......	182—3	......	0.9619$^{14/4}$	1.5053^{14}	al, eth, bz	B6^{3}, 1737
1045	Anisole, 3,4-dimethyl 3,4-$(CH_3)_2C_6H_3OCH_3$	136.19	......	204-5, 96-7^{17}	......	0.9744$^{14/4}$	1.5198^{14}	al, eth, bz	B6^{3}, 1727
1046	Anisole, 3,5-dimethyl 3,5-$(CH_3)_2C_6H_3OCH_3$	136.19	......	194.5, 89^{15}	......	0.9627$^{15/4}$	1.5110^{20}	al, eth, bz, aa	B6^{3}, 1756
1047	Anisole, 2,3-dinitro 2,3-$(NO_2)_2C_6H_3OCH_3$	198.14	nd (al), pl (to)	......	119	1.2290^{137}	......	al, lig	B6^{4}, 1369
1048	Anisole, 3,4-dinitro 3,4-$(NO_2)_2C_6H_3OCH_3$	198.14	ye nd (dil al)	......	71	1.3332^{110}	......	al, MeOH	B6^{4}, 1384
1049	Anisole, 2,6-dinitro 2,6-$(NO_2)_2C_6H_3OCH_3$	198.14	nd (al)	......	118	1.3000^{128}	......	al	B6^{3}, 868
1050	Anisole, 2,4-dinitro 2,4-$(NO_2)_2C_6H_3OCH_3$	198.14	nd (al or w)	206-7^{12}, sub	94.5-5.5	1.3364^{131}	......	al, eth, ace, bz, to	B6^{4}, 1372
1051	Anisole, 2,5-dinitro 2,5-$(NO_2)_2C_6H_3OCH_3$	198.14	nd (bz-lig)	136-8^{2}	97	1.476^{18}	......	ace, bz, al	B6^{4}, 1383
1052	Anisole, 3,5-dinitro 3,5-$(NO_2)_2C_6H_3OCH_3$	198.14	nd (al)	......	105.5	1.558^{12}	......	ace, bz, MeOH	B6^{4}, 1385
1053	Anisole, 2-ethyl 2-$C_2H_5C_6H_4OCH_3$	136.19	......	186-8^{755}, 80^{14}	......	0.9636$^{19/4}$	1.5142^{20}	eth, bz	B6^{3} 1656
1054	Anisole, 3-ethyl 3-$C_2H_5C_6H_4OCH_3$	136.19	......	196-7^{758}, 74^{10}	......	0.9575$^{18/4}$	1.5102	eth, bz	B6^{3}, 1662
1055	Anisole, 4-ethyl 4-$C_2H_5C_6H_4OCH_3$	136.19	......	195-6, 83-4^{16}	......	0.9624$^{15/4}$	1.5120^{20}	eth, bz	B6^{3}, 1665
1056	Anisole, 2-fluoro 2-$FC_6H_4OCH_3$	126.13	......	154-5, 59^{11}	-39	1.5489^{17}	1.4969$^{17.5}$	eth	B6^{4}, 771
1057	Anisole, 4-fluoro 4-$FC_6H_4OCH_3$	126.13	......	154	-45	1.1781$^{18/4}$	1.4886^{18}	eth	B6^{4}, 773
1058	**Anisole, *o*-(1-hydroxyethyl)** **2-$[CH_3CH(OH)]C_6H_4OCH_3$**	152.19	......	125^{768}	-85.1	0.9647$^{20/4}$	1.4024^{20}	al, **eth**, ace, **bz**	B6^{3}, 4563
1059	Anisole, *m*-(1-hydroxyethyl) 3-$[CH_3CH(OH)]C_6H_4OCH_3$	152.19	......	133^{15}	......	1.0781$^{19/4}$	1.5325^{20}	al, eth	B6^{3}, 4564
1060	Anisole, *p*-(1-hydroxyethyl) 4-$[CH_3CH(OH)]C_6H_4OCH_3$	152.19	......	310d, 140-1^{17}	......	1.0794$^{20/4}$	1.5310^{25}	al, eth	B6^{3}, 4565
1061	Anisole, 2-iodo 2-$IC_6H_4OCH_3$	234.04	......	239-40^{730}, 91-2^{2}	......	1.8^{20}	......	al, eth, ace, bz, lig	B6^{4}, 1070
1062	Anisole, 3-iodo 3-$IC_6H_4OCH_3$	234.04	......	244-5, 123^{14}	......	......	......	al, eth	B6^{4}, 1073
1063	Anisole, 4-iodo 4-$IC_6H_4OCH_3$	234.04	lf (al), nd (MeOH)	237^{726}, 139^{25}	53	......	......	eth, al	B6^{4}, 1075
1064	Anisole, 4-(3-methylbutyl) or *p*-Isopentyl anisole 4-*i*-$C_5H_{11}C_6H_4OCH_3$	178.28	......	121^{14}	......	......	......	bz	B6^{3}, 1960
1065	Anisole, 2-nitro 2-$NO_2C_6H_4OCH_3$	153.14	......	276.8^{752}, 144^{4}	10.5	1.2540$^{20/4}$	1.5161^{20}	**al, eth**	B6^{4}, 1249
1066	Anisole, 3-nitro 3-$NO_2C_6H_4OCH_3$	153.14	nd (al), pl (bz-lig)	258	38-9	1.373^{18}	......	al, eth	B6^{4}, 1270
1067	Anisole, 4-nitro 4-$NO_2C_6H_4OCH_3$	153.14	pr (al), nd (dil al)	274	54	1.2192$^{60/4}$	1.5070^{60}	al, eth	B6^{4}, 1282
1068	Anisole, 4-propenyl or Anethole 4-$(CH_3CH{=}CH)_2$-$C_6H_4OCH_3$	148.20	lf (al)	234.5^{763}, 115^{12}	21.3	0.9882$^{20/4}$	1.5615^{20}	al, eth, ace, bz	B6^{3}, 2395
1069	Anisole, 4-propyl 4-$C_3H_7C_6H_4OCH_3$	150.22	......	210, 98^{17}	......	0.94718$^{20/4}$	1.5045^{20}	al, eth, ace, bz, chl	B6^{3}, 1789

PHYSICAL CONSTANTS OF ORGANIC COMPOUNDS (Continued)

No.	Name, Synonyms, and Formula	Mol. wt.	Color, crystalline form, specific rotation and λ_{max} (log ε)	b.p. °C	m.p. °C	Density	n_D	Solubility	Ref.
1070	Anisole, 2,3,4,6-tetrabromo 2,3,4,6-$Br_4C_6HOCH_3$	423.72	nd (dil al or aa)	340	113-4			al	B6^1, 766
1071	Anisole, 2,3,4,5-tetrachloro 2,3,4,5-$Cl_4C_6HOCH_3$	245.92	cr (MeOH)		83			al, bz	B6^2, 182
1072	Anisole, 2,3,5,6-tetrachloro 2,3,5,6-$Cl_4C_6HOCH_3$	245.92	nd (al)		89-90			al, eth, bz	B6^2, 182
1073	Anisole, 2,3,4,6-tetrachloro 2,3,4,6-$Cl_4C_6HOCH_3$	245.92	nd (MeOH), pr (al)		64-5			al, eth, bz	B6^2, 182
1074	Anisole, 2,3,4-tribromo 2,3,4-$Br_3C_6H_2OCH_3$	344.83	nd (al)		106			al, ace, bz	B6^4, 1066
1075	Anisole, 2,3,5-tribromo 2,3,5-$Br_3C_6H_2OCH_3$	344.83	pr (dil al)	305-12	82			al, ace, bz	B6^2, 192
1076	Anisole, 1,3,5-tribromo 1,3,5-$Br_3C_6H_2OCH_3$	344.83	nd (al)	297-9	88	2.491		ace, bz	B6^4, 1067
1077	Anisole, 2,4,5-tribromo 2,4,5-$Br_3C_6H_2OCH_3$	344.83	nd (al)	306-9^{775}	105			al, ace, bz	B6^4, 1067
1078	Anisole, 3,4,5-tribromo 3,4,5-$Br_3C_6H_2OCH_3$	344.83	cr (al)	300-10	91-4			al, ace, bz	B6^2 195
1079	Anisole, 2,3,5-trichloro 2,3,5-$Cl_3C_6H_2OCH_3$	211.48	nd (al)		84			ace, al	B6^2, 180
1080	Anisole, 2,4,5-trichloro 2,4,5-$Cl_3C_6H_2OCH_3$	211.48	nd (dil al)	252-5^{742}	77.5			al, ace	B6^2, 180
1081	Anisole, 2,3,6-trichloro 2,3,6-$Cl_3C_6H_2OCH_3$	211.48	pr (al)	227-9^{754}	45			al, ace	B6^2, 180
1082	Anisole, 2,4,6-trichloro 2,4,6-$Cl_3C_6H_2OCH_3$	211.48	mcl nd (al)	240^{738}	61-2 (65)	1.640		al, ace	B6^1, 723
1083	Anisole, 2,4,6-triido 2,4,6-$I_3C_6H_2OCH_3$	485.83	lf (bz), nd (eth or al)		98-9			al, ace	B6^2, 204
1084	Anisole, 2,3,4-trinitro 2,3,4-$(NO_2)_3C_6H_2OCH_3$	243.13	pa ye lf (al)	exp	155			ace, aa, al	B6^1, 129
1085	Anisole, 2,3,5-trinitro 2,3,5-$(NO_2)_3C_6H_2OCH_3$	243.13	ye nd (al)		106.8	1.618^{15}		al, bz, ace	B6^3, 873
1086	Anisole, 2,4,5-trinitro 2,4,5-$(NO_2)_3C_6H_2OCH_3$	243.13	ye (al)		106-7			eth, bz, w	B6^2, 253
1087	Anisole, 2,4,6-trinitro 2,4,6-$(NO_2)_3C_6H_2OCH_3$	243.13	nd (dil MeOH)		(i) 69, (ii) 56-7	1.4947^{80}		al, eth, bz	B6^4, 1456
1088	Anisole, 2-vinyl or 2-Methoxystyrene 2-$(CH_2{=}CH)C_6H_4OCH_3$	134.18	nd	195-200, 83-4^{12}	29	1.0049$^{17/4}$	1.5388^{20}	al, eth, ace, bz	B6^3, 2383
1089	Anisole, 3-vinyl or 3-Methoxystyrene 3-$(H_2C{=}CH)C_6H_4OCH_3$	134.18		114-6$^{16-7}$, 90-3^{15}		0.999$^{16/4}$	1.5586^{23}	al, eth, bz	B6^3, 2385
1090	Anisole, 4-vinyl or 4-Methoxystyrene 4-$(CH_2{=}CH)C_6H_4OCH_3$	134.18		204-5^{756}, 91^{13}		1.0001$^{13/4}$	1.5642^{13}	al, eth, bz	B6^3, 2386
1091	*o*-Anisonitrile or *o*-Anisic acid nitrile 2-$NCC_6H_4OCH_3$	133.15		255-6, 146^{20}	24.5	1.1063$^{20/4}$		al, eth	B10^3, 159
1092	*p*-Anisonitrile or *p*-Anisic acid nitrile 4-$CH_3OC_6H_4CN$	133.15	nd (w), lf (al)	256-7	61-2			al, eth, bz	B10^3, 344
1093	*o*-Anisyl chloride 2-$CH_3OC_6H_4COCl$	170.60		254, 128^{11}					B10^3, 151
1094	*m*-Anisyl chloride 3-$CH_3OC_6H_4COCl$	170.60		243-4, 123-5^{15}				eth, ace, bz	B10^3, 253
1095	*p*-Anisyl chloride 4-$CH_3OC_6H_4COCl$	170.60	nd	262-3, 91^{1}	24-5	1.261$^{20/4}$	1.580^{20}	eth, ace, bz	B10^3, 337
1095a	*o*-Anisyl ether or bis(2-methoxy phenyl) ether $(2\text{-}CH_3OC_6H_4)_2O$	230.26	pl (lig)	330-1	79-80			al, eth	
1096	Anthracene $C_{14}H_{10}$	178.23	ta or mcl pr (al)	340 (cor), 226.5^{53} sub	216-4	1.283$^{25/4}$		ace, bz	B5^3, 2123
1097	Anthracene, 1-acetyl 1-$CH_3COC_{14}H_9$	220.27	pa ye (al)		107.5-9			al	B7^3, 2538
1098	Anthracene, 2-acetyl 2-$CH_3COC_{14}H_9$	220.27	ye (al), cr (AcOEt-peth)		190-2			al	B7^3, 2538
1099	Anthracene, 9-acetyl 9-$CH_3COC_{14}H_9$	220.27	pa ye (al)		76			al	B7^3, 2539
1100	Anthracene, 1-amino or 1-Anthrylamine 1-$H_2NC_{14}H_9$	193.25	gold-ye nd (al)		130			al	B12^3, 3335

No.	Name, Synonyms, and Formula	Mol. wt.	Color, crystalline form, specific rotation and λ_{max} (log ε)	b.p. °C	m.p. °C	Density	n_D	Solubility	Ref.
1101	Anthracene, 2-amino or 2-Anthrylamine 2-$H_2NC_{14}H_9$	193.25	ye lf (al)	sub	238-41			al	B12³, 3355
1102	Anthracene, 9-amino or 9-Anthrylamine 9-$H_2NC_{14}H_9$	193.25	ye lf (dil al), br (bz)		145-50			al, eth, bz, chl	B7³, 2361
1103	Anthracene, 9-benzoyl or 9-Anthraphenone 9-$C_6H_5COC_{14}H_9$	282.35	ye nd (bz, aa)		148			ace, bz, aa	B7², 502
1104	Anthracene, 1-chloro 1-$ClC_{14}H_9$	212.68	lf (aa)		83.5	$1.1707^{100/4}$	1.6959^{100}	al, eth, bz	B5³, 2132
1105	Anthracene, 2-chloro 2-$ClC_{14}H_9$	212.68	nd or lf		215 (223)			CCl_4	B5³, 2133
1106	Anthracene, 9-chloro 9-$ClC_{14}H_9$	212.68	gold-ye nd (al)		106			al, eth, bz	B5³, 2133
1107	Anthracene, 9,10-diamino 9,10-$(NH_2)_2C_{14}H_8$	208.26	red cr		196				B14³, 291
1108	Anthracene, 9,10-dibenzyl 9,10$(C_6H_5CH_2)_2C_{14}H_8$	358.48			245			lig, bz	B5³, 2640
1109	Anthracene, 9,10-dibromo 9,10-$Br_2C_{14}H_8$	336.03	ye nd (to or xyl)	sub	226			chl, bz	B5³, 2135
1110	Anthracene, 9,10-dichloro 9,10-$Cl_2C_{14}H_8$	247.12	ye nd (MeCOEt or CCl_4)		212			bz	B5³, 2134
1111	Anthracene, 9,10-dihydro or 9,10-Dihydro anthracene $C_{14}H_{12}$	180.25	ta or pr	305, $165\text{-}70^{12}$, sub	111	$0.8976^{11/4}$		al, eth, bz	B5³, 1987
1112	Anthracene, 9,10-dihydro-10-chloro-9-NO_2 10-Cl-9-NO_2-$C_{14}H_{10}$	259.69	nd (bz)		163			bz	B5³, 1989
1113	Anthracene, 9,10-dihydro-9-ethyl 9-$C_2H_5C_{14}H_{11}$	208.30		320-3 (cor)		$1.049^{18/18}$		**al, eth, bz**, aa	B5², 560
1114	Anthracene, 9,10-dihydro-1-hydroxy 1-$HOC_{14}H_{11}$	196.25	sl grsh flr lf or nd (bz, peth)		94			al, eth, aa	B6², 660
1115	**Anthracene, 9,10-dihydro-2-hydroxy** **2-$HOC_{14}H_{11}$**	196.25	(bz-peth), lf (dil al)		129				B6², 660
1116	Anthracene, 9,10-dihydro-9-hydroxy or 9-Hydroanthrol 9-HO-$C_{14}H_{11}$	196.25	nd (peth)		76			al, eth, bz	B6³, 3504
1117	Anthracene, 9,10-dihydro-9-oxo or 9-Anthrone $C_{14}H_{10}O$	194.23	nd (bz-lig, aa)		155			ace, bz	B7³, 2359
1118	Anthracene, 9,10-dihydro-9-oxo-10-nitro $C_{14}H_9NO_3$	239.23	cr (bz-lig), nd (CS_2)		140 (148d)			al, bz	B7³, 2371
1119	Anthracene, 1,2-dihydroxy or 1,2-Anthradiol 1,2-$(HO)_2C_{14}H_8$	210.23	pa gr lf		160-2			al, eth, aa	B6², 998
1120	Anthracene, 1,5-dihydroxy or 1,5-Anthradiol, Ruful 1,5-$(HO)_2C_{14}H_8$	210.23	ye nd		265d			al, eth, bz	B6³, 5683
1121	Anthracene, 1,8-dihydroxy or 1,8-Anthradiol, Chrysazol 1,8-$(HO)_2C_{14}H_8$	210.23	ye nd (dil al), lf (al-aa)		225d			al, eth, bz, AcOEt	B6, 1033
1122	Anthracene, 2,6-dihydroxy or 2,6-Anthradiol, Flavol 2,6-$(HO)_2C_{14}H_8$	210.23	pa ye lf (al)		295-300d			al, eth, aa	B6³, 5684
1123	Anthracene, 9,10-dihydroxy or 9,10-Anthradiol, Anthraquinol 9,10-$(HO)_2C_{14}H_8$	210.23	br or ye nd		180			al, eth	B6³, 5685
1124	Anthracene, 1,3-dimethyl 1,3-$(CH_3)_2C_{14}H_8$	206.29	pa bl fir lf (eth), cr (al)	$140\text{-}5^{2}$	83			al, eth	B5³, 2165
1125	Anthracene, 2,3-dimethyl 2,3-$(CH_3)_2C_{14}H_8$	206.29	bl gr fir lf (bz)		252			al, bz	B5³, 2166
1126	Anthracene, 9-ethyl 9-$C_2H_5C_{14}H_9$	206.29	bl flr lf (al or MeOH)		59	$1.0413^{99/4}$	1.6762^{99}	al, eth	B5³, 2165
1127	Anthracene, 1,2,3,4,5,6-hexahydro $C_{14}H_{16}$	184.28	lf (MeOH), cr (al)	160^{15}	67 (70)			bz	B5², 472
1128	Anthracene, 1-hydroxy or 1-Anthrol 1-$(HO)C_{14}H_9$	194.23	cr (bz), br nd or lf (al or aa)	234^{15}	158			al, eth	B6³, 3551
1129	Anthracene, 2-hydroxy or 2-Anthrol 2-$(HO)C_{14}H_9$	194.23	ye (bz), br lf or nd (dil al)		253			al, eth, ace, bz	B6³, 3552

No.	Name, Synonyms, and Formula	Mol. wt.	Color, crystalline form, specific rotation and λ_{max} (log ε)	b.p. °C	m.p. °C	Density	n_D	Solubility	Ref.
1130	Anthracene, 9-hydroxy or 9-Anthrol $9\text{-}HOC_{14}H_9$	194.23	ye red lf (dil al), pa ye nd (aa)		160-4			al, bz	B6^{3}, 3554
1131	Anthracene, 1-methyl $1\text{-}CH_3C_{14}H_9$	192.26	bl nd (MeOH), lf (al)	199—200	85-6	$1.0471^{99/4}$	$1.6802^{99/4}$	al, eth, bz, aa	B5^{3}, 2149
1132	Anthracene, 2-methyl $2\text{-}CH_3C_{14}H_9$	192.26	gr-bl flr lf (sub)	sub	209	$1.81^{0/4}$		bz, chl	B5^{3}, 2149
1133	Anthracene, 9-methyl $9\text{-}CH_3C_{14}H_9$	192.26	yesh nd (dil al), pr (bz, al)	$196\text{-}7^{12}$	81.5	$1.065^{99/4}$	1.6959^{99}	al, eth, ace, bz	B5^{3}, 2150
1134	Anthracene, 9-nitro $9\text{-}O_2NC_{14}H_9$	223.23	ye nd (al), pr (aa or xyl)	ca 275^{17}	146			bz, al	B5^{3}, 2136
1135	Anthracene, octahydro or Octhracene $C_{14}H_{18}$	186.30	pl (al)	293-5, 167^{12}	78	$0.9703^{80/4}$	1.5372^{80}	al, bz, aa	B5^{4}, 1584
1136	Anthracene, 2-phenyl $2\text{-}C_6H_5C_{14}H_9$	254.33			207				B5^{3}, 2462
1137	Anthracene, 9-phenyl $9\text{-}C_6H_5C_{14}H_9$	254.33	bl flr in sol, lf (al) (aa)		207			al, ace, bz, chl	B5^{3}, 2462
1138	Anthracene, tetradecahydro $C_{14}H_{24}$	192.34		128^{11}	93				B5^{4}, 492
1139	Anthracene, 1,2,3,4-tetrahydro $C_{14}H_{14}$	182.27			106-7				B5^{4}, 1909
1140	Anthracene, 1,2,9-trihydroxy $1,2,9\text{-}(HO)_3C_{14}H_7$	226.23	og ye lf		149-51				B6^{3}, 2728
1141	Anthracene, 1,2,10-trihydroxy $1,2,10\text{-}(HO)_3C_{14}H_7$	226.23	ye lf, nd (al-w)		208			al, eth, ace, bz, aa	B8^{2}, 372
1142	Anthracene, 1,4,9-trihydroxy $1,4,9\text{-}(HO)_3C_{14}H_7$	226.23	og-red nd (al)		156			al	B6^{3}, 2800
1143	Anthracene, 1,5,9-trihydroxy $1,5,9\text{-}(HO)_3C_{14}H_7$	226.23	gold lf (al)		200d without melt			al	B6^{3}, 2801
1144	Anthracene, 1,8,9-trihydroxy or Anthralin $1,8,9\text{-}(HO)_3C_{14}H_7$	226.23	ye pl or nd (lig)		178-80			al, eth, ace, bz	B6^{3}, 2802
1145	Anthracene, 1,9,10-trihydroxy (enol form) $1,9,10\text{-}(HO)_3C_{14}H_7$	226.23	gr nd (eth)		204-6			al, eth	B8^{2}, 372
1146	Anthracene, 1,9,10-trihydroxy (keto form) $1,9,10\text{-}(HO)_3C_{14}H_7$	226.23	ye nd (lig)		135-7			al	B8^{2}, 372
1147	Anthracene, 2,3,9-trihydroxy $2,3,9\text{-}(HO)_3C_{14}H_7$	226.23	ye br nd (al)		288-9			al, eth, ace, aa	B6^{3}, 2803
1148	Anthracene-2-vinyl $2\text{-}CH_2{=}CHC_{14}H_9$	204.27			125				
1149	Anthracene-9-vinyl $9\text{-}CH_2{=}CHC_{14}H_9$	204.27			64-5				
1150	9-Anthraldehyde or 9-Anthracene carboxaldehyde $9\text{-}C_{14}H_9CHO$	206.24	og nd (dil al)		104-5			bz, aa	B7^{3}, 2527
1151	Anthranil or 3,4-Benzo isoxazol C_7H_5NO	119.12		215d, 99^{13}	−18	$1.8127^{20/4}$	1.5845^{20}	al, ace	B27^{2}, 17
1152	Anthranil amide $2\text{-}H_2NC_6H_4CONH_2$	136.15	lf (chl, w)	300	110-1			al, AcOEt	B14^{3}, 889
1153	Anthranilic acid or 2-Aminobenzoic acid $2\text{-}H_2NC_6H_4CO_2H$	137.14	lf (al)	sub	146-7	1.412^{20}		w, al, eth, chl	B14^{3}, 879
1154	Anthranilic acid, *N*-acetyl $2\text{-}CH_3CONHC_6H_4CO_2H$	179.18	nd (aa)		185			al, eth, ace, bz, aa	B14^{3}, 922
1155	Anthranilic acid, *N*-acetyl-4-ethoxy $2\text{-}CH_3CONH\text{-}4\text{-}C_2H_5O\text{-}C_6H_3CO_2H$	223.23	nd (al or MeOH)		199d			al	B14^{1}, 657
1156	Anthanilic acid, butyl ester or Butyl anthranilate $2\text{-}H_2NC_6H_4CO_2C_4H_9$	193.25		182					B14^{2}, 209
1157	Anthranilic acid, iso--Butyl ester or iso-Butyl anthranilate $2\text{-}H_2NC_6H_4CO_2\text{-i-}C_4H_9$	193.25		$156\text{-}7^{13.5}$					B14^{2}, 209
1158	Anthranilic acid, 3,4-dichloro $3,5\text{-}Cl_2\text{-}2\text{-}H_2NC_6H_2CO_2H$	206.03	nd (aa)		237-8			al, eth, chl	B14^{1}, 549
1159	Anthranilic acid, 3,5-dichloro $3,5\text{-}Cl_2\text{-}2\text{-}H_2NC_6H_2CO_2H$	206.03	nd or lf (al)		231-2d			al, ace, eth, bz	B14^{3}, 966
1160	Anthranilic acid, 3,6-dichloro $3,6\text{-}Cl_2\text{-}2\text{-}H_2NC_6H_2CO_2H$	206.03	nd (w or aa)	sub	155			al, eth, ace, w, bz, aa	B14, 367

No.	Name, Synonyms, and Formula	Mol. wt.	Color, crystalline form, specific rotation and λ_{max} (log ε)	b.p. °C	m.p. °C	Density	n_D	Solubility	Ref.
1161	Anthranilic acid, 4,5-dichloro $4,5-Cl_2-2-H_2NC_6H_4CO_2H$	206.03	nd (aa)		213-4			al, eth, aa	$B14^{1}$, 549
1162	Anthranilic acid, 5,6-dichloro $5,6-Cl_2-2-H_2NC_6H_2CO_2H$	206.03	nd (MeOH)		176-7d			al, eth, aa	B14, 368
1163	Anthranilic acid, 3,5-diido $3,5-I_2-2-H_2NC_6H_2CO_2H$	388.93	pr (al)		232-3			eth	$B14^{3}$, 973
1164	Anthranilic acid, 3,5-diido,ethyl ester $3,5-I_2-2-H_2NC_6H_2CO_2C_2H_5$	416.98	pr (al)		101				$B14^{1}$, 555
1165	Anthranilic acid, 4,5-diido $4,5-I_2-2-H_2NC_6H_2CO_2H$	388.93	cr (dil NH_3)		210-2d			al, eth, bz	$B14^{1}$, 555
1166	Anthranilic acid, 4,5-diido,ethyl ester $4,5-I_2-2-H_2NC_6H_2CO_2C_2H_5$	416.99	pr (al)		137			al, eth, bz	$B14^{1}$, 555
1167	Anthranilic acid, *N,N*-dimethyl $2-(CH_3)_2NC_6H_4CO_2H$	165.19	pr nd (eth)	sub d	72			w, al, eth	$B14^{3}$, 896
1168	Anthranilic acid, *N*-ethyl $2-(C_2H_5NH)C_6H_4CO_2H$	165.19	pr or nd (dil al)		154			al, eth, ace	$B14^{3}$, 897
1169	Anthranilic acid, ethyl ester or Ethyl anthranilate $2-H_2NC_6H_4CO_2C_2H_5$	165.19		268, $145-7^{15}$	13	$1.1374^{20/4}$	1.5646^{20}	al, eth	$B14^{3}$, 885
1170	Anthranilic acid hydrochloride or 2-Aminobenzoic acid hydrochloride $2-H_2NC_6H_4CO_2H-HCl$	173.60						w	$B14^{2}$, 207
1171	Anthranilic acid, 3-hydroxy or 2-Amino-3-hydroxy benzoic acid $3-HO-2-H_2NC_6H_3CO_2H$	153.14	lf(w)		164			al, eth, chl	$B14^{3}$, 1463
1172	Anthranilic acid, 3 hydroxy or 2-Amino-3-hydroxy benzoic acid $3-HO-2-H_2NC_6H_3CO_2H$	153.14	lf (w)		164			al, eth, chl	$B14^{2}$, 1463
1173	Anthranilic acid, 4-hydroxy or 2-Amino-4-hydroxy benzoic acid $4-HO-2-H_2NC_6H_3CO_2H$	153.14			148^{4}			w, al, eth, ace	$B14^{3}$, 1475
1174	Anthranilic acid, 5-hydroxy $5-HO-2-H_2NC_6H_3CO_2H$	153.14	vt pr (w)		252d			al, eth, ace, bz	$B14^{3}$, 1468
1175	Anthranilic acid, *N*-methyl $2-CH_3NHC_6H_4CO_2H$	151.16	pl (al or lig)	$80^{0.01}$	179			al, eth, bz, chl, lig	$B14^{3}$, 895
1176	Anthranilic acid, *N*-methyl,ethyl ester or Ethyl-*N*-methyl anthranilate $2-CH_3NHC_6H_4CO_2C_2H_5$	179.22		266	39			eth	$B14^{2}$, 213
1177	Anthranilic acid, *N*-methyl,methyl ester $2-CH_3NHC_6H_4CO_2CH_3$	165.01	cr (peth)	255, $130-1^{13}$	19	1.120^{15}	1.5839^{15}	al, eth	$B14^{3}$, 895
1178	Anthranilic acid, methyl ester or Methyl anthranilate $2H_2NC_6H_4CO_2CH_3$	151.16		256, 135.5^{15}	24—5	$1.1682^{10/4}$	1.5810	al, eth	$B14^{3}$, 894
1179	Anthranilic acid, 3-methyl or 2-Amino-3-methyl benzoic acid $3-CH_3-2-H_2NC_6H_3CO_2H$	151.16	nd (al), pr (w)		172			al, eth	$B14^{2}$, 290
1180	Anthranilic acid, 5-methyl $5-CH_3-2-H_2NC_6H_3CO_2H$	151.16	lf (al), nd (w)		175			al, eth	$B14^{2}$, 291
1181	Anthranilic acid, 6-methyl $6-CH_3-2-H_2NC_6H_3CO_2H$	151.16	nd (MeOH)		125-6d			MeOH	$B14^{3}$, 1502
1182	Anthranilic acid, 3-nitro $3-NO_2-2-H_2NC_6H_3CO_2H$	182.14	ye nd (w)		208-9	1.558^{15}		al, eth	$B14^{3}$, 973
1183	Anthranilic acid, 4-nitro $4-NO_2-2-H_2NC_6H_3CO_2H$	182.14	og pr (dil al)		269			al, eth, ace, xyl	$B14^{3}$, 975
1184	Anthranilic acid, 5-nitro $5-NO_2-2-H_2NC_6H_3CO_2H$	182.14	lf (al), ye nd (w, dil al)			268-70, (280)		al, eth	$B14^{3}$, 980
1185	Anthranilic acid, 6-nitro $6-NO_2-2-H_2NC_6H_3CO_2H$	182.14	ye nd or lf (w)		184			w, al, eth, ace, aa	$B14^{3}$, 988
1186	Anthranilic acid, *N*-phenyl $2-C_6H_5NHC_6H_4CO_2H$	213.24	lf, nd or pr (al)		184d			al	$B14^{3}$, 898
1187	Anthranilic acid, phenyl ester or Phenyl anthranilate $2-H_2NC_6H_4CO_2C_6H_5$	213.24	nd (al)		70			al, eth	$B14^{3}$, 885
1188	Anthranilic acid, propyl ester or Propyl anthranilate $2-H_2NC_6H_4CO_2C_3H_7$	179.22		270				al, eth	$B14^{2}$, 209
1189	Anthranilonitrile or *o*-Cyanoaniline $2-H_2NC_6H_4CN$	118.14	ye pr (CS_2), nd (peth)	263^{751}	51			al, eth, ace, bz	$B14^{2}$, 210
1190	Anthranilonitrile, 4-methyl $4-CH_3-2-H_2NC_6H_3CN$	132.16	lf (dil al)		94			al, ace, bz, chl	$B14^{3}$, 1208

No.	Name, Synonyms, and Formula	Mol. wt.	Color, crystalline form, specific rotation and λ_{max} (log ε)	b.p. °C	m.p. °C	Density	n_D	Solubility	Ref.
1191	Anthanilonitrile, 5-methyl 5-CH_3-2-$H_2NC_6H_3CN$	132.16	cr (dil al)		63			al, eth, ace, bz	B14, 482
1192	Anthanilonitrile, 6-methyl 6-CH_3-2-$H_2NC_6H_3CN$	132.16	ye pr (bz) (w)		128				B14^{2}, 290
1193	Anthranylamide 2-$H_2NC_6H_4CONH_2$	136.15	lf (chl or w)	300	110-1			al, AcOEt, w	B14^{3}, 889
1194	β-Anthraquinoline or Naphtho(2,3:5,6)quinoline $C_{17}H_{11}N$	229.28	lf or ta (al)	446	170			al, eth, bz	B20^{4}, 4290
1195	Anthraquinone or 9,10-Dioxoanthracene $C_{14}H_8O_2$	208.22	ye rh nd (al, bz)	379.8	286 (sub)	1.438^{4}			B7^{3}, 4059
1196	Anthraquinone, 1-amino 1-$H_2NC_{14}H_7O_2$	223.23	red nd (al), gl (aa)	sub	253-4			eth, ace, bz, aa, chl	B14^{3}, 407
1197	Anthraquinone, 1-amino-2-benzoyl 1-NH_2-2-$(C_6H_5CO)C_{14}H_6O_2$	327.34	red nd (aa)		190			aa	B14, 482
1198	Anthraquinone, 1-amino-2-bromo 1-NH_2-2-$BrC_{14}H_6O_2$	302.13	ye, red nd (aa), nd (xyl)		182			to, aa	B14^{3}, 423
1199	Anthraquinone, 1-amino-3-bromo 3-Br-1-$H_2NC_{14}H_6O_2$	302.13	red nd (to)		243				B14^{3}, 423
1200	Anthraquinone, 1-amino-4-bromo 1-H_2N-4-Br-$C_{14}H_6O_2$	302.13			170				B14^{3}, 423
1201	Anthraquinone, 1-amino-4-bromo-2-methyl $C_{15}H_{10}NO_2Br$	316.15			232			al, eth	B14^{3}, 494
1202	Anthraquinone, 1-amino-4-hydroxy $C_{14}H_9NO_3$	239.23			215			al, ace	B14^{3}, 652
1203	Anthraquinone, 1-amino-2-methyl $C_{15}H_{11}NO_2$	237.26			205-6			al, eth, bz, chl, aa	B14^{3}, 492
1204	Anthraquinone, 1-amino-4-methyl $C_{15}H_{11}NO_2$	237.26			181				B14^{2}, 123
1205	Anthraquinone, 1-amino-4-nitro $C_{14}H_8N_2O_4$	268.23			298-300			al	B14^{3}, 427
1206	Anthraquinone, 2-amino 2-$NH_2C_{14}H_8O_2$	223.23	red nd (al aa)	sub	303-6			ace, bz, chl	B14^{3}, 429
1207	Anthraquinone, 2-amino-3-benzoyl 2-NH_2-3$(C_6H_5CO)C_{14}H_6O_2$	327.34	ye pl (Py)		331				B14^{1}, 482
1208	Anthraquinone, 2-amino-3-chloro $C_{14}H_8NO_2Cl$	257.68			280-3				B14^{3}, 435
1209	Anthraquinone, 3-amino-1,2-dihydroxy or 3-Aminoalizarin 3-NH_2-1,2-$(HO_2)_2C_{14}H_5O_2$	255.23	dk red pr (aa)	sub d	>300			aq NH_3	B14^{2}, 185
1210	Anthraquinone, 2-amino-1-hydroxy 2-NH_2-1-$HOC_{14}H_6O_2$	239.23	red br nd (al)	sub	226-7			al, eth, bz	B14^{3}, 651
1211	Anthraquinone, 4-amino-1,2-dihydroxy or 4-Aminoalizarin 4-NH_2-1,2-$(HO)_2C_{14}H_5O_2$	255.23	gr bl nd (al)		d			al	B14^{2}, 185
1212	Anthraquinone, 1-bromo 1-$BrC_{14}H_7O_2$	287.11	ye nd (bz)	sub	188			al, bz	B7^{3}, 4076
1213	Anthraquinone, 1-bromo-4-methylamino 1-Br-4$(CH_3NH)C_{14}H_6O_2$	316.15	br-red nd (Py)		194				B14^{3}, 424
1214	Anthraquinone, 2-bromo 2-$BrC_{14}H_7O_2$	287.11		sub	204-5			bz	B7^{3}, 4076
1215	Anthraquinone, 2-bromo-1-methylamino 2-Br-1-$(CH_3NH)C_{14}H_6O_2$	316.15	br nd (aa)		170-2				B14^{1}, 446
1216	Anthraquinone, 3-bromo-1,2-dihydroxy or 3-Bromoalizarin 3-Br-1,2-$(HO)_2C_{14}H_5O_2$	319.11	br-red nd (to)	sub	260-1			w	B8^{3}, 3772
1217	Anthraquinone, 1-chloro 1-$ClC_{14}H_7O_2$	242.66	ye nd (to or al)	sub	162			eth, bz, aa	B7^{3}, 4064
1218	Anthraquinone, 1-chloro-2-methyl 1-Cl-2-$CH_3C_{14}H_6O_2$	256.69			170-1				B7^{3}, 4105
1219	Anthraquinone, 2-chloro 2-$ClC_{14}H_7O_2$	242.66	pa ye nd (aa or al)	sub	211			bz, al	B7^{3}, 4066
1220	Anthraquinone, 1,2-diamino 1,3-$(NH_2)_2C_{14}H_6O_2$	238.25	vt nd		303-4				B14^{3}, 438
1221	Anthraquinone, 1,3-diamino 1,3-$(NH_2)_2C_{14}H_6O_2$	238.25	red		290			Py	B14^{3}, 439

No.	Name, Synonyms, and Formula	Mol. wt.	Color, crystalline form, specific rotation and λ_{max} (log ε)	b.p. °C	m.p. °C	Density	n_D	Solubility	Ref.
1222	Anthraquinone, 1,4-diamino $1,4-(NH_2)_2C_{14}H_6O_2$	238.25	dk vt nd (Py), vt cr		268			al, bz, Py	B14[3], 439
1223	Anthraquinone, 1,5-diamino $1,5-(NH_2)_2C_{14}H_6O_2$	238.25	dk red nd (al, aa)	sub	319 (cor)				B14[3], 466
1224	Anthraquinone, 1,6-diamino $1,6-(NH_2)_2C_{14}H_6O_2$	238.25	red nd (aa)		297				B14[2], 119
1225	Anthraquinone, 1,7-diamino $1,7-(NH_2)_2C_{14}H_6O_2$	238.25	red, nd		290				B14[1], 470
1226	Anthraquinone, 1,8-diamino $1,8-(NH_2)_2C_{14}H_6O_2$	238.25	red (al or aa)	265				al, aa	B14[1], 478
1227	Anthraquinone, 2,3-diamino $2,3-(NH_2)_2C_{14}H_6O_2$	238.25	red		353			Py	B14[1], 480
1228	Anthraquinone, 2,6-diamino $2,6-(NH_2)_2C_{14}H_6O_2$	238.25	red-br pr (aq Py)		320d			Py, xy, al	B14[1], 480
1229	Anthraquinone, 2,7-diamino $2,7-(NH_2)_2C_{14}H_6O_2$	238.25	og-ye nd (al), dk red nd (sub)	sub	>320				B14[3], 483
1230	Anthraquinone, 2,3-dibromo $2,3-Br_2C_{14}H_6O_2$	366.01	ye nd (to)	sub	283			bz, chl	B7[3], 4077
1231	Anthraquinone, 2,7-dibromo $2,7-Br_2C_{14}H_6O_2$	366.01	lt ye lf	sub	248			bz, aa	B7[2], 718
1232	Anthraquinone, 1,3-dichloro $1,3-Cl_2C_{14}H_6O_2$	277.11	ye nd (aa)		209-10			aa	B7[3], 4068
1233	Anthraquinone, 1,4-dichloro $1,4-Cl_2C_{14}H_6O_2$	277.11	og-ye nd (aa)		187-8			Py, aa	B7[3], 4068
1234	Anthraquinone, 1,5-dichloro $1,5-Cl_2C_{14}H_6O_2$	277.11	yesh (to), ye nd		252			aa	B7[3], 4068
1235	Anthraquinone, 1,6-dichloro $1,6-Cl_2C_{14}H_6O_2$	277.11	pa ye nd (aa)		203-4			ace, to	B7[3], 4070
1236	Anthraquinone, 1,7-dichloro $1,7-Cl_2C_{14}H_6O_2$	277.11	ye nd (aa)		213-4			bz, to	B7[3], 4070
1237	Anthraquinone, 1,8-dichloro $1,8-Cl_2C_{14}H_6O_2$	277.11	ye nd (aa)		202-3			bz	B7[3], 4070
1238	Anthraquinone, 2,3-dichloro $2,3-Cl_2C_{14}H_6O_2$	277.11	ye nd (aa)		271			bz, aa	B7[3], 4071
1239	Anthraquinone, 2,6-dichloro $2,6-Cl_2C_{14}H_6O_2$	277.11	ye nd (aa or al)		291			bz, aa, al	B7[3], 4071
1240	Anthraquinone, 2,7-dichloro $2,7-Cl_2C_{14}H_6O_2$	277.11	yesh nd		212 (231)			eth	B7[3], 4072
1241	Anthraquinone, 1,2-dihydroxy or Alizarin $1,2-(HO)_2C_{14}H_6O_2$	240.22	og or red tcl nd or pr (al, sub)	430 (sub)	289-90 (cor)			al, eth, ace, bz, Py	B8[3], 3767
1242	Anthraquinone, 1,2-dihydroxy-3-iodo or β-iodo alizarin $1,2-(HO)_2-3-IC_{14}H_5O_2$	366.11	og-red nd (xyl)		229			w	B8[3], 3773
1243	Anthraquinone, 1,2-dihydroxy-3-methyl or β-methylalizarin $1,2(HO)_2-3-CH_3-C_{14}H_5O_2$	254.24	og nd	sub	245			al, eth, ace	B8[3], 3808
1244	Anthraquinone, 1,2-dihydroxy-3-nitro or β-nitroalizarin, Alizarin orange $1,2-(HO)_2-3(NO_2)C_{14}H_5O_2$	285.21	og-ye nd (bz), ye, pl (gl aa, al)	sub, d	244d			al, bz, aa	B8[3], 3774
1245	Anthraquinone, 1,2-dihydroxy-4-nitro or 4-Nitroalizarin $1,2-(HO)_2-3(NO_2)C_{14}H_5O_2$	285.21	gold-ye nd (aa or al)	sub, d	289d			al, bz, aa, chl	B8[2], 491
1246	Anthraquinone, 1,3-dihydroxy or Purpuro xanthin $1,3-(HO)_2C_{14}H_6O_2$	240.22	ye-red nd (sub)		268-70			al, ace, bz, aa	B8[3], 3774
1247	Anthraquinone, 1,4-dihydroxy or Quinizarin $1,4-(HO)_2C_{14}H_6O_2$	240.22	ye red lf (eth), dk red nd		200			w, al, bz, eth	B8[3], 3775
1248	Anthraquinone, 1,4-dihydroxy-5,6,7,8-tetrachloro or 5,6,7,8-Tetrachloroquinizarin $1,4-(HO)_2-5,6,7,8-Cl_4C_{14}H_2O_2$	378.00	red pl (aa)		270			bz, aa	B8[3], 3783
1249	Anthraquinone, 1,5-dihydroxy or Anthrarufin $1,5-(HO)_2C_{14}H_6O_2$	240.22	pa ye pl (gl aa)	sub	280			bz	B8[3], 3787
1250	Anthraquinone, 1,6-dihydroxy $1,6-(HO)_2C_{14}H_6O_2$	240.22	og-ye nd (gl aa)		276				B8[3], 3791
1251	Anthraquinone, 1,7-dihydroxy $1,7-(HO)_2C_{14}H_6O_2$	240.22	ye nd (sub)	sub	292-3			al, eth, bz, chl	B8[3], 3791

No.	Name, Synonyms, and Formula	Mol. wt.	Color, crystalline form, specific rotation and λ_{max} (log ε)	b.p. °C	m.p. °C	Density	n_D	Solubility	Ref.
1252	Anthraquinone, 1,8-dihydroxy or Chrysazin 1,8-$(HO)_2C_{14}H_6O_2$	240.22	red or redsh-ye nd or lf (al)	sub	193	...	...	al, eth, ace, chl	$B8^3$, 3792
1253	Anthraquinone, 1,8-dihydroxy-3-hydroxymethyl or Aloe emedin 1,8-$(HO)_2$-3$(HOCH_2)C_{14}H_5O_2$	270.24	og-ye nd (to, al)	sub	223-4	...	...	al, eth, bz	$B8^3$, 4160
1254	Anthraquinone, 1,8-dihydroxy-3-methyl or Chrysophenol, Chrysophanic acid 1,8-$(HO)_2$-3-$CH_3C_{14}H_5O_2$	254.24	ye hex or mcl nd (sub)	sub	196	0.92	...	ace, bz, aa	$B8^2$, 510
1255	Anthraquinone, 1,8-dihydroxy-2,4,5,7-tetrabromo or 2,4,5,7-Tetrabromoquinizarin 1,8-$(HO)_2$-2,4,5,7-$Br_4C_{14}H_2O_2$	555.80	og-ye nd (bz)	...	312	...	...	...	$B8^1$, 722
1256	Anthraquinone, 1,8-dihydroxy-2,4,5,7-tetranitro or Chrysammic acid, Chrysamminic acid 1,8-$(HO)_2$-2,4,5,7-$(NO_2)_4C_{14}H_2O_2$	420.21	ye pl or lf	d	exp	...	...	al, eth	$B8^1$, 723
1257	Anthraquinone, 2,3-dihydroxy or Hystazin 2,3-$(HO)_2C_{14}H_6O_2$	240.23	ye-br nd (aa), ye nd (sub)	sub	>330	...	...	...	$B8^3$, 3794
1258	Anthraquinone, 2,6 dihydroxy or Anthraflavin 2,6-$(HO)_2C_{14}H_6O_2$	240.22	ye nd (al)	...	360d	...	...	...	$B8^3$, 3796
1259	Anthraquinone, 2,7-dihydroxy or Isoanthraflavin 2,7-$(HO)_2C_{14}H_6O_2$	240.22	ye nd (+1 w, dil al), nd (sub)	sub	350-5	...	...	al, aa	$B8^3$, 3798
1260	Anthraquinone, 1,8-dimethoxy 1,8$(CH_3O)_2C_{14}H_6O_2$	268.27	...	...	221	...	...	al, bz, lig	$B8^3$, 3793
1261	Anthraquinone, 2,6-dimethoxy 2,6-$(CH_3O)_2C_{14}H_6O_2$	268.27	...	...	256	...	...	...	$B8^3$, 3797
1262	Anthraquinone, 1,2-dimethyl 1,2-$(CH_3)_2C_{14}H_6O_2$	236.27	nd (ace or aa)	...	156	...	...	al, eth, ace, bz, aa	$B7^3$, 4130
1263	Anthraquinone, 1,3-dimethyl 1,3-$(CH_3)_2C_{14}H_6O_2$	236.27	nd (aa)	...	162	...	...	aa	$B7^3$, 4130
1264	Anthraquinone, 1,4-dimethyl 1,4-$(CH_3)_2C_{14}H_6O_2$	236.27	ye nd (al, sub)	sub	140-1	...	...	bz, xyl, aa	$B7^3$, 4131
1265	Anthraquinone, 2,3-dimethyl 2,3-$(CH_3)_2C_{14}H_6O_2$	236.27	ye nd (al, to or xyl), cr (aa)	sub	210	...	...	al, bz, xyl	$B7^3$, 4131
1266	Anthraquinone, 2,6-dimethyl 2,6-$(CH_3)_2C_{14}H_6O_2$	236.27	ye nd (aa or al)	sub	242	...	...	to	$B7^3$, 4132
1267	Anthraquinone, 2,7-dimethyl 2,7-$(CH_3)_2C_{14}H_6O_2$	236.27	yesh nd (al)	...	170	...	...	al	$B7^3$, 4133
1268	Anthraquinone, 1,3-dinitro 1,3-$(NO_2)_2C_{14}H_6O_2$	298.21	ye nd (HNO_3)	...	246-50	...	...	...	E13, 436
1269	Anthraquinone, 1,5-dinitro 1,5-$(NO_2)_2C_{14}H_6O_2$	298.21	pa ye nd (xyl), ye-cr (sub)	sub	422 (385)	...	...	...	$B7^3$, 4081
1270	Anthraquinone, 1,6-dinitro 1,6-$(NO_2)_2C_{14}H_6O_2$	298.21	...	...	257-9	...	...	...	$B7^3$, 4082
1271	Anthraquinone, 1,7-dinitro 1,7$(NO_2)_2C_{14}H_6O_2$	298.21	...	...	295	...	...	...	$B7^3$, 4082
1272	Anthraquinone, 1,8-dinitro 1,8-$(NO_2)_2C_{14}H_6O_2$	298.21	...	...	312	...	...	...	$B7^3$, 4082
1273	Anthraquinone, 1,5-disulfonic acid 1,5-$(HO_3S)_2C_{14}H_6O_2$	368.33	ye nd (HCl + 4w), pl (dil aa + 4w)	...	310-1d	...	...	w, al, aa	$B11^3$, 634
1274	Anthraquinone, 1,6-disulfonic acid 1,6-$(HO_3S)_2C_{14}H_6O_2$	368.33	ye nd (HCl + 5w), gold pr (dil aa + 5w)	...	215-7d	...	...	w, al, aa	$B11^3$, 634
1275	Anthraquinone, 1,7-disulfonic acid 1,7-$(HO_3S)_2C_{14}H_6O_2$	368.33	ye hyg pw (dil aa + 4w)	...	d at 120	...	...	w, al, aa	$B11^3$, 634
1276	Anthraquinone, 1,8-disulfonic acid 1,8-$(HO_3S)_2C_{14}H_6O_2$	368.33	ye nd (+5w)	...	293-4d	...	...	w, al	$B11^3$, 635
1277	Anthraquinone, 2-ethyl-1-nitro 1-(NO_2)-2$(C_2H_5)C_{14}H_6O_2$	281.27	yesh br (aa)	...	226	...	...	...	$B7^2$, 743

No.	Name, Synonyms, and Formula	Mol. wt.	Color, crystalline form, specific rotation and λ_{max} (log ε)	b.p. °C	m.p. °C	Density	n_D	Solubility	Ref.
1278	Anthraquinone, 1,2,3,5,6,7-hexahydroxy or Rufigallol 1,2,3,5,6,7-$(HO)_6C_{14}H_2O_2$	304.21	red rh, red, ye nd (sub)		sub, d			ace	B8^3, 4401
1279	Anthraquinone, 1-hydroxy or Erythroxyanthraquinone 1-$HOC_{14}H_7O_2$	224.22	red-og nd (al)	sub	194-5			al, eth, bz	B8^3, 2906
1280	Anthraquinone, 2-hydroxy 2-$HOC_{14}H_7O_2$	224.22	ye pl or nd (al or aa)	sub	306			al, eth	B8^3, 2921
1281	Anthraquinone, 2-methoxy 2-$CH_3OC_{14}H_7O_2$	238.24			196			bz, al	B8^3, 2922
1282	Anthraquinone, 2-methyl 2-$CH_3C_{14}H_7O_2$	222.24	yesh nd (al aa)	sub	182-3			al, bz, aa	B7^3, 4104
1283	Anthraquinone, 1-methylamino 1-$(CH_3NH)C_{14}H_7O_2$	237.26	ye-red nd		170			al, to, aa	B14^3, 408
1284	Anthraquinone, 2-methylamino 2-$(CH_3NH)C_{14}H_7O_2$	237.26	red nd (aa)		226-7			al, eth, aa, to	B14^3, 430
1285	Anthraquinone, 2-methyl-1-nitro 2-CH_3-1-$NO_2C_{14}H_6O_2$	267.24	pa ye nd (aa)		270-1				B7^3, 4110
1286	Anthraquinone, 6-methyl-1,2,5-trihydroxy or Morindone 6-CH_3-1,2,5-$(HO)_3C_{14}H_4O_2$	270.24	og-red nd (to)		282			al, eth, bz, Py	B8^3, 4151
1287	Anthraquinone, 6-methyl-1,3,8-trihydroxy or Emodin 6-CH_3-1,3,8-$(HO)_3C_{14}H_4O_2$	270.24	og-red mcl nd (aa), cr (dil aa + 1w)	sub	256-7			al, eth	B8^3, 4154
1288	Anthraquinone, 1-nitro 1-$NO_2C_{14}H_7O_2$	253.21	yesh pr (ace), nd (aa)	270-1^7	232-3			ace, bz, aa	B7^3, 4080
1289	Anthraquinone, 2-nitro 2-$NO_2C_{14}H_7O_2$	253.21	ye nd (aa or al)	sub	184-5			bz, chl	B7^3, 4080
1290	Anthraquinone, 1,2,4,5,8-pentahydroxy or Alizarin cyanine R 1,2,4,5,8-$(HO)_5C_{14}H_3O_2$	288.21							B8^3, 4379
1291	Anthraquinone, 2-sulfonamide 2-$(H_2NO_2S)C_{14}H_7O_2$	287.29	ye nd (aa)		261				B11, 339
1292	Anthraquinone, 1-sulfonic acid 1-$HO_3SC_{14}H_7O_2$	288.27	lf (aa), ye lf (con HCl + 3w)		214 (cor) anh			w, al	B11^3, 626
1293	Anthraquinone, 1-sulfonic acid-5-chloro 1-HO_3S-5-Cl-$C_{14}H_6O_2$	322.72	ye rh pr (HCl or aa + 5w)		236-7			w	B11^3, 627
1294	Anthraquinone, 2-sulfonic acid 2-$HO_3SC_{14}H_7O_2$	288.27	ye lf (+ 3w)					w, al	B11^3, 628
1295	Anthraquinone-2-sulfonic acid-5-nitro 2-HO_3S-5-$NO_2C_{14}H_6O_2$	333.27	yesh pl (dil HNO_3)		255d			w	B11^3, 629
1296	Anthraquinone, 1,2-4,6-tetrahydroxy or Hydroxyflavopurpurin 1,2,4,6-$(HO)_4C_{14}H_4O_2$	272.21	dk red nd (sub)					al, Py	B8^3, 4288
1297	Anthraquinone, 1,2,4,7-tetrahydroxy or 4-Hydroxyanthrapurpurin 1,2,4,7-$(HO)_4C_{14}H_4O_2$	272.21	red-ye (al, Py or aa)					al	B8^3, 4288
1298	Anthraquinone, 1,2,5,6-tetrahydroxy or Rufiopin 1,2,5,6-$(HO)_4C_{14}H_4O_2$	272.21	og-red nd (Py)	sub	340			w, al, aa	B8^3, 4288
1299	Anthraquinone, 1,2,5,8-tetrahydroxy or Quinalizarin 1,2,5,8-$(HO)_4C_{14}H_4O_2$	272.21	og nd		>275				B8^3, 4289
1300	Anthraquinone, 1,2,6,7-tetrahydroxy 1,2,6,7-$(HO)_4C_{14}H_4O_2$	272.21			>330				B8^2, 584
1301	Anthraquinone, 1,2,7,8-tetrahydroxy 1,2,7,8-$(HO)_4C_{14}H_4O_2$	272.21	red nd or pr (aa)		318d				B8^2, 585
1302	Anthraquinone, 1,3,5,7-tetrahydroxy or Anthrachrysone 1,3,5,7-$(HO)_4C_{14}H_4O_2$	272.21	yesh nd (al + 2w)	sub	150—60d (+ 2w), <360 (anh)			al, aa, bz, ace	B8^3, 4289
1303	Anthraquinone, 1,4,5,8-tetrahydroxy 1,4,5,8-$(HO)_4C_{14}H_4O_2$	272.21	gr nd (aa), br nd (bz-lig)	sub	>300			al	B8^3, 4290
1304	Anthraquinone, 1,2,3-trihydroxy or Anthragallol 1,2,3-$(HO)_3C_{14}H_5O_2$	256.21	ye nd (dil al), br (aa), og nd	sub (290)	313			al, eth, aa	B8^3, 4140

No.	Name, Synonyms, and Formula	Mol. wt.	Color, crystalline form, specific rotation and λ_{max} (log ε)	b.p. °C	m.p. °C	Density	n_D	Solubility	Ref.
1305	Anthraquinone, 1,2,4-trihydroxy or Purpurin $1,2,4\text{-}(HO)_3C_{14}H_5O_2$	256.21	og red, dk red or og-ye nd (al)	sub	259	....	....	al, eth, bz	$B8^3$, 4141
1306	Anthraquinone, 1,2,5-trihydroxy or 2-Hydroxyanthrarufin $1,2,5\text{-}(HO)_3C_{14}H_5O_2$	256.21	red nd (gl aa)	....	278	....	....	eth	$B8^2$, 554
1307	Anthraquinone, 1,2,6-trihydroxy or Flavopurpurin $1,2,6\text{-}(HO)_3C_{14}H_5O_2$.	256.21	ye nd (al)	459d	330 (sub)	....	....	al, bz, w	$B8^3$, 4144
1308	Anthraquinone, 1,2,7-trihydroxy or Anthrapurpurin $1,2,7\text{-}(HO)_3C_{14}H_5O_2$	256.21	og nd (al)	462	374	....	....	al, bz, aa	$B8^3$, 4145
1309	Anthraquinone, 1,2,8-trihydroxy or 2-Hydroxychrysazin $1,2,8\text{-}(HO)_3C_{14}H_5O_2$	256.21	red nd (aa-sub)	sub	239-40	....	....	....	$B8^3$, 4145
1310	Anthraquinone, 1,3,8-trihydroxy $1,3,8\text{-}(HO)_3C_{14}H_5O_3$	256.21	bt red-br nd (bz), gold-ye pl (AcOEt)	....	287-8	....	....	....	$B8^3$, 4145
1311	Anthraquinone, 1,4,5-trihydroxy or 5-Hydroxyquinizarin $1,4,5\text{-}(HO)_3C_{14}H_5O_2$	256.21	red-br nd or lf dk red nd (Py)	....	271	....	....	....	$B8^3$, 4146
1312	Anthraquinone, 1,4,6-trihydroxy or 6-Hydroxyquinizarin $1,4,6\text{-}(HO)_3C_{14}H_5O_2$	256.21	bt red nd (al), br pw	....	> 300	....	....	al, Py	$B8^2$, 558
1313	Anthraquinone-2-carboxylic acid $2\text{-}HO_2CC_{14}H_7O_2$	252.23	ye nd (aa)	sub	290-2	....	....	ace	$B10^3$, 3640
1314	Anthraquinone-2-carboxylic acid, 1,3-dihydroxy or Munjiston $1,3\text{-}(HO)_2C_{14}H_5O_2(CO_2H)\text{-}2$	284.24	ye nd (al-w + 1w), lf anh	....	......	....	....	eth	$B10^3$, 4787
1315	Anthraquinone-2-carboxylic acid, 1,4-dihydroxy $1,4\text{-}(HO)_2C_{14}H_5O_2(CO_2H)\text{-}2$	284.24	ye br or red nd	....	249-50	....	....	al, ace	$B10^3$, 4787
1316	Anthraquinone-2-carboxylic acid, 4,5-dihydroxy or Cassic acid $4,5\text{-}(HO)_2C_{14}H_5O_2(CO_2H)\text{-}2$	284.24	ye or og nd (MeOH, Py)	sub	321	....	....	Py	$B10^3$, 4789
1317	Anthraquinone-2-carboxylic acid, 4,5-dihydroxy-7-methoxy or Parietinic acid $4,5\text{-}(HO)_2\text{-}7\text{-}(CH_3O)C_{14}H_4O_2(CO_2H)\text{-}2$	314.25	red-br or ye nd (sub)	....	300	....	....	....	$B10^3$, 4829
1318	Anthraquinone-2-carboxylic acid, 1-nitro $1\text{-}(NO_2)C_{14}H_6O_2(CO_2H)\text{-}2$	297.22	nd (al or aa)	....	288d	....	....	al, ace	$B10^3$, 3650
1319	Anthraquinone-2-carboxylic acid, 5-nitro $5\text{-}(NO_2)\text{-}C_{14}H_6O_2(CO_2H)\text{-}2$	297.22	yesh nd (aa)	....	......	....	....	aa	$B10^3$, 3653
1321	1-Anthroic acid or 1-Anthracene carboxylic acid $1\text{-}HO_2CC_{14}H_9$	222.24	ye nd (aa), ye pr (al, AcOEt)	sub	251-2	....	....	eth	$B9^3$, 3492
1322	2-Anthroic acid or 2-Anthracene carboxylic acid $2\text{-}HO_2CC_{14}H_9$	222.24	ye lf (al), nd, lf (sub)	sub	281	....	....	aa	$B9^3$, 3493
1323	9-Anthroic acid or 9-Anthracene carboxylic acid $9\text{-}HO_2CC_{14}H_9$	222.24	pa ye nd (bz, al)	sub	217d	....	....	al	$B9^3$, 3494
1324	Anthrone or 9-Oxodihydroanthracene $C_{14}H_{10}O$	194.23	nd (bz-lig, aa)	....	155	....	....	ace, bz	$B7^3$, 2359
1325	Anthrone-10-nitro $10\text{-}NO_2\text{-}C_{14}H_9O$	239.23	cr (bz-lig), nd (CS_2)	....	140(148 d)	....	....	al, bz	$B7^3$, 2371
1326	Antimalarine or Plasmocid $C_{17}H_{25}N_3O$	287.41	....	$182^{1.0}$	......	$1.0569^{24/4}$	1.5855^{24}	dil HCl	$B22^4$, 5787
1327	Antipyrine (α - form) or Analgesine, Phenazone $C_{11}H_{12}N_2O$	188.23	mcl lf or sc (w, bz, eth)	319^{741}, $211\text{-}2^{10}$	114	$1.0747^{20/4}$	1.5697	w, al, bz, chl	$B24^2$, 11
1328	Antipyrene (β) $C_{11}H_{12}N_2O$	188.23	cr (unst)	....	109($\beta \rightarrow \alpha$ at 18)	....	....	bz	$B24^2$, 11
1329	Antipyrine, 4-acetamido $4\text{-}(CH_3CONH)C_{11}H_{11}N_2O$	245.28	....	....	201	....	....	w, al	$B24^2$, 152
1330	Antipyrene, *o*-amino $C_{11}H_{13}N_3O$	203.24	nd (AcOEt-eth)	....	165	....	....	w, al	$B24^1$, 210
1331	Antipyrene, *m*-amino $C_{11}H_{13}N_3O$	203.24	redsh in air (bz)	....	148	....	....	w, al, chl	$B24^1$, 210
1332	Antipyrine, *p*-amino $C_{11}H_{13}N_3O$	203.24	....	....	109	....	....	w, al, eth, bz	$B24^2$, 151

PHYSICAL CONSTANTS OF ORGANIC COMPOUNDS (Continued)

No.	Name, Synonyms, and Formula	Mol. wt.	Color, crystalline form, specific rotation and λ_{max} (log ε)	b.p. °C	m.p. °C	Density	n_D	Solubility	Ref.
1333	Antipyrene, *p*-bromo $C_{11}H_{11}BrN_2O$	267.13	nd (w)	300^{9}	122			w, al, eth, to	B24, 33
1334	Antipyrene, *p*-dimethylamino $C_{13}H_{17}N_3O$	231.30	pr or pl (lig or AcOEt)		134—5			w, al, bz	B24, 46
1335	Antipyrene, 2-hydroxybenzoate or Salazolon $C_{11}H_{12}N_2O\text{-}C_7H_6O_2$	310.35	pw		92			al, chl	$B24^{2}$, 16
1336	Aphanin $C_{40}H_{54}O$	550.87	bl blk lf (bz-MeOH)		178			chl	$B7^{3}$, 2858
1337	Aphyiline $C_{15}H_{24}N_2O$	248.37	$[\alpha]^{20}_D$ + 10.3 (MeOH, c = 2)	200d	52-7			al, eth, ace, bz	C26, 2742
1338	Apiol or 2,5-Dimethoxy saffrole parsley camphor $C_{12}H_{14}O_4$	222.24	nd	294, 179^{35}	29.55	$1.015^{20/4}$	1.5360^{20}	al, eth, ace, bz, lig	$B19^{4}$, 1033
1339	Apoatropine or Atropamine $C_{17}H_{21}NO_2$	271.36	pr (chl)		62			al, eth, ace, bz, chl	$B21^{4}$, 175
1340	Apoatropine hydrochloride $C_{17}H_{21}NO_2\text{-}HCl$	307.82	lf (w)		239			w	$B21^{4}$, 175
1341	Apocinchonidine $C_{19}H_{22}N_2O$	294.40	lf (al), $[\alpha]^{20}_D$ − 139.3 (chl-al, c = 2)		252			al, chl	$B23^{1}$, 131
1342	Apocinchonine or Allocinchonine $C_{19}H_{22}N_2O$	294.40	pr (al), $[\alpha]^{20}_D$ + 167.4 (abs al, c=3)		219			al, bz	$B23^{2}$, 367
1343	Apocinchonine hydrochloride $C_{19}H_{22}N_2O\text{-}HCl\text{-}2H_2O$	366.89	nd (+ 2w), $[\alpha]^{16}_D$ + 139 (w, c=0.006)					w	$B23^{2}$, 367
1344	Apocodeine or Apomorphine-3-methyl ether $C_{18}H_{19}NO_2$	281.35	pr (MeOH), $[\alpha]^{23}_D$ − 90 (abs al, c=0.449)		123.4(anh)			al, eth, aa, lig	$B21^{4}$, 2420
1345	Apocodeine, ethanol solvate $C_{18}H_{19}NO_2\text{-}C_2H_5OH$	327.43	lf (al + 1)		104.5-6.5			al	$B21^{4}$, 2420
1346	Apocodeine, methanol solvate $C_{18}H_{19}NO_2\text{-}CH_3OH$	313.40	nd (MeOH + 1)		85			MeOH	$B21^{4}$, 2420
1347	Apocyclene C_9H_{14}	122.21	cr (al)	$138\text{-}9^{764}$	42.5-4.3	$0.8710^{40/4}$	1.4514^{40}	al, eth, bz	$B5^{4}$, 430
1348	Apocynin or 4-Hydroxy-3-methoxy acetophenone $4\text{-}HO\text{-}3\text{-}(CH_3O)C_6H_3COCH_3$	166.18	pr (w)	295-300, $233\text{-}5^{15-20}$	115			al, eth, ace, bz, chl	$B8^{3}$, 2108
1349	Apofenchocamphoric acid or 4,4-Dimethyl-1,3-cyclopentane dicarboxylic acid $4,4\text{-}(CH_3)_2\text{-}1,3\text{-}(HO_2C)_2C_5H_6$	186.21	mcl		144-5			al, eth	$B9^{3}$, 3849
1350	Apomorphine $C_{17}H_{17}NO_2$	267.33	hex pl (chl-peth), rods (eth + l)		195d			al, eth, ace, bz	$B21^{4}$, 2419
1351	Apomorphine-hydrochloride $C_{17}H_{17}NO_2\text{-}HCl\text{-}1/2H_2O$	312.80	gr in air, mcl pr, $[\alpha]^{25}_D$ − 48 (w, c = 1.2)		200-10				$B2^{4}$, 2419
1352	Aponal or *tert*-Pentylcarbamate $H_2NCOOCH_2C(CH_3)_3$	131.17	nd (dil al)		86—7			ace, bz	$B3^{1}$, 14
1353	Apoquinine (α) or Apocupreine $C_{19}H_{22}N_2O_2$	310.40	pr (eth), $[\alpha]^{20}_D$ − 214.8 (al)		190d			al	$B23^{2}$, 412
1354	Apoquinine(β) $C_{19}H_{22}N_2O_2\text{-}2H_2O$	346.43	$[\alpha]^{20}_D$ − 194 (al)		190d			al	$B23^{2}$, 412
1355	Aposafranone or Benzeneindone,10-phenyl-2-phenazinone $C_{18}H_{12}N_2O$	272.31	br, gr (red in sl), nd (al)						$B23^{2}$, 364

No.	Name, Synonyms, and Formula	Mol. wt.	Color, crystalline form, specific rotation and λ_{max} (log ε)	b.p. °C	m.p. °C	Density	n_D	Solubility	Ref.
1356	D-Arabinose (α-anomer) $C_5H_{10}O_5$	150.13	(cr, MeOH)		155.5—6.5 (cor)	1.585		w	B1^4, 4215
1357	D-Arabinose (β-anomer) $C_5H_{10}O_5$	150.13	cr (MeOH), $[\alpha]^{20}{}_D$ − 175→-108 (mut)		155.5—6.5 (cor)	1.625		w	B1^4, 4215
1358	L-Arabinose or pectinose $C_5H_{10}O_5$	150.13	pr nd (al)		164—5 (cor)	1.585$^{20/4}$		w	B1^4, 4223
1359	L-Arabinose (α-anomer) $C_5H_{10}O_5$	150.13	orth cr (pr), cr (dil al)		159-60	1.585$^{20/4}$			B1^4, 4218
1360	L-Arabinose (β-anomer) $C_5H_{10}O_5$	150.13	orth, $[\alpha]^{20}{}_D$ + 190.5→ + 104.5 (c = 3) (mut)		159-60	1.625$^{20/4}$		w	B1^4, 4218
1361	D-Arabinose-diphenylhydrazone $HOCH_2(CHOH)_3CH = NN(C_6H_5)_2$	316.36	orth pr (dil al)		207				B31, 33
1362	DL-Arabinose-diphenylhydrazone $HOCH_2(CHOH)_3CH = NN(C_6H_5)_2$	316.36	nd (aq Py)		206			Py	B31, 47
1363	L-Arabinose-diphenylhydrazone $HOCH_2(CHOH)_3CH = NN(C_6H_5)_2$	316.36	orth pr, nd (dil al), $[\alpha]^{20}{}_D$ + 18.5 Py		204.5			al	B31, 44
1364	D-Arabitol or Arabite, D-Lyxitol 1,2,3,4,5-pentanepentol $HOCH_2(CHOH)_3CH_2OH$	152.15	pr (dil al), $[\alpha]^{20}{}_D$ + 11.8 (borax sol, c = 9.5)					w	B1^4, 2832
1365	DL-Arabitol $HOCH_2(CHOH)_3CH_2OH$	152.15	pr (90% al), cr (al-ace)		106			w	B1^4, 2832
1366	L-Arabitol $HOCH_2(CHOH)_3CH_2OH$	152.15	$[\alpha]_D$ − 5.4 (borax sol)		102-3			w	B1^4, 2832
1367	D-Arabonic acid $HOCH_2(CHOH)_3CO_2H$	166.13	(dil aa), $[\alpha]^{25}{}_D$ + 10.5 (c=6)		114-16			w, al	B3^4, 1205
1368	DL-Arabonic acid $HOCH_2(CHOH)_3CO_2H$	166.13						w	B3^2, 474
1369	L-Arabonic acid $HOCH_2(CHOH)_3CO_2H$	166.13	cr (al), $[\alpha]^{20}{}_D$ − 9.6 (w, c=2.5)						B3^1, 979
1370	Arachidic acid or Eicosanoic acid $CH_3(CH_2)_{18}CO_2H$	312.54	pl (al)	328d, 203—5^1	77	0.8240$^{100/4}$	1.425^{100}	eth, bz, chl	B2^4, 1275
1371	Arachidic acid-ethyl ester or Ethyl arachidate $CH_3(CH_2)_{18}CO_2C_2H_5$	340.69		295-7^{100}, 186-7^2	50			al, eth, bz, chl	B2^3, 1067
1372	Arachidic acid, methyl ester or methyl arachidate $CH_3(CH_2)_{18}CO_2CH_3$	326.56	lf (MeOH)	215-6^{10}, 188^2	54.5		1.4317^{60}	al, eth, bz, chl	B2^4, 1276
1373	Arachidic alcohol or l-Eicosanol $CH_3(CH_2)_{18}CH_2OH$	298.55	wx (al), cr (chl)	309, 220-5^3	72-3	0.8405$^{20/4}$	1.4550^{20}	ace, bz	B1^4, 1900
1374	Arachidonic acid $CH_3(CH_2)_3(CH_2CH{=}CH)_4(CH_2)_3CO_2H$	304.44		d	− 49.5		1.4824^{20}	al, eth, ace	B2^4, 1802
1375	Aramite or Niagaramite $C_{15}H_{23}ClO_4S$	334.86		195^2	− 37.3	1.145$^{20/20}$	1.5100^{20}	al, eth, ace, bz	M, 97
1376	Arbutin or Hydroquinone-β-d-glucopyranoside $C_{12}H_{16}O_7$	272.24	nd (w + l), $[\alpha]^{25}{}_D$ − 64 (w, c=3)		199.5-200(st, anh)			w, al	B17^4, 2983
1377	Arbutin hydrate $C_{12}H_{16}O_7 \cdot H_2O$	290.27	$[\alpha]^{17}{}_D$ − 60.3 (w, p=5)		142			al	B31, 210
1378	Arecaidine or l-Methyl-l,2,5,6-tetrahydroricotinic acid $C_7H_{11}NO_2$	141.17	pl (dil al), ta (dil al + lw)		232 (anh)			w	B22^3, 184
1379	Arecoline or Arecaidine methyl ester $C_8H_{13}NO_2$	155.20		209 (220), 94^1		1.0504$^{20/20}$	1.4860^{20}	w, al, eth chl	B22^3, 185
1380	Arecoline hydrobromide $C_8H_{13}NO_2HBr$	236.11	mcl pr (al)		172 (177)			w, al	B22^3, 185

PHYSICAL CONSTANTS OF ORGANIC COMPOUNDS (Continued)

No.	Name, Synonyms, and Formula	Mol. wt.	Color, crystalline form, specific rotation and λ_{max} (log ε)	b.p. °C	m.p. °C	Density	n_D	Solubility	Ref.
1381	Arecoline hydrochloride $C_8H_{13}NO_2$-HCl	191.66	nd (al)		157-8			w, al	$B22^3$, 185
1382	Arginine (*DL*) or *DL*-2-Amino-5-guanido pentanoic acid $H_2NC(=NH)NH(CH_2)_3CH(NH_2)CO_2H$	174.20			217-8			w	$B4^3$, 1359
1383	Arginine (*L*) $C_6H_{14}N_4O_2$	174.20			244d			w	$B4^3$, 1348
1384	Arginine, benzylidene (*L*) $C_{13}H_{18}N_4O_2$	262.31	lf (w)		204-5			MeOH	$B7^3$, 837
1385	Arginine diflavinate (*L*) $C_6H_{14}N_4O_2(C_{10}H_6N_2O_8S)_2$	802.66	ye nd		202d				$B11^3$, 547
1386	Arginine dipicrate (*DL*) $C_6H_{14}N_4O_2(C_6H_3N_3O_7)_2$	632.41			196d				$B6^4$, 1450
1387	Arginine dipicrate (*L*) $C_6H_{14}N_4O_2(C_6H_3N_3O_7)_2$	632.41			200d				$B6^4$, 1450
1388	Arginine picrate (*D*) $C_6H_{14}N_4O_2 \cdot C_6H_3N_3O_7 \cdot 2H_2O$	439.34	ye-og lf		258-60d				$B11^3$, 547
1389	Arginine picrate (*D*) $C_6H_{14}N_4O_2$-$C_6H_3N_3O_7 \cdot 2H_2O$	439.34	nd (w)		217-8d				$B6^4$, 1450
1390	Arginine picrate (*DL*) $C_6H_{14}N_4O_2$-$C_6H_3N_3O_7 \cdot 2H_2O$	439.34	pr (w)		200 (223d)				$B6^4$, 1450
1391	Arginine picrate (*L*) $C_6H_{14}N_4O_2$-$C_6H_3N_3O_7$-$2H_2O$	439.34	ye nd (w)		217-8d				$B6^4$, 1450
1392	Arsenic acid, diphenyl $(C_6H_5)_2AsO_2H$	262.14	nd pr		178			al, chl	$B16^2$, 443
1393	Arsenic acid, triethyl ester or Triethyl arsenate $(C_2H_5O)_3AsO$	226.10		$235-8, 118-20^{15}$		$1.3023^{20/0}$	1.4343^{20}		$B1^4$, 1358
1394	Arsenobenzene $C_{36}H_{30}As_6$	912.16			212			bz, al	$B16^3$, 1157
1395	Arsenous acid, triethyl ester or Treiethyl arsenite $(C_2H_5O)_3As$	210.10		$165-6, 66-7^{12}$		$1.2239^{20/4}$	1.4369^{13}		$B1^4$, 1358
1396	Arsine, *bis*(trifluoromethyl)iodo $(CF_3)_2AsI$	339.84	ye oil	92			1.425^{23}	eth	$B3^4$, 267
1397	Arsine, chloro,diphenyl $(C_6H_5)_2AsCl$	264.59	rh pl (peth)	$337, 193^{20}$	44	$1.48204^{16/4}$	1.6332^{56}	al, eth, ace, bz	$B16^3$, 944
1398	Arsine, dibromo,trifluoromethyl CF_3AsBr_2	303.75		118^{745}		1.528^{20}			$B3^4$, 267
1399	Arsine, dichloro,trifluoromethyl CF_3AsCl_2	214.83		$133, 37^{25}$	− 42.5	1.8358^{20}	1.5677^{15}	al, eth	$B3^4$, 267
1400	Arsine, dichloro,phenyl $C_6H_5AsCl_2$	222.93		$254-7, 131^{14}$		$1.6516^{20/4}$	1.6386^{15}	al, eth, ace, bz	$B16^3$, 958
1401	Arsine, dichloro (trifluoromethyl) CF_3AsCl_2	214.83		71		1.431^{20}		al, ace	$B3^4$, 268
1402	Arsine, diethyl $(C_2H_5)_2AsH$	134.05		105 (97)	$1.1338^{24/4}$	1.4709		al, eth, ace, bz	$B4^3$, 1797
1403	Arsine, difluoro,ethyl $C_2H_5AsF_2$	141.98	fume in air	$94.3, 74^{100}$	-38.7	1.708^{17}			$B4^3$, 1799
1404	Arsine, difluoro,methyl CH_3AsF_2	127.95	fume in air	76.5	-29.7	1.924^{18}			$B4^3$, 1796
1405	Arsine, difluoro,phenyl $C_6H_5AsF_2$	190.03	wax	110^{48}	42				$B16^3$, 959
1406	Arsine, diido,trifluoromethyl F_3CAsI_2	397.74		$183d, 100^{48}$					$B3^4$, 268
1407	Arsine, dimethyl or Cacodylhydride $(CH_3)_2AsH$	106.00	ign in air	36		$1.213^{29/29}$		**al, eth, ace, bz**	$B4^2$, 978
1408	Arsine, diphenyl $(C_6H_5)_2AsH$	230.14	oil	174^{25}		$1.30^{25/25}$		al, eth	$B16^3$, 918
1409	Arsine, ethyl $C_2H_5AsH_2$	106.00		36		$1.217^{22/22}$		al, eth	$B4^3$, 1797
1410	Arsine, 4-methoxyphenyl,oxide $(4\text{-}CH_3OC_6H_4)AsO$	198.05	cr (chl-eth)- or bz-peth		114-6 (anh)			bz, chl	$B16^2$, 448
1411	Arsine, methyl CH_3AsH_2	91.97		2	-143			**al, eth, ace**	$B4^3$, 1795
1412	Arsine, methyl, oxide CH_3AsO	105.96	pr (CS_2) (al)	275d	95			al, bz, chl	$B4^3$, 1817
1413	Arsine, pheny,oxide C_6H_5AsO	168.03	cr (bz-eth or chl-eth)		144-6			bz, chl	$B16^3$, 956

PHYSICAL CONSTANTS OF ORGANIC COMPOUNDS (Continued)

No.	Name, Synonyms, and Formula	Mol. wt.	Color, crystalline form, specific rotation and λ_{max} (log ε)	b.p. °C	m.p. °C	Density	n_D	Solubility	Ref.
1414	Arsine, triethyl $(C_2H_5)_3As$	162.11		138-9		$1.150^{20/4}$	1.467^{20}	al, eth, ace	B4[3], 1798
1415	Arsine, trifluoromethyl CF_3AsH_2	145.94		-11.6^{781}					B3[4], 266
1416	Arsine, trimethyl $(CH_3)_3As$	120.03		52	-87.3	1.144^{15}		al, eth, bz	B4[3], 1795
1417	Arsine, triphenyl $(C_6H_5)_3As$	306.24			61	$1.2634^{18/4}$	1.6888^{21}	eth, bz, chl	B16[3], 921
1418	Arsine, triphenyl, oxide $(C_6H_5)_3AsO$	322.24			192				B16[3], 1021
1419	Arsine, triphenyl, sulfide $(C_6H_5)_3AsS$	338.20			162				B16[3], 1026
1420	Arsine, *tris*-(pentafluoroethyl) $(C_2F_5)_3As$	431.96		96					C49, 14635
1421	Arsine, *tris*-(trifluoromethyl) $(CF_3)_3As$	281.94		33.3				eth	B3[4], 266
1422	Artemsic acid $C_{15}H_{16}O_3$	244.29	nd (dil aa)		135-6			al, eth, aa, lig	B10[2], 220
1423	Ascaridole $C_{10}H_{16}O_2$	168.24	unst, $[\alpha]_D$ −4.14	exp^{760}, 115^{15}, 39—$40^{0.2}$	3.3	$1.0103^{20/4}$	1.4769^{20}	al, ace, bz	B19[4], 164
1424	D-Ascorbic acid $C_6H_8O_6$	176.13	pl or mcl nd, $[\alpha]^{18}_D$ − 48 (MeOH, c=1), $[\alpha]^{20}_D$ − 23.8 (w, c=3)		192d			w, al	B18[4], 3046
1425	DL-Ascorbic acid $C_6H_8O_6$	176.13			168-9			w, al	B18[4], 3047
1426	L-Ascorbic acid or Vitamin C $C_6H_8O_6$	176.13	pl or mcl nd, $[\alpha]^{20/}_D$ + 24 (w, c = 1)		192d	1.65			B18[4], 3038
1427	L-Ascorbic acid, 6-desoxy $C_6H_8O_5$	160.13	pr (AcOEt), $[\alpha]^{22/}_D$ + 36.7 (O.1N HCl, c = 1)	sub, $160^{0.001}$	168			w, al, ace	B18[4], 2301
1428	β-Asparagine (*D*) $H_2NCOCH_2CH(NH_2)CO_2H$-H_2O	150.13	$[\alpha]^{20/}_D$ + 5.4 (w, c = 1.3)		234.5 (anh)	$1.543^{15/4}$		w	B4[3], 1522
1429	β-Asparagine (*DL*) $H_2NCOCH_2CH(NH_2)CO_2H$-H_2O	150.13	tcl cr (w + 1), pr (w al)	213-5d	182-3	$1.4540^{15/4}$		w	B4[3], 1524
1430	β-Asparagine (*L*) $H_2NCOCH_2CH(NH_2)CO_2H$	132.12	rh (w + 1), $[\alpha]^{20/}_D$ −5.42 (w, c = 1.3)		236 (anh)	$1.543^{15/4}$		w	B4[3], 1513
1431	Aspartic acid (*D*) or Aminosuccinic acid (*D*) $HO_2CCH_2CH(NH_2)CO_2H$	133.10	$[\alpha]^{20/}_D$ −25.5 (HCl)		269-71	$1.6613^{13/13}$		w, dil, HCl	B4[3], 1522
1432	Aspartic acid (*DL*) or Aminosuccinic acid $HO_2CCH_2CH(NH_2)CO_2H$	133.10	mcl pr (w)		338-9 (275 sealed tube)	$1.6632^{13/13}$		w	B4[3], 1523
1433	Aspartic acid (*L*) or Aminosuccinic acid $C_4H_7NO_4$	133.10	rh lf (w)		324d, (270 sealed tube)	$1.6613^{13/13}$		w, dil, HCl	B4[3], 1506
1434	Aspartic acid, *N*-benzoyl $HO_2CCH_2CH(NHCOC_6H_5)CO_2H$	237.21	nd or lf, $[\alpha]^{20/}_D$ + 37 (0.76N NaOH)		171-3				B9[3], 1189
1435	Aspidospermine or Vallesine $C_{22}H_{30}N_2O_2$	354.49	nd or pr (al), nd (peth)					al, bz, chl	J1940, 1051

No.	Name, Synonyms, and Formula	Mol. wt.	Color, crystalline form, specific rotation and λ_{max} (log ε)	b.p. °C	m.p. °C	Density	n_D	Solubility	Ref.
1436	Atabrin or Quinacrine dihydrochloride (*DL*) $C_{23}H_{30}ClN_3O\text{-}2HCl\text{-}2H_2O$	508.92	yesh nd (w), ye cr pw		248-50d			w, al, MeOH	B21[4], 6248
1437	Atisine or Anthorine $C_{22}H_{33}NO_2$	343.51	rh bipym		57-60			al, eth, chl	Am78, 4139
1438	Atisine-hydrochloride $C_{22}H_{33}NO_2HCl$	377.97	nd (dil al), $[\alpha]_D$ + 28, (w, c = 1.1)		340			w, al	C36, 4826
1439	Atrolactic acid or 2-hydroxy-2-phenyl propanoic acid $CH_3C(OH)(C_6H_5)CO_2H$	166.18	pr (w), $[\alpha]^{16.5/}{}_D$ + 37.7 (al, c = 3.5)		116.5-7			w, ace, bz, al	B10[3], 560
1440	Atropic acid or 2-phenylpropenoic acid $CH_2{=}C(C_6H_5)CO_2H$	148.16	lf (al), nd (w)	267d	106-7			al, eth, bz, chl	B9[3], 2751
1441	Atropine or (*DL*)-hyoscyamine $C_{17}H_{23}NO_3$	289.37	rh nd (dil al), orth pr (ace)	sub(vac) 93-100	118-9			al, eth, bz, chl	B21[4], 183
1442	Atropine hydrochloride $C_{17}H_{23}NO_3\text{-}HCl$	325.84	nd (al)		165			w, al	B21[4], 184
1443	Atropine pentanoate $C_{17}H_{23}NO_3\text{-}C_5H_{10}O_2\text{-}\frac{1}{2}H_2O$	400.52	cr		42			w, al, eth	M, 111
1444	Atropine sulfate, hydrate $(C_{17}H_{23}NO_3)_2 \cdot H_2SO_4 \cdot H_2O$	694.82	nd (al eth or al ace)	sub	194 (anh)			w, al	B21[4], 182
1445	Auramine or *bis*(*p*-dimethylaminophenyl)methylene imine $C_{17}H_{21}N_3$	267.37	ye or col pl (al)		136			al	B14[3], 227
1446	Auramine hydrochloride or Auramine O $C_{17}H_{21}N_3\text{-}HCl$	303.83	ye nd (w)		267			al, chl	B14[2], 58
1447	Auramine-*N* $C_{18}H_{23}N_3$	281.40	ye cr (al)		133			al, ace, aa	B14, 93
1448	Aurin or Rosolic acid $C_{19}H_{14}O_3$	290.32	dk red lf or rh		308-10d			al, eth	B8[3], 3024
1449	Auroemycin or Chlorotetracycline $C_{22}H_{23}N_2O_8Cl$	478.89	gold ye, ye flr, $[\alpha]^{25/}{}_D$ −275 (MeOH)		168-9				B14[3], 1710
1450	Auroemycin hydrochloride $C_{22}H_{23}N_2O_8Cl \cdot HCl$	515.35	ye or th, $[\alpha]^{20/}{}_D$ −106.5 (dil al)		216d				B14[3], 1710
1451	Auxin A or Auxenetriolic acid $C_{18}H_{32}O_5$	328.45	hex (al-lig), $[\alpha]^{20/}{}_D$ −3.19 (al)		196	$1.292^{19/4}$		al	B10[3], 2045
1452	Auxin B or Auxenolonic acid $C_{18}H_{30}O_4$	310.43	cr (al-lig), $[\alpha]^{20/}{}_D$ −2.8 (al)		183	$1.269^{20/4}$		al, eth	B10[3], 4194
1453	Azelaic acid or Nonanedioic acid $HO_2C(CH_2)_7CO_2H$	188.22	lf or nd	287^{100}, 225^{10}	106.5	$1.225^{25/4}$	1.4303^{111}	al, eth	B2[4], 2055
1454	Azelaic acid, diethylester or Diethyl azelate $C_2H_5O_2C(CH_2)_7CO_2H_5$	244.33		291-2, $174\text{-}5^{20}$	-18.5	$0.97294^{20/4}$	1.43509^{20}	al, eth	B2[3], 1787
1455	Azelaic acid, di(2-ethylbutyl) ester $H_2C[(CH_2)_3{'}CO_2CH_2CH(C_2H_5)]_2$	356.54		230^5	-45	$0.928^{25/4}$	1.443^{25}	al, ace, bz	B2[3], 1787
1456	Azelaic acid, di-(2-ethylhexyl) ester $CH_2[(CH_2)_3CO_2CH_2CH(C_2H_5)C_4H]_2$	412.65		237^5	-78	1.446^{25}		al, ace, bz	B2[3], 1787
1457	Azelaic acid, dimethylester or Dimethyl azelate $CH_3O_2C(CH_2)_7CO_2CH_3$	216.28		156^{20}		$1.0082^{20/4}$	1.4367^{20}	al, ace, bz	B2[3], 1786
1458	Azelaic acid, diphenylester or Diphenyl azelate $C_6H_5O_2C(CH_2)_7CO_2C_6H_5$	340.42	nd (al)		59-60			eth, bz	B6[3], 606
1459	Azelanitrile or Azelaic acid, dinitrile $NC(CH_2)_7CN$	150.22		$198\text{-}9^{25}$		0.9200^{19}	1.4518^{19}	al, eth, bz	B2[4], 2059
1460	Azelayl chloride $ClOC(CH_2)_7COCl$	225.11		166^{18}		1.4680^{20}		eth, bz	B2[3], 1789
1461	l-Azacyclooctane, 2-methyl or α-Methyl heptamethyleni-mine $C_8H_{17}N$	139.24	nd (ace)	$162\text{-}3^{746}$	156-7	0.853^{20}	1.4620^{21}		B20[1], 30
1462	Azetidine or Trimethyleneimine C_3H_7N	57.10		63^{748}		0.8436^{20}	1.4287^{25}	w, al, eth, ace, bz	B20[4], 53

PHYSICAL CONSTANTS OF ORGANIC COMPOUNDS (Continued)

No.	Name, Synonyms, and Formula	Mol. wt.	Color, crystalline form, specific rotation and λ_{max} (log ε)	b.p. °C	m.p. °C	Density	n_D	Solubility	Ref.
1463	Aziridine or Ethyleneimine C_2H_5N	43.07	...	56^{756}	...	$0.8321^{20/4}$	...	w, al, eth, ace, bz	B20^{4}, 3
1464	Azobenzene (*cis*) $C_6H_5N=NC_6H_5$	182.22	og-red pl (peth)	...	71	...	...	al, eth, bz, aa	B16^{3}, 4
1465	Azobenzene (*trans*) $C_6H_5N=NC_6H_5$	182.22	og red mcl lf (al)	29.3	68.5	$1.203^{20/4}$	1.6266^{78}	al, eth, bz, aa	B16^{3}, 5
1466	Azobenzene, 2-acetamide-4′,5′-dimethyl $2(CH_3CONH)\text{-}5\text{-}CH_3\text{-}C_6H_3N=N(C_6H_4CH_3\text{-}4)$	267.33	ye nd (al-aa)	...	157	...	...	al, eth, chl	B16^{2}, 182
1467	Azobenzene, acetamido $4\text{-}(CH_3CONH)C_6H_4N=NC_6H_5$	239.28	gold-ye nd (al)	...	144-6	...	...	...	B16^{3}, 346
1468	Azobenzene, acetamido-2′,3-dimethyl $4(CH_3CONH)\text{-}3\text{-}CH_3\text{-}C_6H_3N=N(C_6H_4CH_3\text{-}2)$	267.33	red nd (al)	...	186-7	...	...	eth, chl	B16^{3}, 387
1469	Azobenzene, 4-acetoxy-2′-methyl $4(CH_3CO_2)C_6H_4N=N(C_6H_4\text{-}CH_3\text{-}2)$	254.29	red-ye lf (al)	...	68	...	...	al	B16, 105
1470	Azobenzene, 4-acetoxy-3-methyl $4\text{-}(CH_3CO_2)\text{-}3\text{-}(CH_3)C_6H_3N=NC_6H_5$	254.29	ye pl (dil al)	...	81-2	...	...	al, eth, bz, chl	B16, 130
1471	Azobenzene, 4-acetoxy-4′-methyl $4\text{-}(CH_3CO_2)C_6H_4N=N(C_6H_4CH_3\text{-}4)$	254.29	og nd (al, bz)	...	98	...	...	al	B16^{2}, 42
1472	Azobenzene, 2-amino or 2-Benzeneazoaniline $2\text{-}H_2NC_6H_4N=N\text{-}C_6H_5$	197.24	red nd or pr (al)	...	59	...	...	eth, ace, bz	B16^{3}, 334
1473	Azobenzene, 3-amino $3\text{-}H_2NC_6H_4N=NC_6H_5$	197.24	(i) og-ye nd (peth), (ii) br-red cr	...	(i) 69-70, (ii) 90-d1	...	...	al, eth, ace, bz, chl	B16^{3}, 335
1474	Azobenzene, amino $4\text{-}H_2NC_6H_4N=NC_6H_5$	197.24	og mcl nd (al)	>360	127	...	...	al, eth, bz, chl	B16^{2}, 149
1475	Azobenzene, 4-amino-2,2′-dimethyl $4(H_2N)\text{-}2\text{-}(CH_3)C_6H_3N=N(C_6H_4CH_3\text{-}2)$	225.29	ye nd (liq)	...	116-7	...	...	al, lig	B16^{2}, 180
1476	Azobenzene, 4-amino-2,3′-dimethyl $4\text{-}(H_2N)\text{-}2\text{-}(CH_3)\text{-}C_6H_3N=N(C_6H_4CH_3\text{-}3)$	225.29	ye-gold nd (al), ye-br nd (lig)	...	80	...	...	al	B16^{3}, 392
1477	Azobenzene, 4-amino-2′,3-dimethyl $4\text{-}(H_2N)\text{-}3\text{-}CH_3C_6H_3N=N(C_6H_4CH_3\text{-}2)$	225.29	ye lf (al)	...	101.5-3	...	...	al, eth	B16^{3}, 386
1478	Azobenzene, 4-amino-2,4′-dimethyl $4(H_2N)\text{-}2CH_3C_6H_3N=N(C_6H_4CH_3\text{-}4)$	225.29	ye pl (al)-gold-ye (lig)	...	127	...	...	al, lig	B16, 348
1479	Azobenzene, 4-amino-3,3′-dimethyl $4(H_2N)\text{-}3\text{-}(CH_3)C_6H_3N=N(C_6H_4CH_3\text{-}3)$	225.29	ye br lf or nd (lig)	...	124	...	...	al, lig	B16, 345
1480	Azobenzene, 4-amino-3,4′-dimethyl $4(H_2N)\text{-}3\text{-}(CH_3)C_6H_3N=N(C_6H_4CH_3\text{-}4)$	225.29	og-ye nd (lig), ye pl (al)	...	128	...	...	...	B16^{3}, 389
1481	Azobenzene, 4-amino, hydrobromide $4\text{-}H_2NC_6H_4N=NC_6H_5\text{-}HBr$	278.15	bk-vt nd (dil al)	...	206-7	...	...	...	B16, 307
1482	Azobenzene, 4-amino, hydrochloride $4\text{-}H_2NC_6H_4N=NC_6H_5\text{-}HCl$	233.70	bl-vt or pa red, nd or pw	...	240	...	...	w, al	B16^{2}, 149
1483	Azobenzene, 4-benzoxy-4′-methyl $4\text{-}(C_6H_5CH_2O)C_6H_4N=N(C_6H_5CH_3\text{-}4)$	302.38	pa ye lf (lig)	...	128	...	...	al, eth, bz	B16, 107
1484	Azobenzene, 3,3′-*bis*(dimethylamino) $3-(CH_3)_2NC_6H_4N=N(C_6H_4N(CH_3)_2-3)$	268.36	red nd (al)	...	118	...	...	al, bz	B16, 305
1485	Azobenzene, 4,4′-*bis*-(dimethylamino) $4-(CH_3)_2NC_6H_4N=N[C_6H_4N(CH_3)_2-4]$	268.36	og nd (bz)	sub	273	...	...	eth, bz, chl	B16^{3}, 375
1486	Azobenzene, 4-bromo $4\text{-}BrC_6H_4N=NC_6H_5$	261.12	og nd (lig)	...	90-1	...	...	al, eth, ace, lig	B16^{3}, 29
1487	Azobenzene, 2,2′-diamino $2\text{-}H_2NC_6H_4N=N(C_6H_4NH_2\text{-}2)$	212.25	red pl (al or bz)	...	134	...	...	eth	B16^{2}, 148
1488	Azobenzene, 2,4-diamino or Chrysoidine $2,4\text{-}(H_2N)_2C_6H_3N=NC_6H_5$	212.25	pa ye nd (w)	...	117.5	...	...	al, eth, bz	B16^{3}, 436
1489	Azobenzene, 4,4′-diamino or *o*-Azodianiline $4\text{-}H_2NC_6H_4N=N(C_6H_4NH_2\text{-}4)$	212.25	gold-ye nd (al), og-ye pr (al)	...	250-1	...	...	al, bz, chl	B16^{3}, 375
1490	Azobenzene, 2,2′-diethoxy or *o*-Azophenetole $2\text{-}C_2H_5OC_6H_4N=N(C_6H_4OC_2H_5\text{-}2)$	270.33	red pr (al)	240d	131	...	...	al, eth	B16^{3}, 83
1491	Azobenzene, 4,4′-diethoxy or *p*-Azophenetole $4\text{-}C_2H_5OC_6H_4N=N(C_6H_4OC_2H_5\text{-}4)$	270.33	ye-lf (al)	d	162	...	...	eth, bz, chl, aa	B16^{3}, 93

No.	Name, Synonyms, and Formula	Mol. wt.	Color, crystalline form, specific rotation and λ_{max} (log ε)	b.p. °C	m.p. °C	Density	n_D	Solubility	Ref.
1492	Azobenzene, 2,2′-dihydroxy or *o*-Azophenol 2-HOC$_6$H$_4$N=N(C$_6$H$_4$OH-2)	214.22	gold-ye lf (bz), nd (al)	sub	172			eth, al, bz	B16^2, 33
1493	Azobenzene, 2,4-dihydroxy 2,4(HO)$_2$C$_6$H$_3$N=NC$_6$H$_5$	214.22	dk red nd (dil al)		170 (anh)			al, eth, bz, aa	B16^1, 166
1494	Azobenzene, 2,4-dihydroxy-4′-nitro 2,4-(HO)$_2$C$_6$H$_3$N=N(C$_6$H$_4$NO$_2$-4)	259.22	red pw (al or MeOH)		200				B16^1, 162
1495	Azobenzene, 2,5-dihydroxy 2,5(HO)$_2$C$_6$H$_3$N=NC$_6$H$_5$	214.22			149			eth, ace, bz	B16^1, 176
1496	Azobenzene, 3,3′-dihydroxy or *m*-Azophenol 3-HOC$_6$H$_4$N=N(C$_6$H$_4$OH-3)	214.22	ye lf (dil al)		207			eth, ace	B16^3, 85
1497	Azobenzene, 4-dimethylamino or Butteryellow 4-(CH$_3$)$_2$NC$_6$H$_4$=NC$_6$H$_5$	225.30	ye lf (al)	d	117			al, eth, aa, Py	B16^1, 340
1498	Azobenzene, 4,4′-dihydroxy or *p*-Azophenol 4-HOC$_6$H$_4$N=N(C$_6$H$_4$OH-4)	214.22	og-ye pl (dl al, + 1w)		216-6.5 anh			al, eth, ace	B16^1, 92
1499	Azobenzene, 2,2′-dimethyl or *o*-Azotoluene 2-CH$_3$C$_6$H$_4$N=N(C$_6$H$_4$CH$_3$-2)	210.28	dk red mcl pr (eth)		55-6	1.0215$^{65/4}$	1.6180^{65}	al, eth, bz	B16^2, 19
1500	Azobenzene, 2,2′-dimethyl-4-ethoxy 2-CH$_3$-4-C$_2$H$_5$OC$_6$H$_3$N=N(C$_6$H$_4$CH$_3$-2)	254.34	dk red nd (al)		64			al, eth, ace, bz	B16, 134
1501	Azobenzene, 2,2′-dimethyl-4-hydroxy or 4-*o*-Tolueneazo-*m*-cresol 2-CH$_3$-4-HOC$_6$H$_3$N=N(C$_6$H$_4$CH$_3$-2)	226.28	ag-red pl (bz), red cr (w + l)		113 (anh)			al, eth, bz	B16^2, 61
1502	Azobenzene, 2,3′-dimethyl-4-ethoxy 2-CH$_3$-4-C$_2$H$_5$OC$_6$H$_3$N=N(C$_6$H$_4$CH$_3$-3)	254.33	red pr (al)	73				al, eth, ace, bz	B16, 135
1503	Azobenzene, 2′,3-dimethyl-4-ethoxy 3-CH$_3$-4-C$_2$H$_5$OC$_6$H$_3$N=N(C$_6$H$_4$CH$_3$-2)	254.33	red cr (lig)		35-7			al, bz, lig	B16, 131
1504	Azobenzene, 2,3-dimethyl-4-hydroxy or 4-Benzeneazo-*o*-xylenol 2,3(CH$_3$)$_2$-4-HO-C$_6$H$_2$N=NC$_6$H$_5$	226.28	red pr (al, lig)		132			al, eth, ace, bz	B16^2, 63
1505	Azobenzene, 2,3′-dimethyl-4-hydroxy or 4-*m*-Tolueneazo *m*-cresol 2-CH$_3$-4-HOC$_6$H$_3$N=N(C$_6$H$_4$CH$_3$-3)	226.28	og-ye pl (bz)		106-7			eth, bz, lig	B16^3, 107
1506	Azobenzene, 2,4′-dimethyl-4-ethoxy 2-CH$_3$-4-C$_2$H$_5$OC$_6$H$_3$N=N(C$_6$H$_4$CH$_3$-4)	254.33	og-red pl (al)		64			al, eth, bz, lig	B16, 135
1507	Azobenzene, 2,4-dimehyl-5-hydroxy or 5-benzeneazo-2,4-xylenol 2,4-(CH$_3$)$_2$-5-HOC$_6$H$_3$N=NC$_6$H$_5$	226.28	og-ye nd (lig-peth)		114			al, eth, ace, bz, aa	B16, 146
1508	Azobenzene, 2,4′dimethyl-4-hydroxy or 4-*p*-Tolueneazo-*m*-cresol 2-CH$_3$-4-HOC$_6$H$_3$N=N(C$_6$H$_4$CH$_3$-4)	226.28	og-ye pr (bz)		135			al, eth, bz	B16^2, 62
1509	Azobenzene, 3,3′-dimethyl *cis* or *m*-Azotoluene 3-CH$_3$C$_6$H$_4$N=N(C$_6$H$_4$CH$_3$-3)	210.28	red (peth)		46				B16^3, 47
1510	Azobenzene, 3,3′-dimethyl *trans* or *m*-Azotoluene 3-CH$_3$C$_6$H$_4$N=N(C$_6$H$_4$CH$_3$-3)	210.28	og-red orh		84-4.5	1.0123^{66}	1.6152^{66}	al, eth, bz	B16^3, 48
1511	Azobenzene, 3,3′-bis(dimethylamino) 3*(CH$_3$)$_2$N*-C$_6$H$_4$N=NC$_6$H$_4$[N(CH$_3$)$_2$]-3	268.36	og nd (bz)	sub	273			eth, bz, chl	B16^3, 339
1512	Azobenzene, 3,3′-dimethyl-4-ethoxy 3-CH$_3$-4-C$_2$H$_5$OC$_6$H$_3$N=N(C$_6$H$_4$CH$_3$-3)	254.33	red-ye pl (al)		46-7			al, eth, ace, bz, lig	B16, 131
1513	Azobenzene, 3,3′-dimethyl-4-hydroxy or 4-*m*-Tolueneazo *o*-cresol 3-CH$_3$-4-HOC$_6$H$_3$N-N(C$_6$H$_4$CH$_3$-3)	226.28	gold-ye nd (bz)		115			al, eth, bz	B16^2, 60
1514	Azobenzene, 3,4′-dimethyl-4-ethoxy 3-CH$_3$-4-C$_2$H$_5$OC$_6$H$_3$N=N(C$_6$H$_4$CH$_3$-4)	254.33	og-ye nd (al)	251^{42}	73-4			al, eth, ace, bz	B16, 131
1515	Azobenzene, 3,4′-dimethyl-4-hydroxy or 4-*p*-Tolueneazo-*o*-cresol 3-CH$_3$-4-HOC$_6$H$_3$N=N(C$_6$H$_4$CH$_3$-4)	226.28	og nd (bz)		163			al, eth, bz	B16^3, 106
1516	Azobenzene, 3,5-dimethyl-2-hydroxy or 6-Benzeneazo-*m*-4-xylenol 3,5-(CH$_3$)$_2$-2-HOC$_6$H$_2$N=NC$_6$H$_5$	226.28	dk red nd (al, lig)		90			al, eth, bz, lig	B16, 145
1517	Azobenzene, 3,5-dimethyl-4-hydroxy or 4-Benzeneazo-2,6-xylenol 3,5(CH$_3$)$_2$-4-HOC$_6$H$_2$N=NC$_6$H$_5$	226.28	og-red cr (lig)		150-1			al, eth, bz	B16, 145
1518	Azobenzene, 4,4′-dimethyl *cis* or *p*-Azotoluene 4-CH$_3$C$_6$H$_4$N=N(C$_6$H$_4$CH$_3$-4)	210.28	dk red		104			bz, lig	B16^3, 49
1519	Azobenzene, 4,4′-dimethyl *trans* or *p*-Azotoluene 4-CH$_3$C$_6$H$_4$N=N(C$_6$H$_4$CH$_3$-4)	210.28	og ye nd (al)		144			eth, bz, lig	B16^3, 49
1520	Azobenzene, 4,4′-bis(dimethyl amino) 4-(CH$_3$)$_2$N-C$_6$H$_4$-N=N*C$_6$H$_4$[N(CH$_3$)$_2$-4]*	268.36			273			eth, bz	B16^3, 375

PHYSICAL CONSTANTS OF ORGANIC COMPOUNDS (Continued)

No.	Name, Synonyms, and Formula	Mol. wt.	Color, crystalline form, specific rotation and λ_{max} (log ε)	b.p. °C	m.p. °C	Density	n_D	Solubility	Ref.
1521	Azobenzene, 4,4′-dimethyl-2-hydroxy or 6-*p*-Tolueneazo-*m*-cresol .... 4-CH_3-2-HOC_6H_3N=N($C_6H_4CH_3$-4)	226.28	og red cr (lig)	....	150-1	....	....	al, eth, bz, lig	B16³, 106
1522	Azobenzene, 4′,5-dimethyl-2-ethoxy .... 5-CH_3-2-$C_2H_5OC_6H_3N$=N($C_6H_4CH_3$-4)	254.33	pa red nd (abs al)	$253\text{-}5^{63}$	43	....	....	al	B16, 141
1523	Azobenzene, 4′,5-dimethyl-2-hydroxy or 2-*p*-Tolueneazo-P-cresol .... 5-CH_3-2-HOC_6H_3N=N($C_6H_4CH_3$-4)	226.28	red cr, ye pl (to)	....	112-3	....	....	al, bz, to, chl	B16³, 109
1524	Azobenzene, 2-ethoxy or *o*-Benzeneazo phenetole .... 2-$C_2H_5OC_6H_4N$=NC_6H_5	226.28	red pl or mcl pr (peth)	....	44	....	....	al, eth, ace, bz, peth	B16³, 81
1525	Azobenzene, 3-ethoxy or *m*-Benzeneazo phenetole .... 3-$C_2H_5OC_6H_4N$=NC_6H_5	226.28	pl (peth)	208^{22}	63.5-4	....	....	al, eth, ace, bz	B16, 95
1526	Azobenzene, 4-ethoxy or *p*-Benzenazo phenetole .... 4-$C_2H_5OC_6H_4N$=NC_6H_5	226.28	og nd (60-70% al)	339-40	85	$1.0400^{100}/_{4}$	$1.6419^{100/}_{5875}$	al, eth, ace, bz	B16², 40
1527	Azobenzene, 4-ethoxy-2-methyl .... 2-CH_3-4-$C_2H_5OC_6H_3N$=NC_6H_5	240.30	or red nd (dil al)	....	51.5-2.0	....	....	al, eth, lig	B16, 134
1528	Azobenzene, 4-ethoxy-2′-methyl .... 4-$C_2H_5OC_6H_4N$=N($C_6H_4CH_3$-2)	240.30	og pl (al)	....	53	....	....	al, eth, bz, chl	B16, 105
1529	Azobenzene, 4-ethoxy-3-methyl .... 3-CH_3-4-$C_2H_5OC_6H_3N$=NC_6H_5	240.30	og nd or mcl pr (al)	....	60	....	....	al, eth, bz	B16, 130
1530	Azobenzene, 4-ethoxy-3′-methyl .... 4-$C_2H_5OC_6H_4N$=N($C_6H_4CH_3$-3)	240.30	og-red pr (al)	....	65	....	....	al, eth, bz	B16, 106
1531	Azobenzene, 4-ethoxy-4′-methyl .... 4-$C_2H_5OC_6H_4N$=($C_6H_4CH_3$-4)	240.30	red lf (al)	....	121-2	....	....	al, bz, chl	B16, 107
1532	Azobenzene, 2-hydroxy or *o*-Benzeneazo phenol .... 2-HOC_6H_4N=N-C_6H_5	198.22	og-red nd (eth)	....	82.5-3	....	....	al, eth, ace, bz	B16³, 80
1533	Azobenzene, 2-hydroxy, benzoate .... 2.$C_6H_5CO_2C_6H_4N$=NC_6H_5	302.33	og-red nd (lig)	....	93	....	....	al, bz	B16³, 81
1534	Azobenzene, 2-hydroxy-4-methyl .... 4-CH_3-2-HOC_6H_3N=NC_6H_5	212.25	red pl (lig)	....	122	....	....	al, ace, bz, lig	B16¹, 241
1535	Azobenzene, 2-hydroxy-4-methyl,benzoate .... 2-$C_6H_5CO_2$-4-$CH_3C_6H_3N$=NC_6H_5	316.36	....	....	98	....	....	bz	B16¹, 241
1536	Azobenzene, 2-hydroxy-5-methyl or 2-benzeneazo-*p*-cresol .... 2-HO-5-$CH_3C_6H_3N$=NC_6H_5	212.25	og-ye lf (bz, bz-lig, gold lf) (w-al)		108-9	....	....	al, eth, bz, chl	B16, 108
1537	Azobenzene, 3-hydroxy or *m*-Benzeneazo phenol .... 3-HOC_6H_4N=NC_6H_5	198.22	ye nd (w, lig), pr (bz)	....	116.5-7	....	....	al, eth, ace, bz	B16³, 85
1538	Azobenzene, 3-hydroxy, acetate .... 3-$CH_3CO_2C_6H_4N$=NC_6H_5	240.26	og pl (peth)	....	67.5	....	....	peth	B16, 95
1539	Azobenzene, 3-hydroxy, benzoate .... 3-$C_6H_5CO_2$-C_6H_4N=N-C_6H_5	302.33	og-red pl (peth)	....	92	....	....	peth	B16, 95
1540	Azobenzene, 4-hydroxy or *p*-Benzeneazo phenol .... 4-HOC_6H_4N=NC_6H_5	198.22	ye lf (bz), og pr (al)	$220\text{-}30^{20/d}$	155-7	....	....	al, eth, bz	B16², 38
1541	Azobenzene, 4-hydroxy acetate .... 4-$CH_3CO_2C_6H_4N$=N-C_6H_5	240.26	og lf (al), nd (al, lig)	> 360 d	89	....	....	lig, al	B16¹, 236
1542	Azobenzene, 4-hydroxy,benzoate .... 4-$C_6H_5CO_2C_6H_4N$=NC_6H_5	302.33	og-ye nd (al, lig), ye-red pr (eth, al)	....	138	....	....	to, lig, al, eth	B16³, 89
1543	Azobenzene, 4-hydroxy-2′-methoxy-2-methyl .... 2-CH_3-4-HOC_6H_3N=N($C_6H_4OCH_3$-2)	242.28	og	....	161	....	....	al, eth	B16, 135
1544	Azobenzene, 4-hydroxy-2′-methoxy-3-methyl .... 3-CH_3-4-HO-C_6H_3N=N($C_6H_4OCH_3$-2)	242.28	og red pl (dil al)	....	68	....	....	al, eth, bz, lig	B16, 131
1545	Azobenzene, 4-hydroxy-2-methyl or 4-Benzeneazo-*m*-cresol .... 4-HO-2-$CH_3C_6H_3N$=NC_6H_5	212.25	ye nd (lig)	....	109	....	....	al, eth, bz, chl	B16², 61
1546	Azobenzene, 4-hydroxy-2′-methyl or *o*-Toluenwazo-*p*-phenol .... 4-HOC_6H_4N=N($C_6H_4CH_3$-2)	212.25	red pl or lf (bz-lig), og-ye nd (bz)	....	107-8	....	....	al, eth, bz, chl	B16³, 91
1547	Azobenzene, 4-hydroxy-3-methyl or 4-Benzeneazo-*o*-cresol .... 4-HO-3-$CH_3C_6H_3N$=NC_6H_5	212.25	gold-ye lf or nd (al)	....	128-30	....	....	al, eth, bz, lig	B16³, 104
1548	Azobenzene, 4-hydroxy-3-methyl, benzoate .... 4-$C_6H_5CO_2$-3-$CH_3C_6H_3N$=NC_6H_5	316.36	ye nd (al)	....	110-1	....	....	eth, ace, chl	B16, 130

No.	Name, Synonyms, and Formula	Mol. wt.	Color, crystalline form, specific rotation and λ_{max} (log ε)	b.p. °C	m.p. °C	Density	n_D	Solubility	Ref.
1549	Azobenzene, 4-hydroxy-3′-methyl or *m*-Tolueneazo-*p*-phenol $4-HO-C_6H_4N=N(C_6H_4CH_3-3)$	212.25	ye-pl (al)		144-5			al, bz	$B16^{2}$, 41
1550	Azobenzene, 4-hydroxy-4′-methyl or *p*-Tolueneazo-*p*-phenol $4-HOC_6H_4N=N(C_6H_4CH_3-4)$	212.25	og red mcl (bz-lig)		152			al, eth, bz	$B16^{2}$, 42
1551	Azobenzene, 4-hydroxy-4′-methyl, benzoate $4-C_6H_5CO_2C_6H_4N=N(C_6H_4CH_3-4)$	316.36	og-red pr (bz), redsh-ye nd (lig)		178			al, bz, eth	$B16^{1}$, 237
1552	Azobenzene, 2-methoxy or *o*-Benzeneazo anisole $2-CH_3OC_6H_4N=NC_6H_5$	212.25	og-red nd (dil al)	195-7^{14}	41			al, bz	$B16^{3}$, 81
1553	Azobenzene, 3-methoxy or *m*-Benzeneaz anisole $3-CH_3OC_6H_4N=NC_6H_5$	212.25	og-red pl (MeOH)	193^{15}	32-3			al, ace	$B16^{3}$, 85
1554	Azobenzene, 4-methoxy or *p*-Benzeneazo anisole $4-CH_3OC_6H_4N=NC_6H_5$	212.25	og-red pl, lf (al) (peth)	340	56 (64)	1.12^{75}		al, eth, ace	$B16^{2}$, 40
1555	Azobenzene, 4-methoxy-4′-methyl $4-CH_3OC_6H_4N-N(C_6H_4CH_3-4)$	226.28	og-ye pr (al)		110-1			al, eth, ace, bz	$B16^{2}$, 236
1556	Azobenzene, 4-nitro $4-NO_2C_6H_4N=NC_6H_5$	227.22	lf, nd, (al, lig)		135 (155)			al, ace, bz, lig	$B16^{2}$, 17
1557	Azobenzene, 4-phenylamino or 4-Anilineazobenzene $4-C_6H_5NHC_6H_4N=NC_6H_5$	273.34	ye pl or pr		82			al, eth, lig	$B16^{3}$, 343
1558	Azobenzene, 2,4,3′-triamino or Bismark's brown $2,4(H_2N)_2C_6H_3N=N(C_6H_4NH_2-3)$	226.23	og lf (w), red (bz)		143-5			al,eth	B16, 386
1559	2-Azobenzene carboxylic acid or *o*-Benzeneazobenzoic acid $2-HO_2C-C_6H_4N=NC_6H_5$	226.24	og-red nd or pl (al)		97-8			al, eth, ace, bz, aa	$B16^{3}$, 213
1560	2-Azobenzene carboxylic acid, 4′-dimethylamino or Methyl red $2-HO_2CC_6H_4N=N-[C_6H_4N(CH_3)_2-4]$	269.30	vt or red pr (to or bz), nd (ag aa)		183			al, ace, bz, aa	$B16^{2}$, 164
1561	3-Azobenzene carboxylic acid or *m*-Benzeneazobenzoic acid $3-HO_2CC_6H_4N=NC_6H_5$	226.23	og-red pl (w), red lf (al)		170-1			al, eth, bz, chl	B16, 229
1562	3-Azobenzene carboxylic acid, 4-hydroxy or 5-Benzeneazosalicylic acid $3-HO_2C-4-HOC_6H_3NC_6H_5$	242.23	nd (bz)		230d			al, eth, ace	$B16^{3}$, 237
1563	3-Azobenzene carboxylic acid, 4-hydroxy-2′-nitro or 5-*o*-Nitrobenzeneazosalicylic acid $3-(HO_2C)-4-HOC_6H_3N=N(C_6H_4NO_2-2)$	287.23	br-red cr (al)		215-7			al, aa	$B16^{2}$, 101
1564	3-Azobenzene carboxylic acid, 4-hydroxy-3′-nitro or 5-*m*-Nitrobenzeneazosalicylic acid $3-HO_2C-4-HOC_6H_3N=N[C_6H_4NO_2-3]$	287.23	red-br nd (al)					al, eth, bz, aa	$B16^{3}$, 238
1565	3-Azobenzene carboxylic acid, 4-hydroxy, 4′-nitro or 5-*p*-nitro benzeneazosalicylic acid $3-(HO_2C)-4-HOC_6H_3N=N(C_6H_4NO_2-4)$	287.23	og br nd (dil aa)		257			al, aa	$B16^{2}$, 101
1566	3-Azobenzene carboxylic acid, 3′-sulfo $3(HO_2C)C_6H_4N=(C_6H_4SO_3H-3)$								B16268
1567	4-Azobenzene carboxylic acid or *p*-Benzeneazobenzoic acid $4-HO_2CC_6H_4N=NC_6H_5$	226.23	red pl or lf (al)		249			al, eth, ace, bz	$B16^{3}$, 218
1568	2,2′-Azobenzenedicarboxylic acid or *o*-Azobenzoic acid $2-HO_2CC_6H_4N=N(C_6H_4CO_2H-2)$	270.24	dk red nd (al)		245 (cor)			eth, al	$B16^{3}$, 216
1569	3,3′-Azobenzenedicarboxylic acid or *m*-Azobenzoic acid $3-HO_2CC_6H_4N=NC_6H_4CO_2H-3)$	270.24	ye nd (aa)		340d			al	$B16^{3}$, 217
1570	4,4′-Azobenzeenedicarboxylic acid or *p*-Azobenzoic acid $4-HO_2CC_6H_4N=N(C_6H_4CO_2H-4)$	270.24	dk ye nd (al), red nd (aa)		330d				$B16^{3}$, 228
1571	3,3′-Azobenzenedisculfonamide $3-(H_2NO_2S)C_6H_4N=N(C_6H_4SO_2NH_2-3)$	340.37	ye nd (al)		305			al	B16, 268
1572	3,4′-Azobenzene disulfonamide $3-H_2NO_2SC_6H_4N=N(C_6H_4SO_2NH_2-4)$	340.37	ye nd (al)		288			al	B16, 279
1573	4,4′-Azobenzene disulfonamide $4-H_2NO_2SC_6H_4N=N(C_6H_4SO_2NH_2-4)$	340.37	og nd (al)		> 250d			al, eth	$B16^{3}$, 299
1574	3,3′ Azobenzene disulfonic acid or *p*-Azobenzoic acid $3-HO_3SC_6H_4N=N(C_6H_4SO_3H-3)$	342.34	ye lf (w + 5)					w, al, eth	$B16^{3}$, 281
1575	3,4′-Azobenzne disulfonic acid $3-HO_3SC_6H_4N=N(C_6H_4SO_3H-4)$	342.34	syr						$B16^{3}$, 299

PHYSICAL CONSTANTS OF ORGANIC COMPOUNDS (Continued)

No.	Name, Synonyms, and Formula	Mol. wt.	Color, crystalline form, specific rotation and λ_{max} (log ε)	b.p. °C	m.p. °C	Density	n_D	Solubility	Ref.
1576	4,4′-Azobenzene disulfonic acid $4-HO_3SC_6H_4N=N(C_6H_4SO_3H-4)$	342.34	red nd (w + 5)	....	169d (anh)	....	....	w	$B16^3$, 299
1577	3,3′-Azobenzene disulfonyl chloride $3-(ClO_2S)C_6H_4N=N(C_6H_4SO_2Cl-3)$	379.23	red nd (eth)	....	166-7	....	....	eth	B16, 268
1578	3,4′-Azobenzene disulfonyl chloride $3-ClO_2SC_6H_4N=N(C_6H_4SO_2Cl-4)$	379.23	red nd (eth)	....	123-5	....	....	eth	B16, 279
1579	4,4′-Azobenzene disulfonyl chloride $4-ClO_2SC_6H_4N=N(C_6H_4SO_2Cl-4)$	379.23	br-red nd (eth-bz)	....	......	....	....	bz, chl	B16, 280
1580	3-Azobenzene sulfonic acid-4′-hydroxy-3′-nitro $3-NO_2-4-HOC_6H_3N=N(C_6H_4SO_3H-3)$	323.28	gold-ye pr (dil HCl + w)	....	235 (anh), 116 (+ w)	....	....	w, al	B16, 267
1581	4-Azobenzene sulfonic acid or 4-phenyl azobenzene sulfonic acid $4-HO_3SC_6H_4N=N(C_6H_5)$	262.28	og-red lf (w + 3)	....	127 (hyd)	....	....	w	$B16^2$, 281
1582	4-Azobenzene sulfonylchloride $4-ClO_2SC_6H_4N=NC_6H_5$	280.73	og-red nd or pl (bz)	....	82	....	....	al, bz	$B16^3$, 282
1583	*p,p′*-Azobiphenyl or Di-*p*-xenyldiimide $4-C_6H_5C_6H_4N=NC_6H_4C_6H_5-4$	334.42	og-red pl (bz)	....	256	....	....	eth	$B16^3$, 65
1584	Azomethane or Dimethyl diimide $CH_3N=NCH_3$	58.08	col or pa ye gas	1.5	-78	$0.744^{0/15}$	1.4199^{19}	al, eth, ace	$B4^2$, 1747
1585	Azomethane, hexafluoro $F_3CN=NCF_3$	166.03	pa grsh gas	-31.6	-133	....	....	....	$B3^4$, 246
1586	*α,α′*-Azonaphthalene or Di-*α*-naphthyldiimide $\alpha-C_{10}H_7N=NC_{10}H_7-\alpha$	282.34	red nd (gl aa)	sub	190	....	....	ace, bz	$B16^2$, 26
1587	*α,α′*-Azonaphthalene, 4-amino $4-H_2NC_{10}H_6N=NC_{10}H_7$	297.36	red-br nd		183 (cor)	....	....	....	
1588	1,2′-Azonaphthalene or *α,β*-Naphthyldiimide $\alpha-C_{10}H_7N=N-\beta-C_{10}H_7-(2)$	282.34	red nd, br lf (aa)	....	136	....	....	al, bz, aa	B16, 80
1589	2,2′-Azonaphthalene or Di-*β*-Naphthyldiimide $\beta-C_{10}H_7N=N-\beta-C_{10}H_7$	282.34	red lf (bz), og-ye nd (chl, al)	sub	208	....	....	bz	$B16^2$, 26
1590	Azoxybenzene *cis* $C_6H_5N(O)=NC_6H_5$	198.22	....	....	87	$1.166^{20/4}$	1.633^{20}	....	$B16^3$, 580
1591	Azoxybenzene *trans* $C_6H_5N(O)=C_6H_5$	198.22	bt ye nd	d	36	$1.1590^{26/4}$	1.652^{20}	al, eth, lig	$B16^3$, 579
1592	Azoxybenzene, 4-bromo $4-BrC_6H_4N(O)=NC_6H_5$	277.12	dk ye nd	....	93.5-4.5	1.4138^{100}	....	al, bz	$B16^3$, 584
1593	Azoxybenzene, 2,2′-dimethoxy or *o*-Azoxyanisole $2-CH_3OC_6H_4N(O)=N(C_6H_4OCH_3-2)$	258.28	ye-og cr (al)	....	81	....	....	al, eth, ace, bz, chl	B16, 635
1594	Azoxybenzene, 3,3′-dimethoxy or *m*-Azoxyanisole $3-CH_3OC_6H_4N(O)=N(C_6H_4OCH_3-3)$	258.28	ye cr (al)	....	51-2	....	....	al, eth	$B16^2$, 325
1595	Azoxybenzene, 4,4′-dimethoxy or *p*-Azoxyanisole $4-CH_3OC_6H_4N(O)=N(C_6H_4OCH_3-4)$	258.28	ye mcl nd (al)	....	119-20	$1.1711^{115/4}$	....	ace, bz, al	$B16^3$, 597
1596	Azoxybenzene, 2,2′-dimethyl *trans* $2-CH_3C_6H_4N(O)=N(C_6H_4CH_3-2)$	226.28	....	....	60	$1.0215^{65/4}$	1.61804^{65}	al, eth, bz	$B16^3$, 587
1597	Azoxybenzene, 3,3′-dimethyl *trans* $3-CH_3C_6H_4N(O)=N(C_6H_4CH_3-3)$	226.28	....	....	39	1.0123^{66}	1.6152^{66}	al, eth, bz, lig	$B16^3$, 588
1598	2,2-Azoxybenzene dicarboxylic acid or *o*-Azoxybensoic acid $2-HO_2CC_6H_4N(O)=N(C_6H_4CO_2H-2)$	286.24	ye pl or pr (al)	....	254.5	....	....	al, ace, aa	$B16^3$, 609
1599	3,3′-Azoxybenzene dicarboxylic acid or *m*-Azoxybenzoic acid $3-HO_2CC_6H_4N(O)=N(C_6H_4CO_2H-3)$	286.24	pa ye nd or lf (gl aa)	....	320d	....	....	....	$B16^3$, 610
1600	4,4′-Azoxybenzene dicarboxylic acid or *p*-Azoxybenzoic acid $4-HO_2CC_6H_4N(O)-N(C_6H_4CO_2H-4)$	286.24	ye am or pw	....	360d	....	....	Py	$B16^3$, 615
1601	1,1′-Azoxynaphthalene $\alpha-C_{10}H_7N(O)=N-\alpha-C_{10}H_7$	298.34	ye or red orh pl (al)	....	127	....	....	al	$B16^3$, 591
1602	2,2′-Azoxynaphthalene $\beta-C_{10}H_7N(O)=N-\beta-C_{10}H_7$	298.34	ye or red nd (al, eth, gl aa)	....	167-8	....	....	bz, aa, chl	$B16^3$, 592
1603	Azulene $C_{10}H_8$	128.17	bl or grsh-bk lf (al)	270d, $115-35^{10}$	99-100	....	....	al, eth, ace	B541636
1604	Azulene, 1,4-dimethyl-7-isopropyl or Guaiazulene $C_{15}H_{18}$	198.31	bl-vt pl (al)	$167-8^{12}$	31.5	....	....	al, eth	$B5^4$, 1751
1605	Azulene, 1,5-dimethyl-8-isopropyl or Chamazulene $C_{15}H_{18}$	198.31	bl lig	161^{12}	$0.9883^{20/4}$	....	....	al, eth	$B5^2$, 474

No.	Name, Synonyms, and Formula	Mol. wt.	Color, crystalline form, specific rotation and λ_{max} (log ε)	b.p. °C	m.p. °C	Density	n_D	Solubility	Ref.
1606	Azulene, 4,8-dimethyl-2-isopropyl or Elamazulene, Vetivazulene .... $C_{15}H_{18}$	198.31	red-vt nd (al)	140-60[2]	31-2	....	....	al, eth	B5[4], 1754
1607	Azulene, 4,6,8-trimethyl .... $C_{13}H_{14}$	170.25	....	....	79-81	....	....	chl	B5[4], 1721

No.	Name, Synonyms, and Formula	Mol. wt.	Color, crystalline form, specific rotation and λ_{max} (log ε)	b.p. °C	m.p. °C	Density	n_D	Solubility	Ref.
1608	Barbituric acid or Pyrimidinetrione $C_4H_4N_2O_3$	128.09	rh pr (w + 2)	260d	248	...	...	eth, w	$B24^2$, 292
1609	Barbituric acid, 5,allyl-5-butyl $C_{11}H_{16}N_2O_3$	224.26	(w, dil al)	...	128	...	...	w, al	$B24^2$, 292
1610	Barbituric acid, 5-allyl-5(2-cyclopenten-l-yl) or Cycclopal $C_{12}H_{14}N_2O_3$	234.25	(w, dil al)	...	139-40	...	...	w, al	C24, 5308
1611	Barbituric acid, 5-allyl-5-*iso*-butyl or Sandoptal $C_{11}H_{16}N_2O_3$	224.26	(w, dil al)	...	138	...	...	w, al, eth, ace, chl	$B24^2$, 292
1612	Barbituric acid, 5- allyl-5-*iso*-propyl $C_{10}H_{14}N_2O_3$	210.23	...	...	140-1	...	...	...	$B24^2$, 290
1613	Barbituric acid, 5-allyl-5-*iso*-propyl-l-methyl or Narconumal $C_{11}H_{16}N_2O_3$	224.26	(w, dil al)	$176-8^{12}$	56-7	...	...	al, eth, chl, ace, bz	C32, 1052
1614	Barbituric acid, 5- allyl-5-phenyl or Alphenal $C_{13}H_{12}N_2O_3$	244.25	...	...	156-7	...	...	al, eth, chl	C26, 4828
1615	Barbituric acid, 5- amino or Uramil $C_4H_5N_3O_3$	143.10	nd or pl (w)	...	> 400	...	...	w, chl	$B25^1$, 704
1616	Barbituric acid, 5-benzylidene $C_{11}H_8N_2O_3$	216.20	pr (aa)	...	256	...	...	ace	$B24^2$, 299
1617	Barbituric acid, 5-(2-bromoallyl)-5-*iso*-propyl or Propallylonal $C_{10}H_{13}BrN_2O_3$	289.13	(dil aa, dil al)	...	181	...	...	al, eth, ace, aa	$B24^2$, 291
1618	Barbituric acid, 5-butyl-5-ethyl $C_{10}H_{16}N_2O_3$	212.25	...	...	128-9	...	...	...	$B24^2$, 285
1619	Barbituric acid, 5-(l-cyclohexenyl)-5-ethyl or Cyclobarbitone $C_{12}H_{16}N_2O_3$	236.27	lf (w)	...	172-4	...	...	al, eth	$B24^2$, 294
1620	Barbituric acid, 5,5-diallyl or Allobarbitone $C_{10}H_{12}N_2O_3$	208.22	pl (w or 50% al)	...	174	...	...	al, eth, bz	$B24^2$, 293
1621	Barbituric acid, 5,5-dibromo or Dibromin $C_4H_2Br_2N_2O_3$	285.88	pl (MeOH-bz), lf (dil HNO_3)	d	235	...	...	al, eth	$B24^2$, 272
1622	Barbituric acid, 5,5-diethyl or barbital, veronal $C_8H_{12}N_2O_3$	184.19	(i) trig (w) (ii) mcl pr (iii) mcl nd (iv) tcl	...	(i) 190 (ii) 183 (iii) 181 (iv) 176	1.220	...	al, eth, ace, aa	$B24^2$, 279
1623	Barbituric acid, 5,5-diethyl-l-methyl or Metharbital $C_9H_{14}N_2O_3$	198.22	nd	...	150-1	...	...	w	$B24^2$, 281
1624	Barbituric acid, 5,5-dipropyl or Proponal $C_{10}H_{16}N_2O_3$	212.25	pl (w)	...	146 (166)	...	...	al, eth, bz, chl	$B24^2$, 286
1625	Barbituric acid, 5-ethyl-5-hexyl $C_{12}H_{20}N_2O_3$	240.30	nd (w)	...	112-3	...	...	al, eth, bz	$B24^2$, 288
1626	Barbituric acid, 5-ethyl-5(2-methallyl)-2-thio $C_{10}H_{14}N_2O_2S$	226.29	...	...	160-1	...	...	...	C35, 5599
1627	Barbituric acid, 5-ethyl-5-(1-methylbutyl) $C_{11}H_{18}N_2O_3$	226.28	...	...	130	...	...	al, eth	$B24^2$, 287
1628	Barbituric acid, 5-ethyl-5-(1-methylbutyl)-2-thio $C_{11}H_{18}N_2O_2S$	242.34	...	...	158-9	...	...	...	C29, 8237
1629	Barbituric acid, 5-ethyl-5-(3-methylbutyl) $C_{11}H_{18}N_2O_3$	226.28	...	...	157-9.5	...	...	...	...
1630	Barbituric acid, 5-ethyl-(3-methylbutyl) or Amytal $C_{11}H_{18}N_2O_3$	226.28	lf (w, dil al)	...	156-8	...	...	al, eth, bz, chl	$B24^2$, 287
1631	Barbituric acid, 5-ethyl-l-methyl-5-phenyl or Methobarbital, Prominal $C_{13}H_{14}N_2O_3$	246.27	wh cr (w)	...	176	...	...	al, chl	C28, 281
1632	Barbituric acid, 5-ethyl-5-pental $C_{11}H_{18}N_2O_3$	226.28	cr (dil al)	...	135-6	...	...	al, eth, chl	$B24^2$, 286
1633	Barbituric acid, 5-ethyl-5-(2-pentyl) or Nembutal, Pentobarbital $C_{11}H_{18}N_2O_3$	226.28	nd (w)	...	130	...	...	al, eth	$B24^2$, 287
1634	Barbituric acid, 5-ethyl-5-phenyl or Phenobarbital, Luminal $C_{12}H_{12}N_2O_3$	232.24	pl (w)	...	174	...	...	al, eth	$B24^2$, 297
1635	Barbituric acid, 5-ethyl-5-(1-piperidyl) or Eldoral $C_{11}H_{17}N_3O_3$	239.28	wh cr (dil al)	...	215	...	...	al, eth, ace	C36, 3160
1636	Barbituric acid, 5-ethyl-5-*iso*-propyl or Ipral, Probarbital $C_9H_{14}N_2O_3$	198.22	nd (w)	...	203	...	...	eth	$B24^2$, 284

PHYSICAL CONSTANTS OF ORGANIC COMPOUNDS (Continued)

No.	Name, Synonyms, and Formula	Mol. wt.	Color, crystalline form, specific rotation and λ_{max} (log ε)	b.p. °C	m.p. °C	Density	n_D	Solubility	Ref.
1637	Barbituric acid, 5-(2-furfurylidine) $C_9H_6N_2O_4$	206.16			315				$B27^{1}$, 607
1638	Barbituric acid, 5-(2-furfurylidine)-2-thio $C_9H_6N_2O_3S$	222.22	ye pl		> 280d				$B27^{1}$, 607
1639	Barbituric acid, 5-hydroxy or Dialuric acid $C_4H_4N_2O_4$	144.09	pr or pl nd (+ lw, w)		224 (anh)			al, ace, w, aa	$B25^{2}$, 61
1640	Barbituric acid, l-methyl $C_5H_6N_2O_3$	142.11			132-3			al, ace	$B24^{2}$, 270
1641	Barbituric acid, 5-methyl-5-phenyl or Rufinal $C_{11}H_{10}N_2O_3$	218.21	cr		220			al, eth	$B24^{2}$, 296
1642	Barbituric acid, 5-nitro $C_4H_3N_3O_5$	173.08	pr, lf (w + 3)		180-1			al, w	$B24^{2}$, 273
1643	Barbituric acid, 5-(2-phenylethyl) $C_{12}H_{12}N_2O_3$	232.24	cr		212-3			al	$B24^{2}$, 296
1644	Barbituric acid, 2-thio $C_4H_4N_2O_2S$	144.15	pl (w)		235d			al, w	$B24^{2}$, 275
1645	Bebeerine *d* or Chondrodendrin $C_{36}H_{38}N_2O_6$	594.71	cr (bz, eth, chl-MeOH), $[\alpha]^{25}_{D}$ + 297 (al)		221			al, eth, ace, chl	$B27^{2}$, 896
1646	Berbeerine *l* or Curine $C_{36}H_{38}N_2O_6$	594.71	pr, nd (chl-MeOH), $[\alpha]^{20}_{D}$ −328 p_4, $[\alpha]^{28}_{D}$ −298 (al)		221			ace, bz, Py	$B27^{2}$, 894
1647	Behenic acid or Docosanoic acid $CH_3(CH_2)_{20}CO_2H$	340.59	nd	306^{60}	80	0.8223^{90}	1.4270^{100}		$B2^{4}$, 1290
1648	Behenic acid, ethyl ester or Ethyl behenate $CH_3(CH_2)_{20}CO_2C_2H_5$	368.64	nd (al), or (ace)	$240\text{-}2^{10}$	50			al, eth	$B2^{4}$, 1292
1649	Behenic acid, methyl ester or Methylbehenate $CH_3(CH_2)_{20}CO_2CH_3$	354.62	nd (ace)	$224\text{-}5^{15}$	54		1.4339^{60}	al, eth	$B2^{4}$, 1291
1650	Benzal acetate, 2-acetoxy 2-$CH_3CO_2C_6H_4CH(O_2CCH_3)_2$	266.25	nd or pl (al), pr (al)		107			eth, bz, chl	$B8^{3}$, 148
1651	Benzal acetate, 3- acetoxy 3-$CH_3CO_3C_6H_4CH(O_2CCH_3)_2$	266.25	lf (w-al)		76			al, eth	B8, 60
1652	Benzal acetate, 4-acetoxy 4-$CH_3CO_2C_6H_4CH(O_2CCH_3)_2$	266.25	pr (eth or lig)		94			eth, lig, al	$B8^{1}$, 530
1653	Benzalacetophenone or *trans* Chalcone $C_6H_5COCH{=}CHC_6H_5$	208.26	pa ye lf, pr, nd (peth)	345-8d, 208^{25}	(i) 59 (ii) 57 (iii) 49	$1.0712^{62/4}$		eth, bz, chl	$B7^{3}$, 2380
1654	Benzaldehyde C_6H_5CHO	106.12		178, 62^{10}	-26(fr-56)	$1.0415^{10/4}$	1.5463^{20}	**al, eth**, ace, bz, lig	$B7^{3}$, 805
1655	Benzaldehyde, 2-acetamido 2-$CH_3CONHC_6H_4CHO$	163.18			70-1			w, eth, ace	$B14^{3}$, 51
1656	Benzaldehyde, 3-acetamido 3-$CH_3CONHC_6H_4CHO$	163.18	pl (bz)		84			al, eth, bz	B14, 29
1657	Benzaldehyde, 4-acetamido 4-$CH_3CONHC_6H_4CHO$	163.18	pr (w)		156			w, bz	$B14^{2}$, 25
1658	Benzaldehyde, 3-allyl-2-hydroxy 3-$(CH_2{=}CHCH_2)$-2-HOC_6H_3CHO	162.19		$245\text{-}6^{755}$	1.098^{15}				$B8^{1}$, 559
1659	Benzaldehyde, 2-amino or Anthranil aldehyde 2-$H_2NC_6H_4CHO$	121.14	silv lf		39-42			al, eth, bz, chl	$B14^{3}$, 47
1660	Benzaldehyde, 2-amino, oxime 2-$H_2NC_6H_4CH{=}NOH$	136.17	nd (bz)		135-6			al, eth, bz, aa	$B14^{3}$, 50
1661	Benzaldehyde, 3-amino 3-$H_2NC_6H_4CHO$	121.14	nd (AcOEt)		28-30			eth	$B14^{3}$, 53
1662	Benzaldehyde, 3-amino, oxime 3-$H_2NC_6H_4CH{=}NOH$	136.15	nd (bz)		88			al, eth	B14, 28
1663	Benzaldehyde, 4-amino 4-$H_2NC_6H_4CHO$	121.14	pl (w)		71-2			w, al, eth	$B14^{3}$, 57
1664	Benzaldehyde, 4-amino, oxime 4-$H_2NC_6H_4CH{=}NOH$	136.15	ye cr (w)		124			al, eth	$B14^{3}$, 58
1665	Benzaldehyde azine or Dibenzal hydrazine $C_6H_5CH{=}NN{=}CHC_6H_5$	208.26	ye pr (al)		93			al, eth, ace, bz	$B7^{3}$, 844
1666	Benzaldehyde, 2-bromo 2-BrC_6H_4CHO	185.02		230, 118^{12}	21-2		1.5925^{20}	al, bz	$B7^{3}$, 882

No.	Name, Synonyms, and Formula	Mol. wt.	Color, crystalline form, specific rotation and λ_{max} (log ε)	b.p. °C	m.p. °C	Density	n_D	Solubility	Ref.
1667	Benzaldehyde, 2-bromo, diacetate $2\text{-}BrC_6H_4CH(O_2CCH_3)_2$	287.11			84-6			al, eth	B7^{2}, 181
1668	Benzaldehyde, 2-bromo-4-hydroxy $2\text{-}Br\text{-}4\text{-}HOC_6H_3CHO$	201.02	pa ye nd (w)		159.5				B8^{2}, 74
1669	Benzaldehyde, 2-bromo-5-hydroxy $2\text{-}Br\text{-}5\text{-}HOC_6H_3CHO$	201.02	nd (w)		135			al, eth, ace, bz	B8^{2}, 57
1670	Benzaldehyde, 3-bromo $3\text{-}BrC_6H_4CHO$	185.02		233-6			1.5915^{20}	al, eth	B7^{3}, 882
1671	Benzaldehyde, 3-bromo-2-hydroxy or 3-Bromosalicylaldehyde $3\text{-}Br\text{-}2\text{-}HOC_6H_3CHO$	201.02	nd (dil al)		49			al, eth, ace, bz	B8, 54
1672	Benzaldehyde, 3-bromo-4-hydroxy $3\text{-}Br\text{-}4\text{-}HOC_6H_3CHO$	201.02	lf (w)		124			al, eth, ace, bz	B8^{3}, 290
1673	Benzaldehyde, 4-bromo $4\text{-}BrC_6H_4CHO$	185.02	lf (dil al)	66-8^{2}	67			al, bz	B7^{3}, 883
1674	Benzaldehyde, 4-bromo, diacetate $4\text{-}BrC_6H_4CH(O_2CCH_3)_2$	287.11	ye		94-5			eth, al	B7^{2}, 182
1675	Benzaldehyde, 4-bromo-2-hydroxy or 4-Bromosalicylaldehyde $4\text{-}Br\text{-}2\text{-}HOC_6H_3CHO$	201.02	nd (dil al)		52			al, eth, ace, bz	B8, 54
1676	Benzaldehyde, 4-bromo-3-hydroxy $4\text{-}Br\text{-}3\text{-}HOC_6H_3CHO$	201.02			131.5			ace, bz, chl	B8^{2}, 56
1677	Benzaldehyde, 5-bromo-2-hydroxy or 5-Bromosalicylaldehyde $5\text{-}Br\text{-}2\text{-}HOC_6H_3CHO$	201.02	nd (al), lf (eth)		105-6			al, eth	B8^{3}, 269
1678	Benzaldehyde, 5-bromo-4-hydroxy-3-methoxy or 5-Bromovanillan $5\text{-}Br\text{-}4\text{-}HO\text{-}3\text{-}CH_3OC_6H_2CHO$	231.05	pl (aa), nd, pl (al)		164-6			al	B8^{3}, 2050
1679	Benzaldehyde, 2-chloro $2\text{-}ClC_6H_4CHO$	140.57	nd	211.9, 84.3^{10}	12.4	1.2483$^{20/4}$	1.5662^{20}	al, eth, ace, bz	B7^{3}, 864
1680	Benzaldehyde, 2-chloro-3-hydroxy $2\text{-}Cl\text{-}3\text{-}HOC_6H_3CHO$	156.57	cr (ag aa)		139.5			al, aa	B8^{2}, 54
1681	Benzaldehyde, 2-chloro-4-hydroxy $2\text{-}Cl\text{-}4\text{-}HOC_6H_3CHO$	156.57	nd (w or aa)		147-8			al, eth, aa	B8, 81
1682	Benzaldehyde, 2-chloro-5-hydroxy $2\text{-}Cl\text{-}5\text{-}HOC_6H_3CHO$	156.57	nd (w or aa)		110-1			al, aa	B8^{1}, 526
1683	Benzaldehyde, 2-chloro, phenyl hydrazone $2\text{-}ClC_6H_4CH{=}NNHC_6H_5$	230.70			84				B15^{3}, 86
1684	Benzaldehyde, 3-chloro $3\text{-}ClC_6H_4CHO$	140.57	pr	213-4, 55^{1}	17-8	1.2410$^{20/4}$	1.5650^{20}	al, eth, ace, bz	B7^{3}, 869
1685	Benzaldehyde, 3-chloro-2-hydroxy or 3-Chlorosalicylaldehyde $3\text{-}Cl\text{-}2\text{-}HOC_6H_3CHO$	156.57	nd (MeOH)		55			al, eth, ace, bz	B8^{3}, 179
1686	Benzaldehyde, 3-chloro-4-hydroxy $3\text{-}Cl\text{-}4\text{-}HOC_6H_3CHO$	156.57	nd (w)	149-50^{14}	139			al, eth	B8^{3}, 286
1687	Benzaldehyde, 4-chloro $4\text{-}ClC_6H_4CHO$	140.57	pl	213-4, 72-5^{3}	47.5	1.196$^{61/4}$	1.5552^{61}	w, al, eth, ace, bz	B7^{3}, 872
1688	Benzaldehyde, 4-chloro-2-hydroxy or 4-chlorosalicylaldehyde $4\text{-}Cl\text{-}2\text{-}HOC_6H_3CHO$	156.57	nd (al or ag aa)		52.5			w, eth, ace, bz	B8^{2}, 44
1689	Benzaldehyde, 4-chloro-3-hydroxy $4\text{-}Cl\text{-}3\text{-}HOC_6H_3CHO$	156.57	nd (al)		121			w, al, eth, ace, bz	B8^{2}, 55
1690	Benzaldehyde, 4-chloro-3-nitro $4\text{-}Cl\text{-}3\text{-}NO_2C_6H_3CHO$	185.57			64.5			chl	B7^{3}, 919
1691	Benzaldehyde, 5-chloro-2-hydroxy or 5-Chlorosalicylaldehyde $5\text{-}Cl\text{-}2\text{-}HOC_6H_3CHO$	156.57	pl (al)	105^{12}	99-100			al, eth	B8^{2}, 45
1692	Benzaldehyde, 5-chloro-2-hydroxy, oxime $5\text{-}Cl\text{-}2\text{-}HOC_6H_3CH{=}NOH$	171.58	nd (w)		128			al	B8, 53
1693	Benzaldehyde, 5-chloro-4-hydroxy-3-methoxy or 5-Chlorovanillan $5\text{-}Cl\text{-}4\text{-}HO\text{-}3\text{-}CH_3OC_6H_2CHO$	186.60	tetr		165			al, aa	B8^{3}, 2063
1694	Benzaldehyde diacetate or Benzylidene diacetate $C_6H_5CH(O_2CCH_3)_2$	208.22	pl (eth)	220, 154^{20}	46	1.11^{20}		al, eth, bz	B7^{3}, 827
1695	Benzaldehyde, 3,5-dibromo-2-hydroxy or 3,5-Dibromosalicylaldehyde $3,5\text{-}Br_2\text{-}2\text{-}HOC_6H_2CHO$	279.92			82.0-3.5				B8^{3}, 269
1696	Benzaldehyde, 2,3-dichloro $2,3\text{-}Cl_2\text{-}C_6H_3CHO$	175.01	cr (dil al)		65-7			al, eth	B7^{3}, 878

No.	Name, Synonyms, and Formula	Mol. wt.	Color, crystalline form, specific rotation and λ_{max} (log ε)	b.p. °C	m.p. °C	Density	n_D	Solubility	Ref.
1697	Benzaldehyde, 2,4-dichloro 2,4-$Cl_2C_6H_3CHO$	175.01	pr		72			al, eth, bz	$B7^{3}$, 878
1698	Benzaldehyde, 2,4-dichloro-3-hydroxy 2,4-Cl_2-3-HOC_6H_2CHO	191.01	cr (aa)		141			lig	$B8^{2}$, 55
1699	Benzaldehyde, 2,5-dichloro 2,5-$Cl_2C_6H_3CHO$	175.01	nd (al)	231-3	58			al, eth, bz, chl	$B7^{3}$, 878
1700	Benzaldehyde, 2,6-dichloro 2,6-Cl_2-C_6H_3CHO	175.01	nd (lig)		71			al, eth, lig	$B7^{3}$, 878
1701	Benzaldehyde, 2,6-dichloro-3-hydroxy 2,6-Cl_2-3-HOC_6H_2CHO	191.01	cr (w)		142			aa	$B8^{3}$, 204
1702	Benzaldehyde, 3,4-dichloro 3,4-$Cl_2C_6H_3CHO$	175.01		247-8	44			al, eth	$B7^{3}$, 880
1703	Benzaldehyde, 3,5-dichloro 3,5-$Cl_2C_6H_3CHO$	175.01	nd or lf (dil al)	$235-40^{748}$	65			al, eth, ace, bz	$B7^{3}$, 880
1704	Benzaldehyde, 3,5-dichloro-2-hydroxy or 3,5-dichlorsalicylaldehyde 3,5-Cl_2-2-HOC_6H_2CHO	191.01	ye rh (aa)		95				$B8^{3}$, 183
1705	Benzaldehyde, 3,5-dichloro-2-hydroxy, oxime 3,5-Cl_2-2-$HOC_6H_2CH{=}NOH$	206.03	nd (dil al)		195-6			al, eth, bz	B8, 54
1706	Benzaldehyde, 3,5-dichloro-4-hydroxy 3,5-Cl_2-4-HOC_6H_2CHO	191.01	nd (chl, dil al)		158-9			al, eth	B8, 81
1707	Benzaldehyde, 4,6-dichloro-3-hydroxy 4,6-Cl_2-3-HOC_6H_2CHO	191.01	nd		130			w, al, eth, bz	$B8^{1}$, 526
1708	Benzaldehyde, 2,3-diethoxy 2,3-$(C_2H_5O)_2C_6H_3CHO$	194.23		169^{37}				al, eth	$B8^{3}$, 1982
1709	Benzaldehyde, 3,4-diethoxy 3,4-$(C_2H_5O)_2C_6H_3CHO$	194.23		278-30				al	$B8^{2}$, 2025
1710	Benzaldehyde, 4-diethylamino 4-$(C_2H_5)_2NC_6H_4CHO$	177.23	ye nd (w)	174^{7}	41			w, al, eth, bz	$B14^{3}$, 71
1711	Benzaldehyde, 2,4-dihydroxy or β-Resorcyl aldehyde 2,4-$(HO)_2C_6H_3CHO$	138.12	nd (eth-lig)	$220-8^{22}$	201-2			w, al, eth, chl	$B8^{3}$, 1989
1712	Benzaldehyde, 2,4-dihydroxy, oxime 2,4-$(HO)_2C_6H_3CH{=}NOH$	153.14			192			w, al, eth	$B8^{2}$, 274
1713	Benzaldehyde, 2,5-dihydroxy or Gentisic aldehyde 2,5-$(HO)_2C_6H_3CHO$	138.12	ye nd (bz)		99			w, al, eth, chl	$B8^{3}$, 2003
1714	Benzaldehyde, 2,6-dihydroxy-4-methyl or Atranol 2,6-$(HO)_2$-4-$CH_3C_6H_2CHO$	152.16	ye nd (w)		124			w, al, eth, chl	$B8^{3}$, 2006
1715	Benzaldehyde, 3,4-dihydroxy or Protocatechualdehyde 3,4-$(HO)_2C_6H_3CHO$	138.12	lf (w, to)		153-4			w, al, eth	$B8^{3}$, 2009
1716	Benzaldehyde, 3,4-dihydroxy, oxime 3,4-$(HO)_2C_6H_3CH{=}NOH$	153.14	nd (w)		157d			w, al	$B8^{2}$, 285
1717	Benzaldehyde, 3,5-diido-4-hydroxy 3,5-I_2-4-HOC_6H_2CHO	373.91	nd (al)		206.5			al, ace, aa	$B8^{3}$, 251
1718	Benzaldehyde, 2,4-dimethoxy 2,4-$(CH_3O)_2C_6H_3CHO$	166.18	nd (al or lig)	165^{10}	72			al, eth, bz, lig	$B8^{3}$, 1992
1719	Benzaldehyde, 2,5-dimethoxy 2,5-$(CH_3O)_2C_6H_3CHO$	166.18		146^{10}	52			al, eth	$B8^{3}$, 2004
1720	Benzaldehyde, 2,4-dimethoxy-6-hydroxy 2,4-$(CH_3O)_2$-6-HOC_6H_2CHO	182.18	nd, pl (dil al)	$190-5^{25}$, 165^{10}	71			al, eth, bz, aa	$B8^{3}$, 3364
1721	Benzaldehyde, 2,6-dimethoxy-4-hydroxy 2,6-$(CH_3O)_2$-4-HOC_6H_2CHO	182.18	nd or pl (MeOH), pr (bz)		70-1				$B8^{3}$, 3365
1722	Benzaldehyde, 3,4-dimethoxy or Veratraldehyde 3,4-$(CH_3O)_2C_6H_3CHO$	166.18	nd (eth, lig, to)	258, $172-5^{18}$	44 (58)			al, eth	$B8^{3}$, 2020
1723	Benzaldehyde, 3,4-dimethoxy-5-hydroxy 3,4-$(CH_3O)_2$-5-HOC_6H_2CHO	182.18		$177-80^{12}$	62-3			ual, bz, aa, lig	$B8^{2}$, 437
1724	Benzaldehyde, 3,5-dimethoxy 3,5-$(CH_3O)_2C_6H_3CHO$	166.18		151^{16}	45.5			al, bz	$B8^{3}$, 2073
1725	Benzaldehyde, 3,5-dimethoxy-4-hydroxy 3,5-$(CH_3O)_2$-4-HOC_6H_2CHO	182.18	br nd (lig)	$192-3^{14}$	113			al, eth, bz, aa	$B8^{3}$, 3368
1726	Benzaldehyde, 2,4-dimethyl 2,4-$(CH_3)_2C_6H_3CHO$	134.18		218, 99^{10}	-9			al, eth, ace, bz	$B7^{3}$, 1072
1727	Benzaldehyde, 2,5-dimethyl 2,5-$(CH_3)_2C_6H_3CHO$	134.18		220, 100^{10}				w, al, eth, ace, bz	$B7^{3}$, 1072
1728	Benzaldehyde, 3,4-dimethyl 3,4-$(CH_3)_2C_6H_3CHO$	134.18		223-5				al, eth, ace	$B7^{3}$, 1073
1729	Benzaldehyde, 3,5-dimethyl 3,5-$(CH_3)_2C_6H_3CHO$	134.18		220-2	9			al, eth, ace, bz	$B7^{3}$, 1073

No.	Name, Synonyms, and Formula	Mol. wt.	Color, crystalline form, specific rotation and λ_{max} (log ε)	b.p. °C	m.p. °C	Density	n_D	Solubility	Ref.
1730	Benzaldehyde, 4-dimethylamino 4-$(CH_3)_2NC_6H_4CHO$	149.19	lf (w)	$176\text{-}7^{17}$	74	....	....	al, eth, ace, bz	$B14^{3}$, 58
1731	Benzaldehyde, 2,4-dinitro 2,4-$(NO_2)_2C_6H_3CHO$	196.12	pa ye pr (al), pl (bz)	$190\text{-}210^{10\text{-}20}$	72	....	....	al, eth, bz	$B7^{3}$, 923
1732	Benzaldehyde, 2,6-dinitro 2,6-$(NO_2)_2C_6H_3CHO$	196.12	lf (dil al)	....	123	....	....	w, al, eth, bz, aa	$B7^{2}$, 206
1733	Benzaldehyde, 2-ethoxy 2-$C_2H_5OC_6H_4CHO$	150.18	....	247-9, $143\text{-}7^{25}$	20-2	....	....	**al, eth**	$B8^{3}$, 144
1734	Benzaldehyde, 3-ethoxy 3-$C_2H_5OC_6H_4CHO$	150.18	....	245.5	....	$1.0768^{20/4}$	1.5408^{20}	al, eth, bz	B8, 60
1735	Benzaldehyde, 3-ethoxy-2-hydroxy or 3-Ethoxysalicylaldehyde 3-C_2H_5O-2-HOC_6H_3CHO	166.18	....	$263\text{-}4^{740}$	64-5	....	....	....	$B8^{3}$, 1981
1736	Benzaldehyde, 3-ethoxy-4-hydroxy or Ethylvanillan 3-C_2H_5O-4-HOC_6H_3CHO	166.18	....	....	77-8	....	....	al, eth, bz, chl	$B8^{3}$, 2022
1737	Benzaldehyde, 4-ethoxy 4-$C_2H_5OC_6H_4CHO$	150.18	....	249, 140^{20}	13-4	$1.08^{21/21}$	....	al, eth, bz	$B8^{3}$, 223
1738	Benzaldehyde, 4-ethoxy-2-hydroxy or 4-Ethoxysalicylaldehyde 4-C_2H_5O-2-HOC_6H_3CHO	166.18	nd (dil al)	....	35	....	....	al	$B8^{3}$, 1992
1739	Benzaldehyde, 4-ethoxy-3-methoxy 4-C_2H_5O-3-$CH_3OC_6H_3CHO$	180.21	mcl pr	sub	64-5 (73-4)	....	....	**al, eth**, bz, aa	$B8^{3}$, 2024
1740	Benzaldehyde, 5-ethoxy-2-hydroxy or 5-Hydroxysalicylaldehyde 5-C_2H_5O-2-HOC_6H_3CHO	166.18	ye pr	230	51.5	....	....	al, eth, chl	B8, 245
1741	Benzaldehyde hydrazone $C_6H_5CH{=}NNH_2$	120.15	lf	140^{14}	16	....	....	al	$B7^{3}$, 863
1742	Benzaldehyde, 2-hydroxy or Salicylaldehyde 2-HOC_6H_4CHO	122.12	....	197, 93^{25}	-7	$1.1674^{20/4}$	1.5740^{20}	**al, eth**, ace, bz	$B8^{3}$, 135
1743	Benzaldehyde-2-hydroxy-3-methoxy or o-Vanillan 2-HO-3-$CH_3OC_6H_3CHO$	152.15	lt ye-lt gr nd (w, lig)	256-6, 128^{10}	44-5	....	....	al, eth, lig	$B8^{3}$, 1979
1744	Benzaldehyde, 2-hydroxy-4-methoxy or 4-Methoxysalicylaldehyde 2-HO-4-$CH_3OC_6H_3CHO$	152.15	nd (w), cr (al)	....	40-2	....	....	al, eth, bz, lig	$B8^{3}$, 1991
1745	Benzaldehyde, 2-hydroxy-5-methoxy or 5-Methoxysalicylaldehyde 2-HO-5-$CH_3OC_6H_3CHO$	152.15	ye (w)	247-8	4	....	....	al, eth	$B8^{3}$, 2004
1746	Benzaldehyde, 2-hydroxy-3-nitro or 3-Nitrosalicylaldehyde 2-HO-3-NO_2 C_6H_3CHO	167.12	nd (aa)	....	109-10	....	....	al, bz	$B8^{3}$, 190
1747	Benzaldehyde, 2-hydroxy-5-nitro or 5-Nitrosalicylaldehyde 5-NO_2-2--HO-C_6H_3CHO	167.12	(dil aa)	....	128	....	....	al, eth	$B8^{3}$, 192
1748	Benzaldehyde, 2-hydroxy-5-nitro, oxime 2-HO-5-$NO_2C_6H_3CH{=}NOH$	182.14	....	....	225	....	....	....	$B8^{3}$, 193
1749	Benzaldehyde, 3-hydroxy 3-HOC_6H_4CHO	122.12	nd (w)	240, 161^{20}	108	....	....	al, eth, ace, bz	$B8^{3}$, 197
1750	Benzaldehyde, 3,hydroxy, azine 3-$HOC_6H_4CH{=}NN{=}CHC_6H_4OH$-(3)	240.26	ye pl (al)	....	162	....	....	....	$B8^{3}$, 203
1751	Benzaldehyde, 3-hydroxy-4-methoxy or Isovanillan 3-HO-4-$CH_3OC_6H_3CHO$	152.15	pl (w)	179^{15}	116-7	1.198	....	al, eth, bz, chl, aa	$B8^{3}$, 2019
1752	Benzaldehyde, 3-hydroxy-5-methoxy 3-HO-5-$CH_3OC_6H_3CHO$	152.15	....	110-20, sub	130-1	....	....	al, bz, aa	$B8^{3}$, 2073
1753	Benzaldehyde, 3-hydroxy-2-nitro 3-HO-2-$NO_2C_6H_3CHO$	167.12	nd or pl (bz, lig), d	....	152	....	....	al, bz	$B8^{3}$, 210
1754	Benzaldehyde, 3-hydroxy-2-nitro, oxime 3-HO-2-$NO_2C_6H_3CH{=}NOH$	182.14	pa ye nd	....	172.5	....	....	....	$B8^{2}$, 58
1755	Benzaldehyde, 3-hydroxy-4-nitro 4-NO_2-3-HOC_6H_3CHO	167.12	ye lf	....	128	....	....	al, eth, bz	$B8^{3}$, 211
1756	Benzaldehyde, 3-hydroxy-4-nitro, oxime 4-NO_2-3-$HOC_6H_3CH{=}NOH$	182.14	ye nd (chl)	....	164	....	....	al, chl	$B8^{2}$, 58
1757	Benzaldehyde, 3-hydroxy, oxime 3-$HOC_6H_4CH{=}NOH$	137.14	(bz)	....	90	....	....	w, al, eth	B8, 61
1758	Benzaldehyde, 4-hydroxy 4-HOC_6H_4CHO	122.12	nd (w)	....	117	$1.129^{130/4}$	1.5705^{130}	al, eth, bz	$B8^{3}$, 215

PHYSICAL CONSTANTS OF ORGANIC COMPOUNDS (Continued)

No.	Name, Synonyms, and Formula	Mol. wt.	Color, crystalline form, specific rotation and λ_{max} (log ε)	b.p. °C	m.p. °C	Density	n_D	Solubility	Ref.
1759	Benzaldehyde, 4-hydroxy, azine 4-$HOC_6H_4CH=NN=CHC_6H_4OH$(-4)	240.26	ye (al)	...	239-40, 268 (d)	...	...	al, bz	B8^{1}, 531
1760	Benzaldehyde, 4-hydroxy-2-iodo-3-methoxy or 2-Iodo vanillan 4-HO-3-CH_3O-2IC_6H_2CHO	278.05	...	...	155-6	...	...	al	B8^{3}, 2058
1761	Benzaldehyde, 4-hydroxy-5-iodo-3-methoxy or 5-Iodo vanillan 5-I-4-HO-3-$CH_3OC_6H_2CHO$	278.05	pa ye	...	180	...	...	...	B8^{3}, 2059
1762	Benzaldehyde, 4-hydroxy-2-methoxy 4-HO-2-$CH_3OC_6H_3CHO$	152.15	lf (bz), nd (w)	...	153	...	...	al, eth, chl	B8^{3}, 1990
1763	Benzaldehyde, 4-hydroxy-3-methoxy or Vanillan 4-HO-3-CH_3O-C_6H_3CHO	152.15	(i) nd (w,lig) (ii) tetr (w, lig)	285, 170^{15}	(i) 77—9 (ii) 81—2	1.056	...	al, eth, ace, bz, chl	B8^{3}, 2011
1764	Benzaldehyde, 4-hydroxy-2-nitro 4-HO-2-$NO_2C_6H_3CHO$	167.12	ye nd	...	67	...	...	al, eth, bz	B8^{3}, 253
1765	Benzaldehyde, 4-hydroxy-3-nitro 4-HO-3-$NO_2C_6H_3CHO$	167.12	dk ye nd (al, w)	...	144.3	...	...	al, w	B8^{3}, 253
1766	Benzaldehyde, 4-hydroxy-3-nitro, oxime 4-HO-3-$NO_2C_6H_3CH=NOH$	182.14	pr or nd (al-chl)	...	189	...	...	al, eth, aa	B8^{3}, 254
1767	Benzaldehyde imine, *N*-ethyl or *N*-Benzalethylamine $C_6H_5CH=NC_2H_5$	133.19	...	195^{745}	...	$0.937^{20/4}$	1.5378^{15}	al, eth	B7^{3}, 831
1768	Benzaldehyde imine,2-hydroxy-*N*-phenyl 2-$HOC_6H_4CH=NC_6H_5$	197.24	...	...	47-9	1.087	...	al	B12^{3}, 369
1769	Benzaldehyde imine, *N*-methyl $C_6H_5CH=NCH_3$	119.17	...	185^{760}, 90-1^{30}	...	$0.9672^{14/2}$	1.5526^{20}	al, eth, ace, chl	B7^{3}, 831
1770	Benzaldehyde imine, *N*-phenyl $C_6H_5CH=NC_6H_5$	181.24	...	310	54	$1.038^{55/4}$	1.600^{99}	al, eth, chl	B12^{3}, 319
1771	Benzaldehyde imine, *N*-(2-tolyl) $C_6H_5CH=NC_6H_4CH_3$-(2)	195.26	...	314, 176^{15}	...	$1.041^{20/4}$	1.6310^{25}	ace	B12^{3}, 1848
1772	Benzaldehyde imine, *N*-(3-tolyl) $C_6H_5CH=N(C_6H_4CH_3$-3)	195.26	...	315^{775}	30-2	...	1.6353^{2}	ace	B12^{3}, 1958
1773	Benzaldehyde imine, *N*-(4-tolyl) or *N*-Benzal-p-toluidine $C_6H_5CH=N(C_6H_4CH_3$-4)	195.26	ye cr	318^{755}, 178^{11}	35	...	...	ace	B12^{3}, 2096
1774	Benzaldehyde, 2-methoxy or *o*-Anisaldehyde 2-$CH_3OC_6H_4CHO$	136.15	pr	243-4, 124-5^{18}	37-8	$1.1326^{20/4}$	1.5600^{20}	al, eth, ace, bz, chl	B8^{3}, 143
1775	Benzaldehyde, 2-methoxy-3-nitro or 3-Nitro-*o*-anisaldehyde 2-CH_3O-3-$NO_2C_6H_3CHO$	181.15	ye pr (dil al), nd (bz)	...	89-90	...	...	al, eth	B8^{3}, 190
1776	Benzaldehyde, 3-methoxy or *m*-Anisaldehyde 3-$CH_3OC_6H_4CHO$	136.15	...	230, 62^{1}	...	$1.1187^{20/4}$	1.5330^{20}	al, eth, ace, bz	B8^{3}, 199
1777	Benzaldehyde, 4-methoxy or *p*-Anisaldehyde *p*-$CH_3OC_6H_4CHO$	136.15	...	249.5, 83^{2}	0	$1.1191^{15/4}$	1.5730^{20}	al, eth, ace, bz	B8^{3}, 218
1778	Benzaldehyde, 4-methoxy-3-methyl or 4-methoxy-*m*-tolualdehyde 4-CH_3O-3-$CH_3C_6H_3CHO$	150.18	...	80-5^{1}	...	...	1.5670^{20}	...	B8^{3}, 367
1779	Benzaldehyde, 4-methoxy, oxime 4-$CH_3OC_6H_4CH=NOH$	151.16	nd (bz)	...	64—5 (anti), 133 (syn)	...	...	...	B8^{3}, 235
1780	Benzaldehyde, 2-methyl or *o*-Tolualdehyde 2-$CH_3C_6H_4CHO$	120.15	...	200, 94^{10}	...	$1.0386^{19/4}$	1.5481^{20}	al eth, ace, bz, chl	B7^{3}, 1011
1781	Benzaldehyde, 3-methyl or *m*-Tolualdehyde 3-$CH_3C_6H_4CHO$	120.15	...	199, 93-4^{17}	...	$1.0189^{21/4}$	1.5413^{21}	**al, eth,** ace, bz, chl	B7^{3}, 1013
1782	Benzaldehyde, 4-methyl or *p*-Tolualdehyde 4-$CH_3C_6H_4CHO$	120.15	...	204-5, 106^{10}	...	$1.0194^{17/4}$	1.5454^{20}	**al, eth, ace,** chl	B7^{3}, 1016
1783	Benzaldehyde, 2-nitro 2-$NO_2C_6H_4CHO$	151.12	ye nd (w)	153^{23}	43.5-4	$1.2844^{20/4}$	...	al, eth, ace, bz	B7^{3}, 889
1784	Benzaldehyde, 2-nitro, diacetate 2-$NO_2C_6H_4CH(O_2CCH_3)_2$	253.21	pr (lig)	...	90	...	...	eth, ace, bz	B7^{3}, 892
1785	Benzaldehyde, 2-nitro, dimethyl acetal 2-$NO_2C_6H_4CH(OCH_3)_2$	197.19	gr-ye	274-6, 138-9^{11}	...	...	...	bz	B7^{3}, 891
1786	Benzaldehyde, 3-nitro 3-$NO_2C_6H_4CHO$	151.12	lt ye nd (w)	164^{23}	58	$1.2792^{20/4}$	...	al, eth, ace, bz	B7^{3}, 897
1787	Benzaldehyde, 3-nitro, diacetate 3-$NO_2C_6H_4CH(O_2CCH_3)_2$	253.21	pr or nd (al)	...	72	1.393	...	al, eth, ace, bz	B7^{3}, 900

No.	Name, Synonyms, and Formula	Mol. wt.	Color, crystalline form, specific rotation and λ_{max} (log ε)	b.p. °C	m.p. °C	Density	n_D	Solubility	Ref.
1788	Benzaldehyde, 3-nitro, dimethyl acetal 3-$NO_2C_6H_4CH(OCH_3)_2$	197.19		162-4^{19}		1.209^{15}		bz	$B7^3$, 899
1789	Benzaldehyde, 4-nitro 4-$NO_2C_6H_4CHO$	151.12	lf, pr (w)	sub	106	1.496		al, bz, aa	$B7^3$, 907
1790	Benzaldehyde, 4-nitro, diacetate 4-$NO_2C_6H_4CH(O_2CCH_3)_2$	253.21	pr (al)		127			al, eth, ace, bz	$B7^3$, 909
1791	Benzaldehyde, 4-nitro, dimethylacetal 4-$NO_2C_6H_4CH(OCH_3)_2$	197.19		294-6^{774}	23-5			bz	$B7^3$, 909
1792	Benzaldehydeoxime *anti* or *anti*-Benzaldoxime $C_6H_5CH{=}NOH$	121.14	nd (eth)		130	1.145$^{20/4}$		al, eth, w	$B7^3$, 840
1793	Benzaldehydeoxime *syn* or *syn*-Benzaldoxime $C_6H_5CH{=}NOH$	121.14	pr	200,118-9^{10}	36-7	1.1111$^{20/4}$	1.5908^{20}	al, eth, bz	$B7^3$, 840
1794	Benzaldehyde, pentachloro C_6Cl_5CHO	278.35	nd (bz, al)		202.5			eth, ace, bz	$B7^3$, 881
1795	Benzaldehyde, phenylhydrazone $C_6H_5CH = NNHC_6H_5$	196.26	nd (lig), pr		156			ace, bz	$B15^2$, 57
1796	Benzaldehyde, 4-isopropyl or Cumaldehyde 4-$(CH_3)_2CHC_6H_4CHO$	148.21		235-6, 103-4^{10}		0.9755$^{20/4}$	1.5301^{20}	al, eth	$B7^3$, 1095
1797	Benzaldehyde, semicarbazone $C_6H_5CH{=}NNHCONH_2$	163.18			224				$B7^3$, 854
1798	Benzaldehyde, 2,3,4,5-tetrachloro 2,3,4,5Cl_4C_6HCHO	243.90			106				$B7^2$, 181
1799	Benzaldehyde, 2,3,4-trichloro 2,3,4-$Cl_3C_6H_2CHO$	209.46	nd (al)		91			al	B7, 238
1800	Benzaldehyde, 2,3,5- trichloro 2,3,5-$Cl_3C_6H_2CHO$	209.46	nd (al)		56			al, eth, ace, bz	$B7^2$, 180
1801	Benzaldehyde, 2,3,6-trichloro 2,3,6-Cl_3-C_6H_2CHO	209.46	nd (lig)		86-7			eth, ace, bz	$B7^3$, 881
1802	Benzaldehyde, 2,4,5-trichloro 2,4,5-$Cl_3C_6H_2CHO$	209.46	nd (al)		112-3			al, eth, ace, bz	$B7^2$, 180
1803	Benzaldehyde, 2,4,6, trichloro 2,4,6-Cl_3-C_6H_2CHO	209.46			58-9			eth, lig	$B7^3$, 881
1804	Benzaldehyde, 3,4,5-trichloro 3,4,5-Cl_3-C_6H_2CHO	209.46	nd (al)		90-1			al, eth, ace, bz, peth	$B7^2$, 180
1805	Benzaldehyde, 2,4,5-trimethoxy 2,4,5-$(CH_3O)_3C_6H_2CHO$	196.20		140^4-sub	114			w, eth, chl, lig	$B8^3$, 3361
1806	Benzaldehyde, 3,4,5-trimethoxy 3,4,5-$(CH_3O)_3C_6H_2CHO$	196.20			71-4			chl	$B8^3$, 3368
1807	Benzaldehyde, 2,4,6-trimethyl or Mesityl aldehyde 2,4,6-$(CH_3)_3C_6H_2CHO$	148.20		237-40, 192^{50}	14			al, eth, ace, bz	$B7^3$, 1111
1808	Benzaldehyde, 2,4,6-trinitro 2,4,6-$(NO_2)_3C_6H_2CHO$	241.12	dk gr pl (bz)		119			al, eth, ace, bz,aa	$B7^3$, 926
1809	Benzamide or Benzoic acid amide $C_6H_5CONH_2$	121.14	mcl pr or pl (w)	290	132-3	1.0792$^{130/4}$, 1.341^4		al, bz, w	$B9^3$, 1064
1810	Benzamide, 2-amino or Anthranilamide 2-$H_2NC_6H_4CONH_2$	136.15	lf (chl or w)	300	110-1.5			w, al	$B14^3$, 889
1811	Benzamide, 3-amino 3-$H_2NC_6H_4CONH_2$	136.15	ye mcl nd (+ lw) nd		79—80 (hyd), 113—4 (anh)			w, al, eth	$B14^3$, 998
1812	Benzamide, 4-amino 4-$H_2NC_6H_4CONH_2$	136.15	ye cr (+ ¼w)		183			al, eth	$B14^3$, 1061
1813	Benzamide, 2-benzoyl 2-$C_6H_5COC_6H_4COCNH_2$	225.25	nd (to)		165 (cor)			al, bz, to	$B10^3$, 3295
1814	Benzamide, 2-bromo 2-$BrC_6H_4COCNH_2$	200.03	nd (w)	sub	160-1			al	$B9^3$, 1387
1815	Benzamide, 3-bromo 3-$BrC_6H_4CONH_2$	200.03	lf (w or al)	sub	155.3			al	$B9^3$, 1399
1816	Benzamide, 4-bromo 4-$BrC_6H_4CONH_2$	200.03	nd or pl (w)		192			al, w, aa	$B9^3$, 1418
1817	Benzamide, 2-chloro 2-$ClC_6H_4CONH_2$	155.58	rh nd (w)		142.4			al, eth	$B9^3$, 1338
1818	Benzamide, 3-chloro 3-$ClC_6H_4CONH_2$	155.58	nd		135-7			al, eth	$B9^3$, 1349

No.	Name, Synonyms, and Formula	Mol. wt.	Color, crystalline form, specific rotation and λ_{max} (log ε)	b.p. °C	m.p. °C	Density	n_D	Solubility	Ref.
1819	Benzamide, 4-chloro $4\text{-}ClC_6H_4CONH_2$	155.58			179-80			al, eth, w	B9^{3}, 1363
1820	Benzamide, 2,5-dimethoxy $2,5(CH_3O)_2C_6H_3CONH_2$	181.19	lf (hz-peth), nd (w)		141-2			al, eth, ace, bz, chl	B10^{2}, 258
1821	Benzamide, 3,4-dimethoxy or Veratramide $3,4\text{-}(CH_3O)_2C_6H_3CONH_2$	181.19	cr (w)		164			w, eth, bz	B10^{2}, 261
1822	Benzamide, 3,5-dimethoxy or 3,5-Dimethoxy benzamide $3,5\text{-}(CH_3O)_2C_6H_3CONH_2$	181.19	nd (bz)		148—9			w, al, eth, bz, lig	B10^{3}, 1449
1823	Benzamide, *N,N*-diphenyl or *N*-Benzoyl diphenylamine $C_6H_5CON(C_6H_5)_2$	273.33	rh pr (al), nd		180				B12^{3}, 511
1824	Benzamide, 2-ethoxy $2\text{-}C_2H_5OC_6H_4CONH_2$	165.19	nd (w, al)		132-4			al, eth	B10^{2}, 58
1825	Benzamide, 3-ethoxy $3\text{-}C_2H_5OC_6H_4CONH_2$	165.19	nd (w)		139			al, eth, ace, chl	B10^{1}, 64
1826	Benzamide, 4-ethoxy $4\text{-}C_2H_5OC_6H_4NH_2$	165.19	pr (dil al)		206			al	B10^{3}, 342
1827	Benzamide, 3-hydroxy $3\text{-}HOC_6H_4CONH_2$	137.14	pl (w)		170.5			al, eth	B10^{3}, 254
1828	Benzamide, 3-hydroxy $3\text{-}HOC_6H_4CONH_2$	137.14	pl (w)		170.5			al, eth	B10^{2}, 82
1829	Benzamide, 3-hydroxy-*N*-phenyl $3\text{-}HOC_6H_4CONHC_6H_5$	213.24	nd or pl (w, dil al)		156			al	B12^{1}, 269
1830	Benzamide, 4-hydroxy $4\text{-}HOC_6H_4CONH_2$	137.14	nd (w + l)		162 (hyd)			al, eth, chl	B10^{3}, 340
1831	Benzamide, 4-hydroxy-*N*-phenyl $4\text{-}HOC_6H_4CONHC_6H_5$	213.24	pl or nd (w)		201-2			al	B12^{3}, 947
1832	Benzamide, 2-methoxy or *o*-Anisamide $2\text{-}CH_3OC_6H_4CONH_2$	151.16	nd (bz), pr (eth), pl (w)		129			eth, bz	B10^{2}, 58
1833	Benzamide, 4-methoxy $4\text{-}CH_3OC_6H_4CONH_2$	151.16	nd or ta (w)	295	166-7			al	B10^{2}, 100
1834	Benzamide, *N*-methyl $C_6H_5CONHCH_3$	135.17		291^{765}, 167^{12}	82			al, ace	B9^{3}, 1068
1835	Benzamide, *N*-α-naphthyl $C_6H_5CONHC_{10}H_7\text{-}\alpha$	247.30	nd (al or aa)		161-2			bz, aa	B12^{3}, 2870
1836	Benzamide, 2-nitro $2\text{-}NO_2C_6H_4CONH_2$	166.14		317	176.6			al, eth	B9^{3}, 1477
1837	Benzamide, 2-nitro-*N*-phenyl $2\text{-}NO_2C_6H_4CONHC_6H_5$	242.23	nd (al or bz)		155			al, eth, bz, chl	B12^{3}, 506
1838	Benzamide, 3-nitro $3\text{-}NO_2C_6H_4CONH_2$	166.14		310—5	142.7			w, al, eth	B9^{3}, 1515
1839	Benzamide, 4-nitro $4\text{-}NO_2C_6H_4CONH_2$	166.14			201.4			al, eth	B9^{3}, 1710
1840	Benzamide, *N*-(2-nitrophenyl) $C_6H_5CONH(C_6H_4NO_2\text{-}2)$	242.23	gold-ye nd (al)		98			al, eth	B12^{3}, 1525
1841	Benzamide, *N*-(3-nitrophenyl) $C_6H_5CONH(C_6H_4NO_2\text{-}3)$	242.23	pl		157			eth, chl	B12^{3}, 1552
1842	Benzamide, *N*-(4-nitrophenyl) $C_6H_5CONH(C_6H_4NO_2\text{-}4)$	242.23	ye nd (AcOEt)		199				B12^{3}, 1597
1843	Benzamide, *N*-phenyl or Benzanilide $C_6H_5CONHC_6H_5$	197.24	lf (al)	$117\text{-}9^{10}$, sub	163	1.315			B17^{3}, 502
1844	Benzamide, thiono, *N*-phenyl $C_6H_5CSNHC_6H_5$	213.30	ye pl or pr (al)	d	10^{2}			al, eth, bz	B12^{3}, 507
1845	Benzamide, *N*-(2-tolyl) $C_6H_5CONH(C_6H_4CH_3\text{-}2)$	211.26	rh nd (AcOEt-ace)		145-6	1.205^{15}		al, eth	B12^{3}, 1858
1846	Benzamide, *N*-(3-tolyl) $C_6H_5CONH(C_6H_4CH_3\text{-}3)$	211.26	mcl pr (dil al)		125	1.170^{15}		al	B12^{3}, 1964
1847	Benzamide, *N*-(4-tolyl) $C_6H_5CONH(C_6H_4CH_3\text{-}4)$	211.26	rh nd (al)	232	158	1.202^{15}		al, eth	B12^{3}, 2062
1848	Benzamide, 3,4,5-trihydroxy or Gallamide $3,4,5\text{-}(HO)_3C_6H_2CONH_2$	169.14	lf (w + 1)		244-5d			al, w, chl	B10^{2}, 346
1849	Benzamide, 3,4,5-trihydroxy-*N*-phenyl or Gallanilide $3,4,5\text{-}(HO)_3C_6H_4CONHC_6H_5$	245.24	lf (+ 2w, dil al)		207			al, aa, w	B12^{2}, 263
1850	Benzamidine $C_6H_5C(=NH)NH_2$	120.15	lf (al)		80			w, al	B9^{3}, 1264

No.	Name, Synonyms, and Formula	Mol. wt.	Color, crystalline form, specific rotation and λ_{max} (log ε)	b.p. °C	m.p. °C	Density	n_D	Solubility	Ref.
1851	Benzamidine hydrochloride $C_6H_5C(:NH)NH_2 \cdot HCl$	156.61	rh pr (w + 2)		169			w, al	B9[2]199
1852	Benzamidine, *N,N*-diphenyl $C_6H_5C(=NH)N(C_6H_5)_2$	272.35	rh pl (eth)		112			al, eth, bz	B12[3], 511
1853	Benzamidine, *N,N*-diphenyl, hydrochloride $C_6H_5C(=NH)N(C_6H_5)_2.HCl$	308.81	mcl pr or nd (al)		223d			w, al	B12, 270
1854	Benzamidine, *N,N'*-diphenyl $C_6H_5C(=NC_6H_5)NHC_6H_5$	272.35	nd (bz or al), pr (al)	146-7				al, eth	B12[3], 514
1856	Benzamidine, *N*-(1-naphthyl) $C_6H_5C(=NH)NHC_{10}H_7$-(*a*)	246.31	pl (al)		141			al, eth	B12, 1233
1857	Benzamidoxime $C_6H_5C(NH_2)=NOH$	136.15	mcl pr (w)		80			al, eth, bz, chl	B9[3], 1308
1858	1,2-Benzanthracene or Naphthanthracene $C_{18}H_{12}$	228.29	ye-br flr pl or lf (al-aa)	435 (sub)	162			al, eth, ace, bz	B5[3], 2374
1859	1,2-Benzanthracene, 9,10-dihydro $C_{18}H_{14}$	230.31	lf (bz, al, aa)		112				B5[3], 2305
1860	1,2-Benzanthracene, 9,10-dimethyl $9,10(CH_3)_2C_{18}H_{10}$	256.35	pa ye pl (al, aa)		122-3			ace, bz	B5[3], 2413
1861	1,2-Benzanthracene, 4'-hydroxy $4\text{-}HOC_{18}H_{11}$	244.29	pa ye or og nd (bz)		225			bz, to, aa	B6[3], 3725
1862	1,2-Benzanthracene, 3-methyl $3\text{-}CH_3C_{18}H_{11}$	242.32	pl (bz, al), nd (bz-lig)	$160^{0.01}$, (sub)	156-7			al, ace, bz, aa	B5[3], 2388
1863	1,2-Benzanthracene, 4-methyl $4\text{-}CH_3C_{18}H_{11}$	242.32	nd (al)	1	126-7			al, eth	B5[3], 2388
1864	1,2-Benzanthracene, 5-methyl $5\text{-}CH_3C_{18}H_{11}$	242.32	ye mcl pl (bz-al)		160			al, eth, bz, xyl	B5[3], 2388
1865	1,2-Benzanthracene, 6-methyl $6\text{-}CH_3C_{18}H_{11}$	242.32	pl (al)		150-1			al, eth, chl	B5[3], 2388
1866	1,2-Benzanthracene, 7-methyl $7\text{-}CH_3C_{18}H_{11}$	242.32	ye pl (al)		183-3.6			al, aa	2389
1867	1,2-Benzanthracene, 8-methyl $8\text{-}CH_3C_{18}H_{11}$	242.32	pl or nd (al, aa)		118			al, aa	B5[3], 2391
1868	1,2-Benzanthracene, 9-methyl $9\text{-}CH_3C_{18}H_{11}$	242.32	pa ye nd or pl (aa, MeOH)		138			al, aa	B5[3], 2391
1869	1,2-Benzanthracene, 10-methyl $10\text{-}CH_3C_{18}H_{11}$	242.32			141			al, eth, ace	B5[3], 2392
1870	1,2-Benzanthracene, 10-carboxaldehyde $10\text{-}OHCC_{18}H_{11}$	256.30	ye pr or nd		148			al, eth, ace, bz	B7[3], 2807
1871	1,2-Benz-3,4-anthraquinone $C_{18}H_{10}O_2$	258.28	red nd (to)		262-3d				B7[2], 759
1872	1,2-Benz-9,10-anthraquinone $C_{18}H_{10}O_2$	258.28	ye pr (aa)	sub	169			ace, bz, chl	B7[3], 4278
1873	Benzanthrene $C_{17}H_{12}$	216.28	lf (al)		84			al	B5[3], 2289
1874	1,9-Benzanthr-10-one $C_{17}H_{10}O$	230.27	ye nd (xyl or al)		170				B7[3], 2694
1875	1,9-Benzanthr-10-one, *bz*-1-bromo $C_{17}H_9BrO$	309.17	ye nd (aa)		178			al, MeOH	B7[3], 2706
1876	1,9-Benzanthr-10-one, *bz*-1-chloro $C_{17}H_9ClO$	264.71	ye nd (aa)		182-3				B7[3], 2700
1877	1,9-Benzanthr-10-one, *bz*-2-hydroxy $C_{17}H_{10}O_2$	246.27	ye nd		294				B8[3], 1629
1878	1,9-Benzanthr-10-one, 4-hydroxy $C_{17}H_{10}O_2$	246.27	ye nd (aa)		178-9			eth, bz	B8[3], 1638
1879	1,2-Benzazulene $C_{14}H_{10}$	178.23			176				B5[3], 2122
1880	Benzedrine or Amphetamine *dl* $C_6H_5CH_2CH(NH_2)CH_3$	135.21		203, 97^{20}		$0.9306^{25/4}$	1.518^{26}	al, chl	B12[3], 2665
1881	Benzestrol or 2,4-Bis(4-hydroxyphenyl)-3-ethyl hexane $C_{20}H_{26}O_2$	298.43	cr (al)		162-6			al, eth, ace, aa	B6[3], 5538
1882	Benzene C_6H_6	78.11	rh pr	80.1	5.5	$0.8765^{20/4}$	1.5011^{20}	**al, eth, ace, aa**	B5[4], 583

No.	Name, Synonyms, and Formula	Mol. wt.	Color, crystalline form, specific rotation and λ_{max} (log ε)	b.p. °C	m.p. °C	Density	n_D	Solubility	Ref.
1883	Benzene, allyl or Allylbenzene $CH_2=CHCH_2C_6H_5$	118.18	...	156, 47^{13}	-40	$0.8920^{20/4}$	1.5131^{20}	al, eth, bz	B5[4], 1362
1884	Benzene, 1-allyl-4-bromo $4\text{-}BrC_6H_4CH_2CH=CH_2$	197.07	...	222-2, 96^{12}	...	$1.324^{15/4}$	1.559^{20}	eth, bz, chl	B5[4], 1363
1885	Benzene, 1-allyl-3,4-dimethoxy $3,4\text{-}(CH_3O)_2C_6H_3CH_2CH=CH_2$	178.23	cr (hx)	254.7, 104^{2}	-4	$1.0396^{20/4}$	1.5340^{20}	al, eth	B6[3], 5024
1886	Benzene, 1-allyl-2,3,4,5-tetramethoxy $2,3,4,5\text{-}(CH_3)C_6HCH_2CH=CH_2$	238.28	...	...	25	1.087^{25}	1.51462^{25}	...	B6[3], 6691
1887	Benzene, (3-aminobutyl) $C_6H_5(CH_2CH_2CH(NH_2)CH_3)$	149.24	...	$221\text{-}3^{750}$, $101\text{-}2^{14}$	148	$0.9289^{15/4}$	1.51520^{20}	al	B12[3], 2717
1888	Benzene, (1-aminoethyl) *d* $C_6H_5CH(NH_2)CH_3$	121.18	$[\alpha]^{25}/_D$ + 39.3 (MeOH)	187	...	09651^{15}	...	w, al, eth	B12[3], 2386
1889	Benzene, (1-aminoethyl) *dl* $C_6H_5CH(NH_2)CH_3$	121.18	...	187, 87^{24}	...	0.9395^{15}	1.5238^{25}	w, al, eth	B12[3], 2386
1890	Benzene, (1-aminoethyl) *l* $C_6H_5CH(NH_2)CH_3$	121.18	$[\alpha]^{18}/_D$ −38 (MeOH)	184-6, 77^{16}	...	$0.9520^{20/4}$	...	w, al, eth	B12[3], 2386
1891	Benzene, (2-aminoethyl) or β-Phenethylamine $C_6H_5CH_2CH_2NH_2$	121.18	...	197-8	...	$0.9580^{24/4}$	1.5290^{25}	w, al, eth	B12[3], 2408
1892	Benzene, 1-(2-aminoethyl)-3,4-dimethoxy $3,4\text{-}(CH_3O)_2C_6H_3CH_2NH_2$	181.23	...	$163\text{-}5^{14}$	...	...	1.5464^{20}	...	B13[3], 2207
1893	Benzene, 1-benzyl-4-ethyl or *p*-Ethyl diphenylmethane $4\text{-}C_2H_5C_6H_4CH_2C_6H_5$	196.29	...	297, $85^{0.2}$	−24	0.9777^{20}	1.5616^{20}	al, eth, chl	B5[4], 1926
1894	Benzene, 1,2-bis (bromomethyl) or *o*-Xylyene bromide $1,2\text{-}(BrCH_2)_2C_6H_4$	263.96	rh (chl)	$128\text{-}30^{4.5}$	95	1.988^{0}	...	al, eth, peth	B5[4], 929
1895	Benzene, 1,3-bis (bromomethyl) or *m*-Xylyene bromide $1,3\text{-}(BrCH_2)_2C_6H_4$	263.96	nd (chl), pr (ace)	$135\text{-}40^{20}$	77	1.959^{0}	...	al, eth, lig	B5[4], 946
1896	Benzene, 1,4-bis (bromomethyl) or *p*-Xylyene bromide $1,4\text{-}(BrCH_2)_2C_6H_4$	263.96	mcl pr (al), cr (chl, bz)	245, $155\text{-}8^{14}$	145-7	...	...	al, bz, chl	B5[4], 970
1897	Benzene, 1,3,-bis (bromomethyl)-5-methyl $1,3\text{-}(BrCH_2)_2\text{-}5\text{-}CH_3C_6H_3$	278.99	pr	...	66	...	...	al, eth, bz, lig	B5[4], 1027
1898	Benzene, 1,3-bis(carboethoxy) $1,3\text{-}C_6H_4(OCH_2CO_2H_5)_2$	226.19	nd (aa, w)	...	195	...	...	w, aa	B6[3], 4324
1899	Benzene, 1,3-bis (carbethoxymethoxy) $1,3\text{-}C_6H_4(OCH_2CO_2H_5)_2$	282.30	nd (eth)	278^{12}	42	...	...	...	B6, 818
1900	Benzene, 1,2-bis (chloromethyl) or *o* Xylyene chloride $1,2\text{-}(ClCH_2)_2C_6H_4$	175.06	mcl (lig)	239-41, $130\text{-}5^{10}$	55	1.393^{0}	...	al, eth, lig	B5[4], 927
1901	Benzene, 1,3-bis (chloromethyl or *m*-Xylyene chloride $1,3\text{-}(ClCH_2)_2C_6H_4$	175.06	...	250-5	34.2	...	...	al, eth	B5[4], 944
1902	Benzene, 1,4-bis (chlormethyl) or *p*-Xylyene chloride $1,4\text{-}(ClCH_2)_2C_6H_4$	175.06	pl (al)	240-50d, 135^{16}	100	1.417^{0}	...	al, eth, ace, chl	B5[4], 967
1903	Benzene, 1,2-bis (cyanomethyl) or *o*-Xylyene dicyanide $1,2\text{-}(NCCH_2)_2C_6H_4$	156.19	(i) nd or pr (MeOH), (ii) pr (al)	...	(i) 18 (unst), (ii) 60 (st)	...	...	al, eth	B9[3], 4293
1904	Benzene, 1,3-bis (cyanomethyl) or *m* - Xylyene dicyanide $1,3\text{-}(NCCH_2)_2C_6H_4$	156.19	nd (w), pr (eth)	$305\text{-}10^{100}$, $170^{20\text{-}30}$	30	...	...	al, eth, bz, chl	B9[3], 4294
1905	Benzene, 1,4-bis (cyanomethyl) or *p*-Xylyene dicyanide $1,4\text{-}(NCCH_2)_2C_6H_4$	156.19	pr (al), nd (w)	...	98	...	...	al, eth, chl	B9[3], 4295
1906	Benzene, 1,2-bi(dibromomethyl) $1,2\text{-}(Br_2CH)_2C_6H_4$	421.75	mcl	...	116-7	...	...	chl	B5[4], 929
1907	Benzene, 1,3-bis (dibromomethyl) $1,3\text{-}(Br_2CH)_2C_6H_4$	421.75	nd (al), pr (chl)	...	107	...	...	al, bz, chl, lig	B5[3], 839
1908	Benzene, 1,4-bis (dibromomethyl) $1,4\text{-}(Br_2CH)_2C_6H_4$	421.75	mcl pr (chl)	...	172	...	...	bz	B5[3], 860
1909	Benzene, 1,2-bis (diethylamino) $1,2\text{-}[(C_2H_5)_2N]_2C_6H_4$	220.36	...	119^{10}	...	$0.9267^{13/4}$	1.5213^{13}	al, eth	B13[2], 12
1910	Benzene, 1,3-bis (diethylamino) $1,3[(C_2H_5)_2N]_2C_6H_4$	220.36	...	148^{9}	...	$0.9522^{12/4}$	1.5537^{12}	al, eth	B13[2], 26
1911	Benzene, 1,4-bis (diethylamino) $1,4\text{-}[(C_2H_5)_2N]_2C_6H_4$	220.36	mcl pr, pl (al-w)	280	52	...	...	al, eth, bz, lig	B13[2], 40
1912	Benzene, 1,4-bis (dimethylamino) $1,4\text{-}[(CH_3)_2N]_2C_6H_4$	164.25	lf (dil al or lig)	260	51	...	...	al, eth, bz, lig	B13[3], 111
1913	Benzene, 1,2-bis (hydroxymethyl) or *o* -Xylylene glycol $1,2\text{-}C_6H_4(CH_2OH)_2$	138.17	pl (eth, peth)	...	65-6	...	...	w, al, eth	B6[1], 4587
1914	Benzene, 1,3-bis (hydroxymethyl) or Isophthalyl glycol $1,3\text{-}(HOCH_2)_2C_6H_4$	138.17	nd (bz)	$154\text{-}9^{13}$	57	...	...	w, al, eth	B6[3], 4600

No.	Name, Synonyms, and Formula	Mol. wt.	Color, crystalline form, specific rotation and λ_{max} (log ε)	b.p. °C	m.p. °C	Density	n_D	Solubility	Ref.
1915	Benzene, 1,4-bis (hydroxymethyl) or Terephthalyl glycol. $1,4\text{-}(HOCH_2)_2C_6H_4$	138.17	nd (w)		115-6			w, al, eth, ace	B6^{3}, 4608
1916	Benzene, 1,2-bis (methylamino)-4-chloro $4\text{-}Cl\text{-}1,2(CH_3NH)_2C_6H_3$	170.64	pr (lig)		61			w	B13, 25
1917	Benzene, 1,3-bis (3-methylbutoxy) $[(CH_3)_2CHCH_2CH_2O]_2C_6H_4$	250.39			47				B6, 815
1918	Benzene, 1,3-bis((4-tolylamino) $1,3\text{-}(4\text{-}CH_3C_6H_4NH)_2C_6H_4$	288.40	nd (al)		139-9				B13, 42
1919	Benzene, 1,2-bis (trimethylsiloxy) $1,2\text{-}[(CH_3)_3SiO]_2C_6H_4$	254.48		235^{760}			$1.4685^{20/}{}_D$		C^{2}, 249
1920	Benzene, 1,3-bis (trimethylsiloxy) $1,3\text{-}[(CH_3)_3SiO)]_2C_6H_4$	254.48		$237\text{-}40^{740}$		$0.950^{20/4}$	$1.4748^{20/}{}_D$		C^{2}, 249
1921	Benzene, bromo or phenylbromide C_6H_5Br	157.01		$156, 43^{18}$	-30.8	$1.4950^{20/4}$	1.5597^{20}	al, eth, bz	B5^{4}, 670
1922	Benzene, (4-bromobutoxy) $Br(CH_2)_4OC_6H_5$	229.12	cr (al)	$153\text{-}6^{18}$	41			al	B6^{4}, 558
1923	Benzene, 1-bromo-4-*tert* butyl $4\text{-}[(CH_3)_3C]C_6H_4Br$	213.12		$231\text{-}2, 103^{10}$	19	$1.2286^{20/4}$	1.5436^{20}	eth, bz, chl	B5^{4}, 1050
1924	Benzene, 1-bromo-2-chloro $2\text{-}ClC_6H_4Br$	191.45		204^{765}	-12.3	$1.6387^{25/4}$	1.5809^{20}	bz	B5^{4}, 680
1925	Benzene, 1-bromo-3-chloro $3\text{-}ClC_6H_4Br$	191.45		196	-21.5	$1.6302^{20/4}$	1.5771^{20}	al, eth	B5^{4}, 680
1926	Benzene, 1 bromo-4-chloro $4ClC_6H_4Br$	191.45	nd or pl (al, eth)	196^{756}	68	$1.576^{71/4}$	1.5531^{70}	eth, bz, chl	B5^{4}, 681
1927	Benzene, 1-bromo-4-cyclohexyl $4\text{-}C_6H_{11}\text{-}C_6H_4Br$	239.16		160^{25}		$1.283^{25/4}$	1.5584^{20}	bz, chl	B5^{3}, 1258
1928	Benzene, 1-bromo-2,3-dichloro $2,3\text{-}Cl_2C_6H_3Br$	225.90	pl or lf (al)	243^{705}	60			eth, bz, chl	B5, 209
1929	Benzene, 1-bromo-3,5-dichloro $3,5\text{-}Cl_2C_6H_3Br$	225.90	pr (al)	232^{757}	82-4			eth, bz, chl, al	B5^{2}, 162
1930	Benzene, 1-bromo-2,6-dichloro $2,6\text{-}Cl_2C_6H_3Br$	225.90	pr (al)	242^{765}	65			eth, bz, chl	B5, 210
1931	Benzene, 1-bromo-2,5-dichloro $2,5\text{-}Cl_2C_6H_3Br$	225.90	pr or nd (al)	$235^{731}, 119^{20}$	35			al, eth, bz, lig	B5^{3}, 564
1932	Benzene, 1-bromo-3,4-dichloro $3,4\text{-}Cl_2C_6H_3Br$	225.90	pr	$237, 124^{33}$				al, eth, bz, chl	B5^{3}, 564
1933	Benzene, 1-bromo-2,3-dimethyl $2,3\text{-}(CH_3)_2C_6H_3Br$	185.06		$214, 83^{11}$		$1.365^{20/4}$		eth, ace, bz	B5^{4}, 928
1934	Benzene, 1-bromo-2,4-dimethyl $2,4\text{-}(CH_3)_2C_6H_3Br$	185.06		$205, 84^{13}$	0		1.5501^{20}	al, eth, bz	B5^{4}, 945
1935	Benzene, 1-bromo-2,5-dimethyl $2,5\text{-}(CH_3)_2C_6H_3Br$	185.06	lf or pl	$199\text{-}200, 88\text{-}9^{13}$	9	1.3582^{18}	1.5514^{18}	al, bz	B5^{4}, 969
1936	Benzene, 1-bromo-2,6-dimethyl $2,6(CH_3)_2C_6H_3Br$	185.06		$203\text{-}4, 98\text{-}9^{20}$			1.5552^{20}	eth, ace, bz	B5^{4}, 945
1937	Benzene, 1-bromo-3,4-dimethyl $3,4\text{-}(CH_3)_2C_6H_3Br$	185.06		214.5	-0.2	$1.3708^{20/4}$	1.5530^{20}	al, eth	B5^{4}, 928
1938	Benzene, 1-bromo-3,5-dimethyl $3,5\text{-}(CH_3)_2C_6H_3Br$	185.06		$204, 83\text{-}9^{12}$		1.342^{20}	1.5462^{22}	eth, ace, bz	B5^{4}, 945
1939	Benzene, 1-bromo-2,3-dinitro $2,3\text{-}(NO_2)_2C_6H_3Br$	247.00	pa ye pl (al)	320	101-2			al	B5^{4}, 749
1940	Benzene, 1-bromo-2,4-dinitro $2,4\text{-}(NO_2)_2C_6H_3Br$	247.00	ye nd (al)		75			al	B5^{4}, 749
1941	Benzene, 1-bromo-2,5-dinitro $2,5\text{-}(NO_2)_2C_6H_3Br$	247.00	nd (al), pr (al-eth)		70			eth, al	B5^{4}, 750
1942	Benzene, 1-bromo-2,6-dinitro $2,6\text{-}(NO_2)_2C_6H_3Br$	247.00	ye pr (al)		107			al	B5^{4}, 749
1943	Benzene, 1-bromo-3,4-dinitro $3,4\text{-}(NO_2)_2C_6H_3Br$	247.00	nd (al), pl (al-eth)		34-5 (unst)			eth, al	B5^{3}, 640
1944	Benzene, 1-bromo-2-ethoxy $2\text{-}C_2H_5OC_6H_4Br$	201.06		222-6				al, eth	B6^{4}, 1038
1945	Benzene, 1-bromo-3-ethoxy $3\text{-}C_2H_5OC_6H_4Br$	201.06		222				al, eth	B6^{3}, 739
1946	Benzene, 1-bromo-4-ethoxy $4\text{-}C_2H_5OC_6H_4Br$	201.06		$230\text{-}2, 109^{17}$	4	$1.4071^{25/4}$	1.5517^{20}	al, eth	B6^{4}, 1045
1947	Benzene, (1-bromoethyl) *d* $C_6H_5(CHBrCH_3)$	185.06	$[\alpha]^{14/}{}_D$ + 1.5	$203, 86\text{-}8^{15}$		1.3108^{23}	1.5612^{20}	al, eth	B5^{4}, 906

No.	Name, Synonyms, and Formula	Mol. wt.	Color, crystalline form, specific rotation and λ_{max} (log ε)	b.p. °C	m.p. °C	Density	n_D	Solubility	Ref.
1948	Benzene, (1-bromoethyl) *dl* $C_6H_5(CHBrCH_3)$	185.06	...	202-3, 85^{13}	...	$1.3605^{20/4}$	1.5612^{20}	al, eth, bz	$B5^4$, 906
1949	Benzene, (2-bromoethyl) $C_6H_5CH_2CH_2Br$	185.06	...	$217\text{-}8^{734}$, 92^{11}	...	$1.3587^{20/4}$	1.5572^{20}	eth, bz	$B5^4$, 907
1950	Benzene, 1-bromo-2-ethyl $2\text{-}C_2H_5C_6H_4Br$	185.06	...	199	...	...	$1.5473^{20/}{}_D$	...	$B5^4$, 906
1951	Benzene, 1-bromo-4-ethyl $4\text{-}C_2H_5C_6H_4Br$	185.06	...	204	...	...	1.5445^{20}	...	$B5^4$, 906
1952	Benzene, 1-bromo-4-fluoro $4\text{-}F\text{-}C_6H_4\text{-}Br$	175.00	...	152^{764}	-8 (fr-17)	$1.4946^{20/4}$	1.5604^{20}	al, eth	$B5^4$, 678
1953	Benzene, (2-bromo-1-hydroxyethyl) $C_6H_5CH(OH)CH_2Br$	201.06	...	$109\text{-}10^{2}$	...	$1.4994^{20/4}$	$1.5800^{17/}{}_D$	...	$B6^3$, 1690
1954	Benzene, 1-bromo-2-iodo $2\text{-}I\text{-}C_6H_4\text{-}Br$	282.91	...	257^{754}, $120\text{-}1^{15}$	9-10	$2.2571^{25/4}$	1.6618^{25}	ace	$B5^4$, 698
1955	Benzene, 1-bromo-3-iodo $3\text{-}I\text{-}C_6H_4Br$	282.91	...	252^{754}, 120^{18}	-9.3	...	...	...	$B5^4$, 698
1956	Benzene, 1-bromo-4-iodo $4\text{-}I\text{-}C_6H_4Br$	282.91	pr or pl (eth-al)	252^{754}	92	...	...	eth	$B5^4$, 698
1957	Benzene, 1-(bromomethyl)-2-methyl or *o*-Xylylbromide $2\text{-}CH_3C_6H_4CH_2Br$	185.06	pr	$216\text{-}7^{742}$, 108^{16}	21	1.3811^{21}	1.5730^{20}	al, eth, ace, bz	$B5^4$, 928
1958	Benzene, 1-(bromomethyl)-3-methyl or *m*-Xylylbromide $3\text{-}CH_3C_6H_4CH_2Br$	185.06	...	212-3, 105^{11}	...	1.3711^{21}	1.5660^{20}	al, eth, chl	$B5^2$, 293
1959	Benzene, 1-(bromomethyl)-4-methyl or *p*-Xylylbromide $4\text{-}CH_3C_6H_4CH_2Br$	185.06	nd (al)	$218\text{-}20^{740}$	35	1.324	...	al, eth, chl	$B5^3$, 858
1960	Benzene, 1-(bromomethyl)-3,5-dimethyl or Mesityl bromide $3,5\text{-}(CH_3)_2C_6H_3CH_2Br$	199.09	nd (eth)	$229\text{-}31^{740}$(d), 118^{22}	40	...	...	al, eth, bz, chl	$B5^4$, 1027
1961	Benzene, 1-bromo-2-nitro $2\text{-}NO_2C_6H_4Br$	202.01	pa, ye (al)	258^{756}	43	$1.6245^{80/4}$	...	al, eth, ace, bz	$B5^4$, 728
1962	Benzene, 1-bromo-3-nitro $3\text{-}NO_2C_6H_4Br$	202.01	rh	265	17 (unst), 56 (st)	$1.7036^{20/4}$	1.5979^{20}	al, eth, bz	$B5^4$, 729
1963	Benzene, 1-bromo-4-nitro $4\text{-}NO_2C_6H_4Br$	202.01	rh or mcl pr (al)	256	127	1.948	...	al, eth, bz	$B5^4$, 729
1964	Benzene, 1-bromo-2-nitroso $2\text{-}ONC_6H_4Br$	186.01	nd	...	98	...	...	bz, chl	$B5^4$, 706
1965	Benzene, 1-bromo-4-nitroso $4\text{-}ONC_6H_4Br$	186.01	nd (al)	...	95	...	...	al, bz, chl	$B5^4$, 706
1966	Benzene, 3-bromo-1-propenyl) or Cinnamyl bromide $BrCH_2CH{=}CHC_6H_5$	197.07	nd (al, eth)	130^{10}	34	1.3428^{30}	1.610—1.613^{20}	al	$B5^3$, 1188
1967	Benzene, (3-bromopropoxy) or 2-Phenoxypropyl bromide $C_6H_5\text{-}OCH_2CH_2CH_2Br$	215.09	...	127^{18}	7-8	$1.365^{16/16}$	...	eth	$B6^3$, 549
1968	Benzene, (1-bromopropyl)-(*DL*) $C_6H_5CHBrCH_2CH_3$	199.09	$d[\alpha]_D$ +5.9 (al, c = 1), $l[\alpha]_D$ −5.7 (eth, c = 10)	105^{17}	...	$1.3098^{19/4}$	1.5517^{19}	eth, bz	$B5^4$, 981
1969	Benzene, (2-bromopropyl) $C_6H_5CH_2CHBrCH_3$	199.09	...	$107\text{-}9^{16}$	...	1.2908^{16}	1.5450^{20}	chl	$B5^4$, 981
1970	Benzene, (3-bromopropyl) $C_6H_5CH_2CH_2CH_2Br$	199.09	...	110^{12}	...	$1.3106^{25/4}$	1.5440^{25}	eth	$B5^4$, 982
1971	Benzene, (3-bromoisopropyl) $BrCH_2CH(CH_3)C_6H_5$	199.09	$[\alpha]^{25/}{}_D(d)$ + 15.6	188-9, $106\text{-}8^{18}$	...	1.3155^{20}	1.5548^{28} (d)	bz, chl	$B5^4$, 995
1972	Benzene, 1-bromo-4-isopropyl $4\text{-}i\text{-}C_3H_7C_6H_4Br$	199.09	...	218.7, $97\text{-}8^{5}$	-22.5	$1.3145^{20/4}$	1.5569^{20}	eth, bz, chl	$B5^4$, 994
1973	Benzene, 2-bromo-4-*iso*-propyl-1-methyl $4\text{-}i\text{-}C_3H_7\text{-}2\text{-}BrC_6H_3CH_3$	213.12	...	234.3, 99^{9}	...	$1.2689^{18/4}$	1.5360^{20}	al, eth, chl	$B5^4$, 1063
1974	Benzene, 1-bromo-4-triazo $4\text{-}N_3C_6H_4Br$	198.02	pl	105^{10}	20	...	...	eth, bz	$B5^4$, 761
1975	Benzene, 1-bromo-2,3,5-trimethyl $2,3,5\text{-}(CH_3)_3C_6H_2Br$	199.09	...	238, 117^{17}	<-15	...	1.5516^{20}	bz	$B5^4$, 1014
1976	Benzene, 1-bromo-2,4,5-trimethyl $2,4,5\text{-}(CH_3)_3C_6H_2Br$	199.09	nd (al)	233-5	73	...	...	al	$B5^4$, 1014
1977	Benzene, 1-bromo-2,4,6-trimethyl $2,4,6\text{-}(CH_3)_3C_6H_2Br$	199.09	...	225, 117^{25}	-1	1.3191^{10}	1.5510^{20}	eth, bz	$B5^4$, 1027
1978	Benzene, butoxy or Butyl phenyl ether $C_6H_5OC_4H_9$	150.22	...	210, 95^{17}	-19.4	$0.9351^{20/4}$	1.4969^{20}	ace, bz	$B6^4$, 558

No.	Name, Synonyms, and Formula	Mol. wt.	Color, crystalline form, specific rotation and λ_{max} (log ε)	b.p. °C	m.p. °C	Density	n_D	Solubility	Ref.
1979	Benzene, isobutoxy or Isobutyl phenyl ether $i\text{-}C_4H_9OC_6H_5$	150.22	...	196	...	0.9240$^{24/15}$	1.4932^{14}	ace, bz, peth	B6^4, 559
1980	Benzene, 1-*iso*-butoxy-2-nitro $2\text{-}NO_2C_6H_4O\text{-}i\text{-}C_4H_9$	195.22	ye nd	275-80	...	1.1361^{20}	...	eth	B6, 218
1981	Benzene, *sec*-butoxy $C_6H_5OCH(CH_3)CH_2CH_3$	150.22	...	194.5, 70-2^5	...	0.9415$^{20/4}$	1.4926^{25}	eth	B6^4, 558
1982	Benzene, butyl $C_6H_5C_4H_9$	134.22	...	183, 62.2^{10}	-88	0.8601$^{20/4}$	1.4898^{20}	al, eth, ace, bz	B5^4, 1033
1983	Benzene, 1-butyl-2-methyl $2\text{-}CH_3C_6H_4C_4H_9$	148.25	...	208, 81^{10}	...	0.8710$^{20/4}$	1.4960^{20}	eth, ace, bz	B5^3, 999
1984	Benzene, 1-butyl-3-methyl $3\text{-}CH_3C_6H_4C_4H_9$	148.25	...	205, 79^{10}	...	0.8590$^{20/4}$	1.4910^{20}	eth, ace, bz	B5^3, 999
1985	Benzene, 1-butyl-4-methyl $4\text{-}CH_3C_6H_4C_4H_9$	148.25	...	207, 80.6^{10}	-85	0.8586$^{20/4}$	1.4916^{20}	eth, ace, bz	B5^4, 1094
1986	Benzene, *iso*-butyl $C_6H_5CH_2CH_2(CH_3)_2$	134.22	...	172.8, 53.1^{10}	-51.5	0.8532$^{20/4}$	1.4866^{20}	**al, eth, ace, bz, peth**	B5^4, 1042
1987	Benzene, *sec*-butyl(*d*) $C_6H_5CH(CH_3)CH_2CH_3$	134.22	$[\alpha]^{20/}{}_D$ + 27.3 (undil)	173, 63^{15}	-75	0.8621$^{20/4}$	1.4895^{20}	**al, eth, ace, bz**	B5^4, 1038
1988	Benzene, *sec*-butyl(*dl*) $C_6H_5CH(CH_3)CH_2CH_3$	134.22	...	173, 53.6^{10}	-75.5	0.8621$^{20/4}$	1.4902^{20}	**al, eth, ace, bz**	B5^4, 1038
1989	Benzene, *sec*-butyl *l*) $C_6H_5CH(CH_3)CH_2CH_3$	134.22	$[\alpha]^{25/}{}_D$ −17.9 (al, c = 5)	61^{18}	...	0.868$^{24/4}$	1.4891^{20}	**al, eth, ace, bz**	B5^4, 1038
1990	Benzene, 1-*sec*-butyl-4-methyl $4\text{-}CH_3C_6H_4CH(CH_3)C_2H_5$	148.25	...	197, 73^{10}	...	0.8640^{19}	1.493^{20}	eth, bz, chl	B5^4, 1095
1991	Benzene, *tert*-butyl $C_6H_5C(CH_3)_3$	134.22	...	169, 50.7^{10}	-57.8	0.8665$^{20/4}$	1.4927^{20}	al, eth, **ace, bz**	B5^4, 1045
1992	Benzene, 1-*tert*-butyl-3,5-dimethyl,2,4,6-trinitro or Musk xylene $3,5\text{-}(CH_3)_2\text{-}2,4,6\text{-}(NO_2)_3C_6\text{-}tert\text{-}C_4H_9$	297.27	pl, nd (al)	...	110	...	...	eth, al	B5^4, 1132
1993	Benzene, 2-*tert*-butyl-3,5-dinitro-1-*iso*-propyl-4-methyl or Moskene $2\text{-}tert\text{-}C_4H_9\text{-}3,5(NO_2)_2\text{-}4\text{-}CH_3C_6H(i\text{-}C_3H_7)$	280.32	Pa ye cr	...	132-3	...	...	...	C49, 3044
1994	Benzene, 2-*tert*-butyl-1,3-dinitro-4,5,6-trinitro or Musk tebetine $2\text{-}(CH_3)_3C\text{-}1,3\text{-}(NO_2)_2\text{-}4,5,6\text{-}(CH_3)_3C_6$	266.30	pa ye pr (al)	...	135-6	...	...	eth	B5^3, 1055
1995	Benzene, 5-*tert*-butyl-1,3-dinitro-4-methoxy-2-methyl or Musk ambrette $5\text{-}(CH_3)_3C\text{-}1,3\text{-}(NO_2)_2\text{-}4\text{-}CH_3O\text{-}2\text{-}CH_3\text{-}C_6H$	268.25	pa ye lf (al)	135^{16}	84-6	...	...	eth	B6^3, 1984
1996	Benzene, 1-*tert*-butyl-2-methyl $2\text{-}CH_3\text{-}C_6H_4C(CH_3)_3$	148.25	...	170^{743}	...	1.49423$^{17/D}$	...	...	B5^4, 1096
1997	Benzene, 1-*tert*-butyl-3-methyl-2,4,6-trinitro or Artificial musk $3\text{-}CH_3\text{-}2,4,6\text{-}(NO_2)_3C_6H\text{-}tertC_4H_9$	283.24	ye nd (al)	...	96-7	...	...	al, eth, bz, chl	B5^3, 1003
1998	Benzene, 1-*tert*-butyl-4-methyl $4\text{-}CH_3C_6H_4C(CH_3)_3$	148.25	...	193, 70^{10}	-52	0.8612$^{20/4}$	1.4918^{20}	eth, ace, bz	B5^4, 1097
1999	Benzene, 2-*tert*-butyl-4-methyl-1,3,5-trinitro or Musk baur $2\text{-}(CH_3)_3C\text{-}4\text{-}CH_3\text{-}1,3,5\text{-}(NO_2)_3C_6H$	283.24	ye nd (al)	...	97	...	...	al, eth, bz, peth	B5^3, 1003
2000	Benzene, chloro or Phenyl chloride C_6H_5Cl	112.56	...	132, 22^{10}	-45.6	1.1058$^{20/4}$	1.5241^{20}	**al, eth**, bz	B5^4, 640
2001	Benzene, (chloro-*tert*-butyl) $C_6H_5C(CH_3)_2CH_2Cl$	168.67	...	222^{741}, 104-5^{18}	...	1.047$^{20/4}$	1.5247$^{20/D}$	**al, eth, ace, bz**	B5^4, 1048
2002	Benzene, 1-chloro-2,5-diacetamido $2,5\text{-}(CH_3CONH)_2\text{-}C_6H_3Cl$	226.66	nd	...	196-7	...	...	al	B13, 118
2003	Benzene, 1-chloro-2,4-diamino $2,4\text{-}(NH_2)_2C_6H_3Cl$	142.59	pl or nd	...	91	...	...	al	B13^3, 97
2004	Benzene, 1-chloro-2,5-diamino $2,5\text{-}(NH_2)_2C_6H_3Cl$	142.59	nd (bz-lig)	...	64	...	...	w, bz	B13^2, 58
2005	Benzene, 1-chloro-2,6-diamino $2,6\text{-}(NH_2)_2C_6H_3Cl$	142.29	...	...	85-6	...	...	eth	B13^1, 15
2006	Benzene, 1-chloro-3,4-diamino $3,4\text{-}(NH_2)_2C_6H_3Cl$	142.59	pl (bz-lig), lf (w)	...	76	...	...	al, eth, bz, lig	B13^3, 20
2007	Benzene, 1-chloro-3,5-diamino $3,5\text{-}(NH_2)_2C_6H_3Cl$	142.59	rh pr (al), nd (to-al)	...	105-6	...	...	al, eth, ace	B13^2, 29
2008	Benzene, 1-chloro-2,4-dimethyl or 4-Chloro-*m*-xylene $2,4\text{-}(CH_3)_2\text{-}C_6H_3Cl$	140.61	...	187-8, 89^{24}	...	1.0598$^{20/20}$	1.5230^{25}	ace, bz	B5^4, 943

PHYSICAL CONSTANTS OF ORGANIC COMPOUNDS (Continued)

No.	Name, Synonyms, and Formula	Mol. wt.	Color, crystalline form, specific rotation and λ_{max} (log ε)	b.p. °C	m.p. °C	Density	n_D	Solubility	Ref.
2009	Benzene, 1-chloro-2,5-dimethyl or 2-Chloro-*p*-xylene 2,5-$(CH_3)_2C_6H_3Cl$	140.61		187 (192)	1.6	1.0589$^{15/4}$		ace, bz	B5^{4}, 965
2010	Benzene, 1-chloro-2,6-dimethyl 2,6-$(CH_3)_2C_6H_3Cl$	140.61		185-7, 62-3^{12}		1.053^{20}	1.526^{20}	ace, bz, chl	B5^{4}, 943
2011	Benzene, 1-chloro-3,4-dimethyl 3,4-$(CH_3)_2C_6H_3Cl$	140.61		194^{755}	-6	1.0692$^{15/15}$		ace, bz	B5^{3}, 815
2012	Benzene, 1-chloro-3,5-dimethyl or 5-Chloro-*m*-xylene 3,5-$(CH_3)_2C_6H_3Cl$	140.61		187-8, 66^{12}				ace, bz	B5^{3}, 834
2013	Benzene, 1-chloro-2,3-dinitro 2,3-$(NO_2)C_6H_3Cl$	202.55	pr (al), cr (MeOH)		78			eth	B5^{4}, 744
2014	Benzene, 1-chloro-2,4-dinitro 2,4-$(NO_2)_2C_6H_3Cl$	202.55	α: ye rh (eth), β: ye rh (eth), nd (al)	315, 158-60^{2}	α: 53 (st), β: 43 (unst)	1.4982$^{75/4}$	1.5857^{60}	eth, bz	B5^{4}, 744
2015	Benzene, 1-chloro-2,6-dinitro 2,6-$(NO_2)_2C_6H_3Cl$	202.55	ye nd (al, aa)	315	88	1.6867^{16}		al, eth, to	B5^{4}, 744
2016	Benzene, 1-chloro-3,4-dinitro 3,4-$(NO_2)C_6H_3Cl$	202.55	α-cr (al), β-pr (lig), γ-mcl or rh nd (eth or lig)	315d, 160^{4}	α 36, β 37, γ 40-1	1.6867$^{20/}{}_{4}$		eth, bz	B5^{4}, 744
2017	Benzene, 1-chloro,3,5-dinitro 3,5-$(NO_2)_2C_6H_3Cl$	202.55	nd (al, peth)		59			al, eth	B5^{3}, 637
2018	Benzene, 1-chloro-2-ethoxy 2-$C_2H_5OC_6H_4Cl$	156.61		210, 97-8^{15}		1.1288$^{15/4}$	1.5284^{25}	al, eth, bz	B6^{4}, 785
2019	Benzene, 1-chloro-3-ethoxy 3-$C_2H_5OC_6H_4Cl$	156.61		204-5^{747}		1.1712$^{20/4}$		al, eth, bz	B6^{4}, 811
2020	Benzene, 1-chloro-4-ethoxy 4-$C_2H_5OC_6H_4Cl$	156.61		212-4, 98^{17}	21	1.1254$^{20/4}$	1.5252^{20}	al, eth, bz, aa	B6^{4}, 823
2021	Benzene, (2-chlorethoxy) $C_6H_5OCH_2CH_2Cl$	156.61		217-20, 100-2^{12}	28			al, eth, ace, bz	B6^{4}, 556
2022	Benzene, (1-chloroethyl)(*d*) or *d*-α-Phenethyl chloride $C_6H_5CHClCH_3$	140.61	$[\alpha]^{20/}{}_{D}$ + 50.6 (undil)	85^{20}		1.0631$^{20/4}$	1.5250^{25}	al, eth, bz	B5^{4}, 898
2023	Benzene, (1-chloroethyl)(*dl*) or α-Phenethyl chloride $C_6H_5CHClCH_3$	140.61		81-2^{17}		1.0620$^{20/4}$	1.5276^{20}	al, eth, bz	B5^{4}, 898
2024	Benzene, (1-chloroethyl)(*l*) or *l*-α-Phenethyl chloride $C_6H_5CHClCH_3$	140.61	$[\alpha]^{20/}{}_{D}$ −30.1 (undil)	85^{20}		1.0632$^{20/4}$		al, eth, bz	B5^{4}, 898
2025	Benzene, (2-chloroethyl) or β-Phenethyl chloride $C_6H_5CH_2CH_2Cl$	140.61		197-8, 92^{20}		1.069$^{25/4}$	1.5276^{20}	eth, al, ace, bz, lig	B5^{4}, 899
2026	Benzene, (2-chloroethylthio) 2,$C_6H_5SCH_2CHCl$	172.67		117-8^{11}		1.1769$^{25/4}$	1.5828^{20}	eth, chl	B6^{4}, 1469
2027	Benzene, 1-chloro-2-ethyl 2-$C_2H_5C_6H_4Cl$	140.61		178.4	-82.7	1.0569$^{20/4}$	1.5218^{20}	ace, bz, chl	B5^{4}, 897
2028	Benzene, 1-chloro-4-ethyl 4-$C_2H_5C_6H_4Cl$	140.61		184.4, 63.5^{10}	-62.6	1.0455$^{20/4}$	1.5175^{20}	al, eth, ace, peth, bz	B5^{4}, 897
2029	Benzene, 1-chloro-2-fluoro 2-FC_6H_4Cl	130.55		137.6	-43	1.2233$^{30/4}$	1.4968^{30}	ace, bz	B5^{4}, 652
2030	Benzene, 1-chloro-3-fluoro 3-FC_6H_4Cl	130.55		127.6		1.221^{25}	1.4911		B5^{4}, 652
2031	Benzene, 1-chloro-4-fluoro 4-FC_6H_4Cl	130.55		130^{757}	-26.8	1.4990^{15}		al, eth, bz	B5^{4}, 652
2032	Benzene, 1-chloro-3-hydroxylamino 3-$(HONH)C_6H_4Cl$	143.57	pl (bz)		49			w, bz, peth	B15^{2}, 8
2033	Benzene, 1-chloro-4-hydroxylamino 4-$(HONH)C_6H_4Cl$	143.57	lf (dil al)		87-8			al, bz, chl	B15^{2}, 9
2034	Benzene, 1-chloro-2-iodo 2-IC_6H_4Cl	238.46		234-5, 110^{18}		1.9515^{25}	1.6331^{25}	ace	B5^{4}, 695
2035	Benzene, 1-chloro-3-iodo 3-IC_6H_4Cl	238.46		230				bz	B5^{4}, 695
2036	Benzene, 1-chloro-4-iodo 4-IC_6H_4Cl	238.46	lf (ace) (al)	227, 108^{13}	57	1.886$^{27/4}$		al	B5^{4}, 696
2037	Benzene, 1-(chloromethyl)-2,4-dimethyl 2,4-$(CH_3)_2C_6H_3CH_2Cl$	154.64		215-6, 86-7^{12}				al, eth, bz	B5^{4}, 1012

No.	Name, Synonyms, and Formula	Mol. wt.	Color, crystalline form, specific rotation and λ_{max} (log ε)	b.p. °C	m.p. °C	Density	n_D	Solubility	Ref.
2038	Benzene, 1-(chloromethyl)-4-ethyl or 4-Ethylbenzyl chloride $4-C_2H_5C_6H_4CH_2Cl$	154.64	....	$95-6^{15}$	......	....	1.5290^{25}	al, bz, chl	$B5^4$, 1004
2039	Benzene, 1-(chloromethyl)-2-methyl or *o*-xylylchloride $2-CH_3C_6H_4CH_2Cl$	140.61	....	$197-9, 80^{12}$	......	....	1.5410^{25}	**al, eth**	$B5^4$, 926
2040	Benzene, 1-chloromethyl-3-methyl or *m*-xylylchloride $3-CH_3C_6H_4CH_2Cl$	140.61	....	$195-6, 101-2^{10}$	......	1.064^{20}	1.5345^{20}	al, eth	$B5^4$, 943
2041	Benzene, 1-(chloromethyl)-4-methyl or *p*-xylylchloride $4-CH_3C_6H_4CH_2Cl$	140.61	....	$200-2, 81^{15}$	......	$1.0512^{20/4}$	1.5380	al, **eth**	$B5^4$, 966
2042	Benzene, 1-chloroisopropyl $C_6H_5CCl(CH_3)_2$	154.64	....	98^{1}	......	1.192^{25}	1.5290^{25}	....	$B5^4$, 992
2043	Benzene, 1-chloro-2-nitro $2-NO_2C_6H_4Cl$	157.56	mcl nd	$246, 119^{8}$	34-5	....	....	al, eth, ace, bz	$B5^4$, 721
2044	Benzene, 1-chloro-3-nitro $3-NO_2C_6H_4Cl$	157.56	pa ye rh pr (al)	235-6	24 (unst) 46 (st)	$1.343^{50/4}$	$1.5374^{80/\alpha}$	al, eth, bz	$B5^4$, 722
2045	Benzene, 1-chloro-4-nitro $4-NO_2-C_6H_4Cl$	157.56	mcl pr	$242, 113^{8}$	83.6	$1.2979^{90.5}$	$1.5376^{100/\alpha}$	al, ace	$B5^4$, 723
2046	Benzene, 1-chloro-2-nitroso $2-ONC_6H_4Cl$	141.56	nd (al)	....	65-6	....	....	al, eth, bz, peth	$B5^4$, 705
2047	Benzene, 1-Chloro-3-nitroso $3-ONC_6H_4Cl$	141.56	nd (bz)	....	72 (77)	....	....	al, eth, ace, bz, chl	$B5^4$, 705
2048	Benzene, 1-chloro-4-nitroso $4-ONC_6H_4Cl$	141.56	(aa or al)	....	92-3	....	....	al	$B5^4$, 705
2049	Benzene, chloro-pentafluoro C_6ClF_5	202.51	....	$122-3^{750}$	......	1.568^{25}	1.4256^{20}	....	$B5^4$, 654
2050	Benzene, (3-chloropropoxy) $C_6H_5OCH_2CH_2CH_2Cl$	170.64	....	$245-55, 139^{25}$	12	1.1167^{20}	1.5235^{25}	eth, ace	$B6^4$, 557
2051	Benzene, (3-chloropropylthio) $C_6H_5SCH_2CH_2CH_2Cl$	186.70	....	$116-7^{4}$	......	$1.1536^{20/4}$	1.5752^{20}	ace, Py	$B6^3$, 981
2052	Benzene, 1-chloro-4-*iso*-propyl or 4-Chlorocumene $4-(CH_3)_2CHC_6H_4Cl$	154.64	....	$198.3, 74^{10}$	-12.3	$1.0208^{20/4}$	1.5117^{20}	**al, eth, ace,** bz	$B5^4$, 992
2053	Benzene, 1-chloro-4-triazo $4-N_3C_6H_4Cl$	153.57	....	96^{20}	20	....	....	eth	$B5^4$, 760
2054	Benzene, 1-chloro 2,4,6-trimethyl $2,4,6(CH_3)_3C_6H_2Cl$	154.64	....	$204-6, 104^{25}$	<-20	1.0337^{10}	1.5212^{10}	al, eth	$B5^4$, 1026
2055	Benzene, 1-chloro-2,4,5-trinitro $2,4,5-(NO_2)_3C_6H_2Cl$	247.55	ye lf (al)	....	116	....	....	al, bz, aa	$B5^4$, 757
2056	Benzene, 1-chloro-2,4,6-trinitro or Picryl chloride $2,4,6-(NO_2)_3C_6H_2Cl$	247.55	wh nd or pl (chl, al-lig)	....	83	1.797^{20}	....	al, ace, bz, to	$B5^4$, 757
2057	Benzene, cyclohexyl or Phenylcyclohexane $C_6H_5C_6H_{11}$	160.26	pl	$235-6, 127-8^{10}$	7-8	$0.9502^{20/4}$	1.5329^{20}	al, eth	$B5^4$, 1424
2058	Benzene, cyclopentyl or Phenylcyclopentane $C_6H_5C_5H_9$	146.23	....	$219, 102^{18}$	......	$0.9462^{20/4}$	1.5280^{20}	eth	$B5^4$, 1409
2059	Benzene, 1,2-diacetamido $1,2-(CH_3CONH)_2C_6H_4$	192.22	nd (w)	....	185-6	....	....	al, ace, aa, w	$B13^3$, 42
2060	Benzene, 1,4-diacetyl $1,4-(CH_3CO)_2C_6H_4$	162.19	pr (al, eth)	sub	114	....	....	al	$B7^3$, 3504
2061	Benzene, 1,2-diamino or *o*-Phenylenediamine $1,2-(H_2N)_2C_6H_4$	108.14	brsh ye lf (w), pl (chl)	256-8	102-3	....	....	al, eth, bz	$B13^3$, 28
2062	Benzene, 1,3-diamino or *m*-Phenylenediamine $1,3,-(H_2N)_2C_6H_4$	108.14	rh (al)	282-4	63-4	$1.0696^{58/4}$	1.6339^{58}	w, al, eth, bz	$B13^3$, 65
2063	Benzene, 1,4-diamino or *p*-Phenylenediamine $1,4-(H_2N)_2C_6H_4$	108.14	wh pl (bz, eth)	267	140	....	....	eth, chl, w, al	$B13^3$, 104
2064	Benzene, 1,2-dibenzoxy or Catechol dibenzyl ether $1,2-(C_6H_5CH_2O)_2C_6H_4$	290.36	yesh nd or pr (al)	....	63-4	....	....	al, eth, peth	$B6^3$, 4219
2065	Benzene, 1,4-dibenzoxy or Hydroquinone dibenzyl ether $1,4-(C_6H_5CH_2O)_2C_6H_4$	290.36	pl (al)	130	......	....	....	al, eth, aa	$B6^3$, 4402
2066	Benzene, 1,4-dibenzoyl $1,4-(C_6H_5CO)_2C_6H_4$	286.33	....	....	161	....	....	....	$B7^3$, 4299
2067	Benzene, 1,4-dibenzyl $1,4-(C_6H_5CH_2)_2C_6H_4$	258.36	....	225^{18}	87-8	....	....	al, bz	$B5^3$, 2334
2068	Benzene, 1,2-dibromo or *o*-Dibromobenzene $1,2-Br_2C_6H_4$	235.91	....	$225, 92^{10}$	7.1	$1.9843^{20/4}$	1.6155^{20}	al, **eth, ace, bz**	$B5^4$, 682

No.	Name, Synonyms, and Formula	Mol. wt.	Color, crystalline form, specific rotation and λ_{max} (log ε)	b.p. °C	m.p. °C	Density	n_D	Solubility	Ref.
2069	Benzene, (1,2-dibromoethyl) $C_6H_5CHBrCH_2Br$	263.96	....	139-41^{15}	74	....	....	al, eth, bz, lig	B5^4, 909
2070	Benzene, 1,2-dibromo-3-nitro $1,2\text{-}Br_2\text{-}3\text{-}NO_2C_6H_3$	280.90	mcl pr	....	85	....	....	eth, ace, chl	B5^4, 731
2071	Benzene, 1,2-dibromo-4-nitro $1,2\text{-}Br_2\text{-}4\text{-}NO_2C_6H_3$	280.90	nd (al, aa) mcl pr	296, 180^{20}	58-9	2.354^{8}	1.9835^{111}	al, bz, aa	B5^4, 732
2072	Benzene, 1,3-dibromo or *m*-Dibromobenzene $1,3\text{-}Br_2C_6H_4$	235.91	....	218, 66^{5}	-7	1.9523$^{20/4}$	1.6083^{17}	al, eth	B5^4, 682
2073	Benzene, 1,3-dibromo-2-nitro $1,3\text{-}Br_2\text{-}2\text{-}NO_2C_6H_3$	280.91	nd (al), mcl pr (al)	sub	84	1.9211^{111}, 2.211^{8}	....	al, ace, bz	B5^3, 621
2074	Benzene, 1,3-dibromo-5-nitro $1,3\text{-}Br_2\text{-}5\text{-}NO_2C_6H_3$	280.91	pl or pr (eth), lf or nd (al)	....	106	1.9341^{111}, 2.363^{8}	....	al, eth, bz	B5^3, 621
2075	Benzene, 1,4-dibromo or *p*-Dibromobenzene $1,4\text{-}Br_2C_6H_4$	235.91	pl	218-9	87.3	1.5742	....	al, eth, ace, bz	B5^4, 683
2076	Benzene, 1,5-dibromo-2,4-dimethyl or 4,6-Diamino-*m*-xylene $1,5\text{-}Br_2\text{-}2,4\text{-}(CH_3)_2C_6H_2$	263.98	pl (al)	255-6	68	....	....	al	B5^3, 839
2077	Benzene, 1,4-dibromo-2-nitro $1,4\text{-}Br_2\text{-}2\text{-}NO_2C_6H_3$	280.90	yesh pl (ace)	....	85-6	1.9416^{111}, 2.368^{8}	....	bz, ace	B5^4, 739
2078	Benzene, 2,3-dibromo 1,4,5-trimethyl $2,3\text{-}Br_2\text{-}1,4,5\text{-}(CH_3)_3C_6H$	277.99	nd (al)	....	63-4	....	....	al, eth, bz	B5^3, 910
2079	Benzene, 2,4-dibromo-1-nitro $2,4\text{-}Br_2C_6H_3NO_2$	280.90	ye pl or pr (al)	....	62	1.9581^{111}, 2.356^{8}	....	ace, bz	B5^3, 620
2080	Benzene, 2,4-dibromo-1,3,5-trimethyl $2,4\text{-}Br_2\text{-}1,3,5\text{-}(CH_3)_3C_6H$	277.99	nd (al)	285	65.5	....	....	al, bz	B5^3, 921
2081	Benzene, 1,4-di-*tert*-butyl or *p*-Di-*tert*-butyl benzene $1,4\text{-}[(CH_3)_3C]_2C_6H_4$	190.33	nd (MeOH)	237^{743}, 109^{15}	80-1	....	....	al, eth	B5^4, 1163
2082	Benzene, 1,2-dichloro or *o*-Dichlorobenzene $1,2\text{-}Cl_2\text{-}C_6H_4$	147.00	....	180.5, 86^{18}	-17	1.3048$^{20/4}$	1.5515^{20}	al, eth, ace, bz	B5^4, 654
2083	Benzene, (1,2-dichloroethyl) $C_6H_5CHClCH_2Cl$	175.06	....	233-4, 115^{15}	......	1.240$^{15/4}$	1.5544^{15}	ace, bz	B5^4, 901
2084	Benzene, 1,2-dichloro-3-nitro $1,2\text{-}Cl_2\text{-}3NO_2C_6H_3$	192.00	mcl nd (peth, aa)	257-8	61-2	1.721^{14}	....	al, eth, ace, bz	B5^4, 725
2085	Benzene, 1,2-dichloro-4-nitro $1,2\text{-}Cl_2\text{-}4\text{-}NO_2C_6H_3$	192.00	nd (al)	255-6, 189^{100}	43	1.4558$^{75/4}$	....	al, eth	B5^4, 726
2086	Benzene, 1,3-dichloro or *m*-Dichlorobenzene $1,3\text{-}Cl_2C_6H_4$	147.00	....	173, 53^{10}	-24.7	1.2884$^{20/4}$	1.5459^{20}	al, eth, ace, bz	B5^4, 657
2087	Benzene, 1,3-dichloro-2-nitro $1,3\text{-}Cl_2\text{-}2\text{-}NO_2C_6H_3$	192.00	nd or pr (al, CS_2)	130^{8}	72.5	1.603^{17}, 1.4094^{80}	....	eth, al	B5^4, 726
2088	Benzene, 1,3-dichloro-5-nitro $1,3\text{-}Cl_2\text{-}5\text{-}NO_2C_6H_3$	192.00	mcl pr or lf (aa, al)	....	65.4	1.692^{14}, 1.4000^{100}	....	eth, al	B5^4, 727
2089	Benzene, 1,4-dichloro or *p*-Dichlorobenzene $1,4\text{-}Cl_2\text{-}C_6H_4$	147.00	mcl pr, lf (ace)	174, 55^{10}	53.1	1.2475$^{20/4}$	1.5285^{20}	al, eth, ace, bz	B5^4, 685
2090	Benzene, 1,4-dichloro-2-iodo $1,4\text{-}Cl_2\text{-}2\text{-}IC_6H_3$	272.90	pl (al)	255-6, 134^{12}	23	....	....	al, eth, bz, chl	B5^4, 580
2091	Benzene, 1,4-dichloro-2-nitro $1,4\text{-}Cl_2\text{-}2\text{-}NO_2C_6H_3$	192.00	pl or pr (al), pl (AcOEt)	267	56	1.669^{22}, 1.439^{75}	....	al, eth, bz, chl	B5^4, 726
2092	Benzene, 2,4-dichloro-1-ethoxy $2,4\text{-}Cl_2C_6H_3OC_2H_5$	191.06	....	237	......	....	....	al, eth, bz	B6^4, 885
2093	Benzene, 2,4-dichloro-1-nitro $2,4\text{-}Cl_2\text{-}C_6H_3NO_2$	192.00	nd (al)	258.5, 100-1^{4}	34	1.4790^{80}	1.5512$^{78/}{}_{o}$	al, eth	B5^4, 726
2094	Benzene, 2,4-dichloro-1-triazo $2,4\text{-}Cl_2\text{-}C_6H_3N_3$	188.02	ye nd (al), pr (ace or bz)	....	54	....	....	al, eth, bz, peth	B5^2, 208
2095	Benzene, 2,4-dichloro-1,3,5-trimethyl $2,4\text{-}Cl_2\text{-}1,3,5\text{-}(CH_3)_3C_6H$	189.08	....	243.4	59	....	....	al	B5^3, 919
2096	Benzene, 2,5-dichloro-1-ethyl $2,5\text{-}Cl_2\text{-}C_6H_3C_2H_5$	175.06	....	213.5	......	1.239^{0}	....	bz	B5^4, 901
2097	Benzene, 1-dichlorophosphino-4-isopropryl $C_6H_4PCl_2\text{-}4\text{-}(CH_3)_2CH$	221.07	....	268-70, 129-30^{16}	......	1.1917^{25}	1.5677^{25}	ace	B16, 773
2098	Benzene, 1,2-diethyl or *o*-Diethyl benzene $1,2\text{-}(C_2H_5)_2C_6H_4$	134.22	....	183.4, 63^{10}	-31.2	0.8800^{20}	1.5035^{20}	al, eth, ace, bz, lig	B5^4, 1065
2099	Benzene, 1,3-diethyl or *m*-Diethyl benzene $1,3\text{-}(C_2H_5)_2C_6H_4$	134.22	....	181, 60^{10}	-83.9	0.8602$^{20/4}$	1.4955^{20}	al, eth, ace, bz	B5^4, 1066
2100	Benzene, 1,3-diethyl-5-methyl $1,3\text{-}(C_2H_5)_2\text{-}5\text{-}CH_3C_6H_3$	148.25	....	205, 79^{10}	-74.1	0.8748$^{20/4}$	1.5027^{20}	al, eth, ace, bz	B5^4, 1106

No.	Name, Synonyms, and Formula	Mol. wt.	Color, crystalline form, specific rotation and λ_{max} (log ε)	b.p. °C	m.p. °C	Density	n_D	Solubility	Ref.
2101	Benzene, 1,4-diethyl or *p*-Diethylbenzene $1,4\text{-}(C_2H_5)_2C_6H_4$	134.22	...	$183.8, 63^{10}$	−42.8	$0.8620^{20/4}$	1.4967^{20}	**al, eth, ace, bz**	$B5^4$, 1067
2102	Benzene, 1,2-difluoro or *o*-Difluorobenzene $1,2\text{-}F_2C_6H_4$	114.09	...	$91\text{-}2^{751}$	−34	$1.1599^{18/4}$	1.44506^{18}	ace, bz, chl	$B5^4$, 637
2103	Benzene, 1,2-dihydroxy or Catechol $2\text{-}HOC_6H_4OH$	110.11	cr	245	105	1.1493^{22}	1.604	w, al, eth, ace, bz	$B6^2$, 4187
2104	Benzene, 1,3-dihydroxy or Resorcinal $1,3\text{-}(HO)_2C_6H_4$	110.11	nd (bz), pl (w)	178^{16}	111	1.2717	...	w, al, eth, bz	$B6^3$, 4292
2105	Benzene, 1,4-dihydroxy or Hydroquinone $1,4\text{-}(HO)_2C_6H_4$	110.11	mcl pr (sub), nd (w), Pr (MeOH)	285^{730}	173-4	1.328^{15}	...	w, al, eth, ace	$B6^3$, 4374
2106	Benzene, 1,4-dicyclohexyl or *p*-Dicyclohexyl benzene $1,4\text{-}(C_6H_{11})_2C_6H_4$	242.40	...	120-3	...	...	...	...	$B5^4$, 1607
2107	Benzene, 1,4-difluoro or *p*-Difluorobenzene $1,4\text{-}F_2C_6H_4$	114.09	...	89^{760}	-13	$1.1688^{20/4}$	1.4422^{20}	ace, bz	$B5^4$, 637
2108	Benzene, 1,3-difluoro-4-nitro $1,3\text{-}F_2\text{-}4\text{-}NO_2C_6H_3$	159.09	...	207^{760}	9.8	1.4571^{14}	1.5149^{14}	chl	$B5^4$, 720
2109	Benzene, 1,2-diiodo or *o*-Diidobenzene $1,2\text{-}I_2C_6H_4$	329.91	pl or pr (lig)	$286^{750}, 109^{3}$	27	2.54^{20}	1.7179^{20}	eth	$B5^4$, 700
2110	Benzene, 1,3-diiodo or *m*-Diidobenzene $1,3\text{-}I_2C_6H_4$	329.91	rh pl or pr (eth-al)	285	40.4	2.47^{25}	...	al, eth, chl	$B5^4$, 700
2111	Benzene, 1,4-diiodo or *p*-Diidobenzene $1,4\text{-}I_2C_6H_4$	329.91	rh lf (al)	285 sub	131-2	...	...	al, eth	$B5^4$, 700
2112	Benzene, 1,2-dimethoxy-4,5-dinitro $4,5\text{-}(NO_2)_2\text{-}1,2\text{-}(CH_3O)_2C_6H_2$	228.16	ye nd (al)	...	130-2	$1.3164^{140/4}$	...	...	$B6^3$, 4274
2113	Benzene, 1,2-dimethoxy-4-iodo or 4-Iodoveratrole $1,2\text{-}(CH_3O)_2\text{-}4\text{-}IC_6H_3$	264.06	nd (dil MeOH)	$163\text{-}4^{26}, 90^{0.2}$	35	...	...	al, bz	$B6^4$, 4262
2114	Benzene, 1,2-dimethoxy-3-nitro $1,2\text{-}(CH_3O)_2\text{-}3\text{-}NO_2C_6H_3$	183.16	nd (al)	...	64-5	$1.1404^{112/4}$	...	al, eth, bz, aa	$B6^3$, 4263
2115	Benzene, 1,2-dimethoxy-4-nitro $1,2\text{-}(CH_3O)_2\text{-}4\text{-}NO_2C_6H_3$	183.16	ye nd (al-w)	$230^{15\text{-}20}$	98	$1.1888^{133/4}$	...	al, eth, chl	$B6^3$, 4264
2116	Benzene, 1,2-dimethoxy-4-propenyl(*cis*) or Isoengenol methyletha $1,2\text{-}(CH_3O)_2\text{-}4\text{-}CH_3CH{=}CH\text{-}C_6H_3$	178.23	...	$270.5, 138\text{-}40^{12}$	...	$1.0521^{20/4}$	1.5616^{20}	ace, bz	$B6^3$, 4995
2117	Benzene, 1,3-dimethoxy-2-nitro $1,3\text{-}(CH_3O)_2\text{-}2\text{-}NO_2C_6H_3$	183.16	ye nd (al or aa)	...	131	...	...	bz, chl, al	$B6^3$, 4344
2118	Benzene, 1,3-dimethoxy 5-nitro $1,3\text{-}(CH_3O)_2\text{-}5\text{-}NO_2C_6H_3$	183.16	pa ye nd (AcOEt)	...	89	$1.1693^{133/4}$	...	al, bz	$B6^3$, 4347
2119	Benzene, 1,4-dimethoxy-2-nitro $1,4\text{-}(CH_3O)_2\text{-}2\text{-}NO_2C_6H_3$	183.16	gold-ye nd (dil al)	169^{13} sub	72-3	$1.1666^{112/4}$	...	al, bz, chl	$B6^3$, 4442
2120	Benzene, 1,4-dimethoxy-2-*iso*-propyl $1,4\text{-}(CH_3O)_2\text{-}2\text{-}(CH_3)_2CHC_6H_3$	180.25	...	$114\text{-}6^{15}$	...	$1.0129^{17/4}$	1.5103^{17}	eth, bz	B6, 929
2121	Benzene, 2,4-dimethoxy-l-nitro $2,4\text{-}(CH_3O)_2\text{-}1\text{-}NO_2C_6H_3$	183.16	nd (al)	...	76-7	$1.1876^{132/4}$	...	al	$B6^2$, 822
2122	Benzene, 1,2-dimethyl-4-*iso*-propyl $1,2(CH_3)_2\text{-}4\text{-}(CH_3)_2CHC_6H_3$	148.25	...	$199, 86\text{-}7^{16}$	...	$0.8710^{20/4}$	$1.4951^{20/D}$	...	$B5^4$, 1103
2123	Benzene, 1,3-dimethyl-5-*iso*-propyl $1,3\text{-}(CH_3)_2\text{-}5\text{-}(CH_3)_2CHC_6H_3$	148.25	...	$83\text{-}5^{17}$	...	...	$1.4935^{25/D}$	...	$B5^4$, 1105
2124	Benzene, 1,2-dinitro or *o*-Dinitrobenzene $1,2\text{-}(NO_2)_2C_6H_4$	168.11	nd (bz) pl (al)	$319^{775}, 194^{30}$	118.5	$1.3119^{120/4}$, 1.565^{17}	...	al, bz, chl	$B5^4$, 738
2125	Benzene, 1,3-dinitro or *m*-Dinitrobenzene $1,3\text{-}(NO_2)_2C_6H_4$	168.11	rh pl (al)	$291^{756}, 167^{14}$	90	$1.5751^{18/4}$	...	al, eth, ace, bz, Py	$B5^4$, 379
2126	Benzene, 1,4-dinitro or *p*-Dinitro benzene $1,4\text{-}(NO_2)_2C_6H_4$	168.11	nd (al)	$298^{777}, 183^{34}$, sub	174	$1.625^{18/4}$	...	ace, bz, to, aa	$B5^4$, 741
2127	Benzene, 1,4-dinitro-2,3,5,6-tetramethyl or Dinitrodurene $1,4\text{-}(NO_2)_2\text{-}2,3,5,6\text{-}(CH_3)_4C_6$	224.22	pr (al)	sub	211-2	...	...	eth, bz	$B5^4$, 1080
2128	Benzene, 2,4-dinitro-1-fluoro $2,4\text{-}(NO_2)_2C_6H_3F$	186.10	...	$296, 178^{25}$	25.8	1.4718^{84}	...	al	$B5^4$, 742
2129	Benzene, 2,4-dinitro-1,3,5-trimethyl $2,4\text{-}(NO_2)_2\text{-}1,3,5\text{-}(CH_3)_3C_6H$	210.19	rh (al)	418 exp	86	...	...	al	$B5^4$, 1029
2130	Benzene, 1,4-dipropyl or *p*-Dipropyl benzene $1,4\text{-}(C_3H_7)_2C_6H_4$	162.27	...	109^{23}	...	$0.8563^{10.4/4}$	$1.4917^{19.4}$	...	$B5^4$, 1124
2131	Benzene, 1,2-diisopropyl or *o*-Diisopropyl benzene $1,2\text{-}(i\text{-}C_3H_7)_2C_6H_4$	162.27	...	$204, 115.4^{50}$	-57	$0.8701^{20/4}$	1.4960^{20}	**al, eth, ace, bz**	$B5^4$, 1125

No.	Name, Synonyms, and Formula	Mol. wt.	Color, crystalline form, specific rotation and λ_{max} (log ε)	b.p. °C	m.p. °C	Density	n_D	Solubility	Ref.
2132	Benzene, 1,3-di*iso*propyl or *m*-Diisopropyl benzene 1,3-$(i\text{-}C_3H_7)_2C_6H_4$	162.27	...	203.2, 75^{9}	-61	$0.8559^{20/4}$	1.4883^{20}	**al, eth, ace, bz**	$B5^4$, 1125
2133	Benzene, 1,4-di*iso*-propyl or *p*-Diisopropyl benzene 1,4-$(i\text{-}C_3H_7)_2C_6H_4$	162.27	...	210.3, 120^{50}	-17	$0.8568^{20/4}$	1.4898^{20}	**al, eth, ace, bz**	$B5^4$, 1126
2134	Benzene, 1,2-divinyl or *o*-Divinyl benzene 1,2$(CH_2{=}CH)_2Cl_6H_4$	130.19	...	76^{14}	...	$0.9325^{22/4}$	1.5767^{20}	ace, bz	$B5^4$, 1540
2135	Benzene, 1,3-divinyl or *m*-Divinyl benzene 1,3-$(CH_2{=}CH)_2C_6H_4$	130.19	...	121^{76}, 52^{3}	-52.3	$0.9294^{20/4}$	1.5760^{20}	ace, bz	$B5^2$, 1367
2136	Benzene, 1,4-divinyl or *p*-Divinyl benzene 1,4-$(CH_2{=}CH)_2C_6H_4$	130.19	...	95-6^{18}, $34^{0.2}$	31	0.913^{40}	1.5835^{25}	ace, bz	$B5^4$, 1541
2137	Benzene, dodecyl $C_6H_5C_{12}H_{25}$	246.44	...	331, 185-8^{15}	-7	$0.8551^{20/4}$	1.4824^{20}	...	$B5^4$, 1200
2138	Benzene, (epoxy isopropyl) $C_6H_5OCH(CH_3)_2$	136.19	...	83-5^{15}	...	1.0280^{20}	1.5232^{20}	...	$B17^4$, 410
2139	Benzene, ethoxy or Phenetole $C_6H_5OC_2H_5$	122.17	...	170, 60^{9}	-29.5	$0.9666^{20/4}$	1.5076^{20}	al, eth	$B6^4$, 554
2140	**Benzene, ethyl or ethylbenzene** $C_6H_5C_2H_5$	106.17	...	136.2, 25.8^{10}	-95	$0.8670^{20/4}$	1.4959^{20}	**al, eth**	$B5^4$, 885
2141	Benzene, l-ethyl-4-*iso*-butyl 4-$i\text{-}C_4H_9C_6H_4C_2H_5$	162.27	...	211	...	...	...	eth, ace	$B5^4$, 1123
2142	Benzene, l-ethyl-2-Iodo 2-$IC_6H_4C_2H_5$	232.06	...	226	...	$1.6189^{10/4}$	1.5941^{22}	ace, bz	$B5^4$, 970
2143	Benzene, l-ethyl-4-iodo 4-$IC_6H_4C_2H_5$	232.06	...	209	-17	$1.6095^{10/4}$	1.5909^{22}	ace, bz	$B5^4$, 801
2144	Benzene, l-ethyl-2-methyl or 2-Ethyl toluene 2-$CH_3C_6H_4C_2H_5$	120.19	...	165.2, 62.3^{20}	-80.8	0.8807^{20}	1.5046^{20}	**al, eth, ace, bz**	$B5^4$, 999
2145	Benzene, l-ethyl-3-methyl or 3-Ethyl toluene 3-$CH_3C_6H_4C_2H_5$	120.19	...	161.3, 45.6^{10}	-95.5	$0.8645^{20/4}$	1.4966^{20}	al, eth, **ace, bz** peth	$B5^4$, 1001
2146	Benzene, l-ethyl-4-methyl or 4-Ethyl toluene 4-$CH_3C_6H_4C_2H_5$	120.19	...	162, 45.6^{10}	-62.3	$0.8614^{20/4}$	1.4959^{20}	al, eth, **ace, bz** peth	$B5^4$, 1003
2147	Benzene, l-ethyl-2-nitro 2-$NO_2C_6H_4C_2H_5$	151.16	...	228, 116^{22}	-23	1.1345^{0}	...	al, eth, ace	$B5^4$, 911
2148	Benzene, l-ethyl-3-nitro 3-$NO_2C_6H_4C_2H_5$	151.16		242-3	...	1.1345^{0}	...	al, eth, ace	$B5^4$, 911
2149	Benzene, l-ethyl-4-nitro 4-$NO_2C_6H_4C_2H_5$	151.16	...	245-6, 134-6^{23}	-12.3	$1.1192^{20/4}$	1.5455^{20}	al, eth, ace	$B5^4$, 912
2150	Benzene, l-ethyl-4-propyl 4-$C_3H_7C_6H_4C_2H_5$	148.25	...	205	...	$0.8594^{20/4}$	1.4921^{20}	al, eth, ace, bz	$B5^4$, 1100
2151	Benzene, l-ethyl-3-*iso*-propyl 3-$i\text{-}C_3H_7C_6H_4C_2H_5$	148.25	...	192, 69.2^{10}	<-20	$0.859^{22/4}$	1.4921^{20}	al, eth, bz	$B5^3$, 1105
2152	Benzene, l-ethyl-4-*iso*-propyl 4-$i\text{-}C_3H_7C_6H_4C_2H_5$	148.25	...	136.6, 73^{10}	<-20	$0.8585^{20/4}$	1.4923^{20}	eth, bz	$B5^4$, 1100
2153	Benzene, fluoro or Phenyl fluoride C_6H_5F	96.10	...	85.1	-41.2	$1.0225^{20/4}$	1.4684^{40}	**al, eth, ace, bz, lig**	$B5^4$, 632
2154	Benzene, l-fluoro-2-iodo 2-IC_6H_4F	222.00	...	188.6, 78^{20}	...	...	1.5910^{20}	ace, bz, chl	$B5^4$, 693
2155	Benzene, l-fluoro-4-iodo 4-IC_6H_4F	222.00	...	182-4, 67-9^{11}	(i)-27 (ii)-18	1.9523^{15}	1.5270^{22}	al, eth, ace	$B5^4$, 694
2156	Benzene, l-fluoro-2-nitro 2-$NO_2C_6H_4F$	141.10	ye	214.6d, 86-7^{11}	-6	$1.3285^{18/4}$	1.5489^{17}	al, eth	$B5^4$, 718
2157	Benzene, l-fluoro-3-nitro 3-$NO_2C_6H_4F$	141.10	ye	198-200	41	$1.3254^{19/4}$	1.5262^{15}	al, eth	$B5^4$, 719
2158	Benzene, l-fluoro-4-nitro 4-$NO_2C_6H_4F$	141.10	ye nd	206-7, 87^{14}	(i) 27 (st), (ii) 3 (unst)	$1.3300^{20/4}$	1.5316^{20}	al, eth	$B5^4$, 719
2159	Benzene, l-fluoro-2,4,6-trimethyl 2,4,6-$(CH_3)_3C_6H_2F$	138.18	...	168.7	-36.7	$0.9745^{25/4}$	1.4809^{25}	ace, bz, chl	$B5^4$, 1225
2160	Benzene, hexabromo C_6Br_6	551.49	mcl nd (bz)	...	327	...	...	bz, chl, peth	$B5^4$, 687
2161	Benzene, hexachloro C_6Cl_6	284.78	nd (bz-al)	322 sub	230	$1.5691^{23.6}$	...	eth, bz, chl	$B5^4$, 670
2162	Benzene, hexaethyl $C_6(C_2H_5)_6$	246.44	mcl pr (al or bz)	298	129	0.8305^{130}	1.4736^{130}	al, eth, bz	$B5^4$, 1208
2163	Benzene, hexafluoro C_6F_6	186.06	...	80.5	5.3	1.6184^{20}	$1.37774^{20/D}$	...	$B5^4$, 640
2164	Benzene, heptyl $C_6H_5C_7H_{15}$	176.30	...	245.5, 116^{12}	−48	$0.8567^{20/4}$	1.4865^{20}	bz, chl	$B5^4$, 1143

No.	Name, Synonyms, and Formula	Mol. wt.	Color, crystalline form, specific rotation and λ_{max} (log ε)	b.p. °C	m.p. °C	Density	n_D	Solubility	Ref.
2165	Benzene, hexahydroxy or Hexaphenol $C_6(OH)_6$	174.11	nd (w)		>300				$B6^{3}$, 6922
2166	Benzene, hexaiodo C_6I_6	833.49	red-br nd or mcl pr (bz)		350d				$B5^{3}$, 585
2167	Benzene, hexamethyl or Mellitene $C_6(CH_3)_6$	162.27	rh pr or nd (al)	265	166-7	1.0630^{25}		al, eth, ace, bz, aa	$B5^{4}$, 1137
2168	Benzene, 1-hydroxymethyl-4-*iso*-propyl or Cumic alcohol 4-i-$C_3H_7C_6H_4CH_2OH$	150.22		249	28	$0.9818^{20/4}$	1.5210^{20}	al, eth	$B6^{3}$, 1911
2169	Benzene, iodo or Phenyl iodide C_6H_5I	204.01		188.3, 75^{10}	−31.3	$1.8308^{20/4}$	1.6200^{20}	al, eth, ace, bz	$B5^{4}$, 688
2170	Benzene, (2-iodoethoxy) $C_6H_5OCH_2CH_2I$	248.06	cr (dil al)		31-2				$B6^{3}$, 548
2171	Benzene, 1-iodomethyl-2-methyl 2-$CH_3C_6H_4CH_2I$	232.06	nd (eth, (peth)		33-4			eth	$B5^{2}$, 286
2172	Benzene, 1-iodomethyl-4-methyl 4-$CH_3C_6H_4CH_2I$	232.06	nd (eth) (peth)		46-7			bz, eth, peth	$B5^{4}$, 970
2173	Benzene, 1-iodo-2-nitro 2-$NO_2C_6H_4I$	249.01	ye rh nd (al)	$288\text{-}9^{729}$, 162^{18}	54	1.9186^{75}		al, eth	$B5^{2}$, 190
2174	Benzene, 1-iodo-3-nitro 3-$NO_2C_6H_4I$	249.01	mcl pr	280, 153^{14}	(i) 38.5 (st), (ii)10 (unst)	$1.9477^{50/4}$		al, eth	$B5^{4}$, 733
2175	Benzene, 1-iodo-4-nitro 4-$NO_2C_6H_4I$	249.01	ye nd (al)	289^{772}	174			al, aa	$B5^{2}$, 191
2176	Benzene, 1-iodo-2-triazo 2-$N_3C_6H_4I$	245.02	ye	$90\text{-}100^{0.9}$		$1.8893^{25/4}$	1.6631^{25}	al, ace	B5, 278
2177	Benzene, iodoso C_6H_5OI	220.01	ye pw		210 exp			al, w	$B5^{4}$, 692
2178	Benzene, iodoxy $C_6H_5IO_2$	236.01	nd (w)		236-7 exp			w, aa	$B5^{4}$, 693
2179	Benzene, *isocyano or Phenylisocyanate* C_6H_5NCO	119.12		$162\text{-}3^{751}$, 55^{13}		$1.0956^{19.6/4}$	$1.5368^{19.6}$	eth, ace	$B12^{3}$, 903
2180	Benzene, methoxy or Anisole $C_6H_5OCH_3$	108.14		155	−37.5	$0.9961^{20/4}$	1.5179^{20}	al, eth, ace, bz	$B6^{4}$, 548
2181	Benzene, (3-methylbutoxy) or Phenyl isopentyl ether C_6H_5O-*i*-C_5H_{11}	164.25		225		$0.9198^{22/4}$	1.4872^{20}	w	$B6^{3}$, 553
2182	Benzene, (1-methylbutyl) $C_6H_5CH(CH_3)C_3H_7$	148.25		$198\text{-}9^{757}$		$0.8594^{21/4}$	1.4875^{25}	al, eth	$B5^{4}$, 1087
2183	Benzene, (2-methylbutyl) $C_6H_5CH_2CH(CH_3)CH_2CH_3$	148.25		193.8, 102^{15}		$0.8584^{20/4}$	1.4873^{20}		$B5^{4}$, 1089
2184	Benzene, (2-methyl-1-propenyl) $C_6H_5CH{=}C(CH_3)_2$	132.21		$99^{43.5}$		$0.9029^{20/4}$	$1.5388^{20/D}$		$B5^{4}$, 1380
2185	Benzene, (2-methyl-2-butyl) $C_6H_5C(CH_3)_2CH_2CH_3$	148.25		192-4, $71\text{-}2^{12}$		$0.8587^{20/4}$	1.4934^{20}	**al, eth**	$B5^{4}$, 1090
2186	Benzene, (3-methylbutyl) or isopentyl benzene i-$C_5H_{11}C_6H_5$	148.25		199		$0.8558^{20/4}$	1.4853^{20}	al, eth, bz	$B5^{4}$, 1089
2187	Benzene, (4-methylpentyl) or Isohexyl benzene i-$C_6H_{13}C_6H_5$	162.27		$214\text{-}5^{740}$		0.8568^{16}		al, eth, bz	$B5^{4}$, 1116
2188	Benzene, 1-methyl-4-(nitromethyl) or 4-Nitromethyl toluene 4-$(O_2NCH_2)C_6H_4CH_3$	151.16		$150\text{-}1^{35}$	11-2	$1.1234^{20/4}$	1.5278^{20}	bz	$B5^{4}$, 972
2189	Benzene, 1-methyl-2-*iso*-propenyl or 2-*iso*-Propenyl toluene 2-[$CH_2{=}C(CH_3)$]$C_6H_4CH_3$	132.21		175, $59\text{-}62^{11}$		$0.9181^{15/0}$	$1.5112^{30/0}$		$B5^{3}$, 1214
2190	Benzene, 1-methyl-2-*iso*-propyl or l-cymene 2-i-$C_3H_7C_6H_4CH_3$	134.22		178.1, 57.3^{10}	−71.5	0.8766^{2014}	1.5006^{20}	al, eth, ace, bz	$B5^{4}$, 1383
2191	Benzene, 1-methyl-3-*iso*-propyl or *m*-Cymene 3-i-$C_3H_7C_6H_4CH_3$	134.22		175.1, 55^{10}	-63.7	$0.8610^{20/4}$	1.4930^{20}	**al, eth, ace, bz**	$B5^{4}$, 1383
2192	Benzene, 1-methyl-4-*iso*-propyl or *p*-Cymene 4-i-$C_3H_7C_6H_4CH_3$	134.22		177.1, 56.3^{10}	-68	$0.8573^{20/4}$	1.4909^{20}	**al, eth, ace, bz**	$B6^{4}$, 1466
2193	Benzene, methylthio or Thioanisole $CH_3SC_6H_5$	124.20		193, 74^{10}		$1.0579^{20/4}$	1.5868^{20}	al, bz	$B5^{4}$, 708
2194	Benzene, nitro $C_6H_5NO_2$	123.11		210.8	5.7	$1.2037^{20/4}$	1.5562^{20}	al, eth, ace, bz	$B5^{4}$, 998
2195	Benzene, (α-nitro-*iso*-propyl) $C_6H_5C(NO_2)(CH_3)_2$	165.19		224d, $125\text{-}7^{15}$		1.1025^{20}	1.5209^{20}	ace, bz	$B5^{4}$, 728

PHYSICAL CONSTANTS OF ORGANIC COMPOUNDS (Continued)

No.	Name, Synonyms, and Formula	Mol. wt.	Color, crystalline form, specific rotation and λ_{max} (log ε)	b.p. °C	m.p. °C	Density	n_D	Solubility	Ref.
2196	Benzene, nitro,pentachloro or Brassicol $C_6Cl_5NO_2$	295.34	cr (al)	....	144	$1.718^{25/4}$	....	bz, chl	B5^4, 728 6
2197	Benzene, 1-nitro-2,3,5,6-tetrachloro 2,3,5,6-$Cl_4C_6HNO_2$	260.89	....	....	99-100	$1.744^{25/4}$	....	al, bz, chl	B5^4, 728
2198	Benzene, 1-nitro-2-triazo 2-$N_3C_6H_4NO_2$	164.12	ye nd (bz-al), pr (al)	....	53-5	....	....	al, bz, aa, chl	B5^4, 761
2199	Benzene, 1-nitro-3-triazo 3-$N_3C_6H_4NO_2$	164.12	wh nd (dil al or peth)	....	56	....	....	al, eth, bz	B5^4, 761
2200	Benzene, 1-nitro-4-triazo 4-$N_3C_6H_4NO_2$	164.12	wh pl (dil al)	....	75	....	....	al, eth, bz, aa	B5^4, 762
2201	Benzene, 1-nitro-2,3,4-tribromo 2,3,4-$Br_3C_6H_2NO_2$	359.80	cr (al)	....	85.4	....	....	al, eth, bz	B5, 251
2202	Benzene, 1-nitro-2,3,5-tribromo 2,3,5-$Br_3C_6H_2NO_2$	359.80	nd	....	119	....	....	eth, bz	B5, 251
2203	Benzene, 1-nitro-2,3,6-tribromo 2,3,6-$Br_3C_6H_2NO_2$	359.80	pl or pr (eth-al)	....	185 (sub)	....	....	al, eth, bz	B5, 251
2204	Benzene, 1-nitro-2,4,5-tribromo 2,4,5-$Br_3C_6H_2NO_2$	359.80	nd (al)	....	95 (sub)	....	....	al, eth	B5^3, 621
2205	Benzene, 1-nitro-2,4,6-tribromo 2,4,6-$Br_3C_6H_2NO_2$	359.80	mcl pr (chl)	177^{11}	125	....	....	eth, chl, aa	B5^4, 732
2206	Benzene, 1-nitro-3,4,5-tribromo 3,4,5-$Br_3C_6H_2NO_2$	359.80	tcl (eth-al)	sub	112	2.645	....	eth, bz	B5^4, 732
2207	Benzene, 1-nitro-2,3,4-trichloro 2,3,4-$Cl_3C_6H_2NO_2$	226.45	nd (al)	....	55.5	....	....	CS_2	B5^2, 186
2208	Benzene, 1-nitro-2,4,5-trichloro 2,4,5-$Cl_3C_6H_2NO_2$	226.45	pr (al), nd (al)	288	57-8	1.790^{23}	....	al, eth, bz	B5^4, 728
2209	Benzene, 1-nitro-2,3,6-trichloro 2,3,6-$Cl_3C_6H_2NO_2$	226.45	nd (al)	....	89	....	....	al	B5^2, 187
2210	Benzene, 1-nitro-2,4,6-trichloro 2,4,6-$Cl_3C_6H_2NO_2$	226.45	nd (al)	....	71	....	....	al, lig	B5^4, 728
2211	Benzene, 1-nitro-3,4,5-trichloro 3,4,5-$Cl_3C_6H_2NO_2$	226.45	pa ye tcl (al)	....	72.5	1.807	....	al	B5^2, 187
2212	Benzene, 1-nitro-2,3,5-triiodo 2,3,5-$I_3C_6H_2NO_2$	500.80	ye pr	....	124	....	....	bz	B5, 256
2213	Benzene, 1-nitro-2,3,6-triiodo 2,3,6-$I_3C_6H_2NO_2$	500.80	nd (aa)	....	137	....	....	aa	B5, 256
2214	Benzene, 1-nitro-2,4,5-triiodo 2,4,5-$I_3C_6H_2NO_2$	500.80	pa ye nd (CS_2)	....	178	....	....	....	B5, 256
2215	Benzene, 1-nitro-3,4,5-triiodo 3,4,5-$I_3C_6H_2NO_2$	500.80	ye pr (chl), nd (al)	....	167	3.256	....	eth, bz, aa, chl	B5^3, 626
2216	Benzene, 1-nitro-2,3,5-trimethyl 2,3,5-$(CH_3)_3$-$C_6H_2NO_2$	165.19	pr (al)	$139\text{-}40^{7}$	20	....	....	al	B5^4, 1015
2217	Benzene, 1-nitro-2,4,5-trimethyl 2,4,5-$(CH_3)_3C_6H_2NO_2$	165.19	ye nd (al)	265	71	....	....	al, peth	B5^4, 1015
2218	Benzene, 1-nitro-2,4,6-trimethyl 2,4,6-$(CH_3)_3C_6H_2NO_2$	165.19	rh pr (al)	255	44	1.51	....	al	B5^4, 1028
2219	Benzene, nitroso C_6H_5NO	107.11	rh or mcl (al-eth)	$57\text{-}9^{18}$	68-9	....	....	al, eth, bz, lig	B5^4, 702
2220	Benzene, nonyl $C_6H_5C_9H_{19}$	204.36	....	280.1	......	$0.8584^{20/4}$	$1.4816^{20/}$	....	B5^3, 1075
2221	Benzene, pentamino $C_6H(NH_2)_5$	153.19	dk red nd	....	228d	....	....	w	B13^2, 155
2222	Benzene, pentabromo C_6HBr_5	472.59	wh nd (aa or al)	sub	160-1	....	....	bz, chl	B5^4, 687
2223	Benzene, pentachloro C_6HCl	250.34	nd (al)	277	86	$1.8342^{16.5}$	....	....	B5^4, 669
2224	Benzene, pentaethyl $C_6H(C_2H_5)_5$	218.38	....	277	<−20	$0.895^{19/19}$	1.5127^{20}	....	B5^4, 1197
2225	Benzene, pentaiodo C_6HI_5	707.60	nd (al)	sub	172	....	....	bz, chl, aa	B5^4, 702
2226	Benzene, pentamethyl $C_6H(CH_3)_5$	148.25	pr (al)	232, 100^{10}	54.5	$0.917^{20/4}$	1.527^{20}	al, bz	B5^4, 1109
2227	Benzene, pentoxy or Phenyl pentyl ether $C_6H_5OC_5H_{11}$	164.25	....	200^{756}, 111^{17}	......	$0.9270^{20/4}$	1.4947^{20}	al, ace	B6^3, 552
2228	Benzene, pentyl $C_6H_5C_5H_{11}$	148.25	....	205.4, 80.6^{10}	-75	$0.8585^{20/4}$	1.4878^{20}	al, eth, ace, bz	B5^4, 1085

No.	Name, Synonyms, and Formula	Mol. wt.	Color, crystalline form, specific rotation and λ_{max} (log ε)	b.p. °C	m.p. °C	Density	n_D	Solubility	Ref.
2229	Benzene, (2-pentyl) $C_6H_5CH(CH_3)C_3H_7$	148.25	...	$198\text{-}9^{757}$	...	$0.8594^{21/4}$	1.4875^{21}	al, eth	$B5^4$, 1087
2230	Benzene, propenyl (*cis*) $C_6H_5CH{=}CHCH_3$	118.18	...	69^{28}	-60.5	$0.9088^{20/4}$	1.5420^{20}	al, eth, ace, bz	$B5^4$, 1359
2231	Benzene, propenyl (*trans*) $C_6H_5CH{=}CHCH_3$	118.18	...	175.6	-27^1	0.9019^{25}	1.5508^{20}	al, eth, ace, bz	$B5^4$, 1360
2232	Benzene, 1-propenyl-2,4,5-trimethoxy (*α* form) or *α* Asaron $2,4,5\text{-}(CH_3O)_3\text{-}C_6H_2CH{=}CHCH_3$	208.34	mcl nd (w)	296, $167\text{-}8^{12}$	67	$1.165^{20/4}$	1.5683^{20}	al, eth, peth	$B6^3$, 6441
2233	Benzene, *iso*-propyl $C_6H_5C(CH_3){=}CH_2$	118.18	...	165.4, 48.5^{10}	-23.2	$0.9106^{20/4}$	1.5386^{20}	al, eth, ace, bz	$B5^4$, 1364
2234	Benzene, propoxy or Phenyl propyl ether $C_6H_5OC_3H_7$	136.19	...	189.9	-27	$0.9474^{20/4}$	1.5014^{20}	al, eth	$B6^4$, 556
2235	Benzene, *iso*-propoxy or Phenyl-*iso*-propyl ether $C_6H_5OCH(CH_3)_2$	136.19	...	176.8	-33	0.9408	1.4975^{20}	w, al, ace, bz	$B6^3$, 549
2236	Benzene, propyl $C_6H_5C_3H_7$	120.19	...	159.2, 43.3^{10}	-99.5	$0.8620^{20/4}$	1.4920^{20}	al, eth, ace, bz	$B5^4$, 977
2237	Benzene, *iso*-propyl or Cumene $(CH_3)_2CHC_6H_5$	120.19	...	152.4, 38.2^{10}	-96	$0.8618^{20/4}$	1.4915^{20}	al, eth, ace, bz	$B5^4$, 985
2238	Benzene, 1-*iso*-propyl-2,4,5-trimethyl $2,4,5\text{-}(CH_3)_3C_6H_2\text{-}i\text{-}C_3H_7$	162.27	...	218-20	...	$0.8795^{21/4}$	$1.50648^{21/D}$	...	$B5^4$, 1038
2239	Benzene, 1,2,3,5-tetrabromo $1,2,3,5\text{-}Br_4C_6H_2$	393.70	nd (al)	329	99-100	...	...	eth, bz, al	$B5^4$, 686
2240	Benzene, 1,2,4,5-tetrabromo $1,2,4,5\text{-}Br_4C_6H_2$	393.70	mcl pr (CS_2)	...	182	3.072^{20}	...	eth	$B5^4$, 687
2241	Benzene, 1.2.3.4-tetrachloro $1,2,3,4\text{-}Cl_4C_6H_2$	215.89	nd (al)	254	47.5	...	...	eth, lig, aa	$B5^4$, 667
2242	Benzene, 1,2,3,5-tetrachloro $1,2,3,5\text{-}Cl_4C_6H_2$	215.89	nd (al)	246	54.5	...	...	eth, bz, lig, al	$B5^4$, 686
2243	Benzene, 1,2,4,5-tetrachloro $1,2,4,5\text{-}Cl_4C_6H_2$	215.89	nd mcl pr (eth, al or bz)	243-6	139-40	...	...	eth, bz, chl	$B5^4$, 686
2244	Benzene, 1,2,3,4-tetraethyl $1,2,3,4\text{-}(C_2H_5)_4C_6H_2$	190.33	...	251^{714}, 121.7^{14}	11.8	$0.8875^{20/4}$	1.5125^{20}	al, eth	$B5^4$, 1168
2245	Benzene, 1,2,4,5-tetraethyl $1,2,4,5\text{-}(C_2H_5)_4C_6H_2$	190.33	...	250	10	$0.8788^{20/4}$	1.5054^{20}	al, eth	$B5^4$, 1168
2246	Benzene, 1,2,3,4-tetrafluoro $1,2,3,4\text{-}F_4C_6H_2$	150.08	...	93-4	...	...	$1.4054^{20/D}$	...	$B5^4$, 639
2247	Benzene, 1,2,4,5-tetrafluoro $1,2,4,5\text{-}F_4C_6H_2$	150.08	...	89-90	4-5	1.4255^{20}	1.4075^{20}	...	$B5^4$, 639
2248	Benzene, 1,2,3,5-tetrahydroxy $1,2,3,5\text{-}(HO)_4C_6H_2$	142.11	nd (w), lf (eth-AcOEt)	...	165-7	...	...	w, al, ace	$B6^3$, 6652
2249	Benzene, 1,2,4,5-tetrahydroxy $1,2,4,5\text{-}(HO)_4C_6H_2$	142.11	lf (w or aa)	...	232.5	...	...	w, al, ace	$B6^3$, 6655
2250	Benzene, 1,2,3,4-tetraiodo $1,2,3,4\text{-}I_4C_6H_2$	581.70	pr (eth-aa)	sub	136	...	...	al, eth, aa	B5, 229
2251	Benzene, 1,2,3,5-tetraiodo $1,2,3,5\text{-}I_4C_6H_2$	581.70	pr (eth, aa)	sub	148	...	...	aa	B5, 229
2252	Benzene, 1,2,4,5-tetraiodo $1,2,4,5\text{-}I_4C_6H_2$	581.70	pr (bz), nd (eth)	sub (vac)	254 (165)	...	...	aa	B5, 229
2253	Benzene, 1,2,3,4-tetramethyl or Prebnitene $1,2,3,4\text{-}(CH_3)_4C_6H_2$	134.22	cr (peth)	205, 79.4^{10}	-6.2	$0.9052^{20/4}$	1.5203^{20}	al, eth, ace, bz	$B5^4$, 1072
2254	Benzene, 1,2,3,5-tetramethyl or Isodurene $1,2,3,5\text{-}(CH_3)_4C_6H_2$	134.22	...	198, 74.4^{10}	-23.7	$0.8903^{20/4}$	1.5130^{20}	al, eth, ace, bz	$B5^4$, 1073
2255	Benzene, 1,2,4,5-tetramethyl or Durene $1,2,4,5\text{-}(CH_3)_4C_6H_2$	134.22	...	196.8, 73.5^{10}	79.2	$0.8380^{81/4}$	1.4790^{81}	al, eth, ace, bz	$B5^4$, 1076
2256	Benzene, 1,2,4,5-tetra-*iso*propyl $1,2,4,5\text{-}(i\text{-}C_3H_7)_4C_6H_2$	246.44	...	260^{775}, 133^{17}	118.4	$0.758^{150/4}$	...	...	$B5^4$, 1207
2257	Benzene, 1,3,5-triacetyl $1,3,5\text{-}(CH_3CO)_3C_6H_3$	204.23	nd (w,aa or al)	...	163	...	...	aa	$B7^3$, 4579
2258	Benzene, 1,2,3-triamino $1,2,3\text{-}(H_2N)_3C_6H_3$	123.16	cr (dil HCl)	336	103	...	...	w, al, eth	$B13^3$, 551
2259	Benzene, 1,2,4-triamino $1,2,4\text{-}(H_2N)_3C_6H_3$	123.16	pl or lf (chl)	340	95-8	...	...	w, al, chl	$B13^3$, 552
2260	Benzene, triazo or Phynylazide $C_6H_5N_3$	119.13	pa ye oil	70^{11}	-27.5	$1.0880^{20/20}$	1.5589^{25}	...	$B5^3$, 759

No.	Name, Synonyms, and Formula	Mol. wt.	Color, crystalline form, specific rotation and λ_{max} (log ε)	b.p. °C	m.p. °C	Density	n_D	Solubility	Ref.
2261	Benzene, 1,2,3-tribromo $1,2,3\text{-}Br_3C_6H_3$	314.80	pl (al)	...	87.8	2.658	...	eth	$B5^3$, 569
2262	Benzene, 1,2,4-tribromo $1,2,4\text{-}Br_3C_6H_3$	314.80	nd (al or eth)	275	44-5	...	...	al, eth, ace	$B5^4$, 685
2263	Benzene, 1,3,5-tribromo $1,3,5\text{-}Br_3C_6H_3$	314.80	nd or pr (al)	271^{765}	121-2	...	...	eth, bz, chl	$B5^4$, 685
2264	Benzene, 2,4,6-tribromo-1,3,5-trimethyl $1,3,5\text{-}(CH_3)_3C_6(Br_3)$	356.90	tcl nd (al or bz)	...	227-8	...	...	bz	$B5^3$, 921
2265	Benzene, 1,2,3-trichloro $1,2,3\text{-}Cl_3C_6H_3$	181.45	pl (al)	218-9	53-4	...	...	eth, bz	$B5^4$, 664
2266	Benzene, 1,2,4-trichloro $1,2,4\text{-}Cl_3C_6H_3$	181.45	rh	213.5, 84.8^{10}	17	$1.4542^{20/4}$	1.5717^{20}	eth	$B5^4$, 664
2267	Benzene, 1,3,5-trichloro $1,3,5\text{-}Cl_3C_6H_3$	181.45	nd	208^{763}	63-4	...	...	eth, bz, lig	$B5^4$, 666
2268	Benzene, 1,2,3-trichloro-4,5,6-trihydroxy $1,2,3\text{-}Cl_3C_6(OH)_3$	229.45	nd (al or bz), nd (+3w, w)	...	185	...	...	w, al, eth	B6, 1084
2269	Benzene, 1,2,4-trichloro-3,5,6-trihydroxy $1,2,4\text{-}Cl_3C_6(OH)_3$	229.45	nd (bz or aa)	...	160	...	...	al, eth	$B6^2$, 1072
2270	Benzene, 1,3,5-trichloro-2,4,6-trihydroxy $1,3,5\text{-}Cl_3C_6(OH)_3$	229.45	cr (al)	sub	136	...	...	al	B6, 1104
2271	Benzene, 1,3,5-triethoxy or Phloroglucinol triethylether $1,3,5\text{-}(C_2H_5O)_3C_6H_3$	210.27	cr (al, dil al)	175^{24}	43.5	...	...	al, eth	$B6^3$, 6306
2272	Benzene, 1,2,4-triethyl $1,2,4\text{-}(C_2H_5)_3C_6H_3$	162.27	...	217.5^{755}, 99^{15}	...	$0.8738^{20/4}$	1.5024^{20}	al, eth	$B5^4$, 1133
2273	Benzene, 1,3,5-triethyl $1,3,5\text{-}(C_2H_5)_3C_6H_3$	162.27	...	216	-66.5	1.4969^{20}	...	al, eth	$B5^4$, 1133
2274	Benzene, 1,2,3-trihydroxy or Pyrogallol $1,2,3\text{-}(HO)_3C_6H_3$	126.11	lf or nd (bz)	309, 171^{12}	133-4	$1.453^{4/4}$	1.561^{134}	w, al, eth	$B6^3$, 2620
2275	Benzene, 1,2,5-trihydroxy, triacetate $1,2,5\text{-}(CH_3CO_2)_3C_6H_3$	252.22	pr (al)	...	165	...	...	al	$B6^3$, 6282
2276	Benzene, 1,2,4-trihydroxy or Hydroxyquinol $1,2,4\text{-}(HO)_3C_6H_3$	126.11	pl (eth), lf or pl (w)	...	140-1	...	...	w, al, eth	$B6^3$, 6272
2277	Benzene, 1,2,4-trihydroxy,triacetate $1,2,4\ (CH_3CO_2)C_6H_3$	252.22	nd (MeOH)	>300	97-8	...	...	al	$B6^3$, 6282
2278	Benzene, 1,3,5-trihydroxy or Phloroglucinol $1,3,5\text{-}(HO)_3C_6H_3$	126.11	lf or pl (w + 2)	sub	117 (hyd), 218-9 (anh)	1.46	...	al, eth, bz	$B6^3$, 6301
2279	Benzene, 1,3,5-trihydroxy, triacetate or Phloroglucinol triacetate $1,3,5\text{-}(CH_3CO_2)C_6H_3$	252.22	pr (w), hd (dil al)	...	105-6	...	...	al	$B6^3$, 6306
2280	Benzene, 1,2,3-triiodo $1,2,3\text{-}I_3C_6H_3$	455.80	nd (al), pr (bz)	sub	116	...	...	al, eth, chl	$B5^1$, 122
2281	Benzene, 1,2,4-triiodo $1,2,4\text{-}I_3C_6H_3$	455.80	nd (al)	sub	91.5	...	...	eth, chl	$B5^3$, 585
2282	Benzene, 1,3,5-triiodo $1,3,5\text{-}I_3C_6H_3$	455.80	nd (aa)	sub	184.2	...	...	aa	$B5^4$, 702
2283	Benzene, 1,2,3-trimethoxy or Pyrogallol trimethyl ether $1,2,3\text{-}(CH_3O)_3C_6H_3$	168.19	rh nd (al)	235, 140^{12}	48-9	$1.1118^{45/45}$	...	al, eth, bz	$B6^3$, 6265
2284	Benzene, 1,3,5-trimethoxy or Phloroglucinol trimethyl ether $1,3,5\text{-}(CH_3O)_3C_6H_3$	168.19	pr (al), lf (peth)	255.5	54-5	...	...	al, eth, bz	$B6^3$, 6305
2285	Benzene, 1,2,3-trimethyl or Hemimellitene $1,2,3\text{-}(CH_3)_3C_6H_3$	120.19	...	176.1, 56.7^{10}	-25.4	$0.8944^{20/4}$	1.5139^{20}	al, eth, ace, bz	$B5^4$, 1007
2286	Benzene, 1,2,4-trimethyl or Pseudocumene $1,2,4\text{-}(CH_3)_3C_6H_3$	120.19	...	169.3, 51.6^{10}	-43.8	$0.8758^{20/4}$	1.5048^{20}	al, eth, ace, bz	$B5^4$, 1010
2287	Benzene, 1,3,5-trimethyl or Mesitylene $1,3,5\text{-}(CH_3)_3C_6H_3$	120.19	...	164.7, 48.7^{10}	-44.7	$0.8652^{20/4}$	1.4994^{20}	al, eth, ace, bz	$B5^4$, 1016
2288	Benzene, 1,2,3-trimethyl-4,5,6-trinitro $4,5,6\text{-}(NO_2)_3C_6(CH_3)_3$	255.19	pr (al)	...	209	...	...	al, ace, bz	B5, 400
2289	Benzene, 1,2,4-trimethyl-3,5,6-trinitro $3,5,6\text{-}(NO_2)_3C_6(CH_3)_3$	255.19	rh pr (al)	...	185	...	...	bz, to	$B5^4$, 1015
2290	Benzene, 1,3,5-trimethyl-2,4,6-trinitro or Trinitromesitylene $2,4,6\text{-}(NO_2)_3C_6(CH_3)_3$	255.19	tcl nd (al), pr (ace)	415 exp	238.2	...	...	ace, bz	$B5^4$, 1029
2291	Benzene, 1,2,3-trinitro $1,2,3\text{-}(NO_2)_3C_6H_3$	213.11	ye nd or pr (MeOH)	...	127.5	...	...	al	$B5^4$, 754

PHYSICAL CONSTANTS OF ORGANIC COMPOUNDS (Continued)

No.	Name, Synonyms, and Formula	Mol. wt.	Color, crystalline form, specific rotation and λ_{max} (log ε)	b.p. °C	m.p. °C	Density	n_D	Solubility	Ref.
2292	Benzene, 1,2,4-trinitro $1,2,4\text{-}(NO_2)_3C_6H_3$	213.11	lf (eth) pa ye pr (al)		61.2			al, eth, ace, bz, chl	$B5^4$, 754
2293	**Benzene, 1,3,5-trinitro** $1,3,5\text{-}(NO_2)_3C_6H_3$	213.11	rh pl (bz), lf (w)	315, 175^{2}	(i) 121-2, (ii) 61			ace, bz	$B5^4$, 755
2294	Benzene, 1,3,5-triphenyl $1,3,5\text{-}(C_6H_5)_3C_6H_3$	306.41	rh nd (al or aa)	459^{717}	176 (cor)	$1.199^{30/4}$		al, eth, bz	$B5^3$, 2563
2295	Benzene, 1,3,5-tri*iso*-propyl $1,3,5(i\text{-}C_3H_7)_3C_6H_3$	204.36		238	-7.4	0.8545^{20}	1.4882^{20}	ace, bz, chl	$B5^4$, 1178
2296	Benzene, 1,3,5-tris(phenyl amino) or *Sym*-trianilino benzene $1,3,5\text{-}(C_6H_5NH)_3C_6H_3$	351.45	nd (al)		193			eth	$B13^2$, 147
2297	Benzene, 1,3,5-tris-(4-tolylamino) $1,3,5\text{-}(4\text{-}CH_3C_6H_4NH)_3C_6H_3$	393.53	nd (al)		186-7				B13, 299
2298	Benzene arsonic acid or Phenylarsonic acid $C_6H_5AsO(OH)_2$	202.04	cr (w)		158-62d			w, al	$B16^3$, 1057
2299	Benzene arsonic acid-2-amino $2\text{-}H_2NC_6H_4AsO(OH)_2$	217.06	nd (al-eth)		153-4			w, al, aa	$B16^3$, 1099
2300	Benzene arsonic acid-3-amino $3\text{-}H_2NC_6H_4AsO(OH)_2$	217.06	pr (w)		213			w	$B16^3$, 1104
2301	Benzene arsonic acid, 4-amino or Arsanilic acid $4H_2NC_6H_4AsO(OH)_2$	217.06	mcl, nd, (w or al)		232	1.9571^{10}		w, al, eth	$B16^3$
2302	Benzenearsonic acid, 2-chloro $2\text{-}ClC_6H_4AsO(OH)_2$	236.49	nd (w or dil al)		186-7			al, w	$B16^2$, 457
2303	Benzenearsonic acid, 3-chloro $3\text{-}ClC_6H_4AsO(OH)_2$	236.49	cr (w)		175			w, al	$B16^2$, 457
2304	Benzenearsonic acid, 4-chloro $4\text{-}ClC_6H_4AsO(OH)_2$	236.49	wh nd (al)		283-5d			al	$B16^3$, 1058
2305	Benzenearsonic acid, 2-hydroxy $2\text{-}HOC_6H_4AsO(OH)_2$	218.04	nd (w)		190-1			w, al, bz, ace	$B16^2$, 464
2306	Benzenearsonic acid, 3-hydroxy $3\text{-}HOC_6H_4AsO(OH)_2$	218.04	cr (w)		162-73d			w, al	$B16^1$, 454
2307	Benzenearsonic acid, 4-hydroxy $4\text{-}HOC_6H_4AsO(OH)_2$	218.04	nd (aa)		170-4			w, al	$B16^3$, 1070
2308	Benzenearsonic acid, 2-nitro $2\text{-}NO_2C_6H_4AsO(OH)_2$	246.03			233d			aa	$B16^3$, 1059
2309	Benzenearsonic acid, 3-nitro $3\text{-}NO_2C_6H_4AsO(OH)_2$	246.03	lf ye lf (w)		200				$B16^3$, 1059
2310	Benzenearsonic acid, 4-nitro $4\text{-}NO_2C_6H_4AsO(OH)_2$	246.03	lf or nd (w)		>310				$B16^3$, 1059
2311	Benzenearsonic acid, 4-ureido or Carbarsone $C_7H_9AsN_2O_4$	260.08	nd (w)		174			w	$B16^2$, 497
2312	Benzeneazoethane or Ethylphenyl diimide $C_6H_5N{=}NC_2H_5$	134.18	bt ye oil	175-85, $82\text{-}3^{20}$		$0.9628^{22/4}$	α-1.5313 β-1.5579	al, eth, bz	$B16^3$, 4
2313	Benzeneazomethane or Methylphenyl diimide $C_6H_5N{=}NCH_3$	120.15	ye oil	150d, 60^{15}				al, eth	$B16^3$, 3
2314	Benzeneazo-α-naphthylene $C_6H_5N{=}N\text{-}\alpha\text{-}C_{10}H_7$	232.28	dk red lf (al)		70			al, eth, bz, lig	$B16^3$, 57
2315	Benzene-azo-α-naphthalene, 2′-amino $C_6H_5N{=}N\text{-}(\alpha\text{-}C_{10}H_6NH_2\text{-}2)$	247.30	red pl (al)		102-4			al, aa	$B16^3$, 417
2316	Benzeneazo, α-naphthalene, 4-amino or Naphthyl red $C_6H_5N{=}N\text{-}(\alpha\text{-}C_{10}H_6NH_2\text{-}4)$	247.30	red lf (dil al), nd (dil al)		125-6			al, eth, bz	$B16^3$, 406
2317	Benzenazo-α-naphthalene, 4-amino,hydrochloride $C_6H_5N{=}N\text{-}(\alpha\text{-}C_{10}H_6NH_2\text{-}4).HCl$	283.76	gr nd, pr (al, aa)		205-6			al, aa	$B16^1$, 324
2318	Benzeneazo-α-naphthalene, 2,3-dimethyl-2′-hydroxy $2,3\text{-}(CH_3)_2C_6H_3N{=}N\text{-}(\alpha\text{-}C_{10}H_6OH\text{-}2)$	276.34	pa ye amor (al-bz)		125-30			al, to	$B16^2$, 71
2319	Benzeneazo-α-naphthalene, 2,4-dimethyl-2′-hydroxy $2,4\text{-}(CH_3)_2C_6H_3N{=}\text{-}(\alpha C_{10}H_6OH\text{-}2)$	276.34	red nd (al)		166			al, eth	$B16^2$, 72
2320	Benzeneazo-α-naphthalene, 2,5-dimethyl-2′-hydroxy $2,5\text{-}(CH_3)_2C_6H_3N{=}N\text{-}(\alpha\text{-}C_{10}H_6OH\text{-}2)$	276.34	nd (al)		153			al	$B16^3$, 137
2321	Benzeneazo-α-naphthalene, 3,4-dimethyl,2′-hydroxy $3,4\text{-}(CH_3)_2C_6H_3N{=}N(\alpha\text{-}C_{10}H_6OH\text{-}2)$	276.34	red nd (al)		146			bz, chl	$B16^1$, 260
2322	Benzeneazo-α-naphthalene, 2′-hydroxy or Sudan yellow $C_6H_5N{=}N\text{-}(\alpha\text{-}C_{10}H_6OH\text{-}2)$	248.28	red-gold lf or nd (al)	133-4				al, eth, bz	$B16^3$, 129
2323	Benzeneazo-α-naphthylene, 4′-hydroxy $C_6H_5N{=}N\text{-}(\alpha\text{-}C_{10}H_6OH\text{-}4)$	248.28	vt-br lf (bz)		205-6d			al, bz	$B16^3$, 125

No.	Name, Synonyms, and Formula	Mol. wt.	Color, crystalline form, specific rotation and λ_{max} (log ε)	b.p. °C	m.p. °C	Density	n_D	Solubility	Ref.
2324	Benzeneazo-N = N-α-naphthalene, 2-nitro-2′-hydroxy 2-$NO_2C_6H_4N{=}N(\alpha$-$C_{10}H_6OH$-2)	293.28	og-red nd (gl aa)		294			aa	B16^2, 70
2325	Benzeneazo-α-naphthalene, 4-nitro-2′-hydroxy or Para red Paranitroaniline red 4-$NO_2C_6H_4N{=}N(\alpha$-$C_{10}H_6OH$-2)	293.28	br-og pl (to or bz)		257			al, bz	B16^2, 70
2326	Benzene azo-β-naphthalene C_6H_5-$N{=}N\beta$-$C_{10}H_7$	232.28	ye (al)		131			al, eth, bz, lig	B16^3, 60
2327	Benzeneazo-β-naphthalene, 2,4-dimethyl, 1′-hydroxy 2,4-$(CH_3)_2C_6H_3N{=}N(\beta$-$C_{10}H_6OH$-1)	276.34	gold-red lf or nd (al-chl)		186			bz, chl	B16^1, 249
2328	Benzeneazo-β-naphthalene, 4-dimethylamino, 1′-hydroxy 4-$(CH_3)_2NC_6H_4N = N(\beta$-$C_{10}H_6OH$-1)	275.33	ye-br (bz-lig)		174			bz	B16, 323
2329	Benzeneazo-β-naphthalene, 1′-hydroxy $C_6H_5N{=}N(\beta C_{10}H_6OH$-1)	248.28	red nd (al)	sub	138			al, aa	B16^3, 123
2330	Benzene diazonium chloride $C_6H_5N_2Cl$	140.57	nd (al)		exp			w, al, ace, aa	B16^3, 506
2331	Benzene diazonium cyanide $C_6H_5N_2CN$	131.14	ye pr (w)		69				B16, 432
2332	Benzene diazonium nitrate $C_6H_5N_2ONO_2$	167.12	nd (al-eth)		90 exp			w, al	B16^2, 268
2333	Benzene hexacarboxylic acid or Mellitic acid $C_6(CO_2H)_6$	342.17	nd (al)		286-8d			w, al	B9^3, 4914
2334	Benzene pentacarboxylic acid $C_6H(CO_2H)_5$	298.16	nd (w + 5)		228-30 (hyd), 238 (anh)			w, al	B9^3, 4908
2335	Benzene phosphinic acid $C_6H_5P(OH)_2$	142.09			82-3			w, al	B16^3, 874
2336	Benzene phosphinic acid, diethylester $C_6H_5P(OC_2H_5)_2$	198.09		235		1.032^{16}			B16^3, 847
2337	Benzene phosphonic acid $C_6H_5P(O)(OH)_2$	158.09	lf (w)		160			w, al, eth	B16^3, 883
2338	Benzenephosphonyl chloride $C_6H_5POCl_2$	194.98		258, $137\text{-}8^{15}$		1.197^{25}	1.5581^{25}		B16^3, 885
2339	Benzenephosphonic acid, tetrachloride $C_6H_5PCl_4$	249.89			73				B16^3, 885
2340	Benzenephosphothionic acid, ethyl-4-nitro phenylester or EPN $C_6H_5PS(OC_2H_5)OC_6H_4NO_2$-4	323.31			36	$1.27^{25/4}$	1.5978^{30}	al, eth, bz	C48, 2742
2341	Benzeneseleninic acid $C_6H_5SeO_2H$	189.07	pl (w)		124-5	$1.652^{125/4}$			B11^3, 716
2342	Benzenesiliconic acid $C_6H_5SiO_2H$	138.20	glassy (eth)		92			al, eth	B16, 911
2343	Benzene stibonic acid $C_6H_5SbO(OH)_2$	248.87	nd (aa)		139				B16^3, 1178
2344	**Benzene sulfenyl chloride** C_6H_5SCl	144.62	red liq	$73\text{-}5^{9}$				bz	B6^4, 1564
2345	Benzene sulfenyl chloride, 2,4-dinitro 2,4-$(NO_2)_2C_6H_3SCl$	234.61	ye pr (bz-peth)		99			dz, chl, aa	B6^4, 1772
2346	Benzene sulfenyl chloride, 2-nitro 2-$NO_2C_6H_4SCl$	189.62	ye nd (bz)		75			eth, bz, chl	B6^4, 1676
2347	Benzene sulfenyl chloride, 4-nitro 4-$NO_2C_6H_4SCl$	189.62	ye lf (peth)	$125^{0.1}$	52			bz	B6^4, 1717
2348	Benzenesulfinic acid $C_6H_5SO_2H$	142.17	pr (w)	d at 100	84			al, eth, bz	B11^3, 3
2349	Benzenesulfinic acid, 3-acetamido 3-$(CH_3CONH)C_6H_4SO_2H$	199.22	cr (w)		145			w	B14^2, 427
2350	Benzenesulfinic acid, 4-acetamido 4-$(CH_3CONH)C_6H_4SO_2H$	199.22	cr (w)		160			w	B14^3, 1893
2351	Benzenesulfinic acid, 4-bromo 4-$BrC_6H_4SO_2H$	221.07	nd (w)		114			al, eth	B11^3, 5
2352	Benzenesulfinic acid, 4-chloro 4-$ClC_6H_4SO_2H$	176.62	lf or nd (w)		99			al, eth, w	B11^3, 5
2353	Benzenesulfinic acid, ethyl ester $C_6H_5SO_2C_2H_5$	170.23	liq	d				**al, eth, bz**	B11^3, 4
2354	Benzenesulfinic acid, 3-nitro 3-$NO_2C_6H_4SO_2H$	187.17	nd		98			w, al, eth, ace	B11^3, 5

No.	Name, Synonyms, and Formula	Mol. wt.	Color, crystalline form, specific rotation and λ_{max} (log ε)	b.p. °C	m.p. °C	Density	n_D	Solubility	Ref.
2355	Benzenesulfinic acid, 4-nitro $4\text{-}NO_2C_6H_4SO_2H$	187.17	pr or nd (w)		159			al, eth, aa	B11³, 6
2356	Benzenesulfinyl chloride C_6H_5SOCl	160.62	pl (peth)	71-2$^{1.5}$	38	1.3469^{25}	1.3470^{25}	eth, chl	B11², 4
2357	Benzenesulfonamide $C_6H_5SO_2NH_2$	157.19	lf, nd (w)		156			al, eth, w	B11³, 52
2358	Benzenesulfonamide, *N,N*-dichloro or Dichloramine B $C_6H_5SO_2NCl_2$	226.08	ye mcl or pl		76			al	B11³, 79
2359	Benzenesulfonamide, *N*-hydroxy $C_6H_5SO_2NHOH$	173.19	pl (w), rh		126d			w, al, eth, ace, aa	B11³, 83
2360	Benzenesulfonamide, *N*-phenyl $C_6H_5SO_2NHC_6H_5$	233.28	tetr pr (al)		110			al, eth	B12³, 1079
2361	Benzene sulfonamido, 3-acetamido $3\text{-}(CH_3CONH)C_6H_4SO_2NH_2$	214.26	(aa)		216-9			ace, aa	B14¹, 718
2362	Benzenesulfonamido, 4-acetamido $4\text{-}(CH_3CONH)C_6H_4SO_2NH_2$	214.24	nd (aa)		219-20			w, al, ace	B14, 702
2363	Benzenesulfonamide, 2-amino $2\text{-}H_2NC_6H_4SO_2NH_2$	172.20	mcl pl, nd or pr (w)		152-3			al, ace, aa	B14³, 1898
2364	Benzenesulfonamide, 3-amino $3\text{-}H_2NC_6H_4SO_2NH_2$	172.20	lf or nd (w)		142			al	B14³, 1908
2365	Benzenesulfonamide, 4-amino or Sulfanilamide $4\text{-}H_2NC_6H_4SO_2NH_2$	172.20	lf (aq al)		165-6	1.08		w, al, eth, ace, MeOH	B14³, 1919
2366	Benzenesulfonamide, 2-bromo or 2-bromo benzenesulfonamide $2\text{-}BrC_6H_4SO_2NH_2$	236.08	nd (w), pr (al)		186				B11, 56
2367	Benzenesulfonamide, 3-bromo or 3-bromo sulfonamide $3\text{-}Br\text{-}C_6H_4SO_2NH_2$	236.08	nd or lf (w), pr (al)		154			w, al	B11, 57
2368	Benzenesulfonamide, 4-bromo or 4-bromo sulfonamide $4\text{-}Br\text{-}C_6H_4SO_2NH_2$	236.08	nd (w or dil al)		166			al	B11³, 104
2369	Benzene sulfonamide, 2-carboxy or *o*-sulfamyl benzoic acid $2\text{-}HO_2CC_6H_4SO_2NH_2$	201.20	pl or nd (w)		165-7			w, al, eth, bz	B11², 216
2370	Benzenesulfonamide, 3-carboxy or *m*-sulfamyl benzoic acid $3\text{-}HO_2CC_6H_4SO_2NH_2$	201.20	pl (w)		246			w, al	B11³, 663
2371	Benzenesulfonamide, 4-carboxy or *p*-sulfamyl benzoic acid $4\text{-}HO_2CC_6H_4SO_2NH_2$	201.20	pr or lf (w)		290-2d			al	B11³, 672
2372	Benzenesulfonamide, 2-chloro $2\text{-}Cl\text{-}C_6H_4SO_2NH_2$	191.63	lf (al)		188			al	B11³, 88
2373	Benzenesulfonamide, 3-chloro $3\text{-}ClC_6H_4SO_2NH_2$	191.63	lf (dil al)		148			w, al, eth	B11³, 88
2374	Benzenesulfonamide, 4-chloro $4\text{-}ClC_6H_4SO_2NH_2$	191.63	pr or pl (eth)	141^{15}	55			eth, bz	B11³, 90
2375	Benzenesulfonamide, 4-fluoro $4\text{-}FC_6H_4SO_2NH_2$	175.18	pl or nd (w, al)		126			al, eth, ace	B11³,88
2376	Benzenesulfonamide, 4-hydroxy $4\text{-}HOC_6H_4SO_2NH_2$	173.19	cr (al or w)		176-7			w, al	B11³, 506
2377	Benzenesulfonamide, 4-ureido,amide $4\text{-}H_2NCONHC_6H_4SO_2NH_2$	215.23	nd (al)		206-7			w, al	B14³, 2142
2378	Benzenesulfonic acid $C_6H_5SO_3H$	158.17	nd (bz)		65-6 (anh)			w, al, aa	B11³, 32
2379	Benzenesulfonic acid, hydrate $C_6H_5SO_3H.1.5H_2O$	185.19	pl (w)		45-6			w, al	B11², 18
2380	Benzenesulfonic acid, 2-amino or orthanilic acid $2\text{-}H_2NC_6H_4SO_3H$	173.19	pr (+ ½ w)		>320d				B14³, 1896
2381	Benzenesulfonic acid, 2-amino-4-chloro $2\text{-}H_2N\text{-}4\text{-}Cl\text{-}C_6H_3SO_3H$	207.63	nd or pl (w)		310-30d				B14³, 1900
2382	Benzenesulfonic acid, 2-amino-5-chloro $2\text{-}H_2N\text{-}5\text{-}ClC_6H_3SO_3H$	207.63	nd (w)		280d			w	B14³, 1901
2383	Benzenesulfonic acid, 2-amino-3,4-dimethyl $2\text{-}H_2N\text{--}3,4\text{-}(CH_3)_2C_6H_2SO_3H$	201.24	pr (w + 1)		300d				B14¹, 731
2384	Benzenesulfonic acid, 2-amino-3,5,-dimethyl $2\text{-}H_2N\text{-}3,5\text{-}(CH_3)_2C_6H_2SO_3H$	201.24		d					B14², 452
2385	Benzenesulfonic acid, 2-amino-3,6-dimethyl $2\text{-}H_2N\text{-}3,6\text{-}(CH_3)_2C_6H_2SO_3H$	201.24	amor		260				B14¹, 732
2386	Benzenesulfonic acid, 2-amino-4,5-dimethyl $2\text{-}H_2N\text{-}4,5\text{-}(CH_3)_2C_6H_2SO_3H$	201.24	pl		>300				B14¹, 731

PHYSICAL CONSTANTS OF ORGANIC COMPOUNDS (Continued)

No.	Name, Synonyms, and Formula	Mol. wt.	Color, crystalline form, specific rotation and λ_{max} (log ε)	b.p. °C	m.p. °C	Density	n_D	Solubility	Ref.
2387	Benzenesulfonic acid, 3-amino or Metanilic acid $3\text{-}H_2NC_6H_4SO_3H$	173.19	nd, pr (w + l)		d				B14^3, 1907
2388	Benzenesulfonic acid, 3-amino-2,5-dimethyl $3\text{-}H_2N\text{-}2,5\text{-}(CH_3)_2C_6H_2SO_3H$	201.24	nd (w + l)						B14^3, 2237
2389	Benzenesulfonic acid, 3-amino-4,5-dimethyl $3\text{-}H_2N\text{-}4,5\text{-}(CH_3)_2C_6H_2SO_3H$	201.24	red nd (w + l)		315				B14^1, 731
2390	Benzenesulfonic acid, 3-amino-4-hydroxy $3\text{-}H_2N\text{-}4\text{-}HOC_6H_3SO_3H$	189.19	rh (w + l)		>300				B14^3, 2275
2391	Benzenesulfonic acid, 4-amino or Sulfanilic acid $4\text{-}H_2NC_6H_4SO_3H$	173.19	rh pl or mcl (w + 2)		288	$1.485^{25/4}$		w	B14^3, 1916
2392	Benzenesulfonic acid, 4-amino-2,3-dimethyl $4\text{-}H_2N\text{-}2,3\text{-}(CH_3)_2C_6H_2SO_3H$	201.24	nd		305				B14^2, 452
2393	Benzenesulfonic acid, 4-amino-2,5-dimethyl $4\text{-}H_2N\text{-}2,5\text{-}(CH_3)_2C_6H_2SO_3H$	201.24	pl or nd		>300			w	B14^3, 2236
2394	Benzenesulfonic acid, 5-amino-2,3-dimethyl $5\text{-}H_2N\text{-}2,3\text{-}(CH_3)_2C_6H_2SO_3H$	201.24	pl (w + 2)		294d			w	B14^1, 731
2395	Benzenesulfonic acid, 5-amino-2,4-dimethyl $5\text{-}H_2N\text{-}2,4\text{-}(CH_3)_2C_6H_2SO_3H$	201.24	pr or nd		290d				B14^3, 2235
2396	Benzenesulfonic acid, 2-bromo $2\text{-}BrC_6H_4SO_3H$	237.07	nd					w, al	B11, 56
2397	Benzenesulfonic acid, 4-bromo $4\text{-}BrC_6H_4SO_3H$	237.07	nd (al)	155^{25}	102-3			w, al	B11^3, 97
2398	Benzenesulfonic acid, 2-carboxy or 2-sulfobenzoic acid $2\text{-}HO_2CC_6H_4SO_3H$	202.18	nd (w + 3)		141 (anh), 70 (+ 3w)			w, al	B11^3, 658
2399	Benzenesulfonic acid, 2-carboxamide $2\text{-}H_2NCOC_6H_4SO_3H$	201.20	pr (w + 1)		193-4 (anh)			w, al, eth	B11^2, 215
2400	Benzenesulfonic acid, 2-carboxy, anhydride (endo) $C_7H_4SO_4$	184.17	nd or pr (bz)	$184\text{-}6^{18}$	129.5			eth, bz, chl	B19^3, 1641
2401	Benzenesulfonic acid, 2-carboxy, imide or Saccharin, 2-Sulfobenzoic acid imide $C_7H_5NSO_3$	183.18	mcl (ace), pr (al), lf (w)	sub(vac)	228-9d	0.828		al, ace	B27^2, 217
2402	Benzenesulfonic acid, 3-carboxy or 3-Sulfobenzoic acid $3\text{-}HO_2CC_6H_4SO_3H$	202.18	cr (w + 2)		191 (anh), 98 (hyd)			w, al, eth	B11^1, 662
2403	Benzenesulfonic acid, 4-carboxy or 4-Sulfobenzoic acid $4\text{-}HO_2CC_6H_4SO_3H$	202.18	nd (w + 3)		259-60 (anh)			w, al, eth	B11^3, 671
2404	Benzenesulfonic acid, 4-chloro $4\text{-}ClC_6H_4SO_3H$	192.62	nd (w + 1)	$147\text{-}8^{25}$	67			w, al	B11^3, 89
2405	Benzenesulfonic acid, 4-chloro,phenylester or Ovotran $4\text{-}ClC_6H_4SO_3C_6H_5$	268.71			62			ace	C51, 1264
2406	Benzenesulfonic acid, 3,4-diamino $3,4\text{-}(H_2N)_2C_6H_3SO_3H$	188.20	nd		d			w	B14, 717
2407	Benzenesulfonic acid, 3,5-diamino $3,5\text{-}(NH_2)_2C_6H_3SO_3H$	188.20			d				B14^2, 446
2408	Benzenesulfonic acid, 2,5-dichloro, dihydrate $2,5\text{-}Cl_2C_6H_3SO_3H\text{-}2H_2O$	263.09	nd (w + 2)		<100			w, al	B11^2, 30
2409	Benzenesulfonic acid, 3,4-dichloro, dihydrate $3,4\text{-}Cl_2C_6H_3SO_3H\text{-}2H_2O$	263.09	nd		71-2			w, al, eth	B11, 55
2410	Benzenesulfonic acid, 3,5-diido-4-hydroxy, trihydrate or Sozoiodolic acid $3,5\text{-}I_2\text{-}4\text{-}HO\text{-}C_6H_2SO_3H\text{-}3H_2O$	480.02	nd (eth), pr (w + 3)		120 (hyd), 190d			w, al, eth	B11^3, 514
2411	Benzenesulfonic acid, 3,4-dimethyl $3,4\text{-}(CH_3)_2C_6H_3SO_3H$	186.23	pl or pr (chl + 2w)		63-4			w	B11^3, 337
2412	Benzenesulfonic acid, 2,4-dinitro $2,4\text{-}(NO_2)_2C_6H_3SO_3H$	248.17	nd (w + 3)		130 (anh), 108			w, al	B11^3, 159
2413	Benzenesulfonic acid, 3,5-dinitro $3,5\text{-}(NO_2)_2C_6H_3SO_3H$	248.17	ye red (dil al)		235			al, w	B11^2, 36
2414	Benzenesulfonic acid, 3-ethylamino $3\text{-}(C_2H_5NH)C_6H_4SO_3H$	201.24	nd (w)		294d			w	B14^2, 435
2415	Benzenesulfonic acid, 4-ethylamino $4\text{-}(C_2H_5NH)C_6H_4SO_3H$	201.24	pl (w)		258d			w	B14^3, 2025
2416	Benzenesulfonic acid, ethyl ester $C_6H_5SO_3C_2H_5$	186.23		156^{15}		$1.2167^{20/2}$	1.5081^{20}	al, eth, chl	B11^3, 36

No.	Name, Synonyms, and Formula	Mol. wt.	Color, crystalline form, specific rotation and λ_{max} (log ε)	b.p. °C	m.p. °C	Density	n_D	Solubility	Ref.
2417	Benzenesulfonic acid, 2-hydroxy or *o*-phenolsulfonic acid $2\text{-}HOC_6H_4SO_3H$	174.18	cr (w + 1)		145d			w, al	$B11^{3}$, 488
2418	Benzenesulfonic acid, 4-hydroxy or *p*-phenolsulfonic acid $4\text{-}HOC_6H_4SO_3H$	174.18	nd					w, al	$B11^{3}$, 498
2419	Benzenesulfonic acid, 4-hydroxy-3-nitro $4\text{-}HO\text{-}3\text{-}NO_2C_6H_3SO_3H$	219.17	ye pl (AcOEt-bz), nd (w + 3)		141-2 (anh), 51.5 (hyd)			w, al	$B11^{3}$, 515
2420	Benzenesulfonic acid, 4-hydrazino $4\text{-}H_2NNHC_6H_4SO_3H$	188.22	nd or lf (w)		286				$B15^{3}$, 865
2421	Benzenesulfonic acid, methyl ester $C_6H_5SO_3CH_3$	172.20							$B11^{3}$, 36
2422	Benzenesulfonic acid, 2-methyl-4-isopropyl $4\text{-}i\text{-}C_3H_7\text{-}2\text{-}CH_3C_6H_3SO_3H$	214.28	pl or pr		88-90			w	B11, 139
2423	Benzenesulfonic acid, 2-methyl-5-isopropyl $5\text{-}i\text{-}C_3H_7\text{-}2\text{-}CH_3C_6H_3SO_3H$	214.28	pl (+ 2w) mcl pr		220, 78-9 (+ 2w)			w	$B11^{3}$, 349
2424	Benzenesulfonic acid, 5-methyl-2-isopropyl $2\text{-}i\text{-}C_3H_7\text{-}5\text{-}CH_3C_6H_3SO_3H$	214.28			130-1			w, al	$B11^{3}$, 350
2425	Benzenesulfonic acid, 2-nitro $2\text{-}NO_2C_6H_4SO_3H$	203.17			85			w, al	$B11^{2}$, 31
2426	Benzenesulfonic acid, 3-nitro $3\text{-}NO_2C_6H_4SO_3H$	203.17	pl		48			w, al	$B11^{3}$, 118
2427	Benzenesulfonic acid, 4-nitro $4\text{-}NO_2C_6H_4SO_3H$	203.17	(+ 2w)		95, (109-11)			w	$B11^{3}$, 134
2428	Benzenesulfonic acid, 4-phenylamino or 4-Sulfodiphenylamine $4\text{-}C_6H_5NHC_6H_4SO_3H$	249.29	pl (al-eth)		206			w, al	$B14^{1}$, 721
2429	Benzenesulfonic acid, propylester $C_6H_5SO_3C_3H_7$	200.26		$162\text{-}3^{15}$		$1.1804^{17/4}$	1.5035^{25}	al, eth, chl	$B11^{2}$, 20
2430	Benzenesulfonic acid, *iso*-propyl ester $C_6H_5SO_3CH(CH_3)_2$	200.26					1.5003^{20}	al, eth	$B11^{2}$, 20
2431	Benzenesulfonyl chloride $C_6H_5SO_2Cl$	176.62		251-2d, 120^{10}	14.5	$1.3482^{15/15}$		al, eth	$B11^{3}$, 51
2432	Benzenesulfonyl chloride, 3-acetamido $3\text{-}(CH_3CONH)C_6H_4SO_2Cl$	233.67	nd (bz-peth)		88			al, eth, aa	$B14^{2}$, 435
2433	Benzenesulfonyl chloride, 4-acetamide $4\text{-}(CH_3CONH)C_6H_4SO_2Cl$	233.67	nd (bz), pr (bz-chl)		149			al, eth	$B14^{3}$, 2043
2434	Benezenesulfonyl chloride, 2-bromo $2\text{-}BrC_6H_4SO_2Cl$	225.51	pr (eth)		51				B11, 56
2435	Benzenesulfonyl chloride, 4-bromo $4\text{-}BrC_6H_4SO_2Cl$	255.51	tcl or mcl pr (eth)	153^{15}	76			eth	$B11^{3}$, 104
2436	Benzenesulfonyl chloride, 3-carboxy $3\text{-}HO_2CC_6H_4SO_2Cl$	220.63	pr (bz)		133-4			eth, bz	$B11^{3}$, 663
2437	Benzenesulfonyl chloride, 4-carboxy $4\text{-}HO_2CC_6H_4SO_2Cl$	220.63	nd (ace)		237-8d			ace, to	$B11^{3}$, 672
2438	Benzenesulfonyl chloride, 4-chloro $4\text{-}ClC_6H_4SO_2Cl$	211.06	pr or pl (eth)	141^{15}	55			eth, bz	$B11^{3}$, 90
2439	Benzenesulfonyl chloride, 2,5-dichloro or 2,5-dichlorobenzene sulfonyl chloride $2,5\text{-}Cl_2C_6H_3SO_2Cl$	245.51	mcl pr (bz)		38				$B11^{3}$, 93
2440	Benzenesulfonyl chloride, 3,4-dichloro or 3,4-Dichlorobenzenesulfonyl chloride $3,4\text{-}Cl_2C_6H_3SO_2Cl$	245.51	mcl pr		fr. p. 22.4				$B11^{3}$, 94
2441	Benzenesulfonyl chloride, 2,3-dimethyl or 2,3-Dimethylsulfonyl chloride $2,3\text{-}(CH_3)_2C_6H_3SO_2Cl$	204.67	pr (peth)		47			peth	B11, 120
2442	Benzenesulfonyl chloride, 3,4-dimethyl or 3,4-Dimethylbenzenesulfonyl chloride $3,4\text{-}(CH_3)_2C_6H_3SO_2Cl$	204.67	pr (peth)		51-2			eth	$B11^{3}$, 338
2443	Benzenesulfonyl chloride, 3,5-dimethyl or 3,5-Dimethylbenzenesulfonyl chloride $3,5\text{-}(CH_3)_2C_6H_3SO_2Cl$	204.67	nd (peth or bz)		94			eth	$B11^{1}$, 34
2444	Benzenesulfonyl chloride, 4-fluoro or 4-Fluorobenzenesulfonyl chloride $4\text{-}FC_6H_4SO_2Cl$	194.61	pl or nd	$95\text{-}6^{2\ 2}$	36			eth, bz, chl	$B11^{3}$, 88

No.	Name, Synonyms, and Formula	Mol. wt.	Color, crystalline form, specific rotation and λ_{max} (log ε)	b.p. °C	m.p. °C	Density	n_D	Solubility	Ref.
2445	Benzenesulfonyl chloride, 2-methoxy or 2-Methoxybenzenesulfonyl chloride $2\text{-}CH_3OC_6H_4SO_2Cl$	206.64	nd (peth)	$126\text{-}9^{0.1}$	56			eth, peth	B11, 235
2446	Benzenesulfonyl chloride, 4-methoxy or 4-Methoxybenzenesulfonyl chloride $4\text{-}CH_3OC_6H_4SO_2Cl$	206.64	nd or pr (bz)		42-3			al, eth, bz	$B11^3$, 504
2447	Benzenesulfonyl chloride, 2-nitro or 2-Nitrobenzenesulfonyl chloride $2\text{-}NO_2C_6H_4SO_2Cl$	221.62	pr (lig, eth-peth)		68-9			eth	$B11^3$, 114
2448	Benzenesulfonyl chloride, 3-nitro or 3-Nitrobenzenesulfonyl chloride $3\text{-}NO_2C_6H_4SO_2Cl$	221.62	mcl pr (eth), nd (lig)		64			al	$B11^3$, 126
2449	Benzenesulfonyl chloride, 4-nitro or 4-Nitrobenzenesulfonyl chloride $4\text{-}NO_2C_6H_4SO_2Cl$	221.62	mcl pr (peth, lig)		79-80			peth	$B11^3$, 136
2450	Benzenesulfonyl fluoride $C_6H_5SO_2F$	160.16		203-4, $90\text{-}1^4$		$1.3286^{20/4}$	1.4932^{18}	al, eth	$B11^3$, 51
2451	1,3-Benzenedisulfonic acid, 4-amino $4\text{-}H_2NC_6H_3(SO_3H)_2\text{-}(1,3)$	253.24	nd (w + 2)		120d			w, al	$B14^2$, 470
2452	1,3-Benzenedisulfonic acid, 4-hydroxy $4\text{-}HOC_6H_3(SO_3H)_2\text{-}(1,3)$	254.23	nd (w)		>100d			w, al	$B11^3$, 522
2453	1,2,3,4-Benzenetetracarboxylic acid or Prehnitic acid $1,2,3,4\text{-}C_6H_2(COOH)_4$	254.15	pr (+ 6w)		241d			w, ace	$B9^3$, 4872
2454	1,2,3,4-Benzenetetracarboxylic acid, tetramethyl ester $1,2,3,4\text{-}C_6H_2(CO_2CH_3)_4$	310.26	nd (MeOH or w)		133.5			al, bz	$B9^3$, 4872
2455	1,2,3,5-Benzenetetracarboxylic acid, tetramethyl ester or Tetramethyl mellophanate $1,2,3,5\text{-}C_6H_2(CO_2CH_3)_4$	310.26	nd (MeOH, al) MeOH		111			al	$B9^3$, 4872
2456	1,2,4,5-Benzenetetracarboxylic acid or Puromellitic acid $1,2,4,5\text{-}C_6H_2(CO_2H)_4$	254.15	tcl pr (w + 2)		276 (anh), 242 (+ 2w)			al	$B9^3$, 4873
2457	1,2,4,5-Benzenetetracarboxylic acid, tetraethyl ester $1,2,4,5\text{-}C_6H_2(CO_2C_2H_5)_4$	366.37	nd (al)	sub	54			al	$B9^2$, 731
2458	1,2,4,5-Benzenetetracarboxylic acid, tetramethyl ester $1,2,4,5\text{-}C_6H_2(CO_2CH_3)_4$	310.26	lf (al)		143-4				$B9^3$, 4873
2459	1,2,3-Benzenetricarboxylic acid or Hemimellitic acid $1,2,3\text{-}C_6H_3(CO_2H)_3$	210.14	tcl pl (+ 2w), nd (w)		197d (anh), 223-4 (hyd)	1.546^{20}		w	$B9^3$, 4791
2460	1,2,4-Benzenetricarboxylic acid or Trimellitic acid $1,2,4\text{-}C_6H_3(CO_2H)_3$	210.24	nd (w), cr (aa or al)		238d			w, al, eth	$B9^3$, 4792
2461	1,2,4-Benzenetricarboxylic acid, 6-hydroxy $6\text{-}HOC_6H_2(CO_2H)_3\text{-}(1,2,4)$	226.14	pr (w + 2)		278-81d			w, al	B10, 580
2462	1,3,5-Benzenetricarboxylic acid or Trimesic acid $1,3,5\text{-}C_6H_3(CO_2H)_3$	210.14	pr or nd (w + 1)		380 (anh)			al, eth	$B9^3$, 4793
2463	1,3,5-Benzenetricarboxylic acid, 2-chloro $2\text{-}Cl\text{-}C_6H_2(CO_2H)_3\text{-}(1,3,5)$	244.59	nd or pl (w + 1)	sub	285 (anh)			al, eth, w	$B9^2$, 713
2464	1,3,5-Benzenetricarboxylic acid, 2-hydroxy $2\text{-}HOC_6H_2(CO_2H)_3\text{-}(1,3,5)$	226.14	pr (w + 1), nd (w + 2)		306 (anh)			al, w	B10, 580
2465	1,3,5-Benzenetricarboxylic acid, triethyl ester $1,3,5\text{-}C_6H_3(CO_2C_2H_5)_3$	294.30	pr or nd (al)		133-4			al, eth, bz, aa	$B9^3$, 4794
2466	1,3,5-Benzenetricarboxylic acid, trimethyl ester $1,3,5\text{-}C_6H_3(CO_2CH_3)_3$	252.22	nd (dil al)		144			al	$B9^3$, 4794
2467	1,3,5-Benzenetrisulfonic acid $1,3,5\text{-}C_6H_3(SO_3H)_3$	318.29	cr (w + 3)		d>100			w, al	$B11^3$, 483
2468	Benzhydrol or Diphenylmethanol $(C_6H_5)_2CHOH$	184.24	nd (lig)	$297\text{-}8^{748}$, 180^{20}	69			al, eth	$B6^3$, 3364
2469	Benzhydrol, *p*-amino $4\text{-}H_2NC_6H_4CH(OH)C_6H_5$	199.25	nd (w, bz)		121			al, ace	B13, 696
2470	Benzhydrol, 4,4′-dimethyl $(4\text{-}CH_3C_6H_4)_2CHOH$	212.39	nd (al)		69			al, eth, ace, chl	$B6^3$, 3422
2471	Benzhydrol, bis(2-dimethylamino)-α-phenyl $[2\text{-}(CH_3)_2NC_6H_4]_2C(OH)C_6H_5$	346.47	pr (lig)		105				B13, 741
2472	Benzhydrol, bis(3-dimethylamino)-α-phenyl $[3\text{-}(CH_3)_2NC_6H_4]_2C(OH)C_6H_5$	346.47	cr (eth)		128-9			eth	B13, 696

No.	Name, Synonyms, and Formula	Mol. wt.	Color, crystalline form, specific rotation and λ_{max} (log ε)	b.p. °C	m.p. °C	Density	n_D	Solubility	Ref.
2473	Benzhydrol, bis(4-dimethylamino)-α-phenyl $[4-(CH_3)_2NC_6H_4]_2C(OH)C_6H_5$	346.48	cr (eth, bz, lig, MeOH)		121-3			eth, bz	$B13^3$, 1957
2474	Benzhydrol, 4-hydroxy or Benzaurin $4-HOC_6H_4CH(OH)C_6H_5$	200.24	ye red pw		110-20				$B8^3$, 1644
2475	Benzhydrol, 4-methoxy $4-CH_3OC_6H_4CH(OH)C_6H_5$	214.27	nd (w, lig, dil, al)		66-8			al, bz, chl	$B6^3$, 5417
2476	Benzhydrol, α-naphthyl $(C_6H_5)_2C(OH)$-α-$C_{10}H_7$	310.40	cr (lig, bz)	d	136.5			al, eth, bz	$B6^3$, 3817
2477	Benzhydrol, β-naphthyl $(C_6H_5)_2C(OH)$-β-$C_{10}H_7$	310.40	pr (eth-lig)		118			al, eth, ace, bz	$B6^3$, 3818
2478	Benzhydryl amine $(C_6H_5)_2CHNH_2$	183.25	hex pl	304^{763}, 176^{23}	34	$1.0635^{20/20}$	1.5963	bz	$B12^3$, 3221
2479	o-Benzidine or 2,2′Diamino biphenyl $2-H_2NC_6H_4-C_6H_4NH_2-2$	184.24	mcl pr or nd (al)	162^4	81			w, bz	$B13^3$, 410
2480	m-Benzidine or 3,3′-Diamino biphenyl $3-H_2NC_6H_4C_6NH_2-3$	184.24	nd (w), pr (bz)	205	93—4			eth, bz	$B13^3$, 422
2481	p-Benzidine or 4,4′-Diamino biphenyl $(4-H_2NC_6H_4-)_2$	189.24	nd (w)	400^{740}	128			al	$B12^3$, 425
2482	Benzil or Diphenylglyoxal $C_6H_5COCOC_6H_5$	210.23	ye pr (al)	346—8d, 188^{12}	95-6	$1.084^{102/4}$		al, eth, ace, bz	$B7^3$, 3804
2483	Benzil dioxime (anti) or α Benzil dioxime $C_6H_5C=(NOH)C=(NOH)C_6H_5$	240.26	lf (al), or (ace)		238d				$B7^2$, 680
2484	Benzil dioxime (syn) or β-Benzil dioxime $C_6H_5C=(NOH)C=(NOH)C_6H_5$	240.26	nd (al)		207d			al, eth	$B7^2$, 681
2485	Benzil dioxime (amphi) or α-Benzil dioxime $C_6H_5-C=(NOH)C=(NOH)C_6H_5$	240.26	nd (al + 1), (aa)		164-6			al, ace, bz	$B7^1$, 681
2486	Benzil monoxime (α) $C_6H_5C=(NOH)COC_6H_5$	225.25	lf (dil al, bz)	200d	137-8			al, eth, aa, chl	$B7^2$, 678
2487	Benzil monoxime (β) $C_6H_5(C=NOH)COC_6H_5$	225.25			70 (+ ½ bz), 113-4			al, eth, ace, bz, chl	$B7^2$, 679
2488	Benzil osazone (anti) $C_{26}H_{22}N_4$	390.49	ye nd (al)		230-2			bz, chl	$B15^3$, 114
2489	Benzil osazone (syn) $C_{26}H_{22}N_4$	390.49	ye nd (bz-al)		210			al, bz	$B15^3$, 114
2490	Benzilic acid, α-hydroxydiphenyl acetic acid $(C_6H_5)_2C(OH)CO_2H$	228.25	mcl nd (w)	d180	151			al, eth	$B10^3$, 1168
2491	Benzilic acid, ethyl ester or Ethyl benzilate $(C_6H_5)_2C(OH)COOC_2H_5$	256.30	pr or nd	201^{21}	34		1.5620^{20}	al, eth	$B10^3$, 1171
2492	Benzilic acid, methyl ester or Methyl benzilate $(C_6H_5)_2C(OH)COOCH_3$	242.27	mcl or tcl cr (al)	187^{13}	75				$B10^3$, 1170
2493	Benzimidazole or 1,3-Benzodiazole $C_7H_6N_2$	118.14	rh bipym pl (w)	>360	170.5			al	$B23^2$, 151
2494	Benzimidazole, 2-amino $2-H_2N-(C_7H_5N_2)$	133.15	pl (w)		224			w, al, ace	$B24^1$, 116
2495	Benzimidazole, 5,6-dimethyl $5,6-(CH_3)_2(C_7H_4N_2)$	146.19	(eth)	140^2 sub	205-6			w, eth, al, chl	C50, 1087
2496	Benzimidazole, 2-hydroxy or o-Phenylene urea $2-HO(C_7H_5N_2)$	134.14	lf (w or al)		318d			al, ace	$B24^2$, 62
2497	Benzimidazole, 2-mercapto $2-HS(C_7H_5N_2)$	150.20	pl (dil al, aq NH_3)		298			al	$B24^2$, 65
2498	Benzimidazole, 1-methyl $1-CH_3(C_7H_5N_2)$	132.16	nd (peth), pl (al)	286^{756}	66	$1.1254^{20/4}$	1.6013^7	al, peth	$B23^2$, 152
2499	Benzimidazole, 2-methyl $2-CH_3(C_7H_5N_2)$	132.16	pr or nd (w)		176-7			w	$B23^2$, 160
2500	Benzimidazole, 2-methyl-5-nitro-1-phenyl $1-C_6H_5-2-CH_3-5-NO_2(C_7H_3N_2)$	253.26	nd (al)		170-1			al	$B23^2$, 161
2501	Benzimidazole, 4-methyl $4-CH_3(C_7H_5N_2)$	132.16	pl (w), nd (bz)		145			al, w	$B23^1$, 38
2502	Benzimidazole, 5-methyl $5-CH_3(C_7H_5N_2)$	132.16	(w)		114			w	$B23^2$, 157
2503	Benzimidazole, 6-nitro $6-NO_2(C_7H_5N_2)$	163.14	nd (w)		209—10			al	$B23^2$, 154
2504	Benzimidazole, 1-phenyl $1-C_6H_5(C_7H_5N_2)$ $C_{13}H_{10}N_2$	194.24		$210-2^{14}$	98			al, w	$B23^2$, 153

No.	Name, Synonyms, and Formula	Mol. wt.	Color, crystalline form, specific rotation and λ_{max} (log ε)	b.p. °C	m.p. °C	Density	n_D	Solubility	Ref.
2505	Benzimidazole, 2-phenyl 2-$C_6H_5(C_7H_5N_2)$	194.24	pl (aa), (al-w), nd (bz w)		293			al, aa	B23², 238
2506	4,5-Benzindane or 1,2-Cyclopentanonaphthalene $C_{13}H_{12}$	168.24	oil	294-5, 170[5]		1.066[20/4]	1.6290[20]	aa	B5⁴, 1865
2507	1,2-Benzisothiazole C_7H_5NS	135.18		116-8[18]					B27², 16
2508	1,2-Benzisothiazole, 5-hydroxy-3-phenyl 3-C_6H_5-5-$HO(C_7H_3NS)$	227.28	nd (dil al, aa)		159-60			al, ace, bz, aa	B27², 87
2509	Benzisoxazole, 3-methyl C_8H_7NO	133.15		108-10[16]				eth	B27², 19
2510	1,2-Benzocarbazole, 3,4-dihydro $C_{16}H_{13}N$	219.29	(lig or MeOH)		163-4			MeOH, lig	B20⁴, 4187
2511	1,2-Benzo-1-cycloactene-3-one $C_{12}H_{14}O$	174.24		87-8[0.001]			1.5577[25]		AM76, 5462
2512	Benzodichlorofluoride $C_6H_5CCl_2F$	179.02		178-80		1.3138[11]	1.5180[11]	al	B5⁴, 818
2513	5,6-Benzoflavone or β-Naphthoflavone $C_{19}H_{12}O_2$	272.30	nd (al)		167-8			eth, bz	B17⁴, 5555
2514	7,8-Benzoflavone or α-Naphthoflavone $C_{19}H_{12}O_2$	272.30	ye pl (al), lf or nd (dil al)		157-9 (167)				B17⁴, 5551
2515	3,4-Benzofluoranthene $C_{20}H_{12}$	252.32	nd (bz)		168				B5³, 2516
2516	10,11-Benzofluoranthene $C_{20}H_{12}$	252.32	ye pl (al), nd (aa)		166				B5³, 2517
2517	11,12-Benzofluoranthene $C_{20}H_{12}$	252.32	pa ye nd (bz)	480	217			al, bz, aa	B5³, 2516
2518	1,2-Benzofluorene or Chrysofluorene $C_{17}H_{12}$	216.28	pl (ace or aa)	413	189-90			eth, chl	B5³, 2288
2519	1,2-Benzofluorene, 9-phenyl 9-$C_6H_5(C_{17}H_{11})$	292.38	nd (aa)		195.5			eth, bz	B5³, 2558
2520	Benzofuran or Coumaron C_8H_6O	118.14		174, 62-3[15]	<−18	1.0913[25]	1.5615[17]	al, eth	B17⁴, 478
2521	Benzofuran, 2-acetyl 2-$CH_3CO(C_8H_5O)$	160.17		136[11]	76			w	B17⁴, 5080
2522	Benzofuran, 2-benzoyl 2-$C_6H_5CO(C_8H_5O)$	222.24		360	91				B17⁴, 5426
2523	Benzofuran, 2-(chloromethyl)-2,3-dihydro 2-$ClCH_2$-2,3-H_2-(C_8H_5O)	168.62		118-9[11]	41-2	1.2196[7.5/16]	1.5620[7.5]	eth, ace	B17⁴, 421
2524	Benzofuran, 2-methyl 2-$CH_3(C_8H_5O)$	132.16		197-8, 93-4[20]		1.0540[20/4]	1.5495[22]	al, eth	B17⁴, 497
2525	Benzofuran, 3-methyl 3-$CH_3(C_8H_5O)$	132.16		196-7[742], 86[20]		1.0540[25/4]	1.5536[16]	al, eth	B17⁴, 500
2526	Benzofuran, 5-methyl 5-$CH_3(C_8H_5O)$	132.16		197-9, 83.5[17]		1.0603[19]	1.5570[19]	al, eth	B12⁴, 502
2527	Benzofuran, 7-methyl 7-$CH_3(C_8H_5O)$	132.16		190-1		1.0400[19]	1.5525[19]	al, eth	B17⁴, 503
2528	Benzoguananine or 2,4-Diamino-6-phenyl-1,3,5-triazine 2,4-$(NH_2)_2$-6-$C_6H_5(C_3N_3)$	187.20	nd or pl (al)		226.4			al, eth	B26′, 69
2529	Benzohydroxamic acid $C_6H_5CO(NHOH)$	137.14	rh ta, lf (eth)	exp	121—2			al, w	B9³, 1304
2530	Benzohydroxamic acid, 2-hydroxy or Salicylhydroxamic acid 2-$HO(C_6H_4CO(NHOH)$	153.14	nd (aa)		168			al, eth	B10³, 160
2531	Benzoic acid $C_6H_5CO_2H$	122.12	mcl lf or nd	249, 133[10]	122.13	1.0749[130], 1.2659[15/4]	1.504[12]	al, eth, ace, bz, chl	B9³, 360
2532	Benzoic acid, 2-acetamido 2-$(CH_3CONH)C_6H_4CO_2H$	179.18	nd (aa)		185			eth, ace, bz	B14³, 922
2533	Benzoic acid, 3-acetimido 3-$CH_3CONHC_6H_4CO_2H$	179.18	nd (al)		248-50			al	B14², 241
2534	Benzoic acid, 4-acetamido 4-$(CH_3CONH)C_6H_4CO_2H$	179.18	nd (aa)		256.5			al	B14³, 1112
2535	Benzoic acid, (4-acetimidophenyl) ester $C_6H_5CO_2(C_6H_4NHCOCH_3$-4)	255.28	nd (al)		171			al, bz, aa	B13³, 1065
2536	Benzoic acid, 2-acetyl 2-CH_3CO-$C_6H_4CO_2H$	164.16	nd (w), pr (bz)	110-2[2]	114-5			al, w	B10³, 102

No.	Name, Synonyms, and Formula	Mol. wt.	Color, crystalline form, specific rotation and λ_{max} (log ε)	b.p. °C	m.p. °C	Density	n_D	Solubility	Ref.
2537	Benzoic acid (-2-acetylphenyl) ester $C_6H_5CO_2(C_6H_4COCH_3$-2)	240.26	nd (al)		88			al, eth, aa	B9^3, 729
2538	Benzoic acid, 3-acetyl 3-$CH_3COC_6H_4CO_2H$	164.16			210			w	B10^3, 248
2539	Benzoic acid, 4-acetyl 4-$CH_3COC_6H_4CO_2H$	164.16	nd (w)	sub	210			w	B10^3, 294
2540	Benzoic acid, 4-acetyl, methyl ester 4-$CH_3COC_6H_4CO_2CH_3$	178.19	nd (w)	140-5^4 sub	95			w	B10, 695
2541	Benzoic acid, (4-acetylphenyl) ester $C_6H_5CO_2(C_6H_4COCH_3$-4)	240.26	nd (al, aq MeOH)		135-6			al, eth, bz, aa	B9^3, 730
2542	Benzoic acid, allyl ester or Allylbenzoate $C_6H_5CO_2CH_2CH=CH_2$	162.19		242		$1.0578^{15/15}$	1.5178^{20}	al, eth, ace	B9^3, 402
2543	Benzoic acid, 5-allyl-2-hydroxy-3-methoxy or Eugenic acid 5-($CH_2=CHCH_2$)-2-HO-3-$CH_3OC_6H_2CO_2H$	208.21	pl (w + l)		85—8 (+ lw), 127 (anh)			al, eth	B10^3, 1855
2544	Benzoic acid, 2-amino or Anthranilic acid 2-$H_2NC_6H_4CO_2H$	137.14	lf (al)	sub	146-7	1.412^{20}		w, al, eth, chl	B14^3, 879
2545	Benzoic acid, anhydride or Benzoic anhydride $(C_6H_5CO)_2O$	226.23	pr (eth)	360	42-3	$1.989^{15/4}$	1.5767^{15}	al, eth	B9^3, 852
2546	Benzoic acid, 2-benzamido 2-$(C_6H_5CONH)C_6H_4CO_2H$	241.25	nd (al or dz)		181			al, eth	B14^1, 925
2547	Benzoic acid, 3-Benzamido 3-$(C_6H_5CONH)C_6H_4CO_2H$	241.25	red pr (al)		252-3	$1.510^{4/4}$		al	B14^1, 562
2548	Benzoic acid, 4-benzamido 4-$(C_6H_5CONH)C_6H_4CO_2H$	241.25	nd (al)		278			al, eth	B14^1, 577
2549	Benzoic acid, 2-Benzoyl or 2-Benzophenone carboxylic acid 2-$(C_6H_5CO)C_6H_4CO_2H$	226.23	tcl nd (w + l)		127-9 (anh)			al, eth, bz	B10^3, 3289
2550	Benzoic acid, 2-benzoyl-4-chloro 2-C_6H_5CO-4-$ClC_6H_3CO_2H$	260.68	nd (MeOH)		92			al	B10^3, 3296
2551	Benzoic acid, 2-benzoyl, ethyl ester 2-$(C_6H_5CO)C_6H_4CO_2C_2H_5$	254.28	rh pl (dil al)		58	$1.221^{64/4}$	1.560^{64}	al, eth	B10^3, 3297
2552	Benzoic acid, 2-benzoyl, methyl ester 2-$(C_6H_5CO)C_6H_4CO_2CH_3$	240.26	pl or mcl pr (dil al)	350-2	52	$1.1903^{19/4}$	1.591^{20}	al, eth	B10^3, 3291
2553	Benzoic acid, benzoylmethyl ester $C_6H_5CO_2(CH_2COC_6H_5)$	240.26	pl (dil al)		118.5			eth, bz, chl, al	B9^3, 730
2554	Benzoic acid, 3-benzoyl or 3-Benzophenonecarboxylic acid 3-$(C_6H_5CO)C_6H_4CO_2H$	226.23	nd (w), fl (dil al)	sub	161-2			al, eth	B10^3, 3304
2555	Benzoic acid, 4-Benzoyl or 4-Benzophenonecarboxylic acid 4-$(C_6H_5CO)C_6H_4CO_2H$	226.23	nd (dil aa), pl (al), mcl lf (w)	sub	198—200 (226—7)			al, eth, aa	B10^3, 3305
2556	Benzoic acid, benzyl ester or Benzyl benzoate $C_6H_5CO_2CH_2C_6H_5$	212.25	nd or lf	323—4 (cor), 170—1^{11}	21	$1.1121^{25/4}$	1.5680^{20}	al, eth, ace, bz	B9^3, 428
2557	Benzoic acid, 2-benzyl 2-$(C_6H_5CH_2)C_6H_4CO_2H$	212.25	nd (dil al)	sub	118			al, eth, bz, chl	B9^3, 3317
2558	Benzoic acid, 3-benzyl 3-$(C_6H_5CH_2)C_6H_4CO_2H$	212.25	nd (w), lf (dil al)	sub	157-8			al, eth, bz, chl	B9^3, 676
2559	Benzoic acid, 4-benzyl 4-$(C_6H_5CH_2)C_6H_4CO_2H$	212.25	nd (w), lf (dil al)	sub	157-8			al, eth, bz, chl	B9^3, 3319
2560	Benzoic acid, 4-(benzylsulfonamido) or Caronamide 4-$(C_6H_5CH_2SO_2NH)C_6H_4CO_2H$	291.33			229-30			al	C45, 3418
2561	Benzoic acid, 2-bromo or 2-Bromobenzoic acid 2-$BrC_6H_4CO_2H$	201.02	mcl pr (w), nd	sub	150	$1.929^{25/4}$		al, eth, bz, chl	B9^3, 1383
2562	Benzoic acid, 2-bromo, anhydride (2-$BrC_6H_4CO)_2O$	384.02	nd (al)		79.6			al, bz, chl	B9^1, 142
2563	Benzoic acid, 2-bromo-3-chloro 2-Br-3-$ClC_6H_3CO_2H$	235.46	cr (bz)		224-6			bz	B9, 355
2564	Benzoic acid, 2-bromo-5-chloro 2-Br-5-$ClC_6H_3CO_2H$	235.46	cr (bz)		153			al, bz	B9^3, 1426
2565	Benzoic acid, 2-bromo-3,5-dinitro 3,5-$(NO_2)_2$-2-$BrC_6H_2CO_2H$	291.01	ye nd (w)		213			al, bz, aa	B9^3, 1954
2566	Benzoic acid, 2-bromo, ethyl ester 2-$BrC_6H_4CO_2C_2H_5$	229.07		254-5, 135^{15}		$1.4438^{15/4}$	1.5455^{15}	al, eth, ace, bz	B9^3, 1385
2567	Benzoic acid, 2-bromo, methyl ester 2-$BrC_6H_4CO_2CH_3$	215.05		244, 122^{17}				al	B9^3, 1385

No.	Name, Synonyms, and Formula	Mol. wt.	Color, crystalline form, specific rotation and λ_{max} (log ε)	b.p. °C	m.p. °C	Density	n_D	Solubility	Ref.
2568	Benzoic acid, 2-bromo-3-nitro $3-NO_2-2-BrC_6H_3CO_2H$	246.02	(dil al)		191			al	B9[2], 277
2569	Benzoic acid, 2-bromo-4-nitro $4-NO_2-2-BrC_6H_3CO_2H$	246.02	nd (w or dil al)	sub>155	166-7			al, eth	B9[3], 1771
2570	Benzoic acid, 2-bromo-5-nitro $5-NO_2-2-BrC_6H_3CO_2H$	246.02	nd (w)	sub	180-1			al, eth	B9[3], 1771
2571	Benzoic acid, 3-bromo $3-BrC_6H_4CO_2H$	201.02	mcl nd (dil al)	>280	155	1.845^{20}	k	al, eth	B9[3], 1392
2572	Benzoic acid, 3-bromo-2-chloro $3-Br-ClC_6H_3CO_2H$	235.46	cr (al)		165			al	B9[3], 1426
2573	Benzoic acid, 3-bromo-4-chloro $3-Br-4-ClC_6H_3CO_2H$	235.46	pl (dil aa), cr (al)		215-6			al, aa	B9[3], 1427
2574	Benzoic acid, 3-bromo-6-chloro $3-Br-6-ClC_6H_3CO_2H$	235.46	nd (w), cr (aa)		155-6			al, aa	B9[3], 1427
2575	Benzoic acid, 3-bromo, ethyl ester or Ethyl-3-bromobenzoate $3-BrC_6H_4CO_2C_2H_5$	229.07		261, 133^{15}		$1.4308^{19/4}$	1.5430^{19}	al, eth, ace, bz	B9[3], 1393
2576	Benzoic acid, 3-bromo, methyl ester $3-BrC_6H_4CO_2CH_3$	215.05	pl	122.5^{15}	32			al, eth	B9[3], 1393
2577	Benzoic acid, 3-bromo-2-nitro $2-NO_2-3-BrC_6H_3CO_2H$	246.02	(eth)		250			eth, bz	B9[3], 1770
2578	Benzoic acid, 3-bromo-4-nitro $4-NO_2-3-BrC_6H_3CO_2H$	246.02	nd (dil al)		197			al, eth, chl	B9, 408
2579	Benzoic acid, 3-bromo-5-nitro $5-NO_2-3-BrC_6H_3CO_2H$	246.02	nd (w, bz or eth), pl (al)		159-60			al, eth, bz, chl	B9[3], 1771
2580	Benzoic acid, 4-bromo $4-BrC_6H_4CO_2H$	201.02	nd (eth), lf (w), mcl pr		254.5	1.894^{20}		al, eth	B9[3], 1403
2581	Benzoic acid, 4-bromo-3-chloro $4-Br-3-ClC_6H_3CO_2H$	235.46	pl (dil aa), cr (al)		218			al, aa	B9[2], 236
2582	Benzoic acid, 4-bromo, ethyl ester or Ethyl-4-bromo benzoate $4-BrC_6H_4CO_2C_2H_5$	229.07		262^{737}, 125^{15}		$1.4332^{17/4}$	1.5438^{17}	al, eth, ace, bz	B9[3], 1405
2583	Benzoic acid, 4-bromo, methyl ester or Methyl-4-bromobenzoate $4-BrC_6H_4CO_2CH_3$	215.05	lf (dil al), nd (eth)		81	1.689		al, eth, ace, bz, chl	B9[3], 1405
2584	Benzoic acid, 4-bromo-2-nitro $2-NO_2-4-BrC_6H_3CO_2H$	246.02	nd (w)		163			al, eth, bz, chl	B9[3], 1770
2585	Benzoic acid, 4-bromo-3-nitro $3-NO_2-4-BrC_6H_3CO_2H$	246.02	nd (dil aa)	sub	203-4			al	B9[3], 1770
2586	Benzoic acid, 5-bromo-2-nitro $2-NO_2-5-BrC_6H_3CO_2H$	246.02	cr (w, al, bz, to)		140	1.920^{18}		al, bz	B9[1], 165
2587	Benzoic acid, butyl ester or Butylbenzoate $C_6H_5CO_2C_4H_9$	178.23		250.3	-22.4	1.000^{20}	1.4940^{25}	al, eth, ace	B9[3], 392
2588	Benzoic acid, *iso*-butyl ester or Isobutyl benzoate $C_6H_5CO_2$-*i*-C_4H_9	178.23		242		$0.9990^{20/4}$		al, eth, ace	B9[3], 394
2589	Benzoic acid, 2-*tert*-butyl $2-[(CH_3)_3C]C_6H_4CO_2H$	178.23	pl (dil al)		80-1			al	B9[2], 365
2590	Benzoic acid, 3-*tert*-butyl $3-[(CH_3)_3C]C_6H_4CO_2H$	178.23	nd (peth)		127-8			al, peth	B9[3], 2525
2591	Benzoic acid, 4-*tert*-butyl $4-[(CH_3)_3C]C_6H_4CO_2H$	178.23	nd (dil al)		164-5			al, bz	B9[3], 2525
2592	Benzoic acid, (1-chloroethyl) ester or (1-Chloroethyl) benzoate $C_6H_5CO_2CHClCH_3$	184.62		134^{30}		1.172^{20}		al, eth	B9[2], 127
2593	Benzoic acid, (2-chloroethyl) ester or (2-Chloroethyl) benzoate $C_6H_5CO_2CH_2CH_2Cl$	184.62		254^{729}	$118-20^{2}$			al, eth	B9[3], 388
2594	Benzoic acid, 2-chloro $2-ClC_6H_4CO_2H$	156.57	mcl pr (w)	sub	142	1.544^{20}		al, eth, ace, bz	B9[3], 1330
2595	Benzoic acid, 2-chloro, ethyl ester $2-ClC_6H_4CO_2C_2H_5$	184.62		243, $122-5^{15}$		$1.1942^{15/4}$	1.5247^{15}	al, eth	B9[3], 1333
2596	Benzoic acid, 2-chloro, methyl ester $2-ClC_6H_4CO_2CH_3$	170.60		$234-5^{762}$				al	B9[3], 1333
2597	Benzoic acid, 2-chloro-4-methyl $2-Cl-4-CH_3C_6H_3CO_2H$	170.60	nd (al)		155-6			al, eth, bz, chl	B9, 497

No.	Name, Synonyms, and Formula	Mol. wt.	Color, crystalline form, specific rotation and λ_{max} (log ε)	b.p. °C	m.p. °C	Density	n_D	Solubility	Ref.
2598	Benzoic acid, 2-chloro-5-methyl $2\text{-Cl-5-CH}_3C_6H_3CO_2H$	170.60	nd (w or al)		167			w, al	$B9^{3}$, 2327
2599	Benzoic acid, 2-chloro-6-methyl $2\text{-Cl-6-CH}_3C_6H_3CO_2H$	170.60	nd (w)		102				$B9^{3}$, 2310
2600	Benzoic acid, 2-chloro-4-nitro $4\text{-NO}_2\text{-2-ClC}_6H_3CO_2H$	201.57	nd (w)		140-2			al, eth, w	$B9^{3}$, 1766
2601	Benzoic acid, 2-chloro-5-nitro $5\text{-NO}_2\text{-2-ClC}_6H_3CO_2H$	201.57	nd or pr (w)		165	1.608^{18}		al, eth, bz	$B9^{3}$, 1765
2602	Benzoic acid, 3-chloro $3\text{-ClC}_6H_4CO_2H$	156.57	pr (w)	sub	158	$1.496^{25/4}$		al, eth	$B9^{3}$, 1345
2603	Benzoic acid, 3-chloro, anhydride $(3\text{-ClC}_6H_4CO)_2O$	295.14	nd (al or peth)		95.5			al, bz, chl	$B9^{2}$, 224
2604	Benzoic acid, 3-chloro, ethyl ester $3\text{-ClC}_6H_4CO_2C_2H_5$	184.62		$243, 121^{20}$		$1.1859^{15/4}$	1.5223^{20}	al, eth	$B9^{3}$, 1347
2605	Benzoic acid, 3-chloro, methyl ester $3\text{-ClC}_6H_4CO_2CH_3$	170.60		$231, 114^{18}$	21			al	$B9^{3}$, 1346
2606	Benzoic acid, 3-chloro-2-methyl $3\text{-Cl-2-CH}_3C_6H_3CO_2H$	170.60	nd (al)		159			al, eth	$B9^{3}$, 2309
2607	Benzoic acid, 3-chloro-4-methyl $3\text{-Cl-4-CH}_3C_6H_3CO_2H$	170.60	nd or lf (dil al)		200—2			al	$B9^{3}$, 2355
2608	Benzoic acid, 3-chloro-5-methyl $3\text{-Cl-5-CH}_3C_6H_3CO_2H$	170.60	nd (dil al)		178			al	B9, 479
2609	Benzoic acid, 3-chloro-2-nitro $2\text{-NO}_2\text{-3-ClC}_6H_3CO_2H$	201.57	nd or pl (w)		237-9	1.566^{18}		al, eth	B9, 400
2610	Benzoic acid, 3-chloro-5-nitro $5\text{-NO}_2\text{-3-ClC}_6H_3CO_2H$	201.57	nd (w)		147			al, eth, aa	$B9^{1}$, 165
2611	Benzoic acid, 4-chloro $4\text{-ClC}_6H_4CO_2H$	156.57	tcl pr (al-eth)		243			al	$B9^{3}$, 1354
2612	Benzoic acid, 4-chloro, anhydride $(4\text{-ClC}_6H_4CO)_2O$	295.14	nd or lf (bz)		193-4				$B9^{3}$, 1361
2613	Benzoic acid, 2-(4-chlorobenzoyl) $2\text{-(4-ClC}_6H_4CO)\text{-C}_6H_4CO_2H$	260.68	cr (bz, aa)		150			al, eth, bz	$B10^{2}$, 518
2614	Benzoic acid, 4-chloro, ethyl ester $4\text{-ClC}_6H_4CO_2C_2H_5$	184.62		$237\text{-}8, 122^{15}$				al	$B9^{3}$, 1356
2615	Benzoic acid, 4-chloro, methyl ester $4\text{-ClC}_6H_4CO_2CH_3$	170.60	nd or mcl pr		44	1.382^{20}		al	$B9^{3}$, 1356
2616	Benzoic acid, 4-chloro-2-methyl $4\text{-Cl-2-CH}_3C_6H_3CO_2H$	170.60	nd (w, al, dil aa, bz)		173			al, aa	$B9^{2}$, 321
2617	Benzoic acid, 4-chloro-3-methyl $4\text{-Cl-3-CH}_3C_6H_3CO_2H$	170.60	nd (w)		209-10				$B9^{3}$, 2327
2618	Benzoic acid, 4-chloro-2-nitro $2\text{-NO}_2\text{-4-ClC}_6H_3CO_2H$	201.57	pl (bz-lig), pr (bz), nd (w)		142.3			al, eth, bz, w	$B9^{3}$, 1763
2619	Benzoic acid, 4-chloro-3-nitro $3\text{-NO}_2\text{-4-ClC}_6H_3CO_2H$	201.57	nd or pl (w)		181-2	1.645^{18}			$B9^{3}$, 1764
2620	Benzoic acid, 4-chloro-3-nitro, ethyl ester $3\text{-NO}_2\text{-4-ClC}_6H_3CO_2C_2H_5$	229.60	ye nd (al)		59			al, bz, aa	B9, 402
2621	Benzoic acid, 4-chloro-3-nitro, methyl ester $3\text{-NO}_2\text{-4-ClC}_6H_3CO_2CH_3$	215.57	nd (MeOH)		83	1.522^{18}		al	B9, 402
2622	Benzoic acid, 5-chloro-2-methyl $5\text{-Cl-2-CH}_3C_6H_3CO_2H$	170.60	nd (al)		168-9			al	$B9^{2}$, 320
2623	Benzoic acid, 5-chloro-2-nitro, methyl ester $2\text{-NO}_2\text{-5-ClC}_6H_3CO_2CH_3$	215.57	pl (MeOH)		48.5	1.453^{18}		MeOH	B9, 401
2624	Benzoic acid, cyclohexyl ester or Cyclohexylbenzoate $C_6H_5CO_2C_6H_{11}$	204.27		285	<-10	1.0429^{20}	1.5200^{20}	al, eth	$B9^{3}$, 404
2625	Benzoic acid, 3-cyano or Isophthalic acid mononitrile $3\text{-NCC}_6H_4CO_2H$	147.13	nd (w)	sub d	217			al, eth	$B9^{3}$, 4243
2626	Benzoic acid, 2,3-diamino $2,3\text{-(H}_2N)_2C_6H_3CO_2H$	152.15	nd (dil al)	d	190-1d			al, aa	$B14^{3}$, 1172
2627	Benzoic acid, 2,4-diamino $2,4\text{-(H}_2N)_2C_6H_3CO_2H$	152.15		>200d	140			al, aa	B14, 448
2628	Benzoic acid, 2,5-diamino $2,5\text{-(H}_2N)_2C_6H_3CO_2H$	152.15	br pr (w)		darkens				B14, 448
2629	Benzoic acid, 3,4-diamino $3,4\text{-(H}_2N)_2C_6H_3CO_2H$	152.15	lf (w)		215-8d				$B14^{3}$, 1177

No.	Name, Synonyms, and Formula	Mol. wt.	Color, crystalline form, specific rotation and λ_{max} (log ε)	b.p. °C	m.p. °C	Density	n_D	Solubility	Ref.
2630	Benzoic acid, 3,5-diamino 3,5-$(H_2N)_2C_6H_3CO_2H$	152.15	nd (+ lw)		240 (rapid)			al, eth	B14^{3}, 1179
2631	Benzoic acid, 2,3-dibromo 2,3-$Br_2C_6H_3CO_2H$	279.92	nd (w)		149-50				B9^{1}, 146
2632	Benzoic acid, 2,4-dibromo 2,4-$Br_2C_6H_3CO_2H$	279.92	lf (w)	sub	174			al, eth	B9^{2}, 237
2633	Benzoic acid, 2,5-dibromo 2,5-$Br_2C_6H_3CO_2H$	279.92	nd (al or w)	sub	157			al, eth, chl, aa	B9^{3}, 1428
2634	Benzoic acid, 2,6-dibromo 2,6-$Br_2C_6H_3CO_2H$	279.92	nd (w), (lig)	209—10^{16}	150-1			al, eth, ace, chl	B9^{3}, 1428
2635	Benzoic acid, 2,6-dibromo-3,4,5-trihydroxy or Gallobromol 2,6-$Br_2C(OH)_3CO_2H$	327.91	nd, pr or lf (w + 1)		139 (hyd)			w, al, eth	B10^{2}, 347
2636	Benzoic acid, 3,4-dibromo 3,4-$Br_2C_6H_3CO_2H$	279.92	nd (w), pl (al)		234-5			al, eth, MeOH	B9^{3}, 1428
2637	Benzoic acid, 2-(*N,N*-dichlorosulfamyl) 2-$(Cl_2NO_2S)C_6H_4CO_2H$	270.09	yesh gr pl (chl)		146-8 exp				B11, 377
2638	Benzoic acid, 4-(*N,N*-dichlorosulfamyl) or Halazone 4-$(Cl_2NO_2S)C_6H_4CO_2H$	270.09	pr (aa)		213			aa	B11^{1}, 230
2639	Benzoic acid, 2,3-dichloro 2,3-$Cl_2C_6H_3CO_2H$	191.01	nd (w)		168.3			al, eth, w	B9^{2}, 228
2640	Benzoic acid, 2,4-dichloro 2,4-$Cl_2C_6H_3CO_2H$	191.01	nd (w or bz)	sub	164.2			al, eth, bz, chl	B9^{3}, 1374
2641	Benzoic acid, 2,5-dichloro 2,5-$Cl_2C_6H_3CO_2H$	191.01	nd (w)	301	154.4			al, eth, w	B9^{3}, 1376
2642	Benzoic acid, 2,6-dichloro 2,6-$Cl_2C_6H_3CO_2H$	191.01	nd (al), pr (w)	sub	144			al, eth, bz, w	B9^{3}, 1377
2643	Benzoic acid, 3,4-dichloro 3,4-$Cl_2C_6H_3CO_2H$	191.01	nd (w, al, bz)		208-9			al, eth, w	B9^{3}, 1378
2644	Benzoic acid, 3,5-dichloro 3,5-$Cl_2C_6H_3CO_2H$	191.01	nd (al or w)	sub	188			al, eth	B9^{3}, 1379
2645	Benzoic acid, 2,3-dihydroxy or *o*-Pyrocatechuic acid 2,3$(HO)_2C_6H_3CO_2H$	154.12	pr or nd (w + 1)		204 (anh)	1.542$^{20/4}$		w, al, eth	B10^{3}, 1363
2646	Benzoic acid, 2,4-dihydroxy or *β*-Resorcylic acid 2-4-$(HO)_2C_6H_3CO_2H$	154.12	cr + w (w)					al, eth, bz	B10^{3}, 1370
2647	Benzoic acid, 2,4-dihydroxy-6-methyl or 4,6-dihydroxy-*o*-toluic acid 2,4$(HO)_2$-6-$CH_3C_6H_2CO_2H$	168.15	nd (dil aa + 1w)		176d			al, eth	B10^{2}, 272
2648	Benzoic acid, 2,4-dihydroxy-6-methyl, ethyl ester 6-CH_3-2,4-$(HO)_2C_6H_2CO_2C_2H_5$	196.20	lf (aa), pr (al)	sub	132			al, eth	B10^{3}, 1482
2649	Benzoic acid, 2,4-dihydroxy-6-pentyl 6-C_5H_{11}-2,4-$(HO)_2C_6H_2CO_2H$	224.26	wh nd		147			al, eth	B10^{3}, 1577
2650	Benzoic acid, 2,5-dihydroxy or Gentisic acid 2,5-$(HO)_2C_6H_3CO_2H$	154.12	nd or pr (w)		205			w, al, eth	B10^{3}, 1384
2651	Benzoic acid, 2,6-dihydroxy or *γ*-Resorcylic acid 2,6-$(HO)_2C_6H_3CO_2H$	154.12	nd (+ w)		167d			al, eth, w	B10^{3}, 1401
2652	Benzoic acid, 3,4-dihydroxy or Protocatechuic acid 3,4-$(HO)_2C_6H_3CO_2H$	154.12	mcl nd (w + 1)		200-2d	1.524^{4}		w, al, eth	B10^{3}, 1403
2653	Benzoic acid, 3,5-dihydroxy or *α*-Resorcylic acid 3,5-$(HO)_2C_6H_3CO_2H$	154.12	pr or nd		238-40			al, eth, w	B10^{3}, 1446
2654	Benzoic acid, 3,4-dimethoxy or Veratric acid 3,4-$(CH_3O)_2C_6H_3CO_2H$	182.18	nd (w or aa), rh (sub)	sub	181-2 (sub)			al, eth	B10^{3}, 1404
2655	Benzoic acid, 2,3-dimethyl or Hemimellitic acid 2,3-$(CH_3)_2C_6H_3CO_2H$	150.18	pr (al)		144			al, eth	B9^{3}, 2434
2656	Benzoic acid, 2(dimethylamino) or *N,N*-Dimethylanthranilic acid 2-$(CH_3)_2NC_6H_4CO_2H$	165.19	pr nd (eth)	sub d	72			w, al, eth	B14^{3}, 896
2657	Benzoic acid, 3-(dimethylamino) 3-$(CH_3)_2NC_6H_4CO_2H$	165.19	nd (w)		151			w, al, eth	B14^{3}, 1002
2658	Benzoic acid, 4-(dimethylamino) 4 $(CH_3)_2NC_6H_4CO_2H$	165.19	nd (al)		242-3			al	B14^{3}, 1082
2659	Benzoic acid, 2,4-dimethyl or 2,4-Xylylic acid 2,4-$(CH_3)_2C_6H_3CO_2H$	150.18	mcl or tcl nd (w)	267^{727}, sub	127 (anh), 90 (hyd)			al, ace, bz	B9^{2}, 350
2660	Benzoic acid, 2,5-dimethyl or Isoxylylic acid 2,5-$(CH_3)_2C_6H_3CO_2H$	150.18	nd (al)	268 sub	132	1.069$^{21/4}$		al, eth, ace, bz	B9^{1}, 210

No.	Name, Synonyms, and Formula	Mol. wt.	Color, crystalline form, specific rotation and λ_{max} (log ε)	b.p. °C	m.p. °C	Density	n_D	Solubility	Ref.
2661	Benzoic acid, 2,6-dimethyl or 2,6-Xylylic acid 2,6-$(CH_3)_2C_6H_3CO_2H$	150.18	nd (lig)	274-5	116	...	...	al, eth	$B9^3$, 2435
2662	Benzoic acid, 3,4-dimethyl or Paraxylylic acid 3,4-$(CH_3)_2C_6H_3CO_2H$	150.18	pr (al)	...	166	...	...	al, eth, bz	$B9^3$, 2441
2663	Benzoic acid, 3,5-dimethyl or Mesitylenic acid 3,5-$(CH_3)_2C_6H_3CO_2H$	150.18	nd (w or al)	sub	170-1	...	...	al, eth	$B9^3$, 2444
2664	Benzoic acid, 2,4-dinitro 2,4-$(NO_2)_2C_6H_3CO_2H$	212.12	nd (w)	...	...	...	...	...	$B9^3$, 1776
2665	Benzoic acid, 2,4-dinitro-3-hydroxy 2,4-$(NO_2)_2$-3-$HOC_6H_2CO_2H$	228.12	(w)	...	204	...	...	al, eth	$B10^3$, 264
2666	Benzoic acid, 2,5-dinitro 2,5-$(NO_2)_2C_6H_3CO_2H$	212.12	pr (w)	...	177	...	...	al, eth	$B9^3$, 1778
2667	Benzoic acid, 2,6-dinitro 2,6-$(NO_2)_2C_6H_3CO_2H$	212.12	nd (w)	...	202.3	...	...	al, eth	$B9^3$, 1778
2668	Benzoic acid, 3,4-dinitro 3,4-$(NO_2)_2C_6H_3CO_2H$	212.12	nd (w)	...	165	...	...	al, eth	$B9^3$, 177
2669	Benzoic acid, 3,5-dinitro 3,5-$(NO_2)_2C_6H_3CO_2H$	212.12	mcl pr (al)	...	205	...	...	al, aa	$B9^3$, 1779
2670	Benzoic acid, 3,5-dinitro, benzyl ester 3,5-$(NO_2)_2C_6H_3CO_2CH_2C_6H_5$	302.25	nd (lig)	...	112	...	...	...	$B9^3$, 1848
2671	Benzoic acid, 3,5-dinitro, butyl ester 3,5-$(NO_2)_2C_6H_3CO_2C_4H_9$	268.23	mcl nd (al)	...	62.5	...	...	al	$B9^3$, 1782
2672	Benzoic acid, 3,5-dinitro, isobutyl ester 3,5-$(NO_2)_2C_6H_3CO_2$-i-C_4H_9	268.23	...	...	87-8	...	...	w, al	$B9^2$, 281
2673	Benzoic acid, 3,5-dinitro, ethyl ester 3,5-$(NO_2)_2C_6H_3CO_2C_2H_5$	240.18	nd (al)	...	72.9	1.295^{111}	1.560	al	$B9^3$, 1781
2674	Benzoic acid, 3,5-dinitro-2-hydroxy 3,5-$(NO_2)_2$-2-$HOC_6H_2CO_2H$	228.12	ye nd or pl (+1w)	...	182 (anh)	...	...	w, al, eth, bz	$B10^3$, 207
2675	Benzoic acid, 3,5-dinitro-4-hydroxy 3,5-$(NO_2)_2$-4-$HOC_6H_2CO_2H$	228.12	ye lf (al)	...	248-9	...	...	al, eth	$B10^3$, 383
2676	Benzoic acid, 3,5-dinitro, methyl ester 3,5-$(NO_2)_2C_6H_3CO_2CH_3$	226.15	nd (w)	...	112	...	...	w, al	$B9^3$, 1781
2677	Benzoic acid, 3,5-dinitro, pentyl ester 3,5-$(NO_2)_2C_6H_3CO_2C_5H_{11}$	282.25	...	...	46.4	...	...	al	$B9^2$, 281
2678	Benzoic acid, 3,5-dinitro, phenyl ester 3,5-$(NO_2)_2C_6H_3CO_2C_6H_5$	288.22	rods (al)	...	145-6	...	...	al, bz	$B9^3$, 1846
2679	Benzoic acid, 3,5-dinitro, propyl ester 3,5-$(NO_2)_2C_6H_3CO_2C_3H_7$	254.20	mcl pl (al)	...	73	...	...	al	$B9^3$, 1781
2680	Benzoic acid, 3,5-dinitro, isopropyl ester 3,5-$(NO_2)_2C_6H_3CO_2$-i-C_3H_7	254.20	nd (al)	...	122	...	...	al	$B9^3$, 1782
2681	Benzoic acid, 3,5-dinitro, tetrahydrofurfuryl ester 3,5-$(NO_2)_2C_6H_3CO_2CH_2(C_4H_7O)$	296.24	nd (al)	...	83-4	...	...	al	$B17^4$, 1106
2682	Benzoic acid, 2-ethoxy 2-$C_2H_5OC_6H_4CO_2H$	166.18	...	$211\text{-}2^{39}$	20.7	...	...	...	$B10^3$, 98
2683	Benzoic acid, (2-ethoxyethyl)ester or Ethylcellosolve benzoate $C_6H_5CO_2CH_2CH_2OCH_2CH_3$	194.23	...	$260\text{-}1^{739}$	...	$1.0585^{25/25}$	1.4969^{25}	al, eth, ace, bz	$B9^3$, 533
2684	Benzoic acid, 2-ethoxy, ethyl ester 2-$C_2H_5OC_6H_4CO_2C_2H_5$	194.23	...	251, $180\text{-}5^{111}$	...	1.005^{20}	...	al, eth	$B10^3$, 117
2685	Benzoic acid, 3-ethoxy 3-$C_2H_5OC_6H_4CO_2H$	166.18	nd (w)	sub	137	...	...	w, al, eth, bz	$B10^3$, 245
2686	Benzoic acid, 3-ethoxy, ethyl ester 3-$C_2H_5OC_6H_4CO_2C_2H_5$	194.23	...	264, $172\text{-}3^{50}$	...	$1.0725^{20/20}$	...	al, eth	$B10^3$, 251
2687	Benzoic acid, 4-ethoxy 4-$C_2H_5OC_6H_4CO_2H$	166.18	nd (w)	...	198.5	...	...	al, eth, bz	$B10^3$, 282
2688	Benzoic acid, 4-ethoxy, ethyl ester 4-$C_2H_5OC_6H_4CO_2C_2H_5$	194.23	...	275, $148\text{-}9^{14}$	...	1.076^{12}	...	al, eth	$B10^3$, 301
2689	Benzoic acid, 4-ethoxy-2-hydroxy or 4-Ethoxysalicylic acid 4-C_2H_5O-2-$HOC_6H_3CO_2H$	182.18	nd (w or bz)	...	154	...	...	al, eth, bz	B10, 379
2690	Benzoic acid, ethyl ester or Ethyl benzoate $C_6H_5CO_2C_2H_5$	150.18	...	213, 87^{10}	-34.6	$1.0468^{20/4}$	1.5007^{20}	al, eth, ace, bz, peth	$B9^3$, 384
2691	Benzoic acid, 2-ethyl 2-$C_2H_5C_6H_4CO_2H$	150.18	nd (w)	259	68	$1.0431^{100/4}$	1.5099^{100}	al, eth	$B9^3$, 2425
2692	Benzoic acid, 3-ethyl 3-$C_2H_5C_6H_4CO_2H$	150.18	nd (w or dil al)	...	47	1.042^{100}	1.5345^{100}	al, eth	$B9^3$, 2429

No.	Name, Synonyms, and Formula	Mol. wt.	Color, crystalline form, specific rotation and λ_{max} (log ε)	b.p. °C	m.p. °C	Density	n_D	Solubility	Ref.
2693	Benzoic acid, 4-ethyl $4\text{-}C_2H_5C_6H_4CO_2H$	150.18	pr (al), pr or lf (w)	....	113.5	....	....	al, bz, chl	B9³, 2430
2694	Benzoic acid, 2-ethylamino $2\text{-}(C_2H_5NH)C_6H_4CO_2H$	165.19	pr or nd (dil al)	....	154	....	....	al, eth, ace	B14³, 897
2695	Benzoic acid, 3-ethylamino $3\text{-}(C_2H_5NH)C_6H_4CO_2H$	165.19	nd or pl (dil al)	sub	112	....	....	al, eth, ace	B14, 393
2696	Benzoic acid, 4-ethylamino $4\text{-}(C_2H_5NH)C_6H_4CO_2H$	165.19	cr (bz)	....	177-8	....	....	al, eth, ace, bz	B14³, 1086
2697	Benzoic acid, 2-fluoro $2\text{-}FC_6H_4CO_2H$	140.11	nd (w)	....	1.265	$1.460^{25/4}$	....	al, eth, chl	B9³, 1324
2698	Benzoic acid, 3-fluoro $3\text{-}FC_6H_4CO_2H$	140.11	lf (w)	....	124	$1.474^{25/4}$	....	eth	B9³, 1327
2699	Benzoic acid, 4-fluoro $4\text{-}FC_6H_4CO_2H$	140.11	pr (w)	....	185	$1.479^{25/4}$	....	al, eth	B9³, 1327
2700	Benzoic acid, 2-formamido $2\text{-}(HCONH)C_6H_4CO_2H$	165.15	nd (w + 1)	....	169	....	....	al, eth	B14³, 921
2701	Benzoic acid, 2-formyl or Phthalaldehydic acid $2\text{-}(HCO)C_6H_4CO_2H$	150.13	lf (+ w)	....	98-9	1.404	....	w, al, eth	B10¹, 2986
2702	Benzoic acid, 3-formyl or Isophthalaldehydic acid $3\text{-}(HCO)C_6H_4CO_2H$	150.13	nd (w)	....	175	....	....	al, eth, w	B10¹, 2988
2703	Benzoic acid, 4-formyl or Terephthalaldehydic acid $4\text{-}(HCO)C_6H_4CO_2H$	150.13	nd (w)	sub	256	....	....	al, eth, chl	B10¹, 2989
2704	Benzoic acid, hexyl ester or Hexylbenzoate $C_6H_5CO_2C_6H_{13}$	206.28	....	272^{770}, $139\text{-}140^{8}$	113-7	....	....	al, ace	B9³, 398
2705	Benzoic acid, hydrazide or Benzoyl hydrazine $C_6H_5CONHNH_2$	136.15	pl (w)	....	113-7	....	....	w, al	B9³, 1312
2706	Benzoic acid, 2-hydrazino $2\text{-}H_2NNHC_6H_4CO_2H$	152.15	nd (w)	....	250-1	....	....	w, al	B15³, 831
2707	Benzoic acid, 2-hydrazino, hydrochloride $2\text{-}H_2NNHC_6H_4CO_2H.HCl$	188.61	nd (w)	....	194-5d	....	....	w	B15², 295
2708	Benzoic acid, 3-hydrazino $3\text{-}H_2NNHC_6H_4CO_2H$	152.15	pa ye lf (w)	....	186d	....	....	....	B15³, 836
2709	Benzoic acid, 4-hydrazino $4\text{-}H_2NNHC_6H_4CO_2H$	152.15	ye nd or pl (w)	....	220-5d	....	....	....	B15³, 837
2710	Benzoic acid, 2-hydroxy or Salicylic acid $2\text{-}HOC_6H_4CO_2H$	138.12	nd (w), nd pr (al)	211^{20}, sub	158	$.443^{20/4}$	1.565	al, eth, ace	B10¹, 87
2711	Benzoic acid, 3-hydroxy $3\text{-}HOC_6H_4CO_2H$	138.12	nd (w), pl, pr (al)	....	202-3	....	....	eth, ace	B10¹, 2421
2712	Benzoic acid, 3-hydroxy, ethyl ester $3\text{-}HOC_6H_4CO_2C_2H_5$	166.18	pl (bz)	295, 211^{65}	73.5	....	....	al, eth	B10¹, 250
2713	Benzoic acid, 3-hydroxy, methyl ester $3\text{-}HOC_6H_4CO_2CH_3$	152.15	nd (bz-peth)	280^{709}, 178^{17}	71.5	....	....	al, bz	B10¹, 249
2714	Benzoic acid, 4-hydroxy $4\text{-}HOC_6H_4CO_2H$	138.12	pr or pl (w, al), cr (dil al, ace)	....	214-5	....	....	al, eth, ace	B10¹, 277
2715	Benzoic acid, 4-hydroxy, butyl ester $4\text{-}HOC_6H_4CO_2C_4H_9$	194.23	....	....	68-9	....	....	al	B10³, 307
2716	Benzoic acid, 4-hydroxy, ethyl ester $4\text{-}HOC_6H_4CO_2C_2H_5$	166.18	cr (dil al)	297-8	116-8	....	....	al, eth	B10¹300
2717	Benzoic acid, 4-hydroxy, methyl ester $4\text{-}HOC_6H_4CO_2H_3$	152.15	nd (dil al)	270-80	131	....	....	al, eth, ace	B10³, 297
2718	Benzoic acid, 4-hydroxy, propyl ester $4\text{-}HOC_6H_4CO_2C_3H_7$	180.20	pr (eth)	....	96-8	$1.0630^{102/4}$	1.5050^{102}	al, eth	B10³, 306
2719	Benzoic acid, 4-(α-hydroxybenzyl) $4\text{-}[C_6H_5CH(OH)]C_6H_4CO_2H$	228.25	nd (w)	....	164-5	....	....	al, eth, w	B10³, 1184
2720	Benzoic acid, 2-hydroxyethyl $2\text{-}(HOCH_2CH_2)C_6H_4CO_2H$	166.18	....	$260\text{-}1^{739}$	$1.0585^{25/25}$	1.4969^{25}	....	al, eth, ace, bz	B10³, 572
2721	Benzoic acid, 2-hydroxyethyl ester or Ethylene glycol monobenzoate $C_6H_5CO_2CH_2CH_2OH$	166.18	....	$150\text{-}1^{10}$	45	....	....	al	B9³, 532
2722	Benzoic acid, 2-(hydroxymethyl) $2\text{-}(HOCH_2)C_6H_4CO_2H$	152.15	nd (w)	....	128	....	....	al, eth, w	B10³, 500
2723	Benzoic acid, 2-iodo $2\text{-}IC_6H_4CO_2H$	248.02	nd (w)	233 exp	163	....	....	al, eth	B9³, 1432
2724	Benzoic acid, 2-iodo, methyl ester $2\text{-}IC_6H_4CO_2CH_3$	262.05	....	$277\text{-}8^{720}$, 146^{16}	....	....	1.6052^{20}	al	B9³, 1434
2725	Benzoic acid, 2-iodo-4-nitro $4\text{-}NO_2\text{-}2\text{-}IC_6H_3CO_2H$	293.02	pa ye pr (w)	....	146-7	....	....	al, eth	B9³, 1774

No.	Name, Synonyms, and Formula	Mol. wt.	Color, crystalline form, specific rotation and λ_{max} (log ε)	b.p. °C	m.p. °C	Density	n_D	Solubility	Ref.
2726	Benzoic acid, 3-iodo $3-IC_6H_4C_2H$	248.02	mcl pr (ace)	sub	187-8			al	B9^3, 1437
2727	Benzoic acid, 3-iodo, methyl ester $3-IC_6H_4CO_2CH_3$	262.05	nd (dil al)	$276-7^{739}$, 50^{18}	54-5			al, eth, ace	B9^3, 1438
2728	Benzoic acid, 4-iodo $4-IC_6H_4CO_2H$	248.02	mcl pr (dil al), lf (sub)	sub	270	2.184^{20}			B9^3, 1442
2729	Benzoic acid, 4-iodo, methyl ester $4-IC_6H_4CO_2CH_3$	262.05	nd (eth-al)	sub	114			al, eth	B9^3, 1443
2730	Benzoic acid, 4-iodo-2-nitro $2-NO_2-4-IC_6H_3CO_2H$	293.02	ye lf or pr (dil al)		192-3			al, eth, bz	B9^2, 278
2731	Benzoic acid, 4-iodo-3-nitro $3-NO_2-4-IC_6H_3CO_2H$	293.02	ye pr (al)		213			al	B9^2, 278
2732	Benzoic acid, 5-iodo-3-nitro $3-NO_2-5-IC_6H_3CO_2H$	293.02	nd (al), pr (peth)		167			al, w	B9^2, 278
2733	Benzoic acid, 2-iodoso $2-IOC_6H_4CO_2H$	264.02	lf (w)		223-5d				B9^3, 1433
2734	Benzoic acid, 3-iodoso $3-IOC_6H_4CO_2H$	264.02	ye amor		175-80				B9, 365
2735	Benzoic acid, 4-iodoso $4-IOC_6H_4CO_2H$	264.02	amor		212d				B9, 366
2736	Benzoic acid, 2-mercapto or Thiosalicylic acid $2-HSC_6H_4CO_2H$	154.18	lf or nd (al, w, aa)	sub	168-9			al, eth, w, aa	B10^3, 212
2737	Benzoic acid, (2-methoxyethyl) ester $C_6H_5CO_2CH_2CH_2OCH_3$	180.20		254-6		$1.0891^{25/25}$	1.5040^{25}	al, eth, ace, bz	B9^3, 533
2738	Benzoic acid, 3-methoxy-2-methylamino or Damascenine $3-CH_3O-2-CH_3NHC_6H_3CO_2H$	181.19	pr (al)	270^{750}d, $147-8^{10}$	27-9			al, eth, bz, lig	B14^1, 654
2739	Benzoic acid, (2-methoxyphenyl) ester or Guaiacyl benzoate $C_6H_5CO_2C_6H_4OCH_3-2$	228.25			58			al, eth, ace, chl	B9^3, 551
2740	Benzoic acid, methyl ester or Methyl benzoate $C_6H_5CO_2CH_3$	136.15		199.6, $96-8^{24}$	−12.3	$1.0888^{20/4}$	1.5164^{20}	al, **eth,** MeOH	B9^3, 381
2741	Benzoic acid, 3-methylbutyl ester or *iso*Pentyl benzoate $C_6H_5CO_2(CH_2)_2CH(CH_3)_2$	192.26		262.3, 133^{14}		$1.0040^{20/4}$	1.4950^{20}	al, **eth**	B9^{3}397
2742	Benzoic acid, methylene diester or Methanediol dibenzoate $(C_6H_5CO_2)_2CH$	256.26	nd or pr (eth)	225d	99	1.275^{22}		eth, ace, bz	B9^3, 715
2743	Benzoic acid, 2-methyl or *o*-Toluic acid $2-CH_3C_6H_4CO_2H$	136.15	pr or nd (w)	$258-9^{751}$	107-8	1.062^{115}	1.512^{115}	al, eth, chl	B9^3, 2298
2744	Benzoic acid, 2-methylamino $2-CH_3NHC_6H_4CO_2H$	151.16	pl (al or lig)	$80^{0.01}$	179			al, eth, bz	B14^3, 895
2745	Benzoic acid, 2-methylamino, ethyl ester $2-CH_3NHC_6H_4CO_2C_2H_5$	179.22		266, $141-3^{15}$	39			eth	B14^2, 213
2746	Benzoic acid, 2-methylamino, methyl ester $2-(CH_3NH)C_6H_4CO_2CH_3$	165.19	cr (peth)	255, $130-1^{13}$	18-19	1.120^{15}	1.5839^{12}	al, eth	B14^3, 895
2747	Benzoic acid, 3-methylamino $3-(CH_3NH)C_6H_4CO_2H$	151.16	pl (peth)		127			al, ace, bz, chl	B14^1, 559
2748	Benzoic acid, 3-methylamino, methyl ester $3-(CH_3NH)C_6H_4CO_2CH_3$	165.19	cr (al)		72			al, eth	B14, 392
2749	Benzoic acid, 4-methylamino $4-(CH_3NH)C_6H_4CO_2H$	151.16	nd (bz, w or dil al)		168			al, eth, w, bz	B14^3, 1081
2750	Benzoic acid, 4-methylamino, methyl ester $4-(CH_3NH)C_6H_4CO_2CH_3$	165.19	pl (dil al or lig)		93.5			al, eth	B14^3, 1081
2751	Benzoic acid, 3-(3-methylbutoxy) $3-[(CH_3)_2CH(CH_2)_2O]C_6H_4CO_2H$	208.26	cr (al)		74-5				B10^1, 64
2752	Benzoic acid, 4-(3-methylbutoxy) $4-[(CH_3)_2CH(CH_2)_2O]C_6H_4CO_2H$	208.26	nd		141-2				B10^1, 70
2753	Benzoic acid, 2-(α naphthoyl) $2-(\alpha-C_{10}H_7CO)C_6H_4CO_2H$	276.29	pl (al-w)		176.4			al, ace, bz, chl	B10^3, 3426
2754	Benzoic acid, 2-(β-naphthoyl) $2-(\beta-C_{10}H_7CO)C_6H_4CO_2H$	276.29	nd (to)		168			al, eth, ace, bz	B10^3, 3429
2755	Benzoic acid, α napthyl ester $C_6H_5CO_2-\alpha-C_{10}H_7$	248.28	pl or pr (al-eth)		56			al, eth	B9^3, 491
2756	Benzoic acid, β-naphthyl ester $C_6H_5CO_2-\beta-C_{10}H_7$	248.28	nd or pr (al)		107			al	B9^3, 492
2757	Benzoic acid, 2-nitro or 2-Nitrobenzoic acid $2-NO_2C_6H_4CO_2H$	167.12	tcl nd (w)		147-8	$1.575^{20/4}$		al, eth, ace	B9^3, 1466

No.	Name, Synonyms, and Formula	Mol. wt.	Color, crystalline form, specific rotation and λ_{max} (log ε)	b.p. °C	m.p. °C	Density	n_D	Solubility	Ref.
2758	Benzoic acid, (2-nitrobenzyl) ester $C_6H_5CO_2(CH_2C_6H_4NO_2\text{-}2)$	257.25	nd (dil al)	...	101-2	...	...	al, eth, bz, aa	B9, 121
2759	Benzoic acid, 2-nitro, ethyl ester $2\text{-}NO_2C_6H_4CO_2C_2H_5$	195.17	tcl (dil al)	275, 173^{18}	30	...	...	al, eth	$B9^3$, 1469
2760	Benzoic acid, 3-nitro or 3-Nitrobenzoic acid $3\text{-}NO_2C_6H_4CO_2H$	167.12	mcl pr (w)	...	140-2	1.494^{20}	...	al, eth, ace, chl	$B9^3$, 1489
2761	Benzoic acid, (3-nitrobenzyl) ester $C_6H_5CO_2(CH_2C_6H_4NO_2\text{-}3)$	257.25	...	...	71-2	...	...	al, eth	$B9^3$, 431
2762	Benzoic acid, 4-nitro $4\text{-}NO_2C_6H_4CO_2H$	167.12	mcl lf (w)	sub	242	1.610^{20}	...	al, eth, chl	$B9^3$, 1537
2763	Benzoic acid, (4-nitrobenzyl) ester $C_6H_5CO_2(CH_2C_6H_4NO_2\text{-}4)$	257.25	...	...	94-5	...	...	al, eth	$B9^3$, 431
2764	Benzoic acid, 2-nitroso $2\text{-}ONC_6H_4CO_2H$	151.12	cr (al or aa)	...	210d	...	...	...	$B9^3$, 1462
2765	Benzoic acid, 3-nitroso $3\text{-}ONC_6H_4CO_2H$	151.12	cr	...	23d	...	...	al	B9, 369
2766	Benzoic acid, 4-nitroso $4\text{-}ON_6H_4CO_2H$	151.12	ye pw	...	<350	...	...	al	$B9^3$, 1465
2767	Benzoic acid, 4-octyl $4\text{-}C_8H_{17}C_6H_4CO_2H$	234.34	lf (al)	...	139	...	...	al	$B9^3$, 2611
2768	Benzoic acid, pentachloro $C_6Cl_5CO_2H$	294.35	nd or pl (bz or dil aa)	sub (vac)	208	...	...	al, to	$B9^3$, 1383
2769	Benzoic acid, pentamethyl $C_6(CH_3)_5CO_2H$	192.26	nd (w), lf or nd (dil al)	sub	210.5	...	...	al	$B9^3$, 2564
2770	Benzoic acid, per or Perbenzoic acid C_6H_5COOOH	138.12	mcl pl (peth)	97-$110^{13.5}$ sub	41-3	...	...	al, eth, ace, bz	$B9^3$, 1049
2771	Benzoic acid, 2-phenoxy $2\text{-}C_6H_5OC_6H_4CO_2H$	214.22	lf (dil al)	355d	113-4	...	...	al, eth, chl	$B10^3$, 99
2772	Benzoic acid, 3-phenoxy $3\text{-}C_6H_5OC_6H_4CO_2H$	214.22	...	...	145	...	...	al, eth	$B10^3$, 247
2773	Benzoic acid, 4-phenoxy $4\text{-}C_6H_5OC_6H_4CO_2H$	214.22	...	...	161	...	...	al, eth	$B10^3$, 289
2774	Benzoic acid, phenyl ester or Phenylbenzoate $C_6H_5CO_2C_6H_5$	198.22	mcl pr (eth-al)	314	71	$1.235^{20/4}$	...	al, eth	$B9^3$, 415
2775	Benzoic acid, (1-phenylethyl) ester $C_6H_5CO_2CH(C_6H_5)CH_3$	226.27	...	189^{21}	...	1.1108^{18}	1.5588^{21}	al	$B9^3$, 431
2776	Benzoic acid, 2-phoshono $2\text{-}(HO_3P)C_6H_4CO_2H$	202.10	nd (w or al)	...	>300	...	...	w	B16, 820
2777	Benzoic acid, propyl ester or Propyl benzoate $C_6H_5CO_2C_3H_7$	164.20	...	211	-51.6	$1.0230^{20/4}$	...	**al, eth**	$B9^3$, 389
2778	Benzoic acid, *iso*-propyl ester or Isopropyl benzoate $C_6H_5CO_2CH(CH_3)_2$	164.20	...	218	...	$1.0172^{15/15}$	1.4890^{20}	al, eth, ace	$B9^3$, 391
2779	Benzoic acid, 2-propyl $2\text{-}C_3H_7C_6H_4CO_2H$	164.20	lf (dil al)	272^{739}, 164-5^{20}	58	...	...	al, eth, w	$B9^1$, 213
2780	Benzoic acid, 2-*iso*-propyl or *o*-Cuminic acid $2\text{-}[(CH_3)_2CH]C_6H_4CO_2H$	164.20	pr (w or peth)	160-1^{25}	64	...	...	al, eth, bz, peth	$B9^3$, 2481
2781	Benzoic acid, 4-propyl $4\text{-}C_3H_7C_6H_4CO_2H$	164.20	pr or lf (w)	...	141	...	...	al, eth, w, bz, lig	$B9^3$, 2479
2782	Benzoic acid, 4-*iso*-propyl or Cumic acid $4\text{-}[(CH_3)_2CH]C_6H_4CO_2H$	164.20	tcl pl (al)	sub	117-8	1.162^4	...	al, eth, peth	$B9^3$, 2482
2783	Benzoic acid, 2,3,4,5-tetrachloro $2,3,4,5\text{-}Cl_4C_6HCO_2H$	259.90	nd (al), cr (ace-w)	...	194-5	...	...	al, eth	$B9^3$, 1381
2784	Benzoic acid, tetrahydrofurfuryl ester $C_6H_5CO_2(C_4H_7O)$	192.21	...	300-2^{750}, 138-40^2	...	$1.137^{20/4}$	...	al, eth, chl	$B17^4$, 1105
2785	Benzoic acid, 2,3,4,5-tetrahydroxy $2,3,4,5\text{-}(HO)_4C_6HCO_2H$	186.12	pr	...	84-5	...	...	w	C50, 14644
2786	Benzoic acid, thiolo $C_6H_5C(=O)SH$	138.18	ye pl (aa)	85-7^{10}	24	...	1.6040^{20}	al, eth, ace, bz	$B9^3$, 1961
2787	Benzoic acid, 2-tolylester $C_6H_5CO_2C_6H_4CH_3\text{-}2$	212.25	...	307-8^{728}, 154-$6^{8/5}$	...	1.114^{19}	...	al, eth	$B9^3$, 423
2788	Benzoic acid, 3-tolylester $C_6H_5CO_2C_6H_4CH_3\text{-}3$	212.25	...	314, 168-70^8	55-6	...	...	al, eth	$B9^3$, 424
2789	Benzoic acid, 4-tolylester $C_6H_5CO_2C_6H_4CH_3\text{-}4$	212.25	pl (eth-al)	316	71.5	...	...	al, eth	$B9^3$, 427
2790	Benzoic acid, 2-(2-tolyl) $2\text{-}(2\text{-}CH_3C_6H_4)C_6H_4CO_2H$	212.25	nd (w + l), cr (bz)	...	130-2 (anh)	...	...	al, eth, w	$B9^3$, 3323
2791	Benzoic acid, 2-(3-tolyl) $2\text{-}(\text{-}3\text{-}CH_3C_6H_4)C_6H_4CO_2H$	212.25	nd (w + l)	...	162	...	...	eth, ace, w	$B9^3$, 3326

No.	Name, Synonyms, and Formula	Mol. wt.	Color, crystalline form, specific rotation and λ_{max} (log ε)	b.p. °C	m.p. °C	Density	n_D	Solubility	Ref.
2792	Benzoic acid, 2-(4-tolyl) 2-(4-$CH_3C_6H_4$)$C_6H_4CO_2H$	212.25	pr (+ lw, al-to), nd (al)		146			al, eth, ace, bz	B9, 677
2793	Benzoic acid, 2-(4-tolyl), methyl ester 2-(4-$CH_3C_6H_4$)$C_6H_4CO_2CH_3$	226.27	pl (MeOH)		66			al, bz	B10, 759
2794	Benzoic acid, 4-(2-tolyl) 4-(2-$CH_3C_6H_4$)$C_6H_4CO_2H$	212.25			177				$B9^3$, 3328
2795	Benzoic acid, 4-(4-tolyl) 4-(4-$CH_3C_6H_4$)$C_6H_4CO_2H$	240.26	nd (MeOH or ace)		228			al, ace	B9, 677
2796	Benzoic acid, 2,3,5-triamino 2,3,5-$(H_2N)_3C_6H_2CO_2H$	167.17	cr (w)		d			w	B14, 455
2797	Benzoic acid, 3,4,5-triamino 3,4,5-$(H_2N)_3C_6H_2CO_2H$	167.17	nd (w + ½)		d			w	B14, 455
2798	Benzoic acid, 2,3,4-tribromo 2,3,4-$Br_3C_6H_2CO_2H$	358.81	nd (bz)		197-8			al, eth, bz	$B9^1$, 147
2799	Benzoic acid, 2,3,5-tribromo 2,3,5-$Br_3C_6H_2CO_2H$	358.81	nd (al)		193-4			al, eth, ace, bz	$B9^1$, 147
2800	Benzoic acid, 2,4,5-tribromo 2,4,5-$Br_3C_6H_2CO_2H$	358.81	nd (al or bz)		195-6			al, eth	$B9^3$, 1430
2801	Benzoic acid, 2,4,6-tribromo 2,4,6-$Br_3C_6H_2CO_2H$	358.81	pr (w)		198			al, eth, bz	$B9^3$, 1430
2802	Benzoic acid, 3,4,5-tribromo 3,4,5-$Br_3C_6H_2CO_2H$	358.81	nd (bz or al)		240			al, eth	$B9^3$, 1431
2803	Benzoic acid, 2,3,4-trichloro 2,3,4-$Cl_3C_6H_2CO_2H$	225.46	nd (w)		187-8				B9, 345
2804	Benzoic acid, 2,3,5-trichloro 2,3,5-$Cl_3HC_6H_2CO_2H$	225.46	nd (w)		163			al, eth, ace, bz	$B9^3$, 1380
2805	Benzoic acid, 2,3,6-trichloro 2,3,6-$Cl_3C_6H_2CO_2H$	225.46			124-5			eth	$B9^3$, 1380
2806	Benzoic acid, 2,4,5-trichloro 2,4,5-$Cl_3C_6H_2CO_2H$	225.46	nd (w or sub)	sub	168			al, eth, w	$B9^3$, 1380
2807	Benzoic acid, 2,4,6-trichloro 2,4,6,$Cl_3C_6H_2CO_2H$	225.46	nd (w)		164			al, eth, chl	$B9^3$, 1381
2808	Benzoic acid, 3,4,5-trichloro 3,4,5-$Cl_3C_6H_2CO_2H$	225.46	nd (dil al)		210			al, eth, ace, bz	$B9^2$, 230
2809	Benzoic acid, 2,3,4-trihydroxy 2,3,4-$(HO)_3C_6H_2CO_2H$	170.12	nd (+ w)	sub	207-8d			al, eth, ace	$B10^3$, 2057
2810	Benzoic acid, 2,3,4-trihydroxy, ethyl ester 2,3,4-$(HO)_3C_6H_2CO_2C_2H_5$	198.18	cr (w + l)		102, (anh), 86 (+ w)			al, eth, w	B10, 467
2811	Benzoic acid, 2,3,4-trihydroxy, methyl ester 2,3,4-$(HO)_3C_6H_2CO_2CH_3$	184.15	nd (w + 2)		151-2			al, w	$B10^3$, 2058
2812	Benzoic acid, 2,4,5-trihydroxy 2,4,5-$(HO)_3C_6H_2CO_2H$	170.12	nd (w + ½), nd		217-8d			al, w	$B10^3$, 2065
2813	Benzoic acid, 2,4,6-trihydroxy 2,4,6-$(HO)_3C_6H_2CO_2H$	170.12	(w + l)		100d			al, eth, w	$B10^2$, 334
2814	Benzoic acid, 2,4,6-trihydroxy, ethyl ester 2,4,6-$(HO)_3C_6H_2CO_2H_5$	198.18	pr or nd (w + l), pr (lig)		129			al, eth, w, bz	$B10^1$, 236
2815	Benzoic acid, 2,4,6-trihydroxy, methyl ester 2,4,6-$(HO)_3C_6H_2CO_2CH_3$	184.15	cr (dil al)		174-6			al, eth	$B10^3$, 2069
2816	Benzoic acid, 3,4,5-trihydroxy or Gallic acid 3,4,5-$(HO)_3C_6H_2CO_2H$	170.12	pr (w + l)		253d	$1.694^{6/4}$		al, ace	$B10^3$, 2070
2817	Benzoic acid, 3,4,5-trihydroxy, ethyl ester 3,4,5-$(HO)_3C_6H_2CO_2C_2H_5$	198.18	mcl pr (w + 2½), nd (chl)		160-2			al, eth, w	$B10^2$, 343
2818	Benzoic acid, 3,4,5-trihydroxy, methyl ester or Gallicin 3,4,5-$(HO)_3C_6H_2CO_2CH_3$	184.15	mcl pr (MeOH)		202			al	$B10^3$, 2076
2819	Benzoic acid, 3,4,5-trihydroxy, propylester 3,4,5-$(HO)_3C_6H_2CO_2C_3H_7$	212.20	nd (w)		130				$B10^3$, 2078
2820	Benzoic acid, 3,4,5-trihydroxy, *iso*-propylester 3,4,5-$(HO)_3C_6H_2CO_2CH(CH_3)_2$	212.20			123-4			w, al, eth	$B10^2$, 343
2821	Benzoic acid, 2,3,5-triido 2,3,5-$I_3C_6H_2CO_2H$	499.81	pr (al)		224-6			al, eth	$B9^3$, 1456
2822	Benzoic acid, 2,4,5-triido 2,4,5-$I_3C_6H_2CO_2H$	499.81	nd (al)		248			eth, al	$B9^1$, 150

No.	Name, Synonyms, and Formula	Mol. wt.	Color, crystalline form, specific rotation and λ_{max} (log ε)	b.p. °C	m.p. °C	Density	n_D	Solubility	Ref.
2823	Benzoic acid, 3,4,5-triido 3,4,5-$I_3C_6H_2CO_2H$	499.81	pr (al)	....	292-3	....	....	al	$B9^3$, 1457
2824	Benzoic acid, 2,3,4-trimethoxy 2,3,4-$(CH_3O)_3C_6H_2CO_2H$	212.20	cr (w or peth)	....	100	....	....	w, al, eth	$B10^3$, 2058
2825	Benzoic acid, 2,4,5-trimethoxy or Asaronic acid 2,4,5-$(CH_3O)_3C_6H_2CO_2H$	212.20	nd (al or bz-peth)	300	144	....	....	w, al, bz, peth	$B10^3$, 2065
2826	Benzoic acid, 3,4,5-trimethoxy 3,4,5-$(CH_3O)_3C_6H_2CO_2H$	212.20	mcl nd (w)	$225\text{-}7^{10}$	171-2	....	....	al, eth, chl	$B10^3$, 2073
2827	Benzoic acid, 2,3,4-trimethyl or Prehnitylic acid 2,3,4-$(CH_3)_3C_6H_2CO_2H$	164.20	pr (al)	....	167.5	....	....	w, al, eth	$B9^3$, 2489
2828	Benzoic acid, 2,3,5-trimethyl or α-Isodurylic acid 2,3,5-$(CH_3)_3C_6H_2CO_2H$	164.20	pl (lig)	....	127	....	....	al	$B9^3$, 2489
2829	Benzoic acid, 2,3,6-trimethyl 2,3,6-$(CH_3)_3C_6H_2CO_2H$	164.20	nd (w or peth)	....	110-1	....	....	w, al, eth	$B9^3$, 2489
2830	Benzoic acid, 2,4,5-trimethyl or Durylic acid 2,4,5-$(CH_3)_3C_6H_2CO_2H$	164.20	nd (bz)	....	152-3	....	....	al, eth	$B9^3$, 2501
2831	Benzoic acid, 2,4,6-trimethyl or Mesitoic acid 2,4,6-$(CH_3)_3C_6H_2CO_2H$	164.20	pr (lig)	....	155	....	....	al, eth, ace, chl	$B9^3$, 2489
2832	Benzoic acid, 3,4,5-trimethyl or α-Isodurylic acid 3,4,5-$(CH_3)_3C_6H_2CO_2H$	164.20	nd (w)	....	215-6	....	....	al, eth	B9, 554
2833	Benzoic acid, 2,3,6-trinitro 2,3,6-$(NO_2)_3C_6H_2CO_2H$	257.12	wh nd (w + 2)	....	160d, 55 (+ 2w)	....	....	al	$B9^1$, 168
2834	Benzoic acid, 2,4,5-trinitro 2,4,5-$(NO_2)_3C_6H_2CO_2H$	257.12	ye lf or pl (w)	....	194.5d	....	....	al, eth, bz, w	$B9^3$, 1956
2835	Benzoic acid, 2,4,6-trinitro 2,4,6-$(NO_2)_3C_6H_2CO_2H$	257.12	rh (w)	....	228d	....	....	al, eth, ace	$B9^3$, 1956
2836	Benzoic acid, 3,4,5-trinitro 3,4,5-$(NO_2)_3C_6H_2CO_2H$	257.12	ye nd (eth + l)	....	168d	....	....	eth	$B9^1$, 168
2837	Benzoin *(d)* or α-Hydroxybensyl phenyl ketone $C_6H_5CH(OH)COC_6H_5$	212.25	$[\alpha]^{15/}_D$ +92.8 (Py. c=1)	....	133-4	....	....	al, ace, Py	$B8^3$, 1272
2838	Benzoin *(dl)* $C_6H_5CH(OH)COC_6H_5$	212.25	....	344^{768}, 194^{12}	137	$1.310^{20/4}$	....	al, chl, aa	$B8^3$, 1273
2839	Benzoin *(l)* $C_6H_5CH(OH)COC_6H_5$	212.25	nd (MeOH), $[\alpha]^{12/}_D$ −117.5 (ace, c=1.25)	....	133-4	....	....	al, bz	$B8^3$, 1272
2840	Benzoin, acetate *(dl)* $C_6H_5CH(O_2CCH_3)COC_6H_5$	254.29	pr or pl (eth)	....	83	....	....	al, eth	$B8^2$, 196
2841	Benzoin, 4,4′-dimethoxy or Anisoin 4-$CH_3OC_6H_4CH(OH)COC_6H_4OCH_3$-4	272.30	pr (dil al)	....	113	....	....	al, ace	$B8^3$, 3655
2842	Benzoin, ethyl ether $C_6H_5CH(COC_2H_5)COC_6H_5$	252.31	nd (lig)	$194\text{-}5^{20}$	62	$1.1016^{17/4}$	1.5727^{17}	al, eth, bz, lig	$B8^2$, 195
2843	Benzoin hydrazone *(dl)* $C_6H_5CH(OH)C(=NNH_2)C_6H_5$	226.28	pr (al)	....	75	....	....	al	B8, 176
2844	Benzoin methyl ether *(dl)* $C_6H_5CH(OCH_3)COC_6H_5$	226.28	nd (lig)	$188\text{-}9^{15}$	49-50	$1.1278^{14/4}$	....	al, eth, bz	$B8^2$, 195
2845	Benzoin oxime *(l)* $C_6H_5CH(OH)C(=NOH)C_6H_5$	227.26	amor or pr, (bz) $[\alpha]^{24/}_D$ -3.2 (chl, c=0.85)	....	163-4	....	....	al, eth, ace	B8, 167
2846	Benzoin oxime *(dl, anti)* $C_6H_5CH(OH)C(=NOH)C_6H_5$	227.26	pr (bz)	....	151-2	....	....	al, eth, ace	$B8^2$, 196
2847	Benzoin oxime *(dl, syn)* $C_6H_5CH(OH)C(=NOH)C_6H_5$	227.26	pr (eth)	....	99	....	....	al, eth, ace	$B8^2$, 196
2848	Benzonitrile or Benzoic acid nitrile C_6H_5CN	103.12	....	190.7	−13	$1.0102^{15/15}$	1.5289^{20}	**al, eth**, ace, bz	$B9^3$, 1255
2849	Benzonitrile, 2-amino 2-$H_2NC_6H_4CN$	118.14	....	263^{751}	51	....	....	al, eth, ace, bz, chl	$B14^2$, 210
2850	Bensonitrile, 3-amino 3-$H_2NC_6H_4CN$	118.14	....	288-90	53-4	....	....	al, eth, ace, chl	$B14^3$, 1001
2851	Benzonitrile, 4-amino 4-$H_2NC_6H_4CN$	118.14	....	....	86	....	....	al, eth, ace, bz, aa	$B14^3$, 1079
2852	Benzonitrile, 2-bromo 2-BrC_6H_4CN	182.02	nd (w)	$251\text{-}3^{754}$	55.5	....	....	al	$B9^3$, 1387
2853	Benzonitrile, 3-bromo 3-BrC_6H_4CN	182.03	(al)	225	39-40	....	....	al, eth	$B9^3$, 1399

PHYSICAL CONSTANTS OF ORGANIC COMPOUNDS (Continued)

No.	Name, Synonyms, and Formula	Mol. wt.	Color, crystalline form, specific rotation and λ_{max} (log ε)	b.p. °C	m.p. °C	Density	n_D	Solubility	Ref.
2854	Benzonitrile, 4-bromo ... $4\text{-}BrC_6H_4CN$	182.03	nd (w or al)	235-7	114	...	...	al, eth	$B9^{1}$, 1420
2855	Benzonitrile, 2-chloro ... $2\text{-}ClC_6H_4CN$	137.57	nd	232	43-6	...	...	al, eth	$B9^{1}$, 1339
2856	Benzonitrile, 3-chloro ... $3\text{-}ClC_6H_4CN$	137.57	...	$99\text{-}100^{15}$	40-2	...	...	al, eth	$B9^{3}$, 1349
2857	Benzonitrile, 4-chloro ... $4\text{-}ClC_6H_4CN$	137.57	nd (al)	223^{750}, 95^{5}	94-6	...	...	al, eth, bz, chl	$B9^{3}$, 1366
2858	Benzonitrile, 3 chloromethyl or 3-Chloromethyl benzonitrile ... $3\text{-}ClCH_2C_6H_4CN$	151.60	pr (al)	258-60	67	...	...	...	$B9^{3}$, 2328
2859	Benzonitrile, 4-chlormethyl or 4-Chloromethyl benzonitrile ... $4\text{-}ClCH_2C_6H_4CN$	151.60	pr (al)	263^{756}	79.5	...	...	...	$B9^{3}$, 2357
2860	Benzonitrile, 2,4-dimethoxy or 2,4-Dimethoxy benzonitrile ... $2,4\text{-}(CH_3O)_2C_6H_3CN$	163.18	cr (al, lig)	...	96	...	...	aa	$B10^{1}$, 1366
2861	Benzonitrile, 2,5-dimethoxy ... $2,5\text{-}(CH_3O)_2C_6H_3CN$	163.18	nd (al)	...	82	...	...	al, bz, chl	$B10^{1}$, 1388
2862	Benzonitrile, 2,6-dimethoxy or 2,6-Dimethoxy benzonitrile ... $2,6\text{-}(CH_3O)_2C_6H_3CN$	163.18	nd or pl	310	118-20	...	...	al, ace, bz, chl	$B10^{2}$, 260
2863	Benzonitrile, 3,4-dimethoxy or Veratronitrile ... $3,4\text{-}(CH_3O)_2C_6H_3CN$	163.18	nd (w)	...	67-8	...	...	bz, al, w	$B10^{2}$, 264
2864	Benzonitrile, 2-ethoxy ... $2\text{-}C_2H_5OC_6H_4CN$	147.18	...	260.7, 153^{15}	5	...	...	al, eth, lig	B10, 97
2865	Benzonitrile, 4-ethoxy ... $4\text{-}C_2H_5OC_6H_4CN$	147.18	nd (lig)	258	61-2	...	...	al, eth, lig	$B10^{1}$, 344
2866	Benzonitrile, 4-fluoro or 4-Fluoro benzonitrile ... $4\text{-}FC_6H_4CN$	121.11	nd (peth)	188.8	34.8	1.1070^{55}	1.4925^{55}	peth	$B9^{2}$, 221
2867	Benzonitrile, 3-formyl or 3-Cyanobenzaldehyde ... $3\text{-}HCOC_6H_4CN$	131.13	...	210	79-81	...	...	al, eth, w, chl	B10, 671
2868	Benzonitrile, 4-formyl or 4-Cyanobenzaldehyde ... $4\text{-}HCOC_6H_4CN$	131.13	nd (w), pr (eth, or dil al)	133^{12}	101-2	...	...	al, eth, w, chl	$B10^{1}$, 2990
2869	Benzonitrile, 2-hydroxy ... $2\text{-}HOC_6H_4CN$	119.12	...	149^{14}	98	$1.1052^{100/4}$	1.5372^{100}	al, eth, bz, chl	$B10^{2}$, 60
2870	Benzonitrile, 3-hydroxy ... $3\text{-}HOC_6H_4CN$	119.12	pr (al, eth), lf (w)	...	83 4	...	...	al, eth, bz, chl	$B10^{3}$, 255
2871	Benzonitrile, 4-hydroxy ... $4\text{-}HOC_6H_4CN$	119.12	lf (w)	...	113	...	...	al, eth, chl	$B10^{3}$, 344
2872	Benzonitrile, 2-methoxy ... $2\text{-}CH_3OC_6H_4CN$	133.15	...	255-6, 146^{20}	24.5	$1.1063^{20.5/4}$	...	al, eth	$B10^{2}$,,60
2873	Benzonitrile, 4-methoxy ... $4\text{-}CH_3C_6H_4CN$	133.15	nd (w), lf (al)	256-7, $106\text{-}8^{6}$	61-2	...	...	al, eth, bz	$B10^{2}$, 101
2874	Benzonitrile, 2-methyl ... $2\text{-}CH_3C_6H_4CN$	117.15	...	205, 90^{15}	-13.5	$0.9955^{20/4}$	1.5279^{20}	al, eth	$B9^{3}$, 2307
2875	Benzonitrile, 2-methyl-5-nitro ... $5\text{-}NO_2\text{-}2CH_3C_6H_3CN$	162.15	nd (95% al)	$174\text{-}5^{18}$	106	...	...	w, al, eth, ace, bz	$B9^{3}$, 2314
2876	Benzonitrile, 3-methyl ... $3\text{-}CH_3C_6H_4CN$	117.15	...	213, 84.5^{10}	-23	$1.0316^{20/4}$	1.5252^{20}	al, eth	$B9^{3}$, 2324
2877	Benzonitrile, 4-methyl ... $4\text{-}CH_3C_6H_4CN$	117.15	...	217.6, 91^{11}	29.5	$0.9805^{30/30}$	...	al, eth	$B9^{3}$, 2348
2878	Benzonitrile, 3-(trifluoromethyl) ... $3\text{-}CF_3C_6H_4CN$	171.12	...	189	14.5	1.28126^{20}	1.4508^{20}	...	$B9^{3}$, 2327
2879	3,4-Benzophenanthrene or Benzo (c) phenanthrene ... $C_{18}H_{12}$	228.29	...	...	68	...	...	...	$B9^{2}$, 2379
2880	9,10-Benzophenanthrene or Triphenylene ... $C_{18}H_{12}$	228.29	nd (al, chl, bz)	425	199	...	...	al, bz, chl	$B9^{3}$, 2380
2881	1,2-Benzophenazine ... $C_{16}H_{10}N_2$	230.27	pr (al), ye nd (bz)	> 360	142.5	...	...	al, eth, ace, bz, aa	$B23^{2}$, 259
2882	Benzophenone or Diphenyl ketone ... $C_6H_5COC_6H_5$	182.22	(α) rh pr (al, eth); (β) mcl pr	305.9	(α) 48.1, (β) 26	(α) 1.146^{20}, (β) 1.1076	α 1.6077^{19} β 1.6059^{23}	al, eth, ace, bz	$B7^{3}$, 2048
2883	Benzophenone, 2-amino ... $2\text{-}H_2NC_6H_4COC_6H_5$	197.24	pa ye lf or pr al	...	110-1	...	...	al, eth	$B14^{3}$, 213
2884	Benzophenone, 2-amino-5-chloro ... $5\text{-}Cl\text{-}2\text{-}H_2NC_6H_3COC_6H_5$	231.68	...	...	97-98	...	...	w, al, chl, peth	$B14^{3}$, 214

No.	Name, Synonyms, and Formula	Mol. wt.	Color, crystalline form, specific rotation and λ_{max} (log ε)	b.p. °C	m.p. °C	Density	n_D	Solubility	Ref.
2885	Benzophenone, 2-amino-4′-methyl $2\text{-}H_2NC_6H_4CO(C_6H_4CH_3\text{-}4)$	211.26	ye pr or pl (al)		96			al, eth, bz	B14^3, 248
2886	Benzophenone, 2-amino-5-methyl $2\text{-}H_2N\text{-}5\text{-}CH_3C_6H_3COC_6H_5$	211.26	ye nd or pl (al)		66			al, eth, ace, aa, lig	B14^3, 246
2887	Benzophenone, 2-amino-5-nitro $5\text{-}NO_2\text{-}2H_2NC_6H_3COC_6H_5$	242.23			161.5			al, aa	B14,79
2888	Benzophenone, 3-amino $3\text{-}H_2NC_6H_4COC_6H_5$	197.24	ye nd (w)					al, eth	B14^1, 388
2889	Benzophenone, 3-amino-4-methyl $3\text{-}H_2N\text{-}4\text{-}CH_3C_6H_3COC_6H_5$	211.26	pa ye nd (MeOH)		109			al, eth, ace, bz	B14^3, 247
2890	Benzophenone, 3-amino-4′-methyl $3\text{-}H_2NC_6H_4CO(C_6H_4CH_3\text{-}4)$	211.26	pr (al)		111			al, eth	B14, 107
2891	Benzophenone, 4-amino $4\text{-}H_2NC_6H_4COC_6H_5$	197.24	lf (dil al)		124			al, eth, aa	B14^3, 217
2892	Benzophenone, 4-amino-3-methyl $4\text{-}H_2N\text{-}3\text{-}CH_3C_6H_3COC_6H_5$	211.26	ye pr (w)		112			al, eth	B14^3, 245
2893	Benzophenone, 4-amino-4′-methyl $4\text{-}H_2NC_6H_4CO(C_6H_4CH_3\text{-}4)$	211.26	nd (bz)		186-7			al, eth, chl	B14^3, 248
2894	Benzophenone, 4,4′-bis(diethylamino) $[4\text{-}(C_2H_5)_2NC_6H_4]_2CO$	324.47	lf (al)		95-6				B14^2, 59
2895	Benzophenone, 4,4′-bis(dimethylamino) or Michler's ketone $[4\text{-}(CH_3)_2NC_6H_4]_2CO$	268.36	lf (al), nd (bz)	>360d	179			bz, Py	B14^3, 226
2896	Benzophenone, 2-bromo $2\text{-}BrC_6H_4COC_6H_5$	261.12	pl (al), nd (lig)	345	42			ace, al, lig	B7^3, 2079
2897	Benzophenone, 3-bromo $3\text{-}BrC_6H_4COC_6H_5$	261.12	nd (al)	185-7^{5}	81			al	B7^3, 2079
2898	Benzophenone, 4-bromo $4\text{-}BrC_6H_4COC_6H_5$	261.12	lf (al)	350^{757}	82.5				B7^3, 2079
2899	Benzophenone, 5-*tert* butyl-2-methoxy $5\text{-}t\text{-}C_4H_9\text{-}2CH_3OC_6H_3COC_6H_5$	268.35			62-3			chl	B8,187
2900	2-Benzophenone Carboxylic acid $2\text{-}HO_2CC_6H_4COC_6H_5$	226.23			127-9			al, eth, bz	B10^3, 3289
2901	2-Benzophenone carboxylic acid, ethyl ester $2\text{-}(C_2H_5O_2C)C_6H_4COC_6H_5$	254.28			58	1.221$^{64/4}$	1.560^{64}	al, eth	B10^2, 517
2902	2-Benzophenone carboxylic acid, methyl ester $2\text{-}(CH_3O_2C)C_6H_4COC_6H_5$	240.26		350-2	52	1.1903$^{19/4}$	1.591^{20}	al, eth	B10^2, 517
2903	Benzophenone, 2-chloro $2\text{-}ClC_6H_4COC_6H_5$	216.67	pl (chl-lig)	330,185-8^{11}	52-6				B7^3, 2071
2903a	Benzophenone, 2-chloro-3,5-dinitro $2\text{-}Cl\text{-}3,5\text{-}(NO_2)_2C_6H_2COC_6H_5$	306.66	ye nd (aa)		148			chl, aa	B7, 428
2904	Benzophenone, 3-chloro $3\text{-}ClC_6H_4COC_6H_5$	216.67	nd		82-3			bz	B7^3, 2071
2905	Benzophenone, 4-chloro $4\text{-}ClC_6H_4COC_6H_5$	216.67	nd (al)	332^{771}	77-8			w, al, eth	B7^3, 2072
2906	Benzophenone, 2,2′-diamino $(2H_2NC_6H_4)_2CO$	212.25	lf (dil al), pr (bz)		134-5			al, eth	B14,87
2907	Benzophenone, 3,3′-diamino $(3\text{-}H_2NC_6H_4)_2CO$	212.25	nd (al)	285^{11}	173-4			al, eth	B14^3, 225
2908	Benzophenone, 4,4′-diamino $(4\text{-}H_2NC_6H_4)_2CO$	212.25	nd (al)		244-5			al, eth	B14^3, 226
2909	Benzophenone, 3,3′-dibromo $(3\text{-}BrC_6H_4)_2CO$	340.01			141			al, eth	B7^2, 361
2910	Benzophenone, 4,4′-dibromo $(4\text{-}BrC_6H_4)_2CO$	340.01	pl (al)	395	177			al, ace, bz, chl	B7^3, 2081
2911	Benzophenone, 4,4′-dicarboxylic acid $(4\text{-}HO_2CC_6H_4)_2CO$	270.24	nd (al)	sub	>360			aa	B10^3, 4010
2912	Benzophenone, 2,4′-dichloro $2\text{-}ClC_6H_4CO(C_6H_4Cl\text{-}4)$	251.11		214-5^{22}	67	1.393^{14}		al, chl	B7^3, 2075
2913	Benzophenone, 3,3′-dichloro $(3\text{-}ClC_6H_4)_2CO$	251.11		160-6^{2}	124			al, eth	B7^3, 2076
2914	Benzophenone, 3,4′-dichloro $3\text{-}ClC_6H_4CO(C_6H_4Cl\text{-}4)$	251.11			112-3				B7^3, 2076
2915	Benzophenone, 4,4′-dichloro $(4\text{-}ClC_6H_4)_2CO$	251.11	pl (al)	353^{757}	147-8			al, eth, ace, chl, aa	B7^3, 2076
2916	Benzophenone, 3,5-dichloro-2-hydroxy $3,5\text{-}Cl_2\text{-}2HOC_6H_2COC_6H_5$	267.11			116			al, bz	B8^3, 1233

No.	Name, Synonyms, and Formula	Mol. wt.	Color, crystalline form, specific rotation and λ_{max} (log ε)	b.p. °C	m.p. °C	Density	n_D	Solubility	Ref.
2917	Benzophenone, 2,2′-dihydroxy $(2\text{-}HOC_6H_4)_2CO$	214.22	lf or pr (lig)	330-40	59.5	...	...	al, eth, chl	$B8^3$, 2644
2918	Benzophenone, 2,3′-dihydroxy $2\text{-}HOC_6H_4CO(C_6H_4OH\text{-}3)$	214.22	nd (w)	...	126	...	...	al, eth	B8, 315
2919	Benzophenone, 2,4-dihydroxy $2,4\text{-}(HO)_2C_6H_3COC_6H_5$	214.22	nd (w)	...	144	...	...	al, eth, aa	$B8^3$, 2640
2920	Benzophenone, 2,4-dihydroxy--6-methoxy or Isocotoin $2,4\text{-}(HO)_2\text{-}6\text{-}CH_3OC_6H_2COC_6H_5$	244.25	ye nd (lig)	...	162	...	...	w, ace	$B8^2$, 467
2921	Benzophenone, 2,4′-dihydroxy $2\text{-}HOC_6H_4CO(C_6H_4OH\text{-}4)$	214.22	pl (w)	...	150-1	...	...	al, eth, bz	$B8^3$, 2646
2922	Benzophenone, 2,5-dihydroxy $2,5\text{-}(HO)_2C_6H_3COC_6H_5$	214.22	ye nd (dil al)	...	125-6	...	...	al, eth, bz, w	$B8^3$, 2643
2923	Benzophenone, 2,6-dihydroxy-4-methoxy or Cotoin $2,6\text{-}(HO)_2\text{-}4\text{-}CH_3OC_6H_2COC_6H_5$	244.25	yesh pr (chl), lf or nd (w)	...	130-1	...	...	al, eth, ace, bz	$B8^3$, 3638
2924	Benzophenone, 3,3′-dihydroxy $(3\text{-}HOC_6H_4)_2CO$	214.22	nd (w)	...	170	...	...	al, w	$B8^3$, 2647
2925	Benzophenone, 3,4′-dihydroxy $3\text{-}HOC_6H_4CO(C_6H_4OH\text{-}4)$	214.22	nd (w)	...	208	...	...	al, eth, w	$B8^3$, 2648
2926	Benzophenone, 2,4′-dihydroxy $(4\text{-}HOC_6H_4)_2CO$	214.22	nd (lig), cr (w)	...	210	1.133^{131}	...	al, eth, ace	$B8^3$, 2646
2927	Benzophenone, 4,4′-diido $(4\text{-}IC_6H_4)_2CO$	434.01	pl (to), nd (bz)	281^{12}	238.5	...	...	...	B7,425
2928	Benzophenone, 2,4-dimethoxy $2,4\text{-}(CH_3O)_2C_6H_3COC_6H_5$	242.27	pr (dil al)	218^{10}	87-8	...	...	al, chl	$B8^3$, 2641
2929	Benzophenone, 2,4′-dimethoxy $2\text{-}CH_3OC_6H_4CO(C_6H_4OCH_3\text{-}4)$	242.27	nd (al)	...	100	...	...	al, eth, bz, aa	$B8^1$, 640
2930	Benzophenone, 2,5-dimethoxy $2,5(CH_3O)_2C_6H_3COC_6H_5$	242.27	(lig)	225^{38}	51	...	...	...	$B8^3$, 2644
2931	Benzophenone, 3,4-dimethoxy $3,4\text{-}(CH_3O)_2C_6H_3COC_6H_5$	242.27	nd or pl (al)	...	103-4	...	...	al	$B8^2$, 354
2932	Benzophenone, 3,4′-dimethoxy $3\text{-}CH_3OC_6H_4CO(C_6H_4OCH_3\text{-}4)$	242.27	pr (al)	...	58-9	...	...	al	$B8^2$, 354
2933	Benzophenone, 4,4′-dimethoxy $(4\text{-}CH_3OC_6H_4)_2CO$	242.27	nd (al)	...	148	...	...	al, eth, ace, bz	$B8^2$, 355
2934	Benzophenone, 2,2′-dimethyl $(2\text{-}CH_3C_6H_4)_2CO$	210.28	...	310^{740}, $175\text{-}80^{17}$	72	...	...	al, ace, bz	$B7^3$, 2178
2935	Benzophenone, 4,4′-dimethyl $(4\text{-}CH_3C_6H_4)_2CO$	210.28	rh (al)	333^{725}	95	...	...	al, eth, ace, bz	$B7^3$, 2181
2936	Benzophenone, 3-(dimethylamino) $3\text{-}(CH_3)_2NC_6H_4COC_6H_5$	225.29	pa ye pl (al)	216^{15}	47	...	...	al	$B14^1$, 388
2937	Benzophenone, 4-(dimethylamino) $4\text{-}(CH_3)_2NC_6H_4COC_6H_5$	225.29	ye lf (al), nd (peth)	...	92-3	...	...	al, eth, peth, chl	$B14^3$, 218
2938	Benzophenone, 2,2′-dinitro $(2\text{-}NO_2C_6H_4)_2CO$	272.21	nd (to or aa)	...	188-9	...	...	to, aa	$B7^3$, 2085
2939	Benzophenone, 3,3′-dinitro $(3\text{-}NO_2C_6H_4)_2CO$	272.22	...	...	157	...	...	ace	$B7^3$, 2085
2940	Benzophenone, 2-hydroxy $2\text{-}HOC_6H_4COC_6H_5$	198.22	pl (dil al)	250^{560}	39	...	...	al, eth, bz, aa	$B8^3$, 1227
2941	Benzophenone, 3-hydroxy $3\text{-}HOC_6H_4COC_6H_5$	198.22	lf or pl (al)	...	116	...	...	al, eth	$B8^3$, 1235
2942	Benzophenone, 4-hydroxy $4\text{-}HOC_6H_4COC_6H_5$	198.22	nd (al), pr (dil al)	...	135 (st), 122 (unst)	...	...	al, eth, aa, w	$B8^3$, 1237
2943	Benzophenone, 2-hydroxy-4-methoxy $4\text{-}CH_3O\text{-}2\text{-}HOC_6H_3COC_6H_5$	228.25	...	...	65.6	...	...	...	$B8^3$, 2640
2944	Benzophenone, 2-hydroxy-4′-methoxy $2\text{-}HOC_6H_4CO(C_6H_4OCH_3\text{-}4)$	228.25	...	...	98	...	...	...	...
2945	Benzophenone, 2-hydroxy-5-methoxy $5\text{-}CH_3O\text{-}2\text{-}HOC_6H_3COC_6H_5$	228.25	...	...	84	...	...	...	$B8^3$, 2644
2946	Benzophenone, 2-hydroxy-4′-methyl $2\text{-}HOC_6H_4CO(C_6H_4CH_3\text{-}4)$	212.25	...	...	40-1	...	...	...	$B8^3$, 1301
2947	Benzophenone, 2-hydroxy-4-methyl $4\text{-}CH_3\text{-}2\text{-}HOC_6H_3COC_6H_5$	212.25	...	...	63	...	...	...	$B8^3$, 1300
2948	Benzophenone, 2-hydroxy-5-methyl $5\text{-}CH_3\text{-}2\text{-}HOC_6H_3COC_6H_5$	212.25	...	...	84	...	...	...	$B8^3$, 1295

No.	Name, Synonyms, and Formula	Mol. wt.	Color, crystalline form, specific rotation and λ_{max} (log ε)	b.p. °C	m.p. °C	Density	n_D	Solubility	Ref.
2949	Benzophenone, 2-hydroxy-5-nitro 5-NO_2-2-$HOC_6H_3COC_6H_5$	243.22			124-5				B8, 157
2950	Benzophenone, 4-hydroxy-3-methyl 3-CH_3-4-$HOC_6H_3COC_6H_5$	212.25			173-4				B8^3, 1294
2951	Benzophenone, 4-hydroxy-4′-nitro 4-$HOC_6H_4CO(C_6H_4NO_2$-4)	243.22			190-2			al, eth, aa	B8,163
2952	Benzophenone imine $(C_6H_5)_2C{=}NH$	181.24		282,158^{12}		1.0847$^{19/4}$	1.6191^{19}	eth	B7^3, 2061
2953	Benzophenone, 2-methoxy 2-$CH_3OC_6H_4COC_6H_5$	212.25		194-6^{18}	41			al, bz, aa	B8^3, 1227
2954	Benzophenone, 3-methoxy 3-$CH_3OC_6H_4COC_6H_5$	212.25		342-3^{730}, 201^{17}	44			al, bz, aa	B8^3, 1236
2955	Benzophenone, 4-methoxy 4-$CH_3OC_6H_4COC_6H_5$	212.25	pr (eth)	354-5^{729}, 168^{12}	61-2			al, eth, ace, bz, aa	B8^3, 1238
2956	Benzophenone, 2-methyl 2-$CH_3C_6H_4COC_6H_5$	196.25		309.5^{762}, 128^{12}	< −18			al	B7^3, 2123
2957	Benzophenone, 3-methyl 3-$CH_3C_6H_4COC_6H_5$	196.25	oil	314-5^{725}, 170^{9}		1.088$^{17.5}$		al, eth, bz, chl, aa	B7^3, 2126
2958	Benzophenone, 4-methyl 4-$CH_3C_6H_4COC_6H_5$	196.25	mcl pr	327-8	59-60			eth, bz, chl	B7^3, 2127
2959	Benzophenone, 4-methyl, diphenyl acetal 4-$CH_3C_6H_4C(OC_6H_5)_2C_6H_5$	366.47	(dil al, eth-peth)		134				B7^2, 372
2960	Benzophenone, 4-methyl, imine 4-$CH_3C_6H_4C({=}NH)C_6H_5$	195.26		147^{5}	37	1.0617$^{20/4}$	1.6097^{20}		B7^3, 2129
2961	Benzophenone, 4-methyl-2-nitro 2-NO_2-4-$CH_3C_6H_3COC_6H_5$	241.25	nd or pl (al)	sub	126-7			al, bz, chl	B7, 442
2962	Benzophenone, 4-methyl-2′-nitro 4-$CH_3C_6H_4CO(C_6H_4NO_2$-2)	241.25	pr (aa or al)		155			bz, chl, ace	B7^3, 2132
2963	Benzophenone, 4-methyl-3-nitro 3-NO_2-4-$CH_3C_6H_3COC_6H_5$	241.25	pa ye pl (al or aa)		130-2			al, eth, ace, bz, aa	B7^3, 2131
2964	Benzophenone, 4-methyl-3′-nitro 4-$CH_3C_6H_4CO(C_6H_4NO_2$-3)	241.25	lf (al)		111			eth, bz, chl	B7^2, 375
2965	Benzophenone, 4-methyl-4′-nitro 4-$CH_3C_6H_4CO(C_6H_4NO_2$-4)	241.25	nd (al)	sub	122-4			al, eth, bz, chl, aa	B7^2, 375
2966	Benzophenone, 4-methyl-4′-nitro, oxime 4-$CH_3C_6H_4C({=}NOH)(C_6H_4NO_2$-4)	256.26	nd (eth-lig)		145			al, eth, bz	B7, 443
2967	Benzophenone, 2-nitro 2-$NO_2C_6H_4COC_6H_5$	227.22	mcl (al)		105			al	B7^3, 2082
2968	Benzophenone, 3-nitro 3-$NO_2C_6H_4COC_6H_5$	227.22	ye nd (al)	234^{18}	95			al	B7^3, 2082
2969	Benzophenone, 4-nitro 4-$NO_2C_6H_4COC_6H_5$	227.22	nd or lf (al)		138	1.406		bz	B7^3, 2082
2970	Benzophenone oxime $(C_6H_5)_2C{=}NOH$	197.24	nd (al)		144			al, eth, ace, chl	B7^3, 2063
2971	Benzophenone phenyl hydrazone $(C_6H_5)_2C{=}NNHC_6H_5$	272.35	pr or nd (al)		137			eth, bz, aa	B15^2, 63
2972	Benzophenone, 2,2′, 3,4-tetrahydroxy 2,3,4-$(HO)_3C_6H_2CO(C_6H_4OH$-2)	246.22	ye lf or pl (w + l)		149, 102 (+ w)			al, eth, w, aa	B8^2, 539
2973	Benzophenone, 2,2′,4,4′-tetrahydroxy [2,4-$(HO)_2C_6H_3]_2CO$	246.22	ye nd (w + l)		196-8			al, eth, ace, w, bz, aa	B8^3, 4064
2974	Benzophenone, 2,2′,4,6′-tetrahydroxy or Isoeuxanthonic acid 2,4-$(HO)_2C_6H_3CO[C_6H_3(OH)_2$-2,6]	246.22	(w + l)		200d			w, al, eth	B8, 496
2975	Benzophenone, 2,2′,5,6′-tetrahydroxy or Euxanthoic acid 2,5-$(HO)_2C_6H_3CO[C_6H_3(OH)_2$-2,6]	246.22	ye nd (w)		200-2d			w, al	B8^2, 541
2976	Benzophenone, 2,3′,4,4′-tetrahydroxy 2,4-$(HO)_2C_6H_3CO[C_6H_3(OH)_2$-3,4]	246.22	nd (w + 2)		202 (anh)			w, al, eth, ace, aa	B8^2, 541
2977	Benzophenone, 2,3′,4,6-tetrahydroxy 2,4,6$(HO)_3C_6H_2CO(C_6H_4OH$-3)	246.22	pa ye lf (w)		246d			w, al	B8^2, 540
2978	Benzophenone, 2,4,4′,6-tetrahydroxy 2,4,6-$(HO)_3C_6H_2CO(C_6H_4OH$-4)	246.22	pr or nd (w + 2)		210			w, al, eth	B8^2, 540
2979	Benzophenone, 3,3′,4,4′-tetrahydroxy [3,4-$(HO)_2C_6H_3]_2CO$	246.22	(w)		227-8			w, al, bz	B8^2, 541
2980	Benzophenone, 2,2′,4,4′-tetramethoxy (2,4-$(CH_3O)_2C_6H_3)_2CO$	302.33			130			al, bz	B8^3, 4064
2981	Benzophenone, 2,2′,5,5′-tetramethoxy [2,5-$(CH_3O)_2C_6H_3]_2CO$	302.33	ye (aa or al)		109			al, eth, bz, chl, aa	B8^2, 541

No.	Name, Synonyms, and Formula	Mol. wt.	Color, crystalline form, specific rotation and λ_{max} (log ε)	b.p. °C	m.p. °C	Density	n_D	Solubility	Ref.
2982	Benzophenone, 2,2′,6,6′-tetramethoxy $[2,6\text{-}(CH_3O)_2C_6H_3]_2CO$	302.33	pl (bz)	....	204	....	....	chl	B8^{3}, 4065
2983	Benzophenone, 2,3′,4,4′-tetramethoxy $2,4\text{-}(CH_3O)_2C_6H_3CO[C_6H_3(OCH_3)_2\text{-}3,4]$	302.33	nd, lf or pr (al)	....	126	....	....	al	B8^{3}, 4065
2984	Benzophenone, 2,3′,4,5′-tetramethoxy $2,4\text{-}(CH_3O)_2C_6H_3CO[C_6H_3(OCH_3)_2\text{-}3,5]$	302.33	nd (bz-peth)	....	73-4	....	....	al, eth	B8^{1}, 735
2985	Benzophenone, 2,3′4′,5-tetramethoxy $2,5\text{-}(CH_3O)_2C_6H_3CO[C_6H_3\text{-}(OCH_3)_2\text{-}3,4]$	302.33	pr (dil al)	....	101-2	....	....	al, eth	B8, 497
2986	Benzophenone, 2,3,4,6-tetramethoxy $2,3,4,6\text{-}(CH_3O)_4C_6HCOC_6H_5$	302.33	nd (lig)	....	125-6	....	....	al, ace, bz	B8^{1}, 734
2987	Benzophenone, 2,4,4′,5-tetramethoxy $2,4,5\text{-}(CH_3O)_3C_6H_2CO(C_6H_4OCH_3\text{-}4)$	302.33	ye pw (al)	....	122-4	....	....	al, ace, bz, chl	B8^{1}, 734
2988	Benzophenone, 2,4,4′,6-tetramethoxy $2,4,6(CH_3O)_3C_6H_2CO(C_6H_4OCH_3\text{-}4)$	302.33	pr (al)	....	146	....	....	al, eth	B8, 496
2989	Benzophenone, 3,3′,4,4′-tetramethoxy or Veratrophenone $[3,4\text{-}(CH_3O)_2C_6H_3]_2CO$	302.33	pr (al)	....	145	....	....	w, bz	B8^{3}, 4066
2990	Benzophenone, 3,3′,4,5′-tetramethoxy $3,4\text{-}(CH_3O)_2C_6H_3CO[C_6H_3(OCH_3)_2\text{-}3,5]$	302.33	nd (bz)	....	114-5	....	....	al, eth, bz, chl	B8^{1}, 735
2991	Benzophenone, 2,2′,4,4′-tetramethyl $[2,4\text{-}(CH_3)_2C_6H_3]_2CO$	238.33	....	190^{10}	1.043^{15}	1.5790^{25}	....	....	B7^{3}, 2247
2992	Benzophenone, thio or Thiobenzophenone $C_6H_5CSC_6H_5$	198.28	....	174^{14}	53-4	....	....	bz	B7^{3}, 2087
2993	Benzophenone, 2,2′,6-trihydroxy $2,6\text{-}(HO)_2C_6H_3CO(C_6H_4OH\text{-}2)$	230.22	ye nd (dil al)	....	133-4	....	....	al, eth, bz	B8^{2}, 468
2994	Benzophenone, 2,3,4-trihydroxy or Callobenzophenone $2,3,4\text{-}(HO)_3C_6H_2COC_6H_5$	230.22	ye nd (dil al)	....	140-1	....	....	w, al, eth, ace, aa	B8^{3}, 3635
2995	Benzophenone, 2,4,4′-trihydroxy $2,4\text{-}(HO)_2C_6H_3CO(C_6H_4OH\text{-}4)$	230.22	ye nd (w + 2)	....	200-1	....	....	w, al	B8^{3}, 3639
2996	Benzophenone, 2,4,6-trihydroxy $2,4,6\text{-}(HO)_3C_6H_2COC_6H_5$	230.22	ye nd (w + 1, dil al)	....	165	....	....	w, al, eth	B8^{3}, 3637
2997	Benzophenone, 3,4,5-trihydroxy $3,4,5\text{-}(HO)_3C_6H_2COC_6H_5$	230.22	ye pl (+lw), col (chl)	....	177-8 (anh)	....	....	w, al, eth, ace	B8, 422
2998	Benzophenone, 2,3,4-trimethoxy $2,3,4\text{-}(CH_3O)_3C_6H_2COC_6H_5$	272.30	pr (dil al)	....	55	....	....	al, eth	B8,418
2999	Benzophenone, 2,4,4′-trimethoxy $2,4\text{-}(CH_3O)_2C_6H_3CO(C_6H_4OCH_3\text{-}4)$	272.30	nd (al)	....	73-4	....	....	al, eth, bz, aa	B8^{1}, 702
3000	Benzophenone, 2,4,5-trimethoxy $2,4,5\text{-}(CH_3O)_3C_6H_2COC_6H_5$	272.30	ye nd (w)	....	97	....	....	w, al, ace, bz	B8^{1}, 701
3001	Benzophenone, 2,4,6-trimethoxy $2,4,6\text{-}(CH_3O)_3C_6H_2COC_6H_5$	272.30	mcl pr or rh pl (al)	....	115	....	....	al, eth, chl	B8^{3}, 3638
3002	Benzophenone, 3,3′,4-trimethoxy $3,4\text{-}(CH_3O)_2C_6H_3CO(C_6H_4OCH_3\text{-}3)$	272.30	nd (MeOH)	....	83-4	....	....	al, ace, bz	B8^{2}, 468
3003	Benzophenone, 3,4,4′-trimethoxy $3,4\text{-}(CH_3O)_2C_6H_3CO(C_6H_4OCH_3\text{-}4)$	272.30	nd (al)	....	98-9	....	....	al, ace, bz	B8^{2}, 469
3004	Benzophenone, 3,4′,5-trimethoxy $3,5\text{-}(CH_3O)_2C_6H_3CO(C_6H_4OCH_3\text{-}4)$	272.30	nd (bz)	....	97-8	....	....	al, eth, bz, chl	B8^{1}, 702
3005	Benzopinacolone (α-form) or α,α,α-triphenylacetophenone $C_6H_5COC(C_6H_5)_3$	348.44	nd	....	206-7	....	....	bz, chl	C55, 22234
3006	Benzopinacolone (β form) or α,α,α-triphenylacetophenone $C_6H_5COC(C_6H_5)_3$	348.44	nd (al)	....	182	....	....	eth, bz, chl	B7^{3}, 2941
3007	1,2-Benzopyrene or Benzo[a]pyrene $C_{20}H_{12}$	252.32	ye pl (bz-lig)	$310\text{-}12^{10}$	179-179.3	....	....	....	B5^{3}, 2520
3009	3,4-Benzopyrene $C_{20}H_{12}$	252.31	....	$310\text{-}12^{10}$	176.5-7.5	....	....	....	B5^{3}, 2517
3010	3,4-Benzopyrene, 5-amino $C_{20}H_{13}N$	267.33	ye pl (bz-lig)	....	239-41	....	....	al, ace, bz, chl	B12^{3}, 3404
3011	3,4-Benzopyrene, 5-hydroxy $C_{20}H_{12}O$	268.31	nd (eth-lig)	....	207-9	....	....	al, bz, aa	B6^{3}, 3810
3012	3,4-Benzopyrene, 8-hydroxy $C_{20}H_{12}O$	268.31	ye nd (bz-peth)	....	226-7d	....	....	al, eth, ace, bz	E14s, 704
3013	5,6-Benzoquinoline or β-Naphthoquinoline $C_{13}H_9N$	179.22	lf (peth or w)	350^{721}, $202\text{—}5^{8}$	94	....	....	al, eth, ace, bz	B20^{4}, 4009

No.	Name, Synonyms, and Formula	Mol. wt.	Color, crystalline form, specific rotation and λ_{max} (log ε)	b.p. °C	m.p. °C	Density	n_D	Solubility	Ref.
3014	7,8-Benzoquinoline or α-Naphthoquinoline $C_{13}H_9N$	179.22	lf (eth), pl (peth)	338^{719}, 233^{47}	52			al, eth, ace, bz	$B20^4$, 4003
3015	7,8-Benzoquinoline, 2-methyl $C_{14}H_{11}N$	193.25		324-6		1.1464^{20}	1.6738^{20}	al	$B20^4$, 4043
3016	1,2-Benzoquinone or o-Quinone $1,2\text{-}O{=}C_6H_4{=}O$	108.10	red pl or pr		60-70d			eth, ace, bz	$B7^3$, 3352
3017	1,2-Benzoquinone, 3-chloro $3\text{-}Cl\text{-}(1,2\text{-}O{=}C_6H_3{=}O)$	142.54	pa ye-red pr (hx)		68d			al, eth	$B7^1$, 338
3018	1,2-Benzoquinone, 4-chloro $4\text{-}Cl\text{-}(1,2\text{-}O{=}C_6H_3{=}O)$	142.54	pa ye-red nd (hx)		78			eth	$B7^3$, 3354
3019	1,2-Benzoquinone, 4,5-dichloro $4,5\text{-}Cl_2\text{-}(1,2\text{-}O{=}C_6H_2{=}O)$	176.99	yesh red pr or pl		94d			eth, bz	$B7^1$, 338
3020	1,2-Benzoquinone, 4-methoxy,1-oxime $4\text{-}(H_3O\text{-}(1\text{-}HON{=}C_6H_3{=}O\text{-}2)$	153.14	ye pr		158-9			al, bz, aa	$B8^3$, 1966
3021	1,2-Benzoquinone, 3-methoxy $3\text{-}CH_3O(1,2\text{-}O{=}C_6H_3{=}O)$	138.12	br-red pl, pr, nd		115-20			w, al, eth, bz, chl	$B8^3$, 1964
3022	1,2-Benzoquinone, 5-isopropyl-4-methyl-1-oxime $C_{10}H_{13}NO_2$	179.22	nd (bz-chl)		165-7d			bz, chl	$B7^2$, 595
3023	1,4-Benzoquinone or p-Quinone $1,4\text{-}O{=}C_6H_4{=}O$	108.10	ye mcl pr (w)	sub	115-7	1.318^{20}		al, eth	$B7^3$, 3356
3024	1,4-Benzoquinone, 2-bromo-6-methyl $2\text{-}Br\text{-}6\text{-}CH_3\text{-}(1,4\text{-}O{=}C_6H_2{=}O)$	201.02	ye nd (al), pr (eth or lig)	sub	95			al, eth, chl	$B7^3$, 3393
3025	1,4-Benzoquinone, 5-bromo-2-methyl $5\text{-}Br\text{-}2\text{-}CH_3(1,4\text{-}O{=}C_6H_2{=}O)$	201.02	ye lf (lig)		106			al	$B7^2$, 591
3026	1,4-Benzoquinone, 2-chloro $2\text{-}Cl\text{-}(1,4\text{-}O{=}C_6H_3{=}O)$	142.54	ye-red rh (hx)		57			w, al, eth, chl	$B7^3$, 3373
3027	1,4-Benzoquinone, 2-chloro,oxime $2\text{-}Cl\text{-}1,4(1\text{-}HON{=}C_6H_3{=}O)$	157.56	gr-ye nd (bz-aa)		184d			al, eth	$B7^3$, 3374
3028	1,4-Benzoquinone, 2-chloro-3-methyl $2\text{-}Cl\text{-}3\text{-}CH_3(1,4\text{-}O{=}C_6H_2{=}O)$	156.57	cr (lig)		55			al, chl, lig	$B7^3$, 3391
3029	1,4-Benzoquinone, 2-chloro-5-methyl $2\text{-}Cl\text{-}5\text{-}CH_3\text{-}(1,4\text{-}O{=}C_6H_2{=}O)$	156.57	ye nd (w or al)		105			al, eth, chl	$B7^3$, 3391
3030	1,4-Benzoquinone, 2-chloro-6-methyl $2\text{-}Cl\text{-}6\text{-}CH_3(1,4\text{-}O{=}C_6H_2{=}O)$	156.57	ye nd (w)		90			al, eth, chl	$B7^3$, 3392
3031	1,4-Benzoquinone, 2,5-diaminoanil or Bandrowski's base $C_{18}H_{18}N_6$	318.38	dk red or br lf		238			dil HCl	$B14^3$, 363
3032	1,4-Benzoquinone, 2,6-dibromo-1-imine, N-chloro $C_6H_2Br_2ClNO$	299.36	ye pr (aa, al)		85-6				$B7^2$, 584
3033	1,4-Benzoquinone, 2,6-dibromo-4-imine- N-chloro $2,6\text{-}Br_2\text{-}(1\text{-}O{=}C_6H_2{=}NCl\text{-}4)$	299.36	ye pr (al or aa)		83			al	$B7^2$, 584
3034	1,4-Benzoquinone, 3,5-dibromo,2,6-dimethyl $2,6\text{-}(CH_3)_2(1,4\text{-}O{=}C_6Br_2{=}O)$	293.94	ye lf (al)	sub	176			al	$B7^3$, 3401
3035	1,4-Benzoquinone, 3,5-dibromo-2-methyl $3,5\text{-}Br_2\text{-}2\text{-}CH_3(1,4\text{-}O{=}C_6H{=}O)$	279.92	ye		117			al, eth, chl	$B7^3$, 3393
3036	1,4-Benzoquinone, 2,5-di-*tert*-butyl $2,5\text{-}t\text{-}C_4H_9(1,4\text{-}O{=}C_6H_2{=}O)$	220.31	ye (al)		152.5			eth, bz, aa	$B7^3$, 3428
3037	1,4-Benzoquinone, 2,6-di-*tert*-butyl $2,6\text{-}di\text{-}t\text{-}C_4H_9(1,4\text{-}O{=}C_6H_2{=}O)$	220.31			102-4				
3038	1,4-Benzoquinone, 2,3-dichloro $2,3\text{-}Cl_2\text{-}(1,4\text{-}O{=}C_6H_2{=}O)$	176.99	ye lf		100-1			eth, bz	$B7^3$, 3375
3039	1,4-Benzoquinone, 2,5-dichloro $2,5\text{-}Cl_2(1,4\text{-}O{=}C_6H_2{=}O)$	176.99	pa ye mcl pr (al)		161-2			eth, chl	$B7^3$, 3376
3040	1,4-Benzoquinone, 2,5-dichloro-3,6-dihydroxy or Chloranilic acid $3,6\text{-}(HO)_2(1,4\text{-}O{=}C_6Cl_2{=}O)$	208.99	red lf (w + 2)		283-4			w	$B8^3$, 3350
3041	1,4-Benzoquinone, 2,6-dichloro $2,6\text{-}Cl_2(1,4\text{-}O{=}C_6H_2{=}O)$	176.99	ye rh (lig, bz)		120-1			chl	$B7^3$, 3376
3042	1,4-Benzoquinone, 2,6-dichloro-1-imine, N-chloro $2,6\text{-}Cl_2\text{-}(1\text{-}ClN{=}C_6H_2{=}O\text{-}4)$	210.45	ye nd (al)		67-8			eth, chl	$B7^2$, 581
3043	1,4-Benzoquinone, 2,5-dihydroxy $2,5\text{-}(HO)_2\text{-}(1,4\text{-}O{=}C_6H_2{=}O)$	140.10	dk ye nd	215d sub	211			aa	$B8^3$, 3348
3044	1,4-Benzoquinone, 2,5-dihydroxy-3-methoxy-6-methyl or Spinulosin $C_8H_8O_5$	184.16	red-bl	120^1 sub	202-3				$B8^3$, 3983
3045	1,4-Benzoquinone, 2,5-dihydroxy-3-undecyl or Embelin $3\text{-}C_{11}H_{23}\text{-}2,5\text{-}(HO)_2\text{-}(1,4\text{-}O{=}C_6H{=}O)$	294.39	og red pl (al-bz)	sub	143				$B8^2$, 452

No.	Name, Synonyms, and Formula	Mol. wt.	Color, crystalline form, specific rotation and λ_{max} (log ε)	b.p. °C	m.p. °C	Density	n_D	Solubility	Ref.
3046	1,4-Benzoquinone, diimine 1,4-$HN{=}C_6H_4{=}NH$	106.13	ye nd		124			bz, chl	$B7^{2}$, 574
3047	1,4-Benzoquinone, diimine, *N,N*-dichloro 1,4-$ClN{=}C_6H_4{=}NCl$	175.02	nd (w)		126d			al, eth, bz	$B7^{2}$, 574
3049	1,4-Benzoquinone, 2,6-dimethoxy 2,6-$(CH_3O)_2$-(1,4-$O{=}C_6H_2{=}O$)	168.15	ye mcl pr (aa)	sub	256			aa	$B8^{3}$, 3354
3050	1,4-Benzoquinone, 2,3-dimethyl or *o*-Xyloquinone 2,3-$(CH_3)_2$-(1,4-$O{=}C_6H_2{=}O$)	136.15	ye nd	sub	55			al, eth	$B7^{3}$, 3397
3051	1,4 Benzoquinone, 2,5-dimethyl 2,5-$(CH_3)_2$-(1,4-$O{=}C_6H_2{=}O$)	136.15			125			eth, bz	$B7^{3}$, 3397
3052	1,4-Benzoquinone, 2,6-dimethyl 2,6-$(CH_3)_2$-(1,4-$O{=}C_6H_2{=}O$)	136.15	ye nd	sub	72-3	$1.0479^{78/4}$			$B7^{3}$, 3399
3053	1,4-Benzoquinone dioxime 1,4-$C_6H_4({=}NOH)_2$	138.13	pa ye nd (w)		240d				B7,627
3054	1,4-Benzoquinone, 2,5-diphenyl 2,5-$(C_6H_5)_2$-(1,4-$O{=}C_6H_2{=}O$)	260.30	og-ye pl (bz or aa)		214			bz, aa	$B7^{3}$, 4251
3055	1,4-Benzoquinone, 2-hydroxy-3,5,6-trimethyl $C_9H_{10}O_3$	166.18			90-2			chl	$B8^{3}$, 2183
3056	1,4-Benzoquinone, 3-hydroxy-2-methoxy-5-methyl or Fumigatin $C_8H_8O_4$	168.16	br nd or pl (peth)		116			al, eth, ace, bz, chl	$B8^{3}$, 3374
3057	1,4-Benzoquinone, 2-methoxy 2-CH_3O-(1,4-$O{=}C_6H_3{=}O$)	138.12	ye nd (w)	sub	145			al	$B8^{3}$, 1969
3058	1,4-Benzoquinone, 2-methyl or Toluquinone 2-CH_3-(1,4-$O{=}C_6H_3{=}O$)	122.12	ye pl or nd	sub	69	$1.08^{75.5/4}$		al, eth	$B7^{3}$, 3387
3059	1,4-Benzoquinone, 2-methyl-6-propyl-4-oxime 2-CH_3-6-C_3H_7-(1,4-$O{=}C_6H_2{=}NOH$)	179.22	br nd (lig)		93-4			al, bz, aa	$B7^{2}$, 595
3060	1,4-Benzoquinone, 2-methyl-3,5,6-tribromo 2-CH_3-(1,4-$O{=}C_6Br_3{=}O$)	358.81	ye pl (al)		235-6			eth, bz	$B7^{2}$, 592
3061	1,4-Benzoquinone monoimine, *N*-chloro 1-$ClN{=}C_6H_4{=}O$-4	141.56	ye nd (peth)		85			w, al, eth	B7,619
3062	1,4-Benzoquinone, 2-phenyl 2-C_6H_5-(1,4-$O{=}C_6H_3{=}O$)	184.21	ye lf (peth, al)		114			al, bz, chl	$B7^{3}$, 3764
3063	1,4-Benzoquinone, 2-isopropyl-5-methyl 2-i-C_3H_7-5-CH_3[1,4-$O{=}C_6H_2{=}O$]	164.20			45-7			chl	$B7^{3}$, 3411
3064	1,4-Benzoquinone, 2-isopropyl-5-methyl-dioxime or Thymoquinone dioxime 2-i-C_3H_7-5-CH_3[1,4-$HON{=}C_6H_2{=}NOH$]	179.22	nd (bz-chl)		165-7d			bz, chl	$B7^{3}$, 3414
3065	1,4-Benzoquinone, tetrachloro or Chloranil 1,4-$O{=}C_6Cl_4{=}O$	245.88	ye mcl pr (bz), ye lf (aa)	sub	290 (sealed tube)			eth	$B7^{3}$, 3378
3066	1,4-Benzoquinone, tetrahydroxy 1,4-$O{=}C_6(OH)_4{=}O$	172.09	bl-bk cr					al, w	$B8^{3}$, 4204
3067	1,4-Benzoquinone, tetramethyl or Duroquinone 1,4-$O{=}C_6(CH_3)_4{=}O$	164.20	ye nd (al or lig)		111-2			al, eth, ace, bz, aa	$B7^{3}$, 3417
3068	1,4-Benzoquinone, trichloro 1,4-$O{=}C_6Cl_3H{=}O$	211.43	ye pl (al)		169-70			al, eth	$B7^{3}$, 3377
3069	*α*-Benzosuberone $C_{11}H_{12}O$	160.22		$124\text{-}5^{7}$, 108^{1}		$1.0780^{20/4}$	1.5698^{20}	al	$B7^{3}$, 1442
3070	Benzotrichloride $C_6H_5CCl_3$	195.48		220.6, 150^{100}	− 4.75	1.3723^{20}	1.5580^{20}	al, eth, bz	$B5^{4}$, 820
3071	Benzotrichloride, 2-chloro 2-$ClC_6H_4CCl_3$	229.92		264.3, 129.5^{13}	30	$1.5187^{20/4}$	1.5836^{20}	eth, ace	$B5^{4}$, 823
3072	Benzotrichloride, 3-chloro 3-$ClC_6H_4CCl_3$	229.92		255		1.495^{14}	1.4461^{20}	eth, ace	$B5^{4}$, 823
3073	Benzotrichloride, 4-chloro 4-$ClC_6H_4CCl_3$	229.92		245, $108\text{-}12^{8}$			1.4463^{20}	eth, ace	$B5^{4}$, 823
3074	**Benzotrifluoride** $C_6H_5CF_3$	146.11		102, 10^{10}	− 29.1	1.1884^{20}	1.4146^{20}	**al, eth, ace, bz**	$B5^{4}$, 802
3075	Benzotrifluoride, 3-amino 3-$H_2NC_6H_4CF_3$	161.13		187.5^{764}, $74\text{-}5^{10}$			1.4787^{20}	al, eth	$B12^{3}$, 1988
3076	Benzotrifluoride, 2-nitro 2-$NO_2C_6H_4CF_3$	191.11	cr (al)	216.3^{765}	32.5			al, bz, aa	$B5^{3}$, 744
3077	Benzotrifluoride, 3-nitro 3-$NO_2C_6H_4CF_3$	191.11		202.8, 81.6^{10}	− 2.4	$1.4357^{15/4}$	1.4719^{20}	al, eth	$B5^{3}$, 744
3078	Benzothiazole C_7H_5NS	135.18		231, 131^{34}	2	$1.2460^{20/4}$	1.6379^{20}	al, eth, ace, bz	$B27^{2}$, 17

No.	Name, Synonyms, and Formula	Mol. wt.	Color, crystalline form, specific rotation and λ_{max} (log ε)	b.p. °C	m.p. °C	Density	n_D	Solubility	Ref.
3079	Benzothiazole, 2-amino $C_7H_6N_2S$	150.20	pl (w)	....	132	....	....	al, eth, chl	B27², 225
3080	Benzothiazole, 2-amino-6-chloro $C_7H_5N_2ClS$	184.64	....	....	199-201	....	....	....	B27², 230
3081	Benzothiazole, 2-amino-6-ethoxy $C_9H_{10}N_2OS$	194.25	nd (al)	....	163-4	....	....	al	B27², 335
3082	Benzothiazole, 2-amino-4-methyl $C_8H_8N_2S$	164.22	nd (w), pl (al)	....	145	....	....	....	B27², 237
3083	Benzothiazole, 2-amino-5-methyl $C_8H_8N_2S$	164.22	pl (dil al)	....	171-2	....	....	al	B27², 240
3084	Benzothiazole, 2-amino-6-methyl $C_8H_8N_2S$	164.22	nd (w), pr (dil al)	....	142	....	....	al, w	B27², 241
3085	Benzothiazole, 5-amino-2-mercapto $C_7H_6N_2S_2$	182.26	nd	....	216	....	....	aniline	B27², 475
3086	Benzothiazole, 6-amino $C_7H_6N_2S$	150.20	pr (w)	....	87	....	....	al	B27,366
3087	Benzothiazole, 6-amino-2-mercapto $C_7H_6N_2S_2$	182.26	....	....	263	....	....	....	B27², 475
3088	Benzothiazole, 2-chloro C_7H_4ClNS	169.63	....	248, 136-6[28]	24	$1.3715^{10/}{}_4$	1.6338^{10}	al, eth, ace	B27², 18
3089	Benzothiazole, 5-chloro-2-methyl C_8H_6ClNS	183.66	pl (eth)	70-90[0.5] sub	69	....	....	al, peth	B27², 22
3090	Benzothiazole, 6-dimethylamino-2-mercapto $C_9H_{10}N_2S_2$	210.32	ye nd (bz)	....	230d	....	....	ace	B27², 475
3091	Benzothiazole, 2-(2,4-dinitrophenylthio) $C_{13}H_7N_3O_4S_2$	333.35	ye	....	162	$1.24^{20/4}$	....	....	C48,1443
3092	Benzothiazole, 2-hydroxy C_7H_5NOS	151.18	pr (dil al) nd	360	138	....	....	al, eth	B27², 225
3093	Benzothiazole, 2-(2-hydroxyphenyl) $C_{13}H_9NOS$	227.28	nd or lf (al)	175-93[3]	132-3	....	....	al	B27², 91
3094	1,2-Benzisothiazole, 5-hydroxy-3-phenyl $C_{13}H_9NOS$	227.28	nd (dil al or aa)	....	159-60	....	....	al, ace, bz, aa	B27², 87
3095	Benzothiazole, 6-hydroxy-2-phenyl $C_{13}H_9NOS$	227.28	wh nd (dil al)	....	227 (cor)	....	....	al, ace, bz, aa	B27², 88
3096	Benzothiazole, 2-mercapto $C_7H_5NS_2$	167.24	nd (al or dil MeOH)	....	180-2	$1.42^{20/4}$	....	al	B27, 185
3097	Benzothiazole, 2-mercapto,benzoate $C_{14}H_9NOS_2$	271.35	ye	....	132	....	....	....	C51,3491
3098	Benzothiazole, 2-mercapto-4-methyl $C_8H_7NS_2$	181.27	nd (dil aa)	....	186	....	....	aa	B27², 240
3099	Benzothiazole, 2-mercapto-5-methyl $C_8H_7NS_2$	181.27	nd (to)	....	171-3	....	....	al, ace, bz	B27², 241
3100	Benzothiazole, 2-mercapto-6-methyl $C_8H_7NS_2$	181.27	....	....	181	....	....	al, ace, bz	B27², 242
3101	Benzothiazole, 2-mercapto-7-methyl $C_8H_7NS_2$	181.27	....	....	184	....	....	ace, bz, to	B27², 242
3102	Benzothiazole, 2-mercapto-6-nitro $C_7H_4N_2O_2S_2$	212.24	ye nd (aa)	....	255-7	....	....	ace, aa	B27², 234
3103	Benxothiazole, 2-methyl C_8H_7NS	149.21	....	238, 150-1[15]	14	$1.1763^{19/4}$	1.6092^{19}	al	B27², 21
3104	Benzothiazole, 2-methyl-6-nitro $C_8H_6N_2O_2S$	194.21	....	....	175	....	....	....	B27, 47
3105	Benzothiazole, 2-(methylthio) $C_8H_7NS_2$	181.28	pr (dil al)	....	52	....	....	al	B27², 71
3106	Benzothiazole, 2-phenyl $C_{13}H_9NS$	211.28	nd (dil al)	>360	114	....	....	al, eth	B27², 37
3107	Benzothiazole, 2-phenylamino $C_{13}H_{10}N_2S$	226.30	nd (al)	....	161	....	....	AcOEt	B27², 226
3108	Benzothiazoline, 3-methyl-2-imino $C_8H_8N_2S$	164.22	pl (w), nd (al)	....	128	....	....	al, eth, chl	B27², 228
3109	2-Benzothiazolinethione, 3-methyl $C_8H_7NS_2$	181.27	nd (al), pr (aa)	335[757]	90	....	....	al, bz, chl	B27², 233
3110	Benzothiophene or Thionaphthene C_8H_6S	134.20	lf	221, 103-5[20]	32	$1.1484^{32/4}$	1.6374^{37}	al, eth, ace, bz	B17³, 482
3111	Benzothiophene, 3-hydroxy C_8H_6OS	150.20	nd (w)	....	71	....	....	al, eth, ace, bz	B17⁴, 1458

PHYSICAL CONSTANTS OF ORGANIC COMPOUNDS (Continued)

No.	Name, Synonyms, and Formula	Mol. wt.	Color, crystalline form, specific rotation and λ_{max} (log ε)	b.p. °C	m.p. °C	Density	n_D	Solubility	Ref.
3112	Benzothiophene, 4-hydroxy C_8H_6OS	150.20	nd (sub), cr (peth)	sub	78-9			al	B17^{4}, 1467
3113	2,3-Benzothiophene, 5-methyl C_9H_8S	148.22		111-5^{12}	19-22	1.111$^{22/D}$	1.615$^{22/D}$		B17^{4}, 502
3114	2,3-Benzothiophene quinone or Thioisatin $C_8H_4O_2S$	164.18	gold-ye pr (al)	247	121			al, bz, aa	B17^{4}, 6129
3115	1,2,3-Benzotriazole or Azimidobenzene $C_6H_5N_3$	119.13	nd (chl or bz)	204^{15}	100			al, bz, chl	B26^{2}, 17
3116	1,2,3-Benzoxadiazole, 5,7-dinitro $C_6H_2N_4O_5$	210.11	ye pl (al)		158			al	B16^{3}, 549
3117	2,3-Benzoxazin-1-one $C_8H_5NO_2$	147.13	cr (bz)		120d				B27^{2}, 249
3118	Benzoxazole C_7H_5NO	119.12	pr (dil al)	182.5, 45^{4}	31		1.5594^{20}	al	B27^{2}, 17
3119	Benzoxazole, 2-chloro C_7H_4ClNO	153.57		201-2	7	1.3453$^{18/4}$	1.5678^{20}		B27^{2}, 17
3120	Benzoxazole, 2,5-dimethyl C_9H_9NO	147.18		218-9			1.5412^{20}		B27^{2}, 25
3121	Benzoxazole, 2-hydroxy or 2(3)-Benzoxazolone $C_7H_5NO_2$	135.12	ng (bz or w + 1)	230^{30}	141—2 (anh), 97—8 (+w)			al, eth	B27^{2}, 223
3122	Benzoxazole, 2-(2-hydroxyphenyl) $C_{13}H_9NO_2$	211.22	pink nd (al or aa)	338	123-4			al, eth, ace, bz	B27^{2}, 91
3123	Benzoxazole, 4-hydroxy-2-phenyl $C_{13}H_9NO_2$	211.22	nd (bz)		138-9			al, eth, aa	B27^{2}, 88
3124	Benzoxazole, 5-hydroxy-2-phenyl $C_{13}H_9NO_2$	211.22	nd (lig or dil al)		175			al, bz, chl, aa	B27^{2}, 88
3125	Benzoxazole, 6-hydroxy-2-phenyl $C_{13}H_9NO_2$	211.22	nd		216-7			ace, bz, aa	B27^{2}, 88
3126	Benzoxazole, 7-hydroxy-2-phenyl $C_{13}H_9NO_2$	211.22	nd (bz or dil al)		191-2			al, eth, bz, chl, lig	B27^{2}, 91
3127	Benzoxazole, 2-mercapto C_7H_5NOS	151.18	nd (w)		196			eth, aa	B27^{2}, 224
3128	Benzoxazole, 2-methyl C_8H_7NO	133.15		200—1, 59-60^{12}	8.5-10	1.1211$^{20/4}$	1.5497^{20}	al, eth	B27^{2}, 20
3129	Benzoyl acetamide or α-Benzoyl acetanilide $C_6H_5COCH_2CONHC_6H_5$	239.27	lf (bz)		108			al	B12^{3}, 1005
3130	Benzoyl acetic acid $C_6H_5COCH_2CO_2H$	164.16	nd (bz-peth)		103-4d			al, eth, bz	B10^{3}, 2990
3131	Benzoyl acetic acid, ethyl-ester or Ethylbenzoyl acetate. $C_6H_5COCH_2CO_2C_2H_5$	192.21		265-70, 165^{14}	<0	1.1220$^{20/4}$	1.5312^{16}	al, eth	B10^{3}, 2991
3132	Benzoyl acetic acid, methyl-ester $C_6H_5COCH_2CO_2CH_3$	178.19	pa ye	265d, 151.5^{12}		1.158$^{29/4}$	1.537^{20}	**al, eth,** ace	B10^{3}, 2991
3133	Benzoyl acetone or 1-Phenyl-1,3-butanedione $C_6H_5COCH_2COCH_3$	162.19	pr	261-2	56	1.0599$^{74/4}$	1.5678^{78}	eth	B7^{3}, 34821
3134	Benzoyl acetonitrite $C_6H_5COCH_2CN$	145.16	pr or lf (w)	160^{10}	80-1			al, eth, bz	B10^{3}, 2994
3135	Benzoyl azide $C_6H_5CON_3$	147.14	pl (ace)	exp	32			al, eth	B9^{3}, 1324
3136	Benzoyl bromide C_6H_5COBr	185.02		218-9, 48-50$^{0.05}$	−24	1.570^{15}	1.5868^{25}	**eth**	B9^{3}, 1064
3137	Benzoyl chloride C_6H_5COCl	140.57		197.2, 71^{9}		1.2120$^{20/4}$	1.5537^{20}	**eth**	B9^{3}, 1058
3138	Benzoyl chloride, 2-bromo $2\text{-}BrC_6H_4COCl$	219.47	nd	245, 118^{10}	11		1.5963^{20}		B9^{3}, 1387
3139	Benzoyl chloride, 4-bromo $4\text{-}Br\text{-}C_6H_4COCl$	219.47	nd (peth)	245-7, 123-6^{15}	42			al, eth, bz, lig	B9^{3}, 1418
3140	Benzoyl chloride, 2-chloro $2\text{-}ClC_6H_4COCl$	175.01		238, 110^{15}	−4		1.5726^{20}		B9^{3}, 1338
3141	Benzoyl chloride, 3-chloro $3\text{-}ClC_6H_4COCl$	175.01		225, 103-4^{14}			1.5677^{20}		B9^{3}, 1348
3142	Benzoyl chloride, 4-chloro $4\text{-}ClC_6H_4COCl$	175.01		222, 111^{18}	16	1.3770$^{20/4}$	1.5756^{20}		B9^{3}, 1362
3143	Benzoyl chloride, 2,4-dichloro $2,4\text{-}Cl_2C_6H_3COCl$	209.46	lig	150^{34}, 111$^{7.5}$	15-8		1.5895^{20}		B9^{3}, 1375
3144	Benzoyl chloride, 3,4-dichloro $3,4\text{-}Cl_2C_6H_3COCl$	209.46	lig	242, 160^{42}	24-6				B9^{3}, 1379

No.	Name, Synonyms, and Formula	Mol. wt.	Color, crystalline form, specific rotation and λ_{max} (log ε)	b.p. °C	m.p. °C	Density	n_D	Solubility	Ref.
3145	Benzoyl chloride, 3,5-dinitro $3,5\text{-}(NO_2)_2C_6H_3COCl$	230.56	ye nd (bz)	...	74	...	...	eth	$B9^3$, 1779
3146	Benzoyl chloride, 3-hydroxy $3\text{-}HOC_6H_4COCl$	156.57	...	$110\text{-}2^{0.5}$	<−15	...	...	chl	$B10^2$, 82
3147	Benzoyl chloride, 2-methoxy or o-Anisoyl chloride $2\text{-}CH_3OC_6H_4COCl$	170.60	...	254, 128^{11}	...	...	...	...	$B10^3$, 151
3148	Benzoyl chloride, 3-methoxy $3\text{-}CH_3OC_6H_4COCl$	170.60	...	243-4, $123\text{-}5^{15}$	...	...	...	eth, ace, bz	$B10^3$, 252
3149	Benzoyl chloride, 4-methoxy $4\text{-}CH_3OC_6H_4COCl$	170.60	nd	262-3, 91^{1}	24-5	$1.261^{20/4}$	1.580^{20}	eth, ace, bz	$B10^3$, 337
3150	Benzoyl chloride, 2-nitro $2\text{-}NO_2C_6H_4COCl$	185.57	...	275-8, 154^{18}	...	...	...	eth	$B9^3$, 1477
3151	Benzoyl chloride, pentachloro C_6Cl_5COCl	312.79	pl (al)	5	87	...	...	al	$B9^2$, 230
3152	Benzoyl chloride, 2,4,6-trimethyl or Mesitoyl chloride $2,4,6\text{-}(CH_3)_3C_6H_2COCl$	182.65	...	$143\text{-}6^{60}$	...	...	...	ace	$B9^3$, 2469
3153	Benzoyl fluoride C_6H_5COF	124.11	...	154-5	...	>1	...	al, eth	$B9^3$, 1058
3154	Benzoyl glycolic acid $C_6H_5COCH(OH)CO_2H$	180.17	lo pr (lig)	...	112	...	...	al, eth, chl	$B10^2$, 677
3155	Benzoyl iodide C_6H_5COI	232.02	nd	128^{20}	3	$1.748^{18/18}$	1.137^{20}	al, eth	$B9^3$, 1064
3156	Benzoyl peroxide or Dibenzoyl peroxide $C_6H_5CO\text{-}OO\text{-}COC_6H_5$	242.23	rh (eth), pr	exp	106-8	...	1.543	al, eth, ace, bz	$B9^3$, 1052
3157	Benzoyl peroxide, 3,3′-dinitro $3\text{-}O_2NC_6H_4CO\text{-}OO(COC_6H_4NO_2\text{-}3)$	332.23	nd (al)	...	139-40d	...	...	eth, ace, bz	$B9^2$, 252
3158	Benzoyl peroxide, 4,4′-dinitro $4\text{-}O_2NC_6H_4CO\text{-}OO(COC_6H_4NO_2\text{-}4)$	332.23	ye cr (ace), nd (to)	...	156d	...	...	to	$B9^2$, 270
3159	β-Benzoyl propionic acid $C_6H_5COCH_2CH_2CO_2H$	178.19	lf (dil al)	...	116	...	...	al, eth, bz, chl	$B10^2$, 482
3160	Benzyl alcohol or α-Hydroxy toluene $C_6H_5CH_2OH$	108.14	...	205.3, 93^{10}	−15.3	$1.0419^{24/4}$	1.5396^{20}	w, al, eth, ace, bz	$B6^4$, 2222
3161	Benzyl alcohol, 2-amino $2\text{-}H_2NC_6H_4CH_2OH$	123.15	...	270-80, $160^{5\text{-}10}$	83-4	...	...	w, al, eth, bz, chl	$B13^3$, 1615
3162	Benzyl alcohol, 2-amino-3-methyl $3\text{-}CH_3\text{-}2\text{-}H_2NC_6H_3CH_2OH$	137.18	nd (bz)	$135\text{-}45^{12}$	71	...	...	...	$B13^2$, 367
3163	Benzyl alcohol, 2-amino-4-methyl $4\text{-}CH_3\text{-}2\text{-}H_2NC_6H_3CH_2OH$	137.18	nd (bz)	$140\text{-}50^{13}$	141	...	...	...	$B13^2$, 369
3164	Benzyl alcohol, 2-amino-5-methyl $2\text{-}NH_2\text{-}5\text{-}CH_3C_6H_3CH_2OH$	137.18	nd (bz)	$145\text{-}50^{12}$	123	...	...	...	$B13^2$, 367
3165	Benzyl alcohol, 3-amino $3\text{-}H_2NC_6H_4CH_2OH$	123.15	...	...	97	...	...	w, al, eth, bz, chl	$B13^3$, 1617
3166	Benzyl alcohol, 4-amino $4\text{-}H_2NC_6H_4CH_2OH$	123.15	...	...	65	...	...	al, eth, bz	$B13^3$, 1619
3167	Benzyl alcohol, 2-bromo $2\text{-}BrC_6H_4CH_2OH$	187.04	nd (lig)	...	80	...	...	w, al, eth	$B6^4$, 2600
3168	Benzyl alcohol, 4-bromo $4\text{-}BrC_6H_4CH_2OH$	187.04	nd (lig)	...	77	...	...	al, eth, bz	$B6^4$, 2602
3169	Benzyl alcohol, 5-bromo-2-hydroxy or 5-Bromosaligenin $5\text{-}Br\text{-}2\text{-}HOC_6H_3CH_2OH$	203.04	lf (bz)	...	113	...	...	al, eth, bz, chl	$B6^3$, 4541
3170	Benzyl alcohol, 2-chloro $2\text{-}ClC_6H_4CH_2OH$	142.58	lf or nd (dil al)	230, $100\text{-}5^{28}$	74	...	...	al, eth, lig	$B6^4$, 2589
3171	Benzyl alcohol, 4-chloro $4\text{-}ClC_6H_4CH_2OH$	142.58	nd (w), pr (bz or bz-lig)	235	75	...	...	al, eth, bz	$B6^4$, 2593
3172	Benzyl alcohol, 2,5-dihydroxy or Gentisyl alcohol $2,5\text{-}(HO)_2C_6H_3CH_2OH$	140.14	nd (chl)	sub vac	100	...	...	w, al, eth, chl	$B6^3$, 6322
3173	Benzyl alcohol, 2,3-dimethoxy $2,3\text{-}(CH_3O)_2C_6H_3CH_2OH$	168.19	...	257-8, $155\text{-}60^{17}$	50	...	...	...	$B6^2$, 1082
3174	Benzyl alcohol, 3,4-dimethoxy $3,4\text{-}(CH_3O)_2C_6H_3CH_2OH$	168.19	...	$135\text{-}8^{0.1}$	...	$1.179^{17/17}$	1.555^{17}/o	al, w	$B6^3$, 6324
3175	Benzyl alcohol, α,α-dimethyl $C_6H_5C(OH)(CH_3)_2$	136.19	pr	202, 93^{11}	35-7	$0.9735^{20/4}$	1.5325^{20}	al, eth, bz, aa	$B6^3$, 1813
3176	Benzyl alcohol, α-ethyl $C_6H_5CH(OH)C_2H_5$	136.19	...	213-5, 98^{10}	...	$0.9938^{22/4}$	1.5210^{22}	al, eth	$B6^3$, 1792
3177	Benzyl alcohol, hexahydro $C_6H_{11}CH_2OH$	114.19	...	183, 83^{14}	−43	$0.9297^{20/4}$	1.4644^{20}	al, eth	$B6^4$, 106

PHYSICAL CONSTANTS OF ORGANIC COMPOUNDS (Continued)

No.	Name, Synonyms, and Formula	Mol. wt.	Color, crystalline form, specific rotation and λ_{max} (log ε)	b.p. °C	m.p. °C	Density	n_D	Solubility	Ref.
3178	Benzyl alcohol, 2-hydroxy or Saligenin Salicyl alcohol $2\text{-}HOC_6H_4CH_2OH$	124.14	lf (bz), nd or pl (w, eth)	sub	87	1.1613^{25}		w, al, eth, bz, chl	$B6^3$, 4537
3179	Benzyl alcohol, 2-hydroxy, glucoside or Saligenin-β-*D*-glucoside $C_{13}H_{18}O_7$	286.28	rh nd or lf (w), $[\alpha]^{20}{}_D$ +62.6 (w, c=3)	240d	205—9	1.434^{26}		w, al, aa	$B6^3$, 4537
3180	Benzyl alcohol, 3-hydroxy $3\text{-}HO\text{-}C_6H_4CH_2OH$	124.14	nd (bz), cr (CCl_4)	300d	73	1.161^{25}		al, eth, w	$B6^3$, 4545
3181	Benzyl alcohol, 4-hydroxy $4\text{-}HOC_6H_4CH_2OH$	124.14	pr or nd (w)	252	124-5			w, al, eth	$B6^3$, 4546
3182	Benzyl alcohol, 4-hydroxy-3-methoxy or Vanillyl alcohol $3\text{-}CH_3O\text{-}4\text{-}HOC_6H_3CH_2OH$	154.17	pr (w), nd (bz)	d	115				$B6^3$, 6323
3183	Benzyl alcohol, 2-methoxy or Saligenin, 2-methylether $2\text{-}CH_3OC_6H_4CH_2OH$	138.17		249, 119^8		$1.0395^{25/15}$	1.5455^{20}	al, **eth**	$B6^3$, 4538
3184	Benzyl alcohol, 4-methoxy or Anisyl alcohol $4\text{-}CH_3OC_6H_4CH_2OH$	138.17	nd	259.1, $134\text{-}5^{12}$	25	$1.109^{26/4}$	1.5420^{25}	al, eth, w	$B6^3$, 4547
3185	Benzyl alcohol, 2-methyl $2\text{-}CH_3C_6H_4CH_2OH$	122.17	nd	223^{750}, $117\text{-}9^{20}$	37-9	1.023^{40}		al, eth, chl	$B6^3$, 1733
3186	Benzyl alcohol, 3-methyl $3\text{-}CH_3C_6H_4CH_2OH$	122.17		215-6	<−20	0.9157^{17}		al, eth	$B6^3$, 1768
3187	Benzyl alcohol, 4-methyl $4\text{-}CH_3C_6H_4CH_2OH$	122.17	nd (hp)	217, $116\text{-}8^{20}$	61-2	$0.978^{22/4}$		al, eth	$B6^3$, 1779
3188	Benzyl alcohol, α-1-naphthyl $C_6H_5CH(OH)\text{-}1\text{-}C_{10}H_7$	234.30	cr (al, lig)	ca 360	86.5			al, eth, bz	$B6^3$, 3608
3189	Benzyl alcohol, α-2-naphthyl $C_6H_5CH(OH)\text{-}2\text{-}C_{10}H_7$	234.30	nd (al, lig)		87-8			al, eth, bz	$B6^3$, 3609
3191	Benzyl alcohol, 2-nitro $2\text{-}NO_2C_6H_4CH_2OH$	153.14	nd (w)	270, 168^{20}	74			al, eth	$B6^3$, 1563
3192	Benzyl alcohol, 3-nitro $3\text{-}NO_2C_6H_4CH_2OH$	153.14	rh nd (w)	$175\text{-}80^3$	30.5	$1.296^{19/15}$		al, eth, w	$B6^3$, 1565
3193	Benzyl alcohol, 4-nitro $4\text{-}NO_2C_6H_4CH_2OH$	153.14	nd (w)	250-60d, 185^{22}	96-7			al, eth	$B6^3$, 1567
3194	Benzyl alcohol, α-isopropyl $C_6H_5CH(OH)CH(CH_3)_2$	150.22	*d*: $[\alpha]^{20}{}_D$ +47.7, *l*: $[\alpha]^{20}{}_D$ -25.2	222.4, $112\text{-}3^{15}$		0.9869^{14}	1.5193^{14}	al, ace	$B6^3$, 1859
3195	Benzyl alcohol, 4-isopropyl $4\text{-}i\text{-}C_3H_7C_6H_4CH_2OH$	150.22		246, 122.5^{13}	28	$0.9401^{20/4}$	1.519^{20}	**al, eth**, bz	$B6^3$, 1911
3196	Benzyl amine or α-Aminotoluene $C_6H_5CH_2NH_2$	107.16		185^{770}, 90^{12}		$0.9813^{20/4}$	1.5401^{20}	w, **al, eth**, ace, bz	$B12^3$, 2194
3197	Benzyl amine, 3-bromo $3\text{-}BrC_6H_4CH_2NH_2$	186.05		244-5, 84^{15}				eth	$B12^3$, 2349
3198	Benzyl amine, 4-bromo $4\text{-}BrC_6H_4CH_2NH_2$	186.05		127^{15}	20			eth	$B12^3$, 2351
3199	Benzyl amine, N-*tert*-butyl or Benzyl *tert*-butyl amine $C_6H_5NHC(CH_3)_3$	163.26		73.7^{44}			1.4951^{25}		$B12^3$, 2208
3200	Benzyl amine, *N*-(3-butynyl)-*N*-methyl $C_6H_5CH_2CH_2N(CH_3)CH_2CH_2C{=}CH$	173.25		127^{16}		0.9372^{20}	1.5202^{20}		C51,7295
3201	Benzyl amine, 2-chloro $2\text{-}ClC_6H_4CH_2NH_2$	141.60		$72\text{-}3^2$			$1.5594^{25/}{}_D$		$B12^3$, 2340
3202	Benzyl amine, 3-chloro $3\text{-}ClC_6H_4CH_2NH_2$	141.60		89^2			$1.5570^{25/}{}_D$		$B12^3$, 2342
3203	Benzyl amine, 4-chloro $4\text{-}ClC_6H_4CH_2NH_2$	141.60		$109\text{-}10^{13}$			1.5566^{25}		$B12^3$, 2343
3204	Benzyl amine, *N*-cyclopropyl $C_6H_5CH_2NHC_3H_5$	147.22		$80\text{-}1^5$		1.5222^{25}			C59,509
3205	Benzyl amine, 2,4-dichloro $2,4\text{-}Cl_2C_6H_3CH_2NH_2$	176.05		$124\text{-}6^{13}$			$1.5762^{25/}{}_D$	chl	
3206	Benzyl amine, 3,4-dimethoxy $3,4\text{-}(CH_3O)_2C_6H_3CH_2NH_2$	167.21		$154\text{-}8^{12}$	1.43^0			chl	$B13^3$, 2183
3207	Benzyl amine, *N,N*-dimethyl or Benzyl dimethyl amine $C_6H_5CH_2N(CH_3)_2$	135.21		180-2, $73\text{-}4^{15}$		$0.915^{0/4}$	1.5011^{20}	**al, eth**	$B12^3$, 2203
3208	Benzyl amine, 2,4-dimethyl $2,4\text{-}(CH_3)_2C_6H_3CH_2NH_2$	135.21		$86\text{-}90^{10}$					$B12^3$, 2709
3209	Benzyl amine, 2,5-dimethyl $2,5\text{-}(CH_3)_2C_6H_3CH_2NH_2$	135.21		225-6			$1.5377^{20/}{}_D$		ALD 12695-0

No.	Name, Synonyms, and Formula	Mol. wt.	Color, crystalline form, specific rotation and λ_{max} (log ε)	b.p. °C	m.p. °C	Density	n_D	Solubility	Ref.
3210	Benzylamine, *N,N*-diphenyl or Benzyl diphenyl amine $C_6H_5CH_2N(C_6H_5)_2$	259.36	nd (al)		95			eth, ace, bz	$B12^3$, 2220
3211	Benzylamine, *N*-ethyl or Benzyl ethyl amine $C_6H_5CH_2NHC_2H_5$	135.21		194, 82^{15}		$0.9350^{17/15}$	1.5117^{20}	al, eth, bz	$B12^3$, 2204
3212	Benzyl amine, *N*-ethyl-*N*-phenyl or Benzyl ethyl phenyl amine $C_6H_5CH_2N(C_2H_5)C_6H_5$	211.31	pa ye	285-6d	34-6	$1.034^{19/4}$	1.5930^{20}	al, eth	$B12^3$, 2218
3213	Benzyl amine hydrochloride $C_6H_5CH_2NH_2$-HCl	143.62			255-8			w, al	$B12^3$, 2197
3214	Benzyl amine, 2-hydroxy--5-nitro $5\text{-}NO_2\text{-}2\text{-}HOC_6H_3CH_2NH_2$	168.16	ye nd or lf (w)		253d			w, al	B13,587
3215	Benzyl amine, 4-hydroxy-3-nitro $3\text{-}NO_2\text{-}4\text{-}HOC_6H_3CH_2NH_2$	168.16	og red nd (w + l)		225d			w	B13, 610
3216	Benzyl amine, *N*-(3-hydroxy propyl)-*N*-methyl $C_6H_5CH_2N(CH_3)CH_2CH_2CH_2OH$	179.26		$132\text{-}5^4$			1.425^{20}	CS_2	$B12^3$, 2234
3217	Benzyl amine, 3-methoxy $3\text{-}CH_3OC_6H_4CH_2NH_2$	137.18		141^{30}				al, eth	$B13^3$, 1569
3218	Benzyl amine, 4-methoxy $4\text{-}CH_3OC_6H_4CH_2NH_2$	137.18		236-7, $133\text{-}4^{33}$		1.050^{15}	1.5462^{20}	w, al, eth	$B13^3$, 1594
3219	Benzyl amine, 2-methyl $2\text{-}CH_3C_6H_4CH_2NH_2$	121.18		$66\text{-}8^2$			1.5408^{25}		$B12^3$, 2460
3220	Benzyl amine, *N*-nitroso-*N*-phenyl $C_6H_5CH_2N(NO)C_6H_5$	212.25	ye nd (al)		58			al, eth, chl, lig	$B12^3$, 2335
3221	Benzyl amine, *N*-phenyl or *N*-Benzyl aniline $C_6H_5CH_2NHC_6H_5$	183.25		306-7, 171.5^{10}	37-8	$1.0298^{65/4}$	1.6118^{25}	al, eth	$B12^3$, 2215
3222	Benzyl amine, *N*-2-tolyl $C_6H_5CH_2NH(C_6H_4CH_3\text{-}2)$	197.28		300-5, 176^{10}	60	$1.0142^{65/4}$	1.5861^{65}	al, ace, chl	$B12^3$, 2220
3223	Benzyl arsonic acid $C_6H_5CH_2AsO(OH)_2$	216.07	nd (al)		167-8			chl	$B16^3$, 1057
3224	Benzyl azide $C_6H_5CH_2N_3$	133.15		108^{23}, 74^{11}		$1.0655^{25/4}$	1.53414^{25}	**al, eth**	$B5^4$, 759
3225	Benzyl boric acid $C_6H_5CH_2B(OH)_2$	135.96	cr (w, bz)		140, 104 (hyd)			al, bz	$B16^3$, 1278
3226	Benzyl bromide $C_6H_5CH_2Br$	171.04	pr	201, 114^{15}	-3	1.4380^{25}	1.5752^{20}	**al, eth**	$B5^4$, 829
3227	Benzyl bromide, 2-bromo $2\text{-}BrC_6H_4CH_2Br$	249.93	cr (al, lig)	129^{19}	31			**al, eth, aa**	$B5^4$, 836
3228	Benzyl bromide, 3-bromo $3\text{-}BrC_6H_4CH_2Br$	249.93	nd or lf		41			al, eth, aa	$B5^4$, 836
3229	Benzyl bromide, 4-bromo $4\text{-}BrC_6H_4CH_2Br$	249.93	nd (al)		63			al, eth, bz, aa	$B5^4$, 836
3230	Benzyl bromide, 2-chloro $2\text{-}ClC_6H_4CH_2Br$	205.48		120^{10}				bz, chl	$B5^4$, 833
3231	Benzyl bromide, 2,4-dibromo $2,4\text{-}Br_2C_6H_3CH_2Br$	328.83			40-1			al	$B5^2$, 240
3232	Benzyl bromide, 3,5-dibromo $3,5\text{-}Br_2C_6H_3CH_2Br$	328.83	pl or nd (al)	173^{19}	96			al	$B5^3$, 719
3233	Benzyl bromide, 2-nitro $2\text{-}O_2NC_6H_4CH_2Br$	216.03	pl (dil al)		46-7			al, eth, bz, lig	$B5^3$, 752
3234	Benzyl bromide, 3-nitro $3\text{-}O_2NC_6H_4CH_2Br$	216.03	nd or pl (al)	$153\text{-}4^8$	58-9			al	$B5^4$, 860
3235	Benzyl bromide, 4-nitro $4\text{-}O_2NC_6H_4CH_2Br$	216.03	nd (al)		99-100			al, eth, aa	$B5^4$, 861
3236	Benzyl butyl ether $C_6H_5CH_2C_4H_9$	164.25		223, 92^{10}		$0.9227^{20/4}$	1.4833^{20}	**al, eth**, ace	$B6^4$, 2231
3237	Benzyl-*iso*-butyl ether $C_6H_5CH_2O\text{-}i\text{-}C_4H_9$	164.25		$211\text{-}2^{743}$		$0.9233^{20/4}$	1.4826^{20}	eth, chl	$B6^3$, 1457
3238	Benzyl chloride or α-chloro toluene $C_6H_5CH_2Cl$	126.59		179.3, 66^{11}	-39	$1.1002^{20/20}$	1.5391^{20}	**al, eth, chl**	$B5^4$, 809
3239	Benzyl chloride, 2-bromo $2\text{-}BrC_6H_4CH_2Cl$	205.48		$124\text{-}6^{20}$				al, eth	$B5^4$, 832
3240	Benzyl chloride, 3-bromo $3\text{-}BrC_6H_4CH_2Cl$	205.48	nd (al or peth)	119^{18}	22-3			al	$B5^4$, 832
3241	Benzyl chloride, 4-bromo $4\text{-}BrC_6H_4CH_3$	205.48	nd (al or peth)	236, $110\text{-}1^9$	50			al, eth, peth	$B5^4$, 832
3242	Benzyl chloride, 2-chloro $2\text{-}ClC_6H_4CH_2Cl$	161.03		217, $94\text{-}5^{10}$	-17	$1.2699^{0/4}$	1.5530^{20}		$B5^4$, 816

PHYSICAL CONSTANTS OF ORGANIC COMPOUNDS (Continued)

No.	Name, Synonyms, and Formula	Mol. wt.	Color, crystalline form, specific rotation and λ_{max} (log ε)	b.p. °C	m.p. °C	Density	n_D	Solubility	Ref.
3243	Benzyl chloride, 3-chloro $3\text{-}ClC_6H_4CH_2Cl$	161.03	....	215-6^{753}, 110-1^{25}	......	1.2695$^{15/4}$	1.5554^{20}	al	B5^4, 816
3244	Benzyl chloride, 4-chloro $4\text{-}ClC_6H_4CH_2Cl$	161.03	nd (dil al)	222, 117^{20}	31	....	....	eth, bz, aa	B5^4, 816
3245	Benzyl chloride, 2,6-dichloro $2,6\text{-}Cl_2C_6H_3CH_2Cl$	195.48	cr (lig, eth, al-eth)	117-9^{14}	39-40	....	....	al, eth, lig	B5^4, 820
3246	Benzyl chloride, 3,4-dichloro $3,4\text{-}Cl_2C_6H_3CH_2Cl$	195.48	....	241	37.5	....	....	al	B5^4, 820
3247	Benzyl chloride, 3,5-dichloro $3,5\text{-}Cl_2C_6H_3CH_2Cl$	195.48	cr(MeOH)	60$^{0.35}$	36	....	....	al	B5^3, 699
3248	Benzyl chloride, 4-hydroxy-3-nitro $3\text{-}NO_2\text{-}4\text{-}HOC_6H_3CH_2Cl$	187.58	ye nd (lig or al), lf (peth)	....	75	....	....	al, bz	B6, 413
3249	Benzyl chloride, 4-methoxy $4\text{-}CH_3OC_6H_4CH_2Cl$	156.61	....	116-20^{15}, 83-4^{2}	......	1.159^{20}	1.553	....	B6^4, 2137
3250	Benzyl chloride, 2-nitro $2\text{-}NO_2C_6H_4CH_2Cl$	171.58	cr (lig)	....	50-2	....	1.5557^{62}	al, eth, ace, bz, aa	B5^4, 854
3251	Benzyl chloride, 3-nitro $3\text{-}NO_2C_6H_4CH_2Cl$	171.58	pa ye nd (lig)	175-83$^{30-5}$	45-7	....	1.5577^{62}	al, eth, ace, bz, aa	B5^4, 855
3252	Benzyl chloride, 4-nitro $4\text{-}NO_2C_6H_4CH_2Cl$	171.58	pl or nd (al)	....	71	....	1.5647^{62}	al, eth, ace, bz	B5^4, 856
3253	Benzyl chloride, 2,4,5-trichloro $2,4,5\text{-}Cl_3C_6H_2CH_2Cl$	229.92	....	273	......	1.547^{20}	....	al, eth, ace	B5^4, 822
3254	Benzyl chloromethyl ether $C_6H_5CH_2OCH_2Cl$	156.61	....	103^{13}	......	1.1350$^{20/4}$	1.5192^{20}	....	B6^4, 2253
3255	Benzyl ether or Dibenzyl ether $(C_6H_5CH_2)_2O$	198.26	....	298, 160^{11}	3.6	1.0428$^{20/4}$	1.5168^{20}	al, eth	B6^4, 2240
3256	Benzyl diphenyl methanol $(C_6H_5)_2C(OH)CH_2C_6H_5$	274.36	nd (bz-lig), pr (peth)	222^{11}	89-90	....	....	al	B6^3, 3680
3257	Benzyl ethyl ether $C_6H_5CH_2OC_2H_5$	136.29	....	185, 70^{15}	......	0.9490$^{20/4}$	1.4955^{20}	al, eth	B6^4, 2229
3258	Benzyl ethyl sulfide $C_6H_5CH_2SC_2H_5$	152.25	....	218-20	......	....	....	....	B6^4, 2635
3259	Benzyl fluoride $C_6H_5CH_2F$	110.13	nd (fr)	139.8^{753}, 40^{14}	-35	1.0228$^{25/4}$	1.4892^{25}	....	B5^4, 800
3260	*N*-Benzyl hydroxy, 1-amine $C_6H_5CH_2NHOH$	123.15	nd (peth or lig)	....	57	....	....	al, lig	B15^3, 20
3261	Benzyl iodidie or α-Iodo toluene $C_6H_5CH_2I$	218.04	col or ye nd (MeOH)	93^{10}	24.5	1.7335^{25}	1.6334^{25}	al, eth, bz	B5^4, 842
3262	Benzyl isocyanide $C_6H_5CH_2NC$	117.15	....	198-200d, 93-4^{55}	......	0.972^{15}	....	....	B12^3, 2241
3263	Benzyl isothiocyanate or Benzyl mustard oil $C_6H_5CH_2NCS$	149.21	ye oil	243, 124-5^{12}	......	1.1246$^{16/4}$	1.6049^{15}	al, eth	B12^2, 567
3264	Benzylmercapton or α- Mercapto toluene $C_6H_5CH_2SH$	124.20	....	194-5	......	1.058^{20}	1.5751^{20}	al, eth	B6^4, 2632
3265	Benzyl methyl ether $C_6H_5CH_2OCH_3$	122.17	....	170, 59-60^{12}	-52.6	0.9634$^{20/4}$	1.5008^{20}	al, eth, bz	B6^4, 2229
3266	Benzyl (2-methylbutyl) ether $C_6H_5CH_2OCH_2CH(CH_3)C_2H_5$	178.28	....	231^{722}	......	0.911$^{22/4}$	1.4854^{22}	al, eth	B6, 341
3267	Benzyl methyl sulfide $C_6H_5CH_2SCH_3$	138.23	....	195-8	......	....	1.5620$^{20/}{}_D$	....	B6^4, 2633
3268	Benzyl α-naphthyl ketone $\alpha\text{-}C_{10}H_7COCH_2C_6H_5$	246.31	ta, lf (al)	194-6$^{0.05}$	66-7	....	....	eth, chl	B7^2, 461
3269	Benzyl β-naphthyl ketone $\beta\text{-}C_{10}H_7COCH_2C_6H_5$	246.31	nd (al)	....	99.5	....	....	al, eth, bz, chl	B7^2, 461
3270	Benzyl phenyl ketone $C_6H_5COCH_2C_6H_5$	196.25	pl (al)	320, 177^{20}	60	1.201$^{0/4}$	....	al, eth, chl	B7^2, 368
3271	Benzyl isopentyl ether $C_6H_5CH_2O\text{-}i\text{-}C_5H_{11}$	178.27	....	236-7^{748}, 117-9^{19}	......	0.9098$^{20/4}$	1.4792^{20}	al, eth	B6^2, 410
3272	Benzyl phenyl methanol *d* $C_6H_5CH_2CH(OH)C_6H_5$	198.26	nd (eth-peth or dil al), $[\alpha]^{25/}{}_D$ + 53 (al)	167-70^{10}	67-8	1.0358$^{70/4}$	....	al, eth	B6^3, 3390
3273	Benzyl phenyl methanol *dl* $C_6H_5CH_2CH(OH)C_6H_5$	198.26	nd (bz-peth)	177^{15}	69	....	....	al, eth	B6^3, 3390
3274	Benzyl phenyl methanol *l* $C_6H_5CH_2CH(OH)C_6H_5$	198.26	$[\alpha]^{20/}{}_D$ -9.4 (w, c=10)	....	67	1.0358$^{70/4}$	....	al, eth	B6^3, 3390

PHYSICAL CONSTANTS OF ORGANIC COMPOUNDS (Continued)

No.	Name, Synonyms, and Formula	Mol. wt.	Color, crystalline form, specific rotation and λ_{max} (log ϵ)	b.p. °C	m.p. °C	Density	n_D	Solubility	Ref.
3275	Benzyl phenyl ketone or α-Phenylacetophenone $C_6H_5CH_2COC_6H_5$	196.25	pl (al)	320, 177^{20}	60	$1.201^{0/4}$	...	al, eth, chl	$B7^2$, 368
3276	Benzyl phenyl ketone, α-chloro or Desyl chloride $C_6H_5COCHClC_6H_5$	230.69	nd (al)	d	68.5	...	...	...	$B7^2$, 369
3277	Benzyl phenyl sulfide $C_6H_5CH_2SC_6H_5$	200.30	lf (al)	197^{27}	42-3.5	...	...	al, eth	$B6^4$, 2644
3278	Benzyl phenyl sulfone $C_6H_5CH_2SO_2C_6H_5$	232.30	nd (al)	...	146	$1.1261^{153/4}$	...	...	$B6^4$, 2647
3279	Benzyl-4-tolyl sulfone $(4\text{-}CH_3C_6H_4)SO_2CH_2C_6H_5$	246.32	nd (al)	...	144-5	...	...	al, bz, aa	$B6^4$, 2649
3280	Benzyl sulfide or Dibenzyl sulfide $(C_6H_5CH_2)_2S$	214.33	pl (eth or chl)	d	49-50	$1.071^{50/50}$	...	al, eth	$B6^4$, 2649
3281	Benzyl sulfonamide $C_6H_5CH_2SO_2NH_2$	171.21	pr or nd (w), nd (al)	...	105	...	...	w, al	$B11^3$, 331
3282	Benzyl sulfonamide, *N*-methyl $C_6H_5CH_2SO_2NHCH_3$	185.25	nd or lf (aa-lig)	...	108-9	...	...	al, eth	$B11^2$, 73
3283	Benzyl sulfonamide, *N*-2-tolyl $C_6H_5CH_2SO_2NH(C_6H_4CH_3\text{-}2)$	261.34	cr (dil al)	...	83	...	...	al	$B12^2$, 452
3284	Benzyl sulfonamide, *N*-3-tolyl $C_6H_5CH_2SO_2NH(C_6H_4CH_3\text{-}3)$	261.34	cr (al)	...	75	...	...	al	$B12^2$, 473
3285	Benzyl sulfonamide, *N*-4-tolyl $C_6H_5CH_2SO_2NH(C_6H_4CH_3\text{-}4)$	261.35	pr (al)	...	113	...	...	al, eth	$B12^2$, 528
3286	Benzyl sulfonyl chloride $C_6H_5CH_2SO_2Cl$	190.64	pr (eth), nd (bz)	...	93	...	...	eth, bz	$B11^3$, 331
3287	Benzyl sulfoxide or Dibenzyl sulfoxide $(C_6H_5CH_2)_2SO$	230.32	lf (al, w)	210d	134-5	...	...	al, eth	$B6^4$, 2651
3288	Benzyl thiocyanate $C_6H_5CH_2SCN$	149.21	pr (al)	256	43	...	...	al, eth	$B6^4$, 2680
3289	Benzyl 4-tolyl sulfide $C_6H_5CH_2S(C_6H_4CH_3\text{-}4)$	214.33	nd (al)	...	144-5	...	...	al, bz, aa	$B6^4$, 2649
3290	Benzyltrimethylammonium bromide $C_6H_5CH_2N^+(CH_3)_3Br^-$	230.15	pl (al-lig), (w)	...	235	...	...	w, al	$B12^3$, 2204
3291	Benzyltrimethylammonium chloride $C_6H_5CH_2N^+(CH_3)_3Cl^-$	185.70	(ace)	...	243	...	...	w	$B12^3$, 2203
3292	Benzyltrimethylammonium iodide $C_6H_5CH_2N^+(CH_3)_3I^-$	277.15	(al)	...	180	...	...	al	$B12^3$, 2204
3293	Benzyltrimethylammonium nitrate $C_6H_5CH_2N^+(CH_3)_3NO_3$	212.25	...	...	151-60	...	...	w, al	$B12^2$, 546
3294	Benzyldimethylphenylammonium chloride $C_6H_5CH_2N^+(CH_3)_2C_6H_5Cl^-$	247.77	...	...	134-8	...	...	w	$B12^3$, 2218
3295	Benzyltriethylammonium bromide $C_6H_5CH_2N^+(C_2H_5)_3Br^-$	272.23	...	...	194d	...	...	w	$B12^2$, 547
3296	Benzylidene bromide or α,α-Dibromotoluene, Benzal bromide $C_6H_5CHBr_2$	249.93	...	156^{23}	...	1.51^{15}	1.6147^{20}	al, eth	$B5^4$, 836
3297	Benzylidene bromide, 4-nitro $4\text{-}NO_2C_6H_4CHBr_2$	294.93	nd (al)	...	84	...	...	al, eth	$B5^4$, 862
3298	Benzylidene chloride or α,α-Dichlorotoluene $C_6H_5CHCl_2$	161.03	...	205.2	-16.4	1.2557^{14}	1.5502^{20}	al, eth	$B5^4$, 817
3299	Benzylidene chloride, 2-chloro $2\text{-}ClC_6H_4CHCl_2$	195.48	...	228.5	...	1.399^{15}	...	...	$B5^3$, 699
3300	Benzylidene chloride, 2,5-dichloro $2,5\text{-}Cl_2C_6H_3CHCl_2$	229.92	cubic cr (chl)	...	42	...	...	al, eth, bz	$B5^3$, 702
3301	Benzylidene chloride, 3,4-dichloro $3,4\text{-}Cl_2C_6H_3CHCl_2$	229.92	...	257	...	$1.518^{22/22}$	...	al, eth, bz, aa	$B5^3$, 702
3302	Benzylidene chloride, 3,5-dichloro $3,5\text{-}Cl_2C_6H_3CHCl_2$	229.92	cr (MeOH or dil aa)	...	36.5	...	...	eth, ace	$B5^3$, 703
3303	Benzylidene chloride, 3-nitro $3\text{-}NO_2C_6H_4CHCl_2$	206.03	mcl (al)	...	65	...	...	al, eth	$B5^3$, 750
3304	Benzylidene chloride, 4-nitro $4\text{-}NO_2C_6H_4CHCl_2$	206.03	pr (al)	...	46	...	...	al, eth	$B5^4$, 859
3305	Benzylidene chloride, pentachloro $C_6Cl_5CHCl_2$	333.26	lf (al)	334, 199^{11}	119.5	...	...	...	$B5^4$, 824
3306	Benzylidene chloride, 2,3,4-trichloro $2,3,4\text{-}Cl_3C_6H_2CHCl_2$	264.37	cr (lig)	275-85	84	...	...	bz	$B5^1$, 153

No.	Name, Synonyms, and Formula	Mol. wt.	Color, crystalline form, specific rotation and λ_{max} (log ε)	b.p. °C	m.p. °C	Density	n_D	Solubility	Ref.
3307	Benzylidene chloride, 2,3,6-trichloro $2,3,6\text{-}Cl_3C_6H_2CHCl_2$	264.37	nd (MeOH)	$145\text{-}50^{12}$	83	...	...	bz	$B5^4$, 823
3308	Benzylidene chloride, 2,4,5-trichloro $2,4,5\text{-}Cl_3C_6H_2CHCl_2$	264.37	...	$280\text{-}1, 153\text{-}5^{15}$	<0	$1.5956^{20/4}$	1.5992^{20}	bz	$B5^4$, 824
3309	Benzylidene chloride, 2,4,6-trichloro $2,4,6\text{-}Cl_3C_6H_2CHCl_2$	264.37	cr (MeOH)	158^{15}	27	...	...	...	$B5^3$, 704
3310	Benzylidene diacetate or Benzaldehyde diacetate $C_6H_5CH(O_2CCH_3)_2$	208.22	pl (eth)	$220, 154^{20}$	46	1.11^{20}	...	al, eth, bz	$B7^2$, 161
3311	Benzylidene ethyl amine $C_6H_5CH{=}NC_2H_5$	133.20	...	$195^{740}, 117^{12}$	...	$0.9370^{20/4}$	1.5365^{20}	al, eth, ace	$B7^2$, 163
3312	Benzylidene methylamine $C_6H_5CH{=}N\text{-}CH_3$	119.17	...	$185, 90\text{-}1^{30}$	...	$0.9672^{14/2}$	1.5526^{20}	al, eth, ace	$B7^2$, 162
3313	Benzylidene fluoride or α,α-Difluoro toluene $C_6H_5CHF_2$	128.12	...	139.9	...	1.1357^{20}	1.4577^{20}	al	$B5^4$, 801
3314	Berbamine $C_{37}H_{40}N_2O_6$	608.73	lf (+2w, al), cr (peth), $[\alpha]^{20/}{}_D$ +109 (chl)	...	197-200 (anh), 156 (hyd)	...	...	al, eth, peth, chl	$B27^2$, 891
3315	Berberine $C_{20}H_{19}NO_5$	353.37	red-ye nd (w+6), cr (chl+1)	...	145 (anh), 110 (+6w)	...	...	al, eth	$B27^2$, 567
3316	Berberine hydrochloride $C_{20}H_{20}ClNO_5$	389.84	ye cr (w+2), nd (w+4)	...	...	...	...	w	$B27^1$, 514
3317	Berberine nitrate $C_{20}H_{18}N_2O_7$	398.37	red-ye nd (al)	...	155d	...	...	w	B27, 500
3318	Berberine sulfate,trihydrate $C_{40}H_{42}N_2O_{15}S$	822.84	red-ye nd	...	...	...	...	w, al, chl	B27, 500
3319	Berberine, tetrahydro *(dl)* $C_{20}H_{21}NO_4$	339.39	mcl nd (al)	...	173-4	...	...	al, chl	$B27^2$, 557
3320	Berbine *(dl)* $C_{17}H_{17}N$	235.33	nd (eth or MeOH)	...	89	...	...	al, eth, ace	$B20^4$, 4108
3321	Betaine $(CH_3)_3N^+CH_2COO^-$	117.15	(w+1), pr or lf (al)	...	293d	...	...	w, al	$B4^3$, 1127
3322	Betonicine $C_7H_{13}NO_3$	159.19	pr (dil al +1w), $[\alpha]^{21/}{}_D$ -37 (w, c=4.8)	...	252d	...	...	al	$B22^4$, 2054
3323	Betulin or Lupenediol $C_{30}H_{50}O_2$	442.73	nd (al+1), $[\alpha]^{15/}{}_D$ +20 (Py, c=2)	$170\text{-}80^{0.08/d}$	251-2	...	...	eth, aa	$B6^3$, 5234
3324	Betulinic acid or Betulic acid $C_{30}H_{48}O_3$	456.71	pr or nd (al+1), $[\alpha]^{22/546}$ +7.9 (Py)	...	...	...	...	Py	$B10^3$, 1059
3325	Biacene or Biacenaphthene $C_{24}H_{16}$	304.39	red-ye pl or nd (bz)	...	277	...	...	bz	$B5^3$, 2595
3326	Biacetyl or 2,3-Butanedione $CH_3COCOCH_3$	86.09	...	88	−2.4	$0.9808^{18.5/4}$	1.3951^{20}	w, **al, eth,** ace, bz	$B1^4$, 3644
3327	Biacetyl dioxime or Dimethyl glyoxime $CH_3C(:NOH)C(:NOH)CH_3$	116.12	nd (to or dil al)	sub 234-5	245-6	...	...	al, eth	$B1^4$, 3647
3328	Biacetyl monoxime $CH_3COC(:NOH)CH_3$	101.11	pr (chl), lf (w)	185-6	77-8	...	...	al, eth, chl	$B1^4$, 3646
3329	10,10′-Bianthronyl or Bianthrone $C_{28}H_{18}O_2$	386.45	pl (ace)	...	256-8d	...	...	chl	$B7^3$, 4486
3330	Biarsine, tetraethyl or Ethylcacodyl $(C_2H_5)_2As\text{-}As(C_2H_5)_2$	266.09	...	185-7	...	$1.1388^{24/4}$	1.4709	al, eth	$B4^3$, 1832
3331	Biarsine, tetrakis(trifluoromethyl) $(CF_3)_2AsAs(CF_3)_2$	425.87	...	106-7	...	...	1.372^{19}	...	$B3^4$, 269
3332	Bibenzyl, 4,4′-diamino $4\text{-}H_2NC_6H_4CH_2CH_2(C_6H_4NH_2\text{-}4)$	212.29	pl (w)	sub	135-6 sub	...	...	al	$B13^3$, 470
3333	Bicyclo[2,2,1]-hepta-2,5-diene C_7H_8	92.14	...	89.5	−19.1	$0.9064^{20/4}$	1.4702^{20}	al, eth, ace, bz **lig**	$B5^4$, 879

No.	Name, Synonyms, and Formula	Mol. wt.	Color, crystalline form, specific rotation and λ_{max} (log ϵ)	b.p. °C	m.p. °C	Density	n_D	Solubility	Ref.
3334	Bicyclo[2,2,1]heptane or Norbornane C_7H_{12}	96.17	...	sub	87.5	...	...	al, eth, ace, bz	B5^{4}, 258
3335	Bicyclo[4,1,0]-heptane, 7-azo $C_6H_{11}N$	97.16	...	48-51^{22}	20-2	...	...	al, eth, bz	B20^{4}, 1937
3336	Bicyclo[2,2,1]-heptane-2,3-dicarboxylic acid or 2,3-Norcamphane dicarboxylic acid $C_9H_{12}O_4$	184.19	...	...	192-3	...	...	...	B9^{3}, 3970
3337	Bicyclo[2,2,1]-heptane-2-carboxaldehyde $C_8H_{10}O$	122.17	cr (w)	70-2^{22}	...	1.0227$^{19/4}$	1.4760^{25}	eth	B7^{3}, 267
3338	Bicycloheptyl $C_{14}H_{26}$	194.36		290-1^{728}	...	0.9069$^{20/0}$	...	...	B5^{4}, 344
3339	Bicyclo[3,1,0]hex-2-ene-4-one, 5-*iso*-propyl-2-methyl $C_{10}H_{14}O$	150.22	$[\alpha]_D$ − 36.5	219.20^{749}	...	0.9581$^{15/15}$	1.48325	...	B7^{3}, 582
3340	Bicyclohexyl or Dodecahydrobiphenyl *(cis,cis)* C_6H_{11}-C_6H_{11}	166.31	...	238	4	0.8914$^{20/4}$	1.4766^{20}	al, eth	B5^{3}, 273
3341	Bicyclohexyl *(trans,trans)* $C_{12}H_{22}$	166.31	...	217-8, 95-6^{9}	4.2	0.8592$^{20/4}$	1.4663^{20}	al, eth	B5^{4}, 334
3342	Bicyclo[3,1,1]-heptane, 2,4,6-trimethyl $C_{10}H_{18}$	138.25	...	169-70^{768}	...	0.8467$^{21/4}$	1.4605^{21}	...	PCHE2, 238
3343	Bicyclo[3,3,1]-nonane C_9H_{16}	124.23	cr (MeOH)	169-70 sub	145-6	...	...	al, aa	B5^{4}, 293
3344	Bicyclo[2,2,2]-octane C_8H_{14}	110.20	...	...	169-71	...	...	...	B5^{4}, 279
3345	Bicyclo[3,2,1]octane C_8H_{14}	110.20	...	...	139-41	...	...	...	B5^{4}, 278
3346	Bicyclo[3,3,0]octane *(cis)* C_8H_{14}	110.20	...	137^{765}	< − 80	0.8638$^{25/4}$	1.4595^{25}	al	B5^{4}, 277
3347	Bicyclo[3,3,0]octane *(trans)* C_8H_{14}	110.20	...	132^{755}	− 30	0.8624$^{11/4}$	1.4625^{18}	al	B5^{4}, 277
3348	Bicyclo[3,3,0]octane, 2,6-dione $C_8H_{10}O_2$	138.17	...	86-8$^{0.2}$	45	1.1290$^{60/4}$	1.4877^{54}	...	B7^{3}, 3279
3349	Bicyclo[2,2,2]octane, 2-methyl $C_8H_{13}CH_3$	124.23	...	158^{740}	33-4	0.8664$^{40.5/4}$	1.4608$^{40.5}$	...	B5^{4}, 295
3350	Bicyclo[3,3,0]octane, 2-one *(cis)* $C_8H_{12}O$	124.18	...	72^{13}	...	1.0097$^{20/4}$	1.4790^{20}	al, ace	B7^{3}, 264
3351	9.9′-Bifluorenyl $C_{26}H_{18}$	330.43	nd (bz-al)	...	247	...	...	aa, Py	B5^{3}, 2626
3352	9,9′-Bifluorenyl, 9,9′-diphenyl $C_{38}H_{26}$	482.62	pl (bz)	...	256 (under CO_2)	1.266$^{0/4}$	...	...	B5^{3}, 2782
3353	Bifluorenylidene $C_{26}H_{16}$	328.41	red nd (bz)	...	194-5	...	...	eth, bz, chl	B5^{3}, 2652
3354	Biguanide $H_2NC(:NH)NHC(:NH)NH_2$	101.11	pr or nd (al)	d 142	136	...	...	w, al	B3^{2}, 76
3355	Biguanide, 1-phenyl $C_6H_5NHC(=NH)NHC(=NH)NH_2$-HCl	177.21	...	...	143	...	...	...	B12^{3}, 807
3356	Biguanide, 1-(2-tolyl) $C_9H_{13}N_5$	191.24	nd or pl (w + 1)	...	144	...	...	al, ace	B12^{3}, 1873
3357	2-2′-Biindane, 1,1′,3,3′-tetraoxo or Bisdiketohydrindene $C_{18}H_{10}O_4$	290.28	red nd (bz)	...	297	...	...	bz	B7^{2}, 863
3358	Bikhaconitine $C_{36}H_{51}NO_{11}$	673.80	$[\alpha]_D$ + 12 (al)	...	118-23	...	...	al, eth, chl	B21^{4}, 2869
3359	Bilifucsin $C_{16}H_{20}N_2O_4$	304.35	dk br pw	...	183	...	...	al, ace	C30,1936
3360	Bilirubin or Haematoidine $C_{33}H_{36}N_4O_6$	584.67	red mcl pr or pl (chl)	...	...	...	...	bz, chl	C38,1230
3361	Biliverdin or Dehydrobilirubin $C_{33}H_{34}N_4O_6$	582.66	dk gr pl or pr (MeOH)	...	>300	...	...	al, bz	J 1961,2284
3362	α,α'-Binaphthyl or α,α'-Dinaphthyl α-$C_{10}H_7$-α-$C_{10}H_7$, $C_{10}H_{14}$	254.33	(i) pl (aa), (ii) rh (peth)	>360, 240-2^{12}	(i) 144, (ii) 160	...	...	eth, ace, bz	B5^{3}, 2465
3363	α,α'-Binaphthyl, 4,4′-diamino-3,3′-dimethyl $(3\text{-}CH_3\text{-}4\text{-}H_2N\text{-}\alpha\text{-}C_{10}H_5)_2$	312.41	...	...	213	...	...	al, bz	B13^{3}, 542
3364	α,α'-Binaphthyl, 2,2′-dihydroxy or β-Dinaphthol $[2\text{-}HO\text{-}\alpha\text{-}C_{10}H_6]_2$	286.33	nd (al), cr (w)	...	220	...	...	al, eth	B6^{3}, 5877
3365	α,α'-Binaphthyl, 4,4′-dihydroxy or α-Dinaphthol $[4\text{-}HO\text{-}\alpha\text{-}C_{10}H_6]_2$	286.33	pl	sub	300	...	...	al, eth	B6^{3}, 5878

No.	Name, Synonyms, and Formula	Mol. wt.	Color, crystalline form, specific rotation and λ_{max} (log ε)	b.p. °C	m.p. °C	Density	n_D	Solubility	Ref.
3366	β,β′-Binaphthyl or β,β′-Dinaphthyl β-$C_{10}H_7$-β-$C_{10}H_7$	254.33	bl fluor pl (al)	452^{753} sub	187-8			eth, bz	B5³, 2467
3367	Biotin or Vitamin H-Coenzyme R $C_{10}H_{16}N_2O_3S$	244.31	nd (w)		232d			w	Am67, 2096
3368	Biotin, methyl ester $C_{11}H_{18}N_2O_3S$	258.34	pl (MeOH-eth), $[\alpha]^{22}_D$ + 57 chl	sub	166.7			al, ace, chl	C43,1810
3369	Biphenyl or Phenylbenzene C_6H_5-C_6H_5	154.21	lf (dil al)	255.9, 145^{22}	71	$0.8660^{20/4}$	1.475^{20}, 1.588^{75}	al, eth, bz	B5⁴, 1807
3370	Biphenyl, 2-acetamide 2-$(CH_3CONH)C_6H_4$-C_6H_5	211.26	pr or nd (dil al or peth)	355	121			al, eth	B12³, 3125
3371	Biphenyl, 3-acetamido 3-$(CH_3CONH)C_6H_4$-C_6H_5	211.26	nd (al)		149			al, bz	B12², 751
3372	Biphenyl, 4-acetamido-3-NO_2 3-NO_2-4-$(CH_3CONH)C_6H_3$-C_6H_5	211.27	cr (dil MeOH)		172			al, ace	B12³, 3156
3373	Biphenyl, 4-acetamido,3-nitro 4-(CH_3CONH)-3-$NO_2C_6H_3C_6H_5$	256.26	ye nd (al)		132			al, eth, aa	B12³, 3199
3374	Biphenyl, 4-acetyl or 4-Phenylacetophenone 4-$(CH_3CO)C_6H_4$-C_6H_5	196.25	pr (ace), cr (al)	325-7	121			al, bz	B7², 377
3375	Biphenyl, 2-amino 2-$H_2NC_6H_4$-C_6H_5	169.23	lf (dil al)	299, 170^{15}	51-3			al, eth, bz	B12³, 3124
3376	Biphenyl, 2-amino-4′-nitro 2-$H_2NC_6H_4$-$C_6H_4NO_2$-4	214.22	og-red nd (al)		159			al	B12³, 3141
3377	Biphenyl, 2-amino-5-nitro 5-NO_2-2-$H_2NC_6H_3$-C_6H_5	214.22	ye nd (al)		125			al	B12³, 3140
3378	Biphenyl, 3-amino 3-$H_2NC_6H_4$-C_6H_5	169.23	nd	254^{135}	30			al, eth, ace, bz	B12³, 3146
3379	Biphenyl, 3-amino-4-hydroxy 3-H_2N-4-HO-C_6H_3-C_6H_5	185.23	nd (chl)		208			al, eth, bz	B13³, 1946
3380	Biphenyl, 3-amino-4-nitro 4-NO_2-3-$H_2NC_6H_3$-C_6H_5	214.22	og nd (dil al)		116			al	B12³, 3149
3381	Biphenyl, 3-amino-4′-nitro 3-$H_2NC_6H_4$-$C_6H_4NO_2$-4	214.22	og nd (al)		137			al, aa	B12², 753
3382	Biphenyl, 4-amino or Xenylamine 4-$H_2NC_6H_4$-C_6H_5	169.23	lf (dil al)	302, 191^{15}	53-4			al, eth, chl	B12³, 3152
3383	Biphenyl, 4-amino-2′-hydroxy 4-$H_2NC_6H_4$-C_6H_4OH-2	185.23	nd (to)		181-2			al, aa	B13³, 1943
3384	Biphenyl, 4-amino-4′-hydroxy 4-$H_2NC_6H_4$-C_6H_4OH-4	185.23	pl (dil al)		275				B13², 420
3385	Biphenyl, 4-amino-2′-nitro 4-$H_2NC_6H_4$-$C_6H_4NO_2$-2	214.22	red mcl pr (al)		99			al	B12³, 3199
3386	Biphenyl, 4-amino-3-nitro 3-NO_2-4-$H_2NC_6H_3$-C_6H_5	214.22	red nd (al)		170-1			al, eth, chl, aa	B12³, 3198
3387	Biphenyl, 4-amino-4′-nitro 4-$H_2NC_6H_4$-$C_6H_4NO_2$-4	214.22	red nd (al)		203-4			al, aa	B12³, 3200
3388	Biphenyl, 5-amino-2-hydroxy 5-H_2N-2-HOC_6H_3-C_6H_5	185.23	nd (al or bz)		201			al, bz	B13³, 1940
3389	Biphenyl, 2-benzyl or o-Biphenylphenyl methane 2-$(C_6H_5CH_2)C_6H_4$-C_6H_5	244.34	mcl nd (al)	283-7^{110}	54-6			al, eth, bz	B5³, 2323
3390	Biphenyl, 4-benzyl or p-Biphenylphenyl methane 4-$(C_6H_5CH_2)C_6H_4$-C_6H_5	244.34	lf	285-6^{110}	85	$1.171^{0/4}$		al, eth, bz	B5³, 2324
3391	Biphenyl, 4,4′-bis(diethylamino) $[4-(C_2H_5)_2NC_6H_4-]_2$	296.46	nd (al)		85			al, eth	B13³, 430
3392	Biphenyl, 2,4′-bis(dimethylamino) 2-$(CH_3)_2N$-$C_6H_4C_6H_4[N(CH_3)_2$-4]	240.35	pl (al)	206-7^{11}	51-2			al, eth	B13², 88
3393	Biphenyl, 4,4′-bis(dimethylamino) 4-$(CH_3)_2N$-C_6H_4-$C_6H_4[N(CH_3)_2$-4]	240.35	nd (al or bz-lig)	>360	198			bz, chl	B13³, 429
3394	Biphenyl, 4,4′-bis(ethylamino) 4-$C_2H_5NHC_6H_4$-$C_6H_4(NHC_2H_5$-4)	240.35	nd or pl (al)		120.5			al, eth, bz	B13, 222
3395	Biphenyl, 4,4′-bis(methylamino) 4-$CH_3NHC_6H_4$-$C_6H_4(NHCH_3$-4)	212.29	lf (al, w, or lig)		91			al, lig	B13², 97
3396	Biphenyl, 4,4′-bis(phenylamino) 4-$C_6H_5NHC_6H_4$-$C_6H_4(NHC_6H_5$-4)	336.24	lf (to)		244-5			aa, to	B13³, 431
3397	Biphenyl, 2-bromo 2-BrC_6H_4-C_6H_5	233.11		296-8, 160^{11}	1-2	1.2175^{26}	1.6248^{25}	al, eth	B5⁴, 1818
3398	Biphenyl, 3-bromo 3-BrC_6H_4-C_6H_5	233.11		299-301, 169-73^{17}			1.6411^{20}		B5⁴, 1818

No.	Name, Synonyms, and Formula	Mol. wt.	Color, crystalline form, specific rotation and λ_{max} (log ε)	b.p. °C	m.p. °C	Density	n_D	Solubility	Ref.
3399	Biphenyl, 3-bromo-4-hydroxy $3\text{-Br-4-HOC}_6H_3\text{-}C_6H_5$	249.11	nd (chl-peth)		96			al, chl, aa	B6^3, 3333
3400	Biphenul, 4-bromo $4\text{-BrC}_6H_4\text{-}C_6H_5$	233.11	pl (al)	310	91.2	$0.9327^{25/4}$		al, eth, bz, aa	B5^4, 1819
3401	Biphenyl, 4-(bromoacetyl) $4\text{-(BrCH}_2\text{CO)}C_6H_4\text{-}C_6H_5$	275.15	nd (95% al)		127				B7^3, 2137
3402	Biphenyl, 4-bromo-4′-hydroxy $4\text{-BrC}_6H_4\text{-}C_6H_4\text{OH-4)}$	249.11	pl (al)		164-6			al, eth, ace, bz	B6^3, 3334
3403	Biphenyl, 2-chloro $2\text{-ClC}_6H_4\text{-}C_6H_5$	188.66	nd (dil al)	274, 154^{12}	34	$1.1499^{32.5}$		al, eth, lig	B5^4, 1816
3404	Biphenyl, 3-chloro $3\text{-ClC}_6H_4\text{-}C_6H_5$	188.66		284-5, $150\text{-}60^6$	16	$1.1579^{25/4}$	1.6181^{25}	al, eth, ace	B5^4, 1816
3405	Biphenyl, 3-chloro-2-hydroxy $3\text{-Cl-2-HOC}_6H_3\text{-}C_6H_5$	204.66		317-8d	6	$1.24^{25/4}$	1.6237^{30}	al, eth, ace, bz	B6^3, 3297
3406	Biphenyl, 4-chloro $4\text{-ClC}_6H_4\text{-}C_6H_5$	188.66	lf (lig or al)	291, $180\text{-}95^{20\text{-}30}$	77.7			al, eth, lig	B5^4, 1816
3407	Biphenyl, 4-(chloroacetyl) $4\text{-(ClCH}_2\text{CO)}C_6H_4\text{-}C_6H_5$	230.69	pl (al)		125			al	B7, 443
3408	Biphenyl, 4-chloro-4′-hydroxy $4\text{-Cl-}C_6H_4\text{-(}C_6H_4\text{OH-4)}$	204.66	cr (dil al)		146-7			al, eth, ace, bz	B6^3, 3332
3409	Biphenyl, 5-chloro-2-hydroxy $5\text{-Cl-2-HOC}_6H_3C_6H_5$	204.66		319^{745}, $128\text{-}30^2$	11				B6^3, 3299
3410	Biphenyl, 2,2′-diacetamido $[2\text{-(CH}_3\text{CONH)}C_6H_4\text{-}]_2$	268.32	pr (al)		164-5			bz, aa	B13^3, 412
3411	Biphenyl, 2,4-diacetamido $2,4\text{-(CH}_3\text{CONH)}_2C_6H_3\text{-}C_6H_5$	268.32	nd (al)		202			al	B13^3, 417
3412	Biphenyl, 4,4′-diacetamido $[4\text{-(CH}_3\text{CONH)}C_6H_4\text{-}]_2$	268.32	nd (aa)		328.3				B13^3, 437
3413	Biphenyl, 2,2′-diamino or *o*-Benzidine $2\text{-H}_2NC_6H_4\text{-}C_6H_4NH_2\text{-2}$	184.24	mcl pr or nd (al)	162^4	81			w, bz	B13^3, 410
3414	Biphenyl, 2,4′-diamino or Diphenyline $2\text{-H}_2NC_6H_4\text{-}C_6H_4NH_2\text{-4}$	184.24	nd (dil al)	363	54.5			al, eth	B13^3, 416
3415	Biphenyl, 3,3′-diamino or *m*-Benzidine $3\text{-H}_2NC_6H_4\text{-}C_6H_4NH_2\text{-3}$	184.24	nd (w), pr (bz)	205	93-4			eth, bz	B13^3, 422
3416	Biphenyl, 3,4-diamino $3,4\text{-(H}_2\text{N)}_2C_6H_3\text{-}C_6H_5$	184.24	lf (eth or al)		103			al, eth	B13^2, 89
3417	Biphenyl, 4,4′-diamino or Benzidine $4\text{-H}_2NC_6H_4\text{-}C_6H_4NH_2\text{-4}$	184.24	nd (w)	400^{740}	125			al	B13^3, 425
3418	Biphenyl, 4,4′-diamino-3,3′-dimethoxy $[3\text{-CH}_3\text{O-4-H}_2NC_6H_3\text{-}]_2$	244.29	lf or nd (w)		137			al, eth, ace, bz, chl	B13^3, 2310
3419	Biphenyl, 4,4′-diamino-2,2′-dimethyl $[2\text{-CH}_3\text{-4-H}_2NC_6H_3\text{-}]_2$	212.29	pr (w)		108-9			al, eth	B13, 255
3420	Biphenyl, 4,4′-diamino-3,3′-dimethyl $[3\text{-CH}_3\text{-4-H}_2NC_6H_3\text{-}]_2$	212.29	lf (dil al)		131-2			al, eth	B13^3, 484
3421	Biphenyl, 4,4′-diamino-3-ethoxy $3\text{-C}_2H_5\text{O-4-H}_2NC_6H_3\text{-}C_6H_4\text{-NH}_2\text{-4}$	228.29	nd (w)		134			al	B13^2, 419
3422	Biphenyl, 4,4′-dibromo $4\text{-BrC}_6H_4\text{-}C_6H_4\text{Br-4}$	312.00	mcl pr (MeOH)	355-60	164			bz	B5^4, 1820
3423	Biphenyl, 3,3′-dichloro $3\text{-ClC}_6H_4\text{-}C_6H_4\text{Cl-3}$	223.10		322-4	29			al, eth, bz	B5^3, 1739
3424	Biphenyl, 4,4′-dichloro $4\text{-ClC}_6H_4\text{-}C_6H_4\text{Cl-4}$	223.11	pr or nd (al or to-peth)	315-9	148-9			bz	B5^4, 1817
3425	Biphenyl, 4,4′-dichloro-2,2′-dinitro $[4\text{-Cl-2-NO}_2C_6H_3\text{-}]_2$	313.10	ye cr (al)		140			bz, aa	B5^4, 1828
3426	Biphenyl, 2,2′-diethoxy-3,3′-dimethyl $[3\text{-CH}_3\text{-2-C}_2H_5\text{O-}C_6H_3\text{-}]_2$	270.37	lf (al)		85				B6^2, 974
3427	Biphenyl, 2,4′-diethoxy-3,3′-dimethyl $3\text{-CH}_3\text{-2-C}_2H_5\text{O-}C_6H_3\text{-}C_6H_3\text{-OC}_2H_5\text{(4)-CH}_3\text{-3}$	270.37	nd (al)		53				B6^2, 974
3428	Biphenyl, 4,4′-diethoxy-3,3′-diethyl $[3\text{-C}_2H_5\text{-4-C}_2H_5\text{OC}_6H_3\text{-}]_2$	298.43	lf (al)		120				B6, 1015
3429	Biphenyl, 4,4′-diethoxy-3,3′-dimethyl $[3\text{-CH}_3\text{-4-C}_2H_5\text{OC}_6H_3\text{-}]_2$	270.37	pl (al)		156				B6, 1010
3430	Biphenyl, 3,3′-diethyl-6,6′-dihydroxy $[3\text{-C}_2H_5\text{-6-HOC}_6H_3\text{-}]_2$	242.32	nd (dil al)		131				B6^2, 981
3431	Biphenyl, 2,2′-difluoro $2\text{-F-}C_6H_4\text{-}C_6H_4\text{-F-2}$	190.19			118.5-9.5	$1.393^{20/4}$		al	B5^3, 1735

No.	Name, Synonyms, and Formula	Mol. wt.	Color, crystalline form, specific rotation and λ_{max} (log ε)	b.p. °C	m.p. °C	Density	n_D	Solubility	Ref.
3432	Biphenyl, 3,3′-difluoro $3\text{-}FC_6H_4C_6H_4F\text{-}3$	190.19	....	130^{14}	8	$1.192^{25/4}$	1.5678^{20}	....	$B5^3$, 1736
3433	Biphenyl, 4,4′-difluoro $4\text{-}FC_6H_4\text{-}C_6H_4F\text{-}4$	190.19	mcl pr (al), lf (w)	254—5, 119^{14}	94-5	....	....	al, eth, ace, bz, chl	$B5^3$, 1736
3434	Biphenyl, 2,2′-dihydroxy or *o,o′*-Biphenol $2\text{-}HOC_6H_4\text{-}C_6H_4OH\text{-}2$	186.21	lf (w + 1), pr (to)	325-6	110-2 (anh)	....	....	al, eth, ace, bz, aa	$B6^3$, 5374
3435	Biphenyl, 2,2′-dihydroxy-3,3′-dimethyl $[3\text{-}CH_3\text{-}2\text{-}HOC_6H_3\text{-}]_2$	214.26	nd (peth)	sub	113	....	....	al, eth, bz	$B6^3$, 5445
3436	Biphenyl, 2,2′-dihydroxy-5,5′-dimethyl $[5\text{-}CH_3\text{-}2\text{-}HOC_6H_3\text{-}]_2$	214.26	nd (bz or w)	sub	153.5	....	....	al, eth, ace, bz	$B6^3$, 5447
3437	Biphenyl, 2,2′-dihydroxy-6,6′-dimethyl $[6\text{-}CH_3\text{-}2\text{-}HOC_6H_3\text{-}]_2$	214.26	pl (dil al)	....	164	....	....	al	$B6^3$, 5445
3438	Biphenyl, 2,2′-dihydroxy-3,3′,5,5′-tetramethyl $[3,5\text{-}(CH_3)_2\text{-}2\text{-}HOC_6H_2\text{-}]_2$	242.32	nd or pl (eth or lig)	$140\text{-}60^{0.05}$	137-8	....	....	al, eth, lig	$B6^3$, 5483
3439	Biphenyl, 2,4′-dihydroxy or *o,p′*-Biphenol $2\text{-}HOC_6H_4\text{-}C_6H_4OH\text{-}4$	186.21	mcl pr or nd (dil al)	342, $206\text{-}10^{11}$	162-3	....	....	eth	$B6^3$, 5387
3440	Biphenyl, 2,5-dihydroxy $2,5\text{-}(HO)_2C_6H_3C_6H_5$	186.21	nd (dil al)	....	97-8	....	....	al	$B6^3$, 5371
3441	Biphenyl, 2,5′-dihydroxy-2′,5-dimethyl $5\text{-}CH_3\text{-}2\text{-}HOC_6H_3C_6H_3OH(5)\text{-}CH_3\text{-}2$	214.26	....	....	158	....	....	al, eth	$B6^3$, 5445
3442	Biphenyl, 3,3′-dihydroxy or *m,m′*-Biphenol $3\text{-}HOC_6H_4\text{-}C_6H_4OH\text{-}3$	186.21	nd (w)	247^{18}	123-4	....	....	al, eth, bz, chl	$B6^3$, 5388
3443	Biphenyl, 3,4-dihydroxy $3\text{-}HOC_6H_4\text{-}C_6H_4OH\text{-}4$	186.21	....	>360	145	....	....	al, eth, ace, bz, chl	$B6^3$, 5387
3444	Biphenyl, 4,4′-dihydroxy or *p,p′*-Biphenol $4\text{-}HOC_6H_4\text{-}C_6H_4OH\text{-}4$	186.21	nd or pl (al)	sub	274-5	....	....	al, eth	$B6^3$, 5389
3445	Biphenyl, 4,4′-dihydroxy-3,3′-diethyl $[3\text{-}C_2H_5\text{-}4\text{-}HOC_6H_3\text{-}]_2$	242.32	nd (aa)	....	148	....	....	al, ace, bz, aa	$B6^3$, 5482
3446	Biphenyl, 4,4′-dihydroxy-3,3′-dimethyl $[3\text{-}CH_3\text{-}4\text{-}HOC_6H_3\text{-}]_2$	214.26	lf (w), nd	....	161	....	....	al, eth	$B6^3$, 5445
3447	Biphenyl, 4,4′-diacetoxy-3,3′-dimethyl $[3\text{-}CH_3\text{-}4\text{-}CH_3CO_2C_6H_3\text{-}]_2$	298.34	wh nd (al, aa)	....	135.3	....	....	....	$B6^3$, 5446
3448	Biphenyl, 4,4′-dihydroxy-3,3′,5,5′-tetramethoxy or Hydrocerulignone $[3,5\text{-}(CH_3)_2\text{-}4\text{-}HOC_6H_2\text{-}]_2$	306.32	mcl pr (al)	....	190	....	....	al	$B6^1$, 593
3449	Biphenyl, 4,4′-dihydroxy-3,3′,5,5′-tetramethyl $[3,5\text{-}(CH_3)_2\text{-}4\text{-}HOC_6H_2\text{-}]_2$	242.32	pa ye nd or pr (aa)	sub	222-3	....	....	al, aa	B6, 1015
3450	Biphenyl, 5,5′-dihydroxy-2,2′-dimethyl $[2\text{-}CH_3\text{-}5\text{-}HOC_6H_3\text{-}]_2$	214.26	pr (al)	....	229	....	....	al, eth	$B6^3$, 5444
3451	Biphenyl, 4,4′-dihydroxy-3,3′,5,5′-tetranitro $[3,5\text{-}(NO_2)_2\text{-}4\text{-}HOC_6H_2\text{-}]_2$	366.20	ye nd	....	223	....	....	....	$B6^3$, 5399
3452	Biphenyl, 2,2′-dimethoxy or *o,o′*-Bianisole $(2\text{-}CH_3OC_6H_4\text{-})_2$	214.26	rh bipyr pr (al)	$307\text{-}8^{766}$	155	1.268	....	al, bz, chl	$B6^3$, 5375
3453	Biphenyl, 3,3′-dimethoxy or *m, m′* Bianisole $(3\text{-}CH_3OC_6H_4\text{-})_2$	214.26	nd (dil al)	328, $211\text{-}20^{15}$	36	....	....	aI, eth, ace, bz, chl	$B6^3$, 5388
3454	Biphenyl, 4,4′-dimethoxy or *p, p′*-Bianisole $(4\text{-}CH_3OC_6H_4\text{-})_2$	214.26	lf (bz)	sub	173	....	....	al, bz, chl	$B6^3$, 5391
3455	Biphenyl, 2,2′-dimethoxy-5,5′-dimethyl $[5\text{-}CH_3\text{-}2\text{-}CH_3OC_6H_3\text{-}]_2$	242.32	nd (dil al)	188^{12}	71	....	....	al, eth, ace, bz	$B6^3$, 5447
3456	Biphenyl, 2,5′-dimethoxy-2′,5-dimethyl $5\text{-}CH_3\text{-}2\text{-}CH_3OC_6H_3\text{-}C_6H_3OCH_3(5)\text{-}CH_3\text{-}2$	242.32	pr (al)	168^4	86	....	....	bz, peth	$B6^2$, 973
3457	Biphenyl, 4,4′-dimethoxy-3,3′-dimethyl $[3\text{-}CH_3\text{-}4\text{-}CH_3OC_6H_3\text{-}]_2$	242.32	pr (al)	....	145.5	....	....	....	$B6^3$, 5445
3458	Biphenyl, 2,2′-dimethyl or *o, o′*-Bitolyl $(2\text{-}CH_3C_6H_4\text{-})_2$	182.27	cr (al)	256	19—20	0.9906^{20}	1.5752^{20}	al, eth, ace, bz	$B5^4$, 1897
3459	Biphenyl-2,3-dimethyl $2,3\text{-}(CH_3)_2C_6H_3C_6H_5$	182.26	....	141^{14}	42	....	1.5845^{23}	eth	$B5^4$, 1897
3460	Biphenyl, 2,3′-dimethyl or *o,m′*-Bitolyl $2\text{-}CH_3C_6H_4\text{-}C_6H_4CH_3\text{-}3$	182.27	....	270	....	0.9924^{20}	1.5810^{20}	al, eth, ace, bz	$B5^4$, 1902
3461	Biphenyl, 2,4-dimethyl $2,4\text{-}(CH_3)_2C_6H_3\text{-}C_6H_5$	182.27	....	$270\text{-}6^{767}$	0.9947 20/4	1.5844^{20}	....	....	$B5^4$, 1897
3462	Biphenyl, 2,4′-dimethyl or *o,p′*-Bitolyl $2\text{-}CH_3C_6H_4\text{-}C_6H_4CH_3\text{-}4$	182.27	....	273-6, $137^{12.5}$	....	0.9924^{20}	1.5826^{20}	al, eth, ace, bz	$B5^4$, 1903
3463	Biphenyl, 2,5-dimethyl $2,5\text{-}(CH_3)_2C_6H_3\text{-}C_6H_5$	182.27	....	$140^{14.5}$	....	$0.9931^{20/4}$	1.5819^{20}	....	$B5^4$, 1897
3464	Biphenyl, 2,6-dimethyl $2,6\text{-}(CH_3)_2C_6H_3\text{-}C_6H_5$	182.27	....	260-5, $132^{16.5}$	− 5	$0.9907^{20/4}$	1.5745^{20}	....	$B5^4$, 1897

No.	Name, Synonyms, and Formula	Mol. wt.	Color, crystalline form, specific rotation and λ_{max} (log ε)	b.p. °C	m.p. °C	Density	n_D	Solubility	Ref.
3465	Biphenyl, 3,3′-dimethyl or *m*, *m*′-Bitolyl $(3\text{-}CH_3C_6H_4\text{-})_2$	182.27		280, 150^{18}	9	0.9995$^{20/4}$	1.5946^{20}	al, eth, ace, bz	B5^4, 1903
3466	Biphenyl, 3,3′-dimethyl-4,4′-dipropoxy $(3\text{-}CH_3\text{-}4\text{-}C_3H_7OC_6H_3\text{-})_2$	298.43	lf		115				B6, 1010
3467	Biphenyl, 3,4-dimethyl $3,4\text{-}(CH_3)_2C_6H_3C_6H_5$	182.27		281-3, 139-40^{8}	29.2-9.7	1.0087$^{20/4}$	1.6036^{20}	bz	B5^4, 1903
3468	Biphenyl, 3,4′-dimethyl $3\text{-}CH_3C_6H_4\text{-}C_6H_4CH_3\text{-}4$	182.27		288-9^{752}, 153^{15}	14-15	0.9978$^{20/4}$	1.5968^{20}	bz	B5^4, 1905
3469	Biphenyl, 3,5-dimethyl $3,5\text{-}(CH_3)_2C_6H_3\text{-}C_6H_5$	182.27		273-6	22-3	0.9990$^{20/4}$	1.5952^{20}		B5^4, 1903
3470	Biphenyl, 4,4′-dimethyl $(4\text{-}CH_3C_6H_4\text{-})_2$	182.27	mcl pr (eth)	295	125	0.917$^{121/4}$		eth, ace, bz	B5^4, 1906
3471	Biphenyl, 2,2′-dinitro $(2\text{-}NO_2C_6H_4\text{-})_2$	244.21	ye mcl pr or nd (al)		127-8	1.45 (sol)		al, eth, bz, aa	B5^4, 1826
3472	Biphenyl, 2,3′-dinitro $2\text{-}NO_2C_6H_4\text{-}C_6H_4NO_2\text{-}3$	244.21			118-9.5			al	B5^4, 1826
3473	Biphenyl, 2,4′-dinitro $2\text{-}NO_2C_6H_4\text{-}C_6H_4NO_2\text{-}4$	244.21	mcl pr (al)		93-4	1.474		al, eth, bz, aa	B5^4, 1827
3474	Biphenyl, 3,3′-dinitro $(3\text{-}NO_2C_6H_4\text{-})_2$	244.21	ye og nd (al or aa)		200			bz, aa	B5^4, 1827
3475	Biphenyl, 4,4′-dinitro $(4\text{-}NO_2C_6H_4\text{-})_2$	244.21	nd (al)		240-3			bz, aa	B5^4, 1827
3476	Biphenyl, 2-ethoxy $2\text{-}C_2H_5OC_6H_4\text{-}C_6H_5$	198.26	pr (peth)	276, 132^{6}	34			al, eth, ace, bz, chl	B6^3, 3284
3477	Biphenyl, 3-ethoxy $3\text{-}C_2H_5OC_6H_4\text{-}C_6H_5$	198.26	(peth)	305, 158^{8}	35			al, eth, bz, ace	B6^3, 3313
3478	Biphenyl, 3-ethyl $3\text{-}C_2H_5C_6H_4\text{-}C_6H_5$	182.27		283-4^{763}		1.043^{0}			B5^4, 1896
3479	Biphenyl, 2-fluoro $2\text{-}F\text{-}C_6H_4\text{-}C_6H_5$	172.20		248	73.5	1.2452$^{25/4}$		al, eth	B5^4, 1815
3480	Biphenyl, 4-fluoro $4\text{-}F\text{-}C_6H_4\text{-}C_6H_5$	172.20		253	74.5			eth	B5^4, 1815
3481	Biphenyl, 2,2′,4,4′,6,6′-hexamethyl $[2,4,6\text{-}(CH_3)_3C_6H_2\text{-}]_2$	238.37		296^{735}	103—4	1.023^{50}		eth, bz	B5^4, 1989
3482	Biphenyl, 2-hydroxy-2′-methoxy-5,5′-dimethyl $C_{15}H_{16}O_2$	228.29		205^{12}				al, eth, bz, chl	B6^2, 974
3483	Biphenyl, 2-iodo $2\text{-}IC_6H_4\text{-}C_6H_5$	280.11		189-92^{36}		1.6038$^{25/25}$	1.6620^{20}	al, eth, bz, aa	B5^4, 1820
3484	Biphenyl, 4-iodo $4\text{-}IC_6H_4\text{-}C_6H_5$	280.11	nd (al or aa)	320d, 183^{11}	113-4			al, eth, bz, aa	B5^4, 1821
3485	Biphenyl, 2-methoxy or 2-Phenylanisole $2\text{-}CH_3OC_6H_4\text{-}C_6H_5$	184.24	pr (peth)	274, 150^{13}	29	1.0233$^{99/4}$	1.5641^{99}	al, peth	B6^3, 3284
3486	Biphenyl, 4-methoxy or 4-Phenylanisole $4\text{-}CH_3OC_6H_4\text{-}C_6H_5$	184.24	pl (al)	157^{10}	90	1.0278$^{100/4}$	1.5744^{100}	al, eth	B6^3, 3321
3487	Biphenyl, 2-methyl or 2-Phenyl toluene $2\text{-}CH_3C_6H_4\text{-}C_6H_5$	168.24		255.5, 130-6^{27}	-0.2	1.010$^{22/4}$	1.5914^{20}	al, eth	B5^4, 1855
3488	Biphenyl, 3-methyl $3\text{-}CH_3C_6H_4C_6H_5$	168.24		272.7, 148.50^{20}	4.5	1.0182$^{17/4}$	1.5972^{20}	al, eth	B5^4, 1858
3489	Biphenyl, 4-methyl $4\text{-}CH_3C_6H_4C_6H_5$	168.24	pl (lig, MeOH)	267-8, 134-6^{15}	49-50	1.015^{27}		al, eth	B5^4, 1860
3490	Biphenyl, 2-nitro $2\text{-}NO_2C_6H_4\text{-}C_6H_5$	199.21	pl (al or MeOH)	320, 201^{30}	37.2	1.44		al, eth	B5^4, 1823
3491	Biphenyl, 3-nitro $3\text{-}NO_2C_6H_4\text{-}C_6H_5$	199.21	ye pl or nd (dil al)	225-30^{35}	62			al, eth, aa, lig	B5^4, 1823
3492	Biphenyl, 4-nitro $4\text{-}NO_2C_6H_4\text{-}C_6H_5$	199.21	ye nd (al)	340, 224^{30}	114			eth, bz, chl, aa	B5^4, 1823
3493	Biphenyl, 2,2′,4,4′-tetrahydroxy $[2,4\text{-}(HO)_2C_6H_3\text{-}]_2$	218.21			226-7			w, al, eth, ace	B6^3, 6705
3494	Biphenyl, 3,3′,5,5′-tetrahydroxy or Diresorcinol $[3,5\text{-}(HO)_2C_6H_3\text{-}]_2$	218.21	pl or nd (w + 2)		310 (anh)			al, eth, w	B6^2, 1129
3495	Biphenyl, 2,2′,4,4′-tetramethyl $[2,4\text{-}(CH_3)_2C_6H_3\text{-}]_2$	210.32			41			al	B5^3, 1891
3496	Biphenyl, 2,2′,5,5′-tetramethyl $[2,5\text{-}(CH_3)_2C_6H_3\text{-}]_2$	210.32		284^{712}	50			eth, bz	B5^3, 1892
3497	Biphenyl, 2,2′,4,4′-tetranitro $[2,4\text{-}(NO_2)_2C_6H_3]_2$	334.20	ye pr (bz)		165-6			bz, aa	B5^3, 1772

PHYSICAL CONSTANTS OF ORGANIC COMPOUNDS (Continued)

No.	Name, Synonyms, and Formula	Mol. wt.	Color, crystalline form, specific rotation and λ_{max} (log ε)	b.p. °C	m.p. °C	Density	n_D	Solubility	Ref.
3498	Biphenyl, 2,4,4′-triamino $2,4\text{-}(H_2N)_2C_6H_3\text{-}C_6H_4NH_2\text{-}4$	199.26	nd		134				B13^{3}, 560
3499	Biphenyl, 4-vinyl $4\text{-}(CH_2{=}CH)C_6H_4\text{-}C_6H_5$	180.25		136-8^{6}	119				B5^{3}, 1987
3500	2-Biphenyl carboxylic acid $2\text{-}(HO_2C)C_6H_4\text{-}C_6H_5$	198.22	lf (dil al)	343-4, 199^{10}	113-4			al, bz, aa	B9^{3}, 3268
3501	2-Biphenylcarboxylonitrile $2\text{-}NCC_6H_4\text{-}C_6H_5$	179.22	nd	170-2^{15}	41			al, eth	B9^{3}, 3269
3502	3-Biphenylcarboxylic acid $3\text{-}(HO_2C)C_6H_4C_6H_5$	198.22	lf (al)		165—6			al, eth, bz, aa	B9^{3}, 3274
3503	4-Biphenylcarboxylic acid $4\text{-}(HO_2C)C_6H_4C_6H_5$	198.22	nd (bz or al)	sub	228			al, eth, bz	B9^{3}, 3276
3504	3-Biphenylcarboxylic acid, 2-hydroxy $2\text{-}HO\text{-}3\text{-}(HO_2C)C_6H_3\text{-}C_6H_6$	214.22	cr (bz)		186-7			al	B10^{3}, 1159
3505	2,2′-Biphenyldicarboxylic acid or Diphenic acid $[2\text{-}(HO_2C)C_6H_4\text{-}]_2$	242.23	mcl pr or lf (w), cr (aa)	sub	233.5			al, eth	B9^{3}, 4496
3506	2,2′-Biphenyldicarboxylic acid anhydride $C_{14}H_8O_3$	224.22	nd (aa or bz)	sub	217				B17^{4}, 6425
3507	2,2′-Biphenyldicarboxylyl chloride $[2\text{-}(ClCO)C_6H_4\text{-}]_2$	279.14		sub	94 (97)				B9^{2}, 657
3508	2,2′-Biphenyldicarboxylic acid, diethyl ester $[2\text{-}(C_2H_5O_2C)C_6H_4\text{-}]_2$	298.34			42			eth	B9^{2}, 656
3509	2,2′-Biphenyldicarboxylic acid, dimethyl ester $[2\text{-}(CH_3O_2C)C_6H_4\text{-}]_2$	270.28	mcl pr (MeOH)	204-6^{14}	74			al, eth, bz	B9^{3}, 4497
3510	2,2′-Biphenyldicarboxylic acid, imide $C_{14}H_9NO_2$	223.23	nd (al)		219-20			chl	B21^{4}, 5601
3511	2,3′-Biphenyldicarboxylic acid or Isodiphenic acid $2\text{-}(HO_2C)\text{-}C_6H_4\text{-}C_6H_4(CO_2H)\text{-}3$	242.22	nd (w or dil aa)		216			al	B9^{2}, 663
3512	2,4′-Biphenyldicarboxylic acid $2\text{-}(HO_2C)\text{-}C_6H_4\text{-}C_6H_4(CO_2H)\text{-}4$	242.23	lf (al)		272-3			al, bz, aa	B9^{3}, 4514
3513	3,3′-Biphenyldicarboxylic acid $[3\text{-}(HO_2C)\text{-}C_6H_4\text{-}]_2$	242.23	lf (al)		356-7			chl	B9^{3}, 4517
3514	3,3′-Biphenyl dicarboxylic acid, dimethyl ester $[3\text{-}CH_3O_2C\text{-}C_6H_4\text{-}]_2$	270.29	lf (MeOH)		104			al, eth, bz	B9^{3}, 4517
3515	3,4′-Biphenyl dicarboxylic acid $[3\text{-}(HO_2C)\text{-}C_6H_4\text{-}C_6H_4\text{-}(CO_2H)\text{-}4]$	242.23	nd		334-5				B9^{3}, 4518
3516	3,4′-Biphenyldicarboxylic acid, dimethyl ester $3\text{-}(CH_3O_2C)\text{-}C_6H_4\cdot C_6H_4CO_2CH_3\text{-}4$	270.28	nd (lig or MeOH)		98-9			lig, MeOH	B9, 927
3517	3,5-Biphenyldicarboxylic acid $3,5\text{-}(HO_2C)_2C_6H_3C_6H_5$	242.23	lf (aa)		>310			al, eth, ace, bz	B9, 926
3518	3,5-Biphenyldicarboxylic acid, dimethyl ester $3\text{-}5\text{-}(CH_3O_2C)_2C_6H_3C_6H_5$	270.28	lf (MeOH)		214				B9^{2}, 665
3519	2,2′-Biphenyldicarboxylic acid, 3,3′-dimethyl-5-nitro $C_{16}H_{13}NO_6$	315.28	lf (w or dil al)		267			al, eth, bz	B9^{2}, 659
3520	2,2′-Biphenyldicarboxylic acid, 3,3′-dimethyl-6-nitro *dl* $C_{16}H_{13}NO_6$	315.28	lf (w)		248-50d			al, eth, ace, aa	B9^{2}, 659
3521	2,2′-Biphenyldicarboxylic acid, 3,3′-dimethyl-4-nitro $C_{16}H_{13}NO_6$	315.28	lf, pr or wh nd (w)		217 (250)			al, eth	B9^{2}, 659
3522	2,2′-Biphenyldicarboxylic acid, 3,3′-dimethyl $[3\text{-}CH_3\text{-}2\text{-}(HO_2C)C_6H_3\text{-}]_2$	270.28			230			al, eth, bz	B9^{1}, 407
3523	2,2′-Biphenyl disulfonyl chloride $[2\text{-}ClO_2S\text{-}C_6H_4\text{-}]_2$	351.24	pr (chl), cr (aa)		142-4			eth, bz, chl	B11^{3}, 469
3524	3,3′-Biphenyl disulfonamide $[3\text{-}(H_2NO_2S)\text{-}C_6H_4\text{-}]_2$	312.37	nd (ace)		285			al	B11, 219
3525	3,3′-Biphenyl disulfonyl chloride $[3\text{-}(ClO_2S)\text{-}C_6H_4\text{-}]_2$	351.24	nd (chl)		128			eth, bz	B11^{3}, 471
3526	4,4′-Biphenyl disulfonic acid $[4\text{-}(HO_3S)\text{-}C_6H_4\text{-}]_2$	314.34	pr	>200	72.5			w	B11^{3}, 472
3527	4,4′-Biphenyl disulfonamide $[4\text{-}(H_2NO_2S)\text{-}C_6H_4\text{-}]_2$	312.37	nd (w)		300			w, eth	B11, 220
3528	4,4′-Biphenyl disulfonyl chloride $[4\text{-}(ClO_2S)C_6H_4\text{-}]_2$	351.24	pr (aa)		205—7			eth, bz, aa	B11^{3}, 472
3529	2,2′-Biphenyl disulfonic acid, 4,4′-diamino $[2\text{-}(HO_3S)\text{-}4\text{-}H_2N\text{-}C_6H_3\text{-}]_2$	344.37	lf		175d				B14^{3}, 2264
3530	2,2′-Bipyridyl $2\text{-}NC_5H_4\text{-}C_5H_4N\text{-}2$	156.19	pr (peth)	272-5	71-3			al, eth, bz, chl, lig	B23^{2}, 211

No.	Name, Synonyms, and Formula	Mol. wt.	Color, crystalline form, specific rotation and λ_{max} (log ε)	b.p. °C	m.p. °C	Density	n_D	Solubility	Ref.
3531	2,3′-Bipyridyl or Isonicoteine $2\text{-}NC_5H_4\text{-}C_5H_4N\text{-}(3)$	156.19	....	295-6	......	$1.140^{20/4}$	1.6223^{20}	al, eth, bz, chl	$B23^2$, 212
3532	2,4′-Bipyridyl $2\text{-}NC_5H_4\text{-}C_5H_4N\text{-}(4)$	156.19	....	280-2, $148\text{-}50^1$	61.5	....	....	al, eth, chl	B23, 200
3533	3,3′-Bipyridyl $3\text{-}NC_5H_4\text{-}C_5H_4N\text{-}(3)$	156.19	....	291-2, $190\text{-}2^{25}$	68	$1.1635^{20/20}$	....	w, al	$B23^2$, 212
3534	3,4′-Bipyridyl $3\text{-}NC_5H_4\text{-}C_5H_4N\text{-}(4)$	156.19	lf (peth)	297	62	....	....	w, al, peth	B23, 212
3534a	4,4′-Bipyridyl $4\text{-}NC_5H_4\text{-}C_5H_4N\text{-}(4)$	156.19	nd (w + 2)	305 sub	114, 171-2	....	....	al, eth, bz, chl, lig	$B23^2$, 212
3535	2,2′-Biquinolyl $(C_9H_6N)_2$	256.31	pl or lf (al)	....	196	....	....	al, eth, ace, bz	$B23^2$, 267
3536	2,3′-Biquinolyl $(C_9H_6N)_2$	256.31	lf (al), ye pl or nd (bz)	>400	176-7	....	....	al, eth, bz, chl	$B23^2$, 267
3537	2,6′-Biquinolyl $(C_9H_6N)_2$	256.31	pl (al)	....	144	....	....	al, ace, bz	B23, 294
3538	2,7′-Biquinolyl(higher melting) $(C_9H_6N)_2$	256.31	mcl pl (al)	....	160	....	....	al	B23, 294
3539	2,7′-Biquinolyl(lower melting) $C_{18}H_{12}N_2$	256.31	tcl	....	115	....	....	al, eth, bz	B23, 294
3540	3,4′-Biquinolyl $(C_9H_6N)_2$	256.31	pw (peth)	....	83-4	....	....	al, bz	$B23^2$, 267
3541	3,7′-Biquinolyl $(C_9H_6N)_2$	256.31	lf or nd (al or bz)	....	190	....	....	al, bz, chl	$B23^2$, 268
3542	4,4′-Biquinolyl $(C_9H_6N)_2$	256.31	pr (peth)	....	171	....	....	al, ace, bz	$B23^2$,268
3543	4,6′-Biquinolyl $(C_9H_6N)_2$	256.31	cr (bz)	....	122	....	....	al, bz, chl	B23,294
3544	6,6′-Biquinolyl $(C_9H_6N)_2$	256.31	lf (al)	....	181	....	....	al, eth, bz	B23, 295
3545	6,8′-Biquinolyl $(C_9H_6N)_2$	256.31	lf (al)	....	148	....	....	al, bz	B23, 296
3546	8,8′-Biquinolyl $(C_9H_6N)_2$	256.31	lf or pl (al or aa)	....	205-7	....	....	al, ace, bz, chl	$B23^2$, 268
3547	Bismuthine, triphenyl $(C_6H_5)_3Bi$	440.30	....	242^{14}	77.6	$1.715^{75/4}$	1.7040^{75}	eth, ace, bz, lig	$B16^3$, 1188
3548	Biquinone, 3,3′-dihydroxy-5,5′-dimethyl or Phenicin $(C_7H_5O_3)_2$	274.23	yesh-br (al)	....	230-1	....	....	al, chl, aa	$B8^3$, 4251
3549	2,2-Bithiophene or 2,2-Bithienyl $(C_4H_3S)_2$	166.26	lf (al)	260, 103^3	33	....	....	al, eth, aa	$B19^4$, 265
3550	2,2-Bithiophene, hexabromo $C_8Br_6S_2$	639.63	nd (bz)	....	257-8	....	....	bz	$B19^4$, 266
3551	3,3′-Bithiophene $(3\text{-}C_4H_3S\text{-}3)_2$	166.26	....	....	132	....	....	al, eth, bz, chl, lig	$B19^4$, 267
3552	Biuret or Carbamoylurea $H_2NCONHCONH_2$	103.08	pl (al), nd (w + 1)	....	190d	....	....	al, w	$B3^4$, 141
3553	Biuret, acetyl $CH_3CONHCONHCONH_2$	145.12	nd (w or al)	....	193-4	....	....	w, al	$B3^4$, 142
3554	Biuret, 1,5-diamino $H_2NNHCONHCONHNH_2$	133.11	pr (dil al), nd (aa)	....	199-200d	....	....	w, aa	$B3^4$, 178
3555	Bixin $C_{25}H_{30}O_4$	394.51	vt pr (ace)	....	198	....	....	al, ace, chl	$B2^3$, 2020
3556	Boric acid, tributyl ester or Tributoxyborine $B(OC_4H_9)_3$	230.15	oil	230—1, $114\text{-}5^{25}$	......	$0.8567^{20/4}$	1.4106^{18}	al, eth, bz	$B1^4$, 1544
3557	Boric acid, isobutyl $i\text{-}C_4H_9B(OH)_2$	101.94	lo pl (w)	....	112	....	....	al, eth	$B4^3$, 1965
3558	Boric acid, isopentyl $i\text{-}C_5H_{11}B(OH)_2$	115.97	pl (w)	....	169	....	....	al, eth, ace	$B4^2$, 1023
3559	Boric acid, triethyl ester or Triethoxyborine $B(OC_2H_5)_3$	146.00	....	120	......	$0.8546^{20/4}$	1.3749^{20}	al, eth	$B1^4$, 1365
3560	Boric acid, trimethyl ester or Trimethoxyborine $B(OCH_3)_3$	103.92	....	67-9	−29.3	0.915^{20}	1.3568^{20}	al, eth, bz	$B1^4$, 1269
3561	Boric acid, tri isopentyl ester or Triisopentyl borate $B\text{-}(O\text{-}i\text{-}C_5H_{11})_3$	272.24	....	254-5, $132\text{-}3^{17}$	......	$0.8518^{20/4}$	1.4156^{20}	al, eth	$B1^4$, 1688
3562	Boric acid, tripropyl ester or Tripropoxyborine $B(OC_3H_7)_3$	188.07	....	179-80, 64^9	......	$0.8576^{20/4}$	1.3948^{20}	al, eth	$B1^4$, 1436
3563	Boric acid, tri-*iso*-propyl ester or Tri-*iso*-propoxyborine $B\text{-}(O\text{-}i\text{-}C_3H_7)_3$	188.07	....	164, $52\text{-}3^{32}$	......	$0.8251^{20/4}$	1.3772^{20}	al, eth, bz	$B1^4$, 1488

No.	Name, Synonyms, and Formula	Mol. wt.	Color, crystalline form, specific rotation and λ_{max} (log ε)	b.p. °C	m.p. °C	Density	n_D	Solubility	Ref.
3564	Boric acid, tri-isopentyl ester $B\text{-}(O\text{-}i\text{-}C_5H_{11})_3$	272.24	....	254-5, 132-3^{12}	......	$0.8518^{20/4}$	1.4156^{20}	al, eth	B1[4], 1688
3565	Borine, bis(dimethylamino)-fluoro $[(CH_3)_2N]_2BF$	117.96	....	106	−44.3	....	....	eth, ace	B4[4], 303
3566	Borine, bis(methylthio) methyl $(CH_3S)_2BCH_3$	120.05	....	100^{147}	−59	....	....	eth, ace	Am78,1523
3567	Borine, tri-(methylthio) $B(SCH_3)_3$	152.11	....	218.2	5	1.126^{20}	1.5788^{20}	eth, ace	C55,9346
3568	Borine, difluoro-(dimethyl amino) $(CH_3)_2NBF_3$	92.89	rh	sub 132	165-8d	....	....	....	B4[4]303
3569	Borine, phenyl-difluoro $C_6H_5BF_2$	125.91	....	97-8^{747}	−36.2	1.087^{25}	1.4441^{25}	eth, bz	B16[2], 638
3570	Borine, 4-tolyl, difluoro $(4\text{-}CH_3C_6H_4)BF_2$	139.94	....	127-8^{747}	......	1.055^{25}	1.4535^{25}	eth, bz	C18,992
3571	Borine, dimethyl-(dimethyl amino) $(CH_3)_2BN(CH_3)_2$	84.96	....	65	−92	....	....	eth, ace	B4[3], 1960
3572	Borine, dimethyl,methoxy $(CH_3)_2BOCH_3$	71.91	....	21	......	....	....	eth, ace	Am75,3872
3573	Borine, (methylthio)-dimethyl $CH_3SB(CH_3)_2$	87.97	....	71	−84	....	....	eth, ace	Am76,3307
3574	Borine, triethyl $B(C_2H_5)_3$	98.00	....	95 6	−92.9	0.6961^{23}	....	al, eth	B4[3], 1957
3575	Borine, tri-*iso*-butyl $B(i\text{-}C_4H_9)_3$	182.16	....	188^{766}, 86^{20}	......	$0.7380^{25/4}$	1.4188^{23}	al, eth, bz	B4[3], 1958
3576	Borine, trimethyl $B(CH_3)_3$	55.91	....	20	−161.5	....	....	al, eth	B4[3], 1955
3577	Borine, triisopentyl $(i\text{-}C_5H_{11})_3B$	224.24	....	119.14	......	$0.7600^{25/4}$	1.4321	al, eth, ace	B4[2], 1023
3578	Borine, triphenyl $B(C_6H_5)_3$	242.13	wh cr	245—50^{15}	142	....	....	bz, lig	B16[3], 1271
3579	Borine, tri-isopentyl $(i\text{-}C_5H_{11})_3B$	224.24	....	119.14	......	$0.7600^{23/4}$	1.4321	al, eth, ace, w	B4[2], 1023
3580	Borine, tripropyl $B(C_3H_7)_3$	140.08	....	159, 43-4^{17}	−56	$0.7204^{25/4}$	$1.4135^{22.5}$	....	B4[3], 1957
3581	3-Bornanone-*(d)* or Epicamphor $C_{10}H_{16}O$	152.24	$[\alpha]^{17}{}_D$ + 45.4 (bz)	....	182	....	....	al, eth, peth	B7[3], 420
3582	3-Bornanone *(dl)* $C_{10}H_{16}O$	152.24	cr (peth)	....	175-7	....	....	al, eth, peth	B7[3], 421
3583	3-Bornanone oxime-*(d)* $C_{10}H_{17}NO$	167.25	nd (MeOH), $[\alpha]_D$ −98.9	....	103	....	....	eth, ace	B7[1], 86
3584	3-Bornanone oxime-*(dl)* $C_{10}H_{17}NO$	167.25	nd (dil al)	....	98-100	....	....	eth, ace	B7[1], 87
3585	3-Bornanone oxime-*(l)* $C_{10}H_{17}NO$	167.25	nd (dil MeOH), $[\alpha]_D$ + 100.5	....	103-4	....	....	eth, ace	B7[1], 86
3586	3-Bornanone semicarbazide *(d)* $C_{11}H_{19}N_3O$	209.29	nd (al)	....	237-8	....	....	....	B7[3], 421
3587	3-Bornanone semicarbazide *(l)* $C_{11}H_{19}N_3O$	209.29	nd (al), $[\alpha]^{20/}{}_D$ + 145 (MeOH, c=0.73)	....	237-8d	....	....	....	B7[3], 421
3588	Borneol *(d)* $C_{10}H_{18}O$	154.25	lf or hex pl (peth), $[\alpha]^{20/}{}_D$ + 37.7 (al)	....	208	$1.011^{20/4}$	....	al, eth, bz, lig	B6[4], 281
3589	Borneol *(dl)* $C_{10}H_{18}O$	154.25	lf (lig)	sub	210.5	$1.011^{20/4}$	....	al, eth, bz	B6[4], 281
3590	Borneol *(l)* $C_{10}H_{18}O$	154.25	hex pl, $[\alpha]^{20/}{}_D$ −37.74 (al)	210^{779}	208.6	$1.1011^{20/4}$	....	al, eth, ace, bz	B6[4], 281
3591	Borneol acetate *(d)* $C_{12}H_{20}O_2$	196.29	rh, $[\alpha]^{20/}{}_D$ + 44.4 (al)	223-4, 107^{15}	29	$0.9920^{20/4}$	....	....	B6[3], 302
3592	Borneal acetate *(dl)* $C_{12}H_{20}O_2$	196.29	....	223-4	<−17	1.4630^{20}	....	....	B6[3],303

PHYSICAL CONSTANTS OF ORGANIC COMPOUNDS (Continued)

No.	Name, Synonyms, and Formula	Mol. wt.	Color, crystalline form, specific rotation and λ_{max} (log ε)	b.p. °C	m.p. °C	Density	n_D	Solubility	Ref.
3593	Borneol acetate *(l)* $C_{12}H_{20}O_2$	196.29	$[\alpha]^{20/D}$ −44.45 (undil)	223-4, 107^{15}	29	$0.9920^{20/4}$	1.4634^{20}	al, eth	$B6^{3}$303
3594	Borneol formate $C_{11}H_{18}O_2$	182.26	$[\alpha]_D$ +48.75 (undil)	90^{10}	...	1.009^{22}	1.4700^{15}	...	$B6^{3}$, 301
3595	Bornylamine *(d)* $C_{10}H_{17}NH_2$	153.27	$[\alpha]^{20/}{}_D$ +47.2 (al)	200 sub	163	...	...	al, eth, ace, bz	$B12^{3}$, 193
3596	Bornyl chloride *(d)* $C_{10}H_{17}Cl$	172.70	nd	207-8 sub	132	...	...	al, eth, bz, peth	$B5^{4}$, 319
3597	Bornylene *(d)* or 2-Bornene $C_{10}H_{16}$	136.24	cr (al), $[\alpha]_D$ + 30.5 (to)	146^{750} sub	109-10	...	...	...	$B5^{4}$, 460
3598	Bornylene *(l)* $C_{10}H_{16}$	136.24	cr (al), $[\alpha]^{19/}{}_D$ −23.9 (bz)	146^{746}	113	...	...	al, eth, bz	$B5^{4}$, 460
3599	Brazilein $C_{16}H_{12}O_5$	284.27	red-br nd or lf (w + 1)	...	250	...	...	al, w, bz, chl, aa	$B18^{4}$, 2770
3600	Bromcresolgreen or 3,3′,5,5′-Tetrabromo-*m*-Cresol sulfophthalein $C_{21}H_{14}O_5BrS$	698.02	wh or red (+7w), ye (aa)	...	218-9	...	...	al, eth, bz, aa	$B19^{4}$, 1133
3601	Bromo acetamide or Bromo acetamide $BrCH_2CONH_2$	137.96	nd (al or bz)	...	91	...	...	w, bz	$B2^{4}$, 530
3602	Bromoacetic acid $BrCH_2CO_2H$	138.95	hex or rh	208, 127.5^{30}	50	$1.9335^{50/4}$	1.4804^{50}	w, **al, eth,** ace, bz	$B2^{4}$, 526
3603	Bromoacetic acid, *iso*-butyl ester or *iso*-Butyl bromoacetate $BrCH_2CO_2\text{-}i\text{-}C_4H_9$	195.06	...	188^{752}, 74.5^{10}	...	$1.3269^{20/4}$	...	al, eth, ace	$B2^{3}$, 482
3604	Bromoacetic acid, *tert*-butyl ester or *tert*-Butyl bromoacetate $BrCH_2CO_2C(CH_3)_3$	195.06	...	$73\text{-}4^{25}$	...	1.4430^{20}	...	al, eth	$B2^{3}$, 482
3605	Bromoacetic acid, ethyl ester or Ethyl bromoacetate $BrCH_2CO_2C_2H_5$	167.00	...	168-9, 58.9^{15}	...	$1.5059^{20/20}$	1.4489^{20}	**al, eth,** ace	$B2^{4}$, 527
3606	Bromoacetic acid, methyl ester or Methyl bromoacetate $BrCH_2CO_2CH_3$	152.98	...	144d, 64^{33}	...	...	...	al, eth, ace	$B2^{4}$, 527
3607	Bromoacetic acid, phenyl ester $BrCH_2CO_2C_6H_5$	215.05	pl (al)	140^{20}	32	...	...	al, eth	$B6^{1}$, 87
3608	Bromoacetic acid, propyl ester or Propyl bromoacetate $BrCH_2CO_2C_3H_7$	181.03	...	176^{762}	...	$1.4099^{20/4}$	1.4518^{20}	al, eth, ace	$B2^{4}$, 528
3609	Bromoacetonitrile $BrCH_2CN$	119.95	pa ye	$150\text{-}1^{752}$, 46^{13}	...	...	...	eth	$B2^{4}$, 531
3610	Bromo acetyl bromide $BrCH_2COBr$	201.85	...	130	...	$2.317^{22/22}$	1.5449^{20}	ace	$B2^{4}$, 530
3611	Bromochloro acetic acid $BrClCHCO_2H$	173.39	...	215d, $103\text{-}4^{11}$	38	$1.9848^{31/4}$	1.5014^{31}	w, al, eth, ace	$B2^{4}$, 532
3612	Bromochloro acetic acid, ethyl ester or Ethyl bromochloro acetate $BrClCHCO_2C_2H_5$	201.45	...	174d	...	$1.5890^{22/4}$	1.4639^{24}	al, eth	$B2^{4}$, 532
3613	Bromo difluoro acetic acid BrF_2CCO_2H	174.93	lf (chl)	145-6, 87^{82}	40	...	...	w, al, chl	$B2^{4}$, 532
3614	Bromo diphenyl acetic acid $(C_6H_5)_2CBrCO_2H$	291.14	(chl-peth)	...	133-4	...	...	chl, to	$B9^{2}$, 471
3615	Bromodiphenyl acetyl bromide $(C_6H_5)_2CBrCOBr$	354.04	nd (lig)	...	65-6	...	...	al, eth, bz, chl	$B9^{1}$, 283
3616	bis-(2-Bromoethyl) ether $(BrCH_2CH_2)_2O$	231.91	...	115^{32}	...	$1.8222^{27/4}$	$1.5131^{27/D}$	...	$B1^{4}$, 1386
3617	Bromofluoro acetamide $BrCHFCONH_2$	155.96	nd (CCl_4)	...	44	...	...	w, al, eth	B2,217
3618	Bromofluoro acetic acid $BrCHFCO_2H$	156.94	...	183, 102^{20}	48	...	...	w, al, chl	$B2^{4}$531
3619	Bromofluoro acetic acid, ethyl ester $BrFCHCO_2C_2H_5$	184.99	...	154	...	1.5587^{17}	...	...	$B2^{4}$, 531
3620	Bromophenol blue or 3,3′,5,5′-Tetrabromophenol sulfonphthalein $C_{19}H_{10}O_5Br_4S$	669.96	hex pr (aa-ace)	279d	...	...	...	al, bz, aa	$B19^{4}$, 1129
3621	bis-(4-Bromophenyl) ether $(4\text{-}BrC_6H_4)_2O$	328.00	lf (al)	338-40, 210^{11}	60.5	1.8 (sol)	...	al, eth, bz	$B6^{4}$, 1048

PHYSICAL CONSTANTS OF ORGANIC COMPOUNDS (Continued)

No.	Name, Synonyms, and Formula	Mol. wt.	Color, crystalline form, specific rotation and λ_{max} (log ε)	b.p. °C	m.p. °C	Density	n_D	Solubility	Ref.
3622	4-Bromophenyl isocyanate $(4-BrC_6H_4)NCO$	198.02	nd	226, 158^{14}	......	....	....	eth	$B12^{1}$, 321
3623	4-Bromphenyl isothiocyanate $4-BrC_6H_4NCS$	214.08	nd	....	60-1	....	....	al	$B12^{3}$, 1463
3624	Brucine $C_{23}H_{26}N_2O_4$	394.47	mcl pr (w + 4), $[\alpha]^{20/}{}_{546.1}$ −149.5 (chl, c=1	....	178 (anh) 105 (hyd)	....	....	al, chl	$B27^{2}$, 797
3625	Brucine hydrochloride $C_{23}H_{26}N_2O_4 \cdot HCl$	430.93	pr	....	......	....	....	w, al	$B27^{2}$, 801
3626	Brucine nitrate, dihydrate $C_{23}H_{26}N_2O_4.HNO_3.2H_2O$	493.51	pr	sub	230d	....	....	....	$B27^{2}$, 797
3627	Brucine sulfate, heptahydrate $C_{46}H_{54}N_4O_{16}S \cdot 7H_2O$	1013.12	nd $[\alpha]_D$ −24.4 (w)	....	......	....	....	w, MeOH	$B27^{2}$, 797
3628	Bufotalin $C_{26}H_{36}O_6$	444.57	cr (+ 1 al), $[\alpha]^{20/}{}_D$ + 5.4 (chl, c=0.5)	....	223d	....	....	al, chl	$B18^{4}$, 2557
3629	Bulbocapnine *(d)* $C_{19}H_{19}NO_4$	325.36	pr (al), $[\alpha]_D$ + 237.1 (chl, c=4)	....	......	....	....	al, chl	$B27^{2}$, 554
3630	Bulbocapnine *(l)* $C_{19}H_{19}NO_4$	325.36	....	209-10	......	....	....	....	$B27^{2}$, 554
3631	1,2-Butadiene or Methylallene $CH_2=C=CHCH_3$	54.09	....	10.8	−136.2	$0.676^{0\ 4}$	$1.4205^{1\ 3}$	**al, eth**, bz	$B1^{4}$, 975
3632	1,2-Butadiene, 4-bromo $CH_2=C=CH-CH_2Br$	132.99	....	109-11	......	$1.4255^{20/4}$	1.5248^{20}	ace	$B1^{3}$, 929
3633	1,2-Butadiene, 4-chloro $CH_2=C=CH-CH_2Cl$	88.54	....	88	......	$0.9891^{20/4}$	1.4775^{20}	eth, ace, bz	$B1^{4}$, 975
3634	1,2-Butadiene, 4-iodo $CH_2=CCCH-CH_2I$	179.99	....	130	......	$1.7129^{20/4}$	1.5709^{20}	....	$B1^{3}$, 929
3635	1,2-Butadiene, 4-methoxy $CH_2=C=CH-CH_2OCH_3$	84.12	....	87-9	......	$0.8286^{20/4}$	1.435^{20}	al	$B1^{4}$, 2221
3636	1,2-Butadiene, 3-methyl $CH_2=C=C(CH_3)_2$	68.12	....	40	−120	$0.6804^{20/4}$	1.4166^{20}	**al, eth, ace, bz, peth**	$B1^{4}$, 1006
3637	1,2-Butadiene, 4-ol $CH=C=CH-CH_2OH$	70.09	....	126-8, $68-9^{45}$	......	$0.9164^{20/4}$	1.4759^{20}	w, al, ace, eth, bz	$B1^{4}$, 2221
3638	1,3-Butadiene or Bivinyl $CH_2=CH-CH=CH_2$	54.09	....	−4.4	−108.9	$0.6211^{20/4}$	1.4292^{-25}	al, eth, ace, bz	$B1^{4}$, 976
3639	1,3-Butadiene, 2-bromo or Bromoprene $CH_2=CBr-CH=CH_2$	132.99	....	$42-3^{165}$	......	$1.397^{20/4}$	1.4988^{20}	al, eth	$B1^{4}$, 989
3640	1,3-Butadiene, 1-chloro $CH_2=CH-CH=CHCl$	88.54	....	68	......	$0.9606^{20/4}$	1.4712^{20}	al, eth, chl	$B1^{3}$, 949
3641	1,3-Butadiene, 1-chloro-2-methyl $ClCH=C(CH_3)CH=CH_2$	102.56	....	107, 50.4^{100}	......	$0.9710^{20/4}$	1.4792^{20}	al, ace	$B1^{3}$, 974
3642	1,3-Butadiene, 1-chloro-3-methyl $ClCH=CH-C(CH_3)=CH_2$	102.56	....	99-100	......	$0.9543^{20/4}$	1.4719^{20}	al, eth, ace, chl	$B1^{3}$, 975
3643	1,3-Butadiene, 2-chloro or Chloroprene $CH_2=C(Cl)CH=CH_2$	88.54	....	59.4, 6.4^{100}	......	$0.9583^{20/4}$	1.4583^{20}	**eth, ace, bz**	$B1^{4}$, 984
3644	1,3-Butadiene, 2-chloro-3-methyl $CH_2=CCl-C(CH_3)=CH_2$	102.56	....	93	......	$0.9593^{20/4}$	1.4686^{20}	al, eth, ace, chl	$B1^{4}$, 1004
3645	1,3-Butadiene, 4-cyano or 2,4-Pentadieno nitrile $CH_2=CH-CH=CHCN$	79.10	....	135-8	0.8444^{20}	1.4880	....	....	$B2^{4}$, 1692
3646	1,3-Butadiene, 1,1-dichloro $Cl_2C=CH-CH=CH_2$	122.98	....	$42-3^{90}$	......	$1.1831^{20/4}$	1.5022^{20}	eth, bz	$B1^{4}$, 985
3647	1,3-Butadiene, 1,2-dichloro $ClCH=CCl-CH=CH_2$	122.98	....	$60-5^{105}$, 35^{40}	......	$1.1991^{20/44}$	1.4960^{20}	CCl_4	$B1^{4}$, 985
3648	1,3-Butadiene, 2,3-dichloro $CH_2=CCl-CCl=CH_2$	122.98	....	98	......	$1.1829^{20/4}$	1.4890^{20}	chl	$B1^{4}$, 986
3649	1,3-Butadiene, 2,3-dimethyl *(cis,cis)* or Biisopropenyl $CH_2=C(CH_3-C)CH_3=CH_2$	82.15	....	68-78	−76	$1.4394^{20/4}$	....	....	$B1^{4}$, 1023
3650	Butadiene dioxide $CH_2CHCHCH_2$ (with two epoxide O bridges)	86.09	....	144	4	1.113^{20}	1.435^{20}	w, al	$B19^{4}$, 111
3651	1,3-Butadiene, 1,4-diphenyl *(cis,cis)* or *cis,cis*-Bistyryl $C_6H_5CH=CH-CH=CHC_6H_5$	206.29	lf or nd (al or MeOH)	....	70.5	0.9697^{100}	1.6183^{100}	al, eth, bz, chl, peth	$B5^{3}$, 2159

No.	Name, Synonyms, and Formula	Mol. wt.	Color, crystalline form, specific rotation and λ_{max} (log ε)	b.p. °C	m.p. °C	Density	n_D	Solubility	Ref.
3652	1,3, Butadiene, 1,4-diphenyl *(trans,trans)* $C_6H_5CH{=}CH{-}CH{=}CHC_6H_5$	206.29	lf (al or aa)	350^{720}	152.5	....	....	al, eth, bz, chl, peth	B5³, 2159
3653	1,3-Butadiene, 1-ethoxy $C_2H_5OCH{=}CH{-}CH{=}CH_2$	98.15	....	109-12	......	$0.8154^{20/4}$	1.4529^{20}	al, eth, ace, bz, chl	B1⁴, 2222
3654	1,3-Butadiene, 2-ethoxy $CH_2{=}C(OC_2H_5){-}CH{=}CH_2$	98.14	....	94-5	......	$0.8177^{20/4}$	1.4400^{20}	al, eth, ace, bz	B1⁴, 2223
3655	1,3-Butadiene, 2-fluoro or Fluoroprene $CH_2{=}CF{-}CH{=}CH_2$	72.08	....	12	......	$0.843^{4/4}$	1.4004	....	B1⁴, 982
3656	1,3-Butadiene, hexachloro $CCl_2{=}CCl{-}CCl{=}CCl_2$	260.76	....	215, 101^{20}	−21	1.5542^{20}	....	al, eth	B1¹, 955
3657	1,3-Butadiene, hexafluoro $CF_2{=}CF{-}CF{=}CF_2$	162.03	....	6	−132	$1.553^{20/4}$	1.378^{-20}	....	B1⁴, 983
3658	1,3-Butadiene, 2-iodo or Iodoprene $CH_2{=}CI{-}CH{=}CH_2$	179.99	....	111-3	......	$1.7278^{20/4}$	1.5616	....	B1³, 956
3659	1,3-Butadiene, 1-methoxy $CH_3OCH{=}CH{-}CH{=}CH_2$	84.12	....	91-2	......	$0.8296^{20/4}$	1.4594^{20}	w, al	B1⁴, 2221
3660	1,3-Butadiene, 2-methoxy $CH_2{=}C(OCH_3){-}CH{=}CH_2$	84.12	....	75	......	$0.8272^{20/4}$	1.4442^{20}	al, eth, ace, bz	B1⁴, 2223
3661	1,3-Butadiene, 2-methyl or Isoprene $CH_2{=}C(CH_3){-}CH{=}CH_2$	68.12	....	34	−146	$0.6810^{20/4}$	1.4219^{20}	al, eth, ace, bz	B1⁴, 1001
3662	1,3-Butadiene, pentafluoro-2-trifluoro methyl $CF_2{=}C(CF_3){-}CF{=}CF_2$	212.04	....	39	......	$1.527^{0/4}$	1.3000^{0}	eth, ace, bz	C49,2479
3663	1,3-Butadiene, 1-phenyl *(trans)* $C_6H_5CH{=}CH{-}CH{=}CH_2$	130.19	....	76^{11}	4.5	$0.9286^{20/4}$	1.6089^{25}	al, eth, ace, bz	B5⁴, 1536
3664	1,3-Butadiene, 2-phenyl $CH_2{=}C(C_6H_5){-}CH{=}CH_2$	130.19	....	$60\text{-}1^{17}$	......	$0.9266^{20/4}$	1.5489^{20}	eth, bz	B5⁴, 1539
3665	1,3-Butadiene, 1,2,3,4-tetrachloro (liquid) $ClCH{=}CCl{-}CCl{=}CHCl_2$	191.87	....	188, 67^{10}	−4	$1.516^{15/15}$	1.5455^{20}	al, eth, ace, bz	B1⁴, 987
3666	1,3-Butadiene, 1,2,3,4-tetrachloro (solid) $ClCH{=}CCl{-}CCl{=}CHCl$	191.87	....	....	52	1.4961^{20}	1.5438^{20}	al, eth, ace, bz, chl	B1⁴, 987
3667	1,3-Butadiene, 1,2,3-trichloro $ClCH{=}CCl{-}CCl{=}CH_2$	157.43	....	$33\text{-}4^{7}$	......	$1.4060^{20/4}$	1.5262^{20}	eth, chl	B1³, 954
3668	Butadiyne or Biacetylene $CH{\equiv}C{-}C{\equiv}CH$	50.06	....	10.3	−36.4	$0.7364^{0/4}$	1.4189^{5}	al, eth, ace, chl	B1⁴, 1116
3669	Butadiyne, 1,4-dichloro $ClC{\equiv}C{-}C{\equiv}CCl$	118.95	nd	....	1-3	....	....	chl	B1³, 1057
3670	Butadiyne, 1,4-bis(1-hydroxycyclohexyl) $(C_8H_{11}O)_2$	246.35	cr (bz)	174	......	....	....	MeOH	B6¹, 5178
3671	Butanal or Butyraldehyde $CH_3CH_2CH_2CHO$	72.11	....	75.7	−99	$0.8170^{20/4}$	1.3843^{20}	w, **al, eth, ace,** bz	B1⁴, 3229
3672	Butane C_4H_{10}	58.12		−0.5	−138.4	$0.6012^{0/4}$, $0.5788^{20/4}$	1.3543^{-13}, 1.3326^{20}	al, eth, chl, w	B1⁴, 236
3673	Butane, 1-amino or *n*-Butylamine $CH_3CH_2CH_2CH_2NH_2$	73.13	....	77.8	−49.1	$0.7414^{20/4}$	1.4031^{20}	w, al, eth	B4⁴, 540
3674	Butane, 1-amino-3-methyl or *iso*-Pentylamine $(CH_3)_2CHCH_2CH_2NH_2$	87.16	....	$95\text{-}7^{761}$	......	$0.7505^{20/4}$	1.4083^{20}	**w, al, eth,** ace, chl	B4⁴, 696
3675	Butane, 2-amino or *sec*-Butylamine *(d)* $CH_3CH_2CH(NH_2)CH_3$	73.13	$[\alpha]^{20}_{D}$ + 7.4 (w)	63	−104.5	$0.724^{20/4}$	1.344^{20}	w, **al, eth,** ace, chl	B4⁴, 617
3676	Butane, 2-amino *(dl)* or *sec*-Butylamine $CH_3CH_2CH(NH_2)CH_3$	73.13	....	63.5^{764}	<-72	$0.7246^{20/4}$	1.3932^{20}	w, **al, eth,** ace, chl	B4⁴, 617
3677	Butane, 2-amino *(l)* or *sec*-Butylamine $CH_3CH_2CH(NH_2)CH_3$	73.13	$[\alpha]^{20}_{D}$ -7.4 (w, c=4.7)	63	......	$0.7205^{20/4}$	....	w, al, eth, ace, chl	B4⁴, 617
3678	Butane, 2-amino-2,3-dimethyl $(CH_3)_2CH{-}C(NH_2)(CH_3)_2$	101.19	....	104-5	......	$0.7683^{0/4}$	1.4096^{17}	....	B4⁴, 733
3679	Butane, 2-amino-2-methyl $CH_3CH_2C(CH_3)_2NH_2$	87.16	....	77	-105	$0.731^{25/4}$	1.3954^{25}	**w, al, eth,** ace	B4⁴, 694
3680	Butane, 2-amino-3-methyl $(CH_3)_2CH{-}CH(CH_3)NH_2$	87.16	....	84-7	......	0.7574^{19}	1.4096^{18}	w, al	B4⁴, 695
3681	Butane, 3-amino-2,2 dimethyl $CH_3CH(NH_2){-}C(CH_3)_2CH_3$	101.19	....	102	-20	....	....	w	B4⁴, 730
3682	Butane, 3-amino 2,2-dimethyl, hydrochloride $(CH_3)_3CCH(NH_2)CH_3.HCl$	137.65	nd	sub 245	300-1 (cor)	....	....	w	B4⁴, 730
3683	Butane, 1-bromo $CH_3CH_2CH_2CH_2Br$	137.02	....	101.6, 18.8^{30}	-112.4	$1.2758^{20/4}$	1.4401^{20}	**al, eth, ace,** chl	B1⁴, 258
3684	Butane, 1-bromo-4-chloro $BrCH_2CH_2CH_2CH_2Cl$	171.46	....	$174\text{-}5^{756}$, $63\text{-}4^{10}$	......	$1.488^{20/4}$	1.4885^{20}	al, eth, chl	B1⁴, 264

PHYSICAL CONSTANTS OF ORGANIC COMPOUNDS (Continued)

No.	Name, Synonyms, and Formula	Mol. wt.	Color, crystalline form, specific rotation and λ_{max} (log ε)	b.p. °C	m.p. °C	Density	n_D	Solubility	Ref.
3685	Butane, 1-bromo-3,3-dimethyl $BrCH_2CH_2C(CH_3)_3$	165.07		$138, 54^{40}$		$1.556^{20/4}$	1.4440^{20}	al, eth, chl	$B1^3$, 409
3686	Butane, 1-bromo-2,4-diphenyl *(d)* $C_6H_5CH_2CH_2CH(C_6H_5)CH_2Br$	289.21	$[\alpha]^{20}$ 16.8 (chl, c=10)	$122\text{-}30^{0.02}$			1.5812^{25}		
3687	Butane, 1-bromo-2,4-diphenyl *(dl)* $C_6H_5CH_2CH_2CH(C_6H_5)CH_2Br$	289.21		$126\text{-}8^{0.01}$			1.5812^{25}		
3688	Butane, 1-bromo-4-fluoro $Br(CH_2)_4F$	155.01		$134\text{-}5^{740}$			1.4370^{25}	al, eth	$B1^4$, 263
3689	Butane, 1-bromo-2-methyl *(d)* or *act*-Amylbromide $BrCH_2CH(CH_3)CH_2CH_3$	151.05	$[\alpha]^{20}$ + 3.68	121.6		$1.2234^{20/4}$	1.4451^{20}	al, eth, chl	$B1^4$, 327
3690	Butane, 1-bromo-3-methyl *(d)* $BrCH_2CH(CH_3)CH_2CH_3$	151.05		$120.5, 12.3^{10}$		$1.2205^{20/4}$	1.4452^{20}	al, eth, chl	$B1^4$, 327
3691	Butane, 1-bromo-3-methyl or Isoamylbromide $BrCH_2CH_2CH(CH_3)_2$	151.05		$120.4, 12.3^{10}$	-112	$1.2071^{20/4}$	1.4420^{20}	al, eth, chl	$B1^4$, 328
3692	Butane, 2-bromo *(dl)* or *sec*-butylbromide $CH_3CH_2CHBrCH_3$	137.02		91.2	-111.9	$1.2585^{20/4}$	1.4366^{20}	**eth, ace,** chl	$B1^4$, 261
3693	Butane, 2-bromo *(l)* or *sec*-butylbromide $CH_3CH_2CHBrCH_3$	137.02	$[\alpha]^{22/}{}_D$ -23.13 (undil)	90-1		$1.2536^{25/4}$	1.4359^{19}	al, **eth, ace,** chl	$B1^4$, 261
3694	Butane, 2-bromo-1-chloro $ClCH_2CHBrCH_2CH_3$	171.46		$146\text{-}7^{758}$		$1.468^{20/4}$	1.4880^{20}	al, eth, bz, chl	$B1^4$, 264
3695	Butane, 2-bromo-2,3-dimethyl $CH_3CHBr(CH_3)CH(CH_3)_2$	165.07		$132\text{-}3^{742}, 87^{180}$	24-5	1.1772^{10}	1.4517	eth, chl	$B1^4$, 374
3696	Butane, 2-bromo-2-methyl $(CH_3)_2CBrCH_2CH_3$	151.05		108^{765}		$1.198^{18}f^{5/15}$	1.44207		$B1^4$, 327
3697	Butane, 1,4-bis(dicarbethoxy amino) $C_2H_5O_2CNH\text{-}(CH_2)_4NHCO_2C_2H_5$	232.28	nd (lig)		85-6			al, eth, bz, chl	$B4^4$, 1292
3698	Butane, 1-chloro or *n*-Butylchloride $CH_3CH_2CH_2CH_2Cl$	92.57		78.4	-123.1	$0.8862^{20/4}$	1.4021^{20}	**al, eth**	$B1^4$, 246
3699	Butane, 1-chloro-2,3-dimethyl $ClCH_2CH(CH_3CH(CH_3)_2$	120.62		116^{735}		1.4200^{20}		al, eth, chl	$B1^4$, 373
3700	Butane, 1-chloro-3,3-dimethyl $ClCH_2CH_2C(CH_3)_3$	120.62		$115,41^{50}$		$0.8670^{20/4}$	1.4161^{20}	al, eth, chl	$B1^4$, 369
3701	Butane, 1-chloro-4-fluoro $Cl(CH_2)_4F$	110.56		114.7		$1.0627^{25/4}$	1.4020^{25}	al, eth	$B1^4$, 249
3702	Butane, 1 chloro 2 methyl *(d)* or *act* Amyl chloride $ClCH_2CH(CH_3)CH_2CH_3$	106.60	$[\alpha]^{20/}{}_{5892}$ + 1.64	$100.5, 43^{100}$		$0.8857^{20/4}$	1.4126^{20}	al, eth	$B1^4$, 324
3703	Butane, 1-chloro-2-methyl *(dl)* $ClCH_2CH(CH_3)CH_2CH_3$	106.60		$99.9, 52.2^{50}$		$0.8818^{15/15}$	1.4102^{25}	al, eth	$B1^4$, 324
3704	Butane, 1-chloro-3-methyl or Isopentyl chloride $(CH_3)_2CHCH_2CH_2Cl$	106.60		98.5	-104.4	$0.8704^{20/4}$	1.4084^{20}	**al, eth,** chl	$B1^4$, 325
3705	Butane, 1-chloro-2,2,3,3-tetramethyl $(CH_3)_3CC(CH_3)_2CH_2Cl$	148.68		$80\text{-}1^{40}$	52-3			eth	$B1^3$, 502
3706	Butane, 2-chloro *(d)* or *sec*-Butyl chloride $CH_3CH_2CHClCH_3$	92.57		68.2	-131.3	$0.8732^{20/4}$	1.3971^{20}	**al, eth,** bz, chl	$B1^4$, 248
3707	Butane, 2-chloro *(l)* $CH_3CH_2CHClCH_3$	92.57	$[\alpha]^{20/}{}_D$ -8.48	68	-140.5	$0.8950^{0/4}$		**al, eth,** bz, chl	$B1^4$, 248
3708	Butane, 2-chloro-2,3-dimethyl $CH_3CCl(CH_3)CH(CH_3)_2$	120.62		112	-10.4	$0.8780^{20/4}$	1.4191^{20}	al, ace	$B1^4$, 373
3709	Butane, 2-chloro-2-methyl or *tert*-Amyl chloride $CH_3CH_2CCl(CH_3)_2$	106.60		85.6	-73.5	$0.8653^{20/4}$	1.4055^{20}	al, eth	$B1^4$, 324
3710	Butane, 2-chloro-3-methyl *(dl)* $CH_3CHClCH(CH_3)_2$	106.60		$92.8, 25.7^{60}$		$0.8620^{20/4}$	1.4020^{20}	al, eth	$B1^3$, 358
3711	Butane, 2-(chloromethyl)-1,3-dichloro *(dl)* $CH_3CHClCH(CH_2Cl)_2$	175.49		$79\text{-}81^{15}$		$1.2793^{15/4}$		al, eth, chl	C31, 1003
3712	Butane, 2-(chloromethyl)-1,2,3-trichloro $CH_3CHClCCl(CH_2Cl)_2$	209.93		$102\text{-}3^{13}$		$1.3977^{18/4}$	1.5012^{18}	eth, chl	$B1^3$, 362
3713	Butane, 2-chloro-2,3,3-trimethyl $(CH_3)_2CClC(CH_3)_3$	134.65		sub	136			eth	$B1^4$, 411
3714	Butane, 3-chloro-2,3-dimethyl or Pinacolyl chloride $(CH_3)_3CCHClCH_3$	120.62		$111, 7^{10}$	0.9	$0.8767^{20/4}$	1.4182^{20}	eth	$B1^4$, 373
3715	Butane, decafluoro or Perfluoro butane C_4F_{10}	238.03		3.96	-128	1.6484		bz, chl	$B1^4$, 245
3716	Butane, 1,4-diamino or Putrescine $H_2N(CH_2)_4NH_2$	88.15	lf	158-9	27-8	0.877^{25}	1.4969^{20}	w	$B4^4$, 1283
3717	Butane, 1,4-diamino, dihydrochloride $H_2N(CH_2)_4NH_2 \cdot 2HCl$	161.07	nd or lf (al or w)	sub	315d			w, al	$B4^4$, 1284

No.	Name, Synonyms, and Formula	Mol. wt.	Color, crystalline form, specific rotation and λ_{max} (log ε)	b.p. °C	m.p. °C	Density	n_D	Solubility	Ref.
3718	Butane, 1,2-dibromo $CH_3CH_2CHBrCH_2Br$	215.92		166.3	-65.4	$1.7915^{20/4}$	1.4025^{20}	eth, chl	$B1^4$, 266
3719	Butane, 1,3-dibromo $CH_3CHBrCH_2CH_2Br$	215.92		$174, 72^{20}$		1.800^{20}	1.507^{20}	eth, chl	$B1^4$, 266
3720	Butane, 1,4-dibromide or Tetramethylene dibromide $Br(CH_2)_4Br$	215.92		$197, 79^{10}$	-16.5	$1.7890^{20/4}$	1.5190^{20}	chl	$B1^4$, 267
3721	Butane, 2,3-dibromo $CH_3CH_2BrCH_2BrCH_3$	215.92		161	<-80	$1.7893^{22/4}$	1.5133^{22}	eth, chl	$B1^4$, 268
3722	Butane, 2,3-dibromo *(meso)* $CH_3CHBrCHBrCH_3$	215.92		157.3		$1.7913^{15/4}$	$1.5132^{15.4}$		$B1^4$, 268
3723	Butane, 1,1-dichloro or Butylidene dichloride $CH_3CH_2CH_2CHCl_2$	127.01		113.8		$1.0863^{20/4}$	1.4355^{20}	chl	$B1^4$, 250
3724	Butane, 1,1-dichloro-3-methyl *(dl)* $(CH_3)_2CHCH_2CHCl_2$	141.04		$130, 48\text{-}9^{40}$		1.0473^{20}	1.4344^{20}	al, eth	$B1^4$, 326
3725	Butane, 1,2-dichloro $CH_3CH_2CHClCH_2Cl$	127.01		124		$1.1116^{25/4}$	1.4450^{20}	eth, chl	$B1^4$, 250
3726	Butane, 1,2-dichloro-2-methyl $CH_3CH_2ClCl(CH_3)CH_2Cl$	141.04		$133\text{-}5, 71.5^{100}$		$1.0785^{20/4}$	$1.4432^{21.5}$	al, eth, chl	$B1^4$, 325
3727	Butane, 1,3-dichloro $CH_3CHClCH_2CH_2Cl$	127.01		134		$1.1158^{20/4}$	1.4445^{20}	eth, chl	$B1^4$, 250
3728	Butane, 1,3-dichloro-3-methyl *(d)* $(CH_3)_2CClCH_2CH_2Cl$	141.04		$145\text{-}6, 39^{10}$		$1.0654^{20/4}$	1.4455^{20}	eth, chl	$B1^4$, 326
3729	Butane, 1,4-dichloro or Tetramethylene dichloride $Cl(CH_2)_4Cl$	127.01		$153.9, 39.7^{10}$	-37.3	$1.1408^{20/4}$	1.4542^{20}	chl	$B1^4$, 250
3730	Butane, 1,4-dichloro-2-methyl $ClCH_2CH_2CH(CH_3)CH_2Cl$	141.04		$168\text{-}9, 50^{12}$		$1.1003^{25/4}$	1.4562^{21}	chl	$B1^3$, 360
3731	Butane, 2,2-dichloro $CH_3CH_2CCl_2CH_3$	127.01		104	-74		1.4295	chl	$B1^4$, 251
3732	Butane, 2,2-dichloro-3,4-dimethyl $(CH_3)_3CCCl_2CH_3$	155.07			151-2			al, eth	$B1^4$, 369
3733	Butane, 2,3-dichloro $CH_3CHClCHClCH_3$	127.01		$116, 49.5^{80}$	-80	$1.1134^{20/4}$	1.4420^{20}	chl	$B1^4$, 251
3734	Butane, 2,3-dichloro-2,3-dimethyl $(CH_3)_2CClCCl(CH_3)_2$	155.07	pr (dil al)		164			al, eth	$B1^4$, 374
3735	Butane, 2,3-dichloro-2-methyl $CH_3CHClCHClCH(CH_3)_2$	141.04		$129, 37.5^{20}$		$1.0696^{15/4}$	1.4450^{18}	al, eth	$B1^4$, 325
3736	Butane, 1,1-dicyclohexyl $CH_3CH_2CH_2CH(C_6H_{11})_2$	222.41		280.2		$0.8842^{16/0}$	1.485^{16}		$B5^4$, 359
3737	Butane, 1,2-dicyclohexyl $CH_3CH_2CH(C_6H_{11})CH_2C_6H_{11}$	222.41		276.8		$0.9084^{18/0}$	1.475^{21}	chl, eth	$B5^4$, 358
3738	Butane, 1,2,3,4-diepoxy *(dl)* or Butadiene dioxide $CH_2CHCHCH_2$ (with O bridges)	86.09		144	4	1.113^{20}	1.435^{20}	w, al	$B19^4$, 111
3739	Butane, 1,2,3,4-diepoxy *(meso)*	86.09		138^{767}	-16	$1.1157^{20/4}$	1.4330^{20}	w, al	$B19^4$, 110
3740	Butane, 1,1-diethoxy $CH_3CH_2CH_2CH(OC_2H_5)_2$	146.23		143		$0.841^{25/4}$			$B1^4$, 3232
3741	Butane, 1,4-diethoxy $C_2H_5O(CH_2)_4OC_2H_5$	146.23		$155\text{-}7^{730}$					$B1^4$, 2517
3742	Butane, 1,4-difluoro-octachloro $FCCl_2CCl_2CCl_2CCl_2F$	369.66		152.5^{20}	4-5	$1.9272^{20/4}$	1.5256^{20}		$B1^4$, 258
3743	Butane, 2,2-di(2-furyl) $CH_3CH_2C(C_4H_3O)_2CH_3$	190.24		$64\text{-}6^{1}$		$1.0330^{20/4}$	1.4970^{20}	al, eth	$B19^4$, 287
3744	Butane, 1,4-diido or Tetramethylene diiodide $I(CH_2)_4I$	309.92		$125\text{-}6^{15/}{}_{d}$	5.8	$2.349^{26/4}$	1.619^{25}		$B1^4$, 276
3745	Butane, 2,2-dimethyl or *neo*-Hexane $CH_3CH_2C(CH_3)_3$	86.18		49.7	-99.9	$0.6485^{20/4}$	1.3688^{20}	al, eth, ace, bz	$B1^4$, 367
3746	Butane, 2,3-dimethyl $(CH_3)_2CH\text{-}CH(CH_3)_2$	86.18		58	-128.5	$0.6616^{20/4}$	1.3750^{20}	al, eth, ace, bz	$B1^4$, 371
3747	Butane, 2,3-dimethyl-2,3-epoxy $(CH_3)_2C\text{-}C(CH_3)_2$ (with O bridge)	100.16		$90\text{-}4^{745}$		$0.8156^{16/4}$	$1.3984^{16.4/}{}_{D}$	w	$B17^4$, 89
3748	Butane, 1,4-bis (dimethylamino) or Tetramethyl putrescine $(CH_3)_2N(CH_2)_4N(CH_3)_2$	144.26		$168, 78\text{-}80^{28}$		0.7942^{15}	1.4621^{25}	w, al, eth	$B4^4$, 1284
3749	Butane, 1,4-dinitro $O_2N(CH_2)_4NO_2$	148.12	pl (al)	$176\text{-}8^{13}$	33-4			eth, bz	$B1^4$, 280
3750	Butane, 2,3-dinitro *(dl)* $CH_3CH(NO_2)CH(NO_2)CH_3$	148.12			48-9			eth	$B1^4$, 280

No.	Name, Synonyms, and Formula	Mol. wt.	Color, crystalline form, specific rotation and λ_{max} (log ε)	b.p. °C	m.p. °C	Density	n_D	Solubility	Ref.
3751	Butane, 2,3-dinitro *(meso)* $CH_3CH(NO_2)CH(NO_2)CH_3$	148.12	pr (eth)	76-7^{1}	......	....	....	eth	B1^{4}, 280
3752	Butane, 1,1-diphenyl $(C_6H_5)_2CH\text{-}CH_2CH_2CH_3$	210.32	....	286-8, 161-3^{20}	27	0.9928$^{20/4}$	1.5664^{20}	al, eth, bz, chl	B5^{4}, 1944
3753	Butane, 1,2-diphenyl $C_6H_5CH_2CH(C_6H_5)CH_2CH_3$	210.32	....	289^{750}, 152^{11}	......	0.9777^{20}	1.5554^{20}	al, eth, bz, chl	B5^{4}, 1939
3754	Butane, 1,3-diphenyl *(l)* $C_6H_5CH_2CH_2CH(C_6H_5)CH_3$	210.32	$[\alpha]^{20/}{}_D$ -15.6 (chl, c=10)	68-70$^{0.01}$	......	....	1.5503^{25}	chl	B5^{4}, 1938
3755	Butane, 1,4-diphenyl $C_6H_5(CH_2)_4C_6H_5$	210.32	....	317, 108-9$^{0.1}$	52.5	....	....	al, eth, chl	B5^{4}, 1937
3756	1,4-Butane dithiol $HS(CH_2)_4SH$	122.24	....	195-6	-53.9	1.0621$^{0/4}$	1.5290^{20}	al	B1^{4}, 2523
3757	Butane, 1,2-epoxy or 1,2-Butylene oxide $CH_3CH_2CHCH_2$	72.11	....	63.3	......	0.837$^{17/4}$	1.3851^{20}	al, **eth**, ace	B17^{4}, 45
3758	Butane, 2,3-epoxy *(cis)* $CH_3CHCHCH_3$	72.11	....	59.7^{742}	-80	0.8226$^{25/4}$	1.3802^{20}	eth, ace, bz	B17^{4}, 48
3759	Butane, 2,3-epoxy *(trans)* $CH_3CHCHCH_3$	72.11	....	56-7	-85	0.8010$^{25/4}$	1.3736^{20}	eth, ace, bz	B17^{4}, 49
3760	Butane, 2,2-bis(ethylsulfonyl) $CH_3CH_2C(SO_3C_2H_5)_2CH_3$	274.35	pl (w)	d	76	1.199$^{85/4}$	....	al, eth, bz, peth, lig	B1^{3}, 2790
3761	Butanc, 1-fluoro or *n*-Butyl fluoride $CH_3CH_2CH_2CH_2F$	76.11	....	32.5	-134	6.7789$^{20/4}$	1.3396^{20}	al	B1^{4}, 244
3762	Butane, 1,1,2,2,3,3,4,4,-heptachloro $Cl_2CHCHClCCl_2CHCl_2$	299.24	....	137.5$^{13.5}$	......	1.742$^{20/20}$	1.5407^{20}	ace, CCl_4	B1^{4}, 257
3763	Butane, 1,1,2,3,4,4-hexachloro (liquid) $Cl_2CHCHClCHClCHCl_2$	264.79	....	111^{10}	......	1.6460$^{20/4}$	1.5258^{20}	bz, CCl_4	B1^{3}, 288
3764	Butane, 1,1,2,3,4,4-hexachloro (solid) $Cl_2CHCHClCHClCHCl_2$	264.79	nd (al), pr (bz, aa)	....	109-10	....	....	bz, CCl_4	B1^{3}, 288
3765	Butane, 1,1-bis(4-hydroxyphenyl) $(4\text{-}HOC_6H_4)_2CHC_3H_7$	242.32	nd (to)	270^{12}	137	....	....	al, bz, to	B6^{3}, 5476
3766	Butane, 2,2-bis(4-hydroxyphenyl) $(4\text{-}HOC_6H_4)_2CH(CH_3)C_2H_5$	242.32	nd or pr (w)	250-3	133-4	....	....	al, eth, ace, bz	B6^{3}, 5477
3767	Butane, 1-iodo or *n*-Butyl iodide $CH_3CH_2CH_2CH_2I$	184.02	....	130.5, 19.2^{10}	-103	1.6154$^{20/4}$	1.5001^{20}	al, eth, chl	B1^{4}, 271
3768	Butane, 1-iodo-2-methyl *(d)* or *act*-Amyliodide $CH_3CH_2CH(CH_3)CH_2I$	198.05	$[\alpha]^{15}{}_D$ + 5.78 (undil)	148^{20}, 47.1^{20}	......	1.5253$^{20/4}$	1.4977^{20}	al, eth	B1^{3}, 366
3769	Butane, 1-iodo-2-methyl *(dl)* $CH_3CH_2CH(CH_3)CH_2I$	198.05	....	144-7	......	....	1.497^{20}	al, eth	B1^{3}, 366
3770	Butane, 1-iodo-3-methyl or Isopentyl iodide $(CH_3)_2CHCH_2CH_2I$	198.05	....	147	......	1.5118$^{20/4}$	1.4939^{20}	**al, eth**	B1^{4}, 331
3771	Butane, 2-iodo *(dl)* or *sec*-Butyl iodide $CH_3CH_2CHICH_3$	184.02	....	120, 33^{45}	-104.2	1.5920$^{20/4}$	1.4991^{20}	**al, eth**, chl	B1^{4}, 272
3772	Butane, 2-iodo *(l)* $CH_3CH_2CHICH_3$	184.02	$[\alpha]_D$ -12.15 (al, c=20)	117-8	......	1.585$^{20/4}$	1.4945^{19}	**al, eth**, chl	B1^{4}, 272
3773	Butane, 2-iodo-2-methyl or *tert*-Amyliodide $(CH_3)_2CICH_2CH_3$	198.05	....	124.5	......	1.4937$^{20/4}$	1.4981^{20}	**al, eth**	B1^{4}, 331
3774	Butane, 2-iodo-3-methyl $(CH_3)_2CHCHICH_3$	198.05	....	138-9	......	....	....	al, eth, ace	B1^{3}, 367
3775	Butane, 1-isocyano or Butyl carbylamine $CH_3CH_2CH_2CH_2NC$	83.13	....	118	......	1.4061^{20}	....	**al, eth**	B4^{4}, 562
3776	Butane, 1-isocyano-3-methyl or Isoamylcarbylamine $(CH_3)_2CHCH_2CH_2NC$	97.16	....	140	......	0.806$^{20/4}$	1.406^{20}	**al, eth**	B4, 184
3777	Butane, 2-methyl or Isopentane $CH_3CH_2CH(CH_3)_2$	72.15	....	27.8	-159.9	0.6201$^{20/4}$	1.3537^{20}	al, eth	B1^{4}, 320
3778	Butane, 3-methyl-2-phenyl $CH_3CH(C_6H_5)CH(CH_3)_2$	148.25	....	186-8	......	0.8672$^{16/4}$	1.4972^{16}	....	B5^{4}, 1118
3779	Butane, 2-methyl-1,2,3-trichloro $CH_3CHClCCl(CH_3)CH_2Cl$	175.49	....	183-5^{762}, 65.5^{11}	......	1.2527$^{20/4}$	....	chl	B1^{3}, 361
3780	Butane, 2-methyl-2,3,3-trichloro $(CH_3)_2CClCCl_2CH_3$	175.49	....	182-3	......	1.215$^{15/4}$	1.472^{21}	chl, aa	B1^{3}, 361
3781	Butane, 1-nitro $CH_3CH_2CH_2CH_2NO_2$	103.12	....	153	......	0.9710$^{20/4}$	1.4303^{20}	**al, eth**	B1^{4}, 277
3782	Butane, 2-nitro *(dl)* $CH_3CH_2CH(NO_2)CH_3$	103.12	....	140	-132	0.9854$^{17/4}$	1.4044^{20}	....	B1^{4},,278
3783	Butane, 1,1,2,2,3,3,3,4,4-octachloro $Cl_2CHCCl_2CCl_2CHCl_2$	333.68	cr (al)	....	81	....	....	al, eth, ace, bz	B1^{4}, 257

No.	Name, Synonyms, and Formula	Mol. wt.	Color, crystalline form, specific rotation and λ_{max} (log ε)	b.p. °C	m.p. °C	Density	n_D	Solubility	Ref.
3784	Butane, 1,1,2,3,4-pentachloro (liquid) $ClCH_2(CHCl)_2CHCl_2$	230.35	....	95.5^{11}	......	1.561^{18}	1.5140^{18}	al, CCl_4	$B1^4$, 256
3785	Butane, 1,1,2,3,4-pentachloro (solid) $ClCH_2CHClCHClCHCl_2$	230.35	lf (al)	230, 102^{11}	49	1.539^{53}	1.5065^{53}	al, CCl_4	$B1^4$, 256
3786	Butane, 1,2,2,3,4-pentachloro $ClCH_2CHClCCl_2CH_2Cl$	230.35	....	85^{10}	......	$1.5543^{20/4}$	1.5157^{20}	ace, chl	$B1^4$, 256
3787	Butane, 1,1,4,4-tetrabromo $Br_2CHCH_2CH_2CHBr_2$	373.71	....	$138\text{-}45^{10}$	......	$2.529^{20/4}$	1.6077^{20}	eth, bz, chl	$B1^3$, 298
3788	Butane, 1,2,2,3-tetrabromo $CH_3CHBrCBr_2CH_2Br$	373.71	....	$128\text{-}30^{14}$	-2	$2.5100^{20/4}$	1.6070^{20}	ace	$B1^3$, 298
3789	Butane, 1,2,2,4-tetrabromo $BrCH_2CH_2CBr_2CH_2Br$	373.71	nd (lig)	....	72-3	....	....	lig	$B1^3$, 298
3790	Butane, 1,2,3,4-tetrabromo *(dl)* $BrCH_2CHBrCHBrCH_2Br$	373.71	lf (peth)	....	40-1	....	....	al, eth, ace, lig	$B1^4$, 271
3791	Butane, 1,2,3,4-tetrabromo *(meso)* $BrCH_2CHBrCHBrCH_2Br$	373.71	nd (al or lig)	$180\text{-}1^{60}$	118-9	....	....	al, ace, chl	$B1^4$, 271
3792	Butane, 2,2,3,3-tetrabromo $CH_3CBr_2CBr_2CH_3$	373.71	lf (lig), pr (eth-lig)	....	243	....	....	eth, ace, bz, chl lig	$B1^3$, 298
3793	Butane, 1,1,1,2-tetrachloro $CH_3CH_2CHClCCl_3$	195.90	....	$134\text{-}5^{742}$	......	$1.3932^{20/20}$	1.4920^{25}	ace, chl	$B1^4$, 254
3794	Butane, 1,2,2,3-tetrachloro $CH_3CHClCCl_2CH_2Cl$	195.90	....	182, 85^{10}	-48	$1.4276^{18/4}$	1.491^{20}	ace, chl	$B1^3$, 286
3795	Butane, 1,2,3,3-tetrachloro $CH_3CCl_2CHClCH_2Cl$	195.90	....	90^{32}, $55\text{-}7^{10}$	......	$1.4204^{20/4}$	1.4958^{20}	eth, ace, chl	$B1^4$, 254
3796	Butane, 2,2,3,3-tetramethyl $CH_3C(CH_3)_2C(CH_3)_2CH_3$	114.23	lf (eth)	106.5, 13.1^{10}	100.7	0.8242^{20}, 0.6485^{110}	1.4695^{20}	eth	$B1^4$, 447
3797	Butane, 1,1,2-tribromo $CH_3CH_2CHBrCHBr_2$	294.81	....	216.2, 98^{14}	......	$2.1836^{20/4}$	1.5626^{17}	al, eth, chl	$B1^2$, 84
3798	Butane, 1,2,2-tribromo $CH_3CH_2CBr_2CH_2Br$	294.81	....	213.8, 90.1^{14}	......	$2.1692^{20/4}$	1.5624^{10}	al, eth, chl	$B1^2$, 85
3799	Butane, 1,2,3-tribromo $CH_3CHBrCHBrCH_2Br$	294.81	....	220, 97^{10}	-19	$2.1908^{20/4}$	1.5680^{20}	al, eth, chl	$B1^4$, 271
3800	Butane, 1,2,4-tribromo $BrCH_2CH_2CHBrCH_2Br$	294.81	....	215, 93^{10}	-18	$2.170^{20/4}$	1.5608^{20}	al, eth, chl	$B1^4$, 271
3801	Butane, 1,3,3-tribromo $CH_3CBr_2CH_2CH_2Br$	294.81	....	200-5, 70^8	......	$2.1446^{20/4}$	1.5564^{20}	al, eth, chl	$B1^3$, 298
3802	Butane, 2,2,3-tribromo $CH_3CHBrCBr_2CH_3$	294.81	....	200, 86^{10}	1.8	$2.1724^{20/4}$	1.5602^{20}	al, eth, chl	$B1^3$, 298
3803	Butane, 1,1,3-trichloro $CH_3CHClCH_2CHCl_2$	161.46	....	152^{753}	......	1.317^{15}	1.4600^{15}	al, eth, chl	$B1^4$, 252
3804	Butane, 1,2,3-trichloro $CH_3CHClCHClCH_2Cl$	161.46	....	$165\text{-}8^{725}$, 63^{28}	......	$1.3164^{20/4}$	1.4790^{20}	al, eth, chl	$B1^3$, 285
3805	Butane, 1,2,4-trichloro $ClCH_2CHClCH_2CH_2Cl$	161.46	....	$61\text{-}2.5^{10}$	......	1.3175^{20}	1.4820^{20}	bz, chl	$B1^4$, 253
3806	Butane, 2,2,3-trichloro $CH_3CHClCCl_2CH_3$	161.46	....	143-5	......	$1.2699^{20/4}$	1.4645^{20}	chl	$B1^3$, 285
3807	Butane, 2,2,3-trimethyl or Triptane $(CH_3)_2CHC(CH_3)_3$	100.20	....	80.9	-24.2	$0.6901^{20/4}$	1.3864^{20}	al, eth, ace, bz	$B1^4$, 410
3808	Isobutane or 2-Methylpropane $(CH_3)_2CHCH_3$	58.12	....	-11.7	-159.4	0.549^{30}	....	al, eth, chl	$B1^4$, 282
3809	Isobutane, 1,2-dibromo or 1,2-dibromo-2-methylpropane $(CH_3)_2CBrCH_2Br$	215.92	....	149-51, 61^{40}	9-12	$1.759^{20/4}$	1.509	al, eth, bz, chl	$B1^4$, 298
3810	Isobutane, 1,1-dichloro or 1,1-dichloro-2-methyl propane $(CH_3)_2CHCHCl_2$	127.01	....	105-6	......	$1.0111^{12/12}$	1.4330^{25}	al, eth, bz, chl	$B1^4$, 292
3811	Isobutane, 1,2-dichloro or 1,2-dichloro-2-methyl propane $(CH_3)_2CClCH_2Cl$	127.01	....	108, $38\text{-}9^{70}$	......	$1.093^{20/4}$	1.4370^{20}	al, eth, ace, bz	$B1^4$, 292
3812	Isobutane, 1,3-dichloro or 1,3-dichloro-2-methyl propane $CH_3CH(CH_2Cl)_2$	127.01	....	134.6, 60^{49}	......	$1.1325^{25/4}$	1.4488^{25}	al, eth, bz	$B1^4$, 293
3813	Isobutane, 1,2-epoxy or Isobutylene oxide $(CH_3)_2C\text{-}CH_2$ (O bridge)	72.11	....	52	......	0.8650^0	1.3712^{22}	al, eth	$B17^4$, 46
3814	Isobutane, 1,2-epoxy-3-chloro $ClCH_2C(CH_3)CH_2$ (O bridge)	106.55	....	12^2, 51^{55}	......	$1.1011^{20/4}$	1.4340^{20}	eth, w	$B17^4$, 47
3815	Isobutane, 1-nitro $(CH_3)_2CHCH_2NO_2$	103.12	....	140.5, $61\text{-}2^{45}$	......	$0.9625^{25/25}$	1.4066^{20}	al, eth	$B1^4$, 301
3816	Isobutane, 2-nitro $(CH_3)_3CNO_2$	103.12	....	127.2	26.23	0.9501^{28}	1.4015^{20}	al, eth, ace, bz	$B1^4$, 301

PHYSICAL CONSTANTS OF ORGANIC COMPOUNDS (Continued)

No.	Name, Synonyms, and Formula	Mol. wt.	Color, crystalline form, specific rotation and λ_{max} (log ε)	b.p. °C	m.p. °C	Density	n_D	Solubility	Ref.
3817	Isobutane, 1,1,1,2,3-pentachloro $ClCH_2CCl(CH_3)CCl_3$	230.35	(al)	215^{757}, $90\text{-}3^{10}$	73.5	$1.5686^{25/4}$	1.5165^{25}	chl	$B1^4$, 294
3818	Isobutane, 1,1,1,2-tetrabromo $(CH_3)_2CBrCBr_3$	373.71	lf	...	217	...	...	al, eth	B1, 128
3819	Isobutane, 1,1,2,3-tetrabromo $BrCH_2CBr(CH_3)CHBr_2$	373.71	...	134^{11}	...	2.4545^{20}	1.5990^{20}	chl	$B1^3$, 325
3820	Isobutane, 1,1,1,2-tetrachloro $(CH_3)_2CClCCl_3$	195.90	cr (al)	$192^{117.5}$ sub	178-9	...	...	al, eth, chl	$B1^4$, 293
3821	Isobutane, 1,1,2,3-tetrachloro $ClCH_2CCl(CH_3)CHCl_2$	195.90	...	190-1, 69^{12}	-46	$1.4393^{25/4}$	1.4963^{20}	eth, bz, chl	$B1^4$, 294
3822	Isobutane, 1,1,2-tribromo $(CH_3)_2CBrCHBr_2$	294.81	...	208-15d, 96^{14}	...	$2.0169^{20/4}$	...	eth, chl	$B1^3$, 325
3823	Isobutane, 1,2,3-tribromo $(BrCH_2)_2CBrCH_3$	294.81	...	88.5^{9}	...	$2.1750^{20/4}$	1.5652^{20}	eth, chl	$B1^4$, 299
3824	Isobutane, 1,1,2-trichloro $(CH_3)_2CClCHCl_2$	161.46	...	145-6, $46\text{-}7^{18}$	6	$1.2588^{20/4}$	1.4666^{20}	eth, chl	$B1^4$, 293
3825	Isobutane, 1,2,3-trichloro $(ClCH_2)_2CClCH_3$	161.46	...	162-3, 81^{50}	...	$1.3012^{25/4}$	1.4765^{20}	chl	$B1^4$, 293
3826	1,2-Butanediol-*(d)* $CH_3CH_2CHOHCH_2OH$	90.12	$[\alpha]^{20}/_D$ + 14.5 (al, c=6)	192.4, $68^{0.4}$	...	$1.0059^{17.5/0}$	1.4375^{20}	w, al, ace	$B1^4$, 2507
3827	1,2-Butanediol *(dl)* or α-Butylene glycol $CH_3CH_2CHOHCH_2OH$	90.12	...	190.5, 96.5^{10}	...	$1.0024^{20/4}$	1.4378^{20}	w, al, ace	$B1^4$, 2507
3828	1,2-Butanediol *(l)* $CH_3CH_2CHOHCH_2OH$	90.12	$[\alpha]^{22}/_D$ -7.4 (al, c=4)	$94\text{-}6^{12}$	...	...	...	w, al, ace	$B1^4$, 2507
3829	1,3-Butanediol *(d)* or β-Butyleneglycol $CH_3CH(OH)CH_2CH_2OH$	90.12	$[\alpha]^{22}/_D$ + 18.5 (al, c=4)	204, $60\text{-}5^{0.8}$	...	$1.0053^{20/4}$	1.4418^{20}	w, al	$B1^4$, 2508
3830	1,3-Butanediol *(dl)* $CH_3CH(OH)CH_2CH_2OH$	90.12	...	207.5, $103\text{-}4^{8}$	...	$1.0053^{20/4}$	1.4410^{20}	w, al	$B1^4$, 2508
3831	1,3-Butanediol *(l)* $CH_3CH(OH)CH_2CH_2OH$	90.12	$[\alpha]^{25}/_D$ -18.8 (al, c=4)	$107\text{-}10^{23}$	...	$1.005^{20/4}$	...	w, al	$B1^4$, 2508
3832	1,3-Butanediol sulfite $C_4H_8O_3S$	136.17	...	185, $76\text{-}7^{17}$	5	$1.2352^{20/4}$	1.4661^{20}	al, eth, ace, bz	$B1^4$, 2510
3833	1,4-Butanediol or Tetramethylene glycol $HOCH_2CH_2CH_2CH_2OH$	90.12	...	235, 120^{10}	20.1	$1.0171^{20/4}$	1.4460^{20}	w, al	$B1^4$, 2515
3834	1,4-Butanediol diacetate $CH_3CO_2(CH_2)_4CO_2CH_3$	174.20	...	229^{765}	12	1.0479^{15}	1.4251^{15}	...	$B2^4$, 224
3835	1,4-Butanediol dibenzoate $C_6H_5CO_2(CH_2)_4CO_2C_6H_5$	298.34	...	...	81-2	...	...	eth	$B9^3$, 540
3836	1,4-Butanediol, 2-hexyl $HOCH_2CH_2CH(C_6H_{13})CH_2OH$	174.28	...	$118^{0.05}$	...	...	...	...	...
3837	1,4-Butanediol, 2-methyl $HOCH_2CH_2CH(CH_3)CH_2OH$	104.15	$[\alpha]^{22}/_D$ + 11.65	131-3	...	$0.9929^{22/4}$	...	...	$B1^4$, 2546
3838	2,3-Butanediol *(d)* $CH_3CHOHCHOHCH_3$	90.12	$[\alpha]^{25}/_D$ + 12.5 (undil)	180-2	34 (anh), 16.86 (+sw)	0.9872^{25}	1.4306^{25}	**w, al**, eth, ace	$B1^4$, 2524
3839	2,3-Butyldiol *(dl)* $CH_3CHOHCHOHCH_3$	90.12	cr (i-$Pr_2$0)	182.5, 86^{10}	7.6	$1.0033^{20/4}$	1.4310^{25}	**w, al**, eth, ace	$B1^4$, 2524
3840	2,3-Butanediol *(l)* $CH_3CHOHCHOHCH_3$	90.12	$[\alpha]^{25}/_D$ -13.0 (undil)	178-81, 77.5^{10}	19.7	$0.9869^{25/4}$	1.4340^{18}	**w, al**, eth, ace	$B1^4$, 2524
3841	2,3-Butanediol *(meso)* $CH_3CHOHCHOHCH_3$	90.12	cr (i-$Pr_2$0)	181.7, 83.5^{10}	34.4	$1.0003^{20/4}$	1.4367^{20}	w, al, eth	$B1^4$, 2524
3842	2,3-Butanediol, 2,3-dimethyl or Pinacol $(CH_3)_2C(OH)C(OH)(CH_3)_2$	118.18	nd (al or eth)	174.4	43	...	...	al, eth	$B1^4$, 2575
3843	2,3-Butanediol, 2,3-dimethyl, hexahydrate or Pinacol hydrate $(CH_3)_2C(OH)C(OH)(CH_3)_2 \cdot 6H_2O$	226.27	pl (w + 6)	...	47	0.967^{15}	...	al, eth, w	$B1^4$, 2575
3844	2,3-Butanediol, 2,3-diphenyl or Acetophenone Pinacol $C_6H_5COH(CH_3)COH(CH_3)C_6H_5$	242.32	pr (dil al)	...	121-2	...	...	al, eth	$B6^3$, 5474
3845	1,2-Butanediol, 3-methyl or α-Isopentylene glycol $(CH_3)_2CHCHOHCH_2OH$	104.15	...	206, $81\text{-}3^{5}$	...	$0.9987^{0/4}$	...	al, eth	$B1^4$, 2549
3846	1,3-Butanediol, 3-methyl or α-Isopentylene glycol $(CH_3)_2C(OH)CH_2CH_2OH$	104.15	...	202-3, 108^{16}	...	$0.9448^{20/4}$	1.4452^{20}	w, al	$B1^4$, 2549
3847	2,3-Butanediol, 2-methyl or β-Isopentyl glycol $CH_3CHOHC(CH_3)OHCH_3$	104.15	...	175, 68^{5}	...	$0.9920^{25/4}$	1.4375^{20}	w, al, eth	$B1^4$, 2547

No.	Name, Synonyms, and Formula	Mol. wt.	Color, crystalline form, specific rotation and λ_{max} (log ε)	b.p. °C	m.p. °C	Density	n_D	Solubility	Ref.
3848	2,3-Butanedione or Biacetyl, Dimethyl glycol $CH_3COCOCH_3$	86.09	....	88	-2.4	$0.9808^{18.5/4}$	1.3951^{20}	w, **al**, **eth**, ace, bz	$B1^4$, 3644
3849	1,4-Butanedithiol $HS(CH_2)_4SH$	122.24	....	195-6, $110\text{-}12^{50}$	-53.9	$1.0621^{0/4}$	1.5290^{20}	al	$B1^4$, 2523
3850	1,2,3,4-Butane tetracarboxylic acid *(dl)* $HO_2CCH_2CH(CO_2H)CH(CO_2H)CH_2CO_2H$	234.16	lf (w), cr (ace)	....	236-7	....	....	w, al	$B2^4$, 2419
3851	1-Butanethiol or *n*-Butyl mercaptan $CH_3CH_2CH_2CH_2SH$	90.18	....	98.4	-115.7	$0.8337^{20/4}$	1.4440^{20}	al, eth	$B1^4$, 1555
3852	1-Butanethiol, 2-methyl-*(d)* or *act*-Amyl mercaptan $CH_3CH_2CH(CH_3)CH_2SH$	104.21	$[\alpha]^{23/}_D$ + 3.21	118.2	......	$0.8420^{20/4}$	1.4440^{20}	....	$B1^4$, 1668
3853	1-Butanethiol, 3-methyl or Isopentyl mercaptan $(CH_3)_2CHCH_2CH_2SH$	104.21	....	118	......	$0.8350^{20/4}$	1.4418^{20}	**al, eth**	$B1^4$, 1688
3854	2-Butanethiol, *(d)* or *sec*-Butyl mercaptan $CH_3CH_2CHSHCH_3$	90.18	$[\alpha]^{20/}_D$ + 15.7	85-95, 37.4^{134}	......	$0.8299^{20/4}$	1.43385^{25}	al, eth, bz, peth	$B1^3$, 1549
3855	2-Butanethiol, *(dl)* or *sec*-Butyl mercaptan $CH_3CH_2CHSHCH_3$	90.18	$[\alpha]^{17/}_D$ - 17.35	85	......	$0.8295^{20/4}$	1.4366^{20}	al, eth, bz, peth	$B1^4$, 1584
3856	2-Butanethiol, *(l)* $CH_3CH_2CHSHCH_3$	90.18	$[\alpha]^{17/}_D$ - 17.35	83-4	......	$0.8300^{17/4}$	....	al, eth, bz, peth	$B1^3$, 1549
3857	2-Butanethiol, 2-methyl $CH_3CH_2C(SH)(CH_3)_2$	104.21	....	99-100	......	....	1.4385^{20}	....	$B1^4$, 1674
3858	1,2,3-Butanetriol or 1-Methyl glycerol $CH_3CHOHCHOHCH_2OH$	106.12	....	170^{20}	......	....	1.4462^{20}	w, al	$B1^4$, 2774
3859	1,2,4-Butanetriol $HOCH_2CH_2CH(OH)CH_2OH$	106.12	....	$172\text{-}4^{12}$	......	1.018^{20}	1.4688^{20}	w, al	$B1^4$, 2775
3860	1-Butanol or *n*-Butyl alcohol $CH_3CH_2CH_2CH_2OH$	74.12	$[\alpha]^{20/}_D$ + 9.8 (w)	117.2	-89.5	$0.8098^{20/4}$	1.3993^{20}	w, **al**, **eth**, ace, bz	$B1^4$, 1506
3861	1-Butanol, 2-amino-*(d)* $CH_3CH_2CHNH_2CH_2OH$	89.14	$[\alpha]^{20/}_D$ + 9.8 (w)	80^{11}	......	0.947^{20}	1.4518^{20}	**w, al, eth**	$B4^4$, 1705
3862	1-Butanol, 2-amino-*(dl)* $CH_3CH_2CHNH_2CH_2OH$	89.14	....	178	-2	0.9162^{20}	1.4489^{25}	**w, al, eth**	$B4^4$, 1705
3863	1-Butanol, 2-amino-1-phenyl $CH_3CH_2CH(NH_2)CH(OH)C_6H_5$	165.24	pl (bz-eth)	....	79-80	....	....	al, bz, chl	$B13^3$, 1791
3864	1-Butanol, 3-amino $CH_3CHNH_2CH_2CH_2OH$	89.14	....	$82\text{-}5^{19}$	......	....	1.4534^{25}	w, al	$B4^4$, 1710
3865	1-Butanol, 4-amino $H_2NCH_2CH_2CH_2CH_2OH$	89.14	....	206^{776}	......	0.967^{12}	1.4625^{20}	w, al	$B4^4$, 1711
3866	1-Butanol, 2-chloro $CH_3CH_2CHClCH_2OH$	108.57	....	$74\text{-}6^{25}$	......	$1.062^{25/4}$	1.4438^{20}	al, eth	$B1^4$, 1548
3867	1-Butanol, 3-chloro $CH_3CHClCH_2CH_2OH$	108.57	....	170-80, 73^{20}	......	$1.0883^{20/4}$	1.4518^{20}	al, eth	$B1^4$, 1549
3868	1-Butanol, 4-chloro $ClCH_2CH_2CH_2CH_2OH$	108.57	....	$84\text{-}5^{16}$	......	$1.0883^{20/4}$	1.4518^{20}	al, eth	$B1^4$, 1550
3869	1-Butanol, 2,2-dimethyl $CH_3CH_2C(CH_3)_2CH_2OH$	102.18	....	136.7	<-15	$0.8283^{20/4}$	1.4208^{20}	al, eth	$B1^4$, 1726
3870	1-Butanol, 2,3-dimethyl-*(d)* $CH_3CH(CH_3)CH(CH_3)CH_2OH$	102.18	$[\alpha]^{25}_D$ + 1.9	142	......	$0.823^{25/4}$	....	al, eth, ace	$B1^4$, 1729
3871	1-Butanol, 2,3-dimethyl-*(dl)* $CH_3CH(CH_3)CH(CH_3)CH_2OH$	102.18	....	144-5	......	$0.8297^{20.5/4}$	$1.4195^{20.5}$	al, eth, ace	$B1^4$, 1729
3872	1-Butanol, 3,3-dimethyl $(CH_3)_3CCH_2CH_2OH$	102.18	....	143	-60	1.4323^{15}	....	al, eth, ace	$B1^4$, 1729
3873	1-Butanol, 2,4-diphenyl $C_6H_5CH_2CH_2CH(C_6H_5)CH_2OH$	226.32	$[\alpha]^{20}_D$ −17.5 (chl, c=10)	$145\text{-}6^{0.1}$	51-2	....	1.5686^{25}	chl	$B6^3$, 3428
3874	1-Butanol, 2-diethyl $(C_2H_5)_3ChCH_2OH$	130.23	....	146.3	<-15	$0.8326^{20/4}$	1.4220^{20}	al, eth	$B1^4$, 1725
3875	1-Butanol, 4-fluoro $FCH_2CH_2CH_2CH_2OH$	92.11	....	58^{15}	......	....	1.3942^{15}	al, eth, ace	$B1^4$, 1546
3876	1-Butanol, 2,2,3,3,4,4,4-heptafluoro $CF_3CF_2CF_2CH_2OH$	200.06	....	95	......	$1.600^{20/4}$	1.294^{20}	al, ace	$B1^4$, 1547
3877	1-Butanol, 3-methoxy $CH_3CH(OCH_3)CH_2CH_2OH$	104.15	....	160	......	0.923^{23}	1.4148^{25}	al, eth, ace	$B1^4$, 2509
3878	1-Butanol, 2-methyl-*(d)* or *act*-Amyl alcohol-*(d)* $CH_3CH_2CH(CH_3)CH_2OH$	88.15	....	128, 65.7^{50}	......	$1.8191^{20/4}$	1.4102^{20}	**al, eth**, ace	$B1^4$, 1666
3879	1-Butanol, 2-methyl-*(dl)* $CH_3CH_2CH(CH_3)CH_2OH$	88.15	....	127-8, 70^{60}	......	$0.8152^{25/4}$	1.4092^{20}	**al, eth**, ace	$B1^4$, 1666
3880	1-Butanol, 2-methyl-*(l)* $CH_3CH_2CH(CH_3)CH_2OH$	88.15	$[\alpha]^{18/}_D$ + 3.75	129	......	$0.816^{18/4}$	1.4098^{20}	**al, eth**, ace	$B1^4$, 1666

No.	Name, Synonyms, and Formula	Mol. wt.	Color, crystalline form, specific rotation and λ_{max} (log ε)	b.p. °C	m.p. °C	Density	n_D	Solubility	Ref.
3881	1-Butanol, 2-methyl-4-phenyl *(l)* $C_6H_5CH_2CH_2CH(CH_3)CH_2OH$	164.25	....	135^{11}	......	$0.9719^{20/4}$	1.5173^{16}	al **eth**, ace, bz	B6³, 1962
3882	1-Butanol, 3-methyl or *iso*-Amylalcohol $(CH_3)_2CHCH_2CH_2CH_2OH$	88.15	....	128.5^{750}	−117.2	$0.8092^{20/4}$	1.4053^{20}	**al, eth**, ace	B1⁴, 1677
3883	1-Butanol, 3-methyl-1-phenyl or α-Hydroxy isopentyl benzene $(CH_3)_2CHCH_2CH(OH)C_6H_5$	164.25	....	$235\text{-}6^{746}$, 112^{9}	......	$0.9537^{19/4}$	1.5080^{18}	al, eth, ace, bz	B6², 505
3884	1-Butanol, 3-methyl -2-phenyl $(CH_3)_2CHCH(C_6H_5)CH_2OH$	164.25	....	130^{15}	......	$0.9694^{25/4}$	1.5137^{20}	al, eth, ace, bz	B6², 506
3885	1-Butanol, 2-nitro $CH_3CH_2CH(NO_2)CH_2OH$	119.12	....	105^{10}	-47	$1.1332^{25/4}$	1.4390^{20}	w, al, eth, ace, aa	B1⁴, 1555
3886	1-Butanol, 1-phenyl $CH_3CH_2CH_2CH(C_6H_5)OH$	150.22	....	232, $113\text{-}5^{17}$	16	$0.9740^{20/4}$	1.5139^{20}	al, eth	B6³, 1845
3887	1-Butanol, 2,2,3-trichloro $CH_3CHClCCl_2CH_2OH$	177.46	pr (dil al)	199-200, $97\text{-}8^{18}$	62	....	....	al, eth, ace	B1³, 1518
3888	2-Butanol-*(d)* or *sec*-Butyl alcohol $CH_3CH_2CH(OH)CH_3$	74.12	$[\alpha]^{20}/_D$ + 13.9	99.5	......	$0.8080^{20/4}$	1.3954^{20}	**al, eth**, ace, bz	B1⁴, 1566
3889	2-Butanol-*(dl)* or *sec*-Butyl alcohol $CH_3CH_2CHOHCH_3$	74.12	....	99.5, 45.5^{60}	......	$0.8063^{20/4}$	1.3978^{20}	**al, eth**, ace, bz	B1⁴, 1567
3890	2-Butanol-*(l)* $CH_3CH_2CHOHCH_3$	74.12	$[\alpha]^{20}/_D$ + 13.9	99.5	......	$0.8070^{20/4}$	1.3975^{20}	**al, eth**, ace, bz	B1⁴, 1566
3891	2-Butanol, 3-amino $CH_3CHNH_2CHOHCH_3$	89.14	....	$159\text{-}60^{745}$, 70^{20}	18-20 (44)	$0.9299^{25/4}$	1.4502^{20}	w, al, ace	B1⁴, 1726
3892	2-Butanol, 3-bromo-*(dl)* $CH_3CHBrCHOHCH_3$	153.02	....	154, $46\text{-}50^{8}$	......	$1.4550^{20/4}$	1.4786^{20}	al, eth	B1⁴, 1580
3893	2-Butanol, 1-chloro $CH_3CH_2CHOHCH_2Cl$	108.57	....	141, 52^{15}	......	$1.068^{25/4}$	1.4400^{20}	al, eth	B1⁴, 1578
3894	2-Butanol, 1-chloro-2-methyl $CH_3CH_2C(CH_3)OHCH_2Cl$	122.59	....	150-2	......	1.0161^{20}	1.4469^{20}	al	B1⁴, 1673
3895	2-Butanol, 3-chloro $CH_3CHClCHOHCH_3$	108.57	....	138-40, $52\text{-}4^{30}$	......	$1.0669^{20/4}$	1.4432^{20}	al, eth	B1³, 1538
3896	2-Butanol, 3-chloro-*(erythro, dl)* $CH_3CHClCHOHCH_3$	108.57	....	135.4^{748}, 56.1^{30}	......	$1.0610^{25/4}$	1.4397^{25}	al, eth, chl	B1³, 1538
3897	2-Butanol, 3-chloro-*(threo)* $CH_3CH(OH)CHClCH_3$	108.57	....	130.8^{748}	......	$1.0586^{25/4}$	1.4386^{25}	....	B1³, 1538
3898	2-Butanol, 3-chloro-2-methyl $CH_3CHClC(CH_3)OHCH_3$	122.59	....	141-2, $55\text{-}6^{30}$	......	$1.0295^{20/4}$	1.4436^{20}	al, eth	B1³, 1627
3899	2-Butanol, 4-chloro $ClCH_2CH_2CHOHCH_3$	108.57	....	67^{20}	......	....	1.4408^{20}	al, eth	B1⁴, 1578
3900	2-Butanol, 4-cyclohexyl $C_6H_{11}CH_2CH_2CH(OH)CH_3$	156.27	....	112^{14}	......	0.903^{21}	1.464^{21}	....	B6³, 123
3901	2-Butanol, 1,4-dibromo $BrCH_2CH_2CH(OH)CH_2Br$	231.91	....	$114\text{-}5^{15}$	......	2.023^{0}	1.544^{20}	....	B1⁴, 1581
3902	2-Butanol, 1,3-dichloro $CH_3CHClCHOHCH_2Cl$	143.01	....	$63\text{-}4^{10}$	......	$1.2860^{15/4}$	1.4766^{20}	al, eth	B1³, 1541
3903	2-Butanol, 2,3-dimethyl $(CH_3)_2C(CH_3)(OH)CH_3$	102.18	....	118.4	-14	$0.8236^{20/4}$	1.4176^{20}	**al, eth**	B1⁴, 1729
3904	2-Butanol, 3,3-dimethyl or Pinacolyl alcohol $(CH_3)_3CCHOHCH_3$	102.18	....	120.4	5.6	0.8122^{25}	1.4148^{20}	al, **eth**	B1⁴, 1727
3905	2-Butanol, 2-methyl $CH_2CH_2C(CH_3)(OH)CH_3$	88.15	....	102, 50^{60}	-8.4	$0.8059^{25/4}$	1.4052^{20}	**al, eth**, ace, bz	B1⁴, 1668
3906	2-Butanol, 2-methyl-1-phenyl-*(dl)* $CH_3CH_2C(CH_3)(OH)CH_2C_6H_5$	164.25	....	$215\text{-}25^{747}$, $103\text{-}5^{11}$	......	$0.9754^{20/0}$	1.5182^{20}	al, eth, lig	B6³, 1958
3907	2-Butanol, 2-methyl-3-phenyl $CH_3CH(C_6H_5)C(OH)(CH_3)_2$	164.25	....	196-8, 118^{24}	......	$0.9794^{20/4}$	1.5193^{20}	al, eth, ace, bz	B6³, 1971
3908	2-Butanol, 2-methyl-4-phenyl $C_6H_5CH_2CH_2CH(OH)(CH_3)_2$	164.25	nd	121^{13}	24.5	$0.9626^{21/4}$	1.5077^{21}	al, eth, ace, bz	B6³, 1962
3909	2-Butanol, 3-methyl-*(d)* $(CH_3)_2CHCHOHCH_3$	88.15	$[\alpha]^{20}/_D$ + 5.34 (al)	112^{734}	......	$0.8225^{16/4}$	1.4089^{20}	al, eth, ace, bz, chl	B1¹, 196
3910	2-Butanol, 3-methyl-*(dl)* $(CH_3)_2CHCHOHCH_3$	88.15	....	112.9	......	$0.8180^{20/4}$	1.4089^{20}	al, eth, ace, bz	B1⁴, 1675
3911	2-Butanol, 3-methyl-3-phenyl $(CH_3)_2C(C_6H_5)CHOHCH_3$	164.25	....	196-8	......	$0.9653^{13/4}$	1.5161^{13}	al, eth, ace, bz	B6¹, 269
3912	2-Butanol, l-nitro $CH_3CH_2CH(OH)CH_2NO_2$	119.12	....	204^{767}, 75^{2}	......	$1.1353^{20/4}$	1.4435^{20}	al, eth, ace, bz	B1⁴, 1582
3913	2-Butanol, 3-nitro $CH_3CH(NO_2)CH(OH)CH_3$	119.12	....	$55^{0.5}$	......	....	1.4414^{20}	....	B1⁴, 1583

No.	Name, Synonyms, and Formula	Mol. wt.	Color, crystalline form, specific rotation and λ_{max} (log ε)	b.p. °C	m.p. °C	Density	n_D	Solubility	Ref.
3914	2-Butanol, 2-phenyl-*(d)* $CH_3CH_2C(OH)(C_6H_5)CH_3$	150.22	$[\alpha]^{22}_{D}$ + 17.5	112-4^{21}	-13	0.984$^{25/4}$	1.5185^{20}	al, eth	B6^{3}, 185
3915	2-Butanol, 2-phenyl-*(dl)* $CH_3CH_2C(OH)(C_6H_5)CH_3$	150.22		211-2, 90-1^{4}		0.984$^{25/4}$	1.5150^{20}	al, eth	B6^{3}, 1854
3916	2-Butanol, 1,1,1-trichloro $CH_3CH_2CHOHCCl_3$	177.46		169-71^{738}, 82-4^{22}		1.3670$^{25/25}$	1.4800^{20}	al, eth, ace, bz, chl	B1^{4}, 1579
3917	2-Butanol, 2,3,3-trimethyl $CH_3C(CH_3)_2C(CH_3)(OH)CH_3$	116.20	cr (dil al + ½ w)	131-2, 40-1^{15}	83-4	0.8380$^{25/4}$	1.4233^{22}	al, eth, ace	B1^{4}, 1755
3918	2-Butanone or Methyl ethyl ketone $CH_3CH_2COCH_3$	72.11		79.6, 30^{119}	-86.3	0.8054$^{20/4}$	1.3788^{20}	w, al, eth, ace, bz	B1^{4}, 3243
3919	2-Butanone, 1-chloro $CH_3CH_2COCH_2Cl$	106.55		137-8, 34-5^{10}		1.0850$^{20/4}$	1.4372^{20}	MeOH	B1^{4}, 3255
3920	2-Butanone, 3-chloro $CH_3CHClCOCH_3$	106.55		115, 40^{30}		1.0554^{0}	1.4219^{20}	al, eth	B1^{4}, 3256
3921	2-Butanone, 3-chloro-3-methyl $(CH_3)_2CHClCOCH_3$	120.58		117.2^{758}		1.0083$^{20/4}$	1.4204^{20}	al, eth	B1^{3}, 2818
3922	2-Butanone, 4-chloro $ClCH_2CH_2COCH_3$	106.55		120-1d, 48^{15}		1.0680^{23}	1.4284^{23}	al, eth	B1^{4}, 3256
3923	2-Butanone, 3,4-dibromo-4-phenyl or Benzalacetone dibromide $C_6H_5CHBrCHBrCOCH_3$	307.00	nd (al)		124-5			al, chl	B7^{3}, 1085
3924	2-Butanone, 1,3-dichloro $CH_3CHClCOCH_2Cl$	141.00		166-7, 55.5^{10}		1.3116$^{20/4}$	1.4686^{20}	al, eth, ace, bz	B1^{3}, 2786
3925	2-Butanone, 4-(diethylamino) $(C_2H_5)_2NCH_2CH_2COCH_3$	143.23		84^{30}		0.8630$^{20/4}$	1.4333^{24}	al, eth, ace, bz	B4^{4}, 1930
3926	2-Butanone, 3,3-dimethyl or Pinacolone $(CH_3)_3CCOCH_3$	100.16		106	-49.8	0.8012$^{25/4}$	1.3952^{20}	al, eth, ace	B1^{4}, 3310
3927	2-Butanone, 3,3-diphenyl $CH_3C(C_6H_5)_2CH_2COCH_3$	224.30	pr (al)	310-1, 176^{14}	41	1.069$^{20/4}$	1.5748^{20}	al, eth, chl, aa	B7^{3}, 2209
3928	2-Butanone, 1-hydroxy $CH_3CH_2COCH_2OH$	88.11		160, 48^{9}		1.0272$^{20/4}$	1.4189^{20}	w, al, eth	B1^{4}, 3989
3929	2-Butanone, 3-hydroxy or Acetoin $CH_3CHOHCOCH_3$	88.11		143, 37^{11}	-72	1.0062$^{20/20}$	1.4171^{20}	w, ace	B1^{4}, 3991
3930	2-Butanone, 3-methyl or Methyl isopropyl ketone $(CH_3)_2CHCOCH_3$	86.13		94-5	-92	0.8051$^{20/4}$	1.3880^{20}	al, eth, ace	B1^{4}, 3287
3931	2-Butanone, 3-methyl, oxime or Methyl isopropyl ketoxime $(CH_3)_2CHC(:NOH)CH_3$	101.15			157-8			al, eth	B1^{4}, 3289
3932	2-Butanone oxime $CH_3CH_2C(:NOH)CH_3$	87.12		152-3, 59-60^{15}	-29.5	0.9232$^{20/4}$	1.4410^{20}	w, al, eth	B1^{4}, 3250
3933	2-Butanone, 1-phenyl or Ethyl benzyl ketone $CH_3CH_2COCH_2C_6H_5$	148.20		230^{755}		1.002$^{20/4}$		al, eth, ace	B7^{3}, 1080
3934	2-Butanone, 4-phenyl or Benzylacetone $C_6H_5CH_2CH_2COCH_3$	148.20		233-4, 115^{13}		0.9849$^{22/4}$	1.511^{22}	al, eth, ace	B7^{3}, 1081
3935	2-Butenal or Crotonaldehyde $CH_3CH{=}CHCHO$	70.09		104-5	-74	0.8495$^{25/4}$	1.4366^{20}	al, eth, ace, bz	B1^{4}, 3447
3936	2-Butenal diethylacetal or Crotonaldehyde diethylacetal $CH_3CH{=}CHCH(OC_2H_5)_2$	144.21		147-8		0.8473$^{18/4}$	1.4097^{20}	al, eth, ace, bz	B1^{4}, 3450
3937	2-Butenal, 2-bromo, diethyl acetal $CH_3CH{=}CBrCH(OC_2H_5)_2$	223.11		86^{15}		1.2255^{21}	1.4565^{21}		B1^{4}, 3452
3938	2-Butenal, 2-chloro $CH_3CH{=}CClCHO$	104.54		147-8, 53-4^{20}		1.1404$^{23/4}$	1.4780^{25}	al, eth, chl	B1^{3}, 2981
3939	2-Butenal, 3-ethoxy, diethylacetal or 1,1,3-triethoxy-2-butene $CH_3C(OC_2H_5){=}CH\text{-}CH(OC_2H_5)_2$	188.27		190-5, 79-82^{10}		0.908$^{21/0}$	1.430^{21}	al	B1^{4}, 4082
3940	2-Butenal, 2-methyl or Tiglaldehyde $CH_3CH{=}C(CH_3)CHO$	84.12		116.7^{738}, 63-5^{110}		0.8710$^{20/4}$	1.4475^{20}	w, al, eth	B1^{4}, 3464
3941	2-Butenal, 3-methyl or β,β-Dimethylacrolein $(CH_3)_2C{=}CHCHO$	84.12		133^{730}		0.8722$^{20/4}$	1.4528^{20}	w, al, eth	B1^{4}, 3464
3942	1-Butene $CH_3CH_2CH{=}CH_2$	56.11		-6.3	-185.3	0.5951$^{20/4}$	1.3962^{20}	al, eth, bz	B1^{4}, 765
3943	1-Butene, 1-bromo-*(cis)* $CH_3CH_2CH{=}CHBr$	135.00		86.15		1.3265$^{15/4}$	1.4536^{20}	eth, ace, bz	B1^{2}, 174
3944	1-Butene, 1-bromo-*(trans)* $CH_3CH_2{=}CHBr$	135.00		94.7	-100.3	1.3209$^{18/4}$	1.4527^{20}	eth, ace, bz	B1^{3}, 726
3945	1-Butene, 2-bromo $CH_3CH_2CBr{=}CH_2$	135.00		88, 25^{10}	-133.4	1.3209$^{15/4}$	1.4527^{20}	eth, ace, bz	B1^{4}, 775

PHYSICAL CONSTANTS OF ORGANIC COMPOUNDS (Continued)

No.	Name, Synonyms, and Formula	Mol. wt.	Color, crystalline form, specific rotation and λ_{max} (log ε)	b.p. °C	m.p. °C	Density	n_D	Solubility	Ref.
3946	1-Butene, 2-bromo-3-methyl $(CH_3)_2CHCBr{=}CH_2$	149.03	...	105^{757}	...	$1.2328^{20/4}$	1.4504^{20}	eth, bz, chl	$B1^{3}$, 800
3947	1-Butene, 2-bromo-4-phenyl $C_6H_5CH_2CH_2CBr{=}CH_2$	211.10	...	$117\text{-}8^{21}$, $90\text{-}1^{5}$	...	$1.2907^{20/4}$	1.5450^{20}	ace	$B5^{3}$, 1209
3948	1-Butene, 4-bromo $BrCH_2CH_2CH{=}CH_2$	135.00	...	98.5	...	$1.3230^{20/4}$	1.4622^{20}	al, eth, bz	$B1^{4}$, 775
3949	1-Butene, 1-chloro-*(cis)* $CH_3CH_2CH{=}CHCl$	90.55	...	63.5	...	$0.9153^{15/4}$	1.4194^{15}	al, eth, ace, chl	$B1^{3}$, 723
3950	1-Butene, 1-chloro-*(trans)* $CH_3CH_2\text{-}CH{=}CHCl$	90.55	...	68	...	$0.9205^{15/4}$	1.4223^{15}	al, eth, ace, chl	$B1^{3}$, 723
3951	1-Butene, 1-chloro-2-methyl $CH_3CH_2C(CH_3){=}CHCl$	104.58	...	96-7	...	$0.9170^{20/4}$	1.4141^{20}	eth, ace	$B1^{3}$, 787
3952	1-Butene, 1-chloro-3-methyl $(CH_3)_2CH\text{-}CH{=}CHCl$	104.58	...	$86\text{-}8^{756}$	...	...	1.4229^{20}	eth, ace, chl	$B1^{3}$, 799
3953	1-Butene, 2-chloro $CH_3CH_2CCl{=}CH_2$	90.55	...	58.5	...	$0.9107^{15/4}$	1.4165^{21}	al, eth, ace, bz	$B1^{4}$, 769
3954	1-Butene, 2(chloromethyl)-1,3-dichloro $CH_3CHClC(CH_2Cl){=}CHCl$	174.48	...	$68\text{-}70^{8}$	...	$1.2775^{19/4}$	...	CCl_4	$B1^{3}$, 788
3955	1-Butene, 3-chloro $CH_3CHClCH{=}CH_2$	90.55	...	64-5	...	$0.8978^{20/4}$	1.4149^{20}	eth, ace, chl	$B1^{4}$, 769
3956	1-Butene, 3-chloro-2-chloromethyl $CH_3CHClC(CH_2Cl){=}CH_2$	139.02	...	155, $31\text{-}3^{7}$	...	$1.1233^{20/4}$	1.4724^{20}	ace, chl	$B1^{3}$, 787
3957	1-Butene, 3-chloro-2-methyl $CH_3CHClC(CH_3){=}CH_2$	104.58	...	94	...	$0.9088^{20/4}$	1.4304^{20}	eth, ace, chl	$B1^{4}$, 819
3958	1-Butone, 4-chloro $ClCH_2CH_2CH{=}CH_2$	90.55	...	75^{773}	...	$0.9211^{20/4}$	1.4233^{20}	eth, ace, chl	$B1^{4}$, 771
3959	1-Butene, 1,3-dichloro $CH_3CHClCH{=}CHCl$	125.00	...	125, $58\text{-}60^{25}$	...	$1.1341^{24/4}$	1.4647^{20}	al, eth, ace, chl	$B1^{4}$, 772
3960	1-Butene, 2,3-dichloro $CH_3CHClCCl{=}CH_2$	125.00	...	112	...	$1.1340^{20/4}$	1.4580^{20}	eth, ace, chl	$B1^{4}$, 772
3961	1-Butene, 3,3-dichloro-2-methyl $CH_3CCl_2C(CH_3){=}CH_2$	139.02	...	151-3	...	$1.1276^{18/4}$	1.4737^{18}	eth, ace, chl	$B1^{3}$, 787
3962	1-Butene, 3,4-dichloro $ClCH_2CHClCH{=}CH_2$	125.00	...	115-7, 45.5^{40}	...	1.1170^{20}	1.4475^{20}, 1.4641^{20}	al, eth, bz	$B1^{4}$, 772
3963	1-Butene, 2,3-dimethyl $CH_3CH(CH_3)C(CH_3){=}CH_2$	84.16	...	55.67	-157.3	$0.6803^{20/4}$	1.3995^{20}	al, eth, ace	$B1^{4}$, 852
3964	1-Butene, 3,3-dimethyl $(CH_3)_3CCH{=}CH_2$	84.16	...	41.2	-115.2	$0.6529^{20/4}$	1.3763^{20}	al, eth	$B1^{4}$, 850
3965	1-Butene-3,4-diol $HOCH_2CHOHCH{=}CH_2$	88.11	...	196.5, 98^{16}	...	$1.0470^{20/4}$	1.4628^{21}	w, al	$B1^{4}$, 2658
3966	1-Butene, 1,3-diphenyl $CH_3CH(C_6H_5)CH{=}CHC_6H_5$	208.30	...	$175\text{-}6^{14}$	...	1.016^{15}	1.590	...	$B5^{4}$, 2205
3967	1-Butene, 3,4-epoxy or Butadiene monoxide $CH_2CHCH{=}CH_2$ (with O bridging CH_2 and CH)	70.09	...	70	...	0.9006^{0}	1.4168^{20}	al, eth, bz	$B17^{4}$, 145
3968	1-Butene, 2-ethyl $CH_3CH_2C(C_2H_5){=}CH_2$	84.16	...	64.7	-131.5	$0.6894^{20/4}$	1.3969^{20}	eth, ace, bz, chl	$B1^{4}$, 850
3969	1-Butene, 2-ethyl-3-methyl $(CH_3)_2CH\text{-}C(C_2H_5){=}CH_2$	98.19	...	89	...	$0.7150^{20/4}$	1.410^{20}	eth, ace, bz, chl	$B1^{4}$, 871
3970	1-Butene, 2-methyl $CH_3CH_2CCH_3{=}CH_2$	70.13	...	31.2	-137.5	$0.6504^{20/4}$	1.3778^{20}	al, eth, bz	$B1^{4}$, 818
3971	1-Butene, 3-methyl $(CH_3)_2CHCH{=}CH_2$	70.13	...	20	-168.5	$0.6272^{20/4}$	1.3643^{20}	al, eth, bz	$B1^{4}$, 825
3972	1-Butene, perfluoro $F_3CCF_2CF{=}CF_2$	200.03	...	4.8	...	$1.615^{-20/4}$, 1.5443^{0}	...	...	$B1^{4}$, 769
3973	1-Butene, 1-phenyl *(cis)* $CH_3CH_2CH{=}CHC_6H_5$	132.21	...	196.2^{755}, $84\text{-}5^{23}$	...	$0.9106^{20/4}$	1.5381^{16}	al, eth, bz	$B5^{3}$, 1205
3974	1-Butene, 1-phenyl-*(trans)* $CH_3CH_2CH{=}CHC_6H_5$	132.21	...	198.7, $91\text{-}2^{21}$	-43.1	$0.9019^{20/4}$	1.5420^{20}	al, eth, bz	$B5^{4}$, 1374
3975	1-Butene, 4-phenyl $C_6H_5CH_2CH_2CH{=}CH_2$	132.21	...	177, 64^{10}	-70	$0.8831^{20/4}$	1.5059^{20}	eth, bz	$B5^{4}$, 1378
3976	1-Butene, 1,3,4,4-tetrachloro $Cl_2CHCHClCH{=}CHCl$	193.89	...	88^{20}	...	$1.0711^{20/4}$	1.4773^{20}	chl	$B1^{4}$, 774
3977	1-Butene, 2,3,3,4-tetrachloro $ClCH_2CCl_2CCl{=}CH_2$	193.89	...	$41\text{-}2^{7}$	...	$1.4602^{20/4}$	1.5135^{20}	ace, chl	$B1^{3}$, 726
3978	1-Butene, 2,3,4-trichloro $ClCH_2CHClCCl{=}CH_2$	159.44	...	60^{20}	...	$1.3430^{20/4}$	1.4944^{20}	ace, chl	$B1^{3}$, 725

PHYSICAL CONSTANTS OF ORGANIC COMPOUNDS (Continued)

No.	Name, Synonyms, and Formula	Mol. wt.	Color, crystalline form, specific rotation and λ_{max} (log ε)	b.p. °C	m.p. °C	Density	n_D	Solubility	Ref.
3979	1-Butene, 2,3,3-trimethyl or Triptene $(CH_3)_3CC(CH_3)=CH_2$	98.19	....	77.9	-109.9	0.7050$^{20/4}$	1.4025^{20}	eth, bz, MeOH	B1^{4}, 873
3980	2-Butene *(cis)* $CH_3CH=CHCH_3$	56.11	....	3.7	-138.9	0.6213$^{20/4}$	1.3931^{-25}	al, eth, bz	B1^{4}, 778
3981	2-Butene *(trans)* $CH_3CH=CHCH_3$	56.11	....	0.9	-105.5	0.6042$^{20/4}$	1.3848^{-25}	al, eth, bz	B1^{4}, 781
3982	2-Butene, 1-bromo $CH_3CH=CHCH_2Br$	135.00	....	103-6, 13^{10}	......	1.3371$^{25/4}$	1.4822^{20}	al, eth, bz	B1^{4}, 789
3983	2-Butene, 1-bromo-3-methyl $(CH_3)_2C=CH\text{-}CH_2Br$	149.03	....	129-33d, 50-1^{40}	......	1.2819$^{20/0}$	1.4930^{15}	al, eth, ace, bz	B1^{4}, 824
3984	2-Butene, 1-bromo-4-phenyl $C_6H_5CH_2CH=CHCH_2Br$	211.10	....	126-30$^{10.5}$	......	1.2660$^{20/4}$	1.5678^{20}	eth	B5^{4}, 1377
3985	2-Butene, 2-bromo *(cis)* $CH_3CH=CBrCH_3$	135.00	....	93.9	-111.5	1.3416$^{15/4}$	1.4631^{19}	al, eth, bz	B1^{4}, 790
3986	2-Butene, 2-bromo-3-methyl $(CH_3)_2C=CBrCH_3$	149.03	....	119-20	......	1.2773$^{20/4}$	1.4738^{20}	eth, ace, chl	B1^{4}, 824
3987	2-Butene, 2-bromo-3-phenyl $CH_3C(C_6H_5)=CBr\text{-}CH_3$	211.10	....	120-30^{11}	......	1.3348$^{20/4}$	1.5811^{20}	eth, bz, chl	B5^{4}, 1379
3988	2-Butene, 1-chloro *(cis)* $CH_3CH=CHCH_2Cl$	90.55	....	84.1	......	0.9426$^{20/4}$	1.4390^{20}	al, ace, chl	B1^{4}, 783
3989	2-Butene, 1-chloro *(trans)* $CH_3CH=CHCH_2Cl$	90.55	....	84.8^{752}	......	0.9295$^{20/4}$	1.4350^{20}	ace, chl	B1^{4}, 784
3990	2-Butene, 1-chloro-2,3-dimethyl $(CH_3)_2C=C(CH_3)CH_2Cl$	118.61	....	111-2^{756}	......	0.9355$^{20/4}$	1.4605^{20}	eth, ace, chl	B1^{4}, 856
3991	2-Butene, 1-chloro-2-methyl $CH_3CH=C(CH_3)CH_2Cl$	104.58	....	110, 26.4^{25}	......	0.9327$^{20/4}$	1.4481^{20}	al, eth, ace, chl	B1^{4}, 822
3992	2-Butene, 1-chlor-3-methyl $CH_3C(CH_3)=CHCH_2Cl$	104.58	....	109, 54.3^{95}	......	0.9273$^{20/4}$	1.4485^{20}	al, eth, ace, chl	B1^{4}, 823
3993	2-Butene, 2-chloro *(cis)* $CH_3CH=CClCH_3$	90.55	....	70.6	-117.3	0.9239$^{20/4}$	1.4240^{20}	al, ace, chl	B1^{4}, 785
3994	2-Butene, 2-chloro *(trans)* $CH_3CH=CClCH_3$	90.55	....	62.8	-105.8	0.9138$^{20/4}$	1.4190^{20}	al, ace, chl	B1^{4}, 785
3995	2-Butene, 2-chloro-3-methyl $(CH_3)_2C=CClCH_3$	104.58	....	94	......	0.9324$^{20/4}$	1.4320^{20}	al, eth, ace, chl	B1^{3}, 794
3996	2-Butene, 1,4-dibromo *(trans)* $BrCH_2CH=CHCH_2Br$	213.90	pl (peth)	203, 85^{10}	53.4	....	....	al, ace, peth	B1^{4}, 791
3997	2-Butene, 1,1-dichloro $CH_3CH=CHCHCl_2$	125.00	....	125-7	......	1.1310^{20}	1.466^{18}	eth, ace, chl	B1^{4}, 787
3998	2-Butene, 1,2-dichloro (high b.p.) $CH_3CH=CClCH_2Cl$	125.00	....	130-1	......	1.1601$^{20/4}$	1.4734^{20}	eth, ace, chl	B1^{3}, 741
3999	2-Butene, 1,2-dichlor (low b.p.) $CH_3CH=CClCH_2Cl$	125.00	....	116-8^{765}	......	1.1544$^{20/4}$	1.4642^{20}	bz, CCl_4	B1^{3}, 741
4000	2-Butene, 1,3-dichloro *(cis)* $CH_3CCl=CHCH_2Cl$	125.00	....	129.9^{745}, 34^{20}	......	1.1605$^{20/4}$	1.4735^{20}	al, eth, ace, bz	B1^{4}, 786
4001	**2-Butene, 1,3-dichloro *(trans)*** $CH_3CCl=CHCH_2Cl$	125.00	....	130^{745}, 53^{50}	......	1.1585$^{20/4}$	1.4719^{20}	al, eth, ace, bz	B1^{4}, 786
4002	**2-Butene, 1,3-dichloro-2-methyl** $CH_3CCl=C(CH_3)CH_2Cl$	139.02	....	151-3	......	1.1293$^{20/4}$	....	ace, chl	B1^{3}, 795
4003	**2-Butene, 1,4-dichloro *(cis)*** $ClCH_2.CH=CHCH_2Cl$	125.00	....	152.5, 22.5^{3}	-48	1.188$^{25/4}$	1.4887^{25}	al, eth, ace, bz	B1^{4}, 787
4004	**2-Butene, 1,4-dichloro *(trans)*** $ClCH_2CH=CHCH_2Cl$	125.00	....	155.5^{758}, 55.5^{20}	1-3	1.183$^{25/4}$	1.4871^{25}	al, eth, ace, bz	B1^{4}, 787
4005	**2-Butene, 1,4-dichloro-2-methyl** $ClCH_2CH=C(CH_3)CH_2Cl$	139.02	....	93^{50}	1	1.1526$^{20/4}$	1.4932^{20}	ace, chl	B1^{3}, 795
4006	**2-Butene, 2,3-dichloro *(cis)*** $CH_3CCl=CClCH_3$	125.00	....	125-6^{758}	......	1.1618$^{20/4}$	1.4590^{20}	al, eth, ace, bz	B1^{4}, 787
4007	**2-Butene, 2,3-dichloro *(trans)*** $CH_3CCl=CClCH_3$	125.00	....	101-3^{758}	......	1.1416$^{20/4}$	1.4582^{20}	al, eth, ace, bz	B1^{4}, 787
4008	**2-Butene, 2,3-dimethyl** $(CH_3)_2=C(CH_3)_2$	84.16	....	73.2	-74.3	0.7080$^{20/4}$	1.4122^{20}	al, eth, ace, chl	B1^{4}, 853
4009	**2-Butene-1,4-diol *(cis)*** $HOCH_2CH=CHCH_2OH$	88.11	....	235, 132^{16}	4	1.0698$^{20/4}$	1.4782^{20}	w, al	B1^{4}, 2660
4010	**2-Butene-1,4-diol *(trans)*** $HOCH_2CH=CHCH_2OH$	88.11	....	131^{13}	25	1.0700$^{20/4}$	1.4755^{20}	w, al	B1^{4}, 2660
4011	**2-Butene, 1,1,2,3,4,4-hexachloro (liquid)** $Cl_2CHCCl=CClCHCl_2$	262.78	....	97-8^{10}	-19	1.651$^{15/15}$	1.5331	chl	B1^{4}, 789

No.	Name, Synonyms, and Formula	Mol. wt.	Color, crystalline form, specific rotation and λ_{max} (log ε)	b.p. °C	m.p. °C	Density	n_D	Solubility	Ref.
4012	2-Butene, 1,1,2,3,4,4-hexachloro (solid) $Cl_2CHCCl=CClCHCl_2$	262.78	lf (al)	...	80	...	...	al, eth, bz, chl	B1^4, 789
4013	2-Butene, 2-methyl $CH_3CH=C(CH_3)_2$	70.13	...	38.6	-133.8	0.6623$^{20/4}$	1.3874^{20}	al, eth, bz, lig	B1^4, 820
4014	2-Butene, perfluoro $CF_3CF=CFCF_3$	200.03	...	0—3	-129	1.5297^0	...	...	B1^4, 783
4015	2-Butene, 1,1,1,4,4-pentachloro $Cl_2CHCH=CHCCl_3$	228.33	...	78-80^{11}	...	1.612$^{21/21}$	1.5538^{21}	al, chl	B1^3, 744
4016	2-Butene, 1,2,4-trichloro $ClCH_2CH=CClCH_2Cl$	159.44	...	67-9^{10}	...	1.3843$^{20/4}$	1.5175^{20}	al, eth, bz, chl	B1^3, 744
4017	2-Butenoic acid *(cis)* or Isocrotonic acid $CH_3CH=CHCO_2H$	86.09	nd or pr (peth)	169.3, 74^{15}	15.5	1.0267$^{20/4}$	1.4483^{14}	w, al	B2^4, 1497
4017a	2-Butenoic acid (*trans*) or Crotonic acid $CH_3CH=CHCO_2H$	86.09	mcl pr or nd (w or lig)	185	71.5	1.1018$^{15/4}$	1.4249^{77}	w, al, eth, ace	B2^4, 1509
4018	2-Butenoic acid, 2-chloro *(cis)* $CH_3CH=CClCO_2H$	120.54	nd (w)	...	67	...	...	w, al, lig	B2^4, 1510
4019	2-Butenoic acid, 2-chloro, ethyl ester *(cis)* $CH_3CH=CClCO_2C_2H_5$	148.59	...	75^{30}	...	1.1021$^{18/4}$	...	al, eth	B2^4, 1510
4020	2-Butenoic acid, 3-chloro *(cis)* $CH_3CCl=CHCO_2H$	120.54	...	195 sub	61	1.1995$^{66/4}$	1.4704^{66}	al, peth	B2^4, 1511
4021	2-Butenoic acid, 3-chloro, ethyl ester *(cis)* $CH_3CCl=CHCO_2C_2H_5$	148.59	...	161.4, 50^{10}	...	1.0860$^{20/4}$	1.4542$^{19/4}$	al, eth	B2^4, 1510
4022	2-Butenoic acid, 3-chloro, methyl ester *(cis)* $CH_3CCl=CHCO_2CH_3$	134.56	...	142.4, 42-3^{13}	...	1.138$^{20/4}$	1.4573^{19}	eth, MeOH	B2^4, 1500
4023	2-Butenoic acid, ethyl ester *(cis)* or Ethyl isocrotonate $CH_3CH=CHCO_2C_2H_5$	114.14	...	136	...	0.9182$^{20/4}$	1.4242^{20}	al, eth, ace	B2^4, 1551
4024	2-Butenoic acid, 2-methyl *(cis)* or Angelic acid $CH_3CH=C(CH_3)CO_2H$	100.12	mcl pr or nd	185, 88-9^{10}	45-6	0.983$^{49/4}$, 0.9539$^{76/4}$	1.4434^{47}	al, eth	B2^4, 1497
4026	2-Butenoic acid, allyl ester *(trans)* $CH_3CH=CHCO_2CH_2CH=CH_2$	126.16	...	88-9^{70}	...	0.9440$^{20/4}$	1.4465^{20}	...	B2^4, 1503
4027	2-Butenoic acid, 3-amino, ethyl ester *(trans)* $CH_3C(NH_2)=CHCO_2C_2H_5$	129.16	mcl pr	210-5d, 105^{15}	34 (st), 20-1 (unst)	1.0219$^{19/4}$	1.4988^{22}	al, eth, bz, chl, lig	B3^3, 1199
4028	2-Butenoic acid, 3-bromo *(trans)* or 3-Bromo-crotonic acid $CH_3CBr=CHCO_2H$	164.99	nd (lig), lf (w)	...	97	...	...	al, eth, bz, aa	B2^3, 1275
4029	2-Butenoic acid, 4-bromo, ethyl ester *(trans)* or Ethyl-4-bromocrotonate $BrCH_2CH=CHCO_2C_2H_5$	193.04	...	97-8^{15}	...	1.402$^{16/4}$	1.4925^{20}	al	B2^4, 1517
4030	2-Butenoic acid, 2-chloro *(trans)* or 2-Chlorocrotonic acid $CH_3CH=CClCO_2H$	120.54	nd (w or peth)	212, 111-2^{14}	100.5	...	...	al, eth	B2^4, 1509
4031	2-Butenoic acid, 2-chloro, ethyl ester *(trans)* or Ethyl 2-chlorocrotonate $CH_3CH=CHCClCO_2C_2H_5$	148.59	...	176-8, 61^{10}	...	1.1135$^{20/4}$	1.4538^{20}	al, eth	B2^4, 1510
4032	2-Butenoic acid, 2-chloro, methyl ester *(trams)* or Methyl 2-chlorocrotonate $CH_3CH=CClCO_2CH_3$	134.56	...	161.5^{762}, 59.5^{16}	...	1.160$^{20/4}$	1.4569^{23}	eth	B2^4, 1510
4033	2-Butenoic acid, 3-chloro *(trans)* or 3-Chlorocrotonic acid $CH_3CCl=CHCO_2H$	120.54	...	206-11d	94-5	...	...	al, CS_2	B2^4, 1510
4034	2-Butenoic acid, 3-chloro, ethyl ester *(trans)* $CH_3CCl=CHCO_2C_2H_5$	148.59	...	184, 66^{10}	...	1.1062$^{20/4}$	1.4592^{20}	al, eth	B2^4, 1511
4035	2-Butenoic acid, 3-chloro, methyl ester *(trans)* $CH_3CCl=CHCO_2CH_3$	134.56	...	64-7^{14}	...	1.157$^{20/4}$	1.4630^{20}	al, eth	B2^4, 1510
4036	2-Butenoic acid, 4-chloro-*(trans)* $ClCH_2CH=CHCO_2H$	120.54	cr (peth-eth)	117-8^{13}	83	...	...	eth	B2^4, 1511
4037	2-Butenoic acid, ethyl ester *(trans)* or Ethyl crotonate $CH_3CH=CHCO_2C_2H_5$	114.14	...	136.5, 58-9^{48}	...	0.9175$^{20/4}$	1.4243^{20}	al, eth	B2^4, 1500
4038	2-Butenoic acid, 2-ethyl *(trans)* $CH_3CH=C(C_2H_5)CO_2H$	114.14	mcl pr (peth)	209,109^{13}	45-6	0.9578$^{50/4}$	1.4475^{50}	al, eth	B2^3, 1329
4039	2-Butenoic acid, methyl ester *(trans)* or Methyl crotonate $CH_3CH=CHCO_2CH_3$	100.12	...	121	-42	0.9444$^{20/4}$	1.4242^{20}	al, eth	B2^4, 1500
4040	2-Butenoic acid, 2-methyl *(trans)* or Tiglic acid $CH_3CH=C(CH_3)CO_2H$	100.12	ta (w)	198.5	64.5-5	0.9641$^{76/4}$	1.4330^{76}	al, eth	B2^4, 1552
4041	2-Butenoic acid, 2-methyl, ethyl ester *(trans)* or Ethyl tiglate $CH_3CH=C(CH_3)CO_2C_2H_5$	128.17	...	156, 55.5^{11}	...	0.9200$^{20/4}$	1.4340^{20}	al, bz	B2^4, 1553

No.	Name, Synonyms, and Formula	Mol. wt.	Color, crystalline form, specific rotation and λ_{max} (log ε)	b.p. °C	m.p. °C	Density	n_D	Solubility	Ref.
4042	3-Butenoic acid or Vinyl acetic acid $CH_2=CHCH_2CO_2H$	86.09		$169, 69\text{-}70^{12}$	-35	$1.0091^{20/4}$	1.4239^{20}	w, **al, eth**	$B2^4$, 1491
4043	3-Butenoic acid-ethyl ester or Ethyl, 3-butenoate $CH_2=CHCH_2CO_2C_2H_5$	114.14		119		0.9122^{20}	1.4105^{20}	al	$B2^4$, 1491
4044	3-Butenoic acid, 2-hydroxy, ethyl ester $CH_2=CHCHOHCO_2C_2H_5$	130.14		$173d, 68^{15}$		$1.0470^{15/4}$	1.436^{13}	w, al, eth	$B2^3$, 685
4045	3-Butenoic acid, 2-hydroxy-4-phenyl or Benzollactic acid $C_6H_5CH=CHCHOHCO_2H$	178.19	nd (w)		137				$B10^3$, 862
4046	3-Butenoic acid, 4-phenyl or Styrylacetic acid $C_6H_5CH=CHCH_2CO_2H$	162.19	nd (w), pr (CS_2)	302	87			al, eth	$B9^3$, 2756
4047	3-Butenonitrile or Allyl cyanide $CH_2=CHCH_2CN$	67.09		119	-84	$0.8329^{20/4}$	1.4060^{20}	**al, eth**	$B2^4$, 1491
4048	2-Butene-1-ol or Crotyl alcohol $CH_3CH=CHCH_2OH$	72.11		121.2	<-30	$0.8521^{20/4}$	1.4288^{20}	**al, eth**	$B1^4$, 2107
4049	2-Butene-1-ol, 2-chloro $CH_3CH=CClCH_2OH$	106.55		159		$1.1180^{20/4}$	1.4682^{20}	w, al	$B1^3$, 1900
4050	2-Butene-1-ol, 4-chloro $ClCH_2CH=CHCH_2OH$	106.55		$64\text{-}5^{2}$			1.4845^{20}	al, eth	$B1^4$, 2110
4051	3-Butene-1-ol $CH_2=CHCH_2CH_2OH$	72.11		113.5		$0.8424^{20/4}$	1.4224^{20}	w, **al, eth,** ace	$B1^4$, 2105
4052	3-Buten-1-ol, 2-chloro $CH_2=CHCHClCH_2OH$	106.55		$66\text{-}7^{30}$		$1.1044^{20/4}$	1.4665^{20}	al, eth	$B1^4$, 2106
4053	3-Buten-2-ol *(d)* $CH_2=CH\text{-}CHOHCH_3$	72.11	$[\alpha]^{20/}{}_D$ + 33.9 (undil)	96.5^{745}		$0.8362^{15/4}$	1.4120^{20}		$B1^4$, 2102
4054	3-Buten-2-ol *(dl)* $CH_2=CH\text{-}CHOHCH_3$	72.11		97.3	<-100	$0.8318^{20/4}$	1.4137^{20}		$B1^4$, 2102
4055	3-Buten-2-ol, 1-chloro $CH_2=CH\text{-}CHOHCH_2Cl$	106.55		$144\text{-}7, 63.3^{30}$		$1.111^{20/4}$	1.4643^{20}	chl	$B1^3$, 1893
4056	3-Buten-2-ol, 3-chloro $CH_2=CClCHOHCH_3$	106.55		$53\text{-}7^{19}$		$1.1138^{23/4}$		al, eth	$B1^3$, 1893
4057	3-Buten-2-one or Methyl vinyl ketone $CH_2=CHCOCH_3$	70.09		$81.4, 33\text{-}4^{130}$		$0.8636^{20/4}$	1.4081^{20}	w, al, eth, ace, bz	$B1^4$, 3444
4058	3-Butene-2-one, 4-bromo-4-phenyl $C_6H_5CBr=CHCOCH_3$	225.08		$150\text{-}1^{10}$				bz	$B7^3$, 1407
4059	3-Butene-2-one, 4-(2-hydroxyphenyl) $2\text{-}HOC_6H_4CH=CHCOCH_3$	162.19	nd (al or lig), pr (bz)		140			eth, al, bz	$B8^3$, 810
4060	3-Butene-2-one, 4-(3-hydroxyphenyl) $3\text{-}HOC_6H_4CH=CHCOCH_3$	162.19	ye pr (bz)		97-8			bz	$B8^3$, 812
4061	3-Butene-2-one, 4-(4-hydroxyphenyl) $4\text{-}HOC_6H_4CH=CHCOCH_3$	162.19	nd (w)		114-5			al, aa	$B8^3$, 812
4062	3-Butene-2-one, 4-(4-methoxy phenyl) or Anisylidene acetone $4\text{-}CH_3OC_6H_4CH=CHCOCH_3$	176.22	lf (al, eth, aa)		73			al, eth, bz, aa	$B8^3$, 812
4063	3-Butene-2-one, 3-methyl or Isopropenyl methyl ketone $CH_2=C(CH_3)COCH_3$	84.12		98	-54	$0.8527^{20/4}$	1.4220^{20}	**al**	$B1^4$, 3462
4064	3-Butene-2-one, 4-phenyl *(trans)* or Benzal acetone $C_6H_5CH=CHCOCH_3$	146.19	pl	$26^{2}, 140^{16}$	42	$1\ 097^{45}$	1.5836^{45}	al, eth, ace, bz, chl	$B7^3$, 1399
4065	1-Butene-3-yne or Vinyl acetylene $CH_2=CH\text{-}C\equiv CH$	52.08		5.1		$0.7095^{0/0}$	1.4161^{1}	bz	$B1^4$, 1083
4066	1-Butene-3-yne, 4-chloro $CH_2=CH\text{-}C\equiv CCl$	86.52		55-7		$1.0022^{20/4}$	1.4656^{20}	chl	$B1^4$, 1085
4067	1-Butene-3-yne, 4-methoxy $CH_3OCH=CH\text{-}C\equiv CH$	82.10		$122\text{-}5d, 30\text{-}2^{15}$		$0.906^{20/4}$	1.4818^{20}		$B1^4$, 2300
4068	1-Butene-3-yne, 2-methyl $CH_2=C(CH_3)C\equiv CH$	66.10		34		$0.6801^{11/4}$	1.4105^{20}		$B1^4$, 1089
4069	*n*-Butyl alcohol or 1-Butanol $CH_3CH_2CH_2CH_2OH$	74.12		117.2	-89.5	$0.8098^{20/4}$	1.3993^{20}	w, **al, eth,** ace, bz	$B1^4$, 1506
4070	Isobutyl alcohol or 2-methyl-1-propanol $(CH_3)CHCH_2OH$	74.12		108.1		$0.8018^{20/4}$	1.3955^{20}	al, eth, ace	$B1^4$, 1588
4071	Isobutyl alcohol, 2-amino $(CH_3)_2C(NH_2)CH_2OH$	89.14		$165.5, 69\text{-}70^{10}$	25-6	$0.934^{20/4}$	1.449^{20}	w	$B4^4$, 1740
4072	Isobutyl alcohol, 2-chloro or β Isobutylene chlorohydrin $(CH_3)_2CClCH_2OH$	108.57		$132\text{-}3d, 59\text{-}61^{50}$		$1.0472^{20/4}$	1.4388^{20}		$B1^4$, 1603
4073	Isobutyl alcohol, 2-nitro $(CH_3)_2C(NO_2)CH_2OH$	119.12	nd or pl (MeOH)	$94\text{-}5^{10}$	89-90			al, eth	$B1^4$, 1604

No.	Name, Synonyms, and Formula	Mol. wt.	Color, crystalline form, specific rotation and λ_{max} (log ε)	b.p. °C	m.p. °C	Density	n_D	Solubility	Ref.
4074	Isobutyl alcohol, 3-chloro $ClCH_2CH(CH_3)CH_2OH$	108.57	....	$76\text{-}8^{11}$	......	$1.083^{25/4}$	1.4460^{25}	al, eth	$B1^3$, 1564
4075	*sec*-Butyl alcohol or 2-Butanol $CH_3CH_2CH(OH)CH_3$	74.12	....	$99.5, 45.5^{60}$	......	$0.8063^{20/4}$	1.3978^{20}	**al, eth,** ace, bz	$B1^4$, 1566
4076	*tert*-Butyl alcohol or 2-Methyl-2-propanol $(CH_3)_3COH$	74.12	....	$82.3, 20^{31}$	25.5	$0.7887^{20/4}$	1.3878^{20}	**w, al, eth**	$B1^4$, 1609
4077	*tert*-Butyl alcohol, methoxy $(CH_3)_2C(OH)CH_2OCH_3$	104.15	....	116.6^{747}	......	$0.9021^{15/15}$	....	....	$B1^4$, 1615
4078	*tert*-Butyl alcohol, 1,1,1-tribromo or Brometone $(CH_3)_2C(OH)CBr_3$	310.81	nd (lig), cr (dil al)	sub	168-70	....	....	al, eth	$B1^3$, 1588
4079	*tert*-Butyl alcohol, 1,1,1-trichloro or Chloreton $(CH_3)_2C(OH)CCl_3$	177.46	hyg nd (w + 1)	167	98-9, 77 (hyd)	....	....	al, eth, ace, bz	$B1^4$, 1629
4080	*n*-Butyl amine $CH_3CH_2CH_2CH_2NH_2$	73.14	....	77.8	-49.1	$0.7414^{20/4}$	1.4031^{20}	**w**, al, eth	$B4^4$, 540
4081	Butyl dimethyl amino $C_4H_9N(CH_3)_2$	101.19	....	95	......	$0.7206^{20/4}$	1.3970^{20}	**w, al, eth, ace, bz**	$B4^4$, 546
4082	Butyl ethyl amine $C_4H_9NHC_2H_5$	101.19	....	108-9	......	$0.7398^{20/4}$	1.4040^{20}	**al, eth, ace, bz**	$B4^4$, 547
4083	Butyl bis(2-hydroxyethyl) amine $C_4H_9N(CH_2CH_2OH)_2$	161.25	....	$273\text{-}5^{741}, 80^{35}$	......	$0.9692^{20/4}$	1.4625^{20}	**w, al, eth, ace**	$B4^4$, 1520
4084	Butyl phenyl amine or *N*-butylanilene $C_4H_9NHC_6H_5$	149.24	....	$241.6, 118\text{-}20^{15}$	-14.4	$0.9322^{20/4}$	1.5341^{20}	al, eth	$B12^3$, 267
4085	*sec*-Butyl amine *(d)* or 2-Aminobutane *(d)* $CH_3CH_2CH(NH_2)CH_3$	73.14	$[\alpha]^{20/}{}_D$ + 7.4 (w)	63	-104.5	$0.724^{20/4}$	1.344^{20}	w, **al, eth**, ace	$B4^4$, 617
4086	*sec*-Butyl amine *(dl)* or 2-Amino butane *(dl)* $CH_3CH_2CH(NH_2)CH_3$	73.14	....	63.5^{764}	<-72	$0.7246^{20/4}$	1.3932^{20}	**al, eth**, ace	$B4^4$, 618
4087	*sec*-Butyl amine *(l)* or 2-Amino butane *(l)* $CH_3CH_2CH(NH_2)CH_3$	73.14	$[\alpha]^{20/}{}_D$ -7.4 (w, c=4.7)	63	......	$0.7205^{20/4}$	....	w, **al, eth**, ace	$B4^4$, 617
4088	*sec*-Butyl ethyl amine *(d)* *sec*-$C_4H_9NC_2H_5$	101.19	$[\alpha]^{15/}{}_D$ + 18	98	......	$0.7396^{15/4}$	1.4043^{15}	**al, eth, ace, bz**	$B4^3$, 307
4089	*sec*-Butyl ethyl amine *(dl)* or Ethyl-*sec*-butyl lamine *(dl)* *sec*-$C_4H_9NHC_2H_5$	101.19	....	$97\text{-}8^{741}$	-104.3	$0.7358^{20/4}$	....	**al, eth, ace, bz**	$B4^2$, 636
4090	*sec*-Butyl phenyl amine or *N*- *sec*-Butylaniline $CH_3CH_2CH(CH_3)NHC_6H_5$	149.24	....	$223, 112\text{-}4^{22}$	......	....	1.5333^{20}	....	$B12^3$, 269
4091	*tert*-Butyl amine $(CH_3)_3CNH_2$	73.14	....	44.4	-67.5	$0.6958^{20/4}$	1.3784^{20}	**w, al, eth**	$B4^4$, 657
4092	*tert*-Butyl phenyl amine or *N*- *tert*-Butylaniline $(CH_3)_3CNHC_6H_5$	149.24	....	$214\text{-}6^{763}, 93\text{-}8^{19}$	......	....	1.5270^{20}	al, ace, bz, chl	$B12^3$, 270
4093	Butyl 4-aminophenyl ketone $(4\text{-}H_2NC_6H_4)COC_4H_9$	177.25	cr (bz-peth)	$160\text{-}3^3$	74-5	....	....	al, eth	$B14^2$, 43
4094	Butylarsonic acid $C_4H_9AsO(OH)_2$	182.05	....	....	160	....	....	w, al	$B4^3$, 1824
4095	*n*-Butyl bromide or 1-Bromobutane $CH_3CH_2CH_2CH_2Br$	137.02	....	$101.6, 18.8^{30}$	-112.4	$1.2758^{20/4}$	1.4401^{20}	**al, eth, ace,** chl	$B1^4$, 258
4096	Isobutylbromide or 1-Bromo-2-methyl propane $(CH_3)_2CHCH_2Br$	137.02	....	$91.7, 41\text{-}3^{135}$	-117.4	$1.2532^{20/4}$	1.4348^{20}	al, eth, ace, bz	$B1^4$, 294
4097	*sec*-Butyl bromide *(dl)* or 2-Bromobutane $CH_3CH_2CHBrCH_3$	137.02	....	91.2	-111.9	$1.2585^{20/4}$	1.4366^{20}	eth, **ace**, chl	$B1^4$, 261
4098	*tert*-Butyl bromide or 2-Bromo-2-methylpropane $(CH_3)_3CBr$	137.02	....	73.25	-16.2	$1.2209^{20/4}$	1.4278^{20}	....	$B1^4$, 295
4099	*n*-Butyl chloride or 1-Chlorobutane $CH_3CH_2CH_2CH_2Cl$	92.57	....	78.44	-123.1	$0.8862^{20/4}$	1.4021^{20}	**al, eth**	$B1^4$, 246
4100	Isobutyl chloride or 1-Chloro-2-methylpropane $(CH_3)_2CHCH_2Cl$	92.57	....	68-70	-130.3	$0.8810^{20/4}$	1.39841^{20}	eth, ace, chl	$B1^4$, 287
4101	*sec*-Butyl chloride or 2-Chlorobutane $CH_3CH_2CHClCH_3$	92.57	....	68.2	-131.3	$0.8732^{20/4}$	1.3971^{20}	**al, eth,** bz, chl	$B1^4$, 248
4102	*tert*-Butyl chloride or 2-Chloro-2-methyl propane $(CH_3)_3CCl$	92.57	....	....	-25.4	$0.8420^{20/4}$	1.3857^{20}	**al, eth,** bz, chl	$B1^4$, 288
4103	Butyl (β-chloroethyl) ether $C_4H_9OCH_2CH_2Cl$	136.62	....	$154.5d, 49\text{-}50^{11}$	......	$0.9335^{20/4}$	1.4155^{20}	eth	$B1^4$, 1519
4103a	1,2-Butylene oxide $CH_3CH_2\underbrace{CH\text{-}CH_2}_{O}$	72.11	....	63.3	......	$0.837^{17/4}$	1.3851^{20}	al, **eth, ace**	$B1^4$, 796
4104	Isobutylene or 2-methylpropene $(CH_3)_2C{=}CH_2$	56.11	....	-6.9	-140.3	$0.5942^{20/4}$	1.3926^{-25}	al, eth, bz	$B1^3$, 762
4105	Isobutylene (trimer) or Trisobutylene $(C_4H_8)_3$	168.32	....	$174\text{-}81, 56^{10}$	-76	$0.7590^{20/4}$	1.4314^{20}	....	$B1^3$, 763

No.	Name, Synonyms, and Formula	Mol. wt.	Color, crystalline form, specific rotation and λ_{max} (log ε)	b.p. °C	m.p. °C	Density	n_D	Solubility	Ref.
4106	Isobutylene (tetramer) $(C_4H_8)_4$	224.44		242-6, 109.5^{15}	-98	0.7944$^{20/4}$	1.4482^{20}		B1^{4}, 803
4107	Isobutylene, 1-chloro or 1-chloro-2-methylpropene $(CH_3)_2C{=}CHCl$	90.55		68^{754}		0.9186$^{20/4}$	1.4221^{20}	**al, eth, ace**, chl	B1^{4}, 803
4108	Isobutylene, 3-chloro or 3-chloro-2-methyl propene $ClCH_2C(CH_3){=}CH_2$	90.55		71-2		0.9165$^{20/4}$	1.4291^{20}	**al, eth,** ace, chl	B1^{4}, 804
4109	Isobutylene, 1,1-dichloro or 1,1-dichloro-2-methylpropene $(CH_3)_2C{=}CCl_2$	125.00		108-9, 42-3^{75}		1.1449$^{20/0}$	1.4580^{20}	eth, bz, chl	B1^{4}, 2533
4110	Isobutylene glycol or 2-methyl-1,2-propane diol $(CH_3)_2C(OH)CH_2OH$	90.12		176,79-80^{12}		1.0024$^{20/4}$	1.4350^{20}	w, al, eth	B1^{4}, 2536
4111	Isobutylene glycol, 3-chloro or 3-chloro-2-methyl propylene glycol $ClCH_2C(OH)(CH_3)CH_2OH$	124.57		114-7^{20}		1.2362$^{20/4}$	1.4788^{20}	**w, al, eth**	B1^{4}, 805
4112	Isobutylene, 3,3-dichloro or 3,3-dichloro-2-methyl propene $Cl_2CHC(CH_3){=}CH_2$	125.00		108-12, 49-50^{120}		1.3631$^{24/4}$	1.4523^{24}	eth, bz, chl	B1^{3}, 1904
4113	Isobutylene, 3-methoxy or 3-methoxy-2-methyl propene $CH_3OCH_2C(CH_3){=}CH_2$	86.13		68^{773}		0.7698$^{20/4}$	1.3964^{20}		B17^{4}, 45
4115	Isobutylene, 1,1,3-trichloro or 1,1,3-trichloro-2-methyl propene $ClCH_2C(CH_3){=}CCl_2$	159.44		156, 45-6^{12}		1.346^{20}	1.4990^{20}	ace, bz, chl	B1^{4}, 805
4116	Isobutylene, 3,3,3-trichloro or 3,3,3-trichloro-2-methyl propene $Cl_3CC(CH_3)_2{=}CH_2$	159.44			132-4	1.293^{20}	1.4770^{20}	ace, bz, chl, aa	B1^{4}, 805
4117	Butyl ether or Dibutylether $C_4H_9OC_4H_9$	130.23		142	-95.3	0.7689$^{20/4}$	1.3992^{20}	**al, eth**, ace	B1^{4}, 1520
4118	*sec*-Butyl ether *(dl)* or Di-*sec*-Butyl ether $(sec\text{-}C_4H_9)_2O$	130.23		120-1		0.756^{25}	1.393^{25}	**al, eth**, ace	B1^{3}, 1533
4119	Isobutyl ether α,β-dichloro or α,β-Diclorodiisobutyl ether $(CH_3)_2CCCHCl\text{-}O\text{-}i\text{-}C_4H_9$	199.12		192.5, 83^{15}		1.031$^{5/4}$		eth, ace	B1, 675
4120	Butyl isobutyl ether $(CH_3)_2CHCH_2OC_4H_9$	130.23		148-52^{730}		0.7980$^{22/4}$	1.4077^{21}	al, **eth**, ace	B1^{4}, 1594
4121	Butyl ethyl ether $C_4H_9OC_2H_5$	102.18		96		0.7490$^{20/4}$	1.3818^{20}	**al, eth**, ace	B1^{4}, 1518
4122	*sec*-Butyl ethyl ether $sec\text{-}C_4H_9OC_2H_5$	102.18		81		0.7503$^{20/4}$	1.3802^{20}	al, eth	B1^{4}, 1572
4123	*tert*-Butyl ethyl ether $(CH_3)_3COC_2H_5$	102.18		73.1	-94	0.7519^{25}	1.3794^{20}	al, eth	B1^{4}, 1615
4124	Butyl ethynyl ether or Butoxy acetylene $C_4H_9OC{\equiv}CH$	98.14		102-4, exp ca 100		0.8200$^{20/4}$	1.4020^{20}	al, eth	B1^{4}, 2213
4125	Butyl ethyl sulfide $C_4H_9SC_2H_5$	118.24		144.2, 33.3^{10}	-95.1	0.8376$^{20/4}$	1.4491^{20}	al, chl	B1^{4}, 1558
4126	Butyl furfuryl ether $C_4H_9OCH_2(C_4H_3O)$	154.21		189.9^{765}		0.9516$^{20/4}$	1.4522^{20}	al, eth	B17^{2}, 115
4127	*tert*-Butyl hydro peroxide $(CH_3)_3COOH$	90.12		d at89, 35-7^{17}	6	0.8960^{20}	1.4015^{20}	w, al, eth, chl	B1^{4}, 1616
4128	*tert*-Butyl hypochlorite $(CH_3)_3COCl$	108.57	ye lig	77-8		0.9583$^{18/4}$	1.403^{20}	eth, ace, bz	B1^{4}, 1621
4129	*tert*-Butyl iodide or 2-Iodo-2-methyl propane $(CH_3)_3CI$	184.02		100, 20.8^{30}	-38.2	1.5445$^{20/4}$	1.4918^{20}	al, eth	B1^{4}, 300
4130	Isobutyl isocyanate $i\text{-}C_4H_9NCO$	99.13		106					B4^{4}, 653
4131	*tert*-Butyl isocyanate $(CH_3)_3CNCO$	99.13		85.5		0.8670^{9}	1.4061^{20}		B4^{4}, 669
4132	*tert*-Butyl isocyanide or 2-Isocyano-2-methyl propane $(CH_3)_3CNC$	83.13		167-70, 91^{38}				**eth, al**	B4^{4}, 562
4133	Butyl isothiocyanate C_4H_9NCS	115.19		168, 64-6^{12}		0.9546$^{20/4}$	1.501^{20}	al, eth	B4^{4}, 596
4134	*sec*-Butyl isothiocyanate *(d)* or *sec*-Butyl mustard oil $sec\text{-}C_4H_9NCS$	115.19	$[\alpha]^{20/}{}_D$ +61.88	159		0.943$^{20/4}$		al, eth	B4^{1}, 372
4135	*sec*-Butyl isothiocyanate *(dl)* $sec\text{-}C_4H_9NCS$	115.19		159.5		0.944^{12}		al, eth	B4^{4}, 624
4136	*sec*-Butyl isothiocyanate *(l)* $sec\text{-}C_4H_9NCS$	115.19	$[\alpha]^{20/}{}_D$ -61.8	159		0.942$^{20/4}$		al, eth	B4, 161
4137	Isobutyl isothiocyanate or Isobutyl mustard oil $i\text{-}C_4H_9NCS$	115.19		160 (cor)		0.9638$^{14/4}$	1.5005^{14}	al, eth	B4^{4}, 653

No.	Name, Synonyms, and Formula	Mol. wt.	Color, crystalline form, specific rotation and λ_{max} (log ε)	b.p. °C	m.p. °C	Density	n_D	Solubility	Ref.
4138	*tert*-Butyl isothiocyanate $(CH_3)_3CNCS$	115.19		140^{770}	10-11	$0.9187^{10/4}$		eth	$B4^4$, 669
4139	*n*-Butyl mercaptan or l-Butanethiol $CH_3CH_2CH_2CH_2SH$	90.18		98.4	-115.7	$0.8337^{20/4}$	1.4440^{20}	al, eth	$B1^4$, 1555
4140	Isobutyl mercaptan or 2-Methyl-l-propanethiol $(CH_3)_2CHCH_2SH$	90.18		88.7	<-70	$0.8339^{20/4}$	1.4387^{20}	al, eth, ace	$B1^4$, 1605
4141	*sec*-Butyl mercaptan *(d)* or 2-Butanethiol $CH_3CH_2CHSHCH_3$	90.18	$[\alpha]^{20/}_D$ + 15.7	85-95, 37.4^{134}		$0.8299^{20/4}$	1.4338^{25}	al, eth, bz, peth	$B1^3$, 1549
4142	*sec*-Butyl mercaptan *(dl)* $CH_3CH_2CH(SH)CH_3$	90.18		85		$0.8295^{20/4}$	1.4366^{20}	al, eth, bz, peth	$B1^4$, 1584
4143	*sec*-Butyl mercaptan *(l)* $CH_3CH_2CH(SH)CH_3$	90.18	$[\alpha]^{17/}_D$ - 17.35	83-4		$0.8300^{17/4}$		al, eth, bz, peth	$B1^3$, 1549
4144	*tert*-Butyl mercaptan or 2-Methyl-2-propanethiol $(CH_3)_3CSH$	90.18		64.2	1.11	$0.8002^{20/4}$	1.4232^{20}		$B1^4$, 1634
4145	Butyl methyl ether $C_4H_9OCH_3$	88.15		71	-115.5	$0.7443^{20/4}$	1.3736^{20}	**al, eth**, ace	$B1^4$, 1518
4146	Isobutyl methyl ether *i*-$C_4H_9OCH_3$	88.15		58		$0.7311^{20/4}$		al, eth	$B1^4$, 1593
4147	*sec*-Butyl methyl ether *sec*-$C_4H_9OCH_3$	88.15		60		$0.7415^{20/4}$	1.3680^{25}	al, eth, ace	$B1^4$, 1572
4148	*tert*-Butyl methyl ether $(CH_3)_3COCH_3$	88.15		55.2	109	$0.7405^{20/4}$	1.3690^{20}	al, eth	$B1^4$, 1615
4149	Butyl methyl sulfide $C_4H_9SCH_3$	104.21		123.2	-97.8	$0.8426^{20/4}$	1.4477^{20}	al	$B1^4$, 1557
4150	Isobutyl methyl sulfide *i*-$C_4H_9SCH_3$	104.21		112.5		$0.8335^{20/4}$	1.4433^{20}	al, eth, ace	$B1^4$, 1606
4151	*n*-Butyl nitrate $C_4H_9ONO_2$	119.12		135.5, $70\text{-}1^{86}$		1.0228^{30}	1.4013^{23}	al, eth	$B1^4$, 1524
4152	*sec*-Butyl nitrate *sec*-$C_4H_9ONO_2$	119.12		124, 59^{80}		$1.0264^{20/4}$	1.4015^{20}	**al, eth**	$B1^4$, 1573
4153	iso-butyl nitrite *i*-C_4H_9ONO	103.12		67		$0.8699^{22/4}$	1.3715^{22}	al, eth	$B1^4$, 1595
4154	*tert*-Butyl nitrite $(CH_3)_3CONO$	103.12	pa ye	63, 34^{250}		$0.8670^{20/4}$	1.368^{20}	al, eth, chl	$B1^4$, 1622
4155	*tert*-Butyl peroxide or Di-*tert*-butyl peroxide $(CH_3)_3COOC(CH_3)_3$	146.23		111, 70^{197}	-40	0.704^{20}	1.3890^{20}	ace, lig	$B1^4$, 1619
4156	Butyl phenyl ether $C_4H_9OC_6H_5$	150.22		210, 95^{17}	-19.4	$0.9351^{20/4}$	1.4969^{20}	eth, ace	$B6^4$, 558
4157	*sec*-Butyl phenyl ether *sec*-$C_4H_9OC_6H_5$	150.22		194-5,$70\text{-}2^{5}$		$0.9415^{20/4}$	1.4926^{25}	eth	$B6^4$, 558
4158	*tert*-Butyl phenyl ether $(CH_3)_3COC_6H_5$	150.22		185-6		0.9214^{20}			$B6^4$, 559
4159	Butyl phenyl ketone or Valerophenone $C_4H_9COC_6H_5$	162.23		248.5, $131\text{-}3^{13}$		$0.988^{20/20}$	1.5158^{20}	al, eth	$B7^3$, 1114
4160	Butylphosphonic acid $C_4H_9PO(OH)_2$	138.10	pl (bz)	d	106			w, al, eth	$B4^3$, 1782
4161	Isobutyl phosphonic acid *i*-$C_4H_9PO(OH)_2$	138.10	pl (xyl)	d	119			w, al, eth, xyl	$B4^1$, 573
4162	*sec*-Butyl phosphonic acid *sec*-$C_4H_9PO(OH)_2$	138.10	lf (eth-lig)		48			w, al, eth, bz	Am75, 3379
4163	*tert*-Butyl phosphonic acid *tert*-$C_4H_9PO(OH)_2$	138.10	wh nd (xyl, aa-lig)	d	192			w, al	Am75, 3379
4164	Butyl propyl ether $C_4H_9OC_3H_7$	116.20		117.1		$0.7773^{0/0}$		al, eth	$B1^4$, 1519
4165	Isobutyl propyl ether *i*-$C_4H_9OC_3H_7$	116.20		$105\text{-}6^{720}$		0.7549^{20}	1.3852^{25}	al, **eth**	$B1^4$, 1594
4166	Butyl isopropyl ether *i*-$C_3H_7OC_4H_9$	116.20		108^{738}		$0.7594^{15/4}$	1.3870^{15}	al, eth, ace	$B1^4$, 1519
4167	Butyl sulfate $(C_4H_9O)_2SO_2$	210.29		109.5^{4}	1.4192^{20}	1.0616^{20}			$B1^4$, 1523
4168	Butyl sulfide (α-form) or α-Dibutyl ssulfide $(C_4H_9)_2S$	146.29		185	-79.7	$0.8386^{20/4}$	1.4530^{20}	al, eth, chl	$B1^4$, 1559
4169	Butyl sulfide (β-form) or β-Dibutyl sulfide $(C_4H_9)_2S$	146.29		190-230d				al, eth, ace	Ber62, 2168
4170	Butyl sulfide-2,2′-dimethyl *(d)* $[C_2H_5CH(CH_3)CH_2]_2S$	146.29		165		$0.8348^{20/4}$	1.4506^{20}	**al, eth**	B1, 387

No.	Name, Synonyms, and Formula	Mol. wt.	Color, crystalline form, specific rotation and λ_{max} (log ε)	b.p. °C	m.p. °C	Density	n_D	Solubility	Ref.
4171	Isobutyl sulfide or Di isobutyl sulfide $(i\text{-}C_4H_9)_2S$	146.29		170.5^{752}	-105.5	0.8363^{10}			B1^4, 1607
4172	sec-Butyl sulfide or Di-*sec*-Butyl sulfide $(sec\text{-}C_4H_9)_2S$	146.29		165		0.8348$^{20/4}$	1.4506^{20}	al, eth	B1^4, 1586
4173	Butyl sulfite $(C_4H_9O)_2SO$	194.30		230, 116^{19}		0.9957$^{20/4}$	0.4310^{20}	al, eth	B1^4, 1522
4174	Isobutyl sulfite $(i\text{-}C_4H_9O)_2SO$	194.30		209^{741}, 92-4^{13}		0.9862$^{20/4}$	1.4268^{20}		B1^4, 1607
4175	*n*-Butyl sulfonamide $C_4H_9SO_2NH_2$	137.20	lf (eth-lig)		48			w, al, eth, bz	B4^4, 45
4176	*n*-Butyl sulfonyl chloride $C_4H_9SO_2Cl$	156.63		75^{10}			1.4559^{20}		B4^4, 45
4177	Isobutyl sulfonyl chloride $i\text{-}C_4H_9SO_2Cl$	156.63		189-91, 87^{15}			1.4520^{25}		B4^4, 49
4178	Butyl sulfoxide or Di butyl sulfoxide $(C_4H_9)_2SO$	162.29	nd (dil al)	d	32.6	0.8317$^{21/4}$	1.4669^{20}	al, eth	B1^4, 1561
4179	Butyl thiocyanate C_4H_9SCN	115.19		185^{743}		0.9563^{15}	1.4360^{20}	al, eth	B3^4, 329
4180	Isobutyl thiocyanate $(CH_3)_2CHCH_2SCN$	115.19		175.4, 66^{15}	− 59			al, eth	B3^4, 330
4181	*tert*-butyl thiocyanate $(CH_3)_3CSCN$	115.20		140^{770}d, 39-40^{10}	10.5	0.9187^{10}			B3^4, 330
4182	Butyl 2-tolyl ether $(2\text{-}CH_3C_6H_4)OC_4H_9$	164.25		223		0.943$^{0/0}$			B6^3, 1247
4183	Butyl 3-tolyl ether or 3-Butoxy toluene $(3\text{-}CH_3C_6H_4)OC_4H_9$	164.25		229.2		0.9407$^{70/0}$	1.4970^{20}	eth	B6^4, 2040
4184	Butyl 4-tolyl ether or 4-Butoxy toluene $(4\text{-}CH_3C_6H_4)OC_4H_9$	164.25		229.5, 88^3		0.9232$^{25/25}$	1.4970^{20}	eth	B6^3, 1354
4185	Butyl vinyl ether $C_4H_9OCH{=}CH_2$	100.16		93.8	− 92	6.7888$^{20/4}$	1.4026^{20}	al, **eth**, ace, bz	B1^4, 2052
4186	Isobutyl vinyl ether $i\text{-}C_4H_9OCH{=}CH_2$	100.16		83	− 112	0.7645$^{20/4}$	1.3966^{20}	al, **eth**, ace, bz	B1^4, 2054
4187	Butyl nitrite C_4H_9ONO	103.12		− 77.8, 27^{88}		0.8823$^{20/4}$	1.3762^{20}	**al, eth**	B1^4, 1523
4188	*sec*-Butyl nitrite $sec\text{-}C_4H_9ONO$	103.12		68-9, 28^{180}		0.8726$^{20/4}$	1.3710^{20}	al, eth, chl	B1^4, 1573
4189	Isobutylene or 2-Methylpropene $(CH_3)_2C{=}CH_2$	56.11	gas	− 6.9	− 140.3	0.5942$^{20/4}$	1.3926^{25}	al, eth, bz	B1^4, 796
4190	Isobutylene trimer or Triisobutylene $(C_4H_8)_3$	168.32		179.8, 56^{16}	− 76	0.7590^{204}	1.4314^{20}		B1^3, 762
4191	Isobutylene tetramer or Tetraiso butylene $(C_4H_8)_4$	224.43		243-6, 109.5^{22}	− 98	0.7444^{20}	1.4482^{20}		B1^3, 763
4192	Isobutylene, l-chloro $(CH_3)_2C{=}CHCl$	90.55		68^{754}		0.9186$^{20/4}$	1.4221^{20}	**al, eth,** ace, chl	B1^4, 803
4193	Isobutylene, 3-chloro or 3-Chloro-2-methylpropene $ClCH_2C(CH_3){=}CH_2$	90.55		71-2		0.9165$^{20/4}$	1.4291^{20}	**al, eth,** ace, chl	B1^4, 803
4194	Isobutylene oxide or l,2-Epoxy-2-methylpropane $(CH_3)_2CCH_2$ (O bridge)	72.11		52		0.8650^0	1.3712^{22}	al, eth	B17^4, 46
4195	l-Butyne $CH_3CHC{\equiv}CH$	54.09		8.1	− 125.7	0.6784$^{0/0}$	1.3962^{20}	al, eth	B1^4, 969
4196	l-Butyne, 3-chloro $CH_3CHClC{\equiv}CH$	88.54		68.5		0.9466^{25}	1.4218^{25}		B1^4, 970
4197	l-Butyne, 3-chloro-3-methyl $(CH_3)_2CClC{\equiv}CH$	102.56		77-9		0.9061$^{20/4}$			B1^4, 1000
4198	l-Butyne, 3,3-dimethyl or *tert*-Butyl acetylene $(CH_3)_3CC{\equiv}CH$	82.15		39-40	− 81.2	0.6695$^{20/4}$	1.3738^{20}		B1^4, 1022
4199	l-Butyne, 3-methyl or Isopropyl acetylene $(CH_3)_2CHC{\equiv}CH$	68.12		29.5	− 89.7	0.6660$^{20/4}$	1.3723^{20}	**al, eth**	B1^4, 999
4200	l-Butyne, 4-phenyl $C_6H_5CH_2CH_2C{\equiv}CH$	130.19		190		0.9258$^{20/4}$	1.5208^{20}		B5^4, 1535
4201	2-Butynal $CH_3C{\equiv}CCHO$	68.08		106-7, 27-8^{34}	− 26	0.9265$^{17/0}$	1.446^{19}	eth, ace	B1^4, 3540
4202	2-Butyne or Dimethyl acetylene $CH_3C{\equiv}CCH_3$	54.09		27	− 32.2	0.6910$^{20/4}$	1.3921^{20}	al, eth	B1^4, 971
4203	2-Butyne, l-chloro $CH_3C{\equiv}CCH_2Cl$	88.54		104-6		1.0152^{20}	1.4581^{20}	al, eth, ace	B1^4, 973

No.	Name, Synonyms, and Formula	Mol. wt.	Color, crystalline form, specific rotation and λ_{max} (log ε)	b.p. °C	m.p. °C	Density	n_D	Solubility	Ref.
4204	2-Butyne, 1,4-dibromo $BrCH_2C{\equiv}CCH_2Br$	211.88		92^{15}		2.014^{18}	1.588^{18}	eth, ace, chl	B1⁴, 974
4205	2-Butyne, 1,4-dichloro $ClCH_2C{\equiv}CCH_2Cl$	122.98		165-6, 73^{24}		$1.258^{20/4}$	1.5058^{20}	eth, ace, chl	B1⁴, 973
4206	2-Butyne, 1,4-diiodo $ICH_2C{\equiv}CCH_2I$	305.88	nd (al)	$70\text{-}2^{0.1}$	53			al, eth, ace, chl	B1⁴, 975
4207	2-Butyne dinitrile $NCC{\equiv}CCN$	76.06		$76\text{-}6.5^{753}$	20.5-1	$0.9703^{25/4}$	1.46471^{25}		B2⁴, 2295
4208	2-Butyne, 1,4-diol $HOCH_2C{\equiv}CCH_2OH$	86.09	pl (bz, AcOEt)	238, 145^{15}	58		1.4804^{20}	w, al, ace	B1⁴, 2687
4209	2-Butyne, 1,4-diol,diacetate $CH_3CO_2CH_2\text{-}C{\equiv}C\text{-}CH_2O_2CCH_3$	170.17		$122\text{-}3^{10}$			1.4611^{20}		B2⁴, 244
4210	3-Butyne, 1,2-diol $HC{\equiv}CCHOHCH_2OH$	86.09		$64\text{-}6^{0.2}$	40			w, al	B1⁴, 2689
4211	2-Butyne, perfluoro $CF_3C{\equiv}CCF_3$	162.03		− 24.6	− 117.4			al, eth, ace, aa	B1⁴, 972
4212	2-Butynedioic acid or Acetylene dicarboxylic acid $HO_2CC{\equiv}CCO_2H$	114.06	pl (eth)		179			w, al, eth	B2⁴, 2290
4213	2-Butynedioic acid, diethyl ester $H_5C_2O_2CC{\equiv}CCO_2C_2H_5$	170.17		184^{200}	1-2	$1.0675^{20/4}$	1.4425^{20}	al, eth	B2⁴, 2294
4214	2-Butynedioic acid, dimethyl ester $CH_3O_2CC{\equiv}CCO_2CH_3$	142.11		195-8d, 98^{20}		$1.1564^{20/4}$	1.4434^{20}	al, eth	B2⁴, 2291
4215	2-Butynoic acid or Tetrolic acid $CH_3C{\equiv}CCO_2H$	84.07	pl (eth, peth)	203, $99\text{-}100^{18}$	78	$0.9641^{20/4}$		w, al, eth, chl	B2⁴, 1690
4216	2-Butynoic acid, ethyl ester $CH_3C{\equiv}CCO_2C_2H_5$	112.13		163, 105^{190}		$0.9641^{20/4}$	1.4372^{20}		B2⁴, 1691
4217	2-Butyn-1-ol $CH_3C{\equiv}CCH_2OH$	70.09		143, $52\text{-}3^{14}$	− 2.2	$0.9370^{20/4}$	1.4530^{20}	al, **eth**	B1⁴, 2220
4218	3-Butyn-1-ol $HC{\equiv}CCH_2CH_2OH$	70.09		129	− 63.6	$0.9257^{20/4}$	1.4409^{20}	w, al	B1⁴, 2219
4219	3-Butyn-2-ol $HC{\equiv}C\text{-}CHOHCH_3$	70.09		107		0.8858^{20}	1.4265^{20}	w, al, eth	B1⁴, 2218
4220	3-Butyn-2-ol-2, methyl $(CH_3)_2COHC{\equiv}CH$	84.12		104, 56^{97}	+ 3	$0.8618^{20/4}$	1.4207^{20}	w, al	B1⁴, 2229
4221	2-Butynyl methyl ether $CH_3C{\equiv}CCH_2OCH_3$	84.12		99-100, 33^{27}		$0.8496^{20/4}$	1.4262^{20}	al, eth, bz	B1³, 1973
4222	Butyraldehyde or Butanal $CH_3CH_2CH_2CHO$	72.11		75.7	− 99	$0.8170^{20/4}$	1.3843^{20}	w, **al**, **eth**, ace, bz	B1⁴, 3229
4223	Butyraldehyde phenyl hydrazone $C_3H_7CH{=}NNHC_6H_5$	162.23		$190\text{-}5^{80}$, 152^{14}	93-5				B15³, 80
4224	Butyraldehyde oxime or Butyraldoxime $CH_3CH_2CH_2CH{=}NOH$	87.12		152^{715}	− 29.5	$0.923^{20/4}$		al, eth, ace, bz	B1⁴ 3234
4225	Butyraldehyde, 2-bromo $CH_3CH_2CHBrCHO$	151.00		33^{17}		1.469^{20}	1.4683^{20}	eth, ace, bz	B1⁴, 3241
4226	Butyraldehyde, 3-chloro, diethyl acetal $CH_3CHClCH_2CH(OC_2H_5)_2$	180.67		$70\text{-}1^{12}$		$0.9709^{20/4}$	1.4210^{20}	**al**, eth, ace, bz	B1⁴, 3239
4227	Butyraldehyde, 4-chloro $ClCH_2CH_2CH_2CHO$	106.55		$50\text{-}1^{13}$		$1.107^{8\ 15/15}$	$1.4466^{8.5}$	al, eth, ace	B1⁴, 3240
4228	Butyraldehyde, 2,3-dichloro $CH_3CHClCHClCHO$	141.00		$58\text{-}60^{20}$		$1.2666^{21/4}$	1.4618^{21}	al, eth, ace, chl	B1⁴, 3240
4229	Butyraldehyde, 2-ethyl $(C_2H_5)_2CHCHO$	100.16		$117\text{-}9^{160}$		$0.8110^{20/4}$	1.4025^{20}	**al**, **eth**	B1⁴, 3310
4230	Butyraldehyde, 3-hydroxy or Aldol $CH_3CH(OH)CH_2CHO$	88.11		83^{20}, d85		$1.103^{20/4}$	1.4238^{20}	**w**, **al**, eth, ace	B1⁴, 3984
4231	Butyraldehyde, 2-methyl *(dl)* $CH_3CH_2CH(CH_3)CHO$	86.14		92-3, 54^{200}		$0.8029^{20/4}$	1.3869^{20}	al, eth, ace	B1⁴, 3286
4232	Butyraldehyde, 3-methyl or Isovaleraldehyde $(CH_3)_2CH_2CH_2CHO$	86.14		92.5	− 51	$0.7977^{20/4}$	1.3902^{20}	al, eth	B1⁴, 3291
4233	Butyraldehyde, 3-methyl,oxime or Isovaleraldoxime $(CH_3)_2CH_2CH_2CH{=}NOH$	101.15		161.3	48.5	$0.8934^{20/4}$	1.4367^{20}	al, eth, ace	B1⁴, 3293
4234	Butyraldehyde, 2,2,3-trichloro or *n*-Butyl chloral $CH_3CHClCCl_2CHO$	175.44		163-5, 49^{8}		$1.3956^{20/4}$	1.4755^{20}	w, al eth	B1⁴, 3241
4235	Butyraldehyde, 2,2,3-trichloro,hydrate $CH_3CHClCCl_2CH(OH)_2$	192.45	rh pl or lf (w)	d	78	$1.694^{20/4}$		w, al, eth	B1³, 2768
4236	Isobutyraldehyde $(CH_3)_2CHCHO$	72.11		64.2-4.6		$0.7938^{20/4}$	1.3730^{20}	w, eth, ace, chl	B1⁴, 3262

PHYSICAL CONSTANTS OF ORGANIC COMPOUNDS (Continued)

No.	Name, Synonyms, and Formula	Mol. wt.	Color, crystalline form, specific rotation and λ_{max} (log ε)	b.p. °C	m.p. °C	Density	n_D	Solubility	Ref.
4237	Isobutyraldehyde, 2-chloro $(CH_3)_2CClCHO$	106.55		90		$1.053^{15/4}$	1.4160^{16}	al, eth	$B1^4$, 3267
4238	Isobutyraldehyde, 3-(4-isopropylphenyl) or Cyclamenaldehyde $4\text{-}(CH_3)_2CHC_6H_4CH_2CH(CH_3)CHO$	190.29		$133\text{-}7^{99}$, 115^9		0.951^{15}	1.5068^{20}	al, eth, bz	$B7^3$, 1200
4239	Isobutyraldehyde, 2-chloro $(CH_3)_2CClCHO$	106.55							$B1^4$, 3267
4240	Butyramide or Butanoic acid,amide $CH_3CH_2CH_2CONH_2$	87.12	lf (bz)	216	114.8	0.8850^{120}	1.4087^{130}	al	$B2^4$, 804
4241	Butyramide, α-bromo $CH_3CH_2CHBrCONH_2$	166.02	lf (bz), nd (ace)		112—3			w, al, eth, ace, bz	$B2^4$, 834
4242	Butyramide, 2-bromo-2-ethyl or Neuronal $(C_2H_5)_2CBrCONH_2$	194.07			67			al, eth, bz	$B2^3$, 755
4243	Butyramide, 2-bromo-3-methyl *(dl)* $(CH_3)_2CHCHBrCONH_2$	180.04	lf (bz)		133			w, lig	$B2^4$, 906
4244	Butyramide, 2-bromo-2-*iso*-propyl or Neodorme $(CH_3)_2CHCBr(C_2H_5)CONH_2$	208.10	nd (sub)	sub	50-1			al, eth, ace, bz	$B2^4$, 979
4244a	Butyramide, 2-bromo-*N*-methyl-*N*-phenyl $CH_3CH_2CHBrCON(CH_3)C_6H_5$	256.14	cr (lig)		44			to, lig	B12, 254
4245	Butyramide, *N,N*-diethyl $CH_3CH_2CH_2CON(C_2H_5)_2$	143.23		206, 97^{16}			1.4403^{25}	w, al	$B4^4$, 354
4246	Butyramide, *N,N*-dimethyl $CH_3CH_2CH_2CON(CH_3)_2$	115.18		185-8, $124\text{-}5^{100}$	− 40	$0.9064^{25/4}$	1.4391^{25}	w, al, **eth, ace, bz**	$B4^4$, 185
4247	Butyramide, 2,3-dimethyl $(CH_3)_2CHCH(CH_3)CONH_2$	115.18	pl (ace-peth)		130.9			al, eth	$B2^3$, 761
4248	Butyramide, 3,3-dimethyl or *tert*-Butyl acetamide $(CH_3)_3CCH_2CONH_2$	115.18	lf (w or ace-peth)		134			al	$B2^4$, 956
4249	Butyramide, 2-ethyl-*N,N*-diethyl $(C_2H_5)_2CHCON(C_2H_5)_2$	171.28		220-1, 108^{12}				al, eth, bz	B4, 111
4250	Butyramide, 3-methyl $(CH_3)_2CHCH_2CONH_2$	101.15	mcl lf (al)	224-8	137			w, al, eth, peth	$B2^4$, 902
4251	Butyramide, 3-methyl-2-phenyl $(CH_3)_2CHCH(C_6H_5)CONH_2$	177.25	nd (dil al)	$180—2^{14}$	111-2			al, eth	$B9^3$, 2518
4252	Butyramide, 2-phenoxy $CH_3CH_2CH(OC_6H_5)CONH_2$	179.22	nd (w or al)		123			al, eth, ace, chl	$B6^4$, 645
4253	Butyramide, 4-phenoxy $C_6H_5OCH_2CH_2CH_2CONH_2$	179.22	lf (dil al), nd (bz)		80			al	$B6^3$, 617
4254	Butyramide, *N*-phenyl or Butyranilide $C_3H_7CONHC_6H_5$	163.22	mcl pr (al, bz, eth)	189^{15}	97	1.134		al, eth	$B12^3$, 474
4255	Butyramide, 2-phenyl $CH_3CH_2CH(C_6H_5)CONH_2$	163.22	cr	185^{16}	86			al	$B9^3$, 2465
4256	Butyramide, 3-phenyl $CH_3CH(C_6H_5)CH_2CONH_2$	163.22	nd (dil al)		106-7			al	$B9^3$, 2459
4257	Butyramide, 4-phenyl $C_6H_5CH_2CH_2CH_2N$	163.22	pl (w)		84.5			al, eth	$B9^3$, 2453
4258	Butyramide, 2,2,3,3-tetramethyl $CH_3C(CH_3)_2C(OH_3)_2CONH_2$	143.23	nd (peth-al)		201-2			al	$B2^4$, 1018
4259	Isobutyramide $(CH_3)_2CHCONH_2$	87.12			127-9			chl	$B2^4$, 852
4260	Isobutyramide, 2-bromo or 2-Bromo-2-methylpropionamide $(CH_3)_2CBrCONH_2$	166.02	pr (chl)	145^{17}	148			al, chl	$B2^4$, 863
4262	Isobutyramide, *N*-phenyl $(CH_3)_2CHCONHC_6H_5$	163.22	mcl pr (al, eth), nd (lig)		106-7			al, eth	$B12^3$, 475
4263	Butyric acid or Butanoic acid $CH_3CH_2CH_2CO_2H$	88.11		165.5	− 4.5	$0.9577^{20/4}$	1.3980^{20}	al, eth	$B2^4$, 779
4264	Butyric acid, 2-acetyl,ethyl ester $CH_3CH_2CH(OCCH_3)CO_2C_2H_5$	158.20		$85\text{-}7^{13}$		$0.9924^{15/15}$	1.4237^{15}	al, eth	$B2^4$, 1592
4265	Butyric acid, 2-acetyl-3-oxo-ethyl ester or Ethyl-α-acetyl aceto acetate $(CH_3CO)_2CHCO_2C_2H_5$	172.18		209^{11}, 104^{16}		$1.1045^{20/4}$	1.4690^{20}	al, eth, bz	$B2^4$, 1781
4266	Butyric acid, allyl ester or Allylbutyrate, allylbutanoate $CH_3CH_2CH_2CO_2CH_2CH{=}CH_2$	128.17		$142\text{-}3^{772}$, 44.5^{15}		$0.9017^{20/4}$	1.4158^{20}	al, eth	$B2^4$, 793
4267	Butyric acid, 2-amino- *(d)* or 2-Aminobutanoic acid $C_2H_5CHNH_2CO_2H$	103.12	lf (dil al), $[\alpha]^{16}{}_D$ + 8.4		292d			w	$B4^3$, 1294

No.	Name, Synonyms, and Formula	Mol. wt.	Color, crystalline form, specific rotation and λ_{max} (log ε)	b.p. °C	m.p. °C	Density	n_D	Solubility	Ref.
4268	Butyric acid, 2-amino- *(dl)* $C_2H_5CHNH_2CO_2H$	103.12	lf (w)	sub	304d			w	$B4^3$, 1296
4269	Butyric acid, 2-amino-*(l)* $C_2H_5CHNH_2CO_2H$	103.12	lf (w-al), cr (al), $[\alpha]^{20}_D$ − 14.9		292d			w	$B4^3$, 1296
4270	Butyric acid, 3-amino- *(d)* $CH_3CH(NH_2)CH_2CO_2H$	103.12	pr (MeOH), $[\alpha]^{20}_D$ + 35.3		d at 220				$B4^3$, 1312
4271	Butyric acid, 3-amino-*(dl)* $CH_3CH(NH_2)CH_2CO_2H$	103.12	nd (al)		193-4			w	$B4^3$, 1312
4272	Butyric acid, 3-amino-*(l)* $CH_3CHNH_2CH_2CO_2H$	103.12	pr (MeOH), $[\alpha]^{20}_D$ − 35.2 (w, c=10)		d at 220			w	$B4^3$, 1312
4273	Butyric acid, 3-amino-2-hydroxy or 3-Methyl *iso*-serine $CH_3CHNH_2CHOHCO_2H$	119.12	pr (dil al)		200d			w	B4, 513
4274	Butyric acid, 3-amino-3-methyl $(CH_3)_2CNH_2CO_2H$	117.15	pr (w + 1), cr (dil al), nd (eth-al)		217			w	$B4^3$, 1364
4275	Butyric acid, 4-amino or Piperidinic acid $H_2NCH_2CH_2CH_2CO_2H$	103.12	pr or nd (dil al), lf (MeOH-eth)		203d			w	$B4^3$, 1316
4276	Butyric acid, 4-amino-2-hydroxy $H_2NCH_2CH_2CHOHCO_2H$	119.12	pr (w or dil al)		214			w	$B4^1$, 548
4277	Butyric acid, 4-amino-3-hydroxy *(dl)* $H_2NCH_2CHOHCH_2CO_2H$	119.12	pr (w), cr (dil al)		218				$B4^3$, 1635
4278	Butyric acid anhydride or Butyric anhydride $(C_3H_7CO)_2O$	158.20		199-201	− 75	$0.9668^{20/4}$	1.4070^{20}	eth	$B2^4$, 802
4279	Butyric acid, 2-benzoyl,ethyl ester $CH_3CH_2CH(COC_6H_5)CO_2C_2H_5$	220.27		152^{7}		$1.0706^{15/4}$	1.509^{15}	eth	$B10^3$, 3065
4280	Butyric acid, 4-benzoyl $C_6H_5CO(CH_2)_3CO_2H$	192.21	pl (w)		128-9			w	$B10^3$, 3060
4281	Butyric acid, benzyl ester or Benzyl butyrate $C_3H_5CO_2CH_2C_6H_5$	178.23		238-40, 105^{7}		$1.0111^{20/4}$	1.4920^{20}	al, eth	$B6^4$, 2266
4282	Butyric acid, 2-bromo *(d)* or α-Bromobutyric acid $CH_3CH_2CHBrCO_2H$	167.00	$[\alpha]^{20}_D$ + 35.2 (eth, c=20)	$105\text{-}7^{15}$		$1.568^{20/4}$	1.4483^{20}	al	$B2^3$, 630
4283	Butyric acid, 2-bromo *(dl)* $CH_3CH_2CHBrCO_2H$	167.00		217d, 108^{13}	− 4	$1.5669^{20/20}$		al, eth	$B2^4$, 833
4284	Butyric acid, 2-bromo,ethyl ester $CH_3CH_2CHBrCO_2C_2H_5$	195.06		177.5^{765}, $43\text{-}4^{5.5}$		$1.3297^{20/20}$	1.4475^{20}	al, eth	$B2^4$, 834
4285	Butyric acid, 2-bromo,methyl ester $CH_3CH_2CHBrCO_2CH_3$	181.03		170-2, $75\text{-}8^{18}$		1.4528^{20}	1.4029^{25}	al	$B2^4$, 834
4286	Butyric acid, 2-bromo-3-methyl-*(d)* $(CH_3)_2CHCHBrCO_2H$	181.03	pr (peth), $[\alpha]^{20}_D$ + 22.8 (bz, p=4)	230, $95\text{-}100^{2}$	44-5			al, eth, ace, bz	$B2^3$, 705
4287	Butyric acid, 2-bromo-3-methyl-*(dl)* $(CH_3)_2CHCHBrCO_2H$	181.03	pr (eth or chl)	230d, $136\text{-}40^{25}$	44	1.459^{20}		al, eth, ace, bz	$B2^4$, 905
4288	Butyric acid, 2-bromo-3-methyl-*(l)* $(CH_3)_2CHCHBrCO_2H$	181.03	cr (peth), $[\alpha]^{20}_D$ − 21.6 (bz, p=4)	150^{40}, $119\text{-}20^{14}$	43-4			al, eth, ace, bz	$B2^4$, 905
4289	Butyric acid, 2-bromo-3-methyl, ethyl ester $(CH_3)_2CHCHBrCO_2C_2H_5$	209.08		186, $73\text{-}4^{12}$		$1.2760^{20/4}$	1.4496^{20}	al, eth	$B2^4$, 906
4290	Butyric acid, 2-bromo-3-methyl, methyl ester $(CH_3)_2CHCHBrCO_2CH_3$	195.06		176-8, $64\text{-}5^{11}$		$1.353^{13/13}$	1.4530^{20}	al, eth	$B2^3$, 706
4291	Butyric acid, 3-bromo-3-methyl $(CH_3)_2CBrCH_2CO_2H$	181.03	nd (lig)		73.5			al, eth, bz	$B2^4$, 905
4292	Butyric acid, 4-bromo $BrCH_2CH_2CH_2CO_2H$	167.00		$124\text{-}7^{7}$	33				$B2^4$, 835
4293	Butyric acid, 4-bromo, methyl ester $BrCH_2CH_2CH_2CO_2CH_3$	181.03		186-7, 86^{15}		1.371^{25}	1.4567^{25}	al	$B2^4$, 835
4294	Butyric acid, butyl ester or Butyl butyrate $C_3H_7CO_2C_4H_9$	144.21		166.6, 55^{11}	− 91.5	$0.8700^{20/4}$	1.4075^{20}	al, eth	$B2^4$, 789

PHYSICAL CONSTANTS OF ORGANIC COMPOUNDS (Continued)

No.	Name, Synonyms, and Formula	Mol. wt.	Color, crystalline form, specific rotation and λ_{max} (log ε)	b.p. °C	m.p. °C	Density	n_D	Solubility	Ref.
4295	Butyric acid, *sec*-butyl ester-*(d)* or *sec*-Butyl butyrate.... $C_3H_7CO_2$-*sec*-C_4H_9	144.21	$[\alpha]^{20}_D + 22$	151.5^{747}, 54^{18}	− 91.5	$0.8737^{11/4}$	1.4011^{20}	al, bz, Py	$B2^3$, 600
4296	Butyric acid, *iso*-butyl ester or *iso*-Butyl butyrate....... $C_3H_7CO_2$-*i*-C_4H_9	144.21	....	157	......	$0.8364^{18/4}$	1.4032^{20}	al, eth	$B2^4$, 790
4297	Butyric acid, *sec*-butyl ester-*(dl)* .... $C_3H_7CO_2$-*sec*-C_4H_9	144.21	....	152.5, 52^{16}	......	$0.8609^{20/4}$	1.4019^{20}	....	$B2^4$, 790
4298	Butyric acid, *tert*-butyl ester or *tert*-Butyl butyrate...... $C_3H_7CO_2C(CH_3)_3$	144.21	....	145-7	......	....	$1.4007^{17.5}$	al, eth, ace	$B2^4$, 790
4299	Butyric acid, 2-chloro .... $CH_3CH_2CHClCO_2H$	122.55	....	189^{627}, 101^{15}	......	$1.1796^{20/4}$	1.4411^{20}	al, eth	$B2^4$, 821
4300	Butyric acid, 2-chloro, ethyl ester .... $CH_3CH_2CHClCO_2C_2H_5$	150.61	....	163-4, 63^{70}	......	1.0560^{20}	1.4248^{20}	al, eth	$B2^4$, 822
4301	Butyric acid, 2-chloro, methyl ester or Methyl 2-chloro-butyrate .... $CH_3CH_2CHClCO_2CH_3$	136.58	....	$145\text{-}6^{756}$	......	1.0979^{14}	1.4247^{20}	al, eth	$B2^4$, 821
4302	Butyric acid, 2-chloro-2-methyl .... $CH_3CH_2CCl(CH_3)CO_2H$	136.58	....	$200\text{-}5^{754/d}$	......	$1.1204^{20/4}$	1.4445^{20}	al, eth	$B2^3$, 688
4303	Butyric acid, 2-chloro-2-methyl, ethyl ester .... $CH_3CH_2C(CH_2)ClCO_2C_2H_5$	164.63	....	175^{747}	......	1.069^{14}	1.4388^{11}	al, eth	$B2^3$, 688
4304	Butyric acid, 2-chloro-2-methyl-3-oxo, ethyl ester or Ethyl *α*-chloro-*α*-acetopropionate.......................... $CH_3COC(CH_3)ClCO_2C_2H_5$	164.63	....	$178\text{-}9^{756}$	......	1.021^{11}	....	al, eth	$B3^2$, 433
4305	Butyric acid, 2-chloro-3-methyl or 2-chloroisovaleric acid $(CH_3)_2CHCHClCO_2H$	136.58	....	$210\text{-}2^{756}$, 126^{32}	20—2	1.135^{13}	1.4450^{11}	al, eth	$B2^4$, 904
4306	Butyric acid, 3-chloro *(dl)* .... $CH_3CHClCH_2CO_2H$	122.55	cr (eth)	116^{22}	16	$1.1898^{20/4}$	1.4221^{20}	al, eth	$B2^4$, 823
4307	Butyric acid, 3-chloro, ethyl ester .... $CH_3CHClCH_2CO_2C_2H_5$	150.61	....	109, 65^{15}	......	$1.0517^{20/4}$	1.4246^{20}	al	$B2^4$, 824
4308	Butyric acid, 3-chloro, methyl ester .... $CH_3CHClCH_2CO_2CH_3$	136.58	....	155-6	......	$1.0996^{20/4}$	1.4258^{20}	eth	$B2^4$, 824
4309	Butyric acid, 4-chloro .... $ClCH_2CH_2CH_2CO_2H$	122.55	....	196^{22}, $68^{0.2}$	16	$1.2236^{20/4}$	1.4642^{20}	eth	$B2^4$, 825
4310	Butyric acid, 4-chloro, ethyl ester .... $ClCH_2CH_2CH_2CO_2C_2H_5$	150.61	....	186,77^{10}	......	$1.0756^{20/4}$	1.4311^{20}	al, eth, ace	$B2^4$, 825
4311	Butyric acid, 4-chloro, methyl ester .... $ClCH_2CH_2CH_2CO_2CH_3$	136.58	....	$175\text{-}6^{764}$, 55^4	......	$1.1201^{20/4}$	1.4321^{20}	al, eth, ace	$B2^4$, 825
4312	Butyric acid, cyclohexyl ester or Cyclohexyl butyrate..... $C_3H_7CO_2C_6H_{11}$	170.25	....	212^{750}	......	$0.9572^{0/4}$	....	al	$B6^4$, 37
4313	Butyric acid, 2,3-dibromo (high m.p.) .... $CH_3CHBrCHBrCO_2H$	245.90	nd (eth)	$100\text{-}10^{20}$	87	....	....	al, eth, bz	$B2^4$, 837
4314	Butyric acid, 2,3-dibromo (low m.p.) .... $CH_3CHBrCHBrCO_2H$	245.90	nd (lig)	....	59-60	....	....	al, eth	$B2^4$, 837
4315	Butyric acid, 2,3-dibromo, ethyl ester .... $CH_3CHBrCHBrCO_2C_2H_5$	273.95	nd	113^{10}	58-9	....	....	al, eth	$B2^4$, 837
4316	Butyric acid, 2,4-dibromo, ethyl ester .... $BrCH_2CH_2CHBrCO_2C_2H_5$	273.95	....	$149\text{-}50^{52}$	......	$1.6990^{20/0}$	1.4960^{20}	al, eth	$B2^4$, 838
4317	Butyric acid, 2,2-dichloro .... $CH_3CH_2CCl_2CO_2H$	157.00	....	$107\text{-}10^{14}$	......	$1.389^{20/4}$	....	al, eth	$B2^2$, 254
4318	Butyric acid, 2,3-dichloro (high m.p.) .... $CH_3CHClCHClCO_2H$	157.00	pr (dil al)	131.5^{20}	78	....	....	al, eth	$B2^4$, 827
4319	Butyric acid, 2,3-dichloro (low m.p.) .... $CH_3CHClCHClCO_2H$	157.00	pr (dil al)	$124\text{-}5^{20}$	63	....	....	al, eth, bz, chl	$B2^4$, 827
4320	Butyric acid, 2,2-diethyl-3-oxo, ethyl ester .... $CH_3COC(C_2H_5)_2CO_2C_2H_5$	186.25	....	$215\text{-}6^{744}$, 64^3	......	$0.9717^{18/4}$	1.4326^{17}	al, eth	$B3^4$, 1625
4321	Butyric acid, 2,2-dimethyl .... $CH_3CH_2C(CH_3)_2CO_2H$	116.16	....	186, 80^{11}	− 14	$0.9276^{20/4}$	1.4145^{20}	....	$B2^4$, 954
4322	Butyric acid, (2,2-dimethylpropyl) ester or *neo*-pentyl butyrate .... $C_3H_7CO_2CH_2C(CH_3)_3$	158.24	....	165-6	......	0.8719^0	....	....	B2, 272
4323	Butyric acid, 2,2-dimethyl-3-oxo, ethyl ester .... $CH_3COC(CH_3)_2CO_2C_2H_3$	158.20	....	184, $40\text{-}1^3$	......	$0.9773^{20/20}$	1.4180^{20}	al, eth	$B3^4$, 1594
4324	Butyric acid, 2,3-dimethyl .... $(CH_3)_2CH\text{-}CH(CH_3)CO_2H$	116.16	....	191.7	− 1.5	$0.9275^{20/4}$	1.4146^{20}	al, eth	$B2^4$, 958
4325	Butyric acid, 3,3-dimethyl or *tert*-Butylacetic acid...... $(CH_3)_3CCH_2CO_2H$	116.16	....	190, 96^{26}	6-7	$0.9124^{20/4}$	1.4096^{20}	al, eth	$B2^4$, 955
4326	Butyric acid, 3,3-dimethyl-2-oxo .... $(CH_3)_3CCOCO_2H$	130.14	....	189^{747}, 80^{15}	90-1	....	....	eth, bz, chl	$B3^4$, 1595

No.	Name, Synonyms, and Formula	Mol. wt.	Color, crystalline form, specific rotation and λ_{max} (log ε)	b.p. °C	m.p. °C	Density	n_D	Solubility	Ref.
4327	Butyric acid, 2,3-epoxy-3-phenyl, ethyl ester $CH_3C(C_6H_5)CH\text{-}CO_2C_2H_5$	206.24	...	272—5, 147-9^{22}	...	1.0442^{20}	1.5182^{20}	...	B18^2, 275
4328	Butyric acid, ethyl ester or Ethylbutyrate $C_3H_7CO_2C_2H_5$	116.16	...	121-6, 48.8^{50}	− 100.8	0.8785$^{20/4}$	1.4000^{20}	al, eth	B2^4, 787
4329	Butyric acid, 2-ethyl or Diethyl acetic acid $(C_2H_5)_2CHCO_2H$	116.16	...	194, 90^{13}	− 31.8	0.9239$^{20/4}$	1.4132^{20}	**al, eth**	B2^4, 951
4330	Butyric acid, 2-ethyl-2-methyl $(C_2H_5)_2C(CH_3)CO_2H$	130.19	...	208.5, 104^{13}	<−20	...	1.4250^{20}	al	B2^4, 981
4331	Butyric acid, 4-fluoro $FCH_2CH_2CH_2CO_2H$	106.10	...	76-8^5	...	...	1.3993^{25}	al, eth	B2^4, 809
4332	Butyric acid, furfuryl ester $C_3H_7CO_2CH_2(C_4H_3O)$	168.19	...	212-3^{764}	...	1.0530$^{20/4}$	...	al, eth	B17^4, 1247
4333	Butyric acid, heptafluoro $F_3CCF_2CF_2CO_2H$	214.04	...	120^{735}	− 17.5	1.651$^{20/4}$	1.295^{15}	w, eth, to	B2^4, 810
4334	Butyric acid, heptafluoro, ethyl ester $F_3CCF_2CF_2CO_2C_2H_5$	242.09	...	95	...	...	1.3011^{20}	eth, ace	B2^4, 813
4335	Butyric acid, heptafluoro, methyl ester $CF_3CF_2CF_2CO_2CH_3$	228.07	...	80	...	1.483$^{20/4}$	1.295^{20}	eth, ace	B2^4, 812
4336	Butyric acid, heptyl ester or Heptyl butyrate $C_3H_7CO_2C_7H_{15}$	186.29	...	225.8, 105^{10}	− 57.5	0.8637^{20}	1.4231^{20}	al	B2^4, 791
4337	Butyric acid, hexyl ester or Hexyl butyrate $C_3H_7CO_2C_6H_{13}$	172.27	...	208	− 78	0.8652^{20}	1.4160^{15}	al	B2^4, 791
4338	Butyric acid, 2-hydroxy *(dl)* $CH_3CH_2CHOHCO_2H$	104.11	nd (CCl_4)	266d, 140^{14}	44-4.5	1.125^{20}	...	w, al, eth	B3^4, 754
4339	Butyric acid, 2-hydroxy, ethyl ester *(d)* $CH_3CH_2CHOHCO_2C_2H_5$	132.16	$[\alpha]^{22}{}_D + 8.4$	165-70	...	0.978^{15}	1.4101	al	B3^4, 756
4340	Butyric acid, 2-hydroxy, ethyl ester *(dl)* $CH_3CH_2CHOHCO_2C_2H_5$	132.16	...	167, 74.5^{24}	...	1.0069$^{20/4}$	1.4179^{20}	al	B3^4, 756
4341	Butyric acid, 2-hydroxy-3-methyl *(d)* $(CH_3)_2CHCHOHCO_2H$	118.13	cr (eth-peth), $[\alpha]^{20}{}_D$ − 1.81 (w, c=12)	124-5^{13}	69.5	...	...	w, al, eth, ace	B3^4, 830
4342	Butyric acid, 2-hydroxy-3-methyl *(dl)* $(CH_3)_2CHOHCO_2H$	118.13	rh bipyr	...	86	...	...	w, al, eth, ace	B3^4, 830
4343	Butyric acid, 3-hydroxy-*(dl)* $CH_3CHOHCH_2CO_2H$	104.11	...	130$^{12-14}$, 94-6$^{0.1}$	48-50	...	1.4424^{20}	w, al, eth	B3^4, 760
4344	Butyric acid, 3-hydroxy-*(l)* $CH_3CHOHCH_2CO_2H$	104.11	$[\alpha]^{25}{}_D$ − 24.5 (w, c=5)	...	49-50	...	...	w, al, eth	B3^4, 760
4345	Butyric acid, 3-hydroxy, ethyl ester *(dl)* $CH_3CHOHCH_2CO_2C_2H_5$	132.16	...	184-5^{755}, 76-7^{15}	...	1.017$^{20/4}$	1.4182^{20}	al, w	B3^4, 762
4346	Butryic acid, 3-hydroxy, methyl ester *(l)* $CH_3CHOHCH_2CO_2CH_3$	118.13	$[\alpha]^{20}{}_D$ − 21.09	76-7^{20}	...	1.058$^{20/20}$	...	w, al, eth, bz	B3^3, 569
4347	Butyric acid, 3-hydroxy-3-methyl $(CH_3)_2COHCH_2CO_2H$	118.13	...	162^{12}	< − 32	0.9384$^{20/4}$	1.5081^{20}	w, al, eth	B3^4, 827
4348	Butryic acid, 4-hydroxy $4HOCH_2CH_2CH_2CO_2H$	104.11	...	d at 178-80	< − 17	...	...	...	B3^4, 774
4349	Butyric acid, 4-hydroxy, lactone or γ-Butyrolactone $CH_2CH_2CH_2CO$ (ring via O)	86.09	...	206, 89^{12}	− 42	1.1286$^{16/0}$	1.4341^{20}	w, al, eth, ace, bz	B17^4, 4159
4350	Butyric acid, 4-hydroxy-2-methylene, lactone or γ-Butyrolactone $CH_2CH_2C(=CH_2)CO$ (ring via O)	98.10	...	85-6^{10}	...	1.1206^{20}	1.4650^{20}	w, al, eth, ace, bz	B17^4, 4304
4351	Butyric acid, methyl ester or Methyl butyrate $C_3H_7CO_2CH_3$	102.13	...	102.3	− 84.8	0.8984$^{20/4}$	1.3878^{20}	**al, eth**	B2^4, 786
4352	Butyric acid, 2-methyl *(d)* $CH_3CH_2CH(CH_3)CO_2H$	102.13	$[\alpha]^{15}{}_D$ + 19.2 (w)	176, 77^{12}	...	0.9419$^{20/4}$	1.4058^{20}	**al, eth**	B2^4, 888
4353	Butyric acid, 2-methyl *(dl)* $CH_3CH_2CH(CH_3)CO_2H$	102.13	...	177	<− 80	0.9410$^{20/4}$	1.4051^{20}	**al, eth**	B2^4, 889
4354	Butyric acid, 2-methyl *(l)* $CH_3CH_2CH(CH_3)CO_2H$	102.13	$[\alpha]^{20}{}_D$ − 24 (w, c=0.9)	176-7, 71-2^{12}	...	0.9340$^{20/4}$	1.4042^{25}	**al, eth**	B2^4, 888
4355	Butytic acid, 2-methyl butyl ester *(d)* $C_3H_7CO_2CH_2CH(CH_3)C_2H_5$	158.24	$[\alpha]^{20}{}_D + 3.5$	179^{765}	...	0.8620$^{20/4}$	1.4135^{20}	...	B2^3, 304
4356	Butyric acid, 2-methyl, ethyl ester *(d)* $CH_3CH_2CH(CH_3)CO_2C_2H_5$	130.19	$[\alpha]^{26}{}_{5892}$ + 5.16	131-3^{730}, 35^{16}	...	0.8689$^{25/4}$	1.3964^{20}	al, bz	B2^4, 890
4357	Butyric acid, 2-methyl butyl ester *(dl)* $C_3H_7CO_2CH_2CH(CH_3)C_2H_5$	158.24	...	166-7	...	0.862$^{20/4}$	1.4100^{25}	...	B2, 304

No.	Name, Synonyms, and Formula	Mol. wt.	Color, crystalline form, specific rotation and λ_{max} (log ε)	b.p. °C	m.p. °C	Density	n_D	Solubility	Ref.
4358	Butyric acid, 2-(2-methyl butyl) ester or *tert*-Pentyl butyrate $C_3H_7CO_2C(CH_3)_2C_2H_5$	158.24		164		$0.8646^{15/0}$			$B2^3$, 602
4359	Butyric acid, 3-methyl or Isovaleric acid $(CH_3)_2CHCH_2CO_2H$	102.13		176.7	− 29.3	$0.9286^{20/4}$	1.4033^{20}	**al, eth, chl**	$B2^4$, 895
4360	Butyric acid, 3-methyl,anhydride $[(CH_3)_2CHCH_2CO]_2O$	186.25		215^{762}, $102\text{-}3^{15}$		$0.9327^{20/4}$	1.4043^{20}	eth	$B2^4$, 901
4361	Butyric acid, 3-methyl, isopropyl ester $(CH_3)_2CHCH_2CO_2CH(CH_3)_2$	144.21		142^{756}, $69\text{-}70^{55}$		0.8538^{17}	1.3960^{20}	al, eth, ace	$B2^3$, 698
4362	Butyric acid, 3-methylbutyl ester or Isopentyl butyrate $C_3H_7CO_2CH_2CH_2CH(CH_3)_2$	158.24		178.5, $65\text{-}8^{12}$		$0.8651^{20/4}$	1.4110^{20}	al, eth	$B2^4$, 791
4363	Butyric acid, 3-methyl, isobutyl ester $(CH_3)_2CHCH_2CO_2\text{-}i\text{-}C_4H_9$	158.24		171.4, $60\text{-}2^{12}$		$0.8736^{20/4}$	1.4057^{20}	**al, eth,** ace	$B2^4$, 790
4364	Butyric acid, 3-methyl, ethyl ester or Ethyl isovalerate $(CH_3)_2CHCH_2CO_2C_2H_5$	130.19		134.7	− 99.3	$0.8656^{20/4}$	1.3962^{20}	al, eth	$B2^4$, 898
4365	Butyric acid, 3-methyl, methyl ester or Methyl isovalerate $(CH_3)_2CHCH_2CO_2CH_3$	116.16		116.7		$0.8808^{20/4}$	1.3927^{20}	al, eth, ace	$B2^4$, 897
4366	Butyric acid, 3-methyl-2-oxo $(CH_3)_2CHCOCO_2H$	116.12		170.5, 73^{11}	31.0 — 1.5	$0.9968^{20/4}$	1.3850^{16}	w, al, eth	$B3^4$, 1577
4367	Butyric acid, 3-methyl, propyl ester or Propyl isovalerate $(CH_3)_2CHCH_2CO_2C_3H_7$	144.21		155.7, 40.5^{13}		$0.8617^{20/4}$	1.4031^{20}	al, eth	$B2^4$, 898
4368	Butyric acid, 3-methyl-2-phenyl $(CH_3)_2CHCH(C_6H_5)CO_2H$	178.23	pr (lig)	$159\text{-}60^{14}$	63			al, eth	$B9^3$, 2518
4369	Butyric acid, octyl ester or Octyl butyrate $C_3H_7CO_2C_8H_{17}$	200.32		244.1	− 55.6	0.8629^{20}	$1.4267^{15/}{}_{H9}$	al	$B2^4$, 791
4370	Butyric acid, 2-oxo or α-Ketobutyric acid $CH_3CH_2COCO_2H$	102.09	pl	$80\text{-}2^{16}$	31-2	$1.200^{17/4}$	1.3972^{20}	w, al	$B3^4$, 1524
4371	Butyric acid, 2-oxo,oxime $CH_3CH_2C(NOH)CO_2H$	117.10	nd (w), tcl (diox)		164			al	$B3^4$, 1525
4372	Butyric acid, pentyl ester or Pentyl butyrate, amyl butyrate $C_3H_7CO_2C_5H_{11}$	158.24		186.4	− 73.2	$0.8713^{15/4}$	1.4123^{20}	al, eth	$B2^4$, 790
4373	Butyric acid, 1-methylpentyl ester *(d)* or *sec*-Hexyl butyrate $C_3H_7CO_2CH(CH_3)C_4H_9$	172.27	$[\alpha]_D$ + 10.16	85^{20}		$0.8744^{21/4}$			$B2^1$, 120
4374	Butyric acid, 2-phenoxy $CH_3CH_2CH(OC_6H_5)CO_2H$	180.20	nd (w), pl (lig)	258	98			al, eth, ace, bz	$B6^4$, 644
4375	Butyric acid, 2-phenoxy, ethyl ester $CH_3CH_2CH(OC_6H_5)CO_2C_2H_5$	208.26		$250\text{-}1^{749}$, $87\text{-}90^{2}$		1.0388^{21}		**al, eth,** ace, chl	$B6^4$, 645
4376	Butyric acid, 4-phenoxy $C_6H_5OCH_2CH_2CH_2CO_2H$	180.20	pl (lig), cr (w)	$192\text{-}7^{15}$	64-5			al, eth, ace, bz	$B6^4$, 645
4377	Butyric acid, 4-phenoxy, ethyl ester $C_6H_5OCH_2CH_2CH_2CO_2C_2H_5$	208.26		$170\text{-}3^{25}$		$1.045^{35/25}$	1.491^{33}	al, eth	$B6^2$,159
4378	Butyric acid, phenyl ester or Phenyl butyrate $C_3H_7CO_2C_6H_5$	164.20		227-8, 85^{8}		$1.0382^{15/4}$	$1.0267^{15/}{}_{15}$	al, eth	$B6^4$, 615
4379	Butyric acid, 2-phenyl $CH_3CH_2CH(C_6H_5)CO_2H$	164.20	pl (eth)	270-2, $145\text{-}50^{14}$	47.5			eth, bz	$B9^3$, 2461
4380	Butyric acid, 2-phenyl, methyl ester $CH_3CH_2CH(C_6H_5)CO_2CH_3$	178.23	nd (dil al)	228	77-8			al, eth	$B9^3$, 2461
4381	Butyric acid, 3-phenyl *(dl)* $CH_3CH(C_6H_5)CH_2CO_2H$	164.20		$140\text{-}5^{3}$	46-7	1.0701^{20}	1.5155^{20}		$B9^3$, 2457
4382	Butyric acid, 4-phenyl $C_6H_5CH_2CH_2CH_2CO_2H$	164.20	lf (w)	290, 171^{15}	52			al, eth	$B9^3$, 2451
4383	Butyric acid, propyl ester or Propyl butyrate $C_3H_7CO_2C_3H_7$	130.19		143, 39.2^{14}	− 97.2	$0.8730^{20/4}$	1.4001^{20}	**al, eth**	$B2^4$, 788
4384	Butyric acid, *iso*-propyl ester or *iso*-Propyl butyrate $C_3H_7CO_2CH(CH_3)_2$	130.19		130-1		$0.8588^{20/4}$	1.3936^{20}	al	$B2^4$, 789
4385	Butyric acid, 3-isopropyl-3-oxo,ethyl ester $CH_3COCH(i\text{-}C_3H_7)CO_2C_2H_5$	172.22		201^{758}, $97\text{-}8^{20}$		$0.9648^{18/4}$	$1.4256^{18.5}$	**al, eth**	$B3^4$, 1608
4386	Butyric acid, 3-thioxo, ethyl ester $CH_3CSCH_2CO_2C_2H_5$	146.20	dk red	75^{15}		$1.0554^{31/4}$	1.4712^{26}	al, eth	$B3^4$, 1552
4387	Butyric acid, 2,2,3-trichloro $CH_3CHClCCl_2CO_2H$	191.44	lf or nd (peth)	236-8	60			eth	$B2^3$, 629
4388	Butyric acid, 2,2,3-trichloro, ethyl ester $CH_3CHClCCl_2CO_2C_2H_5$	219.50		212, 101.9^{17}		$1.3138^{20/20}$		al, eth	B2, 281
4389	Butyric acid, 2,2,4-trichloro $ClCH_2CH_2CCl_2CO_2H$	191.44	cr (peth)		73-5			al, eth	B2, 281

No.	Name, Synonyms, and Formula	Mol. wt.	Color, crystalline form, specific rotation and λ_{max} (log ε)	b.p. °C	m.p. °C	Density	n_D	Solubility	Ref.
4390	Butyric acid, 2,3,3-trichloro $CH_3CCl_2CHClCO_2H$	191.44	pl (lig)	...	52	...	...	al, eth, ace, bz	B2, 281
4391	Butyric acid, 4,4,4-trichloro $Cl_3CCH_2CH_2CO_2H$	191.44	nd (w)	...	35	...	...	al, eth, chl	$B2^4$, 830
4392	Isobutyric acid or 2-Methyl propionic acid $(CH_3)_2CHCO_2H$	88.11	...	153.2, 53.7^{10}	− 46.1	$0.9681^{20/4}$	1.3930^{20}	al, eth, w	$B2^4$, 843
4393	Isobutyric acid, allyl ester or Allyl isobutyrate $(CH_3)_2CHCO_2CH_2CH=CH_2$	128.17	...	$133\text{-}5^{755}$	...	...	...	**al, eth,** ace	$B2^3$, 650
4394	Isobutyric acid anhydride or Isobutyric anhydride $[(CH_3)_2CHCO]_2O$	158.20	...	181.5^{734}, 89-90^{32}	− 53.5	$0.9535^{20/4}$	1.4061^{19}	eth, chl	$B2^4$, 851
4395	Isobutyric acid, 2-amino $(CH_3)_2C(NH_2)CO_2H$	103.12	ta or pr (w)	sub 280	337 (cor)	...	...	w	$B4^3$, 1322
4396	Isobutyric acid, benzyl ester $(CH_3)_2CHCO_2CH_2C_6H_5$	178.23	...	$114\text{-}5^{20}$	...	1.0075^{15}	1.4883^{20}	...	$B6^4$, 2267
4397	Isobutyric acid, 2-bromo or 2-Bromo-2-methyl propionic acid $(CH_3)_2CBrCO_2H$	167.00	cr (peth)	198-200, 115^{24}	48-9	$1.5225^{60/60}$	...	al, eth	$B2^4$, 862
4398	Isobutyric acid, 2-bromo, ethyl ester $(CH_3)_2C(Br)CO_2C_2H_5$	195.06	...	164^{762}, 70^{20}	...	$1.3182^{20/4}$	1.4446^{20}	al, eth	$B2^4$, 862
4399	Isobutyric acid, isobutyl ester or Isobutyl isobutyrate $(CH_3)_2CHCO_2CH_2CH(CH_3)_2$	144.21	...	148.6, $36\text{-}40^{11}$	− 80.6	$0.8750^{0/4}$	1.3999^{20}	al, eth, ace	$B2^4$, 847
4400	Isobutyric acid, *tert*-butyl ester or *tert*-Butyl isobutyrate $(CH_3)_2CHCO_2C(CH_3)_3$	144.21	...	126.7	...	...	1.3921^{20}	al, eth, ace	$B2^3$, 648
4401	Isobutyric acid, 2-chloro $(CH_3)_2CClCO_2H$	122.55	...	118^{50}	31	...	1.450^{20}	al	$B2^4$, 858
4402	Isobutyric acid, 2-chloro,ethyl ester $(CH_3)_2CClCO_2C_2H_5$	150.61	...	148-9	...	1.062^{0}	1.4109^{16}	al, eth	$B2^4$, 859
4403	Isobutyric acid, 2-chloro, methyl ester $(CH_3)_2CClCO_2CH_3$	136.58	...	135, $42\text{-}4^{17}$	...	$1.0893^{15/15}$	1.4122^{21}	eth	$B2^4$, 859
4404	Isobutyric acid, 3-chloro $ClCH_2CH(CH_3)CO_2H$	122.55	...	$128\text{-}33^{50}$	...	1.0153^{20}	1.4310^{20}	al, eth	$B2^4$, 860
4405	Isobutyric acid, cyclohexyl ester or Cyclohexyl isobutyrate $(CH_3)_2CHCO_2C_6H_{11}$	170.25	...	204^{750}	...	$0.9489^{0/4}$	...	al, eth	$B6^3$, 24
4406	Isobutyric acid, 3,3-dichloro-2-hydroxy $Cl_2CHC(CH_3)(OH)CO_2H$	173.00	pr (al-eth)	d	82-3	...	...	w, al, eth	$B3^2$, 224
4407	Isobutyric acid, ethyl ester or Ethyl isobutyrate $(CH_3)_2CHCO_2C_2H_5$	116.16	...	111.0	− 88.2	$0.8693^{20/4}$	1.3869^{18}	**al, eth,** ace	$B2^4$, 846
4408	Isobutyric acid, furfuryl ester or Furfuryl isobutyrate $(CH_3)_2CHCO_2CH_2(C_4H_3O)$	168.19	...	$85\text{-}6^{15}$	...	$1.0313^{20/4}$	...	al, eth	$B17^2$, 115
4409	Isobutyric acid, 2-hydroxy or Acetonic acid $(CH_3)_2C(OH)CO_2H$	104.11	hyg pr (eth), nd (bz)	212, $108\text{-}11^{8}$	82-3	...	...	al, eth, w	$B3^4$, 782
4410	Isobutyric acid, 2-hydroxy, ethyl ester $(CH_3)_2C(OH)CO_2C_2H_5$	132.16	...	150 (cor), 46^{14}	...	0.987^{20}	1.4080^{20}	w, al	$B3^4$, 783
4411	Isobutyric acid, 2-hydroxy, methyl ester $(CH_3)_2C(OH)CO_2CH_3$	118.13	...	137, $62\text{-}4^{12}$	...	...	1.4056^{20}	w, al	$B3^4$, 783
4412	Isobutyric acid, methyl ester or Methyl isobutyrate $(CH_3)_2CHCO_2CH_3$	102.13	...	92.3	− 84.7	$0.8906^{20/4}$	1.3840^{20}	**al, eth,** ace	$B2^4$, 846
4413	Isobutyric acid, 2-nitropentyl ester or 2-Nitropentyl isobutyrate $(CH_3)_2CHCO_2CH_2CH(NO_2)C_3H_7$	202.23	...	248-51, 122^{10}	...	$1.0329^{20/20}$	1.4315^{20}	**eth,** ace	$B2^3$, 648
4414	Isobutyric acid, isopentyl ester or Isopentyl isobutyrate $(CH_3)_2CHCO_2\text{-i-}C_5H_{11}$	158.24	...	168.9	...	0.8627^{20}	...	al, eth, ace	$B2^3$, 649
4415	Isobutyric acid, 3-phenyl $C_6H_5CH_2CH(CH_3)CH_2CO_2H$	164.20	pl (dil al)	272, $155\text{-}6^{11}$	36.5	...	...	al, eth	$B9^3$, 2472
4416	Isobutyric acid, propyl ester or Propyl isobutylrate $(CH_3)_2CHCO_2C_3H_7$	130.19	...	135-6	...	$0.8843^{0/4}$	1.3955^{20}	al, eth, ace	$B2^4$, 847
4417	Isobutyric acid, isopropyl ester or Isopropyl isobutylrate $(CH_3)_2CHCO_2CH(CH_3)_3$	130.19	...	120.7	...	$0.8471^{21/4}$	...	al, eth, ace	$B2^4$, 847
4418	Isobutyric acid, vinyl ester or Vinyl isobutyrate $(CH_3)_2CHCO_2CH=CH_2$	114.14	...	104-5	...	0.8932^{20}	1.4061^{20}	...	$B2^4$, 848
4419	γ-Butyrolactone $CH_2CH_2CH_2CO$ (ring closed through O)	86.09	...	206, 89^{12}	− 42	$1.1286^{16/0}$	1.4341^{20}	w, al, eth, ace, bz	$B17^4$, 4159

No.	Name, Synonyms, and Formula	Mol. wt.	Color, crystalline form, specific rotation and λ_{max} (log ε)	b.p. °C	m.p. °C	Density	n_D	Solubility	Ref.
4420	γ-Butyrolactone, 2-methylene or Butanoic acid-4-hydroxy-2-methylene, lactone $CH_2CH_2C(=CH_2)CO$	98.10		85-6^{10}		1.1206^{20}	1.4650^{20}	w, al, eth, ace, bz	B17^4, 4304
4421	Butyronitrile $CH_3CH_2CH_2CN$	69.11		118	− 112	0.7936$^{20/4}$	1.3842^{10}	**al, eth**, bz	B2^4, 806
4422	Butyronitrile, 4-bromo $BrCH_2CH_2CH_2CN$	148.00		205-7, 91^{12}		1.4967$^{20/4}$	1.4818^{20}	al, eth	B2^4, 836
4423	Butyronitrile, 4-chloro $ClCH_2CH_2CH_2CN$	103.55		189-91, 75^{11}		1.0934^{15}	1.4413^{20}	al, eth	B2^4, 827
4424	Butyronitrile, 3,4-epoxy or Epicyanohydrin CH_2CHCH_2CN	83.09	pr		162			al	B18^4, 3822
4425	Butyronitrile, 2-ethyl $(C_2H_5)_2CHCN$	97.16		145-6			1.3891^{24}	**al, eth**	B2^4, 953
4426	Butyronitrile, 2-methyl $CH_3CH_2CH(CH_3)CN$	83.13		125		0.7913$^{15/4}$	1.3933^{20}	al, eth	B2^4, 892
4427	Butyronitrile, 3-methyl or Isovaleronitrile $(CH_3)_2CHCH_2CN$	83.13		130.5, 53^{50}	− 100.8 f.p.	0.7914$^{20/4}$	1.3927^{20}	**al, eth**, ace	B2^4, 902
4428	Butyronitrile, 3-methyl-2-phenyl $(CH_3)_2CHCH(C_6H_5)CN$	159.23		245-9^{765}		0.967$^{15.5}$	1.5038^{25}	al, bz	B9^2, 364
4429	Butyronitrile, 2-phenoxy $CH_3CH_2CH(OC_6H_5)CN$	161.20		228-30^{748}, d				al, eth	B6, 164
4430	Butyronitrile, 4-phenoxy $C_6H_5OCH_2CH_2CH_2CN$	161.20	nd	287-9^{765}, 170.5^{22}	45-6			eth	B6^4, 646
4431	Butyronitrile, 2-phenyl $CH_3CH_2CH(C_6H_5)CN$	145.20		238-40^{763}, 141-3^{8}				al, eth, bz	B9^3, 2468
4432	Butyronitrile, 4-phenyl $C_6H_5CH_2CH_2CH_2CN$	145.20		142-5^{16}					B9^3, 2454
4433	Isobutyronitrile $(CH_3)_2CHCN$	69.11		103.8	− 71.5	0.7608$^{30/4}$	1.3720^{20}	al, eth, ace, chl	B2^4, 853
4434	Isobutyronitrile, 2-bromo $(CH_3)_2CBrCN$	148.00		139-40, 61-2^{5}		1.4796$^{15/4}$	1.4739^{15}	al, eth, bz	B2^4, 863
4435	Isobutyronitrile, 2-hydroxy or Acetone cyanohydrin $(CH_3)C(OH)CN$	85.11		82^{23}	− 19	0.932$^{20/4}$	1.3996^{20}	w, al, eth, ace, bz	B3^4, 785
4436	Isobutyronitrile, 2-hydroxy-3-chloro $ClCH_2C(OH)(CH_3)CN$	119.55		110^{27}		1.2027^{15}	1.4356^{11}	w, al, ace	B3^3, 599
4437	Butyrophenone or *n*-Propyl phenyl ketone $CH_3CH_2CH_2COC_6H_5$	148.20		228-9	11-3	0.988$^{20/4}$	1.5203^{20}	**al, eth**, ace	B7^3, 1075
4438	Butyrophenone, *p*-methyl or *p*-Totyl-propyl ketone $(4\text{-}CH_3C_6H_4)COC_3H_7$	162.23		251.5^{758}	12	0.9745$^{20/4}$	1.5232^{20}	al, eth	B7^3, 1127
4439	Isobutyrophenone $C_6H_5COCH(CH_3)_2$	148.20		221, 86^{4}		0.9863$^{11/4}$	1.5172^{20}	al, eth	B7^3, 1088
4440	Butyryl bromide C_3H_7COBr	151.00		128		1.4162$^{17/4}$	1.1596^{17}		B2^4, 804
4441	Butyrylbromide, 2-bromo $CH_3CH_2CHBrCOBr$	229.90		172-4, 57-60^{10}				eth	B2^4, 834
4442	Butyrylbromide, 3-methyl or Isovaleryl bromide $(CH_3)_2CHCH_2COBr$	165.03	mcl pl (al)	143				eth	B2^4, 902
4443	Isobutyryl bromide $(CH_3)_2CHCOBr$	151.00		116-8		1.4067$^{15/4}$	1.4552^{15}		B2^4, 852
4444	Isobutyryl bromide, 2-bromo $(CH_3)_2CBrCOBr$	229.90		162-4, 91-8^{100}		1.4067$^{14/4}$	1.4552^{14}	ace, CS_2	B2^3, 661
4445	Butyryl chloride $CH_3CH_2CH_2COCl$	106.55		102	− 89	1.0277$^{20/4}$	1.4121^{20}	**eth**	B2^4, 803
4446	Butyryl chloride, 2-bromo $CH_3CH_2CHBrCOCl$	185.45		150-2, 41^{12}		1.5320^{20}		eth	B2^4, 834
4447	Butyryl chloride, 2-chloro $CH_3CH_2CHClCOCl$	141.00		130-1, 51-2^{41}		1.2360^{17}	1.4475^{20}	eth	B2^4, 822
4448	Butyryl chloride, 2-chloro-2-methyl $CH_3CH_2C(CH_3)ClCOCl$	155.02		144^{750}		i.187^{14}		eth	B2^4, 893
4449	Butyryl chloride, 2-chloro-3-methyl $(CH_3)_2CHCHClCOCl$	155.02		148-9		1.135^{11}		eth	B2^4, 904
4450	Butyryl chloride, 3-chloro $CH_3CHClCH_2COCl$	141.00		40-1^{12}		1.2163$^{20/4}$	1.4509^{20}	CS_2	B2^4, 824
4451	Butyryl chloride, 4-chloro $ClCH_2CH_2CH_2COCl$	141.00		173-4, 60-1^{12}		1.2581$^{20/4}$	1.4616^{20}	eth	B2^4, 826
4452	Butyryl chloride, 4-chloro-3-oxo $ClCH_2COCH_2COCl$	154.98		117-9^{17}		1.4397$^{20/4}$	1.4860^{20}	bz	B3^3, 1207

No.	Name, Synonyms, and Formula	Mol. wt.	Color, crystalline form, specific rotation and λ_{max} (log ε)	b.p. °C	m.p. °C	Density	n_D	Solubility	Ref.
4453	Butyryl chloride, 2,2-dimethyl $CH_3CH_2C(CH_3)_2COCl$	134.61	....	132, 27[11]	......	$0.9801^{20/4}$	1.4245^{20}	eth	B2[4], 955
4454	Butyryl chloride, 2,3-dimethyl $(CH_3)_2CHCH(CH_3)COCl$	134.61	....	135-6[751], 38-9[18]	......	$0.9795^{20/4}$	....	eth	B2[3], 761
4455	Butyryl chloride, 3,3-dimethyl $(CH_3)_3CCH_2COCl$	134.61	....	128-30[745], 68[100]	......	$0.9696^{20/4}$	1.4210^{20}	eth	B2[4], 956
4456	Butyryl chloride, 2-ethyl or Diethylacetyl chloride $(C_2H_5)_2CHCOCl$	134.61	....	140, 40[20]	......	$0.9825^{20/4}$	1.4234^{20}	eth	B2[4], 952
4457	Butyryl chloride, 3-methyl or Isovaleryl chloride $(CH_3)_2CHCH_2COCl$	120.58	....	114-5[771]	......	$0.9844^{20/4}$	1.4149^{20}	eth	B2[4], 901
4458	Butyryl chloride, 2-methyl-*(dl)* $CH_3CH_2CH(CH_3)COCl$	120.58	....	116	......	$0.9917^{20/4}$	1.4170^{20}	....	B2[4], 891
4459	Butyryl chloride, 2,2,3-trimethyl $(CH_3)_2CHC(CH_3)_2COCl$	148.63	....	148-50	......	....	....	eth	B2[4], 982
4460	Isobutyryl chloride $(CH_3)_2CHCOCl$	106.55	....	92	− 90	$1.0174^{20/4}$	1.4079^{20}	eth	B2[4], 852
4461	Isobutyryl chloride, 2-chloro $(CH_3)_2CClCOCl$	141.00	....	126-7	......	....	1.4369^{20}	eth	B2[4], 859
4462	Isobutyryl chloride. 3-chloro $CH_3CH_2ClCHCOCl$	141.00	....	92	− 90.0	$1.0174^{20/4}$	1.4079^{20}	eth	B2[4], 860

No.	Name, Synonyms, and Formula	Mol. wt.	Color, crystalline form, specific rotation and λ_{max} (log ε)	b.p. °C	m.p. °C	Density	n_D	Solubility	Ref.
4463	Cacodyl or Tetramethyl biarsine $(CH_3)_2AsAs(CH_3)_2$	209.98	pl	165	− 6	1.447^{15}	...	al, eth	B4^{3}, 1831
4464	Cacodyl chloride or Dimethyl chloraisine $(CH_3)_2AsCl$	140.45	...	109	< − 45	$1.5046^{12/4}$	1.5203^{12}	al	B4^{3}, 1797
4465	Cacodyl oxide or bis-Dimethyl arsenous oxide $(CH_3)_2AsOAs(CH_3)_2$	225.98	...	150	− 25	1.4816^{15}	1.5225^{9}	al, eth	B4^{3}, 1814
4466	β-Cadinene-*(l)* or 3,9-Cadinadiene $C_{15}H_{24}$	204.36	$[\alpha]^{27}{}_D$ − 15.9 (chl, c=1)	274, 149^{20}	...	$0.9230^{20/4}$	1.5059^{20}	eth, lig	B5^{3}, 1086
4467	Caffeine or 1,3,7-Trimethylxanthine $C_8H_{10}N_4O_2$	194.19	wh nd (w + 1), hex pr (sub)	sub 178, sub 89^{15}	238 (anh)	1.23^{19}	...	Py, chl	B26^{2}, 266
4468	Caffeine benzoate $C_8H_{10}N_4O_2$-$C_6H_5CO_2H$	316.32	wh so pw	...	...	...	...	al, w	B26^{2}, 268
4469	Caffeine citrate $C_8H_{10}N_4O_2$-$C_6H_8O_7$	386.32	mcl cr	...	...	...	...	...	B26^{2}, 269
4470	Caffeine hydrobromide $C_8H_{10}N_4O_2$-HBr-$2H_2O$	311.14	ye	...	d 80-100	...	...	w	B26^{1}, 137
4471	Caffeine hydrochloride $C_8H_{10}N_4O_2 \cdot HCl \cdot 2H_2O$	266.68	mcl pr	...	d 80-100	...	...	...	B26^{2}, 268
4472	Caffeine, 2-hydroxybenzoate or Caffeine salicylate $C_8H_{10}N_4O_2$-$C_7H_6O_3$	322.32	wh nd (w)	...	137	...	...	al	B26^{2}, 269
4473	Caffeine, 3-methylbutanoate or Caffeine isovalerate $C_{10}H_8N_4O_2$-$C_5H_{10}O_2$	296.33	unst nd	...	...	...	...	w	B26, 467
4474	Caffeine sulfate $C_8H_{10}N_4O_2$-H_2SO_4	292.27	wh nd	...	...	...	...	...	B26, 466
4475	Caffeine, 8-ethoxy or 1,3,7-Trimethyl-2,6-dioxo-8-ethoxy purine $C_{10}H_{14}N_4O_3$	238.25	wh or yesh nd (w)	...	143	...	...	al	B26^{2}, 322
4476	Caffeine, 8-methoxy $C_9H_{12}N_4O_3$	224.22	wh nd (al or w)	...	176	$1.399^{25/4}$	...	al, bz	B26^{2}, 322
4477	Calciferol or Vitamin D2 $C_{28}H_{44}O$	396.66	pr (ace), $[\alpha]^{20}{}_D$ + 102.5 (al)	sub	115—8	...	...	al, eth, ace	B6^{3}, 3089
4478	Camphane or Bornane $C_{10}H_{18}$	138.25	hex pl (al), pr (MeOH)	sub 161	158-9	...	...	al, eth	B5^{4}, 319
4479	Camphane-3-carboxylic acid $C_{10}H_{17}CO_2H$	182.26	pl (dil aa), $[\alpha]^{20}{}_D$ + 56 (al, p=8)	153^{13}	90-1	...	...	al, aa	B9^{3}, 244
4480	Camphene-*(d)* $C_{10}H_{16}$	136.24	nd $[\alpha]^{17}{}_D$ + 103.9 (eth, c=4)	160-2, 52^{17}	52	$0.8450^{50/4}$	1.4570^{25}	eth	B5^{4}, 461
4481	Camphene *(dl)* $C_{10}H_{16}$	136.24	nd (sub)	158-9	51-2	$0.879^{20/4}$	1.4551^{54}	al, eth	B5^{4}, 462
4482	Camphene-*(l)* $C_{10}H_{16}$	136.24	$[\alpha]^{19}{}_D$ − 106.1 (eth, c=4)	158	52	$0.8446^{50/4}$	1.4564^{54}	eth	B5^{4}, 461
4483	Camphenilone $C_9H_{14}O$	138.21	$[\alpha]^{20}{}_D$ + 70.4 (al)	193^{751}, 76^{12}	41	...	...	eth	B7^{3}, 306
4484	3-Camphanol or Epiborneol $C_{10}H_{15}OH$	154.24	nd $[\alpha]^{17}$ + 11.1 (al)	213^{742}	181-2	...	...	peth	B6^{4}, 288
4485	3-Camphanol acetate $C_{12}H_{20}O_2$	196.29	$[\alpha]^{10}{}_D$ + 15.63	101^{11}	< − 15	$0.9872^{14/4}$	1.4651^{14}	...	B6^{2}, 92
4486	Campholic acid-*(d)* $C_{10}H_{18}O_2$	170.25	pr $[\alpha]^{20}{}_D$ + 59.3 (bz)	255^{768}, 146^{12}	106	...	...	al, eth	B9^{3}, 98
4487	Campholic acid-*(dl)* $C_{10}H_{18}O_2$	170.25	tcl pr	...	109	...	...	al, eth	B9^{3}, 99
4488	Campholic acid-*(l)* $C_{10}H_{18}O_2$	170.25	pr (dil al), $[\alpha]^{15}{}_D$ − 49.1 (al)	250	106-7	...	...	al, eth	B9^{3}, 99
4489	Campholytic acid-*(α,l)* $C_9H_{14}O_2$	154.21	$[\alpha]^{13}{}_D$ − 60.4	240-3, 140^{15}	...	1.0145^{18}	1.4712^{17}	lig	B9^{3}, 187
4490	Camphor-*(d)* or 2-Camphanone *(d)* $C_{10}H_{16}O$	152.24	pl $[\alpha]^{20}{}_D$ + 44.26 (al)	sub 204	179.8	$0.990^{25/4}$	1.5462	al, eth, ace, bz	B7^{3}, 400

No.	Name, Synonyms, and Formula	Mol. wt.	Color, crystalline form, specific rotation and λ_{max} (log ε)	b.p. °C	m.p. °C	Density	n_D	Solubility	Ref.
4491	Camphor-*(dl)* $C_{10}H_{16}O$	152.24	wh	sub	178.8			al, eth, ace, bz	B7^3, 406
4492	Camphor-*(l)* $C_{10}H_{16}O$	152.24	$[\alpha]^{16}_D$ − 44.2 (al, c=16.5)	204 sub	178.6	0.9853^{18}		al, eth, ace	B7^3, 405
4493	Camphor oxime-*(d)* $C_{10}H_{16}NOH$	167.25	pr (lig-eth), $[\alpha]^{22}_D$ + 42.5 (al)		115			al, eth	B7^3, 408
4494	Camphor oxime-*(dl)* $C_{10}H_{16}NOH$	167.25	cr (peth)		118			al, eth	B7^3, 409
4495	Camphor oxime-*(l)* $C_{10}H_{16}NOH$	167.25	mcl nd or pr (dil al), $[\alpha]^{20}_D$ − 42.4 (al)	249-54d	118	$1.01^{116/4}$		al, eth	B7^3, 408
4496	Camphor, 3-amino-*(d)* or 3-Camphoryl amine $C_{10}H_{17}NO$	167.25	wx	244	110-15d			al, eth	B14^3, 15
4497	Camphor, 3-bromo-(*d*,α) $C_{10}H_{15}BrO$	231.13	pr (al), $[\alpha]^{20}_D$ + 129.3 (MeOH, c=4.6)	274(d) sub	76			al, eth, bz, chl	B7^3, 414
4498	Camphor, 3-bromo-(*d*,α′) $C_{10}H_{15}BrO$	231.13	$[\alpha]^{20}_D$ 29.4	d265	78	1.484^{14}		al, chl	B7^3, 415
4499	Camphor, 3-bromo-(*dl*,α) $C_{10}H_{15}BrO$	231.13			51			al, eth, bz, chl	B7^3, 415
4500	Camphor, 3-bromo-(*l*,α) $C_{10}H_{15}BrO$	231.13	mcl nd (al), $[\alpha]^{18}_D$ − 138.8 (ace, c=6)		76			al, eth, bz, chl	B7^3, 415
4501	Camphor, 3-bromo-(*d*,α′) $C_{10}H_{15}BrO$	231.13	nd (dil al)	265d	78	1.484^{14}		al, chl, aa	B7^2, 101
4502	Camphor, 5-bromo-*(exo)* $C_{10}H_{15}BrO$	231.13	$[\alpha]^{16}_D$ bz 2%	100^{15}	114			al, eth, ace, bz	B7^3, 415
4503	Camphor, 8-bromo-*(d)* $C_{10}H_{15}BrO$	231.13	tetr pr (lig), $[\alpha]^{19/}_D$ + 122.2 (chl)	sub	93			al, eth, ace, bz	B7^3, 416
4504	Camphor, 8-bromo-(dl) $C_{10}H_{15}BrO$	231.13	pr (eth-peth), pym (eth)	sub	92.7			al, eth, ace, bz	B7^3, 417
4505	Camphor, 10-bromo-*(d)* $C_{10}H_{15}BrO$	231.13	pr (peth), $[\alpha]^{20}_D$ + 19.2 (abs al), + 15.7 (bz)	265d	78			al, eth, ace, bz	B7^3, 416
4506	Camphor, 10-bromo-*(dl)* $C_{10}H_{15}BrO$	231.13	cr		77				B7^3, 416
4507	Camphor, 3-chloro-(*d*,α) $C_{10}H_{15}ClO$	186.68	lf $[\alpha]^{20}_D$ + 71.1	244-7	94			al, eth, ace, bz	B7^3, 411
4508	Camphor, 3-chloro-(*d*,α′) $C_{10}H_{15}ClO$	186.68	$[\alpha]^{20}_D$ + 35 (al, c=5)	231d	118			eth, chl	B7^3, 411
4509	Camphor, 8-chloro-*(d)* $C_{10}H_{15}ClO$	186.68	pr (al), $[\alpha]_D$ + 99.9 (chl)		139				B7^3, 414
4510	Camphor, 8-chloro-*(dl)* $C_{10}H_{15}ClO$	186.68		sub	138				B7, 136
4511	Camphor, 10-chloro-*(d)* $C_{10}H_{15}ClO$	186.68	pr (al), $[\alpha]^{14}_D$ + 40.7 (al)		132-5			al, eth, bz, chl	B7^2, 100
4512	Camphor, 3,3-dibromo-*(d)* $C_{10}H_{14}Br_2O$	310.03	wh ye rh pr (al, peth), $[\alpha]^{20}_D$ + 40 (chl) + 39.2 (al)	sub	64	$1.8954^{21.6/4}$		al, eth, bz, chl	B7^2, 101
4513	Camphor, 3-nitro-*(l)* $C_{10}H_{15}NO_3$	197.23	mcl pr (bz), $[\alpha]^{13}_D$ - 26→-9 (al), (mut)					al, eth, bz, chl	B7^3, 419

No.	Name, Synonyms, and Formula	Mol. wt.	Color, crystalline form, specific rotation and λ_{max} (log ε)	b.p. °C	m.p. °C	Density	n_D	Solubility	Ref.
4514	Camphor-3-carboxylic acid-*(d)* $C_{11}H_{16}O_3$	196.25	pr (eth, 50% al), $[\alpha]^{20}_D$ +34.9 (bz)		128d			w, al, eth, bz	B10^3, 2925
4515	Camphor-3-carboxylic acid-*(dl)* $C_{11}H_{16}O_3$	196.25	cr (bz)		136-7			al, eth, bz	B10^3, 2927
4516	Camphor-3-carboxylic acid $C_{11}H_{16}O_3$	196.25	pr (eth), cr (bz), $[\alpha]^{20}_D$ -64 (al)		127-8d			al, eth, bz	B10^3, 2927
4517	β-Camphor-*(l)* or *l*-EpiCamphor, 1,3-Bornanone $C_{10}H_{16}O$	152.24	$[\alpha]^{19}_D$ -58.21 (bz, c=13)	213	184			al, eth, peth	B7^3, 421
4518	Camphoric acid-*(d)* or 1,2,2-Trimethylcyclopentane-1,3-dicarboxylic acid $C_{10}H_{16}O_4$	200.23	Pr, lf (w), $[\alpha]^{20}_D$ +47.7 (al)		188.2	1.186$^{20/4}$		al, eth, ace	B9^3, 3876
4519	Camphoric acid-*(dl)* $C_{10}H_{16}O_4$	200.23	pr (al, aa), mcl nd		208	1.228$^{20/4}$		al, ace	B9^3, 3878
4519a	Camphoric acid-*(l)* $C_{10}H_{16}O_4$	200.23	Cr (w), $[\alpha]^{18/}_D$ -48.1 (abs. al, c=8)	223	270d	1.190		al, eth, ace	B9^3, 3878
4520	Camphoric anhydride-*(d)* $C_{10}H_{14}O_3$	182.22	rh (al), pr (bz)	7270d	223.5	1.194^{20}		bz	B17^4, 5957
4521	Camphoric anhydride-*(dl)* $C_{10}H_{14}O_3$	182.22	rh (al)	270	221	1.194$^{20/4}$		bz	B17^1, 238
4522	Camphoric anhydride-*(l)* $C_{10}H_{14}O_3$	182.22	$[\alpha]_D$ -77 (bz)	>270	221	1.194$^{20/4}$		bz	B17^4, 5958
4523	Camphoric acid, diethyl ester-*(d)* $C_{14}H_{24}O_4$	256.34	$[\alpha]^{15/}_D$ +7.5 (al), +9 (bz)	286^{752}, 164^{20}		1.0298$^{20/4}$	1.4535^{20}	al, eth, bz	B9^2, 536
4524	Camphoric acid, dimethyl ester-*(d)* $C_{12}H_{20}O_4$	228.29	$[\alpha]_D$ +49.07 (al)	264^{758}, 155^{15}	< -16	1.0747$^{20/4}$	1.4627^{19}	al, eth	B9^3, 3880
4525	Camphoric acid, 1-monoamide or β-Camphoramic acid $C_{10}H_{17}NO_3$	199.25	pl (nd), $[\alpha]^{20}_D$ +74 (al)		183			al, ace	B9^3, 3886
4526	Camphoric acid, 3-monoamide-*(d)* or α-Camphoramic acid $C_{10}H_{17}NO_3$	199.25	nd or lf (w), $[\alpha]^{20/}_D$ +25 (al)		176-7				B9^3, 3887
4527	Camphoric acid, 3-monoamide-*(dl)* or α-Camphoramic acid $C_{10}H_{17}NO_3$	199.25	nd (w)		198				B9, 761
4528	Camphoronic acid-*(d)* or 2,4-Dimethylpentane-1,2,3-tricarboxylic acid $(CH_3)_2C(CO_2H)C(CH_3)(CO_2H)CH_2CO_2H$	218.21	nd (w), $[\alpha]^{19}$ +27.05 (w)					w, al, eth, ace	B2, 837
4529	Camphoronic acid-*(dl)* $(CH_3)_2C(CO_2H)C(CH_3)(CO_2H)CH_2CO_2H$	218.21	nd or pr (w)		172d			al, eth, ace	B2^3, 2045
4530	Camphoronic acid-*(l)* $(CH_3)_2C(CO_2H)C(CH_3)(CO_2H)CH_2CO_2H$	218.21	nd (w), $[\alpha]^{19/}_D$ -26.9		164-5d			w, al, eth, ace	B2^3, 2045
4531	Camphorpinacol-*(l)* or 2,2′-Bicamphane-2,2′-diol $(C_{10}H_{17}O)_2$	306.49	rh $[\alpha]_D$ −27.2 (bz)		158			al, eth	B6^3, 4767
4532	Camphor quinone or 2,3-Camphor dione $C_{10}H_{14}O_2$	166.22	ye nd (dil al, w), pr (eth) $[\alpha]^{20/}_D$ −113.2 (bz)	sub	199			al, eth, bz, chl	B7^3, 3297
4533	Camphor-3-sulfonic acid, methyl ester-*(d)* $C_{11}H_{18}SO_4$	246.32	cr (MeOH), nd (peth), $[\alpha]^{20/}_D$ +98.6 (chl, c=5)					al	B11^2, 179
4534	Camphor-10-sulfonic acid-*(d)* $C_{10}H_{16}O_4S$	232.29	pr (aa), $[\alpha]^{20/}_D$ +32.8 (AcOEt, + c=3), +24 (w)		195d			w	B11^3, 585

No.	Name, Synonyms, and Formula	Mol. wt.	Color, crystalline form, specific rotation and λ_{max} (log ε)	b.p. °C	m.p. °C	Density	n_D	Solubility	Ref.
4535	Camphor-10-sulfonic acid-*(dl)* $C_{10}H_{16}O_4S$	232.29	cr (aa)		202d			w	B11[3], 587
4536	Camphor-10-sulfonic acid-*(l)* $C_{10}H_{16}O_4S$	232.29	cr (aa), nd (AcoEt), $[\alpha]^{20}_D$ −20.75 (w)		194-5d			w	B11[3], 585
4537	α-Camphyl amine or 4-(2-Aminoethyl)-1,5,5-trimethyl cyclopentene $C_{10}H_{19}N$	153.27	$[\alpha]_D$ + 6	194-6, 95^{12}		0.8688^{20}	1.4728^{18}		B12[3], 176
4538	β-Camphyl amine or 2-(2-Aminoethyl)-1,5,5-trimethyl cyclopentene $C_{10}H_{19}N$	153.27		206		$0.8697^{20/20}$			B12, 40
4539	Canadine-*(d)* or *d*-Tetrahydrobeberine $C_{20}H_{21}NO_4$	339.39	ye nd (dil al), $[\alpha]^{29/}_D$ + 299 (chl, c=1)		132 (140)			al, eth, bz, chl	B27[2], 557
4540	Canadine-*(dl)* $C_{20}H_{21}NO_4$	339.39	mcl nd (al)		174			chl	B27[2], 557
4541	Canadine-*(l)* $C_{20}H_{21}NO_4$	339.39	ye nd (al) $[\alpha]^{20/}_D$-299.2 (chl)		134			al, eth, bz, chl	B27[2], 557
4542	Canaline $H_2NOCH_2CH_2CH(NH_2)CO_2H$	134.14	nd (al), $[\alpha]^{23/}_D$ -8.31 (w)		214d			w, al	B4[3], 1636
4543	Canavanine-*(L)* $HN{:}C(NH_2)NHOCH_2CH_2CH(NH_2)CO_2H$	176.18	cr (al), $[\alpha]^{20/D}$ + 7.9 (w, C=2)					w	B4[3], 1636
4544	Cannabidiol $CH_3C_6H_7CCH_3CH_2C_6H_2(OH)_2(CH_2)_4CH_3$	314.47	rods (peth)	$187\text{-}90^{2}$	67			al, eth, bz, chl	B6[3], 5362
4545	Cannabinol $C_{21}H_{26}O_2$	310.44	pl, lf (peth) $[\alpha]^{20/}_D$ -148 (al)	$185^{0.05}$	77			al, eth, ace, bz	B17[4], 1652
4546	Cantharidin $C_{10}H_{12}O_4$	196.20	rh pl or sc	sub 84	218				B19[4], 1958
4547	Capraldehyde or Decanal $CH_3(CH_2)_8CHO$	156.27		208-9, 81^{9}	−5	$0.830^{15/4}$	1.4287^{20}	al, eth, ace	B1[4], 3366
4548	Capradehyde oxime or Capraldoxime $CH_3(CH_2)_8CHOH$	171.28	lf (dil MeOH)		69			al, eth	B1, 711
4549	Capramide $CH_3(CH_2)_8CONH_2$	171.28	lf (eth)		108	$0.999^{20/4}$	1.4261^{110}	al, eth, ace	B2[4], 1050
4550	Capric acid or Decanoil acid $CH_3(CH_2)_8CO_2H$	172.27	nd	270, $148\text{-}50^{11}$	fr 31.5	$0.8858^{40/4}$	1.4288^{40}	al, eth, ace, bz, peth	B2[4], 1041
4551	Capric acid, 2-acetyl,ethyl ester $C_8H_{15}CH(COCH_3)CO_2C_2H_5$	242.36		280-2		$0.9354^{18.5/17.5}$			B3[3], 1278
4552	Capric anhydride $(C_9H_{19}CO)_2O$	326.52	lf		24.7	$0.8865^{25/4}$	1.400^{25}	al, eth	B2[4], 1049
4553	Capric acid, 2-bromo $C_8H_{17}CHBrCO_2H$	251.16		$140\text{-}1^{2}$	4	1.1912^{24}	1.4595^{24}	eth	B2[4], 1054
4554	Capric acid, decyl ester $C_9H_{19}CO_2C_{10}H_{21}$	312.54		219^{15}	9.7	0.8586^{20}	1.4423^{20}	eth	B2[4], 1045
4555	Capric acid, ethyl ester or Ethyl caprate $C_9H_{19}CO_2C_2H_5$	200.32		241.5, $122\text{-}4^{11}$	−20	$0.8650^{20/4}$	1.4256^{20}	al, eth, chl	B2[4], 1044
4556	Capric acid, 10-fluoro $F(CH_2)_9CO_2H$	190.26		$135\text{-}8^{10}$	49			al, eth, lig	B2[4], 1051
4557	Capric acid, methyl ester or Methyl caprate $C_9H_{19}CO_2CH_3$	186.29		224, 114^{15}	−18	$0.8730^{20/4}$	1.4259^{20}	al, eth, chl	B2[4], 1044
4558	Capric acid, 2-octyl or 9-Heptadecane carboxylic acid $(C_8H_{17})_2CHCO_2H$	284.48	nd or lf (al)	$212\text{-}8^{13}$	38.5			al, eth	B2[4], 1254
4559	Capric acid, 4-oxo or γ-Ketocapric acid $C_6H_{13}COCH_2CH_2CO_2H$	186.25	(dil al)		70-1			al	B3[4], 1642
4560	Capric acid, propyl ester or Propyl caprate $C_9H_{19}CO_2C_3H_7$	214.35		128.5^{10}	0.8623^{20}	1.4280^{20}			B2[4], 1045
4561	Capric acid, isopropyl ester or Isoporpyl caprate $C_9H_{19}CO_2CH(CH_3)_2$	214.35		121^{10}		0.8543^{20}	1.4221^{25}		B2[4], 1045
4562	Caprinitrile $C_9H_{19}CN$	153.27		243, 106^{10}	fr−17.9	$0.8199^{20/4}$	1.4296^{20}	**al, eth, ace, chl**	B2[4], 1051

PHYSICAL CONSTANTS OF ORGANIC COMPOUNDS (Continued)

No.	Name, Synonyms, and Formula	Mol. wt.	Color, crystalline form, specific rotation and λ_{max} (log ε)	b.p. °C	m.p. °C	Density	n_D	Solubility	Ref.
4563	Capryl chloride $C_9H_{19}COCl$	190.71	...	232, 114^{15}	−34.5	$0.973^{8/4}$	...	eth	$B2^4$, 1050
4564	Caproaldehyde or Hexanal $CH_3(CH_2)_4CHO$	100.16	...	128, 28^{12}	−56	$0.8139^{20/4}$	1.4039^{20}	al, eth, ace, bz	$B1^4$, 3296
4565	Caproaldehyde oxime or Capraldoxime, Hexanaldoxime $CH_3(CH_2)_4CH{=}NOH$	115.18	cr (MeOH)	...	51	...	...	...	$B1^2$, 745
4566	Caproaldehyde, 2-ethyl $CH_3(CH_2)_3CH(C_2H_5)CHO$	128.21	...	163, 65^{15}	<−100	0.8540^{20}	1.4142^{20}	al, eth	$B1^4$, 3345
4567	Caproaldehyde, 3-methyl or 3-Methylhexanal $CH_3(CH_2)_2CH(CH_3)CH_2CHO$	114.19	...	$142\text{-}3^{755}$	...	$0.8203^{20/4}$	1.4122^{20}	al, eth	$B1^2$, 756
4568	Caproamide or Hexananoamide $CH_3(CH_2)_4CONH_2$	115.18	cr (ace)	255	101	$0.999^{20/4}$	1.4200^{110}	al, eth, bz, chl	$B2^4$, 929
4569	Caproamide, 2-ethyl $C_4H_9CH(C_2H_5)CONH_2$	143.23	nd (w)	...	102-3	...	...	w	$B2^4$, 1007
4570	Caproamide, *N*-phenyl or Capranilide $CH_3(CH_2)_4CONHC_6H_5$	191.27	nd (peth), pr (al)	...	95	1.112	...	al, eth	$B12^3$, 478
4571	Caproic acid or Hexanoic acid $CH_3(CH_2)_4CO_2H$	116.16	...	205	−2	$0.9274^{20/4}$	1.4163^{20}	al, eth	$B2^4$, 917
4572	Caproic acid, 2-acetyl, ethyl ester or Ethyl 2-acetyl caproate $C_4H_9CH(COCH_3)CO_2C_2H_5$	186.25	...	219-24, 104^{12}	...	$0.9523^{20/4}$	1.4301^{20}	eth, ace	$B3^4$, 1616
4573	Caproic acid, 2-amino $C_4H_9CH(NH_2)CO_2H$	131.17	...	...	297-300	...	...	w	$B4^3$, 1386
4574	Caproic acid, 6-amino $H_2NCH_2(CH_2)_4CO_2H$	131.17	lf (eth)	202-3	...	...	...	w	$B4^3$, 1393
4575	Caproic acid, 6-amino-ε-lactam or ε-Caprolactam $C_6H_{11}NO$	113.16	lf (lig)	139^{12}	69-71	...	...	w, al, bz, chl	$B21^4$, 3196
4576	Caproic acid, 6-amino-3-methyl, lactam-*(l)* $CH_2(CH_2)_2CH(CH_3)CH_2CO$ NH (ring)	127.19	cr (bz-peth), $[\alpha]^{20/}_{D}$ −36.1	...	105-6	...	...	w, bz	B21, 243
4577	Caproic acid, 6-amino-5-methyl, lactam-*(l)* $CH_2CH(CH_3)(CH_2)_3CO$ NH (ring)	127.19	cr (peth, bz-lig), $[\alpha]^{20/}_{D}$ −22.2	...	68-9	...	...	w, eth	B21, 242
4578	Caproic acid anhydride $(C_5H_{11}CO)_2O$	214.30	...	254-7d, 143^{15}	−41	$0.9240^{15/4}$	1.4297^{20}	al, eth	$B2^4$, 928
4579	Caproic acid, 6-benzoylamino $C_6H_5CONH(CH_2)_5CO_2H$	235.28	nd (al-eth)	...	79-80	...	...	AcOEt	$B9^3$, 1157
4580	Caproic acid, 6-benzoylamino-2-bromo-*(dl)* $C_6H_5CONH(CH_2)_4CHBrCO_2H$	314.18	cr (dil al)	...	166	...	...	al	$B9^3$, 1158
4581	Caproic acid, 6-benzoylamino-2-bromo-*(l)* $C_6H_5CONH(CH_2)_4CHBrCO_2H$	314.18	nd (dil al), $[\alpha]^{18/D}$ −29.2 (al)	...	129	...	...	al	$B9^3$, 1157
4582	Caproic acid, 2-bromo-*(dl)* $CH_3(CH_2)_3CHBrCO_2H$	195.06	...	240, $140\text{-}2^{23}$	4	...	...	al, eth	$B2^4$, 938
4583	Caproic acid, 2-bromo-*(l)* $CH_3(CH_2)_3CHBrCO_2H$	195.06	$[\alpha]^{20/D}$ −27 (eth, C=5)	129^{14}	...	...	...	al, eth	$B2^3$, 736
4584	Caproic acid, 2-bromo, ethyl ester-*(dl)* $CH_3(CH_2)_3CHBrCO_2C_2H_5$	223.12	...	205-10, $95\text{-}6^{9}$	...	...	...	al	$B2^4$, 938
4585	Caproic acid, 3-bromo $C_3H_7CHBrCH_2CO_2H$	195.06	nd (dil al)	...	35	...	...	al, bz, chl, lig	$B2^4$, 939
4586	Caproic acid, 6-bromo $Br(CH_2)_5CO_2H$	195.06	cr (peth)	$165\text{-}70^{20}$	35	...	...	peth	$B2^4$, 940
4587	Caproic acid, butyl ester or Butyl caproate $C_5H_{11}CO_2C_4H_9$	172.27	...	208	−64.3	$0.8653^{20/4}$	1.4152^{20}	al, eth	$B2^4$, 922
4588	Caproic acid, 6-cyclohexyl $C_6H_{11}(CH_2)_5CO_2H$	198.31	...	180^{11}	33-5	$0.9626^{20/4}$	1.4750^{20}	eth	$B9^3$, 112
4589	Caproic acid, ethyl ester or Ethyl caproate $C_5H_{11}CO_2C_2H_5$	144.21	...	168	−67	$0.8710^{20/4}$	1.4073^{20}	al, eth	$B2^4$, 921
4590	Caproic acid, 2-ethyl $C_4H_9CH(C_2H_5)CO_2H$	144.21	...	228^{755}, 120^{13}	...	$0.9031^{25/4}$	1.4241^{20}	eth	$B2^4$, 1003
4591	Caproic acid, 6-fluoro $F(CH_2)_5CO_2H$	134.15	...	138^{28}, $67\text{-}8^{0.6}$	...	...	1.4166^{25}	al, eth	$B2^4$, 932
4592	Caproic acid, heptyl ester $C_5H_{11}CO_2C_7H_{15}$	214.35	...	261	−34.4	0.8611^{20}	1.4293^{15}	al, eth, ace, bz	$B2^4$, 923

No.	Name, Synonyms, and Formula	Mol. wt.	Color, crystalline form, specific rotation and λ_{max} (log ε)	b.p. °C	m.p. °C	Density	n_D	Solubility	Ref.
4593	Caproic acid, hexyl ester or Hexyl caproate $C_5H_{11}CO_2C_6H_{13}$	200.32		246	−55	0.865^{18}	1.4264^{15}	al, eth, ace, bz	$B2^4$, 922
4594	Caproic acid, 2-hydroxy-*(d)* $C_4H_9CH(OH)CO_2H$	132.16	$[\alpha]^{20/D}$ +0.7 (w, C=14)		60			w, al, eth	$B3^4$, 838
4595	Caproic acid, 2-hydroxy-*(dl)* $CH_3(CH_2)_3CH(OH)CO_2H$	132.16	pr (eth-al, peth)		60-1			w, al, eth, chl	$B3^4$, 838
4596	Caproic acid, 2-hydroxy-*(l)* $C_4H_9CH(OH)CO_2H$	132.16	pr (eth), $[\alpha]^{20/D}$ −3.8 (w, C=4.5)		60-1			w, al, eth, chl	$B3^4$, 838
4597	Caproic acid, 4-hydroxy, lactone or γ-Caprolactone $CH_3CH_2CHCH_2CH_2CO$ (O-bridged)	114.14		215-6, 103^{14}	−18		1.4495^{20}	w, al	$B17^4$, 4194
4598	Caproic acid, 6-hydroxy-ε-lactone $C_6H_{10}O_2$	114.14		108^{10}	−1.3	$1.0693^{20/4}$	1.4611^{20}	al, eth, ace	$B17^4$, 4186
4599	Caproic acid, methyl ester or Methyl caproate $C_5H_{11}CO_2CH_3$	130.19		151, 52^{15}	−71	$0.8846^{20/4}$	1.4049^{20}	al, eth ace, bz	$B2^4$, 921
4600	Caproic acid, 2-methyl-(d) or 2-methyl hexanoic acid $C_4H_9CH(CH_3)CO_2H$	130.19	$[\alpha]^{22/D}$ + 19.6 (eth)	105^5		0.909^{25}	1.4189^{20}	al, eth, ace, bz	$B2^3$, 773
4601	Caproic acid, 2-methyl-*dl* or 2-Methylhexanoic acid $C_4H_9CH(CH_3)CO_2H$	130.19		215-6, 100^{11}		$0.9612^{20/4}$	1.4195^{20}	al, eth, ace, bz	$B2^4$, 969
4602	Caproic acid, 2-methyl-*(l)* or 2-Methylhexanoic acid $C_4H_9CH(CH_3)CO_2H$	130.19	$[\alpha]^{25/}_D$ -4.3 (w, c=26)	121^{20}		$0.909^{25/4}$	1.4189^{25}	al, eth, ace, bz	$B2^3$, 773
4603	Caproic acid, 4-methyl-*(d)* or 4-Methylhexanoic acid $C_2H_5CH(CH_3)CH_2CH_2CO_2H$	130.19	$[\alpha]^{20/}_D$ + 7.6 (MeOH)	221, 115^{16}		$0.9228^{20/4}$	1.4198^{20}	al, ace, bz	$B2^4$, 973
4604	Caproic acid, 4-methyl-*(dl)* or 4-Methylhexanoic acid $C_2H_5CH(CH_3)(CH_2)_2CO_2(CH_2)_2CO_2H$	130.19		217-8, 85^2	-80	$0.9215^{20/4}$	1.4211^{20}	al, eth, ace, bz	$B2^4$, 974
4605	Caproic acid, 5-methyl or 5-Methylhexanoic acid $(CH_3)_2CH(CH_2)_3CO_2H$	130.19		216, 109^{16}	<-25	$0.9138^{21/4}$	1.4220^{20}	al, eth, ace, bz	$B2^4$, 970
4606	Caproic acid, octyl ester or Octyl caproate $C_5H_{11}CO_2C_8H_{17}$	228.38		275	-28	0.8603^{20}	1.4326^{15}	al, eth, ace, bz	$B2^3$, 727
4607	Caproic acid, 4-oxo or Homolevulinic acid $C_2H_5COCH_2CH_2CO_2H$	130.14	hyg ta or lf (eth-peth)	183^{20}, $89^{0.4}$	41-2			w, al, eth	$B3^4$, 1581
4608	Caproic acid, pentyl ester or Pentyl caproate $C_5H_{11}CO_2C_5H_{11}$	186.29		226, 116.6^{20}	-47	$0.8612^{25/4}$	1.4202^{25}	al, eth, ace	$B2^4$, 922
4609	Caproic acid, isopentyl ester or Isopentyl caproate $C_5H_{11}CO_2CH_2CH_2CH(CH_3)_2$	186.29		224-7		$0.861^{20/4}$		al, eth	$B2^3$, 727
4610	Caproic acid, propyl ester or Propyl caproate $C_5H_{11}CO_2C_3H_7$	158.24		187	-68.7	$0.8672^{20/4}$	1.4170^{20}	al, eth	$B2^4$, 922
4611	Caprolactam or Hexanoic acid-6-amino-ε-lactam $CH_2(CH_2)_4CO$ (NH-bridged)	113.16	lf (lig)	139^{12}	69-71			w, al, bz, chl	$B21^4$, 3196
4612	Caprolactam, 6-amino-3-methyl $C_7H_{13}NO$	127.19	cr (bz-peth), $[\alpha]^{20/}_D$ -36.1		105-6			w, bz	B21, 243
4613	Caprolactam, 6-amino-5-methyl $C_7H_{13}NO$	127.19	cr (peth, bz lig), $[\alpha]^{20/}_D$ -22.2		68-9			w, eth	B21, 242
4614	2-Caprolactone $C_6H_{10}O_2$	114.14		215-6, 103^{14}	-18		1.4495^{20}	w, al	$B17^4$, 4194
4615	Capronitrile $CH_3(CH_2)_4CN$	97.16		163.6, 47.3^{10}	-80.3	$0.8051^{20/4}$	1.4068^{20}	al, eth, chl	$B2^4$, 930
4616	Caprophenone or Pentyl phenyl ketone $C_5H_{11}COC_6H_5$	176.26	fl	265, $122\text{-}4^{15}$	27	$0.9576^{20/4}$	1.5027^{25}	al, eth, ace	$B7^3$, 1151
4617	Caproyl chloride $C_5H_{11}COCl$	134.61		153	-87	$0.9754^{20/4}$	1.4264^{20}	eth, ace	$B2^4$, 928
4618	Caproyl chloride, 3-methyl $C_4H_9CH(CH_3)CH_2COCl$	148.63		163^{751}, 82^{50}		$0.967^{20/4}$	1.4293^{25}	bz	$B2^3$, 777
4619	Caproyl chloride, 4-methyl $C_2H_5CH(CH_3)(CH_2)_2COCl$	148.63		167^{767}		$0.9677^{20/4}$		eth	$B2^2$, 299
4620	Caproyl chloride, 5-methyl $(CH_3)_2CH(CH_2)_3COCl$	148.63		168^{739}, $76\text{-}82^{34}$				eth	$B2^4$, 971
4621	Caprylaldehyde or Octanal $CH_3(CH_2)_6CHO$	128.21		171, 72^{20}		$0.8211^{20/4}$	1.4217^{20}	al, eth, ace, bz	$B1^4$, 3337
4622	Caprylaldehyde oxime or Caprylaldoxime $CH_3(CH_2)_6CH{=}NOH$	143.23	nd (peth, dil al)	112^9	60			al, ace	$B1^2$, 758

PHYSICAL CONSTANTS OF ORGANIC COMPOUNDS (Continued)

No.	Name, Synonyms, and Formula	Mol. wt.	Color, crystalline form, specific rotation and λ_{max} (log ε)	b.p. °C	m.p. °C	Density	n_D	Solubility	Ref.
4623	Caprylamide $C_7H_{15}CONH_2$	143.23	lf, pl	239	fr 106-10	0.8450^{110}		al, eth, ace	$B2^4$, 992
4624	Caprylic acid or Octanoic acid $CH_3(CH_2)_6CO_2H$	144.21		239.3, 140^{23}	16.5	0.9088^{20}	1.4285^{20}	al, chl	$B2^4$, 982
4625	Caprylic acid-2-amino-*(d)* $C_6H_{13}CH(NH_2)CO_2H$	159.23	$[\alpha]^{26}{}_D$ +23.5 (6N HCl, c=1)					aa	$B4^2$, 886
4626	Caprylic acid, 2-amino-*(dl)* $C_6H_{13}CH(NH_2)CO_2H$	159.23	lf (w)	(sub, d)	270			aa	$B4^3$, 1472
4627	Caprylic acid, 2-amino-*(l)* $C_6H_{13}CH(NH_2)CO_2H$	159.23	$[\alpha]_D$ -23 (5N HCl)		276			aa	$B4^2$, 886
4628	Caprylic acid, 8-amino $H_2N(CH_2)_7CO_2H$	159.23			172			al	$B4^1$, 527
4629	Caprylic anhydride $(C_7H_{15}CO)_2O$	270.41		280-5, 186^{15}	-1	$0.9065^{18/4}$	1.4358^{18}	al, eth, ace	$B2^3$, 796
4630	Caprylic acid, 2-bromo $C_6H_{13}CHBrCO_2H$	223.11		140^5		1.2785^{24}	1.4613^{24}		$B2^4$, 1000
4631	Caprylic acid, butyl ester or Butyl caprylate $C_7H_{15}CO_2C_4H_9$	200.32		240.5, $121\text{-}2^{20}$	-42.9	0.8628^{20}	1.4232^{25}	al, eth, ace	$B2^4$, 987
4632	Caprylic acid, ethyl ester or Ethyl caprylate $C_7H_{15}CO_2C_2H_5$	172.27		208.5, 104^{80}	-43.1	$0.8693^{20/4}$	1.4178^{20}	al, eth	$B2^4$, 987
4633	Caprylic acid, 8-fluoro $F(CH_2)_7CO_2H$	162.21		$132\text{-}3^4$	35			al, eth	$B2^4$, 994
4634	Caprylic acid, heptyl ester or Heptyl caprylate $C_7H_{15}CO_2C_7H_{15}$	242.40		290.5, 160^{14}	-10.6	0.8596^{20}	1.4340^{20}	al, eth, ace	$B2^3$, 794
4635	Caprylic acid, hexyl ester or Hexyl caprylate $C_7H_{15}CO_2C_6H_{13}$	228.38		277.4	-30.6	0.8603^{20}	1.4323^{25}	al, eth, ace	$B2^3$, 794
4636	Caprylic acid, 2-hydroxy $C_6H_{13}CH(OH)CO_2H$	160.21	pl	$160\text{-}5^{10}$	70			al, eth	$B3^4$, 874
4637	Caprylic acid, 4-hydroxy, lactone or 2-Caprylolactone $C_4H_9CH\text{-}CH_2CH_2CO$ (ring O)	142.20		$132\text{-}3^{20}$		$0.9796^{19/4}$	1.4451^{19}	al	$B17^4$, 4228
4638	Caprylic acid, methyl ester or Methyl caprylate $C_7H_{15}CO_2CH_3$	158.24		192.9, 83^{15}	-40	$0.8775^{20/4}$	1.4170^{20}	al, eth	$B2^4$, 986
4639	Caprylic acid, 2-methyl-3-oxo,ethyl ester $C_5H_{11}COCH(CH_3)CO_2C_2H_5$	200.28		$128\text{-}9^{12}$		$0.963^{0/4}$		al	B3, 713
4640	Caprylic acid, 2-methallyl ester $C_7H_{15}CO_2CH_2(CH_3C{=}CH_2$	198.31		147.8^{50}		0.8703	1.4308	al	$B2^3$, 795
4641	Caprylic acid, octyl ester or Octyl caprylate $C_7H_{15}CO_2C_8H_{17}$	256.43		306.8, 192.5^{30}	-18.1	$0.8554^{20/4}$	1.4352^{20}	al, eth, ace	$B2^4$, 988
4642	Caprylic acid, pentyl ester or Pentyl caprylate $C_7H_{15}CO_2C_5H_{11}$	214.35		260.2, $124\text{-}6^{20}$	-34.8	0.8613^{20}	1.4262^{25}	al, eth, ace	$B2^3$, 794
4643	Caprylic acid, perfluoro $CF_3(CF_2)_6CO_2H$	414.07		187-9	53				$B2^4$, 994
4644	Caprylic acid, perfluoro, methyl ester $C_7H_{15}CO_2CH_3$	428.10		158		$1.684^{20/4}$	1.304^{27}		$B2^4$, 995
4645	Caprylic acid, propyl ester or Propyl caprylate $C_7H_{15}CO_2C_3H_7$	186.29	(peth)	226.4, 112^{20}	-46.2	0.8659^{20}	1.4191^{25}	al, eth, ace	$B2^4$, 987
4646	Caprylic acid, ispropyl ester or Isopropyl caprylate $C_7H_{15}CO_2CH(CH_3)_2$	186.29		93.8^{10}		0.8555^{20}	1.4147^{25}		$B2^4$, 987
4647	Caprylonitrile $C_7H_{15}CN$	125.21		205.2, $77\text{-}8^{10}$	-45.6	$0.8136^{20/4}$	1.4203^{20}	eth	$B2^4$, 993
4648	Caprylyl chloride $C_7H_{15}COCl$	162.66		195.6, 89^{20}	-63	$0.9535^{15/4}$	1.4335^{20}	eth	$B2^4$, 992
4649	Capsaicin $C_{18}H_{27}NO_3$	305.42	mcl pr or sc (peth)	$210\text{-}20^{0.01}$	65			al, eth, bz, peth	$B13^3$, 2192
4650	Carbamic acid, benzyl ester $H_2NCO_2CH_2C_6H_5$	151.16	pl (to), lf (w)	220d	91			al	$B6^3$, 1485
4651	Carbamic acid, *N*-benzyl, ethyl ester $C_6H_5CH_2NHCO_2C_2H_5$	179.22	lf (lig)	230d	49			al, eth, bz, chl	$B12^3$, 2271
4652	Carbamic acid, N-nitro, benzyl ethylester $C_6H_5CH_2N(NO_2)CO_2C_2H_5$	224.22	ye	d		$1.213^{20/20}$	1.5203^{20}	**al, eth**	C55, 24616
4653	Carbamic acid, butyl ester $H_2NCO_2C_4H_9$	117.15	pr	204d	54			al	$B3^4$, 54
4654	Carbamic acid, *N*-butyl, butyl ester $C_4H_9NHCO_2C_4H_9$	173.26		88^3		$0.9238^{20/20}$	1.4359^{20}	**al, eth**	$B4^4$, 577

No.	Name, Synonyms, and Formula	Mol. wt.	Color, crystalline form, specific rotation and λ_{max} (log ε)	b.p. °C	m.p. °C	Density	n_D	Solubility	Ref.
4655	Carbamic acid, *N*-butyl-*N*-nitro, butyl ester $C_4H_9N(NO_2)CO_2C_4H_9$	218.26	...	98^{1}	...	$1.048^{20/20}$	1.4359^{20}	**al, eth**	Am73, 5449
4656	Carbamic acid, isobutyl ester or Isobutyl carbamate $H_2NCO_2CH_2CH(CH_3)_2$	117.15	lf	207	67	...	1.4098^{76}	al, eth	B3^{4}, 56
4657	Carbamic acid, *N*-isobutyl, ethyl ester or Isobutyl urethane $(CH_3)_2CHCH_2NHCO_2C_2H_5$	145.20	...	110^{30}	>-65	$0.9432^{20/4}$	1.4288^{20}	al, eth	B4^{4}, 647
4658	Carbamic acid, *N*-*tert*-butyl-*N*-nitro, ethyl ester $(CH_3)_3CN(NO_2)CO_2C_2H_5$	190.20	...	56^{2}d	...	$1.051^{20/20}$	1.4331^{20}	al, eth	Am83, 1191
4659	Carbamic acid, *N,N*-diethyl $(C_2H_5)_2NCO_2H$	117.15	nd (eth)	171	-15d	$0.9276^{20/4}$	1.4206^{20}	w, al, eth	B4^{3}, 222
4660	Carbamic acid, *N,N*-diphenyl, ethyl ester or Diphenyl urethane $(C_6H_5)_2NCO_2C_2H_5$	241.29	pr (lig)	360	72	...	...	w, eth, bz	B12^{3}, 888
4661	Carbamic acid, ethyl ester or Urethane $H_2NCO_2C_2H_5$	89.09	pr (bz, to)	185	48-50	$0.9862^{21/4}$	1.4144^{51}	al, w, eth, bz	B3^{4}, 40
4662	Carbamic acid, *N*-ethyl,butyl ester $C_2H_5NHCO_2C_4H_9$	145.20	...	66^{3}	...	$0.9413^{20/20}$	1.4301^{20}	w, al, eth	C55, 2334
4663	Carbamic acid, *N*-ethyl, ethyl ester or Ethyl urethane $C_2H_5NHCO_2C_2H_5$	117.15	...	176, 75^{14}	...	$0.9813^{20/4}$	1.4215^{20}	w, al, eth	B4^{4}, 365
4664	Carbamic acid, *N*-ethyl-*N*-nitro,butyl ester $C_2H_5N(NO_2)CO_2C_4H_9$	190.20	...	79^{3}	...	$1.091^{20/20}$	1.4455^{20}	w, **al, eth**	Am73, 5043
4665	Carbamic acid, *N*-ethyl-*N*-nitro,ethyl ester $C_2H_5N(NO_2)CO_2C_2H_5$	162.15	...	107^{31}	...	$1.163^{20/20}$	1.4432^{20}	**al, eth**	Am73, 5449
4666	Carbamic acid, *N*-ethyl-*N*-nitro,methyl ester $C_2H_5N(NO_2)CO_2CH_3$	148.12	...	72^{11}	...	$1.233^{20/20}$	1.4483^{20}	**al, eth**	Am73, 5449
4667	Carbamic acid, *N*-ethylidine, diethyl ester or Ethylidine diurethane $CH_3CH(NHCO_2C_2H_5)_2$	204.23	nd (eth)	$170\text{-}8^{20}$	126	...	...	w, al, ace, chl	B3^{1}, 11
4668	Carbamic acid, methyl ester or Urethylan $H_2NCO_2CH_3$	75.07	nd	177, 82^{14}	54	$1.1361^{56/4}$	1.4125^{56}	w, al, eth	B3^{4}, 37
4669	Carbamic acid, 2-methyl-2-butyl ester or *tert*-Pentylcarbamate, Aponal $H_2NCO_2C(CH_3)_2CH_2CH_3$	131.17	nd (dil al)	...	85-7	...	...	ace, bz	B3^{1}, 14
4670	Carbamic acid, 3-methylbutyl ester or Isopentyl carbamate $H_2NCO_2CH_2CH_2CH(CH_3)_2$	131.17	nd (w)	220, $114\text{-}5^{16}$	64	$0.9438^{71/4}$	1.4175^{71}	al, eth	B3^{4}, 58
4671	Carbamic acid, *N*-methyl, ethyl ester or Methyl urethane $CH_3NHCO_2C_2H_5$	103.12	...	170, 80^{15}	...	$1.0115^{20/4}$	1.4183^{20}	w, al	B4^{4}, 200
4672	Carbamic acid, *N*-nitro,ethyl ester $O_2NNHCO_2C_2H_5$	134.09	pl (eth, lig)	140d	64	$1.0074^{20/4}$	...	w, al, ace, lig	B3^{4}, 247
4673	Carbamic acid, *N*-phenyl,ethyl ester or Ethylcarbanilate $C_6H_5NHCO_2C_2H_5$	165.19	wh nd (w), pl (dil al)	237d	53	$1.1064^{30/4}$	1.5376^{30}	al, eth, bz	B12^{3}, 612
4674	Carbamic acid, *N*-phenyl,isobutyl ester $C_6H_5NHCO_2CH_2CH(CH_3)_2$	193.25	nd (dil al)	216	86	...	...	al, eth, bz	B12^{3}, 614
4675	Carbamic acid, *N*-phenyl,propyl ester $C_6H_5NHCO_2C_3H_7$	179.22	wh nd (dil al)	...	57-9	...	...	al, eth, bz	B12^{3}, 613
4676	Carbamic acid, *N*-phenyl,*iso*propyl ester $C_6H_5NHCO_2CH(CH_3)_2$	179.22	wh nd (al)	...	90	1.09^{20}	1.4989^{91}	al, bz	B12^{3}, 613
4677	Carbamic acid, propyl ester $H_2NCO_2C_3H_7$	103.12	pr	196, 92^{12}	60	...	...	w, al, eth, ace	B3^{4}, 52
4678	Carbamic acid, isopropyl ester $H_2NCO_2CH(CH_3)_2$	103.12	nd	181^{711}	92-4	0.9951^{66}	...	...	B3^{4}, 53
4679	Carbamic acid, *N*-propyl,ethyl ester $C_3H_7NHCO_2C_2H_5$	131.17	...	192^{751}, 92^{22}	...	0.9921^{15}	...	al	B4^{4}, 480
4680	Carbamic acid, *N*-propyl-*N*-nitro, ethyl ester $C_3H_7N(NO_2)CO_2C_2H_5$	176.17	...	66^{3}	...	$1.123^{20/20}$	1.4431^{20}	**al, eth**	Am73, 5449
4681	Carbamic acid, *N*-propyl-*N*-nitro,methyl ester $C_3H_7N(NO_2)CO_2CH_3$	162.15	...	...	...	$1.2585^{15/15}$	...	al, **w, eth**	B4, 146
4682	Carbamic acid, *N*-*iso*propyl, ethyl ester or Isopropylurethane $(CH_3)_2CHNHCO_2C_2H_5$	131.17	...	79^{15}	...	$0.9548^{20/20}$	1.4229^{20}	**al, eth**	B4^{4}, 520
4683	Carbamic acid, *N*-isopropyl-*N*-nitro, ethyl ester $(CH_3)_2CHN(NO_2)CO_2C_2H_5$	176.17	...	72^{7}	...	$1.112^{20/20}$	1.4381^{20}	**al, eth**	Am73, 5449
4684	Carbamic acid, thiolo, ethyl ester $H_2NCOSC_2H_5$	105.15	pl (w)	sub d	109	...	...	al, eth	B3^{4}, 294
4685	Carbamic acid, thiono, ethyl ester or Thiourethane $H_2NCSOC_2H_5$	105.15	nd lf or pyr	d	41	$1.069^{20/4}$	1.520^{20}	**al, eth**, chl	B3^{4}, 294

No.	Name, Synonyms, and Formula	Mol. wt.	Color, crystalline form, specific rotation and λ_{max} (log ε)	b.p. °C	m.p. °C	Density	n_D	Solubility	Ref.
4686	Carbamonitrile, *N*-ethyl-*N*-phenyl $C_6H_5N(CH_2CH_3)CN$	146.19		$271,153^{19}$					$B12^{3}423$
4687	Carbamyl chloride H_2NCOCl	79.49		62d					$B3^{3}$, 65
4688	Carbamyl chloride, *N,N*-diethyl $(C_2H_5)_2NCOCl$	135.59		186					$B4^{4}$, 379
4689	Carbamyl chloride, *N,N*-diphenyl $(C_6H_5)_2NCOCl$	231.68	lf (al)		85				$B12^{3}$, 893
4690	Carbamyl chloride, *N*-methyl-*N*-phenyl $C_6H_5N(CH_3)COCl$	169.61	pl (al)	280	88-9			al, eth	$B12^{3}$, 874
4691	Carbazic acid, methyl ester $H_2NNHCO_2CH_3$	90.08		108^{12}	73			w, al, bz	$B3^{2}$, 78
4692	Carbazide $H_2NNHCONHNH_2$	90.08	nd (dil al)		154	1.616^{20}		w, al	$B3^{4}$, 240
4693	Carbazide, 1,5-diphenyl $(C_6H_5NHNH)_2CO$	242.28	cr (al + 1), cr (aa)	d	170			bz, aa	$B15^{3}$, 187
4694	Carbazide, 1-phenyl $C_6H_5NHNHCONHNH_2$	166.18	nd (al)		151			w	$B15^{3}$, 187
4695	Carbazide, 1,1,5,5-tetraphenyl $[(C_6H_5)_2NNH]_2CO$	394.48	(al), nd (aa)		242			al, aa	$B15^{2}$, 115
4696	Carbazide, 3-thio $(H_2NNH)_2CS$	106.15	nd, pl (w)		170d			w	$B3^{3}$, 319
4697	Carbazole or Dibenzopyrrole $C_{12}H_9N$	167.21	pl or lf	$355, 200^{147}$	247-8			ace	$B20^{4}$, 3824
4698	Carbazole, 9-acetyl $C_{14}H_{11}NO$	209.25	(eth), nd (w)	190^{6}	69	$1.161^{100/24}$	1.640^{100}		$B20^{4}$, 3836
4699	Carbazole, 9-benzoyl $C_{19}H_{13}NO$	271.32	nd or pr (al)		98.5			al, bz	$B20^{4}$, 3838
4700	Carbazole, 9-benzyl $C_{19}H_{15}N$	257.33	nd (al)	$267\text{-}8^{24}$	118-20			bz	$B20^{4}$, 3831
4701	Carbazole, 9-butyl $C_{16}H_{17}N$	223.32	nd (al)	$218\text{-}9^{19}$	58			eth	$B20^{4}$, 3829
4702	Carbazole, 9-ethenyl $C_{14}H_{11}N$	193.25	cr (al)		66			eth	$B20^{4}$, 3830
4703	Carbazole, 9-ethyl $C_{14}H_{13}N$	195.27	nd (al)	190^{10}	68	$1.059^{80/4}$	1.6394^{80}	al, eth	$B20^{4}$, 3829
4704	Carbazole, 9-methyl $C_{13}H_{11}N$	181.24	nd, lf (al)	195^{12}	88			eth	$B20^{4}$, 3828
4705	Carbazole, 1-nitro $C_{12}H_8N_2O_2$	212.21	ye nd (aa)		187			aa	$B20^{4}$, 3863
4706	Carbazole, 3-nitro $C_{12}H_8N_2O_2$	212.21	ye		214				$B20^{4}$, 3865
4707	Carbazole, 3-nitro-9-nitroso $C_{12}H_7N_3O_3$	241.21	ye nd (al)		169			chl	$B20^{4}$, 3868
4708	Carbazole, 1-oxo-1,2,3,4-tetrahydro $C_{12}H_{11}NO$	185.23	nd (dil al)		170			al, bz, aa	$B21^{4}$, 4075
4709	Carbazole, 9-phenyl $C_{18}H_{13}N$	243.31	nd or pl (al)		95			al, eth, bz, aa	$B20^{4}$, 3830
4710	Carbazole, 9-propionyl $C_{15}H_{13}NO$	223.27			90			al, eth	$B20^{4}$, 3837
4711	Carbazole, 9-propyl $C_{15}H_{15}N$	209.29	nd (al)		50			eth	$B20^{4}$, 3829
4712	Carbazole, 1,2,3,4-tetrahydro $C_{12}H_{13}N$	171.24	lf (dil al)	$325\text{-}30, 190^{10}$	120			al, eth, bz	$B20^{4}$, 3566
4713	Carbazone, 1,5-diphenyl $C_6H_5N{=}NCONHNHC_6H_5$	240.26	og nd (bz), pr (al)		157d			al, bz, chl	$B16^{3}$, 18
4714	Carbodiimide, diphenyl $C_6H_5N{=}C{=}NC_6H_5$	194.24		$331, 218^{21}$	168-70			bz	$B12^{3}$, 906
4715	Carbon dioxide CO_2	44.01		−78.6 sub	−56.6 (5.2 atm)	1.0310^{-20}			B3, 4
4716	Carbon diselenide CSe_2	169.93	ye	$126, 46^{50}$	−45.5	$2.6824^{20/4}$	1.8454^{20}		$B3^{4}$, 436
4717	Carbon disulfide CS_2	76.13		46.2	−111.5	$1.2632^{20/4}$	1.6319^{20}	al, eth, chl	$B3^{4}$, 395
4718	Carbonic acid, bis(2-chloroethyl) ester or bis-(2-Chloroethyl) carbonate $(ClCH_2CH_2O)_2CO$	187.02		241	8	$1.3506^{20/4}$	1.461^{20}		$B3^{4}$, 6

No.	Name, Synonyms, and Formula	Mol. wt.	Color, crystalline form, specific rotation and λ_{max} (log ε)	b.p. °C	m.p. °C	Density	n_D	Solubility	Ref.
4719	Carbonic acid, bis (3-chloropropyl) ester $(ClCH_2CH_2CH_2O)_2CO$	215.08		265-70^{740}					B3^3, 8
4720	Carbonic acid, bis(2-ethoxyethyl) ester $(C_2H_5OCH_2CH_2O)_2CO$	206.24		245-6^{758}, 112-3^5		1.0439$^{20/4}$	1.4227^{20}	al, eth, ace	B3^3, 17
4721	Carbonic acid, bis(2 methoxyethyl) ester $(CH_3OCH_2CH_2O)_2CO$	178.19		230-2, 99-100^5		1.0988$^{20/4}$	1.4204^{20}	al, eth, ace	B3^3, 17
4722	Carbonic acid, bis(2-methoxyphenyl) ester or Duotal $(2\text{-}CH_3OC_6H_4O)_2CO$	274.28	cr (al)		89			eth, chl	B6^3, 4233
4723	Carbonic acid, bis(trichloromethyl) ester $[Cl_3CO]_2CO$	296.75	cr (eth, peth)	203d	79				B3^4, 33
4724	Carbonic acid, dibutyl ester or Dibutyl carbonate $(C_4H_9O)_2CO$	174.24		207, 96-7^{16}		0.9251$^{20/4}$	1.4117^{20}	al, eth	B3^4, 8
4725	Carbonic acid, diisabutyl ester $(i\text{-}C_4H_9O)_2CO$	174.24		190, 85^{16}		0.9138$^{20/4}$	1.4072^{20}	al, eth	B3^4, 9
4726	Carbonic acid, di-*tert*-butyl ester or di-*tert*-Butyl carbonate $[(CH_3)_3CO]_2CO$	174.24	cr (al)	158^{767}	40 (sub)			al	B3^4, 9
4727	Carbonic acid, diethyl ester or Diethyl carbonate $(C_2H_5O)_2CO$	118.13		126	-43	0.9752$^{20/4}$	1.3845^{20}	al, eth	B3^4, 5
4728	Carbonic acid, dimethyl ester or Dimethyl carbonate $(CH_3O)_2CO$	90.08		90-1	2-4	1.0694$^{20/4}$	1.3687^{20}		B3^4, 3
4729	Carbonic acid, diphenyl ester or Diphenyl carbonate $(C_6H_5O)_2CO$	214.22	nd (al, bz)	306, 168^{15}	83 (88)	1.1215$^{87/4}$		eth	B6^4, 629
4730	Carbonic acid, di-isopentyl ester or Diisopentyl carbonate $(i\text{-}C_5H_{11}O)_2CO$	202.29		232-5^{751}, 122^{16}		0.9067$^{20/4}$	1.4174^{20}		B3^3, 11
4731	Carbonic acid, di-propyl ester or Dipropyl carbonate $(C_3H_7O)_2CO$	146.19		168, 59.5^{15}		0.9435$^{20/4}$	1.4008^{20}	al, eth	B3^4, 6
4732	Carbonic acid, Diisopropyl ester $(i\text{-}C_3H_7O)_2CO$	146.19		147, 43^{12}		0.9162$^{20/4}$	1.3932^{20}	al	B3^3, 9
4733	Carbonic acid, dithiolo, diethyl ester $(C_2H_5S)_2CO$	150.25	ye	197, 85-7^{19}		1.085^{20}	1.5237^{18}	al, eth	B3^3, 339
4734	Carbonic acid, di-2-tolyl ester $(2\text{-}CH_3C_6H_4O)_2CO$	242.27	nd (al)	144-5$^{0.5}$	60			aa	B6^3, 1256
4735	Carbonic acid, di-3-tolyl ester $(3\text{-}CH_3C_6H_4O)_2CO$	242.27	cr (al)		50-1			bz, chl	B6^3, 1307
4736	Carbonic acid, di-4-tolyl ester $(4\text{-}CH_3C_6H_4O)_2CO$	242.27			115			chl	B6^3, 1366
4737	Carbonic acid, ethyl-2-butoxyethyl ester $C_2H_5O\text{-}CO\text{-}OCH_2CH_2OC_4H_9$	190.24		224^{759}		0.9756$^{25/4}$	1.4143^{25}	al, eth, ace	B3^3, 14
4738	Carbonic acid, ethyl-methyl ester $CH_3O\text{-}CO\text{-}OC_2H_5$	104.11		107-8	−14	1.012$^{20/4}$	1.3778^{20}	al, eth	B3^4, 4
4739	Carbonic acid, trithio $(HS)_2CS$	110.21	red	57d	−30	1.47$^{17/4}$		al, chl, to	B3^4, 428
4740	Carbon suboxide or 1,3-Dioxoallene $OC=C=CO$	68.03	gas	6.8	−107	1.114$^{0/4}$	1.4538^0	eth, bz	B1^4, 3764
4741	Carbonyl fluoride COF_2	66.01		−83	−114	1.139^{-114}			B3^4, 21
4742	Carbonyl sulfide COS	60.08		−50	−138	1.028$^{17/4}$		al	B3^4, 271
4743	Carbothialdine $C_5H_{10}N_2S_2$	162.27	cr (al)		120d			al	B27^2, 687
4744	Δ^3-Carene-*(dl)* $C_{10}H_{16}$	136.24		167^{732}, 44-5^8		0.8602$^{20/4}$	1.4759^{20}	eth, ace, bz, aa	B5^2, 362
4745	Δ^3-Carene-*(l)* $C_{10}H_{16}$	136.24	$[\alpha]^{20/}_{D}$ −5.72	168-9^{705}, 123-4^{200}		0.8586$^{30/30}$	1.4684^{30}	eth, ace, bz, aa	B5^4, 449
4746	Δ^4-Carene-*(l)* $C_{10}H_{16}$	136.24	$[\alpha]^{20/}_{D}$ + 62.2	167^{707}, 64^{20}		0.8441$^{30/4}$	1.4740^{30}	eth, ace, bz, aa	B5^4, 451
4747	Carminic acid $C_{22}H_{20}O_{13}$	492.39	red mcl pr (ag MeOH)		d 136			al	B10^3, 4874
4748	Carnaubyl alcohol $CH_3(CH_2)_{29}OH$	438.82	lf (al)		69			al, ace	B1^2, 472
4749	Carnosine-*(D,-)* or Ignotine. β-Alanyl-*(D,-)*-histidine $C_9H_{14}N_4O_3$	226.24	$[\alpha]^{18/}_{D}$ −20.4 (w, c=1.5)		260			w	B25^2, 408

No.	Name, Synonyms, and Formula	Mol. wt.	Color, crystalline form, specific rotation and λ_{max} (log ε)	b.p. °C	m.p. °C	Density	n_D	Solubility	Ref.
4750	Carnosine-*(L +)* $C_9H_{14}N_4O_3$	226.24	nd (w-al), $[\alpha]^{20/}{}_D$ +24.1 (w, c=1.5)		246-50d			w	$B25^2$, 408
4751	α-Carotene or α-Carotin $C_{40}H_{56}$	536.88	red pl or pr (peth, bz-MeOH)		187.5	$1.00^{20/20}$		eth, bz, chl	$B5^3$, 2457
4752	β-Carotene or Provitamin A $C_{40}H_{56}$	536.88	red br hex pr (bz-MeOH)		184	$1.00^{20/20}$		eth, ace, bz, peth	$B5^3$, 2453
4753	γ-Carotene or γ-Carotin $C_{40}H_{56}$	536.88	red br (bz-MeOH) vt pr (bz-eth)		178			bz, chl	$B5^3$, 2451
4754	Carpaine-*(d)* $C_{28}H_{50}N_2O_4$	478.72	mcl pr (al or ace) $[\alpha]^{21/}{}_D$ +24.7 (al, c=1.07)	sub $120^{0.01}$	121			al, eth, ace, bz, chl	$B27^2$, 209
4755	Carpaine hydrochloride $C_{28}H_{50}N_2O_4$-HCl	515.18	wh mcl nd or pl		225d			w, al, eth	$B27^2$, 210
4756	Carpiline or Carpidine, Pilosine $C_{16}H_{18}N_2O_3$	286.33	pl (al), pr (dil al or w), $[\alpha]^{20/}{}_D$ +35.9 (al)		187			al	$B27^1$, 612
4757	Carvacrol or 2-Methyl-5-*iso*-propyl phenol 2-CH_3-5-$(CH_3)_2CHC_6H_3OH$	150.22	nd	237.7, $101\text{-}2^{10}$	1	$0.9772^{20/4}$	1.5230^{20}	al, eth, ace	$B6^3$, 1885
4758	Carvacrol-acetate or 2-Methyl-5-*iso*-propyl phenyl acetate 2-CH_3-5-$(CH_3)_2CHC_6H_3O_2CCH_3$	192.24		245-8		0.9896^{25}	1.4913^{28}	al, eth	$B6^2$, 494
4759	Carvacrol, 4-amino or 2-Methyl-5-*iso*-propyl-4-amino-phenol 2-CH_3-4-H_2N-5-$CH(CH_3)_2$-C_6H_2OH	165.24	cr (MeOH)		134			MeOH	$B13^3$, 1801
4760	Carvenone-*(dl)* $C_{10}H_{16}O$	152.24		$235\text{-}6^{762}$, 104^{10}		$0.9263^{20/4}$	1.4826^{20}	ace	$B7^3$, 332
4761	Carvenone-*(l)* $C_{10}H_{16}O$	152.24	$[\alpha]_D$ −2.08	232-4		$0.9290^{20/4}$	1.4805	ace	$B7^1$, 66
4762	Carveol, dihydro-*(d)* $C_{10}H_{18}O$	154.25	$[\alpha]^{18/}{}_D$ +34.2	225, 107^{15}		$0.9274^{20/4}$	1.4780^{20}	eth	$B6^3$, 256
4763	Carveol, dihydro-*(l)* $C_{10}H_{18}O$	154.25	$[\alpha]^{20/}{}_D$ −33.3	107^{14}		0.9368^{15}	1.4836^{20}	ace	$B6^3$, 256
4764	β-Carveol, dihydro $C_{10}H_{18}O$	154.25	$[\alpha]_D$ +7.64	130^{20}		$0.9266^{20/4}$	1.4809^{20}		B6, 64
4765	Carvomenthane-*(d)* $C_{10}H_{18}$	138.25	$[\alpha]_{5780}$ +118	175-7, 77^{24}		$0.8246^{18/4}$	1.4563^{18}	al, bz, peth	$B5^4$, 301
4766	Carvomenthol-*(d)* or Hexahydrocarvacrol $C_{10}H_{20}O$	156.27	$[\alpha]^{21/}{}_D$ 31.4	222, 102^{14}		$0.8995^{20/4}$	1.4617^{20}	al, eth	$B6^4$, 148
4767	Carvomenthol-*(l-neo)* $C_{10}H_{20}O$	156.27	$[\alpha]^{21/}{}_D$ −41.7	217-8, 102^{18}		$0.9012^{20/4}$	1.4632^{20}	al, eth	$B6^4$, 148
4768	Carvomenthone $C_{10}H_{18}O$	154.25	$[\alpha]^{21/}{}_D$ +17.15	$218\text{-}20^{745}$, $95\text{-}9^{15}$		0.9075^{15}	1.4544^{20}	al, ace, chl	$B7^3$, 146
4769	Carvone-*(d)* or Carvol $C_{10}H_{14}O$	150.22	$[\alpha]^{20/}{}_D$ +69.1	23^1, 104^{11}		$0.9608^{20/4}$	1.4999^{18}	al, eth, chl	$B7^3$, 561
4770	Carvone-*(dl)* $C_{10}H_{14}O$	150.22		231, 85^5		$0.9645^{15/15}$	1.5003^{20}	al, eth, chl	$B7^3$, 564
4771	Carvone-*(l)* $C_{10}H_{14}O$	150.22	$[\alpha]^{20/}{}_D$ −62.46	231, 98^9		$0.9593^{20/4}$	1.4988^{20}	al, eth, chl	$B7^3$, 561
4772	Carvone oxime-*(α,d)* or *D*-Carotime-*(α)* $C_{10}H_{15}NO$	165.24	lf (al) $[\alpha]^{17/}{}_D$ +39.71 (al, p=8.45)		72			al, eth, ace	$B7^3$, 564
4773	Carvone oxime-*(l,α)* or *L*-Carvoximine-*(α)* $C_{10}H_{15}NO$	165.24	mcl (dil al) $[\alpha]^{18/}{}_D$ −39.43 (al, p=4.33)		73.5	1.0140^{73}		al, eth, ace	$B7^3$, 564

No.	Name, Synonyms, and Formula	Mol. wt.	Color, crystalline form, specific rotation and λ_{max} (log ε)	b.p. °C	m.p. °C	Density	n_D	Solubility	Ref.
4776	Carvone oxime-*(dl)* or *dl*-Carvoxime $C_{10}H_{15}NO$	165.24			93-4			al, eth, ace	$B7^3$, 564
4777	Carvone, dihydro-*(d)* $C_{10}H_{16}O$	152.24	$[\alpha]_D$ + 17.5	221-2		0.928^{19}	1.4724	eth, ace	$B7^3$, 337
4778	Carvone, dihydro-*(l)* $C_{10}H_{16}O$	152.24	$[\alpha]^{20/}_D -19$	221-2, 104^{18}		$0.9253^{20/4}$	1.4717^{20}	eth, ace	$B7^3$, 337
4779	α-Caryophyllene $C_{15}H_{24}$	204.36	$[\alpha]^{20/4}$ + 1±0.3 (chl, c=9.26)	123^{10}		$0.8905^{20/4}$	1.5038^{20}		$B5^4$, 1171
4780	β-Caryophyllene $C_{15}H_{24}$	204.36	$[\alpha]^{20/}_D$ −9.08	$122^{13.5}$		$0.9075^{20/4}$	1.4988^{20}	bz	$B5^4$, 1182
4781	γ-Caryophyllene or Isocaryophyllene $C_{15}H_{24}$	204.36	$[\alpha]^{19/}_D$ −26.2	130-1^{24}		$0.8953^{20/4}$	1.4967^{19}	bz	$B5^3$, 1083
4782	Caryophyllenic acid-*(l,cis)* $C_9H_{14}O_4$	186.21	pr (w) $[\alpha]_{546} -7.4$		77-8			w, al, eth, ace, bz	$B9^3$, 3850
4783	Caryophyllenic acid-*(d,trans)* $C_9H_{14}O_4$	186.21	nd $[\alpha]^{20/}_D + 35.3$ (bz)		81-2			w, al, eth, ace, bz	$B9^3$, 3850
4784	Caryophyllin or Oleanolic acid $C_{30}H_{48}O_3$	456.71	nd or pr (al) $[\alpha]^{20/}_D$ + 83.3 (chl, c=0.9)	280-308, sub (vac)	310d			aa, Py	$B10^3$, 1049
4785	Catechin-*(cis'd)* or 3,5,7,3′,4′-Flavanpentol $C_{15}H_{14}O_6$	290.27	nd (w + 4) $[\alpha]^{18/}_D$ + 18.4 (w, c=0.9)	240-5	96 (hyd) 177 (anh)	1.344^{44}		al, ace, aa	$B17^4$, 3841
4786	Catechin-*(cis,dl)* $C_{15}H_{14}O_6$	290.27	nd (w + 3)		212-4d			al, ace	$B17^4$, 3842
4787	Catechin-*(cis,l)* $C_{15}H_{14}O_6$	290.27	nd (+ 4, w), $[\alpha]_{5780}$ (w ace, p=3)		96 (hyd) 177 (anh)			al, ace, aa	$B17^2$, 255
4788	Catechol or 1,2-Dihydroxybenzene 2-HOC_6H_4OH	110.11	cr	245^{750}	105	1.1493^{21}	1.604	w, al, eth, ace	$B6^3$, 4187
4789	Catechol, 3-acetyl or 2,3-Dihydroxyacetophenone 2, 3-$(HO)_2C_6H_3COCH_3$	152.15	ye pr (bz-lig)		97-8			al	$B8^3$, 2080
4790	Catechol, 4-acetyl or 3,4-Dihydroxyacetophenone 3,4$(HO)_2C_6H_3COCH_3$	152.15	nd (w or chl)		115-6				$B8^3$, 2108
4791	Catechol, 4-bromo or 4-Bromo-1,2-dihydroxybenzene 1,2-$(HO)_2$-4-Br-C_6H_3	189.01	pr or nd (chl)		87			w, al, eth, bz	$B6^3$, 4253
4792	Catechol, 3-chloro or 3-Chloro-1,2-dihydroxybenzene 3-Cl-1,2-$(HO)_2C_6H_3$	144.56	cr (lig)	110-1^{11}	46-8			lig	$B6^3$, 4249
4793	Catechol, 4-chloro 4-Chloro-1,2-$(HO)_2C_6H_3$	144.56	lf (bz-peth)	$139^{10.5}$	90-1			w, al, eth, ace, aa	$B6^3$, 4249
4794	Catechol, diacetate 2-CH_3CO_2-C_6H_4-O_2CCH_3	194.19	nd (al)	142-3^9	64-5			al, eth, chl	$B6^3$, 4228
4795	Catechol, dibenzoate 1,2-$(C_6H_5CO_2)C_6H_4$	318.33	lf (eth-al)		86			al, eth, bz	$B9^3$, 552
4796	Catechol, dibenzyl ether 1,2-$(C_6H_5CH_2O)_2C_6H_4$	290.37	yesh nd or pr (al)		63-4			eth, peth	$B6^3$, 4219
4797	Catechol, dibutyl ether or 1,2-Dibutoxybenzene 1,2-$(C_4H_9O)_2C_6H_4$	222.23	ye	241^{765}, 135-8^{12}					$B6^3$, 4210
4798	Catechol, 3,5-dichloro 3,5-$Cl_2C_6H_2(OH)_2$-1,2	179.00	pr		83-4			al, ace	B6, 783
4799	Catechol, 4,5-dichloro 4,5-$Cl_2C_6H_2(OH)_2$-1,2	179.00	pr (chl-CS_2), nd (bz-peth)		116-7			w, al, bz	$B6^3$, 4252
4800	Catechol, diethyl ether or 1,2-Diethoxybenzene 1,2-$(C_2H_5O)_2C_6H_4$	166.22	pr (peth, dil al)	219	43-5	$1.0075^{20/4}$	1.5083^{25}	al, eth	$B6^3$, 4208
4801	Catechol, dimethylether or Veratrole-1-2 dimethoxy benzene 1,2-$(CH_3O)_2C_6H_4$	138.17	cr (lig)	206^{750}, 90^{10}	22.5	$1.0842^{25/25}$	$1.5827^{21/}_4$	al, eth	$B6^3$, 4205
4802	Catechol, 3,5-dimethyl 3,5-$(CH_3)_2C_6H_2(OH)_2$-1,2	138.17	pr (w), nd (peth-bz)		73-4			w, al, eth	$B6^3$, 4591
4803	Catechol, 4,5-dimethyl 4,5-$(CH_3)_2C_6H_2(OH)_2$-1,2	138.17	mcl pr or nd (peth)	sub	87-8			w, al, eth	$B6^3$, 4584

PHYSICAL CONSTANTS OF ORGANIC COMPOUNDS (Continued)

No.	Name, Synonyms, and Formula	Mol. wt.	Color, crystalline form, specific rotation and λ_{max} (log ε)	b.p. °C	m.p. °C	Density	n_D	Solubility	Ref.
4804	Catechol, dipropylether or 1,2-Dipropoxy benzene $1,2(C_3H_7O)_2C_6H_4$	194.27		234-7, $117\text{-}20^{12}$		$0.9554^{13/4}$	1.4950^{27}		$B6^3$, 4209
4805	Catechol, dithio or 1,2-Dimercaptobenzene $1,2\text{-}(HS)_2C_6H_4$	142.25		238-9, 120^{17}	28-9			al, eth, bz	$B6^3$, 4286
4806	Catechol, 4-iodo $4\text{-}IC_6H_3(OH)_2\text{-}1,2$	236.01	lf (CCl_4)	sub	92			al, eth, ace, bz	$B6^3$, 4262
4807	Catechol, 3-methoxy $3\text{-}CH_3OC_6H_4(OH)_2\text{-}1,2$	140.14	nd	129^{10}	43-4				$B6^3$, 6264
4808	Catechol, 3-methyl $3\text{-}CH_3C_6H_3(OH)_2\text{-}1,2$	124.14	lf (bz)	241, 127^{12}	68			w, al, bz, chl	$B6^3$, 4492
4809	Catechol, monoacetate $2\text{-}(CH_3CO_2)C_6H_4OH$	152.15	pl	$189\text{-}91^{102}$, 148^{25}	57-8			w, al, ace, peth	$B6^3$, 4227
4810	Catechol, monobenzoate $2\text{-}(C_6H_5CO_2)C_6H_4OH$	214.22	nd (w)		130-1			al	$B9^3$, 551
4811	Catechol, monobenzylether $2\text{-}(C_6H_5CH_2O)C_6H_4OH$	200.24		$173\text{-}4^{13}$		1.154^{22}	1.5906^{18}	al, eth	$B6^3$, 4218
4812	Catechol, monobutyl ether $2\text{-}(C_4H_9O)C_6H_4OH$	166.22		231-4, 159^{69}		1.026^{25}	1.5113^{25}		$B6^3$, 4209
4813	Catechol, monoethyl ether or 2-Ethoxy phenol $2\text{-}(C_2H_5O)C_6H_4OH$	138.17		217, 68^4	29			al, eth	$B6^3$, 4207
4814	Catechol, monomethyl ether or 2-Methoxy phenol, Guaiacol $2\text{-}(CH_3O)C_6H_4OH$	124.14	hex pr	205	32	$1.1287^{21/4}$	1.5429^{20}	al, eth, chl	$B6^3$, 4200
4815	Catechol, monomethyl ether, acetate $2\text{-}CH_3OC_6H_4OCOCH_3$	166.18		$123\text{-}4^{13}$		$1.1285^{25/4}$	1.5101^{25}	al, eth	$B6^3$, 4227
4816	Catechol, monopropryl ether or 2-Propoxy phenol $2\text{-}(C_3H_7O)C_6H_4OH$	152.19		228-9, $80\text{-}3^4$		1.0523^{25}	1.5176^{25}	al	$B6^3$, 4209
4817	Catechol, 4-nitro $4\text{-}NO_2C_6H_3(OH)_2\text{-}1,2$	155.11		174				w, al, eth, chl	$B6^3$, 4263
4818	Catechol, 3-*iso*-propyl-6-methyl $3\text{-}i\text{-}C_3H_7\text{-}6\text{-}CH_3\text{-}C_6H_2(OH)_2\text{-}1,2$	166.22		270	48			eth, ace	$B6^3$, 4673
4819	Catechol, 4-propyl $4\text{-}C_3H_7C_6H_3(OH)_2\text{-}1,2$	152.19	pr (w, bz)	152^{13}	60	$1.100^{18/4}$	1.4440^{18}	al, eth, ace	$B6^3$, 4613
4820	Catechol, 4-*iso*-propyl $4\text{-}i\text{-}C_3H_7C_6H_3(OH)_2\text{-}1,2$	152.29	lf (lig)	270-2, 168^{26}	78				$B6^3$, 4632
4821	Catechol, tetrabromo $Br_4C_6(OH)_2\text{-}1,2$	425.70	nd (bz, al) lf (bz-peth)		192-3			al, bz	$B6^3$, 4261
4822	Catechol, tetrachloro $Cl_4C_6(OH)_2\text{-}1,2$	247.89	cr (dil al, bz) cr (+3w) (aq aa)		110 (anh), 94 (+3w)				$B6^3$, 4253
4823	Catechol, 3,4,5-trichloro $3,4,5\text{-}Cl_3C_6H(OH)_2\text{-}1,2$	213.45	(i) pr (+1w, aa), (ii) pr (+½w, bz)	115, 134-5				al, eth, aa	$B6^1$, 389
4824	Cedrene $C_{15}H_{24}$	204.36	$[\alpha]^{20}{}_D$ −91.3	262-3, $124\text{-}6^{12}$		$0.9342^{20/4}$	1.5034^{20}	bz, lig	$B5^3$, 1095
4825	Cedrol $C_{15}H_{26}O$	222.37	$[\alpha]$ +10.08 (chl, c=10)		86	$0.9496^{90/20}$	$1.4824^{90/0}$		$B6^3$, 424
4826	β-Cellobiose or 4-*o*-β-D-Glucopyranosyl-β-D-glucose $C_{12}H_{22}O_{11}$	342.30	cr (dil al), $[\alpha]^{20}{}_D$ +14.2→ +34.6 (mut) (w, c=8, 15 hr)		d 225			w	$B17^4$, 3061
4827	Cellobiose, octa-acetate (α-anomer) $C_{28}H_{38}O_{19}$	678.60	nd (al), $[\alpha]^{20/D}$ +43.6 (chl, c=6)		229.5			chl, aa	$B17^4$, 3589
4828	Cellobiose, octa-acetate (β-anomer) $C_{28}H_{38}O_{19}$	678.60	nd (al), $[\alpha]^{20}{}_D$ −14.7 (chl, c=5)		202			chl	$B17^4$, 3590
4829	Cellulose or Polycellobioso $(C_6H_{10}O_5)_x$	$(162.14)_x$	wh amor		260-70d	1.27-1.60			

PHYSICAL CONSTANTS OF ORGANIC COMPOUNDS (Continued)

No.	Name, Synonyms, and Formula	Mol. wt.	Color, crystalline form, specific rotation and λ_{max} (log ε)	b.p. °C	m.p. °C	Density	n_D	Solubility	Ref.
4830	Cellulose, hexanitrate or Gun cotton $(C_{12}H_{14}N_6O_{22})_x$	(594.27)	wh amor		160-70 (ign)	1.66		ph NO_2	
4831	Cellulose, pentanitrate $(C_{12}H_{15}N_5O_{20})_x$	(549.28)	wh amor		1.66			eth-al	
4832	Cellulose, tetranitrate or in Collodion $(C_{12}H_{16}N_4O_{18})_x$	(504.28)	wh amor			1.66		eth-al	
4833	Cellulose, triacetate $(C_{12}H_{16}O_8)_x$	(288.25)	yesh fl $[\alpha]_D$ −22.5 (chl)					aa	
4834	Cellulose, triethylether or Ethylcellulose $(C_{12}H_{22}O_5)_x$	(246.30)	wh nd (bz) $[\alpha]^{20}/_D$ + 26.1 (bz)		240-55			eth	
4835	Cellulose, trinitrate or in Collodion $(C_{21}H_{17}N_3O_{16})_x$	(459.28)	wh			1.66		ace, aa	
4836	Cepharanthine $C_{37}H_{38}N_2O_6$	606.72	ye amor pw $[\alpha]^{20}/_D$ + 277 (chl)		145-55			al, eth, ace, bz	C49, 1745
4837	Cerane or Isohexacosane $CH_3(CH_2)_{24}CH_3$	366.71	pl (eth), sc (w)	$207^{0.7}$	61			al, eth	B1[2], 143
4838	Cerulignone $[(C_6H_2O(CH_3O)_2]_2$	304.30	bl gr						B8[2], 573
4839	Cetane or Hexadecane $CH_3(CH_2)_{14}CH_3$	226.45	lf (ace)	287, 149^{10}	18.2	$0.7733^{20/4}$	1.4345	**eth**	B1[4], 537
4840	Cetene or 1-Hexadecene $CH_3(CH_2)_{13}CH=CH_2$	224.43	lf	284.4, 155^{15}	4.1	$0.7811^{20/4}$	1.4412^{20}	al, eth, peth	B1[4], 927
4841	Cetyl alcohol or 1-Hexadecanol $CH_3(CH_2)_{14}CH_2OH$	242.45	fl (AcOEt)	344, 190^{15}	50	$0.8176^{50/4}$	1.4283^{79}	eth, ace, bz, chl	B1[4], 1876
4842	Cetylamine or 1-Amino hexadecane $CH_3(CH_2)_{14}CH_2NH_2$	241.46	lf	322.5, 144^2	46.8	$0.8129^{20/4}$	1.4496^{20}	al, eth, ace, bz, chl	B4[4], 818
4843	Cetyl Phenyl Ether or Hexadecyl phenyl ether $C_{16}H_{33}OC_6H_5$	318.54	lf (al)	200^1	41.8	0.8434^{82}	1.4556^{82}		B6[3], 555
4844	Cetyl sulfate $(C_{16}H_{33}O)_2SO_2$	546.93			66.2			w	B1[4], 1879
4845	Cevagenine $C_{27}H_{43}NO_8$	509.64	nd (MeOH-eth), $[\alpha]_{20/D}$ −47.5 (al)		246-8				B21[4], 6815
4846	Chalcone dibromide-*(threo)* $C_6H_5CHBrCHBrCOC_6H_5$	368.07	nd (al)		122-3			al	B7[3], 2155
4847	Chalcone dibromide-*(erythro)* $C_6H_5CHBrCHBrCOC_6H_5$	368.07	pr or nd (al)		159-60				B7[3], 2154
4848	Chalcone-*(trans)* or Benzalacetophenone $C_6H_5COCH=CHC_6H_5$	208.26	pa ye lf, pr, nd (peth)	345-8d, 208^{25}	(i)59 (ii)57 (iii)49	$1.0712^{62/4}$		eth, bz, chl	B7[3], 2380
4849	Chalcone, 4,4-dimethyl $(4-(CH_3C_6H_4)COCH=CH(C_6H_4CH_3\text{-}4)$	236.32	cr (MeOH)		127-9			al	B7[2], 441
4850	Chalcone, 3,3′-dinitrotro $(3-O_2NC_6H_4)COCH=CH(C_6H_4NO_2\text{-}3)$	298.25	pa ye nd (aa)		210-1			bz	B7[3], 2407
4851	Chalcone, 2-methoxy or 2-Anisylidene acetophenone $(2-CH_3OC_6H_4)CH=CHCOC_6H_5$	238.29	yesh nd (peth or eth-lig)		64-5			al, eth, bz, chl	B8[3], 1456
4852	Chalcone, 3-methoxy $(3-CH_3OC_6H_4)CH=CHCOC_6H_5$	238.29	yesh pl or pr (MeOH)	247^{12}	65			al, eth, ace, bz	B8[3], 1463
4853	Chalcone, 4-methoxy $(4-CH_3OC_6H_4)CH=CHCOC_6H_5$	238.29	ye nd (al)	$187\text{-}8^{18}$	79			al, eth, chl, aa	B8[3], 1464
4854	Chalcone, 3,4-methylene dioxy or Piperonylidine acetophenone $[3,4-(CH_2O_2)C_6H_3]CH=CHCOC_6H_5$	252.27	ye nd (al)		128			al, aa	B19[4], 1866
4855	Chalcone, 2-nitro $(2-O_2NC_6H_4)CH=CHCOC_6H_5$	253.26	pa br nd (al)		125			al, eth, aa	B7[3], 2399
4856	Chalcone, 2′-nitro $C_6H_5CH=CHCO(C_6H_4NO_2\text{-}2)$	253.26	nd (al)		128-9			al, eth	B7[3], 2402
4857	Chalcone, 3-nitro $(3-O_2NC_6H_4)CH=CHCOC_6H_5$	253.26	ye nd (al or bz)		145-6			al, bz, chl, aa	B7[3], 2400
4858	Chalcone, 4-nitro $(4-O_2NC_6H_4)CH=CHCOC_6H_5$	253.26	pa ye nd (al), pl (bz)		164			al, chl	B7[3], 2401

No.	Name, Synonyms, and Formula	Mol. wt.	Color, crystalline form, specific rotation and λ_{max} (log ε)	b.p. °C	m.p. °C	Density	n_D	Solubility	Ref.
4859	Chalcone, α-nitro $C_6H_5C(NO_2)=CHCOC_6H_5$	253.26	ye pl (eth or bz-lig) cr aa		90			eth, ace, bz	B7[3], 2403
4860	Chaulmoogric acid-(d) or d-13(2-cyclopentenyl)tridecanoic acid $C_{18}H_{32}O_2$	280.46	pl or lf (al, aa), $[\alpha]_D$ + 62 (chl)	$247\text{-}8^{20}$	68.5			eth, chl	B9[3], 284
4861	Chaulmoogric acid-(dl) $C_{18}H_{32}O_2$	280.45	cr (peth)	$247\text{-}8^{20}$				eth, chl	B9[3], 285
4862	Chloroacetaldehyde, trimer or 2,4,6-tris (chloromethyl) 1,3,5-trioxane $C_6H_9Cl_3O_3$	235.49	nd (eth)	$142\text{-}4^{10}$	87			eth	B19[4], 4718
4863	Chelerythrine $C_{21}H_{19}NO_5$	365.39	cr (chl-MeOH), cr (al + 1)	207				chl	B27[2], 563
4864	Chelidonic acid or 4-Pyrone-2,6-dicarboxylic acid $C_7H_4O_6$	184.11	rose mcl nd (al-w + 1w)		262				B18[4], 6136
4865	Chelidonic acid, diethyl ester $C_{11}H_{12}O_6$	240.21	pr, nd		69			eth	B18[4], 6137
4866	Chelidonine-(d) $C_{20}H_{19}NO_5$	353.37	mcl pl (al + 1w), $[\alpha]_D$ + 151 (al, 1%)		136-40d			al, eth, chl	B27[2], 615
4867	Chelidonine, hydrochloride-(d) $C_{20}H_{19}NO_5\text{-}HCl$	389.84	wh (w)						B27, 557
4868	Chloral or Trichloroacetaldehyde Cl_3CCHO	147.39		97.8	−57.5	$1.51214^{20/4}$	1.4557^{20}	w, al, eth	B1[4], 3142
4869	Chloral ammonia $Cl_3CCH(OH)NH_2$	164.42	nd (al)	100d	72-4			al, eth, bz	B1[4], 3147
4870	Chloral hydrate $Cl_3CCH(OH)_2$	165.40	mcl pl (w)	$96.3^{764}/_d$	57	$1.9081^{20/4}$		w, al, eth, ace, bz, chl, Py	B1[4], 3143
4871	Chloralide or 2,5-bis(trichloromethyl)1,3-dioxolan-4-one $C_5H_2Cl_6O_3$	322.79	pr (al or eth)	272-3, $147\text{-}8^{12}$	116			eth, aa	B19[4], 1571
4872	Chloroacetamide $ClCH_2CONH_2$	93.51	mcl pr	224.5^{743}	121			w, al	B2[4], 490
4873	Chloroacetamide, *N*-allyl-*N*-phenyl $ClCH_2CON(C_6H_5)CH_2CH=CH_2$	217.74		$119^{0.15}$			1.5079^{25}		Am78, 2556
4874	Chloroacetamide, *N,N*-bis(2-chloroallyl) $ClCH_2CON(CH_2C(Cl)=CH_2)_2$	242.53		$161—3^{12}$			1.5220^{25}		B4[4], 1089
4875	Chloroacetamide-*N,N-bis*(3-chloroallyl) $ClCH_2CON[CH_2C(Cl)=CH_2]_2$	242.53		140.5^{1}			1.5220^{25}		Am78, 2556
4876	Chloroacetamide, *bis*(2-chloropropyl) $ClCH_2CON(CH_2CHClCH_3)_2$	246.57		$134^{0.7}$			1.5018^{25}		B4[4], 499
4877	Chloroacetamide, *N,N-bis*(2-ethylhexyl) $ClCH_2CON[CH_2CH(C_2H_5)C_4H_9]_2$	217.94		$154^{0.8}$			1.4622^{25}		Am78, 2556
4878	Chloroacetamide, *N,N-bis*(2-methylallyl) $ClCH_2CON[CH_2C(CH_3)=CH_2)]_2$	201.70		$133\text{-}5^{20}$			1.4882^{2}		B4[4], 1105
4879	Chloroacetamide, *N,N-bis*(3-Methylbutyl) $ClCH_2CON[CH_2CH_2CH(CH_3)_2]_2$	233.78		$109^{0.6}$			1.4625^{25}		B4[4], 702
4880	Chloroacetamide, *N*-butyl $ClCH_2CONHC_4H_9$	149.62		110^{7}			1.4665^{25}		B4[4], 567
4881	Chloroacetamide, *N*-butyl-*iso*-propyl $ClCH_2CON(C_4H_9)CH(CH_3)_2$	191.70		$101^{1.4}$			1.5078^{25}		B4[4], 567
4882	Chloroacetamide, *N*-*sec*-butyl $ClCH_2CONHCH(CH_3)C_2H_5$	149.62		$68^{0.7}$	45-6			peth	B4[4], 621
4883	Chloroacetamide, *N*-*tert*-butyl $ClCH_2CONHC(CH_3)_3$	149.62	cr (peth)		84			peth	B4[4], 662
4884	Chloroacetamide, *N*-butyl-*N*-ethyl $ClCH_2CON(C_2H_5)C_4H_9$	177.67		$90^{1.5}$			1.4665^{25}		B4[4], 567
4885	Chloroacetamide, *N*-2-chloroallyl $ClCH_2CONHCH_2CCl=CH_2$	168.02		$101^{1.4}$			1.5078^{25}		B4[4], 1089
4886	Chloroacetamide, *N*-2-chloroallyl-*N*-phenyl $ClCH_2CON(C_6H_5)CH_2CCl=CH_2$	243.11		$138^{0.7}$			1.5602^{25}		Am78, 2556
4887	Chloroacetamide, *N*-3-chloroallyl $ClCH_2CONHCH_2CH=CHCl$	168.02	cr (peth)	$112^{0.5}$	52-3			peth	B4[4], 1087
4888	Chloroacetamide, *N*-(2-chloro-4-nitrophenyl) $(4\text{-}NO_2\text{-}2\text{-}Cl\text{-}C_6H_3)NCOCH_2Cl$	248.05	cr (peth)		118-9				B12[3], 1664

No.	Name, Synonyms, and Formula	Mol. wt.	Color, crystalline form, specific rotation and λ_{max} (log ε)	b.p. °C	m.p. °C	Density	n_D	Solubility	Ref.
4889	Chloroacetamide, *N*-(4-chlorophenyl)-*N*-ethyl $(4\text{-}ClC_6H_4)N(C_2H_5)COCH_2Cl$	232.11	cr (peth)		70-1			peth	Am78, 2557
4890	Chloroacetamide, *N*-2-chloropropyl $ClCH_2CONHCH_2CHClCH_3$	170.04		$88^{1.5}$			1.4942^{25}		$B4^4$, 499
4891	Chloroacetamide, *N*-3-chloropropyl $ClCH_2CONHCH_2CH_2CH_2Cl$	170.04			36-7				$B4^4$, 501
4892	Chloroacetamide, *N*-*N*-diallyl $ClCH_2CON(CH_2CH=CH_2)_2$	173.64		$92^{0.7}$			1.4932^{25}		$B4^4$, 1064
4893	Chloroacetamide, *N,N*-dibenzyl $ClCH_2CON(CH_2C_6H_5)_2$	273.76		$190^{1.8}$			1.5837^{25}		Am78, 2556
4894	Chloroacetamide, *N,N-di-iso*-butyl $ClCH_2CON(i\text{-}C_4H_9)_2$	205.73		99^{2}			1.4642^{25}		$B4^4$, 633
4895	Chloroacetamide, *N,N-di-sec*-butyl $ClCH_2CON(sec\text{-}C_4H_9)_2$	205.73		$92^{0.7}$			1.4681^{25}		Am78, 2556
4896	Chloroacetamide, *N*-(2,3-dichloroallyl) $ClCH_2CONHCH_2CCl=CHCl$	202.47		$126\text{-}31^{1.8}$			1.5311^{25}		Am78, 2556
4897	Chloroacetamide, N-(2,4-dichlorobenzyl) $ClCH_2CONH\text{-}(CH_2C_6H_3Cl_2\text{-}2,4)$	252.53	cr (bz)		96-7			bz	Am78,2556
4898	Chloroacetamide, *N*-(2,4-dichlorobenzyl) $ClCH_2CONH\text{-}(CH_2C_6H_3Cl_2\text{-}2,4)$	252.53	cr (dil al)		105-6			al	Am78, 2556
4899	Chloroacetamide, *N*-(2,4-dichlorophenyl) $ClCH_2CONH(C_6H_3Cl_2\text{-}2,4)$	238.50	cr (dil al)		101-2			al, peth	Am78, 2556
4900	Chloroacetamide, *N*-(2,5-dichlorophenyl) $ClCH_2CONH(C_6H_3Cl_2\text{-}2,5)$	238.50	cr (dil al)		116-7			al	Am78, 2556
4901	Chloroacetamide, *N*(2,3-dichloropropyl) $ClCH_2CONHCH_2CHClCH_2Cl$	204.48	cr (peth)		65-6			peth	$B4^4$, 502
4902	Chloroacetamide, *N,N*-diethyl $ClCH_2CON(C_2H_5)_2$	149.62		$190\text{-}5^{25}$					$B4^4$, 350
4903	Chloroacetamide, *N,N*-dihexyl $ClCH_2CON(C_6H_{13})_2$	261.84	cr (peth)		114-5				$B4^4$, 713
4904	Chloroacetamide, *N*-(2,4-dinitrophenyl) $[2,4\text{-}(NO_2)_2C_6H_3]NHCOCH_2Cl$	259.61	nd (al)		114-5				$B9^2$, 315
4905	Chloroacetamide, *N,N*-dipentyl $ClCH_2CON(C_5H_{11})_2$	233.78		126^{1}			1.4651^{25}		$B4^4$, 679
4906	Chloroacetamide, *N,N*-dipropyl $ClCH_2CON(C_3H_7)_2$	177.67		120^{8}, $90\text{-}2^{0.8}$			1.4670^{20}		$B4^4$, 476
4907	Chloroacetamide, *N,N*-di-*iso*-propyl $ClCH_2CON(i\text{-}C_3H_7)_2$	177.67	cr (peth)	$86^{2.7}$	48-9		1.4619^{25}		$B4^4$, 516
4908	Chloroacetamide, *N*-ethyl-*N*-hexyl $ClCH_2CON(C_2H_5)C_6H_{13}$	205.73		120^{11}			1.4978^{25}		Am78, 2556
4909	Chloroacetamide, *N*-furfuryl $ClCH_2CONH(CH_2C_4H_3O)$	173.60	cr(peth)		58			peth	$B18^4$, 7080
4910	Chloroacetamide, *N*-hexyl $ClCH_2CONHC_6H_{13}$	177.67	cr (peth)	$95\text{-}105^{0.2}$	108-9			peth	$B4^4$, 713
4911	Chloroacetamide, *N*-hexyl-*N*-methyl $ClCH_2CON(CH_3)C_6H_{13}$	191.70		$134^{3.8}$			1.5005^{25}		Am78, 2556
4912	Chloroacetamide, 3-methoxypropyl $ClCH_2CONH(CH_2)_3COCH_3$	165.62		$88^{0.5}$	30		1.4712^{25}		$B4^4$, 1645
4913	Chloroacetamide, *N*-(2-methylallyl) $ClCH_2CONHCH_2C(CH_3)=CH_2$	147.61		96^{1}			1.4860^{25}		$B4^4$, 1105
4914	Chloroacetamide, *N*-(3-methylbutyl) $ClCH_2CONH\text{-}CH_2CH_2CH(CH_3)_2$	163.65		$134\text{-}5^{13}$	−15				$B4^2$, 647
4915	Chloroacetamide, *N*-pentyl $ClCH_2CONHC_5H_{11}$	163.65		$82^{0.5}$			1.4665^{25}		$B4^4$, 678
4916	Chloroacetamide, *N*-propyl $ClCH_2CONHC_3H_7$	135.59		$105\text{-}6^{10.5}$	62			al, eth, ace, chl	$B4^4$, 476
4917	Chloroacetamide, *N-iso*-propyl $ClCH_2CONHCH(CH_3)_2$	135.59	cr (peth)		62			peth	$B4^4$, 515
4918	Chloroacetamide, *N*-tetradecyl $ClCH_2CONHC_{14}H_{29}$	289.89	cr(peth)		64-5				$B4^4$, 814
4919	Chloroacetamide, *N*-tetrahydrofurfuryl $ClCH_2CONH(C_4H_7O)$	177.63	cr (peth)		62-3			peth	$B18^4$, 7039
4920	Chloroacetic acid (*α*) $ClCH_2CO_2H$	94.50	mcl pr	187.8, 104^{20}	63	$1.4043^{40/4}$	1.4351^{55}	w, al, eth, bz, chl	$B2^4$, 474
4921	Chloroacetic acid (*β*) $ClCH_2CO_2H$	94.50	mcl pr	187.9, 104^{20}	56.2	$1.4043^{40/4}$	1.4351^{50}	w, al, eth, bz, chl	$B2^4$, 474

No.	Name, Synonyms, and Formula	Mol. wt.	Color, crystalline form, specific rotation and λ_{max} (log ε)	b.p. °C	m.p. °C	Density	n_D	Solubility	Ref.
4922	Chloroacetic acid (γ) $ClCH_2CO_2H$	94.50		187.8, 104^{20}	52.5	$1.4043^{40/4}$	1.4351^{55}	w, al, eth, bz, chl	B2[4], 474
4923	Chloroacetic acid, anhydride $(ClCH_2CO)_2O$	170.98	pr (bz)	203	46	1.5497^{20}			B2[4], 487
4924	Chloroacetic acid, benzyl ester or Benzyl chloracetate $ClCH_2CO_2CH_2C_6H_5$	184.62		147.5^{9}, 84-6$^{0.4}$		$1.2223^{4/4}$	1.5426^{18}	al, eth	B6[3], 1479
4925	Chloroacetic acid, butyl ester or Butyl chloroacetate $ClCH_2CO_2C_4H_9$	150.61		183, 94^{38}		$1.0704^{20/4}$	1.4297^{20}	al, eth	B2[4], 482
4926	Chloroacetic acid, *iso*-butyl ester or *iso*-Butyl chloroacetate $ClCH_2CO_2CH_2CH(CH_3)_2$	150.61		170		$1.0612^{20/4}$	1.4255^{20}	eth, ace	B2[4], 483
4927	Chloroacetic acid, *sec*-butyl ester or *sec*-Butylchloroacetate $ClCH_2CO_2$-*sec*-C_4H_9	150.61		163-4		$1.062^{20/20}$	1.4251^{19}	al, eth	B2[3], 443
4928	Chloroacetic acid, 2-chloroethyl ester or 2-chloroethyl chloroacetate $ClH_2CO_2CH_2CH_2Cl$	157.00		202, 89^{10}		$1.3600^{25/4}$	1.4619^{25}	eth	B2[4], 481
4929	Chloroacetic acid, ethyl ester or Ethyl chloroacetate $ClCH_2CO_2C_2H_5$	122.55		144^{740}, 52^{20}	−26	$1.1585^{20/4}$	1.4215^{20}	**al, eth, ace,** bz	B2[4], 481
4930	Chloroacetic acid, 2-hydroxyethyl ester or 2-Hydroxyethyl chloroacetate $ClCH_2CO_2CH_2CH_2OH$	138.55		240d, $86^{1.5}$		$1.3300^{20/4}$	1.4609^{20}	**w, al**	B2[3], 447
4931	Chloroacetic acid, hydrazide $ClCH_2CONHNH_2$	108.53			93				
4932	Chloroacetic acid, 2-methoxyethyl ester or 2-Methoxyethyl chloroacetate $ClCH_2CO_2CH_2CH_2OCH_3$	152.57		$85\text{-}6^{9}$		$1.2015^{20/4}$	1.4382^{20}	eth	B2[3], 447
4933	Chloroacetic acid, methyl ester or Methyl chloroacetate $ClCH_2CO_2CH_3$	108.52		129.8, 29^{10}	−32.1	$1.2337^{20/4}$	1.4218^{20}	**al, eth, ace, bz**	B2[4], 480
4934	Chloroacetic acid, phenyl ester or Phenyl chloroacetate $ClCH_2CO_2C_6H_5$	170.60	nd or pl (al)	230-5, 114^{8}	44-5	$1.2202^{44/4}$	1.5146^{44}	al, eth	B6[3], 598
4935	Chloroacetic acid, propylester or Propyl chloroacetate $ClCH_2CO_2C_3H_7$	136.58		161^{764}		$1.1033^{20/4}$	1.4261^{20}	eth	B2[4], 482
4936	Chloroacetic acid, *iso*-propyl ester or *iso*-Propyl chloroacetate $ClCH_2CO_2$-*i*-C_3H_7	136.58		150-1		$1.0888^{20/4}$	1.4382^{20}	eth	B2[4], 482
4937	Chloroacetic acid, 4-tolyl ester $ClCH_2CO_2(C_6H_4CH_3\text{-}4)$	184.62	pl	162^{41}	32			al, eth	B6[2], 378
4938	Chloroacetone cyanohydrin $ClCH_2C(CH_3)(OH)CN$	119.55		110^{27}		1.2027^{15}	1.4536^{11}	w, al, ace	B3[3], 599
4939	Chloroacetomitrile $ClCH_2CN$	75.50		126-7, $30\text{-}2^{15}$		1.1930^{20}	1.4202^{25}	al, eth	B2[4], 492
4940	Chloroacetyl chloride $ClCH_2COCl$	112.94		107		$1.4202^{20/4}$	1.4541^{20}	**eth,** ace	B2[4], 488
4941	2-Chloroally isotthiocyanate $CH{=}CClCH_2NCS$	133.60		182		1.27^{12}			B4, 219
4942	Chlorodifluoro acetic acid ClF_2CCOOH	130.48			22.9			chl	B2[4], 497
4943	Chlorodiphenyl acetamide $(C_6H_5)_2CClCONH_2$	245.71	cr (to)		115			al, eth, bz, chl	B9[3], 3308
4944	Chlorodiphenyl acetic acid $(C_6H_5)_2CClCO_2H$	246.79	pl (bz-lig)	118-9d				al, eth, ace, bz	B9[3], 3307
4945	Chlorodiphenyl acetic acid, ethyl ester $(C_6H_5)_2CClCO_2C_2H_5$	274.75	pl (chl), cr (al)	185^{14}	43-4			al, eth	B9[3], 3307
4946	Chlorodiphenyl acetyl chloride $(C_6H_5)_2CClCOCl$	265.14	cr (lig)	180^{14}	50-1				B9[3], 3308
4947	bis-(2-Chloroethyl) methylamine, hydrochloride $(ClCH_2CH_2)_2NCH_3 \cdot HCl$	192.52	hyg nd		111-2			w, al	B4[4], 446
4948	tris-(2-Chloroethyl) amine or Nitrogen mustard gas $(ClCH_2CH_2)_3N$	204.53	Pa ye	$143\text{-}4^{15}$	−4			**al, eth, bz**	B4[4], 447
4949	(α-Chloroethyl)ether or bis-(α-chloroethyl) ether $(CH_3CHCl)_2O$	143.02		116-7		$1.1060^{25/4}5$	1.4186^{25}	al, **eth,** chl	B1[4], 3120
4950	(α-Chloroethyl)(β-Chloroethyl) ether $CH_3CHClOCH_2CH_2Cl$	143.01		d^{760}, $55\text{-}7^{17}$		$1.1867^{20/4}$	1.4473^{20}	al, eth, chl	B1[3], 2655
4951	(β-Chloroethyl)diethyl amminium chloride $ClCH_2CH_2NH(C_2H_5)_2Cl^-$	172.10	nd (al-eth)		210-1			**w, al**	B4[4], 447

No.	Name, Synonyms, and Formula	Mol. wt.	Color, crystalline form, specific rotation and λ_{max} (log ε)	b.p. °C	m.p. °C	Density	n_D	Solubility	Ref.
4952	α-Chloroethyl methyl ether $CH_3OCHClCH_3$	94.54	...	$72\text{-}3^{751}$	...	$0.9902^{20/4}$	1.4004^{20}	eth	$B1^4$, 3119
4953	α-Chloroethyl pentyl ether $CH_3CH_2Cl\text{-}O\text{-}C_5H_{11}$	150.65	...	$63\text{-}6^{8}$	...	$0.9200^{20/4}$	1.4218^{20}	eth	$B1^3$, 2656
4954	α Chloroethyl propyl ether $CH_3CHClOC_3H_7$	122.59	...	$112\text{-}5^{731}$	...	$0.9322^{20/4}$	1.4013^{20}	eth	$B1^3$, 2655
4955	(β-Chloroethyl)ether or bis-(β-Chloroethyl)ether $(ClCH_2CH_2)_2O$	143.01	...	$178, 75^{20}$	−24.5	$1.2199^{20/4}$	1.4575^{20}	al, eth, ace, bz	$B1^4$, 1375
4956	α-Chloroethyl methyl ether $ClCH_2CH_2OCH_3$	94.54	...	92-3	...	$1.0345^{20/4}$	1.4111^{20}	w, eth	$B1^4$, 1375
4957	β-Chloroethyl vinyl ether $ClCH_2CH_2OCH{=}CH_2$	106.55	...	108	...	$1.0475^{20/4}$	1.4378^{20}	al, eth	$B1^4$, 2051
4958	Chloroformic acid, benzyl ester or Carbobenzoxy chloride $ClCO_2CH_2C_6H_5$	170.60	...	103^{20}	...	1.20	1.5150^{20}	eth, ace, bz	$B6^3$, 1485
4959	Chloroformic acid, butyl ester $ClCO_2C_4H_9$	136.58	...	$138^{750}, 35.5^{13}$	...	$1.0513^{20/4}$	1.4121^{20}	eth, ace	$B3^4$, 25
4960	Chloroformic acid, isobutyl ester or Isobutyl chloroformate $ClCO_2CH_2CH(CH_3)_2$	136.58	...	128.8	...	$1.0426^{18/4}$	$1.4071^{18/}{}_{Hc}$	eth, bz, chl	$B3^4$, 26
4961	Chloroformic acid, β-chloroethyl ester $ClCO_2CH_2CH_2Cl$	142.97	...	156	...	$1.3847^{20/4}$	1.4483^{20}	al, eth, ace, bz	$B3^4$, 24
4962	Chloroformic acid, chloromethyl ester $ClCO_2CH_2Cl$	128.54	...	107	...	1.465^{15}	1.4286^{22}	eth, ace	$B3^4$, 30
4963	Chloroformic acid, (3-chloropropyl) ester $ClCO_2CH_2CH_2CH_2Cl$	157.00	...	177	...	$1.2949^{25/20}$	1.4456^{20}	...	$B3^3$, 25
4964	Chloroformic acid, cyclohexyl ester $ClCO_2C_6H_{11}$	162.62	...	87.5^{27}	...	...	...	eth	$B6^4$, 43
4965	Chloroformic acid, dichloromethyl ester $ClCO_2CHCl_2$	163.39	...	110-1	...	...	...	ace	$B3^4$, 31
4966	Chloroformic acid, β-ethoxyethyl ester or Cellosolve chloroformate $ClCO_2CH_2CH_2OC_2H_5$	152.58	...	67.2^{14}	...	1.1341^{25}	1.4169^{25}	al	$B3^2$, 29
4967	Chloroformic acid, ethyl ester or Ethyl chloroformate $ClCO_2C_2H_5$	108.52	...	95	−80.6	$1.1352^{20/4}$	1.3974^{20}	eth, bz, chl	$B3^4$, 23
4968	Chloroformic acid, 2-methoxyethyl ester $ClCO_2CH_2CH_2OCH_3$	138.55	...	58.7^{13}	...	1.1905^{25}	1.4163^{20}	eth	$B3^2$, 29
4969	Chloroformic acid, methyl ester or Methyl chloroformate $ClCO_2CH_3$	94.50	...	70-1	...	$1.2231^{20/4}$	1.3868^{20}	**al, eth**, bz, chl	$B3^4$, 23
4970	Chloroformic acid, isopentyl ester or Isopentyl chloroformate $ClCO_2\text{-}i\text{-}C_5H_{11}$	150.61	...	$154.3, 60^{15}$	...	$1.0288^{17/4}$	1.4176^{20}	**al, eth**	$B3^4$, 26
4971	Chloroformic acid, pentyl ester $ClCO_2C_5H_{11}$	150.61	...	$60\text{-}2^{15}$	...	...	1.4181^{18}	eth	$B3^4$, 26
4972	Chloroformic acid, isopropenyl ester $ClCO_2C(CH_3){=}CH_2$	120.54	...	100	...	$1.103^{20/20}$	...	eth	$B3^3$, 28
4973	Chloroformic acid, propyl ester or Propyl chloroformate $ClCO_2C_3H_7$	122.55	...	115.2	...	$1.0901^{20/4}$	1.4035^{20}	al, eth	$B3^4$, 24
4974	Chloroformic acid, isopropyl ester or Isopropyl chloroformate $ClCO_2CH(CH_3)_2$	122.55	...	$105, 66.3^{200}$	...	...	1.4013^{20}	...	$B3^4$, 24
4975	Chloroformic acid, trichloro methyl ester or Diphosgene $ClCO_2CCl_3$	197.83	...	$128, 49^{50}$	−57	1.6525^{14}	1.4566^{22}	al, eth	$B3^4$, 33
4976	Chloromethyl ether or bis-(Chloromethyl) ether $(ClCH_2)_2O$	114.96	...	104	-41.5	$1.328^{15/4}$	1.435^{21}	al, eth	$B1^4$, 3051
4977	Chloromethyl propyl ether $ClCH_2OC_3H_7$	108.57	...	109	...	$0.9884^{20/4}$	1.4125^{20}	al, eth	$B1^3$, 2589
4978	Chloromethyl thiocyanate $ClCH_2SCN$	107.56	...	185	...	1.37^{15}	...	...	$B3^2$, 124
4979	Chloromycetin or Chloroamphenicol $C_{11}H_{12}Cl_2N_2O_5$	323.14	pa ye pl or nd (w), $[\alpha]^{25/}{}_D$ + 19 (al, c=5) -25.5 (AcOEt)	sub vac	150-1	...	...	al, ace, chl	$B13^3$, 2268
4980	(4-Chlorophenyl) ether or bis-(4-Chlorophenyl) ether $(4\text{-}ClC_6H_4)_2O$	239.10	nd (al)	$312\text{-}4, 168.7^{7}$	30	1.1231^{20}	1.611^{20}	...	$B6^4$, 826

No.	Name, Synonyms, and Formula	Mol. wt.	Color, crystalline form, specific rotation and λ_{max} (log ε)	b.p. °C	m.p. °C	Density	n_D	Solubility	Ref.
4981	4-Chlorophenyl phenyl ether $(4\text{-}ClC_6H_4)OC_6H_5$	204.66		284-5		1.2026^{15}	1.599_D		$B6^4$, 826
4982	2-Chlorophenyl isocyanate $2\text{-}ClC_6H_4NCO$	153.57		$115\text{-}7^{43}$	30-1				$B12^3$, 1296
4983	3-Chlorophenyl isocyanate $3\text{-}ClC_6H_4NCO$	153.57		$113\text{-}4^{43}$					$B12^3$, 1316
4984	4-Chlorophenyl isocyanate $4\text{-}ClC_6H_4NCO$	153.57		$115\text{-}7^{45}$					$B12^3$, 1376
4985	4-Chlorophenyl isothiocyanate $4\text{-}ClC_6H_4NCS$	169.63	nd (al)	249-50	45				$B12^3$, 1376
4986	Chlorophyll-a $C_{55}H_{72}MgN_4O_5$	893.51	bl bk hex pl		150-3			al, eth, lig	M, 245
4987	Chlorophyll b $C_{55}H_{70}MgN_4O_6$	907.49	bl bk gr pw		120-30			al, eth, lig	M, 245
4988	Chloroprene or 2-Chloro-1,3-Butadiene $CH_2{=}CCl\text{-}CH{=}CH_2$	88.54		59.4, 6.4^{100}		$0.9583^{20/4}$	1.4583^{20}	**eth, ace, bz**	$B1^4$, 984
4989	Chlorpromazine or Thorazine $C_{17}H_{19}ClN_2S$	318.86		$200\text{-}5^{0.8}$				al, eth, bz, chl	C50, 1951
4990	Chlorpromazin, hydrochloride $C_{17}H_{19}ClN_2S\text{-}HCl$	355.33		194-7d				w, al, chl	C50, 1931
4991	(β-Chloropropyl) ether or bis-(β-chloropropyl) ether $CH_3CHClCH_2OCH_2CHClCH_3$	171.07		188		$1.109^{20/4}$	1.4467^{20}	al, eth	$B1^4$, 1442
4992	(γ-Chloropropyl) ether or bis-(α-Chloropropyl)ether $(ClCH_2CH_2CH_2)_2O$	171.07		215^{745}, 90.5^{11}		$1.140^{20/20}$	1.4158^{20}	al, eth	$B1^2$, 370
4993	(β-Chloroisopropyl) ether or bis-(β-Chloroisopropyl) ether $[ClCH_2CH(CH_3)]_2O$	171.07		187		$1.103^{20/4}$	1.4505^{20}	**al, eth**, ace, bz	$B1^3$, 1470
4994	Bis-(1,2,2,2-Tetrachloroethyl) ether $(Cl_3CCHCl)_2O$	349.68	cr (al, MeOH)	$130\text{-}1^{11}$	40.2			bz, peth	$B1^3$, 2672
4995	Chlorosulfinic acid, ethyl ester or Ethyl chlorosulfinate $(C_2H_5O)SOCl$	128.57		52.5^{44}, 32^{16}		$1.2766^{25/4}$	1.4550^{25}	eth	$B1^3$, 1316
4996	Chlorosulfonic acid, ethyl ester or Ethyl chlorosulfonate $C_2H_5OSO_2Cl$	144.57		151-4, 52^{14}		$1.3502^{25/4}$	1.416^{20}	eth, chl, lig	$B1^4$, 1326
4997	Chlorosulfonic acid, methyl ester or Methyl chlorosulfon-ate CH_3OSO_2Cl	130.55		133-5, 48^{29}		$1.4805^{25/4}$	1.4138^{18}	eth, ace, bz	$B1^4$, 1252
4998	Cholanic acid or Ursocholamic acid $C_{24}H_{40}O_2$	360.58	nd (al), cr (aa), $[\alpha]^{20}{}_D$ + 21.7 (chl)		163-4			al, chl, aa	$B9^3$, 2656
4999	Cholanic acid, 3α, 6α-dihydroxy or Hyodeoxycholic acid $C_{24}H_{40}O_4$	392.58	cr (AcOEt), $[\alpha]^{20}{}_D$ + 37.2 (MeOH)		198-9			al, aa	$B10^3$, 1631
5000	Cholanic acid, 3α,7α-dihydroxy or Chenodeoxycholic acid $C_{24}H_{40}O_4$	392.58	nd (AcOEt-hp), $[\alpha]^{20}{}_D$ + 11.1 (al, c=2.1)		143			al, eth, ace, aa	$B10^3$, 1635
5001	Cholanic acid, ethyl ester $C_{26}H_{44}O_2$	388.63	lf, nd (dil al), $[\alpha]^{19}{}_D$ + 21 (chl)	273^{12}	93-4			chl	$B9^3$, 2658
5002	Cholanic acid, methyl ester $C_{25}H_{42}O_2$	374.61	nd $[\alpha]_D$ + 23±2 (diox)		87-8			diox	$B9^3$, 2658
5003	Clolanthrene or Benz *(j)* aceanthrylene $C_{20}H_{14}$	254.33	pa ye lf (bz-al)	sub, $210^{0.2}$	174-5d			al, bz, aa, lig	$B5^3$, 2469
5004	Cholanthrene, 6,7-dihydro-20-methyl $C_{21}H_{18}$	270.37	lf (MeOH)		155			MeOH	$B5^3$, 2429
5005	Cholanthrene, 11,14-dihydro-20-methyl $C_{21}H_{18}$	270.37	nd (PrOH)		138-9				$B5^3$, 2429
5006	Cholanthrene, 20-methyl $C_{21}H_{16}$	268.36	yesh nd (bz)		180				$B5^3$, 2484
5007	$\Delta^{2,4}$-Cholestadiene $C_{27}H_{44}$	368.65	cr (eth-ace), $[\alpha]^{23}{}_D$ + 168.5 (eth, c=1.5)		68.5			al, eth, chl	$B5^3$, 1428

No.	Name, Synonyms, and Formula	Mol. wt.	Color, crystalline form, specific rotation and λ_{max} (log ε)	b.p. °C	m.p. °C	Density	n_D	Solubility	Ref.
5008	Δ 3,5-Cholestadiene or Cholesterilene $C_{27}H_{44}$	368.65	wh nd (al), $[\alpha]^{20/}{}_D$ - 129.6 (chl, c=3)	260^{11}	80	$0.925^{100/4}$	....	al, eth, bz, chl	B5⁴, 1620
5009	Δ 3,7-Cholestadiene-3β-ol or 7-Dehydrocholesterol $C_{27}H_{44}O$	384.65	pl (+ 1w, eth-MeOH), $[\alpha]^{25/}{}_D$ - 115 (chl, C=2.5)	....	150-1	....	....	eth, ace	B6³, 2819
5010	Δ 4,6-Cholestadiene-3-one $C_{27}H_{42}O$	382.63	ye pr $[\alpha]^{20/}{}_D$ + 31 (chl)	....	78	....	....	....	B7³, 1760
5011	Cholestane $C_{27}H_{48}$	372.68	sc or pl (eth-al, ace), $[\alpha]^{20/}{}_D$ + 30.2 (chl, c=2)	250^{1}	80	$0.9090^{88/4}$	1.4887^{88}	eth, bz, chl	B5⁴, 1227
5012	3β-Cholestane carboxylic acid $C_{28}H_{48}O_2$	416.69	nd $[\alpha]^{25/}{}_D$ + 28.8 (chl, c=1.7)	....	210-1	....	....	....	B9³, 2668
5013	3,6-Cholestane dione $C_{27}H_{44}O_2$	400.65	nd $[\alpha]^{20/}{}_D$ + 8.9	....	171 2	....	....	....	B7³, 3282
5014	3α-Cholestanol or Epidihydro cholesterol $C_{27}H_{48}O$	388.68	nd (al), $[\alpha]^{20/}{}_D$ + 34 (chl, c=1.1)	....	188	....	....	....	B6³, 2135
5015	3β-Cholestanol or Dihydrocholesterol $C_{27}H_{48}O$	388.68	lf (al + 1w), pr, pl (MeOH), $[\alpha]^{22/}{}_D$ + 24.2 (chl, c=1.3	....	141-3	....	....	al, eth, chl	B6³, 2131
5016	3-Cholestanone or Zymostanone $C_{27}H_{46}O$	386.66	nd or lf (al), $[\alpha]_D$ + 42 (chl, c=2.12)	....	128-30	....	....	al	B7³, 1330
5017	6-Cholestanone, 3β-hydroxy or 6-Keto Cholestanol $C_{27}H_{46}O_2$	402.66	nd (al), $[\alpha]^{\prime}{}_D$ -3.0 (chl)	....	150-1	....	....	al	B8³, 658
5018	7-Cholestanone, 3β-hydroxy or 7-Oxocholestanol $C_{27}H_{46}O_2$	402.66	pl $[\alpha]^{22/}{}_D$ - 34 (chl)	....	165-8	....	....	al	B8³, 665
5019	2-Cholestene or Neocholestene $C_{27}H_{46}$	370.66	nd (eth-ace or al), $[\alpha]_D$ + 66 (c=1.65)	....	75-6	....	....	....	B5⁴, 1507
5020	3-Cholestene $C_{27}H_{46}$	370.66	wh pr or nd (al), $[\alpha]^{18/}{}_D$ - 56.3 (chl)	....	93-4	....	....	....	B5⁴, 1508
5021	5-Cholestene, 3β-bromo or Cholestyryl bromide $C_{27}H_{45}Br$	449.56	mcl lf (al), $[\alpha]^{20/}{}_D$ -19 (bz, c=0.4)	....	100-2	....	....	bz, chl	B5⁴, 1512
5022	5-Cholestene, 3β-chloro or Cholesteryl chloride $C_{27}H_{45}Cl$	405.11	nd (al or ace), $[\alpha]^{20/}{}_D$ - 26.4 (bz) - 33.3 (chl)	....	96	....	....	bz, chl	B5⁴, 1510
5023	5-Cholestene, 3β,7β-diol or 7β-Hydroxycholesterol $C_{27}H_{46}O_2$	402.66	wh nd (eth), $[\alpha]^{20/}{}_D$ + 7.2 (chl, c=2)	sub, $145^{0.005}$	177-8	....	....	....	B6³, 5130
5024	5-Cholestene, 3β-iodo or Cholestyryl iodide $C_{27}H_{45}I$	496.56	nd (ace or AcOEt)	....	106-7	....	....	bz, chl, lig	B5⁴, 1512
5025	1-Cholesten-3-one $C_{27}H_{44}O$	384.65	nd (dil ace or al), $[\alpha]^{25/}{}_D$ + 88.2 (chl)	....	99-101	....	....	bz	B7³, 1592

No.	Name, Synonyms, and Formula	Mol. wt.	Color, crystalline form, specific rotation and λ_{max} (log ε)	b.p. °C	m.p. °C	Density	n_D	Solubility	Ref.
5026	4-Cholestene-3-one $C_{27}H_{44}O$	384.65	nd or pl (al), $[\alpha]^{25}{}_D$ + 92 (chl, c=2.01)		81-2			eth, bz, peth	B7[3], 1594
5027	5-Cholesten-3-one $C_{27}H_{44}O$	384.65	lf (al), $[\alpha]^{20}{}_D$ -4.3 (chl)		127			eth, al	B7[3], 1608
5028	Cholesterol $C_{27}H_{45}OH$	386.66	rh or tcl lf (al + 1w) nd (eth), $[\alpha]^{20}{}_D$ - 31.5 (eth), $[\alpha]^{20}{}_D$ - 31.5 (eth, c=2) -39.5 (chl, c=2)	360d, $233^{0.5}$	148.5 (anh)	$1.067^{20/4}$		eth, bz, chl, aa	B6[2], 2607
5029	Cholesterol, acetate or Cholestyryl acetate $C_{29}H_{48}O_2$	428.70	wh nd (ace or al), $[\alpha]^{20}{}_D$ - 47.7 (chl, c=2)		115-6			eth, ace, bz, chl	B6[3], 2630
5030	Cholesterol, benzoate $C_{34}H_{50}O_2$	490.77	wh nd, $[\alpha]$ - 13.7 (chl, c=0.9)		152-3			eth, chl	B9[3], 460
5031	Cholesterol, hexadecanoate or Cholesteryl palmitate $C_{43}H_{76}O_2$	625.08	wh nd (eth or al), $[\alpha]^{20}$ -25.4 (chl, c=2)		80			bz, chl	B6[3], 2640
5032	Cholestrophane or Dimethylparabanic acid Oxalyldimethylurea $C_5H_6N_2O_3$	142.11	lf or pl (w or al)	275-7, $148\text{-}50^{13}$	155.5			w, eth	B24[2], 265
5033	Cholic acid $C_{24}H_{40}O_5$	408.58	rh (eth), tetr rh (w or dil al), cr (al + 1), $[\alpha]^{20}{}_D$ + 37 (al, c=0.6)		198 (anh)			al, eth, ace, chl, aa	B10[3], 2162
5034	Choline or Trimethyl(2-hydroxymethyl) ammonium hydroxide $(CH_3)_3N^+CH_2CH_2OH \cdot OH^-$	121.18	syr					w, al	B4[4], 1443
5035	Choline, *O*-acetyl, bromide or Acetylcholine bromide $(CH_3)_3N^+CH_2CH_2O_2CCH_3\text{-}Br^-$	226.11	hex pr (dil al)	d	143			w, al	B4[4], 1446
5036	Choline, *O*-acetyl, chloride or Acetyl choline chloride $(CH_3)_3N^+CH_2CH_2O_2CCH_3\text{-}Cl^-$	181.66	yesh nd		153			w, al	B4[4], 1446
5037	Choline, *O*-benzyl, chloride $(CH_3)_3N^+CH_2CHOO_2CC_6H_5Cl^-$	243.73	pr (ace-al)		200			al, ace	B9[3], 877
5038	Choline, carbonate, chloride $(CH_3)_3N^+CH_2CH_2O_2CNH_2\text{-}Cl^-$	182.65	hyg pr or pw		210-2			w	B4[4], 1455
5039	Chroman or Dihydrobenzopyran $C_9H_{10}O$	134.18		$214\text{-}5^{742}$, $98\text{-}9^{18}$		1.0610^{20}	1.5444^{20}		B17[4], 413
5040	Chroman, 2,2-dimethyl $C_{11}H_{14}O$	162.23		225^{769}, 98^{11}		$1.009^{15/15}$			B17[4], 445
5041	Chromanone $C_9H_8O_2$	148.16		160^{50}, 128^{13}	38-9	$1.1291^{100/4}$	1.5460^{100}	al, eth, ace, bz	B17[4], 4957
5042	3-Chromene C_9H_8O	132.16		91^{13}, 50^{1}			1.5879^{20}		B17[4], 496
5043	Chromone or α-Benzopyrone $C_9H_6O_2$	146.15	nd (peth or w)	sub	59			al, eth, bz, chl	B17[4], 5052
5044	*cis*-Chrysanthemucic acid-*(d)* or Cyclopropane carboxylic acid-2,2-dimethyl-3-(2-methyl propenyl) $C_{10}H_{16}O_2$	168.24	pr $[\alpha]^{22}{}_D$ + 83.3 (chl, c=1.6)	$95^{0.1}$	40-2			al, eth	B9[3], 210
5045	*trans*-Chrysanthemucic acid-*(d)* or Cyclopropane carboxylic acid-2,2-dimethyl-3-(2 methyl propenyl*(trans)*. $C_{10}H_{16}O_2$	168.24	pr $[\alpha]^{19}{}_D$ + 25.8 (chl, c=2.5)	245d, 135^{12}	18-21			al, eth, chl	B9[3], 211
5046	Chrysazin or 1,8-Dihydroxyanthraquinone $1,8\text{-}(HO)_2C_{14}H_6O_2$	240.22	red or redsh-ye nd or lf (al)	sub	193			al, eth, ace, chl	B8[3], 3792

No.	Name, Synonyms, and Formula	Mol. wt.	Color, crystalline form, specific rotation and λ_{max} (log ε)	b.p. °C	m.p. °C	Density	n_D	Solubility	Ref.
5047	Chrysene or 1,2-Benzophenanthrene $C_{18}H_{12}$	228.29	red bl flr rh pl (bz-aa)	448	255-6	1.274^{20}	...	...	B5[3], 2380
5048	Chrysene, 5,6-dimethyl $C_{20}H_{16}$	256.35	pl or nd (bz-al)	$200^{0.5}$ (sub 140 vac)	128-9	...	...	al, aa	B5[3], 2418
5049	Chrysene, 1-methyl $C_{19}H_{14}$	242.32	lf (hx, bz, to)	sub 130-140 (vac)	256-7	...	...	al	B5[3], 2395
5050	Chrysene, 2-methyl $C_{19}H_{14}$	242.32	lf (bz-al)	...	229-30	...	...	al, aa	B5[3], 2395
5051	Chrysene, 3-methyl $C_{19}H_{14}$	242.32	lf (bz-peth)	...	172-3	...	...	al	B5[1], 2395
5052	5,6-Chrysoquinone $C_{18}H_{10}O_2$	258.28	red nd (bz or to), lf or pl (aa)	sub	239.5	...	...	al, bz	B7[3], 4285
5053	6,12-Chrysoquinone $C_{18}H_{10}O_2$	258.25	red ye nd (aa)	...	288-90 d	...	...	al, aa	B7[3], 4285
5054	Cinchonamine $C_{19}H_{24}N_2O$	296.41	rh nd (al), orh pr (MeOH), $[\alpha]^{20/}{}_D$ + 123 (al, c=0.66	...	186 (194)	...	...	al, eth, bz, chl	B23[2], 358
5055	Cinchonicine or Cinchotoxine $C_{19}H_{22}N_2O$	294 40	nd or pr (eth), $[\alpha]^{15/}{}_D$ + 48 (al, c=1)	...	58-60	...	...	al, eth, ace, bz, chl	B24[2], 100
5056	Cinchonidine or Cinchovatine $C_{19}H_{22}N_2O$	294.40	orh pl or pr (al), $[\alpha]^{20/}{}_D$ - 109.2 (al, p=1)	sub	210.5	...	...	al	B23[2], 373
5057	Cinchonidine, hydrochloride $C_{19}H_{22}N_2O\text{-}HCl\text{-}H_2O$	348.87	wh pr (w), $[\alpha]^{20/}{}_D$ - 117.6 (w, c=1.2)	...	242d (anh)	...	...	w, al, chl	B23[2], 373
5058	Cinchonidine, sulfate $(C_{19}H_{22}N_2O)_2\text{-}H_2SO_4 \cdot 6H_2O$	794.97	mcl nd (al + 2w), $[\alpha]^{18.5/}{}_D$ - 97.9 (w, c=1.2)	...	205 (anh)	...	...	...	B23[2], 373
5059	β-Cinchonidine $C_{19}H_{22}N_2O$	294.40	pr or lf (al or dil al)	...	241	...	...	al, chl	B23[1], 131
5060	Cinchonine $C_{19}H_{22}N_2O$	294.40	nd or mcl cr (al)	sub	255 (265)	...	...	chl	B23[2], 369
5061	Cinchonine, dihydrochloride $C_{19}H_{22}N_2O \cdot 2HCl$	367.32	pl $[\alpha]^{24/}{}_D$ + 205.5 (w, c=3.6)	...	...	...	...	w, al	B23[1], 133
5062	Cinchonine, hydrochloride $C_{19}H_{22}N_2O\text{-}HCl\text{-}2H_2O$	366.89	mcl $[\alpha]^{25/}{}_D$ + 133.6 (chl, c=1.4)	...	ca 215 (anh)	1.234	...	w, al, chl	B23[2], 370
5063	Cinchonine, sulfate $(C_{19}H_{22}N_2O)_2\text{-}H_2SO_4$	686.87	rh pr (w), $[\alpha]^{15/}{}_D$ + 169 (w, c=1.4)	...	206-7	...	...	al, w	B23[2], 371
5064	Cinchotine or Hydrocinchonine $C_{19}H_{24}N_2O$	296.41	pr $[\alpha]^{21/}{}_D$ + 203.4	...	268-9	...	...	w	B23[2], 356
5065	1,4-Cineole or 1,4-epoxy-p-menthane $C_{10}H_{18}O$	154.24	...	173-4	1	0.8997^{20}	1.4562^{20}	al, eth, bz, lig	B17[4], 213
5066	1,8-Cineole or Eucalyptol $C_{10}H_{18}O$	154.24	...	176.4, 61^{14}	1.5	0.9267^{20}	1.4586^{20}	al, eth, chl	B17[4], 273
5067	Cineolic acid-*(d)* $C_{10}H_{16}O_5$	216.23	cr (w + 1), $[\alpha]^{20/}{}_D$ + 18.6 (w, p=8.21)	...	79 (+ w), 138-9 (anh)	...	...	al	B18, 322
5068	Cineolic acid-*(dl)* $C_{10}H_{16}O_5$	216.23	...	...	(i)197.5 (ii)208	...	...	al, eth	B18[4], 4446
5069	Cineolic acid-*(l)* $C_{10}H_{16}O_5$	234.23	rh (w + 1), $[\alpha]^{20/}{}_D$ - 19.1(w)	...	79 (+ w), 138.9 (anh)	...	...	al	B18, 322

No.	Name, Synonyms, and Formula	Mol. wt.	Color, crystalline form, specific rotation and λ_{max} (log ε)	b.p. °C	m.p. °C	Density	n_D	Solubility	Ref.
5070	Cinnamaldehyde-*(trans)* or β-Phenyl acrolein $C_6H_5CH=CHCHO$	132.16	yesh	253d, 127^{16}	-7.5	$1.0497^{20/4}$	1.6195^{20}	al, eth, chl	$B7^3$, 1364
5071	Cinnamaldehyde, β-bromo-*(cis)* $C_6H_5CBr=CHCHO$	211.06	...	$144\text{-}6^{12}$	...	$1.492^{20/4}$	1.6368^{20}	eth	$B7^3$, 1383
5072	Cinnamaldehyde, α-ethyl $C_6H_5CH=C(C_2H_5)CHO$	160.22	...	$157\text{-}8^5$	...	$1.0201^{22/4}$	1.578^{20}	...	$B7^3$, 1435
5073	Cinnamaldehyde, 4-hydroxy-3-methoxy or Coniferaldehyde $[4\text{-}(HO)\text{-}3\text{-}(CH_3O)C_6H_3]CH=CHCHO$	178.19	cr (bz)	$157^{2.5}$	84	...	...	al, eth, bz	$B8^3$, 2331
5074	Cinnamaldehyde, α-methyl $C_6H_5CH=C(CH_3)CHO$	146.19	ye	150^{100}	...	$1.0407^{17/4}$	1.6057^{17}	...	$B7^3$, 1412
5075	Cinnamaldehyde, 4-methyl $4\text{-}CH_3C_6H_4CH=CHCHO$	146.19	ye lf (dil al)	154^{25}	41.5	...	...	al	$B7^3$, 1414
5076	Cinnamaldehyde, 2-nitro $2\text{-}O_2NC_6H_4CH=CHCO$	177.16	nd (eth or al)	...	127.5	...	...	al, eth, chl	$B7^3$, 1387
5077	Cinnamaldehyde, 3-nitro $3\text{-}O_2NC_6H_4CH=CHCHO$	177.16	ye nd (w), pr (al), cr (aa)	...	116	...	...	bz, aa	$B7^2$, 282
5078	Cinnamaldehyde, 4-nitro $4\text{-}O_2NC_6H_4CH=CHCHO$	177.16	nd (w or al)	...	141-2	...	...	al, eth, ace, bz	$B7^3$, 1388
5079	Cinnamaldehyde, oxime-*(trans)* $C_6H_5CH=CHCH=NOH$	147.18	nd	...	138.5	...	...	bz	$B7^3$, 1376
5080	Cinnamamide-*(trans)* $C_6H_5CH=CHCONH_2$	147.18	nd (bz)	...	148	...	...	al, eth	$B9^3$, 2711
5081	Cinnamic acid-*(cis)*-(1st form) or Phenyl acrylic acid $C_6H_5CH=CHCO_2H$	148.16	mcl pr (w)	...	42	...	...	al, aa, lig	$B9^3$, 2670
5082	Cinnamic acid-*(cis)*-(2nd form) $C_6H_5CH=CHCO_2H$	148.16	mcl pr (lig)	265	58	...	...	al, eth, ace, chl	$B9^3$, 2670
5083	Cinnamic acid-*(cis)*-(3rd form) or Allocinnamic acid $C_6H_5CH=CHCO_2H$	148.16	mcl pr	...	68	...	...	al, eth, lig	$B9^3$, 2670
5084	Cinnamic acid -*(trans)* $C_6H_5CH=CHCO_2H$	148.16	mcl pr (dil al)	300 (cor)	135-6	$1.2475^{4/4}$	...	al, eth, ace, bz, chl	$B9^3$, 2671
5085	Cinnamic acid, α-acetamido $C_6H_5CH=C(NHOCH_3)CO_2H$	205.21	cr (+ 2w)	...	193-4	...	...	w	$B10^3$, 3001
5086	Cinnamic acid, allyl ester or Allylcinnamate $C_6H_5CH=CHCO_2CH_2CH=CH_2$	188.23	...	268d, 163^{17}	...	1.048^{23}	1.530^{20}	al, eth	$B9^2$, 387
5087	Cinnamic acid, 2-amino-*(trans)* $2\text{-}H_2NC_6H_4CH=CHCO_2H$	163.18	ye nd (w)	...	158-9d	...	...	al, eth	$B14^3$, 1304
5088	Cinnamic acid, 3-amino-*(trans)* $3\text{-}H_2NC_6H_4CH=CHCO_2H$	163.18	nd (al)	...	191-3	...	...	aa	$B14^3$, 1305
5089	Cinnamic acid, 4-amino $4\text{-}H_2NC_6H_4CH=CHCO_2H$	163.18	ye nd (w or al)	...	175-6d	...	...	al, eth	$B14^3$, 1305
5090	Cinnamic acid-*(trans)*, anhydride $(C_6H_5CH=CHCO)_2O$	278.31	nd (bz or al) pr, (al)	...	138	...	...	bz	$B9^3$, 2703
5091	Cinnamic acid, 2-benzamido-*(trans)* $2\text{-}(C_6H_5CONH)C_6H_4CH=CHCO_2H$	267.28	nd (al)	...	191-3	...	...	aa	$B14^3$, 1304
5092	Cinnamic acid, 3-benzamido-*(trans)* $3\text{-}(C_6H_5CONH)C_6H_4CH=CHCO_2H$	267.28	nd (AcOEt)	...	229	...	...	ace, aa	$B14^3$, 1305
5093	Cinnamic acid, 4-benzamido-*(trans)* $4\text{-}(C_6H_5CONH)C_6H_4CH=CHCO_2H$	267.28	lf (aa), nd (ace)	...	274d	...	...	ace, bz, aa	$B14^3$, 1312
5094	Cinnamic acid, benzyl ester-*(trans)* or Benzyl cinnamate $C_6H_5CH=CHCO_2CH_2C_6H_5$	238.29	pr	350d, 244^5	39	1.109^{15}	...	al, eth	$B9^3$, 2691
5095	Cinnamic acid, α-bromo-*(cis)* $C_6H_5CH=CBrCO_2H$	227.06	lf (w), pr (chl)	...	120-1	...	...	al, bz	$B9^3$, 2734
5096	Cinnamic acid, α-bromo-*(trans)* $C_6H_5CH=CBrCO_2H$	227.06	nd (w)	...	131-2	...	...	al, eth, bz	$B9^3$, 2734
5097	Cinnamic acid, β-bromo-*(cis)* $C_6H_5CBr=CHCO_2H$	227.06	nd (bz), pl (al)	...	159-60	...	...	eth, chl	$B9^3$, 2732
5098	Cinnamic acid, β-bromo-*(trans)* $C_6H_5CBr=CHCO_2H$	227.06	pa ye nd or pl (w), pr (chl)	...	135	...	...	al, bz	$B9^3$, 2733
5099	Cinnamic acid, 2 carboxy-*(trans)* $(2\text{-}HO_2CC_6H_4)CH=CHCO_2H$	192.17	pr or nd (w)	...	208-9	...	...	al	$B9^3$, 4384
5100	Cinnamic acid, 3-carboxy-*(trans)* $(3\text{-}HO_2CC_6H_4)CH=CHCO_2H$	192.17	nd (ace-lig)	...	275	...	...	aa	$B9^2$, 642
5101	Cinnamic acid, 4-carboxy-*(trans)* $4\text{-}HO_2CC_6H_4CH=CHCO_2H$	192.17	pw	sub> 350	358d	...	...	...	$B9^2$, 642

No.	Name, Synonyms, and Formula	Mol. wt.	Color, crystalline form, specific rotation and λ_{max} (log ε)	b.p. °C	m.p. °C	Density	n_D	Solubility	Ref.
5102	Cinnamic acid, α-chloro-*(cis)* $C_6H_5CH=CClCO_2H$	182.61			111			al	B9^3, 2729
5103	Cinnamic acid, α-chloro,methyl ester-*(trans)* $C_6H_5CH=CHClCO_2CH_3$	196.63		108-9$^{0.5}$	33			ace	B9^2, 396
5104	Cinnamic acid, β-chloro-*(cis)* $C_6H_5CCl=CHCO_2H$	182.61			133			al, eth	B9^3, 2728
5105	Cinnamic acid, β-chloro-*(trans)* $C_6H_5CCl=CHCO_2H$	182.61			143			al, eth	B9^3, 2728
5106	Cinnamic acid, β-chloro,methyl ester $C_6H_5CCl=CHCO_2CH_3$	196.63		113-4$^{0.5}$	29	1.2248$^{21/4}$	1.5781^{21}	ace	B9^2, 396
5107	Cinnamic acid, α,β-dibromo-*(cis)* $C_6H_5CBr=CBrCO_2H$	305.95	ye pr or pl (chl or lig)	124$^{0.5}$	100			eth, chl, aa	B9^3, 2736
5108	Cinnamic acid, 2,4-dihydroxy-*(trans)* or Umbellic acid $2,4\text{-}(HO)_2C_6H_3CH=CHCO_2H$	180.16	ye nd or pl		260d 240 (darkens)			al	B10^3, 1830
5109	Cinnamic acid, 2,5-dihydroxy-*(trans)* $2,5\text{-}(HO)_2C_6H_3CH=CHCO_2H$	180.16	ye cr (w), cr (dil al + 1w)		207d			al	B10^3, 1833
5110	Cinnamic acid, 3,4-dihydroxy-*(trans)* or Caffeic acid $3,4\text{-}(HO)_2C_6H_3CH=CHCO_2H$	180.17	ye pr or pl (w)		225d			al	B10^3, 1834
5111	Cinnamic acid, 3,5-dihydroxy-*(trans)* $3,5\text{-}(HO)_2C_6H_3CH=CHCO_2H$	180.16	nd (w + ½)		245-6			al, eth	B10^2, 297
5112	Cinnamic acid, 2,3-dimethoxy-*(trans)* $2,3\text{-}(CH_3O)_2C_6H_3CH=CHCO_2H$	208.21			180-1				B10^3, 1829
5113	Cinnamic acid, 2,4-dimethoxy-*(cis)* $2,4\text{-}(CH_3O)_2C_6H_3CH=CHCO_2H$	208.21	nd (al)		138			al, eth, bz	B10^3, 1831
5114	Cinnamic acid, 2,4-dimethoxy-*(trans)* $2,4\text{-}(CH_3O)_2C_6H_3CH=CHCO_2H$	208.21	nd (w or dil al)		187-9			al, eth, bz, chl	B10^3, 1831
5115	Cinnamic acid, 2,5-dimethoxy-*(trans)* $2,5\text{-}(CH_3O)_2C_6H_3CH=CHCO_2H$	208.21	pa ye or ye-gr nd (w)		148-9			al, eth	B10^3, 1833
5116	Cinnamic acid, 3,4-dimethoxy-*(trans)* $3\text{-}4\text{-}(CH_3O)_2C_6H_3CH=CHCO_2H$	208.21	nd (w, dil al), pa ye pw (dil aa)		183			al, eth	B10^3, 1835
5117	Cinnamic acid, 3,4-dimethoxy, methyl ester $3,4\text{-}(CH_3O)_2C_6H_3CH=CHCO_2CH_3$	222.24			68-70			chl	B10^3, 1838
5118	Cinnamic acid, 3,5-dimethoxy-*(trans)* $3,5\text{-}(CH_3O)_2C_6H_3CH=CHCO_2H$	208.22	nd (w)		175-6			al, bz	B10^2, 297
5119	Cinnamic acid, 3,5-dimethoxy-4-hydroxy-*(trans)* or 5-Methoxyferulic acid $4\text{-}HO\text{-}3,5\text{-}(CH_3O)_2C_6H_2CH=CHCO_2H$	224.21	pa ye nd (al)		192			al	B10, 508
5120	Cinnamic acid, α-ethyl-*(cis)* or α-Benzal butyric acid $C_6H_5CH=C(C_2H_5)CO_2H$	176.22	nd (w)		82			al, eth, bz	B9^3, 2783
5121	Cinnamic acid, α-ethyl-*(trans)* or α-Benzal butyric acid $C_6H_5CH=C(C_2H_5)CO_2H$	176.22	nd (w)		106 (114)			al, eth	B9^3, 2783
5122	Cinnamic acid, ethyl ester-*(trans)* or Ethyl cinnamate $C_6H_5CH=CHCO_2C_2H_5$	176.22		271.5, 144^{15}	12	1.0491$^{20/4}$	1.5598^{20}	al, eth, ace, bz	B9^3, 2682
5123	Cinnamic acid, α-fluoro $C_6H_5CH=CFCO_2H$	166.15		290	157.6			al, eth	B9^1, 237
5124	Cinnamic acid, 2-hydroxy-*(trans)* or *o*-Coumaric acid $2\text{-}HOC_6H_4CH=CHCO_2H$	164.16	nd (w)		217d			al	B10^3, 833
5125	Cinnamic acid, 3-hydroxy-*(trans)* or *m*-Coumaric acid $3\text{-}HOC_6H_4CH=CHCO_2H$	164.16	pr (w)		193			al, eth, bz	B10^3, 840
5126	Cinnamic acid-4-hydroxy-*(trans)* or *p*-Coumaric acid $4\text{-}HOC_6H_4CH=CHCO_2H$	164.16	nd (w + 1), cr (w)		215d			eth	B10^3, 844
5127	Cinnamic acid-4-hydroxy-3-methoxy or Ferulic acid $4\text{-}(HO)\text{-}3\text{-}CH_3OC_6H_3CH=CHCO_2H$	194.19	pr or nd (w)		171			al, chl	B10^3, 1834
5128	Cinnamic acid-2-methoxyphenyl ester-*(trans)* or *o*-Anisylcinnamate $C_6H_5CH=CHCO_2(C_6H_4OCH_3\text{-}2)$	254.29	wh nd (al)		130			ace, bz, chl	B9^3, 2698
5129	Cinnamic acid-4-methoxy $4\text{-}CH_3OC_6H_4CH=CHCO_2H$	178.19	wh nd (al)		172-5			aa	B10^3, 845
5130	Cinnamic acid-methyl ester-*(trans)* $C_6H_5CH=CHCO_2CH_3$	162.19	cr (peth or dil al)	261.9, 127^{10}	36.5	1.0911$^{20/4}$	1.5766^{22}	al, eth, ace, bz	B9^3, 2680
5131	Cinnamic acid-α-methyl-*(cis)* or α-Benzal propionic acid $C_6H_5CH=C(CH_3)CO_2H$	162.19	nd (bz)	288, 190^{21}	74			al, eth, bz, peth	B9^3, 2764

PHYSICAL CONSTANTS OF ORGANIC COMPOUNDS (Continued)

No.	Name, Synonyms, and Formula	Mol. wt.	Color, crystalline form, specific rotation and λ_{max} (log ε)	b.p. °C	m.p. °C	Density	n_D	Solubility	Ref.
5132	Cinnamic acid-α-methyl-*(trans)* or α-Benzal propionic acid $C_6H_5CH=C(CH_3)CO_2H$	162.19	pr (aa, eth or dil al)	...	81-2	...	...	al, eth, bz, peth	$B9^3$, 2764
5133	Cinnamic acid-α-methyl-2-nitro $2\text{-}O_2NC_6H_4CH=C(CH_3)CO_2H$	207.19	mcl pr (al)	...	164-5	...	...	al, eth	$B9^3$, 2766
5134	Cinnamic acid-α-methyl-3-nitro $3\text{-}O_2NC_6H_4CH=C(CH_3)CO_2H$	207.19	nd or pw	...	203.5	...	...	eth, bz, aa	$B9^3$, 2767
5135	Cinnamic acid-α-methyl-4-nitro $4\text{-}O_2NC_6H_4CH=C(CH_3)CO_2H$	207.19	ye rh (aa), tcl pym (aa, al-eth)	...	208	...	...	al, eth, bz	$B9^3$, 2767
5136	Cinnamic acid-2-methyl-4-nitro $[2\text{-}(CH_3)\text{-}4\text{-}(O_2N)C_6H_3]CH=CHCO_2H$	207.19	nd (al)	...	256	...	...	...	$B9^1$, 256
5137	Cinnamic acid-4-methyl-3-nitro-*(trans)* $[4\text{-}(CH_3)\text{-}3\text{-}(O_2N)C_6H_3]CH=CHCO_2H$	207.19	ye pl or nd (al)	...	173.5	...	...	al, eth	$B9^3$, 2770
5138	Cinnamic acid-2-nitro-*(cis)* $2\text{-}O_2NC_6H_4CH=CHCO_2H$	193.16	yesh (bz or chl)	...	146-7	...	...	al, bz, chl	$B9^1$, 246
5139	Cinnamic acid-2-nitro-*(trans)* $2\text{-}O_2NC_6H_4CH=CHCO_2H$	193.16	nd (al)	sub	242-3	...	...	...	$B9^3$, 2739
5140	Cinnamic acid-2-nitro, ethyl ester-*(trans)* $2\text{-}O_2NC_6H_4CH=CHCO_2C_2H_5$	221.21	ye rh bipym (al)	...	44	...	...	al, eth, bz	$B9^3$, 2739
5141	Cinnamic acid-2-nitro,methyl ester-*(trans)* $2\text{-}O_2NC_6H_4CH=CHCO_2CH_3$	207.19	wh nd (w)	$187\text{-}9^{15}$	73	...	...	al	$B9^3$, 2739
5142	Cinnamic acid-3-nitro-*(cis)* $3\text{-}O_2NC_6H_4CH=CHCO_2H$	193.16	nd	...	158	...	...	...	$B9^1$, 247
5143	Cinnamic acid-3-nitro-*(trans)* $3\text{-}NO_2C_6H_4CH=CHCO_2H$	193.16	nd (al)	...	204-5	...	...	al	$B9^3$, 2741
5144	Cinnamic acid-3-nitro,ethyl ester-*(trans)* $3\text{-}NO_2C_6H_4CH=CHCO_2C_2H_5$	221.21	nd (al), pr (aa)	...	78-9	...	...	...	$B9^3$, 2742
5145	Cinnamic acid-3-nitro,methyl ester-*(trans)* $3\text{-}O_2NC_6H_4CH=CHCO_2CH_3$	207.19	pa ye pr (MeOH)	d	123-4	...	...	eth, bz, chl	$B9^3$, 2742
5146	Cinnamic acid-4-nitro-*(trans)* $4\text{-}O_2NC_6H_4CH=CHCO_2H$	193.16	yesh-wh pr (al)	...	286	...	...	...	$B9^3$, 2744
5147	Cinnamic acid-4-nitro,ethyl ester-*(trans)* $4\text{-}O_2NC_6H_4CH=CHCO_2C_2H_5$	221.21	pl (aa)	...	141-2	...	...	...	$B9^3$, 2744
5148	Cinnamic acid-4-nitro,methyl ester-*(trans)* $4\text{-}O_2NC_6H_4CH=CHCO_2CH_3$	207.19	wh nd (al)	281-6	162	...	...	al	$B9^3$, 2744
5149	Cinnamic acid-2-octyl ester-*(d)* $C_6H_5CH=CHCO_2CH(CH_3)C_6H_{13}$	260.38	$[\alpha]^{17}{}_D$ + 40.2	218^{28}	...	$0.9645^{20/4}$	1.5145^{20}	bz, chl	$B9^3$, 8687
5150	Cinnamic acid-2-octyl ester-*(trans,dl)* $C_6H_5CH=CHCO_2CH(CH_3)C_6H_{13}$	260.38	...	240^{60}	...	$0.9715^{17/4}$	...	bz, chl	$B9^1$, 230
5151	Cinnamic acid, 2-octyl ester*(trans,l)* $C_6H_5CH=CHCO_2CH(CH_3)C_6H_{13}$	260.38	$[\alpha]^{17}{}_D$ - 39.78	211^{28}	...	$0.9692^{17/4}$	...	bz, chl	$B9^3$, 2687
5152	Cinnamic acid, α-phenyl or α-Phenyl cinnamic acid $C_6H_5CH=C(C_6H_5)CO_2H$	224.26	nd (lig, dil al)	sub	134-5	...	...	al, eth	$B9^3$, 3414
5153	Cinnamic acid, phenylester-*(trans)* $C_6H_5CH=CHCO_2H_5$	224.26	...	$205\text{-}7^{15}$	72.5	...	...	eth, bz	$B9^3$, 2689
5154	Cinnamic acid, propyl ester-*(trans)* $C_6H_5CH=CHCO_2C_3H_7$	190.24	...	285	...	$1.0435^{0/0}$	...	...	$B9^3$, 2684
5155	Cinnamic acid, *iso*-propyl ester-*(trans)* or Isopropyl cinnamate $C_6H_5CH=CHCO_2CH(CH_3)_2$	190.24	...	268-70, $153\text{-}5^{20}$	...	1.0320^{20}	1.5455^{20}	al, eth, ace	$B9^3$, 2685
5156	Cinnamic acid, 4-isopropyl or Cumiliden acetic acid $4\text{-}i\text{-}C_3H_7C_6H_4CH=CHCO_2H$	190.24	pr (bz)	...	165	...	...	al	$B9^3$, 2814
5157	Cinnamic acid, 3,4,5-trimethoxy $3,4,5\text{-}(CH_3O)_3C_6H_2CH=CHCO_2H$	238.24	nd (w)	...	126-7	...	...	chl	$B10^3$, 2200
5158	Cinnamonitrile, *(cis)* or Allocinnamonitrile $C_6H_5CH=CHCN$	129.16	...	249, 139^{30}	-4.4	...	1.5843^{20}	al, bz	$B9^3$, 2720
5159	Cinnamonitrile, *(trans)* $C_6H_5CH=CHCN$	129.16	...	263.8, $134\text{-}6^{28}$	22	$1.0304^{20/4}$	1.6013^{20}	al, ace	$B9^3$, 2721
5160	Cinnamylalcohol, (cis) or 3-phenyl-2-propen-2-ol $C_6H_5CH=CHCH_2OH$	134.18	wh nd (eth-peth)	257.5, $127\text{-}8^{10}$	34	$1.0440^{20/4}$	1.5819^{20}	al, eth	$B6^3$, 2401
5161	Cinnamyl alcohol, acetate-*(trans)* or Cinnamyl acetate $C_6H_5CH=CHCH_2O_2CCH_3$	176.22	...	$145\text{-}6^{15}$, 114^1	...	1.0567^{20}	1.5425^{20}	al, eth, ace, bz	$B6^3$, 2406
5162	Cinnamyl chloride, *(trans)* $C_6H_5CH=CHCOCl$	166.61	...	257.5, 131^{11}	37-8	$1.1617^{45/4}$	$1.614^{42.5}$	lig	$B9^3$, 2710

No.	Name, Synonyms, and Formula	Mol. wt.	Color, crystalline form, specific rotation and λ_{max} (log ε)	b.p. °C	m.p. °C	Density	n_D	Solubility	Ref.
5163	Citraconic acid or Methyl maleic acid $HO_2CC(CH_3)=CHCO_2H$	130.10	tcl pr or pl (eth-bz) nd (eth-lig)	...	93-4	1.617	...	w	B2[4], 2230
5164	Citraconic anhydride or Methylmaleic anhydride $C_5H_4O_3$	112.08	...	213-4, 99-100^{15}	7-8	1.2469$^{16/4}$	1.4716^{21}	al, eth, ace	B17[4], 5912
5165	Citraconic acid, diethyl ester or Diethyl citraconate $C_2H_5O_2CC(CH_3)=CHCO_2C_2H_5$	186.21	...	228^{766}, 120^{20}	...	1.0491$^{20/4}$	1.4467^{20}	al, eth, aa	B2[4], 2232
5166	Citraconic acid, dimethyl ester or Dimethyl citraconate $CH_3O_2CC(CH_3)=CHCO_2CH_3$	158.15	...	210.5^{758}, 92.8^{10}	...	1.1153$^{20/4}$	1.4473^{20}	al, eth, ace, aa	B2[4], 2232
5167	Citral a or Geranial $C_{10}H_{16}O$	152.24	...	229, 118-9^{20}	...	0.8888^{20}	1.4898^{20}	al, eth	B1[4], 3569
5168	Citral b or Neral $C_{10}H_{16}O$	152.24	...	120^{20}	...	0.8869^{20}	1.4869^{20}	al, eth	B1[4], 3569
5169	β-Citraurin $C_{30}H_{40}O_2$	432.65	pl (bz-peth), cr (al)	...	147	...	...	al, eth, ace, bz	B8[3], 1599
5170	Citric acid or 2-Hydroxy-1,2,3-propane tricarboxylic acid $HOC(CH_2CO_2H)_2CO_2H$	192.13	rh (w + 1)	d	153 anh	1.665$^{20/4}$	...	w, al, eth	B3[4], 1272
5171	Citric acid, anhydro-methylene or 1,3-Dioxolan-4-one-5,5-diacetic acid $C_9H_{12}O_7$	220.18	(w)	...	298	...	...	bz, chl	B19[2], 324
5172	Citric acid, tribenzyl ester or Benzyl citrate $HOC(CH_2CO_2CH_2C_6H_5)_2CO_2CH_2C_6H_5$	462.50	nd (al)	...	51	...	...	al	B6[3], 1537
5173	Citric acid, triethyl ester or Ethyl citrate $HOC(CH_2CO_2C_2H_5)_2CO_2C_2H_5$	276.29	...	294, 185^{17}	...	1.1369$^{20/4}$	1.4455^{20}	al, eth	B3[4], 1276
5174	Citric acid, trimethyl ester or Methyl citrate $HOC(CH_2CO_2CH_3)_2CO_2CH_3$	234.21	tcl	287d, 176^{16}	78-9	...	...	al, eth	B3[4], 1276
5175	Citric acid, triphenyl ester or Phenyl citrate $HOC(CH_2CO_2C_6H_5)_2CO_2C_6H_5$	420.42	nd (al)	...	124.5	...	...	eth	B6, 170
5176	Citric acid, tripropyl ester or Propyl citrate $HOC(CH_2CO_2C_3H_7)_2CO_2C_3H_7$	318.37	...	198^{18}	...	...	...	al, eth	B3, 568
5177	Citric acid, triethyl ester, acetate $CH_3CO_2C(CH_2CO_2C_2H_5)CO_2C_2H_5$	318.32	...	131-2	...	1.135	1.4380	...	...
5178	Citric triamide $HOC(CH_2CONH_2)_2CONH_2$	189.17	(w)	...	210-5d	...	...	w	B3[1], 197
5179	Citrinin $C_{13}H_{14}O_5$	250.25	ye nd (MeOH), $[\alpha]^{11/}{}_D$ -37	...	178-9d	...	...	ace, bz, chl	B18[4], 6329
5180	Citronellal-*(d)* or *d*-Rhodinal $(CH_3)_2C=CH(CH_2)_2CH(CH_3)CH_2CHO$	154.25	$[\alpha]^{18/}{}_D$ + 13.09	207.8, 92^{14}	...	0.8573$^{20/4}$	1.4456^{20}	al, eth	B1[4], 3515
5181	Citronellal-*(dl)* or *dl*-Rhodinal $C_{10}H_{18}O$	154.25	...	207-8, 79-81^{10}	...	0.8535$^{17/4}$	1.4473^{20}	...	B1[4], 3515
5182	Citronellal-*(l)* or *l*-Rhodinal $C_{10}H_{18}O$	154.25	$[\alpha]^{20/}{}_D$ -2.5	205-6, 87^{10}	...	0.8567$^{17/4}$	1.4479^{20}	**al, eth**	B1[4], 3515
5183	Citronellol-*(d)* or *d*-Rhodinol $(CH_3)_2C=CH(CH_2)_2CH(CH_3)CH_2CH_2OH$	156.27	$[\alpha]^{17/}{}_D$ + 6.8	244.4, 118^{17}	...	0.8590^{20}	1.4565^{20}	al, eth	B1[4], 2188
5184	Citronellol-*(dl)* or Dihydrogeranoil $C_{10}H_{20}O$	156.27	...	99^{10}	...	0.8560$^{20/4}$	1.4543^{20}	**al, eth**	B1[4], 2188
5185	Citronellol-*(l)* or *l*-Rhodinol $C_{10}H_{20}O$	156.27	$[\alpha]^{18/}{}_D$ -5.3	108-9^{10}	...	0.859$^{18/4}$	1.4576^{18}	**al, eth**	B1[4], 2188
5186	Citrulline-*(L)* or α-Amino-δ-ureido valeric acid $H_2NCONH(CH_2)_3CH(NH_2)CO_2H$	175.19	pr (aq MeOH), $[\alpha]^{20/}{}_D$ + 3.7 (w, c=2)	...	234-7	...	...	w	B4[3], 1347
5187	Clovene $C_{15}H_{24}$	204.36	$[\alpha]_D$ + 2.84	259-60	...	0.9241^{18}	1.4999^{18}	...	B5[2], 356
5188	Cocaine-*(d)* or Benzoyl methyl ecgonine $C_{17}H_{21}NO_4$	303.36	mcl pr (eth), $[\alpha]^{20/}{}_D$ + 15.8 (chl, p=10)	...	98	...	...	al, eth, ace, bz	B22[2], 150
5189	Cocaine-*dl* $C_{17}H_{21}NO_4$	303.36	rh bipym pr (peth)	...	79-80	...	...	al, eth, ace, bz, lig	B22[4], 2103

No.	Name, Synonyms, and Formula	Mol. wt.	Color, crystalline form, specific rotation and λ_{max} (log ε)	b.p. °C	m.p. °C	Density	n_D	Solubility	Ref.
5190	Cocaine-*(l)* $C_{17}H_{21}NO_4$	303.36	mcl pr (al), $[\alpha]^{20}/_D$ -16.3 (chl, c=4)	187-8$^{0.1}$, (sub, vac)	98		1.5022^{98}	al, eth, ace, bz, chl	B22^4, 2101
5191	Cocaine, hydrochloride-*(dl)* $C_{17}H_{21}NO_4$-HCl	339.82	pl (al)		187 (cor)			w, al, ace	B22^2, 156
5192	Cocaine, hydrochloride-*(l)* $C_{17}H_{21}NO_4$-HCl	339.82	mcl pr (al), cr (w + 2), $[\alpha]^{20}/_D$ -71.95		197			w, al, ace	B22^4, 2102
5193	Coclaurine-*(l)* $C_{17}H_{19}NO_3$	285.34	pl (al), $[\alpha]^{20}/_D$ -17.01		220-1				B21^4, 2605
5194	Codamine $C_{20}H_{25}NO_4$	343.42	pr (bz or eth)		127			al, eth, chl	B21^4, 2704
5195	Codeine or Morphine-3-methylether $C_{18}H_{21}NO_3$	299.37	rh oct (+ 1w, w or dil al), cr (eth)	250^{22}, sub 140$^{1.5}$	157-8	1.32		al, eth, bz, chl	B27^2, 137
5196	Codeine, hydrate $C_{18}H_{21}NO_3$-H_2O	317.38	rh oct (w, aq al), $[\alpha]^{25}/_D$ -136 (al, c=2.8)			1.31		al, eth, ace, bz, chl	B27^2, 137
5197	Codeine, hydrochloride $C_{18}H_{21}NO_3$-HCl-2H_2O	371.86	nd, $[\alpha]^{15}/_D$ -108.2		287d			w, al	B27^2, 143
5198	Codeine, phosphate $C_{18}H_{21}NO_3 \cdot H_3PO_4 \cdot 1\frac{1}{2}H_2O$	424.39	lf or pr (dil al)		220-35d			eth, chl	B27^2, 144
5199	Codeine, sulfate $(C_{18}H_{21}NO_3)_2$-$H_2SO_4 5H_2O$	786.89	pr, $[\alpha]^{15}/_D$ -100.9 (w, p=3)		278 (anh)			w	B27^2, 144
5200	β-Codiene or Neopine $C_{18}H_{21}NO_3$	299.37	nd (peth), $[\alpha]^{23}/_D$ -28 (chl, c=7.5)		127.5			al, eth, bz, chl	B27^2, 176
5201	Codeine, dihydro-*(d)* or Dihydroneopine $C_{18}H_{23}NO_3$	301.39	cr (+ 1w, dil MeOH)	248^{15}	112-3				B27^2, 103
5202	Colchiceine or *N*-Acetyltrimethyl colchicinic acid $C_{21}H_{23}NO_6$	385.42	pa ye nd (diox)		178-9	1.24		al, chl	B14^3, 692
5203	Colchicine $C_{22}H_{25}NO_6$	399.44	ye pl (w + 1½), pa ye nd (AcOEt), ye cr (bz), $[\alpha]^{17}/_D$ -121 (chl, c=0.9), -429 (w, c=1.72)		155-7			w, al	B14^3, 693
5204	Colchicinic acid, trimethyl $C_{19}H_{21}NO_5$	342.38	pa ye nd (al), $[\alpha]^{25}/_D$-184.5 (chl, c=1)		155-7			w, al	Am75, 5292
5205	Conessine or Neriine $C_{24}H_{40}N_2$	356.60	lf or pl (ace), $[\alpha]^{20}/_D$ + 25.3 (al, c=0.7)	165-7$^{0.1}$	125-6			aa	B22^4, 4382
5206	Congo red $C_{32}H_{22}N_6Na_2O_6S_2$	696.66	pw					al	B16^3, 474
5207	Conhydrine-*(d)* or 2(1-Hydroxypropyl)piperidine $C_8H_{17}NO$	143.23	lf (eth), $[\alpha]_D$ + 10 (w)	226	121			w, al, eth, chl	B21^4, 122
5208	Conhydrine-*(dl)*(lower m.p.) $C_8H_{17}NO$	143.23	nd (peth)	sub	69-70			w, al, eth, bz	B21^4, 122
5209	Conhydrine-*(dl)*(higher m.p.) or 2-(1-Hydroxypropyl) piperidine $C_8H_{17}NO$	143.23	nd (eth)	sub	98-9			w, al, eth, bz	B21^4, 123

No.	Name, Synonyms, and Formula	Mol. wt.	Color, crystalline form, specific rotation and λ_{max} (log ε)	b.p. °C	m.p. °C	Density	n_D	Solubility	Ref.
5210	α-Coniceine-*(d)* or 2-Methylconidine $C_8H_{15}N$	125.21	$[\alpha]^{15}/_D$ + 18.4 (al, c=2)	158	-16	$0.891^{15.5/4}$	...	al	B20, 152
5211	Corticosterone, 11-dehydro $C_{21}H_{28}O_4$	344.45	pr (ace-w, al or ace-eth), $[\alpha]^{25}/_D$ + 258 (al)	...	183-4	...	...	al, ace, bz	B8[3], 3624
5212	α-Coniceine-*(dl)* or 2-Methyl conidine $C_8H_{15}N$	125.21	...	156-9	...	$0.890^{15.5/4}$	...	al	B20, 153
5213	β-Coniceine-*(dl)* or 2-Propenyl piperidine $C_8H_{15}N$	125.21	nd $[\alpha]^{45}/_D$ + 49.9	168-9	38-9	...	...	al, eth	B20, 146
5214	β-Coniceine-*(dl)* or 2-Propenylpiperidine $C_8H_{15}N$	125.21	nd	$168\text{-}70^{753}$	8	$0.8716^{15/4}$	...	al, eth	B20, 146
5215	β-Coniceine-*(l)* or 2-Propenylpiperidine $C_8H_{15}N$	125.21	nd $[\alpha]^{45}/_D$ - 50.5	168-9	41	$0.8520^{50/4}$	...	al, eth	B20, 146
5216	γ-Coniceine $C_8H_{15}N$	125.21	...	173-4, $64\text{-}5^{14}$	...	$0.8720^{20/4}$	1.4607^{18}	al	B20[4], 1970
5217	ε-Coniceine-*(d)* or *d*-2-Methyl coniceine $C_8H_{15}N$	125.21	$[\alpha]^{15}/_D$ + 67.4	152-4	...	$0.8856^{15/4}$	...	...	B20, 151
5218	ε-Coniceine-*(dl)* $C_8H_{15}N$	125.21	...	150-1	...	$0.8836^{15/4}$	...	...	B20, 152
5219	ε-Coniceine-*l* $C_8H_{15}N$	125.21	$[\alpha]^{15}/_D$ - 87.34	151-3.5	...	$0.8642^{15/4}$	...	...	B20, 151
5220	α-Conidendrin or Tsugalactone $C_{20}H_{20}O_6$	356.38	cr (al), $[\alpha]^{20}/_D$ - 54.5 (ace, c=2.1)	...	255-6	...	...	al, eth, bz	B18[4], 3346
5221	α-Conidine, 3-methyl-*(l)* $C_8H_{15}N$	125.21	$[\alpha]^{8}/_D$ + 16	158	...	$0.8856^{15/4}$	...	al, eth	B20, 153
5222	α-Conidine, 3-methyl-*(dl)* $C_8H_{15}N$	125.21	...	158	...	$0.8946^{15/4}$	...	al, eth	B20, 153
5223	Conidine, 3-methyl-*(l)* $C_8H_{15}N$	125.21	$[\alpha]^{17}/_D$ -17.1	158	...	$0.8856^{15/4}$	...	al, eth	B20, 153
5224	Coniferin or 4-(3-hydroxypropenyl)-3-Methoxyphenyl-D-glucoside $C_{16}H_{22}O_8$	342.35	nd (w + 2), $[\alpha]^{20}/_D$ -68 (w, c=0.5)	...	186 (anh)	...	...	...	B17[4], 2999
5225	Coniferyl alcohol $C_{10}H_{12}O_3$	180.20	pr (eth-lig)	$163\text{-}5^{3}$	74	...	...	al, eth	B6[3], 6442
5226	Coniine-*(d)* or 2-Propyl piperidine $C_8H_{17}N$	127.23	$[\alpha]^{20}/_D$ + 15.6	166-7, 64^{18}	-2	0.8440^{20}	1.4512^{22}	al, eth, bz	B20[4], 1611
5227	Coniine-*(dl)* or 2-Propyl piperidine $C_8H_{17}N$	127.23	...	$166\text{-}7^{745}$, $59\text{-}63^{17}$	...	0.8447^{20}	1.4513^{23}	al, eth, bz	B20[4], 1611
5228	Coniine-*(l)* $C_8H_{17}N$	127.23	$[\alpha]^{15}/_D$ -15.6	166, 64^{18}	...	$0.845^{15/4}$	1.4512^{22}	al, eth, bz	B20[2], 62
5229	Coniine, hydrobromide-*(d)* $C_8H_{17}N\text{-}HBr$	208.14	pr	...	211	...	...	w, al, eth, chl	B20, 112
5230	Coniine, hydrochloride-*(d)* $C_8H_{17}N\text{-}HCl$	163.69	orth (w), $[\alpha]^{20}/_D$ + 10.1 (lig, NH_3)	...	221	...	...	w, al	B20[4], 1611
5231	Coniine, hydrochloride-*(dl)* $C_8H_{17}N\text{-}HCl$	163.69	nd (al-eth)	...	216-7	...	...	w, al	B20[1], 31
5232	Coniine, hydrochloride-*(l)* $C_8H_{17}N\text{-}HCl$	163.69	nd	...	220-1	...	...	w, al	B20, 118
5233	Coniine, *N*-methyl-*(d)* $C_9H_{19}N$	141.26	$[\alpha]^{24}/_D$ + 82.4	$173\text{-}4^{757}$	...	$0.8326^{23/4}$	1.4538^{13}	al, ace	B20[4], 1612
5234	Coniine, picrate-*(d)* $C_8H_{17}N\text{-}C_6H_3N_3O_7$	356.34	ye pr (w)	...	75	...	...	al, eth	B20, 112
5235	Coniine, picrate-*(l)* $C_8H_{17}N\text{-}C_6H_3N_3O_7$	356.34	ye pr (w)	...	74	...	...	al, eth	B20[2], 62
5236	Conquinamine $C_{19}H_{24}N_2O_2$	312.41	ye tetr, $[\alpha]^{15}/_D$ + 200 (al, c=0.5)	...	123	...	...	al, eth, chl	B27[2], 667
5237	Copaene $C_{15}H_{24}$	204.36	$[\alpha]_D$ -25.8	246-51, $119\text{-}20^{10}$	...	$0.8996^{20/4}$	1.4894^{20}	eth, ace, aa, lig	B5[4], 1189

No.	Name, Synonyms, and Formula	Mol. wt.	Color, crystalline form, specific rotation and λ_{max} (log ε)	b.p. °C	m.p. °C	Density	n_D	Solubility	Ref.
5238	Coproergostane or Pseudoergostane $C_{28}H_{50}$	386.71	nd (ace), $[\alpha]^{19/}{}_D$ + 25.3 (chl, c=2)		64			eth, chl	B5³, 1143
5239	Coprostane or Pseudocholestane $C_{27}H_{48}$	372.68	orth nd (al, ace or eth-al), $[\alpha]^{20/}{}_D$ + 25.1 (chl, c=2)		72	$0.9119^{87.7/4}$	$1.4884^{88/4}$	eth, chl	B5⁴, 1226
5240	3β-Coprostanol or Coprosterol $C_{27}H_{48}O$	388.68	nd (MeOH), $[\alpha]^{18/}{}_D$ + 28 (chl, c=1.8)		102 (105)			al, eth, bz, chl	B6³, 2128
5241	Coprostenol or Allocholesterol $C_{27}H_{46}O$	386.66	nd (eth-MeOH), $[\alpha]_D$ + 43.7 (bz, c=1)					al, eth, ace, bz, chl	B6³, 2604
5242	Coprostenone or Δ^4-Cholesten-3-one $C_{27}H_{44}O$	384.65	pl (MeOH), $[\alpha]^{/}{}_D$ + 88.6 (chl)		81-2			eth, bz, lig	B7³, 1592
5243	Coronene or Hexabenzobenzene $C_{24}H_{12}$	300.36	ye nd (bz)	525	438-40 (cor)	1.371			B5³, 2651
5244	Corticosterone or 11,21-Dihydroprogesterone $C_{21}H_{30}O_4$	346.47	nd (al), pl (ace), $[\alpha]^{15/}{}_D$ + 223 (al, c=1.1)	sub $190^{0.01}$	180-2			al, eth, ace	B8³, 3574
5245	Corticosterone, 17-hydroxy or Cortisol Hydrocortisone $C_{21}H_{30}O_5$	362.47	pr(al or i-prOH), $[\alpha]^{22/}{}_D$ + 167 (al)		220			aa, al	B8³, 4036
5246	Cortisone-*(d)* $C_{21}H_{28}O_5$	360.45	$[\alpha]^{25/}{}_D$ + 209 E + OH, c=1.2)		220-4			al, ace	B8³, 4057
5247	Cortisone, 21-acetate $C_{23}H_{30}O_6$	402.49	nd (ace), rods (chl), $[\alpha]^{24/}{}_D$ + 164 (ace, c=0.5)		239-40			ace, chl	B8³, 4058
5248	Corybulbine-*(d)* or Corydalis-6 $C_{21}H_{25}NO_4$	355.43	nd (al), $[\alpha]^{20/}{}_D$ 303 (chl, c=1.4)		237-8			ace, chl	B21⁴, 2779
5249	Corybulbine-*(dl)* $C_{21}H_{25}NO_4$	355.43	cr (chl-al)		220-2			al, ace, chl, bz	B21, 217
5250	Corycavamine $C_{21}H_{21}NO_5$	367.40	pr (eth or al), $[\alpha]^{20/}{}_D$ + 166.6 (chl, c=2.2)		149			al, chl	B27², 621
5251	Corycavine $C_{21}H_{21}NO_5$	367.40	orh pl (al)		221-2			chl	B27², 621
5252	Corydaldine $C_{11}H_{13}NO_3$	207.23	mcl pr (w or al)		175			w, al, eth, bz, chl	B21⁴, 6443
5253	Corydaline-*(d)* or Corydolis-A $C_{22}H_{27}NO_4$	369.46	pr (al), $[\alpha]^{20/}{}_D$ + 311 (al, c=0.8)		136			al, eth	B21⁴, 2779
5254	Corydaline-*(dl)* $C_{22}H_{27}NO_4$	369.46	cr (al)		135-6			eth, al, w	B21⁴, 2780
5255	Corydaline-*(meso,d)* $C_{22}H_{27}NO_4$	369.46	pr (eth), $[\alpha]_D$ + 180 (chl, c=3)		155-6			eth, al	B21⁴, 2779
5256	Corydaline-*(meso,dl)* $C_{22}H_{27}NO_4$	369.46	cr (al)		163-4			eth, al, w	B21⁴, 2779

No.	Name, Synonyms, and Formula	Mol. wt.	Color, crystalline form, specific rotation and λ_{max} (log ε)	b.p. °C	m.p. °C	Density	n_D	Solubility	Ref.
5257	Corydaline-*(meso,l)* $C_{22}H_{27}NO_4$	369.46	pr (eth), $[\alpha]^{20}/_D$ -181 (chl, c=3)		155-6			eth, al, w	$B21^1$, 257
5258	Corynantheine $C_{22}H_{26}N_2O_3$	366.46	$[\alpha]^{18}/_D$ + 28.8 (MeOH, c=28.8)		165-6			al	$B25^2$, 212
5259	Cotarnine $C_{12}H_{15}NO_4$	237.26	nd (bz), cr (eth)		132-3d			al, eth, bz, chl	$B27^2$, 543
5260	Cotarnine, chloride or Stypticin $C_{12}H_{14}ClNO_3 \cdot 2H_2O$	291.73	ye pw or nd (al-AcOEt)		197			w, al	$B27^1$, 456
5261	Cotarnine, *O*-phthalate or Styptol $(C_{12}H_{14}NO_3)_2C_6H_4(CO_2)_2$	604.61	og cr or pw		103-5			w	B27, 476
5262	Coumalic acid or α-Pyrone-5-carboxylic acid $C_6H_4O_4$	140.10	pr (MeOH)	218^{120}	205-10d			al, aa	$B18^4$, 5382
5263	Coumaran or 2,3-Dihydrobenzofuran C_8H_8O	120.15		188-9, 76^{14}	-21.5	$1.0576^{24/4}$	1.5426^{20}	al, eth, chl	$B17^4$, 404
5264	3-Coumaranone $C_8H_6O_2$	134.13	red nd (al)	$152\text{-}4^{10}$	102-3			bz	$B17^4$, 1456
5265	Coumarin or 1,2-Benzopyrone $C_9H_6O_2$	146.15	rh pym (eth)	301.7	71	$0.935^{20/4}$		al, eth, chl	$B17^4$, 5055
5266	Coumarin, 6-amino $C_9H_7NO_2$	161.16			168-70			w, al	$B18^4$, 7920
5267	Coumarin, 4-chloro $C_9H_5ClO_2$	180.59	nd (al)		165			al, eth, bz	$B17^4$, 5058
5268	Coumarin, 7-diethylamino-4-methyl $C_{14}H_{17}NO_2$	231.29	cr (al, bz-lig)		89 (al), 135 (bz-lig)			al, eth, ace	B18, 612
5269	Coumarin, 3,4-dihydro or Hydrocoumarin $C_9H_8O_2$	148.16	lf	272, 145^{18}	25	1.169^{18}	1.5563^{20}	chl	$B17^4$, 4956
5270	Coumarin, 5,7-dihydroxy-4-methyl $C_{10}H_8O_4$	192.17	nd (al), lf (aa)		282-4			al	$B18^4$, 1367
5271	Coumarin, 6,7-dihydroxy or Aesculetin $C_9H_6O_4$	178.14	nd (w + 1), pr (aa), lf (sub)	sub	276			al, ace, chl	$B18^4$, 1322
5272	Coumarin, 7,8-dihydroxy or Daphnetin $C_9H_6O_4$	178.14	yesh (dil al)	sub	261-3			al	$B18^4$, 1330
5273	Coumarin, 6,7-dihydroxy-4-methyl or 4-Methyl aesculetin $C_{10}H_8O_4$	192.17	ye nd (dil al)		274-6			al	$B18^4$, 1371
5274	Coumarin, 7,8-dihydroxy-6-methoxy or Fraxetin $C_{10}H_8O_5$	208.17	pl (dil al)		230-2			al	$B18^4$, 2371
5275	Coiumarin, 5,7-dimethoxy or Citropten $C_{11}H_{10}O_4$	206.20	pr or nd (al)	200d	148-50			al, ace, chl, aa	$B18^4$, 1322
5276	Coumarin, 7-ethoxy-4-methyl or Maraniol $C_{12}H_{12}O_3$	204.23	wh		114			al	$B18^4$, 334
5277	Coumarin, 3-hydroxy $C_9H_6O_3$	162.14			154			w, al, eth, ace	$B17^4$, 6152
5278	Coumarin, 4-hydroxy or Benzotetronic acid $C_9H_6O_3$	162.14	nd (w)		213-4 (232)			al, eth	$B17^4$, 6153
5279	Coumarin, 4-hydroxy-3-(1-phenyl-3-oxobutyl) or Coumadin, Warfarin $C_{19}H_{16}O_4$	308.33	cr (al)		161			al, bz, diox	$B17^4$, 6794
5280	Coumarin, 5-hydroxy $C_9H_6O_3$	162.14			224-7			al, eth, bz	$B18^3$, 291
5281	Coumarin, 6-hydroxy $C_9H_6O_3$	162.14	nd (dil HCl)		250	1.25		al	$B18^4$, 293
5282	Coumarin, 7-hydroxy or Umbelliferone $C_9H_6O_3$	162.14	nd (w)	sub	230-1			**al, chl, aa**	$B18^4$, 294
5283	Coumarin, 7-hydroxy-6-methoxy or Chrysatropic acid $C_{10}H_8O_4$	192.17	nd or pr (al)		204			chl, aa	$B18^4$, 1323
5284	Coumarin, 7-hydroxy-4-methyl or 4-Methyl umbelliferone $C_{10}H_8O_3$	176.17	nd (al)		185-7			al, aa	$B18^4$, 332
5285	Coumarin, 8-hydroxy $C_9H_6O_3$	162.14	nd (dil al)		160			al, aa	$B18^4$, 304

No.	Name, Synonyms, and Formula	Mol. wt.	Color, crystalline form, specific rotation and λ_{max} (log ε)	b.p. °C	m.p. °C	Density	n_D	Solubility	Ref.
5286	Coumarin, 7-methoxy or Herniarin $C_{10}H_8O_3$	176.17	lf (w or MeOH)		117-8			al, eth	B18^4, 295
5287	Coumarin, 8-methoxy $C_{10}H_8O_3$	176.17			89			al, eth, bz	B18^4, 304
5288	Coumarin, 3-methyl $C_{10}H_8O_2$	160.17	rh bipym (al)	292.5	91			al	B17^4, 5073
5289	Coumarin, 4-methyl $C_{10}H_8O_2$	160.17	nd (w), pr (bz)		83-4			al, bz	B17^4, 5074
5290	Coumarin, 5-methyl $C_{10}H_8O_2$	160.17		$173\text{-}4^{12}$	65.8			al, eth, bz	B17^4, 5075
5291	Coumarin, 6-methyl $C_{10}H_8O_2$	160.17		303^{725}, 174^{14}	75-6			al, eth, bz	B17^4, 5076
5292	Coumarin, 7-methyl $C_{10}H_8O_2$	160.17		171.5^{11}	128			al, eth	B17^4, 5077
5293	Coumarin, 8-methyl $C_{10}H_8O_2$	160.17		178^{20}	109-10			al, eth, bz	B17^4, 5079
5294	Coumarin, 6,7,8-trimethoxy or Fraxetin dimethyl ether $C_{12}H_{12}O_5$	236.22	rh bipym pl (dil al)	$90\text{-}100^{0.2}$	103-4			al, eth	B18^4, 2372
5295	3-Coumarin carboxylic acid $C_{10}H_6O_4$	190.16	nd (w or bz)		190d			al	B18^4, 5569
5296	2-Coumarone carboxylic acid or Coumarilic acid $C_9H_6O_3$	162.14	nd (w)	310-5d	192-3			al	B18^4, 4247
5297	2-Coumarone carboxylic acid, ethyl ester $C_{11}H_{10}O_3$	190.20		274^{720}, 161^{15}	30-1	$1.1656^{28.5/4}$	$1.564^{27.6}$		B18^1, 442
5298	2-Coumarone carboxylic acid, 3-methyl $C_{10}H_8O_3$	176.18	nd (dil al)		188-9			al	B17^4, 5074
5299	*P*-Coumaryl alcohol or -3(4 Hydroxyphenyl)-2-propene-1-ol 4-$HOC_6H_4CH=CH$-CH_2OH	150.18	pr (dil al)		124			al, eth, ace, bz	C45, 3359
5300	Coumestrol $C_{15}H_8O_5$	268.23	gy micr rods		385d				B19^4, 2870
5301	Coumestrol, diacetate $C_{19}H_{12}O_7$	352.30	pl		234				B19^4, 2872
5302	Creatine or (α-Methylguanido) acetic acid $H_2NC(=NH)N(CH_3)CH_2CO_2H$	131.13	mcl pr (w + 1)		303	1.33		w	B4^3, 1170
5303	Creatinine or l-Methylglylocy amidine $C_4H_7N_3O$	113.12	rh pr (w + 2), lf (w)		ca 300d				B24^2, 128
5304	*o*-Cresol or 2-Hydroxytoluene 2-$CH_3C_6H_4OH$	108.14		191, 74.9^{10}	30.9	$1.0273^{20/4}$	1.5361^{20}	al, eth, **ace, bz**	B6^4, 1940
5305	*o*-Cresol, 3-amino 3-H_2N-2-$CH_3C_6H_3OH$	123.15	nd (w)		129			al	B13, 579
5306	*o*-Cresol, 4-amino 4-H_2N-2-$CH_3C_6H_3OH$	123.15	nd or lf (bz)	sub	175			al, eth	B13^3, 1531
5307	*o*-Cresol, 4-amino-6-nitro 6-NO_2-4-NH_2-2-$CH_3C_6H_2OH$	168.15	br-red nd (al)		118			al	B13, 578
5308	*o*-Cresol, 6-amino 6-H_2N-2-$CH_3C_6H_3OH$	123.15	pl (w)		89			al, eth, ace, bz	B13^3, 1527
5309	*o*-Cresol, 6-amino-4-nitro 4-NO_2-6-H_2N-2-$CH_3C_6H_2OH$	168.15	red-br nd (bz)		176			al, bz	B13^3, 1528
5310	*o*-Cresol, 3-bromo 3-Br-2-$CH_3C_6H_3OH$	187.04	nd (peth)	$55\text{-}7^4$	95			al, eth, ace, bz	B6, 360
5311	*o*-Cresol, 4-bromo 4-Br-2-$CH_3C_6H_3OH$	187.04	nd (al or peth)	235 sub, $137\text{-}43^{18}$	64			al, eth, ace	B6^4, 2006
5312	*o*-Cresol, 5-bromo 5-Br-2-$CH_3C_6H_3OH$	187.04	nd (lig or peth)		80			al, eth, ace	B6^2, 333
5313	*o*-Cresol, 3-chloro 3-Cl-2-$CH_3C_6H_3OH$	142.58	lo nd (w)	225	86			al, eth, bz	B6^4, 2000
5314	*o*-Cresol, 4-chloro 4-Cl-2-$CH_3C_6H_3OH$	142.58	nd (peth)	223	51				B6^4, 1987
5315	*o*-Cresol, 5-chloro 5-Cl-2-$CH_3C_6H_3OH$	142.58	nd (peth)		73-4			al, bz	B6^4, 1986
5316	*o*-Cresol, 6-chloro 6-Cl-2-$CH_3H_6H_3OH$	142.58		$188\text{-}9^{740}$, $80\text{-}1^{20}$			1.5449^{20}	eth	B6^4, 1984
5317	*o*-Cresol, 3,5-dibromo 3,5-Br_2-2-$CH_3C_6H_2OH$	265.93	nd (peth)	$283\text{-}7^{758}$	98-101			peth	B6^2, 334
5318	*o*-Cresol, 3,6-dibromo 3,6-Br_2-2-$CH_3C_6H_2OH$	265.93	cr	255-60	38				B6^1, 176

No.	Name, Synonyms, and Formula	Mol. wt.	Color, crystalline form, specific rotation and λ_{max} (log ε)	b.p. °C	m.p. °C	Density	n_D	Solubility	Ref.
5319	*o*-Cresol, 4,6-dibromo $4,6\text{-}Br_2\text{-}2\text{-}CH_3C_6H_2OH$	265.93	nd (peth)	$263\text{-}6^{745}/_d$	58	...	...	al, eth, bz	B6³, 1271
5320	*o*-Cresol, 4,5-dichloro $4,5\text{-}Cl_2\text{-}2\text{-}CH_3C_6H_2OH$	177.03	nd (peth)	...	101	...	...	al, bz, aa	B6², 333
5321	*o*-Cresol, 4,5dichloro $4,6\text{-}Cl_2\text{-}2\text{-}CH_3C_6H_2OH$	177.03	nd (w or peth)	$266.5, 73\text{-}8^4$	55	...	...	al, eth, chl	B6⁴, 2001
5322	*o*-Cresol, 3,5-dinitro $3,5\text{-}(NO_2)_2\text{-}2\text{-}CH_3C_6H_2OH$	198.14	ye pr (al)	...	85.8	...	...	al, eth, ace	C51, 10414
5323	*o*-Cresol, 4,6-dinitro $4,6\text{-}(NO_2)_2\text{-}2\text{-}CH_3C_6H_2OH$	198.14	ye pr or nd (al)	...	86.5	...	...	al, eth, ace	B6⁴, 2014
5324	*o*-Cresol, 3-nitro $3\text{-}NO_2\text{-}2\text{-}CH_3C_6H_3OH$	153.14	pa ye nd (w)	...	147	...	...	al, eth	B6¹, 178
5325	*o*-Cresol, 4-nitro $4\text{-}NO_2\text{-}2\text{-}CH_3C_6H_3OH$	153.14	ye or col nd (w or aq al)	$186\text{-}90^9$	96 (anh)	...	...	al, eth, bz, aa	B6⁴, 2011
5326	*o*-Cresol, 5-nitro $5\text{-}NO_2\text{-}2\text{-}CH_3C_6H_3OH$	153.14	ye nd (lig)	...	118	...	...	al, eth, bz	B6⁴, 2010
5327	*o*-Cresol, 6-nitro $6\text{-}NO_2\text{-}2\text{-}CH_3C_6H_3OH$	153.14	ye pr (dil al or peth)	$250\text{-}60d, 185^{12}$	70	...	...	al, eth	B6⁴, 2009
5328	*o*-Cresol, 4-nitroso $4\text{-}ON\text{-}2\text{-}CH_3C_6H_3OH$	137.14	nd (w)	...	134-5d	...	...	al, eth, bz, chl	B7³, 3388
5329	*o*-Cresol-4-isopropyl $4\text{-}i\text{-}C_3H_7\text{-}2\text{-}CH_3C_6H_3OH$	150.22	...	$230^{766}, 83^3$	8.6	0.9793^{25}	1.5253^{20}	al, bz, chl	B6³, 1884
5330	*o*-Cresol, 5-isopropyl $5\text{-}i\text{-}C_3H_7\text{-}2\text{-}CH_3\text{-}C_6H_3OH$	150.22	...	117-20	0-3	$0.976^{20/4}$	1.523^{20}	...	B6³, 1885
5331	*o*-Cresol, 6-isopropyl $6\text{-}i\text{-}C_3H_7\text{-}2\text{-}CH_3C_6H_3OH$	150.22	...	$225, 104^{14}$	-14.5	0.9789^{25}	1.5239^{20}	al, bz, chl	B6³, 1882
5332	*o*-Cresol-3,4,5,6-tetrabromo $2\text{-}CH_3C_6Br_4OH$	423.72	ye nd (chl or aa)	d	208	...	...	al, eth, bz, chl, aa	B6³, 1272
5333	*o*-Cresol-3,4,5,6-tetrachloro $2\text{-}CH_3C_6Cl_4OH$	245.92	nd (lig)	...	190	...	...	al, eth, bz, aa	B6³, 333
5334	*o*-Cresol-4,5,6-trinitro $4,5,6\text{-}(NO_2)_3\text{-}2\text{-}CH_3C_6HOH$	243.13	og-ye pr (ace)	...	102	...	...	al, eth, ace, chl	B6, 369
5335	*o*-Cresylphosphate or Tri-*o*-cresylphosphate $(2\text{-}CH_3C_6H_4O)_3PO$	368.37	col or pa ye	$410, 283\text{-}5^{20}$	11	$1.1955^{20/4}$	1.5575^{20}	al, eth, aa	B6⁴, 1979
5336	*m*-Cresol or 3-Hydroxy toluene $3\text{-}CH_3C_6H_4OH$	108.14	...	$202.2, 86^{10}$	11.5	$1.0336^{20/4}$	1.5438^{20}	al, eth, ace, bz	B6⁴, 2035
5337	*m*-Cresol-2-amino $2\text{-}NH_2\text{-}3\text{-}CH_3C_6H_3OH$	123.15	pl (w)	sub	150	...	...	eth	B13², 324
5338	*m*-Cresol-4-amino $4\text{-}H_2N\text{-}3\text{-}CH_3C_6H_3OH$	123.15	pr (dil al), cr (bz)	...	179	...	...	al, eth	B13³, 1559
5339	*m*-Cresol-5-amino $5\text{-}H_2N\text{-}3\text{-}CH_3C_6H_3OH$	123.15	cr (dil MeOH)	245	139	...	...	al	C47, 9300
5340	*m*-Cresol-5-amino-2-nitro $2\text{-}NO_2\text{-}H_2\text{-}5\text{-}H_2N\text{-}3CH_3C_6H_2OH$	168.15	red-br nd (al)	...	201	...	...	al	B13, 595
5341	*m*-Cresol-6-amino $6\text{-}H_2N\text{-}3\text{-}CH_3C_6H_3OH$	123.15	nd (bz, dil al)	...	162d	...	...	al, eth, ace	B13³, 1552
5342	*m*-Cresol-4-bromo $4\text{-}Br\text{-}3\text{-}CH_3C_6H_3OH$	187.04	nd (peth or w)	$137\text{-}43^{16}$	63.5	...	...	eth, Py	B6⁴, 2072
5343	*m*-Cresol-5-bromo $5\text{-}Br\text{-}3\text{-}CH_3C_6H_3OH$	187.04	nd (w)	$161\text{-}2^{28}$	56-7	...	...	al, eth	B6², 357
5344	*m*-Cresol-6-bromo $6\text{-}Br\text{-}3\text{-}CH_3C_6H_3OH$	187.04	cr (peth)	$206\text{-}8^{731}, 81\text{-}2^4$	38	...	...	al, eth, ace	B6³, 1320
5345	*m*-Cresol-2-chloro $2\text{-}Cl\text{-}3\text{-}CH_3C_6H_3OH$	142.58	lf or nd (dil al)	$230, 100\text{-}5^{28}$	74	...	...	al, eth	B6⁴, 2064
5346	*m*-Cresol-4-chloro $4\text{-}Cl\text{-}3\text{-}CH_3C_6H_3OH$	142.58	nd (peth)	235	66-8	...	...	al, eth	B6⁴, 2064
5347	*m*-Cresol-6-chloro $6\text{-}Cl\text{-}3\text{-}CH_3C_6H_3OH$	142.58	pr (peth)	196	45-6	1.215^{15}	...	al, w	B6³, 1315
5348	*m*-Cresol-2,4-dichloro $2,4\text{-}Cl_2\text{-}3\text{-}CH_3C_6H_3OH$	177.03	pr (peth)	241-2	27	...	...	eth, chl	B6³, 1319
5349	*m*-Cresol-2,6-dichloro $2,6\text{-}Cl_2\text{-}3\text{-}CH_3C_6H_3OH$	177.03	...	$235\text{-}6^{745}, 75\text{-}80^4$	58-9	...	...	eth, chl	B6³, 1319
5350	*m*-Cresol-4,6-dichloro $4,6\text{-}Cl_2\text{-}3\text{-}CH_3C_6H_2OH$	177.03	pr (peth)	$235\text{-}6, 110^{18}$	72-4	...	1.572^{20}	chl, peth	B6⁴, 2069
5351	*m*-Cresol, 4-nitro $4\text{-}NO_2\text{-}3\text{-}CH_3C_6H_3OH$	153.14	nd or pr (w)	...	129	...	...	al, eth, bz, chl	B6⁴, 2075

PHYSICAL CONSTANTS OF ORGANIC COMPOUNDS (Continued)

No.	Name, Synonyms, and Formula	Mol. wt.	Color, crystalline form, specific rotation and λ_{max} (log ε)	b.p. °C	m.p. °C	Density	n_D	Solubility	Ref.
5352	*m*-Cresol, 5-nirto 5-NO_2-3-$CH_3C_6H_3OH$	153.14	pa ye cr (bz)		90-1			eth, bz	$B6^{2}$, 361
5353	*m*-Cresol, 6-nitro 6-NO_2-3-$CH_3C_6H_3OH$	153.14	ye mcl nd (eth or bz)		56			al, eth, bz	$B6^{3}$, 1326
5354	*m*-Cresol, 4-nitroso 4-ON-3-$CH_3C_6H_3OH$	137.14	nd (w or bz), pr (aa)		165d			al, eth, bz, aa	$B7^{3}$, 3389
5355	*m*-Cresol, 2,4,5,6-tetrabromo 3-$CH_3C_6Br_4OH$	423.72	nd (chl aa)		194			eth	$B6^{3}$, 1324
5356	*m*-Cresol, 2,4,5,6-tetrachloro 3-$CH_3C_6Cl_4OH$	245.92	nd (peth)		189-90			al, eth, ace, bz	$B6^{4}$, 2071
5357	*m*-Cresol, 2,4,5,6-tribromo 2,4,6-Br_3-3-CH_3C_6HOH	344.83			84				$B6^{3}$, 1324
5358	*m*-Cresol, 2,4,6-trinitro or Methyl picric acid 2,4,6-$(NO_2)_3$-3-CH_3C_6HOH	243.13	pa ye nd (w or al)	150 exp	109-10			al, eth, ace, bz, chl	$B6^{4}$, 2079
5359	*m*-Cresyl phosphate or Tri-*m*-Cresyl phosphate $(3\text{-}CH_3C_6H_4O)_3PO$	368.37	wax	260^{15}	25-6	1.150^{25}	1.5575^{20}	eth, aa	$B6^{4}$, 2057
5360	*p*-Cresol or 4-Hydroxytoluene 4-$CH_3C_6H_4OH$	108.14	pr	201.9, 85.7^{10}	34.8	$1.0178^{20/4}$	1.5312^{20}	**al, eth, ace, bz**	$B6^{4}$, 2093
5361	*p*-Cresol, 2-amino 2-NH_2-4-$CH_3C_6H_3OH$	123.15	cr (w), rh (bz), lf or nd (sub)	sub	137			al, eth, chl	$B13^{3}$, 1576
5362	*p*-Cresol, 2-amino-5-nitro 5-NO_2-2-NH_2-4-$CH_3C_6H_2OH$	168.15	ye-og cr (al)		199-200d			al, dil HCl	$B13^{2}$, 346
5363	*p*-Cresol, 2-amino-6-nitro 6-NO_2-2-NH_2-4-$CH_3C_6H_2OH$	168.15	red-br (al)		119			al	$B13^{2}$, 345
5364	*p*-Cresol, 3-amino 3-H_2N-4-$CH_3C_6H_3OH$	123.15	cr (w or eth), lf (sub)	sub	156-7			eth	$B13^{2}$, 337
5365	*p*-Cresol, 2-bromo 2-Br-4-$CH_3C_6H_3OH$	187.04	nd (peth)	213-4	56-7	$1.5468^{25/25}$	1.5772^{20}	al, bz	$B6^{4}$, 2143
5366	*p*-Cresol, 3-bromo 3-Br-4-CH_3-C_6H_3OH	187.04	nd (peth)	245-7	56			al, eth, ace, bz	$B6^{4}$, 2143
5367	*p*-Cresol, 2-chloro 2-Cl-4-$CH_3C_6H_3OH$	142.58		195-6		$1.1785^{25/4}$	1.5200^{27}	al, eth, bz, aa	$B6^{4}$, 2135
5368	*p*-Cresol, 3-chloro 3-Cl-4-$CH_3C_6H_3OH$	142.58	nd (al)	228	55-6			al, eth, bz, aa	$B6^{3}$, 1374
5369	*p*-Cresol, 2,6-dichloro 2,6-Cl_2-4-$CH_3C_6H_2OH$	177.03	nd (lig)	$138\text{-}9^{28}$	39 (42)			al, eth, aa	$B6^{4}$, 2141
5370	*p*-Cresol, 2,6-dinitro 2,6-$(NO_2)_2$-4-$CH_3C_6H_2OH$	198.14	ye nd (eth or peth)		85			al, eth, bz	$B6^{4}$, 2152
5371	*p*-Cresol, 2-methoxy or Cresolol 2-CH_3O-4-$CH_3C_6H_3OH$	138.17	pr	221, 113.5^{22}	5.5	$1.098^{20/4}$	1.5353^{25}	al, eth	$B6^{2}$, 865
5372	*p*-Cresol, 3-methylamino 3(CH_3NH)-4-$CH_3C_6H_3OH$	137.18	cr (bz-lig)		108			al, eth, bz	B13, 599
5373	*p*-Cresol, 2-nitro 2-NO_2-4-$CH_3C_6H_3OH$	153.14	ye nd (al or w)	125^{22}	36.5	$1.2399^{20/4}$	1.574^{40}	al, eth, ace, bz	$B6^{4}$, 2149
5374	*p*-Cresol, 3-nitro 3-NO_2-4-$CH_3C_6H_3OH$	153.14	ye pr (eth)		79			al, eth	$B6^{3}$, 1384
5375	*p*-Cresol, 2-isopropyl 2-*p*-C_3H_7-4-$CH_3C_6H_3OH$	150.22		228-9, 82^{3}	36-7	$0.9910^{20/4}$	1.5275^{20}	al, bz, chl	$B6^{3}$, 1882
5376	*p*-Cresol, 2,3,5,6-tetrabromo 4-$CH_3C_6Br_4OH$	423.72	nd (al or chl)		198-9			al, eth, chl	$B6^{3}$, 1383
5377	*p*-Cresol, 2,3,5,6-tetrachloro 4-$CH_3C_6Cl_4OH$	245.92	nd (dil al, aa bz-lig)		190			al, bz, chl, aa	$B6^{4}$, 2142
5378	*p*-Cresyl phosphate or Tri-*p*-Cresyl phosphate $(4\text{-}CH_3C_6H_4O)_3PO$	368.37	nd (al), ta (eth)	$224^{3.5}$	77-8	1.247^{25}		al, eth, bz, chl	$B6^{4}$, 2130
5379	Crocetin-*(trans)* or Gardenin $C_{20}H_{24}O_4$	328.41	brick red rh		285-7 (cor)			Py	$B2^{3}$, 2018
5380	Croconic acid or Crocic acid $C_5H_2O_5$	142.07	pa ye nd (+3w al-diox)		sub d> 150			w, al	$B8^{3}$, 3977
5381	Crotonaldehyde or 2-Butenal $CH_3CH=CHCHO$	70.09		104-5	-74	$0.8495^{25/4}$	1.4355^{20}	al, eth, ace, **bz**	$B1^{4}$, 3447
5382	Crotonaldehyde, diethylacetal or 2-Butenaldiethylacetal $CH_3CH=CHCH(OC_2H_5)_2$	144.21		147-8, 49^{17}		$0.8473^{18/4}$	1.4097^{20}	**al, eth, ace, bz**	$B1^{4}$, 3450
5383	Crotonamide $CH_3CH=CHCONH_2$	85.11	nd (ace)	sub at 140^{13}	161.5 (cor)		1.4420^{165}	al, bz	$B2^{4}$, 1506

No.	Name, Synonyms, and Formula	Mol. wt.	Color, crystalline form, specific rotation and λ_{max} (log ε)	b.p. °C	m.p. °C	Density	n_D	Solubility	Ref.
5384	Crotonic acid or *(trans)*-2-Butenoic acid $CH_3CH=CHCO_2H$	86.09	mcl pr or nd (w or lig)	185	71.5	$1.018^{15/4}$	1.4249^{77}	w, al, eth, ace	**$B2^4$, 1498**
5385	Crotonic acid, allyl ester $CH_3CH=CHCO_2CH_2CH=CH_2$	126.15		$88-9^{70}$		$0.9440^{20/4}$	1.4465^{20}		**$B2^4$, 1503**
5386	Crotonic acid, anhydride or Crotonic anhydride $(CH_3CH=CHCO)_2O$	154.17		246-8, 129^{19}		1.0397^{20}	1.4745^{20}	eth	**$B2^4$, 1505**
5387	Crotonic acid, ethyl ester $CH_3CH=CHCO_2C_2H_5$	114.14		136.5, $58-9^{48}$		$0.9175^{20/4}$	1.4243^{20}	al, eth	**$B2^4$, 1500**
5388	Crotonic acid, methyl ester or Methyl crotonate $CH_3CH=CHCO_2CH_3$	100.13		121	-42	$0.9444^{20/4}$	1.4242^{20}	al, eth	**$B2^4$, 1500**
5389	Crotononitrile or 2-Butenonitrile $CH_3CH=CHCN$	67.09		$12-1^{762}$	-51.5	$0.8239^{20/4}$	1.4225^{20}	eth, ace	**$B2^4$, 1507**
5390	Crotonyl acetate $CH_3CO_2CH_2CH=CHCH_3$	114.15		132		$0.9192^{20/4}$	1.4181^{20}	al, eth, ace	**$B2^4$, 183**
5391	Crotonyl chloride $CH_3CH=CHCOCl$	104.54		124-5, 35^{18}		1.0905^{20}	1.460^{18}	ace	**$B2^4$, 1506**
5392	Cryptopine or Cryptocavine $C_{21}H_{23}NO_5$	369.42	pr or pl (bz), nd (chl-MeOH)		223 (cor)	$1.315^{20/4}$		chl, aa	**$B27^2$, 578**
5393	Cryptoxanthin or *β*-Caroten-3-ol $C_{40}H_{56}O$	552.88	garnet red pr (bz-MeOH)		169			bz, chl	**$B6^3$, 3772**
5394	Cumene or *iso*-Propylbenzene $(CH_3)_2CHC_6H_5$	120.19		152.4, 38.2^{10}	-96	$0.8618^{20/4}$	1.4915^{20}	**al, eth, ace, bz**	**$B5^4$, 985**
5395	Cumene, 2-nitro or 1-*iso*-Propyl-2-nitro benzene $2-NO_2C_6H_4CH(CH_3)_2$	165.19	pa ye		103^9	1.101^{12}	1.5259^{20}	ace, bz, aa	**$B5^4$, 997**
5396	Cumene, 4-nitro or 4-Nitro-1-*iso*-propyl benzene $4-NO_2C_6H_4CH(CH_3)_2$	165.19	pa ye oil	122^9		$1.0830^{20/4}$	1.5367^{20}	ace, bz, lig	**$B5^4$, 997**
5397	Cubebin $C_{20}H_{20}O_6$	356.38	nd (al or bz), $[\alpha]^{25}/_D$ -45.6 (chl, c=5)		131-2			al, eth, chl	**$B19^4$, 5967**
5398	Cumic alcohol or 1-Hydroxymethyl-4-isopropylbenzene $4-i-C_3H_7C_6H_4CH_2OH$	150.22		289	28	$0.9818^{18/4}$	1.5210^{20}	al, eth	**$B6^3$, 1911**
5399	Cupreine or Hydroxycinchonine $C_{19}H_{22}N_2O_2$	310.40	pr (eth), $[\alpha]^{17}/_D$ -175.5 (al)		198 (anh)			al	**$B23^4$, 416**
5400	C-Curarine-III-hydroxide or C-flurorcuraninehydroxide $C_{20}H_{28}N_2O_2$	328.46	cr (MeOH-eth)		212			w, al	H36, 102
5401	ar-Curcumene $C_{15}H_{22}$	202.34	$[\alpha]^{18}/_D$ + 35.8	140^{19}		$0.8821^{20/20}$	1.4989^{20}	bz	**$B5^4$, 1465**
5402	Curcumin $C_{21}H_{20}O_6$	368.39	or ye pr, rh pr (MeOH)		183			al, aa	**$B8^3$, 4312**
5403	Cuscohygrine or *α,α'* Bis(*N*-methyl-*α*-pyrrolidyl) acetone $C_{13}H_{24}N_2O$	224.35		185^{22}		$0.9782^{16/4}$		w, al, eth, bz	**$B24^2$, 36**
5404	Cuscohygrine, hydrate $C_{13}H_{24}N_2O.3\frac{1}{2}H_2O$	287.40	nd (a)		40-1			eth, bz	**B24, 78**
5405	Cusparine $C_{19}H_{17}NO_3$	307.35	(i) wh, nd (peth), (ii) ye, nd, (iii) pr		(i) 92, (ii) 92, (iii) 110-22			al, eth, ace, bz, chl	**$B27^2$, 545**
5406	Cyamelide or *sym*-Trioxane triimine $C_3H_3N_3O_3$	129.08	am or pw	d		$1.127^{15/4}$			**$B3^3$, 30**
5407	Cyanamide or Carbamonitrile H_2NCN	42.04	nd	140^{19}	42 (46)	$1.282^{20/4}$	1.4418^{48}	w, al, eth, ace, bz, chl	**$B3^4$, 145**
5408	Cyanamide, benzyl $C_6H_5CH_2NHCN$	132.16	pl (eth)		43			al, eth	**B12, 1051**
5409	Cyanamide, diallyl $(CH_2=CHCH_2)_2NCN$	122.17		$140-5^{90}$, 95^9				al, eth, ace, bz	**$B4^4$, 1078**
5410	Cyanamide, dibutyl $(C_4H_9)_2NCN$	154.26		$187-91^{190}$, $146-51^{35}$				al, eth, ace, bz	**$B4^4$, 592**
5411	Cyanamide, diethyl $(C_2H_5)_2NCN$	98.15		188, 62^{10}		$0.854^{20/4}$	1.4126^{25}	al, eth	**$B4^4$381**
5412	Cyanamide, dimethyl $(CH_3)_2NCN$	70.09		163.5, 56^{15}			1.4089^{19}	al, eth, ace	**$B4^4$, 226**

No.	Name, Synonyms, and Formula	Mol. wt.	Color, crystalline form, specific rotation and λ_{max} (log ε)	b.p. °C	m.p. °C	Density	n_D	Solubility	Ref.
5413	Cyanamide, diphenyl $(C_6H_5)_2NCN$	194.24	pr (al)	$235\text{-}40^{60}$	73-4	...	...	al, lig	B12[3], 895
5414	Cyanamide, methyl-α-naphthyl $\alpha\text{-}C_{10}H_7N(CH_3)CN$	182.22	yesh	$185\text{-}7^{2}$	...	...	...	al, eth	B12[2], 697
5415	Cyanamide, phenyl or Carbanilonitrile C_6H_5NHCN	118.14	cr (w, eth), lf (aa)	...	47 (hyd)	...	...	al, eth	B12[3], 805
5416	Cyanic acid HOCN	43.03	gas	23.5	-81	$1.140^{20/4}$	...	w, eth, bz, chl, aa	B3[4], 80
5417	Cyanic acid, ethyl ester or Ethyl cyanate C_2H_5OCN	71.08	...	$162d, 30^{12}$	...	$0.89^{20/4}$	1.3788^{25}	al, eth	Tet, 964, 2829
5418	Cyano acetamide $NCCH_2CONH_2$	84.08	pl (w)	...	121-2	...	...	w	B2[4], 1891
5419	Cyano acetamide, *N*-phenyl or α-Cyano acetanilide $NCCH_2CONHC_6H_5$	160.18	nd (al)	...	199-200	...	...	...	B12[2], 167
5420	Cyanoacetic acid $NCCH_2CO_2H$	85.06	...	$108^{0.15}$, d	70-1	...	...	w, al, eth	B2[4], 1888
5421	Cyano acetic acid, benzal or α-Cyanocinnamic acid $C_6H_5CH=C(CN)CO_2H$	173.17	cr (al)	...	183	...	...	...	B9[3], 4379
5422	Cyanoacetic acid, benzal, ethyl ester $C_6H_5CH=C(CN)CO_2C_2H_5$	201.22	(i) nd, (al), (ii) oil	(ii) 188^{15}	(i) 51	(ii) 1.1076	(ii) 1.5033	eth, bz, chl	B9[.], 4380
5423	Cyano acetic acid, 1-cyclo hexenyl $C_6H_9CH(CN)CO_2H$	165.19	nd (bz)	...	109-10	...	...	al, ace, bz	B9[2], 560
5424	Cyanoacetic acid, diethyl, ethyl ester $(C_2H_5)_2C(CN)CO_2C_2H_5$	169.22	...	$214\text{-}5, 100\text{-}1^{15}$	...	...	1.4200^{27}	**al, eth**	B2[3], 1761
5425	Cyanoacetic acid, ethyl ester or Ethyl cyanoacetate $NCCH_2CO_2C_2H_5$	113.12	...	$205, 99^{15}$	-22.5	$1.0654^{20/4}$	1.4175^{20}	**al, eth**	B2[4], 1889
5426	Cyanoacetic acid, methyl ester or Methyl cyanoacetate $NCCH_2CO_2CH_3$	99.09	...	$200\text{-}1, 115^{36}$	-22.5	1.1128^{20}	1.4176^{20}	al, eth	B2[4], 1889
5427	Cyanoacetic acid, phenyl, ethyl ester or Ethyl phenyl cyenate $C_6H_5CH(CN)CO_2C_2H_5$	189.21	oil	$275d, 165^{20}$	...	$1.091^{20/4}$	1.5012^{25}	**al, eth**, ace, bz	B9[3], 4262
5428	α-Cyanocaproic acid, ethyl ester $C_4H_9CH(CN)CO_2C_2H_5$	169.22	...	$245\text{-}50^{762}$, 105^{9}	...	0.988^{15}	1.4248^{20}	al, eth	B2[3], 1746
5429	Cyanogen or Oxalodinitrile NCCN	52.04	gas	-21.2	-27.9	0.9537^{-21}	...	w, al, eth	B2[4], 1863
5430	Cyanogen bromide BrCN	105.92	nd	61.4	52	$2.015^{20/4}$	...	w, al, eth	B3[4], 92
5431	Cyanogen chloride ClCN	61.47	gas	12.7	-6	$1.186^{20/4}$	...	w, al, eth	B3[4], 90
5432	Cyanogen iodide ICN	152.92	nd (al or eth)	sub>45	146-7	2.84^{18}	...	al, eth	B3[4], 93
5433	Cyanogen sulfide $S(CN)_2$	84.10	rh pl	sub 30-40	65	...	...	w, al, eth	B3[4], 339
5434	Cyanuric acid, dihydrate or 2,4,6,-Triazinetriol $C_3H_3N_3O_3\cdot 2H_2O$	165.11	mcl (w + 2)	d	>360d	$2.500^{20/4}$	...	w	B26[2], 131
5435	Cyanuric acid, tribenzyl ester or Benzyl cyanurate $C_{24}H_{21}N_3O_3$	399.45	nd (al)	>320	159	...	...	al	B26[1], 76
5436	Cyanuric chloride $C_3N_3Cl_3$	184.41	cr (eth or bz)	190^{720}	154	...	...	al	B26[2], 16
5437	Cyclamen aldehyde $4\text{-}i\text{-}C_3H_7C_6H_4CH_2CH(CH_3)CHO$	190.29	...	$133\text{-}7^{99}, 115^{5}$	...	0.951^{15}	1.5068^{20}	al, eth, bz	B7[3], 1200
5438	Cyclobutane or Tetramethylene C_4H_8	56.11	...	12	-50	$0.720^{5/4}$	1.4260^{20}	al, **eth**, ace, bz	B5[4], 6
5439	Cyclobutane, benzoyl or Cyclobutyl phenyl ketone $C_6H_5COC_4H_7$	160.22	...	$260, 122^{10}$	...	$1.0457^{25/25}$	1.5472^{20}	...	B7, 374
5440	Cyclobutane, ethyl $C_2H_5C_4H_7$	84.16	...	70.7	-142.9	$0.7284^{20/4}$	1.4020^{20}	al, eth, ace, bz	B5[4], 87
5441	Cyclobutane, methyl $CH_3C_4H_7$	70.13	...	36.3	...	$0.6884^{20/4}$	1.3866^{20}	**al, eth,** ace, bz, peth	B5[4], 21
5442	Cyclobutane, octafluoro or Perfluro cyclobutane C_4F_8	200.03	...	-4^{764}	-38.7	...	...	eth	B5[4], 8
5443	Cyclobutane carboxylic acid $C_4H_7CO_2H$	100.12	...	$190^{754}, 74\text{-}5^{2}$	-2	$1.0599^{20/4}$	1.4400^{20}	al, eth	B9[3], 6
5444	1,1-Cyclobutane dicarboxylic acid $1,1\text{-}C_4H_6(CO_2H)_2$	144.13	pr (eth or w)	...	156.6 (cor)	...	...	w, al, eth, bz	B9[3], 3797
5445	1,1-Cyclobutane dicarboxylic acid, diethyl ester $1,1\text{-}C_4H_6(CO_2C_2H_5)_2$	200.23	...	$229^{735}, 104^{12}$	...	$1.0456^{20/4}$	1.4344^{20}	al	B9[3], 3798

PHYSICAL CONSTANTS OF ORGANIC COMPOUNDS (Continued)

No.	Name, Synonyms, and Formula	Mol. wt.	Color, crystalline form, specific rotation and λ_{max} (log ε)	b.p. °C	m.p. °C	Density	n_D	Solubility	Ref.
5446	1,2-Cyclobutane dicarboxylic acid *(cis, dl)* $1,2\text{-}C_4H_6(CO_2H)_2$	144.13	pl (w), pr (bz)		138			w, al, eth	$B9^{3}$, 3798
5447	1,2-Cyclobutane dicarboxylic acid *(trans, d)* $1,2\text{-}C_4H_6(CO_2H)_2$	144.13	$[\alpha]^{30/}{}_D$ + 123.3 (w, c=2)		105			w, al	$B9^{3}$, 3799
5448	1,2-Cyclobutane dicarboxylic acid *(trans-dl)* $1,2\text{-}C_4H_6(CO_2H)_2$	144.13	rh nd (bz)		131			w, al	$B9^{3}$, 3799
5449	1,2-Cyclobutane dicarboxylic acid *(trans-l)* $1,2\text{-}C_4H_6(CO_2H_2$	144.13	nd (HCl), $[\alpha]^{30/}{}_D$ + 123.3 (w, c=0.8)		105			w, al	$B9^{3}$, 3798
5450	1,3-Cyclobutane dicarboxylic acid *(cis)* $1,3\text{-}C_4H_6(CO_2H)_2$	114.13	pr (w)	252	143-4			w, al	$B9^{3}$, 3801
5451	1,3-Cyclobutane dicarboxylic acid *(trans)* $1,3\text{-}C_4H_6(CO_2H)_2$	144.13	pr (w), nd (sub)	sub	171			w, al	$B9^{3}$, 3802
5452	Cyclobutane, acetyl $CH_3COC_4H_7$	98.14		137-9		0.9020^{20}	1.4322^{19}		$B7^{3}$, 45
5453	Cyclobutane, 1,2,bis (amino methyl) $1,2\text{-}(CH_3NH)_2C_4H_6$	114.19			193-4		$1.4778^{27.5}$		
5454	Cyclobutanone C_4H_6O	70.09		96-7		$0.9548^{0/0}$	1.4215^{20}	al, eth, bz, chl	$B7^{3}$, 4
5455	Cyclobutene C_4H_6	54.09			2	$0.733^{0/4}$		ace, bz, peth	$B5^{4}$, 207
5456	Cyclobutene, perfluoro C_4F_6	162.03		3	-60	$1.602^{-20/4}$	1.298^{-20}		$B5^{4}$, 208
5457	Cyclobutyl phenyl ketone $C_4H_7COC_6H_5$	160.22		260,121-2^{10}		$1.0457^{25/25}$	1.5472^{20}		B7, 374
5458	Cyclocamphene or Epicyclene $C_{10}H_{16}$	136.24		150-1	117-8	0.7948^{121}		al, aa	$B5^{3}$, 393
5459	1,6-Cyclodecanediol *(trans)* $C_{10}H_{20}O_2$	172.27	cr (chl, AcOEt)		151-3			eth	$B6^{3}$, 4106
5460	1,6-Cyclodecadione $C_{10}H_{16}O_2$	168.24	cr (eth)		100			ace, aa	$B7^{3}$3246
5461	Cyclodecanol $C_{10}H_{19}OH$	156.27		125^{12}	40-1	$0.9606^{20/4}$	1.4926^{20}	al	$B6^{4}$, 138
5462	Cyclodecanone $C_{10}H_{18}O$	154.25	amor pw	106-7^{13}	28	$0.9654^{20/4}$	1.4806^{20}	eth, bz, chl	$B7^{3}$, 134
5463	Cyclofenchene $C_{10}H_{16}$	136.24		144-6		$0.859^{20/4}$	1.4503^{22}		$B5^{4}$, 468
5464	9-Cycloheptadecen-1-one or Civetone $C_{17}H_{30}O$	250.42		342^{742}, 159^{2}	32.5	$0.9170^{33/4}$	1.4830^{33}	al, bz	$B7^{3}$, 524
5465	Cycloheptane or Suberane C_7H_{14}	98.19		118.5	-12	$0.8098^{20/4}$	1.4436^{20}	al, eth, bz, lig, chl	$B5^{4}$, 92
5466	Cycloheptane, 1-aza $C_6H_{13}N$	99.18		138^{749}		$0.8643^{22/4}$	1.4631^{20}	al, eth	$B20^{4}$, 1406
5467	Cycloheptane, bromo or Suberyl bromide $C_7H_{13}Br$	177.08		101.5^{40}, 75^{12}		$1.2887^{22/4}$	1.4996^{20}	eth, chl	$B5^{4}$, 93
5468	Cycloheptane, methyl $C_7H_{13}CH_3$	112.22		134		$0.8001^{20/4}$	1.4401^{20}	al, eth, bz, peth	$B5^{4}$, 114
5469	1,3-Cycloheptanedione $C_7H_{10}O_2$	126.16	ye	107-9^{17}	-40	$1.0607^{22/22}$	1.4689^{22}	al	C50, 6327
5470	Cycloheptanol or Suberol $C_7H_{13}OH$	114.19		185, 95^{24}	2	0.9554^{20}	1.4705^{20}	al, eth	$B6^{4}$, 94
5471	Cycloheptanone or Suberone $C_7H_{12}O$	112.17		178-9, 71^{9}		$0.9508^{20/4}$	1.4608^{20}	al, eth	$B7^{3}$, 46
5472	Cycloheptane carboxylic acid $C_7H_{13}CO_2H$	142.20		254-8^{711}, 130-1^{8}		$1.0423^{20/4}$	1.4753^{20}	al	$B9^{3}$, 47
5473	Cycloheptasiloxane, tetradicamethyl $C_{14}H_{42}O_7Si_7$	519.08		154^{20}	-26	$0.9703^{20/4}$	1.4040^{20}		$B4^{3}$, 1886
5474	1,,3,5-Cycloheptatriene or Tropilidene C_7H_8	92.14	cubic (at-80)	117, 60.5^{122}	-79.5	$0.8875^{19/4}$	1.5343^{20}	al, eth, bz, chl	$B5^{4}$, 765
5475	2,4,6-Cycloheptatriene-1-one or Tropone C_7H_6O	106.12		113^{15}	-7	$1.095^{22/4}$	1.6172^{22}		Am73, 876
5476	2,4,6-Cycloheptatriene-1-one, 2-amino $2\text{-}H_2N(C_7H_5O)$	121.14	ye pl (bz)		106-7			al, bz, chl	C46, 7559
5477	2,4,6-Cycloheptatriene-1-one, 3-bromo-2-hydroxy $3\text{-}Br\text{-}2\text{-}HO(C_7H_4O)$	201.02	ye pl or nd		107-8			al, eth	C49, 2405

No.	Name, Synonyms, and Formula	Mol. wt.	Color, crystalline form, specific rotation and λ_{max} (log ε)	b.p. °C	m.p. °C	Density	n_D	Solubility	Ref.
5478	2,4,6-Cycloheptatrien-1-one, 2-hydroxy or Tropolone $2\text{-HO}(C_7H_6O)$	122.12	nd	sub 40^{4}	51-2			w, eth, ace	Am74, 4456
5479	2,4,6-Cycloheptatriene-1-one, 2-hydroxy-4-isopropyl $C_{10}H_{12}O_2$	164.20	pa ye (peth)		50-1				$B8^{3}$, 440
5480	2,4,6-Cycloheptatriene-1-one, 2-hydroxy-4-methyl $2\text{-HO-4-}CH_3(C_7H_4O)$	136.15	nd (peth)		75-6			eth, chl	$B8^{3}$, 260
5481	2,4,6-Cycloheptatriene-1-one, 2-methoxy $2\text{-}CH_3O\text{-}(C_7H_5O)$	136.15	pa, ye nd (+ ½ w)	128^{5}	41(+ ½ w)			al, bz	C46, 4521
5482	Cycloheptene or Suberene C_7H_{12}	96.17		115	-56	$0.8228^{20/4}$	1.4552^{20}	al, eth, bz, peth	$B5^{4}$, 244
5483	1,3-Cyclohexadiene or 1,2-Dihydrobenzene C_6H_8	80.13		80.5	-89	$0.8405^{20/4}$	1.4755^{20}	al, eth, bz, chl	$B5^{4}382$
5484	1,3-Cyclohexadiene, 5-methyl *(dl)* $5\text{-}CH_3C_6H_7$	94.16		101.5^{762}		$0.8354^{20/4}$	1.4763^{20}	al, eth, bz, lig	$B5^{3}$, 318
5485	1,3-Cyclohexadiene, perfluoro C_6F_8	224.05			62-4	1.601^{20}	1.3149^{20}		$B5^{4}384$
5486	1,4-Cyclohexadiene or 1,4-Dihydrobenzene C_6H_8	80.13		85.6	-49.2	$0.8471^{20/4}$	1.4725^{20}	**al, eth,** bz, peth, chl	$B5^{4}$, 385
5487	1,4-Cyclohexadiene, octafluoro C_6F_8	224.05			57-8		1.318^{18}		$B5^{4}$, 386
5488	1,4-Cyclohexadiene-1,2-dicarboxylic acid or 3,6-Dihydrophthalic acid $1,4\text{-}(HO_2C)_2C_6H_6$	168.15	mcl pr (w)		153			al	$B9^{3}$, 4047
5489	2,4-Cyclohexadiene-1,2-dicarboxylic acid or 2,3-Dihydrophthalic acid $1,2\text{-}(HO_2C)_2C_6H_6$	168.15	pr (w or al)		179.80			al	$B9^{3}$, 4047
5490	2,6-Cyclohexadiene-1,2-dicarboxylic acid or 4,5-Dihydrophthalic acid $1,2\text{-}(HO_2C)_2C_6H_6$	168.15	tcl (w)		215			al, ace	$B9^{2}$, 575
5491	Cyclohexane C_6H_{12}	84.16		80.7	6.5	$0.7785^{20/4}$	1.4266^{20}	**al, eth, ace, bz, lig**	$B5^{4}$, 27
5492	Cyclohexane, acetyl or Cyclohexyl methyl ketone $CH_3COC_6H_{11}$	126.20		180-1, 69^{12}		$0.9176^{20/4}$	1.4565^{16}	eth	$B7^{3}$, 84
5493	Cyclohexane, allyl $(CH_2{=}CHCH_2)C_6H_{11}$	124.23		131.5^{757}		0.8135^{20}	1.4500^{20}	al, eth, ace, bz, chl	$B5^{4}$, 283
5494	Cyclohexane, amino or Cyclohexyl amine $C_6H_{11}NH_2$	99.18		134.5, 30.5^{15}	-17.7	$0.8191^{20/4}$	1.4372^{20}	w, al, **eth, ace, bz**	$B12^{3}$, 10
5495	Cyclohexane, bromo or Cyclohexyl bromide $C_6H_{11}Br$	163.06		166.2, 45.5^{10}	-56.5	$1.3359^{20/4}$	1.4957^{20}	**al, eth, ace, bz**	$B5^{4}$, 67
5496	Cyclohexane, 1-bromo-1-methyl $1\text{-Br-1-}CH_3(C_6H_{10})$	177.08		156-60, $65\text{-}9^{10}$		1.2510^{20}	1.4866^{20}	al, chl	$B5^{4}$, 100
5497	Cyclohexane, 1-bromo-2-methyl $1\text{-Br-2-}CH_3(C_6H_{10})$	177.08		$90\text{-}2^{10}$					$B5^{2}$, 12
5498	Cyclohexane, 1-bromo-3-methyl *(dl)* $1\text{-Br-3-}CH_3(C_6H_{10})$	177.08		181, 60^{11}		$1.275^{25/4}$	1.4979^{20}	eth, bz	$B5^{3}$, 76
5499	Cyclohexane, 1-bromo-4-methyl $1\text{-Br-4-}CH_3(C_6H_{10})$	177.08		130^{200}, 55^{15}				eth, bz	$B5^{4}$, 100
5500	Cyclohexane, (bromomethyl) $BrCH_2C_6H_{11}$	177.08		$76\text{-}7^{26}$		$1.2763^{25/4}$	1.4907^{20}	eth, bz, chl	$B5^{4}$, 100
5501	Cyclohexane, butyl or Cyclohexyl butane $C_4H_9C_6H_{11}$	140.27		181, 59^{10}	−74.7	$0.7992^{20/4}$	1.4408^{20}		$B5^{4}$, 146
5502	Cyclohexane, isobutyl $i\text{-}C_4H_9C_6H_{11}$	140.27		171.3	−95	$0.7952^{20/4}$	1.4386^{20}	al, eth, ace, bz, chl	$B5^{4}$, 147
5503	Cyclohexane, *sec*-butyl $CH_3CH_2CH(CH_3)C_6H_{11}$	140.27		179.3		$0.8131^{20/4}$	1.4467^{20}	ace	$B5^{4}$, 146
5504	Cyclohexane, *tert*-butyl $(CH_3)_3CC_6H_{11}$	140.27		171.5	−41.2	$0.8127^{20/4}$	1.4469^{20}		$B5^{4}$, 147
5505	Cyclohexane, butylamino $C_4H_9NHC_6H_{11}$	155.28		207				al, eth	$B12^{3}$, 15
5506	Cyclohexane, chloro or Cyclohexyl chloride $C_6H_{11}Cl$	118.61		143	−43.9	$1.000^{20/4}$	1.4626^{20}	**al, eth, ace, bz,** chl	$B5^{4}$, 48
5507	Cyclohexane, cyclopentyl $C_5H_9C_6H_{11}$	152.28		215.1		$0.8758^{20/4}$	1.4725^{20}		$B5^{4}$, 328
5508	Cyclohexane, 1,2-dibromo *(cis)* $1,2\text{-}Br_2C_6H_{10}$	241.95		115^{14}	9.7	$1.803^{25/25}$	1.5514^{25}	eth, ace, bz, chl, lig	$B5^{4}$, 70
5509	Cyclohexane, 1,2-dibromo *(trans, dl)* $1,2\text{-}Br_2C_6H_{10}$	241.95		$145\text{—}6^{100}$, 105^{20}	−4	$1.7759^{20/4}$	1.5445^{19}	al, eth, ace, bz	$B5^{4}$, 71

No.	Name, Synonyms, and Formula	Mol. wt.	Color, crystalline form, specific rotation and λ_{max} (log ε)	b.p. °C	m.p. °C	Density	n_D	Solubility	Ref.
5510	Cyclohexane, 1,3-dibromo *(cis)* 1,3-$Br_2C_6H_{10}$	241.95	rods (al)	...	112	...	...	bz, al	$B5^4$, 71
5511	Cyclohexane, 1,3-dibromo *(trans)* 1,3-$Br_2C_6H_{10}$	241.95	...	116^{16}	1	...	1.5480^{20}	al, bz	$B5^4$, 72
5512	Cyclohexane, 1,4-dibromo *(cis)* 1,4-$Br_2C_6H_{10}$	241.95	...	$137\text{-}8^{25}$	...	$1.7834^{20/4}$	1.5531^{20}	eth	$B5^4$, 72
5513	Cyclohexane, 1,4-dichloro *(cis)* 1,4-$Cl_2C_6H_{10}$	153.05	...	80.3^{25}	18	$1.1900^{20/4}$	1.4942^{20}	...	$B5^4$, 51
5514	Cyclohexane, 1,4-dibromo *(trans)* 1,4-$Br_2C_6H_{10}$	241.95	cr (eth)	...	...	...	...	eth	$B5^4$, 72
5515	Cyclohexane, 1,.2-dichloro *(cis)* 1,2-$Cl_2C_6H_{10}$	153.05	...	$206\text{-}9^{762}$, 91^{20}	−1.5	$1.2021^{20/4}$	1.4967^{20}	bz	$B5^4$, 50
5516	Cyclohexane, 1,2-dichloro-*(trans,dl)* 1,2-$Cl_2C_6H_{10}$	153.05	...	189, 78^{20}	−6.3	$1.1839^{20/4}$	1.4902^{20}	...	$B5^4$, 51
5517	Cyclohexane, (diethyl amino) $(C_2H_5)_2NC_6H_{11}$	155.28	...	$192\text{-}3^{740}$, $85\text{-}6^{20}$	...	$0.872^{0/0}$	...	al	$B12^3$, 14
5518	Cyclohexane, (difluoramino) decafluro $C_6F_{11}NF_2$	333.05	...	75-6	...	$1.787^{25/4}$	1.286^{25}	...	J1950, 1966
5519	Cyclohexane, 1,1-dimethyl 1,1-$(CH_3)_2C_6H_{10}$	112.22	...	119.5, 10^{10}	−33.5	$0.7809^{20/4}$	1.4290^{20}	al, eth, ace, bz, lig	$B5^4$, 117
5520	Cyclohexane, 1,2-dimethyl *(cis)* 1,2-$(CH_3)_2C_6H_{10}$	112.22	...	129.7, 18.3^{10}	−50.1	$0.7963^{20/4}$	1.4360^{20}	al, eth, ace, bz, lig	$B5^4$, 118
5521	Cyclohexane, 1,2-dimethyl *(trans)* 1,2-$(CH_3)_2C_6H_{10}$	112.22	...	123.4, 12.9^{10}	−89.2	$0.7760^{20/4}$	1.4270^{20}	al, eth, ace, bz, lig	$B5^4$, 118
5522	Cyclohexane, 1,2-dimethyl,perfluoro 1,2-$(CF_3)_2C_6F_{10}$	400.06	...	101.5	−56	$1.829^{25/4}$	1.283^{25}	...	$B5^3$, 98
5523	Cyclohexane, 1,3-dimethyl *(cis)* 1,3-$(CH_3)_2C_6H_{10}$	112.22	...	120.1, 11.1^{10}	−75.6	$0.7660^{20/4}$	1.4229^{20}	al, eth, ace, bz	$B5^4$, 121
5524	Cyclohexane, 1,3-dimethyl *(trans, d)* 1,3-$(CH_3)_2C_6H_{10}$	112.22	$[\alpha]_{549}$ + 1.33	124.4, 15^{10}	−90	$0.7847^{20/4}$	1.4309^{20}	al, eth, ace, bz, lig	$B5^4$, 121
5525	Cyclohexane, 1,4-dimethyl *(cis)* 1,4-$(CH_3)_2C_6H_{10}$	112.22	...	124.3, 14.4^{10}	−87.4	$0.7829^{20/4}$	1.4230^{20}	al, eth, ace, bz, lig	$B5^4$, 122
5526	Cyclohexane, 1,4-dimethyl *(trans, dl)* 1,4-$(CH_3)_2C_6H_{10}$	112.22	...	119.3, 10^{10}	−37.0	$0.7626^{20/4}$	1.4209^{20}	al, eth, ace, bz, lig	$B5^4$, 123
5527	Cyclohexane, 1,2-dimethylene $(1,2\text{-}CH_2)_2C_6H_8$	108.18	...	124^{740}, $60\text{—}1^{90}$	...	$0.8229^{25/4}$	1.4718^{25}	al, eth, ace, bz, chl	$B5^4$, 409
5528	Cyclohexane, 1,2-epoxy or Cyclohexene oxide $C_6H_{10}O$	98.14	...	131.5, $54\text{-}5^{10}$	<−10	0.9663^{20}	1.4519^{20}	al, eth, ace, bz	$B17^4$, 164
5529	Cyclohexane, 1,2-epoxy-4 (epoxy ethyl) or 4-Vinylcycloh-exene dioxide $C_8H_{12}O_2$	140.18	...	227, 92^5	<−55	$1.0986^{20/20}$	1.4787^{20}	w	$B19^4$, 161
5530	Cyclohexane, 1,2-epoxy-4-vinyl $C_8H_{12}O$	124.18	...	169, 20^2	<−100	$0.9598^{20/20}$	1.4700^{20}	...	$B17^4$, 314
5531	Cyclohexane, ethyl $C_2H_5C_6H_{11}$	112.22	...	131.8, 20.5^{10}	−111.3	$0.7880^{20/4}$	1.4330^{20}	al, eth, ace, bz, lig	$B5^4$, 115
5532	Cyclohexane, ethylamino $C_2H_5NHC_6H_{11}$	127.23	...	164, $62\text{-}5^{15}$	...	$0.868^{0/0}$	...	al, eth	$B12^3$, 14
5533	Cyclohexane, fluoro or Cyclohexyl fluoride $C_6H_{11}F$	102.15	...	100.2, 48^{100}	13	$0.9279^{20/4}$	1.4146^{20}	Py	$B5^4$, 44
5534	Cyclohexane, 1,1,2,3,4,5,6-heptachloro $C_6H_5Cl_7$	325.28	rods	...	55-6	...	...	...	$B5^4$, 63
5535	Cyclohexane, 1,2,3,4,5,6-hexabromo or Benzene-β-hexa-bromide *(β or cis)* $C_6H_6Br_6$	557.54	pr	...	253d	...	...	...	$B5^4$, 78
5536	Cyclohexane, 1,2,3,4,5,6-hexabromo *(α or trans)* $C_6H_6Br_6$	557.54	mcl pr (xyl)	...	212	...	...	...	$B5^4$, 78
5537	Cyclohexane, 1,2,3,4,5,6-hexachloro *(α dl)* or Benzene-*trans*-hexachloride $C_6H_6Cl_6$	290.83	mcl pr (al or aa)	288	159-60	...	...	al, bz, chl	$B5^4$, 60
5538	Cyclohexane, 1,2,3,4,5,6-hexachloro-*(β)* or Benzene-*cis*-hexachloride $C_6H_6Cl_6$	290.83	cr (bz, al or xyl)	$60^{0.50}$	314-5 sub	1.89^{19}	...	...	$B5^4$, 61
5539	Cyclohexane, 1,2,3,4,5,6-hexachloro -γ- or Benzene-γ-hexachloride, Lindane $C_6H_6Cl_6$	290.83	nd (al)	323.4, 176.2^{10}	112-3	...	...	ace, bz	$B5^4$, 58
5540	Cyclohexane, 1,2,3,4,5,6-hexachloro-(δ) $C_6H_6Cl_6$	290.83	pl	$60^{0.36}$	141-2	...	...	...	$B5^4$, 57

No.	Name, Synonyms, and Formula	Mol. wt.	Color, crystalline form, specific rotation and λ_{max} (log ε)	b.p. °C	m.p. °C	Density	n_D	Solubility	Ref.
5541	Cyclohexane, α-hydroxyethyl $C_6H_{11}[CH(OH)CH_3]$	128.21		$189, 81\text{-}2^{15}$		$0.9250^{20/4}$	1.4677^{20}	al, eth	B6[4], 117
5542	Cyclohexane, β-hydroxy ethyl $C_6H_{11}CH_2CH_2OH$	128.21		$207\text{-}9^{757}, 97\text{-}9^{15}$		$0.9229^{20/4}$	1.4641^{20}	al, eth, bz	B6[4], 119
5543	Cyclohexane, hydroxymethyl $C_6H_{11}CH_2OH$	114.19		$183, 83^{14}$	−43	$0.9297^{20/4}$	1.4644^{20}	al, eth	B6[4], 106
5544	Cyclohexane, iodo or Cyclohexy iodide $C_6H_{11}I$	210.06		$180d, 81.5^{20}$		$1.6244^{20/4}$	1.5477^{20}	al, eth, ace, bz, lig, chl	B5[4], 78
5545	Cyclohexane, methyl or Hexahydrotoluene $CH_3C_6H_{11}$	98.19		$100.9, 16.3^{10}$	−126.6	$0.7694^{20/4}$	1.4231^{20}	al, eth, ace, bz	B5[4], 94
5546	Cyclohexane, methylamino $CH_3NHC_6H_{11}$	113.20		$145\text{-}7, 76\text{-}7^{18}$		0.8660^{23}	1.4530^{23}	al, eth	B12[3], 13
5547	Cyclohexane, 1-methyl-4-ethyl $4\text{-}C_2H_4(C_6H_4)CH_3$	126.24		150-1		$0.791^{20/20}$	1.435^{20}		B5[4], 136
5548	Cyclohexane, 1-methyl-4-methylene $4\text{-}CH_2{=}(C_6H_9)CH_3$	110.20		122		$0.7923^{19/19}$	1.4465^{18}		B5[4], 271
5549	Cyclohexane, 1-methyl-2-pentyl $2\text{-}C_5H_{11}(C_6H_{10})CH_3$	168.32		216-9		$0.816^{20/20}$	1.4487^{20}		B5[4], 171
5550	Cyclohexane, methyl, perfluoro $CF_3C_6F_{11}$ $CF_3C_6F_{11}$	350.05		76.1	−44.7	$1.7878^{25/4}$	1.285^{17}	ace, bz	B5[4], 97
5551	Cyclohexane, methylene $CH_2{:}(C_6H_{10})$	96.17		$102\text{—}3^{764}$	−106.7	$0.8074^{20/4}$	1.4523^{20}	eth, bz, lig	B5[4], 250
5552	Cyclohexane, nitro $O_2NC_6H_{11}$	129.16		$205.5^{768}, 95^{22}$	fr-34	$1.0610^{20/4}$	1.4612^{19}	al, lig	B5[4], 81
5553	Cyclohexane, 1,2,3,4,5-pentahydroxy *(d)* or *d*-Quercitol $1,2,3,4,5\text{-}(HO)_5C_6H_7$	164.16	pr (w or dil al), $[\alpha]^{15}{}_D$ +25 (w,c=10)		235-7	1.5845^{13}		w	B6[3], 6873
5554	Cyclohexane, 1,2,3,4,5-pentahydroxy *(l)* or Viboquercetol $1,2,3,4,5\text{-}(HO)_5C_6H_7$	164.16	pr (w), nd (al), nd (w + 1), $[\alpha]^{20}{}_D$ −50 (w,c=4)		180-1 (anh)			w	B6[3], 6873
5555	Cyclohexane, pentyl $C_5H_{11}C_6H_{11}$	154.30		$202.8, 75.3^{10}$	−57.5	$0.8037^{20/4}$	1.4437^{20}	al, eth, ace, bz	B5[4], 164
5556	Cyclohexane, *iso*-pentyl $i\text{-}C_5H_{11}C_6H_{11}$	154.30		196.5		$0.8023^{20/4}$	1.4420^{20}	bz, lig	B5[3], 143
5557	Cyclohexane, phenyl or Cyclohexyl benzene $C_6H_5C_6H_{11}$	160.26		$235\text{-}6, 127\text{-}8^{30}$	7-8	$0.9502^{20/4}$	1.5329^{20}	al, eth	B5[4], 1424
5558	Cyclohexane, propyl $C_3H_7C_6H_{11}$	126.24		$156.7, 40.1^{10}$	−94.9	$0.7936^{20/4}$	1.4370^{20}	al, eth, ace, bz, peth	B5[4], 134
5559	Cyclohexane, *iso*-propyl or Hexahydrocumene $(CH_3)_2CHC_6H_{11}$	126.24		$154.5, 38.3^{10}$	90	$0.8023^{20/4}$	1.4410^{20}	al, eth, ace, bz	B5[4], 134
5560	Cyclohexane, 1,1,3-trimethyl $1,1,3\text{-}(CH_3)_3C_6H_9$	126.24		138-94		$0.7664^{20/0}$	1.4237^{15}		B5[4], 137
5561	Cyclohexane, 1,3,5-trimethyl *(cis)* or Hexahydromesitylene $1,3,5\text{-}(CH_3)_3C_6H_9$	126.24		138.5	−49.7	$0.7708^{20/4}$	1.4269^{20}	eth, bz, lig	B5[4], 138
5562	Cyclohexane, 1,3,5-trimethyl *(trans)* or Hexahydromesitylene $1,3,5\text{-}(CH_3)_3C_6H_9$	126.24		140.5	−107.4	$0.7794^{20/4}$	1.4307^{20}	eth, bz, lig	B5[4], 138
5563	Cyclohexanecarboxaldehyde or Hexahydrobenzaldehyde $C_6H_{11}CHO$	112.17		$159.3, 36^{10}$		$0.9035^{20/4}$	1.4496^{20}	eth	B7[3], 66
5564	Cyclohexane carboxylic acid or Hexahydrobenzoic acid $C_6H_{11}CO_2H$	128.17	mcl pr	$232\text{-}3, 120\text{-}1^{13}$	31-2	$1.0334^{22/4}$	1.4599^{22}	al, bz, chl	B9[3], 15
5565	Cyclohexane carboxylic acid, ethyl ester $C_6H_{11}CO_2C_2H_5$	156.22		$196, 63^{12}$		$0.9362^{20/4}$	1.4501^{15}	al, eth, ace, chl	B9[3], 17
5566	Cyclohexane carboxylic acid, 2-hydroxy or Hexahydrosalicylic acid $2\text{-}HOC_6H_{10}CO_2H$	144.17	nd (AeOEt)		111			w, al, eth	B10[3], 14
5567	Cyclohexane carboxylic acid, methyl ester $C_6H_{11}CO_2CH_3$	142.20		$183, 73^{15}$		$0.9954^{15/4}$	1.4433^{20}	al, eth, ace, chl	B9[3], 16
5568	Cyclohexane carboxylic acid, propyl ester $C_6H_{11}CO_2C_3H_7$	170.25		215.5		$0.9530^{15/4}$	1.4486^{15}	al, eth, ace, chl	B9[3], 17

PHYSICAL CONSTANTS OF ORGANIC COMPOUNDS (Continued)

No.	Name, Synonyms, and Formula	Mol. wt.	Color, crystalline form, specific rotation and λ_{max} (log ε)	b.p. °C	m.p. °C	Density	n_D	Solubility	Ref.
5569	Cyclohexane carboxylic acid, 1,3,4,5-tetrahydroxy *(d)* or *d*-Quinic acid $1,3,4,5\text{-}(HO)_4C_6H_7CO_2H$	192.17	mcl pr (w), $[\alpha]^{20/D}$ + 44 (w,c=10)	d	164	1.637	...	w	B10, 538
5570	Cyclohexane carboxylic acid, 1,3,4,5-tetra hydroxy *(dl)* or *dl*-Quinic acod $1,3,4,5\text{-}(HO)_4C_6H_7CO_2H$	192.17	pr (w)	...	142	...	...	w	$B10^3$, 2408
5571	Cyclohexane carboxylic acid, 1,3,4,5-tetrahydroxy *(l)* or *l*-Quinic acid $1,3,4,5\text{-}(HO)_4(C_6H_7)CO_2H$	192.17	pr (w), $[\alpha]^{18/D}$ −44.1 (w, c=12)	d	172	1.64	...	w	$B10^3$, 2407
5572	Cyclohexane carboxylonitrile, epoxy C_7H_9NO	123.15	...	244.5, 110^{10}	−33	$1.0929^{20/20}$	1.4763^{20}	w, eth	$B18^4$, 3891
5573	Cyclohexanecarboxylyl chloride $C_6H_{11}COCl$	146.62	...	180, $75\text{-}7^{15}$	...	$1.0962^{15/4}$	1.4711^{20}	...	$B9^3$, 27
5574	1,2-Cyclohexanedicarboxylic acid *(cis)* or Hexahydrophthalic acid $1,2\text{-}C_6H_{10}(CO_2H)_2$	172.18	tcl nd (al)	d	192	...	...	al, eth, ace, bz	$B9^3$, 3812
5575	1,2-Cyclohexane dicarboxylic acid *(trans, d)* $1,2\text{-}C_6H_{10}(CO_2H)_2$	172.18	pw (w), $[\alpha]_D$ + 18.2	...	179-83	...	...	w	$B9^3$, 3812
5576	Cyclohexanehexone, octahydrate $C_6O_6 \cdot 8H_2O$	312.18	mic nd (dil HNO_3)	...	100-1	...	...	...	$B7^3$, 4857
5577	Cyclohexanol $C_6H_{11}OH$	100.16	hyg nd	161.1	25.1	$0.9624^{20/4}$	1.4641^{20}	w, al, ace, eth, bz	$B6^4$, 20
5578	Cyclohexanol, 1-acetyl $1\text{-}CH_3COC_6H_{10}OH$	142.20	...	125-6, 91^{11}	...	$1.0248^{25/4}$	1.4670^{25}	al, eth	$B8^3$, 15
5579	Cyclohexanol, 2-allyl *(trans)* $2\text{-}(CH_2{=}CHCH_2)C_6H_{10}OH$	140.23	...	$94\text{-}6^{15}$	...	$0.947^{20/4}$	1.4778^{20}	aa	$B6^4$, 235
5580	Cyclohexanol, 2-amino *(trans, dl)* $2\text{-}H_2NC_6H_{10}OH$	115.18	hyg	105^{10}	68	...	...	bz, chl, aa	$B13^3$, 704
5581	Cyclohexanol, 2-butyl *(trans)* $2\text{-}C_4H_9C_6H_{10}OH$	156.27	...	$111\text{-}2^{16}$	...	$0.9020^{20/4}$	1.4641^{20}	eth, ace, bz	$B6^3$, 121
5582	Cyclohexanol, 2-chloro *(cis, dl)* $2\text{-}ClC_6H_{10}OH$	134.61	hyg (peth)	$93\text{-}4^{26}$	36-7	1.1261^{25}	1.4894^{25}	al, bz, chl	$B6^4$, 64
5583	Cyclohexanol, 2-chloro *(cis, l)* $2\text{-}ClC_6H_{10}OH$	134.61	hyg $[\alpha]_{549}$ 19.5	87^{15}	...	1.137^{15}	1.4894^{25}	w, al, bz, chl	$B6^3$, 39
5584	Cyclohexanol, 2-chloro *(trans)* $2\text{-}ClC_6H_{10}OH$	134.61	pr (bz-lig)	93^{26}	29	$1.146^{16/4}$	1.4899^{20}	al, eth, bz, chl	$B6^4$, 64
5585	Cyclohexanol, 4-chloro *(trans)* $4\text{-}ClC_6H_{10}OH$	134.16	pl (cy)	106^{14}	82-3	$1.1435^{17/4}$	1.4930^{17}	al, eth, bz, chl	$B6^4$, 68
5586	Cyclohexanol, 3-dimethylamino $3\text{-}(CH_3)_2NC_6H_{10}OH$	143.23	...	231, $126\text{-}7^{22}$	73	$0.9766^{25/25}$	1.4852^{20}	al	$B13^3$, 719
5587	Cyclohexanol, 1-ethyl $1\text{-}C_2H_5C_6H_{10}OH$	128.21	pr	166, 67^{10}	34-5	0.9227^{25}	1.4633^{20}	bz, peth	$B6^4$, 115
5588	Cyclohexanol, 2-ethyl *(cis, dl)* $2\text{-}C_2H_5C_6H_{10}OH$	128.21	...	180-2, 74^{12}	...	$0.9274^{20/4}$	1.4655^{21}	eth, ace, bz, peth	$B6^3$, 85
5589	Cyclohexanol, 2-ethyl *(trans, dl)* $2\text{-}C_2H_5C_6H_{10}OH$	128.21	...	79^{12}	...	$0.9193^{21/4}$	1.4640^{21}	eth, ace, bz, peth	$B6^4$, 117
5590	Cyclohexanol, 1-ethynyl $1\text{-}(HC{\equiv}C)C_6H_{10}OH$	124.18	cr (peth)	174, 73^{12}	31-2	$0.9873^{20/4}$	1.4822^{20}	al, bz, peth	$B6^4$, 348
5591	Cyclohexanol, 2-(l-hydroxyethyl) $2[CH_3CH(OH)]C_6H_{10}OH$	144.21	...	140^{12}	...	$0.976^{20/7}$	1.4900^{20}	...	C50, 3299
5592	Cyclohexanol, 1-methyl $1\text{-}CH_3C_6H_{10}OH$	114.19	...	155, 70^{25}	25	$0.9194^{20/4}$	1.4595^{20}	al, bz, chl	$B6^4$, 95
5593	Cyclohexanol, 2-methyl *(cis, dl)* $2\text{-}CH_3C_6H_{10}OH$	114.19	...	165, 60^{12}	7	$0.9360^{20/4}$	1.4640^{20}	al, eth	$B6^4$, 100
5594	Cyclohexanol, 2-methyl *(trans, d)* $2\text{-}CH_3C_6H_{10}OH$	114.19	$[\alpha]^{20/}{}_D$ + 17.19 (un-dil)	166, 78^{20}	...	0.9454^{20}	1.4610^{20}	al, eth	$B6^3$, 62
5595	Cyclohexanol, 2-methyl *(trans, dl)* $2\text{-}CH_3C_6H_{10}OH$	114.19	...	167-8, 78^{20}	−4	$0.9247^{20/4}$	1.4616^{20}	al, eth	$B6^4$, 100
5596	Cyclohexanol, 2-methyl-*(trans, l)* $2\text{-}CH_3C_6H_{10}OH$	114.19	$[\alpha]^{20}{}_D$ − 35.5 (un-dil)	166, 78^{20}	...	0.9454^{20}	1.4610^{20}	al, eth	$B6^3$, 62
5597	Cyclohexanol, 3-methyl *(cis, l)* $3\text{-}CH_3C_6H_{10}OH$	114.19	$[\alpha]^{22/}{}_D$ − 4.75 (un-dil)	174-5, 94^{12}	−4.7	$0.9155^{20/4}$	1.4574^{20}	al, eth	$B6^3$, 67

PHYSICAL CONSTANTS OF ORGANIC COMPOUNDS (Continued)

No.	Name, Synonyms, and Formula	Mol. wt.	Color, crystalline form, specific rotation and λ_{max} (log ε)	b.p. °C	m.p. °C	Density	n_D	Solubility	Ref.
5598	Cyclohexanol, 3-methyl *(trans, l)* 3-$CH_3C_6H_{10}OH$	114.19	$[\alpha]^{20}/_D - 7.3$ (undil)	174-5, 84^{13}	−1	$0.9214^{20/4}$	1.4590^{20}	al, eth	$B6^3$, 68
5599	Cyclohexanol, 4-methyl *(cis)* 4-$CH_3C_6H_{10}OH$	114.19		173-4, $78\text{-}9^{20}$	−9.2	$0.9170^{20/4}$	1.4614^{20}	al, eth	$B6^4$, 105
5600	Cyclohexanol, 4-methyl *(trans)* $C_7H_{14}O$	114.19		173-4, 54^3		$0.9118^{20/4}$	1.4561^{20}	al, eth	$B6^4$, 105
5601	Cyclohexanol, 1-phenyl 1-$C_6H_5C_6H_{10}OH$	176.26		157.5^{28}, $112\text{-}3^5$	63-3.5	1.035^{16}	1.5415^{16}		$B6^3$, 2510
5602	Cyclohexanol, 2-phenyl *(cis, dl)* 2-$C_6H_5C_6H_{10}OH$	176.26		$140\text{-}1^{16}$	41-2 (56)	1.035^{16}	1.5415^{16}		$B6^3$, 2510
5603	Cyclohexanol, 2-phenyl *(trans, dl)* 2-$C_6H_5C_6H_{10}OH$	176.26	cr (peth)	$152\text{-}5^{16}$	56-7			al, chl	$B6^3$, 2511
5604	Cyclohexanol, 2-isopropyl *(cis)* 2-$(CH_3)_2CHC_6H_{10}OH$	142.24		77^{13}	52-3	0.9223^{25}	1.4665^{25}		$B6^4$, 131
5605	Cyclohexanol, 2,2,6,6-tetrakis (hydroxymethyl) 2,2,6,6,$(HOCH_2)_4C_6H_7OH$	220.27	pl (al)		131			w, al	$B6^3$, 6877
5606	Cyclohexanol, 1,2,2-trimethyl *(dl)* 1,2,2-$(CH_3)_3C_6H_8OH$	142.24	cr (+ ½ w)	$81\text{-}2^{20}$	41 (hyd)	$0.9230^{20/4}$	1.4682^{20}	al, eth, bz	$B6^3$, 114
5607	Cyclohexanol, 1,2,6-trimethyl 1,2,6-$(CH_3)_3C_6H_8OH$	142.24		78^{22}		$0.9126^{15/4}$	1.4598^{15}	al, eth, ace, bz	$B6^1$, 17
5608	Cyclohexanol, 1,3,3-trimethyl 1,3,3-$(CH_3)_3C_6H_8OH$	142.24	pr (dil al)		74			al, eth, ace, bz	$B6^1$, 16
5609	Cyclohexanol, 1,3,5-trimethyl 1,3,5-$(CH_3)_3C_6H_8OH$	142.24		181, $82\text{-}3^{19}$		$0.8876^{17/4}$	$1.454^{16.3}$	al, eth, chl	$B6^3$, 117
5610	Cyclohexanol, 1,4,4-trimethyl 1,4,4-$(CH_3)_3C_6H_8OH$	142.24	hyg nd (dil al)	$79\text{-}80^{15}$	58			al, eth, chl	$B6^1$, 16
5611	Cyclohexanol, 2,2,3-trimethyl 2,2,3-$(CH_3)_3C_6H_8OH$	142.24		$85\text{-}7^{15}$				al, eth	$B6^1$, 16
5612	Cyclohexanol, 2,2,5-trimethyl or Pulenol 2,2,5-$(CH_3)_3C_6H_8OH$	142.24		187-9, $90\text{-}2^{23}$		$0.8955^{22/4}$	1.4569^{20}	al	B6, 22
5613	Cyclohexanol, 2,2,6-trimethyl (liquid) 2,2,6-$(CH_3)_3C_6H_8OH$	142.24		$186\text{-}7^{752}$		$0.9128^{20/4}$	1.4600^{20}	al, eth, chl	$B6^4$, 135
5614	Cyclohexanol, 2,2,6-trimethyl (solid) 2,2,6-$(CH_3)_3C_6H_8OH$	142.24	cr (peth or al)	87^{28}	51			al, eth, chl	$B6^4$, 135
5615	Cyclohexanol, 2,3,3-trimethyl 2,3,3-$(CH_3)_3C_6H_8OH$	142.24	nd	197, 97^{19}	28			al, ace	$B6^1$, 16
5616	Cyclohexanol, 2,3,6-trimethyl 2,3,6-$(CH_3)_3C_6H_8OH$	142.24		$193\text{-}5^{747}$		$0.9117^{17/4}$		al, chl	B6, 22
5617	Cyclohexanol, 2,4,5-trimethyl *(cis)* 2,4,5-$(CH_3)_3C_6H_8OH$	142.24	hyg	191-3, 84^{17}		$0.9120^{20/4}$	1.463^{20}	al, eth, chl	$B6^2$, 36
5618	Cyclohexanol, 2,4,5-trimethyl *(trans)* 2,4,5-$(CH_3)C_6H_8OH$	142.24	hyg	196, 112^{35}		$0.906^{20/4}$	1.461^{20}	al, eth, chl	$B6^2$, 36
5619	Cyclohexanol, 3,3,5-trimethyl *(cis)* 3,3,5-$(CH_3)_3C_6H_8OH$	142.24		$201\text{-}3^{750}$, 92^{12}	37.3	$0.9006^{16/4}$	1.4550^{16}	al, eth, chl	$B6^4$, 135
5620	Cyclohexanol, 3,3,5-trimethyl *(trans)* 3,3,5-$(CH_3)_3C_6H_8OH$	142.24	cr (eth)	189.2	55.8	$0.8647^{60/20}$		al, eth, chl	$B6^4$, 135
5621	Cyclohexanone $C_6H_{10}O$	98.14		155.6, 47^{15}	−16.4	$0.9478^{20/4}$	1.4507^{20}	al, eth, ace, bz, chl	$B7^3$, 14
5622	Cyclohexanone, 2-acetyl 2-$CH_3CO(C_6H_9O)$	140.18		$111\text{-}2^{18}$		1.0782^0	1.5138^{20}		$B7^3$, 3223
5623	Cyclohexanone, 2,6-dibromo *(cis)* 2,6-$Br_2(C_6H_8O)$	255.94	cr (eth, aa)		106-7				$B7^3$, 38
5624	Cyclohexanone, 2-butyl 2-$C_4H_9(C_6H_9O)$	154.25		70^2		$0.905^{20/4}$	1.4545^{20}		$B7^3$, 140
5625	Cyclohexanone, 2-butylidene $C_{10}H_{16}O$	152.24		$98\text{-}100^{10}$		$0.935^{20/4}$	1.4800^{20}	al, eth, ace, bz	C49, 1598
5626	Cyclohexanone, 2-chloro 2-$Cl(Cl_6H_9O)$	132.59		82^{15}	23	$1.161^{20/15}$	1.4825^{20}	eth, bz	$B7^3$, 36
5627	Cyclohexanone, 3-chloro 3-$Cl(C_6H_9O)$	132.59		$91\text{-}2^{14}$				eth	B7, 10
5628	Cyclohexanone, 4-chloro 4-$Cl(C_6H_9O)$	132.59		95^{17}			1.4867^{20}	eth	$B7^2$, 11
5629	Cyclohexanone, 2-β-cyanoethyl 2-$NCCH_2CH_2\text{-}(C_6H_9O)$	151.21		$138\text{-}42^{10}$		$1.0181^{20/4}$	1.4755^{20}		$B10^3$, 2835
5630	Cyclohexanone, cyanohydrin $C_6H_{10}(OH)CN$	125.17		$109\text{-}13^9$	34-6		1.4643^{20}	eth, w	$B10^3$, 11

No.	Name, Synonyms, and Formula	Mol. wt.	Color, crystalline form, specific rotation and λ_{max} (log ε)	b.p. °C	m.p. °C	Density	n_D	Solubility	Ref.
5631	Cyclohexanone, 2-cyclohexyl $2\text{-}C_6H_{11}(C_6H_9O)$	180.29	...	264	−32	$0.9752^{25/25}$	1.4877^{25}	...	$B7^3$, 474
5632	Cyclohexanone, 2,6-dibenzyl $2,6\text{-}(C_6H_5CH_2)_2(C_6H_8O)$	278.39	ye nd (al)	$185\text{-}95^{20}$	117-8	...	...	bz, aa	$B7^3$, 2661
5633	Cyclohexanone, 2,4-dimethyl *(trans, d)* $2,4\text{-}(CH_3)_2(C_6H_8O)$	126.20	$[\alpha]^{24}/_D$ 64.8 al 6%	178.7^{766}, 69^{17}	...	$0.9004^{16/4}$	1.4488^{22}	eth, ace, bz	$B7^3$, 93
5634	Cyclohexanone, 2,5-dimethyl *(d)* $2,5\text{-}(CH_3)_2(C_6H_8O)$	126.20	$[\alpha]^{20}$ + 11.5 (undil)	$172\text{-}4^{750}$, 51^{10}	...	$0.8985^{20/4}$	1.4445^{20}	al, eth	$B7^3$, 97
5635	Cyclohexanone, 2,5-dimethyl *(trans, dl)* $2,5\text{-}(CH_3)_2(C_6H_8O)$	126.20	...	171-3, $76\text{-}7^{27}$	...	0.9025^{20}	1.4446^{20}	al, eth	$B7^3$, 97
5635a	Cyclohexanone, 2-(dimethylaminomethyl) $2\text{-}(CH_3)_2NCH_2(C_6H_9O)$	155.24	...	$92^{10.5}$	...	$0.9504^{20/4}$	1.4672^{20}	al, eth	$B14^3$, 5
5636	Cyclohexanone, 2-ethylidene $(2\text{-}CH_3CH{=})(C_6H_8O)$	124.18	...	92^{20}	...	$0.962^{20/4}$	1.4882^{20}	...	$B7^2$, 58
5637	Cyclohexanone, 2-hydroxy or Adipoin $2\text{-}HO(C_6H_9O)$	114.14	nd (al or MeOH)	...	113	...	1.4785^{21}	w, al	$B8^3$, 4
5638	Cyclohexanone, 2-methyl *(d)* $2\text{-}CH_3(C_6H_9O)$	112.17	$[\alpha]^{25}/_D$ + 14.21 (chl)	$167\text{-}8^{735}$	...	$0.9262^{18/4}$	1.4440^{25}	...	$B7^3$, 49
5639	Cyclohexanone, 2-methyl *(dl)* $2\text{-}CH_3(C_6H_9O)$	112.17	...	165^{757}, 90^{20}	−13.9	$0.9250^{20/4}$	1.4483^{25}	al, eth	$B7^3$, 49
5640	Cyclohexanone, 2-methyl *(l)* $2\text{-}CH_3(C_6H_9O)$	112.17	$[\alpha]^{25}/_D$ − 15.22 (undil)	$59\text{-}60^{20}$	...	$0.9230^{25/4}$	1.4440^{25}	al, eth	$B7^3$, 49
5641	Cyclohexanone, 3-methyl *(d)* $3\text{-}CH_3(C_6H_9O)$	112.17	$[\alpha]^{20}/_D$ + 12.7 (undil)	169	...	$0.9155^{20/4}$	1.4493^{20}	al, eth	$B7^3$, 55
5642	Cyclohexanone, 3-methyl *(dl)* $3\text{-}CH_3(C_6H_9O)$	112.17	...	$168\text{-}9^{738}$, 65^{15}	−73.5	$0.9136^{20/4}$	1.4456^{20}	al, eth	$B7^3$, 57
5643	Cyclohexanone, 4-methyl $4\text{-}CH_3(C_6H_9O)$	112.17	...	170	−40.6	$0.9138^{20/4}$	1.4451^{20}	al, eth	$B7^3$, 63
5644	Cyclohexanone, oxime $C_6H_{10}{=}NOH$	113.16	hex pr (lig)	206-10	90	...	...	w, al, eth	$B7^3$, 32
5645	Cyclohexanone, 2-propyl $2\text{-}C_3H_7(C_6H_9O)$	140.23	...	195, 70^{6}	...	$0.927^{20/4}$	1.4538^{20}	al, eth, ace, bz	$B7^3$, 115
5646	Cyclohexanone, 2-isopropyl $2\text{-}i\text{-}C_3H_7(C_6H_9O)$	140.23	...	...	72-3	$0.922^{16/4}$	1.4564^{15}	al, eth, ace, bz	$B7^3$, 117
5647	Cyclohexanethiol or Cyclohexyl mercaptan $C_6H_{11}SH$	116.22	...	158, 41^{12}	...	$0.9782^{20/4}$	1.4921^{20}	al, eth, ace, bz, chl	$B6^4$, 72
5648	Cyclohexanethione $C_6H_{10}S$	114.21	...	74^{11}	...	...	1.5375^{20}	al, eth, ace	$B7^3$, 39
5649	1,2-Cyclohexane dicarboxylic acid *(trans, dl)* $1,2\text{-}C_6H_{10}(CO_2H)_2$	172.19	lf or pr (w)	...	222	...	...	w	$B9^3$, 3813
5650	1,2-Cyclohexane dicarboxylic acid *(trans, l)* $1,2\text{-}C_6H_{10}(CO_2H)_2$	172.19	pw (w)	...	179-82	...	...	...	B9, 732
5651	1,2-Cyclohexane dicarboxylic acid, diethyl ester *(cis)* $1,2\text{-}C_6H_{10}(CO_2HC_{25})_2$	228.29	...	133^{10}	...	$1.0540^{22/4}$	1.45512^{14}	eth	$B9^3$, 3813
5652	1,2-Cyclohexane dicarboxylic acid,diethyl ester *(trans, dl)* $1,2\text{-}C_6H_{10}(CO_2C_2H_5)_2$	228.29	...	135^{11}	...	$1.040^{20/4}$	1.4522^{13}	eth	$B9^3$, 3813
5653	1,3-Cyclohexane dicarboxylic acid *(cis)* or *cis*-Hexadhydro isophthalic acid $1,3\text{-}C_6H_{10}(CO_2H)_2$	172.18	nd (con HCl), cr (w)	...	167-8	...	...	w, al, eth, bz	$B9^3$, 3817
5654	1,3-Cyclohexane dicarboxylic acid *(trans, d)* $1,3\text{-}C_6H_{10}(CO_2H)_2$	172.18	cr (w), $[\alpha]^{22}/_D$ + 23.8 (w, c=4)	...	134	...	...	w, al, eth	$B9^2$, 523
5655	1,3-Cyclohexane dicarboxylic acid *(trans, dl)* $1,3\text{-}C_6H_{10}(CO_2H)_2$	172.18	nd (w)	...	150.5	...	...	w, al, eth	$B9^3$, 3817
5656	1,3-Cyclohexane dicarboxylic acid *(trans, l)* $1,3\text{-}C_6H_{10}(CO_2H)_2$	172.18	w, $[\alpha]^{22}/_D$ − 23.2 (w, c=2)	...	134	...	...	w, al, eth	$B9^3$, 523
5657	1,3-Cyclohexane dicarboxylic acid, diethyl ester *(cis)* $1,3\text{-}C_6H_{10}(CO_2C_2H_5)_2$	228.29	...	288, 142^{11}	...	$1.0450^{20/4}$	1.4521^{20}	...	$B9^3$, 3817
5658	1,3-Cyclohexane dicarboxylic acid, diethyl ester *(trans, dl)* $1,3\text{-}C_6H_{10}(CO_2C_2H_5)_2$	228.29	...	286^{756}, 142^{12}	...	$1.0485^{21/4}$	1.4530^{20}	...	$B9^3$, 3817
5659	1,4-Cyclohexane dicarboxylic acid *(cis)* or *cis*-Hexahydro terephthalic acid $1,4\text{-}C_6H_{10}(CO_2H)_2$	172.18	lf (w)	...	170-1	...	...	al, eth, chl	$B9^3$, 3818

No.	Name, Synonyms, and Formula	Mol. wt.	Color, crystalline form, specific rotation and λ_{max} (log ε)	b.p. °C	m.p. °C	Density	n_D	Solubility	Ref.
5660	1,4-Cyclohexane dicarboxylic acid *(trans)* $1,4\text{-}C_6H_{10}(CO_2H)_2$	172.19	pr (w), pl (ace)	300 sub	312-3			al, ace	$B9^3$, 3818
5661	1,4-Cyclohexane dicarboxylic acid, diethyl ester *(cis)* $1,4\text{-}C_6H_{10}(CO_2C_2H_5)_2$	228.29		151^{13}		$1.0516^{21/4}$	1.4522^{21}	eth	$B9^2$, 524
5662	1,4-Cyclohexane dicarboxylic acid, diethyl ester *(trans)* $1,4\text{-}C_6H_{10}(CO_2C_2H_5)_2$	228.29	nd		43-4	$1.0110^{20/4}$	1.4337^{64}	eth	$B9^2$, 524
5663	1,2-Cyclohexanediol *(cis)* $1,2\text{-}(HO)_2C_6H_{10}$	116.16	cr (eth), pl (bz)	120^{15}	99-101	$1.0297^{101/4}$		al, ace, bz	$B6^3$, 4058
5664	1,2-Cyclohexanediol *(trans)* $1,2\text{-}(HO)_2C_6H_{10}$	116.16	cr (ace)	117^{13}	105	$1.147^{24/4}$		w, al	$B6^3$, 4060
5665	1,4-Cyclohexanediol *(cis)* or *cis*-Quinitol $1,4\text{-}(HO)_2C_6H_{10}$	116.16	pr (ace)		113-4			w, al	$B6^3$, 4080
5666	1,4-Cyclohexanediol *(trans)* $1,4\text{-}(HO)_2C_6H_{10}$	116.16	mcl pr (ace)		143	$1.18^{20/4}$		w, al	$B6^3$, 4081
5667	1,2-Cyclohexanedione $C_6H_8O_2$	112.13	cr (peth)	193-5, $96\text{-}7^{25}$	38-40		1.4995^{20}	w, al, eth, bz	$B7^3$, 3209
5668	1,2-Cyclohexanedione, 3,5-dimethyl $C_8H_{12}O_2$	140.18	cr (dil MeOH)		71-2			w	$B7^1$, 314
5669	1,2-Cyclohexanedione, 5,5-dimethyl or Dimedone $C_8H_{12}O_2$	140.18	yesh nd (w, aq ace), mcl pr (al-eth)		150			ace, chl, aa	$B7^3$, 3225
5670	1,2-Cyclohexanedione, dioxime or Nioxime $C_6H_{10}N_2O_2$	142.16	nd (w or ace)		191-3			ace	$B7^3$, 3210
5671	1,3-Cyclohexanedione $C_6H_8O_2$	112.13	pr (bz)		105-6	1.0861^{91}	1.4576^{102}	w, al, ace, chl	$B7^3$, 3210
5672	1,3-Cyclohexanedione, 2-bromo $C_6H_7BrO_2$	191.02	micr nd		169-70			al	B7, 556
5673	1,3-Cyclohexanedione, dioxime $C_6H_{10}N_2O_2$	142.16	cr (w)		156-7			w, al, aa	$B7^3$, 3211
5674	1,4-Cyclohexanedione $C_6H_8O_2$	112.13	mcl pl (w), nd (peth)	sub 100	78			w, al, eth, ace, bz	$B7^3$, 3211
5675	1,4-Cyclohexanedione, dioxime $C_6H_{10}N_2O_2$	142.16	cr (w)		188			w	$B7^3$,3212
5676	Cyclohexasiloxane, dodecamethyl $C_{12}H_{36}O_6Si_6$	444.93		245, 128^{20}	−3	0.9672	1.4015^{20}		$B4^3$, 1886
5677	Cyclohexasiloxane, 2,4,6,8,10,12-hexamethyl $C_6H_{24}O_6Si_6$	360.77			−79	$1.006^{20/4}$	1.3944^{20}		$B4^3$, 1874
5678	Cyclohexene C_6H_{10}	82.15		83	−103.5	$0.8102^{20/4}$	1.4465^{20}	**al, eth, ace, bz**	$B5^4$, 218
5679	Cyclohexene, 1-acetyl $1\text{-}(CH_3CO)C_6H_9$	124.18		201-2, $63\text{-}4^6$		$0.9655^{20/4}$	1.4881^{20}	al, eth	$B7^4$, 244
5680	1-Cyclohexene, 1-bromo $1\text{-}BrC_6H_9$	161.04		164-6, 69^{35}		$1.3901^{20/4}$	1.5134^{20}	eth, ace, bz	$B5^4$, 236
5681	Cyclohexene, 3-bromo $3\text{-}BrC_6H_9$	161.04		$80\text{-}2^{40}$		$1.3890^{20/4}$	1.5230^{20}	eth, bz, chl	$B5^4$, 237
5682	Cyclohexene, 1-chloro $1\text{-}ClC_6H_9$	116.59		142-3, 35^{13}		$1.0361^{19/4}$	1.4797^{20}	eth, ace, chl	$B5^4$, 230
5683	Cyclohexene, 1,2-dimethyl $1,2\text{-}(CH_3)_2C_6H_8$	110.20		136		$0.823^{20/4}$	1.4580^{21}		$B5^4$, 268
5684	Cyclohexene, 1-ethyl $1\text{-}C_2H_5C_6H_9$	110.20		$135\text{-}6^{755}$		0.8238^{19}	1.4567^{19}		$B5^4$, 266
5685	Cyclohexene, 1-methyl $1\text{-}CH_3C_6H_9$	96.17		110, 24.6^{30}	−121	$0.8102^{20/4}$	1.4503^{20}	eth, bz	$B5^4$, 245
5686	Cyclohexene, 3-methyl *(d)* $3\text{-}CH_3C_6H_9$	96.17	$[\alpha]^{20}{}_D$ + 110	104		$0.8010^{20/4}$	1.4414^{20}	eth, bz, peth, chl	$B5^3$, 200
5687	Cyclohexene, 3-methyl *(dl)* $3\text{-}CH_3C_6H_9$	96.17		104	−115.5	$0.7990^{20/4}$	1.4414^{20}	eth, bz, peth, chl	$B5^4$, 247
5688	Cyclohexene, 4-methyl $4\text{-}CH_3C_6H_9$	96.17		102.7, 19^{30}	−115.5	$0.7991^{20/4}$	1.4414^{20}	al, eth	$B5^4$, 248
5689	Cyclohexene, 1,3,4,5,6-pentachloro *(γ)* $C_6H_5Cl_5$	254.37		$115\text{-}6^4$			1.5630^{20}		$B5^4$, 234
5690	Cyclohexene, 1,3,4,5,6-pentachloro *(δ)* $C_6H_5Cl_5$	254.37			68-9	1.80		al	$B5^4$, 234
5691	Cyclohexene-per fluoro C_6F_{10}	262.05		$52\text{-}3^{750}$			1.293^{20}		$B5^4$, 229

No.	Name, Synonyms, and Formula	Mol. wt.	Color, crystalline form, specific rotation and λ_{max} (log ε)	b.p. °C	m.p. °C	Density	n_D	Solubility	Ref.
5692	Cyclohexene, 1-phenyl $1\text{-}C_6H_5C_6H_9$	158.24	.....	251-3, $125\text{-}6^{14}$	−11	$0.9939^{20/4}$	1.5718^{20}	MeOH	$B5^4$, 1557
5693	Cyclohexene, 1-isopropyl-4-methylene $i\text{-}C_3H_7\text{-}4(CH_2{=})C_6H_7$	136.24	.....	173-4	.....	0.838^{22}	1.4754^{22}	.....	$B5^4$, 437
5694	Cyclohexene, 4-isopropyl-1-methyl $4\text{-}i\text{-}C_3H_7C_6H_8CH_3$	136.24	.....	174.5	.....	$0.8465^{15.5/15.5}$	1.4735^{20}	.....	$B5^4$, 300
5695	Cyclohexene, 1-vinyl $1\text{-}(CH_2{=}CH)C_6H_9$	108.18	.....	145, $50\text{-}2^{22}$	.....	$0.8623^{15/4}$	1.4915^{20}	eth, bz, MeOH	$B5^4$, 405
5696	Cyclohexene, 4-vinyl $4\text{-}(CH_2{=}CH)C_6H_9$	108.18	.....	128.9, $66\text{-}7^{100}$	.....	$0.8299^{20/4}$	1.4639^{20}	eth, bz, peth	$B5^4$, 406
5697	Cyclohexene, 1-carboxaldehyde C_6H_9CHO	110.16	.....	72^{15}	.....	$0.9694^{20/4}$	1.5005^{20}	al, eth	$B7^3$, 234
5698	3-Cyclohexene, 1-carboxaldehyde C_6H_9CHO	110.16	.....	164, 52^{13}	fr−96.1	$0.9709^{20/4}$	1.4725^{19}	ace, MeOH	$B7^3$, 237
5699	Cyclohexene, 1-carboxylic acid $2\text{-}C_6H_9CO_2H$	126.16	.....	240-2, 138^{14}	38	$1.109^{20/4}$	1.4902^{20}	al, ace	$B9^3$, 144
5700	3-Cyclohexene, 1-carboxylic acid $4\text{-}C_6H_9CO_2H$	126.16	.....	237^{748}, $132\text{-}3^{20}$	17	$1.0815^{20/4}$	1.4812^{20}	w, al, ace	$B9^3$, 148
5701	1-Cyclohexene-1,2-dicarboxylic acid or Δ^1-Tetrahydraphthalic acid $C_8H_{10}O_4$	170.17	nd pr (w)	.....	126	.....	.....	w	$B9^3$, 3939
5702	1-Cyclohexene-1,2-dicarboxylic anhydride $C_8H_8O_3$	152.15	pl (eth)	.....	74	.....	.....	al, eth, ace, chl	$B17^4$, 5995
5703	2-Cyclohexene-1,2-dicarboxylic anhydride $C_8H_8O_3$	152.15	pr (eth)	.....	78-9	.....	.....	al, eth, chl	$B17^4$, 5994
5704	4-Cyclohexene-1,2-dicarboxylic anhydride *(cis)* $C_8H_8O_3$	152.15	pl (eth or lig)	.....	103-4	.....	.....	al, ace, chl	$B17^4$, 5996
5705	4-Cyclohexene-1,2-dicarboxylic anhydride *(trans, d)* $C_8H_8O_3$	152.15	lf $[\alpha]^{25/}{}_D$ + 6.6 (al)	.....	128	.....	.....	al, bz	B17, 462
5706	4-Cyclohexene-1,2-dicarboxylic anhydride *(trans, dl)* $C_8H_8O_3$	152.15	cr (bz-lig)	.....	141	.....	.....	al, bz, chl	$B17^4$, 5996
5707	2-Cyclohexene-1-ol C_6H_9OH	98.14	.....	164-6, $63\text{-}5^{12}$	.....	$0.9923^{15/4}$	1.4790^{22}	al, ace	$B6^4$, 196
5708	2-Cyclohexene-1-ol, 5-methyl *(cis, d)* $CH_3C_6H_8OH$	112.17	$[\alpha]^{30/}{}_D$ + 6.95	83^{25}	.....	$0.9391^{25/4}$	1.4727^{25}	eth, lig	$B6^4$, 206
5709	2-Cyclohexene-1-ol, 5-methyl-*(cis, l)* $CH_3C_6H_8OH$	112.17	$[\alpha]^{25/}{}_D$ − 7	82^{25}	.....	$0.9391^{25/4}$	1.4727^{25}	eth, lig	$B6^4$, 206
5710	2-Cyclohexene-1-ol, 5-methyl *(trans, d)* $CH_3C_6H_8OH$	112.17	$[\alpha]^{27/}{}_D$ + 127 (ace, c=19.4)	$68\text{-}9^{24}$	.....	$0.9430^{20/4}$	1.4737^{25}	eth, lig	$B6^4$, 207
5711	2-Cyclohexene-1-ol, 5-methyl-*(trans, l)* $CH_3C_6H_8OH$	112.17	$[\alpha]^{27/}{}_D$ − 163.9	$82\text{-}3^{24}$	.....	$0.9430^{20/4}$	1.4737^{25}	eth, lig	$B6^4$, 207
5712	3-Cyclohexene-1-ol C_6H_9OH	98.14	.....	164, $68\text{-}9^{16}$	.....	$0.9845^{20/4}$	1.4851^{20}	eth, ace	$B6^4$, 200
5713	2-Cyclohexene-1-one C_6H_8O	96.13	.....	169-71, $61\text{-}2^{10}$	.....	0.9620^{25}	1.4883^{20}	al, ace	$B7^3$, 224
5714	2-Cyclohexene-1-one, 2,3-dimethyl $(CH_3)_2(C_6H_6O)$	124.18	.....	$93\text{-}6^{20}$	.....	0.9695^{20}	1.4995^{20}	al, eth	$B7^3$, 250
5715	2-Cyclohexene-1-one, 2,5-dimethyl $(CH_3)_2(C_6H_6O)$	124.18	.....	189-90	.....	0.938^{22}	1.4753^{22}	al, eth	$B7^1$, 51
5716	2-Cyclohexene-1-one, 3,5-dimethyl $(CH_3)_2(C_6H_6O)$	124.18	.....	208-9, 94^{17}	.....	$0.9400^{20/4}$	1.4812^{20}	al, eth	$B7^3$, 255
5717	2-Cyclohexene-1-one, 3,5-dimethyl-4-carboxyethyl $C_{11}H_{16}O_3$	196.25	.....	$157\text{-}8^{18}$	.....	$1.0493^{20/4}$	1.4773^{20}	ace	$B10^3$, 2902
5718	2-Cyclohexene-1-one, 3,6-dimethyl $(CH_3)_2(C_6H_6O)$	124.18	.....	75^{19}	.....	$1.008^{18/18}$	1.4805^{18}	al, eth	$B7^3$, 257
5719	2-Cyclohexene-1-one, 2-methyl $CH_3(C_6H_7O)$	110.16	.....	178-9, 56^{9}	.....	$0.9667^{20/4}$	1.4833^{20}	bz	$B7^3$, 233
5720	2-Cyclohexene-1-one, 3-methyl $CH_3(C_6H_7O)$	110.16	.....	200-2, $78\text{-}9^{12}$	−21	$0.9693^{20/4}$	1.4947^{20}	bz	$B7^3$, 230
5721	3-Cyclohexene-1-one, 4-methyl $CH_3(C_6H_7O)$	110.16	.....	$169\text{-}72^{755}$, 74^{17}	.....	$0.9551^{20/4}$	1.4652^{20}	al, ace, bz	$B7^3$, 232
5722	2-Cyclohexene-1-one, 5-methyl *(dl)* $CH_3(C_6H_7O)$	110.16	.....	179-83, 60^{8}	.....	$0.947^{20/4}$	1.4739^{25}	al, ace, bz	$B7^3$, 236
5723	2-Cyclohexene-1-one, 5-isopropyl-3-methyl or Hexeton $C_{10}H_{16}O$	152.24	pa ye	244, 124^{15}	.....	0.9340^{21}	1.4865^{21}	al, ace	$B7^3$, 323

No.	Name, Synonyms, and Formula	Mol. wt.	Color, crystalline form, specific rotation and λ_{max} (log ε)	b.p. °C	m.p. °C	Density	n_D	Solubility	Ref.
5724	3-Cyclohexene-1-one, 4,6-dimethyl $(CH_3)_2(C_6H_6O)$	124.18	....	194	......	0.9539^{0}	....	eth	$B7^{3}$, 256
5725	Cycloheximide $C_{15}H_{23}NO_4$	281.35	wh pl (w), $[\alpha]^{29}{}_D$ − 3.4 (al)	....	119-21	....	....	al, eth, ace	$B21^{4}$, 6632
5726	Cyclohexyl acetic acid $C_6H_{11}CH_2CO_2H$	142.20	nd (HCO_2H)	224-6, 135^{13}	33	$1.0423^{18/4}$	1.4775^{20}	al, eth	$B9^{3}$, 47
5727	Cyclohexylamine $C_6H_{11}NH_2$	99.18	....	134.5, 30.5^{15}	−17.7	$0.8191^{20/4}$	1.4372^{20}	al, eth, ace, bz	B12 , 10
5728	Cyclohexylamine, hydrochloride $C_6H_{11}NH_2.HCl$	135.64	nd (w or al-eth)	....	206-7	....	....	w, al	$B12^{3}$, 12
5729	Cyclohexyl bromide $C_6H_{11}Br$	163.06	....	166.2, 45.5^{10}	−56.5	$1.3359^{20/4}$	1.4957^{20}	al, eth, ace, bz, lig	$B5^{4}$, 67
5730	Cyclohexyl chloride $C_6H_{11}Cl$	118.61	....	143	$1.000^{20/4}$	1.4626^{20}	....	al, eth, ace, bz, chl	$B5^{4}$, 48
5731	Cyclohexyl ether or Dicyclohexyl ether $(C_6H_{11})_2O$	182.31	....	242-3	......	$0.9227^{20/4}$	1.4741^{20}	....	$B6^{3}$, 19
5732	Cyclohexyl 2-furyl ether $C_6H_{11}O(C_4H_3O)$	166.22	....	118.9^{28}	......	$1.0200^{28/4}$	1.4861^{28}	al, eth, ace	$B17^{4}$, 1219
5733	Cyclohexyl hydro peroxide $C_6H_{11}OOH$	116.16	....	$42^{0.1}$	−20	$1.019^{20/4}$	1.4645^{25}	al, eth, aa	$B6^{4}$, 53
5734	Cyclohexyl methyl ether or Hexahydro anisole $C_6H_{11}OCH_3$	114.19	....	133	−74.4	$0.8756^{20/4}$	1.4355^{20}	al, eth	$B6^{4}$, 26
5735	Cyclohexyl methyl ether, 2-bromo *(dl, trans)* $2\text{-}BrC_6H_{10}OCH_3$	193.08	....	$78\text{-}9^{12}$	......	$1.3314^{20/4}$	1.4871^{20}	al, eth	$B6^{3}$, 43
5736	Cyclohexyl phenyl ether $C_6H_{11}OC_6H_5$	176.26	....	128^{15}	......	$1.0077^{20/4}$	1.520^{22}	ace, bz	$B6^{4}$, 565
5737	Cyclohexyl isothiocyanate $C_6H_{11}NCS$	141.23	....	219^{746}	......	....	1.5375^{20}	al, eth	$B12^{3}$, 53
5738	Cyclononanone $C_9H_{16}O$	140.23	....	148.5^{24}	34	$0.9560^{20/4}$	1.4729^{20}	al	$B7^{3}$, 111
5739	Cyclononasiloxane, octadecamethyl $C_{18}H_{54}O_9Si_9$	667.39	....	188^{20}	......	....	1.4070^{20}	bz, lig	$B4^{3}$, 1886
5740	Cyclononene *(cis)* C_9H_{16}	124.23	....	167-9, $73\text{-}4^{30}$	......	$0.8671^{20/4}$	1.4805^{20}	bz	$B5^{4}$, 280
5741	Cyclononene-*(trans)* C_9H_{16}	124.23	....	$94\text{-}6^{30}$	......	$0.8615^{20/4}$	1.4799^{20}	bz	$B5^{4}$, 281
5742	1,5-Cyclooctadiene *(cis, cis)* C_8H_{12}	108.18	....	150.8^{757}, $51\text{-}2^{25}$	−70	$0.8818^{25/4}$	1.4905^{25}	bz	$B5^{4}$, 403
5743	1,3-Cyclooctadiene *(cis, cis)* C_8H_{12}	108.18	....	$54\text{-}5^{35}$	−57	$0.8699^{25/4}$	1.4940^{25}	eth, bz, chl	$B5^{4}$, 401
5744	Cyclooctane C_8H_{16}	112.22	....	$148\text{-}9^{749}$, 63^{45}	14.3	$0.8349^{20/4}$	1.4586^{20}	bz, lig	$B5^{4}$, 111
5745	Cycloctanol $C_8H_{15}OH$	128.21	....	99^{16}	25.1	$0.9740^{20/4}$	1.4871^{20}	al	$B6^{4}$, 113
5746	Cycloctanone or Azelaone $C_8H_{14}O$	126.20	....	194-8, 74^{12}	28-30	$0.9581^{20/4}$	1.4694^{20}	al, ace, bz	$B7^{3}$, 77
5747	Cycloctanone, semicarbazone $C_8H_{14}NHCONHNH_2$	183.25	lf (dil MeOH)	....	170-1	....	....	....	$B7^{3}$, 78
5748	Cyclooctasiloxane, hexadecamethyl $C_{16}H_{48}O_8Si_8$	593.24	....	290, 175^{20}	31.5	1.177	1.4060^{20}	bz, lig	$B4^{3}$, 1886
5749	Cyclooctatetraene C_8H_8	104.16	ye or wh	140.5, 29.1^{10}	−4.7	$0.9206^{20/4}$	1.5381^{20}	al, eth, ace, bz	$B5^{4}$, 1331
5750	Cyclooctatetraene, chloro C_8H_7Cl	138.60	....	$50\text{-}1^{5.5}$	......	$1.1199^{25/4}$	1.5542^{25}	ace, bz	$B5^{4}$, 1334
5751	Cyclooctatetraene, methyl $CH_3C_8H_7$	118.18	....	84.5^{67}	......	$0.8978^{25/4}$	1.5249^{25}	al, eth, bz, chl	$B5^{4}$, 1358
5752	Cyclooctene *(cis)* C_8H_{14}	110.20	....	138, 42^{18}	−12	$0.8472^{20/4}$	1.4698^{20}	al, eth	$B5^{4}$, 262
5753	Cyclooctene *(trans)* C_8H_{14}	110.20	....	143, 75^{78}	−59	$0.8483^{20/4}$	1.4741^{25}	al, chl	$B5^{4}$, 263
5754	Cyclopentadecanone or Exhaltone $C_{15}H_{28}O$	224.39	....	$120^{0.3}$	63	0.8895	1.4637^{66}	al, ace	$B7^{3}$, 203
5755	Cyclopentadecanone, 3-methyl or Muscone, muskone $3\text{-}CH_3C_{15}H_{27}O$	238.41	ye $[\alpha]^{17}{}_D$ − 13 (undil)	$327\text{-}30^{752}$, $130^{0.5}$	......	$0.9221^{17/4}$	1.4802^{17}	al, eth, ace	$B7^{3}$, 208

PHYSICAL CONSTANTS OF ORGANIC COMPOUNDS (Continued)

No.	Name, Synonyms, and Formula	Mol. wt.	Color, crystalline form, specific rotation and λ_{max} (log ε)	b.p. °C	m.p. °C	Density	n_D	Solubility	Ref.
5756	Cyclopentadiene C_5H_6	66.10		40.0	−97.2	$0.8021^{20/4}$	1.4440^{20}	**al, eth,** ace, **bz**	B5[4], 377
5757	Cyclopentadiene, perchloro C_5Cl_6	272.77	ye gr lig	239^{753}, $48\text{-}9^{0.3}$	−9	$1.7019^{25/4}$	1.5658^{20}		B5[4], 381
5758	1,3-Cyclopentadiene, 5-isopropylidene $[CH_3{=}C(CH_3)]C_5H_5$	106.17		$49\text{-}50^{11}$		$0.881^{20/4}$	1.5474^{20}		B5[4], 974
5759	Cyclopentadiene benzoquinone $C_{11}H_{10}O_2$	174.20	gr-ye lf (MeOH)		77-8			al, eth, ace, bz	B6[3], 5311
5760	Cyclopentadienone, tetraphenyl or Tetracyclone $(C_6H_5)_4C_5O$	384.48	bk-vt lf, cr (aa or xyl)		220-1			al, bz	B7[3], 2997
5761	Cyclopentasiloxane, 2,4,6,8,10-pentamethyl $C_5H_{20}O_5Si_5$	300.64		169	−108	$0.9985^{20/4}$	1.3912^{20}		B4[3], 1874
5762	Cyclopentane or Pentamethylene C_5H_{10}	70.13		49.2	−93.9	$0.7457^{20/4}$	1.4065^{20}	**al, eth, ace, bz**	B5[4], 14
5763	Cyclopentane, acetyl or Cyclopentyl methyl ketone $CH_3COC_5H_9$	112.17		158-9		$0.918^{20/20}$	1.4409^{20}	eth	B7[3], 71
5764	Cyclopentane, allyl $(CH_2{=}CHCH_2)C_5H_9$	110.20		124-6		$0.793^{25/4}$	1.4412^{20}		B5[4], 272
5765	Cyclopentane, amino or Cyclopentyl amine $H_2NC_5H_9$	85.15		108	−85.7	$0.8689^{20/4}$	1.4778^{25}	ace, bz	B12[3], 5
5766	Cyclopentane, bromo or Cyclopentyl bromide C_5H_9Br	149.03		137-8, 56^{49}		$1.3873^{20/4}$	1.4886^{20}		B5[4], 19
5767	Cyclopentane, butyl $C_4H_9C_5H_9$	126.24		156.7, 41.6^{10}	−108	$0.7846^{20/4}$	1.4316^{20}	**al, eth, ace, bz, peth**	B5[4], 139
5768	Cyclopentane, chloro or Cyclopentyl chloride C_5H_9Cl	104.58		$113\text{-}4^{752}$		$1.0051^{20/4}$	1.4510^{20}	eth, ace, bz	B5[4], 18
5769	Cyclopentane, 1,2-diethyl *(trans)* $1,2\text{-}(C_2H_5)_2C_5H_8$	126.24		153.6	−95.6	$0.7832^{20/4}$	1.4295^{20}	eth, bz, peth	B5[4], 141
5770	Cyclopentane, 1,1-dimethyl $1,1\text{-}(CH_3)_2C_5H_8$	98.19		87.5		$0.7552^{20/0}$	1.4139^{20}		B5[4], 105
5771	Cyclopentane, 1,2-dimethyl *(cis)* $1,2\text{-}(CH_3)_2C_5H_8$	98.19		99.25	−62	$0.7718^{20/4}$			B5[4], 106
5772	Cyclopentane, 1,2-dimethyl *(trans)* $1,2\text{-}(CH_3)_2C_5H_8$	98.19		91.8	−120	$0.7495^{20/4}$			B5[4], 107
5773	Cyclopentance, ethyl $C_2H_5C_5H_9$	98.19		103.5, 19.4^{30}	−138.4	$0.7665^{20/4}$	1.4198^{20}	**al, eth, ace, bz,** peth	B5[4], 104
5774	Cyclopentane, β-hydroxyethyl $C_5H_9CH_2CH_2OH$	114.19		$183\text{-}4^{770}$, $96\text{-}7^{24}$		$0.9180^{20/4}$	1.4577^{20}	eth	B6[4], 110
5775	Cyclopentane, iodo or Cyclopentyl iodide C_5H_9I	196.03		166-7, 52^{12}		$1.7096^{20/4}$	1.5447^{20}	eth, bz	B5[4], 20
5776	Cyclopentane, methyl $CH_3C_5H_9$	84.16		71.8	−142.4	$0.7486^{20/4}$	1.4097^{20}	**al, eth, ace, bz**	B5[4], 84
5777	Cyclopentane, 1-methyl-2-propyl (trans) $1\text{-}CH_3\text{-}2\text{-}C_3H_7(C_5H_8)$	126.24		152.6	−104.9	$0.7921^{20/4}$	1.4321^{20}		B5[4], 140
5778	Cyclopentane, nitro $C_5H_9NO_2$	115.13		$90\text{-}1^{40}$		$1.0776^{23/4}$	1.4538^{20}	bz	B5[4], 20
5779	Cyclopentane, phenyl $C_6H_5C_5H_9$	146.23		219, 102^{18}		$0.9462^{20/4}$	1.5309^{20}		B5[4], 1409
5780	Cyclopentane, propyl $C_3H_7C_5H_9$	112.22		131, 21.2^{10}	−117.3	$0.7763^{20/4}$	1.4266^{20}	**al, eth, ace, bz**	B5[4], 125
5781	Cyclopentane, isopropyl $(CH_3)_2CHC_5H_9$	112.22		126.4, 16.3^{10}	−111.4	$0.7765^{20/4}$	1.4258^{20}	**al, eth, ace, bz**	B5[4], 125
5782	Cyclopentane, 1,1,2-trimethyl $1,1,2\text{-}(CH_3)_3C_5H_7$	112.22		$113\text{-}4^{749}$		$0.7661^{20/D}$	1.4199^{20}		B5[4], 127
5783	Cyclopentane, 1,1,3-trimethyl $1,1,3\text{-}(CH_3)_3C_5H_7$	112.22		115-6		$0.7703^{20/4}$	1.4223^{20}		B5[4], 128
5784	Cyclopentane carboxaldehyde C_5H_9CHO	98.14		133-4		$0.9371^{20/4}$	1.1432^{20}	w, al, eth	B7[3], 43
5785	Cyclopentane carboxylic acid $C_5H_9CO_2H$	114.14		$212\text{-}3^{752}$, 104^{11}	−7	$1.0527^{20/4}$	1.4532^{20}		B9[3], 11
5786	Cyclopentanecarboxylic acid, 3-formyl-2,2,3-trimethyl,methyl ester *(d)* $C_{11}H_{18}O_3$	198.26	$[\alpha]_D$ + 51.4 (al)	$130\text{-}2^{8}$		$1.048^{20/4}$	1.4160^{22}	al	B10[3], 2868
5787	Cyclopentanecarboxylic acid, 2,oxo,ethyl ester $C_8H_{12}O_3$	156.18		218^{704}, 110^{16}		$1.0781^{21/4}$	1.4519^{20}	eth, bz	B10[3], 2808
5788	Cyclopentanecarboxylic acid, 3-oxo $C_6H_8O_3$	128.13	197^{30}	64-5					B10[3], 2812

No.	Name, Synonyms, and Formula	Mol. wt.	Color, crystalline form, specific rotation and λ_{max} (log ε)	b.p. °C	m.p. °C	Density	n_D	Solubility	Ref.
5789	1,2-Cyclopentanedicarboxylic acid *(cis)* 1,2-$C_5H_8(CO_2H)_2$	158.15	nd (w)		140			w	$B9^3$, 3807
5790	1,2-Cyclopentanedicarboxylic acid *(trans d)* $C_7H_{10}O_4$	158.15	cr (w), $[\alpha]_D$ + 87.6 (w, c=0.9)		181			w, al	$B9^3$, 3807
5791	1,2-Cyclopentanedicarboxylic acid *(trans, dl)* $C_7H_{10}O_4$	158.15	cr (w)		162-3			w, al	$B9^3$, 3807
5792	1,2-Cyclopentanedicarboxylic acid *(trans,l)* $C_7H_{10}O_4$	158.15	cr (w)$[\alpha]_D$ −85.9 (w, c=1.2)		180-1			w, al	$B9^3$, 3807
5793	1,3-Cyclopentanedicarboxylic acid *(cis)* or Norcamphoric acid 1,3-$C_5H_8(CO_2H)_2$	158.15	p^r (w)	>300d	121			al, eth, ace, chl	$B9^3$, 3808
5794	1,3-Cyclopentanedicarboxylic acid *(trans, d)* 1,3-$C_5H_8(CO_2H)_2$	158.15	cr (CCl_4), $[\alpha]_D$ + 5.9 (w, c=5)		93.5			w	$B9^2$, 519
5795	1,3-Cyclopentanedicarboxylic acid *(trans, dl)* $C_7H_{10}O_4$	158.15	pr (CCl_4)		88			w	$B9^3$, 3808
5796	1,3-Cyclopentanedicarboxylic acid *(trans, l)* $C_7H_{10}O_4$	158.15	cr (CCl_4), $[\alpha]_D$ − 5.3 (w, c=5)		93			w	$B9^2$, 519
5797	1,3-Cyclopentanedicarboxylic acid, 4,4-dimethyl or Apofenchocamphoric acid 4,4-$(CH_3)_2C_5H_6(CO_2H)_2$-1,3	186.21	mcl		144-5			w, al, eth	$B9^3$, 3849
5798	Cyclopentanol C_5H_9OH	86.13		140.8, 53^{10}	−19	$0.9478^{20/4}$	1.4530^{20}	al, eth, ace	$B6^4$, 5
5799	Cyclopentanol, 2,acetyl-1,3,3,4,4-pentamethyl (α) or Desoxymesityl oxide $C_{12}H_{22}O_2$	198.31	cr (peth-eth)		45			bz	$B8^3$, 32
5800	Cyclopentanol, 1-methyl 1-$CH_3C_5H_8OH$	100.16		$53\text{-}4^{30}$	35-7	$0.9044^{23.5/4}$	$1.4429^{23.5}$		$B6^4$, 86
5801	Cyclopentanone C_5H_8O	84.12		130.6	−51.3	$0.9487^{20/4}$	1.4366^{20}	al, eth, ace	$B7^3$, 5
5802	Cyclopentanone, 2-methyl 2-$CH_3(C_5H_7O)$	98.14		139.5, 44^{18}	−75	0.9139^{20}	1.4364^{20}	al, eth, ace	$B7^3$, 40
5803	Cyclopentanone, 3-methyl *(d)* 3-$CH_3(C_5H_7O)$	98.14	$[\alpha]^{25}{}_D$ + 143.7 (undil)	143.5^{742}, $43\text{-}4^{12}$	−58.4	$0.9140^{19/4}$	1.4340^{19}	w, al, eth, ace, aa	$B7^3$, 42
5804	Cyclopentanone, 3-methyl *(dl)* 3-$CH_3(C_5H_7O)$	98.14		144, 38^{11}		0.913^{22}	1.4329^{20}	w, al, eth, ace	$B7^3$, 43
5805	Cyclopentasiloxane, decamethyl $C_{10}H_{30}O_5Si_5$	370.77		210, 101^{20}	−38	$0.9593^{20/4}$	1.3982^{20}		$B4^3$, 1885
5806	Cyclopentene C_5H_8	68.12		44.2	−135	$0.7720^{20/4}$	1.4225^{20}	al, eth, bz, peth	$B5^4$, 209
5807	Cyclopentene, 3-chloro 3-ClC_5H_7	102.56		$25\text{-}31^{30}$		1.0577^{15}	1.4708^{26}	al, eth, chl	$B5^4$, 212
5808	Cyclopentene, perchloro C_5Cl_8	343.68	nd (al)	283, 140^{10}	41	$1.8200^{50/4}$	1.5660^{50}	al	$B5^4$, 213
5809	Cyclopentene, 1,2-deimthyl $1,2(CH_3)_2C_5H_6$	96.17		103^{757}		$0.7992^{13.5}$	$1.4447^{13.5}$		$B5^4$, 254
5810	Cyclopentene, 2,3-dimethyl-4-isopropyl 2,3-$(CH_3)_2$-4-$(CH_3)_2CHC_5H_5$	138.25		164-6		$0.8085^{22/4}$	1.4503^{22}		PCHE 2, 319
5811	Cyclopentene, 2,4-dimethyl 2,4-$(CH_3)_2C_5H_6$	96.17		93.2			1.4283^{20}		$B5^4$, 255
5812	Cyclopentene, 1-ethyl 1-$C_2H_5C_5H_7$	96.17		108	−123.3	$0.8000^{20/4}$	1.4429^{21}		$B5^4$, 252
5813	Cyclopentene, 3-ethyl 3-$C_2H_5C_5H_7$	96.17		$99\text{-}103^{758}$		$0.7874^{20/4}$	1.4303^{20}		$B5^4$, 252
5814	Cyclopentene, 1,2-dichloro,perfluoro $C_5Cl_2F_6$	244.95		90.7	−105.8	$1.6546^{20/4}$	1.3676^{20}		$B5^4$, 213
5815	Cyclopentene, 1-methyl 1-$CH_3C_5H_7$	82.15		75.5	−127.2	$0.7851^{15/4}$	1.4347^{15}		$B5^4$, 239
5816	Cyclopentene, 3-methyl 3-$CH_3C_5H_7$	82.15		69-71		$0.9705^{20/4}$	1.42476^{20}		$B5^4$, 240
5817	Cyclopentene, 1,2,3-trimethyl $1,2,3(CH_3)_3C_5H_5$	110.20		121.6		$0.8039^{15/4}$	$1.4464^{16.5}$		$B5^3$, 219
5818	2-Cyclopentene-1-one, 3-phenyl $C_{11}H_{10}O$	158.20		234.2	−23	0.9711^{20}	1.5440^{20}	al, ace, chl	$B5^3$, 1654

No.	Name, Synonyms, and Formula	Mol. wt.	Color, crystalline form, specific rotation and λ_{max} (log ε)	b.p. °C	m.p. °C	Density	n_D	Solubility	Ref.
5819	3-Cyclopentene-1-one, 3,4-bis(4-methoxyphenyl) $(CH_3OC_6H_4)_2C_5H_4O$	294.35	ye br	...	129	...	...	al, eth, ace	B8, 355
5820	1,2-Cyclopentenophenanthrene $C_{17}H_{14}$	218.30	nd (al), cr (peth)	...	135-6	...	...	al	$B5^3$, 2241
5821	1,2-Cyclopentenophenanthrene, 3-methyl $C_{18}H_{16}$	232.33	cr (aa)	...	126-7	...	...	...	$B5^3$, 2252
5822	2,3-Cyclopentenophenanthrene $C_{17}H_{14}$	218.30	pl or pr (al), nd (MeOH)	...	84-5	...	...	al	$B5^3$, 2240
5823	9,10-Cyclopentenophenanthrene $C_{17}H_{14}$	218.30	pl (xyl), nd (i-ProH)	...	155-6	...	...	al, bz	$B5^3$, 2242
5824	Cyclopropane C_3H_6	42.08	...	−32.7	−127.6	$0.720^{-79/4}$	$1.3799^{-42.5}$	al, eth, bz, peth	$B5^4$, 3
5825	Cyclopropane, acetyl or Cyclopropyl methyl ketone $CH_3COC_3H_5$	84.12	...	114^{772}	fp − 68.4	$0.8984^{20/4}$	1.4251^{20}	w, al, eth	$B7^3$, 13
5826	Cyclopropane, amino or Cyclopropyl amine $H_2NC_3H_5$	57.10	...	50-1	...	$0.8240^{20/4}$	1.4210^{20}	w, al, eth	$B12^3$, 3
5827	Cyclopropane, 1,1-dimethyl $1,1\text{-}(CH_3)_2C_3H_4$	70.13	...	20.6	−109	$0.6589^{20/4}$	1.3668^{20}	al, eth	$B5^4$, 24
5828	Cyclopropane, 1,2-dimethyl *(cis)* $1,2\text{-}(CH_3)_2C_3H_4$	70.13	...	37	...	0.6928^{20}	1.3822^{20}	al, eth	$B5^4$, 25
5829	Cyclopropane, 1,2-dimethyl *(trans, dl)* $1,2\text{-}(CH_3)_2C_3H_4$	70.13	...	29	...	0.6769^{20}	1.3713^{20}	al, eth	$B5^4$, 26
5830	Cyclopropane, 1,2-dimethyl *(trans, l)* $1,2\text{-}(CH_3)_2C_3H_4$	70.13	$[\alpha]^{22}/_D$ − 2.39 diglyme 13.9%	28-9	...	1.3699^{16}	...	al, eth	TETRA 20, 1965
5831	Cyclopropane, ethyl $C_2H_5C_3H_5$	70.13	...	34.5	...	$0.677^{20/4}$	1.379^{20}	...	$B5^4$, 23
5832	Cyclopropane, methoxy $CH_3OC_3H_5$	72.11	...	44.7	−119	$0.8100^{20/4}$	1.3802^{20}	w, al, eth, bz	$B6^3$, 3
5833	Cyclopropane, methyl $CH_3C_3H_5$	56.11	...	4-5	−117.2	$0.6912^{-20/4}$	...	al, eth	$B5^4$, 13
5834	Cyclopropane, 1-methyl-1-phenyl $1\text{-}CH_3\text{-}1\text{-}C_6H_5C_3H_4$	132.21	...	91^{50}	...	...	1.5160^{20}	ace, bz, chl	$B5^4$, 1388
5835	Cyclopropane, phenyl $C_6H_5C_3H_5$	118.18	...	173.6^{758}, 80^{37}	−31	$0.9317^{20/4}$	1.5285^{20}	eth, ace, chl	$B5^3$, 1200
5836	Cyclopropane, isopropenyl $[CH{=}CH(CH_3)]C_3H_5$	82.15	...	$69.5\text{-}70^{751}$	...	$0.7500^{20/4}$	1.4252^{20}	...	$B5^4$, 243
5837	Cyclopropane, 1,1,2-trimethyl *(dl)* $1,1,2\text{-}(CH_3)_3C_3H_3$	84.16	...	52.6	− 138.3	$0.6974^{20/4}$	1.3864^{20}	eth, bz	$B5^4$, 91
5838	Cyclopropane carbonitrile C_3H_5CN	67.09	...	135, $69\text{-}70^{88}$	...	$0.8946^{20/4}$	1.4229^{20}	eth	$B9^3$, 6
5839	Cyclopropane carboxylic acid $C_3H_5CO_2H$	86.09	...	182-4	18-9	$1.0885^{20/4}$	1.4390^{20}	al, eth	$B9^3$, 3
5840	Cyclopropane carboxylic acid, 2,2-dimethyl-3-(2-methylpropenyl) or *d-cis*-Chrysanthemamic acid $C_{10}H_{16}O_2$	168.24	pr $[\alpha]^{22}/_D$ + 83.3 (chl, c=1.6)	$95^{0.1}$	40-2	...	...	al, eth, chl	$B9^3$, 210
5841	Cyclopropane carboxylic acid, 2,2-dimethyl-3-(2-methylpropenyl)*cis,dl* $C_{10}H_{16}O_2$	168.24	pr (AcOEt), cr (peth)	...	115-6	...	...	al, eth	$B9^3$, 211
5842	Cyclopropane carboxylic acid, 2,2-dimethyl-3-(2-methylpropenyl)*(cis,l)* $C_{10}H_{16}O_2$	168.24	pr $[\alpha]^{19}/_D$ − 83.3 (chl, c=1.6)	$95^{0.1}$	41-3	...	...	al, eth	$B9^3$, 211
5843	Cyclopropane carboxylic acid, 2,2-dimethyl-3-(2-methylpropenyl)*trans,d* or *trans*-Chrysanthemucic acid *(d)* $C_{10}H_{16}O_2$	168.24	pr $[\alpha]^{20}/_D$ + 25.8 (chl, c=2.5)	245d, 135^{12}	18-21	...	...	al, eth, chl	$B9^3$, 211
5844	Cyclopropane carboxylic acid, 2,2-dimethyl-3-(2-methylpropenyl)*trans,dl* $C_{10}H_{16}O_2$	168.24	pr (AcOEt)	$145\text{-}6^{13}$	54	...	...	al, eth, chl	$B9^3$, 212
5845	Cyclopropane carboxylic acid, 2,2-dimethyl-3-(2-methylpropenyl)*trans,l* $C_{10}H_{16}O_2$	168.24	$[\alpha]^{20}/_D$ − 25.8 (chl 2.9%)	$99\text{-}100^{0.2}$	17-21	...	...	al	$B9^3$, 211
5846	Cyclopropane carboxylic acid, methyl ester $C_3H_5CO_2CH_3$	100.12	...	119	...	$0.9848^{20/4}$	1.4144^{19}	...	$B9^3$, 3
5847	1,1-Cyclopropane dicarboxylic acid $1,1\text{-}C_3H_4(CO_2H)_2$	130.10	pr or nd (chl), pr (w + 1)	...	140-1	...	...	w, eth	$B9^3$, 3795

PHYSICAL CONSTANTS OF ORGANIC COMPOUNDS (Continued)

No.	Name, Synonyms, and Formula	Mol. wt.	Color, crystalline form, specific rotation and λ_{max} (log ε)	b.p. °C	m.p. °C	Density	n_D	Solubility	Ref.
5848	1,1-Cyclopropane dicarboxylic acid, diethyl ester $1,1\text{-}C_3H_4(CO_2C_2H_5)_2$	186.21		$214\text{-}6^{748}$, $99\text{-}100^{12}$		$1.0566^{25/25}$	1.4345^{18}	al, eth	$B9^3$, 3795
5849	1,2-Cyclopropane dicarboxylic acid *(cis)* $1,2\text{-}C_3H_4(CO_2H)_2$	130.10	pr (eth or w)		139			w, al, eth	$B9^3$, 3796
5850	1,2-Cyclopropane dicarboxylic acid *(trans, d)* $1,2\text{-}C_3H_4(CO_2H)_2$	130.10	$[\alpha]^{27/}{}_D$ + 84.87 (w)		175			w, al, eth	$B9^3$, 3796
5851	1,2-Cyclopropane dicarboxylic acid *(trans, dl)* $1,2\text{-}C_3H_4(CO_2H)_2$	130.10	nd (eth), pl (ace-bz)	210^{30}	175			w, al, eth	$B9^3$, 3797
5852	1,2-Cyclopropane dicarboxylic acid *(trans, l)* $1,2\text{-}C_3H_4(CO_2H)_2$	130.10	$[\alpha]^{27/}{}_D$ − 84.40 (w)		175			w, al, eth	$B9^3$, 3797
5853	1,2-Cyclopropane dicarboxylic acid, 1-bromo $1\text{-}BrC_3H_3(CO_2H)_2\text{-}1,2$	209.00	pr (eth-chl), cr (ace-bz)		175			eth, ace	$B9^2$, 514
5854	1,2-Cyclopropane dicarboxylic acid, diethyl ester-*(cis)* $1,2\text{-}C_3H_4(CO_2C_2H_5)_2$	186.21		106-7		$1.062^{12/4}$	1.4450^{20}	al, eth	$B9^2$, 513
5855	1,2-Cyclopropane dicarboxylic acid, dimethyl ester $C_7H_{20}O_4$	158.15		219-20, 110^3		$1.1584^{16/4}$	1.4472^{14}	al, eth	$B9^2$, 513
5856	1,2,3-Cyclopropane tricarboxylic acid $C_3H_3(CO_2H)_3$	174.11	nd (w), cr (HCl)		220			w, al	$B9^3$, 4746
5857	Cyclopropyl methyl ketone $CH_3COC_3H_5$	84.12		114^{772}	fp − 68.4	$0.8984^{20/4}$	1.4251^{20}	w, al, eth	$B7^3$, 13
5858	Cyclotetrasiloxane, octamethyl $C_8H_{24}O_4Si_4$	296.62		175.8, 74^{20}	17.5	0.9561^{20}	1.3968^{20}		$B4^3$, 1885
5859	Cyclotetrasiloxane, octaphenyl $(C_6H_5)_8O_4Si_4$	793.19	nd (bz-al or aa)	$330\text{-}4^1$	200-1			bz, aa	$B16^3$, 1213
5860	Cyclotetrasiloxane, 1,3,5,7-tetramethyl-1,3,5,7-tetra-phenyl $C_{28}H_{32}O_4Si_4$	544.90	cr (aa)	$237^{1.5}$	99	$1.1183^{20/4}$	1.5461^{20}	ace	$B16^3$, 1211
5861	Cyclotetrasiloxane, 2,4-6,8-tetramethyl $C_4H_{16}O_4Si_4$	240.51		134-5	−65	$0.9912^{20/4}$	1.3870^{20}		$B4^3$, 1874
5862	Cyclotrisiloxane, hexaphenyl $(C_6H_5)_6O_3Si_3$	594.89	pl (bz-al or aa)	$290\text{-}300^1$	190	$1.23^{25/4}$		bz, aa	$B16^3$, 1213
5863	Cyclotrisiloxane, 1,3,5-triethyl,1,3,5-triphenyl $(C_2H_5)_3(C_6H_5)_3O_3Si_3$	450.76		$166^{0.025}$	177.5	$1.0952^{25/4}$	1.5402^{25}	bz	$B16^3$, 1211
5864	Cymarose or 4,5-Dihydroxy-3-methoxyhexanal or 2,6-didesoxy-3-*O*-methyl-D-allose $CH_3(CHOH)_2CH(OCH_3)CH_2CHO$	162.19	pr (eth-peth), nd (ace), $[\alpha]^{21/}{}_D$ + 53.4 (w, c=2.2)		100-2			w, al, ace	$B1^4$, 4193
5865	*o*-Cymene or 1-Methyl-2-*iso*-propyl benzene $2\text{-}(CH_3)_2CHC_6H_4CH_3$	134.22		178.1, 57.3^{10}	−71.5	$0.8766^{20/4}$	1.5006^{20}	**al, eth, ace, bz**	$B5^4$, 1057
5866	*m*-Cymene or 1-Methyl-3-*iso*-propyl benzene $3\text{-}(CH_3)_2CHC_6H_4CH_3$	134.22		175.1, 55^{10}	−63.7	$0.8610^{20/4}$	1.4930^{20}	**al, eth, ace, bz**	$B5^4$, 1058
5867	*p*-Cymene or 1-Methyl-4-*iso*-propyl benzene $4\text{-}(CH_3)_2C\,HC_6H_4CH_3$	134.22		177.1, 56.3^{10}	−67.9	$0.8573^{20/4}$	1.4909^{20}	**al, eth, ace, bz**	$B5^4$, 1060
5868	*p*-Cymene, 2-chloro or 2-Chloro-4-isopropyltoluene $2\text{-}Cl\text{-}4\text{-}i\text{-}C_3H_7C_6H_3CH_3$	168.67		216-7, 104^{20}		$1.0104^{25/4}$	1.50782^{20}	ace, bz	$B5^4$, 1062
5869	*p*-Cymene, 2-nitro $4\text{-}(CH_3)_2CH\text{-}2\text{-}NO_2\text{-}C_6H_3CH_3$	179.22		126^{10}		$1.0744^{20/4}$	1.5301^{20}	al, eth	$B5^4$, 1064
5870	Cysteic acid *(d)* or 1,2-amino-3-sulfopropanoic acid $HO_2CCH(NH_2)CH_2SO_3H$	169.15	oct cr or nd (dil al), pr or nd (w + 1), $[\alpha]^{20/}{}_D$ + 8.66 D (w)		260d			w	$B5^3$, 1713
5871	Cysteic acid *(dl)* $HO_2CCH(NH_2)CH_2SO_3H$	169.15	pr (w)		272-4d			w	$B4^3$, 1714
5872	Cysteine *(L)* or *L*-β-Mercaptoalanine $HSCH_2CH(NH_2)CO_2H$	121.15	cr (w), $[\alpha]^{30;/}{}_D$ + 9.8 (w, c=1.3)		240d			w, al, aa	$B4^3$, 1580
5873	Cystine *(D)* or *D*-Dicystine $[HO_2CCH(NH_2)CH_2S]_2$	240.29	cr (dil NH3), $[\alpha]^{20/}{}_D$ + 224 (1N HCl, c=1)		247-9				$B4^3$, 1618
5874	Cystine *(DL)* $[HO_2CCH(NH_2)CH_2S\text{-}]_2$	240.29			260				$B4^3$, 1621

No.	Name, Synonyms, and Formula	Mol. wt.	Color, crystalline form, specific rotation and λ_{max} (log ε)	b.p. °C	m.p. °C	Density	n_D	Solubility	Ref.
5875	Cystine *(L)* $[HO_2CCH(NH_2)CH_2S\text{-}]_2$	240.29	hex pl or pr (w), $[\alpha]^{20}{}_D$ − 223.4 (1N HCl, c=1)		260-1d	1.677			B4[1], 1593
5876	Cystine *(meso)* $[HO_2CCH(NH_2)CH_2S\text{-}]_2$	240.29			200-21d				B4[1], 1593
5877	Cytidine or 1-β-D-ribofuranosylcytosine $C_9H_{13}N_3O_5$	243.22	nd (dil al), $[\alpha]^{20.5}{}_D$ + 35.3 (w, c=1)		230-1d			w	B31, 24
5878	Cytidylic acid or 3-Cytosylic acid $C_9H_{14}N_3O_8P$	323.20	orh nd $[\alpha]^{20}{}_D$ + 49.4 (w, c=1)		233-4d			w, al	B31, 25
5879	Cytisine *(l)* or Sophorine, Baptitoxine $C_{11}H_{14}N_2O$	190.24	orh pr (aa or ace)	218[2]	154.5			w, al, ace, bz, chl	B24[2], 70
5880	Cytisine, *N*-methyl or Caulophylline $C_{12}H_{16}N_2O$	204.27	cr (w + 2), nd (al, bz or lig), pr (al)		137			w, al, ace, bz	B24[2], 70
5881	Cytosine or 4-amino-1,2-dihydro-1,3-diazin-2-one $C_4H_5N_3O$	111.10	mcl or tcl pl (w + 1)		320-5d			w	B24, 314
5882	Cytosine, 5-methyl $C_5H_7N_3O$	125.13	pr (w + ½)		270d			w	B24, 355

PHYSICAL CONSTANTS OF ORGANIC COMPOUNDS (Continued)

No.	Name, Synonyms, and Formula	Mol. wt.	Color, crystalline form, specific rotation and λ_{max} (log ε)	b.p. °C	m.p. °C	Density	n_D	Solubility	Ref.
5883	1,3-Decadiene $CH_3(CH_2)_5CH=CHCH=CH_2$	138.25		168-70		0.752^{20}		bz	$B1^4$, 1056
5884	3,7-Decadiene-5-yne, 4,7-dipropyl $C_2H_5CH=C(C_3H_7)C \equiv CCH=C(C_3H_7)C_2H_5$	218.38		$125\text{-}7^{18}$		$0.8131^{19/4}$	1.4890^{20}	bz	$B1^4$, 1134
5885	2,4-Decadienoic acid, ethyl ester $C_5H_{11}CH=CHCH=CHCO_2C_2H_5$	196.29		$130^{0.6}$			1.5020^{26}		$B2^4$, 17314
5886	4,6-Decadiyne or Dipentyne $C_3H_7C \equiv C\text{-}C \equiv CC_3H_7$	134.22		88^{12}		$0.8695^{19/4}$			$B1^3$, 1064
5887	Decalin *(cis)* or Decahydronaphthalene $C_{10}H_{18}$	138.25		195.6, 69.4^{10}	−43	$0.8965^{20/4}$	1.4810^{20}	al, eth, ace, bz, chl	$B5^4$, 310
5888	Decalin *(trans)* or Decahydronaphthalene $C_{10}H_{18}$	138.25		187.2, 63^{10}	−30.4	$0.8699^{20/4}$	1.4695^{20}	al, eth, ace, bz, chl	$B5^4$, 311
5889	Decalin, 1-amino *(cis)* $1\text{-}H_2NC_{10}H_{17}$	153.27		100^{12}	(i) 8, (ii) − 2				$B12^3$, 178
5890	Decalin, 1-amino *(trans)* $1\text{-}H_2NC_{10}H_{17}$	153.27		106^{16}	(i) − 18, (ii) − 1				$B12^2$, 35
5891	Decalin, 1-chloro $C_{10}H_{17}Cl$	172.70		d^{760}, $114\text{-}6^{20}$					$B5^4$, 313
5892	Decalin, 2-methylene *(trans)* $2\text{-}CH_2{:}C_{10}H_{16}$	150.26		$200\text{-}1^{756}$, 82^{10}		$0.8897^{20/4}$	1.4841^{22}		$B5^3$, 397
5893	Decalin-1,3-dione *(cis)* $C_{10}H_{14}O_2$	166.22	nd (bz, ace, dil al)		124-5			al, ace, bz	$B7^3$, 3288
5894	Decalin-1,3-dione *(trans)* $C_{10}H_{14}O_2$	166.22	nd (bz, dil al)		152-3			al	$B7^3$, 3289
5895	Decalin-2,3-dione *(cis)* $C_{10}H_{14}O_2$	166.22	rh (al, ace, lig)		88-9			al	$B7^2$, 552
5896	Decalin-2,3-dione *(trans)* $C_{10}H_{14}O_2$	166.22	nd (w), lf (dil al)		100-1			al, lig	$B7^3$, 3291
5897	2-Decalincarboxylic acid *(cis)* or Decahydro-2-naphthoic acid $2\text{-}C_{10}H_{17}CO_2H$	182.26	cr (hx)	150^{15}	81			al, eth, bz, chl	$B9^3$, 233
5898	Decanal or Capraldehyde $CH_3(CH_2)_8CHO$	156.27		208-9, 81^7	*ca* −5	$0.830^{15/4}$	1.4287^{20}	al, eth, ace	$B1^4$, 3366
5899	Decane $C_{10}H_{22}$	142.28		174.1, 57.6^{10}	−29.7	$0.7300^{20/4}$	1.4102^{20}	al, eth	$B1^4$, 464
5900	Decane, 1-amino or Decylamine $CH_3(CH_2)_8CH_2NH_2$	157.30		220.5, 95.8^{10}	17	$0.7936^{20/4}$	1.4369^{20}	**al, eth, ace, bz, chl**	$B4^4$, 783
5901	Decane, 1-bromo or Decyl bromide $CH_3(CH_2)_8CH_2Br$	221.18		240.6, 110^{10}	−29.2	$1.0702^{20/4}$	1.4557^{20}	eth, chl	$B1^4$, 470
5902	Decane, 1-bromo-10-fluoro $F(CH_2)_{10}Br$	239.17		$131\text{-}2^{11}$		$1.152^{20/4}$	1.4512^{25}		**$B1^4$, 471**
5903	Decane, 2-bromo *(dl)* $CH_3(CH_2)_7CHBrCH_3$	221.18		111^{11}		1.0512^{20}	1.4526^{25}	eth, chl	$B1^3$, 523
5904	Decane, 1-chloro or Decyl chloride $CH_3(CH_2)_8CH_2Cl$	176.73		223.4, 97^{10}	−31.3	$0.8705^{20/4}$	1.4379^{20}	eth, chl	$B1^4$, 469
5905	Decane, 1-chloro-10-fluoro $F(CH_2)_{10}Cl$	194.72		115^9		$0.957^{20/4}$	1.4333^{25}	al, eth	$B1^4$, 469
5906	Decane, 1,10-dibromide or Decamethylene dibromide $Br(CH_2)_{10}Br$	300.08	pl (al)	160^{15}d, $127\text{-}30^4$	28	1.335^{30}	1.4905^{20}	eth	$B4^4$, 1368
5907	Decane, 1,10-dibromide or Decamethyiene dibromide $Br(CH_2)_{10}Br$	300.09	pr (al)	160^{15}d, $127\text{-}30^4$	28	1.335^{30}	1.4905^{20}	eth	$B1^4$, 471
5908	Decane, 2,5-dimethyl $(CH_3)_2CH(CH_2)_2CH(CH_3)C_5H_{11}$	170.34	$[\alpha]^{25/}{}_D$ −0.05	122^{100}		$0.739^{25/4}$			**$B1^4$, 506**
5909	Decane, 1-fluoro or Decyl fluoride $CH_3(CH_2)_8CH_2F$	160.28		186.2, 69^{10}	−35	$0.8194^{20/4}$	1.4085	eth	$B1^4$, 468
5910	Decane, 1-iodo or Decyl iodide $CH_3(CH_2)_8CH_2I$	268.18		132^{15}	−16.3	$1.2546^{20/4}$	1.4858^{20}	al, eth	$B1^4$, 472
5911	Decane, 2-methyl $(CH_3)_2CH(CH_2)_7CH_3$	156.31		189.2	−48.86	0.7368^{20}	1.4154^{20}		**$B1^4$, 491**
5912	Decane, 3-methyl $C_7H_{15}CH(CH_3)C_2H_5$	156.31		188.1, $71\text{-}2^{12}$	−92.9	$0.7422^{20/4}$	1.4177^{20}		**$B1^4$, 491**
5913	Decane, 1-nitro $CH_3(CH_2)_8CH_2NO_2$	187.28		86^1			1.4387^{20}		**$B1^4$, 473** 8536
5914	Decandioic acid or Sebacic acid $HO_2C(CH_2)_8CO_2H$	202.25	lf	295^{100}, 232^{10}	134.5	$1.2705^{20/4}$	1.422^{133}	al, eth	$B2^4$, 2078
5915	1-10-Decanediol or Decamethylene glycol $HO(CH_2)_{10}OH$	174.28	nd (w)	$175\text{-}6^{14}$	72-5			al	$B1^4$, 2613

PHYSICAL CONSTANTS OF ORGANIC COMPOUNDS (Continued)

No.	Name, Synonyms, and Formula	Mol. wt.	Color, crystalline form, specific rotation and λ_{max} (log ε)	b.p. °C	m.p. °C	Density	n_D	Solubility	Ref.
5916	1-Decanethiol or Decylmercaptan $CH_3(CH_2)_8CH_2SH$	174.34		240.6, $125\text{-}7^{19}$	−26	$0.8443^{20/4}$	1.4569^{20}	al, eth	$B1^4$, 1821
5917	Decanoic acid or Capric acid $CH_3(CH_2)_8CO_2H$	172.27	nd	270, $148\text{-}50^{11}$	fr 31.5	$0.8858^{40/4}$	1.4288^{40}	al, eth, ace, bz, chl	$B2^4$, 1041
5918	Decanoic acid, 10-fluoro or 10-Fluoro capric acid $F(CH_2)_9CO_2H$	190.26		$135\text{-}8^{10}$	49			al, eth, lig	$B2^4$, 1052
5919	Decanoic acid, 2-octyl or 9-Heptadecane carboxylic acid $(C_8H_{17})_2CHCO_2H$	284.49	nd or lf (al)	$212\text{-}8^{11}$	38.5			al, eth	$B2^4$, 1254
5920	Decanoic acid, 4-oxo or γ-Ketocapric acid $C_6H_{13}COCH_2CH_2CO_2H$	186.25	(dil al)		70-1			al	$B3^4$, 1642
5921	1-Decanol or Decyl alcohol $CH_3(CH_2)_8CH_2OH$	158.28		229, $107\text{-}8^{7}$	fr 7	$0.8297^{20/4}$	1.4372^{20}	al, eth, ace, bz, chl	$B1^4$,1815
5922	1-Decanol, 10-chloro $Cl\text{-}(CH_2)_{10}OH$	192.73		$185\text{-}9^{15}$	12-3	0.9630^{25}	1.4578^{20}	al, eth	$B1^4$, 1821
5923	1-Decanol, 10-fluoro $F(CH_2)_{10}OH$	176.27		$136\text{-}7^{15}$	*ca* 22	$0.919^{20/4}$	1.4322^{25}	al, eth	$B1^4$, 1821
5924	2-Decanol *(dl)* $CH_3(CH_2)_7CHOHCH_3$	158.28		211, $110\text{-}11^{10}$	−2.4	$0.8250^{20/4}$	1.4326^{25}	al, eth, ace, bz	**$B1^4$, 1823**
5925	4-Decanol $C_3H_7CHOHC_6H_{13}$	158.28		210-1, 96^{13}	−11	$0.8262^{20/0}$	1.4320^{20}	al	**$B1^4$, 1824**
5926	2-Decanone or Methyl-*m*-octyl ketone $CH_3(CH_2)_7COCH_3$	156.27	nd	$210\ 1^{767}$, $95\text{-}7^{12}$	14	0.8248^{20}	1.4255^{20}	al, eth	$B1^4$, 3367
5927	3-Decanone or Ethyl-*M*-heptyl ketone $CH_3(CH_2)_6COCH_2CH_3$	156.27		203^{754}	1-4	$0.8251^{20/4}$	1.4252^{20}	al, eth	$B1^4$, 3368
5928	4-Decanone or Propyl hexyl ketone $C_6H_{13}COC_3H_7$	156.27	nd	206-7, $87\text{-}9^{11}$	-9	$0.824^{20/D}$	1.4240^{21}	al, eth	$B1^4$, 3368
5929	Decasiloxane, dicosamethyl $CH_3[Si(CH_3)_2O]_9Si(CH_3)_3$	755.62		183^{4}		$0.925^{20/4}$	1.3988^{20}	bz, lig	C47, 4679
5930	1,4,9-Decatriene *(trans)* $CH_2{=}CHCH_2CH{=}CH(CH_2)_3CH{=}CH_2$	136.24		164-6			1.4496^{20}		
5931	1-Decene $CH_3(CH_2)_7CH{=}CH_2$	140.27		170.5, 54.3^{10}	fr-66.3	$0.7408^{20/4}$	1.4215^{20}	al, eth	$B1^3$, 858
5932	1-Decene, 2-bromo $CH_3(CH_2)_7CBr{=}CH_2$	219.16		$115\text{-}6^{22}$		$1.0844^{20/4}$	1.4629^{20}		$B1^3$, 859
5933	1-Decene-3-yne $CH_3(CH_2)_5C{\equiv}C\text{-}CH{=}CH_2$	136.24		76^{20}		0.7873^{20}	1.4620^{20}		$B1^4$, 1105
5934	1-Decene-4-yne $CH_3(CH_2)_4C{\equiv}CCH_2CH{=}CH_2$	136.24		$73\text{-}4^{22}$		0.7880^{20}	1.445^{20}		$B1^3$, 1049
5935	2-Decene, 1-bromo $CH_3(CH_2)_6CH{=}CHCH_2Br$	219.16		121^{17}		$1.074^{18/4}$	1.4716^{18}	lig	$B1^4$, 902
5936	2-Decen-4-yne $CH_3(CH_2)_4C{\equiv}C\text{-}CH{=}CHCH_3$	136.24		55^{5}		$0.7850^{25/4}$	1.4609^{25}		$B1^3$, 1049
5937	5-Decene *(cis)* $CH_3(CH_2)_3CH{=}CH(CH_2)_3CH_3$	140.27		170^{739}, 73^{20}	−112	$0.7445^{20/4}$	1.4258^{20}	al, eth	$B1^4$, 902
5938	5-Decene *(trans)* $CH_3(CH_2)_3CH{=}CH(CH_2)_3CH_3$	140.27		170.2^{739}	−73	$0.7401^{20/4}$	1.4243^{20}	al, eth	$B1^4$, 902
5939	*n*-Decyl nitrate $C_{10}H_{21}ONO_2$	203.28		$127\text{-}8^{11}$, $88\text{-}9^{1}$		$0.951^{0/4}$		al, eth	$B1^4$, 1819
5940	Decyl nitrite $C_{10}H_{21}ONO$	187.28	yesh	$105\text{-}8^{12}$			1.4247^{20}	al, eth	$B1^4$, 1819
5941	Decyl sulfate $(C_{10}H_{21}O)_2SO_2$	378.61			37-8				$B1^4$, 1819
5942	1-Decyne or *n*-Octyl acetylene $CH_3(CH_2)_7C{\equiv}CH$	138.25		174, 57^{10}	−36	$0.7655^{20/4}$	1.4265^{20}	al, eth	$B1^4$, 1054
5943	3-Decyne $CH_3CH_2C{\equiv}C(CH_2)_5CH_3$	138.25		175-6		$0.765^{21/4}$	1.433^{21}		$B1^4$, 1055
5944	4-Decyne $C_4H_9C{\equiv}CC_3H_7$	138.25		74.5^{19}		$0.772^{17/4}$	1.436^{17}		$B1^3$, 1017
5945	4-Decyne, 3,3-dimethyl $CH_3C\ H_2C(CH_3)_2C{\equiv}C(CH_2)_4CH_3$	166.31		86^{20}		$0.7731^{20/4}$	1.4399^{20}		$B1^3$, 1026
5946	5-Decyne or Dibutyl acetylene $C_4H_9C{\equiv}CC_4H_9$	138.25		177^{751}, 788^{25}	−73	$0.7690^{20/4}$	1.4331^{20}	al, eth	$B1^4$, 1055
5947	Dehydroacetic acid $C_8H_8O_4$	168.15	nd (w), rh nd or pr (al)	270, $132\text{-}3^{5}$	109			eth, w	$B17^2$, 524

No.	Name, Synonyms, and Formula	Mol. wt.	Color, crystalline form, specific rotation and λ_{max} (log ε)	b.p. °C	m.p. °C	Density	n_D	Solubility	Ref.
5948	Dehydrochloric acid or 3,7,12-Trioxocholanic acid $C_{24}H_{34}O_5$	402.53	(ace), $[\alpha]^{20/}_{D}$ + 26 (al, c=1.4)	....	237	....	....	ace, chl	B10³, 3986
5949	Dehydro ergosterol $C_{28}H_{42}O$	394.64	lf (al + 1w), pl (al), nd (eth), $[\alpha]^{15}_{D}$ + 149.2 (chl,c=1.9	$230^{0.5}$	146	....	....	al, eth, ace, bz, chl	B6³, 3479
5950	Delphinidine chloride or 3,3′,4′,5′,7′-hexahydroxy flavinium chloride $C_{15}H_{11}ClO_7$	338.70	br pr, nd or pl (HCl)	....	>350		....	w, al	B18², 247
5951	Delphinine $C_{33}H_{45}NO_9$	599.72	orh (al), $[\alpha]^{25/}_{D}$ + 25 (al)	....	198-200d	....	....	al, eth, ace, chl	B2⁴, 2867
5952	Demissine, solamine-d $C_{50}H_{30}NO_{20}$	1018.21	nd (al), $[\alpha]$ −20 Py	....	276-9	....	....	al	B21⁴, 844
5953	Derritol $C_{21}H_{22}O_6$	370.40	ye nd (MeOH)	$220\text{-}5^{0.06}$	164	....	....	....	B18⁴, 3352
5954	Deserpidine or Canescine $C_{32}H_{38}N_2O_8$	578.66	nd or pr $[\alpha]^{24.5/}_{D}$ −137 (chl)	....	229-32	....	....	al, chl	Am77, 4335
5955	Desoxycholic acid or 3,12-Dihydroxy cholamic acid $C_{24}H_{40}O_4$	392.58	(al) $[\alpha]^{20/}_{D}$ + 57 (al)	....	176	....	....	al	B10³, 1641
5956	Desoxycorticosterone or Δ^4-Pregnene-3,20-dion-21-ol $C_{19}H_{30}O_3$	306.45	pl (eth), $[\alpha]^{20/}_{D}$ + 178 (al,c=1)	....	141-2	....	....	al, eth, ace	B8³, 2506
5957	Desthiobiotin $C_{10}H_{18}N_2O_3$	214.26	lo nd (w), $[\alpha]^{21/}_{D}$ + 10.7 (w,c=2)	....	156-8	....	....	w	J1948, 1552
5958	Desthiobiotin, methyl ester $C_{11}H_{20}N_2O_3$	228.29	cr (MeOH), $[\alpha]^{28/}_{D}$ + 2.6 (chl,c=2)	$194\text{-}7^{0.03}$	69-70	....	....	al	C44, 4934
5959	Dextrin (starch) or Amylin $(C_6H_{10}O_5)_x$	(162.14)ₓ	amor $[\alpha]'_{D}$ > + 200	....	chars	$1.0384^{20/4}$	....	w	J1925, 636
5960	Dextropimaric acid, methyl ester $C_{21}H_{32}O_2$	316.48	$[\alpha]_D$ + 60.5 (MeOH)	$140^{0.03}$	69	$1.030^{19/4}$	1.5208^{19}	al, eth	B9³, 2912
5961	Diacetone alcohol or 4-Methyl-2-pentanon-4-ol $(CH_3)_2C(OH)CH_2COCH_3$	116.16	....	164, $67\text{-}9^{19}$	−44	$0.9387^{20/4}$	1.4213^{20}	**w, al, eth**	B1⁴, 4023
5962	Diacetyl disulfide $CH_3COSSCOCH_3$	150.21	....	$105\text{-}6^{18}$	20	....	....	al, eth	B2⁴, 564
5963	Diallyl amine $(CH_2{=}CHCH_2)_2NH$	97.16	....	111	....		1.4387^{20}	al, eth	B4⁴, 1060
5964	Diallyl trisulfide $(CH_2{=}CHCH_2)_2S_3$	178.33	....	$112\text{-}22^{16}$	....	1.0845^{15}	....	**eth**	B1, 441
5965	Di(4-aminophenyl)amine $(4\text{-}H_2NC_6H_4)_2NH$	199.26	lf (w)	d	158	....	....	al, eth	B13³, 256
5966	Diaziridine, 3-ethyl-3-methyl $C_4H_{10}N_2$	86.14	....	32^{17}	....	....	1.4390^{20}	....	
5967	Diazoacetic acid, ethyl ester or Ethyl diazo acetate $N_2CH_2CO_2C_2H_5$	114.10	ye rh	$140\text{-}1^{750}{}_{d}$	−22	$1.0852^{18/4}$	1.4605^{20}	**al, eth, bz, lig**	B3⁴, 1495
5968	Diazoamino benzene or 1,3-Diphenyltriazine $C_6H_5N{=}N\text{-}NHC_6H_5$	197.24	ye lf or pr (al)	....	98	....	....	al, eth, bz, Py	B16³, 643
5969	Diazoamino benzene, 2,2′-dimethyl $(2\text{-}CH_3C_6H_4)N{=}N\text{-}NH(C_6H_4CH_3\text{-}2)$	225.29	og (al)	....	51	....	....	al, eth, lig	B16³, 653
5970	Diazoamino benzene, 2′,3-dimethyl $(3\text{-}C_6H_3C_6H_4)N{=}N\text{-}NH(C_6H_4CH_3\text{-}2)$	225.29	ye cr (lig)	....	74	....	....	ace, bz, lig	B16³, 654
5971	Diazoamino benzene, 2,4′-dimethyl $(2\text{-}CH_3C_6H_4)N{=}N\text{-}NH(C_6H_4\text{-}CH_3\text{-}4)$	225.29	ye nd (lig)	....	120	....	....	lig	B16³, 655
5972	Diazoamino benzene, 3,3′-dimethyl $(3\text{-}CH_3C_6H_4)N{=}N\text{-}NH(C_6H_4CH_3\text{-}3)$	225.29	ye nd (peth)	....	52	....	....	eth, ace, bz	B16³, 655

No.	Name, Synonyms, and Formula	Mol. wt.	Color, crystalline form, specific rotation and λ_{max} (log ε)	b.p. °C	m.p. °C	Density	n_D	Solubility	Ref.
5973	Diazoamino benzene, 3,4'-dimethyl (3-$CH_3C_6H_4$)N=NNH($C_6H_4CH_3$-4)	225.29	ye nd (lig)		97				B16^3, 656
5974	Diazoamino benzene, 4,4'-dimethyl (4-$CH_3C_6H_4$)N=N-NH($C_6H_4CH_3$-4)	225.29	red-ye nd (lig) pr (al)		118			al, lig	B16^3, 656
5975	Diazoamino benzene, 4,4'-dinitro (4-$O_2NC_6H_4$)N=NNH($C_6H_4NO_2$-4)	287.23	ye nd (al), lf (bz)		240d			eth	B16^3, 650
5976	Diazoamino benzene, 3-methyl (3-$CH_3C_6H_4$)N=N-NHC_6H_5	211.27	ye nd (lig)		86				B16^1, 654
5977	Diazoamino benzene, 4-methyl (4-$CH_3C_6H_4$)N=$NNHC_6H_5$	211.27	ye pl (lig)		86-7				B16^1, 655
5978	α,α'-Diazoamino naphthalene α-$C_{10}H_7$N=N-NH-α-$C_{10}H_7$	297.36	ye lf (al)		exp > 100				B16, 716
5979	β,β'-Diazoamino naphthalene β-$C_{10}H_7$N=N-NH-β-$C_{10}H_7$	297.36	red nd (xyl)		156				Ber19, 1282
5980	Diazoethane, 1,1,1-trifluoro CF_3CHN_2	110.04	ye	13^{752}				w, eth	B1^3, 2659
5981	1,4-Diazopine, 1-methyl-perhydro $C_6H_{14}N_2$	114.19		152-5^{742}		0.9111^{20}	1.4769^{20}		CAS 59, 7370
5982	1,2:3,4-Dibenzanthracene $C_{22}H_{14}$	278.35	nd (aa or al)		205			bz	B5^3, 2555
5983	1,2:5,6-Dibenzanthracene $C_{22}H_{14}$	278.35	pl (dil ace)		269-70			ace, bz, aa	B5^3, 2553
5984	1,2:5,6-Dibenzanthracene, 4',4'' dihydroxy $C_{22}H_{14}O_2$	310.35	og (bz)	sub	415-8				B6^3, 5914
5985	1,2:6,7-Dibenzanthracene $C_{22}H_{14}$	278.35	ye lf or nd (xyl)	sub 275$^{2-4}$	263-4				B5^3, 2552
5986	1,2:7,8-Dibenzanthracene $C_{22}H_{14}$	278.35	og lf or nd (bz)		197-8			peth	B5^3, 2553
5987	Dibenzanthrone or Violanthrone $C_{34}H_{16}O_2$	456.50	vt-bl or bk nd ($PhNO_2$)		490-5d				B7^3, 4539
5988	2,3:6,7-Dibenzocycloheptadiene-5-one $C_{23}H_{16}O$	308.38		203-4^{7}	30	1.1635^{20}	1.6324^{20}		
5989	1,2:5,6-Dibenzofluorene $C_{21}H_{14}$	266.34	pl (bz-al)	195-200$^{0.1}$	174-5			bz	B5^3, 2527
5990	1,2:7,8-Dibenzo fluorene $C_{21}H_{14}$	266.34	lf (bz), pl (aa)		234			bz	B5^3, 2527
5991	1,2:6,7-Dibenzo-9-fluorenone $C_{21}H_{12}O$	280.33	og-ye pl or pr (aa or xyl)	sub 190-200$^{0.04}$	214				B7^3, 2897
5992	Dibenzofuran or Diphenylene oxide $C_{12}H_8O$	168.19	lf or nd (al)	287	86-7	1.0886$^{99/4}$	1.6079^{99}	al, eth, ace, aa	B17^4, 585
5993	Dibenzofuran, 1-amino 1-$H_2N(C_{12}H_7O)$	183.21	br nd (dil MeOH)		85				B18^{4}7183
5994	Dibenzofuran, 2-amino 2-$H_2N(C_{12}H_7O)$	183.21	pl (dil al)		128			al, eth	B18^{4}7184
5995	Dibenzofuran, 3-amino 3-$H_2N(C_{12}H_7O)$	183.21	(dil al)		94 (99)				B18^4, 7191
5996	Dibenzofuran, 4-amino 4-$H_2N(C_{12}H_7O)$	183.21	(al)		85				B18^4, 7211
5997	Debenzofuran, 2-bromo 2-$Br(C_{12}H_7O)$	247.09	nd (al), lf (aa)	220^{40}	110			al	B17^4, 588
5998	Dibenzofuran, 3-bromo 3-$Br(C_{12}H_7O)$	247.09	lf (al)	220^{40}	120			al, eth	B17^4, 588
5999	Dibenzofuran, 4-bromo 4-$Br(C_{12}H_7O)$	247.09			67				B17^4, 589
6000	Dibenzofuran, 2,8-dibromo 2,8-$Br_2(C_{12}H_6O)$	325.98	lf (al)		199-200			al, eth, bz, aa	B17^4, 589
6001	Dibenzofuran, 3,6-dinitro 3,6-$(NO_2)_2(C_{12}H_6O)$	258.19			245				B17^4, 595
6002	Dibenzofuran, 3,8-dinitro 3,8-$(NO_2)_2(C_{12}H_6O)$	258.19			225-6			ace, bz	B17^4, 594
6003	Dibenzofuran, 1-nitro 1-$O_2N(C_{12}H_7O)$	213.19	lt ye nd (al)		120-1				B17^4, 591
6004	Dibenzofuran, 3-nitro 3-$O_2N(C_{12}H_7O)$	213.19	ye nd (aa)	180-5^{3}	181-2			aa	B17^4, 591

PHYSICAL CONSTANTS OF ORGANIC COMPOUNDS (Continued)

No.	Name, Synonyms, and Formula	Mol. wt.	Color, crystalline form, specific rotation and λ_{max} (log ε)	b.p. °C	m.p. °C	Density	n_D	Solubility	Ref.
6005	Dibenzofuran, 4-nitro 4-$O_2N(C_{12}H_7O)$	213.19	ye nd	190-205^{15}	138-9				B17^4, 591
6006	1-Dibenzofuran carboxylic acid 1-$(C_{12}H_7O)CO_2H$	212.20	nd (50% al)		232-3				B18^4, 4340
6007	2-Dibenzo furan carboxylic acid 2-$(C_{12}H_7O)CO_2H$	212.20	nd (dil al aa)		246-7 (252)			al, eth	B18^4, 4341
6008	4-Dibenzofuran carboxylic acid 4-$(C_{12}H_7O)CO_2H$	212.20	nd (al)		209-10				B18^4, 4346
6009	1,2:6,7-Dibenzophenanthrene $C_{22}H_{14}$	278.35	pa gr-ye lf (xyl)		294			bz, diox	B5^3, 2552
6010	2,3:6,7-Dibenzophenanthrene $C_{22}H_{14}$	278.35	ye-gr nd or lf (xyl)		257			al, bz	B5^3, 2552
6011	1,2:4,5-Dibenzopyrene $C_{24}H_{14}$	302.38	pa ye nd (xyl)		233-4				B5^3, 2621
6012	Dibenzothiophene or Diphenylene sulfide $C_{12}H_8S$	184.26	nd (dil al or lig)	332-3, 152-4^3	99-100			al, bz	B17^4, 601
6013	Dibenzothiophene, 2-amino 2-$H_2N(C_{12}H_7S)$	199.27	lf (dil al)		122-3				B18^4, 7186
6014	Dibenzothiophene, 3-amino 3-$H_2N(C_{12}H_7S)$	199.27	(dil al)		129-31				B18^4, 7204
6015	Dibenzothiophene, 2-bromo 2-$Br(C_{12}H_7S)$	263.15	nd (al)		125-6				B17^4, 605
6016	Dibenzothiophene, 2-bromo-monoxide 2-$Br(C_{12}H_7SO)$	279.15			171-2				B17^2, 71
6017	Dibenzothiophene, 3-bromo, dioxide 3-$Br\ (C_{12}H_7SO_2)$	295.15			224-5				B17^4, 605
6018	Dibenzothiophene, 4-bromo 4-$Br(C_{10}H_7S)$	263.15	(al)		84				B17^4, 605
6019	Dibenzothiophene, 2,8-diamino 2,8-$(NH_2)_2(C_{12}H_6S)$	214.28	nd (al)		194-6			al	B18^4, 7286
6020	Dibenzothiophene, 3,7-diamino 3,7-$(NH_2)_2(C_{12}H_6S)$	214.28	pa ye cr		169-70				B18^4, 7289
6021	Dibenzothiophene, 3,7-diamino, dioxide 3,7-$(NH_2)_2(C_{12}H_6SO_2)$	246.28	ye nd (al)		327-8				B18^4, 7289
6022	Dibenzothiophene, 2,8-dibromo 2,8-$Br_2(C_{12}H_6S)$	342.05	(aa)		229				B17^4, 605
6023	Dibenzothiophene, 2,8-dibromo, dioxide 2,8-$Br_2(C_{12}H_6SO_2)$	374.05	(aa)		361-2				B17^4, 606
6024	Dibenzothiophene, 3,7-dinitro, dioxide 3,7-$(NO_2)_2(C_{12}H_6SO_2)$	306.25	(ace)		273-5				B17^4, 608
6025	Dibenzothiophene, 2-nitro 2-$O_2N(C_{12}H_7S)$	229.25	pa ye nd		186				B17^4, 606
6026	Dibenzothiophene, 2-nitro, dioxide 2-$O_2N(C_{12}H_7SO_2)$	261.25	(ace)		257-8				B17^4, 607
6027	Dibenzothiophene, 3-nitro 3-$O_2N(C_{12}H_7S)$	229.25	pa ye (dil al)		153-4				B17^4,607
6028	Dibenzothiophene, 3-nitro, monoxide 3-$O_2N(C_{12}H_7SO)$	245.25	(al)		210				B17^4, 607
6029	2-Dibenzothiophene carboxylic acid 2-$(C_{12}H_7S)CO_2H$	228.27	(al)		255				B18^4, 4344
6030	4-Dibenzothiophene carboxylic acid 4-$(C_{12}H_7S)CO_2H$	228.27	(dil MeOH)		261-2				B18^4, 4349
6031	Dibenzoyl disulfide $C_6H_5COSSCOC_6H_5$	274.35	pr (al) sc (chl-peth)	d	136				B9^3, 1977
6032	Dibenzoylmethane $(C_6H_5CO)_2CH_2$	224.26			70-1			eth, chl	B7^3, 3838
6033	Dibenzoyl methane, (enol form) $C_6H_5COCH{=}C(OH)C_6H_5$	224.26	rh bipym (eth)	219-21^{18}	78-9			al, eth, chl	B7^3, 3838
6034	Dibenzoylmethane, keto form $(C_6H_5CO)_2CH_2$	224.26	nd or pl (eth)		81			al, eth, chl	B7^3, 3838
6035	Dibenzoylmethane, α,α-dibromo $(C_6H_5CO)_2CBr$	382.05	pr (eth)		95				B7^3, 3846
6036	Dibenzoylmethane, oxo or Diphenyl triketone $(C_6H_5CO)_2CO$	238.24	ye nd (lig)	289^{175}, 248^{60}	68-70			eth	B7^3, 4620
6037	Dibenzyl acetamide $(C_6H_5CH_2)_2CHCONH_2$	239.32	nd (al, w)	259^{18}	129			al, eth	B9^2, 476

PHYSICAL CONSTANTS OF ORGANIC COMPOUNDS (Continued)

No.	Name, Synonyms, and Formula	Mol. wt.	Color, crystalline form, specific rotation and λ_{max} (log ϵ)	b.p. °C	m.p. °C	Density	n_D	Solubility	Ref.
6038	Dibenzylacetic acid $(C_6H_5CH_2)_2CHCO_2H$	240.30	pl (peth, dil aa) nd (w)	235^{18}	89	...	...	al, eth, bz, chl, aa	$B9^3$, 3356
6039	Dibenzyl acetic acid, methyl ester or Methyldibenzyl acetate $(C_6H_5CH_2)_2CHCO_2CH_3$	254.33	nd (al)	...	42-3	...	...	...	$B9^2$, 475
6040	Dibenzyl acetonitrile $(C_6H_5CH_2)_2CHCN$	221.30	lf or pl (al)	...	89-91	...	...	...	$B9^3$, 3359
6041	Dibenzyl amine $(C_6H_5CH_2)_2NH$	197.28	...	300d, 270^{250}	-26	$1.0256^{22/4}$	1.5731^{20}	al, eth	$B12^2$, 2221
6042	Dibenzyl disulfide $C_6H_5CH_2SSCH_2C_6H_5$	246.39	lf (MeOH or al) nd (aa)	...	71-2	...	...	al, eth, bz, MeOH	$B6^4$, 2760
6043	Dibenzylphenyl amine $(C_6H_5CH_2)_2NC_6H_5$	273.38	nd or pr (al)	> 300d, 226^{10}	71-2	$1.0444^{80/4}$	1.6065^{80}	eth, bz	$B12^3$, 2225
6044	Dibenzyl sulfone or Benzylsulfene $(C_6H_5CH_2)_2SO_2$	246.32	nd (al-bz)	290d	155	...	...	ace, bz, aa	$B6^4$, 2651
6045	Dibenzyl sulfoxide $(C_6H_5CH)_2SO$	230.32	lf (al or w)	210d	134-5	...	...	al, eth	$B6^4$, 2651
6046	Diborane, methylthio $CH_3SB_2H_5$	73.75	...	53	-101.5	...	...	ace, bz	$B1^4$, 1287
6047	Dibromo acetamide, N,N-dimethyl $Br_2CHCON(CH_3)_2$	244.91	pr (w or eth)	128^{16}	79-80	...	...	...	B4, 59
6048	Dibromoacetic acid Br_2CHCO_2H	217.84	dlq cr	195^{250}	48	...	...	w, al, eth	$B2^4$, 533
6049	Dibromoacetic acid, ethyl ester or Ethyl dibromo acetate $Br_2CHCO_2C_2H_5$	245.90	...	194, 121^{74}	...	$1.9025^{20/20}$	1.5017^{13}	al, eth	$B2^4$, 533
6050	Dibromoacetic acid, methyl ester or Methyl dibromo acetate $Br_2CHCO_2CH_3$	231.87	...	182-3	...	...	...	al, eth	B2, 219
6051	Dibutyl amine $(C_4H_9)_2NH$	129.25	...	159, 48^{13}	−60	$0.7670^{20/4}$	1.4177^{20}	w, al, eth, ace, bz	$B4^4$, 550
6052	Diisobutyl amine $(i\text{-}C_4H_9)_2NH$	129.25	...	139-40	−73.5	...	1.4090^{20}	al, eth, ace, bz	$B4^4$, 630
6053	Di-*sec*-butyl amine $(Sec\text{-}C_4H_9)_2NH$	129.25	...	135^{765}	...	$0.7534^{20/4}$	1.4162^{20}	w, al	$B4^4$, 620
6054	Dibutyl disulfide $C_4H_9SSC_4H_9$	178.35	...	226, 85^{3}	...	$0.9383^{20/4}$	1.4926^{20}	**al, eth**	$B1^4$, 1562
6055	Dibutyl sulfone or Butyl sulfone $(C_4H_9)_2SO_2$	178.29	pl (w or al)	...	46	...	...	al, eth	$B1^4$, 1561
6056	Dibutyl sulfoxide $(C_4H_9)_2SO$	162.29	nd (dil al)	d	32.6	$0.8317^{23/4}$	1.4669^{20}	al, eth	$B1^4$, 1561
6057	Dichloroacetaldehyde Cl_2CHCHO	112.94	...	90-1	...	...	...	al	$B1^4$, 3140
6058	Dichloroacetaldehyde, diethyl acetal $Cl_2CHCH(OC_2H_5)_2$	187.07	...	183-4, $67\text{-}71^{12}$	...	1.1383^{14}	...	al, eth	$B1^4$, 3140
6059	Dichloroacetamide $Cl_2CHCONH_2$	127.96	mcl pr (w)	$233\text{-}4^{745}$ sub	99.4	...	...	al, eth	$B2^4$, 505
6060	Dichloroacetic acid Cl_2CHCO_2H	128.94	...	194	13.5	$1.5634^{20/4}$	1.4658^{20}	w, **al, eth,** ace	$B2^4$, 498
6061	Dichloroacetic acid, anhydride or Dichloroacetic anhydride $(Cl_2CHCO)_2O$	239.87	...	214-6d	...	1.574^{24}	...	...	$B2^4$, 503
6062	Dichloroacetic acid, butyl ester or Butyl dichloroacetate $Cl_2CHCO_2C_4H_9$	185.05	...	193-4, 102^{37}	...	$1.1820^{20/4}$	1.4420^{20}	al, eth	$B2^4$, 502
6063	Dichloroacetic acid, ethyl ester or Ethyl dichloroacetate $Cl_2CHCO_2C_2H_5$	157.00	...	155.5^{764}, 56^{10}	...	$1.2827^{20/4}$	1.4386^{20}	**al, eth,** ace	$B2^4$, 501
6064	Dichloroacetic acid, 2-hydroxy ethyl ester or (2-Hydroxyethyl) dichloroacetate $Cl_2CHCO_2CH_2CH_2OH$	173.00	...	$81\text{-}2^{0.5}$	...	$1.438^{20/4}$	1.4735^{20}	al	$B2^3$, 460
6065	Dichloroacetic acid, methyl ester or Methyl dichloroacetate $Cl_2CHCO_2CH_3$	142.97	...	142.8, 38^{10}	−51.9	$1.3774^{20/4}$	1.4429^{20}	al	$B2^4$, 501
6066	Dichloroacetic acid, propyl ester or Propyl dichloro acetate $Cl_2CHCO_2C_3H_7$	171.02	...	176	...	1.2240^{20}	1.4398^{20}	al, eth	$B2^4$, 502
6067	Dichloroacetic acid, *iso*-propyl ester or *iso*-Propyl dichloro acetate $Cl_2CHCO_2CH(CH_3)_2$	171.02	...	163-4	...	$1.2053^{20/4}$	1.4328^{20}	al, eth	$B2^4$, 502

No.	Name, Synonyms, and Formula	Mol. wt.	Color, crystalline form, specific rotation and λ_{max} (log ε)	b.p. °C	m.p. °C	Density	n_D	Solubility	Ref.
6068	Dichloro acetonitrile Cl_2CHCN	109.94		112-3		1.369^{20}	1.4391^{25}	al	$B2^{4}$, 506
6069	Dichloroacetyl chloride $Cl_2CHCOCl$	147.39		108-10		$1.5315^{16/4}$	1.4591^{20}	**eth**	$B2^{4}$, 504
6070	Di-(β-chloroethyl) methyl amine, hydrochloride $(ClCH_2CH_2)_2NCH_3HCl$	192.52	hyg nd		111-2			w, al	$B4^{4}$, 446
6071	Dichloro fluoro acetic acid Cl_2FCCO_2H	146.93			162.5				$B2^{4}$, 507
6072	Dicoumarin-(cis) $C_{18}H_{10}O_4$	290.28	lf (aa)		262				$B19^{4}$, 2106
6073	Dicoumarin-(trans) $C_{18}H_{10}O_4$	290.28	nd or pl (aa)		>275				$B19^{4}$, 2106
6074	Dictamnine $C_{12}H_9NO_2$	199.21	pr (al)		133-4			al, eth, chl	$B27^{2}$, 79
6075	Dicumarol or 4,4′-Dihydroxy-3,3′-methylene bis coumarin $C_{19}H_{12}O_6$	336.30	nd		288-92				$B19^{4}$, 2261
6076	Dicyclohexadiene $C_{12}H_{16}$	160.26		229-30, 104^{16}		$0.9950^{20/4}$	$1.5267^{20.5}$	eth, ace, bz, aa	$B5^{3}$, 1267
6077	Dicyclohexylamine $(C_6H_{11})_2NH$	181.32		255.8d, $113\text{-}5^{9}$	−0.1	$0.9123^{20/4}$	1.4842^{20}	al, eth, bz	$B12^{3}$, 19
6078	Dicyclohexyl ketone $(C_6H_{11})_2CO$	194.32		159^{20}		0.986%	1.4860^{20}	eth, ace	$B7^{3}$, 494
6079	Dicyclohexyl methane, 4,4′-diamino *(cis,cis)* $(4\text{-}H_2NC_6H_{10})_2CH_2$	210.36		141^{2}	60-2		1.5014^{27}		Am73, 641
6080	Dicyclohexyl methane, 4,4′-diamino (cis,trans) $(4\text{-}H_2NC_6H_{10})_2CH_2$	210.36	cr	$127\text{-}8^{1.2}$	36-7	$0.9608^{25/4}$	1.5046^{25}		Am73, 741
6081	Dicyclohexyl methane, 4,4′-diamino (trans,trans) $C_{13}H_{26}N_2$	210.36	cr (peth)	$130\text{-}1^{0.3}$	64-5		1.5032^{25}		Am73, 741
6082	α-Dicylopentadiene *(endo form)* $C_{10}H_{12}$	132.21		170d, $64\text{-}5^{14}$	32	$0.9302^{35/4}$	1.5050^{35}	al, eth, aa	$B5^{4}$, 1399
6083	α-Dicyclopentadiene, 3,4,5,6,7,8,8a-heptachloro or Heptachlor $C_{10}H_5Cl_7$	373.32	wh		95-6	1.57^{9}		al, eth, bz, lig	$B5^{3}$, 1236
6084	α-Dicyclopentadiene, tetrahydro or Tricyclodecane $C_{10}H_{16}$	136.24	(al or aa)	193^{769}, $86\text{-}7^{12}$	77	0.9128^{79}	1.4726^{79}	al, aa	$B5^{4}$, 467
6085	Didodecylamine or Dilaurylamine $(C_{12}H_{25})_2NH$	353.68			55-6			al, eth, bz, chl	$B4^{4}$, 801
6086	Dieldrin or Octalox $C_{12}H_8OCl_6$	380.91			175-6	1.75		ace, bz	$B17^{4}$, 526
6087	Diethanol amine $(HOCH_2CH_2)_2NH$	105.14		271, $154\text{-}5^{10}$	28	$1.0966^{20/4}$	1.4776^{20}	w, al	$B4^{4}$, 1514
6088	Diethanol amine, *N*-phenyl $(HOCH_2CH_2)_2NC_6H_5$	181.23	pl (al)	228^{15}	58			al, eth, ace, bz	$B12^{3}$, 299
6089	Diethoxyacetic acid, ethyl ester or Ethyldiethoxy acetate $(C_2H_5O)_2CHCO_2C_2H_5$	176.21		199, $83\text{-}5^{13}$		0.994^{18}	1.4089^{25}	al, eth	$B3^{4}$, 1494
6090	Diethoxy disulfide $C_2H_5OSSOC_2H_5$	154.24		$67\text{-}8^{16}$		$1.0913^{20/4}$	1.4766^{20}		$B1^{4}$, 1324
6091	Diethoxy sulfide $(C_2H_5O)_2S$	122.18		117^{733}		$0.9940^{20/4}$	1.4234^{20}	w, al, eth, ace, bz	$B1^{3}$, 1314
6092	Diethyl amine $(C_2H_5)_2NH$	73.14		56.3	−48	$0.7056^{20/4}$	1.3864^{20}	w, **al**, eth	$B4^{4}$, 313
6093	Diethyl amine, hydrochloride $(C_2H_5)_2NH.HCl$	109.60	lf (al-eth)	320-30	227-30	$1.0477^{22/4}$		w, al	$B4^{4}$, 318
6094	Diethyl amine, *N*-nitro $(C_2H_5)_2N.NO_2$	118.14		206.5^{757}, 93^{16}		1.057^{15}		**al, eth**	$B4^{3}$, 233
6095	Diethyl amine, *N*-nitroso $(C_2H_5)_2N.NO$	102.14	ye	1 76.9		$0.9422^{20/4}$	1.4386^{20}	w, al, eth	$B4^{3}$, 233
6096	Diethyl (B-bromo ethyl) amine, hydrobromide $(C_2H_5)_2NC\ H_2CH_2Br.HBr$	261.00	nd (al-eth)		209			w	$B4^{3}$, 249
6097	Diethyl (2,2-diethoxy ethyl) amine or Diethylamino acetal $(C_2H_5)_2NCH_2CH(OC_2H_5)_2$	189.30		194-5		0.863	1.4189^{20}	w, al, eth	$B4^{4}$, 1919
6098	Diethyl disulfide $C_2H_5SSC_2H_5$	122.24		154	−101.5	$0.9931^{20/4}$	1.5073^{20}	**al, eth**	$B1^{4}$, 1397
6099	Diethyl (methoxymethyl) amine $(C_2H_5)_2NCH_2OCH_3$	117.19		117^{763}				w, al, eth	$B4^{3}$, 203
6100	Diethyl methyl amine $(C_2H_5)_2NCH_3$	87.16		66		$0.703^{25/4}$	1.3879^{25}	w, al, eth	$B4^{4}$, 321

No.	Name, Synonyms, and Formula	Mol. wt.	Color, crystalline form, specific rotation and λ_{max} (log ε)	b.p. °C	m.p. °C	Density	n_D	Solubility	Ref.
6101	Diethyl selenide $(C_2H_5)_2Se$	137.08	pa ye	108	......	$1.2300^{20/4}$	1.4768^{20}	al, eth, bz	$B1^4$, 1411
6102	Diethyl sulfide $(C_2H_5)_2S$	90.18	....	92.1	-103.8	$0.8362^{20/4}$	1.4430^{20}	al, eth	$B1^4$, 1394
6103	Diethyl sulfide, 2,2-diamino or bis(2-Aminoethyl) sulfide $(H_2NCH_2CH_2)_2S$	120.21	ye	$231\text{-}3^{755}$, $118\text{-}20^{17}$	......	....	....	....	$B4^4$, 1577
6104	Diethyl sulfide, 2,2′-dichloro or bis-(2-Chloroethyl) sulfide mustard gas $(ClCH_2CH_2)_2S$	159.07	ye pr	217, 95^{10}	13-4	$1.2741^{20/4}$	1.5312^{20}	al, eth, ace, bz	$B1^4$, 1407
6105	Diethyl sulfide, 2,2′-dihydroxy or bis-(2-Hydroxyethyl) sulfide $(HOCH_2CH_2)_2S$	122.18	....	$164\text{-}6^{20}$	−10	$1.1819^{20/4}$	1.5203^{20}	**w, al**, eth, chl	$B1^4$, 2437
6106	Diethyl sulfide, 2,2′-diphenoxy or bis-(2-Phenoxyethyl) sulfide $(C_6H_5OCH_2CH_2)_2S$	274.38	nd (al)	....	42	....	....	al, eth	$B6^4$, 581
6107	Diethyl sulfone or Ethyl sulfone $(C_2H_5)_2SO_2$	122.18	rh pl	248	73-4	$1.357^{20/4}$	....	w, bz	$B1^4$, 1396
6108	Diethyl sulfoxide $(C_2H_5)_2SO$	106.18	syr	104^{25}	14	....	....	w, al, eth	$B1^4$, 1395
6109	bis (Diethylthiocarbamyl) disulfide or Antabuse $[(C_2H_5)_2NCS]_2S_2$	296.52	(al)	117^{17}	71-2	....	....	al, chl	$B4^4$, 398
6110	Diethyl trisulfide $C_2H_5S_3C_2H_5$	154.30	ye	$96\text{-}7^{26}$	......	1.114^{20}	1.5689^{11}	....	$B1^4$, 1398
6111	Diethylene glycol $HOCH_2CH_2OCH_2CH_2OH$	106.12	....	245, 133^{14}	-10.5	$1.1197^{15/4}$	1.4472^{20}	w, al, eth	$B1^4$, 2390
6112	Diethylene glycol, monobutyl ether or Butylcarbitol $C_4H_9OCH_2CH_2OCH_2CH_2OH$	162.23	....	231, 118^{12}	-68.1	$0.9553^{20/4}$	1.4321^{20}	**w,** al, eth, ace, bz	$B1^4$, 2394
6113	Diethylene glycol, monobutyl ether, acetate $C_4H_9OCH_2CH_2OCH_2CH_2O_2CCH_3$	204.27	....	245	−32	$0.985^{20/4}$	1.4262^{20}	w, **al, eth, ace**	$B2^3$, 308
6114	Diethylene glycol, diacetate $CH_3CO_2CH_2CH_2OCH_2CH_2O_2CCH_3$	190.20	....	245-51, $110\text{-}35^{16}$	......	$1.1078^{15/15}$	1.4348^{20}	w, al, eth	$B2^4$, 216
6115	Diethylene glycol, dibenzyl ether $(C_6H_5CH_2OCH_2CH_2)_2O$	286.37	....	$279\text{-}81^{24}$, 250^1	33.5	$1.1701^{15/15}$	....	w, al	$B9^3$, 535
6116	Diethylene glycol, diisobutyrate $(i\text{-}C_3H_7CO_2CH_2CH_2)_2O$	246.30	....	$178\text{-}80^{40}$	......	....	1.4282^{20}	....	$B2^3$, 652
6117	Diethylene glycol, diethyl ether or Diethyl carbitol $(C_2H_5OCH_2CH_2)_2O$	162.23	....	189	......	$0.9063^{20/4}$	1.4115^{20}	w, al, eth	$B1^4$, 2394
6118	Diethylene glycol, dimethyl ether $(CH_3OCH_2CH_2)_2O$	134.18	....	162	-68	$0.9451^{20/20}$	....	w, al, eth	$B1^4$, 2393
6119	Diethylene glycol, di-octadecanoate or Diethylene glycol disfearate $(C_{17}H_{35}CO_2CH_2CH_2)_2O$	639.07	wax	....	54-5	$0.9333^{20/4}$	....	....	$B2^4$, 1223
6120	Diethylene glycol, dioleate $(C_{17}H_{33}CO_2CH_2CH_2)_2O$	635.04	pa ye oil	....	......	$0.9310^{20/4}$	....	**al, eth**	C47, 4618
6121	Diethylene glycol, monododecanoate or Diethylene glycol monolaurate $C_{11}H_{23}CO_2CH_2CH_2OCH_2CH_2OH$	288.43	lt ye	> 270	17-8	$0.96^{25/25}$	....	**al, eth, ace,** bz	$B2^3$, 887
6122	Diethylene glycol monoethyl ether or Ethylcellosolve $C_2H_5OCH_2CH_2OCH_2CH_2OH$	134.19	hyg lig	195	......	$0.9881^{20/4}$	1.4300^{20}	**w, al**, eth, ace, **bz**	$B1^4$, 2377
6123	Diethylene glycol, monoethyl ether, acetate or Carbitol acetate $C_2H_5OCH_2CH_2OCH_2CH_2O_2CCH_3$	176.21	....	218	−25	$1.0096^{20/4}$	1.4230^{25}	**w, al, eth,** ace	$B2^3$, 308
6124	Diethylene glycol, (2-hydroxypropyl) ether $[(CH_3CH(OH)CH_2]OCH_2CH_2OCH_2CH_2OH$	164.20	....	277-9	......	$1.0789^{20/4}$	1.4498^{20}	**w**, al, bz	C52, 17693
6125	Diethylene glycol, methyl ether or Methyl carbitol $CH_3OCH_2CH_2OCH_2CH_2OH$	120.15	....	193	......	$1.0270^{20/4}$	1.4264^{20}	**w**, al, eth, **ace**	$B1^4$, 2392
6126	Diethylene triamine or bis-(2-Aminoethyl) amine $H_2NCH_2CH_2NHCH_2CH_2NH_2$	103.17	ye hyg liq	207	−39	$0.9586^{20/20}$	1.4810^{25}	**w, al**, lig	$B4^3$, 1238
6127	Difluoroacetic acid F_2CHCO_2H	96.03	....	134.2, $67\text{-}70^{20}$	−0.3	1.5255^{20}	1.3420^{20}	**w, al, eth, ace, bz**	$B2^4$, 455
6128	Difluoroacetic acid, ethyl ester or Ethyl difluoro acetate $F_2CHCO_2C_2H_5$	124.09	....	99.2^{750}	......	$1.1893^{9.8}$	....	....	$B2^4$, 455
6129	Difurfuryl amine $(OC_4H_3CH_2)_2NH$	177.20	....	$135\text{-}42^{15}$	......	$1.1045^{20/4}$	1.5168^{20}	eth	$B18^4$, 7092
6130	Di(2-furfuryl) disulfide $[(2\text{-}OC_4H_3)CH_2]_2S_2$	226.31	....	$112\text{-}3^{0.5}$	10	....	....	al	$B17^4$, 1258
6131	*m*-Digallic acid or Gallicacid-3-monogallate $C_{14}H_{10}O_9$	332.23	nd (dil al + lw)	....	268-70d	....	....	al, ace	$B10^3$, 2086

PHYSICAL CONSTANTS OF ORGANIC COMPOUNDS (Continued)

No.	Name, Synonyms, and Formula	Mol. wt.	Color, crystalline form, specific rotation and λ_{max} (log ε)	b.p. °C	m.p. °C	Density	n_D	Solubility	Ref.
6132	Digitalose or 3-*o*-Methyl-D-fucose $C_7H_{14}O_5$	178.19	nd (AcOEt) $[\alpha]^{22}_D$ + 109→ + 126 (mut)		106→ 119			w	B1[4], 4270
6133	Digitogenin or 5α,22α-Spirostan-2,3,15-triol $C_{27}H_{44}O_5$	448.64	nd (al) $[\alpha]^{20}_D$ −18 (chl, c=1.4)		280-3			chl	B19[4], 1242
6134	Digitoxigenin $C_{23}H_{34}O_4$	374.52	(dil MeOH) $[\alpha]^{20}_D$ + 119.1 (MeOH, c=1.36)		253 (256)			al	B18[4], 1468
6135	Digitoxin $C_{41}H_{64}O_{13}$	764.95	wh (chl-eth) pr (dil al) $[\alpha]^{20}_D$ + 4.8 (diox, c=1.2)		255-6			al, eth, chl	B18[4], 1478
6136	Digitoxose or 2-Deoxy-D-*altro*-Methylose $C_6H_{12}O_4$	148.16	cr (MeOH-eth) $[\alpha]^{25}_D$ + 27.9→ 43.3 (Pyr=1, mut)		112			w, ace, py	B1[4], 4191
6137	Diglycolic acid or Oxydiethanoic acid $O(CH_2CO_2H)_2$	134.09	mcl pr (w + 1)	d	148			w, al, eth	B3[4], 577
6138	Diglycolyl chloride $O(CH_2COCl)_2$	146.96		116^{15}				chl	B3, 240
6139	Diglycolic acid, hydrate $O(CH_2CO_2H)_2 \cdot H_2O$	152.10	mcl pr (w + 1)	d	148			w, al, eth	B3, 234
6140	Digoxigenin $C_{23}H_{34}O_5$	390.52	pr (AcoEt)		222			al, MeOH	B18[4], 2450
6141	Diheptyl amine $(C_7H_{15})_2NH$	213.41	nd	271^{750}, $134\text{-}6^{9}$	30			al, eth	B4[4], 736
6142	Diheptyl sulfide $(C_7H_{15})_2S$	230.45		298, 164^{20}		$0.8416^{20/4}$	1.4606^{20}	eth	B1[4], 1739
6143	Dihexyl amine $(C_6H_{13})_2NH$	185.35		192-5, $112\text{-}4^{12}$			1.4339^{20}	al, eth	B4[4], 711
6144	Dihexyl sulfide $(C_6H_{13})_2S$	202.40		230, 113.5^{4}		$0.8411^{20/4}$	1.4586^{20}		B1[4], 1706
6145	Dihydrosamidin $C_{21}H_{24}O_7$	388.42	$[\alpha]_D$ + 19(al)		117-9			al, eth	B19[4], 2788
6146	3,4-Dihydroxybenzoic acid, 5-bromo $5\text{-}Br\text{-}3,4\text{-}(HO)_2C_6H_2CO_2H$	233.02	nd (w, aa, dil al)		230			aa	B10[1], 192
6147	2,3-Dihydroxybenzoic acid, 5-bromo $5\text{-}Br\text{-}2,3\text{-}(HO)_2C_6H_2CO_2H$	233.02	pr (w + 1) nd (w)		187 (pr) 215 (nd)			al, eth, aa	B10[3], 1367
6148	2,4-Dihydroxybenzoic acid, 3-bromo $3\text{-}Br\text{-}2,4\text{-}(HO)_2C_6H_2CO_2H$	233.02	br or ye nd (w)		202			aa	B10[2], 254
6149	2,4-Dihydroxybenzoic acid, 5-bromo $5\text{-}Br\text{-}2,4\text{-}(HO)_2C_6H_2CO_2H$	233.02	micr pr (w + 1)		212			al, eth	B10[3], 1378
6150	Di-(β-Hydroxyethyl) methyl amine $(HOCH_2CH_2)_2NCH_3$	119.16		$246\text{-}8^{747}$, $123\text{-}5^{4}$		1.0377^{20}	1.4642^{20}	w, al	B4[4], 1571
6151	Diidoacetic acid I_2CHCO_2H	311.85	lf ye cr or wh nd (bz)		110			w, al, eth, bz	B2[4], 537
6152	Diisoeugenol $C_{20}H_{24}O_4$	328.41	nd(bz liq or al)		180-1			al, eth, chl	B6[3], 6765
6153	Dilactic acid $[HO_2CCH(CH_3)]_2O$	162.14	rh		112-3			w, eth	B3[3], 468
6154	Dilauryl amine $(C_{12}H_{25})_2NH$	353.68			55-6			al, eth, bz, chl	B4[4], 801
6155	Dimethisoquin hydrochloride or Quotane $C_{17}H_{24}N_2OHCl$	308.85			146			w, al	B21[4], 1335
6156	Dimethyl amine $(CH_3)_2NH$	45.08		7.4	−93	$0.6804^{0/4}$	1.350^{17}	w, al, eth	B4[4], 128

No.	Name, Synonyms, and Formula	Mol. wt.	Color, crystalline form, specific rotation and λ_{max} (log ε)	b.p. °C	m.p. °C	Density	n_D	Solubility	Ref.
6157	Dimethyl amine hydrochloride or Dimethylammonium chloride $(CH_3)_2NH.HCl$	81.55	rh, nd (al)		171			w, al, chl	B4^4, 132
6158	Dimethyl amine, hexafluoro $(CF_3)_2NH$	153.03		−6.7	−130				B3^4, 78
6159	Dimethylamine, perfluoro $(CF_3)_2NF$	171.02		−37					B3^4, 170
6160	Dimethylamine, *N*-nitro $(CH_3)_2N\text{-}NO_2$	90.08	nd(eth)	187	58	$1.1090^{72/4}$	1.4462^{72}	w, al, eth, ace, bz	B4^3, 167
6161	Dimethyl amine, *N*-nitroso or Dimethylnitrosamine $(CH_3)_2\text{-}N\text{-}NO$	74.08	ye	154		$1.0059^{20/4}$	1.4358^{20}	w, al, eth	B4^3, 166
6162	3,4-Dimethylbenzoic acid, ethyl ester $3,4\text{-}(CH_3)_2C_6H_3CO_2C_2H_5$	178.23		$127\text{-}8^{10}$			1.5144^{20}		B9^2, 353
6163	Dimethyl isobutyl amine $i\text{-}C_4H_9N(CH_3)_2$	101.19		80-1		$0.7097^{20/4}$	1.3907^{20}	w	B4^4, 627
6164	Dimethyl disulfide CH_3SSCH_3	94.19		109.7, 6.4^{10}	−84.7	$1.0625^{20/4}$	1.5289^{20}	**al, eth**	B1^4, 1281
6165	Dimethyl disulfide, hexafluoro CF_3SSCF_3	202.13		34.6	> 1			al, peth	B3^4, 278
6166	Dimethyl ethyl amine $(CH_3)_2NC_2H_5$	73.14		36-7	−36	0.675	1.3705		B4^4, 312
6167	Dimethyl glyoxime $CH_3C(:NOH)C(:NOH)CH_3$	116.12	nd (to or dil al)	sub 234	245-6			al eth	B1^4, 3647
6168	Dimethyl hexesterol or 3,4-bis(4-hydroxy-3-methylphenyl) hexane $(C_{10}H_{13}O)_2$	298.43	cr (dil al)		145				B6^3, 5541
6169	Dimethyloxonium bromide $[(CH_3)_2OH]^+Br^-$	126.98			−13			eth	B1^4, 1247
6170	Dimethyloxonium chloride $[(CH_3)_2OH]^+Cl^-$	82.53	gas	−2	−97			eth, lig, HCl	B1^4, 1247
6171	Dimethyl pentyl amine $(CH_3)_2NC_5H_{11}$	115.22		123		$0.743^{20/4}$	1.4083^{20}	**eth**	B4^4, 675
6172	Dimethyl phosphinic acid, ethyl ester $(CH_3)_2PO_2C_2H_5$	122.10		89^{15}		$1.0278^{25/4}$	1.4281^{25}	al, eth	Am73, 5466
6173	bis-(Dimethyl phosphino) amine or Amino-bis(dimethylphosphine) $[(CH_3)_2P]_2NH$	137.10		$33.5^{5.4}$ sub	39.5				Am75, 3869
6174	Dimethyl sulfone $(CH_3)_2SO_2$	94.13	pr	238	110	$1.1702^{110/0}$	1.4226	w, al, bz	B1^4, 1279
6175	Dimethyl sulfoxide $(CH_3)_2SO$	78.13		189, $85\text{-}7^{20}$	18.4	$1.1014^{20/4}$	1.4770^{20}	w, al, eth, ace	B1^4, 1277
6176	bis-(Dimethylthiocarbamyl) disulfide or Arasan $[(CH_3)_2NCS]_2S_2$	240.41	wh or ye mcl (chl-al)	129^{20}	155.6			chl	B4^4, 242
6177	Di-α-Naphthyl amine $(\alpha\text{-}C_{10}H_7)_2NH$	269.35		$310\text{-}5^{15}$	115			al, eth, ace, bz, chl	B12^3, 2859
6178	Diβ-Naphthyl amine $(\beta\text{-}C_{10}H_7)_2NH$	269.35	lf (bz)	471	172.2			eth	B12^3, 3003
6179	α,α′-Dinaphthyl disulfide $(\alpha\text{-}C_{10}H_7)_2S_2$	318.45	pl (al) nd (lig)		91	1.144^{20}		eth	B6^3, 2947
6180	β,β′-Dinaphthyl disulfide $(\beta\text{-}C_{10}H_7)_2S_2$	318.45	nd		139-40	$0.8409^{20/4}$	1.4555^{20}	al, eth	B6^3, 3013
6181	α,α′-Dinaphthyl ketone $\alpha\text{-}C_{10}H_7CO\text{-}\alpha\text{-}C_{10}H_7$	282.34	wh nd yesh pr (aa)		104			al, eth, bz	B7^2, 503
6182	α,β′-Dinaphthyl ketone $\alpha\text{-}C_{10}H_7CO\text{-}\beta\text{-}C_{10}H_7$	282.34	nd (al bz-lig)	$235^{0.06}$	136-7			chl	B6^3, 2869
6183	β,β′-Dinaphthyl ketone $\beta\text{-}C_{10}H_7CO\text{-}\beta\text{-}C_{10}H_7$	282.34	(i) nd (eth), (ii) lf (chl-eth)		(i) 125.5, (ii) 164.5			chl	B7^2, 504
6184	Di(α-naphthyl)methane $(\alpha\text{-}C_{10}H_7)_2CH_2$	268.36	pr or nd (al)	>360, 270^{14}	109			eth, bz, chl	B5^3, 2480
6185	Di(β-Naphthyl)methane $(\beta\text{-}C_{10}H_7)_2CH_2$	268.36	nd (al eth)		93			bz	B5^1, 360
6186	2,4-Dinitrobenzoic acid $2,4\text{-}(NO_2)_2C_6H_3CO_2H$	212.12	nd(w)		183			bz	B9^3, 1776
6187	2,5-Dinitrobenzoic acid $2,5\text{-}(NO_2)_2C_6H_3CO_2H$	212.12	pr (w)		177			al, eth	B9^3, 1778

No.	Name, Synonyms, and Formula	Mol. wt.	Color, crystalline form, specific rotation and λ_{max} (log ε)	b.p. °C	m.p. °C	Density	n_D	Solubility	Ref.
6188	2,6-Dinitrobenzoic acid $2,6\text{-}(NO_2)_2C_6H_3CO_2H$	212.12	nd(w)		202-3			al, eth	B9[3], 1778
6189	3,4-Dinitrobenzoic acid $3,4\text{-}(NO_2)_2C_6H_3CO_2H$	212.12	nd (w)		165			al, eth	B9[3], 1778
6190	3,5-Dinitrobenzoic acid $3,5\text{-}(NO_2)_2C_6H_3CO_2H$	212.12	mcl pr (al)		205			al, aa	B9[3], 1779
6191	3,5-Dinitrobenzoic acid, benzyl ester $3,5\text{-}(NO_2)_2C_6H_3CO_2CH_2C_6H_5$	302.24	nd (lig)		112				B9[3], 1848
6192	3,5-Dinitrobenzoic acid, 2-bromo $3,5\text{-}(NO_2)_2\text{-}2\text{-}BrC_6H_3CO_2H$	291.01	ye nd (w)		213			al, bz, aa, lig	B9[3], 1954
6193	3,5-Dinitrobenzoic acid, butyl ester $3,5\text{-}(NO_2)_2C_6H_3CO_2C_4H_9$	268.23	mcl nd (al)		62.5		1.488		B9[3], 1782
6194	3,5-Dinitrobenzoic acid, isobutyl ester $3,5\text{-}(NO_2)_2C_6H_3CO_2\text{-i-}C_4H_9$	268.23	mcl pl, nd (al)		87-8			al	B9[3], 1783
6195	3,5-Dinitrobenzoic acid, ethyl ester $3,5\text{-}(NO_2)_2C_6H_3CO_2C_2H_5$	240.17	nd (al)		92.9	1.295^{111}	1.560	al	B9[3], 1781
6196	3,5-Dintrobenzoic acid, furfuryl ester $3,5\text{-}(NO_2)_2C_6H_3CO_2CH_2(C_4H_3O)$	292.20	(bz-py)		78-81				B17[4], 1249
6197	3,5-Dinitrobenzoic acid, methyl ester $3,5\text{-}(NO_2)_2C_6H_3CO_2CH_3$	226.15	nd(w)		112			al, w	B9[3], 1781
6198	3,5-Dinitrobenzoic acid, tetrahydrofurfuryl ester $3,5\text{-}(NO_2)_2C_6H_3CO_2CH_2(C_4H_7O)$	296.24	nd (al)		83-4			al	B17[4], 1106
6199	3,5-Dinitrobenzoic acid, pentyl ester $3,5\text{-}(NO_2)_2C_6H_3CO_2C_5H_{11}$	282.25			46.4			al	B9[2], 281
6200	3,5-Dinitrobenzoic acid, phenyl ester $3,5\text{-}(NO_2)_2C_6H_3CO_2C_6H_5$	288.22	rods (al)		145-6			al, bz	B9[3], 1846
6201	3,5-Dinitrobenzoic acid, propyl ester $3,5\text{-}(NO_2)_2C_6H_3CO_2C_3H_7$	254.20	mcl pl (al)		73			al	B9[3], 1781
6202	3,5-Dintrobenzoic acid, isopropyl ester $3,5\text{-}(NO_2)_2C_6H_3CO_2\text{-i-}C_3H_7$	254.20	nd (al)		122			al	B9[3], 1782
6203	3,5-Dinitrobenzoyl chloride $3,5\text{-}(NO_2)_2C_6H_3COCl$	230.56	ye nd bz	196^{12}	74			eth	B9[3], 1936
6204	Dioctadecyl amine or Distearyl amine $(C_{18}H_{37})_2NH$	522.00			73-4			chl	B4[4], 829
6205	Dioctyl amine $(C_8H_{17})_2NH$	241.46	nd	297-8, 175^{14}	35.6	$0.7968^{26/4}$	1.4415^{26}	al, eth	B4[4], 753
6206	Di-2-Octyl amine $[C_6H_{13}CH(CH_3)]_2NH$	241.46		281.5^{739}		$0.7948^{20/4}$		al, eth	B4[4], 765
6207	1,3-Dioxane or *m*-Dioxane $C_4H_8O_2$	88.11		105^{755}	−42	$1.0342^{20/4}$	1.4165^{20}	**w, al, eth, ace, bz**	B19[4], 8
6208	1,3-Dioxane, 2,4-dimethyl $C_6H_{12}O_2$	116.16		115-8		$0.9392^{20/4}$	1.4136^{20}		B19[4], 61
6209	1,3-Dioxane, 5-ethyl-4-propyl $C_9H_{18}O_2$	158.24		196		$0.9305^{20/4}$	1.4370^{20}		C52, 2795
6210	1,3-Dioxane, 5-hydroxy-2-methyl $C_5H_{10}O_3$	118.13		176		$1.0705^{17/4}$	1.4375^{17}	**w**	B19[4], 624
6211	1,3-Dioxane, 4-methyl $C_5H_{10}O_2$	102.13		114		0.9758^{20}	1.4159^{20}		B19[4], 49
6212	1,3-Dioxane, 4-methyl-4-phenyl $C_{11}H_{14}O_2$	178.23		256, 102^4	35-40	1.0864^{20}	1.5240^{20}		B19[4], 233
6213	1,3-Dioxane, 2-phenyl $C_{10}H_{12}O_2$	164.20	nd (peth)	252-4, $98\text{-}9^6$	41			al, eth	B19[4] 215
6214	1,3-Dioxane, 4-phenyl $C_{10}H_{12}O_2$	164.20		245, $128\text{-}30^{10}$		$1.1038^{20/4}$	1.5306^{18}		B19[4], 218
6215	1,4-Dioxane or Diethylene dioxide $C_4H_8O_2$	88.11		101^{750}	11.8	$1.0337^{20/4}$	1.4224^{20}	**w, al, eth, ace, bz**	B19[4], 9
6216	1,4-Dioxane, -2,3-dichloro $C_4H_6Cl_2O_2$	157.00		$80\text{-}2^{10}$	30	$1.468^{20/4}$	1.4928^{20}	eth, ace, bz	B19[4], 30
6217	1,4-Dioxane, heptachloro $C_4HCl_7O_2$	329.22		$123\text{-}8^8$	54-6			eth, ace, bz, lig	B19[4], 32
6218	1,4-Dioxene $C_4H_6O_2$	86.09		94.1		$1.0836^{20/4}$	1.4372^{20}	eth, ace, bz	B19[4], 108
6219	1,4-Dioxine $C_4H_4O_2$	84.07		74.6^{748}		$1.115^{20/4}$	1.4350^{20}	eth, ace, bz	B19[4], 154
6220	1,3-Dioxolane or Glycol methylene ether $C_3H_6O_2$	74.08		78^{765}	−95	$1.0600^{20/4}$	1.3974^{20}	w, al, eth, ace	B19[4], 5

No.	Name, Synonyms, and Formula	Mol. wt.	Color, crystalline form, specific rotation and λ_{max} (log ϵ)	b.p. °C	m.p. °C	Density	n_D	Solubility	Ref.
6221	1,3-Dioxolane, 4-(hydroxymethyl)-2-methyl or Glycerol-ethylidine ether $C_5H_{10}O_3$	118.13	....	187, 68-70[1]	......	$1.1243^{17/4}$	1.4413^{17}	al	$B19^4$, 631
6222	1,3-Dioxolane, 2-methyl or Glycol ethylidine ether $C_4H_8O_2$	88.11	81-2	....	......	$0.9811^{20/4}$	1.4035^{17}	w, al, eth	$B19^4$, 42
6223	1,3-Dioxlane-4-carboxaldehyde, 2,2-dimethyl $C_6H_{10}O_3$	130.14	....	74^{50}	......	....	1.4189^{25}	w	$B19^4$, 1579
6224	1,3-Dioxolan-2-one or 1,2-Ethanediol carbonate $C_3H_4O_3$	88.06	mcl pl (al)	248^{760}	39-40	$1.3214^{39/4}$	1.4158^{50}	w, al, eth, bz, chl	$B19^4$, 1556
6225	1,3-Dioxolane-2-one, 4-methyl or Propylene carbonate $C_4H_6O_3$	102.09	....	24.2	−48.8	$1.2069^{20/20}$	1.4189^{20}	w, al, eth, ace, bz	$B19^4$, 1564
6226	Dipentyl amine $(C_5H_{11})_2NH$	157.30	....	202-3, 91-3[14]	......	$0.7771^{20?4}$	1.4272^{20}	al, %eth,] ace	$B4^4$, 676
6227	Dipentylamine, 2,2′-dimethyl-2,2′-dihydroxy $[C_3H_7C(OH)(CH_3)CH_2]_2NH$	217.35	....	165-70[15]	......	$0.9264^{20?4}$	1.4585^{20}	al, %eth,] ace	C53, 16829
6228	Diisopentyl amine $(i\text{-}C_5H_{11})_2NH$	157.30	....	188	−44	$0.7672^{21?4}$	1.4235^{20}	al, %eth]	$B4^4$, 699
6229	Dipentyl disulfide or [m] -Amyldi sulfide $C_5H_{11}SSC_5H_{11}$	206.40	....	119[7]	......	$0.9221^{20?4}$	1.4889^{20}	....	$B1^4$, 1654
6230	Di-[iso] -Pentyl disulfide or Isoamylsulfide $([i]\text{-}C_5H_{11})_2S_2$	206.40	....	250	......	$0.9192^{20?4}$	1.4864^{20}	....	$B1^4$, 1689
6231	Diphenadione or 2-Diphenyl acetyl-1,3-indanedione $C_{23}H_{16}O_3$	340.38	pa ye mcl (al)	....	146-7	....	1.670	ace, aa	C49, 3264
6232	Diphenyl acetamide $(C_6H_5)_2CHCONH_2$	211.26	pl (al)	....	167-8	....	....	al	$B9^1$, 3301
6232a	Diphenyl acetamide, α-chloro $(C_6H_5)_2CClCONH_2$	245.71	cr (to)	....	115	....	....	al, eth, bz, chl	....
6233	Diphenyl acetic acid $(C_6H_5)_2CHCO_2H$	212.25	nd (w) lf (al)	194[25] sub	148	$1.258^{15?15}$	....	al, eth, chl	$B9^1$, 3290
6234	Diphenyl acetic acid-anhydride $[(C_6H_5)_2CHCO]_2O$	406.48	nd (eth,)	220-5[15]	98	....	....	bz, chl	$B9^1$, 281
6235	Diphenyl acetic acid, α-bromo $(C_6H_5)_2C(Br)CO_2H$	291.14	(chl-peth)	....	133-4	....	....	chl, to	$B9^2$, 471
6236	Diphenyl acetic acid, α-chloro $(C_6H_5)_2C(Cl)CO_2H$	246.69	pl (bz-liq)	....	118-9d	....	....	al, eth, ace, bz	$B9^1$, 3307
6237	Diphenyl acetic acid, α-chloro, ethyl ester or Ethyl-diphenyl chloroacetate $(C_6H_5)_2C(Cl)CO_2C_2H_5$	274.75	pl (chl) cr (al)	185[14]	43-4	....	....	al, eth	$B9^1$, 3307
6238	Diphenl acetic acid, ethyl ester or Ethyldiphenyl acetate $(C_6H_5)_2CHCO_2C_2H_5$	240.30	nd (al) rh (AcOEt)	195[25]	59	....	....	al, eth	$B9^1$, 3291
6239	Diphenyl acetic acid, α-hydroxy or Benzilic acid $(C_6H_5)_2C(OH)CO_2H$	228.25	mcl nd (w)	d180	151	....	....	al, eth	$B10^1$, 1168
6240	Diphenyl acetic acid, α-hydroxy, ethyl ester $(C_6H_5)_2C(OH)CO_2C_2H_5$	256.30	pr or nd	201[21]	34	....	1.5620^{20}	al, eth	$B10^1$, 1171
6241	Diphenyl acetic acid, α-hydroxy, methyl ester or Methyl benzilate $(C_6H_5)_2C(OH)CO_2CH_3$	242.27	mcl or tcl cr (al)	187[13]	75	....	....	al, eth, aa	$B10^1$, 1170
6242	Diphenyl acetic acid, methyl ester or Methyl diphenyl acetate $(C_6H_5)_2CHCO_2CH_3$	226.27	mcl pl (AcOEt) lf (dil al)	....	60	....	....	al, eth	$B9^1$, 3291
6243	Diphenyl acetic acid, 2,2′,4,4′-tetranitro,ethyl ester $[2,4\text{-}(NO_2)C_6H_3]_2CHCO_2C_2H_5$	420.29	lf (al) nd (bz, aa)	....	154	....	....	....	$B9^1$, 3315
6244	Diphenyl acetic acid, 2,2′,4,4′-tetranitro,methyl ester $[2,4\text{-}(NO_2)C_6H_3]_2CHCO_2CH_3$	406.27	lf (chl-MeOH)	....	159	....	....	chl	B9, 675
6245	Diphenyl acetonitrile $(C_6H_5)_2CHCN$	193.25	pr (eth) lf (dil al)	181-4[12]	72-3	....	....	al, eth	$B9^1$, 3304
6246	Diphenyl acetyl bromide, α-bromo $(C_6H_5)_2CBrCOBr$	354.04	nd (lig)	....	65-6	....	....	al, eth, bz, chl	$B9^1$, 283
6247	Diphenyl acetyl chloride $(C_6H_5)_2CHCOCl$	230.69	pl (lig)	178[15]	56-7	....	....	....	$B9^3$, 3300
6248	Diphenyl acetyl chloride, α-chloro $(C_6H_5)_2CClCOCl$	265.14	cr (lig)	180[14]	50-1	....	....	....	$B9^1$, 3308
6249	Diphenyl amine $(C_6H_5)_2NH$	169.23	mcl lf (dil al)	302, 179[22]	54-5	$1.160^{22?20}$	....	al, eth, ace, bz	$B12^1$, 284
6250	Diphenyl amine, 4,4′-dimethylamino $[4\text{-}(CH_3)_2NC_6H_4]_2NH$	255.36	tetr pl (CS_2)	....	119	....	....	eth	$B13^2$, 56

PHYSICAL CONSTANTS OF ORGANIC COMPOUNDS (Continued)

No.	Name, Synonyms, and Formula	Mol. wt.	Color, crystalline form, specific rotation and λ_{max} (log ε)	b.p. °C	m.p. °C	Density	n_D	Solubility	Ref.
6251	Diphenyl amine, 2,2′-dinitro $(2-NO_2C_6H_4)_2NH$	259.23	lf (al ace) ye cr (al aa)	.....	169	.....	.....	al, ace, aa	B12[3], 1518
6252	Diphenyl amine, 2,4-dinitro $2,4-(NO_2)C_6H_3NHC_6H_4$	259.23	ye red nd (al)	.....	157	.....	.....	al, ace, chl, py	B12[3], 1683
6253	Diphenyl amine, 2,4′-dinitro $2-NO_2C_6H_4NH(C_6H_4NO_2-4)$	259.23	red nd (aa)	.....	222-3	.....	.....	chl, to	B12[3], 1587
6254	Diphenyl amine, 2,6-dinitro $2,6-(NO_2)_2C_6H_3NHC_6H_5$	259.23	og lf (al, aa)	.....	107-8	.....	.....	al, aa	B12[3], 1705
6255	Diphenyl amine, 3,4′-dinitro $3-NO_2C_6H_4NH(C_6H_4NO_2-4)$	249.23	pa ye (chl aq py)	.....	217	.....	.....	ace, chl	B12[3], 1587
6256	Diphenyl amine, 4,4′-dinitro $(4-NO_2C_6H_4)_2NH$	259.23	ye nd (al)	.....	216	.....	.....	ace, aa	B12[3], 1587
6257	Diphenyl amine, 2,4-dinitro-4′-hydroxy $2,4-(NO_2)_2C_6H_3NH(C_6H_4OH-4)$	275.22	red lf	.....	195-6	.....	.....	.....	B13[3],1019
6258	Diphenyl amine, 2,4-dinitro-5-hydroxy $2,4-(NO_2)-5-HOC_6H_2NHC_6H_5$	275.22	dk ye nd (al)	.....	139	.....	.....	al	B13[1], 138
6259	Diphenyl amine, 2,6-dinitro-2′-hydroxy $2,6-(NO_2)C_6H_3NH(C_6H_4OH-2)$	275.22	red vt nd (al)	.....	191	.....	.....	al, eth, bz	B13, 365
6260	Diphenyl amine, 2,6-dinitro-3-hydroxy $2,6-(NO_2)_2-3-HOC_6H_2NHC_6H_5$	275.22	ye br nd (MeOH)	.....	124-5	.....	.....	al	B13[2], 216
6261	Diphenyl amine, 2,2′,4,4′,6,6′-hexanitro or Dipicryl amine $[2,4,6-(NO_2)_3C_6H_2]_2NH$	439.22	pa ye pr (aa)	.....	244d	.....	.....	py	B12[3], 1734
6262	Diphenyl amine, hydrobromide $(C_6H_5)_2NH.HBr$	250.14	pl (dil al)	.....	230d	.....	.....	w	B12, 180
6263	Diphenyl amine, *N*-nitroso $(C_6H_5)_2N-NO$	198.23	ye pl (lig)	.....	66.5	.....	.....	al, bz	B12[3], 1120
6264	Diphenyl amine, 2,2′,4,4′-tetrabromo $(2,4-Br_2C_6H_3)_2NH$	484.81	nd (chl or bz)	.....	187.5	.....	.....	bz, chl	B12[3], 1472
6265	Diphenyl amine, 2,2′,4,4′-tetranitro $[2,4-(NO_2)_2C_6H_3]_2NH$	349.22	ye nd or cr (aa)	.....	201	.....	.....	py	B12[3], 1684
6266	Diphenyl diselenide $(C_6H_5)_2Se_2$	312.13	ye nd	.....	63-4	$1.557^{80/4}$	.....	al, eth, xyl	B6[3], 1110
6267	Diphenyl disulfide $(C_6H_5)_2S_2$	218.33	nd (al) or rh	310, 192^{15}	61-2	$1.353^{20/4}$	.....	al, eth, bz	B6[3], 1027
6268	1,2-Diphenylethane or Bibenzyl $C_6H_5CH_2CH_2C_6H_5$	182.27	nd (al)	285, $95\text{-}6^{1}$	52.2	$0.9583^{60/4}$	1.5478^{60}	al, eth	B5[4], 1868
6269	1,2-Diphenylethane, 4,4′-diamino $(4-H_2NC_6H_4)CH_2CH_2(C_6H_4NH_2-4)$	212.29	pl (w)	sub	135-6 sub	.....	.....	al	B13[3], 470
6270	1,2-Diphenylethane, 2,2′-dibromo $(2-BrC_6H_4)CH_2CH_2(C_6H_4Br-2)$	340.06	pl (al)	$138\text{-}40^{0.012}$	84.5	.....	.....	al	B5[4], 1875
6271	1,2-Diphenylethane, 4,4′-dibromo $(4-BrC_6H_4)CH_2CH_2(C_6H_4Br-4)$	340.06	pr (al)	ca 198^{10}	115	.....	.....	.....	B5[4], 1875
6272	1,2-Diphenyl ethane, α,α′-dinitro*(dl)* $C_6H_5CH(NO_2)CH(NO_2)C_6H_5$	272.26	(al)pr(aa)	.....	154-5	.....	.....	al, eth, ace, bz, chl	B5[4], 1878
6273	1,2-Diphenyl ethane, α,α′-dinitro-*(meso)* $C_6H_5CH(NO_2)CH(NO_2)C_6H_5$	272.26	nd (aa)	.....	235-6	.....	.....	ace	B5[4], 1878
6274	1,2-Diphenyl ethane, 2,2′-dinitro $(2-O_2NC_6H_4)CH_2CH_2(C_6H_4NO_2-2)$	272.26	pr (aa)	.....	127	.....	.....	eth, bz, aa	B5[4], 1878
6275	1,2-Diphenyl ethane, 4,4′-dinitro $(4-O_2NC_6H_4)CH_2CH_2(C_6H_4NO_2-4)$	272.26	yesh nd (al or bz)	.....	180.5	.....	.....	.....	B5[4],1878
6276	1,2-Diphenyl ethane, α-hydroxy-*(l)* $C_6H_5CH(OH)CH_2C_6H_5$	198.26	nd (eth-peth), $[\alpha]^{20}/_D$ -9.4 (w, c=10)	.....	67	$1.0358^{20/4}$	.....	al, eth	B6[2], 637
6277	1,2-Diphenyl ethane, α-hydroxy-*(dl)* $C_6H_5CH(OH)CH_2C_6H_5$	198.26	nd (bz-peth)	177^{15}	69	.....	.....	al, eth	B6[3], 3390
6278	1,2-Diphenyl ethane, α-hydroxy-*(d)* $C_6H_5CH(OH)CH_2C_6H_5$	198.26	nd (eth-peth or dil al), $[\alpha]^{25}/_D$ + 53 (al)	$167\text{-}70^{10}$	67-8	$1.0358^{20/4}$	.....	al, eth	B6[3], 3390
6279	Diphenyl ether, 2,3′-dimethoxy $(2-CH_3OC_6H_4)O(C_6H_4OCH_3-3)$	230.26	pr(bz-peth)	326-9, 152^{2}	54	.....	.....	al, eth, bz	B6[3], 4318
6280	Diphenyl glyoxal or Benzil $C_6H_5COCOC_6H_5$	210.23	ye pr (al)	346-8d, 188^{12}	95-6	$1.084^{102/4}$	.....	al, eth, ace, bz	B7[3], 3804
6281	Diphenyl glyoxime (anti) or Benzildioxime (anti) $C_6H_5C(=NOH)C(=NOH)C_6H_5$	240.26	lf (al or ace)	.....	238d	.....	.....	.....	B7[3], 3816

No.	Name, Synonyms, and Formula	Mol. wt.	Color, crystalline form, specific rotation and λ_{max} (log ε)	b.p. °C	m.p. °C	Density	n_D	Solubility	Ref.
6282	Diphenyl methane or Ditan $(C_6H_5)_2CH_2$	168.24	pr nd	264.3, 125.5^{10}	25.3	$1.0060^{20/4}$	1.5753^{20}	al, eth, chl	$B5^4$, 1841
6283	Diphenyl methane, α-amino $(C_6H_5)_2CHNH_2$	183.25	hex pl	304^{763}, 176^{23}	34	$1.0675^{20/0}$	1.5963	bz	$B12^3$, 3221
6284	Diphenyl methane, 3-amino $(3\text{-}H_2NC_6H_4)CH_2C_6H_5$	183.25	cr(lig)	...	46	...	...	lig	$B12^3$, 3215
6285	Diphenyl methane, 4-amino $(4\text{-}H_2NC_6H_4)CH_2C_6H_5$	183.25	mcl (lig)	300	34-5	1.038^{55}	...	al, eth, lig	$B12^3$, 3215
6286	Diphenyl methane, α-bromo $(C_6H_5)_2CHBr$	247.13	tcl (peth)	193^{26}, $111^{0.3}$	45	...	...	al, bz	$B5^4$, 1850
6287	Diphenyl methane, α-chloro or Benzydryl chloride $(C_6H_5)_2CHCl$	202.68	nd	173^{19}	20.5	$1.1398^{20/4}$	1.5959^{20}	...	$B5^4$, 1847
6288	Diphenyl methane, 4-chloro $(4\text{-}ClC_6H_4)CH_2C_6H_5$	202.68	...	298^{742}, $147\text{-}8^{8}$	7.5	$1.1247^{20/4}$	...	ace	$B5^4$, 1847
6289	Diphenyl methane, 4,4′-diamino or 4,4′-Diamino ditan $(4\text{-}H_2NC_6H_4)_2CH_2$	198.27	pl or nd (w) pl (bz)	$398\text{-}9^{768}$, 257^{18}	92-3	...	...	al, eth, bz	$B13^3$, 454
6290	Diphenyl methane, diazo $(C_6H_5)_2CN_2$	194.24	bl-red nd (peth)	exp	30-2	...	...	al, eth, ace	$B7^3$, 2068
6291	Diphenyl methane, α,α-dichloro or Benzophenone dichloride $(C_6H_5)_2CCl_2$	237.13	...	305d, 190^{21}	...	1.235^{18}	...	eth, bz	$B5^4$, 1848
6292	Diphenyl methane, 4,4′-dichloro $(4\text{-}ClC_6H_4)_2CH_2$	237.13	...	$186\text{-}90^{18}$	55-6	1.365^{17}	...	al	$B5^4$, 1848
6293	Diphenyl methane, 5,5′-dichloro-2,2′-dihydroxy or Dichlorophene $(5\text{-}Cl\text{-}2\text{-}HOC_6H_3)_2CH_2$	269.13	cr (bz peth)	...	177-8	...	...	al, ace	$B6^3$, 5406
6294	Diphenyl methane, 2,4′-dihydroxy $(2\text{-}HOC_6H_4)CH_2(C_6H_4OH\text{-}4)$	200.24	nd (dil al, bz or w)	...	119-20	...	...	al, eth	$B6^3$, 5409
6295	Diphenyl methane, 3,3′-dihydroxy or *m,m′*-Methylene diphenol $[3\text{-}HOC_6H_4]_2CH_2$	200.24	nd (dil aa)	$230\text{-}40^{3}$	102-3	...	...	al, eth, aa	$B6^3$, 5411
6296	Diphenyl methane, 4,4′-dihydroxy or *p,p′*-Methylene diphenol $(4\text{-}HOC_6H_4)_2CH_2$	200.24	lf or nd (w)	sub	162-3	...	...	al, eth, chl	$B6^3$, 5412
6297	Diphenyl methane, 4,4′-dimethylamino $[4\text{-}(CH_3)_2NC_6H_4]_2CH_2$	254.38	pl or ta (al lig)	390d, $182\text{-}5^{3}$	91-2	...	...	eth, bz	$B13^3$, 454
6298	Diphenyl methane, α,α-dinitro $(C_6H_5)_2C(NO_2)_2$	258.23	pl (dil al)	...	79-80	...	...	al, eth, bz, chl	$B5^3$, 1797
6299	Diphenyl methane, 2,2′-dinitro $(2\text{-}O_2NC_6H_4)_2CH_2$	258.23	cr	...	83.5	...	...	al, eth	$B5^3$, 1796
6300	Diphenyl methane, 2,2′-dinitro,4,4′-diamino $[2\text{-}NO_2\text{-}4\text{-}H_2NC_6H_3]_2CH_2$	288.26	og pl (al)	...	205	...	...	al, aa	$B13^2$, 113
6301	Diphenyl methane, 2,4′-dinitro $(2\text{-}O_2NC_6H_4)CH_2(C_6H_4NO_2\text{-}4)$	258.23	ye mcl pr(bz)	...	118	...	...	...	$B5^3$, 1797
6302	Diphenyl methane, 3,3′-dinitro $(3\text{-}O_2NC_6H_4)_2CH_2$	258.23	lf(aa)	...	175.5	...	...	al, bz, aa	$B5^3$, 1797
6303	Diphenyl methane, 3,3′-dinitro-4,4′-diamino $[3\text{-}NO_2\text{-}4\text{-}NH_2C_6H_3]_2CH_2$	288.26	red nd	...	232-3	...	...	...	$B13^3$, 113
6304	Diphenyl methane, 3,4′-dinitro $(3\text{-}O_2NC_6H_4)CH_2(C_6H_4NO_2\text{-}4)$	258.23	nd(al)	...	103-4	...	...	...	$B5^3$, 1797
6305	Diphenyl methane, 4,4′-dinitro $(4\text{-}O_2NC_6H_4)_2CH_2$	258.23	nd (bz, peth, aa)	...	188	...	...	bz, aa	$B5^3$, 1797
6306	Diphenyl methane, 4-ethyl $(4\text{-}C_2H_5\text{-}C_6H_4)CH_2C_6H_5$	196.29	...	297 $85^{0.2}$	−24	0.9777^{20}	1.5618^{20}	al, eth, chl	$B5^3$, 1870
6307	Diphenyl methane, 4-hydroxy or 4-Benzylphenol $(4\text{-}HOC_6H_4)CH_2C_6H_5$	184.24	nd or pl(al)	325-30, $198\text{-}200^{10}$	84	...	...	al, eth, bz, chl, aa	$B6^3$, 3357
6308	Diphenyl methane, 2-methyl or Phenyl-2-tolyl methane $(2\text{-}CH_3H_5)CH_2C_6H_5$	182.27	...	280.5	6.6	...	$1.5763^{20}/_D$	...	$B5^1$, 1855
6309	Diphenyl methane, 3-methyl or Phenyl-3-tolyl methane $(3\text{-}CH_3C_6H_4)CH_2C_6H_5$	182.27	...	279.2, $120^{0.2}$	−28	$0.9913^{20/4}$	1.5712^{20}	al, eth, bz, aa, chl	$B5^4$, 1858
6310	Diphenyl methane, 4-methyl $(4\text{-}CH_3C_6H_4)CH_2C_6H_5$	182.27	...	286, $114\text{-}5^{3}$	−30	$0.9976^{20/4}$	1.5712^{20}	al, eth, bz, chl, aa	$B5^4$, 1860
6311	Diphenyl methane, 4-propyl $(4\text{-}C_3H_7C_6H_4)CH_2C_6H_5$	210.32	...	$152\text{-}5^{10}$	...	$0.9739^{18/4}$	1.5552^{20}	...	$B5^4$, 1950
6312	Diphenyl methane, 2,2′,4,4′-tetramethyl $[2,4\text{-}(CH_3)_2C_6H_3]_2CH_2$	224.35	...	$140\text{-}2^{3}$	...	...	1.5635^{15}	...	$B5^4$, 1971
6313	Diphenyl methane, 2,2′,4,4′-tetranitro $[2,4\text{-}(NO_2)_2C_6H_3]_2CH_2$	348.23	ye pr (aa)	...	181	...	...	...	$B5^3$, 1798

PHYSICAL CONSTANTS OF ORGANIC COMPOUNDS (Continued)

No.	Name, Synonyms, and Formula	Mol. wt.	Color, crystalline form, specific rotation and λ_{max} (log ε)	b.p. °C	m.p. °C	Density	n_D	Solubility	Ref.
6314	Diphenyl methanol or Benzhydrol $(C_6H_5)_2CHOH$	184.24	nd(lig)	$297-8^{748}$, 180^{20}	69			al, eth, chl	B6^{3}, 3364
6315	Diphenyl methanol, 4,4′-dimethyl or di-(4-tolyl)methanol $(4-CH_3C_6H_4)_2CHOH$	212.29	nd(al)		69			al, eth, ace, chl, aa	B6^{3}, 3422
6316	Di-(α-Phenylethyl) amine $[CH_3CH(C_6H_5)]_2NH$	225.23	ye	295-8, 190^{10}		1.018^{13}	1.573		B12^{2}, 589
6317	Di-(β-Phenylethyl)amine $(C_6H_5CH_2CH_2)_2NH$	225.33		$335-7^{603}$, 190^{15}	28-30		1.5550^{25}	al, eth	B12^{2}, 593
6318	Diphenyl ethyl amine $(C_6H_5)_2NC_2H_5$	197.28		295-6, 148^{11}		$1.0396^{20/20}$	1.6095^{20}	al, eth	B12^{3}, 291
6319	Diphenyl methyl amine $(C_6H_5)_2NCH_3$	183.25		293-4, 145^{10}	−7.5	$1.0476^{20/4}$	1.6193^{20}		B12^{3}, 290
6320	Diphenyl propanetrione or Diphenyl triketone $C_6H_5COCOCOC_6H_5$	238.24	ye nd(lig)	289^{175}, 248^{60}	68-70			eth	B7^{3}, 4620
6321	Diphenyl selenide $(C_6H_5)_2Se$	233.17	ye nd(bz)	301-2, $126-7^{5}$	2.5	$1.351^{20/4}$	1.5500^{20}	al, eth, bz, xyl	B6^{4}, 1779
6322	Diphenyl selenonium dichloride $(C_6H_5)_2SeCl_2$	304.08	pa ye pr(xyl, ace)nd(al)		183			w, al, ace	B6^{3}, 1107
6323	Diphenyl sulfide $(C_6H_5)_2SO_2$	218.27	mcl pr (bz) pl (al) nd (w)	379, 232^{18}	128-9	$1.252^{20/4}$		eth, bz	B6^{4}, 1490
6324	Diphenyl sulfoxide $(C_6H_5)_2SO$	202.27	pr(lig)	340d, 210^{15}	70.5			al, eth, bz, aa	B6^{4}, 1489
6325	Diphosphine, tetrakis(trifluoromethyl) $(F_3C)_2PP(CF_3)_2$	337.97		84		>1.0			B3^{4}, 265
6326	Dipicryl amine $[2,4,6-(NO_2)_3C_6H_2]_2NH$	439.21	pa ye pr (aa)		244d			py	B12^{2}, 422
6327	Diploicin $C_{16}H_{10}Cl_4O_5$	424.06	(bz)		232				B19^{4}, 2347
6328	Dipropanol amine, *N*-methyl $(HOCH_2CH_2CH_2)_2NCH_3$	147.22	hyg	$164-5^{12}$				w, al, bz	B4^{4}, 1644
6329	Di-*iso*-Propanol amine $[CH_3CH(OH)CH_2]_2NH$	133.19	cr	$249-50^{745}$, 151^{23}	44-5			w, al	B4^{3}, 761
6330	Di-*iso*-Propanol amine, *N*-2-hydroxy ethyl $[CH_3CH(OH)CH_2]_2NCH_2CH_2OH$	177.24		$155-6^{1}$		$1.0458^{20/4}$	1.4708^{20}	w, al, ace	B4^{3}, 764
6331	Dipropyl amine $(C_3H_7)_2NH$	101.19		109-10	−39.6	$0.7400^{20/4}$	1.4050^{20}	w, al, eth, ace, bz	B4^{4}, 469
6332	Dipropyl amine, *N*-nitroso $(C_3H_7)_2N-NO$	130.19	gold	206, 89^{13}		$0.9163^{20/4}$	1.4437^{20}	al, eth	B4^{3}, 264
6333	Di-*m*-Propyl-*iso*-butyl amine, perfluoro $i-C_4F_9N(C_3F_7)_2$	571.08		146-8		$1.84^{25/4}$	1.283^{25}	al, eth	B2^{4}, 858
6334	Di-*iso*-Propyl amine $(i-C_3H_7)_2NH$	101.19		84	−61	$0.7169^{20/4}$	1.3924^{20}	al, eth, ace, bz	B4^{4}, 510
6335	Di-*iso*-Propyl amine, *N*-nitroso $[(CH_3)_2CH]_2N-NO$	130.19	cr(eth w)	194.5, $76-81^{14}$	48	$0.9422^{20/4}$		al, eth, bz	B4^{3}, 281
6336	Dipropyl disulfide $(C_3H_7)_2S_2$	150.30		193.5		$0.9599^{20/4}$	1.4981^{20}		B1^{4}, 1454
6337	Diisopropyl disulfide $(i-C_3H_7)_2S_2$	150.30		177.2, 56.8^{10}		$0.9435^{20/4}$	1.4916^{20}		B1^{4}, 1503
6338	Dipropylene glycol $(CH_3CH(OH)CH_2)_2O$	134.18		229-32		$1.0224^{20/20}$		w, al	B1^{4}, 2473
6339	Dipropyl sulfide $(C_3H_7)_2SO_2$	150.24	sc		29-30	$1.0278^{50/4}$	1.4456^{30}	al, eth	B1^{4}, 1453
6340	Di-isopropyl sulfone $(i-C_3H_7)_2SO_2$	150.24	eth		36			w, eth	B1^{4}, 1502
6341	Diisopropyl ketone, cyanohydrin $(i-C_3H_7)_2C(OH)CN$	141.21	rh(eth or peth)	111^{18}	59			al, eth, ace, bz	B3^{2}, 239
6342	Di-(α-Pyrryl)methane $[2-C_4H_3NH]_2CH_2$	146.19	lf or nd(al)	$163-7^{12}$	73			al, eth, bz	B23, 167
6343	Disilane, 1,2-dichloro-1,1,2,2-tetramethyl $(CH_3)_2SiClSiCl(CH_3)_2$	187.22		$49-50^{18}$		1.010^{20}	1.4548^{20}		KHOC, 366
6344	Disilane, 1,2-difluoro-1,1,2,2-tetramethyl $(CH_3)_2SiFSiF(CH_3)_2$	154.31		92-9		$0.9120^{20/4}$	1.3837^{20}		BOSC2^{2}, 183
6345	Disilane, 1,2-diphenyl-1,1,2,2-tetramethyl $C_6H_5Si(CH_3)_2Si(CH_3)_2C_6H_5$	270.52		111^{1}, $73^{0.1}$	34.5	$0.9892^{20/4}$	1.5161^{20}		BOSC2^{2}, 285
6346	Disilane, hexamethyl $(CH_3)_3SiSi(CH_3)_3$	116.31		112.5-4.3	12.8-14	$0.7247^{22.5/4}$	1.4229^{20}		KHOC, 597

No.	Name, Synonyms, and Formula	Mol. wt.	Color, crystalline form, specific rotation and λ_{max} (log ε)	b.p. °C	m.p. °C	Density	n_D	Solubility	Ref.
6347	Disiloxane, 1,3-dichlo-1,1,3,3,-tetramethyl $(CH_3)_2SiClOSiCl(CH_3)_2$	203.22	...	138	−37.5	$1.038^{20/4}$	...	...	KHOC, 593
6348	Disiloxane, dichloromethyl-pentamethyl $Cl_2CHSi(CH_3)_2OSi(CH_3)_3$	231.27	...	200-5	...	1.046^{20}	1.4382^{20}	...	KHOC, 596
6349	Disiloxane, 1,3-diethenyl-1,1,3,3-tetramethyl $CH_2{=}CHSi(CH_3)_2\text{-}O\text{-}Si(CH_3)_2CH{=}CH_2$	186.40	...	39	−99.7	0.811^{20}	1.4123^{20}	...	KHOC, 602
6350	Disiloxane, 1,3-dimethoxy-1,1,3,3-tetramethyl $CH_3OSi(CH_3)_2O\text{-}Si(CH_3)_2OCH_3$	194.38	...	139	...	$0.9048^{20/4}$	1.3835^{20}	...	$BOSC2^2$, 197
6351	Disiloxane, 1,3-diphenyl-1,1,3,3-tetramethyl $C_6H_5Si(CH_3)_2\text{-}O\text{-}Si(CH_3)_2C_6H_5$	286.53	...	110^2	...	$0.9763^{20/4}$	$1.5176^{2}3^{0}$	...	Am79, 1437
6352	Disiloxane, hexaethyl $[(C_2H_5)_3Si]_2O$	246.54	...	233^{756}, 129^{30}	...	$0.8590^{0/0}$	1.4340^{20}	...	B4, 627
6353	Disiloxane, hexakis (2-ethylbutoxy) $([C_2H_5)_2CHCH_2O]_3Si)_2O$	679.18	...	220^1	>-54	$0.9219^{20/20}$	1.4330^{20}	eth, bz	C48, 3761
6354	Disiloxane, hexakis(2-ethyhexoxy) $([CH_3(CH_2)_3CH(C_2H_5)CH_2O]_3Si)_2O$	847.50	...	$253^{0.9}$	...	$0.9044^{20/20}$	1.4402^{20}	eth, bz	C48, 3761
6355	Disiloxane, hexamethyl $(CH_3)_3SiOSi(CH_3)_3$	162.38	...	99.5-100	-66	0.7638^{20}	$1.3774^{20/}{}_D$	...	KHOC, 598
6356	Disiloxane, 1,1,3,3-tetramethyl $(CH_3)_2SiH\text{-}O\text{-}SiH(CH_3)_2$	134.33	...	$70.5\text{-}71^{731}$	...	$0.7572^{20/4}$	$1.3700^{20/}{}_D$	...	$BOSC2^2$, 184
6357	1,2-Dithiane $C_4H_8S_2$	120.23	...	89^{14}	32-3	...	1.5981^{25}	eth, bz, chl	$B19^4$, 8
6358	1,4-Dithiane or Diethylene disulfide $C_4H_8S_2$	120.23	mcl pr	199-200	111-2	...	...	al, eth, aa	$B19^4$, 35
6359	1,3,5-Dithiazine, 4,5-dihydro-5-methyl or Methylthioformaldine $C_4H_9NS_2$	135.24	nd(eth)	185d	65	...	...	al, eth, aa	$B27^2$, 524
6360	1,4-Dithiine, 2,5-diphenyl $C_{16}H_{12}S_2$	268.39	ye pr (al)	...	118-9	...	...	al, bz	$B19^4$, 404
6361	1,3-Dithiolane or Trimethylene-1,3-disulfide $C_3H_6S_2$	106.20	...	175	−50	1.259^{17}	1.5975^{15}	al, eth, xyl	$B19^4$, 6
6362	di(2-Thienyl)ketone or Thienone $C_4H_3SCOC_4H_3S$	194.27	nd(al)	326	90	...	...	eth, ace	$B19^4$, 1745
6363	Dithizone or Diphenyl thiocarbazone $C_6H_5N{=}NCSNHNHC_6H_5$	256.33	bl-bk(chl-al)	...	165-9d	...	...	chl	$B16^3$, 19
6364	Dithioacetic acid or Thiolo-thionoacetic acid CH_3CS_2H	92.17	ye-red oil	66^{85}, 37^{15}	...	1.24^{20}	...	w, al, eth, ace, bz	$B2^4$, 572
6365	Di-2-tolylamine $(2\text{-}CH_3C_6H_4)_2NH$	197.28	bl flr, wh cr	312^{727}, 192^{23}	52-3	...	...	...	$B12^2$, 437
6366	Di-3-tolylamine $(3\text{-}CH_3C_6H_4)_2NH$	197.28	pa ye (peth)	319-20	53	...	...	al, eth, peth	$B12^2$, 467
6367	Di-4-tolylamine $(4\text{-}CH_3C_6H_4)_2NH$	197.28	nd(peth)	330.5	79	...	...	eth, peth	$B12^3$, 2033
6368	3,4'-Ditolylamine hydrochloride $(3\text{-}CH_3C_6H_4)NH(C_6H_4CH_3\text{-}4).HCl$	233.74	cr	...	202-3	...	...	al, ace, bz, chl	$B12^1$, 414
6369	Di-2-tolyl disulfide $(2\text{-}CH_3C_6H_4)_2S_2$	246.39	lf(al)	...	38-9	...	...	al, eth, ace	$B6^4$, 2027
6370	Di-4-tolyl disulfide $(4\text{-}CH_3C_6H_4)_2S_2$	246.39	nd or lf(al)	$210\text{-}15^{20}$	47-8	1.114^{51}	...	al, eth, ace	$B6^4$, 2206
6371	2,2'-Ditolylsulfide or *o*-Totylsulfide $(2\text{-}CH_3C_6H_4)_2S$	214.33	pl(al)	285, 174^{15}	64	...	...	al, eth, chl	$B6^4$, 2019
6372	3,4'-Ditolylsulfide $(3\text{-}CH_3C_6H_4)S(C_6CH_3\text{-}4)$	214.33	nd(al)	179^{11}	28	...	...	al, eth	$B6^4$, 2173
6373	4,4'-Ditolylsulfide or *p*-Totylsulfide $(4\text{-}CH_3C_6H_4)_2S$	214.33	nd(al)	>300, 179^{11}	57.3	...	...	al, eth, ace, bz	$B6^4$, 2173
6374	2,4'-Ditolylsulfide $(2\text{-}CH_3C_6H_4)S(C_6H_4CH_3\text{-}4)$	214.33	...	173^{11}	...	$1.0774^{15/4}$	...	al, eth	$B6^4$, 2172
6375	2,2'-Ditolylsulfone or *o*-Tolylsulfone $(2\text{-}CH_3C_6H_4)_2SO_2$	246.32	nd(al)	...	134-5	...	...	al, eth, bz, chl	$B6^4$, 2020
6376	3,4'-Ditolylsulfide $(3\text{-}CH_3C_6H_4)SO_2(C_6H_4CH_3\text{-}4)$	246.32	...	...	116	...	...	...	$B6^3$, 1404
6377	4,4'-Ditolylsulfide $(4\text{-}CH_3C_6H_4)_2SO_2$	246.32	pr(bz)nd(w, al)pl(al)	405^{714}	159	...	...	bz, chl	$B6^4$, 2174
6378	4,4'-Ditolylsulfoxide or bis-(4-tolyl)sulfoxide $(4\text{-}CH_3C_6H_4)_2SO$	230.32	cr(lig)	...	94	...	...	al, eth, bz, chl, aa	$B6^4$, 2173
6379	Diurea or Dicarbamide $H_2NCONHNHCONH_2$	116.08	pr(w)	...	270	...	...	...	$B3^4$, 236

PHYSICAL CONSTANTS OF ORGANIC COMPOUNDS (Continued)

No.	Name, Synonyms, and Formula	Mol. wt.	Color, crystalline form, specific rotation and λ_{max} (log ε)	b.p. °C	m.p. °C	Density	n_D	Solubility	Ref.
6380	Djenkoic acid or β,β′-Methylene dithio dialanine $[HO_2CCH(NH_2)CHS]_2CH_2$	254.32	nd(w)		300-50d			w	$B4^{1}$, 1591
6381	Docosane $CH_3(CH_2)_{20}CH_3$	310.61	pl(to) cr(eth)	368.6, 213^{10}	44.4	$0.7944^{20/4}$	1.4455^{20}	al, eth, chl	$B1^{4}$, 572
6382	Docosanoic acid or Behenic acid $CH_3(CH_2)_{20}CO_2H$	340.59	nd	306^{60}	80	0.8223^{90}	1.4270^{100}		$B2^{4}$, 1290
6383	Docosanoic acid, ethyl ester or Ethyl behenate $C_{21}H_{43}CO_2C_2H_5$	368.64	nd (al) cr (ace)	$240\text{-}2^{10}$	50	0.8820^{51}		al, eth	$B2^{4}$, 1292
6384	Docosanoic acid, methyl ester or Methyl behenate $C_{21}H_{43}CO_2CH_3$	354.62	nd(ace)	$224\text{-}5^{12}$	54		1.4339^{60}	al, eth	$B2^{4}$, 1291
6385	1-Docosanol or Docosyl alcohol $CH_3(CH_2)_{20}CH_2OH$	326.61	(ace chl)	$180^{0.22}$	71(87)			al, chl	$B1^{4}$, 1906
6386	4,7,11-Docosatriene-18-ynoic acid or Clupanodonic acid $C_{22}H_{34}O_2$	330.51	pa ye	236^{5}	<−78	0.9290^{20}	1.4868^{20}	eth	$B1^{3}$, 1528
6387	13-Docosenoic acid-*(cis)* or Erucic acid $CH_3(CH_2)_7CH{=}CH(CH_2)_{11}CO_2H$	338.57	nd(al)	265^{15}	33-4	$0.860^{55/4}$	1.4758^{20}	al, eth	$B2^{4}$, 1676
6388	13-Docosenoic acid-*(trans)* or Brassidic acid $CH_3(CH_2)_7CH{=}CH(CH_2)_{11}CO_2H$	338.57	pl(al)	282^{30}	61.5	$0.8585^{57/4}$	1.4472^{64}		$B2^{4}$, 1677
6389	13-Docosenoic acid, anhydride-*(trans)* or Brassidic anhydride $(C_{21}H_{41}CO)_2O$	659.12	nd (al) pl (eth)		64	$0.835^{70/4}$	1.4366^{100}	eth, ace, peth	$B2^{2}$, 448
6390	13-Docosynoic acid or Behenolic acid $CH_3(CH_2)_7C{\equiv}C(CH_2)_{11}CO_2H$	336.56	mcl pr or nd(al)		59.5			al, eth, chl	$B2^{4}$, 1764
6391	1,11-Dodecadiene $CH_2{=}CH(CH_2)_8CH{=}CH_2$	166.31		$208\text{-}9^{758}$		$0.7702^{20/4}$	1.4400^{20}		$B1^{4}$, 1067
6392	Dodecanal or Lauraldehyde $CH_3(CH_2)_{10}CHO$	184.32	lf	185^{100}, $100^{3.5}$	44.5	$0.8352^{15/4}$	1.435^{22}	al, eth	$B1^{4}$, 3380
6393	Dodecane $C_{12}H_{26}$	170.34		216.3, 91.5^{10}	−9.6	$0.7487^{20/4}$	1.4216^{20}	al, eth, ace, chl	$B1^{4}$, 498
6394	Dodecane, 1-amino or Lauryl amine $CH_3(CH_2)_{10}CH_2NH_2$	185.35		259, 126.5^{10}	28.3	$0.8015^{20/4}$	1.4421^{20}	al, eth, bz, chl	$B4^{4}$, 1794
6395	Dodecane, 1-bromo or Lauryl bromide $CH_3(CH_2)_{10}CH_2Br$	249.23		276, 139^{10}	−9.5	$1.0399^{20/4}$	1.4583^{20}	al, eth, ace	$B1^{4}$, 502
6396	Dodecane, 1-bromo-12-fluoro $F(CH_2)_{12}Br$	267.22		$85.8^{0.15}$			1.4524^{25}	al, eth, ace	$B1^{4}$, 502
6397	Dodecane, 1-chloro or Lauryl chloride $CH(CH_2)_{10}CH_2Br$	204.78		260, 126.4^{10}	fr−9.3	$0.8687^{20/4}$	1.4433^{20}	al, ace, bz, lig	$B1^{4}$, 501
6398	Dodecane, 1,12-dibromo $Br(CH_2)_{12}Br$	328.13	nd(aa al)	215^{15}	41			al, eth, chl, aa	$B1^{4}$, 503
6399	Dodecane, 1-iodo or Lauryl iodide $CH_3(CH_2)_{10}CH_2I$	296.24		298.2, 153^{10}	0.3	$1.1999^{20/4}$	1.4840^{20}	al, eth, ace, chl	$B1^{4}$, 503
6400	6-Dodecanol $C_5H_{11}CH(OH)C_6H_{13}$	186.34	peth	119^{9}	30			al, eth	$B1^{1}$, 1794
6401	2-Dodecanone or *n*-Decyl methyl Ketone $CH_3COC_{10}H_{21}$	184.32		246-7, 144^{11}	21	$0.8198^{20/4}$	1.4330^{20}	al, eth, ace	$B1^{4}$, 3382
6402	6-Dodecanone $C_6H_3COC_5H_{11}$	184.32		112^{9}	9		$1.4302^{20/}{}_{D}$		$B1^{4}$, 3383
6403	1-Dodecene $CH_3(CH_2)_9CH{=}CH_2$	168.32		213.4, 88.7^{10}	−35.2	$0.7584^{20/4}$	1.4300^{20}	al, eth, ace, bz	$B1^{4}$, 914
6404	Dodecanedioic acid $HO_2C(CH_2)_{10}CO_2H$ $HO_2C(CH_2)_{10}CO_2H$	230.30			128				$B2^{4}$, 2126
6405	Dodecanedioic acid, dimethyl ester $CH_3O_2C(CH_2)_{10}CO_2CH_3$	258.36	pr	$167\text{-}9^{9}$	31.3				$B2^{4}$, 2126
6406	1-Dodecanethiol or Lauryl mercapton $CH_3(CH_2)_{10}CH_2SH$	202.40		$142\text{-}5^{15}$		$0.8450^{20/20}$	1.4589^{20}	al, eth	$B1^{4}$, 1851
6407	Dodecanoic acid or Lauric acid $CH_3(CH_2)_{10}CO_2H$	200.32	nd(al)	131^{1}	44	$0.8679^{50/4}$	1.4304^{50}	al, eth, ace, bz, peth	$B2^{4}$, 1082
6408	Dodecanoic acid, 2-bromo $CH_3(CH_2)_9CHBrCO_2H$	279.22	pl	$157\text{-}9^{2}$	32	1.1474^{74}	1.4585^{24}	al, eth, bz, chl, lig	$B2^{4}$, 1106
6409	1-Dodecanol or Lauryl alcohol $CH_3(CH_2)_{10}CH_2OH$	186.34	lf(dil al)	255-9, 150^{20}	26	$0.8309^{24/4}$		al, eth	$B1^{4}$, 1844
6410	2-Dodecanol $CH_3(CH_2)_9CH(OH)CH_3$	186.34		252	19	0.8286^{20}	$1.4400^{20/}{}_{D}$		$B1^{3}$, 1793
6411	2-Dodecenedioic acid-*(cis)* or Traumatic acid $HO_2CCH{=}CH(CH_2)_8CO_2H$	228.29	(al, ace)		67-8			al, eth, bz, chl	$B2^{3}$, 1979
6412	1-Dodecene-3-yne $CH_3(CH_2)_7C{\equiv}C\text{-}CH{=}CH_2$	164.29		78^{4}		$0.7858^{25/4}$	1.4510^{25}		$B1^{4}$, 1112

No.	Name, Synonyms, and Formula	Mol. wt.	Color, crystalline form, specific rotation and λ_{max} (log ε)	b.p. °C	m.p. °C	Density	n_D	Solubility	Ref.
6413	2-Dodenedioic acid-*(trans)* $HO_2CCH=CH(CH_2)_8CO_2H$	228.29	(al, ace)		165-6			al, eth, chl	B2[4], 2279
6414	Dodecyl sulfate $(C_{12}H_{25}O)_2SO_2$	434.72			48.5				B1[4], 1849
6415	Dodecyltrimethyl ammonium chloride $C_{12}H_{25}N^+(CH_3)_3Cl^-$	263.89			246d			w, al, ace, chl	B4[4], 798
6416	1-Dodecyne $CH_3(CH_2)_9C\equiv CH$	166.31		215, 89^{10}	−19	$0.7788^{20/4}$	1.4340^{20}		B1[4], 1066
6417	2-Dodecyne $CH_3(CH_2))_8C\equiv CCH_3$	166.31		105^{15}	−9	$0.7917^{15/4}$	1.4828^{20}		B1, 261
6418	3-Dodecyne $CH_3(CH_2)_7C\equiv CCH_2CH_3$	166.31		95^{12}		$0.7871^{20/4}$	1.4442^{20}	eth, ace	B1[3], 1025
6419	6-Dodecyne $CH_3(CH_2)_4C\equiv C(CH_2)_4CH_3$	166.31		209^{745}, 100^{14}		$0.7871^{20/4}$	1.4442^{20}	al, eth, ace	B1[4], 1067
6420	Dotriacontane or Dicetyl $CH_3(CH_2)_{30}CH_3$	450.88	pl (bz, chl, aa, eth)	467, 292.7^{10}	69.7	$0.8124^{20/4}$	1.4550^{20}	eth, bz	B1[4], 595
6421	1-Dotriacontanol $CH_3(CH_2)_{30}CH_2OH$	466.88	pl(bz)	sub 200-50^1	89.4				B1[4], 1919
6422	Durene or 1,2,4,5-tetramethyl benzene $1,2,4,5\text{-}(CH_3)_4C_6H_2$	134.22		196.8, 73.5^{10}	79.2	$0.8380^{81/4}$, $0.8875^{20/4}$	1.4790^{81}, 1.5116^{20}	al, eth, ace, bz	B5[4], 1076

No.	Name, Synonyms, and Formula	Mol. wt.	Color, crystalline form, specific rotation and λ_{max} (log ε)	b.p. °C	m.p. °C	Density	n_D	Solubility	Ref.
6423	Ecgonidine-*(l)* or Anhydroecgonine $C_9H_{13}NO_2$	167.21	cr(MeOH-eth), $[\alpha]^{14/}{}_D$ -84.6 (w, p=1.7)	...	225d	...	...	al	B22⁴, 284
6424	Ecgonine-*(dl)* $C_9H_{15}NO_3$	185.22	pl(w + 3)	...	203	...	...	w, al	B22², 156
6425	Ecgonine, Ecqonine $C_9H_{15}NO_3$	185.23	mcl pr, $[\alpha]^{15/}{}_D$ -45.5 (w, c=5)	...	205	...	...	w, al	B22⁴, 2097
6426	Ecgonine hydrate-*(l)* $C_9H_{15}NO_3 \cdot H_2O$	203.24	mcl pr (al) eff 120-30	...	198	...	...	...	B22, 196
6427	Ecgonine benzoate-*(l)* or O-Benzoyl ecqonine $C_{16}H_{19}NO_4$	289.34	nd(w), $[\alpha]^{14/}{}_D$ -63.5 (w, p=1.7)	...	195	...	...	al, bz	B22⁴, 2098
6428	Ecgonine benzoate, ethyl ester (l) or Homococaine $C_{19}H_{23}NO_4$	317.38	pr(eth)	...	109	...	...	al, eth	B22, 202
6429	Ecognine benzoate, tetrahydrate (l) $C_{16}H_{19}NO_4 \cdot 4H_2O$	361.39	pr(w)	...	92	...	...	al, bz	B22, 197
6430	Ecgonine hydrochloride (l) $C_9H_{15}NO_3.HCl$	221.68	rh(al), $[\alpha]^{25/}{}_D$ -59 (w, c=10)	...	246	...	...	w, al	B22⁴, 2098
6431	Echinochrome A $C_{12}H_{10}O_7$	266.21	dk red nd (to)	sub 120 ^{0001}	220d	...	...	al, eth, ace	B8³, 4365
6432	Echinochrome A, 3,6,7-trimethyl ether $C_{15}H_{16}O_7$	308.29	dk red lf (aq diox)	...	133 vac	...	...	al	B8³, 4361
6433	Echinopsine or N-Methyl-α-quinoline $C_{10}H_9NO$	159.19	α-nd(bz) β-cr(al)	...	α152 β135	...	...	w, al, bz, chl	B21⁴, 3722
6434	Echitamidine $C_{20}H_{26}N_2O_3$	342.44	pl(eth), $[\alpha]^{16/}{}_D$ -515 (al)	...	244d	...	...	w, al	J1932, 2628
6435	Echitamine or Ditaine $C_{22}H_{28}N_2O_4 \cdot 4H_2O$	456.54	pr (al + 4w) eff −105 (-3w), $[\alpha]^{20/}{}_D$ -29 (al)	...	206 (+ 1w)	...	...	w, al, eth	J1925, 1640
6436	Echitamine, hydrochloride $C_{22}H_{28}N_2O_4.HCl$	420.94	nd(w), $[\alpha]^{15/}{}_D$ -58 (w, c=1)	...	295-300d	...	...	al	J1925, 1640
6437	Echitin $C_{32}H_{52}O_2$	468.76	lf, $[\alpha]_D$ + 73 (eth)	...	170	...	...	al, eth, ace, bz	M, 397
6438	Egonol $C_{19}H_{18}O_5$	326.35	pl(BuOH)	$228\text{-}30^{0.15}$	118	...	...	chl	B19⁴, 4882
6439	Eicosane or Didecyl $C_{20}H_{42}$	282.55	lf(al)	343, 195.7^{10}	36.8	$0.7886^{20/4}$	1.4425^{20}	eth, ace, bz, peth	B1⁴, 563
6440	Eicosane, 1-cyclohexyl $C_{26}H_{52}$	364.70	...	422	48.5	0.8318^{20}	$1.4622^{20/}{}_D$	...	B5⁴, 198
6441	Eicosane, 9-octyl $C_{11}H_{23}CH(C_8H_{15})_2$	394.77	...	257^{10}, $199^{0.5}$	0.5	$0.8075^{20/4}$	$1.4515^{20/}{}_D$	...	B1⁴, 590
6442	Eicosane, 1-phenyl $CH_3(CH_2)_{19}C_6H_5$	358.65	...	212^1	42.3	$0.8235^{60/4}$	$1.4725^{40/}{}_D$	...	B5⁴, 1222
6443	Eicosane, 2-phenyl $CH_3(CH_2)_{17}CH(CH_3)C_6H_5$	358.65	...	204.5^1	29.0	$0.8547^{20/4}$	1.4795^{20}	...	B5⁴, 1222
6444	Eicosone, 3-phenyl $CH_3(CH_2)_{16}CH(C_2H_5)C_6H_5$	358.65	...	202.0^1	29.3	$0.8546^{20/4}$	$1.4796^{20/}{}_D$	...	B5⁴, 1222
6445	Eicosane, 4-phenyl $CH_3(CH_2)_{15}CH(C_3H_7)C_6H_5$	358.65	...	199.0^1	31.4	$0.8546^{20/4}$	$1.4794^{20/}{}_D$	...	B5⁴, 1223
6446	Eicosane, 5-phenyl $CH_3(CH_2)_{14}CH(C_4H_9)C_6H_5$	358.65	...	197^1	30.2	$0.8549^{20/4}$	$1.4796^{20/}{}_D$	...	B5⁴, 1223
6447	Eicosane, 9-phenyl $CH_3(CH_2)_7CH(C_6H_5)(CH_2)_{10}CH_3$	358.65	...	196^{10}	17.9	$0.8534^{20/4}$	1.4790^{20}	...	B5⁴, 1223
6448	Eicosanedioic acid or Octadecane dicarboxylic acid $HO_2C(CH_2)_{18}CO_2H$	342.52	cr(bz or al)	$233\text{-}4^2$	125-6	...	...	eth	B2⁴, 2185
6449	Eicosanedioic acid, diethyl ester or Diethyl eicosanedioate $C_2H_5O_2C(CH_2)_{18}CO_2C_2H_5$	398.63	...	240^{12}	54-5	...	...	al, eth	B2³, 1881
6450	Eicosanoic acid or Arachidic acid $CH_3(CH_2)_{18}CO_2H$	312.54	pl(al)	328d, $203\text{-}5^1$	77	$0.8240^{100/4}$	1.425^{100}	eth, chl	B2⁴, 1275

No.	Name, Synonyms, and Formula	Mol. wt.	Color, crystalline form, specific rotation and λ_{max} (log ε)	b.p. °C	m.p. °C	Density	n_D	Solubility	Ref.
6451	Eicosanoic acid, ethyl ester or Ethyleicosonoate .... $C_{19}H_{39}CO_2C_2H_5$	340.59	....	295-7^{100}, 186-7^{2}	50	....	....	al, eth, bz, chl	B2^{3}, 1067
6452	Eicosanoic acid, methyl ester or Methyl eicosanoate..... $C_{19}H_{39}CO_2CH_3$	326.56	lf(MeOH)	215-6^{10}	54.5	....	1.4317^{60}	al, eth, bz, chl	B2^{4}, 1276
6453	1-Eicosanol or Arachidic alcohol.............. $CH_3(CH_2)_{18}CH_2OH$	298.55	wx(al) cr(chl)	309, 220-5^{3}	72-3	0.8405$^{20/4}$	1.4550^{20}	ace, bz	B1^{4}, 1900
6454	2-Eicosonal .... $CH_3(CH_2)_{17}CH(OH)CH_3$	298.56	cr(MeOH)	357	63-4	0.8378$^{20/4}$	1.4312^{80}	ace, bz	B1^{4}, 1901
6455	2-Eicosanone or Methyl octadecyl ketone.............. $C_{18}H_{35}COCH_3$	296.54	lf(MeOH)	....	58	....	....	al, ace, bz	B1^{4}, 3402
6456	3-Eicosanone or Ethyl heptadecyl ketone.............. $C_{17}H_{35}C(NOH)C_2H_5$	296.54	lf(al)	....	60-1	....	....	eth, ace, bz, chl, aa	B1^{4}, 3403
6457	3-Eicosanone oxime .... $C_{17}H_{35}C(H)C_2H_5$	311.55	nd(al)	....	α55-6, β64-5	....	....	al, eth	B1^{3}, 2932
6458	7-Eicosanone or Hexyl tridecyl ketone .... $C_{13}H_{27}COC_6H_{13}$	296.54	cr	210-1^{11}	52-3	....	1.4258^{20}	eth, ace	B1^{4}, 3403
6459	5,8,11,14-Eicosatetraenoic acid or Arachidonic acid .... $CH_3(CH_2)_3(CH_2CH{=}CH)_4(CH_2)_3CO_2H$	304.47	....	d	−49.5	....	1.4824^{20}	al, eth, ace, chl	B2^{4}, 1802
6460	1-Eicosene .... $CH_3(CH_2)_{17}CH{=}CH_2$	280.54	....	341, 151$^{1.5}$	28.5	0.7882$^{30/4}$	1.440^{30}	bz, peth	B1^{4}, 934
6461	1-Eicosyne .... $CH_3(CH_2)_{17}C{\equiv}CH$	278.52	....	340, 191.8^{10}	36	0.8073$^{20/4}$	1.4501^{20}	bz, peth	B1^{4}, 1077
6462	Elaidamide .... $CH_3(CH_2)_7CH{=}CH(CH_2)_7CONH_2$	281.48	....	....	93-4	....	....	al	B2^{4}, 1668
6463	Elaidic acid or 9-Octadecenoic acid (trans)............. $CH_3(CH_2)_7CH{=}CH(CH_2)_7CO_2H$	282.47	pl(al)	288^{100}, 234^{15}	45	0.8734^{45}	1.4499^{45}	al, eth, bz, chl	B2^{4}, 1647
6464	Elaidic acid, dibromide or 9,10-Dibromo octadecanedioic acid .... $CH_3(CH_2)_7CHBrCHBr(CH_2)_7CO_2H$	442.27	col or ye	....	29-30	1.2458$^{30/4}$	1.4893^{42}	eth	B2^{3}, 1048
6465	Elaidic acid, ethyl ester or Ethylelaidate............... $C_{17}H_{33}CO_2C_2H_5$	310.52	....	217-9^{15}	5.8	0.8664^{25}	1.4480^{25}	al, eth	B2^{4}, 1652
6466	Elaidic acid, methyl ester .... $C_{17}H_{33}CO_2CH_3$	296.49	....	213-5^{15}	......	0.8730^{20}	1.4513^{20}	al, eth	B2^{4}, 1651
6467	Elaidyl alcohol or 9-Octadecen-1-ol *(trans)*............. $C_8H_{17}CH{=}CH(CH_2)_7CH_2OH$	268.48	....	333, 198^{10}	36-7	0.8338$^{40/4}$	1.4552^{40}	al, eth, ace	B1^{4}, 2204
6468	α-Elaterin .... $C_{32}H_{44}O_8$	556.70	cr(chl-MeOH or al), $[\alpha]^{20}{}_{D}$ -64.3 (chl, c=1.6)	....	234	....	....	eth, bz, chl	B8^{3}, 4377
6469	β-Elaterin .... $C_{20}H_{28}O_5$	348.44	nd(al), $[\alpha]^{25}{}_{D}$ + 13.9	....	195.5	....	....	chl	B8^{3}, 4377
6470	Elemane or Dihydroelemene .... $C_{15}H_{30}$	210.40	....	115-9^{10}	......	0.8509$^{20/4}$	1.4640^{20}	eth, bz, peth	B5^{2}, 117
6471	α-Elemene (d) .... $C_{15}H_{24}$	204.36	$[\alpha]_D$ 116 (chl, 14.85%)	120-30^{7}	......	0.8782$^{20/4}$	1.5130^{26}	ace, bz	B5^{3}, 1083
6472	Elemenonic acid or Dihydro-β-elemonic acid........... $C_{30}H_{48}O_3$	456.71	nd(al or AcOEt)	....	249-50	....	....	....	B10^{3}, 3227
6473	α-Elemol .... $C_{15}H_{26}O$	222.37	cr, $[\alpha]_D$ + 43.7 (chl, c=1.9)	142-3^{12}	52-3	0.9345$^{18/4}$	1.4980^{18}	....	B6^{3}, 410
6474	α-Eleostearic acid or 9,11,13-Octadecatrienoic acid (cis) .. $C_4H_9[CH{=}CH]_3(CH_2)_7CO_2H$	278.44	$[\alpha]_D$ -5.8 (chl, c=3.4) nd (al)	235^{12}d, 170^{1}	49	0.9028$^{50/4}$	1.5112^{50}	al, eth	B2^{4}, 1787
6475	β-Eleostearic acid or 9,11,13-Octadecatrienoic acid (trans) $C_4H_9[CH{=}CH]_3(CH_2)_7CO_2H$	278.44	lf (al, MeOH)	188^{1}	71-2	0.8839$^{80/4}$	1.5000^{80}	MeOH, al	B2^{4}, 1787
6476	Eluetherin .... $C_{16}H_{16}O_4$	272.30	$[\alpha]^{15}{}_{D}$ 346 (chl)	....	175	....	....	....	B18^{4}, 1642
6477	Ellagene or Indino-2′,3′:2,3-fluorene .... $C_{20}H_{14}$	254.33	pl(bz)	....	216	....	....	....	E14s, 498
6478	Ellagic acid, dihydrate .... $C_{14}H_6O_8.2H_2O$	338.23	pa ye nd(py)	....	450-80d	....	....	....	B19^{4}, 3164
6479	Elliptic acid .... $C_{20}H_{18}O_8$	386.36	nd(aq al)	....	190	....	....	....	B18^{4}, 3342

No.	Name, Synonyms, and Formula	Mol. wt.	Color, crystalline form, specific rotation and λ_{max} (log ε)	b.p. °C	m.p. °C	Density	n_D	Solubility	Ref.
6480	Emeraldine $C_{48}H_{40}N_8$	728.90	indigo-bl pw					aa(80)	B13¹, 257
6481	Emetine (l) or Cephaline-O-methyl ether $C_{29}H_{40}N_2O_4$	480.65	amor pw $[\alpha]^{20/}{}_D$ -50 (chl, c=2)		74 (cor)			al, eth, ace	B23², 449
6482	Emetine, hydrochloride (l) $C_{29}H_{40}N_2O_4 \cdot 2HCl \cdot 7H_2O$	679.68	nd(w), $[\alpha]_D$ + 11 (w, c=1)		269-70d			w, al	B23², 451
6483	Emicymarin $C_{30}H_{46}O_9$	550.69	nd or pr (+ MeOH), $[\alpha]^{20/}{}_D$ + 12.5 (al, c=2.5)		Ca 207				B18⁴, 2440
6484	Enneaphyllin $C_{90}H_{154}$	1236.21	rods (bz)		295-6			al	C32, 2686
6485	Eosin or 2,4,5,7-tetrabromfluorescein $C_{20}H_8O_5Br_4$	647.90	ye-red		295-6			al	B19⁴, 2917
6486	Ephedrine (d) $C_6H_5CH(OH)CH(CH_3)NHCH_3$	165.24	pl(w)	225	40			w, al, eth, bz, chl	B13³, 1723
6487	Ephedrine, hydrochloride-(d) $C_{10}H_{15}NO \cdot HCl$	201.70	pl(abs al), $[\alpha]^{20/}{}_D$ + 35.8 (w, c=11.5)		218			w, al	B13³, 1723
6488	Ephedrine (dl) or Racephedrine $C_6H_5CH(OH)CH(CH_3)NHCH_3$	165.24	nd(eth or peth)	135-7¹²	76-7			w, al, eth, bz, chl	B13³, 1723
6489	Ephedrine, p-amino dihydrochloride $CH_3CH(NH_2)CH(OH)C_6H_4NH_2$-4.2HCl	239.15	lf(al-eth)		192-3d			w, al	C27, 2762
6490	Ephedrine, hydrochloride (dl) or Ephotonin $C_{10}H_{15}NO.HCl$	201.70	pl(al)		189-90			w, al	B13³, 1724
6491	Ephedrine (l) or Natural ephedrine $C_6H_5CH(OH)CH(CH_3)NHCH_3$	183.25	pl(w + 1)	225	40			w, al, eth, bz, chl	B13³, 1720
6492	Ephedrine, hydrochloride (l) $C_{10}H_{15}NO.HCl$	201.70	orb nd, $[\alpha]^{20/}{}_D$ - 36.6		218-20			w, al	B13³, 1721
6493	Ephedrine, N-methyl $C_6H_5CH(OH)CH(CH_3)N(CH_3)_2$	179.27	nd or pl (al or eth), $[\alpha]_D$ -29.5 (MeOH, c=4.5)		87-8			al, eth, MeOH	B13³, 1726
6494	Ephedrine, N-(4-nitrobenzoyl) (dl) $C_6H_5CH(OH)CH(CH_3)NH(CH_3)COC_6H_4NO_2$-(4)	314.34	pa ye pl(al)		162				B13², 384
6495	Ephedrine sulfate (l) $(C_{10}H_{15}NO)_2.H_2SO_4$	428.54	hex pl or orh nd(w)		245-8d			w	C21, 2169
6496	Epi-β-amyrin acetate $C_{30}H_{50}O$	426.73	cr(MeOH)		225 (cor)				B6³, 2896
6497	Epi-α-amyrin acetate $C_{32}H_{52}O_2$	468.76	nd(chl-MeOH), $[\alpha]_D$ + 39 (chl)		135 (cor)				B6³, 2890
6498	Epiandrosterone $C_{10}H_{17}OH$	290.45	cr(bz-peth ace), $[\alpha]^{20/}{}_D$ + 108 (MeOH)		177-9				B8³, 584
6499	Epiborneol (l) or 3-Camphanol $C_{10}H_{15}OH$	154.25	nd(peth)	213⁷⁴²	181-2				B6⁴, 288
6500	Epiborneol, acetate or 3-Camphanol acetate $C_{12}H_{20}O_2$	196.29	$[\alpha]^{19/}{}_D$ + 15.63	101¹¹	<-15	$0.9872^{14/0}$	1.4651^{14}		B6³, 318
6501	Epibreinonol or Breinonol A $C_{30}H_{48}O_2$	440.71	pl (MeOH) nd (chl), $[\alpha]^{17/}{}_D$ + 37 (chl)		204			chl	B8³, 1093
6502	Epicamphor (d) or 3-Bornanone $C_{10}H_{16}O$	152.24	$[\alpha]^{17/}{}_D$ + 45.4 (bz)		182			al, eth, peth	B7³, 420
6503	Epicamphor, oxime (d) or d-3-Bornanone oxime $C_{10}H_{16}NOH$	167.25	nd(MeOH)		103			eth, ace	B7¹, 86
6504	Epicamphor, semicarbazone (d) $C_{10}H_{16}NNHCONH_2$	209.29	nd(al)		237-8			al	B7³, 421

No.	Name, Synonyms, and Formula	Mol. wt.	Color, crystalline form, specific rotation and λ_{max} (log ε)	b.p. °C	m.p. °C	Density	n_D	Solubility	Ref.
6505	Epicamphor (dl) or 3-Bornanone $C_{10}H_{16}O$	152.24	cr(peth)		175-7			al, eth, peth	B7^{3}, 421
6506	Epicamphor, oxime (dl) $C_{10}H_{16}NOH$	167.25	nd(dil al)		98-100			eth, ace	B7^{1}, 87
6507	Epicamphor-(l) or β-Camphor $C_{10}H_{16}O$	152.24		213	184 (187)			al, eth, bz, peth	B7^{3}, 421
6508	Epicamphor, bromo (l) $BrC_{10}H_{15}O$	231.13	nd or pw (peth), $[\alpha]_D$ -86.6 (AcOGt ,c=3.6)		133-4			bz, chl, lig	B7^{3}, 422
6509	Epicamphor, oxime (l) or β-camphor oxime $C_{10}H_{16}NOH$	167.25	nd(dil MeOH), $[\alpha]_D$ + 100.5 (bz, c=6.3)		103-4			al, eth, ace	B7^{1}, 86
6510	Epicamphor, semicarbazone (l) $C_{10}H_{16}NNHCONH_2$	209.29	nd(al)		237-8			al	B7^{3}, 421
6511	α-Epicamphyl amine $C_{10}H_{17}NH_2$	153.27	$[\alpha]_D$ + 17.6 (bz, c=6.5)	127-8^{100}				w	B12^{3}, 176
6512	Epicatechin (dl) or Epicatechol $C_{15}H_{14}O_6$	290.27	nd(w + 1)pr (w + 4)		224-6d			al, ace	B17^{2}, 258
6513	Epicatechin (l) or Epicatechol $C_{15}H_{14}O_6$	290.27	cr(w + 4), $[\alpha]^{25/}_D$ -69 (al)		245d			al, ace	B17^{4}, 3841
6514	Epicholestan-3-ol or 3-α-hydroxycholestane $C_{27}H_{48}O$	388.68	nd (al, MeOH), $[\alpha]^{20/}_D$ + 34 (chl)		185-6			eth, chl	B6^{3}, 2135
6515	Epicholestan-4-ol or 4α-Hydroxycholestane $C_{27}H_{48}O$	388.68	lf(al MeOH-ace), $[\alpha]^{21/}_D$ + 29.0 (chl)		187-8			eth, chl	B6^{3}, 2155
6516	Epicholesterol or 5-Cholesten-3α-ol $C_{27}H_{46}O$	386.66	cr(al chl MeOH), $[\alpha]^{30/}_D$ - 37.5 (al)		141.5				B6^{3}, 2622
6517	Epicoprostanol or 3-α-Coprostanol $C_{27}H_{48}O$	388.68	cr(al ace), $[\alpha]^{20/}_D$ + 31.6 (chl)		117-8			al, eth, bz, chl	B6^{3}, 2130
6518	Epicoprostenol or 4-Cholesten-3α-ol $C_{27}H_{46}O$	386.66	nd(ace), $[\alpha]^{24/}_D$ + 120.8 (bz)		84			al, eth, bz, chl	B6^{3}, 2605
6519	Epidicentrin (dl) or dl-Domesticine methyl ether $C_{20}H_{21}NO_4$	339.39	pr(MeOH)		142			al, eth, chl	B27^{2}, 553
6520	Epidicentrin (l) or l-Domesticin methyl ether $C_{20}H_{21}NO_4$	339.39	pr(MeOH), $[\alpha]^{18/}_D$ - 101.3 (chl, c=0.5)		138-9			al, eth, chl	B27^{2}, 533
6521	Epiergosterol *(d)* or Δ-7,9,(11),22-Ergostatrien-3α-ol $C_{28}H_{44}O$	396.66	nd(eth-MeOH), $[\alpha]^{19/}_D$ + 36.2 (chl)		203-4				B6^{3}, 3096
6522	D-*epi*-Fucitol or 6-deoxy-D-glucitol $C_6H_{14}O_5$	166.17	cr(eth w), $[\alpha]^{21/}_D$ + 2.2 (w, c=1)		105-7			w, eth	Am74, 4373
6523	L-*epi*-Fucitol $C_6H_{14}O_5$	166.17	cr(w), $[\alpha]^{20/}_D$ -2.3 (w, c=1)		105-7			w, eth	C24, 2431
6524	L-*epi*-Fucose or L-Quinovose $CH_3(CHOH)_4CHO$	164.16	cr(AcOEt), $[\alpha]_D$ -36.9 (w, c=6)		135-45			w	B1^{4}, 4267

PHYSICAL CONSTANTS OF ORGANIC COMPOUNDS (Continued)

No.	Name, Synonyms, and Formula	Mol. wt.	Color, crystalline form, specific rotation and λ_{max} (log ε)	b.p. °C	m.p. °C	Density	n_D	Solubility	Ref.
6525	Epiisofenchol or 4,6,6-Trimethyl-2-norbornanol .... $C_{10}H_{17}OH$	154.25	nd(sub), $[\alpha]^{20}{}_D$ -7.35	....	71-2	....	....	....	B6³, 292
6526	Epiisofenchone or 4,6,6-Trimethyl norbornanone (d) ... $C_{10}H_{16}O$	152.24	$[\alpha]_D$ + 19.6	195	......	$0.934^{20/4}$	1.459^{20}	eth, ace	B7³, 396
6527	Epiisofenchone (dl) or 4,6,6,-Trimethyl norbornanone .. $C_{10}H_{16}O$	152.24	....	195-8, $90\text{-}3^{21}$	......	....	1.4625^{25}	eth, ace	B7³, 397
6528	Epilupinine or d-isolupinine .... $C_{10}H_{19}NO$	169.27	nd(peth), $[\alpha]^{17}{}_D$ + 32 (al, c=1.5)	....	76-8	....	....	eth, ace, bz	B21⁴, 290
6529	Epinine .... $C_9H_{13}NO_2$	167.21	nd(al)	....	188-9	....	....	....	B13³, 2209
6530	Epiquinidine .... $C_{20}H_{24}N_2O_2$	324.42	cr (AcOEt) lf (eth), $[\alpha]^{20}{}_D$ + 103.7 (al, c=1.86)	....	113	....	....	al, eth	B23, 505
6531	Epirhodanhydrin or 2,3-Epoxy propyl rhodanine .... CH_2CHCH_2SCN (epoxide O bridging first two C)	115.15	dk red liq- (garlic odor)	d	......	....	....	al, chl	B17, 106
6532	Episarsapogenin .... $C_{27}H_{44}O_3$	416.64	nd(ace), $[\alpha]_D$ -71	....	2 04-6	....	....	....	B19⁴, 825
6533	Epitruxillic acid or 2,4-*cis*-Diphenyl cyclobutane-1,3-*trans*-dicarboxylic acid .... $C_{18}H_{16}O_4$	296.32	cr(dil al or bz-aa)	....	285-7	....	....	....	B9³, 4625
6534	2,3-Epoxypropyl ethyl ether .... $CH_2CHCH_2OC_2H_5$ (epoxide O bridging first two C)	102.13	....	128	......	0.9700^{20}	1.4320^{20}	w, al, eth	B17⁴, 987
6535	2,3-Epoxypropyl phenyl ether .... $CH_2CHCH_2OC_6H_5$ (epoxide O bridging first two C)	150.18	....	242.5^{755}	......	$1.1109^{21\ 2/4}$	$1.5307^{21}{}_D$	....	B17⁴, 990
6536	Equilenin (d) .... $C_{18}H_{18}O_2$	266.34	$[\alpha]^{16}{}_D$ + 87 (diox)	sub 170-80° ¹	258-9	....	....	....	B8³, 1523
6537	Equilenin (dl) .... $C_{18}H_{18}O_2$	266.34	cr(bz)	....	276-8	....	....	....	B8³, 1525
6538	Equilenin (l) .... $C_{18}H_{18}O_2$	266.34	$[\alpha]^{20}{}_D$ -85 (diox)	sub 170-80¹	258-9	....	....	al	B8³, 1522
6539	Equilin .... $C_{18}H_{20}O_2$	268.36	orh sph pl(AcOEt)	170-200 (sub vac)	238-40	....	....	al, ace	B8³, 1415
6540	Equilin, α-dihydro .... $C_{18}H_{22}O_2$	270.37	cr(ace), $[\alpha]_D$ + 220 (diox, c=1)	....	174-6	....	....	al, ace, diox	B6³, 5530
6541	Equisetrin .... $C_{27}H_{30}O_{16}$	610.53	ye nd (+2w)	....	195-6	....	....	al	B18, 3296
6542	Equol or 4,7-Isoflavandiol .... $C_{15}H_{14}O_3$	242.27	cr(aq al)	....	189-90	....	....	....	B17⁴, 2186
6543	Eremophilol .... $C_{15}H_{24}O$	220.35	visc oil	164.5^{13}	......	....	1.5202^{20}	....	B6³, 2080
6544	Eremophilone .... $C_{15}H_{22}O$	218.34	nd(MeOH)	171^{15}	42-3	$0.9994^{25/25}$	1.5182^{25}	....	B7³, 1266
6545	Eremophilone, 8,9-epoxy .... $C_{15}H_{22}O_2$	234.34	nd(peth)	....	63-4	....	....	al, eth, ace, bz	B17⁴, 4767
6546	Ergine or Ergonovine .... $C_{16}H_{17}N_3O$	267.33	cr (MeOH) Pr (aq ace)	....	135-40	....	....	....	H32, 506
6547	Ergocornine .... $C_{37}H_{39}N_5O_5$	561.68	cr(MeOH), $[\alpha]^{20}{}_D$ -188 (chl, c=1)	....	182-4d	....	....	al, ace, bz, chl	H52, 1549
6548	Ergocorninine .... $C_{31}H_{39}N_5O_5$	561.68	lo pr (al), $[\alpha]^{20}{}_D$ + 409 (chl, c=1)	....	228d	....	....	al, ace, bz, chl	H52, 1549
6549	Ergocristine .... $C_{35}H_{39}N_5O_5$	609.73	rh (bz + 2), $[\alpha]^{20}{}_D$ -183 (chl, c=1) -93 Py	....	175d	....	....	al, ace, chl	H34, 1944

No.	Name, Synonyms, and Formula	Mol. wt.	Color, crystalline form, specific rotation and λ_{max} (log ε)	b.p. °C	m.p. °C	Density	n_D	Solubility	Ref.
6550	Ergocristinine $C_{35}H_{39}N_5O_5$	609.73	pr(al), $[\alpha]^{20/}{}_D$ + 366 (chl, c=0.68)		237-8d				$B27^2$, 860
6551	Ergocryptine $C_{32}H_{41}N_5O_5$	575.71	pr(al), $[\alpha]^{20/}{}_D$ -187 (chl, c=1)		212-4d			al, chl	$B27^2$, 860
6552	Ergocryptinine $C_{32}H_{41}N_5O_5$	575.71	lo pr (al) $[\alpha]^{20/}{}_D$ + 408 (chl, c=1) + 479 Py		245d,			ace, chl	$B27^2$, 860
6553	Ergometrine (l) or Ergobasine $C_{19}H_{23}N_3O_2$	325.41	nd(bz), $[\alpha]^{20/}{}_D$ -89 (w)		159-62d			al, ace	Am60, 1701
6554	Ergometrinine (d) or Ergobasinine $C_{19}H_{23}N_3O_2$	325.41	pr(ace), $[\alpha]^{20/}{}_D$ + 416 (chl, c=0.26)		195-7d			chl, py	Am60, 1701
6555	Ergometrinine (l) $C_{19}H_{23}N_3O_2$	325.41	pr(ace), $[\alpha]^{20/}{}_D$ -41.5 (chl, c=0.26)		196			chl, py	Am60, 1701
6556	Ergopinacol II or Bisergostadienol $C_{56}H_{36}O_2$	791.30	nd(bz-al), $[\alpha]_D$ -155 (Py, c=0.8)		205				$B6^1$, 5897
6557	Ergosine $C_{30}H_{37}N_5O_5$	547.65	pr(MeOH, AcOEt)		228d			ace, chl	H34, 1544
6558	Ergosinine $C_{30}H_{37}N_5O_5$	547.65	Pr (al), (aq ace, bz) nd MeOH, $[\alpha]^{20/}{}_D$ + 420 (chl, c=1)		220d			ace, chl	J1937, 396
6559	$\Delta^{5}:^{6,7}:^{8}$-Ergostadien-3β-ol or Provitamin D_4 $C_{28}H_{46}O$	398.67	nd(MeOH-AcOEt), $[\alpha]^{19/}{}_D$ -109 (chl)		152-3				$B6^1$, 2836
6560	$\Delta^{14,22}$-Ergostadien-3β-ol $C_{28}H_{46}O$	398.67	cr(MeOH), $[\alpha]^{20/}{}_D$ -9 (chl)		116				E14s, 1762
6561	α-Ergostadienone $C_{28}H_{44}O$	396.66	lf(al), $[\alpha]^{19/}{}_D$ + 2 (chl)		182-3				$B7^1$, 1767
6562	Ergostane or 24-Methyl-5α-cholestane $C_{28}H_{50}$	386.71	lf or pl(eth-MeOH or ace), $[\alpha]^{25/}{}_D$ + 21 (chl, c=2)		85			eth, ace, chl	$B5^4$, 1234
6563	Ergostanol or Ergostan-3β-ol $C_{28}H_{50}O$	402.70	nd(MeOH-eth), $[\alpha]_D$ + 15.4 (chl, c=1.8)		144-5			eth, chl	$B6^3$, 2161
6564	$\Delta^{3,5,7,22}$-Ergostatetraene $C_{28}H_{42}$	378.64	pl(al)$[\alpha]^{20/}{}_D$ -40.5 (chl)		104				$B5^3$, 1927
6565	$\Delta^{4,6,22}$-Ergostatrienone or Isoergosterone $C_{28}H_{42}O$	394.64	nd(ace-eth)		110				$B7^3$, 2031
6566	α-Ergostenol or α-Tetrahydroergosterol $C_{28}H_{48}O$	400.69	lf or nd (MeOH) nd (gl aa), $[\alpha]^{16/}{}_D$ + 11 (MeOH, c=0.9)		131 (135)			eth, bz, chl	$B6^3$, 2685

PHYSICAL CONSTANTS OF ORGANIC COMPOUNDS (Continued)

No.	Name, Synonyms, and Formula	Mol. wt.	Color, crystalline form, specific rotation and λ_{max} (log ε)	b.p. °C	m.p. °C	Density	n_D	Solubility	Ref.
6567	β-Ergostenol or β-Tetrahydroergosterol $C_{28}H_{48}O$	400.69	pl or ta(al), $[\alpha]^{20}{}_D$ + 21.2 (chl, c=0.9)		141-2				B6^3, 2689
6568	α-Ergostenol or α-Tetrahydroergosterol $C_{28}H_{48}O$	400.69	nd (MeOH) cr (ProH) lf (w)		148 (152)				B6^3, 2683
6569	d-Δ-^{8}:^{9}-Ergostenol $C_{28}H_{46}O$	400.69	$[\alpha]_D$ + 39		155			chl	B6^3, 2685
6570	Ergosterol or Δ ^{5},^{7},^{22}-Ergostatrien-2β-ol $C_{28}H_{44}O$	396.66	pl (+ w, al) nd (eth), $[\alpha]^{20}{}_D$ -135 (chl, c=1.2)	$250^{0.01}$	168 (+ w)			bz, chl	B6^3, 3099
6571	Ergosterol D $C_{28}H_{44}O$	396.76	nd(al), $[\alpha]^{17}{}_D$ + 24.6		167			bz, chl	B6^3, 3120
6572	Ergosterol-5,6-dihydro or α-Dihydroergosterol $C_{28}H_{46}O$	398.67	lf (ace) pl (chl-MeOH), $[\alpha]^{20}{}_D$ -19 (chl)		176-7				B6^3, 2841
6573	Ergosterone $C_{28}H_{42}O$	394.64	nd(ace-MeOH), $[\alpha]^{20}{}_D$ -4.52 (chl)		132				B7^3, 2032
6574	Ergotamine $C_{33}H_{35}N_5O_5$	581.67	nd (al) pr (bz) pl (aq ace), $[\alpha]^{20}{}_D$ -160 (chl, c=1)		213-4d			eth, bz, chl	B27^2, 860
6575	Ergotaminine $C_{33}H_{35}N_5O_5$	581.67	rh pl (MeOH) pl (al), $[\alpha]^{20}{}_D$ + 369 (chl, c=0.5)		252d			chl, py	B27^2, 860
6576	Ergothioneine or Thiasine $C_9H_{15}N_3O_2S$	229.30	pl(w + 2)nd or lf (dil al), $[\alpha]_D$ + 115 (w, c=1)		290d				B25^2, 413
6577	Erucic acid or *cis*-13-Docosenoic acid $CH_3(CH_2)_7CH{=}CH(CH_2)_{11}CO_2H$	338.57	nd(al)	265^{15}	33-4	$0.860^{55/4}$	1.4758^{20}	al, eth	B2^4, 1676
6578	Erysocine $C_{18}H_{21}NO_3$	299.37	nd(eth), $[\alpha]_D$ + 238.1		162			al, eth, chl	B21^4, 2624
6579	Erysodine $C_{18}H_{21}NO_3$	299.37	nd(al), $[\alpha]^{27}{}_D$ + 248 (al)		204-5			al, eth	B21^4, 2623
6580	Erysonine $C_{17}H_{19}NO_3$	285.34	cr(al)$[\alpha]^{25}{}_D$ + 285 (aq.HCl)		236-7d				B21^4, 2623
6581	Erysopine $C_{17}H_{19}NO_3$	285.34	cr(al), $[\alpha]^{25}{}_D$ + 265.2 (al-glycerol)		241-2				B21^4, 2622
6582	Erysothiopine $C_{10}H_{21}NO_7S$	299.37	cr(al-w), $[\alpha]^{25}{}_D$ + 194 (al)		168-9				B21^4, 2624
6583	Erysovine $C_{18}H_{21}NO_3$	299.37	pr(eth), $[\alpha]_D$ + 252 (al)		178-9			al, eth, chl	B21^4, 2623
6584	Erythraline $C_{18}H_{19}NO_3$	297.35	cr(al), $[\alpha]^{27}{}_D$ + 211.8 (al)		106-7			al, chl	Am73, 589

No.	Name, Synonyms, and Formula	Mol. wt.	Color, crystalline form, specific rotation and λ_{max} (log ε)	b.p. °C	m.p. °C	Density	n_D	Solubility	Ref.
6585	Erythramine or Dihydroerythraline $C_{18}H_{21}NO_3$	299.37	cr(eth-peth), $[\alpha]^{20}_D$ + 228 (al, c=0.19)	$125^{4\times10^{-4}}$	103-4			al, eth, ace, bz	Am73, 589
6586	Erythraline $C_{18}H_{19}NO_4$	315.33	cr(eth-peth), $[\alpha]^{28}_D$ + 145.5 (al)		170				Am73, 589
6587	*meso*-Erythritol $HOCH_2(CHOH)_2CH_2OH$	122.12	bipym tetr pr,	329-31	121.5	$1.451^{20/4}$		w, py	$B1^4$, 2807
6588	Erythritol anhydride $C_4H_6O_2$	86.09		138		$1.113^{18/4}$			
6589	Erythritol, tetranitrate or Cardilate $C_4H_6N_4O_{12}$	302.11	leaflets		61			al, w	
6590	β-Erythroidine $C_{16}H_{19}NO_3$	273.33	cr(abs al), $[\alpha]^{25}_D$ + 88.8 (w)		99-100			w, al, eth, bz, chl	Am80, 3905
6591	D-Erythronic acid-γ-lactone $C_4H_6O_4$	118.09	pr, $[\alpha]^{20}_D$ -73.2 (w, c=4)		104-5			w, al	$B18^4$, 1099
6592	L-Erythronic acid-γ-lactone $C_4H_6O_4$	118.09	nd (AcOEt), $[\alpha]^{20}_D$ + 73 (w, c=4)		105			al	$B18^4$, 1099
6593	D-Erythrose $C_4H_8O_4$	120.11	syr $[\alpha]^{20}_D$ + 1→-14.3 (w, c=11)					w, al	$B1^4$, 4172
6594	L-Erythrose $C_4H_8O_4$	120.11	syr $[\alpha]^{24}_D$ + 11.5→ + 30.5 (w, c=3)						$B1^4$, 4172
6595	Erythrosin or 2,4,5,7-Tetraiodo fluorescein $C_{20}H_8O_5I_4$	835.90	og-ye (eth)					al, eth	$B19^4$, 2923
6596	L-Erythrulose $HOCH_2CH(OH)COCH_2OH$	120.11	syr $[\alpha]D$ + 11.2	d				w, al	$B1^4$, 4176
6597	Escholerine $C_{41}H_{61}NO_{13}$	775.93	pl (ace-w), $[\alpha]^{25}_D$ -30 (Py, c=1)		235d			ace	$B21^4$, 6844
6598	Esculin or Aesculin (αβ-glucoside) $C_{15}H_{16}O_9$	340.29	pr (w + 2) $[\alpha]^{18}$/D-78.4 (50% aq diox)	230d	205d			aa, py	$B18^4$, 1326
6599	α-Estradiol $C_{18}H_{24}O_2$	272.39	nd(+ ½ w) (80% al), $[\alpha]^{20}_D$ + 56 (diox, c=0.9)		220-3			al, ace, peth	$B6^3$, 5332
6600	β-Estradiol $C_{18}H_{24}O_2$	272.39	pr (80% al), $[\alpha]^{25}_D$ + 76 (diox)		178-9			al, ace, diox	$B6^3$, 5337
6601	Estriol $C_{18}H_{24}O_3$	288.39	lf(al) mcl (dil al), $[\alpha]_D$ + 61 (al) + 30 (Py)		288d	1.27			$B6^3$, 6520
6602	Estrone $C_{18}H_{22}O_2$	270.37	α mcl(al) βγ orth(al)		260.2	β-1.236, γ-1.228		diox, py	$B8^3$, 1171
6603	Ethane CH_3CH_3	30.07	gas, hex cr	−88.6	−183.3	$0.572^{-100/4}$	$1.0377^{0/}_{546}$ mm	bz	$B1^4$, 108
6604	Ethane, 1,2-bis (ethylthio) $C_2H_5SCH_2CH_2SC_2H_5$	150.30		217, 95.5^{12}		$0.9815^{20/4}$	1.5118^{20}	al, eth	$B1^4$, 2452

No.	Name, Synonyms, and Formula	Mol. wt.	Color, crystalline form, specific rotation and λ_{max} (log ε)	b.p. °C	m.p. °C	Density	n_D	Solubility	Ref.
6605	Ethane, 2,2-bis (4-methoxyphenyl)-1,1,1-trichloro or Methoxychlor ... $(4\text{-}CH_3OC_6H_4)_2CHCCl_3$	345.65	cr (dil al)	...	94	...	...	al, eth, bz	$B6^3$, 5436
6606	Ethane, 1,2-bis(methyl thio) ... $CH_3SCH_2CH_2SCH_3$	122.24	...	182.5^{750}, $78\text{-}80^{11}$	...	$1.0371^{20/4}$	1.5292^{20}	w, al, eth, ace, chl	$B1^4$, 2457
6607	Ethane, 1,2-bis (phenyl sulfonyl) ... $C_6H_5SO_2CH_2CH_2O_2SC_6H_5$	310.38	nd or lf (al)	...	180	...	...	al, bz, aa	$B6^4$, 1495
6608	Ethane, 1,2-bis (phenylthio) ... $C_6H_5SCH_2CH_2SC_6H_5$	246.39	ta (al)	...	70	...	...	ace	$B6^4$, 1493
6609	Ethane, 2,2-bis (4 chlorophenyl)-1,11-trichloro or DDT, Dichloro diphenyl trichloro ethane ... $C_{14}H_9Cl_5$	354.49	nd (al)	260	108-9	...	...	eth, ace, bz, chl, peth	$B5^3$, 1833
6610	Ethane, bromo or Ethyl bromide ... C_2H_5Br	108.97	...	38.4	−118.6	$1.4604^{20/4}$	1.4239^{20}	al, eth, chl	$B1^4$, 150
6611	Ethane, 1-bromo-2-chloro or Ethylene chloro bromide ... $ClCH_2CH_2Br$	143.41	...	107	−16.7	$1.7392^{20/4}$	1.4908^{20}	al, eth	$B1^4$, 155
6612	Ethane, 1-bromo-2-fluoro ... FCH_2CH_2Br	126.96	...	71-2	...	$1.7044^{25/4}$	1.4236^{20}	al, eth	$B1^4$, 154
6613	Ethane, 1-bromo-2-methoxy ... $BrCH_2CH_2OCH_3$	140.00	...	110.3	...	$1.4623^{20/4}$	1.44753^{20}	...	$B1^4$, 1386
6614	Ethane, chloro or Ethyl chloride ... C_2H_5Cl	64.51	...	12.3	−136.4	$0.8978^{20/4}$	1.3676^{20}	al, eth	$B1^4$, 124
6615	Ethane, 1-chloro-2-fluoro ... FCH_2CH_2Cl	82.51	...	59^{750}	...	$1.1747^{20/4}$	1.3775^{20}	al, eth	$B1^4$,127
6616	Ethane, 1-chloro-2-iodo ... ICH_2CH_2Cl	190.41	...	140	−15.6	2.16439^{0}	...	...	$B1^4$, 167
6617	Ethane, 1-chloro-1-nitro ... $CH_3CHCl(NO_2)$	109.51	...	124-5	...	$1.2860^{20/20}$	...	...	$B1^4$, 172
6618	Ethane, chloro pentafluoro ... $ClCF_2CF_3$	154.47	gas	−38	−106	...	...	al, eth	$B1^4$, 129
6619	Ethane, 2-chloro-1,1,1-trifluoro ... $ClCH_2CF_3$	118.49	...	6.93	−105.5	$1.389^{0/4}$	$1.3090^{0}/_{0}$	...	$B1^4$, 128
6620	Ethane, 1-(2 chloroethoxy)-2-phenoxy ... $ClCH_2CH_2OCH_2CH_2OC_6H_5$	200.67	...	149^{10}	...	$1.149^{15/15}$	...	...	$B6^3$, 568
6621	Ethane, 1,1-dibromo or Ethylidene bromide ... CH_3CHBr_2	187.86	...	108, 9.0^{10}	−63	$2.0555^{20/4}$	1.5128^{20}	al, eth, ace, bz	$B1^4$, 157
6622	Ethane, 1,2-dibromo or Ethylene dibromide ... CH_2BrCH_2Br	187.86	...	131.3, 29.1^{10}	9.8	$2.1792^{20/4}$	1.5387^{20}	al, eth, ace, bz	$B1^4$, 158
6623	Ethane, 1,2-dibromo-1,1-dichloro ... $BrCH_2CBrCl_2$	256.75	...	178.3, 58.8^{10}	−66.8	$2.2623^{20/4}$	1.5567^{20}	al, eth, ace, bz	$B1^4$, 161
6624	Ethane, 1,2-dibromo-1,2-dichloro ... $BrCHCl\text{-}CHBrCl$	256.75	...	195, 84^{45}	−26	$2.135^{20/4}$	1.5662^{20}	al, eth, ace, bz	$B1^4$, 161
6625	Ethane, 1,2-dibromo-tetrafluoro ... $BrCF_2CF_2Br$	259.82	...	46.4	−112	$2.149^{25/4}$	...	...	$B1^4$, 160
6626	Ethane, 1,1-dichloro or Ethylidene chloride ... CH_3CHCl_2	98.96	...	57.3	−97	$1.1757^{20/4}$	1.4164^{20}	al, eth, ace, bz	$B1^4$, 130
6627	Ethane, 1,1-dichloro-1-fluoro ... CH_3CCl_2F	116.95	...	32	−103.5	$1.250^{10/4}$	$1.3600^{10}/_{D}$	...	$B1^4$, 134
6628	Ethane, 1,1-dichloro-1,2,2,2 tetrafluoro ... F_3CCCl_2F	170.92	...	3.6	−94	$1.455^{25/4}$	1.3092^{0}	al, eth, bz, chl	$B1^4$, 136
6629	Ethane, 1,2-dichloro or Ethylene dichloride ... $ClCH_2CH_2Cl$	98.96	...	83.5	−35.3	1.2351^{20}	1.4448^{20}	al, eth, ace, bz	$B1^4$, 131
6630	Ethane, 1,2-dichloro-1,1-difluoro ... $CH_2Cl_2ClCClF_2$	134.94	...	46.8	−1012	$1.4163^{20/4}$	$1.36193^{20}/_{D}$	...	$B1^4$, 135
6631	Ethane, 1,2-dichloro-1-fluoro ... $ClCH_2CHClF$	116.95	...	73.7	...	$1.3814^{20/4}$	$1.41132^{20}/_{D}$	...	$B1^4$, 134
6632	Ethane, 1,1-difluoro or Ethylidene fluoride ... CH_3CHF_2	66.05	gas	−24.7	−117	0.95^{20} (sat pr)	1.3011^{-72}	...	$B1^4$, 120
6633	Ethane, 1,1-difluoro-1,2,2,2-tetrachloro ... Cl_3CCClF_2	203.83	...	91.5	40.6	...	...	al, eth, chl	$B1^4$, 146
6634	Ethane, 1,2-difluoro or Ethylene difluoride ... CH_2FCH_2F	66.05	...	30.7	...	...	...	eth, bz, chl	$B1^4$, 121
6635	Ethane, 1,2-difluoro-1,1,2,2-tetrachloro ... $FCCl_2\text{-}CFCl_2$	203.83	...	93	25	$1.6447^{25/4}$	1.4130^{25}	al, eth, chl	$B1^4$, 146
6636	Ethane, 1,1-diiodo or Ethylidine iodide ... CH_3CHI_2	281.86	...	179-80, $60\text{-}1^{12}$	...	2.84^{0}	1.673^{20}	al, eth, chl, ace	$B1^4$, 169

No.	Name, Synonyms, and Formula	Mol. wt.	Color, crystalline form, specific rotation and λ_{max} (log ε)	b.p. °C	m.p. °C	Density	n_D	Solubility	Ref.
6637	Ethane, 1,2-diido or Ethylene diiodo CH_2ICH_2I	281.86	ye mcl pr or rh (eth) d in lt	200, 74^{10}	83	$3.325^{20/4}$	1.871^{20}	al, eth, ace, chl	$B1^4$, 169
6638	Ethane, 1,2-di-N-morpholyl $C_{10}H_{20}N_2O_2$	200.28	wh-yesh (eth or lig)	$160\text{-}3^{25}$	75			w, al, ace, bz	B27, 7
6639	Ethane, 1,1-dinitro $CH_3CH(NO_2)_2$	120.06	ye mcl (bz or MeOH)	185-6, 72^{12}		$1.3503^{24/24}$		al, eth	$B1^4$, 174
6640	Ethane, 1,2-dinitro-1,1,2,2-tetrafluoro $O_2NCF_2CF_2(NO_2)$	192.03		58-9	−41.5	$1.6024^{25/4}$	1.3265^{25}	ace	$B1^4$, 175
6641	Ethane, 1,1-diphenyl or α-Methylditan $CH_3CH(C_6H_5)_2$	182.27		286, 148^{15}	−21.5	$0.9997^{20/4}$	1.5756^{20}	**al, eth**, bz	$B5^4$, 1880
6642	Ethane, 1,2-diphenyl or Bibenzyl $C_6H_5CH_2CH_2C_6H_5$	182.27	nd (al)	285, $95\text{-}6^{1}$	52.2	$0.9583^{60/4}$	1.5478^{60}	al, eth	$B5^4$, 1868
6643	Ethane, 1,2-di-(4-tolyl) $(4\text{-}CH_3C_6H_4)CH_2CH_2(C_6H_4CH_3\text{-}4)$	210.32	lf(MeOH or dil al) pl (lig)	296-8, 178^{18}	82-3			bz, peth	$B5^4$, 1943
6644	Ethane, 2,2-di-4-tolyl-1,1,1-trichloro $(4\text{-}CH_3C_6H_4)_2CHCCl_3$	313.65	mcl pr (al eth-al)		92			al, eth, ace	$B5^4$, 1949
6645	Ethane, 1-ethoxy-2-methylamino or Ethyl-β-methyl amino ethyl ester $C_2H_5OCH_2CH_2NHCH_3$	103.17		$114\text{-}5^{744}$		$0.8363^{20/4}$	1.4147^{20}	w, al, eth, ace, bz	$B4^3$, 647
6646	Ethane, fluoro or Ethyl fluoride CH_3CH_2F	48.06	gas	−37.7	−143.2	$0.7182^{20/4}$ (liq)	1.2656^{20}	al, eth	$B1^4$, 120
6647	Ethane, 1-fluoro-1,2,2-trichloro $Cl_2CHCHFCl$	151.40		101-3		1.54968^{17}	$1.4390^{20}{}_D$		$B1^4$, 141
6648	Ethane, fluoro penta chloro FCl_2CCCl_3	220.29		134-6	101.3			al, eth	$B1^4$, 148
6649	Ethane, hexabromo or Perbromo ethane C_2Br_6	503.45	rh pr (bz)	d200-10	d	$2.823^{20/4}$	1.863		$B1^3$, 193
6650	Ethane, hexachloro or Perchloro ethane C_2Cl_6	236.74	rh (al-eth)	186^{777}	186-7 (sealed tube)	$2.091^{20/4}$		al, eth, bz	$B1^4$, 148
6651	Ethane, hexafluoro or Perfluoro ethane C_2F_6	138.01	gas	−79	−94	1.590^{-78}			$B1^4$, 123
6652	Ethane, hexaphenyl $(C_6H_5)_3CC(C_6H_5)_3$	486.66	cr (ace)	d	145-7d			eth, ace, chl, McOII	$B5^3$, 2746
6653	Ethane, iodo or Ethyl iodide C_2H_5I	155.97		72.3	−108	$1.9358^{20/4}$	1.5133^{20}	al, eth	$B1^4$, 163
6654	Ethane, isocyano or Ethyl carbylamine CH_3CH_2NC	55.08		79^{775}	<-66	$0.7402^{20/4}$	1.3622^{20}	**al, eth,** ace	$B4^4$, 342
6655	Ethane-1-(4-methoxyphenyl)-1-phenyl or 1-*p*-Anisyl-1-phenyl ethane $CH_3CH(C_6H_5)(C_6H_4OCH_3\text{-}4)$	212.29		$180\text{-}2^{19}$		$1.0473^{20/4}$	1.5725^{20}	eth, ace, bz	$B6^2$, 639
6656	Ethane, nitro $C_2H_5NO_2$	75.07		115	−50	$1.0448^{25/4}$	1.3917^{20}	**al, eth,** ace	$B1^4$, 170
6657	Ethane, nitro-pentafluoro $CF_3CF_2NO_2$	165.02		0				eth	$B1^4$, 172
6658	Ethane, 1,nitro-2,2,2-trifluoro $F_3CCH_2NO_2$	129.04		96		$1.3914^{20/4}$	1.3394^{20}	eth	$B1^4$, 172
6659	Ethane, nitroso-pentafluoro CF_3CF_2NO	149.02		−42					$B1^4$, 169
6660	Ethane, pentabromo Br_2CHBr_3	424.55	mcl pr (dil al)	210^{300}	56-7	$3.312^{20/4}$		al, eth	$B1^3$, 193
6661	Ethane, pentachloro $CHCl_2CCl_3$	202.29		162	−29	$1.6796^{20/4}$	1.5025^{20}	**al, eth**	$B1^4$, 147
6662	Ethane, pentaiodo CHI_2CI_3	659.55	mcl pr (aa)		182-4			al, eth, bz, aa	$B1^1$, 31
6663	Ethane, perfluoro CF_3CF_3	138.01		−79	−100.6	1.590^{-98}			$B1^4$, 123
6664	Ethane, 1,1,1,2-tetrabromo CH_2BrCBr_3	345.65		$112^{18}{}_d$	0.0	$2.8748^{20/4}$	1.6277^{20}	al, eth, ace, bz, chl	$B1^4$, 162
6665	Ethane, 1,1,2,2,-tetrabromo $CHBr_2CHBr_2$	345.65	yesh	243.5, 114.8^{10}	0	$2.9656^{20/4}$	1.6353^{20}	**al, eth,** ace, bz, aa	$B1^4$, 162
6666	Ethane, 1,1,1,2--tetrachloro CH_2ClCCl_3	167.85	yesh red	130.5, 22.1^{10}	-70.2	$1.5406^{20/4}$	1.4821^{20}	**al, eth,** ace, bz, chl	$B1^4$, 143
6667	Ethane, 1,1,2,2-tetrachloro $CHCl_2CHCl_2$	167.85		146.2, 33.9^{10}	−36	$1.5953^{20/4}$	1.4940^{20}	**al, eth,** ace, bz	$B1^4$, 144

No.	Name, Synonyms, and Formula	Mol. wt.	Color, crystalline form, specific rotation and λ_{max} (log ε)	b.p. °C	m.p. °C	Density	n_D	Solubility	Ref.
6668	Ethane, 1,1,1,2-tetrafluoro CH_2FCF_3	102.03		-26.5^{736}				eth	$B1^4$, 123
6669	Ethane, 1,1,1,2-tetraphenyl $C_6H_5CH_2C(C_6H_5)_3$	334.46	mcl (eth-peth)	$277\text{-}80^{21}$	143-4			bz	$B5^3$, 2575
6670	Ethane, 1,1,2,2-tetraphenyl $(C_6H_5)_2CHCH(C_6H_5)_2$	334.46	cr (bz + l) rh nd (chl)	358-62, 260^{16}	214-5			aa	$B5^1$, 2574
6671	Ethane, 1,1,2-tribromo $CH_2BrCHBr_2$	266.76		188.9, 73.1^{10}	−29.3	$2.6211^{20/4}$	1.5933^{20}	al, eth, bz, chl	$B1^4$, 161
6672	Ethane, 1,1,1-trichloro or Methyl chloroform CH_3CCl_3	133.40		74.1	−30.4	$1.3390^{20/4}$	1.4379^{20}	**al, eth,** chl	$B1^4$, 138
6673	Ethane, 1,1,1-trichloro-2,2,2-trifluoro CF_3CCl_3	187.38		45.8	14.2	$1.5790^{20/4}$	1.3610^{25}	al, eth, chl	$B1^4$, 142
6674	Ethane, 1,1,2-trichloro $CH_2ClCHCl_2$	133.40		113.8, 9.5^{10}	−36.5	$1.4397^{20/4}$	1.4714^{20}	al, **eth,** chl	$B1^4$, 139
6675	Ethane, 1,1,2-trichloro-1,2,2-trifluoro CF_2ClCCl_2F	187.38		47.7	−36.4	$1.5635^{25/4}$	1.3557^{25}	al, **eth, bz**	$B1^4$, 142
6676	Ethane, 1,1,1-trifluoro CH_3CF_3	84.04	gas	−47.3	−111.3			eth, chl	$B1^4$, 122
6677	Ethane, 1,1,1-triido CH_3CI_3	407.76	ye oct (al)		95			eth, bz	$B1^1$, 199
6678	Ethane, 1,1,1-triphenyl or α-Methyl tritan $CH_3C(C_6H_5)_3$	258.36	nd (al, eth)	$205\text{-}10^{18}$	95			eth, bz	$B5^1$, 2331
6679	Ethane, 1,1,2-triphenyl $C_6H_5CH_2CH(C_6H_5)_2$	258.36	mcl lf (dil al) nd (al)	$348\text{-}9^{751}$	57			al, eth, bz	$B5^1$, 2329
6680	Ethanediol or Ethylene glycol $HOCH_2CH_2OH$	62.07		198, 93^{13}	−11.5	$1.1088^{20/4}$	1.4318^{20}	**w, al,** eth, **ace**	$B1^4$, 2369
6681	1,2-Ethanediol, 1,2-dicylohexyl (dl) or Cyclohoxanone pinacol $C_6H_{11}CH(OH)\text{-}CH(OH)C_6H_{11}$	226.36	nd		129-30			bz, peth	$B6^1$, 4156
6682	1,2-Ethanediol, 1,2-diphenyl (d) $C_6H_5CH(OH)CH(OH)C_6H_5$	214.26	nd (w) lf or pr (abs al) [α] + 9_2 (abs al c=1.2) + 128 Cbz, c=0.3		148-9d			al	$B6^1$, 5431
6683	1,2-Ethanediol, 1,2-diphenyl (dl) $C_6H_5CH(OH)CH(OH)C_6H_5$	214.26	nd (w or al) ta (eth)	>300, 133^{0} ^{023}	122-3			al, eth, chl	$B6^1$, 5431
6684	1,2-Ethanediol, 1,2-diphenyl (l) $C_6H_5CH(OH)CH(OH)C_6H_5$	214.26	lf (eth abs al or bz) pr(bz or abs al)		148-9			al, eth, ace, bz	$B6^1$, 5431
6685	1,2-Ethanediol, 1,2-diphenyl-*(meso)* $C_6H_5CH(OH)CH(OH)C_6H_5$	214.26	nd or lf (w or bz-peth) mcl lf (al, w)	>300, 139^{0} ^{023}	139-40			al, chl	$B6^3$, 5429
6686	1,2-Ethanediol, phenyl-(dl) or Phenyl ethylene glycol $C_6H_5CH(OH)CH_2OH$	138.17	nd (lig)	$272\text{-}4^{755}$	69-70			w, al, eth, bz	$B6^3$, 4572
6687	1,2-Ethanol, tetraphenyl or Bensopinacol $(C_6H_5)_2C(OH)C(OH)(C_6H_5)_2$	366.46	pr (bz + l) cr (ace)		182			eth, ace, bz, chl	$B6^3$, 5923
6688	Ethanedione, di-(2-furyl) $(OC_4H_3)COCO(C_4H_3O)$	190.16	ye nd (al) cr (bz)		165-6			al, eth, bz, chl	$B19^4$, 2007
6689	1,2-Ethane disulfonic acid $HO_3SCH_2CH_2SO_3H$	190.19	hyg nd (gl aa)		111-2 (+2w) 174 (anh)			w, al, diox	$B4^4$, 78
6690	1,2-Ethanedithiol or Dithioglycol $HSCH_2CH_2SH$	94.19		146, $46\text{-}7^{6}$	−41.2	$1.1243^{20/4}$	1.5590^{20}	al, eth, ace, bz	$B1^4$, 2450
6691	Ethane dithiolic acid, diethyl ester or Diethyl dithio oxalate $C_2H_5SCO\text{-}COSC_2H_5$	178.26	ye nd (eth)	235, $80\text{-}2^{32}$	27	$1.0565^{21/4}$		eth	$B1^1$, 244
6692	Ethane phosphonic acid $C_2H_5PO(OH)_2$	110.05	pl or nd (w)		61-2			w, al, eth	$B4^3$, 1779
6693	Ethane phosphonic acid, diethyl ester $C_2H_5PO(OC_2H_5)_2$	166.16		198, 83^{13}		$1.0259^{20/4}$	1.4163^{20}	al, eth	$B4^3$, 1779
6694	Ethane phosphonic acid, dimethyl ester $C_2H_5PO(OCH_3)_2$	138.10		82^{18}		$1.1029^{30/4}$	1.4128^{30}	w, al, bz	J1954, 3222
6695	Ethane sulfonic acid or Ethyl sulfonic acid $CH_3CH_2SO_3H$	110.13	hyg	123^{1}	−17	1.3341^{25}	1.4335^{20}	w, al	$B4^4$, 33

No.	Name, Synonyms, and Formula	Mol. wt.	Color, crystalline form, specific rotation and λ_{max} (log ε)	b.p. °C	m.p. °C	Density	n_D	Solubility	Ref.
6696	Ethane sulfonyl chloride $C_2H_5SO_2Cl$	128.57 etc.	pa ye	171, 65^{11}		$1.357^{22.5}$	1.4531^{20}	eth	$B4^4$, 34
6697	Ethane sulfonyl chloride, 2-bromo $BrCH_2CH_2SO_2Cl$	207.47	pa ye	102^{11}		1.921^{20}	1.5242^{20}		$B4^4$, 37
6698	Ethane sulfonyl chloride, 1-chloro $CH_3CHClSO_2Cl$	163.02		$80\text{-}1^{22}$			1.4782^{20}		$B1^4$, 3121
6699	Ethane sulfonyl chloride, 2-chloro $ClCH_2CH_2SO_2Cl$	163.02		200-3, $93\text{-}7^{17}$		$1.555^{20/4}$	1.4920^{20}		$B4^4$, 36
6700	Ethane sulfonic acid, 2-hydroxy, dihydrate or Isethionic acid $HOCH_2CH_2SO_3H_2H_2O$	162.16	hyg cr (aa-AC_2O)		111-2			w, al	$B4^4$, 84
6701	1,1,2,2-Ethane tetracarboxylic acid, 1,2-diethyl ester $C_2H_5O_2CCH(CO_2H)CH(CO_2H)CO_2C_2H_5$	262.22	hyg lf (+ 1/2 w)		132-3d			w, al, eth	B2, 858
6702	1,1,2,2-Ethane tetracarboxylic acid, tetraethyl ester $(C_2H_5O_2C)_2CHCH(CO_2C_2H_5)_2$	318.32		d 305	77	1.064^{80}	1.4105^{80}	al	$B2^4$, 2415
6703	1,1,2,2-Ethane tetracarboxylic acid, tetramethyl ester $(CH_3O_2C)_2CHCH(CO_2CH_3)_2$	262.22	cr (eth al bz)		138			al	$B2^3$, 2076
6704	Ethanethiol or Ethyl mercaptan C_2H_5SH	62.13		35	−144.4	$0.8391^{20/4}$	1.4310^{20}	al, eth, ace	$B1^4$, 1390
6705	Ethanethiol, 2-amino or Cysteamine $H_2NCH_2CH_2SH$	77.14	cr (sub)	d^{760} Sub (vac)	99-100			w, al	$B4^4$, 1570
6706	Ethanethiol, 2-chloro $ClCH_2CH_2SH$	96.57		113		$1.1826^{20/4}$	1.4929^{20}	al, eth, diox	$B1^4$, 1406
6707	Ethanethiol, 1-phenyl (l) $CH_3CH(C_6H_5)SH$	138.23	$[\alpha]^{20}_D$ −89 al, c=6	199-200, 83^{10}		$1.022^{20/4}$	1.5593^{20}	al, eth, bz	$B6^3$, 1697
6708	Ethanol or Ethyl alcohol CH_3CH_2OH	46.07		78.5	-117.3	$0.7893^{20/4}$	1.3611^{20}	**w, eth, ace**, bz	$B1^4$, 1289
6709	Ethanol, β-(4-amino phenyl) $(4\text{-}NH_2C_6H_4)CH_2CH_2OH$	137.18	nd (al)		108				$B13^3$, 1679
6710	Ethanol, 2-bromo or Ethylene bromohydrin $BrCH_2CH_2OH$	124.97		$149\text{-}50^{750}$, 51^4		$1.7629^{20/4}$	1.4915^{20}	**w, al, eth**	$B1^4$, 1385
6711	Ethanol-2-chloro or Ethylene chlorohydrin $ClCH_2CH_2OH$	80.51		128, 44^{20}	-67.5	$1.2003^{20/4}$	1.4419^{20}	**w, al**	$B1^4$, 1372
6712	Ethanol, 2-chloro-1-phenyl or Styrene Chlorohydrin $ClCH_2CH(OH)C_6H_5$	156.61		128^{17}		$1.1926^{20/4}$	1.5523^{20}	al, eth	$B6^3$, 1683
6713	Ethanol, 2,2-dichloro Cl_2CHCH_2OH	114.96		146, $37\text{-}8^6$		$1.4040^{25/4}$	1.4626^{25}	al, eth	$B1^4$, 1383
6714	Ethanol, 2,2-diphenyl $(C_6H_5)_2CHCH_2OH$	198.26		195^{20}, $144\text{-}5^1$	64.5			eth, al, ace	$B6^3$, 3397
6715	Ethanol, 2-ethylthio $C_2H_5SCH_2CH_2OH$	106.18		184	Ca-100	$1.0166^{20/4}$	1.4867^{20}	al, ace	$B1^4$, 2430
6716	Ethanol, 2-fluoro or Ethylene fluorohydrin FCH_2CH_2OH	64.06		103.5	-26.4	$1.1040^{20/4}$	1.3647^{18}	**w, al, eth**, ace	$B1^4$, 1366
6717	Ethanol, 2-iodo or Ethylene iodohydrin ICH_2CH_2OH	171.97		176-7d, $85\text{-}8^{25}$		$2.1968^{20/4}$	1.5713^{20}	w, al, eth	$B1^4$, 1387
6718	Ethanol, 2-mercapto or Monothio ethylene glycol $HSCH_2CH_2OH$	78.13		$157\text{-}8^{742}$, 55^{13}		$1.1143^{20/4}$	1.4996^{20}	w, al, eth, bz	$B1^4$, 2428
6719	Ethanol, 2-methylthio $CH_3SCH_2CH_2OH$	92.16		$68\text{-}70^{20}$		$1.6640^{20/20}$	1.4867^{30}	w, al, eth	$B1^4$, 2429
6720	Ethanol, 2-nitro or β-Nitro ethanol $O_2NCH_2CH_2OH$	91.07		194^{765}, 102^{10}	-80	$1.270^{15/4}$	1.4438^{19}	**w, al, eth**	$B1^4$, 1388
6721	Ethanol, 2-triazo $N_3CH_2CH_2OH$	87.08		75^{40}		$1.149^{24/24}$		**w**	$B1^4$, 1389
6722	Ethanol, 2,2,2-tribromo or Avertin, bromethol CBr_3CH_2OH	282.76	nd or pr (peth)	$92\text{-}3^{10}$	81			al, eth, bz	$B1^3$, 1362
6723	Ethanol, 2,2,2-trichloro CCl_3CH_2OH	149.40	hyg rh ta or pl	151^{737}, 52^{11}	19		1.4861^{20}	**al, eth**	$B1^4$, 1383
6724	Ethanol, 2,2,2-trifluoro CF_3CH_2OH	100.04		74	-43.5	1.4680^{20}, $1.3739^{22/4}$	1.2907^{22}	al, eth, ace, bz	$B1^4$, 1370
6725	Ethanol, 1,1,2-triphenyl $C_6H_5CH_2C(OH)(C_6H_5)_2$	274.36	nd (bz-lig) pr (peth)	222^{11}	89-90			al	$B6^3$, 3680
6726	Ethanol, 2,2,2-triphenyl $(C_6H_5)_3CCH_2OH$	274.36	cr (al eth lig)		110.5d			al, eth, bz, lig	$B6^3$, 3683
6727	Ethanolamine or 2-Amino ethanol $H_2NCH_2CH_2OH$	61.08		170, 58^5	10.3	$1.0180^{20/4}$	1.4541^{20}	**w, al**, chl	$B4^4$, 1406

No.	Name, Synonyms, and Formula	Mol. wt.	Color, crystalline form, specific rotation and λ_{max} (log ε)	b.p. °C	m.p. °C	Density	n_D	Solubility	Ref.
6728	Ethanolamine, N-acetyl $CH_3CONHCH_2CH_2OH$	103.12	nd (ace)	$166\text{-}7^{8}$	63-5	$1.1079^{25/4}$	1.4674^{20}	**w**	$B4^{4}$, 1535
6729	Ethanolamine, N-butyl $C_4H_9NHCH_2CH_2OH$	117.19	...	199-200, $91\text{-}2^{11}$	...	$0.8907^{20/4}$	1.4437^{20}	w, al, eth	$B4^{3}$, 682
6730	Ethanolamine, N-isobutyl i-$C_4H_9NHCH_2CH_2OH$	117.19	...	199-200, 90^{16}	...	$0.8818^{20/4}$	1.4402^{20}	w, al, eth	$B4^{3}$, 683
6731	Ethanolamine, N,N-diethyl $(C_2H_5)_2NCH_2CH_2OH$	117.19	hyg	163, $56\text{-}7^{15}$	...	$0.8921^{20/4}$	1.4412^{20}	**w**, al, eth, ace, bz	$B4^{4}$, 1471
6732	Ethanolamine, N,N-dimethyl $(CH_3)_2NCH_2CH_2OH$	89.14	...	134	...	$0.8866^{20/4}$	1.4300^{20}	**w, al, eth**	$B4^{4}$, 1424
6733	Ethanolamine, N-ethyl $C_2H_5NHCH_2CH_2OH$	89.14	...	169-70, $78\text{-}80^{27}$	...	$0.914^{20/4}$	1.444^{20}	w, al, eth	$B4^{4}$, 1465
6734	Ethanolamine, N-methyl $CH_3NHCH_2CH_2OH$	75.11	...	158, 52^{6}	...	0.937^{20}	1.4385^{20}	**w, al, eth**	$B4^{4}$, 1422
6735	Ethanolamine, N-methyl-N-phenyl $C_6H_5N(CH)_3CH_2CH_2OH$	151.21	yesh	150^{14}	...	$0.9995^{15/0}$	...	al, eth, ace, bz	$B12^{3}$, 296
6736	Ethanolamine, N-β-naphthyl β-$C_{10}H_7NHCH_2CH_2OH$	187.24	lf (eth or al)	$197\text{-}8^{3}$	52	...	...	eth	$B12^{2}$, 717
6737	Ethanolamine, N-phenyl $C_6H_5NHCH_2CH_2OH$	137.18	...	286, 167^{17}	...	1.0945^{20}	1.5760^{20}	al, eth, chl	$B12^{3}$, 293
6738	Ethanolamine, 1-phenyl $H_2NCH_2CH(OH)C_6H_5$	137.18	nd (al-eth-peth)	160^{17}	56-7	...	...	w, al	$B13^{4}$, 1657
6739	Ethanolamine, N-isopropyl i-$C_3H_7NHCH_2CH_2OH$	103.16	...	172-4, $76\text{-}7^{15}$	...	$0.8970^{20/4}$	1.4395^{20}	**w, al, eth**	$B4^{3}$, 681
6740	Ethoxyacetic acid $C_2H_5OCH_2CO_2H$	104.11	...	206-7, 111^{25}	...	$1.1021^{20/4}$	1.4194^{20}	w, al, eth	$B2^{4}$, 574
6741	Ethoxyacetic acid, ethyl ester or Ethyl ethoxy acetate $C_2H_5OCH_2CO_2C_2H_5$	132.16	...	158, 52^{12}	...	$0.9702^{20/4}$	1.4029^{20}	al, eth, ace	$B3^{4}$, 581
6742	Ethoxyacetic acid, l-menthyl ester or l-Menthyl ethoxy acetate $C_2H_5OCH_2CO_2C_{10}H_{19}$	242.36	$[\alpha]^{20}_{D}$ −66.35	155^{20}	...	$0.9545^{20/4}$	...	al, eth, chl	$B6^{2}$, 47
6743	Ethoxyacetic acid, methyl ester or Methylethoxy acetate $C_2H_5OCH_2CO_2CH_3$	118.13	...	$147\text{-}8^{734}$	...	1.0112^{15}	...	al, eth, ace	$B3^{4}$, 578
6744	2-Ethoxybenzoic acid 2-$C_2H_5OC_6H_4CO_2H$	166.18	...	$211\text{-}2^{35}$	20.7	...	...	...	$B10^{3}$, 98
6745	2-Ethoxybenzoic acid, ethyl ester 2-$C_2H_5OC_6H_4CO_2C_2H_5$	194.23	...	251, $180\text{-}5^{113}$	...	1.005^{20}	...	al, eth	$B10^{3}$, 117
6746	3-Ethoxybenzoic acid 3-$C_2H_5OC_6H_4CO_2H$	166.18	nd (w or sub)	sub	137	...	...	al, eth, bz	$B10^{3}$, 245
6747	3-Ethoxybenzoic acid, ethyl ester or Ethyl-3-ethoxy benzoate 3-$C_2H_5OC_6H_4CO_2C_2H_5$	194.23	...	264, $172\text{-}3^{50}$	...	$1.0725^{20/20}$	...	al, eth	$B10^{3}$, 251
6748	4-Ethoxybenzoic acid 4-$C_2H_5OC_6H_4CO_2H$	166.18	nd (w)	...	198.5	...	...	al, eth, bz	$B10^{3}$, 282
6749	4-Ethoxybenzoic acid, ethyl ester or Ethyl-4-ethoxy benzoate 4-$C_2H_5OC_6H_4CO_2C_2H_5$	194.23	...	275 $148\text{-}9^{14}$	: lf	1.076^{14}	...	al, eth	$B10^{3}$, 301
6750	Ethoxyl amine or α-Ethyl hydroxyl amine $C_2H_5ONH_2$	61.08	...	...	68	$0.8872^{8/8}$	...	**w, al, eth**	$B1^{4}$, 1326
6751	Ethylaceto acetate $CH_3COCH_2CO_2C_2H_5$	130.14	...	180.4, 74^{14}	<-80	$1.0282^{20/4}$	1.4194^{25}	**al, eth,** bz, chl	$B3^{4}$, 1528
6752	Ethylacetoacetate (enol form) $CH_3C(OH){=}CHCO_2C_2H_5$	130.14	...	...	...	1.0119^{10}	1.4432^{20}	...	$B3^{4}$, 1528
6753	Ethylacetoacetate (keto form) $CH_3COCH_2CO_2C_2H_5$	130.14	...	...	-39	$1.0368^{10/4}$	1.4171^{20}	...	$B3^{4}$, 1528
6754	Ethyl amine $C_2H_5NH_2$	45.08	...	16.6	-81	$0.6829^{20/4}$	1.3663^{20}	**w, al, eth**	$B4^{4}$, 307
6755	Ethyl amine, hydrobromide $C_2H_5NH_2\text{-}HBr$	126.00	mcl nd or pl (al)	...	159.3	...	...	w, al	$B4^{4}$, 310
6756	Ethyl amine, hydrochloride $C_2H_5NH_2.HCl$	81.55	mcl pl (al)	d 315	109-10	...	...	w, al	$B4^{4}$, 310
6757	Ethyl amine, hydroiodide $C_2H_5NH_2HI$	173.00	mcl nd (w)	...	188.5	2.100	...	w	$B4^{4}$, 310
6758	Ethyl amine, 2-bromo, hydrochloride $BrCH_2CH_2NH_2.HCl$	160.44	lf (al-AcOEt)	...	174-5d	...	...	w, al, bz	$B4^{3}$, 248
6759	Ethyl amine, 2-chloro, hydrochloride $ClCH_2CH_2NH_2.HCl$	115.99	hyg cr (al-eth)	...	144 (148)	...	...	w, al, ace	$B4^{3}$, 236

No.	Name, Synonyms, and Formula	Mol. wt.	Color, crystalline form, specific rotation and λ_{max} (log ε)	b.p. °C	m.p. °C	Density	n_D	Solubility	Ref.
6760	Ethyl amine, 2-ethoxy $C_2H_5OCH_2CH_2NH_2$	89.14	...	108^{758}	...	$0.8512^{20/4}$	1.4101^{20}	w, al, eth, ace, bz	$B4^4$, 1411
6761	Ethyl amine, 2-methoxy $CH_3OCH_2CH_2NH_2$	75.11	...	95^{756}	...	...	...	w, al	$B4^4$, 1411
6762	Ethyl amime, perfluoro $CF_3CF_2NF_2$	171.02	...	−35	...	...	...	...	$B2^4$, 473
6763	Ethyl β-amino, ethyl ether $C_2H_5OCH_2CH_2NH_2$	89.14	...	108^{758}	...	$0.8512^{20/4}$	1.4101^{20}	w, al, eth, ace, bz	$B4^4$, 1411
6764	Ethyl benzene $C_6H_5CH_2CH_3$	106.17	...	136.2, 25.8^{10}	-95	$0.8670^{20/4}$	1.4959^{20}	al, eth	$B5^3$, 776
6765	2-Ethyl benzoic acid $2\text{-}C_2H_5C_6H_4CO_2H$	150.18	nd (w)	259	68	$1.0413^{100/4}$	1.5099^{100}	al, eth	$B9^3$, 2425
6766	3-Ethyl benzoic acid $3\text{-}C_2H_5C_6H_4CO_2H$	150.18	nd (w or dil al)	...	47	1.042^{100}	1.5345^{100}	al, eth	$B9^1$, 208
6767	4-Ethyl benzoic acid $4\text{-}C_2H_5C_6H_4CO_2H$	150.18	pr (al) pl or lf (w)	...	113.5	...	...	al, bz, chl	$B9^2$, 349
6768	Ethyl benzyl ketone or 1-Phenyl-2-butanone $CH_3CH_2COCH_2C_6H_5$	148.29	...	230^{755}	...	$1.002^{20/4}$	...	al, **eth**, ace	$B7^3$, 1080
6769	Ethylboric acid $C_2H_5B(OH)_2$	73.89	pl (eth)	d	40 (sub)	...	...	w, al, eth	$B4^3$, 1964
6770	Ethyl bromide CH_3CH_2Br	108.97	...	38.4	-118.6	$1.4604^{20/4}$	1.4239^{20}	al, eth, chl	$B1^4$, 150
6771	Ethyl 2-bromoethyl ether $C_2H_5OCH_2CH_2Br$	153.02	ye nd (al)	$127\text{-}8^{755}$, 40^{24}	...	$1.3572^{20/4}$	1.4447^{20}	al, eth	$B1^4$, 1386
6772	Ethyl bromomethyl ether $C_2H_5OCH_2Br$	138.99	...	109^{746}	...	$1.4402^{20/4}$	1.4515^{20}	eth	$B1^3$, 2594
6773	Ethyl isobutyl ether $C_2H_5\text{-O-}i\text{-}C_4H_9$	102.18	...	81	...	$0.751^{20/4}$	1.3739^{25}	**al, eth**, ace, chl	$B1^4$, 1593
6774	Ethyl isobutyl sulfide $i\text{-}C_4H_9SC_2H_5$	118.24	...	134.2, 24.8^{10}	...	$0.8306^{20/4}$	1.4450^{20}	**al**, eth	$B1^4$, 1606
6775	Ethyl chloride CH_3CH_2Cl	64.51	...	12.3	-136.4	$0.8978^{20/4}$	1.3676^{20}	al, eth	$B1^4$, 124
6776	Ethyl (l-chloroethyl) ether $C_2H_5OCHClCH_3$	108.57	...	92-5	...	$0.950^{20/4}$	1.4053^{20}	...	$B1^4$, 3119
6777	Ethyl β-chloroethyl ether $C_2H_5OCH_2CH_2Cl$	108.57	...	107-8	...	$0.9894^{20/4}$	1.4113^{20}	**eth**, chl	$B1^4$, 1375
6778	Ethyl β-chloroethyl sulfide $C_2H_5SCH_2CH_2Cl$	124.63	...	156, $63\text{-}5^{47}$	...	$1.0663^{20/4}$	1.4878^{20}	chl	$B1^4$, 1707
6779	Ethyl chloromethyl ether $C_2H_5OCH_2Cl$	94.54	...	83^{763}	...	$1.0372^{0/4}$	1.4040^{20}	al, eth	$B1^4$, 3047
6780	Ethyl chloro sulfinate C_2H_5OSOCl	128.57	...	52.5^{44}	...	$1.2766^{25/4}$	1.4550^{25}	eth	$B1^3$, 1316
6781	Ethyl chloro sulfonate $C_2H_5OSO_2Cl$	144.57	...	151-4, 52^{14}	...	$1.3502^{25/4}$	1.416^{20}	eth, chl, lig	$B1^4$, 1326
6782	Ethyl cyanoacetate $NCCH_2CO_2C_2H_5$	113.12	...	205, 99^{15}	-22.5	$1.0654^{20/4}$	1.4175^{20}	**al, eth**	$B2^4$, 1889
6783	Ethyl cyanoformate $NCCO_2C_2H_5$	99.09	...	$116\text{-}8^{765}$	...	$1.0034^{20/4}$	1.3821^{20}	al, eth	$B2^4$, 1862
6784	Ethyl di-benzylamine $C_2H_5N(CH_2C_6H_5)_2$	225.33	...	306, 131^{11}	...	...	...	al, eth	$B12^3$, 2222
6785	Ethyl (1,2-dibromoethyl) ether $C_2H_5OCH_2BrCH_2Br$	231.91	...	80^{20}	...	$1.7320^{20/4}$	1.5044^{20}	al, chl	$B1^4$, 3152
6786	Ethyl (1,2-dichloroethyl) ether $C_2H_5OCHClCH_2Cl$	143.01	...	145, $66\text{-}8^{45}$	...	$1.1370^{20/4}$	1.4435^{20}	al, eth	$B1^4$, 3136
6787	Ethyl (1,2-dichlorovinyl) ether $C_2H_5OCCl{=}CHCl$	141.00	...	128.2	...	$1.1972^{25/4}$	1.4558^{17}	...	$B1^4$, 3425
6788	Ethyl (diethylaminomethyl) ether $(C_2H_5)_2NCH_2OC_2H_5$	131.22	...	136, 76^{11}	...	...	...	al, eth, ace	$B4^4$, 336
6789	Ethyl-di-isopropylamine $C_2H_5N(i\text{-}C_3H_7)_2$	129.25	...	126.5	...	...	1.4138^{20}	...	$B4^4$, 511
6790	Ethyl 2,3-epoxypropyl ether $C_2H_5OCH_2CHCH_2$ (O bridging CH and CH_2)	102.13	...	128	...	0.9700^{20}	1.4320^{20}	w, al, eth	B17, 105
6791	Ethyl ether or Diethyl ether $C_2H_5OC_2H_5$	74.12	...	34.5	fr−116.2	$0.7138^{20/4}$	1.3526^{20}	**al**, ace, **bz, chl**	$B1^4$, 1314
6792	Ethyl ether borofluoride $(C_2H_5)_2O.BF_3$	141.93	...	125-6 60^{20}	-60.4	$1.3572^{20/4}$	1.4447^{20}	**al, eth**	$B1^4$, 1321

PHYSICAL CONSTANTS OF ORGANIC COMPOUNDS (Continued)

No.	Name, Synonyms, and Formula	Mol. wt.	Color, crystalline form, specific rotation and λ_{max} (log ε)	b.p. °C	m.p. °C	Density	n_D	Solubility	Ref.
6793	Ethyl ethynyl ether or Ethoxy acetylene $C_2H_5OC{\equiv}CH$	70.09	...	50 exp 100	...	$0.8000^{20/4}$	1.3796^{20}	...	$B1^4$, 2211
6794	Ethyl furfuryl ether $C_2H_5OCH_2(C_4H_3O)$	126.16	...	$149\text{-}50^{770}$	...	$0.9844^{20/4}$	1.4523^{20}	al, eth	$B17^2$, 114
6795	Ethyl-2-furyl ether $(2\text{-}C_4H_3O)OC_2H_5$	112.13	...	125-6	...	$0.9849^{23/4}$	1.4500^{23}	...	$B17^4$, 1219
6796	Ethyl heptyl ether $C_2H_5OC_7H_{15}$	144.26	...	166.6	...	$0.790^{16/4}$	1.4111^{20}	al, eth	$B1^4$, 1733
6797	Ethyl hexyl ether $C_2H_5OC_6H_{13}$	130.23	...	$142\text{-}3^{773}$, 42^{14}	...	$0.7722^{20/4}$	1.4008^{20}	al, eth	$B1^4$, 1697
6798	Ethyl hydrogensulfate $C_2H_5OSO_3H$	126.13	...	280d	...	$1.3657^{20/4}$	1.4105^{20}	w	$B1^4$, 1324
6799	Ethyl hydroperoxide C_2H_5OOH	62.07	...	93-7 ext >100	-100	$0.9332^{20/4}$	1.3800^{20}	**w, al, eth,** bz	$B1^4$, 1323
6800	Ethyl-bis (2-hydroxyethyl) amine $C_2H_5N(CH_2CH_2OH)_2$	133.19	ye	246-8, 118^1	...	$1.0135^{20/4}$	1.4663^{20}	al	$B4^3$, 693
6801	Ethyl (2-hydroxy--5-methylphenyl) ketone $(5\text{-}CH_3\text{-}2\text{-}HOC_6H_3)COC_2H_5$	164.20	...	$129\text{-}30^{16.5}$	2	$1.0841^{14/4}$	$1.549^{13.8}$	chl	$B8^2$, 120
6802	Ethyl (2-hydroxy--3-methylphenyl) ketone $(3\text{-}CH_3\text{-}2\text{-}HOC_6H_3)COC_2H_5$	164.20	...	$127\text{-}9^{15}$	22-3	...	...	chl	$B8^2$, 119
6803	Ethyl (2-hydroxy--4-methylphenyl) ketone $(4\text{-}CH_3\text{-}2\text{-}HOC_6H_3)COC_2H_5$	164.20	...	$115\text{-}20^{10}$	41.5-2.5	...	...	chl	$B8^3$, 471
6804	Ethyl hypochlorite C_2H_5OCl	80.51	ye lig	36^{732}	...	$1.013^{-6/4}$	...	al, **eth, bz, chl**	$B1^4$, 1324
6805	Ethyl iodide CH_3CH_2I	155.97	...	72.3	-108	$1.9358^{20/4}$	1.5133^{20}	**al,** eth	$B1^4$, 163
6806	Ethyl isocyanate C_2H_5NCO	71.08	...	60	...	$0.9031^{20/4}$	1.3808^{20}	**al, eth**	$B4^4$, 402
6807	Ethyl methyl amine or Methylethylamine $CH_3NHC_2H_5$	59.11	...	36.7	...	...	...	w, al, eth, ace	$B4^4$, 312
6808	Ethyl methyl amine, hydrochloride $C_2H_5NHCH_3.HCl$	95.57	pl (al-eth)	...	126-30	$1.0874^{20/4}$	...	w, al, eth, ace	$B4^2$, 589
6809	Ethyl methyl ether $C_2H_5OCH_3$	60.10	...	10.8	...	$0.7252^{0/0}$	1.3420^4	w, **al, eth,** ace, chl	$B1^4$, 1314
6810	Ethyl methyl sulfide $C_2H_5SCH_3$	76.16	...	66.6	-105.9	$0.8422^{20/4}$	1.4404^{20}	**al,** eth	$B1^4$, 1392
6811	Ethyl α-naphthyl ether or 1-Ethoxynaphthalene $\alpha\text{-}C_{10}H_7OC_2H_5$	172.23	nd	280.5, $136\text{-}8^{14}$	5.5	$1.060^{20/4}$	1.5953^{25}	al, eth	$B6^3$, 2924
6812	Ethyl β-naphthyl ether $\beta\text{-}C_{10}H_7OC_2H_5$	172.23	pl (al)	282, 148^{10}	37-8	$1.0640^{20/20}$	1.5975^{36}	al, eth, lig, to	$B6^3$, 2972
6813	Ethyl nitrate $C_2H_5ONO_2$	91.07	flam	87.2	-94.6	$1.1084^{20/4}$	1.3852^{20}	w, **al, eth**	$B1^4$, 1327
6814	Ethyl nitrite C_2H_5ONO	75.07	yesh	$16\text{-}7^{725}$	...	$0.90^{15/15}$	1.3418^{10}	**al, eth**	$B1^4$, 1327
6815	Ethyl octyl ether $C_2H_5OC_8H_{17}$	158.28	...	186.3, 74^9	12.5	$0.7847^{20/4}$	1.4127^{20}	al	$B1^4$, 1759
6816	Ethyl pentyl ether $C_2H_5OC_5H_{11}$	116.20	...	119-20	...	$0.7622^{20/4}$	1.3927^{20}	**al, eth**	$B1^3$, 1602
6817	Ethyl *iso*pentyl ether $C_2H_5O\text{-}i\text{-}C_5H_{11}$	116.20	...	112-3	...	$0.7695^{21/15}$	...	**al, eth**	$B1^4$, 1681
6818	Ethyl *tert*-pentyl ether $C_2H_5O\text{-}tert\text{-}C_5H_{11}$	116.20	...	101	...	$0.7657^{20/4}$	1.3912^{20}	al, **eth**	$B1^3$, 1626
6819	Ethylarsonic acid $C_2H_5AsO(OH)_2$	154.00	nd (al) rh nd (w)	$209\text{-}11^{12}$	99.5	...	...	w, al	$B4^3$, 1823
6820	Ethyl perchlorate $C_2H_5OClO_3$	128.51	oil	...	89	...	...	**al, eth**	$B1^3$, 1314
6821	Ethyl Diethylperoxide $C_2H_5OOC_2H_5$	90.12	...	65	-70	0.8240^{19}	1.3715^{17}	**al, eth**	$B1^4$, 1323
6822	Ethyl phenyl sulfide or (Ethylthio)benzene $C_6H_5SC_2H_5$	138.23	...	205, 84^{10}	...	$1.0211^{20/4}$	1.5670^{20}	al	$B6^4$, 1468
6823	Ethyl propyl amine $C_2H_5NHC_3H_7$	87.16	...	61-2	...	0.7204^{17}	1.3858^{25}	w, al, ace	$B4^4$, 468
6824	Ethyl isopropyl amine $i\text{-}C_3H_7NHC_2H_5$	87.16	...	70-1	...	...	1.3872^{25}	...	$B4^4$, 508
6825	Ethyl propyl ether $C_2H_5OC_3H_7$	88.15	...	63.6	<-79	$0.7386^{20/4}$	1.3695^{20}	**al, eth, aa**	$B1^4$, 1421

PHYSICAL CONSTANTS OF ORGANIC COMPOUNDS (Continued)

No.	Name, Synonyms, and Formula	Mol. wt.	Color, crystalline form, specific rotation and λ_{max} (log ε)	b.p. °C	m.p. °C	Density	n_D	Solubility	Ref.
6826	Ethyl iso propyl ether $C_2H_5OCH(CH_3)_2$	88.15	...	63-4	...	$0.720^{25/4}$	1.3698^{25}	**al, eth,** ace, chl	B1[4], 1471
6827	Ethyl propyl sulfide $C_2H_5SC_3H_7$	104.21	...	118.5, 13.5^{10}	-117	$0.8370^{20/4}$	1.4462^{20}	al	B1[4], 1451
6828	Ethyl 1-propynyl ether $C_2H_5OC{=}CCH_3$	84.12	...	84	...	$0.8276^{20/4}$	1.4039^{20}	al, eth	B1[3], 1969
6829	Ethyl 2-propynyl ether or Ethyl propargyl ether $C_2H_5OCH_2C{\equiv}CH$	84.12	...	82	...	$0.8326^{20/4}$	1.4039^{20}	al, eth	B1[4], 2215
6830	N-Ethyl quinolinium iodide $C_9H_7N^+(C_2H_5)I^-$	285.13	ye pr (al)	...	158	...	...	w, al, chl	B20[4], 3358
6831	Ethyl pyridinium bromide $(C_5H_5N^+)C_2H_5Br^-$	188.07	cr (al)	...	111-2	...	...	w, al	B20[4], 2309
6832	Ethyl sulfate $(C_2H_5O)_2SO_2$	154.18	...	208d, 96^{15}	-24.5	$1.1774^{20/4}$	1.4004^{20}	**al, eth**	B1[4], 1326
6833	Ethyl sulfite $(C_2H_5O)_2SO$	138.18	...	230, 116^{10}	...	$0.9957^{20/4}$	1.4310^{20}	al, eth	B1[4], 1324
6834	Ethyl sulfoxide or Diethyl sulfoxide $(C_2H_5)_2SO$	106.18	syr	104^{25}	14	...	...	w, al, eth	B1[4], 1395
6835	Ethyl telluride or Diethyl telluride $(C_2H_5)_2Te$	185.72	red-ye	137-8	...	$1.599^{15/4}$	1.5182^{15}	al	B1[4], 1412
6836	Ethyl thioacetic acid $C_2H_5SCH_2CO_2H$	120.17	...	164^{83}, 109^{5}	-8.5	$1.1497^{20/4}$	...	w, al, eth	B3[4], 603
6837	Ethyl thiocyanate C_2H_5SCN	87.14	...	145^{758}	-85.5	$1.0071^{22/4}$	1.4684^{15}	**al, eth**	B3[4], 328
6838	Ethyl isothiocyanate or Ethyl mustard oil C_2H_5NCS	87.14	...	131-2	-5.9	$0.9990^{20/4}$	1.5130^{20}	**al, eth**	B4[4], 403
6839	Ethyl vinyl ether $C_2H_5OCH{=}CH_2$	72.11	...	35-6	-115.8	$0.7589^{20/4}$	1.3767^{20}	al, **eth**	B1[4], 2049
6840	Ethylene $CH_2{=}CH_2$	28.05	gas, mcl pr	-103.7	-169	...	1.363^{100}	eth	B1[4], 677
6841	Ethylene, amino or Vinylamine $CH_2{=}CHNH_2$	43.07	...	$55\text{-}6^{750}$	...	0.8321^{24}	...	**w**, al, **eth**	B4, 203
6842	Ethylene, 1,2-bis(trimethylsilyl)-*(trans)* $(CH_3)_3SiCH{=}CHSi(CH_3)_3$	172.42	...	145.5	...	0.7589^{20}	$1.4310^{20/}{}_{D}$	...	KHOC, 603
6843	Ethylene, l-bromo-2-chloro $BrCH{=}CHCl$	141.39	...	84.6	-86.7	1.7972^{15}	1.4982	...	B1[3], 671
6844	Ethylene, 1-bromo-1,2,2-triphenyl $C_6H_5CBr{=}C(C_6H_5)_2$	335.24	nd (aa)	...	116-7	...	...	aa	B5[3], 2400
6845	Ethylene, 1-chloro-2-dichloroarsino (trans) or Lewisite $ClCH{=}CHAsCl_2$	207.32	...	196d, 93^{20}	0.1	$1.888^{20/4}$	...	al, eth	B4[3], 1810
6846	Ethylene chlorohydrin $ClCH_2CH_2OH$	80.51	...	128, 44^{20}	-67.5	$1.2003^{20/4}$	1.4419^{20}	w, al	B1[4], 1372
6847	Ethylene, 1-chloro-1,2,2-trifluoro $FCCl{=}CF_2$	116.47	...	-26.2	-157.5	$1.54^{-60/4}$	1.38^{0}	bz	B1[4], 704
6848	Ethylene, 1-chloro-1,2,2-triphenyl $(C_6H_5)_2C{=}CClC_6H_5$	290.79	...	...	117	...	...	al, eth, ace, bz, chl	B5[3], 2400
6849	Ethylenediamine $H_2NCH_2CH_2NH_2$	60.10	...	116.5	8.5	$0.8995^{20/20}$	1.4568^{20}	w, **al**	B4[4], 1166
6850	Ethylene diamine, N,N′-dibenzoyl $C_6H_5CONHCH_2CH_2NHCOC_6H_5$	268.32	pr or nd (al)	...	247	...	...	aa	B9[3], 1210
6851	Ethylenediamine, N,N′-diethyl $C_2H_5NHCH_2CH_2NHC_2H_5$	116.21	...	144, $38\text{-}40^{15}$	...	$0.8280^{20/4}$	1.4340^{20}	**w**, al, eth, to	B4[4], 1174
6852	Ethylenediamine, N,N′-dimethyl $CH_3NHCH_2CH_2NHCH_3$	88.15	...	120	...	$0.828^{15/4}$	...	al, eth, dil HCl	B4[4], 1171
6853	Ethylenediamine, N,N′-diphenyl $C_6H_5NHCH_2CH_2NHC_6H_5$	212.29	lf (dil al)	$178\text{-}82^{2}$	74	...	...	al, eth	B12[1], 1042
6854	Ethylenediamine, hydrate $H_2NCH_2CH_2NH_2.H_2O$	78.11	...	118	10	$0.964^{20.5/4}$	$1.4500^{20.5}$	**w**	B4[4], 1168
6855	Ethylenediamine, hydrochloride $H_2NCH_2CH_2NH_2.2HCl$	133.02	mcl pr (w)	sub	300-30 sub	...	1.633	w	B4[4], 1168
6856	Ethylenediamine, N-β-hydroxyethyl $HOCH_2CH_2NHCH_2CH_2NH_2$	104.15	...	238-40, 123^{10}	...	$1.0254^{25/D}$	1.4861^{20}	**w, al,** ace	B4[4], 1558
6857	Ethylene, dibenzoyl *(cis)* $C_6H_5COCH{=}CHCOC_6H_5$	236.27	nd (al)	...	134	...	...	eth, ace, bz, chl	B3[7], 4115
6858	Ethylene, dibenzoyl *(trans)* $C_6H_5COCH{=}CHCOC_6H_5$	236.27	ye nd (al or bz)	...	111	...	...	bz, chl, aa	B3[7], 4116
6859	Ethylene, 1,1-dibromo or Vinylidene bromide $CH_2{=}CBr_2$	185.85	...	92	...	$2.1780^{21/4}$	...	al, eth, ace, bz	B1[4], 720

No.	Name, Synonyms, and Formula	Mol. wt.	Color, crystalline form, specific rotation and λ_{max} (log ε)	b.p. °C	m.p. °C	Density	n_D	Solubility	Ref.
6860	Ethylene, 1,1-dibromo--2-ethoxy $C_2H_5OCH=CBr_2$	229.90	...	$170-2^{747}$, $73-5^{15}$	...	$1.7697^{18/4}$	...	eth	$B1^2$, 473
6861	Ethylene, 1,2-dibromo *(cis)* $CHBr=CHBr$	185.85	...	112.5^{760}	-53	2.2464^{20}	1.5428^{20}	al, eth, ace, bz, chl	$B1^4$, 720
6862	Ethylene, 2,2-dibromo *(trans)* $CHBr=CHBr$	185.85	...	108	-6.5	2.2308^{20}	1.5505^{18}	al, eth, ace, bz, chl	$B1^4$, 721
6863	Ethylene, 1,1-dichloro or Vinylidene chloride $CH_2=CCl_2$	96.94	...	37	-122.1	1.218^{20}	1.4249^{20}	al, eth, ace, bz, chl	$B1^4$, 706
6864	Ethylene, 1,1-dichloro-2-fluoro $CCl_2=CHF$	114.93	...	37.5	...	$1.37324^{16.4}$	$1.4031^{16.4/}{}_D$	...	$B1^4$, 711
6865	Ethylene, 1,2-dichloro *(cis)* $CHCl=CHCl$	96.94	...	60.3	-80.5	$1.2837^{20/4}$	...	**al, eth, ace,** bz, chl	$B1^4$, 707
6866	Ethylene, 1,2-dichloro *(trans)* $CHCl=CHCl$	96.94	...	47.5	-50	$1.2565^{20/4}$	1.4454^{20}	**al, eth, ace,** bz, chl	$B1^4$, 709
6867	Ethylene, 1,2-dichloro-1,2-difluoro $CFCl=CFCl$	132.92	...	21.1	-130.5	$1.4950^{0/4}$	$1.3777^{0/}{}_D$	...	$B1^4$, 712
6868	Ethylene, 1,1-difluoro or Vinylidene fluoride $CH_2=CF_2$	64.03	gas	...	<-84	...	...	al, eth	$B1^4$, 696
6869	Ethylene, 1,2-diiodo $CHI=CHI$	279.85	...	72.5^{16}	-14	3.0625^{20}	...	eth, chl	$B1^4$, 724
6870	Ethylene, 1,1-diphenyl $CH_2=C(C_6H_5)_2$	180.25	...	277, $94-5^{11}$	8.2	$1.0281^{16/4}$	1.6100^{20}	eth, chl	$B5^3$, 1975
6871	Ethylene, fluoro-trichloro $CFCl=CCl_2$	149.38	...	71	-108.9	$1.5460^{20/4}$	$1.4379^{20/}{}_D$	chl	$B1^4$, 715
6872	Ethylene, nitro $CH_2=CHNO_2$	73.05	...	98.5, $38-9^{80}$	-55.5	$1.2212^{14/4}$	1.4282^{20}	al, eth, ace, bz, chl	$B1^4$, 725
6873	Ethylene, tetrabromo $Br_2C=CBr_2$	343.64	pl (dil al) nd (al)	225-7, 100^{15}	56.5	...	...	al, eth, ace, bz	$B1^4$, 722
6874	Ethylene, tetrachloro $Cl_2C=CCl_2$	165.83	...	121, 14^{10}	-19	$1.6227^{20/4}$	1.5053^{20}	**al, eth, bz**	$B1^4$, 715
6875	Ethylene, tetracyano $(NC)_2C=C(CN)_2$	128.09	...	223	198-200	1.348^{25}	1.560^{25}	ace	$B2^4$, 2450
6876	Ethylene, tetrafluoro $F_2C=CF_2$	100.02	gas	-76.3	-142.5	$1.519^{-76.3}$	...	...	$B1^4$, 698
6877	Ethylene, tetraiodo $I_2C=CI_2$	531.64	ye lf pr (eth)	sub	192	2.983^{20}	...	bz, chl	$B1^4$, 724
6878	Ethylene, tetraphenyl $(C_6H_5)_2C=C(C_6H_5)_2$	332.44	mcl or rh (bz-eth or chl-al)	415-25	225	$1.155^{0/4}$	...	bz	$B5^3$, 2598
6879	Ethylene tetracarboxylic acid, tetraethyl ester $(C_2H_5O_2C)_2C=C(CO_2C_2H_5)_2$	316.31	tcl pr (eth)	325-8d, 210^{22}	58	...	...	al, eth	$B2^4$, 2450
6880	Ethylene, tribromo $BrCH=CBr_2$	264.74	...	163-4, 75^{15}	...	$2.708^{20.5/4}$	1.6045^{16}	al, eth, ace, chl	$B1^4$, 722
6881	Ethylene, trichloro $ClCH=CCl_2$	131.39	...	87	−73	$1.4642^{20/4}$	1.4773^{20}	al, eth, ace, chl	$B1^4$, 712
6882	Ethylene, triphenyl $C_6H_5CH=C(C_6H_5)_2$	256.35	lf (al or MeOH)	$220-1^{14}$	72-3	$1.0373^{78/4}$	1.6292^{78}	al, eth	$B5^3$, 2398
6883	Ethylene glycol or Ethanediol $HOCH_2CH_2OH$	62.07	...	198, 93^{13}	-11.5	$1.1088^{20/4}$	1.4318^{20}	**w, al, eth, ace**	$B1^4$, 2369
6884	Ethylene glycol, bis(chloroacetate) $ClCH_2CO_2CH_2CH_2O_2CCH_2Cl$	215.03	pr (eth-peth)	$142-4^{2}$	45-6	...	...	eth	$B2^3$, 448
6885	Ethylene glycol, bis(2-chloroethyl)ether $ClCH_2CH_2OCH_2CH_2OCH_2CH_2Cl$	187.07	...	230, 118^{10}	...	$1.197^{20/20}$	1.4592^{25}	...	$B1^4$, 2379
6886	Ethylene glycol, diacetate or Ethylene glycol diacetate $CH_3CO_2CH_2CH_2O_2CCH_3$	146.14	...	190	-31	$1.1063^{20/20}$	1.4159^{20}	w, **al, eth, ace, bz**	$B2^4$, 217
6887	Ethylene glycol, dibenzoate $C_6H_5CO_2CH_2CH_2O_2CC_6H_5$	270.28	rh pr (eth)	>360d	73-4	...	...	eth	$B9^3$, 536
6888	Ethylene glycol, dibutyrate $C_3H_7CO_2CH_2CH_2O_2CC_3H_7$	202.25	...	240, $118-21^{11}$	...	$1.0005^{20/4}$	1.4262^{20}	al, eth	$B2^4$, 796
6889	Ethane-1,1-di-4-tolyl $(4CH_3C_6H_4)_2CHCH_3$	210.32	...	298-9, $153-6^{11}$	<-20	$0.974^{20/4}$	...	bz	$B5^4$, 1948
6890	Ethylene glycol, diethylether or 1,2-diethoxyethane $C_2H_5OCH_2CH_2OC_2H_5$	118.18	...	123.5	...	0.8484^{20}	1.3860^{20}	al, eth, ace, bz	$B1^4$, 2379
6891	Ethylene glycol, diformate $HCO_2CH_2CH_2O_2CH$	118.09	...	174	...	$1.193^{0/4}$	1.3580	al, eth	$B2^4$, 37
6892	Ethylene glycol, dilaurate or Ethyleneglycoldidocleca-noate $C_{11}H_{23}CO_2CH_2CH_2O_2CC_{11}H_{23}$	426.68	pl (al)	188^{20}	56.6	...	...	al, eth	$B2^4$, 1094

No.	Name, Synonyms, and Formula	Mol. wt.	Color, crystalline form, specific rotation and λ_{max} (log ε)	b.p. °C	m.p. °C	Density	n_D	Solubility	Ref.
6893	Ethylene glycol, dimethylether or 1,2-dimethoxyethane $CH_3OCH_2CH_2OCH_3$	90.12	...	83-4	-58	$0.8628^{20/4}$	1.3796^{20}	w, al, eth, ace, bz	$B1^4$, 2376
6894	Ethylene glycol, dimyristate or Ethylene glycol ditetradecylate $C_{13}H_{27}CO_2CH_2CH_2O_2CC_{13}H_{27}$	482.79	cr (eth or ace)	...	65	...	...	eth, ace, bz	$B2^4$, 1133
6895	Ethylene glycol, dinitrate $O_2NOCH_2CH_2ONO_2$	152.06	ye	197-200	-22.3	$1.4918^{20/4}$	...	al, eth	$B1^4$, 2413
6896	Ethylene glycol, dinitrite $ONOCH_2CH_2ONO$	120.06	...	98	<-15	$1.2156^{0/4}$	...	al, eth	$B1^4$, 2411
6897	Ethylene glycol, dipalmitate or Ethanediol dihexadecylate $C_{15}H_{31}CO_2CH_2CH_2O_2CC_{15}H_{31}$	538.90	lf or nd (al-chl)	226 (vac)	72	0.8594^{78}	...	eth, ace	$B2^4$, 1169
6898	Ethylene glycol, di-phenyl ether $C_6H_5OCH_2CH_2OC_6H_5$	214.26	lf (al)	$180\text{-}5^{12}$	98	...	...	eth, chl	$B6^4$, 573
6899	Ethylene glycol, dipropionate $C_2H_5CO_2CH_2CH_2O_2CC_2H_5$	174.20	...	...	211	1.020^{15}	...	**al, eth**, ace	$B2^4$, 715
6900	Ethylene glycol, distearate or Ethylene glycol di-octadecylate $C_{17}H_{35}CO_2CH_2CH_2O_2CC_{17}H_{35}$	595.00	lf	241^{20}	79	...	...	eth, ace	$B2^4$, 1223
6901	Ethylene glycol, dithiocyanate $NCSCH_2CH_2SCN$	144.21	rh pl or nd (w) ta (al or eth)	d	90	...	...	al, eth, ace	$B3^4$, 333
6902	Ethylene glycol, methyl, ethyl ether $CH_3OCH_2C_2OC_2H_5$	104.15	...	102	...	$0.8529^{20/4}$	1.3868^{20}	**w, al, eth** ace, bz	$B1^3$, 2078
6903	Ethylene glycol, monoacetate or 2-Hydroxyethyl acetate $CH_3CO_2CH_2CH_2OH$	104.11	...	187-9	...	1.108^{15}	...	w, al, eth	$B2^4$, 154
6904	Ethylene glycol, monoallyl ether $HOCH_2CH_2OCH_2CH{=}CH_2$	102.13	...	159^{755},64^{15}	...	$0.9580^{20/4}$	1.4358^{20}	w, al, bz	$B1^4$, 2388
6905	Ethylene glycol, monobenzoate $C_6H_5CO_2CH_2CH_2OH$	166.18	...	$150\text{-}1^{10}$	45	...	...	al	$B9^3$, 532
6906	Ethylene glycol, monobenzyl ether or Benzyl cellusolve $HOCH_2CH_2OCH_2C_6H_5$	152.19	...	256,138^{15}	<-75	$1.0640^{20/4}$	1.5233^{20}	w, al, eth	$B6^4$, 2241
6907	Ethylene glycol, monobutyl ether or Butyl cellosolve $HOCH_2CH_2OC_4H_9$	118.18	...	171, 50^4	...	$0.9015^{20/4}$	1.4198^{20}	**w, al, eth**	$B1^4$, 2380
6908	Ethylene glycol, mono-*iso*-butyl ether or Isobutyl cellosolve $HOCH_2CH_2OCH_2CH(CH_3)_2$	118.18	...	159^{745}	...	$0.8900^{20/4}$	1.4143^{20}	...	$B1^4$, 2382
6909	Ethylene glycol, mono-β-chloroethyl ether or β-Chloroethyl cellosolve $HOCH_2CH_2OCH_2CH_2Cl$	124.57	...	180-85, $91\text{-}2^{13}$	...	...	1.4805^{19}	w, **al, eth**	$B1^3$, 2078
6910	Ethylene glycol, monoethyl ether or Ethyl cellosolve $HOCH_2CH_2OC_2H_5$	90.12	...	135, 35^{10}	...	$0.9297^{20/4}$	1.4080^{20}	**w, al, eth**, ace	$B1^4$, 2377
6911	Ethylene glycol, monohexyl ether or Hexyl cellosolve $HOCH_2CH_2OC_6H_{13}$	146.23	...	208,96^{13}	-45.1	$0.8894^{20/20}$	1.4291^{20}	al, eth	$B1^4$, 2383
6912	Ethylene glycol, monomethyl ether or Methyl cellosolve $HOCH_2CH_2OCH_3$	76.10	...	125^{768}	-85.1	$0.9647^{20/4}$	1.4024^{20}	w, al, **eth**, ace, bz	$B1^4$, 2375
6913	Ethylene glycol, monophenyl ether or Phenyl cellosolve $HOCH_2CH_2OC_6H_5$	138.17	...	237, $134\text{-}5^{18}$	...	1.1020^{20}	1.5340^{20}	al, eth	$B6^4$, 571
6914	Ethylene glycol, monopropyl ether or Propyl cellosolve $HOCH_2CH_2OC_3H_7$	104.15	...	150^{743}	...	$0.9112^{20/4}$	1.4133^{20}	w, al, eth	$B1^4$, 2379
6915	Ethylene glycol, mono isopropyl ether or Isopropyl cellosolve $i\text{-}C_3H_7OCH_2CH_2OH$	104.15	...	144^{743}	...	$0.9030^{20/4}$	1.4095^{20}	**w, al, eth**, ace	$B1^4$, 2380
6916	Ethylene glycol, monostearate $C_{17}H_{35}CO_2CH_2CH_2OH$	328.54	(peth)	$189\text{-}91^3$	60-1	$0.8780^{60/4}$	1.4310^{60}	eth	$B4^2$, 1222
6917	Ethylene glycol sulfate $C_2H_4O_2SO_2$	124.1	nd or pr (bz -lig)	sub	99	...	...	al, eth, ace, bz	$B1^3$, 2110
6918	Ethylene glycol sulfite $C_2H_4O_2SO$	108.11	...	173, $70\text{-}1^{20}$	-11	$1.4402^{20/4}$	1.4463^{20}	w, al, eth, ace, bz	$B1^4$, 2409
6919	Ethylene imine or Aziridine C_2H_5N	43.07	...	50^{756}	...	$0.8321^{20/4}$	...	**w**, al, eth, ace, bz	$B20^4$, 3
6920	Ethylene oxide or Epoxyethane C_2H_4O	44.06	...	13.2^{746}	-111	$0.8824^{10/10}$	1.3597^7	w, al, eth, ace, bz	$B17^4$, 3
6921	Ethylidene diacetate $CH_3CH(O_2CCH_3)_2$	146.14	...	169, $65\text{-}7^{10}$	18.9	1.070^{25}	1.3985^{25}	al, eth	$B2^4$, 282
6922	Ethynyl methyl ether $CH_3OC{\equiv}CH$	56.06	...	50	...	$0.8001^{20/4}$	1.3812^{20}	al, eth	$B1^4$, 2211
6923	Ethynyl phenyl ether C_6H_5OCCH	118.14	...	$61\text{-}2^{25}$	-36	$1.0614^{20/4}$	1.5125^{20}	al, eth	$B6^4$, 565

No.	Name, Synonyms, and Formula	Mol. wt.	Color, crystalline form, specific rotation and λ_{max} (log ε)	b.p. °C	m.p. °C	Density	n_D	Solubility	Ref.
6924	Ethynyl propyl ether $C_3H_7OC{\equiv}CH$	84.12	....	75	......	$0.8080^{20/4}$	1.3935^{20}	al, 6924	B1^{4}, 2213
6925	α -Eucaine $C_{19}H_{27}NO_4$	333.43	pr (eth or al)	....	104-5	....	....	w, al, eth, bz, chl	B22, 194
6926	α-Eucaine, hydrochloride $C_{19}H_{27}NO_4 \cdot HCl$	369.89	pl (w + 1) pr	....	*ca* 200d	....	....	w, al	B22, 194
6927	β-Eucaine-*(d)* $C_{15}H_{21}NO_2$	247.34	pr (peth)	....	57-8	....	....	al, eth, bz, chl, peth	B21^{2}, 14
6928	β-Eucaine *(dl)* or Betacaine $C_{15}H_{21}NO_2$	247.34	pr (peth)	....	57-8	....	....	al, eth, bz, chl, peth	B21^{2}, 13
6929	β-Eucaine *(l)* $C_{15}H_{21}NO_2$	247.34	pr (peth)	....	57-8	....	....	al, eth, bz, chl, peth	B21^{2}, 14
6930	β-Eucaine, hydrochloride-*(dl)* $C_{15}H_{21}NO_2.HCl$	283.80	pl (w)	....	277-9	....	....	w, al, eth, chl	B21^{2}, 13
6931	Eugenol or 5-Allylguaiacol 4-$(CH_2{=}CHCH_2)$-2-$CH_3OC_6H_3OH$	164.20	cr (hx)	253.2, 130.5^{10}	−7.5	$1.0652^{20/4}$	1.5405^{20}	al 6931	B6^{3}, 5021
6932	Eugenol-acetate 4-$(CH_2{=}CHCH_2)$-2-$CH_3OC_6H_3O_2CCH_3$	206.24	pr (al)	$127\text{-}8^{6}$	30-1	$1.0806^{20/4}$	1.5205^{20}	al	B6^{3}, 5029
6933	Eupitone or Eupittonic acid $C_{25}H_{26}O_9$	470.48	nd (al-eth)	....	200	....	....	a: 6933	B8^{3}, 4427
6934	Euxanthic acid $C_{19}H_{16}O_{16}$	404.33	ye nd (w + 1) α−108 (+ lw)	....	130d (+ w), 162d (anh)	....	....	al	B31, 277
6935	Eucarvone $C_{10}H_{14}O$	150.22	....	$99\text{-}100^{22}$, 88^{16}	......	$0.9490^{20/4}$	1.50872^{20}	eth, ace	JOCEA 26, 1609
6936	Evernic acid $C_{17}H_{16}O_7$	332.31	nd (w or ace) pr (al)	....	170	....	....	....	B10^{3}, 1488
6937	Evodiamine *(d)* or Rhetsine $C_{19}H_{17}N_3O$	303.36	yesh lf (al) $[\alpha]^{15}_{D}$ + 352 (ace,c=0.5)	....	278	....	....	....	B26^{2}, 103
6938	Evodiamine, hydrate *(d)* $C_{19}H_{17}N_3O.H_2O$	321.38	pl (al)	....	146-7	....	....	....	B24^{2}, 72

No.	Name, Synonyms, and Formula	Mol. wt.	Color, crystalline form, specific rotation and λ_{max} (log ε)	b.p. °C	m.p. °C	Density	n_D	Solubility	Ref.
6939	Fagaramide $C_{14}H_{17}NO_3$	247.30	nd (bz, dil al or peth)	...	119.5	...	...	al, bz	B19², 299
6940	β-Fagarine or Skimmianine $C_{14}H_{13}NO_4$	259.27	pym oct (al)	...	177	...	...	al, chl	B27, 134
6941	α-Fagarine or Haplophine $C_{13}H_{11}NO_3$	229.24	pr (al)	...	142	...	...	al, eth, bz, chl	C51, 4402
6942	α-Farnesine or 3,7,11-trimethyl-1,3,6,10-dodecatetracene $C_{15}H_{24}$	204.36	...	$129\text{-}32^{12}$	...	$0.8410^{20/4}$	1.4836^{20}	eth, ace, peth, lig	B1³, 1067
6943	β-Farnesine $C_{15}H_{24}$	204.36	...	$121\text{-}2^{9}$	...	$0.8363^{20/4}$	1.4899^{20}	eth, ace, chl, aa	B1⁴, 1133
6944	Farnesol *(trans, trans)* $(CH_3)_2C{=}CH(CH_2)_2C(CH_3){=}CH(CH_2)_2C(CH_3){=}CHCH_2OH$	222.37	...	160^{10}	...	$0.8846^{20/4}$	1.4877^{20}	al, eth, ace	B1⁴, 2335
6945	Farnesol *(cis, trans)* $C_{15}H_{26}O$	222.37	...	$120^{0.3}$	...	$0.8846^{20/4}$	1.4877^{20}	al, eth, ace	B1⁴, 2335
6946	Fenchane *(d)* $C_{10}H_{18}$	138.25	...	$151\text{-}2^{765}$	...	$0.8345^{20/4}$	1.44714^{20}	al, eth	B5³, 256
6947	Fenchane *(dl)* $C_{10}H_{18}$	138.25	...	$151\text{-}2^{265}$	...	$0.8345^{20/4}$	1.4471^{20}	al, eth	B5³, 256
6948	α-Fenchene *(d)* $C_{10}H_{16}$	136.24	$[\alpha]^{14/}{}_D + 29$	155-6	...	$0.8660^{20/4}$	1.4713^{20}	al, eth, ace	B5¹, 86
6949	α-Fenchene *(dl)* or Isopinene $C_{10}H_{16}$	136.24	...	154-6	...	$0.8660^{20/4}$	1.4705^{20}	al, eth, ace	B5³, 389
6950	α-Fenchene *(l)* $C_{10}H_{16}$	136.24	$[\alpha]^{20/}{}_D\text{-}43.8$	158-9	...	$0.8670^{20/4}$	1.4713^{20}	al, eth, ace	B5⁴, 466
6951	β-Fenchene $C_{10}H_{16}$	136.24	$[\alpha]'_D$ 62.91 (al)	151-3	...	$0.8591^{20/4}$	$1.4645^{25/}{}_D$	...	B5⁴, 465
6952	Fenchone *(d)* or Trimethyl norcamphor $C_{10}H_{16}O$	152.24	$[\alpha]^{20/}{}_D$ + 66.9 (al)	193.5, 80^{20}	6	$0.9465^{20/4}$	1.4623^{20}	al, eth, ace	B7³, 392
6953	Fenchone-*(dl)* $C_{10}H_{16}O$	152.24	...	193-4, $72\text{-}3^{12}$	-18	$0.9501^{15/15}$	1.4702^{20}	al, eth, ace	B7³, 393
6954	Fenchone *(l)* $C_{10}H_{16}O$	152.24	$[\alpha]^{23}{}_D$ −66.94 (al)	192-4	5(8.5)	0.948^{20}	1.4636^{20}	al, eth, ace	B7³, 392
6955	Fenchyl alcohol *(dl)* $C_{10}H_{18}O$	154.25	...	α202-301, β201	α38-9, β6	...	...	al, eth	B6;3, 288
6956	Filixic acid BBB or Filicin $C_{36}H_{44}O_{12}$	668.74	cr (AcOEt-ace)	...	172-4	...	...	bz, chl	B8³, 4436
6957	Ferrocene or Dicyclopentadienyl iron $C_{10}H_{10}Fe$	186.04	...	249	172.5-3.0	...	...	...	KHOC, 1525
6958	Flavaniline or 2-(*p*-Aminophenyl) lepidine $C_{16}H_{14}N_2$	234.30	pr (bz)	$133\text{-}41^{15}$	97	...	...	al, bz	B22, 469
6959	Flavanone or 2,3-dihydro-2-phenyl-1,4-benzopyrone $C_{15}H_{12}O_2$	224.26	nd (lig)	...	76	...	...	ace, bz	B17⁴, 5338
6960	Flavanone, 4′-methoxy-3′,5,7-trihydroxy or Hesperetin $C_{16}H_{14}O_6$	302.28	pl (dil al + ½ w)	sub $205^{0.004}$	227-8	...	...	al, eth	B18⁴, 3215
6961	Flavanone, 3′,4′,5,7-tetra hydroxy or Eriodictyol $C_{15}H_{12}O_6$	288.26	Pa br nd (dil al + 1.5 w)	...	267 d	...	...	al, aa	B18⁴, 3214
6962	Flavone or 2-Phenyl-α-benzopyrone $C_{15}H_{10}O_2$	222.24	nd (lig)cr (30% al)	...	100	...	...	al, eth, ace, bz, chl	B17⁴, 5413
6963	Flavone, 6-bromo $C_{15}H_9BrO_2$	301.14	nd (al)	...	191-2	...	...	al	B17⁴, 5417
6964	Flavone, 5,7-dihydroxy or Chrysin $C_{15}H_{10}O_4$	254.24	Pa ye pl or pr (MeOH) nd (sub)	sub	275	...	...	al, ace	B18⁴, 1766
6965	Flavone, 5,7-dihydroxy 4′-methoxy or Acacetin $C_{16}H_{12}O_5$	284.27	Pa ye nd (al)	...	261	...	...	ace	B18⁴, 2683
6966	Flavone, 5,7-dihydroxy-6-methoxy $C_{16}H_{12}O_5$	284.27	ye nd (al)	...	231-2	...	...	al, eth, ace, aa	B18⁴, 2671
6967	Flavone, 3-hydroxy or Flavanol $C_{15}H_{10}O_3$	238.24	Pa ye nd (al)	...	169-70	...	...	al	B17⁴, 6428
6968	Flavone, 2′,3,3′,5,7-pentahydroxy $C_{15}H_{10}O_7$	302.24	ye nd (aa + 1.5 w)	...	300	...	...	al	B18⁴, 3468
6969	Flavone, 2′,3,4′,5,7-pentahydroxy or Morin $C_{15}H_{10}O_4$	302.24	Pa ye nd (+ 1w, dil al)	...	303-4	...	...	al, bz	B18⁴, 3468

No.	Name, Synonyms, and Formula	Mol. wt.	Color, crystalline form, specific rotation and λ_{max} (log ε)	b.p. °C	m.p. °C	Density	n_D	Solubility	Ref.
6970	Flavone, 2′,3,5,5′,7-pentahydroxy $C_{15}H_{10}O_7$	302.24	red ye cr (dil al + 1 w)		306-8				B18⁴, 3470
6971	Flavone, 3,3′,4′,5,7-pentahydroxy or Quercitin $C_{15}H_{10}O_7$	302.24	ye rd (dil al + 2w)	sub	316-7			al, ace, aa	B18⁴, 3470
6972	Flavone, 3,3′,4′,7,8-pentahydroxy $C_{15}H_{10}O_7$	302.24	ye nd (dil al + 1w)		308d			al, ace	B18⁴, 3506
6973	Flavone, 3,3′,5,5′,7-pentahydroxy $C_{15}H_{10}O_7$	302.24	ye nd		>300				B18², 239
6974	Flavone, 3′,4′,5,5′,7-pentahydroxy or Tricetin $C_{15}H_{10}O_7$	302.24	ye nd(dil al + w)		>330d				B18⁴, 3454
6975	Flavone, 2′,3,5,7-tetrahydroxy or Datiscetin $C_{15}H_{10}O_6$	286.24	Pa ye nd (al, aq aa)		277-8			al, eth, ace	B18⁴, 3281
6976	Flavone, 3,3′,4′,7-tetrahydroxy or Fisetin $C_{15}H_{10}O_6$	286.24	lf ye nd (dil al + 1w)	sub^{760}	330			al, ace	B18⁴, 3304
6977	Flavone, 3′,4′,5,7-tetrahydroxy or Luteolin $C_{15}H_{10}O_6$	286.24	ye nd (dil al + 1w)	sub^{760}	329-30 d			al, eth	B18⁴, 3261
6978	Flavone, 3,4′,5,7-tetrahydroxy or Kaempferol $C_{15}H_{10}O_6$	286.24	ye nd (al + 1w) (aa)		276-8			al, ace	B18⁴, 3283
6979	Flavone, 4′,5,7-trihydroxy or Apigenin $C_{15}H_{10}O_5$	270.24	ye nd (py-w) lf (al))	sub^{760}	347-8			al, py	B18⁴, 2682
6980	Flavone, 5,6,7-trihydroxy or Baicalein $C_{15}H_{10}O_5$	270.24	ye pr (al)		264-5d			al, eth, ace	B18⁴, 2671
6981	Floridoside or 2-*o*-α-D-Galactosyl glycerol $C_9H_{18}O_8$	254.24	pr (al) $[\alpha]'_D$ + 15.1 (w)		86-7			w	B17⁴, 2995
6982	Fluoran or 9-Hydroxy-9-xanthene-*o*-benzoic acid acetone $C_{20}H_{12}O_3$	300.31	nd (al + 2 al)		182-3			al	B19⁴, 2903
6983	Fluoran, 1,6-dihydroxy $C_{20}H_{12}O_5$	332.31	ye nd		>260				B19⁴, 2902
6984	Fluoran, 2,6-dihydroxy $C_{20}H_{12}O_5$	332.31	ye nd (al)		177				B19⁴, 2902
6985	Fluoran, 3,5-dihydroxy $C_{20}H_{12}O_5$	332.31	ye gr nd or pl (aa)		179				B19⁴, 2903
6986	Fluoranthene or 1,2-Benzacenaphthene $C_{16}H_{10}$	202.26	pa ye nd or pl (al)	375, 217^{50}		$1.252^{0/4}$		al, eth, bz, aa	B5³, 2276
6987	Fluorene or 2,3-Benzindene $C_{13}H_{10}$	166.22	lf (al)	293-5	116-7	$1.203^{0/4}$		eth, ace, bz	B5³, 1936
6988	Fluorene, 2-acetamido or *N*(2-Fluorenyl) acetamide $C_{15}H_{13}NO$	223.27	nd (50% al or 50% aa)		194			al, eth, aa	B12³, 3287
6989	Fluorene, 9-acetamido $C_{15}H_{13}NO$	223.27	nd (aa)		262				B12³, 3299
6990	Fluorene, 1-amino-9-hydroxy $C_{13}H_{11}NO$	197.24	dk red rd (w)		142			al, eth, bz, aa	B13², 435
6991	Fluorene, 2-amino $C_{13}H_{11}N$	181.24	lo pl or nd (dil al)		131-2			al, eth	B12³, 3285
6992	Fluorene, 2-amino-9-hydroxy $C_{13}H_{11}NO$	197.24	irid nd (al)		200-1			al	B13³, 2023
6993	Fluorene, 4-amino-9-hydroxy $C_{13}H_{11}NO$	197.24	ye (60% al)		183-4			al	B13³, 2023
6994	Fluorene, 9-amino $C_{13}H_{11}N$	181.24	nd (lig)		64-5			al, eth, ace, bz, chl	B12³, 3297
6995	Fluorene, 9-benzhydrylidene or ω, ω-Diphenyl dibenzo fulvene $C_{26}H_{18}$	330.43	ye (bz)		229.5			chl	B5³, 2625
6996	Fluorene, 9-benzylidene or ω-Phenyl dibenzo fulvene $C_{20}H_{14}$	254.33	lf (al)		76			al, bz	B5³, 2464
6997	Fluorene, 2-bromo $C_{13}H_9Br$	245.12	nd or pl (al)	185^{135}	113-4			al, chl, aa	B5³, 1943
6998	Fluorene, 9-bromo $C_{13}H_9Br$	245.12	(lig or al)		104-5			al, ace	B5³, 1944
6999	Fluorene, 9 (3-bromobenzylidene) $C_{20}H_{13}Br$	333.23	ye nd (aa)		92-3			al, MeOH	B5¹, 358
7000	Fluorene, 9 (4-bromobenzylidene) $C_{20}H_{13}Br$	333.23	ye nd (aa, AcOEt-chl)		147-8			aa	B5², 640

No.	Name, Synonyms, and Formula	Mol. wt.	Color, crystalline form, specific rotation and λ_{max} (log ε)	b.p. °C	m.p. °C	Density	n_D	Solubility	Ref.
7001	Fluorene, 9 (2-chlorobenzylidene) $C_{20}H_{13}Cl$	288.78	ye nd (aa, MeOH)	$180^{0.7}$	69-70			al, aa	B5³, 2464
7002	Fluorene, 9-(3-chlorobenzylidene) $C_{20}H_{13}Cl$	288.78	Pa ye pr or pym (MeOH)		90.5			al, aa	B5¹, 358
7003	Fluorene, 9-(4 chlorobenzylidene) $C_{20}H_{13}Cl$	288.78	ye nd (aa,al)		151			al, aa	B5¹, 358
7004	Fluorene, 9-Cinnamylidene *(trans)* $C_{22}H_{16}$	280.37	pa ye nd (aa)		155			al, chl	B5³, 2533
7005	Fluorene, 2,7-diamino $C_{13}H_{12}N_2$	196.25	nd (w) pr (bz) pl (eth)		165-7			al, chl	B13³, 507
7006	Fluorene, 2,7-dichloro $C_{13}H_8Cl_2$	235.11	pl or nd (bz)	sub	128			bz, chl	B5³, 1942
7007	Fluorene, 9,9-dichloro $C_{13}H_8Cl_2$	235.11	rh pr (bz-eth) nd (peth)		103			al, eth, ace, bz	B5³, 1942
7008	Fluorene, 1,8-dimethyl-9-(2-tolyl) $C_{22}H_{20}$	284.40			168-9			al, eth, ace	C50, 11293
7009	Fluorene, 2-hydroxy $C_{13}H_{10}O$	182.22	lf (w) nd (chl)		171-4			al, eth, ace, aa	B6³, 3487
7010	Fluorene, 9-hydroxy or 9-Fluorenol $C_{13}H_{10}O$	182.22	hex nd (w or peth)		154			eth, ace, bz	B6³, 3489
7011	Fluorene, 9-hydroxy-9-phenyl $C_{19}H_{14}O$	258.32	ye or col pr (lig)		108-9			bz, aa	B6³, 3732
7012	Fluorene, 9-methyl $C_{14}H_{12}$	180.25	pr	$154\text{-}6^{15}$	46-7	$1.0263^{66/4}$	1.610^{66}	al, eth, ace, bz, chl	B5³, 1992
7013	Fluorene, 9-(2-methylbenzylidene) $C_{21}H_{16}$	268.36	nd or pr (aa)		109.5			al, bz	B5¹, 359
7014	Fluorene, 9-(4-methylbenzylidene) $C_{21}H_{16}$	268.36	nd (aa) pr (al)		97.5			al, bz	B5¹, 359
7015	Fluorene, 9-methylene or Dibenzofulvene $C_{14}H_{10}$	178.23			53			al, eth, ace, bz	B5³, 2147
7016	Fluorene, 2-nitro $C_{13}H_9NO_2$	211.22	nd (50% aa or ace)		158			ace, bz	B5³, 1948
7017	Fluorene, 3-nitro $C_{13}H_9NO_2$	211.22	pa ye nd (al, chl-peth)		106			al, ace	B5³, 1949
7018	Fluorene, 9-nitro $C_{13}H_9NO_2$	211.22	gr ye nd (al) lf (bz)		181-2d			eth, ace,bz, chl, aa	B5³, 1949
7019	Fluorene, 9-phenyl $C_{19}H_{14}$	242.32	nd or lf (al or bz)		148			al, bz, aa, chl	B5³, 2385
7020	2-Fluorene sulfonic acid $C_{13}H_{10}SO_3$	246.28	nd (aa)		155 (+1w)			w, al, ace, chl	B11³, 440
7021	9-Fluorenone or 9-Oxofluorene $C_{13}H_8O$	180.21	ye rh bipym (al, bz-peth)	341.5	84	$1.1300^{99/4}$	1.6369^{99}	al, eth, ace, bz	B7³, 2330
7022	9-Fluorenone, 1-amino $C_{13}H_9NO$	195.22	ye nd (dil al)		118-20			al, eth, ace	B14³, 285
7023	9-Fluorenone, 2-amino $C_{13}H_9NO$	195.22	red vt pr (al)		163			al, eth, bz, aa	B14³, 286
7024	9-Fluorenone, 3-amino $C_{13}H_9NO$	195.22	ye nd (w or dil al)		158-9			al	B14³, 289
7025	9-Fluorenone, 4-amino $C_{13}H_9NO$	195.22	red nd (al)		145			al, eth, ace, chl, aa	B14³, 289
7026	9-Fluorenone, 2-bromo $C_{13}H_7BrO$	259.10	ye nd (al or aa)		149			ace, bz, chl, aa	B7³, 2340
7027	9-Fluorenone, 2-chloro $C_{13}H_7ClO$	214.65	og ye nd (dil al)	sub	125-6			al	B7³, 2338
7028	9-Fluorenone, 1,8-dimethyl $C_{15}H_{12}O$	208.26	ye		197-8			chl, aa	C62, 7611
7029	9-Fluorenone, 2-nitro $C_{13}H_7NO_3$	225.20	ye nd or lf (aa)	sub	222-3			ace	B7³, 2344
7030	9-Fluorenone, oxime $C_{13}H_9NO$	195.22	nd (chl-peth or bz)		195-6			al, chl	B7³, 2335
7031	Fluorenone, 2,3,7-trinitro $C_{13}H_5N_3O_7$	315.20	Pa ye nd (aa)		180-1			bz, chl	B7³, 2348

No.	Name, Synonyms, and Formula	Mol. wt.	Color, crystalline form, specific rotation and λ_{max} (log ε)	b.p. °C	m.p. °C	Density	n_D	Solubility	Ref.
7032	9-Fluorenone, 2,4,7-trinitro $C_{13}H_5N_3O_7$	315.20	pa ye nd (aa or bz)		176			ace, bz, chl	B7^3, 2348
7033	9-Fluorenone-1-carboxamide $(C_{13}H_7O)CONH_2$	223.23	ye nd (al)		229-30			ace	B10^3, 3369
7034	9-Fluorenone-1-carboxylic acid $(C_{13}H_7O)CO_2H$	224.22	og, red nd (dil al)		192-4			al, eth	B10^3, 3368
7035	9-Fluorenone-1-carboxylic acid, ethyl ester $(C_{13}H_7O)CO_2C_2H_5$	252.27	ye nd (dil al)		84-5			al, eth	B10^3, 3369
7036	9-Fluorenone-1-carboxylyl chloride $(C_{13}H_7O)COCl$	242.66	pa ye nd (bz)		140			al, eth, bz	B10, 774
7037	9-Fluorenone-2-carboxylic acid $(C_3H_7O)_2CO_2H$	224.22	ye nd (al or aa)	sub 340	338			al, aa	B10^3, 3370
7038	9-Fluorenone-2-carboxylic acid, methyl ester $C_{15}H_{10}O_3$ 2-$(C_3H_7O)CO_2CH_3$	238.24	ye nd (MeOH)		181			al, eth, ace	B10^3, 3370
7039	9-Fluorenone-3-carboxylic acid $C_{14}H_8O_3$	224.22	ye (aa, MeOH)		299			al	B10^3, 3372
7040	9-Fluorenone 4-carboxylic acid $C_{14}H_7O_3$	224.22	ye nd (al)		227			al, eth	B10^3, 3372
7041	Fluorescein or 3′,4′-Dehydroxy fluoran $C_{20}H_{12}O_5$	332.31	red rh pr		314-6d (sealed tube)			ace, py, MeOH	B19^4, 2904
7042	Fluorescin or 2 (3,6-Dihydroxyxanthyl) benzoic acid $C_{20}H_{14}O_5$	334.33	col or ye nd (aa or eth) pl (bz)		125-7			al, eth, ace	B19^4, 2904
7043	Fluoroacetic acid FCH_2CO_2H	78.04	nd	165	35.2	1.3693^{36}		al, w	B2^4, 446
7044	Fluorophosphoric acid, diisopropyl ester $(i\text{-}C_3H_7O)_2POF$	184.15		62^9		1.055	1.3830^{25}	eth	B1^4, 1480
7045	Folic acid or Pteroylglutamic acid, Vitamin Bc $C_{19}H_{19}N_7O_6$	441.40	ye og nd (w) $\alpha^{25/}{}_D$ + 23 (0.1 N NaOH, c=0.5)		250d (darkens)			al, aa, py	Am69, 1476
7046	Folinic acid or 5-Formyl-5,6,7,8-tetrahydropteroyl-*L*-glutamic acid $C_{20}H_{23}N_7O_7$	473.45	cr (w + 3) $[\alpha]^{25/}{}_D$ + 16.76 (5% Na_2CO_3, c=3.5)		248-50d				Am73, 1979
7047	Formaldehyde or Methanal HCHO	30.03	gas	−21	−92	0.815$^{20/4}$		w, al, **eth, ace, bz**	B1^4, 3017
7048	Formaldehyde, bis-4-chlorophenyl acetal $H_2C(OC_6H_4Cl\text{-}4)_2$	269.13		189-94^6	70			eth, ace, bz	B6^4, 833
7049	Formaldehyde, dibutyl acetal $H_2C(OC_4H_9)_2$	160.26		179.2	−58.1	0.834$^{20/D}$	1.4072$^{17\ 2/}{}_D$		B1^4, 3029
7050	Formaldehyde, β,β′-(dichloro isopropyl) ethyl acetal $[(ClCH_2)_2CHO]CH_2[OC_2H_5]$	187.07		96-8^{16}		1.182$^{17/17}$	1.4491^{17}	eth	B1^3, 2574
7051	Formaldehyde, diethyl acetal or Ethylal $CH_2(OC_2H_5)_2$	104.15		89	−66.5	0.8319$^{20/4}$	1.3748^{18}	w, **al, eth**, ace bz	B1^4, 3027
7052	Formaldehyde, dimethyl acetal or Methylal $CH_2(OCH_3)_2$	76.10		45.5	−104.8	0.8593$^{20/4}$	1.3513^{20}	w,**al, eth, ace, bz**	B1^4, 3026
7053	Formaldehyde, 2,4-dinitrophenyl hydrazone $[2,4\text{-}(NO_2)_2C_6H_3]NHN{=}CH_2$	210.15	ye cr (al) pr (lig)		167				B15^3, 426
7054	Formaldehyde, dipropyl acetal $CH_2(OC_3H_7)_2$	132.20		140.5	−97.3	0.8345^{20}	1.3939^{19}	w, al, eth, ace, bz	B1^4, 3029
7055	Formaldehyde, fluoro or Formyl fluoride FCHO	48.02	gas	−24					B2^4, 42
7056	Formaldehyde, oxime or Formaldoxime $H_2C{=}NOH$	45.04		109^{15}	2.5	1.133		w, al, eth	B1^4, 3055
7057	Formaldomedone $C_{17}H_{22}O_4$	292.36	nd (al or bz)		189-90				B7^3, 4736
7058	Formamide $HCONH_2$	45.04		111^{20}	2.5	1.1334$^{20/4}$	1.4472^{20}	**w, al**, ace	B2^4, 45
7059	Formamide, *N,N*-diethyl $HCON(C_2H_5)_2$	101.15		177-8, 68^{15}		0.9080^{19}	1.4321^{25}	w, al, eth, **ace, bz**	B4^4, 346
7060	Formamide, *N,N*-dimethyl $HCON(CH_3)_2$	73.09		149-56, 39.9^{10}	−60.5	0.9487$^{20/4}$	1.4305^{20}	**w, al, eth, ace, bz, chl**	B4^4, 171

No.	Name, Synonyms, and Formula	Mol. wt.	Color, crystalline form, specific rotation and λ_{max} (log ε)	b.p. °C	m.p. °C	Density	n_D	Solubility	Ref.
7061	Formamide, *N,N*-diphenyl $HCON(C_6H_5)_2$	197.24	rh (dil al)	337.5, 190^{13}	73-4	...	...	al, eth, bz	$B12^{3}$, 455
7062	Formamide, *N*-ethyl $HCONHC_2H_5$	73.09	...	197-9, 109.6^{10}	...	$0.9552^{20/4}$	1.4320^{20}	**w, al, eth**	$B4^{4}$, 346
7063	Formamide, *N*-(1-hydroxy-2,2,2-trichloro ethyl) or Chloral formamide $HCONH[CH(OH)CCl_3]$	192.43	cr	118	...	...	...	w, al, eth, ace	$B2^{2}$, 37
7064	Formamide, *N*-methyl $HCONHCH_3$	59.07	...	180-5, $102\text{-}3^{20}$	...	1.011^{19}	1.4319^{20}	w, al, **ace**	$B4^{4}$, 170
7065	Formamide, *N*-phenyl or Formanilide $HCONHC_6H_5$	121.14	mcl pr (lig-xyl)	271, 166^{14}	50	$1.1322^{50/50}$	...	al, eth, bz	$B12^{3}$, 453
7066	Formamide, *N*-2-tolyl $HCONH(C_6H_4CH_3\text{-}2)$	135.17	lf (al)	288	62	$1.086^{55/4}$	...	al	$B12^{3}$, 1852
7067	Formamide, *N*-3-tolyl $HCONH(C_6H_4CH_3\text{-}3)$	135.17	...	$278^{724}/_{d}$, $176\text{-}8^{17}$	<-18	...	...	w	$B12^{3}$, 1962
7068	Formamide, *N*-4-tolyl $HCONH(C_6H_4CH_3\text{-}4)$	135.17	nd	...	53	...	...	al, ace	$B12^{3}$, 2050
7069	Formamidine $HN{=}CHNH_2$	44.06	pr	d	81	...	...	w, al	$B2^{4}$, 82
7070	Formamidine, *N,N'*-diphenyl $C_6H_5N{-}CHNHC_6H_5$	196.25	nd (al)	> 250	142	...	...	al, eth, ace, bz, chl	$B12^{3}$, 456
7071	Formamidoxime or Isoretin $HON{=}CHNH_2$	60.06	rh nd (al)	d	114-5	...	...	al	$B2^{4}$, 84
7072	Formic acid or Methanoic acid HCO_2H	46.03	...	100.7, 50^{120}	8.4	$1.220^{20/4}$	1.3714^{20}	**w, al, eth,** ace bz	$B2^{4}$, 3
7073	Formic acid, allyl ester or Allyl formate $HCO_2CH_2CH{=}CH_2$	86.09	...	83.6	...	$0.9460^{20/4}$	...	al, **eth,**	$B2^{3}$, 46
7074	Formic acid, benzyl ester or Benzyl formate $HCO_2CH_2C_6H_5$	136.15	...	$202\text{-}3^{747}$, $84\text{-}5^{10}$	...	$1.081^{20/4}$	1.5154^{20}	al, **eth**, ace	$B6^{4}$, 2262
7075	Formic acid, butyl ester or Butyl formate $HCO_2C_4H_9$	102.13	...	106.8	−91.3	$0.8885^{20/4}$	1.3912^{20}	**al, eth**, ace	$B2^{4}$, 28
7076	Formic acid, *iso*butyl ester or Isobutyl formate $HCO_2\text{-}i\text{-}C_4H_9$	102.13	...	98.4	−95.8	$0.8854^{20/4}$	1.3857^{20}	**al, eth**, ace	$B2^{4}$, 29
7077	Formic acid, *sec*-butyl ester-*(dl)* or *sec*-Butyl formate $HCO_2\text{-}sec\text{-}C_4H_9$	102.13	...	97	...	$0.8846^{20/4}$	1.3865^{20}	**al, eth**, ace	$B2^{4}$, 29
7078	Formic acid, cyclohexyl ester or Cyclohexyl formate $HCO_2C_6H_{11}$	128.17	...	162.5^{750}	...	$1.0057^{0/4}$	1.4430^{20}	al, eth, aa	$B6^{4}$, 35
7079	Formic acid, ethyl ester or Ethyl formate $HCO_2C_2H_5$	74.08	...	54.5	−80.5	0.9168^{20}	1.3598^{10}	w, **al, eth**, ace	$B2^{4}$, 23
7080	Formic acid, heptyl ester or Heptyl formate $HCO_2C_7H_{15}$	144.22	...	178.1, 83^{30}	...	0.8784^{20}	1.4140^{20}	**al, eth**	$B2^{4}$, 31
7081	Formic acid, hexyl ester or Hexyl formate $HCO_2C_6H_{13}$	130.19	...	155.5	−62.6	$0.8813^{20/4}$	1.4071^{20}	**al, eth**	$B2^{4}$, 31
7082	Formic acid, hydrazide $HCONHNH_2$	60.06	ye lf or nd (al)	...	54	...	...	al, eth, bz, chl	$B2^{4}$, 85
7083	Formic acid, methyl ester or Methyl formate HCO_2CH_3	60.05	...	31.5	−99	$0.9742^{20/4}$	1.3433^{20}	w, **al**, eth	$B2^{4}$, 20
7084	Formic acid, octyl ester or Octyl formate $HCO_2C_8H_{17}$	158.24	...	198.8	−39.1	0.8744^{20}	1.4208^{15}	al, **eth**	$B2^{4}$, 31
7085	Formic acid, pentyl ester $HCO_2C_5H_{11}$	116.16	...	132.1	−73.5	$0.8853^{20/4}$	1.3992^{20}	**al, eth**	$B2^{4}$, 30
7086	Formic acid, isopentyl ester $HCO_2\text{-}i\text{-}C_5H_{11}$	116.16	...	124.2	−93.5	$0.8857^{20/4}$	1.3976^{20}	al, **eth**	$B2^{4}$, 30
7087	Formic acid, propyl ester or Propyl formate $HCO_2C_3H_7$	88.11	...	81.3	−92.9	$0.9058^{20/4}$	1.3779^{20}	**al, eth**	$B2^{4}$, 26
7088	Formic acid, isopropyl ester or Isopropyl formate $HCO_2CH(CH_3)_2$	88.11	...	68.2	...	$0.8728^{20/4}$	1.3678^{20}	**al, eth**, ace	$B2^{4}$, 27
7089	Formimidic acid, *N*-phenyl, ethyl ether $C_6H_5N{=}CHOC_2H_5$	149.19	...	213-5	...	$1.0051^{20/4}$	1.5279^{20}	eth, bz	$B12^{3}$, 455
7090	Formonitrolic acid or Methyl nitrolic acid $HON{=}CHNO_2$	90.04	nd (eth or eth-peth)	...	68d	...	...	w, al, eth	$B2^{4}$, 85
7091	Frangulin A (*αβ*-L-Rhamnoside) $C_{21}H_{20}O_9$	416.38	ye or red (al or AcOEt)	...	228	...	...	al, bz, aa	$B17^{4}$, 2535
7092	Fraxin or *αβ*-Glucoside of fraxetin $C_{16}H_{18}O_{10}$	370.31	ye nd (al) (w + 3)	...	205	...	...	...	$B18^{4}$, 2373

No.	Name, Synonyms, and Formula	Mol. wt.	Color, crystalline form, specific rotation and λ_{max} (log ε)	b.p. °C	m.p. °C	Density	n_D	Solubility	Ref.
7093	β-D-Fructose or Levulose $C_6H_{12}O_6$	180.16	pr or nd (w) orh, pr (al) $[\alpha]^{20/}{}_{D}$(mut) - 133→-92 (w, c=2)		103-5d	$1.60^{20/4}$		w, al, ace	B1⁴, 4401
7094	α-L-Fucose or 6-Deoxy-L-galactose $C_6H_{12}O_5$	164.16	nd (al) $[\alpha]^{20/}{}_{D}$(mut) - 124.1→- 75.6 (w, c=9)						B1⁴, 4265
7095	Fucoxanthin $C_{42}H_{58}O_6$	658.92	red br pl (eth-peth) hex pl c + 2w, dil al) $[\alpha]^{18/}{}_{D}$ + 72.5 (chl)		168			al, eth	B18⁴, 2820
7096	Fulvene C_6H_6	78.11		$7\text{-}8^{56}$			1.4920^{20}	bz	B5⁴, 764
7097	Fulvene-6-vinyl *(trans)* 6-$(CH_2{=}CH)C_6H_5$	104.15		45^{12}	−35	0898^{20}		bz, chl	HCACA 47, 1022
7098	Fumagacin $C_{32}H_{41}O_7$	537.68	nd (dil aa) $[\alpha]^{18/}{}_{D}$ - 125 (chl, c=1)		212 d			eth, ace, bz, aa	Am78, 5275
7099	Fumaric acid or *trans*-Butenedioic acid $HO_2CCH{=}CHCO_2H$	116.08	nd mcl pr or lf (w)	$165^{1.7}$ sub	300-2 (sealed tube)	$1.635^{20/4}$		al	B2⁴, 2202
7100	Fumaric acid, bromo or Bromofumaric acid $HO_2CCBr{=}CHCO_2H$	194.97	pr (AcOEt)	d 200	185-6			w, al	B2⁴, 2224
7101	Fumaric acid, dibutyl ester or Dibutyl fumarate $C_4H_9O_2CCH{=}CHCO_2C_4H_9$	228.29		150^{4}		$0.9869^{20/4}$	1.4469^{20}	ace, chl	B2⁴, 2210
7102	Fumaric acid, diisobutyl ester i-$C_4H_9O_2CCH{=}CHCO_2$-c-C_4H_9	228.29		170^{160}, 122^{5}		$0.9760^{20/4}$	1.4432^{20}	al, eth, ace	B2⁴, 2211
7103	Fumaric acid, chloro $HO_2CCCl{=}CHCO_2H$	150.52	pl (aa)	sub	192-3			al, eth	B2⁴, 2221
7104	Fumaric acid, chloro, diethyl ester $C_2H_5O_2CCCl{=}CHCO_2C_2H_5$	206.63		250d, 127^{10}		$1.1880^{20/4}$	1.4571^{20}	al, eth	B2³, 1909
7105	Fumaric acid, chloro, dimethyl ester $CH_3O_2CCCl{=}CHCO_2CH_3$	178.57		224, 108^{15}		$1.2899^{25/4}$	1.4720^{18}	al, eth	B2³, 1909
7106	Fumaric acid, diethyl ester $C_2H_5O_2CCH{=}CHCO_2C_2H_5$	172.18		214, 98^{10}	1-2	$1.0452^{20/4}$	1.4412^{20}	ace, chl	B2⁴, 2207
7107	Fumaric acid, dimethyl $HO_2CC(CH_3){=}C(CH_3)CO_2H$	144.13	nd (w)		241				B2⁴, 2243
7108	Fumaric acid, dimethyl ester $CH_3O_2CCH{=}CHCO_2CH_3$	144.13		193	103-4	$1.37^{20/4}$	1.40625^{111}	ace, chl	B2⁴, 2205
7109	Fumaric acid, diphenyl ester $C_6H_5O_2CCH{=}CHCO_2C_6H_5$	268.27	nd (al)	219^{14}	161-2				B6⁴, 628
7110	Fumaric acid, dipropyl ester $C_3H_7O_2CCH{=}CHCO_2C_3H_7$	200.24		110^{5}		$1.0129^{20/4}$	1.4435^{20}	al, eth	B2⁴, 2209
7111	Fumaric acid, diisopropyl ester $(CH_3)_2CHO_2CCH{=}CHCO_2CH(CH_3)_2$	200.24		225-6				al, eth, ace	B2⁴, 2209
7112	Fumaric acid, methyl or Mesaconic acid $HO_2CC(CH_3){=}CHCO_2H$	130.10	rh nd or mcl pr (eth, AcOEt)	sub	204-5	$1.466^{20/4}$		al, eth	B2⁴, 2231
7113	Fumaronitrile $NCCH{=}CHCN$	78.07	nd (bz-peth)	186	96.8	0.9416^{111}	1.4349^{111}	w, al, eth, ace, bz	B2⁴, 2219
7114	Fumaryl chloride $ClOCCH{=}CHCOCl$	152.97	pa ye lig	158-60, 63^{13}		1.408^{20}	1.5004^{18}		B2⁴, 2217
7115	Fumaryl chloride, chloro $ClOCCCl{=}CHCOCl$	187.41	pa gr	184-7d, $73\text{-}5^{20}$		$1.564^{20/4}$	1.5206^{20}	eth, aa	B2³, 1909
7116	Furan C_4H_4O	68.08		31.4	−85.6	$0.9514^{20/4}$	1.4214^{20}	al, eth, ace, bz	B17⁴, 225
7117	Furan, 2-acetyl or 2-Furyl methyl ketone 2-$CH_3CO(C_4H_3O)$	110.11	cr (lig)	175, 67^{10}	33	1.098^{20}	1.5017^{20}	al, eth	B17⁴, 4500
7118	Furan, 2-benzoyl or 2-Furyl phenyl ketone $C_6H_5CO(2\text{-}C_4H_3O)$	172.18		285, 164^{19}	<-15	1.1732^{20}	1.6055^{20}	al, eth	B17⁴, 5184

PHYSICAL CONSTANTS OF ORGANIC COMPOUNDS (Continued)

No.	Name, Synonyms, and Formula	Mol. wt.	Color, crystalline form, specific rotation and λ_{max} (log ε)	b.p. °C	m.p. °C	Density	n_D	Solubility	Ref.
7119	Furan, 2-bromo or α-Furyl bromide $2\text{-}Br(C_4H_3O)$	146.97	....	102^{744}	......	1.6500^{20}	1.4980^{20}	al, eth, ace, bz	B17^4, 231
7120	Furan, 3-bromo or β Furyl bromide $3\text{-}Br(C_4H_3O)$	146.97	....	103	......	$1.6606^{20/4}$	1.4958^{20}	al, eth, ace, bz	B17^4, 232
7121	Furan, 2-*tert*butyl $2\text{-}(CH_3)_3C(C_4H_3O)$	124.19	....	119-20	......	$0.869^{20/4}$	1.4373^{20}	al, eth, ace	B17^4, 307
7122	Furan, 2-chloro or α -Furyl chloride $2\text{-}Cl(C_4H_3O)$	102.52	....	77.5^{744}	......	$1.1923^{70/4}$	1.4569^{20}	al, eth, ace	B17^4, 230
7123	Furan, 3-chloro or β -Furyl chloride $3\text{-}Cl(C_4H_3O)$	102.52	....	79^{742}	......	$1.2094^{20/4}$	1.4601^{20}	eth, ace	B17^4, 230
7124	Furan, 2,5-dibromo $2,5\text{-}Br_2(C_4H_2O)$	225.87	pl	$164\text{-}5^{764}$, 62^{13}	......	$2.27^{20/20}$	1.5455^{20}	....	B17^4, 232
7125	Furan, 2,5-di-*tert*-butyl $2,5\text{-}[tert\text{-}(C_4H_9)_2](C_4H_2O)$	180.29	....	210, $61\text{-}2^{17}$	......	$0.837^{20/4}$	1.4369^{20}	al, eth,	B17^4, 336
7126	Furan, 2,5-dichloro $2,5\text{-}Cl_2(C_4H_2O)$	136.97	....	115	......	1.371^{25}	....	....	B17^4, 230
7127	Furan, 2,5-dimethoxy $2,5(CH_3O)_2(C_4H_2O)$	128.13	....	145-7	......	....	$1.4168^{20/}{}_{D}$	....	B17^4, 1993
7128	Furan, 2,4-dimethyl $2,4(CH_3)_2(C_4H_2O)$	96.13	....	94	......	$0.8993^{20/4}$	1.4371^{20}	....	B17^4, 287
7129	Furan, 2,5-dimethyl $2,5\text{-}(CH_3)_2(C_4H_2O)$	96.13	....	93-4	−62.8	$0.8883^{20/4}$	1.4363^{20}	al, eth, ace, bz	B17^4, 289
7130	Furan, 2,5-dinitro $2,5\text{-}(NO_2)_2(C_4H_2O)$	158.07	nd (w) pr (al)	....	102	....	....	eth	B17^4, 234
7131	Furan, 2,5-diphenyl $2,5\text{-}(C_6H_5)_2(C_4H_2O)$	220.27	nd or lf (dil al)	343-5	91	....	....	al, eth, ace, bz	B17^4, 682
7132	Furan, 2-ethyl $2\text{-}C_2H_5(C_4H_3O)$	96.14	....	$92\text{-}3^{768}$	......	$0.912^{15/15}$	1.4466^{21}	al, eth, ace, bz	B17^4, 284
7133	Furan, 2-furfuryl $(C_4H_3O)CH_2(C_4H_3O)$	148.16	....	94^{22}	......	$1.102^{20/4}$	1.5049^{20}	al, eth, ace	B19^4, 269
7134	Furan, 2-iodo or α -Furyl iodide $2\text{-}I\text{-}(C_4H_3O)$	193.97	....	$43\text{-}5^{15}$	......	$2.024^{20/4}$	1.5661^{20}	eth	B17^4, 232
7135	Furan, 3-iodo or β Furyl iodide $3\text{-}I(C_4H_3O)$	193.97	....	132.2^{732}, $37\text{-}8^{22}$	......	$2.045^{20/4}$	1.5610^{20}	eth	B17^4, 232
7136	Furan, 2-methyl or Sylvan $2\text{-}CH_3(C_4H_3O)$	82.10	....	63^{737}	......	$0.9132^{20/4}$	1.4342^{20}	al, eth	B17^4, 265
7137	Furan, 3-methyl $3\text{-}CH_3(C_4H_3O)$	82.10	....	65.5^{749}	......	$0.923^{18/4}$	1.4330^{19}	al, eth	B17^4, 276
7138	Furan, 2-nitro $2\text{-}O_2N(C_4H_3O)$	113.07	yesh mcl cr (peth)	$133\text{-}5^{123}$	29	....	....	al, eth	B17^4, 233
7139	Furan, 2-phenyl $2\text{-}C_6H_4(C_4H_3O)$	144.17	....	$107\text{-}8^{18}$	......	$1.083^{20/4}$	1.5920^{20}	ace, bz	B17^4, 542
7140	Furan, 2-propionyl or Ethyl Furyl Ketone $2\text{-}CH_3CH_2CO(C_4H_3O)$	124.14	cr	88^{14}	28	1.0626^{28}	1.4922^{25}	eth	B17^4, 4537
7141	Furan, 2-propyl $2\text{-}C_3H_7(C_4H_3O)$	110.16	....	$114\text{-}5^{750}$	......	$0.8876^{20/4}$	1.4549^{20}	al, eth, ace	B17^4, 296
7142	Furan, tetrahydro C_4H_8O	72.11	....	67	fr-108	$0.8892^{20/4}$	1.4050^{20}	al, eth, ace, bz	B17^4, 24
7143	Furan, tetrahydro-2,2-diethyl $2,2\text{-}(C_2H_5)_2(C_4H_6O)$	128.21	....	146	......	$0.8703^{20/4}$	1.4317^{20}	al, eth, ace,,bz	B17^4, 107
7144	Furan, tetrahydro,2-ethyl $2\text{-}C_2H_5(C_4H_7O)$	100.16	....	109	......	$0.8570^{19/4}$	1.4147^{19}	al, eth, ace, bz	B17^4, 78
7145	Furan, tetrahydro-2-methyl $2\text{-}CH_3(C_4H_7O)$	86.13	....	80	......	$0.8552^{20/4}$	1.4059^{21}	al, eth, ace, bz, chl	B17^4, 60
7146	Furan, tetrahydro-3-methyl $3\text{-}CH_3(C_4H_7O)$	86.13	....	86-7	......	$0.8642^{20/4}$	1.4122^{20}	al, eth, ace, bz	B17^2, 21
7147	Furan, tetrahydro-2-propyl $2\text{-}C_3H_7(C_4H_7O)$	114.19	....	139.4	......	$0.8547^{20/4}$	1.4242^{20}	....	B17^4, 93
7148	Furan, tetraiodo C_4I_4O	571.66	nd	....	165	....	....	al	B17^4, 233
7149	Furan, tetraphenyl $(C_6H_5)_4(C_4O)$	372.47	....	220	175	....	....	al, eth, ace,bz, aa	B17^4, 810
7150	Furan, 2,3,5-trichloro $2,3,5\text{-}Cl_3(C_4HO)$	171.41	....	147	......	1.50^{25}	....	....	B17^4, 231
7151	2-Furancarboxylic acid or α -Furoic acid $2\text{-}(C_4H_3O)CO_2H$	112.09	mcl nd or lf (w)	230-2, $141\text{-}4^{20}$	133-4	....	....	w, al, eth	B18^4, 3914

PHYSICAL CONSTANTS OF ORGANIC COMPOUNDS (Continued)

No.	Name, Synonyms, and Formula	Mol. wt.	Color, crystalline form, specific rotation and λ_{max} (log ε)	b.p. °C	m.p. °C	Density	n_D	Solubility	Ref.
7152	2-Furancarboxylic acid, 5-ethoxy $5\text{-}C_2H_5O(C_4H_2O)CO_2H\text{-}2$	156.14			140-1			al	B18^4, 4827
7153	2-Furancarboxylic acid, 5-iodo $5\text{-}I(C_4H_2O)CO_2H\text{-}2$	237.99			197d				B18^4, 3992
7154	2-Furancarboxylic acid, 5-methoxy $5\text{-}CH_3O(C_4H_2O)CO_2H\text{-}2$	142.11			136-8d			al	B18^4, 4827
7155	3-Furancarboxylic acid or β-Furoic acid $3\text{-}(C_4H_3O)CO_2H$	112.09	nd (w)	105-10^{12} sub	122-3			al, eth	B18^4, 4052
7156	3-Furancarboxylic acid, 4-methyl $4\text{-}CH_3(C_4H_2O)CO_2H\text{-}3$	126.11	nd (bz-peth)		138-9			w	B18^4, 4089
7157	3-Furancarboxylic acid, 5-methyl $5\text{-}CH_3(C_4H_2O)CO_2H\text{-}3$	126.11	(w)	sub	119			w, eth	B18^4, 4076
7158	2,3-Furandicarboxylic acid $2,3\text{-}(C_4H_2O)(CO_2H)_2$	156.10	pr (aa or sub)	sub	226			w, al	B18^4, 4477
7159	2,3-Furandicarboxylic acid, dimethyl ester $2,3\text{-}(C_4H_2O)(CO_2CH_3)_2$	184.15	(MeOH)		39			al, eth	B18^4, 4477
7160	2,4-Furandicarboxylic acid $2,4\text{-}(C_4H_2O)(CO_2H)_2$	156.10	lf (w + l)	sub	266			al, ace	B18^4, 4479
7161	2,4-Furandicarboxylic acid, dimethyl ester $2,4\text{-}(C_4H_2O)(CO_2CH_3)_2$	184.15	pr (MeOH)		109-10				B18^4, 4480
7162	2,5-Furandicarboxylic acid or Dehydromucic acid $2,5\text{-}(C_4H_2O)(CO_2H)_2$	156.10	nd (w) lf (al)	sub	> 320				B18^4, 4481
7163	2,5-Furandicarboxylic acid, dimethyl ester $2,5\text{-}(C_4H_2O)(CO_2CH_3)_2$	184.15	nd (w) cr (MeOH)	154-6^{15}	112			al, eth, chl	B18^4, 4482
7164	3,4-Furandicarboxylic acid $3,4\text{-}(C_4H_2O)(CO_2H)_2$	156.10			217-8				B18^4, 4497
7165	2-Furanone, tetrahydro $C_4H_6O_2$	86.09		206, 89^{12}	−42	1.1286$^{16/0}$	1.4341^{20}	w, al, eth, ace, bz	B17^4, 4159
7166	2-Furanone, 5-methyl (d), tetrahydro $C_5H_8O_2$	100.12	$[\alpha]^{20/}{}_D$ + 13.5 undil	86-90^{14}					B17^2, 288
7167	2-Furanone, 5-methyl (dl), tetrahydro $C_5H_8O_2$	100.12		206, 83-4^{13}	−31	1.0465^{23}	1.4328^{20}		B17^4, 4176
7168	2-Furanone, 5-methyl (l), tetrahydro $C_5H_8O_2$	100.12	$[\alpha]^{20/}{}_D$ −4.6 eth 10%	78-80^{8}			1.4322^{20}		B17^4, 4176
7169	Furazan, 3,4-dimethyl or 3,4-dimethyl-1,2,5-oxadiazole $C_4H_6N_2O$	98.10		156^{764}	−7	1.0528$^{14/4}$	1.4237^{20}	al, eth	B27^2, 628
7170	Furfural or α Furaldehyde $(OC_4H_3)CHO$	96.09		161.7, 90^{65}	−38.7	1.1594$^{20/4}$	1.5261^{20}	al, eth, ace, bz, chl	B17^4, 4403
7171	Furfural acetone $(C_4H_3O)CH{=}CHCOCH_3$	136.15	nd	229d, 112-3^{10}	39-40	1.0496$^{57/4}$	1.5788^{25}	al, eth, chl, peth	B17^4, 4714
7172	Furfural, 5-bromo $5\text{-}Br(OC_4H_2)CHO\text{-}2$	174.98	cr (50% al)	112^{16}	82			al, eth	B17^4, 4456
7173	Furfural, 5-chloro $5\text{-}Cl(C_4H_2O)CHO\text{-}2$	130.53		70^{1}	31-3				B17^4, 4454
7174	Furfural, diacetate or Furfurylidene diacetate $(C_4H_3O)CH(O_2CCH_3)_2$	198.18	nd or pl (eth-peth)	220, 143-4^{20}	52-3			al, eth, bz	B17^4, 4413
7175	Furfural, diethyl acetal $(C_4H_3O)CH(OC_2H_5)_2$	170.21		191-2, 62-4^{5}		0.9994$^{20/20}$	1.4451^{20}	al	B17^4, 4412
7176	Furfural, 5-hydroxymethyl $5\text{-}HOCH_2(C_4H_2O)CHO\text{-}2$	126.11	nd (eth-peth)	114-6$^{0.5}$	35	1.2062$^{25/4}$	1.5627^{18}	w, al, eth, bz, chl	B18^4, 100
7177	Furfural, 5-methyl $5\text{-}CH_3(C_4H_2O)CHO\text{-}2$	110.11		187, 79-81^{12}		1.1072$^{18/4}$	1.5262^{20}	w, al, eth	B17^4, 4523
7178	Furfural, 5-nitro $5\text{-}NO_2(C_4H_2O)CHO\text{-}2$	141.08	pa ye (peth)	128-32^{10}	35-6			peth	B17^4, 4459
7179	Furfural, 5-nitro, semicarbazone $C_6H_6N_4O_4$	198.14	pa ye nd (w) darkens in light		237d				B17^4, 4467
7180	Furfural, oxime, *(anti)* or α-Furfuraldoxime $2\text{-}(C_4H_3O)CH{=}NOH$	111.10	nd (lig)		75-6			al, eth, bz, aa	B17^4, 4430
7181	Furfural, oxime *(syn)* or β Furfuraldoxime $2\text{-}(C_4H_3O)CH{=}NOH$	111.10	nd (lig)	201-8d, 98^{9}	91-2			al, eth, bz, chl	B17^4, 4428
7182	Furfural, phenylhydrazone $2\text{-}(C_4H_3O)CH{=}NNHC_6H_5$	186.21	ye lf (al)		97-8			al, eth	B17^4, 4432
7183	Furfural, tetrahydro $2\text{-}(C_4H_7O)CHO$	100.12		142-3^{779}, 45-7^{29}		1.0727$^{20/4}$	1.4366^{20}	w, eth,	B17^4, 4179
7184	Furfuryl alcohol or 2-Hydroxymethyl furan $2\text{-}(C_4H_3O)CH_2OH$	98.10	col ye	171^{750}, 68-9^{20}		1.1296$^{20/4}$	1.4868^{20}	w, al, eth	B17^4, 1242

PHYSICAL CONSTANTS OF ORGANIC COMPOUNDS (Continued)

No.	Name, Synonyms, and Formula	Mol. wt.	Color, crystalline form, specific rotation and λ_{max} (log ε)	b.p. °C	m.p. °C	Density	n_D	Solubility	Ref.
7185	Furfuryl alcohol, 5- methyl 5-$CH_3(C_4H_2O)CH_2OH$-2	112.13	....	$194\text{-}6^{744}{}_d$, 81^{23}	......	$1.0769^{20/4}$	1.4853^{20}	al, eth	B17[4], 1279
7186	Furfuryl alcohol, tetrahydro 2-$(C_4H_7O)CH_2OH$	102.13	....	$177\text{-}8^{750}$, $80\text{-}2^{20}$	......	$1.0544^{20/4}$	1.4517^{20}	eth, ace	B17[4], 1095
7187	Furfuryl amine 2-$(C_4H_3O)CH_2NH_2$	97.12	....	145-6, 80^{84}	......	$1.0995^{20/4}$	1.4908^{20}	w, al, eth	B18[4], 7068
7188	Furfuryl amine, tetrahydro 2-$(C_4H_7O)CH_2NH_2$	101.15	....	$151\text{-}2^{735}$	......	$0.9770^{20/20}$	1.4551^{20}	w, **al, eth**	B18[4], 7034
7189	Furfuryl bromide 2-$(C_4H_3O)CH_2Br$	161.00	pa ye	$33\text{-}4^{2}$	......	$1.560^{20/20}$	1.5380^{20}	eth	B17[4], 268
7190	2-Furfuryl bromide, tetrahydro 2-$(C_4H_7O)CH_2Br$	165.03	....	$168\text{-}70^{744}$, $69\text{-}70^{22}$	......	$1.3653^{20/4}$	1.4850^{20}	al, eth	B17[4], 62
7191	Furfuryl chloride 2-$(C_4H_3O)CH_2Cl$	116.55	....	49^{26}	......	$1.1783^{20/4}$	1.4941^{20}	al, eth, ace, bz	B17[4], 268
7192	Furfuryl mercaptan 2-$(C_4H_3O)CH_2SH$	114.16	....	155, 47^{12}	......	$1.1319^{20/4}$	1.5329^{20}	....	B17[4], 1255
7193	Furfurin $C_{15}H_{12}N_2O_3$	268.27	lt br nd or rh pr (w or eth)	....	116-7	....	....	al, eth	B27[2], 918
7194	Furfuryl ether or Difurfuryl ether $[(C_4H_3O)CH_2]_2O$	178.19	....	101^{2}	......	$1.1405^{20/4}$	1.5088^{20}	....	B17[2], 116
7196	Furfuryl methyl ether $(C_4H_3O)CH_2OCH_3$	112.14	....	131-3	......	$1.0163^{20/4}$	1.4570^{20}	al, eth	B17[4], 1243
7197	α-Furoic acid or 2-Furancarboxylic acid 2-$(C_4H_3O)CO_2H$	112.09	mcl nd or lf (w)	230-2, $141\text{-}4^{20}$	133-4	....	....	w, al, eth	B18[4], 3914
7198	α-Furoic acid, allyl ester 2-$(C_4H_3O)CO_2(CH_2CH{=}CH_2)$	152.15	....	206-9	......	$1.118^{25/25}$	1.4945^{20}	eth, ace	B18[4], 3919
7199	α-Furoic acid, benzyl ester 2-$(C_4H_3O)CO_2CH_2C_6H_5$	202.21	ye	$179\text{-}81^{18}$	......	$1.1623^{22/4}$	1.5550^{20}	eth, ace	B18[4], 3921
7200	α-Furoic acid, butyl ester or Butyl furoate 2-$(C_4H_3O)CO_2C_4H_9$	168.19	....	233, $83\text{-}4^{1}$	......	$1.0555^{20/4}$	1.4740	al, eth, bz, peth	B18[4], 3918
7201	α-Furoic acid, *iso*butyl ester or Isobutyl α-furoate 2-$(C_4H_3O)CO_2$-i-C_4H_9	168.19	....	221-3, $97^{13.5}$	......	$1.0388^{20/4}$	1.4676^{20}	al, eth, ace, bz	B18[2], 266
7202	α-Furoic acid, *sec*butyl ester 2-$(C_4H_3O)CO_2CH(CH_3)C_2H_5$	168.19	....	$67\text{-}9^{1}$	......	$1.0465^{20/4}$	....	al, eth	B18[2], 266
7203	α-Furoic acid, 5-ethoxy 5-$C_2H_5O(C_4H_2O)CO_2H$-2	156.14	....	....	140-1	....	....	al	B18[4], 4827
7204	α-Furoic acid, ethyl ester or Ethyl-α-furoate 2-$(C_4H_3O)CO_2C_2H_5$	140.14	lf or pr	196.8, 128^{95}	34-5	$1.1174^{21/4}$	1.4797^{21}	al, eth, ace, bz, peth	B18[4], 3917
7205	α-Furoic acid, furfuryl ester 2-$(C_4H_3O)CO_2CH_2(C_4H_3O)$	192.17	dimorphic	122^{2}	27.5	$1.2384^{25/25}$	1.5280^{20}	al, eth, ace, bz, peth	B18[4], 3936
7206	α-Furoic acid, heptyl ester or Heptyl-α-furoate 2-$(C_4H_3O)CO_2C_7H_{15}$	210.27	....	$116\text{-}7^{1}$	......	$1.0005^{20/4}$	....	al, eth	B18[2], 267
7207	α-Furoic acid, hexylester or Hexyl-α-Furoate 2-$(C_4H_3O)CO_2C_6H_{13}$	196.25	....	$105\text{-}7^{1}$	......	$1.0170^{20/4}$	....	al, eth, bz	B18[2], 267
7208	α-Furoic acid, 5-iodo 5-$I(C_4H_2O)CO_2H$-2	237.98	....	....	197d	....	....	....	B18[4], 3992
7209	α-Furoic acid, 5-methoxy 5-$CH_3O(C_4H_2O)CO_2H$-2	142.11	....	....	136-8d	....	....	al	B18[4], 4827
7210	α-Furoic acid, methyl ester or Methyl-α-Furoate 2-$(C_4H_3O)CO_2CH_3$	126.11	....	181.3	......	$1.1786^{21/4}$	1.4860^{20}	al, eth, bz	B18[4], 3916
7211	α-Furoic acid, 3-methyl 3-$CH_3(C_4H_2O)CO_2H$-2	126.11	nd (w)	sub	134	....	....	....	B18[4], 4067
7212	α-Furoic acid, 3-methyl, ethyl ester 3-$CH_3(C_4H_2O)(CO_2C_2H_5)$-2	154.17	pl	205	47-8	....	....	....	B18[4], 4067
7213	α-Furoic acid, 3-methyl, methyl ester 3-$CH_3(C_4H_2O)(CO_2CH_3)$-2	140.14	pl (al)	$72\text{-}6^{2}$	36-8	....	....	....	B18[4], 4067
7214	α-Furoic acid, 4-methyl 4-$CH_3(C_4H_2O)CO_2H$-2	126.11	nd (bz-peth)	....	131-2	....	....	....	B18[4], 4073
7215	α-Furoic acid, 5-methyl 5-$CH_3(C_4H_2O)CO_2H$-2	126.11	pl or nd (w)	105^{1}	109-10	....	....	al, eth, chl	B18[4], 4076
7216	α-Furoic acid, 5-methyl, methyl ester 5-$CH_3(C_4H_2O)CO_2CH_3$-2	140.14	....	205, 98^{15}	......	....	....	....	B18[4], 4076
7217	α-Furoic acid, 5-nitro 5-$O_2N(C_4H_2O)CO_2H$-2	157.09	pa ye pl (w)	sub	184	....	....	al, eth	B18[4], 3992
7218	α-Furoic acid, octyl ester or Octyl-α-furoate 2-$(C_4H_3O)CO_2C_8H_{17}$	224.30	....	$126\text{-}7^{1}$	......	$0.9885^{20/4}$	....	al	B18[2], 267

No.	Name, Synonyms, and Formula	Mol. wt.	Color, crystalline form, specific rotation and λ_{max} (log ε)	b.p. °C	m.p. °C	Density	n_D	Solubility	Ref.
7219	α-Furoic acid, pentyl ester or Pentyl-α-Furoate $2-(C_4H_3O)CO_2C_5H_{11}$	182.22		$95-7^{1}$		$1.0335^{20/4}$		al	$B18^{2}$, 266
7220	α-Furoic acid, isopentyl ester or Isopentyl-α-Furoate $2-(C_4H_3O)CO_2-i-C_5H_{11}$	182.22		282		$1.030^{20/4}$	1.4274^{20}	al, bz, peth	$B18^{2}$, 226
7221	α-Furoic acid, 5-phenoxy $5-C_6H_5O(C_4H_2O)CO_2H-2$	204.19			122-3			al	$B18^{4}$, 4827
7222	α-Furoic acid, propylester or Propyl-α-Furoate $2-(C_4H_3O)CO_2C_3H_7$	154.17		210.9		$1.0745^{20/4}$	1.4737^{20}	al, eth, ace, bz, peth	$B18^{4}$, 3918
7223	α-Furoic acid, isopropyl ester or isopropyl-α-furoate $2-(C_4H_3O)CO_2CH(CH_3)_2$	154.17		198-9		$1.0655^{24/4}$	1.4682^{24}	al, eth, ace, bz	B18, 275
7224	α-Furoic acid, tetrahydro $2-(C_4H_7O)CO_2H$	116.12		145^{25}	21	$1.1933^{20/20}$	1.4612^{20}	w	$B18^{4}$, 3824
7225	α-Furonitrile $2-(C_4H_3O)CN$	93.09		146^{738}		$1.0822^{20/4}$	1.4798^{20}	al, eth	$B18^{4}$, 3964
7226	α-Furoyl chloride or 2-Furan carboxylyl chloride $2-(C_4H_3O)COCl$	130.53		$173, 66^{10}$	-2			eth, chl	$B18^{4}$, 3938
7227	β-Furoic acid or 3-Furan carboxylic acid $3-(C_4H_3O)CO_2H$	112.09	nd (w)	$105-10^{12}$, sub	122-3			al, eth	$B18^{4}$, 4052
7228	β-Furoic acid, 4-bromo $4-Br(C_4H_2O)CO_2H-3$	190.98	nd (w)		129			al, eth, bz, chl	$B18^{4}$, 3980
7229	β-Furoic acid, 5-bromo $5-Br(C_4H_2O)CO_2H-3$	190.98	lf (w)		190-1			al, eth, chl	$B18^{4}$, 3980
7230	β-Furoic acid, 5-bromo, ethyl ester $5-Br(C_4H_2O)(CO_2C_2H_5)-3$	219.04	pr	$235^{767}, 134-6^{34}$	17	1.528^{20}		al, eth	$B18^{4}$, 3981
7231	β-Furoic acid, 4-chloro $4-Cl(C_4H_2O)CO_2H-3$	146.53	pl or pr (w)		149			al, eth, bz	$B18^{4}$, 3977
7232	β-Furoic acid, 5-chloro $5-Cl(C_4H_2O)CO_2H-3$	146.53	lf (w)		179-80			al, eth, bz	$B18^{4}$, 3977
7233	β-Furoic acid, 5-chloro, methyl ester $5-Cl(C_4H_2O)CO_2CH_3-3$	160.56			40-1			eth	$B18^{4}$, 3978
7234	β-Furoic acid, 2,5-dimethyl or Pyrotritaric acid $2,5-(CH_3)_2(C_4HO)CO_2H-3$	140.14	nd (w)	sub	135			al, eth	$B18^{4}$, 4099
7235	β-Furoic acid, methyl ester $3-(C_4H_3O)CO_2CH_3$	126.11		$160, 79^{42}$		$1.1744^{15/15}$	1.4676^{20}	al, ace	$B18^{4}$, 4052
7236	β-Furoic acid, 2-methyl $2-CH_3(C_4H_2O)CO_2H-3$	126.11	cr (w)		102-3			al, eth	$B18^{4}$, 4072
7237	β-Furoic acid, 2-methyl, ethyl ester $2-CH_3(C_4H_2O)CO_2C_2H_5-3$	154.17		$85-7^{20}$		$1.0102^{25/4}$	1.4620^{25}	eth	$B18^{4}$, 4072
7238	β-Furoic acid, 4-methyl $4-CH_3(C_4H_2O)CO_2H-3$	126.11	nd (bz-peth)		138-9				$B18^{4}$, 4089
7239	β-Furoic acid, 5-methyl $5-CH_3(C_4H_2O)CO_2H-3$	126.11	(w)	sub	119			w, eth	$B18^{4}$, 4076
7240	Furoin $C_{10}H_8O_4$	192.17	nd (al)		138-9			eth, MeOH	$B19^{4}$, 2543
7241	2-Furylacetic acid or 2-Furanacetic acid $(2-C_4H_3O)CH_2CO_2H$	126.11	lf (bz peth)	$102-4^{0.4}$	68-9			w, MeOH	$B18^{4}$, 4061
7242	2-Furylacetonitrile or Furfuryl cyanide $(2-C_4H_3O)CH_2CN$	107.11		$75-80^{20}$		$1.0854^{25/4}$	1.4693^{20}	al, eth	$B18^{4}$, 4062
7243	2-Furyl octyl ether $(2-C_4H_3O)OC_8H_{17}$	196.29		$129-30^{18}$		$0.9214^{28/4}$	1.4520^{28}	eth	$B17^{4}$, 1219
7244	2-Furyl phenyl ether $(2-C_4H_3O)OC_6H_5$	160.17		$105-6^{18}$		$1.1010^{23/4}$	1.5418^{23}	eth	$B17^{4}$, 1219
7245	2-Furyl phenyl ketone or 2-Benzoyl furan $(2-C_4H_3O)COC_6H_5$	172.18		$285, 164^{19}$	<-15	1.1732^{20}	1.6055^{20}	al, eth, bz	$B17^{4}$, 5184
7246	2-Furyl phenyl sulfide $(2-C_4H_3O)SC_6H_5$	176.23		$119-20^{8}$		1.1341^{26}	1.5976^{20}	al, eth	$B17^{4}$, 1220
7247	2-Furyl isopropyl ether or 2-isopropoxy furan $(2-C_4H_3O)O-i-C_3H_7$	126.16		135-6		$0.9689^{20/4}$	1.4419^{20}		$B17^{4}$, 1219
7248	2-Furyl methyl ether $(2-C_4H_3O)OCH_3$	98.10		110-1		$1.0646^{25/4}$	1.4468^{25}		$B17^{4}$, 1219

PHYSICAL CONSTANTS OF ORGANIC COMPOUNDS (Continued)

No.	Name, Synonyms, and Formula	Mol. wt.	Color, crystalline form, specific rotation and λ_{max} (log ε)	b.p. °C	m.p. °C	Density	n_D	Solubility	Ref.
7249	Galacitol or D-dulcitol $CH_2OH(CHOH)_4CH_2OH$	182.18	mcl pr (w)	275-8[1] sub	189	1.466	...	w	B1[4], 2844
7250	D-Galactonic acid-γ-lactone $C_6H_{10}O_6$	178.14	nd (w + 1) nd (al or AcOEt) $[\alpha]^{20}{}_D$ −65.5 (w)	...	112 (anh), 66 (hyd)	...	...	w	B18[4], 3026
7251	D-Galactose $C_6H_{12}O_6$	180.16	pl or pr (al) pr or nd (w + 1) $[\alpha]$ + 83.3 (w)	...	...	...	...	w, py	B1[4], 4336
7252	D-Galactose, 3,6-anhydro- $C_6H_{10}O_5$	162.14	lf (PrOH) $[\alpha]_D$ + 24 (w)	...	123-5	...	...	w	B18[4], 2278
7253	α-D-Galactose, 2,3,4,6-tetra-O-methyl $C_{10}H_{20}O_6$	236.27	$[\alpha]^{20}{}_D$ + 150 → + 114 (w)	172^{12}	71-3	...	...	w, al, eth	B1[4], 4371
7254	D-Galacturonic acid $C_6H_{10}O_7$	194.14	nd (w + 1) β-$[\alpha]^{20}{}_D$ + 27 (w) α-$[\alpha]_D$ + 98 → + 53 (w, c=10)	...	α 156, β 150	...	...	w, al	B3[4], 2000
7255	Galegine or 4-Guanidino-2-methyl-2-butene $(CH_3)_2C{=}CHCH_2N{=}C(NH_2)_2$	127.19	hyg	d	60-5	...	...	w, al	B4[3], 465
7256	Galipine or Galipoline methyl ether $C_{20}H_{21}NO_3$	323.39	pr (al eth) nd (peth)	...	115.5	...	...	al, eth, ace, bz	B21[4], 2656
7257	Gallein or 4,5-Dihydro fluorescein $C_{20}H_{12}O_5$	332.31	br red pw (+ 1.5w) red (anh)	...	>300	...	...	al, ace	B19[4], 3147
7258	Gallin or 4,5-Dihydroxyfluorescin $C_{20}H_{14}O_7$	366.33	nd (eth) turns red in air	...	...	...	...	al, ace, aa	B18, 368
7259	Gelsemine (d) or Gelseminine $C_{20}H_{22}N_2O_2$	322.41	cr (ace)	...	178	...	...	al, eth, ace, bz,chl	B27[2], 720
7260	Gelsemine, hydrochloride $C_{20}H_{22}N_2O_2.HCl$	358.87	pr (w)	...	326	...	...	w	B27[2], 720
7261	Geranial or Citral a $C_{10}H_{16}O$	152.24	$[\alpha]^{20}{}_D$ + 2.5 (w)	229, 118-9^{20}	...	0.8888^{20}	1.4898^{20}	**al, eth**	B1[4], 3569
7262	Geraniol or 2,7-Dimethyl-2,6-octadiene-1-ol $(CH_3)_2C{=}CHCH_2CH_2C(CH_3)CH_2CH_2OH$	154.25	...	230, 121^{18}	>-15	0.8894^{20}	1.4766^{20}	al, eth, ace	B1[4], 2277
7263	Geraniol formate $C_{11}H_{18}O_2$	182.26	...	229d, 113-4^{25}	...	$0.9086^{25/4}$	1.4659^{20}	al, eth, ace	B2[4], 35
7264	Geraniol, tetrahydro (d) or 3,7-Dimethyl-1-octanol $(CH_3)_2CH(CH_2)_3CH(CH_3)CH_2CH_2OH$	158.28	$[\alpha]^{20}{}_D$ + 4.09	212-3, 105-6^{10}	...	$0.8285^{20/4}$	1.4355^{20}	eth	B1[4], 1830
7265	Geraniol, tetrahydro or 3,7-Dimethyl-1-octanol $(CH_3)_2CH(CH_2)_3CH(CH_3)CH_2CH_2OH$	158.28	...	202-3, 106^{12}	...	$0.8308^{10/4}$	1.4367^{20}	al, eth, ace, bz	B1[4], 1830
7266	Geraniol, tetrahydro (l) or 3,7-Dimethyl-1-octanol $(CH_3)_2CH(CH_2)_3CH(CH_3)CH_2CH_2OH$	158.28	$[\alpha]^{27}{}_{548}$ − 3.67	212-3, 109^{15}	...	0.830^{18}	1.4370^{15}	eth	B1[3], 1768
7267	Geranyl bromide $(CH_3)_2C{=}CHCH_2CH_2C(CH_3)CH_2CH_2Br$	217.15	...	101-2^{12}, 47-$8^{0.005}$	...	$1.0940^{22/4}$	1.5027^{20}	al, eth	B1[4], 1059
7268	Germanidine $C_{37}H_{57}NO_{10}$	675.87	nd $[\alpha]^{24}{}_D$ − 30 (al)		221-2 (vac)	...	...	...	B21[4], 6806
7269	Germanitrine $C_{39}H_{59}NO_{11}$	717.90	nd (aq aw) $[\alpha]^{24}{}_D$ − 61 (Py, c=1)	...	228-9 (vac)	...	...	...	B21[4],6807
7270	Germerine $C_{37}H_{59}NO_{11}$	693.88	cr (ace) lf (bz) $[\alpha]^{25}{}_D$ + 15.7 (chl, c=1.02)	...	193-5d	...	...	al, ace, bz, chl	B21[4], 6808
7271	Germidine $C_{34}H_{53}NO_{10}$	635.80	pl (aq MeOH) $[\alpha]^{25}{}_D$ + 13 (chl)	...	242-4, 203-3	...	...	al, bz	B21[4], 6802

No.	Name, Synonyms, and Formula	Mol. wt.	Color, crystalline form, specific rotation and λ_{max} (log ε)	b.p. °C	m.p. °C	Density	n_D	Solubility	Ref.
7272	Germine $C_{27}H_{43}NO_8$	509.64	pr or cr (MeOH) $[\alpha]^{25}{}_D$ + 5 (95% al)		220			bz	B21^4, 6796
7273	Germinitrine $C_{39}H_{57}NO_{11}$	715.89	pr (dil ace) $[\alpha]^{24}{}_D$ −36 (Py, c=1.12)		175 (vac)				Am75, 4925
7274	Germitrine $C_{39}H_{61}NO_{12}$	735.92	cr (dil al) $[\alpha]^{25}{}_D$ + 11 (chl)		197-9			bz, chl	B21^4, 6809
7275	Gitogenin or Digine $C_{27}H_{44}O_4$	432.65	lf (bz) nd (eth) $[\alpha]^{20}{}_D$ −75 (chl, c=1)		271-2			al, chl	B19^4, 1050
7276	Gitoxigenin or Hydroxy digitoxin $C_{23}H_{34}O_5$	390.52	pr (AcOEt) pr (+w, dil al) $[\alpha]^{20}{}_{546}$ + 38.5 (MeOH, c=0.7)		234 (anh)			chl	B18^4, 2456
7277	Gitoxin or Anhydrogitalin $C_{41}H_{64}O_{14}$	780.96	pr (chl MeOH) $[\alpha]^{24}{}_D$ + 5 (py, c=1)		285d				B18^4, 2462
7278	D-Glucoascorbic acid or 3-keto-D-*keto*-heptonofuranolactone $C_7H_{10}O_7$	206.16	reds (+w) (ace MeOH Peth) $[\alpha]^{20}{}_D$ −22 (MeOH, c=1)		191 (anh) 140 (hyd)			w, al	B18^4, 3400
7279	D-*gluco*-Heptose $C_7H_{14}O_7$	210.19	rh pl (w) $[\alpha]^{20}{}_D$ −19.7 (w) (mut)		193 (210)			w	B1^4, 4436
7280	D-gluco-Methylose or D-Isorhamnose $C_6H_{12}O_5$	164.16	cr (AcOEt) $[\alpha]^{20}{}_D$ + 73→ + 29.7 (w, mut)		139.40			w, al	B1^4, 4260
7281	D-Gluconic acid or Dextronic acid $C_6H_{12}O_7$	196.16	nd (al eth) $[\alpha]^{25}{}_D$ −3.49→ + 12.95 (w, mut)		131			w	B3^4, 1255
7282	D-Gluconic acid-γ-lactone $C_6H_{10}O_6$	178.14	nd (al) $[\alpha]_D$ + 67.5→ + 6.2 (w)		134-6			al	B18^4, 3024
7283	D-Gluconic acid-δ-lactone $C_6H_{10}O_6$	178.14	nd (al) $[\alpha]^{25}{}_D$ + 63.5→ + 6.2 (w)						B18^4, 3018
7284	D-Gluconitrile $C_6H_{11}NO_5$	177.16	(i) cr (al, aa) (ii) pl (al) (i) $[\alpha]^{24}{}_D$ + 10.0 (w, c=1.8) (ii) $[\alpha]^{21}{}_D$ + 8.8 (w)		(i)146.8, (ii)120.5			w, py	B18^4, 3024
7285	D-Gluconic acid, phenylhydrazide $C_{12}H_{18}N_2O_6$	286.29	pr (w) $[\alpha]$ + 12(w)		204-5			w	B15^2, 122
7286	D-Gluconic acid, 5-oxo or 5-keto-D-gluconic acid $C_6H_{10}O_7$	194.14	cr or syr $[\alpha]_D$ −14.5 (w)		125-6			w, al	B3^4, 1993
7287	Gluconol (d) $C_6H_{10}O_4$	146.15	hyg nd $[\alpha]^{22}{}_D$ −7 (w)		60				B17^4, 2332

PHYSICAL CONSTANTS OF ORGANIC COMPOUNDS (Continued)

No.	Name, Synonyms, and Formula	Mol. wt.	Color, crystalline form, specific rotation and λ_{max} (log ε)	b.p. °C	m.p. °C	Density	n_D	Solubility	Ref.
7288	D-Glucose (equilib mixt) or Dextrose $C_6H_{12}O_6$	180.16	$[\alpha]^{20/}_D$ +52.7(w)	...	146 (150)	...	...	w	B1[4], 4302
7289	α-D-Glucose $C_6H_{12}O_6$	180.16	reds cubes orh nd (al) $[\alpha]^{20/}_D$ +112.2→ +52.7 (w, c=4)	...	146d	$1.5620^{18/4}$	...	w	B1[4], 4304
7290	α-D-Glucose, monohydrate $C_6H_{12}O_6.H_2O$	198.18	lf pl or orh (w) $[\alpha]^{20/}_D$ +102→ +47.9	...	86	$1.54^{25/4}$	...	w	B1[4], 4306
7291	β-D-Glucose $C_6H_{12}O_6$	180.16	nd (al) (w + 1) $[\alpha]^{20/}_D$ +17.5→ +52.7 (w, c=4)	...	150	$1.5620^{18/4}$	...	w	B1[4], 4306
7292	α-D-Glucose, pentaacetate $C_{16}H_{22}O_{11}$	390.35	pl or nd (al) $[\alpha]^{20/}_D$ +100.9 (al, c=0.5)	sub	112-3	...	...	eth	B17[4], 3276
7293	β-D-Glucose, pentaacetate $C_{16}H_{22}O_{11}$	390.35	nd (al) $[\alpha]'_D$ +3.9 (chl, c=6)	sub (vac)	134	...	...	bz, chl	B17[4], 3278
7294	D-Glucose, pentamethyl ether $C_{11}H_{22}O_6$	250.30	$[\alpha]^{20/}_D$ +147.4 (w, p=10)	$180^{0.4}$	...	$1.0944^{20/4}$	1.4466^{20}	w, al, eth, ace	B17[4], 2928
7295	D-Glucose, phenylhydrazone (α type) $C_{12}H_{18}N_2O_5$	270.29	pl (al) $[\alpha]^{25/}_D$ −8.7→−52.2 (w, c=2) (mut)	...	160	...	...	w	B31, 173
7296	D-Glucose, phenylhydrazone (β type) $C_{12}H_{18}N_2O_5$	270.29	pr nd (al) $[\alpha]^{19/}_D$ -4.5→-53.7 (w py) (mut)	...	140-1	...	...	w	B31, 173
7297	D-Glucose, phenylosazone or D-Glucosazone, D-Fructosazone, D-Mannosazone $C_{18}H_{22}N_4O_4$	358.40	ye nd (dil al) $[\alpha]'_D$-41 (MeOH)	d 213	210	...	...	...	B31, 350
7298	DL-Glucose, phenylosazone $C_{18}H_{22}N_4O_4$	358.40	ye nd (al)	...	...	...	...	...	B31, 355
7299	β-D-Glucose, 2-amino or D-Glucosamine $C_6H_{13}NO_5$	179.18	nd (al MeOH) $[\alpha]_D$ +28→ +47.5 (w, c=0.4)	...	110d	...	...	w	B4[4], 2019
7300	D-Glucose, 2-amino, hydrochloride $C_6H_{13}NO_5 \cdot HCl$	215.64	mcl (w or dil al) $[\alpha]_{20D}$ +25→ +72.6 (w)	...	...	...	...	w	B4[4], 2019
7301	D-Glucose, 2-(methylamino) or N-methyl-D-glucosamine $C_7H_{15}NO_5$	193.20	gummy, $[\alpha]^{25/}_4$ -65 (MeOH, c=1)	...	130-2d	...	...	w	B18[4], 7526
7302	D-Glucothiose $C_6H_{12}SO_5$	196.22	hyg pw(w + 1), $[\alpha]^{30/}_D$ +48.7 (w, c=1.4)	...	70	...	...	w	B1[4], 4391
7303	D-Glucoside, α-methyl or Methyl-α-D-glucopyranoside $C_7H_{14}O_6$	194.19	rh nd, (al), $[\alpha]^{20/}_D$ +158.9 (w)	$200^{0.2}$	168	$1.46^{30/4}$	...	w	B17[4], 2909

No.	Name, Synonyms, and Formula	Mol. wt.	Color, crystalline form, specific rotation and λ_{max} (log ε)	b.p. °C	m.p. °C	Density	n_D	Solubility	Ref.
7304	D-Glucoside, β-methyl or Methyl-β-D-glucopyranoside $C_7H_{14}O_6$	194.18	tetr pr (al), $[\alpha]^{20/}_D$ -34.2 (w, p=10)	...	115-6	...	...	w	$B17^4$, 2911
7305	β-D-Glucuronic acid $C_6H_{10}O_7$	194.14	nd(al AcOEt), $[\alpha]^{20/}_D$ + 11.7 + 36.3 (w, c=6) (mut)	...	165	...	...	w, al	$B3^4$, 1996
7306	D-Glucuronic acid, γ-lactone or D-glucurone $C_6H_8O_6$	176.13	mcl pl (w) cr (al), $[\alpha]^{25/}_D$ + 19.8 (w, c=5.2)	...	177-8	$1.76^{20/4}$	...	w	$B18^4$, 3055
7307	Glutaconic acid, diethyl ester or Ethyl glutaconate $C_2H_5O_2CCH_2CH{=}CHCO_2C_2H_5$	186.21	...	236-8, 125^{12}	...	$1.0496^{20/4}$	1.4411^{20}	al, eth	$B2^4$, 2227
7308	Glutamic acid (D) or 2-Aminopentanedioic acid $HO_2CCH_2CH_2CH(NH_2)CO_2H$	147.13	lf (w), $[\alpha]^{25/}_D$ -31.7 (1.7 NHCl)	...	213d	...	...	...	$B4^3$, 1549
7309	Glutamic acid (DL) $HO_2CCH_2CH_2CH(NH_2)CO_2H$	147.13	rh(al w)	...	199d, (225-7)	$1.4601^{20/4}$	...	...	$B4^3$, 1550
7310	Glutamic acid (L +) $HO_2CCH_2CH_2CH(NH_2)CO_2H$	147.13	orh(dil al), $[\alpha]^{22/}_D$ + 31.4 (6N HCl, c=1)	sub 175^{10}	224-5d	$1.538^{20/4}$	...	...	$B4^3$, 1530
7311	Glutamic acid, hydrochloride (L) or Acidulin $HO_2CCH_2CH_2CH(NH_2)CO_2H.HCl$	183.59	rh pl (w), $[\alpha]^{19/}_D$ + 31.1 (dil HCl)	...	214d	...	...	w, al	$B4^3$, 1537
7312	Glutamic acid, N-acetyl (L) $HO_2CCH_2CH_2CH(NHCOCH_3)CO_2H$	189.17	pr (w), $[\alpha]_D$ -15.3 (w, c=2)	...	199	...	...	...	$B4^3$, 1544
7313	Glutamic acid, 3-hydroxy (D) $HO_2CCH_2CH(OH)CH(NH_2)CO_2H$	163.13	hyg pr(w)	d	135	...	...	w, aa	$B4^3$, 1676
7314	Glutamic acid, 3-hydroxy (DL) $HO_2CCH_2CH(OH)CH(NH_2)CO_2H$	163.13	rh pr or nd(w)	d	198d	...	...	w	$B4^3$, 1676
7315	Glutamine (L +) or α-Amino glutaramic acid $HO_2CCH(NH_2)CH_2CH_2CONH_2$	146.15	nd (w or dil al), $[\alpha]^{25/}_D$ + 6.5 (w, c=2)	...	185-6d	...	...	w	$B4^3$, 1540
7316	Glutyraldehyde or 1,5-Pentanedial $OHC(CH_2)_3CHO$	100.12	...	187-9d, 71-2^{10}	...	...	...	**w, al**, bz	$B1^4$, 3111
7317	Glutyraldehyde, dioxime or Glutyraldoxime $HON{=}CH(CH_2)_3CH{=}NOH$	130.15	nd(w or Py)	sub	178	...	...	...	$B1^4$, 3660
7318	Glutaric acid or Pentanedioic acid $HO_2C(CH_2)_3CO_2H$	132.12	nd(bz)	302-4d, 200^{20}	99	$1.424^{25/4}$	1.4188^{106}	w, al, eth, chl	$B2^4$, 1934
7319	Glutaric acid, 2-acetyl, diethyl ester $C_2H_5O_2CCH_2CH_2CH(COCH_3)CO_2C_2H_5$	230.26	...	271-2d, 119-$2^{0.1}$	...	$1.0712^{20/4}$	1.4420^{15}	al, eth	$B3^4$, 1835
7320	Glutaric anhydride, 2-phenyl $C_{11}H_{10}O_3$	190.20	nd(eth)	218-30^{13}	95	...	...	al	$B17^4$, 6185
7321	Glutaric anhydride, 3-phenyl $C_{11}H_{10}O_3$	190.20	cr(bz)	217-9^{15}	105	...	...	eth, bz, chl	$B17^4$, 6185
7322	Glutaric acid, diethyl ester or Diethylglutarate $C_2H_5O_2C(CH_2)_3CO_2C_2H_5$	188.22	syr	236-7, 103-4^7	−24.1	$1.0220^{20/4}$	1.4241^{20}	eth	$B2^4$, 1937
7323	Glutaric acid, dimethyl ester or Dimethyl glutarate $CH_3O_2C(CH_2)_3CO_2CH_3$	160.17	...	214^{751}, 109^{21}	...	$1.0876^{20/4}$	1.4242^{20}	al, eth	$B2^4$, 1937
7324	Glutaric acid, 2,3-dimethyl $HO_2CCH_2CH_2C(CH_3)_2CO_2H$	160.17	nd(bz-lig)	...	85	...	...	al, chl, aa	$B2^4$, 2018
7325	Glutaric acid, 3,3-dimethyl $(CH_3)_2C(CH_2CO_2H)_2$	160.17	mcl pl nd(bz)	126-$7^{4.5}$	103-4	$1.4278^{20/4}$	...	w, al, eth	$B2^4$, 2023
7326	Glutaric acid, diphenyl ester $C_6H_5O_2C(CH_2)_3CO_2C_6H_5$	284.31	nd(lig)	236.5^{15}	54	...	...	al, ace, lig	$B6^4$, 626
7327	Glutaric acid, 2-ethyl-3-methyl $HO_2CCH_2CH(CH_3)CH(C_2H_5)CO_2H$	174.20	(i) pr (w), (ii) pr (chl-lig)	...	(i)100 (ii)88	...	...	w, eth	$B2^4$, 2047

No.	Name, Synonyms, and Formula	Mol. wt.	Color, crystalline form, specific rotation and λ_{max} (log ε)	b.p. °C	m.p. °C	Density	n_D	Solubility	Ref.
7328	Glutaric acid, 2-ethyl-4-methyl (dl) or Paramethyl ethyl glutaric acid $HO_2CCH(CH_3)CH_2CH(C_2H_5)CO_2H$	174.20	nd(w)		107			eth, lig	B2^4, 2042
7329	Glutaric acid, 2-ethyl-4-methyl (meso) $HO_2CCH(CH_3)CH_2CH(C_2H_5)CO_2H$	174.20	nd(al)		83-4			eth	B2^4, 2042
7330	Glutaric acid, 3-ethyl-3-methyl $(HO_2CCH_2)_2C(CH_3)C_2H_5$	174.20	pl(bz-peth)	260^{740}	87			al, eth, bz	B2^3, 1781
7331	Glutaric acid, 2-hydroxy (d) $HO_2CCH_2CH_2CH(OH)CO_2H$	148.12	$[\alpha]^{19}{}_D$ + 1.76 (w, c=1)		72				B3^4, 1146
7332	Glutaric acid, 2-hydroxy (dl) $HO_2CCH_2CH_2CH(OH)CO_2H$	148.12	pr(AcOEt)		72			w, al	B3^4, 1146
7333	Glutaric acid, 2-hydroxy (l) $HO_2CCH_2CH_2CH(OH)CO_2H$	148.12	$[\alpha]_D$ -1.98 (w)		72-3			w, al	B3^4,1146
7334	Glutaric acid, 3-methyl $CH_3CH(CH_2CO_2H)_2$	146.14		165-7$^{0.5}$	87			w, al, eth	B2^4, 1992
7335	Glutaric acid, 2-oxo $HO_2CCH_2CH_2COCO_2H$	146.10	cr(ace-bz)		115-6			w, al, eth, ace	B3^4, 1813
7336	Glutaric acid, 3-oxo or Acetone dicarboxylic acid $(HO_2CCH_2)_2C{=}O$	146.10	nd(al AcOEt) rh(w)	d	135d			w, al	B3^4, 1816
7337	Glutaric acid, 3-oxo, diethyl ester $(C_2H_5O_2CCH_2)_2C{=}O$	202.21		250, 140^{12}		1.113$^{20/4}$		al	B3^4, 1817
7338	Glutaric acid, 2-phenyl $HO_2CCH_2CH_2CH(C_6H_5)CO_2H$	208.21	cr(bz or eth-peth)		82-3				B9^3, 4298
7340	Glutaric acid, 2,3,4-trihydroxy (d) $HO_2C(CHOH)_3CO_2H$	180.11	cr(ace) pl(w), $[\alpha]^{20}{}_D$ + 22.2		128			w, al	B3^4, 1265
7341	Glutaric acid, 2,3,4-trihydroxy (dl) $HO_2C(CHOH)_3CO_2H$	180.11	cr(ace)		154d			w, al, ace	B3, 553
7342	Glutaronitrile $NC(CH_2)_3CN$	94.12		286, 160.4^{22}	−29	0.9911$^{15/4}$	1.4295^{20}	al, chl	B2^4, 1941
7343	Glutaryl chloride $ClCO(CH_2)_3COCl$	169.01		216-8		1.324$^{30/4}$	1.4728^{20}	eth	B2^4, 1939
7344	Glutathione or α-Glutamyl cysteinyl glycine $C_{10}H_{17}N_3O_6S$	307.32	rh (w) cr (50% al), $[\alpha]^{27}{}_D$ -21.3 (w, c=2)		195			w, DMF	B4^3, 1612
7345	D-Glyceraldehyde or D-2,3-Dihydroxy proponal $HOCH_2CH(OH)CHO$	90.08	syr, $[\alpha]_D$ + 14 (w)					w	B1^4, 4114
7346	DL-Glyceraldehyde $HOCH_2CH(OH)CHO$	90.08	nd or pr(40% MeOH)	140-50$^{0.8}$	145	1.455$^{18/18}$		w	B1^4, 4114
7347	L-Glyceraldehyde $HOCH_2CH(OH)CHO$	90.08	$[\alpha]$ -13.8 (w, c=1)					w	B1^4, 4114
7348	D-Glyceraldehyde, diethyl acetal or 3,3-Diethoxy-1,2- $HOCH_2CH(OH)CH(OC_2H_5)_2$	164.20	$[\alpha]^{15}{}_D$ + 21.2 (w, c=18)	127-9^{17}				w, al, eth, ace	B1^2, 888
7349	DL-Glyceric acid or 2,3-Dihydroxy propionic acid $HOCH_2CH(OH)CO_2H$	106.08	syr	d				w, al, eth	B3^4, 1050
7350	D-Glyceric acid, ethyl ester or D-Ethyl glycerate $HOCH_2CH(OH)CO_2C_2H_5$	134.13	$[\alpha]^{11}{}_D$ -22.73					w, al, eth	B3^4, 1052
7351	DL-Glyceric acid, ethyl ester $HOCH_2CH(OH)CO_2C_2H_5$	134.13		230-40, 120-1^{16}		1.1908$^{15/15}$		w, al, eth	B3^4, 1052
7352	DL-Glyceric acid, methyl ester or DL-Methyl glycerate $HOCH_2CH(OH)CO_2CH_3$	120.11		239-44, 119-20^{14}		1.2814$^{15/15}$	1.4502^{20}	w, al	B3^4, 1052
7353	L-Glyceric acid, methyl ester or L-Methyl glycerate $HOCH_2CH(OH)CO_2CH_3$	120.11	$[\alpha]^{15}{}_D$ −6.44	119-20^{14}, 74-5$^{0.2}$		1.2798$^{15/15}$		w, al	B3^4, 1052
7354	Glycerol or Glycerin, 1,2,3-trihydroxy propane $HOCH_2CH(OH)CH_2OH$	92.09	syr rh pl	290d, 182^{20}	20	1.2613$^{20/4}$	1.4746^{20}	w, al	B1^4, 2751
7355	Glycerol, 1-monoacetate or α-Monoacetin $HOCH_2(CHOH)CH_2O_2CCH_3$	134.13		158^{165}, 129-31^3		1.2060$^{20/4}$	1.4157^{20}	w, al	B2^4, 251
7356	Glycerol borate $(C_3H_5BO_3)n$	(99.88)n	glass		Ca 150				B1, 519
7357	Glycerol, 1-monobutyrate (dl) or α-Monobutyrin $HOCH_2CH(OH)CH_2O_2CC_3H_7$	162.19		289-71, 163^{16}		1.129^{18}	1.4531^{20}	w, al	B2^4, 798

PHYSICAL CONSTANTS OF ORGANIC COMPOUNDS (Continued)

No.	Name, Synonyms, and Formula	Mol. wt.	Color, crystalline form, specific rotation and λ_{max} (log ε)	b.p. °C	m.p. °C	Density	n_D	Solubility	Ref.
7358	Glycerol, 1(2-chlorophenyl) ether ...... $(2\text{-ClC}_6H_4)OCH_2CH(OH)CH_2OH$	202.64	nd (bz)	250^{19}	71-2	......	......	eth	B6^{4}, 790
7359	Glycerol, 1-(4-chlorophenyl) ether ...... $(4\text{-ClC}_6H_4)OCH_2CH(OH)CH_2OH$	202.64	nd (eth peth) lf (bz)	$214\text{-}5^{19}$	76	......	......	al, eth, ace	B6^{4}, 831
7360	Glycerol, 1,3-diacetate or Diacetin ...... $HOCH(CH_2O_2CCH_3)_2$	176.17	......	280, $155\text{-}6^{15}$	40	$1.1779^{15/4}$	1.4395^{20}	w, al	B2^{4}, 252
7361	Glycerol, 1,2-dibutyrate (d) or α,β-Dibutyrin ...... $HOCH_2CH(O_2CC_3H_7)CH_2(O_2CC_3H_7)$	232.28	$[\alpha]_D$ + 1.7 (py,c=7)	273.5, 167^{20}	......	......	1.4422^{20}	al	B2^{4}, 799
7362	Glycerol, 1,3-dilaurate or α,γ-Dilarurin ...... $HOCH(CH_2O_2CC_{11}H_{23})_2$	456.71	pl (al) nd (eth-al)	......	[α] −49.5 (unst) β −56.5 (st)	......	......	al, eth, bz, chl, lig	B2^{4}, 1098
7363	Glycerol, 1,2-dimethyl ether (dl) or 2,3-Dimethoxy-1-propanol...... $HOCH_2CH(OCH_3)CH_2OCH_3$	120.15	......	180, 100^{40}	......	$1.016^{25/4}$	1.4200^{20}	w, al, eth	B1^{3}, 2317
7364	Glycerol, 1,3-dimethyl ether or 1,3-Dimethoxy-2-propanol...... $HOCH(CH_2OCH_3)_2$	120.15	......	169, 88^{40}	......	$1.0085^{20/4}$	1.4192^{20}	w, al, eth	B1^{3}, 2318
7365	Glycerol, 1,3-dinitrate ...... $HOCH_2(CH_2ONO_2)_2$	182.09	pr (w) cr (eth + 1w)	148^{15}, $116^{0.6}$	26 (hyd)	$1.523^{20/4}$	1.4715^{20}	w, al, eth	B1^{2}, 591
7366	Glycerol, 1,3-dipalmitate ...... $HOCH(CH_2O_2CC_{15}H_{31})_2$	568.92	cr (al, chl)	......	72-4	......	......	eth	B2^{4}, 1174
7367	Glycerol, 1,3-diphenylether ...... $HOCH(CH_2OC_6H_5)_2$	244.29	lf (al)	$224.5^{17.5}$, 175^{2}	81-2	$1.179^{24/4}$	......	al, eth, bz, chl	B6^{4}, 590
7368	Glycerol, 1,3-diphenylether-2-acetate ...... $CH_3CO_2CH(CH_2OC_6H_5)_2$	286.33	(dil al)	190^{160}	70-1	......	......	al, eth, bz, chl	B6^{3}, 583
7369	Glycerol, 1,3-dipropionate ...... $HOCH(CH_2O_2CC_2H_5)_2$	204.22	......	$170\text{-}3^{10}$	......	......	......	al	B2^{4}, 717
7370	Glycerol, 1,3-distearate or α,γ-Distearin...... $HOCH(CH_2O_2CC_{17}H_{35})_2$	625.03	nd or pl (eth chl lig)	......	79.1	......	......	eth	B2^{4}, 1231
7371	Glycerol, 1-mono (2-hydroxybenzoate) or α-Glyceryl salicylate ...... $(2\text{-HOC}_6H_4)CO_2CH_2CH(OH)CH_2OH$	212.20	nd (eth)	......	76	......	......	al, bz	B10^{3}, 142
7372	Glycerol, 1-monolaurate (dl) or α-Monolaurin (dl)...... $HOCH_2CH(OH)CH_2O_2CC_{11}H_{23}$	274.40	lf (peth)	186^{2}	63	0.9248^{97}	1.4350^{86}	eth, ace, bz, chl	B2^{4}, 1096
7373	Glycerol, 1-monolaurate (l) or α-Monolaurin (l) ...... $HOCH_2CH(OH)CH_2O_2CC_{11}H_{23}$	274.40	cr (eth or peth)	......	54-5	......	......	eth, ace, bz, chl	B2^{4}, 1096
7374	Glycerol, 1-linoleate (dl) or Glycerol-1-(9,12-actadecadienoate) ...... $C_{17}H_{31}CO_2CH_2CH(OH)CH_2OH$	354.53	cr (bz)	......	14-5	......	1.4758^{20}	eth, bz, chl	B2^{4}, 1758
7375	Glycerol, 1-(2-methoxyphenyl) ether or Guaiacol-α-glyceryl ether...... $(2\text{-CH}_3OC_6H_4)OCH_2CH(OH)CH_2OH$	198.22	rh pr (eth, eth-peth)	215^{19}, $126^{0.2}$	78-9	......	......	al, bz, chl, w	B6^{3}, 4224
7376	Glycerol, 1-methyl ether or 3-Methoxy-1,2-propanediol .. $HOCH_2CH(OH)CH_2OCH_3$	106.12	hyg liq	220, $110\text{-}2^{13}$	......	$1.830^{20/4}$	1.442^{25}	w, al, eth, ace	B1^{4}, 2755
7377	Glycerol, 2-methyl ether or 2-Methoxy-1,3-propandiol ... $CH_3OCH(CH_2OH)_2$	106.12	hyg liq	232, $119\text{-}20^{9}$	......	$1.124^{25/4}$	1.4505^{12}	w, al, eth, ace	B1^{3}, 2317
7378	Glycerol, 1-mononitrate ...... $HOCH_2CH(OH)CH_2ONO_2$	137.09	pr (w, al, eth)	155-60, 102^{1}	61	$1.4164^{20/4}$	1.4698^{20}	w, al	B1^{4}, 2761
7379	Glycerol, 2-mononitrate ...... $ONO_2CH(CH_2OH)_2$	137.09	lf (w)	155-60	54	$1.40^{22/4}$	......	w, al, eth	B1^{2}, 591
7380	Glycerol, 1-octadecyl ether (d) or Batyl alcohol...... $HOCH_2CH(OH)CH_2OC_{18}H_{37}$	344.58	pl (bz, aa)	$215\text{-}20^{2}$	70-1	......	......	eth	B1^{4}, 2758
7381	Glycerol, 1-oleate or α-Monoolein Glycerol-1-(9-octadeceneoate) ...... $C_{17}H_{33}CO_2CH_2CH(OH)CH_2OH$	356.55	pl (al)	$238\text{-}40^{3}$	35	$0.9420^{20/4}$	1.4626^{20}	al, eth, chl	B2^{4}, 1657
7382	Glycerol, 1-oleate or α-Monoolein Glycerol- mono hexadicaneate ...... $C_{15}H_{31}CO_2CH_2CH(OH)CH_2OH$	330.51	pl or lf(eth lig)	......	77	......	......	al	B2^{4}, 1170
7383	Glycerol, monopalmitate (l) or α-Monopalmitate (l)...... $HOCH_2CH(OH)CH_2O_2CC_{15}H_{31}$	330.51	$[\alpha]_D$ −4.37 (py)	......	71-2	......	......	......	B2^{4}, 1170
7384	Glycerol, 1-palmityl ether or Glycerol-1-hexadecylether... $HOCH_2CH(OH)CH_2OC_{16}H_{33}$	316.52	lf (hex) $[\alpha]^{20}_D$ + 3 (chl)	$120^{0.005}$	64	......	......	ace, chl, peth	B1^{3}, 2322
7385	Glycerol, 1-phenyl ether or Antodyne...... $HOCH_2CH(OH)CH_2OC_6H_5$	168.19	nd (eth peth)	200^{22}	67-8	$1.225^{20/4}$	......	w, al, eth, bz	B6^{4}, 589

No.	Name, Synonyms, and Formula	Mol. wt.	Color, crystalline form, specific rotation and λ_{max} (log ϵ)	b.p. °C	m.p. °C	Density	n_D	Solubility	Ref.
7386	Glycerol, 1-ricinoleate or 1-Mono (12-hydroxy-9-octadecanoate) .... $HOCH_2CH(OH)CH_2[O_2C(CH_2)_7CH{=}CHCH_2(OH)CHC_6H_{13}]$	372.55	ye	....	......	$1.028^{20/4}$	....	al, eth, ace, bz	$B3^4$, 1030
7387	Glycerol, 1-stearate or α-Monostearin .... $C_{17}H_{35}CO_2CH_2CH(OH)CH_2OH$	358.36	pl (MeOH)	....	81	$0.9841^{20/4}$	1.4400^{86}	lig	$B2^4$, 1225
7388	Glycerol, 1-stearate (l) .... $C_{17}H_{35}CO_2CH(OH)CH_2OH$	358.56	cr (eth or peth) $[\alpha]_D$ −3.58 (py)	....	76-7	....	....	....	$B2^4$, 1225
7389	Glycerol, 1-(2-tolyl) ether .... $(2\text{-}CH_3C_6H_4)OCH_2CH(OH)CH_2OH$	182.22	nd (bz-peth)	....	70-1d	....	....	al	$B6^3$, 1952
7390	Glycerol, triacetate or Triacetin .... $(CH_3CO_2CH_2)_2CHO_2CCH_3$	218.21	cr (al)	258-60, 130.5^7	4.1	$1.1596^{20/4}$	1.4301^{20}	**al, eth,** ace, **bz, chl**	$B2^4$, 253
7391	Glycerol, tribenzoate or Tribenzoin .... $C_6H_5CO_2CH(CH_2O_2CC_6H_5)_2$	404.43	nd (MeOH)	....	76	$1.228^{12/4}$	....	eth, ace, bz, chl	$B9^3$, 666
7392	Glycerol, tributyrate or Tributyrin .... $C_3H_7CO_2CH(CH_2O_2CC_3H_7)_2$	302.37	....	305-10, 190^{15}	−75	$1.0350^{20/4}$	1.4359^{20}	al, eth, ace, bz	$B2^4$, 799
7393	Glycerol, tricaproate or Tricaproin Glycerol trihexanoate $C_5H_{11}CO_2CH(CH_2O_2CC_5H_{11})_2$	386.54	....	>200	−60	$0.9867^{20/4}$	1.4427^{20}	**al, eth,** ace, **bz,** peth	$B2^4$, 926
7394	Glycerol, tricaprylate or Tricuprylin Glycerol trioctoate .. $C_7H_{15}CO_2CH(CH_2O_2CC_7H_{15})_2$	470.70	....	233.1	10 (st) −22 (unst)	$0.9540^{20/4}$	1.4482^{20}	**al,** eth, bz, chl, lig	$B2^4$, 991
7395	Glycerol, trielaidate or Trielaidin Glycerol tri (*trans*-9-octacteceneate) .... $C_{17}H_{33}CO_2CH(CH_2O_2CC_{17}H_{33})_2$	885.47	....	....	α-16.6β-42.8	....	....	eth, bz, chl	$B2^4$, 1664
7396	Glycerol, trilaurate or Trilaurin Glycerol tridodicanoate .. $C_{11}H_{23}CO_2CH(CH_2O_2CC_{11}H_{23})_2$	639.03	nd(al)	....	46	0.8986^{55}	1.4404^{60}	al, eth, ace, bz, chl	$B2^4$, 1098
7397	Glycerol, trimethylether or Trimethoxy propane .... $CH_3OCH(CH_2OCH_3)_2$	134.18	....	....	148	$0.9460^{15/4}$	1.4055^{15}	w, eth, ace, bz	$B1^4$, 2755
7398	Glycerol, trimyristate or Glycerol tritetradecanoate ..... $C_{13}H_{27}CO_2CH(CH_2O_2CC_{13}H_{27})_2$	768.28	poly-morphic (al-eth)	311	56.5 (st) 32 (unst)	$.08848^{60/4}$	1.4428^{60}	eth, ace, bz, chl	$B2^4$, 1134
7399	Glycerol, trinitrate or Nitroglycerin .... $O_2NOCH(CH_2ONO_2)_2$	227.09	pa ye fcl or rh	256 exp, 125^2	13	$1.5931^{20/4}$	1.4786^{12}	al, eth, ace, bz, chl	$B1^4$, 2762
7400	Glycerol, trioleate or Trioleen, Glycerol-(*cis*-9-octodeceneate) .... $C_{17}H_{33}CO_2CH(CH_2O_2CC_{17}H_{33})_2$	885.47	poly-morphic	$235\text{-}40^{18}$	......	0.8988^{40}	1.4621^{40}	eth, chl, peth	$B2^4$, 1664
7401	Glycerol, tripalmatate or Tripalmitin Glycerol trihexadecanoate .... $C_{15}H_{31}CO_2CH(CH_2O_2CC_{15}H_{31}$	807.35	nd (eth)	310-20	66 (st) 44.7 (unst)	$0.8752^{70/4}$	1.4381^{80}	eth, bz, chl	$B2^4$, 1176
7402	Glycerol, tripropionate .... $C_2H_5CO_2CH(CH_2O_2CC_2H_5)_2$	260.29	....	$175\text{-}6^{20}$	......	$1.100^{20/18}$	1.4318^{19}	al, eth, chl	$B2^4$, 717
7403	Glycerol, tristearate or Tristearin Glycerol triactadecanoate .... $C_{17}H_{35}CO_2CH(CH_2O_2CC_{17}H_{35})_2$	891.51	....	....	α-55, β-73	$0.8559^{90/4}$	1.4395^{80}	ace	$B2^4$, 1233
7404	Glycerol, triisovalerate or Glycerol tri (3-methylbutyrate) $i\text{-}C_4H_9CO_2CH(CH_2O_2C\text{-}i\text{-}C_4H_9)$	344.45	....	330-5, 194^{15}	......	$0.9984^{20/4}$	1.4354^{20}	al, eth	$B2^4$, 900
7405	Glycidic acid or 2,3-Epoxypropionic acid .... $CH_2CHCOOH$ (epoxide, O bridging)	88.06	....	....	......	....	....	**w, al, eth**	$B18^1$, 435
7406	Glycidol (d) or 3-Hydroxypropylene oxide .... CH_2CHCH_2OH (epoxide, O bridging)	74.08	α + 15 (undil)	56.5^{11}	......	$1.117^{20/4}$	1.4293^{16}	w, al, eth, ace, bz	$B17^4$, 985
7407	Glycidol dl or 2,3-Epoxy-1-propanol (dl) .... CH_2CHCH_2OH (epoxide, O bridging)	74.08	....	$65\text{-}6^{2.5}$	......	1.1143^{25}	1.4287^{20}	**w, al,** eth, ace, bz	$B17^4$, 985
7408	Glycidol (l) or 3-Hydroxy propylene oxide .... $HOCH_2CHCH_2$ (epoxide, O bridging)	74.08	$[\alpha]^{18}{}_D$ −8.6 (indil)	56^{11}	......	1.1050^{18}	1.4293^{16}	w, al, eth, ace, bz	$B17^4$, 985
7409	Glycidol, phenyl .... $C_6H_5CHCHCH_2OH$ (epoxide, O bridging)	150.18	....	138^3	26.5	1.512^{27}	1.5432^{27}	al, eth	$B6^3$, 1800
7410	Glycine or Aminoacetic acid .... $H_2NCH_2CO_2H$	75.07	mcl or trg pr (dil al)	....	262d	1.607	....	w	$B4^3$, 1097
7411	Glycine, N-Leucyl (DL) or DL-Leucyl Glycine .... $(CH_3)_2CHCH_2\text{-}CH(NH_2)CONHCH_2CO_2H$	188.23	....	....	243d	....	....	w	$B4^3$, 1434
7412	Glycine, -Leucyl (L) or L-Leucyl glycine .... $(CH_3)_2CHCH_2CH(NH_2)CONHCH_2CO_2H$	188.23	lf or nd (w al)	....	248d	....	....	w	$B4^3$, 1414
7413	Glycyl glycine .... $H_2NCH_2CONHCH_2CO_2H$	132.12	$[\alpha]^{20}{}_D$ + 85.8 (w,c=2)	....	dec 215	....	....	....	$B4^3$, 1191

No.	Name, Synonyms, and Formula	Mol. wt.	Color, crystalline form, specific rotation and λ_{max} (log ε)	b.p. °C	m.p. °C	Density	n_D	Solubility	Ref.
7414	Glycocholic acid or Cholylglycine $C_{26}H_{43}NO_6$	465.63	nd (w) $[\alpha]^{23/}_D$ + 32.3 (al,c=1)		165-8 (anh) 132-4 (+ w)				B10[3], 2176
7415	Glycocyamine or N-Guanyl glycine $NH{=}C(NH_2)NHCH_2CO_2H$	117.11	pl or nd (w)		>300				B4[3], 1165
7416	Glycogen or Animal starch $(C_6H_{10}O_5)n$	(162.14) n	wh pw $[\alpha]^{25/}_D$ + 196.5 (w)					w	C25, 4940
7417	Glycolic acid or Hydroxyacetic acid $HOCH_2COOH$	76.05	rh nd (w) lf (eth)	d	80			w, al, eth	B3[4], 571
7418	Glycolic acid, benzoyl $C_6H_5COCH(OH)CO_2H$	180.16	lo pr (lig)		112			al, eth, chl	B9, 954
7419	Glycolide or 2,5-*p*-dioxane dione $C_4H_4O_4$	116.07	lf (al-chl or al)		86-7			ace	B19[4], 1922
7420	Glycoluril or Glyoxaldiurene $C_4H_6N_4O_2$	142.12	nd or pr (w)		300d			eth	B26[2], 260
7421	18α-Glycyrrhetinic acid or Glycyrhetic acid $C_{30}H_{46}O_4$	470.69	α-pl (dil al) β-nd (al peth)	α283 β296				al	B10[3], 4392
7422	Glycyrrhizic acid or Glycyrhizin $C_{42}H_{62}O_{16}$	822.94	pl or pr (aa)		220d			w	B18[4], 5156
7423	Glyoxal or Ethanedial $OHCCHO$	58.04	ye pr	50.4	15	1.14^{20}	1.3826^{20}	w, al, eth	B1[4], 3625
7424	Glyoxal, dioxine $HON{=}CHCH{=}NOH$	88.07	rh pl (w)	sub	178d			w, al, eth	B1[4], 3629
7425	Glyoxal, methyl, phenyl or Methylphenyl glyoxal $CH_3COCOC_6H_5$	148.16	ye oil	228, 101^{12}		$1.0065^{20/4}$	1.537^{10}	w, al, eth	B7[3], 3463
7426	Glyoxal, methylphenyl, dioxime or Methylphenyl glyoxime $CH_3C({=}NOH)C({=}NOH)C_6H_5$	178.19	nd (dil al)		140-1			al	B7[3], 3465
7427	Glyoxal, phenyl or Benzoyl formaldehyde C_6H_5COCHO	134.13	nd (+ w)	142^{125}	91	(hyd)		w, al, eth, ace, bz	B7[3], 3443
7428	Glyoxal, phenyl, hydrate $C_6H_5COCH(OH)_2$	152.15	nd (w, chl, al, lig)	93.4				al, eth, chl	B7[3], 3443
7429	Glyoxal, phenyl-1-oxime $C_6H_5COCH{=}NOH$	149.15	mcl pr or lf (chl, w)		129			chl	B7[3], 3447
7430	Glyoxylic acid or Oxoacetic acid $HCOCO_2H$	74.04	rh pr (w + 1/2)		98			w	B3[4], 1489
7431	Glyoxylic acid, phenyl or Phenylglyoxylic acid, Benzoyl formic acid $C_6H_5COCO_2H$	150.13	pr (CCl_4)	147-51^{12}	66			w, al, eth	B10[3], 2972
7432	Glyoxylic acid, phenyl, methyl ester or Methyl phenyl glyoxylate $C_6H_5COCO_2CH_3$	164.16	ye	246-8, 137^{14}			1.5268^{20}		B10[3], 2973
7433	Glyoxylonitrile, phenyl or Benzoyl cyanide C_6H_5COCN	131.13	ta	206-8, 99^{19}	32-3			al, eth	B10[3], 2976
7434	Gramine or 3-(Dimethylaminomethyl indole) $C_{11}H_{14}N_2$	174.25	lf (eth) nd (ace)		138-9			al, eth, chl	B22[4], 4302
7435	Griseofulvin or Fulvicin $C_{17}H_{17}ClO_6$	352.77	oct or rh (bz) $[\alpha]^{17/}D$ + 376 (chl,Sat sul)		220				B18[4], 3160
7435a	Guaiol or champicol $C_{15}H_{26}O$	222.37	trq Pr (al), $[\alpha]^{-30}$ c = 4	165_{17}	91	$0.9074^{100/4}$	$1.4716^{100/6}$	al, eth	
7436	Guanamine or 2,4-Diamino-1,3,5-triazine $2,4(H_2N)_2(C_3HN_3)$	111.11	nd (w)		329d			w	B26[1], 65
7437	Guanidine or Aminomethanamidine, Carbaniedine $HN{=}C(NH_2)_2$	59.07	cr		Ca 50			al, w	B3[4], 148
7438	Guanidine acetate $HN{=}C(NH_2)_2\cdot CH_3CO_2H$	119.12	nd (al-eth)		229-30			w, al	B3[4], 152
7439	Guanidine, amine or Guanyl hydrazine $HN{=}C(NH_2)NHNH_2$	74.09	cr		d			w, al	B3[4], 236
7440	Guanidine, carbonate $[HN{=}C(NH_2)_2]_2.H_2CO_3$	180.17	oct tetr pr (w)		198	1.24^4		w	B3[4], 152
7441	Guanidine, 1-cyano $HN{=}C(NH_2)NHCN$	84.08	rh lf or pl (al)	d	211-2	1.404^{14}		w, al, ace	B3[4], 160

No.	Name, Synonyms, and Formula	Mol. wt.	Color, crystalline form, specific rotation and λ_{max} (log ε)	b.p. °C	m.p. °C	Density	n_D	Solubility	Ref.
7442	Guanidine, 1,3-diphenyl or Melaniline $HN{=}C(NHC_6H_5)_2$	211.27	mcl nd (al or to)	d 170	150	$1.13^{20/4}$	....	al, eth	B12[3], 805
7443	Guanidine, 1,3-di(2-totyl) $HN{=}C[NH(C_6H_4CH_3\text{-}2)]$	239.32	cr (dil al)	....	179	$1.10^{20/4}$	....	eth, chl	B12[3], 1871
7444	Guanidine, hydrochloride $HN{=}C(NH_2)_2).HCl$	95.53	rh bipym (al)	....	178-85	$1.354^{20/4}$	....	w, al	B3[4], 150
7445	Guanidine, nitrate $HN{=}C(NH_2)_2.HNO_3$	122.08	lf (w)	d	217	....	....	w, al	B3[4], 151
7446	Guanidine, 1-nitro $HN{=}C(NH_2)NHNO_2$	104.07	nd or pr (w)	....	239d	....	....	....	B3[4], 249
7447	Guanidine, picrate $HN{=}C(NH_2)_2.C_6H_3N_3O_7$	288.18	og ye pl or nd (w)	....	333d	....	....	....	B6[3], 960
7448	Guanidine, thiocyanate $HN{=}C(NH_2)_2HCNS$	118.16	lf	....	118	....	....	w	B3[2], 121
7449	Guanidine, tetraphenyl $HN{=}C[N(C_6H_5)_2]_2$	363.46	rh (lig)	....	130-1	....	....	al, eth, bz	B12, 430
7450	Guanidine, 1,1,3-triphenyl $HN{=}C(NHC_6H_5)N(C_6H_5)_2$	287.37	ta (dil al)	....	134	....	....	al, eth	B12[3], 895
7451	Guanidine, 1,2,3-triphenyl $C_6H_5N{=}C(NHC_6H_5)_2$	287.37	nd or pr(al)	d	146-7	$1.163^{20/4}$	....	al	B12[3], 907
7452	Guanidine, 1-ureido or Dicyandiamidine $HN{=}C(NH_2)NHCONH_2$	102.10	pr(al)	d160	105	....	....	py	B3[4], 155
7453	Guanine or 2-Aminohypoxanthine $C_5H_5N_5O$	151.13	nd or pl(aq NH_3)	sub	360d	....	....	....	B26[2], 262
7454	Guanosine or 9-D-Ribosidoguanine $C_{10}H_{13}N_5O_5$	283.24	nd (w), $[\alpha]^{20}{}_D$ -60.5 (0.1 NNaOH, p=3)	....	239d	....	....	aa	B31, 28
7455	Guanylic acid or Guanosine phosphoric acid $C_{10}H_{14}N_5O_8P$	363.22	nd or pr (w + 2), $[\alpha]^{20}{}_D$ -7.5 (w, p=1) $^{25}{}_D$ -65 (5% NaOH, c=2)	....	208d	....	....	w	B31, 29
7456	Guiaicol or Catechol monomethyl ether $2\text{-}CH_3OC_6H_4OH$	124.14	hex pr	205, 106.5[24]	32	$1.1287^{21/4}$	1.5429^{20}	al, eth, chl	B6[3], 4200
7457	Guiaicol, 3-nitro or 2-Methoxy-6-nitro phenol $2\text{-}CH_3O\text{-}6\text{-}NO_2C_6H_3OH$	169.14	og ye nd(sub)	sub	62	....	....	w, al	B6[3], 4263
7458	Guiaicol, 4-nitro or 2-Methoxy-5-nitro phenol $2\text{-}CH_3O\text{-}5\text{-}NO_2C_6H_3OH$	169.14	pa ye nd(w)	....	105	....	....	al, eth	B6[3], 4264
7459	Guiaicol, 5-nitro or 2-Methoxy-4-nitro phenol $2\text{-}CH_3O\text{-}4\text{-}NO_2C_6H_3OH$	169.14	ye nd(w)	....	103-4	....	....	al, eth	B6[3], 4264
7460	Guiaicol, 6-nitro or 2-Methoxy-3-nitro phenol $2\text{-}OCH_3\text{-}3\text{-}NO_2C_6H_3OH$	169.14	yesh rh pr(peth)	....	102-3	....	....	al	B6[3], 4263
7461	Guaiacol, 5-vinyl or 2-Methoxy-4-vinyl phenol $2\text{-}CH_3O\text{-}4\text{-}(CH_2{=}CH)C_6H_3OH$	150.18	cr	....	57	....	....	al, eth	B6[3], 4981
7462	D-Gulonic acid-γ-lactone $C_6H_{10}O_6$	178.14	pr ta (w), $[\alpha]^{20}{}_D$ −57.1 (w)	....	180-1	....	....	w	B18[4], 3025
7463	L-Gulonic acid-γ-lactone $C_6H_{10}O_6$	178.14	rh pr (w), $[\alpha]^{20}{}_D$ + 55.1 (w)	....	185	....	....	w	B18[4], 3026
7464	D-Gulonic acid, phenylhydrazide $C_{12}H_{18}N_2O_6$	286.28	(w), $[\alpha]^{20}{}_D$ + 13.45	d 195	147-9	....	....	w, al	B15[1], 82
7465	D-Gulose $C_6H_{12}O_6$	180.16	syr $[\alpha]^{20}{}_D$ −20.4 (w)	d	......	....	....	w	B1[4], 4333
7466	L-Gulose $C_6H_{12}O_6$	180.16	sy $[\alpha]^{20}{}_D$ + 61.6 (w)	d	......	....	....	w	B1[4], 4334
7467	Guvacine or 1,2,5,6-Tetrahydronicotinic acid $C_6H_9NO_2$	127.14	pr(w) rods(+ 1w dil al)	....	295d	....	....	w	B22[1], 489
7468	β-Gurjunene $C_{15}H_{24}$	204.36	$[\alpha]_D$ + 74.5	120-3[13]	......	0.9348	1.5028	....	B5[3], 1093

No.	Name, Synonyms, and Formula	Mol. wt.	Color, crystalline form, specific rotation and λ_{max} (log ε)	b.p. °C	m.p. °C	Density	n_D	Solubility	Ref.
7469	Halostachine (l) $C_6H_5CH(OH)CH_2NHCH_3$	151.21	[α] -47	....	43-5	....	....	w, al, eth	B13[3], 1658
7470	Harmaline or 3,4-Dihydroharmine $C_{13}H_{14}N_2O$	214.27	ta(MeOH) rh pr(al)	....	250d	....	....	....	B23[2], 345
7471	Harmine or Banisterine, Telepathine $C_{13}H_{12}N_2O$	212.25	rh(al) pr(MeOH)	sub	272-4	....	....	py	B23[2], 348
7472	Hecogenin $C_{27}H_{42}O_4$	430.63	pl (eth), $[\alpha]^{22/}{}_D$ + 7	....	265-8	....	....	al, eth, ace	B19[4], 2581
7473	Hecogenin, acetate $C_{29}H_{44}O_5$	472.67	cr(MeOH)	....	243 (252)	....	....	al	B19[4], 2584
7474	Hederagenin $C_{30}H_{48}O_4$	472.71	pr (al), $[\alpha]^{20/}{}_D$ + 70.1 (chl-MeOH)	....	332-4	....	....	al	B10[3], 1923
7475	Helenine or Alantolactone $C_{15}H_{20}O_2$	232.32	nd	275, 197^{10}	76	....	....	al, eth, bz	B17[4], 5030
7476	Helicin or Salicylaldehyde-β-D- glucoside $C_{13}H_{16}O_7$	284.27	nd (a), $[\alpha]^{20/}{}_D$ -60.4 (w)	....	175	....	....	w, al	B17[4], 3010
7476a	Helvolic acid or Fumagacin $C_{32}H_{44}O_8$	554.68	nd (dil al), $[\alpha]^{18/}{}_D$ -125 (chl, c=1)	....	212d	....	....	eth, ace, bz, aa	Am 78, 5275
7477	Hematein or Haematein $C_{16}H_{12}O_6$	300.27	red br cr	....	250d	....	....	....	B18[4], 3343
7478	Hematin or Ferriporphyrin hydroxide $C_{34}H_{32}N_4O_4.FeOH$	633.51	br pw(py)	....	>200	....	....	....	M, 508
7479	Hematommic acid $C_9H_{14}O_5$	202.21	nd(aa)	....	172-3	....	....	aa	H16, 282
7480	Hematoporphyrin or Photodyn $C_{34}H_{38}N_4O_6$	598.70	red	....	172-3	....	....	aa	M, 509
7481	Hematoxylin $C_{16}H_{14}O_6 \cdot 3H_2O$	362.38	yesh cr, [α] + 11 (w, c=3.7)	....	140	....	....	al	B17[2], 273
7482	Hemimellitic acid or 1,2,3-Benzene tricarboxylic acid 1,2,3-$C_6H_3(CO_2H)_3$	210.14	tcl pl(+ 2w) nd(w)	....	197d	1.546^{20}	....	w	B9[3], 4791
7483	Heneicoscane or Uneicosane $CH_3(CH_2)_{19}CH_3$	296.58	cr (w)	356.5, 203^{10}	40.5	$0.7919^{20/4}$	1.4441^{20}	Peth	....
7484	Heptachlor or 3,4,5,6,7,8,8a Heptachloro-α-Dicyclopentadiene $C_{10}H_5Cl_7$	373.32	wh	....	95-6	1.57^9	....	al, eth, bz, lig	B5[3], 1236
7485	Heptacosane $CH_3(CH_2)_{25}CH_3$	380.74	cr(al bz) lf(AcOEt)	442, 270^{15}	59.5	$0.7796^{60/4}$	1.4345^{65}	....	B1[4], 586
7486	7,10-Heptadecadiyne $CH_3(CH_2)_5C{\equiv}CCH_2C{\equiv}C(CH_2)_5CH_3$	232.41	....	150^6	....	$0.84^{19/4}$	1.4700^{19}	....	B1[3], 1068
7487	Heptadecanal or Margaraldehyde $CH_3(CH_2)_{15}CHO$	254.46	nd (peth) cr (al + 1)	204^{26}	36	....	....	eth, bz, aa	B1[4], 3395
7488	Heptadecane $CH_3(CH_2)_{15}CH_3$	240.47	hex lf	301.8, 161.7^{10}	22	$0.7780^{20/4}$	1.4369^{20}	eth	B1[4], 548
7489	Heptadecane, 1-amino or Heptadecyl amine $CH_3(CH_2)_{15}CH_2NH_2$	255.49	....	336, 189^{10}	49	$0.8510^{20/4}$	1.4510^{20}	al, eth	B4[4], 824
7490	Heptadecane, 1-bromo or Heptadecyl bromide $CH_3(CH_2)_{15}CH_2Br$	319.37	....	349, 199^{10}	32	$0.9916^{20/4}$	1.4625^{20}	chl	B1[4], 549
7491	Heptadecane, 1,17-dibromo $Br(CH_2)_{17}Br$	398.26	lf(al)	$208\text{-}10^3$	38	....	....	chl	B1[3], 564
7492	Heptadecane, 9-hexyl $CH_3(CH_2)_5CH[(CH_2)_7CH_3]_2$	324.63	....	213.0^{10}, $151^{0.5}$	-19.4	$0.7976^{20/4}$	$1.4465^{20/}{}_D$	....	B1[4], 578
7493	Heptadecane-9-octyl $CH[(CH_2)_7CH_3]_3$	352.69	....	231.5^{10}	-13.8	$0.8020^{20/4}$	$1.4487^{20/}{}_D$	....	B1[4], 583
7494	Heptadecane-9-Phenethyl $C_6H_5(CH_2)_2CH(C_8H_{17})_2$	344.62	....	189^{10}	-26.7	$0.8560^{20/4}$	$1.4806^{20/}{}_D$	....	PCHG 3, 180
7495	Heptadecanoic acid or Margaric acid $CH_3(CH_2)_{15}CO_2H$	270.46	pl(peth)	227^{100}	62-3	0.8532^{60}	1.4342^{60}	eth, ace, bz, chl	B2[4], 1193
7496	Heptadecanoic acid, ethyl ester or Ethyl heptadecanoate $CH_3(CH_2)_{15}CO_2C_2H_5$	298.51	pl(dil al)	185^5	28	....	....	al, eth, ace, bz	B2[4], 1194

No.	Name, Synonyms, and Formula	Mol. wt.	Color, crystalline form, specific rotation and λ_{max} (log ε)	b.p. °C	m.p. °C	Density	n_D	Solubility	Ref.
7497	Heptadecanoic acid, methyl ester $CH_3(CH_2)_{15}CO_2CH_3$	284.48	pl(al)	$184\text{-}7^{9}$	30	...	...	al, eth, ace, bz	B2⁴, 1194
7498	Heptadecanonitrile or Margaronitrile $CH_3(CH_2)_{15}CN$	251.46	cr(al)	$349, 183^{10}$	34	$0.8315^{20/4}$	1.4467^{20}	eth	B2⁴, 1195
7499	1-Heptadecanol $CH_3(CH_2)_{15}CH_2OH$	256.47	lf(al) cr(ace)	308	54	$0.8475^{20/4}$	...	al, eth	B1⁴, 1884
7500	2-Heptadecanol $CH_3(CH_2)_{14}CH(OH)CH_3$	256.47	pl(dil al)	$140^{0.5}$	54	...	1.4407^{37}	al, eth	B1⁴, 1885
7501	9-Heptadecanol $(C_8H_{17})_2CHOH$	256.47	pl(dil al)	174^{9}	61	...	1.4262^{80}	al, eth, ace, bz	B1⁴, 1884
7502	2-Heptadecanone or Methyl penta decylketone $C_{15}H_{31}COCH_3$	254.46	pl(dil al)	$320, 246^{110}$	48	$0.8140^{48/48}$	...	eth, ace, bz	B1⁴, 3395
7503	9-Heptadecanone or Pelargone $(C_8H_{17})_2CO$	254.46	pl(MeOH)	$250\text{-}3, 142^{1}$	53	...	...	...	B1⁴, 3396
7504	1-Heptadecene $CH_3(CH_2)_{14}CH=CH_2$	238.46	...	$300, 160^{10}$	11.2	$0.7852^{20/4}$	1.4432^{20}	eth, bz, lig	B1⁴, 927
7505	8-Heptadecene, 9-octyl $C_7H_{15}CH=C(C_8H_{17})_2$	350.67	...	227^{10}	...	$0.8086^{20/4}$	$1.4554^{20}/_{D}$	...	B1⁴, 936
7506	Heptadecyl amine $CH_3(CH_2)_{15}CH_2NH_2$	255.49	...	$336, 189^{10}$	49	$0.8510^{20/4}$	1.4510^{20}	al, eth	B4⁴, 827
7507	1,4-Heptadiene $CH_3CH_2CH=CHCH_2CH=CH_2$	96.17	...	93^{772}	...	$0.7270^{20/4}$	1.4370^{20}	eth, bz, peth	B1³, 999
7508	1,5-Heptadiene $CH_3CH=CHCH_2CH_2CH=CH_2$	96.17	...	94	...	$0.7186^{20/4}$	1.4200^{20}	al, eth, ace, bz	B1³, 999
7509	1-5-Heptadiene-4-ol $CH_3CH=CHCH(OH)CH_2CH=CH_2$	112.17	...	$155\text{-}6^{742}$, 68^{24}	...	$0.8598^{20/4}$	1.4510^{25}	ace	B1⁴, 2248
7510	1,6-Heptadiene-3-yne $CH_2=CHCH_2C\equiv CCH=CH_2$	92.14	...	110^{950}	...	$0.787^{25/4}$	1.4694^{25}	bz, peth	B1³, 1061
7511	2,4-Heptadiene $CH_3CH_2CH=CHCH=CHCH_3$	96.17	...	108	...	$0.7384^{20/4}$	1.4578^{20}	al, eth, ace, bz	B1⁴, 1029
7512	2,5-Heptadien-4-one,2,6-dimethyl or Phorone $CH_3C(CH_3)=CHCOCH=C(CH_3)_2$	138.21	ye gr pr	197.8	28	$0.8850^{20/4}$	1.4998^{20}	al, eth, ace	B1⁴, 3051
7513	3,5-Heptadiene-2-one or Crotonylidene acetone $CH_3CH=CHCH=CHCOCH_3$	110.16	...	88^{28}	...	$0.8946^{19/4}$	1.5177^{19}	eth	B1⁴, 3549
7514	1,5-Heptadiyne $CH_3C\equiv CCH_2CH_2C\equiv CH$	92.14	...	26^{30}	...	$0.8100^{21/4}$	1.4521^{21}	bz, peth	B1², 247
7515	1,6-Heptadiyne $HC\equiv C(CH_2)_3C\equiv CH$	92.14	...	$112, 36^{20}$	−85	$0.8164^{17/4}$	1.451^{17}	bz, aa	B1⁴, 1121
7516	Heptamethyleneimine, α-methyl or 1-Azacyclooctane-2-methyl $C_8H_{17}N$	127.23	nd(ace)	$162\text{-}3^{746}$	156-7	0.853^{30}	1.4620^{21}	w, **al**, eth, ace, bz	B20¹, 30
7517	Heptanal or Enanthaldehyde. Heptaldehyde $CH_3(CH_2)_5CHO$	114.19	...	152.8, 59.6^{30}	−43.3	$0.8495^{20/4}$	1.4113^{20}	**al, eth**	B1⁴, 3314
7518	Heptanal, 2-benzylidene or α-Pentylcinnamaldehyde $C_5H_{11}C(=CHC_6H_5)CHO$	202.30	ye oil	$174\text{-}5^{20}$	80	0.9711^{20}	1.5381^{20}	ace	B7³, 1517
7519	Heptanal, oxime or Heptaldoxime $CH_3(CH_2)_5CH=NOH$	129.20	pl(al)	$195, 100.5^{14}$	57-8	0.8583^{55}	1.4210^{20}	al, eth	B1⁴, 3316
7520	1,6-Heptadiene-4-ol, 4-methyl $CH_3C(OH)(CH_2CH=CH)_2$	126.20	...	158.4, $54\text{-}6^{11}$	...	$0.86258^{20/20}$	1.4500^{23}	al, ace	B1⁴, 2260
7521	1,6-Heptadiene-3-yne $CH_2=CHC\equiv CCH_2CH=CH_2$	92.14	...	110^{750}	...	$0.787^{25/4}$	1.4694^{25}	bz, peth	B1³, 1061
7522	Heptane C_7H_{16}	100.20	...	98.4	−90.6	$0.6837^{20/4}$	1.3878^{20}	al, **eth, ace, chl, peth**	B1⁴, 376
7523	Heptane, 1-amino or Heptyl amine $CH_3(CH_2)_5CH_2NH_2$	115.22	...	156.9, 45.6^{20}	−18	$0.7754^{20/4}$	1.4251^{20}	**al, eth**	B4⁴, 734
7524	Heptane, 2-amino $CH_3(CH_2)_4CHNH_2CH_3$	115.22	...	142	...	0.7665^{19}	1.4199^{19}	al, eth, peth	B4⁴, 743
7525	Heptane, 1-bromo or Heptyl bromide $CH_3(CH_2)_5CH_2Br$	179.11	...	$178\text{-}9, 59.7^{10}$	−56.1	$1.1400^{20/4}$	1.4502^{20}	al, eth, chl	B1⁴, 391
7526	Heptane, 1-bromo-7-fluoro $Br(CH_2)F$	197.10	...	85^{11}	...	...	1.4463^{20}	al, eth	B1⁴, 392
7527	Heptane, 2-bromo $C_5H_{11}CHBrCH_3$	179.10	...	165-7, 63.8^{20}	...	$1.1277^{20/4}$	1.4503^{20}	bz	B1³, 431
7528	Heptane, 3-bromo $CH_3CH_2CHBr(CH_2)_3CH_3$	179.10	...	62^{18}	...	$1.1362^{20/4}$	1.4503^{20}	bz, chl	B1⁴, 392

No.	Name, Synonyms, and Formula	Mol. wt.	Color, crystalline form, specific rotation and λ_{max} (log ε)	b.p. °C	m.p. °C	Density	n_D	Solubility	Ref.
7529	Heptane, 4-bromo $CH_3(CH_2)_2CHBrCH_2CH_2CH_3$	179.10	...	84.6^{72}	...	$1.1351^{20/4}$	1.4495^{20}	bz, chl	$B1^4$, 392
7530	Heptane, 1-chloro or Heptyl chloride $CH_3(CH_2)_5CH_2Cl$	134.65	...	159, 45^{10}	−69.5	$0.8758^{20/4}$	1.4256^{20}	**al, eth**	$B1^4$, 389
7531	Heptane, 1-chloro-7-fluoro $F(CH_2)_7Cl$	152.64	...	70^{10}	...	0.993^{20}	1.4222^{25}	al, eth, bz, chl	$B1^4$, 390
7532	Heptane, 2-chloro $CH_3(CH_2)_4CHClCH_3$	134.65	...	46^{19}	...	0.8672^{20}	1.4221^{20}	eth, bz, chl, aa	$B1^4$, 390
7533	Heptane, 2-chloro-2-methyl $C_5H_{11}CCl(CH_3)_2$	148.68	...	50^{15}	...	$0.8568^{25/4}$	1.4240^{25}	al, eth, bz, chl	$B1^4$, 428
7534	Heptane, 2-chloro-6-methyl $(CH_3)_2CH(CH_2)_3CHClCH_3$	148.68	...	74^{35}	...	...	1.4260^{15}	al, eth, bz, chl	$B1^3$, 472
7535	Heptane, 3-chloro $CH_3(CH_2)_3CHClCH_2CH_3$	134.65	...	144^{751}, 48.3^{20}	...	$0.8960^{20/4}$	1.4228^{20}	eth, bz	$B1^4$, 390
7536	Heptane, 3-chloro-2,3-dimethyl $C_4H_9CCl(CH_3)CH(CH_3)_2$	162.70	...	54^8	...	0.8395^{20}	1.4391^{20}	eth, chl	$B1^3$, 511
7537	Heptane, 3-chloro-3-ethyl $C_4H_9CCl(C_2H_5)_2$	162.70	...	46^3	...	$0.8856^{20/4}$	1.4400^{20}	eth, chl	$B1^3$, 510
7538	Heptane, 3-chloromethyl $CH_3CH_2CH(CH_2Cl)(CH_2)_3CH_3$	148.68	...	174	...	$0.8769^{20/4}$	1.4319^{20}	al, eth, ace, bz	$B1^4$, 430
7539	Heptane, 3-chloro-3-methyl $C_4H_9CCl(CH_3)C_2H_5$	148.68	...	64^{27}	...	$0.8764^{20/4}$	1.4317^{20}	al, eth, bz, chl	$B1^4$, 430
7540	Heptane, 4-chloro $(C_3H_7)_2CHCl$	134.65	...	144^{758}, 48.9^{21}	...	$0.8710^{20/4}$	1.4237^{20}	eth,bz	$B1^4$, 390
7541	Heptane, 4-chloro-4-ethyl $(C_3H_7)_2C(Cl)C_2H_5$	162.70	...	67^{12}	...	$0.8821^{20/4}$	1.4438^{20}	eth, bz, chl	$B1^3$, 511
7542	Heptane, 4-chloro-4-methyl $(C_3H_7)_2C(Cl)CH_3$	148.68	...	50^{12}	...	$0.8690^{20/4}$	1.4310^{15}	al, eth, bz, chl	$B1^3$, 477
7543	Heptane, 5-chloro-2,5-dimethyl $CH_3CH_2CCl(CH_3)CH_2CH_2CH(CH_3)_2$	162.70	...	63^{15}	...	$0.8692^{18/4}$	1.4346^{15}	bz, chl	$B1^1$, 64
7544	Heptane, 1,7-diamino or Heptamethylenediamine $H_2N(CH_2)_7NH_2$	130.23	...	223-5, $104\text{-}5^{12}$	28.9	...	...	al, eth, ace, bz	$B4^4$, 1354
7545	Heptane, 1,7-dibromo $Br(CH_2)_7Br$	258.00	...	263, 132^{11}	−41.7	$1.5306^{20/4}$	1.5034^{20}	eth, ace, bz	$B1^4$, 393
7546	Heptane, 1,1-dichloro $CH_3(CH_2)_5CHCl_2$	169.09	...	187, 82^{20}	...	$1.0008^{20/4}$	1.4440^{20}	eth, bz, chl	$B1^4$, 390
7547	Heptane, 1,2-dichloro $CH_3(CH_2)_4CHClCH_2Cl$	169.09	...	$68\text{-}72^7$	...	$1.064^{20/4}$	1.4490^{20}	eth, bz, chl	$B1^3$, 430
7548	Heptane, 2,2-dichloro $CH_3(CH_2)_4CCl_2CH_3$	169.09	...	77^{25}	...	$1.012^{20/4}$	1.4440^{20}	eth, bz, chl	$B1^1$, 430
7549	Heptane, 2,6-dichloro-2,6-dimethyl $(CH_3)_2CCl(CH_2)_3CCl(CH_3)_2$	197.15	...	93^{16}	43	...	...	bz, chl	$B1^3$, 513
7550	Heptane, 4,4-dichloro $(C_3H_7)_2CCl_2$	169.09	...	86^{27}	...	1.0008^{17}	1.448^{17}	eth, chl	$B1^2$, 117
7551	Heptane, 3,3-di(hydroxymethyl) $C_2H_5C(CH_2OH)_2C_4H_9$	160.26	wh	262, 123^{15}	43.8	$0.929^{50/20}$	1.4587^{25}	al	$B1^3$, 2228
7552	Heptane, 2,2-dimethyl $(CH_3)_3C(CH_2)_4CH_3$	128.26	...	132.7	−113	$0.7105^{20/4}$	1.4016^{20}	eth, ace, bz	$B1^4$, 457
7553	Heptane, 2,3-dimethyl $CH_3(CH_2)_3CH(CH_3)CH(CH_3)_2$	128.26	...	140.5, 29.4^{10}	−116	$0.7260^{20/4}$	1.4088^{20}	**al, eth, ace, bz**	$B1^4$, 457
7554	Heptane, 2,4-dimethyl $C_3H_7CH(CH_3)CH_2CH(CH_3)_2$	128.26	...	133.5, 23.8^{10}	...	$0.7143^{20/4}$	1.4031^{20}	**al, eth, ace, bz**	$B1^4$, 457
7555	Heptane, 2,5-dimethyl (d) $C_2H_5CH(CH_3)(CH_2)_2CH(CH_3)_2$	128.26	...	136	...	$0.7198^{20/4}$	1.4033^{20}	**al, eth, ace, bz**	$B1^4$, 457
7556	Heptane, 2,6-dimethyl $(CH_3)_2CH(CH_2)_3CH(CH_3)_2$	128.26	...	135.2, 25.5^{10}	−102.9	$0.7089^{20/4}$	1.4011^{20}	...	$B1^4$, 458
7557	Heptane, 3,3-dimethyl $C_4H_9C(CH_3)_2CH_2CH_3$	128.26	...	137.3, 26^{10}	...	$0.7254^{20/4}$	1.4087^{20}	**al,** eth, ace, bz	$B1^4$, 458
7558	Heptane, 3,4-dimethyl $CH_3CH_2CH(CH_3)CH(CH_3)C_3H_7$	128.26	...	140.1	...	$0.7314^{20/4}$	1.4108^{20}	eth, ace, bz	$B1^3$, 514
7559	Heptane, 3,5-dimethyl $CH_3CH_2CH(CH_3)CH_2CH(CH_3)CH_2CH_3$	128.26	...	136	...	$0.7225^{20/4}$	1.4083^{20}	eth, ace, bz	$B1^4$, 458
7560	Heptane, 4,4-dimethyl $(C_3H_7)_2C(CH_3)_2$	128.26	...	135.2	...	$0.7221^{20/4}$	1.4076^{20}	eth, ace, bz	$B1^4$, 458
7561	Heptane, 4-ethyl $(C_3H_7)_2CHC_2H_5$	128.26	...	141.2, 31^{10}	...	$0.7270^{20/4}$	1.4096^{20}	**al,** eth, **ace, bz, chl**	$B1^4$, 457

PHYSICAL CONSTANTS OF ORGANIC COMPOUNDS (Continued)

No.	Name, Synonyms, and Formula	Mol. wt.	Color, crystalline form, specific rotation and λ_{max} (log ε)	b.p. °C	m.p. °C	Density	n_D	Solubility	Ref.
7562	Heptane, 1-fluoro or Heptyl fluoride $CH_3(CH_2)_5CH_2F$	118.12	....	117.9	−73	$0.8062^{20/4}$	1.3854^{20}	eth, ace, bz, peth	$B1^4$, 387
7563	Heptane, perfluoro C_7F_{16}	388.05	....	82.4	−78	1.7333^{20}	1.2618^{20}	al, eth, ace, chl	$B1^4$, 388
7564	Heptane, 1-iodo or Heptyl iodide $CH_3(CH_2)_5CH_2I$	226.10	....	204, 76.1^{10}	−48.2	$1.3791^{20/4}$	1.4904^{20}	al, eth, ace, chl	$B1^4$, 393
7565	Heptane, 2-iodo $CH_3CHI(CH_2)_4CH_3$	226.10	....	98^{50}	......	1.304^{20}	1.4826	ace, bz	$B1^4$, 393
7566	Heptane, 2-methyl $C_5H_{11}CH(CH_3)_2$	114.23	....	117.6, 12.3^{10}	−109	$0.6980^{20/4}$	1.3949^{20}	al eth, ace, bz, chl	$B1^4$, 428
7567	Heptane, 2- methylamino $C_5H_{11}CH(CH_3)NHCH_3$	129.25	....	155	......	....	....	....	$B4^4$, 743
7568	Heptane, 3-methyl (d) $C_4H_9CH(CH_3)C_2H_5$	114.23	$[\alpha]^{26}{}_D$ + 9.34	115-8	......	$0.7075^{16/4}$	1.4002^{18}	al, eth, ace, bz, chl	$B1^4$, 429
7569	Heptane, 3-methyl (dl) $C_4H_9CH(CH_3)C_2H_5$	114.23	....	119, 13.3^{10}	−120.5	$0.7058^{20/4}$	1.3985^{20}	al, eth, ace, bz, chl	$B1^4$, 429
7570	Heptane, 3-methyl (l) $C_4H_9CH(CH_3)C_2H_5$	114.23	....	$117\text{-}8^{745}$	......	....	1.3990^{20}	al, eth, ace, bz, chl	$B1^3$, 476
7571	Heptane, 4-methyl $(C_3H_7)_2CHCH_3$	114.23	....	117.7, 12.4^{10}	−121	$0.7046^{20/4}$	1.3979^{20}	al, eth ace, bz, chl	$B1^4$, 431
7572	Heptane, 2,2,4,4,6-pentamethyl $(CH_3)_2CHCH_2CH(CH_3)_2CH_2C(CH_3)_3$	170.34	....	177.8	−67	$0.7463^{20/4}$	$1.4440^{20}{}_D$	....	$B1^4$, 510
7573	Heptane, 2,2,4-trimethyl $(CH_3)_3CCH_2CH(CH_3)(C_3H_7)$	142.28	....	147.7, 32.9^{10}	......	$0.7275^{20/4}$	1.4092^{20}	bz, chl	$B1^4$, 481
7574	Heptane, 3,3,5-trimethyl $CH_3CH_2C(CH_3)_2CH_2CH(CH_3)CH_2CH_3$	142.28	....	155.7, 38.9^{10}	......	$0.7248^{20/4}$	1.4170^{20}	bz, chl	$B1^4$, 483
7575	Heptanedioic acid or Pimelic acid $HO_2C(CH_2)_5CO_2H$	160.17	pr(w)	272^{100}sub, 212^{10}	106	1.329^{15}	....	w, al, eth	$B2^4$, 2003
7576	1,7-Heptanediol or Heptamethylene glycol $HO(CH_2)_7OH$	132.20	....	262, 151^{14}	22	$0.9569^{25/4}$	1.4520^{25}	w, al	$B1^4$, 2580
7577	2,4-Heptanediol, 3-methyl $C_3H_7CH(OH)CH(CH_3)CH(OH)CH_3$	146.23	....	115^3	......	$0.928^{20/4}$	1.4459^{20}	al	B1, 491
7578	2,4-Heptanedione $C_3H_7COCH_2COCH_3$	128.17	....	174, 70^{20}	......	$0.9411^{25/4}$	....	....	$B1^4$, 3698
7579	1-Heptanethiol $CH_3(CH_2)_5CH_2SH$	132.26	....	177	−43	$0.8427^{20/4}$	1.4521^{20}	al, eth	$B1^4$, 1738
7580	1,4,7-Heptanetriol $(HOCH_2CH_2CH_2)_2CHOH$	148.20	....	$230\text{-}2^{25}$, 146^1	−35	1.075^{18}	1.4725^{20}	w, al, ace	$B1^4$, 2787
7581	2,4,6-Heptanetrione or Diacetyl acetone $(CH_3COCH_2)_2CO$	142.15	lf	121^{10}	49	$1.0681^{40/40}$	1.4930^{20}	w, al, eth	$B1^4$, 3783
7582	Heptano amide $CH_3(CH_2)_5CONH_2$	129.20	nd(al)lf(w)	250-8	96	$0.852^{110/4}$	1.4217^{110}	w, al, eth	$B2^4$, 963
7583	Heptanoic acid or Enanthic acid $CH_3(CH_2)_5CO_2H$	130.19	....	223, 116^{11}	−7.5	$0.9200^{20/4}$	1.4170^{20}	al, eth, ace	$B2^4$, 958
7584	Heptanoic acid, 7-amino $H_2N(CH_2)_6CO_2H$	145.20	cr (w, MeOH-peth)	....	195	....	....	w, al	$B4^3$, 1467
7585	Heptanoic anhydride $(C_6H_{13}CO)_2O$	242.36	....	268-71, $164^{12.5}$	−12.4	$0.9321^{20/4}$	1.4335^{15}	al, eth	$B2^4$, 962
7586	Heptanoic acid, 2-bromo $CH_3(CH_2)_4CHBrCO_2H$	209.08	....	250d, 147^{12}	......	1.319^{15}	1.471^{18}	eth, ace	$B2^4$, 967
7587	Heptanoic acid, 7-bromo $Br(CH_2)_6CO_2H$	209.08	wh cr(dil al)	280	31	....	....	al, eth, ace, bz	$B2^4$, 968
7588	Heptanoic acid, butyl ester $CH_3(CH_2)_5CO_2C_4H_9$	186.29	....	226.2	−67.5	0.8638^{20}	1.4204^{20}	al, eth, ace, bz	$B2^3$, 768
7589	Heptanoic acid, *iso*-butyl ester $C_6H_{13}CO_2\text{-}i\text{-}C_4H_9$	186.29	....	208	......	0.8593^{20}	....	al, eth, ace, bz	$B2^1$, 145
7590	Heptanoic acid, ethyl ester or Ethyl heptanoate $CH_3(CH_2)_5CO_2C_2H_5$	158.24	....	187, 78^{14}	−66.1	$0.8817^{20/4}$	1.4100^{20}	al, eth	$B2^4$, 960
7591	Heptanoic acid, 7-fluoro $F(CH_2)_6CO_2H$	148.18	....	133^{10}	......	1.039^{20}	1.4207^{25}	....	$B2^4$, 964
7592	Heptanoic acid, heptyl ester $C_6H_{13}CO_2C_7H_{15}$	228.38	....	276-8	−33	$0.8649^{20/4}$	1.4320^{20}	al, eth	$B2^4$, 961
7593	Heptanoic acid, hexyl ester $C_6H_{13}CO_2C_6H_{13}$	214.35	....	261	−48	0.8611^{20}	1.429^{15}	al, eth, ace, bz	$B2^3$, 768
7594	Heptanoic acid, 7-iodo $I(CH_2)_6CO_2H$	256.08	lf(dil al)	....	49-51	....	....	al, eth, ace, bz	$B2^4$, 969

PHYSICAL CONSTANTS OF ORGANIC COMPOUNDS (Continued)

No.	Name, Synonyms, and Formula	Mol. wt.	Color, crystalline form, specific rotation and λ_{max} (log ε)	b.p. °C	m.p. °C	Density	n_D	Solubility	Ref.
7595	Heptanoic acid, methyl ester $C_6H_{13}CO_2CH_3$	144.21		172	−56	$0.8815^{20/4}$	1.4152^{20}	al, eth, ace	$B2^4$, 960
7596	Heptanoic acid, octyl ester or Octyl heptanoate $C_6H_{13}CO_2C_8H_{17}$	242.40		290	−22.5	0.8596^{20}	1.4349^{15}	al, eth, ace, bz	$B2^1$, 768
7597	Heptanoic acid, 6-oxo $CH_3CO(CH_2)_4CO_2H$	144.17		$250\text{-}3^{280}$, 135^1	40.2		1.4306^{25}	w, al, eth, ace	$B3^4$, 1598
7598	Heptanoic acid, pentyl ester $C_6H_{13}CO_2C_5H_{11}$	200.32		245.4	−50	0.8623^{20}	1.4263^{15}	al, eth, ace, bz	$B2^4$, 960
7599	Heptanoic acid, propyl ester or Propyl heptanoate $C_6H_{13}CO_2C_3H_7$	172.27	cr(peth)	207.9	−63.5	$0.8641^{15/4}$	1.4183^{15}	al, eth, ace, bz	$B2^1$, 767
7600	Heptanonitrile $CH_3(CH_2)_5CN$	111.19		183^{765}, $70\text{-}2^{10}$		$0.8107^{20/0}$	1.4104^{30}	eth, ace, bz, aa	$B2^4$, 963
7601	Heptanoyl chloride $CH_3(CH_2)_5COCl$	148.63		125.2	−83.8	0.9590^{20}	1.4345^{18}	eth, lig	$B2^4$, 963
7602	1-Heptanol $CH_3(CH_2)_5CH_2OH$	116.20		176	−34.1	$0.8219^{20/4}$	1.4249^{20}	al, eth	$B1^4$, 1731
7603	1-Heptanol, 7-chloro $Cl(CH_2)_7OH$	150.65	cr(peth or bz)	150^{20}	11	$0.9998^{15/4}$	1.4537^{25}	al, peth	$B1^4$, 1738
7604	1-Heptanol, 7-fluoro $F(CH_2)_7OH$	134.19		$98\text{-}9^{12}$		$0.956^{20/4}$	1.4197^{25}	al, eth	$B1^4$, 17369
7605	1-Heptanol, 4-methyl $C_3H_7CH(CH_3)CH_2CH_2CH_2OH$	130.23		183, 71.6^{20}		$0.8065^{25/4}$	1.4258^{20}	al	$B1^4$, 1789
7606	1-Heptanol, 6-methyl $(CH_3)_2CH(CH_2)_5OH$	130.23		188^{764}, 95.8^{20}	−106	$0.8176^{25/4}$	1.4251^{25}	al, eth	$B1^4$, 1782
7607	1-Heptanol, 1-phenyl $CH_3(CH_2)_5CH(C_6H_5)OH$	192.30		275, $153\text{-}5^{18}$		0.946	1.5024^{20}		$B6^2$, 513
7608	2-Heptanol (d) $C_5H_{11}CH(OH)CH_3$	116.20	$[\alpha]^{20}{}_D$ + 11.4 (al)	160-2, 73^{20}		$0.8190^{20/4}$	1.4209^{20}	al, eth	$B1^4$, 1740
7609	2-Heptanol (dl) $C_5H_{11}CH(OH)CH_3$	116.20		160, 66^{20}		$0.8167^{20/4}$	1.4210^{20}	al, eth	$B1^4$, 1740
7610	2-Heptanol (l) $C_5H_{11}CH(OH)CH_3$	116.20	$[\alpha]^{12}{}_D$ -10.5	74^{23}		$0.8184^{20/4}$	1.4201^{20}	al, eth	$B1^3$, 1687
7611	2-Heptanol, 6-amino-2-methyl, hydrochloride $CH_3CH(NH_2)(CH_2)_3C(OH)(CH_3)_2 \cdot HCl$	181.71	cr		154-5			w, al	$B4^4$, 1809
7612	2-Heptanol, 1-chloro $C_5H_{11}CH(OH)CH_2Cl$	150.65		93^{13}		$0.9885^{20/4}$	1.4499^{20}	al, eth, ace	$B1^4$, 1741
7613	2-Heptanol, 2-methyl $C_5H_{11}C(OH)(CH_3)_2$	130.23		156, $66\text{-}8^{15}$		$0.8142^{20/4}$	1.4250^{20}	al, eth	$B1^4$, 1780
7614	2-Heptanol, 3-methyl $C_4H_9CH(CH_3)CH(OH)CH_3$	130.23		166.1, 68.1^{20}		$0.8177^{25/4}$	1.4199^{25}	al, eth	$B1^3$, 1729
7615	3-Heptanol (d) $C_4H_9CH(OH)C_2H_5$	116.20		157^{750}, 66^{18}	−70	$0.8227^{20/4}$	1.4201^{20}	al, eth	$B1^4$, 1741
7616	3-Heptanol, 2,6-dimethyl $(CH_3)_2CHCH(OH)CH_2CH_2CH(CH_3)_2$	144.26		175		$0.8212^{20/4}$	1.4246^{20}		$B1^3$, 1753
7617	3-Heptanol, 2-methyl (d) $C_4H_9CH(OH)CH(CH_3)_2$	130.23	$[\alpha]^{20}{}_D$ + 27.7 (al)	72^{12}		$0.8235^{20/4}$	1.4265^{20}	al, eth	$B1^1$, 209
7618	3-Heptanol, 2-methyl (dl) $C_4H_9CH(OH)CH(CH_3)_2$	130.23		167.2, 73^{19}		$0.8235^{20/4}$	1.4265^{20}	al, eth	$B1^4$, 1781
7619	3-Heptanol, 2-methyl (l) $C_4H_9CH(OH)CH(CH_3)_2$	130.23	$[\alpha]_D$ -21.08	87^{36}		$0.8235^{20/4}$	1.4265^{20}	al, eth	$B1^1$, 209
7620	3-Heptanol, 3-methyl $C_4H_9C(OH)(CH_3)CH_2CH_3$	130.23		163, $64\text{-}5^{10}$	−83	$0.8282^{20/4}$	1.4279^{20}	al, eth	$B1^4$, 1783
7621	4-Heptanol $(C_3H_7)_2CHOH$	116.20		161, 63.8^{16}	−41.2	$0.8183^{20/4}$	1.4205^{20}	al, eth	$B1^4$, 1743
7622	4-Heptanol, 2,6-dimethyl $[(CH_3)_2CHCH_2]_2CHOH$	144.26		176-7		$0.809^{21/4}$	1.4242^{20}	al, eth	$B1^4$, 1810
7623	4-Heptanol, 4-ethyl $(C_3H_7)_2C(OH)C_2H_5$	144.26		182		0.8350^{20}	1.4332^{20}	al, eth	$B1^4$, 1809
7624	4-Heptanol, 3-methyl (dl) $C_2H_5CH(CH_3)CH(OH)C_3H_7$	130.23		164.7, 67.3^{20}		$0.8335^{25/4}$	1.4211^{25}	al, eth	$B1^3$, 1730
7625	4-Heptanol, 4-methyl $(C_3H_7)_2C(CH_3)OH$	130.23		161, $61\text{-}3^{12}$	-82	$0.8248^{20/4}$	1.4258^{20}	al, eth	$B1^4$, 1789
7626	4-Heptanol, 4-propyl $(C_3H_7)_3COH$	158.28		190-2, $89\text{-}90^{15}$		$0.8338^{21/0}$	1.4355^{31}	al, eth, bz	$B1^4$, 1831
7627	2-Heptanone or Methyl hexyl ketone $CH_3(CH_2)_4COCH_3$	114.19		151.4, 111^{21}	−35.5	$0.8111^{20/4}$	1.4088^{20}	al, eth	$B1^4$, 3318

PHYSICAL CONSTANTS OF ORGANIC COMPOUNDS (Continued)

No.	Name, Synonyms, and Formula	Mol. wt.	Color, crystalline form, specific rotation and λ_{max} (log ε)	b.p. °C	m.p. °C	Density	n_D	Solubility	Ref.
7628	2-Heptanone, 1-chloro $C_5H_{11}COCH_2Cl$	148.64	...	83^{16}	...	0.802^{20}	1.4371^{20}	**al, eth**	$B1^{4}$, 3320
7629	2-Heptanone, 3-methyl $C_4H_9CH(CH_3)COCH_3$	128.21	...	167	...	$0.8218^{20/4}$	1.4172^{20}	al, eth, ace, bz	$B1^{3}$, 2878
7630	2-Heptanone, 6-methyl $(CH_3)_2CH(CH_2)_3COCH_3$	128.21	...	167, $51\text{-}3^{12}$	...	$0.8151^{20/4}$	1.4162^{20}	al, eth, ace, bz	$B1^{4}$, 3344
7631	2-Heptanone, 4-isopropyl $i\text{-}C_3H_7CH(C_3H_7)CH_2COCH_3$	156.27	...	$82\text{-}4^{14}$	...	...	...	al, eth, ace, bz	$B1^{3}$, 2902
7632	3-Heptanone or Butyl ethyl ketone $C_4H_9COC_2H_5$	114.19	...	147^{765}	−39	$0.8183^{20/4}$	1.4057^{20}	**al, eth**	$B1^{4}$, 3321
7633	3-Heptanone, 6-dimethylamino-4,4-diphenyl (l) or l-Methadone $(CH_3)_2NCH(CH_3)CH_2C(C_6H_5)_2COC_2H_5$	309.45	$[\alpha]^{20}/_D$ -32 (al)	...	99-100	...	...	al	$B14^{3}$, 278
7634	3-Heptanone, 6-dimethylamino-4,4-diphenyl,hydrochloride (dl) or Physopeptone $(CH_3)_2NCH(CH_3)CH_2C(C_6H_5)_2COC_2H_5.HCl$	345.91	pl(al-eth)	...	236	...	...	w, al, eth, chl	$B14^{3}$, 279
7635	3-Heptanone, 6-dimethylamino-4,4-diphenyl hydrochloride (l) $(CH_3)_2NCH(CH_3)CH_2C(C_6H_5)_2COC_2H_5,HCl$	345.91	$[\alpha]^{20}/_D$ -169 (al, c=2.1)	...	245-6	...	...	w, al, chl	$B14^{3}$, 278
7636	3-Heptanone, 2-methyl $C_4H_9COCH(CH_3)_2$	128.21	...	158, $63\text{-}5^{25}$	...	$0.8163^{20/4}$	1.4115^{20}	al, eth, ace	$B1^{4}$, 3343
7637	3-Heptanone, 6-methyl $(CH_3)_2CHCH_2CH_2COC_2H_5$	128.21	...	163^{734}	...	0.8304^{20}	1.4209^{20}	al, eth, bz	$B1^{4}$, 3344
7638	4-Heptanone or Depropyl ketone $(C_3H_7)_2CO$	114.19	...	144	−33	$0.8174^{20/4}$	1.4069^{20}	**al, eth**	$B1^{4}$, 3323
7639	4-Heptanone, 2,6-dimethyl or Isovalerone $(i\text{-}C_4H_9)_2CO$	142.24	...	168, $60\text{-}1^{18}$	...	$0.8053^{20/4}$	1.412^{20}	**al, eth**	$B1^{4}$, 3360
7640	4-Heptanone, 2-methyl $C_3H_7COCH_2CH(CH_3)_2$	128.21	...	155^{750}	...	$0.813^{22/0}$	...	al, eth	$B1^{4}$, 3343
7641	Heptano phenone $C_6H_{13}COC_6H_5$	190.29	lf	283.3, 155^{15}	16.4	$0.9516^{20/4}$	1.5060^{20}	al, eth, ace	$B7^{3}$, 1188
7642	Heptasiloxane, hexadecamethyl $CH_3[Si(CH_3)_2O\text{-}]_6Si(CH_3)_3$	533.15	...	270, 165^{20}	−78	$0.9012^{20/4}$	1.3965^{20}	bz, lig	$B4^{3}$, 1880
7643	1-Heptene $CH_3(CH_2)_4CH=CH_2$	98.19	...	93.6	−119	$0.6970^{20/4}$	1.3998^{20}	al, eth	$B1^{4}$, 857
7644	1-Heptene, 1-chloro $C_5H_{11}CH=CHCl$	132.63	...	155, $78\text{-}82^{75}$	...	0.8948^{20}	1.4380^{20}	eth, ace, bz, chl	$B1^{3}$, 823
7645	1-Heptene, 2-chloro $C_5H_{11}CCl=CH_2$	132.63	...	138^{748}	...	$0.8895^{20/4}$	1.4349^{20}	al, eth, chl	$B1^{3}$, 823
7646	1-Heptene, 2-methyl $C_5H_{11}C(CH_3)=CH_2$	112.22	...	118.2	−90.1	$0.2206^{20/4}$	$1.4120^{20}/_D$	...	$B1^{4}$, 881
7647	2-Heptene *(cis)* $CH_3(CH_2)_3CH=CHCH_3$	98.19	...	98.5	...	$0.708^{20/4}$	1.406^{20}	al, eth, ace, bz, chl	$B1^{3}$, 824
7648	2-Heptene (trans) $CH_3(CH_2)_3CH=CHCH_3$	98.19	...	98	−109.5	$0.7012^{20/4}$	1.4045^{20}	al, eth, ace, bz, chl	$B1^{4}$, 860
7649	2-Heptene, 4-chloro $C_3H_7CHClCH=CHCH_3$	132.63	...	140-5, 49^{21}	...	$0.879^{18/4}$	1.4430^{23}	al, eth, chl	$B1^{3}$, 825
7650	2-Heptene, 6-chloro-2-methyl $CH_3CHCl(CH_2)_2CH=C(CH_3)_2$	146.66	...	$60\text{-}1^{15}$	...	$0.8931^{18/4}$	1.4458^{18}	al, eth, ace, bz	$B1^{2}$, 200
7651	2-Heptene, 2-methyl $C_4H_9CH=C(CH_3)_2$	112.22	...	122.6	...	$0.7241^{20/0}$	1.4170^{20}	eth, bz	$B1^{4}$, 882
7652	2-Heptene, 2-methyl-6-methylamino $CH_3NHCH(CH_3)CH_2CH_2CH=C(CH_3)_2$	141.26	...	176-8, $58\text{-}9^{17}$	...	...	...	al, eth	$B4^{3}$, 467
7653	3-Heptene *(cis)* $C_3H_7CH=CHC_2H_5$	98.19	...	95.8	...	$0.7030^{20/4}$	1.4059^{20}	al, eth, ace, bz, chl	$B1^{4}$, 861
7654	3-Heptene (trans) $C_3H_7CH=CHC_2H_5$	98.19	...	95.7	−136.6	$0.6981^{20/4}$	1.4043^{20}	al, eth, ace, bz, peth	$B1^{4}$, 861
7655	3-Heptene, 4-chloro $C_3H_7CCl=CHC_2H_5$	132.63	...	139	...	0.883^{14}	1.437^{14}	al, eth, chl	$B1^{3}$, 827
7656	3-Heptene, 4-propyl $C_2H_5CH=C(C_3H_7)_2$	140.27	...	160.5	...	$0.7518^{17.8}$	$1.4302^{17.8}$	...	$B1^{4}$, 906
7657	1-Heptene-2-carboxaldehyde, 1-phenyl or Jasminaldehyde $C_5H_{11}C(CHO)=CHC_6H_5$	203.30	...	$174\text{-}5^{20}$, 140^{5}	...	0.9718^{20}	1.5381^{20}	al, eth	$B7^{3}$, 1517
7658	1-Heptene-4-ol, 4-methyl $C_3H_7COH(CH_3)CH_2CH=CH_2$	128.21	...	159-60	...	$0.8345^{20/0}$	1.4479^{18}	al, eth	$B1^{3}$, 1945
7659	2-Heptene-1-ol *(trans)* $C_4H_9CH=CHCH_2OH$	114.19	...	177-9, 75^{10}	...	0.8516^{20}	1.4460^{20}	al, ace	$B1^{3}$, 1936

No.	Name, Synonyms, and Formula	Mol. wt.	Color, crystalline form, specific rotation and λ_{max} (log ε)	b.p. °C	m.p. °C	Density	n_D	Solubility	Ref.
7660	2-Heptene-4-ol (dl) $C_3H_7CH(OH)CH=CHCH_3$	114.19		152-4, 64^{14}		$0.8445^{20/4}$	1.4373^{20}	al, eth	$B1^4$, 2155
7661	3-Heptene-2-one *(trans)* $C_3H_7CH=CHCOCH_3$	112.17		62^{15}		$0.8496^{20/4}$	1.4436^{20}	al, eth	$B1^4$, 3481
7662	5-Heptene-2-one, 6-methyl $(CH_3)_2C=CHCH_2CH_2COCH_3$	126.20		173, 58.6^{10}	−67	$0.8546^{16/4}$	1.4445^{20}	**al, eth**	$B1^4$, 3493
7663	1-Heptene-3-yne- $C_3H_7C\equiv C\text{-}CH=CH_2$	94.16		110, 44^{75}		$0.7603^{20/4}$	1.4520^{25}	al, ace, bz, peth	$B1^4$, 1097
7664	1-Heptene-4-one, 6,6-dimethyl $(CH_3)_3CC\equiv CCH_2CH=CH_2$	122.21		125, 68^{100}		0.758^{20}	1.4312^{20}	al, ace, bz, peth	C55, 23329
7665	6-Heptene-4-one-3-ol, 3-ethyl $CH_2=CCHC\equiv CC(OH)(C_2H_5)_2$	138.21		62^{4}		$0.8875^{20/4}$	1.4800^{20}	al, eth	$B1^3$, 2034
7666	Heptyl amine $CH_3(CH_2)_5CH_2NH_2$	115.22		156.9, 45.6^{10}	−18	$0.7754^{20/4}$	1.4251^{20}	**al, eth**	$B4^4$, 734
7667	Heptyl ether or Diheptyl ether $(C_7H_{15})_2O$	214.39		258.5		$0.8008^{20/4}$	1.4275^{20}	al, eth	$B1^4$, 1733
7668	Heptyl methyl ether $C_7H_{15}OCH_3$	130.23		151		$0.7869^{15/15}$	1.4073^{20}	**al, eth,** ace	$B1^3$, 1682
7669	Heptyl nitrite $C_7H_{15}ONO$	145.20		155-8, 44^{18}		$0.8939^{0/4}$	1.4032^{20}	eth	$B1^4$, 1735
7670	Heptyl phenyl ether $C_7H_{15}OC_6H_5$	192.31		267, $128\text{-}30^{12}$		$0.9178^{15/15}$	1.4912^{20}	**al, eth,** ace	$B6^4$, 560
7671	Heptyl sulfate $(C_7H_{15}O)_2SO_2$	294.45	cr(peth)	$146.6^{1.5}$	13	$0.9819^{25/25}$	1.4362^{25}		$B1^3$, 1683
7672	Heptyl thiocyanate $C_7H_{15}SCN$	157.27		234-6, 136^{28}		0.92^{20}		**al, eth**	$B3^4$, 331
7673	1-Heptyne $CH_3(CH_2)_4C\equiv CH$	96.17		99.7, 6^{10}	−81	$0.7328^{20/4}$	1.4087^{20}	**al, eth,** bz, chl, peth	$B1^4$, 1025
7674	1-Heptyne, 1-bromo $CH_3(CH_2)_4C\equiv CBr$	175.07		164^{755}, 69^{25}		$1.2120^{22/4}$	1.4678^{22}	al, eth, ace, chl	$B1^3$, 998
7675	1-Heptyne, 1-chloro $CH_3(CH_2)_4C\equiv CCl$	130.62		141		$0.9250^{24/4}$	1.4411^{24}	al, eth	$B1^3$, 997
7676	1-Heptyne, 1-iodo $CH_3(CH_2)_4C\equiv CI$	222.07		$90\text{-}2^{17}$		$1.4701^{19/4}$	1.5123^{19}_{D}		$B1^3$, 998
7677	2-Heptyne $CH_3(CH_2)_3C\equiv CCH_3$	96.17		112		$0.7480^{20/4}$	1.4230^{20}	**al, eth,** bz, chl, peth	$B1^4$, 1026
7678	2-Heptyne, 1-bromo $CH_3(CH_2)_3C\equiv CCH_2Br$	175.07		104^{55}, 84^{20}			1.4878^{25}	al, eth, ace	$B1^4$, 1026
7679	2-Heptyne, 1-chloro $CH_3(CH_2)_3C\equiv CCH_2Cl$	130.62		167, 73^{24}			1.4570^{25}		$B1^4$, 1026
7680	2-Heptyne, 7-chloro $Cl(CH_2)_4C\equiv CCH_3$	130.62		166			1.4507^{25}	al, eth	$B1^3$, 998
7681	3-Heptyne $C_3H_7C\equiv CC_2H_5$	96.17		105-6		$0.7527^{20/4}$	1.4220^{20}	**al, eth,** bz, chl, peth	$B1^4$, 1027
7682	3-Heptyne, 1-chloro $C_3H_7C\equiv CCH_2CH_2Cl$	130.62		162, $90\text{-}93^{20}$			1.4520^{25}		$B1^3$, 999
7683	3-Heptyne, 7-chloro $Cl(CH_2)_3C\equiv CC_2H_5$	130.62		164, $74\text{-}5^{31}$			1.4517^{20}	al, eth	$B1^3$, 999
7684	3-Heptyne, 2,6-dimethyl $(CH_3)_2CHCH_2C\equiv CCH(CH_3)_2$	124.23		130-6		$0.785^{20/4}$		eth, ace	$B1^3$, 1015
7685	3-Heptyne, 5,5-dimethyl $CH_3CH_2C(CH_3)_2C\equiv CC_2H_5$	124.23		69^{100}		$0.7610^{20/4}$	1.4360^{20}	eth, ace	$B1^3$, 1015
7686	3-Heptyne, 5-ethyl-5-methyl $(C_2H_5)_2C(CH_3)C\equiv CC_2H_5$	138.25		88^{100}		$0.7714^{20/4}$	1.4386^{20}	eth, ace	$B1^3$, 1021
7687	2-Heptyne-1-ol or Butyl propargyl alcohol $C_4H_9C\equiv CCH_2OH$	112.17		94^{22}			1.4523^{25}		$B1^4$, 2247
7688	Heroin or *o,o*-Diacetyl morphine $C_{21}H_{23}NO_5$	369.42	rh, $[\alpha]^{15}_{D}$ -166 (MeOH)	$272\text{-}4^{12}$	173	1.56-1.61		bz, chl	$B27^2$, 151
7689	Hesperidin or Hesperitin-7-(6-α-L-rhamnopyranosyl)-β-D-glucoside $C_{28}H_{34}O_{15}$	610.57	wh nd(dil MeOH or aa), $[\alpha]^{20}_{D}$ -77.5 (py, c=4.2		261-3			al, py, aa	$B18^4$, 3219
7690	Hexacene or Anthraceno-2′:3′,2:3-anthracene $C_{26}H_{16}$	328.41	dk bl-gr cr(sub)	sub	Ca 380				$B5^3$, 2654

No.	Name, Synonyms, and Formula	Mol. wt.	Color, crystalline form, specific rotation and λ_{max} (log ε)	b.p. °C	m.p. °C	Density	n_D	Solubility	Ref.
7691	Hexachlorophene or Bis(2-hydroxy-3,5,6-trichlorophenyl) methane $(2\text{-}HO\text{-}3,5,6\text{-}Cl_3C_6H)_2CH_2$	406.91	nd(bz)	...	166-7	...	...	al, eth, ace, chl	B6^{1}, 5407
7692	Hexacosane or Cerane $CH_3(CH_2)_{24}CH_3$	366.71	mcl tcl or rh (bz) cr (eth)	412.2, 248.2^{22}	56.4	0.7783^{60}, $0.8032^{20/4}$	1.4357^{60}	bz, lig, chl	B1^{4}, 583
7693	Hexacosane, 13-dodecyl $C_{13}H_{27}CH(C_{12}H_{25})_2$	353.03	...	272^{1}	13.7	$0.8188^{20/4}$	1.4577^{20}	...	B1^{4}, 600
7694	1-Hexacosanol or Ceryl alcohol Cerotin $CH_3(CH_2)_{24}CH_2OH$	382.71	rh pl (dil al)	305^{20} d	80	...	...	al, eth	B1^{4}, 1912
7695	1,15-Hexadecadiyne $HC{\equiv}C(CH_2)_{12}C{\equiv}CH$	218.38	fl(al)	$152\text{-}5^{12}$	44-5	...	...	...	B1^{2}, 249
7696	6,9-Hexadecadiyne $CH_3(CH_2)_5C{\equiv}CCH_2C{\equiv}C(CH_2)_4CH_3$	218.38	...	169^{15}	...	$0.845^{18/4}$	1.4694^{18}	...	B1^{3}, 1067
7697	6,10-Hexadecadiyne $CH_3(CH_2)_4C{\equiv}CCH_2CH_2C{\equiv}C(CH_2)_4CH_3$	218.28	...	157^{10}	...	$0.7907^{20/4}$	1.4523^{20}	...	B1^{3}, 1067
7698	Hexadecanal or Palmitaldehyde $CH_3(CH_2)_{14}CHO$	240.43	pl (eth) nd (peth)	$200\text{-}2^{29}$	34	...	...	al, eth, ace, bz	B1^{4}, 3393
7699	Hexadecanal, dimethyl acetal $CH_3(CH_2)_{14}CH(OCH_3)_2$	286.50	...	144^{2}	10	0.8542^{20}	1.4382^{25}	al, eth, ace	B1^{4}, 3393
7700	Hexadecanal, oxime $CH_3(CH_2)_{14}CH{=}NOH$	255.45	nd(dil al)	...	88	...	...	al, chl	B1^{3}, 2923
7701	Hexadecane or Cetane $CH_3(CH_2)_{14}CH_3$	226.45	lf(ace)	287, 149^{10}	18.2	$0.7733^{20/4}$	1.4345	**eth**	B1^{4}, 537
7702	Hexadecane, 1-amino or Cetyl amine $CH_3(CH_2)_{14}CH_2NH_2$	241.46	lf	322.5, 144^{2}	46.8	$0.8129^{20/4}$	1.4496^{20}	al, eth, ace, bz, chl	B4^{4}, 818
7703	Hexadecane, 1-bromo or Cetyl bromide $CH_3(CH_2)_{14}CH_2Br$	305.34	...	336, 188^{10}	17-9	$0.9991^{20/4}$	1.4618^{25}	...	B1^{4}, 542
7704	Hexadecane, 1-chloro or Cetyl chloride $CH_3(CH_2)_{14}CH_2Cl$	260.89	...	322, 177^{10}	17.9	$0.8652^{20/4}$	1.4505^{20}	...	B1^{4}, 542
7705	Hexadecane, 1,16-dibromo $Br(CH_2)_{16}Br$	384.24	lf(al)	204^{4}	56	...	...	chl	B1^{2}, 138
7706	Hexadecane, 6,11-dipentyl $(C_5H_{11})_2CH(CH_2)_4CH(C_5H_{11})_2$	366.71	...	231^{10}	−16.2	$0.8072^{20/4}$	$1.4502^{20/D}$	...	B1^{4}, 586
7707	Hexadecane, 1-fluoro or Cetyl fluoride $CH_3(CH_2)_{14}CH_2F$	244.44	...	289, 152.6^{10}	18	$0.8321^{20/4}$	1.4317^{20}	eth, lig	B1^{4}, 542
7708	Hexadecane, 1-iodo or Cetyl iodide $CH_3(CH_2)_{14}CH_2I$	352.34	lf(al)	357, 202^{10}	24.7	$1.1257^{20/4}$	1.4818^{20}	eth, ace, **bz**, chl	B1^{4}, 543
7709	Hexadecane, 1-phenyl or Cetylbenzene $CH_3(CH_2)_{14}CH_2C_6H_5$	302.54	...	237^{16}	27	$0.8560^{20/4}$	1.4814^{20}	eth, bz, lig	B5^{4}, 1216
7710	Hexadecanedioic acid or Thapsic acid $HO_2C(CH_2)_{14}CO_2H$	286.41	pl(al, AcOEt)	...	126	...	...	al, ace	B2^{4}, 2162
7711	Hexadecanoic acid or Palmitic acid $CH_3(CH_2)_{14}CO_2H$	256.43	nd(al)	350, 267^{100}	63	$0.8527^{62/4}$	1.4335^{60}	al, **eth**, ace, bz, chl	B2^{4}, 1157
7712	1-Hexadecanol or Cetyl alcohol $CH_3(CH_2)_{14}CH_2OH$	242.45	fl(AcOEt)	344, 190^{15}	50	$0.8176^{50/4}$	1.4283^{20}	al, eth, ace, bz	B1^{4}, 1876
7713	2-Hexadecanol $CH_3(CH_2)_{13}CHOHCH_3$	242.45	...	314	44	0.8338^{20}	$1.4479^{20/D}$	...	B1^{4}, 1882
7714	1-Hexadecanethiol or Cetyl mercapton $CH_3(CH_2)_{14}CH_2SH$	258.51	(lig)	$123\text{-}8^{0.5}$	18-20	...	...	eth	B1^{4}, 1881
7715	1-Hexadecene or Cetene $CH_3(CH_2)_{13}CH{=}CH_2$	224.43	lf	284.4, 155^{15}	4.1	$0.7811^{20/4}$	1.4412^{20}	al, eth, peth	B1^{4}, 927
7716	2-Hexadecenoic acid (form I) or Gaidic acid $CH_3(CH_2)_{12}CH{=}CHCO_2H$	254.41	lf(al)	...	39	...	...	al	B2, 461
7717	2-Hexadecenoic acid (form II) or Δ-α,β-Hypogeic acid $CH_3(CH_2)_{12}CH{=}CHCO_2H$	254.41	fl(al)	...	49	...	...	al, eth, chl, peth	B2^{4}, 1629
7718	7-Hexadecenoic acid $CH_3(CH_2)_7CH{=}CH(CH_2)_5CO_2H$	254.41	cr	230^{10}	33	...	...	al, eth	B2, 460
7719	Hexadecyl ether or Dicetyl ether $(C_{16}H_{33})_2O$	466.88	lf(al)	270d	55	0.978^{19}	...	al, eth	B1^{4}, 1878
7720	Hexadecylpyridenium chloride $(C_5H_5N^+)C_{16}H_{33}Cl^-$	340.00	wh pw	...	77-83	...	...	w, chl	B20^{4}, 2316
7721	1-Hexadecyne $CH_3(CH_2)_{13}C{\equiv}CH$	222.41	...	284, 147.8^{10}	15	$0.7965^{20/4}$	1.4440^{20}	bz	B1^{4}, 1073
7722	2-Hexadecyne $CH_3(CH_2)_{12}C{\equiv}CCH_3$	222.41	fl	160^{15}	20	$0.8039^{20/4}$	...	...	B1^{3}, 1028
7723	7-Hexadecynoic acid $C_8H_{17}C{\equiv}C(CH_2)_5CO_2H$	252.40	nd(w) cr(al)	214^{15}	47	...	...	al, eth	B2^{3}, 1474

PHYSICAL CONSTANTS OF ORGANIC COMPOUNDS (Continued)

No.	Name, Synonyms, and Formula	Mol. wt.	Color, crystalline form, specific rotation and λ_{max} (log ε)	b.p. °C	m.p. °C	Density	n_D	Solubility	Ref.
7724	2,4-Hexadienal or Sorbaldehyde $CH_3CH=CHCH=CHCHO$	96.13	....	$173\text{-}4^{756}$, 76^{30}	......	0.898^{20}	1.5384^{20}	....	B1⁴, 3545
7725	1,2-Hexadiene or Propylallene $CH_3CH_2CH_2CH=C=CH_2$	82.15	....	76	......	0.7149^{20}	1.4282^{20}	eth, chl	B1⁴, 1011
7726	1,3-Hexadiene $CH_3CH_2CH=CH\text{-}CH=CH_2$	82.15	....	73	......	$0.7050^{20/4}$	1.4380^{20}	eth	B1⁴, 1011
7727	1,3-Hexadiene, 3-chloro $CH_3CH_2CH=CClCH=CH_2$	116.59	....	68^{117}	......	$0.9390^{20/4}$	1.4770^{20}	eth, chl	B1⁴, 1012
7728	1,4-Hexadiene $CH_3CH=CHCH_2CH=CH_2$	82.15	....	65	......	$0.7000^{20/4}$	1.4150^{20}	eth	B1⁴, 1013
7729	1,4-Hexadiene, 5-chloro-2-isopropyl $CH_3CCl=CHCH_2C(i\text{-}C_3H_7)=CH_2$	158.67	....	95^{18}	......	$0.9310^{25/4}$	1.4370^{25}	ace, chl	B1³, 1014
7730	1,4-Hexadiene, 3,3,6-trichloro $ClCH_2CH=CHCCl_2CH=CH_2$	185.48	....	$100\text{-}3^{4}$	......	$1.3036^{20/4}$	1.5585^{20}	eth, chl	B1³, 982
7731	1,5-Hexadiene or Biallyl $H_2C=CHCH_2CH_2CH=CH_2$	82.15	....	59.5	−141	$0.6880^{20/4}$	1.4042^{20}	al, eth, bz, chl	B1⁴, 1013
7732	1,5-Hexadiene, 2,5-dimethyl $CH_2=C(CH_3)CH_2CH_2C(CH_3)=CH_2$	110.20	....	134	−75.6	0.7512^{20}	1.4399^{21}	ace, chl	B1⁴, 1042
7733	1,5-Hexadiene, 2-methyl $CH_2=C(CH_3)CH_2CH_2CH=CH_2$	96.17	....	90-3	−128.8	$0.7198^{20/4}$	$1.4183^{20/}{}_D$	....	B1⁴, 1030
7734	1,5-Hexadiene, perchloro $Cl_2C=CClCCl_2CCl_2CCl=CCl_2$	426.60	cr(ace)	$121^{0.03}$	49	$1.905^{52/4}$	1.6012^{51}	eth, ace, chl	B1⁴, 1016
7735	1,5-Hexadeiene, 3,4-diol (d) or Divinyl glycol $CH_2=CHCH(OH)CH(OH)CH=CH_2$	114.14	$[\alpha]^{17/}{}_D$ + 94.8 (al)	198, 97^{13}	−60	$1.006^{20/4}$	1.4700^{20}	al, eth, chl	B1⁴, 2693
7736	1,5-Hexadiene-3,4-diol (dl) $CH_2=CHCH(OH)CH(OH)CH=CH_2$	114.14	hyg	$90\text{-}1^{8}$	21.7	$1.017^{19/4}$	1.4790^{19}	al, eth, chl	B1³, 2272
7737	1,5-Hexadiene-3,4-diol *(meso)* $CH_2=CHCH(OH)CH(OH)CH=CH_2$	114.14	hyg	100^{14}	18	$1.023^{19/4}$	1.4810^{19}	w, al, eth, chl	B1⁴, 2693
7738	2,4-Hexadiene $CH_3CH=CHCH=CHCH_3$	82.15	....	80	−79	$0.7196^{20/4}$	1.4500^{20}	al, eth, chl	B1⁴, 1016
7739	2,4-Hexadiene, 6-chloro-2-methyl $ClCH_2CH=CHCH=C(CH_3)_2$	130.62	....	57^{11}	......	$0.9416^{20/4}$	1.5120^{20}	ace, chl	B1³, 1000
7740	2,4-Hexadiene, 1,3-dichloro $CH_3CH=CHCCl=CHCH_2Cl$	151.04	....	$80\text{-}2^{17}$	......	$1.1456^{20/4}$	1.5271^{20}	bz, chl	B1⁴, 1017
7741	2,4-Hexadiene, 2,5-dimethyl $(CH_3)_2C=CHCH=C(CH_3)_2$	110.20	....	134, 75^{100}	14-5	$0.7625^{20/4}$	1.4785^{20}	al, eth, bz, chl	B1⁴, 1043
7742	2,4-Hexadiene, 1,3,4,6-tetrachloro $ClCH_2CH=CClCCl=CHCH_2Cl$	219.93	....	$84\text{-}9^{2}$	......	$1.4013^{20/4}$	1.5465^{20}	chl, MeOH	B1³, 987
7743	2,4-Hexadienedioic acid or *cis*-Muconic acid $HO_2CCH=CHCH=CHCO_2H$	142.11	....	194-5	......	....	....	aa	B2⁴, 2297
7744	2,4-Hexadienedioic acid *(trans)* or *trans*-Muconic acid $HO_2CCH=CHCH=CHCO_2H$	142.11	nd(al)	320	305d	....	....	aa, AcOEt	B2⁴, 2298
7745	2,4-Hexadienoic acid or Sorbic acid $CH_3CH=CHCH=CHCO_2H$	112.13	nd(dil al)	228d, 153^{50}	134.5	$1.204^{19/4}$	....	al, eth	B2⁴, 1701
7746	2,4-Hexadien-1-ol or Sorbyl alcohol $CH_3CH=CHCH=CHCH_2OH$	98.14	nd	76^{12}	30-1	$0.8967^{23/4}$	1.4981^{20}	al, eth	B1⁴, 2239
7747	3,5-Hexadien-2-ol $CH_2=CHCH=CHCH(OH)CH_3$	98.14	....	$77\text{-}8^{26}$	......	0.8678^{20}	1.4816^{20}	al	B1⁴, 2237
7748	3,5-Hexadiene, 2-one-6-phenyl or Cinnamylidene acetone $C_6H_5CH=CHCH=CHCOCH_3$	172.23	wh lf (eth)	$170\text{-}2^{15}$	68	....	....	al, eth, ace, bz, chl	B7³, 1656
7749	1,5-Hexadiene-3-one or Divinyl acetylene $CH_2=CHC\equiv CCH=CH_2$	78.11	....	85	−88	$0.7851^{20/4}$	1.5035^{20}	bz	B1⁴, 1120
7750	1,5-Hexadiene, 3-one-2,5-dimethyl $CH_2=C(CH_3)C\equiv CC(CH_3)=CH_2$	106.17	ye	123	......	$0.7863^{25/4}$	1.4845^{20}	bz, chl	B1⁴, 1124
7751	3,5-Hexadiene, 1-yne $CH_2=CHCH=CH_2C\equiv CH$	78.11	....	83-4, 32^{100}	......	$0.7806^{20/4}$	1.5095^{20}	bz	B1⁴, 1120
7752	1,4-Hexadiyne $CH_3C\equiv CCH_2C\equiv CH$	78.11	....	78-83	<-80	$0.825^{0/4}$	....	bz, chl	B1³, 1057
7753	1,5-Hexadiyne or Dipropargyl $HC\equiv CCH_2CH_2C\equiv CH$	78.11	....	86 ,20^{46}	−6	$0.8049^{20/4}$	1.4380^{23}	al, eth, ace, bz	B1⁴, 1118
7754	1,5-Hexadiyne, 1,6-diamino $H_2NC\equiv CCH_2CH_2C\equiv CNH_2$	108.14	....	....	104-5	....	....	bz	B4⁴, 1399
7755	2,4-Hexadiyne $CH_3C\equiv C\text{-}C\equiv C\text{-}CH_3$	78.11	pr(sub)	129-30	68.5	....	....	al, eth	B1⁴, 1119
7756	Hexaethyl tetraphosphate $[(C_2H_5O)_2P(O)O]_3PO$	506.26	hyg	>150d	ca−40	$1.2917^{27/4}$	1.4273^{27}	**al, ace, bz**	B1³, 1331

No.	Name, Synonyms, and Formula	Mol. wt.	Color, crystalline form, specific rotation and λ_{max} (log ε)	b.p. °C	m.p. °C	Density	n_D	Solubility	Ref.
7757	Hexamethylene diamine or 1,6-Diamino hexane $H_2N(CH_2)_6NH_2$	116.21	rh bi pym pl	204-5, 100^{20}	41-2	....	....	w, al, bz	$B4^{4}$, 1320
7758	Hexamethylene tetramine or Hexamin. Urotropine $C_6H_{12}N_4$	140.19	rh (al)	sub	285-95 sub	1.331^{-5}	....	w, al, ace, chl	$B26^{2}$, 200
7759	Hexanal or Caproaldehyde $CH_3(CH_2)_4CHO$	100.16	....	128, 28^{12}	−56	$0.8139^{20/4}$	1.4039^{20}	al, eth, ace, bz	$B1^{4}$, 3296
7760	Hexane C_6H_{14}	86.18	....	69	−95	$0.6603^{20/4}$	1.3751^{20}	al, eth, chl	$B1^{4}$, 338
7761	Hexane, 1-amino or *n*-Hexyl amine $CH_3(CH_2)_4CH_2NH_2$	101.19	....	130	−19	0.7660^{20}	1.4180^{20}	**al, eth**	$B4^{4}$, 709
7762	Hexane, 2-amino (d) $CH_3(CH_2)_3CHNH_2CH_3$	101.19	....	114-5, 64^{90}	......	$0.755^{27/4}$	....	al, eth	$B4^{3}$, 361
7763	Hexane, 2-amino (dl) $CH_3(CH_2)_3CHNH_2CH_3$	101.19	....	117-8	−19	$0.7534^{20/0}$	1.4080^{25}	al, eth	$B4^{4}$, 721
7764	Hexane, 2-amino-4-methyl $CH_3CH_2CH(CH_3)CH_2CH(NH_2)CH_3$	115.22	....	130-5	......	0.7655^{20}	1.4150^{25}	al, eth, chl	$B4^{4}$, 747
7765	Hexane, 1-bromo or *n*-Hexyl bromide $CH_3(CH_2)_4CH_2Br$	165.07	....	155.3, 41^{10}	−84.7	$1.1744^{20/4}$	1.4478^{20}	**al, eth**, ace, chl	$B1^{4}$, 352
7766	Hexane, 1-bromo-6-fluoro $F(CH_2)_6Br$	183.06	....	$67\text{-}8^{11}$	......	$1.293^{20/4}$	1.4435^{25}	al, eth, ace, chl	$B1^{4}$, 353
7767	Hexane, 2-bromo or *sec*-Hexyl bromide $CH_3(CH_2)_3CHBrCH_3$	165.07	....	144^{749}, 78^{90}	......	$1.1658^{20/4}$	1.4832^{25}	al, eth, ace, chl	$B1^{4}$, 353
7768	Hexane, 3-bromo $C_3H_7CHBrC_2H_5$	165.07	....	141-3	......	$1.1799^{20/4}$	1.4472^{20}	al, eth, ace, chl	$B1^{4}$, 353
7769	Hexane, 1-chloro or *n*-Hexyl chloride $CH_3(CH_2)_4CH_2Cl$	120.62	....	134.5	−94	$0.8785^{20/4}$	1.4199^{20}	al, eth, ace, bz, chl	$B1^{4}$, 349
7770	Hexane, 1-chloro-3-ethyl (d) $C_3H_7CH(C_2H_5)CH_2CH_2Cl$	148.68	$[\alpha]^{27}{}_D$ + 1.15	85^{40}	......	$0.879^{21/4}$	1.4335^{25}	eth	$B1^{4}$, 432
7771	Hexane, 1-chloro-6-fluoro $F(CH_2)_6Cl$	138.61	....	167^{740}, 62^{15}	......	$1.015^{20/4}$	1.4168^{25}	al, eth, chl	$B1^{4}$, 350
7772	Hexane, 1-chloro-3-methyl $C_3H_7CH(CH_3)CH_2CH_2Cl$	134.65	....	$150\text{-}2^{758}$	......	$0.8766^{20/4}$	1.4274^{20}	al, eth, chl	$B1^{2}$, 119
7773	Hexane, 2-chloro or *sec*-Hexyl chloride $CH_3(CH_2)_3CHClCH_3$	120.62	....	122-3, 61^{100}	......	$0.8694^{21/4}$	1.4142^{22}	al, eth, ace, bz; chl	$B1^{4}$, 349
7774	Hexane, 2-chloro-2,5-dimethyl $(CH_3)_2CHCH_2CH_2CCl(CH_3)_2$	148.68	....	86^{100}	......	$0.8476^{18/4}$	1.4232^{20}	al, eth, ace, bz	$B1^{4}$, 434
7775	Hexane, 2-chloro-2-methyl $C_4H_9CCl(CH_3)_2$	134.65	....	135d, 59.5^{52}	......	$0.8635^{20/4}$	1.4200^{20}	al, eth, chl	$B1^{4}$, 398
7776	Hexane-2-chloro-5-methyl $(CH_3)_2CHCH_2CH_2CHClCH_3$	134.65	....	138^{735}d	......	$0.863^{20/4}$	....	al, eth, chl	$B1^{3}$, 436
7777	Hexane-3-chloro $C_3H_7CHClC_2H_5$	120.62	....	123, 60^{95}	......	$0.8700^{20/20}$	1.4163^{20}	al, eth, ace, bz, chl	$B1^{4}$, 349
7778	Hexane-3-chloro-2,3-dimethyl $C_3H_7CCl(CH_3)CH(CH_3)_2$	148.68	....	$41\text{-}3^{12}$	......	$0.8869^{20/4}$	1.4333^{25}	al	$B1^{3}$, 481
7779	Hexane-3-chloro-3-ethyl $C_3H_7CCl(C_2H_5)_2$	148.68	....	155d, $62\text{-}3^{24}$	......	0.9018	1.4358^{20}	eth, chl	$B1^{3}$, 479
7780	Hexane-3-chloro-3-methyl $C_3H_7CCl(CH_3)C_2H_5$	134.65	....	135	......	$0.8787^{20/4}$	1.4250^{20}	al, eth, chl	$B1^{3}$, 438
7781	Hexane-3-chloro-2,2,3-trimethyl $C_3H_7CCl(CH_3)C(CH_3)_3$	162.70	....	$64\text{-}5^{13}$	......	$0.9010^{20/4}$	1.4465^{20}	al, eth, chl	$B1^{3}$, 515
7782	Hexane-1,6-diamino or Hexamethylene diamine $H_2N(CH_2)_6NH_2$	116.21	rh bipym pl	204-5, 100^{20}	41-2	....	....	w, al, bz	$B4^{4}$, 1320
7783	Hexane-1,6-diamino, dihydrochloride $H_2N(CH_2)_6NH_2.2HCl$	189.13	nd(al-eth)	....	248-50	....	....	w	$B4^{4}$, 1320
7784	Hexane-2,5-diamino $CH_3CH(NH_2)CH_2CH_2CH(NH_2)CH_3$	116.21	....	175	......	....	....	w, al, eth	B4, 269
7785	Hexane-1,2-dibromo $C_4H_9CHBrCH_2Br$	243.97	....	$103\text{-}5^{36}$	......	$1.5774^{20/4}$	1.5024^{20}	al, **eth**, bz, chl	$B1^{3}$, 392
7786	Hexane-1,6-dibromo $Br(CH_2)_6Br$	243.97	....	245-6, 110^{12}	−2.3	1.5948^{15}	1.5037^{20}	eth, ace	$B1^{4}$, 354
7787	Hexane-2,5-dibromo (dl) $CH_3CHBrCH_2CH_2CHBrCH_3$	243.97	....	$108\text{-}9^{30}$	−44.64	1.5788^{20}	1.5007^{20}	eth, ace, chl	$B1^{4}$, 354
7788	Hexane-1,2-dichloro $C_4H_9CHClCH_2Cl$	155.07	....	172-4, $73\text{-}4^{0.03}$	......	1.085^{15}	....	eth, chl	$B1^{4}$, 350
7789	Hexane-1,6-dichloro $Cl(CH_2)_6Cl$	155.07	....	203-5, 94^{22}	......	$1.0677^{20/4}$	1.4572^{20}	eth, chl	$B1^{4}$, 350

No.	Name, Synonyms, and Formula	Mol. wt.	Color, crystalline form, specific rotation and λ_{max} (log ε)	b.p. °C	m.p. °C	Density	n_D	Solubility	Ref.
7790	Hexane-2,2-dichloro $C_4H_9CCl_2CH_3$	155.07	...	68^{49}	...	$1.0150^{25/4}$	1.4353^{25}	eth, chl	B1[3], 390
7791	Hexane-2,3-dichloro $C_3H_7CHClCHClCH_3$	155.07	...	162-5	...	1.0527^{11}	...	eth, chl	B1[4], 350
7792	Hexane-2,5-dichloro (dl) $CH_3CHClCH_2CH_2CHClCH_2$	155.07	...	177^{751}, 106^{91}	fp-38.4	1.0474^{20}	1.4491^{20}	chl	B1[4], 350
7793	Hexane-2,5-dichloro *(meso)* $CH_3CHClCH_2CH_2CHClCH_3$	155.07	...	178^{752}, 109^{99}	19.9	$1.0474^{20/4}$	1.4484^{20}	chl	B1[4], 350
7794	Hexane-2,5-dichloro-2,5-dimethyl $(CH_3)_2CClCH_2CH_2CCl(CH_3)_2$	183.12	lf, nd	...	67-8	0.9543^{70}	...	al, eth, bz, chl	B1[4], 435
7795	Hexane-3,4-dichloro (dl) $CH_3CH_2CHClCHClCH_2CH_3$	155.07	...	167.7, 62^{20}	...	1.0617^{20}	1.4541^{20}	ace, chl	B1[4], 351
7796	Hexane-3,4-dichloro-3,4-dimethyl $C_2H_5CCl(CH_3)CCl(CH_3)C_2H_5$	183.12	...	165, $114\text{-}5^{18}$	...	...	...	chl	B1[4], 436
7797	Hexane-1,6-diido $I(CH_2)_6I$	337.97	nd	$141\text{-}2^{10}$	10	$2.03^{22/4}$	1.585^{20}	al, eth	B1[4], 356
7798	Hexane-2,2-dimethyl $C_4H_9C(CH_3)_3$	114.23	...	106.8	−121	$0.6953^{20/4}$	1.3935^{20}	al, **eth, ace, bz, chl**	B1[4], 432
7799	3-Hexane-2,3-dimethyl-(dl) $C_3H_7CH(CH_3)CH(CH_3)_2$	114.23	...	115.6, 9.9^{0}	...	$0.7121^{20/4}$	1.4011^{20}	**al**, eth, **ace, bz, chl, lig**	B1[4], 432
7800	Hexane-2,3-dimethyl (l) $C_3H_7CH(CH_3)CH(CH_3)_2$	114.23	$[\alpha]^{25/}{}_D$-0.92	113	...	...	...	**al**, eth, **ace, bz, lig, chl**	B1[3], 482
7801	Hexane, 2,4-dimethyl (d) $C_2H_5CH(CH_3)CH_2CH(CH_3)_2$	114.23	$[\alpha]^{30/}{}_D$ + 2.99	111	...	$0.696^{20/4}$	1.3810^{20}	**al**, eth, **ace, bz, chl, lig**	B1[3], 483
7802	Hexane, 2,4-dimethyl (dl) $C_2H_5CH(CH_3)CH(CH_3)_2$	114.23	...	109.4, 5.2^{10}	...	$0.7004^{20/4}$	1.3953^{20}	**al**, eth, **ace, bz, lig, chl**	B1[4], 433
7803	Hexane, 2,4-dimethyl (l) $C_2H_5CH(CH_3)CH(CH_3)_2$	114.23	$[\alpha]^{21/}{}_D$ -10.85	110	...	$0.703^{21/4}$	...	**al**, eth, **ace, bz, chl, lig**	B1[3], 483
7804	Hexane, 2,5-dimethyl $(CH_3)_2CHCH_2CH_2CH(CH_3)_2$	114.23	...	109, 5.3^{10}	−91.2	$0.6935^{20/4}$	1.3925^{20}	**al**, eth, **ace, bz, lig, chl**	B1[4], 434
7805	Hexane, 3,3-dimethyl $C_3H_7C(CH_3)_2C_2H_5$	114.23	...	112, 6.1^{10}	−126.1	$0.7100^{20/4}$	1.4001^{20}	al, eth, ace, bz	B1[4], 435
7806	Hexane, 3,4-dimethyl $C_2H_5CH(CH_3)CH(CH_3)C_2H_5$	114.23	...	117.7, 11.3^{10}	...	$0.7200^{20/4}$	1.4046^{20}	**al**, eth, **ace, bz, chl, lig**	B1[4], 436
7807	Hexane, 1,6-dinitro $O_2N(CH_2)_6NO_2$	176.17	cr(MeOH)	$100\text{-}3^{0.3}$	37.5	...	...	aa	B1[4], 357
7808	Hexane, 3-ethyl $C_3H_7CH(C_2H_5)_2$	114.23	...	118.5, 12.8^{10}	...	$0.7136^{20/4}$	1.4018^{20}	**al, eth, ace, bz, lig**	B1[4], 431
7809	Hexane, 3-ethyl-2-methyl $(CH_3)_2CHCH(C_2H_5)CH_2CH_2CH_3$	128.26	...	138	...	...	$1.4106^{20/}{}_D$	...	B1[4], 459
7810	Hexane, 3-ethyl-3-methyl $(CH_3CH_2)_2C(CH_3)C_3H_7$	128.26	...	140.6	...	...	$1.4140^{20/}{}_D$	...	B1[4], 459
7811	Hexane, 3-ethyl-4-methyl $(CH_3CH_2)_2CHCH(CH_3)C_2H_5$	128.26	...	140.4	...	...	$1.4134^{20/}{}_D$	...	B1[4], 459
7812	Hexane, 4-ethyl-2-methyl $(C_2H_5)_2CHCH_2CH(CH_3)_2$	128.26	...	133.8	...	...	$1.4063^{20/}{}_D$	...	B1[4], 459
7813	Hexane, 1-fluoro or *n*-Hexyl fluoride $CH_3(CH_2)_4CH_2F$	104.17	...	91.5	−103	0.7995^{20}	1.3738^{20}	eth, bz	B1[4], 348
7814	Hexane, 3,4-bis(4-hydroxy-3-methylphenyl) or Dimethylexesterol $[4\text{-}HO\text{-}3\text{-}CH_3C_6H_3CH(C_2H_5)_2\text{-}]_2$	298.43	cr(dil aa)	...	145	...	...	...	B6[3], 5541
7815	Hexane, 2,4-bis(4-hydroxyphenyl)-3-ethyl or Benzestrol $4\text{-}HOC_6H_4CH(C_2H_5)CH(C_2H_5)CH(CH_3)(C_6H_4OH\text{-}4)$	298.43	cr(al)	...	162-6	...	...	al, eth, ace, aa	B6[3], 5538
7816	Hexane, 1-iodo or *n*-Hexyl iodide $CH_3(CH_2)_4CH_2I$	212.07	...	181.3, 58.2^{10}	−75	$1.4397^{20/4}$	1.4929^{20}	...	B1[4], 355
7817	Hexane, 2-iodo (l) $CH_3CHIC_4H_9$	212.07	$[\alpha]^{17/}{}_D$ -38.35	$90\text{-}1^{70}$. 45^{9}	...	$1.4354^{17/4}$	1.4878^{25}	ace, chl	B1[3], 395
7818	Hexane, 2-methyl or Isoheptane $CH_3(CH_2)_3CH(CH_3)_2$	100.20	...	90	−118.3	$0.6787^{20/4}$	1.3848^{20}	al, **eth, ace, bz, lig, chl**	B1[4], 397
7819	Hexane, 3-methyl (d) $C_3H_7CH(CH_3)C_2H_5$	100.20	$[\alpha]^{20/}{}_D + 9.5$	92	−119	$0.6860^{20/4}$	1.3887^{20}	al, **eth, ace, bz, chl, lig**	B1[4], 400
7820	Hexane, 3-methyl (dl) $C_3H_7CH(CH_3)C_2H_5$	100.20	...	92	ca−173	0.6872^{20}	1.3885^{20}	**al, eth, ace, bz, lig, chl**	B1[4], 400
7821	Hexane, 3-methyl (l) $C_3H_7CH(CH_3)C_2H_5$	100.20	$[\alpha]^{21/}{}_D$ -7.75	92	...	$0.687^{21/4}$	1.3854^{25}	al, **eth, ace, bz, chl, lig**	B1[3], 440
7822	Hexane, 1-nitro $CH_3(CH_2)_4CH_2NO_2$	131.17	...	$193\text{-}4^{765}$, 84^{21}	...	$0.9396^{20/4}$	1.4270^{20}	al, eth, ace, bz	B1[4], 356
7823	Hexane, 1,1,1,2,2,pentachloro $CH_3(CH_2)_3CCl_2CCl_3$	258.40	...	$129\text{-}31^{10}$	...	1.370^{25}	1.4872^{25}	eth, chl	B1[3], 390

No.	Name, Synonyms, and Formula	Mol. wt.	Color, crystalline form, specific rotation and λ_{max} (log ε)	b.p. °C	m.p. °C	Density	n_D	Solubility	Ref.
7824	Hexane, perfluoro C_6F_{14}	338.04	....	57.11	−87.1	$1.6995^{20/4}$	1.2515^{20}	eth, bz, chl	B1^4, 348
7825	Hexane, 1-phenyl or Hexylbenzene $CH_3(CH_2)_4CH_2C_6H_5$	162.27	....	227	−62	0.8613^{20}	1.4900^{20}	**eth**, bz, peth	B5^4, 1115
7826	Hexane, 2-phenyl $CH_3(CH_2)_3CH(C_6H_5)CH_3$	162.27	....	208	......	$0.869^{15/4}$	1.492^{15}	....	B5^4, 1116
7827	Hexane, 3-phenyl $CH_3CH_2CH(C_6H_5)C_3H_7$	162.27	....	209-12	......	$0.8254^{25/20}$	$1.4859^{20/}{}_D$	....	B5^4, 1117
7828	Hexane, 1,1,2,2,-tetrachloro $CH_3(CH_2)_3CCl_2CHCl_2$	223.96	....	$99\text{-}101^{14}$	......	$1.3096^{25/4}$	1.488^{25}	ace, bz, chl	B1^3, 390
7829	Hexane, 2,2,3-trimethyl $C_3H_7CH(CH_3)C(CH_3)_3$	128.26	....	131.7	......	....	1.4100^{20}	....	B1^4, 459
7830	Hexane, 2,2,4-trimethyl $C_2H_5CH(CH_3)CH_2C(CH_3)_3$	128.26	....	126.5	120	$0.711^{20/4}$	$1.40328^{20/}{}_D$	....	B1^4, 459
7831	Hexane, 2,2-5-trimethyl $(CH_3)_2CHCH_2CH_2C(CH_3)_3$	128.26	....	124, 16.2^{10}	−105.8	$0.7072^{20/4}$	1.3997^{20}	al, eth, ace, bz, lig	B1^4, 460
7832	Hexane, 2,3,3-trimethyl $C_3H_7C(CH_3)_2CH(CH_3)_2$	128.26	....	137.7	−116.8	....	$1.4141^{20/}{}_D$	....	B1^4, 461
7833	Hexane, 2,3,4-trimethyl $C_2H_5CH(CH_3)CH(CH_3)CH(CH_3)_2$	128.26	....	139	......	....	$1.4144^{20/}{}_D$	....	B1^4, 461
7834	Hexane, 2,3,5-trimethyl $CH_3CH(CH_3)CH(CH_3)CH_2CH(CH_3)_2$	128.26	....	131.3	−s127.9	$0.7818^{20/4}$	$1.4051^{20/}{}_D$	....	B1^4, 461
7835	Hexane, 2,4,4-trimethyl $C_2H_5C(CH_3)_2CH_2CH(CH_3)_2$	128.26	....	126.5	−123.4	$0.711^{20/4}$	$1.40328^{20/}{}_D$	....	B1^4, 461
7836	Hexane, 3,3,-4-trimethyl $CH_3CH_2C(CH_3)_2CH(CH_3)_2CH_2CH_3$	128.26	....	140.5	−101.2	....	$1.4178^{20/}{}_D$	....	B1^4, 462
7837	1,6-Hexanedial or Adipaldehyde $OHC(CH_2)_4CHO$	114.14	....	$92\text{-}4^{9}$	ms;8	$1.003^{19/4}$	1.4350^{20}	al, eth, bz, aa	B1^4, 3686
7838	Hexanedioic acid or Adipic acid $HOOC(CH_2)_4COOH$	146.14	mcl pr(w, ace-lig)	265^{100}, 205^{10}	153	$1.360^{25/4}$	....	al, eth	B2^4, 1956
7839	1,3-Hexanediol, 2-ethyl $C_3H_7CH(OH)CH(C_2H_5)CH_2OH$	146.23	....	244	−40	$0.9325^{22/4}$	1.4497^{20}	al, eth	B1^4, 2597
7840	1,6-Hexanediol or Hexamethylene glycol $HO(CH_2)_6OH$	118.18	nd(w)	250, 132^{9}	43(59)	....	....	w, al, ace	B1^4, 2556
7841	2,3-Hexanediol $C_3H_7CH(OH)CH(OH)CH_3$	118.18	cr	204-6, $102^{0.8}$	......	0.9900^{15}	1.4510^{15}	w, al, eth	B1^4, 2561
7842	2,5-Hexanediol $CH_3CH(OH)CH_2CH_2CH(OH)CH_3$	118.18	cr(eth)	$216\text{-}8^{750}$, $85\text{-}7^{1}$	43	$0\ 9610^{20/4}$	1.4475^{20}	w, al, eth	B1^4, 2562
7843	3,4-Hexanediol, 3,4-diethyl $(C_2H_5)_2C(OH)C(OH)(C_2H_5)_2$	174.28	cr(eth)	230, 112^{10}	28	$0.9630^{13/25}$	1.467^{13}	al, eth	B1^4, 2621
7844	2,5-Hexanediol, 2,5-dimethyl $(CH_3)_2C(OH)CH_2CH_2C(OH)(CH_3)_2$	146.23	pr (AcOEt) fl (peth)	214, 118^{15}	92	0.898^{20}	....	w, al, bz, chl	B1^4, 2600
7845	2,3-Hexandione, 3-oxime $C_3H_7C(=NOH)COCH_3$	129.16	cr(al)	....	60	....	....	al	B1^3, 3127
7846	2,5-Hexanedione or Acetonyl acetone $CH_3COCH_2CH_2COCH_3$	114.14	....	194^{754}, 89^{25}	−5.5	$0.9737^{20/4}$	1.4421^{20}	**w, al, eth**, ace, bz	B1^4, 3688
7847	2,5-Hexanedione, dioxime $CH_3C(=NOH)CH_2CH_2C(=NOH)CH_3$	144.17	pl(bz)	....	137	....	....	al, eth	B1^3, 3130
7848	2,5-Hexanedione, 3-hydroxy $CH_3COCH_2CH(OH)COCH_3$	130.14	....	$62\text{-}7^{0.5}$	......	....	1.4497^{25}	eth, bz	B1^3, 3317
7849	3,4-Hexanedione, 2,2,5,5-tetramethyl or Di-*tert*-butyl glyoxal $(CH_3)_3CCOCOC(CH_3)_3$	170.25	....	168^{745}	−2	0.8776^{20}	1.4157^{20}	**eth**	B1^4, 3726
7850	1,2,3,4,5,6-Hexanehexol or Dulcitol $HOCH_2(CHOH)_4CH_2OH$	182.17	mcl pr	$257\text{-}80^{1}$	189	1.466^{15}	....	w	B1^4, 2844
7851	1,3,4,5-Hexanetetrol or Digitoxit $CH_3(CHOH)_3CH_2CH_2OH$	150.17	pr, $[\alpha]^{15/}{}_D$ -86.2	....	88	....	....	al	B1^2, 603
7852	1-Hexanethiol or Hexyl mercapton $CH_3(CH_2)_4CH_2SH$	118.24	....	151	−81	$0.8424^{20/4}$	1.4496^{20}	al, eth	B1^4, 1705
7853	2-Hexanethiol or *sec*-Hexyl mercaptan $CH_3(CH_2)_3CH(SH)CH_3$	118.24	....	142, $60\text{-}6^{50}$	−147	$0.8345^{20/4}$	1.4451^{20}	al, eth, bz	B1^4, 1711
7854	3-Hexanethiol $C_3H_7CH(SH)C_2H_5$	118.24	....	57^{25}	......	0.9206^{20}	1.4496^{20}	eth	B1^4, 1713
7855	1,2,3-Hexanetriol (threo) $C_3H_7CH(OH)CH(OH)CH_2OH$	134.18	....	$130^{0.05}$	64-5	1.089^{26}	1.472^{26}	w, al	B1^4, 2784
7856	1,2,4-Hexanetriol $C_2H_5CH(OH)CH_2CH(OH)CH_2OH$	134.18	....	$190\text{-}2^{30}$	......	....	....	w, al	B1, 521

No.	Name, Synonyms, and Formula	Mol. wt.	Color, crystalline form, specific rotation and λ_{max} (log ε)	b.p. °C	m.p. °C	Density	n_D	Solubility	Ref.
7857	1,2,5-Hexanetriol $CH_3CH(OH)CH_2CH_2CH(OH)CH_2OH$	134.18		181^{10}		$1.1012^{20/4}$		w, al	$B1^4$, 2784
7858	2,3,4-Hexanetriol $CH_3CH_2(CHOH)_3CH_3$	134.18		256-7, $155\text{-}6^{20}$				w, al	$B1^3$, 2349
7859	Hexanoic acid or Caproic acid $CH_3(CH_2)_4CO_2H$	116.16		205	−2	$0.9274^{20/4}$	1.4163^{20}	al, eth	$B2^4$, 917
7860	Hexanoic acid, 2-acetyl, ethyl ester $C_4H_9CH(COCH_3)CO_2C_2H_5$	186.25		219-24, 104^{12}		$0.9523^{20/4}$	1.4301^{20}	eth, ace	$B3^4$, 1616
7861	Hexanoic acid, 2,4-dioxo or Propionyl pyruvic acid $CH_3CH_2COCH_2COCO_2H$	144.13	cr(al w + 1)		83			w, al, eth	$B3^3$, 1334
7862	Hexanoic acid, 2,4-dioxo, ethyl ester $CH_3CH_2COCH_2COCO_2C_2H_5$	172.18		163-5, $108\text{-}11^{11}$				al, eth, ace	$B3^4$, 1779
7863	Hexanoic acid, 2-ethyl $C_4H_9CH(C_2H_5)CO_2H$	144.21		228^{755}, 120^{13}		$0.9031^{25/4}$	1.4241^{20}	eth	$B2^4$, 1003
7864	Hexanoic acid, 2-methyl (dl) or 2-Methyl caproic acid $C_4H_9CH(CH_3)CO_2H$	130.19		215-6, 100^{12}		$0.9612^{20/4}$	1.4193^{20}	al, eth, ace, bz, chl	B24, 969
7865	Hexanoic acid, 4-methyl (dl) or 4-Methyl caproic acid $C_2H_5CH(CH_3)(CH_2)_2CO_2H$	130.19		217-8, 85^2	−80	$0.9215^{20/4}$	1.4211^{20}	al, eth, ace, bz	$B2^4$, 974
7866	Hexanoic acid, 5-methyl or 5-Methyl caproic acid $(CH_3)_2CH(CH_2)_3CO_2H$	130.19		216	<−25	$0.9138^{21/4}$	1.4220^{20}	al, eth, ace, bz	$B2^4$, 970
7867	Hexanoic acid, 5-oxo $CH_3CO(CH_2)_3CO_2H$	130.14		155^{12}	13-14		1.4445^{20}	w, al, eth	$B3^4$, 1583
7868	1-Hexanol or n-Hexyl alcohol $CH_3(CH_2)_4CH_2OH$	102.18		158	−46.7	$0.8136^{20/4}$	1.4178^{20}	al, eth, ace, bz, chl	$B1^4$, 1694
7869	1-Hexanol, 6-chloro $Cl(CH_2)_6OH$	136.62		107^{12}		$1.0241^{20/4}$	1.4550^{20}	al, eth	$B1^4$, 1704
7870	1-Hexanol, 3,5-dimethoxy $CH_3CH(OCH_3)CH_2CH(OCH_3)CH_2CH_2OH$	162.23		114^{13}		0.9631^{25}	1.4329^{25}		$B1^4$, 2785
7871	3-Hexanol, 2-ethyl $CH_3(CH_2)_3CH(C_2H_5)CH_2OH$	130.23		185, $84\text{-}6^{15}$	<−76	$0.8328^{20/4}$	1.4328^{20}	al, eth, ace, bz	$B1^4$, 1783
7872	1-Hexanol, 6-fluoro $F(CH_2)_6OH$	120.17		$85\text{-}6^{14}$		$0.975^{20/4}$	1.4141^{27}	al, eth	$B1^4$, 1703
7873	1-Hexanol, 2-methyl (d) $C_4H_9CH(CH_3)CH_2OH$	116.20	$[\alpha]^{25}/_D$ + 2.45(al)	164-5, $70\text{-}2^{15}$		$0.8313^{13/4}$	1.4245^{17}	al, eth	$B1^2$, 444
7874	1-Hexanol, 2-methyl (dl) $C_4H_9CH(CH_3)CH_2OH$	116.20		164^{750}		$0.8270^{20/4}$	1.4226^{20}	al, eth	$B1^4$, 1745
7875	1-Hexanol, 3-methyl (dl) $C_3H_7CH(CH_3)CH_2CH_2OH$	116.20		$168\text{-}9^{754}$, 91^{13}		0.8258^{20}	1.4245^{20}	al, eth, ace	$B1^4$, 1748
7876	1-Hexanol, 3-methyl (l) $C_3H_7CH(CH_3)CH_2CH_2OH$	116.20	$[\alpha]^{27}/_D$ -1.67 (chl)	$161\text{-}2^{740}$, 80^{25}		$0.8208^{25/4}$	1.4204^{30}	al, eth, ace	$B1^4$, 1748
7877	1-Hexanol, 4-methyl (d) $C_2H_5CH(CH_3)(CH_2)_3OH$	116.20	$[\alpha]^{28}/_D$ + 2.2	77^{20}		$0.809^{23/4}$	1.4233^{25}	al, eth, ace, bz	$B1^4$, 1749
7878	1-Hexanol, 4-methyl (dl) $C_2H_5CH(CH_3)(CH_2)_3OH$	116.20		173, 83^{24}		$0.8239^{20/4}$	1.4219^{20}	al, eth, ace, bz	$B1^4$, 1749
7879	1-Hexanol, 5-methyl $(CH_3)_2CH(CH_2)_3CH_2OH$	116.20		170^{755}, $53\text{-}5^{15}$		$0.8119^{20/4}$	1.4175^{20}	al, eth	$B1^4$, 1748
7880	1-Hexanol, 1-phenyl $C_5H_{11}CH(OH)C_6H_5$	178.27		170^{50}		$0.9477^{25/4}$	1.5105^{20}	al, eth	$B6^3$, 1994
7881	1-Hexanol, 2-isopropyl-5-methyl $(CH_3)_2CHCH_2CH(i\text{-}C_3H_7)CH_2OH$	158.28		211		$0.8345^{20/4}$	1.4369^{20}	al, eth	$B1^4$, 1833
7882	2-Hexanol (d) $CH_3(CH_2)_3CH(OH)CH_3$	102.18	$[\alpha]^{25}/_D$ + 14.1 (eth, c=1)	138		$0.8104^{25/4}$	1.4126^{25}	al, eth	$B1^3$, 1663
7883	2-Hexanol (dl) $CH_3(CH_2)_3CH(OH)CH_3$	102.18		140		$0.8159^{20/4}$	1.4144^{20}	al, eth	$B1^4$, 1708
7884	2-Hexanol (l) $C_4H_9CH(OH)CH_3$	102.18	$[\alpha]^{18}/_{5780}$ -12.04	$136\text{-}8^{754}$		$0.8178^{18/4}$		al, eth	$B1^3$, 1663
7885	2-Hexanol, 1-chloro $C_4H_9CH(OH)CH_2Cl$	136.62		$73\text{-}5^{12}$		$0.0139^{20/4}$	1.4478^{20}	al, eth, ace	$B1^4$, 1710
7886	2-Hexanol, 2-methyl $C_4H_9C(OH)(CH_3)_2$	116.20		143, $53\text{-}5^{15}$		$0.8119^{20/4}$	1.4175^{20}	al, eth	$B1^4$, 1745
7887	2-Hexanol, 3-methyl $C_3H_7CH(CH_3)CH(OH)CH_3$	116.20		$79\text{-}81^{52}$		$0.8820^{25/4}$	1.4198^{18}	al, eth, ace	$B1^3$, 1693
7888	2-Hexanol, 5-methyl $(CH_3)_2CHCH_2CH_2CH(OH)CH_3$	116.20		150^{744}, 78^{28}		$0.814^{20/4}$	1.4180^{20}	al, eth	$B1^4$, 1747
7889	2-Hexanol, 2,3,4-trimethyl $C_2H_5CH(CH_3)CH(CH_3)C(OH)(CH_3)_2$	144.26		57^5		$0.853^{15/4}$	1.4415^{15}	al, eth, ace, peth	$B1^3$, 1756

No.	Name, Synonyms, and Formula	Mol. wt.	Color, crystalline form, specific rotation and λ_{max} (log ε)	b.p. °C	m.p. °C	Density	n_D	Solubility	Ref.
7890	2-Hexanol, 2,3,5-trimethyl $(CH_3)_2CHCH_2CH(CH_3)C(OH)(CH_3)_2$	144.26		171^{755}, 72^{21}		$0.8271^{20/20}$	1.4321^{20}	al, eth, ace	$B1^1$, 212
7891	3-Hexanol (d) $C_3H_7CH(OH)C_2H_5$	102.18	$[\alpha]^{20/}{}_D$ + 6.8 (chl)	131-3		$0.8213^{20/4}$	1.4150^{20}	al, **eth**, ace	$B1^3$, 1665
7892	3-Hexanol (dl) $C_3H_7CH(OH)C_2H_5$	102.18		135		$0.8182^{20/4}$	1.4167^{20}	al, **eth**, ace	$B1^4$, 1711
7893	3-Hexanol (l) $C_3H_7CH(OH)C_2H_5$	102.18	$[\alpha]^{20/}{}_D$-7.17	135		$0.8213^{20/4}$	1.4140^{20}	al, **eth**, ace	$B1^3$, 1665
7894	3-Hexanol, 1-chloro $C_3H_7CH(OH)CH_2CH_2Cl$	136.62		120^{35}, 78^6		$1.003^{25/4}$	1.446^{25}	al, eth, ace, bz	$B1^4$, 1712
7895	3-Hexanol, 2-chloro $C_3H_7CH(OH)CHClCH_3$	136.62		171		1.0143^{11}		al, eth, ace, bz	$B1^2$, 438
7896	3-Hexanol, 5-chloro $CH_3CHClCH_2CH(OH)C_2H_5$	136.62		$78\text{-}9^{13}$		$1.0012^{15/4}$	1.4433^{19}	al, eth, ace, bz	$B1^4$, 1712
7897	3-Hexanol, 2,4-dimethyl $C_2H_5CH(CH_3)CH(OH)CH(CH_3)_2$	130.23		$61\text{-}2^{18}$		$0.8371^{20/4}$	$1.4309^{20/}{}_D$		$B1^4$, 1791
7898	3-Hexanol, 3-ethyl $C_3H_7C(OH)(C_2H_5)_2$	130.23		160		$0.8373^{20/4}$	1.4300^{20}	al, eth, ace, bz	$B1^4$, 1790
7899	3-Hexanol, 3-ethyl-5-methyl $(CH_3)_2CHCH_2C(OH)(C_2H_5)_2$	144.26		272		$0.8396^{22/4}$	1.4346^{13}	al, eth	$B1^3$, 1755
7900	3-Hexanol, 3-methyl $C_3H_7C(OH)(CH_3)C_2H_5$	116.20		143, 56^{18}		$0.8234^{20/0}$	1.4231^{20}	al, eth	$B1^4$, 1749
7901	3-Hexanol, 5-methyl (d) $(CH_3)_2CHCH_2CH(OH)C_2H_5$	116.20	$[\alpha]^{35/}{}_D$ + 21.2	81^{60}			1.4171^{25}		$B1^3$, 1692
7902	3-Hexanol, 5-methyl (dl) $(CH_3)_2CHCH_2CH(OH)C_2H_5$	116.20		$147\text{-}8^{756}$		0.827^0	1.4128^{20}	al, eth	$B1^4$, 1747
7903	3-Hexanol, 5-methyl (l) $(CH_3)_2CHCH_2CH(OH)C_2H_5$	116.20	$[\alpha]^{23/}{}_D$ -3.88	$93\text{-}6^{105}$, 63^{19}			1.4171^{25}		$B1^3$, 1692
7904	3-Hexanol, 2,2,5,5-tetramethyl $(CH_3)_3CCH_2CH(OH)C(CH_3)_3$	158.28	cr(peth)	166-70	52-3			al, eth, ace, peth	$B1^3$, 1772
7905	3-Hexanol, 2,2,3-trimethyl $C_3H_7C(OH)(CH_3)C(CH_3)_3$	144.26		170		$0.8474^{20/4}$	1.4402^{20}	al, eth, ace	$B1^3$, 1755
7906	3-Hexanol, 2,3,5-trimethyl $(CH_3)_2CHCH_2C(OH)(CH_3)CH(CH_3)_2$	144.26		72^{21}		$0.8271^{20/20}$	1.4321^{20}	al, eth, ace	$B1^4$, 1813
7907	3-Hexanol, 2,4,4-trimethyl $C_2H_5C(CH_3)_2CH(OH)CH(CH_3)_2$	144.26		170		0.8489^{20}	1.4395^{20}	al, eth, ace	$B1^3$, 1756
7908	3-Hexanol, 2,5,5-trimethyl $(CH_3)_3CCH_2CH(OH)CH(CH_3)_2$	144.26		77^{32}		0.8250^{20}	1.4286^{20}	al, eth, ace	$B1^4$, 1813
7909	3-Hexanol, 3,4,4-trimethyl $C_2H_5C(CH_3)_2C(OH)(CH_3)C_2H_5$	144.26		165-6		$0.8323^{21/0}$	1.4341^{21}	al, eth, ace	B1, 425
7910	3-Hexanol, 3,5,5-trimethyl $(CH_3)_3CCH_2C(OH)(CH_3)C_2H_5$	144.26		62^{14}		0.8350^{20}	1.4352^{20}	al, eth, ace	$B1^3$, 1755
7911	2-Hexanone or Methyl butyl ketone $C_4H_9COCH_3$	100.16		128	−57	$0.8113^{20/4}$	1.4007^{20}	**al, eth**, ace	$B1^4$, 3298
7912	2-Hexanone, 3,3-dimethyl $C_3H_7C(CH_3)_2COCH_3$	128.21		149^{765}		$0.838^{0/4}$	1.4098^{20}	al, eth, ace	$B1^4$, 3350
7913	2-Hexanone, 3,4-dimethyl $C_2H_5CH(CH_3)CH(CH_3)COCH_3$	128.21		158		$0.8295^{22/4}$	1.4193^{20}	al, eth, ace	$B1^3$, 2882
7914	2-Hexanone, 3-methyl $C_3H_7CH(CH_3)COCH_3$	114.19		142-5		0.828^{25}	1.4035^{20}	al, eth, ace, bz	$B1^3$, 2863
7915	2-Hexanone, 4-methyl $C_2H_5CH(CH_3)CH_2COCH_3$	114.19		142, $35\text{-}7^{11}$			1.4081^{24}	al, eth, ace, bz	$B1^4$, 3329
7916	2-Hexanone, 5-methyl $(CH_3)_2CH(CH_2)_2COCH_3$	114.19		144		$0.888^{20/4}$	1.4062^{20}	**al, eth**, ace, bz	$B1^4$, 3329
7917	2-Hexanone, 5-methyl, oxime $(CH_3)_2CH(CH_2)_2C(=NOH)CH_3$	129.20		195-6		$0.8881^{20/4}$	1.4448^{20}		B1, 701
7918	3-Hexanone or Ethyl propyl ketone $C_3H_7COC_2H_5$	100.16		125		$0.8118^{20/4}$	1.4004^{20}	al, eth, aee	$B1^4$, 3301
7919	3-Hexanone, 2,2-dimethyl or *tert*-Butyl propyl ketone $C_3H_7COC(CH_3)_3$	128.21		$145\text{-}8^{745}$		$0.8105^{25/4}$	1.4119^{20}	al, eth, ace	$B1^4$, 3347
7920	3-Hexanone, 2,2-dimethyl,oxime $C_3H_7C(=NOH)C(CH_3)_3$	143.23	nd(al)		78				$B1^3$, 2881
7921	3-Hexanone, 2,5-dimethyl $(CH_3)_2CHCH_2COCH(CH_3)_2$	128.21		147-8		$0.8270^{0/0}$	1.4049^{20}	al, eth, ace	$B1^4$, 3349
7922	3-Hexanone, 4,4-dimethyl $C_2H_5C(CH_3)_2COC_2H_5$	128.21		151		0.8285^{20}	1.4203^{25}	al, bz, chl	$B1^4$, 3350

PHYSICAL CONSTANTS OF ORGANIC COMPOUNDS (Continued)

No.	Name, Synonyms, and Formula	Mol. wt.	Color, crystalline form, specific rotation and λ_{max} (log ε)	b.p. °C	m.p. °C	Density	n_D	Solubility	Ref.
7923	3-Hexanone, 6-dimethylamino-4,4-diphenyl-5-methyl (l) or l-Isomethadone $(CH_3)_2NCH_2CH(CH_3)C(C_6H_5)_2COC_2H_5$	309.46	...	162-5$^{0.5}$	...	...	...	al	B14^{3}, 287
7924	3-Hexanone, 4-hydroxy or Propioin $CH_3CH_2CH(OH)COC_2H_5$	116.16	...	132-5^{227}, 73^{30}	...	0.956$^{21/4}$	1.4340^{21}	al, ace	B1^{4}, 4021
7925	3-Hexanone, 4-hydroxy-2,2,5,5-tetramethyl $(CH_3)_3CCH(OH)COC(CH_3)_3$	172.27	...	80^{10}	81(sub)	...	...	eth	B1^{4}, 4060
7926	3-Hexanone, 2-methyl or Propyl isopropyl ketone $C_3H_7COCH(CH_3)_2$	114.19	...	134-6	...	0.8091^{20}	1.4042^{20}	al, eth, ace, chl	B1^{4}, 3328
7927	3-Hexanone, 4-methyl $C_2H_5CH(CH_3)COC_2H_5$	114.19	...	134-5	...	0.8162^{20}	1.4069^{20}	al, eth, ace, bz	B1^{4}, 3329
7928	3-Hexanone, 5-methyl $(CH_3)_2CHCH_2COC_2H_5$	114.19	...	134^{735}	...	0.8090^{20}	1.4047^{20}	**al, eth**	B1^{4}, 3328
7929	Hexaphenyl ethane $(C_6H_5)_3CC(C_6H_5)_3$	486.66	cr(ace)	d	145-7d	...	...	eth, ace, chl, MeOH	B5^{3}, 2746
7930	Hexasiloxane, tetradecamethyl $(CH_3)_3SiO[\text{-}Si(CH_3)_2O\text{-}]_4Si(CH_3)_3$	443.97	...	245.5, 142^{20}	−59	0.8910$^{20/4}$	1.3948^{20}	bz	B4^{3}, 1880
7931	1,2,3,5 Hexatetracene, 4-chloro $CH_2{=}CHCCl{=}C{=}C{=}CH_2$	112.56	...	127d, 55^{54}	...	0.9997$^{20/4}$	1.5280^{20}	eth, ace, chl	B1^{3}, 1061
7932	1,2,4,5-Hexatetracene, 3,4-dichloro $CH_2{=}C{=}CClCCl{=}C{=}CH_2$	147.00	...	38-40^{8}	...	1.1819$^{20/4}$	1.5456^{20}	eth, ace, chl	B1^{3}, 1061
7933	1,2,4-Hexatriene, 3,4,6-trichloro $ClCH_2CH{=}CClCCl{=}C{=}CH_2$	183.46	...	50^{1}	...	1.3132$^{20/4}$	1.5517^{20}	ace, bz, chl	B1^{3}, 1041
7934	1,3,4-Hexatriene, 3,6-dichloro $ClCH_2CH{=}C{=}CClCH{=}CH_2$	149.02	...	45-6^{3}	...	1.1807$^{20/4}$	1.5195^{20}	al	B1^{3}, 1041
7935	1,3,5-Hexatriene *(cis)* or Divinyl ethylene $CH_2{=}CHCH{=}CHCH{=}CH_2$	80.13	...	78	−12	0.7175$^{20/4}$	1.4577^{20}	al, ace, chl, peth	B1^{3}, 1041
7936	1,3,5-Hexatriene *(trans)* $CH_2{=}CHCH{=}CHCH{=}CH_2$	80.13	...	78.5	−12	0.7369$^{15/4}$	1.5135^{20}	al, ace, chl, peth	B1^{4}, 1093
7937	1,3,5-Hexatriene, 2,5-dimethyl $CH_2{=}C(CH_3)CH{=}CHC(CH_3){=}CH_2$	108.18	...	145^{747}	−9	0.7822$^{20/4}$	1.5122^{20}	ace, lig, MeOH	B1^{4}, 1102
7938	1,3,5-Hexatriene, 1,6-diphenyl $C_6H_5CH{=}CHCH{=}CHCH{=}CHC_6H_5$	232.33	lf(ace)	...	200-3	...	...	...	B5^{3}, 2243
7939	2-Hexenal *(trans)* $C_3H_7CH{=}CHCHO$	98.14	...	146-7, 43^{12}	...	0.8491$^{20/4}$	1.4480^{20}	...	B1^{4}, 3468
7940	3-Hexenal $CH_3CH_2CH{=}CHCH_2CHO$	98.14	...	42-3^{28}	...	0.8455$^{22/4}$	1.4275$^{21.5}$	eth, ace	B1^{4}, 3469
7941	1-Hexene $C_4H_9CH{=}CH_2$	84.16	...	63.3	−139.8	0.6731$^{20/4}$	1.3837^{20}	al, eth, bz, chl, peth	B1^{4}, 828
7942	1-Hexene, 5-amino-4-methyl $CH_3CH(NH_2)CH(CH_3)CH_2CH{=}CH_2$	113.20	...	133-6	...	0.793$^{15/0}$	...	w	B4, 226
7943	1-Hexene, 1-chloro $C_4H_9CH{=}CHCl$	118.61	...	121	...	0.8872^{22}	1.4300^{22}	eth, ace, bz, chl	B1^{4}, 831
7944	1-Hexene, 2-chloro $C_4H_9CCl{=}CH_2$	118.61	...	113, 63^{118}	...	0.8886$^{25/4}$	1.4278^{25}	ace, bz, chl	B1^{3}, 803
7945	1-Hexene, 5-chloro $CH_3CHCl(CH_2)_2CH{=}CH_2$	118.61	...	120.7, 28-30^{13}	...	0.8891$^{25/4}$	1.4305^{20}	eth, ace, bz, chl	B1^{4}, 832
7946	1-Hexene, 1,2-dichloro *(cis)* $C_4H_9CCl{=}CHCl$	153.05	...	88^{30}	...	1.0812$^{25/4}$	1.4631^{25}	bz, chl	B1^{3}, 803
7947	1-Hexene, 1,2-dichloro *(trans)* $C_4H_9CCl{=}CHCl$	153.05	...	63-5^{22}	...	1.1167$^{25/4}$	1.4576^{25}	bz, chl	B1^{3}, 803
7948	1-Hexene, 2-ethyl $C_4H_9C(C_2H_5){=}CH_2$	112.22	...	120	...	0.7270$^{20/4}$	1.4157^{20}	eth, bz, peth	B1^{4}, 884
7949	1-Hexene, 2-methyl $C_4H_9C(CH_3){=}CH_2$	98.19	...	91.1	...	0.7000$^{20/4}$	1.4040$^{20/}{}_{D}$	...	B1^{4}, 863
7950	1-Hexene, 3-methyl $C_3H_7CH(CH_3)CH{=}CH_2$	98.19	...	84	...	0.6945$^{20/4}$	1.3970^{20}	...	B1^{4}, 865
7951	1-Hexene, 4-methyl $C_2H_5CH(CH_3)CH_2CH{=}CH_2$	98.19	...	87.5	...	1.6969$^{20/4}$	1.3985$^{20/}{}_{D}$	...	B1^{4}, 867
7952	1-Hexene, perfluoro $C_4F_9CF{=}CF_2$	300.05	...	57	...	...	...	chl	B1^{4}, 831
7953	1-Hexene, 1,1,2-trichloro $C_4H_9CCl{=}CCl_2$	187.50	...	90-93^{10}	...	1.125^{25}	1.4760^{25}	eth	B1^{3}, 803
7954	2-Hexene *(cis)* $C_3H_7CH{=}CHCH_3$	84.16	...	68.8	−141.3	0.6869$^{20/4}$	1.3977^{20}	al, eth, bz, lig, chl	B1^{4}, 833
7955	2-Hexene *(trans)* $C_3H_7CH{=}CHCH_3$	84.16	...	68^{750}	−133	0.6784$^{20/4}$	1.3935^{20}	al, eth, bz, lig, chl	B1^{4}, 834

PHYSICAL CONSTANTS OF ORGANIC COMPOUNDS (Continued)

No.	Name, Synonyms, and Formula	Mol. wt.	Color, crystalline form, specific rotation and λ_{max} (log ε)	b.p. °C	m.p. °C	Density	n_D	Solubility	Ref.
7956	2-Hexene, 4-chloro $C_2H_5CHClCH=CHCH_3$	118.61	....	123, 30^{10}	......	$0.8934^{20/4}$	1.4400^{20}	eth, ace, bz, chl	B1[4], 835
7957	2-Hexene, 2,3-dimethyl $C_3H_7C(CH_3)=C(CH_3)CH_3$	112.22	....	122.1	......	$0.7405^{20/4}$	$1.4269^{20/}{}_D$	....	B1[4], 887
7958	2-Hexene, 2,5-dimethyl $(CH_3)_2CHCH_2CH=C(CH_3)_2$	112.22	....	112.6	......	$0.7182^{20/4}$	$1.4135^{20/}{}_D$	....	B1[4], 888
7959	2-Hexene, 2-methyl $C_3H_7CH=C(CH_3)_2$	98.19	....	95-8	......	....	$1.4040^{20.5/}{}_D$	....	B1[4], 863
7960	3-Hexene *(cis)* $C_2H_5CH=CHC_2H_5$	84.16	....	66.4	−137.8	$0.6796^{20/4}$	1.3947^{20}	al, eth, bz, lig, chl	B1[4], 837
7961	3-Hexene *(trans)* $C_2H_5CH=CHC_2H_5$	84.16	....	67.1	−113.4	$0.6772^{20/4}$	1.3943^{20}	al, eth, bz, lig, chl	B1[4], 237
7962	3-Hexene, 1-chloro $C_2H_5CH=CH(CH_2)_2Cl$	118.61	....	61^{60}	......	$0.900^{24/4}$	1.435^{24}	eth, ace, bz, chl	B1[4], 838
7963	3-Hexene, 1-chloro-4-ethyl $(CH_3CH_2)_2C=CHCH_2CH_2Cl$	146.66	....	173	......	$0.9102^{20/4}$	1.4524^{20}	bz, chl	B1[2], 201
7964	3-Hexene, 2-chloro,2,5-dimethyl $(CH_3)_2CHCH=CHCCl(CH_3)_2$	146.66	....	45-60^{15}	......	....	1.450^{20}	bz, chl	Am63, 3474
7965	3-Hexene, 3-chloro *(cis)* $C_2H_5CH=CClC_2H_5$	118.61	....	119.6	......	$0.9009^{20/4}$	1.4360^{20}	eth, ace, bz, chl	B1[4], 838
7966	3-Hexene, 1,2,3,4,5,6-hexachloro $ClCH_2CHClCCl=CClCHClCH_2Cl$	290.83	cr (peth)	110-12^{2}	58-9	....	....	chl, MeOH	B1[4], 839
7967	3-Hexene, 3-(4-hydroxyphenyl)-4-(4-methoxyphenyl) $CH_3CH_2C(4\text{-}CH_3OC_6H_4)=C(4\text{-}HOC_6H_4)CH_2CH_3$	282.38	nd (bz-lig) lf (70% al)	185-$95^{0.3}$	117-8	....	....	al, eth, ace	B6[3], 5623
7968	3-Hexene, perfluoro $C_2F_5CF=CFC_2F_5$	300.05	....	49	......	....	....	chl	B1[4], 838
7969	2-Hexenoic acid, 2-methyl *(trans)* $C_3H_7CH=C(CH_3)CO_2H$	128.17	....	204-6, 118^{11}	......	$0.9627^{20/4}$	1.4601^{20}	eth, ace	B2[4], 1581
7970	2-Hexenoic acid, 3-phenyl $C_3H_7C(C_6H_5)=CHCO_2H$	190.24	cr (peth)	183-4^{14}	94	....	....	bz	B9[3], 2811
7971	3-Hexenoic acid or Hydrosorbic acid $C_2H_5CH=CHCH_2CO_2H$	114.14	....	208, 81-2^{2}	12	$0.9640^{23/4}$	1.4935^{20}	....	B2[4], 1566
7972	5-Hexenoic acid $CH_2=CH(CH_2)_3CO_2H$	114.14	....	203, 107^{17}	−37	$0.9610^{20/4}$	1.4343^{20}	al, eth	B2[4], 1562
7973	1-Hexene-3-ol $C_3H_7CH(OH)CH=CH_2$	100.16	....	134, 50^{20}	......	$0.834^{22/4}$	1.4297^{18}	al, eth, ace	B1[4], 2136
7974	2-Hexene-1-ol *(cis)* $C_3H_7CH=CHCH_2OH$	100.16	....	156-8, 58-60^{15}	......	$0.8472^{20/4}$	1.4397^{20}	al, eth, ace	B1[4], 2138
7975	3-Hexene-1-ol *(cis)* $C_2H_5CH=CHCH_2CH_2OH$	100.16	....	156-7, 58^{12}	......	$0.8478^{22/4}$	1.4380^{20}	al, eth	B1[4], 2139
7976	3-Hexene-2-one $C_2H_5CH=CHCOCH_3$	98.14	....	140, 36-8^{18}	......	$0.86554^{20/4}$	1.4418^{20}	al, eth, ace	B1[4], 3468
7977	5-Hexene-2-one, 5-methyl $CH_2=C(CH_3)CH_2CH_2COCH_3$	112.17	....	150	......	$0.8475^{20/20}$	1.4348^{20}	al, eth, ace	B1[4], 3482
7978	1-Hexene-3-one, 1-phenyl-5-methyl or Benzal pinacolone $(CH_3)_2CHCH_2COCH=CHC_6H_5$	188.26	cr	154^{25}	43	0.9509^{46}	1.5523^{25}	al, bz, chl	B7[3], 1487
7979	4-Hexene-3-one, 2-methyl $CH_3CH=CHCO(CH_3)_2$	112.17	....	147-8.5^{739}	......	$0.843^{20/4}$	1.4345^{20}	al, eth, ace	B1[4], 3483
7980	2-Hexene-4-one $CH_3CH_2COCH=CHCH_3$	98.14	....	138-9	......	$0.8559^{20/4}$	1.4388^{20}	al, eth, ace	B1[4], 3468
7981	1-Hexene-3-yne $C_2H_5C\equiv CCH=CH_2$	80.13	....	85^{758}	......	$0.7492^{20/4}$	1.4522^{20}	eth, bz, peth, chl	B1[4], 1091
7982	1-Hexene-3-yne, 5-chloro-5-methyl $(CH_3)_2CClC\equiv C\text{-}CH=CH_2$	128.60	....	48^{28}	......	0.9375^{15}	1.4778^{20}	al, eth, ace, bz, peth	B1[4], 1098
7983	1-Hexene-5-yne $HC\equiv CCH_2CH_2CH=CH_2$	80.13	....	70	......	$0.7650^{20/4}$	1.4318^{20}	eth, bz, peth, chl	B1[4], 1092
7984	3-Hexene-1-yne, 3-propyl $C_2H_5CH=C(C_3H_7)C\equiv CH$	122.21	....	136	......	$0.7799^{25/4}$	1.4432^{25}	bz, peth, chl	B1[3], 1049
7985	4-Hexene-1-yn-3-ol (d) $CH_3CH=CHCH(OH)C\equiv CH$	96.13	$[\alpha]^{20/D}$ + 16.06	157-9	......	$0.9090^{20/4}$	1.4645^{17}	al, ace, bz	B1[4], 2307
7986	4-Hexene-1-yn-3-ol (dl) $CH_3CH=CHCH(OH)C\equiv CH$	96.13	....	154-6, 60^{18}	......	$0.9148^{25/4}$	1.4651^{23}	al, ace, bz	B1[4], 2307
7987	Hexylamine or 1-aminohexane $CH_3(CH_2)_4CH_2NH_2$	101.19	....	130	-19	0.7660^{20}	1.4180^{20}	**al, eth**	B4[4], 709
7988	Hexyl nitrite $C_6H_{13}ONO$	131.17	ye	129-30^{774}, 52^{44}	......	$0.8778^{20/4}$	1.3987^{20}	al, eth	B1[4], 1699

No.	Name, Synonyms, and Formula	Mol. wt.	Color, crystalline form, specific rotation and λ_{max} (log ε)	b.p. °C	m.p. °C	Density	n_D	Solubility	Ref.
7989	Hexyl phenyl ether $C_6H_{13}OC_6H_5$	178.27		$240, 130^{22}$	-19	$0.9174^{20/4}$	1.4921^{20}	eth	$B6^4$, 560
7990	Hexyl sulfate $(C_6H_{13}O)_2SO_2$	266.40		125.3^{2}		$1.0036^{21/0}$	1.433^{21}		$B1^4$, 1699
7991	1-Hexyne or *n*-Butyl acetylene $C_4H_9C{\equiv}CH$	82.15		71.3	-131.9	$0.7155^{20/4}$	1.3989^{20}	al, eth, bz, peth, chl	$B1^4$, 1006
7992	1-Hexyne, 5-methyl or isopentyl acetylene $(CH_3)_2CHCH_2CH_2C{\equiv}CH$	96.17		92	-125	$0.7274^{20/4}$	1.4059^{20}	al, eth, bz, peth, chl	$B1^3$, 1000
7993	2-Hexyne $C_3H_7C{\equiv}CCH_3$	82.15		84	-89.6	$0.7315^{20/4}$	1.4138^{20}	al, eth, bz, peth, chl	$B1^4$, 1009
7994	2-Hexyne, 5-methyl $(CH_3)_2CHCH_2C{\equiv}CH_3$	96.17		102.5	-92.9	$0.7378^{20/4}$	1.4176^{20}	eth, ace, bz, peth, chl	$B1^4$, 1030
7995	3-Hexyne or Diethyl acetylene $C_2H_5C{\equiv}CC_2H_5$	82.15		81.5^{764}	-103	$0.7231^{20/4}$	1.4115^{20}	al, eth, bz, peth, chl	$B1^4$, 1009
7996	3-Hexyne, 2,5-dimethyl-2,5-dichloro $(CH_3)_2CClC{\equiv}C\text{-}CCl(CH_3)_2$	181.11		$175\text{-}8^{745}$	29			chl	$B1^4$, 1042
7997	3-Hexyne, 1:2,5:6-diepoxy $CH_2CHC{\equiv}C\text{-}CH\text{-}CH_2$	110.11		98^{20}	-16	1.1189^{23}	1.4871^{23}	chl	$B19^2$, 19
7998	3-Hexyne, 2,5-dimethyl-2,5-diol or Acetylene pinacol $(CH_3)_2C(OH)C{\equiv}CC(OH)(CH_3)_2$	142.20	nd(w)	205	95	$0.949^{20/20}$		al, eth, ace, bz, chl	$B1^4$, 2699
7999	3-Hexyne, 2-methyl $CH_3CH_2C{\equiv}CCH(CH_3)_2$	96.17		95.2	-116.7	$0.7263^{20/4}$	1.4114^{20}	eth, bz, peth, chl	$B1^4$, 1029
8000	1-Hexyn-3-ol, 3-methyl $C_3H_7C(OH)(CH_3)C{\equiv}CH$	112.17		137		$0.8620^{20/4}$	1.4338^{20}	al, eth	$B1^4$, 2252
8001	3-Hexyne-2,5-diol $CH_3CH(OH)C{\equiv}CCH(OH)CH_3$	114.14		120^{11}		1.0180^{20}	1.4691^{20}		$B1^3$, 2271
8002	3-Hexyn-2-ol, 2-methyl $C_2H_5C{\equiv}CC(OH)(CH_3)_2$	112.17		145-7, $46\text{-}7^{7}$		0.962^{0}	1.4392^{25}	al, eth	$B1^4$, 2250
8003	Hippuric acid or N-benzoylaminoacetic acid $C_6H_5CONHCH_2CO_2H$	179.18	pr (w or al)		190-3	$1.371^{20/4}$		al, w	$B9^3$, 1123
8004	Hippuric acid-*p*-amino $(4\text{-}H_2NC_6H_4)CONHCH_2CO_2H$	194.19	pr or nd (w)		198-9			al	$B14^3$, 1069
8005	Hippuric acid, *o*-bromo $(2\text{-}BrC_6H_4)CONHCH_2CO_2H$	258.07	nd (w)		192-3			AcOEt	$B9^3$, 1387
8006	Hippuric acid, *m*-bromo $(3\text{-}BrC_6H_4)CONHCH_2CO_2H$	258.07	nd (w)		146-7			al, MeOH	$B9^2$, 233
8007	Hippuric acid-*p*-bromo $(4\text{-}BrC_6H_4)CONHCH_2CO_2H$	258.07	nd (w)		162			al	$B9^2$, 236
8008	Histamine or 4-Imidoazol ethylamine $C_5H_9N_3$	111.15	wh nd (chl)	209^{18}	86			w, al	$B25^2$, 302
8009	Histamine, dihydrochloride $C_5H_9N_3.2HCl$	184.07	pl (eth-ace) pr (w)		249-52			w, MeOH	$B25^2$, 303
8010	Histidine (d) $C_6H_9N_3O_2$	155.16	ta (w) $[\alpha]^{20}_D$ + 40.2 (w)		287d			w	$B25^2$, 404
8011	Histidine (dl) $C_6H_9N_3O_2$	155.16	ta, tetr pr (w)		285d			w	$B25^2$, 409
8012	Histidine (l) $C_6H_9N_3O_2$	155.16	nd or pl (dil al) $[\alpha]^{20}_D$ −39.7 (w, c=1.13)		287d			w	$B25^2$, 404
8013	Histidine, bis(3,4-dichlorobenzene sulfonate) (d) $C_6H_6N_3O_2\text{-}2Cl_2C_6H_4O_3S$	609.28	rh nd(w)		280d				C51, 5184
8014	Histidine, diflavianate $C_6H_9N_3O_2.2C_{10}H_6N_2O_8S$	783.61	nd(w)		251-4d				$B25^2$, 407
8015	Histidine, dihydrochloride (dl) $C_6H_9N_3O_2.2HCl$	228.08			237d			w, al	$B25^1$, 718
8016	Histidine, dihydrochloride (l) $C_6H_9N_3O_2.2HCl$	228.08	rh pl		252d			w, al	B25, 513
8017	Histidine, monohydrochloride, hydrate $C_6H_9N_3O_2.HCl.H_2O$	209.63	pl (w) $[\alpha]^{26}_D$ + 8.0 (3NHCl, c=2)		259d				$B25^2$, 407
8018	Holocaine or N,N-bis(p-ethoxyphenyl) acetamidine $C_{18}H_{22}N_2O_2$	298.38	nd (al)		117-8			al, eth, ace, bz	$B13^3$, 1069

No.	Name, Synonyms, and Formula	Mol. wt.	Color, crystalline form, specific rotation and λ_{max} (log ε)	b.p. °C	m.p. °C	Density	n_D	Solubility	Ref.
8019	Holocaine, hydrochloride or Phenacaine $C_{18}H_{22}N_2O_2.HCl$	334.85	cr(w + 1)		190-2			w, al, chl	B13³, 1069
8020	Homatropine or Mandelyl tropine $C_{16}H_{21}NO_3$	275.35	pr (al or eth)		99-100			al, eth, ace, chl	B21⁴, 179
8021	Homatropine, hydrobromide $C_{16}H_{21}NO_3 \cdot HB$-	356.26	rh pym or pl(w)		217-8d			w, al	B21⁴, 179
8022	Homatropine, hydrochloride $C_{16}H_{21}NO_3 \cdot HCl$	311.81	wh pr(w)		220-7d			w, al	B21⁴, 179
8023	Homocysteine (dl) $HSCH_2CH_2CH(NH_2)CO_2H$	135.18			270-5 d			w	B4³, 1647
8024	Homocystine (d) $HO_2CCH(NH_2)CH_2CH_2SSCH_2CH_2CH(NH_2)CO_2H$	268.35	$_D,^{26/}{}_D$ −79 in HCl		281-4 d			w	B4³, 1646
8025	Homogentisic acid or 2,5-dihydroxyphenyl acetic acid $(2,5\text{-}(HO)_2C_6H_3)CH_2CO_2H$	168.15	pr (w + 1) lf (al-chl)		152-4			w, al, eth	B10³, 1456
8026	Homoveratric acid or 3,4-dimethoxyphenyl acetic acid $[3,4\text{-}(CH_3O)_2C_6H_3]CH_2CO_2H$	196.20	nd (w + 1) cr (bz-peth)		80-2 (hyd) 98-9 (anh)			w, al, eth	B10³, 1459
8027	Hordenine or 1-(dimethylamino)-2-(4-hydroxyphenyl)ethane $4\text{-}HOC_6H_4CH_2CH_2N(CH_3)_2$	165.24	rh pr (al or bz-peth) nd (w)	173-4¹¹ sub	117-8			al, eth, bz, lig, chl	B13³, 1640
8028	Hordenine, sulfate $2(C_{10}H_{15}NO).H_2SO_4$	428.54	fl		210-11			w	B13³, 1641
8029	Hordenine, sulfate, dihydrate $(C_{10}H_{15}NO)_2.H_2SO_4.2H_2O$	464.57	pr or pl		197			w	B13³, 1641
8030	Humulon or α-Lupulic acid $C_{21}H_{30}O_5$	362.47	yesh cr (eth) $[\alpha]^{20/}{}_D$ −232 (bz)		66.5			al, eth, ace, bz	B8³, 4034
8031	Hydantoic acid or N-Carbomoyl glycine $H_2NCONHCH_2CO_2H$	118.09	mcl pr		180d			w, al	B4³, 1163
8032	Hydantoic acid, ethyl ester or Ethyl hydantoate $H_2NCONHCH_2CO_2C_2H_5$	146.15	nd (w)		135			w, al	B4³, 1167
8033	Hydantoic acid, phenylthio or N-phenylpseudothio hydantoic acid $C_6H_5N{=}C(NH_2)SCH_2CO_2H$	210.26	wh (al)	175	159-61				B12³, 868
8034	Hydantoin or Glycol urea $C_3H_4N_2O_2$	100.08	nd (MeOH) lf (w)		220			w, al	B24², 127
8035	Hydantoin, 1-acetyl-2-thio $C_5H_6N_2O_2S$	158.17	pl (al)		175-6			al	B24¹, 293
8036	Hydantoin, 1-benzoyl-2-thio $C_{10}H_8N_2O_2S$	220.25	pr (al)		165d			w	B24¹, 294
8037	Hydantoin, 5-benzylidene-2-thio $C_{10}H_8N_2OS$	204.25	ye nd (al)		258d			al	B24¹, 355
8038	Hydantoin, 5,5-dimethyl $C_5H_8N_2O_2$	128.13	pr (w-al)	sub	178			al, w, eth, ace, bz	B24², 157
8039	Hydantoin, 5,5-diphenyl $C_{15}H_{12}N_2O_2$	252.27	nd (al)		286			al, ace, aa	B24², 227
8040	Hydantoin, 3,5-diphenyl-2-thio $C_{15}H_{12}N_2OS$	268.33			233			al, eth, bz	B24, 385
8041	Hydantoin, 5-ethyl-5-phenyl (d) or d-Nirvanol $C_{11}H_{12}N_2O_2$	204.23	pl (10 al) $[\alpha]_D$ + 123 (al)		237			al	B24², 206
8042	Hydantoin, 5-ethyl-5-phenyl (dl) or dl-Nirvanol $C_{11}H_{12}N_2O_2$	204.23	pr (dil al)		199-200			al, aa	B24², 206
8043	Hydantoin, 5(2-hydroxybenzylidene)-2-thio $C_{10}H_8N_2O_2S$	220.25	nd (aa)		248			aa	B25¹, 502
8044	Hydantoin, 1-methyl $C_4H_6N_2O_2$	114.10	cr (w) pl (al)		157-9			w, al, chl	B24², 128
8045	Hydantoin, 5-methyl (dl) or α-Lactylurea $C_4H_6N_2O_2$	114.10	pr (w)		145-6 (anh)			w, al	B24², 155
8046	Hydantoin, 5-methyl (l) $C_4H_6N_2O_2$	114.10	(w) $[\alpha]^{20/}{}_D$ −50.6		175			w, al	B24¹, 304
8047	Hydantoin, 5-methyl,hydrate (dl) $C_4H_6N_2O_2.H_2O$	132.12	rh (w)		155-6			w, al, ace	B24², 155
8048	Hydantoin, 5-phenyl $C_9H_8N_2O_2$	176.17			184-5				B24², 201

No.	Name, Synonyms, and Formula	Mol. wt.	Color, crystalline form, specific rotation and λ_{max} (log ε)	b.p. °C	m.p. °C	Density	n_D	Solubility	Ref.
8049	Hydantoin, 2-thio or Glycol thiourea $C_3H_4N_2OS$	116.14	wh nd (w)		229-31d			w, al, eth	B24², 138
8050	β(-)Hydrastine $C_{21}H_{21}NO_6$	383.40	yesh pr (al) $[\alpha]^{17/}_D$ −67.8 (chl,c=2.5)		132 (135)			eth, ba, chl	B27², 603
8051	β(-)Hydrastine, hydrochloride $C_{21}H_{21}NO_6.HCl$	419.86	micr pw $[\alpha]'_D$ +158.0 (w, c=2)		116			w	B27², 604
8052	Hydrastinine $C_{11}H_{13}NO_3$	207.23	nd (lig) cr (eth)		116-7			al, eth, chl	B27², 530
8053	Hydrastinine, bisulfate $C_{11}H_{13}NO_3.H_2SO_4$	305.30	gr-flr ye cr (al)		216d			w, al	B27², 530
8054	Hydrastinine, hydrochloride $C_{11}H_{13}NO_3.HCl$	243.69	pa ye nd		212d			w, al	B27², 530
8055	Hydratropamide or 2-Phenyl propionamide $C_6H_5CH(CH_3)CONH_2$	149.19	lo nd (w dil al) $[\alpha]^{28/}_D$ +57.9 (chl, c=1.6)		100.5			al, chl	B9³, 2420
8056	Hydratropic acid (d) or 2-Phenyl propionic acid $CH_3CH(C_6H_5)CO_2H$	150.18	$[\alpha]^{20/}_D$ +81.1 (al, c=3)	152^{16}					B9³, 2417
8057	Hydratropic acid (dl) or 2-Phenyl propionic acid $C_6H_5CH(CH_3)CO_2H$	150.18		260-2, 160^{25}	<-20	$1.1^{0/4}$	1.5237^{20}		B9³, 2418
8058	Hydratropic acid (l) or 2-Phenyl propionic acid $C_6H_5CH(CH_3)CO_2H$	150.18	$[\alpha]^{20/}_D$ −58 (al)	152^{10}					B9³, 2417
8059	Hydratroponitrile or 2-Phenyl propionitrile $C_6H_5CH(CH_3)CN$	131.18		230-2, $116\text{-}7^{20}$		$0.9854^{20/4}$	1.5095^{25}	al, eth	B9³, 2421
8060	Hydrazine, 1-acetyl-2-phenyl $CH_3CONHNHC_6H_5$	150.18	hex pr (eth)		130-2			w, al, bz, chl	B15², 92
8061	Hydrazine-allyl $CH_2{=}CHCH_2NHNH_2$	72.11		$122\text{-}4^{757}$				w, eth, chl	B4¹, 562
8062	Hydrazine, 1,2-bis-(3-aminophenyl) or m,m′-hydrazino dianiline $3\text{-}H_2NC_6H_4NHNHC_6H_4NH_2\text{-}(3)$	214.27	pym (al)		151				B15³, 876
8063	Hydrazine, 1,2-bis-(4-aminophenyl) or p,p′-Hydrazino dianiline $4\text{-}H_2NC_6H_4NHNHC_6H_4NH_2\text{-}(4)$	214.27	ye cr		145			al, eth	B15³, 879
8064	Hydrazine, benzyl $4\text{-}CH_3C_6H_4NHNHCH_2C_6H_5$	212.29	fl or pr (al)	103^{41}	26			w, **al, eth**	B15³, 699
8065	Hydrazine, 1-benzyl-2-(4-tolyl) $4\text{-}H_2NC_6H_4NHNHCH_2C_6H_5$	212.29		212^{17}					B15, 533
8066	Hydrazine, 2-bromophenyl $2\text{-}BrC_6H_4NHNH_2$	187.04	nd		48				B15¹, 117
8067	Hydrazine, 4-bromophenyl $4\text{-}BrC_6H_4NHNH_2$	187.04	nd (w) lf (lig) cr (al)		108			al, eth, lig	B15³, 289
8068	Hydrazine, 1-butyl-1-phenyl $C_6H_5N(C_4H_9)NH_2$	164.25		250^{763}					B15¹, 28
8069	Hydrazine, 1,2-diallyl $(CH_2{=}CHCH_2)NHNH(CH_2CH{=}CH_2)$	112.17		145^{752}					B4³, 1737
8070	Hydrazine, 1,2-dibenzoyl $C_6H_5CONHNHCOC_6H_5$	240.26	nd (al)		241				B9³, 1318
8071	Hydrazine, 1,2-dibenzoyl-1,2-dimethyl $C_6H_5CONH(CH_3)N(CH_3)COC_6H_5$	268.32	pr (al)		85-6			al	B9², 217
8072	Hydrazine, 1,1-dibenzyl $(C_6H_5CH_2)_2NNH_2$	212.29	cr (peth)		65			al, eth	B15², 245
8073	Hydrazine, 1,2-dibenzyl $C_6H_5CH_2NHNHCH_2C_6H_5$	212.29	lf (dil al)		47				B15³, 701
8074	Hydrazine, 1,2-diisobutyl $i\text{-}C_4H_9NHNII\text{-}i\text{-}C_4H_9$	144.26		170^{735}, 63^{10}		$0.8002^{20/4}$	1.4276	al, eth, ace, bz	B4², 962
8075	Hydrazine, (2,4-dichlorophenyl) $2,4\text{-}Cl_2C_6H_3NHNH_2$	177.03	nd (eth or peth)		94			al, eth, aa	B15², 152
8076	Hydrazine, 1,1-diethyl $(C_2H_5)_2NNH_2$	88.15		$98\text{-}9^{750}$		$0.8804^{20/4}$	1.4214^{20}	w, al, eth, bz, chl	B4², 959

PHYSICAL CONSTANTS OF ORGANIC COMPOUNDS (Continued)

No.	Name, Synonyms, and Formula	Mol. wt.	Color, crystalline form, specific rotation and λ_{max} (log ε)	b.p. °C	m.p. °C	Density	n_D	Solubility	Ref.
8077	Hydrazine, 1,2-diethyl $C_2H_5NHNHC_2H_5$	88.15		85-6		0.797^{26}	1.4204^{20}	al, eth, bz	$B4^1$, 1730
8078	Hydrazine, 1,1-dimethyl $(CH_3)_2NNH_2$	60.10		63^{752}		0.7914^{22}	1.4075^{22}	w, al, eth	$B4^1$, 1726
8079	Hydrazine, 1,2-dimethyl $CH_3NHNHCH_3$	60.10		81^{753}		$0.8274^{20/4}$	1.4209^{20}	w, al, eth	$B4^1$, 1727
8080	Hydrazine, 1,2-dimethyl, dihydrochloride $CH_3NHNHCH_3.2HCl$	133.02	pr (w)		170d			w, al	$B4^1$, 1727
8081	Hydrazine, (2,3-dimethylphenyl) $[2,3\text{-}(CH_3)_2C_6H_3]NHNH_2$	136.20	nd (al)		111-2			al, eth	$B15^3$, 718
8082	Hydrazine, (2,4-dimethylphenyl) $[2,4\text{-}(CH_3)_2C_6H_3]NHNH_2$	136.20	nd (eth)		85			al, eth	$B15^2$, 249
8083	Hydrazine, (2,5-dimethylphenyl) $[2,5\text{-}(CH_3)_2C_6H_3]NHNH_2$	136.20	nd (dil aa)		78			al, eth, ace, bz	$B15^3$, 719
8084	Hydrazine, (2,6-dimethylphenyl) $[2,6\text{-}(CH_3)_2C_6H_3]NHNH_2$	136.20	nd (lig)		46			lig	$B15^2$, 552
8085	Hydrazine, (3,4-dimethylphenyl) $[3,4\text{-}(CH_3)_2C_6H_3]NHNH_2$	136.20	yesh nd (eth)		57				$B15^2$, 249
8086	Hydrazine, 1,2-di-α-naphthyl or 1,1′-Hydrazonaphthalene $\beta\text{-}C_{10}H_7NHNH\text{-}\alpha\text{-}C_{10}H_7$	284.36	lf (bz) pl (peth)		153			eth, bz	$B15^3$, 729
8087	Hydrazine, 1,2-di-β-naphthyl or 2,2′-Hydrazonaphthalene $\beta\text{-}C_{10}H_7NHNH\text{-}\beta\text{-}C_{10}H_7$	284.36	red pl (bz)		140-1				$B15^3$, 734
8088	Hydrazine, (2,4-dinitrophenyl) $[2,4\text{-}(O_2N)_2C_6H_3]NHNH_2$	198.14	blsh-red (al)		194 (198d)				$B15^3$, 425
8089	Hydrazine, (2,6-dinitrophenyl) $[2,6\text{-}(O_2N)_2C_6H_3]NHNH_2$	198.14	red nd (dil al)		145				$B15^2$, 219
8090	Hydrazine, 1,1-diphenyl $(C_6H_5)_2NNH_2$	184.24	ta (lig)	220^{40-50}	49-52	$1.190^{16/4}$		al, eth, bz, chl	$B15^3$, 74
8091	Hydrazine, 1,2-diphenyl or Hydrazobenzene $C_6H_5NHNHC_6H_5$	184.24	ta (al-eth)		131	$1.158^{16/4}$		al	$B15^3$, 76
8092	Hydrazine, 1,2-diisopropyl $(CH_3)_2CHNHNHCH(CH_3)_2$	116.21		125, 63^{84}		$0.7894^{20/4}$	1.4173^{20}	al, eth, ace, bz	$B4^3$, 1732
8093	Hydrazine, 1,1-di-4-tolyl $(4\text{-}CH_3C_6H_4)_2NNH_2$	212.29	lf (al)		93			al	$B15^1$, 154
8094	Hydrazine, 1,2-di-(2-tolyl) or *o*-Hydrazotoluene $(2\text{-}CH_3C_6H_4)NHNH(C_6H_4CH_3\text{-}2)$	212.29	lf (al)		165			al, eth, bz	$B15^3$, 655
8095	Hydrazine, 1,2-di-(3-tolyl) or *m*-Hydrazotoluene $(3\text{-}CH_3C_6H_4)NHNH(C_6H_4CH_3\text{-}3)$	212.29	cr (peth)	224	38			al, eth, bz	$B15^3$, 670
8096	Hydrazine, 1,2-di-(4-tolyl) or *p*=Hydrazotoluene $(4\text{-}CH_3C_6H_4)NHNH(C_6H_4CH_3\text{-}4)$	212.29	lf (lig) cr (bz-al) pl (al-eth)		135	$0.957^{20/4}$		al, eth, bz	$B15^3$, 677
8097	Hydrazine, 1-isobutyl-1-phenyl $i\text{-}C_4H_9N(C_6H_5)NH_2$	164.25			240-5	$0.9633^{15/4}$			$B15^3$, 75
8098	Hydrazine, ethyl $C_2H_5NHNH_2$	60.11		101				w, al, eth, ace, bz	$B4^3$, 1730
8099	Hydrazine, 1-ethyl-1-phenyl $C_2H_5N(C_6H_5)NH_2$	136.20		237, $115\text{-}9^{19}$		$1.0181^{21/4}$	1.5711^{21}	al, eth, ace, bz	$B15^3$, 74
8100	Hydrazine, 1-ethyl-2-phenyl $C_2H_5NHNHC_6H_5$	136.20		240^{750}, 110^{14}		$1.0150^{20/4}$	1.5676^{20}	al, eth, bz, chl	$B15^3$, 74
8101	Hydrazine, methyl CH_3NHNH_2	46.07		87.5^{760}	−52.4	0.874^{25}	$1.4325^{20/D}$	w, al, eth	$B4^3$, 1726
8102	Hydrazine-l-methyl, l-phenyl $C_6H_5N(CH_3)NH_2$	122.17		227^{745}, 131^{35}		$1.0404^{20/4}$	1.5691^{20}	al, eth, bz, chl	$B15^3$, 73
8103	Hydrazine-l-methyl, 2-phenyl $CH_3NHNHC_6H_5$	122.17		230^{728}, 112^{14}		$1.0320^{20/4}$	1.5733^{20}	al, eth, bz, chl	$B15^3$, 73
8104	Hydrazine-l-methyl, 2-(3-tolyl) $(3\text{-}CH_3C_6H_4)NHNHCH_3$	136.20	ye		59-61	$1.0265^{100/4}$		al, bz	$B15^2$, 229
8105	Hydrazine-l-methyl, 2-(4-tolyl) $(4\text{-}CH_3C_6H_4)NHNHCH_3$	136.20	ye nd (eth) pl (lig)		91			al, eth, bz	$B15^2$, 154
8106	Hydrazine, α-naphthyl $\alpha\text{-}C_{10}H_7NHNH_2$	158.20		203^{20}	117			al, eth, bz, chl	$B15^3$, 728
8107	Hydrazine, β-naphthyl $\beta\text{-}C_{10}H_7NHNH_2$	158.20	lf (w)		124-5			al, bz	$B15^3$, 734
8108	Hydrazine, (2-nitrophenyl) $(2\text{-}O_2NC_6H_4)NHNH_2$	153.14	red nd (bz)		90-2			w	$B15^3$, 316

PHYSICAL CONSTANTS OF ORGANIC COMPOUNDS (Continued)

No.	Name, Synonyms, and Formula	Mol. wt.	Color, crystalline form, specific rotation and λ_{max} (log ε)	b.p. °C	m.p. °C	Density	n_D	Solubility	Ref.
8109	Hydrazine, (3-nitrophenyl) $(3\text{-}O_2NC_6H_4)NHNH_2$	153.14	red nd or pr (ace) ye nd (al)		93			chl, aa	B15³, 326
8110	Hydrazine, (4-nitrophenyl) $(4\text{-}O_2NC_6H_4)NHNH_2$	153.14	og-red lf or nd (al)		158d			eth, chl	B15³, 331
8111	Hydrazine, l-pentyl-2-phenyl *(d)* $C_5H_{11}NHNHC_6H_5$	178.28	$[\alpha]^t_D$ + 4.45	$173\text{-}5^{50}$		0.986^{20}	1.5523^{20}	eth	B15, 121
8112	Hydrazine, l-isopentyl-l-phenyl $i\text{-}C_5H_{11}N(C_6H_5)NH_2$	178.28			236	0.9588^{15}			B15, 121
8113	Hydrazine, phenyl $C_6H_5NHNH_2$	108.14	mcl pr or pl	243, 115^{10}	19.8	$1.0986^{20/4}$	1.6084^{10}	al, eth, ace, bz, chl	B15³, 67
8114	Hydrazine, phenyl, hemihydrate $C_6H_5NHNH_2.½H_2O$	117.15	fl (w)	120^{12}	24	$1.0970^{25/25}$	1.6081^{20}		B15, 68
8115	Hydrazine, phenyl, hydrochloride $C_6H_5NHNH_2.HCl$	144.60	lf (al)	sub	243-6d			w, al	B15³, 71
8116	Hydrazine, l-phenyl-2-(2-tolyl) $(2\text{-}CH_3C_6H_4)NHNHC_6H_5$	198.27	pl (al)		101-2			eth, bz	B15³, 655
8117	Hydrazine, l-phenyl-2-(3-tolyl) $(3\text{-}CH_3C_6H_4)NHNHC_6H_5$	198.27	ye cr (peth)		61	$1.0265^{100/4}$		al, bz, lig	B15², 229
8118	Hydrazine, l-phenyl-2-(4-tolyl) $(4\text{-}CH_3C_6H_4)NHNHC_6H_5$	198.27	pl (lig), cr (al)		91			al, bz	B15³, 677
8119	Hydrazine, propyl $C_3H_7NHNH_2$	74.13			119				B4³, 1731
8120	Hydrazine, *iso*-propyl $(CH_3)_2CHNHNH_2$	74.13		107^{750}				w, al, bz	B4³, 1731
8121	Hydrazine, l-isopropyl-2-methyl $(CH_3)_2CHNHNHCH_3$	88.15	cr (eth)	100					B4², 960
8122	Hydrazine, tetraphenyl $(C_6H_5)_2NN(C_6H_5)_2$	336.44	pr (chl-al)		149d			eth, ace, bz, chl	B15³, 77
8123	Hydrazine, (2-tolyl) $(2\text{-}CH_3C_6H_4)NHNH_2$	122.17	nd (dil al)		59			al, eth, chl	B15³, 654
8124	Hydrazine, (3-tolyl) $(3\text{-}CH_3C_6H_4)NHNH_2$	122.17		244d		$1.057^{20/4}$		al, eth, bz	B15³, 669
8125	Hydrazine, (4-tolyl) $(4\text{-}CH_3C'6H_4)NHNH_2$	122.17	lf (w or eth)	244d	66			al, eth, bz	B15³, 676
8126	Hydrazine, (2,4,6-tribromophenyl) $(2,4,6\text{-}Br_3C_6H_2)NHNH_2$	344.83	nd (peth) cr (lig)		146			bz, chl, lig	B15¹, 126
8127	Hydrazine, (2,4,6-trichlorophenyl) $(2,4,6\text{-}Cl_3C_6H_2)NHNH_2$	211.48	cr (bz)		143				B15³, 281
8128	Hydrazine, (2,4,6-trinitrophenyl) $[2,4,6(O_2N)_3C_6H_2]NHNH_2$	243.14	red pl (al)		186			al, aa	B15³, 652
8129	Hydrazine, triphenyl $(C_6H_5)_2NNHC_6H_5$	260.34	nd (bz-peth) cr (al)		142d	$0.869^{70/4}$		al, bz	B15², 54
8130	Hydrazine carboxylic acid, ethyl ester or *N*-Amino urethane $H_2NNHCO_2C_2H_5$	104.11	cr	198d, 93^{9}	46			al, eth	B3⁴, 174
8131	l,l-Hydrazine dicarboxylic acid, diethyl ester $H_2NN(CO_2C_2H_5)_2$	176.17	pr (w)	138^{12}	29			al	B3³, 79
8132	l,2-Hydrazine dicarbon amide $H_2NCONHNHCONH_2$	118.10	pl (w)		257-9				B3⁴, 236
8133	l,2-Hydrazine dicarboxylic acid, diethyl ester $C_2H_5O_2CNHNH\ CO_2C_2H_5$	176.17	nd (chl) pr (w)	250d	135	1.324^{8}		al, eth	B3⁴, 175
8134	Hydrazobenzene or l,2-Diphenyl hydrazine $C_6H_5NHNHC_6H_5$	184.24	ta (al-eth)		131	$1.158^{16/4}$		al	B15³, 76
8135	Hydrazo diformic acid OCHNHNHCHO	88.07	pr (al)		160			w	B2⁴, 86
8136	α,α'-Hydrazo naphthalene $\alpha\text{-}C_{10}H_7NHNH\text{-}\alpha\text{-}C_{10}H_7$	284.36	lf (bz) pl (peth)		153			eth, bz	B15³, 729
8137	β,β'-Hydrazo naphthalene $\beta\text{-}C_{10}H_7NHNH\text{-}\beta\text{-}C_{10}H_7$	284.36	red pl (bz)		140-1				B15³, 734
8138	Hydrindane *(trans, l)* or Hexahydro indane C_9H_{16}	124.23		161, 71.7^{40}		$0.8627^{20/4}$	1.4636^{20}	eth, bz, peth	B5⁴, 292
8139	2-Hydrindanone *(cis)* $C_9H_{14}O$	138.21		225^{754}, 108^{23}	10		1.4830^{20}	al, bz, lig	B7³, 294
8140	2-Hydrindanone *(trans)* $C_9H_{14}O$	138.21		218^{754}	-12	$0.9807^{17/4}$	1.4769^{17}	al, bz, lig	B7³, 294

PHYSICAL CONSTANTS OF ORGANIC COMPOUNDS (Continued)

No.	Name, Synonyms, and Formula	Mol. wt.	Color, crystalline form, specific rotation and λ_{max} (log ε)	b.p. °C	m.p. °C	Density	n_D	Solubility	Ref.
8141	Hydrobenzamide or Tribenzal diamine $C_6H_5CH(N{=}CHC_6H_5)_2$	298.39	nd (bz) cr (al w)	130	110	...	...	al, eth	$B7^3$, 838
8142	Hydroberberine *(d)* or *d*-Canadine $C_{20}H_{21}NO_4$	339.39	$[\alpha]^{20}_D$ + 297.4 (chl, c=1)	...	132	...	...	al, eth, bz, chl	$B27^2$, 556
8143	Hydroberberine *(l)* $C_{20}H_{21}NO_4$	339.39	nd (al) $[\alpha]^{20/}_D$ −298.2 (chl,c=1)	...	154	...	...	al, eth, bz, chl	$B27^2$, 557
8144	Hydrocinchonidine or Cinchamidine $C_{19}H_{24}N_2O$	296.41	lf (al) $[\alpha]^{20/}_D$ −98.4 (al)	(al)	229	...	...	al	$B23^2$, 357
8145	Hydrocinnamaldehyde or 3-Phenylpropionaldehyde $C_6H_5CH_2CH_2CHO$	134.18	mcl	223^{745}, $104\text{-}5^{13}$	47	...	...	al, eth	$B7^3$, 1046
8146	Hydrocinnamide or 3-Phenylpropionamide $C_6H_5CH_2CH_2CONH_2$	149.19	nd (w)	...	106-8	...	...	al, eth	$B9^3$, 2393
8147	Hydrocinnamic acid or 3-Phenylpropionic acid $C_6H_5CH_2CH_2CO_2H$	150.18	pr (peth)	279.8, $169\text{-}70^{28}$	48.6	$1.0712^{49/4}$	...	al, eth, bz	$B9^3$, 2382
8148	Hydrocinnamic acid, 3-amino *(d)* or 3-Amino-3-phenyl-propionic acid $C_6H_5CH(NH_2)CH_2CO_2H$	165.19	pl (w) $[\alpha]^{20/}_D$ + 7.0 (w,p=1)	...	234-5d	...	...	...	$B14^3$, 1218
8149	Hydrocinnamic acid, 3-amino *(dl)* $C_6H_5CH(NH_2)CH_2CO_2H$	165.19	cr (w)	...	231d	...	...	...	$B14^3$, 1218
8150	Hydrocinnamic acid, 3-amino *(l)* $C_6H_5CH(NH_2)CH_2CO_2H$	165.19	cr (w) $[\alpha]^{25/}_D$ −7.5 (w,c=1)	...	234-5d	...	...	...	$B14^3$, 1218
8151	Hydrocinnamic acid, benzyl ester or Benzyl hydrocinnamate $C_6H_5CH_2CH_2CO_2CH_2C_6H_5$	240.30	...	310-40, $198\text{-}9^{20}$	...	1.090^{15}	...	eth	$B9^2$, 339
8152	Hydrocinnamic acid, ethyl ester or Ethyl hydrocinnamate $C_6H_5CH_2CH_2CO_2C_2H_5$	178.23	...	247.2, 123^{16}	...	1.0147^{20}	1.4954^{20}	al, eth	$B9^3$, 2385
8153	Hydrocinnamic acid, methyl ester or Methyl hydrocinnamate $C_6H_5CH_2CH_2CO_2CH_3$	164.20	...	$238\text{-}9^{757}$	...	1.0455^0	...	al, eth, bz	$B9^3$, 2384
8154	Hydrocinnamic acid, propyl ester or Propyl hydrocinnamate $C_6H_5CH_2CH_2CO_2C_3H_7$	192.26	...	262.1, 135^{16}	...	1.008^{12}	...	...	$B9^3$, 2386
8155	Hydrocinnamic acid, isopropyl ester or Isopropyl hydrocinnamate $C_6H_5CH_2CH_2CO_2CH(CH_3)_2$	192.26	...	126^{11}	...	$0.9860^{25/4}$	...	al, eth	$B9^2$, 339
8156	Hydrocinnamonitrile or 3-Phenyl propionitrile $C_6H_5CH_2CH_2CN$	131.18	...	261, $125\text{-}6^{15}$	...	1.0016^{20}	1.5266^{28}	al, eth	$B9^3$, 2395
8157	Hydrocinnamyl chloride or 3-Phenyl propionyl chloride $C_6H_5CH_2CH_2COCl$	168.62	...	225d, 105^{10}	...	1.135^{21}	...	eth	$B9^3$, 2393
8158	Hydroconiferyl alcohol or 3-(4-hydroxy-3-methoxy-phenyl)-l-propanol $(3\text{-}CH_3O\text{-}4\text{-}HOC_6H_3)CH_2CH_2CH_2OH$	182.22	...	197^{15}	65	...	1.5545^{25}	al, eth	$B6^3$, 6347
8159	Hydrocotamine, hemihydrate $C_{12}H_{15}NO_3.\frac{1}{2}H_2O$	230.26	pr (eth)	...	56	...	...	al, eth, ace, chl, aa	$B27^2$, 541
8160	Hydrocupreine $C_{19}H_{24}N_2O_2$	312.41	pl (dil al) $[\alpha]^{23/}_D$ −159.2 (abs, al)	...	230	...	...	eth, chl	$B23^2$, 399
8161	Hydrocyanic acid or Hydrogen cyanide HCN	27.03	...	25.7	−13.2	$0.6876^{20/4}$	1.2614^{20}	w, al, eth	$B2^4$, 50
8162	Hydrofuramide or Furfur amide $C_{15}H_{12}N_2O_3$	268.27	nd (al)	...	117	...	...	al, eth	$B17^4$, 4428
8163	Hydrohydrastinine $C_{11}H_{13}NO_2$	191.23	nd (lig) cr (peth)	303^{752}	66	...	...	al, eth, ace, bz, aa	$B27^2$, 528
8164	Hydrolaphacol $C_{15}H_{16}O_3$	244.29	ye nd (al) cr (peth)	...	94	...	...	al	$B8^3$, 2595
8165	Hydroquinidine $C_{20}H_{26}N_2O_2$	326.44	nd (al) $[\alpha]^{18/}_D$ + 229.26	...	168-9	...	...	al, eth, ace, chl	B23,411
8166	Hydroquinine *(d)* or Quinotine $C_{20}H_{26}N_2O_2$	326.44	$[\alpha]^{18/}_D$ + 143.5	...	171	...	...	al, eth, ace, chl	$B23^2$, 400

PHYSICAL CONSTANTS OF ORGANIC COMPOUNDS (Continued)

No.	Name, Synonyms, and Formula	Mol. wt.	Color, crystalline form, specific rotation and λ_{max} (log ε)	b.p. °C	m.p. °C	Density	n_D	Solubility	Ref.
8167	Hydroquinine *(dl)* $C_{20}H_{26}N_2O_2$	326.44	nd (al, chl)		175-7			al, eth, ace, chl	$B23^2$, 400
8168	Hydroquinine *(l)* $C_{20}H_{26}N_2O_2$	326.44	nd (eth, chl) $[\alpha]^{20/}{}_D$ −142.2		172.3			al, eth, ace, chl	$B23^2$, 400
8169	Hydroquinone or 1,4-Dehydroxy benzene $1,4-(HO)_2C_6H_4$	110.11	mcl pr (sub) nd (w)	285^{750}	173-4	1.328^{15}		w, al, eth, ace	$B6^3$, 4374
8170	Hydroquinone, 2-acetyl or 2,5-Dihydroxy acetophenone $2,5(HO)_2C_6H_3COCH_3$	152.15	ye-gr nd (dil al or w)		204-5			al	$B8^3$, 2100
8171	Hydroquinone, monobenzoate $4-(C_6H_5CO_2)C_6H_4OH$	214.22	nd (al)		163-4			al, eth, bz, chl	$B9^3$, 558
8172	Hydroquinone, 2-bromo or Adurol-1,4-dehydroxy-2-bromobenzene $1,4-(HO)_2-2-Br-C_6H_3$	189.01	lf (lig) cr (chl)	sub	110-1			w, al, eth, bz	$B6^3$, 4436
8173	Hydroquinone, 2-chloro $2-Cl-1,4-(HO)_2C_6H_3$	144.56	red lf (chl) nd (bz)	263	108			w, al, eth, bz	$B6^3$, 4432
8174	Hydroquinone, diacetate $1,4-(CH_3CO_2)_2C_6H_4$	194.19	pl (w, al)		123-4	$0.8731^{25/4}$		al, eth, lig, chl	$B6^3$, 4414
8175	Hydroquinone, dibenzoate $1,4-(C_6H_5CO_2)_2C_6H_4$	318.33	mcl nd (al or to)		204			w	$B9^3$, 559
8176	Hydroquinone, monobenzyl ether $4-C_6H_5CH_2OC_6H_4OH$	200.24	pl (w)		122			al, eth, bz	$B6^3$, 4402
8177	Hydroquinone, dibenzyl ether $1,4-(C_6H_5CH_2O)_2C_6H_4$	290.36	pl (al)		130			eth, aa	$B6^3$, 4402
8178	Hydroquinone, dibenzoyl $1,4-(C_6H_5CO)_2C_6H_4$	286.33			161				$B7^3$, 4299
8179	Hydroquinone, 2,5-di-*tert*-butyl $2,5-(t-C_4H_9)_2C_6H_2(OH)_2-1,4$	222.33	cr (aq aa)		213.4				$B6^3$, 4741
8180	Hydroquinone, 2,3-dichloro $2,3-Cl_2C_6H_2(OH)_2-1,4$	179.00	cr (sub) nd (w + 2)		146-8 (anh)			al	$B6^3$, 4434
8181	Hydroquinone, 2,5-dichloro $2,5-Cl_2-C_6H_2-(OH)_2-1,4$	179.00	nd or pr (w, ace, bz)		172.5			al, eth, ace	$B6^3$, 4434
8182	Hydroquinone, 2,6-dichloro $2,6-Cl_2C_6H_2(OH)_2-1,4$	179.00	nd or lf (w, bz)		164			al, ace	$B6^3$, 4435
8183	Hydroquinone, diethyl ether or 1,4-diethoxybenzene $1,4-(C_2H_5O)_2C_6H_4$	166.22	pl (dil al)	246	72			al, eth, bz, chl	$B6^3$, 4387
8184	Hydroquinone, 2,3-dimethyl $2,3-(CH_3)_2C_6H_2(OH)_2-1,4$	138.17	cr (w)		224-5d			w, al, eth	$B6^3$, 4582
8185	Hydroquinone, 2,5-dimethyl $2,5-(CH_3)_2C_6H_2(OH)_2-1,4$	138.17	lf (w, al, bz)		217 sub			al, eth, chl	$B6^3$, 4601
8186	Hydroquinone, 2,6-dimethyl $2,6-(CH_3)_2-C_6H_2(OH)_2-1,4$	138.17	nd (xyl) cr (w)		153-4			w, al, eth	$B6^3$, 4588
8187	Hydroquinone, dimethyl ether or 1,4-dimethoxybenzene $1,4-(CH_3O)_2C_6H_4$	138.17	lf (w)	212.6, 109^{20}	58-60	$1.0526^{55/55}$		al, eth, bz	$B6^3$, 4385
8188	Hydroquinone, dithio or 1,4-dimercapto benzene $1,4-(HS)_2C_6H_4$	142.23	lf (al)		98			al, bz, aa	$B6^3$, 4472
8189	Hydroquinone, monoethyl ether or 4-ethoxy phenol $4-C_2H_5OC_6H_4OH$	138.17	pr or lf (w)	246-7	66-7			al, eth	$B6^3$, 4387
8190	Hydroquinone, monoheptyl ether $4-C_7H_{15}OC_6H_4OH$	208.30	cr (lig)		60				$B6^3$, 4391
8191	Hydroquinone, monohexyl ether $4-C_6H_{13}OC_6H_4OH$	194.27	cr (lig)		48				$B6^3$, 4390
8192	Hydroquinone, 2-iodo $2,5-(HO)_2C_6H_3I$	236.01			115-6				$B6^3$, 4440
8193	Hydroquinone, 2-methyl $2,5-(HO)_2C_6H_3CH_3$	124.14		sub 163^{11}	128			w, al, eth, ace	$B6^3$, 4498
8194	Hydroquinone, monomethyl ether or 4-methoxyphenol $4-CH_3OC_6H_4OH$	124.14	pl	243	57			w, al, eth, bz	$B6^3$, 4383
8195	Hydroquinone, 2-nitro $2,5-(HO)_2C_6H_3NO_2$	155.11	og-red rh (w)		133-4			al, eth, bz	$B6^3$, 4442
8196	Hydroquinon, monooctyl ether $4-C_8H_{17}OC_6H_4OH$	222.33	cr (lig)		60-1			al	$B6^3$, 4391
8197	Hydroquinone, phenyl or 2,5-dihydroxybiphenyl $2,5-(HO)_2C_6H_3.C_6H_5$	186.21	nd (dil al)		97-8			al	$B6^3$, 5371

No.	Name, Synonyms, and Formula	Mol. wt.	Color, crystalline form, specific rotation and λ_{max} (log ε)	b.p. °C	m.p. °C	Density	n_D	Solubility	Ref.
8198	Hydroquinone, monopropyl ether or 4-propoxyphenol $4\text{-}C_3H_7OC_6H_4OH$	152.19	cr (w, lig, al)		56-7				B6³, 4388
8199	Hydroquinone, 2-*iso*-propyl $2,5\text{-}(HO)_2C_6H_3\text{-}i\text{-}C_3H_7$	152.19	nd (w)		130-1				B6³, 4632
8200	Hydroquinone, 2-*iso*-propyl-5-methyl or Thynohydro quinone $[2,5\text{-}(HO)_2\text{-}4\text{-}CH_3]C_6H_2\text{-}i\text{-}C_3H_7$	166.22	pr (dil al)	290 sub	148			al, eth	B6³, 4673
8201	Hydroquinone, tetrabromo $1,4\text{-}(HO)_2C_6Br_4$	425.70	mcl pr (al-eth)		244	3.023^{21}		al, eth, aa	B6³, 4440
8202	Hydroquinone, tetrachloro $1,4\text{-}(HO)_2C_6Cl_4$	247.89	nd (aa)	sub	232			al, eth	B6³, 4436
8203	Hydroquinone, tetraiodo $1,4\text{-}(HO)_2C_6I_4$	613.70	(aa)		258			eth, chl	B6¹, 417
8204	Hydroquinone, tetra methyl or Durohydroquinone $1,4\text{-}(HO)_2C_6(CH_3)_4$	166.22	nd (al)		233				B6³, 4682
8205	Hydroquinone, 2,3,5-tribromo $2,3,5\text{-}Br_3C_6H(OH)_2\text{-}1,4$	346.80	nd (chl)		136-7			al, eth, bz, aa, chl	B6³, 4439
8206	Hydroquinone, 2,3,5-trimethyl $2,3,5\text{-}(CH_3)_3C_6H(OH)_2\text{-}1,4$	152.19	nd (w)		168-70d			al, eth, bz	B6², 897
8207	Hydroquinonephthalein or 2,7-Dihydroxy fluoran $C_{20}H_{12}O_5$	332.31	nd (eth)		228-9			al, eth, ace, aa	B19², 247
8208	Hydroxyacetamide $HOCH_2CONH_2$	75.07	lf (al) rh (AcOEt)		120	1.415^{13}		w	B3⁴, 597
8209	Hydroxyacetic acid or Glycolic acid $HOCH_2COOH$	76.05	rh, nd (w) lf (eth)	d	80			w, al, eth	B3⁴, 571
8210	Hydroxyacetic acid, acetate or Acetoxyacetic acid $CH_3CO_2CH_2COOH$	118.19	nd (bz)	$144\text{-}5^{12}$	67-8			w, al, eth, ace, chl	B3⁴, 576
8211	Hydroxyacetic acid, anhydride or Glycolic anhydride $(HOCH_2CO)_2O$	134.09	pw	d	128-30				B3¹, 92
8212	Hydroxyacetic acid, 4-bromophenyl ester *(dl)* $(4\text{-}BrC_6H_4)O_2CCH_2OH$	231.05	nd (bz)		118-20			al, eth, bz	B16², 125
8213	Hydroxyacetic acid, ethyl ester or Ethyl glycolate $HOCH_2CO_2C_2H_5$	104.11		$160, 69^{25}$		$1.0826^{23/4}$	1.4180^{20}	al, eth	B3⁴, 580
8214	Hydroxyacetic acid, methyl ester or Methyl glycolate $HOCH_2CO_2CH_3$	90.08		151.1		$1.1677^{18/4}$		w, **al, eth**	B3⁴, 578
8215	Hydroxyacetic acid, propyl ester or Propyl glycolate $HOCH_2CO_2C_3H_7$	118.14		170-1		$1.0631^{18/4}$	1.4231^{18}		B3⁴, 588
8216	Hydroxyacetonitrile or Glycolonitrile $HOCH_2CN$	57.05		183d, 119^{24}	<-72		1.4117^{19}	w, al, eth	B3⁴, 598
8217	Hydroxy amphetamine or Paredrine $C_9H_{13}NO$	151.21			125-6			al, chl	B13³, 1709
8218	Hydroxy amphetamine, hydrobromide $C_9H_{13}NO.HBr$	232.12			189			w, al, ace	B13³, 1709
8219	Hydroxycitronellal $(CH_3)_2COH(CH_2)_3CH(CH_3)CH_2CHO$	172.27		103^{3}		0.9220^{20}	1.4494^{20}	al, ace	B1⁴, 4058
8220	bis [3-Hydroxy-*sec*-butyl] amine $[CH_3CH(OH)CH(CH_3)_2]NH$	161.24	yesh	$112\text{-}5^{3}$		$0.9775^{20/4}$	1.4162^{20}	w, al, ace	B4⁴, 1726
8221	β-Hydroxyethyl 2-tolyl amine $(2\text{-}CH_3C_6H_4)NHCH_2CH_2OH$	151.21		285-6, 149^{4}		$1.0794^{20/4}$	1.5675^{20}	al, eth	B12³, 1846
8222	bis (2-Hydroxyethyl) methyl amine $(HOCH_2CH_2)_2NCH_3$	119.17		$246\text{-}8^{747}$, $123\text{-}5^{4}$		1.0377^{20}	1.4642^{20}	w, **al**	B4⁴, 1517
8223	bis-(2-Hydroxyethyl) ethyl amine $(HOCH_2CH_2)_2NC_2H_5$	133.19	ye	246-8, 118^{3}		$1.0135^{20/4}$	1.4663^{20}	w, al	B4³, 693
8224	β-Hydroxyethyl-4-tolyl amine $(4\text{-}CH_3C_6H_4)NHCH_2CH_2OH$	151.21	pl (eth-lig)	286-8, $153\text{-}5^{4}$	42-3			al, eth, bz, chl	B12³, 2034
8225	2-(Hydroxyethyl)-2-(hydroxybutyl) amine $(HOCH_2CH_2)NH(CH_2CH(OH)CH_2CH_3)$	133.19	yesh	137^{9}		1.0310^{204}	1.4690^{30}	w, al, ace	B4⁴, 1725
8226	(2-Hydroxyethyl)-3-(hydroxybutyl) amine $(HOCH_2CH_2)NH(CH_2CH_2CH(OH)CH_3)$	133.19	yesh	$107\text{-}9^{1}$		$1.0331^{20/4}$	1.4718^{20}	w, ace	C59, 6397
8227	(2-Hydroxypropyl)trimethylammonium chloride or β-Methyl choline chloride $CH_3CH(OH)CH_2N^+(CH_3)_3Cl^-$	153.65	pr (BuOH)	d	165			w, al	B4³, 754
8228	tris-(2-Hydroxyethyl) amine or Triethanolamine $(HOCH_2CH_2)_3N$	149.19	hyg cr	277^{150}	21.2	$1.1242^{20/4}$	1.4852^{20}	w, al, chl	B4⁴, 1524
8229	2-Hydroxybenzoic acid or Salicylic acid $2\text{-}HOC_6H_4CO_2H$	138.12	nd (w), mcl pr (al)	211^{20} sub	159	$1.443^{20/4}$	1.565	al, eth, ace, bz	B10³, 87

PHYSICAL CONSTANTS OF ORGANIC COMPOUNDS (Continued)

No.	Name, Synonyms, and Formula	Mol. wt.	Color, crystalline form, specific rotation and λ_{max} (log ε)	b.p. °C	m.p. °C	Density	n_D	Solubility	Ref.
8230	3-Hydroxybenzomide 3-$HOC_6H_4CONH_2$	137.14	pl (w)		170.5			al, eth	B10^{3}, 255
8231	3-Hydroxybenzamide-*N*-phenyl 3-$HOC_6H_4CONHC_6H_5$	213.24	nd or pl (w, dil al)		156			al	B12^{1}, 269
8232	3-Hydroxybenzoic acid 3-$HOC_6H_4CO_2H$	138.12	nd (w) pl or pr (al)		201-3			eth, ace, MeOH	B10^{3}, 242
8233	3-Hydroxybenzoic acid, 2-bromo 2-Br-3-$HOC_6H_3CO_2H$	217.02	nd (w)		160-1			eth	B10^{2}, 83
8234	3-Hydroxybenzoic acid, 4-bromo 4-Br-3-$HOC_6H_3CO_2H$	217.02	pl, nd (w)		214			al	B10^{3}, 258
8235	3-Hydroxybenzoic acid, 6-bromo 6-Br-3-$HOC_6H_3CO_2H$	217.02	(w)		185d			eth	B10^{3}, 258
8236	3-Hydroxybenzoic acid, 2-chloro 2-Cl-3-HO-$C_6H_3CO_2H$	172.57	lf (w or bz)		157-8			bz	B10^{3}, 257
8237	3-Hydroxybenzoic acid, 4-chloro 4-Cl-3-$HOC_6H_3CO_2H$	172.57	nd (w)		219-20				B10^{2}, 83
8238	3-Hydroxybenzoic acid, 6-chloro 6-Cl-3-$HOC_6H_3CO_2H$	172.57	(w)		178-9			al, ace	B10^{3}, 258
8239	3-Hydroxybenzoic acid, 4,5-dimethoxy 4,5-$(CH_3O)_2$-3-$HOC_6H_2CO_2H$	198.18	nd (aa or w)		197-8			al, aa	B10^{3}, 2073
8240	3-Hydroxybenzoic acid, 5,6-dimethoxy 5,6-$(CH_3O)_2$-3-$HOC_6H_2CO_2H$	198.18	pl or lf (w)		186-8			al	B10^{3}, 2060
8241	3-Hydroxybenzoic acid, 2,4-dinitro 2,4-$(NO_2)_2$-3-$HOC_6H_2CO_2H$	228.12	(w)		204			al, eth	B10^{3}, 264
8242	3-Hydroxybenzoic acid, ethyl ester or Ethyl 3-hydroxybenzoate 3-$HOC_6H_4CO_2C_2H_5$	166.18	pl (bz)	295, 211^{65}	73.8			al, eth	B10^{3}, 250
8243	3-Hydroxybenzoic acid, 4-formyl 4-HCO-3-$HOC_6H_3CO_2H$	166.13	nd (w)	sub	234			al, eth	B10, 954
8244	3-Hydroxybenzoic acid, 2-iodo 2-I-3-$HOC_6H_3CO_2H$	264.02	nd (chl)	158-9				al, eth	B10^{2}, 84
8245	3-Hydroxybenzoic acid, 4-iodo 4-I-3-$HOC_6H_3CO_2H$	264.02	nd (w)		226-8			al, eth	B10^{3}, 261
8246	3-Hydroxybenzoic acid, 6-iodo 6-I-3-$HOC_6H_3CO_2H$	264.02	nd (w)	sub 160	198			al, eth	B10^{3}, 261
8247	3-Hydroxybenzoic acid, 4-methoxy or Isovanillic acid 4-CH_3O-3-$HOC_6H_3CO_2H$	168.15	nd, pr, pl (w)	sub	255-7			al, eth	B10^{3}, 1404
8248	3-Hydroxybenzoic acid, methyl ester 3-$HOC_6H_4CO_2CH_3$	152.15	nd (bz-peth)	280^{709}, 178^{17}	71.5			al	B10^{3}, 249
8249	3-Hydroxybenzoic acid, 2-methyl or 2,3-Cresotic acid 2-CH_3-3-$HOC_6H_3CO_2H$	152.15	nd (w, dil al)		145-6			al, eth	B10^{3}, 494
8250	3-Hydroxybenzoic acid, 4-methyl or 3,4-Cresotic acid 4-CH_3-3-$HOC_6H_3CO_2H$	152.15	nd or pr (w)	sub	208.5			al, eth	B10^{3}, 527
8251	3-Hydroxybenzoic acid, 5-methyl or 3,5-Cresotic acid 5-CH_3-3-$HOC_6H_3CO_2H$	152.15	nd (w)	sub	210			al, eth	B10, 227
8252	3-Hydroxybenzoic acid, 6-methyl or 3,6-Cresotic acid 6-CH_3-3-$HOC_6H_3CO_2H$	152.15	nd or pr (w)		185			al, eth	B10, 215
8253	3-Hydroxybenzoic acid, 2-nitro 2-NO_2-3-$HOC_6H_3CO_2H$	183.12	pl or pr (w + 1)		180-1			al, eth	B10^{2}, 84
8254	3-Hydroxybenzoic acid, 4-nitro 4-NO_2-3-$HOC_6H_3CO_2H$	183.12	ye lf (w)		235			al, eth	B10^{3}, 263
8255	3-Hydroxybenzoic acid, 5-nitro 5-NO_2-3-$HOC_6H_3CO_2H$	183.12	ye lf or pl (25% HCl)		167			al, eth	B10^{2}, 85
8256	3-Hydroxybenzoic acid, 6-nitro 6-NO_2-3-$HOC_6H_3CO_2H$	183.12	ye nd or pr (w + 1)		172			al, eth	B10^{3}, 263
8257	3-Hydroxybenzoic acid, 4-sulfo 4-HO_3S-3-$HOC_6H_3CO_2H$	218.18	ye-gr nd (w + 2)		208 (213)			w, al	B11^{3}, 707
8258	3-Hydroxybenzoic acid, 5-sulfo 5-HO_3S-3-$HOC_6H_3CO_2H$	218.18	nd (w + 1)		120d			w, al, eth	B11^{3}, 708
8259	3-Hydroxybenzonitrile 3-HOC_6H_4CN	119.12	pr (al or eth) lf (w)		83-4			al, eth, bz, chl	B10^{3}, 255
8260	3-Hydroxybenzoyl chloride 3-HOC_6H_4COCl	156.57		110-3$^{0.5}$	<-15			chl	B10^{2}, 82
8261	4-Hydroxybenzamide 4-$HOC_6H_4CONH_2$	137.14	nd (w + 1)		162 (hyd)			al, eth	B10^{3}, 340

PHYSICAL CONSTANTS OF ORGANIC COMPOUNDS (Continued)

No.	Name, Synonyms, and Formula	Mol. wt.	Color, crystalline form, specific rotation and λ_{max} (log ε)	b.p. °C	m.p. °C	Density	n_D	Solubility	Ref.
8262	4-Hydroxybenzamide, *N*-phenyl 4-$HOC_6H_4CONHC_6H_5$	213.24	pl or nd (w)		201-2			al	$B12^3$, 947
8263	4-Hydroxybenzoic acid 4-$HOC_6H_4CO_2H$	138.12	pr or pl (w, al, xyl-al) cr (dil al or ace)		214-5			eth, ace	$B10^3$, 277
8264	4-Hydroxybenzoic acid, 2-bromo 2-Br-4-$HOC_6H_3CO_2H$	217.02	nd (w)		151			eth	$B10^2$, 103
8265	4-Hydroxybenzoic acid, 3-bromo 3-Br-4-$HOC_6H_3CO_2H$	217.02	nd or pr (+w) (w)		177			al, eth, aa	$B10^3$, 363
8266	4-Hydroxybenzoic acid, butyl ester or Butyl 4-hydroxy-benzoate 4-$HOC_6H_4CO_2C_4H_9$	194.23			68-9			al	$B10^3$, 307
8267	4-Hydroxybenzoic acid, 2-chloro 2-Cl-4-$HOC_6H_3CO_2H$	172.57	nd (w)		159			ace	$B10^3$, 360
8268	4-Hydroxybenzoic acid, 3-chloro 3-Cl-4-$HOC_6H_3CO_2H$	172.57	nd (w)	sub	170-2			al, eth, ace	$B10^2$, 102
8269	4-Hydroxybenzoic acid, 3,5-dichloro 3,5-Cl_2-4-$HOC_6H_2CO_2H$	207.01	nd (dil al or dil aa)	sub d	269			al, eth	$B10^3$, 362
8270	4-Hydroxybenzoic acid, 3,5-diido 3,5-I_2-4-$HOC_6H_2CO_2H$	389.92	nd (dil al)	d 260	237			al, eth	$B10^3$, 370
8271	4-Hydroxybenzoic acid, 3,5-diido, ethyl ester 3,5-I_2-4-$HOC_6H_2CO_2C_2H_5$	417.98	nd (dil al)		123			al	$B10^3$, 372
8272	4-Hydroxybenzoic acid, 2,3-dimethoxy 2,3-$(CH_3O)_2$-4-$HOC_6H_2CO_2H$	198.18	pl or lf (w or al)		154-5			al, ace, chl	$B10^2$, 332
8273	4-Hydroxybenzoic acid, 2,6-dimethoxy 2,6-$(CH_3O)_2$-4-$HOC_6H_2CO_2H$	198.18	pl (w)		175			bz, py	$B10^1$, 235
8274	4-Hydroxy benzoic acid, 3,5-dimethoxy 3,5-$(CH_3O)_2$-4-$HOC_6H_2CO_2H$	198.18	nd (w)		204-5			al, eth, ace, chl	$B10^3$, 2073
8275	4-Hydroxybenzoic acid, 3,5-dinitro 3,5-$(NO_2)_2$-4-$HOC_6H_2CO_2H$	228.12	ye lf (al)		248-9			al, eth	$B10^3$, 383
8276	4-Hydroxybenzoic acid, ethyl ester 4-$HOC_6H_4CO_2C_2H_5$	166.18	cr (dil al)	297-8	116-8			al, eth	$B10^3$, 300
8277	4-Hydroxybenzoic acid, 3-formyl 3-HCO-4-$HOC_6H_3CO_2H$	166.13	pr (w)	sub	244			al, eth	$B10^2$, 675
8278	4-Hydroxybenzoic acid, 2-iodo 2-I-4-$HOC_6H_3CO_2H$	264.02	nd (w)		215d			al, eth	$B10^2$, 104
8279	4-Hydroxybenzoic acid, 3-iodo 3-I-4-$HOC_6H_3CO_2H$	264.02	nd (w + ½)	sub	173-4			al, eth, aa	$B10^3$, 367
8280	4-Hydroxybenzoic acid, 3-methoxy or Vanillic acid 3-(CH_3O)-4-$HOC_6H_3CO_2H$	168.15	nd (w)	sub	213-5			eth	$B10^3$, 1403
8281	4-Hydroxybenzoic acid, 3-methoxy, ethyl ester or Ethyl 3-methoxy-4-hydroxybenzoate 3-CH_3O-4-$HOC_6H_3CO_2C_2H_5$	196.20	nd (dil al)	291-3	44			al, eth	B10, 397
8282	4-Hydroxybenzoic acid, 3-methoxy methyl ester 3-CH_3O-4-$HOC_6H_3CO_2CH_3$	182.18	nd (dil al)	285-7, 118^2	64			al, chl	$B10^3$, 1410
8283	4-Hydroxybenzoic acid, methyl ester or Methyl 4-hy-droxy-benzoate 4-$HOC_6H_4CO_2CH_3$	152.15	nd (dil al)	270-80d	131			al, eth, ace	$B10^3$, 296
8284	4-Hydroxybenzoic acid, 2-methyl or 4,2-Cresotic acid 2-CH_3-4-$HOC_6H_3CO_2H$	152.15	nd (w + ½)	236-7 sub	177-8			al, eth	$B10^3$, 494
8285	4-Hydroxybenzoic acid, 3-methyl or 4,3-Cresotic acid 3-CH_3-4-$HOC_6H_3CO_2H$	152.15	nd (w + ½)		174-5 sub			al, eth	$B10^3$, 512
8286	4-Hydroxybenzoic acid, 3-nitro 3-NO_2-4-$HOC_6H_3CO_2H$	183.12	nd or lf (w)		186-7			al, eth	$B10^3$, 376
8287	4-Hydroxybenzoic acid, propyl ester 4-$HOC_6H_4CO_2C_3H_7$	180.20	pr (eth)		96-8	$1.0630^{102/4}$	1.5050^{102}	eth	$B10^3$, 306
8288	4-Hydroxybenzoic acid, 5-*iso*-propyl-2-methyl or *p*-Thymotinic acid 5-$(CH_3)_2CH$-2-CH_3-4-$HOC_6H_2CO_2H$	194.23	pl (dil al)		157			al, eth, bz, chl	$B10^3$, 631
8289	4-Hydroxybenzoic acid, 3-sulfo 3-HO_3S-4-$HOC_6H_3CO_2H$	218.18	nd or lf (w)		d			w, al	$B11^3$, 709
8290	4-Hydroxybenzonitrile 4-HOC_6H_4CN	119.12	lf (w)		113			al, eth, chl	$B10^3$, 344
8291	Hydroxylamine, *N*-ethyl or β-Ethyl hydroxyl amine C_2H_5NHOH	61.08	nd (lig)		59-60d	$0.9079^{20/4}$	1.4152^{66}	w, al	$B4^3$, 1717

No.	Name, Synonyms, and Formula	Mol. wt.	Color, crystalline form, specific rotation and λ_{max} (log ε)	b.p. °C	m.p. °C	Density	n_D	Solubility	Ref.
8292	Hydroxylamine, *N*-methyl CH_3NHOH	47.06	hyg nd	62.5^{15}	87-8	1.0003$^{20/4}$	1.4164^{20}	w, al	B4^3, 1715
8293	Hydroxyquinol or 1,2,4-Trihydroxybenzene 1,2,4-$(HO)_3C_6H_3$	126.11	pl (eth), lf or pl (w)		140-1			w, al, eth	B6^3, 6276
8294	Hydroxyquinol, triacetate or 1,2,4-Triacetoxy benzene 1,2,4-$(CH_3CO_2)_3C_6H_3$	252.22	nd (MeOH)	>300	97-8			al	B6^3, 6282
8295	Hydroxyquinol, 3,5,6-trichloro or 3,5,6-trichloro-1,2,4-trihydroxybenzene $C_6H_3Cl_3O_3$	229.45	nd (bz, aa)		160			al, eth	B6^2, 1072
8296	Hyenic acid $C_{25}H_{50}O_2$	382.67	cr (eth) nd (bz)		77-8			eth	B2^2, 380
8297	Hygrine-*(l)* or 2-acetonyl-1-methylpyrrolidine $C_8H_{15}NO$	141.21	$[\alpha]_D$ −1.3	193-5, 92-4^{20}		0.935$^{17/4}$		al, chl	B21^4, 3257
8298	Hyoscine *(dl)* or Scopalamine $C_{17}H_{21}NO_4$	303.36	syr					al, eth, ace, bz	B27^1, 248
8299	Hyoscine, hydrobromide *(d)* $C_{17}H_{21}NO_4.HBr$	384.27	$[\alpha]_D$ + 26.3 (w)		195			w, al	B27^1, 247
8300	Hyoscine, hydrobromide *(dl)* $C_{17}H_{21}NO_4.HBr$	384.27	eff (ace)		185			w, al	B27^1, 248
8301	Hyoscine, hydrobromide *(l)* $C_{17}H_{21}NO_4.HBr$	384.27	lf (al) $[\alpha]^t_D$ −26		209			w	B27^2, 64
8302	Hyoscine, hydrobromide, trihydrate *(d)* $C_{17}H_{21}NO_4.HBr.\ 3\ H_2O$	438.32	ta (w) $[\alpha]^t_D$ + 26.3 (w,c=3)		55				B27^1, 247
8303	Hyoscine, hydrobromide, trihydrate *(dl)* $C_7H_{21}NO_4.HBr\text{-}3\ H_2O$	438.32			55-8				B27^1, 248
8304	Hyoscine, hydrobromide, trihydrate-*(l)* $C_{17}H_{21}NO_4.HBr\text{-}3\ H_2O$	438.32	ta (w + 3) $[\alpha]_D$-22.8 (w,c=2)					w	B27^2, 64
8305	Hyoscine, hydrochloride, dihydrate *(l)* $C_{17}H_{21}NO_4.HCl.2H_2O$	375.85	pr (w)		80			w, al	B27, 101
8306	Hyoscine, monohydrate *(dl)* $C_{17}H_{21}NO_4.H_2O$	321.37	cr (w + 1)		56-7			al, eth, chl	B27, 102
8307	Hyoscine, monohydrate-*(l)* $C_{17}H_{21}NO_4H_2O$	321.37	cr (w + 1) $[\alpha]^{20/}_D$-28 (w,c=2.7)		59			al, eth, ace, bz, chl	B27^2, 63
8308	Hyoscine, hydrochloride *(l)* $C_{17}H_{21}NO_4.HCl$	339.82	cr (al)		200			w, al	B27^2, 64
8309	Hyoscyamine *(d)* $C_{17}H_{23}NO_3$	289.37	nd (dil al) $[\alpha]^t_D$ + 31.3 (al,c=4)		106			al, eth, bz, chl	B21^4, 181
8310	Hyoscyamine *(l)* or Daturine $C_{17}H_{23}NO_3$	289.37	tetr nd (dil al)$[\alpha]^{20/}_D$-1 (al,c=1)		108.5			al, chl	B21^4, 181
8311	Hyoscyamine, hydrobromide-*(l)* $C_{17}H_{23}NO_3.HBr$	370.29	pr		152			w, al, chl	B21^4, 182
8312	Hyoscyamine, sulfate *(l)* $(C_{17}H_{23}NO_3)_2.H_2SO_4$	676.82	dlq nd (al or w)		206 anh			w, al	B21^4, 182
8313	Hyoscyamine, hydrochloride-*(l)* $C_{17}H_{23}NO_3.HCl$	325.84	$[\alpha]^t_D$ −23.2 (w,c=0.5)		149-51			w, al	B21^4, 182
8314	Hyoscyamine, sulfate dihydrate *(l)* $(C_{17}H_{23}NO_3)_2.H_2SO_4.2H_2O$	712.85	dlq nd (al, w) $[\alpha]^t_D$ −28.3 (w)		206			w, al, bz	B21^4, 182
8315	Hypaphorine *(d)* or *N,N*-Dimethyl-*L*-tryptophane betaine $C_{14}H_{18}N_2O_2$	246.31	cr (dil al) $[\alpha]^{25/}_D$ + 113.4 (w,c=1.6)		255d			w, al	B22^2, 469
8316	Hypnal or Antipyrine chloral hydrate $C_{11}H_{12}N_2O\cdot Cl_3CCH(OH)_2$	353.63	wh		68			w, al	B24^1, 196
8317	Hypoxanthene or 6-hydroxy purine $C_5H_4N_4O$	136.11	oct nd (w)		150 d				B26^2, 252

No.	Name, Synonyms, and Formula	Mol. wt.	Color, crystalline form, specific rotation and λ_{max} (log ε)	b.p. °C	m.p. °C	Density	n_D	Solubility	Ref.
8318	D-Iditol or 1,2,3,4,5,6-Hexane hexol $HOCH_2(CHOH)_4CH_2OH$	182.17	mcl pr (al) $[\alpha]^{20/}_D$ + 3.5 (w)	...	73.5	...	...	w	B1[4], 2843
8319	D-Idonic acid or 2,3,4,5,6-pentahydroxy hexanoic acid $HOCH_2(CHOH)_4CO_2H$	196.16	nd (w, al) $\alpha^{20/}_D$ + 5.2→-13,7 (mut)	...	205d	...	...	w	B3[4], 1257
8320	D-Idonic acid-γ-lactone $C_6H_{10}O_6$	178.14	pl $\alpha^{20/}_D$ 52.6 (w)	...	174	...	...	w	B18[4], 3026
8321	D-Idonic acid, phenylhydrazide $HOCH_2(CHOH)_4CONHNHC_6H_5$	286.28	$[\alpha]^{20/}_D$ −15.´ (w,c=1)	...	115-7	...	...	w, al	B15[3], 208
8322	D-Idose $C_6H_{12}O_6$	180.16	syr $[\alpha]^{23/}_D$ + 16 (w)	...	...	...	...	w	B1[4], 4335
8323	L-Idose $C_6H_{12}O_6$	180.16	syr $[\alpha]^{20/}_D$ −17.4 (w,c = 6.2)	...	...	...	...	w	B1[4], 4336
8324	Imesatin or Isatin-3-imide $C_8H_6N_2O$	146.15	dk ye pr (dil al)	...	175-6	...	...	al	B21[4], 4984
8325	Imidazole or 1,3-Diazole. Glyoxaline $C_3H_4N_2$	68.08	ncl pr (bz)	257, 138.2^{12}	90-1	$1.0303^{101/4}$	1.4801^{101}	w, al, eth, ace, chl	B23[2], 34
8326	Imidazole, 2-mercapto-1-methyl or Methimazole $C_4H_6N_2S$	114.17	lf (al)	280d	142	...	...	w, al, chl	B24, 17
8327	Imidazole, l-ethyl $C_5H_8N_2$	96.13	...	209-10	...	0.999	...	w	B23, 46
8328	Imidazole, 1-methyl or Oxalmethylene $C_4H_6N_2$	82.11	...	195-6, .94-5	−6	1.0325^{20}	1.4970^{20}	w, al, eth, ace	B23[2], 35
8329	Imidazole, 4-methyl $C_4H_6N_2$	82.11	...	263, $120^{0.02}$	56	$1.0416^{14.3/4}$	$1.5037^{14.3}$	w, al	B23[2], 60
8330	Imidazole, 1-phenyl $C_9H_8N_2$	144.18	...	276,$153\text{-}4^{23}$	13	...	1.6025^{25}	eth, ace	B23[2], 36
8331	4,5-Imidazoledicarboxylic acid or 1,3-Diazole-4,5-dicarboxylic acid $C_5H_4N_2O_4$	156.10	pr	288d	...	1.749	...	...	B25[2], 159
8332	2-Imidazolidine thione or *N,N*-Ethylene-thiourea $C_3H_6N_2S$	102.15	nd (al), pr (al)	...	200-3	...	...	w, al	B24[2], 4
8333	2-Imidazolidone or Ethylene urea $C_3H_6N_2O$	86.09	nd (chl)	...	131-3	...	...	w, al	B24[2], 3
8334	2-Imidazoline, 2-methyl or 2-methyl-2-glyoxalidine $C_4H_8N_2$	84.12	hyg	195-8	107	...	...	w, al, chl	B23[2], 26
8335	Imino-di-acetic acid or Diglycolamidic acid $HN(CH_2CO_2H)_2$	133.10	rh pr	...	247.5	...	...	...	B4[3], 1176
8336	Imperatorin or Ammidin $C_{16}H_{14}O_4$	270.28	cr (al)	...	102	...	...	al, eth, ace, bz	B19[4], 2635
8337	Indaconitine or Acetylbenzoyl pseudoaconine $C_{34}H_{47}NO_{10}$	629.75	cr	...	202-3d	...	...	al, eth, chl	B21[4], 2889
8338	Indan or 2,3-dihydroindene C_9H_{10}	118.18	...	178, 73^{13}	−51.4	$0.9639^{20/4}$	1.5978^{20}	al, eth	B5[4], 1371
8339	Indan, 1-amino or *dl*-1-hydrindamine $C_9H_9NH_2$	133.19	oil	220.5^{747}, $96\text{-}7^{8}$	...	$1.038^{15/4}$	1.5613^{20}	eth, ace, bz	B12[3], 2798
8340	Indan-5-amino or 5-Hydrindamine $5\text{-}H_2NC_9H_9$	133.19	nd (peth)	$247\text{-}9^{745}$, 131^{15}	37-8	...	...	eth, ace, bz	B12[3], 2800
8341	Indan, 2,3-dibromo or Indene dibromide $2,3\text{-}Br_2C_9H_8$	275.97	...	144^{10}	31-2	$1.747^{25/4}$	1.6290^{25}	eth	B5[3], 1203
8342	Indan, 2,3-dichloro or Indene dichloride $2,3\text{-}Cl_2C_9H_8$	187.07	...	$87\text{-}90^{2}$	...	$1.254^{25/4}$	1.5715^{23}	...	B5[3], 1202
8343	Indan, 1,1-dimethyl $C_{11}H_{14}$	146.23	...	191	...	0.919^{20}	$1.5135^{25/}_D$	...	B5[4], 1415
8344	Indan, 1,2-dimethyl $1,2\text{-}(CH_3)_2C_9H_8$	146.23	...	$79\text{-}80^{10}$	...	$0.927^{20/4}$	$1.5186^{20/}_D$	...	B5[4], 1415
8345	Indan, 4,6-dimethyl $4,6\text{-}(CH_3)_2C_9H_8$	146.23	...	52.9^{1}	...	...	$1.5325^{20/}_D$	...	B5[2], 395
8346	Indan, 4,7-dimethyl $4,7\text{-}(CH_3)_2C_9H_8$	146.23	...	$94\text{-}7^{10}$	...	$0.949^{20/4}$	$1.5342^{20/}_D$	...	B5[4], 1416

PHYSICAL CONSTANTS OF ORGANIC COMPOUNDS (Continued)

No.	Name, Synonyms, and Formula	Mol. wt.	Color, crystalline form, specific rotation and λ_{max} (log ε)	b.p. °C	m.p. °C	Density	n_D	Solubility	Ref.
8347	Indan, 5,6-dimethyl $5\text{-}6\text{-}(CH_3)_2C_9H_8$	146.23	....	94^{10}	......	0.9449^{20}	$1.5360^{20}/_D$	....	$B5^4$, 1416
8348	Indan, 1-ethyl $1\text{-}C_2H_5C_9H_9$	146.23	....	222	......	0.9348^{25}	$1.5121^{25}/_D$	....	$B5^4$, 1414
8349	Indan, 5-hexyl $5\text{-}C_6H_{13}C_9H_9$	202.34	....	292.1	......	0.9114^{20}	1.5122^{2D}	....	$B5^4$, 1474
8350	Indan, 1-methyl $1\text{-}CH_3C_9H_9$	132.21	....	188-90, 60^{10}	......	$0.9402^{20/4}$	1.5260^{20}	....	$B5^4$, 1397
8351	Indan, 2-methyl $2\text{-}CH_3C_9H_9$	132.21	....	187, 70^{10}	......	$0.9034^{20/4}$	1.5070^{20}	....	$B5^4$, 1397
8352	Indan, 4-methyl $4\text{-}CH_3C_9H_9$	321.21	....	205.5	......	0.9577^{20}	$1.5356^{20}/_D$	....	$B5^4$, 1397
8353	Indan, 5-methyl $5\text{-}CH_3C_9H_9$	132.21	....	$101\text{-}2^{32}$, 82^{17}	......	$0.9494^{20/4}$	1.5316^{20}	....	$B5^4$, 1398
8354	Indan, 4-nitro $4\text{-}O_2NC_9H_9$	163.18	wh cr (al)	139^{10}	44	....	....	....	$B5^3$, 1204
8355	Indan, per hydro C_9H_{16}	124.23	....	165-6	......	0.8334^{20}	$1.4629^{20}/_D$	....	$B5^4$, 292
8356	Indan, 1-phenyl-1,3,3-trimethyl $1\text{-}C_6H_5\text{-}1,3,3\text{-}(CH_3)_3C_9H_6$	236.36	tcl pr (al)	307-10, $161\text{-}5^{12}$	52-3	$1.0009^{20/4}$	1.5681^{20}	bz, MeOH	$B5^4$, 2246
8357	Indan, 1,1,4,7-tetramethyl $1,1,4,7(CH_3)_4C_9H_6$	174.29	....	114.2^{15}	......	0.934^{25}	$1.5216^{25}/_D$	....	$B5^3$, 1282
8358	Indan, 1,1,3-trimethyl $1,1,3\text{-}(CH_3)_3C_9H_7$	160.26	....	204^{748}	......	....	1.5082^{20}	....	$B5^4$, 1433
8359	Indan, 1,1,4-trimethyl $1,1,4\text{-}(CH_3)_3C_9H_7$	160.26	....	52.5^{1}	......	....	$1.5157^{20}/_D$	....	$B5^4$, 1433
8360	Indan, 1,1,5-trimethyl $1,1,5\text{-}(CH_3)_3C_9H_7$	160.26	....	$86.^{10}$	......	0.9119^{20}	1.5126^{20}	....	$B5^4$, 1433
8361	Indan, 1,1,6-trimethyl $1,1,6\text{-}(CH_3)_3C_9H_7$	160.26	....	55^{1}	......	....	1.5134^{20}	....	API 23, 2(35.5202)
8362	Indan, 1,4,7-trimethyl $(CH_3)_3C_9H_7$	160.26	....	95^{10}	......	0.938^{20}	1.5252^{20}	....	$B5^4$, 1434
8363	Indan, 1,5,7-trimethyl $1,5,7\text{-}(CH_3)_3C_9H_7$	160.26	....	106.1^{15}	......	....	1.5231^{25}	....	$B5^3$, 1267
8364	Indan, 4,5,7-trimethyl $4,5,7\text{-}(CH_3)_3C_9H_7$	160.26	....	59.9^{1}	......	....	1.5322^{20}	....	API 23, 2(35.5202)
8365	1,2-Indandione or α, β -Dioxohydrindene $C_9H_6O_2$	146.15	gold ye pl or lf (bz, eth)	....	114-6	....	....	al, chl	$B7^3$, 3593
8366	1,2-Indandione, β -oxime $C_9H_6O(=NOH)$	161.16	nd (al), (bz)	....	215-20d	....	....	al, bz	$B7^3$, 3593
8367	1,3-Indandione or 1,3-Dioxohydrindene $C_9H_6O_2$	146.15	nd (eth, lig)	....	131-2d	1.37^{21}	....	al, eth, bz	$B7^3$, 3594
8368	1,3-Indandione, 2,2-dimethylpropoxy or Pivalyl indandione $C_{14}H_{14}O_3$	230.26	(dil al)	....	108-10	....	....	al, eth, ace	$B7^3$, 4599
8369	1,3-Indandione, dioxime $1,3\text{-}C_9H_6(=NOH)_2$	176.17	nd (w)	....	ca 225d	....	....	....	B7, 695
8370	1,3-Indandione, 2(3-methylbutoxy) or Valone $C_{14}H_{14}O_3$	230.26	ye (dil al)	....	67-8	....	....	al, eth, ace	$B7^3$, 4598
8371	1,3-Indandione, 2-phenyl or Danilone $C_{15}H_{10}O_2$	222.24	lf (al, bz)	....	149-51	....	....	al, eth, ace, bz, chl	$B7^3$, 4100
8372	1-Indanol or -1-hydroxyindan $1\text{-}HOC_9H_9$	134.18	pl (peth)	255, 128^{12}	54	....	....	al, eth, bz, chl	$B6^3$, 2423
8373	4-Indanol or 4-Hydroxy Indan $4\text{-}HOC_9H_9$	134.18	(i) tcl pr (peth) (ii) nd (peth)	120^{12}	(i) 50 (ii) 40	....	....	....	$B6^3$, 2427
8374	5-Indanol or 5-Hydroxy indan $5\text{-}HOC_9H_9$	134.18	nd (peth)	225, 110^{8}	56	....	....	al, eth	$B6^3$, 2428
8375	1-Indanone or α-Hydrindone C_9H_8O	132.16	ta, nd (w + 3)	$241\text{-}2^{739}$, 129^{12}	42	$1.1028^{40/40}$	1.561^{25}	al, eth, ace, chl, lig	$B7^3$, 1392
8376	1-Indanone, 3,3-dimethyl $C_{11}H_{12}O$	160.22	....	$130\text{-}1^{18}$	......	$1.0320^{14.5/4}$	$1.5453^{14.5}$	....	$B7^3$, 1450
8377	1-Indanone, 2-nitro $C_9H_7NO_3$	177.16	ye nd (bz-lig)	....	117d	....	....	w, al, eth, ace, bz	$B7^1$, 192
8378	1-Indanone, 6-nitro $C_9H_7NO_3$	177.16	ye lf or nd (peth, al)	....	74	....	....	al, eth, ace, bz, chl	$B7^2$, 285

No.	Name, Synonyms, and Formula	Mol. wt.	Color, crystalline form, specific rotation and λ_{max} (log ε)	b.p. °C	m.p. °C	Density	n_D	Solubility	Ref.
8379	2-Indanone or β-Hydrindone C_9H_8O	132.16	nd (al or eth)	218d	59(61)	$1.0712^{69/4}$	1.538^{67}	al, eth, ace, chl	B7³, 1397
8380	2-Indanone, 5-nitro $C_9H_7NO_3$	177.16	br nd (al)		141			al, eth, aa	B7, 364
8381	Indanthrene or Dihydroanthraquinonazine-indanthrone $C_{28}H_{14}N_2O_4$	442.43	bl nd		470-500d				B24², 317
8382	Indazole or 1,2-Benzodiazole $C_7H_6N_2$	118.14	nd (al or w)	$267\text{-}70^{743}$	147-9			al, eth	B23², 117
8383	Indazole, 1-benzhydrylidene or ω,ω-Diphenyl benzofulvene $C_{22}H_{16}$	280.37	og-ye (al)		114.5			eth, ace, bz	B5³, 2532
8384	Indazole, 3-chloro $C_7H_5ClN_2$	152.58	nd (w or lig)	sub	148			al, eth, bz	B23², 139
8385	Indazole, 4-chloro $C_7H_5ClN_2$	152.58	nd (to)		156			w, al, eth, ace	B23², 139
8386	Indazole, 2-methyl $C_8H_8N_2$	132.16		261, 135^{16}	56			al, eth, ace	B23², 118
8387	Indazole, 1-methyl $C_8H_8N_2$	132.16		231, 109^{17}	60-1			al, ace, eth	B23², 118
8388	Indazole, 3-methyl $C_8H_8N_2$	132.16		280-1, $169\text{-}74^{9}$	113			al, eth, ace	B23², 155
8389	Indazole, 5 methyl $C_8H_8N_2$	132.16		$293\text{-}4^{747}$	117				B23², 157
8390	Indazole, 4-nitro $C_7H_5N_3O_2$	163.14	nd (w)		205-7			al, eth, ace, bz, aa	B23², 144
8391	Indazole, 5-nitro $C_7H_5N_3O_2$	163.14	yesh nd or col nd (al)		208			al, eth, ace, bz, aa	B23², 145
8392	Indazole, 6-nitro $C_7H_5N_3O_2$	163.14	nd (w, al, aa)		181d			al, eth, ace, bz	B23², 146
8393	Indazole, 7-nitro $C_7H_5N_3O_2$	163.14	cr (al)		188-90 sub			eth, ace	B23², 150
8394	3-Indazolinone or Benzo pyrazolone $C_7H_6N_2O$	134.14	nd or lf (w or MeOH) pl or nd (al)		250-2				B24², 59
8395	Indene or Indonaphthene C_9H_8	116.16	(aa)	182.6	−1.8	$0.9960^{25/4}$	1.5768^{20}	al, eth, ace, bz, py	B5⁴, 1532
8396	Indene, 1,2-Diphenyl 1,2-$(C_6H_5)_2C_9H_6$	268.36	lo nd (aa)		177-8			eth	B5³, 2473
8397	Indene, 1,3-diphenyl 1,3-$(C_6H_5)_2C_9H_6$	268.36	(i) nd (aa), (ii) pym (aa)	230^{15}	(i) 68-9 (ii) 85			eth, ace	B5³, 2474
8398	Indene, 2,3-diphenyl 2,3-$(C_6H_5)_2C_9H_6$	268.36	pr (aa)	$235\text{-}40^{12}$	108-9			eth, ace, bz	B5³, 2473
8399	Indene, 2-methyl 2-$CH_3C_9H_7$	130.19	oil	187, $62\text{-}5^{20}$		$0.9034^{20/4}$	1.5070^{20}	eth, ace, bz	B5⁴, 1545
8400	Indene, 3-methyl 3-$CH_3C_9H_7$	130.19		198.5d, 70^{10}		$0.9640^{20/4}$	1.5591^{27}	eth, ace, bz	B5⁴, 1545
8401	1-Indenecarboxylic acid 1-$C_9H_7CO_2H$	160.17	pa ye nd(bz)	$193\text{-}5^{12}$	161			eth	B9³, 3068
8402	2-Indenecarboxylic acid 2-$C_9H_7CO_2H$	160.17	nd or lf(bz)		234 sub			al, eth	B9³, 3069
8403	1-Indenone, 2,3-diphenyl $C_{21}H_{14}O$	282.34	og-red (lig or al)		153-5			al, ace, bz	B7³, 2861
8404	Indican or Indoxyl-β-glucoside $C_{14}H_{17}NO_6.3H_2O$	349.34	orh nd (w + 3), $[\alpha]^{19/}{}_{546}$ -65.6 (w, c=1)		57-8 (hyd) 178-80d (anh)			w, al, ace	B21⁴, 748
8405	Indigo white or 2,2′-Diindoxyl leucoindigo $C_{16}H_{12}N_2O_2$	264.28	ye cr (dil al)					al, eth	B23², 429
8406	Indigotin $C_{16}H_{10}N_2O_2$	262.27			390-2d	1.35			B24², 233
8407	4,4′-Indigotin dicarboxylic acid $C_{18}H_{10}N_2O_6$	350.29	bl nd						B25, 273
8408	5,5′-Indigotin disulfonic acid, sodium Salt or Indigo carmine $C_{16}H_{18}N_2O_8Na_2S_2$	476.43	dk bl amor or br-red cr					w, al	B25², 298

No.	Name, Synonyms, and Formula	Mol. wt.	Color, crystalline form, specific rotation and λ_{max} (log ε)	b.p. °C	m.p. °C	Density	n_D	Solubility	Ref.
8409	Indigotin sulfonic acid $C_{16}H_{10}N_2S_2O_8$	422.38	amor		200d			w, al	$B25^2$, 246
8410	Indirubin or Indigo red $C_{16}H_{10}N_2O_2$	262.27	red or br rh nd (sub)		sub			eth	$B24^2$, 246
8411	Indole or 1-Benzo [b] pyrrole C_8H_7N	117.15	lf (w, peth) cr (eth)	254, $123\text{-}4^5$	52.5	1.22		al, eth, bz, lig	$B20^4$, 3176
8412	Indole, 1-acetyl $1\text{-}CH_3CO(C_8H_6N)$ $C_{10}H_9NO$	159.19		$152\text{-}3^{14}$ $100^{0.001}$				eth, ace	$B20^4$, 3182
8413	Indole, 3-(2-aminoethyl) or Tryptamine $C_{10}H_{12}N_2$	160.22	nd (al-bz or liq)	$137^{0.15}$	120 (146)			al, ace	$B22^4$, 4319
8414	Indole, 1,3-dimethyl or N-Methyl skatole $1,3\text{-}(CH_3)_2C_8H_5N$	145.20	nd	257-60, 119^7	141-3			eth	$B20^4$, 3208
8415	Indole, 2,3-dimethyl $C_{10}H_{11}N$ $2,3\text{-}(CH_3)_2C_8H_5N$	145.20		285	105-7				$B20^4$, 3226
8416	Indole, 2-hydroxy-3-nitroso or β-Isatoxime $C_8H_6N_2O_2$	162.15	gold-ye nd		214			al	$B21^4$, 4988
8417	Indole, 3-hydroxy or Indoxyl $3\text{-}HOC_8H_6N$	133.15	bt ye pr		85			w, al, eth, ace, bz	$B21^3$, 746
8418	Indole, 1-methyl $1\text{-}CH_3C_8H_6N$	131.18		240-1, $70\text{-}5^2$		1.0707^0		al, eth, bz	$B20^4$, 3180
8419	Indole, 2-methyl $2\text{-}CH_3C_8H_6N$	131.18	pl(dil al)nd or lf(w)	272	61	$1.07^{20/4}$		al, eth, ace	$B20^4$, 3202
8420	Indole, 3-methyl or Skatole $3\text{-}CH_3C_8H_6N$	131.18	lf(liq)	$265\text{-}6^{755}$	97-8			w, al, eth, ace, bz	$B20^4$, 3206
8422	Indole, 3-methyl-2-phenyl $3\text{-}CH_3\text{-}2\text{-}C_6H_5C_8H_5N$	207.27		$280\text{-}90^{120}$	91-2			al, bz	B20, 474
8423	Indole, 2-phenyl $2\text{-}C_6H_5C_8H_6N$	193.25		250^{10}	189			eth, bz	$B20^2$, 302
8424	Indole, 1,2,3-trimethyl $1,2,3\text{-}(CH_3)_3C_8H_4N$	159.23		$283\text{-}4^{750}$	18				$B20^4$, 3227
8425	3-Indolylacetic acid, 2-methyl $C_{11}H_{11}NO_2$	189.21	ace	195-200				al, eth, ace	$B22^4$, 1117
8426	2-Indolecarboxylic acid $C_9H_7NO_2$	161.16	ye pl(bz peth)		205-8			al, eth	$B22^4$, 1059
8427	2-Indolecarboxylic acid, 3-hydroxy or Indoxylic acid $C_9H_7NO_3$	177.16	cr(sub)	122-3 sub					$B22^2$, 168
8428	Indoline or 2,3-Dihydroindole C_8H_9N	119.17		228-30, $70\text{-}5^2$		$1.069^{20/4}$	1.5923^{20}	eth, ace, bz	$B20^4$, 2896
8429	3-Indolylacetic acid or Heteroauxin $(C_8H_6N)CH_2CO_2H$	175.19	lf(bz)pl(chl)		165-6			al, eth, ace	$B20^4$, 1088
8430	Indone-2,3-dibromo $C_9H_4Br_2O$	289.95	og-ye nd(al aa)		123			al, eth, chl	$B7^3$, 1647
8431	Indophenin $C_{24}H_{14}N_2O_2S_2$	426.51	bl nd or pw		d				$B21^2$, 330
8432	Indopenol $C_{12}H_9NO_2$	199.21	red br pl (ace-peth)		160			w, al, eth, bz, chl	$B13^1$, 1047
8433	Indoxazene or 4,5-Benzoisoxazol C_7H_5NO	119.12	oil	100^{26}		$1.1727^{21/4}$	1.5570^{20}	eth	$B27^2$, 15
8434	Inosine or Hyoxanthosine-β- -ribose $C_{10}H_{12}N_4O_5$	268.23	pl (w + 2) nd (80% al)		90 (+ 2w) 218d (anh)			al	B31, 25
8435	D-Inositol or 1,2,3,4,5,6-cyclohexanehexol $C_6H_{12}O_6$	180.16	pr (w + 2) (al), $[\alpha]_D$ + 65.0 (w, 12%)		249-50			w, aa	$B6^3$, 6925
8436	DL-Inositol or Phaseo mannitol $C_6H_{12}O_6$	180.16	mcl pr (w)cr(gl aa)	319(vac)	253	1.752^{15}		w, aa	$B6^3$, 6925
8437	L-Inositol $C_6H_{12}O_6$	180.16	nd(w + 2)	250 vac	247	1.598^{20}		w, aa	$B6^3$, 6925
8438	Inulin or Plant starch $(C_6H_{10}O_5)n$	~7000	wh amor or pw, $[\alpha]^{21}\Sigma_D$ - 38.3		178d	$1.35^{20/4}$			J1952, 2384

PHYSICAL CONSTANTS OF ORGANIC COMPOUNDS (Continued)

No.	Name, Synonyms, and Formula	Mol. wt.	Color, crystalline form, specific rotation and λ_{max} (log ε)	b.p. °C	m.p. °C	Density	n_D	Solubility	Ref.
8439	Iodoacetamide ICH_2CONH_2	184.96	cr(w)		95			w	B2^{4}, 536
8440	Iodoacetic acid ICH_2COOH	185.95	pl(w,peth)	d	83			w, al	B2^{4}, 534
8441	Iodoacetic acid, ethyl ester or Ethyliodoacetate $ICH_2COOC_2H_5$	214.00	oil	178-80, 73^{16}		$1.8173^{13/4}$	1.5079^{13}	al, eth	B2^{4}, 535
8442	Iodogorgoic acid (d) or 3,5-Diidotyrosine(d) $C_9H_9I_2NO_3$	432.98	yesh nd(w or 70% al), $[\alpha]^{20}{}_D$ + 2.9 (HCl, c=5)		213				B14^{3}, 1563
8443	Iodogorgoic acid (dl) $C_9H_9I_2NO_3$	432.98	nd (50%) al, pl (w)		200d				B14^{3}, 1563
8444	Iodogorgoic acid (l) $C_9H_9I_2NO_3$	432.98	nd(w or70%), $[\alpha]^{20}{}_D$ -2.98 (4% HCl, c=5)	213d				B14^{2}, 366	B14^{3}, 1563
8445	Ionene $C_{13}H_{18}$	174.29		$238\text{-}9^{730}$, $90\text{-}1^{4}$		$0.9356^{20/4}$	1.5257^{20}	al, eth, bz, chl	B5^{4}, 1445
8446	α-Ionol $C_{13}H_{22}O$	194.32	oil	127^{15}		$0.9474^{20/4}$	1.4735^{20}	al, eth, ace	B6^{3}, 402
8447	β-Ionol $C_{13}H_{22}O$	194.32		131^{15}, $89^{0.7}$		$0.9243^{20/4}$	1.4969^{20}	al, eth, ace	B6^{3}, 401
8448	α-Ionone (d,trans) or 4-(2,6,6-trimethyl-2-cyclohenyl)-3-buturic-2-one $C_{13}H_{20}O$	192.30	$[\alpha]^{25}{}_D$ + 347				$1,5061^{20}$		B7^{3}, 640
8449	α-Ionone (dl) $C_{13}H_{20}O$	192.30		$146\text{-}7^{28}$		0.9298^{21}	1.5041^{20}	al, eth, ace	B7^{3}, 641
8450	α-Ionone (l) $C_{13}H_{20}O$	192.30	$[\alpha]^{27}{}_D$ -406	$73\text{-}7^{0.15}$			1.5000^{25}	al, eth, ace	B7^{3}, 640
8451	α-Ionone, semicarbazone (dl) $C_{14}H_{23}N_3O$	249.36	(60% al)		(i) 107-8, (ii) 143				B7^{3}, 644
8452	β-Ionone $C_{13}H_{20}O$	192.30		140^{18}, $72\text{-}4^{0.1}$		$0.9462^{20/4}$	1.5198^{20}	al, eth	B7^{3}, 634
8453	β-Ionone, semicarbazone $C_{14}H_{23}N_3O$	249.36	nd(al)		149			al, eth, bz, chl	B7^{3}, 639
8454	Irene or anhydro irone (?) $C_{14}H_{20}$	188.31		$120\text{-}5^{10}$		$0.9332^{20/4}$	1.5217^{20}	ace, bz	B5^{4}, 1460
8455	β-Irone or 4-(2,5,6,6-tetramethyl-1-cyclohexenyl)-3-butene-2-one $C_{14}H_{22}O$	206.33		$85\text{-}90^{0.1}$		$0.9434^{21/4}$	1.5017^{20}	al, eth, bz, chl, lig	B7^{3}, 666
8456	Isatic acid or 2-Aminobenzoyl formic acid $H_2NC_6H_4COCO_2H$	165.15	pw		d			w	B14^{3}, 1650
8457	Isatin or 2,3-Indolinedione $C_8H_5NO_2$	147.13	yesh red pr(sub)	sub	203-5			al, ace, bz	B21^{4}, 4981
8458	Isatin, 1-acetyl $C_{10}H_7NO_3$	189.17	ye pr or nd		144-5			al, ace	B21^{4}, 4998
8459	Isatin, chloride or 2-chloro-3-indolone C_8H_4ClON	165.58	br nd		180d			al, eth, aa	B21^{4}, 3720
8460	Isatin, 1-methyl $C_9H_7NO_2$	161.16	red-ye orh nd (w)		134			al, eth, ace, bz	B21^{4}, 4991
8461	Isatin, 5-methyl $C_9H_7NO_2$	161.16	red pl (w)nd (w or al)		187			al	B21^{4}, 5451
8462	Isatin, 7-methyl $C_9H_7NO_2$	161.16			267				B21^{4}, 5454
8463	Isatin, 5-nitro $C_8H_4N_2O_4$	192.13	ye nd al		254-5d				B21^{4}, 5015
8464	Isatin, 2-oxime $C_8H_6N_2O_2$	162.15	ye-og nd al or w		198-200d			eth, ace, aa	B21^{4}, 4987
8465	Isatin, 3-oxime $C_8H_6N_2O_2$	162.15	gold ye nd		225d			al	B21^{4}, 4988
8466	Isatoic acid, anhydride or N-carboxyanthranilic anhydride $C_8H_5NO_3$	163.13	pr(al or gl aa)cr (al)		243d				B27^{2}, 299

No.	Name, Synonyms, and Formula	Mol. wt.	Color, crystalline form, specific rotation and λ_{max} (log ε)	b.p. °C	m.p. °C	Density	n_D	Solubility	Ref.
8467	d-α-Isatropic acid or d,α-1-phenyl-1,4-tetralindicarboxylic acid $C_{18}H_{16}O_4$	296.32	pr, $[\alpha]^{25/}{}_D$ + 9.44 (al, c=12.6)		239d				B9[1], 417
8468	dl-α-Isatropic acid or dl-α-1-phenyl-1,4-tetralindicarboxylic acid $C_{18}H_{16}O_4$	296.32	cr(chl-peth)		238-9			aa	B9[3], 4635
8469	dl-β-Isatropic acid $C_{18}H_{16}O_4$	296.32	pl w		208-9			al, aa	B9[3], 4635
8470	l,β-Isatropic acid $C_{18}H_{16}O_4$	296.32	$[\alpha]_D$ -8.8 (al, c=5)		197			al, aa	B9[3], 4634
8471	Isoapiol $C_{12}H_{14}O_4$	222.24	mcl pr lf or nd (al)	303-4, 189^{33}	56			al, eth, ace, bz	B19[4], 1033
8472	Isobergaptene $C_{12}H_8O_4$	216.19	cr(al)		222-3			al, diox	B19[4], 2638
8473	Isoborneol (d) or 2-Hydroxybornane $C_{10}H_{18}O$	154.25	cr(peth)		212 (sealed tube)			al, eth, bz, chl	B6[3], 299
8474	Isoborneol (dl) or α,β-Camphol $C_{10}H_{18}O$	154.25	ta (peth)	sub	212(sealed tube)			al, eth, chl	B6[4], 282
8475	Isoborneol (l) or β-Camphol $C_{10}H_{18}$	154.25	(peth), $[\alpha]_D$ + 33.9 (al)		214 (218)			al, eth, chl	B6[4], 282
8476	Isoborneol, acetate (d) $CH_3CO_2C_{10}H_{17}$	196.29	$[\alpha]^{20/}{}_D$ -50.2 (al)	112^{17}		$0.9905^{20/4}$	1.4633^{20}	al, ace	B6[2], 90
8477	Isoborneol, acetate (dl) $CH_3CO_2C_{10}H_{17}$	196.29		$115\text{-}7^{21}$		$0.9841^{20/4}$	1.4640^{20}	al, ace	B6[4], 283
8478	Isoborneol, acetate (l) $CH_3CO_2C_{10}H_{17}$	196.29		225, $123\text{-}7^{35}$	<-50	$1.002^{11/11}$		al, ace	B6, 89
8479	Isoborneol, formate (d) $HCO_2C_{10}H_{17}$	182.26	$[\alpha]^{20/}{}_D$ + 129.5 (al, c=5)	94^{15}		$1.0136^{20/4}$	1.4678^{22}	al, ace	B6[3], 301
8480	Isobornylamine $C_{10}H_{17}NH_2$	153.27	pw, $[\alpha]_D$ -47.7 (4% al)		184			eth, ace	B12[3], 195
8481	Isocalycanthine $C_{22}H_{28}N_4.H_2O$	366.51	rh		235-6			al, eth, ace, chl	Am32, 1305
8482	Isocamphane (d) or Dihydrocamphene $C_{10}H_{18}$	138.25	cr (MeOH), $[\alpha]^{20/}{}_D$ + 8.68 (bz, p=20)	166^{750}	62-3			al, ace	B5[3], 263
8483	Isocamphane (dl) $C_{10}H_{18}$	138.25	cr (MeOH)	$165\text{-}6^{730}$	65-7	$0.8276^{67/4}$	1.4419^{67}	al, ace, bz	B5[3], 263
8484	Isocamphane (l) $C_{10}H_{18}$	138.25	cr (al), $[\alpha]^{20/}{}_D$ -8.5 (al)	$164\text{-}5^{757}$, $62\text{-}3^{17}$	64			al, ace	B5[3], 263
8485	Isocamphoric acid (d) or α-*trans*-1,2,2-Trimethyl-1,3-cyclopentane-dicarboxylic acid $C_{10}H_{16}O_4$	200.23	lf (w), $[\alpha]^{20/}{}_D$ + 48.6		171-2			al, aa	B9, 762
8486	Isocamphoric acid (dl) $C_{10}H_{16}O_4$	200.23	pl(al or gl aa)cr(w)		197	1.249		eth	B9[3], 3879
8487	Isocamphoric acid (l) $C_{10}H_{16}O_4$	200.23	tetr, $[\alpha]^{17/}{}_D$ -48.4 (MeOH, p=9.9)		173			al, aa	B9[3], 3878
8488	Isocarotene or Dehydro-β-carotene $C_{40}H_{54}$	534.87	vt pr(bz-MeOH) vt nd lf(bz)		192-3				B5[3], 2515
8489	Isocarvomenthol (d) $C_{10}H_{20}O$	156.27	$[\alpha]_D$ + 20.2	110^{20}		$0.904^{20/4}$	1.4669^{18}	al, eth, ace	B6[3], 131
8490	Isocarvomenthol (l) $C_{10}H_{20}O$	156.27	$[\alpha]^{16/}{}_D$ -17.7	106^{17}		$0.9109^{20/4}$	1.4662^{20}	al, eth, ace	B6[4], 149
8491	Isocodeine $C_{18}H_{21}NO_3$	299.37	pl (bz) pr (AcoEt or aa), $[\alpha]^{15/}{}_D$ -152 (chl, c=2)	d	171-2	1.87^4	1.675		B27[2], 175
8492	Isocorybulbine $C_{21}H_{25}NO_4$	355.43	lf (al), $[\alpha]^{15/}{}_D$ + 301 (chl, c=1)		187-8	$1.045^{20/4}$		al, chl	B21[4], 4779

PHYSICAL CONSTANTS OF ORGANIC COMPOUNDS (Continued)

No.	Name, Synonyms, and Formula	Mol. wt.	Color, crystalline form, specific rotation and λ_{max} (log ε)	b.p. °C	m.p. °C	Density	n_D	Solubility	Ref.
8493	Isocorydine or Corytuberine methyl ether $C_{20}H_{23}NO_4$	341.41	pl, $[\alpha]^{20/}_D$ + 195.3 (chl)		185			chl	B21^{4}, 2755
8494	Isocoumarin or *o*-(β-Hydroxyvinyl) benzoic acid lactone $C_9H_6O_2$	146.15	pl(bz)	285-6^{719}	47			al, eth, bz	B17^{4}, 5062
8495	Isocyanuric acid, trimethyl ester $C_6H_9N_3O_3$	171.16	mcl pr (w or al)	274	176-7			al	B26^{2}, 134
8496	Isoderritol $C_{21}H_{22}O_6$	370.40	ye lf (MeOH)		150				B18^{4}, 3351
8497	Isodurene or -1,2,3,5-tetramethylbenzene 1,2,3,5-$(CH_3)_4C_6H_2$	134.22		198, 24.4^{10}	-23.7	0.8903$^{20/4}$	1.5130^{20}	**al, eth, ace, bz**	B5^{4}, 1073
8498	8-Isoestradiol $C_{18}H_{24}O_2$	272.39	cr (dil MeOH chl), $[\alpha]^{20/}_D$ + 18 (diox)		181			al, diox	B6^{3}, 5332
8499	8-Isoestrone or 8-Epiestrone $C_{18}H_{22}O_2$	270.37	cr (MeOH), $[\alpha]^{20/}_D$ + 94 (diox)		247			eth, diox	B8^{3}, 1170
8500	Iso-β-eucaine (dl) or 4-Benzoyloxy-2,2,6-trimethyl piperidine $C_{15}H_{21}NO_2$	247.34		188^{19}	<-5	1.0467$^{22/22}$		al	B21^{4}, 131
8501	Iso-β-eucaine, hydrochloride (l) $C_{15}H_{21}NO_2 \cdot HCl$	283.80	nd (w) $[\alpha]_{5461}$ + 17 (w,c=1)		271-3			w	B21^{2}, 16
8502	Iso-β-eucaine, hydrochloride (dl) $C_{15}H_{21}NO_2.HCl$	283.80	ta (w) pl (aq al or al-eth)		269-71			w	B21^{2}, 15
8503	Iso-β-eucaine, hydrochloride (l) $C_{15}H_{21}NO_2.HCl$	283.80	nd α_{5461} −16.3 (w,c=1)		271-3			w	B21^{2}, 15
8504	Isoeugenol *(cis)* or 2-Methoxy-4-propenylphenol 2-CH_3O-4-$(CH_2CH=CH)$-C_6H_3OH	164.20		134-5^{13}, 80-1$^{0.5}$		1.0837$^{20/4}$	1.5726^{20}		B6^{3}, 4992
8505	Isoeugenol (trans) 2-CH_3O-4-$(CH_2CH=CH)C_6H_3OH$	164.20		141-2^{13}	33-4	1.0852$^{20/4}$	1.5784^{20}	al, eth	B6^{3}, 4992
8506	Isoeugenol (cis), acetate 2-CH_3O-4-$(CH_2CH=CH)C_6H_3O_2CCH_3$	206.24		160-2^{13}		1.0947$^{19/4}$	1.5418^{20}	eth	B6^{3}, 5007
8508	Isoeugenol, acetate (trans) 2-CH_3O-4-$(CH_2CH=CH)C_6H_3O_2CCH_3$	206.24	nd (al bz-lig)	282-3	80-1	1.0251$^{100/4}$	1.5052^{100}		B6^{3}, 5008
8509	Isoflavone, 4′,7-dihydroxy or Daidzein $C_{15}H_{10}O_4$	254.24	pa ye pr (50% al)	sub	323d			al, eth	B18^{4}, 1805
8510	Isoflavone, 4′,5,7-trihydroxy or Genistein $C_{15}H_{10}O_5$	270.24	nd (eth) pr (dil al)		301-2d				B18^{4}, 2724
8511	Isofurfurine $C_{15}H_{12}N_2O_3$	268.27	nd (w)		143				B27, 674
8512	Isogeraniolene or 2,6-Dimethyl-1,3-heptadiene $CH_2=C(CH_3)CH=CHCH_2CH(CH_3)_2$	124.23		143-4^{755}, 31^{7}		0.7561$^{20/4}$	1.4520^{20}	eth, bz	B1, 1015
8513	Isoleucine (d) $C_2H_5CH(CH_3)CH(NH_2)CO_2H$	131.17	$[\alpha]^{20/}_D$ −12.2H_2O −(3,2%)		283-4 d			w	B4^{3}, 1458
8514	Isoleucine (l) $C_2H_5CH(CH_3)CH(NH_2)CO_2H$	131.17	$[\alpha]^{25/}_D$ + 12.2 (3.2% in H_2O) 36.7 (INHCl)		285-6 d			w	B4^{3}, 1454
8515	Isoleucine, allo (l) $C_2H_5CH(CH_3)CH(NH_2)CO_2H$	131.17	$[\alpha]^{20/}_D$ −14.2H_2O 2		dec 280-1			w	B4^{3}, 1462
8516	Isolysergic acid (d) $C_{16}H_{16}N_2O_2$	268.32	cr (w + 2) $[\alpha]^{20/}_D$ + 281 (py,c=1)		218d			py	B27^{2}, 860

No.	Name, Synonyms, and Formula	Mol. wt.	Color, crystalline form, specific rotation and λ_{max} (log ε)	b.p. °C	m.p. °C	Density	n_D	Solubility	Ref.
8517	D-Isomannide or 1,4,3,6-Dianhydro-D-mannitol $C_6H_{10}O_4$	146.14	mcl cr $[\alpha]^{26/}_D$ + 62.2 ^{D} (chl) + 91 (w)	274d	87-9			w	$B19^4$, 990
8518	Isomenthol (d) or 2-isopropyl-5-methyl cyclohexanol $C_{10}H_{20}O$	156.27	nd (dil al) $[\alpha]^{20/}_D$ + 26.5 (al,c=4)	218.6, 96.5^{10}	82 (85)			al, eth, aa	$B6^4$, 151
8519	Isomenthol (dl) $C_{10}H_{20}O$	156.27	nd	218.5, $97.4^{16.5}$	53-4	0.9040^{30}	1.4510^{60}	al, eth, aa	$B6^4$, 152
8520	Isomenthol (l) $C_{10}H_{20}O$	156.27	$[\alpha]^{25/}_D$ −24.1		82.5			al, eth, aa	$B6^4$, 152
8521	Isomenthone $C_{10}H_{18}O$	154.25		$89\text{-}90^{15}$		$0.8995^{20/4}$	$1.4527^{20/}_D$		$B7^3$, 151
8522	α-Isomorphine $C_{17}H_{19}NO_3$	285.34	nd (MeOH-AcOEt) $[\alpha]^{15/}_D$ −167 (MeOH,c=3)		248			al, MeOH	$B27^2$, 174
8523	Isonicotine or 4-(4-pyridyl) piperidine $C_{10}H_{14}N_2$	162.23	hyg wh nd	292	80			al, eth, bz, lig	B23, 119
8524	Isonicotinaldehyde or 4-Pyridine carboxaldehyde 4-C_5H_4NCHO	107.11		$77\text{-}8^{12}$			1.5423^{20}	w, al, aa, chl	$B21^4$, 3529
8525	Isonicotinic acid or 4-Pyridine carboxylic acid 4-$C_5H_4NCO_2H$	123.11	nd (w)	sub 260^{15}	319				$B22^4$, 518
8526	Isonicotinic acid-2,6-dihydro or Citrazinic acid 4-$C_5H_6NCO_2H$	155.11	yesh-gr pw (w)		>330d				$B22^4$, 2459
8527	Isonicotinic acid, ethyl betaine $C_8H_9NO_2$	151.16	nd		241d			w, al	B22, 47
8528	Isonicotinic acid, ethyl ester or Ethylisonicotinate 4-$C_5H_4NCO_2H_5$	151.16	nd	220, 110^{15}	23	1.1052^{20}	1.5177^{20}	al, eth, bz, chl	$B22^4$, 521
8530	Isonicotinic acid, hydrazide or Isoniazid 4-$C_5H_4NCONHNH_2$	137.14	nd (al)		171				B22, 47
8531	Isonicotinic acid, methyl ester or Methyl isonicotinate 4-$C_5H_4NCO_2CH_3$	137.14		209d, 104^{21}	8.5	$1.1599^{20/4}$	1.5315^{20}	al, eth, bz	$B22^4$, 545
8532	Isonicotinonitrile 4-C_5H_4NCN	104.11	nd (liq-eth)		83			w, al, eth, bz	$B22^4$, 520
8533	Isopapaverine, N-benzyl $C_{27}H_{27}NO_4$	429.52	ye lf (al)		139-40			eth	$B22^4$, 542
8534	Isopapaverine, N-ethyl $C_{22}H_{25}NO_4$	367.44	pr(al)		ca 101			eth, bz	B21, 229
8535	Isopapaverine, N-methyl $C_{21}H_{23}NO_4$	353.42	ye hg mcl pr (al)		129-31			w	$B21^4$, 2799
8536	Isopelletriene or 2-Acetonyl piperidine $C_8H_{15}NO$	141.21	oil		$91\text{-}2^{14}$	$0.9624^{20/4}$	1.4683^{20}	al, chl	$B21^4$, 3264
8537	Isopelletriene, N-methyl $C_9H_{17}NO$	155.24		$96\text{-}8^{11}$		$0.9478^{20/4}$	1.4674^{20}	w, lig	$B21^4$, 3265
8538	Isophenolphthalein or 3-(*o*-hydroxy phenyl)-3-(*p*-hydroxy phenyl) phthalide $C_{20}H_{14}O_4$	318.33	(dil aa)		189-90			al, ace	$B10^3$, 2013
8539	3-Isophenothiazin, 3-one or Azthione thiazone $C_{12}H_7NO$	181.19	red (dil al)		164			al, bz, chl	$B27^1$, 251
8540	3-Isophenothiazin-3-one, 7-hydroxy or Thionol $C_{12}H_7NO_2S$	229.25	red br pw or nd (aa)		>360			al	$B27^2$, 109
8541	Isophorone or 3,5,5-trimethyl-2-cyclohexene-1-one $C_9H_{14}O$	138.21		214^{754}, 99^{18}		0.9229^{20}	1.4759^{20}	al, eth, ace	$B7^3$, 283
8542	Isophthaldehyde or 1,3-Benzene dicarboxaldehyde 1,3-$(OHC)_2C_6H_4$	134.13	nd(dil al)	$245\text{-}8^{77}$	89-90			al, ace, bz	$B7^3$, 3459
8543	Isophthalamide 1,3-$(H_2NOC)_2C_6H_4$	164.16	pl (w)		280				$B9^1$, 372
8544	Isophthalamide, N,N,N′,N′-tetraethyl 1,3-$C_6H_4[CON(C_2H_5)_2]_2$	276.38		242^{12}	85			al, eth, ace, bz	$B9^3$, 4242
8545	Isophthalic acid or 1,3-Benzene dicarboxylic acid 1,3-$C_6H_4(CO_2H)_2$	166.13	nd (w or al)	sub	348			al, aa	$B9^3$, 4240
8546	Isophthalic acid, 2-amino 2-$H_2NC_6H_3(CO_2H)_2$-1,3	181.15	pl (al), nd (aa)	sub 267	>260			al, eth	$B14^2$, 337

PHYSICAL CONSTANTS OF ORGANIC COMPOUNDS (Continued)

No.	Name, Synonyms, and Formula	Mol. wt.	Color, crystalline form, specific rotation and λ_{max} (log ε)	b.p. °C	m.p. °C	Density	n_D	Solubility	Ref.
8547	Isophthalic acid, 2-amino, dimethyl ester $2\text{-}H_2NC_6H_3(CO_2CH_3)_2\text{-}1,3$	209.20	nd (al)		103-4				B14², 337
8548	Isophthalic acid, 4-amino $4\text{-}H_2NC_6H_3(CO_2H)\text{-}1,3$	181.15	nd (w)		336-7			al, eth, ace, aa	B14¹, 633
8549	Isophthalic acid, 4-amino, dimethyl ester $4\text{-}H_2NC_6H_3(CO_2CH_3)_2\text{-}1,3$	209.20	nd(al)		131.5				B14², 337
8550	Isophthalic acid, 4-amino $5\text{-}H_2NC_6H_3(CO_2H)_2\text{-}1,3$	181.15	pr (al) pl (w)	sub	>360				B14¹, 636
8551	Isophthalic acid, 5-amino, dimethyl ester $5\text{-}H_2NC_6H_3(CO_2CH_3)_2\text{-}1,3$	209.20	lf or pl (MeOH)		176			eth	B14, 556
8552	Isophthalic acid, 4-bromo $4\text{-}BrC_6H_3(CO_2H)_2\text{-}1,3$	245.03	nd (al)		287			al	B9³, 4247
8553	Isophthalic acid, 4-chloro $4\text{-}ClC_6H_3(CO_2H)_2\text{-}1,3$	200.58	nd (w)		295			al	B9³, 4245
8554	Isophthalic acid, 5-chloro $5\text{-}ClC_6H_3(CO_2H)_2\text{-}1,3$	200.58	nd (w + ½)		278 (anh)			al	B9³, 4246
8555	Isophthalic acid, 4,6-dichloro $4,6\text{-}Cl_2C_6H_2(CO_2H)_2\text{-}1,3$	235.02	nd (w, dil al)		280			al, eth, chl	B9³, 4246
8556	Isophthalic acid, 4,5-dimethoxy or Isohemipinic acid $4,5\text{-}(CH_3O)_2C_6H_2(CO_2H)_2\text{-}1,3$	226.19	nd (w)		245-6			al, eth	B10³, 2435
8557	Isophthalic acid, diethyl ester or Diethyl isophthalate $1,3\text{-}C_6H_4(CO_2C_2H_5)_2$	222.24		302, $170^{2\,4}$	11.5	$1.1239^{17/4}$	1.508^{18}		B9³, 4241
8558	Isophthalic acid, dimethyl ester or Dimethylisophthalate $1,3\text{-}C_6H_4(CO_2CH_3)_2$	194.19	nd (dil al)	282, 124^{12}	67-8	$1.194^{20/4}$	1.5168^{20}		B9³, 4241
8559	Isophthalic acid, 4,6-dimethyl or α-Cumidic acid $4,6\text{-}(CH_3)_2C_6H_2(CO_2H)_2\text{-}1,3$	194.19	nd (w) pr (al-bz) lf (sub)	sub	266			al	B9³, 4298
8560	Isophthalic acid, 2-hydroxy $2\text{-}HOC_6H_3(CO_2H)_2\text{-}1,3$	182.13	nd (w + 1)		244-5			al, eth, chl	B10³, 2192
8561	Isophthalic acid, 4-hydroxy $4\text{-}HOC_6H_3(CO_2H)_2\text{-}1,3$	182.13	nd (w) lf (dil al)		310			al, eth	B10³, 2193
8562	Isophthalic acid, 5-hydroxy $5\text{-}HOC_6H_3(CO_2H)_2\text{-}1,3$	182.13	nd (w + 2) cr (aq al)	sub	293			al, eth, bz	B10³, 2195
8563	Isophthalic acid, 5-methyl or Uvitic acid $5\text{-}CH_3C_6H_3(CO_2H)_2\text{-}1,3$	180.16	nd (w)		298			al, eth, ace	B9³, 4274
8564	Isophthalic acid, 2-nitro, dimethyl ester $2\text{-}NO_2\text{-}C_6H_3(CO_2CH_3)_2\text{-}1,3$	239.18	nd (w or al)		135			al, eth	B9¹, 373
8565	Isophthalic acid, 5-nitro $5\text{-}NO_2C_6H_3(CO_2H)_2\text{-}1,3$	211.13	gr lf (+ ³⁄₂w)		260-1 (anh)			al, eth	B9¹, 373
8566	Isophthalic acid, 5-nitro, diethyl ester $5\text{-}NO_2C_6H_3(CO_2C_2H_5)_2\text{-}1,3$	267.24	nd (al)		83.5			al, eth	B9, 840
8567	Isophthalic acid, 5-nitro, dimethyl ester $5\text{-}NO_2C_6H_3(CO_2CH_3)_2\text{-}1,3$	239.18	nd (dil al)		123			al, eth	B9², 611
8568	Isophthalic acid, tetrabromo $1,3\text{-}(HO_2C)_2C_6Br_4$	481.71	nd (w)		288-92				B9, 839
8569	Isophthalonitrile $1,3\text{-}C_6H_4(CN)_2$	128.13	nd (al)		162			al, eth, bz, chl	B9³, 4243
8570	Isophthalylalcohol or *m*-Xylylene glycol $1,3\text{-}(HOCH_2)_2C_6H_4$	138.17	nd (bz)	$154\text{-}9^{13}$	57	1.1359^{53}		w, al, eth	B6³, 4600
8571	Isophthalyl chloride $1,3\text{-}C_6H_4(COCl)_2$	203.02	pr (eth)	276	43-4	$1.3880^{17/4}$	1.570^{47}	eth	B9³, 4242
8572	Isopilocarpine or N-Methyl isopilocarpidine $C_{11}H_{16}N_2O_2$	208.26	pr	261^{10}				w, al, eth, bz, chl	B27², 697
8573	Isopimpinellin $C_{13}H_{10}O_5$	246.22	ye nd (MeOH)		151			MeOH	B19⁴, 2811
8574	Isopomiferin $C_{25}H_{24}O_6$	420.46	nd (dil al)		265d			al	B19⁴, 5231
8575	Isoprene or 2-Methyl-1,3-butadiene $CH_2=C(CH_3)\text{-}CH=CH_2$	68.12		34	−146	$0.6810^{20/4}$	1.4219^{20}	**al, eth, ace, bz**	B1⁴, 1001
8576	Isopropenyl methyl ketone or 3-Methyl-3-buten-2- one $CH_2=C(CH_3)COCH_3$	84.12		98	-54	$0.8527^{20/4}$	1.4220^{20}	**al**	B1⁴, 3462
8577	Isopulegol (d) or $\Delta^{8(9)}$-p-Menthenol-3 $C_{10}H_{18}O$	154.25	$[\alpha]_{5461}$ + 29.3	212, $93\text{-}4^{14}$		$0.9110^{20/4}$	1.4723^{20}	al, eth	B6³, 257
8578	Isopulegol (l) $C_{10}H_{18}O$	154.25	$[\alpha]^{20}_{5461}$ −25.9	212, 94^{14}		$0.9110^{20/4}$	1.4723^{20}	al, eth	B6³, 257

No.	Name, Synonyms, and Formula	Mol. wt.	Color, crystalline form, specific rotation and λ_{max} (log ε)	b.p. °C	m.p. °C	Density	n_D	Solubility	Ref.
8579	α-Isoquinine $C_{20}H_{24}N_2O_2$	324.42	(bz-peth) $[\alpha]^{18/}_D$ −245 (al,c=1)	...	196.5	...	...	al, eth	B23^2, 414, 423
8580	β-Isoquinine $C_{20}H_{24}N_2O_2$	324.42	pr (dil al or amor) $[\alpha]^{17/}_D$ −187 (97% al,c=1)	...	190-1	...	...	al, bz, chl	B23^2, 413
8581	Isoquinoline C_9H_7N	129.16	hyg pl	242.2^{743}, 142^{40}	26.5	1.0986^{20}	1.6148^{20}	al, **eth**, ace, **bz**	B20^4, 3410
8582	Isoquinoline, hydrochloride $C_9H_7N.HCl$	165.62	pr or pl (al)	...	209	...	...	w	B20^4, 3412
8583	Isoquinoline, hydrogen sulfate $C_9H_7N.H_2SO_4$	227.23	pr or pl (al)	...	209	...	...	w	B20^4, 3413
8584	Isoquinoline, 1-amino 1-$H_2N(C_9H_6N)$	144.18	pl (w)	...	123	...	...	al	B22^4, 4736
8585	Isoquinoline, 3-amino 3-$H_2N(C_9H_6N)$	144.18	...	...	178-9	...	...	...	B22^4, 4744
8586	Isoquinoline, 5-amino 5-$H_2N(C_9H_6N)$	144.18	pa ye nd (peth)	sub	128	...	...	...	B22^4, 4747
8587	Isoquinoline, 4-bromo 4-$Br(C_9H_6N)$	208.06	(peth)	280-5	40-3	...	...	eth	B20^4, 3448
8588	Isoquinoline, 1-chloro 1-$Cl(C_9H_6N)$	163.61	...	274-5, 135-40^{10}	37-8	...	...	bz	B20^4, 3444
8589	Isoquinoline, 6,7-dimethoxy-1,2-dimethyl-1,2,3,4-tetrahydro or Carnegine $C_{13}H_{19}NO_2$	221.30	pa br syr $[\alpha]^{25/}_D$ +20 (w)	170^1	...	...	...	w, al, eth, chl	B21^4, 2125
8590	Isoquinoline, 1-hydroxy or Isocarbostyril 1-$HO(C_9H_6N)$	145.16	mcl (bz) nd (bz al w)	240 sub	209-10	...	...	al	B21^4, 1205
8591	Isoquinoline, 7-hydroxy 7-$HO(C_9H_6N)$	145.16	...	...	230	...	...	al	B21^4, 1214
8592	Isoquinoline, 7-methoxy 7-$CH_3O(C_9H_6N)$	159.19	...	182-6^{34}	49	...	...	al, lig	B21^4, 1214
8593	Isoquinoline, 1-methyl 1-$CH_3(C_9H_6N)$	143.19	...	248, 124-5^{10}	10	1.0777$^{20/4}$	1.6095^{20}	eth, ace, bz	B20^4, 3505
8594	Isoquinoline, 3-methyl 3-$CH_3(C_9H_6N)$	143.19	cr(eth)	246	68	...	...	eth, ace	B20^4, 3507
8595	Isoquinoline, 4-methyl 4-$CH_3(C_9H_6N)$	143.19	...	256	...	...	...	eth, ace, bz	B20^4, 3510
8596	Isoquinoline, 6-methyl 6-$CH_3(C_9H_6N)$	143.19	cr	265.5	85-6	...	...	al, eth, ace, bz	B20^2, 247
8597	Isoquinoline, 7-methyl 7-$CH_3(C_9H_6N)$	143.19	...	245	67-8	...	...	eth, ace	B20^4, 3511
8598	Isoquinoline, 8-methyl 8-$CH_3(C_9H_6N)$	143.19	...	258	...	...	...	eth, ace, bz	B20, 404
8599	Isoquinoline, 5-nitro 5-$O_2N(C_9H_6N)$	174.16	nd (w + 1)	sub	110	...	...	al, eth, chl, aa	B20^4, 3450
8600	Isoquinoline, 1,2,3,4-tetrahydro or 2-Azatetralin $C_9H_{11}N$	133.19	...	232-3	<−15	1.0642$^{24/4}$	1.5668^{20}	al	B20^4, 2949
8601	1-Isoquinoline carboxylonitrile or 1-Cyanoisoquinoline 1-$NC(C_9H_6N)$	154.17	nd (peth MeOH)	...	78(93)	...	...	al, eth, bz	B22^4, 1205
8602	5-Isoquinoline carboxylonitrile or 5-Cyanoisoquinoline 5-$NC(C_9H_6N)$	154.17	nd (w or dil al)	sub 100-120	135	...	...	al, eth	B22^4, 1208
8603	Isoraunescine $C_{31}H_{36}N_2O_8$	564.64	wh nd	...	241-2	...	...	chl, aa	C50, 1267
8604	Isoreserpilline $C_{23}H_{28}N_2O_5$	412.49	wh pr$[\alpha]^{20/}_D$ −84 (py)	...	210-2	...	...	...	C53, 14134
8605	Isorubijervine or Δ^5-β-18-Dihydroxy solanidene $C_{27}H_{43}NO_2$	413.64	dr (al) $[\alpha]^{/}_D$ +9.2 (al)	...	241-4	...	...	bz, chl	B21^4, 2312
8606	Isorubijervosine $C_{33}H_{53}NO_7$	575.79	wh nd $[\alpha]^{24/}_D$ −20 (py,c=1.45)	...	279-80	...	...	...	B21^4, 2313
8607	Isosaccharic acid or 3,4-Dihydroxytetrahydro-2,5-furandicarboxylic acid $C_6H_8O_7$	192.13	rh $[\alpha]^{20/}_D$ +46.7 (w,p=4.2)	d	185	...	...	w, al	B18^2, 309

No.	Name, Synonyms, and Formula	Mol. wt.	Color, crystalline form, specific rotation and λ_{max} (log ε)	b.p. °C	m.p. °C	Density	n_D	Solubility	Ref.
8608	Isosafrole *(trans)* $C_{10}H_{10}O_2$	162.19		253, 111-2[6]	fp 6.8	$1.1224^{20/4}$	1.5782^{20}	al, eth, ace, bz, chl	B19[4], 273
8609	Isoserine (l) $H_2NCH_2CH(OH)CO_2H$	105.09	cr (dil al w) $[\alpha]^{20}{}_D$ −32.6		199-201d			w	B4[1], 1566
8610	Isothebaine (d) $C_{19}H_{21}NO_3$	311.38	rh cr (al) $[\alpha]^{18}{}_D$ +285 (al,c=2)		203-4			al, chl	B21[4], 2646
8611	Isothebaine, sulfate $(C_{19}H_{21}NO_3)_2.H_2SO_4$	720.83	nd		120-1d			w	B21[1], 250
8612	Isovaleric acid or 3-Methylbutyric acid $(CH_3)_2CHCH_2CO_2H$	102.13		176.7	−29.3	$0.9286^{20/4}$	1.4033^{20}	al, eth, chl	B2[4], 895
8613	Isovalerophenone or Isobutyl phenyl ketone $(CH_3)_2CHCH_2COC_6H_5$	162.23		236.5, 137-8[38]		$0.9701^{16.4/4}$	$1.5139^{15.3}$	al, eth, ace	B7[3], 1121
8614	Isovaline (d) or 2-Amino-2-methyl butyric acid $CH_3CH_2C(CH_3)(NH_2)CO_2H$	117.15	nd (aq al) $[\alpha]^{t}{}_D$ +13 (w,c=2)	sub	ca 300			w	B7[3], 1361
8615	Isovaline (dl) or 2-Amino-2-methyl butyric acid $CH_3CH_2CH(CH_3)(NH_2)CO_2H$	117.15	rh nd (al-eth) mcl pr	sub 300	315 (sealed tube)			w, al	B4[3], 1361
8616	Isovaline (l) or 2-Amino-2-methyl butyric acid $CH_3CH_2CH(CH_3)(NH_2)CO_2H$	117.15	lo nd (w, ace) $[\alpha]^{20/}{}_D$ −9.1 (w,c=2)					w, al	B4[2], 851
8617	Isoxanthen-3-one, 9-phenyl-2,6,7-trihydroxy or 9-Phenyl-2,6,7-trihydroxy-fluorone $C_{19}H_{12}O_5$	320.30	og red (al-HCl)		>300				B18[4], 2824
8618	Isoxazole C_3H_3NO	69.06		95-6		$1.078^{20/4}$	1.4298^{17}		B27[2], 9
8619	Isoxazole, 5-methyl 5-$CH_3(C_3H_2NO)$	83.09		122		$1.4386^{20/0}$			B27[2], 9
8620	Itaconic acid or Methylene succinic acid $CH_2{=}C(CO_2H)CH_2CO_2H$	130.10	rh (bz)	d	175	1.632		w, al, ace, chl	B2[4], 2228
8621	Itaconic anhydride or Methylene succinic anhydride $C_5H_4O_3$	112.08	rh bipym pr (eth chl)	139-40[30], 114-5[18]	68-70			chl	B17[4], 5913
8622	Itaconic acid, diethyl ester or Diethylitaconate $CH_2{=}C(CO_2C_2H_5)CH_2CO_2C_2H_5$	186.21		228, 111[13]	58-9	$1.0467^{20/4}$	1.4377^{20}	al, eth, ace, bz	B2[4], 2230
8623	Itaconic acid, dimethyl ester or Dimethylitaconate $CH_2{=}C(CO_2CH_3)CH_2CO_2CH_3$	158.15	hyg mcl (MeOH)	208, 108[11]	38	$1.1241^{18/4}$	1.4457^{20}	al, eth, ace	B2[4], 2229
8624	Iticonyl chloride or Methylene succienyl chloride $CH_2{=}C(COCl)CH_2COCl$	166.99		89[17]			1.4919^{20}	ace	B2[3], 1934

No.	Name, Synonyms, and Formula	Mol. wt.	Color, crystalline form, specific rotation and λ_{max} (log ε)	b.p. °C	m.p. °C	Density	n_D	Solubility	Ref.
8625	Jacareubin $C_{18}H_{14}O_6$	326.33	ye pr (MeOH)	...	256-7d	...	...	al, ace	$B19^4$, 3041
8626	Jaconecic acid $C_{10}H_{16}O_6$	232.23	nd (eth) $[\alpha]^{25}_D$ + 28.1 (95% al)	...	183-4	...	...	...	$B18^4$, 5079
8627	Japaconitine-A or Acetyl benzoyl aconitine $C_{34}H_{47}NO_{11}$	645.75	rh $[\alpha]^{25}_D$ + 20.7 (chl)	...	202-3d	...	...	ace, chl	$B21^4$, 2901
8628	Japaconitine-Al $C_{34}H_{47}NO_{11}$	645.75	rh (MeOH) $[\alpha]^{22}_D$ + 26.4 (chl)	...	208-9	...	...	al, eth, chl	Ber57, 1462
8629	Japaconitine-B $C_{34}H_{47}NO_{11}$	645.76	rh (MeOH) $[\alpha]^{21}_D$ + 26.9	...	208-9d	...	...	al, eth, chl	Ber57, 1462
8630	Jasmin aldehyde $C_5H_{11}C(CHO)=CHC_6H_5$	203.30	...	$174-5^{20}$, 140^{5}	...	0.9718^{20}	1.5381^{20}	al, eth	$B7^3$, 1517
8631	Jasmone or 3-Methyl-2-(2-pentenyl)-2-cyclopenten-1-one $C_{11}H_{16}O$	164.25	ye oil	$257-8^{755}$, $134-5^{12}$	...	$0.9437^{22/4}$	1.4979^{22}	al, eth, lig	$B7^3$, 601
8632	Javanicin $C_{15}H_{14}O_6$	290.27	red (al)	...	208d	...	...	...	$B8^3$, 4231
8633	Jervine $C_{27}H_{39}NO_3.2H_2O$	461.64	nd (w + 2) (MeOH w) $[\alpha]^{23}_D$ −158.5 (al,c=0.99)	...	243-4d	...	...	al, ace, chl	Am73, 2970
8634	Julolidine $C_{12}H_{15}N$	173.26	...	280d, $155-6^{17}$	40	1.003^{20}	1.568^{25}	...	$B20^4$, 3281
8635	Julolidine-1,6-dioxo or 1,6-Diketo julolidine $C_{12}H_{11}NO_2$	201.22	ye (al)	$190-210^{0.3}$	145-6	...	...	al, MeOH	$B21^4$, 5525
8636	Junipal or 5-(α-propynyl)-2-formyl thiophene C_8H_6OS	150.20	nd (peth or dil al)	...	80	...	...	...	$B17^4$, 4946
8637	Juniperol or Macrocarpol $C_{15}H_{24}O$	220.35	tcl (al) $[\alpha]^{20}_D$ + 25.4 (al)	286-8d	112	$1.0460^{20/20}$	1.519	...	$B6^3$, 426
8638	Junipic acid or 5-(α-propynyl)-2-thiophencarboxylic acid $C_8H_6O_2S$	166.19	ye nd (peth) cr (aq al)	...	180	...	...	al	$B18^4$, 4198
8639	Junipic acid, methyl ester or Methyl janipate $C_9H_8O_2S$	180.22	(aq MeOH)	sub 50-5	62	...	...	ace	$B18^4$, 4199

No.	Name, Synonyms, and Formula	Mol. wt.	Color, crystalline form, specific rotation and λ_{max} (log ε)	b.p. °C	m.p. °C	Density	n_D	Solubility	Ref.
8640	Ketene CH_2=CO	42.04	....	−56	−151	....	....	....	B1[4], 3418
8641	Ketene, diethyl acetal or 1,1-diethoxyethylene CH_2=C$(OC_2H_5)_2$	116.16	....	68^{100}	....	$0.7932^{20/4}$	1.3643^{21}	....	B1[4], 3420
8642	Ketene, dimethyl $(CH_3)_2$C=C=O	70.09	ye	34	−97.5	....	....	....	B1[4], 3453
8643	Ketene, diphenyl $(C_6H_5)_2$C=C=O	194.23	ye-red	265-70d, 146^{12}	....	$1.1107^{14/4}$	1.615^{14}	eth, bz	B7[3], 2356
8644	Ketene, methyl CH_3CH=C=O	56.06	....	....	−80	....	....	eth	B1[4], 3433
8645	Khellin $C_{14}H_{12}O_5$	260.25	(MeOH or eth)	180-200$^{0.05}$	154-5 (vac)	....	....	ace, MeOH	B19[4], 2816
8646	Kynurenine or 3-Anthranyloyl alanine $C_{10}H_{12}N_2O_3$	208.22	lf (+ ½w) $[\alpha]^{20}{}_D$ −29 (w,c=4)	....	191d	....	....	....	B14[4], 1656

PHYSICAL CONSTANTS OF ORGANIC COMPOUNDS (Continued)

No.	Name, Synonyms, and Formula	Mol. wt.	Color, crystalline form, specific rotation and λ_{max} (log ε)	b.p. °C	m.p. °C	Density	n_D	Solubility	Ref.
8647	Lactamide (d) $CH_3CH(OH)CONH_2$	89.09	cr (AcOEt) $[\alpha]^{18/}_{578}$	...	49-51	...	...	w, al	B3³, 450
8648	Lactamide (dl) $CH_3CH(OH)CONH_2$	89.09	pl (AcOEt)	...	75.5	$1.1381^{80/4}$	...	w, al	B3⁴, 674
8649	Lactamide, N-(4-ethoxyphenyl) or N-Lactyl-β-phenetidide $CH_3CH(OH)CONH(C_6H_5OC_2H_5\text{-}4)$	209.25	nd (w)	...	118	...	...	al	B13³, 1135
8650	D-Lactic acid or D-2-Hydroxypropionic acid $CH_3CH(OH)CO_2H$	90.08	pl (chl aa) $[\alpha]^{/}_{D}$ −2.26 (w,c=1.24)	103^{2}	53	...	...	w, al	B3⁴, 633
8651	DL-Lactic acid $CH_3CH(OH)CO_2H$	90.08	ye	122^{15}	18	$1.2060^{21/4}$	1.4392^{20}	w, al, eth	B3⁴, 633
8652	L-Lactic acid $CH_3CH(OH)CO_2H$	90.08	hyg pr (eth), $[\alpha]^{15/}_{D}$ + 3.8 (w, c=10.5)	...	...	...	...	w, al	B3⁴, 633
8653	DL-Lactic acid acetate $CH_3CO_2CH(CH_3)CO_2H$	132.13	dlq	$167\text{-}70^{78}$, 127^{11}	57-60	$1.1758^{20/4}$	1.4240^{20}	al, bz	B3⁴, 638
8654	DL-Lactic acid, allyl ester or Allyl lactate $CH_3CH(OH)CO_2CH_2CH=CH_2$	130.14	...	$56\text{-}60^{8}$	...	$1.0452^{20/4}$	1.4369^{20}	py	B3³, 486
8655	DL-Lactic acid anhydride $[CH_3CH(OH)CO]_2O$	162.14	pa ye amor or syr	250d	...	...	...	al, eth	B3³, 494
8656	D-Lactic acid, butyl ester or D-Butyl lactate $CH_3CH(OH)CO_2C_4H_9$	146.19	$[\alpha]^{27/}_{D}$ + 13.6	77^{10}	...	$0.9744^{27/4}$	...	al, eth	B3², 188
8657	DL-Lactic acid, butyl ester or DL-Butyl lactate $CH_3CH(OH)CO_2C_4H_9$	146.19	...	83^{13}	−49	$0.9807^{22/4}$	1.4217^{20}	al, eth	B3⁴, 649
8658	D-Lactic acid, ethyl ester or D-Ethyl lactate $CH_3CH(OH)CO_2C_2H_5$	118.13	$[\alpha]^{19/}_{D}$ + 14.5	58^{20}	...	$1.0324^{20\ 4/4}$	1.4125^{20}	w, al, eth	B3³, 449
8659	DL-Lactic acid, ethyl ester $CH_3CH(OH)CO_2C_2H_5$	118.13	...	154.5, 58^{19}	...	$1.0302^{20/4}$	1.4124^{20}	w, al, eth	B3⁴, 643
8660	L-Lactic acid, ethyl ester $CH_3CH(OH)CO_2C_2H_5$	118.13	$[\alpha]^{19/}_{D}$ -11.3	$69\text{-}70^{36}$	...	$1.0314^{20/4}$	1.4156^{20}	w, al, eth	B3², 446
8661	D-Lactic acid, methyl ester or D-Methyl lactate $CH_3CH(OH)CO_2CH_3$	104.11	$[\alpha]^{20/}_{D}$ + 7.5	40^{11}	...	$1.0857^{25/4}$	...	w, al, eth	B3³, 449
8662	DL-Lactic acid, Methyl ester $CH_3CH(OH)CO_2CH_3$	104.11	...	144.8	...	$1.0928^{20/4}$	1.4141^{20}	w, al, eth	B3⁴, 640
8663	L-Lactic acid, Methyl ester $CH_3CH(OH)CO_2CH_3$	104.11	$[\alpha]^{20/}_{D}$ -8.3	58^{19}	...	$1.0895^{20/4}$	1.4139^{20}	w, al, eth	B3³, 445
8664	Lactic acid, isopentyl ester or Isopentyl lactate $CH_3CH(OH)CO_2\text{-i-}C_5H_{11}$	160.21	...	202.4, 82^{7}	...	$0.9617^{25/25}$	1.4240^{25}	al, eth	B3³, 483
8665	D-Lactic acid phenyl or Atrolatic acid $CH_3C(OH)(C_6H_5)CO_2H$	166.18	pr (w), $[\alpha]^{16.5/}_{D}$ + 37.7 (al, c=3.5)	...	116-7	...	...	al, ace, bz	B10³, 560
8666	DL-Lactic acid, 2-phenyl $CH_3C(OH)(C_6H_5)CO_2H$	166.18	nd pl(liq)	...	93-5	...	...	w, al, ace, bz	B10³, 560
8667	L-Lactic acid, 2-phenyl $CH_3C(OH)(C_6H_5)CO_2H$	166.18	nd (bz, w), $[\alpha]^{13.8/}_{D}$ -37.7 (al, c=3.4)	...	116-7	...	...	w, al, ace, bz	B10³, 560
8668	Lactic acid, 3-phenyl (d) or α-Hydroxy hydrocinnamic acid $C_6H_5CH_2CH(OH)CO_2H$	166.18	nd (w), $[\alpha]^{20/}_{D}$ + 22.2 (w, c=2.2)	...	124-6	...	...	w, al, ace	B10³, 554
8669	Lactic acid, 3-phenyl (dl) $C_6H_5CH_2CH(OH)CO_2H$	166.18	cr(chl bz) pr (w)	$148\text{-}50^{15}$	98	...	...	al, eth, ace	B10³, 554
8670	Lactic acid, 3-phenyl (l) $C_6H_5CH_2CH(OH)CO_2H$	166.18	nd (w), $[\alpha]^{20/}_{D}$ -19.9 (w, c=3.2)	...	124-5	...	...	al, eth, ace	B10³, 554
8671	Lactic acid, isopropyl ester or Isopropyl lactate $CH_3CH(OH)CO_2CH(CH_3)_2$	132.16	...	166-8, $75\text{-}80^{12}$	...	$0.9980^{20/4}$	1.4082^{25}	w, al, eth, bz	B3³, 479
8672	Lactic acid, 3,3,3-trichloro $CCl_3CH(OH)CO_2H$	193.41	pr (eth)	$140\text{-}70^{45}$	125	...	...	w, al, eth, chl	B3⁴, 680

No.	Name, Synonyms, and Formula	Mol. wt.	Color, crystalline form, specific rotation and λ_{max} (log ε)	b.p. °C	m.p. °C	Density	n_D	Solubility	Ref.
8673	Lactide (d) or 2,5-Dimethyl, 3,6-dioxo-1,4-dioxane $C_6H_8O_4$	144.13	hyg rh (eth), $[\alpha]^{18}_{D}$ -298 (bz, c=1.17)	150^{25}	95				$B19^4$, 1927
8674	Lactide (dl) $C_6H_8O_4$	144.13	pa ye tcl pr or nd (al)	255^{757}, $138\text{-}42^{8}$	124.5	$0.862^{10/4}$		ace, bz	$B19^4$, 1927
8675	Lactide (l) $C_6H_8O_4$	144.13	rh (eth), $[\alpha]^{26}_{D}$ +281.6 (bz, c=0.82)	150^{25}	95				$B19^4$, 1927
8676	D-Lactobionic acid $C_{12}H_{22}O_{12}$	358.30	syr					w	$B17^4$, 3392
8677	D-Lactonitrile or Acetaldehyde cyanohydrin $CH_3CH(OH)CN$	71.08	ye lig	182-4d, 102^{30}	−40	$0.9877^{20/4}$	1.4058^{18}	**w, al**, eth	$B3^4$, 675
8678	D-Lactonitrile, acetate $CH_3CO_2CH(CH_3)CN$	113.12		172-3, $76\text{-}7^{25}$		$1.0278^{20/4}$	1.4027^{20}	w	$B3^4$, 676
8679	D-Lactonitrile-3,3,3-trichloro $Cl_3CCH(OH)CN$	174.41	pl(w)	215-20d	61			w, al, eth	$B3^4$, 680
8680	Lactose or Milk sugar (α-anomer) $C_{12}H_{22}O_{11}$	342.30	pw, $[\alpha]^{20}_{D}$ +92.6→ +52.3 (w, c=4.5)		222.8			w	$B17^4$, 3066
8681	Lactose (β-anomer) $C_{12}H_{22}O_{11}$	342.30	$[\alpha]^{20}_{D}$ +34.2→ +52.3 (mut, w)		253	1.59^{20}		w	$B17^4$, 3068
8682	Lactose, monohydrate (α-anomer) $C_{12}H_{22}O_{11}.H_2O$	360.32	mcl (w), $[\alpha]^{20}_{D}$ +83.5 (w, 10min)	d	201-2	1.525^{20}		w	$B17^4$, 3067
8683	Lactyl chloride, acetate (dl) $CH_3CO_2CH(CH_3)COCl$	150.56		150d, 56^{11}		$1.1920^{17/4}$	1.4241^{17}		$B3^4$, 674
8684	Lactyl chloride, acetate (d+) $CH_3CO_2CH(CH_3)COCl$	150.56	$[\alpha]^{18}_{578}$ +32.4	$51\text{-}3^{11}$		1.177^{20}			$B3^2$, 189
8685	Lanosterol or Isocholesterol $C_{30}H_{50}O$	426.73	nd (eth) cr (MeOH-ace), $[\alpha]^{20}_{D}$ +62 (chl, c=1)		140-1			al, eth, chl	$B6^3$, 2880
8686	Lanosterol, benzoate or Isocholesterol benzoate $C_{34}H_{50}O_2$	490.77	pw or nd (eth), $[\alpha]^{17}_{D}$ +72.2		191.5			al, eth	$B9^3$, 485
8687	Lanthionine (D) or d-β,β-Thio di-alanine $[HO_2CCH(NH_2)CH_2]_2S$	208.23	hex pl, $[\alpha]^{21}_{D}$ -8.0 (2.4N NaOH, c=5)		293-5d darkens 245				$B4^3$, 1618
8688	Lanthionine (DL) $[HO_2CCH(NH_2)CH_2]_2S$	208.23	hex pl		282-95d chars 240				$B4^3$, 1620
8689	Lanthionine (L) $[HO_2CCH(NH_2)CH_2]_2S$	208.23	hex pl, $[\alpha]^{22}_{D}$ +8.6 (2.4N NaOH, c=5)		293-5d darkens 245				$B4^3$, 1593
8690	Lanthionine *(meso)* $[HO_2CCH(NH_2)]_2S$	208.23	hex pl (aq NH_3)		304d softens 270				$B4^3$, 1620
8691	Lanthopine $C_{23}H_{25}NO_4$	379.46	pw		200			chl	C20, 2715
8692	Lapachol or Taiguic acid $C_{15}H_{14}O_3$	242.27	ye pr (eth,bz) pl (aa,al)		139-40			al, eth, bz, aa	$B8^2$, 365
8693	Lapachol-∂-hydroxy $C_{15}H_{14}O_4$	258.27	ye nd (bz,w)		127			al, eth, bz	$B8^3$, 3661

No.	Name, Synonyms, and Formula	Mol. wt.	Color, crystalline form, specific rotation and λ_{max} (log ε)	b.p. °C	m.p. °C	Density	n_D	Solubility	Ref.
8694	Lappaconitine $C_{32}H_{44}N_2O_9$	600.71	hex pl (al), $[\alpha]^{18/}_{D}$ + 27 (chl)		223			bz, chl	B21^4, 2850
8695	Laudanidine (d) $C_{20}H_{25}NO_4$	343.42	(MeOH), $[\alpha]^{17/}_{D}$ + 93.5 (chl, c=1)		184-5				B21^4, 2703
8696	Laudanidine (l) or Tritopine $C_{20}H_{25}NO_4$	343.42	hex pr (al), $[\alpha]^{17/}_{D}$ -94.8 (chl, c=2)		184-5			w, bz	B21^4, 2703
8697	Laudanine (dl) $C_{20}H_{25}NO_4$	343.42	ye wh pr (dil al or al-chl)		167	$1.26^{20/4}$		al, bz, chl	B21^4, 2703
8698	Laudanosine (d) or N-Methyl tetrahydro papaverine $C_{21}H_{27}NO_4$	357.45	nd (peth) pr (al), $[\alpha]^{16/}_{D}$ + 106 (al, c=1.6)		89			al, eth, ace, chl	B21^4, 2704
8699	Laudanosine (dl) $C_{21}H_{27}NO_4$	357.45	nd(al)		115-6			al, eth, ace, bz, chl	B21^4, 2704
8700	Laudanosine (l) $C_{21}H_{27}NO_4$	357.45	cr (al), $[\alpha]^{15/}_{D}$ -105.4 (al, c=31)		89				B21^4, 2704
8701	Laureline $C_{19}H_{19}NO_3$	309.36	ta (al) cubes (peth), $[\alpha]^{18/}_{D}$ -99 (abs al,c=0.7)		114			al, eth	B27^1, 461
8702	Lauraldehyde or Dodecanal $CH_3(CH_2)_{10}CHO$	184.32	lf	185^{100}, $100^{3.5}$	44.5	$0.8352^{15/4}$	1.435^{22}	al, eth	B1^4, 3380
8703	Lauraldehyde, dimethylacetal $CH_3(CH_2)_{10}CH(OCH_3)_2$	230.39		$132\text{-}4^{5}$			1.4310^{25}	al, eth	B1^4, 3381
8704	Lauramide $CH_3(CH_2)_{10}CONH_2$	199.34	nd	199^{12}	110		1.4287^{110}	al, ace	B2^4, 1103
8705	Lauramide, N-Phenyl $CH_3(CH_2)_{10}CONHC_6H_5$	275.43	nd (dil al)		78			al, eth, ace, bz, chl	B12^3, 485
8706	Lauric acid or Dodecanoic acid $CH_3(CH_2)_{10}CO_2H$	200.32	nd (al)	131^{1}	44	$0.8679^{50/4}$	1.4304^{50}	al, eth, ace, bz, peth	B2^4, 1082
8707	Lauric anhydride $(C_{11}H_{23}CO)_2O$	382.63	lf (al,eth)		41.8	$0.8533^{70/4}$	1.4292^{70}	al	B2^4, 1100
8708	Lauric acid, benzyl ester or Benzyl laurate $C_{11}H_{23}CO_2CH_2C_6H_5$	290.45		$209\text{-}11^{12}$	8.5	$0.9457^{25/25}$	1.4812^{24}	al, eth, bz, chl, peth	B6^4, 2267
8709	Lauric acid, 2-bromo $CH_3(CH_2)_9CHBrCO_2H$	279.22	pl	$157\text{-}9^{2}$	32	1.1474^{24}	1.4585^{24}	al, eth, bz, chl, lig	B2^4, 1106
8710	Lauric acid, ethyl ester or Ethyl laurate $C_{11}H_{23}CO_2C_2H_5$	228.38		273^{764}, 154^{15}	fr−1.8	$0.8618^{20/4}$	1.4311^{20}	al, **eth**	B2^4, 1092
8711	Lauric acid, 12-fluoro $F(CH_2)_{11}CO_2H$	218.31			60-1			al, eth	B2^4, 1105
8712	Lauric acid, methyl ester or Methyl laurate $C_{11}H_{23}CO_2CH_3$	214.35		262^{766}, 141^{15}	fr 5.2	$0.8702^{20/4}$	1.4319^{20}	**al, eth, ace, bz**	B2^4, 1090
8713	Lauric acid, phenyl ester or Phenyl laurate $C_{11}H_{23}CO_2C_6H_5$	276.42	lf (al)	210^{15}	24.5			al, eth, ace	B6^4, 618
8714	Lauric acid, propyl ester or Propyl laurate $C_{11}H_{23}CO_2C_3H_7$	242.40		205^{60}, 124^{2}		0.8600^{20}	1.4335^{20}		B2^4, 1092
8715	Lauric acid, isopropyl ester or *iso*-Propyl laurate $C_{11}H_{23}CO_2CH(CH_3)_2$	242.40		196^{60}, 117^{2}		0.8536^{20}	1.4280^{25}	al, eth	B2^4, 1092
8716	Laurone or 12-Tricosanone Diundecyl ketone $(C_{11}H_{23})_2CO$	338.62	lf (al)		69.3	$0.8086^{69/4}$	1.4283^{80}	eth, bz, chl	B1^4, 3408
8717	Lauronitrile $C_{11}H_{23}CN$	181.32		277, 131^{10}	fr 4	$0.8240^{20/4}$	1.4361^{20}	**al, eth, ace, bz, chl**	B2^4, 1104
8718	Laurophenone or n-undecyl phenyl ketone $C_{11}H_{23}COC_6H_5$	260.42	og cr	$222\text{-}3^{21}$	46-7	$0.8969^{52/4}$	1.4850^{52}	ace	B7^3, 1291
8719	Lauroyl peroxide $C_{11}H_{23}CO\text{-}OO\text{-}COC_{11}H_{23}$	398.63	wh pl		49				B2^4, 1102
8720	Lauryl alcohol or 1-Dodecanol $CH_3(CH_2)_{10}CH_2OH$	186.34	lf (dil al)	255-9, 150^{20}	26	$0.8309^{24/4}$		al, eth	B1^4, 1844

No.	Name, Synonyms, and Formula	Mol. wt.	Color, crystalline form, specific rotation and λ_{max} (log ε)	b.p. °C	m.p. °C	Density	n_D	Solubility	Ref.
8721	Lauryl amine or 1-Amino dodecane $CH_3(CH_2)_{10}CH_2NH_2$	185.35	...	259, 126.5^{10}	28.3	$0.8015^{20/4}$	1.4421^{20}	**al, eth, bz, chl**	$B4^{4}$, 794
8722	Lauryl amine, acetate $C_{12}H_{25}NH_2.CH_3CO_2H$	245.41	...	...	fr 69.5	...	...	w, al	$B4^{4}$, 797
8723	Lauryl amine, hydrochloride $C_{12}H_{25}NH_2.HCl$	221.81	...	...	98	...	...	w, al	$B4^{4}$, 795
8724	Lauroyl chloride $C_{11}H_{23}COCl$	218.77	...	145^{18}	−17	...	1.4458^{20}	eth	$B2^{4}$, 1103
8725	Lauryl mercaptan $CH_3(CH_2)_{11}SH$	202.40	...	$142\text{-}5^{15}$	...	$0.8450^{20/20}$	1.4589^{20}	al, eth	$B1^{4}$, 1851
8726	Lauryl sulfate $(C_{12}H_{25}O)_2SO_2$	434.72	...	...	48.5	...	...	...	$B1^{4}$, 1849
8727	Lead, tetraphenyl $Pb(C_6H_5)_4$	515.61	...	126^{13}	227-8	...	...	...	$B16^{3}$, 1252
8728	Lecithin $RCO_2CH_2(CHOCOR)CH_2OP(O^-)(=O)OCH_2CH_2\overset{+}{N}(CH_3)_3$	...	$[\alpha]^{24}{}_D$ + 7.0	...	236-7	...	...	eth, chl, peth	$B4^{4}$, 1462
8729	Ledol $C_{15}H_{26}O$	222.37	nd (al), $[\alpha]^{20}{}_D$ + 28 (chl, c=10)	292 sub	105-6 sub	$0.9094^{100/20}$	1.4667^{110}	al, eth, ace	$B6^{4}$, 426
8730	*p*-Leucaniline, N,N,N′,N′-tetramethyl or bis(4-dimethylaminophenyl)4-aminophenylmethane $C_{23}H_{27}N_3$	345.49	(al)	...	151-2	...	...	...	$B13^{3}$, 566
8731	Leucic acid (D) or 2-Hydroxy-4-methyl valeric acid $(CH_3)_2CHCH_2CH(OH)CO_2H$	132.16	nd (bz) pr (eth-peth), $[\alpha]^{125}{}_D$ + 10.7 (w, c=5)	...	80-1	...	...	w, al, eth	$B3^{4}$, 850
8732	Leucic acid-(DL) $(CH_3)_2CHCH_2CH(OH)CO_2H$	132.16	pl (eth-peth)	...	77	...	...	w, al, eth	$B3^{4}$, 851
8733	Leucic aicd (l) $(CH_3)_2CHCH_2CH(OH)CO_2H$	132.16	rh (eth), $[\alpha]^{20}{}_D$ -11.3 (w, c=1)	...	81-2	...	...	w, al, eth	$B3^{4}$, 851
8734	Leucin amide (dl) or α-Amino isocaproamide $(CH_3)_2CHCH_2CH(NH_2)CONH_2$	130.19	pr (bz)	...	106-7	...	...	w, al, ace	$B4^{3}$, 1434
8735	Leucine (D) or D-α-Amino isocaproic acid $(CH_3)_2CHCH_2CH(NH_2)CO_2H$	131.17	pl (al), $[\alpha]^{20}{}_D$ + 10.34	sub	293 (sealed tube)	...	...	...	$B4^{3}$, 1424
8736	Leucine (DL) $(CH_3)_2CHCH_2CH(NH_2)CO_2H$	131.17	lf (w)	sub	293-5 (sealed tube)	$1.293^{18/4}$	...	w	$B4^{3}$, 1430
8737	Leucine (L) $(CH_3)_2CHCH_2CH(NH_2)CO_2H$	131.17	hex pl (dil al), $[\alpha]^{25}{}_D$ -10.4 (w, p=22)	sub	293-5 (sealed tube)	$1.293^{18/4}$	...	...	$B4^{3}$, 1408
8738	Leucine, N-acetyl-(dl) $(CH_3)_2CHCH_2CH(NHCOCH_3)CO_2H$	173.21	nd (dil al)	...	161	...	...	al	$B4^{3}$, 1439
8739	Leucine, N-benzoyl $(CH_3)_2CHCH_2CH(NHCOC_6H_5)CO_2H$	235.28	nd (dil al)	...	137-41	...	...	w, al, eth, chl	$B9^{3}$, 1159
8740	Leucine, N-glycyl (dl) $(CH_3)_2CHCH_2CH(NHOCH_2NH_2)CO_2H$	188.23	tetr (dil al)	...	242d	1.181	...	w	$B4^{3}$, 1443
8741	Leucine, N-glycyl (l) $(CH_3)_2CHCH_2CH(NHCONH_2)CO_2H$	188.23	pl (dil al), $[\alpha]^{20}{}_D$ -35.2 (w)	...	256d	...	...	w	$B4^{3}$, 1420
8742	Leucomethylene blue $C_{16}H_{19}N_3S$	285.41	ye nd (eth,al)	...	185	...	...	al	$B27^{2}$, 448
8743	Levopimaric acid, methyl ester $C_{21}H_{32}O_2$	316.48	(MeOH or eth), $[\alpha]_D$ -190.4 (al) -268 (eth)	$166\text{-}9^{0.5}$	63-4	$1.0312^{22/4}$	1.5232^{22}	al	$B9^{3}$, 2904
8744	Levulin or Fructosin $(C_6H_{10}O_5)n$	(162.14)n	amor, [α] -52.1	...	140-5d	...	...	w, al	B1, 925
8745	Levulinaldehyde or 4-Oxovaleraldehyde $CH_3COCH_2CH_2CHO$	100.12	...	186-8d, 70^{12}	<−21	$1.0184^{21/4}$	1.4257^{22}	**w, al, eth**, ace, bz	$B1^{4}$, 3659
8746	Levulinic acid or 4-Oxovalericacid $CH_3COCH_2CH_2CO_2H$	116.12	lf or pl	245-6d, $139\text{-}40^{8}$	37.2	$1.1335^{20/4}$	1.4396^{20}	w, al, eth	$B3^{4}$, 1560

PHYSICAL CONSTANTS OF ORGANIC COMPOUNDS (Continued)

No.	Name, Synonyms, and Formula	Mol. wt.	Color, crystalline form, specific rotation and λ_{max} (log ε)	b.p. °C	m.p. °C	Density	n_D	Solubility	Ref.
8747	Levulinic acid, benzyl ester or Benzyl levulinate $CH_3COCH_2CH_2CO_2CH_2C_6H_5$	206.24	...	$132\text{-}4^{2}$	...	$1.0935^{20/4}$	1.5090^{20}	to	B6[4], 2481
8748	Levulinic acid, butyl ester or Butyl levulinate $CH_3COCH_2CH_2CO_2C_4H_9$	172.22	...	237.8	...	$0.9735^{20/4}$	1.4290^{20}	al, eth, ace, bz	B3[4], 1563
8749	Levulinic acid, iso butyl ester or Butyl levulinate $CH_3COCH_2CH_2CO_2\text{-i-}C_4H_9$	172.22	...	231	...	$0.9705^{20/4}$	1.4268^{20}	al, eth, ace, bz	B3[4], 1563
8750	Levulinic acid, *Sec*-butyl ester or *Sec*-Butyl levulinate $CH_3COCH_2CH_2CO_2CH(CH_3)C_2H_5$	172.22	...	225.8	...	$0.9669^{20/4}$	1.4249^{20}	al, eth, ace	B3[3], 1221
8751	Levulinic acid, methyl ester or Methyl levulinate $CH_3COCH_2CH_2CO_2CH_3$	130.14	...	$196, 85\text{-}6^{14}$	...	$1.0511^{20/4}$	1.4233^{20}	al, eth, ace, bz	B3[4], 1562
8752	Levulinic acid-propyl ester or Propyl levulinate $CH_3COCH_2CH_2CO_2C_3H_7$	158.20	...	221.2	...	$0.9896^{20/4}$	1.4258^{20}	al, eth, ace, bz	B3[4], 1563
8753	Levulinic acid-isopropyl ester or Isopropyl levulinate $CH_3COCH_2CH_2CO_2CH(CH_3)_2$	158.20	...	209.3	...	$0.9842^{20/4}$	1.4420^{20}	al, eth, ace, bz	B3[4], 1563
8754	Limonene (d) or 4-isopropenyl-1-methyl cyclohexene $C_{10}H_{16}$	136.24	$[\alpha]^{20/}_{D}$ + 125.6 (undil)	$178, 61^{12}$	−74.3	$0.8411^{20/4}$	1.4730^{20}	al, eth	B5[4], 438
8755	Limonene (dl) or Dipentene $C_{10}H_{16}$	136.24	...	$178, 64.4^{15}$	−95.5	$0.8402^{21/4}$	1.4727^{20}	...	B5[4], 440
8756	Limonene (l) $C_{10}H_{16}$	136.24	$[\alpha]^{20/}_{D}$ -122.1 (undil)	$177\text{-}8^{755}$, 64.4^{15}	...	$0.8422^{20/4}$	1.4746^{20}	al, eth	B5[4], 440
8757	Linalool (d) or 3,7-Dimethyl-1,6-octadien-3-ol $CH_2{=}CHC(OH)(CH_3)CH_2CH_2CH{=}C(CH_3)_2$	154.25	$[\alpha]^{20/}_{D}$ + 19.18	$198\text{-}200, 87\text{-}8^{12}$	...	$0.8700^{20/4}$	1.4636^{20}	al, eth	B1[3], 2012
8758	Linalool, acetate (l) or Linalyl acetate, Bergamol $C_{12}H_{20}O_2$	196.29	$[\alpha]^{20/}_{D}$ -9.45	220^{762}	...	$0.8951^{20/4}$	1.4544^{21}	al, eth	B2[4], 204
8759	Linalool, formate or Linalyl formate $CH_2{=}CHC(CH_3)(O_2CH)CH_2CH_2CH{=}C(CH_3)_2$	182.26	...	$100\text{-}3^{10}$	...	$0.915^{25/4}$	1.456^{20}	al	B2[4], 35
8760	Linalool, tetrahydro-(dl) or 3,7-dimethyl-3-octanol-(dl) $(CH_3)CH(CH_2)_3C(OH)(CH_3)CH_2CH_3$	158.28	...	196-7, $87\text{-}8^{10}$	31-2	0.8280^{20}	1.4335^{20}	al	B1[4], 1829
8761	Linaloolen or 2,6-Dimethyl-2,7-octadiene $(CH_3)_2C{=}CHCH_2CH_2CH(CH_3)CH{=}CH_2$	138.25	...	$168, 58^{12}$	...	0.7882^{20}	1.4561^{20}	al	B1, 261
8762	β-Linaloolene or Dihydromyrcene $CH_2{=}CHCH(CH_3)CH_2CH_2CH{=}C(CH_3)_2$	138.25	...	165-8	...	$0.7601^{20/4}$	1.4362^{20}	...	B1[4], 1060
8763	Linamarin or Acetone cyanohydrin-β-d-glycopyranoside $C_{10}H_{17}NO_6$	247.25	nd (w al), $[\alpha]^{18/}_{D}$ -29.1 (w, p=5)	...	145	...	...	ace	B17[4], 3340
8764	Linoleic acid or 9,12-Octadecadienoic acid $CH_3(CH_2)_4CH{=}CHCH_2CH{=}CH(CH_2)_7CO_2H$	280.45	...	$229\text{-}30^{16}$	−5	0.9022^{204}	1.4699^{20}	al, eth, ace, bz, chl	B2[4], 1754
8765	Linoleic acid, ethyl ester or Ethyl linoleate $CH_3(CH_2)_4CH{=}CH\text{-}CH_2\text{-}CH{=}CH\text{-}(CH_2)_7CO_2C_2H_5$	308.50	ye or col	$270\text{-}5^{180}$, 212^{12}	...	$0.8865^{20/4}$	...	al, eth	B2[4], 1757
8766	Linoleic acid, methyl ester or Methyl linoleate $CH_3(CH_2)_4CH{=}CHCH_2CH{=}CH(CH_2)_7CO_2CH_3$	294.48	...	215^{20}	−35	$0.8886^{18/4}$	1.4638^{20}	al, eth	B2[4], 1756
8767	10,12-Linoleic acid or 10,12-Octadecadienoic acid $CH_3(CH_2)_4CH{=}CHCH{=}CH(CH_2)_8CO_2H$	280.45	...	...	56-7	$0.8686^{70/4}$	1.4689^{60}	...	B2[4], 1752
8768	α-Linolenic acid or 9,12,15-Octadecatrienoic acid *(cis,cis,cis)* $CH_3[CH_2CH{=}CH]_3(CH_2)_7CO_2H$	278.44	...	$230\text{-}2^{17}$, $129^{0.05}$	−11.3	$0.9164^{20/4}$	1.4800^{20}	al, eth	B2[4], 1781
8769	α-Linolenic acid, ethyl ester or Ethyl linolenate $CH_3[CH_2CH{=}CH]_3(CH_2)CO_2C_2H_5$	306.49	...	218^{15}	...	$0.8919^{20/4}$	1.4694^{20}	...	B2[4], 1782
8770	α-Lipoic acid (d) or 6,8-Epidithiooctanoic acid $C_8H_{14}O_2S_2$	206.32	pa ye pl, $[\alpha]^{25/}_{D}$ + 96.7	...	47.5	...	...	bz, MeOH	B19[4], 3459
8771	Lithocholic acid or 3α-Hydroxycholanic acid $C_{24}H_{40}O_3$	376.58	hex lf (al) pr (dil al or aa), $[\alpha]^{20/}_{D}$ + 32.14 (al)	...	186	...	...	al, chl, aa	B10[3], 687
8772	Lobelanidine $C_{22}H_{29}NO_2$	339.48	sc(al eth)	distb vac	150	...	...	al, ace, bz, chl, py	B21[4], 2380
8773	Lobelanine $C_{22}H_{25}NO_2$	335.45	nd(peth or eth)	...	99	...	...	al, ace, bz, aa, chl	B21[4], 5630
8774	Lobeline (dl) $C_{22}H_{27}NO_2$	337.46	ye pl (eth) pr (al)	...	110	...	...	al, eth, bz, chl	B21[4], 6346

PHYSICAL CONSTANTS OF ORGANIC COMPOUNDS (Continued)

No.	Name, Synonyms, and Formula	Mol. wt.	Color, crystalline form, specific rotation and λ_{max} (log ε)	b.p. °C	m.p. °C	Density	n_D	Solubility	Ref.
8775	Lobeline (l) $C_{22}H_{27}NO_2$	337.46	nd (al eth bz), $[\alpha]^{15/}_{D}$ -42.8 (al, c=1)	......	130.1	......	......	al, eth, ace, bz, chl	$B21^4$, 6345
8776	Longifolene (d) $C_{15}H_{24}$	204.36	$[\alpha]^{18/}_{D}$ + 42.7	$254\text{-}6^{706}$, $126\text{-}7^{15}$	......	$0.9319^{10/4}$	1.5040^{20}	bz	$B5^4$, 1192
8777	Lophine or 2,4,5-Triphenyl imidazole $C_{21}H_{16}N_2$	296.37	nd (al)	sub	275	......	......	al, eth	$B23^2$, 280
8778	Luciculine $C_{22}H_{38}NO_3$	361.52	cr (+ 1w ace), $[\alpha]^{11.5/}_{D}$ -11.4 (al)	$165^{0.02}$	148-50	......	......	al	$B21^4$, 2584
8779	Luciculine, hydrochloride $C_{22}H_{38}NO_3HCl$	397.99	cr (w + $^3/_2$), $[\alpha]_D$ -9.4 (w)	......	198-203	......	......	al	$B21^4$, 2584
8780	Luminol or 3-Aminophthalhydrazide $C_8H_7N_3O_2$	177.16	ye nd (al)	......	329-32	......	......	......	$B25^2$, 389
8781	Lumisterol $C_{28}H_{44}O$	396.66	nd (ace MeOH), $[\alpha]^{18/}_{D}$ + 197 (chl)	......	118	......	......	al, eth, ace, bz, aa	$B6^3$, 3108
8782	Lupanine (d) $C_{15}H_{24}N_2O$	248.37	hyg nd, $[\alpha]^{20/}_{D}$ + 82.4 (w, c=3)	$190\text{-}3^3$	40-4	......	1.544^{26}	w, al, eth, chl	$B24^2$, 53
8783	Lupanine (dl) $C_{15}H_{24}N_2O$	248.37	nd(peth) rh pr(ace)	$233\text{-}4^{18}$	98-9	......	......	w, al, eth, chl, peth	$B24^2$, 55
8784	Lupeol $C_{30}H_{50}O$	426.73	nd (al or ace), $[\alpha]^{20/}_{D}$ + 27.2 (chl)	......	215-7	$0.9457^{218/4}$	1.4910^{218}	al, eth, ace, bz, chl	$B6^3$, 2901
8785	Lupinine (l) $C_{10}H_{19}NO$	169.27	rh (peth), $[\alpha]^{17/}_{D}$ -20.3 (al)	$269\text{-}70^{754}$	70	......	......	w, al, eth, bz, chl	$B21^4$, 291
8786	Lupinine, hydrochloride $C_{10}H_{19}NO.HCl$	205.73	pr (dil al), $[\alpha]_D$ -14 (w)	......	212-3	......	......	w, al	$B21^4$, 292
8787	Lupulone or β-Lupulinic acid $C_{26}H_{38}O_4$	414.59	pr (MeOH)	......	93	......	......	al, peth	$B7^3$, 4753
8788	2,3-Lutidine or 2,3-dimethyl pyridine 2,3-$(CH_3)_2(C_5H_3N)$	107.16	......	163-4	......	$0.9319^{25/4}$	1.5057^{20}	w, al, eth	$B20^4$, 2765
8789	2,4-Lutidine or 2,4-dimethyl pyridine 2,4-$(CH_3)_2(C_5H_3N)$	107.16	......	159	......	$0.9309^{20/4}$	1.5010^{20}	w, al, eth, ace	$B20^4$, 2768
8790	2,5-Lutidine or 2,5-dimethyl pyridine 2,5-$(CH_3)_2(C_5H_3N)$	107.16	......	157-9	−16	$0.9297^{20/4}$	1.5006^{20}	al, **eth**, ace	$B20^4$, 2774
8791	2,6-Lutidine or 2,6-dimethyl pyridine 2,6-$(CH_3)_2(C_5H_3N)$	107.16	......	145.7	−6.1	$0.9226^{20/4}$	1.4953^{20}	w, eth, ace	$B20^4$, 2776
8792	3,4-Lutidine or 3,4-dimethyl pyridine 3,4-$(CH_3)_2(C_5H_3N)$	107.16	......	163-4	......	$0.9281^{20/4}$	1.5096^{20}	al, eth, ace, chl	$B20^4$, 2787
8793	3,5-Lutidine or 3,5-dimethyl pyridine 3,5-$(CH_3)_2(C_5H_3N)$	107.16	......	171.6	......	$0.9419^{20/4}$	1.5061^{20}	w, al, eth, ace	$B20^4$, 2788
8794	Lycaconitine $C_{34}H_{34}N_2O_6$	566.65	amor, $[\alpha]_D$ + 31.5	......	111-4	......	......	al, bz, chl, peth	$B21^4$, 4564
8795	Lycomarasmine $HO_2CCH_2CH(CO_2H)NHCH_2CH(CO_2H)NHCH_2CONH_2$	277.24	$[\alpha]^{20/}_{D}$ -48 (w)	......	227-9d	......	......	......	$B4^3$, 1521
8796	Lycopene $C_{40}H_{56}$	536.88	red pr or nd (peth)	......	175	......	......	eth, bz, chl	$B1^4$, 1166
8797	Lycorine (l) or Narcissine $C_{16}H_{17}NO_4$	287.32	pr (al py), $[\alpha]^{16/}_{D}$ -129 (al, c=0.16)	sub	280	......	......	......	$B27^2$, 547
8798	Lycoxanthin $C_{40}H_{56}O$	552.88	red pl (bz-MeOH)	......	168	......	......	bz	$B1^4$, 2368

PHYSICAL CONSTANTS OF ORGANIC COMPOUNDS (Continued)

No.	Name, Synonyms, and Formula	Mol. wt.	Color, crystalline form, specific rotation and λ_{max} (log ε)	b.p. °C	m.p. °C	Density	n_D	Solubility	Ref.
8799	Lysergic acid $C_{16}H_{16}N_2O_2$	268.32	lf or hex sc(w), $[\alpha]^{20/}_D$ + 40 (py, c=0.5)		240d			al, py	J1955, 1626
8800	Lysine(L) or L-α-ω-Diaminocaproic acid $H_2N(CH_2)_4CH(NH_2)CO_2H$	146.19	nd (w, dil al), $[\alpha]^{20/}_D$ + 14.6 (w, c=6)		224-5d darkens 210			w	B4[3], 1400
8801	Lysine, dihydro chloride (L) $H_2N(CH_2)_4CH(NH_2)CO_2H.2HCl$	219.11	(al-eth or aq HCl), $[\alpha]^{20/}_D$ + 15.3 (w)		201-2			w, al	B4[3], 1402
8802	Lysine, α-N-benzoyl (dl) $H_2N(CH_2)_4CH(NHCOC_6H_5)CO_2H$	250.30	nd (w)		235			w	B9[3], 1235
8803	Lysine, ω-N-benzoyl (dl) $C_6H_5CONH(CH_2)_4CH(NH_2)CO_2H$	250.30	cr (w)		268			w	B9[3], 1238
8804	Lysine, ω-N-benzoyl-(1) $C_6H_5CONH(CH_2)_4CH(NH_2)CO_2H$	250.30	lf (w), $[\alpha]^{19/}_D$ + 20.1 (aq HCl)		240d			w	B9[3], 1237

PHYSICAL CONSTANTS OF ORGANIC COMPOUNDS (Continued)

No.	Name, Synonyms, and Formula	Mol. wt.	Color, crystalline form, specific rotation and λ_{max} (log ε)	b.p. °C	m.p. °C	Density	n_D	Solubility	Ref.
8805	Malathion or S(1,2-Dicarboxymethyl)-0,0-dimethyldithiophosphate $C_{10}H_{19}PS_2O_6$	330.35	ye-br	$156\text{-}7^{0.7}/_d$	2.8	1.2076^{20}	1.4960^{20}	al, eth, bz	$B3^4$, 1136
8806	Maleicacid-*cis*-butenedioic acid $HO_2CCH=CHCO_2H$	116.07	mcl pr (w)	...	139-40	1.590^{20}	...	w, al, eth, ace, aa	$B2^4$, 2199
8807	Maleic acid, bromo or Bromo maleicacid $HO_2C\ CH=CHCO_2H$	194.97	nd or pr	d	136-40	...	...	w, al, eth	$B2^4$, 2224
8808	Maleic acid, chloro $HO_2CCl=CHCO_2H$	150.52	pr(eth-chl)	...	108 (114)	...	...	al, eth, ace, aa	$B2^4$, 2221
8809	Maleic acid, chloro,diethyl ester or Diethyl chloromaleate $C_2H_5O_2CCCl=CHCO_2C_2H_5$	206.63	...	235d, 125^{19}	...	$1.1741^{20/4}$	...	al, eth, ace, aa	$B2^3$, 1928
8810	Maleic acid, chloro, dimethyl ester $CH_3O_2CCCl=CHCO_2CH_3$	178.57	...	106.5^{18}	...	$1.2775^{25/4}$	...	al, eth	$B2^3$, 1928
8811	Maleic acid, dichloro $HO_2CCCl=CClCO_2H$	184.96	nd (liq eth)	...	119-20	...	...	w, al, eth	$B2^4$, 2223
8812	Maleic acid, diethyl ester or Diethyl maleate $C_2H_5O_2CCH=CHCO_2H_5$	172.18	...	223, $105\text{-}6^{14}$	−88	$1.0662^{20/4}$	1.4416^{20}	al, eth	$B2^4$, 2207
8813	Maleic acid, dihydroxy $HO_2CC(OH)=C(OH)CO_2H$	148.07	pl (w + 2)	...	155 (anh)	...	...	...	$B3^4$, 1975
8814	Maleic acid, dimethyl ester or Dimethyl maleate $CH_3O_2CCH=CHCO_2CH_3$	144.13	...	202, 102^{17}	−19	$1.1606^{20/4}$	1.4416^{20}	eth	$B2^4$, 2204
8815	Maleic acid, diphenyl ester $C_6H_5O_2CCH=CHCO_2C_6H_5$	286.27	pl(lig)	226.15	73	...	...	al, eth, ace, bz, chl	$B6^4$, 628
8816	Maleic acid, dipropyl ester or Dipropylmaleate $C_3H_7O_2CCH=CHCO_2C_3H_7$	200.23	...	126^{12}	...	$1.0245^{20/4}$	1.4434^{20}	al, eth, ace, bz	$B2^4$, 2209
8817	Maleic acid, hydrazide $C_4H_4N_2O_2$	112.09	cr (w)	...	>300d	...	...	...	$B2^4$, 2220
8818	Maleic acid, methyl or Citraconic acid $HO_2CC(CH_3)=CHCO_2H$	130.10	tcl pr or pl (eth-bz) nd (eth-lig)	...	93-4	1.617	...	w	$B2^4$, 2230
8819	Maleic acid, monoamide or Maleamic acid $HO_2CCH=CHCONH_2$	115.09	lf(w)	...	178	...	...	w, al	$B2^4$, 2218
8820	Maleic anhydride $C_4H_2O_3$	98.06	nd(chl eth)	197-9, 82^{14}	60	1.314^{60}	...	eth, ace, chl	$B17^4$, 5897
8821	Maleic anhydride, dimethyl or Pyrocinchonic anhydride $OC(CH_3)=C(CH_3)CO$	126.11	pl or lf(dil al)	223, 105^{12}	96	$1.107^{100/4}$	...	al, eth, bz, chl	$B17^4$, 5919
8822	Maleic anhydride, methyl or Citraconic anhydride $C_5H_4O_3$	112.08	...	213-4, $99\text{-}100^{15}$	7-8	$1.2469^{16/4}$	1.4710^{21}	al, eth, ace	$B17^4$, 5912
8823	Maleic anhydride, phenyl $C_{10}H_7O_3$	174.16	ye nd (CS_2)	...	122	...	...	al, eth	$B17^4$, 6279
8824	Maleimide or Maleic acid, imide OCHC=CH−(O·NH) └─N──┘	97.07	pl (bz)	sub	93-5	...	...	w, al, eth	$B21^4$, 4627
8825	Maleimide, N-ethyl $C_6H_7NO_2$	125.13	cr (bz)	...	45.5	...	...	al, eth	$B21^4$, 4629
8826	Maleimide, N-phenyl or Maleanil $C_{10}H_7NO_2$	173.17	ye nd (bz-lig)	162^{12}	90-1	...	...	al, eth, bz	$B21^4$, 4631
8827	Malonamide $H_2C(CONH_2)_2$	102.09	mcl pr (w)	...	171-2	...	...	w	$B2^4$, 1887
8828	Malonamide, benzyl $C_6H_5CH_2CH(CONH_2)_2$	192.22	nd (al)	...	225	...	...	...	$B9^3$, 4285
8829	Malonamide, N,N′-diphenyl or Malonanilide $H_2C(CONHC_6H_5)_2$	254.29	nd (al)	...	226	...	...	al, bz, aa	$B12^3$, 558
8830	Malonic acid or Propanedioic acid $HO_2CCH_2CO_2H$	104.06	tcl(al)	d140	135.6	1.619^{16}	...	al, eth, py	$B2^4$, 1874
8831	Malonic acid, acetamide,diethyl ester $CH_3CONHCH(CO_2C_2H_5)_2$	217.22	cr(al bz-peth)	185^{20}	95-6	...	...	al	$B4^3$, 1503
8832	Malonic acid, acetyl, diethyl ester $(CH_3CO)CH(CO_2C_2H_5)_2$	202.21	...	232, 120^{17}	...	$1.0834^{26/4}$	1.4435^{25}	ace	$B3^4$, 1819
8833	Malonic acid, allyl or 3-Butene-1,1-dicarboxylic acid $(CH_2=CHCH_2)CH(CO_2H)_2$	144.13	tcl (eth)	>180d	105	...	...	w, al, eth	$B2^3$, 1945
8834	Malonic acid, allyl,diethyl ester $(CH_2=CHCH_2)CH(CO_2C_2H_5)_2$	200.23	...	222-3, 93^6	...	$1.0098^{20/4}$	1.4305^{20}	al, eth	$B2^4$, 2240
8835	Malonic acid, amino,monohydrate $H_2NCH(CO_2H)_2.H_2O$	137.09	pr (w + 1)	...	112d	...	...	...	$B4^3$, 1501
8836	Malonic acid, amino,diethyl ester $H_2NCH(CO_2C_2H_5)_2$	175.18	...	$122\text{-}3^{16}$	...	$1.100^{16/4}$	1.4353^{16}	w, al, eth, ace, bz	$B4^3$, 1501

No.	Name, Synonyms, and Formula	Mol. wt.	Color, crystalline form, specific rotation and λ_{max} (log ε)	b.p. °C	m.p. °C	Density	n_D	Solubility	Ref.
8837	Malonic acid, benzoyl amino,diethyl ester ... $(C_6H_5CONH)CH(CO_2C_2H_5)_2$	279.29	nd(peth)	...	61	...	...	al, eth	$B9^3$, 1186
8838	Malonic acid, benzyl ... $C_6H_5CH_2CH(CO_2H)_2$	194.19	pr(bz eth chl-peth)	...	121-2	...	...	w, al, eth, bz	$B9^3$, 4283
8839	Malonic acid, benzyl,diethyl ester ... $C_6H_5CH_2CH(CO_2C_2H_5)_2$	250.29	...	$300, 169^{212}$	...	$1.0750^{20/4}$	1.4872^{20}	...	$B9^3$, 4284
8840	Malonic acid, benzyl,hydroxy ... $C_6H_5CH_2C(OH)(CO_2H)_2$	210.19	pr	...	147d	...	...	w, al, eth	$B10^3$, 2216
8841	Malonic acid, benzylidene ... $C_6H_5CH{=}C(CO_2H)_2$	192.17	pr(w)	...	195-6d	...	...	al, ace	$B9^3$, 4378
8842	Malonic acid, benzylidene,diethyl ester ... $C_6H_5CH{=}C(CO_2C_2H_5)_2$	248.28	...	308-12d, 180^{10}	32	$1.1045^{20/4}$	1.5389^{20}	al, eth, ace, bz	$B9^3$, 4379
8843	Malonic acid, bromo ... $BrCH(CO_2H)_2$	182.96	nd(eth) pl(ace-bz)	...	113d	...	...	al, eth	$B2^4$, 1904
8844	Malonic acid, bromo,diethyl ester ... $BrCH(CO_2C_2H_5)_2$	239.07	...	253-5d, 123^{20}	−54	$1.4022^{25/4}$	1.4521^{20}	**al, eth**, ace	$B2^4$, 1904
8845	Malonic acid, butyl or 1,1-Pentane dicarboxylic acid ... $C_4H_9CH(CO_2H)_2$	160.17	pr(w)	...	104-5	...	...	w, al, eth	$B2^4$, 2011
8846	Malonic acid, butyl,diethyl ester ... $C_4H_9CH(CO_2C_2H_5)_2$	216.28	...	235-40, 122^{12}	...	...	1.4250^{20}	al, eth	$B2^4$, 2011
8847	Malonic acid, isobutyl ... $i\text{-}C_4H_9CH(CO_2H)_2$	160.17	cr(bz)	d	115d	...	...	w, al, eth	$B2^4$, 2023
8848	Malonic acid, isobutyl,diethyl ester ... $i\text{-}C_4H_9CH(CO_2C_2H_5)_2$	216.28	...	255, $119\text{-}20^{16}$	...	$0.9804^{20/4}$	1.4236^{20}	al, eth	$B2^3$, 1756
8849	Malonic acid, sec-Butyl,diethyl ester ... $C_2H_5CH(CH_3)CH(CO_2C_2H_5)_2$	216.28	...	245-50, 105^{9}	...	0.988^{15}	1.4248^{20}	al, eth	$B2^4$, 2019
8850	Malonic acid, cetyl or Hexadecyl malonic acid ... $C_{16}H_{33}CH(CO_2H)_2$	328.69	nd(lig) lf(aa)	...	121-2	...	...	eth, bz, aa	$B2^4$, 2183
8851	Malonic acid, cetyl, diethyl ester ... $C_{16}H_{33}CH(CO_2C_2H_5)_2$	384.60	amor	$238\text{-}40^{14}$	(i)25.1 (ii)12.7	...	1.4433^{20}	al	$B2^4$, 2183
8852	Malonic acid, chloro ... $ClCH(CO_2H)_2$	138.51	pr (w)	d	133	...	...	w, sl, eth	$B2^3$, 1637
8853	Malonic acid, chloro, diethyl ester or Ethyl chloromalonate ... $ClCH(CO_2C_2H_5)_2$	194.61	...	222, 118^{16}	...	$1.2040^{20/4}$	1.4327^{20}	**al, eth, chl**	$B2^4$, 1903
8854	Malonic acid, (3-chloropropyl), diethyl ester ... $Cl(CH_2)_3CH(CO_2C_2H_5)_2$	236.70	...	$147\text{-}9^{10}$	...	...	1.4429^{20}	al, eth, chl	$B2^4$, 1992
8855	Malonic acid, cinnamylidene-*(trans)* or Cinnamal malonic acid ... $C_6H_5CH{=}CHCH{=}C(CO_2H)_2$	218.21	dk ye nd	...	212d	...	...	al, chl	$B9^3$, 4432
8856	Malonic acid, cyclohexyl, diethyl ester ... $C_6H_{11}CH(CO_2C_2H_5)_2$	242.32	...	$163\text{-}5^{20}$	...	$1.0281^{19/4}$	1.4478^{25}	al, eth, ace, bz	$B9^3$, 3834
8857	Malonic acid, 2-cylopentenyl, diethyl ester ... $(2\text{-}C_5H_7)CH(CO_2C_2H_5)_2$	226.27	...	141^{10}	...	$1.0507^{20/4}$	1.4536^{20}	eth, ace	$B9^3$, 3951
8858	Malonic acid, cyclopentylidene, diethyl ester ... $(C_5H_8{=})C(CO_2C_2H_5)_2$	226.27	...	140^{10}	...	$1.0616^{20/4}$	1.4724^{20}	eth, ace	$B9^3$, 3952
8859	Malonic acid, diallyl, diethyl ester ... $(CH_2{=}CHCH_2)_2C(CO_2C_2H_5)_2$	240.30	...	243-4, 128^{16}	...	$0.9943^{20/4}$	1.4445^{22}	al, eth, ace	$B2^3$, 2003
8860	Malonic acid, dibenzyl, diethyl ester ... $(C_6H_5CH_2)_2C(CO_2C_2H_5)_2$	340.42	...	$234\text{-}5^{23}$	14	$1.093^{20/4}$	...	al, eth	$B9^3$, 4553
8861	Malonic acid, dibromo, diethyl ester ... $Br_2C(CO_2C_2H_5)_2$	317.96	...	250-6d, 154^{28}	...	...	...	al, eth, ace	$B2^4$, 1905
8862	Malonic acid, dibutyl ester ... $CH_2(CO_2C_4H_9)_2$	216.28	...	251-2, 137^{14}	−83	$0.9824^{20/4}$	1.4262^{20}	al, eth, ace, bz, aa	$B2^4$, 1884
8863	Malonic acid, dibutyl,diethyl ester ... $(C_4H_9)_2C(CO_2C_2H_5)_2$	272.38	...	$153\text{-}4^{14}$	...	$0.9457^{20/4}$	1.4341^{20}	al, eth	$B2^4$, 2119
8864	Malonic acid, diethyl or 3,3-Pentanedicarboxylic acid ... $(C_2H_5)_2C(CO_2H)_2$	160.17	pr (w bz)	d	127	...	...	w, al, eth	$B2^4$, 2026
8865	Malonic acid, diethyl, diethyl ester ... $(C_2H_5)_2C(CO_2C_2H_5)_2$	216.28	...	230, 100^{12}	...	0.9643^{30}	1.4240^{20}	**al, eth**	$B2^4$, 2026
8866	Malonic acid, diethyl ester or Diethyl malonate ... $CH_2(CO_2C_2H_5)_2$	160.17	...	199.3, 96^{22}	−48.9	$1.0551^{20/4}$	1.4139^{20}	**al, eth**, ace, bz, chl	$B2^4$, 1881
8867	Malonic acid, dimethyl ester or Dimethyl malonate ... $CH_2(CO_2CH_3)_2$	132.12	...	181.4, 78.4^{15}	−61.9	1.156^{20}	1.4135^{20}	**al, eth**, ace, bz, chl	$B2^4$, 1880
8868	Malonic acid, dimethyl, diethyl ester ... $(CH_3)_2C(CO_2C_2H_5)_2$	188.22	...	197, $97\text{-}8^{22}$	−30.4	$0.9964^{20/4}$	1.4129^{20}	**al, eth**	$B2^4$, 1955

No.	Name, Synonyms, and Formula	Mol. wt.	Color, crystalline form, specific rotation and λ_{max} (log ε)	b.p. °C	m.p. °C	Density	n_D	Solubility	Ref.
8869	Malonic acid, dipropyl ester $CH_2(CO_2C_3H_7)_2$	188.22	glass	229, 113^{13}	−77.1	$1.0097^{20/4}$	1.4206^{20}	al, eth, ace, bz	$B2^{3}$, 1619
8870	Malonic acid, ethoxymethylene),diethyl ester $(C_2H_5OCH{=})C(CO_2C_2H_5)_2$	216.23		279-81d, 165^{19}			1.4600^{20}	al, eth	$B3^{4}$, 1192
8871	Malonic acid, ethyl or 1,1-Propanedicarboxylic acid $C_2H_5CH(CO_2H)_2$	132.12	pr (w + 1)	160d	114			w, al, eth, bz, chl	$B2^{4}$, 1952
8872	Malonic acid, ethyl, diethyl ester $C_2H_5CH(CO_2C_2H_5)_2$	188.22		$207\text{-}9^{755}$, $98\text{-}9^{12}$		$1.0047^{20/4}$	1.4166^{20}	al, eth, ace, chl	$B2^{4}$, 1953
8873	Malonic acid, ethyl (methyl) or 2,2-Butane dicarboxylic acid $C_2H_5C(CH_3)(CO_2H)_2$	146.14	pr or nd (eth)		122			w, al, eth, ace	$B2^{4}$, 1977
8874	Malonic acid, ethyl (phenyl), diethyl ester $C_6H_5C(C_2H_5)(CO_2C_2H_5)_2$	264.32		170^{19}		$1.071^{20/4}$	1.4896^{25}	al, eth	$B9^{3}$, 4304
8875	Malonic acid, ethyl (isopropyl), diethyl ester $(CH_3)_2CHC(C_2H_5)(CO_2C_2H_5)_2$	230.30		$232\text{-}4^{742}$, $108\text{-}10^{11}$			1.4280^{25}	al, eth, ace	$B2^{4}$, 2052
8876	Malonic acid, ethylidene,diethyl ester $(CH_3CH{=})C(CO_2C_2H_5)_2$	186.21		$115\text{-}8^{17}$		$1.0194^{17/4}$	1.4308^{17}	al, eth	$B2^{4}$, 2234
8877	Malonic acid, formamido,diethyl ester $HCONHCH(CO_2C_2H_5)_2$	203.20	pl (MeOH)	$173\text{-}4^{11}$	48-9			al	$B4^{3}$, 1503
8878	Malonic acid, heptyl or 1,1-Octanedicarboxylic acid $C_7H_{15}CH(CO_2H)_2$	202.25	pr (bz-peth)		96-8			al, eth, ace	$B2^{4}$, 2094
8879	Malonic acid, hexyl,diethyl ester $C_6H_{13}CH(CO_2C_2H_5)_2$	244.33		268-70, 143^{15}		$0.9577^{21/4}$	1.4278^{21}	al, eth, ace, bz	$B2^{3}$, 1791
8880	Malonic acid, hydroxy or Tartronic acid $HOCH(CO_2H)_2$	120.06	pr (w + 1)	sub	156-8			w, al	$B3^{4}$, 1120
8881	Malonic acid, hydroxy,diethyl ester $HOCH(CO_2C_2H_5)_2$	176.17		222-5, 121^{15}	−2.5			al, eth, ace, bz	$B3^{3}$, 905
8882	Malonic acid, hydroxy,dimethyl ester $HOCH(CO_2CH_3)_2$	148.12	cr (eth-peth)	122^{19}	45			w, al, eth, ace, bz	$B3^{3}$, 905
8883	Malonic acid, (2-hydroxycyclohexyl) lactone,ethyl ester $C_{11}H_{16}O_4$	212.25	pa ye	199^{30}		1.0735^{19}			$B18^{4}$, 5351
8884	Malonic acid, hydroxy-(methyl) or α-Isomalic acid $CH_3C(OH)(CO_2H)_2$	134.09		170d	142d			w, al, eth	$B3^{3}$, 927
8885	Malonic acid, methyl or 1,1-Ethane dicarbozylic acid $CH_3CH(CO_2H)_2$	118.09	nd (AcOEt-bz) pr(eth-bz)		135d	$1.455^{20/4}$		w, al, eth, aa	$B2^{4}$, 1932
8886	Malonic acid, methyl,diethyl ester or Diethyl methylmalonate $CH_3CH(CO_2C_2H_5)_2$	174.20		201, 94^{16}		$1.0225^{20/4}$	1.4126^{20}	al, eth, ace, chl	$B2^{4}$, 1932
8887	Malonic acid, methyl,dimethyl ester $CH_3CH(CO_2CH_3)_2$	146.14		176.5		$1.0977^{20/4}$	1.4128^{20}	al, eth, ace, chl	$B2^{3}$, 1681
8888	Malonic acid, monoamide-N-Phenyl or Malonanilic acid $HO_2CCH_2CONHC_6H_5$	179.18	cr(w eth al)		132			al, eth	$B12^{3}$, 557
8889	Malonic acid, monochloride,ethyl ester or Carbethoxy acetyl chloride $ClCOCH_2CO_2C_2H_5$	150.56		170-80d, $75\text{-}7^{15}$				eth	$B2^{4}$, 1887
8890	Malonic acid, monochloride,methyl ester $ClCOCH_2CO_2CH_3$	136.53		71^{15}				eth	$B2^{4}$, 1886
8891	Malonic acid, octyl or 1,1-Nonane dicarboxylic acid $C_8H_{17}CH(CO_2H)_2$	216.28	pr(bz-peth)		116	1.173^{17}		al, ace	$B2^{4}$, 2114
8892	Malonic acid, pentyl,diethyl ester $C_5H_{11}CH(CO_2C_2H_5)_2$	230.30		$134\text{-}6^{14}$		$0.9652^{20/4}$	1.4253^{20}	al, eth	$B2^{4}$, 2034
8893	Malonic acid, isopentyl,diethyl ester $i\text{-}C_5H_{11}CH(CO_2C_2H_5)_2$	230.30		240-2, 137.5^{19}		$0.9580^{25/4}$	1.4255^{25}	al, eth, ace	$B2^{3}$, 1777
8894	Malonic acid, (3-pentyl), diethyl ester $[(C_2H_5)_2CH]CH(CO_2C_2H_5)_2$	230.30		130^{16}			1.4291^{20}	al, eth	$B2^{3}$, 1781
8895	Malonic acid, phenyl,diethyl ester $C_6H_5CH(CO_2C_2H_5)_2$	236.27		205d, 168^{12}	16-7	$1.0950^{20/4}$	1.4977^{20}	al, ace	$B9^{3}$, 4260
8896	Malonic acid, phenyl,dimethyl ester $C_6H_5CH(CO_2CH_3)_2$	208.21	cr(lig)	$145\text{-}7^{13}$	51			al, eth	$B9^{3}$, 4260
8897	Malonic acid, phenyl amino,diethyl ester $C_6H_5NHCH(CO_2C_2H_5)_2$	251.28	cr(al or lig)		45			al, eth, bz, chl	$B12^{3}$, 971
8898	Malonic acid, phthalimido,diethyl ester $C_{15}H_{15}NO_6$	305.29	pr(al)		74			al, eth, ace, bz, chl	$B21^{4}$, 5264
8899	Malonic acid, propyl $C_3H_7CH(CO_2H)_2$	146.14	pl(bz)	d	96-7			w, al, eth, chl	$B2^{4}$, 1991
8900	Malonic acid, propyl,diethyl ester $C_3H_7CH(CO_2C_2H_5)_2$	202.25		221^{767}, 114^{22}		0.9873^{20}	1.4197^{20}	al, eth	$B2^{4}$, 1991

No.	Name, Synonyms, and Formula	Mol. wt.	Color, crystalline form, specific rotation and λ_{max} (log ε)	b.p. °C	m.p. °C	Density	n_D	Solubility	Ref.
8901	Malonic acid, isopropyl,diethyl ester $(CH_3)_2CHCH(CO_2C_2H_5)_2$	202.25	....	235, 107-9^{18}	......	0.9970$^{20/15}$	1.4188^{21}	al, eth, chl	B2^4, 2011
8902	Malonic acid, isopropylidene $(CH_3)_2C{=}C(CO_2H)_2$	144.13	cr (ace-chl)	d	170-1	....	....	al	B2^3, 1948
8903	Malonic acid, isopropylidene,diethyl ester $(CH_3)_2C{=}C(CO_2C_2H_5)_2$	200.23	....	175-8, 140-1^{20}	......	1.0282$^{18/4}$	1.4486^{17}	al, ace	B2^4, 2244
8904	Malononitrile $CH_2(CN)_2$	66.06	....	218-9, 109^{20}	32	1.1910$^{20/4}$	1.4146^{34}	w, al, eth, ace, bz	B2^4, 1892
8905	Malononitrile, benzyl $C_6H_5CH_2CH(CN)_2$	156.19	pl(al) nd(w lig)	174^{23}	91	....	....	al, eth, bz	B9, 870
8906	Malononitrile, phenyl $C_6H_5CH(CN)_2$	142.16	cr (dil al)	152-3^{21}	70-1	....	....	al	B9^3, 4263
8907	Malonyl chloride $CH_2(COCl)_2$	140.95	....	58^{26}	......	1.4509$^{20/4}$	1.4639^{20}	eth	B2^4, 1887
8908	Maltose (pyranose form, α-anomer, anhydrous) or 4-*O*-α-D-glucopyranosyl-D-glucose $C_{12}H_{22}O_{11}$	342.30	nd (abs al), $[\alpha]_D$ +140.7 (w, c=10)	....	160-5	....	....	w	B17^4, 3057
8909	Maltose, monohydrate (β-anomer) $C_{12}H_{22}O_{11}\cdot H_2O$	360.32	nd (w), $[\alpha]^{20}{}_D$ +111.7→ +130.4 (c=4, mut)	....	102-3	1.54	....	w	B17^4, 3057
8910	Malvidine chloride or Syringidine chloride $C_{17}H_{15}ClO_7$	366.75	rh ta or pr(al-aq HCl + 1w)	....	>300	....	....	al	B18^4, 3561
8911	Mandelic acid-(D) or α-Hydroxyphenyl acetic acid $C_6H_5CH(OH)CO_2H$	152.15	ta, $[\alpha]^{20}{}_D$ -158 (w, c=2.5)	....	133-5	1.341	....	w, al, eth, aa, chl	B10^3, 445
8912	Mandelic Acid (DL) or α-Hyroxyphenyl acetic acid $C_6H_5CH(OH)COOH$	152.15	pl (wh) rh (bz)	d	121.3	1.300$^{20/4}$	....	w, al, eth	B10^3, 448
8913	Mandelic acid (L+) or L+-α-Hydroxyphenylacetic acid $C_6H_5CH(OH)CO_2H$	152.15	pl(w), $[\alpha]^{20}{}_D$ +156.6 (w, c=2.9)	....	......	....	....	w, al, eth, chl	B10^3, 447
8914	Mandelic acid, 2,4-dimethyl or (2,4-Dimethylphenyl)hydroxyacetic acid $[2,4\text{-}(CH_3)_2C_6H_3]CH(OH)CO_2H$	180.20	rh(w) nd(bz) lf(to peth al)	....	119	....	....	al, eth, chl	B10^3, 613
8915	Mandelic acid, 2,5-dimethyl or (2,5-Dimethylphenyl)-α-Hydroxyacetic acid $[2,5\text{-}(CH_3)_2C_6H_3]CH(OH)CO_2H$	180.20	nd or pr(bz)	....	116-7	....	....	al, eth, chl	B10^3, 612
8916	Mandelic acid, 3,4-dimethyl or (3,4-Dimethylphenyl)-α-hydroxyacetic acid $[3,4\text{-}(CH_3)_2C_6H_3]CH(OH)CO_2H$	180.20	lf(bz)	....	135	....	....	w, bz	B10^3, 611
8917	Mandelic acid, *p*-iodo-*(dl)* $4{=}IC_6H_4CH(OH)CO_2H$	278.05	....	....	135	....	....	w, al, eth	B10^3, 481
8918	Mandelic acid, *p*-isopropyl-*(dl)* $[4\text{-i-}C_3H_7C_6H_4]CH(OH)CO_2H$	194.23	nd(w)	....	159-60	....	....	al, eth	B10^3, 629
8919	Mandelic acid, *p*-isopropyl-*(dl)* $[4\text{-i-}C_3H_7C_6H_4]CH(OH)CO_2H$	194.23	lf (w), $[\alpha]^{17}{}_D$ +134.9 (abs, al, c=4)	....	153-4	....	....	al, eth	B10, 279
8920	Mandelic acid, *p*-isopropyl-*(l)* $[4\text{-i-}C_3H_7C_6H_4]CH(OH)CO_2H$	194.23	ta (20% al), $[\alpha]^{17}{}_D$ -135 (abs, al, c=4.09)	....	153-4	....	....	al, eth	B10, 279
8921	D-Mannitane or 3,6-(1,4)Anhydro-D-mannitol $C_6H_{12}O_5$	164.16	amor pl	....	146-7	....	....	w, al	B17^4, 2641
8922	D-Mannitol or D-Mannite $C_6H_{14}O_6$	182.17	rh nd or pr (w), $[\alpha]^{25}{}_D$ -0.49 (w)	295$^{3.5}$	168	1.489$^{20/4}$	1.3330	w	B1^4, 2841
8923	DL-Mannitol or α-Acritol $C_6H_{14}O_6$	182.17	cr(al)	....	168	....	....	w	B1^3, 2405
8924	L-Mannitol or L-Mannite $C_6H_{14}O_6$	182.17	nd(al)	....	163-4	....	....	w	B1^4, 2843
8925	Mannitol, hexanitrate or Nitro mannitol $O_2NOCH_2(CHONO_2)_4CH_2ONO_2$	452.16	nd (al), $[\alpha]^{25}{}_{546}$ +46.8	120 exp	112.3	1.8$^{20/4}$	....	al, eth, bz, aa	B1^4, 2849

No.	Name, Synonyms, and Formula	Mol. wt.	Color, crystalline form, specific rotation and λ_{max} (log ε)	b.p. °C	m.p. °C	Density	n_D	Solubility	Ref.
8926	α-D-Mannoheptose $C_7H_{14}O_7$	210.18	nd (al) $[\alpha]^{20}_D$ +85→ +68.6 (w,c=1)		134-5			w	$B1^4$, 4438
8927	β-D-Mannoheptose, monohydrate $C_7H_{14}O_7 \cdot H_2O$	228.20	cr (w+1) $[\alpha]^{20}_D$ +45.7		83			w	$B1^4$, 4441
8928	D-Mannoic acid-γ-lactone $C_6H_{10}O_6$	178.14	pr (abs al) $[\alpha]^{20}_D$ +54 (w,c=2)		151			w	$B18^4$, 3018
8929	D-Mannoheptose (β-anomer), hydrate $C_7H_{14}O_7$	210.18	rh nd (w) $[\alpha]_D$ −51.8		151			w	$B18^2$, 199
8930	D-Mannosaccharic acid, γ,γ′-dilactone $C_6H_6O_6$	174.11	nd (w+2) $[\alpha]^{23}_D$ +202 (w)		190d			al	$B19^4$, 2962
8931	L-Mannosaccharic acid, γ,γ′-dilactone $C_6H_6O_6$	174.11	nd (w al) $[\alpha]^{20}_D$ −202.5 (w)		183-5d, 68 (hyd)			w	$B19^2$, 266
8932	D-Mannose (β-anomer) $C_6H_{12}O_6$	180.16	nd or orh pr (al or aa) $[\alpha]^{20}_D$ −17→ +14.6 (w,p=3)		132d	$1.539^{20/4}$			$B1^4$, 4328
8933	DL-Mannose $C_6H_{12}O_6$	180.16	cr (al)		132-3			w	B31, 294
8934	L-Mannose (β-anomer) $C_6H_{12}O_6$	180.16	cr (al) $[\alpha]_D$ +14→ −14		132			w	$B1^4$, 4333
8935	D-Mannose, phenylhydrazone $C_{12}H_{18}N_2O_5$	270.29	ye pr (w) nd (al) $[\alpha]_D$ +26.3→ +33.8 (py)		199-200				B31, 290
8936	L-Mannose, phenylhydrazone $C_{12}H_{18}N_2O_5$	270.29	ye pr (w)		195				B31, 294
8937	D-Mannuronic acid (α-anomer), monohydrate $C_6H_{10}O_7$	194.14	hyg nd (al eth) $[\alpha]^{25}_D$ +16→ −6 (w,mut)		120-30d			w	$B3^4$, 1998
8938	D-Mannuronic acid (β-anomer) $C_6H_{10}O_7$	194.14	cr (w ace eth) $[\alpha]^{25}_D$ −47.9→ −23.9 (w)		165-7			w	$B3^4$, 1999
8939	Margaric acid or Heptadecanoic acid $CH_3(CH_2)_{15}CO_2H$	270.46	pl (peth)	227^{100}	62-3	0.8532^{60}	1.4342^{60}	eth, ace, bz, chl	$B2^4$, 1193
8940	Matrine or Sophocarpidine $C_{15}H_{24}N_2O$	248.37	α, nd or pl, β, orh prγ, lig;δ,pr or lf (peth), $[\alpha]^{15}_D$ α40.9 (w), β −28.7 (w)	γ-223^6	α,76-7 β, 87 δ, 84	γ-$1.088^{20/4}$	γ-1.5286^{85}	al, eth, ace, bz	$B24^2$, 58
8941	Meconic acid or 3-hydroxy-γ-pyrone-2,6-dicarboxylic acid $C_7H_4O_7$	200.10	rh pl (w dilHCl) (+3w)	d at 120	-w at 100			al, bz	$B18^4$, 6203
8942	Meconidine $C_{21}H_{23}NO_4$	353.42	ye amor		58			al, eth, ace, bz, chl	M, 641
8943	Meconin or 6,7-Dimethoxyphthalide $C_{10}H_{10}O_4$	194.19	wh nd (w)	155sub	102-3			al, eth, ace, bz, chl	$B18^4$, 1226
8944	Medicagenic acid $C_{30}H_{46}O_6$	502.69	pr or nd		352-3				Am76, 2271
8945	Medicagenic acid, diacetate $C_{34}H_{50}O_8$	586.77	mcl $[\alpha]^{22}_D$ +92 chl		210-2	$1.190^{25/25}$			Am79, 5292
8946	Melam $C_6H_9N_{11}$	235.21	pw						$B3^4$, 319
8947	Melamine or 2,4,6-triamino-1,3,5-triazine $C_3H_6N_6$	126.12	mcl pr (w)	sub	345d	1.573^{16}	1.872^{20}		$B26^2$, 132

PHYSICAL CONSTANTS OF ORGANIC COMPOUNDS (Continued)

No.	Name, Synonyms, and Formula	Mol. wt.	Color, crystalline form, specific rotation and λ_{max} (log ε)	b.p. °C	m.p. °C	Density	n_D	Solubility	Ref.
8948	Melene $C_{30}H_{60}$	420.81	nd (ace) cr (peth)	380	62-3	$0.9037^{25/25}$	1.4228^{90}	...	B1[3], 885
8949	Melissic acid $CH_3(CH_2)_{29}CO_2H$	466.83	sc or nd (al or ace)	...	93	...	...	bz	B2[3], 1097
8950	Melezitose $C_{18}H_{32}O_{16}$	504.44	cr (w + 2) $[\alpha]^{20}_D$ + 88 (w,c=4)	...	153-4 (anh)	1.5565^{0}	...	w	B17[4], 3815
8951	Melibiose (α-anomer) or 6-*O*-α-D-galactopyranosyl-α-D-glucose $C_{12}H_{22}O_{11}$	342.30	amor (anh) $[\alpha]^{15}$ + 145.8 →141.6 (w,c=2,mut)	...	...	...	...	w	B17[4], 3075
8952	Melibiose (β-anomer), dihydrate $C_{12}H_{22}O_{11}.2H_2O$	378.33	mcl cr (w + 2) $[\alpha]^{20}_D$ + 111.7 → + 129.5 (w,c=4)	...	84-5	...	...	w, MeOH	B17[4], 3075
8953	Mellitic acid or Benzene hexacarboxylic acid $C_6(CO_2H)_6$	342.17	nd (al)	...	286-8	...	...	w, al	B9[3], 4914
8954	P-Menthane *(cis)* or 1-isopropyl-4-methyl cyclohexane $C_{10}H_{20}$	140.27	...	170.9^{725}	−89.9	$0.8039^{20/4}$	1.4431^{20}	al, eth, bz, peth	B5[4], 151
8955	P-Menthane *(trans)* $C_{10}H_{20}$	140.27	...	170.6, 58.5^{15}	...	$0.7928^{20/4}$	1.4366^{20}	al, eth, bz, lig	B5[4], 151
8956	*p*-Menthane, 1,8-epoxy $C_{10}H_{18}O$	154.25	...	176-7	1.5	0.9267^{20}	1.4584^{15}_D	al, eth, chl	B17[4], 213
8957	Δ^3-*p*-Menthene(d) (one form) or 1-isopropyl-4-methyl cyclohexene-(3) $C_{10}H_{18}$	138.25	$[\alpha]^{20}_D$ + 115.6 (undil)	168	...	$0.8118^{20/4}$	1.4524^{20}	al, eth, bz, aa	B5[4], 301
8958	Δ^1-*p*-methene(d) (one form) $C_{10}H_{18}$	138.25	$[\alpha]_D$ + 29.6 → + 54.4	167-8	...	$0.8078^{20/4}$	...	al, eth, bz, peth	B5[4], 299
8959	Δ^3-*p*-methene(dl) $C_{10}H_{18}$	138.25	...	168^{754}, 60.5^{12}	...	$0.8069^{20/4}$	1.4503^{15}	al, eth, bz, peth	B5[4], 301
8960	Δ^3-*p*-Menthene $C_{10}H_{18}$	138.25	$[\alpha]_D$ −13.5 (al)	167-8	1	$0.8112^{19/19}$	1.4511^{20}	al, eth, bz, peth	B5[4], 301
8961	Menthol-(d) or 3-*p*-Menthol, Hexahydro thymol $C_{10}H_{20}O$	156.27	$[\alpha]_D$ + 49.2 (al,c=5)	$103\text{-}4^{9}$	42-3	...	...	al, eth, ace, bz	B6[4], 150
8962	Menthol-(dl) $C_{10}H_{20}O$	156.27	nd (peth)	216, $103\text{-}5^{16}$	(i)28 (ii)38	$0.904^{15/15}$	1.4615^{20}	al, eth, ace, bz	B6[4], 151
8963	Menthol-(l) $C_{10}H_{20}O$	156.27	nd (MeOH) $[\alpha]^{20}_D$ −48 (al,c=2.5)	216.4, 111^{20}	44	$0.904^{15/15}$	1.460^{22}	al, eth, ace, chl	B6[4], 151
8964	Menthol, 3-isovalerate $C_{15}H_{28}O_2$	240.39	$[\alpha]^{20}_D$ −64 (bz,c=10)	129^{9}	...	0.9089^{15}	1.4486^{20}	al, ace	B6[3], 145
8965	Menthol, acetate(l) $C_{12}H_{22}O_2$	198.31	$[\alpha]^{20}_D$ −79.42	109^{10}	...	$0.9185^{20/4}$	1.4469^{20}	...	B6[3], 142
8966	Menthone(d) or 2-Isopropyl-5-methyl cyclohexanone $C_{10}H_{18}O$	154.25	$[\alpha]^{18}_D$ + 24. 8	204^{750}, 85^{14}	...	$0.8963^{20/20}$	1.4503^{20}	al, eth, ace, **bz**, aa	B7[3], 154
8967	Menthone (dl) $C_{10}H_{18}O$	154.25	...	210.5	...	0.911^{0}	...	al, eth, ace, **bz**, aa	B7[3], 154
8968	Menthone (l) $C_{10}H_{18}O$	154.25	$[\alpha]^{20}_D$-29.6	209.6, 96^{20}	...	$0.8954^{20/4}$	1.4505^{20}	**al, eth**, ace, bz	B7[3], 152
8969	Menthoxy acetic acid (l) $C_{12}H_{22}O_3$	214.30	cr (eth) $[\alpha]^{20}_D$ −92.9 (MeOH)	171^{11}	53-5	...	...	al, eth, ace	B6[3], 155
8970	Menthoxyacetyl chloride (l) $C_{12}H_{21}ClO_2$	232.75	$[\alpha]^{13}_D$ −84.8 (chl)	$128\text{-}31^{11}$	...	...	...	eth	B6[3], 156
8971	Mercaptoacetamide, N-Phenyl $HSCH_2CONHC_6H_5$	167.23	nd (al)	...	110-1	...	...	al, eth	B12[3], 925
8972	Mercaptoacetamide, N-β-naphthyl or Thionalide $HSCH_2CONH\text{-}\beta\text{-}C_{10}H_7$	217.29	nd	...	111-2	...	...	al	B12[3], 3052

PHYSICAL CONSTANTS OF ORGANIC COMPOUNDS (Continued)

No.	Name, Synonyms, and Formula	Mol. wt.	Color, crystalline form, specific rotation and λ_{max} (log ε)	b.p. °C	m.p. °C	Density	n_D	Solubility	Ref.
8973	Mercaptoacetic acid or Thioglycolic acid $HSCH_2COOH$	92.11		120^{20}	−16.5	1.3253^{20}	1.5030^{20}	**w, al, eth**	$B3^4$, 600
8974	Mercaptoacetic acid, acetate or Acetyl thioglycolic acid $(CH_3COS)CH_2CO_2H$	134.15	yo	$158\text{-}9^{17}$, $115\text{-}8^{2.5}$				w	$B3^4$, 610
8975	Mercaptoacetic acid, ethyl ester or Ethylmercapto acetate $HSCH_2CO_2C_2H_5$	120.17		156-8, 55^{17}		1.0964^{15}	1.4582^{20}	al, eth	$B3^4$, 617
8976	Mercaptoacetic acid, butyl ester or Butyl mercapto acetate $HSCH_2CO_2C_4H_9$	148.22		$85\text{-}8^{16}$		$1.03^{20/4}$			$B3^4$, 620
8977	Mercaptoacetic acid, dodecyl ester or Lauryl mercapto-acetate $HSCH_2CO_2C_{12}H_{25}$	260.44			3	$0.43^{20/4}$			$B3^3$, 432
8978	Mercaptoacetic acid, 2-ethylhexyl ester or 2-Ethylhexyl mercaptoacetate $HSCH_2CO_2[CH_2CH(C_2H_5)(CH_2)_3CH_3]$	204.33		133.5		$0.97^{20/4}$			
8979	Mercaptoacetic acid, hexadecyl ester $HSCH_2CO_2C_{16}H_{33}$	316.54			20-5			chl	$B3^4$, 622
8980	Mercaptoacetic acid, isopropyl ester or isopropyl mercap-toacetate $HSCH_2CO_2CH(CH_3)_2$	134.19		$80\text{-}8^{45}$		$1.05^{20/4}$			$B3^3$, 430
8981	Mercaptoacetic acid, methyl ester or Methyl-mercaptoace-tate $HSCH_2CO_2CH_3$	106.14		$42\text{-}3^{10}$			1.4657^{20}	al, eth	$B3^4$, 614
8982	Mercaptoacetic acid, octyl ester or Octyl mercaptoacetate $HSCH_2CO_2C_8H_{17}$	204.33		125^{17}			1.4606^{21}		$B3^3$, 431
8983	Mesaconic acid or Methylfumaric acid $HO_2CC(CH_3){=}CHCO_2H$	130.10	rh nd or mcl pr (eth, AcOEt)	sub	204-5	$1.466^{20/4}$		al, eth	$B2^4$, 2231
8984	Mesaconic acid, diethyl ester or Diethyl mesaconate $C_2H_5O_2CC(CH_3){=}CCO_2C_2H_5$	186.21		229, $93\text{-}5^{10}$		$1.0453^{20/20}$	1.4488^{20}	al, eth, ace, bz	$B2^4$, 2232
8985	Mesaconic acid, dimethyl ester or Dimethyl mesaconate $CH_3O_2CC(CH_3){=}CHCO_2CH_3$	158.15		203.5, 100^{10}		$1.0914^{20/4}$	1.4512^{20}	al, eth, ace	$B2^4$, 2232
8986	Mescaline or 3,4,5-trimethoxy-β-phenethylamine $C_{11}H_{17}NO_3$	211.26	cr	180^{12}	35-6			w, al, bz, chl	$B13^3$, 2375
8987	Mesitylene or 1,3,5-trimethylbenzene $1,3,5\text{-}(CH_3)_3C_6H_3$	120.19		164.7, 48.7^{10}	−44.7	$0.8652^{20/4}$	1.4994^{20}	**al, eth, ace, bz**	$B5^4$, 1016
8988	Mesitylene, acetyl or 2,4,6-trimethyl acetophenone $2,4,6\text{-}(CH_3)_3C_6H_2COCH_3$	162.23		240.5^{735}, 120^{12}		$0.9754^{20/4}$	1.5175^{20}	al, eth, ace, bz	$B7^3$, 1137
8989	Mesitylene, γ-bromo isobutylryl $2,4,6\text{-}(CH_3)_3C_6H_2COCBr(CH_3)_2$	269.18	gold-ye oil	$160\text{-}70^{24}$	27			eth	$B7^3$, 1209
8990	Mesitylene, vinyl $2,4,6(CH_3)_3C_6H_2CH{=}CH_2$	146.23		208-10, 83^{12}		$0.9057^{20/4}$	1.5296^{20}		$B5^4$, 1408
8991	Mesityl oxide or 4-Methyl-3-penten-2-one $(CH_3)_2C{=}CHCOCH_3$	98.14		129.7, 41^{25}	−52.9	$0.8653^{20/4}$	1.4440^{20}	w, **al, eth**, ace	$B1^4$, 3471
8992	Mesoxalic acid, hydrate or Dihydroxymalonic acid $(HO)_2C(CO_2H)_2$	136.06	dlq nd (w)		120-1			w, al, eth	$B3^3$, 1355
8993	Mesoxalic acid, diethyl ester $O{=}C(CO_2C_2H_5)_2$	174.15	pa ye gr oil	208^{220}, 105^{19}	ca −30	$1.1419^{16/4}$	1.4310^{22}	w, al, eth, chl	$B3^3$, 1356
8994	Mesoxalic acid, diethyl ester,hydrate or Ethyl dihydroxy malonate $(HO)_2C(CO_2C_2H_5)_2$	192.17	pl (bz)	ca 200	57			w, al, eth, ace, bz	$B3^3$, 1356
8995	Mesoxalic acid, diethyl ester,oxime $HON{=}C(CO_2C_2H_5)_2$	189.17		172^{12}		$1.1821^{18/4}$	1.4544^{18}	al, eth, ace, bz	$B3^4$, 1805
8996	Mesoxalonitrile or Oxomalonic acid, dinitrile $O{:}C(CN)_2$	80.05		65.5	−36	$1.124^{20/4}$	1.3919^{20}	eth, ace	$B3^4$, 1806
8997	Mestilbol or 3-(4-hydroxyphenyl)-4(4-methoxyphenyl)-3-Hexene $CH_3CH_2C(4\text{-}CH_3OC_6H_5){=}C(4\text{-}HOC_6H_5)CH_2CH_3$	282.38	nd (bz-lig) lf (70% al)	$185\text{-}95^{0.3}$	117-8			al, eth, ace	$B6^3$, 5623
8998	Metacrolein or 2,4,6-triethynyl-1,3,5-trioxane $C_9H_{12}O_3$	168.19	pl (al)	170	50			al, eth	$B1^4$, 3435
8999	Metaldehyde or Metacetaldehyde $(C_2H_4O)_{4\text{-}6}$	(44.05)	tetr nd or pr (al)	sub 115	246.2 (sealed tube)				$B19^4$, 5643
9000	Metaldehyde II(tetramer) $(C_2H_4O)_4$	176.21		110, 65^{15}	47			al, eth, ace, bz	$B19^4$, 5643
9001	Metameconin or 5,6-Dimethoxy phthalide $C_{10}H_{10}O_4$	194.19	cr (dil al)		155-7				$B18^4$, 1223

No.	Name, Synonyms, and Formula	Mol. wt.	Color, crystalline form, specific rotation and λ_{max} (log ε)	b.p. °C	m.p. °C	Density	n_D	Solubility	Ref.
9002	Methacrolein $CH_2{=}C(CH_3)CHO$	70.09		68.4		$0.837^{20/4}$	1.4144^{20}	w, al, eth,	$B1^4$, 3455
9003	Methacryl amide $CH_2{=}C(CH_3)CONH_2$	85.11	cr (bz)		110-1			al	$B2^4$, 1538
9004	Methacrylic acid or 2-Methylpropenoic acid $CH_2{=}C(CH_3)COOH$	86.09	pr	$162\text{-}3^{757}$, 60^{12}	16	$0.0153^{20/4}$	1.4314^{20}	w, al, eth	$B2^4$, 1518
9005	Methacrylic acid, anhydride $[CH_2{=}C(CH_3)CO]_2O$	154.17		89^5			1.4540^{20}	al, eth	$B2^4$, 1537
9006	Methacrylic acid, butyl ester or Butyl methacrylate $CH_2{=}C(CH_3)CO_2C_4H_9$	142.20		160, 52^{11}		$0.9836^{20/4}$	1.4240^{20}	al, eth	$B2^4$, 1525
9007	Methacrylic acid, isobutyl ester or isobutyl methacrylate $CH_2{=}C(CH_3)CO_2\text{-}i\text{-}C_4H_9$	142.20		155, 45^{11}		$0.8858^{20/4}$	1.4199^{20}	al, eth	$B2^4$, 1526
9008	Methacrylic acid, ethyl ester or Ethyl methacrylate $CH_2{=}C(CH_3)CO_2C_2H_5$	114.14		117, 30^{18}		$0.9135^{20/4}$	1.4147^{20}	al, eth	$B2^4$, 1523
9009	Methacrylic acid, methyl ester or Methyl methacrylate $CH_2{=}C(CH_3)CO_2CH_3$	100.12		100-1, 24^{32}	−48	$0.9440^{20/4}$	1.4142^{20}	al, eth, ace	$B2^4$, 1519
9010	Methacrylic acid, propyl ester or Propyl methacrylate $CH_2{=}C(CH_3)CO_2C_3H_7$	128.17		141		$0.9022^{20/4}$	1.4190^{20}	al, eth	$B2^4$, 1524
9011	Methacrylic acid, isopropyl ester or Isopropyl methacrylate $CH_2{=}C(CH_3)CO_2\text{-}i\text{-}C_3H_7$	128.17		125		$0.8847^{20/4}$	1.4122^{20}	al, eth, ace, bz	$B2^4$, 1525
9012	Methacrylonitrile $CH_2{=}C(CH_3)CN$	67.09		90.3	−35.8	$0.7998^{20/4}$	1.4190^{20}	al, eth	$B2^4$, 1539
9013	Methacrylyl chloride $CH_2{=}C(CH_3)COCl$	104.54		96, 50^{135}		$1.0871^{20/4}$	1.4435^{20}	eth, ace, chl	$B2^4$, 1537
9014	Methallyl alcohol or 2-Methyl-2-propen-1-ol $CH_2{=}C(CH_3)CH_2OH$	72.11		114.5		0.8515^{20}	1.4255^{20}	w, al, eth	$B2^4$, 2114
9015	Methane CH_4	16.04	gas	−164	−182	0.466^{-164}, 0.5547^0		al, eth, bz	$B1^4$, 3
9016	Methane, bis (dimethyl amino) $[(CH_3)_2N]_2CH_2$	102.18		82.5		$0.7491^{18.7}$		w	$B4^4$, 153
9017	Methane, bis-(2-hydroxy-3,5,6-trichlorophenyl) or Hexachlorophene $(2\text{-}HO\text{-}3,5,6\text{-}Cl_3C_6H)_2CH_2$	406.91	nd (bz)		166-7			al, eth, ace, chl	$B6^3$, 5407
9018	Methane, bis (trichlorosilyl) $(Cl_3Si)_2CH_2$	282.92		182.7^{745},64^{10}		$1.5567^{20/4}$	1.4740^{20}	al, eth	$B1^4$, 3077
9019	Methane, bis (4-chlorophenoxy) or Oxythane $(4\text{-}ClC_6H_4O)_2CH_2$	269.13		$189\text{-}94^6$	69-70			eth, ace, bz	$B6^4$, 833
9020	Methane, bromo or Methyl bromide CH_3Br	94.94		3.6	−93.6	$1.6755^{20/4}$	1.4218^{20}	al, eth	$B1^4$, 68
9021	Methane, bromochloro CH_2BrCl	129.39		68.1	−86.5	$1.9344^{20/4}$	1.4838^{20}	al, eth, ace, bz	$B1^4$, 74
9022	Methane, bromochloro-dinitro $BrCCl(NO_2)_2$	219.38		$75\text{-}6^{15}$	9.3	$2.0394^{20/4}$	1.4793	al	$B1^3$, 115
9023	Methane, bromo-chloro-fluoro $BrCHClF$	147.37		36.1^{756}	−115	$1.9771^{0/4}$	1.4144^{25}	eth, ace, chl	$B1^4$, 75
9024	Methane, bromo-dichloro $BrCHCl_2$	163.83		90	−57.1	$1.980^{20/4}$	1.4964^{20}	al, eth, ace, bz, chl	$B1^4$, 76
9025	Methane, bromo-difluoro $BrCHF_2CHF_2$	130.92			−14.5	1.55^{16}		w, al	$B1^4$, 72
9026	Methane, bromo difluoro nitroso BrF_2CNO	159.92	bl gas	−12					$B1^4$, 100
9027	methane, bromo-diido $BrCHI_2$	346.73	ye (peth)	110^{25}	60				B1, 72
9028	Methane, bromo-diphenyl or Benzydryl bromide $(C_6H_5)_2CHBr$	247.13	tcl (peth)	193^{26}d, $111^{0.3}$	45			al, bz	$B5^4$, 1850
9029	Methane, bromo fluoro $BrCH_2F$	112.93		18-20				al, chl	$B1^4$, 72
9030	Methane, bromo-iodo $BrCH_2I$	220.84		138-41		2.926^{17}	1.6410^{20}	chl	$B1^4$, 95
9031	Methane, bromo-nitro $BrCH_2NO_2$	139.94		$148\text{-}9^{742}$			1.4880^{20}	al	$B1^4$, 106
9032	Methane, bromo-trichloro $BrCCl_3$	198.27		104.7, 0.6^{10}	−5.6	$2.0122^{20/4}$	1.5063^{20}	al, eth	$B1^4$, 77
9033	Methane, bromo-trifluoro $BrCF_3$	148.91	gas	-59^{740}				chl	$B1^4$, 73
9034	Methane, bromo-trinitro $BrC(NO_2)_3$	229.93		56^{10}	17-8	$2.0313^{20/4}$	1.4808^{20}	al, chl	$B1^3$, 116

PHYSICAL CONSTANTS OF ORGANIC COMPOUNDS (Continued)

No.	Name, Synonyms, and Formula	Mol. wt.	Color, crystalline form, specific rotation and λ_{max} (log ε)	b.p. °C	m.p. °C	Density	n_D	Solubility	Ref.
9035	Methane, chloro or Methyl chloride CH_3Cl	50.49	gas	−24.2	−97.1	$0.9159^{20/4}$	1.3389^{20}	al,eth, ace, bz, chl	$B1^{4}$, 28
9036	Methane, chloro-dibromo $ClCHBr_2$	208.28		$119\text{-}20^{748}$		$2.451^{20/4}$	1.5482^{20}	al, eth, ace, bz	$B1^{4}$, 81
9037	Methane, chloro-difluoro or Freon 22 $ClCHF_2$	86.47	gas	−40.8	−146			eth, ace, chl	$B1^{4}$, 32
9038	Methane, chloro difluoro nitro ClF_2CNO_2	131.47		25				chl	$B1^{4}$, 106
9039	Methane, chloro difluoro nitroso ClF_2CNO	115.47	bl gas	*ca* -35					$B1^{4}$, 99
9040	Methane, chloro diiodo $ClCHI_2$	302.28		200d, 88^{30}	−4			eth, ace, chl	$B1^{4}$, 97
9041	Methane, chloro dinitro $ClCH(NO_2)_2$	140.48		$34\text{-}6^{13}$		1.6125^{20}	1.4575^{20}		$B1^{3}$, 115
9042	Methane, chloro diphenyl or Benzhydryl chloride $(C_6H_5)_2CHCl$	202.68	nd	173^{19}	20.5	$1.1398^{20/4}$	1.5959^{20}		$B5^{4}$, 1847
9043	Methane, chloro-fluoro $ClCH_2F$	68.48	gas		−9.1			chl	$B1^{4}$, 32
9044	Methane, chloro-iodo $ClCH_2I$	176.38		109		$2.422^{20/4}$	1.5822^{20}	al, eth, ace, bz, chl	$B1^{4}$, 94
9045	Methane, chloro-nitro $ClCH_2NO_2$	95.49		122-3		1.466^{15}		w	$B1^{4}$, 106
9046	Methane, chloro tribromo $ClCBr_3$	287.18	lf (eth)	158-9	55	2.71^{15}		eth	$B1^{4}$, 85
9047	Methane, chloro trifluoro or Freon 13 $ClCF_3$	104.46	gas	−81.1	−181				$B1^{4}$, 34
9048	Methane, chloro trinitro $ClC(NO_2)_3$	185.48		133-5d, 56^{40}	4.5	$1.6769^{20/4}$	1.4500^{20}	al, eth, chl	$B1^{3}$, 116
9049	Methane, chloro triphenyl or Trityl chloride $(C_6H_5)_3CCl$	278.78	nd or pr (bz-peth)	310, $230\text{-}5^{20}$	113-4			eth, ace, bz, chl	$B5^{3}$, 2315
9050	Methane, deutero-trichloro $CDCl_3$	120.38		61-2	-64.1	$1.5004^{20/4}$	1.4450^{20}		$B1^{4}$, 54
9051	Methane, diazo or Acomethylene CH_2N_2	42.04	ye gas	*ca* -0	−145			eth	$B1^{4}$, 3056
9052	Methane, diazo diphenyl $(C_6H_5)_2CN_2$	194.24	bl-red nd (peth)	exp	30-2			al, eth, ace, bz	$B7^{3}$, 2068
9053	Methane, dibromo or Methylene bromide CH_2Br_2	173.83		97	−52.5	$2.4970^{20/4}$	1.5420^{20}	al, eth, ace	$B1^{4}$, 78
9054	Methane, dibromo chloro fluoro Br_2CClF	226.27		80.3		2.3173^{22}	$1.4570^{20/D}$		$B1^{4}$, 82
9055	Methane, dibromo dichloro Br_2CCl_2	242.73		150.2	38			al, eth, ace, bz	$B1^{4}$, 82
9056	Methane, dibromo difluoro Br_2CF_2	209.82		24.5				al, eth, ace, bz	$B1^{4}$, 80
9057	Methane, dibromo dinitro $Br_2C(NO_2)_2$	263.83	nd	158d, 77^{21}	5.5	$2.4440^{20/4}$	1.5280^{25}	al	$B1^{3}$, 115
9058	Methane, dibromo fluoro $FCHBr_2$	191.83		64.9^{757}		$2.421^{20/4}$	1.4685^{20}	al, eth, ace, bz, chl	$B1^{4}$, 80
9059	Methane, dibromo iodo $ICHBr_2$	299.73	pl (peth)	91^{42}	22.5				$B1^{3}$, 99
9060	Methane, dichloro or Methylene chloride CH_2Cl_2	84.93		40	−95.1	$1.3266^{20/4}$	1.4242^{20}	al, eth	$B1^{4}$, 35
9061	Methane, dichloro difluoro or Freon 12 Cl_2CF_2	120.91		−29.8	−158	1.75^{-115}, 1.1834^{57}		al, eth, aa	$B1^{4}$, 40
9062	Methane, dichloro-dinitro $Cl_2C(NO_2)_2$	174.93		121-2, 46^{20}		$1.6124^{20/4}$	1.4575^{20}	al, eth, bz, chl	$B1^{3}$, 115
9063	Methane, dichloro diphenyl $(C_6H_5)_2CCl_2$	237.13		305d, 190^{21}		1.235^{18}			$B5^{4}$, 1848
9064	Methane, dichloro-fluoro or Freon 21 Cl_2CHF	102.92		9	−135	1.405^{9}	1.3724^{9}	al, eth, chl, aa	$B1^{4}$, 39
9065	Methane, dichloro-iodo Cl_2CHI	210.83		132, 40^{30}		$2.392^{20/4}$	1.5840^{20}	al, eth, ace, bz, chl	$B1^{4}$, 95
9066	Methane, dichloro-nitro Cl_2CHNO_2	129.93		107					$B1^{3}$, 113
9067	Methane, difluoro or Methylene fluoride CH_2F_2	52.02		−51.6		0.909^{20}	1.190^{20}	al	$B1^{4}$, 24

No.	Name, Synonyms, and Formula	Mol. wt.	Color, crystalline form, specific rotation and λ_{max} (log ε)	b.p. °C	m.p. °C	Density	n_D	Solubility	Ref.
9068	Methane, difluoro-iodo F_2CHI	177.92		21.6	−122	3.238^{-19}			B1^4, 92
9069	Methane, di-2-furyl $(2\text{-}C_4H_3O)_2CH_2$	148.16		$94^{22.5}$		$1.102^{20/4}$	1.5049^{20}	al, eth, ace	B19^4, 269
9070	Methane, diiodo or Methylene iodide CH_2I_2	267.84	ye nd or lf	182; 60^{10}	6.1	$3.3254^{20/4}$	1.7425^{20}	al, eth, bz, chl	B1^4, 96
9071	Methane, diiodo-fluoro $FCHI_2$	285.83	pa ye	100-1, 50^{50}	−34.5	3.1969^{22}		al, eth	B1^4, 97
9072	Methane, dinitro $CH_2(NO_2)_2$	106.04	ye nd	100 exp	<−15			al, eth	B1^4, 107
9073	Methane, dinitro diphenyl $(C_6H_5)_2C(NO_2)_2$	258.23	pl (dil al)		79-80			al, eth, bz, chl	B5^3, 1797
9074	Methane, diphenyl or Diphenyl nethane $(C_6H_5)_2CH_2$	168.24	pr nd	264.3, 125.5^{10}	25.3	$1.0060^{20/4}$	1.5753^{20}	al, eth, chl	B5^4, 1841
9075	Methane, diphenyl 3-tolyl $(3\text{-}CH_3C_6H_4)CH(C_6H_5)_2$	258.36	pr (al, MeOH)	354	62	1.07^{16}		eth, bz, chl, aa	B5^3, 2332
9076	Methane, di(α-pyrryl) or 2, 2′-Methylene dipyrrole $(2\text{-}C_4H_4N)_2CH_2$	146.19	lf or nd (al)	$163\text{-}7^{12}$	73			al, eth, bz	B23^2, 167
9077	Methane, disilano $(SiH_3)_2CH_2$	76.25		14.7^{754}		$0.6979^{4/4}$	1.4115^4		B1^4, 3072
9078	Methane, di-(4-tolyl) phenyl $(4\text{-}CH_3C_6H_4)_2CHC_6H_5$	272.39	nd (MeOH)	$218\text{-}20^{12}$	56			al, eth, bz, chl	B5^3, 2342
9079	Methane, fluoro or Methyl fluoride CH_3F	34.03		−78.4	−141.8	0.8428^{-60}	1.1727^{20}	al, eth, bz, chl	B1^4, 22
9080	Methane, fluoro-iodo FCH_2I	159.93		53.4		$2.366^{20/4}$	1.5256^{20}	eth, ace, bz, chl	B1^4, 92
9081	Methane, fluoro-tribromo $FCBr_3$	270.72	hyg nd	62.5^{15}	42	$1.0003^{20/4}$	1.4164^{20}	al	B1^4, 85
9082	Methane, iodo or Methyl iodide CH_3I	141.94		42.4	−66.4	$2.279^{20/4}$	1.5380^{20}	**al, eth** ace, bz	B1^4, 87
9083	Methane, iodo-trichloro $ICCl_3$	245.27		142		$2.355^{20/4}$	1.5854^{20}	eth, ace, bz, chl	B1^4, 95
9084	Methane, iodo-trifluoro CF_3I	195.91		−22.5		$2.3608^{-32/4}$	$1.3790^{-42}/_D$		B1^4, 92
9085	Methane, nitro CH_3NO_2	61.04		100.8	fr -17	$1.1371^{20/4}$	1.3817^{20}	al,eth,ace	B1^4, 100
9086	Methane, nitro-tribromo or Bromopicrin Br_3CNO_2	297.73		exp^{760}, $89\text{-}90^{20}$	10.2	$2.7930^{20/4}$	1.5790^{20}	al, eth, ace, bz, aa	B1^4, 106
9087	Methane, nitro-trichloro or Chloropicrin Cl_3CNO_2	164.38		111.8	−64.5	$1.6566^{20/4}$	1.4622^{20}	al, ace, bz, aa	B1^4, 106
9088	Methane, nitro-trifluoro or Fluoropicrin F_3CNO_2	115.01		−31.1					B1^4, 105
9089	Methane, nitroso-trifluoro F_3CNO	99.01		−84	−197				B1^4, 99
9090	Methane, (pentafluorothio)trifluoro $F_3C(SF_5)$	196.06		−20					B1^4, 35
9091	Methane, tetrabromo or Carbon tetrabromide CBr_4	331.63	mcl ta (dil al)	189-90, 102^{50}	90-4	$2.9609^{100/4}$	1.5942^{100}	al, eth	B1^4, 85
9092	Methane, tetrachloro or Carbon tetrachloride CCl_4	153.82		76.5	−23	$1.5940^{20/4}$	1.4601^{20}	al,**eth, ace, bz, chl**	B1^4, 56
9093	Methane, tetrafluoro or Carbon tetrafluoride CF_4	88.00		-129^{754}	−150	3.034^0		bz, chl	B1^4, 26
9094	Methane, tetraiodo or Carbon tetraiodide CI_4	519.63	red lf (bz,chl)	$130\text{-}40^{1\text{-}2}$	171d	4.23^{20}		chl, py	B1^4, 98
9095	Methane, tetranitro $C(NO_2)_4$	196.03		126, $21\text{-}3^{22}$	14.2	$1.6380^{20/4}$	1.4384^{20}	al, eth	B1^4, 107
9096	Methane, tetraphenyl $C(C_6H_5)_4$	320.43	rh nd (bz,sub)	431 sub					B5^3, 2568
9097	Methane, tribromo or Bromoform $CHBr_3$	252.73	hex sc	149.5, 46^{15}	8.3	2.8899	1.5976^{20}	**al, eth,** bz, chl, lig	B1^4, 82
9098	Methane, trichloro or Chloroform $CHCl_3$	119.38		61.7	−63.5	$1.4832^{20/4}$	1.4459^{20}	**al, eth,** ace, **bz, lig**	B1^4, 42
9099	Methane, tricyclohexyl $(C_6H_{11})_3CH$	262.48		322-9	48	$0.9274^{50/0}$	$1.4986^{40}/_D$	eth, bz	B5^4, 504
9100	Methane, trifluoro or Fluroform CHF_3	70.01		−82.2	−160	1.52^{-100}		al, ace, bz	B1^4, 24

No.	Name, Synonyms, and Formula	Mol. wt.	Color, crystalline form, specific rotation and λ_{max} (log ε)	b.p. °C	m.p. °C	Density	n_D	Solubility	Ref.
9101	Methane, triido or Iodoform CHI_3	393.73	ye hex pr or nd(ace)	*ca* 218	123	4.008$^{20/4}$	...	eth, ace, chl, aa	B1^4, 97
9102	Methane, trinitro or Nitroform $CH(NO_2)_3$	151.04	...	exp^{760}, 45-7^{22}	19	1.479$^{20/4}$	1.4451^{24}	al, ace	B1^4, 107
9103	Methane, triphenyl or Triphenyl methane $(C_6H_5)_3CH$	244.34	rh (al)	358-9^{754}, 190-215^{10}	94	1.014$^{99/4}$	1.5839^{99}	eth, ba, chl	B5^3, 2307
9104	Methane, tris (4-aminophenyl) or *p*-Leucaniline $(4\text{-}NH_2C_6H_4)_3CH$	289.38	lf (w,al,bz)	...	208	...	...	al, eth	B13^3, 566
9105	Methane, tris (4-dimethylaminophenyl) or Leucocrystal violet $[4\text{-}(CH_3)_2NC_6H_4]_3CH$	373.54	lf (al) nd (bz, lig)	...	175	...	...	eth, bz, chl, aa	B13^3, 566
9106	Methane, tris (2-tolyl) $(2\text{-}CH_3C_6H_4)_3CH$	286.42	nd (al)	...	130-1	...	...	al, eth	B5^3, 2347
9107	Methane arsonic acid $CH_3AsO(OH)_2$	139.97	lf (al)	...	160-1	...	...	w, al	B4^3, 1822
9108	Methanedisulfonic acid, dihydrate or Methionic acid $CH_2(SO_3H)_2.2H_2O$	212.18	hyg nd (w + 2)	220-70$^{15-20}$ d	...	...	...	w, al	B1^4, 3054
9109	Methane phosphonic acid or Methyl phosphonic acid $CH_3PO(OH)_2$	96.02	hyg pl	d	108-9	...	...	w, al, eth	B4^3, 1778
9110	Methanephosphonic acid, diethyl ester or Diethyl methylphosphonate $CH_3PO(OC_2H_5)_2$	152.13	...	194, 85^{15}	...	1.0406$^{30/4}$	1.4101^{30}	w, al, eth	B4^5, 1778
9111	Methanephosphonic acid, dimethyl ester or Dimethyl methylphosphonate $CH_3PO(OCH_3)_2$	124.08	...	181^{754}, 79.5^{20}	...	...	...	w	B4^3, 1778
9112	Methanesulfenyl chloride, trichloro or Perchloromethylmercaptan Cl_3CSCl	185.88	ye oil	147-8, 51^{25}	...	1.6947$^{20/4}$	1.5484^{20}	eth	B3^4, 290
9113	Methanesulfenylchloride, trifluoro or Perfluoromethylmercaptan F_3CSCl	136.52	...	-0.7	...	...	...	...	B3^4, 289
9114	Methane sulfinic acid, amino, imino $HN{=}C(NH_2)SO_2H$	108.11	nd (al)	...	144d	...	...	w	B3^4, 145
9115	Methane sulfonamide, trifluoro $F_3CSO_2NH_2$	149.09	...	...	119	...	...	w, chl	B3^4, 35
9116	Methanesulfonic acid CH_3SO_3H	96.10	...	167^{10}	20	1.4812$^{18/4}$	1.4317^{18}	w, al, eth	B4^4, 10
9117	Methanesulfonic acid, trifluoro F_3CSO_3H	150.07	hyg liq	162, 81$^{37.5}$	...	...	...	**eth**	B3^4, 34
9118	Methane sulfonamide, trifluoro-*N,N*-diethyl $F_3CSO_2N(C_2H_5)_2$	205.20	...	55^7	...	...	...	eth	B4^4, 415
9119	Methane sulfonyl chloride, trifluoro F_3CSO_2Cl	168.52	...	31.6	...	...	...	...	B3^4, 35
9120	Methane sulfonic acid, trifluoro,ethyl ester $F_3CSO_3C_2H_5$	178.13	...	115, 42^{40}	...	...	...	eth	B3^4, 34
9121	Methane sulfonyl chloride CH_3SO_2Cl	114.55	...	161^{730}, 55^{11}	...	1.4805$^{18/4}$	1.4573^{20}	al, eth	B4^4, 27
9122	Methane sulfonyl chloride, trichloro Cl_3CSO_2Cl	217.88	cr (al-w)	170 (sub)	140-1	...	...	al, eth	B3^4, 36
9123	Methane sulfonyl fluoride, trifluoro F_3CSO_2F	152.06	...	−21.7	...	...	...	...	B3^4, 34
9124	Methanethiol or Methyl mercaptan CH_3SH	48.10	...	6.2	−123	0.8665$^{20/4}$	...	al, eth	B1^4, 1273
9125	Methanetricarboxylic acid, trimethyl ester $HC(CP_2CH_3)_3$	190.15	pr (MeOH)	242.7, 128^{15}	46-7	...	...	al, eth, bz, chl	B2^3, 2023
9126	Methantheline bromide or Banthine bromide $C_{21}H_{26}NO_3Br$	420.35	cr (i-prOH)	...	172-7	...	...	w, al, eth	B18^4, 4352
9127	Methanol or Methyl alcohol CH_3OH	32.04	...	65, 15^{73}	−93.9	0.7914$^{20/4}$	1.3288^{20}	**w, al, eth, ace,** bz, chl	B1^4, 1227
9128	Methanol-*d* or *o*-Deutero methanol CH_3OD	33.05	...	65.5	−100	0.8127$^{20/4}$	...	**w, al, eth,** ace, bz	B1^4, 1244
9129	Methapyrilene (base) or Histadyl base $C_{14}H_{19}N_3S$	261.38	...	173-5^3	...	...	1.5915^{20}	...	B22^4, 3950
9130	Methionine *(DL)* or *dl*-2-amino-4-(methylthio) butyric acid $CH_3SCH_2CH_2CH(NH_2)CO_2H$	149.21	pl	...	281d	1.340	...	w	B4^3, 1647

No.	Name, Synonyms, and Formula	Mol. wt.	Color, crystalline form, specific rotation and λ_{max} (log ε)	b.p. °C	m.p. °C	Density	n_D	Solubility	Ref.
9131	Methionine *(L)* $CH_3SCH_2CH_2CH(NH_2)CO_2H$	149.21	hex pl (dil al $[\alpha]^{25/}_D$ −8.2 (w,c=1) +22.5 (INHCl)	sub 186	283d	...	...	w	B4[3], 1639
9132	Methoxyacetic acid or Methyl glycolic acid $CH_3OCH_2CO_2H$	90.08	hyg	203-4, 96^{13}	...	$1.1768^{20/4}$	1.4168^{20}	w, al, eth	B3[4], 574
9133	Methoxyacetic acid, ethyl ester or Ethyl methoxy acetate $CH_3OCH_2CO_2C_2H_5$	118.13	...	142, 44-5^{9}	...	1.0118^{15}	1.4050^{20}	al, eth	B3[3], 381
9134	Methoxyacetic acid, methyl ester or Methyl methoxy acetate $CH_3OCH_2CO_2CH_3$	104.11	...	131^{763}, 57^{50}	...	$1.0511^{20/4}$	1.3962^{20}	al, eth, ace	B3[4], 578
9135	2-Methoxybenzoic acid or *o*-Anisic acid $2\text{-}CH_3OC_6H_4CO_2H$	152.15	pl (w), fl (al)	200	101	...	...	al, eth, bz, chl	B10[3], 97
9136	2-Methoxybenzoic acid, 5-chloro $5\text{-}Cl\text{-}2\text{-}CH_3OC_6H_3CO_2H$	186.59	nd (w)	...	81-2	...	...	al, ace	B10[3], 165
9137	2-Methoxybenzoic acid, ethyl ester $2\text{-}CH_3OC_6H_4CO_2C_2H_5$	180.20	...	261, 135-6^{12}	...	$1.1124^{20/4}$	1.5224^{20}	al, eth	B10[3], 116
9138	2-Methoxybenzoic acid, methyl ester $2\text{-}CH_3OC_6H_4CO_2CH_3$	166.18	...	245, 127^{11}	...	$1.1571^{19/4}$	$1.534^{19.5}$	al	B10[3], 109
9139	3-Methoxybenzoic acid or *m*-Anisic acid $3\text{-}CH_3OC_6H_4CO_2H$	152.15	nd (w)	170-2^{10}	110	...	...	al, eth, bz, chl	B10[3], 244
9140	3-Methoxybenzoic acid, ethyl ester $3\text{-}CH_3OC_6H_4CO_2C_2H_5$	180.20	...	260-1, 110^{5}	...	$1.0993^{20/4}$	1.5161^{20}	al, eth	B10[3], 250
9141	3-Methoxybenzoic acid, methyl ester $3\text{-}CH_3OC_6H_4CO_2CH_3$	166.18	...	252, 121-4^{10}	...	$1.1310^{20/4}$	1.5224^{20}	al	B10[3], 250
9142	4-Methoxybenzoic acid or *p*-Anisic acid $4\text{-}CH_3OC_6H_4CO_2H$	152.15	pr or nd (w)	275-80	185	...	...	al, eth, chl	B10[3], 280
9143	4-Methoxybenzoic acid, butyl ester $4\text{-}CH_3OC_6H_4CO_2C_4H_9$	208.26	...	183^{40}	...	$1.054^{16.5}$	$1.5141^{18.5}$	...	B10[3], 307
9144	4-Methoxybenzoic acid, ethyl ester $4\text{-}CH_3OC_6H_4CO_2C_2H_5$	180.20	...	269-70	7-8	$1.1038^{20/4}$	1.5254^{20}	al, eth	B10[3], 300
9145	4-Methoxybenzoic acid, methyl ester $4\text{-}CH_3OC_6H_4CO_2CH_3$	166.18	fl (al,eth)	256, 160^{20}	49	...	...	al, eth	B10[3], 297
9146	Methoxylamine, hydrochloride or *o*-Methyl hydroxylamine hydrochloride $CH_3ONH_2.HCl$	83.52	pr	149	...	...	...	w, al	B1[4], 1252
9147	2-Methoxyphenyl isothiocyanate $2\text{-}CH_3OC_6H_4NCS$	165.21	...	131-2^{11}	...	1.1878^{20}	$1.6458^{20/}{}_D$	...	B13[3], 823
9148	Methylamine or Amino methane CH_3NH_2	31.06	gas	−6.3	−93.5	0.699^{-4}, 0.6628^{20}	...	w, al, **eth**, ace, bz	B4[4], 118
9149	Methylamine, hydrochloride $CH_3NH_2.HCl$	67.52	diq tetr ta (al)	sub 30^{15}	227-8	...	...	w, al	B4[4], 122
9150	Methyl isobutyl ketone or 4-methyl-2-pentanone $i\text{-}C_4H_9COCH_3$	100.16	...	116.8, 35-40^{16}	−84.7	0.7978^{20}	1.3962^{20}	**al, eth, ace, bz,** chl	B1[4], 3305
9151	*β*-Methyl chalcone $C_6H_5C(CH_3){=}CHCOC_6H_5$	222.29	...	340-5d, 225^{22}	...	$1.108^{20/0}$	1.6312^{20}	eth	B7[3], 2422
9152	Methyl *β*-chloroethyl sulfide $CH_3SCH_2CH_2Cl$	110.60	...	140, 44^{20}	...	$1.1097^{25/4}$	1.4902^{20}	al, eth, ace	B1[4], 1406
9153	Methyl chlorosulfonate CH_3OSO_2Cl	130.55	...	133-5, 48^{29}	...	$1.4805^{25/4}$	1.4138^{18}	eth, ace, bz	B1[4], 1252
9154	*β*-Methyl choline chloride or (2-hydroxypropyl) trimethyl ammonium chloride $CH_3CH(OH)CH_2N^+(CH_3)_3\ Cl^-$	153.65	pr (BROH)	d	165	...	...	w, al	B4[3], 754
9155	Methyl ether or Dimethyl ether CH_3OCH_3	46.07	gas	−25	−138.5	...	...	w, al, **eth** ace, chl	B1[4], 1245
9156	Methyl ether, boro fluoride $(CH_3)_2O.BF_3$	113.88	...	127d	−14	$1.2410^{20/4}$	1.302^{20}	...	B1[4], 1248
9157	Methyl chloromethyl ether CH_3OCH_2Cl	80.51	...	59.1	−103.5	$1.0605^{20/4}$	1.3974^{20}	al,, eth, ace, chl	B1[4], 3046
9158	Methyl (2,3-dibromopropyl) ether $CH_3OCH_2CHBrCH_2Br$	231.91	...	185, 84^{15}	...	$1.8320^{12/4}$	1.5123^{20}	eth	B1[4], 1447
9159	Methyl (2,3-epoxypropyl) ether or Epimethylin $CH_3OCH_2\underset{O}{CHCH_2}$	89.11	...	115-8	...	0.9890^{20}	1.4320^{20}	w, al, eth, ace	B17[4], 986
9160	Methyl ethyl amine or Ethyl methyl amine $CH_3NHC_2H_5$	59.11	...	36.7	...	...	...	w, al, eth, ace	B4[4], 312

PHYSICAL CONSTANTS OF ORGANIC COMPOUNDS (Continued)

No.	Name, Synonyms, and Formula	Mol. wt.	Color, crystalline form, specific rotation and λ_{max} (log ε)	b.p. °C	m.p. °C	Density	n_D	Solubility	Ref.
9161	Methyl ethyl ketone or 2-Butanone $CH_3CH_2COCH_3$	72.11		$79.6, 30^{119}$	−86.3	$0.8054^{20/4}$	1.3788^{20}	w, **al, eth, ace, bz**	$B1^4$, 3243
9162	Methyl ethynyl ether or Methoxy acetylene $CH_3OC{\equiv}CH$	56.06		50		$0.8001^{20/4}$	1.3812^{20}	al, eth	$B1^4$, 2211
9163	Methyl glyoxal or 2-oxopropionaldehyde CH_3COCHO	72.06	ye hyg liq	72		1.0455^{24}	1.4002^{18}	al, eth, bz	$B1^4$, 3631
9164	Methylglyoxal, dioxime or Methyl glyoxime $CH_3C(NOH)CH{=}NOH$	102.09	nd (w, sub) Pr (al)	sub	157			al, eth	$B1^4$, 3633
9165	Methylglyoxal, 1-oxime $CH_3COCH{=}NOH$	87.08	nd (CCl_4) lf (eth-peth)	sub	69	1.0744^{67}		w, eth	$B1^4$, 3632
9166	Methyl green or Heptamethyl pararosaniline chloride $C_{26}H_{33}N_3Cl_2$	458.47	gr pw (al)					w	$B13^3$, 2074
9167	Methyl hydrogen sulfate CH_3OSO_3H	112.10		130-40d	<−30			w, al, **eth**	$B1^4$, 1250
9168	Methyl hydroperoxide CH_3OOH	48.04		$38\text{-}40^{65}$		$1.9967^{15/4}$	1.3641^{15}	**w, al, eth,** bz	$B1^4$, 1249
9169	Methyl isocyanate CH_3NCO	57.05		39.1-40.1	−45	$0.9230^{27/4}$	1.3419^{18}	w	$B4^4$, 247
9170	Methyl isothiocyanate or Methyl mustard oil CH_3NCS	73.11		119^{758}	36	$1.0691^{37/4}$	1.5258	**al**, eth	$B4^4$, 248
9171	Methyl α-naphthyl ether $CH_3O\text{-}\alpha\text{-}C_{10}H_7$	158.20		$269, 135^{10}$	<−10	$1.0964^{14/2}$	1.6227^{22}	al, eth, bz, chl	$B6^3$, 2922
9172	Methyl β-napthyl ether $CH_3O\text{-}\beta\text{-}C_{10}H_7$	158.20	lf (eth) pl (peth)	$274, 138^{10}$ sub	73-4			eth, bz, chl	$B6^3$, 2969
9173	Methyl nitrate CH_3ONO_2	77.04	exp.vapor	64.6 exp	−82.3	$1.2075^{20/4}$	1.3748^{20}	al, eth	$B1^4$, 1254
9174	Methyl nitrite CH_3ONO	61.04	gas	−12	−16	0.991^{15} (liq)		al, eth	$B1^4$, 1253
9175	Methyl nitro amine CH_3NHNO_2	76.05		$80\text{-}5^{10}$	38	1.2433^{49}	1.4616^{49}	w, al, eth, bz	$B4^3$, 1753
9176	Methyl (2-mitrophenyl) sulfide $(2\text{-}O_2NC_6H_4SCH_3$	169.20			64.5	$1.2628^{78/4}$	1.6246^{78}	al, bz, chl	$B6^4$, 1661
9177	Methyl (4-nitrophenyl) sulfide $(4\text{-}O_2NC_6H_4)SCH_3$	169.20		$135\text{-}40^2$	72	$1.2391^{80/4}$	1.64008^{20}		$B6^4$, 1687
9178	Methyl (2-octyn-1-yl) ether $CH_3O[CH_2C{\equiv}C(C_5H_{11})]$	140.23		77^{19}		$0.8370^{25/4}$	1.4380^{20}	eth	$B1^3$, 1996
9179	Methyl orange or Sodium 4′-dimethylamino azobenzene-4-sulfonate $C_{14}H_{14}N_3O_3NaS$	327.33	og,ye pl or sc (w)		d				$B16^3$, 371
9180	Methyl pentyl ether $CH_3OC_5H_{11}$	102.18		99-100		0.767^{19}	1.3855^{19}	al, eth, ace	$B1^4$, 1643
9181	Methyl *iso*pentyl ether $CH_3O\text{-}i\text{-}C_5H_{11}$	102.18		91^{765}		0.7517^{20}	1.3830^{20}	al, eth	$B1^4$, 1681
9182	Methyl-*tert*-pentyl ether $CH_3OC(CH_3)_2CH_2CH_3$	102.18		86.3		$0.7703^{20/4}$	1.3885^{20}	al, **eth**	$B1^4$, 1671
9183	Methyl pentyl sulfide $CH_3SC_5H_{11}$	118.24		145	−94	$0.8431^{20/4}$	1.4506^{20}	al, eth, ace, bz, chl	$B1^3$, 1608
9184	Methyl perchlorate CH_3OClO_3	114.49	oil	*ca* 52				**al, eth**	$B1^4$, 1249
9185	Methyl phenyl glyoxal $C_6H_5COCOCH_3$	148.16	ye oil	$222, 101^{12}$		$1.0065^{20/4}$	1.537^{10}	w, al, eth	$B7^3$, 3463
9186	Methylphenylglyoxal, dioxime or Methylphenylglyoxime $C_6H_5C(=NOH)C(=NOH)CH_3$	178.19	nd (dil al)		238-40			al	$B7^3$, 3465
9187	Methylphenylglyoxal, 1-oxime $C_6H_5C(=NOH)COCH_3$	163.18	pa ye nd or lf (aa, al)		166-7				$B7^3$, 3464
9188	Methylphenylglyoxal, 2-oxime $C_6H_5COC(=NOH)CH_3$	163.18	wh nd (w)		115				$B7^3$, 3464
9189	Methyl phenyl sulfide $CH_3SC_6H_5$	124.20		$193, 74^{10}$		$1.0579^{20/4}$	1.5868^{20}	al, bz	$B6^4$, 1466
9190	Methyl propyl ether $CH_3OC_3H_7$	74.12		38-9		$0.738^{20/4}$	1.3579^{25}	w,**al, eth,** ace	$B1^4$, 1421
9191	Methyl isopropyl ether $(CH_3)_2CHOCH_3$	74.12		32.5^{777}		$0.7237^{15/4}$	1.3576^{20}	al, eth	$B1^4$, 1471
9192	Methyl isopropyl ketone or 3-methyl-2-butanone $i\text{-}C_3H_7COCH_3$	86.13		94-5	−92	$0.8051^{20/4}$	1.3880^{20}	**al, eth**, ace	$B1^4$, 3287
9193	Methyl propyl sulfide $CH_3SC_3H_7$	90.18		$95.5, -4^{10}$	−113	0.8424^{20}	1.4442^{20}	w, al, eth, ace	$B1^4$, 1450

PHYSICAL CONSTANTS OF ORGANIC COMPOUNDS (Continued)

No.	Name, Synonyms, and Formula	Mol. wt.	Color, crystalline form, specific rotation and λ_{max} (log ε)	b.p. °C	m.p. °C	Density	n_D	Solubility	Ref.
9194	Methyl isopropyl sulfide i-$C_3H_7SCH_3$	90.18		84.7	−101.5	$0.8291^{20/4}$	1.4932^{20}	al, eth, ace	$B1^4$, 1500
9195	*N*-Methyl quinolinium chloride $C_{10}H_9NCl$	178.64	cr (+w, al)		126			w, chl	$B20^4$, 3357
9196	Methyl red or 4′ dimethylamino azobenzene-2-carboxylic acid $C_{15}H_{15}N_3O_2$	269.30	vt or red pr (to, bz) nd (aq, aa) lf (dil al)		183			al, ace, bz, chl, aa	$B16^3$, 367
9197	Methyl selenide or Dimethyl selenide $(CH_3)_2Se$	109.03		$54\text{-}5^{753}$		$1.4077^{15/4}$		al, eth, chl	$B1^4$, 1288
9198	Methyl sulfate or Dimethyl sulfate $(CH_3O)_2SO_2$	126.13		188.5d, 76^{15}	−31.7	1.3283^{20}	1.3874^{20}	w, al, eth, bz	$B1^4$, 1251
9199	Methyl sulfide or Dimethyl sulfide $(CH_3)_2S$	62.13		37.3	−98.3	$0.8483^{20/4}$	1.4438^{20}	al, eth	$B1^4$, 1275
9200	Methyl sulfite or Dimethyl sulfite $(CH_3O)_2SO$	110.13		126, 52^{45}		$1.2129^{20/4}$	1.4093^{20}	al, eth	$B1^4$, 1250
9201	Methyl sulfone or Dimethyl sulfone $(CH_3)_2SO_2$	94.13	pr	238	110	$1.1707^{110/0}$	1.4226	w, al, bz	$B1^4$, 1279
9202	Methyl sulfoxide or Dimethyl sulfoxide $(CH_3)_2SO$	78.13		189, $85\text{-}7^{20}$	18.4	$1.1014^{20/4}$	1.4770^{20}	w, al, eth, ace	$B1^4$, 1277
9203	Methyl telluride or Dimethyl telluride $(CH_3)_2Te$	157.67	pa ye	93.5^{749}		> 1		al	$B1^4$, 1288
9204	Methyl thiocyanate CH_3SCN	73.11		132.9^{757}	−51	$1.0678^{25/4}$	1.4669^{25}	al, eth	$B3^4$, 327
9205	Methyl vinyl ether $CH_3OCH{=}CH_2$	58.08		12	−122	$0.7725^{0/4}$	1.3730^{0}	al, eth, ace, bz	$B1^4$, 2049
9206	Methyl vinyl ketone or 3-buten-2-one $CH_2{=}CH\text{-}COCH_3$	70.09		81.4, $33\text{-}4^{130}$		$0.8636^{20/4}$	1.4081^{20}	w, al, eth, ace, bz	$B1^4$, 3444
9207	Methyl vinyl sulfide $CH_3SCH{=}CH_2$	74.14		69-70		$0.9026^{20/4}$	1.4837^{20}		$B1^4$, 2065
9208	Methyl vinyl sulfone $CH_3SO_2CH{=}CH_2$	106.14		$122\text{-}4^{24}$		$1.2117^{20/4}$	1.4636^{20}	eth, ace	$B1^4$, 2065
9209	Methylene blue or 3,9-bis Dimethylamino phenazothionium chloride $C_{16}H_{18}N_3ClS$	319.85	dk gr cr or pw (chl-eth)					w, al, chl	$B27^2$, 448
9210	5,5′-Methylene disalicylic acid or Bis-(3-carboxy-4 hydroxy phenyl)methane $[4\text{-}HO\text{-}3HO_2CC_6H_3]_2CH_2$	288.26	nd (bz)		243-4			al, eth, ace	$B10^3$, 2507
9211	Methysticin or Kavatin $C_{15}H_{14}O_5$	274.27	nd (MeOH) pr (ace) $[\alpha]^{20}_{D}$ +94.34 (aa,p=5)		137			al, ace, bz, chl	$B19^4$, 5161
9212	Metrazol or Cardiazole Leptazolo $C_6H_{10}N_4$	138.17	cr (bz-liq)	194^{12}	59-60			w, al, eth, ace, bz	$B26^2$, 213
9213	Metycaine or Piperocaine hydrochloride $C_{16}H_{19}NO_2Cl$	292.79			172-5			w, al, chl	$B20^4$, 1460
9214	Michler's hydrol $(4\text{-}(CH_3)_2NC_6H_4)_2CHOH$	270.38	lt gr lf or pr (bz)		98			al, eth, bz, aa	$B13^3$, 1958
9215	Mimosine (*l*) or *l*-Leucenol $C_8H_{10}N_2O_4$	198.18	ta (w) $[\alpha]^{22}_{D}$ −21 (w,c=0.5)	228-9d					$B21^4$, 4648
9216	Morphine $C_{17}H_{19}NO_3$	285.34	pr		254-6			py, MeOH	$B27^2$, 118
9217	Morphine, acetate,trihydrate (*l*) $C_{17}H_{19}NO_3.CH_3CO_2H.3H_2O$	398.43	cr (dil al) $[\alpha]^{15}_{D}$ −77 (w)		200d			w	$B27^2$, 134
9218	Morphine, hydrate $C_{17}H_{19}NO_3.H_2O$	303.36	orh pr (dil al) $[\alpha]^{25}_{D}$ −132 (MeOH, c=1)	d	254-6 anh	$1.32^{20/4}$	1.56-1.64		$B27^2$, 122
9219	Morphine, hydrochloride,trihydrate $C_{17}H_{19}NO_3.HCl.3H_2O$	375.85	nd or fl (dil HCl) $[\alpha]^{25}_{D}$ −113.5 (w,c=2.2)		200d			w, al	$B27^2$, 132

PHYSICAL CONSTANTS OF ORGANIC COMPOUNDS (Continued)

No.	Name, Synonyms, and Formula	Mol. wt.	Color, crystalline form, specific rotation and λ_{max} (log ε)	b.p. °C	m.p. °C	Density	n_D	Solubility	Ref.
9220	Morphine, *N*-oxide or Genomorphine $C_{17}H_{19}NO_4$	301.34	pr (50% al)		274-5				$B27^2$, 159
9221	Morphine, sulfate,pentahydrate $2(C_{17}H_{19}NO_3).H_2SO_4.5H_2O$	758.84	pw or cubes $[\alpha]^{25}_D$ −107.8 (w,c=4)		*ca* 250d			w	$B27^2$, 133
9222	Morphine, *o,o*-diacetyl or Heroin $C_{21}H_{23}NO_5$	369.42	rh $[\alpha]^{15}_D$ −166 (MeOH)	$272\text{-}4^{12}$	173	1.56-1.61		bz, chl	$B27^2$, 151
9223	Morphine-*o,o*-diacetyl, hydrochloride, monohydrate $C_{21}H_{23}NO_5.HCl.H_2O$	423.89	$[\alpha]^{20}_D$ −153 (w,c=1.17)		231-2			w, al, chl	$B27^2$, 153
9224	Morphine-3-ethyl ether hydrochloride, dihydrate or Dionin $C_{19}H_{23}NO_3.HCl.2H_2O$	385.89	cr		123-5d, 170 (anh)			w, al	$B27^2$, 148
9225	Morpholine or Tetrahydro-1,4-isoxazine C_4H_9NO	87.12	hyg	128.3, 24.8^{10}	−4.7	$1.0005^{20/4}$	1.4548^{20}	w, al, eth, ace, bz	$B27^2$, 3
9226	Morpholine, 4-acetyl $4\text{-}CH_3CO(C_4H_8NO)$	129.16		152^{50}, 118^{12}	14.5	$1.1165^{20/20}$	1.4827^{20}	w, al, ace	C50, 7112
9227	Morpholine, 4-(2-aminoethyl) $4\text{-}(H_2NCH_2CH_2)(C_4H_8NO)$	130.19		116^{50}	25.6	$0.9915^{20/20}$	1.4715^{20}	**w, al,** ace, **bz, lig**	C42, 6747
9228	Morpholine, 4-(3 aminopropyl) $4\text{-}[H_2N(CH_2)_3](C_4H_8NO)$	144.22		219^{733}, 134^{50}	−15	$0.9872^{20/20}$	1.4762^{20}	**w, al,** ace, **bz, lig**	Am66, 725
9229	Morpholine, 4-benzyl $4\text{-}C_6H_5CH_2(C_4H_8NO)$ $C_{11}H_{15}NO$	177.25		260-1, $128\text{-}9^{13}$		$1.0387^{20/4}$	1.5302^{20}	ace, bz	$B27^1$, 203
9230	Morpholine, 4-butyl $4\text{-}C_4H_9(C_4H_8NO)$ $C_8H_{17}NO$	143.23		213-4, $67\text{-}8^{10}$	−57.1	$0.9068^{20/4}$	1.4451^{20}	w, al, ace, bz	Am61, 171
9231	Morpholine, 2,6,-dimethyl $2,6\text{-}(CH_3)_2(C_4N_7NO)$	115.18		146.6, 58^{30}	fr -85	$0.9346^{20/20}$	1.4460^{20}	**w, al,** ace **bz, lig**	Am80, 1257
9232	Morpholine, 4-(2-ethoxy ethyl) $4\text{-}(C_2H_5OCH_2CH_2)(C_4H_8NO)$ $C_8H_{17}NO_2$	159.23		206, $93\text{-}7^{14}$	−100	0.963^{20}		**w,** eth, ace, bz	Am63, 298
9233	Morpholine, 4-ethyl $4\text{-}C_2H_5(C_4H_8NO)$ $C_6H_{13}NO$	115.18		$138\text{-}9^{763}$		$0.9886^{20/4}$	1.4400^{20}	**w, al, eth,** ace, bz	$B27^1$, 203
9234	Morpholine, 4-(2-hydroxy ethyl) or 4-(β-morpholinol) ethanol $C_6H_{13}NO_2$	131.17		227^{757}		$1.0710^{20/4}$	1.4763^{20}	w, al	B27, 7
9235	Morpholine, 4-(2-hydroxy propyl) $C_7H_{15}NO_2$	145.20		$92\text{-}4^{13}$		$1.0174^{20/4}$	1.4638^{20}	w, al, eth, ace, bz	Am64, 970
9236	Morpholine, 4-methyl $C_5H_{11}NO$ $4\text{-}CH_3(C_4H_8NO)$	101.15		$115\text{-}6^{750}$		$0.9051^{20/4}$	1.4332^{20}	w, al, eth	$B27^1$, 203
9237	Morpholine, 4-phenyl $4\text{-}C_6H_5(C_4H_8NO)C_{10}H_{13}NO$	163.22	cr (al-eth)	$259\text{-}60^{745}$, $165\text{-}70^{45}$	57-8			eth	$B27^2$, 3
9238	Morpholine, 4(4-tolyl) $4(4\text{-}CH_3C_6H_4)(C_4H_8NO)$ $C_{11}H_{15}NO$	177.25	cr (dil al)	167^{30}	51			al, eth	$B27^2$, 4
9239	Mucic acid or 2,3,4,5-Tetrahydroxy hexanedioic acid or Galactaric acid $HO_2C(CHOH)_4CO_2H$	210.14	pr (w)		255 (rapid htng)				$B3^4$, 1292
9240	*cis*-Muconic acid or 2,4-Hexadiendioic acid $HO_2CCH=CHCH=CHCO_2H$	142.11			194-5			aa	$B2^4$, 2297
9241	*trans*-Muconic acid $HO_2CCH=CHCH=CHCO_2H$	142.11	nd (al)	320	305d				$B2^4$, 2298
9242	Murexide or Ammonium purpurate $C_8H_{10}N_6O_7$	302.20	red-gr pr (aq.NH_4Cl)						$B25^1$, 709
9243	Mycophenolic acid $C_{17}H_{20}O_6$	320.34	nd (w)		141			al, eth, chl	$B18^4$, 6513
9244	Myrcene, dihydro or β-Linaloolene $CH_2=CHCH(CH_3)CH_2CH_2CH=C(CH_3)_2$	138.25		165-8		$0.7601^{20/4}$	1.4362^{20}		$B1^3$, 1020
9245	Myrcene or -7-methyl-3-methylene-1,6-octadiene $(CH_3)_2C=CH(CH_2)_2\text{-}C(=CH_2)CH=CH_2$	136.24		167, 65^{20}		$0.8013^{15/4}$	1.4722^{20}	al, eth, bz, chl, aa	$B1^4$, 1108
9246	Myricyl alcohol or 1-Triacontanol $CH_3(CH_2)_{28}CH_2OH$	438.82	nd (eth) pl (bz)		88	0.777^{95}		al, eth, bz	$B1^3$, 1850
9247	Myristaldehyde or Tetradecanal $C_{13}H_{27}CHO$	212.38	lf	166^{24}	30			al, eth, ace	$B1^4$, 3389
9248	Myristaldehyde, dimethyl acetal $C_{13}H_{27}CH(OCH_3)_2$	258.43		$134\text{-}6^4$			1.4342^{25}	al, eth	$B1^4$, 3389
9249	Myristaldehyde, oxime $C_{13}H_{27}CH=NOH$	227.39	lf or nd (al)		82-3			al	$B1^2$, 770
9250	Myristamide $C_{13}H_{27}CONH_2$	227.39	lf (ace)	217^{12}	105-7			al	$B2^4$, 1138

No.	Name, Synonyms, and Formula	Mol. wt.	Color, crystalline form, specific rotation and λ_{max} (log ϵ)	b.p. °C	m.p. °C	Density	n_D	Solubility	Ref.
9251	Myristic acid or Tetradecanoic acid $C_{13}H_{27}CO_2H$	228.38	lf (eth, 80% aa)	250.5^{100}, 149.3^{1}	58	$0.8439^{80/4}$	1.4305^{60}	al, ace, bz, chl	$B2^4$, 1126
9252	Myistic acid, anhydride $(C_{13}H_{27}CO)_2O$	438.73	lf (peth)	vac distb	53.4	$0.8502^{70/4}$	1.4335^{70}	al, eth	$B2^4$, 1138
9253	Myristic acid, benzyl ester or Benzyl myristate $C_{13}H_{27}CO_2CH_2C_6H_5$	318.50	....	229.3^{11}	20.5	$0.9321^{25/25}$	....	al, eth, bz, chl	$B6^2$, 417
9254	Myristic acid, ethyl ester or Ethyl myristate $C_{13}H_{27}CO_2H_5$	256.43	....	295, 162.5^{9}	12.3	$0.8573^{25/4}$	1.4362^{20}	al	$B2^4$, 1131
9255	Myristic acid, methyl ester or Methyl myristate $C_{13}H_{27}CO_2CH_3$	242.40	....	295^{751}, $155\text{-}7^{7}$	19	....	1.425^{45}	**al, eth, ace, bz, chl**	$B2^4$, 1131
9256	Myristic acid, propyl ester or Propyl myristate $C_{13}H_{27}CO_2C_3H_7$	270.46	....	147^{2}	....	0.8592^{20}	1.4356^{25}	al, eth, ace, bz	$B2^4$, 1132
9257	Myristic acid, isopropyl ester or Isopropyl myristate $C_{13}H_{27}CO_2CH(CH_3)_2$	270.46	....	192.6^{20}, 140.2^{2}	....	0.8532^{20}	1.4325^{25}	al, eth, ace, bz, chl	$B2^4$, 1132
9258	Myristicin or 1,3-Benzodioxide,4-methoxy-6-(2-propenyl) $C_{11}H_{12}O_3$	192.21	....	276-7, 157^{21}	<−20	$1.1437^{20/20}$	1.5403^{20}	eth, bz	$B19^4$, 801
9259	Myristonitrile $C_{13}H_{27}CN$	209.38	....	226.5^{100}, 119^{1}	19.2	$0.8281^{19/4}$	1.4392^{23}	**al, eth, ace, bz, chl**	$B2^4$, 1139
9260	Myristoyl chloride $C_{13}H_{27}COCl$	246.82	....	174^{16}	−1	....	....	eth	$B2^4$, 1138
9261	Myristyl alcohol or 1-Tetradecanol $C_{13}H_{27}CH_2OH$	214.39	lf	263.2, 167^{15}	39-40	$0.8236^{38/4}$	....	al, eth, ace, bz, chl	$B1^4$, 1864

No.	Name, Synonyms, and Formula	Mol. wt.	Color, crystalline form, specific rotation and λ_{max} (log ε)	b.p. °C	m.p. °C	Density	n_D	Solubility	Ref.
9262	Naphthacene or 2,3-Benzanthracene $C_{18}H_{12}$	228.29	og-ye lf (bz, xyl)	sub	357				$B5^{2}$, 2372
9263	Naphthacene, 9,10,-dihydro $C_{18}H_{14}$	230.31	nd (xyl) lf (bz)	*ca* 400	212			bz, aa	$B5^{3}$, 2304
9264	Naphthacene, 9,10,-diphenyl $C_{30}H_{20}$	380.49	og (eth)		207-8			eth	$B5^{3}$, 2703
9265	Naphthacene, 9,11-diphenyl $C_{30}H_{20}$	380.49	ye		301-2			eth, bz	$B5^{3}$, 2702
9266	Naphthacene, 9,10,11-triphenyl $C_{36}H_{24}$	456.59	og (eth) (bz)		236-7, 177 (+1 bz)			eth, bz	$B5^{3}$, 2766
9267	9,10-Naphthacene quinone or 2,3-Benzanthraquinone $C_{18}H_{10}O_2$	258.28	ye nd (aa)	sub	294			al	$B7^{3}$, 4273
9268	9,11-Naphthacene quinone $C_{18}H_{10}O_2$	258.28	dk red (aa,xyl)		322				$B7^{3}$, 4276
9269	1-Naphthaldehyde α-$C_{10}H_7CHO$	156.18	pa ye	292, 160^{15}	33-4	$1.1503^{20/4}$	1.6507^{20}	al, eth, ace, bz	$B7^{3}$, 1953
9270	1-Naphthaldehyde, 2-ethoxy 2-C_2H_5O-α-$C_{10}H_6CHO$	200.24	yesh nd (al,aa)	$185\text{-}7^{25}$	115			al, aa	$B8^{3}$, 1110
9271	1-Naphthaldehyde, 4-ethoxy 4-C_2H_5O-α-$C_{10}H_6CHO$	200.24	yesh cr (aa)		75			al, eth	$B8^{2}$, 174
9272	1-Naphthaldehyde, 2-hydroxy 2-HO-α-$C_{10}H_6CHO$	172.18	pr (al) nd (AcOEt)	192^{27}	82			al, eth, peth	$B8^{3}$, 1108
9273	2-Naphthaldehyde β-$C_{10}H_7CHO$	156.18	lf (w)	160^{19}	61-3	$1.0775^{99/4}$	1.6211^{99}	al, eth, ace	$B7^{3}$, 1957
9274	2-Naphthaldehyde, 1-hydroxy 1-HO-β-$C_{10}H_6CHO$	172.18	grsh-ye nd (dil al,dil aa,lig)		60			al, eth, aa	$B8^{3}$, 1118
9275	Naphthalene $C_{10}H_8$	128.17	mcl pl (al)	218, 87.5^{10}	80.5	$0.9625^{100/4}$, $1.0253^{2}0$	1.5898^{85}, 1.4003^{24}	al, eth, ace, bz	$B5^{3}$, 1549
9276	Naphthalene, 1-acetoxy or α-Naphthyl acetate α-$CH_3CO_2C_{10}H_7$	186.21	nd or pl (al)		49			al, eth	$B6^{3}$, 2928
9277	Naphthalene, 1-acetoxy-2-methyl 1-CH_3CO_2-2-$CH_3C_{10}H_6$	200.24	nd(eth-peth)		81-2			eth	$B6^{3}$, 3027
9278	Naphthalene, 1-acetoxy-4-methyl 1-CH_3CO_2-4-$CH_3C_{10}H_6$	200.24	nd(eth-peth)		86-8				C64, 19519
9279	Naphthalene, 1-acetoxy-7-methyl 1-CH_3CO_2-7-$CH_3C_{10}H_6$	200.24		188^{15}	38-41				$B6^{3}$, 3030
9280	Naphthalene, 2-acetoxy or β-Naphthyl acetate β-$CH_3CO_2C_{10}H_7$	186.21	nd (al)		70			al, eth, chl	$B6^{3}$, 2982
9281	Naphthalene, 5-acetoxy-2-methyl 5-CH_3CO_2-2-$CH_3C_{10}H_6$	200.24		124^{2}					$B6^{3}$, 3028
9282	Naphthalene, 1-acetyl or Methyl-α-naphthyl ketone α-$CH_3COC_{10}H_7$	170.21		296-8, 170^{20}	34	$1.1171^{21.5/4}$	1.6280^{22}	al, eth, ace	$B7^{3}$, 1960
9283	Naphthalene, 2-acetyl or Methyl-β-naphthyl ketone β-$CH_3COC_{10}H_7$	170.21	nd (liq, dil al)	301-3, $171\text{-}3^{11}$	56				$B7^{3}$, 1967
9284	Naphthalene, 1-allyl α-$(CH_2{=}CHCH_2)C_{10}H_7$	168.24		265-7, $129\text{-}30^{10}$		$1.0228^{20/4}$	1.6140^{20}	al, bz, chl	$B5^{4}$, 1740
9285	Naphthalene, 1(2-aminoethyl) α-$(H_2NCH_2CH_2)C_{10}H_7$	171.24		$182\text{-}3^{18}$				al, xyl	$B12^{3}$, 3112
9286	Naphthalene, β-(2-aminoethyl) β-$(H_2NCH_2CH_2)C_{10}H_7$	171.24		174^{25}				al, aa	$B12^{3}$, 3114
9287	Naphthalene, 1-aminomethyl α-$(H_2NCH_2)C_{10}H_7$	157.22	yesh turns red in air	294-5, $162\text{-}3^{12}$				al, eth	$B12^{3}$, 3097
9288	Naphthalene, 2-aminomethyl β-$(H_2NCH_2)C_{10}H_7$	157.22	pr (eth)	180^{24}	59-60			al, eth	$B12^{3}$, 3109
9289	Naphthalene, α-benzyl α-$(C_6H_5CH_2)C_{10}H_7$	218.30	mcl lf or ta (al)	350, $217\text{-}20^{19}$	59-60	1.166^{17}		al, eth, bz, chl	$B5^{3}$, 2236
9290	Naphthalene, β-benzyl or β-Benzyl naphthalene β-$(C_6H_5CH_2)C_{10}H_7$	218.30	mcl pr (al, MeOH)	350	58	1.176^{0}		eth, bz, chl,	$B5^{3}$, 2237
9291	Naphthalene, α-bromo α-$BrC_{10}H_7$	207.07	pr (β form)	281, 139^{16}	α-6.2, β2-3	$1.4826^{20/4}$	1.658^{20}	al, eth, ace, bz, chl	$B5^{4}$, 1665
9292	Naphthalene, 1-bromo-2-(bromomethyl) 2-$BrCH_2$-1-$BrC_{10}H_6$	299.99	nd (al), cr (peth)		107-8			al, bz, liq	$B5^{3}$, 1633
9293	Naphthalene, α-(bromo methyl) α-$(BrCH_2)C_{10}H_7$	221.10	cr (peth, al)	183^{18}, $145\text{-}50^{2}$	56			al, eth, ace, bz	$B5^{4}$, 1693

No.	Name, Synonyms, and Formula	Mol. wt.	Color, crystalline form, specific rotation and λ_{max} (log ε)	b.p. °C	m.p. °C	Density	n_D	Solubility	Ref.
9294	Naphthalene, β-bromo β-$BrC_{10}H_7$	207.07	pl or rh lf (al)	287-2, 147^{18}	59	1.605^{0}	1.6382^{60}	al, eth, bz, chl	$B5^4$, 1667
9295	Naphthalene, β-bromomethyl β-$(BrCH_2)C_{10}H_7$	221.10	lf(al)	213^{100}, $165\text{-}9^{14}$	56			al, eth, chl,	$B5^4$, 1698
9296	Naphthalene, α-butyl α-$C_4H_9C_{10}H_7$	184.28		289.34, $151\text{-}2^{14}$	−19.76	$0.9738^{20/4}$	1.5819^{20}	al, eth, ace, bz,	$B5^4$, 1737
9297	Naphthalene, β-butyl β-$C_4H_9C_{10}H_7$	184.28		292, 146^{12}	−5	$0.9673^{20/4}$	1.57774^{20}	al, ace, bz	$B5^4$, 1738
9298	Naphthalene, β-*tert*-butyl β-$[(CH_3)_3C]C_{10}H_7$	184.28		$274\text{-}7^{756}$	−4	$0.9674^{20/4}$	$1.5685^{20}/_D$	al, ace, bz	$B5^3$, 1805
9299	Naphthalene, α-chloro α-$ClC_{10}H_7$	162.62	cr (al, ace)	258.8^{753}, 106.5^{5}	−2.3	$1.1938^{20/4}$	1.6326^{20}	al, eth, bz	$B5^4$, 1658
9300	Naphthalene, 1-chloro-2-nitro 1-Cl-2-$(O_2N)C_{10}H_6$	207.62	pa ye nd (al, liq)		81			al	$B5^4$, 1677
9301	Naphthalene, 1-chloro-3-nitro 1-Cl-3-$(NO_2)C_{10}H_6$	207.62	ye rd (al)		129.5			al	$B5^4$, 1677
9302	Naphthalene, 1-chloro-4-nitro 1-Cl-4$(NO_2)C_{10}H_6$	207.62	br ye nd (peth,al)		87			al, eth	$B5^4$, 1676
9303	Naphthalene, 1-chloro-5-nitro 1-Cl-5-$(NO_2)C_{10}H_6$	207.62	nd (dil al,aa)	>360, 181^{2}	111			al	$B5^3$, 1597
9304	Naphthalene, 1-chloro-6-nitro 1-Cl-6-$(NO_2)C_{10}H_6$	207.62	ye nd (dil al)		131			al, eth, ace, bz, chl	$B5^3$, 1598
9305	Naphthalene, 1-chloro-8-nitro 1-Cl-8-$(NO_2)C_{10}H_6$	207.62	lt ye nd (gl aa bz liq)	175^{2}	94-5			al, aa	$B5^4$, 1676
9306	Naphthalene, β-chloro β-$ClC_{10}H_7$	162.62	pl (dil al)lf	256, $121\text{-}2^{12}$	61	$1.1377^{71/4}$	1.6079^{13}	al, eth, bz	$B5^4$, 1660
9307	Naphthalene, 2-chloro-1-nitro 1-NO_2-2-$ClC_{10}H_6$	207.62	nd (peth) ye nd (al)	>360	99-100			al, eth, ace, bz	$B5^4$, 1676
9308	Naphthalene, 2-chloro-3-nitro 2-Cl-3$(NO_2)C_{10}H_6$	207.62	br cr or nd (al)		94.5				$B5^4$, 1677
9309	Naphthalene, 2-chloro-6-nitro 2-Cl-6-$(NO_2)C_{10}H_6$	207.62	ye nd	$180\text{-}90^{15}$	170				$B5^3$, 1598
9310	Naphthalene, 2-chloro-7-nitro 2-Cl-7$(NO_2)C_{10}H_6$	207.62	pa ye nd		136				$B5^4$, 1677
9311	Naphthalene, 2-chloro-8-nitro 8-(NO_2)-2-$ClC_{10}H_6$	207.62	ye nd (al)		116			al, eth	$B5^4$, 1676
9312	Naphthalene, 3-chloro-1-nitro 3-Cl-1-$(NO_2)C_{10}H_6$	207.62	grsh br nd		105				$B5^4$, 1676
9313	Naphthalene, 3-chloro-8-nitro 8-(NO_2)-3-$ClC_{10}H_6$	207.62	nd (aq ace)		100.5			ace	$B5^4$, 1676
9314	Naphthalene, α-(chloromethyl) α-$(ClCH_2)C_{10}H_7$	176.65	pr	291-2, $135\text{-}6^{6}$	32			al, bz	$B5^4$, 1692
9315	Naphthalene, β-(chloromethyl) β-$(ClCH_2)C_{10}H_7$	176.65	lf (al)	170^{20}	48-9			al, bz	$B5^4$, 1697
9316	Naphthalene, decahydro *(cis)* or Decalin $C_{10}H_{18}$	138.25		155.6, 69.4^{10}	−43	$0.8965^{22/4}$	1.4810^{20}	**al**, eth, ace, **bz**, chl	$B5^4$, 310
9317	Naphthalene, 1,2-diamino or 1,2-Naphthalenediamine 1,2-$(H_2N)_2C_{10}H_6$	158.20	lf (w) (red in air)	214^{13}	98.5			al, eth, chl	$B13^3$, 377
9318	Naphthalene, 1,3-diamino-2-phenyl 2-C_6H_5-1,3-$(CH_2N)_2C_{10}H_5$	234.30	pl, (MeOH, bz) (red in air)		116			al, bz	$B13^2$, 131
9319	Naphthalene, 1,4-diamino or 1,4-Nahthylene diamine 1,4-$(H_2N)_2C_{10}H_6$	158.20	ye nd (w)		120		1.6441^{18}		$B13^3$, 383
9320	Naphthalene, 1,4-diamino-2-methyl 2-CH_3-1,4-$(H_2N)_2C_{10}H_5$	172.23	ye cr (peth)		113-4				$B13^3$, 406
9321	Naphthalene, 1,4-diamino-2-methyl,dihydrochloride or Vitamin K_6 2-CH_3-1,4-$(H_2N)_2C_{10}H_5$.2HCl	245.15	cr (dil HCl)		300d			w	$B13^3$, 406
9322	Naphthalene, 1,5-diamino or 1,5-Naphthylenediamine 1,5-$(H_2N)_2C_{10}H_6$	158.20	pr (eth,al,w)	sub	190	1.4		al, eth, chl	$B13^3$, 390
9323	Naphthalene, 1,5-diamino-2,methyl 2-CH_3-1,5-$(H_2N)_2C_{10}H_5$	172.23	red-ye lf (dil al)		136			al, eth, bz	$B13^3$, 407
9324	Naphthalene, 1,6-diamino or 1,6-Naphthylene diamine 1,6-$(H_2N)_2C_{10}H_6$	158.20	nd (w,eth)		85-6	$1.1477^{99/4}$	1.7083^{99}	al, bz	$B13^2$, 85
9325	Naphthalene, 1,7-diamino or 1,7-Naphthylenediamine 1,7-$(H_2N)_2C_{10}H_6$	158.20	lf (bz)nd (w)		117.5			al, bz	$B13^3$, 398

PHYSICAL CONSTANTS OF ORGANIC COMPOUNDS (Continued)

No.	Name, Synonyms, and Formula	Mol. wt.	Color, crystalline form, specific rotation and λ_{max} (log ε)	b.p. °C	m.p. °C	Density	n_D	Solubility	Ref.
9326	Naphthalene, 1,8-diamino or 1,8-Naphthylenediamine 1,8-$(CH_2N)_2C_{10}H_6$	158.20	nd(dil al)	205^{12},sub	66.5	$1.1265^{99/4}$	1.6828^{99}	al, eth	B13³, 398
9327	Naphthalene, 2,3-diamino or 2,3-Naphthylenediamine 2,3-$(H_2N)_2C_{10}H_6$	158.20	lf (eth or w)		199	$1.0968^{26/4}$	1.6342^{26}	al, eth	B13³, 402
9328	Naphthalene, 2,6-diamino or 2,6-naphthylenediamine 2,6-$(H_2N)_2C_{10}H_6$	158.20	nd or lf (w) lf (al)		222d				B13³, 402
9329	Naphthalene, 1,2-dichloro 1,2-$Cl_2C_{10}H_6$	197.06	pl (al)	295-8, $151\text{-}3^{19}$	35-7	$1.3147^{49/4}$	1.5338^{49}	al, eth	B5⁴, 1661
9330	Naphthalene, 1,3-dichloro 1,3-$Cl_2C_{10}H_6$	197.06	nd or pr (al)	291^{775}	61.5			al	B5⁴, 1661
9331	Naphthalene, 1,4-dichloro 1,4-$Cl_2C_{10}H_6$	197.06	nd or pr (al,aa,ace)	$286\text{-}7^{740}$, 147^{12}	68	$1.2997^{76/4}$	1.6228^{76}	eth, ace, bz, aa	B5⁴, 1661
9332	Naphthalene, 1,5-dichloro 1,5-$Cl_2C_{10}H_6$	197.06	nd or lf (al,aa) pr (sub)	sub	107				B5⁴, 1662
9333	Naphthalene, 1,6-dichloro 1,6-$Cl_2C_{10}H_6$	197.06	nd or pr (al,peth,sub)	sub	49				B5⁴, 1662
9334	Naphthalene, 1,7-dichloro 1,7-$Cl_2C_{10}H_6$	197.06	nd or pr (al, aa)	285-6	63-4	$1.2611^{100/4}$	1.6092^{100}	al, eth, bz, aa	B5⁴, 1662
9335	Naphthalene, 1,8-dichloro 1,8-$Cl_2C_{10}H_6$	197.06	rh pl (hx) nd (al,sub)	d sub	89	$1.2924^{100/4}$	1.6236^{100}	al, peth	B5⁴, 1662
9336	Naphthalene, 2,3-dichloro 2,3-$Cl_2C_{10}H_6$	197.06	rh lf (al)		120			eth	B5⁴, 1662
9337	Naphthalene, 2,6-dichloro 2,6-$Cl_2C_{10}H_6$	197.06	pr (aa), nd or lf (al) pl (eth,bz)	285	140-1			eth, bz, chl, aa	B5⁴, 1662
9338	Naphthalene, 2,7-dichloro 2,7-$Cl_2C_{10}H_6$	197.06	pl or lf (al)		114			al	B5⁴, 1662
9339	Naphthalene, 2,3-dihydrazino 2,3-$(H_2NNH)_2C_{10}H_6$	188.23	red-br (al,w) col nd (bz)		167-8d			al	B15, 583
9340	Naphthalene, 1,2-dihydro $C_{10}H_{10}$	130.19	lf, pl	206-7, 78^{9}	−8	$0.9974^{20/4}$	1.5814^{20}		B5⁴, 1543
9341	Naphthalene, 1,2-dihydro-3-methyl $C_{11}H_{12}$	144.22		105^{13}		$0.9837^{20/20}$	1.5751^{20}	eth, bz	B5⁴, 1552
9342	Naphthalene, 1,2-dihydro-4-methyl $C_{11}H_{12}$	144.22		112^{18}		$0.9895^{20/4}$	1.5758^{20}	eth, bz	B5⁴, 1551
9343	Naphthalene, 1,4-dihydro $C_{10}H_{10}$	130.19	pl	211-2, 94^{17}	25(30)	$0.9928^{33/4}$	1.5577^{20}	aa	B5⁴, 1544
9344	Naphthalene, 2,3-dihydro-1,4-dimethyl $C_{12}H_{16}$	160.26		234		0.940^{20}	$1.528^{20/}{}_{D}$		35.5214) 35.5214)
9345	Naphthalene, 1,2-dihydroxy or 1,2-Naphthalenediol 1,2-$(HO)_2C_{10}H_6$	160.17	lf or nd (CS_2) lf (w + 1) nd (liq)		103-4 (anh), 58-60 (+ 1w)			eth	B6³, 5240
9346	Naphthalene, 1,4-dihydroxy or 1,4-Naphthalenediol 1,4-$(HO)_2C_{10}H_6$	160.17	mcl nd (bz,w)		192			al, eth, aa	B6³, 5260
9347	Naphthalene, 1,4-diacetoxy-2-methyl 2-CH_3-1,4-$(CH_3CO_2)_2C_{10}H_5$	258.28	pr (al)		113			al	B6³, 5302
9348	Naphthalene, 1,5-dihydroxy or 1,5-Napthalene diol 1,5$(HO)_2C_{10}H_6$	160.17	pr (w) nd (sub)	sub	265d			eth, ace, aa	B6³, 5265
9349	Naphthalene, 1,5-diacetoxy 1,5-$(CH_3CO_2)_2C_{10}H_6$	244.25	nd (bz)		161			bz,aa	B6³, 5267
9350	Naphthalene, 1,6-dihydroxy or 1,6-Naphthylene diol 1,6-$(HO)_2C_{10}H_6$	160.17	pr (bz)	sub	138			eth, ace, bz, MeOH	B6³, 5279
9351	Naphthalene, 1,7-dihydroxy 1,7$(HO)_2C_{10}H_6$	160.17	nd (bz or sub)	sub	178-81			al, eth, bz, aa	B6³, 5281
9352	Naphthalene, 1,8-dihydroxy or 1,8-Napthalenediol 1,8-$(HO)_2C_{10}H_6$	160.17	lf or nd (w)		144			al, eth, bz	B6³, 5283
9353	Naphthalene, 2,3-dihydroxy or 2,3 -Napthalenediol 2,3-$(HO)_2C_{10}H_6$	160.17	lf(w)		163-4			al, eth, bz, aa, lig	B6³, 5287
9354	Naphthalene, 2,6-dihydroxy or 2,6-Naphthalenediol 2,6-$(HO)_2C_{10}H_6$	160.17	rh pl (w)	sub	222			al, eth, ace, aa	B6³, 5287
9355	Naphthalene, 2,7-dihydroxy or 2,7-naphthalenediol 2,7-$(HO)_2C_{10}H_6$	160.17	nd (w,dil al) pl (dil al)	sub	190-4			al, eth, bz, chl	B6³, 5291

No.	Name, Synonyms, and Formula	Mol. wt.	Color, crystalline form, specific rotation and λ_{max} (log ε)	b.p. °C	m.p. °C	Density	n_D	Solubility	Ref.
9356	Naphthalene, 1,5-dimercapto or 1,5-naphthylenedithiol 1,5-$(HS)_2C_{10}H_6$	192.29	ye lf (al,eth,bz)	...	119	...	...	al, eth, bz	B6^{3}, 5276
9357	Naphthalene, 1,2-dimethyl 1,2-$(CH_3)_2C_{10}H_6$	156.23	...	266-7, 132^{12}	−1.6	1.0179$^{20/4}$	1.61656^{20}	eth, bz	B5^{4}, 1708
9358	Naphthalene, 1,3-dimethyl 1,3-$(CH_3)_2C_{10}H_6$	156.23	...	263, 138-40^{12}	−6	1.0144$^{20/4}$	1.6140^{20}	eth, bz	B5^{4}, 1708
9359	Naphthalene, 1,4-dimethyl 1,4-$(CH_3)_2C_{10}H_6$	156.23	...	268, 129^{10}	7.6	1.0166$^{20/4}$	1.6127^{20}	al, **eth, ace, bz, lig**	B5^{4}, 1709
9360	Naphthalene, 1,5-dimethyl 1,5-$(CH_3)_2C_{10}H_6$	156.23	...	265	82	...	...	eth, bz	B5^{4}, 1710
9361	Naphthalene, 1,6-dimethyl 1,6-$(CH_3)_2C_{10}H_6$	156.23	...	264, 126^{13}	−16.9	1.0021^{20}	1.61656^{20}	eth, bz	B5^{4}, 1711
9362	Naphthalene, 1,6-dimethyl-4-isopropyl or Cadalene 1,6-$(CH_3)_2C_{10}H_5$-*i*-C_3H_7-4	198.31	...	291-2^{720}, 165^{20}	...	0.9792$^{19/4}$	1.5851^{19}	...	B5^{4}, 1758
9363	Naphthalene, 1,7-dimethyl 1,7-$(CH_3)_2C_{10}H_6$	156.23	...	263, 148^{15}	−13.9	1.0115$^{20/4}$	1.60831^{20}	eth, bz	B5^{4}, 1711
9364	Naphthalene, 1,8-dimethyl 1,8-$(CH_3)_2C_{10}H_6$	156.23	...	270, 140^{18}	65	...	...	eth, bz	B5^{4}, 1712
9365	Naphthalene, 2,3-dimethyl or Guaiene 2,3-$(CH_3)_2C_{10}H_6$	156.23	lf (al)	268, 128.7^{10}	105	1.003$^{20/4}$	1.5060^{20}	eth, bz	B5^{4}, 1713
9366	Naphthalene, 1,2-dinitro 1,2-$(NO_2)_2C_{10}H_6$	218.17	...	...	162-3	...	...	...	B5^{3}, 1605
9367	Naphthalene, 1,3-dinitro 1,3-$(NO_2)_2C_{10}H_6$	218.17	ye nd (bz,py-w)	sub	147-9	...	...	al, ace	B5^{4}, 1680
9368	Naphthalene, 1,5-dinitro 1,5-$(NO_2)_2C_{10}H_6$	218.17	hex nd (aa,ace)	sub	219	...	...	eth, bz	B5^{4}, 1680
9369	Naphthalene, 1,7-dinitro 1,7-$(NO_2)_2C_{10}H_6$	218.17	...	...	156	...	...	ace, bz	B5^{4}, 1680
9370	Naphthalene, 1,8-dinitro 1,8-$(NO_2)_2C_{10}H_6$	218.17	ye rh pl (chl)	445d	173	...	...	ace, py	B5^{4}, 1681
9371	Naphthalene, 2,3-dinitro 2,3-$(NO_2)_2C_{10}H_6$	218.17	...	...	172-4	...	...	al, bz	B5^{4}, 1681
9372	Naphthalene, 2,4-dinitro-1-triazo 1-N_3-2,4-$(NO_2)_2C_{10}H_5$	259.18	ye rh nd (al)	...	105d	...	...	eth, bz, chl	B5^{2}, 460
9373	Naphthalene, 2,6-dinitro 2,6-$(NO_2)_2C_{10}H_6$	218.17	...	...	278	...	...	al, bz	B5^{3}, 1609
9374	Naphthalene, 2,7-dinitro 2,7$(NO_2)_2C_{10}H_6$	218.17	...	...	234	...	...	al, eth, bz	B5^{4}, 1681
9375	Naphthalene, 1-ethoxy or Ethyl α-naphthyl ether α-$C_{10}H_7OC_2H_5$	172.23	nd	280.5, 136-8^{14}	5.5	1.060$^{20/4}$	1.5953^{25}	al, eth	B6^{3}, 2924
9376	Naphthalene, 2-ethoxy or Ethyl β-naphthyl ether β-$C_{10}H_7OC_2H_5$	172.23	pl (al)	282, 148^{10}	37-8	1.0640$^{20/20}$	1.5975^{36}	al, eth, lig, to	B6^{3}, 2972
9377	Naphthalene, α-ethyl α-$C_2H_5C_{10}H_7$	156.23	...	258.6, 120^{10}	−13.9	1.0082$^{20/4}$	1.6062^{20}	**al, eth**	B5^{4}, 1705
9378	Naphthalene, β-ethyl β-$C_2H_5C_{10}H_7$	156.23	...	258, 119^{10}	−7.4	0,9922$^{20/4}$	1.5999^{20}	**al, eth**	B5^{4}, 1707
9379	Naphthalene, α-fluoro α-$FC_{10}H_7$	146.16	...	215^{756}, 80^{11}	−9	1.1322^{20}	1.5939^{20}	al, eth, bz, chl, aa	B5^{4}, 1657
9380	Naphthalene, β-fluoro β-$FC_{10}H_7$	146.16	nd (al)	211.5^{737}, 90^{16}	61	...	...	al, eth, bz, chl, aa	B5^{4}, 1658
9381	Naphthalene, 1,2,3,4,9,10-hexahydro $C_{10}H_{14}$	134.22	...	200, 82^{2}	...	0.934^{23}	1.5260^{16}	eth, bz	B5^{4}, 1082
9382	Naphthalene, 1-(α-hydroxyethyl)-*(dl)* α-$C_{10}H_7CH(OH)CH_3$	172.23	nd (peth)	178^{15}	66	1.1190$^{14/4}$	1.6188^{25}	al, ace, bz, chl	B6^{3}, 3034
9383	Naphthalene, 1-(α-hydroxy ethyl *(l)* α-$C_{10}H_7$-$CH(OH)CH_3$	172.23	$[\alpha]^{20}{}_{D}$ -78.9 (al, c=5)	166^{11}	47	1.1190$^{14/4}$	1.6180^{25}	al, ace, bz, chl	B6^{3}, 3034
9384	Naphthalene, α-hydroxymethyl α-$(HOCH_2)C_{10}H_7$	158.20	nd (w,al) cr (bz-liq)	301^{715},163^{12}	64	1.1039$^{80/4}$	...	al, eth	B6^{3}, 3024
9385	Naphthalene, α-iodo α-$IC_{10}H_7$	254.07	...	302	4.2	1.7399$^{20/4}$	1.7026^{20}	**al, eth, bz**	B5^{4}, 1670
9386	Naphthalene, β-iodo β-$IC_{10}H_7$	254.07	lf (dil al)	308, 172^{21}	54.5	1.6319$^{99/4}$	1.6662^{99}	al, eth, aa	B5^{4}, 1671
9387	Naphthalene, α-mercapto or Naphthalene thiol α-$C_{10}H_7SH$	160.23	...	285d, 161^{20}	...	1.1607$^{20/4}$	1.6802^{20}	al, eth	B6^{3}, 2943
9388	Naphthalene, β-mercapto or β-Naphthalene thiol β-$C_{10}H_7SH$	160.23	pl(al)	288, 162.7^{20}	81	1.550	...	al, eth, lig	B6^{3}, 3007

No.	Name, Synonyms, and Formula	Mol. wt.	Color, crystalline form, specific rotation and λ_{max} (log ε)	b.p. °C	m.p. °C	Density	n_D	Solubility	Ref.
9389	Naphthalene, α-methoxy or Methyl α-naphthyl ether α-$C_{10}H_7OCH_3$	158.20	...	269, 135^{10}	<−10	$1.0964^{14/2}$	1.6940^{25}	al, eth, bz, chl	B6³, 2922
9390	Naphthalene, β-methoxy or Methyl β-naphthyl ether β-$C_{10}H_7OCH_3$	158.20	lf (eth) pl (peth)	274, 138^{10}	73-4	...	...	eth, bz, chl	B6³, 2969
9391	Naphthalene, α-methyl α-$CH_3C_{10}H_7$	142.20	...	244.6, 107.4^{10}	−22	$1.0202^{20/4}$	1.6170^{20}	al, eth, bz	B5⁴, 1687
9392	Naphthalene, 1-methyl-2-nitro 1-CH_3-2-$NO_2C_{10}H_6$	187.20	lt ye nd (al)	...	58-9	...	...	al	B5⁴, 1693
9393	Naphthalene, 1-methyl-3-nitro 1-CH_3-3$NO_2C_{10}H_6$	187.20	ye nd (al)	...	81-2	...	...	al	B5³, 1624
9394	Naphthalene, 1-methyl-4-nitro 1-CH_3-4-$NO_2C_{10}H_6$	187.20	pa ye nd (al)	$182\text{-}3^{18}$	71-2	...	...	al, eth, ace	B5³, 1624
9395	Naphthalene, 1-methyl-5-nitro 1-CH_3-5-$NO_2C_{10}H_6$	187.20	brsh nd (al)	...	82-3	...	...	al, ace	B5³, 1624
9396	Naphthalene, 1-methyl-6-nitro 1-CH_3-6$NO_2C_{10}H_6$	187.20	ye nd (dil al)	...	76-7	...	...	al	B5³, 1625
9397	Naphthalene, 1-methyl-7-nitro 1-CH_3-7-$NO_2C_{10}H_6$	187.20	ye nd (al)	...	98-9	...	...	al	B5³, 1625
9398	Naphthalene, 1-methyl-8-nitro 1-CH_3-8-$NO_2C_{10}H_6$	187.20	br lf (al)	...	65	...	...	al	B5³, 1625
9399	Naphthalene, 1-methyl-7-isopropyl 1-CH_3-7-*i*-$C_3H_7C_{10}H_6$	184.28	...	152^{18}	...	$0.9740^{20/4}$	1.5833^{20}	eth, ace, bz	B5⁴, 1741
9400	Naphthalene, β-methyl β-$CH_3C_{10}H_7$	142.20	mcl (al)	241, 104.7^{10}	34.6	$1.0058^{20/4}$	1.6015^{40}	al, eth, bz	B5⁴, 1693
9401	Naphthalene, 2-methyl-1-nitro 2-CH_3-1-$NO_2C_{10}H_6$	187.20	yesh pr or nd (al)	188^{20}	81-2	...	...	al, ace	B5⁴, 1698
9402	Naphthalene, 2-methyl-3-nitro 2-CH_3-3-$NO_2C_{10}H_6$	187.20	yesh pl (al)	...	117-8	...	...	al	B5⁴, 1698
9403	Naphthalene, 2-methyl-4-nitro 2-CH_3-4-$NO_2C_{10}H_6$	187.20	pa ye nd (al)	...	49-50	...	...	al	B5³, 1634
9404	Naphthalene, 2-methyl-5-nitro 2-CH_3-5-$NO_2C_{10}H_6$	187.20	yd nd (al)	...	61-2	...	...	al	B5³, 1634
9405	Naphthalene, 2-methyl-6-nitro 2-CH_3-6-$NO_2C_{10}H_6$	187.20	ye nd (al)	...	119	...	...	al	B5³, 1634
9406	Naphthalene, 2-methyl-7-nitro 2-CH_3-7-$NO_2C_{10}H_6$	187.20	yesh pl (al)	...	105	...	...	al	B5³, 1634
9407	Naphthalene, 2-methyl-8-nitro 2-CH_3-8-$NO_2C_{10}H_6$	187.20	ye nd (al)	...	36-8	...	...	al	B5³, 1635
9408	Naphthalene, α-nitramino α-$(O_2NNH)C_{10}H_7$	188.19	lt ye nd (w)	...	123-4	...	...	bz	B16², 346
9409	Naphthalene, β-nitramino β-$(O_2NNH)C_{10}H_7$	188.19	lf or nd	...	131-6	...	...	...	B16, 675
9410	Naphthalene, α-nitro α-$NO_2C_{10}H_7$	173.17	ye nd (al)	304 sub, $30\text{-}40^{0.01}$	61.5	$1.332^{20/4}$	...	al, eth, bz, py	B5⁴, 1673
9411	Naphthalene, 1-nitro-5-triazo 1-NO_2-5-$N_3C_{10}H_6$	214.18	gold-ye nd (al)	...	121	...	...	al, ace	B5², 459
9412	Naphthalene, β-nitro β-$NO_2C_{10}H_7$	173.17	ye rh nd or pl (al)	312.5^{734}, 165^{15}	79	...	...	al, eth	B5⁴, 1675
9413	Naphthalene, 2-nitro-1-triazo 2-NO_2-1-$N_3C_{10}H_6$	214.18	ye nd (dil ace)	...	103-4d	...	...	al, ace, bz, ace	B5, 565
9414	Naphthalene α-(nitrosohydroxyl amino) α-$C_{10}H_7(N(NO)OH)$	188.19	nd (peth)	...	54-5	...	...	chl	B16³, 639
9415	Naphthalene, β-(nitrosohydroxylamino) β-$C_{10}H_7(N(NO)OH)$	188.19	nd (AcOEt-peth)	...	88-92	...	...	eth, aa	B16¹, 396
9416	Naphthalene, octachloro or Perchloro naphthalene $C_{10}Cl_8$	403.73	nd (bz-CCl_4)	$440\text{-}2^{7\ 4}$	197-8	...	...	bz, chl, lig	B5⁴, 1665
9417	Naphthalene, α-isopentoxy or Isopentyl-α-naphthyl ether α-$C_{10}H_7O$-*i*-C_5H_{11}	214.31	...	$317\text{-}9^{742}$, $148\text{-}53^{3}$	...	$1.0069^{14/4}$	1.5705^{16}	...	B6³, 2925
9418	Naphthalene, β-Isopentoxy or Isopentyl-β-naphthyl ether *i*-$C_5H_{11}O$-β-$C_{10}H_7$	214.31	lf	323-6d, $155\text{-}60^{6}$	26.5	$1.0155^{12/4}$	1.5768^{12}	al, eth	B6³, 2974
9419	Naphthalene, α-phenyl α-$C_6H_5C_{10}H_7$	204.27	cr	334, 190^{12}	ca 45	$1.096^{20/4}$	1.6664^{20}	al, eth, bz, aa	B5³, 2230
9420	Naphthalene, β-phenyl β-$C_6H_5C_{10}H_7$	204.27	lf (al)	345-6, $185\text{-}90^{5}$	103-4	...	...	al, eth, bz, aa, chl	B5³, 2231

PHYSICAL CONSTANTS OF ORGANIC COMPOUNDS (Continued)

No.	Name, Synonyms, and Formula	Mol. wt.	Color, crystalline form, specific rotation and λ_{max} (log ε)	b.p. °C	m.p. °C	Density	n_D	Solubility	Ref.
9421	Naphthalene, picrate ... $C_{10}H_8.C_6H_3N_3O_7$	357.28	ye pr or pl (aa) cr (eth)	...	152	1.53	...	eth, bz	$B5^{3}$, 1568
9422	Naphthalene, 1-propoxy or Propyl α-naphthyl ether ... α-$C_{10}H_7OC_3H_7$	186.25	...	293.5, 167^{18}	...	$1.0447^{18/4}$	1.5928^{18}	...	$B6^{3}$, 2924
9423	Naphthalene, 2-propoxy or Propyl β-naphthyl ether... β-$C_{10}H_7OC_3H_7$	186.25	nd (al)	305, 144^{10}	41	...	...	al	$B6^{3}$, 2973
9424	Naphthalene, α-propyl ... α-$C_3H_7C_{10}H_7$	170.25	...	274-5	−8.6	...	...	...	$B5^{4}$, 1721
9425	Naphthalene, β-isopropyl ... β-i-$C_3H_7C_{10}H_7$	170.25	...	268^{2}, $129\text{-}30^{14}$	...	$0.9753^{20/4}$	1.58482^{20}	al, eth, bz	$B5^{4}$, 1722
9426	Naphthalene, β-isopropyl ... β-i-$C_3H_7C_{10}H_7$	170.25	...	...	225	...	...	w, al, ace, bz	$B6^{3}$, 6699
9427	Naphthalene, 1,2,3,4-tetrahydroxy ... 1,2,3,4-$(HO)_4C_{10}H_4$	192.17	...	...	106-7	...	...	al, ace, bz	$B5^{4}$, 1745
9428	Naphthalene, 1,2,5,6-tetramethyl ... 1,2,5,6-$(CH_3)_4C_{10}H_4$	184.28	...	$150\text{-}5^{12}$	118	...	...	al, bz	$B5^{4}$, 1746
9429	Naphthalene, 1,3,6,8-tetramethyl ... 1,3,6,8-$(CH_3)_4C_{10}H_4$	184.28	...	$115\text{-}6^{2}$	84-5	...	...	al	$B5^{4}$, 1746
9430	Naphthalene, 1,3,5,8-tetranitro or γ-Tetranitro naphthalene... 1,3,5,8-$(NO_2)_4C_{10}H_4$	308.16	lt ye tetr (ace)	...	194-5	...	...	ace	$B5^{4}$, 1683
9431	Naphthalene, 1,3,6,8-tetranitro or β-tetranitro naphthalene... 1,3,6,8$(NO_2)_4C_{10}H_4$	308.16	ye nd (al,bz)	exp	207	...	...	bz, aa	$B5^{4}$, 1683
9432	Naphthalene, α-triazo or α-Naphthyl azide ... α-$C_{10}H_7N_3$	169.19	pa ye pr	d	12	1.1713^{25}	1.6550^{25}	al, **eth, ace**	$B5^{4}$, 1684
9433	Naphthalene, β-triazo or β-Napthyl azide... β-$C_{10}H_7N_3$	169.19	pl (al nd (peth) ye in air	...	33	...	...	al, eth, ace. bz	$B5^{3}$, 1614
9434	Naphthalene, 1,2,3-trimethyl ... 1,2,3-$(CH_3)_3C_{10}H_5$	170.25	...	$125\text{-}30^{12}$	27-8	...	...	...	$B5^{4}$, 1725
9435	Naphthalene, 1,2,4-trimethyl ... 1,2,4-$(CH_3)_3C_{10}H_5$	170.25	...	146^{12}	55-6	...	...	al, eth, bz, chl	$B5^{4}$, 1725
9436	Naphthalene, 1,2,5-trimethyl ... 1,2,5-$(CH_3)_3C_{10}H_5$	170.25	nd (al)	140^{12}	33.5	$1.0103^{22/4}$	16093^{22}	eth, bz	$B5^{4}$, 1726
9437	Naphthalene, 1,2,6-trimethyl ... 1,2,6-$(CH_3)_3C_{10}H_5$	170.25	lf	146^{10}	14	...	1.6010^{20}	eth, bz	$B5^{4}$, 1726
9438	Naphthalene, 1,2,7-trimethyl ... 1,2,7-$(CH_3)_3C_{10}H_5$	170.25	...	$147\text{-}8^{16}$	...	$1.0087^{20/4}$	1.6097^{20}	eth, bz	$B5^{4}$, 1727
9439	Naphthalene, 2,3,6-trimethyl ... 2,3,6-$(CH_3)_3C_{10}H_5$	170.25	...	263-4, $146\text{-}8^{14}$	100-2	...	...	eth, bz	$B5^{4}$, 1730
9440	Naphthalene, 1,2,5-trinitro ... 1,2,5-$(NO_2)_3C_{10}H_5$	263.17	lt ye nd (al)	...	112-3	...	...	al	$B5^{3}$, 1614
9441	Naphthalene, 1,3,5-trinitro ... 1,3,5-$(NO_2)_3C_{10}H_5$	263.17	ye rh (chl)	364exp	122	...	...	al, ace, aa, chl	$B5^{4}$, 1682
9442	Naphthalene, 1,3,8-trinitro ... 1,3,8-$(NO_2)_3C_{10}H_5$	263.17	yesh mcl pr (al,ace,aa)	...	218	...	...	ace, py	$B5^{4}$, 1682
9443	Naphthalene, 1,4,5-trinitro ... 1,4,5-$(NO_2)_3C_{10}H_5$	263.17	ye lf or rh pl (al,aa,bz)	...	154	...	...	ace	$B5^{2}$, 458
9444	Naphthalene, β-vinyl ... β-$(CH_2{=}CH)C_{10}H_7$	154.21	...	$136\text{-}8^{17}$	66	...	...	al, ace, bz	$B5^{4}$, 1833
9445	1,2-Naphthalene dicarboxylic acid ... 1,2-$C_{10}H_6(CO_2H)_2$	216.19	nd (al) cr (w)	...	175d	...	...	al, eth, aa	$B9^{3}$, 4462
9446	1,2-Naphthalene dicarboxylic acid, 3,4-dihydro, anhydride 1,2-$C_{10}H_8(CO)_2O$	200.19	pa ye nd (lig,al)	$227\text{-}30^{23}$	126-7	...	...	bz, aa, MeOH	$B9^{4}$, 650
9447	1,4-Naphthalene dicarboxylic acid ... 1,4-$C_{10}H_6(CO_2H)_2$	216.19	rods (aa)	...	309 (320)	...	...	al, aa	$B9^{3}$, 4463
9448	1,5-Naphthalene dicarboxylic acid ... 1,5-$C_{10}H_6(CO_2H)_2$	216.19	nd	...	320-2d	...	...	...	$B9^{3}$, 4465
9449	1,6-Naphthalene dicarboxylic acid ... 1,6-$C_{10}H_6(CO_2H)_2$	216.19	nd (aa)	...	310 sinters	...	...	al,aa	$B9^{2}$, 651
9450	1,7-Naphthalene dicarboxylic acid ... 1,7-$C_{10}H_6(CO_2H)_2$	216.19	ye pw (dil al,aa)	...	308d	...	...	al, eth, ace, aa	$B9^{3}$, 4466
9451	2,3-Naphthalene dicarboxylic acid ... 2,3-$C_{10}H_6(CO_2H)_2$	216.19	pr (aa,w) pr (sub)	...	246	...	...	...	$B9^{3}$, 4470

PHYSICAL CONSTANTS OF ORGANIC COMPOUNDS (Continued)

No.	Name, Synonyms, and Formula	Mol. wt.	Color, crystalline form, specific rotation and λ_{max} (log ε)	b.p. °C	m.p. °C	Density	n_D	Solubility	Ref.
9452	2,6-Naphthalene dicarboxylic acid $2,6\text{-}C_{10}H_6(CO_2H)_2$	216.19	nd (al or sub)	...	>300d	...	...	al	$B9^3$, 4471
9453	2,7-Naphthalene dicarboxylic acid $2,7\text{-}C_{10}H_6(CO_2H)_2$	216.19	nd (w,al,dil HCl)	...	>300d	...	...	al	$B9^2$, 653
9454	1,3-Naphthalene disulfonic acid, 7-amino or Amino-G-acid $7\text{-}NH_2\text{-}1,3\text{-}C_{10}H_5(SO_3H)_2$	303.30	mcl pr or nd (w + 2)	...	273-5	...	...	w, al	$B14^3$, 2263
9455	1,3-Naphthalene disulfonic acid, 7-hydroxy or G-acid $7\text{-}HO\text{-}1,3\text{-}C_{10}H_5(SO_3H)_2$	304.29	...	...	...	...	...	w	$B11^3$, 560
9456	1,5-Naphthalene disulfonic acid $1,5\text{-}C_{10}H_6(SO_3H)_2$	288.29	pl (+ 4w,dil aa)	...	240-5d	1.493	...	w, al	$B11^3$, 17
9457	1,6-Naphthalene disulfonic acid $1,6\text{-}C_{10}H_6(SO_3H)_2$	288.29	og pr (+ 4w,aa or w)	...	125d (anh)	...	...	w, al	$B11^3$, 467
9458	2,7-Naphthalene disulfonic acid $2,7\text{-}C_{10}H_6(SO_3H)_2$	288.29	hyg nd (con HCl)	...	199	...	...	w	$B11^3$, 468
9459	2,7-Naphthalene disulfonic acid, 4-amino-5-hydroxy or H acid $4\text{-}NH_2\text{-}5\text{-}HO\text{-}2,7\text{-}C_{10}H_4(SO_3H)_2$	319.30	...	...	...	...	...	...	$B14^3$, 2292
9460	2,7-Naphthalene disulfonic acid, 4,5-dihydroxy or Chromotropic acid $4,5\text{-}(HO)_2\text{-}2,7\text{-}C_{10}H_4(SO_3H)_2$	320.29	nd or lf (w + 2)	...	...	...	...	w	$B11^3$, 576
9461	2,7-Naphthalene disulfonic acid, 3-hydroxy or R-acid $3\text{-}HO\text{-}2,7\text{-}C_{10}H_5(SO_3H)_2$	304.29	dlq nd	...	d	...	...	w, al	$B11^3$, 559
9462	α-Naphthalene phosphonic acid $\alpha\text{-}C_{10}H_7PO(OH)_2$	208.15	cr (w)	...	189	...	...	al	$B16^3$, 896
9463	α-Naphthalene phosphonyl chloride $\alpha\text{-}C_{10}H_7PO(Cl)_2$	245.04	...	...	*ca* 60	...	...	...	$B16^2$, 392
9464	α-Naphthalene sulfinic acid $\alpha\text{-}C_{10}H_7SO_2H$	192.23	nd (w)	...	104	...	...	w, al	$B11^3$, 12
9465	β-Naphthalene sulfinic acid $\beta\text{-}C_{10}H_7SO_2H$	192.23	nd (w)	...	98(105)	...	...	w, al, eth	$B11^3$, 14
9466	α-Naphthalene sulfonic acid $\alpha\text{-}C_{10}H_7SO_3H$	208.23	pr (+ 2w,dil HCl)	...	90 (hyd) 139-40 (anh)	...	...	w, al	$B11^3$, 382
9467	α-Naphthalene sulfonic acid, 4-amino or Naphthionic acid $4\text{-}H_2N\text{-}\alpha\text{-}C_{10}H_6SO_3H$	232.25	wh nd (w + ½) red-br cr	...	d	$1.6703^{25/4}$	...	py, MeOH	$B14^3$, 2241
9468	α-Naphthalene sulfonic acid, 4-amino-5-hydroxy or S-acid $4\text{-}H_2N\text{-}5\text{-}HO\text{-}\alpha\text{-}C_{10}H_5SO_3H$	239.25	nd	...	...	...	...	...	$B14^3$, 2290
9469	α-Naphthalene sulfonic acid, 7-amino or Bayer's acid, Cassela's acid $7\text{-}H_2N\text{-}\alpha\text{-}C_{10}H_6SO_3H$	223.25	nd (w + 1) pl (aq ace)	...	...	...	...	aa	$B14^3$, 2245
9470	α-Naphthalene sulfonic acid, 4-hydroxy or Nenile-winther acid $4\text{-}HO\text{-}\alpha\text{-}C_{10}H_6SO_3H$	224.23	ta or pl (w)	...	170d (rapid htg)	...	...	...	$B11^3$, 540
9471	α-Naphthalene sulfonic acid, 5-hydroxy or α-Naphtholsulfonic acid L $5\text{-}HO\text{-}\alpha\text{-}C_{10}H_6SO_3H$	224.23	dlq	...	120	...	...	w, aa	$B11^3$, 541
9472	α-Naphthalene sulfonic acid, 7-hydroxy or Croceic acid $7\text{-}HO\text{-}\alpha\text{-}C_{10}H_6SO_3H$	224.23	...	...	...	...	...	w	$B11^3$, 556
9473	α-Naphthalene sulfonic acid, 8-hydroxy or α-Naphthol sulfonic acid S $8\text{-}HO\text{-}\alpha\text{-}C_{10}H_6SO_3H$	224.23	cr (w + 1)	...	106-7	...	...	w	$B11^2$, 157
9474	α-Naphthalene sulfonic acid, 8-hydroxy,lactone or 1-Naphthol-8-sulfonic acid sultone $C_{10}H_6O_3S$	206.22	pr (bz)	>360	154	...	...	bz, chl	$B19^4$, 323
9475	α-Naphthalene sulfonyl chloride $\alpha\text{-}C_{10}H_7SO_2Cl$	226.68	lf (eth)	195^{11}, $147.5^{0.9}$	68	...	...	al, eth, bz	$B11^3$, 383
9476	β-Naphthalene sulfonic acid $\beta\text{-}C_{10}H_7SO_3H$	208.23	dlq pl (+ 1w) cr (+ 3w,dil HCl)	d	124-5 (+ 1w) 83(+ 3 w)	$1.441^{25/4}$	...	w, al, eth	$B11^3$, 397
9477	β-Naphthalene sulfonic acid, 4-amino or Cleve's acid $4\text{-}H_2N\text{-}\beta\text{-}C_{10}H_6SO_3H$	223.25	md (w + 1)	...	...	...	...	...	$B14^3$, 2248
9478	β-Naphthalene sulfonic acid, 5,7-dinitro-8-hydroxy or Flavianic acid $5,7\text{-}(NO_2)_2\text{-}8\text{-}HO\text{-}\beta\text{-}C_{10}H_4SO_3H$	314.23	pa ye nd (con HCl, + 3w) cr (w)	...	100 (+ 3w) 151 (anh)	...	...	w, al	$B11^3$, 542

No.	Name, Synonyms, and Formula	Mol. wt.	Color, crystalline form, specific rotation and λ_{max} (log ε)	b.p. °C	m.p. °C	Density	n_D	Solubility	Ref.
9479	β-Naphthalene sulfonic acid, 1-hydroxy or α-Naphthol sulfonic acid $1\text{-HO-}\beta\text{-}C_{10}H_6SO_3H$	224.23	pl (w)	....	>250	....	....	w, al	B11^{3}, 540
9480	β-Naphthalene sulfonic acid, 6-hydroxy or Schaeffer acid $6\text{-HO-}\beta\text{-}C_{10}H_6SO_3H$	224.23	lf cr (w + 1)	....	167 (anh) 129 (+1w) 118 (+2w)	....	....	w, al, aa	B11^{3}, 553
9481	β-Naphthalene sulfonic acid, 7-hydroxy or F acid $7\text{-HO-}\beta\text{-}C_{10}H_6SO_3H$	224.23	nd (HCl) cr (w + 1,2 or 4)	....	115-6 (anh) 108-9 (+1w)	....	....	w, al	B11^{3}, 555
9482	β-Naphthalene sulfonyl chloride $\beta\text{-}C_{10}H_7SO_2Cl$	226.68	pw or lf (bz-peth)	201^{13}, $148^{0.5}$	79	....	....	al, eth, bz, chl	B11^{3}, 399
9483	1,4,5,8-Naphthalene tetracarboxylic acid $1,4,5,8\text{-}C_{10}H_4(CO_2H)_4$	304.21	lf or nd (w,dil HCl)	....	320	....	....	ace, aa	B9^{3}, 4889
9484	1,4,5,8-Naphthalene tetracarboxylic acid, 1,8: 4,5-dianhydride $C_{14}H_4O_6$	268.18	nd (al)	sub 320^{3}	>300	....	....	....	B19^{4}, 2258
9485	1,2,5-Naphthalene tricarboxylic acid $1,2,5\text{-}C_{10}H_5(CO_2)_3$	260.20	nd (MeOH or sub)	....	270-2	....	....	w, MeOH	B9^{4}, 4835
9486	1,4,5-Naphthalene tricarboxylic acid $1,4,5\text{-}C_{10}H_5(CO_2)_3$	260.20	cr (eth or con HCl)	....	266-8	....	....	al, eth	B9^{3}, 4835
9487	Naphthalic acid or 1,8-Naphthalene dicarboxylic acid $1,8\text{-}C_{10}H_6(CO_2H)_2$	216.19	nd (al)	....	260	....	....	....	B9^{3}, 4466
9488	Naphthalic anhydride $1,8\text{-}C_{10}H_6(CO)_2O$	198.18	nd (al) pr (aa) lf (sub)	sub	274	....	....	aa	B17^{4}, 6392
9489	Naphthalic acid, diethyl ester or Diethyl naphthalate $1,8\text{-}C_{10}H_6(CO_2C_2H_5)_2$	272.30	yesh mcl or lf (dil al)	$238\text{-}9^{19}$	59-60	$1.1399^{70/4}$	1.5586^{70}	al, eth	B9^{3}, 4467
9490	Naphthalic acid, dimethyl ester or Dimethyl naphthalate $1,8\text{-}C_{10}H_6(CO_2CH_3)_2$	244.25	nd (al) pr MeOH)	....	104	....	....	al, aa, MeOH	B9^{3}, 4467
9491	Naphthalic acid, 3,6-dinitro or 3,6-Dinitronaphthalic acid $3,6\text{-}(NO_2)_2\text{-}1,8\text{-}C_{10}H_4(CO_2H)_2$	306.19	silvery lf (w)	....	212	....	....	al, aa	B9^{3}, 4470
9492	Naphthalic acid, imide $1,8\text{-}C_{10}H_6(CO)_2NH$	197.19	nd (chl-al)	....	300	....	....	....	B21^{4}, 5557
9493	Naphthalic anhydride, 3-nitro $3\text{-}NO_2\text{-}1,8\text{-}C_{10}H_5(CO)_2O$	243.18	yesh nd (aa)	....	252-3	....	....	aa	B17^{4}, 6396
9494	Naphthalic acid, 4-nitro $4\text{-}NO_2\text{-}1,8\text{-}C_{10}H_5(CO_2H)_2$	261.19	ye nd	....	d140-50	....	....	aa	B9^{2}, 653
9495	Naphthaloyl chloride $1,8\text{-}C_{10}H_6(COCl)_2$	253.08	pr (CS_2)	$195\text{-}200^{0.2}$	84-6	....	....	bz, chl	B9^{3}, 4468
9496	1-Naphthamidine $C_{11}H_{10}N_2$	170.21	lf (dil al)	....	154	....	....	al, ace, chl	B9^{3}, 3147
9497	2-Naphthamidine $C_{11}H_{10}N_2$	170.21	cr (bz)	....	133-6	....	....	al, bz, aa	B9^{3}, 3187
9498	α-Naphthamide $\alpha\text{-}C_{10}H_7CONH_2$	171.20	nd or pl (al,gl aa)	sub	204-5	....	....	aa	B9^{3}, 3145
9499	Naphthocaine, hydrochloride $C_{17}H_{23}N_2O_2Cl$	322.83	pa ye	....	212-6	....	....	....	B14^{3}, 1333
9500	α-Naphthoic acid or α-Naphthalene carboxylic acid $\alpha\text{-}C_{10}H_7CO_2H$	172.18	nd (aa-w,w,al)	>300, 231^{50}	161	1.398	....	al, eth, chl	B9^{3}, 3136
9501	α-Naphthoic acid-2-amino $2\text{-}H_2N\text{-}\alpha\text{-}C_{10}H_6CO_2H$	187.20	nd (dil al)	....	126d	....	....	al, eth, aa	B14^{3}, 1330
9502	α-Naphthoic acid-3-amino $3\text{-}H_2N\text{-}\alpha\text{-}C_{10}H_6CO_2H$	187.20	ye or pksh nd (eth)	....	181-2	....	....	al, MeOH	B14^{3}, 1330
9503	α-Naphthoic acid-4,amino $4\text{-}H_2N\text{-}\alpha\text{-}C_{10}H_6CO_2H$	187.20	brsh nd (w,al)	....	177	....	....	al, eth, ace, aa	B14^{3}, 1332
9504	α-Naphthoic acid-5-amino $5\text{-}H_2N\text{-}\alpha\text{-}C_{10}H_6CO_2H$	187.20	og nd (w, dil al) (sub) nd	211-2, 196 (sub)	......	....	....	al, aa	B14, 553
9505	α-Naphthoic acid-6-amino $6\text{-}H_2N\text{-}\alpha\text{-}C_{10}H_6CO_2H$	187.20	pa ye nd (al, w)	....	205-6	....	....	eth, ace, aa	B14^{3}, 1340
9506	α-Naphthoic aicd-7-amino $7\text{-}H_2N\text{-}\alpha\ C_{10}H_6CO_2H$	187.20	pa br pr (al)	....	223-4	....	....	al, eth, ace, aa	B14^{3}, 1340

PHYSICAL CONSTANTS OF ORGANIC COMPOUNDS (Continued)

No.	Name, Synonyms, and Formula	Mol. wt.	Color, crystalline form, specific rotation and λ_{max} (log ε)	b.p. °C	m.p. °C	Density	n_D	Solubility	Ref.
9507	α-Naphthoic anhydride $(\alpha\text{-}C_{10}H_7CO)_2O$	326.35	pr (bz)	...	145-6	...	...	eth, bz	$B9^{2}$, 450
9508	α-Naphthoic acid-4-bromo $4\text{-Br-}\alpha\text{-}C_{10}H_6CO_2H$	251.08	nd (aa, dil al, xyl)	...	220	...	...	al, eth, ace, bz	$B9^{3}$, 3153
9509	α-Naphthoic acid-5-bromo $5\text{-Br-}\alpha\text{-}C_{10}H_6CO_2H$	251.08	nd (aa, al)	sub	261	...	...	bz	$B9^{3}$, 3153
9510	α-Naphthoic acid-8-Bromo $8\text{-Br-}\alpha\text{-}C_{10}H_6CO_2H$	251.08	pr (w, bz)	...	178	...	...	al, eth, bz, aa	$B9^{3}$, 3154
9511	α-Naphthoic acid-2-chloro $2\text{-Cl-}\alpha\text{-}C_{10}H_6CO_2H$	206.63	cr (w, bz)	...	153	...	...	al, eth	$B9^{3}$, 3148
9512	α-Naphthoic acid-4-chloro $4\text{-Cl-}\alpha\text{-}C_{10}H_6CO_2H$	206.63	nd (al)	...	210 (223)	...	...	al, aa	$B9^{3}$, 3148
9513	α-Naphthoic acid-5-chloro $5\text{-Cl-}\alpha\text{-}C_{10}H_6CO_2H$	206.63	nd (dil al)	sub	245	...	...	al	$B9^{3}$, 3149
9514	α-Naphthoic acid-8-chloro $8\text{-Cl-}\alpha\text{-}C_{10}H_6CO_2H$	206.63	pl (al, w, bz)	sub	171-2	...	...	ace, aa	$B9^{3}$, 3151
9515	α-Naphthoic acid-1,2-dihydro $\alpha\text{-}C_{10}H_9CO_2H$	174.20	cr (50% al)	...	138	...	...	w	$B9^{3}$, 3076
9516	α-Naphthoic acid-1,4-dihydro $\alpha\text{-}C_{10}H_9CO_2H$	174.20	nd or pl (lig)	...	91	...	...	al, eth, ace	$B9^{3}$, 3075
9517	α-Naphthoic acid-3,4-dihydro $\alpha\text{-}C_{10}H_9CO_2H$	174.20	nd (w, AcOEt) cr (peth)	$305\text{-}6^{748}$	125	...	...	al, chl	$B9^{3}$, 3076
9518	α-Naphthoic acid-4,5-dinitro $4,5\text{-}(NO_2)_2\text{-}\alpha\text{-}C_{10}H_5CO_2H$	262.18	yesh nd or pl (al)	...	267 sub	...	...	al, eth	$B9^{3}$, 3168
9519	α-Naphthoic acid-ethyl ester or Ethyl α-naphthoate $\alpha\text{-}C_{10}H_7CO_2C_2H_5$	200.24	...	310, $183\text{-}6^{70}$	...	$1.1274^{15/15}$	1.5966^{15}	al	$B9^{3}$, 3139
9520	α-Naphthoic acid-5-hydroxy $5\text{-HO-}\alpha\text{-}C_{10}H_6CO_2H$	188.18	pr or nd (w) cr (bz)	...	236	...	...	al, eth, chl, aa	$B10^{3}$, 1070
9521	α-Naphthoic acid-6-hydroxy $6\text{-HO-}\alpha\text{-}C_{10}H_6CO_2H$	188.18	nd or pr (w)	...	212-3	...	...	al, eth, ace	$B10^{3}$, 1070
9522	α-Naphthoic acid-7-hydroxy $7\text{-HO-}\alpha\text{-}C_{10}H_6CO_2H$	188.18	nd (w, al aa)	...	256	...	...	al, aa	$B10^{3}$, 1072
9523	α-Naphthoic acid-8-iodo $8\text{-I-}\alpha\text{-}C_{10}H_6CO_2H$	298.08	br pr (w)	...	164-5	...	...	al, eth, bz, aa	$B9^{3}$, 3157
9524	α-Naphthoic acid-methyl ester or Methyl α-naphthoate $\alpha\text{-}C_{10}H_7CO_2CH_3$	186.21	...	$167\text{-}9^{20}$, $100\text{-}2^{0.04}$	59.5	$1.1290^{20/4}$	1.6086^{20}	al, bz	$B9^{3}$, 3138
9525	α-Naphthoic acid-3-nitro $3\text{-}NO_2\text{-}\alpha\text{-}C_{10}H_6CO_2H$	217.18	cr (al)	...	271-5	...	...	al	$B9^{3}$, 3158
9526	α-Naphthoic acid, 4-nitro $4\text{-}NO_2\text{-}\alpha\text{-}C_{10}H_6CO_2H$	217.18	yesh nd (al,aa)	...	225-6	...	...	al, chl, aa	$B9^{3}$, 3160
9527	α-Naphthoic acid, 5-nitro $5\text{-}NO_2\text{-}\alpha\text{-}C_{10}H_6CO_2H$	217.18	yesh nd (al)	sub	241-2	...	...	aa	$B9^{3}$, 3162
9528	α-Naphthoic acid, 8-nitro $8\text{-}NO_2\text{-}\alpha\text{-}C_{10}H_6CO_2H$	217.18	nd (w) pr (al)	...	217	...	...	al	$B9^{3}$, 3164
9529	α-Naphthoic acid, 1,2,3,4-tetrahydro $\alpha\text{-}C_{10}H_{11}CO_2H$	176.22	tcl pr (AcOEt)	...	85	...	...	al, eth, ace, bz	$B9^{3}$, 2801
9530	α-Naphthoic acid, 5,6,7,8-tetrahydro $\alpha\text{-}C_{10}H_{11}CO_2H$	176.22	pr (w) wh nd (dil aa)	...	150	...	...	al, bz, chl	$B9^{3}$, 2794
9531	α-Naphthonitrile $\alpha\text{-}C_{10}H_7CN$	153.18	nd (lig)	299, 148^{12}	37.5	$1.1113^{25/25}$	1.6298^{18}	al, eth, lig	$B9^{3}$, 3146
9532	α-Naphthoyl chloride $\alpha\text{-}C_{10}H_7COCl$	190.63	...	297.5, 172^{15}	20	...	...	...	$B9^{3}$, 3145
9533	β-Naphthamide $\beta\text{-}C_{10}H_7CONH_2$	171.20	lf (al)	...	195	...	...	al, eth, ace, chl, lig	$B9^{2}$, 454
9534	β-Naphthoic acid or 2-Naphthalene carboxylic acid $\beta\text{-}C_{10}H_7CO_2H$	172.20	nd (lig, chl, sub) pl (ace)	>300	185.5	$1.077^{100/4}$	...	al, eth, chl	$B9^{3}$, 3174
9535	β-Naphthoic acid, 4-acetyl-3-hydroxy $4\text{-}CH_3CO\text{-}3\text{-HO-}\beta\text{-}C_{10}H_5CO_2H$	230.22	ye pr (aa)	...	194	...	...	al	$B10^{3}$, 4408
9536	β-Naphthoic acid, 1-amino $1\text{-}H_2N\text{-}\beta\text{-}C_{10}H_6CO_2H$	187.20	nd (dil al,aa)	...	205 (rapid htq)	...	...	al, eth, bz	$B14^{3}$, 1340
9537	β-Naphthoic acid, 3-amino $3\text{-}H_2N\text{-}\beta\text{-}C_{10}H_6CO_2H$	187.20	ye lf (dil al)	...	216-7	...	...	al, eth	$B14^{3}$, 1341
9538	β-Naphthoic acid, 4-amino $4\text{-}H_2N\text{-}\beta\text{-}C_{10}H_6CO_2H$	187.20	nd (dil al)	...	215-6	...	...	al, eth, ace, bz	$B14^{3}$, 1345

No.	Name, Synonyms, and Formula	Mol. wt.	Color, crystalline form, specific rotation and λ_{max} (log ε)	b.p. °C	m.p. °C	Density	n_D	Solubility	Ref.
9539	β-Naphthoic acid, 5-amino 5-H_2N-β-$C_{10}H_6CO_2H$	187.20	ye lf (al)		291-2			al, eth, ace, bz	B14[3], 1345
9540	β-Naphthoic acid, 6-amino 6-H_2N-β-$C_{10}H_6CO_2H$	187.20	pa ye nd (w,dil al)		225			al, eth, ace, bz, aa	B14[2], 324
9541	β-Naphthoic acid, 7-amino 7-H_2N-β-$C_{10}H_6CO_2H$	187.20	pa ye nd or lf (al)		243			al, eth, ace, aa	B14[2], 324
9542	β-Naphthoic acid, 8-amino 8-H_2N-β-$C_{10}H_6CO_2H$	187.20	grsh-ye nd (aa)		220			al, eth, ace	B14[2], 324
9543	β-Naphthoic anhydride (β-$C_{10}H_7CO)_2O$	326.35	nd (eth)		135			bz, aa	B9[3], 3184
9544	β-Naphthoic acid, 1-bromo 1-Br-β-$C_{10}H_6CO_2H$	251.08	nd (aa,bz)		191			al, bz, aa	B9[3], 3195
9545	β-Naphthoic acid, 5-bromo 5-Br-β-$C_{10}H_6CO_2H$	251.08	nd (al,sub)	sub	270			al, eth, bz, aa	B9[3], 3196
9546	β-Naphthoic acid, 1-chloro 1-Cl-β-$C_{10}H_6CO_2H$	206.63	nd (bz)		196			al, ace	B9[3], 3191
9547	β-Naphthoic acid, 3-chloro 3-Cl-β-$C_{10}H_6CO_2H$	206.63	cr (dil MeOH)		216.5			al, eth, ace, bz, chl	B9[3], 3192
9548	β-Naphthoic acid, 5-chloro 5-Cl-β-$C_{10}H_6CO_2H$	206.63	nd (al,aa)		270			bz, aa	B9[3], 3192
9549	β-Naphthoic acid, 1,2-dihydro β-$C_{10}H_9CO_2H$	174.20	nd or pr (dil al,w,peth)		105-6			al, chl, aa	B9[3], 3076
9550	β-Naphthoic acid, 1,4-dihydro β-$C_{10}H_9CO_2H$	174.20	pl (bz,w,dil al)		162-3			al, eth	B9[3], 3076
9551	β-Naphthoic acid, 3,4-dihydro β-$C_{10}H_9CO_2H$	174.20	nd (dil aa,dil al)		120			bz, chl, aa	B9[3], 3076
9552	β-Naphthoic acid, 1,3-dihydroxy,ethyl ester 1,3-$(HO)_2$-β-$C_{10}H_5CO_2C_2H_5$	232.24	nd (dil al, dil aa)		83-4			al, eth, lig	B10[3], 1932
9553	β-Naphthoic acid, ethyl ester or Ethyl β-naphthoate β-$C_{10}H_7CO_2C_2H_5$	200.24		308-9, 224^{74}	32	$1.1143^{23/4}$	1.5951^{23}	al, eth, chl, aa	B9[3], 3177
9554	β-Naphthoic acid, 1-hydroxy 1-HO-β-$C_{10}H_6CO_2H$	188.18	cr (dil al,w,aa) nd (al,eth,bz)		195			al, th, bz	B10[3], 1075
9555	β-Naphthoic acid, 1-hydroxy-4-chloro 1-HO-4-Cl-β-$C_{10}H_5CO_2H$	222.63	nd (al,aa)		234			al, ace	B10[3], 1079
9556	β-Naphthoic acid, 1-hydroxy,phenyl ester 1-HO-β-$C_{10}H_6CO_2C_6H_5$	264.28			96			al, bz	B10, 332
9557	β-Naphthoic acid, 3-hydroxy 3-HO-β-$C_{10}H_6CO_2H$	188.18	ye lf (dil al) nd (dil al)		222.3			al, eth, bz, chl	B10[3], 1084
9558	β-Naphthoic acid, 3-hydroxy-4-chloro 4-Cl-3-HO-β-$C_{10}H_5CO_2H$	222.63	ye nd		231d				B10[3], 1091
9559	β-Naphthoic acid, 3-hydroxy,ethyl ester 3-HO-β-$C_{10}H_6CO_2C_2H_5$	216.24	nd or mcl pr (aa)	291	85			ace, chl	B10[3], 1086
9560	β-Naphthoic acid, 3-hydroxy,hexadecyl ester 3-HO-β-$C_{10}H_6CO_2C_{16}H_{33}$	412.61	grsh-wh pr		72-3			bz, lig	B10[3], 1087
9561	β-Naphthoic acid, 3-hydroxy,methyl ester 3-HO-β-$C_{10}H_6CO_2CH_3$	202.21	pa ye rh nd (dil MeOH)	205-7	75-6			al	B10[3], 1086
9562	β-Naphthoic acid, 5-hydroxy 5-HO-β-$C_{10}H_6CO_2H$	188.18	wh nd (w,dil al)		213			al, eth, ace, aa	B10[3], 1098
9563	β-Naphthoic acid, 7-hydroxy 7-HO-β-$C_{10}H_6CO_2H$	188.18	lf or pl (dil al) pa ye nd (al)		269-70			al, eth, ace, aa	B10[3], 1101
9564	β-Naphthoic acid, methyl ester or Methyl β-naphthoate β-$C_{10}H_7CO_2CH_3$	186.21	lf (MeOH)	290, $141\text{-}3^{4}$	77			al, eth, bz, chl	B9[3], 3176
9565	β-Naphthoic acid, 5-nitro 5-NO_2-β-$C_{10}H_6CO_2H$	217.18	yesh nd (al)		295			ace	B9[3], 3203
9566	β-Naphthoic acid, 8-nitro 8-NO_2-β-$C_{10}H_6CO_2H$	217.18	yesh nd (al)	sub	295				B9[2], 455
9567	β-Naphthoic acid, 1,2,3,4-tetra hydro β-$C_{10}H_{11}CO_2H$	176.22	nd (dil al)	$168\text{-}70^{15}$	97			al, eth, bz, chl	B9[3], 2805
9568	β-Naphthoic acid, 5,6,7,8-tetra hydro β-$C_{10}H_{11}CO_2H$	176.22	nd (al) cr (aa,bz)	216^{14}	154			al, bz	B9[3], 2802
9569	β-Naphthonitrile β-$C_{10}H_7CN$	153.18	lf (lig)	306.5, $156\text{-}8^{12}$	66	$1.0939^{60/60}$		al, eth, lig	B9[3], 3186
9570	β-Naphthoyl chloride β-$C_{10}H_7COCl$	190.63	cr (peth)	304-6, $142\text{-}3^{5}$	51			eth, bz, chl aa	B9[3], 3185

PHYSICAL CONSTANTS OF ORGANIC COMPOUNDS (Continued)

No.	Name, Synonyms, and Formula	Mol. wt.	Color, crystalline form, specific rotation and λ_{max} (log ε)	b.p. °C	m.p. °C	Density	n_D	Solubility	Ref.
9571	α-Naphthol or 1-Hydroxy naphthalene α-$C_{10}H_7OH$	144.17	ye mcl nd (w)	288 sub	96	$1.0989^{99/4}$	1.6224^{99}	al, eth, ace, bz, chl	$B6^3$, 2193
9572	α-Naphthol, 2-acetyl 2-CH_3CO-α-$C_{10}H_6OH$	186.21	(i)pr(bz,lig) (ii)gr-ye nd(al)	325d	(i)98 (ii)103	....	....	bz, aa	$B8^3$, 1130
9573	α-naphthol, 2-acetyl-4-bromo 2-CH_3CO-4-Br-α-$C_{10}H_5OH$	265.11	ye nd (al)	....	126-7	....	....	al, eth, bz, chl	$B8^3$, 1134
9574	α-Naphthol, 2-acetyl-4-nitro 2-CH_3CO-4-NO_2-α-$C_{10}H_5OH$	231.21	ye nd (al)	....	159	....	....	eth, bz	$B8^3$, 1135
9575	α-Naphthol, 3-acetyl 3-CH_3CO-α-$C_{10}H_6OH$	186.21	nd (bz)	....	173-4	....	....	al, aa	B8, 150
9576	α-naphthol, 4-acetyl 4-CH_3CO-$C_{10}H_6OH$	186.21	yesh pr (aa,al,to)	....	198	....	....	al, bz, aa	$B8^3$, 1124
9577	α-Naphthol, 2-benzyl 2-$C_6H_5CH_2$-α-$C_{10}H_6OH$	234.30	nd (lig) pr (bz)	$237\text{-}40^{12}$	73-4	....	....	....	$B6^3$, 3608
9578	α-Naphthol, 4-benzyl 4-$C_6H_5CH_2$-α-$C_{10}H_6OH$	234.30	pl or nd (aa-lig) pr (bz)	237^{10}	125-6	....	....	....	$B6^2$, 680
9579	α-Naphthol, 4-bromo 4-Br-α-$C_{10}H_6OH$	223.07	nd (dil al,hex)	....	128	....	....	chl, aa	$B6^3$, 2935
9580	α-Naphthol, 5-bromo 5-Br-α-$C_{10}H_6OH$	223.07	nd (w)	....	137	....	....	eth	$B6^3$, 2936
9581	α-Naphthol, 6-bromo 6-Br-α-$C_{10}H_6OH$	223.07	nd (w)	....	129-30	....	....	w	$B6^3$, 2936
9582	α-Naphthol, 7-bromo 7-Br-α-$C_{10}H_6OH$	223.07	cr (w)	....	105-6	....	....	w	$B6^2$, 583
9583	α-Naphthol, 8-bromo 8-Br-α-$C_{10}H_6OH$	223.07	pl (peth)	....	61	....	....	al	B6, 614
9584	α-Naphthol, 4-methyl 4-CH_3-α-$C_{10}H_6OH$	158.20	....	$165\text{-}7^{13}$	86-7	....	....	al, eth, ace, bz	$B6^3$, 3022
9585	α-Naphthol, 8-nitro or 8-nitro-1-naphthol 8-NO_2-α-$C_{10}H_6OH$	189.17	grsh ye nd (al,chl,bz-hx)	....	130-3	....	....	al, eth, ace, bz	$B6^3$, 2939
9586	α-Napthol, 2-propionyl 2-CH_3CH_2CO-α-$C_{10}H_6OH$	200.24	grsh ye lf or pl (al)	....	81	....	....	al, eth	$B8^3$, 1144
9587	α-Naphthol, 2,3,4-trichloro 2,3,4-Cl_3-α-$C_{10}H_4OH$	247.51	nd (aa or lig)	....	168	....	....	eth, al	$B6^2$, 582
9588	β-Naphthol or 2-Hydroxy naphthalene β-$C_{10}H_7OH$	144.17	mcl lf (w)	295	123-4	1.28^{20}	....	al, eth, bz, chl	$B6^3$, 2955
9589	β-Naphthol, 1-acetoamido 1-CH_3CONH-β-$C_{10}H_6OH$	201.22	lf (w,dil al)	sub	235d	....	....	al, eth, ace, bz	$B13^2$, 414
9590	β-Naphthol, acetate or 2-Acetoxy naphthalene β-$C_{10}H_7(O_2CCH_3)$	186.21	nd (al)	$132\text{-}4^2$	71-2	....	....	al, eth, chl	$B6^3$, 2982
9591	β-Naphthol, 1-acetyl 1-CH_3CO-β-$C_{10}H_6OH$	186.21	pa ye lf (peth) rh (lig)	....	64-5	....	....	al, eth, bz	$B8^3$, 1122
9592	β-Naphthol, 3-acetyl 3-CH_3CO-β-$C_{10}H_6OH$	186.21	ye lf or nd (al,peth)	....	112	....	....	ace, bz	$B8^3$, 1135
9593	β-Napthol, 1-amino 1-H_2N-β-$C_{10}H_6OH$	159.19	silvery lf (bz, eth)	....	150d	....	....	al	$B13^3$, 1891
9594	β-Naphthol, 3-amino 3-H_2N-β-$C_{10}H_6OH$	159.19	silvery lf (bz) nd (al,w)	....	235	....	....	al	$B13^3$, 1897
9595	β-Naphthol, 5-amino 5-H_2N-β-$C_{10}H_6OH$	159.19	nd or og pr (w)	....	190.6	....	....	al, eth, ace, w	$B13^3$, 1901
9596	β-Naphthol, 6-amino 6-H_2N-β-$C_{10}H_6OH$	159.19	pr (w)	....	192-4, 212d	....	....	al, w	$B13^3$, 1902
9597	β-Naphthol, 7-amino 7-H_2N-β-$C_{10}H_6OH$	159.19	nd or lf (al)	....	201 (208)	....	....	al, eth	$B13^2$, 416
9598	β-Naphthol, 8-amino 8-H_2N-β-$C_{10}H_6OH$	159.19	nd (w,al)	sub	205-7	....	....	eth	$B13^3$, 1907
9599	β-Naphthol, benzoate or 2-Benzoyloxy naphthalene β-$C_{10}H_7(O_2CC_6H_5)$	248.28	nd or pr (al) cr (lig)	....	108	....	....	al, eth	$B9^3$, 492
9600	β-Naphthol, 1-benzyl 1-$C_6H_5CH_2$-β-$C_{10}H_6OH$	234.30	nd (bz)	$247\text{-}50^{13}$	115	....	....	al, eth, ace, bz, chl	$B6^3$, 3607
9601	β-Naphthol, 1-bromo 1-Br-β-$C_{10}H_6OH$	223.07	rh pr(bz-lig) nd(aa lig)	d130	84	....	....	al, eth, bz, aa, lig	$B6^3$, 2994

No.	Name, Synonyms, and Formula	Mol. wt.	Color, crystalline form, specific rotation and λ_{max} (log ε)	b.p. °C	m.p. °C	Density	n_D	Solubility	Ref.
9602	β-Naphthol, 3-bromo 3-Br-β-$C_{10}H_6OH$	223.07	nd(lig)		84-5			al, bz	B6^3, 2995
9603	β-Naphthol, 5-Bromo 5-Br-β-$C_{10}H_6OH$	223.07	nd(w)		105			al	B6^3, 2996
9604	β-Naphthol, 6-bromo 6-Br-β-$C_{10}H_6OH$	223.07	nd(bz)		127			al, bz	B6^3, 3020
9605	β-Naphthol, 6-bromo-1-methyl 6-Br-1-CH_3-β-$C_{10}H_5OH$	262.31	nd(bz)		129			al, eth, ace, bz, chl	B6^3, 3020
9606	β-Naphthol, 7-bromo 7-Br-β-$C_{10}H_6OH$	223.07	cr(peth)		132-3			eth, ace	B6^2, 605
9607	β-Naphthol, 1-chloro 1-Cl-β-$C_{10}H_6OH$	178.62	nd(lig) pr(chl) pl(w)		71			al, bz, chl	B6^3, 2990
9608	β-Naphthol, 1,6-dibromo 1,6-Br_2-β-$C_{10}H_5OH$	301.97	nd(peth bz) lf(al)		106			al, eth, ace	B6^3, 2998
9609	β-Naphthol, 1,6-dinitro 1,6-$(NO_2)_2$-β-$C_{10}H_5OH$	234.17	pa ye nd(chl)		195d			al, eth, chl, py	B6^3, 3005
9610	β-Naphthol, 1-methyl 1-CH_3-β-$C_{10}H_6OH$	158.20	nd(w bz-lig dil aa)	180^{12} sub	112			al, eth, ace, bz, aa	B6^3, 3019
9611	β-Naphthol, 1-nitro or 1-nitro-2-naphthol 1-NO_2-β-$C_{10}H_6OH$	189.17	ye nd lf or pr(al)	$115^{0.05}$	104			al, eth	B6^3, 3002
9612	β-Naphthol, 5-nitro 5-NO_2-β-$C_{10}H_6OH$	189.17	lt ye nd(w)		147-9			eth, ace	B6^3, 3004
9613	β-Naphthol, 1,3,6-tribromo 1,3,6-Br_3-β-$C_{10}H_4OH$	380.86	nd(aa al)		133			al, bz	B6^3, 3000
9614	β-Naphthol, 1,4,6-tribromo 1,4,6-Br_3-β-$C_{10}H_4OH$	380.86	nd(bz)		157-8			al, bz, chl, aa	B6^3, 3000
9615	β-Naphthol, 3,4,6-tribromo 3,4,6-Br_3-β-$C_{10}H_4OH$	380.86	nd(bz)		127-8			al, bz	B6^2, 607
9616	β-Naphthol, 1,3,4-trichloro 1,3,4-Cl_3-β-$C_{10}H_4OH$	247.51	nd		162			al, aa	B6^2, 604
9618	1,2-Naphthoquinone $C_{10}H_6O_2$	158.16	ye-red nd (eth) og lf (bz)		146	1.450		w, a;, eth	B7^3, 3686
9619	1,2-Naphthoquinone, 3-bromo $C_{10}H_5BrO_2$	237.05	red nd or pl(aa al)	sub	178			al, bz	B7^3, 3692
9620	1,2-Naphthoquinone, 4-bromo $C_{10}H_5BrO_2$	237.05	red nd(bz-lig)		154			al, bz, aa	B7^3, 3691
9621	1,2-Naphthoquinone, 6-bromo $C_{10}H_5BrO_2$	237.05	og-red or ye pr or pl (bz) nd (w)		168d			al, ace	B7^3, 3691
9622	1,2-Naphthoquinone, 3-chloro $C_{10}H_5ClO_2$	192.60	red nd(al aa bz chl)		172d			bz	B7^3, 3690
9623	1,2-Naphthoquinone, 4-chloro $C_{10}H_5O_2Cl$	192.60			134-6			al	B7^3, 3690
9624	1,2-Naphthoquinone, 3,4-dibromo $C_{10}H_4Br_2O_2$	315.95	red lf or pl(aa bz)		172-4			bz	B7^3, 3693
9625	1,2-Naphthoquinone, 3,6-dibromo $C_{10}H_4Br_2O_2$	315.95	red pr(AcOEt)		176			al	B7^3, 3692
9626	1,2-Naphthoquinone, 4,6-dibromo $C_{10}H_4Br_2O_2$	315.95	og-red pr (bz peth) nd (AcOEt)		153			bz, aa	B7^3, 3692
9627	1,2-Naphthoquinone, 3,4-dichloro $C_{10}H_4Cl_2O_2$	227.05	red lf or pl(aa bz) nd(bz chl)	sub	184			bz, chl	B7^3, 3691
9628	1,2-Naphthoquinone, dioxime $C_{10}H_8N_2O_2$	188.19	ye nd(bz-lig dil al)		169			al, bz, diox	B7^3, 3690
9629	1,2-Naphthoquinone, 6-hydroxy $C_{10}H_6O_3$	174.16	red lf(ace)		165d			al, eth, ace, aa	B8^3, 2542
9631	1,2-Naphthoquinone, 7-hydroxy $C_{10}H_6O_3$	174.16	br nd		194			al	B8^3, 2542
9632	1,2-Naphthoquinone, 3-methyl $C_{11}H_8O_2$	172.18	red or og nd(abs al) lf(bz peth)		116 (122)			al	B8^3, 3709
9633	1,2-Naphthoquinone, 4-methyl $C_{11}H_8O_2$	172.18	og nd (MeOH) nd (aa)		248-50d			al	B7^3, 3708

No.	Name, Synonyms, and Formula	Mol. wt.	Color, crystalline form, specific rotation and λ_{max} (log ε)	b.p. °C	m.p. °C	Density	n_D	Solubility	Ref.
9634	1,2-Naphthoquinone, 3-nitro $C_{10}H_5NO_4$	203.15	red pl(aa)	....	158	....	....	bz, aa	B7^{2}, 651
9635	1,2-Naphthoquinone, 1-oxime $C_{10}H_7NO_2$	173.17	ye nd(bz) og pr or pl(al)	....	112	....	....	al, eth, ace, bz	B7^{1}, 3688
9636	1,4-Naphthoquinone $C_{10}H_6O_2$	158.16	bt ye nd(al peth) ye(sub)	sub	128.5	....	....	al, eth, bz, aa	B7^{1}, 3696
9637	1,4-Naphthoquinone, 2-amino $C_{10}H_7NO_2$	173.17	....	....	207	....	....	al, eth	B14^{1}, 388
9638	1,4-Naphthoquinone, 4-anil,2-anilino $C_{22}H_{16}N_2O$	324.38	ye red nd(bz al)	....	182-3	....	....	aa, bz	B14^{1}, 390
9639	1,4-Naphthoquinone, 2-bromo $C_{10}H_5BrO_2$	237.05	ye pl or nd (al, dil aa)	....	132	....	....	aa, bz, chl, aa	B7^{1}, 3705
9640	1,4-Naphthoquinone, 2-bromo-3-methyl $C_{11}H_7BrO_2$	251.08	ye-br nd(al)	sub 100	151	....	....	al, eth, ace, bz, chl	B7^{1}, 3714
9641	1,4-Naphthoquinone, 5-bromo-2,3-dichloro $C_{10}H_3BrCl_2O_2$	305.94	ye pr(al)	....	180	....	....	ace, bz	B7^{2}, 655
9642	1,4-Naphthoquinone, 6-bromo $C_{10}H_5BrO_2$	237.05	og-red or ye pr(bz AcOEt) nd(w)	....	168d	....	....	al, ace	B7, 722
9643	1,4-Naphthoquinone, 2-chloro $C_{10}H_5ClO_2$	192.60	ye nd(w al aa)	....	117-8	....	....	al, ace, bz	B7^{1}, 3702
9644	1,4-Naphthoquinone, 5-chloro $C_{10}H_5ClO_2$	192.60	ye nd(lig)	sub	163	....	....	al, aa, lig	B7^{3}, 3701
9645	1,4-Naphthoquinone, 6-chloro $C_{10}H_5ClO_2$	192.60	red-br(al eth) ye cr(dil MeOH)	....	109-10	....	....	bz	B7^{1}, 3702
9646	1,4-Naphthoquinone, 2,3-dibromo $C_{10}H_4Br_2O_2$	315.95	ye nd(aa)	....	218	....	....	aa	B7^{1}, 3705
9647	1,4-Naphthoquinone, 5,8-dibromo $C_{10}H_4Br_2O_2$	315.95	ye nd(al)	....	171-3	....	....	....	B7, 732
9648	1,4-Naphthoquinone, 2,3-dichloro $C_{10}H_4Cl_2O_2$	227.05	ye nd(al)	....	195	....	....	chl	B7^{1}, 3703
9649	1,4-Naphthoquinone, 2,6-dichloro $C_{10}H_4Cl_2O_2$	227.05	dk ye nd(al)	....	148-9	....	....	al	B7, 730
9650	1,4-Naphthoquinone, 5,6-dichloro $C_{10}H_4Cl_2O_2$	227.05	ye nd(al)	sub	181	....	....	eth	B7, 730
9651	1,4-Naphthoquinone, 5,8-dichloro $C_{10}H_4Cl_2O_2$	227.05	ye nd(al)	sub	173-4	....	....	eth	B7^{1}, 3702
9651a	1,4-Naphthoquinone, 2,3-dihydro $C_{10}H_8O_2$	160.17	lf (hx) nd (peth)	....	98-9	....	....	al	E12B, 2806
9652	1,4-Naphthoquinone, 2,3-dihydroxy or Isonaphthazarin $C_{10}H_6O_4$	190.16	red og nd or lf (sub)	sub	282	....	....	ace	B8^{1}, 3596
9653	1,4-Naphthoquinone, 3,5-dihydroxy-2-methyl or Proserone $C_{11}H_8O_4$	204.18	og-ye nd(al aa)	sub 100^{3}	181	....	....	al, eth, peth	B8^{3}, 3604
9654	1,4-Naphthoquinone, 5,8-dihydroxy or Naphthazarin $C_{10}H_6O_4$	190.16	dk red mcl pr (bz) red-br nd (al)	sub	276-80	....	....	aa	B8^{1}, 3600
9655	1,4-Naphthoquinone, 5,8-dihydroxy-2-methyl $C_{11}H_8O_4$	204.18	gr pl	....	173	....	....	al	B8^{3}, 3605
9656	1,4-Naphthoquinone, 2,3-dimethyl $C_{12}H_{10}O_2$	186.21	ye pr(al)	....	127	....	....	bz, aa	B7^{3}, 3717
9657	1,4-Naphthoquinone, 2,5-dimethyl $C_{12}H_{10}O_2$	186.21	ye nd(peth eth)	....	95	....	....	diox	B7^{3}, 3716
9658	1,4-Naphthoquinone, 2,6-dimethyl $C_{12}H_{10}O_2$	186.21	ye pr or nd (AcOEt)	....	136-7	....	....	al, eth, bz	B7^{3}, 3720
9659	1,4-Naphthoquinone, 2,8-dimethyl $C_{12}H_{10}O_2$	186.21	pr (peth) nd (MeOH)	....	135-6	....	....	al	B7^{3}, 3716
9660	1,4-Naphthoquinone, dioxime $C_{10}H_8N_2O_2$	188.19	nd(dil al)	....	207d	....	....	w, al	B7^{2}, 653
9661	1,4-Naphthoquinone, 2-ethyl $C_{12}H_{10}O_2$	186.21	pr(al) nd aa peth MeOH	....	88-9	....	....	al, aa	B7^{3}, 3715

No.	Name, Synonyms, and Formula	Mol. wt.	Color, crystalline form, specific rotation and λ_{max} (log ε)	b.p. °C	m.p. °C	Density	n_D	Solubility	Ref.
9662	1,4-Naphthoquinone, 2-ethyl-3-hydroxy $C_{12}H_{10}O_3$	202.21	ye nd(dil MeOH)		141			eth, ace	B8³, 2586
9663	1,4-Naphthoquinone, 2-ethyl-3,5,6,7,8-pentahydroxy or Echinochrome A $C_{12}H_{10}O_7$	266.21	red nd(diox-w)	sub 120¹⁰	220d			w, al, eth, ace, bz	B8³, 4360
9664	1,4-Naphthoquinone, 2-hydroxy or Lawsone $C_{10}H_6O_3$	174.16	redsh-br(aa)		192d			al, aa	B8³, 2543
9665	1,4-Naphthoquinone, 2-hydroxy,acetate $C_{12}H_8O_4$	216.19	ye lf(al)		131			al, eth, chl, aa	B8³, 2547
9666	1,4-Naphthoquinone, 2-hydroxy-3-methyl or Phthiocol $C_{11}H_8O_3$	188.18	ye pr(eth-peth)	sub	173-4			eth, ace	B8³, 2569
9667	1,4-Naphthoquinone, 2-hydroxy-3-phenyl $C_{16}H_{10}O_3$	250.25	gold-ye or og pr or nd(al bz MeOH)		147			al, eth, bz, chl, lig	B8³, 2981
9668	1,4-Naphthoquinone, 3-hydroxy-2-bromo $C_{10}H_5BrO_3$	253.05	ye mcl pr (al) nd (al, w)	sub	202			ace	B8³, 2552
9669	1,4-Naphthoquinone, 3-hydroxy-2-chloro $C_{10}H_5ClO_3$	208.60	ye nd(al aa)	sub	213			al, eth, bz	B8², 347
9670	1,4-Naphthoquinone, 5-hydroxy or Juglon, Nucin $C_{10}H_6O_3$	174.16	redsh-ye nd or pr(chl bz)	sub	154 (161)			al, eth, bz, chl, aa	B8³, 2558
9671	1,4-Naphthoquinone, 5-hydroxy-2-methyl or Plumbagin $C_{11}H_8O_3$	188.18	gold pr or og-ye nd(dil al)	sub	78-9			al, eth, ace, bz, chl	B8³, 2576
9672	1,4-Naphthoguinone, 6-hydroxy $C_{10}H_6O_3$	174.16	gold-ye or red ye nd(w bz al)		170d			al, eth, ace, MeOH	B8², 348
9673	1,4-Naphthoquinone, 2-methyl or Menadione, Vitamin K_3 $C_{11}H_8O_2$	172.18	ye nd(al peth)		107			ace, bz	B7³, 3709
9674	1,4-Naphthoquinone, 2-phenyl $C_{16}H_{10}O_2$	234.25	gold-ye nd(al)		111			al, eth, bz, chl	B7³, 4209
9675	1,4-Naphthoquinone, 2-phenylamino or Lawsone anilide $C_{16}H_{11}NO_2$	249.27	red nd(dil al)	sub	193			eth, bz	B14³, 389
9676	1,4-Naphthoquinone, 5,6,7,8-tetrahydro $C_{10}H_{10}O_2$	162.19	gold-ye nd(peth)		55-6			al, eth	B7³, 3506
9677	1,4-Naphthoquinone, 2,5,8-trihydroxy $C_{10}H_6O_5$	206.15	red nd(bz MeOH)		195			al, aa	B8³, 4043
9678	1,4-Naphthoquinone, 2-oxime $C_{10}H_7NO_2$	173.17	pa ye nd(bz) nd(dil al)		198			al, eth, ace	B7³, 3700
9679	2,6-Naphthoquinone $C_{10}H_6O_2$	158.16	ye-red pr(bz bz-peth)		135d			al, MeOH	B7³, 3706
9680	2,6-Naphthoquinone, 1,5-dichloro $C_{10}H_4Cl_2O_2$	227.05	og pr(chl) gold-ye nd(al)		206d			ace, bz, chl, aa	B7³, 3706
9681	α-Naphthoxy acetic acid α-$C_{10}H_7OCH_2CO_2H$	202.21	pr		190			al, eth	B6³, 2930
9682	β-Naphthoxy acetic acid β-$C_{10}H_7OCH_2CO_2H$	202.21	pr(w)		156			al, eth, aa	B6³, 2985
9683	α-Naphthy acetamide α-$C_{10}H_7CH_2CONH_2$	185.23	nd(w al)		180-1 sub			eth, bz, aa	B9³, 3208
9684	α-Naphthyl acetic acid α-$C_{10}H_7CH_2CO_2H$	186.21	wh nd(w)	d	133			eth, ace, bz, chl, aa	B9³, 3206
9685	α-Naphthyl acetonitrile α-$C_{10}H_7CH_2CN$	167.21	wx	162-4¹²	32-3		1.6192²⁰		B9³, 3209
9686	β-Naphthyl acetamide β-$C_{10}H_7CH_2CONH_2$	185.23	lf(w)		202-4d			al, eth	B9³, 3212
9687	β-Naphthyl acetonitrile β-$C_{10}H_7CH_2CN$	167.21	nd or lf(dil al)	145.50²	85-6			eth, bz, chl	B9³, 3212
9688	α-Naphthyl amine α-$H_2NC_{10}H_7$	143.19	nd(dil al eth)	300.8, 160¹² sub	50	1.1229²⁵/²⁵	1.6703⁵¹	al, eth	B12³, 2846
9689	α-Naphthyl amine, 2-acetyl,hydrochloride 2-CH_3CO-α-$C_{10}H_6NH_2.HCl$	221.69	cr(w)		220d			w, aa	B14³, 208
9690	α-Naphthyl amine, N-acetyl α-$C_{10}H_7NHCOCH_3$	185.23	cr(al)		160				B12³, 2866

No.	Name, Synonyms, and Formula	Mol. wt.	Color, crystalline form, specific rotation and λ_{max} (log ε)	b.p. °C	m.p. °C	Density	n_D	Solubility	Ref.
9691	α-Naphthyl amine, N-acetyl-2-nitro 2-O_2N-α-$C_{10}H_6NHCOCH_3$	230.22	lt ye nd(aa al)	....	200	....	....	al, aa	B12^3, 2969
9692	α-Naphthyl amine, N-acetyl-4-nitro 4-O_2N-α-$C_{10}H_6NHCOCH_3$	230.22	pa ye nd(ace)	....	192-3	....	....	al, ace	B12^3, 2972
9693	α-Naphthyl amine, N-acetyl-5-nitro 5-O_2N-α-$C_{10}H_6NHCOCH_3$	230.22	br pr(aa) ye cr(al)	....	220	....	....	al	B12^3, 2973
9694	α-Naphthyl amine, N-acetyl-8-nitro 8-NO_2-α-$C_{10}H_6NHCOCH_3$	230.22	nd(w)	....	191	....	....	....	B12^3, 2976
9695	α-Naphthyl amine, N-2-aminoethyl α-$C_{10}H_7(NHCH_2CH_2NH_2)$	186.26	ye	320d, 204^{9}	......	$1.114^{25/4}$	1.6648^{25}	al, ace	B12^3, 2955
9696	α-Naphthyl amine, N-benzylidene α-$C_{10}H_7N{=}CHC_6H_5$	231.30	ye lf(al)	....	73.5	....	....	al, eth, bz, MeOH	B12^3, 2861
9697	α-Naphthyl amine, 4-bromo 4Br-α-$C_{10}H_6NH_2$	222.08	nd(al bz peth)	....	102	....	....	al, ace, bz, lig	B12^3, 2964
9698	α-Naphthyl amine, 4-bromo-2-nitro 4-Br-2-NO_2-α-$C_{10}H_5NH_2$	267.08	og cr(al aa)	....	200	....	....	al, bz, chl	B12^3, 2979
9699	α-Naphthyl amine, 5-bromo 5-Br-α-$C_{10}H_6NH_2$	222.08	lf or pl(w lig)	sub	69	....	....	al, eth, ace, bz, lig	B12^3, 2966
9700	α-Naphthyl amine, 2-chloro 2-Cl-α-$C_{10}H_6NH_2$	177.63	nd(peth dil al)	....	60	....	....	al, ace	B12^3, 2960
9701	α-Naphthyl amine, 4-chloro 4-Cl-α-$C_{10}H_6NH_2$	177.63	nd(al bz lig)	....	99-100	....	....	al, eth, bz	B12^3, 2961
9702	α-Naphthyl amine, 2,4-dibromo 2,4-Br_2-α-$C_{10}H_5NH_2$	300.98	nd or pl(dil al)	....	118-9	....	....	al, eth, bz, chl, lig	B12^3, 2966
9703	α-Naphthyl amine, 2,4-dichloro 2,4-Cl_2-α-$C_{10}H_5NH_2$	212.08	nd(al)	....	83-4	....	....	al	B12^3, 2963
9704	α-Naphthyl amine, N,N-diethyl or 1-Diethylamino naphthalene α-$(C_2H_5)_2NC_{10}H_7$	199.30	....	285, $155\text{-}65^{30}$	......	$1.015^{20/20}$	1.5961^{20}	al, eth, aa	B12^3, 2855
9705	α-Naphthyl amine, 5,8-dihydro α-$C_{10}H_9NH_2$	145.20	pl or nd (to) (pink in air)	247^{408}	37.5	....	....	al, chl	B12^3, 2837
9706	α-Naphthyl amine, N,N-dimethyl α-$C_{10}H_7N(CH_3)_2$	171.24	vt flr	274.5^{711}, $139\text{-}40^{13}$	......	$1.0423^{20/4}$	1.624^{15}	al, eth	B12^3, 2854
9707	α-Naphthyl amine, 2,4-dinitro 2,4-$(NO_2)_2$-α-$C_{10}H_5NH_2$	233.18	ye nd(al) pr(aa) cr(ace)	....	242	....	....	ace	B12^3, 2983
9708	α-Naphthyl amine, N-ethyl α-$C_{10}H_7NHC_2H_5$	171.24	....	303^{723}, 191^{16}	......	1.060^{20}	1.6477^{15}	....	B12^3, 2854
9709	α-Naphthyl amine, 4-fluoro 4-F-α-$C_{10}H_6NH_2$	161.18	lt ye	162^{16}	48	....	....	....	B12^3, 2960
9710	α-Naphthyl amine, N-formyl α-$C_{10}H_7NHCHO$	171.20	nd(w)	....	137.5	....	....	al, eth, ace	B12^3, 2866
9711	α-Naphthyl amine, hydrochloride α-$C_{10}H_7NH_2$.HCl	179.65	nd	sub	......	....	....	al, w	B12^3, 2849
9712	α-Naphthyl amine, N-methyl α-$C_{10}H_7NHCH_3$	157.22	oil	293, $165\text{-}7^{15}$	......	....	1.6722^{20}	al, eth	B12^3, 2854
9713	α-Naphthyl amine, 2-methyl 2-CH_3-α-$C_{10}H_6NH_2$	157.22	nd (peth) (turns red in air)	....	32	....	....	eth, lig	B12^3, 3702
9714	α-Naphthyl amine, 3-methyl 3-CH_3-α-$C_{10}H_6NH_2$	157.22	cr(peth)	....	51-2	....	....	al, eth, lig	B12^3, 3106
9715	α-Naphthyl amine, 4-methyl 4-CH_3-α-$C_{10}H_6NH_2$	157.22	nd(peth)	176^{12}	51-2	....	....	eth	B12^3, 3093
9716	α-Naphthyl amine, 2-nitro 2-O_2N-α-$C_{10}H_6NH_2$	188.19	ye-red mcl pr(al)	....	144	....	....	al	B12^3, 2968
9717	α-Naphthyl amine, 3-nitro 3-O_2N-α-$C_{10}H_6NH_2$	188.19	og-ye nd(50% al)	....	137	....	....	al, bz, chl	B12^3, 2969
9718	α-Naphthyl amine, 4-nitro 4-O_2N-α-$C_{10}H_6NH_2$	188.19	og-ye nd(al)	....	195	....	....	al, aa	B12^3, 2971
9719	α-Naphthyl amine, 5-nitro 5-O_2N-α-$C_{10}H_6NH_2$	188.19	red nd(w)	....	118-9	....	....	eth, aa	B12^3, 2973
9720	α-Naphthyl amine, 6-nitro 6-O_2N-α-$C_{10}H_6NH_2$	188.19	og-red nd(chl)	....	172-3	....	....	al	B12^3, 2974
9721	α-Naphthyl amine, 8-nitro 8-O_2N-α-$C_{10}H_6NH_2$	188.19	red lf(peth)	....	96-7	....	....	eth	B12^3, 2975

No.	Name, Synonyms, and Formula	Mol. wt.	Color, crystalline form, specific rotation and λ_{max} (log ε)	b.p. °C	m.p. °C	Density	n_D	Solubility	Ref.
9722	β-Naphthyl amine β-$C_{10}H_7NH_2$	143.19	lf(w)	306.1	113	$1.0614^{98/4}$	1.6493^{98}	al, eth	$B12^3$, 2989
9723	β-Naphthyl amine, N-acetyl β-$C_{10}H_7(NHCOCH_3)$	185.23	lf(al w)		134				$B12^3$, 3014
9724	β-Naphthyl amine, N-acetyl-1-nitro 1-NO_2-β-$C_{10}H_6NHCOCH_3$	230.22	ye nd(al)		126			al, eth, bz, aa	$B12^2$, 731
9725	β-Naphthyl amine, N-acetyl-5-nitro 5-NO_2-β-$C_{10}H_6NHCOCH_3$	230.22	ye-br rh(al) ye nd(bz)		186			al, aa	$B12^3$, 732
9726	β-Naphthyl amine, N-acetyl-6-nitro 6-NO_2-β-$C_{10}H_6NHCOCH_3$	220.22	lf ye nd(dil al)		224			al, bz, aa	$B12^2$, 733
9727	β-Naphthyl amine, N-acetyl-8-nitro 8-NO_2-β-$C_{10}H_6NHCOCH_3$	230.22	ye nd(al)		195.5			aa	$B12^2$, 733
9728	β-Naphthyl amine, N-benzylidene β-$C_{10}H_7N{=}CHC_6H_5$	231.30	yesh nd(al)		103			al, bz, chl, aa	$B12^3$, 3006
9729	β-Naphthyl amine, 1-bromo 1-Br-β-$C_{10}H_6NH_2$	222.08	rh nd(dil al lig)		63-4			al, eth, bz, chl	$B12^3$, 3070
9730	β-Naphthyl amine, 3-bromo 3-Br-β-$C_{10}H_6NH_2$	222.08	pl(al)		169			al, aa	$B12^3$, 3072
9731	β-Naphthyl amine, 4-bromo 4-Br-β-$C_{10}H_6NH_2$	222.08	nd(bz-peth or 90% aa)		72			al, eth, bz	$B12^3$, 3073
9732	β-Naphthyl amine, 5-bromo 5-Br-β-$C_{10}H_6NH_2$	222.08	cr	$207\text{-}10^{16}$	38			al, ace	$B12^3$, 3074
9733	β-Naphthyl amine, 6-bromo 6-Br-β-$C_{10}H_6NH_2$	222.08	lf(al w peth)		128			al, ace, bz	$B12^3$, 3074
9734	β-Naphthyl amine, 1-chloro 1-Cl-β-$C_{10}H_6NH_2$	177.63	nd(al peth)		60			al, ace	$B12^3$, 3065
9735	β-Naphthyl amine, 1,4-dibromo 1,4-Br_2-β-$C_{10}H_5NH_2$	300.98	nd(al bz)		106-7			al, eth, bz	B12, 1311
9736	β-Naphthyl amine, 1,6-dibromo 1,6-Br_2-β-$C_{10}H_5NH_2$	300.98	nd(al peth)		121			al, bz, aa	$B12^1$, 544
9737	β-Naphthyl amine, N,N-dimethyl β-$C_{10}H_7N(CH_3)_2$	171.24	dk red nd	305, $160\text{-}2^{12}$	52-3	$1.0455^{60/60}$	1.6443^{53}	al, eth	$B12^3$, 2995
9738	β-Naphthyl amine, 1,4-dimethyl 1,4-$(CH_3)_2$-β-$C_{10}H_5NH_2$	171.24		333	75			al, eth	B12, 1317
9739	β-Naphthyl amine, 1,6-dinitro 1,6-$(NO_2)_2$-β-$C_{10}H_5NH_2$	233.18	gold-ye nd(al aa) ye pw(py)		248				$B12^3$, 3087
9740	β-Naphthyl amine, N-ethyl β-$C_{10}H_7NHC_2H_5$	171.24		316-7, 191^{25}	<15	1.0545^{21}	1.6544^{21}		$B12^3$, 2996
9741	β-Naphthyl amine, N-formyl β-$C_{10}H_7NHCHO$	171.20	lf(bz-peth)		129			al, bz	$B12^3$, 3013
9742	β-Naphthyl amine, hydrochloride β-$C_{10}H_7NH_2.HCl$	179.65	lf		254			w, al	$B12^3$, 2992
9743	β-Naphthyl amine, 3-iodo 3-I-β-$C_{10}H_6NH_2$	269.08	cr(al)		137			al, bz, chl, aa	$B12^3$, 3078
9744	β-Naphthyl amine, N-methyl β-$C_{10}H_7NHCH_3$	157.22	dk in air	317, $165\text{-}70^{12}$			1.6722^{20}		$B12^3$, 2995
9745	β-Naphthyl amine, 1-methyl 1-CH_3-β-$C_{10}H_6NH_2$	157.22	nd(lig) pr(peth)		51			al, eth, ace, bz	$B12^3$, 3091
9746	β-Naphthyl amine, 6-methyl 6-CH_3-β-$C_{10}H_6NH_2$	157.22	lf(peth w)(turns red in air)		129-30				$B12^2$, 743
9747	β-Naphthyl amine, 1-nitro 1-NO_2-β-$C_{10}H_6NH_2$	188.19	og-ye nd(al)		126-7			al, ace, aa	$B12^3$, 3079
9748	β-Naphthyl amine, 5-nitro 5-NO_2-β-$C_{10}H_6NH_2$	188.19	red nd(al)		143.5			al, bz, aa	$B12^3$, 3082
9749	β-Naphthyl amine, 6-nitro 6-NO_2-β-$C_{10}H_6NH_2$	188.19	lt og pl(al) ye pl(aa)		207.5			bz, aa, diox	$B12^3$, 3082
9750	β-Naphthyl amine, 8-nitro 8-NO_2-β-$C_{10}H_6NH_2$	188.19	red nd(aa)		104.5			al, eth, bz	$B12^3$, 3084
9751	β-Naphthyl amine, 1-nitroso 1-ON-β-$C_{10}H_6NH_2$	172.19	gr nd(dil al bz)		150.2			al	$B7^3$, 3690
9752	β-Naphthyl amine, 1,3,6-tribromo 1,3,6-Br_3-β-$C_{10}H_4NH_2$	379.88	pa red cr (chl al-eth)		143			eth, chl	$B12^3$, 3078
9753	α-Naphthyl ether or α-Naphthyl ether $(\alpha$-$C_{10}H_7)_2O$	270.33	lf(al al-eth)	$280\text{-}5^{22}$	110			eth, bz	$B6^3$, 2926

No.	Name, Synonyms, and Formula	Mol. wt.	Color, crystalline form, specific rotation and λ_{max} (log ε)	b.p. °C	m.p. °C	Density	n_D	Solubility	Ref.
9754	β-Naphthyl ether or bis-(2-Naphthyl) ether $(\beta\text{-}C_{10}H_7)_2O$	270.33	nd or lf(al)	250d	105			eth, bz	$B6^3$, 2976
9755	α-β-Naphthyl ether $\alpha\text{-}C_{10}H_7O\text{-}\beta\text{-}C_{10}H_7$	270.33	lf(al or al-eth)	264^{15}	81			eth, bz	$B6^2$, 600
9756	α-Naphthyl isocyanate $\alpha\text{-}C_{10}H_7NCO$	169.18		269-70		$1.1774^{20/4}$		eth, bz	$B12^3$, 2948
9757	β-Naphthyl isocyanate $\beta\text{-}C_{10}H_7NCO$	169.18	lf		55-6			eth, bz	$B12^3$, 3051
9758	α-Naphthyl isothiocyanate or α-Naphthyl mustard oil $\alpha\text{-}C_{10}H_7NCS$	185.24	wh nd(al)		58			al, eth, ace, bz, chl	$B12^3$, 2948
9759	β-Naphthyl isothiocyanate $\beta\text{-}C_{10}H_7NCS$	185.24	yesh nd(al)		62-3			al, eth, bz, chl	$B12^3$, 3052
9760	α-Naphthyl pentyl ether $\alpha\text{-}C_{10}H_7OC_5H_{11}$	214.31	nd(al)	322	30			al, eth, bz, chl	$B6^3$, 2925
9761	β-Naphthyl pentyl ether $\beta\text{-}C_{10}H_7OC_5H_{11}$	214.31	lf(al)	335	24.5		1.5587^{30}	al, eth, bz, chl	$B6^3$, 2973
9762	α-Naphthyl phenyl amine $\alpha\text{-}C_{10}H_7NHC_6H_5$	219.29	lf(lig) pr or nd(al)	335^{528}, 226^{8}	62			al, eth, bz, chl, aa	$B12^3$, 2856
9763	α-Naphthyl propyl amine $\alpha\text{-}C_{10}H_7NHC_3H_7$	185.27	ye	$316\text{-}8^{771}$					B12, 1224
9764	β-Naphthyl phenyl amine $\beta\text{-}C_{10}H_7NHC_6H_5$	219.29	nd(MeOH)	395-9, 237-13	108			al, eth, bz, aa	$B12^3$, 2999
9765	α-Naphthyl sulfide $(\alpha\text{-}C_{10}H_7)_2S$	286.39	nd or pr(al)	$289\text{-}90^{15}$	110			bz, aa	$B6^3$, 2945
9766	β-Naphthyl sulfide $(\beta\text{-}C_{10}H_7)_2S$	286.39	pl(al) lf(bz)	$295\text{-}6^{15}$	151			bz	$B6^3$, 3009
9767	α-Naphthyl thiocyanate $\alpha\text{-}C_{10}H_7SCN$	185.24	cr(peth)		55			peth	$B6^2$, 588
9768	β-Naphthyl thiocyanate $2\text{-}C_{10}H_7SCN$	185.24			35				$B6^2$, 611
9769	α-Naphthyl (2-tolyl)amine $(2\text{-}CH_3C_6H_4)NH\text{-}\alpha\text{-}C_{10}H_7$	233.31	nd(lig)	$198\text{-}202^{9}$	94-5			al, eth, bz	$B12^3$, 2857
9770	α-Naphthyl (4-tolyl) amine $(4\text{-}CH_3C_6H_4)NH\text{-}\alpha\text{-}C_{10}H_7$	233.31	pr(al)	360^{528}, 236^{15}	79			eth, bz	$B12^3$, 2857
9771	β-Naphthyl 2-tolyl amine or Yellow OB $(2\text{-}CH_3C_6H_4)NH\text{-}\beta\text{-}C_{10}H_7$	233.31	lf(lig)	400-5, $235\text{-}7^{14}$	95-6 (105)			al, eth, ace, bz, lig	$B12^3$, 3001
9772	β-Naphthyl 4-tolyl amine $(4\text{-}CH_3C_6H_4)NH\text{-}\beta\text{-}C_{10}H_7$	233.31	red lf(al)		103			eth, bz	$B12^3$, 3001
9773	α-Naphthyl 2-tolyl ketone $\alpha\text{-}C_{10}H_7CO(C_6H_4CH_3\text{-}2)$	246.31	(al)	365	64			al, eth	$B7^3$, 2639
9774	α-Naphthyl 3-tolyl ketone $\alpha\text{-}C_{10}H_7CO(C_6H_4CH_3\text{-}3)$	246.31			74-5			al	$B7^3$, 2639
9775	Narceine or Pseudonarcene $C_{23}H_{27}NO_8.3H_2O$	499.52	nd or pr (w + 3)		145.2 (+3w) 176-7 (anh)			al	$B19^4$, 4382
9776	Narceine, bisulfate,decahydrate $C_{23}H_{27}NO_8.H_2SO_4.10H_2O$	723.70	nd(sulf)		d			w, al, eth	B19, 372
9777	Narceine, hydrochloride,trihydrate $C_{23}H_{27}NO_8.HCl.3H_2O$	535.98	pr(HCl)		192 (anh)			al	$B19^4$, 4383
9778	α-Narcotine (dl) $C_{22}H_{23}NO_7$	413.43	nd(al chl MeOH)		232-3				$B27^2$, 607
9779	α-Narcotine (l) $C_{22}H_{23}NO_7$	413.43	pr or nd (al), [α] -200 (chl)		176			al, ace, bz, chl	$B27^2$, 605
9780	α-Narcotine, hydrochloride $C_{22}H_{23}NO_7\cdot HCl$	449.89	(w + 3) [α] + 100		193 anh			w	$B27^2$, 606
9781	β-Narcotine-(dl) or β-Gnoscopine $C_{22}H_{23}NO_7$	413.43	nd pr(MeOH al)		180			al	$B27^1$, 559
9782	β-Narcotine, hydrochloride $C_{22}H_{23}NO_7.HCl$	449.89	pr		86-8, 224 (on standing)				$B27^1$, 559
9783	Naringin $C_{27}H_{32}O_{14}\cdot 2H_2O$	616.57	nd (w + 8), $[\alpha]^{11}{}_D$ -82.1 (al)		82 (+8w), 17 (+2w)				$B18^4$, 2637

No.	Name, Synonyms, and Formula	Mol. wt.	Color, crystalline form, specific rotation and λ_{max} (log ε)	b.p. °C	m.p. °C	Density	n_D	Solubility	Ref.
9784	Neoabietic acid, methyl ester or Methyl neoabietate $C_{21}H_{32}O_2$	316.48	cr(MeOH)	......	61-2	......	......	MeOH	$B9^{2}$, 433
9785	Neoamygdalin or L-Mandelonitrile-β-gentobioside $C_{20}H_{27}NO_{11}$	457.43	cr (al), $[\alpha]^{25/}{}_D$ -61.4 (w, c=8.5)	......	212	......	......	w, al	B31, 404
9786	Neobornyl amine or Isobornyl amine $C_{10}H_{19}N$	153.27	pw, $[\alpha]_D$ -47.7 (4%al)	......	184	......	......	eth, ace	$B12^{3}$, 195
9787	Neocarvomenthol (dl) $C_{10}H_{20}O$	156.27	......	......	......	......	1.4637^{20}	al	$B6^{3}$, 130
9788	Neocarvomenthol (l) $C_{10}H_{20}O$	156.27	$[\alpha]^{21/}{}_D$ -41.7	102^{18}	......	$0.9012^{20/4}$	1.4632^{20}	al	$B6^{4}$, 148
9789	Neoergesterol $C_{27}H_{40}O$	380.61	pr or nd (al), $[\alpha]^{17/}{}_D$ -12 (chl, c=2)	......	155-7	......	......	eth, ace	$B6^{3}$, 3474
9790	Neogermitrine $C_{36}H_{55}NO_{11}$	677.83	wh nd, $[\alpha]^{25/}{}_D$ -79.2 (py)	......	237-9	......	......	chl	$B21^{4}$, 6803
9791	Neoisocarvomenthol (l) $C_{10}H_{20}O$	156.27	$[\alpha]^{17/}{}_D$ -34.7	$87\text{-}8^{4}$	<-25	$0.9102^{20/4}$	1.4676^{20}	al, ace	$B6^{4}$, 148
9792	Neoisamenthol (d) or p-menthol-3 $C_{10}H_{20}O$	156.27	$[\alpha]^{15/}{}_D$ + 2.2 (al, c=2)	214.6, 91.5^{11}	−8	$0.9131^{18/4}$	1.4670^{20}	al, ace	$B6^{4}$, 149
9793	Neoisomenthol (dl) $C_{10}H_{20}O$	156.27	......	214.5, 81^{6}	14	$0.8854^{55/4}$	1.4649^{20}	al, ace	$B6^{4}$, 149
9794	Neomenthol (d) or *d*,β-Pulegomenthol $C_{10}H_{20}O$	156.27	$[\alpha]^{20/}{}_D$ + 19.6 (al)	211.7, 95^{12}	−15	$0.897^{22/4}$	1.4600^{20}	al, ace	$B6^{4}$, 149
9795	Neomenthol (dl) $C_{10}H_{20}O$	156.27	pl or pr(peth)	211.7, $103\text{-}5^{16}$	52	$0.903^{15/15}$	1.4600^{20}	al, ace	$B6^{4}$, 150
9796	Neomenthol-(l) $C_{10}H_{20}O$	156.27	$[\alpha]^{18/}{}_D$ -19.6 (al)	211.7, 97.6^{10}	......	......	1.4603^{20}	al, ace	$B6^{3}$, 140
9797	Neopentane or 2,2-Dimethylpropane $(CH_3)_4C$	72.15	gas	9.5	−16.5	0.6135^{20}	1.3476^{6}	al, eth	$B1^{4}$, 333
9798	Neopentyl alcohol or 2,2-Dimethyl-1-propanol $(CH_3)_3CCH_2OH$	88.15	......	113-4	52-3	0.812	......	al, eth	$B1^{4}$, 1690
9799	Neopentyl alcohol, 1-phenyl $(CH_3)_3CCH(OH)C_6H_5$	164.25	nd	$114\text{-}6^{16}$	45	......	......	al, eth	$B6^{3}$, 1972
9800	Neopentyl alcohol, 3-phenyl $C_6H_5CH_2C(CH_3)_2CH_2OH$	164.25	nd	$125\text{-}6^{14.5}$	34-5	......	......	al, eth	$B6^{2}$, 507
9801	Neopentyl amine or 1-amino-2,2-dimethyl propane $(CH_3)_3CCH_2NH_2$	87.16	......	$81\text{-}2^{741}$	......	$0.7455^{20/4}$	1.4023^{20}	eth	$B4^{4}$, 707
9802	Neopentyl bromide or 1-bromo-2,2-dimethyl propane $(CH_3)_3CCH_2Br$	151.05	......	106, 34.6^{100}	......	$1.1997^{20/4}$	1.4370^{20}	al, eth, ace, bz, chl	$B1^{4}$, 337
9803	Neopentyl chloride or 1-Chloro-2,2-dimethyl propane $(CH_3)_3CCH_2Cl$	106.60	......	84.3	−20	$0.86604^{20/4}$	1.4044^{20}	al, eth, bz, chl	$B1^{4}$, 336
9804	Nepentylene glycol or 2,2-Dimethyl-1,3-propanediol $(HOCH_2)_2C(CH_3)_2$	104.15	nd (bz)	206^{747}, $120\text{-}30^{15}$	130	......	......	w, al, eth	$B1^{4}$, 2551
9805	Neopentyl iodide or 1-Iodo-2,2-dimethyl propane $(CH_3)_3CCH_2I$	198.05	......	127-9d, $42\text{-}4^{20}$	......	1.4940^{20}	1.4890^{20}	al, eth	$B1^{4}$, 338
9806	Neral or Citral b $C_{10}H_{16}O$	152.24	......	120^{20}	......	0.8869^{20}	1.4869^{20}	al, eth	$B1^{4}$, 3569
9807	Nerol or 3,7-Dimethyl-2,6-octadiene-1-ol $(CH_3)_2C=CHCH_2CH_2C(CH_3)=CHCH_2OH$	154.25	......	$224\text{-}5^{745}$, 125^{25}	<-15	$0.8756^{20/4}$	1.4746^{20}	al	$B1^{4}$, 2276
9808	Nerolidol (d) or α-3,7,11-trimethyl-1,6,10-dodecatriene-3-ol $(CH_3)_2C=CHCH_2CH_2C(CH_3)=CH(CH_2)_2C(CH_3)(OH)CH=CH_2$	222.37	$[\alpha]^{20/}{}_D$ + 15.5 (undil)	276, $128\text{-}9^{6}$	......	$0.8778^{20/4}$	1.4898^{20}	al, eth, ace, aa	$B1^{4}$, 2336
9809	Nerolidol (dl) $C_{15}H_{26}O$	222.37	......	$145\text{-}6^{12}$, $75\text{-}6^{0.1}$	......	$0.8756^{19/4}$	1.4801^{16}	al, eth, ace	$B1^{4}$, 2336
9810	Nerolidol (l) $C_{15}H_{26}O$	222.37	$[\alpha]_D$-6.5 (undil)	$124\text{-}6^{3}$	......	$0.8881^{15/15}$	1.4799^{20}	al, eth, ace, aa	$B1^{3}$, 2042
9811	Neurine or Trimethyl ethenyl ammonium hydroxide $CH_2=CHN^+(CH_3)_3OH^-$	103.16	syr	......	......	......	......	w, al, eth	$B4^{4}$, 1053
9812	Nicotine (d) or α-N-Methyl-*d*-β-pyridylpyrrolidine $C_{10}H_{14}N_2$	162.23	hyg $[\alpha]^{20/}{}_D$ + 163.2	$245\text{-}6^{729}$	......	$1.0094^{20/4}$	1.5280^{20}	w, al, eth, chl, lig	B23, 117

No.	Name, Synonyms, and Formula	Mol. wt.	Color, crystalline form, specific rotation and λ_{max} (log ε)	b.p. °C	m.p. °C	Density	n_D	Solubility	Ref.
9813	Nicotine (dl) or Tetrahydronicotyrine $C_{10}H_{14}N_2$	162.23		242-3		$1.0082^{20/4}$	1.5289^{20}	w, al, eth, chl, lig	B23[2], 111
9814	Nicotine (l) $C_{10}H_{14}N_2$	162.23	hyg (br in air) $[\alpha]^{20}{}_D$ −169	246.7^{745}, $124\text{-}5^{18}$	−79	$1.0097^{20/4}$	1.5282^{20}	w, al, eth, chl, lig	B23[2], 107
9815	Nicotine-hydrochloride (d) $C_{10}H_{14}N_2.HCl$	198.70	diq $[\alpha]^{20}{}_D$ + 104 (w, p=10)			1.0337		w	B23, 114
9816	Nicotine amide or Niacin amide $3\text{-}C_5H_4NCONH_2$	122.13	wh pw nd (bz)	$150\text{-}60^{0.0005}$	129-31	1.400	1.466	w, al	B22[4], 389
9817	Nicotinamide-N,N-diethyl or Coramine $3\text{-}C_5H_4NCON(C_2H_5)_2$	178.23	yesh	280d, 175^{25}	24-6	$1.060^{25/4}$	1.525^{20}	w, al, eth, ace, chl	B22[4], 393
9818	Nicotinic acid or Niacin 3-Pyridine carboxylic acid $3\text{-}C_5H_4NCO_2H$	123.11	nd (w al)	sub	236-7	1.473			B22[4], 348
9819	Nicotinic acid-6-amino $6\text{-}N_2N\text{-}(C_5H_3N)CO_2H\text{-}3$	138.13	cr (dil aa + 2w) aa		312				B22[4], 6726
9820	Nicotinic acid-ethyl betaine $C_8H_9NO_2$	151.16	hyg pl		84-6			w	B22, 43
9821	Nicotinic acid-ethyl ester or Ethyl nicotinate $3\text{-}(C_5H_4N)CO_2C_2H_5$	151.16		224, $103\text{-}5^{5}$	8-9	1.1070^{20}	1.5024^{20}	w, al, eth, bz	B22[4], 357
9823	Nictinic acid-hydrochloride $3\text{-}(C_5H_4N)CO_2H.HCl$	159.57	pr or pl rh bipym(w)		274			w, al	B22[4], 354
9824	Nicotinic acid-2-hydroxy $2\text{-}HO(C_5H_3N)CO_2H\text{-}3$	139.11	nd (w)		α259-61d, β301-2d				B22[4], 2139
9825	Nicotinic acid-4-hydroxy $4\text{-}HO(C_5H_3N)CO_2H\text{-}3$	139.11	nd (w + 2) cr (al)		254-5				B22[4], 2145
9826	Nicotinic acid, 6-hydroxy $6\text{-}HO\text{-}3(C_5H_3N)CO_2H\text{-}3$	139.11	nd (w)	sub	304d				B22[4], 2147
9827	Nicotinic acid, N-methyl or Trigoneiline $C_7H_7NO_2$	137.14	pr (aq al + 1w)		218d anh			w	B22[4], 462
9828	Nicotinic acid, methyl ester or Methyl nicotinate $3\text{-}(C_5H_4N)CO_2CH_3$	137.14	cr	204, 118.5^{23}	42-3			w, al, bz	B22[4], 356
9829	Nicotinonitrile $3\text{-}(C_5H_4N)CN$	104.11	nd (lig peth-eth)	240-5	50-2			w, al, eth, bz	B22[4], 434
9830	2,2′-Nicotyrine (solid) or N-Methyl-2-(2-pyridyl)pyrrole $C_{10}H_{10}N_2$	162.19	cr, br in air		43-4			al, eth, bz, dil HCl	B23[2], 192
9831	2,2′-Nicotyrine (liquid) $C_{10}H_{10}N_2$	162.19		273^{764}, $149\text{-}50^{22}$	−28			ace, bz	B23[2], 191
9832	3,2′-Nicotyrine or N-Methyl-2-(3-pyridyl)pyrrole $C_{10}H_{10}N_2$	162.19	br in air	$280\text{-}1^{744}$, 150^{5}		$1.2111^{20/4}$	1.6057^{20}	al, eth, ace	B23[2], 192
9833	Ninhydrin or 1,2,3-triketo hydrindene monohydrate $C_9H_6O_4$	178.14	pr (w)		241-3d			w, al	B7[3], 4592
9834	Nitranilic acid or 2,5-Dihydroxy-3,6-dinitro-p-benzoquinone $C_6H_2N_2O_8$	230.09	gold-ye pl (+ w dil HNO_3)	exp	100d			w, al	B8[3], 3351
9835	Nitro acetic acid $O_2NCH_2CO_2H$	105.05	nd (chl)		92-3d			al, eth, bz, chl	B2[4], 537
9836	Nitro acetic acid, ethyl ester or Ethyl nitro acetate $O_2NCH_2CO_2C_2H_5$	133.10		$105\text{-}7^{25}$		$1.1953^{20/4}$	1.4250^{20}	al, eth	B2[4], 537
9837	2-Nitrobenzamide $2\text{-}NO_2C_6H_4CONH_2$	166.14	nd (dil al)	317	176-6			al, eth	B9[3], 1477
9838	2-Nitrobenzamide, N-phenyl or N-Phenyl-2-nitrobenzamide $2\text{-}NO_2C_6H_4CONHC_6H_5$	242.23	nd (al bz)		155			al, bz, chl	B12[3], 506
9839	2-Nitrobenzoic acid $2\text{-}NO_2C_6H_4CO_2H$	167.12	tcl nd (w)		147-8	$1.575^{20/4}$		al, eth, ace	B9[3], 1466
9840	2-Nitrobenzoic acid, azide $2\text{-}NO_2C_6H_4CON_3$	192.13	ye pr (eth)		37.5			eth, bz, chl	B9[3], 1489
9841	2-Nitrobenzoic acid, 3-bromo $3\text{-}Br\text{-}2\text{-}NO_2C_6H_3CO_2H$	246.02	(eth)		250			eth, bz	B9[3], 1770
9842	2-Nitrobenzoic acid, 4-bromo $4\text{-}Br\text{-}2\text{-}NO_2C_6H_3CO_2H$	246.02	nd (w)		163			al, eth, bz, chl	B9[3], 1770
9843	2-Nitrobenzoic acid, 5-bromo $5\text{-}Br\text{-}2\text{-}NO_2C_6H_3CO_2H$	246.02	cr (w al bz)		140	1.920^{18}		al, bz	B9[1], 165
9844	2-Nitrobenzoic acid, 3-chloro $3\text{-}Cl\text{-}2\text{-}NO_2C_6H_3CO_2H$	201.57	nd or pl (w)		237-9	1.566^{18}		al, eth	B9, 400

No.	Name, Synonyms, and Formula	Mol. wt.	Color, crystalline form, specific rotation and λ_{max} (log ε)	b.p. °C	m.p. °C	Density	n_D	Solubility	Ref.
9845	2-Nitrobenzoic acid, 4-chloro $4\text{-Cl-2-NO}_2C_6H_3CO_2H$	201.57	pl (bz-lig) pr (bz) nd (w)	....	142.3	....	....	al, eth	B9[3], 1763
9846	2-Nitrobenzoic acid, 5-chloro,methyl ester $5\text{-Cl-2-NO}_2C_6H_3CO_2CH_3$	215.59	pl (MeOH)	....	48.5	1.453[18]	....	MeOH	B9, 401
9847	2-Nitrobenzoic acid, ethyl ester or Ethyl 2-nitrobenzoate $2\text{-NO}_2C_6H_4CO_2C_2H_5$	195.17	tcl (dil al)	275, 173[18]	30	....	....	al, eth	B9[3], 1469
9848	2-Nitrobenzoic acid, hydrazide $2\text{-NO}_2C_6H_4CONHNH_2$	181.15	ye-br pr (w)	....	123	....	....	al	B9[3], 1481
9849	2-Nitrobenzoic acid, 4-iodo $4\text{-I-2-NO}_2C_6H_3CO_2H$	293.02	ye lf or pr (dil al)	....	192-3	....	....	al, eth	B9[2], 278
9850	2-Nitrobenzoic acid, methyl ester $2\text{-NO}_2C_6H_4CO_2CH_3$	181.15	....	275, 176[21]	−13	1.2855[20]	....	al, eth, bz, chl	B9[3], 1469
9851	2-Nitrobenzonitrile $2\text{-NO}_2C_6H_4CN$	148.12	nd (w, aa)	sub	111	....	....	al, eth, ace, bz, aa	B9[3], 1479
9852	2-Nitrobenzonitrile, 4-chloro $4\text{-Cl-2-NO}_2C_6H_3CN$	182.57	nd (w)	....	100-1	....	....	al, eth	B9[3], 1763
9853	3-Nitrobenzamide $3\text{-NO}_2C_6H_4CONH_2$	166.14	ye mcl nd (w)	310-5	142.7	....	....	al, eth	B9[3], 1515
9854	3-Nitrobenzamide, N-phenyl or N-Phenyl-3-nitrobenzamide $3\text{-NO}_2C_6H_4CONHC_6H_5$	242.23	lf (w, al)	sub	153-4	....	....	al, eth, bz	B12[3], 506
9855	3-Nitrobenzoic acid $3\text{-NO}_2C_6H_4CO_2H$	167.12	mcl pr (w)	....	140-2	1.494[20/4]	....	al, eth, ace	B9[3], 1489
9856	3-Nitrobenzoic acid, azide $3\text{-NO}_2C_6H_4CON_3$	192.13	pl (dil al)	....	68	....	....	al, eth, bz, aa	B9[3], 1536
9857	3-Nitrobenzoic acid, 2-bromo $2\text{-Br-3-NO}_2C_6H_3CO_2H$	246.02	(dil al)	....	191	....	....	al	B9[2], 277
9858	3-Nitrobenzoic acid, 4-bromo $4\text{-Br-3-NO}_2C_6H_3CO_2H$	246.02	nd (dil aa)	sub	203-4	....	....	al	B9[3], 1770
9859	3-Nitrobenzoic acid, 5-bromo $5\text{-Br-3-NO}_2C_6H_3CO_2H$	246.02	nd (w, bz eth), pl (al)	....	159-60	....	....	al, eth, bz, aa, chl	B9[3], 1771
9860	3-Nitrobenzoic acid, 6-bromo $6\text{-Br-3-NO}_2C_6H_3CO_2H$	246.02	nd (w)	sub	180-1	....	....	al, eth, chl	B9[3], 1771
9861	3-Nitrobenzoic acid, 2-chloro $2\text{-Cl-3-NO}_2C_6H_3CO_2H$	201.57	....	....	185	1.662	....	al	B9[2], 275
9862	3-Nitrobenzoic acid, 4-chloro $4\text{-Cl-3-NO}_2C_6H_3CO_2H$	201.57	nd or pl (w)	....	181-2	1.645[18]	....	....	B9[3], 1764
9863	3-Nitrobenzoic acid, 4-chloro, ethyl ester $4\text{-Cl-3-NO}_2C_6H_3CO_2C_2H_5$	229.62	ye nd (al)	....	59	....	....	al, bz, aa	B9, 402
9864	3-Nitrobenzoic acid, 4-chloro, methyl ester $4\text{-Cl-3-NO}_2C_6H_3CO_2CH_3$	215.59	nd (MeOH)	....	83	1.522[18]	....	al	B9, 402
9865	3-Nitrobenzoic acid, 5-chloro $5\text{-Cl-3-NO}_2C_6H_3CO_2H$	201.57	nd (w)	....	147	....	....	al, eth, aa	B9[1], 165
9866	3-Nitrobenzoic acid, 6-chloro $6\text{-Cl-3-NO}_2C_6H_3CO_2H$	201.57	nd or pr (w)	....	165	1.608[18]	....	al, eth, bz	B9[3], 1765
9867	3-Nitrobenzoic acid, ethyl ester or Ethyl-3-nitrobenzoate $3\text{-NO}_2C_6H_4CO_2C_2H_5$	195.17	mcl pr	296-8, 156[10]	47	....	....	al, eth	B9[3], 1493
9868	3-Nitrobenzoic acid hydrazide $3\text{-NO}_2C_6H_4CONHNH_2$	181.15	nd (w)	....	153-4	....	....	....	B9[3], 1524
9869	3-Nitrobenzoicacid, 2-iodo $2\text{-I-3-NO}_2C_6H_3CO_2H$	293.02	pr (w)	....	206	....	....	al, eth	B9[2], 278
9870	3-Nitrobenzoic acid, 4-iodo $4\text{-I-3-NO}_2C_6H_3CO_2H$	293.02	ye pr (al)	....	213	....	....	al	B9[2], 278
9871	3-Nitrobenzoic acid, 5-iodo $5\text{-I-3-NO}_2C_6H_3CO_2H$	293.02	nd (al), pr (peth)	....	167	....	....	al	B9[2], 278
9872	3-Nitrobenzoic acid, methyl ester $3\text{-NO}_2C_6H_4CO_2CH_3$	181.15	nd	....	78	....	....	....	B9[3], 1493
9873	3-Nitrobenzonitrile $3\text{-NO}_2C_6H_4CN$	148.12	nd (w)	sub	118	....	....	al, eth, ace, aa	B9[3], 1521
9874	3 Nitrobenzoyl chloride $3\text{-NO}_2C_6H_4COCl$	185.57	ye cr	275-8, 154-5[18]	35	....	....	eth	B9[3], 1514
9875	4-Nitrobenzamide $4\text{-NO}_2C_6H_4CONH_2$	166.14	nd (w)	....	201-4	....	....	al, eth	B9[3], 1710

No.	Name, Synonyms, and Formula	Mol. wt.	Color, crystalline form, specific rotation and λ_{max} (log ε)	b.p. °C	m.p. °C	Density	n_D	Solubility	Ref.
9876	4-Nitrobenzamide, N-phenyl or N-Phenyl-4-nitrobenzamide .... $4-NO_2C_6H_4CONHC_6H_5$	242.23	lf (eth)	....	211	....	....	al, eth	$B12^3$, 506
9877	4-Nitrobenzoic acid .... $4-NO_2C_6H_4CO_2H$	167.12	mcl lf (w)	sub	242	1.610^{20}	....	al, eth, chl	$B9^3$, 1537
9878	4-Nitrobenzoic acid, 2-bromo .... $2-Br-4-NO_2C_6H_3CO_2H$	246.02	nd (w, dil al)	sub >155	166-7	....	....	al, eth	$B9^3$, 1771
9879	4-Nitrobenzoic acid, 3-bromo .... $3-Br-4-NO_2C_6H_3CO_2H$	246.02	nd (dil al)	....	197	....	....	al, eth, chl	B9, 408
9880	4-Nitrobenzoic acid, butyl ester or Butyl-4-nitrobenzoate $4-NO_2C_6H_4CO_2C_4H_9$	223.23	nd	160^8	35.3	....	....	eth, bz	$B9^3$, 1544
9881	4-Nitrobenzoic acid, 2-chloro .... $2-Cl-4-NO_2C_6H_3CO_2H$	201.57	nd (w)	....	140-2	....	....	al, eth	$B9^3$, 1766
9882	4-Nitrobenzoic acid, ethyl ester .... $4-NO_2C_6H_4CO_2C_2H_5$	195.17	tcl lf (al)	186.3	57	....	....	al, eth	$B9^3$, 1541
9883	4-Nitrobenzoic acid, hydrazide .... $4-NO_2C_6H_4CONHNH_2$	181.15	yesh nd (w)	....	214	....	....	....	$B9^3$, 1751
9884	4-Nitrobenzoic acid, 2-iodo .... $2-I-4-NO_2C_6H_3CO_2H$	293.02	pa ye pr (w)	....	146-7	....	....	al, eth	$B9^2$, 278
9885	4-Nitrobenzoic acid, methyl ester .... $4-NO_2C_6H_4CO_2CH_3$	181.15	ye mcl lf	....	96	....	....	al, eth, chl	$B9^3$, 1541
9886	4-Nitrobenzoic acid, 2,2,2-trichloroethyl ester .... $4-NO_2C_6H_4CO_2CH_2C(Cl)_3$	298.51	pr (al)	$106-7^1$	71	1.5343^{26}	....	....	$B9^3$, 1542
9887	4-Nitrobenzonitrile .... $4-NO_2C_6H_4CN$	148.12	lf (al), nd (bz)	sub	149	....	....	chl	$B9^3$, 1748
9888	4-Nitrobenzoyl chloride .... $4-NO_2C_6H_4COCl$	185.57	ye nd (lig)	$202-5^{105}$, $150-2^{15}$	75	....	....	eth	$B9^3$, 1709
9889	2-Nitrophenyl isocyanate .... $2-O_2NC_6H_4NCO$	164.12	wh nd (peth)	135^7	41	....	....	eth, bz, chl	$B12^3$, 1535
9890	3-Nitrophenyl isocyanate .... $3-O_2NC_6H_4NCO$	164.12	wh lf (lig)	$130-1^{11}$	51	....	....	eth, bz, chl	$B12^3$, 1573
9891	4-Nitrophenyl isocyanate .... $4-O_2NC_6H_4NCO$	164.12	pa ye nd	$137-8^{11}$	57	....	....	eth, bz, chl	$B12^3$, 1630
9892	bis-(2-nitrophenyl)trisulfide .... $(2-NO_2C_6H_4)_2S_3$	340.39	ye nd (al)	....	175-6	....	....	....	$B6^3$, 1062
9893	2-Nonanol (d) .... $CH_3CH(OH)C_7H_{15}$	144.26	$[\alpha]^{20}/_D$	105^{19}	......	$0.8230^{20/4}$	1.4299^{20}	al, eth	$B1^1$, 211
9894	2-Nonanol (dl) .... $CH_3CH(OH)C_7H_{15}$	144.26	....	193-4, 91^{12}	......	$0.84708^{20/4}$	1.43533^{20}	al, eth	$B1^4$, 1803
9895	3-Nonanol (s) .... $C_2H_5CH(OH)C_6H_{13}$	144.26	$[\alpha]^{25}/_D$ 7.08	97^{17}	......	$0.8281^{17/4}$	1.4308^{20}	al, eth	$B1^3$, 1746
9896	3-Nonanol (dl) .... $C_2H_5CH(OH)C_6H_{13}$	144.26	....	195^{750}, 93^{18}	−22	$0.8250^{20/4}$	1.4289^{20}	al, eth	$B1^4$, 1803
9897	4-Nonanol (s) .... $C_3H_7CH(OH)C_5H_{11}$	144.26	$[\alpha]^{25}/_D$ 0.57 eth 0.9%	192-3, $94-5^{18}$	......	$0.8282^{20/4}$	1.41971^{20}	al, eth	$B1^3$, 1747
9898	5-Nonanol .... $(C_4H_9)_2CHOH$	144.26	....	193^{750}, 97^{20}	......	$0.8356^{20/4}$	1.4289^{20}	al	$B1^4$, 1803
9899	Nitron or 4,5-Dihydro-1,4-diphenyl-3,5-phenylamino-1,2,4-triazole .... $C_{19}H_{16}N_4$	300.36	ye lf (al), nd (+ chl)	189d	......	....	....	al, ace, bz, chl	$B26^2$, 76,199
9900	Nonacosane .... $CH_3(CH_2)_{27}CH_3$	408.80	rh cr (peth)	440.8, 271.4^{10}	63.7	$0.8083^{20/4}$, $0.7630^{100/4}$	1.4529^{20}	al, eth, ace, bz	$B1^4$, 591
9901	Nonacosane, 2-methyl .... $(CH_3)_2CH(CH_2)_{26}CH_3$	422.82	pl (lig)	$222^{0.3}$	73-4	....	....	eth, ace, bz	$B1^1$, 72
9902	1-Nonacosanol .... $CH_3(CH_2)_{27}CH_2OH$	424.79	cr (al)	sub $200-50^1$	84-5	....	....	....	$B1^4$, 1916
9904	Nonadecane .... $CH_3(CH_2)_{17}CH_3$	268.53	wax	329.7, 193^{15}	32.1	$0.7855^{20/4}$	1.4409^{20}	eth, ace	$B1^4$, 560
9905	1,2,3-Nonadecane tricarboxylic acid, 2-hydroxy or Agaric acid, Laricic acid .... $CH_3(CH_2)_{15}CH(CO_2H)C(OH)(CO_2H)CH_2CO_2H$	416.56	lf (+ $^3/_2$ w, dil al), $[\alpha]^{19}/_D$ -8.8 (NaOH)	....	142d	....	....	....	$B3^4$, 1284
9906	1,2,3-Nonadecane tricarboxylic acid, 2-hydroxy,triethyl ester .... $C_{28}H_{52}O_7$	500.72	nd	36-7	......	....	....	al, bz	$B3^2$, 373
9907	1,2,3-Nonadecane tricarboxylic acid, 2-hydroxy,trimethyl ester .... $C_{25}H_{46}O_7$	458.64	nd (al)	....	63-4	....	....	bz	$B3^2$, 373

PHYSICAL CONSTANTS OF ORGANIC COMPOUNDS (Continued)

No.	Name, Synonyms, and Formula	Mol. wt.	Color, crystalline form, specific rotation and λ_{max} (log ε)	b.p. °C	m.p. °C	Density	n_D	Solubility	Ref.
9908	Nonadecanoic acid or *n*-Nonadecylic acid $C_{18}H_{37}CO_2H$	298.51	lf (al)	297-8^{100}, 227-30^{10}	69.4	...	...	al, eth, bz, chl, lig	B2^{4}, 1256
9909	1-Nonadecanol $C_{18}H_{37}CH_2OH$	284.53	cr (ace)	166-7$^{0.32}$	62-3	...	1.4328^{75}	eth, ace	B1^{4}, 1898
9910	2-Nonadecanone or Methyl-n-heptadecyl ketone $C_{17}H_{35}COCH_3$	282.51	pr (al)	266.5^{110}, 165^{2}	57	0.8108^{56}	...	eth, ace, chl	B1^{4}, 3400
9911	2-Nonadecanone oxime $C_{17}H_{35}C(=NOH)CH_3$	297.50	cr (al)	...	76-7	...	...	al	B1, 718
9912	4-Nonadecanone or Propyl pentadecyl ketone $C_{15}H_{31}COC_3H_7$	282.51	lf (al)	211^{11}d	50.5	...	...	eth, ace	B1^{4}, 3401
9913	10-Nonadecanone or Caprinone-dinonyl ketone $(C_9H_{19})_2CO$	282.51	lf (al)	>350, 155.6^{11}	65.5	...	...	eth, ace, bz, chl, lig	B1^{4}, 3402
9914	1-Nonadecyne $C_{17}H_{35}C{\equiv}CH$	264.49	...	327, 181.6^{10}	37-8	0.8054$^{20/4}$	1.4488^{20}	eth, ace, bz	B1^{3}, 1030
9915	1,8-Nonadiyne $HC{\equiv}C(CH_2)_5C{\equiv}CH$	120.19	...	162, 55^{13}	−27.3	0.8158$^{20/4}$	1.4490^{20}	eth, ace	B1^{4}, 1125
9916	Nonanal $C_8H_{17}CHO$	142.24	...	190-2, 93.5^{23}	...	0.8264$^{22/4}$	1.4273^{20}	eth	B1^{4}, 3352
9917	Nonanal oxime $C_8H_{17}CH=NOH$	157.26	lf (dil al)	...	64	...	...	al, eth, ace	B1^{2}, 761
9918	Nonane $CH_3(CH_2)_7CH_3$	128.26	...	150.8, 39^{10}	−51	0.7176$^{20/4}$	1.4054^{20}	al, eth, **ace, bz, chl**	B1^{4}, 447
9919	Nonane, 1-amino or Nonyl amine $C_8H_{17}CH_2NH_2$	143.27	...	202.2, 80.8^{10}	−1	0.7886$^{20/4}$	1.4336^{20}	al, eth	B4^{4}, 777
9920	Nonane, 1-bromo $C_9H_{19}Br$	207.15	...	88^{4}	...	1.0183$^{20/20}$	1.4533^{20}	...	B1^{4}, 451
9921	Nonane, 5-butyl $(C_4H_9)_3CH$	184.37	...	217-8	...	0.7635$^{18.5/4}$	1.4273$^{18.5}$	...	B1^{4}, 517
9922	Nonane, 1-chloro or Nonyl chloride $CH_3(CH_2)_7CH_2Cl$	162.70	...	203.4, 80.5^{10}	−39.4	0.8720$^{20/4}$	1.4345^{20}	eth, chl	B1^{4}, 450
9923	Nonane, 1-chloro-9-fluoro $F(CH_2)_9Cl$	180.69	...	102^{11}	...	0.966$^{20/4}$	1.4301^{25}	al, eth	B1^{4}, 450
9924	Nonane, 2-chloro $CH_3(CH_2)_6CHClCH_3$	162.70	...	190^{764}	...	0.8790^{20}	1.4420^{20}	chl	B1, 166
9925	Nonane, 5-chloro $C_4H_9CHClC_4H_9$	162.70	...	85-7^{14}	...	0.8639$^{15/4}$	1.4314^{15}	eth	B1^{2}, 128
9926	Nonane, 2-methyl $CH_3(CH_2)_6CH(CH_3)_2$	142.28	...	166.8	−74.3	0.7281$^{20/4}$	1.4099^{20}	eth, bz, chl	B1^{4}, 473
9927	Nonane, 3-methyl (dl) $C_2H_5CH(CH_3)(CH_2)_5CH_3$	142.28	...	167.8	−84.6	0.7354$^{20/4}$	1.4125^{20}	eth, bz, chl	B1^{4}, 474
9928	Nonane, 4-methyl $C_3H_7CH(CH_3)(CH_2)_4CH_3$	142.28	...	165.7	−101.6	0.7323$^{20/4}$	1.4123^{20}	eth, bz, chl	B1^{4}, 475
9929	Nonane, 5-methyl $CH_3CH[(CH_2)_3CH_3]_2$	142.28	...	165.1	−86.5	0.7326$^{20/4}$	1.4116^{20}	eth, bz, chl	B1^{4}, 475
9930	Nonanedioic acid or Azelaic acid $HO_2C(CH_2)_7CO_2H$	188.22	lf or nd	>360d, 287^{100}, 225^{10}	106.5	1.225$^{25/4}$	1.4303^{111}	al	B2^{4}, 2055
9931	1,9-Nonanediol or Nonamethylene glycol $HOCH_2(CH_2)_7CH_2OH$	160.26	cr (bz)	173-5^{20}	45.8	...	...	al, eth, bz	B1^{4}, 2607
9932	Nonanoic acid or Pelargonic acid $CH_3(CH_2)_7CO_2H$	158.24	...	255, 150^{20}	15	0.9057$^{20/4}$	1.4343^{19}	al, eth, chl	B2^{4}, 1018
9933	1-Nonanol or Nonyl alcohol $CH_3(CH_2)_7CH_2OH$	144.26	...	213.5, 118^{15}	−5.5	0.8273$^{20/4}$	1.4333^{20}	al, eth	B1^{4}, 1798
9934	5-Nonanol, 5-butyl or Tri-*n*-butyl carbinol $(C_4H_9)_3COH$	200.36	...	230-5d, 118-20^{17}	20	0.8408$^{20/4}$	1.4445^{20}	al	B1^{4}, 1863
9935	1-Nonanol, 9-chloro $Cl(CH_2)_8CH_2OH$	178.70	...	146-8^{14}	28	...	1.4575^{20}	al, eth	B1^{4}, 1802
9936	1-Nonanol, 9-fluoro $F(CH_2)_8CH_2OH$	162.25	...	125-6^{15}	...	0.928$^{20/4}$	1.4279^{25}	al, eth	B1^{4}, 1801
9937	2-Nonanone or Heptyl methyl ketone $C_7H_{15}COCH_3$	142.24	...	195.3, 73.8^{10}	−7.5	0.8208$^{20/4}$	1.4210^{20}	al, eth, ace, bz, chl	B1^{4}, 3353
9938	4-Nonanone or Propyl pentyl ketone $C_5H_{11}COC_3H_7$	142.24	...	187-8, 75-6^{20}	...	0.8190$^{25/4}$	1.4189^{20}	al, eth, ace, chl	B1^{4}, 3354
9939	5-Nonanone or Dibutyl ketone $(C_4H_9)_2CO$	142.24	...	188.4, 88^{22}	−4.8	0.8217$^{20/4}$	1.4195^{20}	al, eth, chl	B1^{4}, 3355
9940	Nonasiloxane, eicosa methyl $CH_3[Si(CH_3)_2O]_8Si(CH_3)_3$	681.46	...	307.5, 198.8^{16}	...	0.9173^{20}	1.3980^{20}	bz	B4^{3}, 1881

PHYSICAL CONSTANTS OF ORGANIC COMPOUNDS (Continued)

No.	Name, Synonyms, and Formula	Mol. wt.	Color, crystalline form, specific rotation and λ_{max} (log ε)	b.p. °C	m.p. °C	Density	n_D	Solubility	Ref.
9941	1,3,6,8-Nonatetraen-5-one, 1,9-diphenyl or Dicinnamylidene acetone $(C_6H_5CH{=}CHCH{=}CH)_2CO$	286.33	ye nd (abs al)		144				$B7^3$, 2756
9942	1-Nonene $C_7H_{15}CH{=}CH_2$	126.24		146		0.730^{21}	1.414^{21}		$B1^4$, 894
9943	3-Nonene *(trans)* $C_2H_5CH{=}CHC_5H_{11}$	126.24		147-8		$0.732^{21/4}$	1.4181^{21}	eth, bz, chl	$B1^4$, 895
9944	3-Nonene, 2-methyl $CH_3CH(CH_3)CH{=}CHC_5H_{11}$	140.27		161.0		$0.7340^{20/4}$	1.4202^{20}		$B1^4$, 903
9945	4-Nonene *(trans)* $C_3H_7CH{=}CHC_4H_9$	126.24				$0.7318^{20/}{}_4$	1.4205^{20}	eth, bz, chl	$B1^4$, 896
9946	4-Nonene, 5-butyl $C_3H_7CH{=}C(C_4H_9)_2$	182.35		215-6		$0.7745^{20/4}$	1.4375^{20}		$B1^4$, 922
9947	1-Nonene-3-yne $C_5H_{11}C{\equiv}CCH{=}CH_2$	122.21		$27\text{-}8^4$		$0.7602^{25/4}$	1.4487^{25}	eth, ace	$B1^4$, 1103
9948	1-Nonene-4-yne $C_4H_9C{\equiv}C\text{-}CH_2CH{=}CH_2$	122.21		58^{22}		$0.777^{25/4}$	1.4413^{25}	eth, ace	$B1^4$, 1103
9949	2-Nonene-4-yne $C_4H_9C{\equiv}C\text{-}CH{=}CHCH_3$	122.21		70^{20}		$0.7832^{25/4}$	1.4590^{25}	eth, ace	$B1^3$, 1048
9950	1-Nonyne $C_7H_{15}C{\equiv}CH$	124.23		150.8, 33.3^{10}	−50	$0.7568^{20/4}$	1.4217^{20}	eth, bz	$B1^4$, 1047
9951	1-Nonyne, 1-chloro $C_7H_{15}C{\equiv}CCl$	158.67		$75\text{-}7^{15}$		0.906^{20}	1.450^{20}	eth	$B1^3$, 1012
9952	2-Nonyne $C_6H_{13}C{\equiv}C{-}CH_3$	124.23		158-9		$0.7690^{20/4}$	1.4337^{20}	eth, lig	$B1^3$, 1013
9953	3-Nonyne $C_5H_{11}C{\equiv}CC_2H_5$	124.23		$153\text{-}5^{740}$, 92^{97}		$0.7616^{20/4}$	1.4299^{20}	eth, lig	$B1^3$, 1013
9954	4-Nonyne $C_4H_9C{\equiv}CC_3H_7$	124.23		$150\text{-}4^{750}$		$0.757^{25/4}$	1.4296^{25}	eth, ace	$B1^4$, 1047
9955	4-Nonyne, 3,3-dimethyl $C_4H_9C{\equiv}C\text{-}C(CH_3)_2C_2H_5$	152.28		82^{40}		$0.7667^{20/4}$	1.4317^{20}	eth, bz	$B1^3$, 1024
9956	4-Nonyne, 8-methyl $(CH_3)_2CHCH_2CH_2C{\equiv}CC_3H_7$	138.25		104.5^{97}		$0.7681^{20/4}$	1.4311^{20}	eth, bz	$B1^3$, 1018
9957	*n*-Nonyl amine or 1-Amino nonane $C_8H_{17}CH_2NH_2$	143.27		202.2, 80.8^{10}	−1	$0.7886^{20/4}$	1.4336^{20}	al, eth	$B4^4$, 777
9958	Nopinone (d) $C_9H_{14}O$	138.21	$[\alpha]^{20/}{}_D$ + 34 (chl)	209, $87\text{-}8^{14}$	0	$0.9807^{20/4}$	1.4787^{20}	w, al, eth	$B7^3$, 303
9959	Noradrenaline-(l) or l-Norepinephrine $C_8H_{11}NO_3$	169.18	$[\alpha]^{25/}{}_D$ −37.5 (dil HCl)		216-8d			dil HCl	$B13^3$, 2382
9960	Norbornane or Norcamphane, bicyclo [2,2,1] heptane C_7H_{12}	96.17		sub	87-8			al, eth, ace, bz	$B5^4$, 258
9961	Norbornyl amine or Isobornyl amine $C_{10}H_{17}NH_2$	153.27	pw $[\alpha]'_D$ −47.7 (4% al)		184			eth, ace	$B12^3$, 195
9962	Norcamphane-2-carboxaldehyde $C_8H_{10}O$	122.17	cr (w)	$70\text{-}2^{22}$		$1.0227^{19/4}$	1.4760^{25}	eth	$B7^3$, 545
9963	2,3-Norcomphane dicarboxylic acid or Bicyclo [2,2,1]-heptane dicarboxylic acid $C_9H_{12}O_4$	184.19			192-5				$B9^3$, 3971
9964	2,3-Norcamphane dicarboxylic anhydride $C_9H_{10}O_3$	166.18			165-7			bz	$B17^4$, 6005
9965	Nordihydroguaiaretic acid or NDGA $C_{18}H_{22}O_4$	302.37	nd (w, al, aa)		185-6			al, eth, ace	$B6^3$, 6731
9966	Norephedrine hydrochloride-(dl) $C_6H_5CH(OH)CH(CH_3)NH_2.HCl$	187.67	pl (abs al), cr (dil HCl, al)		194			w, al	$B13^2$, 371
9967	Norephedrine, N,N-diethyl, hydrochloride $C_6H_5CH(OH)CH(CH_3)N(C_2H_5)_2 \cdot HCl$	243.78	cr (al-ace)		205.6			al	$B13^2$, 380
9968	Norephedrine, N-ethyl $C_6H_5CH(OH)CH(CH_3)NHC_2H_5$	179.26	cr (lig)	143^{18}	51.5			bz	$B9^3$, 1728
9969	Normorphine or Desmethylmorphine $C_{16}H_{17}NO_3$	271.32	(w + 3/2)		273 (+ 3/2 w), 263-4 (anh)				$B27^2$, 117

No.	Name, Synonyms, and Formula	Mol. wt.	Color, crystalline form, specific rotation and λ_{max} (log ε)	b.p. °C	m.p. °C	Density	n_D	Solubility	Ref.
9970	Nornicotine-(l) or *l*-3-(2-Pyrrolidyl) pyridine $C_9H_{12}N_2$	148.21	hyg $[\alpha]^{22}{}_D$ −88.8 (undil)	270, 130-1^{11}	......	1.0737$^{19.5}$	1.5378$^{18.5}$	w, al, eth, ace, chl	B23^2, 107
9971	Ocimene or 3,7-Dimethyl-1,3,7-octatriene $CH_2=C(CH_3)CH_2CH_2CH=C(CH_3)CH=CH_2$	136.24	....	176-8d, 73-4^{21}	......	0.8000^{20}	1.4862^{20}	al, eth, bz, chl, aa	B1^4, 1108
9972	Ocimene, dihydro or 2,6-Dimethyl-2,6-octadiene $CH_3CH=C(CH_3)CH_2CH_2CH=C(CH_3)_2$	138.25	....	168, 75^{10}	......	0.775$^{20/4}$	1.4498^{20}	al, eth, ace, aa	B1^4, 1058
9973	Octacosane $CH_3(CH_2)_{26}CH_3$	394.77	mcl or rh (bz-al)	431.6, 264^{10}	64.5	0.8067$^{20/4}$, 0.7750^{70}	1.4520^{20}, 1.4330^{70}	ace, bz, chl	B1^4, 588
9974	Octacosanoic acid $CH_3(CH_2)_{26}CO_2H$	424.75	(ace or aa)	....	90.4	0.8191^{100}	1.4313^{100}	....	B2^4, 1318
9975	1-Octacosanol $CH_3(CH_2)_{26}CH_2OH$	410.77	(ace or peth)	sub, 200-50^1	83.3	....	....	....	B1^4, 1915
9976	9,12-Octadecadienoic acid or -Linoleic acid *(cis,cis)* $CH_3(CH_2)_4CH=CHCH_2CH=CH(CH_2)_7CO_2H$	280.45	....	229.30^{16}	−5	0.9022$^{20/4}$	1.4699^{20}	al, eth, ace, bz, chl	B2^4, 1754
9977	10,12-Octadecadienoic acid *(trans,trans)* or 10,12-Linoleic acid $CH_3(CH_2)_4CH=CH\text{-}CH=CH(CH_2)_8CO_2H$	280.45	(bz or al)	....	56-7	0.8686$^{70/4}$	1.4689^{60}	....	B2^4, 1752
9978	7,11-Octadecadiyne $CH_3(CH_2)_5C\equiv CCH_2CH_2C\equiv C(CH_2)_5CH_3$	246.44	....	167-8^7	......	0.841$^{19/4}$	1.4698^{19}	....	B1^3, 1068
9979	Octadecanal or Stearaldehyde $CH_3(CH_2)_{16}CHO$	268.48	nd (peth)	261, 212-13^{12}	55	....	....	....	B1^4, 3397
9980	Octadecanal, dimethyl acetal or 1,1-Dimethyoxy octadecane $CH_3(CH_2)_{16}CH(OCH_3)_2$	314.55	....	167-70^3	......	....	1.4410^{25}	al, eth	B1^4, 3397
9981	Octadecane $CH_3(CH_2)_{16}CH_3$	254.50	nd (al, eth-MeOH)	316.1, 173.5^{20}	28.2	0.7768$^{20/4}$	1.4390^{20}	eth, ace, lig	B1^4, 553
9982	Octadecane, 1-amino or Stearyl amine $C_{18}H_{37}NH_2$	269.51	(w)	348.8, 199.5^{20}	fr 52.9	0.8618$^{20/4}$	1.4522^{20}	al, eth, bz, chl	B4^4, 825
9983	Octadecane, 1-amino, hydro chloride $C_{18}H_{37}NH_2.HCl$	305.98	orh pl (al)	....	162-3	....	....	....	B4^4, 826
9984	Octadecane, 1-bromo $CH_3(CH_2)_{16}CH_2Br$	333.40	cr (al)	210^{10}	28.2	0.9848$^{20/4}$	1.4631^{20}	al, eth, peth	B1^4, 555
9985	Octadecane, 1-chloro $CH_3(CH_2)_{16}CH_2Cl$	288.96	....	348, 199^{10}	28.6	0.8641^{20}	1.4531^{20}	....	B1^4, 554
9986	Octadecane, 1,18-dibromo $Br(CH_2)_{18}Br$	412.29	nd or lf (al)	205-7^{15}	64	....	....	chl	B1^4, 556
9987	Octadecane, 1-iodo $CH_3(CH_2)_{16}CH_2I$	380.40	lf (lig), nd (ace, al-ace)	383, 223^{10}	34	1.0994$^{20/4}$	1.4810^{20}	....	B1^4, 556
9988	Octadecane, 9-(4-tolyl) $C_8H_{17}CH(4\text{-}C_6H_4CH_3)C_9H_{19}$	344.62	....	185^{10}	......	0.8549$^{20/4}$	1.4811^{20}	....	B5^4, 1221
9989	Octadecandioic acid, diethyl ester or Diethyl eicosanedioate $C_2H_5O_2C(CH_2)_{16}CO_2C_2H_5$	370.57	....	240^{12}	54-5	....	....	al, eth	B2^4, 2176
9990	1,18-Octadecanediol $HO(CH_2)_{18}OH$	286.50	lf (al, bz), nd (bz, diox)	210-1^2	97-9	....	....	....	B1^4, 2639
9991	1-Octadecanethiol or *n*-Octadecyl mercaptan $CH_3(CH_2)_{16}CH_2SH$	286.56	....	188$^{1-2}$	24-8	0.8475^{20}	1.4645^{20}	eth	B1^4, 1894
9992	Octadecanoic acid or Stearic acid $CH_3(CH_2)_{16}CO_2H$	284.48	mcl lf (al)	360d, 232^{15}	71-2	0.9408$^{20/4}$	1.4299^{80}	eth, ace, chl	B1^4, 1206
9993	1-Octadecanol or Stearyl alcohol $CH_3(CH_2)_{16}CH_2OH$	270.50	lf (al)	210.5^{15}	59-60	0.8124$^{59/4}$	....	al, eth, chl	B1^4, 1888
9994	3-Octadecanone or Ethyl pentadecyl ketone $C_{15}H_{31}COC_2H_5$	268.48	....	48-50	......	....	....	....	B1^4, 3398
9995	9,11,13-Octadecatrienoic acid *(cis)* or α-Eleostearic acid $C_4H_9[CH=CH]_3(CH_2)_7CO_2H$	278.44	nd (al)	235$^{12}{}_d$, 170^1	49	0.9028$^{50/4}$	1.5112^{50}	al, eth	B2^4, 1787
9996	9,12-15-Octadecatrienoic acid *(cis,cis,cis)* or α-Linolenic acid $CH_3[CH_2CH=CH]_3(CH_2)_7CO_2H$	278.44	....	230-2^{17}, 125$^{0.05}$	−11.3	0.9164$^{20/4}$	1.4800^{20}	al, eth	B2^4, 1781
9997	9-Octadecenal or Olealdehyde $CH_3(CH_2)_7CH=CH(CH_2)_7CHO$	266.47	ye nd	168-9^3	......	0.8509$^{20/4}$	1.4558^{20}	....	B1^4, 3533
9998	1-Octadecene $C_{16}H_{33}CH=CH_2$	252.48	....	179^{15}, 145^8	17.5	0.7891$^{20/4}$	1.4448^{20}	bz	B1^4, 930

No.	Name, Synonyms, and Formula	Mol. wt.	Color, crystalline form, specific rotation and λ_{max} (log ε)	b.p. °C	m.p. °C	Density	n_D	Solubility	Ref.
9999	9-Octadecene $C_8H_{17}CH=CHC_8H_{17}$	252.48	...	162^{9}	−30.5	$0.7916^{20/4}$	1.4470^{20}	...	$B1^{4}$, 932
10000	9-Octadecenoic acid (trans) or Elaidic acid $CH_3(CH_2)_7CH=CH(CH_2)_7CO_2H$	282.47	pl (al)	288^{100}, 234^{15}	45	0.8734^{41}	1.4499^{45}	al, eth, bz, chl	$B2^{4}$, 1647
10001	6-Octadecenoic acid, 6,7-diiodo $CH_3(CH_2)_{10}CI=CI(CH_2)_4CO_2H$	534.26	nd (al)	d	48.5	...	...	eth, bz, chl	$B2^{4}$, 1638
10002	9-Octadecenoic acid *(cis)*, 12-hydroxy or Ricinoleic acid $C_6H_{13}CH(OH)CH=CH(CH_2)_8CO_2H$	298.47	$[\alpha]^{22}{}_{D}$ + 5.05	$226\text{-}8^{10}$	α:7.7 β:16 γ:5.5	$0.9450^{21/4}$	1.4716^{21}	al, eth	$B3^{4}$, 1026
10003	11-Octadecenoic acid *(trans)* or Vaccenic acid $C_6H_{13}CH=CH(CH_2)_9CO_2H$	282.47	...	...	44	...	1.4439^{60}	ace	$B2^{4}$, 1640
10004	9-Octadecene, 1-ol *(cis)* or Oleylalcohol $CH_3(CH_2)_7CH=CH(CH_2)_7CH_2OH$	268.48	...	$205\text{-}10^{15}$	6-7	$0.8489^{20/4}$	1.4606^{20}	al, eth	$B1^{4}$, 2204
10005	9-Octadecenl-1-ol *(trans)* or Elaidyl alcohol $CH_3(CH_2)_7CH=CH(CH_2)_7CH_2OH$	268.48	(al or ace)	198^{10}	36-7	$0.8338^{40/4}$	1.4552^{40}	al, eth	$B1^{4}$, 2204
10006	Octadecyl sulfate $(C_{18}H_{37}O)_2SO_2$	603.04	...	...	70.5	...	...	...	$B1^{4}$, 1892
10007	1-Octadecyne $C_{16}H_{33}C\equiv CH$	250.47	(al)	313, 180^{15}	22.5	$0.8025^{20/4}$	1.4774^{20}	...	$B1^{4}$, 1075
10008	2-Octadecyne $C_{15}H_{31}C\equiv CCH_3$	250.47	(al)	184^{15}	30	$0.8016^{30/4}$	...	...	$B1^{4}$, 1075
10009	9-Octadecyne $C_8H_{17}C\equiv CC_8H_{17}$	250.47	...	$163\text{-}4^{7}$	3	$0.8012^{20/4}$	1.4488^{25}	...	$B1^{4}$, 1076
10010	9-Octadecynoic acid or Stearolic acid $CH_3(CH_2)_7C\equiv C(CH_2)_7CO_2H$	280.45	pr (al, peth), nd (dil al)	$189\text{-}90^{1.8}$	48	...	1.4510^{54}	eth	$B2^{4}$, 1751
10011	1,3-Octadiene, 3-chloro $C_4H_9CH=CClCH=CH_2$	144.64	...	$64\text{-}5^{18}$	...	$0.9366^{20/4}$	1.4794^{20}	eth	$B1^{4}$, 1038
10012	1,6-Octadiene, 7-methyl-3-methylene $(CH_3)_2C=CHCH_2CH_2C(=CH_2)CH=CH_2$	136.24	...	$56\text{-}7^{12}$	...	$0.7982^{20/4}$	1.47065^{20}	...	$B1^{4}$, 1108
10013	1,7-Octadiene $CH_2=CH(CH_2)_4CH=CH_2$	110.20	...	113-8	...	$0.735^{20/20}$	1.424^{20}	...	$B1^{4}$, 1038
10014	2,4-Octadiene, 7-methyl $(CH_3)_2CHCH_2CH=CHCH=CHCH_3$	124.23	...	149	...	$0.7521^{18/4}$	1.4543^{18}	...	$B1^{2}$, 239
10015	2-6-Octadiene $CH_3CH=CHCH_2CH_2CH=CHCH_3$	110.20	...	118-20	...	0.748^{17}	1.4292^{17}	...	$B1^{4}$, 1039
10016	2,6-Octadiene, 2,6-dimethyl *(cis,cis)* $CH_3CH=C(CH_3)CH_2CH_2CH=C(CH_3)_2$	138.25	...	168, 75^{30}	...	$0.775^{21/4}$	1.4498^{20}	al, eth, ace	$B1^{4}$, 1058
10017	2,6-Octadıene, 2,6-dimethyl or Dihydroocimene $CH_3CH=C(CH_3)CH_2CH_2CH=C(CH_3)_2$	138.25	...	168, 75^{30}	...	$0.775^{21/4}$	1.4498^{20}	al, eth, ace, aa	$B1^{4}$, 1058
10018	2,7-Octadiene, 2,6-dimethyl or Linaloolen $(CH_3)_2C=CHCH_2CH_2CH(CH_3)CH=CH_2$	138.25	...	168, 58^{12}	...	0.7882^{20}	1.4561^{20}	al	B1, 261
10019	1,5-Octadien, 3-yn-5-propyl $C_2H_5CH=C(C_3H_7)C\equiv CCH=CH_2$	148.25	...	$57\text{-}8^{6}$	...	$0.8047^{20/4}$	1.4949^{20}	eth, ace	$B1^{4}$, 1129
10020	2,6-Octadien-4-yne, 3,6-diethyl $CH_3CH=C(C_2H_5)C\equiv CC(C_2H_5)=CHCH_3$	162.27	...	169-71, 99^{12}	...	$0.8196^{20/4}$	1.4965^{20}	eth, ace	$B1^{3}$, 1066
10021	2,6-Octadien-4-yne, 3,6-dimethyl $CH_3CH=C(CH_3)C\equiv CC(CH_3)=CHCH_3$	134.22	...	170	−45	$0.8071^{22/4}$	1.4998^{20}	eth, ace	$B1^{4}$, 1129
10022	1,7-Octadiyne $HC\equiv C(CH_2)_4C\equiv CH$	106.17	...	135-6, $93\text{-}5^{16}$	...	$0.8169^{21/4}$	1.4521^{18}	eth	$B1^{4}$, 1122
10023	2,6-Octadiyne $CH_3C\equiv CCH_2CH_2C\equiv CCH_3$	106.17	...	62^{19}	27	$0.828^{80/4}$	1.4658^{30}	eth	$B1^{4}$,1123
10024	3,5-Octadiyne $CH_3CH_2C\equiv CC\equiv CCH_2CH_3$	106.17	...	163-4, 78^{34}	...	$0.826^{0/4}$	1.4968^{0}	eth	$B1^{4}$, 1123
10025	3,5-Octadiyne, 2,7-dimethyl $(CH_3)_2CHC\equiv CC\equiv CCH(CH_3)_2$	134.22	...	74^{12}	...	$0.8090^{20/4}$	...	eth	$B1^{1}$, 128
10026	Octanal or Caprylaldehyde $CH_3(CH_2)_6CHO$	128.21	...	171, 72^{20}	...	$0.8211^{20/4}$	1.4217^{20}	al, eth, ace, bz	$B1^{4}$, 3337
10027	Octane $CH_3(CH_2)_6CH_3$	114.23	...	125.7, 19.2^{10}	−56.8	$0.7025^{20/4}$	1.3974^{20}	al, eth, ace, bz, chl, peth	$B1^{4}$, 412
10028	Octane, 1-amino or Octyl amine $CH_3(CH_2)_6CH_2NH_2$	129.25	...	179.6, 63.2^{10}	0	$0.7826^{20/4}$	1.4924^{20}	al, eth	$B4^{4}$, 751
10029	Octane, 2-amino (d) $C_6H_{13}CH(NH_2)CH_3$	129.25	$[\alpha]^{17}{}_{D}$ + 8.6 (undil)	70^{25}	...	$0.771^{25/4}$	1.4220^{25}	al, eth	$B4^{4}$, 762
10030	Octane, 2-amino (dl) $C_6H_{13}CH(NH_2)CH_3$	129.25	...	163-5, $58\text{-}9^{13}$	...	$0.7745^{20/0}$	1.4232^{25}	al, eth	$B4^{4}$, 763

No.	Name, Synonyms, and Formula	Mol. wt.	Color, crystalline form, specific rotation and λ_{max} (log ε)	b.p. °C	m.p. °C	Density	n_D	Solubility	Ref.
10031	Octane, 1-bromo or Octyl bromide $CH_3(CH_2)_6CH_2Br$	193.13	...	$200.8, 77.3^{10}$	−55	$1.1122^{20/4}$	1.4524^{20}	al, eth	B1⁴, 422
10032	Octane, 1-Bromo-8-fluoro $Br(CH_2)_8F$	211.12	...	$118\text{-}20^{22\,5}$	...	...	1.4500^{20}	al, eth	B1⁴, 424
10033	Octane, 2-bromo (d) $C_6H_{13}CHBrCH_3$	193.13	$[\alpha]^{25/}_D$ + 34.2	71^{14}	...	$1.0982^{25/4}$	1.4500^{20}	al, eth	B1⁴, 423
10034	Octane, 2-bromo (dl) $C_6H_{13}CHBrCH_3$	193.13	...	$188\text{-}9, 72^{14}$	...	$1.0878^{25/4}$	1.4442^{25}	**al, eth**	B1⁴, 423
10035	Octane, 2-bromo (l) $C_6H_{13}CH(Br)CH_3$	193.13	$[\alpha]^{25/}_D$ −37.5	$72\text{-}3^{18}, 46^{1}$	...	$1.0920^{20/4}$	1.4475^{25}	**al, eth**	B1⁴, 423
10036	Octane, 1-chloro or Octyl chloride $CH_3(CH_2)_6CH_2Cl$	148.68	...	$182, 78^{15}$	−57.8	$0.8738^{20/4}$	1.4305^{20}	...	B1⁴, 419
10037	Octane, 1-chloro-8-fluoro $F(CH_2)_8Cl$	166.67	...	87^{10}	...	$0.978^{20/4}$	1.4266^{25}	al, eth, ace	B1⁴, 420
10038	Octane, 2-chloro (d) $C_6H_{13}CHClCH_3$	148.68	$[\alpha]^{20/}_D$ + 33.7	$171\text{-}3, 75^{28}$	...	$0.8658^{17/4}$	1.4273^{21}	al, eth	B1⁴, 419
10039	Octane, 3-chlor-3-methyl $C_5H_{11}CCl(CH_3)CH_2CH_3$	162.70	...	$73\text{-}4^{15}$	...	$0.8680^{25/4}$	1.4351^{20}	eth, ace, bz, chl	B1³, 508
10040	Octane, 4-chloro (d) $C_4H_9CHClC_3H_7$	148.68	$[\alpha]^{25/}_D$ + 0.28 (undil)	92^{50}	...	...	...	al, eth, chl	B1³, 466
10041	Octane, 4-chloro-4-methyl $C_4H_9CCl(CH_3)C_3H_7$	162.70	...	$71^{14\,5}$	...	$0.8723^{20/4}$	1.4360^{20}	eth, chl	B1³, 509
10042	Octane, 1-cyclohexyl $C_8H_{17}C_6H_{11}$	196.38	...	263.6	−19.7	0.8138^{20}	1.4504^{20}	...	B5⁴, 178
10043	Octane, 1,2-dibromo $C_6H_{13}CHBrCH_2Br$	272.02	...	$240\text{-}2, 118.5^{15}$	...	$1.4580^{20/4}$	1.4970^{20}	...	B1³, 468
10044	Octane, 1,8-dibromo $Br(CH_2)_8Br$	272.02	...	270-2, $92\text{-}3^{0\,45}$	15-6	$1.4594^{25/4}$	1.4971^{25}	eth, chl	B1⁴,424
10045	Octane, 2,3-dimethyl $C_5H_{11}CH(CH_3)CH(CH_3)_2$	142.28	...	164.7	...	$0.7377^{20/4}$	1.4146^{20}	...	B1⁴, 476
10046	Octane, 2,6-dimethyl $CH_3CH_2CH(CH_3)(CH_2)_3CH(CH_3)_2$	142.28	...	$160\text{-}1, 90\text{-}3^{26}$	...	$0.7313^{20/4}$	1.4097^{20}	...	B1⁴, 477
10047	Octane, 2,7-dimethyl $(CH_3)_2CH(CH_2)_4CH(CH_3)_2$	142.28	...	159.6	−54.6	$0.7240^{20/4}$	1.4092^{20}	eth	B1⁴, 478
10048	Octane, 1-fluoro or Octyl fluoride $CH_3(CH_2)_6CH_2F$	132.22	...	142-3	...	$0.8103^{20/4}$	1.3935^{20}	...	B1⁴, 418
10049	Octane, 1-iodo or Octyl iodide $CH_3(CH_2)_7CH_2I$	240.13	...	$225.5, 86.5^{5}$	−45.7	$1.3297^{20/4}$	1.4889^{20}	al, eth	B1⁴, 425
10050	Octane, 2-iodo (D-) $C_6H_{13}CHICH_3$	240.13	$[\alpha]^{28/}_D$ −45.5	92^{12}	...	$1.3219^{20/4}$	1.4863^{25}	al, eth, lig	B1⁴, 425
10051	Octane, 2-iodo (dl) $C_6H_{13}CHICH_3$	240.13	...	$210, 95\text{-}6^{16}$	...	$1.3251^{20/4}$	1.4896^{20}	al, eth, lig	B1⁴, 425
10052	Octane, 2-iodo (l, +) $C_6H_{13}CHICH_3$	240.13	$[\alpha]^{26/}_D$ + 46.3	101^{22}	...	$1.3314^{17/4}$	1.4877^{22}	al, eth, lig	B1⁴, 425
10053	Octane, 2-methyl $C_6H_{13}CH(CH_3)_2$	128.26	...	142.8	−80.1	$0.7107^{20/4}$	1.4029^{20}	al, eth, lig	B1⁴, 454
10054	Octane, 3-methyl (d) $C_5H_{11}CH(CH_3)C_2H_5$	128.26	$[\alpha]^{17/}_D$ + 9.4	143-4	−107.6	0.7206^{17}	1.4068^{20}	ace, bz	B1⁴, 455
10055	Octane, 3-methyl (l) $C_5H_{11}CH(CH_3)C_2H_5$	128.26	$[\alpha]^{27/}_D$ −8.5	143	...	$0.714^{27/4}$	1.4052^{25}	ace, bz	B1³, 509
10056	Octane, 4-methyl (dl) $C_4H_9CH(CH_3)C_3H_7$	128.26	...	$142.4, 32^{10}$	−113.2	$0.7199^{20/4}$	1.4061^{20}	al, eth, ace, bz	B1⁴, 456
10057	Octane, 4-methyl (l) $C_4H_9CH(CH_3)C_3H_7$	128.26	$[\alpha]^{19/}_D$ −1.06	141	...	$0.717^{19/4}$	...	al, eth, ace, bz	B1³, 510
10058	Octane, 1-phenyl or Octylbenzene $C_7H_{15}CH_2C_6H_5$	190.33	...	$264\text{-}5, 131\text{-}4^{12}$	−7	$0.8582^{20/4}$	1.4851^{20}	**bz, eth**	B5⁴, 1157
10059	Octane, 2-phenyl $C_6H_{13}CH(C_6H_5)CH_3$	190.33	...	$123\text{-}5^{20}$	...	$0.8611^{20/4}$	1.4837^{20}	...	B5⁴, 1157
10060	Octanedial or Suberaldehyde $OHC(CH_2)_6CHO$	142.20	...	230-40d, $96\text{-}8^{3}$	...	...	1.4439^{20}	w, al	B1⁴, 3706
10061	Octanedioic acid or Suberic acid $HO_2C(CH_2)_6CO_2H$	174.20	lo nd or pl (w)	300 sub, 219.5^{10}	144	...	...	al	B2⁴, 2028
10062	1,7-Octanediol, 3,7-dimethyl $(CH_3)_2C(OH)(CH_2)_3CH(CH_3)CH_2CH_2OH$	174.28	...	265	...	0.937^{20}	1.4599^{20}	...	B1³, 2233
10063	1,8-Octanediol $HO(CH_2)_8OH$	146.23	nd (bz-lig), pr	172^{20}	63	...	...	al, eth, ace	B1⁴, 2592

PHYSICAL CONSTANTS OF ORGANIC COMPOUNDS (Continued)

No.	Name, Synonyms, and Formula	Mol. wt.	Color, crystalline form, specific rotation and λ_{max} (log ε)	b.p. °C	m.p. °C	Density	n_D	Solubility	Ref.
10064	4,5-Octanediol (dl) or 1,2-Dipropyl ethylene glycol $C_3H_7CHOHCHOHC_3H_7$	146.23	....	110^{8}	28	....	1.4419^{25}	....	$B1^{4}$, 2593
10065	4,5-Octanediol *(meso)* $C_3H_7CHOHCHOHC_3H_7$	146.23	....	....	123-4	....	....	....	$B1^{4}$, 2593
10066	2,3-Octanedione $C_5H_{11}COCOCH_3$	142.20	....	$172\text{-}3^{733}$	......	....	....	al	$B1^{3}$, 3138
10067	2,3-Octanedione dioxime or Methyl pentyl glyoxime $C_5H_{11}C(=NOH)C(=NOH)CH_3$	172.23	nd (dil al)	....	173	....	....	al, ace	$B1^{4}$, 3706
10068	2,3-Octanedione, 3-oxime $C_5H_{11}C(=NOH)COCH_3$	157.21	(lig)	133^{15}	59	....	....	eth	B1, 795
10069	2,7-Octanedione $CH_3CO(CH_2)_4COCH_3$	142.20	pl (bz)	114^{10}	44	....	....	al	$B1^{4}$, 3707
10070	2,7-Octanedione dioxime $CH_3C(=NOH)(CH_2)_4C(=NOH)CH_3$	172.23	(al)	....	158	....	....	al	B1, 795
10071	3,6-Octanedione $CH_3CH_2COCH_2CH_2COCH_2CH_3$	142.20	pl (al)	98^{14} sub	35-6	....	....	al	$B1^{4}$, 3708
10072	3,6-Octanedione, 2,2,7,7-tetramethyl $(CH_3)_3CCOCH_2CH_2COC(CH_3)_3$	198.31	....	$115\text{-}7^{17}$, $55\text{-}60^{0.5}$	2.5	$0.900^{27/4}$	1.4400^{20}	al	$B1^{4}$, 3734
10073	4,5-Octanedione $C_3H_7COCOC_3H_7$	142.20	ye oil	168, 60^{12}	......	$0.934^{0/4}$	....	al, eth, ace	$B1^{4}$, 3708
10074	4,5-Octanedione dioxime $C_3H_7C(=NOH)C(=NOH)C_3H_7$	172.23	....	sub	186-7	....	....	al, eth	$B1^{2}$, 846
10075	1-Octanethiol or Octyl mercaptan $CH_3(CH_2)_6CH_2SH$	146.29	....	199.1, 86^{15}	−49.2	$0.8433^{20/4}$	1.4540^{20}	al	$B1^{4}$, 1767
10076	2-Octanethiol (dl) $C_6H_{13}CHSHCH_3$	146.29	....	186.4, 88.9^{30}	−79	$0.8366^{20/4}$	1.4504^{20}	al, eth, bz	$B1^{4}$, 1777
10077	2-Octanethiol (l) $C_6H_{13}CHSHCH_3$	146.29	$[\alpha]^{25/}_{546}$ −36.4	$78\text{-}80^{22}$	......	$0.830^{25/4}$	....	al, eth, bz	$B1^{4}$, 1777
10078	Octanoic acid or Caprylic acid $CH_3(CH_2)_6CO_2H$	144.21	....	239.3, 140^{23}	16.5	0.9088^{20}	1.4285^{20}	**al, chl**	$B1^{4}$, 982
10079	1-Octanol or Octyl alcohol $CH_3(CH_2)_6CH_2OH$	130.23	....	194.4, 98^{19}	−16.7	$0.8270^{20/4}$	1.4295^{20}	**al, eth**	$B1^{4}$, 1756
10080	1-Octanol, 8-chloro $Cl(CH_2)_8OH$	164.68	....	139^{19}	......	....	1.4563^{25}	al, eth	$B1^{4}$, 1766
10081	1-Octanol, 3,7-dimethyl (d) or Tetrahydrogeraniol $(CH_3)_2CH(CH_2)_3CH(CH_3)CH_2CH_2OH$	158.28	$[\alpha]^{20/}_{D}$ +4.1	212-3, $105\text{-}6^{10}$	......	$0.8285^{20/4}$	1.4355^{20}	eth	$B1^{4}$, 1830
10082	1-Octanol, 3,7-dimethyl (l) $(CH_3)_2CH(CH_2)_3CH(CH_3)CH_2CH_2OH$	158.28	$[\alpha]^{11/}_{546}$ −3.7	212-3, 109^{15}	......	0.830^{18}	1.4370^{15}	eth	$B1^{3}$, 1768
10083	1-Octanol, 8-fluoro $F(CH_2)_8OH$	148.22	....	$106\text{-}7^{10}$	......	$0.945^{20/4}$	1.4248^{25}	al, eth	$B1^{4}$, 1766
10084	2-Octanol (d) $C_6H_{13}CH(OH)CH_3$	130.23	$[\alpha]^{17/}_{D}$ +9.9	86^{20}	......	$0.8216^{20/4}$	1.4264^{20}	al, eth, ace	$B1^{4}$, 1770
10085	2-Octanol (dl) $C_6H_{13}CH(OH)CH_3$	130.23	....	180, 87^{20}	−31.6	$0.8193^{20/4}$	1.4203^{20}	al, eth, ace	$B1^{4}$, 1770
10086	2-Octanol (l) $C_6H_{13}CH(OH)CH_3$	130.23	$[\alpha]^{17/}_{D}$ −9.9	86^{20}	......	$0.8201^{20/4}$	1.4264^{20}	al, eth, ace	$B1^{3}$, 1721
10087	2-Octanol, 2-methyl $C_6H_{13}C(OH)(CH_3)_2$	144.26	....	178, $81\text{-}3^{16}$	......	$0.8210^{20/4}$	1.4280^{20}	al, eth	$B1^{4}$, 1805
10088	3-Octanol $C_2H_5CH(OH)C_5H_{11}$	130.23	....	166-72	......	....	....	....	$B1^{4}$, 1779
10089	3-Octanol, 3,6-dimethyl $C_2H_5CH(CH_3)CH_2CH_2C(OH)(CH_3)C_2H_5$	158.28	....	202.2	−67.5	....	1.4370^{20}	....	$B1^{1}$, 214
10090	3-Octanol, 3,7-dimethyl (dl) or Tetrahydrolinalool $(CH_3)_2CH(CH_2)_3C(OH)(CH_3)CH_2CH_3$	158.28	....	196-7, $87\text{-}8^{10}$	31-2	0.8280^{20}	1.4335^{20}	al	$B1^{4}$, 1829
10091	3-Octanol, 3-ethyl $C_5H_{11}C(OH)(C_2H_5)_2$	158.28	....	199, 84^{12}	......	$0.8361^{25/4}$	1.4390^{20}	al	$B7^{3}$, 1766
10092	4-Octanol (d) $C_3H_7CH(OH)C_4H_9$	130.23	$[\alpha]^{22/}_{D}$ 0.74	79^{16}	......	$0.8159^{25/4}$	1.4275^{25}	al	$B1^{2}$, 452
10093	4-Octanol (dl) $C_3H_7CH(OH)C_4H_9$	130.23	....	176.3, 81.3^{20}	−40.7	$0.8186^{20/4}$	1.4248^{20}	al	$B1^{4}$, 1779
10094	2-Octanone or Methyl hexyl ketone $C_6H_{13}COCH_3$	128.21	....	173, $59\text{-}60^{11}$	−16	$0.8202^{20/4}$	1.4151^{20}	**al, eth**	$B1^{4}$, 3339
10095	3-Octanone or Ethyl pentyl ketone $C_5H_{11}COC_2H_5$	128.21	....	167^{749}	......	$0.8221^{20/4}$	1.4153^{20}	**al, eth**	$B1^{4}$, 3341
10096	4-Octanone or Butyl propyl ketone $C_4H_9COC_3H_7$	128.21	....	163, 70^{26}	......	$0.8146^{25/4}$	1.4173^{14}	**al, eth**	$B1^{4}$, 3342

No.	Name, Synonyms, and Formula	Mol. wt.	Color, crystalline form, specific rotation and λ_{max} (log ε)	b.p. °C	m.p. °C	Density	n_D	Solubility	Ref.
10097	4-Octanone, 5-hydroxy or Butyroin $C_3H_7CH(OH)COC_3H_7$	144.21		180-90, 95^{20}	−10	0.9231^{20}	1.4290^{20}	al, eth, ace	$B1^4$, 4042
10098	4-Octanone, 7-methyl or Propyl-*iso*-pentyl ketone $(CH_3)_2CHCH_2CH_2COC_3H_7$	142.24		177-9		$0.8239^{20/4}$	1.4210^{20}	al, eth	$B1^4$, 3356
10099	Octasiloxane, octadecamethyl $CH_3[-Si(CH_3)_2O-]_7Si(CH_3)_3$	607.31		$153^{5\ 1}$		0.913	1.3970^{20}	bz, lig, peth	$B4^3$, 1881
10100	1,3,5,7-Octatetraene $CH_2=CHCH=CHCH=CHCH=CH_2$	106.17	(bz)	sub	ca 50			aa	$B1^4$, 1124
10101	2,4,6-Octatriene *(trans,trans,trans)* $CH_3CH=CHCH=CHCH=CHCH_3$	108.18	lf	147-8, 43^{10}	52	$0.7961^{23/4}$	1.5131^{27}	al, chl, lig	$B1^4$, 1101
10102	1,3,7-Octatriene, 3,7-dimethyl or Ocimene $CH_2=C(CH_3)CH_2CH_2CH=C(CH_3)CH=CH_2$	136.24		176-8d, $73\text{-}4^{21}$		0.8000^{20}	1.4862^{20}	al, eth, bz, chl, aa	$B1^4$, 1108
10103	2,4,6-Octatriene, 2,6-dimethyl (4-*trans*,6-*trans*) or Allocimene A $CH_3CH=C(CH_3)CH=CH-CH=C(CH_3)_2$	136.24		188^{750}, 91^{20}	−35.4	$0.8118^{20/4}$	1.5446^{20}		$B1^4$, 1106
10104	2,4,6-Octatriene, 2,6-dimethyl (4-*trans*,6-*cis*) or Allocimene B $CH_3CH=C(CH_3)CH=CHCH=C(CH_3)_2$	136.24		89^{20}	−20.6	0.8060^{20}	1.5446^{20}		$B1^4$, 1106
10105	1-Octene $C_6H_{13}CH=CH_2$	112.22		121.3, 15.4^{10}	−101.7	$0.7149^{20/4}$	1.4087^{20}	al, eth, ace, bz, chl	$B1^4$, 874
10106	1-Octene, 2-chloro $C_6H_{13}CCl=CH_2$	146.66		168-70		$0.9274^{0/0}$		eth, ace, bz	$B1^1$, 840
10107	1-Octene, 3,7-dimethyl $(CH_3)_2CH(CH_2)_3CH(CH_3)CH=CH_2$	140.27		154		$0.7396^{20/4}$	1.4212^{20}		$B1^4$, 905
10108	1-Octene, 2-methyl $C_6H_{13}C(CH_3)=CH_2$	126.24		114.8	−77.8	$0.7343^{20/4}$	1.4184^{20}		$B1^4$, 896
10109	2-Octene *(cis)* $C_5H_{11}CH=CHCH_3$	112.22		125.6, 16.5^{10}	−100.2	$0.7243^{20/4}$	1.4150^{20}	al, eth, ace, bz, chl	$B1^4$, 878
10110	2-Octene *(trans)* $C_5H_{11}CH=CHCH_3$	112.22		125, 16^{10}	−87.7	$0.7199^{20/4}$	1.4132^{20}	al, eth, ace, bz, chl	$B1^4$, 879
10111	2-Octene, 2-chloro $C_5H_{11}CH=CClCH_3$	146.66		167-8		$0.8923^{16/16}$	1.4424^{16}	al, eth, ace, bz	$B1^4$, 879
10112	2-Octene, 4-chloro $C_4H_9CHClCH=CHCH_3$	146.66		153, $65\text{-}6^{15}$		$0.8924^{20/4}$	1.4452^{20}	eth, ace, bz, chl	$B1^1$, 840
10113	2-Octene, 2,6-dimethyl $C_2H_5CH(CH_3)CH_2CH_2CH_2C(CH_3)=CH_2$	140.27		163		$0.746^{22/4}$	1.425^{22}		$B1^4$, 904
10114	3-Octene *(cis)* $C_4H_9CH=CHC_2H_5$	112.22		122.9, 14.3^{10}	−126	$0.7189^{20/4}$	1.4135^{20}	al, eth, ace, bz, lig	$B1^4$, 880
10115	3-Octene *(trans)* $C_4H_9CH=CHC_2H_5$	112.22		123.3, 14.6^{10}	−110	$0.7152^{20/4}$	1.4126^{20}	al, eth, ace, bz, lig	$B1^4$, 880
10116	4-Octene *(cis)* $C_3H_7CH=CHC_3H_7$	112.22		122.5, 14^{10}	−118.7	$0.7212^{20/4}$	1.4148^{20}	al, eth, ace, bz, lig	$B1^4$, 880
10117	4-Octene *(trans)* $C_3H_7CH=CHC_3H_7$	112.22		122.3, 13.7^{10}	−93.8	$0.7141^{20/4}$	1.4114^{20}	al, eth, ace, bz, lig	$B1^4$, 880
10118	4-Octene, 4-chloro *(cis)* $C_3H_7CH=CClC_3H_7$	146.66		165.3		$0.8912^{20/4}$	1.4447^{20}	eth	$B1^4$, 881
10119	1-Octen-3-yne $C_4H_9C≡C-CH=CH_2$	108.18		62^{60}		$0.7830^{20/4}$	1.4592^{20}	eth	$B1^4$, 1100
10120	1-Octen-3yne-5-ol, 5-methyl $C_3H_7C(OH)(CH_3)C≡C-CH=CH_2$	138.21		80^{13}		$0.8851^{15/4}$	1.4735^{20}	al	$B1^3$, 2034
10121	Octyl ether or Dioctyl ether $(C_8H_{17})_2O$	242.45		286-7		$0.8063^{20/4}$	1.4327^{20}	al, eth	$B1^4$, 1760
10122	Octyl nitrate $C_8H_{17}ONO_2$	175.23		$110\text{-}2^{20}$		$0.8419^{17/17}$		al, eth	$B1^4$, 1762
10123	Octyl nitrite $C_8H_{17}ONO$	159.23	ye	174-5, 60^{10}		0.862^{17}	1.4127^{20}	al, eth	$B1^4$, 762
10124	Octyl phenyl ether $C_8H_{17}OC_6H_5$	206.33		285, $164\text{-}7^{20}$	8	$0.9319^{15/15}$	1.4875^{20}	al, eth	B6, 144
10125	Octyl thiocyanate $C_8H_{17}SCN$	171.30		$141\text{-}2^{19}$	105	$0.9149^{25/4}$	1.4649^{20}	al, eth	$B3^1$, 282
10126	1-Octyne or Hexyl acetylene $C_6H_{13}C≡CH$	110.20		125.2, 19.7^{10}	−79.3	0.7461^{20}	1.4159^{20}	al, eth	$B1^4$, 1034
10127	1-Octyne, 1-chloro $C_6H_{13}C≡CCl$	144.64		$61\text{-}2^{17}$		0.912^{20}	1.445^{20}	al, eth	$B1^3$, 1005
10128	1-Octyne-3-ol, 3-methyl $C_5H_{11}C(OH)(CH_3)C≡CH$	140.23		75^{10}		$0.863^{10/10}$	1.443^{10}		$B1^4$, 2268

PHYSICAL CONSTANTS OF ORGANIC COMPOUNDS (Continued)

No.	Name, Synonyms, and Formula	Mol. wt.	Color, crystalline form, specific rotation and λ_{max} (log ε)	b.p. °C	m.p. °C	Density	n_D	Solubility	Ref.
10129	2-Octyne $C_5H_{11}C{\equiv}CCH_3$	110.20	...	138	−61.6	$0.7596^{20/4}$	1.4278^{20}	al, eth	$B1^4$, 1035
10130	2-Octyne-1-ol $C_5H_{11}C{\equiv}CCH_2OH$	126.20	...	$98-9^{15}$	−18	$0.8805^{20/4}$	1.4556^{20}	**eth**	$B1^4$, 2256
10131	3-Octyne $C_4H_9C{\equiv}CC_2H_5$	110.20	...	$133, 85^{169}$	−103.9	$0.7529^{20/4}$	1.4250^{20}	al, eth	$B1^4$, 1036
10132	3-Octyne, 2-chloro-2-methyl $C_4H_9C{\equiv}CCCl(CH_3)_2$	158.67	...	68^{15}	...	$0.8929^{20/4}$	1.4480^{20}	eth	$B1^4$, 1048
10133	3-Octyne, 2,2-dimethyl $C_4H_9C{\equiv}CC(CH_3)_3$	138.25	...	79^{60}	...	$0.7491^{20/4}$	1.4270^{20}	eth	$B1^3$, 1018
10134	3-Octyne, 7-methyl $(CH_3)_2CHCH_2CH_2C{\equiv}CC_2H_5$	124.23	...	87^{99}	...	$0.7599^{20/4}$	1.4280^{20}	eth	$B1^3$, 1014
10135	4-Octyne $C_3H_7C{\equiv}CC_3H_7$	110.20	...	131.5	−102.5	$0.7509^{20/4}$	1.4248^{20}	al, eth	$B1^4$, 1037
10136	Olealdehyde or 9-Octadecenal $CH_3(CH_2)_7CH{=}CH(CH_2)_7CHO$	266.47	ye nd	$168-9^{3}$	...	$0.8509^{20/4}$	1.4558^{20}	...	$B1^4$,3533
10137	Oleamide $CH_3(CH_2)_7CH{=}CH(CH_2)_7CONH_2$	281.48	...	...	76	...	...	eth	$B2^4$, 1668
10138	Oleamide, N-phenyl or Oleanilide $CH_3(CH_2)_7CH{=}CH(CH_2)_7CONHC_6H_5$	357.58	nd	143.5^{10}	41	...	...	eth, bz, aa	$B12^2$, 150
10139	Oleic acid or 9,10-Octadecenoic acid (*cis*) $CH_3(CH_2)_7CH{=}CH(CH_2)_7CO_2H$	282.47	...	$286^{100}, 228-9^{15}$	16.3	$0.8935^{20/4}$	1.4582^{20}	**al, eth, ace, bz, chl**	$B2^4$, 1641
10140	Oleic acid, benzyl ester or Benzyl oleate $C_{17}H_{33}CO_2CH_2C_6H_5$	372.59	...	237^{1}	...	$0.9330^{25/25}$	1.4875^{25}	al, eth	$B6^4$, 2269
10141	Oleic acid, butyl ester or Butyl oleate $C_{17}H_{33}CO_2C_4H_9$	338.57	ye	$227-8^{15}$	−26.4	0.8704^{15}	1.4480^{25}	al	$B2^4$, 1653
10142	Oleic acid, ethyl ester or Ethyl oleate $C_{17}H_{33}CO_2C_2H_5$	310.52	...	207^{13}	0.8720^{2}	1.4515^{20}	...	**al, eth**	$B2^4$, 1651
10143	Oleic acid, methyl ester or Methyl oleate $C_{17}H_{33}CO_2CH_3$	296.49	...	218.5^{20}	−19.9	0.8739^{20}	1.4522^{20}	...	$B2^4$, 1649
10144	Oleic acid, isopentyl ester or Isopentyl oleate $C_{17}H_{33}CO_2\text{-i-}C_5H_{11}$	352.60	...	$223-4^{10}$	...	0.897^{15}	...	al, eth	$B2^3$, 1414
10145	Oleonitrile $C_{17}H_{33}CN$	263.47	...	$330-5d, 204^{12}$	−1	$0.848^{17/17}$	1.4566^{20}	...	$B2^4$, 1668
10146	Oleyl alcohol or 9-Octadecen-1-ol $CH_3(CH_2)_7CH{=}CH(CH_2)_7CH_2OH$	268.48	...	$205-10^{15}$	6-7	$0.8489^{20/4}$	1.4606^{20}	al, eth	$B1^4$, 2204
10147	Orcein or Orcin $C_{28}H_{24}N_2O_7$	500.51	br-red pw	...	...	...	...	al, ace	$B6^3$, 4531
10148	Ornithine (L) or L-2,5-Diamino valeric acid $H_2N(CH_2)_3CH(NH_2)CO_2H$	132.16	cr (al-eth), $[\alpha]^{25}{}_D$ + 11.5 (w, c=6.5)	...	140	...	...	w, al	$B4^3$, 1346
10149	Ornithine monohydrochloride (L) $H_2N(CH_2)_3CH(NH_2)CO_2H.HCl$	168.62	nd $[\alpha]^{25}{}_D$ + 11.0 (w, c=5.5)	...	215 (230)	...	...	w	$B4^3$, 1347
10150	Ornithine-sulfate (L) $H_2N(CH_2)_3CH(NH_2)CO_2H.H_2SO_4$	230.24	$[\alpha]^{25}{}_D$ + 8.4 (w)	...	234d	...	...	w	$B4^3$, 1347
10151	Orthoacetic acid, triethyl ester or 1,1,1-triethoxyethane, triethyl ortho acetate $CH_3C(OC_2H_5)_3$	162.23	...	$144-6, 66.5^{41}$	...	$0.8847^{25/4}$	1.3980^{20}	**al, eth, chl**	$B2^4$, 137
10152	Orthoacetic acid, trimethyl ester or 1,1,1-trimethoxy ethane,trimethyl ortho acetate $CH_3C(OCH_3)_3$	120.15	...	107-9	...	$0.9438^{25/4}$	1.3859^{25}	al, eth	$B2^4$, 127
10153	Orthocarbonic acid, tetraethyl ester or Ethyl ortho carbonate,Tetraethoxy methane $C(OC_2H_5)_4$	192.26	...	$160-1, 62^{28}$	...	$0.9186^{20/4}$	1.3928^{20}	**al, eth**	$B3^4$, 6
10154	Orthocarbonic acid, tetrapropyl ester or Propyl ortho carbonate $C(OC_3H_7)_4$	248.36	...	224.2	...	$0.897^{20/4}$	1.4100^{20}	al, eth	$B3^4$, 7
10155	Orthoformic acid, triisobutyl ester or Isobutyl ortho formate $HC(O\text{-i-}C_4H_9)_3$	232.36	...	224-6	...	$0.8582^{20/4}$	1.4120^{20}	al, eth	$B2^4$, 29
10156	Orthoformic acid, triethyl ester or Ethyl ortho formate,Triethoxy methane $HC(OC_2H_5)_3$	148.20	...	$143^{765}, 60^{20}$	...	$0.8909^{20/4}$	1.3922^{20}	al, eth	$B2^4$, 25
10157	Orthoformic acid, trimethyl ester or Methyl ortho formate $HC(OCH_3)_3$	106.12	...	103-5	...	$0.9676^{20/4}$	1.3793^{20}	al, eth	$B2^4$, 22

PHYSICAL CONSTANTS OF ORGANIC COMPOUNDS (Continued)

No.	Name, Synonyms, and Formula	Mol. wt.	Color, crystalline form, specific rotation and λ_{max} (log ε)	b.p. °C	m.p. °C	Density	n_D	Solubility	Ref.
10158	Orthoformic acid, triisopentyl ester or isopentyl ortho formate $HC(O\text{-}i\text{-}C_5H_{11})_3$	274.44		267-9d, 166^{25}		0.8628$^{20/4}$	1.4233^{20}		B2^{4}, 30
10159	Orthoformic acid, triphenyl ester or Phenyl ortho formate $HC(OC_6H_5)_3$	292.33		269-70^{50} d	76-7			al, eth	B6^{4}, 611
10160	Orthoformic acid, tripropyl ester or Propyl ortho formate $HC(OC_3H_7)_3$	190.28		190-1^{745}, 93^{30}		0.8805$^{20/4}$	1.4072^{20}	al, eth	B2^{4}, 27
10161	Orthoformic acid, triisopropyl ester or isopropyl ortho formate $HC[OCH(CH_3)_2]_3$	190.28		166-8		0.8621$^{20/4}$	1.4000^{20}	al, eth	B2^{4}, 28
10162	Orthoformic acid, trithio, triethyl ester or Ethyl orthothio formate $HC(SC_2H_5)_3$	196.38		235d, 127-8^{12}		1.053$^{20/4}$	1.5410^{15}	al, eth	B2^{4}, 93
10163	Orthopropionic acid, triethyl ester or Ethyl ortho propionate $CH_3CH_2C(OC_2H_5)_3$	176.26		171, 44^{9}			1.4000^{25}	al, eth	B2^{4}, 707
10164	Orthosilicic acid, tetraethyl ester or Ethyl ortho silicate $Si(OC_2H_5)_4$	208.33	liq	168.8	−82.5	0.9320$^{20/4}$	1.3928^{20}	al, eth	B1^{4}, 1360
10165	Orthosilicic acid, tetrakis(2-ethylbutyl)ester $Si[OCH_2CH(C_2H_5)_2]_4$	432.76	liq	358		0.8920$^{20/4}$	1.4307^{20}	eth, bz	B1^{4}, 1725
10166	Orthosilicic acid, tetrakis(2-ethylhexyl)ester $Si[OCH_2CH(C_2H_5)C_4H_9]_4$	544.98		419, 227^{5}	−90	0.8803$^{20/4}$	1.4388^{20}	eth, bz	B1^{4}, 1787
10167	Orthosilicic acid, tetramethyl ester or Methyl ortho silicate $Si(OCH_3)_4$	152.22	nd	121, 25-7^{12}	−2	1.0232^{20}	1.3683^{20}	al	B1^{4}, 1266
10168	1,3,4-Oxadiazole, 2,5-dimethyl $C_4H_6N_2O$	98.10		178-9				w, al, eth	B27, 565
10169	Oxalic acid or Ethanedoic acid HO_2CCO_2H	90.04	mcl ta or pr (+ 2w, w), orh (anh)	157 sub	a:189.5, β:182 (anh), 101.5 (hyd)	α:1.900$^{17/4}$ β:1.895		w, al	B2^{4}, 1819
10170	Oxalic acid, diallyl ester or Diallyl oxalate $C_3H_5O_2CCO_2C_3H_5$	170.17		217, 86^{2}		1.1582^{20}	1.4481^{20}	al, ace, bz	B2^{4}, 1851
10171	Oxalic acid, dibutyl ester or Dibutyl oxalate $C_4H_9O_2CCO_2C_4H_9$	202.25		242^{773}, 96^{2}	−30.5	0.9873$^{20/4}$	1.4234^{20}	al, eth	B2^{4}, 1850
10172	Oxalic acid, diisobutyl ester $(i\text{-}C_4H_9O_2C\text{-})_2$	202.25		229, 143^{20}		0.9737$^{20/4}$	1.4180^{20}	al, eth, ace	B2^{4}, 1850
10173	Oxalic acid, di(2-chloroethyl)ester $[ClCH_2CH_2O_2C\text{-}]_2$	215.03	lf (dil al)	132^{3}	45			al, bz	B2^{4}, 1849
10174	Oxalic acid, dicyclohexyl ester or Dicyclohexyl oxalate $(C_6H_{11}O_2C\text{-})_2$	254.33	(MeOH)	190-1^{73}	42 (47)			al, eth	B6^{3}, 26
10175	Oxalic acid, diethyl ester or Diethyl oxalate $(C_2H_5O_2I\text{-})_2$	146.14		185.7, 97^{20}	−38.5	1.0785$^{20/4}$	1.4101^{20}	al, eth, ace	B2^{4}, 1848
10176	Oxalic acid dihydrazide $H_2NNHCOCONHNH_2$	118.10	nd (w)		243d	1.458$^{22.5}$			B2^{4}, 1868
10177	Oxalic acid, dimethyl ester or Dimethyl oxalate $CH_3O_2CCO_2CH_3$	118.09	mcl ta	164.5	54	1.148^{15}, 1.1716^{60}	1.379^{82}	al, eth, ace	B2^{4}, 1847
10178	Oxalic acid, di-isopentyl ester or Diisopentyl oxalate $(i\text{-}C_5H_{11}O_2C\text{-})_2$	230.30		267-8, 144^{14}		0.968$^{11/11}$		al, eth	B2^{4}, 1851
10179	Oxalic acid, dipropyl ester or Dipropyl oxalate $C_3H_7O_2CCO_2C_3H_7$	174.20		211, 78-80^{3}	−44.3	1.0188$^{20/4}$	1.4158^{20}	al, eth	B2^{4}, 1849
10180	Oxalic acid, diisopropyl ester $(i\text{-}C_3H_7O_2C\text{-})_2$	174.20		191^{765}		1.0010$^{20/4}$	1.4100^{20}	al, eth	B2^{4}, 1850
10181	Oxalic acid, di(2-tolyl)ester $[(2\text{-}CH_3C_6H_4)O_2C\text{-}]_2$	270.28	nd (al)	distb	91			al, eth, ace, bz, chl	B6^{2}, 330
10182	Oxalic acid, di-(3-tolyl)ester $[(3\text{-}CH_3C_6H_4)O_2C\text{-}]_2$	270.28	nd (al)	distb	106			al, eth, ace, bz, chl	B6^{2}, 353
10183	Oxalic acid, di-(4-tolyl)ester $[(4\text{-}CH_3C_6H_4)O_2C\text{-}]_2$	270.28	lf or pl, (al-eth)		148			al, eth, ace, bz, chl	B6^{4}, 2114
10184	Oxalic acid, monochloride, monoethyl ester $ClCOCO_2C_2H_5$	136.53	hyg	137, 30^{10}		1.2226$^{20/4}$		eth, bz	B2^{4}, 1853
10185	Oxalic acid, monoethyl, monomethyl ester $CH_3O_2CCO_2C_2H_5$	132.12		173.7		1.5505$^{0/0}$		al, eth	B2^{1}, 232
10186	Oxaluric acid or Oxalic acid ureide $H_2NCONHCOCO_2H$	132.08	cr		208-10d			w	B3^{4}, 121
10187	Oxalyl chloride $ClCOCOCl$	126.93	nd (eth, peth)	63-4	−16	1.4785$^{20/4}$	1.4316^{20}	eth	B2^{4}, 1853

No.	Name, Synonyms, and Formula	Mol. wt.	Color, crystalline form, specific rotation and λ_{max} (log ϵ)	b.p. °C	m.p. °C	Density	n_D	Solubility	Ref.
10188	Oxamic acid or Oxalic acid monoamide HO_2CCONH_2	89.05	cr (w)	...	210d	...	...	...	B2[4], 1857
10189	Oxamic acid, N-acetyl, ethyl ester $CH_3CONHCOCO_2C_2H_5$	159.14	pl (eth)	...	54-5	...	...	al, eth	B2[2], 509
10190	Oxamic acid, N-sec-butyl Sec-$C_4H_9NHCOCO_2H$	145.16	cr (eth)	...	88-9	...	...	eth	B4, 162
10191	Oxamide, dithio or Dithiooxamide $H_2NCSCSNH_2$	120.19	og-red	sub	...	...	...	...	B2[4], 1871
10192	Oxamic acid, ethyl ester or Ethyl oxamate $H_2NCOCO_2C_2H_5$	117.10	...	...	114-5	...	...	w, eth	B2[4], 1857
10193	Oxamic acid, N-phenyl or Oxanilic acid $HO_2CCONHC_6H_5$	165.15	nd (bz)	...	150	...	...	al, eth, chl	B12[3], 550
10194	Oxamic acid, N-phenyl, ethyl ester $C_6H_5NHCOCO_2C_2H_5$	193.20	pl or pr (al), nd (w)	260-300	65-6	...	...	al, eth, ace, bz	B12[3], 550
10195	Oxamide $H_2NOCCONH_2$	88.07	nd (w)	...	419d	...	...	...	B2[4], 1860
10196	Oxamide, N,N′-diethyl $C_2H_5NHCOCONHC_2H_5$	144.17	nd (al)	...	175 (180)	1.169^{4}	...	al	B4[4], 359
10197	Oxamide, N,N′-dimethyl $CH_3NHCOCONHCH_3$	116.12	pl or nd (al)	sub	217	$1.3^{4/4}$	...	w, chl	B4[4], 246
10198	Oxamide, N,N′-dephenyl or Oxanilide $C_6H_5NHCONHC_6H_5$	240.26	lf (bz)	>360	254	...	...	bz	B12[3], 551
10199	Oxamide, N,N′-diisopropyl [i-C_3H_7NHCO-]$_2$	172.23	nd (al)	...	212	...	...	al	B4[4], 518
10200	1,4-Oxathiane C_4H_8OS	104.17	...	147^{755}	...	$1.1174^{20/4}$	...	...	B19[4], 33
10201	Oxazole C_3H_3NO	69.06	...	69-70	...	...	$1.4285^{17.5}$	...	B27[2], 9
10202	Oxazole, 2,4-dimethyl 2,4-$(CH_3)_2(C_3HNO)$	97.12	...	108	...	$0.9352^{15/4}$	1.4166^{15}	w, al, eth	B27[2], 10
10203	Oxazole, 2,5-dimethyl 2,5-$(CH_3)_2(C_3HNO)$	97.12	...	117-8	...	$0.9958^{21/4}$	1.4385^{21}	w	B27[2], 10
10204	Oxazole, 2,4-diphenyl 2,4-$(C_6H_5)_2(C_3HNO)$	221.26	lf (al)	338-40	103	...	...	al, eth, bz	B27, 78
10205	Oxazole, 2,5-diphenyl 2,5-$(C_6H_5)_2(C_3HNO)$	221.26	nd (lig)	360	74	$1.0940^{100/4}$	1.6231^{100}	al, eth	B27[2], 43
10206	Oxazole, 4,5-diphenyl 4,5-$(C_6H_5)_2(C_3HNO)$	221.26	pl or pr (lig)	$192\text{-}5^{15}$	44	...	1.6283^{100}	...	B27, 79
10207	Oxazole, 2,4,5-triphenyl or Azobenzil 2,4,5-$(C_6H_5)_3(C_3NO)$	297.36	pr	...	116	...	...	bz	B27[2], 56
10208	2,4-Oxazolidenedione, 5,5-dipropyl $C_9H_{15}NO_3$	185.22	...	$148\text{-}50^{3}$	42-3	...	...	...	Am67, 522
10209	Oxetane or Trimethylene oxide $\underbrace{CH_2CH_2CH_2}_{O}$	58.08	...	47.8	...	$0.8930^{25/4}$	1.3961^{20}	w, al, eth, ace	B17[4], 13
10210	Oxetane, 3-chloro or 2-Chloro-1,3-epoxy propane C_3H_5ClO	92.53	...	...	132-4	...	...	al, eth, chl	B17[4], 14
10211	Oximide $O{=}C{-}C{=}O$ (bridged by NH) NH	71.04	pr (al)	...	...	...	...	...	B21, 368
10212	Oxindole C_8H_7NO	133.15	nd (w)	227^{23}	127	...	...	al, eth	B21[4], 3611
10213	Oxindole, 1-ethyl 1-$C_2H_5(C_8H_6NO)$	161.20	nd (ace, w)	...	97-8	...	...	ace	B21[4], 3615
10214	Oxindole, 3-hydroxy or Diozindole 3-HO(C_8H_6NO)	149.15	cr (w, al)	...	180	...	...	...	B21[4], 6076
10215	Oxindole, 1-methyl-3-ethyl 1-CH_3-3-$C_2H_5(C_8H_5NO)$	175.23	...	$280\text{-}5^{745}$, $103\text{-}7^{0.5}$	...	...	1.557^{25}	al, eth	B21[2], 258
10216	Oxacanthine or Vinetine $C_{37}H_{40}N_2O_6$	608.73	nd (al, eth), $[\alpha]^{20}{}_D$ +131.5 (chl, c=1)	...	216-7	...	...	al, eth, bz, chl	B27[2], 892
10217	Oxyacanthine hydrochloride $C_{37}H_{40}N_2O_6.HCl$	645.20	nd, $[\alpha]^{15}{}_D$ +163.8 (w, c=3)	...	270-1	...	...	w	B27[2], 893
10218	Oxyacanthine nitrate,dihydrate $C_{37}H_{40}N_2O_6.HNO_3.2H_2O$	707.78	nd	...	195-200	...	...	...	B27[2], 893

No.	Name, Synonyms, and Formula	Mol. wt.	Color, crystalline form, specific rotation and λ_{max} (log ε)	b.p. °C	m.p. °C	Density	n_D	Solubility	Ref.
10219	Oxynarcotine or α-Narcotine-*N*-oxide $C_{22}H_{23}NO_8$	429.43	hyg nd, $[\alpha]_D$ + 135 (chl)					w, al, chl	B27[2], 607
10220	Oxysparteine or Isolupanine $C_{15}H_{24}N_2O$	248.37	ye to col hyg nd (peth), $[\alpha]^{18}/_D$ −10.0 (al, c=18)	209^{12}	111			w, al, eth, chl	B24[2], 56
10221	Oxysparteine monohydrochloride, tetrahydrate $C_{15}H_{24}N_2.HCl.4H_2O$	356.89	wh cr (w)		48-50			w, al	B24[2], 57

No.	Name, Synonyms, and Formula	Mol. wt.	Color, crystalline form, specific rotation and λ_{max} (log ε)	b.p. °C	m.p. °C	Density	n_D	Solubility	Ref.
10222	Palmitaldehyde or Hexadecanal $CH_3(CH_2)_{14}CHO$	240.43	pl (eth), nd (peth)	$200\text{-}2^{29}$	34	...	...	al, eth, ace, bz	$B1^4$, 3393
10223	Palmitamide $CH_3(CH_2)_{14}CONH_2$	255.44	lf	236^{12}	107	...	...	...	$B2^4$, 1182
10224	Palmitamide, *N*-phenyl or Palmitanilide $CH_3(CH_2)_{14}CONHC_6H_5$	331.54	nd (al)	$282\text{-}4^{17}$	90.5	...	...	al, ace, bz, chl	$B12^1$, 486
10225	Palmitic acid or Hexadecanoic acid $CH_3(CH_2)_{14}CO_2H$	256.43	nd (al)	350, 267^{100}	63	$0.8527^{62/4}$	1.4335^{60}	al, **eth**, ace, bz, chl	$B2^4$, 1157
10226	Palmitic acid anhydride $(C_{15}H_{31}CO)_2O$	494.84	lf (peth)	...	64	$0.8383^{81/4}$	1.4364^{68}	eth	$B2^4$, 1181
10227	Palmitic acid, benzyl ester or Benzyl palmitate $C_{15}H_{31}CO_2CH_2C_6H_5$	346.55	cr (al)	...	36	$0.9136^{35/25}$	1.4689^{50}	al, eth, bz, chl	$B6^1$, 1481
10228	Palmitic acid, butyl ester or Butyl palmitate $C_{15}H_{31}CO_2C_4H_9$	312.54	cr (dil al)	...	16.9	...	1.4312^{50}	al, eth	$B2^4$, 1167
10229	Palmitic acid, ethyl ester or Ethyl palmitate $C_{15}H_{31}CO_2C_2H_5$	284.48	nd	191^{10}	α: 24, β: 19.3	$0.8577^{25/4}$	1.4347^{34}	al, eth, ace, bz, chl	$B2^4$, 1165
10230	Palmitic acid, hexadecyl ester or Cetyl palmitate $C_{15}H_{31}CO_2C_{16}H_{33}$	480.86	pl (eth)	360	53-4	$0.8324^{50/4}$	1.4425^{50}	eth, bz, chl	$B2^4$, 1168
10231	Palmitic acid, 16-hydroxy $HO(CH_2)_{15}CO_2H$	272.43	cr (bz-eth)	...	95	...	...	al, ace	$B3^1$, 664
10232	Palmitic acid, (2-hydroxy ethyl) ester or β-hydroxy ethyl palmitate $C_{15}H_{31}CO_2CH_2CH_2OH$	300.48	...	$173\text{-}4^{3}$	51	$0.8768^{60/4}$	...	al	$B2^4$, 1168
10233	Palmitic acid, methyl ester or Methyl palmitate $C_{15}H_{31}CO_2CH_3$	270.46	...	$415\text{-}8^{747}$, 148^{2}	30	...	...	al, eth, ace, bz, chl	$B2^4$, 1165
10234	Palmitic acid, myricyl ester or Myricyl palmitate $C_{15}H_{31}CO_2C_{31}H_{63}$	691.26	cr (eth)	...	72	...	...	eth	B2, 373
10235	Palmitic acid, propyl ester $C_{15}H_{31}CO_2C_3H_7$	298.51	nd	190^{12}	20.4	0.8455^{88}	1.4392^{25}	...	$B2^4$, 1167
10236	Palmitic acid, isopropyl ester $C_{15}H_{31}CO_2CH(CH_3)_2$	298.51	...	160^{2}	13-4	0.8404^{38}	1.4364^{25}	al, eth, ace, bz, chl	$B2^4$, 1167
10237	Palmitic acid, 9,10,16-trihydroxy or Aleuritic acid $HO(CH_2)_6CH(OH)CH(OH)(CH_2)_7CO_2H$	304.43	lf (dil al), nd (w)	...	102	...	...	...	$B2^4$, 1118
10238	Palmitonitrile $C_{15}H_{31}CN$	237.43	hex	333, 251^{100}	31	$0.8303^{20/4}$	1.4450^{20}	al, eth, ace, bz, chl	$B2^4$, 1183
10239	Palmityl chloride $C_{15}H_{31}COCl$	274.87	...	199^{20}	12	...	1.4514^{20}	eth	$B2^4$, 1182
10240	Palmitone $C_{15}H_{31}COC_{15}H_{31}$	450.83	lf (al)	...	83	$0.7947^{91/4}$	1.4297^{94}	eth	$B1^4$, 3413
10241	Paludrine or Proganil $C_{11}H_{16}ClN_5$	253.73	pl (dil al)	...	130-1	...	...	al	C40, 2931
10242	Paludrine hydrochloride $C_{11}H_{16}ClN_5 \cdot HCl$	290.20	nd (w)	...	245	...	...	al	J1946, 729
10243	Pamelonitrile $NC(CH_2)_5CN$	122.17	...	175^{14}	−31.4	0.949^{18}	1.4472^{20}	**al, eth, chl**	$B2^4$, 2006
10244	Panthesin $C_{18}H_{32}N_2O_5S$	388.52	pa ye pw (al)	...	157-9	...	...	w, al	$B3^{14}$, 1058
10245	Pantothenic acid *(d)* $HOCH_2C(CH_3)_2CH(OH)CONH(CH_2)_2CO_2H$	219.24	ye visc oil, $[\alpha]^{25}/_D$ + 37.5 (w)	...	...	...	...	w, eth, bz	$B4^3$, 1283
10246	Pantothenic acid, calcium salt *(d)* or Calcium pantothenate $(C_9H_{16}NO_5)_2Ca$	476.56	wh (MeOH), $[\alpha]^{26}/_D$ + 28.2 (w)	...	195-6	...	...	w	$B4^3$, 1286
10247	Pantothenic acid, calcium salt-*(l)* or Calcium pantothenate $(C_9H_{16}NO_5)_2Ca$	476.54	cr (MeOH), $[\alpha]^{26}/_D$ -27.8 (w)	...	187-9	...	...	w	$B4^3$, 1288
10248	Pantothenyl alcohol or Panthenol $HOCH_2C(CH_3)_2CH(OH)CONH(CH_2)_3OH$	205.26	hyg oil	d, $118\text{-}20^{0.02}$	...	$1.2^{20/20}$	1.497^{20}	w, al	$B4^4$, 1652
10249	Papaveraldine or Xanthaline $C_{20}H_{19}NO_5$	353.37	nd (al) cr, (bz, peth)	210-1	...	...	...	bz, chl, aa	$B21^4$, 6738
10250	Papaverine or Papaveroline tetramethyl ether $C_{20}H_{21}NO_4$	339.39	wh pr (aleth) nd, (chl-peth)	d, sub, $135\text{-}40^{11}$	147-8	$1.337^{20/4}$	1.625	al, ace, chl, Py	$B21^4$, 2788
10251	Papaverine hydrochloride $C_{20}H_{21}NO_4.HCl$	375.85	wh mcl pr (w)	...	224-5	...	...	w, al	$B21^4$, 2790

No.	Name, Synonyms, and Formula	Mol. wt.	Color, crystalline form, specific rotation and λ_{max} (log ε)	b.p. °C	m.p. °C	Density	n_D	Solubility	Ref.
10252	Parabanic acid or oxalurea $C_3H_2N_2O_3$	114.06	mcl nd (w)	sub 100	243-5d	...	...	w, al	B24^{2}, 263
10253	Parabutyraldehyde or 2,4,6-Tripropyl-1,3,5-trioxane $C_{12}H_{24}O_3$	216.32	...	98-100^{35}	...	0.918	...	...	B19^{4}, 4725
10254	Para isobutyraldehyde or 2,4,6-Triisopropyl-1,3,5-trioxane $C_{12}H_{24}O_3$	216.32	nd (al)	195 sub	59-60	...	...	al, eth	B19^{4}, 4726
10255	Paraconic acid or Hydroxymethyl succinic acid-γ-lactone $C_5H_6O_4$	130.10	dlq	...	57-8	...	...	w	B18^{4}, 5264
10256	Paracyanogen $(CN)_x$	...	br pw	...	sub	...	...	...	B2^{4}, 1864
10257	Paraldehyde or 2,4,6-Trimethyl-1,3,5-trioxane $C_6H_{12}O_3$	132.16	...	128	12.6	0.9943$^{20/4}$	1.4049^{20}	**al, eth, chl**	B19^{4}, 4715
10258	Paraldol $C_8H_{16}O_4$	176.21	wh tcl pr	90^{15}	89-91	1.116^{20}	1.4610^{20}	w, al, eth	B1^{4}, 3987
10259	Parasorbic acid $C_6H_8O_2$	112.13	oily liq, $[\alpha]^{19/}_{D}$ + 210 (al, c=3)	100^{15}	...	1.079$^{18/4}$	1.4730^{20}	w, al, eth	B17^{4}, 4305
10260	Parathion or Diethyl-*p*-nitro phenyl mono thiophosphate $C_{10}H_{14}NO_5PS$	291.26	ye liq	375, 157-62$^{0.5}$	6.1	1.2704$^{20/20}$	1.5370^{25}	**al,** eth, ace, chl	B6^{4}, 1337
10261	Patulin or Clavacin $C_7H_6O_4$	154.12	pl or pr (eth, chl)	...	111	...	...	w, al, eth, ace, bz	B18^{4}, 1184
10262	Patchouli alcohol $C_{15}H_{26}O$	222.37	...	...	56	0.9924$^{65/20}$	1.5029^{65}	al, eth	B6^{3}, 426
10263	Paucin, hydrate $C_{27}H_{39}N_5O_5.6\frac{1}{2}H_2O$	630.74	ye lf	...	126d	...	...	...	B10^{3}, 1841
10264	Pelargonaldehyde $C_8H_{17}CHO$	142.24	...	190-2, 93.5^{23}	...	0.8264$^{22/4}$	1.4275^{20}	eth	B1^{4}, 3352
10265	Pelargonaldehyde oxime or Pelargonaldoxime $C_8H_{17}CH{=}NOH$	157.26	lf (dil al)	...	64	...	...	al, eth, ace	B1^{2}, 761
10266	Pelargonamide $CH_3(CH_2)_7CONH_2$	157.26	...	sub	99-100	0.8394^{110}	1.4248^{110}	...	B2^{4}, 1023
10267	Pelargonic acid or Nonanoic acid $CH_3(CH_2)_7CO_2H$	158.24	...	255, 150^{20}	fp 12.2	0.9057$^{20/4}$	1.4343^{19}	al, eth	B2^{4}, 1018
10268	Pelargonic acid, 9-amino $H_2N(CH_2)_8CO_2H$	173.26	...	185-7	...	...	...	w, al	B4^{3}, 1479
10269	Pelargonic acid, ethyl ester or Ethyl pelargonate $CH_3(CH_2)_7CO_2C_2H_5$	186.29	...	227, 96-8^{10}	−36.7	0.8657$^{20/4}$	1.4220^{20}	al, eth, ace	B2^{4}, 1019
10270	Pelargonic acid, 9-fluoro $F(CH_2)_8CO_2H$	176.23	...	88-90$^{0.2}$	*ca* 18	...	1.4289^{25}	al, eth	B2^{4}, 1024
10271	Pelargonic acid, methyl ester or Methyl pelargonate $CH_3(CH_2)_7CO_2CH_3$	172.27	...	213-4^{757}, 104-6^{23}	...	0.8799^{15}	1.4214^{20}	al, eth	B2^{4}, 1019
10272	Pelargononitrile $CH_3(CH_2)_7CN$	139.24	...	224.4, 91.9^{10}	−34.2	0.8178$^{20/4}$	1.4255^{20}	al, eth	B2^{4}, 1024
10273	Pelargonyl chloride $CH_3(CH_2)_7COCl$	176.69	...	215.3, 98^{15}	−60.5	0.9463$^{15/4}$	...	eth, ace	B2^{4}, 1023
10274	Pelargonidin chloride $C_{15}H_{11}ClO_5$	306.70	red br hyg (anh) pr, or pl (dil HCl)	...	>350 (anh)	...	...	w, al	B18^{4}, 3198
10275	Pellotine or *N*-Methyl anhalonidine $C_{13}H_{19}NO_3$	237.30	pl (al, peth)	...	111.5	...	...	al, eth, ace, chl, peth	B21^{4}, 2525
10276	Penicilic acid $C_8H_{10}O_4$	170.17	rh or hex pl (+1w), nd (peth)	...	64-5 (+1w) 8-7 (anh)	...	...	w, al, eth, ace, bz	B3^{2}, 519
10277	Pentacene or 2,3,6,7-Dibenzanthracene $C_{22}H_{14}$	278.35	deep vt-bl nd or lf (ph NO_2), cr (bz)	290-300 sub (vac)	270-1	...	...	...	B5^{3}, 2551
10278	Pentacene, 6,13-diphenyl $C_{34}H_{22}$	430.55	vt-bl nd	318-20	...	...	...	bz, aa	B5^{3}, 2738
10279	Pentacosane $CH_3(CH_2)_{23}CH_3$	352.69	...	401.9, 239.9^{10}	...	0.8012$^{20/4}$	1.4491^{20}	bz, chl	B1^{4}, 582
10280	Pentacosane, 13-phenyl $C_6H_5CH(C_{12}H_{25})_2$	428.78	...	235^{10}	31.7	0.8537$^{20/4}$	1.4787$^{20}/_{4}$	...	B5^{4}, 1241
10281	Pentacosane, 13-undecyl $(C_{12}H_{25})_2CHC_{11}H_{23}$	506.98	...	307^{10}	9.7	0.8168$^{20/4}$	1.4567^{20}	...	B1^{4}, 600

No.	Name, Synonyms, and Formula	Mol. wt.	Color, crystalline form, specific rotation and λ_{max} (log ε)	b.p. °C	m.p. °C	Density	n_D	Solubility	Ref.
10282	6,9-Pentadecadiyne $H_2C(C{\equiv}C\text{-}C_5H_{11})_2$	204.36		$135\text{-}6^{4}$		$0.840^{21/4}$	1.4693^{21}		$B1^{4}$, 1132
10283	Pentadecanal $CH_3(CH_2)_{13}CHO$	226.40	nd	185^{25}	24-5			al, eth, ace	$B1^{4}$, 3391
10284	Pentadecanal oxime $CH_3(CH_2)_{13}CH{=}NOH$	241.42	nd (dil al)		86			eth	$B1^{2}$, 770
10285	Pentadecane $CH_3(CH_2)_{13}CH_3$	212.42		270.6, 136.10	10	$0.7685^{20/4}$	1.4315^{20}	al, eth	$B1^{4}$, 529
10286	Pentadecane, 1-amino or Pentadecylamine $CH_3(CH_2)_{13}CH_2NH_2$	227.43	fl	307.6, 165.8^{10}	37.3	$0.8104^{20/4}$	1.4480^{20}	al, eth	$B4^{4}$, 817
10287	Pentadecane, 1-bromo or Pentadecyl bromide $CH_3(CH_2)_{13}CH_2Br$	291.32		322, 177^{10}	19	$1.0675^{20/4}$	1.4611^{20}	ace, chl	$B1^{4}$, 531
10288	Pentadecane, 1,15-dibromo $Br(CH_2)_{15}Br$	370.21	lf (al)	$215\text{-}25^{15}$, 192^{2}	27			chl	$B1^{4}$, 531
10289	Pentadecane, 1(2,3-dihydroxyphenyl) or Tetra hydra urushiol $(2,3\text{-}(HO)_2C_6H_3)CH_2(CH_2)_{13}CH_3$	320.52	nd (to, eth, peth)		59			al, eth, bz, aa, chl	$B6^{3}$, 4771
10290	Pentadeconoic acid or Pentadecylic acid $CH_3(CH_2)_{13}CO_2H$	242.40	pl (aq, al, aa), cr (peth)	257^{100}, 158^{1}	53-4	0.8423^{80}	1.4254^{80}	al, eth, ace, bz, chl	$B2^{4}$, 1147
10291	Pentadecanoic acid, methyl ester or Methyl pentadeconoate $CH_3(CH_2)_{13}CO_2CH_3$	256.43	nd (dil al)	153.5	18.5	$0.8618^{25/4}$	1.4390^{25}	al, eth	$B2^{3}$, 936
10292	2-Pentadecanol $C_{13}H_{27}CH(OH)CH_3$	228.42		299	35	0.8328^{20}	1.4463^{20}		$B1^{4}$, 1871
10293	2-Pentadecanone $C_{13}H_{27}COCH_3$	226.40		294	39.5	0.8182^{39}			$B1^{4}$, 3391
10294	8-Pentadecanone or Diheptyl ketone $(C_7H_{15})_2CO$	226.40	cr (al)	291, 178^{20}	43			al, eth, bz, chl	$B1^{4}$, 3392
10295	1-Pentadecene $CH_3(CH_2)_{12}CH{=}CH_2$	210.40		268.2, 133.7^{10}	2-8	$0.7764^{20/4}$	1.4389^{20}	ace	$B1^{4}$, 926
10296	1-Pentadecyne $CH_3(CH_2)_{12}C{\equiv}CH$	208.39		268, 129.8^{10}	10	$0.7928^{20/4}$	1.4419^{20}	ace	$B1^{4}$, 1072
10297	2,4-Pentadienal, 5-phenyl or Cinnamylidene acetaldehyde $C_6H_5CH{=}CHCH{=}CHCHO$	158.20		$155\text{-}65^{3}$, $92\text{-}5^{0.05}$	42-3			al, eth, bz	$B7^{3}$, 1653
10298	1,2-Pentadiene or Ethyl allene $CH_3CH_2CH{=}C{=}CH_2$	68.12		44.9	−137.3	$0.6926^{20/4}$	1.4209^{20}	al, eth, ace, bz	$B1^{4}$, 993
10299	1,2-Pentadiene, 1-chloro-3-ethyl $(C_2H_5)_2C{=}C{=}CHCl$	130.62		$85\text{-}8^{100}$		$0.9297^{19/4}$		eth	$B1^{3}$, 1002
10300	1,2-Pentadiene, 1-chloro-3-methyl $CH_3CH_2C(CH_3){=}C{=}CHCl$	116.59		$68\text{-}70^{100}$		$0.9562^{20/4}$		eth	$B1^{3}$, 990
10301	1,3-Pentadiene or Piperylene $CH_3CH{=}CHCH{=}CH_2$	68.12		42	−87.5	$0.6760^{20/4}$	1.4301^{20}	al, eth, ace, bz	$B1^{4}$, 994
10302	1,3-Pentadiene, 1-chloro-3-methyl $CH_3CH{=}C(CH_3)CH{=}CHCl$	116.59		$62\text{-}3^{100}$		$0.9574^{20/4}$		eth, chl	$B1^{3}$, 990
10303	1,3-Pentadiene, 2-chloro-3-methyl $CH_3CH{=}C(CH_3)CCl{=}CH_2$	116.59		$57\text{-}60^{95}$		$0.9437^{20/4}$	1.4671^{20}	eth, chl	$B1^{3}$, 991
10304	1,3-Pentadiene, 3-chloro or Methyl chloroprene $CH_3CH{=}C(Cl)CH{=}CH_2$	102.56		99-101		$0.9576^{20/4}$	1.4785^{20}	eth, ace, bz, chl	$B1^{3}$, 962
10305	1,3-Pentadiene, 2,4-dimethyl $(CH_3)_2C{=}CHC(CH_3){=}CH_2$	96.17		93^{758}		$0.7343^{23/4}$	1.43904^{23}		$B1^{4}$, 1034
10306	1,3-Pentadiene, 4-methyl $(CH_3)_2C{=}CHCH{=}CH_2$	82.15		76.5		$0.7181^{20/4}$	1.4532^{20}		$B1^{4}$, 1020
10307	1,4-Pentadiene $CH_2{=}CHCH_2CH{=}CH_2$	68.12		26	−148.3	$0.6608^{20/4}$	1.3888^{20}	al, eth, ace, bz	$B1^{4}$, 998
10308	2,3-Pentadiene $CH_3CH{=}C{=}CHCH_3$	68.12		48.2	−125.6	$0.6950^{20/4}$	1.4284^{20}	al, eth, ace, bz	$B1^{4}$, 999
10309	2,4-Pentadienoic acid or β-Vinyl acrylic acid $CH_2{=}CHCH{=}CHCO_2H$	98.10	hyg pr (eth)	d110-5	80			w, al, eth, bz, peth	$B2^{4}$, 1694
10310	2,4-Pentadienoic acid, 4-hydroxy-γ-lactone $CH_2{=}C\text{-}CH{=}CHCO$ (ring via O)	96.09	pa ye oil	73^{11}				chl	$B17^{4}$, 4495
10311	2,4-Pentadienoic acid, 5-(3,4-methylene dioxy)phenyl or Piperic acid $(3,4\text{-}CH_2O_2)C_6H_3CH{=}CHCH{=}CHCO_2H$	218.21	ye in light nd (al), ye nd (sub)	sub	215			al	$B19^{4}$, 3565
10312	2,4-Pentadienoic acid, 5-phenyl or Cinnamylidenacetic acid $C_6H_5CH{=}CHCH{=}CHCO_2H$	174.20	pl (al), pr (bz)		166-7			eth, bz	$B9^{3}$, 3070

No.	Name, Synonyms, and Formula	Mol. wt.	Color, crystalline form, specific rotation and λ_{max} (log ε)	b.p. °C	m.p. °C	Density	n_D	Solubility	Ref.
10313	2,4-pentadienoic acid, 5-phenyl,ethyl ester $C_6H_5CH=CHCH=CHCO_2C_2H_5$	202.25	ye oil	$149\text{-}50^{4}$	25-7	$1.0469^{20/4}$	1.5768^{80}	al, eth	$B9^{2}$, 441
10314	2,4-Pentadienoic acid, 5-phenyl,methyl ester $C_6H_5CH=CHCH=CHCO_2CH_3$	188.23	lf or pl	185^{20}	71	...	...	al	$B9^{3}$, 3071
10315	2,4-Pentadienonitrile *(cis)* $CH_2=CHCH=CHCN$	79.10	...	49.5^{32}, 32.5^{13}	−64	$0.8541^{26/4}$	1.4855^{20}	eth, ace	$B2^{4}$, 1695
10316	2,4-Pentadienonitrile *(trans)* $CH_2=CHCH=CHCN$	79.10	...	41^{13}	−43	$0.8576^{20/4}$	1.4986^{20}	eth, ace	$B2^{4}$, 1696
10317	1,4-Pentadien-3-one, 1,5-bis-(2-ethoxyphenyl) $[(2\text{-}C_2H_5OC_6H_4)CH=CH]_2CO$	322.40	ye lf (dil al)	...	89	...	...	al	$B8^{1}$, 666
10318	1,4-Pentadiene-3-one, 1,5-bis(2-hydroxyphenyl) $[(2\text{-}HOC_6H_4)CH=CH]_2CO$	266.30	ye nd (dil al)	...	168d	...	...	al, eth , ace, bz, Py	$B8^{3}$, 2954
10319	1,4-Pentadien-3-one, 1,5-bis-(4-hydroxyphenyl) or Bis-(4-hydroxystyryl)ketone $[(4\text{-}HOC_6H_4)CH=CH]_2CO$	266.30	ye-og nd or lf (dil al)	...	237-8	...	...	ace, al	$B8^{3}$, 2958
10320	1,4-Pentadien-3-one, 1,5-bis-(2-methoxyphenyl) or Bis (2-methoxystyryl)ketone $[(2\text{-}CH_3OC_6H_4)CH=CH]_2CO$	294.35	ye nd or lf (al)	...	127	...	...	...	$B8^{2}$, 405
10321	1,4-Pentadien-3-one, 1,5-bis-(3-methoxyphenyl) or Bis (3-methoxystyryl)ketone $[(3\text{-}CH_3OC_6H_4)CH=CH]_2CO$	294.35	nd (chl-MeOH)	...	55-6	...	...	ace, chl	B81, 666
10322	1,4-Pentadien-3-one, 1,5-bis-(4-methoxyphenyl) or Dianisal acetone $[(4\text{-}CH_3OC_6H_4)CH=CH]_2CO$	294.35	ye lf (aa)	...	129-30	...	...	bz, aa, chl	$B8^{3}$, 2958
10323	1,4-Pentadien-3-one, 1,5-bis(3,4-methylene dioxyphenyl) or Dipiperonylidene acetone $C_{19}H_{14}O_5$	322.32	ye nd (bz)	...	185	...	...	ace, chl	$B19^{4}$, 5891
10324	1,4-Pentadien-3-one, 1,5-bis-(4-nitrophenyl) or Bis (2-nitrostyryl)ketone $[(2\text{-}O_2NC_6H_4)CH=CH]_2CO$	324.29	ye nd (aa)	...	170-1	...	...	chl	$B7^{2}$, 455
10325	1,4-Pentadien-3-one, 1,5-bis-(3-nitrophenyl) $[(3\text{-}O_2NC_6H_4)CH=CH]_2CO$	324.29	ye br (Ac_2O)	...	238	...	...	ace	$B7^{2}$, 455
10326	1,4-Pentadien-3-one, 1,5-bis-(4-nitrophenyl) or Bis(4-nitrosytyryl) ketone $[(4\text{-}O_2NC_6H_4)CH=CH]_2CO$	324.29	ye (Ac_2O)	...	254	...	...	bz	$B7^{2}$, 455
10327	1,4-Pentadiene-3-one, 1(2-chlorphenyl)-5-(3-chlorophenyl) $(2\text{-}ClC_6H_4)CH=CH\text{-}CO\text{-}CH=CH(C_6H_4Cl\text{-}3)$	303.19	ye nd (dil al)	...	67-8	...	...	...	$B7^{2}$, 454
10328	1,4-Pentadiene-3-one, 1(2-chlorophenyl)-5-(4-chlorophenyl) $(2\text{-}ClC_6H_4)CH=CH\text{-}CO\text{-}CH=CH(C_6H_4Cl\text{-}4)$	303.19	ye nd (al)	...	109	...	...	...	$B7^{2}$, 454
10329	1,4-Pentadien-3-one, 1,5-di(2-furyl) $[2\text{-}(C_4H_3O)CH=CH]_2CO$	214.22	dlq pr (peth), ye pr (lig)	d	60-1	...	...	al, eth, chl	$B19^{4}$, 1822
10330	1,4-Pentadien-3-one, 1,5-diphenyl or Dibenzal acetone $[C_6H_5CH=CH]_2CO$	234.30	pl or lf (ace, AcOEt)	d	113d	...	...	ace, chl	$B7^{3}$, 2559
10331	1,3-Pentadiyne $CH_3C\equiv CC\equiv CH$	64.09	...	55-6	...	$0.7375^{20/4}$	1.4431^{21}	eth, bz, chl	$B1^{4}$, 1117
10332	Pentaerythritol or Tetramethylol methane $C(CH_2OH)_4$	136.15	cr (dil HCl)	sub	269	...	1.548	w	$B1^{4}$, 2812
10333	Pentaerythritol tetra acetate $(CH_3CO_2CH_2)_4C$	304.30	tetr nd (w, bz)	...	83-4	$1.273^{18/4}$	...	w, al, eth	$B1^{4}$, 264
10334	Pentaerythrityl tetrabromide or 2,2-bis(Bromoethyl)-1,3-dibromo propane $C(CH_2Br)_4$	387.73	cr (ace), nd (lig)	305-6	163	2.596^{15}	...	al, bz	$B1^{4}$, 337
10335	Pentaerythrityl tetrachloride or 2,2-bis(chloromethyl)-1,3-dichloropropane $C(CH_2Cl)_4$	209.93	...	110^{12}	97	...	...	eth, chl	$B1^{4}$, 336
10336	Pentaerythrityl tetraiodide or 2,2-bis(iodomethyl)-1,3-diiodo methane $C(CH_2I)_4$	575.74	nd (to)	...	225	...	...	...	$B1^{4}$, 338
10337	Pentaerythritol tetranitrate or PETN $C(CH_2ONO_2)_4$	316.14	tetr (ace), pr (ace-al)	...	140-1	$1.773^{20/4}$	...	ace, bz	$B1^{4}$, 2816
10338	Pentamethylene sulfide or Tetrahydro thiopyran $C_5H_{10}S$	102.19	...	141.7, 93^{82}	19	$0.9861^{20/4}$	1.5067^{20}	al, eth, ace, bz	$B17^{4}$, 55
10339	Pentanal or Valeraldehyde $CH_3(CH_2)_3CHO$	86.13	...	103	−91.5	$0.8095^{20/4}$	1.3944^{20}	al, eth	$B1^{4}$, 3268

No.	Name, Synonyms, and Formula	Mol. wt.	Color, crystalline form, specific rotation and λ_{max} (log ε)	b.p. °C	m.p. °C	Density	n_D	Solubility	Ref.
10340	Pentane $CH_3(CH_2)_3CH_3$	72.15	....	36.1	−130	$0.6262^{20/4}$	1.3575^{20}	**al, eth, ace, bz, chl**	B1^{4}, 303
10341	Pentane, 1-amino or Pentylamine $CH_3(CH_2)_3CH_2NH_2$	87.16	....	104.4, 5.9^{10}	−55	$0.7547^{20/4}$	1.4118^{20}	**al, eth,** ace, bz	B4^{4}, 675
10342	Pentane, 2-amino $C_3H_7CH(NH_2)CH_3$	87.16	....	91.5^{755}	......	0.7384^{20}	1.4027^{20}	w, **al, eth,** ace, bz	B4^{4}, 689
10343	Pentane, 3-amino $H_2NCH(C_2H_5)_2$	87.16	....	91	......	$0.7487^{20/4}$	1.4063^{20}	al	B4^{4}, 692
10344	Pentane, 3,3-bis(ethyl sulfonyl) $(C_2H_5)_2C(SO_2C_2H_5)_2$	256.38	lf (dil al)	....	85	....	....	al, eth, w	B4, 681
10345	Pentane, 1-bromo or Pentyl bromide $CH_3(CH_2)_3CH_2Br$	151.05	....	129.6, 21^{10}	−87.9	$1.2182^{20/4}$	1.4447^{20}	al, **eth**, bz, chl	B1^{4}, 312
10346	Pentane, 1-bromo-5-fluoro $F(CH_2)_5Br$	169.04	....	162	......	1.3604^{25}	1.4406^{25}	al, eth	B1^{4}, 313
10347	Pentane, 1-bromo-2-methyl $C_3H_7CH(CH_3)CH_2Br$	165.07	....	142-5, $51\text{-}3^{25}$	......	$1.1624^{20/4}$	1.4495^{20}	eth, chl	B1^{4}, 361
10348	Pentane, 1-bromo-3-methyl $C_2H_5CH(CH_3)CH_2CH_2Br$	165.07	....	$148\text{-}9^{766}$	......	$1.1829^{20/4}$	1.4496^{20}	eth, chl	B1^{4}, 365
10349	Pentane, 1-bromo-4-methyl or Isohexyl bromide $(CH_3)_2CH(CH_2)_3Br$	165.07	....	147-8	......	1.1683^{20}	1.4490	eth, chl	B1^{4}, 361
10350	Pentane, 2-bromo $C_3H_7CHBrCH_3$	151.05	....	117.4, 58.4^{100}	−95.5	$1.2075^{20/4}$	1.4413^{20}	al, eth, bz, chl	B1^{4}, 312
10351	Pentane, 2-bromo-2-methyl $C_3H_7CBr(CH_3)_2$	165.07	....	142-3, 70^{100}	......	....	1.442^{23}	eth, chl	B1^{4}, 361
10352	Pentane, 3-bromo $BrCH(C_2H_5)_2$	151.05	....	118.6, 10.8^{10}	−126.2	$1.2124^{20/4}$	$1.4441^{20/4}$	al, eth, bz, chl	B1^{4}, 313
10353	Pentane, 3-bromo-3-methyl $(C_2H_5)_2C(Br)CH_3$	165.07	....	129-31, $82\text{-}3^{145}$	......	1.1835^{20}	1.4525^{20}	eth, chl	B1^{4}, 365
10354	Pentane, 1-chloro or Pentyl chloride $CH_3(CH_2)_3CH_2Cl$	106.60	....	107.8, 5^{10}	−99	$0.8818^{20/4}$	1.4127^{20}	**al, eth**, bz, chl	B1^{4}, 309
10355	Pentane, 1-chloro-5-fluoro $F(CH_2)_5Cl$	124.59	....	143.2	......	1.0325^{25}	1.4120^{23}	al, eth	B1^{4}, 309
10356	Pentane, 2-chloro $C_3H_7CHClCH_3$	106.60	$[\alpha]_D$ + 34.1	96.9	−137	$0.8698^{20/4}$	1.4069^{20}	al, eth, bz, chl	B1^{4}, 309
10357	Pentane, 2-chloro-2,3-dimethyl $C_2H_5CH(CH_3)CCl(CH_3)_2$	134.65	....	$38\text{-}9^{20}$	......	....	1.4264^{20}	eth, bz	B1^{4}, 406
10358	Pentane, 2-chloro-2,4-dimethyl $(CH_3)_2CHCH_2CCl(CH_3)_2$	134.65	....	$127\text{-}8^{733}$, $33\text{-}4^{20}$	......	$0.861^{20/4}$	1.4180^{20}	eth	B1^{4}, 408
10359	Pentane, 2-chloro-3-ethyl $(C_2H_5)_2CHCH(Cl)CH_3$	134.65	....	83.5^{100}	......	$0.8951^{25/25}$	1.4318^{20}	eth, chl	B1^{2}, 120
10360	Pentane, 2-chloro-2-methyl $C_3H_7CCl(CH_3)_2$	120.62	....	$110\text{-}1^{734}{}_d$, $36\text{-}7^{15}$	......	$0.863^{20/4}$	1.4126^{20}	eth	B1^{4}, 360
10361	Pentane, 2-chloro-4-methyl $(CH_3)_2CHCH_2CHClCH_3$	120.62	....	$111\text{-}2^{733}$	......	$0.8610^{20/4}$	1.4113^{20}	eth	B1^{3}, 399
10362	Pentane, 2-chloro-2,4,4-trimethyl $(CH_3)_2CClCH_2C(CH_3)_3$	148.68	....	145-50d, 44^{16}	−26	0.8746^{20}	1.4308^{20}	al	B1^{4}, 444
10363	Pentane, 3-chloro $ClCH(C_2H_5)_2$	106.60	....	97.8, $38\text{-}9^{20}$	−105	$0.8731^{20/4}$	1.4082^{20}	al, eth, bz, chl	B1^{4}, 309
10364	Pentane, 3-chloro-2,2-dimethyl-3-ethyl $(C_2H_5)_2CClC(CH_3)_3$	162.70	....	d $53\text{-}4^{6}$	......	....	1.4528^{25}	eth, chl	B1^{4}, 462
10365	Pentane, 3-chloro-2,3-dimethyl $C_2H_5C(CH_3)ClCH(CH_3)_2$	134.65	....	$135\text{-}8^{757}{}_d$, $41\text{-}2^{20}$	......	$0.884^{22/22}$	1.4318^{20}	eth, chl	B1^{4}, 406
10366	Pentane, 3-chloro-3-ethyl $(C_2H_5)_3CCl$	134.65	....	143-4, $43\text{-}4^{20}$	......	$0.8856^{20/4}$	1.4400^{20}	eth	B1^{4}, 403
10367	Pentane, 3-chloro-3-ethyl-2-methyl $(C_2H_5)_2CClCH(CH_3)_2$	148.68	....	150-5d, $98\text{-}100^{98}$	......	1.0325^{25}	1.4120^{23}	al, eth	B1^{4}, 437
10368	Pentane, 3-chloro-2-methyl $C_2H_5CHClCH(CH_3)_2$	120.62	....	$115\text{-}7^{752}{}_d$	......	....	....	eth, chl	B1^{2}, 111
10369	Pentane, 3-chloro-3-methyl $(C_2H_5)_2C(Cl)CH_3$	120.62	....	116, 35^{25}	......	$0.8900^{20/4}$	1.4210^{20}	eth, bz, chl	B1^{4}, 365
10370	Pentane, 3-(chloromethyl) $(C_2H_5)_2CHCH_2Cl$	120.62	....	125-7	......	$0.8914^{20/4}$	1.4222^{20}	eth, bz, chl	B1^{3}, 403
10371	Pentane, 4-chloro-2,2-dimethyl $CH_3CHClCH_2C(CH_3)_3$	134.65	....	93^{250}	......	$0.855^{20/4}$	1.4180^{20}	eth, chl	B1^{4}, 404
10372	Pentane, 1,5-diamino or Pentamethylendiamine Cadavarine $H_2N(CH_2)_5NH_2$	102.18	....	178-80	9	0.867^{25}	1.4561^{25}	w, al	B4^{4}, 1310

No.	Name, Synonyms, and Formula	Mol. wt.	Color, crystalline form, specific rotation and λ_{max} (log ε)	b.p. °C	m.p. °C	Density	n_D	Solubility	Ref.
10373	Pentane, 1,5-dibromo $Br(CH_2)_5Br$	229.94	....	$222.3, 98.6^{20}$	−39.5	$1.7018^{20/4}$	1.5126^{20}	bz, chl	$B1^{4}$, 314
10374	Pentane, 1,2-dichloro $C_3H_7CHClCH_2Cl$	141.04	....	$148\text{-}9, 58\text{-}9^{20}$	......	$1.0872^{20/4}$	1.4485^{20}	al, chl	$B1^{1}$, 341
10375	Pentane, 1,2-dichloro-4,4-dimethyl $(CH_3)_3CCH_2CHClCH_2Cl$	169.09	....	$173\text{-}5^{745}, 58\text{-}9^{12}$	......	1.0259^{20}	1.4489^{20}	bz, chl	$B1^{1}$, 445
10376	Pentane, 1,3-dichloro $C_2H_5CHClCH_2CH_2Cl$	141.04	....	80.4^{60}	......	$1.0834^{20/4}$	1.4485^{20}	ace, chl	$B1^{1}$, 342
10377	Pentane, 1,4-dichloro $CH_3CHClCH_2CH_2CH_2Cl$	141.04	....	161-3, $58\text{-}60^{15}$	......	$1.0840^{20/4}$	1.4503^{20}	ace, chl	$B1^{4}$, 309
10378	Pentane, 1,5-dichoro $Cl(CH_2)_5Cl$	141.04	....	$180, 59^{10}$	−72.8	$1.1006^{22/4}$	1.4564^{20}	al, eth, bz, chl	$B1^{4}$, 310
10379	Pentane, 1,5-dichloro-3,3-dimethyl $(ClCH_2CH_2)_2C(CH_3)_2$	169.09	....	$135^{80}, 58\text{-}9^{8}$	......	$1.0563^{20/4}$	1.4652^{20}	chl	$B1^{4}$, 409
10380	Pentane, 2,2-dichloro $C_3H_7CCl_2CH_3$	141.04	....	$128\text{-}9, 36\text{-}7^{20}$	......	1.040^{20}	1.434^{20}	eth, bz, chl	$B1^{1}$, 342
10381	Pentane, 2,4-dichloro $CH_2(CHClCH_3)_2$	141.04	....	147-50, 62^{12}	......	$1.0634^{15/4}$	1.447^{18}	eth, bz, chl	$B1^{1}$, 343
10382	Pentane, 2,4-dichloro-2,4-dimethyl $CH_2[CCl(CH_3)_2]_2$	169.09	....	$51\text{-}7^{8}$	23-4	$1.0292^{20/4}$	1.4537^{20}	....	$B1^{4}$, 408
10383	Pentane, 3,3-dichloro $(C_2H_5)_2CCl_2$	141.04	....	$131\text{-}2^{750}, 32^{14}$	......	1.053^{20}	1.442^{20}	bz, chl	$B1^{1}$, 343
10384	Pentane, 3,3-dichloro-2,4-dimethyl $[(CH_3)_2CH]_2CCl_2$	169.09	....	118-20d	......	0.9513^{9}	....	eth, chl	B1, 158
10385	Pentane, 3,3-diethyl or Tetraethyl methane $C(C_2H_5)_4$	128.26	....	146.2, 30.7^{10}	−33.1	$0.7536^{20/4}$	1.4206^{20}	eth, bz	$B1^{4}$, 462
10386	Pentane, 1,5-diido $I(CH_2)_5I$	323.94	....	$149^{20}, 101\text{-}2^{3}$	9	2.1903^{15}	1.6046^{15}	eth, chl	$B1^{4}$, 317
10387	Pentane, 2,2-dimethyl $C_3H_7C(CH_3)_3$	100.20	....	79.2	−123.8	$0.6739^{20/4}$	1.3822^{20}	al, eth, **ace**, **bz**, **chl**	$B1^{4}$, 403
10388	Pentane, 2,2-dimethyl-3-ethyl $(CH_3)_3CCH(C_2H_5)_2$	128.26	....	133.83	−99.3	$0.74378^{20/4}$	1.41227^{20}	....	$B1^{4}$, 462
10389	Pentane, 2,3-dimethyl $C_2H_5CH(CH_3)CH(CH_3)_2$	100.20	....	89.8	......	$0.6951^{20/4}$	1.3919^{20}	al, eth, **ace**, **bz**, **chl**	$B1^{4}$, 405
10390	Pentane, 2,3-dimethyl-3-ethyl $(CH_3)_2CHC(C_2H_5)_2CH_3$	128.26	....	141.6	......	....	1.4186^{20}	....	$B1^{4}$, 462
10391	Pentane, 2,4-dimethyl $CH_2[CH(CH_3)_2]_2$	100.20	....	80.5	−119.2	$0.6727^{20/4}$	1.3815^{20}	al, eth, **ace**, **bz**, **chl**	$B1^{4}$, 406
10392	Pentane, 2,4-dimethyl-3-ethyl $CH_3CH(CH_3)CH(C_2H_5)CH(CH_3)_2$	128.26	....	136.73	−122.4	$0.7365^{20/4}$	1.4131^{20}	....	$B1^{4}$, 462
10393	Pentane, 2,4-dimethyl-3-phenyl $(CH_3)_2CHCH(C_6H_5)CH(CH_3)_2$	176.29	....	220-5	......	$0.8822^{10/0}$	1.512	....	$B5^{1}$, 214
10394	Pentane, 3,3-dimethyl $(C_2H_5)_2C(CH_3)_2$	100.20	....	86.1	−134.4	$0.6936^{20/4}$	1.3909^{20}	al, eth, **ace**, **bz**, **chl**	$B1^{4}$, 409
10395	Pentane, 1,5-dinitro $O_2N(CH_2)_5NO_2$	162.15	....	$134^{1.2}$	......	....	1.461^{20}	bz	$B1^{4}$, 319
10396	Pentane, 1,1-diphenyl $C_4H_9CH(C_6H_5)_2$	224.35	....	307.9	−12	0.9659^{20}	1.5511^{20}	....	$B5^{4}$, 1966
10397	Pentane, 1,5-diphenyl $C_6H_5(CH_2)_5C_6H_5$	224.35	....	330.6, $187\text{-}9^{10}$	......	$0.9814^{19/0}$	1.559^{19}	....	$B5^{4}$, 1963
10398	Pentane, 1,2-epoxy-2,4,4-trimethyl $(CH_3)_3CCH_2C(CH_3)CH_2$ (C and CH_2 bridged by O)	128.21	....	$140.9, 20^{5.4}$	−64	$0.8287^{20/20}$	1.4097^{20}	eth, bz	$B17^{4}$, 112
10399	Pentane, 3-ethyl or Triethyl methane $(C_2H_5)_3CH$	100.20	....	93.5	−118.6	$0.6982^{20/4}$	1.3934^{20}	al, eth, **ace**, **bz**, **chl**	$B1^{4}$, 402
10400	Pentane, 3-ethyl-2-methyl $(C_2H_5)_2CHCH(CH_3)_2$	114.23	....	115.6, 9.5^{10}	−115	$0.7193^{20/4}$	1.4040^{20}	al, eth, ace, bz, chl	$B1^{4}$, 437
10401	Pentane, 3-ethyl-3-methyl $(C_2H_5)_3CCH_3$	114.23	....	$118.2, 9.9^{10}$	−90.9	$0.7274^{20/4}$	1.4078^{20}	**al**, eth, **ace**, **bz**, **chl**	$B1^{4}$, 438
10402	Pentane, 1-fluoro or Pentyl fluoride $C_5H_{11}F$	90.14	....	62.8	−120	$0.7907^{20/4}$	1.3591^{20}	al, eth	$B1^{4}$, 308
10403	Pentane, 1-iodo or Pentyl iodide $C_5H_{11}I$	198.05	....	$157, 39.3^{10}$	−85.6	$1.5161^{20/4}$	1.4959^{20}	al, eth	$B1^{4}$, 315
10404	Pentane, 2-iodo $C_3H_7CHICH_3$	198.05	....	144-5	......	$1.5096^{20/4}$	1.4961^{20}	eth, ace, bz	$B1^{4}$, 316
10405	Pentane, 3-iodo $(C_2H_5)_2CHI$	198.05	....	145-6, 68^{50}	......	$1.5176^{20/4}$	1.4974^{20}	eth, ace, bz	$B1^{1}$, 349

No.	Name, Synonyms, and Formula	Mol. wt.	Color, crystalline form, specific rotation and λ_{max} (log ε)	b.p. °C	m.p. °C	Density	n_D	Solubility	Ref.
10406	Pentane, 2-methyl $C_3H_7CH(CH_3)_2$	86.18		60.3	−153.7	$0.6532^{20/4}$	1.3715^{20}	al, eth, **ace, bz, chl**	B1^4, 358
10407	Pentane, 2-methyl-2-phenyl $C_6H_5C(CH_3)_2CH_2CH_2CH_3$	162.27		205-6		$0.8796^{10/4}$	$1.4955^{16.5}$		B5^3, 1019
10408	Pentane, 3-methyl $(C_2H_5)_2CHCH_3$	86.18		63.3		$0.6645^{20/4}$	1.3765^{20}	al, **eth, ace, bz**	B1^4, 363
10409	Pentane, 3-methyl-1-phenyl $C_6H_5CH_2CH_2CH(CH_3)C_2H_5$	162.27		220^{757}		$0.8644^{14.5/4}$	1.4896		B5^4, 1116
10410	Pentane, 3-nitro $(C_2H_5)_2CHNO_2$	117.15		153-5		$0.957^{0/4}$		al, eth, ace	B1^4, 318
10411	Pentane, perfluoro C_5F_{12}	288.04		57.73, -30.5^{10}		$1.7326^{20/4}$	1.2564^{22}	bz	B1^4, 308
10412	Pentane, 3-phenyl $C_6H_5CH(C_2H_5)_2$	148.25		187.5, $83\text{-}5^{22}$		$0.8649^{20/4}$	1.4880^{20}		B5^4, 1090
10413	Pentane, 1,1,1,5-tetrachloro $CCl_3(CH_2)_3CH_2Cl$	209.93		112^{24}		1.3416^{25}	1.4859^{25}		B1^4, 311
10414	Pentane, 2,2,3,3,-tetramethyl $(CH_3)_3CC(CH_3)_2C_2H_5$	128.26		140.27^{760}			1.4236^{20}		B1^4, 463
10415	Pentane, 2,2,3,4-tetramethyl $(CH_3)_3CCH(CH_3)CH(CH_3)_2$	128.26		133, 70.6^{104}		$0.7389^{20/4}$	1.4147^{20}		B1^4, 463
10416	Pentane, 2,2,4,4-tetramethyl $CH_2[C(CH_3)_3]_2$	128.26		122.7, 12.5^{10}	−66.5	$0.7195^{20/4}$	1.4069^{20}	al, bz	B1^4, 464
10417	Pentane, 2,3,3,4-tetramethyl $(CH_3)_2C[CH(CH_3)_2]_2$	128.26		141.5, 78^{104}	-102.14	$0.7547^{20/4}$	1.4222^{20}		B1^4, 464
10418	Pentane, 2,2,3-trimethyl $C_2H_5CH(CH_3)C(CH_3)_3$	114.23		110, 3.9^{10}	−112.3	$0.7161^{20/4}$	$1.4030^{20/4}$	**al, eth, ace, bz, chl**	B1^4, 438
10419	Pentane, 2,2,4-trimethyl or Isooctane $(CH_3)_2CHCH_2C(CH_3)_3$	114.23		99.2, 4.3^{10}	−107.4	$0.6919^{20/4}$	1.3915^{20}	**al,** eth, **ace, bz, chl**	B1^4, 439
10420	Pentane, 2,3,3-trimethyl $C_2H_5C(CH_3)_2CH(CH_3)_2$	114.23		114.7, 6.9^{10}	−100.7	$0.7262^{20/4}$	1.4075^{20}	al, **eth, ace, bz, chl**	B1^4, 445
10421	Pentane, 2,3,4-trimethyl $CH_3CH[CH(CH_3)_2]_2$	114.23		113.4, 7.1^{10}	−109.2	$0.7191^{20/4}$	1.4042^{20}	al, **eth, ace, bz, chl**	B1^4, 446
10422	Isopentane or 2-Methyl butane $CH_3CH_2CH(CH_3)_2$	72.15		27.8	−159.9	$0.6201^{20/4}$	1.3537^{20}	**al, eth**	B1^4, 320
10423	1,5-Pentanedial or Glutyraldehyde $OHC(CH_2)_3CHO$	100.12		187-9d, $71\text{-}2^{10}$				**w, al,** bz	B1^4, 3659
10424	Pentanedioic acid or Glutaric acid $HO_2C(CH_2)_3CO_2H$	132.12	nd (bz)	302-4d, 200^{20}	99	$1.424^{25/4}$	1.4188^{106}	w, al, eth, chl	B2^4, 1934
10425	1,2-Pentanediol *(d)* or 1,2-Pentylene glycol $C_3H_7CHOHCH_2OH$	104.15	[α] +0.95	$210\text{-}12^{751}$, $99\text{-}102^{13}$		$0.9802^{20/20}$	1.4412^{19}		B1^4, 2538
10426	1,2-Pentanediol, 2,4,4-trimethyl $(CH_3)_3CCH_2C(OH)(CH_3)CH_2OH$	146.23	pr or pl (bz)		62-3			w, al, eth	B1^4, 2604
10427	1,3-Pentanediol, 2,2-dimethyl $C_2H_5CH(OH)C(CH_3)_2CH_2OH$	132.20	(eth)	212-4, 119^{21}	60-3			al	B1, 490
10428	1,3-Pentanediol, 2,2,4-trimethyl $(CH_3)_2CHCHOHC(CH_3)_2CH_2OH$	146.23	pl (bz)	234^{737}, $81\text{-}2^{1}$	51-2	$0.937^{15/15}$	1.4513^{15}	al, eth	B1^4, 2604
10429	1,4-Pentanediol or γ-Pentylene glycol $CH_3CHOHCH_2CH_2CH_2OH$	104.15		220^{715}, $124\text{-}6^{10}$		$0.9883^{20/4}$	1.4452^{23}	**w, al, chl**	B1^4,2539
10430	1,4-Pentanediol, 2,2,4-trimethyl $(CH_3)_2C(OH)CH_2C(CH_3)_2CH_2OH$	146.23	cr (eth)	209.11, $114\text{-}5^{13}$	86			al, eth	B1, 493
10431	1,5-Pentanediol or Pentamethylene glycol $HO(CH_2)_5OH$	104.15		260, $137\text{-}8^{12}$	−18	$0.9939^{20/20}$	1.4494^{20}	w, al	B1^4, 2540
10432	1,5-Pentanediol, diacetate or 1,5-Diacetoxy pentane $CH_3CO_2(CH_2)_5O_2CCH_3$	188.22		$122\text{-}3^{3}$		1.0296^{20}	1.4261^{19}		B2^4, 226
10433	1,5-Pentanediol, 2,2-dimethyl $HOCH_2CH_2CH_2C(CH_3)_2CH_2OH$	132.20		130^{12}				al, eth	B1^1, 254
10434	2,3-Pentanediol $C_2H_5CHOHCHOHCH_3$	104.15		187.5, 97^{17}		$0.9800^{19/0}$	1.4412^{25}	w, al	B1^4, 2543
10435	2,3-Pentanediol, 2,4,4-trimethyl $(CH_3)_3CCH(OH)C(CH_3)(OH)CH_3$	146.23	mcl pr (lig)		65-6			al, eth	B1^4, 2604
10436	2,4-Pentanediol, 2-methyl $CH_3CH(OH)CH_2C(OH)(CH_3)_2$	118.18		197		$0.9254^{17/4}$	1.4250^{20}	w, al, eth	B1^4, 2565
10437	2,4-Pentanediol, 3-methyl $CH_3CH[CH(OH)CH_3]_2$	118.18		211-2, 91^{3}		0.9640^{20}	1.4433^{20}	w, al	B1^4, 2572
10438	2,3-Pentanedione $C_2H_5COCOCH_3$	100.12	dk ye liq	108		$0.9565^{19/4}$	1.4014^{19}	w, **al, eth, ace**	B1^4, 3660

No.	Name, Synonyms, and Formula	Mol. wt.	Color, crystalline form, specific rotation and λ_{max} (log ε)	b.p. °C	m.p. °C	Density	n_D	Solubility	Ref.
10439	2,3-Pentanedione, dioxime $C_2H_5C(NOH)C(NOH)CH_3$	130.15	ye nd (al), pl (to,al)	sub	172-3			al	B1[4], 3661
10440	2,3-Pentanedione, 2-oxime $C_2H_5COC)NOH)CH_3$	115.13	lf (dil al)		69-72			al, eth	B1[4], 3661
10441	2,3-Pentanedione, 3-oxime $C_2H_5C(NOH)COCH_3$	115.13	pl (lig)	183-7d	58-9			al, eth, chl	B1[4], 3661
10442	2,4-Pentanedione or Acetylacetone $CH_3COCH_2COCH_3$	100.12		139^{746}	−23	$0.9721^{25/4}$	1.4494^{20}	w, **al, eth, ace, chl**	B1[4], 3662
10443	2,4-Pentanedione, 3,3-dimethyl $(CH_3CO)_2C(CH_3)_2$	128.17		173, 58^{10}	19	$0.9575^{20/4}$	1.4306^{20}	eth	B1[4], 3705
10444	2,4-Pentanedione, dioxime $CH_2[C(=NOH)CH_3]_2$	130.15	pr (eth)		149-50			al	B1[3], 3123
10445	2,4-Pentanedione, 3-ethyl $(CH_3CO)_2CHC_2H_5$	128.17		177-80, $69\text{-}70^{13}$		$0.9531^{19/4}$	1.4408^{19}	al, eth, chl	B1[4], 3703
10446	2,4-Pentanedione, monoimide $CH_3COCH_2C(NH)CH_3$	99.13		209	43			w, eth	B1[4], 3678
10447	1-Pentanethiol or Pentylmercaptan $CH_3(CH_2)_3CH_2SH$	104.21		126.6^{460}, 99.5^{10}	−75.7	$0.8421^{20/4}$	1.4469^{20}	**al, eth**	B1[4], 1653
10448	2-Pentanethiol $C_3H_7CH(SH)CH_3$	104.21		112.9, 63.9^{150}	−169	$0.8327^{20/4}$	1.4412^{20}	al, lig	B1[4], 1662
10449	3-Pentanethiol $(C_2H_5)_2CHSH$	104.21		105	−110.8	$0.8410^{20/4}$	1.4447^{20}	al	B1[4], 1665
10450	3-Pentanecarboxylic acid-1,5-dinitrile, 3-acetyl, ethyl ester $CH_3COC(CH_2CN)_2CO_2C_2H_5$	236.27	cr	$190\text{-}200^{2}$	83				B3[2], 512
10451	1,2,3-Pentanetriol $C_2H_5(CHOH)_2CH_2OH$	120.15	syr	192^{63}		$1.0851^{34/0}$		w, al, eth	B1[4], 2778
10452	1,2,5-Pentanetriol $HOCH_2CHOH(CH_2)_3OH$	120.15		$190\text{-}1^{13}$		$1.136^{20/15}$	1.4730^{20}	w, al	B1[4], 2779
10453	1,3,5-Pentanetriol $(HOCH_2CH_2)_2CHOH$	120.15		$188\text{-}9^{11}$		$1.1291^{20/4}$	1.4785^{20}	w, al, eth, ace	B1[4], 2779
10454	Pentanoic acid or Valeric acid $CH_3(CH_2)_3CO_2H$	102.13		186, 82.7^{10}	−33.8	$0.9391^{20/4}$	1.4085^{20}	w, al, eth	B2[4], 868
10455	1-Pentanol or *n*-Pentyl alcohol $CH_3(CH_2)_3CH_2OH$	88.15		137.3^{748}, 50^{13}	−79	$0.8144^{20/4}$	1.4101^{20}	**al, eth**, ace	B1[4], 1640
10456	1-Pentanol, 5-amino $HO(CH_2)_5NH_2$	103.16		221-2, $126\text{-}8^{22}$	38-9	$0.9488^{17/4}$	1.4618^{17}	w, al, ace	B4[4], 1750
10457	1-Pentanol, 5-chloro $Cl(CH_2)_5OH$	122.59		112^{12}			1.4518^{20}	al, eth	B1[4], 1650
10458	1-Pentanol, 2,4-dimethyl(*(dl)* $(CH_3)_2CHCH_2CH(CH_3)CH_2OH$	116.20		160-2, $65\text{-}7^{16}$		$0.793^{20/4}$	1.427^{20}	al, eth	B1[4], 1753
10459	1-Pentanol, 2,4-dimethyl *(l)* $(CH_3)_2CHCH_2CH(CH_3)CH_2OH$	116.20	$[\alpha]^{22}{}_D$ −1.1 (undil)	157		$0.816^{25/4}$		al, eth	B1[4], 1753
10460	1-Pentanol, 5-fluoro $F(CH_2)_5OH$	106.14		$70\text{-}1^{11}$			1.4057^{25}	al, eth	B1[4], 1647
10461	1-Pentanol, 2-methyl $C_3H_7CH(CH_3)CH_2OH$	102.18		148		$0.8263^{20/4}$	1.4182^{20}	al, eth, ace	B1[4], 1713
10462	1-Pentanol, 3-methyl(-*(dl)* $C_2H_5CH(CH_3)CH_2CH_2OH$	102.18		152.4, $51\text{-}3^{8}$		$0.8242^{20/4}$	1.4112^{23}	al, eth	B1[4], 1722
10463	1-Pentanol, 4-methyl or Isohexyl alcohol $(CH_3)_2CH(CH_2)_3OH$	102.18		151.6		$0.8131^{20/4}$	1.4134^{25}	al, eth	B1[4], 1721
10464	1-Pentanol, 1-phenyl $C_4H_9CH(OH)C_6H_5$	164.25	$[\alpha]^{20}{}_D$ +40.8	$140\text{-}2^{25}$		$0.9672^{20/20}$	1.4086^{25}	al, eth, ace	B6[4], 1952
10465	1-Pentanol, 5-phenyl $C_6H_5(CH_2)_5OH$	164.25		155^{20}		0.9725^{20}	1.5156^{20}	al, eth	B6[4], 1954
10466	2-Pentanol $C_3H_7CH(OH)CH_3$	88.15		118.9, 62^{60}		$0.8103^{20/4}$	1.4053^{20}	w, al, eth	B1[4], 1655
10467	2-Pentanol, 1-chloro $C_3H_7CH(OH)CH_2Cl$	122.59		$157\text{-}60^{735}$, $59\text{-}62^{14}$		$1.037^{20/20}$	1.4404^{25}	al, eth	B1[3], 1613
10468	2-Pentanol, 2,4-dimethyl $(CH_3)_2CHCH_2C(OH)(CH_3)_2$	116.20		133.1, $53\text{-}4^{25}$	>−20	$0.8103^{20/4}$	1.4172^{20}		B1[4], 1753
10469	2-Pentanol, 2-methyl $C_3H_7C(OH)(CH_3)_2$	102.18	col	120-2, $49.5^{27.5}$	−103	$0.8350^{16/4}$	1.4100^{20}	al, eth	B1[4], 1714
10470	2-Pentanol, 3-methyl $C_2H_5CH(CH_3)CH(OH)CH_3$	102.18		134.3, 75.6^{50}		$0.8307^{20/4}$	1.4182^{20}	al, eth	B1[3], 1672
10471	2-Pentanol, 4-methyl $(CH_3)_2CHCH_2CH(OH)CH_3$	102.18		133, $50\text{-}5^{25}$		$0.8075^{20/4}$	1.4100^{20}	al, eth	B1[4], 1717

PHYSICAL CONSTANTS OF ORGANIC COMPOUNDS (Continued)

No.	Name, Synonyms, and Formula	Mol. wt.	Color, crystalline form, specific rotation and λ_{max} (log ε)	b.p. °C	m.p. °C	Density	n_D	Solubility	Ref.
10472	2-Pentanol, 4-methyl,acetate or 4-Methyl-2-pentyl acetate $(CH_3)_2CHCH_2CH(CH_3)O_2CCH_3$	144.21	....	147-8, 76.5^{47}	......	$0.8805^{0/0}$	1.4066^{20}	ace, bz	B2,133
10473	2-Pentanol, 2-phenyl $C_3H_7C(OH)(CH_3)C_6H_5$	164.25	....	216, 112^{14}	......	$0.9723^{22/4}$	....	al	$B6^2$, 505
10474	3-Pentanol $(C_2H_5)_2CHOH$	88.15	....	116.1, 30^{12}	......	$0.8212^{20/4}$	1.4104^{20}	al, eth, ace	$B1^4$, 1662
10475	2-Pentanol, 2,4,4-trimethyl $(CH_3)_3CCH_2C(OH)(CH_3)_2$	130.23	....	147.5, $42\text{-}4^{7}$	−20	$0.8225^{20/4}$	1.4284^{20}	eth	$B1^4$, 1796
10476	3-Pentanol, 4-amino-2-methyl $CH_3CH(NH_2)CH(OH)CH(CH_3)_2$	117.19	....	174^{745}	35-6	....	....	al, eth, ace, bz	$B4^4$, 796
10477	3-Pentanol, 1-chloro $C_2H_5CH(OH)CH_2CH_2Cl$	122.59	....	173, 77^{20}	......	$1.0327^{25/4}$	1.448^{25}	al, eth	$B1^4$, 1665
10478	3-Pentanol, 2,2-dimethyl $C_2H_5CH(OH)C(CH_3)_3$	116.20	....	135, $44\text{-}5^{15}$	−5	$0.8253^{20/4}$	1.4223^{20}	al, eth	$B1^4$, 1751
10479	3-Pentanol, 2,3-dimethyl $C_2H_5C(OH)(CH_3)CH(CH_3)_2$	116.20	....	139.7, $44\text{-}5^{14}$	<−30	$0.833^{20/4}$	1.4287^{20}	al, eth	$B1^4$, 1752
10480	3-Pentanol, 2,4-dimethyl $(i\text{-}C_3H_7)_2CHOH$	116.20	....	138.7, 87.5^{125}	<−70	$0.8288^{20/4}$	1.4250^{20}	al, eth	$B1^4$, 1754
10481	3-Pentanol, 2,4-dimethyl-3-phenyl $(i\text{-}C_3H_7)_2C(OH)C_6H_5$	192.30	ye	229, 157^{60}	......	$0.9755^{20/4}$	1.5239^{20}	eth	$B6^1$, 273
10482	3-Pentanol, 3-ethyl $(C_2H_5)_3COH$	116.20	....	143.1, 73^{52}	......	$0.8407^{22/4}$	1.4294^{20}	al, eth	$B1^4$, 1750
10483	3-Pentanol, 3-ethyl-2-methyl $(C_2H_5)_2C(OH)CH(CH_3)_2$	130.23	....	$159\text{-}61^{750}$, $55\text{-}7^{48}$	......	$0.8295^{20/20}$	1.4372^{10}	al, eth	$B1^4$, 1794
10484	3-Pentanol, 2-methyl $C_2H_5CH(OH)CH(CH_3)_2$	102.18	....	126.7	......	$0.8243^{20/4}$	1.4175^{20}	**al, eth**	$B1^4$,1716
10485	3-Pentanol, 3-methyl $(C_2H_5)_2C(OH)CH_3$	102.18	$[\alpha]^{20}_D$ + 40.8	122.4	−23.6	$0.8286^{20/4}$	4186^{20}	**al, eth**	$B1^4$, 1723
10486	3-Pentanol, 1-phenyl *(d)* $C_2H_5CH(OH)CH_2CH_2C_6H_5$	164.25	$[\alpha]^{20/}_D$ + 26.8 (al)	143^{19}	38	$0.9687^{20/4}$	....	al	$B6^4$, 1953
10487	3-Pentanol, 3-phenyl $(C_2H_5)_2C(OH)C_6H_5$	164.25	....	$223\text{-}4^{762}$, 110^{12}	<−17	$0.9831^{20/4}$	1.5165^{20}	vs	$B6^4$, 1963
10488	3-Pentanol, 2,3,4-trimethyl $(CH_3)_2CHC(OH)(CH_3)CH(CH_3)_2$	130.23	....	156.5	......	$0.8507^{20/20}$	1.4353^{20}	....	$B1^4$, 1798
10489	1-Pentanol-2-one $C_3H_7COCH_2OH$	102.13	....	152, $62\text{-}4^{18}$	......	$0.9860^{20/4}$	1.4234^{12}	w, al, eth	$B1^4$, 4004
10490	1-Pentanol-4-one $HO(CH_2)_3COCH_3$	102.13	....	208^{730}, $116\text{-}8^{33}$	......	$1.0071^{20/4}$	1.4390^{20}	al, eth, **w**	$B2^4$, 704
10491	2-Pentanol-4-one, 2-methyl or Diacetone alcohol $(CH_3)_2C(OH)CH_2COCH_3$	116.16	....	164, $67\text{-}9^{19}$	−44	$0.9387^{20/4}$	1.4213^{20}	**w, al, eth**	$B1^4$, 4023
10492	3-Pentanol-2-one $C_2H_5CH(OH)COCH_3$	102.13	....	147-8, 59^{27}	......	$0.9500^{20/4}$	1.4350^{10}	**al**, eth, ace, bz	$B1^3$, 3220
10493	2-Pentanol-3-one $C_2H_5COCH(OH)CH_3$	102.13	....	152.5, 63^{20}	......	$0.9742^{20/4}$	1.4128^{20}	al, eth	$B1^4$, 4008
10494	4-Pentanol-2-one $CH_3CH(OH)CH_2COCH_3$	102.13	....	177, $62\text{-}4^{12}$	......	$1.0071^{20/4}$	1.4265^{20}	al, eth	$B1^4$, 4005
10495	2-Pentanone or Methyl propyl ketone $C_3H_7COCH_3$	86.13	....	102	−77.8	$0.8089^{20/4}$	1.3895^{20}	**al, eth**	$B1^4$, 3271
10496	2-Pentanone, 4-amino-4-methyl $(CH_3)_2C(NH_2)CH_2COCH_3$	115.18	....	$25^{0.14}$	......	<1	....	**al, eth**	$B4^3$, 894
10497	2-Pentanone, 3-benzylidene $C_6H_5CH{=}C(C_2H_5)COCH_3$	174.24	....	$136\text{-}8^{12}$	......	$1.0005^{22/4}$	1.5650^{22}	....	$B7^1$, 198
10498	2-Pentanone, 1-chloro $C_3H_7COCH_2Cl$	120.58	....	154-6d, $58\text{-}9^{17}$	......	....	....	MeOH	$B1^4$, 3276
10499	2-Pentanone, 5-chloro $Cl(CH_2)_3COCH_3$	120.58	....	76^{34}	......	$1.0523^{20/4}$	1.4375^{20}	eth, bz	$B1^4$, 3277
10500	2-Pentanone, 3-ethyl-4-methyl $(CH_3)_2CHCH(C_2H_5)COCH_3$	128.21	....	154-5	......	$0.812^{20/4}$	1.4105^{20}	**al, eth, bz, chl,** aa	$B1^3$, 2883
10501	2-Pentanone, 3-methyl (dl) or *sec* butyl methyl ketone $C_2H_5CH(CH_3)COCH_3$	100.16	....	118^{758}	......	$0.8130^{30/4}$	1.4002^{20}	**al, eth**, chl	$B1^4$, 3309
10502	2-Pentanone, 4-methyl or Methyl isobutyl Ketone $(CH_3)_2CHCH_2COCH_3$	100.16	....	116.8, $35\text{-}40^{16}$	−84.7	0.7978^{20}	1.3962^{20}	**al, eth, ace, bz,** chl	$B1^4$, 3305
10503	2-Pentanone oxime or Methyl propyl ketoxime $C_3H_7C({=}NOH)CH_3$	101.15	....	167^{748}	......	$0.9095^{20/4}$	1.4450^{20}	w, **al, eth**	$B1^4$, 3274
10504	3-Pentanone or Diethyl ketone $C_2H_5COC_2H_5$	86.13	....	101.7	−39.8	$0.8138^{20/4}$	1.3924^{20}	w, **al, ace**	$B1^4$, 3282

No.	Name, Synonyms, and Formula	Mol. wt.	Color, crystalline form, specific rotation and λ_{max} (log ε)	b.p. °C	m.p. °C	Density	n_D	Solubility	Ref.
10505	3-Pentanone, 1-chloro $C_2H_5COCH_2CH_2Cl$	120.58	...	68^{20}	...	...	1.4361^{20}	al, eth	$B1^{4}$, 3284
10506	3-Pentanone, 2-chloro $C_2H_5COCHClCH_3$	120.58	...	135	...	...	...	al, eth	$B1^{4}$, 3284
10507	3-Pentanone, 2,2-dimethyl or Ethyl *tert*-butyl ketone $C_2H_5COC(CH_3)_3$	114.19	...	124.5^{730}	−45	$0.8125^{20/4}$	1.4065^{20}	al, eth, ace, chl	$B1^{4}$, 3331
10508	3-Pentanone, 2,4 dimethyl or Diisopropyl ketone $(i\text{-}C_3H_7)_2CO$	114.19	...	124-5	−69	$0.8108^{20/4}$	1.3999^{20}	**al, eth**, bz	$B1^{4}$, 3334
10509	3-Pentanone, 2-methyl or Ethyl isopropyl ketone $(CH_3)_2CHCOC_2H_5$	100.16	...	$114\text{-}5^{745}$	...	$0.830^{0/0}$	1.3975^{20}	al, **eth, ace**, bz, chl	$B1^{4}$, 3304
10510	3-Pentanone, 2,2,4,4-tetramethyl or Pivalone $[(CH_3)_3C]_2CO$	142.24	...	152, 70^{43}	...	0.8240^{18}	1.4194^{20}	al, eth, ace, chl, aa	$B1^{4}$, 3334
10511	Pentaguine $C_{18}H_{27}N_3O$	301.43	...	$165\text{-}70^{0.02}$	...	...	1.5785^{25}	dil HCl	Am68, 1524
10512	Pentasiloxane, dodecamethyl $CH_3[Si(CH_3)_2O\text{-}]_4Si(CH_3)_3$	384.84	...	229^{710}, $103\text{-}7^{12}$	−80	0.8755^{20}	1.3925^{20}	bz, lig	Am68, 2284
10513	Pentatriacontane $CH_3(CH_2)_{33}CH_3$	492.96	cr (al)	490, 311^{10}	75	$0.8157^{20/4}$	1.4568^{20}	ace	$B1^{4}$, 598
10514	18-Pentatriacontanone or Stearone $C_{17}H_{35}COC_{17}H_{35}$	506.94	lf (lig)	...	88.4	$0.793^{95/4}$	...	...	$B1^{4}$, 3413
10515	2-Pentenal, 2-methyl $C_2H_5CH{=}C(CH_3)CHO$	98.14	...	136-7, $38\text{-}9^{18}$	...	$0.8581^{20/4}$	1.4488^{20}	al, eth, bz	$B1^{4}$, 3471
10516	4-Pentenal $CH_2{=}CHCH_2CH_2CHO$	84.12	...	96	...	$0.852^{20/4}$	1.4191^{20}	eth, ace	$B1^{4}$, 3459
10517	1-Pentene $CH_3CH_2CH_2CH{=}CH_2$	70.13	...	30	−138	$0.6405^{20/4}$	1.3715^{20}	al, eth, bz	$B1^{4}$, 808
10518	1-Pentene, 1-bromo $C_3H_7CH{=}CHBr$	149.03	...	121-2, 43.5^{30}	...	$1.2606^{20/4}$	1.4572^{20}	eth, bz, chl	$B1^{3}$, 775
10519	1-Pentene, 2-bromo $C_3H_7CBr{=}CH_2$	149.03	...	107-8	...	1.228^{20}	1.4535^{20}	eth, bz, chl, lig	$B1^{3}$, 775
10520	1-Pentene, 3-bromo $CH_3CH_2CHBrCH{=}CH_2$	149.03	...	30.5^{30}	...	$1.2417^{25/4}$	1.4626^{25}	ace, bz, chl	$B1^{3}$, 775
10521	1-Pentene, 2-chloro $CH_3CH_2CH_2CCl{=}CH_2$	104.58	...	95-7	...	0.872^{5}	...	al, eth	$B1^{3}$, 774
10522	1-Pentene, 2-chloro-3-ethyl-3-methyl $(C_2H_5)_2C(CH_3)CCl{=}CH_2$	146.66	...	147^{742}, 53^{20}	...	$0.9147^{20/4}$	1.4450^{25}	bz, chl	$B1^{3}$, 847
10523	1-Pentene, 3-chloro $CH_3CH_2CHClCH{=}CH_2$	104.58	...	$93\text{-}4^{764}$	...	$0.8978^{20/4}$	1.4254^{20}	al, eth, ace	$B1^{3}$, 774
10524	1-Pentene, 3-chloro-2-methyl $C_2H_5CHClC(CH_3){=}CH_2$	118.61	...	121-4	...	...	1.4422^{20}	eth, bz, chl	$B1^{3}$, 809
10525	1-Pentene, 4-chloro $CH_3CHClCH_2CH{=}CH_2$	104.58	...	97-100	...	0.934^{15}	1.417^{15}	eth, chl	$B1^{3}$, 774
10526	1-Pentene, 5-Chloro $ClCH_2CH_2CH_2CH{=}CH_2$	104.58	...	$103\text{-}4^{773}$	...	$0.9125^{20/4}$	1.4297^{20}	eth, ace	$B1^{4}$, 810
10527	1-Pentene, 2,3-dimethyl $C_2H_5CH(CH_3)C(CH_3){=}CH_2$	98.19	...	84.3	−134.8	$0.7051^{20/4}$	1.4033^{20}	**al, eth**	$B1^{4}$, 870
10528	1-Pentene, 2,4-dimethyl $(CH_3)_2CHCH_2C(CH_3){=}CH_2$	98.19	...	81.6	−123.8	$0.6943^{20/4}$	1.3986^{20}	**al, eth**, bz, chl	$B1^{4}$, 870
10529	1-Pentene, 3,3-dimethyl $C_2H_5C(CH_3)_2CH{=}CH_2$	98.19	...	77.5	−134.3	$0.6974^{20/4}$	1.3984^{20}	**al, eth**, bz, chl	$B1^{4}$, 873
10530	1-Pentene, 4,4-dimethyl $(CH_3)_3CCH_2CH{=}CH_2$	98.19	...	72.5	−136.6	$0.6827^{20/4}$	1.3918^{20}	**al, eth**, bz	$B1^{4}$, 869
10531	1-Pentene, 2-ethyl $C_3H_7C(C_2H_5){=}CH_2$	98.19	...	94	...	$0.7079^{20/4}$	1.405^{20}	al, eth, bz	$B1^{4}$, 867
10532	1-Pentene, 3-ethyl $(C_2H_5)_2CHCH{=}CH_2$	98.19	...	85	...	0.6948^{22}	1.3966^{23}	...	$B1^{4}$, 867
10533	1-Pentene, 2-methyl $C_3H_7C(CH_3){=}CH_2$	84.16	...	60.7	−135.7	$0.6799^{20/4}$	1.3920^{20}	al, bz, chl, peth	$B1^{4}$, 841
10534	1-Pentene, 3-methyl $C_2H_5CH(CH_3)CH{=}CH_2$	84.16	...	51.1	−153	$0.6675^{20/4}$	1.3841^{20}	al, bz, chl, peth	$B1^{4}$, 847
10535	1-Pentene, 4-methyl $(CH_3)_2CHCH_2CH{=}CH_2$	84.16	...	53.9	−153.6	$0.6642^{20/4}$	1.3828^{20}	al, bz, chl, peth	$B1^{4}$, 846
10536	1-Pentene, 2-methyl, perfluoro $F_3C(CF_2)_2C(CF_3){=}CF_2$	300.05	...	60	...	...	...	bz	$B1^{4}$, 842
10537	1-Pentene, perfluoro $CF_3CF_2CF_2CF{=}CF_2$	250.04	...	$29\text{-}30^{740}$	...	...	1.2571^{25}	chl	$B1^{4}$, 810

PHYSICAL CONSTANTS OF ORGANIC COMPOUNDS (Continued)

No.	Name, Synonyms, and Formula	Mol. wt.	Color, crystalline form, specific rotation and λ_{max} (log ε)	b.p. °C	m.p. °C	Density	n_D	Solubility	Ref.
10538	1-Pentene, 2,4,4-trimethyl or Diisobutylene $(CH_3)_3CCH_2C(CH_3)=CH_2$	112.22	...	101.4	−93.5	$0.7150^{20/4}$	1.4086^{20}	eth, bz, lig, chl	$B1^{4}$, 892
10539	2-Pentene (cis) $CH_3CH_2CH=CHCH_3$	70.13	...	36.9	−151.4	$0.6556^{20/4}$	1.3830^{20}	**al, eth**, bz	$B1^{4}$, 814
10540	2-Pentene-*(trans)* $CH_3CH_2CHCHCH_3$	70.13	...	36.3	−136	$0.6482^{20/4}$	1.3793^{20}	**al, eth**, bz	$B1^{4}$, 814
10541	2-Pentene, 1-bromo $CH_3CH_2CH=CHCH_2Br$	149.03	...	123-4, 35^{25}	...	1.2545^{20}	1.4731^{20}	ace, bz, chl	$B1^{4}$, 816
10542	2-Pentene, 2-bromo $CH_3CH_2CH=CBrCH_3$	149.03	...	110.5^{750}	...	$1.277^{20/20}$	1.4580^{20}	ace, bz, chl	$B1^{3}$, 784
10543	2-Pentene, 3-bromo $CH_3CH_2CBr=CHCH_3$	149.03	...	115.2^{750}	...	$1.273^{20/20}$	1.4628^{20}	ace, bz, chl	$B1^{3}$, 784
10544	2-Pentene, 4-bromo $CH_3CHBrCH=CHCH_3$	149.03	...	117d, 22^{9}	...	1.2312^{21}	1.4752^{21}	ace, bz, chl	$B1^{4}$, 816
10545	2-Pentene, 5-bromo $BrCH_2CH_2CH=CHCH_3$	149.03	...	121.7^{621}	...	$1.2715^{20/4}$	1.4695^{20}	ace, bz, chl	$B1^{4}$, 816
10546	2-Pentene, 1-chloro $CH_3CH_2CH=CHCH_2Cl$	104.58	...	109.5, 62^{148}	...	$0.908^{22/4}$	1.4352^{22}	al, eth, ace, chl	$B1^{3}$, 781
10547	2-Pentene, 2-chloro $CH_3CH_2CH=CClCH_3$	104.58	...	95-7, 45^{130}	...	$0.9067^{20/4}$	1.4261^{20}	eth, ace	$B1^{3}$, 782
10548	2-Pentene, 3-chloro $CH_3CH_2CCl=CHCH_3$	104.58	...	90-2	...	1.423^{24}	0.9125^{20}	eth, ace	$B1^{3}$,782
10549	2-Pentene, 3-chloro-2,4-dimethyl $(CH_3)_2CHCCl=C(CH_3)_2$	132.63	...	118-20, 44-5^{30}	...	$0.9513^{9/9}$	...	eth, bz, chl	B1, 221
10550	2-Pentene, 4-chloro $CH_3CHClCH=CHCH_3$	104.58	...	103, 18-20^{12}	...	$0.9004^{20/20}$	1.4322^{20}	eth, ace, chl	$B1^{4}$, 815
10551	2-Pentene, 5-chloro $ClCH_2CH_2CH=CHCH_3$	104.58	...	107-8^{755}	...	$0.9043^{20/4}$	1.4310^{20}	ace, bz, chl	$B1^{4}$, 816
10552	2-Pentene, 5-chloro-2-methyl $ClCH_2CH_2CH=C(CH_3)_2$	118.61	...	132-3^{756}	...	0.9135^{20}	...	ace, chl	$B1^{4}$, 843
10553	2-Pentene, 2,5-dichloro $ClCH_2CH_2CH=CClCH_3$	139.02	...	40-1^{8}	...	$1.1182^{15/4}$	...	chl	$B1^{4}$, 816
10554	2-Pentene, 2,3-dimethyl $C_2H_5C(CH_3)=C(CH_3)_2$	98.19	...	97.5	−118.3	0.7277^{20}	1.4208^{20}	al, eth, bz, chl	$B1^{4}$, 870
10555	2-Pentene, 2,4-dimethyl $(CH_3)_2CHCH=C(CH_3)_2$	98.19	...	83.4	−127.7	$0.6954^{20/4}$	1.4040^{20}	al, eth, bz, chl	$B1^{4}$, 872
10556	2-Pentene, 3,4-dimethyl $(CH_3)_2CH(CH_3)=CHCH_3$	98.19	...	87	...	$0.7126^{20/4}$	1.4070^{20}	...	$B1^{4}$, 871
10557	2-Pentene, 4,4-dimethyl *(trans)* $(CH_3)_3CCH=CHCH_3$	98.19	...	76.7	−115.2	$0.6889^{20/4}$	1.3982^{20}	al, eth, bz, chl	$B1^{4}$, 868
10558	2-Pentene, 3-ethyl $(C_2H_5)_2C=CHCH_3$	98.19	...	94	...	$0.7079^{20/4}$	1.405^{20}	al, eth, bz, chl	$B1^{4}$, 868
10559	2-Pentene, 2-methyl $C_2H_5CH=C(CH_3)_2$	84.16	...	67.3	−135	$0.6863^{20/4}$	1.4004^{20}	al, bz, chl, peth	$B1^{4}$, 842
10560	2-Pentene, 3-methyl (cis) $C_2H_5C(CH_3)=CHCH_3$	84.16	...	67.6	−138.4	$0.6986^{20/4}$	1.4045^{20}	al, bz, chl, peth	$B1^{4}$, 848
10561	2-Pentene, 3-methyl *(trans)* $C_2H_5C(CH_3)=CHCH_3$	84.16	...	70.4	−134.8	$0.6942^{20/4}$	1.4016^{20}	al, bz, chl, peth	$B1^{4}$, 848
10562	2-Pentene, 4-methyl *(cis)* $(CH_3)_2CHCH=CHCH_3$	84.16	...	56.3	−134.4	$0.6690^{20/4}$	1.3800^{20}	al, bz, chl, peth	$B1^{4}$, 844
10563	2-Pentene, 4-methyl *(trans)* $(CH_3)_2CHCH=CHCH_3$	84.16	...	58.5	−140.8	$0.6686^{20/4}$	1.3889^{20}	al, bz, chl, peth	$B1^{4}$, 844
10564	2-Pentene, 2,3-dimethyl. $(CH_3)_2CHC(CH_3)=C(CH_3)_2$	98.19	...	116.5	-113.38	0.74342^{204}	1.4274^{20}	...	$B1^{4}$, 894
10565	2-Pentene, 2,4,4-trimethyl $(CH_3)_3CCH=C(CH_3)_2$	112.22	...	104.9	−106.3	$0.7218^{20/4}$	1.4160^{20}	eth, bz, lig, chl	$B1^{4}$, 891
10566	2-Pentenedioic acid, diethyl ester (*trans*) or Ethyl glutaconate $C_2H_5O_2CCH_2CH=CHCO_2C_2H_5$	186.21	...	236-8, 125^{12}	...	$1.0496^{20/4}$	1.4411^{20}	al, eth	$B2^{4}$, 2227
10567	2-Pentenoic acid, 4-hydroxy, lactone or β-Angelica lactone $CH_3CHCH=CHCO$ (ring closed through O)	98.10	...	208-9^{751}, 98^{15}	<−17	$1.0810^{20/4}$	1.4454^{20}	w, al, eth	$B17^{4}$, 4302
10568	2-Pentenoic acid, 2-methyl (trans) $C_2H_5CH=C(CH_3)CO_2H$	114.14	...	213^{750}, 112^{12}	24.4	0.9751^{20}	1.4513^{20}	eth, chl	$B2^{4}$, 1568
10569	2-Pentenoic acid, 4-methyl $(CH_3)_2CHCH=CHCO_2H$	114.14	...	217, 115-6^{20}	35	$0.9529^{21/4}$	1.4489^{21}	al, eth, ace	$B2^{4}$, 1569

PHYSICAL CONSTANTS OF ORGANIC COMPOUNDS (Continued)

No.	Name, Synonyms, and Formula	Mol. wt.	Color, crystalline form, specific rotation and λ_{max} (log ε)	b.p. °C	m.p. °C	Density	n_D	Solubility	Ref.
10570	3-Pentenoic acid, 4-hydroxy, γ-lactone or γ-Angelica lactone $CH_3\text{-}C{=}CHCH_2CO$	98.10	...	167, 53^{12}	18	$1.084^{20/4}$	1.4476^{20}	w, al, eth	$B17^4$, 4300
10571	3-Pentenonitrile, 2-hydroxy (*cis*) $CH_3CH{=}CHCH(OH)CN$	97.12	...	139^{70}	...	$0.9675^{15/4}$	1.4460^{21}	al, eth, bz, chl	$B3^4$, 1001
10572	4-Pentenamide, 2,2-diethyl or Novonal $CH_2{=}CHCH_2C(C_2H_5)_2CONH_2$	155.24	wh pw, cr (eth-peth)	155^{10}	75-6	...	...	al, eth	$B2^3$, 1351
10573	4-Pentenoic acid or Allyl acetic acid $CH_2{=}CHCH_2CH_2CO_2H$	100.12	...	188-9, 93^{20}	−22.5	$0.9809^{20/4}$	1.4281^{20}	al, eth	$B2^4$, 1542
10574	4-Pentenoic acid, 2-acetyl ethyl ester $CH_2{=}CHCH_2CH(COCH_3)CO_2C_2H_5$	170.21	...	211-2, 102^{12}	...	$0.9898^{20/4}$	1.4388^{18}	al, eth, bz	B3, 738
10575	4-Pentenonitrile $CH_2{=}CHCH_2CH_2CN$	81.12	...	140, $60\text{-}1^{40}$	...	0.8239^{24}	1.4213^{14}	al, eth	$B2^4$, 1593
10576	1-Penten-3-ol $C_2H_5CH(OH)CH{=}CH_2$	86.13	...	114-6, 37^{20}	...	$0.8935^{22/4}$	1.4239^{20}	**al, eth**	$B1^4$, 2117
10577	2-Penten-1-ol (cis) $C_2H_5CH{=}CHCH_2OH$	86.13	...	138, 41.2^7	...	$0.8529^{20/4}$	1.4354^{20}	al, eth, ace	$B1^4$, 2121
10578	2-Penten-1-ol (*trans*) $C_2H_5CH{=}CHCH_2OH$	86.13	...	139.5, 42^7	...	$0.8471^{20/4}$	1.4341^{20}	al, eth, ace	$B1^4$, 2121
10579	3-Pentene-2-ol (d) $CH_3CH{=}CHCH(OH)CH_3$	86.13	$[\alpha]_D$ + 0.413	118-21, 33^{10}	...	$0.8354^{20/4}$	$1.4250^{25/20}$	al, eth, ace	$B1^4$, 2122
10580	3-Pentene-2-ol (dl) $CH_3CH{=}CHCH(OH)CH_3$	86.13	...	121.6, 60.6^{55}	...	0.8328^{25}	1.4280^{20}	al, eth, ace	$B1^4$, 2122
10581	3-Penten-2-ol (l) $CH_3CH{=}CHCH(OH)CH_3$	86.13	$[\alpha]_D$ -3.3 (H_2O 1.5%)	119-21	...	$0.8354^{20/4}$	1.4280^{20}	al, eth, ace	$B1^4$, 2124
10582	3-Penten-2-ol, 2-methyl $CH_3CH{=}CHC(OH)(CH_3)_2$	100.16	...	$121\text{-}2^{757}$	...	$0.8347^{20/4}$	1.4302^{20}	**al, eth**	$B1^4$, 2145
10583	4-Pentene-1-ol $CH_2{=}CHCH_2CH_2CH_2OH$	86.13	...	140-2	...	0.8457^{204}	1.4309^{20}	eth	$B1^4$, 2119
10584	4-Penten-2-ol $CH_2{=}CHCH_2CH(OH)CH_3$	86.13	...	$115\text{-}6^{750}$	...	$0.8367^{20/4}$	1.4225^{20}	**al, eth**	$B1^4$, 2118
10585	4-Penten-2-ol, 2-methyl $CH_2{=}CHCH_2C(OH)(CH_3)_2$	100.16	...	119.5	...	$0.8300^{20/4}$	1.4263^{20}	al, eth	$B1^4$, 2146
10586	1-Penten-3-one or Ethyl vinyl ketone $C_2H_5COCH{=}CH_2$	84.12	...	102^{740}, 44^{90}	...	$0.8468^{20/4}$	1.4192^{20}	al, eth, ace, bz	$B1^4$, 3457
10587	1-Penten-3-one, 5,5-dimethyl-1-phenyl or Benzalpinacolone $(CH_3)_2CHCH_2COCH{=}CHC_6H_5$	188.27	cr	154^{25}	43	0.9508^{46}	1.5523^{25}	al, bz, chl	$B7^3$, 1487
10588	1-Penten-3-one, 1-(4 methoxy phenyl) or Ethyl-4-methoxy styryl ketone $C_2H_5COCH{=}CH(C_6H_4OCH_3\text{-}4)$	190.24	col-lt ye pl (eth-peth)	...	60	...	...	al, eth	$B8^3$, 832
10589	1-Penten-3-one, 2-methyl or Ethyl isopropenyl ketone $C_2H_5COC(CH_3){=}CH_2$	98.14	...	118.5	−69.5	$1.8530^{20/4}$	1.4289^{20}	al, ace	$B1^4$, 3470
10590	1-Penten-3-one, 1-Phenyl $C_2H_5COCH{=}CHC_6H_5$	160.22	lf (lig)	142^{12}	38-9	$0.8697^{20/4}$	1.5684^{20}	al, eth, bz	$B7^2$, 298
10591	3-Penten-2-one (trans) or Methyl propenyl ketone $CH_3CH{=}CHCOCH_3$	84.12	...	122	...	$0.8624^{20/4}$	1.4350^{20}	eth, ace	$B1^4$, 3460
10592	3-Penten-2-one, 4-methyl or Mesityl oxide $(CH_3)_2C{=}CHCOCH_3$	98.14	...	129.7, 41^{23}	−52.8	$0.8653^{20/4}$	1.4440^{20}	w, **al, eth**, ace	$B1^4$, 3471
10593	1-Penten-3-yne or Puryline $CH_3C{\equiv}C\text{-}CH{=}CH_2$	66.10	...	59-60	...	$0.7401^{20/4}$	1.4496^{20}	eth, bz	$B1^4$, 1087
10594	1-Penten-3-yne, 2-methyl $CH_3C{\equiv}CC(CH_3){=}CH_2$	80.13	...	$81\text{-}2^{100}$	...	...	1.4002^{20}	eth, bz	$B1^4$, 1095
10595	1-Penten-4-yne or Allyl acetylene $CH_2{=}CHCH_2C{\equiv}CH$	66.10	...	42-3	...	$0.777^{22/22}$	1.3653^{22}	eth, bz	$B1^4$, 998
10596	3-Penten-1-yne, 3-ethyl $CH_3CH{=}C(C_2H_5)C{\equiv}CH$	94.16	...	96.5, $41\text{-}3^{100}$	...	0.7886^{25}	1.4338^{25}	eth, bz	$B1^4$, 1099
10597	3-Penten-1-yne, 3-methyl $CH_3CH{=}C(CH_3)C{\equiv}CH$	80.13	...	66-7	...	0.789^{20}	1.4332^{20}	eth, bz	$B1^4$, 1096
10598	Peucedanin $C_{15}H_{14}O_4$	258.27	pr or pl (bz-peth), yesh cr (eth)	$276\text{-}81^{17}$	109	...	...	eth, chl, aa	$B19^4$, 2647
10599	Pentyl amine or 1-Amino pentane $CH_3(CH_2)_3CH_2NH_2$	87.16	...	104.4, 5.9^{10}	−55	$0.7547^{20/4}$	1.4118^{20}	**al, eth**, ace, bz	$B4^4$, 674
10600	iso-Pentyl amine $(CH_3)_2CHCH_2CH_2NH_2$	87.16	...	95-7	...	$0.7505^{20/4}$	1.4083^{20}	**w, al, eth**, ace, chl	$B4^4$, 696

PHYSICAL CONSTANTS OF ORGANIC COMPOUNDS (Continued)

No.	Name, Synonyms, and Formula	Mol. wt.	Color, crystalline form, specific rotation and λ_{max} (log ε)	b.p. °C	m.p. °C	Density	n_D	Solubility	Ref.
10601	Isopentyl boric acid $i\text{-}C_5H_{11}B(OH)_2$	115.97	pl (w)	...	169	...	...	w, al, eth, ace	B4^{2}, 1023
10602	Pentyl ether or Dipentyl ether $C_5H_{11}OC_5H_{11}$	158.28	...	190, 70^{12}	−69	$0.7833^{20/4}$	1.4119^{20}	**al, eth**	B1^{4}, 1643
10603	Isopentyl ether or Diisopentyl ether $(i\text{-}C_5H_{11})_2O$	158.28	...	172-3, 60^{10}	...	$0.7777^{20/4}$	1.4085^{20}	**al, eth, chl**	B1^{4}, 1682
10604	Pentyl isocyanide $C_5H_{11}NC$	97.16	...	155.5, 50^{45}	−51.1	$0.806^{20/4}$	...	al	B4^{3}, 331
10605	Pentyl isothiocyanate $C_5H_{11}NCS$	129.22	br ye	...	193.4	...	...	al, eth	B4^{4}, 685
10606	Isopentyl isothiocyanate $(CH_3)_2CHCH_2CH_2NCS$	129.22	...	182-4	...	0.9419^{17}	...	al, eth	B4^{4}, 707
10607	Isopentyl α-naphtyl ether $i\text{-}C_5H_{11}O\text{-}\alpha\text{-}C_{10}H_7$	214.31	...	317^{742}, $148\text{-}53^{1}$	...	$1.0069^{14/4}$	1.5705^{14}	...	B6^{4}, 2925
10608	Isopentyl β-naphthyl ether $i\text{-}C_5H_{11}O\text{-}\beta\text{-}C_{10}H_7$	214.31	lf	323-6d, $155\text{-}60^{6}$	26.5	$1.0155^{12/4}$	1.5768^{12}	al, eth	B6^{4}, 2974
10609	Isopentyl nitrate $(CH_3)_2CHCH_2CH_2ONO_2$	133.15	...	147-8	...	$0.9961^{22/4}$	1.4122^{22}	**al, eth**	B1^{4}, 1683
10610	Pentyl nitrite $C_5H_{11}ONO$	117.15	ye	104-5, 29^{60}	...	$0.8817^{20/4}$	1.3851^{20}	**al, eth**	B1^{4}, 1644
10611	Isopentyl nitrite $(CH_3)_2CHCH_2CH_2ONO$	117.15	...	99.2, 30^{60}	...	$0.8828^{20/4}$	1.3918^{20}	**al, eth**	B1^{4}, 1683
10612	Pentyl sulfide or Dipentyl sulfide $(C_5H_{11})_2S$	174.34	...	230, 84.5^{4}	−51.3	$0.8409^{20/4}$	1.4556^{20}	eth	B1^{4}, 1654
10613	Isopentyl sulfide or Disopentyl sulfide $(i\text{-}C_5H_{11})_2S$	174.34	col pa-ye	216, 85.5^{6}	...	$0.8323^{20/4}$	1.4520^{20}	**al**, eth	B1^{4}, 1689
10614	Pentyl sulfate $(C_5H_{11}O)_2SO_2$	238.34	...	$117^{3.5}$	14	$1.029^{20/0}$	1.4290^{20}	...	B1^{2}, 418
10615	Isopentyl sulfate $(i\text{-}C_5H_{11}O)_2SO_2$	238.34	...	$139\text{-}41^{12}$	−20	...	...	...	B1^{2}, 434
10616	Isopentyl thiocyanate $(CH_3)_2CHCH_2CH_2SCN$	129.22	...	197	...	...	...	al, eth	B3^{1}, 282
10617	*Iso*-Pentyl vinyl ether $i\text{-}C_5H_{11}OCH{=}CH_2$	114.19	...	112-3	...	$0.7826^{20/4}$	1.4072^{20}	al, eth	B1^{4}, 2055
10618	1-Pentyne $C_3H_7C{\equiv}CH$	68.12	...	40.2	−90	$0.6901^{20/4}$	1.3852^{20}	al, **eth**, bz, chl	B1^{4}, 990
10619	1-Pentyne, 1-bromo $C_3H_7C{\equiv}CBr$	147.01	...	$44\text{-}6^{57}$	...	$1.281^{13/4}$	1.4579^{13}	...	B1^{3}, 958
10620	1-Pentyne, 3-chloro-3-ethyl $(C_2H_5)_2C(Cl)C{\equiv}CH$	130.62	...	$73\text{-}6^{100}$	...	$0.9230^{19/4}$	1.4437^{19}	eth, bz, chl	B1^{4}, 1032
10621	1-Pentyne, 3-chloro-3-methyl $C_2H_5CCl(CH_3)C{\equiv}CH$	116.59	...	102.3, 55^{130}	...	$0.9163^{20/4}$	1.4330^{20}	eth, bz, chl	B1^{4}, 1021
10622	1-Pentyne, 4,4-dimethyl $(CH_3)_3CCH_2C{\equiv}CH$	96.17	...	76.1	−75.7	$0.7142^{20/4}$	1.3983^{20}	eth, bz, chl	B1^{4}, 1033
10623	1-Pentyne, 3-ethyl $(C_2H_5)_2CHC{\equiv}CH$	96.17	...	87-9	...	$0.7246^{25/4}$	1.4043^{25}	eth, bz, chl	B1^{3}, 1002
10624	1-Pentyne, 3-ethyl-3-methyl $(C_2H_5)_2C(CH_3)C{\equiv}CH$	110.20	...	101-2	...	$0.7422^{20/4}$	1.4110^{20}	eth, bz, chl	B1^{4}, 1046
10626	1-Pentyne, 1-iodo $C_3H_7C{\equiv}CI$	194.02	...	54^{23}	...	$1.6127^{19/4}$	1.5148^{19}	...	B1^{4}, 992
10627	1-Pentyne, 4-methyl $(CH_3)_2CHCH_2C{\equiv}CH$	82.15	...	61-2	−105.1	$0.7092^{15/4}$	1.3936^{15}	bz, chl	B1^{4}, 1019
10628	2-Pentyne $C_2H_5C{\equiv}CCH_3$	68.12	...	56	−101	$0.7107^{20/4}$	1.4039^{20}	al, **eth**, bz, chl	B1^{4}, 992
10629	2-Pentyne, 4-chloro-4-methyl $(CH_3)_2C(Cl)C{\equiv}CCH_3$	116.59	...	55^{70}	...	...	1.4143^{20}	ace, bz, chl	B1^{4}, 1018
10630	2-Pentyne, 4,4-dimethyl $(CH_3)_3CC{\equiv}CCH_3$	96.17	...	83	−82.4	$0.7176^{20/4}$	1.4071^{20}	eth, bz, chl	B1^{4}, 1032
10631	2-Pentyne, 4-methyl $(CH_3)_2CHC{\equiv}CCH_3$	82.15	...	72.5	−110.4	$0.716^{19/4}$	1.4078^{19}	bz, chl	B1^{4}, 1018
10632	2-Pentynoic acid $CH_3CH_2C{\equiv}CCO_2H$	98.10	cr (peth)	122^{10}	50	0.978^{20}	1.4619^{20}	w	B2^{4}, 1693
10633	4-Pentynoic acid $HC{\equiv}CCH_2CH_2CO_2H$	98.10	...	102^{17}	57.7	...	...	w, al, eth	B2^{4}, 1693

No.	Name, Synonyms, and Formula	Mol. wt.	Color, crystalline form, specific rotation and λ_{max} (log ε)	b.p. °C	m.p. °C	Density	n_D	Solubility	Ref.
10634	1-Pentyn-3-ol, 3,4-dimethyl $(CH_3)_2CHC(OH)(CH_3)C{\equiv}CH$	112.17	...	133^{735}	...	$0.8691^{20/4}$	1.4372^{20}	w, al, eth	B1^{4}, 2255
10635	1-Pentyn-3-ol, 3-methyl $C_2H_5C(OH)(CH_3)C{\equiv}CH$	98.14	...	120-1, 61^{70}	30-1	$0.8688^{20/4}$	1.4310^{20}	...	B1^{4}, 2242
10636	Perbenzoic acid C_6H_5COOOH	138.12	mcl pl (peth)	97-110^{13-5} sub	41-3	...	...	al, eth, ace, bz, chl	B9^{3}, 1049
10637	Pereirine $C_{19}H_{26}N_2O$	298.43	pa ye amor pw, [α] + 137.5 (al)	...	135d	...	...	al, eth, chl	C28, 5459
10638	Perimidine or Peri-naphthimidazole $C_{11}H_8N_2$	168.20	gr cr (dil al)	...	222	...	...	al, eth, ace, bz	B23^{2}, 209
10639	Perseitol or α-Mannoheptitol $C_7H_{16}O_7$	212.20	nd, $[\alpha]^{20}$ + 4.53 (w)	...	188	...	...	w	B1^{4}, 2854
10640	Perylene or *peri*-Dinaphthalene $C_{20}H_{12}$	252.32	gold-br ye pl (bz, aa)	sub 350-400	277-9	1.35	...	ace, bz, chl	B5^{3}, 2521
10641	Perylene-3-carboxylic acid $3\text{-}C_{20}H_{11}CO_2H$	296.33	og-br nd ($PhNO_2$)	...	330	...	...	...	B9^{3}, 3664
10642	α-Phellandrene (d) or 5-isopropyl-2-methyl-1,3-cyclohexadiene $C_{10}H_{16}$	136.24	$[\alpha]_D$ + 49.1 (undil)	175-6, 61^{11}	...	$0.8463^{25/4}$	1.4777^{22}	eth	B5^{4}, 436
10643	β-Phellandrene or 3-isopropyl-6-methylene cyclohexene $C_{10}H_{16}$	136.24	$[\alpha]^{20}{}_D$ + 65.2 (undil)	171-2, 57^{11}	...	$0.8520^{20/4}$	1.4788^{20}	eth	B5^{4}, 436
10644	Phenacyl acetate or α-Hydroxy acetophene acetate $C_6H_5COCH_2O_2CCH_3$	178.19	rh pl (eth, lig)	270, 150-2^{10}	49	$1.1169^{65/4}$	...	al, eth, bz, chl	B8^{3}, 301
10645	Phenacyl alcohol or α-Hydroxy acetophenone $C_6H_5COCH_2OH$	136.15	hex pl (al, eth), pl (w, dil al), pr (lig)	124-6^{12} (sub 56^{1})	90.5 (anh)	$1.0963^{99/4}$	...	al, eth, chl	B8^{3}, 298
10646	Phenacyl bromide or α-Bromo acetophenone $C_6H_5COCH_2Br$	199.05	nd (al), rh pr (dil al), pl (peth)	135^{18}	50-1	$1.647^{20/4}$	...	al, eth, bz, chl	B7^{3}, 979
10647	Phenacyl chloride or α-Chloro acetophenone $C_6H_5COCH_2Cl$	154.60	pl (dil al), rh, lf (peth)	247, 139-41^{14}	56-5	$1.324^{15/4}$	...	al, eth, ace, bz	B7^{3}, 967
10648	Phenacyl chloride-2,4-dimethyl $2,4(CH_3)_2C_6H_3COCH_2Cl$	182.65	nd	...	62	...	...	al, eth, bz, chl	B7^{2}, 172
10649	Phenanthrene $C_{14}H_{10}$	178.23	mcl pl (al), lf (sub)	340, 210-15^{12}	101	0.9800^{4}	1.5943	al, eth, ace, bz, aa	B5^{4}, 2297
10650	Phenanthrene, 2-acetyl $2\text{-}CH_3COC_{14}H_9$	220.27	nd (MeOH)	...	144-5	...	...	al, bz	B7^{3}, 2543
10651	Phenanthrene, 3-acetyl $3\text{-}CH_3COC_{14}H_9$	220.27	nd (MeOH)	...	72	...	...	al, bz, chl, aa	B7^{3}, 2544
10652	Phenanthrene, 9-acetyl $9\text{-}CH_3COC_{14}H_9$	220.27	nd (MeOH)	...	74.5	...	...	al, eth, bz	B7^{3}, 2549
10653	Phenanthrene, 2-amino or 2-Phenanthryl amine $2\text{-}NH_2C_{14}H_9$	193.25	lt ye (lig)	...	85	...	...	...	B12^{3}, 3339
10654	Phenanthrene, 3-amino or 3-Phenanthryl amine $3\text{-}NH_2C_{14}H_9$	193.25	α:lf (lig) β:cr (lig)	...	α:143, β:87.5	...	...	al	B12^{3}, 3339
10655	Phenanthrene, 4-amino $4\text{-}H_2NC_{14}H_9$	193.25	...	...	104-5	...	...	al, eth, bz, chl	B12^{3}, 3341
10656	Phenanthrene, 9-amino (a form) or 9-Phenanthryl amine $9\text{-}NH_2C_{14}H_9$	193.25	lt ye cr (al)	sub	137-8	...	...	al, eth, bz, chl	B12^{3}, 3341
10657	Phenanthrene, 9-amino (b form) $9\text{-}NH_2C_{14}H_9$	193.25	lt ye nd (al)	sub	104	...	...	al, eth, bz, chl	B12^{1}, 555
10658	Phenanthrene, 9-bromo $9\text{-}BrC_{14}H_9$	257.13	pr (al)	>360, 190^{12} sub	64-5	$1.4093^{10/4}$	...	al, eth	B5^{3}, 2145
10659	Phenanthrene, 9,10-diamino $9,10\text{-}(NH_2)_2C_{14}H_8$	208.26	pa ye lf	...	166	...	...	...	B13^{3}, 524
10660	Phenanthrene, 9,10-dihydro $C_{14}H_{12}$	180.25	nd (MeOH)	168-9^{15}	34-5	$1.0757^{40/4}$	$1.6415^{20/}{}_4$	al, eth	B5^{3}, 1989
10661	Phenanthrene, 3,4-dihydroxy or 3,4 Phenanthrenediol $3,4\text{-}(HO)_2C_{14}H_8$	210.23	col nd (peth), dk in air	sub 130 (vac)	143	...	...	al, eth	B6^{3}, 5689
10662	Phenanthrene, 3,4-dimethoxy $3,4\text{-}(CH_3O)_2C_{14}H_8$	238.29	bt ye lf (MeOH)	298-303^{112}	45	...	...	al, eth	B6^{3}, 5689

No.	Name, Synonyms, and Formula	Mol. wt.	Color, crystalline form, specific rotation and λ_{max} (log ε)	b.p. °C	m.p. °C	Density	n_D	Solubility	Ref.
10663	Phenanthrene, 4,5-dimethyl 4,5-$(CH_3)_2C_{14}H_8$	206.29	pr (MeOH)		76-7			chl	B5³, 2175
10664	Phenanthrene, 9,10-dimethyl 9,10-$(CH_3)_2C_{14}H_8$	206.29	lt red pr (aa), nd (MeOH)	sub	144			bz, chl, aa	B5³, 2175
10665	Phenanthrene, 9,10-diphenyl 9,10-$(C_6H_5)_2C_{14}H_8$	330.43	nd (eth, bz)	sub 270	240			eth, bz	B5³, 2624
10666	Phenanthrene, 1,2,3,4,5,6,7,8,9,10,11,12-dodecahydro $C_{14}H_{22}$	190.33		$81\text{-}2^{1.5}$		$0.9674^{20/4}$	1.5102^{20}	ace, bz	B5³, 1074
10667	Phenanthrene, 9-ethyl 9-$C_2H_5C_{14}H_9$	206.29		$199\text{-}200^{1}$	62-3 (66)	$1.0603^{78/4}$	1.6582^{78}	al, bz	B5³, 2170
10668	Phenanthrene, 1,2,3,4,11,12-hexadydro $C_{14}H_{16}$	184.28		307	−3	1.045^{20}	1.5810^{15}	eth, bz, aa, chl, peth	B5³, 1676
10669	Phenanthrene, 1-hydroxy or 1-Phenanthrol 1-$HOC_{14}H_9$	194.23	nd (peth, bz-lig, eth)		157				B6³, 3557
10670	Phenanthrene, 2-hydroxy or 2-Phenanthrol 2-$HOC_{14}H_9$	194.23	pl, lf (al, eth, lig)		168			al, eth, bz	B6³, 3557
10671	Phenanthrene, 3-hydroxy or 3-Phenanthrol 3-$HOC_{14}H_9$	194.23	nd (al, lig)		122-3			al, eth	B6³, 3558
10672	Phenanthrene, 9-hydroxy or 9-Phenanthrol 9-$HOC_{14}H_9$	194.23	nd (lig, bz)		158			al, eth, bz, chl, lig	B6³, 3560
10673	Phenanthrene, 1-methyl 1-$CH_3C_{14}H_9$	192.26	lf, pl (dil al)		123			al	B5³, 2151
10674	Phenanthrene, 3-methyl 3-$CH_3C_{14}H_9$	192.26	pr or nd (al)	$140\text{-}50^{6}$	65			ace	B5³, 2154
10675	Phenanthrene, 4-methyl 4-$CH_3C_{14}H_9$	192.26		$175\text{-}80^{10}$	52.5			al	B5³, 2154
10676	Phenanthrene, 9-methyl 9-$CH_3C_{14}H_9$	192.26			90-1				B5³, 2154
10677	Phenanthrene, 2-nitro 2-$O_2NC_{14}H_9$	223.23			119-20			al, eth, ace	B5³, 2146
10678	Phenanthrene, 3-nitro 3-$O_2NC_{14}H_9$	223.23			172-4			ace, bz, chl	B5³, 2146
10679	Phenanthrene, 9-nitro 9-$NO_2C_{14}H_9$	223.23	ye nd (al)		116-7			al, eth, bz	B5³, 2147
10680	Phenanthrene, 1,2,3,4,5,6,7,8-octahydro or Octathrene $C_{14}H_{18}$	186.30		$295, 169^{15}$	16.7	$1.026^{20/4}$	1.5569^{17}	ace, bz, aa	B5⁴, 1585
10681	Phenanthrene, 1,2,8-trimethyl 1,2,8-$(CH_3)_3C_{14}H_7$	220.31		$210\text{-}20^{15}$	144-5				B5³, 2187
10682	Phenanthrene, 1,4,7-trimethyl 1,4,7-$(CH_3)_3C_{14}H_7$	220.31			72.3				B5³, 2188
10683	Phenanthrene, 1,2,3,4,9,10,11,12-octahydro (cis) $C_{14}H_{18}$	186.30		$129^{6}, 88\text{-}90^{0.1}$		$1.0072^{25/4}$	1.5549^{21}	bz	B5⁴, 1585
10684	Phenanthrene, 1,2,3,4,9,10,11,12-octahydro (*trans*) $C_{14}H_{18}$	186.30	nd	$94\text{-}5^{15}$	23-4	$1.0060^{20/4}$	1.5528^{21}	bz	B5³, 1402
10685	Phenanthrene, perhydro $C_{14}H_{24}$	192.34		$86\text{-}9^{2}$		$0.9447^{20/4}$	1.5011^{20}	eth, ace, bz	B5⁴, 492
10686	Phenanthrene, 7-isopropyl-1-methyl 1-CH_3-7-*i*-$C_3H_7C_{14}H_8$	234.34	pl (al)	$390, 290^{10}$	100-1	1.035		bz, lig	B5³, 2199
10687	Phenanthrene, 1,2,3,4-tetrahydro or Tetranthrene $C_{14}H_{14}$	182.27	lf (MeOH)	173^{11}	33-4	$1.0601^{40/4}$		al, eth, ace, bz, chl, lig	B5⁴, 1909
10688	Phenanthrene, 3,4,5-trihydroxy or 3,4,5-Phenanthrene-triol 3,4,5-$(HO)_3C_{14}H_7$	226.23	lf or pl (w)		148			al, eth, chl	B6¹, 1411
10689	Phenanthrene, 1,6,7-trimethyl 1,6,7-$(CH_3)_3C_{14}H_7$	220.31			123-4				B5³, 2188
10690	1-Phenanthrene carboxylic acid or 1-Phenanthroic acid 1-$C_{14}H_9CO_2H$	222.24	nd (al)		232-3			al, bz	B9³, 3496
10691	2-Phenanthrene carboxylic acid or 2-Phenanthroic acid 2-$C_{14}H_9CO_2H$	222.24	nd (aa)		258-60			al, bz, aa	B9³, 3497
10692	2-Phenanthronitrile 2-$C_{14}H_9CN$	203.24	cr (bz-lig, al)		108-10			al, eth, ace	B9³, 3498
10693	3-Phenanthrene carboxylic acid or 3-Phenanthroic acid 3-$C_{14}H_9CO_2H$	222.24	nd (aa)	sub	270			al, eth, aa	B9³, 3498
10694	3-Phenanthronitrile 3-$C_{14}H_9CN$	203.24	nd (abs al)		102			al, eth	B9³, 3499

PHYSICAL CONSTANTS OF ORGANIC COMPOUNDS (Continued)

No.	Name, Synonyms, and Formula	Mol. wt.	Color, crystalline form, specific rotation and λ_{max} (log ε)	b.p. °C	m.p. °C	Density	n_D	Solubility	Ref.
10695	9-Phenanthrene carboxylic acid or 9-Phenanthroic acid 9-$C_{14}H_9CO_2H$	222.24	nd (aa), lf (sub)	sub	256-7			al, eth, bz, aa	B9^3, 3501
10696	9-Phenanthronitrile 9-$C_{14}H_9CN$	203.24	nd (al)		103			eth, ace	B9^3, 3502
10697	2-Phenanthrene sulfonic acid 2-$C_{14}H_9SO_3H$	258.29	cr (bz), cr (w + 1)		ca 150			w, al, bz	B11^3, 445
10698	3-Phenanthrene sulfonic acid 3-$C_{14}H_9SO_3H$	258.29	lf (bz), cr (w)		88-9 (+ 2w), 120-1 (+ 1w), N175-6 (anh)			w	B11^3, 445
10699	9-Phenanthrene sulfonic acid 9-$C_{14}H_9SO_3H$	258.29	lf or nd (bz, w + 2)		134 hyd, 174 sub			w, al, aa	B11^2, 111
10700	Phenanthridine or 3-4-Benzoguinoline $C_{13}H_9N$	179.22	nd (dil al)	349	106-7			al, eth, ace, bz, chl	B20^4, 4016
10701	Phenanthridine, 6-hydroxy $C_{13}H_9NO$	195.22	nd (al, sub)	sub	293-4				B21^4, 1572
10702	1,7-Phenanthrolene or 1,7-Diazaphenanthrene $C_{12}H_8N_2$	180.21	pl (anh), nd (w + 2)	>360	78 (anh) 65.5 (+ 2w)			al	B23^1, 61
10703	1,10-Phenanthroline or 4,5-Diazaphenanthrene $C_{12}H_8N_2$	180.21	wh nd (bz), cr (w + 1)	>300	117 (anh)			al, ace, bz	B23, 227
10704	4,7-Phenanthroline or 1,8-Dizaphenanthrene $C_{12}H_8N_2$	180.21	nd (w)	sub 100	177			al, chl	B23^1, 61
10705	4,7-Phenanthroline hydrate $C_{12}H_8N_2.H_2O$	198.22	wh nd (w)		100-3				B23^2, 235
10706	1,7-Phenanthroline, 9-nitro $C_{12}H_7N_3O_2$	225.21	nd (dil al)		168				B23^2, 236
10707	9,10-Phenanthroquinone $C_{14}H_8O_2$	208.22	og nd (to), og-red pl (sub)	>360 sub	208-10	1.405$^{22/4}$		eth	B7^3, 4084
10708	9,10-Phenanthroquinone-2-bromo $C_{14}H_7BrO_2$	287.11	red-ye cr (aa)		233-4				B7^3, 4093
10709	9,10-Phenanthroquinone, 3-bromo $C_{14}H_7BrO_2$	287.11	ye nd (aa)		268-9			bz	13, 925
10710	9,10-Phenanthraquinone, 1-chloro $C_{14}H_7O_2Cl$	242.66			229			al, bz	B7^3, 4092
10711	9,10-Phenanthraquinone, 2-Chloro $C_{14}H_7ClO_2$	242.66	ye-red nd (aa)		252-3			al	B7^3, 4093
10712	9,10-Phenanthraquinone, 3-chloro $C_{14}H_7ClO_2$	242.66	og-ye nd (aa, bz-al)		264-5			al, bz, aa	B7^3, 4093
10713	9,10-Phenanthraquinone, 1,2-dihydroxy $C_{14}H_8O_4$	240.22	dk red nd (ace)		d			al, ace, aa	B8^2, 506
10714	9,10-Phenanthraquinone, 2,5-dihydroxy $C_{14}H_8O_4$	240.22	dk red nd (w)		400d				B8^2, 507
10715	9,10-Phenanthraquinone, 2,7-dihydrocy $C_{14}H_8O_4$	240.22	dk red or br nd		>400d			al, eth, ace, bz, aa	B8^2, 507
10716	9,19-Phenanthraquinone, 4,5-dihydroxy $C_{14}H_8O_4$	240.22	dk red (al), nd (w)	d>400				al	B8^2, 508
10717	9,10-Phenanthroquinone, 2,5-dinitro $C_{14}H_6N_2O_6$	298.21	red ye pr (aa)		228			aa	B7^3, 4096
10718	9,10-Phenanthroquinone, 2,7-dinitro $C_{14}H_6N_2O_6$	298.21	gold ye nd (aa)		301-3			aa	B7^2, 730
10719	9,10-Phenanthroquinone, 2-hydroxy $C_{14}H_8O_3$	224.22	br-red or bk-vt nd (aa)	sub	283			w	B8^3, 2929
10720	9,10-Phenanthroquinone, 3-hydroxy $C_{14}H_8O_3$	224.22	ye-red or red nd (aa MeOH)	sub	330d			w, al	B8^3, 2930
10721	9,10-Phenanthroquinone, 2-nitro $C_{14}H_7NO_4$	253.21	ye lf or nd (aa)		260				B7^3, 4095

No.	Name, Synonyms, and Formula	Mol. wt.	Color, crystalline form, specific rotation and λ_{max} (log ε)	b.p. °C	m.p. °C	Density	n_D	Solubility	Ref.
10722	9,10-Phenanthroquinone, 7-isopropyl-1-methyl or Retenoquinone $C_{18}H_{16}O_2$	264.32	og nd (chl-al)	sub	197-8	...	...	al, eth	$B7^3$, 4165
10723	9,10-Phenanthroquinone, 1,2,4-trihydroxy $C_{14}H_8O_5$	256.21	red (al + 1)	...	d	...	...	al	$B8^2$, 558
10724	9,10-Phenanthroquinone, 2,3,4-trihydroxy $C_{14}H_8O_5$	256.21	red br pw	...	185d	...	...	w	$B8^2$, 559
10725	Phenazine or Azophenylene $C_{12}H_8N_2$	180.21	ye-red nd (aa)	>360 sub	176-7	...	...	bz	$B23^2$, 233
10726	Phenazine, 1,4-dihydroxy-di-(N-oxide) $C_{12}H_8N_2O_4$	244.21	purp (chl)	...	236d	...	...	chl	C50, 358
10727	Phenazine, 9,10-dihydro or Hydrazophenylene $C_{12}H_{10}N_2$	182.22	rh lf	...	317 (sealed tube)	...	...	bz	$B23^2$, 225
10728	Phenazine, 1-hydroxy or 1-Phenazinol $C_{12}H_8N_2O$	196.21	ye nd (bz, dil MeOH) lf (dil MeOH)	sub	158	...	...	Py	$B23^2$, 360
10729	Phenazine, 2-methyl $C_{13}H_{10}N_2$	194.24	lt ye nd or pr	350d	117	...	...	al, eth, chl	$B23^7$, 239
10730	β-Phenethyl amine $C_6H_5CH_2CH_2NH_2$	121.18	...	197-8	...	$0.9580^{24/4}$	1.5290^{25}	al, eth	$B12^3$, 2408
10731	β-Phenethyl amine, hydrochloride $C_6H_5CH_2CH_2NH_2.HCl$	157.64	pl or lf (al)	...	218-9	...	...	w, al	$B12^3$, 2409
10732	*o*-Phenetidine or 2-Ethoxy aniline $2\text{-}C_2H_5OC_6H_4NH_2$	137.18	...	232.5, $127\text{-}8^{14}$	<−20	...	1.5560^{20}	al, eth	$B13^3$, 756
10733	o-Phenetidine, 4-nitro or 2-Ethoxy-4-nitro aniline $4\text{-}NO_2\text{-}2\text{-}C_2H_5OC_6H_3NH_2$	182.18	ye nd (dil al)	...	91	...	...	al, eth, ace	$B13^3$, 888
10734	*o*-Phenetidine, 5-nitro or 2-Ethoxy-5-nitro aniline $5\text{-}NO_2\text{-}2\text{-}C_2H_5OC_6H_3NH_2$	182.18	ye nd (dil al)	$205\text{-}6^{14}$	96-7	...	...	al, eth	$B13^3$, 878
10735	*o*-Phenetidine, 6-nitro or 2-Ethoxy-6-nitro aniline $6\text{-}NO_2\text{-}2\text{-}C_2H_5OC_6H_3NH_2$	182.18	ye or og cr (w)	...	60	...	...	w	$B13^3$, 876
10736	*m*-Phenetidine or 3-Ethoxy aniline $3\text{-}C_2H_5OC_6H_4NH_2$	137.18	...	248, $127\text{-}8^{11}$	...	...	...	al, eth	$B13^3$, 932
10737	*m*-Phenetidine, 4-nitro or 3-Ethoxy-4-nitro aniline $4\text{-}NO_2\text{-}3\text{-}C_2H_5OC_6H_3NH_2$	182.18	nd (dil al)	...	122-3	...	...	al, eth, ace, bz	$B13^3$978
10738	*m*-Phenetidine, 5-nitro or 5-Ethoxy-3-nitro aniline $3\text{-}NO_2\text{-}5\text{-}C_2H_5OC_6H_3NH_2$	182.18	ye og red nd (al)	...	115	...	...	al, ace, bz	$B13^3$, 977
10739	*m*-Phenetidine, 6-nitro or 5-Ethoxy-2-nitro aniline $5\text{-}(C_2H_5O)\text{-}2\text{-}(NO_2)C_6H_3NH_2$, $2\text{-}NO_2\text{-}5\text{-}\ C_2H_5OC_6H_3NH_2$	182.18	ye nd (dil al)	...	105-6	...	...	ace, bz	$B13^3$975
10740	*p*-Phenetidine or 4-Ethoxy aniline $4\text{-}C_2H_5OC_6H_4NH_2$	137.18	...	254, 125^{12}	2.4	$1.0652^{16/4}$	1.5528^{20}	al, eth	$B13^3$, 996
10741	*p*-Phenetidine, 2-nitro or 4-Ethoxy-2-nitro aniline $2\text{-}NO_2\text{-}4\text{-}C_2H_5OC_6H_3NH_2$	182.18	red pr (al)	...	113	...	...	eth, chl	$B13^3$, 1203
10742	*p*-Phenetidine, 3-nitro or 4-Ethoxy-3-nitro aniline $3\text{-}NO_2\text{-}4\text{-}C_2H_5OC_6H_3NH_2$	182.18	og-ye nd (bz, dil al)	...	41	...	...	ace, bz	$B13^2$, 284
10743	Phenetole or Ethoxy benzene $C_6H_5OC_2H_5$	122.17	...	170, 60^9	−29.5	$0.9666^{20/4}$	1.5076^{20}	al, eth	$B6^4$, 554
10744	Phenetole, 2-bromo or 1-Bromo-2-ethoxybenzene $2\text{-}BrC_6H_4OC_2H_5$	201.06	...	222-6	...	...	...	al, eth	$B6^4$, 1038
10745	Phenetole, 3-bromo $3\text{-}BrC_6H_4OC_2H_5$	201.06	...	222	...	...	...	al, eth	$B6^3$, 739
10746	Phenetole, 4-bromo $4\text{-}BrC_6H_4OC_2H_5$	201.06	...	230-2, 109^{17}	4	$1.4071^{25/4}$	1.5517^{20}	al, eth	$B6^4$, 1045
10747	Phenetole, 2-chloro or 1-Chloro-2-ethoxy benzene $2\text{-}Cl\text{-}C_6H_4OC_2H_5$	156.61	...	210, $97\text{-}8^{15}$	...	$1.1288^{25/4}$	1.5284^{25}	al, eth, bz	$B6^4$, 785
10748	Phenetole, 3-chloro or 1-Chloro-3-ethoxy benzene $3\text{-}ClC_6H_4OC_2H_5$	156.61	...	$204\text{-}5^{717}$	...	$1.1712^{20/4}$	...	al, eth, bz, aa	$B6^4$, 811
10749	Phenetole, 4-chloro or 4-Chloro-1-ethoxy benzene $4\text{-}ClC_6H_4OC_2H_5$	156.61	...	212-4, 98^{17}	21	$1.1254^{20/4}$	1.5252^{20}	al, eth, bz, aa	$B6^4$, 823
10750	Phenetole, 2,4-dichloro $2,4\text{-}Cl_2\text{-}C_6H_3OC_2H_5$	191.06	...	237	...	...	...	al, eth, bz	$B6^4$, 885
10751	Phenetole, 2,4-dinitro $2,4\text{-}(NO_2)_2C_6H_3OC_2H_5$	212.16	nd or lf (al)	...	86-7	...	...	ace	$B6^4$, 1373
10752	Phenetole, 2,5-dinitro $2,5\text{-}(NO_2)_2C_6H_3OC_2H_5$	212.16	lf (al)	...	96-8	...	...	...	$B6^2$, 245

No.	Name, Synonyms, and Formula	Mol. wt.	Color, crystalline form, specific rotation and λ_{max} (log ε)	b.p. °C	m.p. °C	Density	n_D	Solubility	Ref.
10753	Phenetole, 2,6-dinitro $2,6\text{-}(NO_2)C_6H_3OC_2H_5$	212.16	nd (eth)	137-9^{3}	60-1			eth	B6^{3}, 868
10754	Phenetole, 3,5 dinitro $3,5\text{-}(NO_2)_2C_6H_3OC_2H_5$	212.16	nd (al)		97.5				B6^{3}, 869
10755	Phenetole, 4-ethyl $4\text{-}C_2H_5C_6H_4OC_2H_5$	150.22		211, 92-3^{12}		0.9385$^{17/4}$		al, ace, bz	B6^{3}, 1665
10756	Phenetole, 2-fluoro $2\text{-}FC_6H_4OC_2H_5$	140.16		171.4, 64^{11}	−16.7	1.0874^{17}	1.4932^{17}	ace, bz	B6^{4}, 771
10757	Phenetole, 3-fluoro $3\text{-}FC_6H_4OC_2H_5$	140.16		171.4^{755}, 65.2^{15}	−27.5	1.0716$^{16/4}$	1.4847^{17}	bz	B6^{3}, 669
10758	Phenetole, 4-fluoro $4\text{-}FC_6H_4OC_2H_5$	140.16		173^{766}, 54^{7}	−8.5	1.0715^{18}	1.4826^{18}	bz, chl	B6^{4}, 774
10759	Phenetole, 2-iodo $2\text{-}IC_6H_4OC_2H_5$	248.06		245^{736}, 121-31^{18}				al, eth, ace, bz	B6^{3}, 769
10760	Phenetole, 4-iodo $4\text{-}IC_6H_4OC_2H_5$	248.06	cr (dil MeOH)	249-50^{729}	29			al, eth, bz, chl	B6^{4}, 1077
10761	Phenetole, 3-mercapto $3\text{-}HSC_6H_4OC_2H_5$	154.23		238-9				al, eth, ace	B6, 833
10762	Phenetole, 4-merapto $4\text{-}HSC_6H_4OC_2H_5$	154.23		238	1.6			al, eth, ace, bz	B6^{2}, 852
10763	Phenetole, 2-nitro $2\text{-}NO_2C_6H_4OC_2H_5$	167.16	br ye	267, 149^{15}	2.1	1.1903^{15}	1.5425^{20}	al, eth	B6^{4}, 1250
10764	Phenetole, 3-nitro $3\text{-}NO_2C_6H_4OC_2H_5$	167.16	br ye	284d, 169^{70} d	36			al, eth	B6^{4}, 1271
10765	Phenetole, 4-nitro $4\text{-}NO_2C_6H_4OC_2H_5$	167.16	pr (dil al, eth)	283, 168^{15}	60	1.1176$^{100/4}$		eth, **ace, bz**	B6^{4}, 1283
10766	Phenicin or 3,3′-Dihydroxy-5,5′-dimethyl biguinone $C_{14}H_1li0O_6$	274.23	yesh-br (al)		230-1			al, chl, aa	B8^{3}, 4251
10767	Phenol C_6H_5OH	94.11		181.7, 70.9^{10}	43	1.0576$^{20/4}$	1.5408^{41}	w, al, eth, **ace**, **bz**, chl	B6^{4}, 531
10768	Phenol, 2-acetyl or 2-Hydroxy acetophenone $2\text{-}CH_3COC_6H_4OH$	136.15		218, 106^{17}	4-6	1.1307$^{20/4}$	1.5584^{20}	**al, eth, aa**	B8^{3}, 261
10769	Phenol, 3-acetyl or 3-Hydroxy acetophenone $3\text{-}CH_3COC_6H_4OH$	136.15	nd or lf	296^{756}, 153^{5}	96	1.0992^{109}	1.5348^{109}	al, eth, bz, chl	B8^{3}, 272
10770	Phenol, 4-acetyl or 4-Hydroxy acetophenone $4\text{-}CH_3COC_6H_4OH$	136.15	nd (eth, dil al)	147-8^{3}	109-10	1.1090^{109}	1.5577^{109}		B8^{3}, 276
10771	Phenol, 2-allyl $2\text{-}(CH_2{=}CHCH_2)C_6H_4OH$	134.18		220, 93-4^{8}	−6	1.0255$^{15/15}$	1.5181^{20}	eth	B6^{2}, 528
10772	Phenol, 2-allyl-4-chloro $2\text{-}(CH_2{=}CHCH_2)\text{-}4\text{-}ClC_6H_3OH$	168.62	pr (lig)	130-2^{15}	48	1.171^{15}		al, bz	B6^{1}, 282
10773	Phenol, 2-allyl-6-methoxy $2\text{-}C_3H_4\text{-}6\text{-}(CH_3O)C_6H_3OH$	164.20		250-1, 115^{9}		1.2090$^{20/4}$	1.5545^{20}	ace, bz	B6^{3}, 5013
10774	Phenol, 4-allyl or Chavicol $4\text{-}(CH_2{=}CHCH_2)C_6H_4OH$	134.18		235-6, 120^{12}	16	1.033$^{18/4}$	1.5448^{20}	**al, eth, chl, peth**	B6^{3}, 2415
10775	Phenol, 5-allyl-2-methoxy or Chavibetol $5\text{-}(CH_2{=}CHCH_2)\text{-}2\text{-}(CH_3O)C_6H_3OH$	164.20		253-4, 111^{8}	8.5	1.0613$^{25/4}$	1.5413^{20}	al, eth	B6^{3}, 5024
10776	Phenol, *o*-amino $2\text{-}H_2NC_6H_4OH$	109.13	wh rh bipym nd (bz)	sub 153^{11}	174	1.328		al, eth, w	B13^{3}, 752
10777	Phenol, 2-amino-3-chloro $3\text{-}Cl\text{-}2\text{-}H_2NC_6H_3OH$	143.57	nd		122			w	B13^{2}, 182
10778	Phenol, 2-amino-4-chloro-5-nitro $5\text{-}NO_2\text{-}4\text{-}Cl\text{-}2\text{-}H_2NC_6H_2OH$	188.57	ye nd		225d (dk at 200)			al	B13^{2}, 196
10779	Phenol, 2-amino-4-chloro-6-nitro $6\text{-}NO_2\text{-}4\text{-}Cl\text{-}2H_2NC_6H_2OH$	188.57			152			al	B13^{2}, 196
10780	Phenol, 2-amino-6-chloro-4-niro $4\text{-}NO_2\text{-}6\text{-}Cl\text{-}2\text{-}H_2NC_6H_2OH$	188.57	ye nd (w + 1)		160				B13^{3}, 895
10781	Phenol, 2-amino-5-chloro $5\text{-}Cl\text{-}2\text{-}H_2NC_6H_3OH$	143.57	nd (al), pr (dil al)		154-5			al, eth	B13^{3}, 850
10782	Phenol, 2-amino-3,5-dibromo $3,5\text{-}Br_2\text{-}2\text{-}H_2NC_6H_2OH$	266.92	nd (lig)		145			w	B13^{2}, 187
10783	Phenol, 2-amino-4,6-dibromo $4,6\text{-}Br_2\text{-}2\text{-}H_2NC_6H_2OH$	266.92	ye nd (dil al)		99			al, eth, bz, chl	B13^{2}, 188
10784	Phenol, 2-amino-3,5-dichloro $3,5\text{-}Cl_2\text{-}2\text{-}H_2NC_6H_2OH$	178.02	nd (bz, w)		132-3			al, ace, bz	B13^{2}, 185

PHYSICAL CONSTANTS OF ORGANIC COMPOUNDS (Continued)

No.	Name, Synonyms, and Formula	Mol. wt.	Color, crystalline form, specific rotation and λ_{max} (log ε)	b.p. °C	m.p. °C	Density	n_D	Solubility	Ref.
10785	Phenol, 2-amino-4,6-dinitro or Picramic acid $4,6\text{-}(NO_2)_2\text{-}2\text{-}H_2NC_6H_2OH$	199.12	dk red nd (al), pr (chl)		169			al, bz, aa	B13^3, 899
10786	Phenol, 2-(2-aminoethyl) $2\text{-}(H_2NCH_2CH_2)C_6H_4OH$	137.18	rh (al-eth)		152-3				B13^3, 1624
10787	Phenol, 2-amino-3-nitro $3\text{-}NO_2\text{-}2\text{-}H_2NC_6H_3OH$	154.13	red nd (w)	sub	216-7			w	B13^3, 875
10788	Phenol, 2-amino-4-nitro $4\text{-}NO_2\text{-}2\text{-}H_2NC_6H_3OH$	154.13	og pr (+ w)		80-90 (+ w), 145-7 (anh)			al, eth, aa	B13^3, 877
10789	Phenol, 2-amino-5-nitro $5\text{-}NO_2\text{-}2\text{-}NH_2C_6H_3OH$	154.13			207-8			w, al, bz	B13^3, 887
10790	Phenol, 2-amino-6-nitro $6\text{-}NO_2\text{-}2\text{-}H_2NC_6H_3OH$	154.13	red nd (dil al)		111-2			al, eth, bz, chl, aa	B13^2, 195
10791	Phenol, 3-amino $3\text{-}H_2NC_6H_4OH$	109.13	pr (to)	164^{11}	123			al, eth	B13^3, 931
10792	Phenol, *m*-amino,hydrobromide $3\text{-}H_2NC_6H_4OH.HBr$	190.04	pr (w)		224			w	B13, 403
10793	Phenol, *m*-amino,hydrochloride $3\text{-}H_2NC_6H_4OH.HCl$	145.59	pr (w)		229			w	B13^3, 932
10794	Phenol, *m*-amino,hydroiodide $3\text{-}H_2NC_6H_4OH.HI$	237.04	pr (w)		209			w	B13, 403
10795	Phenol, 3-(2-aminoethyl), hydrochloride $3\text{-}(H_2NCH_2CH_2)C_6H_4OH.HCl$	173.64	cr (al-eth)		145				B13^3, 1630
10796	Phenol, 3-amino-2-chloro $2\text{-}Cl\text{-}3\text{-}H_2NC_6H_3OH$	143.57			85-7			al, eth	B13, 420
10797	Phenol, 3-amino-4,6-dichloro $4,6\text{-}Cl_2\text{-}3\text{-}H_2NC_6H_2OH$	178.02	ye br pr (w)		135-6			al, eth, ace, bz, chl	B13^3, 970
10798	Phenol, 3-amino-4-nitro $4\text{-}NO_2\text{-}3\text{-}H_2NC_6H_3OH$	154.13	og nd (w)		185-6			al, eth, bz, chl	B13^1, 136
10799	Phenol, *p*-amino $4\text{-}H_2NC_6H_4OH$	109.13	wh pl (w)	110$^{0.3}$	186-7			al	B13^3, 991
10800	Phenol, 4-(2 aminoethyl) or Tyramine $4\text{-}(H_2NCH_2CH_2)C_6H_4OH$	137.18	pl or nd (bz), cr (al), nd (w)	205-7^{25}	164-5			al, xyl	B13^3, 1637
10801	Phenol, 4-(2 aminopropyl) (l) or L-Paredrine $4\text{-}[CH_3CH(NH_2)CH_2]C_6H_4OH$	151.21	$[\alpha]^{17}{}_D$ −52 (al)		111			al, eth, chl	H34, 2202
10802	Phenol, 4-amino-2-chloro $2\text{-}Cl\text{-}4\text{-}H_2NC_6H_3OH$	143.57	nd (al, eth, w)		153			al, eth	B13^3, 1180
10803	Phenol, 4-amino-2-chloro-6-nitro $6\text{-}NO_2\text{-}2\text{-}Cl\text{-}4\text{-}H_2NC_6H_2OH$	188.57			130			al	B13, 524
10804	Phenol, 4-amino-3-chloro $3\text{-}Cl\text{-}4\text{-}H_2NC_6H_3OH$	143.57	nd		160			al, eth	B13^3, 1184
10805	Phenol, 4-amino-2,6-dibromo $2,6\text{-}Br_2\text{-}4\text{-}H_2NC_6H_2OH$	266.92	nd (al, bz)		192-3			al, bz	B13^3, 1195
10806	Phenol, 4-amino 2,5-dichloro $2,5\text{-}Cl_2\text{-}4\text{-}H_2NC_6H_2OH$	178.02	cr (bz)		178-9			al, eth, aa	B13^2, 274
10807	Phenol, 4-amino -2,6-dichloro $2,6\text{-}Cl_2\text{-}4\text{-}H_2NC_6H_2OH$	178.02	nd or lf (w, bz)	sub	167			al, eth	B13^2, 274
10808	Phenol, 4-amino-3,5-dichloro $3,5\text{-}Cl_2\text{-}4\text{-}H_2NC_6H_2OH$	178.02	nd (w, bz)		154			al, eth, chl, aa	B13^3, 1189
10809	Phenol, 4-amino-2-nitro $2\text{-}NO_2\text{-}4\text{-}H_2NC_6H_3OH$	154.13	dk red pl or nd (w, al)		131			al, eth	B13^2, 284
10810	Phenol, 4-amino-3-nitro $3\text{-}NO_2\text{-}4\text{-}H_2NC_6H_3OH$	154.13	dk red pr (eth)		154			w, al, eth, chl	B13^3, 1203
10811	Phenol, 4-(2-aminopropyl) $4\text{-}CH_3CH(NH_2)CH_2C_6H_4OH$	151.21	cr (bz)		125-6			w, al, chl	B13^3, 1709
10812	Phenol, 2-benzyl $2\text{-}(C_6H_5CH_2)C_6H_4OH$	184.24	cr (peth)	312, 159-62^{12}	51-3				B6^3, 3349
10813	Phenol, 4-benzyl $4\text{-}(C_6H_5CH_2)C_6H_4OH$	184.24	nd (al)	320-2, 198-200^{10}	84			al, eth, bz, chl, aa	B6^3, 3357
10814	Phenol, 2-bromo $2\text{-}BrC_6H_4OH$	173.01		194-5, 87.3^{13}	5.6	1.4924$^{20/4}$	1.589^{20}	al, eth	B6^4, 1037
10815	Phenol, 2-bromo-3-nitro $2\text{-}Br\text{-}3\text{-}NO_2C_6H_3OH$	218.01	pa ye nd (HCl)	sub	147-8			al, eth	B6^3, 844

No.	Name, Synonyms, and Formula	Mol. wt.	Color, crystalline form, specific rotation and λ_{max} (log ε)	b.p. °C	m.p. °C	Density	n_D	Solubility	Ref.
10816	Phenol, 2-bromo-4-nitro $2\text{-Br-4-NO}_2C_6H_3OH$	218.01	cr (to, w), nd (chl, eth, dil al)		114			al, eth, chl	B6³, 845
10817	Phenol, 2-bromo-5-nitro $2\text{-Br-5-NO}_2C_6H_3OH$	218.01	pa ye nd (w), cr (peth)		129-30			al, eth, ace, bz	B6⁴, 1365
10818	Phenol, 2-bromo-6-nitro $2\text{-Br-6-NO}_2C_6H_3OH$	218.01	pa ye nd (chl, al)		68			al, aa	B6³, 844
10819	Phenol, 3-bromo 3-BrC_6H_4OH	173.01		236.5, $135\text{-}40^{12}$	33			al, eth, chl	B6⁴, 1042
10820	Phenol, 3-bromo-5-chloro 3-Br-5-ClC_6H_3OH	207.45	nd (peth)	$256\text{-}60^{756}$	70			al, aa	B6², 187
10821	Phenol, 3-bromo-2,4-dinitro $2,4\text{-}(NO_2)_2\text{-3-BrC}_6H_2OH$	263.02	pa ye nd (w)		175			eth	B6², 249
10822	Phenol, 3-bromo-2,6-dinitro $2,6\text{-}(NO_2)_2\text{-3-BrC}_6H_2OH$	263.00	nd (peth)		131			al, eth	B6², 250
10823	Phenol, 3-bromo-2-nitro $3\text{-Br-2-NO}_2C_6H_3OH$	218.01	ye nd (peth), col nd (+w)		65-7 (anh)			al	B6³, 842
10824	Phenol, 3-bromo-4-nitro $3\text{-Br-4-NO}_2C_6H_3OH$	218.01	pa ye nd (bz, peth)		129-30			al, eth, bz	B6⁴, 1365
10825	Phenol, 3-bromo-5-nitro $3\text{-Br-5-NO}_2C_6H_3OH$	218.01	cr (w)		145			al, eth	B6², 233
10826	Phenol, 4-bromo 4-BrC_6H_4OH	173.01		238, 118.2^{11}	66.4	1.840^{15}		w, al, eth, chl	B6⁴, 1043
10827	Phenol, 4-bromo-2-chloro 4-Br-2-ClC_6H_3OH	207.45	nd (lig, bz, aa)	233-4, $127\text{-}30^{12}$	50-1	$1.6170^{20/4}$	1.5859^{20}	al, eth, ace, bz	B6³, 750
10828	Phenol, 4-bromo-2,6-dinitro $2,6\text{-}(NO_2)_2\text{-4-BrC}_6H_2OH$	263.00	pa ye nd (w), nd (al), pr (aa)	sub	78			al, eth, bz, chl	B6³, 872
10829	Phenol, 4-bromo-2-nitro $4\text{-Br-2-NO}_2C_6H_3OH$	218.01	ye nd or lf (al), pr (eth)	sub	92			al, eth, bz, chl	B6⁴, 1363
10830	Phenol, 4-bromo-3-nitro $4\text{-Br-3-NO}_2C_6H_3OH$	218.01	ye nd (w)		147			eth, bz	B6⁴, 1364
10831	Phenol, 5-bromo-2,4-dinitro $2,4(NO_2)_2\text{-5-BrC}_6H_2OH$	263.00	pr (al, eth)		92			al, eth	B6³, 871
10832	Phenol, 5-bromo-2-nitro $5\text{-Br-2-NO}_2C_6H_3OH$	218.01	ye pr or nd (lig)		44			al, eth, lig	B6³, 844
10833	Phenol, 6-bromo-2,4-dinitro $2,4\text{-}(NO_2)_2\text{-6-BrC}_6H_2OH$	263.00	pa ye cr (al), pr (eth)	sub	118-9			eth, bz, lig	B6³, 871
10834	Phenol, *o*-butyl $2\text{-}C_4H_9C_6H_4OH$	150.22		235, 106.5^{10}		$0.975^{20/4}$	$1.5180^{25.5}$	al, eth	B6³, 1843
10835	Phenol, *o*-*sec*-butyl $2\text{-}CH_3CH_2CH(CH_3)C_6H_4OH$	150.22		$227\text{-}8^{751}$, 116^{21}	16	0.9804^{25}	1.5200^{25}		B6³, 1852
10836	Phenol, *o*-*tert*-butyl or 2-*tert*-butyl phenol $2\text{-}(CH_3)_3CC_6H_4OH$	150.22		221, 99^{10}		$0.9783^{20/4}$	1.5160^{20}	al, eth	B6³, 1861
10837	Phenol, *m*-butyl or 3-Butyl phenol $3\text{-}C_4H_9C_6H_4OH$	150.22		248, 123^{10}		$0.974^{20/4}$		al, eth	B6², 485
10838	Phenol, *m*-*tert*-butyl $3\text{-}(CH_3)_3CC_6H_4OH$	150.22	nd (peth)	240, 132.5^{20}	41-2			al, eth	B6³, 1862
10839	Phenol, *p*-butyl or 4-Butyl phenol $4\text{-}C_4H_9C_6H_4OH$	150.22		248, $138\text{-}9^{18}$	22	$0.978^{20/4}$	$1.5165^{25.5}$	al, eth	B6³, 1844
10840	Phenol, *p*-isobutyl $4\text{-i-}C_4H_9C_6H_4OH$	150.22		235-9	51-2	$0.9796^{20/20}$	1.5319^{25}	al, eth, ace	B6³, 1859
10841	Phenol, *p*-*sec*-butyl-(d) $4\text{-}sec\,C_4H_9C_6H_4OH$	150.22	nd $[\alpha]^{20}_{d}$ + 13.3 (xyl)	240-2	61-2	0.9883^{20}	1.5182^{21}	al, eth	B6², 487
10842	Phenol, *p*-*tert*-butyl or 4-*tert*-butyl phenol $4\text{-}(CH_3)_3CC_6H_4OH$	150.22	nd (lig)	239.5, 114^{10}	101	$0.908^{80/4}$	1.4787^{114}	al, eth	B6³, 1862
10843	Phenol, *o*-chloro 2-ClC_6H_4OH	128.56		174.9, 56.4^{10}	9.0	$1.2634^{20/4}$	1.5524^{20}	al, eth, bz	B6⁴, 782
10844	Phenol, 2-chloro-4-cyclohexyl $2\text{-Cl-4-}C_6H_{11}C_6H_3OH$	210.70		$122\text{-}3^{3}$	39				B6³, 2508

No.	Name, Synonyms, and Formula	Mol. wt.	Color, crystalline form, specific rotation and λ_{max} (log ε)	b.p. °C	m.p. °C	Density	n_D	Solubility	Ref.
10845	Phenol, 2-chloro-3,4-dimethyl $3,4\text{-}(CH_3)_2\text{-}2\text{-}Cl\text{-}C_6H_2OH$	156.61	cr (peth)	187-9	27	1.5538^{20}	....	peth	$B6^{2}$, 456
10846	Phenol, 2-chloro-4,6-dinitro $2\text{-}Cl\text{-}4,6\text{-}(NO_2)_2C_6H_2OH$	218.55	....	....	114-6	....	....	al, eth, ace, bz	$B6^{4}$, 1385
10847	Phenol, 2-chloro-4-nitro $2\text{-}Cl\text{-}4\text{-}NO_2C_6H_3OH$	173.56	wh nd (50% al)	....	111	....	....	al, eth, chl	$B6^{4}$, 1353
10848	Phenol, 2-chloro-5-nitro $2\text{-}Cl\text{-}5\text{-}NO_2C_6H_3OH$	173.56	ye nd or pr (w)	....	121-2	....	....	....	$B6^{4}$, 1353
10849	Phenol, 2-chloro-6-nitro $2\text{-}Cl\text{-}6\text{-}NO_2C_6H_3OH$	173.56	....	....	70	....	....	al, chl	$B6^{1}$, 837
10850	Phenol, *m*-chloro $3\text{-}ClC_6H_4OH$	128.56	....	214	33	1.268^{25}	1.5565^{40}	al, eth, bz	$B6^{4}$, 810
10851	Phenol, 3-chloro-4-nitro $3\text{-}Cl\text{-}4\text{-}NO_2C_6H_3OH$	173.56	....	....	121-2	....	....	al, bz	$B6^{4}$, 1357
10852	Phenol, 4-Chloro or *p*-Chloro phenol $4\text{-}ClC_6H_4OH$	128.56	....	219.7, 125^{18}	43-4	$1.2651^{40/4}$	1.5579^{40}	al, eth, bz	$B6^{4}$, 820
10853	Phenol, 4-chloro-2-allyl $2(CH_2{=}CHCH_2)\text{-}4\text{-}ClC_6H_3OH$	168.62	pr (lig)	$130\text{-}2^{15}$	48	1.171^{15}	....	al, bz	$B6^{1}$, 282
10854	Phenol, 4-chloro-2,6-dimethyl $2,6\text{-}(CH_3)_2\text{-}4\text{-}ClC_6H_2OH$	156.61	nd (w)	....	83	....	....	al, bz, aa	$B6^{1}$, 1738
10855	Phenol, 4-chloro-2,3-dinitro $4\text{-}Cl\text{-}2,3(NO_2)_2C_6H_2OH$	218.55	pr	....	127	....	....	al, eth, bz, chl	B6, 259
10856	Phenol, 4-chloro-2-nitro $4\text{-}Cl\text{-}2\text{-}NO_2C_6H_3OH$	173.56	ye mcl pr (al)	....	88-9	....	....	al, eth, chl	$B6^{4}$, 1349
10857	Phenol, 5-chloro-2,4-dinitro $5\text{-}Cl\text{-}2,4\text{-}(NO_2)_2C_6H_2OH$	218.55	nd (al, peth)	....	92	1.74^{22}	....	al, eth, chl, peth	$B6^{4}$, 1385
10858	Phenol, 5-chloro-2-nitro $5\text{-}Cl\text{-}2\text{-}NO_2C_6H_3OH$	173.56	ye pr or nd (w)	sub	41	....	....	al, eth, aa	$B6^{1}$, 836
10859	Phenol, *o*-cyclohexyl $2\text{-}C_6H_{11}C_6H_4OH$	176.26	nd (lig)	283, 147^{17}	56-7	....	....	al, aa	$B6^{1}$, 2492
10860	Phenol, *p*-cyclohexyl $4\text{-}C_6H_{11}C_6H_4OH$	176.26	nd (bz)	$293\text{-}5^{752}$, $132\text{-}5^{4}$	133	....	....	al, eth, bz	$B6^{1}$, 2501
10861	Phenol, 2,4-diamino $2,4\text{-}(H_2N)_2C_6H_3OH$	124.14	lf	....	78-80d	....	....	al, ace, w	$B13^{3}$, 1338
10862	Phenol, 2,4-diamino, dihydrochloride or Amidol $2,4\text{-}(H_2N)_2C_6H_3OH\cdot 2HCl$	197.06	nd	....	230-40d	....	....	w	$B13^{3}$, 1338
10863	Phenol, 2,5-diamino $2,5\text{-}(H_2N)_2C_6H_3OH$	124.14	nd	....	68	....	....	w	$B13^{2}$, 312
10864	Phenol, 3,4-diamino $3,4\text{-}(H_2N)_2C_6H_3OH$	124.14	nd	....	170-2	....	....	w	$B13^{1}$, 210
10865	Phenol, 3,5-diamino $3,5\text{-}(H_2N)_2C_6H_3OH$	124.14	nd or pr (chl)	....	168-70 (180)	....	....	w	$B13^{1}$, 1370
10866	Phenol, 2,4-dibromo $2,4\text{-}Br_2C_6H_3OH$	251.91	nd(peth)	238-9, 177^{17}	40	....	....	al, eth, bz	$B6^{4}$, 1061
10867	Phenol, 2,4-dibromo-6-nitro $2,4\text{-}Br_2\text{-}6\text{-}NO_2C_6H_2OH$	296.90	....	....	118	....	....	eth, bz, chl	$B6^{1}$, 848
10868	Phenol, 2,6-dibromo $2,6\text{-}Br_2C_6H_3OH$	251.91	nd (w)	162^{21} (sub)	56-7	....	....	al, eth	$B6^{4}$, 1064
10869	Phenol, 2,6-dibromo-4-nitro $2,6\text{-}Br_2\text{-}4\text{-}NO_2C_6H_2OH$	296.90	pa ye pr or lf (al)	d>144	145-6	....	....	al, eth	$B6^{4}$, 1366
10870	Phenol, 3,5-dibromo $3,5\text{-}Br_2C_6H_3OH$	251.91	....	$120\text{-}2^{3}$	81	....	....	al, eth	$B6^{4}$, 1065
10871	Phenol, 2,4-di-*tert*-butyl $2,4\text{-}[(CH_3)_3C]_2\text{-}C_6H_3OH$	206.33	....	263.5, 146^{20}	56.5	....	1.5080^{20}	....	$B6^{3}$, 2062
10872	Phenol, 2,4-di-*tert*-butyl-5-methyl $2,4(t\text{-}C_4H_9)_2\text{-}5\text{-}CH_3C_6H_2OH$	220.35	....	282, 167^{20}	62.1	$0.912^{80/4}$	....	al, eth, ace, bz	$B6^{3}$, 2072
10873	Phenol, 2,4-di-*tert*-butyl-6-methyl $2,4\text{-}(t\text{-}C_4H_9)_2\text{-}6\text{-}CH_3C_6H_2OH$	220.35	....	269, 138.5^{10}	51	$0.891^{80/4}$	....	....	$B6^{3}$, 2073
10874	Phenol, 2,6-di-*sec*-butyl $2,6\text{-di-}(sec\text{-}C_4H_9)_2C_6H_3OH$	206.33	....	255-60	−42	....	1.5080^{20}	....	ALD 11971, 7
10875	Phenol, 2,6-di-*tert*-butyl $2,6\text{-}[(CH_3)_3C]_2C_6H_3OH$	206.33	pr (al)	133^{20}	39	....	1.5001^{20}	al	$B6^{3}$, 2061
10876	Phenol, 2,6-di-*tert*-butyl-4-ethyl $2,6\text{-}(t\text{-}C_4H_9)_2\text{-}4\text{-}C_2H_5C_6H_2OH$	234.38	....	272, 140^{10}	44	....	....	....	$B6^{3}$, 2087
10877	Phenol, 3,6-di-*tert*-butyl-4-(2-methyl butyl) $4\text{-}[CH_3CH_2CH(CH_3)CH_2]\text{-}2,6\text{-}(t\text{-}C_4H_9)_2C_6H_2OH$	276.46	....	$135\text{-}8^{6}$	47	....	....	....	$B6^{3}$, 2097

PHYSICAL CONSTANTS OF ORGANIC COMPOUNDS (Continued)

No.	Name, Synonyms, and Formula	Mol. wt.	Color, crystalline form, specific rotation and λ_{max} (log ε)	b.p. °C	m.p. °C	Density	n_D	Solubility	Ref.
10878	Phenol, 2,6-di-*tert*-butyl-4-methyl or Ionol $2,6\text{-}(t\text{-}C_4H_9)_2\text{-}4\text{-}CH_3C_6H_2OH$	220.35		$265, 136^{10}$	71	$0.8937^{75/4}$	1.4859^{75}	al, ace, bz, chl	$B6^3$, 2073
10879	Phenol, 2,3-dichloro $2,3\text{-}Cl_2C_6H_3OH$	163.00	cr (lig, bz)		57-9			al, eth	$B6^4$, 883
10880	Phenol, 2,3-dichloro-4,5-dimethyl $2,3\text{-}Cl_2\text{-}4,5\text{-}(CH_3)_2C_6HOH$	191.06	nd (peth)		102.5			eth, ace, bz	$B6^2$, 456
10881	Phenol, 2,3-dichloro-5,6-dimethyl $2,3\text{-}Cl_2\text{-}5,6\text{-}(CH_3)_2C_6HOH$	191.06	cr (peth)		90			eth, ace, bz, chl	$B6^2$, 454
10882	Phenol, 2,4-dichloro $2,4\text{-}Cl_2C_6H_3OH$	163.00	hex nd (bz)	$210, 145\text{-}7^{110}$	45			al, eth, bz, chl	$B6^4$, 885
10883	Phenol, 2,4-dichloro-3,5-dimethyl $2,4\text{-}Cl_2\text{-}3,5\text{-}(CH_3)_2C_6HOH$	191.06			83			eth	$B6^3$, 1760
10884	Phenol, 2,5-dichloro $2,5\text{-}Cl_2C_6H_3OH$	163.00	pr (bz, peth)	211^{764}	59			al, eth, bz	$B6^4$, 942
10885	Phenol, 2,5-dichloro-3,4-dimethyl $2,5\text{-}Cl_2\text{-}3,4\text{-}(CH_3)_2C_6HOH$	191.06	nd (peth)		84			eth	$B6^3$, 1730
10886	Phenol, 2,6-dichloro $2,6\text{-}Cl_2C_6H_3OH$	163.00	nd (peth)	$219\text{-}20^{740}$	68-9			al, eth, bz	$B6^4$, 949
10887	Phenol, 2,6-dichloro-3,4-dimethyl $2,6\text{-}Cl_2\text{-}3,4\text{-}(CH_3)_2C_6HOH$	191.06	cr (peth)		52			eth	$B6^2$, 456
10888	Phenol, 2,6-dichloro-3,5-dimethyl $2,6\text{-}Cl_2\text{-}3,5\text{-}(CH_3)_2C_6HOH$	191.06	cr (peth)	$105\text{-}10^{1}$	87-8			eth, chl	$B6^4$, 3157
10889	Phenol, 3,4-dichloro $3,4\text{-}Cl_2C_6H_3OH$	163.00	nd (bz-peth)	253.5^{767}	68			al, eth, bz	$B6^4$, 952
10890	Phenol, 3,5-dichloro $3,5\text{-}Cl_2C_6H_3OH$	163.00	pr (peth)	233^{757}, $122\text{-}4^{8}$	68			al, eth	$B6^4$, 957
10891	Phenol, 2,6-dichloro-4-nitro $2,6\text{-}Cl_2\text{-}4\text{-}NO_2C_6H_2OH$	208.00	br nd (w)		127d (exp <100)	1.822		eth, chl	$B6^4$, 1361
10892	Phenol, 3-(diethyl amino) $3\text{-}(C_2H_5)_2NC_6H_4OH$	165.24	rh bipym (CS_2-lig)	$276\text{-}80, 170^{15}$	78			al, eth, w	$B13^3$, 942
10893	Phenol, 2,4-diido $2,4\text{-}I_2C_6H_3OH$	345.91	nd (w)	sub 100	72-3			al, eth	$B6^4$, 1082
10894	Phenol, 2,3-dimethoxy or Pyrogallol-1,2-dimethyl ether $2,3(CH_3O)_2C_6H_3OH$	154.17		$232\text{-}4, 124\text{-}5^{17}$			1.5392^{20}		$B6^3$, 6264
10895	Phenol, 3-(dimethyl amino) $3\text{-}(CH_3)_2NC_6H_4OH$	137.18	nd (lig)	$265\text{-}8, 152\text{-}3^{15}$	87		1.5895^{26}	al, eth, ace, bz	$B13^3$, 934
10896	Phenol, 4-(dimethyl amino) $4\text{-}(CH_3)_2NC_6H_4OH$	137.18		165^{30}	76			al, eth	$B13^3$, 1009
10897	Phenol, 2,3-dimethyl or *o*-3-Xylenol $2,3\text{-}(CH_3)_2C_6H_3OH$	122.17	nd (w, dil al)	$218, 95.4^{10}$	75		1.5420^{20}	al, eth	$B6^3$, 1722
10898	Phenol, 2,3,dimethyl-4-*tert*-butyl $2,3\text{-}(CH_3)_2\text{-}4\text{-}(CH_3)_3C\text{-}C_6H_2OH$	178.27		$259, 145^{20}$					$B6^3$, 2019
10899	Phenol, 2,3-dimethyl-4-chloro $2,3\text{-}(CH_3)_2\text{-}4\text{-}Cl\text{-}C_6H_2OH$	156.61	nd (peth)		85			al, eth, ace	$B6^3$, 1724
10900	Phenol, 2,3-dimethyl-6-chloro $2,3\text{-}(CH_3)_2\text{-}6\text{-}Cl\text{-}C_6H_2OH$	156.61		$221\text{-}3$, 100^{17}				al, ace, bz	$B6^3$, 1747

No.	Name, Synonyms, and Formula	Mol. wt.	Color, crystalline form, specific rotation and λ_{max} (log ε)	b.p. °C	m.p. °C	Density	n_D	Solubility	Ref.
10901	Phenol, 2,3 dimethyl-5-nitro 2,3-$(CH_3)_2$-5-$NO_2C_6H_2OH$	167.16	og-ye nd (bz, w)		109 (120)			al, ace, aa	B6^2, 455
10902	Phenol, 2,3-dimethyl-4,5,6-trichloro 2,3-$(CH_3)_2$-4,5,6-Cl_3C_6OH	225.50	nd (dil al, peth)		180-1			al	B6^2, 454
10903	Phenol, 2,4-dimethyl 2,4-$(CH_3)_2C_6H_3OH$	122.17	nd (w)	210, 89.3^{10}	27-8	$0.9650^{20/4}$	1.5420^{14}	**al, eth**	B6^3, 1741
10904	Phenol, 2,4-dimethyl, acetate or 2,4-Dimethyl phenyl acetate 2,4-$(CH_3)_2C_6H_3O_2CCH_3$	164.20		226, 108^{13}		$1.0298^{15.5/4}$	1.4990^{15}	al, eth	B6^2, 459
10905	Phenol, 2,4-dimethyl-6-*tert*-butyl 2,4-$(CH_3)_2$-6-$(CH_3)_3C$-C_6H_2OH	178.27		249, 115^{10}	22.3	$0.917^{80/4}$	1.5183^{20}		B6^3, 2020
10906	Phenol, 2,4-dimethyl-5-chloro 2,4-$(CH_3)_2$-5-Cl-C_6H_2OH	156.61	nd (lig, w)		90-1			al, eth, ace	B6^3, 1747
10907	Phenol, 2,4-dimethyl-6-(hydroxy methyl) 2,4-$(CH_3)_2$-6-$(HOCH_2)C_6H_2OH$	152.19	nd (bz-peth)	57-8				al, eth	B6^3, 4653
10908	Phenol, 2,4-dimethyl-6-nitro 2,4$(CH_3)_2$-6-$NO_2C_6H_2OH$	167.16	ye nd (al)		73				B6^3, 1750
10909	**Phenol, 2,4-dimethyl-3,5,6-trichloro** **2,4-$(CH_3)_2$-3,5,6-Cl_3C_6OH**	225.50	pa ye nd		174			eth	B6^2, 460
10910	Phenol, 2,5-dimethyl 2,5-$(CH_3)_2C_6H_3OH$	122.17	nd (w), pr (al-eth)	211.5	75			al, eth	B6^3,1769
10911	Phenol, 2,5-dimethyl, acetate or 2,5-Dimethyl phenyl acetate 2,5$(CH_3)_2C_6H_3O_2CCH_3$	164.20		237^{768}	<−20	1.0624^{15}		al, eth	B6^2, 467
10912	Phenol, 2,5-dimethyl-4-*tert*-butyl 2,5$(CH_3)_2$-4-$(CH_3)_3C$-C_6H_2OH	178.27		264, 136^{10}	71-2	$0.939^{80/4}$, $1.001^{27/4}$	1.5311^{20}		B6^3, 2019
10913	Phenol, 2,5-dimethyl-4-chloro 2,5$(CH_3)_2$-4-Cl-C_6H_2OH	156.61	silv-gr nd (lig)		74-5			al, bz, aa, peth	B6^3, 1773
10914	Phenol, 2,5-dimethyl-3-nitro 2,5-$(CH_3)_2$-3-$NO_2C_6H_2OH$	167.16	ye lf (peth)		91			al, eth	B6, 497
10915	Phenol, 2,5-dimethyl-6-nitro 2,5-$(CH_3)_2$-6-$NO_2C_6H_2OH$	167.16	nd (peth)	236d, 150^{15}	34.5			al, eth, ace, bz	B6^2, 246
10916	Phenol, 2,5-dimethyl-4-nitro 2,5-$(CH_3)_2$-4-$NO_2C_6H_2OH$	167.16	pa ye nd (dil al)		122-3			al, eth, ace	B6^3, 1775
10917	Phenol, 2,5-dimethyl-2,4,6-trichloro 2,5-$(CH_3)_2$-2,4,6-Cl_3C_6HOH	225.50	pa gr nd (al, aq MeOH)		175-6			al, eth, bz, chl	B6^2, 467
10918	Phenol, 2,6-dimethyl 2,6-$(CH_3)_2C_6H_3OH$	122.17	lf or nd (al)	212, 91.2^{10}	49			al, eth	B6^3, 1735
10919	Phenol, 2,6-dimethyl-4-*tert*-butyl 2,6-$(CH_3)_2$-4-$(CH_3)_3C$-C_6H_2OH	178.27	pr (peth)	248, 119^{10}	82.4	$0.916^{80/4}$			B6^3, 2019
10920	Phenol, 2,6-dimethyl-3-nitro 2,6-$(CH_3)_2$-3-$NO_2C_6H_2OH$	167.16	lf pr (bz), nd (lig)		99-100			al, chl	B6, 485
10921	**Phenol, 2,6-dimethyl-4-nitro** **2,6-$(CH_3)_2$-4-NO_2-C_6H_2OH**	167.16	pr (MeOH)		171			al, ace, chl	B6, 486
10922	**Phenol, 2,6-bis(1,1-dimethylpropyl)-4-methyl** **4-CH_3-2,6-$[CH_3CH_2C(CH_3)_2]_2C_6H_2OH$**	248.41		283, 165^{20}		$0.931^{25/4}$	1.4950^{20}		B6^3, 2071
10923	Phenol, 3,4-dimethyl 3,4-$(CH_3)_2C_6H_3OH$	122.17	nd (w)	225, 106-8^{10}	66-8	$0.9830^{20/4}$		al, **eth**	B6^3, 1725
10924	Phenol, 3,4-dimethyl, acetate or 3,4-Dimethyl phenyl acetate 3,4-$(CH_3)_2C_6H_3O_2CCH_3$	164.20		235, 140^{80}	22			al, eth, bz	B6^3, 1728
10925	Phenol, 3,4-dimethyl-6-*tert*-butyl 3,4-$(CH_3)_2$-6-$(CH_3)_3C$-C_6H_2OH	178.27		145^{20}	46.0	$0.920^{80/4}$, $0.973^{27/4}$	1.5222^{20}		B6^3, 2019

PHYSICAL CONSTANTS OF ORGANIC COMPOUNDS (Continued)

No.	Name, Synonyms, and Formula	Mol. wt.	Color, crystalline form, specific rotation and λ_{max} (log ε)	b.p. °C	m.p. °C	Density	n_D	Solubility	Ref.
10926	Phenol, 3,4 dimethyl-5-chloro $3,4-(CH_3)_2-5-Cl-C_6H_2OH$	156.61	nd (peth)	......	98	......	......	al, ace, chl	B6[2], 456
10927	Phenol, 3,4-dimethyl-6-chloro $3,4-(CH_3)_2-6-Cl-C_6H_2OH$	156.61	nd (peth)	......	72	......	......	al, bz, aa	B6[3], 1729
10928	Phenol, 3,4-dimethyl-6-nitro $3,4-(CH_3)_2-6-NO_2C_6H_2OH$	167.16	ye rh (al)	......	87-9	......	......	al, eth, bz, chl	B6[3], 1731
10929	Phenol, 3,4-dimethyl-2,5,6-tribromo $3,4-(CH_3)_2-2,5,6-Br_3-C_6-OH$	358.85	nd (al)	......	173-4	......	......	al	B6[3], 1730
10930	Phenol, 3,4-dimethyl-2,5,6-trichloro $3,4-(CH_3)_2-2,5,6-Cl_3C_6OH$	225.50	nd (peth)	......	182.5	......	......	al	B6[2], 456
10931	Phenol, 3,5-dimethyl $3,5-(CH_3)_2C_6H_3OH$	122.17	nd (w, peth)	219.5 sub, 102-3[10]	68	0.9680[20/4]	......	al	B6[3], 1753
10932	Phenol, 3,5-dimethyl-4-chloro $3,5-(CH_3)_2-4-Cl-C_6H_2OH$	156.61	......	246	115-8	......	......	al, eth, bz	B6[3], 1759
10933	Phenol, 3,5-dimethyl-6-chloro $3,5-(CH_3)_2-6-Cl-C_6H_2OH$	156.61	nd	......	49-50	......	......	eth	B6[2], 464
10934	Phenol, 3,5-dimethyl-2-nitro $3,5-(CH_3)_2-2-NO_2C_6H_2OH$	167.16	ye nd (lig, dil MeOH)	......	66	......	......	al, eth, ace, bz	B6[3], 1765
10935	Phenol, 3,5-dimethyl-2,4,6-tribromo $3,5-(CH_3)_2-2,4,6-Br_3C_6OH$	358.85	nd (al)	......	166	......	......	al	B6[3], 1763
10936	Phenol, 3,5-dimethyl-2,4,6-trichloro $3,5-(CH_3)_2-2,4,6-Cl_3C_6OH$	225.50	ye nd (peth)	......	117-8	......	......	peth	B6[3], 1761
10937	Phenol, 2,3-dinitro $2,3-(NO_2)_2C_6H_3OH$	184.11	ye nd (w)	......	144-5	1.681[20]	......	al, eth, bz	B6[4], 1369
10938	Phenol, 2,4-dinitro $2,4-(NO_2)_2C_6H_3OH$	184.11	pa ye pl or lf (w)	sub	115-6	1.683[24]	......	al, eth, ace, bz, chl	B6[4], 1369
10939	Phenol, 2,5-dinitro $2,5(NO_2)_2C_6H_3OH$	184.11	ye mcl pr or nd (dil al, w, lig)	......	108	......	......	eth, bz	B6[4], 1383
10940	Phenol, 2,6-dinitro $2,6-(NO_2)_2C_6H_3OH$	184.11	pa ye rh nd or lf (dil al)	......	63-4			al, eth, ace, bz, Py	B6[4], 1383
10941	Phenol, 3,4-dinitro $3,4-(NO_2)_2C_6H_3OH$	184.11	tcl nd (w)	......	134	1.672	......	al, eth, bz	B6[4], 1384
10942	Phenol, 3,5-dinitro $3,5-(NO_2)_2C_6H_3OH$	184.11	lf (w)	......	126.1	1.702	......	al, eth, bz, chl	B6[4], 1385
10943	Phenol, 2,6-dipropyl $2,6(C_3H_7)_2C_6H_3OH$	178.27	......	256, 114-6[5]	28	......	......	eth	B6[3], 2013
10944	Phenol, 2-ethyl $2-C_2H_5C_6H_4OH$	122.17	......	207, 84.1[10]	<−18	1.0371[0]	1.5367[20]	al, eth, ace, bz	B6[3], 1655
10945	Phenol, 2-ethylamino $2-C_2H_5NHC_6H_4OH$	137.18	pl (bz)	......	113-4	......	......	al	B13[3], 763
10946	Phenol, 2-ethyl-4-*tert*-butyl $2-C_2H_5-4-(CH_3)_3C-C_6H_3OH$	178.27	......	257, 141[20]	......	......	......	......	B6[3], 2012

PHYSICAL CONSTANTS OF ORGANIC COMPOUNDS (Continued)

No.	Name, Synonyms, and Formula	Mol. wt.	Color, crystalline form, specific rotation and λ_{max} (log ε)	b.p. °C	m.p. °C	Density	n_D	Solubility	Ref.
10947	Phenol, 3-ethyl 3-$C_2H_5C_6H_4OH$	122.17	....	214, 99.3^{10}	−4	$1.0283^{20/4}$	....	al, eth	$B6^3$, 1660
10948	Phenol, 3-(ethylamino) 3-$C_2H_5NHC_6H_4OH$	137.18	cr (bz-peth)	176^{12}	62	....	....	al, eth, bz, chl	B13, 408
10949	Phenol, 4-(ethylamino) 4-$C_2H_5NHC_6H_4OH$	137.18	nd (w)	....	110-2	....	....	al, eth	$B13^3$, 1012
10950	Phenol, 4-ethyl 4-$C_2H_5C_6H_4OH$	122.17	nd	219, 99.5^{10}	47-8	....	0.5239^{25}	al, eth, ace, bz	$B6^3$, 1663
10951	Phenol, 4-ethyl-2-*tert*-butyl 4-C_2H_5-2-$(CH_3)_3C$-C_6H_3OH	178.27	....	250, 123^{10}	23	....	....	al	$B6^2$, 2012
10952	Phenol, *o*-fluoro 2-F-C_6H_4OH	112.10	....	151-2	16.1	....	....	w	$B6^4$, 770
10953	Phenol, *p*-fluoro 4-F-C_6H_4OH	112.10	....	185.5, 87^{23}	48	....	....	ace, peth	$B6^4$, 773
10954	Phenol, 4-fluoro-2-nitro 4-F-2-$NO_2C_6H_3OH$	157.10	....	....	73.7	....	....	al	$B6^1$, 121
10955	Phenol, 4-[bis-(2-hydroxyethyl) amino 4[$(HOCH_2CH_2)_2N]C_6H_4OH$	197.23	cr (w)	....	140	....	....	w	$B13^3$, 1033
10956	Phenol, 2-hydroxymethyl-4-methyl or Homosaligenine 2-$HOCH_2$-4-$CH_3C_6H_3OH$	138.17	lf (w, chl)	....	106-7	....	....	w, al, eth	$B6^3$, 4598
10957	Phenol, *o*-iodo 2-IC_6H_4OH	220.01	nd	$186\text{-}7^{160}$, $91\text{-}2^{2}$	43	1.8757^{80}	....	al, eth	$B6^4$, 1070
10958	Phenol, *m*-iodo 3-IC_6H_4OH	220.01	nd (lig)	d	118	....	....	al, eth	$B6^4$, 1073
10959	Phenol, *p*-iodo 4-IC_6H_4OH	220.01	nd (w or sub)	$138\text{-}40^{5}$d	93-4	1.8573^{112}	....	al, eth	$B6^4$, 1074
10960	Phenol, 2-mercapto 2-HSC_6H_4OH	126.17	oil	$216\text{-}7^{751}$, $88\text{-}90^{8}$	5-6	$1.2373^{0/0}$	....	eth	$B6^3$, 4276
10961	Phenol, 3-mercapto 3-HSC_6H_4OH	126.17	cr	168^{35}	16-7	....	....	al	$B6^2$, 827
10962	Phenol, 4-mercapto 4-HSC_6H_4OH	126.17	cr	$166\text{-}8^{45}$, $133\text{-}7^{11}$	29-30	$1.1285^{25/4}$	1.5101^{25}	w, al	$B6^3$, 4445
10963	Phenol, 2-methoxy-3-nitro or 6-Nitroguiaicol 2-CH_3O-3-$NO_2C_6H_3OH$	169.14	yesh rh pr (peth)	....	102-3	....	....	al	$B6^3$, 4263
10964	Phenol, 2-methoxy-4-nitro or 5-Nitroguiaicol 2-CH_3O-4-$NO_2C_6H_3OH$	169.14	ye nd (w)	....	103-4	....	....	al, eth	$B6^3$, 4264
10965	Phenol, 2-methoxy-5-nitro or 4-Nitroguiaicol 2-CH_3O-5-$NO_2C_6H_3OH$	169.14	pa ye nd (w)	....	105	....	....	al, eth	$B6^3$, 4264
10966	Phenol, 2-methoxy-6-nitro or 3-Nitroguiaicol 2-CH_3O-6-$NO_2C_6H_3OH$	169.14	og-ye nd (sub)	sub	62	....	....	w, al	$B6^2$, 789
10967	Phenol, 2-methoxy-4-propenyl (cis) or Isoeugenol (cis) 2-CH_3O-4-$(CH_3CH{=}CH)C_6H_3OH$	164.20	....	$134\text{-}5^{13}$, $80\text{-}1^{0.5}$	......	$1.0837^{20/4}$	1.5726^{20}	al, eth	$B6^3$, 4992
10968	Phenol, 3-methoxy-4-nitro 3-CH_3O-4-$NO_2C_6H_3OH$	169.14	ye nd (al)	....	144	....	....	al	$B6^2$, 822
10969	Phenol, 3-methoxy-5-nitro 3-CH_3O-5-$NO_2C_6H_3OH$	169.14	ye cr (al)	....	144	....	....	al	$B6^3$, 4347
10970	Phenol, 4-methoxy-2-nitro 4-CH_3O-2-$NO_2C_6H_3OH$	169.14	og ye nd or mcl cr (al, lig)	....	80 (83)	....	....	al	$B6^3$, 4442
10971	Phenol, 4-methoxy-3-nitro 4-CH_3O-3-$NO_2C_6H_3OH$	169.14	pa ye nd (w), cr (bz)	....	98-100	....	....	al	$B6^2$, 848
10972	Phenol, 5-methoxy-2-nitro 5-CH_3O-2-$NO_2C_6H_3OH$	169.14	yesh nd (al)	....	95	....	....	al	$B6^3$, 4345
10973	Phenol, 2-(methylamino) 2-$CH_3NHC_6H_4OH$	123.15	pl (bz-peth)	....	96-7	....	....	al, bz	$B13^3$, 761
10974	Phenol, 4-methyl amino,sulfate or Metol,Pictol $(4\text{-}CH_3NHC_6H_4OH)_2H_2SO_4$	344.37	wh nd(w)	....	250-60d	....	....	w, al	$B13^3$, 1007
10975	Phenol, 4(methyl amino)-2-nitro 2-NO_2-4-$(CH_3NH)C_6H_3OH$	168.15	dk red-br nd (al)	....	113-4	....	....	al	$B13^1$, 186
10976	Phenol, 4-(2-methylaminopropyl) 4-$[CH_3CH(NHCH_3)CH_2]C_6H_4OH$	165.24	cr (MeOH)	....	161	....	....	al, eth	$B13^3$, 1710
10977	Phenol, 4-(2-methyl-2-butyl) 4-$[H_5C_2\text{-}C(CH_3)_2]$-$C_6H_4OH$	164.25	nd	262.5, 138^{15}	94-6	....	....	....	$B6^3$, 1965
10978	Phenol, 2-methyl-4-*tert*-butyl 2-CH_3-4-$(CH_3)_3C$-C_6H_3OH	164.25	yesh	$235\text{-}7^{740}$, 132^{20}	27-8	$0.965^{20/4}$	1.5230^{20}	eth, ace, bz	$B6^3$, 1979

No.	Name, Synonyms, and Formula	Mol. wt.	Color, crystalline form, specific rotation and λ_{max} (log ε)	b.p. °C	m.p. °C	Density	n_D	Solubility	Ref.
10979	Phenol, 2-methyl-4(3-methyl-3-pentyl) $2-CH_3-4-[(C_2H_5)_2C(CH_3)]-C_6H_3OH$	192.30	....	145-6^{11}	35	....	1.5200^{25}	ace	C48, 1328
10980	Phenol, 2-methyl-5-isopropyl or Carvacrol $2-CH_3-5-i-C_3H_7C_6H_3OH$	150.22	....	237.7, 101-3^{10}	1	0.9772$^{20/4}$	1.5230^{20}	al, eth, ace	B6³, 1885
10981	Phenol, 2-methyl-5-isopropyl-3-nitro $2-CH_3-5-i-C_3H_7-3-ONC_6H_2OH$	195.22	yesh pr (bz), nd (dil al)	....	153	....	....	al, eth, bz, chl	B7³,3412
10982	Phenol, 2-methyl-3,4,5-trichloro $2-CH_3-3,4,5-Cl_3C_6HOH$	211.48	nd (lig)	....	77	....	....	ace	B6³, 1268
10983	Phenol, 3-methyl-6-*tert*-butyl $3-CH_3-6-(CH_3)_3C-C_6H_3OH$	164.25	....	224, 127^{11}	46-7	0.922$^{80/4}$	1.5250^{20}	al, eth, ace	B6³, 1982
10984	Phenol, 4-methyl-2-*tert*-butyl $4-CH_3-2-(CH_3)_3C-C_6H_3OH$	164.25	nd (peth)	237, 111^{10}	55	0.9247$^{75/4}$	1.4969^{75}	eth, ace, bz	B6³, 1978
10985	Phenol, 4-(3-methyl-1-butyl) $4-(CH_3)_2CHCH_2CH_2C_6H_4OH$	164.25	nd (w)	255, 126^{14}	93	0.9579$^{23/20}$	1.5050^{27}	al, eth	B6³, 1960
10986	Phenol, 3-methyl-2,4,6-trichloro $3-CH_3-2,4,6-Cl_3-C_6HOH$	211.48	nd (w), pl (peth)	265, 162-3^{28}	47	....	....	al, eth, chl	B6⁴, 2070
10987	Phenol, 4-methyl-2,3,5-trichloro $4-CH_3-2,3,5-Cl_3C_6HOH$	211.48	nd (aa, lig)	....	66-7	....	....	ace	B6¹, 204
10988	Phenol, *o*-nitro $2-O_2NC_6H_4OH$	139.11	ye nd or pr (eth, al)	216, 96-7^{10}	45-6	1.2942^{40}, 1.485^{14}	1.5723^{50}	al, eth, ace, bz, chl	B6⁴, 1246
10989	Phenol, *m*-nitro $3-O_2NC_6H_4OH$	139.11	ye mcl (eth, aq HCl)	194^{70}	97	....	....	al, eth, ace, bz	B6⁴, 1269
10990	Phenol, *p*-nitro $4-O_2NC_6H_4OH$	139.11	ye mcl pr (to)	279d sub	114-6	1.479^{20}	....	al, eth, ace, Py	B6⁴, 1279
10991	Phenol, *p*-nitroso $4-ONC_6H_4OH$	123.11	pa ye rh nd (ace, bz)	....	d 144	....	....	al, eth, ace, bz	B7³, 3367
10992	Phenol, *o*-octyl $4-C_8H_{17}C_6H_4OH$	206.33	....	169^{10}	41-2	....	....	....	B6³, 2046
10993	Phenol, penta bromo Br_5C_6OH	488.59	mcl pr (aa), nd (al)	sub	229.5	....	....	al, bz	B6⁴, 1069
10994	Phenol, penta chloro Cl_5C_6OH	266.34	mcl pr (al + 1w), nd (bz)	309-10^{754}d	174 (+1w), 191 (anh)	1.978$^{22/4}$	....	al, eth, bz	B6⁴, 1025
10995	Phenol, penta fluoro C_6F_5OH	184.07	....	72-3^{48}	......	....	1.4263^{26}	....	B6⁴, 782
10996	Phenol, penta methyl $(CH_3)_5C_6OH$	164.25	nd (al, peth, ace)	267	128	....	....	al	B6³, 1991
10997	Phenol, *p*-pentyl $4-C_5H_{11}C_6H_4OH$	164.25	....	250.5, 119-20^{3}	23	0.960$^{20/4}$	1.5272^{25}	al, eth	B6³, 1950
10998	Phenol, *o*-phenyl or 2-Hydroxy biphenyl $2-HOC_6H_4C_6H_5$	170.21	nd (peth)	286, 145^{14}	58-60	1.213$^{25/4}$	....	al, eth, ace, bz, lig	B6³, 3281
10999	Phenol, *m*-phenyl or 3-hydroxy biphenyl $3-HOC_6H_4C_6H_5$	170.21	nd (w or peth)	>300	78	....	....	al, eth, bz, peth	B6³, 3311
11000	Phenol, *p*-phenyl or 4-Hydroxy biphenyl $4-HOC_6H_4C_6H_5$	170.21	nd or pl (dil al)	305-8 sub	165-7	....	....	al, eth, chl	B6³, 3319
11001	Phenol, 4-propenyl $4-(CH_3CH=CH)C_6H_4OH$	134.18	pl (w), lf	250d, 140-5^{15}	93-4	....	....	al, eth	B6³, 2394
11002	Phenol, *o*-propionyl $2(CH_3CH_2CO)C_6H_4OH$	150.18	....	150^{80}, 115^{15}	......	....	1.5501^{20}	al, eth	B8³, 373
11003	Phenol, *p*-propionyl $4-(CH_3CH_2CO)C_6H_4OH$	150.18	wh nd or pr (w)	....	149	....	....	al, eth	B8³, 379
11004	Phenol, 2-propyl $2-C_3H_7C_6H_4OH$	136.19	....	220, 106.7^{10}	......	1.015$^{20/4}$	....	al, eth	B6³, 1784
11005	Phenol, 3-propyl $3-C_3H_7C_6H_4OH$	136.19	....	228, 111.2^{10}	26	0.987^{20}	1.5223^{20}	al, eth	B6³, 1787
11006	Phenol, 4-propyl $4-C_3H_7C_6H_4OH$	136.19	....	232.6, 111.7^{10}	22	1.009$^{20/4}$	1.5379^{25}	al	B6³, 1783
11007	Phenol, 2-isopropyl or *o*-Cumenol $2-i-C_3H_7C_6H_4OH$	136.19	....	213-4	15-6	1.012$^{20/4}$	1.5315^{20}	al, eth, bz	B6³, 1807
11008	Phenol, 2-isopropyl-5-methyl-4-nitroso or 4 Nitroso carvacrol $2-i-C_3H_7-5-CH_3-4-ON-C_6H_2OH$	179.22	yesh red	....	175	....	....	al, eth, chl	B7³, 3412
11009	Phenol, 3-isopropyl or *m*-Cumenol $3-i-C_3H_7C_6H_4OH$	136.19	....	228	26	....	1.5261^{20}	eth	B6³, 1810

No.	Name, Synonyms, and Formula	Mol. wt.	Color, crystalline form, specific rotation and λ_{max} (log ε)	b.p. °C	m.p. °C	Density	n_D	Solubility	Ref.
11010	Phenol, 4-isopropyl or *p*-Cumenol 4-*i*-$C_3H_7C_6H_4OH$	136.19	nd (peth)	228-30^{745}, 109-10^{10}	62-3	0.990^{20}	1.5228^{20}	al	$B6^3$, 1810
11011	Phenol, seleno or Selenyl benzene C_6H_5SeH	157.07	...	183.6	...	1.4865^{15}	...	al, eth	$B6^4$, 1777
11012	Phenol, 2,3,4,6-tetrabromo 2,3,4,6-Br_4CHOH	409.70	nd (al, aa)	sub	113-4 (120)	...	...	al, bz	$B6^3$, 766
11013	Phenol, 2,3,4,5-tetrachloro 2,3,4,5-Cl_4C_6HOH	231.89	nd (peth, sub)	sub	116-7	...	...	al	$B6^4$, 1020
11014	Phenol, 2,3,4,6-tetrachloro 2,3,4,6-Cl_4C_6HOH	231.89	nd (lig, aa)	150^{15}	70	...	...	al, bz, chl, lig	$B6^4$, 1021
11015	Phenol, 2,3,5,6-tetrachloro 2,3,5,6-Cl_4C_6HOH	231.89	lf (lig)	...	115	...	...	bz	$B6^4$, 1025
11016	Phenol, 2,3,4,5-tetramethyl 2,3,4,5-$(CH_3)_4C_6HOH$	150.22	...	266	86-7	...	...	al, eth	$B6^3$, 1918
11017	Phenol, 2,3,4,6-tetramethyl or Isodurenol 2,3,4,6-$(CH_3)_4C_6HOH$	150.22	cr (peth)	230-50	80-1	...	...	al	$B6^3$, 1919
11018	Phenol, 2,3,5,6-tetramethyl or Durenol 2,3,5,6-$(CH_3)_4C_6HOH$	150.22	nd (lig), pr (al)	247	118-9	...	...	aa, peth	$B6^3$, 1919
11019	Phenol, 2,3,4,6-tetranitro 2,3,4,6-$(NO_2)_4C_6HOH$	274.10	lt ye nd (chl)	exp	140d	...	...	w	$B6^3$, 973
11020	Phenol, 2,4,6-triamino 2,4,6-$(H_2N)_3C_6H_2OH$	139.16	...	257	...	...	...	w, al, eth	$B13^3$, 1373
11021	Phenol, 2,3,4-tribromo 2,3,4-$Br_3C_6H_2OH$	330.80	nd or pl (w, lig)	...	94-5	...	...	al, eth, ace, lig	$B6^2$, 192
11022	Phenol, 2,4,5-tribromo 2,4,5-$Br_3C_6H_2OH$	330.80	nd (lig)	...	87	...	...	al, lig	$B6^3$, 760
11023	Phenol, 2,4,6-tribromo or Bromol 2,4,6-$Br_3C_6H_2OH$	330.80	nd (al), pr (bz), cr (aa + l)	282-90^{764} (sub)	95-6	2.55$^{20/20}$	...	al, eth	$B6^4$, 1067
11024	Phenol, 3,4,5-tribromo 3,4,5-$Br_3C_6H_2OH$	330.80	tcl (bz-lig)	...	129	...	...	al, bz, aa	$B6^2$, 195
11025	Phenol, 2,4,6-tri-*tert*-butyl 2,4,6-$(t\text{-}C_4H_9)_3C_6H_2OH$	262.44	cr (al, peth)	278, 130^{15}	131	0.864$^{27/4}$	...	ace	$B6^3$, 2094
11026	Phenol, 2,3,4-trichloro 2,3,4-$Cl_3C_6H_2OH$	197.45	nd (bz, lig, sub)	sub	83.5	...	...	al, eth, bz, aa	$B6^3$, 716
11027	Phenol, 2,3,5-trichloro 2,3,5-$Cl_3C_6H_2OH$	197.45	nd (al)	248-9^{250}	62	...	...	al, eth	$B6^3$, 716
11028	Phenol, 2,3,6-trichloro 2,3,6-$Cl_3C_6H_2OH$	197.45	nd (dil al, lig)	...	58	...	...	al, eth, bz, aa, lig	$B6^4$, 962
11029	Phenol, 2,4,5-trichloro 2,4,5-$Cl_3C_6H_2OH$	197.45	nd (al, peth)	sub	68-70	...	...	al, lig	$B6^4$,962
11030	Phenol, 2,4,6-trichloro 2,4,6-$Cl_3C_6H_2OH$	197.45	rh nd (aa)	246	69.5	1.4901$^{75/4}$	...	al, eth	$B6^4$, 1005
11031	Phenol, 3,4,5-trichloro 3,4,5-$Cl_3C_6H_2OH$	197.45	nd(lig)	271-7^{746}	101	...	...	eth	$B6^3$, 729
11032	Phenol, 2,3,5-triido 2,3,5-$I_3C_6H_2OH$	471.80	nd (peth, bz-lig)	...	114	...	...	al, eth, ace, bz	$B6^4$, 1085
11033	Phenol, 2,4,6-triido 2,4,6-$I_3C_6H_2OH$	471.80	nd (dil al)	sub, d	158-9	...	...	eth, ace	$B6^4$, 1085
11034	Phenol, 2,4,5-trimethyl or Psuedocumenol 2,4,5-$(CH_3)_3C_6H_2OH$	136.19	nd (lig)	232	72	...	...	al, eth	$B6^3$, 1831
11035	Phenol, 2,4,5-trimethyl,acetate 2,4,5-$(CH_3)_3C_6H_2(O_2CCH_3)$	178.23	nd (peth)	245-6	34	...	...	al, eth	$B6^2$, 482
11036	Phenol, 2,4,6-trimethyl or Mesitol 2,4,6-$(CH_3)_3C_6H_2OH$	136.19	nd (peth, MeOH)	221 sub	72	...	...	al, eth	$B6^3$, 1835
11037	Phenol, 2,3,6-trinitro 2,3,6-$(O_2N)_3C_6H_2OH$	229.11	ye nd (w)	...	119	...	...	al, eth, bz, aa	$B6^2$, 253
11038	Phenol, 2,4,5-trinitro 2,4,5-$(O_2N)_3C_6H_2OH$	229.11	wh nd (w, dil al)	...	96	...	...	al, eth, bz, aa	$B6^4$, 1388
11039	Phenol, 2,4,6-trinitro or Picric acid 2,4,6-$(O_2N)_3C_6H_2OH$	229.11	ye lf (w), pr (eth), pl (al)	sub exp>300	122-3	...	1.763	al, eth, ace, bz, aa	$B6^4$, 1388
11040	Phenol, *o*-vinyl 2-$(CH_2{=}CH)C_6H_4OH$	120.15	nd	101^{14}	29-30	1.0609$^{18/4}$	1.5851^{20}	al, eth	$B6^3$, 2383
11041	Phenol, *m*-vinyl 3-$(CH_2{=}CH)C_6H_4OH$	120.15	...	114-6^{16}	...	1.0353$^{21/4}$	1.5804^{21}	...	$B6^3$, 2385

No.	Name, Synonyms, and Formula	Mol. wt.	Color, crystalline form, specific rotation and λ_{max} (log ε)	b.p. °C	m.p. °C	Density	n_D	Solubility	Ref.
11042	Phenolphthalein or 2,2-bis(4-hydroxyphenyl)phthalide $C_{20}H_{14}O_4$	318.33	wh rh nd	...	262-3	$1.277^{32/4}$	$1.277^{32/}{}_4$	al, eth, ace, chl, Py	B18[4], 1945
11043	Phenolphthalein-3′,3″,5′,5″-tetrabromo $C_{20}H_{10}Br_4O_4$	633.91	nd (al, eth, aa)	...	295-7	...	...	eth	B18[4], 1948
11044	Phenolphthalein-3′,3″,5′,5″-tetrachloro $C_{20}H_{10}Cl_4O_4$	456.11	cr (bz, aa)	...	225	...	...	al, ace, bz	B18[2], 123
11045	Phenolphthalein-3′,3″,5′,5″-tetraiodo or Iodophen $C_{20}H_{10}I_4O_4$	821.92	amor	...	227-9d	$2.0246^{22/22}$	...	...	B18[4], 1949
11046	Phenolphthalin $C_{20}H_{16}O_4$	320.34	nd (w)	...	229-32	...	...	al	B10[3], 2013
11047	Phenolsulfonphthalein or Phenol red $C_{19}H_{14}O_5S$	354.38	dk red nd or pl	...	>300	...	...	...	B19[4], 1128
11048	Phenothiazine $C_{12}H_9NS$	199.27	ye pr (al), ye lf or pl (to)	371, 290^{40} (sub 130^{1})	186-9	...	...	al, eth, ace, bz	B27[2], 32
11049	Phenothiazine, 2,8-dinitro,5-oxide $C_{12}H_7N_3O_5S$	305.26	ye-red lf (aa)	...	d	...	...	...	B27[1], 229
11050	Phenothiazine, 10-(2-diethylaminoethyl) $C_{18}H_{22}N_2S$	298.45	ye oil	$195\text{-}208^{4-5}$	...	...	...	dil HCl	Am66, 888
11051	Phenothiazine, 10-(2-dimethylaminopropyl) or Promethazine $C_{17}H_{20}N_2S$	284.42	...	$190\text{-}3^{0.5}$	60	...	...	...	C42, 575
11052	Phenothiazine, 10-(2-dimethylaminopropyl), hydrochloride $C_{17}H_{20}N_2S.HCl$	320.88	...	...	230-2	...	...	w, al, chl	C42, 575
11053	Phenothiazine, 10-phenyl $C_{18}H_{13}NS$	275.37	ye pr (al)	...	89-90	...	...	al	B27[1], 227
11054	Phenothiazine, 10-octadecyl $C_{30}H_{45}NS$	451.75	...	$290\text{-}300^{0.5}$	$51.0^{-2.5}$	...	...	...	CAS38, 3985
11056	Phenothiazine, N-propyl $C_{15}H_{15}NS$	241.35	...	$155\text{-}65^{0.8}$	48-9	...	...	...	CAS53, 18045
11057	Phenoxanthin $C_{12}H_8OS$	200.25	nd, cr (MeOH)	311^{745}, $183\text{-}4^{12}$	59-60	...	...	al, eth, ace	B19[4], 341
11058	Phenoxazine or Dibenzoxazine $C_{12}H_9NO$	183.21	lf (dil al, bz)	d	156	...	...	al, eth, bz, aa	B27[1], 223
11059	Phenoxy acetamide $C_6H_5OCH_2CONH_2$	151.16	nd (w al)	...	101.5	...	...	al	B6[4], 638
11060	Phenoxyacetic acid $C_6H_5OCH_2CO_2H$	152.15	nd or pl (w)	285d	98-9	...	...	w, al, eth, bz, aa	B6[4], 634
11061	Phenoxyacetic acid, anhydride $(C_6H_5OCH_2CO)_2O$	286.29	lf (eth)	...	67-9	...	...	bz	B6, 162
11062	Phenoxyacetic acid, 2-bromo $2\text{-}BrC_6H_4OCH_2CO_2H$	231.05	nd (dil al), cr (w)	...	142.5	...	...	al, eth	B6[4], 1040
11063	Phenoxyacetic acid, 4-bromo $4\text{-}Br\text{-}C_6H_4OCH_2COOH$	231.05	pr (al), cr (w)	...	161-2	...	...	al, eth	B6[4], 1052
11064	Phenoxyacetic acid, 2-carboxy or Salicyl acetic acid $(2\text{-}HO_2CC_6H_4O)CH_2CO_2H$	196.16	nd (w)	...	190-2	...	...	al, eth, ace, aa, chl	B10[3], 106
11065	Phenoxyacetic acid, 3-carboxy $(3\text{-}HO_2CC_6H_4O)CH_2CO_2H$	196.16	...	...	206-7	...	...	...	B10[1], 65
11066	Phenoxyacetic acid, 4-carboxy $(4\text{-}HO_2CC_6H_4O)CH_2CO_2H$	196.16	nd (ace, w)	...	280-2	...	...	al, eth, ace, bz, chl	B10[2], 94
11067	Phenoxyacetic acid, 2-chloro $2\text{-}ClC_6H_4OCH_2CO_2H$	186.59	nd (w, al)	...	148-9	...	...	w, al	B6[4], 796
11068	Phenoxyacetic acid, 3-chloro $3\text{-}ClC_6H_4OCH_2CO_2H$	186.59	cr (w)	...	110	...	...	...	B6[4], 816
11069	Phenoxyacetic acid, 4-chloro $4\text{-}ClC_6H_4OCH_2COOH$	186.59	pr or nd (w)	...	156-7	...	...	...	B6[4], 845
11070	Phenoxyacetic acid, 4-chloro-2-methyl $[4\text{-}Cl\text{-}2(CH_3)C_6H_3O]CH_2CO_2H$	200.62	...	...	120	...	...	al, eth, bz	B6[4], 1991
11071	Phenoxyacetic acid, 2,4-dichloro $(2,4\text{-}Cl_2\text{-}C_6H_3O)CH_2CO_2H$	221.04	cr (bz)	$160^{0.4}$	140-1	...	...	al	B6[4], 908
11072	Phenoxyacetic acid, 2,4-dinitro $(2,4\text{-}(NO_2)_2C_6H_3O)CH_2CO_2H$	242.14	pa ye pr (w)	...	147-8	...	...	al	B6[2], 244
11073	Phenoxyacetic acid, 3,5-dinitro $(3,5\text{-}(NO_2)_2C_6H_3O)CH_2CO_2H$	242.14	pa br cr pw	...	207	...	...	al	B6, 259

PHYSICAL CONSTANTS OF ORGANIC COMPOUNDS (Continued)

No.	Name, Synonyms, and Formula	Mol. wt.	Color, crystalline form, specific rotation and λ_{max} (log ε)	b.p. °C	m.p. °C	Density	n_D	Solubility	Ref.
11074	Phenoxyacetic acid, ethyl ester or Ethyl phenoxy acetate $C_6H_5OCH_2CO_2C_2H_5$	180.20		250-1, 136^{19}		1.104^{18}		al, eth	$B6^4$, 635
11075	Phenoxyacetic acid, 3-hydroxy $(3\text{-}HOC_6H_4O)CH_2CO_2H$	168.15	nd or pr (w, to)		158-9			al	B6, 817
11076	Phenoxy acetic acid, 3-hydroxy, ethyl ester or Ethyl-3-hydroxy phenoxy acetate $(3\text{-}HOC_6H_4O)CH_2CO_2C_2H_5$	196.20	pr (w, bz)	274d, $170\text{-}3^{11}$	55				B6, 817
11077	Phenoxy acetic acid, 4-hydroxy $(4\text{-}HOC_6H_4O)CH_2CO_2H$	168.15	nd (to), pr (+ ½w)		154 (hyd)				B6, 847
11078	Phenoxy acetic acid, 2-methoxy $(2\text{-}CH_3OC_6H_4O)CH_2CO_2H$	182.18	nd (w)		123-5 (129)			w, al, eth, bz, aa	$B6^3$, 4234
11079	Phenoxy acetic acid, 3-methoxy $(3\text{-}CH_3OC_6H_4O)CH_2CO_2H$	182.18	nd (w)		118				$B6^3$, 4323
11080	Phenoxy acetic acid, methyl ester or Methyl phenoxy acetate $C_6H_5OCH_2CO_2CH_3$	166.18		245, 130^{14}		$1.1493^{20/4}$	1.5155^{20}	**al, eth**	$B6^4$, 635
11081	Phenoxy acetic acid, 2-methyl or o-Cresoxyacetic acid $(2\text{-}CH_3C_6H_4O)CH_2CO_2H$	166.18	lf (w)		157			al	$B6^2$, 331
11082	Phenoxy acetic acid, 3-methyl or m-Cresoxyacetic acid $(3\text{-}CH_3C_6H_4O)CH_2CO_2H$	166.18	nd (w)		103-4			al, bz	$B6^2$, 353
11083	Phenoxy acetic acid, 4-methyl or p-Cresoxyacetic acid $(4\text{-}CH_3C_6H_4O)CH_2CO_2H$	166.18	nd (w)		136			al, bz	$B6^2$, 380
11084	Phenoxy acetic acid, 2-nitro $(2\text{-}O_2NC_6H_4O)CH_2CO_2H$	197.15	pr (w)		158			al, eth	$B6^3$, 804
11085	Phenoxy acetic acid, 2-*iso*-propyl-5-methyl or Thymoxy acetic acid $(2\text{-}i\text{-}C_3H_7\text{-}5\text{-}CH_3C_6H_3O)CH_2CO_2H$	208.26	nd (dil al), cr (bz)		149-50			al, eth	$B6^3$, 1902
11086	Phenoxy acetic acid, 2,4,6-tribromo $[2,4,6\text{-}(Br)_3C_6H_2O]CH_2CO_2H$	388.84	nd (dil al)		200			al, eth	$B6^4$, 1068
11087	Phenoxy acetic acid, 2,4,5-trichloro $[2,4,5\text{-}(Cl)_3\text{-}C_6H_2O]CH_2CO_2H$	255.48	cr (bz)		157-8			al	$B6^4$, 973
11088	Phenoxy acetic acid, 2,4,6-trichloro $[2,4,6\text{-}(Cl)_3\text{-}C_6H_2O]CH_2CO_2H$	255.48	cr (al)		177			al	$B6^4$, 1011
11089	Phenoxy acetonitrile $C_6H_5OCH_2CN$	133.15		239-40, 128^{17}		$1.0991^{20/4}$	1.5246^{20}	al, eth	$B6^4$, 640
11090	Phenoxy acetyl chloride $C_6H_5OCH_2COCl$	170.60		225-6, 111^{13}				eth	$B6^3$, 613
11091	β-Phenoxy ethyl ether $(C_6H_5OCH_2CH_2)_2O$	258.32	nd (dil al)		66-7			al	$B6^4$, 573
11092	Phenyl acetaldehyde or α-Tolualdehyde $C_6H_5CH_2CHO$	120.15	(w)	195, 88^{18}	33-4	$1.0272^{20/4}$	1.5255^{20}	**al, eth**, ace	$B7^3$, 1003
11093	Phenyl acetamide $C_6H_5CH_2CONH_2$	135.17	pl or lf (w)		157				$B9^3$, 2193
11094	Phenyl acetamide, 3,4-dimethoxy $3,4\text{-}(CH_3O)_2C_6H_3CH_2CONH_2$	195.22	cr (w)		145-7			al, eth	$B10^3$, 1463
11095	Phenyl acetamide, α-hydroxy (dl) or Mandelamide $C_6H_5CH(OH)CONH_2$	151.16	pl (bz, al)		134-5			al, bz	$B10^3$, 469
11096	Phenyl acetamide, α-hydroxy-N-ethyl (d) or *l*-N-Ethoxy mandelamide $C_6H_5CH(OH)CONHC_2H_5$	179.22	pl (chl-peth), $[\alpha]^{18}{}_D$ −103.6 (ace)		65-6			al, eth, ace, bz, chl	$B10^2$, 117
11097	Phenyl acetamide, α-hydroxy-N-ethyl (dl) or dl-N-Ethyl mandelamide $C_6H_5CH(OH)CONHC_2H_5$	179.22	pl (bz-peth)		53-4			w, al, eth, ace, bz	$B10^1$, 89
11098	Phenyl acetamide, 2-hydroxy $2\text{-}HOC_6H_4CH_2CONH_2$	151.16	lf (al-chl)		118				B10, 188
11099	Phenyl acetamide, N-methyl $C_6H_5CH_2CONHCH_3$	149.20	cr (bz)		58			al, eth, chl	$B9^3$, 2195
11100	Phenyl acetamide, N-phenyl or α-Phenyl acetamide $C_6H_5CH_2CONHC_6H_5$	211.27	pr (al)		117-8			al, eth	$B12^3$, 521
11101	Phenyl acetamide, 4-isopropyl $4\text{-}i\text{-}C_3H_7C_6H_4CH_2CONH_2$	177.25	pl (bz)		170			al	$B9^3$, 2529
11102	Phenyl acetic acid $C_6H_5CH_2CO_2H$	136.15	lf, pl (peth)	265.5, $144\text{-}5^{12}$	77	$1.091^{77/4}$, 1.228^6		al, eth, ace	$B9^3$, 2169
11103	Phenyl acetic acid, anhydride $(C_6H_5CH_2CO)_2O$	254.29	pr or nd (eth)	$195\text{-}8^{12}$	71-2			eth, chl	$B9^3$, 2190

PHYSICAL CONSTANTS OF ORGANIC COMPOUNDS (Continued)

No.	Name, Synonyms, and Formula	Mol. wt.	Color, crystalline form, specific rotation and λ_{max} (log ε)	b.p. °C	m.p. °C	Density	n_D	Solubility	Ref.
11104	Phenyl acetic acid, 4-amino or Aminophenyl acetic acid p-$H_2NC_6H_4CH_2CO_2H$	151.16	pl (w)		199-200d				B14^3, 1182
11105	Phenyl acetic acid, α-bromo or α-Bromophenyl acetic acid $C_6H_5CH(Br)CO_2H$	215.05			82-3				B9^3, 2276
11106	Phenyl acetic acid, α-bromo, ethyl ester or Ethyl-α-bromophenyl acetate $C_6H_5CH(Br)CO_2C_2H_5$	243.10		145-55^{18}					B9^3, 2276
11107	Phenyl acetic acid, 2-bromo or 2-Bromotoluic acid $2\text{-}BrC_6H_4CH_2CO_2H$	215.05	cr (aa)		105-6			al, eth, aa	B9^3, 2273
11108	Phenyl acetic acid, 3-bromo or 3-Bromotoluic acid $3\text{-}BrC_6H_4CH_2CO_2H$	215.05	nd (w)		100-1				B9^3, 2274
11109	Phenyl acetic acid, 4-bromo $4\text{-}BrC_6H_4CH_2CO_2H$	215.05	nd (w)	sub	116			al, eth	B9^3, 2275
11110	Phenyl acetic acid, *iso*-butyl ester or *iso*-Butyl phenyl acetate, Eglantine $C_6H_5CH_2CO_2\text{-}i\text{-}C_4H_9$	192.26		247, 123-5^{14}		0.999^{18}		al, eth	B9^3, 2180
11111	Phenyl acetic acid, 2-carboxy $2\text{-}(HO_2C)C_6H_4CH_2CO_2H$	180.16	cr (w, eth)		185-7			al, w	B9^3, 4266
11112	Phenyl acetic acid, 3-carboxy $3\text{-}(HO_2C)C_6H_4CH_2CO_2H$	180.16	nd or pl (w)		184-5			al, eth	B9^3, 4269
11113	Phenyl acetic acid, 4-carboxy $4\text{-}(HO_2C)C_6H_4CH_2CO_2H$	180.16	cr (dil al)		239-41			al, eth, bz	B9^3, 4269
11114	Phenyl acetic acid, α-chloro (d) d-$C_6H_5CHClCOOH$	170.60	cr (peth), $[\alpha]^{20}_{D}$ + 192 (bz)		60-1			al, eth, bz, chl	B9^3, 2265
11115	Phenyl acetic acid, α-chloro (dl) dl-$C_6H_5CHClCOOH$	170.60	lf (peth)		78			al, eth	B9^3, 2265
11116	Phenyl acetic acid, α-chloro (l) l-$C_6H_5CHClCOOH$	170.60	nd (peth), $[\alpha]^{18}_{D}$ −191.3 (bz)		61			al, eth, bz, chl	B9^2, 307
11117	Phenyl acetic acid, 2-chloro $2\text{-}ClC_6H_4CH_2CO_2H$	170.60	nd (w)		96			al	B9^3, 2262
11118	Phenyl acetic acid, 3-chloro or 3-Chloro phenyl acetic acid $3\text{-}ClC_6H_4CH_2CO_2H$	170.60	pl (dil al), nd (hep)		77-8			eth	B9^3, 2263
11119	Phenyl acetic acid, 4-chloro $4\text{-}ClC_6H_4CH_2CO_2H$	170.60	nd (w)		105-6			al, eth, bz	B9^3, 2264
11120	Phenyl acetic acid, 2,5-dihydroxy or Homogentisic acid $2,5\text{-}(HO)_2C_6H_3CH_2CO_2H$	168.15	pr (w + 1), lf (al-chl)		152-4			al, eth, w	B10^3, 1456
11121	Phenyl acetic acid, 3,4-dihydroxy $3,4\text{-}(HO)_2C_6H_3CH_2CO_2H$	168.15			131-2			w, al, eth	B10^3, 1458
11122	Phenyl acetic acid, 3,4-dimethoxy or Homoveratric acid $3,4\text{-}(CH_3O)_2C_6H_3CH_2CO_2H$	196.20	nd (w + 1), cr (bz-peth)		98-9 (anh), 80-2 (+ w)			w, al, eth	B10^3, 1459
11123	Phenyl acetic acid, 2,4-dimethyl-α-hydroxy or 2,4-Dimethyl mandelic acid $2,4\text{-}(CH_3)_2C_6H_3\text{-}CH(OH)CO_2H$	180.20	rh (w), lf (peth-chl)		119			al, eth, chl, w	B10^3, 613
11124	Phenyl acetic acid, 2,5-dimethyl-α-hydroxy or 2,5-Dimethyl mandelic acid $2,5\text{-}(CH_3)_2C_6H_3CH(OH)CO_2H$	180.20	nd or pr (bz)		116-7			al, eth, chl	B10^3, 612
11125	Phenyl acetic acid, 3,4-dimethyl-α-hydroxy or 3,4-Dimethyl mandelic acid $3,4\text{-}(CH_3)_2C_6H_3CH(OH)CO_2H$	180.20	lf (bz)		135			al, bz	B10^3, 611
11126	Phenyl acetic acid, 2,4-dinitro or 2,4-Dinitrophenyl acetic acid $2,4\text{-}(NO_2)_2C_6H_3CH_2CO_2H$	226.15			179-80d			al, eth	B9^3, 2292
11127	Phenyl acetic acid, 2,4-dinitro, ethyl ester $2,4\text{-}(NO_2)_2C_6H_3CH_2CO_2C_2H_5$	254.20	nd (w)		37			al, eth	B9^3, 2292
11128	Phenyl acetic acid, 2,6-dinitro $2,6\text{-}(NO_2)_2C_6H_3CH_2CO_2H$	226.15	ye lf (aa)		201-2d			al	B9^1, 185
11129	Phenyl acetic acid, 2-ethoxy $2\text{-}(C_2H_5O)C_6H_4CH_2CO_2H$	180.20	nd (lig), cr (w)		103-4				B10^1, 82
11130	Phenyl acetic acid, ethyl ester or Ethyl phenyl acetate $C_6H_5CH_2CO_2C_2H_5$	164.20		227, 120-1^{20}		1.0333$^{20/4}$	1.4980^{20}	**al, eth**	B9^3, 2176
11131	Phenyl acetic acid, α-hydroxy-(D) or *l*-Mandelic acid $C_6H_5CH(OH)CO_2H$	152.15	ta $[\alpha]^{20}_{D}$ −158 (w, c=2.5)		133-5	1.341		w, al, eth, aa, chl	B10^3, 447

No.	Name, Synonyms, and Formula	Mol. wt.	Color, crystalline form, specific rotation and λ_{max} (log ε)	b.p. °C	m.p. °C	Density	n_D	Solubility	Ref.
11132	Phenyl acetic acid, α-hydroxy (DL) or dl-Mandelic acid $C_6H_5CH(OH)CO_2H$	152.15	pl (w), rh (bz)	d	121-3	$1.300^{20/4}$	....	al, eth	B10[3], 448
11133	Phenyl acetic acid, α-hydroxy (L +) or d-Mandelic acid $C_6H_5CH(OH)CO_2H$	152.15	pl (w), $[\alpha]^{20/}{}_D$ +156.6 (w, c=2.9)	....	134-5	....	....	w, al, eth, chl	B10[3], 445
11134	Phenyl acetic acid, α-hydroxy,acetate(D) or l-Acetyl mandelic acid $C_6H_5CH(O_2CCH_3)CO_2H$	194.19	nd (w + 1) $[\alpha]^{20/}{}_D$ −156−4 (ace)	....	96-8 (anh)	....	....	al, eth, ace, bz, chl	B10[3], 453
11135	Phenyl acetic acid, α-hydroxy,acetate (DL) or dl-acetyl mandelic acid $C_6H_5CH(CO_2CH_3)COOH$	194.19	cr (bz), amor (chl-peth), cr (w + 1)	....	38-9 (+1w), 79-80 (anh)	....	....	al, eth, bz, chl	B10[3], 453
11136	Phenyl acetic acid, α-hydroxy-4-bromo or p-Bromo mandelic acid 4-Br-$C_6H_4CH(OH)CO_2H$	231.05	nd (bz)	....	118-20	....	....	al, eth, bz, chl	B10[3], 480
11137	Phenyl acetic acid, α-hydroxy-4-chloro or 4-Chloro mandelic acid 4-Cl-$C_6H_4CH(OH)CO_2H$	186.59	nd (bz)	....	119-20	....	....	w, al	B10[3], 479
11138	Phenyl acetic acid, α-hydroxy, ethyl ester (D-) or l-Ethyl mandelate $C_6H_5CH(OH)CO_2C_2H_5$	180.20	(peth), $[\alpha]^{20/}{}_D$ −128.4 (chl, c=6.7)	150^{20}	35	$1.1270^{20/4}$	....	al, eth	B10[3], 456
11139	Phenyl acetic acid, α-hydroxy, ethyl ester (DL) or dl-Ethyl mandelate $C_6H_5CH(OH)CO_2C_2H_5$	180.20	nd (peth)	253-5, 141^{15}	37	....	....	al, eth, lig	B10[3], 457
11140	Phenyl acetic acid, α-hydroxy, ethyl ester (L +) or d-Ethyl mandelate $C_6H_5CH(OH)CO_2C_2H_5$	180.20	(peth), $[\alpha]^{20/}{}_D$ +205 (CS_2, c=0.7)	....	33	$1.1270^{20/4}$	....	al, eth	B10[3], 456
11141	Phenyl acetic acid, α-hydroxy, methyl ester (D-) or l-Methyl mandelate $C_6H_5CH(OH)CO_2CH_3$	166.18	cr (peth), $[\alpha]^{20/}{}_D$ −131.5 (w)	160^{22}	55	1.1756^{20}	....	w, al, ace, bz, chl	B10[3], 454
11142	Phenyl acetic acid, α-hydroxy, methyl ester (DL) or dl-Methyl mandelate $C_6H_5CH(OH)CO_2CH_3$	166.18	pl (bz-lig)	250d, 144^{20}	58	1.1756^{20}	....	al, chl	B10[3], 454
11143	Phenyl acetic acid, α-hydroxy, methyl ester (L +) or d-Methyl mandelate $C_6H_5CH(OH)CO_2CH_3$	166.18	cr (peth), $[\alpha]^{20/}{}_D$ −133.6 (w)	160^{32}	55.5	1.1756^{20}	....	w, al, ace, bz, chl	B10[3], 454
11144	Phenyl acetic acid, α-hydroxy-4-*iso*-propyl (d) or d-4-*iso*-Propyl mandelic acid 4-i-$C_3H_7C_6H_4CH(OH)CO_2H$	194.23	lf (w), $[\alpha]^{17/}{}_D$ +135 (abs al, c = 4)	....	153-4	....	....	al, eth	B10, 279
11145	Phenyl acetic acid, α-hydroxy-4-*iso*-propyl (dl) or dl-4-iso-Propyl mandelic acid 4-i-$C_3H_7C_6H_4CH(OH)CO_2H$	194.23	nd (w)	....	159-60	....	....	al, eth	B10[3], 629
11146	Phenyl acetic acid, α-hydroxy-4-*iso*propyl (l) or l-4-*iso*-propyl mandelic acid 4-i-$C_3H_7C_6H_4CH(OH)CO_2H$	194.23	ta (20% al), $[\alpha]^{17/}{}_D$ −135 (abs al, c=4)	....	153-4	....	....	al, eth	B10, 279
11147	Phenyl acetic acid, 2-hydroxy 2-$HOC_6H_4CH_2CO_2H$	152.15	....	240-3d	147-9	....	....	eth	B10[3], 422
11148	Phenyl acetic acid, 2-hydroxy-5-nitro 2-(HO)-5-$(NO_2)C_6H_3CH_2CO_2H$	197.15	nd	....	160-2	....	....	w, al, eth	B10, 189
11149	Phenyl acetic acid, 2-hydroxy-5-nitro, ethyl ester or Ethyl-2-hydroxy-2-nitro phenyl acetate 5-NO_2-2-$HOC_6H_3CH_2CO_2C_2H_5$	225.20	pr or pl (al)	....	154-5	....	....	bz, chl	B10, 189
11150	Phenyl acetic acid, 2-hydroxy, hydrazide 2-HO-C_6H_4-$CH_2CONHNH_2$	166.18	lf (chl, bz), nd (al)	....	154	....	....	bz, al	B10[2], 112
11151	Phenyl acetic acid, 3-hydroxy 3-$HOC_6H_4CH_2CO_2H$	152.15	nd (bz-lig)	190^{11}	131-4	....	....	w, al, eth, bz, chl	B10[3], 428
11152	Phenyl acetic acid, 4-hydroxy 4-$HOC_6H_4CH_2CO_2H$	152.15	nd (w)	sub	149-51	....	....	al, eth	B10[3], 430
11153	Phenyl acetic acid, 3-hydroxy-4-methoxy or Homoisovanillic acid 3-(HO)-4-$(CH_3O)C_6H_3CH_2CO_2H$	182.18	....	....	130-1	....	....	w, al, eth	B10[3], 1458

No.	Name, Synonyms, and Formula	Mol. wt.	Color, crystalline form, specific rotation and λ_{max} (log ε)	b.p. °C	m.p. °C	Density	n_D	Solubility	Ref.
11155	Phenyl acetic acid, 2-mercapto $2\text{-}HSC_6H_4CH_2CO_2H$	168.21	pl (w, bz-lig)		96-7			al, eth, bz	B10^{1}, 1782
11156	Phenyl acetic acid, α-methoxy (D-) $C_6H_5CH(OCH_3)CO_2H$	166.18	nd (peth), $[\alpha]^{13/}{}_D$ -150 (al)		63-4			w, al, ace	B10^{3}, 451
11157	Phenyl acetic acid, α-methoxy (DL) $C_6H_5CH(OCH_3)CO_2H$	166.18	pl (lig)		71-2			al, eth	B10^{3}, 452
11158	Phenyl acetic acid, 2-methoxy $2\text{-}CH_3OC_6H_4CH_2CO_2H$	166.18	nd (w)	100-1^{2}	123			al, eth, ace, bz, chl	B10^{3}, 422
11159	Phenyl acetic acid, 4-methoxy or Homoanisic acid $4\text{-}CH_3OC_6H_4CH_2CO_2H$	166.18	pl (w)	138-40$^{2\text{-}3}$	86			al, eth	B10^{3}, 431
11160	Phenyl acetic acid, 4-methoxy-2-nitro $4\text{-}(CH_3O)\text{-}2\text{-}(NO_2)C_6H_3CH_2CO_2H$	211.17	ye nd (50% al)		157-8d			al, aa	B10^{2}, 113
11161	Phenyl acetic acid, methyl ester or Methyl phenyl acetate $C_6H_5CH_2CO_2CH_3$	150.18		218, 131-2^{50}		1.0633$^{16/16}$	1.5075^{20}	al, eth, ace	B9^{3}, 2175
11162	Phenyl acetic acid, 2-methyl or 2-Tolylacetic acid $(2\text{-}CH_3C_6H_4)CH_2CO_2H$	150.18	nd (w)		88-90				B9^{3}, 2426
11163	Phenyl acetic acid, 3-methyl or 3-Methyl-α-toluic acid $3\text{-}CH_3C_6H_4CH_2CO_2H$	150.18	nd (w)	120-3^{26}	62				B9^{3}, 2429
11164	Phenyl acetic acid, 4-methyl or 4-Methyl phenyl acetic acid $4\text{-}CH_3C_6H_4CH_2CO_2H$	150.18	nd or pl (al, w)	265-7 sub	91-3			al, eth, bz, chl	B9^{3}, 2432
11165	Phenyl acetic acid, 2-nitro $2\text{-}NO_2C_6H_4CH_2CO_2H$	181.15	nd (w), pl (dil al)		141-2			al	B9^{3}, 2282
11166	Phenyl acetic acid, 3-nitro $3\text{-}NO_2C_6H_4CH_2CO_2H$	181.15	nd (w)		122			al	B9^{3}, 2283
11167	Phenyl acetic acid, 4-nitro $4\text{-}NO_2C_6H_4CH_2CO_2H$	181.15	pa ye nd (w)		153-5			al, eth, bz	B9^{3}, 2284
11168	Phenyl acetic acid, phenethyl ester $C_6H_5CH_2CO_2CH_2CH_2C_6H_5$	240.30		177-8$^{4.5}$	26.5	1.080$^{25/25}$		al	B9^{3}, 2183
11169	Phenyl acetonitrile $C_6H_5CH_2CN$	117.15		234, 107^{12}	−23.8	1.0157$^{20/4}$	1.5230^{20}	**al, eth**, ace	B9^{3}, 2252
11170	Phenyl acetonitrile, α-amino (dl) $C_6H_5CH(NH_2)CN$	132.16	hyg lf (lig)		55				B14^{3}, 1189
11171	Phenyl acetonitrile, 2-amino $2\text{-}H_2NC_6H_4CH_2CN$	132.16	lf (dil al)		72			al, bz	B14^{3}, 1181
11172	Phenyl acetonitrile, 4-amino or 4-Amino phenyl acetonitrile $4\text{-}H_2NC_6H_4CH_2CN$	132.16	lf (w)	312, 177^{11}	46			al	B14^{3}, 1182
11173	Phenyl acetonitrile, α-bromo $C_6H_5CH(Br)CN$	196.05	yesh cr (dil al)	242d, 132-4^{12}	29	1.539$^{29/4}$		al, eth, ace, bz	B9^{3}, 2278
11174	Phenyl acetonitrile, 2-bromo or 2-Bromo phenyl acetonitrile $2\text{-}Br\text{-}C_6H_4CH_2CN$	196.05		145-7^{14}	1			al	B9^{3}, 2274
11175	Phenyl acetonitrile, 4-bromo or 4-Bromo phenyl acetonitrile $4\text{-}BrC_6H_4CH_2CN$	196.05	pa ye cr (al)		47			al, bz	B9^{3}, 2275
11176	Phenyl acetonitrile, 2-carboxy $2\text{-}(HO_2C)C_6H_4CH_2CN$	161.16	cr (aa), nd (w, bz)		116d			al, eth, bz	B9^{3}, 4267
11177	Phenyl acetonitrile, 2-chloro $2\text{-}ClC_6H_4CH_2CN$	151.60	grsh-ye nd	251, 170^{120}	24	1.1737$^{18/4}$	1.534^{18}		B9^{3}, 2263
11178	Phenyl acetonitrile, α-hydroxy (D+) or d-Mandelonitrile $C_6H_5CH(OH)CN$	133.15	nd, $[\alpha]^{25/}{}_{546\cdot1}$ +46.9 (bz)		28-9			al, eth	B10^{3}, 473
11179	Phenyl acetomitrile, α-hydroxy (DL) or dl-Mandelonitrile $C_6H_5CH(OH)CN$	133.15	ye pr	170d	22	1.1165$^{20/4}$	1.5201^{20}	al, eth	B10^{3}, 474
11180	Phenyl acetonitrile, α-hydroxy (L-) or l-Mandelonitrile $C_6H_5CH(OH)CN$	133.15						al, eth	B10^{1}, 86
11181	Phenyl acetonitrile, 2-hydroxy $2\text{-}HOC_6H_4CH_2CN$	133.15	nd (bz-lig)		117-9			w, eth, ace, bz	B10^{3}, 425
11182	Phenyl acetonitrile, 3-hydroxy $3\text{-}HOC_6H_4CH_2CN$	133.15	pl (w)		52-3			w, al, eth	B10, 189
11183	Phenyl acetonitrile, 4-hydroxy $4\text{-}HOC_6H_4CH_2CN$	133.15	pl (w), mcl pr	330^{756}, 210^{10}	69-70			al, eth	B10^{3}, 438
11184	Phenyl acetonitrile, 2-methoxy or 2-Methoxy phenyl acetonitrile $2\text{-}CH_3OC_6H_4CH_2CN$	147.18	pr (bz-lig)	141-3^{11}	68.5			bz	B10^{3}, 425

No.	Name, Synonyms, and Formula	Mol. wt.	Color, crystalline form, specific rotation and λ_{max} (log ε)	b.p. °C	m.p. °C	Density	n_D	Solubility	Ref.
11185	Phenyl acetonitrile, 4-methoxy $4-CH_3OC_6H_4CH_2CN$	147.18	...	286-7, 152^{16}	...	$1.0845^{20/4}$	1.5309^{20}	al, eth	$B10^3$, 439
11186	Phenyl acetonitrile, 2-methyl or 2-Totyl acetonitrile $2-CH_3C_6H_4CH_2CN$	131.18	...	244	...	1.0156^{22}	1.5252^{20}	al, eth, bz	$B9^3$, 2427
11187	Phenyl acetonitrile, 3-methyl or 3-Totyl acetonitrile $3-CH_3C_6H_4CH_2CN$	131.18	...	$245-7^{745}$d, 133^{15}	...	1.0022^{22}	1.5233^{20}	al, eth, bz	$B9^2$, 349
11188	Phenyl acetonitrile, 4-methyl or 4-Totyl acetonitrile $4-CH_3C_6H_4CH_2CN$	131.18	...	242-3, 122^{13}	18	0.9922^{22}	1.5167^{20}	al, eth, bz	$B9^3$, 2433
11189	Phenyl acetonitrile, α-nitro $C_6H_5CH(NO_2)CN$	162.15	...	...	39-40	...	...	al	$B9^3$, 2291
11190	Phenyl acetonitrile, 2-nitro $2-NO_2C_6H_4CH_2CN$	162.15	nd (al-w), pr (aa, al)	178^{12}	84	...	...	al, eth, ace, bz, chl	$B9^3$, 2283
11191	Phenyl acetonitrile, 3-nitro $3-NO_2C_6H_4CH_2CN$	162.15	(eth-lig)	$180^{1.5}$	63	...	...	al, eth, bz, chl	$B9^2$, 312
11192	Phenyl acetonitrile, 4-nitro $4-NO_2C_6H_4CH_2CN$	162.15	pl	$195-7^{12}$	116-7	...	...	al, eth, bz, chl	$B9^3$, 2291
11193	Phenyl acetyl chloride $C_6H_5CH_2COCl$	154.60	...	170^{250}, $104-5^{24}$	...	$1.1682^{20/4}$	1.5325^{20}	eth	$B9^3$, 2192
11194	Phenyl acetyl chloride, α-acetyl (DL) or dl-Acetyl mandelyl chloride $C_6H_5CH(CO_2CH_3)COCl$	196.63	...	$150-5^{33}$	...	...	...	eth, bz, chl	$B10^1$, 89
11195	Phenyl acetyl chloride, 2,4-dinitro $2,4-(NO_2)_2C_6H_3CH_2COCl$	244.59	ye lf (CS_2)	...	77	...	...	eth, bz, chl	$B9^1$, 185
11196	Phenyl acetylene $C_6H_5C{\equiv}CH$	102.14	...	142-4, 44^{18}	−44.8	$0.9281^{20/4}$	1.5485^{20}	**al, eth**, ace	$B5^4$, 1525
11197	α-Phenyl acrylic acid or Atropic acid $CH_2{=}C(C_6H_5)CO_2H$	148.16	lf (al), nd (w)	267d	106-7	...	...	al, eth, bz, chl	$B9^3$, 2751
11198	α-Phenyl acrylic acid, 3-(2-nitro phenyl)(*trans*) $(3-NO_2C_6H_4)CH{=}C(C_6H_5)CO_2H$	269.26	ye cr (al)	...	196-7	...	...	eth, bz	$B9^3$, 3423
11199	α-Phenyl acrylic acid, 3-(3-nitro phenyl)-(*cis*) $(3-NO_2C_6H_4)CH{=}C(C_6H_5)CO_2H$	269.26	nd (al, aa)	...	195-6	...	...	...	$B9^3$, 3423
11200	α-Phenyl acrylic acid, 3-(3-nitrophenyl)-(*trans*) $(3-NO_2C_6H_4)CH{=}C(C_6H_5)CO_2H$	269.26	ye pr (eth), nd (dil al)	...	182	...	...	al, ace, bz, chl	$B9^3$, 3424
11201	α-Phenyl acrylic acid, 3-(4-nitrophenyl)-cis $(4-NO_2C_6H_4)CH{=}C(C_6H_5)CO_2H$	269.26	ye cr pr (dil al + 1w), lf (bz + ½)	...	144	...	...	al, eth	$B9^3$, 3424
11202	α-Phenyl acrylic acid, 3-(4-nitrophenyl)-trans $(4-NO_2C_6H_4)CH{=}C(C_6H_5)CO_2H$	269.26	ye pr or nd (al)	...	213-4	...	...	al	$B9^3$, 3425
11203	Phenylalanine $C_6H_5CH_2CH(NH_2)CO_2H$	165.19	pr (w), $[\alpha]^{18/}{}_D$ + 70 (w)	sub at 295	283-4d	...	...	w	$B14^3$, 1228
11204	Phenylalanine (D) $C_6H_5CH_2CH(NH_2)CO_2H$	165.19	nd or pr (w), $[\alpha]^{20/}$ -69.5 (w)	sub at 295	283d	...	1.600	w	$B14^3$, 1228
11205	Phenylalanine (DL) $C_6H_5CH_2CH(NH_2)CO_2H$	165.19	red-br lf (dil al), pr or nd, (w)	sub d	284-8d	...	...	w	$B14^3$, 1229
11206	Phenylalanine, N-acetyl (d,t) $C_6H_5CH_2CH(NHCOCH_3)CO_2H$	207.23	cr (w), $[\alpha]^{20/}{}_D$ -50.9	...	172	...	...	al	$B14^2$, 297
11207	Phenylalanine, N-acetyl (dl) $C_6H_5CH_2CH(NHCOCH_3)CO_2H$	207.23	nd or pl (w), hex tab (ace)	...	152-3	...	...	w, al	$B14^3$, 1238
11208	Phenylalanine, N-acetyl (l) $C_6H_5CH_2CH(NHCOCH_3)CO_2H$	207.23	$[\alpha]^{26/}{}_D$ + 35.1	...	172	...	...	al	$B14^2$, 298
11209	Phenylalanine, ethyl ester $C_6H_5CH_2CH(NH_2)CO_2C_2H_5$	193.25	...	148^{15}	...	1.065^{15}	...	...	$B14^3$, 1232
11210	Phenylboric acid or Benzeneboronic acid $C_6H_5B(OH)_2$	121.93	nd (w)	...	218-20	...	...	al, eth, bz	$B16^2$, 638
11211	*o*-Phenylenediamine or 1,2-Diamino benzene $1,2-(H_2N)_2C_6H_4$	108.14	brsh-ye lf (w), pl (chl)	256-8	102-3	...	...	al, eth, bz, chl	$B13^3$, 28
11212	*o*-Phenylenediamine, 3,5-dichloro $1,2-(H_2N)_2-3,5-Cl_2C_6H_2$	177.03	nd (al)	...	60.5	...	...	...	B13, 27
11213	*o*-Phenylenediamine, 3,6-dichloro $1,2-(H_2N)_2-3,6-Cl_2C_6H_2$	177.03	nd (50% al)	...	100	...	...	w, al, eth, ace	$B13^2$, 20

No.	Name, Synonyms, and Formula	Mol. wt.	Color, crystalline form, specific rotation and λ_{max} (log ε)	b.p. °C	m.p. °C	Density	n_D	Solubility	Ref.
11214	*o*-Phenylenediamine, 4,5-dimethoxy $1,2\text{-}(H_2N)_2\text{-}4,5\text{-}(CH_3O)_2C_6H_2$	168.20	bl pr	....	131-2	....	....	w, al	B13[3], 2127
11215	*o*-Phenylenediamine, 3-methoxy-6-methyl $1,2\text{-}(H_2N)_2\text{-}3\text{-}CH_3O\text{-}6\text{-}CH_3\text{-}C_6H_2$	152.20	pr (eth-bz)	....	75-6	....	....	al, eth	B13[2], 349
11216	*o*-Phenylenediamine, 4-methoxy $1,2\text{-}(H_2N)_2C_6H_3OCH_3\text{-}4$	138.17	gr pl	167-70[11]	50-2	....	....	eth	B13[3], 1362
11217	*o*-Phenylenediamine, 4-methyl $1,2\text{-}(H_2N)_2\text{-}4\text{-}CH_3C_6H_3$	122.17	pl (lig)	265, 92[1]	89-90	....	....	w	B13[3], 292
11218	*o*-Phenylene diamine, 3-nitro $1,2\text{-}(H_2N)_2\text{-}3\text{-}NO_2\text{-}C_6H_3$	153.14	dk red nd (dil al)	....	158-9	....	....	....	B13[3], 61
11219	*o*-Phenylenediamine, 4-nitro $1,2\text{-}(H_2N)_2\text{-}4\text{-}NO_2\text{-}C_6H_3$	153.14	dk red nd	....	199-200	....	....	....	B13[3], 62
11220	*o*-Phenylenediamine, N-phenyl $2\text{-}H_2NC_6H_4NHC_6H_5$	184.24	nd (w)	312.5[744]	80-1	....	....	ace, bz, chl	B13[3], 34
11221	*m*-Phenylenediamine or 1,3-Diamino benzene $1,3\text{-}(H_2N)_2C_6H_4$	108.14	rh (al)	282-4	63-4	1.0696[58/4]	1.6339[58]	w, al, eth, bz	B13[3], 65
11222	Phenylenediamine, N-acetyl $3\text{-}(CH_3CONH)C_6H_4NH_2$	150.18	lf (ace-bz), nd or pl (bz)	d at 100	87-9	....	....	w, al, eth, ace	B13[3], 75
11223	*m*-Phenylenediamine, 2,5-dichloro $2,5\text{-}Cl_2\text{-}1,3\text{-}(H_2N)_2C_6H_2$	177.03	nd (w)	....	100	....	....	w, al, bz	B13[2], 29
11224	*m*-Phenylenediamine, 4,6-dichloro $4,6\text{-}Cl_2\text{-}1,3\text{-}(H_2N)_2C_6H_2$	177.03	nd (dil al)	....	136-7	....	....	al	B13[3], 98
11225	*m*-Phenylenediamine, 6-methoxy $6\text{-}CH_3O\text{-}1,3\text{-}(NH_2)_2C_6H_3$	138.17	nd (eth)	....	67-8	....	....	al, eth	B13[2], 308
11226	*p*-Phenylenediamine or 1,4-Diamino benzene $1,4\text{-}(H_2N)_2C_6H_4$	108.14	wh pl (bz, eth)	267	140	....	....	eth, chl	B13[3],104
11227	*p*-Phenylenediamine, N-acetyl,sulfate $4\text{-}(CH_3CONH)C_6H_4NH_2.H_2SO_4$	248.25	nd (eth-al)	....	285 d	....	....	w, al	B13, 95
11228	*p*-Phenylenediamine, 2,5-dichloro $2,5\text{-}Cl_2\text{-}1,4\text{-}(H_2N)_2C_6H_2$	177.03	pr (w)	....	170	....	....	....	B13, 118
11229	*p*-Phenylenediamine, 2,6-dichloro $2,6\text{-}Cl_2\text{-}1,4\text{-}(H_2N)_2C_6H_2$	177.03	nd, pr (dil al)	....	124-6	....	....	al, eth, ace, bz	B13[3], 269
11230	*p*-Phenylenediamine, 2-methoxy-5-methyl $2\text{-}CH_3O\text{-}5\text{-}CH_3\text{-}1,4\text{-}(H_2N)_2C_6H_2$	152.20	....	....	166	....	....	al, eth	B13[2], 349
11231	*p*-Phenylenediamine, N-phenyl $4\text{-}H_2NC_6H_4NHC_6H_5$	184.24	nd (al), cr (lig)	354, 155[0.026]	66 (al), 75 (lig)	....	....	al, eth, lig	B13[3], 115
11232	*p*-Phenylenediamine, 2,3,5,6-tetramethyl or Diamino durene $1,4\text{-}(H_2N)_2\text{-}2,3,5,6\text{-}(CH_3)_4C_6$	164.25	nd (w)	....	148	....	....	al, eth, chl	B13[3], 360
11233	bis-(1-phenylethyl) amine $[C_6H_5CH(CH_3)]_2NH$	225.13	ye	295-8, 190[10]	....	1.018[15]	1.573	....	B12[2], 589
11234	Phenyl-2-aminophenyl sulfide $2\text{-}H_2NC_6H_4SC_6H_5$	201.29	pl (al)	257.5[100]	35-6	....	....	al	B13[3], 904
11235	Phenyl-2-aminophenyl sulfone $(2\text{-}H_2NC_6H_4)SO_2C_6H_5$	233.28	lf (dil al)	....	122-4	....	....	al, bz, aa	B3[3], 905
11236	Phenyl-4-aminophenyl sulfide $(4\text{-}H_2NC_6H_4)SC_6H_5$	201.22	nd (dil al), cr (lig)	242.5[29]	95.8	....	....	al, eth	B13[3], 1224
11237	Phenyl-4-aminophenyl sulfone $(4\text{-}H_2NC_6H_4)SO_2C_6H_5$	233.28	nd (al)	....	176	....	....	al, bz, aa	B13[3], 1226
11238	Phenyl-4-aminophenyl sulfoxide $(4\text{-}H_2NC_6H_4)SOC_6H_5$	217.29	nd (w)	....	152	....	....	al, eth	B13[3], 1226
11239	Phenyl-4-bromophenyl ether $4\text{-}BrC_6H_4OC_6H_5$	249.11	....	310.1, 163[10]	18.7	1.4208[20/4]	1.6084[20]	eth	B6[4], 1047
11240	Phenyl *tert*-butyl ketone or Pivalophenone $C_6H_5COC(CH_3)_3$	162.23	....	219-21, 97-8[16]	....	0.963[26]	1.5086[19]	ace	B7[3], 1125
11241	Phenyl-α-chlorobenzyl ketone or Desylchloride $C_6H_5CO(CHClC_6H_5)$	230.69	nd (al)	d	68.5	....	....	al	B7[3], 2106
11242	Phenyl-4-chlorophenyl sulfone or Sulphenone $(4\text{-}ClC_6H_4)SO_2C_6H_5$	252.72	cr (al)	....	98	....	....	eth, ace, bz	B6[4], 1587
11243	Phenyl-2,4-diaminophenyl sulfone $[2,4\text{-}(NH_2)_2C_6H_3]SO_2C_6H_5$	248.30	nd (al)	....	188	....	....	al	B13, 553
11244	Phenyl-2,5-dihydroxyphenyl sulfone $[2,5\text{-}(HO)_2C_6H_3]SO_2C_6H_5$	250.27	pr (w), nd (dil al)	....	196	....	....	al	B6[2], 1072
11245	Phenyl-2,4-dinitrophenyl ether $[2,4\text{-}(NO_2)_2C_6H_3]OC_6H_5$	260.21	pl (al), nd (al-ace)	230-50[27]	71	....	....	al, eth	B6[4], 1375

PHYSICAL CONSTANTS OF ORGANIC COMPOUNDS (Continued)

No.	Name, Synonyms, and Formula	Mol. wt.	Color, crystalline form, specific rotation and λ_{max} (log ε)	b.p. °C	m.p. °C	Density	n_D	Solubility	Ref.
11246	Phenyl-2,4-dinitrophenyl sulfide $[2,4\text{-}(O_2N)_2C_6H_3]SC_6H_5$	276.27	pa ye nd, (ace, bz-al)		121			ace, bz, aa	$B6^4$, 1746
11247	Phenyl-2,6-dinitrophenyl ether $[2,6\text{-}(NO_2)_2C_6H_3]OC_6H_5$	260.21	lf (al)		99-100				$B6^2$, 245
11248	Phenyl-3,4-dinitrophenyl ether $[3,4\text{-}(NO_2)_2C_6H_3]OC_6H_5$	260.21	lf (al)		99-100				$B6^1$, 127
11249	Phenyl ether or Diphenyl ether $(C_6H_5)_2O$	170.21		$257.9, 121^{10}$		1.0748^{20}	1.5787^{25}	al, eth, bz, aa	$B6^4$, 568
11250	α-Phenyl ethyl alcohol (d) $C_6H_5CH(OH)CH_3$	122.17	$[\alpha]^{19/}_D$ +42.9 (undil)	$203, 100^{18}$		$1.0129^{20/4}$	1.5272^{20}	al, chl	$B6^3$, 1671
11252	α-Phenyl ethyl alcohol (dl) $C_6H_5CH(OH)CH_3$	122.17	glassy	$203.4, 87.2^{10}$	20	$1.0135^{20/4}$	1.5275^{20}	**al, eth**	$B6^3$, 1673
11253	α-Phenyl ethyl alcohol (l) $C_6H_5CH(OH)CH_3$	122.17	$[\alpha]^{20/}_D$ -45.5 (MeOH, c=5)	$202\text{-}4, 93^{14}$		$1.0129^{20/4}$	1.5272^{20}	al, eth	$B6^3$, 1672
11254	α-Phenyl ethyl alcohol, 3-methyl $[3\text{-}CH_3C_6H_4]CH(OH)CH_3$	136.19		112^{12}		$0.9974^{15/4}$	1.5240^{20}	al, eth	$B6^3$, 1823
11255	α-Phenyl ethyl alcohol, 4-methyl $(4\text{-}CH_3C_6H_4)CH(OH)CH_3$	136.19		$219^{756}, 120^{19}$		$0.9944^{20/4}$	1.5246^{20}	al, eth	$B6^3$, 1826
11256	β-Phenyl ethyl alcohol $C_6H_5CH_2CH_2OH$	122.17	glass	218.2, 97.4^{10}	fr−27	$1.0202^{20/4}$	1.5325^{20}	**al, eth**	$B6^3$, 1703
11257	β-Phenyl ethyl alcohol, 2-amino $(2\text{-}H_2NC_6H_4)CH_2CH_2OH$	137.18	ye in air	$152\text{-}3^6$			1.5849^{19}	w	$B13^3$, 1679
11258	β-Phenyl ethyl alcohol, 4-amino $(4\text{-}H_2NC_6H_4)CH_2CH_2OH$	137.18	nd (al)		108				$B13^3$, 1679
11259	β-Phenyl ethyl alcohol, 2-methoxy or 2-Anisyl methyl carbinol $(2\text{-}CH_3OC_6H_4)CH_2CH_2OH$	152.19		128^{17}		$1.0862^{15/4}$	1.5312^{25}	al, eth	$B6^3$, 4569
11260	β-Phenyl ethyl alcohol, 3-methoxy or 3-Anisyl methyl carbinol $(3\text{-}CH_3OC_6H_4)CH_2CH_2OH$	152.19		133^{15}		$1.0781^{19/4}$	1.5325^{20}	al, eth	$B6^3$, 4570
11261	β-Phenyl ethyl alcohol, 4-methoxy or 4-Anisyl methyl carbinol $(4\text{-}CH_3OC_6H_4)CH_2CH_2OH$	152.19		$140\text{-}1^{17}$		$1.0794^{20/4}$	1.5310^{25}	al, eth	$B6^3$, 4571
11262	bis-(2-Phenyl ethyl)amine $[C_6H_5CH_2CH_2]_2NH$	225.33		$335\text{-}7^{603}, 190^{15}$	28-30		1.5550^{25}	al, eth	$b12^3$, 2415
11263	bis-(α-Phenyl ethyl) ether (dl) $[C_6H_5CH(CH_3)]_2O$	226.32		280.2, $167\text{-}8^{23}$		$1.0058^{15/4}$	1.5454^{21}	eth, chl	$B6^3$, 1677
11264	bis-(β-Phenyl ethyl) ether $(C_6H_5CH_2CH_2)_2O$	226.32	vt-bl flr	317-20, 194.5^{20}		$1.0141^{18/4}$	1.5488^{18}	eth, chl	$B6^3$, 1707
11265	Phenyl ethynyl ether or Phenoxy acetylene $C_6H_5OC{\equiv}CH$	118.14		$61\text{-}2^{25}$	−36	$1.0614^{20/4}$	1.5125^{20}	al, eth	$B6^4$, 565
11266	Phenyl glycidol $C_6H_5CHCHCH_2OH$ (O bridging CHCH)	150.18		138^1	26.5	1.512^{27}	1.5432^{27}	al, eth	$B17^4$, 1349
11267	Phenyl glyoxal C_6H_5COCHO	134.13	nd (+w)	$142^{125}, 95\text{-}7^{25}$	91 (+w)			al, eth, ace, bz, chl	$B7^3$, 3443
11268	**Phenyl glyoxime** $C_6H_5C(NOH)C(NOH)H$	164.16	nd (chl)		180			w, al, eth	$B7^3$, 3448
11269	Phenyl glyoxylic acid, 2-nitro $(2\text{-}NO_2C_6H_4)COCO_2H$	195.13	pr (w + l)		123			w, al, ace, aa	$B10^1$, 315
11270	Phenyl glyxylonitrile C_6H_5COCN	131.13	ta	$206\text{-}8, 99^{19}$	32-3			al, eth	$B10^3$, 2976
11271	Phenylhydrazine, 1-Benzoyl $C_6H_5NHNHCOC_6H_5$	212.25	pr (al), nd (w), lf (dil al)	314	168			bz, chl	$B15^3$, 163
11272	Phenylhydrazine, *o*-bromo $2\text{-}BrC_6H_4NHNH_2$	187.04	nd		48				$B15^1$, 117
11273	Phenyl hydrazine, *p*-bromo $4\text{-}BrC_6H_4NHNH_2$	187.04	nd (w), lf (lig), cr (al)		108			al, eth	$B15^3$, 289
11274	Phenyl hydrazine, 1-butyl $H_2NN(C_4H_9)C_6H_5$	164.25		250^{763}					$B15^1$, 28
11275	Phenylhydrazine, 2,4,6-tribromo $(2,4,6\text{-}Br_3C_6H_2)NHNH_2$	344.83	nd (peth), cr (lig)		146			bz, chl, lig	$B15^1$, 126

No.	Name, Synonyms, and Formula	Mol. wt.	Color, crystalline form, specific rotation and λ_{max} (log ε)	b.p. °C	m.p. °C	Density	n_D	Solubility	Ref.
11276	Phenylhydrazine, 2,4,6-trichloro $(2,4,6\text{-}Cl_3C_6H_2)NHNH_2$	211.48	cr (bz)		143				B15³, 281
11277	Phenylhydroxylamine C_6H_5NHOH	109.13	nd (w, bz, peth)		83-4			al, eth, bz, chl	B15³, 5
11278	Phenylhydroxylamine, N-nitroso $C_6H_5N(NO)OH$	138.13	nd (lig)		59			al, eth	B16³, 638
11279	Phenyl, 2-hydroxyphenyl amine $(2HOC_6H_4)NHC_6H_5$	185.23	pr (w)	$180\text{-}9^{20}$	69-70			al, eth, aa	B13³, 764
11280	Phenyl, 3-hydroxyphenyl amine $(3\text{-}HOC_6H_4)NHC_6H_5$	185.23	lf (w)	340	81-2			al, eth, ace, bz	B13³, 931
11281	Phenyl, 4-hydroxyphenyl amine $(4\text{-}HOC_6H_4)NHC_6H_5$	185.23	lf (w)	330, $215\text{-}6^{12}$	73			al, eth, bz, chl	B13³, 991
11282	Phenyl isocyanate C_6H_5NCO	119.12		$162\text{-}3^{750}$, 55^{13}		$1.0956^{20/4}$	1.5368^{20}	eth	12³, 903
11283	Phenyl isothiocyanate or Phenyl mustard oil C_6H_5NCS	135.18		221, 95^{12}	−21	$1.1303^{20/4}$	1.6492^{23}	al, eth	12³, 908
11284	Phenyl, 2-methoxyphenyl ether $(2\text{-}CH_3OC_6H_4)OC_6H_5$	200.24	cr (MeOH), nd (lig)	288^{745}, $91\text{-}2^{7}$	79			al, eth, bz	B6³, 4215
11285	Phenyl nitromethane or α-Nitrotoluene $C_6H_5CH_2NO_2$	137.14	ye liq	225-7, $118\text{-}9^{16}$		$1.1598^{20/0}$	1.5323^{20}	eth, ace	B5⁴, 850
11286	Phenyl 2-nitrophenyl amine $(2\text{-}NO_2C_6H_4)NHC_6H_5$	214.22	og pl (dil al), rh bipym		75.5			al	B12³, 1517
11287	Phenyl 3-Nitrophenyl amine $(3\text{-}NO_2C_6H_4)NHC_6H_5$	214.22	red nd or pl (dil al)		114			al, eth, bz	B9³, 3585
11288	Phenyl 4-nitrophenyl amine $(4\text{-}NO_2C_6H_4)NHC_6H_5$	214.22	ye nd, tab (CCl_4)	211^{20}	133-4			al, aa	B12³, 1586
11289	Phenyl 4-nitrosophenyl amine $(4\text{-}ONC_6H_4)NHC_6H_5$	198.22	ye pw pl (bz)		143			al, eth, bz, chl	B12³, 347
11290	Phenyl 2-nitrophenyl ether $(2\text{-}O_2NC_6H_4)OC_6H_5$	215.21	ye liq	235^{60}, $183\text{-}5^{8}$	<−20	1.2539^{22}	1.575^{20}	al, eth, bz, chl, aa	B6⁴, 1252
11291	Phenyl 4-nitrophenyl ether $(4\text{-}O_2NC_6H_4)OC_6H_5$	215.21	pl (peth, MeOH)	$188\text{-}90^{8}$	61			eth, bz	B6⁴, 1287
11292	Phenyl 2-nitrophenyl sulfide $(2\text{-}NO_2C_6H_4)SC_6H_5$	231.27	ye-og nd (lig, al-eth)	210^{15}	82			al, eth	B6⁴, 1663
11293	Phenyl 4-nitrophenyl sulfide $(4\text{-}NO_2C_6H_4)SC_6H_5$	231.27	pa ye mcl pr (lig)	240^{25}	55			al, eth	B6⁴, 1694
11294	Phenyl isopentyl ketone or Isocaprophenone $i\text{-}C_5H_{11}COC_6H_5$	176.26		255-6, $145\text{-}7^{30}$	−2	$0.9623^{15/4}$	1.533^{20}	al, eth, ace, bz, chl	B7³, 1158
11295	Phenyl phosphinic acid $C_6H_5P(OH)_2$	142.09			82-3			w, al	B16³, 874
11296	2-Phenyl propionic acid, 2-hydroxy (D) or 2-Hydroxy hydratropic acid $C_6H_5C(OH)(CH_3)CO_2H$	166.18	$[\alpha]^{10.5/}_{D}$ +37.7 (al, c=3.5)		116-7			ace, bz	B10³, 560
11297	2-Phenyl propionic acid, 2-hydroxy (DL) or 2-Hydroxy hydratropic acid $C_6H_5C(OH)(CH_3)CO_2H$	166.18	nd pl (lig)		93-5			ace, bz	B10³, 560
11298	2-Phenyl propionic acid, 3-hydroxy (*d*) or 3-Hydroxy hydratropic acid $HOCH_2CH(C_6H_5)CO_2H$	166.18	nd (w, bz), pr (eth, w)		130			al, eth	B10², 158
11299	2-Phenyl propionic acid, 3-hydroxy (*dl*) or 3-Hydroxy hadratropic acid $HOCH_2CH(C_6H_5)CO_2H$	166.18	nd, pl (al, bz, w)	d	118			w, al, eth	B10³, 564
11300	2-Phenyl propionic acid, 3-hydroxy-(*l*) or 3-Hydroxy hydratropic acid $HOCH_2CH(C_6H_5)CO_2H$	166.18	pl (AcOEt), nd (w), $[\alpha]^{15/}_{D}$ -81.2 (w, c=1.5)					al, eth, AcOEt	B10³, 564
11301	2-Phenyl propionic acid, 2-hydroxy (L) or 2-Hydroxy hydratropic acid $C_6H_5C(OH)(CH_3)CO_2H$	166.18	nd (bz, w), $[\alpha]^{13.8/}_{D}$ -37.7 (al, c=3.4)		116-7			w, al, ace, bz	B10³, 560
11302	3-Phenyl propionic acid, 2-hydroxy (d) or 2-Hydroxy hydrocinnamic acid $C_6H_5CH_2CH(OH)CO_2H$	166.18	nd (w), $[\alpha]^{20/}_{D}$ +22.2 (w, c=2.2)		124-6			w, al, ace	B10¹, 554

No.	Name, Synonyms, and Formula	Mol. wt.	Color, crystalline form, specific rotation and λ_{max} (log ε)	b.p. °C	m.p. °C	Density	n_D	Solubility	Ref.
11303	3-Phenyl propionic acid, 2-hydroxy-(dl) or 2-Hydroxy hydrocinnamic acid $C_6H_5CH_2CH(OH)CO_2H$	166.18	cr (chl, bz) pr (w)	148-50[15]	98	...	...	al, eth, ace	B10[3], 554
11304	3-Phenyl propionic acid, 2-hydroxy (l) or 2-Hydroxy hydrocinnamic acid $C_6H_5CH_2CH(OH)CO_2H$	166.18	nd (w), $[\alpha]^{20/}{}_D$ -19.9 (w, c=3.2)	...	124.5	...	...	al, eth, ace	B10[3], 554
11305	3-Phenyl propionic acid, 3-hydroxy-(d) or 3-Hydroxy hydrocinnamic acid $C_6H_5CH(OH)CH_2CO_2H$	166.18	cr (bz), $[\alpha]^{18/}{}_D$ + 20.6 (MeOH, c=5)	...	116	...	...	al, eth	B10[3], 545
11306	3-Phenyl propionic acid, 3-hydroxy-(dl) or 3-Hydroxy hydrocinnamic acid $C_6H_5CH(OH)CH_2CO_2H$	166.18	pr (w)	...	96	...	...	al, ace, chl	B10[3], 546
11307	3-Phenyl propionic acid, 3-hydroxy (l) or 3-Hydroxy hydrocinnamic acid $C_6H_5CH(OH)CH_2CO_2H$	166.18	nd (bz), $[\alpha]^{18/}{}_D$ -19.8 (al, c=4.7)	...	115-6	...	...	...	B10[3], 546
11308	Phenyl sulfide or Diphenyl sulfide $(C_6H_5)_2S$	186.27	...	296, 145[8]	−25.9	$1.1136^{20/4}$	1.6334^{20}	**eth, bz**	B6[4], 1488
11309	Phenyl sulfide, 2,4′-diamino $(2\text{-}H_2NC_6H_4)S(C_6H_4NH_2\text{-}4)$	216.30	nd (w, al), pr (dil al)	...	62.5	...	...	al, eth, bz	B13[3], 1245
11310	Phenyl sulfide, 4,4′-diamino or bis-(4-Amino phenyl)sulfide $(4\text{-}H_2NC_6H_4)_2S$	216.30	nd (w)	...	108-9	...	...	al, eth, bz	B13[3], 1246
11311	Phenyl sulfide, 4,4′-dibromo $(4\text{-}BrC_6H_4)_2S$	344.06	...	268.5[40]	115	1.84	...	eth, chl	B6[4], 1651
11312	Phenyl sulfide, 4,4′-dichloro-2,2′-dinitro $(4\text{-}Cl_2\text{-}2\text{-}NO_2C_6H_3)_2S$	345.16	br-ye nd (90% aa)	...	149-50	...	...	bz	B6[2], 312
11313	Phenyl sulfide, 4,4′-dihydroxy or bis-(4-Hydroxy phenyl)sulfide $(4\text{-}HOC_6H_4)_2S$	218.27	mcl pr or lf (al)	...	151	...	...	al, eth	B6[3], 4455
11314	Phenyl sulfide, 2,2′-dimethoxy or *o*-Anisyl disulfide $(2\text{-}CH_3OC_6H_4)_2S$	246.32	lf (al)	252-3[10]	73	...	...	al, eth, bz	B6, 794
11315	Phenyl sulfide, 4,4′-dinitro or bis-(4-Nitro phenyl)sulfide $(4\text{-}O_2NC_6H_4)_2S$	276.27	og pl (aa)	...	160-1	...	...	...	B6[4], 1696
11316	Phenyl sulfide, 2,2′-dinitro or bis-(2-Nitro phenyl)sulfide $(2\text{-}O_2NC_6H_4)_2S$	276.27	gold-ye pl (aa)	sub d	122-3	...	...	...	B6[4], 1665
11317	Phenyl sulfide, 2,2′,4,4′-tetranitro or bis-(2,4-Dinitro phenyl)sulfide $[2,4\text{-}(NO_2)_2C_6H_3]_2S$	366.26	ye nd or pl (aa)	...	197	...	...	...	B6[4], 1748
11318	Phenyl sulfone $(C_6H_5)_2SO_2$	218.27	mcl pr (bz), pl (al), nd (w)	379, 232[18]	128-9	$1.252^{20/4}$	...	eth, bz	B6[4], 1490
11319	Diphenyl sulfone, 4,4′-diacetamide $(4\text{-}CH_3CONHC_6H_4)_2SO_2$	332.37	pa ye nd (eth, dil aa), lf (dil al)	...	282-5	...	...	al	B13[3], 1286
11320	Phenyl sulfone, 3,3′-diamino or bis-(3-Amino phenyl)sulfone $(3\text{-}H_2NC_6H_4)_2SO_2$	248.30	pr	...	168	...	...	w, al	B13[3], 984
11321	Phenyl sulfone, 4,4′-diamino or bis-(4-Amino phenyl)sulfone $(4\text{-}H_2NC_6H_4)_2SO_2$	248.30	lf (dil al)	...	178	...	...	al	B13[3], 1246
11322	Phenyl sulfone, 4,4′-dichloro or bis-(4-Chloro phenyl)sulfone $(4\text{-}ClC_6H_4)_2SO_2$	287.16	mcl	sub	148-9	...	...	...	B6[4], 1587
11323	Phenyl sulfone, 4,4′-diethoxy or bis-(4-Ethoxy phenyl)sulfone $(4\text{-}C_2H_5OC_6H_4)_2SO_2$	306.38	pl (al, aa)	...	163	...	...	al, eth	B6[3], 4459
11324	Phenyl sulfone, 4,4′-diethyl $(4\text{-}C_2H_5C_6H_4)_2SO_2$	274.38	...	...	102	...	...	eth, bz	B6, 475
11325	Phenyl sulfone, 2,2′-dihydroxy or bis-(2-Hydroxy phenyl)sulfone $(2\text{-}HOC_6H_4)_2SO_2$	250.27	nd (bz)	...	164-5 (179)	...	...	w, al, eth, aa	B6[3], 4278
11326	Phenyl sulfone, 3,3′-dihydroxy or bis-(3-Hydroxy phenyl)sulfone $(3\text{-}HOC_6H_4)_2SO_2$	250.27	190-1	...	...	...	...	al, eth	B6[3], 4365

No.	Name, Synonyms, and Formula	Mol. wt.	Color, crystalline form, specific rotation and λ_{max} (log ε)	b.p. °C	m.p. °C	Density	n_D	Solubility	Ref.
11327	Phenyl sulfone, 4,4′-dihydroxy or bis-(4-Hydroxy phenyl)sulfone $(4\text{-}HOC_6H_4)_2SO_2$	250.27	nd (w), rh bipym		240-1	1.3663^{15}		al, eth	$B6^3$, 4456
11328	Phenyl sulfone, 2,2′-dimethoxy or bis-(2-Methoxy phenyl)sulfone $(2\text{-}CH_3OC_6H_4)_2SO_2$	278.32	nd (bz)		157-8			al, aa	$B6^3$, 4278
11329	Phenyl sulfone, 4,4′-dimethoxy or bis-(4-Methoxy phenyl)sulfone $(4\text{-}CH_3OC_6H_4)_2SO_2$	278.32	lf or pr (al), nd (al-eth)	sub	130			al	$B6^3$, 4458
11330	Phenyl-o-tolyl sulfide $(2\text{-}CH_3C_6H_4)SC_6H_5$	200.30		304.5^{224}, 164^{12}		$1.0893^{20/4}$		ace, bz	$B6^4$, 2018
11331	Phenyl-m-tolyl sulfide $(3\text{-}CH_3H_4)SC_6H_5$	200.30		309.5, 164.5^{11}	−6.5	$1.0937^{15/4}$		ace, bz	$B6^4$, 2080
11332	Phenyl-*p*-tolyl sulfide $(4\text{-}CH_3C_6H_4)SC_6H_5$	200.30		317, 167.5^{11}	15.7	$1.0986^{25/4}$	1.6225^{25}	ace, bz	$B6^4$, 2169
11333	Phenyl-2-tolyl sulfone $(2\text{-}CH_3C_6H_4)SO_2C_6H_5$	232.30	pl (al)		81			al, eth, bz	$B6^4$, 2018
11334	Phenyl-4-tolyl sulfone $(4\text{-}CH_3C_6H_4)SO_2C_6H_5$	232.30	pl (al)		127-8				$B6^4$, 2171
11335	Phenyl sulfoxide or Diphenyl sulfoxide $(C_6H_5)_2SO$	202.27	pr (lig)	340d, 210^{15}	70.5			al, eth, bz, aa	$B6^4$, 1489
11336	Phenyl sulfoxide, 4-amino-4′-nitro $(4\text{-}H_2NC_6H_4)SO(C_6H_4NO_2\text{-}4)$	262.28	ye (al)		132			al	$B13^3$, 1226
11337	Phenyl sulfoxide, 4,4′-diamino or bis(4-Amino phenyl)sulfoxide $(4\text{-}H_2NC_6H_4)_2SO$	232.30	pr (w, al)		175d				$B13^3$, 1246
11338	Phenyl sulfoxide, 4,4′-dibromo $(4\text{-}BrC_6H_4)_2SO$	360.06			153-4			al, bz, chl	$B6^4$, 1651
11339	Phenyl sulfoxide, 4,4′-dichloro or bis-(4-Chloro phenyl)sulfoxide $(4\text{-}ClC_6H_4)_2SO$	271.16			143-4			chl	$B6^4$, 1587
11340	Phenyl sulfoxide, 4,4′-dihydroxy or bis-(4-Hydroxy phenyl)sulfone $(4\text{-}HOC_6H_4)_2SO$	234.27	nd (ace)		195			al, ace	$B6^3$, 4456
11341	Phenyl thiocyanate C_6H_5SCN	135.18		232-3, $71\text{-}3^{15}$		$1.155^{18/18}$		al, eth	$B6^4$, 1536
11342	Phenyl thiocyanate, 4-amino $(4\text{-}H_2NC_6H_4)SCN$	150.20	nd (w), cr (dil al)		57-8			al, eth, bz	$B13^3$, 1239
11343	Phenyl thiocyanate, 4-chloro $(4\text{-}ClC_6H_4)SCN$	169.63	nd (al)		35-6			al	$B6^4$, 1601
11344	Phenyl thiocyanate, 4-(dimethyl amino) $[4\text{-}(CH_3)_2NC_6H_4]SCN$	178.25	nd (lig, w, al)		73-4			eth	$B13^3$, 1251
11345	Phenyl trimethyl ammonium bromide $C_6H_5N^+(CH_3)_3Br^-$	216.12	hygm pr (al, al-eth)		213-4			w	$B12^3$, 254
11346	Phenyl trimethyl ammonium iodide $C_6H_5N^+(CH_3)_3I^-$	263.12	lf (al)		224			w, al, aa	$B12^3$, 254
11347	Phenyl vinyl ether $C_6H_5OCH{=}CH_2$	120.15		155-6		$0.9770^{20/4}$	1.5224^{20}	eth	$B6^4$, 561
11348	Phloretin or Dihydronaringenin $C_{15}H_{14}O_5$	274.27	nd (dil al), cr (dil ace)		262-4d			al, bz	$B8^3$, 4076
11349	Phlorhizin, dihydrate or Asebotin (a β-glucoside) $C_{21}H_{24}O_{10}\cdot 2H_2O$	472.45	nd (w), $[\alpha]^{25}{}_D$ -52 (96% al)		108 (+2w), 170d (anh)	1.4298		al, ace, Py	$B17^4$, 3042
11350	Phloroglucinol or 1,3,5-Trihydroxybenzene $1,3,5\text{-}(HO)_3C_6H_3$	126.11	lf or pl (w + 2)	sub	218-9 (anh)	1.46		al, eth, bz, Py	$B6^3$, 6301
11351	Phloroglucinol, 2-acetyl or 2,4,6-Trihydroxy acetophenone $2,4,6(HO)_3C_6H_2COCH_3$	168.15	nd (w + 1)		222-4 (anh)			al, eth, ace, aa	$B8^3$, 3386
11352	Phloroglucinol diacetate $C_{10}H_{10}O_5$	210.19	pl (w)		104			al, eth, ace	$B6^1$, 547
11353	Phloroglucinol dimethyl ether or 3,5-Dimethoxy phenol $3,5\text{-}(CH_3O)_2C_6H_3OH$	154.17	cr (bz-lig)	$172\text{-}5^{17}$	36-8			eth, bz	$B6^3$, 6305
11354	Phloroglucinol triacetate $1,3,5(CH_3CO_2)_3C_6H_3$	252.22	pr (w), nd (dil al)		105-6			al	$B6^3$, 6306
11355	Phloroglucinol, 2,4,6-trichloro or 2,4,6-Trichloro-1,3,5-trihydroxy benzene $C_6H_3Cl_3O_3$	229.45	cr (al)	sub	136			al	B6, 1104

No.	Name, Synonyms, and Formula	Mol. wt.	Color, crystalline form, specific rotation and λ_{max} (log ε)	b.p. °C	m.p. °C	Density	n_D	Solubility	Ref.
11356	Phloroglucinol, triethyl ether or 1,3,5-Triethoxy benzene $1,3,5-(C_2H_5O)_3C_6H_3$	210.27	cr (al, dil al)	175^{24}	43.5	....	....	al, eth	$B6^3$, 6306
11357	Phloroglucinol-trimethyl ether or 1,3,5-Trimethoxy benzene $1,3,5-(CH_3O)_3C_6H_3$	168.19	pr (al), lf (peth)	255.5	54-5	....	....	al, eth, bz	$B6^3$,6305
11358	Phorone or 2,6-Dimethyl, 2,5-heptadiene-4-one $CH_3C(CH_3)=CHCOCH=C(CH_3)_2$	138.21	ye gr pr	197.8	28	$0.8850^{20/4}$	1.4998^{20}	al, eth, ace	B^4, 3564
11359	Phosgene or Carbonyl chloride $COCl_2$	98.92	....	7.6	−118	$1.381^{20/4}$	....	bz, chl, aa	$B3^4$, 31
11360	Phosphine, bis trifluoromethyl $(F_3C)_2PH$	169.99	spont flam	1	−137	....	....	....	$B3^4$, 255
11361	Phosphine, bis(trifluoromethyl) chloro $(F_3C)_2PCl$	204.44	spont flamm	21	......	....	....	....	$B3^4$, 257
11362	Phosphine, bis(trifluoromethyl) cyano $(F_3C)_2PCN$	195.00	spont. flamm	48	......	....	1.3248^{20}	....	$B3^4$, 256
11363	Phosphine, bis(trifluoromethyl) iodo $(F_3C)_2PI$	295.89	....	73	......	....	1.403^{15}	....	$B3^4$, 257
11364	Phosphine, dichloro-2,4-(dimethyl phenyl) $[2,4-(CH_3)_2C_6H_3]PCl_2$	207.04	....	256-8	......	....	....	....	B16, 773
11365	Phosphine, dichloro-(2,5-dimethyl phenyl) $[2,5-(CH_3)_2C_6H_3]PCl_2$	207.04	....	253-4	−30	$1.25^{18/18}$	....	....	$B16^3$, 848
11366	Phosphine, dichloro(4-ethyl phenyl) $[4-C_2H_5C_6H_4]PCl_2$	207.04	....	250-2, $85^{0.4}$	......	$1.237^{20/4}$	1.584^{20}	....	$B16^3$, 848
11367	Phosphine, dichloro-phenyl $C_6H_5PCl_2$	178.99	....	224-6, $99\text{-}101^{11}$	......	$1.356^{20/4}$	1.6030^{20}	bz	$B16^3$, 847
11368	Phosphine, dichloro-trifluoromethyl F_3CPCl_2	170.89	....	37	......	....	....	....	$B3^4$, 258
11369	Phosphine, diethyl $(C_2H_5)_2PH$	90.11	....	85	......	....	....	....	$B4^3$, 1761
11370	Phosphine, diido-trifluoromethyl F_3CPI_2	353.79	ye fum	d 760, 73^{37}	......	....	1.630^{20}	....	$B3^4$, 258
11371	Phosphine, dimethyl $(CH_3)_2PH$	62.05	....	25	<1	....	....	al, eth	$B4^3$, 1759
11372	Phosphine, diphenyl-ethyl $(C_6H_5)_2PC_2H_5$	214.25	....	293	......	....	....	....	$B16^3$, 833
11373	Phosphine, ethyl $C_2H_5PH_2$	62.05	....	25	<1	....	....	....	$B4^3$, 1761
11374	Phosphine, methyl CH_3PH_2	48.02	gas	−14	......	....	....	eth	$B4^3$, 1759
11375	Phosphine, phenyl $C_6H_5PH_2$	110.10	....	160-1, 40^{10}	......	1.001^{15}	1.5796^{20}	....	$B16^3$, 831
11376	Phosphine, triethyl $(C_2H_5)_3P$	118.16	....	129^{762}	−88	$0.8006^{19/4}$	1.458^{15}	al, eth	$B4^3$, 1761
11377	Phosphine, triethyl, oxide $(C_2H_5)_3PO$	134.16	wh hyg nd	243	50	....	....	al, eth, w	$B4^3$, 1775
11378	Phosphine, triethyl, sulfide $(C_2H_5)_3PS$	150.22	cr (al)	....	94	....	....	w	$B4^3$, 1775
11379	Phosphine, trifluoromethyl F_3CPH_2	102.00	spont. flamm	−26.5	......	....	....	....	$B3^4$, 255
11380	Phosphine, trimethyl $(CH_3)_3P$	76.08	....	37.8	−85	<1	....	eth	$B4^3$, 1759
11381	Phosphine, triphenyl $(C_6H_5)_3P$	262.29	....	188^1	80	$1.0749^{80/4}$	1.6358^{80}	al, eth, bz, chl	$B16^3$, 833
11382	Phosphine, triphenyl, oxide $(C_6H_5)_3PO$	278.29	pr	>360	156-7	$1.2124^{23/4}$	....	al, bz	$B16^3$, 864
11383	Phosphine, tris (chloromethyl) $(ClCH_2)_3P$	179.41	....	100^7	......	1.414^{20}	....	al, eth, ace, bz	$B1^3$, 2608
11384	Phosphine, tris (trichloromethyl) $(Cl_3C)_3P$	386.08	....	....	53	....	....	....	$B3^4$, 259
11385	Phosphine, tris(trifluoromethyl) $(F_3C)_3P$	237.99	spont, flamm.	17.3	−112	....	....	....	$B3^4$, 255
11386	Phosphine, tris(trifluoromethyl) oxide $(F_3C)_3PO$	253.99	....	23.6	......	....	....	....	J1955, 574
11387	Phosphinic acid, diethyl $(C_2H_5)_2PO(OH)$	122.10	....	320, $134^{0.7}$	18.5	....	....	w, al, eth	$B4^3$, 1776
11388	Phosphinic acid, diethyl, anhydride $[(C_2H_5)_2PO]_2O$	226.19	....	188^{14}	......	$1.1053^{20/4}$	1.4647^{20}	....	Am73, 5466

No.	Name, Synonyms, and Formula	Mol. wt.	Color, crystalline form, specific rotation and λ_{max} (log ε)	b.p. °C	m.p. °C	Density	n_D	Solubility	Ref.
11389	Phosphinic acid, diethyl, ethyl ester $(C_2H_5)_2PO(OC_2H_5)$	150.16		95^{14}		$0.9908^{20/4}$	1.4337^{20}	al, eth	AM73, 5466
11390	Phosphinic acid, dimethyl $(CH_3)_2PO(OH)$	94.05	cr (bz)	377	92			w, al, eth	$B4^{3}$, 1776
11391	Phosphinic acid, dimethyl, anhydride $[(CH_3)_2PO]_2O$	170.09	nd (bz)	192^{15}	119-21				Am73, 5466
11392	Phosphinic acid, bis(trifluoromethyl) $(F_3C)_2PO(OH)$	201.99	visc liq	182, $137\text{-}8^{238}$		<1.9		w	J1955, 563
11393	Phosphinyl chloride dimethyl $(CH_3)_2POCl$	112.50	hyg nd (bz, peth)	204	67-8				Am73, 5466
11394	Phosphinyl chloride, diethyl $(C_2H_5)_2POCl$	140.55		104^{15}		$1.1394^{20/4}$	1.4647^{20}		$B4^{3}$, 1776
11395	Phosphonic acid, acetyl,diethyl ester $CH_3COPO(OC_2H_5)_2$	180.14		$114\text{-}5^{20}$			1.4200^{26}		$B2^{4}$, 440
11396	Phosphonic acid, benzyl,diethyl ester $C_6H_5CH_2PO(OC_2H_5)_2$	228.23		110^{2}			1.4930^{20}		$B16^{3}$, 889
11397	Phosphonic acid, carboxy methyl $HO_2CCH_2PO(OH)_2$	140.03			142-3				$B4^{2}$, 975
11398	Phosphonic acid, diethyl ester $(C_2H_5O)_2P(O)H$	138.10		$53.0\text{-}55^{6}$					$B1^{4}$, 1329
11399	Phosphonic acid, diisobutyl ester $(i\text{-}C_4H_9O)_2P(O)H$	194.21		235-6		$0.9759^{20/4}$			$B1^{4}$, 1596
11400	Phosphonic acid, diisopropyl ester $(i\text{-}C_3H_7O)_2\text{-}P(O)H$	166.16		$76\text{-}7^{10}$		$0.9972^{18/0}$			$B1^{4}$, 1475
11401	Phosphonic acid, diphenyl ester $(C_6H_5O)_2P(O)H$	234.19		$218\text{-}9^{26}$	12		1.5564^{25}		$B6^{4}$, 703
11402	Phosphonic acid, methyl, diisopropyl ester $CH_3PO(O\text{-}i\text{-}C_3H_7)_2$	180.18		66^{3}			$1.4120^{16.5}$		$B4^{3}$, 1778
11403	Phosphonic acid, methyl, diphenyl ester $CH_3P(O)(OC_6H_5)_2$	248.22		205^{13}	35	$1.2051^{20/0}$			$B6^{2}$, 164
11404	Phosphoric acid, methyl,ethyl ester $CH_3P(O)(OC_2H_5)(OH)$	124.08		$106\text{-}7^{0.1}$		1.1800^{20}	1.4258^{20}		CAS58, 5720
11405	Phosphonic acid, diethyl ester or Diethyl phosphate $(C_2H_5O)_2PO(OH)$	154.10	syr	203d		$1.186^{25/4}$	1.4170^{20}	eth	$B1^{4}$, 1339
11406	Phosphoric acid, dimethyl ester or Dimethyl phosphate $(CH_3O)_2PO(OH)$	126.05		172-6d		1.335^{25}	1.408^{25}	w, al, ace	$B1^{4}$, 1259
11407	Phosphoric acid, monoethyl ester $C_2H_5OPO(OH)_2$	126.05	hyg cr	d		$1.430^{25/4}$	1.427	w, al, eth, ace	$B1^{4}$, 1338
11408	Phosphoric acid, phenyl ester or Phenyl phosphate $C_6H_5OPO(OH)_2$	174.09	pl (chl), nd (w)		99.5			w, al, eth, bz, chl	$B6^{4}$, 708
11409	Phosphoric acid triamide, hexamethyl $[(CH_3)_2N]_3PO$	179.20		$98\text{-}100^{6}$		$1.024^{25/25}$	1.4579^{20}	al, eth	$B4^{4}$, 284
11410	Phosphoric acid, tributyl ester or Tributyl phosphate $(C_4H_9O)_3PO$	266.32		289, $160\text{-}2^{15}$		$0.9727^{25/4}$	1.4224^{25}	w, al, eth, bz	$B1^{4}$, 1531
11411	Phosphoric acid, tris (2,4-dimethylphenyl) ester $[2,4\text{-}(CH_3)_2C_6H_3O]_3PO$	410.46	glassy	232-5		$1.142^{38/4}$	1.5550^{20}	bz, hx	B6, 488
11412	Phosphoric acid, tris (-2,5-dimethylphenyl) ester $[2,5\text{-}(CH_3)_2C_6H_3O]_3PO$	410.46	cr (dil al)	$260\text{-}5^{8}$	78-81	1.197^{25}		eth, bz	$B6^{3}$, 1773
11413	Phosphoric acid, tris(2,6-dimethylphenyl) ester $[2,6\text{-}(CH_3)_2C_6H_3O]_3PO$	410.46	wax	$262\text{-}4^{6}$	136-8			bz	J1956, 3043
11414	Phosphoric acid, tris (3,4-dimethylphenyl) ester $[3,4\text{-}(CH_3)_2C_6H_3O]_3PO$	410.46	wax	$260\text{-}3^{7}$	71-2			bz	B6, 482
11415	Phosphonic acid, tris (3,5-dimethylphenyl) ester $[3,5\text{-}(CH_3)_2C_6H_3O]_3PO$	410.46	wax	290^{10}	46			aa	C31, 187
11416	Phosphoric acid, triisobutyl ester $(i\text{-}C_4H_9O)_3PO$	266.32		264, 138^{10}		$0.9681^{20/4}$	1.4193^{20}	w, al, eth, bz	$B1^{4}$, 1598
11417	Phosphoric acid, triethyl ester or Triethyl phosphate $(C_2H_5O)_3PO$	182.16		215-6, 103^{25}	−56.4	$1.0695^{20/4}$	1.4053^{20}	al, eth, bz	$B1^{4}$, 1339
11418	Phosphoric acid, trimethyl ester or Trimethyl phosphate $(CH_3O)_3PO$	140.08		197.2, 85^{26}	α −46 (st), β-62	$1.2144^{20/4}$	1.3967^{20}	w, eth	$B1^{4}$, 1259
11419	Phosphoric acid, tripentyl ester or Tripentyl phosphate $(C_5H_{11}O)_3PO$	308.40		225^{50}, 167^{5}		$0.9608^{20/4}$	1.4319^{20}	al, eth, to	$B1^{4}$, 1645
11420	Phosphoric acid, triphenyl ester or Triphenyl phosphate $(C_6H_5O)_3PO$	326.29	cr (abs al-lig), pr, (al), nd (eth-lig)	245^{11}	50-1	$1.2055^{50/4}$		al, eth, bz, chl	$B6^{4}$, 720

No.	Name, Synonyms, and Formula	Mol. wt.	Color, crystalline form, specific rotation and λ_{max} (log ε)	b.p. °C	m.p. °C	Density	n_D	Solubility	Ref.
11421	Phosphoric acid, tripropyl ester or Tripropyl phosphate $(C_3H_7O)_3PO$	224.24		252, 107.5^{5}		1.0121$^{20/4}$	1.4165^{20}	al, eth, to	B1^{4}, 1428
11422	Phosphoric acid, triisopropyl ester or Triisopropyl phosphate $(i\text{-}C_3H_7O)_3PO$	224.24		218-20, 95-6^{8}		0.9867$^{20/4}$	1.4057^{20}	al	B1^{4}, 1478
11423	Phosphoric acid, tri(2-tolyl) ester or Tri-*o*-cresyl phosphate $(2\text{-}CH_3C_6H_4O)_3PO$	368.37	col or pa ye	410, 283-5^{20}	11	1.955$^{20/4}$	1.5575^{20}	al, eth, aa, to	B6^{4}, 1979
11424	Phosphoric acid, tri(3-tolyl) ester or Tri-*m*-cresyl phosphate $(3\text{-}CH_3C_6H_4O)_3PO$	368.37	wax	260^{15}	25-6	1.150^{25}	1.5575^{20}	eth	B6^{4}, 2057
11425	Phosphoric acid, tri (4-tolyl) ester or Tri-*p*-cresyl phosphate T.P.C. $(4\text{-}CH_3C_6H_4O)_3PO$	368.37	nd (al), ta (eth)	244$^{3.5}$	77-8	1.247^{25}		al, eth, bz, chl, aa	B6^{4}, 2130
11426	Phosphoric acid, thiono, tri(2-tolyl) ester $(2\text{-}CH_3C_6H_4O)_3PS$	384.43	nd (al)	260-5^{1}	45-6			al, aa	B6^{4}, 1980
11427	Phosphonic acid, thiono, tri-3-tolyl ester $(3\text{-}CH_3C_6H_4O)_3PS$	384.43	nd (al)	270-2^{1}	40-1			aa	B6^{4}, 2058
11428	Phosphonic acid, thiono, tri-4-tolyl ester $(4\text{-}CH_3C_6H_4O)_3PS$	384.43	nd (al)		93-4				B6^{4}, 2132
11430	Phosphorous acid, diethyl ester or Diethyl phosphite $(C_2H_5O)_2P(OH)$	138.10		87^{20}		1.0720$^{20/4}$	1.4101^{20}	al, eth	B1^{4}, 1329
11431	Phosphorous acid, dimethyl ester or Dimethyl phosphite $(CH_3O)_2P(OH)$	110.05		170-1, 70^{25}		1.2004$^{20/0}$	1.4036^{20}	al, Py	B1^{4}, 1255
11432	Phosphorous acid, tributyl ester or Tributyl phosphite $(C_4H_9O)_3P$	250.32		122^{12}		0.9259$^{20/4}$	1.4321^{19}	al, eth	B1^{4}, 1527
11433	Phosphorous acid, triethyl ester or Triethyl phosphite $(C_2H_5O)_3P$	166.16		157.9, 49^{12}		0.9629$^{20/4}$	1.4127^{20}		B1^{4}, 1333
11434	Phosphorous acid, trimethyl ester or Trimethyl phosphite $(CH_3O)_3P$	124.08		111-2, 22^{23}		1.0520$^{20/0}$	1.4095^{20}	al, eth	B1^{4}, 1256
11435	Phosphorous acid, triphenyl ester or Triphenyl phosphite $(C_6H_5O)_3P$	310.29		360, 200-1^{5}	*ca* 25	1.1844$^{20/0}$	1.5900^{20}	al	B6^{4}, 695
11436	Phosphorous acid, tripropyl ester or Tripropyl phosphite $(C_3H_7O)_3P$	208.24		206-7, 92^{14}		0.9417^{20}	1.4282^{20}	al, eth	B1^{4}, 1426
11437	Phosphorous acid, trisopropyl ester or Trisopropyl phosphite $(i\text{-}C_3H_7O)_3P$	208.24		60-1^{10}		0.9687$^{18.5/0}$	1.4085^{25}	al, eth	B1^{4}, 1476
11438	Phosphorous acid, tris (2,2,2-trichloro ethyl) ester $(Cl_3CCH_2O)_3P$	476.16		263, 127-31$^{0.1}$			1.5174^{20}		B1^{4}, 1384
11439	Phosphorous acid, tri(2-tolyl) ester or Tri-*o*-cresyl phosphite $(2\text{-}CH_3C_6H_4O)_3P$	352.38	ye	238^{11}		1.1423$^{20/4}$	1.5740^{28}	eth	B6^{4}, 1977
11440	Phosphorous acid, tri(4-tolyl) ester or Tri-*p*-cresyl phosphite $(4\text{-}CH_3C_6H_4O)_3P$	352.37	pa ye	250-5^{10}		1.1313$^{25/25}$	1.5703^{28}	eth	B6^{4}, 2128
11441	Phthaladehyde or *o*-Phthalic aldehyde $1,2\text{-}C_6H_4(CHO)_2$	134.13	ye cr or nd (lig)		53-6			al, eth	B7^{3}, 3457
11442	Phthalamic acid or Phthalic acid monoamide $2\text{-}HO_2CC_6H_4CONH_2$	165.15	pr		148-9			al	B9^{3}, 4191
11443	Phthalamic acid, *N*-(1-naphthyl) or Alanap-1 $2\text{-}HO_2CC_6H_4CONH\text{-}\alpha\text{-}C_{10}H_7$	291.31			185				B12^{3}, 2876
11444	Phthalamide $1,2\ C_6H_4(CONH_2)_2$	164.16	cr	d	222				B9^{3}, 4197
11445	Phthalmide, *N,N,N'N'*-tetraethyl $1,2\text{-}C_6H_4[CON(C_2H_5)_2]_2$	276.38		204^{16}	36				B9^{3}, 4198
11446	Phthalazine $C_8H_6N_2$	130.15		175^{17}	90-1			w, al, bz	B23, 174
11447	Phthalazine, 1,2-dihydro-1-oxo or Phthalazone $C_8H_6N_2O$	146.15	nd (w), pr (sub)	337^{755}, sub	184-5			al, bz	B24^{2}, 70
11448	Phthalhydrazide or Phthalic acid hydrazide $C_8H_6N_2O_2$	162.15	mcl nd (w, dil al, aa)		342-4			aa	B24^{2}, 194
11449	Phthalic acid or 1,2 Benzene dicarboxylic acid $1,2\ C_6H_4(CO_2H)_2$	166.13	pl (w)	d	210-11d, 191 (sealed tube)	1.593		al	B9^{3}, 4094
11450	Phthalic acid, 3-amino $3\text{-}H_2NC_6H_3(CO_2H)_2\text{-}1,2$	181.15	nd		231-2				B14^{3}, 1393

No.	Name, Synonyms, and Formula	Mol. wt.	Color, crystalline form, specific rotation and λ_{max} (log ε)	b.p. °C	m.p. °C	Density	n_D	Solubility	Ref.
11451	Phthalic acid, 4-amino, dimethyl ester $4\text{-}H_2NC_6H_3(CO_2CH_3)_2\text{-}1,2$	209.20	pl (al, bz), pr (w)		54			al, chl, Py	B14^3, 1397
11452	Phthalic acid, 3-benzoyl or 2,3-Benzophenone dicarboxylic acid $3\text{-}C_6H_5COC_6H_3(CO_2H)_2\text{-}1,2$	270.26	pl, nd (w + 1)	d	140-1			al, bz	B10^3, 4007
11453	Phthalic acid, 4-benzoyl or 3,4-Benzophenone dicarboxylic acid $4\text{-}C_6H_5COC_6H_3(CO_2H)_2\text{-}1,2$	270.26	lf (xyl)		177			al	B10^3, 4010
11454	Phthalic acid, 3-bromo $3\text{-}BrC_6H_3(CO_2H)_2\text{-}1,2$	245.03	nd		188d			al, eth	B9^3, 4212
11455	Phthalic acid, 4-bromo $4\text{-}BrC_6H_3(CO_2H)_2\text{-}1,2$	245.03				173-5		al, eth	B9^3, 4213
11456	Phthalic acid, butyl ester *(mono)* or *mono*-Butyl phthalate $2\text{-}HO_2CC_6H_4CO_2C_4H_9$	222.24	pl (ace, al)		73-4			al, chl	B9^3, 4101
11457	Phthalic acid, *sec*-butyl ester *(mono) (d)* or *mono-sec*-Butyl phthalate $2\text{-}sec\text{-}C_4H_9O_2CC_6H_4CO_2H$	222.24	$[\alpha]^{20/}{}_D$ + 38.4 (al)		48			al, chl	B9^3, 4104
11458	Phthalic acid, *sec*-butyl ester *(mono) (dl)* or mono-*sec*-butyl phthalate $2\text{-}HO_2CC_6H_4CO_2\text{-}sec\text{-}C_4H_9$	222.24	cr (peth)		63			al, chl	B9^3, 4104
11459	Phthalic acid, 3-chloro $3\text{-}ClC_6H_3(CO_2H)_2\text{-}1,2$	200.58	nd (w)		186-7			al, eth	B9^3, 4202
11460	Pathalic acid, 4-chloro $4\text{-}ClC_6H_3(CO_2H)_2\text{-}1,2$	200.58	nd (dil al)		157			al, eth	B9^3, 4202
11461	Phthalic acid, dibenzyl ester or Dibenzyl phthalate $1,2\text{-}C_6H_4(CO_2CH_2C_6H_5)_2$	346.39	pr (al)	277^{15}	42-3			al, eth	B9^3, 4158
11462	Phthalic acid, dibutyl ester or Dibutyl phthalate $1,2\text{-}C_6H_4(CO_2C_4H_9)_2$	278.35		340, 206^{20}		$1.047^{20/20}$	1.4911^{20}	al, eth, bz	B9^3, 4102
11463	Phthalic acid, diisobutyl ester or Diisobutyl phthalate $1,2\text{-}C_6H_4(CO_2\text{-}i\text{-}C_4H_9)_2$	278.35		295-8, $182\text{-}4^{10}$		1.0490^{15}			B9^3, 4105
11464	Phthalic acid, 3,4-dichloro $3,4\text{-}Cl_2C_6H_2(CO_2H)_2\text{-}1,2$	235.02	pl (w)		195			al, eth	B9, 817
11465	Phthalic acid, 3,5-dichloro $3,5\text{-}Cl_2C_6H_2(CO_2H)_2\text{-}1,2$	235.02	nd, ta (aq HCl)	sub	164d			al, eth, ace	B9^3, 4204
11466	Phthalic acid, 3,6-dichloro $3,6\text{-}Cl_2C_6H_2(CO_2H)_2\text{-}1,2$	235.02	pl (w)		d 100			al, eth	B9^3, 4204
11467	Phthalic acid, 4,5-dichloro $4,5\text{-}Cl_2C_6H_2(CO_2H)_2\text{-}1,2$	235.02	nd (w)		*ca* 200d			eth	B9^3, 4205
11468	Phthalic acid, dicyclohexyl ester or Dicyclohexyl phthalate $1,2\text{-}C_6H_4(CO_2C_6H_{11})_2$	330.42	pr (al)		66	$1.383^{20/4}$	1.451^{20}	al, eth	B9^3, 4123
11469	Phthalic acid, di-(2-ethoxyethyl) ester $1,2\text{-}C_6H_4(CO_2CH_2CH_2OC_2H_5)_2$	310.35		345, $233\text{-}5^{23}$	34	1.1229^{21}			B9^2, 597
11470	Phthalic acid, diethyl ester or Diethylphthalate $1,2\text{-}C_6H_4(CO_2C_2H_5)_2$	222.24		298, 172^{12}		$1.1175^{20/4}$	1.5000^{21}	**al, eth**, ace, bz	B9^3, 4099
11471	Phthalic acid, 4,5-dimethoxy or *m*-Hemipic acid $4,5\text{-}(CH_3O)_2C_6H_2(CO_2H)_2\text{-}1,2$	226.19	nd (w), pr (w + 2)		174-5				B10^3, 2431
11472	Phthalic acid, di-(2-methoxyethyl) ester $1,2\text{-}C_6H_4(CO_2CH_2CH_2OCH_3)_2$	282.29		230^{10}		1.1708^{15}			B9^3, 4173
11473	Phthalic acid, dimethyl ester or Dimethyl phthalate $1,2\text{-}C_6H_4(CO_2CH_3)_2$	194.19	pa ye	283.8	0-2	$1.1905^{20/4}$	1.5138^{20}	**al, eth**, bz	B9^3, 4098
11474	Phthalic acid, diisopentyl ester or Diisopentylphthalate $1,2\text{-}C_6H_4(CO_2\text{-}i\text{-}C_5H_{11})_2$	306.40		330-8d, 225^{40}		$1.0220^{16/16}$	1.4871^{20}	al	B9^3, 4107
11475	Phthalic acid, diphenyl ester or Diphenyl phthalate $1,2\text{-}C_6H_4(CO_2C_6H_5)_2$	318.33	pr (al, lig)	$250\text{-}7^{14}$, sub	73				B9^3, 4157
11476	Phthalic acid, dipropyl ester or Dipropyl phthalate $1,2\text{-}C_6H_4(CO_2C_3H_7)_2$	250.29		304-5				al, eth	B9^3, 4101
11477	Phthalic acid, ethyl ester or Mono-ethylphthalate $2\text{-}HO_2CC_6H_4CO_2C_2H_5$	194.19		d	2	$1.1877^{22/4}$	1.509^{22}	al, eth	B9^3, 4099
11478	Phthalic acid, 3-hydroxy $3\text{-}HOC_6H_3(CO_2H)_2\text{-}1,2$	182.13	nd, pr (eth-peth, w)	sub	150d			al, eth	B10^3, 2189
11479	Phthalic acid, 4-hydroxy $4\text{-}HOC_6H_3(CO_2H)_2\text{-}1,2$	182.13	rosettes (w)		204-5d			al, eth	B10^3, 2190
11480	Phthalic acid mono amide or Phthalamic acid $2\text{-}HO_2CC_6H_4\text{-}CONH_2$	165.15	pr		148-9			al	B9^3, 4191
11481	Phthalic acid mononitrile $2\text{-}NCC_6H_4CO_2H$	147.13	nd (al)		187d			ace, al	B9^3, 4199

PHYSICAL CONSTANTS OF ORGANIC COMPOUNDS (Continued)

No.	Name, Synonyms, and Formula	Mol. wt.	Color, crystalline form, specific rotation and λ_{max} (log ε)	b.p. °C	m.p. °C	Density	n_D	Solubility	Ref.
11482	Phthalic acid mononitrile, monoamide or Phthalamonitrile $2\text{-}NCC_6H_4CONH_2$	146.15	nd (MeOH), cr (aa)		175			al, ace	B9^3, 4199
11483	Phthalic acid, 3-nitro $3\text{-}NO_2C_6H_3(CO_2H)_2\text{-}1,2$	211.13	pa ye pr (w)		218			al	B9^3, 4215
11484	Phthalic acid, 3-nitro, diethyl ester $3\text{-}NO_2C_6H_3(CO_2C_2H_5)_2\text{-}1,2$	267.24	pr (al), nd (peth)		46			al, eth	B9^3, 4216
11485	Phthalic acid, 4-nitro $4\text{-}NO_2 1,2\text{-}C_6H_3(CO_2H)_2$	211.13	pa ye nd (w, eth)		165-6			al, w	B9^3, 4234
11486	Phthalic acid, 4-nitro, dimethyl ester $4\text{-}NO_2C_6H_3(CO_2CH_3)_2\text{-}1,2$	239.19	cr (dil al)		69-71			al, MeOH	B9^2, 607
11487	Phthalic acid, 2-octyl ester *(D)* or 2-Octyl phthalate (mono) $2\text{-}HO_2CC_6H_4CO_2[CH(CH_3)C_6H_{13}]$	278.35	pr (peth), $[\alpha]^{20}{}_D$ + 48.7 (al)		75	$1.027^{72/4}$		al, bz, chl	B9^3, 4112
11488	Phthalic acid, 2-octyl ester-mono *(dl)* $2\text{-}HO_2CC_6H_4CO_2[CH(CH_3)C_6H_{13}]$	278.35	pr (peth)		55			al, ace, bz, chl	B9^1, 358
11489	Phthalic acid, 2-octyl ester-mono *(l)* $2\text{-}HO_2CC_6H_4CO_2[CH(CH_3)C_6H_{13}]$	278.35	pr (peth)		75			al, bz, chl	B9^3, 4112
11490	Phthalic acid, tetrabromo $C_6Br_4(CO_2H)_2\text{-}1,2$	481.72	nd (w)		266d (anh)				B9^1, 367
11491	Phthalic acid, tetrachloro $C_6Cl_4(CO_2H)_2\text{-}1,2$	303.91	pl (w)	2	250d (anh)			ace	B9^3, 4205
11492	Phthalic acid, tetrachloro, monoethyl ester $HO_2CC_6Cl_4(CO_2C_2H_5)$	331.97	pr (dil al)	d 250	94-5			al, eth	B9^3, 4206
11493	Phthalic anhydride $1,2\text{-}C_6H_4(CO)_2O$	148.12	wh nd (al, bz)	295 sub	131.6			al	B17^4, 6135
11494	Phthalic anhydride, 3-chloro $3\text{-}Cl\text{-}1,2\text{-}C_6H_3(CO)_2O$	182.56	nd (sub)	sub	124-5				B17^4, 6142
11495	Phthalic anhydride, 4-chloro $4\text{-}Cl\text{-}1,2\text{-}C_6H_3(CO)_2O$	182.56	pr	$294\text{-}5^{720}$	98.5			al, eth	B17^4, 6142
11496	Phthalic anhydride, 3,4-dichloro $3,4\text{-}Cl_2\text{-}1,2\text{-}C_6H_2(CO)_2O$	217.01	pl	329	121			al, chl	B17^4, 6142
11497	Phthalic anhydride, 3,5-dichloro $3,5\text{-}Cl_2\text{-}1,2\text{-}C_6H_2(CO)_2O$	217.01	nd		89			bz, chl	B17^4, 6143
11498	Phthalic anhydride, 3,6-dichloro $3,6\text{-}Cl_2\text{-}1,2\text{-}C_6H_2(CO)_2O$	217.01	nd	339	194.5				B17^4, 6143
11499	Phthalic anhydride, 4,5-dichloro $4,5\text{-}Cl_2\text{-}1,2\text{-}C_6H_2(CO)_2O$	217.01	ta or pr (to)	313	187-9			al, eth, to	B17^4, 6143
11500	Phthalic anhydride, 3-nitro $3\text{-}(NO_2)\text{-}1\text{-},2\text{-}C_6H_3(CO)_2O$	193.12	nd (aa, ace, al)		164			ace, al	B17^4, 6149
11501	Phthalic anhydride, tetra bromo $1,2\text{-}C_6Br_4(CO)_2O$	463.70	nd (aa-xyl)		279-81				B17^4, 6147
11502	Phthalic anhydride, tetrachloro $1,2\text{-}C_6Cl_4(CO)_2O$	285.90	pr, nd (sub)	sub	255-7				B17^4, 6144
11503	Phthalic anhydride, tetra iodo $1,2\text{-}C_6I_4(CO)_2O$	651.70	ye pr, nd (aa), nd (sub)	sub	327-8			al	B17^4, 6149
11504	Phthalimide $1,2\text{-}C_6H_4(CO)_2NH$	147.13	nd (w), pr (aa), lf (sub)		238			bz	B21^4, 5017
11505	Phthalimide, *N*-acetyl $1,2\text{-}C_6H_4(CO)_2NCOCH_3$	189.17	nd (bz), cr (al, aa)		135-6			eth, chl	B21^4, 5171
11506	Phthalimide, *N*-benzyl $1,2\text{-}C_6H_4(CO)_2NCH_2C_6H_5$	237.26	ye nd (al), cr (aa)		116	1.343^{18}		al	B21^4, 5053
11507	Phthalimide, *N*(2-bromo-isobutyl) $1,2\text{-}C_6H_4(CO)_2N[CH_2CBr(CH_3)_2]$	282.14	nd (al), lf (chl)		97			al, bz, chl	B21^2, 349
11508	Phthalimide, *N*(2-bromo ethyl) $1,2\text{-}C_6H_4(CO)_2NCH_2CH_2Br$	254.08	nd (w)		82-4			eth	B21^4, 5033
11509	Phthalimide, *N*-bromo methyl $1,2\text{-}C_6H_4(CO)_2NCH_2Br$	240.06	pr (chl, bz, aa)		151.5			ace	B21^4, 5110
11510	Phthalimide, *N*-(2-bromo propyl) $1,2\text{-}C_6H_4(CO)_2N(CH_2CHBrCH_3)$	268.11	nd (al, MeOH)		110-1			al, eth	B21^2, 349
11511	Phthalimide, *N*-(3-bromo propyl) $1,2\text{-}C_6H_4(CO)_2N[(CH_2)_3Br]$	268.11	nd (lig)		72-3			al, eth	B21^4, 5033
11512	Phthalimide, *N*-isobutyl $1,2\text{-}C_6H_4(CO)_2N\text{-}i\text{-}C_4H_9$	203.24		293-5	93				B21^4, 5036
11513	Phthalimide, 3,6-dihydroxy $3,6\text{-}(HO)_2\text{-}1,2\text{-}C_6H_2(CO)_2NH$	179.13	gr-ye nd (w + 3)		273-4			w	B21^1, 478

No.	Name, Synonyms, and Formula	Mol. wt.	Color, crystalline form, specific rotation and λ_{max} (log ε)	b.p. °C	m.p. °C	Density	n_D	Solubility	Ref.
11514	Phthalimide, *N*-ethyl $1,2\text{-}C_6H_4(CO)_2NC_2H_5$	175.19	nd (al)	285.6	79	...	...	eth	B21^4, 5032
11515	Phthalimide, *N*-(2-hydroxy ethyl) $1,2\text{-}C_6H_4(CO)_2NCH_2CH_2OH$	191.19	nd (al), lf (w)	...	129.5	...	...	...	B21^4, 5063
11516	Phthalimide, *N*-(hydroxy methyl) $1,2\text{-}C_6H_4(CO)_2NCH_2OH$	177.16	lf, pr (to)	...	141-2	...	...	to	B21^4, 5108
11517	Phthalimide, *N*-methyl $1,2\text{-}C_6H_4(CO)_2NCH_3$	161.16	nd (al), lf (sub)	285-7	134	...	...	...	B21^4, 5030
11518	Phthalimide, *N*-α-naphthyl $1,2\text{-}C_6H_4(CO)_2N\text{-}\alpha\text{-}C_{10}H_7$	273.29	pr, pl (al-aa)	...	180-1	...	...	...	B21^4, 5059
11519	Phthalimide, *N*-β-naphthyl $1,2\text{-}C_6H_4(CO)_2N\text{-}\beta\text{-}C_{10}H_7$	273.29	nd (aa)	...	216	...	...	al, aa	B21^4, 5060
11520	Phthalimide, 4-nitro $4\text{-}NO_2\text{-}1,2\text{-}C_6H_3(CO)_2NH$	192.13	col nd (w), ye lf (al-ace)	...	202	...	...	ace, aa	B21^2, 373
11521	Phthalimide, *N*-(4-nitrophenyl) $1,2\text{-}C_6H_4(CO)_2N\text{-}(C_6H_4NO_2\text{-}4)$	268.23	cr (aa)	...	271-2	...	...	...	B21^4, 5049
11522	Phthalimide, *N*-phenyl $1,2\text{-}C_6H_4(CO)_2NC_6H_5$	223.23	wh nd (al)	sub	210	...	...	chl	B21^4, 5047
11523	Phthalonitrile $1,2\text{-}C_6H_4(CN)_2$	128.13	nd (w, lig)	...	141	...	...	al, eth, ace, bz	B9^3, 4199
11524	Phthalonitrile, 3,6-dihydroxy $3,6\text{-}(HO)_2C_6H_2(CN)_2\text{-}1,2$	160.13	yesh lf (w + 2)	...	230	...	...	al, eth	B10^3, 2430
11525	Phthalyl alcohol or *o*-Xylylene glycol $1,2\text{-}C_6H_4(CH_2OH)_2$	138.17	pl (eth, peth)	...	65-6	...	...	w, al, eth	B6^3, 4587
11526	Phthalyl chloride $1,2\text{-}C_6H_4(COCl)_2$	203.02	...	281.1, 131-3^9	15-6	$1.4089^{20/4}$	1.5684^{20}	...	B9^3, 4190
11527	Phthalyl fluoride $1,2\text{-}C_6H_4(COF)_2$	170.12	...	227-8, 84^{13}	42-3	...	...	peth	B9^3, 4190
11528	Phthalide or α-Hydroxy-*o*-toluic acid lactone $C_8H_6O_2$	134.13	nd or pl (w)	290	75	$1.1636^{99/4}$	1.536^{99}	al, eth	B17^4, 4948
11529	Phthalide, 3-benzylidene *(trans)* $C_{15}H_{10}O_2$	222.24	mcl pr	...	108	...	...	al	B17^4, 5433
11530	Phthalide, 3,3-diphenyl or Phthalophenone $C_{20}H_{14}O_2$	286.33	lf (al)	235^{15}	120	...	...	...	B17^4, 5561
11531	Phthalide, 6-nitro $C_8H_5NO_4$	179.13	ye nd (al, aa)	...	145	...	...	al, eth, bz, aa	B17^4, 4953
11532	Phthalocyanine or Tetrabenzoporphyrazine $C_{32}H_{18}N_8$	514.55	grsh-bl mcl (guino-line)	sub 55d (vac)	...	...	...	...	J1938, 1151
11533	Physostigmine or Eserine $C_{15}H_{21}N_3O_2$	275.35	orth pr (eth, bz), $[\alpha]^{17}_D$ -82 (chl, c=1.3)	...	(i) 105-6 (st), (ii) 86-7 (unst)	...	...	al, eth, bz, chl	B23^2, 330
11534	Physostigmine, 2-hydroxybenzoate or Eserin salicylate $C_{22}H_{27}N_3O_5$	413.47	pr (al)	...	185-7	...	...	al, chl	B23^2, 332
11535	Physostigmine sulfate $(C_{15}H_{21}N_3O_2)_2.H_2SO_4$	648.77	dlq sc (ace-eth), $[\alpha]_D$ -130 (w)	...	140-2 (anh)	...	...	w, al, ace	B23^2, 332
11536	Phytadiene-*(d)* or 2,6,10,14-Tetramethyl-13,15-hexadeca-diene $C_{20}H_{38}$	278.52	$[\alpha]_D$ + 0.89 (undil)	186-8^{14}	...	$0.826^{0/4}$	...	peth, aa, MeOH	B1^4, 1078
11537	Phytol *(dl)* or 3,7,11,15-Tetramethyl-2-hexadecene-1-ol $C_{20}H_{40}O$	296.54	...	202-4^{10}, 140-1$^{0.03}$	...	$0.8497^{25/4}$	1.4595^{25}	...	B1^4, 2208
11538	Picein or *p*-Hydroxyacetophenone-*D*-glucoside $C_{14}H_{18}O_7$	298.29	nd (w + 1), nd (MeOH), $[\alpha]_D$ -86.5	...	195-6	...	...	al, eth, al	B17^4, 3013
11539	Picene or 1,2-Benzochrysene $C_{22}H_{14}$	278.35	lf, pl (xyl, Py, sub)	518-20, sub 300^2	367-9	...	...	...	B5^3, 2555
11540	α-Picoline or 2-Methyl pyridine $2\text{-}CH_3(C_5H_4N)$	93.13	...	128.8	−66.8	$0.9443^{20/4}$	1.4957^{20}	w, al, eth, ace	B20^4, 2679
11541	α-Picoline, 6-amino or 6-Amino-2-methylpyridine $6\text{-}NH_2\text{-}2\text{-}CH_3(C_5H_3N)$	108.14	hyg (lig)	208-9	41	...	...	w, al, eth, ace, bz	B22^4, 4133
11542	α-Picoline, 6-dimethyl amino or 6-Dimethyl amino-2-methyl pyridine $6\text{-}(CH_3)_2N\text{-}2\text{-}CH_3(C_5H_3N)$	136.20	...	198-200, 88^{15}	...	...	...	al, eth	B22^4, 4134

No.	Name, Synonyms, and Formula	Mol. wt.	Color, crystalline form, specific rotation and λ_{max} (log ε)	b.p. °C	m.p. °C	Density	n_D	Solubility	Ref.
11543	α-Picoline, 4-ethyl $4\text{-}C_2H_5\text{-}2\text{-}CH_3(C_5H_3N)$	121.18		179		$0.9130^{25/4}$		w, al, eth, ace, bz	$B20^4$, 2798
11544	α-Picoline, 5-ethyl or 5-ethyl-2-methyl pyridine $5\text{-}C_2H_5\text{-}2\text{-}CH_3(C_5H_3N)$	121.18		178.3, 65-6^{17}		$0.9219^{20/20}$	1.4971^{20}	al, eth, ace, bz	$B20^4$, 2798
11545	α-Picoline, 6-ethyl $6\text{-}C_2H_5\text{-}2\text{-}CH_3(C_5H_3N)$	121.18		160-1, 73-6^{12}		$0.9207^{25/4}$	1.4920^{25}	al, eth, ace	$B20^4$, 2803
11546	β-Picoline or 3-Methylpyridine $3\text{-}CH_3(C_5H_4N)$	93.13		144.1	−18.3	$0.9566^{20/4}$	1.5040^{20}	**w, al, eth,** ace	$B20^4$, 2710
11547	β-Picolene, 2-amino or 2, Amino-3-methyl pyridine $2\text{-}NH_2\text{-}3\text{-}CH_3(C_5H_3N)$	108.14	hyg	221.5^{748}, 95^{8}	33.5			w, al, eth, ace, bz	$B22^4$, 4154
11548	γ-Picoline or 4-Methylpyridine $4\text{-}CH_3(C_5H_4N)$	93.13		144.9	3.6	$0.9548^{20/4}$	1.5037^{20}	**w, al, eth,** ace	$B20^4$, 2732
11549	γ-Picoline, 2-amino or 2-Amino-4-methyl pyridine $2\text{-}NH_2\text{-}4\text{-}CH_3(C_5H_3N)$	108.14	lf or pl (lig)	115-7^{11}	100			w, al, eth, ace, bz	$B22^4$, 4172
11550	γ-Picoline, 3-Amino or 3-Amino-4-methyl pyridine $3\text{-}NH_2\text{-}4\text{-}CH_3(C_5H_3N)$	108.14	pr (bz-peth)	254^{735}	106			w, al, eth, ace, bz	$B22^4$, 4181
11551	γ-Picoline-2-ethyl or 2-Ethyl-4-methylpyridine $2\text{-}C_2H_5\text{-}4\text{-}CH_3(C_5H_3N)$	121.18		173-5^{748}		$0.9239^{20/0}$		al, eth, ace	$B20^4$, 2798
11552	γ-Picoline-3-ethyl or β-Collidine $3\text{-}C_2H_5\text{-}4\text{-}CH_3(C_5H_3N)$	121.18		198, 76^{12}		$0.9286^{17/4}$		al, eth, ace, chl	$B20^4$, 2804
11553	Picolinamide or 2-Pyridine carboxamide $2\text{-}(C_5H_4N)CONH_2$	122.13	mcl pr (w)		107-8			al, bz	$B22^4$, 311
11554	Picolinic acid or 2-Pyridine carboxylic acid $2\text{-}(C_5H_4N)CO_2H$	123.11	nd (w, al, bz)	sub	136-7			al, aa	$B22^4$, 303
11555	Picolinic acid, ethyl ester or Ethyl picolinate $2\text{-}(C_5H_4N)CO_2C_2H_5$	151.16	ye in air	243, 122^{13}	0-2	1.1194^{20}	1.5104^{20}	**w, al, eth**	$B22^4$, 308
11556	Picolinic acid, *N*-Methyl or Trigonelline $C_7H_7NO_2$	137.14	pr (aq al + 1 w)		218d (anh), 130 (+1w)			w	$B22^4$, 462
11557	Picolinic acid, 2,3,6-tetrahydro or Baikiain $2\text{-}(C_5H_8N)CO_2H$	127.14	pr (MeOH), $[\alpha]^{20}{}_D$ −201.6		274d			w	$B22^4$, 182
11558	Picolino nitrile $2\text{-}(C_5H_4N)CN$	104.11	nd or pr (eth)	222-7	29	$1.0810^{25/4}$	1.5242^{25}	w, al, eth, bz chl	$B22^4$, 320
11559	Picric acid or 2,4,6-Trinitro phenol $2,4,6\text{-}(NO_2)_3C_6H_2OH$	229 11	ye lf (w), pr (eth), pl (al)	sub exp >300	122-3		1.763	al, eth, ace, bz, aa, Py	$B6^4$, 1388
11560	Picrolonic acid $C_{10}H_8N_4O_5$	264.20	ye nd (al)		116-7, (d 125)				$B24^2$, 25
11561	Picropodophyllin $C_{22}H_{22}O_8$	414.41	col nd (al, bz)		228			al, eth, ace, bz, chl	$B19^4$, 5298
11562	Picrotoxin or Cocculin $C_{30}H_{34}O_{13}$	602.59	rh lf, $[\alpha]^{16}{}_D$ −29.3 (abs al, c=4)		203-4			al, Py	$B19^4$, 5245
11563	Pilocarpidine *(d)* $C_{10}H_{14}N_2O_2$	194.23	syr, $[\alpha]^{20}{}_D$ +81.3 (w, al)					w, al	$B27^2$, 694
11564	Pilocarpidine nitrate $C_{10}H_{14}N_2O_2.HNO_3$	257.25	pr (w), $[\alpha]^{20}{}_D$ +73.2 (w)		137			w, al	$B27^2$, 694
11565	Pilocarpine $C_{11}H_{16}N_2O_2$	208.26	nd $[\alpha]^{20}{}_D$ +100.5 (w)	260^{5}	34			w, al, chl	$B27^2$, 694
11566	Pilocarpine hydrochloride $C_{11}H_{16}N_2O_2.HCl$	244.72	hyg cr		204-5			w, al	$B27^2$, 695
11567	Pilocarpine, 2-hydroxybenzoate or Pilocarpine salicylate $C_{18}H_{22}N_2O_5$	346.38	nd or lf (al), $[\alpha]^{t}{}_D$ +63		120			w, al, eth	$B27^2$, 696
11568	Pilocarpine nitrate $C_{11}H_{16}N_2O_2.HNO_3$	271.27	wh pw or cr (al), $[\alpha]_D$ + 80 (w, c=4)		178			w	$B27^2$, 695

No.	Name, Synonyms, and Formula	Mol. wt.	Color, crystalline form, specific rotation and λ_{max} (log ε)	b.p. °C	m.p. °C	Density	n_D	Solubility	Ref.
11569	Pilocarpine, sulfate $2(C_{11}H_{16}N_2O_2)_2H_2SO_4$	514.59	hyg cr (al-eth), $[\alpha]^{20/}{}_D$ + 85 (w, c=7)		132			w, al	B27, 635
11570	Pimaric acid *(d)* or Dextropimaric acid $C_{20}H_{30}O_2$	302.46	orh (ace), pr (al) $[\alpha]^{20/}{}_D$ + 87.3 (chl)	282^{18}	218-9			al, eth, Py	$B9^3$, 2911
11571	Pimelic acid or Heptanedioic acid $HO_2C(CH_2)_5CO_2H$	160.17	pr (w)	272^{100} (sub), 212^{10}	106	1.329^{15}		w, al, eth	$B2^4$, 2003
11572	Pimelic acid, diethyl ester or Diethyl pimelate $C_2H_5O_2C(CH_2)_5CO_2C_2H_5$	216.28		$252\text{-}5^{745}$, $139\text{-}41^{15}$	−24	0.9945^{20}	1.4305^{20}	al, eth	$B2^4$, 2004
11573	Pimelic acid, dimethyl ester or Pimethyl pamelate $CH_3O_2C(CH_2)_5CO_2CH_3$	188.22		120^{10}, 80^1	−21	$1.0625^{20/4}$	1.4309^{20}	al, eth	$B2^4$, 2004
11574	Pimelic acid, ethyl ester, mono $HO_2C(CH_2)_5CO_2C_2H_5$	188.22	cr (eth)	182^{18}, 160^4	10		1.4415^{20}		$B2^3$, 1742
11575	Pimelic acid, 4-oxo $OC(CH_2CH_2CO_2H)_2$	174.15	rh pl (w)		143			al, w	$B3^3$, 1380
11576	Pimelonitrile or 1,5-Dicyanopentane $NC(CH_2)_5CN$	122.17		175^{14}	−31.4	0.949^{18}	1.4472^{20}	**al, eth, chl**	$B2^4$, 2006
11577	Pimelyl chloride or Heptanedioyl chloride $ClOC(CH_2)_5COCl$	197.06		137^{15}					$B2^4$, 2005
11578	Pinacolone or 3,3 Dimethyl-2-butanone $(CH_3)_3CCOCH_3$	100.16		106	−49.8	$0.8012^{25/4}$	1.3952^{20}	al, eth, ace	$B1^3$, 354
11580	Pinacolyl alcohol or 3,3-Dimethyl-2-butanol $(CH_3)_3CCH(OH)CH_3$	102.18		120.14	5.6	0.8122^{25}	1.4148^{20}	al, **eth**	$B1^4$, 1727
11581	Pinane *(d-cis)* $C_{10}H_{18}$	138.25	$[\alpha]^{20/}{}_D$ + 23.3 (undil)	169, 60.1^{18}	−53	$0.8560^{20/4}$	1.4629^{20}	eth, bz	$B5^4$, 318
11582	Pinane *(dl)* or Pincocamphane $C_{10}H_{18}$	138.25		164-5		$0.8551^{20/4}$	1.4609^{20}	eth, bz	$B5^1$, 48
11583	Pinane *(l, cis)* $C_{10}H_{18}$	138.25	$[\alpha]^{20}{}_D$ −47 (undil)	$167\text{-}8^{757}$		0.8556^{21}	1.4645^{20}	eth, bz	$B5^4$, 318
11584	α-Pinene *(dl)* $C_{10}H_{16}$	136.24	oil	156.2, 51.4^{20}	−55	$0.8582^{20/4}$	1.4658^{20}	**al, eth,** chl	$B5^4$, 456
11585	β-Pinene *(d)* or Nopinene $C_{10}H_{16}$	136.24	$[\alpha]_D$ + 28.6	164-6, 59.7^{20}		$0.8654^{20/4}$	1.4789^{20}	al, eth, bz, chl	$B5^4$, 456
11586	β-Pinene *(l)* $C_{10}H_{16}$	136.24	$[\alpha]^{25/}{}_D$ −21.5	164, 59.7^{20}		$0.8694^{20/4}$	1.4762^{20}	al, eth, bz, chl	$B5^4$, 457
11587	Pinic acid *(dl)*(α-form) or 3-Carboxy-2,2-dimethyl cyclobutyl acetic acid $C_9H_{14}O_4$	186.21	wh pr (w)	$214\text{-}6^9$	101-2	$1.0925^{109/4}$	1.4458^{109}		$B9^3$, 3852
11588	Pinic acid-*(dl)*(β form) $C_9H_{14}O_4$	186.21		$214\text{-}6^9$	68				$B9^3$, 3852
11589	Pinic acid *(dl)*(c form) $C_9H_{14}O_4$	186.21	pl	$214\text{-}6^9$	58				$B9^3$, 3852
11590	Pinic acid *(l)* $C_9H_{14}O_4$	186.21	nd (w)		135-6				$B9^3$, 3853
11591	Pinol *(dl)* or *(dl)*-Sobrerone-6,8-epoxy-1-*p*-menthene $C_{10}H_{16}O$	152.24	$[\alpha]_D$ −7.1 (ace)	183-4, $76\text{-}7^{14}$		$0.9515^{20/4}$	1.4695^{20}	al, eth	$B17^4$, 327
11592	Pinol hydrate *(trans, dl)* $C_{10}H_{16}O.H_2O$	170.25	pl or nd (w)	270-1, $157\text{-}8^{12}$	132			w, al, eth	$B6^3$, 4136
11593	Piperitenone or 3-Terpinolenone $C_{10}H_{14}O$	150.22	$[\alpha]_{546}$ −0.1	$120\text{-}2^{14}$		0.9774^{20}	1.5294^{20}	al, eth	$B7^3$, 559
11594	Piperazine or Hexahydropyrazine $C_4H_{10}N_2$	86.14	hyg pl or lf (al)	146	106		1.446^{111}	w, al	$B23^2$, 3
11595	Piperazine, *N*-benzyl $C_{11}H_{16}N_2$	176.26		$145\text{-}7^{12}$			1.5430^{28}	w, al, eth	Am72, 753
11596	Piperazine, *N,N′*-bis-(4 methoxybenzoyl) or *N,N′*-Dianisoylpiperazine $C_{20}H_{22}N_2O_4$	354.41	wh		192-4				Am56, 150
11597	Piperazine, *N,N′*-bis-(phenylacetyl) or *N,N′*-Di-(α-toluyl) piperazine *N,N′*-$(C_6H_5CH_2CO)_2(C_4H_8N_2)$	322.41	wh		150-1				Am56, 150

No.	Name, Synonyms, and Formula	Mol. wt.	Color, crystalline form, specific rotation and λ_{max} (log ε)	b.p. °C	m.p. °C	Density	n_D	Solubility	Ref.
11598	Piperazine, *N,N′*-bis-(3-phenyl propionyl) or *N,N′*-Bis-(hydrocinnanmyl)-piperazine $N,N'\text{-}(C_6H_5CH_2CH_2CO)_2(C_4H_8N_2)$	350.46	wh		122-3				Am56, 150
11599	Piperazine, dihydrobromide $C_4H_{10}N_2.2HBr$	247.96	wh nd		d			w	J1957, 1881
11600	Piperazine, dihydrochloride, monohydrate $C_4H_{10}N_2 \cdot 2HCl \cdot H_2O$	177.07	nd (dil al)		82-3			w	B23, 5
11601	Piperazine, *N,N′*-dimethyl or *N,N′*-Dimethylpiperazine $C_6H_{14}N_2$	114.19		131-2^{764}		0.8600$^{20/4}$	1.4474^{20}	w, al, eth	B23^2, 5
11602	Piperazine, 2,5-dimethyl *(cis)* $2,5\text{-}(CH_3)_2C_4H_8N_2$	114.19	rh bipym nd or pr (chl)	162 sub	114		1.4720^{20}	w, al, chl	B23^2, 21
11603	Piperazine, 2,5-dimethyl *(trans)* $2,5\text{-}(CH_3)_2(C_4H_8N_2)$	114.19	mcl pl or pr (bz, chl)	162 sub	118-9			w, al, chl	B23^2, 19
11604	Piperazine, 2,6-dimethyl *(cis)* $2,6\text{-}(CH_3)_2(C_4H_8N_2)$	114.19	lf or pl (bz)	162	111-3			w, al, peth, chl	B23^1, 8
11605	Piperazine, *N,N′*-dinitroso or *N,N′*-Dinitrosopiperazine $C_4H_8N_4O_2$	144.13	pa ye pl (w)		158			al	B23^1, 7
11606	Piperazine, *N,N′*-diphenyl $C_{16}H_{18}N_2$	238.33		230-5^{112}, dec 300	164				B23^2, 5
11607	Piperazine, *N*-ethyl $C_6H_{14}N_2$	114.19		154^{753}				w, al, eth	B23^2, 5
11608	Piperazine hexahydrate $C_4H_{10}N_2 \cdot 6H_2O$	194.23		125-3	44-5			w, al	CAS47, 10014
11609	Piperazine, *N*-methyl $N\text{-}CH_3(C_4H_9N_2)$	100.16		138			1.4378^{20}	w, al, eth	C51, 10538
11610	Piperazine, 2-methyl $2\text{-}CH_3(C_4H_9N_2)$	100.16	hyg lf (al)	155^{763}	62			w, al, eth, bz, chl	B23^2, 16
11611	Piperazine, *N*-phenyl $N\text{-}C_6H_5(C_4H_9N_2)$	162.23	pa ye oil	286.5, 156-7^{10}		1.0621$^{20/4}$	1.5875^{20}	**al, eth**	C49, 11662
11612	Piperazine, 1,2,4-trimethyl $1,2,4\text{-}(CH_3)_3C_4H_7N_2$	128.22		149-50			1.4433^{20}		
11613	*N*-Piperazine carboxaldehyde $N\text{-}OHC(C_4H_9N_2)$	114.15		94-7$^{0.5}$			1.5094$^{20}/_D$		
11614	*N*-Piperazine carboxylic acid, ethyl ester $C_7H_{14}N_2O_2$	158.20		237, 116-7^{12}			1.4760^{25}	w, al, eth	B23^2, 9
11615	*N,N′*-Piperazine dicarboxylic acid-diethyl ester $N,N'\text{-}(C_2H_5O_2C)_2C_4H_8N_2$	230.26	nd (hex)	315, 131-3^3	49			al, eth	
11616	2,5-Piperazinedione $C_4H_6N_2O_2$	114.10	ta or pl (w)		318-20d, sub 260			al	B24^2, 141
11617	Piperidine or hexahydropyridine $C_5H_{10}NH$	85.15		106, 17.7^{20}	−9	0.8606$^{20/4}$	1.4530^{20}	**w, al,** eth, ace, bz, chl	B20^4, 287
11618	Piperidine, *N*-acetyl $N\text{-}CH_3CO(NC_5H_{10})$	127.19		226-7, 109^{18}		1.011^9	1.4790^{25}	w, al	B20^4, 965
11619	Piperidine, 2-allyl $2\text{-}CH_2{=}CHCH_2(C_5H_9NH)$	125.21			170-1	0.8823$^{15/4}$			B20, 147
11620	Piperidine, *N*-benzoyl $C_6H_5CO(NC_5H_{10})$	189.26	tcl	320-1, 180^{15}	49			al, eth	B20^4, 972
11621	Piperidine, 4-benzoyl-*N*-methyl $C_{13}H_{17}NO$	203.28	nd	160-3^{13}, 130-7^2	35-7		1.5430^{23}	al, eth, ace, bz	B21^4, 3689
11622	Piperidine, 4-benzyl $4\text{-}C_6H_5CH_2(C_5H_9NH)$	175.27		279, 150-2^{17}	6-7	0.9972$^{20/0}$	1.5337^{25}	al, eth	B20^4, 3055
11623	Piperidine, *N*-butyl $C_4H_9(NC_5H_{10})$	141.26		175-7, 47-8^{20}		0.8245$^{20/4}$	1.4467^{20}		B20^4, 311
11624	Piperidine, 2-butyl-*(d)* $2\text{-}C_4H_9(C_5H_9NH)$	141.26	$[\alpha]^t_D$ +15.7			0.8512			B20, 127
11625	Piperidine, 2-butyl *(dl)* $2\text{-}C_4H_9(C_5H_9NH)$	141.26		191-3, 75^{14}		0.8529$^{15/4}$			B20^4, 1633
11626	Piperidine, 2-butyl *(l)* $2\text{-}C_4H_9(C_5H_9NH)$	141.26	$[\alpha]^{16}_D$ −18.7			0.8533			B20, 127
11627	Piperidine, 3-butyl $3\text{-}C_4H_9(C_5H_9NH)$	141.26		196-7				al, eth, chl	B20^4, 1635
11628	Piperidine, *N*-isobutyl $i\text{-}C_4H_9(NC_5H_{10})$	141.26		160-1, 87^{43}		0.8161^{25}	1.4382^{25}	w	B20^4, 314
11629	Piperidine, 2-isobutyl $2\text{-}i\text{-}C_4H_9(C_5H_9NH)$	141.26		181-2		0.8510$^{22/4}$	1.4553^{22}	al, eth	B20^4, 1636

No.	Name, Synonyms, and Formula	Mol. wt.	Color, crystalline form, specific rotation and λ_{max} (log ε)	b.p. °C	m.p. °C	Density	n_D	Solubility	Ref.
11630	Piperidine, *N*-*sec*-butyl *(dl)* $CH_3CH_2CH(CH_3)(NC_5H_{10})$	141.26	...	175-6	...	0.8378$^{20/4}$	1.4506^{20}	al, eth	B20^4, 313
11631	Piperidine, *N*-*sec*-butyl-*(l)* $CH_3CH_2CH(CH_3)(NC_5H_{10})$	141.26	$[\alpha]^{25/}{}_D$ −54.6	175	...	0.835$^{25/4}$	1.4486^{21}	al, eth	B20^4, 313
11632	Piperidine, *N*-*tert*-butyl $(CH_3)_3C(NC_5H_{10})$	141.26	...	166	...	0.8465$^{20/4}$	1.4532^{20}	al, eth	B20^4, 315
11633	Piperidine, 2,4-diethyl $2,4\text{-}(C_2H_5)_2(C_5H_8NH)$	141.26	...	174-9	...	0.8722^0	...	...	B20, 128
11634	Piperidine, 2,5-diethyl $2,5(C_2H_5)_2(C_5H_8NH)$	141.26	...	190, 100-5^{22}	...	0.8722^0	...	eth, chl	B20, 128
11635	Piperidine, 3,4-diethyl *(cis)* $3,4\text{-}(C_2H_5)_2(C_5H_8NH)$	141.26	$[\alpha]^{22/}{}_D$ + 26.0 (90% al, c=4.35)	70^{12}	...	...	...	...	B20^4, 1637
11636	Piperidine, 3,4-diethyl *(trans)* $3,4\text{-}(C_2H_5)_2(C_5H_8NH)$	141.26	...	193^{720}	...	...	...	...	B20^4, 1637
11637	Piperidine, *N*, 2-dimethyl *(d)* or *N*-Methyl-α-pipecoline $N\text{-}2\text{-}(CH_3)_2(NC_5H_9)$	113.20	$[\alpha]^{15/}{}_D$ + 68.8 (undil)	127	...	0.825^{16}	1.4395^{20}	...	B20^4, 1444
11638	Piperidine, *N*,2-dimethyl-*(dl)* $N,2\text{-}(CH_3)_2(NC_5H_9)$	113.20	...	127.5	...	0.824$^{15/4}$	1.4395^{20}	w, al, eth	B20^4, 1444
11639	Piperidine, *N*, 3-dimethyl-*(dl)* or *N*-Methyl)-β-pipecoline $N,3\text{-}(CH_3)_2(NC_5H_9)$	113.20	...	124-6	...	0.818^{15}	...	...	B20^4, 1499
11640	Piperidine, 2,3-dimethyl or α,β-Lupetidine $2,3\text{-}(CH_3)_2(C_5H_8NH)$	113.20	...	138-40^{720}	...	...	...	w, al	B20^4, 1573
11641	Piperidine, 2,4-dimethyl *(d)* or α,γ-Lupetidine $2,4\text{-}(CH_3)_2(C_5H_8NH)$	113.20	$[\alpha]_D$ + 23.2	140-2	...	0.845	...	...	B20, 108
11642	Piperidine, 2,4-dimethyl *(dl)* $2,4\text{-}(CH_3)_2(C_5H_8NH)$	113.20	...	140-2	...	0.8615^0	1.4366^{25}	al, eth	B20^4, 1574
11643	Piperidine, 2,5-dimethyl or α,β'-Lupetidine $2,5\text{-}(CH_3)_2(C_5H_8NH)$	113.20	...	138-40	...	...	1.4452^{25}	al	B20^4, 1574
11644	Piperidine, 2,6-dimethyl or α,α'-Lupetidine $2,6\text{-}(CH_3)_2(C_5H_8NH)$	113.20	...	127-8^{768}	...	0.8158$^{25/4}$	1.4377^{20}	w, al, eth	B20^4, 1579
11645	Piperidine, 3,3-dimethyl or β,β'-Lupetidine $3,3\text{-}(CH_3)_2(C_5H_8NH)$	113.20	...	137, 45-6^{20}	...	...	1.4452^{25}	al	B20^4, 1595
11646	Piperidine, 4,4-dimethyl or γ,γ-Lupetidine $4,4\text{-}(CH_3)_2(C_5H_8NH)$	113.20	...	145-6, 30-2^{12}	...	...	1.4489^{25}	w	B20^4, 1596
11647	Piperidine, *N*-(2,2-dimethyl propyl) or *N*-Neopentyl piperdine $CH_3C(CH_3)_2CH_2(NC_5H_{10})$	155.28	...	188	...	0.8608$^{20/4}$	1.4593^{20}	al, eth	B20^2, 13
11648	Piperidine, *N*-dodecyl $C_{12}H_{25}(NC_5H_{10})$	253.47	pa ye	161^5, 114-6$^{0.6}$	...	0.8378$^{20/4}$	1.4588^{20}	...	B20^4, 321
11649	Piperidine, *N*-ethyl $C_2H_5(NC_5H_{10})$	113.20	...	130.8	...	0.8237$^{20/4}$	1.4480^{20}	...	B20^4, 307
11650	Piperidine, 2-ethyl *(d)* $2\text{-}C_2H_5(C_5H_9NH)$	113.20	$[\alpha]_D$ + 17.1	...	142-4	0.8680^4	...	...	B20^4, 1564
11651	Piperidine, 2-ethyl *(dl)* $2\text{-}C_2H_5(C_5H_9NH)$	113.20	...	142-3,73-5^{52}	...	0.8650$^{0/0}$	1.4494^{21}	...	B20^1, 28
11652	Piperidine, 2-ethyl *(l)* $2\text{-}C_2H_5(C_5H_9NH)$	113.20	$[\alpha]_D$ −14.9	143^{720}	...	0.8680^4	1.4544^{20}	...	B20^4, 1564
11653	Piperidine, 3-ethyl *(dl)* $3\text{-}C_2H_5(C_5H_9NH)$	113.20	fum in air	152.6	...	0.8565$^{23/4}$	1.4531^{20}	ace	B20^4, 1567
11654	Piperidine, 3-ethyl *(l)* $3\text{-}C_2H_5(C_5H_9NH)$	113.20	$[\alpha]^{15/}{}_D$ −4.5	155	...	...	...	ace	B20^4, 1567
11655	Piperidine, 4-ethyl $4\text{-}C_2H_5(C_5H_9NH)$	113.20	...	156-8	...	0.8759^0	1.4503^{25}	...	B20^4, 1569
11656	Piperidine, *N*-heptyl $C_7H_{15}(NC_5H_{10})$	183.34	...	259.5, 100-3^9	...	0.8316^{20}	1.4531^{20}	...	B20^4, 317
11657	Piperidine, *N*-hexyl $C_6H_{13}(NC_5H_{10})$	169.31	...	219.2, 103-4^{20}	...	0.8292$^{20/4}$	1.4522^{20}	...	B20^4, 317
11658	Piperidine, 2-(1-hydroxyethyl) $2\text{-}[CH_3CH(OH)](C_5H_9NH)$	129.20	...	106-10^{18}	...	...	...	...	B21^4, 85
11659	Piperidine, *N*-(2-hydroxyethyl) $HOCH_2CH_2(NC_5H_{10})$	129.20	...	200-2^{742}, 90^{12}	...	0.9732$^{25/25}$	1.4749^{20}	w, al	B20^4, 387
11660	Piperidine, 2-(2-hydroxyethyl) $2\text{-}(HOCH_2CH_2)(C_5H_9NH)$	129.20	...	234.5, 145-6^{36}	39-40	1.01^{27}	...	w, al, eth	B21^4, 85
11661	Piperidine, 3-(2-hydroxyethyl) $3\text{-}(HOCH_2CH_2)(C_5H_9NH)$	129.20	...	121-3^6	...	1.0106$^{25/4}$	1.4888^{25}	w, al, eth	B21^4, 92

No.	Name, Synonyms, and Formula	Mol. wt.	Color, crystalline form, specific rotation and λ_{max} (log ε)	b.p. °C	m.p. °C	Density	n_D	Solubility	Ref.
11662	Piperidine, 4-(2-hydroxyethyl) $4-(HOCH_2CH_2)(C_5H_9NH)$	129.20	syr	227-8, $120-5^{15}$	132-3	$1.0059^{15/4}$	1.4907^{20}	w, al, eth	$B21^4$, 93
11663	Piperidine, *N*-methyl $CH_3(NC_5H_{10})$	99.18		107		$0.8159^{20/4}$	1.4355^{20}	w, **al, eth**	$B20^4$, 305
11664	Piperidine, 2-methyl *(d)* or *d*-α-Pipecoline $2-CH_3(C_5H_9NH)$	99.18	$[\alpha]^{22/}{}_D$ + 5.6 (al)	117^{745}			1.4459^{20}		$B20^4$, 1444
11665	Piperidine, 2-methyl *(dl)* or α-Pipecoline $2-CH_3(C_5H_9NH)$	99.18		$117-8^{747}$	−4.9	$0.8436^{24/4}$	1.4459^{20}	w, al, eth	$B20^4$, 1444
11666	Piperidine, 3-methyl *(dl)* or β-Pipecoline $3-CH_3(C_5H_9NH)$	99.18		$125-6^{763}$	0-5	$0.8446^{26/4}$	1.4470^{20}	w	$B20^4$,1499
11667	Piperidine, 3-methyl *(l)* $3-CH_3(C_5H_9NH)$	99.18	$[\alpha]^{25/}{}_D$ −4	124				w	$B20^4$, 1498
11668	Piperidine, 4-methyl or γ-Pipecoline $4-CH_3(C_5H_9NH)$	99.18		132-4		0.8674^0	1.4458^{20}	w	$B20^4$, 1511
11669	Piperidine, *N*-methyl-4-benzoyl $C_{13}H_{17}NO$	203.28		$160-3^{13}$	35-7		1.5430^{25}	al, eth, ace, bz	$B21^4$, 3689
11670	Piperidine, *N*-(2-methyl-2-pentyl) $C_{11}H_{23}N$	169.31		205-7		$0.8517^{20/4}$	1.4592^{20}	ace	$B20^2$, 14
11671	Piperidine, *N*-(3-methyl-3-pentyl) $C_{11}H_{23}N$	169.31		214^{752}		$0.8614^{20/4}$	1.4637^{20}		$B20^2$, 14
11672	Piperidine, *N*-nitroso $N-ON(C_5H_{10}N)$	114.15	pa ye	217^{721}, 109^{20}		$1.0631^{18.5/4}$	$1.4933^{18.5}$	w	$B20^4$, 1371
11673	Piperidine, *N*-nonyl $C_9H_{19}(NC_5H_{10})$	211.39		$135-7^{11}$		0.8313^{25}	1.4538^{25}		$B20^4$, 319
11674	Piperidine, *N*-octyl $C_8H_{17}(NC_5H_{10})$	197.36		$136-8^{13}$, 89^1		$0.8324^{20/4}$	1.4544^{20}		$B20^4$, 318
11675	Piperidine, *N*-pentyl $C_5H_{11}(NC_5H_{10})$	155.28		198.2, 80^8		0.8282^{20}	1.4498^{20}		$B20^4$, 315
11676	Piperidine, *N*-isopentyl $i-C_5H_{11}(NC_5H_{10})$	155.28		188-9, $76-9^{20}$				ace	$B20^4$, 316
11677	Piperidine, *N*-phenyl $C_6H_5(NC_5H_{10})$	161.25		$257-8^{752}$, 126.5^{15}				al, eth, bz, chl	$B20^4$, 339
11678	Piperidine, *N*-phenyl-4-dimethyl amino or Irenal $4-(CH_3)_2N(C_5H_9NC_6H_5)$	204.32	lf (bz)	$123-6^{0.5}$	47-8			bz	$B22^4$, 3751
11679	Piperidine, 2-propenyl *(d)* or β-Coniceine $2-(CH_2{=}CHCH_2)C_5H_9NH$	125.21	nd, $[\alpha]^{45/}{}_D$ + 49.9	168-9	39			al, eth	B20, 146
11680	Piperidine, 2-propenyl-*(dl)* $2-(CH_2{=}CHCH_2)C_5H_9NH$	125.21	nd	$168-70^{753}$	8	$0.8716^{25/4}$		al, eth	B20, 146
11681	Piperidine, 2-propenyl *(l)* $2-(CH_2{=}CHCH_2)C_5H_9NH$	125.21	nd, $[\alpha]^{15/}{}_D$ -50.5	168-9	39-40	$0.8672^{15/4}$			B20, 146
11682	Piperidine, *N*-propyl $C_3H_7(NC_5H_{10})$	127.23		151.2		0.8231^{20}	1.4446^{20}	w, al, eth	$B20^4$, 309
11683	Piperidine, 3-propyl *(d)* $3-C_3H_7(C_5H_9NH)$	127.23	$[\alpha]^{16/}{}_D$ + 5.9	174^{752}		$0.8517^{19/4}$		w, al	B20, 120
11684	Piperidine, 3-propyl *(dl)* $3-C_3H_7(C_5H_9NH)$	127.23		174^{758}		$0.8475^{26/4}$		al, w	B20, 119
11685	Piperidine, 3-propyl *(l)* $3-C_3H_7(C_5H_9NH)$	127.23	oil, $[\alpha]^{16/}{}_D$ -6.6	174^{752}		$0.8517^{19/4}$		w, al	B20, 120
11686	Piperidine, 4-propyl $4-C_3H_7(C_5H_9NH)$	127.23		172^{748}		0.864^{22}	1.4465^{23}	al	$B20^4$, 1616
11687	Piperidine, *N*-isopropyl $(CH_3)_2CH(NC_5H_{10})$	127.23		$149-50^{757}$		$0.8389^{20/4}$	1.4491^{20}	w	$B20^4$, 311
11688	Piperidine, 2-isopropyl *(dl)* $2-(CH_3)_2CH(C_5H_9NH)$	127.23		162		0.8668^0			$B20^4$, 1620
11689	Piperidine, 2-isopropyl *(l)* $2-(CH_3)_2CH(C_5H_9NH)$	127.23	$[\alpha]_D$ -13.1	161.5		0.8503^{19}			B20, 121
11690	Piperidine, 4-isopropyl $4-(CH_3)_2CH(C_5H_9NH)$	127.23		168-71, $66-70^{15}$					$B20^4$, 1620
11691	Piperidine, 2,2,6,6-tetramethyl $2,2,6,6-(CH_3)_4(C_5H_6NH)$	141.26		155-7		$0.8367^{16/4}$	1.4455^{20}	eth	$B20^4$, 1639
11692	Piperidine, 2,2,4-trimethyl $2,2,4-(CH_3)_3(C_5H_7NH)$	127.23		148		0.832^{15}	1.4458^{20}	al, eth	$B20^4$, 1624
11693	Piperidine, 2,3,6-trimethyl $2,3,6-(CH_3)_3(C_5H_7NH)$	127.23		36^5		$0.8302^{20/4}$	1.4434^{20}		$B20^4$, 1625
11694	Piperidine, 2,4,6-trimethyl $2,4,6-(CH_3)_3(C_5H_7NH)$	127.23		165-6		$0.8315^{19/4}$	1.4412^{20}	**al, eth**	$B20^4$, 1625
11695	*N*-Piperidine carboxaldehyde or *N*-Formylpiperidine $(C_5H_{10}N)CHO$	113.16	pa ye	222, $106-10^{17}$		$1.0205^{25/4}$	1.4700^{20}	**w, al, eth, bz, chl**, lig	$B20^4$, 964

No.	Name, Synonyms, and Formula	Mol. wt.	Color, crystalline form, specific rotation and λ_{max} (log ε)	b.p. °C	m.p. °C	Density	n_D	Solubility	Ref.
11696	4-Piperidine carboxylic acid or Hexahydroisonicotinic acid 4-$(HNC_5H_9)CO_2H$	129.16	nd (w)		ca 326			w	B22^4, 128
11697	4-Piperidine carboxylic acid, ethyl ester 4-$(HNC_5H_9)CO_2C_2H_5$	157.21	col oil	100-1^{10}			1.4591^{20}	w, al, eth, bz	B22^4, 128
11698	4-Piperidine carboxylic acid, methyl ester 4-$(HNC_5H_9)CO_2CH_3$	143.19	col oil	107-10^{22}			1.4635^{25}	w, al, eth, bz	B22^4, 128
11699	4-Piperidine carboxylic acid, *N*-methyl, methyl ester *N*-$CH_3(NC_5H_9)CO_2CH_3$-4	157.21		96-100^{20}			1.4539^{24}	w, al, eth, bz	B22^4, 131
11700	4-Piperidine carboxylic acid, *N*, methyl-4-phenyl,ethyl ester hydrochlo or Demerol hydrochloride *N*-$CH_3(NC_5H_8)(C_6H_5)(CO_2C_2H_5)$-4,4	283.80	cr (al)		186-9			w, ace	B22^4, 1004
11701	2,4-Piperidine dione, 3,3-diethyl or Piperidone $C_9H_{15}NO_2$	169.22	nd (w)		103-5			w, al	B21^4, 4614
11702	2-Piperidone or 5-Amino pentanoic acid lactone C_5H_9NO	99.13	hyg	256, 137^{14}	39-40			w, al, eth	B21^4, 3170
11703	2-Piperidone, 3-hydroxy *(dl)* $C_5H_9NO_2$	115.13	nd or pr (AcOEt)		141-2			w, eth	B21^4, 6022
11704	2-Piperidone, 5-hydroxy $C_5H_9NO_2$	115.13	(al, al-AcOEt)		145-6			w, al	B21^4, 6023
11705	4-Piperidone hydrochloride $C_5H_9NO.HCl$	135.59	cr (+ $^3/_2$ al, al-eth)		147-9			w	B21^4, 3183
11706	4-Piperidone, 2,3,6,6-tetramethyl or Triacetone amine $C_9H_{17}NO$	155.24	rh pl (moist eth), nd (eth)	205	58			w, al, eth	B21^4, 3278
11707	Piperine or 1-Piperylpiperidine $C_{17}H_{19}NO_3$	285.34	pr (AcOEt), pl or mcl pr (al), cr (bz-lig)		130-3			al, bz, chl, aa, Py	B20^4, 1341
11708	Piperitone *(d)* or 1-Methyl-4-isopropyl-3-cyclohexenone $C_{10}H_{16}O$	152.24	yesh in air, $[\alpha]^{20}{}_D$ +49.1 (undil)	222-30, 116-8^{20}		0.9344$^{20/4}$	1.4843^{20}	eth	B7^3, 326
11709	Piperitone- *(dl)* $C_{10}H_{16}O$	152.24		232-3, 113^{18}		0.9331$^{20/4}$	1.4845^{20}		B7^3, 326
11710	Piperitone *(l)* $C_{10}H_{16}O$	152.24	$[\alpha]^{20}{}_D$ -51.5 (undil)	235, 109-10^{15}		0.9324$^{20/4}$	1.4848^{20}		B7^3, 324
11711	Piperoin $C_{16}H_{12}O_6$	300.27			120			al, chl	B19^4, 5958
11712	Piperolidine *(dl)* or δ-Coniceine $C_8H_{15}N$	125.21		161.5^{750}, 65-7^{10}		0.9012$^{15/4}$	1.4748	al, eth	B20^4, 1989
11713	Piperolidine *(l)* $C_8H_{15}N$	125.21	$[\alpha]^{27}{}_D$ -7.9	158^{729}, 65-7^{18}		0.8976$^{20/4}$	1.4748^{20}	al, eth	B20^4, 1989
11714	Piperonal or 3,4-Methylenedioxy benzaldehyde $(3,4\text{-}CH_2O_2)C_6H_3CHO$	150.13	wh-ye (w)	263, 140^{15}	37			al, **eth**, ace	B19^4, 1649
11715	Piperonal oxime $(3,4\text{-}CH_2O_2)C_6H_3CH{=}NOH$	165.15			146			eth	B19^4, 1666
11716	Piperonyl alcohol or 3,4-Methylenedioxy benzyl alcohol $(3,4\text{-}CH_2O_2)C_6H_3CH_2OH$	152.15	nd (peth)	157^{16}	58			al, eth, bz, chl	B19^4, 734
11717	Piperonylic acid or 3,4-Methylene dioxy benzoic acid $(3,4\text{-}CH_2O_2)C_6H_3CO_2H$	166.13	nd (al), pr (sub)	sub	229-31				B19^4, 3493
11718	Piperonylic acid, ethyl ester or Ethyl piperonylate $(3,4\text{-}CH_2O_2)C_6H_3CO_2C_2H_5$	194.19	pr	285-6, 164-5^{11}	18.5			al, eth, peth	B19^2, 293
11719	Piperonylic acid, methyl ester or Methyl piperonylate $(3,4\text{-}CH_2O_2)C_6H_3CO_2CH_3$	180.16	nd or lf (peth)	273-4d	53			al, eth	B19^4, 3493
11720	Piperonyl chloride $(3,4\text{-}CH_2O_2)C_6H_3COCl$	184.58	cr	155^{25}	80				B19^4, 3497
11721	Piperylene or 1,3,-Pentadiene $CH_3CH{=}CHCH{=}CH_2$	68.12		42	−87.5	0.6760$^{20/4}$	1.4301^{20}	**al, eth, ace, bz**	B1^4, 994
11722	Pivaldehyde or 2,2-Dimethyl propionaldehyde $(CH_3)_3CCHO$	86.13		77-8	6	0.7923^{17}	1.3791^{20}	al, eth	B1^4, 3295
11723	Pivaldehyde oxime or Pivaldoxime $(CH_3)_3CCH{=}NOH$	101.15		65^{20}	48				B1^3, 2824
11724	Pivalamide, *N,N*-diethyl $(CH_3)_3CCON(C_2H_5)_2$	157.26		203		0.891^{15}		al, eth	B4^3, 211
11725	Pivalic acid or 2,2-Dimethyl propionic acid $(CH_3)_3CCO_2H$	102.13	nd	164, 70^{14}	35	0.905^{50}	1.3931$^{36.5}$	al, eth	B2^4, 908

PHYSICAL CONSTANTS OF ORGANIC COMPOUNDS (Continued)

No.	Name, Synonyms, and Formula	Mol. wt.	Color, crystalline form, specific rotation and λ_{max} (log ε)	b.p. °C	m.p. °C	Density	n_D	Solubility	Ref.
11726	Pivalic acid, chloro or 3-Chloro-2,2-dimethyl propionic acid $ClCH_2C(CH_3)_2CO_2H$	136.58		$108\text{-}12^{10}$	41-2				$B2^4$, 914
11728	Pivalic acid, ethyl ester or Ethyl pivalate $(CH_3)_3CCO_2C_2H_5$	130.19		118	−89.5	$0.856^{20/4}$	1.3906^{20}	al, eth	$B2^4$, 910
11729	Pivalic acid, methyl ester or Methyl pivalate $(CH_3)_3CCO_2CH_3$	116.16		101		$0.891^{0/4}$	1.3880^{20}	al, eth	$B2^4$, 909
11730	Pivalonitrile or *tert*-Butyl cyanide $(CH_3)_3CCN$	83.13		105-6	15-6	$0.7586^{25/4}$	1.3774^{20}		$B2^4$, 913
11731	Pivalophenone or *tert*-Butyl phenyl ketone $C_6H_5COC(CH_3)_3$	162.23		219-21, $97\text{-}8^{16}$		0.963^{26}	1.5086^{19}	ace	$B7^3$, 1125
11732	Pivalyl chloride or 2,2-Dimethyl propionyl chloride $(CH_3)_3CCOCl$	120.58		107, 48^{100}		1.003^{20}	1.4139^{20}	eth	$B2^4$, 912
11733	Pivalyl chloride-chloro $ClCH_2C(CH_3)_2COCl$	155.02		$85\text{-}6^{60}$			1.4539^{20}		$B2^4$, 914
11734	Podophyllotoxin $C_{22}H_{22}O_8$	414.41	$[\alpha]^{20}{}_D$ -132.7 ($CHCl_3$, 2%)		114-8			al, ace, bz, chl	$B19^4$, 5299
11735	Polyglycolid $[O\text{-}CH_2CO]_n$				223				$B19^1$, 679
11736	Polyporic acid $C_{18}H_{12}O_4$	292.29			310-2				$B8^3$, 3854
11737	Popalin or Salicinbenzoate (an α-glucoside) $C_{20}H_{22}O_8$	390.39	nd (w + 2), pr (al), $[\alpha]_D$ -2.0 (Py, c=5)		180			eth, aa	$B17^4$, 3298
11738	Porphin or Tetramethene tetrapyrrole $C_{20}H_{14}N_4$	310.36	red or og lf (chl-MeOH)	sub 300^{12}	darkens 360	1.336		bz, diox	$B26^2$, 228
11739	5-α-Pregnane $C_{21}H_{36}$	288.52	$[\alpha]^{19}{}_D$ ($CHCl_3$, c=1.69)		84-5				$B5^4$, 1215
11740	5α-Pregnane-20β-ol-3 one $C_{21}H_{34}O_2$	318.50	$[\alpha]_D$ (20 $CHCl_3$, c=1.2)		185				$B8^3$, 614
11741	5β-Pregnane or 17β-Ethyletiocholane $C_{21}H_{36}$	288.52	mcl sc or pl (MeOH), $[\alpha]^{19}{}_D$ + 21.2 (chl, c=2)		83.5	$1.032^{15/4}$		chl	$B5^4$, 1215
11742	5β-Pregnane-3α, 20 α-diol $C_{21}H_{36}O_2$	320.52	pl (ace), $[\alpha]^{20}{}_D$ + 27.4 (al, c=0.7)		243-4	1.15			$B6^3$, 4779
11743	5 β-Pregnane 3 α-20 β-diol $C_{21}H_{36}O_2$	320.52	cr (al), $[\alpha]^{20}{}_D$ + 10		244-6			al	$B6^3$, 4779
11744	5 β-Pregnane-3 β-20 α-diol $C_{21}H_{36}O_2$	320.52	cr (al, ace)		182			al	$B6^3$, 4778
11745	5 β-Pregnane-3 β-20 β-diol $C_{21}H_{36}O_2$	320.52	cr (AcOEt-peth, dil al)		174-6			AcOEt	$B6^3$, 4778
11746	5 β-Pregnane-3,20-dione $C_{21}H_{32}O_2$	316.48	nd (dil al), cr (dil ace)		123			al, eth, ace	$B7^3$, 3568
11747	5 β-Pregnan-3 α-ol-20-one $C_{21}H_{34}O_2$	318.50	nd (bz), cr (dil al)		149.5			al	$B8^3$, 618
11748	5 β-Pregnan-3 β-ol-20-one $C_{21}H_{34}O_2$	318.50	cr (dil al)		149			al	$B8^3$, 617
11749	5 β-Pregnan-20 α-ol-3-one $C_{21}H_{34}O_2$	318.50	pr (ace)		152				$B8^3$, 614
11750	5 β-Pregnan-20 β-ol-3-one $C_{21}H_{34}O_2$	318.50	cr (dil MeOH)		172			MeOH	$B8^3$, 614
11751	5-Pregnen-3 β-ol-20-one $C_{21}H_{32}O_2$	316.48	nd (dil al), $[\alpha]^{20}{}_D$ + 28 (al)		192				$B8^3$, 949

No.	Name, Synonyms, and Formula	Mol. wt.	Color, crystalline form, specific rotation and λ_{max} (log ε)	b.p. °C	m.p. °C	Density	n_D	Solubility	Ref.
11752	4-Pregnen-11 β, 17 α, 20 β, 21-tetrol-3-one or Reichstein's substance E $C_{21}H_{32}O_5$	364.48	cr (aq, ace)		ca 125d			al, ace	B8³, 4027
11753	4-Pregnen-17 α, 20 β, 21-triol-3-one $C_{21}H_{32}O_4$	348.48	cr (MeOH), $[\alpha]_D$ + 63 (diox, c=1)		190			diox, chl, MeOH	B8³, 3533
11754	Prehnitene or 1,2,3,4-Tetramethyl benzene 1,2,3,4-$(CH_3)_4C_6H_2$	134.22	cr (peth)	205, 79.4¹⁰	−6.2	$0.9052^{20/4}$	1.5203^{20}	al, eth, ace, bz	B5⁴, 1072
11755	Primeverose or 6-O-β-D-xylopyranosyl-α-D-glucose $C_{11}H_{20}O_{10}$	312.27	cr (MeOH), $[\alpha]^{20}{}_D$ + 23→-3.2 (w, c=5) (mut)		210			w, MeOH	B17⁴, 2447
11756	α-Progesterone or 17 α-Progesterone $C_{21}H_{30}O_2$	314.47	orh pr (dil al), $[\alpha]_D$ + 192		129-31	1.166^{21}			B7³, 3648
11757	β-Progesterone or 17 β-Progesterone $C_{21}H_{30}O_2$	314.47	nd (peth), $[\alpha]^{20}{}_D$ + 172 (diox, c=2)		121-2	1.171^{20}		al, ace, diox	B7³, 3648
11758	Progesterone dioxime $C_{21}H_{32}N_2O_2$	344.50	pl (dil al)		243			al	B7³, 3654
11759	Proline *(D)* or *d*-2-Pyrrolidene carboxylic acid $C_5H_9NO_2$	115.13	hyg pr (al-eth), $[\alpha]^{20}{}_D$ + 81.9 (w)		215-20d			w, al	B22⁴, 8
11760	Proline-*(DL)* $C_5H_9NO_2$	115.13	hyg nd (al-eth), cr (+w)		205d (anh)			w, al	B22⁴, 12
11761	Proline *(L)* $C_5H_9NO_2$	115.13	nd (al-eth), pr (w), $[\alpha]^{20}{}_D$ -80.9 (w, c=1)		220-2d			w	B22⁴, 8
11762	Proline, 4-hydroxy *(D-cis)* or *Allo*-4-hydroxyproline $C_5H_9NO_3$	131.13	nd (w + 1), $[\alpha]^{18}{}_D$ + 58.6 (w, p=5)		237-41			w	B22⁴, 2045
11763	Proline-4-hydroxy *(DL, cis)* or *Allo*-4-hydroxyproline $C_5H_9NO_3$	131.13	cr (w, dil al)		250			w	B22⁴, 2046
11764	Proline-4-hydroxy *(L, cis)* or *Allo*-4-hydroxyproline $C_5H_9NO_3$	131.13	nd (w + 1), $[\alpha]^{18}{}_D$ -58.1 (w, p=5.2)		238-41			w, al	B22⁴, 2046
11765	Proline-4-hydroxy *(D-trans)* or α-4-Hydroxyproline $C_5H_9NO_3$	131.13	lf (dil al), $[\alpha]^{21}{}_D$ + 75.2 (w)		274			w	B22⁴, 2047
11766	Proline-4-hydroxy *(DL,trans)* or α-4-Hydroxyproline $C_5H_9NO_3$	131.13	pl (MeOH)		270			w	B22⁴, 2049
11767	Proline-4-hydroxy-*(L,trans)* or α-4-Hydroxyproline $C_5H_9NO_3$	131.13	lf (dil al), pr (w), $[\alpha]_D$ -76.5 (w, c=2.5)		274			w	B22⁴, 2047
11768	Proline, 4-hydroxy, betaine *(d)* $C_7H_{13}NO_3$	159.19	pr (w + 1), $[\alpha]^{21}{}_D$ + 36		249d				B22⁴, 2053
11769	Prontosil $C_{12}H_{14}N_5O_2ClS$	327.79	og-red pw		248-51			al, ace, oils	B16³, 439
11770	Propadiene or Allene $CH_2=C=CH_2$	40.06	gas	−34.5	−136	0.787	1.4168	bz, peth	B1⁴, 966
11771	Propanal or Propionaldehyde CH_3CH_2CHO	58.08		48.8	−81	$0.8058^{20/4}$	1.3636^{20}	w, al, eth	B1⁴, 3165
11772	Propane $CH_3CH_2CH_3$	44.10		−42.1	−189.7	$0.5853^{-45/4}$	1.2898^{20}	al, eth, bz, chl	B1⁴, 176
11773	Propane, 1-amino or *n*-Propyl amine $CH_3CH_2CH_2NH_2$	59.11		47.8	−83	$0.7173^{20/4}$	1.3870^{20}	w, al, eth, ace, bz, chl	B4⁴, 464

No.	Name, Synonyms, and Formula	Mol. wt.	Color, crystalline form, specific rotation and λ_{max} (log ε)	b.p. °C	m.p. °C	Density	n_D	Solubility	Ref.
11774	Propane, 1-amino-2,2-dimethyl or Neopentyl amine $(CH_3)_3CCH_2NH_2$	87.16	....	$81-2^{741}$	......	$0.7455^{20/4}$	1.4023^{20}	eth	B4[4], 707
11775	Propane, 1-amino-3-dodecyloxy $C_{12}H_{25}O(CH_2)_3NH_2$	243.43	....	140^{1}	......	$0.8439^{20/4}$	1.4487^{20}	ace, bz, chl, MeOH	C55, 22136
11776	Propane, 1-amino-3-methoxy $CH_3O(CH_2)_3NH_2$	89.14	....	116-9	......	$0.8727^{20/4}$	1.4391^{20}	w, al, bz, MeOH, chl	B4[4], 1623
11777	Propane, 2-amino or Isopropyl amine $(CH_3)_2CHNH_2$	59.11	....	32.4	−95.2	0.6891^{20}	1.3742^{20}	w, **al**, **eth**, ace, bz, chl	B4[4], 504
11778	Propane, 2,2,bis(4-aminophenyl) $(4\text{-}H_2NC_6H_4)_2C(CH_3)_2$	226.32	....	$200-31^{5.5}$	128.2-9.6	....	....	....	B13[3], 495
11779	Propane, 1,3-bis (dimethylamino) $CH_2[CH_2N(CH_3)_2]_2$	130.23	....	144	......	$0.7837^{18/4}$	....	w, al, eth	B4[4], 1259
11780	Propane, 1,1-bis(4-hydroxyphenyl) or 4,4′-Propylidenediphenol $CH_3CH_2CH(C_6H_4OH\text{-}4)_2$	228.29	nd (w)	275^{20}	130	....	....	al, eth, aa	B6[3], 5457
11781	Propane, 2,2-bis(4-hydroxyphenyl) or Bis-phenol A -4,4′-isopropylidene diphenol $(4\text{-}HOC_6H_4)_2C(CH_3)_2$	228.29	pr (dil aa), nd (w)	$250-2^{13}$	152-3	....	....	al, eth, bz, aa	B6[3], 5459
11782	Propane, 1-bromo or *n*-Propylbromide $CH_3CH_2CH_2Br$	122.99	....	71	−110	$1.3537^{20/4}$	1.4343^{20}	al, eth, ace, bz	B1[4], 205
11783	Propane, 1-bromo-2-chloro $CH_3CHClCH_2Br$	157.44	....	118	......	$1.531^{10/4}$	1.4745^{20}	al, eth, ace, bz	B1[4], 212
11784	Propane, 1-bromo-3-chloro $Cl(CH_2)_3Br$	157.44	....	143.3, 32.4^{10}	−58.9	$1.5969^{20/4}$	1.4864^{20}	al, eth, chl	B1[4], 212
11785	Propane, 1-bromo 2,3-dimethyl or Neopentyl bromide $(CH_3)_3CCH_2Br$	151.05	....	106, 34.6^{100}	......	$1.1997^{20/4}$	1.4370^{20}	al, eth, ace, bz, chl	B1[4], 337
11786	Propane, 1-bromo-3-fluoro $F(CH_2)_3Br$	140.98	....	101.4	......	$1.542^{25/4}$	1.4290^{25}	al, eth, bz, chl	B1[4], 210
11787	Propane, 1-bromo-1-nitro $CH_3CH_2CH(Br)NO_2$	167.99	....	160-5, 82.5^{50}	......	....	....	al, eth	B1[3], 260
11788	Propane, 2-bromo or Isopropyl bromide $(CH_3)_2CHBr$	122.99	....	59.4	−89	$1.3140^{20/4}$	1.4251^{20}	**al**, **eth**, ace, bz, chl	B1[4], 208
11789	Propane, 2-bromo-1-chloro $CH_3CHBrCH_2Cl$	157.44	....	118^{756}	......	$1.537^{20/4}$	1.4795^{20}	al, eth, ace, bz, chl	B1[4], 212
11790	Propane, 2-bromo-2-chloro $(CH_3)_2CClBr$	157.44	....	$93-5^{745}$	......	1.474^{22}	1.4575^{20}	al, eth, ace, bz, chl	B1[4], 213
11791	Propane, 3-bromo-1,2-epoxy *(d)* or Epibromohydrin $BrCH_2CHCH_2$ (epoxy O bridge)	136.98	$[\alpha]^{16}{}_D$ + 45.4 (undil)	134 6^{50}	......	....	....	eth, bz, chl	B17[4], 22
11792	Propane, 3-bromo-1,2-epoxy *(dl)* or Epibromohydrin $BrCH_2CHCH_2$	136.98	....	138-40, $61-2^{50}$	......	1.615^{14}	1.4841^{20}	eth, bz, chl	B17[4] 22
11793	Propane, 2-bromo-2-nitro $(CH_3)_2CBrNO_2$	167.99	....	152^{745}, $73-5^{50}$	......	1.6562^{0}	....	al, eth	B1[4], 233
11794	Propane, 2-(bromomethyl)-1,2,3-tribromo $(BrCH_2)_3CBr$	373.71	....	$150-1^{14}$	25	$2.5595^{20/4}$	1.6246^{20}	eth, ace	B1[2], 91
11795	Propane, 1-chloro or *n*-Propyl chloride $CH_3CH_2CH_2Cl$	78.54	....	46.6	−122.8	$0.8909^{20/4}$	1.3879^{20}	**al**, **eth**, bz, chl	B1[4], 189
11796	Propane, 1-chloro-2,2, difluoro $CH_3CF_2CH_2Cl$	114.52	....	55	−56.2	$1.2001^{20/4}$	1.3520^{20}	eth, bz, chl	B1[4], 193
11797	Propane, 1-chloro-2,2-dimethyl or Neopentyl chloride $(CH_3)_3CCH_2Cl$	106.60	....	84.3	−20	$0.8660^{20/4}$	1.4044^{20}	al, eth, bz, chl	B1[4], 336
11798	Propane, 1-chloro-3-fluoro $F(CH_2)_3Cl$	96.53	....	79.5^{740}	......	....	1.3871^{25}	al, eth, bz, chl	B1[4], 193
11799	Propane, 1-chloro-3-iodo $ICH_2CH_2CH_2Cl$	204.44	....	170-2, 57.10	......	1.904^{20}	1.5472^{20}	eth, bz, chl	B1[4], 226
11800	Propane, 1-chloro-1-nitro $CH_3CH_2CH(Cl)NO_2$	123.54	....	141-3, 67^{56}	......	$1.209^{20/20}$	1.4251^{20}	al, eth	B1[4], 232
11801	Propane, 1-chloro-2-nitro $CH_3CH(NO_2)CH_2Cl$	123.54	....	172-3, 94^{46}	......	1.245^{22}	1.4432^{25}	al, eth, chl	B1[4], 233
11802	Propane, 1-chloro-3-nitro $ClCH_2CH_2CH_2NO_2$	123.54	....	197d, $115-6^{40}$	......	1.267^{20}	....	al, eth	B1[3], 259
11803	Propane, 1-chloro-2-phenyl *(dl)* $CH_3CH(C_6H_5)CH_2Cl$	154.64	....	85^{13}	......	1.0484^{20}	1.5245^{20}	....	B5[4], 992
11804	Propane, 1-chloro-3-phenyl $C_6H_5(CH_2)_3Cl$	154.64	....	219-20, 110^{21}	......	$1.056^{21/4}$	1.5160^{25}	....	B5[4], 980
11805	Propane, 2-chloro or Isopropyl chloride $(CH_3)_2CHCl$	78.54	....	35.7	−117.2	$0.8617^{20/4}$	1.3777^{20}	**al**, **eth**, bz, chl	B1[4], 191

PHYSICAL CONSTANTS OF ORGANIC COMPOUNDS (Continued)

No.	Name, Synonyms, and Formula	Mol. wt.	Color, crystalline form, specific rotation and λ_{max} (log ε)	b.p. °C	m.p. °C	Density	n_D	Solubility	Ref.
11806	Propane, 2-chloro-1-nitro $CH_3CHClCH_2NO_2$	123.54		$172, 75^{15}$		1.2361^{15}	1.4447^{20}	al, eth	$B1^4$, 232
11807	Propane, 2-chloro-2-nitro $(CH_3)_2CClNO_2$	123.54		$134d, 57^{50}$		1.230^{19}	1.4378^{19}	al, eth	$B1^4$, 233
11808	Propane, 2-chloro-1-phenyl *(d)* $CH_3CHClCH_2C_6H_5$	154.64	$[\alpha]_{4359}$, + 21.2	94^{17}		$1.038^{19/4}$	1.5198^{20}	ace, bz, chl	$B5^4$, 980
11809	Propane, 2-chloro-1-phenyl *(dl)* $CH_3CHClCH_2C_6H_5$	154.64		79^{10}		$1.0367^{17/4}$	1.5134^{20}	ace, bz, chl	$B5^4$, 980
11810	Propane, 2-chloro-1-Phenyl *(l)* $CH_3CHClCH_2C_6H_5$	154.64	$[\alpha]_{5461}$ -24.9	94^{17}		$1.038^{19/4}$	1.5198^{22}	ace, bz, chl	$B5^4$, 980
11811	Propane, 2-(chloromethyl)-1,1,2,3-tetrachloro $(ClCH_2)_2CClCHCl_2$	230.35		$226^{737}, 95^{9}$		$1.5686^{25/4}$	1.5165^{25}	al, bz, chl	$B1^3$, 321
11812	Propane, 2-(chloro methyl)-1,2,3-trichloro $(ClCH_2)_3CCl$	195.90		$209\text{-}10^{737}$ 87^{9}		$1.5036^{25/4}$	1.508^{20}	chl	$B1^3$, 321
11813	Propane, 3-chloro-1,2-epoxy *(dl)* or α- Epichlorohydrin $ClCH_2CHCH_2$ (O bridge)	92.53		116.5, $60\text{-}1^{100}$	−48	$1.1801^{20/4}$	1.4361^{20}	al, eth, bz	$B17^4$, 22
11814	Propane, 3-chloro-1,2,-epoxy *(l)* or α -Epichlorohydrin $ClCH_2CHCH_2$ (O bridge)	92.53	$[\alpha]^{18/}{}_D$ -25.6	$92\text{-}3^{360}$		1.2007		al, eth, bz	$B17^2$, 12
11815	Propane, 3-chloro-2-methyl-1,2-epoxy $ClCH_2C(CH_3)CH_2$ (O bridge)	106.55		$122, 51^{55}$		$1.1011^{20/4}$	1.4340^{20}	w, eth	$B17^4$, 47
11816	Propane, 1,2-diamino *(d)* or 1,2-Propanediamine $CH_3CH(NH_2)CH_2NH_2$	74.13	$[\alpha]^{25/}{}_D$ + 29.8	120.5		$0.8584^{25/4}$		w, chl	$B4^4$, 1255
11817	Propane, 1,2-diamino-*N,N'*-diacetyl or 1,2-Diacetamido propane $CH_3CH(NHCOCH_3)CH_2NHCOCH_3$	158.20	nd (bz)	190^{18}	138-9			w, al, chl	$B4^3$, 552
11818	Propane, 1,3-diamino or Trimethylene diamine $H_2N(CH_2)_3NH_2$	74.13		135.5		$0.884^{25/4}$	1.4600^{20}	w, al, eth	$B4^4$, 1258
11819	Propane, 1,1-dibromo or Propylidine bromide $CH_3CH_2CHBr_2$	201.89		$133.5, 28.4^{10}$		$1.982^{20/4}$	1.5100^{20}	al, eth, chl	$B1^4$, 215
11820	Propane, 1,1-dibromo-2,2-dimethyl or Neopentylidene bromide $(CH_3)_3CCHBr_2$	229.94		$180, 66^{20}$	14	$1.6695^{20/4}$	1.5047^{20}	eth, bz, chl	$B1^3$, 371
11821	Propane , 1,2-dibromo $CH_3CHBrCH_2Br$	201.89		$140, 35.7^{10}$	−55.2	$1.9324^{20/4}$	1.5201^{20}	al, eth, chl	$B1^4$, 215
11822	Propane, 1,2-dibromo-2-methyl $(CH_3)_2CBrCH_2Br$	215.92		$149\text{-}50, 61^{40}$	9-12	$1.759^{20/4}$	1.509	al, eth, bz, chl	$B1^4$, 298
11823	Propane, 1,3,-dibromo $Br(CH_2)_3Br$	201.89		$167.3, 56.6^{20}$	fr−34.2	$1.9822^{20/4}$	1.5232^{20}	al, eth	$B1^4$, 216
11824	Propane, 1,3-dibromo-2,2-dimethyl $(CH_3)_2C(CH_2Br)_2$	229.94		$185\text{-}90d, 72^{14}$		$1.6934^{20/4}$	1.5050^{20}	al, eth, bz, chl	$B1^4$, 337
11825	Propane, 2,2-dibromo $(CH_3)_2CBr_2$	201.89		$114\text{-}5^{740}$		$1.7825^{20/4}$		al, eth, chl	$B1^4$, 217
11826	Propane, 1,1-dichloro or Propylidene dichloride $CH_3CH_2CHCl_2$	112.99		88.1		$1.1321^{20/4}$	1.4289^{20}	al, eth, bz, chl	$B1^4$, 195
11827	Propane, 1,1-dichloro-2-methyl or Isobutylidene chloride $(CH_3)_2CHCHCl_2$	127.01		105-6		$1.0111^{12/12}$	1.4330^{15}	al, eth, bz, chl	$B1^4$, 292
11828	Propane, 1,2-dichloro $CH_3CHClCH_2Cl$	112.99		$96.4, -3.7^{10}$	−100.4	$1.1560^{20/4}$	1.4394^{20}	al, eth, bz, chl	$B1^4$, 195
11829	Propane, 1,2-dichloro-2-fluoro $CH_3CClFCH_2Cl$	130.98		88.6	−91.7	$1.2624^{20/4}$	1.4099^{20}	ace, bz	$B1^4$, 197
11830	Propane, 1,2-dichloro-2-methyl or Isobutylene chloride $(CH_3)_2CClCH_2Cl$	127.01		$108, 38\text{-}9^{70}$		$1.093^{20/4}$	1.4370^{20}	al, eth, ace, bz	$B1^4$, 292
11831	Propane, 1,3-dichloro $Cl(CH_2)_3Cl$	112.99		120.4, 14^{10}	−99.5	$1.1876^{20/4}$	1.4487^{20}	al, eth, bz, chl	$B1^4$, 196
11832	Propane, 1,3-dichloro-2-methyl $CH_3CH(CH_2Cl)_2$	127.01		134.6, 60^{49}		$1.1325^{25/4}$	1.4488^{25}	al, eth, bz	$B1^4$, 293
11833	Propane, 2,2-dichloro $(CH_3)_2CCl_2$	112.99		69.3	−33.8	$1.1136^{20/4}$	1.4148^{20}	al, eth, bz, chl	$B1^4$, 196
11834	Propane, 1,1-dicyclohexyl $CH_3CH_2CH(C_6H_{11})_2$	208.39		270-1		$0.8887^{23/0}$	1.485^{23}		$B5^4$, 350
11835	Propane, 1,2-dicyclohexyl $CH_3CH(C_6H_{11})CH_2C_6H_{11}$	208.39		272-3		$0.8725^{21/0}$	1.479^{21}		$B5^4$, 350
11836	Propane, 1,3-dicyclohexyl $CH_2(CH_2C_6H_{11})_2$	208.39		291-2	−17	$0.8752^{24/24}$	1.4736^{24}		$B5^4$, 349
11837	Propane, 1,3-dicyclohexyl-2-ethyl $C_2H_5CH(CH_2C_6H_{11})_2$	236.44		296		$0.8846^{21/0}$	1.483^{21}		$B5^4$, 363
11838	Propane, 1,3-dicylohexyl-2-methyl $CH_3CH(CH_2C_6H_{11})_2$	222.41		295.2	0.6	0.8715^{20}	1.4756^{20}		$B5^4$, 358

PHYSICAL CONSTANTS OF ORGANIC COMPOUNDS (Continued)

No.	Name, Synonyms, and Formula	Mol. wt.	Color, crystalline form, specific rotation and λ_{max} (log ε)	b.p. °C	m.p. °C	Density	n_D	Solubility	Ref.
11839	Propane, 2,2-dicyclohexyl $(C_6H_{11})_2C(CH_3)_2$	308.39	...	273-4	...	$0.9002^{23/0}$	1.490^{23}	...	$B5^4$, 350
11840	Propane, 2,2-di(ethyl sulfonyl) or Sulfonal $(CH_3)_2C(SO_2C_2H_5)_2$	228.32	mcl (w), pr (al)	300d	125.8	...	...	al, bz, chl	$B1^3$, 2754
11841	Propane, 1,1-difluoro-1,2,2,3,3-pentachloro $CClF_2CCl_2CCl_2H$	252.30	...	168.4	...	$1.73162^{20/4}$	1.46241^{20}	...	$B1^4$, 204
11842	Propane, 1,3-difluoro $F(CH_2)_3F$	80.08	...	41.6	...	$1.0057^{25/4}$	1.3190^{26}	bz	$B1^3$, 218
11843	Propane, 2,2-difluoro $(CH_3)_2CF_2$	80.08	...	−0.4	−104.8	$0.9205^{20/4}$	1.2904^{20}	...	$B1^4$, 187
11844	Propane, 1,2-diido CH_3CHICH_2I	295.89	...	...	...	$2.490^{18.5}$	...	al, eth	B1, 115
11845	Propane, 1,3-diido $I(CH_2)_3I$	295.89	...	227d, 110^{19}	fr−20	$2.5755^{20/4}$	1.6423^{20}	eth, chl	$B1^4$, 228
11846	Propane, 2,2-diido $(CH_3)_2CI_2$	295.89	...	173, 53^{10}	...	$2.5755^{20/4}$	1.651^{20}	eth, chl	$B1^3$, 255
11847	Propane, 2,2-dimethyl or Neopentane $(CH_3)_4C$	72.15	gas	9.5	−16.5	0.6135^{20}	1.3476^{0}	al, eth	$B1^4$, 333
11848	Propane, 1,1-dinitro $CH_3CH_2CH(NO_2)_2$	134.09	...	184	−42	1.2610^{25}	1.4339^{20}	...	$B1^4$, 234
11849	Propane, 1,3-dinitro $O_2N(CH_2)_3NO_2$	134.09	...	103^{1}	−21.4	$1.353^{26/4}$	1.4654^{20}	eth	$B1^4$, 234
11850	Propane, 2,2-dinitro $(CH_3)_2C(NO_2)_2$	134.09	...	185.5, $48\text{-}50^{2}$	53	1.30^{25}	...	...	$B1^4$, 234
11851	Propane, 1,2-diphenoxy $CH_3CH(OC_6H_5)CH_2OC_6H_5$	228.29	rh (MeOH)	$175\text{-}8^{12}$	32	$1.0748^{33.3/4}$	$1.5542^{33.3}$	al, eth, ace, bz, chl	$B6^2$, 151
11852	Propane, 1,3-diphenoxy $C_6H_5O(CH_2)_3OC_6H_5$	228.29	lf (al)	338-40, 160^{25}	61	...	...	al, eth	$B6^4$, 577
11853	Propane, 1,1-diphenyl $CH_3CH_2CH(C_6H_5)_2$	196.29	...	280	...	$0.9951^{14/4}$	1.5681^{14}	...	$B5^4$, 1920
11854	Propane, 1,2-diphenyl $CH_3CH(C_6H_5)CH_2C_6H_5$	196.29	...	$280\text{-}1^{758}$, 109^{2}	52	$0.9807^{20/4}$	1.5700^{20}	...	$B5^4$, 1914
11855	Propane, 1,3-diphenyl $C_6H_5(CH_2)_3C_6H_5$	196.29	...	300.3, 124^{2}	6	$1.007^{20/4}$	1.5760^{20}	...	$B5^4$, 1913
11856	Propane, 2,2-diphenyl $(C_6H_5)_2C(CH_3)_2$	196.29	...	282-3	29	...	...	...	$B5^4$, 1925
11857	Propane, 1,2-epoxy or Propylene oxide CH_3CHCH_2	58.08	...	34.3	...	$0.859^{0/4}$	1.3670^{20}	**w, al, eth**	$B17^4$, 16
11858	Propane, 1,2-epoxy-3-fluoro FCH_2CHCH_2	76.07	...	85.0-6.5	...	$1.090^{20/20}$	1.3730^{20}	...	$B17^4$, 20
11859	Propane, 1,2-epoxy-3 iodo or Epiiodohydrin ICH_2CHCH_2	183.98	...	160-2	...	1.982^{24}	...	al, eth	$B17^4$, 23
11860	Propane, 1,2-epoxy,3,3,3-trichloro or 3,3,3-Trichloro propylene oxide Cl_3CCHCH_2	161.42	...	149, $44\text{-}5^{13}$	...	$1.495^{20/4}$	1.4737^{25}	eth	$B17^4$, 22
11861	Propane, 1-ethoxy-3-phenoxy $C_6H_5O(CH_2)_3OC_2H_5$	180.25	oil	328-30	...	...	...	al, eth	B6, 147
11862	Propane, 1-fluoro or Propyl fluoride $CH_3CH_2CH_2F$	62.09	...	2.5	−159	$0.7956^{20/4}$	1.3115^{20}	al, eth	$B1^4$, 187
11863	Propane, 1,1,1,2,2,3,3-heptachloro $Cl_2CHCCl_2CCl_3$	285.21	amor	247-8, 132^{30}	29.4	$1.8048^{24/4}$	...	chl	$B1^4$, 205
11864	Propane, 1,1,1,2,3,3,3,-heptachloro $(Cl_3C)_2CHCl$	258.21	...	249, 93^{2}	11	$1.7921^{34/4}$	1.5427^{21}	chl	$B1^4$, 205
11865	Propane, heptafluoro-1-nitro $CF_3CF_2CF_2NO_2$	215.03	...	25	...	...	...	...	$B1^4$, 232
11866	Propane, heptafluoro-1-nitroso $CF_3CF_2CF_2NO$	199.03	deep bl	−12	−150	...	...	...	$B1^4$, 229
11867	Propane, 1,1,1,2,2,3-hexafluoro $CH_2FCF_2CF_3$	152.04	...	1.2	...	...	...	...	$B1^4$, 188
11868	Propane, 1-iodo or *n*-Propyl iodide $CH_3CH_2CH_2I$	169.99	...	102.4	−101	$1.7489^{20/4}$	1.5058^{20}	**al, eth, bz, chl**	$B1^4$, 222
11869	Propane, 1-iodo-2,2-dimethyl or Neopentyl iodide $(CH_3)_3CCH_2I$	198.05	...	127-9d, $42\text{-}4^{20}$	...	1.4940^{20}	1.4890^{20}	al, eth	$B1^4$, 338
11870	Propane, 2-iodo or Isopropyl iodide $(CH_3)_2CHI$	169.99	...	89.4	−90.1	$1.7033^{20/4}$	1.5028^{20}	**al, eth, bz, chl**	$B1^4$, 223
11871	Propane, 1-methoxy-3-phenoxy $C_6H_5O(CH_2)_3OCH_3$	166.22	oil	230-1	...	...	...	...	B6, 147

PHYSICAL CONSTANTS OF ORGANIC COMPOUNDS (Continued)

No.	Name, Synonyms, and Formula	Mol. wt.	Color, crystalline form, specific rotation and λ_{max} (log ε)	b.p. °C	m.p. °C	Density	n_D	Solubility	Ref.
11872	Propane, 2-methyl or Isobutane $(CH_3)_2CHCH_3$	58.12		−11.633	−159.4	0.549^{20}		al, eth, chl	$B1^{4}$, 282
11873	Propane, 1-nitro $CH_3CH_2CH_2NO_2$	89.09		130-1	−108	$1.0081^{24/4}$	1.4016^{20}	**al, eth** chl	$B1^{4}$, 229
11874	Propane, 2-nitro $(CH_3)_2CHNO_2$	89.09		120	−93	$0.9876^{20/4}$	1.3944^{20}	chl	$B1^{4}$, 230
11875	Propane, 3-nitro-1,1,1-trifluoro $F_3CCH_2CH_2NO_2$	143.07		135		$1.4203^{20/20}$	1.3549^{20}	eth	$B1^{4}$, 232
11876	Propane, 1,1,1,2,3-pentachloro $CH_2ClCHClCCl_3$	216.32		179-80			1.5130^{20}	al, eth	$B1^{4}$, 203
11877	Propane, 1,1,2,3,3-pentachloro $Cl_2CHCHClCHCl_2$	216.32		198-200, $78\text{-}100^{20}$		$1.6086^{14/4}$	1.5131^{17}	al, eth, chl	$B1^{4}$, 204
11878	Propane, 1,1,2,3,3-pentafluoro-1,2,3-trichloro $CF_2ClCClFC.ClF_2$	237.38		73.7	−72	$1.6631^{20/4}$	1.3512^{20}		$B1^{4}$, 200
11879	Propane, perchloro C_3Cl_8	319.66		$268\text{-}9^{734}$	160			al, eth	$B1^{4}$, 205
11880	Propane, perfluoro C_3F_8	188.02		−36	−183				$B1^{4}$, 189
11881	Propane, 1,1,1,2-tetrachloro $CH_3CHClCCl_3$	181.88		150, 37^{10}		$1.473^{20/4}$	1.4867^{20}	al, eth, chl	$B1^{4}$, 201
11882	Propane, 1,1,1,3-tetrachloro $CH_2ClCH_2CCl_3$	181.88		159, 59^{24}		$1.4510^{20/6}$	1.4825^{20}	al, eth, bz, chl	$B1^{4}$, 201
11883	Propane, 1,1,2,2-tetrachloro $CH_3CCl_2CHCl_2$	181.88		153		1.47^{12}	1.4850^{25}	**al, eth**, chl	$B1^{4}$, 201
11884	Propane, 1,1,2,3-tetrachloro $CH_2ClCHClCHCl_2$	181.88		179-80		1.513^{17}	1.5037^{17}	al, eth	$B1^{4}$, 202
11885	Propane, 1,2,2,3-tetrachloro $CH_2ClCCl_2CH_2Cl$	181.88		165, 51^{12}		1.500^{18}	1.4940^{18}	al, eth, chl	$B1^{4}$, 202
11886	Propane, 1,2,2,3-tetrachloro,-1,1,3,3,-tetrafluoro $CF_2ClCCl_2CF_2Cl$	253.84		112	−42.9	$1.7199^{20/4}$	1.39584^{20}		$B1^{4}$, 203
11887	Propane, 1,1,2-tribromo $CH_3CHBrCHBr_2$	280.78		200-1, 83^{6}		$2.3548^{20/4}$	1.5790^{20}	al, eth, chl, aa	$B1^{3}$, 251
11888	Propane, 1,2,2-tribromo $CH_3CBr_2CH_2Br$	280.78		190-1, 81^{20}		$2.2985^{20/4}$	1.5670^{20}	al, eth, chl, aa	$B1^{2}$, 77
11889	Propane, 1,2,3-tribromo $CH_2BrCHBrCH_2Br$	280.78		222.1, 98.5^{10}	16.9	$2.4209^{20/4}$	1.5862^{20}	al, eth	$B1^{4}$, 221
11890	Propane, 1,1,1-trichloro $CH_3CH_2CCl_3$	147.43		107-9		$1.287^{23/4}$		al, eth, chl	$B1^{4}$, 198
11891	Propane, 1,1,2- trichloro $CH_3CHClCHCl_2$	147.43		140		1.372^{15}		al, eth, chl	$B1^{4}$, 199
11892	Propane, 1,1,3- trichloro $CH_2ClCH_2CHCl_2$	147.43		145.5, 33.4^{10}	−59	$1.3557^{20/4}$	1.4718^{20}	al, eth, chl, aa	$B1^{4}$, 199
11893	Propane, 1,2,2-trichloro $CH_3CCl_2CH_2Cl$	147.43		123-5		1.318^{25}	1.4609^{20}	al, eth, chl	$B1^{4}$, 199
11894	Propane, 1,2,3- trichloro $CH_2ClCHClCH_2Cl$	147.43		156.8, 41.9^{10}	−14.7	$1.3889^{20/4}$	1.4852^{20}	al, eth, chl	$B1^{4}$, 199
11895	Propane, 1,1,1-triphenyl $CH_3CH_2C(C_6H_5)_3$	272.39	pr		51				$B5^{3}$, 2340
11896	1-Propanearsonic acid $CH_3CH_2CH_2AsO_3H_2$	168.02	nd (al), pl (w)		134-5			w, al	$B4^{3}$, 1824
11897	1-Propane boronic acid or *n*-Propylboric acid $CH_3CH_2CH_2B(OH)_2$	87.91	wh nd	d	107			w, al, eth	$B4^{3}$, 1964
11898	1,2-Propanediol or Propylene glycol $CH_3CH(OH)CH_2OH$	76.10		189, $96\text{-}8^{21}$		$1.0361^{20/4}$	1.4324^{20}	**w, al,** eth, bz	$B1^{4}$, 2468
11899	1,2-Propanediol, 3-amino or 3-Amino-1,2-propylene glycol $H_2NCH_2CH(OH)CH_2OH$	91.11		265d, 145^{9}		$1.1752^{20/4}$	1.4910^{25}	w, al	$B4^{4}$, 1865
11900	1,2-Propanediol carbonate or 1,2-Propylene glycol carbonate $C_4H_6O_3$	102.09		240, 110^{10}	−48.8	$1.2041^{20/4}$	1.4189^{20}	w, al, ace, bz	$B19^{4}$, 1564
11901	1,2-Propanediol, 3-chloro $ClCH_2CH(OH)CH_2OH$	110.54	yesh liq	213d, 116^{11}		$1.326^{18/15}$	1.4809^{20}	w, al, eth	$B1^{4}$, 2484
11902	1,2-Propanediol, 3-chloro, diacetate $ClCH_2CH(OCOCH_3)CH_2OCOCH_3$	194.61		245, 116^{22}		$1.199^{25/4}$	1.4407^{20}	al, eth	$B2^{3}$, 313
11903	1,2, Propanediol, 3-chloro-2-methyl $ClCH_2C(OH)(CH_3)CH_2OH$	124.57		$114\text{-}7^{20}$		$1.2362^{20/4}$	1.4748^{20}	**w, al, eth**	$B1^{4}$, 2536
11904	1,2-Propanediol diacetate $CH_3CH(O_2CCH_3)CH_2O_2CCH_3$	160.17		$190\text{-}1^{762}$		$1.059^{20/4}$	1.4173^{20}	w, al, eth	$B2^{4}$, 220

No.	Name, Synonyms, and Formula	Mol. wt.	Color, crystalline form, specific rotation and λ_{max} (log ε)	b.p. °C	m.p. °C	Density	n_D	Solubility	Ref.
11905	1,2-Propane diol, 3-(diethyl amino) .... $(C_2H_5)_2NCH_2CH(OH)CH_2OH$	147.22	syr	233-5	......	....	....	w, al, eth, chl	$B4^3$, 840
11906	1,2-Propanediol, 3-(dimethylamino) .... $(CH_3)_2NCH_2CH(OH)CH_2OH$	119.16	....	220^{749}	......	....	....	w, al, eth, chl	$B4^3$, 840
11907	1,2-Propanediol, 3-mercapto or 1-Thioglycerol........ $HSCH_2CH(OH)CH_2OH$	108.16	visc	$100\text{-}1^{1}$	......	1.2455^{20}	1.5268^{20}	w, al, ace	$B1^3$, 2339
11908	1,2-Propanediol, 2-methyl or Isobutylene glycol .... $(CH_3)_2C(OH)CH_2OH$	90.12	....	176, $79\text{-}80^{12}$	......	$1.0024^{20/4}$	1.5268^{20}	w, al, eth	$B1^4$, 2533
11909	1,2-Propanediol-sulfite .... $C_3H_6O_3S$	122.14	....	175, 85^{25}	<−60	$1.2960^{20/4}$	1.4370^{20}	w, al, eth, ace, bz	$B1^4$, 2476
11910	1,3-Propanediol or Trimethylene glycol .... $HOCH_2CH_2CH_2OH$	76.10	....	213.5, 110^{12}	......	$1.0597^{20/4}$	1.4398^{20}	**w, al,** eth	$B1^4$, 2493
11911	1,3-Propanedioll, 2-amino-2-ethyl .... $C_2H_5C(CH_2OH)_2NH_2$	119.16	ye	$143\text{-}5^{10}$	37-8	$1.099^{20/4}$	1.490^{20}	**w**	$B4^4$, 1883
11912	1,3-Propanediol, 2-amino-2-(hydroxymethyl) .... $(HOCH_2)_3CNH_2$	121.14	nd or fl (MeOH)	$219\text{-}20^{18}$	170-1	....	....	w	$B4^4$, 1903
11913	1,3,-Propanediol, 2-amino-2-methyl .... $(HOCH_2)_2C(CH_3)NH_2$	105.14	....	151.2^{10}	109-11	....	....	w, al	$B4^4$, 1881
11914	1,3-Propanediol, 2-butyl-2-ethyl or 3,3-Di(hydroxymethyl) heptane .... $(HOCH_2)_2C(C_2H_5)(C_4H_9)$	160.26	wh	262, 123^{15}	43.8	$0.929^{20/20}$	1.4587^{25}	al	$B1^4$, 2611
11915	1,3-Propanediol, 2-chloro .... $(HOCH_2)_2CHCl$	110.54	....	146^{18}	......	$1.3219^{20/4}$	1.4831^{20}	w, al, ace	$B1^4$, 2499
11916	1,3 Propanediol diacetate .... $CH_3CO_2CH_2CH_2CH_2O_2CCH_3$	160.17	....	209-10, 84.5^{10}	......	1.070^{14}	1.4192	w, al	$B2^4$, 221
11917	1,3-Propanediol, 2,2-diethyl .... $(HOCH_2)_2C(C_2H_5)_2$	132.20	wh	240-1, 131^{13}	61-2	$1.052^{20/20}$	....	w, al, eth	$B1^4$, 2589
11918	1,3-Propanediol, 2,2-dimethyl or Neopentylene glycol.... $(HOCH_2)_2C(CH_3)_2$	104.15	nd (bz)	206^{747}, $120\text{-}30^{15}$	130	....	....	w, al, eth	$B1^4$, 2551
11919	1,3-Propanediol, 2,2-dinitro .... $(HOCH_2)_2C(NO_2)_2$	166.09	wh pl (bz)	....	142	....	....	....	$B1^4$, 2501
11920	1,3-Propanediol, 2-ethyl-2-hydroxymethyl or TMP Trimethylol propane .... $CH_3CH_2C(CH_2OH)_3$	134.18	wh pw or pl	160^{5}	58	....	....	**w, al**	$B1^4$, 2786
11921	1,3-Propanediol, 2-ethyl-2-nitro .... $(HOCH_2)_2C(NO_2)C_2H_5$	149.15	nd (w)	d	57-8	....	....	w, al, eth	$B1^4$, 2550
11922	1,3-Propanediol, 2-hydroxymethyl-2-nitro .... $(HOCH_2)_3CNO_2$	151.12	nd or pr	d	165	....	....	w, al, eth	$B1^4$, 2777
11923	1,3,Propanediol, 2-methyl-2-(hydroxy methyl) or Trimethylol ethane .... $CH_3C(CH_2OH)_3$	120.15	wh pw or nd (al)	$135\text{-}7^{15}$	204	....	....	**w, al**	$B1^4$, 2780
11924	1,3-Propanediol, 2-methyl-2-nitro .... $(HOCH_2)_2C(CH_3)NO_2$	135.12	mcl	d	149-50	....	....	w, al	$B1^4$, 2537
11925	1,3-Propane diol, 2-methyl-2-propyl or 2,2-Bis-(hydroxymethyl) pentane........ $(HOCH_2)_2C(CH_3)C_3H_7$	132.20	cr (hx)	234	62-3	....	....	w	$B1^4$, 2585
11926	1,2 Propanedithiol .... $CH_3CH(SH)CH_2SH$	108.22	....	$41\text{-}3^{11}$	......	$1.08^{20/4}$	1.532^{20}	chl	$B1^4$, 2492
11927	1,3-Propanedithol .... $HSCH_2CH_2CH_2SH$	108.22	....	172.9, 63^{15}	−79	$1.0783^{20/4}$	1.5392^{20}	**al, eth, bz, chl**	$B1^4$, 2503
11928	1-Propane phosphonic acid or *n*-Propylphosphonic acid $CH_3CH_2CH_2PO(OH)_2$	124.08	pl (bz)	d	23	....	....	w, al, eth	$B4^3$, 1781
11929	2-Propane phosphonic acid or Isopropyl phosphonic acid $(CH_3)_2CHPO_3H_2$	124.08	pl (bz)	d	74-5	....	....	w, al, eth	$B4^3$, 1781
11930	1-Propane sulfonamide .... $C_3H_7SO_2NH_2$	123.17	pr (eth), cr (bz)	....	53.5	....	....	w, al	$B4^4$, 39
11931	1-Propanesulfonyl chloride .... $C_3H_7SO_2Cl$	142.60	....	180d, 77^{12}	......	$1.2826^{15/4}$	1.452^{20}	....	$B4^4$, 39
11932	2-Proponesulfonamide or Isopropyl sulfonamide........ $(CH_3)_2CHSO_2NH_2$	123.17	cr (eth-peth)	....	67.5	....	....	w, al, eth	$B4^4$, 42
11933	2-Propanesulfonic acid or Isopropyl sulfonic acid........ $(CH_3)_2CHSO_3H$	124.15	....	$159^{1\ 4}$	−37	1.187^{25}	1.4332^{20}	w	$B4^4$, 42
11934	1,1,2,3-Propane tetra carboxylic acid, tetra ethyl ester .... $C_2H_5O_2CCH_2CH(CO_2C_2H_5)CH(CO_2C_2H_5)_2$	332.35	....	$203\text{-}4^{18}$	......	$1.1184^{20/4}$	1.4395^{20}	al	$B2^3$, 2077
11935	1,1,3,3-Propane tetra carboxylic acid, tetra ethyl ester or Ethyl methane dimalonate........ $[(C_2H_5O_2C)_2CH]_2CH_2$	332.35	....	300-10d, 195^{8}	−30	1.116^{20}	1.4398^{20}	al	$B2^4$, 2417

No.	Name, Synonyms, and Formula	Mol. wt.	Color, crystalline form, specific rotation and λ_{max} (log ε)	b.p. °C	m.p. °C	Density	n_D	Solubility	Ref.
11936	1-Propanethiol or *n*-Propyl mercaptan ... $CH_3CH_2CH_2SH$	76.16	...	67-8	−113.3	0.8411^{20}	1.4380^{20}	al, eth, ace, bz	$B1^4$, 1449
11937	2- Propane thiol or Isopropyl mercaptan ... $(CH_3)_2CHSH$	76.16	...	52.5	−130.5	$0.8143^{20/4}$	1.4255^{20}	**al, eth,** ace	$B1^4$, 1498
11938	1,2,3-Propane tricarboxylic acid or Tricarballic acid ... $(HO_2CCH_2)_2CHCO_2H$	176.13	orh (w, eth)	...	166	...	...	w, al	$B2^4$, 2366
11939	1,2,3-Propanetricarboxylic acid-1,2-dihydroxy *(l)* ... $HO_2CCH_2C(OH)(CO_2H)CH(OH)(CO_2H)$	208.12	nd, $[\alpha]_D$ 17.7 (acid)	...	159-60	1.39^{25}	...	w, eth	$B3^4$, 1298
11940	1,2,3-Propanetricarboylic acid, 1-hydroxy or Isocitric acid $HO_2CCH_2CH(CO_2H)CH(OH)CO_2H$	192.13	yesh syr	...	105	...	...	...	$B3^4$, 1270
11941	Propanoic acid or Propionic acid ... $CH_3CH_2CO_2H$	74.08	...	141, 41.6^{10}	−20.8	0.9930^{20}	1.3869^{20}	**w, al,** eth	$B2^4$, 695
11942	1-Propanol or *n*-Propyl alcohol ... $CH_3CH_2CH_2OH$	60.10	...	97.4	−126.5	$0.8035^{20/4}$	1.3850^{20}	**w, al, eth,** ace, bz	$B1^4$, 1413
11943	1-Propanol, 2-amino *(dl)* ... $CH_3CH(NH_2)CH_2OH$	75.11	...	173-6, 80^{18}	...	...	1.4502^{20}	w, al, eth	$B4^4$, 1615
11944	1-Propanol, 3-amino or Propanolamine... $H_2NCH_2CH_2CH_2OH$	75.11	...	187-8	...	$0.9824^{26/4}$	1.4617^{20}	w, al, eth	$B4^4$, 1623
11945	1-Propanol, 3-benzyloxy ... $C_6H_5CH_2OCH_2CH_2CH_2OH$	166.22	...	172^{43}		$1.0474^{20/4}$	...	al, eth	$B6^4$, 2243
11946	1-Propanol, 3-bromo ... $BrCH_2CH_2CH_2OH$	138.99	...	$98\text{-}112^{185}$, 62^5	...	$1.5374^{20/4}$	1.4834^{25}	w, **al, eth**	$B1^4$, 1446
11947	1-Propanol, 2-chloro or Propylene chlorohydrin ... $CH_3CHClCH_2OH$	94.54	...	133-4	...	1.103^{20}	1.4390^{20}	w, al, eth	$B1^4$, 1440
11948	1-Propanol, 2-chloro-2-methyl or 2-Chloro isobutyl alcohol... $(CH_3)_2CClCH_2OH$	108.57	visc	132-3d, $59\text{-}61^{50}$	...	$1.0477^{20/4}$	1.4388^{20}	...	$B1^4$, 1603
11949	1-Propanol, 3-chloro or Trimethylene chlorohydrin... $ClCH_2CH_2CH_2OH$	94.54	...	165, 53^6	...	$1.1309^{20/4}$	1.4459^{20}	w, al, eth	$B1^4$, 1441
11950	1-Propanol, 2,3-dibromo *(d)* ... $CH_2BrCHBrCH_2OH$	217.89	$[\alpha]_D$ + 7.3	219d	...	2.11	...	**al, eth, ace, bz**	$B1^2$, 371
11951	1-Propanol, 2,3-dibromo *(dl)* ... $CH_2BrCHBrCH_2OH$	217.89	...	219d, 118^{17}	...	$2.0739^{20/4}$	1.5466^{20}	**al, eth, ace, bz**	$B1^4$, 1446
11952	1-Propanol, 2,3-dichloro ... $CH_2ClCHClCH_2OH$	128.99	visc	183-5, $70\text{-}80^{17}$	...	$1.3607^{20/4}$	1.4819^{20}	**al, eth, ace, bz**	$B1^4$, 1442
11953	1-Propanol, 2-diethylamino ... $(C_2H_5)_2NCH(CH_3)CH_2OH$	131.22	...	$166\text{-}9^{749}$, 78^{12}	...	$0.8665^{27/20}$	1.4332^{20}	al, eth, ace, bz	$B4^4$, 1618
11954	1-Propanol, 3-diethylamino ... $(C_2H_5)_2NCH_2CH_2CH_2OH$	131.22	...	189.5, 87^{16}	...	$0.8600^{20/4}$	1.4439^{20}	al, eth, ace, bz	$B4^4$, 1633
11955	1-Propanol, 2,3-dimercapto or 1,2 Dithioglycerol... $HSCH_2CH(SH)CH_2OH$	124.22	visc liq	120^{15}	...	$1.2463^{20/4}$	1.5733^{20}	al, eth, oils	$B1^4$, 2770
11956	1-Propanol, 3-(3,5-dimethoxy-4-hydroxyphenyl) or 2-Syringyl ethanol. Hydrosinapyl alcohol... $(3,5\text{-}(CH_3O)_2\text{-}4\text{-}HOC_6H_2)CH_2CH_2CH_2OH$	212.25	wh nd (eth-peth)	...	75-6	...	...	w	$B6^3$, 6669
11957	1-Propanol, 2,2-dimethyl or Neopentyl alcohol... $(CH_3)_3CCH_2OH$	88.15	...	113-4	52-3	0.812	...	al, eth	$B1^4$, 1690
11958	1-Propanol, 2,3-epoxy or Glycidol ... CH_2CHCH_2OH (epoxy O bridging CH_2 and CH)	74.08	...	166-7d, $65\text{-}6^{2.5}$	...	1.1143^{25}	1.4287^{20}	**w, al,** eth, ace. bz, chl	$B17^4$, 985
11959	1-Propanol, 2-ethoxy ... $CH_3CH(OC_2H_5)CH_2OH$	104.15	...	140-1	...	$0.9044^{20/4}$	1.4122^{20}	w, al, eth	$B1^3$, 2147
11960	1-Propanol, 3-fluoro ... $FCH_2CH_2CH_2OH$	78.09	...	127.8	...	$1.0390^{25/4}$	1.3771^{25}	w, al, eth	$B1^4$, 1437
11961	1-Propanol, 3-(4-hydroxy-3-methoxyphenyl) or Hydroconiferyl alcohol... $(3\text{-}CH_3O\text{-}4\text{-}HOC_6H_3)CH_2CH_2CH_2OH$	182.22	...	197^{15}	65	...	1.5545^{25}	al, eth	$B6^3$, 6347
11962	1-Propanol, 2-methoxy ... $CH_3CH(OCH_3)CH_2OH$	90.12	...	130^{758}	...	$0.938^{20/4}$	1.4070^{20}	...	$B1^4$, 2471
11963	1-Propanol, 2-methyl or Isobutyl alcohol... $(CH_3)_2CHCH_2OH$	74.12	...	108	−108	$0.8018^{20/4}$	1.3955^{20}	w, al, eth, ace	$B1^4$, 1588
11964	1-Propanol, 2-methyl-2-nitro ... $(CH_3)_2C(NO_2)CH_2OH$	119.12	nd or pl (MeOH)	$94\text{-}5^{10}$	89-90	...	...	al, eth	$B1^4$, 1604
11965	1-Propanol, 2-nitro ... $CH_3CH(NO_2)CH_2OH$	105.09	...	100^{12}	...	$1.1841^{25/4}$	1.4379^{20}	al, eth	$B1^4$, 1448
11966	1-Propanol, 2-phenoxy ... $CH_3CH(OC_6H_5)CH_2OH$	152.19	...	244, $124\text{-}6^{20}$	...	$0.9830^{25/25}$	1.4760^{25}	al, eth	$B6^4$, 582
11967	1-Propanol, 3-phenoxy ... $C_6H_5OCH_2CH_2CH_2OH$	152.19	oil	$249\text{-}50^{764}$, 170^{60}	...	...	1.491^{20}	...	$B6^4$, 584

No.	Name, Synonyms, and Formula	Mol. wt.	Color, crystalline form, specific rotation and λ_{max} (log ε)	b.p. °C	m.p. °C	Density	n_D	Solubility	Ref.
11968	1-Propanol, 1-phenyl- *(dl)* $CH_3CH_2CH(OH)C_6H_5$	136.19	...	$213-5^{740}$, 98^{10}	...	$0.9938^{23/4}$	1.5210^{23}	al, eth	B6^{3}, 1793
11969	1-Propanol, 3-phenyl $C_6H_5CH_2CH_2CH_2OH$	136.19	...	$236-7^{750}$, 132^{21}	<−18	$1.008^{20/4}$	1.5278^{20}	w, al, eth	B6^{3}, 1880
11970	1-Propanol, 2,2,3,3-tetrafluoro $CHF_2CF_2CH_2OH$	132.06	...	109-1	−15	$1.4853^{20/4}$	1.3197^{20}	al, ace, chl	B1^{4}, 1438
11971	2-Propanol or Isopropyl alcohol $(CH_3)_2CHOH$	60.10	...	82.4	−89.5	$0.7855^{20/4}$	1.3776^{20}	w, al, eth, ace, bz	B1^{4}, 1461
11972	2-Propanol, 1-amino *(dl)* or 150 Propanol amine $CH_3CH(OH)CH_2NH_2$	75.11	...	159.4, 59.5^{10}	1.7	$0.9611^{20/4}$	1.4479^{20}	w, al, eth, ace, bz	B4^{4}, 1665
11973	2-Propanol, 1-amino *(l)* $CH_3CH(OH)CH_2NH_2$	75.11	...	$156-8^{758}$	...	0.973^{18}	...	w, al, eth, ace, bz	B4^{4}, 1664
11974	2-Propanol, 1-amino-3-(diethylamino) $(C_2H_5)_2NCH_2CH(OH)CH_2NH_2$	146.23	...	223, $116-8^{25}$	...	$1.937^{20/4}$	1.465^{20}	w	B4^{3}, 767
11975	2-Propanol, 2-benzyl $C_6H_5CH_2C(OH)(CH_3)_2$	150.22	...	214-6, $104-5^{10}$	24	$0.9790^{20/25}$	1.5174^{20}	al	B6^{3}, 1860
11976	2-Propanol, 1,3-bis (dimethylamino) $[(CH_3)_2NCH_2]_2CHOH$	146.23	...	178-85, $79-81^{18}$	...	$0.8788^{20/4}$	1.4418^{20}	w	B4^{4}, 1695
11977	2-Propanol, 1-bromo $CH_3CH(OH)CH_2Br$	138.99	...	145-8, 49.6^{12}	...	1.5585^{10}	1.4801^{20}	w, al, eth	B1^{4}, 1495
11978	2-Propanol, 1-butoxy or 1,2-Propylene glycol-1-monobutyl ether $CH_3CH(OH)CH_2OC_4H_9$	132.20	...	168.75	...	$1.0035^{20/4}$	1.4168^{20}	al, eth, bz	B1^{4}, 2471
11979	2-Propanol, 1-chloro $CH_3CH(OH)CH_2Cl$	94.54	...	$126-7^{750}$	...	$1.115^{20/20}$	1.4392^{20}	w, al, eth	B1^{4}, 1490
11980	2-Propanol, 1-chloro-2-methyl $(CH_3)_2C(OH)CH_2Cl$	108.57	...	128-9, 71^{100}	−20	$1.0628^{20/4}$	1.4380^{24}	w, al	B1^{4}, 1628
11981	2-Propanol, 1-chloro-3-propoxy $C_3H_7OCH_2CH(OH)CH_2Cl$	152.62	...	$92-5^{15}$	...	$1.0526^{25/4}$	1.4378^{25}	eth, ace	B1^{3}, 2153
11982	2-Propanol, 1-chloro-3-isopropoxy $(CH_3)_2CHOCH_2CH(OH)CH_2Cl$	152.62	...	87^{20}	...	1.0530^{25}	1.4370^{25}	al, eth	B1^{4}, 2485
11983	2-Propanol, 1,3-diamino $(H_2NCH_2)_2CHOH$	90.13	cr	235, $93-5^{2}$	42-5	...	...	w	B4^{4}, 1694
11984	2-Propanol, 1-3 diamino, dihydrochloride $(H_2NCH_2)_2CHOH.2HCl$	163.05	hyg pr or nd (dil al)	...	184.5	...	...	w	B4^{2}, 739
11985	2-Propanol, 1,3-dibromo $(BrCH_2)_2CHOH$	217.89	yesh liq	219d, 105^{16}	...	$2.1202^{25/4}$	1.5495^{25}	al, eth, ace	B1^{4}, 1496
11986	2-Propanol, 1-1,dichloro $CH_3CH(OH)CHCl_2$	128.99	...	$146-8^{765}$	...	$1.3334^{22/4}$	...	al, eth, ace	B1^{2}, 383
11987	2-Propanol, 1,1-dichloro-2-methyl $(CH_3)_2C(OH)CHCl_2$	143.01	...	150-5, 52^{10}	8	$1.2363^{19/4}$	1.4598^{19}	al, bz	B1^{4}, 1629
11988	2-Propanol, 1,3-dichloro $(ClCH_2)_2CHOH$	128.99	...	176, 69^{12}	...	$1.3506^{17/4}$	1.4837^{20}	w, al, eth, ace	B1^{4}, 1491
11989	2-Propanol, 1,3-dichloro-2-methyl $(ClCH_2)_2C(OH)CH_3$	143.01	...	174-5, $55-6^{10}$	...	$1.2745^{20/4}$	1.4744^{21}	al, ace	B1^{4}, 1629
11990	2-Propanol, 1-(diethylamino) $CH_3CH(OH)CH_2N(C_2H_5)_2$	131.22	...	$158-9^{756}$, 63^{22}	...	$0.8511^{20/0}$	1.4255^{20}	w, al, ace	B4^{3}, 759
11991	2-Propanol, 1-ethoxy $CH_3CH(OH)CH_2OC_2H_5$	104.15	...	131	...	$0.9028^{20/4}$	1.4075^{20}	w, al, eth	B1^{4}, 2471
11992	2-Propanol, 1-methoxy $CH_3CH(OH)CH_2OCH_3$	90.12	...	118	...	$0.9620^{20/4}$	1.4034^{20}	...	B1^{4}, 2471
11993	2-Propanol, 1-methoxy-2-methyl $CH_3OCH_2C(OH)(CH_3)_2$	104.15	...	116.6^{747}	...	$0.9021^{15/15}$	...	...	B1^{4}, 2533
11994	2-Propanol, 2-methyl or *tert*-Butyl alchol $(CH_3)_3COH$	74.12	...	82.3, 20^{31}	25.5	$0.7887^{20/4}$	1.3878^{20}	w, al, eth	B1^{4}, 1609
11995	2-Propanol, 1-nitro $CH_3CH(OH)CH_2NO_2$	105.09	...	112^{13}, 68^{1}	...	1.1906^{20}	1.4383^{20}	w, al, eth, bz	B1^{4}, 1496
11996	2-Propanol, 1-nitro-3,3,3-trichloro $O_2NCH_2CH(OH)CCl_3$	208.43	pr or pl (chl)	$105-6^{15}$	45-6	...	...	al, eth	B1^{4}, 1497
11997	2-Propanol, 1-phenoxy $CH_3CH(OH)CH_2OC_6H_5$	152.19	...	134.5^{20}	...	$1.0622^{20/4}$	1.5210^{22}	al, eth	B6^{4}, 582
11998	2-Propanol, 2-phenyl $(CH_3)_2C(OH)C_6H_5$	136.19	pr	202, 93^{13}	35-7	$0.9735^{20/4}$	1.5325^{20}	al, eth, bz, aa	B6^{3}, 1813
11999	2-Propanol, 1-propoxy $CH_3CH(OH)CH_2OC_3H_7$	118.18	...	$148-9^{730}$	...	$0.8886^{20/4}$	1.4130^{20}	...	B1^{3}, 2147
12000	2-Propanol, 1-isopropoxy $CH_3CH(OH)CH_2OCH(CH_3)_2$	118.18	...	137-8	...	$0.879^{20/4}$	1.4070^{20}	...	B1^{4}, 2471

No.	Name, Synonyms, and Formula	Mol. wt.	Color, crystalline form, specific rotation and λ_{max} (log ε)	b.p. °C	m.p. °C	Density	n_D	Solubility	Ref.
12001	2-Propanol, 1,1,1,3-tetrachloro $CH_2ClCH(OH)CCl_3$	197.88		95-6^{17}		1.610$^{20/4}$	1.5145^{20}	eth	B1^3, 1474
12002	2-Propanol, 1,1,3,3-tetrachloro $(Cl_2CH)_2CHOH$	197.88		80-90^{14}		1.612$^{20/4}$	1.5133^{20}	eth	B1^3, 1474
12003	2-Propanol, 1,1,1-trichloro $CH_3CH(OH)CCl_3$	163.43		161.8^{773}, 53-5^{12}	50-1			al, eth, ace, bz	B1^3, 1474
12004	2-Propanol, 1,1,1-trifluoro (dl) $F_3CCH(OH)CH_3$	114.07		77.7^{754}	−52	1.2799$^{15/4}$	1.3172^{15}	al, eth, ace, bz	B1^4, 1489
12005	2-Propanone or Acetone, Dimethyl Ketone CH_3COCH_3	58.08		56.2	−95.3	0.7899$^{20/4}$	1.3588^{20}	w, al, eth, bz, chl	B1^4, 3180
12006	Propargyl aldehyde or Propynal $HC{\equiv}CCHO$	54.05		59-61			1.4033^{25}	w, al, eth, ace, bz	B1^4, 3537
12007	Propargyl aldehyde, diethyl acetyl $HC{\equiv}CCH(OC_2H_5)_2$	128.17		130.4, 37^{11}		0.8942$^{22/4}$	1.4140^{20}	al, eth, ace, chl	B1^4, 3538
12008	Propargyl aldehyde, phenyl $C_6H_5C{\equiv}CCHO$	130.15		127-8^{28} 65$^{0.1}$		1.0639$^{16/4}$	1.6079^{18}		B7^3, 1644
12009	Propargyl bromide or 3-Bromo propyne $BrCH_2C{\equiv}CH$	118.96		88-90, 33^{130}		1.579^{19}	1.4922^{20}	al, eth, bz, chl	B1^4, 964
12010	Propargyl alcohol or 2-Propyn-1-ol $HC{\equiv}CCH_2OH$	56.06		113.6, 30^{21}	−48	0.9485^{20}	1.4322^{20}	w, al, eth	B1^4, 2214
12011	Propargyl alcohol, acetate or Propargyl acetate $CH_3CO_2CH_2C{\equiv}CH$	98.10		124-5		1.0082$^{20/4}$	1.4205^{20}	al, eth	B2^4, 197
12012	Propargyl alcohol, 3-phenyl $C_6H_5C{\equiv}CCH_2OH$	132.16		137-8^{15}		1.07$^{18/18}$	1.5873^{18}	eth, ace, bz	B6^3, 2736
12013	Propargylic acid or Propynoic acid $HC{\equiv}CCO_2H$	70.05	cr (CS_2)	144d, 83-4^{50}	18 (anh)	1.1380$^{20/4}$	1.4306^{20}	w, al, eth, ace, chl	B2^4, 1687
12014	Propargylic acid, ethyl ester or Ethyl propargylate $HC{\equiv}CCO_2C_2H_5$	98.10		119^{745}		0.9583$^{25/25}$	1.4105^{20}	al, eth, chl	B2^4, 1688
12015	Propargylic acid, (2-chlorophenyl) $(2\text{-}ClC_6H_4)C{\equiv}CCO_2H$	180.59	cr (bz, 50% aa)		133-4			aa	B9^3, 3066
12016	Propargylic acid, 3-chlorophenyl $(3\text{-}ClC_6H_4)C{\equiv}CCO_2H$	180.59	cr (aa, bz-peth)		144-5			aa	B9^3, 3066
12017	Propargylic acid, (4-chlorophenyl) $(4\text{-}ClC_6H_4)C{\equiv}CCO_2H$	180.59	cr (aa), pl (bz)		192-3			aa	B9^3, 3066
12018	Propargylic acid, (2-nitrophenyl) $(2\text{-}O_2NC_6H_4)C{\equiv}CCO_2H$	191.14		exp	157d			al, eth	B9^3, 3067
12019	Propargylic acid, (4-nitrophenyl) $(4\text{-}O_2NC_6H_4)C{\equiv}CCO_2H$	191.14	nd (eth, al)		181			eth, aa, chl	B9^3, 3067
12020	**Propargylic acid, phenyl** $C_6H_5C{\equiv}CCO_2H$	146.15		sub	137			al, eth	B9^3, 3061
12021	Propargylic acid, phenyl, ethyl ester $C_6H_5C{\equiv}C\text{-}CO_2C_2H_5$	174.20		260-70d, 144^{11}		1.0550$^{25/4}$	1.5535^{20}	eth	B9^3, 3063
12022	Propargylonitrile or Propynonitrile $HC{\equiv}CCN$	51.05		42.5	5	0.8167$^{17/4}$	1.3868^{25}	al	B2^4, 1689
12023	**Propenal or Acrolein** $CH_2{=}CHCHO$	56.06		52-3	−87	0.8410$^{20/4}$	1.4017^{20}	w, al, eth, ace	B1^4, 3435
12024	2-Propen-1-arsonic acid or Allylarsonic acid $CH_2{=}CHCH_2AsO_3H_2$	166.01			130-1			w, al	B4^3, 1826
12025	Propene or Propylene $CH_3CH{=}CH_2$	42.08	gas	−47.4	−185.2	0.5193$^{20/4}$ liq	1.3567^{-70}	al, aa	B1^4, 725
12026	1-Propene-3,3, diol, diacetate $(CH_3CO_2)_2CHCH{=}CH_2$	158.15		180, 76^{13}	−37.6	1.0760$^{20/4}$	1.4193^{20}	al, eth, ace, bz, lig	B2^4, 291
12027	1,2,3-Propene tricarboxylic acid (*cis*) or *cis*-Aconitic acid $HO_2CCH_2C(CO_2H){=}CHCO_2H$	174.11	nd (w)		130			w	B2^4, 2405
12028	1,2,3-Propene tricarboxylic acid (*trans*) or *trans*-Aconitic acid $HO_2CCH_2C(CO_2H){=}CHCO_2H$	174.11	lf (w), nd (w, eth)		198-9			w, al	B2^4, 2405
12029	Propenyl benzene (*cis*) $(CH_3CH{=}CH)C_6H_5$	118.18		69^{28}	−60.5	0.9088$^{20/4}$	1.5420^{20}	al, eth, ace, bz	B5^4, 1359
12030	Propenyl benzene-(*trans*) $(CH_3CH{=}CH)C_6H_5$	118.18		175-6	−27.1	0.9019^{25}	1.5508^{20}	al, eth, ace, bz	B5^4, 1359
12031	Propenyl benzene, α-chloro $CH_3CH{=}CClC_6H_5$	152.62		90.5$^{0.9}$		1.085$^{20/4}$	1.5635^{15}	ace, bz	B5^2, 372
12032	Propenyl benzene, β-chloro $C_6H_5CH{=}CClCH_3$	152.62		118-23^{20}, 61-2^{1}		1.0738$^{19/4}$	1.5565^{19}	ace, bz, chl	B5^3, 1186
12033	Propenyl benzene, γ-chloro-(*trans*) or *trans*-Cinnamyl chloride $C_6H_5CH{=}CHCH_2Cl$	152.62		106-7^{13}	8-9	1.0926$^{20/4}$	1.5851^{20}	al, eth, ace, bz	B5^4, 1360

PHYSICAL CONSTANTS OF ORGANIC COMPOUNDS (Continued)

No.	Name, Synonyms, and Formula	Mol. wt.	Color, crystalline form, specific rotation and λ_{max} (log ε)	b.p. °C	m.p. °C	Density	n_D	Solubility	Ref.
12034	Propenyl benzene, γ-ethoxy $C_2H_5OCH_2CH=CHC_6H_5$	162.23		127-8^{22}		0.970$^{15/4}$	1.547^{15}		B6^{3}, 2404
12035	Propenyl benzene, β-nitro $C_6H_5CH=C(NO_2)CH_3$	163.18	ye nd (peth)		65-6			al	B5^{4}, 1360
12036	Isopropenyl methyl ketone $CH_2=C(CH_3)COCH_3$	84.12		98	−54	0.8527$^{20/4}$	1.4220^{20}	al	B1^{4}, 3462
12037	Propenoic acid or Acrylic acid $CH_2=CHCO_2H$	72.06		141.6, 48.5^{15}	13	1.0511$^{20/4}$	1.4224^{20}	w, al, eth, ace, bz	B2^{4}, 1455
12038	Propenoic acid, 2-methyl or Methacrylic acid $CH_2=C(CH_3)CO_2H$	86.09	pr	162-3^{757}, 60^{12}	16	1.0153$^{20/4}$	1.4314^{20}	w, al, eth	B2^{4}, 1518
12039	2-Propene-1-ol, 2-methyl or Methallyl alcohol $CH_2=C(CH_3)CH_2OH$	72.11		114.5		0.8515^{20}	1.4255^{20}	w, al, eth	B1^{4}, 2114
12040	2-Propene-1-ol, 1-phenyl or α-Vinyl benzyl alcohol $CH_2=CH\text{-}CH(OH)C_6H_5$	134.18		215-6, 111^{18}		1.0251$^{21/0}$	1.5406^{20}	al, eth, bz, chl	B6^{3}, 2417
12041	β-Propiolactone CH_2CH_2CO (ring via O)	72.06		162d, 51^{10}	−33.4	1.1460$^{20/5}$	1.4105^{20}	eth, chl	B17^{4}, 4157
12042	Propionaldehyde or Propanal CH_3CH_2CHO	58.08		48.8	−81	0.8058$^{20/4}$	1.3636^{20}	w, al, eth	B1^{4}, 3165
12043	Propionaldehyde, 2-Bromo $CH_3CHBrCHO$	136.98		109-10, 52-4^{80}		1.592^{20}	1.4813^{20}	eth	B1^{4}, 3177
12044	Propionaldehyde, 2-chloro $CH_3CHClCHO$	92.53		86		1.182$^{15/4}$	1.431^{17}	eth, bz	B1^{3}, 2691
12045	Propionaldehyde, 3-chloro $ClCH_2CH_2CHO$	92.53		130-1, 40^{19}		1.268^{15}	1.475^{25}	al, eth	B1^{4}, 3174
12046	Propionaldehyde, 3-chloro, diethyl acetal $ClCH_2CH_2CH(OC_2H_5)_2$	166.65		84^{25}		0.9951$^{19/4}$	1.4268^{20}	ace, bz	B1^{3}, 2692
12047	Propionaldehyde, 2,3-dibromo $BrCH_2CHBrCHO$	215.87	pa ye fum liq	73-5^{10}		2.198^{15}	1.5082^{20}	eth	B1^{4}, 3178
12048	Propionaldehyde, 2,2-dichloro CH_3CCl_2CHO	126.97	(peth)	80	38-9				B1^{4}, 3174
12049	Propionaldehyde, 2,3-dichloro $ClCH_2CHClCHO$	126.97	hyd (w)	73^{50}		1.400^{20}	1.4762^{20}		B1^{4}, 3174
12050	Propionaldehyde diethyl acetal or 1,1-Diethoxypropane $CH_3CH_2CH(OC_2H_5)_2$	132.20		122.8^{744}		0.8232$^{20/4}$	1.3924^{19}	w, al, eth, ace, bz	B1^{4}, 3168
12051	Propionaldehyde, 2,3-epoxy CH_2CHCHO (ring via O)	72.06		112-3	−62		1.4265^{20}		B17^{4}, 4159
12052	Propionaldehyde, 3-ethoxy, diethyl acetal or 1,1,3-Triethoxy propane $C_2H_5OCH_2CH_2CH(OC_2H_5)_2$	176.26		184-6d, 78^{14}		0.898$^{19/4}$	1.4067^{20}	al	B1^{4}, 3971
12053	Propionaldehyde, 3-hydroxy-2-oxo (enol form) or Reductone $HOCH=C(OH)CHO$	88.06	ye nd (w)		200-20d			w, al	B1^{4}, 4145
12054	Propionaldehyde oxime or Propionaldoxime $CH_3CH_2CH=NOH$	73.09		131.5	40	0.9258$^{20/4}$	1.4287^{20}		B1^{4}, 3170
12055	Propionaldehyde, 2-phenoxy $CH_3CH(OC_6H_5)CHO$	150.18		229-30, 99-101^{19}				al, eth, bz	B6, 151
12056	Propionaldehyde, 2-phenoxy, oxime $CH_3CH(OC_6H_5)CH=NOH$	165.19	nd (dil al)		110			al, eth	B6, 151
12057	Propionaldehyde, 2-phenyl or Hydratropic aldehyde $CH_3CH(C_6H_5)CHO$	134.18		202-5, 92.5^{10}		1.0089$^{20/4}$	1.5176^{20}	al	B7^{3}, 1050
12058	Propionaldehyde, 3-phenyl or Hydrocinnamaldehyde $C_6H_5CH_2CH_2CHO$	134.18	mcl	223^{745}, 104-5^{13}	47			al, eth	B7^{3}, 1046
12059	Propionaldehyde, 3-(4-tolyl) $(4\text{-}CH_3C_6H_4)CH_2CH_2CHO$	148.20		122^{15}		0.999$^{14/4}$	1.525^{14}		B7^{2}, 247
12060	Propionaldehyde, 2,2,3-trichloro $ClCH_2CCl_2CHO$	161.42		63-5^{45}		1.470^{25}	1.473^{25}	eth	B1^{3}, 2693
12061	Propionamide $CH_3CH_2CONH_2$	73.09	rh, pl (bz)	213	81.3	0.9262^{110}	1.4180^{110}	w, al, ace, chl	B2^{4}, 725
12062	Propionamide, 3-bromo $CH_2BrCH_2CONH_2$	151.99	cr (w)		111			al, eth, ace	B2^{4}, 766
12063	Propionamide, 3-chloro-N-(2-tolyl) $ClCH_2CH_2CONH(C_6H_4CH_3\text{-}2)$	197.66	(w, dil al)		78			al	B12^{2}, 441
12064	Propionamide, N,N-diethyl $CH_3CH_2CON(C_2H_5)_2$	129.20		191, 81-5^{20}			1.4425^{20}	al	B4^{4}, 353
12065	Propionamide, 2-hydroxy $CH_3CH(OH)CONH_2$	89.09			75.5	1.1381$^{80/4}$		w, al	B3^{4}, 674
12066	Propionamide, 2-phenoxy $CH_3CH(OC_6H_5)CONH_2$	165.19	nd or pl (to, w)		132-3			al, eth, aa	B6^{4}, 643

No.	Name, Synonyms, and Formula	Mol. wt.	Color, crystalline form, specific rotation and λ_{max} (log ε)	b.p. °C	m.p. °C	Density	n_D	Solubility	Ref.
12067	Propionamide, 3-phenoxy $C_6H_5OCH_2CH_2CONH_2$	165.19	nd (w)	...	119	...	...	al, eth	$B6^3$, 616
12068	Propionamide, N-phenyl or Propionanilide $CH_3CH_2CONHC_6H_5$	149.19	pl (eth, al, bz)	222.2	105-6	1.175	...	al, eth	$B12^3$, 472
12069	Propionamide, N-propionyl or Dipropionamide $(CH_3CH_2CO)_2NH$	129.16	nd (w, eth)	210-20 sub	154	...	...	...	$B2^4$, 727
12070	Propionamide, N-(2-tolyl) $CH_3CH_2CONH(C_6H_4CH_3\text{-}2)$	163.22	nd (bz)	298-9	89.5	...	...	al, eth, chl, aa	$B12^2$, 440
12071	Propionamide, N-(3-tolyl) $CH_3CH_2CONH(C_6H_4CH_3\text{-}3)$	163.22	nd (eth)	...	81	...	...	al, eth, lig	B12, 861
12072	Propionic acid or Propanoic acid $CH_3CH_2CO_2H$	74.08	...	141, 41.6^{10}	−20.8	0.9930^{20}	1.3809^{20}	**w, al, eth**	$B2^4$, 695
12073	Propionic acid, allyl ester or Allyl propionate $CH_3CH_2CO_2CH_2CH{=}CH_2$	114.14	...	$124\text{-}5^{774}$	...	0.9140^{20}	1.4105^{20}	al, eth, ace	$B2^4$, 711
12074	Propionic anhydride or Propionic anhydride $(CH_3CH_2CO)_2O$	130.14	...	$168^{71\ 2}$, 67.5^{18}	−45	$1.0110^{20/4}$	1.4038^{20}	**eth**	$B2^4$, 722
12075	Propionic acid, β-benzoyl or β-Benzoyl propionic acid $C_6H_5COCH_2CH_2CO_2H$	178.19	lf (dil al)	...	116	...	...	al, eth, bz, chl	$B10^3$, 3035
12076	Propionic acid, 2-bromo-(dl) $CH_3CHBrCO_2H$	152.98	pr	203.5, 96^{10}	25.7	$1.7000^{20/4}$	1.4753^{20}	w, al, eth	$B2^4$, 761
12077	Propionamide, 2-bromo-N-(2-tolyl) $CH_3CHBrCONH(C_6H_4CH_3\text{-}2)$	242.12	nd	...	131	...	...	al, eth, bz, chl	B12, 794
12078	Propionic acid, 2-bromo,ethyl ester $CH_3CHBrCO_2C_2H_5$	181.03	...	159-61d, 71^{26}	...	$1.4135^{20/4}$	1.4490^{20}	**al, eth**, chl	$B2^4$, 762
12079	Propionic acid, 2-bromo,methyl ester(d) $CH_3CHBrCO_2CH_3$	167.00	$[\alpha]^{17/}_{D}$ +42.65	144, $61\text{-}2^{26}$	...	1.482^{17}	...	al, eth, chl	B2, 253
12080	Propionic acid, 2-bromo,methyl ester (l) $CH_3CHBrCO_2CH_3$	167.00	$[\alpha]^{20/}_{578}$ -55.5	$61\text{-}3^{32}$	...	1.484^{20}	...	al, eth	$B2^2$, 229
12080a	Propionic acid, 2-bromo,methyl ester-(dl) $CH_3CHBrCO_2CH_3$	167.00	...	143-5, 51.5^{19}	...	$1.4966^{25/4}$	1.4451^{22}	al, eth	$B9^3$, 2423
12081	Propionic acid, 2-bromo-2-phenyl (dl) $CH_3CBr(C_6H_5)CO_2H$	229.07	pl (CS_2)	...	93-4	...	...	bz	$B2^4$, 764
12082	Propionic acid, 3-bromo $BrCH_2CH_2CO_2H$	152.98	pl (CCl_4)	$140\text{-}2^{45}$	62.5	1.48	...	al, eth, bz, chl	$B2^4$, 765
12083	Propionic acid, 3-bromo, butyl ester $CH_2BrCH_2CO_2C_4H_9$	209.09	...	130^{26}	...	1.4549^{20}	1.3051^{20}	al, eth, ace	$B2^4$, 765
12084	Propionic acid, 3-bromo, ethyl ester $CH_2BrCH_2CO_2C_2H_5$	181.03	...	179, 70^{12}	...	...	1.4516^{20}	al, eth, ace	$B2^4$, 765
12085	Propionic acid, 3-bromo,methyl ester $CH_2BrCH_2CO_2CH_3$	167.00	...	105.5^{60}	...	1.4897^{15}	1.4542^{20}	al, eth, ace	$B2^4$, 762
12087	Propionic acid, 3-bromo,isopentyl ester $CH_2BrCH_2CO_2\text{-i-}C_5H_{11}$	223.11	...	$110\text{-}1^{11}$	...	$1.2217^{15/4}$	1.4556^{9}	al, eth	$B2^2$, 231
12088	Propionic acid, 3-bromo-2-phenyl $CH_2BrCH(C_6H_5)CO_2H$	229.07	pr (CS_2)	...	93-4	...	...	al, eth, bz	B9, 526
12089	Propionic acid, 3-bromo-3-phenyl $C_6H_5CHBrCH_2CO_2H$	229.07	mcl pr (al)	...	137	...	...	al	$B9^3$, 2401
12090	Propionic acid, 3-bromo-3-phenyl,methyl ester $C_6H_5CHBrCH_2CO_2CH_3$	243.10	pr	...	37-8	...	...	al	$B9^1$, 201
12091	Propionic acid, butyl ester or Butyl propionate $CH_3CH_2CO_2C_4H_9$	130.19	...	145.5	−89.5	$0.8754^{20/4}$	1.4014^{20}	**al, eth**	$B2^4$, 708
12092	Propionic acid, isobutyl ester $CH_3CH_2CO_2\text{-i-}C_4H_9$	130.19	...	136.8, 66.5^{60}	−71.4	$0.8687^{20/4}$	1.3973^{20}	**al, eth**	$B2^4$, 709
12093	Propionic acid, *sec*-butyl ester or *sec*-Butyl propionate $C_2H_5CO_2CH(CH_3)C_2H_5$	130.19	...	132-1	...	0.8657^{20}	1.3952^{20}	al, eth	$B2^4$, 709
12094	Propionic acid, 2-chloro $CH_3CHClCO_2H$	108.52	...	186, 84^{12}	...	$1.2585^{20/4}$	1.4380^{20}	w, **al, eth**, ace	$B2^4$, 745
12095	Propionyl chloride, 2-chloro-(d) $CH_3CHClCOCl$	126.97	$[\alpha]^{18/}_{D}$ +0.2	110^{744}	...	$1.2394^{7.5}$	...	...	$B2^4$, 747
12096	Propionic acid, 2-chloro, butyl ester $CH_3CHClCO_2C_4H_9$	164.63	...	183-5, $72\text{-}3^{10}$	...	$1.0253^{20/4}$	1.4263^{20}	eth	$B2^4$, 746
12097	Propionic acid, 2-chloro, ethyl ester $CH_3CHClCO_2C_2H_5$	136.58	...	147-8, $52\text{-}4^{18}$	...	$1.0793^{20/4}$	1.4178^{20}	**al, eth**	$B2^4$, 746
12098	Propionic acid, 2-chloro, isobutyl ester (d) $CH_3CHClCO_2\text{-i-}C_4H_9$	164.63	$[\alpha]_D$ +5.2	175-7	...	$1.0312^{20/4}$	1.4247^{20}	al, eth, ace	B2, 248
12099	Propionic acid, 2-chloro, isopropyl ester $CH_3CHClCO_2CH(CH_3)_2$	150.61	...	151-2, $46\text{-}7^{12}$	...	$1.0315^{20/4}$	1.4149^{20}	al, eth	$B2^4$, 746

PHYSICAL CONSTANTS OF ORGANIC COMPOUNDS (Continued)

No.	Name, Synonyms, and Formula	Mol. wt.	Color, crystalline form, specific rotation and λ_{max} (log ε)	b.p. °C	m.p. °C	Density	n_D	Solubility	Ref.
12100	Propionic acid, 2-chloro, methyl ester (d) ... $CH_3CHClCO_2CH_3$	122.55	$[\alpha]_D$ +19.9 (undil)	133-4, 50^{35}	...	$1.1815^{20/4}$	...	al	B2, 248
12101	Propionic acid, 2-chloro, methyl ester-(dl) ... $CH_3CHClCO_2CH_3$	122.55	...	132.5	...	$1.209^{20/4}$	1.4182^{20}	al	B2[4], 746
12102	Propionic acid, 2-chloro, methyl ester-(l) ... $CH_3CHClCO_2CH_3$	122.55	$[\alpha]_D$ -26.9 (undil)	$79\text{-}80^{120}$	...	$1.158^{5/4}$	...	al	B2[3], 553
12103	Propionamide, 2-chloro-N-(2-tolyl) ... $CH_3CHClCONH(C_6H_4CH_3\text{-}2)$	197.66	nd (abs al)	...	111	...	...	al	B12, 794
12104	Propionic acid, 3-chloro ... $CH_2ClCH_2CO_2H$	108.52	lf (w), hyg cr (lig)	204d	41 (61)	...	...	w, al, eth	B2[4], 748
12105	Propionic acid, 3-chloro, butyl ester ... $CH_2ClCH_2CO_2C_4H_9$	164.63	...	104^{22}	...	$1.0370^{20/4}$	1.4321^{20}	w, eth	B2[4], 749
12106	Propionic acid, 3-chloro, isobutyl ester ... $CH_2ClCH_2CO_2$-i-C_4H_9	164.63	...	191.3	...	$1.0323^{20/4}$	1.4295^{20}	al, eth	B2[4], 749
12107	Propionic acid, 3-chloro, ethyl ester ... $CH_2ClCH_2CO_2C_2H_5$	136.58	...	162, 56^{11}	...	$1.1086^{20/4}$	1.4254^{20}	al, eth	B2[4], 749
12108	Propionic acid, 3-chloro, methyl ester ... $CH_2ClCH_2CO_2CH_3$	122.55	...	155-7, $40\text{-}2^{10}$	...	$1.1861^{15/4}$	1.4263^{20}	al	B2[4], 748
12109	Propionic acid, 3-chloro, isopentyl ester ... $CH_2ClCH_2CO_2$-i-C_5H_{11}	178.66	...	$207\text{-}8^{740}$, 87^{12}	...	$1.0171^{20/4}$	1.4343^{20}	al, eth	B2[4], 749
12110	Propionic acid, 3-chloro, propyl ester ... $CH_2ClCH_2CO_2C_3H_7$	150.61	...	180, $77\text{-}8^{12}$	...	$1.0656^{20/4}$	1.4290^{20}	al, eth	B2[4], 749
12111	Propionic acid, 3-cyclohexyl ... $C_6H_{11}CH_2CH_2CO_2H$	156.22	...	275-8, 143^{11}	16	$0.9966^{20/4}$	1.4634^{20}	eth	B9[3], 64
12112	Propionic acid, cyclohexyl ester or Cyclohexyl propionate ... $CH_3CH_2CO_2C_6H_{11}$	156.22	...	193^{750}, 93^{35}	...	$0.9359^{20/4}$	1.4403^{20}	al, eth, ace	B6[4], 37
12113	Propionic acid, 2,3-diamino ... $H_2NCH_2CH(NH_2)CO_2H$	104.11	hyg rosettes	...	ca 110-20	...	...	...	B4[3], 1292
12114	Propionic acid, 2,3-dibromo ... $CH_2BrCHBrCO_2H$	231.87	...	220-40d, 160^{20}	66-7	...	...	al, eth, bz	B2[4], 767
12115	Propionic acid, 2,3-dibromo, ethyl ester ... $CH_2BrCHBrCO_2C_2H_5$	259.93	...	214-5, 112^{23}	...	$1.7966^{20/4}$	1.5007^{20}	al, eth	B2[4], 767
12116	Propionic acid, 2,3-dibromo, methyl ester ... $CH_2BrCHBrCO_2CH_3$	245.90	...	206, 115^{25}	...	$1.9333^{20/4}$	1.5127^{20}	al	B2[4], 767
12117	Propionic acid, 2,3-dibromo-3-phenyl (d) or *d*-Cinnamic acid dibromide ... $C_6H_5CHBrCHBrCO_2H$	307.97	pr (chl), $[\alpha]^{15}{}_D$ +45.8 (abs al)	182	...	...	...	al, eth	B9[3], 2404
12118	Propionic acid, 2,3-dibromo-3-phenyl (dl) ... $C_6H_5CHBrCHBrCO_2H$	307.97	mcl pr (chl)	sub	240d	...	...	al, eth	B9[3], 2405
12119	Propionic acid, 2,3-dibromo-3-phenyl (*meso*) ... $C_6H_5CHBrCHBrCO_2H$	307.97	nd	...	91.3	...	...	bz	B9[3], 2406
12120	Propionic acid, 2,3-dibromo-3-phenyl, ethyl ester-(d) ... $C_6H_5CHBrCHBrCO_2C_2H_5$	336.02	cr (CS_2)	...	71	...	...	al	B9, 518
12121	Propionic acid, 2,3-dibromo-3-phenyl, ethyl ester-(dl) ... $C_6H_5CHBrCHBrCO_2C_2H_5$	336.02	mcl pr or pl	...	15-6	...	...	...	B9[3], 2406
12122	Propionic acid, 2,2-dichloro ... $CH_3CCl_2CO_2H$	142.97	...	185-90, $90\text{-}2^{14}$	...	$1.389^{12/4}$	...	al, eth	B2[4], 753
12123	Propionic acid, 2,2-dichloro-3,3,3-trifluoro, methyl ester $CF_3CCl_2CO_2CH_3$	210.97	...	$116\text{-}7^{625}$	...	1.5092^{20}	1.3806^{20}	...	B2[4], 758
12124	Propionic acid, 2,2-dichloro-3,3,3-trifluoro, propyl ester $CF_3CCl_2CO_2C_3H_7$	239.02	...	144.5^{625}	...	1.3531^{20}	1.3888^{20}	...	B2[4], 758
12125	Propionic acid, 2,3-dichloro ... $CH_2ClCHClCO_2H$	142.97	hyg nd (peth)	210, 113^{12}	50	...	1.4650^{20}	al, eth, w	B2[4], 756
12126	Propionic acid, 2,3-dichloro, ethyl ester ... $CH_2ClCHClCO_2C_2H_5$	171.02	...	183-4, $76\text{-}7^{15}$	...	$1.2461^{20/4}$	1.4482^{20}	al, eth	B2[4], 756
12127	Propionic acid, 2,3-dichloro, methyl ester-(d) ... $CH_2ClCHClCO_2CH_3$	157.00	$[\alpha]^{20}{}_D$ +1.7	92^{50}	...	$1.3282^{20/4}$	...	al, eth, ace	B2[1], 111
12128	Propionic acid, 2,3-dichloro, isopropyl ester ... $CH_2ClCHClCO_2CH(CH_3)_2$	185.05	...	$61\text{-}2^{5}$	...	1.2010	1.4470	...	B2[4], 756
12129	Propionic acid, 3,3-dichloro ... $CHCl_2CH_2CO_2H$	142.97	pr	...	56	...	...	al, ace, bz, chl	B2[4], 758
12130	Propionic acid, 2,3-dihydroxy or Glyceric acid ... $HOCH_2CH(OH)CO_2H$	106.08	syr	d	...	...	...	**w, al, ace**	**B3[4], 1050**

PHYSICAL CONSTANTS OF ORGANIC COMPOUNDS (Continued)

No.	Name, Synonyms, and Formula	Mol. wt.	Color, crystalline form, specific rotation and λ_{max} (log ε)	b.p. °C	m.p. °C	Density	n_D	Solubility	Ref.
12131	Propionic acid, 2,3-dihydroxy, propyl ester $HOCH_2CH(OH)CO_2C_3H_7$	148.16		$132\text{-}4^{3}$		$1.1537^{20/4}$	1.4503^{20}		B3^{3}, 854
12132	Propionic acid, 2,2-dimethyl or Pivalic acid $(CH_3)_3CCO_2H$	102.13	nd	164, 70^{14}	35	0.905^{50}	$1.3931^{36.5}$	al, eth	B2^{4}, 908
12133	Propionic acid, 2,2-diphenyl $CH_3C(C_6H_5)_2CO_2H$	226.27	pl (bz-peth), nd (w), lf (dil al)	sub >300	175-7			al, eth, bz, chl	B9^{3}, 3342
12134	Propionic acid, 2,3-diphenyl (d) $C_6H_5CH_2CH(C_6H_5)CO_2H$	226.27	cr (dil al), $[\alpha]^{20}{}_D$ + 94 (bz)		83-9			al, eth, bz	B9^{3}, 3333
12135	Propionic acid, 2,3-diphenyl (dl) $C_6H_5CH_2CH(C_6H_5)CO_2H$	226.27	(i) pr (chl) (ii) pl (chl) (iii) cr (MeOH)	330-40	(i)88-9 (ii)95-6 (iii)82			al, eth, bz	B9^{3}, 3333
12136	Propionic acid, 2,3-diphenyl (l) $C_6H_5CH_2CH(C_6H_5)CO_2H$	226.27	nd (dil al), $[\alpha]^{20}{}_D$ -85.1 (bz)		83-9			al, eth, bz	B9^{3}, 3333
12137	Propionic acid, 2,3-diphenyl-2-hydroxy or α-Benzylmandelic acid $C_6H_5CH_2C(C_6H_5)(OH)CO_2H$	242.27	nd (bz), cr (dil al)		165.6			al, eth, aa	B10^{3} 1193
12138	Propionic acid, 3,3-diphenyl or Benzhydryl acetic acid $(C_6H_5)_2CHCH_2CO_2H$	226.27	nd (dil al)		155			al, eth	B9^{3}, 3338
12139	Propionic acid, 3,3-diphenyl-2-hydroxy or β,β-Diphenyl lactic acid $(C_6H_5)_2CHCH(OH)CO_2H$	242.27	nd (w)		159d			al, eth	B10^{2}, 228
12140	Propionic acid, 3,3-diphenyl-3-hydroxy $(C_6H_5)_2C(OH)CH_2CO_2H$	242.27	nd (dil al)		212			al, ace, aa	B10^{3}, 1196
12141	Propionic acid, 3,3-diphenyl-3-hydroxy, ethyl ester $(C_6H_5)_2C(OH)CH_2CO_2C_2H_5$	270.33	pr (dil al)		87			al	B10^{2}, 228
12142	Propionic acid, 2,3-epoxy or Glycidic acid CH_2CHCO_2H (–O–)	88.06						**w, al, eth**	B18^{1}, 435
12143	Propionic acid, ethyl ester or Ethyl propionate $CH_3CH_2CO_2C_2H_5$	102.13		99.1	−73.9	0.8917^{20}	1.3839^{20}	**al, eth**, ace	B2^{4}, 705
12144	Propionic acid, 3-fluoro $FCH_2CH_2CO_2H$	92.07		$83\text{-}4^{14}$			1.3889^{25}	w, al, eth	B2^{4}, 734
12145	Propionic acid, 3-(2-furyl) or Furfuryl acetic acid $(2\text{-}C_4H_3O)CH_2CH_2CO_2H$	140.14	cr (chl-lig, w, peth)	229, $108\text{-}10^{10}$	58			w, eth, chl	B18^{4}, 4090
12146	Propionic acid, 3-(2-furyl), ethyl ester $(2\text{-}C_4H_3O)CH_2CH_2CO_2C_2H_5$	168.19		212, $108\text{-}10^{10}$			1.4812^{25}		B18^{4}, 4090
12147	Propionic acid, furfuryl ester or Furfuryl propionate $CH_3CH_2CO_2CH_2(C_4H_3O)$	154.17		$195\text{-}6^{762}$		$1.1085^{20/4}$		al, **eth**, ace	B17^{2}, 115
12148	Propionic acid, heptyl ester or Heptyl propionate $CH_3CH_2CO_2C_7H_{15}$	172.27		210, $124\text{-}5^{16}$	−50.9	$0.8679^{20/4}$	1.4201^{15}	al, eth, ace	B2^{4}, 710
12149	Propionic acid, hexyl ester or Hexyl propionate $CH_3CH_2CO_2C_6H_{13}$	158.24		190, $73\text{-}4^{10}$	−57.5	0.8698^{20}	1.4162^{15}	al, eth, ace	B2^{4}, 709
12150	Propionic acid, 2-hydroxy (D) or D-Lactic acid $CH_3CH(OH)CO_2H$	90.08	pl (chl aa), $[\alpha]_D$ -2.3 (w, c=1.24)	103^{2}	53			w, al	B3^{4}, 633
12151	Propionic acid, 3-hydroxy or Hydracrylic acid $HOCH_2CH_2CO_2H$	90.08	syr	d			1.4489^{20}	w, al, **eth**	B3^{4}, 689
12152	Propionic acid, 3-hydroxy, lactone or β-Propiolactone CH_2CH_2CO (–O–)	72.06		162d, 51^{10}	−33.4	$1.1460^{20/5}$	1.4105^{20}	**eth**, chl	B17^{4}, 4157
12153	Propionic acid, 3-Hydroxy, methyl ester $HOCH_2CH_2CO_2CH_3$	104.11		121^{94}		1.118^{25}	1.4306^{23}	w, al, eth	B3^{3}, 525
12154	Propionic acid, 3-(2-hydroxyphenyl) or o-Hydrocoumaric acid $(2\text{-}HOC_6H_4)CH_2CH_2CO_2H$	166.18	pr (w)		82-4			w, al, eth	B10^{3}, 534
12155	Propionic acid, 3-(4-hydroxy phenyl) $(4\text{-}HOC_6H_4)CH_2CH_2CO_2H$	166.18		$208\text{-}10^{14}$	129-30			w, al, eth, bz	B10^{3}, 539
12156	Propionic acid, 3-(4-hydroxy-3-methoxy phenyl) or Hydroferulic acid $(3\text{-}CH_3O\text{-}4\text{-}HOC_6H_3)CH_2CH_2CO_2H$	196.20	pl (w)		89-90			al, eth	B10^{3}, 1517
12157	Propionic acid, 2-hydroxy-3,3,3-trichloro $Cl_3CCH(OH)CO_2H$	193.41	pr (eth)	$140\text{-}7^{49}$	125			al, eth, chl	B3^{4}, 680

PHYSICAL CONSTANTS OF ORGANIC COMPOUNDS (Continued)

No.	Name, Synonyms, and Formula	Mol. wt.	Color, crystalline form, specific rotation and λ_{max} (log ε)	b.p. °C	m.p. °C	Density	n_D	Solubility	Ref.
12158	Propionic acid, 2,2′-imino-di-(dl) $HN[CH(CH_3)CO_2H]_2$	161.16	(i) nd (w), pr (eth) (ii) cr	....	(i) 234-5d (ii) 254-5	....	....	w	B4^{3}, 1251/1252
12159	Propionic acid, 3,3-imino-di, N-methyl, diethyl ester $CH_3N[CH_2CH_2CO_2C_2H_5]_2$	231.29	....	136-8^{4}	....	1.0190$^{20/20}$	1.4421^{20}	al, eth	B4^{2}, 829
12160	Propionic acid, β-(3-indolyl) $C_{11}H_{11}NO_2$	189.21	pl (w)	....	134	....	....	al, eth, ace, bz, chl	B22^{4}, 1112
12161	Propionic acid, β-(3-indolyl), methyl ester $C_{12}H_{13}NO_2$	203.24	pr (MeOH)	....	79-80	....	....	al	B22^{2}, 53
12162	Propionic acid, 2-iodo (dl) CH_3CHICO_2H	199.98	nd (bz, w, al)	93-6$^{0.2}$	45-7	2.073$^{18/4}$	....	al, eth	B2^{3}, 573
12163	Propionic acid, 3-iodo $ICH_2CH_2CO_2H$	199.98	lf (w)	....	85	....	....	al, eth, ace	B2^{4}, 770
12164	Propionic acid, 3-iodo, methyl ester $ICH_2CH_2CO_2CH_3$	214.00	....	188^{756}	....	1.8408^{7}	....	al	B2^{3}, 574
12165	Propionic acid, 2-mercapto $CH_3CH(SH)CO_2H$	106.14	....	106-7^{15}	10-4	1.1938$^{20/4}$	1.4810^{20}	w, al, eth	B3^{4}, 682
12166	Propionic acid, 3-mercapto $HSCH_2CH_2CO_2H$	106.14	amor	110-12^{15}	17-9	1.218^{21}	1.4911^{20}	w, al, eth	B3^{4}, 726
12167	Propionitrile, 2-methoxy CH_3-$CH(OCH_3)CN$	85.11	....	118^{740}	....	0.8928$^{20/4}$	1.3818^{20}	al	B3^{4}, 675
12168	Propionitrile, 3-methoxy $CH_3OCH_2CH_2CN$	85.11	....	163.5^{761}, 85.5^{49}	....	0.9379$^{20/4}$	1.4043^{20}	al, eth	B3^{4}, 708
12169	Propionic acid, methyl ester or Methyl propionate $CH_3CH_2CO_2CH_3$	88.11	....	79.9	−87.5	0.9150$^{20/4}$	1.3775^{20}	....	B2^{4}, 704
12170	Propionic acid, 2-methyl or Isobuturic acid $(CH_3)_2CHCO_2H$	88.11	....	153.2, 53.7^{10}	−46.1	0.9681$^{20/4}$	1.3930^{20}	**al, eth**, w	B2^{4}, 843
12171	Propionic acid, 3-(α-naphthyl) $(\alpha$-$C_{10}H_7)CH_2CH_2CO_2H$	200.24	cr (bz), nd (al)	179^{11}	156-7	....	....	al	B9^{3}, 3219
12172	Propionic acid, 3-(β-naphthyl) $(\beta$-$C_{10}H_7)CH_2CH_2CO_2H$	200.24	lf or nd (w, al)	....	135	....	....	al	B9^{3}, 3221
12173	Propionic acid, octyl ester or Octyl propionate $CH_3CH_2CO_2C_8H_{17}$	186.29	....	228	−42.6	0.8663^{20}	1.4221$^{15}{}_{D}$	al, eth, bz	B2^{4}, 710
12174	Propionic acid, 2-oxo or Pyruvic acid CH_3COCO_2H	88.06	....	165d, 54^{10}	13.8	1.2272$^{20/4}$	1.4280^{20}	**w, al, eth**, ace	B3^{4}, 1505
12175	Propionic acid, 3-oxo-2-phenyl, ethyl ester $C_6H_5CH(CHO)CO_2C_2H_5$	192.21	pl (chl)	136^{16}	70-1	1.1204$^{20/20}$	1.532^{21}	**al**, eth	B10^{3}, 3023
12176	Propionic acid, pentachloro $CCl_3CCl_2CO_2H$	246.30	cr (CCl_4)	....	200-15d	....	....	....	B2^{2}, 228
12177	Propionic acid, pentyl ester or Pentyl propionate $CH_3CH_2CO_2C_5H_{11}$	144.21	....	168.6	−73.1	0.8761$^{25/4}$	1.4096^{15}	**al, eth**, bz	B2^{4}, 709
12178	Propionic acid, isopentyl ester or Isopentyl propionate $CH_3CH_2CO_2$-i-C_5H_{11}	144.21	....	160.7	....	0.8697$^{20/4}$	1.4069^{20}	....	B2^{4}, 709
12179	Propionic acid, 2-phenoxy (D) $CH_3CH(OC_6H_5)CO_2H$	166.18	nd (w), $[\alpha]^{17}_{D}$ + 39.3 (al,c=1.2)	265-6^{758}	87	....	....	al, eth	B6^{4}, 642
12180	Propionic acid, 2-phenoxy (DL) $CH_3CH(OC_6H_5)CO_2H$	166.18	nd (w)	265-6, 105-6^{5}	115-6	1.1865$^{20/4}$	1.5184^{20}	al, eth	B6^{4}, 642
12181	Propionic acid, 2-phenoxy, ethyl ester $CH_3CH(OC_6H_5)CO_2C_2H_5$	194.23	....	243-4, 120-5^{6}	....	1.360$^{17/4}$	....	....	B6^{4}, 643
12182	Propionic acid, 3-phenoxy $C_6H_5OCH_2CH_2CO_2H$	166.18	nd (w), lf (lig)	234-45^{771}, 188-9^{24}	97-8	....	....	....	B6^{4}, 643
12183	Propionic acid, 3-phenoxy, ethyl ester $C_6H_5OCH_2CH_2CO_2C_2H_5$	194.23	nd (peth)	170^{40}	24	1.0821$^{25/25}$	1.5007^{18}	al, eth	B6^{4}, 644
12184	Propionic acid, phenyl ester or Phenyl propionate $CH_3CH_2CO_2C_6H_5$	150.18	pr	211, 100^{16}	20	1.0467$^{25/25}$	1.4980^{20}	al, eth, bz	B6^{4}, 615
12185	Propionic acid, 2-phenyl (d) or Hydratropic acid $CH_3CH(C_6H_5)CO_2H$	150.18	$[\alpha]^{20}_{D}$ + 81.1 (al,c=3)	152^{16}	....	....	....	....	B9^{3}, 2417
12186	Propionic acid, 3-phenyl or Hydrocinnamic acid $C_6H_5CH_2CH_2CO_2H$	150.18	nd (w)	....	106-8	....	....	al, eth	B9^{3}, 2382
12187	Propionic acid, 3-(3-pyrenyl) $(3$-$C_{16}H_9)CH_2CH_2CO_2H$	274.32	pl (aa)	....	180	....	....	ace, bz	E14, 441
12188	Propionic acid, propyl ester or Propyl propionate $CH_3CH_2CO_2C_3H_7$	116.16	....	122.3	−75.9	0.8809$^{20/4}$	1.3935^{20}	**al, eth**, ace, chl	B2^{4}, 707

No.	Name, Synonyms, and Formula	Mol. wt.	Color, crystalline form, specific rotation and λ_{max} (log ε)	b.p. °C	m.p. °C	Density	n_D	Solubility	Ref.
12189	Propionic acid, isopropyl ester or Isopropyl propionate $CH_3CH_2CO_2CH(CH_3)_2$	116.16		109-10		$0.8660^{20/4}$	1.3872^{20}	al, eth	B2[4], 708
12190	Propionic acid, 2,2,3,3-tetrachloro $Cl_2CHCCl_2CO_2H$	211.86	cr (CS_2-chl)		76			w	B2[4], 760
12191	Propionic acid, 2,3,3,3 tetrafluoro, ethyl ester $CF_3CHFCO_2C_2H_5$	174.10		108-9		$1.289^{20/4}$	1.3260^{20}		B2[4], 737
12192	Propionic acid, tetrahydrofurfuryl ester $CH_3CH_2CO_2CH_2(C_4H_7O)$	158.20		$204\text{-}7^{756}$, $85\text{-}7^{3}$		$1.044^{20/0}$		al, eth, chl	B17[4], 1104
12193	Propionic acid, 3-(2-tetrahydrofuryl) $(2\text{-}C_4H_7O)CH_2CH_2CO_2H$	144.17		263, $118\text{-}20^{2}$		$1.1155^{20/20}$	1.4578^{25}	al, ace	B18[4], 3846
12194	Propionic acid, 3-(2-tetrahydrofuryl), ethyl ester $(2\text{-}C_4H_7O)CH_2CH_2CO_2C_2H_5$	172.22		$221\text{-}2^{750}$, 73^{2}		$1.024^{7/15}$	1.440^{20}	al, ace	B18[4], 3846
12195	Propionic acid, 2,2,3-trichloro $CH_2ClCCl_2CO_2H$	177.41	hyg pr (CS_2)	140^{40}	65-6			al, bz	B2[4], 759
12196	Propionic acid, 2,3,3-triphenyl $(C_6H_5)_2CHCH(C_6H_5)CO_2H$	302.37	nd (dil al, peth)		222-3			al, eth	B9[3], 3598
12197	Propionic acid, 3,3,3-triphenyl $(C_6H_5)_3CCH_2CO_2H$	302.37	pr (al)		179-80			al, eth	B9[3], 3603
12198	Propionitrile CH_3CH_2CN	55.08		97.3	−92.9	$0.7818^{20/4}$	1.3655^{20}	w, al, eth, ace, bz	B2[4], 728
12199	Propionitrile, 2-bromo $CH_3CHBrCN$	133.98		59^{24}		$1.5505^{20/4}$	1.4585^{20}	eth, ace	B2[4], 764
12200	Propionitrile, 3-bromo $BrCH_2CH_2CN$	133.98		92^{25}, 69^{7}		$1.6152^{20/4}$	1.4800^{20}	al, eth	B2[4], 766
12201	Propionitrile, 2-chloro $CH_3CHClCN$	89.52		123-4, 73^{144}		1.0792^{10}			B2[4], 748
12202	Propionitrile, 3-chloro $ClCH_2CH_2CN$	89.52		175-6, $85\text{-}7^{20}$		1.1573^{20}	1.4360^{20}		B2[4], 751
12203	Propionitrile, 2,3-dichloro $CH_2ClCHClCN$	123.97		$58\text{-}9^{8}$		1.3500^{20}	1.4640^{20}		B2[4], 757
12204	Propionitrile, 2,2-dimethyl $(CH_3)_3CCN$	83.13		105-6	15-6	$0.7586^{25/4}$	1.3774^{20}		B2[4], 913
12205	Propionitrile, 2-(dimethylamino) $[(CH_3)_2N]CH(CH_3)CN$	98.15		144				al, eth	B4[3], 1236
12206	Propionitrile, 2-ethoxy $CH_3CH(OC_2H_5)CN$	99.13		136^{765}		$0.8743^{20/4}$	1.3890^{22}	al, eth, aa	B3[4], 675
12207	Propionitrile, 3-ethoxy $C_2H_5OCH_2CH_2CN$	99.13		171, 65^{15}		$0.9285^{15/4}$	1.4068^{20}	al, eth	B3[4], 708
12208	Propionitrile, 2-hydroxy or Acetaldehyde cyanohydrin $CH_3CH(OH)CN$	71.08		182-4, 102^{30}	−40	$0.9877^{20/4}$	1.4058^{18}	w, al, eth, chl	B3[4], 675
12209	Propionitrile, 2-hydroxy-3,3,3-trichloro or Chloral cyanohydrin $Cl_3CCH(OH)CN$	174.41	pl (w)	215-20d	61			w, al, eth	B3[4], 680
12210	Propionitrile, 3-hydroxy or Hydracrylonitrile $HOCH_2CH_2CN$	71.08		230, 110^{15}		1.0588^{0}	1.4240^{20}	w, al	B3[4], 708
12211	Propionitrile, 2,2′-imino-di- or 2,2′-imino dipropionitrile $HN[CH(CH_3)CN]_2$	123.16	nd (eth)		68			al, eth	B4[3], 1251
12212	Propionitrile, 3-(2-oxocyclohexyl) $C_9H_{13}NO$	151.21		$138\text{-}42^{10}$		$1.0181^{20/4}$	1.4755^{20}		B10[3], 2835
12213	Propionitrile, 2-propoxy $CH_3CH(OC_3H_7)CN$	113.16		150^{727}		$0.866^{20/4}$	1.398^{20}		B3, 285
12214	Propionitrile, 3-phenyl $C_6H_5CH_2CH_2CN$	131.18		261, $125\text{-}6^{15}$		1.0016^{20}	1.5266^{20}	al, eth	B9[3], 2395
12215	Propionyl bromide CH_3CH_2COBr	136.98		$103\text{-}1^{770}$		$1.5210^{16/4}$	1.4578^{16}	eth	B2[4], 724
12216	Propionyl bromide, 2-bromo $CH_3CHBrCOBr$	215.87		152-4		$2.0612^{16/4}$			B2[4], 764
12217	Propionyl chloride CH_3CH_2COCl	92.53		80	−94	1.0646^{20}	1.4032^{20}	eth	B2[4], 724
12218	Propionyl chloride, 2-bromo $CH_3CHBrCOCl$	171.42		131-3		1.697^{11}		eth, chl	B2[4], 764
12219	Propionyl chloride, 3 chloro CH_2ClCH_2COCl	126.97	yesh	$143\text{-}5^{763}$, 82^{102}		1.3307^{13}	1.4549^{20}	al, eth, chl	B2[4], 750
12220	Propionyl chloride, 2,2-dichloro CH_3CCl_2COCl	161.42		$117\text{-}8^{750}$, 68-72		$1.4062^{20/4}$	1.4524^{20}		B2[4], 755
12221	Propionyl chloride, 2,3-dichloro $CH_2ClCHClCOCl$	161.42		$52\text{-}4^{16}$		$1.4757^{20/4}$	1.4764^{20}	eth	B2[4], 757

PHYSICAL CONSTANTS OF ORGANIC COMPOUNDS (Continued)

No.	Name, Synonyms, and Formula	Mol. wt.	Color, crystalline form, specific rotation and λ_{max} (log ε)	b.p. °C	m.p. °C	Density	n_D	Solubility	Ref.
12222	Propionyl chloride, 3,3-dichloro $CHCl_2CH_2COCl$	161.42		$43\text{-}4^{10}$		$1.4557^{20/4}$	1.4738^{20}	eth, diox	$B2^3$, 562
12223	Propionyl chloride, pentachloro CCl_3CCl_2COCl	264.75	nd		42			bz	$B2^4$, 761
12224	Propionyl chloride, 2-phenoxy $CH_3CH(OC_6H_5)COCl$	184.62		$146\text{-}7^{55}$, $115\text{-}7^{10}$		$1.1865^{20/4}$	1.5178^{20}	eth	$B6^4$, 643
12225	Propionyl fluoride CH_3CH_2COF	76.07		44		$0.972^{15/4}$	1.329^{11}		$B2^4$, 724
12226	Propionyl iodide CH_3CH_2COI	183.98		127-8					$B2^3$, 542
12227	Propioryl urea $CH_3CH_2CONHCONH_2$	116.12	cr (w)		210-11			w, al	$B3^3$, 121
12228	Propiophenone or Ethyl phenyl ketone $C_6H_5COC_2H_5$	134.18		217.5, 91.6^{10}	18.6	1.0096^{20}	1.5269^{20}	al, eth	$B7^3$, 1022
12229	Propiophenone, α-amino, hydrochloride $C_6H_5COCH(NH_2)CH_3.HCl$	185.65	nd (al-eth)		187			w	$B14^3$, 147
12230	Propiophenone, o-amino $(2\text{-}H_2NC_6H_4)COCH_2CH_3$	149.19	pa ye lf (peth) pl (dil al)	$93^{0.8}$	46-7			w, al, eth, ace	$B14^3$, 143
12231	Propiophenone, [p] -amino $(4\text{-}H_2NC_6H_4)COCH_2CH_3$	149.19	pl (al, w), nd (w)		140			w, al, chl	$B14^3$, 146
12232	Propiophenone, α-bromo $C_6H_5COCHBrCH_3$	213.07		245-50, $134\text{-}5^{18}$		$1.4298^{20/4}$	1.5720^{20}	al, eth, ace, bz, chl	$B7^3$, 1033
12233	Propiophenone, [p] -bromo $(4\text{-}BrC_6H_4)COC_2H_5$	213.07	nd	169^{15}	48			al, eth, ace	$B7^3$, 1032
12234	Propiophenone, β-chloro $C_6H_5COCH_2CH_2Cl$	168.62	lf (eth), cr (peth al)		49-50				$B7^3$, 1030
12235	Propiophenone, [p]- chloro $(4\text{-}ClC_6H_4)COC_2H_5$	168.62		$134\text{-}7^{31}$	36-7			al	$B7^3$, 1029
12236	Propiophenone, [p] -chloro, oxime $4\text{-}ClC_6H_4C(=NOH)C_2H_5$	183.64	pl (al)		62-3			al	B7, 301
12237	Propiophenone, α,β-dibromo-β-phenyl or [threo] -Chalone dibromide $C_6H_5COCHBrCHBrC_6H_5$	368.07	nd (al)		122-3			al	$B7^3$, 2155
12238	Propiophenone, [m] -methoxy $(3\text{-}CH_3OC_6H_4)COCH_2CH_3$	164.20		258-60, $95\text{-}7^{0.7}$		1.0812^{0}	1.5230^{25}	al, eth	B8, 106
12239	Propiophenone, β,β-diphenyl $C_6H_5COCH_2CH(C_6H_5)_2$	286.37	nd (al)		96			ace, bz, chl	$B7^3$, 2756
12240	Propiophenone, [p] -methoxy $(4\text{-}CH_3OC_6H_4)COCH_2CH_3$	164.20		267-9, 142^{14}	<−15	$1.0670^{18/4}$	1.5253^{20}	al, eth	$B8^3$, 381
12241	Propiophenone, [m] -methyl or Ethyl-3-tolyl ketone $(3\text{-}CH_3C_6H_4)COC_2H_5$	148.20		234^{745}, $130\text{-}5^{33}$		$1.0059^{0/4}$		al, eth, ace, bz	$B7^3$, 1092
12242	Propiophenone, [p] -methyl or Ethyl-4-tolyl ketone $(4\text{-}CH_3C_6H_4)COC_2H_5$	148.20		238-9, 120^{18}		$0.9926^{20/4}$	1.5278^{20}	al, eth, ace, bz	$B7^3$, 1093
12243	Propiophenone oxime or Propiophenoxime $C_6H_5C(=NOH)C_2H_5$	149.19	pl (peth)	245-6d, 165^{38}	53-5			al, eth	$B7^3$, 1025
12244	n-Propyl alcohol or 1-Propcenol $CH_3CH_2CH_2OH$	60.10		97.4	−126.5	$0.8035^{20/4}$	1.3850^{20}	w, **al, eth**, ace, bz	$B1^4$, 1413
12245	Isopropyl alcohol or 2-Propanol $(CH_3)_2CHOH$	60.10		82.4	−89.5	$0.7855^{20/4}$	1.3776^{20}	**w, al, eth**, ace, bz	$B1^4$, 1461
12246	n-Propyl amine or 1-Amino propane $CH_3CH_2CH_2NH_2$	59.11		47.8	−83	$0.7173^{20/4}$	1.3870^{20}	w, al, eth, ace, bz, chl	$B4^4$, 464
12247	Propyl amine, 3-(diethyl amino) $(C_2H_5)_2NCH_2CH_2CH_2NH_2$	130.23		165-72	$0.825^{20/20}$	1.443^{20}_D			$B4^4$, 1260
12248	n-Propyl amine, N,N-dimethyl or -1-(dimethyl amino)propane $CH_3CH_2CH_2N(CH_3)_2$	87.16		65.5^{752}		$0.7152^{20/4}$	1.3860^{20}	al, eth, bz, chl	$B4^4$, 467
12249	Propyl amine, N-nitro $CH_3CH_2CH_2NHNO_2$	104.11		$128\text{-}9^{40}$	−21	1.1046^{15}	1.4610^{20}	**al, eth**	$B4^1$, 569
12250	Isopropyl amine or 2-Amino propane $(CH_3)_2CHNH_2$	59.11		32.4	−95.2	0.6891^{20}	1.3742^{20}	**w, al, eth**, ace, bz, chl	$B4^4$, 504
12251	n-Propyl bromide or 1-Bromo propane $CH_3CH_2CH_2Br$	122.99		71	−110	$1.3537^{20/4}$	1.4343^{20}	al, eth, ace, bz	$B1^4$, 205
12252	Isopropyl bromide or 2-Bromo propane $(CH_3)_2CHBr$	122.99		59.4	−89	$1.3140^{20/4}$	1.4251^{20}	**al, eth**, ace, bz, chl	$B1^4$, 208
12253	n-Propyl chloride or 1-Chloro propane $CH_3CH_2CH_2Cl$	78.54		46.6	−122.8	$0.8909^{20/4}$	1.3879^{20}	**al, eth**, bz, chl	$B1^4$, 189

No.	Name, Synonyms, and Formula	Mol. wt.	Color, crystalline form, specific rotation and λ_{max} (log ε)	b.p. °C	m.p. °C	Density	n_D	Solubility	Ref.
12254	Isopropyl chloride or 2-Chloro propane $(CH_3)_2CHCl$	78.54	...	35.7	−117.2	$0.8617^{20/4}$	1.3777^{20}	**al, eth**, bz, chl	$B1^4$, 191
12255	Propyl, 1,2-dichloro propyl ether $C_3H_7O(CHClCHClCH_3)$	171.07	...	176	...	$1.129^{15/4}$	1.447^{16}	eth	$B1^3$, 2691
12256	Propyl, 1,3-dichloro propyl ether $C_3H_7O(CHClCH_2CH_2Cl)$	171.07	...	65^{12}	...	1.112^{20}	1.4476^{20}	al, eth	$B1^2$, 690
12257	Propylene or Propene $CH_3CH=CH_2$	42.08	gas	−47.4	−185.2	$0.5193^{20/4}$ liq	1.3567^{-70}	al, aa	$B1^4$, 725
12258	Propylene, 3-amino or Allyl amine $H_2NCH_2CH=CH_2$	57.10	...	58	...	$0.7621^{20/4}$	1.4205^{20}	**w, al, eth**, chl	$B4^4$, 1057
12259	Propylene, 1-bromo (cis) or 1-Bromopropene, Propenyl bromide $CH_3CH=CHBr$	120.98	...	57.8	−113	$1.4291^{20/4}$	1.4560^{20}	eth, ace, chl	$B1^4$, 754
12260	Propylene, 2-bromo or Isopropenyl bromide $CH_3CBr=CH_2$	120.98	...	48.4^{748}	−124.8	$1.362^{20/4}$	1.4440^{20}	eth, ace, chl	$B1^4$, 754
12261	Propylene, 2-bromo-3-cyclohexyl $C_6H_{11}CH_2CBr=CH_2$	203.12	...	90^{15}	...	1.215^{17}	1.495^{17}	eth	$B5^3$, 224
12262	Propylene, 3-bromo or Allyl bromide $BrCH_2CH=CH_2$	120.98	...	70^{752}	−119.4	$1.398^{20/4}$	1.4697^{20}	**al, eth**	$B1^4$, 754
12263	Propylene, 3-bromo-3,3-difluoro $BrCF_2CH=CH_2$	156.96	...	42	...	$1.543^{25/4}$	1.3773^{25}	...	$B1^4$, 756
12264	Propylene, 1-chloro or [cis] -Propenyl chloride $CH_3CH=CHCl$	76.53	...	32.8	−134.8	$0.9347^{20/4}$	1.4055^{20}	eth, ace, bz, chl	$B1^4$, 737
12265	Propylene, 1-chloro-([trans]) or [trans] -Propenyl chloride $CH_3CH=CHCl$	76.53	...	37.4	−99	$0.9350^{20/7}$	1.4054^{20}	eth, ace, bz, chl	$B1^4$, 737
12266	Propylene, 2-chloro or Isopropenyl chloride $CH_3CCl=CH_2$	76.53	...	22.6	−137.4	$0.9014^{20/4}$	1.3973^{20}	eth, ace, bz, chl	$B1^4$, 737
12267	Propylene, 3-chloro or Allyl chloride $ClCH_2CH=CH_2$	76.53	...	45	−134.5	0.9376^{20}_{D}	1.4157^{20}	**al, eth, ace, bz, lig**	$B1^4$, 738
12268	Propylene, 3-chloro-2-(chloro methyl) $(ClCH_2)_2C=CH_2$	125.00	...	138, $30\text{-}1^{9}$	−15	$1.1782^{20/3}$	1.4754^{20}	al, chl	$B1^4$, 805
12269	Propylene, 2-cyclopropyl $CH_2=C(C_3H_5)CH_3$	82.15	...	$69.5\text{-}70^{751}$	...	$0.7500^{20/4}$	1.4252^{20}	...	$B5^4$, 243
12270	Propylene, 1,1-dibromo $CH_3CH=CBr_2$	199.87	...	127.4, 41.5^{30}	...	$1.9803^{20/20}$	1.5260^{20}	chl, bz	$B1^4$, 759
12271	Propylene, 2,3-dibromo or 2,3-Dibromo propene $CH_2BrCBr=CH_2$	199.87	...	141, $37\text{-}7^{11}$	...	$2.0346^{25/4}$	1.5416^{25}	eth, ace, chl	$B1^4$, 760
12272	Propylene, 1,1-dichloro or 1,1-Dichloro propene $CH_3CH=CCl_2$	110.97	...	76-7	...	$1.1864^{25/4}$	1.4430^{25}	eth, ace, chl	$B1^4$, 742
12273	Propylene, 1,1-dichloro-2-fluoro $CH_3CF=CCl_2$	128.96	...	77.7^{745}	...	$1.3026^{25/4}$	1.4196^{25}	...	$B1^4$, 745
12274	Propylene, 1,1-dichloro-3-phenyl or Cinnamylidene chloride $C_6H_5CH_2CH=CCl_2$	187.07	cr (eth, chl), pl (peth)	$142\text{-}3^{30}$	59	...	...	eth, bz, chl	$B5^4$, 1363
12275	Propylene, 1,2-dichloro ([cis]) $CH_3CCl=CHCl$	110.97	...	92.5^{742}	...	...	1.4549^{20}	ace, bz, chl	$B1^4$, 742
12276	Propylene, 1,2-dichloro ([trans]) or 1,2-Dichloro propene $CH_3CCl=CHCl$	110.97	...	77^{757}	...	$1.1818^{20/4}$	1.4471^{20}	al	$B1^4$, 742
12277	Propylene, 1,2-dichloro-1,3,3,3-tetrafluoro $CF_3CCl=CFCl$	182.93	...	47.3	−137	1.5468	1.3511^{20}	...	$B1^4$, 747
12278	Propylene, 1,2-dichloro-3,3,3-trifluoro $CF_3CCl=CHCl$	164.94	...	53.7	−109.2	$1.4653^{20/4}$	1.3670^{20}	...	$B1^4$, 746
12279	Propylene, 1,3-dichloro ([cis]) or 1,3-Dichloro propene $ClCH_2CH=CHCl$	110.97	...	104.3	...	$1.217^{20/4}$	1.4730^{20}	eth, bz, chl	$B1^4$, 743
12280	Propylene, 1,3-dichloro-([trans]) $ClCH_2CH=CHCl$	110.97	...	112	...	$1.224^{20/4}$	1.4682^{20}	eth, bz, chl	$B1^4$, 744
12281	Propylene, 2,3-dichloro $CH_2ClCCl=CH_2$	110.97	...	94	...	$1.211^{20/4}$	1.4603^{20}	**al,** eth, bz, chl	$B1^4$, 744
12282	Propylene, 3,3-dichloro or 3,3-Dichloro propene $CHCl_2CH=CH_2$	110.97	...	84.4	...	$1.175^{20/4}$	1.4510^{20}	al, eth, bz, chl	$B1^4$, 745
12283	Propylene, 1,1-diphenyl or 1,1-Diphenyl propene $CH_3CH=C(C_6H_5)_2$	194.28	lf (al)	280-1, 149^{11}	52	$1.0250^{20/4}$	1.5880^{20}	al, bz	$B5^3$, 1998
12284	Propylene, 2,3-diphenyl $C_6H_5CH_2C(C_6H_5)=CH_2$	194.28	...	289^{757}	48	$1.10143^{20/4}$	1.5903^{20}	...	$B5^2$, 553
12285	Propylene, 1,2-epoxy or Methyl oxirene $CH_3C=CH$ (O)	56.06	...	63	...	...	...	**al, eth**	B17, 20
12286	Propylene, 3-fluoro or Allyl fluoride $FCH_2CH=CH_2$	60.07	gas	−3	...	...	...	al, eth, chl	$B1^4$, 733

PHYSICAL CONSTANTS OF ORGANIC COMPOUNDS (Continued)

No.	Name, Synonyms, and Formula	Mol. wt.	Color, crystalline form, specific rotation and λ_{max} (log ε)	b.p. °C	m.p. °C	Density	n_D	Solubility	Ref.
12287	Propylene glycol or 1,2-Propanediol $CH_3CH(OH)CH_2OH$	76.10	...	189, 96-8^{21}	...	1.0361$^{20/4}$	1.4324^{20}	**w, al, eth**, bz	B1^4, 2468
12288	Propylene, 3-iodo or Allyl iodide $CH_2{=}CHCH_2I$	167.98	...	102	−99.3	1.8494$^{20/4}$	1.5530^{20}	al, eth, chl	B1^4, 761
12289	Propylene, 1-nitro $CH_3CH{=}CHNO_2$	87.08	...	59-60^{34}	...	1.0661$^{20/4}$	1.4527^{20}	eth, ace, chl	B1^4, 763
12290	Propylene, 2-nitro $CH_2{=}C(NO_2)CH_3$	87.08	...	57^{100}, 32^{30}	...	1.0643$^{20/4}$	1.4358^{20}	eth, ace, chl	B1^4, 764
12291	Propylene, 1,1,2,3,3-pentachloro $CHCl_2CCl{=}CCl_2$	214.31	...	185, 116^{9}	...	1.6317$^{34/4}$	1.5313^{20}	eth	B1^4, 753
12292	Propylene, perchloro $CCl_3CCl{=}CCl_2$	248.75	...	209-10, 140^{100}	...	1.7652$^{20/4}$	1.5455^{20}	chl	B1^4, 753
12293	Propylene, perfluoro $CF_3CF{=}CF_2$	150.02	gas	−29.4	−156.2	1.583$^{-40/4}$	...	...	B1^4, 735
12294	Propylene, 2-phenyl or α-Methyl styrene $C_6H_5C(CH_3){=}CH_2$	118.18	...	163-4, 60^{17}	24.3	0.9082$^{20/4}$	1.5303^{20}	eth, bz, chl	B5^4, 364
12295	Propylene, 1,2,3,3-tetrachloro $CHCl_2CCl{=}CHCl$	179.86	...	165, 50.6^{12}	...	1.537$^{19/4}$	1.5121^{19}	bz, chl	B1^4, 752
12296	Propylene, 1,1,2-trichloro $CH_3CCl{=}CCl_2$	145.42	...	118, 41^{52}	...	1.382$^{20/4}$	1.4827^{20}	al, eth, bz, chl	B1^4, 747
12297	Propylene, 1,1,2-trichloro-3,3,3-trifluoro $CF_3CCl{=}CCl_2$	199.39	...	88.1	−114.6	1.617$^{20/4}$	...	...	B1^4, 751
12298	Propylene, 1,2,3-trichloro $CH_2ClCCl{=}CHCl$	145.42	...	142, 32-3^{14}	...	1.414$^{20/20}$	1.5030^{20}	al, eth, bz, chl	B1^4, 748
12299	Propylene, 3,3,3-trichloro $CCl_3CH{=}CH_2$	145.42	...	114-5, 57^{102}	−30	1.369$^{20/20}$	1.4827^{20}	al, eth, bz, chl	B1^4, 749
12300	Propylene oxide or 1,2-Epoxy propane CH_3CHCH_2 (epoxide O bridging CH and CH₂)	58.08	...	34.3	...	0.859$^{0/4}$	1.3670^{20}	**w, al, eth**	B17^4, 16
12301	Propyl ether or Dipropyl ether $C_3H_7OC_3H_7$	102.18	...	91	fr−122	0.7360$^{20/4}$	1.3809^{20}	**al, eth**	B1^4, 1422
12302	Isopropyl ether or Diisopropyl ether $(CH_3)_2CHOCH(CH_3)_2$	102.18	...	68	−85.9	0.7241$^{20/4}$	1.3679^{20}	**al, eth**, ace	B1^4, 1471
12303	Propyl ethynyl ether $C_3H_7OC{\equiv}CH$	84.12	...	75	...	0.8080$^{20/4}$	1.3935^{20}	al, eth	B1^4, 2213
12304	*n*-Propyl fluoride or 1-Fluoro propane $CH_3CH_2CH_2F$	62.09	...	250	−159	0.7956$^{20/4}$	1.3115^{20}	al, eth	B1^4, 187
12305	Propyl hexedrine or 1-Cyclohexyl-2-methyl amino propane $C_{10}H_{21}N$	135.28	...	205, 92-3^{20}	...	0.8501$^{20/4}$	1.4600^{20}	al	B12^3, 108
12306	*n*-Propyl iodide or 1-Iodo propane $CH_3CH_2CH_2I$	169.99	...	102.4	−10.3	1.7489$^{20/4}$	1.5058^{20}	al, eth, bz, chl	B1^4, 222
12307	Isopropyl iodide or 2-Iodo propane $(CH_3)_2CHI$	169.99	...	891.4	−90.1	1.7033$^{20/4}$	1.5028^{20}	**al, eth, bz, chl**	B1^4, 223
12308	Propyl isocyanide $CH_3CH_2CH_2NC$	69.11	...	99.5	...	...	...	**al, eth**	B4^4, 474
12309	Isopropyl isocyanide $i\text{-}C_3H_7NC$	69.11	...	87	...	0.7596^{0}	...	**al, eth**	B4, 154
12310	Propyl isothiocyanate C_3H_7NCS	101.17	...	153	...	0.9781$^{16/4}$	1.5085^{16}	**al, eth**	B4^4, 491
12311	Isopropyl isothiocyanate or Isopropyl mustard oil $(CH_3)_2CHNCS$	101.17	...	138, 29-30^{10}	...	...	...	al, eth	B4^4, 532
12312	Isopropyl methyl ketone or 3-Methyl-2-butanone $i\text{-}C_3H_7COCH_3$	86.13	...	94-5	−92	0.8051$^{20/4}$	1.3880^{20}	**al, eth**, ace	B1^4, 3287
12313	Propyl-α-naphthyl ether $C_3H_7O\text{-}\alpha\text{-}C_{10}H_7$	186.25	...	293.5, 167^{18}	...	1.0447$^{18/4}$	1.5928^{18}	...	B6^3, 2924
12314	Propyl-β-naphthyl ether $C_3H_7O\text{-}\beta\text{-}C_{10}H_7$	186.25	nd (al)	305, 144^{10}	41	...	...	al	B6^3, 2973
12315	Propyl nitrate $C_3H_7ONO_2$	105.09	...	110^{762}	...	1.0538$^{20/4}$	1.3973^{20}	al, eth	B1^4, 1424
12316	Isopropyl nitrate $(CH_3)_2CHONO_2$	105.09	...	100-1	...	1.036$^{19/19}$	1.3912^{16}	al, eth	B1^4, 1475
12317	Propyl nitrite C_3H_7ONO	89.09	...	79	...	0.935^{21}	1.3604^{20}	al, eth	B1^4, 1424
12318	Isopropyl nitrite $(CH_3)_2CHONO$	89.09	...	45	...	0.8684$^{15/4}$	...	al, eth	B1^4, 1474
12319	Propyl isopropyl ether $i\text{-}C_3H_7OC_3H_7$	102.18	...	83	...	0.7370$^{20/4}$	1.376^{21}	al, eth, ace	B1^4, 1471

PHYSICAL CONSTANTS OF ORGANIC COMPOUNDS (Continued)

No.	Name, Synonyms, and Formula	Mol. wt.	Color, crystalline form, specific rotation and λ_{max} (log ε)	b.p. °C	m.p. °C	Density	n_D	Solubility	Ref.
12320	Propyl red or 4′-Dipropyl aminoazobenzene-2-carboxylic acid $C_{19}H_{23}N_3O_2$	325.41	vt-bl or purp-red cr (al)					w, al	$B16^3$, 368
12321	Propyl sulfide or Dipropyl sulfide $(C_3H_7)_2S$	118.24		142.4, 32.3^{10}	−102.5	$0.8377^{20/4}$	1.4487^{20}	al, eth	$B1^4$, 1452
12322	Propyl sulfide, 2,2′-dichloro or bis(2-chloropropyl)sulfide $(CH_3CHClCH_2)_2S$	187.13		122^{23}	−40	$1.1569^{25/4}$	1.5020^{20}	al	$B1^3$, 1437
12323	Propyl sulfide, 3,3′-dichloro or bis(3-chloro propyl)sulfide $(ClCH_2CH_2CH_2)_2S$	187.13		162^{43}		$1.1774^{25/4}$	1.5075^{20}	al, eth, to	$B1^3$, 1438
12324	Isopropyl sulfide or Diisopropyl sulfide $[(CH_3)_2CH]_2S$	118.24		120	−78.1	$0.8142^{20/4}$	1.4438^{20}	al, eth	$B1^4$, 1502
12325	Propyl sulfone or Dipropyl sulfone $(C_3H_7)_2SO_2$	150.24	sc		29-30	$1.0278^{50/4}$	1.4456^{30}	al, eth	$B1^4$, 1453
12326	Isopropyl sulfone or Diisopropyl sulfone $(i\text{-}C_3H_7)_2SO_2$	150.24	(eth)		36			w	$B1^4$, 1502
12327	Propyl sulfate or Dipropyl sulfate $(C_3H_7O)_2SO_2$	182.23		121^{20}		$1.1064^{20/4}$	1.4135^{20}	peth	$B1^4$, 1423
12328	Propyl sulfoxide or Dipropyl sulfoxide $(C_3H_7)_2SO$	134.24	nd	80^2	22-3	$0.9654^{20/4}$	1.4663^{20}	al, eth	$B1^4$, 1453
12329	Propyl thiocyanate C_3H_7SCN	101.17		163				al, eth	$B3^4$, 329
12330	Isopropyl thiocyanate $(CH_3)_2CHSCN$	101.17		$152\text{-}3^{754}$		0.9784^{20}		**al, eth**	$B3^4$, 329
12331	Isopropyl vinyl ether $i\text{-}C_3H_7OCH{=}CH_2$	86.13		55-6		$0.7534^{20/4}$	1.3840^{20}	al, eth, ace, bz	$B1^4$, 2052
12332	Propynal or Propargyl aldehyde $HC{\equiv}CCHO$	54.05		59-61			1.4033^{25}	**w**, al, eth, ace, bz	$B1^4$, 3537
12333	Propyne or Methyl acetylene $CH_3C{\equiv}CH$	40.06	gas	−23.2	−101.5	0.7062^{-50}	1.3863^{-40}	al, bz, chl	$B1^4$, 958
12334	Propyne, 3-bromo or Propargyl bromide $BrCH_2C{\equiv}CH$	118.96		88-90, 33^{130}		1.579^{18}	1.4922^{20}	al, eth, bz, chl	$B1^4$ 964
12335	Propyne, 3-chloro or Propargyl chloride $ClCH_2C{\equiv}CH$	74.51		65		$1.0297^{20/4}$	1.4320^{20}	**al, eth**, bz, chl	$B1^4$, 963
12336	Propyne, 3-cyclohexyl $C_6H_{11}CH_2C{\equiv}CH$	122.21		157-8, 48^{11}		$0.8449^{20/4}$	1.4605^{20}	eth	$B5^4$, 419
12337	Propyne, 1,3-dibromo $BrCH_2C{\equiv}CBr$	197.86		$73\text{-}4^{30}$		2.1894^{20}	1.5690^{20}	eth, chl	$B1^4$, 965
12338	Propyne, 1-iodo $CH_3C{\equiv}CI$	165.96		110		2.08^{22}		eth, al	$B1^4$, 965
12339	Propyne, 3-iodo or Propargyl iodide $ICH_2C{\equiv}CH$	165.96		115		2.0177^0		eth	$B1^4$, 965
12340	Propyne, 3-methoxy or Methyl propargyl ether $CH_3OCH_2C{\equiv}CH$	70.09		63		0.83^{12}	1.5035^{20}	**al, eth**	$B1^4$, 2215
12341	Propyne, 1-phenyl $CH_3C{\equiv}CC_6H_5$	116.16		183, 77^{17}		$0.9388^{20/4}$	1.5650^{20}	eth	$B5^4$, 1530
12342	Propyne, 3,3,3-trifluoro $F_3CC{\equiv}CH$	94.04	gas	−48.3					$B1^4$, 962
12343	Propynoic acid or Propargylic acid $HC{\equiv}CCO_2H$	70.05	cr (CS_2)	144d, $83\text{-}4^{50}$	18 (anh)	$1.1380^{20/4}$	1.4306^{20}	**w, al, eth**, ace, **chl**	$B2^4$, 1687
12344	2-Propyn-1-ol or Propargyl alcohol $HC{\equiv}CCH_2OH$	56.06		113.6, 30^{21}	−48	0.9485^{20}	1.4322^{20}	w, **al, eth**	$B1^4$, 2214
12345	Prostigmine bromide or Neostigmine bromide $C_{12}H_{19}O_2N_2Br$	303.20	(al-eth)		167d			w, al	$B13^3$, 939
12346	Protopine or Fumarine $C_{20}H_{19}NO_5$	353.37	mcl pr (al-chl)		208			chl	$B27^2$, 625
12347	Protoveratridine $C_{32}H_{51}NO_9$	593.76	cr (al-chl), $[\alpha]^{20/}{}_D$ −14 (Py, c=1)		272-3			al	$B21^4$, 6801
12348	Protoveratrine $C_{39}H_{61}NO_{12}$	751.91	pl (ace), $[\alpha]^{24/}{}_D$ −8.3 (chl)		225d				$B21^4$, 6845
12349	Protoveratrine A $C_{41}H_{63}NO_{14}$	793.95	lf (al), $[\alpha]^{20/}{}_D$ −44.1 (Py, c=1.12)		305d				$B21^4$, 6845

PHYSICAL CONSTANTS OF ORGANIC COMPOUNDS (Continued)

No.	Name, Synonyms, and Formula	Mol. wt.	Color, crystalline form, specific rotation and λ_{max} (log ε)	b.p. °C	m.p. °C	Density	n_D	Solubility	Ref.
12350	Protoveratrine B or Neoprotoveratrine $C_{41}H_{63}NO_{15}$	809.95	lf (chl-al), $[\alpha]^{20/}{}_D$ −39.8 (Py, c=1.24)	....	285-90d	....	....	....	B21^{4}, 6847
12351	Protoverine $C_{27}H_{43}NO_9$	525.64	nd (MeOH)	....	220-2	....	....	al, bz	B21^{4}, 6841
12352	Prulaurasin or dl-Mandelonitrile glucoside $C_{14}H_{17}NO_6$	295.29	wh nd or pl (al), $[\alpha]_D$ −54	....	122	....	....	w, al	B31, 240
12353	Prunasin or d-Mandelonitrile-β-d-glucoside $C_{14}H_{17}NO_6$	295.29	$[\alpha]^{22/}{}_D$ −27.0	....	149-50	....	....	w, al, ace	B17^{4}, 3356
12354	Pseudoaconitine $C_{36}H_{49}NO_{12}$	687.78	tcl (MeOH), $[\alpha]^{15/}{}_D$ + 18.4 (al)	....	214	....	....	al, eth	B21^{4}, 2890
12355	Pseudocodeine or Neoisocodeine $C_{18}H_{21}NO_3$	299.37	wh nd, $[\alpha]_D$ −96.6 (al)	....	181-2	1.290^{80}	1.574	al	B27^{2}, 112
12356	Pseudoconhydrine or 5-Hydroxyconine $C_8H_{17}NO$	143.23	hyg nd (eth), $[\alpha]^{20/}{}_D$ + 11.0 (al, c=10)	236	106 (anh), 60 (+ 1 w)	....	....	w, al, eth	B21^{4}, 121
12357	Psuedoconiceine (L) or γ-Coniceine $C_8H_{15}N$	127.23	hyg, $[\alpha]^{15/}{}_D$ + 122.6	171-2	....	$0.8776^{15/4}$	1.4607^{10}	al, eth, chl	B20, 146
12358	Pseudocumene or 1,2,4-Trimethyl benzene 1,2,4-$(CH_3)_3C_6H_3$	120.19	....	169.3, 51.6^{10}	−43.8	$0.8758^{20/4}$	1.5048^{20}	al, eth, ace, bz, peth	B5^{4}, 1010
12359	Pseudocumene-5-acetyl or 2,4,5-Trimethyl acetophenone 2,4,5-$(CH_3)_3C_6H_2COCH_3$	162.23	....	246-7, 137-8^{20}	10-1	$1.0039^{15/4}$	1.541^{15}	al, eth, bz, aa	B7^{3}, 1145
12360	Pseudoephedrine(d) or Isoephedrine $C_{10}H_{15}NO$	165.24	pr or lf (eth), $[\alpha]^{20/}{}_D$ + 51.9 (abs al, c=0.6)	....	118.7	....	....	al, eth, bz	B13^{1}, 1719
12361	Pseudoephedrine (dl) $C_{10}H_{15}NO$	165.24	nd (eth)	130^{16}	118.2	....	....	al, eth, bz	B13^{1}, 1720
12362	Pseudoephedrine (l) $C_{10}H_{15}NO$	165.24	lf or pr (eth), $[\alpha]^{20/}{}_D$ −51.9 (abs al, c=0.6)	....	118.7	....	....	al, eth, bz	B13^{1}, 1719
12363	Pseudoephedrine hydrochloride (d) $C_{10}H_{15}NO \cdot HCl$	201.70	rh pl or nd (al), $[\alpha]^{20/}{}_D$ + 62 (w, c=0.8)	....	181-2	....	....	....	B13^{1}, 1719
12364	Pseudoephedrine hydrochloride (dl) $C_{10}H_{15}NO \cdot HCl$	201.70	nd (abs al)	....	164	....	....	w, al, eth	B13^{1}, 1720
12365	Pseudoephedrine hydrochloride (l) $C_{10}H_{15}NO.HCl$	201.70	nd (abs al or AcOEt), $[\alpha]^{20/}{}_D$ −62.1 (w, c=1.8)	....	182	....	....	w	B13^{2}, 377
12366	Pseudohyoscyamine or Norhyoscyanine $C_{16}H_{21}NO_3$	275.35	nd, $[\alpha]_D$ −22 (al)	....	140.5	....	....	al, chl	B21^{4}, 167
12367	Pseudoionone or Citrylidenacetone $(CH_3)_2CCH_2CH_2C(CH_3)COCH_3$	140.23	pa ye oil	$143\text{-}5^{12}$	....	0.8984^{20}	1.5335^{20}	al, eth, chl	B1^{4}, 3598
12368	Pseudojervine $C_{33}H_{49}NO_8$	587.75	wh nd or hex pl, $[\alpha]_D$ −133 (al-chl, ⅓)	....	304-5d	....	....	....	C39, 1413
12369	Pseudomorphine or 2,2′-Bimorphine $C_{34}H_{36}N_2O_6$	568.67	cr (aq NH_3 + 3w), $[\alpha]^{24/}{}_D$ + 44.8 (in HCl, c=0.86)	....	282-3	....	....	Py	B27^{2}, 886

No.	Name, Synonyms, and Formula	Mol. wt.	Color, crystalline form, specific rotation and λ_{max} (log ε)	b.p. °C	m.p. °C	Density	n_D	Solubility	Ref.
12370	Pseudomorphine hydrochloride $C_{24}H_{36}N_2O_6.2HCl.2H_2O$	677.62	pw, $[\alpha]_D$ −114.76						B27[2], 886
12371	Pseudopelletierine or Pseudoplenicine $C_9H_{15}NO$	153.22	orh pr (peth)	246	54	1.001^{100}	1.4760^{100}	w, al, eth, chl	B21[4], 3315
12372	Pseudoreserpine $C_{32}H_{38}N_2O_9$	594.66	$[\alpha]^{24}/_D$ -N-65 (chl)		257-8			ace	C52, 2876
12373	Pseudothiohydantiol or 2-Imino-4-thiazolidine $C_3H_4N_2OS$	116.14	pr or nd (w)		255-8d				B27[2], 284
12374	Pseudotropine or Pseudotropanol $C_8H_{15}NO$	141.21	rh ta or pr (eth), rh bipym (peth-bz)		109			w, al, bz, chl	B21[4], 169
12375	D-Psicose or D-Allulose $C_6H_{12}O_6$	180.16	$[\alpha]^{25}/_D$ + 4.7 (w, c=4.3)		58			w, al	B1[4], 4400
12376	Pteridine or Pyrimido[(4,5)]pyrazine $C_6H_4N_4$	132.12	ye pl (bz, sub)	sub 125-30^{20}	139-40			w, al	J1951, 474
12377	Pteridine, 2-amino-4,6-dihydroxy or Uropterin $C_6H_5N_5O_2$	179.14	og-ye (w + 1)		>410d dark-ens 360				B26[2], 313
12378	Pteridine, 2-amino-4-hydroxy or 2-Amino-4-pteridol $C_6H_5N_5O$	163.14	ye		>360				C47, 5945
12379	Pukateine (l) or 4-Hydroxy-5,6-methylenedioxyaporphin $C_{18}H_{17}NO_3$	295.34	cr (al, eth), $[\alpha]^{15}/_D$ −200 (al, c=6)	210-5^{2}	200				B27[1], 461
12380	Pulegenone or 4-Methyl-1-isopropyl cyclopentinen-5-one $C_9H_{14}O$	138.21		188-9		$0.9144^{20/0}$	1.4660^{20}	al, eth, ace	B7[3], 291
12381	Pulegone or 4(8)p-Menthen-3-one $C_{10}H_{16}O$	152.24	$[\alpha]^{20}/_D$ + 23.4 (undil)	224, 103^{17}		0.9346^{45}	1.4894^{20}	**al, eth, chl**	B7[3], 334
12382	Purine or 7-Imidazo[4,5]pyrimidine $C_5H_4N_4$	120.11	nd (to al)	sub	216-7			w, al, ace	B26, 354
12383	Purine, 6-amino or Adenine $C_5H_5N_5$	135.13	nd or lf (w + 3)	sub at 220	360-5d (anh)				B26[2], 252
12384	Purine, 6-mercapto or 6-Purinethiol $C_5H_4N_4S$	152.17	ye pr (w + 1)		313-4d				Am74, 411
12385	Purine, 2,6,8-trichloro $C_5HCl_3N_4$	223.45	nd (al)		159-61			al, eth, ace, bz	B26, 356
12386	Pyocyanine $C_{13}H_{10}N_2O$	210.24	dk bl nd (w + 1), (chl-peth)	sub	133d			ace, chl, Py	B23[2], 361
12387	Pyraconitine $C_{32}H_{43}O_9N$	585.69	nd		171			al	B21[4], 6794
12388	γ-Pyran C_5H_6O	82.10	col oil	80			1.4559^{20}	al, eth, bz	B17, 36
12389	γ-Pyran, 2,3-dihydro C_5H_8O	84.12		86-7		$0.922^{19/15}$	1.4402^{19}	w, al	B17[4], 148
12390	γ-Pyran, tetrahydro or Pentamethylene oxide $C_5H_{10}O$	86.13		88		$0.8810^{20/4}$	1.4200^{20}	al, eth, bz	B17[4], 51
12391	Pyranthrone or 8,16-Pyranthenedione $C_{30}H_{14}O_2$	406.44	red-ye or red-br nd ($PhNO_2$)	sub, vac	d				B7[3], 4514
12392	Pyrazine or 1,4-Diazine $C_4H_4N_2$	80.09	pr (w)	115-6^{768}	54	$1.0311^{61/4}$	1.4953^{61}	w, al, eth, ace	B23[2], 80
12393	Pyrazine, 2,3-dimethyl $2,3(CH_3)_2C_4H_2N_2$	108.14		156		$1.0281^{0/4}$		w, al, eth	B23[2], 80
12394	Pyrazine, 2,5-dimethyl-3-ethyl $2,5-(CH_3)_2-3-C_2H_5(C_4HN_2)$	136.20		180-1		$0.9657^{24/4}$	1.5014^{24}		B23, 99
12395	Pyrazine, 2,5-dimethyl or Ketine $2,5-(CH_3)_2(C_4H_2N_2)$	108.15		155	15	$0.9887^{20/4}$	1.4980^{20}	**w, al, eth**, ace	B23[2], 80
12396	Pyrazine, 2,6-dimethyl $2,6-(CH_3)_2(C_4H_2N_2)$	108.14		155.6	47-8	$0.9647^{50/4}$		w, al, eth	B23, 97
12397	Pyrazine, 2-methyl $2-CH_3(C_4H_3N_2)$	94.12		136-7		$1.0290^{20/4}$	1.5067^{19}	**w, al, eth**, ace	B23, 94
12398	Pyrazine carboxamide $(C_4H_3N_2)CONH_2$	123.11	wh nd (w, al)	sub	191-3				C48, 2074

No.	Name, Synonyms, and Formula	Mol. wt.	Color, crystalline form, specific rotation and λ_{max} (log ε)	b.p. °C	m.p. °C	Density	n_D	Solubility	Ref.
12399	2,3-Pyrazine dicarboxylic acid $2,3-(C_4H_2N_2)(CO_2H)_2$	168.11	pr (w + 2)		193d			w, ace, MeOH	B25², 164
12400	2,3-Pyrazine dicarboxylic acid, 5-methyl $5-CH_3-2,3-(C_4HN_2)(CO_2H)_2$	182.14	(w, ace, dil al-eth)		174-5			w, al	B25², 165
12401	2,5-Pyrazine dicarboxylic acid $2,5-(C_4H_2N_2)(CO_2H)_2$	168.11	nd (w + 2)		255-6d (sealed tube)				B25², 164
12402	2,6-Pyrazine dicarboxylic acid $2,6-(C_4H_2N_2)(CO_2H)_2$	168.11	mcl nd (w + 2)		217-8d			al	B25, 168
12403	Pyrazole or 1,2-Diazole $C_3H_4N_2$	68.08	nd or pr (lig)	186-8	69-70		1.4203	w, al, eth, bz	B23², 33
12404	Pyrazole, 4-bromo-1,3-dimethyl $4-Br-1,3-(CH_3)_2(C_3HN_2)$	175.03		$76-7^{10}$		$1.4975^{15/4}$	1.5214^{15}	al, eth, ace	B23², 54
12405	Pyrazole, 4-bromo-1,5-dimethyl $4-Br-1,5-(CH_3)_2(C_3HN_2)$	175.03	cr	85^{10}	38-9				B23², 54
12406	Pyrazole, 4-bromo-3,5-dimethyl $4-Br-3,5-(CH_3)_2(C_3HN_2)$	175.03	nd (dil al)		123			eth, ace	B23², 68
12407	Pyrazole, 4-bromo-3-methyl $4-Br-3-CH_3(C_3H_2N_2)$	161.00	cr (dil al)		76-7	$1.5638^{100/4}$	1.5182^{100}	al, eth, ace	B23², 54
12408	Pyrazole, 3-chloro-1,5-dimethyl $3-Cl-1,5-(CH_3)_2(C_3HN_2)$	130.58	pl (w)	210-2, 138^{72}	47-8	$1.0823^{100/4}$	1.4648^{100}	al, eth, ace, bz	B23², 49
12409	Pyrazole, 4-chloro-3,5-dimethyl $4-Cl-3,5-(CH_3)_2(C_3HN_2)$	130.58	pr (w), cr (al)	220-2	117-8			al, eth, ace, bz	B23², 67
12410	Pyrazole, 5-chloro-1,3-dimethyl $5-Cl-1,3-(CH_3)_2(C_3HN_2)$	130.58		157-8		$1.1367^{18/4}$	1.4877^{18}	w	B23², 49
12411	Pyrazole, 5-chloro-3-methyl $5-Cl-3-CH_3(C_3H_2N_2)$	116.55	cr (eth, lig)	258, 138^{15}	118-9			w	B23², 49
12412	Pyrazole, 1,3-dimethyl $1,3-(CH_3)_2(C_3H_2N_2)$	96.13		136-8, 31^{1}		$0.9561^{17/4}$	1.4734^{15}	w	B23², 44
12413	Pyrazole, 1,5-dimethyl $1,5-(CH_3)_2(C_3H_2N_2)$	96.13		153		$0.9813^{17/4}$	1.4782^{16}	w, al, eth	B23², 44
12414	Pyrazole, 3,4-dimethyl $3,4-(CH_3)_2(C_3H_2N_2)$	96.13	(peth)	111^{10}	58			al, eth, ace, bz	B23², 64
12415	Pyrazole, 3,5-dimethyl $3,5-(CH_3)_2(C_3H_2N_2)$	96.13	cr (peth, al)	218^{758}	107-8	$0.8839^{16/4}$		al, eth, ace, bz, chl	B23², 65
12416	Pyrazole, 1-ethyl $1-C_2H_5(C_3H_3N_2)$	96.13		139		$0.9537^{20/4}$	1.4700^{20}	al, eth, ace, bz, Py	C55, 22291
12417	Pyrazole, 3-methyl $3-CH_3(C_3H_3N_2)$	82.11		204^{752}, 108^{25}	36-7	$1.0203^{16/4}$	1.4915^{20}	w, al, eth	B23², 44
12418	Pyrazole, 1-isopropyl $1-i-C_3H_7(C_3H_3N_2)$	110.16		143				w, al	C53, 21043
12419	2-Pyrazoline or 4,5-Dihydropyrazole $C_3H_6N_2$	70.09		144		$1.0200^{17/4}$	1.4796^{17}	w, al, eth	B23², 24
12420	2-Pyrazoline, 1-phenyl $1-C_6H_5(C_3H_5N_2)$	146.19	pl (lig)	273^{754}, 151^{17}	52	$1.0689^{58/4}$	1.6015^{58}	al, eth, bz	B23², 25
12421	5-Pyrazolinone, 3-methyl $C_4H_6N_2O$	98.10			215			w	B24², 8
12422	5-Pyrazolinone, 1-methyl-3-phenyl $C_{10}H_{10}N_2O$	174.20		330-40, 235^{68}	213-7			al, eth	B24, 148
12423	5-Pyrazolinone, 3-methyl-1-phenyl $C_{10}H_{10}N_2O$	174.20	[α] + 1.637	287^{105}, 191^{17}	127			w, al	B24², 9
12424	5-Pyrazolone or 2-Pyrazolin-5-on $C_3H_4N_2O$	84.08	nd (to, w)	sub	165			w, al	B24², 6
12425	5-Pyrazolone, 3-methyl $C_4H_6N_2O$	98.10	pr (w), nd (al), lf (sub)	sub	215 (219)			w	B24², 8
12426	5-Pyrazolone, 3-methyl-1(nitrophenyl) $C_{10}H_9N_3O_3$	219.20	ye amor (aa)		185				B24¹, 191
12427	5-Pyrazolone, 3-methyl-1-phenyl $C_{10}H_{10}N_2O$	174.20	mcl pr (w)	287^{105}, 191^{17}	127		1.637		B24², 9
12428	5-Pyrazolone, 3-methyl-1-(4-sulfophenyl) $C_{10}H_{10}N_2O_4S$	254.26	nd (w + 1)		290-320d				B24², 20
12429	5-Pyrazolone-3-carboxylic acid, 1-phenyl $C_{10}H_8N_2O_3$	204.19	nd (w, al)		261			al	B25², 219
12430	Pyrene or Benzo[d,e,f]phenanthrene $C_{16}H_{10}$	202.26	pa ye pl (to, sub)	393, 260^{60}	156	$1.271^{23/4}$		al, eth, bz, lig	B5⁴, 2467

No.	Name, Synonyms, and Formula	Mol. wt.	Color, crystalline form, specific rotation and λ_{max} (log ε)	b.p. °C	m.p. °C	Density	n_D	Solubility	Ref.
12431	Pyrene, 1-acetyl 1-$(CH_3CO)C_{16}H_9$	244.29	gr-ye lf (al, MeOH)	...	89-90	...	...	eth, bz	$B7^3$, 2726
12432	Pyrene, 1-amino 1-$NH_2C_{16}H_9$	217.27	ye nd (hx), lf (dil al)	...	117-8	...	...	al, ace	$B12^3$, 3368
12433	Pyrene, 1,2-dihydro $C_{16}H_{12}$	204.27	...	...	132	...	...	...	BCHE, 4239
12434	Pyrene, 2,7-dimethyl $C_{18}H_{14}$	230.31	...	...	228-32	...	...	...	$B5^3$, 2307
12435	Pyrene perhydro (3a,8a-*cis*) $C_{16}H_{26}$	218.38	...	318.5	90	...	...	...	$B5^4$, 1199
12436	3-Pyrene carboxylic acid 3-$(C_{16}H_9)CO_2H$	246.27	ye nd (eth-al, sub)	sub	274	...	...	eth, chl	$B9^3$, 3575
12437	4-Pyrene carboxylic acid 4-$(C_{16}H_9)CO_2H$	246.27	gr lf or nd (Ph NO_2)	...	327-8	...	...	chl	$C5^2$, 11081
12438	Pyrethrin I or Chrysatnthemum monocarboxylic acid $C_{21}H_{28}O_3$	328.45	visc liq, $[\alpha]^{25}{}_D$ −32.3 (eth, c=5.66)	$170^{0.1}{}_d$	...	...	1.5192^{18}	al, eth, peth	$B9^3$, 215
12439	Pyrethrin II or Chrysanthemum dicarboxylic acid, methyl ester $C_{22}H_{28}O_5$	372.46	visc liq, $[\alpha]^{20}{}_D$ −6 (eth, c=5)	$200^{0.1}{}_d$	...	...	1.5258^{20}	al, eth, peth	$B9^3$, 3988
12440	Pyridazine or 1,2-Diazine $C_4H_4N_2$	80.09	...	208, $47\text{-}8^1$	−8	$1.1035^{20/4}$	1.5218^{20}	**w, al**, eth, ace, bz	$B23^1$, 28
12441	Pyridine C_5H_5N	79.10	...	115.5	−42	0.9819^{20}	1.5095^{20}	**w, al, eth, ace, bz**	$B20^4$, 2205
12442	Pyridine, 3-acetamido 3-$CH_3CONH(C_5H_4N)$	136.15	...	326.7	133	...	...	w, al	$B22^4$, 4073
12443	Pyridine, 2-acetyl or Methyl 2-pyridyl ketone 2-$CH_3CO(C_5H_4N)$	121.14	ye in air	192, 78^{12}	...	...	1.5203^{20}	al, eth	$B21^4$, 3544
12444	Pyridine, 3-acetyl or Methyl-3-pyridyl ketone 3-$CH_3CO(C_5H_4N)$	121.14	ye in air	220, 106^{12}	13-4	...	1.5341^{20}	w, al, eth	$B21^4$, 3548
12445	Pyridine, 3-acetyl,oxime 3-$CH_3C(=NOH)(C_5H_4N)$	136.15	cr (al, bz)	...	113 (130.5)	...	...	...	$B21^4$, 3549
12446	Pyridine, 2-allyl 2-$(CH_2=CHCH_2)(C_5H_4N)$	119.17	...	190, 58^{10}	...	0.959^{20}	...	**al, eth**	$B20^4$, 2890
12447	Pyridine, 2-amino or α-Pyridyl amine 2-$H_2N(C_5H_4N)$	94.12	lf (liq)	204 (sub), $104\text{-}6^{20}$	57-8	...	...	al, eth, ace, bz	$B22^4$, 3840
12448	Pyridine, 2-amino-5-bromo 5-Br-2-$H_2N(C_5H_3N)$	173.01	cr (bz)	...	137	...	...	al, bz	$B22^4$, 4031
12449	Pyridine, 2-amino-5-chloro 5-Cl-2-$H_2N(C_5H_3N)$	128.56	pl	$127\text{-}8^{11}$	136-8	...	...	w, al	$B22^4$, 4019
12450	Pyridine, 2-amino-3,5-dibromo 3,5-Br_2-2-$H_2N(C_5H_2N)$	251.92	nd (dil al, peth)	...	104	...	...	al, eth	$B22^4$, 4041
12451	Pyridine, 2-amino-3,5-dichloro 3,5-Cl_2-2-$H_2N(C_5H_2N)$	163.01	nd or pr (dil al)	...	84-5	...	...	al, ace, lig	$B22^4$, 4028
12452	Pyridine, 2-amino-5-iodo 5-I-2-$H_2N(C_5H_3N)$	220.01	nd or lf (dil al)	...	129	...	...	al, eth	$B22^4$, 4045
12453	Pyridine, 2-amino-3-nitro 3-NO_2-2-$H_2N(C_5H_3N)$	139.11	ye nd (dil al)	...	165-7	...	...	al	$B22^4$, 4053
12454	Pyridine, 2-amino-5-nitro 5-NO_2-2-$H_2N(C_5H_3N)$	139.11	ye lf (dil al)	...	188	...	...	al	$B22^4$, 4054
12455	Pyridine, 3-amino or β-Pyridyl amine 3-$H_2N(C_5H_4N)$	94.12	lf (bz-lig)	252, $131\text{-}2^{12}$	64-5	...	...	w, al, eth	$B22^4$, 4067
12456	Pyridine, 4-amino or γ-Pyridyl amine 4-$H_2N(C_5H_4N)$	94.12	nd (bz)	180^{13}	158-9	...	...	w, al, eth, bz	$B22^4$, 4098
12457	Pyridine, 4-amino-2,6-dichloro 2-6-Cl_2-4-$H_2N(C_5H_2N)$	163.01	nd (dil al)	...	176	...	...	al, eth, ace, bz	$B22^4$, 4119
12458	Pyridine, 4-amino-3,5-dinitro 3,5-$(NO_2)_2$-4-$H_2N(C_5H_2N)$	184.11	ye pl (dil al)	...	170-1	...	...	...	$B22^4$, 4128
12459	Pyridine, 4-amino-3-nitro 3-NO_2-4-$H_2N(C_5H_3N)$	139.11	ye nd (w)	...	200	...	...	al	$B22^4$, 4122
12460	Pyridine, 5-amino-2-butoxy 5-NH_2-2-$C_4H_9O(C_5H_3N)$	166.22	...	$148\text{-}50^{12}$	...	1.037^{25}	1.5373^{20}	al, eth	$B22^4$, 5574
12461	Pyridine, 2-benzoyl or Phenyl-2-pyridyl ketone 2-$C_6H_5CO(C_5H_4N)$	183.21	...	$114\text{-}5^{0.01}$	41-3	...	...	chl	$B21^4$, 4119

PHYSICAL CONSTANTS OF ORGANIC COMPOUNDS (Continued)

No.	Name, Synonyms, and Formula	Mol. wt.	Color, crystalline form, specific rotation and λ_{max} (log ε)	b.p. °C	m.p. °C	Density	n_D	Solubility	Ref.
12462	Pyridine, 4-benzoyl or Phenyl-4-pyridyl ketone $4\text{-}C_6H_5CO(C_5H_4N)$	183.21	nd (peth), pl (w)	314^{742}, $170\text{-}2^{10}$	71-3	...	...	al, eth, bz	$B21^4$, 4125
12463	Pyridine, 2-benzyl $2\text{-}C_6H_5CH_2(C_5H_4N)$	169.23	nd	276^{742}, 149^{16}	11-4	$1.067^{0/0}$	1.5785^{20}	al, eth	$B20^4$, 3649
12464	Pyridine, 3-benzyl $3\text{-}C_6H_5CH_2(C_5H_4N)$	169.23	nd	286^{740}	34	1.061	...	al, eth	$B20^4$, 3651
12465	Pyridine, 4-benzyl $4\text{-}C_6H_5CH_2(C_5H_4N)$	169.23	...	287^{742}, $180\text{-}1^{31}$	...	$1.0614^{20/0}$	1.5818^{20}	al, eth	$B20^4$, 3652
12466	Pyridine, 2-bromo $2\text{-}Br(C_5H_4N)$	158.00	...	$193\text{-}4^{764}$, $74\text{-}5^{13}$	...	1.657^{15}	1.5734^{20}	al, eth	$B20^4$, 2503
12467	Pyridine, 3-bromo $3\text{-}Br(C_5H_4N)$	158.00	nd (al)	$172\text{-}3^{751}$, $68\text{-}70^{18}$	142-3	$1.645^{0/4}$	1.5694^{20}	w, al, eth	$B20^4$, 2505
12468	Pyridine, 3-bromo-6-hydroxy $3\text{-}Br\text{-}6\text{-}HO(C_5H_3N)$	174.00	pr (w, al)	...	177-8	...	...	al	$B21^4$, 362
12469	Pyridine, 4-bromo $4\text{-}Br(C_5H_4N)$	158.00	...	$172\text{-}3^{752}$, $68\text{-}70^{18}$	0-1	$1.6450^{0/4}$	1.5694^{20}	w, al, eth	$B20^4$, 2512
12470	Pyridine, 5-bromo-2-hydroxy $2\text{-}HO\text{-}5\text{-}Br(C_5H_3N)$	174.00	pr (w, al)	...	177-8	...	...	al	$B21^4$, 2512
12471	Pyridine, 2-chloro $2\text{-}Cl(C_5H_4N)$	113.55	oil	170, $54\text{-}8^{10}$	...	1.205^{15}	1.5320^{20}	al, eth	$B20^4$, 2493
12472	Pyridine, 2-chloro-5-nitro $2\text{-}Cl\text{-}5\text{-}NO_2(C_5H_3N)$	158.54	...	...	108-10	...	...	...	$B20^4$, 2533
12473	Pyridine, 3-chloro $3\text{-}Cl(C_5H_4N)$	113.55	...	148^{764}, $85\text{-}7^{100}$	...	...	1.5304^{20}	...	$B20^4$, 2497
12474	Pyridine, 4-chloro $4\text{-}Cl(C_5H_4N)$	113.55	...	147-8, $85\text{-}7^{100}$	−42.5	...	...	w, al	$B20^4$, 2499
12475	Pyridine, 5-chloro-2-trichloro methyl $5\text{-}Cl\text{-}2\text{-}Cl_3C(C_5H_3N)$	230.91	...	$139\text{-}42^{25}$	45.5-6.5	...	...	...	$B20^4$, 2704
12476	Pyridine, 2,3-diamino-(one form) $2,3\text{-}(NH_2)_2(C_5H_3N)$	109.13	lf or pl (dil al)	$148\text{-}50^5$	122	...	...	al	$B22^2$, 395
12477	Pyridine, 2,3-diamino (one form) $2,3\text{-}(NH_2)_2(C_5H_3N)$	109.13	nd (bz)	sub	116	...	...	w, al, bz	$B22^2$, 394
12478	Pyridine, 2,4-diamino $2,4\text{-}(NH_2)_2(C_5H_3N)$	109.13	hyg lf or nd	...	107	...	...	w	$B22^2$, 394
12479	Pyridine, 2,5-diamino $2,5\text{-}(NH_2)_2(C_5H_3N)$	109.13	nd	$180\text{-}5^{12}$	109-10	...	...	w, al	$B22^2$, 394
12480	Pyridine, 3,4-diamino $3,4\text{-}(NH_2)_2(C_5H_3N)$	109.13	nd or lf	...	218-9	...	...	...	$B22^2$, 395
12481	Pyridine, 3,5-diamino $3,5\text{-}(NH_2)_2(C_5H_3N)$	109.13	hyg nd or lf	...	119-20	...	...	w	$B22^1$, 648
12482	Pyridine, 2,6-dibromo $2,6\text{-}Br_2(C_5H_3N)$	236.89	...	255	118	...	...	...	$B20^4$, 2515
12483	Pyridine, 3,5-dibromo $3,5\text{-}Br_2(C_5H_3N)$	236.89	nd (al)	222, sub 100	112	...	...	eth	$B20^4$, 2516
12484	Pyridine, 1,2-dihydro-1-methyl-2-oxo C_6H_7NO	109.13	nd	250, 121^{10}	7	...	...	w	$B21^4$, 3348
12485	Pyridine, 2,4-dihydroxy or 2,4-Pyridinediol $2,4\text{-}(HO)_2(C_5H_3N)$	111.10	rh bipym (al, w)	...	260-5d	...	...	w, al	$B21^4$, 2058
12486	Pyridine, 2,5-dihydroxy $2,5\text{-}(HO)_2(C_5H_3N)$	111.10	...	...	248	...	...	w, al	$B21^4$, 2062
12487	Pyridine, 3,5-diido-2-hydroxy $3,5\text{-}I_2\text{-}2\text{-}HO(C_5H_2N)$	346.89	br nd (al)	...	261-2	...	...	...	$B21^4$, 365
12488	Pyridine, 2,3-dimethyl or 2,3-Lutidine $2,3\text{-}(CH_3)_2(C_5H_3N)$	107.16	...	163-4	...	$0.9319^{25/4}$	1.5057^{20}	w, al, eth	$B20^4$, 2765
12489	Byridine, 2,6-dimethyl-3-ethyl $2,6\text{-}(CH_3)_2\text{-}3\text{-}C_2H_5(C_5H_2N)$	135.21	...	$83\text{-}5^{23}$	...	0.9120^{18}	...	...	$B20^4$, 2828
12490	Pyridine, 2,6-dimethyl-4-ethyl $2,6\text{-}(CH_3)_2\text{-}4\text{-}C_2H_5(C_5H_2N)$	135.21	...	$187.5\text{-}8^{758}$	...	$0.9089^{25/4}$	1.4964^{25}	...	$B20^4$, 2828
12491	Pyridine, 3,5-dimethyl-2-ethyl or α-Parvoline $3,5\text{-}(CH_3)_2\text{-}2\text{-}C_2H_5(C_5H_2N)$	135.21	...	$198\text{-}9^{764}$, $85\text{-}7^{15}$	...	0.9338^0	...	al, eth, ace	$B20^4$, 2828
12492	Pyridine, 2-(dimethylamino) $2\text{-}(CH_3)_2N(C_5H_4N)$	122.17	...	196, 88^{15}	...	$1.0157^{14/14}$	1.5663^{20}	al, eth, bz	$B22^4$, 3847
12493	Pyridine, 4-(dimethylamino) $4\text{-}(CH_3)_2N(C_5H_4N)$	122.17	pl (eth)	...	114	...	...	w, al, bz, chl	$B22^4$, 4101

No.	Name, Synonyms, and Formula	Mol. wt.	Color, crystalline form, specific rotation and λ_{max} (log ε)	b.p. °C	m.p. °C	Density	n_D	Solubility	Ref.
12495	Pyridine, 2-ethyl 2-$C_2H_5(C_5H_4N)$	107.16	...	148.6	−63.1	0.9502^{0}	1.4964^{20}	**al**, eth, ace, w	B20⁴, 2755
12496	Pyridine, 2-ethyl-4-methyl or 2-Ethyl-γ-picoline 2-C_2H_5-4-$CH_3(C_5H_3N)$	121.18	...	$173\text{-}5^{748}$	...	$0.9239^{20/0}$	...	al, eth, ace	B20⁴, 2798
12497	Pyridine, 3-ethyl 3-$C_2H_5(C_5H_4N)$	107.16	...	165	−76.9	0.9539^{0}	1.5021^{20}	w, al, eth, ace	B20⁴, 2758
12498	Pyridine, 4-ethyl 4-$C_2H_5(C_5H_4N)$	107.16	...	167.7	−90.5	0.9417^{20}	1.5009^{20}	w, al, eth, ace	B20⁴, 2761
12499	Pyridine, 2-fluoro 2-$F(C_5H_4N)$	97.09	...	125^{758}	...	...	1.4574^{20}	...	B20⁴, 2491
12500	Pyridine hydrochloride $C_5H_5N.HCl$	115.56	hyg pl or sc (al)	281-9	82	...	...	w, al, chl	B20⁴, 2230
12501	Pyridine, 2-hydroxy or 2-Pyridol 2-$HO(C_5H_4N)$	95.10	nd (bz)	280-1	106-7	...	...	w, al, bz, chl	B21⁴, 344
12502	Pyridine, 2-(β-hydroxyethyl) 2-$(HOCH_2CH_2)(C_5H_4N)$	123.15	hyg	$118\text{-}21^{15}$	...	$1.1111^{4/0}$	1.5368^{20}	w, al, chl	B21⁴, 512
12503	Pyridine, 2-hydroxy-4-methyl 4-CH_3-2-$HO(C_5H_3N)$	109.13	...	307-9	130	...	...	w, al, chl	B21⁴, 505
12504	Pyridine, 2-hydroxymethyl 2-$HOCH_2(C_5H_4N)$	109.13	...	$112\text{-}3^{16}$, 102.5^{8}	...	$1.1317^{20/4}$	1.5444^{20}	w, al, eth, ace, bz	B21⁴, 487
12505	Pyridine, 2-(1-hydroxypropyl) 2-$[CH_3CH_2CH(OH)](C_5H_4N)$	137.18	pa ye	213-6, $112\text{-}3^{12}$	...	$1.0501^{20/4}$	1.5197^{20}	w, al	B21⁴, 544
12506	Pyridine, 2-(β-hydroxypropyl) 2-$[CH_3CH(OH)CH_2](C_5H_4N)$	137.18	pr	123.5^{20}	32	...	...	w, al, chl	B21⁴, 545
12507	Pyridine, 2-(β-hydroxyisopropyl) 2-$[HOCH_2CH(CH_3)](C_5H_4N)$	137.18	...	$128\text{-}31^{17}$	...	...	...	w, al	B21, 57
12508	Pyridine, 3-hydroxy or 3-Pyridol 3-HOC_5H_4N	95.10	nd (bz)	...	129	...	...	w, al	B21⁴, 402
12509	Pyridine, 3-(α-hydroxy isopropyl) 3-$[(CH_3)_2C(OH)](C_5H_4N)$	137.18	cr	$140\text{-}1^{12}$	58	...	...	w, al, chl	B21⁴, 552
12510	Pyridine, 4-hydroxy or 4-Pyridol 4-HOC_5H_4N	95.10	pr or nd (w + 1)	>350, $257\text{-}60^{10}$	148.5 (anh), 65 (hyd)	...	...	w, al	B21⁴, 446
12511	Pyridine, 4-(hydroxymethyl) 4-$HOCH_2(C_5H_4N)$	109.13	...	$140\text{-}2^{12}$	51-5	...	...	chl	B21⁴, 507
12512	Pyridine, 2-iodo 2-I-C_5H_4N	205.00	...	93^{13}	...	1.9735^{20}	1.6366^{20}	al, eth, ace, bz	B20⁴, 2522
12513	Pyridine, 2-mercapto 2-$HS(C_5H_4N)$	111.16	...	...	128	...	...	w, al, bz, chl	B21⁴, 373
12514	Pyridine, 2-methoxy 2-$CH_3O(C_5H_4N)$	109.13	...	142-3	...	...	$1.5042.^{20}$	...	B21⁴, 345
12515	Pyridine, 4-methoxy 4-$CH_3O(C_5H_4N)$	109.13	...	191^{738}, 95^{45}	...	...	...	**w**	B21⁴, 447
12516	Pyridine, 2-methyl or α-Picoline 2-$CH_3(C_5H_4N)$	93.13	...	128.8	−66.8	$0.9443^{20/4}$	1.4957^{20}	w, **al**, **eth**, ace	B20⁴, 2679
12517	Pyridine, 2-methylamino 2-$CH_3NH(C_5H_4N)$	108.14	...	200-1, 90^{9}	15	$1.052^{29/29}$	...	w, al, eth, bz, aa	B22⁴, 3847
12518	Pyridine, 4-methylamino 4-$CH_3NH(C_5H_4N)$	108.14	pl (eth)	...	117-8	...	...	w, al, eth, ace, bz	B22⁴, 4100
12519	Pyridine-N-oxide C_5H_5NO	95.10	...	$146\text{-}7^{13}$	65-6	...	...	...	B20⁴, 2305
12520	Pyridine, pentachloro C_5Cl_5N	251.33	...	sub 280	125-6	...	...	al, bz, lig	B20⁴, 2503
12521	Pyridine, 2-phenyl or α-Pyridyl benzene 2-$C_6H_5(C_5H_4N)$	155.20	...	270-2, 146^{15}	...	1.0833^{25}	1.6210^{20}	**al, eth**	B20⁴, 3639
12522	Pyridine, 3-phenyl or β-Pyridyl benzene 3-$C_6H_5(C_5H_4N)$	155.20	pa ye oil	273-4	...	...	1.6123^{25}	al, eth	B20⁴, 3643
12523	Pyridine, 4-phenyl 4-$C_6H_5(C_5H_4N)$	155.20	pl (w)	280-2	77-8	...	...	al, eth	B20⁴, 3645
12524	Pyridine picrate $C_5H_5N.C_6H_3N_3O_7$	308.21	ye nd (al)	...	167-8	...	...	...	B20⁴, 2287
12525	Pyridine, 2-propyl or Conyrine 2-$C_3H_7(C_5H_4N)$	121.18	...	166-8, 60^{11}	2	$0.9119^{20/4}$	1.4925^{20}	**al, eth**, ace	B20⁴, 2790
12526	Pyridine, 4-propyl 4-$C_3H_7(C_5H_4N)$	121.18	...	184-6, 80^{2}	...	0.9381^{15}	1.4966^{20}	al, eth	B20⁴, 2792

PHYSICAL CONSTANTS OF ORGANIC COMPOUNDS (Continued)

No.	Name, Synonyms, and Formula	Mol. wt.	Color, crystalline form, specific rotation and λ_{max} (log ε)	b.p. °C	m.p. °C	Density	n_D	Solubility	Ref.
12527	Pyridine, 2-isopropyl $2\text{-i-}C_3H_7(C_5H_4N)$	121.18		159.8		0.9342^{0}	1.4915^{20}	al, eth, ace	$B20^4$, 2794
12528	Pyridine, 4-isopropyl $4\text{-i-}C_3H_7(C_5H_4N)$	121.18		178	−54.9	$0.9382^{25/4}$	1.4962^{20}	al, eth, ace	$B20^4$, 2795
12529	**Pyridine, 2,3,4,6-tetramethyl or β-Parvoline** $2,3,4,6\text{-}(CH_3)_4(C_5HN)$	135.21		203^{750}		$0.9322^{25/4}$	1.5087^{25}	al, eth	$B20^4$, 2830
12530	Pyridine, 2-(trichloromethyl) $2\text{-}Cl_3C(C_5H_4N)$	196.46		$125\text{-}6^{25}$	−10	1.4526^{25}	1.5596^{25}		$B20^4$, 2703
12531	Pyridine, 2,4,6-trihydroxy $2,4,6(HO)_3(C_5H_2N)$	127.10	ye nd or pw		230d			al, eth, ace	$B21^4$, 2504
12532	Pyridine, 2,3,4-trimethyl $2,3,4\text{-}(CH_3)_3(C_5H_2N)$	121.18		192-3, $79\text{-}82^{14}$		$0.9127^{15/4}$	1.5150^{20}	w, al, eth	$B20^4$, 2806
12533	Pyridine, 2,3,5-trimethyl $2,3,5\text{-}(CH_3)_3(C_5H_2N)$	121.18		186.7		$0.9352^{19/4}$	1.5057^{25}	al, eth, ace, bz	$B20^4$, 2807
12534	Pyridine, 2,3,6-trimethyl $2,3,6\text{-}(CH_3)_3(C_5H_2N)$	121.18		176-8		$0.9220^{25/4}$	1.5053^{20}	al, eth, ace, bz	$B20^4$, 2808
12535	Pyridine, 2,4,5-trimethyl $2,4,5\text{-}(CH_3)_3(C_5H_2N)$	121.18		188		$0.9330^{25/4}$	1.5054^{25}	al, eth, ace, bz	$B20^4$, 2809
12536	Pyridine, 2,4,6-trimethyl or γ-Collidine $2,4,6\text{-}(CH_3)_3(C_5H_2N)$	121.18		170.5^{762}	−44.5	$0.9166^{22/4}$	1.4959^{25}	w, al, eth, ace	$B20^4$, 2810
12537	Pyridine, 2 vinyl $2\text{-}CH_2{=}CH(C_5H_4N)$	105.14		159-60, $50\text{-}5^{4}$		$0.9985^{20/0}$	1.5495^{20}	al, eth, ace, chl	$B20^4$, 2884
12538	Pyridine, 3-vinyl $3\text{-}CH_2{=}CH(C_5H_4N)$	105.14		160, $67\text{-}8^{18}$			1.5530^{20}	al, chl	$B20^4$, 2886
12539	Pyridine, 4-vinyl $4\text{-}CH_2{=}CH(C_5H_4N)$	105.14	red to dk-br	65^{15}		$0.9800^{20/4}$	1.5449^{20}		$B20^4$, 2887
12540	2-Pyridine carboxaldehyde $2\text{-}(C_5H_4N)CHO$	107.11		180^{750}, 81.5^{25}		$1.1201^{20/4}$	1.53653^{20}	w, al, eth	$B21^4$, 3495
12541	2-Pyridine carboxylic acid or Picolinic acid $2\text{-}(C_5H_4N)CO_2H$	123.11	nd (w, al, bz)	sub	136-7			al, aa	$B22^4$, 303
12542	3-Pyridine carboxaldehyde $3\text{-}(C_5H_4N)CHO$	107.11		89.5^{14}		$1.415^{20/4}$		w, al, ace, chl	$B21^4$, 3517
12543	3-Pyridine carboxylic acid or Nicotinic acid, Niacin $3\text{-}(C_5H_4N)CO_2H$	123.11	nd (w, al)	sub	236-7	1.473			$B22^4$, 348
12544	4-Pyridine carboxaldehyde or Isonicotinaldehyde $4\text{-}OHC(C_5H_4N)$	107.11		$77\text{-}8^{12}$			1.5423^{20}	w, eth	$B21^4$, 3529
12545	4-Pyridine carboxylic acid or Isonicotinic acid $4\text{-}(C_5H_4N)CO_2H$	123.11	nd (w)	sub at 260^{15}	319				$B22^4$, 518
12546	4-Pyridine carboxylic acid anhydride $C_{12}H_8N_2O_3$	228.21	nd		103-4				$B22^4$, 526
12547	4-Pyridine carboxylic acid, ethyl betaine $C_8H_9O_2N$	151.16	nd		241d			w, al	B22, 47
12548	2,3-Pyridine dicarboxylic acid or Quinolinic acid $2,3\text{-}(C_5H_3N)(CO_2H)_2$	167.12	mcl pr (w)		228-9				$B22^4$, 1618
12549	2,4-Pyridine dicarboxylic acid or Lutidinic acid $2,4\text{-}(C_5H_3N)(CO_2H)_2$	167.12	lf (w + 1)		248-50 (anh)	0.942			$B22^4$, 1630
12550	2,4-Pyridine dicarboxylic acid, 6-methyl or Uvitonic acid $6\text{-}CH_3\text{-}2,4\text{-}(C_5H_2N)(CO_2H)_2$	181.15	cr (w)		282d				$B22^2$, 107
12551	2,5-Pyridine dicarboxylic acid or Isocinchomeronic acid $2,5\text{-}(C_5H_3N)(CO_2H)_2$	167.12	lf (w + 1)	sub	256-8d				$B22^4$, 1632
12552	2,6-Pyridine dicarboxylic acid or Dipicolinic acid $2,6\text{-}(C_5H_3N)(CO_2H)_2$	167.12	nd (w + 3/2)		252 (anh)				$B22^4$, 1635
12553	3,4-Pyridine dicarboxylic acid or Cinchomeronic acid $3,4\text{-}(C_5H_3N)(CO_2H)_2$	167.12	pr, nd or lf (w)	sub	262d				$B22^4$, 1641
12554	3,4-Pyridine dicarboxylic acid, 4,5-dihydro-2,6-dimethyl,diethyl ester $C_{13}H_{19}NO_4$	253.30	bl flr pl (eth)		85			al, ace	$B22^2$, 98
12555	3,4-Pyridine dicarboxylic acid, 2,6-dimethyl,diethyl ester $2,6\text{-}(CH_3)_2(C_5HN)(CO_2C_2H_5)_2\text{-}3,4$	251.28		270d, 163^{13}	16				$B22^2$, 109
12556	3,4-Pyridine dicarboxylic acid, 5-methoxy-6-methyl $6\text{-}CH_3\text{-}5\text{-}CH_3O\text{-}(C_5HN)(CO_2H)_2\text{-}3,4$	211.17	cr (w-ace)		213-5				$B22^4$, 2591
12557	3,5-Pyridine dicarboxylic acid or Dinicotinic acid $3,5\text{-}(C_5H_3N)(CO_2H)_2$	167.12	cr (w)	sub	325d				$B22^4$, 1643
12558	3,5-Pyridine dicarboxylic acid, 1,4-dihydro-2,6-dimethyl,diethyl ester $C_{13}H_{19}NO_4$	253.30	ye nd or lf (al)		184-5			chl	$B22^4$, 1582

No.	Name, Synonyms, and Formula	Mol. wt.	Color, crystalline form, specific rotation and λ_{max} (log ε)	b.p. °C	m.p. °C	Density	n_D	Solubility	Ref.
12559	3,5-Pyridine dicarboxylic acid, 1,4-dihydro-2,4,6-trimethyl,diethyl est $C_{14}H_{21}NO_4$	267.33	lt bl flr pl (al)		131			chl	B22^4, 1594
12560	3,5-Pyridine dicarboxylic acid, 2,6-dimethyl,diethyl ester $2,6\text{-}(CH_3)_2(C_5HN)(CO_2C_2H_5)_2\text{-}3,5$	251.28	nd (al), pr (eth)	301-2, 180^{16}	75-6			al, eth, bz, chl, lig	B22^4, 1656
12561	2,6-Pyridine diol hydrate $C_5H_5NO_2.H_2O$	129.12	ye pr		202-3			w, al	B21^4, 2063
12562	Pyridine sulfate $C_5H_5NH_2SO_4$	177.17	cr (al)					w, al	B20^4, 2233
12563	4-Pyridine sulfonic acid $4\text{-}C_5H_4NSO_3H$	159.16			333-4			w, al	B22^4, 3465
12564	Pyridine pentacarboxylic acid-dihydrate $C_5N(CO_2H)_5 \cdot 2H_2O$	335.18	cr (eth, w)		220d			w	B22^2, 142
12565	3-Pyridine sulfonic acid $3\text{-}C_5H_4NSO_3H$	159.16	orh		375d	$1.718^{25/25}$		w	B22^4, 3458
12566	2,3,4,5-Pyridine tetracarboxylic acid $2,3,4,5\text{-}(C_5HN)(CO_2H)_4$	255.14	cr		160d			w	B22, 188
12567	2,3,4,6-Pyridine tetracarboxylic acid $2,3,4,6\text{-}(C_5HN)(CO_2H)_4$	255.14	nd (w + 3)		236d			w, aa	B22^2, 142
12568	2,3,5,6-Pyridine tetracarboxylic acid $2,3,5,6\text{-}(C_5HN)(CO_2H)_4$	255.14	cr (w + 2)		200d			w	B22^1, 544
12569	2,3,4-Pyridine tricarboxylic acid or γ-Carbocinchomeronic acid $2,3,4\text{-}(C_5H_2N)(CO_2H)_3$	211.13	lf (w + $^3/_2$)		250 (anh)				B22^4, 1777
12570	2,3,5-Pyridine tricarboxylic acid or Carbodinicotinic acid $2,3,5\text{-}(C_5H_2N)(CO_2H)_3$	211.13	pl, lf or nd (w, dil al)		323 (anh)			w	B22^2, 136
12571	2,4,5-Pyridine tricarboxylic acid or Berberonic acid $2,4,5(C_5H_2N)(CO_2H)_3$	211.13	tcl pr (dil HCl + 2w)		243 (anh)				B22^4, 1778
12572	2,4,6-Pyridine tricarboxylic acid or Trimesic acid $2,4,6\text{-}(C_5H_2N)(CO_2H)_3$	211.13	nd (w + 2)	sub	227d				B22^4, 1778
12573	3,4,5-Pyridine tricarboxylic acid or β-Carbocinchomeronic acid $3,4,5\text{-}(C_5H_2N)(CO_2H)_3$	211.13	lf or pl (w + 3), ta (w)		261d −w115			al	B22^4, 1779
12574	2,4,6-Pyridine triol $2,4,6\text{-}(HO)_3(C_5H_2N)$	127.10	ye nd or pw		230d			al, eth, ace	B21^4, 2504
12575	Pyridoxal hydrochloride $C_8H_9NO_3 \cdot HCl$	203.63	rh		165d			w	B21^4, 6419
12576	Pyridoxal oxime $C_8H_{10}N_2O_3$	182.18	cr (al)		205-6d			al	B21^4, 6427
12577	Pyridoxamine dihydrochloride $C_8H_{12}N_2O_2 \cdot 2HCl$	241.12	pr (al)		226-7d			w	C51, 14833
12578	Pyrimidine or 1,3-Diazin $C_4H_4N_2$	80.09		123-4	22		1.4998^{20}	w, al	B23, 89
12579	Pyrimidine, 2-amino $2\text{-}H_2N(C_4H_3N_2)$	95.11	nd (AcOEt)	sub	127-8			w	B24, 80
12580	Pyrimidine, 2-amino-4,6-dihydroxy $4,6\text{-}(HO)_2\text{-}2\text{-}H_2N(C_4HN_2)$	127.10	pr (w + 1)		>330				B24, 468
12581	Pyrimidine, 2-amino-4,5-dimethyl $4,5\text{-}(CH_3)_2\text{-}2\text{-}H_2N(C_4HN_2)$	123.16	nd (w)	sub	214-5			al, ace, bz, chl	B24, 91
12582	Pyrimidine, 2-amino-4,6-dimethyl $4,6\text{-}(CH_3)_2\text{-}2\text{-}H_2N(C_4HN_2)$	123.16			153-4			w, al, ace, bz, chl	B24^1, 234
12583	Pyrimidine, 2-amino-4-methyl $4\text{-}CH_3\text{-}2\text{-}H_2N(C_4H_2N_2)$	109.13	pl (w), nd (sub)	sub	159-60			al	B24, 84
12584	Pyrimidine, 2-amino-5-methyl $5\text{-}CH_3\text{-}2\text{-}H_2N(C_4H_2N_2)$	109.13	pl (sub), pr (w)	sub	193.5			al	B24, 87
12585	Pyrimidine, 2-amino-5-nitro $5\text{-}NO_2\text{-}2\text{-}H_2N(C_4H_2N_2)$	140.10	nd (al)		236-7			al, ace	B24^1, 231
12586	Pyrimidine, 4-amino $4\text{-}H_2N(C_4H_3N_2)$	95.10	pl (AcOEt)		151-2			w, al	B24, 81
12587	Pyrimidine, 4-amino-2,6-dimethyl or Kyanmethin $2,6\text{-}(CH_3)_2\text{-}4\text{-}H_2N(C_4HN_2)$	123.16	nd (al), pl (bz)	sub	183 (192)				B24^2, 46
12588	Pyrimidine, 4-amino-2-methyl $2\text{-}CH_3\text{-}4\text{-}H_2N(C_4H_2N_2)$	109.13	rh (ace)		205			w, al, ace, bz	B24, 84
12589	Pyrimidine, 4-amino-5-methyl $5\text{-}CH_3\text{-}4\text{-}H_2N(C_4H_2N_2)$	109.13	pl (al, AcOEt)		176			al, w	B24, 87
12590	Pyrimidine, 4-amino-6-methyl $6\text{-}CH_3\text{-}4\text{-}H_2N(C_4H_2N_2)$	109.13	pr (w), nd, lf (sub)	sub	197			w, al	B24, 85

PHYSICAL CONSTANTS OF ORGANIC COMPOUNDS (Continued)

No.	Name, Synonyms, and Formula	Mol. wt.	Color, crystalline form, specific rotation and λ_{max} (log ε)	b.p. °C	m.p. °C	Density	n_D	Solubility	Ref.
12591	Pyrimidine, 6-amino-2,4-dihydroxy or 4-Amino uracil $2,4\text{-}(HO)_2\text{-}6\text{-}H_2N(C_4HN_2)$	127.10	cr (w)	......	d	......	......	w	B24, 469
12592	Pyrimidine, 6-amino-4,5-dimethyl $4,5\text{-}(CH_3)_2\text{-}6\text{-}NH_2(C_4HN_2)$	123.16	nd (w)	......	230	......	......	bz, ace, chl	B24, 92
12593	Pyrimidine, 2-chloro-4-dimethylamino-6-methyl $4\text{-}[(CH_3)_2N]\text{-}6\text{-}CH_3\text{-}2\text{-}Cl(C_4HN_2)$	171.63	br wax	$140\text{-}7^{4}$	87	......	......	al	C36, 911
12594	Pyrimidine, 2,4-dichloro-5-methyl $5\text{-}CH_3\text{-}2,4\text{-}Cl_2(C_4HN_2)$	163.01	pl (al)	235	25-7	......	......	al, eth, bz, chl	B23, 93
12595	Pyrimidine, 2,4-dichloro-6-methyl $6\text{-}CH_3\text{-}2,4\text{-}Cl_2(C_4HN_2)$	163.01	nd (lig)	219	46-7	......	......	al, eth, bz, chl	B23, 92
12596	Pyrimidine, 2,4-dichloro-5-nitro $2,4\text{-}Cl_2\text{-}5\text{-}NO_2(C_4HN_2)$	193.98	......	$153\text{-}5^{58}$	29.3	......	......	......	B23, 90
12597	Pyrimidine, 2-hydroxy,hydrochloride $2\text{-}HO(C_4H_3N_2).HCl$	132.55	rods (al)	......	205	......	......	w	B24^{1}, 231
12598	Pyrimidine, 2-methyl $2\text{-}CH_3(C_4H_3N_2)$	94.12	......	138^{758}	−4	......	......	w	B23, 92
12599	Pyrimidine, 4-methyl $4\text{-}CH_3(C_4H_3N_2)$	94.12	......	142	32	$1.031^{16/16}$	1.5000^{20}	w	B23, 92
12600	Pyrimidine, 5-methyl $5\text{-}CH_3(C_4H_3N_2)$	94.12	......	152-4	30.5	......	......	w	B23, 93
12601	Pyrimidine, 4-methylamino $4\text{-}CH_3NH(C_4H_3N_2)$	109.13	lf (al), nd (peth)	$142\text{-}4^{16}$	74-5	......	......	w	B24^{2}, 38
12602	Pyrocalciferol or 9α-Lumisterol $C_{28}H_{44}O$	396.66	nd (MeOH), $[\alpha]^{20/}{}_D$ + 512 (al)	......	93-5	......	......	al	B6^{3}, 3098
12603	Pyrocoll $C_{10}H_6N_2O_2$	186.17	ye mcl pl (aa)	sub	268	......	......	al, eth, bz, chl	B24^{1}, 360
12604	Pyrogallol or 1,2,3-Trihydroxy benzene $1,2,3\text{-}(HO)_3C_6H_3$	126.11	lf or nd (bz)	309, 171^{12}	133-4	$1.453^{4/4}$	1.561^{134}	w, al, eth	B6^{3}, 6260
12605	Pyrogallol, 4-acetyl or 2,3,4-Trihydroxy acetophenone $2,3,4\text{-}(HO)_3C_6H_2(COCH_3\text{-}1)$	168.15	pa ye nd or lf (w)	......	173	......	......	al, eth, ace, aa	B8^{3}, 3376
12606	Pyrogallol, 1,2-dimethyl ether or 2,3-Dimethoxy phenol $2,3(CH_3O)_2C_6H_3OH$	154.17	......	232-4, $124\text{-}5^{17}$	......	1.5392^{20}	......	......	B6^{3}, 6264
12607	Pyrogallol, 1,3-dimethyl ether or 2,6-Dimethoxy phenol $2,6\text{-}(CH_3O)_2C_6H_3OH$	154.17	mcl pr (w)	262-7	56-7	......	......	al, eth	B6^{3}, 6264
12608	Pyrogallol triacetate $1,2,3\text{-}(CH_3CO_2)_3C_6H_3$	252.22	pr (al)	......	165	......	......	al	B6^{3}, 6269
12609	Pyrogallol, 4,5,6-trichloro or 4,5,6-Trichloro-1,2,3-hydroxy benzene $C_6H_3Cl_3O_3$	229.45	nd (al, bz)	......	185	......	......	w, al, eth	B6, 1084
12610	Pyrogallol trimethyl ether or 1,2,3-Trimethoxy benzene $1,2,3\text{-}(CH_3O)_3C_6H_3$	168.20	rh nd (al)	235, 140^{12}	48-9	$1.1118^{45/45}$	......	al, eth, bz	B6^{3}, 6265
12611	2-Pyrone, 4,6-dimethyl $4,6\text{-}(CH_3)_2(C_5H_2O_2)$	124.14	lf (eth)	245, 126^{11}	51.5	......	......	w, al, eth	B17^{4}, 4531
12612	2-Pyrone, 3-hydroxy $3\text{-}HO\text{-}(C_5H_3O_2)$	112.08	nd (w + 2)	112^{20}	95 (anh), 80-5 (hyd)	......	......	w, al, eth, chl	B17^{4}, 5908
12613	4-Pyrone or 4H-Pyran-4-one $C_5H_4O_2$	96.09							
12614	4-Pyrone, 2,6-dimethyl $2,6\text{-}(CH_3)_2(C_5H_2O_2)$	124.14	hyg cr	$215\text{-}7^{742}$, 105^{23}	32.5	1.190	1.5238	w, al, eth, bz, chl	B17^{4}, 4399
12615	4-Pyrone, 3-Hydroxy-2-methyl or Larixinic acid, Maltol $2\text{-}CH_3\text{-}3\text{-}HO(C_5H_2O_2)$	126.11	pl, nd (sub)	$248\text{-}9^{713}$, $139\text{-}40^{25}$ sub	132	......	......	w, al, eth, ace	B17^{4}, 4532
12616	4-Pyrone, 5-hydroxy-2-(hydroxymethyl) or Kojic acid $2\text{-}HOCH_2\text{-}5\text{-}HO(C_5H_2O_2)$	142.11	red mcl pr (chl)	sub at 93	162.4	......	......	chl	B17^{4}, 5916
12617	4-Pyrone-2-carboxylic acid-5,6-dihydro-6,6-dimethyl, butyl ester $6,6\text{-}(CH_3)_2\text{-}2\text{-}C_5H_3O_2CO_2C_4H_9$	226.27	ye or pa red-br	256-70, $113\text{-}4^{14}$	......	1.052^{25}	1.4745^{29}	**al, eth, bz, chl**	B18^{4}, 5345
12618	4-Pyrone-2-carboxylic acid-5-hydroxy or Comenic acid $C_6H_4O_5$	156.09	ye cr	......	>270d	......	......	......	B18^{4}, 5985
12619	Pyrophosphoramide, octamethyl $[(CH_3)_2N]_4P_2O_3$	286.25	......	$118\text{-}22^{0.3}$	......	1.1343^{25}	1.462^{25}	w, al, chl	B4^{4}, 288
12620	Pyrophosphoric acid, tetraethyl ester or Tetraethyl pyrophosphate $(C_2H_5O)_4P_2O_3$	290.19	......	155^{3}	......	$1.1847^{20/4}$	1.4180^{20}	**al, eth, ace, chl**	B1^{4}, 1340

PHYSICAL CONSTANTS OF ORGANIC COMPOUNDS (Continued)

No.	Name, Synonyms, and Formula	Mol. wt.	Color, crystalline form, specific rotation and λ_{max} (log ε)	b.p. °C	m.p. °C	Density	n_D	Solubility	Ref.
12621	Pyrophosphoric acid, dithiono, tetraethyl ester or Sulfotep $(C_2H_5O)_4P_2OS_2$	322.31	col oil	$136\text{-}9^{2}$		$1.196^{25/4}$	1.4753^{25}	al	$B1^4$, 1351
12622	Pyrotartaric acid (dl) or Methyl succinic acid $HO_2CCH(CH_3)CH_2CO_2H$	132.12	pr	d	115		1.4303	w, al, eth	$B2^4$, 1948
12623	Pyrrole C_4H_4NH	67.09		130-1		$0.9691^{20/4}$	1.5085^{20}	al, eth, ace, bz	$B20^4$, 2072
12624	Pyrrole, N-acetyl $C_4H_4N(COCH_3)$	109.13		181-2					$B20^4$, 2089
12625	Pyrrole, 2-acetyl $2\text{-}CH_3CO(C_4H_3NH)$	109.13	mcl nd (w)	220	90			w, al, eth	$B21^4$, 3437
12626	Pyrrole, 2-acetyl-N-methyl $2\text{-}CH_3CO(C_4H_3N)CH_3$	123.15		$200\text{-}2^{252}$, $92\text{-}4^{22}$		$1.0445^{15/4}$	1.5403^{15}	al, bz, chl	$B21^4$, 3439
12627	Pyrrole, N-benzyl $(C_4H_4N)CH_2C_6H_5$	157.22		247, 138^{27}	15		1.5655^{24}	al, eth	$B20^4$, 2086
12628	Pyrrole, N-butyl $(C_4H_4N)C_4H_9$	123.20	oil	170-1, $53\text{-}4^{11}$			1.4727^{20}		$B20^4$, 2082
12629	Pyrrole, 2,3-dimethyl-4-ethyl or Hemopyrrole $2,3\text{-}(CH_3)_2\text{-}4\text{-}C_2H_5(C_4HNH)$	123.20		198^{725}, 113^{16}	16-7	$0.915^{20/4}$		w	$B20^4$, 2152
12630	Pyrrole, 2,4-dimethyl $2,4\text{-}(CH_3)_2(C_4H_2NH)$	95.14	pa bl flr	171, $62\text{-}3^{10}$		$0.9236^{20/4}$	1.5048^{20}	al, eth, bz	$B20^4$, 2107
12631	Pyrrole, 2,4-dimethyl-3-ethyl or Krypto pyrrole $2,4\text{-}(CH_3)_2\text{-}3\text{-}C_2H_5(C_4HNH)$	123.20	pr	197^{710}, 96^{16}	0	$0.913^{20/4}$	1.4961^{20}	al, eth, bz, chl	$B20^4$, 2153
12632	Pyrrole, 2,5-dimethyl $2,5\text{-}(CH_3)_2(C_4H_2NH)$	95.14		$170\text{-}2^{765}$, $50\text{-}3^{8}$		$0.9353^{20/4}$	1.5036^{20}	al, eth	$B20^4$, 2152
12633	Pyrrole, 1,3-diphenyl $3\text{-}C_6H_5(C_4H_3N)C_6H_5$	219.29	pl (al)		122-3			eth, bz, chl	$B20^1$, 148
12634	Pyrrole, 2,5-diphenyl $2,5\text{-}(C_6H_5)_2(C_4H_2NH)$	219.29	lf (aa, dil al)		143.5			al, eth, bz, aa	$B20^4$, 4162
12635	Pyrrole, N-ethyl $(C_4H_4N)C_2H_5$	95.14		$129\text{-}30^{762}$		$0.9009^{20/4}$	1.4841^{20}	al	$B20^4$, 2082
12636	Pyrrole, 2-ethyl $2\text{-}C_2H_5(C_4H_3NH)$	95.14		163-5, $59\text{-}60^{15}$		$0.9042^{20/4}$	1.4942^{20}	al	$B20^4$, 2106
12637	Pyrrole, 3-ethyl-4-methyl or Opsopyrrole $3\text{-}C_2H_5\text{-}4\text{-}CH_3(C_4H_2NH)$	109.17	ye oil	70^{11}	3	$0.9059^{20/4}$	1.4913^{20}	al, eth	$B20^4$, 2145
12638	Pyrrole, 3-ethyl-2,4,5-trimethyl or Phyllopyrrole $3\text{-}C_2H_5\text{-}2,4,5\text{-}(CH_3)_3(C_4NH)$	137.22	pl (sub), lf (eth), wh lf (peth)	213^{725}, $92\text{-}3^{12}$	67-8			al, eth	$B20^4$, 2174
12639	Pyrrole, N-methyl $(C_4H_4N)CH_3$	81.12		$114\text{-}5^{747}$		$0.9145^{15/4}$	1.4875^{20}	**al, eth**	$B20^4$, 2080
12640	Pyrrole, 2-methyl $2\text{-}CH_3(C_4H_3NH)$	81.12		$147\text{-}8^{750}$		$0.9446^{15/4}$	1.5035^{16}	**al, eth**	$B20^4$, 2103
12641	Pyrrole, 3-methyl $3\text{-}CH_3(C_4H_3NH)$	81.12		$142\text{-}3^{743}$, 45^{11}			1.4970^{20}	**al, eth**	$B20^4$, 2105
12642	Pyrrole, N-Phenyl $(C_4H_4N)C_6H_5$	143.19	pl (sub), red in air	234, 140^{38} sub	62			al, eth, ace, bz, peth	$B20^4$, 2084
12643	Pyrrole, 2-phenyl $2\text{-}C_6H_5(C_4H_3NH)$	143.19	pl (al, sub)	$271\text{-}2^{726}$	129			al, eth, bz, chl	$B20^4$, 3452
12644	Pyrrole, N-Propyl $(C_4H_4N)C_3H_7$	109.17		145-6		$0.8833^{20/4}$		al, eth	$B20^4$, 3082
12645	Pyrrole, 3-Propyl $3\text{-}C_3H_7(C_4H_3NH)$	109.17		90^{30}, 49^{2}			1.4900^{25}	eth, ace	$B20^4$, 2143
12646	Pyrrole, 2-isopropyl $2\text{-}i\text{-}C_3H_7(C_4H_3NH)$	109.17		$171\text{-}2^{741}$		$0.908^{25/4}$	1.491^{25}	al, eth	$B20^4$, 2143
12647	Pyrrole, tetraiodo or Iodol C_4I_4NH	570.68	ye nd (al)		150d			eth, ace, chl, aa	$B20^4$, 2098
12648	Pyrrole, tetramethyl $C_4(CH_3)_4NH$	123.20	lf (dil al, peth)	130^{7}	111			al, eth, ace, bz	$B20^4$, 2156
12649	2-Pyrrole carboxaldehyde $(HNC_4H_3)CHO\text{-}(2)$	95.10	rh pr (peth)	217-9, 114^{15}	46-7		1.5939^{16}		$B21^4$, 3419
12650	2-Pyrrole carboxylic acid $(HNC_4H_3)CO_2H\text{-}2$	111.10	lf (w)		208d			w, al, eth	$B22^4$, 225
12651	2-Pyrrole carboxylic acid, 3,5-dimethyl, ethyl ester $3,5\text{-}(CH_3)_2\ (C_4HNH)CO_2C_2H_5\text{-}2$	167.21	cr (al)		125			al, ace	$B22^4$, 260
12652	3-Pyrrole carboxylic acid $(HNC_4H_3)CO_2H\text{-}3$	111.10	nd (lig)		161-2				$B22^4$, 235

No.	Name, Synonyms, and Formula	Mol. wt.	Color, crystalline form, specific rotation and λ_{max} (log ε)	b.p. °C	m.p. °C	Density	n_D	Solubility	Ref.
12653	3-Pyrrole carboxylic acid, 2,4-dimethyl, ethyl ester $2,4\text{-}(CH_3)_2\text{-}3\text{-}(C_4HNH)CO_2C_2H_5$	167.21	cr (eth-lig, peth)	291, 181-2^{35}	78-9			al, eth	B22^4, 255
12654	3-Pyrrole carboxylic acid, 2,5-dimethyl, ethyl ester $2,5\text{-}(CH_3)_2\text{-}3\text{-}(C_4HNH)CO_2C_2H_5$	167.21	rh (al)	290^{731}, 130^{15}	117-8			al	B22^4, 267
12655	3-Pyrrole carboxylic acid, 4,5-dimethyl, ethyl ester $4,5\text{-}(CH_3)_2\text{-}3\text{-}(C_4HNH)CO_2C_2H_5$	167.21	cr (dil al)		110-1			al, eth, chl	B22^4, 260
12656	2,4-Pyrrole dicarboxylic acid $2,4\text{-}(C_4H_2NH)(CO_2H)_2$	155.11	cr (w)		295d			w	B22^4, 1542
12657	2,4-Pyrrole dicarboxylic acid, 3,5-dimethyl, diethyl ester $3,5\text{-}(CH_3)_2\text{-}2,4\text{-}(C_4NH)(CO_2C_2H_5)_2$	239.27	nd (dil al)		136-7			ace, bz, chl	B22^4, 1568
12658	2,4-Pyrrole dicarboxylic acid, 3,5-dimethyl-N-ethyl, diethyl ester $C_{14}H_{21}NO_4$	267.33	cr (dil al)		40-1			al	B22^4, 1574
12659	2,4-Pyrrole dicarboxylic acid, 5-formyl-3-methyl, diethyl ester $3\text{-}CH_3\text{-}5\text{-}HCO\text{-}2,4\text{-}(C_4NH)(CO_2C_2H_5)_2$	253.25	nd		124-5			al, to	B22^4, 3265
12660	2-Pyrone $C_5H_4O_2$	96.09		206-9, 120^{30}	8-9	1.2000$^{20/4}$	1.5270^{25}	w, ace	B17^4, 4399
12661	2-Pyrone, tetrahydro $C_5H_8O_2$	100.12		218-20, 113-4^{14}	−12.5	1.0794^{20}	1.4503^{20}	w, **al, eth**	B17^4, 4169
12662	Pyrrolidine or Tetrahydropyrrol C_4H_8NH	71.12		88-9		0.8520$^{22/4}$	1.4431^{20}	al, eth, chl	B20^4, 61
12663	Pyrrolidine, N-butyl $(C_4H_8N)C_4H_9$	127.23		154-5		0.816^{25}	1.4373^{25}	al	B20^4, 67
12664	Pyrrolidine, 2-butyl $2\text{-}C_4H_9(C_4H_7NH)$	127.23		173-4^{741}, 67^{18}		0.8277$^{20/4}$	1.4490^{20}	w, al	B20^4, 1627
12665	Pyrrolidine, N-(chloroacetyl) $(C_4H_8N)COCH_2Cl$	147.60		112$^{0.5}$	44-6				B20^4, 186
12666	Pyrrolidine, N,2-dimethyl $2\text{-}CH_3(C_4H_7N)CH_3$	99.18		96		0.7994$^{20/4}$	1.4252^{20}	w, al, eth	B20^4, 1384
12667	Pyrrolidine, 2,4-dimethyl $2,4\text{-}(CH_3)_2(C_4H_6NH)$	99.18		115-7		0.8297$^{20/4}$	1.4325^{20}	w, al, eth	B20^4, 1537
12668	Pyrrolidine, 2,5-dimethyl (*cis*) $2,5\text{-}(CH_3)_2(C_4H_6NH)$	99.18			106-7	0.8205$^{20/4}$	1.4299^{20}	**w, al, eth**	B20^4, 1538
12669	Pyrrolidine, N-methyl $(C_4H_8N)CH_3$	85.15		81.3		0.8188$^{20/4}$	1.4247^{20}	w, eth	B20^4, 64
12670	Pyrrolidine, N-phenyl $(C_4H_8N)C_6H_5$	147.22		119-20^{12}		1.0260^{25}	1.5813^{20}	eth	B20^4, 78
12671	2-Pyrrolidone or γ-Butyrolactam C_4H_7NO	85.11	cr (peth)	250.5^{742}, 133^{12}	24.6	1.120$^{20/4}$	1.4806^{30}	w, al, eth, bz	B21^4, 3142
12672	2-Pyrrolidone, N,5-dimethyl $C_6H_{11}NO$	113.16		215-7^{743}, 87.5^{10}			1.4650^{20}	**w, eth**	B21^4, 3192
12673	2-Pyrrolidone, 3,3-dimethyl $C_6H_{11}NO$	113.16	lf (bz)	237	65-7			w, al, eth, ace	B21, 242
12674	2-Pyrrolidone, N-methyl C_5H_9NO	99.13		202, 84-5^{14}	−23	1.0260$^{25/25}$	1.4684^{20}	w, eth, ace	B21^4, 3145
12675	3-Pyrroline or Dihydropyrrole C_4H_6NH	69.11		90-1		0.9097$^{20/4}$	1.4664^{20}	**al**, w, **eth**, ace	B20^4, 1906
12676	Pyruvic acid or 2-oxo-Propionic acid CH_3COCO_2H	88.06		165d, 54^{10}	13.8	1.2272$^{20/4}$	1.4280^{20}	w, **al, eth**, ace	B3^4, 1505
12677	Pyruvic acid-acetyl $CH_3COCH_2COCO_2H$	130.10	nd (bz-chl), pr (bz)	130^{37} sub	101 (anh) 55-63 (hyd)			w, al, eth, ace, bz, chl	B3^3, 1331
12678	Pyruvic acid, acetyl, ethyl ester $CH_3COCH_2COCO_2C_2H_5$	158.15		213-5, 111-2^{16}	18	1.1251^{20}	1.4757^{17}	al, eth	B3^4, 1777
12679	Pyruvic acid, 3-(2,5-dimethoxyphenyl) $[2,5\text{-}(CH_3O)_2C_6H_3]CH_2COCO_2H$	224.21	yesh cr (aa)		166-70d			ace, MeOH	B10^2, 723
12680	Pyruvic acid, 3-(3,4-dimethoxy phenyl) $[3,4\text{-}(CH_3O)_2C_6H_3]CH_2COCO_2H$	224.21	lf (aa)		ca 187d			al, ace	**B10^3, 4521**
12681	Pyruvic acid, ethyl ester or Ethyl pyruvate $CH_3COCO_2C_2H_5$	116.12		155, 69-71^{42}	−50	1.0596$^{15.6/4}$	1.4052^{20}	**al, eth**, ace	**B3^4, 1513**
12682	Pyruvic acid, methyl ester or Methyl pyruvate $CH_3COCO_2CH_3$	102.09		134-7, 53^{15}		1.154$^{0/4}$	1.4046^{25}	**al, eth**, ace	**B3^3, 1160**
12683	Pyruvic acid, 3-phenyl $C_6H_5CH_2COCO_2H$	164.16	lf (chl, bz)	157-8				al, eth	**B10^3, 3000**
12684	Pyruvonitrile CH_3COCN	69.06	rh	92-3		0.9745$^{20/4}$	1.3764^{20}	eth, ace	**B3^4, 1515**
12685	Pyruvic acid, 3-(2-nitro phenyl) $(2\text{-}NO_2C_6H_4)CH_2COCO_2H$	209.16	ye nd or lf (w, al, bz)		130			al, eth, aa	**B10^3, 3017**

No.	Name, Synonyms, and Formula	Mol. wt.	Color, crystalline form, specific rotation and λ_{max} (log ε)	b.p. °C	m.p. °C	Density	n_D	Solubility	Ref.
12686	o,o′-Quaterphenyl or o,o′-Diphenyl biphenyl $C_{24}H_{18}$	306.41	pr (al)	420	118-9	....	....	al, eth, ace, bz, chl	B5[3], 2561
12687	p,p′-Quaterphenyl or Tetraphenyl $C_{24}H_{18}$	306.41	lf (bz)	428[18]	320	....	....	....	B5[3], 2562
12688	Quercitrin or 3-D-α-L-rhamnopyranosylquercitrin $C_{21}H_{20}O_{11}$	448.38	pa ye nd or pl (+2w dil al), $[\alpha]_{578}$ −73.5 (al, c=4)	....	182-5 (hyd), 250-2 (anh)	....	....	al	B18[4], 3491
12689	Quillaic acid or Quillaja sapogenin $C_{30}H_{46}O_5$	486.69	nd (dil al), $[\alpha]^{20}{}_D$ +56.1 (py, c=2.91)	....	294	....	....	al, eth, ace, aa, Py	B10[3], 4652
12690	Quinacrine dihydrochloride (dl) or Atebrine $C_{23}H_{30}ClN_3O.2HCl.2H_2O$	532.94	yesh nd (w), ye cr pw	....	248-50d	....	....	w, MeOH	B21[4], 6247
12691	Quinamine $C_{19}H_{24}N_2O_2$	312.41	pr (bz), nd (80%al), $[\alpha]^{15}{}_D$ +93 (chl, c=2)	....	185-6	....	....	eth, ace, bz, lig	B27[2], 667
12692	Quinazoline or 1,3-Benzodiazine $C_8H_6N_2$	130.15	ye pl (peth)	241.5, 117-20[15]	48	....	....	w, al, eth, ace, bz	B23[2], 177
12693	Quinazoline, 3,4-dihydro-4-oxo or Quinazolinone $C_8H_6N_2O$	146.15	nd (dil aa)	360	216-8	....	....	....	B24[2], 71
12694	Quinazoline, 3,4-dihydro-2-phenyl $C_{14}H_{12}N_2$	208.26	lf (dil al)	....	142	....	....	al, eth, chl, aa	B23, 239
12695	Quinazoline, 3,4-dihydro-3-phenyl or Phenzoline $C_{14}H_{12}N_2$	208.26	pl (eth-lig)	d	96-7	1.290[4]	....	al, eth, bz, chl	B23[2], 155
12696	Quinazoline, 3,4-dihydro-4-phenyl $C_{14}H_{12}N_2$	208.26	pl (al, AcOEt)	....	166-7	....	....	al, eth	B23, 239
12697	Quinazoline, 2,4-dihydroxy-6,8-dinitro $C_8H_4N_4O_6$	252.14	ye gr pr (aa)	....	274-5d	....	....	....	B24[1], 344
12698	2,4-Quinazolinedione or Benzoylene urea $C_8H_6N_2O_2$	162.15	nd (w, al), lf (aa)	....	356	....	....	....	B24[2], 197
12699	Quinhydrone or Benzoquinhydrone $C_{12}H_{10}O_4$	218.21	red br nd	sub	171	1.401[20/4]	....	al, eth	B7[3], 3363
12700	Quinicine or Quinotoxin $C_{20}H_{24}N_2O_2$	324.42	red ye amor, $[\alpha]^{15}{}_D$ +44.1 (chl)	....	ca 60	....	....	al, eth, chl	B25[2], 20
12701	Quinicine oxalate (d) $(C_{20}H_{24}N_2O_2)_2H_2C_2O_4.9H_2O$	901.02	pr (chl), nd (al), $[\alpha]^{15}{}_D$ +19.5 (al-chl), (+9.5w)	....	149	....	....	w, al, chl	B25[2], 20
12702	Quinidine $C_{20}H_{24}N_2O_2$	324.42	cr (+2.5w, dil al)	....	174-5 (anh)	....	....	al, bz, chl	B23[2], 414
12703	Quinidine hydrate or Conquinine $C_{20}H_{24}N_2O_2.2\frac{1}{2}H_2O$	369.46	cr (dil al)	....	171d	....	....	al, eth, bz, chl	B23[2], 414
12704	Quinidine hydrochloride $C_{20}H_{24}N_2O_2.HCl.H_2O$	378.90	pr (w), $[\alpha]^{20}{}_D$ +200 (w, c=1)	....	258-9d (anh)	....	....	chl	B23[2], 415
12705	Quinidine sulfate (d) $(C_{20}H_{24}N_2O_2)_2.H_2SO_4.2H_2O$	782.95	pr, nd (w), $[\alpha]^{25}{}_D$ +212 (al)	....	....	....	....	w, al, chl	B23[2], 415
12706	Quinine $C_{20}H_{24}N_2O_2.3H_2O$	378.47	cr (+3w, eth), nd (+3w, al), rh nd (abs al), $[\alpha]^{15}{}_D$ −145.2 (al)	....	177 (anh), 57 (hyd)	....	1.625	al, eth, chl, Py	B23[2], 416

PHYSICAL CONSTANTS OF ORGANIC COMPOUNDS (Continued)

No.	Name, Synonyms, and Formula	Mol. wt.	Color, crystalline form, specific rotation and λ_{max} (log ε)	b.p. °C	m.p. °C	Density	n_D	Solubility	Ref.
12707	Quinine bisulfate $C_{20}H_{24}N_2O_2.H_2SO_4.7H_2O$	548.60	pr (+w, w), pr (+5w, al), $[\alpha]^{15}/_D$ +168.4 (al)		160d			chl	B23^{2}, 416
12708	Quinine, *o*-ethyl carbonate $C_{23}H_{28}N_2O_4$	396.49	nd (w), cr (dil al)		95			al, eth, chl	B23^{2}, 424
12709	Quinine formate or Quinoform $C_{20}H_{24}N_2O_2.HCO_2H$	370.45	nd $[\alpha]^{20}/_D$ −144.2 (w, c=1)		149-50 (anh), 126 (+1w)			w, al, chl	B23^{1}, 169
12710	Quinine hydrobromide $C_{20}H_{24}N_2O_2.HBr.H_2O$	423.35	silky efflor nd (w)		ca 208 softens 152			w, al, chl	B23^{1}, 168
12711	Quinine hydrochloride $C_{20}H_{24}N_2O_2.HCl$	360.88	silky efflor nd (w), $[\alpha]^{15}/_D$ −145 (al)		158-60 (anh)			w, al, chl	B23^{2}, 420
12712	Quinine hydrochloride, hydrate $C_{20}H_{24}N_2O_2.HCl.2H_2O$	396.91	silky efflor nd (w), $[\alpha]^{20}/_D$ −149.8 (w, c=1.3)		156-90			w, al, chl	B23^{2}, 420
12713	Quinine salicylate or Quinine-2-hydroxy benzoate $C_{20}H_{24}N_2O_2.C_7H_6O_3.H_2O$	480.56	wh pr (al), cr (+2w, w)		195			al, chl	B23^{2}, 422
12714	Quinine sulfate $2(C_{20}H_{24}N_2O_2).H_2SO_4$	746.92	silky nd (w)		235.2			al	B23^{2}, 420
12715	Quinine sulfate, dihydrate $2(C_{20}H_{24}N_2O_2)_2.H_2SO_4.2H_2O$	782.95	silky nd (w), $[\alpha]^{15}/_D$ −220 (0.5N HCl)		205			al	B23^{2}, 420
12716	Quinine valerate $C_{20}H_{24}N_2O_2.C_4H_9CO_2H.H_2O$	444.57	wh		ca 96			al, eth	M, 892
12717	Quininone $C_{20}H_{22}N_2O_2$	322.41	nd lf (eth), $[\alpha]^{20}/_D$ +75.5 (al, c=2)		108 (rapid htng)			al, eth, bz, chl	B25^{2}, 23
12718	Quinizarin or 1,4-Dihydroxy anthraquinone $1,4-(HO)_2C_{14}H_6O_2$	240.22	ye red lf (eth), dk red nd (al), red cr (to, aa)		200-2 (aa), 194 (to)				B8^{1}, 3775
12719	Quinoline or Benzo[*b*]-pyridine C_9H_7N	129.16		238, 114^{17}	fr −15.6	$1.0929^{20/4}$	1.6268^{20}	al, eth, ace, bz	B20^{4}, 3334
12720	Quinoline, 3-acetamido $3-(CH_3CONH)(C_9H_6N)$	186.21	cr (w)		166-7			al	B22^{4}, 4607
12721	Quinoline, 4-acetamide $4-(CH_3CONH)(C_9H_6N)$	186.21	nd (w + 1)	sub	176			al	B22^{4}, 4614
12722	Quinoline, 5-acetamido $5-(CH_3CONH)(C_9H_6N)$	186.21	pl (w), pr (dil al)		178			al	B22^{4}, 4671
12723	Quinoline, 6-acetamido $6-(CH_3CONH)(C_9H_6N)$	186.21	nd (w)		138			w, al	B22^{2}, 355
12724	Quinoline, 8-acetamido $8-(CH_3CONH)(C_9H_6N)$	186.21	cr (w, al)		167.5				B22^{2}, 356
12725	Quinoline, 3-acetamido $8-(CH_3CONH)(C_9H_6N)$	186.21	nd (al)		103			al	B22^{1}, 640
12726	Quinoline, 2-amino or α-Quinolyl amine $2-H_2N(C_9H_6N)$	144.18	lf (w)	sub	131.5			al, eth, ace, chl, aa	B22^{4}, 4587
12727	Quinoline, 2-amino-4-hydroxy $4-HO-2-H_2N(C_9H_5N)$	160.18	nd (w + 1), rh (al)		303-4				B22^{4}, 5718
12728	Quinoline, 2-amino-4-methyl or 2-Aminolepidine $4-CH_3-2-H_2N(C_9H_5N)$	158.20	cr pw (bz)	320	133			al, eth, chl, aa	B22^{4}, 4801
12729	Quinoline, 3-amino or β-Quinolyl amine $3-H_2N(C_9H_6N)$	144.18	rh (w, dil al)		94			al, eth, chl	B22^{4}, 4605

PHYSICAL CONSTANTS OF ORGANIC COMPOUNDS (Continued)

No.	Name, Synonyms, and Formula	Mol. wt.	Color, crystalline form, specific rotation and λ_{max} (log ε)	b.p. °C	m.p. °C	Density	n_D	Solubility	Ref.
12730	Quinoline, 3-amino-2-methyl or 3-Amino quinaldine 2-CH_3-3-$H_2N(C_9H_5N)$	158.20	pa ye nd (eth, peth)	278d, 198^{16}	160			al, eth, bz, chl	B22^{4}, 4756
12731	Quinoline, 4-amino or γ-Quinolyl amine 4-$H_2N(C_9H_6N)$	144.18	nd (w + 1), nd (bz, dil al)	180^{12}	154 (anh), 70 (+ w)			w, al, eth, chl	B22^{4}, 4611
12732	Quinoline, 4-amino-2-methyl or 4-Amino quinaldine 2-CH_3-4-$H_2N(C_9H_5N)$	158.20	nd (bz-lig), pr (eth-bz)	333	168			al, eth, ace, bz	B22^{4}, 4559
12733	Quinoline, 5-amino or 5-Quinalyl amine 5-$H_2N(C_9H_6N)$	144.18	ye nd (al), lf (eth)	310 sub, 184^{10}	110			al, eth, bz	B22^{4}, 4669
12734	Quinoline, 5-amino-6-hydroxy 6-HO-5-$H_2N(C_9H_5N)$	160.18	gr nd (w + 2)		185			al	B22, 501
12735	Quinoline, 5-amino-8-hydroxy 8-HO-5-$H_2N(C_9H_5N)$	160.18	nd (bz)		143				B22, 5865
12736	Quinoline, 5-amino-2-methyl or 5-Amino quinaldine 2-CH_3-5-$H_2N(C_9H_5N)$	158.20	grsh pl or nd (w + 1)		117-8			w, al, bz, lig	B22^{2}, 360
12737	Quinoline, 5-amino-8-methyl 8-CH_3-5-$H_2N(C_9H_5N)$	158.20	yesh nd (w, dil al)		143			al	B22^{4}, 4828
12738	Quinoline, 6-amino or 6-Quinolyl amine 6-$H_2N(C_9H_6N)$	144.18	cr (w + 2), pr (eth)	187^{12}	114 (anh)			al	B22^{4}, 4681
12739	Quinoline, 6-amino-2-methyl or 6-Amino quinaldine 2-CH_3-6-$H_2N(C_9H_5N)$	158.20	pa br (w, dil al)		187.5			w, al, chl	B22^{4}, 4780
12740	Quinoline, 6-amino-4-methyl 4-CH_3-6-$H_2N(C_9H_5N)$	158.20	nd (w)		169-70			al, eth, chl	B22^{4}, 4812
12741	Quinoline, 7-amino or 7-Quinolyl amine 7-$H_2N(C_9H_6N)$	144.18	ye nd (+ 1w)		93-4 (anh), 74-5 (+ w)			al	B22^{4}, 4704
12742	Quinoline, 7-amino-8-hydroxy 8-HO-7-$H_2N(C_9H_5N)$	160.18	br pr (eth, dil al)		124			al, eth, bz, chl	B22^{4}, 5873
12743	Quinoline, 7-amino-2-methyl 2-CH_3-7-$H_2N(C_9H_5N)$	158.20	pr (dil al)	304	129			al, eth, ace, bz	B22^{4}, 4784
12744	Quinoline, 7-amino-8-methyl 8-CH_3-7-$H_2N(C_9H_5N)$	158.20	pr (dil al)	304	125			al, eth, ace, bz	B22, 456
12745	Quinoline, 8-amino or 8-Quinolyl amine 8-$H_2N(C_9H_6N)$	144.18	pa ye nd (sub), cr (al, lig)	157-8^{19}	70			w, al	B22^{4}, 4708
12746	Quinoline, 8-amino-6-methoxy 6-CH_3O-8-$H_2N(C_9H_5N)$	174.20	cr	137-8^{1}	51				B22^{4}, 5747
12747	Quinoline, 8-amino-2-methyl or 8-Amino quinaldine 2-CH_3-8-$H_2N(C_9H_5N)$	158.20	pr (lig)		57-8			al, eth, ace, bz	B22^{4}, 4785
12748	Quiniline, 8-amino-6-methyl 6-CH_3-8-$H_2N(C_9H_5N)$	158.20	nd	sub	73			w, al, eth, ace, bz	B22^{4}, 4822
12749	Quinoline, 2-bromo 2-$Br(C_9H_6N)$	208.06	nd (al)		49			al, eth, bz, chl	B20^{4}, 3387
12750	Quinoline, 3-bromo 3-$Br(C_9H_6N)$	208.06	ye oil	274-6, $98^{0.5}$	13-5		1.6641^{20}	aa	B20^{4}, 3388
12751	Quinoline, 3-bromo-2-hydroxy or β-Bromo carbostyril 3-Br-2-$HO(C_9H_5N)$	224.06	pr (al)	sub	253				B21^{4}, 1063
12752	Quiniline, 4-bromo 4-$Br(C_9H_6N)$	208.06	cr	270d	29-30				B20^{4}, 3389
12753	Quinoline, 4-bromo-2-hydroxy or γ-Bromo carbostyril 4-Br-2-$HO(C_9H_5N)$	224.06	nd (al)	sub	266-7			al	B21^{4}, 1063
12754	Quinoline, 5-bromo 5-$Br(C_9H_6N)$	208.06	nd	280^{756}, 105-$7^{1.2}$	52				B20^{4}, 3389
12755	Quinoline, 5-bromo-2-hydroxy or 5-Bromo carbostyril 5-Br-2-$HO(C_9H_5N)$	224.06	nd (al)		300				B21^{4}, 1063
12756	Quinoline, 5-bromo-6-hydroxy 5-Br-6-$HO(C_9H_5N)$	224.06	nd (dil al)		186				B21^{4}, 1119
12757	Quinoline, 5-bromo-8-hydroxy 5-Br-8-$HO(C_9H_5N)$	224.06	nd (al), nd or lf (sub)	sub	124			w, al, bz, chl	B21^{4}, 1184
12758	Quinoline, 6-bromo 6-$Br(C_9H_6N)$	208.06		278, 155-6^{15}	24			al, eth	B20^{4}, 3390
12759	Quinoline, 6-bromo-2-hydroxy 6-Br-2-$HO(C_9H_5N)$	224.06	ye nd (al)		269			w, al, eth, chl	B21^{4}, 1063
12760	Quinoline, 7-bromo 7-$Br(C_9H_6N)$	208.06	nd	290	34 (52)			al, eth	B20^{4}, 3390

PHYSICAL CONSTANTS OF ORGANIC COMPOUNDS (Continued)

No.	Name, Synonyms, and Formula	Mol. wt.	Color, crystalline form, specific rotation and λ_{max} (log ε)	b.p. °C	m.p. °C	Density	n_D	Solubility	Ref.
12761	Quinoline, 7-bromo-2-hydroxy or 7-Bromo carbostyril 7-Br-2-HO(C_9H_5N)	224.06	nd (aa), pl (al)	sub	228	...	...	al, eth, chl	$B21^4$, 1064
12762	Quinoline, 8-Bromo 8-Br(C_9H_6N)	208.06	...	302-4, 165-6^{20}	<−10 (80)	...	...	al	$B20^4$, 3391
12763	Quinoline, 8-bromo-5-hydroxy 8-Br-5-HO(C_9H_5N)	224.06	nd (al)	...	190d	...	...	al, chl	B21, 85
12764	Quinoline, 2-chloro 2-Cl(C_9H_6N)	163.61	nd (aq al)	265-6^{750}, 153-4^{22}	38	1.2464$^{25/4}$	1.6342^{25}	al, eth, bz, lig	$B20^4$, 3376
12765	Quinoline, 2-chloro-4-methyl or 2-Chloro lepidine 4-CH_3-2-Cl(C_9H_5N)	177.63	nd (dil al)	296	59	...	...	al, eth, chl	$B20^4$, 3482
12766	Quinoline, 2-chloro-6-methyl 6-CH_3-2-Cl(C_9H_5N)	177.63	nd (dil al)	...	112 (116)	...	...	al, eth, bz, chl	$B20^4$, 3492
12767	Quinoline, 2-chloro-8-methyl 8-CH_3-2-Cl(C_9H_5N)	177.63	nd (eth)	286^{734}	61	...	...	al, eth, bz, chl	$B20^1$, 152
12768	Quinoline, 3-chloro 3-Cl(C_9H_6N)	163.61	hyg cr	255, 141^{15}	...	...	...	...	$B20^4$, 3377
12769	Quinoline, 3-chloro-2-methyl or 3-Chloro quinaldine 2-CH_3-3-Cl(C_9H_5N)	177.63	nd (dil al)	...	71-2	...	...	al, eth	$B20^4$, 3461
12770	Quinoline, 3-chloro-4-methyl or 3-Chloro lepidine 4-CH_3-3-Cl(C_9H_5N)	177.63	nd (dil al)	...	55	...	...	al	$B20^1$, 150
12771	Quinoline, 3-chloro-6-methyl 6-CH_3-3-Cl(C_9H_5N)	177.63	nd (dil MeOH)	...	85.5	...	...	al, MeOH	$B20^2$, 246
12772	Quinoline, 4-chloro 4-Cl(C_9H_6N)	163.61	cr	261^{764}, 130^{15}	34-5	1.251	...	al, eth	$B20^4$, 3377
12773	Quinoline, 4-chloro-2-methyl or 4-Chloro quinaldine 2-CH_3-4-Cl(C_9H_5N)	177.63	nd (+1w)	269-70	42-3	...	...	al, eth, bz, chl	$B20^4$, 3461
12774	Quinoline, 5-chloro 5-Cl(C_9H_6N)	163.61	cr (al)	256-7^{756}	45	...	...	al	$B20^4$, 3379
12775	Quinoline, 5-chloro-7-iodo-8-hydroxy or Vioform 5-Cl-7-I-8-HO(C_9H_4N)	305.50	ye br nd (al, aa)	...	178-9	...	...	...	$B21^4$, 1190
12776	Quinoline, 6-chloro 6-Cl(C_9H_6N)	163.61	pr (eth), nd (al)	262-4	44-5	...	1.6110^{56}	...	$B20^4$, 3379
12777	Quinoline, 6-chloro-2-methyl or 6-Chloro quinaldine 2-CH_3-6-Cl(C_9H_5N)	177.63	lf or nd (dil al)	...	91	...	...	eth	$B20^4$, 3462
12778	Quinoline, 6-chloro-4-methyl or 6-Chloro lepidine 4-CH_3-6-Cl(C_9H_5N)	177.63	nd (al)	...	70-2	...	...	al, eth, ace, bz	$B20^4$, 3482
12779	Quinoline, 7-chloro 7-Cl(C_9H_6N)	163.61	nd or pr	267-8	31-2	1.2158$^{58/4}$	1.6108$^{58/}{}_4$	al, eth, ace, bz, chl	$B20^4$, 3381
12780	Quinoline, 7-chloro-4-hydroxy 7-Cl-4-HO(C_9H_5N)	179.61	nd (al-w)	...	276-80	...	...	...	$B21^4$, 1084
12781	Quinoline, 7-chloro-2-methyl or 7-Chloro quinaldine 2-CH_3-7-Cl(C_9H_5N)	177.63	nd (eth), cr (lig)	87$^{0.5}$	75-6	...	...	...	$B20^4$, 3462
12782	Quinoline, 8-chloro 8-Cl(C_9H_6N)	163.61	...	288-9	fr−20	1.2834$^{14/4}$	1.6408$^{14.3}$	al, eth, ace, bz, chl	$B20^4$, 3377
12783	Quinoline, 8-chloro-5-methyl 5-CH_3-8-Cl(C_9H_5N)	177.63	nd (w)	...	49	...	...	al, eth, bz	$B20^2$, 246
12784	Quinoline, 8-(chloromethyl) 8-$ClCH_2$(C_9H_6N)	177.63	nd (w)	...	49	...	...	al, eth, bz	$B20^4$, 3502
12785	Quinoline, decahydro (*cis*) $C_9H_{17}N$	139.24	...	205^{735}, 90^{20}	−40	0.9426$^{20/4}$	1.4926^{20}	al, eth	$B20^4$, 2017
12786	Quinoline, decahydro (*trans,d*) $C_9H_{17}N$	139.24	$[\alpha]^{25/}{}_D$ +4.8 (al, c=3)	200-2	75	...	...	al, eth, ace, bz	$B20^4$, 2017
12787	Quinoline, decahydro (*trans,dl*) $C_9H_{17}N$	139.24	pr (lig, sub)	203^{735} sub	48	0.9610^{22}	1.4692^{56}	al, eth, ace, bz	$B20^4$, 2017
12788	Quinoline, decahydro (*trans,l*) $C_9H_{17}N$	139.24	$[\alpha]^{25/}{}_D$ −4.5 (al, c=3)	200-1	74-5	...	...	...	$B20^4$, 2017
12789	Quinoline, 5,7-dibromo-8-hydroxy 5,7-Br_2-8-HO(C_9H_4N)	302.95	nd (al)	sub	196	...	...	al, ace, bz, aa, chl	$B21^4$, 1180
12790	Quinoline, 6,8-dibromo-2-hydroxy or 6,8-Dibromo carbostyril 6,8-Br_2-2-HO(C_9H_4N)	302.95	nd (dil al)	...	230	...	...	al	$B21^4$, 1064
12791	Quinoline, 2,3-dichloro 2,3-Cl_2(C_9H_5N)	198.05	(dil al)	...	104-5	...	...	al, eth, bz	$B20^4$, 3382
12792	Quinoline, 2,4-dichloro 2,4-Cl_2(C_9H_5N)	198.05	nd (dil al)	280-2	67-8	...	...	al, eth, bz, chl	$B20^4$, 3382

No.	Name, Synonyms, and Formula	Mol. wt.	Color, crystalline form, specific rotation and λ_{max} (log ε)	b.p. °C	m.p. °C	Density	n_D	Solubility	Ref.
12793	Quinoline, 2,6-dichloro $2,6\text{-}Cl_2(C_9H_5N)$	198.05	nd (eth)		156 (161)			al, eth	B20^{4}, 3383
12794	Quinoline, 2,7-dichloro $2,7\text{-}Cl_2(C_9H_5N)$	198.05	nd (al)	sub 100^{2}	120			al, eth	B20^{4}, 3383
12795	Quinoline, 3,4-dichloro $3,4\text{-}Cl_2(C_9H_5N)$	198.05	peth		69-70			al, eth	B20^{4}, 3383
12796	Quinoline, 4,5-dichloro $4,5\text{-}Cl_2(C_9H_5N)$	198.05		134$^{5.5}$	118			al, eth	B20^{4}, 3383
12797	Quinoline, 4,6-dichloro $4,6\text{-}Cl_2(C_9H_5N)$	198.05	(peth)		104			al, eth	B20^{4}, 3383
12798	Quinoline, 4,7-dichloro $4,7\text{-}Cl_2(C_9H_5N)$	198.05	cr (MeOH), nd (80% al)	148^{10}	93				B20^{4}, 3384
12799	Quinoline, 4,8-dichloro $4,8\text{-}Cl_2(C_9H_5N)$	198.05			155-6			al, eth	B20^{4}, 3385
12800	Quinoline, 5,6-dichloro $5,6\text{-}Cl_2(C_9H_5N)$	198.05	nd (al)		85			al, eth, peth	B20, 361
12801	Quinoline, 5,7-dichloro $5,7\text{-}Cl_2(C_9H_5N)$	198.05	nd (al)		117			al, eth	B20^{4}, 3385
12802	Quinoline, 5,7-dichloro-8-hydroxy $5,7\text{-}Cl_2\text{-}8\text{-}HO(C_9H_4N)$	214.05	nd (al)		183			bz, peth	B21^{4}, 1180
12803	Quinoline, 5,8-dichloro $5,8\text{-}Cl_2(C_9H_5N)$	198.05	nd (al), pl (eth)	sub	97-8			al, eth	B20^{4}, 3385
12804	Quinoline, 6,8-dichloro $6,8\text{-}Cl_2(C_9H_5N)$	198.05	nd (al)		104-5			eth	B20^{4}, 3385
12805	Quinoline, 7,8-dichloro $7,8\text{-}Cl_2(C_9H_5N)$	198.05	nd		85.5			al, eth	Prak 48, Prak 279
12806	Quinoline, 1,2-dihydro-1-methyl-2-oxo or N-Methyl carbostyril $C_{10}H_9NO$	159.19	nd (lig)	324^{728}	74			al, eth, ace, bz, chl	B21^{4}, 3737
12807	Quinoline, 2,4-dihydroxy $2,4\text{-}(HO)_2(C_9H_5N)$	161.16			355				B21^{4}, 2219
12808	Quinoline, 5,7-diido-8-hydroxy or Diidoquin $5,7\text{-}I_2\text{-}8\text{-}HO(C_9H_4N)$	396.95	yesh nd (aa, xyl)		210d			al	B21^{4}, 1191
12809	Quinoline, 5,8-diido-6-hydroxy $5,8\text{-}I_2\text{-}6\text{-}HO(C_9H_4N)$	396.95	yesh		191			al, eth	B21^{2}, 54
12810	Quinoline, 2,3-dimethyl or 3-Methyl quinaldine $2,3\text{-}(CH_3)_2(C_9H_5N)$	157.22	ye rh (eth)	261^{729}	68-9	1.1013		al, eth, lig	B20^{4}, 3521
12811	Quinoline, 2,4-dimethyl or 4-Methyl quinaldine $2,4\text{-}(CH_3)_2(C_9H_5N)$	157.22	rh pr (eth)	264-6, 143^{15}		1.0611^{15}	1.6075^{20}	al, eth	B20^{4}, 3522
12812	Quinoline, 2,4-dimethyl-6-hydroxy $2,4\text{-}(CH_3)_2\text{-}6\text{-}HO(C_9H_4N)$	173.21	pr or pl (al)	360d	214			al, ace	B21^{2}, 68
12813	Quinoline, 2,4-dimethyl-7-hydroxy $2,4\text{-}(CH_3)_2\text{-}7\text{-}HO(C_9H_4N)$	173.21	nd (al)		218			al	B2^{1}, 116
12814	Quinoline, 2,4-dimethyl-8-hydroxy $2,4\text{-}(CH_3)_2\text{-}8\text{-}HO(C_9H_4N)$	173.21	pr (eth)	281 sub	65			al, eth, ace, bz, chl	B21^{4}, 1292
12815	Quinoline,2,3-dimethyl-4-hydroxy $2,3\text{-}(CH_3)_2\text{-}4\text{-}HO(C_9H_4N)$	173.21	pr (w + 1)	sub	319-20				B21^{4}, 1293
12816	Quinoline, 2,6-dimethyl or 6-Methyl quinaldine $2,6\text{-}(CH_3)_2(C_9H_5N)$	157.22	rh pr (eth)	266-7, 152-5^{13}	60			bz	B20^{4}, 3525
12817	Quinoline, 2,6-dimethyl-4-hydroxy $2,6\text{-}(CH_3)_2\text{-}4\text{-}HO(C_9H_4N)$	173.21	nd (w + 1)		279			w	B21^{4}, 1294
12818	Quinoline, 2,7-dimethyl $2,7\text{-}(CH_3)_2(C_9H_5N)$	157.22		264-5, 115-6^{7}	61			al, eth, chl	B20^{4}, 3528
12819	Quinoline, 2,8-dimethyl or *o*-Toluquinaldine $2,8\text{-}(CH_3)_2(C_9H_5N)$	157.22		255.3, 103-4^{5}	27	1.0394$^{20/4}$	1.6022^{20}	al, eth	B20^{4}, 3529
12820	Quinoline, 2,8-dimethyl-4-hydroxy $2,8\text{-}(CH_3)_2\text{-}4\text{-}HO(C_9H_4N)$	173.21	lf or pl (w + 1)	sub	260-1			al	B21^{4}, 1296
12821	Quinoline, 3,4-dimethyl or 3-Methyl lepidine $3,4\text{-}(CH_3)_2(C_9H_5N)$	157.22	cr (eth)	290^{737}	73-4			al, eth	B20^{4}, 3530
12822	Quinoline, 4,6-dimethyl-2-hydroxy $4,6\text{-}(CH_3)_2\text{-}2\text{-}HO(C_9H_4N)$	173.21	pr (al)		249-50				B21^{4}, 1301
12823	Quinoline, 4,7-dimethyl-2-hydroxy $4,7\text{-}(CH_3)_2\text{-}2\text{-}HO(C_9H_4N)$	173.21	cr (aa)		220			al, aa	B21^{4}, 1303
12824	Quinoline, 4,8-dimethyl-2-hydroxy $4,8\text{-}(CH_3)_2\text{-}2\text{-}HO(C_9H_4N)$	173.21	pl (aq aa)		217-8			al	B21^{4}, 1304
12825	Quinoline, 5,8-dimethyl $5,8\text{-}(CH_3)_2(C_9H_5N)$	157.22	nd	265^{736}	4-5	1.070^{21}		al, eth	B20^{4}, 3538

PHYSICAL CONSTANTS OF ORGANIC COMPOUNDS (Continued)

No.	Name, Synonyms, and Formula	Mol. wt.	Color, crystalline form, specific rotation and λ_{max} (log ε)	b.p. °C	m.p. °C	Density	n_D	Solubility	Ref.
12826	Quinoline, 6,8-dimethyl or β-Cytisolidine 6,8-$(CH_3)_2(C_9H_5N)$	157.22	...	268-9, $133\text{-}4^{14}$	...	1.0665^{4}	...	al, eth	B20^{4}, 3539
12827	Quinoline, 6,8-dimethyl-2-hydroxy 6,8-$(CH_3)_2$-2-$HO(C_9H_4N)$	173.21	nd (al)	...	201-2	...	...	al	B21^{1}, 225
12828	Quinoline, 6,8-dimethyl-5-hydroxy or Cytisoline 6,8-$(CH_3)_2$-5-$HO(C_9H_4N)$	173.21	pl (chl), cr (al)	sub	197-8	...	...	al, bz, chl	B21, 117
12829	Quinoline, 6-(dimethylamino)-2-methyl or 6-Dimethyl amino quinaldine 2-CH_3-6-$[(CH_3)_2N](C_9H_5N)$	186.26	ye pr (aa, AcOEt)	319	101	...	...	al, eth, bz	B22^{2}, 361
12830	Quinoline, 3-ethyl-2-hydroxy or 3-Ethyl carbostyril 3-C_2H_5-2-$HO(C_9H_5N)$	173.21	(dil HCl)	...	168	...	...	bz, chl	B21, 115
12831	Quinoline, 2-hydrazino 2-$H_2NNH(C_9H_6N)$	159.19	(bz)	...	142-3	...	...	al	B22^{1}, 690
12832	Quinoline, 5-hydrazino 5-$H_2NNH(C_9H_6N)$	159.19	ye nd (w)	...	150-1	...	...	al	B22, 565
12833	Quinoline hydrochloride or Quinilirium chloride $C_9H_7N.HCl$	165.62	pr (w)	...	134.5	...	...	w, al, chl	B20^{2}, 226
12834	Quinoline hydrogen sulfate $C_9H_7N.H_2SO_4$	227.23	cr (al, aa)	...	164	...	...	w, aa	B20^{2}, 226
12835	Quinoline, 2-hydroxy or 2-Quinolinol 2-$HO(C_9H_6N)$	145.16	pr (al, dil al + 1w), nd (sub)	sub	199-200 (anh)	...	...	al, eth	B21^{4}, 1057
12836	Quinoline, 2-hydroxy-3-methyl or 3-Methyl carbostyril 3-CH_3-2-$HO(C_9H_5N)$	159.19	yesh nd (ace, dil al)	sub	234-5	...	...	al	B21^{4}, 1246
12837	Quinoline, 2-hydroxy-4-methyl or 2-Hydroxy lepidine 4-CH_3-2-$HO(C_9H_5N)$	159.19	nd (w)	>360, 270^{17}	245	...	...	al	B21^{4}, 1252
12838	Quinoline, 2-hydroxy-6-methyl or 6-Methyl carbostyril 6-CH_3-2-$HO(C_9H_5N)$	159.19	nd (al)	$240\text{-}1^{12}$	237	...	...	al, eth, ace, bz	B21^{4}, 1271
12839	Quinoline, 3-hydroxy or 3-Quinolinol 3-$HO(C_9H_6N)$	145.16	cr (bz, dil al)	...	200-1	...	...	al, bz	B21^{4}, 1075
12840	Quinoline, 3-hydroxy-2-methyl or 3-Hydroxy quinaldine 2-CH_3-3-$HO(C_9H_5N)$	159.19	nd (al)	...	203-5	...	...	al, eth, chl	B21^{4}, 1216
12841	Quinoline, 4-hydroxy or 4-Quinolinol 4-$HO(C_9H_6N)$	145.16	nd (w + 3)	...	210 (anh), 100 (hyd)	...	...	al	B21^{4}, 1079
12842	Quinoline, 4-hydroxy-2-methyl or 4-Hydroxy quinaldine 2-CH_3-4-$HO(C_9H_5N)$	159.19	pr (w + 2)	300d	232	...	...	al	B21^{4}, 1218
12843	Quinoline, 4-hydroxy-2-phenyl 2-C_6H_5-4-$HO(C_9H_5N)$	221.26	pl or pr (al), nd (aa)	...	256-7	...	...	al	B21^{4}, 1625
12844	Quinoline, 5-hydroxy or 5-Quinolinol 5-$HO(C_9H_6N)$	145.16	nd (al), pl	sub	224d	...	...	...	B21^{4}, 1103
12845	Quinoline, 5-hydroxy-2-methyl or 5-Hydroxy quinaldine 2-CH_3-5-$HO(C_9H_5N)$	159.19	pl (al)	...	246-7	...	...	eth	B21^{2}, 63
12846	Quinoline, 5-hydroxy-6-methyl 6-CH_3-5-$HO(C_9H_5N)$	159.19	nd (al, sub)	sub	230	...	...	al, eth, ace	B21, 111
12847	Quinoline, 5-hydroxy-8-methyl 8-CH_3-5-$HO(C_9H_5N)$	159.19	nd (dil al, sub)	sub	762-3	...	...	...	B21, 112
12848	Quinoline, 6-hydroxy or 6-Quinolinol 6-$HO(C_9H_6N)$	145.16	pr (al, eth)	>360	193	...	...	...	B21^{2}, 53
12849	Quinoline, 6-hydroxy-2-methyl or 6-Hydroxy quinaldine 2-CH_3-6-$HO(C_9H_5N)$	159.19	(w)	304-5, 186^{15}	213	1.1665^{0}	...	al, eth	B21, 106
12850	Quinoline, 6-hydroxy-4-methyl or 6-Hydroxy lepidine 4-CH_3-6-$HO(C_9H_5N)$	159.19	nd (w, dil al)	...	222-4	...	...	al, ace, chl	B21^{4}, 1265
12851	Quinoline, 6-hydroxy-8-methyl 8-CH_3-6-$HO(C_9H_5N)$	159.19	nd (dil al)	...	200	...	...	al	B21, 113
12852	Quinoline, 7-hydroxy or 7-Quinolinol 7-$HO(C_9H_6N)$	145.16	pr (al), nd (dil al-eth)	sub	238-40	...	...	al	B21^{4}, 1130
12853	Quinoline, 7-hydroxy-6-methyl 6-CH_3-7-$HO(C_9H_5N)$	159.19	nd (al)	240^{22} sub	244	...	...	al, eth, bz	B21, 111
12854	Quinoline, 8-hydroxy or 8-Quinolinol 8-$HO(C_9H_6N)$	145.16	nd (dil al)	266.6^{752} sub	75-6	1.034^{209}	...	al, ace, bz, chl	B21^{4}, 1135
12855	Quinoline, 8-hydroxy-2-methyl or 8-Hydroxy quinaldine 2-CH_3-8-$HO(C_9H_5N)$	159.19	pr (dil al)	267, $145\text{-}60^{22}$	74-5	...	...	eth, bz	B21^{4}, 1232
12856	Quinoline, 8-hydroxy-4-methyl or 8-Hydroxy lepidine 4-CH_3-8-$HO(C_9H_5N)$	159.19	nd (lig)	...	141	...	...	ace, bz, chl, aa	B21^{4}, 1266

PHYSICAL CONSTANTS OF ORGANIC COMPOUNDS (Continued)

No.	Name, Synonyms, and Formula	Mol. wt.	Color, crystalline form, specific rotation and λ_{max} (log ε)	b.p. °C	m.p. °C	Density	n_D	Solubility	Ref.
12857	Quinoline, 8-hydroxy-5-methyl 5-CH_3-8-HO(C_9H_5N)	159.19	nd (dil al)	sub 100	122-4	....	....	....	$B21^4$, 1269
12858	Quinoline, 8-hydroxy-6-methyl 6-CH_3-8-HO(C_9H_5N)	159.19	nd (chl, bz)	sub	95-6	....	....	al	$B21^4$, 1274
12859	Quinoline, 8-hydroxy-7-methyl 7-CH_3-8-HO(C_9H_5N)	159.19	nd (dil al)	sub 100	72-4	....	....	al	$B21^4$, 1279
12860	Quinoline, 8-hydroxy-5-nitroso 5-NO-8-HO(C_9H_5N)	174.16	nd (al)	....	245d	....	....	....	$B21^1$, 405
12861	Quinoline, 8-hydroxy, sulfate or Chinosol $(8\text{-}HOC_9H_6N)_2 \cdot H_2SO_4$	388.39	....	....	177.5	....	....	w, al	B21, 92
12862	Quinoline, 2-iodo 2-I(C_9H_6N)	255.06	nd (dil al)	....	52-3	....	....	al, eth, ace	$B20^4$, 3393
12863	Quinoline, 4-iodo 4-I(C_9H_6N)	255.06	nd or pr	sub	97 (100)	....	....	al, eth	B20, 370
12864	Quinoline, 5-iodo 5-I(C_9H_6N)	255.06	nd (al, eth)	sub	101-2	....	....	....	$B20^1$, 141
12865	Quinoline, 6-iodo 6-I(C_9H_6N)	255.06	lf (w), nd (sub)	sub	91	....	....	al, eth	B20, 370
12866	Quinoline, 8-iodo 8-I(C_9H_6N)	255.06	lo nd (al)	....	36	....	....	al, eth, ace, bz, lig	$B20^1$, 141
12867	Quinoline, 2-methoxy-6-nitro 6-NO_2-2-CH_3O(C_9H_5N)	204.19	nd (dil aa, bz, sub)	....	189-90	....	....	al, eth, bz, chl, lig	B21, 81
12868	Quinoline, 2-methoxy-8-nitro 8-NO_2-2-CH_3O(C_9H_5N)	204.19	nd (dil al)	....	124-5	....	....	al, bz	B21, 82
12869	Quinoline, 4-methoxy 4-CH_3O(C_9H_6N)	159.19	....	245, 167^{20}	41	....	....	al, eth, bz	$B21^4$, 1080
12870	Quinoline, 5-methoxy 5-CH_3O(C_9H_6N)	159.19	....	282^{750}	....	....	....	al, eth	$B21^4$, 1103
12871	Quinoline, 6-methoxy or *p*-Quinanisole 6-CH_3O(C_9H_6N)	159.19	hyg lf	305^{740}, 153^{12}	26.5	$1.154^{20/20}$, 1.000^{209}	....	al, eth	$B21^4$, 1108
12872	Quinoline, 6-methoxy-5-nitro 5-NO_2-6-CH_3O(C_9H_5N)	204.19	cr (al)	....	104-5	....	....	ace	$B21^4$, 1121
12873	Quinoline, 6-methoxy-8-nitro 8-NO_2-6-CH_3O(C_9H_5N)	204.19	yesh nd (al)	....	159-60	....	....	chl	$B21^4$, 1121
12874	Quinoline, 6-methoxy-1,2,3,4-tetrahydro or Thalline 6-CH_3O($C_9H_{10}N$)	163.22	pr (peth, al), rh pym (w)	283^{735}, $127\text{-}30^{1}$	42-3	....	1.5718^{20}	al, eth, bz	B22, 61
12875	Quinoline, 8-methoxy or *o*-Quinanisole 8-CH_3O(C_9H_6N)	159.19	nd (peth)	282^{750}, 164^{14}	49-50	1.034^{29}	....	al, eth, bz, peth	$B21^4$, 1154
12876	Quinoline, 8-methoxy-5-nitro 5-NO_2-8-CH_3O(C_9H_5N)	204.19	cr (al)	....	151.5	....	....	al	$B21^4$, 1193
12877	Quinoline, 1-methyl-1,2,3,4-tetrahydro or Kairoline ($C_9H_{10}N$)CH_3	147.22	....	247-50, $123\text{-}6^{14}$	....	$1.022^{20/4}$	1.5802^{23}	al	$B20^2$, 174
12878	Quinoline, 2-methyl or Quinaldine 2-CH_3(C_9H_6N)	143.19	....	247.6, 118^{10}	−2	$1.0585^{20/4}$	1.8116^{20}	al, eth, ace, chl	$B20^4$, 3454
12879	Quinoline, 2-methyl-5-nitro or 5-Nitro quinaldine 5-NO_2-2-CH_3(C_9H_5N)	188.19	nd (dil al)	....	82	....	....	al, eth	$B20^4$, 3468
12880	Quinoline, 2-methyl-6-nitro or 6-Nitro quinaldine 6-NO_2-2-CH_3(C_9H_5N)	188.19	cr (80% MeOH)	....	165 (173)	....	....	al	$B20^4$, 3468
12881	Quinoline, 2-methyl-8-nitro or 8-Nitro quinaldine 8-NO_2-2-CH_3(C_9H_5N)	188.19	pa ye nd (dil al)	....	137	....	....	al, eth, bz	$B20^4$, 3468
12882	Quinoline, 3-methyl or β-Methyl quinoline 3-CH_3(C_9H_6N)	143.19	pr	259.8, $140\text{-}2^{25}$	16-7	$1.0673^{20/4}$	1.6171^{20}	al, eth, ace	$B20^4$, 3472
12883	Quinoline, 4-methyl or Lepidine 4-CH_3(C_9H_6N)	143.19	red br	264.2, 133^{15}	9-10	$1.0862^{20/4}$	1.6206^{20}	al, eth, ace, bz, lig	$B20^4$, 3477
12884	Quinoline, 4-methyl-3-nitro or 3-Nitro lepidine 3-NO_2-4-CH_3(C_9H_5N)	188.19	pr (w)	....	118	....	....	al	$B20^4$, 3485
12885	Quinoline, 4-methyl-8-nitro 8-NO_2-4-CH_3(C_9H_5N)	188.19	lf (abs al)	....	126-7	....	....	eth	$B20^4$, 3485
12886	Quinoline, 5-methyl 5-CH_3(C_9H_6N)	143.19	....	262.7	19	$1.0832^{20/4}$	1.6219^{20}	al, eth, ace	$B20^4$, 3488
12887	Quinoline, 6-methyl or *p*-Toluquinoline 6-CH_3(C_9H_6N)	143.19	....	258.6, 130^{15}	ca -22	$1.0654^{20/4}$	1.6157^{20}	al, eth, ace	$B20^4$, 3489
12888	Quinoline, 6-methyl-5-nitro 5-NO_2-6-CH_3(C_9H_5N)	188.19	pa ye nd (al)	....	116-7	....	....	al, eth, ace, bz	$B20^4$, 3495
12889	Quinoline, 6-methyl-8-nitro 8-NO_2-6-CH_3(C_9H_5N)	188.19	pa ye nd (w)	....	122	....	....	al, eth, ace, bz	B20, 400

No.	Name, Synonyms, and Formula	Mol. wt.	Color, crystalline form, specific rotation and λ_{max} (log ε)	b.p. °C	m.p. °C	Density	n_D	Solubility	Ref.
12890	Quinoline, 7-methyl or *m*-Toluquinoline $7\text{-}CH_3(C_9H_6N)$	143.19	ye	$257.6, 144^{18}$	39	$1.0609^{20/4}$	1.6150^{20}	al, eth, ace	$B20^4$, 3497
12891	Quinoline, 8-methyl or *o*-Toluquinoline $8\text{-}CH_3(C_9H_6N)$	143.19		$247^{8}, 143^{34}$		$1.0719^{20/4}$	1.6164^{20}	**al, eth**, ace	$B20^4$, 3500
12892	Quinoline, 8-methyl-5-nitro $5\text{-}NO_2\text{-}8\text{-}CH_3(C_9H_5N)$	188.19	pa ye nd (al)		93			al, eth, ace, bz	$B20^4$, 3504
12893	Quinoline, 8-methyl-6-nitro $6\text{-}NO_2\text{-}8\text{-}CH_3(C_9H_5N)$	188.19	cr (al)		129			al	$B20^4$, 3504
12894	Quinoline, 3-nitro $3\text{-}NO_2(C_9H_6N)$	174.16	nd (dil al)		128			al, ace	$B20^4$, 3395
12895	Quinoline, 4-nitro, oxide $4\text{-}NO_2(C_9H_6NO)$	190.16	ye nd pl (ace)		154				$B20^4$, 3396
12896	Quinoline, 5-nitro $5\text{-}NO_2(C_9H_6N)$	174.16	pl (w or al), nd (+ w, w)	sub	73-5 (sub)			bz	$B20^4$, 3397
12897	Quinoline, 6-nitro $6\text{-}NO_2(C_9H_6N)$	174.16	ye pl (HCl-aa)	sub	153-4			bz	$B20^4$, 3397
12898	Quinoline, 7-nitro $7\text{-}NO_2(C_9H_6N)$	174.16	nd or lf, w al pl (sub)	sub	132-3			eth, chl	$B20^4$, 3398
12899	Quinoline, 8-nitro $8\text{-}NO_2(C_9H_6N)$	174.16	mcl pr (al)		91-2			al, eth	$B20^4$, 3399
12900	Quinoline, 2-oxo-1,2,3,4-tetrahydro C_9H_9NO	147.18	pr (al, eth)	201^{45}	163-4			al, eth	$B21^4$, 3638
12901	Quinoline, 2-phenyl $2\text{-}C_6H_5(C_9H_6N)$	205.26	nd (dil al)	$363, 310^{187}$	86			al, eth, ace, bz	$B20^4$, 4137
12902	Quinoline, 3-phenyl $3\text{-}C_6H_5(C_9H_6N)$	205.26	pl (eth)	$205\text{-}7^{12}$	52			al, eth, ace, bz, chl	$B20^4$, 4145
12903	Quinoline, 4-phenyl $4\text{-}C_6H_5(C_9H_6N)$	205.26		260^{77}	61-2			al, ace, bz, chl	$B20^4$, 4149
12904	Quinoline, 5-phenyl $5\text{-}C_6H_5(C_9H_6N)$	205.26	nd (dil al)		83			al, eth, ace	$B20^4$, 4150
12905	Quinoline, 6-phenyl $6\text{-}C_6H_5(C_9H_6N)$	205.26	pl (al, bz)	260^{77}	110	1.945^{20}		al, ace, bz, chl	$B20^4$, 4151
12906	Quinoline, 8-phenyl $8\text{-}C_6H_5(C_9H_6N)$	205.26	ye gr oil	283^{187}				al, ace, eth, bz, chl	$B20^4$, 4153
12907	Quinoline, 1,2,3,4-tetrahydro $C_9H_{11}N$	133.19	nd	251	20	$1.0588^{20/4}$	1.6062^{19}	**al, eth**	$B20^4$, 2923
12908	Quinoline, 5,6,7,8-tetrahydro $C_9H_{11}N$	133.19		$222, 92\text{-}5^{12}$		1.0304^{13}	1.5435^{20}	al, eth, ace, bz	$B20^4$, 2922
12909	Quinoline, 2,3,4-trimethyl or 3,4-Dimethyl quinaldine $2,3,4\text{-}(CH_3)_3(C_9H_4N)$	171.24		$285, 156\text{-}8^{12}$	92				$B20^4$, 3557
12910	Quinoline, 2,3,8-trimethyl $2,3,8\text{-}(CH_3)_3(C_9H_4N)$	171.24		285	86-7			al, eth, chl	$B20^4$, 3557
12911	Quinoline, 2,4,6-trimethyl or 4,6-Dimethyl quinaldine $2,4,6\text{-}(CH_3)_3(C_9H_4N)$	171.24	nd (w, dil al + 1w)	$281\text{-}2, 146\text{-}8^{13}$	65.5 (anh), 40 (+ w)			al, eth, ace, bz, chl	$B20^4$, 3558
12912	Quinoline, 2,4,7-trimethyl or 4,7-Dimethyl quinaldine $2,4,7\text{-}(CH_3)_3(C_9H_4N)$	171.24	nd (w)	200-1	63-4	$1.0337^{20/4}$	1.5973^{24}	al, eth, ace, bz	$B20^4$, 3558
12913	Quinoline, 2,4,8-trimethyl or 4,8-Dimethyl quinaldine $2,4,8\text{-}(CH_3)_3(C_9H_4N)$	171.24		287	50-1		1.5855^{50}		$B20^4$, 3559
12914	Quinoline, 2,5,6-trimethyl $2,5,6(CH_3)_3(C_9H_4N)$	171.24	nd		69-70			al, eth, bz	B20, 415
12915	Quinoline, 2,5,7-trimethyl or Tetracoline $2,5,7\text{-}(CH_3)_3(C_9H_4N)$	171.24	pr	$286.6^{746}, 107\text{-}8^{719}$				al, eth, bz, peth	$B20^4$, 3559
12916	Quinoline, 2,6,8-trimethyl $2,6,8\text{-}(CH_3)_3(C_9H_4N)$	171.24	pr (peth), lf (dil al)	260^{719}	46			al, eth, bz, peth	$B20^4$, 3560
12917	Quinoline, 4,5,8-trimethyl $4,5,8\text{-}(CH_3)_3(C_9H_4N)$	171.24		155^{13}	73-4			al, eth, bz	$B20^4$, 3562
12918	2-Quinoline carbonitrite $2\text{-}NC(C_9H_6N)$	154.17	nd (lig, chl)	$100\text{-}70^{20\text{-}23}$	94			al, eth, bz, chl, lig	$B22^4$, 1157
12919	2-Quinoline carboximide $2\text{-}H_2NCO(C_9H_6N)$	172.19	nd (dil al, bz-lig)		133			al, bz, chl	$B22^4$, 1154
12920	2-Quinoline carboxylic acid or Quinaldinic acid $2\text{-}HO_2C(C_9H_6N)$	173.17	nd (+ 3w), (bz)		157 (anh)			w, bz	$B22^4$, 1149
12921	2-Quinoline carboxylic acid, 4,8-dihydroxy or Xanthurenic acid $2\text{-}HO_2C\text{-}4,8\text{-}(HO)_2(C_9H_4N)$	205.17	ye micr cr (w)		289 (297)			al	$B22^4$, 2513

No.	Name, Synonyms, and Formula	Mol. wt.	Color, crystalline form, specific rotation and λ_{max} (log ε)	b.p. °C	m.p. °C	Density	n_D	Solubility	Ref.
12922	2-Quinoline carboxylic acid, 4-hydroxy or Kynurenic acid 4-HO-2-$HO_2C(C_9H_5N)$	189.17	ye nd (+w dil aa)		282-3				B22^4, 2245
12923	2-Quinoline carboxylic acid, methyl ester 2-$CH_3O_2C(C_9H_6N)$	187.20	nd (lig)		86			al, lig	B22^4, 1152
12924	2-Quinoline carboxylyl chloride 2-$ClCO(C_9H_6N)$	191.62	nd (eth, lig)		(i)97-8, (ii)175-6			eth, bz	B22^4, 509
12925	3-Quinoline carbonitrile or 3-Cyanoquinoline 3-$NC(C_9H_6N)$	154.17	(al or sub)		108			eth, ace, bz	B22^4, 1171
12926	3-Quinoline carboxylic acid 3-$HO_2C(C_9H_6N)$	173.17	pl (al, dil al)		275d			al	B22^4, 1167
12927	3-Quinoline carboxylic acid, 2-phenyl 2-C_6H_5-3-$HO_2C(C_9H_5N)$	249.27	nd (al)		230d			al, ace, aa	B22^4, 1355
12928	4-Quinoline carbonitrile or Cinchoninonitrile 4-$NC(C_9H_6N)$	154.17	cr (chl, lig, eth) nd(sub)	240-5	103-4			al, eth, ace, bz	B22^4, 1181
12929	4-Quinoline carboxylic acid or Cinchoninic acid 4-$HO_2C(C_9H_6N)$	173.17	mcl pr (w), nd (+1w), mcl or tcl (+2w)		257-8 (anh)				B22^4, 1177
12930	4-Quinoline carboxylic acid, 2-(3-carboxy-4-hydroxy phenyl) or Hexophan $C_{17}H_{11}NO_5$	309.28	vesh pw		283-4d			chl	B22^2, 206
12931	4-Quinoline carboxylic acid, 6-methoxy or Quininic acid 6-CH_3O-4-$HO_2C(C_9H_5N)$	203.20	pa ye pr (dil al)	sub	285d				B22^4, 2297
12932	4-Quinoline carboxylic acid, 8-methoxy-2-phenyl or Isatophan 8-CH_3O-2-C_6H_5-4-$HO_2C(C_9H_4N)$	279.30	ye nd (al)		216			al, chl	B22^1, 559
12933	4-Quinoline carboxylic acid, 6-methyl-2-phenyl,ethyl ester or Novatophan 6-CH_3-2-C_6H_5-4-$(C_2H_5O_2C)(C_9H_4N)$	291.35	ye cr (al)		75-6			eth, ace, bz, chl	B22^4, 1395
12934	4-Quinoline carboxylic acid, 2-phenyl or Cinchophene 2-C_6H_5-4-$HO_2C(C_9H_5N)$	249.27	nd (MeOH, dil al), ye in air		218			eth	B22^4, 1358
12935	4-Quinoline carboxylic acid, 2-phenyl,allyl ester or Atoquinol 2-C_6H_5-4-$(C_3H_5O_2C)(C_9H_5N)$	289.33	nd (dil al)	265^{15}	30			al, eth, ace, bz	B22^4, 1361
12936	5-Quinoline carbonitrile 5-$NC(C_9H_6N)$	154.17	nd (lig), nd(+3/2 w, dil al)		89, 70 (+w)			al, bz	B22^4, 1195
12937	5-Quinoline carboxylic acid 5-$HO_2C(C_9H_6N)$	173.17	cr (aa, sub)	sub<338	342				B22^4, 1195
12938	6-Quinoline carboxylic acid 6-$HO_2C(C_9H_6N)$	173.17	nd, pr or pl (sub)	sub<290	291-2				B22^4, 1196
12939	7-Quinoline carboxylic acid 7-$HO_2C(C_9H_6N)$	173.17	nd (w al)	sub	249-50			al	B22^4, 1199
12940	8-Quinoline carbonitrile or 8-Cyano quinoline 8-$NC(C_9H_6N)$	154.17	nd (dil al)		84			al	B22^4, 1203
12941	8-Quinoline carboxylic acid 8-$HO_2C(C_9H_6N)$	173.17	nd (w)	sub	187			al	B22^4, 1200
12942	5-Quinoline sulfonic acid, 8-hydroxy-7-iodo $C_9H_6NO_4IS$	351.12			260d				B22^4, 3497
12943	5-Quinoline sulfonic acid, 6-hydroxy 6-HO-5-$(HO_3S)(C_9H_5N)$	225.22	ye nd (+½ w, w, al)		270d			w	B22, 407
12944	5-Quinoline sulfonic acid, 8-hydroxy 8-HO-5-$HO_3S(C_9H_5N)$	225.22	ye lf, nd (+1w, dil HCl)		322-3				B22^4, 3493
12945	5-Quinoline sulfonic acid, 8-hydroxy-7-iodo or Loretin 8-HO-7-I-5-$HO_3S(C_9H_4N)$	351.12	ye pr or lf (al)		260d				B22^4, 3497
12946	7-Quinoline sulfonic acid, 8-hydroxy 8-HO-7-$HO_3S(C_9H_5N)$	225.22	ye nd (al)		314-5			w, al	B22, 408
12947	8-Quinoline sulfonic acid, 5-hydroxy-6-iodo or Lorenite 5-HO-6-I-8-$HO_3S(C_9H_4N)$	351.12	ye nd or lf		210-30d				B22, 406
12948	2-Quinolinone, 1-methyl $C_{10}H_9NO$	159.19		324^{728}	74			al, eth, ace, bz, chl	B20^4, 3505
12949	Quinovic acid or Quinovaic acid $C_{30}H_{46}O_5$	486.69	pl or nd $[\alpha]^{16}{}_D$ +87 (aq KOH)		298d				B10^3, 2307

PHYSICAL CONSTANTS OF ORGANIC COMPOUNDS (Continued)

No.	Name, Synonyms, and Formula	Mol. wt.	Color, crystalline form, specific rotation and λ_{max} (log ε)	b.p. °C	m.p. °C	Density	n_D	Solubility	Ref.
12950	Quinoxaline or 1,4-Benxodiazine $C_8H_6N_2$	130.15	cr (peth)	229.5, 108-11^{12}	28	1.1334$^{48/4}$	1.6231^{48}	w, **al, eth, ace, bz**	B23^2, 177
12951	Quinoxaline, 6-amino 6-$H_2N(C_8H_5N_2)$	145.16	ye nd (eth)	sub	159	....	....	w, al, eth, chl	B25, 326
12952	Quinoxaline, 6-chloro 6-$Cl(C_8H_5N_2)$	164.59	nd (w)	117-9^{10} sub	64	....	....	....	B23^2, 177
12953	Quinoxaline, 2,3-dichloro 2,3-$Cl_2(C_8H_4N_2)$	199.04	cr (al, bz)	....	151-3	....	....	al, bz, chl, aa	B23^2, 177
12954	Quinoxaline, 2,3-dihydroxy 2,3$(HO)_2(C_8H_4N_2)$	162.15	nd (w)	....	410	....	....	w	B24^2, 200
12955	Quinoxaline, 2,3-dimethyl 2,3-$(CH_3)_2(C_8H_4N_2)$	158.20	nd (w + 3) (ace)	....	106 (anh), 85 (+ w)	....	....	al, eth, ace, bz	B23^2, 197
12956	Quinoxaline, 2,6-dimethyl 2,6-$(CH_3)_2(C_8H_4N_2)$	158.20	....	267-9	54	....	....	w, al, ace, bz	B23, 192
12957	Quinoxaline, 2-hydroxy or 2-Quinoxalinol 2-$HO(C_8H_5N_2)$	146.15	lf (al)	sub 200$^{0.5}$	271	....	....	....	B24^2, 72
12958	Quinoxaline, 6-methoxy 6-$CH_3O(C_8H_5N_2)$	160.18	nd (w)	128^7 sub	57.5	....	....	eth, ace	B23, 387
12959	Quinoxaline, 2 methyl 2-$CH_3(C_8H_5N_2)$	144.18	ye	245-7, 118^{16}	180-1	....	....	w, al, **eth, ace, bz**	B23^2, 190
12960	Quinoxaline, 6-methyl 6-$CH_3(C_8H_5N_2)$	144.18	....	248^{748}, 141.5^{29}	218-9	1.1164$^{20/4}$	1.6211$^{18.4}$	**w, al, eth, ace, bz**	B23, 184
12961	Quinoxaline, 1,2,3,4-tetrahydro $C_8H_{10}N_2$	134.18	lf (w, eth, peth)	289, 153-4^{14}	99	....	....	al, eth, bz, chl	B23^2, 106
12962	Quinoxaline, 5-hydroxy or 5-Quinoxalol 5-$HO(C_8H_5N_2)$	146.15	(eth)	184^7, sub 90$^{2.5}$	101-2	....	....	....	C48, 8232
12963	Quinuclidine or 1,4-Ethylene piperidine $C_7H_{13}N$	111.19	(eth)	....	158 (sealed tube)	....	....	w, al, eth, ace, bz	B20^4, 1966
12964	Quinuclidine, 3-hydroxy $C_7H_{13}NO$	127.19	cr (bz)	....	221-1	....	....	ace	B21^4, 237
12965	2-Quinuclidine carboxylic acid $C_8H_{13}NO_2$	155.20	col hyg cr (al-ace)	....	280d	....	....	....	B22^4, 200

PHYSICAL CONSTANTS OF ORGANIC COMPOUNDS (Continued)

No.	Name, Synonyms, and Formula	Mol. wt.	Color, crystalline form, specific rotation and λ_{max} (log ε)	b.p. °C	m.p. °C	Density	n_D	Solubility	Ref.
12966	Raffinose or Melitriose $C_{18}H_{32}O_{14}\cdot 5H_2O$	562.52	wh pw, pr or nd (+5w, dil al) $[\alpha]^{20/}{}_D$ +101, +123	d 130	118-9 (anh), 80 (hyd)	1.465^0		w, MeOH, Py	$B17^4$, 3801
12967	Raunescine hydrate $C_{11}H_{16}H_2O_8\cdot H_2O$	582.65	wh hex pr (90% MeOH), $[\alpha]^{25/}{}_D$ −74 (chl)		160-70			chl, aa	Am79, 250
12968	Reductone or 3-Hydroxy-2-oxo-propionaldehyde (enol-form) HOCH=C(OH)CHO	88.06	ye nd (w)	200-20d				w, al	$B1^4$, 4145
12969	Resazurin $C_{12}H_7NO_4$	229.19	dk red to grsh pr or pl (aa, AcOEt)	sub vac	d				B27, 128
12970	Rescinnamine or Moderil $C_{35}H_{42}N_2O_9$	634.73	nd (bz) $[\alpha]^{24/}{}_D$ −98 (chl, c=0.1)		238-9			ace, chl	Am77, 2241
12971	Reserpic acid or Reserpinolic acid $C_{22}H_{28}N_2O_5$	400.47	(MeOH), $[\alpha]^{23/}{}_D$ −81 (of hydro-chloride)		241-3				Am78, 2023
12972	Reserpine or Rivasin-Serparsin $C_{33}H_{40}N_2O_9$	608.69	lo pr (dil ace), $[\alpha]'_D$ −117.7 (chl, c=1)		264-5, 277 (sealed tube)			bz, chl	H37, 67
12973	Reserpinine or Raubasinine $C_{22}H_{26}N_2O_4$	382.46	pa ye pl (aq ace), $[\alpha]^{20/}{}_D$ −131 (chl, c=1.18)		243-4d			al, chl	C49, 11672
12974	Resorcinol or 1,3-Dihydroxy benzene $1,3\text{-}(HO)_2C_6H_4$	110.11	nd (bz), pl (w)	178^{16}	111	1.2717		w, al, eth, aa	$B6^3$, 4292
12975	Resorcinol monoacetate $3\text{-}(CH_3CO_2)C_6H_4OH$	152.18	ye	283d				al	$B6^3$, 4319
12976	Resorcinol, 4-acetyl or 2,4-Dihydroxy acetophenone. Resacetophenone $2,4\text{-}(HO)_2C_6H_3COCH_3$	152.15	nd or lf		147	1.1800^{141}		aa, Py	$B8^3$, 2082
12977	Resorcinol, 5-acetyl or 3,5-Dihydroxy acetophenone $3,5\text{-}(HO)_2C_6H_3COCH_3$	152.15	cr (w)		147-8			w, al, eth, ace	$B8^2$, 301
12978	Resorcinol benzoate (mono) $3\text{-}C_6H_5CO_2C_6H_4OH$	214.22	pr (dil al)		135-6			chl, aa	$B9^3$, 555
12979	Resorcinol, 4-(2-aminoethyl), hydrochloride $4(H_2NCH_2CH_2)\text{-}1,3\text{-}(HO)_2C_6H_3.HCl$	189.64	nd (w)		237d			w, al	$B13^2$, 486
12980	Resorcinol acetate (mono) or 3-Hydroxy phenyl acetate $3\text{-}HO\text{-}C_6H_4O_2CCH_3$	152.15	ye	283d				al	$B6^3$, 4319
12981	Resorcinol, monobenzyl ether $3\text{-}C_6H_5CH_2OC_6H_4OH$	200.24	cr (CCl_4)	200^5	69.2			w	$B6^3$, 4314
12982	Resorcinol, 2-bromo or 1,3-Dihydroxy-2-bromo benzene $1,3\text{-}(HO)_2\text{-}2\text{-}BrC_6H_3$	189.01	nd (chl), cr (peth)		102-3			w, al, eth, chl	$B6^3$, 4336
12983	Resorcinol, 4-bromo or 1,3-Dihydroxy-4-bromo benzene $1,3\text{-}(HO)_2\text{-}4\text{-}BrC_6H_3$	189.01		$138\text{-}41^{12}$	103			w, eth	$B6^3$, 4336
12984	Resorcinol, 5-bromo or 1,3-Dihydroxy-5-bromo benzene $1,3\text{-}(HO)_2\text{-}5\text{-}BrC_6H_3$	189.01	nd (bz), pr(+1)		87, 79 (+w)			w, eth, bz	$B6^2$, 820
12985	Resorcinol, 4-butyl $4\text{-}C_4H_9\text{-}1,3\text{-}(HO)_2C_6H_3$	166.22		$164\text{-}6^{6\text{-}7}$	47-8				$B6^3$, 4657
12986	Resorcinol, 4-isobutyl $4\text{-}i\text{-}C_4H_9\text{-}1,3\text{-}(HO)_2C_6H_3$	166.22		$166\text{-}8^{6\text{-}7}$	62-3			al, eth	$B6^2$, 899
12987	Resorcinol, 2-chloro or 2-Chloro-1,3-dihydroxy benzene $2\text{-}Cl\text{-}1,3\text{-}(HO)_2C_6H_3$	144.56			97-8				$B6^3$, 4333
12988	Resorcinol, 4-chloro or 4-Chloro-1,3-dihydroxy benzene $4\text{-}Cl\text{-}1,3\text{-}(HO)_2C_6H_3$	144.56		259, 147^{18}	(i)89 (ii)105			w, al, eth, ace, bz	$B6^3$, 4333
12989	Resorcinol, 5-chloro or 5-Chloro-1,3-dihydroxy benzene $5\text{-}Cl\text{-}1,3\text{-}(HO)_2C_6H_3$	144.56	hyg nd (sub)	sub	117			w, eth, ace	$B6^2$, 819
12990	Resorcinol, 4-cyclohexyl or 4-Cyclohexyl-1,3-dihydroxy benzene $4\text{-}C_6H_{11}C_6H_3(OH)_2(1,3)$	192.26	nd (bz chl), cr (bz-peth)		127-8				$B6^3$, 5057

No.	Name, Synonyms, and Formula	Mol. wt.	Color, crystalline form, specific rotation and λ_{max} (log ε)	b.p. °C	m.p. °C	Density	n_D	Solubility	Ref.
12991	Resorcinol diacetate $1,3\text{-}(CH_3CO_2)_2C_6H_4$	194.19		278, 153-4^{12}				al, eth	$B6^3$, 4320
12992	Resorcinol, 4,6-diacetyl or Resodiacetophenone $4,6\text{-}(CH_3CO)_2C_6H_2(OH)_2\text{-}(1,3)$	194.19	nd (al)		185			eth, ace, bz	$B8^3$, 3511
12993	Resorcinol dibenzoate $3\text{-}(C_6H_5CO_2)C_6H_4(O_2CC_6H_5)$	318.33	pl (dil al)		117			al, eth	$B9^3$, 555
12994	Resorcinol, 4,6-dichloro $4,6\text{-}Cl_2\text{-}1,3\text{-}(HO)_2C_6H_2$	179.00		254 sub	113			w, al, eth, ace	$B6^3$, 4335
12995	Resorcinol diethyl ether or 1,3-Diethoxy benzene $1,3\text{-}(C_2H_5O)_2C_6H_4$	166.22	pr	235^{756}	12.4			al, eth	$B6^3$, 4307
12996	Resorcinol, 2,5-dimethyl $2,5\text{-}(CH_3)_2\text{-}1,3\text{-}(HO)_2C_6H_2$	138.17	nd (bz), pr (w)	277-80	163			al, eth	$B6^3$, 4606
12997	Resorcinol, 4,5-dimethyl $4,5\text{-}(CH_3)_2\text{-}1,3\text{-}(HO)_2C_6H_2$	138.17	nd (bz), pr (w + 1)	sub	136-7, 115-7 (+ w)			w, al, eth, aa	$B6^3$, 4581
12998	Resorcinol, 4,6-dimethyl $4,6\text{-}(CH_3)_2\text{-}1,3\text{-}(HO)_2C_6H_2$	138.17	mcl pr (w + 1), nd (sub)	276-9 sub	125			w, al, eth	$B6^3$, 4595
12999	Resorcinol dimethyl ether or 1,3-Dimethoxy benzene $1,3\text{-}(CH_3O)_2C_6H_4$	138.17		217-8	−52	1.0552$^{25/25}$	1.5231^{20}	al, eth, bz	$B6^3$, 4305
13000	Resorcinol, 2,4-dinitro $2,4\text{-}(NO_2)_2\text{-}1,3\text{-}(HO)_2C_6H_2$	200.11	ye lf (al)		147 8				$B6^3$, 4351
13001	Resorcinol, 2,4-dinitroso $2,4\text{-}(ON)_2\text{-}1,3\text{-}(HO)_2C_6H_2$	168.11	ye rh pl (50% al), lf (aq MeOH)		168			w, al	$B7^3$, 4732
13002	Resorcinol dipropyl ether or 1,3-Dipropoxy benzene $1,3\text{-}(C_3H_7O)_2C_6H_4$	194.27		251, 127-8^{12}		1.035$^{20/21}$	1.5138^{83}		$B6^3$, 4308
13003	Resorcinol, dithio or 1,3-Dimercapto benzene $1,3\text{-}(HS)_2C_6H_4$	142.23	lf (dil al)	245, 123^{17}	27			al, eth, ace, bz	$B6^3$, 4366
13004	Resorcinol, 4-ethyl $4\text{-}C_2H_5\text{-}1,3\text{-}(HO)_2C_6H_3$	138.17	pr (chl, bz)	131^{15}	98-9				$B6^3$, 4554
13005	Resorcinol ethyl ether(mono) or 3-Ethoxy phenol $3\text{-}C_2H_5OC_6H_4OH$	138.17	ye	246-7, 117$^{5.5}$		1.0705$^{4/4}$		al, eth, bz	$B6^3$, 4307
13006	Resorcinol, 4-hexyl $4\text{-}C_6H_{13}\text{-}1,3\text{-}(HO)_2C_6H_3$	194.27	pa ye nd (lig)	333, 178^{8}	68-9			al, eth, ace, bz, chl	$B6^3$, 4712
13007	Resorcinol, 4-iodo $4\text{-}I\text{-}1,3\text{-}(HO)_2C_6H_3$	236.01	pr, nd (+ 1w, w)		67			al, eth, ace, chl	$B6^3$, 4341
13008	Resorcinol, 5-iodo $5\text{-}I\text{-}1,3\text{-}(HO)_2C_6H_3$	236.01	nd (bz, sub)	sub	92.3			al	$B6^2$, 821
13009	Resorcinol, 2-methoxy $2\text{-}CH_3O\text{-}1,3\text{-}(HO)_2C_6H_3$	140.14	cr (bz)	154-5^{24}	85-7				$B6^3$, 6264
13010	Resorcinol, 5-methoxy $5\text{-}CH_3O\text{-}1,3\text{-}(HO)_2C_6H_3$	140.14	pl (bz)	213^{16}	78-81			al, eth	$B6^3$, 6304
13011	Resorcinol, 2-methyl $2\text{-}CH_3\text{-}1,3\text{-}(HO)_2C_6H_3$	124.14	pr (bz)	264, 168^{16}	119-21			w, al, eth, bz	$B6^3$, 4512
13012	Resorcinol methyl ether(mono) or 3-Methoxy phenol $3\text{-}CH_3OC_6H_4OH$	124.14		244.3, 144^{25}	<−17		1.5520^{20}	al, eth	$B6^3$, 4303
13013	Resorcinol, 4-(4-methyl pentyl) $4[(CH_3)_2CH(CH_2)_3]\text{-}1,3\text{-}(HO)_2C_6H_3$	194. 27	lf (peth)	190^{12}	71		1.5292^{15}	al, eth	$B6^3$, 4717
13014	Resorcinol, 2-nitro $2\text{-}NO_2\text{-}1,3\text{-}(HO)_2C_6H_3$	155.11	og-red pr (al)		87-8			al	$B6^3$, 4343
13015	Resorcinol, 4-nitro $4\text{-}NO_2\text{-}1,3\text{-}(HO)_2C_6H_3$	155.11			122			al, eth, bz	$B6^3$, 4344
13016	Resorcinol, 4-nitroso or 2-Hydroxy-4-benzoquinone oxime $4\text{-}ON\text{-}1,3\text{-}(HO)_2C_6H_3$	139.11	ye nd (+ w), br nd (chl)		150d			w, al, eth, ace, chl	$B8^3$, 1965
13017	Resorcinol, 4-pentyl $4\text{-}C_5H_{11}\text{-}1,3\text{-}(HO)_2C_6H_3$	180.25		168-70$^{6-7}$	72-3			al, eth, bz	$B6^3$, 4693
13018	Resorcinol, 4-isopentyl $4\text{-}i\text{-}C_5H_{11}\text{-}1,3\text{-}(HO)_2C_6H_3$	180.25		177-8^{8}	68-70			al, eth, bz	$B6^3$, 4700
13019	Resorcinol, 5-pentyl or Olivetol $5\text{-}C_5H_{11}\text{-}1,3\text{-}(HO)_2C_6H_3$	180.25	nd (+ w), pr (bz-lig)	164^{5}	49			w, al, eth, ace, bz	$B6^3$, 4695
13020	Resorcinol, 4-propionyl $4\text{-}(CH_3CH_2CO)C_6H_3(OH)_2\text{-}1,3$	166.18	ye nd (al)	176-8^{6}	97 (anh)			al, eth, bz, aa	$B8^3$, 2144
13021	Resorcinol monopropyl ether or 3-Propoxy phenol $3\text{-}C_3H_7OC_6H_4OH$	152.19	cr (w, al)	256-7, 120^{5}	55				$B6^3$, 4308

No.	Name, Synonyms, and Formula	Mol. wt.	Color, crystalline form, specific rotation and λ_{max} (log ε)	b.p. °C	m.p. °C	Density	n_D	Solubility	Ref.
13022	Resorcinol, 4-propyl $4\text{-}C_3H_7\text{-}1,3\text{-}(HO)_2C_6H_3$	152.19	pr (bz), nd (peth)	172^{14}	82-3			w, al, eth	B6^{3}, 4611
13023	Resorcinol, 4-iso-propyl $4\text{-}(CH_3)_2CH\text{-}1,3\text{-}(HO)_2C_6H_3$	152.19	cr (aq aa)	265-81, $114^{0.2}$	105			al, eth	B6^{2}, 896
13024	Resorcinol, 5-propyl or Divarinol $5\text{-}C_3H_7\text{-}1,3\text{-}(HO)_2C_6H_3$	152.19	lf (+1w), lf (bz)	$148\text{-}9^{3}$	83-4, 51 (+w)			al, eth, ace, bz, aa	B6^{3}, 4622
13025	Resorcinol, tetrachloro $1,3\text{-}(HO)_2C_6Cl_4$	247.89	nd (w)		141			al, eth, bz, aa	B6^{2}, 819
13026	Resorcinol, 2,4,6-tribromo $2,4,6\text{-}Br_3C_6H(OH)_2\text{-}1,3$	346.80	nd (w)		112			al, eth	B6^{3}, 4340
13027	Resorcinol, 2,4,6-trichloro $2,4,6\text{-}Cl_3C_6H(OH)_2\text{-}1,3$	213.45	nd (w)		83			al, eth	B6^{3}, 4336
13028	Resorcinol, 2,4,5-trimethyl $2,4,5\text{-}(CH_3)_3C_6H(OH)_2\text{-}1,3$	152.19	cr		156				B6^{3}, 4643
13029	Resorcinol, 2,4,6-trimethyl or Mesorcinol $2,4,6\text{-}(CH_3)_3\text{-}C_6H(OH)_2\text{-}1,3$	152.19	pl (al), lf	275	149-50			al, eth, bz	B6^{3}, 4653
13030	Resorcinol, 4,5,6-trimethyl $4,5,6\text{-}(CH_3)_3C_6H(OH)_2\text{-}1,3$	152.19	nd or lf		163-4			al, eth, ace, bz, chl	B6^{3}, 4643
13031	Resorcinol, 2,4,6-trinitro or Styphnic acid $2,4,6\text{-}(NO_2)_3C_6H(OH)_2\text{-}1,3$	245.11	hex (al), ye cr (aa)	sub	179-80			al, eth	B6^{3}, 4354
13032	Resorufin or 7-Hydroxy-2-phenoxazone $C_{12}H_7NO_3$	213.19	br nd ($PhNH_2$), pr (HCl)						B27^{2}, 108
13033	Retene or 7-isopropyl-1-methyl phenanthrene $C_{18}H_{18}$	234.34	pl (al)	390, 200^{10}	100-1	1.035		bz, lig	B5^{3}, 2199
13034	Retromecine or Senecifolinene $C_8H_{13}NO_2$	155.20	pr (ace) $[\alpha]^{t}{}_{D}$ +27.4 (w), $^{26}{}_{D}$ +50.2 (al)	$80^{0.01}$	121-2 (130)			w, al, ace	B21^{4}, 2048
13035	Rhamnetin $C_{16}H_{12}O_7$	316.27	ye nd (al)		294-6			ace	B18^{4}, 3474
13036	D-Rhamnitol or Rhamnite $C_6H_{14}O_5$	166.17	pr (ace), $[\alpha]^{20}{}_{D}$ −12.4 (w,c = 0.5)		123			w, al, Py	B1^{4}, 2837
13037	D-Rhamnose, hydrate (α-anomer) $C_6H_{12}O_5.H_2O$	182.17	(wh), $[\alpha]^{18.5}{}_{D}$ −8.25 (w)		90-1			w	B1^{4}, 4260
13038	DL-Rhamnose $C_6H_{12}O_5$	164.16	(w)		151-3 (anh)			w, al	B1^{4}, 4261
13039	L-Rhamnose, hydrate $C_6H_{12}O_5.H_2O$	182.17	mcl pl (al w+1) $[\alpha]^{20}{}_{D}$ −77→ +8.9 (mut)	105^{2} sub	92	1.4708^{20}		w, al	B1^{4}, 4261
13040	L-Rhamnose (β-anomer) $C_6H_{12}O_5$	164.16	nd (ace), $[\alpha]^{20}{}_{D}$ +38.4 +8.9 (w,mut)		122-6			w, al	B1^{4}, 4262
13041	Rheadin or Rhoedine $C_{21}H_{21}NO_6$	383.40	nd (chl, eth, al), $[\alpha]^{17.5}{}_{D}$ +232 (al)	sub	256-8				C38, 6060
13042	Rhizopterin or 12-Formylpteroic acid $C_{15}H_{12}N_6O_4$	340.30	lt ye pl (w)		>300 darkens 285				Am69, 2751
13043	Rhodamine β or Tetraethyl rhodamine $C_{28}H_{30}N_2O_3$	442.56	gr lf (w+4), col pr (al, xyl)		165 (anh)			w, al, eth, bz	B19^{4}, 4301
13044	Rhodamine, β-hydrochloride $C_{28}H_{30}N_2O_3.HCl$	479.02	gr or red vt pw (w), lf (dil HCl)					w, al, bz	B19^{4}, 4301
13045	Rhodamine or 4-Thioxo-4-thiazolidone $C_3H_3NOS_2$	133.18	lt ye pr (al, w, aa)		170	0.868		al, eth	B27^{2}, 288

No.	Name, Synonyms, and Formula	Mol. wt.	Color, crystalline form, specific rotation and λ_{max} (log ε)	b.p. °C	m.p. °C	Density	n_D	Solubility	Ref.
13046	Rhodanine, 3-amino $C_3H_4N_2OS_2$	148.20			100				B27[2], 289
13047	Rhodanine, 3-amino-5-benzylidene $C_{10}H_8N_2OS_2$	236.31			194-7				C61, 7190
13048	Rhodanine, 5-(dimethylamino) amino-benzylidene $C_{12}H_{12}N_2OS_2$	264.36	red or og nd (al, Py-w)		296			eth, ace	B27[2], 484
13049	Rhodanine, 3-ethyl $C_5H_7NOS_2$	161.24			37-9				B27[1], 309
13050	Rhodanind, 5-ethyl $C_5H_7NOS_2$	161.24	ye amor (dil al)		105			al, eth, ace, aa	B27[1], 313
13051	Rhodanine, 3-methyl $C_4H_5NOS_2$	147.21			69-71				B27[2], 288
13052	Rhodanine, 3-phenyl $C_9H_7NOS_2$	209.28	ye pl (aa), nd or pr (al)		194-5				B27[2], 288
13053	Rhodanine, 5-isopropylidene $C_6H_7NOS_2$	173.25			197-8				C44, 2979
13054	Rhodizonic acid or 5,6-Dihydroxy-5-cyclohexene-1,2,3,4-tetrone $C_6H_2O_6$	170.08	dk og nd (sub)		155-60			al	B8[3], 4214
13055	Riboflavin or Lactoflavin Vitamin B2 $C_{17}H_{20}N_4O_6$	376.37	ye or og-ye nd (w, dil aa), $[\alpha]^{25}/_D$ −112→ −122 (0.02N NαOH, c = 0.5), (mut)		280d				C28, 2036
13056	D-Ribose $C_5H_{10}O_5$	150.13	pl (abs al), $[\alpha]_D$ −21.5 (w)		95			w	B1[4], 4211
13057	Ricinidine or 1-Methyl, 2-pyridone-3-carbonitrile $C_7H_6N_2O$	134.14	nd (sub), (chl, al)	243^{28}	140			al	B22[2], 222
13058	Ricinine or Ricidine $C_8H_8N_2O_2$	164.16	pr or lf (w al)	sub $170\text{-}80^{20}$	201.5			Py	B22[4], 3354
13059	Ricinoleic acid or 12-Hydroxy-9-octadecenoic acid (*cis*) $C_6H_{13}CH(OH)CH_2CH=CH(CH_2)_7CO_2H$	298.47	$[\alpha]^{12}/_D$ + 5.05	226.8^{10}	α:7.7 β:16 γ:5.5	$0.9450^{21/4}$	1.4716^{21}	al, eth	B3[4], 1026
13060	Ricinoleic acid, butyl ester (*cis*) or Butyl ricinoleate $C_6H_{13}CH(OH)CH_2CH=CH(CH_2)_7CO_2C_4H_9$	354.57	$[\alpha]^{25}/_D$ + 3.7	278^{12}		0.9058^{22}	1.4566^{22}	eth	B3[3], 711
13061	Ricinoleic acid, isobutyl ester or Isobutyl ricinoleate $C_6H_{13}CHOHCH_2CH=CH(CH_2)_7CO_2\text{-i-}C_4H_9$	354.57	[α] + 4.01	282^{9}		0.9078^{22}	1.4538^{22}	eth	B3[1], 138
13062	Ricinoleic acid, ethyl ester (*cis*) or Ethyl ricinoleate $C_6H_{13}CH(OH)CH_2CH=CH(CH_2)_7CO_2C_2H_5$	326.52	$[\alpha]^{22}/_D$ + 5.3	258^{13}		0.9045^{22}	1.4618^{22}		B3[4], 1029
13063	Rosaniline $C_{20}H_{21}N_3O$	319.41	nd (w)		186d			al	B13[3], 2078
13064	Rosinduline, anhydro base $C_{22}H_{15}N_3$	321.38	red-br lf (eth-bz)		198-9			al, eth, bz	B25[2], 322
13065	Rotenone or Tubotoxine $C_{23}H_{22}O_6$	394.42	nd or lf (al, aq ace), $[\alpha]^{29.5}/_D$ −225.2 (bz)	$210\text{-}20^{0.5}$	(i)163 (ii) 176			al, ace, bz, chl, aa	B19[4], 5227
13066	Rubicene $C_{26}H_{14}$	326.40	red nd (xyl)		306			CS_2	B5[3], 2673
13067	Rubignol $C_5H_4O_4$	128.08			203.5			w, al, eth, ace	B17[4], 6679

No.	Name, Synonyms, and Formula	Mol. wt.	Color, crystalline form, specific rotation and λ_{max} (log ε)	b.p. °C	m.p. °C	Density	n_D	Solubility	Ref.
13068	Rubijervine $C_{27}H_{43}NO_2$	413.64	nd (+1w dil al), $[\alpha]^{25}/_D$ +19 (al, c=1)	....	242	....	....	al, bz, chl	B21[4], 2310
13069	Rubixanthin or 3-Hydroxy-γ-carotene $C_{40}H_{56}O$	552.88	dk red nd (bz-MeOH), og-red (bz-peth)	....	160	....	....	bz, chl	B6[3], 3772
13070	Rubrene or 9,10,11,12-Tetraphenyl naphthacene $C_{42}H_{28}$	532.68	og red (bz-lig)	....	331	....	....	bz	B5[3], 2803
13071	Rufol or 1,5-Dihydroxy anthracene 1,5-$(HO)_2C_{14}H_8$	210.23	ye nd	....	265d	....	....	al, eth, bz	B6, 1032
13072	Rutaecarpine $C_{18}H_{13}N_3O$	287.32	yesh nd (al, AcOEt)	....	259-60	....	....	....	B26[2], 104
13073	Rutinose or 6-D-β-L-rhamnopyransoyl-D-glucose $C_{12}H_{22}O_{10}$	326.30	hyg pw (al, eth), $[\alpha]^{20}/_D$ −10 (al)	....	189-92d	....	....	w, al	B17[4], 2536
13074	Sabadine $C_{29}H_4$li9NO_8	541.70	nd (eth), $[\alpha]_D$ −11 (al)	....	256-60	....	....	al, ace	B21[4], 2883
13075	Sabinane or d-Thujane $C_{10}H_{18}$	138.25	$[\alpha]_D$ +73.1	157^{758}	....	$0.8139^{20/4}$	1.4376^{20}	al, eth, ace, bz	B5[4], 317

No.	Name, Synonyms, and Formula	Mol. wt.	Color, crystalline form, specific rotation and λ_{max} (log ε)	b.p. °C	m.p. °C	Density	n_D	Solubility	Ref.
13076	Sabinene (d) or 4(10)-Thujene $C_{10}H_{16}$	136.24	$[\alpha]^{20/}{}_D$ + 101.4	$163\text{-}5^{758}$, 49^{13}		$0.8437^{20/4}$	1.4676^{20}	al, eth, bz, chl	B5^3, 365
13077	Sabinene (l) $C_{10}H_{16}$	136.24	$[\alpha]^{15/}{}_D$ −42.5	162-6		0.8468^{20}	1.4674^{17}	al, eth, bz	B5^3, 365
13078	Sabinol (d) or 4(10)Thujene-3-01 $C_{10}H_{16}O$	152.24	$[\alpha]^{18/}{}_D$ + 3.94	208, 90^{11}		$0.9488^{19/4}$	1.4871^{25}	eth	B6^4, 382
13079	Sabinol acetate(d) $C_{12}H_{18}O_2$	194.27	$[\alpha]_D$ + 79	$81\text{-}2^3$		0.972^{15}			B6^3, 384
13080	D-Saccaharic acid $C_6H_{10}O_8$	210.14	nd (95%al), $[\alpha]_D$ + 6.9→ + 20.6 (w, c=2.5, mut)		125-6			w, al	B3^4, 1291
13081	Saccharin or 2-Sulfobenzoic acid imide $C_7H_5NO_3S$	183.18	mcl (ace), pr (al), lf (w)	sub vac	229d	0.828		al, ace	B27^2, 214
13082	Safrole or 1-Allyl-3,4-(methylene dioxy) benzene $C_{10}H_{10}O_2$	162.19	mcl	234.5, $104\text{-}5^6$	11.2	$1.000^{20/4}$	1.5381^{20}	al, eth, chl	B19^4, 275
13083	Salicylaldehyde or 2-Hydroxybenzaldehyde $2\text{-}HOC_6H_4CHO$	122.12		197, 93^{25}	−7	$1.1674^{20/4}$	1.5740^{20}	al, eth, ace, bz	B8^3, 135
13084	Salicylaldehyde azine or Salazine $C_{14}H_{12}N_2O_2$	240.26	ye nd or lf (al)		214			al, bz, chl	B8^2, 43
13085	Salicylaldehyde oxime or Salicylaldoxime $2\text{-}HOC_6H_4CH{=}NOH$	137.14	pr (bz-peth)		63			al, eth, bz	B8^3, 168
13086	Salicylamide or 2-Hydroxy benzamide $2\text{-}HOC_6H_4CONH_2$	137.14	ye lf (dil al)	181.5^{14}	142	$1.175^{140/4}$		al	B10^3, 152
13087	Salicylamide, N-phenyl or N-Phenyl salicylamide $2\text{-}HOC_6H_4CONHC_6H_5$	213.24	pr (w, al)		136-7			al	B12^3, 944
13088	Salicylamide, N-(2-tolyl) or N-(2-Tolyl) salicylamide $2\text{-}HOC_6H_4CONH(C_6H_4CH_3\text{-}2)$	227.26	nd (al)		144			al	B12^3, 1888
13089	Salicylic acid or 2-Hydroxybenzoic acid $2\text{-}HOC_6H_4CO_2H$	138.12	nd (w), mcl pr (al)	211^{20} sub	159	$1.443^{20/4}$	1.565	al, eth, ace	B10^3, 87
13090	Salicylic acid, allyl ester or Allyl salicylate $2\text{-}HOC_6H_4CO_2CH_2CH{=}CH_2$	178.19		247-50, $105\text{-}6^5$		1.1000^{15}			B10^3, 125
13091	Salicylic acid, 4-acetamido, phenyl ester or Salophene $4\text{-}CH_3CONH\text{-}2\text{-}HOC_6H_3CO_2C_6H_5$	271.27	pl (w), lf (al)		187-8			al, eth, bz	B13^3, 1066
13092	Salicylic acid acetate or Aspirin $2\text{-}CH_3CO_2C_6H_4CO_2H$	180.16	nd (w), mcl ta (w)		135			al, eth, chl	B10^3, 102
13093	Salicylic acid, 3-amino or 3-Amino-2-hydroxybenzoic acid $3\text{-}NH_2\text{-}2\text{-}HOC_6H_3CO_2H$	153.14			235d				B14^3, 1434
13094	Salicylic acid, 4-amino or 4-Amino-2-hydroxybenzoic acid $4\text{-}H_2N\text{-}2\text{-}HOC_6H_3CO_2H$	153.14			150-1d			w, al, eth, ace	B14^3, 1436
13095	Salicylic acid, 5-amino or 5-Amino-2-hydroxybenzoic acid $5\text{-}H_2N\text{-}2\text{-}HOC_6H_3CO_2H$	153.14			283				B14^3, 1456
13096	Salicylic acid anhydride $(2\text{-}HOC_6H_4CO)_2O$	258.23	amor		255d			al, eth	B19^4, 2059
13097	Salicylic acid, benzyl ester or Benzyl salicylate $2\text{-}HOC_6H_4CO_2CH_2C_6H_5$	228.25		320, 170^5		$1.1799^{20/4}$	1.5805^{20}	al, eth	B10^3, 132
13098	Salicylic acid, benzyl ester, acetate $2\text{-}(CH_3CO_2)C_6H_4CO_2CH_2C_6H_5$	270.28		$197\text{-}200^7$	26			al, eth, ace, bz	B10^2, 52
13099	Salicylic acid, 3-bromo or Hydroxy-3-bromobenzoic acid $3\text{-}Br\text{-}2\text{-}HOC_6H_3CO_2H$	217.02	nd (dil al)		184-5			al, eth, ace	B10^3, 174
13100	Salicylic acid, 4-bromo $4\text{-}Br\text{-}2\text{-}HO\text{-}C_6H_3CO_2H$	217.02	pl, nd (w)		214			al	B10^2, 63
13101	Salicylic acid, 5-bromo $5\text{-}Br\text{-}2\text{-}HOC_6H_3CO_2H$	217.02	nd (w, dil al)	sub>100	168-9			w, al, eth	B10^3, 176
13102	Salicylic acid, 5-bromo,acetate $5\text{-}Br\text{-}2(CH_3CO_2)C_6H_3CO_2H$	259.06	nd (al)		60			al, eth	B10^2, 64
13103	Salicylic acid, butyl ester or Butyl salicylate $2\text{-}HOC_6H_4CO_2C_4H_9$	194.23		270-2, $136\text{-}8^{10}$	−5.9	$1.0728^{20/4}$	1.5115^{20}		B10^3, 121
13104	Salicylic acid, *iso*butyl ester or *Iso*Butyl salicylate $2\text{-}HOC_6H_4CO_2\text{-}i\text{-}C_4H_9$	194.23		260-2, $136\text{-}8^{10}$	5.9	$1.0639^{20/4}$	1.5087^{20}	al, eth	B10^3, 121
13105	Salicylic acid, 3-chloro $3\text{-}Cl\text{-}2\text{-}HOC_6H_3CO_2H$	172.57	nd(w)		180			al, chl, aa	B10^3, 163
13106	Salicylic acid, 3-chloro,ethyl ester or Ethyl 3-chloro salicylate $3\text{-}Cl\text{-}2\text{-}HOC_6H_3CO_2C_2H_5$	200.62	nd	269-70, 147^{12}	21			al	B10, 101

No.	Name, Synonyms, and Formula	Mol. wt.	Color, crystalline form, specific rotation and λ_{max} (log ε)	b.p. °C	m.p. °C	Density	n_D	Solubility	Ref.
13107	Salicylic acid, 3-chloro,methyl ester or Methyl-3-chloro salicylate 3-Cl-2-$HOC_6H_3CO_2CH_3$	186.59	nd (MeOH or al)	260d	38				B10^2, 61
13108	Salicylic acid, 4-chloro or 4-Chloro-2-hydroxy benzoic acid 4-Cl-2-$HOC_6H_3CO_2H$	172.57	nd (w)	sub (d)	207			al, bz, chl	B10^3, 164
13109	Salicylic acid, 5-chloro 5-Cl-2-$HOC_6H_3CO_2H$	172.57	nd (w, or al)		173-4			al, eth, bz, chl	B10^3, 165
13110	Salicylic acid, 5-chloro,ethyl ester 5-Cl-2-$HOC_6H_3CO_2C_2H_5$	200.62	nd (al)		25			al	B10, 103
13111	Salicylic acid, 5-chloro, methyl ester 5-Cl-2-$HOC_6H_3CO_2CH_3$	186.59	nd (al)	249d	50			al	B10^3, 166
13112	Salicylic acid, 6-chloro 6-Cl-2-$HOC_6H_3CO_2H$	172.57	nd (w)		166			al, eth, ace, bz	B10, 104
13113	Salicylic acid, 3-diazo or 3-Diazo-2-hydroxy benzoic acid 3-N_2-2-$HOC_6H_3CO_2H$	164.12	ye nd (ace)		155d			ace	B16, 553
13114	Salicylic acid, 3,5-dichloro 3,5-Cl_2-2-$HOC_6H_2CO_2H$	207.01	nd (dil al), rh pr	sub d	220-1			al, eth	B10^3, 169
13115	Salicylic acid, 3,5-diido 3,5-I_2-2-$HOC_6H_2CO_2H$	389.92	nd (al, aa)		235-6			al, eth	B10^3, 189
13116	Salicylic acid, 3,5-diido, ethyl ester 3,5-I_2-2-$HOC_6H_2CO_2C_2H_5$	417.97	lf (al)		133			al, eth, bz	B10, 114
13117	Salicylic acid, 3,4-dimethoxy 3,4-$(CH_3O)_2$-2-$HOC_6H_2CO_2H$	198.18	nd (w)		169-72			chl	B10^3, 2058
13118	Slaicylic acid, 4,5-dimethoxy 4,5$(CH_3O)_2$-2-$HOC_6H_2CO_2H$	198.18	nd (w)		202d			al	B10^3, 2065
13119	Salicylic acid, 4,5-dimethoxy,methyl ester 4,5-$(CH_3O)_2$-2-$HOC_6H_2CO_2CH_3$	212.20	nd (w)		95				B10^3, 2067
13120	Salicylic acid, 4,6-dimethoxy 4,6-$(CH_3O)_2$-2-$HOC_6H_2CO_2H$	198.18	nd (eth-bz)		152-4d			al, eth, ace	B10^3, 2069
13121	Salicylic acid, 2,4-dinitrobenzyl ester 2-$HOC_6H_4CO_2[CH_2C_6H_3(NO_2)_2$-2,4]	318.24	ye pl (aa)		168				B10^2, 52
13122	Salicylic acid, 3,5-dinitro 3,5$(NO_2)_2$-2-$HOC_6H_2CO_2H$	228.12	ye nd or pl (+1w)		182			al, eth, bz	B10^3, 207
13123	Salicylic acid, dithio or Dithiosalicylic acid 2-$HOC_6H_4CS_2H$	170.24	og-ye nd (peth)		48-50			w, al, eth, bz	B10^2, 78
13124	Salicylic acid, 4-ethoxy 4-C_2H_5O-2-$HOC_6H_3CO_2H$	182.18	nd (w, bz)		154			al, eth, bz	B10, 379
13125	Salicylic acid, ethyl ester or Ethyl salicylate 2-$HOC_6H_4CO_2C_2H_5$	166.18		234, 132.8^{27}	2-3	1.1326$^{20/4}$	1.5296^{20}	**al**, eth	B10^3, 115
13126	Salicylic acid, ethyl ester, acetate or Ethyl aspirin 2$(CH_3CO_2)C_6H_4CO_2C_2H_5$	208.21		272, 148-50^{15}		1.1566^{15}		al, eth, ace, bz	B10^2, 48
13127	Salicylic acid, 3-formyl 3-HCO-2-$HOC_6H_3CO_2H$	166.13	nd (w + l)		179			al	B10^3, 4207
13128	Salicylic acid, 5-formyl 5-HCO-2-$HOC_6H_3CO_2H$	166.13	nd (w)		250d			eth	B10^3, 4209
13129	Salicylic acid hydrazide 2-$HOC_6H_4CONHNH_2$	152.15	pl (al), pr (w)		147			al, bz	B10^3, 161
13130	Salicylic acid, β-hydroxyethyl ester 2-$HOC_6H_4CO_2CH_2CH_2OH$	182.18		173^{15}	37	1.2537$^{15/15}$		al, eth, bz, chl	B10^3, 138
13131	Salicylic acid, 3-iodo 3-I-2-$HOC_6H_3CO_2H$	264.02	nd (w)		199			al, eth	B10^3, 186
13132	Salicylic acid, 4-iodo 4-I-2-$HOC_6H_3CO_2H$	264.02	pr or nd (al)		230d			al, eth	B10^2, 65
13133	Salicylic acid, 5-iodo 5-I-2-$HOC_6H_3CO_2H$	264.02	nd (w)		197			al, eth	B10^3, 188
13134	Salicylic acid, menthyl ester or Menthyl salicylate $C_{17}H_{24}O_3$	276.38		190^{15}, 143$^{0.5}$		1.0467^{20}	1.5198^{26}	**al**, eth, ace, **bz**	B10^3, 126
13135	Salicylic acid, 4-methoxy-6-methyl or Everninic acid 4-CH_3O-6-CH_3-2-$HOC_6H_2CO_2H$	182.18	nd (peth, w)		171-2			al, ace, aa	B10^2, 273
13136	Salicylic acid, 4-methoxy-6-methyl,methyl ester or Sparassol 4-CH_3O-6-CH_3-2-$HOC_6H_2CO_2CH_3$	196.20	pr (w), lf (MeOH)		67-8			al, eth, ace, chl, peth	B10^2, 273
13137	Salicylic acid, (methoxymethyl) ester or Mesotan 2-$HOC_6H_4CO_2CH_2OCH_3$	182.18		162^{742}		1.2^{15}		**al, eth, ace, bz, chl**	B10^3, 143
13138	Salicylic acid, (2-methoxyphenyl) ester 2-$HOC_6H_4CO_2(C_6H_4OCH_3$-2)	244.25			70			al, eth, chl	B10, 81

PHYSICAL CONSTANTS OF ORGANIC COMPOUNDS (Continued)

No.	Name, Synonyms, and Formula	Mol. wt.	Color, crystalline form, specific rotation and λ_{max} (log ε)	b.p. °C	m.p. °C	Density	n_D	Solubility	Ref.
13139	Salicylic acid, methyl ester $2\text{-}HOC_6H_4CO_2CH_3$	152.15	....	223.3	−8	$1.1738^{20/4}$	1.5360^{20}	al, eth	B10³, 107
13140	Salicylic acid, methyl ester, benzoate $2\text{-}(C_6H_5CO_2)C_6H_4CO_2CH_3$	256.26	pr (al, eth)	385d, $270\text{-}80^{120}$	92	....	....	al, eth, bz, chl	B10³, 112
13141	Salicylic acid, 3-methyl or 2,3-Cresotic acid $3\text{-}CH_3\text{-}2\text{-}HOC_6H_3CO_2H$	152.15	nd (w, dil al)	....	169-70	....	....	al, eth, chl	B10³, 505
13142	Salicylic acid, 4-methyl or 2,4-Cresotic acid $4\text{-}CH_3\text{-}2\text{-}HOC_6H_3CO_2H$	152.15	nd (w), pr (al), pl (chl)	sub	177.8	....	....	al, eth, chl	B10³, 521
13143	Salicylic acid, 5-methyl or 2,5-Cresotic acid $5\text{-}CH_3\text{-}2\text{-}HOC_6H_3CO_2H$	152.15	nd (w, peth)	....	153	....	....	al, eth, bz, chl	B10³, 516
13144	Salicylic acid, 6-methyl or 2,6-Cresotic acid $6\text{-}CH_3\text{-}2\text{-}HOC_6H_3CO_2H$	152.15	nd (chl)	....	173 (184)	....	....	al, eth, chl	B10³, 496
13145	Salicylic acid, α-naphthyl ester or Alphol $2\text{-}HOC_6H_4CO_2\text{-}\alpha\text{-}C_{10}H_7$	264.28	....	....	83	....	....	eth	B10², 52
13146	Salicylic acid, β-naphthyl ester or Betol $2HOC_6H_4CO_2\text{-}\beta\text{-}C_{10}H_7$	264.28	cr (al)	....	95.5	1.11^{116}	....	eth, bz	B10³, 136
13147	Salicylic acid, 3-nitro $3\text{-}NO_2\text{-}2\text{-}HOC_6H_3CO_2H$	183.12	yesh nd (aa, w + 1)	....	148-9 (anh), 128-9 (+ w)	....	....	al, eth, ace, bz, chl	B10³, 190
13148	Salicylic acid, 4-nitro $4\text{-}NO_2\text{-}2\text{-}HOC_6H_3CO_2H$	183.12	ye nd (w, dil al)	....	235	....	....	al, ace, chl, aa	B10³, 194
13149	Salicylic acid, 5-nitro $5\text{-}NO_2\text{-}2\text{-}HOC_6H_3CO_2H$	183.12	nd (w)	....	229-30	1.650^{20}	....	al, eth, ace, bz	B10³, 197
13150	Salicylic acid, 6-nitro $6\text{-}NO_2\text{-}2\text{-}HOC_6H_3CO_2H$	183.12	pa ye nd	....	....	....	....	w, al	B10³, 205
13151	Salicylic acid, 4-nitrobenzyl ester $2\text{-}HOC_6H_4CO_2(CH_2C_6H_4NO_2\text{-}4)$	273.25	cr (dil al)	....	97-8	....	....	al, ace	B10², 52
13152	Salicylic acid, pentyl ester $2\text{-}HOC_6H_4CO_2C_5H_{11}$	208.26	....	148-54	....	$1.065^{15/15}$	1.506^{20}	**al, eth**	B10³, 122
13153	Salicylic acid, isopentyl ester or Isopentyl salicylate $2\text{-}HOC_6H_4CO_2\text{-}i\text{-}C_5H_{11}$	208.26	....	$276\text{-}7^{741}$, $151\text{-}2^{15}$	....	$1.0535^{20/4}$	1.5080^{20}	al, eth, chl	B10³, 123
13154	Salicylic acid, phenyl ester or Phenyl salicylate Salol $2\text{-}HOC_6H_4CO_2C_6H_5$	214.22	pl (MeOH)	173.12	43	$1.2614^{30/4}$	....	al, eth, ace, bz, aa	B10³, 127
13155	Salicylic acid, propyl ester $2\text{-}HOC_6H_4CO_2C_3H_7$	180.20	....	239	96-8	$1.0979^{20/4}$	1.5161^{20}	**al, eth**	B10³, 119
13156	Salicylic acid, *iso*-propyl ester $2\text{-}HOC_6H_4CO_2CH(CH_3)_2$	180.20	....	240-2, 118^{17}	....	$1.0729^{20/4}$	1.5065^{20}	**al, eth**	B10³, 120
13157	Salicylic acid, 3-*iso*-propyl-6-methyl or *o*-Thymotinic acid $3\text{-}i\text{-}C_3H_7\text{-}6\text{-}CH_3\text{-}2\text{-}HOC_6H_2CO_2H$	194.23	nd (w, bz, lig)	sub	127	....	....	al, eth, bz, aa	B10³, 629
13158	Salicylic acid, 5-*iso*-propyl-3-methyl $5\text{-}i\text{-}C_3H_7\text{-}3\text{-}CH_3\text{-}2\text{-}HOC_6H_2CO_2H$	194.23	nd (w)	....	147	....	....	al	B10, 282
13159	Salicylic acid, 5-*iso*-propyl-4-methyl $5\text{-}i\text{-}C_3H_7\text{-}4\text{-}CH_3\text{-}2\text{-}HOC_6H_2CO_2H$	194.23	nd (bz)	....	189-90	....	....	....	B10², 171
13160	Salicylic acid, 6-*iso*-propyl-3-methyl or *o*-Carvacrotinic acid $6\text{-}i\text{-}C_3H_7\text{-}3\text{-}CH_3\text{-}2\text{-}HOC_6H_2CO_2H$	194.23	nd (w)	sub	136	....	....	al, eth	B10, 282
13161	Salicylic acid, 3-sulfo $3\text{-}SO_3H\text{-}2\text{-}HOC_6H_3CO_2H$	218.18	pr (w + 2)	....	120	....	....	w, al, eth, aa	B11³, 701
13162	Salicylic acid, 5-sulfo $5\text{-}SO_3H\text{-}2\text{-}HOC_6H_3CO_2H$	218.28	hyg nd (w + 2)	....	198d 120 (+ 2w)	....	....	**w, al,** eth	B11³, 704
13163	Salicylonitrile $2\text{-}HOC_6H_4CN$	119.12	pr (bz)	149^{14}	98	$1.052^{100/4}$	1.5372^{100}	al, eth, bz, chl	B10³, 159
13164	Salicyl chloride $2\text{-}HOC_6H_4COCl$	156.57	....	92^{15}	19	1.3112^{20}	1.5812^{20}	eth	B10³, 150
13165	Salicyl chloride, 3-methyl or 3-Methyl salicyl chloride $3\text{-}CH_3\text{-}2\text{-}HOC_6H_3COCl$	170.60	....	$87\text{-}9^{16}$	27-8	....	....	....	B10², 132
13166	Salsoline $C_{11}H_{15}NO_2$	193.25	pw or cr (al), $[\alpha]^{20}{}_D$ + 34.5 (0.1 NHCl, c=1)	....	221-2	....	....	al, chl	B21⁴, 2121

No.	Name, Synonyms, and Formula	Mol. wt.	Color, crystalline form, specific rotation and λ_{max} (log ε)	b.p. °C	m.p. °C	Density	n_D	Solubility	Ref.
13167	Salvarsan or 606. Arspenamine $C_{12}H_{12}N_2As_2O_2 \cdot 2HCl \cdot 2H_2O$	475.04	pw		185-95d			w	$B16^{2}$, 560
13168	Sambunigrin or d-Mandelonitrile glucoside $C_6H_5CH(CN)OC_6H_{11}O_5$	295.29	nd (bz-peth), $[\alpha]^{18}{}_D$ −75.1		151-2			w, al	$B17^{4}$, 3356
13169	Samidin $C_{21}H_{22}O_7$	386.40	cubic, $[\alpha]^{21}{}_D$ +49.1 (chl, c=1.59)		138-9			al, eth	$B19^{4}$, 2788
13170	Sanquinarine $C_{20}H_{15}NO_5$	349.34	cr (eth al)		266			al, eth, ace, bz	$B27^{2}$, 614
13171	Santalic acid or Guerbet's acid $C_{15}H_{22}O_2$	234.34	red		β:202 γ:189		β:1.5136^{20} γ:1.5055^{2}		$B9^{3}$, 2619
13172	α-Santalol or Arheol $C_{15}H_{24}O$	220.35	$[\alpha]^{20}{}_D$ +17	301-2, 167^{14}		$0.9679^{20/4}$	1.5023^{20}	al	$B6^{3}$, 2083
13173	β-Santalol $C_{15}H_{24}O$	230.35	$[\alpha]^{20}{}_D$ −90.5	$167\text{-}8^{10}$		$0.9750^{20/4}$	1.5115^{20}		$B6^{3}$, 2080
13174	Santene or 2,3-Dimethyl-2-norbornene C_9H_{14}	122.21		140-1, 35^{15}		$0.8698^{17/4}$	1.4688^{17}	eth, ace, bz	$B5^{3}$, 333
13175	Santenic acid (*cis,d*) or π-Norcamphenic acid $C_9H_{14}O_4$	186.21	pl (w), $[\alpha]^{33}{}_D$ + 38.3 (al)		170-1			al	$B9^{3}$, 3847
13176	β-Santenol (*cis,exo*) or 2,3-Dimethyl-2-norbornano $C_9H_{16}O$	140.23	nd (al), tab (lig)	19^{2}	101-2			eth	$B6^{4}$, 243
13177	α-Santenone (*cis*) or 1,7-Dimethyl-1,2-norbarnanone $C_9H_{14}O$	138.21	pl $[\alpha]^{22}{}_D$ + 11.4 (al)	191	55			al, ace	$B7^{3}$, 305
13178	Santonin or Santonic acid $C_{15}H_{18}O_3$	246.31	orh (w, aa, eth) $[\alpha]^{18}{}_D$ −173 (al, c=2)		174-6	$1.187^{66/4}$	1.590	bz, chl, Py	$B17^{4}$, 6232
13179	Sarmentogenin $C_{23}H_{34}O_5$	390.52	pr (95% al, MeOH-eth)	270-5				al, Py	$B18^{4}$, 2443
13180	Sarpagine or Raupine $C_{19}H_{22}N_2O_2$	310.40	nd $[\alpha]^{20}{}_D$ + 54 (Py)		320			al	Am84, 622
13181	Sarsaspogenin or Parigenin $C_{27}H_{44}O_3$	416.64	lo pr nd (ace), $[\alpha]^{25}{}_D$ −75 (chl, c=0.5)		200-1			al, ace, bz, chl	$B19^{4}$, 824
13182	Scarlet red $C_{24}H_{20}N_4O$	380.45	dr br pw or nd	d 260	186d			chl, peth	B16, 172
13183	Scilliroside (a β-glucoside) $C_{32}H_{44}O_{12}$	620.69	lo pr (dil MeOH) $[\alpha]^{20}{}_D$ -60 (MeOH)	d	168-70			al, diox	$B18^{4}$, 3178
13184	Scopoline or Oscin $C_8H_{13}NO_2$	155.20	hyg nd (lig, eth, chl, peth)	248	108-9	$1.0891^{134/4}$		w, al, ace	$B27^{2}$, 61
13185	Scyllitol $C_6H_{12}O_6$	180.16	pr (+ 3w)		353d	$1.659^{19/4}$			$B6^{3}$, 6926
13186	Sebacamide or Decandiamide $H_2NOC(CH_2)_8CONH_2$	200.28	pr or pl (aa)		210			aa	$B2^{4}$, 2088
13187	Sebacic acid or Decanedioic acid $HO_2C(CH_2)_8CO_2H$	202.25	lf	295^{100}	134.5	$1.2705^{20/4}$	1.422^{133}	al, eth	$B2^{4}$, 2078
13188	Sebacic acid, dibutyl ester or di-Butyl sebacate $C_4H_9O_2C(CH_2)_8CO_2C_4H_9$	314.47		344-5, 227^{17}	−10	0.9405^{15}	1.4433^{15}	eth	$B2^{4}$, 2081
13189	Sebacic acid, diethyl ester or Diethyl sebacate $C_2H_5O_2C(CH_2)_8CO_2C_2H_5$	258.36		306^{773}, 188^{19}	5	$0.9646^{20/4}$	1.4366^{20}	al, ace	$B2^{4}$, 2080
13190	Sebacic acid, di(2-ethyl butyl) ester $(C_2H_5)_2CHCH_2O_2C(CH_2)_8CO_2CH_2CH(C_2H_5)_2$	370.57		344-6	−22	$0.920^{25/4}$		al, ace, bz	$B2^{4}$, 2082
13191	Sebacic acid, di(2-ethyl hexyl) ester $C_4H_9CH(C_2H_5)CH_2O_2C(CH_2)_8CO_2CH_2CH(C_2H_5)C_4H_9$	426.68		256^{5}	−48	$0.912^{25/4}$	1.451^{25}	al, ace, bz	$B2^{4}$, 2083
13192	Sebacic acid, dimethyl ester or Dimethyl sebacate $CH_3O_2C(CH_2)_8CO_2CH_3$	230.30	lo pr	175^{20}, 144^{5}	38	0.9882^{28}	1.4355^{28}	al, eth, ace	$B2^{4}$, 2080
13193	Sebacic acid, monomethyl ester, mononitrile $NC(CH_2)_8CO_2CH_3$	197.28		178^{16}	3-4	0.934^{20}	1.4398^{25}	eth	$B2^{4}$, 2079

No.	Name, Synonyms, and Formula	Mol. wt.	Color, crystalline form, specific rotation and λ_{max} (log ε)	b.p. °C	m.p. °C	Density	n_D	Solubility	Ref.
13194	Sebaconitrile $NC(CH_2)_8CN$	164.25		204^{16}		$0.9313^{20/4}$	1.4474^{20}		$B2^{4}$, 2089
13195	Sebacyl chloride $ClOC(CH_2)_8COCl$	239.14		220^{75}, 165^{11}	−2.5	$1.1212^{20/4}$	1.4684^{18}		$B2^{4}$, 2088
13196	Sedarmide $H_2NCONHCOCH(CH_2CH{=}CH_2)CH(CH_3)_2$	184.24	nd (al)		194			al, chl	$B3^{2}$, 53
13197	Selanthrone $C_{12}H_8Se_2$	310.12	pr (al), nd (aa)	223^{11}	181			CS_2	$B19^{4}$, 352
13198	Seleno urea $H_2NCSeNH_2$	123.02	pr or nd (w)	200d				w	$B3^{4}$, 435
13199	Seleno urea, 1-ethyl $C_2H_5NHCSeNH_2$	151.07	nd (al, peth, w)		ca 125			al	$B4^{2}$, 610
13200	α-Selinene $C_{15}H_{24}$	204.36	$[\alpha]^{t}_{D}$ + 49.5	268-72, $128\text{-}32^{11}$		$0.9196^{20/15}$	1.5048^{20}		$B5^{3}$, 1090
13201	B-Selinene $C_{15}H_{24}$	204.36	$[\alpha]^{20}_{D}$ + 38.2	$121\text{-}2^{6}$		$0.9170^{18/15}$	1.4956^{21}		$B5^{3}$, 1090
13202	Semicarbazide or Aminourea $H_2NCONHNH_2$	75.07	pr (al)		96			w, al	$B3^{4}$, 177
13203	Semicarbazide, 1,1-diphenyl $(C_6H_5)_2NNHCONH_2$	227.27	nd (al, bz)		195			al, bz	B15, 304
13204	Semicarbazide, 1,1-diphenyl-3-thio $(C_6H_5)_2NNHCSNH_2$	243.33	cr (al, bz)		202			al, ace, bz, chl	$B15^{3}$, 193
13205	Semicarbazide, 1,4-diphenyl $C_6H_5NHNHCONHC_6H_5$	227.27	nd or lf (al,bz)		177			al	$B15^{3}$, 184
13206	Semicarbazide, 1,4-diphenyl-3-thio $C_6H_5NHCSNHNHC_6H_5$	243.33			176-7			al, ace, chl	$B15^{3}$, 190
13207	Semicarbazide, 2,4-diphenyl $H_2NN(C_6H_5)CONHC_6H_5$	227.27	lf (al)		165.5			al, eth, bz, chl	B15, 277
13208	Semicarbazide, 2,4-diphenyl-3-thio $H_2NN(C_6H_5)CSNHC_6H_5$	243.33	lf (al)		139			ace, bz	$B15^{3}$, 182
13209	Semicarbazide, 4,4-diphenyl $(C_6H_5)_2NCONHNH_2$	227.27			154			al, eth, chl	$B12^{3}$, 896
13210	Semicarbazide hydrochloride $H_2NCONHNH_2.HCl$	111.52	pr (dil al)		175-7d			w	$B3^{4}$, 177
13211	Semicarbazide, 4-methyl-3-thio $CH_3NHCSNHNH_2$	105.16			135-8			w, al	$B4^{4}$, 220
13212	Semicarbazide, 1-phenyl $H_2NCONHNHC_6H_5$	151.17			172			al, ace, MeOH	$B15^{3}$, 184
13213	Semicarbazide, 1-phenyl-3-thio $C_6H_5NHNHCSNH_2$	167.23	pr (al)		200-1d				$B15^{3}$, 190
13214	Semicarbazide, 2-phenyl $H_2NN(C_6H_5)CONH_2$	151.17	nd (bz, al)		120			w, al, chl	$B15^{2}$, 103
13215	Semicarbazide, 2-phenyl-3-thio $H_2NN(C_6H_5)CSNH_2$	167.23	w		153			al	$B15^{1}$, 70
13216	Semicarbazide, 4-phenyl $H_2NNHCONHC_6H_5$	151.17	nd (bz), pl (w)		128			al, chl	$B12^{3}$, 822
13217	Semicarbazide, 4-phenyl-3-thio $H_2NNHCSNHC_6H_5$	167.23	pl (al)		140d				$B12^{2}$, 232
13218	Semicarbazide, 3-thio or Thiosemicarbazide $H_2NNHCSNH_2$	91.13	lo nd (w)		183			w, al	$B3^{4}$, 374
13219	Semicarbazide, 1-(3-tolyl) or Maretine $(3\text{-}CH_3C_6H_4)NHNHCONH_2$	165.19	lf (w, dil al)		183-4			al	B15, 508
13220	Semicarbazide, 1-(4-tolyl) $(4\text{-}CH_3C_6H_4)NHNHCONH_2$	165.19			190-1			w	$B15^{2}$, 239
13221	Semicarbazide, 1,1,4-triphenyl $(C_6H_5)_2NNHCONHC_6H_5$	303.36	nd (al)		206-7				$B15^{2}$, 115
13222	Semicarbazide, 1,4,4,-triphenyl $C_6H_5NHNHCON(C_6H_5)_2$	303.36	pl (al)		151-2			al	$B15^{3}$, 185
13223	Semicarbazide, 2,4,4-triphenyl $H_2NN(C_6H_5)CON(C_6H_5)_2$	303.36	cr (dil al)		128			al	$B15^{3}$, 181
13224	Serine *(D)* or *D*-2-Amino-3-hydroxy propionic acid $HOCH_2CH(NH_2)CO_2H$	105.09	nd or hex pr (w), $[\alpha]^{20}_{D}$ + 6.9 (w, c=10)	d	228d			w	$B4^{3}$, 1572
13225	Serine *(DL)* or *DL*-2-Amino-3-hydroxy propionic acid $HOCH_2CH(NH_2)CO_2H$	105.10	mcl pr or lf (w)		246d (sealed tube)	1.603^{22}		w	$B4^{3}$, 1573

No.	Name, Synonyms, and Formula	Mol. wt.	Color, crystalline form, specific rotation and λ_{max} (log ε)	b.p. °C	m.p. °C	Density	n_D	Solubility	Ref.
13226	Serine *(L)* or *L*-2-Amino-3-hydroxy propionic acid $HO\text{-}CH_2CH(NH_2)CO_2H$	105.09	hex pl or pr (w), $[\alpha]^{20}{}_D$ −6.8 (w, c=10)	sub150^{10-4}	228d			w	$B4^3$, 1568
13227	Serpentine $C_{21}H_{20}N_2O_3$	348.40	ye rods or lf (al), $[\alpha]^{25}{}_D$ +292 (MeOH)		175			al, eth, ace, MeOH	Am76, 2843
13228	Sesamin or Asaranin $C_{20}H_{18}O_6$	354.36	nd (al), $[\alpha]^{20}{}_D$ +68.4 (chl, c=24)		123-4			al, ace, bz, chl	$B19^4$, 6236
13229	Shikonine or *d*-Alkanin $C_{16}H_{16}O_5$	288.30	br-red nd (bz), $[\alpha]^{20}{}_{644}$ +135 (bz, c=1.3)		147			al, eth, ace, bz	$B8^3$, 4088
13230	Silane, allyl-trimethyl $(CH_3)_3SiCH_2CH{=}CH_2$	114.26		85.4^{752}		$0.7193^{20/4}$	1.4074^{20}		$B4^3$, 1854
13231	Silane, benzyl-trimethyl $(CH_3)_3SiCH_2C_6H_5$	164.32		$191\text{-}2^{748}$, 93^{15}		$0.8933^{20/4}$	1.5042^{20}		$B16^3$, 1201
13232	Silane, 4-bromophenoxy-trimethyl $4\text{-}BrC_6H_4OSi(CH_3)_3$	245.19		126^{25}		$1.2619^{20/4}$	1.5145^{20}		$B6^4$, 1056
13233	Silane, butyl-chloro-dimethyl $C_4H_9Si(Cl)(CH_3)_2$	245.19		138.4^{747}		$0.8751^{20/4}$	1.5145^{20}		$BOSC2^1$, 180
13234	Silane, butyl-trichloro $C_4H_9SiCl_3$	191.56		148.9		$1.1606^{20/4}$	1.4363^{20}	bz, eth	$B4^1$, 582
13235	Silane, isobutyl-trichloro $i\text{-}C_4H_9SiCl_3$	191.56			141-6	$1.154^{20/4}$			$B4^1$, 582
13236	Silane, butyl-trimethyl $C_4H_9Si(CH_3)_3$	130.31		115		$0.7353^{0/4}$			$B4^3$, 1851
13237	Silane, isobutyl-trimethyl $i\text{-}C_4H_9Si(CH_3)_3$	130.31			108-9	$0.7330^{0/4}$			$B4^1$, 580
13238	Silane, chloro-dimethyl-ethyl $(CH_3)_2Si(Cl)C_2H_5$	122.67		89-90		0.8675^{20}	1.4105^{20}		$B4^3$, 1865
13239	Silane, chloro-dimethyl-phenyl $(CH_3)_2Si(Cl)C_6H_5$	170.71		196		1.0646^{20}	1.5184^{20}		$B16^3$, 1206
13240	Silane, chloro-dimethyl-vinyl $(CH_3)_2Si(Cl)CH{=}CH_2$	120.65		83-4		$0.8744^{20/4}$	1.4141^{20}		KHOC, 240
13241	Silane, chloro-ethyl-methyl-phenyl $C_6H_5SiCl(CH_3)C_2H_5$	184.74						CS_2	
13242	Silane, chloro-phenyl $C_6H_5SiH_2Cl$	142.66		162-3		$1.0683^{20/4}$	1.5340^{20}		BOSC, 2,, 147
13243	Silane, chloro-triethoxy $(C_2H_5O)_3SiCl$	198.72		156, 69^{12}	−51	$1.032^{20/20}$	1.3999^{20}	al	$B1^3$, 1336
13244	Silane, chloro-triethyl $(C_2H_5)_3SiCl$	150.72		144-5		$0.8967^{20/4}$	1.4314^{20}		$B4^3$, 1867
13245	Silane, chloro-trimethyl $(CH_3)_3SiCl$	108.64		57.7, $0^{61.5}$	−57.7	$0.8580^{20/4}$	1.3885^{20}		$B4^3$, 1857 334, $BOSC2^1$, 71
13246	Silane, chloro-triphenyl $(C_6H_5)_3SiCl$	294.86		$240\text{-}3^{15}$					$B16^3$, 1208
13247	Silane, chloromethyl-dimethyl-phenyl $(CH_3)_2Si(C_6H_5)CH_2Cl$	184.74		225				CS_2	$B16^3$, 1200
13248	Silane, (4-chlorophenoxy)-trimethyl $4\text{-}ClC_6H_4OSi(CH_3)_3$	200.74		214^{758}		$1.0320^{20/4}$	1.4930^{20}		$B6^4$, 878
13249	Silane, (2-chlorophenoxy)-trimethyl $2\text{-}ClC_6H_4OSi(CH_3)_3$	200.74		$95\text{-}6^{15}$					$B6^4$, 809
13250	Silane, 3-chlorophenyl-trimethyl $3\text{-}ClC_6H_4Si(CH_3)_2$	200.74		206-7, $84\text{-}5^{9}$		$1.0071^{20/4}$	1.5108^{20}		$B16^3$, 1200
13251	Silane, 3-chloropropyl-trimethyl $Cl(CH_2)_3Si(CH_3)_3$	150.72		148		0.8825^{20}	$1.4310^{20}{}_D$		KHOC, 263
13252	Silane, dibenzyl-diethoxy $(C_6H_5CH_2)_2Si(OC_2H_5)_2$	300.47		$139^{0.5}$			1.5250^{20}		$BOSC2^1$, 674
13253	Silane, dichloro-diisobutyl $(i\text{-}C_4H_9)_2SiCl_2$	213.22		93^{16}		$1.00^{20/4}$			$B4^3$, 1893
13254	Silane, dichloro-diethoxy $(C_2H_5O)_2SiCl_2$	189.11		135.9, 51.6^{12}	−130	$1.1290^{20/4}$			$B1^4$, 1364

No.	Name, Synonyms, and Formula	Mol. wt.	Color, crystalline form, specific rotation and λ_{max} (log ε)	b.p. °C	m.p. °C	Density	n_D	Solubility	Ref.
13255	Silane, dichloro-diethyl $(C_2H_5)_2SiCl_2$	157.11		d128-30	−96.5	1.0504^{20}	1.4309^{20}		$B4^3$, 1890
13256	Silane, dichloro-dimethyl $(CH_3)_2SiCl_2$	129.06		70.5	−76	$1.070^{25/25}$	$1.405^{25/D}$		$B4^3$, 1877
13257	Silane, dichloro-diphenyl $(C_6H_5)_2SiCl_2$	253.20		302-5, $163\text{-}5^{10}$		$1.2216^{20/4}$	1.5819^{20}	al, eth, ace, bz	$B16^3$, 1214
13258	Silane, dichloro-ethyl-methyl $C_2H_5SiCl_2CH_3$	143.09		100^{747}		$1.0630^{20/4}$	1.4197^{20}		$B4^3$, 1889
13259	Silane, dichloro-methyl-phenyl $C_6H_5SiCl_2CH_3$	191.13		206-7		1.1866^{20}	1.5180^{20}		$B16^3$, 1211
13260	Silane, dichloro-methyl-isopropyl $i\text{-}C_3H_7SiCl_2CH_3$	157.11		121-2		1.0385^{20}	1.4270^{20}		C52, 6159
13261	Silane, dichloro methyl vinyl $CH_2=CHSiCl_2CH_3$	141.07		93		$1.085^{27/27}$	$1.4295^{20/D}$		$B4^3$, 1894
13262	Silane, dichloro-phenyl $C_6H_5SiHCl_2$	177.11		180-2		$1.225^{25/25}$			$B16^3$, 1210
13263	Silane, dichloro-isopropyl-methyl $CH_3SiCl_2CH(CH_3)_2$	157.11		121-2		1.0385^{20}	1.4270^{20}		C52, 6159
13264	Silane, diethoxy-difluoro $(C_2H_5O)_2SiF_2$	156.20		83	−122				$B1^3$, 1336
13265	Silane, diethoxy-dimethyl $(C_2H_5O)_2Si(CH_3)_2$	148.28		113.8, 69.5^{179}	−87	$0.8395^{20/4}$	$1.3805^{20/D}$		$B4^3$, 1874
13266	Silane, diethoxy-diphenyl $(C_2H_5O)_2Si(C_6H_5)_2$	272.42		$302\text{-}4^{767}$, $151\text{-}3^{6}$		$1.0329^{20/4}$	$1.52695^{20/D}$		$B16^2$, 608
13267	Silane, diethoxy-methyl $(C_2H_5O)_2SiHCH_3$	134.25		$135\text{-}8^{731}$		0.829^{25}	$1.3724^{25/D}$		KHOC, 358
13268	Silane, diethoxy-methyl-vinyl $(C_2H_5O)_2Si(CH_3)CH=CH_2$	160.29		133		0.8620^{20}	$1.4001^{20/D}$		KHOC, 388
13269	Silane, diethyl $(C_2H_5)_2SiH_2$	88.22		56	−134.4	$0.6832^{20/4}$	1.3920^{20}		$B4^3$, 1846
13270	Silane, diethyl-difluoro $(C_2H_5)_2SiF_2$	124.21		60.9^{755}	−78.7	$0.9348^{20/4}$	1.3385^{20}		$B4^3$, 1890
13271	Silane, difluoro-diphenyl $(C_6H_5)_2SiF_2$	220.29		252, 158^{50}		$1.145^{17/4}$	1.5221^{25}	bz	$B16^3$, 1213
13272	Silane, dimethoxy-dimethyl $(CH_3O)_2Si(CH_3)_2$	120.22		80.5-82		0.8535^{20}	$1.3679^{20/D}$		KHOC, 346
13273	Silane, dimethoxy-diphenyl $(CH_3O)_2Si(C_6H_5)_2$	244.37		$90\text{-}5^{13}$		1.0771^{20}	$1.5447^{20/D}$		KHOC, 489
13274	Silane, dimethyl $(CH_3)_2SiH_2$	60.17		−20.1	−150.2	0.68^{-80}			$B4^1$, 579
13275	Silane, dimethyl-diphenoxy $(C_6H_5O)_2Si(CH_3)_2$	244.37		$130\text{-}2^{5}$	−23	1.0599^{25}	1.5330^{20}		$B6^4$, 764
13276	Silane, dimethyl-diphenyl $(C_6H_5)_2Si(CH_3)_2$	212.37		276^{740}, 173^{45}		$0.9867^{20/4}$	1.5644^{20}		$B16^2$, 605
13277	Silane, dimethylphenyl $C_6H_5SiH(CH_3)_2$	136.27		157		0.8891^{20}	1.4995^{20}		KHOC, 396
13278	Silane, dimethylpropyl $C_3H_7SiH(CH_3)_2$	102.25		73-4					$B4^3$, 1850
13279	Silane, dimethylamino-trimethyl $(CH_3)_3SiN(CH_3)_2$	117.27		86, $0^{22.5}$			1.4379^{24}		BOSC 2^1, 138
13280	Silane, diphenyl $(C_6H_5)_2SiH_2$	184.31		75				CS_2	
13281	Silane, diphenyl-methyl $(C_6H_5)_2SiHCH_3$	198.34		266.8, 93.5^{1}		$0.9973^{20/4}$	1.5694^{20}		BOSC 2^2, 522
13282	Silane, diphenyl-divinyl $(C_6H_5)_2Si(CH=CH_2)_2$	236.39		$130\text{-}1^{0.05}$		$1.0092^{25/4}$	$1.5350^{25/4}$		BOSC 2^1, 614
13283	Silane, ethoxy-trichloro $C_2H_5OSiCl_3$	179.51		101.9	−135	1.2274^{20}	1.4045^{20}	al	$B1^4$, 1364
13284	Silane, ethoxy-triethyl $(C_2H_5)_3SiOC_2H_5$	160.33		154-5		0.8160^{20}	1.4140^{20}	al, eth	$B4^3$, 1866
13285	Silane, ethoxy-trifluoro $C_2H_5OSiF_3$	130.14	gas	−7	−122				$B1^3$, 1336
13286	Silane, ethoxy-trimethyl $C_2H_5OSi(CH_3)_3$	118.25		75^{745}		$0.7573^{20/4}$	$1.3741^{20/D}$	al, eth, ace	$B4^3$, 1856
13287	Silane, ethoxy-triphenyl $(C_6H_5)_3SiOC_2H_5$	304.46		344, 207^{3}	65			chl	$B16^3$, 1207

No.	Name, Synonyms, and Formula	Mol. wt.	Color, crystalline form, specific rotation and λ_{max} (log ε)	b.p. °C	m.p. °C	Density	n_D	Solubility	Ref.
13288	Silane, ethyl trichloro $C_2H_5SiCl_3$	163.51	...	97.9	−105.6	$1.2381^{20/4}$	1.4257^{20}	...	$B4^{3}$, 1900
13289	Silane, ethyl-triethoxy $(C_2H_5O)_3SiC_2H_5$	192.33	...	158.9	...	$0.8594^{20/4}$	1.3955^{20}	al, eth	$B4^{3}$, 1899
13290	Silane, ethyl-trifluoro $C_2H_5SiF_3$	114.14	...	−4.4	−105	1.227^{-76}	...	...	$B4^{3}$, 1900
13291	Silane, ethyl-trimethoxy $(CH_3O)_3SiC_2H_5$	150.25	...	124.3	...	$0.9488^{20/4}$	1.3838^{20}	al	$B4^{3}$, 1899
13292	Silane, ethyl-triphenyl $(C_6H_5)_3SiC_2H_5$	288.46	...	...	76	...	...	chl	...
13293	Silane, fluoro-triethoxy $(C_2H_5O)_3SiF$	182.27	...	134.6	...	...	...	...	$B1^{3}$, 1336
13294	Silane, fluoro-triethyl $(C_2H_5)_3SiF$	134.27	...	110	...	$0.8354^{25/4}$	1.3900^{25}	peth	$B4^{3}$, 1866
13295	Silane, fluoro-triphenyl $(C_6H_5)_3SiF$	278.40	...	$245\text{-}52^{30}$, $160\text{-}70^{0.5}$	64	...	...	...	$B16^{3}$, 1208
13296	Silane, methyl CH_3SiH_3	46.14	...	−57	−156.5	...	...	...	$B4^{1}$, 579
13297	Silane, methyl-phenyl $C_6H_5SiH_2CH_3$	122.24	...	140, $46\text{-}7^{20}$	...	$0.8895^{20/4}$	1.5058^{20}	...	BOSC 2^{1}, 209
13298	Silane, methyl-tribromo CH_3SiBr_3	282.83	...	131.3^{744}	−28.4	2.2130^{25}	1.5152^{25}	...	$B4^{3}$, 1897
13299	Silane, methyl-trichloro CH_3SiCl_3	149.48	...	66.4	−77.8	$1.273^{25/23}$	...	...	$B4^{3}$, 1896
13300	Silane, methyl-triethoxy $(C_2H_5O)_3SiCH_3$	178.30	...	143	...	0.8923^{20}	1.3835^{20}	al	$B4^{3}$, 1895
13301	Silane, methyl-trimethoxy $(CH_3O)_3SiCH_3$	136.22	...	102	...	$0.951^{25/4}$	1.3687^{25}	chl	BOSC 2^{1}, 109
13302	Silane, methyl-triphenoxy $(C_6H_5O)_3SiCH_3$	322.44	...	$267\text{-}71^{100}$, $210\text{-}1^{12}$	...	1.135^{20}	1.5599^{20}	...	$B6^{4}$, 765
13303	Silane, methyl-triphenyl $(C_6H_5)_3SiCH_3$	274.44	...	$196\text{-}200^{9}$	69.5	...	...	CS_2	$B16^{3}$, 1199
13304	Silane, 2-methylphenoxy-trimethyl $2\text{-}CH_3C_6H_4OSi(CH_3)_3$	180.32	...	192	...	$0.9287^{20/4}$	1.4830^{20}	...	BOSC 2^{1}, 386
13305	Silane, 3-methylphenoxy-trimethyl $3\text{-}CH_3C_6H_4OSi(CH_3)_3$	180.32	...	198^{3}	...	$0.9186^{20/4}$	1.4791^{20}	...	BOSC 2^{1}, 386
13306	Silane, 4-methylphenoxy-trimethyl $4\text{-}CH_3C_6H_4OSi(CH_3)_3$	180.32	...	199	...	$0.9183^{20/4}$	1.4790^{20}	...	BOSC 2^{1}, 386
13307	Silane, pentyl-trichloro $C_5H_{11}SiCl_3$	205.59	...	171^{742}, 60.5^{15}	...	$1.1330^{20/4}$	1.4503^{20}	...	$B4^{3}$, 1905
13308	Silane, pentyl-triethoxy $C_5H_{11}Si(OC_2H_5)_3$	234.41	...	95^{13}	...	0.8862^{20}	1.4059^{20}	...	$B4^{3}$, 1905
13309	Silane, phenyl $C_6H_5SiH_3$	108.22	...	118-20	...	$0.8681^{20/4}$	1.5125^{20}	...	$B16^{3}$, 1198
13310	Silane, phenyl-tri-butyl $(C_4H_9)_3SiC_6H_5$	276.54	...	$140\text{-}2^{4}$	...	0.8753^{20}	1.4915^{20}	...	$B16^{3}$, 1199
13311	Silane, phenyl-trichloro $C_6H_5SiCl_3$	211.55	...	201	...	...	...	chl	$B16^{3}$, 1216
13312	Silane, phenyl-triethoxy $C_6H_5Si(OC_6H_5)_3$	240.37	...	235-8	...	$0.9961^{20/4}$	1.4718^{20}	...	$B16^{3}$, 1215
13313	Silane, phenyl-triethyl $C_6H_5Si(C_2H_5)_3$	192.38	...	283	...	$0.8915^{20/4}$	1.4999^{20}	...	$B16^{3}$, 1198
13314	Silane, phenyl-trifluoro $C_6H_5SiF_3$	162.19	...	101.5, 16.5^{24}	...	$1.2169^{20/4}$	1.4110^{20}	al, bz	$B16^{3}$, 1216
13315	Silane, phenyl-trimethyl $C_6H_5Si(CH_3)_3$	150.30	...	170	...	...	...	CS_2	$B16^{3}$, 1198
13316	Silane, phenyl-tripropyl. $(C_3H_7)_3SiC_6H_5$	234.46	...	$146\text{-}7^{18}$	...	$0.8799^{20/4}$	1.4950^{20}	...	BOSC 2^{1}, 601
13317	Silane, propyl-trichloro $C_3H_7SiCl_3$	177.53	...	122.2^{740}	...	$1.1851^{20/4}$	1.4290^{20}	...	$B4^{3}$, 1902
13318	Silane, propyl-trichloro $(CH_3)_2CHSiCl_3$	177.53	...	120.3^{748}	−87.7	$1.1934^{20/4}$	$1.4319^{20}/_D$	...	$B4^{3}$, 1903
13319	Silane, tetra(allyloxy) $(C_3H_5O)_4Si$	256.37	...	$134\text{-}5^{34}$	...	$0.9824^{20/4}$	1.4349^{20}	...	$B1^{3}$, 1886
13320	Silane, tetraethoxy $(C_2H_5O)_4Si$	208.33	...	165.5-8	−77	$0.9356^{20/20}$	1.3832^{20}	...	$B1^{4}$, 1360
13321	Silane, tetraethyl $(C_2H_5)_4Si$	144.33	...	153	...	0.7658^{20}	1.4268^{20}	...	$B4^{3}$, 1847

PHYSICAL CONSTANTS OF ORGANIC COMPOUNDS (Continued)

No.	Name, Synonyms, and Formula	Mol. wt.	Color, crystalline form, specific rotation and λ_{max} (log ε)	b.p. °C	m.p. °C	Density	n_D	Solubility	Ref.
13322	Silane, tetramethoxy $(CH_3O)_4Si$	152.22	....	121-2	......	1.032^{20}	1.3683^{20}	....	B1⁴, 1266
13323	Silane, tetramethyl $(CH_3)_4Si$	88.22	....	26.5	−102.2	$0.648^{19/4}$	1.3587^{20}	al, eth	B4³, 1843
13324	Silane, tetraphenyl $(C_6H_5)_4Si$	336.51	....	228^5	236-7	$1.078^{20/4}$	....	CS_2	B16³, 1199
13325	Silane, tetrapropoxy $(C_3H_7O)_4Si$	264.44	....	$225\text{-}7^{757}$, 94^5	......	$0.9158^{20/4}$	1.4012^{20}	....	B1⁴, 1435
13326	Silane, tetravinyl $(CH_2{=}CH)_4Si$	136.27	....	130.2	......	0.7999^{20}	1.4625^{20}	....	KHOC, 278
13327	Silane, 2-tolyl-trichloro $(2\text{-}CH_3C_6H_4)SiCl_3$	225.58	....	226.7	−26	$1.306^{25/4}$	1.5336^{25}	....	BOSC 2¹, 204
13328	Silane, tributyl $(C_4H_9)_3SiH$	200.44	....	215-20 $88\text{-}9^5$	......	$0.7794^{20/4}$	$1.4380^{20}{}_D$	....	BOSC 2¹, 509
13329	Silane, triethoxy $(C_2H_5O)_3SiH$	164.28	....	132-5	......	$0.8745^{20/4}$	....	....	B1⁴, 1359
13330	Silane, triethoxy-vinyl $(C_2H_5O)_3SiCH{=}CH_2$	190.31	....	....	......	....	....	....	BOSC 2¹, 294
13331	Silane, triethyl $(C_2H_5)_3SiH$	116.28	....	109^{755}	......	0.7302^{20}	1.4117^{20}	....	B4³, 1847
13332	Silane, trimethyl vinyl $(CH_3)_3SiCH{=}CH_2$	100.24	....	55.5^{767}	......	0.6910^{20}	$1.3914^{20}{}_D$	....	KHOC, 354
13333	Silane, trioctyl $(C_8H_{17})_3SiH$	368.76	....	$163\text{-}5^{0.15}$	......	$0.8207^{20/20}$	1.4545^{20}	....	BOSC 2², 81
13334	Silane, triphenyl $(C_6H_5)_3SiH$	260.41	....	152-61	36-7	....	....	....	B16³, 1199
13335	Silane, tripropyl $(C_3H_7)_3SiH$	158.36	....	171-3	......	$0.7723^{0/4}$	1.4280^{20}	....	B4³, 1850
13336	Silane, vinyl-trichloro $CH_2{=}CHSiCl_3$	161.49	....	91-2	−95	$1.2426^{20/4}$	1.4295^{20}	chl	B4³, 1908
13337	Silanol, dimethyl-ethyl $(CH_3)_2Si(OH)C_2H_5$	104.22	....	120^{774}, 58^{50}	......	$0.8332^{20/4}$	1.4070^{20}	....	BOSC 2¹, 108
13338	Silanol, diphenyl-methyl $(C_6H_5)_2Si(OH)CH_3$	214.34	....	148^3	167	....	....	CS_2	....
13339	Silanol, triethyl $(C_2H_5)_3Si(OH)$	132.28	....	154, 46.6^9	......	$0.8647^{20/4}$	1.4329^{20}	al, eth	B4³, 1866
13340	Silanediol, dimethyl, diacetate $(CH_3CO_2)_2Si(CH_3)_2$	176.24	....	165^{750}, $44\text{-}5^3$	......	$1.0540^{20/4}$	1.4030^{20}	....	B4³, 1876
13341	Silanetriol, methyl-triacetate $(CH_3CO_2)_3SiCH_3$	220.25	....	$110\text{-}2^{17}$	40.5	$1.1677^{25/4}$	1.4083^{20}	....	B4³, 1896
13342	Silin, 1,1-diphenyl-perhydro $C_{17}H_{20}Si$	252.43	....	$193\text{-}8^5$	......	1.0319^{25}	1.5779^{20}	....	BOSC2¹, 647
13343	Sinapine hydrogen sulfate $C_{16}H_{23}NO_5.H_2SO_4$	407.44	lf (al)	....	186-7d	....	....	w	B10², 354
13344	Sinapine, thiocyanate, monohydrate $C_{16}H_{23}NO_5.HSCN.H_2O$	386.46	ye nd	....	180-1	....	....	....	B10², 354
13345	Sinomenine or Cuculine $C_{19}H_{23}NO_4$	329.40	nd (bz), $[\alpha]^{26}{}_D$ −71 (al, c=2.1)	....	162	....	....	al, ace, chl	B21⁴, 6670
13346	α-Sitosterol $C_{29}H_{48}O$	412.70	nd (al), $[\alpha]^{28}{}_D$ −1.7 (chl,c=2)	....	166	....	....	al, chl	B6³, 2876
13347	α_2-Sitosterol $C_{30}H_{50}O$	426.73	cr (al-peth), $[\alpha]^{25}{}_D$ + 3.5 (chl,c=2)	....	156	....	....	al	B6³, 2697
13348	α_3-Sitosterol $C_{29}H_{48}O$	412.70	pl (al), $[\alpha]^{20}{}_D$ + 5.2 (chl)	....	142-3	....	....	al	B6³, 2707
13349	β-Sitosterol or Verosterol $C_{29}H_{50}O$	414.72	pl (al), nd (MeOH), $[\alpha]^{25}{}_D$ −37 (chl,c=2)	....	140	....	....	al, eth, aa	B6³, 2696
13350	Skimmin or 7-Hydroxycoumarin-β-D-glucose $C_{15}H_{16}O_8$	324.29	cr (w + 1), $[\alpha]^{18}{}_D$ −80	....	219-21	....	....	al	B18⁴, 301

No.	Name, Synonyms, and Formula	Mol. wt.	Color, crystalline form, specific rotation and λ_{max} (log ε)	b.p. °C	m.p. °C	Density	n_D	Solubility	Ref.
13351	Smilagenin or Isosapogenin $C_{27}H_{44}O_3$	416.64	silky nd (ace), $[\alpha]^{25/}{}_D$ −69 (chl,c=0.5)		185			al, ace, bz, chl	B19^4, 826
13352	Solanidine-T or Solatubine $C_{27}H_{43}NO$	397.64	lo nd (chl-MeOH)	sub d	218-9			bz, chl	B21^4, 1398
13353	Solanine-S or Purapurine $C_{45}H_{73}NO_{16}$	884.07	nd (al), fl (diox) $[\alpha]^{20/}{}_D$ −69 (al)		*ca* 190				J1963, 745
13354	Solasodine or Purapuridine $C_{27}H_{43}NO_2$	413.64	hex pl (sub), $[\alpha]^{20/}{}_D$ −92.4 (bz)		202			ace, bz, chl, MeOH, Py	J1942, 13
13355	Sorbaldehyde or 2,4-Hexadienal $CH_3CH{=}CHCH{=}CHCHO$	96.13		173-4^{756}, 76^{30}		0.898^{20}	1.5384^{20}		B1^4, 3545
13356	Sorbamide $CH_3CH{=}CHCH{=}CHCONH_2$	111.14	nd (w)		171-2			w, al	B2^4, 1705
13357	Sorbic Acid or 2,4-Hexadienoic acid $CH_3CH{=}CHCH{=}CHCO_2H$	112.13	nd (dil al)	228d, 153^{50}	134.5	1.204$^{19/4}$		al, eth	B2^4, 1701
13358	Sorbic acid, ethyl ester or Ethyl sorbate $CH_3CH{=}CHCH{=}CHCO_2C_2H_5$	140.18		195.5, 85^{20}		0.9506$^{20/4}$	1.4951^{20}	al, eth, chl	B2^4, 1703
13359	Sorbic acid, methyl ester or Methyl sorbate $CH_3CH{=}CHCH{=}CHCO_2CH_3$	126.16	lf	180, 70^{20}	15	0.9777$^{20/4}$	1.5025^{22}	al, eth	B2^4, 1703
13360	Sorbylalcohol or 2,4-Hexadien-1-ol $CH_3CH{=}CHCH{=}CHCH_2OH$	98.14	nd	76^{12}	30-1	0.8967$^{20/4}$	1.4981^{20}	al, eth	B1^4, 2240
13361	Sorbyl chloride $CH_3CH{=}CHCH{=}CHCOCl$	130.57		78^{15}		1.0666$^{19/4}$	1.5545^{20}	ace	B2^3, 1460
13362	Sorbierite or L-Iditol $HOCH_2(CHOH)_4CH_2OH$	182.17	pr (al), $[\alpha]^{/}{}_D$ −3.5 (w,p=2)		73-4			w	B1^4, 2843
13363	D-Sorbitol or D-Gulcitol $C_6H_{14}O_6$, $HOCH_2(CHOH)_4CH_2OH$	182.17	nd (w + 0.5), $[\alpha]^{25/}{}_D$ −1.98 (w)	295$^{3.5}$	110-2 (anh), 75 (hyd)	1.489$^{20/4}$	1.3330^{20}	w, ace, aa	B1^4, 2839
13364	L-Sorbitol or D-Gulitol $C_6H_{14}O_6$,$HOCH_2(CH_2OH)_4CH_2OH$	182.17	nd (w + 0.5), $[\alpha]$ + 1.7 (w)		77			w, ace, aa	B1^3, 2405
13365	D-Sorbitol, hexaacetate $C_{18}H_{26}O_{12}$	434.40	pr (w), $[\alpha]^{18/}{}_D$ + 6.8 (aa)		99.5 (120)			al	B2^4, 275
13366	D-Sorbose $C_6H_{12}O_6$	180.16	rh (al), $[\alpha]^{20/}{}_D$ + 42.9 (w)		165	1.612^{17}		w	B1^4, 4411
13367	DL-Sorbose or β-Acrose $C_6H_{12}O_6$	180.16	rh (dil al)		162-3	1.638^{17}		w, MeOH	B31, 348
13368	L-Sorbose $C_6H_{12}O_6$	180.16	rh (al), $[\alpha]^{20/}{}_D$ −43.2 (w,c=5)		165	1.612^{17}		w	B1^4, 4412
13369	Sparteine *(D)* or Lupinidine $C_{15}H_{26}N_2$	234.38	$[\alpha]^{20/}{}_D$ −19.5 (al)	325^{754}, 173^8	30-1	1.0196$^{20/4}$	1.5312^{20}	al, eth, chl	B23^2, 97
13370	Sparteine sulfate, pentahydrate $C_{15}H_{26}N_2 \cdot H_2SO_4 \cdot 5H_2O$	422.53	pr $[\alpha]^{17/}{}_D$ −15.3		242		1.5289	al	B23^2, 99
13371	Spiro [5,5]-hendecane-2,4,8,10-tetraoxa or Pentacrythritol dimethylene ether $C_7H_{12}O_4$	160.17		147^{53}	46.5			w, al, eth, ace, bz	B19^4, 5650
13372	Spiropentane C_5H_8	68.12		39-40^{746}		0.7266$^{20/4}$	1.4120^{20}		B5^4, 218
13373	Squalene or Spinacene $C_{30}H_{50}$	410.73		280^{17}	<−20	0.8584$^{20/4}$	1.4990^{20}	eth, ace	B1^4, 1146
13374	Squalene, perhydro or 2,6,10,15,19,23-Hexamethyltetracosane $C_{30}H_{62}$	422.82	oil	*ca* 350, 263^{10}	−38	0.8125$^{15/4}$	1.4525^{20}	eth, bz, peth, chl	B1^3, 585

PHYSICAL CONSTANTS OF ORGANIC COMPOUNDS (Continued)

No.	Name, Synonyms, and Formula	Mol. wt.	Color, crystalline form, specific rotation and λ_{max} (log ε)	b.p. °C	m.p. °C	Density	n_D	Solubility	Ref.
13375	Stachydrine *(L)* or Hygric acid methylbetaine $C_7H_{13}NO_2$	143.19	cr (w + 1), $[\alpha]^{25/}_D$ −40.2 (w,c=4)		235d (anh), 116-8 (+ w)			w, al	$B22^4$, 26
13376	Stachydrine-*(L)*-oxalate $C_7H_{13}NO_2.H_2C_2O_4$	233.22	nd		105-7				$B22^1$, 484
13377	Starch or Amylum $(C_6H_{10}O_5)n$		amor pw		d				C51, 11746
13378	Starch triacetate $C_{216}H_{288}O_{144}$	5188.6	pw $[\alpha]^{25/}_D$ + 72.5		180d			ace, bz	C46, 8401
13379	Stearamide $CH_3(CH_2)_{16}CONH_2$	283.50	lf (al)	250-1^{12}	109			eth, chl	$B2^4$, 1240
13380	Stearamide, *N*-phenyl or Stearanilide $C_{17}H_{35}CONHC_6H_5$	359.60	nd (al)	153.5^{10}	94			al, eth, ace, bz, chl	$B12^3$, 486
13381	Stearic acid or Octadecanoic acid $CH_3(CH_2)_{16}CO_2H$	284.48	mcl lf (al)	360d, 232^{15}	71.2	0.9408$^{20/4}$	1.4299^{20}	eth, ace, chl	$B2^4$, 1206
13382	Stearic acid anhydride or Stearic anhydride $[C_{17}H_{35}CO]_2O$	550.95			72	0.8365$^{82/4}$	1.4362^{80}		$B2^4$, 1239
13383	Stearic acid, benzyl ester or Benzyl stearate $C_{17}H_{35}CO_2CH_2C_6H_5$	374.61	pa ye		28 (45)	0.9075$^{50/25}$	1.4663^{50}		$B6^3$, 1481
13384	Stearic acid, butyl ester $C_{17}H_{35}CO_2C_4H_9$	340.59		343,223^{15}	27.5	0.855$^{20/4}$	1.4328^{50}	al, ace	$B2^4$, 1219
13385	Stearic acid, isobutyl ester $C_{17}H_{35}CO_2\text{-}i\text{-}C_4H_9$	340.59	wax	199^5	(i)22.5 (ii)28.9	0.8498$^{20/4}$		eth	$B2^3$, 1017
13386	Stearic acid, cyclohexyl ester or Cyclohexyl stearate $C_{17}H_{35}CO_2C_6H_{11}$	366.63			44	0.890$^{15/15}$		eth	$B6^4$, 38
13387	Stearic acid, 9,10-dibromo *(trans)* $CH_3(CH_2)_7CHBrCHBr(CH_2)_7CO_2H$	442.27	col or ye		29-30	1.2458$^{10/4}$	1.4893^{42}	eth	$B2^3$, 1048
13388	Stearic acid, 2,3-dihydroxy $C_{15}H_{31}CHOHCHOHCO_2H$	316.48	nd (aa)		α: 107 β: 126				$B3^4$, 1089
11388a	Stearic acid, 9,10-dihydroxy $C_8H_{17}CH(OH)CH(OH)(CH_2)_7CO_2H$	316.48	lf		95				$B2^4$, 1218
13389	Stearic acid, ethyl ester or Ethyl stearate $C_{17}H_{35}CO_2C_2H_5$	312.54		199^{10}	31-3	1.057$^{20/4}$	1.4349^{40}	al, eth, ace	$B2^4$, 1249
13390	Stearic acid, 9,10,12,13,15,16-hexabromo $CH_3[CH_2CHBrCHBr]_3(CH_2)_7CO_2H$	757.86	cr (diox)		182				$B2^4$, 1220
13391	Stearic acid, hexadecyl ester or Cetyl stearate $C_{17}H_{35}CO_2C_{16}H_{33}$	508.91	lf or pl (eth, aa)		57		1.4410^{70}	eth, ace, chl	$B3^4$, 934
13392	Stearic acid, 2-hydroxy $C_{16}H_{33}CH(OH)CO_2H$	300.48	flat nd (chl, AcOEt)		93			al, eth, ace, bz, chl	$B3^4$, 935
13393	Stearic acid, 3-hydroxy $C_{15}H_{31}CH(OH)CH_2CO_2H$	300.48	pl (chl)		89-90			al, eth, chl	$B3^4$, 940
13394	Stearic acid, 10-hydroxy $C_8H_{17}CHOH(CH_2)_8CO_2H$	300.48	pl		84			al	$B3^4$, 941
13395	Stearic acid, 11-hydroxy $C_7H_{15}CH(OH)(CH_2)_9CO_2H$	300.48	ta or pl (al)		81-2				$B3^4$, 942
13396	Stearic Acid, 12-hydroxy $C_6H_{13}CH(OH)(CH_2)_{10}CO_2H$	300.48	(al)		82			al, eth, chl	$B2^4$, 1222
13397	Stearic acid, 2-hydroxyethyl ester or Glycolmonostearate $C_{17}H_{35}CO_2CH_2CH_2OH$	328.54	(peth)	189-91^3	60-1	0.8780$^{60/4}$	1.4310^{60}		$B2^4$, 1216
13398	Stearic acid, methyl ester or Methylstearate $C_{17}H_{35}CO_2CH_3$	298.51		442-3^{747}, 215^{15}	39.1	0.8498$^{40/4}$	1.4367^{40}	al, eth, ace, chl	$B2^4$, 1272
13399	Stearic acid, 10-methyl *(D)* or Tuberculostearic acid $C_8H_{17}CH(CH_3)(CH_2)_8CO_2H$	298.51	$[\alpha]^{22/}_D$ −0.11 (ace)	175-8$^{0.7}$	12-3	0.8771$^{25/4}$	1.4512		$B3^4$, 1093
13401	Stearic acid, 9,10-dioxo or Stearoxylic acid $CH_3(CH_2)_7COCO(CH_2)_7CO_2H$	312.45	ye pl (al)		86			al, eth, ace, lig	$B3^4$, 1796
13402	Stearic acid, 9,10,12,13,15,16-hexabromo or α-Linolenic acid hexabromide $CH_3[CH_2CHBrCHBr]_3(CH_2)_7CO_2H$	757.86	cr (diox)		182				$B2^4$, 1249
13403	Stearic acid, 3-oxo or Palmitoyl acetic acid $C_{15}H_{31}COCH_2CO_2H$	298.47			102-3			al	$B3^4$, 1687
13404	Stearic acid, 3-oxo, ethyl ester $C_{15}H_{31}COCH_2CO_2C_2H_5$	326.52	(al or peth)		37-8			al, peth	$B3^4$, 1687
13405	Stearic acid, 6-oxo or Lactaric acid $C_{12}H_{25}CO(CH_2)_4CO_2H$	298.47	pl (al, peth)		87			eth, chl	$B3^4$, 1688

No.	Name, Synonyms, and Formula	Mol. wt.	Color, crystalline form, specific rotation and λ_{max} (log ε)	b.p. °C	m.p. °C	Density	n_D	Solubility	Ref.
13406	Stearic acid, 6-oxo, ethyl ester $C_{12}H_{25}CO(CH_2)_4CO_2C_2H_5$	326.52	cr	...	47	...	...	al, eth	B3^{1}, 253
13407	Stearic acid, 10-oxo $C_8H_{17}CO(CH_2)_8CO_2H$	298.47	pl (al)	...	76 (82)	...	...	...	B3^{4}, 1689
13408	Stearic acid, 10-oxo, ethyl ester $C_8H_{17}CO(CH_2)_8CO_2C_2H_5$	326.53	pl (al)	...	41	...	...	al	B3, 725
13409	Stearic acid, 12-oxo $C_6H_{13}CO(CH_2)_{10}CO_2H$	298.47	lf (aa), cr (lig)	...	82	...	...	al	B3^{4}, 1690
13410	Stearic acid, 12-oxo, ethyl ester $C_6H_{13}CO(CH_2)_{10}CO_2C_2H_5$	326.52	lf (al)	$199\text{-}200^{3}$	38	...	...	al	B3^{3}, 1294
13411	Stearic acid, pentyl ester or Pentylo, stearate Amylstearate $C_{17}H_{35}CO_2C_5H_{11}$	354.62	pl	...	30	...	1.4342^{50}	al, eth	B2^{4}, 1220
13412	Stearic acid, isopentyl ester or Isoamyl stearate $C_{17}H_{35}CO_2CH_2CH_2CH(CH_3)_2$	354.62	...	192^{2}	25.5	$0.855^{20/4}$	1.433^{50}	eth, ace	B2^{3}, 1017
13413	Stearic acid, phenyl ester or Phenyl stearate $C_{17}H_{35}CO_2C_6H_5$	360.59	...	267^{15}	51-3	...	...	al, eth	B6^{4}, 618
13414	Stearic acid, propyl ester or Propyl stearate $C_{17}H_{35}CO_2C_3H_7$	326.56	cr (eth), pr (peth)	186.8^{2}	fr 28.9	0.8452^{38}	1.4400^{30}	al, eth, ace	B2^{4}, 1219
13415	Stearic acid, isopropyl ester or Isopropyl stearate $C_{17}H_{35}CO_2CH(CH_3)_2$	326.56	...	207^{6}	28	0.8403^{38}	...	al, eth, ace, chl	B2^{4}, 1219
13416	Stearic acid, 9,10,11,12-tetrabromo or Linoleic acid tetrabromide $C_5H_{11}[CHBrCHBrCH_2]_2(CH_2)_6CO_2H$	600.07	pl or lf (aa)	...	114-5	...	...	al, eth, bz, chl, aa	B2^{4}, 1247
13417	Stearic acid, 9,10,12,13-tetrabromo, ethyl ester $C_5H_{11}[CHBrCHBrCH_2]_2(CH_2)_6CO_2C_2H_5$	628.12	nd	...	63	...	...	...	B2^{3}, 1049
13418	Stearic acid, 9,10,12,13-tetrabromo, methyl ester $C_5H_{11}[CHBrCHBrCH_2]_2(CH_2)_6CO_2CH_3$	614.09	lf	215^{15}	63	...	1.4346^{45}	al, eth, chl	B2^{4}, 1248
13419	Stearic acid, tetrahydrofurfuryl ester $C_{17}H_{35}CO_2CH_2(C_4H_70)$	368.60	...	...	22	$0.917^{25/25}$	...	al, eth	Am50, 134
13420	Stearolic acid or 9-Octadecynoic acid $CH_3(CH_2)_7C{\equiv}C(CH_2)_7CO_2H$	280.45	pr (al, peth), nd (dil al)	260, $189\text{-}90^{1.8}$	48	1.4510^{54}	...	eth	B2^{4}, 1751
13421	Stearolic acid, ethyl ester or Ethylstearolate $CH_3(CH_2)_7C{\equiv}C(CH_2)_7CO_2C_2H_5$	308.50	...	$180^{2.5}$	...	...	1.4555^{20}	...	B2^{4}, 1752
13422	Stearolic acid, methyl ester or Methylstearolate $CH_3(CH_2)_7C{\equiv}C(CH_2)_7CO_2CH_3$	294.48	...	175^{3}	...	...	1.4562^{20}	...	B2^{4}, 1751
13423	Stearone or 18-Pentatriacontanone $(C_{17}H_{35})_2CO$	506.94	lf (lig)	...	88.4	$0.793^{95/4}$	...	...	B1^{4}, 3413
13424	Stearonitrile $C_{17}H_{35}CN$	265.48	...	362, 193^{10}	41	$0.8325^{20/4}$	1.4389^{45}	al, eth, ace, chl	B2^{4}, 1242
13425	Stearophenone $C_{17}H_{35}COC_6H_5$	344.58	...	...	56	...	...	ace	B7^{3}, 1317
13426	Stearoxylic acid or 9,10-Dioxostearic acid $C_8H_{17}COCO(CH_2)_7CO_2H$	312.45	yr pl (al)	...	86	...	...	al, eth, ace, lig	B3^{4}, 1796
13427	Stearoyl chloride $C_{17}H_{35}COCl$	302.93	...	215^{15}	23	...	1.4523^{24}	al	B2^{4}, 1240
13428	Stearyl alcohol or 1-Octadecanol $CH_3(CH_2)_{16}CH_2OH$	270.50	lf (al)	210.5^{15}	59-60	$0.8124^{59/4}$	...	al, eth, chl	B1^{4}, 1888
13429	Stearyl sulfate or Octadecyl sulfate $(C_{18}H_{37}O)_2SO_2$	603.04	...	...	70.5	...	...	...	B1^{4}, 1892
13430	Stibine, bis (trifluoromethyl)-bromo $(F_3C)_2SbBr$	339.67	...	113	...	...	...	...	J1957, 3708
13431	Stibene, bis (trifluoromethyl)-chloro $(F_3C)_2SbCl$	295.22	...	*ca* 88, 17^{20}	...	...	...	...	J1957, 3708
13432	Stibene, bis(trifluoromethyl)-iodo $(F_3C)_2SbI$	386.67	...	*ca* 129, 16^{8}	−42	...	...	...	J1957, 3708
13433	Stibene, dibromo-trifluoromethyl CF_3SbBr_2	350.56	...	*ca* 157, $34^{2.5}$	...	...	...	...	J1957, 3708
13434	Stibene, dichloro-phenyl $C_6H_5SbCl_2$	269.76	nd or ta	$110\text{-}5^{10}$	69-70	...	...	al, ace, aa	B16^{3}, 1167
13435	Stibene, diido-trifluoromethyl F_3CSbI_2	444.57	bt ye	>200d	4-8	...	...	...	J1957, 3708
13436	Stibene, triethyl $(C_2H_5)_3Sb$	208.94	...	161.4	−98	1.3224^{15}	...	al, eth	B4^{3}, 1834
13437	Stibene, trimethyl $(CH_3)_3Sb$	166.85	...	80.6	−62	1.523^{15}	1.42^{15}	al, eth	B4^{3}, 1834

No.	Name, Synonyms, and Formula	Mol. wt.	Color, crystalline form, specific rotation and λ_{max} (log ε)	b.p. °C	m.p. °C	Density	n_D	Solubility	Ref.
13438	Stibene, triphenyl $(C_6H_5)_3Sb$	353.07	pr (peth)	>360, >220^1	53.5	$1.4343^{25/4}$	1.6948^{42}	eth, ace, bz, chl, aa	B16^3, 1159
13439	Stibene, tris(trifluoromethyl) $(CF_3)_3Sb$	328.77		73	−58				J1957, 3708
13440	Stibene, tri-(2-tolyl) $(2\text{-}CH_3C_6H_4)_3Sb$	395.15	(al)		102			eth, bz, chl, peth	B16, 892
13441	Stibene, tri-(3-tolyl) $(3\text{-}CH_3C_6H_4)_3Sb$	395.15	(peth)		72	$1.3957^{16/4}$		al, eth, bz, chl, aa	B16, 892
13442	Stibene, tri(4-tolyl) $(4\text{-}CH_3C_6H_4)_3Sb$	395.15	rh (eth, MeOH)		127-8	1.3595^{16}		eth, bz, chl	B16^3, 1161
13443	Stigmastanol or β-Sitostanol $C_{29}H_{52}O$	416.74	pl (al), cr (+1 w), $[\alpha]^{20}{}_D$ +24.8 (chl, c=1.1)		144.5			chl	B6^3, 2172
13444	Stigmasterol $C_{29}H_{48}O$	412.70	cr (al, +1 w), $[\alpha]^{22}{}_D$ −51 (chl,c=2)		170			eth, ace, bz, chl	B6^3, 2857
13445	2-Stilbazole *(cis)* $C_{13}H_{11}N$	181.24		141^{10}	−50			al, eth, bz	J1958, 2202
13446	2-Stilbazole *(trans)* $C_{13}H_{11}N$	181.24		324-5^{750}, 194^{14}	91-2			al, eth, bz, lig	B20^4, 3874
13447	4-Stilbazole-*(trans)* $C_{13}H_{11}N$	181.24		208^{35}	131			al, eth, chl	B20^4, 3880
13448	Stilbene-*(cis)* $C_6H_5CH{=}CHC_6H_5$	180.25		141^{12}	5-6	$1.0143^{20/4}$	1.6130^{20}	al, eth, ace, bz, chl	B5^3, 1958
13449	Stilbene *(trans)* or *trans*-1,2-Diphenylthylene $C_6H_5CH{=}CHC_6H_5$	180.25	cr (al)	305^{720}, 166-7^{12}	124-5	0.9707	1.6264^{17}	eth, bz	B5^3, 1953
13450	Stilbene, 2-bromo *(cis)* $2\text{-}BrC_6H_4CH{=}CHC_6H_5$	259.15		$121^{0.5}$			1.6404^{25}	al, lig	Am78, 475
13451	Stilbene, 2-bromo *(trans)* $2\text{-}BrC_6H_4CH{=}CHC_6H_5$	259.15		$145^{0.5}$	34		1.6822^{25}	al, lig	B5^3, 1964
13452	Stilbene, α-chloro *(cis)* $C_6H_5CCl{=}CHC_6H_5$	214.69		160-2^{12}			1.6281^{19}	al, eth, bz, chl	B5^3, 1961
13453	Stilbene, α-chloro-*(trans)* $C_6H_5CCl{=}CHC_6H_5$	214.69		320-4	53-4			al, eth, bz, chl	B5^3, 1961
13454	Stilbene, 2-chloro *(trans)* $2\text{-}ClC_6H_4CH{=}CHC_6H_5$	214.69		208-10^{30}, 138-40^2	39-40			al, chl	B5^3, 1961
13455	Stilbene, 3-chloro *(trans)* $3\text{-}ClC_6H_4CH{=}CHC_6H_5$	214.69		175-$80^{0.2}$	73-4			al, chl	B5^3, 1961
13456	Stilbene, 2,2′-diamino *(cis)* $2\text{-}H_2NC_6H_4CH{=}CHC_6H_4NH_2\text{-}2$	210.28	red nd (w)		123			ace	B13^3, 510
13457	Stilbene, 2,2′-diamino *(trans)* $2\text{-}H_2NC_6H_4CH{=}CH(C_6H_4NH_2\text{-}2)$	210.28	gold-ye pr (al)		176			ace	B13^3, 510
13458	Stilbene, 4,4′-diamino (trans) $4\text{-}H_2NC_6H_4CH{=}CH(C_6H_4NH_2\text{-}4)$	210.28	ye nd or lf (al)	sub	231			MeOH	B13^3, 513
13459	Stilbene, 2,2′-dibromo *(trans)* $2\text{-}BrC_6H_4CH{=}CH(C_6H_4Br\text{-}2)$	338.04			215-6			al, bz	B5^3, 1965
13460	Stilbene, α,β-dichloro *(cis)* $C_6H_5CCl{=}CClC_6H_5$	249.14		180^{18}	67.5-8			al, eth	B5^3, 1963
13461	Stilbene, α β-dichloro-*(trans)* $C_6H_5CCl{=}CClC_6H_5$	249.14		180^{18}	143-4			al, eth	B5^3, 1963
13462	Stilbene, 2,2′-dihydroxy (α-form) $2\text{-}HOC_6H_4CH{=}CH(C_6H_4OH\text{-}2)$	212.25	nd (al)		95			al, eth	B6, 1022
13463	Stilbene, 2,2′-dihydroxy-(β form) $2\text{-}HOC_6H_4CH{=}CH(C_6H_4OH\text{-}2)$	212.25	flat nd (al)		197			eth, bz	B6^3, 5574
13464	Stilbene, 3,5-dihydroxy *(trans)* or Pinosylvin $3,5\text{-}(HO)_2C_6H_3CH{=}CHC_6H_5$	212.25	nd (aa)		156			ace, bz, chl, aa	B6^3, 5577
13465	Stilbene, 4,4′-dihydroxy *(trans)* or Stilbestrol $4\text{-}HOC_6H_4CH{=}CH(C_6H_4OH\text{-}4)$	212.25	nd (aa), ta (al)		284			ace, bz	B6^3, 5581
13466	Stilbene, 4,4′-dimethoxy or Bianisal $4\text{-}CH_3OC_6H_4CH{=}CH(C_6H_4OCH_3\text{-}4)$	240.30	lf (bz, aa)	sub	214-5			ace	B6^3, 5582
13467	Stilbene, α,β-dimethyl *(cis)* $C_6H_5C(CH_3){=}C(CH_3)C_6H_5$	208.30			67-8	$0.9537^{78/4}$	1.5612^{78}	al	B5^3, 2009

No.	Name, Synonyms, and Formula	Mol. wt.	Color, crystalline form, specific rotation and λ_{max} (log ε)	b.p. °C	m.p. °C	Density	n_D	Solubility	Ref.
13468	Stilbene, α, β -dimethyl *(trans)* $C_6H_5C(CH_3)=C(CH_3)C_6H_5$	208.30	...	...	107	0.987^{20}	1.6173^{20}	eth	$B5^3$, 2010
13469	Stilbene, α, β -dinitro *(cis)* $C_6H_5C(NO_2)=C(NO_2)C_6H_5$	270.24	ye pym pr (al)	d	108-9	...	...	al, eth, ace, bz, chl	$B5^3$, 1973
13470	Stilbene, α, β dinitro-*(trans)* $C_6H_5C(NO_2)=C(NO_2)C_6H_5$	270.24	pa ye nd or pr (al)	...	187-8	...	...	ace, bz, chl, aa	$B5^3$, 1973
13471	Stilbene, 2,2′-dinitro *(cis)* $2\text{-}O_2NC_6H_4CH=CH(C_6H_4NO_2\text{-}2)$	270.24	ye nd (aa)	...	126	...	...	...	B5, 637
13472	Stilbene, 2,2′-dinitro *(trans)* $2\text{-}O_2NC_6H_4CH=CH(C_6H_4NO_2\text{-}2)$	270.24	pa ye nd (chl)	420 exp	199	...	...	...	$B5^3$, 1970
13473	Stilbene, 2,4-dinitro *(cis)* $2,4\text{-}(NO_2)_2C_6H_3CH=CHC_6H_5$	270.24	ye pl (aa)	...	127	...	...	bz, chl, aa	$B5^2$, 541
13474	Stilbene, 2,4-dinitro *(trans)* $2,4\text{-}(NO_2)_2C_6H_3CH=CHC_6H_5$	270.24	pa ye (aa)	412 exp	143-5	...	...	chl, xyl	$B5^3$, 1970
13475	Stilbene, 2,6-dinitro *(trans)* $2,6\text{-}(NO_2)_2C_6H_3CH=CHC_6H_5$	270.24	ye nd (bz, aa)	...	114	...	...	chl	$B5^3$, 1970
13476	Stilbene, 3,4′-dinitro *(cis)* $3\text{-}O_2NC_6H_4CH=CH(C_6H_4NO_2\text{-}4)$	270.24	ye nd (aa)	...	155	...	...	ace, bz, chl	$B5^2$, 541
13477	Stilbene, 3,4′-dinitro *(trans)* $3\text{-}O_2NC_6H_4CH=CH(C_6H_4NO_2\text{-}4)$	270.24	ye nd (aa, Py)	...	220-2	...	...	...	$B5^3$, 1971
13478	Stilbene, 4,4′-dinitro *(cis)* $4\text{-}O_2NC_6H_4CH=CH(C_6H_4NO_2\text{-}4)$	270.25	ye nd (aa, chl)	...	186	...	...	bz, chl	$B5^3$, 1972
13479	Stilbene, 4,4′-dinitro *(trans)* $4\text{-}O_2NC_6H_4CH=CH(C_6H_4NO_2\text{-}4)$	270.25	pa ye lf (aa), nd (aa)	...	303-4	...	...	al, eth, ace, bz, chl	$B5^3$, 1971
13480	Stilbene, α-ethyl $C_6H_5C(C_2H_5)=CHC_6H_5$	208.30	...	296-7	57	...	...	...	$B5^3$, 2009
13481	Stilbene, 2-methoxy *(trans)* $2\text{-}CH_3OC_6H_4CH=CHC_6H_5$	210.28		...	59	...	...	...	$B6^3$, 3494
13482	Stilbene, 3-methoxy *(trans)* $3\text{-}CH_3OC_6H_4CH=CHC_6H_5$	210.28	...	$173\text{-}4^{4}$	34	...	...	al, eth, ace	$B6^3$, 3497
13483	Stilbene, 4-methoxy *(trans)* $4\text{-}CH_3OC_6H_4CH=CHC_6H_5$	210.28	...	142.5^{15}	136-6.5	...	...	al, eth, ace, bz	$B6^3$, 3498
13484	Stilbene, α -methyl $C_6H_5CH=C(CH_3)C_6H_5$	194.28	...	$195\text{-}200^{45}$	82-3	$0.9565^{100/4}$	1.5635^{17}	al, eth, bz	$B5^3$, 1995
13485	Stilbene, 2-nitro *(cis)* $2\text{-}O_2NC_6H_4CH=CHC_6H_5$	225.25	...	187^{11}	62.5-3.5	...	...	al, eth	$B5^3$, 1967
13486	Stilbene, 2-nitro *(trans)* $2\text{-}O_2NC_6H_4CH=CHC_6H_5$	225.25	...	209^{11}	73	...	...	al, eth	$B5^3$, 1966
13487	Strophanthiden or Corchorgenin $C_{23}H_{32}O_6$	404.50	orh ta (MeOH-w), lf (w + 2)	...	235 (anh), 171-5 (+ w)	...	...	al, ace, bz, chl, aa	$B18^4$, 3127
13488	g-Strophanthin or Quabain $C_{29}H_{44}O_{12}$	584.66	hyg pl (+ aw), $[\alpha]^{25/}_{D}$ −34	...	200	...	...	al	$B18^4$, 3554
13489	Strychnine $C_{21}H_{22}N_2O_2$	334.42	orh pr (al), $[\alpha]^{18/}_{D}$ −139.3 (chl, c=1)	270^{5}	286-8	$1.36^{20/4}$	...	chl	$B27^2$, 723
13490	Strychnine hydrochloride $C_{21}H_{22}N_2O_2.HCl.2H_2O$	406.91	nd (w), $[\alpha]_D$ −28.3 (w, c=0.7)	...	...	...	...	w	$B27^2$, 730
13491	Strychnine nitrate $C_{21}H_{22}N_2O_2.HNO_3$	397.44	nd (w)	...	280-310	1.627	...	w, MeOH	$B27^2$, 732
13492	Strychnine sulfate $(C_{21}H_{22}N_2O_2)_2.H_2SO_4.5H_2O$	856.99	wh mcl $[\alpha]_D$ + 13.2	...	200d	...	...	w, MeOH	$B27^2$, 731
13493	Styracin or Cinnamyl cinnamate $C_6H_5CH=CHCO_2(CH_2CH=CHC_6H_5)$	264.32	nd	...	44	1.1565^{4}	...	al, eth	$B9^3$, 2693
13494	Styracitol or 1,5-anhydro-D-mannitol $C_6H_{12}O_5$	164.16	pr (90% al), $[\alpha]^{20/}_{D}$−49.9, −71.7 (w)	...	157	...	...	w	$B17^4$, 2579
13495	Styrene or Vinyl benzene $C_6H_5CH=CH_2$	104.15	...	145.2, 33.6^{10}	−30.6	$0.9060^{20/4}$	1.5468^{20}	al, eth, ace, bz, peth	$B5^4$, 1334

PHYSICAL CONSTANTS OF ORGANIC COMPOUNDS (Continued)

No.	Name, Synonyms, and Formula	Mol. wt.	Color, crystalline form, specific rotation and λ_{max} (log ε)	b.p. °C	m.p. °C	Density	n_D	Solubility	Ref.
13496	Styrene, 2-amino or 2-Vinyl aniline $2\text{-}H_2NC_6H_4CH{=}CH_2$	119.17	....	112^{20}	......	$1.0181^{20/4}$	1.6124^{20}	ace, bz	$B12^3$, 2785
13497	Styrene, 3-amino or 3-Vinyl aniline $3\text{-}H_2NC_6H_4CH{=}CH_2$	119.17	....	$112\text{-}5^{12}$	......	$1.0216^{20/20}$	1.6069^{26}	ace, bz	$B12^3$, 2786
13498	Styrene, 4-amino or 4-Vinyl aniline $4\text{-}H_2NC_6H_4CH{=}CH_2$	119.17	....	116^{9}	23.5	$1.012^{20/20}$	1.6250^{22}	ace, bz	$B12^3$, 2786
13499	Styrene, α-bromo $C_6H_5CBr{=}CH_2$	183.05	....	$86\text{-}7^{14}$	−44	1.4025^{23}	1.5881^{20}	....	$B5^4$, 1349
13500	Styrone, β-bromo-(*trans*) $C_6H_5CH{=}CHBr$	183.05	....	219d, 108^{20}	7	1.4269^{16}	1.6093^{20}	al, eth	$B5^4$, 1349
13501	Styrene, 2-bromo $2\text{-}BrC_6H_4CH{=}CH_2$	183.05	....	206.2, 98^{20}	fr−52.8	$1.4160^{20/4}$	1.5927^{20}	....	$B5^4$, 1349
13502	Styrene, 3-bromo $3\text{-}BrC_6H_4CH{=}CH_2$	183.05	....	$90\text{-}4^{20}$	......	$1.4059^{20/4}$	1.5933^{20}	....	$B5^3$, 1176
13503	Styrone, 4-bromo $4\text{-}BrC_6H_4CH{=}CH_2$	183.05	....	103^{20}, $50^{2.5}$	4.5	$1.3984^{20/4}$	1.5947^{20}	chl, aa	$B5^4$, 1349
13504	Styrene, α-chloro $C_6H_5CCl{=}CH_2$	138.60	....	199, 73^{16}	−23	$1.1016^{18/4}$	1.5612^{20}	al, eth	$B5^4$, 1345
13505	Styrene, β-chloro-(*trans*) $C_6H_5CH{=}CHCl$	138.60	....	199, 90^{18}	......	$1.1095^{18/4}$	1.5648^{20}	al, eth, ace	$B5^4$, 1346
13506	Styrene, 2-chloro $2\text{-}ClC_6H_4CH{=}CH_2$	138.60	....	188.7, 64.6^{10}	−63.1	1.1000^{20}	1.5649^{20}	al, eth, ace, **peth**, aa	$B5^4$, 1345
13507	Styrene, 3-chloro $3\text{-}ClC_6H_4CH{=}CH_2$	138.60	....	$62\text{-}3^{6}$	......	$1.1168^{20/4}$	1.5625^{20}	al, eth	$B5^4$, 1345
13508	Styrene, 4-chloro $4\text{-}ClC_6H_4CH{=}CH_2$	138.60	....	192, 66.3^{10}	−15.9	$1.0868^{20/4}$	1.5660^{20}	al, eth, **ace, bz, peth**	$B5^4$, 1345
13509	Styrene, ββ-dichloro $C_6H_5CH{=}CCl_2$	173.04	....	225, $85\text{-}6^{5}$	......	$1.2531^{20/4}$	1.5852^{20}	ace, chl	$B5^4$, 1348
13510	Styrene, 2,4-dimethyl $2,4\text{-}(CH_3)_2C_6H_3CH{=}CH_2$	132.20	....	$79\text{-}80^{12}$	......	$0.9022^{21.5}$	$1.5214^{21.5}$	....	$B5^4$, 1386
13511	Styrene, 3-ethyl $3\text{-}C_2H_5C_6H_4CH{=}CH_2$	132.20	....	190	−101	0.8945^{20}	1.5351^{20}	....	$B5^3$, 1217
13512	Styrene, 4-ethyl $4\text{-}C_2H_5C_6H_4CH{=}CH_2$	132.20	....	192.3	−49.7	0.8925^{20}	1.5376^{70}	....	$B5^4$, 1384
13513	Styrene, 2-fluoro $2\text{-}F\text{-}C_6H_4CH{=}CH_2$	122.14	....	46^{32}	......	1.0282^{20}	1.5200^{20}	al, eth, bz	$B5^4$, 1342
13514	Styrene, 3-fluoro $3\text{-}FC_6H_4CH{=}CH_2$	122.14	....	$30\text{-}1^{4}$	......	1.0177^{20}	1.5170^{20}	al, eth, bz	$B5^4$, 1343
13515	Styrene, 4-fluoro $4\text{-}FC_6H_4CH{=}CH_2$	122.14	....	67.4^{50}, $29\text{-}30^{4}$	−34.5	$1.0220^{20/4}$	1.5150^{20}	al, eth, bz	$B5^4$, 1343
13516	Styrene, α-methyl $C_6H_5C(CH_3){=}CH_2$	118.18	....	163-4, 60^{17}	24.3	$0.9082^{20/4}$	1.5303^{20}	eth, bz, chl	$B5^3$, 1192
13517	Styrene, 2-methyl or 2-Vinyl toluene $2\text{-}CH_3C_6H_4CH{=}CH_2$	118.18	....	171	......	$0.9106^{20/4}$	1.5450^{20}	bz, chl	$B5^4$, 1367
13518	Styrene, 3-methyl or 3-Vinyl toluene $3\text{-}CH_3C_6H_4CH{=}CH_2$	118.18	....	168, $61\text{-}2^{18}$	−70	$0.9028^{20/20}$	1.5410^{20}	al, eth, bz	$B5^4$, 1367
13519	Styrene, 4-methyl or 4-Vinyl toluene $4\text{-}CH_3C_6H_4CH{=}CH_2$	118.18	....	169, 63^{15}	−37.8	$0.8760^{20/4}$	1.5428^{20}	bz	$B5^4$, 1369
13520	Styrene, β-nitro-(*trans*) $C_6H_5CH{=}CHNO_2$	149.15	ye pr (peth, al)	250-60, 150^{14}	60	....	....	al, eth, ace, chl, peth	$B5^4$, 1352
13521	Styrene, 3-nitro $3\text{-}O_2NC_6H_4CH{=}CH_2$	149.15	....	$120\text{-}1^{11}$	−1	$1.1552^{32/4}$	1.5836^{20}	al, eth, bz, chl, lig	$B5^4$, 1351
13522	Styrene, 4-nitro $4\text{-}O_2NC_6H_4CH{=}CH_2$	149.15	pr (lig)	d	29	....	....	al, eth, chl, lig, aa	$B5^4$, 1351
13523	Styrene oxide or (1,2-Epoxy ethyl) benzene $C_6H_5\text{-}\underbrace{CH\text{-}CH_2}_{O}$	120.15	....	194.1, $84\text{-}5^{25}$	−35.6	$1.0523^{16/4}$	1.5342^{20}	al, eth, bz	$B17^4$, 398
13524	Styrene, 2,4,6-trimethyl or Vinyl mesitylene $2,4,6\text{-}(CH_3)_3C_6H_2CH{=}CH_2$	146.23	....	208-10, 83^{12}	......	$0.9057^{20/4}$	1.5296^{20}	....	$B5^4$, 1408
13525	β-Styrene sulfonyl chloride $C_6H_5CH{=}CHSO_2Cl$	202.66	fl (bz)	....	88	....	....	....	$B11^3$, 370
13526	Subboric acid, tetramethyl ester $(CH_3O)_2BB(OCH_3)_2$	145.76	....	93	−24.3	....	....	....	$B1^4$, 1273
13527	Suberaldehyde or Octanedial $OHC(CH_2)_6CHO$	142.20	....	230-40d, $96\text{-}8^{3}$	......	....	1.4439^{20}	w, al	$B1^4$, 3706
13528	Suberaldehyde oxime or Suberaldoxime $HON{=}C(CH_2)_6C{=}NOH$	172.23	pr (al)	....	155-6	....	....	al	$B1^4$, 3706
13529	Suberic acid or Octanedioic acid $HO_2C(CH_2)_6CO_2H$	174.20	lo nd or pl (w)	300 sub, 219.5^{10}	144	....	....	al	$B2^4$, 2028

No.	Name, Synonyms, and Formula	Mol. wt.	Color, crystalline form, specific rotation and λ_{max} (log ε)	b.p. °C	m.p. °C	Density	n_D	Solubility	Ref.
13530	Suberic acid, diethyl ester or Diethyl suberate $C_2H_5O_2C(CH_2)_6CO_2C_2H_5$	230.30	....	282-6, $140\text{-}1^{8}$	5.9	$0.9811^{20/4}$	1.4328^{20}	al, eth	$B2^4$, 2029
13531	Suberic acid, dimethyl ester $CH_3O_2C(CH_2)_6CO_2CH_3$	202.25	....	268, 120^{6}	−3.1	$1.0217^{20/4}$	1.4341^{20}	al, eth, ace	$B2^4$, 2029
13532	Succinaldehyde or Butanedial $OHCCH_2CH_2CHO$	86.09	....	169-70d, 58.5^{9}	......	$1.064^{20/4}$	1.4262^{18}	w, al, eth, ace, bz	$B1^4$, 3642
13533	Succinamicacid or Succinic acid monoamide $HO_2CCH_2CH_2CONH_2$	117.10	nd (w, ace)	....	156-8	....	....	w	$B2^4$, 1922
13534	Succinamic acid, N-phenyl or Succinanilic acid $HO_2CCH_2CH_2CONHC_6H_5$	194.19	nd (w)	....	148.5	....	....	al, eth	B12, 295
13535	Sucinamide $H_2NCOCH_2CH_2CONH_2$	116.12	orh nd (w)	125.5 (sub)	268-70d	....	....	....	$B2^4$, 1922
13536	Succinic acid or Butanedioic acid $HO_2CCH_2CH_2CO_2H$	118.09	tcl or mcl pr	235d	188	$1.572^{25/4}$	1.450	al, eth, ace	$B2^4$, 1908
13537	Succinic acid, acetyl,diethyl ester or Diethyl acetyl succinate $C_2H_5O_2CCH(COCH_3)CH_2CO_2C_2H_5$	216.23	....	254-6, 139^{12}	......	$1.081^{20/4}$	1.438^{16}	al, eth, bz	$B3^4$, 1826
13538	Succinic acid anhydride or Succinic anhydride $C_2H_4(CO)_2O$	100.07	nd (al), rh pym (chl)	261, 139^{15}	119.6	$1.2340^{20/4}$	....	al, chl	$B17^4$, 5820
13539	Sucinic acid, benzyl ester-(mono) $HO_2CCH_2CH_2CO_2CH_2C_6H_5$	208.21	cr (bz)	....	59	....	....	....	$B6^4$, 2271
13540	Succinic acid, benzyl-(dl) $C_6H_5CH_2CH(CO_2H)CH_2CO_2H$	208.21	nd or lf	....	163-4	....	....	al, eth	$B9^3$, 4300
13541	Succinic acid, bromo $HO_2CCHBrCH_2CO_2H$	196.99	(w)	....	161	2.073	....	w, al	$B2^4$, 1929
13542	Succinic acid, bromo,dimethyl ester $CH_3O_2CCHBrCH_2CO_2CH_3$	225.04	....	$132\text{-}6^{30}$	......	1.5094^{15}	....	....	$B2^4$, 1930
13543	Succinic acid, dibenzyl ester $C_6H_5CH_2O_2CCH_2CH_2CO_2CH_2C_6H_5$	298.34	....	245^{15}	49-50	1.256	1.596	al, eth, bz, chl	$B6^4$, 2271
13544	Succinic acid, 2,3-dibromo (d) $HO_2CCHBrCHBrCO_2H$	275.88	pl (aa-CCl_4), $[\alpha]^{24/}{}_D$ + 147.8 (AcOEt)	....	157-8	....	....	w, al, ace	$B2^3$, 1679
13545	Succinic acid, 2,3-dibromo(dl) $HO_2CCHBrCHBrCO_2H$	275.88	(w or AcOEt)	....	171	....	....	w, al, bz	$B2^4$, 1930
13546	Succinic acid, 2,3-dibromo(l) $HO_2CCHBrCHBrCO_2H$	275.88	nd (bz), $[\alpha]^{18/}{}_D$ −148 (AcOEt, c=5.8)	....	157-8d	....	....	w, al, ace	$B2^3$, 1679
13547	Succinic acid, 2,3-dibromo (*meso*) $HO_2CCHBrCHBrCO_2H$	275.88	....	sub 275	255 (sealed tube)	....	....	al, eth	$B2^4$, 1930
13548	Succinic acid, dibutyl ester or Dibutyl succinate $C_4H_9O_2CCH_2CH_2CO_2C_4H_9$	230.30	....	274.5, 145^{4}	−29.2	$0.9752^{20/4}$	1.4299^{20}	al, eth, bz	$B2^4$, 1916
13549	Succinic acid, di-*sec*-butyl ester sec-$C_4H_9O_2CCH_2CH_2CO_2$-*sec*-C_4H_9	230.30	....	256	......	$0.9735^{20/4}$	1.4238^{25}	al, eth, bz	$B2^3$, 1665
13550	Succinic acid, di(2-carboxy phenyl) ester or Diaspirin. Succinyl salicylic acid	358.30	nd (aa, al)	....	176-8	....	....	....	$B10^2$, 43
13551	Succinic acid, 2,3-dichloro(d) $HO_2CCHClCHClCO_2H$	186.98	mcl, $[\alpha]^{19/}{}_D$ + 3.6 (w, c=3)	....	168	1.820^{15}	....	w, eth, ace, chl	$B2^4$, 1928
13552	Succinic acid, 2,3-dichloro (dl) $HO_2CCHClCHClCO_2H$	186.98	pr (w, eth-peth)	....	175d	1.844^{15}	....	eth, ace, chl	$B2^4$, 1929
13553	Succinic acid, 2,3-dichloro-(meso) $HO_2CCHClCHClCO_2H$	186.98	hex pr (w)	....	221d	....	....	w, al, eth, ace, chl	$B2^4$, 1928
13554	Succinic acid, 2,3-dichloro,diethyl ester(dl) $C_2H_5O_2CCHClCHClCO_2C_2H_5$	243.09	....	132^{15}	......	$1.1963^{77/4}$	1.4512^{20}	eth	$B2^2$, 558
13555	Succinic acid, 2,3-dichloro,diethyl ester(*meso*) $C_2H_5O_2CCHClCHClCO_2C_2H_5$	243.09	nd (dil al)	$125.5^{12.5}$	63	$1.490^{99/4}$	1.4266^{65}	al, eth	$B2^2$, 558
13556	Succinic acid, diethyl ester or Diethyl succinate $C_2H_5O_2CCH_2CH_2CO_2C_2H_5$	174.20	....	216.5, 105^{15}	−20.6	$1.0402^{20/4}$	1.4198^{20}	**al, eth**, ace	$B2^4$, 1914
13557	Succinic acid, 2,3-dimethoxy,dimethyl ester $CH_3O_2CCH(OCH_3)CH(OCH_3)CO_2CH_3$	206.20	pr, $[\alpha]^{18/}{}_D$ + 79.9 (MeOH, c=2)	$130\text{-}2^{12}$	53-4	$1.1317^{60/4}$	1.4340^{20}	MeOH	$B3^3$, 1019
13558	Succinic acid, dimethyl ester or Dimethyl succinate $CH_3O_2CCH_2CH_2CO_2CH_3$	146.14	....	196.4, 80^{11}	19	$1.1198^{20/4}$	1.4197^{20}	al, eth, ace	$B2^4$, 1913

PHYSICAL CONSTANTS OF ORGANIC COMPOUNDS (Continued)

No.	Name, Synonyms, and Formula	Mol. wt.	Color, crystalline form, specific rotation and λ_{max} (log ε)	b.p. °C	m.p. °C	Density	n_D	Solubility	Ref.
13559	Succinic acid, diphenyl ester or Diphenyl succinate $C_6H_5O_2CCH_2CH_2CO_2C_6H_5$	270.28	lf (al)	222.5[15]	121			eth, ace, bz	B6[3], 605
13560	Succinic acid, dipropyl ester $C_3H_7O_2CCH_2CH_2CO_2C_3H_7$	202.25		250.8, 101.5[3]	−5.9	1.0020[20/4]	1.4250[20]	eth, ace, bz	B2[4], 1916
13561	Succinic acid, di-(4-tolyl)ester $(4\text{-}CH_3C_6H_4)O_2CCH_2CH_2CO_2(C_6H_4CH_3\text{-}4)$	298.34	nd or lf (al)		121			al, eth, ace, bz, aa	B6[2], 379
13562	Succinic acid, ethyl ester(mono) *dl* $HO_2CCH_2CH_2CO_2C_2H_5$	146.14	pr or nd	172[42], 119[3]	8	1.1466[20/4]	1.4327[20]	w, al, eth	B2[4], 1915
13563	Succinic acid, ethyl,methyl ester $C_2H_5O_2CCH_2CH_2CO_2CH_3$	160.17		208.2, 90-5[3]	<−20	1.076[20/4]		al, eth	B2, 609
13564	Succinic acid, ethyl-(d) $HO_2CCH(C_2H_5)CH_2CO_2H$	146.14	pr or nd, $[\alpha]^{18/}{}_D$ +20.6 (aa, c=3.7)	180-3	96	1.0017[20/4]		w, al, eth	B2[4], 1995
13565	Succinic acid, ethyl (l) $HO_2CCH(C_2H_5)CH_2CO_2H$	146.14	pr or nd, $[\alpha]^{24/}{}_D$ −20.8 (ace, c=4.6)	180-3	96	1.0018[20]		w, al, eth	B2[4], 1995
13566	Succinic acid, formyl,diethyl ester $C_2H_5O_2CCH(CHO)CH_2CO_2C_2H_5$	202.21		130-4[15]			1.4486[25]	**al, eth**	B3[4], 1819
13567	Succinic acid, 2-hydroxy-2-methyl (*dl*) $HO_2C(CH_3)(OH)CH_2CO_2H$	148.12	mcl pr	sub	123			w, al, ace	B3[4], 1149
13568	Succinic acid, 2-hydroxy-3-methyl (d) or d-Citramalic acid $HO_2CC(CH_3)CH(OH)CH_2CO_2H$	148.12	$[\alpha]^{14/}{}_D$ +34.7 (w)		108-9			w, al, ace	B3[4], 1151
13569	Succinic acid, mercapto (d) or d-Thiomalic acid $HO_2CCH(SH)CH_2CO_2H$	150.15	cr (AeOEt-bz), $[\alpha]^{17/}{}_D$ +64.4 (al)		154			w, al, ace	B3[4], 1130
13570	Succinic acid, mercapto (dl) $HO_2CCH(SH)CH_2CO_2H$	150.15	cr (eth)		151			w, al, eth, ace	B3[4], 1130
13571	Succinic acid, mercapto (l) $HO_2CCH()SH)CH_2CO_2H$	150.15	$[\alpha]^{17/}{}_D$ −64.8 (al)		152-3			w, al, ace	B3[2], 287
13572	Succinic acid, methyl ester (mono) $HO_2CCH_2CH_2CO_2CH_3$	132.12		121-3[4]	56			w	B2[4], 1913
13573	Succinic acid, methyl (dl) or Pyrotartaric acid $HO_2CCH(CH_3)CH_2CO_2H$	132.12	pr	d	115		1.4303	w, al, eth	B2[4], 1948
13574	Succinic acid, monochloride,monomethyl ester $CH_3O_2CCH_2CH_2COCl$	150.56		102[35], 53-4[1]			1.4412[20]		B2[4], 1921
13575	Succinic acid, oxo-diethyl ester or Diethyl oxaloacetate $C_2H_5O_2CCOCH_2CO_2C_2H_5$	188.18		131-2[24]		1.131[20/4]	1.4561[17]	**al, eth**, ace, **bz**	B3[4], 1809
13576	Succinic acid, 2-oxo-3-methyl,diethyl ester or Diethyl methyl oxaloacetate $C_2H_5O_2CCH(CH_3)COCO_2C_2H_5$	202.21		137-8[23], 75-8[2]		1.0970[20/4]	1.4313[20]	**al, eth**	B3[4], 1818
13577	Succinic acid, 2-oxo-3-phenyl-1-ethyl ester-4-nitrile or Ethyl phenyl cyanopyruvate $C_2H_5O_2CCOCH(C_6H_5)CN$	217.23	(eth-lig)	206[20]	120			al, chl	B10[3], 3957
13578	Succinic acid, phenyl (d) $HO_2CCH(C_6H_5)CH_2CO_2H$	194.19	pr (w), $[\alpha]^{16.5/}{}_D$ +148.3 (al, c=1.5)		173-4			al, eth, ace	B9[3], 4276
13579	Succinic acid, phenyl (dl) $HO_2CCH(C_6H_5)CH_2CO_2H$	194.19	lf or nd (w)	d	168			al, eth, ace	B9[3], 4276
13580	Succinic acid, phenyl (l) $HO_2CCH(C_6H_5)CH_2CO_2H$	194.19	$[\alpha]^{15/}{}_D$ −173.3 (ace)		173-4			al, eth, ace	B9[3], 4276
13581	Succinic acid, phenyl,anhydride (d) $C_{10}H_8O_3$	176.17	nd (bz-peth), $[\alpha]^{15/}{}_D$ +100.9 (bz)		83-4			al, bz, chl	B17[1], 259
13582	Succinic acid, phenyl,anhydride (dl) $C_{10}H_8O_3$	176.17	mcl pr or nd (eth)	204-6[22]	54			al, eth, ace, bz	B17[4], 6173
13583	Succinic acid, phenyl,anhydride (l) $C_{10}H_8O_3$	176.17	$[\alpha]^{14/}{}_D$ −100.9 (bz)		83-4			bz, chl	B17[4], 6173
13584	Succinic acid, (3-phenyl propenyl) $(C_6H_5CH_2CH{=}CH)CH(CO_2H)CH_2CO_2H$	234.25	lf (eth, bz)		112			eth, ace, bz	B9, 909
13585	Succinic acid, isopropylidene or Tetraconic acid $(CH_3)_2C{=}C(CO_2H)CH_2CO_2H$	158.15	tcl nd (eth)		160-1			al, eth	B2[4], 2251

No.	Name, Synonyms, and Formula	Mol. wt.	Color, crystalline form, specific rotation and λ_{max} (log ε)	b.p. °C	m.p. °C	Density	n_D	Solubility	Ref.
13586	Succinic acid, tetrahydroxy $HO_2CC(OH)_2C(OH)_2CO_2H$	182.09			114-5				B3^4, 1883
13587	Succinic acid, tetramethyl $HO_2CC(CH_3)_2C(CH_3)_2CO_2H$	174.20	tcl (60% MeOH, lig), mcl (eth, ace)	sub	200	1.30		al, bz, chl	B2^4, 2054
13588	Succinimide $C_4H_5NO_2$	99.09	pl (+ 1w, al), rh (ace)	287-8d	126-7	1.418		w	B21^4, 4539
13589	Succinimide, N-benzyl $C_{11}H_{11}NO_2$	189.21	nd (w), pr (al)	390-400	103			al, eth, bz, chl	B21^4, 4550
13590	Succinimide, N-bromo $C_4H_4BrNO_2$	177.99	cr (bz)	173.5d		2.098		ace, AcOEt	B21^4, 4575
13591	Succinimide, N-(2-bromophenyl) $C_{10}H_8BrNO_2$	254.08	(dil al)		91			al	B21^2, 304
13592	Succinimide, N-(3-bromophenyl) $C_{10}H_8BrNO_2$	254.08	nd (al)		118			al	B21^2, 304
13593	Succinimide, N-(4-bromophenyl) $C_{10}H_8BrNO_2$	254.08	pr (al)		172			ace, bz	B21^4, 4547
13594	Succinimide, N-chloro $C_4H_4ClNO_2$	133.53	plCCl_4)		150	1.65		ace, aa	B21^4, 4575
13595	Succinimide, N-chloromethyl $C_5H_6ClNO_2$	147.56		158-60^{12}	58			w, ace, bz	B21^2, 305
13596	Succinimide, N-(4-chlorophenyl) $C_{10}H_8ClNO_2$	209.63	nd (dil al)		170			al, ace, bz	B21^4, 4547
13597	Succinimide, N-(ethoxymethyl) $C_7H_{11}NO_3$	157.17	nd	262, 151-2^{14}	31-2			w, al	B21^2, 304
13598	Succinimide, N-(4-ethoxyphenyl) $C_{12}H_{13}NO_3$	219.24	pr(al)		155				B21^4, 4554
13599	Succinimide, N-ethyl $C_6H_9NO_2$	127.14	(eth)	236	26			w, al, eth	B21^4, 4544
13600	Succinimide, N-(hydroxymethyl) $C_5H_7NO_3$	129.12	lf (bz)		66			w	B21^2, 304
13601	Succinimide, N-iodo $C_4H_4INO_2$	224.99	(ace)	234	135			w, al, eth	B21^4, 4576
13602	Succinimide, N-methyl $C_5H_7NO_2$	113.12	nd (eth-peth, al, ace)	234	71			w, al, eth	B21^4, 4544
13603	Succinimide, N-α-naphthyl $C_{14}H_{11}NO_2$	225.25	nd (dil al)		153			al	B21, 376
13604	Succinimide, N-β-naphthyl $C_{14}H_{11}NO_2$	225.25	nd (al)		183			al, bz	B21^4, 4551
13605	Succinimide, N-phenyl or Succinanil $C_{10}H_9NO_2$	175.19	mcl pr or nd (w, al)	ca 400	156	1.356		eth	B21^4, 4547
13606	Succinonitrile $NCCH_2CH_2CN$	80.09		265-7, 124^5	57	0.9867$^{60/4}$	1.4173^{60}	al, ace, bz, chl	B2^4, 1923
13607	Succinonitrile, tetramethyl $NCC(CH_3)_2C(CH_3)_2CN$	136.20	mcl pl, lf and pr (dil al)		170-1	1.070		al	B2^4, 2054
13608	Succinyl chloride $ClCOCH_2CH_2COCl$	154.98	pl or lf	193.3, 88.5^{19}	20	1.3748$^{20/4}$	1.4683^{20}	eth, ace, bz	B2^4, 1921
13609	Succinyl chloride, 2,3-dichloro-dl) $ClCOCHClCHClCOCl$	223.87		78.5^7	39			eth	B2^2, 558
13610	Succinyl chloride, 2,3-dichloride (*meso*) $ClCOCHClCHClCOCl$	223.87		105-6^{45}				CCl_4	B2^2, 558
13611	Sucrose or Saccharose $C_{12}H_{22}O_{11}$	342.30	mcl, $[\alpha]^{20/}{}_D$ + 66.37 (w)		185-6	1.5805$^{17.5}$	1.5376	w, Py	B31, 424
13612	Sucrose octaacetate $C_{28}H_{38}O_{19}$	678.60	nd, (al), $[\alpha]^{20/}{}_D$ + 59.6 (chl)	d 285, 260^1	86-87	1.27^{16}	1.4660	eth, ace, bz, chl	B31, 453
13613	Sudan III or Tetrazobenzene-β-naphthol $C_{22}H_{16}N_4O$	352.40	br lf with gr lustre(aa)		195			al, eth, ace, bz	B16^3, 148
13614	Sulfadiazine or 2-Sulfanilamido pyrimidine $C_{10}H_{10}N_4O_2S$	250.28	cr (w), wh pw		255-6d				C55, 25956
13615	Sulfaguanidine $C_7H_{10}N_4O_2S$	214.24	nd (w)		190-3				B14^3, 1970

No.	Name, Synonyms, and Formula	Mol. wt.	Color, crystalline form, specific rotation and λ_{max} (log ε)	b.p. °C	m.p. °C	Density	n_D	Solubility	Ref.
13616	Sulfaguanidine, monohydrate $C_7H_{10}N_4O_2S.H_2O$	232.26	nd (w)		143 (sealed tube)				B14[3], 1971
13617	Sulfamerazine or Sulfamethyldiazine $C_{11}H_{12}N_4O_2S$	264.30	cr		234-8				C55, 5501
13618	Sulfomethazine $C_{12}H_{14}N_4O_2S$	278.33	pa ye(w + ½) cr (diox-w)		198-9 (205-7)			w	Am64, 567
13619	Sulfamethylthiazole $C_{10}H_{11}N_3O_2S_2$	269.34			236-8			al	Am64, 2905
13620	Sulfamic acid, N-cyclohexyl $C_6H_{11}NHSO_2OH$	179.23	(al)		169-70				B12[3], 1594
13621	Sulfamide, tetramethyl $[(CH_3)_2N]_2SO_2$	152.21	pl or nd (dil al)	225	73			al	B4[4], 270
13622	Sulfanilamide or 4-Amino sulfonamide $4\text{-}H_2NC_6H_4SO_2NH_2$	172.20	lf (aq al)		165-6	1.08		al, eth, ace	B14[3], 1920
13623	Sulfanilamide, N-acetyl $4\text{-}(CH_3CO)NHC_6H_4SO_2NH_2$	214.24	pr (w)		182-4			al, ace	B14[3], 2042
13624	Sulfanilic acid or 4-Aminobenzene sulfonic acid $4\text{-}H_2NC_6H_4SO_3H$	173.19	rh pl or mcl (w + 2)		288	$1.485^{25/4}$			B14[3], 1916
13625	Sulfapyrazine or N-(2-Pyrazinyl)sulfanilamide $C_{10}H_{10}N_4O_2S$	250.28	nd ($PhNO_2$)		251			Py	Am63, 3153
13626	Sulfapyridine $C_{11}H_{11}N_3O_2S$	249.29	ye og (al)		191-3				C34, 1814
13627	Sulfaquinoxaline $C_{14}H_{12}N_4O_2S$	300.33			247-8				C49, 5525
13628	Sulfathiadiazole $C_8H_8N_4S_2O_2$	256.30			218			al, Py	C40, 5411
13629	Sulfathiazole $C_9H_9N_3O_2S_2$	255.31	br pl, rods or pw (45 al)		(i) 202.5 (ii) 125				C40, 7518
13630	Sulfathiazole, 4-nitro $C_9H_8N_4O_4S_2$	300.31	pa ye pw		258-62				C53, 12281
13631	Sulfathiazole, phthalyl, $C_{17}H_{13}N_3O_5S_2$	403.43			272-4				C51, 9689
13632	Sulfathiazole, succinyl $C_{13}H_{13}N_3O_5S_2$	355.38	cr		192-5				C51, 12148
13633	Sulfoacetic acid or Sulfoethanoic acid $HO_3SCH_2CO_2H$	140.11	hyg ta (w + 1)	245d	84-6			w, al, ace	B4[4], 102
13634	Sylvestrene (d) or Carvestrene $C_{10}H_{16}$	136.24	$[\alpha]^{18}{}_D$ + 83.2 (chl, c=4.3), (undil), + 66.3	175^{751}		$0.8479^{15/4}$	1.4760^{18}	al, eth	B14[3], 1916
13635	Syringenin or Sinapyl alcohol. 3-(3,5-dimethoxy-4-hydroxyphenyl)-2-propen-1-ol. $[3,5\text{-}(CH_3O)_2\text{-}4\text{-}HOC_6H_2]CH{=}CHCH_2OH$	210.23	nd (eth-peth)		66-7			eth	B6[3], 6690
13636	Syringin or Methoxy coniferine $C_{17}H_{24}O_9$	372.38	cr (w), nd (al), $[\alpha]_D$ −17.1		192			al	B31, 222
13637	Syringoyl methyl ketone $C_{11}H_{12}O_5$	224.21	ye nd		80-1			al, bz	Am62, 986

PHYSICAL CONSTANTS OF ORGANIC COMPOUNDS (Continued)

No.	Name, Synonyms, and Formula	Mol. wt.	Color, crystalline form, specific rotation and λ_{max} (log ε)	b.p. °C	m.p. °C	Density	n_D	Solubility	Ref.
13638	Tachysterol $C_{28}H_{44}O$	396.66	$[\alpha]^{18/}{}_D$ −70 (bz)	220 vac	......	....	....	al, eth, ace, bz	B6³, 3087
13639	D-Tagatose $C_6H_{12}O_6$	180.16	cr (dil al), $[\alpha]^{20/}{}_D$ −5 (w, c=1)	....	134-5	....	....	w	B31, 348
13640	D-Talitol or D-Altritol, D-Talaite $C_6H_{14}O_6$	182.17	pr (al)$[\alpha]^{20}{}_D$ + 3.1	....	87-8	....	....	w, al	B1⁴, 2839
13641	D-Talonic acid, hemihydrate $C_6H_{12}O_6 \cdot 1/2H_2O$	205.16	cr (aq al + ½ w), $[\alpha]^{25/}{}_D$ + 16.7→ −21.6 (w, c=4, mut)	....	138-9	....	....	w	B3³, 1068
13642	D-Talonic acid, γ-lactone $C_6H_{10}O_6$	178.14	pr (al), $[\alpha]^{25/}{}_D$ −34.6→ −28.4 (w), (mut)	....	135-7	....	....	....	B18⁴, 30273
13643	D-Talose $C_6H_{12}O_6$ HOCH(CHOH)₄CHO	180.16	[α]cr (al) β:cr (MeOH), $[\alpha]^{27/}{}_D$ + 29→ + 19.7 (w, c=1, mut)	....	α:130-5 β:120-1	....	....	w	B31, 283
13644	Tannic acid or Gallotannic acid Tannin $C_{76}H_{52}O_{46}$	1701.22	pa ye br amor or fl	....	210-5d	....	....	w, al, ace	B31, 133
13645	Taraxanthin $C_{40}H_{56}O_4$	600.88	ye pr (MeOH), $[\alpha]^{20/}{}_{Cd}$ + 200 (AcOEt)	....	185-6	....	....	al, ace, peth	B17⁴, 2239
13646	L-Tartaric acid or L-Threoic acid or L-2,3-Dihydroxy butanedoic acid $HO_2CCH(OH)CH(OH)CO_2H$	150.09	mcl (anh), rh pr (w + 1), $[\alpha]^{20/}{}_D$ + 12.7 (w, c=17.4)	....	171-4	1.7598^{20}	1.4955	w, al, ace	B3⁴, 1229
13647	DL-Tartaric acid or Racemic acid $HO_2CCH(OH)CH(OH)CO_2H$	150.09	mcl pr (w, al + 1w)	....	206	1.788	....	w	B3⁴, 1229
13648	Tartaric acid (*meso*) $HO_2CCH(OH)CH(OH)CO_2H$	150.09	tcl pl (w)	....	146-8	$1.666^{20/4}$	1.5	w, al	B3⁴, 1218
13649	Tartaric acid anhydride,diacetate (d) or α,α-Diacetoxy succinic anhydride $C_8H_8O_7$	216.16	nd (bz)	....	135	....	....	al, eth, ace	B18⁴, 2296
13650	Tartaric acid dibenzyl ester(d) or d-Dibenzyl tartrate $C_6H_5CH_2O_2C(CHOH)_2CO_2CH_2C_6H_5$	330.34	$[\alpha]^{15/}{}_D$ + 19.3	$250\text{-}70^4$	50	1.2036^{72}	....	al, Py	B6³, 1537
13651	Tartaric acid, dibutyl ester (*d*) or d-Dibutyl tartrate $C_4H_9O_2C(CHOH)_2CO_2C_4H_9$	262.30	pr $[\alpha]^{20/}{}_D$ + 11.3 (al)	320, 178^{12}	22	$1.0909^{20/4}$	1.4451^{20}	w, al, ace	B3⁴, 1232
13652	DL-Tartaric acid, dibutyl ester $C_4H_9O_2C(CHOH)_2CO_2C_4H_9$	262.30	....	320^{765}, 185^{12}	......	$1.0879^{25/4}$	1.4474^{15}	w, al, ace	B3³, 1032
13653	Tartaric acid, diisobutyl ester-(d) or d-Diisobutyl tartrate $i\text{-}C_4H_9O_2C(CHOH)_2CO_2\text{-}i\text{-}C_4H_9$	262.30	$[\alpha]_D$ + 11.8 (al)	323-5, 183^{11}	73	$1.0265^{81/4}$	....	al	B3³, 1021
13654	DL-Tartaric acid, diiosbutyl ester or DL-Diisobutyl tartrate $i\text{-}C_4H_9O_2C(CHOH)_2CO_2\text{-}i\text{-}C_4H_9$	262.30	cr (bz)	311^{768}, 195^{13}	63	$1.0386^{68/4}$	....	al	B3⁴, 1232
13655	Tartaric acid, diethyl ester (d) or d-Diethyl tartrate $C_2H_5O_2C(CHOH)_2CO_2C_2H_5$	206.20	$[\alpha]^{16/}{}_D$ + 7.9 (undil)	280, 142^8	18.7	$1.2036^{20/4}$	1.4468^{20}	w, al, eth, ace, aa	B3³, 1025
13656	DL-Tartaric acid, diethyl ester or DL-Diethyl racemate $C_2H_5O_2C(CHOH)_2CO_2C_2H_5$	206.20	....	281^{765}, 158^{14}	18.7	$1.2046^{20/4}$	1.4438^{20}	**al, eth**, ace	B3⁴, 1232
13657	Tartaric acid, diethyl ester-(*l*) $C_2H_5O_2C(CHOH)_2CO_2C_2H_5$	206.20	$[\alpha]^{20/}{}_D$ −7.55 (undil)	280, 162^{15}	......	$1.2054^{20/4}$	1.4468^{20}	al, eth, ace	B3³, 1020
13658	Tartaric acid, diethyl ester-(*meso*) $C_2H_5O_2C(CHOH)_2CO_2C_2H_5$	206.20	....	157.5^{14}	60	$1.1350^{99/4}$	1.4315^{65}	eth, ace	B3³, 1031
13659	Tartaric acid, diethyl ester, diacetate-(*d*) $C_2H_5O_2C(CHO_2CCH_3)_2CO_2C_2H_5$	290.27	mcl cr (lig), $[\alpha]^{100/}{}_D$ + 6.3 (undil)	296^{764}, 163^{10}	67	$1.1149^{66/4}$	....	al, eth	B3³, 1021

No.	Name, Synonyms, and Formula	Mol. wt.	Color, crystalline form, specific rotation and λ_{max} (log ε)	b.p. °C	m.p. °C	Density	n_D	Solubility	Ref.
13660	Tartaric acid, dimethyl ester (d) $CH_3O_2C(CHOH)_2CO_2CH_3$	178.14	(i)cr (bz), (ii)cr (bz), (iii)cr (w), $[\alpha]^{50/}_D$ +6.7 (MeOH, c=16)	280, 166^{12}	(i)48(ii) 50(iii) 61	1.306^{45}	.....	w, al, eth, ace, chl	B3[3], 1018
13661	DL-Tartaric acid, dimethyl ester or DL-Dimethyl tartrate $CH_3O_2C(CHOH)_2CO_2CH_3$	178.14	orh nd (bz), ta (chl)	280, 169^{20}	90	1.2604^{90}	.....	w, al, eth, ace	B3[4], 1232
13662	Tartaric acid, dimethyl ester-(*meso*) or *meso*-Dimethyl tartrate $CH_3O_2C(CHOH)_2CO_2CH_3$	178.14	nd (chl), cr (MeOH)	$98^{0.01}$ sub	114	.....	.....	w, al, ace	B3[3], 1031
13663	Tartaric acid, di(2-methyl butyl)ester-(*d*) $C_5H_{11}O_2C(CHOH)_2CO_2C_5H_{11}$	290.36	$[\alpha]^{20/}_D$ +14.1	208^{20}	.....	$1.0636^{20/4}$	.....	al, ace	B3, 519
13664	Tartaric acid, dinitrate (d) $HO_2C(CHONO_2)_2CO_2H$	240.08	nd (eth-bz), $[\alpha]^{20/}_D$ +13.7 (MeOH, p=9)	.....	d	.....	.....	al, eth, ace	B3[3], 1018
13665	Tartaric acid, dipropyl ester-(d) or d-Dipropyl tartrate $C_3H_7O_2C(CHOH)_2CO_2C_3H_7$	234.25	$[\alpha]^{20/}_D$ +12.4 (w)	303, 181^{23}	.....	$1.1390^{20/4}$	.....	w, al, eth, ace	B3[2], 331
13666	DL-Tartaric acid, dipropyl ester $C_3H_7O_2C(CHOH)_2CO_2C_3H_7$	234.25	pr (al-eth)	286^{765}, 167^{11}	.....	$1.1256^{20/4}$	.....	w, al, eth, ace	B3[2], 337
13667	Tartaric acid, diisopropyl ester-(d) or d-Diisopropyl tartrate i-$C_3H_7O_2C(CHOH)_2CO_2$-i-C_3H_7	234.25	$[\alpha]^{20/}_D$ +14.9	275^{765}, 152^{12}	.....	$1.1300^{20/4}$	.....	al, eth, ace	B3[2], 331
13668	DL-Tartaric acid, diisopropyl ester or DL-Diisopropyl tartrate i-$C_3H_7O_2C(CHOH)_2CO_2$-i-C_3H_7	234.25	.....	275^{765}, 154^{12}	34	$1.1166^{20/4}$	.....	al, eth, ace	B3[2], 337
13669	Tartaric acid, ethyl ester-(mono) $HO_2C(CHOH)_2CO_2C_2H_5$	178.14	$[\alpha]^{/}_D$ +21.8 (w)	.....	90	.....	.....	w, al	B3[3], 1020
13670	Taurine or 2-Amino ethane sulfonic acid $H_2NCH_2CH_2SO_3H$	125.14	mcl pr (w)	.....	328	.....	.....	w	B4[3], 1697
13671	Taurine, N,N-dimethyl or 2-(N,N-dimethylamino)ethane sulfonic acid $(CH_3)_2NCH_2CH_2SO_2H$	153.20	pr (MeOH), pl (w + 1)	.....	315-6 (anh), 270-80 d (+ w)	.....	.....	w, aa	B4[3], 1699
13672	Taruine, N-methyl or 2-(N-Methyl)ethane sulfonic acid $CH_3NHCH_2CH_2SO_3H$	139.17	pr	.....	241-2	.....	.....	w	B4[3], 1699
13673	Taurine, N-methyl-N-phenyl $C_6H_5NC(CH_3)CH_2CH_2SO_3H$	215.27	pa vt (al)	.....	239-40	.....	.....	al	B12[2], 285
13674	Taurine, N-phenyl $C_6H_5NHCH_2CH_2SO_3H$	201.24	lf (w), pr (al)	.....	277-80d	.....	.....	w	B12[2], 284
13675	Taurocholic acid or Cholaic acid. Cholyl taurine $C_{26}H_{45}NO_7S$	515.71	pr (al-eth), $[\alpha]^{18/}_D$ +38.8 (al, c=2)	.....	ca 125d	.....	.....	w, al	E14, 195
13676	Tephrosin or Hydroxy deguelin $C_{23}H_{22}O_7$	410.42	pr (chl-MeOH)	.....	198 (218)	.....	.....	eth, ace, chl	B19[4], 5271
13677	Terebic acid-(dl) or 3.3-Dimethyl paraconic acid $C_7H_{10}O_4$	158.15	mcl pr (al)	.....	176	$0.8155^{24/4}$	.....	.....	B18[4], 5287
13678	Terephthaldehyde 1,4-$(OHC)_2C_6H_4$	134.13	nd (w)	247^{771}	116	.....	.....	al, eth	B7[3], 3460
13679	Terephthalamide 1,4-$C_6H_4(CONH_2)_2$	164.16	nd (w), pl (aa)	.....	>250	.....	.....	.....	B9[3], 4253
13680	Terephthalamide, N,N,N′,N′-tetraethyl 1,4-$C_6H_4[CON(C_2H_5)_2]_2$	276.38	cr (eth-al)	.....	127	.....	.....	al, bz	B9[3], 4253
13681	Terephthalic acid or 1,4-Benzene dicarboxylic acid 1,4-$C_6H_4(CO_2H)_2$	166.13	nd (sub)	>300 (sub without melting)	sub	.....	.....	.....	B9[3], 4249
13682	Terephthalic acid, 2-amino 2-$H_2NC_6H_3(CO_2H)_2$-1,4	181.15	ye cr (w)	.....	324-5d	.....	.....	.....	B14[1], 637
13683	Terephthalic acid, 2-amino, dimethyl ester 2-$H_2NC_6H_3(CO_2CH_3)_2$-1,4	209.20	nd (bz), cr (al)	.....	134	.....	.....	ace, chl	B14[2], 338
13684	Terephthalic acid, 2-benzyl or Benzophenone-2,5-dicarboxylic acid 2-$(C_6H_5CO)C_6H_3(CO_2H)_2$-1,4	270.24	nd	.....	291-2	.....	.....	al, eth	B10[3], 4008
13685	Terephthalic acid, 2-bromo 2-$BrC_6H_3(CO_2H)_2$-1,4	245.03	nd (w, al)	.....	299	.....	.....	al	B9[3], 4258

No.	Name, Synonyms, and Formula	Mol. wt.	Color, crystalline form, specific rotation and λ_{max} (log ε)	b.p. °C	m.p. °C	Density	n_D	Solubility	Ref.
13686	Terephthalic acid, 2-chloro 2-$ClC_6H_3(CO_2H)_2$-(1,4)	200.58	cr (w)	...	320	...	...	al, eth	B9[3], 4256
13687	Terephthalic acid, 2,5-dichloro 2,5-$Cl_2C_6H_2(CO_2H)_2$-1,4	235.02	nd (w)	sub	306	...	...	al, eth	B9[3], 4257
13688	Terephthalic acid, diethyl ester or Diethyl terephthalate 1,4-$C_6H_4(CO_2C_2H_5)_2$	222.24	nd pr (al, peth)	302, 142[2]	44	$1.1098^{45/45}$	...	al, eth	B9[3], 4250
13689	Terephthalic acid, 2,5-dihydroxy 2,5-$(HO)_2C_6H_2(CO_2H)_2$-1,4	198.13	ye cr (al, w)	...	d	...	...	...	B10[3], 2438
13690	Terephthalic acid, dimethyl ester 1,4-$C_6H_4(CO_2CH_3)_2$	194.19	nd (eth)	sub	141-2	...	...	eth	B9[3], 4250
13691	Terephthalic acid, monoamide or Terephthalamic acid 4-$HO_2CC_6H_4CONH_2$	165.15	...	sub 250	>300	...	...	...	B9[3], 4253
13692	Terephthalic acid, mononitrile or 4-Cyanobenzoic acid 4-$NCC_6H_4CO_2H$	147.13	pl or lf (w)	...	219	...	...	al, eth	B9[3], 4254
13693	Terephthalic acid, 2-nitro 2-$NO_2C_6H_3(CO_2H)_2$-1,4	211.13	nd (w)	...	270-5	...	...	al	B9[3], 4258
13694	Terephthalic acid, tetrabromo $Br_4C_6(CO_2H)_2$-1,4	481.72	nd (w)	...	266 (anh)	...	...	...	B9[1], 367
13695	Terephthalonitrile 1,4-$C_6H_4(CN)_2$	128.13	nd (w, MeOH)	sub	222	...	...	bz, aa	B9[3], 4255
13696	Terephthalyl chloride 1,4-$C_6H_4(COCl)_2$	203.02	nd or pl (lig)	125-7[9]	83-4	...	...	eth	B9[3], 4252
13697	Terephthalyl alcohol or *p*-Xylylene glycol 1,4-$(HOCH_2)_2C_6H_4$	138.17	nd (w)	...	115-6	...	...	w, al, eth, ace	B6[3], 4608
13698	Terpenolic acid or Terpenylic acid $C_8H_{12}O_4$	172.18	lf or pr (w + 1)	sub 130-40	90 (anh), 57 (+ w)	...	...	w	B18[4], 5302
13699	*o*-Terphenyl or 1,2-Diphenyl benzene 1,2-$(C_6H_5)_2C_6H_4$	230.31	mcl pr (MeOH)	332, 160-70[2]	58	...	...	ace, bz, chl, MeOH	B5[3], 2292
13700	*m*-Terphenyl or 1-3-Diphenyl benzene 1,3-$(C_6H_5)_2C_6H_4$	230.31	ye nd (al)	365	89	...	...	al, eth, bz, aa	B5[3], 2294
13701	*p*-Terphenyl 1,4-$(C_6H_5)_2C_6H_4$	230.31	...	sub 250[45]	213	...	...	eth, bz	B5[3], 2296
13702	α-Terpinene or p-Mentha-1,3-diene $C_{10}H_{16}$	136.24	...	177.2, 68-70[12]	...	$0.8502^{20/4}$	1.4784^{10}	**al, eth**	B5[4], 435
13703	β-Terpinene $C_{10}H_{16}$	136.24	...	173-4	...	0.838^{23}	1.4754^{22}	...	B5[4], 437
13704	γ-Terpinene $C_{10}H_{16}$	136.24	...	183	...	$0.849^{20/4}$	$1.4765^{14.5}$	...	B5[4], 436
13705	α-Terpineol (dl) or *dl*-p-Menth-1-en-8-ol $C_{10}H_{18}O$	154.25	cr (peth)	220, 85[2]	40-1	$0.9337^{20/4}$	1.4831^{20}	al, eth, ace, bz, chl	B6[4], 251
13706	α-Terpineol acetate $C_{12}H_{20}O_2$	196.29	*d*: $[\alpha]_D$ + 52.5 (undil), *l*: $[\alpha]_D$ −73 (undil)	140[40], 104-6[11]	...	$0.9659^{20/4}$	1.4689^{21}	al, eth, bz	B6[4], 252
13707	Terpin hydrate (*cis*) or *cis*-Terpino hydrate $C_{10}H_{20}O_2.H_2O$	214.30	rh cr	sub 100	123d	...	1.51	al	B6[3], 4113
13708	Terpinolene $C_{10}H_{16}$	136.24	...	185, 76[10]	...	$0.8623^{20/4}$	1.4883^{20}	**al, eth**, bz	B5[4], 437
13709	3-Terpinolenone or Piperitenone $C_{10}H_{14}O$	150.22	$[\alpha]_{546}$ −0.1	120-2[14]	...	$0.9774^{20/4}$	1.5294^{20}	al, eth	B7[3], 559
13710	Terramycin or Oxytetracycline $C_{22}H_{22}N_2O_9$	460.44	...	...	184-5	1.634^{20}	...	...	AM87, 134
13711	Testosterone or 17β-Hydroxy-4-androsten-3-one $C_{19}H_{28}O_2$	288.43	nd (dil ace), $[\alpha]^{24}{}_D$ + 109 (al, c=4)	...	155	...	...	al, eth, ace	B8[3], 892
13712	Testosterone, 4,5-dihydro-17-methyl $C_{20}H_{32}O_2$	304.47	cr (AcOEt)	...	192-3	...	...	...	B8[3], 609
13713	Testosterone, 17-ethenyl or 17-Vinyl testosterone $C_{21}H_{30}O_2$	314.47	pr nd (peth-eth), $[\alpha]_D$ + 87.6	...	140-1	...	...	eth, ace, chl, MeOH	B8[3], 1067
13714	Testosterone, 17-ethyl $C_{21}H_{32}O_2$	316.48	nd (AcOEt)	...	143-4	...	...	...	B8[3], 946

No.	Name, Synonyms, and Formula	Mol. wt.	Color, crystalline form, specific rotation and λ_{max} (log ε)	b.p. °C	m.p. °C	Density	n_D	Solubility	Ref.
13715	Testosterone, 17-ethynyl $C_{21}H_{28}O_2$	312.45	cr (chl-MeOH, AcOEt), $[\alpha]^{20/}{}_D$ + 22.5 (diox)	(sub, vac)	270-2	....	....	diox, Py	B8³, 1206
13716	Testosterone, 17-methyl $C_{20}H_{30}O_2$	302.46	nd (bz), $[\alpha]^{20/}{}_D$ + 82 (al)	....	165-6	....	....	al, eth	B8³, 939
13717	Testosterone propionate $C_{22}H_{32}O_3$	344.49	$[\alpha]^{25/}{}_D$ + 83-90 (diox, c=1)	....	118-22	....	....	al, eth, Py	B8³, 897
13718	Tetrabenzotriazaporphyrin $C_{32}H_{18}N_8$	514.55	purple nd and pl (quinoline)	....	......	....	....	Py	J1939, 1809
13719	Tetrabutyl ammonium iodide $(C_4H_9)_4N^+I^-$	369.37	lf(w bz)	....	148	....	....	al, w	B4⁴, 558
13720	Tetracosane $CH_3(CH_2)_{22}CH_3$	338.66	cr (eth)	391.3, 231.3^{10}	54	$0.7665^{70/4}$, 0.7991^{20}	1.4283^{70}, 1.4480^{20}	eth	B1⁴, 578
13721	Tetracoscene, 11-decyl $(C_{10}H_{21})_2CHC_{13}H_{27}$	478.93	....	301^{10}	10.8	$0.8161^{20/4}$	1.4556^{20}	....	B1⁴, 598
13722	Tetracosane, 3-ethyl $(C_2H_5)_2CHC_{21}H_{43}$	366.71	....	255.5^{10}	30.1	$0.7949^{40/4}$	1.4436^{40}	....	B1⁴, 584
13723	Tetracyclone or Tetraphenyl cyclopentadienone $C_{29}H_{20}O$	384.48	bk-vt lf, cr (aa, xyl)	....	220-1	....	....	al, bz	B7³, 2997
13724	6,8-Tetradecadiyne $C_5H_{11}C{\equiv}CC{\equiv}CC_5H_{11}$	190.33	....	$118\text{-}9^{4}$	2	$0.8699^{16/4}$	....	eth	B1³, 1067
13725	Tetradecanal or Myristaldehyde $C_{13}H_{27}CHO$	212.38	lf	166^{24}	30	....	....	al, eth, ace	B1⁴, 3389
13726	Tetradecane $CH_3(CH_2)_{12}CH_3$	198.39	....	253.7, 121.9^{10}	5.9	$0.7628^{20/4}$	1.4290^{20}	al, eth	B1⁴, 520
13727	Tetradecane, 1-amino or Myristyl amine $C_{13}H_{27}CH_2NH_2$	213.41	....	291.2, 162^{15}	83.1	$0.8079^{20/4}$	1.4463^{20}	al, eth, ace, bz, chl	B4⁴, 812
13728	Tetradecane, bromo or Myristyl bromide $C_{13}H_{27}CH_2Br$	277.29	....	307, 181^{21}	5.6	$1.0170^{20/4}$	1.4603^{20}	al, **ace, bz**, chl	B1⁴, 523
13729	Tetradecane, 1-chloro or Myristyl chloride $C_{13}H_{27}CH_2Br$	232.84	....	292, 153^{10}	4.9	$0.8665^{20/4}$	1.4473^{20}	al, ace, bz, chl	B1³, 550
13730	Tetradecane, 1,14-dibromo $Br(CH_2)_{14}Br$	356.18	lf (al-eth), cr (al)	$190\text{-}2^{8}$	50.4	....	....	al, eth, chl	B1⁴, 523
13731	Tetradecane, 1,1-dicyclohexyl $C_{13}H_{27}CH(C_6H_{11})_2$	362.68	....	406	37.6	0.8735^{20}	1.4799^{20}	....	B5³, 300
13732	Tetradecane, 1-phenyl or Tetradecyl benzene $C_{13}H_{27}CH_2C_6H_5$	274.49	....	358.9, 210^{12}	16.1	$0.8559^{20/4}$	1.4818^{20}	....	B5⁴, 1212
13733	Tetradecanedioic acid, dimethyl ester $CH_3O_2C(CH_2)_{12}CO_2CH_3$	286.41	nd (MeOH)	$191\text{-}2^{10}$	43-5	....	....	....	B2⁴, 2149
13734	1,14-Tetradecanediol or Tetradecamethylene glycol $HO(CH_2)_{14}OH$	230.39	nd (bz)	200^{9}	84.5	....	....	al, eth	B1⁴, 2631
13735	1-Tetradecane sulfonic acid $C_{13}H_{27}CH_2SO_3H$	278.45	....	....	65.5 (anh), 55-6 (+1w)	$0.9996^{25/4}$	....	w	B4⁴, 66
13736	1-Tetradecanethiol or Myristyl mercaptan $C_{13}H_{27}CH_2SH$	230.45	....	$176\text{-}80^{22}$	......	$0.8484^{20/20}$	1.4597^{20}	al, eth	B1⁴, 1867
13737	Tetradecanoic acid or Myristic acid $C_{13}H_{27}CO_2H$	228.38	lf (eth 80% aa)	250.5^{100}, 149.3^{2}	58	$0.8439^{80/4}$	1.4305^{80}	al, ace, bz, chl	B2⁴, 1126
13738	1-Tetradecanol or Myristyl alcohol $C_{13}H_{27}CH_2OH$	214.40	lf	263.2, 167^{11}	39-40	$0.8236^{38/4}$	....	al, eth, ace, bz, chl	B1⁴, 1864
13739	2-Tetradecanone or n-Dodecyl methyl ketone $C_{12}H_{25}COCH_3$	214.39	cr (dil al)	$205\text{-}6^{100}$, 134^{13}	33-4	....	....	al, ace	B1⁴, 3389
13740	3-Tetradecanone or Ethyl undecyl ketone $C_{11}H_{23}COC_2H_5$	212.38	cr (MeOH)	152^{16}	34	....	....	al, ace	B1⁴, 3389
13741	5,9-Tetradecadien-7-yne, 6,9-dimethyl $C_4H_9CH{=}C(CH_3)C{\equiv}CC(CH_3){=}CHC_4H_9$	218.38	....	$95\text{-}8^{0.3}$	......	$0.8241^{20/4}$	1.4866^{20}	eth, bz	B1⁴, 1133
13742	1-Tetradecene $C_{12}H_{25}CH{=}CH_2$	196.38	....	232-4, 125^{15}	−12	$0.7745^{20/4}$	1.4351^{20}	al, eth, bz	B1⁴, 924
13743	2-Tetradecyne $C_{11}H_{23}C{\equiv}CCH_3$	194.36	....	252.5, 134^{15}	6.5	$0.8000^{20/4}$	....	al, eth	B1⁴, 1070

PHYSICAL CONSTANTS OF ORGANIC COMPOUNDS (Continued)

No.	Name, Synonyms, and Formula	Mol. wt.	Color, crystalline form, specific rotation and λ_{max} (log ε)	b.p. °C	m.p. °C	Density	n_D	Solubility	Ref.
13744	7-Tetradecyne $C_6H_{13}C{\equiv}CC_6H_{13}$	194.36	....	144^{30}	......	$0.7991^{20/4}$	1.4330^{25}	al, eth	$B1^3$, 1067
13745	Tetraethyl ammonium bromide $(C_2H_5)_4N^+Br^-$	210.16	dlq (al)	....	......	$1.3970^{20/4}$	....	w, al, chl	$B4^4$, 332
13746	Tetraethyl ammonium hydroxide, hydrate $(C_2H_5)_4N^+OH^-.4H_2O$	219.32	nd (w + 4)	d (vac)	49-50	....	....	w, al	$B4^4$, 331
13747	Tetraethylene glycol $HOCH_2(CH_2OCH_2)_3CH_2OH$ $C_8H_{18}O_5$	194.23	....	328, 198^{14}	−6.2	$1.1285^{15/4}$	1.4577^{20}	w, al, eth	$B1^4$, 2403
13748	Tetraethylene glycol, dimethyl ether $CH_3OCH_2(CH_2OCH_2)_3CH_2OCH_3$	222.29	....	275.8	......	$1.0132^{20/20}$	....	w, al, eth	$B1^4$, 2404
13749	Tetraethylene glycol, mono stearate $C_{17}H_{35}CO_2CH_2(CH_2OCH_2)_3CH_2OH$	476.69	....	328	40	$1.1285^{15/4}$	1.4593^{20}	....	C37, 3202
13750	Tetraethylene pentamine $H_2NCH_2(CH_2NHCH_2)_3CH_2NH_2$	189.31	....	340.3, $186\text{-}92^{14}$	......	....	1.5042^{20}	....	$B4^1$, 1244
13751	Tetraethyl hypophosphate $(C_2H_5O)_2POOP(OC_2H_5)_2$	274.19	....	$116\text{-}7^2$	......	1.1283^{18}	1.4284^{20}	bz	$B1^4$, 1358
13752	Tetralin or 1,2,3,4-Tetrahydronaphthalene $C_{10}H_{12}$	132.21	....	207.6, 79.4^{10}	−35.8	$0.9702^{20/4}$	1.5413^{20}	al, eth	$B5^4$, 1388
13753	Tetralin, 6-acetyl $6\text{-}(CH_3CO)C_{10}H_{11}$	174.24	....	289-91d, 182^{20}	......	....	....	....	$B7^3$, 1473
13754	Tetralin, 2-amino-(*dl*) $2\text{-}H_2NC_{10}H_{11}$	147.22	....	249^{710}, 140^{20}	38	$1.0295^{22/4}$	1.5604^{22}	al, eth, ace, bz	$B12^3$, 2811
13755	Tetralin, 5-amino $5\text{-}H_2NC_{10}H_{11}$	147.22	....	276.8, 155^{22}	......	1.0625^{16}	1.6050^{20}	al, eth	$B12^3$, 2805
13756	Tetralin, 1,1-dimethyl $1,1\text{-}(CH_3)_2.C_{10}H_{10}$	160.26	....	221	......	0.950^{20}	1.5292^{20}	....	$B5^4$, 1430 (35.5214)
13757	Tetralin, 1-ethyl $1\text{-}C_2H_5C_{10}H_{11}$	160.26	....	239-46	......	0.95285^{20}	1.5318^{20}	....	$B5^4$, 1429
13758	Tetralin, 2-ethyl $2\text{-}C_2H_5C_{10}H_{11}$	160.26	....	237, $63\text{-}5^{0.5}$	......	$0.9401^{15.5}$	$1.5250^{15.5}$	....	$B5^3$, 1261
13759	Tetralin, 6-ethyl $6\text{-}C_2H_5C_{10}H_{11}$	160.26	....	244	......	$0.9632^{17/4}$	1.5414^{16}	....	$B5^4$, 1429
13761	Tetralin, 1-hydroxy $1\text{-}HOC_{10}H_{11}$	148.20	....	$102\text{-}4^2$	......	....	1.5658^{20}	....	$B6^3$, 2457
13762	Tetralin, 1-methyl $1\text{-}CH_3C_{10}H_{11}$	146.23	....	220.59	......	0.9583^{20}	1.5353^{20}	....	$B5^4$, 1412 (35.5210)
13763	Tetralin, 2-methyl $2\text{-}CH_3C_{10}H_{11}$	146.23	....	220-2, $99\text{-}101^{13}$	......	....	....	....	$B5^4$, 1414
13764	Tetralin, 6-methyl $6\text{-}CH_3C_{10}H_{11}$	146.23	....	220-2	......	$0.9541^{15/4}$	1.5332^{15}	....	$B5^4$, 1413
13765	Tetralin, 1,2,3,4-tetrabromo or Naphthalene tetrabromide $1,2,3,4\text{-}Br_4C_{10}H_8$	447.79	mcl pr (chl)	....	111d	....	....	CS_2, bz	$B5^3$, 1229
13766	Tetralin, 5,6,7,8-tetrachloro $5,6,7,8\text{-}Cl_4C_{10}H_8$	269.99	....	180^{26}	172	....	....	....	$B5^3$, 1227
13767	1,4-Tetralin dicarboxylic acid, 1-phenyl-(*d*,*α*) or d-*α*-Isatropic acid $C_{18}H_{16}O_4$	296.33	pr $[\alpha]^{20}{}_D$ + 9.4 (al, c=12.6)	....	239d	....	....	....	$B9^1$, 416
13768	5-Tetralin sulfonamide $5\text{-}C_{10}H_{11}SO_2NH_2$	211.28	lf (al, 30% aa)	....	139-40	....	....	al	$B11^2$, 87
13769	5-Tetralin sulfonic acid $5\text{-}C_{10}H_{11}SO_3H$	212.26	cr (chl + 1w)	....	105-10 (+ 1w)	....	....	w	$B11^2$, 87
13770	5-Tetralin sulfonyl chloride $5\text{-}C_{10}H_{11}SO_2Cl$	230.71	pl (peth)	....	70.5	....	....	eth	$B11^2$, 87
13771	6-Tetralin sulfonic acid $6\text{-}C_{10}H_{11}SO_3H$	212.26	cr (chl, dil sulf)	....	75	....	....	w, eth, chl	$B11^2$, 88
13772	6-Tetralin sulfonyl chloride $6\text{-}C_{10}H_{11}SO_2Cl$	230.71	pl (eth)	$197\text{-}200^{18}$	58	....	....	eth	$B11^2$, 88
13773	1-Tetralone or 1-oxo-1,2,3,4-tetrahydronaphthalene $1\text{-}C_{10}H_{10}O$	146.19	....	255-7, 129^{12}	8	$1.0989^{16/4}$	1.5672^{20}	....	$B7^3$, 1416
13773a	1-Tetralone, 7-ethyl $7\text{-}C_2H_5\text{-}(1\text{-}C_{10}H_9O)$	174.24	....	$152\text{-}3^{12}$	......	$1.0556^{17/4}$	1.5599^{17}	al, eth, bz	$B7^3$, 1473
13774	1-Tetralone, 4-methyl $4\text{-}CH_3(C_{10}H_9O)$	160.22	....	$133\text{-}4^{12}$	......	$1.0779^{19/4}$	1.5620^{19}	....	$B7^3$, 1444
13775	2-Tetralone $2\text{-}C_{10}H_{10}O$	146.19	....	234-40, 138^{16}	18	$1.1055^{27/4}$	1.5598^{20}	eth, bz	$B7^3$, 1424

No.	Name, Synonyms, and Formula	Mol. wt.	Color, crystalline form, specific rotation and λ_{max} (log ε)	b.p. °C	m.p. °C	Density	n_D	Solubility	Ref.
13776	Tetramethyl ammonium bromide $(CH_3)_4N^+Br^-$	154.05	dlq ditetr bipym	(360 sub,vac)	230d	1.56	....	w, MeOH	B4⁴, 145
13777	Tetramethyl ammonium chloride $(CH_3)_4N^+Cl^-$	109.60	dlq ditetr bipym (dil)	....	420 (sealed tube)	$1.169^{20/4}$	....	w, MeOH	B4⁴, 145
13778	Tetramethyl ammonium hydroxide, monohydrate $(CH_3)_4N^+OH^-.H_2O$	109.17	....	....	d 130-5	....	....	w	B4, 50
13779	Tetramethyl ammonium hydroxide, trihydrate $(CH_3)_4N^+OH^-.3H_2O$	145.20	....	....	60	....	....	w	B4, 50
13780	Tetramethyl ammonium hydroxide, pentahydrate $(CH_3)_4N^+OH^-.5H_2O$	181.23	....	d	62-3	....	....	w, al	B4², 557
13781	Tetramethyl ammonium iodide $(CH_3)_4NI$	201.05	....	....	>230d	1.829	....	....	B4⁴, 146
13782	Tetraphenylene or Tetrabenzocyclooctatetraene $C_{24}H_{16}$	304.39	cr (al, AcOEt)	$200^{0.2}$ sub	233	....	....	al	B5³, 2595
13783	Tetrapropyl ammonium bromide $(C_3H_7)_4N^+Br^-$	266.27	....	....	252	....	....	w, chl	B4⁴, 471
13784	Tetrapropyl ammonium iodide $(C_3H_7)_4N^+I^-$	313.27	rh bipym	....	280d	$1.3138^{25/4}$	....	w, al, chl, aa	B4⁴, 472
13785	Tetrasiloxane, decamethyl $(CH_3)_3Si(OSi(CH_3)_2)_3CH_3$	310.69	....	194, 88^{20}	−76	$0.8536^{20/4}$	1.3895^{20}	bz, peth	B4³, 1879
13786	Tetrasiloxane, 1,1,1,3,5,7,7,7, octamethyl $[(CH_3)_3SiOSi(CH_3)_2]_2O$	310.69	....	170	......	$0.8559^{20/4}$	1.3854^{20}	....	B4³, 1874
13787	1-Tetratriacontanol or *n*-Carnatyl alcohol $CH_3(CH_2)_{33}OH$	494.93	nd (ace)	....	913	....	....	ace	B1⁴, 1921
13788	1,2,4,5-Tetrazine or *s*-Tetrazine $C_2H_2N_4$	82.06	dk red pr	sub	99	....	....	w, al, eth	B26², 212
13789	1,2,3,4-Tetrazole CH_2N_4	70.05	pl (al)	sub	156	....	....	w, al, ace, aa	B26², 196
13790	Thebaine or Paramorphine $C_{19}H_{21}NO_3$	311.38	pl (eth), pr (dil al), $[\alpha]^{25/}{}_D$ −218.5 (al, p=2)	sub $91^{0.01}$	193	$1.305^{20/4}$	....	al, bz, chl, Py	B27², 177
13791	Thebaine, dihydro $C_{19}H_{23}NO_3$	313.40	$[\alpha]^{20/}{}_D$ −267 (bz, c=1.02)	....	162-3	....	....	al, bz, AcOEt	B27², 110
13792	Thebaine, hydrochloride monohydrate $C_{19}H_{21}NO_3.HCl.H_2O$	365.86	orh pr (al), $[\alpha]^{20/}{}_D$ −157 (al)	....	......	....	....	w, al, eth	B27², 181
13793	Thebainone A $C_{18}H_{21}NO_3$	299.37	nd (dil al + ½ w), nd or pr (al, aa)	....	151-2	....	....	ace, bz, chl	B21⁴, 6548
13794	Theobromine or 3,7-Dimethylxanthine $C_7H_8N_4O_2$	180.17	rh or mcl nd (w)	sub 290	351 (357)	....	....	....	B26², 264
13795	Theophylline or 1,3-Dimethylxanthine $C_7H_8N_4O_2$	180.17	nd or pl (w + 1)	....	272-4	....	....	w	B26², 263
13796	Thevetin $C_{42}H_{66}O_{18}$	858.98	nd (al), pl or nd (i-proh)	....	210	....	....	al, Py, MeOH	B18⁴, 1493
13797	Thiacyclobutane or Trimethylene sulfide C_3H_6S	74.14	....	94.7, 14^{30}	−73.2	$1.0200^{20/4}$	1.5102^{20}	al, ace, bz	B17⁴, 14
13798	1,3,4-Thiadiazole, 2.5-dimethyl $C_4H_6N_2S$	114.17	....	202-3	64	....	....	....	B27, 565
13799	Thialdine or Thioacetaldehyde ammonia $C_6H_{13}NS_2$	163.30	pl (al-eth)	d	46	$1.0632^{50/20}$	....	al, eth	B27², 525
13800	Thiane $C_5H_{10}S$	102.19	....	141.8, 93^{82}	19	$0.9861^{20/4}$	1.5067^{20}	al, eth, ace, bz, chl	B17⁴, 55
13801	Thianthrene or Dephenylene disulfide $C_{12}H_8S_2$	216.32	mcl pr or pl (al)	336, 204^{11}	158-9	....	....	eth, bz	B19⁴, 347
13802	Thiazole C_3H_3NS	85.12	....	116.8	......	$1.998^{17/4}$	1.5969^{20}	al, eth, ace	B27², 9
13803	Thiazole, 2-acetamido $2\text{-}CH_3CONH(C_3H_2NS)$	142.18	....	....	208	....	....	....	C35, 5110
13804	Thiazole, 2-amino or Abadol $2\text{-}H_2N(C_3H_2NS)$	100.14	ye pl (al)	140^{11}	93	....	....	....	B27², 205
13805	Thiazole, 2-amino-4-methyl $4\text{-}CH_3\text{-}2\text{-}H_2N(C_3HNS)$	114.17	hyg cr	281-2d, 136^{30-40}	45-6	....	....	w, al, eth	B27², 206

No.	Name, Synonyms, and Formula	Mol. wt.	Color, crystalline form, specific rotation and λ_{max} (log ε)	b.p. °C	m.p. °C	Density	n_D	Solubility	Ref.
13806	Thiazole, 2-amino-5-methyl $5\text{-}CH_3\text{-}2\text{-}H_2N(C_3HNS)$	114.17	pl (w)	...	96	...	...	al, eth	B27, 162
13807	Thiazole, 2-amino-5-sulfanilyl $C_9H_9N_3O_2S_2$	255.31	nd (al)	...	219-21d	...	...	al, eth, ace, diox	C41, 447
13808	Thiazole, 2,4-dimethyl $2,4\text{-}(CH_3)_2(C_3HNS)$	113.18	...	144-5^{719}, 70-3^{50}	...	1.0562$^{15/4}$	1.5091^{20}	al, eth	B27^{2}, 10
13809	Thiazole, 4,5-dimethyl $4,5\text{-}(CH_3)_2(C_3HNS)$	113.18	...	158, 81-3^{59}	83-4	...	...	al, eth	AM74, 5778
13810	Thiazole, 2,4-diphenyl $2,4\text{-}(C_6H_5)_2(C_3HNS)$	237.32	lf (al)	>360	92-3	1.1554$^{98/4}$	...	al, eth	B27^{2}, 43
13811	Thiazole, 5-(2-hydroxyethyl)-4-methyl $4\text{-}CH_3\text{-}5\text{-}(HOCH_2CH_2)(C_3HNS)$	143.20	col to pa ye	135^{7}	...	1.196$^{24/4}$	...	w, al, eth, bz, chl	AM71, 2931
13812	Thiazole, 2-mercapto-4-methyl $4\text{-}CH_3\text{-}2\text{-}HS(C_3HNS)$	131.21	ye (dil al)	188^{3}	88-9	...	...	al	B27^{2}, 208
13813	Thiazole, 2-methyl $2\text{-}CH_3(C_3H_2NS)$	99.15	...	128-9, 65-70^{80}	...	...	1.510	w, al, ace	B27, 16
13814	Thiazole, 4-methyl $4\text{-}CH_3(C_3H_2NS)$	99.15	...	132^{745},70^{90}	...	1.112^{25}	...	w, al, eth	B27^{2}, 9
13815	5-Thiazole carboxylic acid $5\text{-}(C_3H_2NS)\text{-}CO_2H$	129.13	lf ye nd (dil HCl)	...	217-8	...	...	eth	C48, 2688
13816	5-Thiazole carboxylic acid, 2-amino,ethyl ester $2\text{-}H_2N\text{-}(C_3HNS)CO_2C_2H_5\text{-}5$	172.20	...	213-5d	163-4	...	...	...	C47, 7453
13817	5-Thiazole carboxylic acid, 4-methyl $4\text{-}CH_3\text{-}(C_3HNS)CO_2H\text{-}5$	143.16	pr or pl (w), nd (al)	sub >250	280d	...	...	...	B27, 316
13818	5-Thiazole carboxylic acid, 4-methyl,ethyl ester $4\text{-}CH_3\text{-}(C_3HNS)CO_2C_2H_5\text{-}5$	171.21	pr	232-3^{735}, 110-5^{15}	28	...	...	al, eth, ace	B27, 316
13819	5-Thiazole carboxylic acid, 4-methyl,ethyl ester hydrochloride $C_7H_9NO_2S.HCl$	207.67	nd (al)	...	155d	...	...	w, al	J1939, 443
13820	5-Thiazole carboxylic acid, 4-methyl-2-sulfanilamide $C_{11}H_{11}N_3O_4S_2$	313.35	cr	...	241-2	...	...	...	C38, 2250
13821	Thiazolidine or Tetrahydrothiazole C_3H_7NS	89.16	...	164-5	...	1.131$^{25/4}$	1.551^{20}	w, al, eth, ace	AM59, 200
13822	2,4-Thiazolidinedione or 2,4-Dioxothiazolidine $C_3H_3O_2NS$	117.12	pl (w), pr (al)	179^{19}	128	...	...	eth	B27^{2}, 284
13823	2-Thiazolidine thione or 2-thiothiazolidone $C_3H_5NS_2$	119.20	nd (w, MeOH)	...	106-7	...	...	...	B27^{2}, 198
13824	2-Thiazoline or 4,5-Dihydrothiazole C_3H_5NS	87.14	...	138^{750}	...	...	...	eth, ace, bz	B27^{1}, 206
13825	2-Thiazoline, 2-amino $C_3H_6N_2S$	102.15	nd or lf (bz)	d	84-5	...	...	w, al, bz, chl	B27^{2}, 194
13826	2-Thiazoline-2,5-dimethyl C_5H_9NS	115.19		152	...	...	...	...	B27, 14
13827	2-Thiazoline, 5-ethyl-2-methyl $5\text{-}C_2H_5\text{-}2\text{-}CH_3(2\text{-}C_3H_3NS)$	129.22	...	62-3^{16}	...	...	...	...	...
13828	2-Thiazoline, 4-hexyl-2-methyl $4\text{-}C_6H_{13}\text{-}2\text{-}CH_3(2\text{-}C_3H_3NS)$	185.33	...	92-3^{2}	...	...	...	...	...
13829	2-Thiazoline, 5-hexyl-2-methyl $5\text{-}C_6H_{13}\text{-}2\text{-}CH_3(2\text{-}C_3H_3NS)$	185.33	...	82-4^{1}	...	...	...	...	...
13830	2-Thiazoline, 2-mercapto or 2-Thiazolinethiol $2HS\text{-}(2\text{-}C_3H_4NS)$	119.20	nd (w, MeOH)	...	106-7	...	...	w, al	B27^{2}, 198
13831	2-Thiazoline, 2,4,4-trimethyl $2,4,4\text{-}(CH_3)_3(2\text{-}C_3H_2NS)$	129.22	...	146-8	...	0.969^{25}	1.4825^{25}	...	C53, 11368
13832	Thienone or Di-(2-thienyl)ketone $(C_4H_3S)_2CO$	194.27	nd (al)	326	90	...	...	eth, ace	B19^{4}, 1745
13833	Thiepane $C_6H_{12}S$	116.22	...	169-71^{747}	...	0.9883$^{20/4}$	1.5138^{20}	eth, ace, chl	B17^{4}, 81
13834	Thietane C_3H_6S	74.14	...	94.7, 14^{30}	−73.25	1.0200$^{20/4}$	1.5102^{20}	al, ace, bz	B17^{4}, 14
13835	Thietane, 1,1-dioxide $C_3H_6SO_2$	106.14	...	91.2^{14}	75.5-6	...	1.5156^{20}	w, al, eth, lig	B17^{4}, 16
13836	Thietane, 2-methyl $2\text{-}CH_3(C_3H_5S)$	88.17	...	105.5-7.5^{747}	...	0.9571$^{20/4}$	1.4852^{20}	al, eth, ace, bz, chl	B17^{4}, 44
13837	Thiirane or Ethylene sulfide C_2H_4S	60.11	...	55-6d	...	1.0368$^{0/4}$	1.4935^{20}	ace, chl	B17^{4}, 11
13838	Thioacetamide CH_3CSNH_2	75.13	cr (al), pl (eth)	...	115-6	...	...	w, al	B2^{4}, 565

No.	Name, Synonyms, and Formula	Mol. wt.	Color, crystalline form, specific rotation and λ_{max} (log ε)	b.p. °C	m.p. °C	Density	n_D	Solubility	Ref.
13839	Thioacetamide, *N*-phenyl or Thioacetanilide $CH_3CSNHC_6H_5$	151.23	nd (w)	d	75-6				B12^{2}, 142
13840	Thioacetic acid CH_3COSH	76.11	ye	87, 26-7^{35}	<−17	1.064$^{20/4}$	1.4648^{20}	w, al, eth, ace	B2^{4}, 542
13841	Thioacetic acid, ethyl ester or Ethyl thioacetate $CH_3COSC_2H_5$	104.17		116.4		0.9792$^{20/4}$	1.4583^{21}	al, eth	B2^{4}, 543
13842	Thioanisole or Methyl phenyl sulfide $CH_3SC_6H_5$	124.20		193, 74^{10}		1.0579$^{20/4}$	1.5868^{20}	al, bz	B6^{4}, 1466
13843	Thiobenzophenone $(C_6H_5)_2CS$	198.28	bl nd (peth)	174^{14}	53-4			bz, chl	B7^{3}, 2087
13844	Thiobenzophenone, 4,4′ bis(dimethyl amino) $[4\text{-}(CH_3)_2NC_6H_4]_2CS$	284.42	pl		204			bz, chl, aa	B14^{3}, 233
13845	Thiocarbamoyl chloride, *N,N*-diethyl $(C_2H_5)_2NCSCl$	151.65	pr	108^{10}	48-51				B4^{4}, 389
13846	Thiocarbamoylchloride, *N,N*-dimethyl $(CH_3)_2NCSCl$	123.60	pr	98^{10}	42-3			eth, chl	B4^{3}, 147
13847	Thiochrome $C_{12}H_{14}N_4OS$	262.33	ye pr (chl)	sub, vac	227-8			w, MeOH	C29, 6242
13848	Thiocyanuric acid or Trithiocyanuric acid $C_3S_3N_3H_3$	177.26	ye pr		200d				B26, 256
13849	Thiodiglycolic acid $S(CH_2CO_2H)_2$	150.15	cr (AcOEt-bz, w)		129			w, al	B3^{4}, 612
13850	Thioglycolamide, *N*-β-naphthyl $HSCH_2CONH\text{-}(\beta\text{-}C_{10}H_7)$	217.29	nd		111-2			al	C29, 3330
13851	Thioglycolic acid $HSCH_2CO_2H$	92.11		120^{20}	−165	1.3253^{20}	1.5030^{20}	**w, al, eth**	B3^{4}, 600
13852	Thioglycolic acid, acetate or Acetylthioglycolic acid $CH_3COSCH_2CO_2H$	134.15	ye	158-9^{17}				w	B3^{4}, 610
13853	Thioindigo $C_{16}H_8O_2S_2$	296.36	br-red nd (xyl), red mcl nd (bz)	sub	359			xyl	B19^{4}, 2091
13854	Thiolane or Tetrahydrothiophene C_4H_8S	88.17		121.1, 14.5^{10}	−96.2	0.9987$^{20/4}$	1.5048^{20}	al, **eth, ace, bz,** chl	B17^{4}, 34
13855	Thiolane, 1,1-dioxide $C_4H_8SO_2$	120.17		285	27		1.4840^{20}	chl	B17^{4}, 37
13856	Thiolane, 2-methyl $2\text{-}CH_3(C_4H_7S)$	102.19		132.5^{750}		0.9541$^{20/4}$	1.4900^{20}	al, eth, ace, bz, chl	B17^{4}, 62
13857	Thiolane, 3-methyl $3\text{-}CH_3(C_4H_7S)$	102.19		137^{742}		0.9625$^{20/4}$	1.4917^{20}	al, eth, ace, bz, chl	B17^{4}, 64
13858	Thiomorpholine or 1,4-Thiazan $C_{14}H_9NS$	103.18		174^{746}, 110^{100}		1.0882$^{20/4}$	1.5386^{20}	w, al, eth, ace, bz	B27^{2}, 4
13859	Thionamide, tetramethyl $[(CH_3)_2N]_2SO$	136.21		209, 70^{16}	31			eth	B4^{4}, 269
13860	Thiomin hydrochloride or 7-Amino-3-imino-3H-2-phenothiazine hydrochloride Lauth's violet $C_{12}H_9N_3S\text{-}HCl$	263.74	dk br or gr pl or nd					bz, chl	B27^{2}, 447
13861	Thiophene or Thiofuran C_4H_4S	84.14		84.2	−38.2	1.0649$^{20/4}$	1.5289^{20}	**al, eth, ace, bz, Py**	B17^{4}, 234
13862	Thiophene, acetamido $2\text{-}CH_3CONH(C_4H_3S)$	141.19	lf (w)		161-2			al, ace	B17^{1}, 136
13863	Thiophene, 2-acetyl or Methyl-2-thienyl ketone $2\text{-}CH_3CO(C_4H_3S)$	126.17		213.5, 94-6^{13}	10-11	1.1679$^{20/4}$	1.5667^{20}	**al, eth**	B17^{4}, 4507
13864	Thiophene, 2-acetyl-5-bromo $5\text{-}Br\text{-}2\text{-}CH_3CO(C_4H_2S)$	205.07	nd (al)	105-7$^{4.5}$	94-5				B17^{4}, 4512
13865	Thiophene, 2-acetyl-5-chloro $5\text{-}Cl\text{-}2\text{-}CH_3CO(C_4H_2S)$	160.62	ta (al, eth)	88-9$^{4.5}$	52			w, al, eth	B17^{4}, 4510
13866	Thiophene, 2-acetyl-3-hydroxy $2\text{-}CH_3CO\text{-}3\text{-}HO(C_4H_2S)$	142.17		47-9$^{0.2}$	51.5-2.5		1.5795^{20}	al	TETRA21, 3331
13867	Thiophene, 2-acetyl-5-methyl $2\text{-}CH_3CO\text{-}5\text{-}CH_3(C_4H_2S)$	140.20		232-3, 98-100^{8}	27-8	1.1185$^{25/4}$	1.5604	eth, ace, bz	B17^{4}, 4550
13868	Thiophene, 2-amino or Thiophenine $2\text{-}H_2N(C_4H_3S)$	99.15	pa ye tab (al)	77-9^{11}				w, al	B17^{2}, 296
13869	Thiophene, 2-benzoyl or Phenyl 2-thienyl ketone $(C_4H_3S)COC_6H_5$	188.24	nd (dil al)	300	56-7	1.1890$^{54/4}$	1.6181^{54}	al, eth, ace	B17^{4}, 5187
13870	Thiophene, 3-benzoyl or Phenyl 3-thienyl ketone $3\text{-}C_6H_5CO(C_4H_3S)$	188.24		129-30^{3}	63-4			al, eth	B17^{4}, 5193

No.	Name, Synonyms, and Formula	Mol. wt.	Color, crystalline form, specific rotation and λ_{max} (log ε)	b.p. °C	m.p. °C	Density	n_D	Solubility	Ref.
13871	Thiophene, 2-bromo 2-Br-(C_4H_3S)	163.03		149.51, 42-6^{13}		$1.684^{20/4}$	1.5868^{20}	eth, ace	B17^4, 245
13872	Thiophene, 2-bromo-5-chloro 5-Cl-2-Br-(C_4H_2S)	197.48		70^{18}	−20	$1.803^{25/25}$	1.5925^{25}	eth, ace	B17^4, 246
13873	Thiophene, 2-bromo-5-iodo 5-I-2-Br-(C_4H_2O)	288.93		116^{13}					B17^4, 252
13874	Thiophene, 2-bromo-3-methyl 2-Br-3-CH_3(C_4H_2S)	177.06		175^{729}, 27$^{1.8}$		$1.5844^{18/4}$	1.5714^{20}	eth, bz	B17^4, 279
13875	Thiophene, 2-bromo-5-methyl 2-Br-5-CH_3(C_4H_2S)	177.06	col to pa ye	177^{740}, 29$^{1.8}$		$1.5529^{20/4}$	1.5673^{20}	eth, bz	B17^4, 272
13876	Thiophene, 3-bromo 3-Br(C_4H_3S)	163.03		159-60, 66-8^{31}		$1.735^{20/4}$	1.5919^{20}	ace, bz	B17^4, 245
13877	Thiophene, 2-butyl 2-C_4H_9(C_4H_3S)	140.24		181-2		$0.9537^{20/4}$	1.5090^{20}		B17^4, 305
13878	Thiophene, 2-chloro 2-Cl(C_4H_3S)	118.58		128.3	−71.9	$1.2863^{20/4}$	1.5487^{20}	**al, eth**	B17^4, 241
13879	Thiophene, 2-chloro-5-butyl 2-Cl-5-C_4H_9(C_4H_2S)	174.69		117-8^{88}		$1.0842^{17/4}$	1.5162^{20}	bz	B17, 44
13880	Thiophene, 2-chloro-5-iodo 2-Cl-5-I(C_4H_2S)	244.48		95-6^{14}	−25				B17^4, 252
13881	Thiophene, 2-chloro-5-methyl 2-Cl-5-CH_3(C_4H_2S)	132.61		154-5^{742}, 55^{19}		$1.2147^{25/4}$	1.5372^{20}	al, eth, ace, bz	B17^4, 271
13882	Thiophene, 2,5-dibromo 2,5-Br_2(C_4H_2S)	241.93		210.3, 76-80^{10}	−6	$2.147^{23/23}$	1.6288^{20}	al, eth	B17^4, 248
13883	Thiophene, 2,5-dibromo 3,4-dinitro 3,4-$(NO_2)_2$-2,5-Br_2(C_4S)	331.92	cr (al)		139-40				B17^4, 261
13884	Thiophene, 2,3-dichloro 2,3-Cl_2(C_4H_2S)	153.03		173-4	−26.2	$1.4605^{20/4}$	1.5651^{20}	**al**, eth, ace, bz	B17^4, 242
13885	Thiophene, 2,4-dichloro 2,4-Cl_2(C_4H_2S)	153.03		167.6	−34	$1.4553^{20/4}$	1.5660^{20}	al, eth, ace, bz	B17^4, 242
13886	Thiophene, 2,5-dichloro 2,5-Cl_2(C_4H_2S)	153.03		162	−40.5	$1.4422^{20/4}$	1.5626^{20}	**al, eth**	B17^4, 243
13887	Thiophene, 2,5-diido 2,5-I_2(C_4H_2S)	335.93	lf (al)	139-40^{15}	41-2			al	B17^4, 253
13888	Thiophene, 2,3-dimethyl or 2,3-Thioxene 2,3-$(CH_3)_2$(C_4H_2S)	112.19		141.6, 29.3^{10}	−49	$1.0021^{20/4}$	1.5192^{20}	al, eth, bz	B17^4, 287
13889	Thiophene, 2,4-dimethyl or 2,4-Thioxene 2,4$(CH_3)_2$(C_4H_2S)	112.19		140.7, 29.9^{10}		$0.9956^{20/20}$	1.5104^{20}	al, eth, bz	B17^4, 288
13890	Thiophene, 2,5-dimethyl or α,α′-Thioxene 2,5-$(CH_3)_2$(C_4H_2S)	112.19		136.7, 26.2^{10}	−62.6	$0.985^{20/4}$	1.5129^{20}	al, eth, bz	B17^4, 290
13891	Thiophene, 3,4-dimethyl 3,4-$(CH_3)_2$(C_4H_2S)	112.19		144-6^{762}, 70-1^{55}		$0.994^{25/15}$	1.5206^{20}	al, eth	B17^4, 293
13892	Thiophene, 2,5-dinitro 2,5-$(NO_2)_2C_4H_2S$)	174.13	(i) ye lf (al) (ii) ye nd (al, w)	290	(i)52 (ii)80-2			al, eth	B17^4, 260
13893	Thiophene, 2,4-diphenyl 2,4-$(C_6H_5)_2$(C_4H_2S)	236.33			119-20			al, ace, chl	B17^4, 680
13894	Thiophene, 2-ethyl 2-C_2H_5(C_4H_3S)	112.19		104, 24^{10}		$0.9930^{20/4}$	1.5122^{20}	al, eth	B17^4, 285
13895	Thiophene, 3-ethyl 3-C_2H_5(C_4H_3S)	112.19		136, 26^{10}	−89.1	$0.9980^{20/4}$	1.5146^{20}	al, eth	B17^4, 286
13896	Thiophene, 2-hydroxy 2-HO(C_4H_3S)	100.14		217-9, 91-3^{13}		$1.255^{20/4}$	1.5644^{20}	w, ace, chl	B17^4, 4286
13897	Thiophene, 2-hydroxy-5-methyl or 2,5-Thiotenol 5-CH_3-2-HO(C_4H_2S)	114.16		85^{40}				al, eth	B17^4, 4301
13898	Thiophene, 2-hydroxymethyl 2-$HOCH_2$(C_4H_3S)	114.16		207, 96^{12}		$1.2053^{16/4}$	1.5280^{20}	al, eth, ace	B17^4, 1242
13899	Thiophene, 2-iodo 2-I-(C_4H_3S)	210.03		180-2, 73^{15}	−40		1.6465^{25}	al, eth	B17^4, 250
13900	Thiophene, 2-iodo-5-nitro 5-NO_2-2-I(C_4H_2S)	255.03	ye pr (al)		74			al	B17^4, 259
13901	Thiophene, 2-methyl or α-Thiololene 2-CH_3(C_4H_3S)	98.16		112.6, 9.2^{10}	−63.4	$1.0193^{20/4}$	1.5203^{20}	**al, eth, ace, bz**	B17^4, 269
13902	Thiophene, 3-methyl or β-Thiololene 3-CH_3(C_4H_3S)	98.16		115.4, 11^{10}	−69	$1.0218^{20/4}$	1.5204^{20}	**al, eth, ace, bz,** chl	B17^4, 277
13903	Thiophene, 2-methyl-5-phenyl 2-CH_3-5-C_6H_5(C_4H_2S)	174.26	nd	270-2	51			al, eth, lig	B17^4, 551

PHYSICAL CONSTANTS OF ORGANIC COMPOUNDS (Continued)

No.	Name, Synonyms, and Formula	Mol. wt.	Color, crystalline form, specific rotation and λ_{max} (log ε)	b.p. °C	m.p. °C	Density	n_D	Solubility	Ref.
13904	Thiophene, 2-methylamino or *N*-Methyl-2-thiophenine $2\text{-}CH_3NH(C_4H_3S)$	113.18		$88\text{-}92^{15}$				eth, ace	B17[1], 136
13905	Thiophene, 2-nitro $2\text{-}O_2N(C_4H_3S)$	129.13	lf ye mcl nd (peth)	224-5	46.5	$1.3644^{43/4}$		al	B17[4], 255
13906	Thiophene, 3-nitro $3\text{-}3NO_2(C_4H_3S)$	129.13		$225, 95^{12}$	78-9			al, lig	B17[4], 256
13907	Thiophene, 3-nitro-2,4,5-trichloro $2,4,5\text{-}Cl_3\text{-}3\text{-}O_2N(C_4S)$	232.47	red-ye nd (al)		86			al, eth, bz	B17, 35
13908	Thiophene, 4-nitro-2,3,5-tribromo $2,3,5\text{-}Br_3\text{-}4\text{-}O_2N(C_4S)$	365.82	red-ye nd (al)		106			eth	B17, 35
13909	Thiophene, 2-octyl $2\text{-}C_8H_{17}(C_4H_3S)$	196.35		257-9		$0.8118^{20.5}_{20.5}$		eth	B17[4], 335
13910	Thiophene, 2-isopentyl $2\text{-}i\text{-}C_5H_{11}(C_4H_3S)$	154.27		$74\text{-}5^{11}$		$0.9481^{12/4}$	$1.5014^{12/}_{587.6}$		B17[4], 316
13911	Thiophene, 2-propyl $2\text{-}C_3H_7(C_4H_3S)$	126.22		157-9, 55.5^{20}		$0.9683^{20/4}$	1.50757^{20}	al, eth, bz	B17[4], 297
13912	Thiophene, 3-propyl $3\text{-}C_3H_7(C_4H_3S)$	126.22		163.2		$0.97377^{20/4}$	1.5057^{20}	al, eth, bz	B17[2], 42
13913	Thiophene, tetrabromo C_4Br_4S	399.72	nd (al)	326, $170\text{-}3^{14}$	117-8			eth	B17[4], 250
13914	Thiophene, tetrachloro C_4Cl_4S	221.92	nd (dil al)	233.4, $75\text{-}7^{2}$	30-1	$1.7036^{30/4}$	1.5915^{30}	al, **eth**	B17[4], 244
13915	Thiophene, tetrahydro or Thiolane C_4H_8S	88.17		121.1, $14\text{-}5^{10}$	−96.6	$0.9987^{20/4}$	1.5018^{20}	**al, eth, ace, bz, chl**	B17[4], 34
13916	Thiophene, tetraphenyl $(C_6H_5)_4(C_4S)$	388.53		400	189-90			eth, bz	B17[4], 810
13917	Thiophene, 2,3,5-tribromo $2,3,5\text{-}Br_3(C_4HS)$	320.82	nd (al)	260	29			al, eth	B17[4], 249
13918	Thiophene, 2,3,5-trichloro $2,3,5\text{-}Cl_3(C_4HS)$	187.47		198.7	−16.1	$1.5856^{20/4}$	1.5791^{20}	al, eth	B17[4], 244
13919	Thiophene, 2,3,5-trimethyl $2,3,5\text{-}(CH_3)_3(C_4HS)$	126.22		164.5, 46.8^{10}		$0.9753^{20/4}$	1.5112^{20}	al, eth, ace, bz	B17[4], 301
13920	2-Thiophenacetic acid $2\text{-}HO_2CH_2C(C_4H_8S)$	142.17	cr (w)		76			al, eth	B18[4], 4062
13921	2-Thiophene carboxaldehyde or 2-Thiophenealdehyde $2\text{-}OHC(C_4H_3S)$	112.15	pa ye liq	$197, 85\text{-}6^{16}$		$1.215^{21/21}$	1.5920^{20}	al, eth	B17[4], 4477
13922	2-Thiophene carboxaldehyde oxime or 2 Thiophenaldoxime $2\text{-}(C_4H_3S)CH{=}NOH$	127.16	nd		133			eth	B17[4], 4482
13923	2-Thiophene carboxaldehyde phenylhydrazone $2\text{-}(C_4H_3S)CH{=}NNHC_6H_5$	202.27	ye nd (al)		134.5			al	B17[4], 4482
13924	2-Thiophene carboxaldehyde, 5-bromo $5\text{-}Br\text{-}(C_4H_2S)\text{-}CHO\text{-}2$	191.04		$105\text{-}7^{11}$			1.6378^{20}		B17[4], 4487
13925	2-Thiophene carboxaldehyde, 5-chloro $5\text{-}Cl\text{-}(C_4H_2S)CHO\text{-}2$	146.59		77.5^{5}			1.6036^{25}		B17[4], 4485
13926	2-Thiophene carboxaldehyde, 5-methyl $5\text{-}CH_3\text{-}(C_4H_2S)CHO\text{-}2$	126.17		114^{25}			1.5825^{20}	chl	B17[4], 4529
13927	3-Thiophene carboxaldehyde $3\text{-}(C_4H_3S)CHO$	112.15		$86\text{-}7^{20}$			1.5855^{20}	al, eth	B17[4], 4497
13928	2-Thiophene carboxylic acid or α-Thiophenic acid $2\text{-}(C_4H_3S)CO_2H$	128.15	nd (w)	260d	129-30			w, al, eth, chl	B18[4], 4011
13929	2-Thiophene carboxylic acid, ethyl ester or Ethyl-α-thiophene carboxylate $2\text{-}(C_4H_3S)CO_2C_2H_5$	156.20		218, 94^{10}		$1.1623^{16/4}$	1.5248^{20}	al, ace	B18[4], 4012
13930	2-Thiophene carboxylic acid hydrazide $2\text{-}(H_2NNHCO)(C_4H_3S)$	142.17			136				B18[4], 4024
13931	2-Thiophene carboxylic acid, 3-methyl $3\text{-}CH_3(C_4H_2S)CO_2H\text{-}2$	142.17	nd (dil al, w)		144			al, eth, ace	B18[4], 4068
13932	2-Thiophene carboxylic acid, 5-methyl $5\text{-}CH_3(C_4H_3S)CO_2H\text{-}2$	142.17			138-9			al, eth	B18[4], 4086
13933	2-Thiophene (carboxylyl) chloride $2\text{-}(C_4H_3S)COCl$	146.59		$201, 77^{10}$					B18[4], 4016
13934	3-Thiophene carboxylic acid $3\text{-}HO_2C(C_4H_3S)$	128.15			138.4			w	B18[4], 4054
13935	2,3-Thiophene dicarboxylic acid $2,3\text{-}(C_4H_2S)(CO_2H)_2$	172.16	pr or nd (w)		272-4			eth	B18[4], 4478
13936	2,4-Thiophene dicarboxylic acid $2,4\text{-}(C_4H_2S)(CO_2H)_2$	172.16	cr	sub>200	280d				B18[4], 4480

No.	Name, Synonyms, and Formula	Mol. wt.	Color, crystalline form, specific rotation and λ_{max} (log ε)	b.p. °C	m.p. °C	Density	n_D	Solubility	Ref.
13937	2,5-Thiophene dicarboxylic acid $2,5-(C_4H_2S)(CO_2H)_2$	172.16	...	sub 150-300	359 (sealed tube)	...	...	al, eth	B18[4], 4495
13938	2,5-Thiophene dicarboxylic acid, diethyl ester or Diethyl-2,5-thiophene dicarboxylate $2,5-(C_4H_2S)(CO_2C_2H_5)_2$	228.26	nd (al)	...	51.5	...	...	al	B18, 331
13939	2-Thiophene sulfonamide $2-(C_4H_3S)SO_2NH_2)$	163.21	nd (w)	...	147	...	...	...	B18[4], 6706
13940	2-Thiophene sulfonyl chloride $2-(C_4H_3S)(SO_2Cl)$	182.64	...	99-101^{6}, sub	28	...	...	eth	B18[4], 6706
13941	3-Thiophene sulfonamide $3-(C_4H_3S)(SO_2NH_2)$	163.21	pl (w)	...	152-3	...	...	...	B18[4], 6714
13942	3-Thiophene sulfonyl chloride $3-(C_4H_3S)(SO_2Cl)$	182.64	cr (eth)	98-9$^{0.5}$	43	...	...	eth	B18[4], 6713
13943	Thiophenol or Mercaptobenzene C_6H_5SH	110.17	...	168.7, 46.4^{10}	−14.8	1.0766$^{20/4}$	1.5893^{20}	al, eth, bz	B6[4], 1463
13944	Thiophenol, *o*-amino $2H_2NC_6H_4SH$	125.19	nd	234, 125-7^{6}	26	...	1.4606^{20}	al, eth	B13[3], 902
13945	Thiophenol, *m*-amino $3-H_2N-C_6H_4SH$	125.19	pa ye oil	180-90^{16}	...	...	...	al, aa, eth	B13[3], 982
13946	Thiophenol, *m*-amino, hydrochloride $3-H_2NC_6H_4SH.HCl$	161.65	...	sub	232	...	...	w, al	B13, 425
13947	Thiophenol, *p*-amino $4H_2NC_6H_4SH$	125.19	cr	140-5^{15}	46	...	...	w, al	B13[3], 1221
13948	Thiophenol, *p*-bromo $4-BrC_6H_4SH$	189.07	lf (al)	230-1	73	...	...	eth, chl	B6[4], 1650
13949	Thiophenol, *o*-chloro $2-Cl_6H_4SH$	144.62	...	205-6, 117^{15}	...	1.2752^{10}	...	...	B6[4], 1570
13950	Thiophenol, *m*-chloro $3-ClC_6H_4SH$	144.62	...	205-7	...	1.2637^{13}	...	al, eth, peth, chl	B6[4], 1576
13951	Thiophenol, *p*-chloro $4-ClC_6H_4SH$	144.62	...	198-9	61	1.1911$^{20/4}$	1.5480^{20}	al, eth	B6[4], 1581
13952	Thiophenyl, *m*-ethoxy $3-C_2H_5OC_6H_4SH$	154.23	...	238-9	...	...	...	al, eth, ace, bz	B6, 833
13953	Thiophenol, *p*-ethoxy $4-C_2H_5OC_6H_4SH$	154.23	...	238	1.6	...	...	al, eth, ace, bz	B6[3], 4446
13954	Thiophenol, *m*-methoxy $3-CH_3OC_6H_4SH$	140.20	...	223-6	...	...	1.5874^{20}	chl	B6[3], 4363
13955	Thiophenol, *p*-methoxy or 4-Mercaptoanisole $4-CH_3OC_6H_4SH$	140.20	...	227-9, 89-90^{5}	...	1.1313$^{25/4}$	1.5801^{25}	al, eth, bz	B6[3], 4445
13956	Thiophenol, *o*-nitro $2-NO_2C_6H_4SH$	155.17	...	...	58.5	...	...	al, eth, bz	B6[4], 1661
13957	Thiophenol, *p*-nitro $4-NO_2C_6H_4SH$	155.18	(eth, chl, ace)	...	79	...	...	al, eth, ace	B6[4], 1687
13958	Thiophenol, 2,4,6-trinitro or Thiopicric acid $2,4,6-(NO_2)_3C_6H_2SH$	245.17	ye nd	...	114	...	...	al, eth, ace, bz, chl	B6[2], 316
13959	Thiophosgene or Thiocarbonyl chloride $CSCl_2$	114.98	red	73	...	1.508^{15}	1.5442^{20}	al, eth	B3[4], 281
13960	Thiophthene (solid) or Thieno[3,2-b]thiophene $C_6H_4S_2$	140.22	bipym orh (lig)	221-4	56	...	...	eth	B19[4], 189
13961	Thiophthene (liquid) or Thieno[3,2-b]thiophene $C_6H_4S_2$	140.22	...	224-6, 106^{16}	6.5	...	...	eth	B19[4], 189
13962	Thiopyran, tetrahydro or Pentamethylene sulfide $C_5H_{10}S$	102.19	...	141.7, 93^{82}	19	0.9861$^{20/4}$	1.5067^{20}	al, eth, ace, bz	B17[4], 55
13963	Thiopyrine or 1,5-Dimethyl-2-phenyl-3-thio-3-pyrazolone $C_{11}H_{12}N_2S$	204.29	cr (w)	...	166	...	...	al, eth	B24[2], 28
13964	Thiosalicylic acid or 2-Mercaptobenzoic acid $2-HSC_6H_4CO_2H$	154.18	lf or nd (al, w, aa)	sub	168-9	...	...	al, eth, aa	B10[3], 265
13965	Thiosemicarbazide or Semicarbazide-3-thio $H_2NNHCSNH_2$	91.13	lo nd (w)	...	183	...	...	w, al	B3[4], 374
13966	2-Thiouracil $C_4H_4N_2OS$	128.15	pr (w, al)	...	>340d	...	...	...	B24[2], 171
13967	2-Thiouracil, 6-methyl $C_5H_6N_2OS$	142.18	pl (w)	...	299-303d	...	...	...	B24[2], 183
13968	2-Thiouracil, 6-propyl $C_7H_{10}N_2OS$	170.23	pw (w)	...	219	...	...	...	C43, 674

No.	Name, Synonyms, and Formula	Mol. wt.	Color, crystalline form, specific rotation and λ_{max} (log ε)	b.p. °C	m.p. °C	Density	n_D	Solubility	Ref.
13969	4-Thiouracil $C_4H_4N_2OS$	128.15	yesh nd or pr (w)		328d				B24, 323
13970	4-Thiouracil, 6-methyl $C_5H_6N_2OS$	142.18	ye pr (w)		>250d				B24, 352
13971	Thiourea H_2NCSNH_2	76.12	rh (al)		182	1.405		w, al	B3^{4}, 342
13972	Thiourea, 1-acetyl $CH_3CONHCSNH_2$	118.15	pr (w), rh (al)		165			al	B3^{4}, 354
13973	Thiourea, *S*-acetyl, hydrochloride $H_2NC(=NH)SCOCH_3.HCl$	154.61	pl		109d			w	B3^{3}, 314
13974	Thiourea, 1-allyl $(CH_2=CHCH_2)NHCSNH_2$	116.18	mcl or rh, pr (w)		(i)71, (ii)78.4	$1.219^{20/20}$	1.5936^{78}	w, al	B4^{4}, 1072
13975	Thiourea, 1-allyl-3-phenyl $(CH_2=CHCH_2)NHCSNHC_6H_5$	192.28			98				B12^{3}, 855
13976	Thiourea, 1-(4-aminobenzene sulfonyl) or 1-Sulfanyl thiourea $(4\text{-}H_2NC_6H_4SO_2)NHCSNH_2$	231.29	pw, pl (w)		182				B14^{3}, 1975
13977	Thiourea, 1-benzoyl $C_6H_5CONHCSNH_2$	180.22	pr (dil al)		171			al	B9^{3}, 1120
13978	Thiourea, 1-benzyl $C_6H_5CH_2NHCSNH_2$	166.24	pr (w)		164-5				B12^{3}, 2277
13979	Thiourea, 1-benzyl 3-phenyl $C_6H_5CH_2NHCSNHC_6H_5$	242.34			153-4			al, eth	B12^{2}, 564
13980	Thiourea, 1-butyl-3-phenyl $C_4H_9NHCONHC_6H_5$	208.32	pr		85			bz	B12^{3}, 853
13981	Thiourea, 1-(2-chlorophenyl) $2\text{-}ClC_6H_4NHCSNH_2$	186.66	nd or pl		146			al, bz	B12^{3}, 1296
13982	Thiourea, 1-(4-chlorophenyl) $4\text{-}ClC_6H_4NHCSNH_2$	186.66	pl or nd (al)		178			al, bz	B12^{3}, 1368
13983	Thiourea, 1,3-dibenzyl $(C_6H_5CH_2NH)_2CS$	256.37			147-8			al, eth	B12^{3}, 2278
13984	Thiourea, 1,3-diethyl $(C_2H_5NH)_2CS$	132.22		d	144			w, al, eth	B4^{4}, 375
13985	Thiourea, 1,3-diethyl-1,3-diphenyl $[C_6H_5N(C_2H_5)]_2CS$	284.42	rh pl (lig), pr (al)		75.5			al	B12^{3}, 884
13986	Thiourea, 1,3-dimethyl $(CH_3NH)_2CS$	104.17	dlq pl		62			w, al, ace, chl	B12^{4}, 217
13987	Thiourea, 1,3-di(α-naphthyl) $(\alpha\text{-}C_{10}H_7NH)_2CS$	328.43	lf (to)		207.5			to	B12^{2}, 696
13988	Thiourea, 1,3-di(β-naphthyl) $(\beta\text{-}C_{10}H_7NH)_2CS$	328.43	lf (aa)		203				B12^{3}, 3048
13989	Thiourea, 1,1-diphenyl $(C_6H_5)_2NCSNH_2$	228.31	pr (al)		210d, (218)			al	B12^{3}, 898
13990	Thiourea, 1,3-diphenyl $(C_6H_5NH)_2CS$	228.31	lf (al)		154, (189)			al, eth, chl	B12^{3}, 858
13991	Thiourea, 1,1-dipropyl $(C_3H_7)_2NCSNH_2$	160.28			67				B4, 144
13992	Thiourea, 1,3-dipropyl $(C_3H_7NH)_2CS$	160.28	lf (w)		71			eth	B4, 143
13993	Thiourea, 1,3-diisopropyl $(i\text{-}C_3H_7NH)_2CO$	160.28			141			chl	B4^{4}, 526
13994	Thiourea, 1,3-di(2-tolyl) $(2\text{-}CH_3C_6H_4NH)_2CS$	256.37	nd (al, sub)	216-8 (sub)	165-6			al, bz, chl, aa	B12^{3}, 1881
13995	Thiourea, 1,3-di(4-tolyl) $(4\text{-}CH_3C_6H_4NH)_2CS$	256.37	rh bipym pr		176			eth	B12^{3}, 2099
13996	Thiourea, 1-ethyl-3-phenyl $C_2H_5NHCSNHC_6H_5$	180.27			107			al, bz	B12^{3}, 853
13997	Thiourea, ethylidene $CH_3CH=NCSNH_2$	102.15	(al)		*ca* 212d				B3^{1}, 76
13998	Thiourea, *S*- methyl $HN=C(SCH_3)NH_2$	90.14	lf (ace)		79			w, al, ace	B3^{4}, 358
13999	Thiourea, *S*-methyl, hydroiodide or *S*-Methyl isothiouronium iodide $HN=C(SCH_3)NH_2.HI$	218.06	pr		117			w, al	B3^{1}, 78
14000	Thiourea, *S*-methyl, nitrate or *S*-Methyl isothiouronium nitrate $HN=C(SCH_3)NH_2.HNO_3$	153.16			109-10			w, al	B3^{1}, 78

No.	Name, Synonyms, and Formula	Mol. wt.	Color, crystalline form, specific rotation and λ_{max} (log ε)	b.p. °C	m.p. °C	Density	n_D	Solubility	Ref.
14001	Thiourea, *S*-methyl, sulfate or *S*-Methyl isothiouronium sulfate $[HN{=}C(SCH_3)NH_2]_2.H_2SO_4$	278.36	nd (al, w)		244d				$B3^4$, 358
14002	Thiourea, *S*-methyl-1,3-diphenyl $C_6H_5N{=}C(SCH_3)NHC_6H_5$	242.34	nd (al)		109-10			al	$B12^3$, 911
14003	Thiourea, 1-methyl $CH_3NHCSNH_2$	90.14	pr		120-1			w, al, ace	$B4^4$, 216
14004	Thiourea, 1-methyl-3-(α-naphthyl) α-$C_{10}H_7NHCSNHCH_3$	216.30	pl (al)		198			al	$B12^2$, 696
14005	Thiourea, 1-methyl-3-phenyl $CH_3NHCSNHC_6H_5$	166.24	ta, pl		112-3			al	$B12^3$, 853
14006	Thiourea, 1-(α-naphthyl) α-$C_{10}H_7NHCSNH_2$	202.27	pr (al)		198				$B12^3$, 2941
14007	Thiourea, 1-(α-naphthyl)-3-phenyl α-$C_{10}H_7NHCSNHC_6H_5$	278.37	lf		162-3				B12, 1241
14008	Thiourea, 1-β-naphthyl β-$C_{10}H_7NHCSNH_2$	202.27	lf (al)		186			al	$B12^3$, 3045
14009	Thiourea, *S*-phenyl $HN{=}C(SC_6H_5)NH_2$	152.22	nd (bz)		96-7d			bz	$B6^1$, 146
14010	Thiourea, 1-phenyl $C_6H_5NHCSNH_2$	152.22	nd (w), pr (al)		154			al	$B12^3$, 852
14011	Thiourea, 1,1,3,3,-tetramethyl $[(CH_3)_2N]_2CS$	132.22		245	78-9			w, al	$B4^4$, 232
14012	Thiourea, 1,1,3,3-tetraphenyl $[(C_6H_5)_2N]_2CS$	380.51	nd (al, MeOH)		194-5			eth, bz	$B12^3$, 898
14013	Thiourea, 1-(2-tolyl) $(2\text{-}CH_3C_6H_4)NHCSNH_2$	166.24	nd (dil al, w)		162			w, al	$B12^3$, 1878
14014	Thiourea, 1-(3-tolyl) $(3\text{-}CH_3C_6H_4)NHCSNH_2$	166.24	pr (al)		110-1			w	$B12^3$, 1975
14015	Thiourea, 1-(4-tolyl) $(4\text{-}CH_3C_6H_4)NHCSNH_2$	166.24	pl (al)		188-9			al	$B12^3$, 2097
14016	Thiourea, 1,1,3-trimethyl $(CH_3)_2NCSNHCH_3$	118.20	pr (bz-lig)		87-8			w, al, bz, chl	$B4^4$, 232
14017	Thiourea, 1,1,3-triphenyl $(C_6H_5)_2NCSNHC_6H_5$	304.41	nd (al)		152				B12, 432
14018	Thioxanthene or Dibenzthiopyran $C_{13}H_{10}S$	198.28	nd (al-chl)	340^{710}, sub	128-9			chl, peth	$B17^4$, 615
14019	Thioxanthone or 9-Oxothioxanthene $C_{13}H_8OS$	212.27	ye nd (chl)	$371\text{-}3^{715}$, sub	209 (212)			bz, aa	$B17^4$, 5303
14020	D-Threonic acid or 2,3,4-Trihydroxy butyric acid $HOCH_2(CHOH)_2CO_2H$	136.10	nd (al), $[\alpha]_D$ −30 (w)		197-8			w	$B3^4$, 1112
14021	DL-Threonic acid $HOCH_2(CHOH)_2CO_2H$	136.10	cr (al, ace)		99			w	$B3^3$, 895
14022	L-Threonic acid $HOCH_2(CHOH)_2CO_2H$	136.10	nd (al-eth), $[\alpha]_D$ + 9.5 (w)		169-70			w, ace	$B3^4$, 1112
14023	Threonine *(D)* or *D*-2-Amino-3-hydroxy butyric acid $CH_3CH(OH)CH(NH_2)CO_2H$	119.12	cr (80% al), $[\alpha]^{26}/_D$ −28.3 (w, c=1.1)		255-7d			w	$B4^3$, 1625
14024	Threonine *(DL)* or *DL*-2-Amino-3-hydroxy butyric acid $CH_3CH(OH)CH(NH_2)CO_2H$	119.12	cr (dil al)		234-5d, 229-30 (+ ½ w)			w	$B4^3$, 1625
14025	D-Threose or D-Trihydroxy butyraldehyde $HOCH_2(CHOH)_2CHO$	120.11	hyg syr or nd(w), $[\alpha]^{22}/_D$ + 29.1→ 19.6 (w, mut)		126-32			w	$B1^4$, 4173
14026	Thujane *(d)* or Sabinane $C_{10}H_{18}$	138.25	$[\alpha]_D$ + 73.1	157^{758}		$0.8139^{20/4}$	1.4376^{20}	al, eth, ace, bz	$B5^4$, 317
14027	3-Thujene $C_{10}H_{16}$	136.24	$[\alpha]^t/_D$ −37.20	151		$0.8301^{20/4}$	1.4515^{20}		$B5^4$, 451
14028	4(10)-Thujene *(d)* or Sabinene $C_{10}H_{16}$	136.24	$[\alpha]_D$ + 80.2	163-5, 66^{30}		0.842^{20}	1.4678^{20}	al, eth, ace, bz, chl	$B5^4$, 451
14029	4(10)Thujene *(l)* or *l*-Sabinene $C_{10}H_{16}$	136.24	$[\alpha]^{15}/_D$ −42.5	163-5		$0.8464^{20/4}$	1.4515^{20}	al, eth, ace, bz, chl	$B5^4$, 452

PHYSICAL CONSTANTS OF ORGANIC COMPOUNDS (Continued)

No.	Name, Synonyms, and Formula	Mol. wt.	Color, crystalline form, specific rotation and λ_{max} (log ε)	b.p. °C	m.p. °C	Density	n_D	Solubility	Ref.
14030	3-Thujone $C_{10}H_{16}O$	152.24	$[\alpha]^{18/}{}_D$ −19.9	200-1, 75^{9}	......	0.9152^{20}	1.4490^{25}	**al, eth**, ace	B7^3, 379
14031	Thymidine or Thymine-β-D-2-desoxy-riboside $C_{10}H_{14}N_2O_5$	242.23	nd (AcOEt), $[\alpha]^{25/}{}_D$ + 30.6 (w, c=1.03)	......	186-7	......	......	w, MeOH, aa, Py	C48, 226
14032	Thymol or 2-Methyl-5-isopropyl phenol 2-CH_3-5-*i*-$C_3H_7C_6H_3OH$	150.22	pl (aa, ace)	233, 92^{2}	52	$0.925^{80/4}$	1.5227^{20}	al, eth	B6^3, 1893
14033	Thymol acetate 5-CH_3-2-$(CH_3)_2CHC_6H_3O_2CCH_3$	192.26	......	245, 131^{21}	......	1.009^{9}	......	**al, eth, bz, chl**	B6^3, 1901
14034	Thymol, 4-amino or 2-Amino-3-methyl-6-*iso*-propyl phenol 2-H_2N-3-CH_3-6-*i*-C_3H_7-C_6H_2OH	165.24	nd, sc (bz)	......	178-9	......	......	al, bz	B13^3, 1803
14035	Thymol carbonate or Thymatol Tyranol 2-CH_3-5-*i*-$C_3H_7C_6H_3OH.H_2CO_3$	212.25	nd or pr	......	49	......	......	eth, chl	B6^2, 499
14036	Thymol, 6-chloro or 4-Chloro-2-*iso*-propyl-5-methyl phenol 4-Cl-2-*i*-C_3H_7-5-$CH_3C_6H_2OH$	184.67	nd or pl (lig)	259-63	59-61	......	......	w, al, eth, bz, peth	B6^3, 1906
14037	Thymol, 2,6-dinitro or 2,4-Dinitro-3-methyl-6-*iso*-propyl phenol 2,4$(NO_2)_2$ 3 CH_3-6-$[(CH_3)_2CH]C_6HOH$	240.22	pr (peth)	......	55.5	......	......	al, eth	B6^3, 1911
14038	α-Thymolsulfonic acid or 4-Hydroxy-5-*iso*-propyl-2 methyl benzene sulfonic acid 4-HO-5$(CH_3)_2CH$-2-$CH_3C_6H_2SO_3H$	230.28	pl (+ 1w)	......	91-2	......	......	w	B11^3, 537
14039	Thymol blue or Thymolsulfonephthalein $C_{27}H_{30}O_5S$	466.59	gr-red (al, eth, aa)	......	221-4d	......	......	al, aa	B19^4, 1135
14040	Thymolphthalein $C_{28}H_{30}O_4$	430.56	pr or nd (al)	......	253	......	......	eth, ace	B18^4, 1955
14041	Thyroxine *(d)* $C_{15}H_{11}I_4NO_4$	776.87	nd $[\alpha]^{25/}{}_{546}$ + 3 (al-NaOH)	......	237d	......	......	......	B14^3, 1566
14042	Thyroxine-*(l)* $C_{15}H_{11}I_4NO_4$	776.87	nd $[\alpha]^{20/}{}_D$ −4.4 (al-NaOH)	......	235-6	......	......	......	B14^3, 1566
14043	Tiglaldehyde or 2-Methyl-2-butenal $CH_3CH{=}C(CH_3)CHO$	84.12	......	116-7^{738}, 63-5^{119}	......	$0.8710^{20/4}$	1.4475^{20}	w, **al, eth**	B1^4, 3464
14044	Tiglic acid or *trans*-2-Methyl -2-butenoic acid $CH_3CH{=}C(CH_3)CO_2H$	100.12	ta (w)	198.5	64-5	$0.9641^{20/4}$	1.4330^{76}	al, eth	B2^4, 1552
14045	Tiglyl chloride or *trans*-2-Methyl-2-butenyl chloride $CH_3CH{=}C(CH_3)COCl$	118.56	......	146-8, 45^{12}	......	......	......	eth	B2^4, 1554
14046	Tigogenin or 5-α,22α-Spirostan-3β-ol $C_{27}H_{44}O_3$	416.64	lf (al + 1w), pr (ace), $[\alpha]^{25/}{}_D$ −49 (Py, c=1)	......	205-6	......	......	eth, ace, MeOH, peth	B19^4, 828
14047	α-Tocopherol or Vitamin E. 5,7,8-Trimethyl tocol $C_{29}H_{50}O_2$	430.71	pa ye visc oil $[\alpha]^{25/}{}_D$ + 0.65 (al)	350d, 140^{10-6}	2-3	......	......	al, eth, ace	B17^4, 1436
14048	α-Tocopherol acetate $C_{31}H_{52}O_3$	472.75	......	$184^{0.01}$	−27.5	$0.9533^{21.5}$	1.495-1.497^{20}	eth, ace, chl	B17^4, 1439
14049	α-Tocopherolquinone or α-tocoquinone $C_{29}H_{50}O_3$	446.71	ye oil	$120^{0.002}$	......	......	......	eth, peth	**Am73, 5148**
14050	β-Tocopherol or Vitamin E. 5,8-Dimethyl tocol $C_{28}H_{48}O_2$	416.69	pa ye visc oil $[\alpha]^{20/}{}_D$ + 6.4	200-$10^{0.1}$	......	......	......	**al, eth, ace, chl**	B17^4, 1427
14051	γ-Tocopherol or Vitamin E. 7,8-Dimethyltocol $C_{28}H_{48}O_2$	416.69	pa ye visc oil $[\alpha]^{20/}{}_D$ −2.4 (al)	200-$10^{0.1}$	−3	......	......	**al, eth, ace, chl**	B17^4, 1429
14052	γ-Tocopherol or γ-Methyltocol $C_{27}H_{46}O_2$	402.66	pa ye visc oil $[\alpha]^{25/}{}_{546}$ + 3.4 (al, c=1.5)	$150^{00.001}$	......	......	......	al, eth, ace, chl	B17^4, 1411

No.	Name, Synonyms, and Formula	Mol. wt.	Color, crystalline form, specific rotation and λ_{max} (log ε)	b.p. °C	m.p. °C	Density	n_D	Solubility	Ref.
14053	Tolbutamide or 1-Butyl-3(*p*-tolylsulfonyl)urea $C_{12}H_{18}N_2O_3S$	270.35	orth cr		128-9	1.245^{25}		al, eth, chl	C53, 13084
14054	*o*-Tolualdehyde or 2-Methylbenzaldehyde $2\text{-}CH_3C_6H_4CHO$	120.15		200, 94^{10}		1.0386$^{19/4}$	1.5481^{20}	al, eth, ace, bz, chl	B7^{3}, 1011
14055	*m*-Tolualdehyde or 3-Methylbenzaldehyde $3\text{-}CH_3C_6H_4CHO$	120.15		199, 93-4^{17}		1.0189^{21}		**al, eth**, ace, bz, chl	B7^{3}, 1013
14056	*p*-Tolualdehyde or 4-Methylbenzaldehyde $4\text{-}CH_3C_6H_4CHO$	120.15		204-5, 106^{10}		1.0194$^{17/4}$	1.5454^{20}	**al, eth, ace**, chl	B7^{3}, 1016
14057	*o*-Toluamide or 2-Methylbenzamide $2\text{-}CH_3C_6H_4CONH_2$	135.17	pl or nd (w)		147			al	B9^{3}, 2304
14058	*m*-Toluamide or 3-Methylbanzamide $3\text{-}CH_3C_6H_4CONH_2$	135.17	nd (eth)		97			al	B9^{3}, 2322
14059	*p*-Toluamide or 4-Methylbenzamide $4\text{-}CH_3C_6H_4CONH_2$	135.17	nd or pl (w)		160			al, eth	B9^{3}, 2343
14060	Toluene or Methyl benzene $C_6H_5CH_3$	92.14		110.6, 14.5$^{14.5}$	−95	0.8669$^{20/4}$	1.4961^{20}	**al, eth**, ace, bz, lig	B5^{4}, 766
14061	Toluene, 2-allyl $2\text{-}CH_2{=}CHCH_2C_6H_4CH_3$	132.21		182-3^{757}, 88-90^{25}		0.9005$^{20/4}$	1.5187^{20}		B5^{3}, 1212
14062	Toluene, 2-amino or *o*-Toluidine $2\text{-}CH_3C_6H_4NH_2$	107.16		200.2, 80.1^{10}	−14.7	0.9984$^{20/4}$	1.5725^{20}	**al, eth**	B12^{3}, 1837
14063	Toluene, 3-amino or *m*-Toluidine $3\text{-}CH_3\text{-}C_6H_4NH_2$	107.16		203.3, 82.3^{10}	−30.4	0.9889$^{20/4}$	1.5681^{20}	**al, eth, ace, bz**	B12^{3}, 1949
14064	Toluene, 4-amino or *p*-Toluidine $4\text{-}CH_3C_6H_4NH_2$	107.16	lf (w + l)	200.5, 79.6^{10}	44-5	0.9619$^{20/4}$	1.5534^{45}, 1.5636^{20}	al, eth, ace, Py	B12^{3}, 2017
14065	Toluene, 2-bromo or *o*-Tolylbromide $2\text{-}Br\text{-}C_6H_4CH_3$	171.04		181.7	−27.8	1.4232$^{20/4}$	1.5565^{20}	al, eth, bz	B5^{4}, 825
14066	Toluene, 2-bromo-3-nitro $2\text{-}Br\text{-}3\text{-}NO_2C_6H_3CH_3$	216.03	ye pr (al)	157^{22}	41^{2}			al	B5^{4}, 860
14067	Toluene, 2-bromo-4-nitro $2\text{-}Br\text{-}4\text{-}NO_2C_6H_3CH_3$	216.03	nd (al)	150-1^{20}	78			eth	B5^{4}, 861
14068	Toluene, 2-bromo-5-nitro $2\text{-}Br\text{-}5\text{-}NO_2C_6H_3CH_3$	216.03	cr (al)	140-3^{17}	78			eth	B5^{4}, 860
14069	Toluene, 2-bromo-6-nitro $2\text{-}Br\text{-}6\text{-}NO_2C_6H_3CH_3$	216.03	pe ye nd (dil al)	143^{22}	42			al	B53, 751
14070	Toluene, 3-bromo or *m*-Tolylbromide $3\text{-}BrC_6H_4CH_3$	171.04		183.7	−39.8	1.4099$^{20/4}$	1.5510^{20}	al, **eth**, ace, chl	B5^{4}, 827
14071	Toluene, 3-bromo-2-nitro $3\text{-}Br\text{-}2\text{-}NO_2C_6H_3CH_3$	216.03	pa ye nd	129-30^{10}	28			al, eth	B5^{4}, 860
14072	Toluene, 3-bromo-4-nitro $3\text{-}Br\text{-}4\text{-}NO_2C_6H_3CH_3$	216.03	pa ye pr or nd (MeOH)	154-5^{20}	37			al, eth	B5^{4}, 861
14073	Toluene, 3-bromo-5-nitro $3\text{-}Br\text{-}5\text{-}NO_2C_6H_3CH_3$	216.03	pa ye nd or pr (MeOH)	269-70	84			al, eth	B5^{3}, 752
14074	Toluene, 4-bromo or *p*-Tolylbromide $4\text{-}Br\text{-}C_6H_4CH_3$	171.04	cr (al)	184.3, 61.9^{10}	28.5	1.3995$^{20/4}$	1.5477^{20}	al, eth, ace, bz, chl	B5^{4}, 827
14075	Toluene, 4-bromo-2-nitro $4\text{-}Br\text{-}2\text{-}NO_2C_6H_3CH_3$	216.03	pa ye nd (dil al)	256-7, 130-2	47			al, eth	B5^{4}, 860
14076	Toluene, 4-bromo-3-nitro $4\text{-}Br\text{-}3\text{-}NO_2C_6H_3CH_3$	216.03	pa ye nd (MeOH)		35		1.5682^{20}		B5^{4}, 860
14077	Toluene, 5-bromo-2-nitro $5\text{-}Br\text{-}2\text{-}NO_2C_6H_3CH_3$	216.03	cr (al)	267, 143^{10}	56				B5^{4}, 860
14078	Toluene, 2-chloro or *o*-Tolylchloride $2\text{-}ClC_6H_4CH_3$	126.59		159.1,42.6^{10}	-35.1	1.0825$^{20/4}$	1.5268^{20}	al, **eth, ace** bz chl	B5^{4}, 805
14079	Toluene, 2-chloro-4-nitro $2\text{-}Cl\text{-}4\text{-}NO_2C_6H_3CH_3$	171.58	nd (al)	260	68		1.5470^{69}	al, eth	B5^{4}, 855
14080	Toluene, 2-chloro-6-nitro $2\text{-}Cl\text{-}6\text{-}NO_2C_6H_3CH_3$	171.58	nd (dil al)	238	37-40		1.5377^{69}	al	B5^{4}, 854
14081	Toluene, 3-chloro or *m*-Tolyl chloride $3\text{-}Cl\text{-}C_6H_4CH_3$	126.59		162	−47.8	1.0722$^{20/4}$	1.5214^{19}	al, **eth**, bz, chl	B5^{4}, 806
14082	Toluene, 3-chloro-4-nitro $3\text{-}Cl\text{-}4\text{-}NO_2C_6H_3CH_3$	171.58	pa ye nd	146^{19}	24				B5^{4}, 856
14083	Toluene, 3-chloro-5-nitro $3\text{-}Cl\text{-}5\text{-}NO_2C_6H_3CH_3$	171.58	ye nd (al)		61		1.5404^{69}	al	B5^{4}, 855
14084	Toluene, 4-chloro or *p*-Tolyl chloride $4\text{-}ClC_6H_4CH_3$	126.59		162, 44^{10}	7.5	1.0697$^{20/4}$	1.5150^{20}	al, **eth**, chl, aa	B5^{4}, 806
14085	Toluene, 4-chloro-3-nitro $4\text{-}Cl\text{-}2\text{-}NO_2C_6H_3CH_3$	171.58	mcl nd	240^{720}, 115.5^{11}	38	1.2559^{80}		eth	B5^{4}, 853

No.	Name, Synonyms, and Formula	Mol. wt.	Color, crystalline form, specific rotation and λ_{max} (log ε)	b.p. °C	m.p. °C	Density	n_D	Solubility	Ref.
14086	Toluene, 4-chloro-3-nitro $4\text{-Cl-3-NO}_2C_6H_3CH_3$	171.58	...	260^{745}, 118^{11}	7	...	1.5572^{20}	...	B5[4], 855
14087	Toluene, 5-chloro-2-nitro $5\text{-Cl-2-NO}_2C_6H_3CH_3$	171.58	ye	...	24.9	...	1.5496^{65}	...	B5[4], 854
14088	Toluene, 2,3-diamino $2,3\text{-}(NH_2)_2C_6H_3CH_3$	122.17	cr	255	63-4	...	...	w, al, eth	B13[3], 277
14089	Toluene, 2,4-diamino $2,4\text{-}(NH_2)_2C_6H_3CH_3$	122.17	nd (w), cr (al)	29^2, $148\text{-}50^8$	99	...	...	w, al, eth, bz	B13[3], 278
14090	Toluene, 2,5-diamino $2,5\text{-}(NH_2)_2C_6H_3CH_3$	122.17	pl (bz)	273-4	64	...	...	w, al, eth	B13[3], 284
14091	Toluene, 2,6-diamino $2,6\text{-}(NH_2)_2C_6H_3CH_3$	122.17	pr (bz, w)	...	106	...	...	w, al	B13[3], 297
14092	Toluene, 3,4-diamino $3,4\text{-}(NH_2)_2C_6H_3CH_3$	122.17	lf (lig)	265 sub	89-90	...	...	w	B13[3], 292
14093	Toluene, 3,5-diamino $3,5\text{-}(NH_2)_2C_6H_3CH_3$	122.17	...	283-5	<0	...	...	w, al, eth	B13[3], 299
14094	Toluene, 3,5-diamino, hydrochloride $3,5\text{-}(NH_2)_2C_6H_3CH_3.2HCl$	195.09	nd (dil al)	...	255-60d	...	...	w	B13, 164
14095	Toluene, 2,5-dibromo $2,5\text{-}Br_2C_6H_3CH_3$	249.93	...	236, $135\text{-}6^{35}$	fr 5.6	1.8127^{19}	1.5982^{18}	...	B5[4], 835
14096	Toluene, 3,5-dibromo $3,5\text{-}Br_2C_6H_3CH_3$	249.93	nd	246	39	...	...	...	B5[4], 836
14097	Toluene, 2,3-dichloro $2,3\text{-}Cl_2C_6H_3CH_3$	161.03	...	207-8, $61\text{-}2^3$	...	...	1.5511^{20}	bz	B5[4], 815
14098	Toluene, 2,4-dichloro $2,4\text{-}Cl_2C_6H_3CH_3$	161.03	...	196-7	−13.5	$1.2498^{20/20}$	1.5511^{20}	...	B5[4], 815
14099	Toluene, 2,5-dichloro $2,5\text{-}Cl_2C_6H_3CH_3$	161.03	...	200^{770}	5	1.2535^{20}	1.5449^{20}	bz	B5[3], 694
14100	Toluene, 2,6-dichloro $2,6\text{-}Cl_2C_6H_3CH_3$	161.03	...	198	...	...	1.5507^{20}	chl	B5[4], 815
14101	Toluene, 3,4-dichloro $3,4\text{-}Cl_2C_6H_3CH_3$	161.03	...	208.9, 81.8^{10}	−15.2	$1.2564^{20/4}$	1.5471^{20}	**al, eth, ace, bz**	B5[4], 815
14102	Toluene, 3,5-dichloro $3,5\text{-}Cl_2C_6H_3CH_3$	161.03	...	201-2	26	...	...	...	B5[4], 816
14103	Toluene, 2,5-diethoxy $2,5\text{-}(C_2H_5O)_2C_6H_3CH_3$	180.25	nd (lig)	247-9	24-5	1.0134^{15}	...	**al, eth, bz, chl**	B6[3], 4499
14104	Toluene, 2,3-dihydroxy or Isohomocatechol $2,3\text{-}(HO)_2C_6H_3CH_3$	124.14	lf (bz)	241, 127^{12}	68	...	...	w, al, eth, bz, chl	B6[3], 4492
14105	Toluene, 2,4-dihydroxy or Cresorcinol $2,4\text{-}(HO)_2C_6H_3CH_3$	124.14	cr (bz-peth)	267-70	105-7	...	...	w, al, eth	B6[3], 4495
14106	Toluene, 2,5-dihydroxy or Toluhydroquinone $2,5\text{-}(HO)_2C_6H_3CH_3$	124.14	rh pl (bz)	163^{11} sub	126-7	...	...	w, al, eth	B6[3], 4498
14107	Toluene, 2,5-dihydroxy, diacetate or 2.5-Diacetoxy toluene $2.5\text{-}(CH_3CO_2)_2C_6H_3CH_3$	208.21	nd (aa), pr (lig), nd or pr (w)	...	52	...	...	eth, al, aa	B6[3], 4501
14108	Toluene, 3,4-dihydroxy or 4-Homocatechol $3,4\text{-}(HO)_2C_6H_3CH_3$	124.14	lf (bz-lig), pr (bz)	251, $143\text{-}6^{26}$ sub	65	$1.1287^{74/4}$	1.5425^{74}	w, al, eth, ace	B6[3], 4514
14109	Toluene, 3,5-dihydroxy or Orcinol $3.5\text{-}(HO)_2C_6H_3CH_3$	124.14	pr (w + l), lf (chl)	289-90, 147^5	107-8 (anh), 58 (hyd)	1.290^4	...	w, al, eth, bz	B6[3], 4531
14110	Toluene, 2,4-dimercapto or Dithiocresorcinol $2,4\text{-}(HS)_2C_6H_3CH_3$	156.26	...	263	36.7	...	...	...	B6, 873
14111	Toluene, 3,4-dimercapto $3,4\text{-}(HS)_2C_6H_3CH_3$	156.26	...	...	28-30	...	...	chl	B6[3], 4530
14112	Toluene, 2,3-dimethoxy or Isohomoveratrol $2,3\text{-}(CH_3O)_2C_6H_3CH_3$	152.19	...	202-3, $92\text{-}3^{18}$	...	$1.0335^{20/4}$	1.5121^{25}	al, eth	B6[3], 4493
14113	Toluene, 2,4-dimethoxy $2,4\text{-}(CH_3O)_2C_6H_3CH_3$	152.19	...	211, $110\text{-}20^{30}$	...	...	...	al, eth	B6[3], 4496
14114	Toluene, 3,4-dimethoxy or Homoveratrol $3,4\text{-}(CH_3O)_2C_6H_3CH_3$	152.19	pr (eth)	219-21, $122\text{-}4^{27}$	24	$1.0509^{25/4}$	1.5257^{25}	...	B6[3], 4516
14115	Toluene, 3,5-dimethoxy $3,5\text{-}(CH_3O)_2C_6H_3CH_3$	152.19	...	244, 102^8	...	1.0478^{15}	1.5234^{20}	al, eth, bz, aa	B6[3], 4533
14116	Toluene, 2,4-dinitro $2,4\text{-}(NO_2)_2C_6H_3CH_3$	182.14	ye nd or mcl pr	300d	71	1.3208^{71}	1.442	al, eth, ace, bz, Py	B5[4], 865
14117	Toluene, 2,5-dinitro $2,5\text{-}(NO_2)_2C_6H_3CH_3$	182.14	nd (al)	...	52.5	1.282^{111}	...	al, bz	B5[4], 866

No.	Name, Synonyms, and Formula	Mol. wt.	Color, crystalline form, specific rotation and λ_{max} (log ε)	b.p. °C	m.p. °C	Density	n_D	Solubility	Ref.
14118	Toluene, 2,6-dinitro $2,6-(NO_2)_2C_6H_3CH_3$	182.14	rh nd (al)	....	66	1.2833^{111}	1.479	al	B5[4], 866
14119	Toluene, 3,4-dinitro $3,4-(NO_2)_2C_6H_3CH_3$	182.14	ye nd (CS_2)	....	58.3	1.2594^{111}	....	al	B5[4], 866
14120	Toluene, 3,5-dinitro $3,5-(NO_2)_2C_6H_3CH_3$	182.14	ye rh nd (aa)	sub	93	1.2772^{111}	....	al, eth, bz, chl	B5[4], 867
14121	Toluene, 4,6-Dinitro-3-Fluoro $4,6-(NO_2)_2-3-FC_6H_2CH_3$	200.13	....	$129-30^{0.5}$	78	....	....	....	B5[4], 867
14122	Toluene, 2,4-diisopropyl $2,4-(i-C_3H_7)_2C_6H_3CH_3$	176.30	....	82^{7}	....	0.8636^{20}	1.4912^{20}	CS_2	B5[4], 1152
14123	Toluene, 3,5-diisopropyl $3,5-(i-C_3H_7)_2C_6H_3CH_3$	176.30	....	215-8	....	0.8668^{20}	1.4950^{20}	....	B5[4], 1153
14124	Toluene, 2-ethoxy or Ethyl o-tolyl ether $(2-CH_3C_6H_4)OC_2H_5$	136.19	....	184, 70^{12}	....	$0.9592^{13/4}$	1.508^{13}	al, eth	B6[4], 1944
14125	Toluene, 3-ethoxy or Ethyl *m*-tolyl ether $(3-CH_3C_6H_4)OC_2H_5$	136.19	....	192	....	0.949^{20}	1.513^{20}	al, eth	B6[4], 2039
14126	Toluene, 4-ethoxy or Ethyl *p*-tolyl ether $(4-CH_3C_6H_4)OC_2H_5$	136.19	....	188-9	....	$0.9509^{18/4}$	1.5058^{18}	al, eth	B6[4], 2099
14127	Toluene, 2-fluoro or *o*-Tolyl fluoride $2-FC_6H_4CH_3$	110.13	....	114, 30^{26}	−62	$1.0041^{13/4}$	1.4704^{20}	al, eth	B5[4], 799
14128	Toluene, 3-fluoro or *m*-Tolyl fluoride $3-FC_6H_4CH_3$	110.13	....	116	−87.7	0.9986^{20}	1.4691^{20}	al, eth	B5[4], 799
14129	Toluene, 4-fluoro or 4-Tolyl fluoride $4-FC_6H_4CH_3$	110.13	....	116.6	−56.8	$1.0007^{16/4}$	1.4699^{20}	al, eth	B5[4], 800
14130	Toluene, 3-(α-hydroxy ethyl) $3-[CH_3CH(OH)]C_6H_4CH_3$	136.19	....	112^{12}	....	$0.9974^{15/4}$	1.5240^{20}	al, eth	B6[3], 1823
14131	Toluene, 4-(α-hydroxy ethyl) $4-[CH_3CH(OH)]C_6H_4CH_3$	136.19	....	219^{756}, 120^{19}	....	$0.9944^{20/4}$	1.5246^{20}	al, eth	B6[3], 1826
14132	Toluene, 2-hydroxylamino or *o*-Toyl hydroxyl amine $2-HONHC_6H_4CH_3$	123.15	nd (eth-bz)	....	44	....	....	al, eth, bz	B15[3], 15
14133	Toluene, 2-hydroxylamino-6-nitro $2-HONH-6-NO_2C_6H_3CH_3$	168.15	col or ye (bz)	....	120-1	....	....	al, eth	B15[2], 14
14134	Toluene, 3-hydroxylamino or *m*-Tolyl hydroxyl amine $3-(HONH)C_6H_4CH_3$	123.15	lf (bz-peth)	....	68.5	....	....	al, eth, chl	B15[3], 17
14135	Toluene, 4-hydroxylamino or *p*-Tolyl hydroxyl amine $4-(HONH)C_6H_4CH_3$	123.15	lf (bz)	115-20d	98	....	....	al, eth, chl	B15[3], 18
14136	Toluene, 4-hydroxylamino-2-nitro $4-(HONH)-2-NO_2C_6H_3CH_3$	168.15	ye (bz)	....	108-9	....	....	bz	B15[2], 16
14137	Toluene, 2-iodo or *o*-Tolyl iodide $2-IC_6H_4CH_3$	218.04	....	211-2, $73-5^{7}$	....	$1.713^{20/4}$	1.6079^{20}	**al, eth**	B5[4], 838
14138	Toluene, 3-iodo or *m*-Tolyl iodide $3-IC_6H_4CH_3$	218.04	....	213, $80-2^{10}$	−27.2	1.705^{20}	1.6053^{20}	al, eth	B5[4], 839
14139	Toluene, 4-iodo or *p*-Tolyl iodide $4-IC_6H_4CH_3$	218.04	lf (al)	211 sub	36-7	$1.678^{20/4}$	....	al, eth	B5[4], 840
14140	Toluene, 2-mercapto or *o*-Thiocresol $2-CH_3C_6H_4SH$	124.20	pl or lf	194.2, 67.1^{10}	15	$1.041^{20/4}$	1.570^{20}	al, eth	B6[4], 2014
14141	Toluene, 3-mercapto or *m*-Thiocresol $3-CH_3C_6H_4SH$	124.20	....	195.1, 67.8^{10}	<−20	$1.044^{20/4}$	1.572^{20}	al, **eth**	B6[4], 2079
14142	Toluene, 4-mercapto or *p*-Thiocresol $4-CH_3C_6H_4SH$	124.20	lf (eth, dil al)	195, 67.6^{10}	44	....	....	al, eth	B6[4], 2153
14143	Toluene, 2-methoxy or Methyl-*o*-tolyl ether $2-CH_3OC_6H_4CH_3$	122.17	....	171.3	....	$0.9851^{15/15}$	1.5161^{20}	al, eth, ace	B6[4], 1943
14144	Toluene, 3-methoxy or Methyl-*m*-tolyl ether $3-CH_3OC_6H_4CH_3$	122.17	....	177.2	....	$0.9697^{25/25}$	1.5164^{13}	al, eth, ace, bz	B6[4], 2039
14145	Toluene, 4-methoxy or Methyl-*p*-tolyl ether $4-CH_3OC_6H_4CH_3$	122.17	....	176.5	....	$0.9689^{25/25}$	1.5124^{19}	al, eth	B6[4], 2098
14146	Toluene, 2-nitro $2-NO_2C_6H_4CH_3$	137.14	(i) nd, (ii) cr	221.7, 118^{16}	(i)-9.5 (ii)-2.9	1.1629^{20}	1.5450^{20}	**al, eth**	B5[4], 845
14147	Toluene, 3-nitro $3-NO_2C_6H_4CH_3$	137.14	pa ye	232.6	16	$1.1571^{20/4}$	1.5466^{20}	al, **eth**, bz	B5[4], 847
14148	Toluene, 3-nitro-4-triazo $4-N_3-3-NO_2-C_6H_3CH_3$	178.15	ye nd or pl (lig, dil al)	....	38	....	....	al	B5[3], 773
14149	Toluene, 4-nitro $4-NO_2C_6H_4CH_3$	137.14	orth cr (al, eth)	238.3, 105^{9}	54.5	$1.1038^{75/4}$	....	al, eth, ace, bz	B5[4], 848

No.	Name, Synonyms, and Formula	Mol. wt.	Color, crystalline form, specific rotation and λ_{max} (log ε)	b.p. °C	m.p. °C	Density	n_D	Solubility	Ref.
14150	Toluene, 2-nitroso 2-ONC$_6$H$_4$CH$_3$	121.14	nd or pr		72.5			al, eth, chl	B5^3, 728
14151	Toluene, 3-nitroso 3-ONC$_6$H$_4$CH$_3$	121.14	nd		53.5			eth, chl	B5^4, 844
14152	Toluene, 4-nitroso 4-ONC$_6$H$_4$CH$_3$	121.14	nd (lig)		48.5			bz, chl	B5^4, 845
14153	Toluene, 4-isopropenyl 4-[CH$_2$=C(CH$_3$)]C$_6$H$_4$CH$_3$	132.21		185.3		$0.8936^{23/4}$	1.52832^{23}		B5^4, 1383
14154	Toluene, 4-isopropyl-2-nitro 4-i-C$_3$H$_7$-2-NO$_2$C$_6$H$_3$CH$_3$	179.22					1.5280^{20}		B5^4, 1064
14155	Toluene, 2,3,4,5,6-pentabromo C$_6$Br$_5$CH$_3$	486.62			288	2.97^{17}		bz	B5^3, 720
14156	Toluene, 2,3,4,5,6-pentachloro C$_6$Cl$_5$CH$_3$	264.37	nd (bz, peth)	301	224			to	B5^4, 823
14157	Toluene, 2-phenyl or 2-Methyl biphenyl 2-C$_6$H$_5$-C$_6$H$_4$CH$_3$	168.24		255.3, 130-6^{27}	−0.2	$1.010^{22/4}$	1.5914^{20}	al, eth	B5^4, 1855
14158	Toluene, 3-phenyl or 3-Methyl biphenyl 3-C$_6$H$_5$-C$_6$H$_4$CH$_3$	168.24		272.7, 148-50^{20}	4.5	$1.0182^{17/4}$	1.5972^{20}	al, eth	B5^4, 1858
14159	Toluene, 4-phenyl 4-C$_6$H$_5$C$_6$H$_4$CH$_3$	168.24	pl (lig, MeOH)	267-8, 134-6^{15}	49-50	1.015^{27}		al, eth	B5^4, 1860
14160	Toluene, 2-*iso*-propenyl 2-[CH$_2$=C(CH$_3$)]C$_6$H$_4$CH$_3$	132.21		175, 59-62^{11}		$0.9181^{15/0}$	1.5112^{30}		B5^3, 1214
14161	Toluene, 2-propoxy or Propyl-*o*-tolyl ether 2-C$_3$H$_7$OC$_6$H$_4$CH$_3$	150.22		204.1		$0.9517^{0/0}$			B6^3, 1246
14162	Toluene, 3-propoxy or Propyl-*m*-tolyl ether 3-C$_3$H$_7$OC$_6$H$_4$CH$_3$	150.22		210.6		$0.9484^{0/0}$			B6^4, 2040
14163	Toluene, 4-propoxy or Propyl-*p*-tolyl ether 4-C$_3$H$_7$OC$_6$H$_4$CH$_3$	150.22		210.4		$0.9497^{0/0}$			B6^3, 1354
14164	Toluene, 2-propyl or 1-Methyl-2-propyl benzene 2-C$_3$H$_7$C$_6$H$_4$CH$_3$	134.22		185, 63.5^{10}	−60.2	0.8744^{20}	1.4998^{20}		B5^4, 1056
14165	Toluene, 3-propyl or 1-Methyl-3-propyl benzene 3-C$_3$H$_7$C$_6$H$_4$CH$_3$	134.22		182, 61.4^{10}		0.8610^{20}	1.4936^{20}		B5^4, 1056
14166	Toluene, 4-propyl or 1-Methyl-4-propyl benzene 4-C$_3$H$_7$C$_6$H$_4$CH$_3$	134.22		183, 61.9^{10}	−63.6	0.8584^{20}	1.4919^{20}	al, eth	B5^4, 1057
14167	Toluene, 2,3,5,6-tetrabromo 2,3,5,6-Br$_4$C$_6$HCH$_3$	407.72	nd		116-7			ace, bz	B5, 310
14168	Toluene, 2,3,4,5-tetrachloro 2,3,4,5-Cl$_4$C$_6$HCH$_3$	229.92	nd (MeOH, dil al)		98.1			al, eth, ace	B5^4, 233
14169	Toluene, 2,3,4,6-tetrachloro 2,3,4,6-Cl$_4$C$_6$HCH$_3$	229.92	nd (al, eth)	276.5	96			al, bz	B5^3, 702
14170	Toluene, 2,3,5,6-tetrachloro 2,3,5,6--Cl$_4$C$_6$HCH$_3$	229.92	nd (MeOH)	sub	93-4			al, eth	B5^3, 702
14171	Toluene, 2,4,5-triacetoxy 2,4,5-(CH$_3$CO$_2$)$_3$C$_6$H$_2$CH$_3$	266.25	cr (al)		114-5			w, al, bz	B6, 1109
14172	Toluene, 2,4,6-triacetoxy 2,4,6-(CH$_3$CO$_2$)$_3$C$_6$H$_2$CH$_3$	266.25	(i) cr (eth-peth) (ii) nd (peth)		(i) 76 (ii) 58			al, bz	B6^3, 6320
14173	Toluene, 2,4,6-triamino 2,4,6-(NH$_2$)$_3$C$_6$H$_2$CH$_3$	137.18	col nd (bz), red in air		105			w, al, bz	B13^3, 557
14174	Toluene, 2-triazo or *o*-Tolyl azide 2-N$_3$C$_6$H$_4$CH$_3$	133.15	pa ye liq	90.5^{30}	<-10			eth	B5^4, 876
14175	Toluene, 3-triazo or *m*-Tolyl azide 3-N$_3$C$_6$H$_4$CH$_3$	133.15		92.5^{31}				eth	B5^4, 876
14176	Toluene, 4-triazo or *p*-Tolyl azide 4-N$_3$C$_6$H$_4$CH$_3$	133.15		d 180, 80^{10}		$1.0527^{23/4}$		al, eth	B5^4, 876
14177	Toluene, 2,3,4-tribromo 2,3,4-Br$_3$C$_6$H$_2$CH$_3$	328.83	rh pl (aa, lig-CS$_2$)		45-6	2.456^{20}			B5^1, 156
14178	Toluene, 2,3,5-tribromo 2,3,5-Br$_3$C$_6$H$_2$CH$_3$	328.83	mcl pr (eth-to)		53-4	2.467^{17}		eth	B5^3, 719
14179	Toluene, 2,3,6-tribromo 2,3,6-Br$_3$C$_6$H$_2$CH$_3$	328.83	mcl pr or lf (lig, chl)		60.5	2.471^{17}			B5^1, 156
14180	Toluene, 2,4,5-tribromo 2,4,5-Br$_3$C$_6$H$_2$CH$_3$	328.83	mcl pr (eth-al), nd (al)		113.5	2.472^{17}			B5^3, 719
14181	Toluene, 2,4,6-tribromo 2,4,6-Br$_3$C$_6$H$_2$CH$_3$	328.83	lo nd or mcl pr (eth-AcOEt)	290	70	2.479^{17}			B5^3, 719

No.	Name, Synonyms, and Formula	Mol. wt.	Color, crystalline form, specific rotation and λ_{max} (log ε)	b.p. °C	m.p. °C	Density	n_D	Solubility	Ref.
14182	Toluene, 2,3,4-trichloro $2,3,4\text{-}Cl_3C_6H_2CH_3$	195.48	nd (al, MeOH)	244	43-4			al, eth, ace	B5[4], 819
14183	Toluene, 2,3,5-trichloro $2,3,5\text{-}Cl_3C_6H_2CH_3$	195.48	nd(al)	229-31[757]	45-6			al, ace	B5, 299
14184	Toluene, 2,3,6-trichloro $2,3,6\text{-}Cl_3C_6H_2CH_3$	195.48	nd (al)		45-6			al, ace	B5[4], 819
14185	Toluene, 2,4,5-trichloro $2,4,5\text{-}Cl_3C_6H_2CH_3$	195.48	nd or lf (al)	229-30[716]	82.4			al, ace	B5[4], 819
14186	Toluene, 2,4,6-trichloro $2,4,6\text{-}Cl_3C_6H_2CH_3$	195.48	nd (al)		38			al, ace	B5[4], 819
14187	Toluene, 3,4,5-trichloro $3,4,5\text{-}Cl_3C_6H_2CH_3$	195.48		246-7[768]	45-6			al, ace	B[5], 299
14188	Toluene, 2,4,6-trihydroxy $2,4,6\text{-}(HO)_3C_6H_2CH_3$	140.14			222-3			w, al, eth	B6[3], 6318
14189	Toluene, 3,4,5-trihydroxy or 5-Methyl pyrogallol $3,4,5\text{-}(HO)_3C_6H_2CH_3$	140.14	pa br nd (bz)	sub	129			bz	B6[3], 6320
14190	Toluene, 2,3,4-triido $2,3,4\text{-}I_3C_6H_2CH_3$	469.83	pa br nd (al)		92			al, bz	B5[1], 157
14191	Toluene, 2,3,5-triido $2,3,5\text{-}I_3C_6H_2CH_3$	469.83	og pl (al)		72-3				B5[1], 157
14192	Toluene, 2,3,6-triido $2,3,6\text{-}I_3C_6H_2CH_3$	469.83	nd (al)		80.5				B5[1], 158
14193	Toluene, 2,4,5-triido $2,4,5\text{-}I_3C_6H_2CH_3$	469.83	pa br nd (al)		118-20				B5[1], 158
14194	Toluene, 2,4,6-triido $2,4,6\text{-}I_3C_6H_2CH_3$	469.83	nd (al, bz)	300d	118-9			al	B5[1], 158
14195	Toluene, 3,4,5-triido $3,4,5\text{-}I_3C_6H_2CH_3$	469.83	nd(al)		122-3			al	B5, 317
14196	Toluene, 2,3,4-trinitro $2,3,4\text{-}(NO_2)_3C_6H_2CH_3$	227.13	tcl lf (al), pr (ace)		112	1.62		eth, ace, bz	B5[4], 872
14197	Toluene, 2,4,5-trinitro $2,4,5\text{-}(NO_2)_3C_6H_2CH_3$	227.13	yesh pl (ace), pa ye rh bipym (al)	exp 290-310	104			eth, ace, bz	B5[4], 872
14198	Toluene, 2,4,6-trinitro or TNT $2,4,6\text{-}(NO_2)_3C_6H_2CH_3$	227.13	orh (al)	240 exp	82	1.654		eth, ace, bz, Py	B5[4], 873
14199	Toluene, 2-vinyl $2\text{-}(CH_2{=}CH)C_6H_4CH_3$	118.18		171		$0.9106^{20/4}$	1.5450^{20}	bz, chl	B5[4], 1367
14200	Toluene, 3-vinyl $3\text{-}(CH_2{=}CH)C_6H_4CH_3$	118.18		168, 61-2[18]	−70	$0.9028^{20/20}$	1.5410^{20}	al, eth, bz	B5[4], 1367
14201	Toluene, 4-vinyl $4\text{-}(CH_2{=}CH)C_6H_4CH_3$	118.18		169, 63[15]	−37.8	$0.8760^{20/4}$	1.5428^{20}	bz	B5[4], 1369
14202	*o*-Toluene arsonic acid or *o*-Tolyl arsinic acid $2\text{-}CH_3C_6H_4AsO(OH)_2$	216.07	nd (w)		163-4			al	B16[3], 1060
14203	*m*-Toluene arsonic acid or *m*-Tolyl arsinic acid $3\text{-}CH_3C_6H_4AsO(OH)_2$	216.07	nd (w)		150			al	B16[3], 1061
14204	*p*-Toluene arsonic acid or *p*-Tolyl arsinic acid $4\text{-}CH_3C_6H_4AsO(OH)_2$	216.07	nd (w)		3 (−w, 105-10)			al	B16[3], 1061
14205	*o*-Toluene boronic acid or *o*-Tolyl boronic acid $2\text{-}CH_3C_6H_4B(OH)_2$	135.96	nd or pl (w)	d	165-8			al, eth, bz	B16[3], 1277
14206	*m*-Toluene boronic acid or *m*-Tolyl boronic acid $3\text{-}CH_3C_6H_4B(OH)_2$	135.96	cr (w)		137-40			al, eth	B16[3], 1277
14207	*p*-Toluene boronic acid or *p*-Tolyl boronic acid $4\text{-}CH_3C_6H_4B(OH)_2$	135.96	nd (w)		245			eth	B16[3], 1277
14208	*o*-Toluene sulfinic acid or *o*-Tolyl sulfinic acid $2\text{-}CH_3C_6H_4SO_2H$	156.20	nd (w)	d	80			w, al, eth, ace	B11[3], 6
14209	*p*-Toluene sulfinic acid or *p*-Tolyl sulfinic acid $4\text{-}CH_3C_6H_4SO_2H$	156.20	rh pl or lo nd (w)		86-7			al, eth	B11[3], 8
14210	*p*-Toluene sulfinyl chloride $4\text{-}CH_3C_6H_4SOCl$	174.64	nd	115-20[4]	54-8			chl	B11[3], 11
14211	*o*-Toluene sulfonamide or 2-Toluene sulfonamide $2\text{-}CH_3C_6H_4SO_2NH_2$	171.21	oct (al), pr (w)		156.3			al	B11[3], 167
14212	*o*-Toluene sulfonamide, 4-amino $4\text{-}NH_2\text{-}2\text{-}CH_3C_6H_3SO_2NH_2$	186.23	nd or pl (w)		164			al	B14[3], 2208
14213	*o*-Toluene sulfonamide, N-methyl $2\text{-}CH_3C_6H_4SO_2NHCH_3$	185.24	pl (bz-lig)		74-5			al, ace, chl	B11, 87

No.	Name, Synonyms, and Formula	Mol. wt.	Color, crystalline form, specific rotation and λ_{max} (log ε)	b.p. °C	m.p. °C	Density	n_D	Solubility	Ref.
14214	*o*-Toluene sulfonamide, N-phenyl $2\text{-}CH_3C_6H_4SO_2NHC_6H_5$	247.31	pr (dil al)	....	136	....	....	al	B12, 566
14215	*o*-Toluene sulfonamide, 4-propyl $4\text{-}C_3H_7\text{-}2\text{-}CH_3C_6H_3SO_2NH_2$	213.29	pl (bz, dil al)	....	101-2	....	....	w, al	B11, 138
14216	*o*-Toluene sulfonamide, 5-isopropyl $5\text{-}i\text{-}C_3H_7\text{-}2\text{-}CH_3C_6H_3SO_2NH_2$	213.29	pl (al), pr or lf (dil al)	....	75	....	....	lig	B11, 140
14217	*o*-Toluene sulfonic acid or 2-Toluene sulfonic acid $2\text{-}CH_3C_6H_4SO_3H$	172.21	hyg pl (w + 2)	128.8^{25}	67.5	....	....	w, al	B11^{3}, 167
14218	*o*-Toluene sulfonic acid, 4-amino $4\text{-}H_2N\text{-}2\text{-}CH_3C_6H_3SO_3H$	187.21	mcl pr (w + l)	....	d	....	....	....	B14^{3}, 2208
14219	*o*-Toluene sulfonic acid, 5-amino $5\text{-}H_2N\text{-}2\text{-}CH_3C_6H_3SO_3H$	187.21	pl (w + l)	....	>275	....	....	....	B14^{3}, 2210
14220	*o*-Toluene sulfonic acid, 6-amino $6\text{-}H_2N\text{-}2\text{-}CH_3C_6H_3SO_3H$	187.21	nd (+ ½w)	....	130d	....	....	....	B14^{1}, 723
14221	*o*-Toluene sulfonic acid, 4-nitro, dihydrate $4\text{-}NO_2\text{-}2\text{-}CH_3C_6H_3SO_3H.2H_2O$	253.23	pl (w)	....	133.5	....	....	w, al, eth, chl	B11^{3}, 173
14222	*o*-Toluene sulfonic acid, 5-nitro $5\text{-}NO_2\text{-}2\text{-}CH_3C_6H_3SO_3H$	217.20	pr or pl (w + 2)	....	130-3	....	....	w, al, eth, chl	B11^{3}, 173
14223	*o*-Toluene sulfonic acid, 2-tolyl ester or (2-Tolyl)-*o*-toluene sulfonate $2\text{-}CH_3C_6H_4SO_2(OC_6H_4CH_3\text{-}2)$	262.32	cr (al)	....	50-1	....	....	bz, al	B11, 85
14224	2-Toluene sulfonic acid, 3-tolyl ester $2\text{-}CH_3C_6H_4SO_2(OC_6H_4CH_3\text{-}3)$	262.32	cr (al)	....	60	....	....	al, bz	B11, 85
14225	2-Toluene sulfonic acid, 4-tolyl ester $2\text{-}CH_3C_6H_4SO_2(OC_6H_4CH_3\text{-}4)$	262.32	cr (al)	....	70-1	....	....	al, bz	B11, 85
14226	*o*-Toluene sulfonyl bromide or 2-Toluene sulfonyl bromide $2\text{-}CH_3C_6H_4SO_2Br$	235.10	cr	138^{10}	13	....	....	....	B11, 86
14227	*o*-Toluene sulfonyl chloride or 2-Toluene sulfonyl chloride $2\text{-}CH_3C_6H_4SO_2Cl$	190.65	....	154^{36}	10.2	$1.3383^{20/4}$	1.5565^{20}	eth, bz	B11^{3}, 167
14228	*m*-Toluene sulfonamide or 3-Toluene sulfonamide $3\text{-}CH_3C_6H_4SO_2NH_2$	171.21	pl (w), mcl pr	....	108	....	....	al	B11^{1}, 23
14229	*m*-Toluene sulfonamide, N-phenyl $3\text{-}CH_3C_6H_4SO_2NHC_6H_5$	247.31	pr (al)	....	96	....	....	al, eth	B11^{1}, 23
14230	*m*-Toluene sulfonamide, 4-isopropyl $4\text{-}i\text{-}C_3H_7\text{-}3\text{-}CH_3C_6H_3SO_2NH_2$	213.29	fl (dil al)	....	149.9	....	....	al, eth	B11^{3}, 350
14231	*m*-Toluene sulfonic acid or 3-Toluene sulfonic acid $3\text{-}CH_3C_6H_4SO_3H$	172.21	oil	....	....	....	....	w, al, eth	B11^{3}, 176
14232	*m*-Toluene sulfonic acid, 4-amino $4\text{-}NH_2\text{-}3\text{-}CH_3C_6H_3SO_3H$	187.21	lf ye nd (w + ½)	....	132d (hyd)	....	....	w	B14^{3}, 2213
14233	*m*-Toluene sulfonyl chloride-6-acetamido $6\text{-}(CH_3CONH)\text{-}3\text{-}CH_3C_6H_3SO_2Cl$	247.70	nd (bz)	....	159	....	....	....	B14^{2}, 448
14234	*p*-Toluene sulfonamide or 4-Toluene sulfonamide $(4\text{-}CH_3C_6H_4)SO_2NH_2$	171.21	mcl pl (w + 2)	....	138-9 (anh), 105 (hyd)	....	....	al	B11^{3}, 266
14235	*p*-Toluene sulfonamide, 2-amino $2\text{-}NH_2\text{-}4\text{-}CH_3C_6H_3SO_2NH_2$	186.23	pr (w)	....	176	....	....	....	B14^{3}, 2220
14236	*p*-Toluene sulfonamide, N,N-dichloro or Dichloramine T $4\text{-}CH_3C_6H_4SO_2NCl_2$	240.10	pa ye pr (peth-chl)	....	83	....	....	al, eth, bz, aa, chl	B11^{3}, 301
14237	*p*-Toluene sulfonamide, N-ethyl $4\text{-}CH_3C_6H_4SO_2NHC_2H_5$	199.27	pl (dil al, lig)	....	64	....	....	al	B11^{3}, 268
14238	*p*-Toluene sulfonamide, N-methyl $4\text{-}CH_3C_6H_4SO_2NHCH_3$	185.24	pl (dil al)	....	78-9	1.340	....	al, eth	B11^{3}, 267
14239	*p*-Toluene sulfonamide, N-methyl-N-nitroso $4\text{-}CH_3C_6H_4SO_2N(NO)CH_3$	214.24	cr	....	60	....	....	al, eth	B11^{1}, 29
14240	*p*-Toluene sulfonamide, N-methyl-N-phenyl $4\text{-}CH_3C_6H_4SO_2N(CH_3)C_6H_5$	261.34	pl or mcl pr (AcOEt)	....	95	....	....	al, eth	B12^{3}, 1102
14241	*p*-Toluene sulfonamide, N-methyl-N-(2-tolyl) $4\text{-}CH_3C_6H_4SO_2N(CH_3)(C_6H_4CH_3\text{-}2)$	275.37	pr (al)	....	119-20	....	....	al	B12^{1}, 388
14242	*p*-Toluene sulfonamide, N-methyl-N-(4-tolyl) $4\text{-}CH_3C_6H_4SO_2N(CH_3)(C_6H_4CH_3\text{-}4)$	275.37	pr	....	60	....	....	....	B12^{3}, 2145
14243	*p*-Toluene sulfonamide, N-phenyl or 4-Toluene sulfoanilide $4\text{-}CH_3C_6H_4SO_2NHC_6H_5$	247.33	dimorphic α: tcl, β: mcl pr (dil al, bz)	....	103-4	....	....	al, aa	B12^{3}, 1081

No.	Name, Synonyms, and Formula	Mol. wt.	Color, crystalline form, specific rotation and λ_{max} (log ε)	b.p. °C	m.p. °C	Density	n_D	Solubility	Ref.
14244	*p*-Toluene sulfonamide, 3-isopropyl $3\text{-i-}C_3H_7\text{-}4\text{-}CH_3C_6H_3SO_2NH_2$	213.29	nd (w)	....	162	....	....	al	$B11^{3}$, 349
14245	*p*-Toluene sulfonamide, N-(2-tolyl) $4\text{-}CH_3C_6H_4SO_2NH(C_6H_4CH_3\text{-}2)$	261.34	rh bipym (al), nd (dil aa)	....	110	....	....	al, eth, ace, bz, chl	$B12^{2}$, 452
14246	*p*-Toluene sulfonamide, N-(4-tolyl) $4\text{-}CH_3C_6H_4SO_2NH(C_6H_4CH_3\text{-}4)$	261.34	tcl pr or nd (aa)	....	118-9	....	....	al, bz, chl	$B12^{3}$, 2143
14247	*p*-Toluene sulfonic acid or 4-Toluene sulfonic acid $4\text{-}CH_3C_6H_4SO_3H$	172.20	hyg pl (w + 1), mcl lf or pl	140^{20}	104-5	....	....	w, al, eth	$B11^{3}$, 183
14248	*p*-Toluene sulfonic acid, 2-amino $2\text{-}NH_2\text{-}4\text{-}CH_3C_6H_3SO_3H$	187.21	pl, nd or pr (w)	....	......	....	....	....	$B14^{3}$, 2219
14249	*p*-Toluene sulfonic acid, butyl ester $4\text{-}CH_3C_6H_4SO_2(OC_4H_9)$	228.31	....	$164\text{-}6^{6}$	......	$1.1319^{20/4}$	1.5050^{20}	eth	$B11^{3}$, 189
14250	*p*-Toluene sulfonic acid, 2-chloro ethyl ester or 2-Chloro ethyl-*p*-toluene sulfonate $4\text{-}CH_3C_6H_4SO_2(OCH_2CH_2Cl)$	234.70	....	210^{21}	......	....	....	....	$B11^{3}$, 188
14251	*p*-Toluene sulfonic acid, 2-chloro propyl ester or 2-Chloro propyl-*p*-toluene sulfonate $4\text{-}CH_3C_6H_4SO_2(OCH_2CHClCH_3)$	248.72	....	$216\text{-}7^{17}$	......	$1.2674^{20/4}$	1.5225^{21}	....	$B11^{2}$, 45
14252	*P*-Toluene sulfonic acid, ethyl ester or Ethyl-*p*-toluene sulfonate $4\text{-}CH_3C_6H_4SO_2(OC_2H_5)$	200.25	mcl pr (aa, AcOEt)	173^{15}	34-5	$1.166^{48/4}$	....	al, eth	$B11^{3}$, 188
14253	4-Toluene sulfonic acid, ethylene ester or Ethylene-*p*-toluene sulfonate $(4\text{-}CH_3C_6H_4SO_2)CH_2CH_2(O_2SC_6H_4CH_3\text{-}4)$	370.44	cr (bz)	....	128	....	....	....	$B11^{3}$, 225
14254	*p*-Toluene sulfonic acid, methyl ester or Methyl-*p*-toluene sulfonate $(4\text{-}CH_3C_6H_4)SO_2(OCH_3)$	186.23	mcl lf or pr (eth-lig)	292, $168\text{-}70^{13}$	28-9	....	....	al, eth, bz, chl	$B11^{3}$, 187
14255	*p*-Toluene sulfonic acid, propyl ester or Propyl-*p*-toluene sulfonate $(4\text{-}CH_3C_6H_4)SO_2(OC_3H_7)$	214.28	....	189^{9}	<−20	$1.144^{20/4}$	1.4998^{20}	....	$B11^{3}$, 188
14256	*p*- Toluene sulfonic acid, isoprtopyl ester $(4\text{-}CH_3C_6H_4)SO_2(O\text{-}i\text{-}C_3H_7)$	214.28	....	....	21	$^{1}.5065^{20}$	....	....	$B11^{3}$, 189
14257	*p*-Toluene sulfonic acid, 2-tolyl ester or 2-Tolyl-*p*-toluene sulfonate $4\text{-}CH_3C_6H_4SO_2(OC_6H_4CH_3\text{-}2)$	262.32	nd	....	54-5	....	....	al	$B11^{3}$, 204
14258	*p*-Toluene sulfonic acid, (2,4,6-tribromo phenyl)ester $(2,4,6\text{-}Br_3C_6H_2O)SO_2C_6H_4CH_3\text{-}4$	484.98	cr (al)	....	113	....	....	al	$B11^{2}$, 47
14259	*p*-Toluene sulfonyl chloride $4\text{-}CH_3C_6H_4SO_2Cl$	190.64	tcl (eth, peth)	$145\text{-}6^{15}$	71	....	....	al, eth, bz	$B11^{3}$, 265
14260	*o*-Toluic acid or 2-Methyl benzoic acid $2\text{-}CH_3C_6H_4CO_2H$	136.15	pr or nd (w)	$258\text{-}9^{751}$	107-8	1.062^{115}	1.512^{115}	al, eth, chl	$B9^{3}$, 2298
14261	*o*-Toluic acid anhydride $(2\text{-}CH_3C_6H_4CO)_2O$	254.29	....	>325	38-9	....	....	eth	$B9^{3}$, 2303
14262	*o*-Toluic acid, 3-chloro $3\text{-}Cl\text{-}2\text{-}CH_3C_6H_3CO_2H$	170.60	nd (al)	....	159	....	....	al, eth	$B9^{3}$, 2309
14263	*o*-Toluic acid, 4-chloro or 4-Chloro-2-methyl benzoic acid $4\text{-}Cl\text{-}2\text{-}CH_3C_6H_3CO_2H$	170.60	nd (w, al, bz)	....	173	....	....	w, al, aa	$B9^{2}$, 320
14264	*o*-Toluic acid, 5-chloro or 5-Chloro-2-methyl benzoic acid $5\text{-}Cl\text{-}2\text{-}CH_3C_6H_3CO_2H$	170.60	nd (al)	....	168-9	....	....	al	$B9^{2}$, 320
14265	*o*-Toluic acid, 4,6-dihydroxy or *o*-Orsellinic acid $4,6\text{-}(HO)_2\text{-}2\text{-}CH_3C_6H_2CO_2H$	168.15	nd (dil aa, + 1w)	....	176d	....	....	al, eth	$B10^{3}$, 1479
14266	*o*-Toluic acid, 4,6-dihydroxy, ethyl ester $4,6\text{-}(HO)_2\text{-}2CH_3C_6H_2CO_2C_2H_5$	196.20	wh nd	....	147	....	....	al, eth	$B10^{3}$, 1482
14267	*o*-Toluic acid, ethyl ester or Ethyl-*o*-toluate $2\text{-}CH_3C_6H_4CO_2C_2H_5$	164.21	....	227, 102.5^{12}	<−10	$1.0325^{21/4}$	1.507^{22}	al, eth	$B9^{3}$, 2301
14268	*o*-Toluic acid, methyl ester or Methyl-*o*-toluate $2\text{-}CH_3C_6H_4CO_2CH_3$	150.18	....	215, 97^{15}	<−15	$1.068^{20/4}$	....	al, eth	$B9^{3}$, 2301
14269	*o*-Toluic acid, 6-methyl $6\text{-}CH_3\text{-}2\text{-}CH_3C_6H_3CO_2H$	150.18	nd (w)	....	102	....	....	....	$B9^{3}$, 2310
14270	*o*-Toluonitrile, or 2-methyl benzonitrile $2\text{-}CH_3C_6H_4CN$	117.15	....	205, 90^{15}	−14	$0.9955^{20/4}$	1.5279^{20}	al, eth	$B9^{3}$, 2307
14271	*o*-Toluonitrile, 4-nitro $4\text{-}NO_2\text{-}2\text{-}CH_3C_6H_3CN$	162.15	lf (sub)	sub	100 (105)	....	....	ace, bz, chl	$B9^{1}$, 188
14272	*o*-Toluonitrile, 5-nitro $5\text{-}NO_2\text{-}2\text{-}CH_3C_6H_3CN$	162.15	nd (95% al)	$174\text{-}5^{18}$	105	....	....	al, eth, ace, bz, chl	$B9^{3}$, 2314

No.	Name, Synonyms, and Formula	Mol. wt.	Color, crystalline form, specific rotation and λ_{max} (log ε)	b.p. °C	m.p. °C	Density	n_D	Solubility	Ref.
14273	*o*-Toluonitrile, 6-nitro or 6-Nitro-*o*-toluonitrile $6-NO_2-2-CH_3C_6H_3CN$	162.15	pl (bz)		109-10			al, ace, bz	$B9^{2}$, 323
14274	*o*-Toluyl chloride or 2-Methyl benzoyl chloride $2-CH_3C_6H_4COCl$	154.60		213-4, $88-90^{12}$			1.5549^{20}	eth	$B9^{3}$, 2304
14275	*m*-Toluic acid or 3-Methyl benzoic acid $3-CH_3C_6H_4CO_2H$	136.15	pr (w, al)	263 sub	111-3	1.054^{112}	1.509	al, eth	$B9^{3}$, 2318
14276	*m*-Toluic acid, 2-amino or 2-Amino-3-methyl benzoic acid $2-H_2N-3-CH_3C_6H_3CO_2H$	151.16	nd (al), pr (w)		172			al, eth	$B14^{2}$, 291
14277	*m*-Toluic acid, 4-amino or 4-Amino-3-methyl benzoic acid $4-H_2N-3-CH_3C_6H_3CO_2H$	151.16	nd (w)		170			w	$B14^{2}$, 290
14278	*m*-Toluic acid anhydride $(3-CH_3C_6H_4CO)_2O$	254.29	cr (peth)	230^{17}	71			al, eth, ace, bz, chl	$B9^{3}$, 2321
14279	*m*-Toluic acid, 4-chloro or 4-Chloro-3-methyl benzoic acid $4-Cl-3-CH_3C_6H_3CO_2H$	170.60	nd (w)		209-10				$B9^{3}$, 2327
14280	*m*-Toluic acid, 5-chloro or 5-Chloro-3-methyl benzoic acid $5-Cl-3-CH_3C_6H_3CO_2H$	170.60	nd (dil al)		178			al	B9, 479
14281	*m*-Toluic acid, 6-chloro $6-Cl-3-CH_3C_6H_3CO_2H$	170.60	nd (w al)		167			w, al	$B9^{3}$, 2327
14282	*m*-Toluic acid, ethyl ester or Ethyl-*m*-toluate $3-CH_3C_6H_4CO_2C_2H_5$	164.20		234, $103-5^{10}$		$1.0265^{21/4}$	1.5052^{22}	**al, eth**	$B9^{3}$, 2320
14283	*m*-Toluic acid, methyl ester or Methyl-*m*-toluate $3-CH_3C_6H_4CO_2CH_3$	150.18		221^{758}		$1.061^{20/4}$		al	$B9^{3}$, 2320
14284	*m*-Toluonitrile $3-CH_3C_6H_4CN$	117.15		213, 84.5^{10}	−23	$1.0316^{20/4}$	1.5252^{20}	al, eth	$B9^{3}$, 2324
14285	*m*-Toluonitrile, 2-nitro $2-NO_2-3-CH_3C_6H_3CN$	162.15	nd (al)		84			ace, bz, chl	$B9^{3}$, 2330
14286	*m*-Toluonitrile, 4-nitro $4-NO_2-3-CH_3C_6H_3CN$	162.15	pr (al), nd		80			al	$B9^{3}$, 2331
14287	*m*-Toluonitrile, 5-nitro $5-NO_2-3-CH_3C_6H_3CN$	162.15	nd (lig)		104.5			al	$B9^{3}$, 2331
14288	*m*-Toluonitrile, 6-nitro $6-NO_2-3-CH_3C_6H_3CN$	162.15	nd (al)		93-4			al, bz, aa	$B9^{3}$, 2332
14289	*m*-Toluyl chloride or 3-Methyl benzoyl chloride $3-CH_3C_6H_4COCl$	154.60		219.20, 105^{20}	−23	$1.0265^{21/4}$	1.505^{22}	**al, eth**	$B9^{3}$, 2321
14290	*p*-Toluic acid or 4-Methyl benzoic acid $4-CH_3C_6H_4CO_2H$	136.15	nd (w)	275 (sub)	182			al, eth	$B9^{3}$, 2334
14291	*p*-Toluic acid anhydride $(4-CH_3C_6H_4CO)_2O$	254.29	pl (MeOH), nd (al)		95			eth, ace, bz, chl	$B9^{3}$, 2341
14292	*p*-Toluic acid, 2-chloro or 2-Chloro-4-methyl benzoic acid $2-Cl-4-CH_3C_6H_3CO_2H$	170.60	nd (al)		155.6			al, eth, bz, chl	B9, 497
14293	*p*-Toluic acid, 3-chloro $3-Cl-4-CH_3C_6H_3CO_2H$	170.60	nd or lf (dil al)		200-2			al	$B9^{3}$, 2355
14294	*p*-Toluic acid, ethyl ester or Ethyl-*p*-toluate $4-CH_3C_6H_4CO_2C_2H_5$	164.20		235.7, 110^{12}		$1.0269^{18/4}$	1.5089^{18}	**al, eth**	$B9^{3}$, 2337
14295	*p*-Toluic acid, methyl ester or Methyl-*p*-toluate $4-CH_3C_6H_4CO_2CH_3$	150.18	cr (aq MeOH, peth)	222.5	33.2			al, eth	$B9^{3}$, 2337
14296	*p*-Toluonitrile or 4-Methyl benzonitrile $4-CH_3C_6H_4CN$	117.15	nd (al)	217.6, 91^{11}	29.5	$0.9805^{30/30}$		al, eth	$B9^{3}$, 2348
14297	*p*-Toluonitrile, 2-nitro $2-NO_2-4-CH_3C_6H_3CN$	162.15	nd (w)		101			al, bz, chl	$B9^{2}$, 334
14298	*p*-Toluonitrile, 3-nitro $3-NO_2-4-CH_3C_6H_3CN$	162.15	pa ye nd (w)	171^{12}	107-8			al, eth, ace, bz, chl	$B9^{3}$, 2359
14299	*p*-Toluyl chloride or 4-Methyl benzoyl chloride $4-CH_3C_6H_4COCl$	154.60		225-7, 102^{15}	−2	$1.1686^{20/4}$	1.5547^{20}		$B9^{3}$, 2342
14300	*o*-Toluidine or 2-Amino toluene $2-H_2NC_6H_4CH_3$	107.16		200.2, 80.1^{10}	−14.7	$0.9984^{20/4}$	1.5725^{20}	**al, eth**	$B12^{3}$, 1837
14301	*o*-Toluidine, N-acetyl-4-bromo $4-Br-2-CH_3C_6H_3(NHCOCH_3)$	228.09	nd (dil al, lig)		159-60			al	$B12^{2}$, 456
14302	*o*-Toluidine, N-acetyl-5-bromo $5-Br-2-CH_3 \cdot C_6H_3(NHCOCH_3)$	228.09	nd (bz)		165.5			al, bz	$B12^{3}$, 1923
14303	*o*-Toluidine, hydrochloride $2-CH_3C_6H_4NH_2.HCl$	143.62	mcl pr (w)	242	215			w, al	$B12^{3}$, 1840

PHYSICAL CONSTANTS OF ORGANIC COMPOUNDS (Continued)

No.	Name, Synonyms, and Formula	Mol. wt.	Color, crystalline form, specific rotation and λ_{max} (log ε)	b.p. °C	m.p. °C	Density	n_D	Solubility	Ref.
14304	*o*-Toluidine, N-acetyl-6-bromo 6-Br-2-$CH_3C_6H_3(NHCOCH_3)$	228.09	nd (bz)		166			al	$B12^3$, 1922
14305	*o*-Toluidine, N-benzyl 2-$CH_3C_6H_4NHCH_2C_6H_5$	197.28	cr (al, eth)	300-5, 176^{10}	60	$1.0142^{65/4}$	1.5861^{65}	al, ace, chl	$B12^3$, 2220
14306	*o*-Toluidine, N-benzyl-N-ethyl 2-$CH_3C_6H_4N(C_2H_5)CH_2C_6H_5$	225.33	ye	230^{20-5}					B12, 1033
14307	*o*-Toluidine, N-benzyl-N-methyl 2-$CH_3C_6H_4N(CH_3)CH_2C_6H_5$	211.31	ye	167^{13}				al	B12, 1033
14308	*o*-Toluidine, 4-bromo 4-Br-2-CH_3-$C_6H_3NH_2$	186.05	cr (al)	240	59.5			al, eth, aa	$B12^2$, 456
14309	*o*-Toluidine, 5-bromo 5-Br-2-$CH_3C_6H_3NH_2$	186.05	lf	253-7d, 139^{17}	33			al, eth	$B12^3$, 1923
14310	*o*-Toluidine, 6-bromo 6-Br-2-$CH_3C_6H_3NH_2$	186.05		130^{16}				eth	$B12^3$, 1922
14311	*o*-Toluidine, 3-chloro 3-Cl-2-$CH_3C_6H_3NH_2$	141.60		245, $96\text{-}9^{10}$	0-2		1.5880^{20}	al	$B12^3$, 1919
14312	*o*-Toluidine, 4-chloro 4-Cl-2-$CH_3C_6H_3NH_2$	141.60	lf (al)	241	29-30			al	$B12^3$, 1914
14313	*o*-Toluidine, 5-chloro 5-Cl-2-$CH_3C_6H_3NH_2$	141.60		237^{722}, 140^{38}	26			al	$B12^3$, 1910
14314	*o*-Toluidine, N,N-diethyl 2-$CH_3C_6H_4N(C_2H_5)_2$	163.26		$208\text{-}9^{755}$				al, eth	$B12^3$, 1844
14315	*o*-Toluidine, N,N-dimethyl 2-$CH_3C_6H_4N(CH_3)_2$	135.21		185.3, $70\text{-}2^{15}$	−60	$0.9286^{20/4}$	1.5152^{20}	**al, eth**	$B12^3$, 1843
14316	*o*-Toluidine, N-ethyl 2-$CH_3C_6H_4NHC_2H_5$	135.21		218, 95.5^{10}	<−15	$0.948^{25/4}$	1.5456^{20}	al, eth	$B12^3$, 1843
14317	*o*-Toluidine, 4-iodo 4-I-2-$CH_3C_6H_3NH_2$	233.05	nd (dil al), pr (lig)		91-2			al, eth, bz, aa, lig	$B12^3$, 1927
14318	*o*-Toluidine, 5-iodo 5-I-2-$CH_3C_6H_3NH_2$	233.05	nd (aq al)	273d	48-9			al, aa	$B12^1$, 391
14319	*o*-Toluidine, 4-methoxy or 2-Methyl-*p*-anisidine 4-CH_3O-2-$CH_3C_6H_3NH_2$	137.18	cr (lig)	248-9, $146\text{-}7^{23}$	29-30		1.5647^{20}	al	$B13^3$, 1560
14320	*o*-Toluidine, 5-methoxy or 6-Methyl-*m*-anisidine 5-CH_3O-2-$CH_3C_6H_3NH_2$	137.18	nd (w)	253, 140^{20}	47			eth	$B13^3$, 1573
14321	*o*-Toluidine, 6-methoxy or 6-Methyl-*o*-anisidine 6-CH_3O-2-$CH_3C_6H_3NH_2$	137.18	nd (w)	$119\text{-}21^{16}$	31			al	$B13^3$, 1538
14322	*o*-Toluidine, N-Methyl 2-$CH_3C_6H_4NHCH_3$	121.18		207-8, 99^{17}		$0.9769^{20/4}$	1.5649^{20}	**al, eth**, ace	$B12^3$, 1842
14323	*o*-Toluidine, 3-nitro 3-NO_2-2-$CH_3C_6H_3NH_2$	152.15	ye rh nd (w), ye lf (al)	305d	92 (97)			al, eth, bz, chl	$B12^3$, 1944
14324	*o*-Toluidine, 4-nitro 4-NO_2-2-$CH_2C_6H_3NH_2$	152.15	ye mcl pr or nd (w, al, lig)		134-5	$1.1586^{140/4}$		al, bz, aa	$B12^3$, 1938
14325	*o*-Toluidine, 5-nitro 5-NO_2-2-$CH_3C_6H_3NH_2$	152.15	ye mcl pr (al)		107-8			al, eth, ace, bz, chl	$B12^3$, 1932
14326	*o*-Toluidine, 6-nitro 6-NO_2-2-$CH_3C_6H_3NH_2$	152.15	og-ye pr (dil al)		97	$1.1900^{100/4}$		al, eth, bz, chl	$B12^3$, 1929
14327	*m*-Toluidine or 3-Amino toluene 3-$CH_3C_6H_4NH_2$	107.16		203.3, 82.3^{10}	−30.4	$0.9889^{20/4}$	1.5681^{20}	**al, eth, ace, bz**	$B12^3$, 1949
14328	*m*-Toluidine, N-benzyl 3-$CH_3C_6H_4NHCH_2C_6H_5$	197.28	pa ye oil	312, 180^{10}		$1.0083^{65/4}$	1.5845^{65}	eth, ace, bz, chl	$B12^2$, 552
14329	*m*-Toluidine, 4-bromo 4-Br-3-$CH_3C_6H_3NH_2$	186.05	pl (50% al), cr (al)	240	81			al	$B12^3$, 1999
14330	*m*-Toluidine, 5-bromo 5-Br-3-$CH_3C_6H_3NH_2$	186.05		255-60, $150\text{-}1^{15}$	37-8	1.1422^{19}		al	$B12^2$, 474
14331	*m*-Toluidine, 6-bromo 6-Br-3-CH_3-$C_6H_3NH_2$	186.05	pr	$129\text{-}30^{15}$	46	$1.474^{25/25}$	1.5990^{25}	al, eth	$B12^3$, 1998
14332	*m*-Toluidine, 4-chloro 4-Cl-3-$CH_3C_6H_3NH_2$	141.60	nd (peth)	241	83-4			al, ace, bz	$B12^3$, 1993
14333	*m*-Toluidine, 6-chloro 6-Cl-3-$CH_3C_6H_3NH_2$	141.60	pl	228-30	29-30			al	$B12^3$, 1992
14334	*m*-Toluidine, N,N-dimethyl 3-$CH_3C_6H_4N(CH_3)_2$	135.21		212		$0.9410^{20/4}$	1.5492^{20}	**al, eth**	$B12^3$, 1953
14335	*m*-Toluidine, N-ethyl 3-$CH_3C_6H_4NHC_2H_5$	135.21		221, $111\text{-}2^{20}$			1.5451^{20}	al, eth	$B12^3$, 1954
14336	*m*-Toluidine hydrochloride 3-$CH_3C_6H_4NH_2.HCl$	143.62	lf (w)	250	228			w, al	$B12^3$, 1951

PHYSICAL CONSTANTS OF ORGANIC COMPOUNDS (Continued)

No.	Name, Synonyms, and Formula	Mol. wt.	Color, crystalline form, specific rotation and λ_{max} (log ε)	b.p. °C	m.p. °C	Density	n_D	Solubility	Ref.
14337	*m*-Toluidine, 2-iodo $2\text{-I-3-}CH_3C_6H_3NH_2$	233.05	pr	...	41-2	...	...	al, eth, ace	$B12^{3}$, 2003
14338	*m*-Toluidine, 4-iodo $4\text{-I-3-}CH_3C_6H_3NH_2$	233.05	lf or pl (al, peth)	...	46	...	...	al, eth, bz, aa, lig	$B12^{2}$, 475
14339	*m*-Toluidine, 5-iodo $5\text{-I-3-}CH_3C_6H_3NH_2$	233.05	nd (peth)	...	78.5	...	...	al, eth, ace	$B12^{1}$, 406
14340	*m*-Toluidine, 6-iodo $6\text{-I-3-}CH_3C_6H_3NH_2$	233.05	nd (dil al), br in air	...	48	...	...	al, eth, chl	$B12^{2}$, 475
14341	*m*-Toluidine, 5-methoxy or 3-Methyl-*p*-anisidine $5\text{-}CH_3O\text{-3-}CH_3C_6H_3NH_2$	137.18	cr (dil al)	...	59-60	...	...	al, eth, ace, bz	$B13^{2}$, 320
14342	*m*-Toluidine, 6-methoxy or 5-Methyl-*o*-anisidine $6\text{-}CH_3O\text{-3-}CH_3C_6H_3NH_2$	137.18	nd or lf (al, lig, peth)	235	93-4	...	...	al, eth, bz	$B13^{3}$, 1577
14343	*m*-Toluidine, N-methyl $3\text{-}CH_3C_6H_4NHCH_3$	121.18	...	206-7, $120\text{-}1^{40}$	...	...	1.5557^{25}	**al, eth**, ace	$B12^{3}$, 1953
14344	*m*-Toluidine, 2-nitro $2\text{-}NO_2\text{-3-}CH_3C_6H_3NH_2$	152.15	ye-og pr or nd (bz-peth)	...	108	...	...	al, eth	$B12^{3}$, 2004
14345	*m*-Toluidine, 4-nitro $4\text{-}NO_2\text{-3-}CH_3C_6H_3NH_2$	152.15	lt ye nd (w, dil al)	...	138	...	...	al, eth	$B12^{3}$, 2008
14346	*m*-Toluidine, 5-nitro $5\text{-}NO_2\text{-3-}CH_3C_6H_3NH_2$	152.15	ye-red or red-br nd (al)	...	98	...	...	al, eth, bz	$B12^{3}$, 2007
14347	*m*-Toluidine, 6-nitro $6\text{-}NO_2\text{-3-}CH_3C_6H_3NH_2$	152.15	ye lf (w), pl (dil al)	...	112	...	...	al, eth, bz, chl	$B12^{3}$, 2004
14348	*p*-Toluidine or 4-Amino toluene $4\text{-}CH_3C_6H_4NH_2$	107.16	lf (w + 1)	200.5, 79.6^{10}	44-5	$0.9619^{20/4}$	1.5534^{45}, 1.5636^{20}	al, eth, ace, Py	$B12^{3}$, 2017
14349	*p*-Toluidine, N-acetyl-2-bromo $2\text{-Br-4-}CH_3C_6H_3NH(COCH_3)$	228.09	nd (al)	...	118	...	...	al	$B12^{3}$, 2159
14350	*p*-Toluidine, N-acetyl-3-bromo $3\text{-Br-4-}CH_3C_6H_3NH(COCH_3)$	228.09	nd (bz, dil al)	...	113	...	...	w, al, bz	$B12^{3}$, 2157
14351	*p*-Toluidine, N-benzyl $4\text{-}CH_3C_6H_4NHCH_2C_6H_5$	197.28	ye lf	319^{765}, 181^{10}	19-20	$1.0064^{65/4}$	1.5832^{65}	al, eth, bz, chl	$B12^{3}$, 2220
14352	*p*-Toluidine, 2-bromo $2\text{-Br-4-}CH_3C_6H_3NH_2$	186.05	lf	240, $120\text{-}2^{30}$	26	1.510^{20}	1.5999^{20}	al, eth	$B12^{3}$, 2158
14353	*p*-Toluidine, 2-bromo-5-nitro $2\text{-Br-5-}NO_2\text{-4-}CH_3C_6H_2NH_2$	231.05	br to pa ye nd (al, aa)	...	121	...	...	...	$B12^{1}$, 441
14354	*p*-Toluidine, 3-bromo $3\text{-Br-4-}CH_3C_6H_3NH_2$	186.05	...	254-7	26	...	...	eth	$B12^{3}$, 2157
14355	*p*-Toluidine, 2-chloro $2\text{-Cl-4-}CH_3C_6H_3NH_2$	141.60	...	219^{732}	7	1.151^{20}	1.5748^{22}	...	$B12^{3}$, 2152
14356	*p*-Toluidine, 3-chloro $3\text{-Cl-4-}CH_3C_6H_3NH_2$	141.60	...	242-4, $112\text{-}3^{3}$	26	...	...	al	$B12^{3}$, 2151
14357	*p*-Toluidine, N,N-diethyl $4\text{-}[(C_2H_5)_2N]C_6H_4CH_3$	163.26	...	229^{770}	...	0.9242^{16}	...	**al, eth**	$B12^{3}$, 2028
14358	*p*-Toluidine, N,N-dimethyl $4\text{-}CH_3C_6H_4N(CH_3)_2$	135.21	...	211	...	$0.9366^{20/4}$	1.5366^{20}	**al, eth**	$B12^{3}$, 2026
14359	*p*-Toluidine, 3,5-dinitro $3,5\text{-}(NO_2)_2\text{-4-}CH_3C_6H_2NH_2$	197.15	ye nd (w, aa)	...	171	...	...	al, eth, ace, bz, chl	$B12^{3}$, 2184
14360	*p*-Toluidine, N-ethyl $4\text{-}CH_3C_6H_4NHC_2H_5$	135.21	...	217	...	0.9391^{16}	...	al, eth	$B12^{3}$, 2027
14361	*p*-Toluidine hydrochloride $4\text{-}CH_3C_6H_4NH_2.HCl$	143.62	mcl nd (aa-eth)	258	243	...	...	w, al, aa	$B12^{3}$, 2021
14362	*p*-Toluidine, 2-iodo $2\text{-I-4-}CH_3C_6H_3NH_2$	233.05	pr	d	40	...	...	al, eth, ace, bz, chl, peth	$B12^{2}$, 533
14363	*p*-Toluidine, 3-iodo $3\text{-I-4-}CH_3C_6H_3NH_2$	233.05	nd (dil al, peth)	...	39-40	...	...	al, eth, ace, aa	$B12^{2}$, 533
14364	*p*-Toluidine, 2-methoxy or 4-Methyl-*o*-anisidine $2\text{-}CH_3O\text{-4-}CH_3C_6H_3NH_2$	137.18	pa ye	237-9, $179\text{-}80^{46}$	...	...	...	al, eth, ace	$B13^{3}$, 1552
14365	*p*-Toluidine, 3-methoxy or 4-Methyl-*m*-anisidine $3\text{-}CH_3O\text{-4-}CH_3C_3H_3NH_2$	137.18	...	250-2	58	...	...	al, eth, bz, lig	$B13^{1}$, 213
14366	*p*-Toluidine, N-methyl $4\text{-}CH_3C_6H_4NHCH_3$	121.18	...	209-11, 102^{20}	...	$0.9348^{55/4}$	1.5568^{20}	**al, eth**, ace	$B12^{3}$, 2025
14367	*p*-Toluidine, 2-nitro $2\text{-}NO_2\text{-4-}CH_3C_6H_3NH_2$	152.15	red lf (dil al), mcl pr (al)	...	117	$1.164^{121/4}$	...	al	$B12^{3}$, 2174
14368	*p*-Toluidine, 3-nitro $3\text{-}NO_2\text{-4-}CH_3C_6H_3NH_2$	152.15	ye nd (w)	...	78-9	...	...	eth, bz	$B12^{3}$, 2165

No.	Name, Synonyms, and Formula	Mol. wt.	Color, crystalline form, specific rotation and λ_{max} (log ε)	b.p. °C	m.p. °C	Density	n_D	Solubility	Ref.
14369	*o*-Tolyl acetic acid or *o*-Methyl-α-toluic acid $2\text{-}CH_3C_6H_4CH_2CO_2H$	150.18	nd (w)		88-90			w	B9³, 2426
14370	2-Tolyl acetonitrile or *o*-Methyl-α-tolunitrile $2\text{-}CH_3C_6H_4CH_2CN$	131.18		244		1.0156^{22}	1.5252^{20}	al, eth, bz	B9³, 2427
14371	3-Tolyl acetic acid or *m*-Methyl-α-toluic acid $3\text{-}CH_3C_6H_4CH_2CO_2H$	150.18	nd (w)	$120\text{-}3^{26}$	62				B9³, 2429
14372	3-Tolyl acetonitrile or *m*-Methyl-α-tolunitrile $3\text{-}CH_3C_6H_4CH_2CN$	131.18		$245\text{-}7^{745}/_d$, 133^{15}		1.0022^{22}	1.5233^{20}	al, eth, bz	B9², 349
14373	4-Tolyl acetic acid or *p*-Methyl-α-toluic acid $4\text{-}CH_3C_6H_4CH_2CO_2H$	150.18	nd or pl (al, w)	265-7 sub	91-3			al, eth, bz, chl	B9³
14374	2-Tolyl ether or Di-2-Tolyl ether $(2\text{-}CH_3C_6H_4)_2O$	198.26		$146\text{-}7^{17}$		1.047^{24}		al, eth, bz	B6⁴, 1947
14375	3-Tolyl ether or Di-3-Tolyl ether $(3\text{-}CH_3C_6H_4)_2O$	198.26		284, $135\text{-}7^{14}$		1.0323^{21}		al, eth, bz	B6⁴, 2043
14376	4-Tolyl ether or Di-4-Tolyl ether $(4\text{-}CH_3C_6H_4)_2O$	198.26		285	51			al, eth, bz	B6⁴, 2103
14377	2-Tolyl hydrazine $2\text{-}CH_3C_6H_4NHNH_2$	122.17	nd (dil al)		59			al, eth	B15³, 654
14378	2-Tolyl isocyanate $2\text{-}CH_3C_6H_4NCO$	133.15		184-6			1.5282^{20}	eth, bz	B12³, 1886
14379	3-Tolyl isocyanate $3\text{-}CH_3C_6H_4NCO$	133.15		195-8				eth, bz	B12³, 1979
14380	4-Tolyl isocyanate $4\text{-}CH_3C_6H_4NCO$	133.15		187^{751}				eth, bz	B12³, 2110
14381	2-Tolyl selenide or Di-(2-Tolyl)selenide $(2\text{-}CH_3C_6H_4)_2Se$	261.23	pr or lf (al)	186^{16}	65				B6², 343
14382	4-Tolyl selenide or Di-(4-Tolyl)selenide $(4\text{-}CH_3C_6H_4)_2Se$	261.23	rods or nd (al)	196^{16}	69-70				B6⁴, 2218
14383	2-Tolyl sulfone or 2,2′-Ditolyl sulfone $(2\text{-}CH_3C_6H_4)_2SO_2$	246.32	nd (al)		134-5			al, eth, bz, chl	B6⁴, 2020
14384	4-Tolyl sulfone or Di-4-Tolyl sulfone $(4\text{-}CH_3C_6H_4)_2SO_2$	246.32	pr (bz), nd (w, al), pl (al)	405^{714}	159			bz, chl	B6⁴, 2174
14385	4-Tolyl sulfoxide or Di-4-Tolyl sulfoxide $(4\text{-}CH_3C_6H_4)_2SO$	230.32	cr (lig-peth)		94			al, eth, bz, chl, aa	B6⁴, 2173
14386	4-Tolyl thiocyanate $4\text{-}CH_3C_6H_4SCN$	149.21		240-5, $116\text{-}8^{10}$				al, bz, chl	B6³, 1421
14387	Tomatidine $C_{27}H_{45}NO_2$	415.66	pl, $[\alpha]^{20}/_D$ + 5 (MeOH)		210-1			eth	Am7³, 4018
14388	Tomatine or Lycopersicin $C_{50}N_{83}NO_{21}$	1034.20	nd (MeOH), $[\alpha]^{20}/_D$ −30 (Py)		270			al, diox	C41, 3502
14389	Torularhodin or Torulene $C_{37}H_{48}O_2$	524.79	red nd (MeOH-eth), vt-bk (bz-MeOH)		201-3d			ace, chl, Py	B9³, 3661
14390	α-Toxicarol (dl) or Hydroxydequelin $C_{23}H_{22}O_7$	410.42	gr ye pl (al)		219-23 (231)		1.580	chl	B19⁴, 5270
14391	α-Toxicarol (l) $C_{23}H_{22}O_7$	410.42	gr-yesh pl or nd (AcOEt-al)		125-7			ace, AcOEt	B19⁴, 5270
14392	α-Toxicarol, dihydro - (l) $C_{23}H_{24}O_7$	412.44	pa ye pl or rods, $[\alpha]^{20}/_D$ −30 (bz, c=5)		179 (206)				B19⁴, 5262
14393	β-Toxicarol-(dl) $C_{23}H_{22}O_7$	410.42	pa ye pl (al)		169-70			al	B19⁴, 5270
14394	Trasentin hydrochloride or Adephenine. 2-(Diethyl amino)ethyl diphenyl acetate hydrochloride $(C_6H_5)_2CHCO_2CH_2CH_2N(C_2H_5)_2.HCl$	347.88	nd		114-5			w	B9³, 3297
14395	α,α-Trehalose or Mycose $C_{12}H_{22}O_{11}$	342.30	orh cr, $[\alpha]^{20}/_D$ + 199 (w, c=6)		214-6 (anh), 97 (hyd)		$1.58^{24/24}$	w	B17⁴, 3505

No.	Name, Synonyms, and Formula	Mol. wt.	Color, crystalline form, specific rotation and λ_{max} (log ε)	b.p. °C	m.p. °C	Density	n_D	Solubility	Ref.
14396	α,α-Trehalose, dihydrate $C_{12}H_{22}O_{11}\cdot 2H_2O$	378.33	cr (dil al), $[\alpha]^{20/}{}_D$ + 178.3 (w, c=7)	...	103	...	...	w	B17^4, 3505
14397	Triacetamide or N,N-Diacetyl acetamide $(CH_3CO)_3N$	143.14	nd (eth)	...	79	...	...	eth	B2^4, 416
14398	Triacetin or Glycerol triacetate $CH_3CO_2CH(CH_2O_2CCH_3)_2$	218.21	cr (al)	258-60, 130.5^7	4.1	1.1596$^{20/4}$	1.4301^{20}	**al, eth**, ace, **bz, chl**	B2^4, 253
14399	Triacontane $CH_3(CH_2)_{28}CH_3$	422.82	orh (eth bz)	449.7, 304^{15}	65.8	0.7750$^{78/4}$	1.4352^{70}, 1.4536^{20}	eth, bz	B1^4, 592
14400	1-Triacontanol or Myricyl alcohol $CH_3(CH_2)_{28}CH_2OH$	438.82	nd (eth), pl (bz)	...	88	0.777^{95}	...	al, eth, bz	B1^4, 1918
14401	Triallylamine $(CH_2{=}CHCH_2)_3N$	137.22	...	155-6	...	0.809$^{20/4}$	1.4502^{20}	al, eth, ace, bz	B4^4, 1061
14402	1,3,5-Triazine $C_3H_3N_3$	81.08	...	114	86	1.38	...	al, eth	Am76, 5646
14403	1,3,5-Triazine, 2,4-diamino or Guanamine $2,4\text{-}(NH_2)_2C_3HN_3$	111.11	nd (w)	...	329d	...	...	w	B26^1, 65
14404	1,3,5-Triazine, 2,4-diamino-6-phenyl or Benzoguanamine $2,4\text{-}(NH_2)_2\text{-}6\text{-}C_6H_5C_3N_3$	187.20	nd or pr (al)	...	226-8	...	...	al, eth	B26^1, 69
14405	1,3,5-Triazine, perhydro, 1,3,5-tributyl $1,3,5\text{-}(C_4H_9)_3C_3H_6N_3$ $C_{15}H_{33}N_3$	255.45	...	130 2^3	...	...	1.4602^{25}	...	B26, 3
14406	1,3,5-Triazine, perhydro, 1,3,5-triethyl $1,3,5\text{-}(C_2H_5)_3C_3H_6N_3$	171.29	...	78-9^6	...	...	1.4580^{25}	...	B26^2, 3
14407	1,3,5-Triazine, perhydro, 1,3,5-trinitro or Cyclonite . Hexogen $1,3,5\text{-}(NO_2)_3C_3H_6N_3$	222.12	orh cr (ace)	...	205-6	1.82^{20}	...	ace, aa	B26^2, 5
14408	1,3,5-Triazine-perhydro-1,3,5-triphenyl $1,3,5\text{-}(C_6H_5)_3C_3H_6N_3$	315.42	nd (lig), pr (eth, chl-al)	185	143	...	...	eth, ace, bz, chl	B26^2, 3
14409	1,3,5-Triazino-2,4,6-tricarboxylic acid or Cyanuric tricarboxylic acid $2,4,6\ C_3N_3(CO_2H)_3$	213.11	pw	...	>250d	...	...	...	B26^2, 168
14410	1,3,5-Triazino-2,4,6-tricarboxylic acid, triethyl ester $2,4,6\text{-}C_3N_3(CO_2C_2H_5)_3$	297.27	nd (al)	d	168	...	...	...	B26^2, 168
14411	1,3,5-Triazino-2,4,6-tricarbonitrile $2,4,6\text{-}C_3N_3(CN)_3$	156.11	mcl pr (bz)	262^{771}, 119^1	119	...	...	bz	B26^1, 91
14412	1,2,3-Triazole or Osotriazole $C_2N_3H_3$	69.07	hyg cr	203^{739}	23	1.1861$^{25/4}$	1.4854^{25}	w, eth, ace	B26^1, 5
14413	1,2,4-Triazole or Pyrrodiazole $C_2N_3H_3$	69.07	pr (w), nd (al, eth, chl, bz)	260	120-1	1.132^{153}	1.4854^{25}	w, al	B26^2, 7
14414	1,2,4-Triazole, 3-amino or Amizol . ATA $3\text{-}NH_2C_2N_3H_2$	84.08	cr (w, al, AcOEt)	...	159	...	...	w, al, chl	B26^2, 76
14415	1,2,4-Triazole, 4-amino $4\text{-}NH_2C_2N_3H_2$	84.08	hyg nd (al, chl)	...	82-3	...	...	w, al	B26^2, 7
14416	1,2,4-Triazole, 3,5-diamino $C_2H_5N_5$	99.10	...	...	211-2	...	...	w, al	B26^1, 57
14417	Tribenzylamine $(C_6H_5CH_2)_3N$	287.40	pl, (eth), mcl (al)	380-90, 230^{13}	91-2	0.9912$^{95/4}$	...	eth	B12^3, 2226
14418	Triabromo acetamide Br_3CCONH_2	295.77	mcl pr (al)	sub	121-2	...	...	w, al, eth	B2^3, 485
14419	Tribromoacetic acid Br_3CCOOH	296.74	mcl	245	135	...	...	w, al, eth	B2^4, 534
14420	Tribromoacetic acid, ethyl ester or Ethyl tribromo acetate $Br_3CCOOC_2H_5$	324.79	...	225, 148^{73}	...	2.2300$^{20/20}$	1.5438^{13}	al, eth	B2^3, 485
14421	Tribromoacetyl bromide Br_3CCOBr	359.64	...	210-5	...	...	...	eth, bz, chl	B2^3, 485
14422	Bis-(tribromomethyl) trisulfide $(Br_3C)_2S_3$	599.63	rh pr (eth)	d	125d	...	...	bz, chl, peth	B3^2, 107
14423	Tributylamine $(C_4H_9)_3N$	185.35	hyg	213, 91-2^9	...	0.7771$^{20/0}$	1.4297^{20}	al, eth, ace, bz	B4^4, 554
14424	Tributylamine, perfluoro $(C_4F_9)_3N$	671.10	...	179	...	1.873$^{25/4}$	1.291^{25}	ace	B2^4, 819
14425	Tri-*iso*-butyl amine $(i\text{-}C_4H_9)_3N$	185.35	...	191.5, 84^{15}	−21.8	0.7684$^{20/4}$	1.4252^{17}	al, **eth**	B4^4, 631
14426	Tributyrin or Glycerol tributyrate $C_4H_9CO_2CH(CH_2O_2CC_4H_9)_2$	302.37	...	305-10, 190^{15}	−75	1.0350$^{20/4}$	1.4359^{20}	al, eth, ace, bz	B2^4, 799

PHYSICAL CONSTANTS OF ORGANIC COMPOUNDS (Continued)

No.	Name, Synonyms, and Formula	Mol. wt.	Color, crystalline form, specific rotation and λ_{max} (log ε)	b.p. °C	m.p. °C	Density	n_D	Solubility	Ref.
14427	Trichloro acetamide Cl_3CCONH_2	162.40	mcl pr (w)	238-9^{746}	142			al, eth	B2^4, 520
14428	Trichloro acetamide, *N,N*-diethyl $Cl_3CCON(C_2H_5)_2$	218.52	pr	109^{9}	27		1.4900^{24}		B4^4, 351
14429	Trichloro acetamide, *N,N*-dimethyl or *N,N*-Dimethyl trichloroacetamide $Cl_3CON(CH_3)_2$	190.46		230-3d, 84^{4}	12	1.390^{20}	1.5017^{25}	bz, chl	B4^4, 182
14430	Trichloro acetamide, *N*-phenyl or α-Trichloroacetanilide $Cl_3CCONHC_6H_5$	238.50	lf (dil al)	168-70	95-7			al	B12^3, 464
14431	Trichloroacetic acid Cl_3CCO_2H	163.39	dlq cr	197.5, 141-2^{25}	α:58 β:49.6	1.62$^{25/4}$, 1.6218$^{64/1}$	1.4603^{61}	al, eth, w	B2^4, 508
14432	Trichloroacetic acid anhydride or Trichloroacetic anhydride $(Cl_3CCO)_2O$	308.76		222-4d, 98-100^{12}		1.6908^{20}		eth, aa	B2^4, 518
14433	Trichloroacetic acid butyl ester or Butyl trichloroacetate $Cl_3CCO_2C_4H_9$	219.50		203-5, 97-9^{19}		1.2778$^{20/4}$	1.4525^{25}	al, eth, ace, bz	B2^4, 515
14434	Trichloroacetic acid, *iso*-butyl ester or *iso*-Butyl trichloroacetate $Cl_3CCO_2\text{-}i\text{-}C_4H_9$	219.50		187-9, 93-4^{20}		1.2636$^{20/4}$	1.4483^{20}	al, eth, bz	B2^4, 515
14435	Trichloroacetic acid, *sec*-butyl ester or *sec*-Butyl trichloroacetate Cl_3CCO_2 *sec*-C_4H_9	219.50		93-4^{24}		1.2636$^{20/4}$	1.4483^{20}	al, eth, bz	B2^3, 472
14436	Trichloro acetic acid, *tert*-butyl ester or *tert*-Butyl trichloroacetate $Cl_3CCO_2C(CH_3)_3$	219.50	cr (MeOH)	54-5^{7}	25.5	1.2363$^{25/4}$	1.4398^{25}	al, eth	B2^4, 516
14437	Trichloroacetic acid, 2-chloroethyl ester or 2-Chloroethyl trichloro acetate $Cl_3CCO_2CH_2CH_2Cl$	225.89		217, 100^{14}		1.5357$^{20/4}$	1.4813^{20}	eth	B2^4, 515
14438	Trichloroacetic acid, ethyl ester or Ethyl trichlorro acetate $Cl_3CCO_2C_2H_5$	191.44		167-8, 62^{12}		1.3836$^{20/4}$	1.4505^{20}	al, eth, bz	B2^4, 514
14439	Trichloroacetic acid, 2-hydroxyethyl ester $Cl_3CCO_2CH_2CH_2OH$	207.44		130-4^{12}		1.532$^{20/4}$	1.4775^{20}	al	B2^3, 474
14440	Trichloroacetic acid, 2-methoxyethyl ester or 2-Methoxyethyl trichloro acetate $Cl_3CCO_2CH_2CH_2OCH_3$	221.47		98-9^{17}	14.5	1.3826$^{20/4}$	1.4563^{20}	al, eth, bz	B2^3, 474
14441	Trichloroacetic acid, methyl ester or Methyl trichloro acetate $Cl_3CCO_2CH_3$	177.41		153.8, 44.5^{12}	−17.5	1.4874$^{20/4}$	1.4572^{20}	al, eth	B2^4, 513
14442	Trichloroacetic acid, 2-(2-methyl butyl) ester or 2-(2-Methylbutyl) trichloroacetate $Cl_3CCO_2CH_2CH(CH_3)CH_2CH_3$	233.52		217, 92-5^{12}		1.2314$^{20/4}$	1.4521^{20}	al, eth	B2^3, 473
14443	Trichloroacetic acid, pentyl ester or Pentyltrichloroacetate $Cl_3CCO_2C_5H_{11}$	233.52		220-2, 118^{30}		1.2475$^{20/20}$		al, eth	B2^3, 473
14444	Trichloroacetic acid, isopentyl ester or Isopentyl trichloro acetate $Cl_3CO_2\text{-}i\text{-}C_5H_{11}$	233.52		217, 92-5^{11}		1.2314$^{20/4}$	1.4521^{20}	al, eth	B2^4, 516
14445	Trichloroacetic acid, propyl ester or Propyl trichloro acetate $Cl_3CCO_2C_3H_7$	205.47		187, 69^{10}		1.3221$^{20/4}$	1.4501^{20}	al, eth	B2^4, 515
14446	Trichloroacetic acid, *iso*-propyl ester or *iso*-Propyl trichloroacetate $Cl_3CCO_2CH(CH_3)_2$	205.47		173.5^{747}, 65-7^{15}		1.3034$^{20/4}$	1.4428^{20}	al, eth, bz	B2^4, 515
14447	Trichloroacetic acid, trichlomethyl ester or Trichloromethyl trichloro acetate $Cl_3CCO_2CCl_3$	280.75		191-2, 73-4^{10}	34			eth, bz, chl, lig	B3^3, 35
14448	Trichloroacetonitrile Cl_3CCN	144.39		84.6^{741}	−42	1.4403$^{25/4}$	1.4409^{20}		B2^4, 524
14449	Trichloroacetyl bromide Cl_3CCOBr	226.28		143		1.900$^{15/15}$		eth, bz	B2^3, 476
14450	Trichloroacetyl chloride Cl_3CCOCl	181.83		118		1.6202$^{20/4}$	1.4695^{20}	eth	B2^4, 519
14451	Trichloromethyl perchlorate Cl_3COClO_3	217.82		exp d 40	−55			eth	B3^3, 37
14452	Tricosane $CH_3(CH_2)_{21}CH_3$	324.63	lf (eth-al)	380, 243^{15}	47.6	0.7785$^{48/4}$, 0.7969$^{20/4}$	1.4468^{20}	eth	B1^4, 576
14453	Tricosane, 2-methyl $(CH_3)_2CH(CH_2)_{20}CH_3$	338.66		205^{3}	37.6	0.7539$^{90/4}$	1.4201^{90}		B1^3, 576
14454	Tricosanoic acid, methyl ester or Methyltricosonate $CH_3(CH_2)_{21}CO_2CH_3$	368.64			55.6				B1^3, 1078

PHYSICAL CONSTANTS OF ORGANIC COMPOUNDS (Continued)

No.	Name, Synonyms, and Formula	Mol. wt.	Color, crystalline form, specific rotation and λ_{max} (log ε)	b.p. °C	m.p. °C	Density	n_D	Solubility	Ref.
14455	1-Tricosanol $C_{22}H_{43}CH_2OH$	340.63	...	191-3$^{0.7}$	74	...	...	...	B1^4, 1908
14456	Tricyclene $C_{10}H_{16}$	136.24	...	153.5	67.5	0.8268$^{80/4}$	1.4389^{20}	...	B5^4, 468
14457	1,12-Tridecadiyne $HC{\equiv}C(CH_2)_9C{\equiv}CH$	176.30	...	115.5^{12}	−3	0.8262$^{21/4}$	1.454^{20}	...	B1^2, 248
14458	Tridecanal $C_{12}H_{25}CHO$	198.35	...	156^{13}	14	0.8356$^{18/4}$	1.4384^{18}	al	B1^4, 3386
14459	Tridecanal oxime $C_{12}H_{25}CH{=}NOH$	213.36	nd (dil al)	...	80.5	...	...	eth, chl	B1^2, 769
14460	Tridecane $CH_3(CH_2)_{11}CH_3$	184.37	...	235.4, 107^{10}	−5.5	0.7564$^{20/4}$	1.4256^{20}	al, eth	B1^4, 512
14461	Tridecane, 1-amino or Tridecyl amine $C_{12}H_{25}CH_2NH_2$	199.38	...	275.8, 140.1^{10}	27.4	0.8049$^{20/4}$	1.4443^{20}	al, eth	B4^4, 810
14462	Tridecane, 1-bromo or Tridecyl bromide $C_{12}H_{25}CH_2Br$	263.26	...	296, 162^{16}	6.2	1.0177$^{20/4}$	1.4593^{20}	chl	B1^4, 514
14463	Tridecane, 1,13-dibromo $Br(CH_2)_{13}Br$	342.16	...	188.92^{12}	8-10	1.276^{15}	1.4880^{27}	eth, chl	B1^3, 548
14464	Tridecane, 7-hexyl $(C_6H_{13})_3CH$	268.53	...	170.5^{10}	−28.3	0.7877$^{20/4}$	1.4409^{20}	...	B1^4, 562
14465	Tridecane, 7-methyl $(C_6H_{13})_2CHCH_3$	198.39	...	115.5^{10}	−37.2	0.7634$^{20/4}$	1.4291^{20}	...	B1^4, 525
14466	Tridecane, 7-phenyl $(C_6H_{13})_2CHC_6H_5$	260.46	...	183-4^{20}	...	0.8723$^{20/4}$	1.49307^{18}	...	B5^4, 1210
14467	Tridecanedioic acid $HO_2C(CH_2)_{11}CO_2H$	244.33	...	...	114	...	...	al, eth, chl	B2^4, 2141
14468	1,12-Tridecanediol $HO(CH_2)_{11}CH(OH)CH_3$	216.36	cr (dil al)	188-90^8	60-1	...	...	al	B1^2, 563
14469	1,13-Tridecanediol or Tridecamethylene glycol $HO(CH_2)_{13}OH$	216.36	cr (bz)	195-7^{10}	76.5	...	...	al, aa	B1^4, 2630
14470	Tridecan amide $C_{12}H_{25}CONH_2$	213.36	lf (al)	...	100	...	...	al, eth	B2^4, 1119
14471	Tridecanoic acid $C_{12}H_{25}CO_2H$	214.35	cr (peth, ace)	236^{100}, 140.5^1	44-5	...	...	al, eth, ace, aa	B2^4, 1117
14472	Tridecanoic acid, methyl ester or Methyl tridecanoate $C_{12}H_{25}CO_2CH_3$	228.38	...	90-5^1	fr 6.5	...	1.4405^{20}	**al**	B2^4, 1118
14473	Tridecanonitrile $C_{12}H_{25}CN$	195.35	...	293, 142^{10}	9.7	0.8257$^{20/4}$	1.4378^{20}	al, eth	B2^3, 906
14474	1-Tridecanol or Tridecyl alcohol $C_{12}H_{25}CH_2OH$	200.36	cr (al)	152^{14}	32-3	0.8223$^{31/4}$	...	al, eth	B1^4, 1860
14475	2-Tridecanol $C_{11}H_{23}CH(OH)CH_3$	200.36	...	161^{30}, 95-6$^{0.5}$	23	...	1.4188^{70}	al, eth, bz	B1^4, 1861
14476	2-Tridecanone or Undecyl methyl ketone $C_{11}H_{23}COCH_3$	198.35	...	263, 160^{16}	30.5	0.8217$^{30/4}$	1.4318^{20}	al, eth, ace, bz	B1^4, 3386
14477	3-Tridecanone or Ethyl decyl ketone $C_{10}H_{21}COC_2H_5$	198.35	pl	140^{17}	31	...	...	ace	B1^4, 3387
14478	7-Tridecanone or Dihexyl ketone $(C_6H_{13})_2CO$	198.35	lf (al)	261, 138^{12}	33	0.825^{30}	...	al, eth, chl, lig	B1^4, 3387
14479	1-Tridecene $C_{11}H_{23}CH{=}CH_2$	182.35	...	232.8, 104^{11}	−13	0.7658$^{20/4}$	1.4340^{20}	al, eth, bz	B1^4, 921
14480	1-Tridecyne $C_{11}H_{23}C{\equiv}CH$	180.33	...	94.5^{25}	...	0.7729$^{20/4}$	1.4309^{20}	eth, bz	B1^4, 1069
14481	Triethanol amine or Tris (2-hydroxyethyl) amine $(HOCH_2CH_2)_3N$	149.19	hyg cr	277^{150}	21-2	1.1242$^{20/4}$	1.4852^{20}	**w, al,** chl	B4^4, 1524
14482	Triethanol amine hydrochloride $(HOCH_2CH_2)_3N.HCl$	185.65	cr (al)	...	179-80	...	...	...	B4^4, 1525
14483	Triethylamine $(C_2H_5)_3N$	101.19	...	89.3	−114.7	0.7275$^{20/4}$	1.4010^{20}	w, al, eth, ace, bz	B4^4, 322
14484	Triethylamine hydrochloride $(C_2H_5)_3N.HCl$	137.65	hex (al)	sub 245	260d	1.0689$^{21/4}$	...	w, al, chl	B4^4, 327
14485	Triethylamine hydroiodide $(C_2H_5)_3N \cdot HI$	229.10	pr (95% al)	...	181	1.924	...	w, al, chl	B4^4, 328
14486	Triethylamine, perfluoro $(C_2F_5)_3N$	371.05	...	70.3	...	1.736$^{20/4}$	1.262^{25}	...	B2^4, 471
14487	Triethylene glycol $HO(CH_2CH_2O)_2CH_2CH_2OH$	150.17	hyg liq	278.3, 165^{14}	−5	1.1274$^{15/4}$	1.4531^{20}	**w, al, bz**	B1^4, 2400
14488	Triethylene glycol, 3-amino propyl ether $HO(CH_2CH_2O)_2CH_2CH_2OCH_2CH_2CH_2NH_2$	207.27	glassy	184^{10}	−50	1.0682$^{20/20}$	1.4668^{20}	**w, al**	C55, 5935

No.	Name, Synonyms, and Formula	Mol. wt.	Color, crystalline form, specific rotation and λ_{max} (log ε)	b.p. °C	m.p. °C	Density	n_D	Solubility	Ref.
14489	Triethylene glycol diacetate $CH_3CO_2(CH_2CH_2O)_2CH_2CH_2O_2CCH_3$	234.25		300				**w, al, eth**	B2[4], 215
14490	Triethylene glycol, monobutyl ether $HO(CH_2CH_2O)_2CH_2CH_2OC_4H_9$	206.28		278		$0.9890^{20/4}$	1.4389^{20}	**w**, al, MeOH	B1[4], 2402
14491	Triethylene tetramine $H_2NCH_2(CH_2NHCH_2)_2CH_2NH_2$	146.24		266-7, 157^{20}	12		1.4971^{20}	w, al	B4[4], 1242
14492	Trifluoro acetamide, *N*-phenyl or Trifluoro acetanilide $F_3CCONHC_6H_5$	189.14	(60% al)		87.6			al	B12[2], 141
14493	Trifluoroacetic acid F_3CCO_2H	114.02		72.4	−15.2	1.5351^{0}		w, al, eth, ace	B2[4], 458
14494	Trifluoroacetic acid anhydride $(F_3CCO)_2O$	210.03		39-40	−65	$1.490^{25/4}$	1.269^{25}	eth, aa	B2[4], 469
14495	Trifluoro acetic acid, butyl ester or Butyltrifluoro acetate $F_3CCO_2C_4H_9$	170.13		100.2		$1.0268^{22/4}$	1.353^{20}	chl	B2[4], 464
14496	Trifluoro acetic acid, *tert*-butyl ester $F_3CCO_2C(CH_3)_3$	170.13		83			1.3300^{25}	chl	B2[4], 465
14497	Trifluoroacetic acid, ethyl ester or Ethyltrifluoroacetate $F_3CCO_2C_2H_5$	142.08		60-2		$1.19^{20/4}$	1.308^{20}		B2[4], 463
14498	Trifluoroacetic acid, methyl ester or Methyl trifluoro acetate $F_3CCO_2CH_3$	128.05				$1.28^{20/4}$			B2[4], 463
14499	Trifluoroacetic acid, propyl ester or Propyltrifluoro acetate $F_3CCO_2C_3H_7$	156.10		82.5		$1.1285^{25/4}$	$1.3233^{22.5}$	chl	B2[4], 464
14500	Trifluoroacetonitrile F_3CCN	95.02		−64					B2[4], 472
14501	Trifloromethyl peroxide $CF_3\text{-}OO\text{-}CF_3$	170.01		−32					B3[4], 22
14502	Trifluoromethyl sulfide $(CF_3)_2S$	170.07		−22.2					B3[4], 278
14503	Trifurfuryl amine $(C_5H_5O)_3N$	257.29		$136\text{-}8^{1}$				eth	**B18[4], 7094**
14504	Triglycine $H_2N(CH_2CONH)_2CH_2CO_2H$	189.17	nd (dil al)		246d			w	**B4[3], 1198**
14505	Triglycine-*N*-phthalyl $C_{14}H_{13}N_3O_6$	319.27	nd (al)		234-5d			w, al	**B21[4], 5186**
14506	Tri-heptyl amine $(C_7H_{15})_3N$	311.60		330^{762}, $151\text{-}4^{1}$				al, eth	**B4[4], 736**
14507	Tri-hexyl amine $(C_6H_{13})_3N$	269.51		263-4, 119^{15}				al, eth	**B4[4], 711**
14508	Triidoacetic acid I_3CCO_2H	437.74	ye lf		150d			w, al, eth	**B2[4], 537**
14509	Trilaurin or Glycerol trilaurate $C_{11}H_{23}CO_2CH(CH_2O_2CC_{11}H_{23})_2$	639.01	nd (al)		46.4	0.8986^{55}	1.4404^{60}	al, eth, ace, bz, chl	**B2[4], 1098**
14510	Trilauryl amine or Tridodecylamine $(C_{12}H_{25})_3N$	522.00		$220\text{-}8^{0.03}$	fr 15.7		1.4567^{25}	**eth, bz, chl**	Am74, 428
14511	Trimellitic acid or 1,2,4-Benzene tricarboxylic acid $1,2,4\text{-}C_6H_3(CO_2H)_3$	210.14	nd (w), cr (aa, al)		238d			w, al, eth	**B9[3], 4792**
14512	Trimesic acid or 1,3,5-Benzene tricarboxylic acid $1,3,5\text{-}C_6H_3(CO_2H)_3$	210.14	pr or nd (w + 1)		380 (anh)			al, eth	**B9[3], 4793**
14513	Trimethadione or 3,5,5-trimethyl 2,4-oxazoldinedione $C_6H_9NO_3$	143.13	cr (50% MeOH)	$78\text{-}80^{5}$	46			al, eth, ace, bz, chl	**B4[4], 8013**
14514	Trimethyl amine $(CH_3)_3N$	59.11		2.9	−117.2	$0.6356^{20/4}$	1.3631^{0}	w, al, eth, bz, chl	**B4[4], 134**
14515	Trimethyl amine hydrochloride $(CH_3)_3N.HCl$	95.57	mcl dlq nd (al)	sub at 200	277-8			w, al, chl	**B4[4], 138**
14516	Trimethylamine oxide $(CH_3)_3NO$	75.11	hyg nd (w + 2)		255-7			w, al	**B4[4], 144**
14517	Trimethylene oxide, hydrochloride $(CH_3)_3NO.HCl$	111.57	nd (al)		218-20d			w, al	**B4[4], 144**
14518	Trimethylamine oxide, hydroiodide $(CH_3)_3NO.HI$	203.02	pr (al)		130d			w, al	**B4, 50**
14519	Trimethylamine, perfluoro $(CF_3)_3N$	221.03			−7				**B3[4], 79**
14520	Tris-(2-methylbutyl) amine $[CH_3CH_2CH(CH_3)CH_2]_3N$	227.43		230-7, 94^{4}		0.7964^{13}	1.4330^{20}	al, eth, ace, bz	**B4, 179**
14521	Trimethylene sulfide or Thiacyclobutane C_3H_6S	74.14		94.7, 14^{30}	−73.2	$1.0200^{20/4}$	1.5102^{20}	al, ace, bz	**B17[4], 14**

No.	Name, Synonyms, and Formula	Mol. wt.	Color, crystalline form, specific rotation and λ_{max} (log ε)	b.p. °C	m.p. °C	Density	n_D	Solubility	Ref.
14522	Trimethylol ethane $CH_3C(CH_2OH)_3$	120.15	wh pw or nd (al)	135-7^{15}	204			w, al	B1^{4}, 2780
14523	1,1,1-Trimethylol propane or TMP $CH_3CH_2C(CH_2OH)$	134.18	wh pw or pl	160^{5}	58			w, al	B1^{4}, 2786
14524	Trimethyloxonium fluoborate $[(CH_3)_3O]^+BF_4^-$	147.91	hyg nd		148d			ace, chl	B1^{4}, 1248
14525	Bis-(trimethylsilyl) amine $[(CH_3)_3Si]_2NH$	161.39		126.2		0.7741^{25}	1.4090^{20}		B4^{3}, 1861
14526	Trinitro acetonitrile $(NO_2)_3CCN$	176.05	wax	220 exp	41.5			eth	B2, 229
14527	Tri-octylamine $(C_8H_{17})_3N$	353.68		365, 182^{5}			1.4510^{19}		B4^{4}, 754
14528	Triolein or Glycerol trioleate $C_{17}H_{33}CO_2CH(CH_2O_2CC_{17}H_{33})_2$	885.45	poly-morphic	235-40^{18}	−5.5	0.8988^{40}	1.4621^{40}	eth, chl, peth	B2^{4}, 1664
14529	1,3,5-Trioxane or Metaformaldehyde $C_3H_6O_3$	90.08	rh nd (eth)	114.5, sub 46^{1}	64	1.17^{65}		w, al, eth, bz, chl	B19^{4}, 4710
14530	Tripalmitin or Glycerol tripalmitate $C_{15}H_{31}CO_2CH(CH_2O_2CC_{15}H_{31})_2$	807.34	nd (eth)	310-20	66.4	0.8752$^{70/4}$	1.4381^{80}	eth, bz, chl	B2^{3}, 971
14531	Tripentylamine or Triamylamine $(C_5H_{11})_3N$	227.43		240-5, 130^{14}		0.7907$^{20/4}$	1.4366^{20}	al, eth	B4^{4}, 676
14532	Tri-isopentylamine $(i\text{-}C_5H_{11})_3N$	227.43		235, 94^{4}		0.7848$^{20/4}$	1.4331^{20}	al, **eth, bz**	B4^{4}, 700
14533	Triphenyl acetic acid $(C_6H_5)_3CCO_2H$	288.35	mcl pr (al), lf (aa)		271			al, aa, lig	B9^{3}, 3585
14534	Triphenyl amine $(C_6H_5)_3N$	245.32	mcl (MeOH, AcOEt, bz)	365	127	0.774$^{0/0}$	1.353^{16}	eth, bz	B12^{3}, 292
14535	Triphenylene $C_{18}H_{12}$	228.29		425	199			al, bz, chl	B5^{3}, 2384
14536	Triphenylethylene, chloro $(C_6H_5)_2C{=}C(Cl)C_6H_5$	290.79			117			al, eth, ace, bz, chl	B5^{3}, 2400
14537	Triphenyl methane or Tritan $(C_6H_5)_3CH$	244.34	rh (al)	358-9^{754}, 190-215^{10}	94	1.014$^{99/4}$	1.5839^{99}	eth, bz, chl	B5^{3}, 2307
14538	Triphenyl methane, 3-amino or 3-Amino tritan $3\text{-}H_2NC_6H_4CH(C_6H_5)_2$	259.35	nd (eth)		120			al, eth	B12^{2}, 790
14539	Triphenylmethane, 4-amino or 4-Aminotritan $4\text{-}H_2NC_6H_4CH(C_6H_5)_2$	259.35	pr or lf (eth, lig), ta (Peth)	ca 248^{12}	84-5			eth, bz, lig	B12^{2}, 790
14540	Triphenyl methane, α-chloro or Trityl chloride $(C_6H_5)_3CCl$	278.78	nd or pr (bz-peth)	310, 230—5^{20}	113-4			eth, ace, bz, chl	B5^{3}, 2315
14541	Triphenylmethane, 4,4′-diamino $(4\text{-}H_2NC_6H_4)_2CHC_6H_5$	274.37	pr (bz, eth)		139-40			al, eth, chl, lig	B13^{3}, 529
14542	Triphenyl methane, 4,4′-dimethyl $(4\text{-}CH_3C_6H_4)_2CHC_6H_5$	272.39	nd (MeOH)	218-20^{12}	56			al, eth, bz, chl, lig	B13^{3}, 2342
14543	Triphenylmethane, 2,4-bis (dimethylamino) $2,4\text{-}[(CH_3)_2N]_2C_6H_3CH(C_6H_5)_2$	330.47	pl (peth)		122-3			al, eth, ace, bz	B13, 273
14544	Triphenyl methane, 4,4′-bis(dimethyl amino) $(4\text{-}(CH_3)_2NC_6H_4)_2CHC_6H_5$	330.47	nd or lf (al, bz)		102			eth, bz	B13^{3}, 529
14545	Triphenyl methane, 4,4′-bis(dimethyl amino)-4″-amino $[4\text{-}(CH_3)_2NC_6H_4]_2CHC_6H_4NH_2\text{-}4$	345.49	(al)		151-2				B13^{3}, 566
14546	Triphenyl methane, 4,4′-bis(dimethyl amino)-2″-hydroxy $[4\text{-}(CH_3)_2NC_6H_4]_2CH(C_6H_4OH\text{-}2)$	346.47	nd (al)		127-8			bz, lig	B13^{3}, 2063
14547	Triphenyl methane, 4,4′-bis(dimethyl amino)-3″-hydroxy $[4\text{-}(CH_3)_2NC_6H_4]_2CH(C_6H_4OH\text{-}3)$	346.47	cr (al)		149			bz	B13^{3}, 440
14548	Triphenyl methane, 4,4′-bis(dimethyl amino)-4″-hydroxy $[4\text{-}(CH_3)_2NC_6H_4]_2CH(C_6H_4OH\text{-}4)$	346.47	cr (al)		165			al, bz	B13^{3}, 440
14549	Triphenylmethane, 3-methyl or 3-Methyltritan $(3\text{-}CH_3C_6H_4)CH(C_6H_5)_2$	258.36	pr (al)	354^{706}	62			eth, bz, chl, aa, lig	B5^{3}, 2332
14550	Triphenylmethane, 2,4′,4″-triamino $(4\text{-}H_2NC_6H_4)_2CH(C_6H_4NH_2\text{-}2)$	289.38	cr (al)		165			al	B13^{3}, 565
14551	Triphenylmethane, 3,4′,4″-triamino $(4\text{-}H_2NC_6H_4)_2CH(C_6H_4NH_2\text{-}3)$	289.38	nd (eth, eth-lig)		150			al	B13^{3}, 565
14552	Triphenyl methane, 4,4′,4″-triamino or *p*-Leucaniline $(4\text{-}H_2NC_6H_4)_3CH$	289.38	lf (w, al, bz)		208			al, eth	B13^{3}, 566
14553	Triphenyl methane, 4,4′,4″-tris(dimethylamino) or Leuco crystal violet $[(CH_3)_2NC_6H_4]_3CH$	373.54	lf (al), nd (bz, lig)		175			eth, bz, chl, aa	B13^{3}, 566

PHYSICAL CONSTANTS OF ORGANIC COMPOUNDS (Continued)

No.	Name, Synonyms, and Formula	Mol. wt.	Color, crystalline form, specific rotation and λ_{max} (log ε)	b.p. °C	m.p. °C	Density	n_D	Solubility	Ref.
14554	Triphenyl methane, 4,4′,4″-trihydroxy or Leucoaurin $(4\text{-}HOC_6H_4)_3CH$	292.33	pr (aa, dil al)		240			al, chl, aa	B6³, 6578
14555	Triphenylmethane, 2,2′,2″-trimethyl or Tri-2-tolylmethane $[2\text{-}CH_3C_6H_4]_3CH$	286.42	nd (al)		130-1			al, eth	B5³, 2347
14556	Triphenylmethane, 4,4′,4″-trinitro $(4\text{-}O_2NC_6H_4)_3CH$	379.33	sc (bz)		212.5				B5⁴, 2501
14557	Triphenylmethanol or Tritanol $(C_6H_5)_3COH$	260.34	pl (al), trg (bz)	380	164.2	$1.199^{0/4}$		al, eth, ace, bz, aa	B6³, 3640
14558	Triphenylmethanol, 2,2′-bis-(dimethylamino) $[2\text{-}(CH_3)_2NC_6H_4]_2C(OH)C_6H_5$	346.47	pr (lig)		105				B13, 741
14559	Triphenylmethanol, 2,3′-bis(dimethylamino) $2,3'[(CH_3)_2NC_6H_4]_2C(OH)C_6H_5$	346.47	pl (bz)		183-4			bz, HCl	B13, 742
14560	Triphenylmethanol, 2,4′-bis (dimethylamino) $C_{23}H_{26}N_2O$	346.47	(al)		169-70			bz	B13, 742
14561	Triphenyl methanol, 3,3′-bis(dimethylamino) $[3\text{-}(CH_3)_2NC_6H_4]_2$	346.47	cr (eth)		128-9			eth, aa	B13, 742
14562	Triphenylmethanol, 3,4′-bis(dimethyl amino) $C_{23}H_{26}N_2O$	346.47	pr (bz-al)		140			bz	B13, 742
14563	Triphenylmethanol, 4,4′-bis(dimethylamino) or Michler's hydrol $[4\text{-}(CH_3)_2NC_6H_4]_2C(OH)C_6H_5$	346.47	cr (eth, bz, lig, MeOH)		121-3			eth, bz, lig	B13³, 2068
14564	Triphenyl methanol, 4,4′-dihydroxy or Benzaurin $[4\text{-}HOC_6H_4]_2C(OH)C_6H_5$	292.33	ye-red pw		110-20				B6³, 6582
14565	Triphenyl methanol, 4,4,4″-triamino or Pararosaniline $(4\text{-}H_2NC_6H_4)_3COH$	305.38	col to red lf		189-205			al	B13³, 2072
14566	Triphenylmethanol, 3,3′,3″-trinitro $(3\text{-}NO_2C_6H_4)_3COH$	395.33	rh (MeOH, chl)		167			bz, aa	B6¹, 352
14567	Triphenylmethanol, 4,4′,4″-trinitro $[\text{-}O_2NC_6H_4]_3COH$	395.33	mcl pr (bz, aa)		(i) 190, (ii) 167			bz, aa	B6³, 3673
14568	Triphenylselenonium chloride $(C_6H_5)_3SeCl$	345.73	orh (AcOEt)		230d			w, al, chl	B6⁴, 1780
14569	Triphenylselenonium fluoride $(C_6H_5)_3SeF$	329.28	oct deliq		145d			w, al, ace, chl	B6³, 1108
14570	Bis-(2 nitrophenyl) trisulfide $(2\text{-}NO_2C_6H_4)_2S_3$	340.39	ye nd (al)		175-6				B6³, 1062
14571	2,3,5-Triphenyl,1,2,3,4-tetrazolium chloride or Tetrazolium salt . T.T.C. $C_{19}H_{15}N_4Cl$	334.81	nd (al, chl)		243d			w, al, ace	B26, 363
14572	Triisopropanolamine $[CH_3CH(OH)CH_2]_3N$	191.27		$170\text{-}80^{10}$	45	$1.0^{20/4}$		w, al	B4⁴, 1680
14573	Tripropylamine $(C_3H_7)_3N$	143.27		156	−93.5	$0.7558^{20/4}$	1.4181^{20}	al, eth	B4⁴, 470
14574	Tripropylamine, perfluoro $(C_3F_7)_3N$	521.07		130		$1.822^{4/4}$	1.279^{25}		B2⁴, 742
14575	Triptane or 2,2,3-Trimethyl butane $(CH_3)_2CHC(CH_3)_3$	100.20		80.9	−24.2	$0.6901^{20/4}$	1.3894^{20}	al, eth, ace, bz	B1⁴, 410
14576	Trisiloxane, 1,5-dichloro-1,1,3,3,5,5-hexamethyl $ClSi(CH_3)_2OSi(CH_2)_2OSi(CH_2)_2Cl$	277.37		184	−53	$1.018^{20/4}$			B4³, 1884
14577	Trisiloxane, 1,5-dihydroxy-1,1,3,3,5,5-hexamethyl $HOSi(CH_2)_2OSi(CH_2)_2OSi(CH_2)_2OH$	240.48			−23	0.9950^{20}	$1.4090^{20/D}$		KHOC, 653
14578	Trisiloxane, 1,1,1,3,5,5,5-heptamethyl $(CH_3)_2SiOSiH(CH_3)OSi(CH_3)_3$	222.51				$0.8194^{20/4}$	1.3818^{20}		B4³, 1874
14579	Trisiloxane, octamethyl $(CH_3)_3SiOSi(CH_3)_2OSi(CH_3)_3$	236.53		153^{774}, $50\text{-}2^{17}$	−80	$0.8200^{20/4}$	1.3840^{20}	bz, peth	B4³, 1879
14580	Trisiloxane, 1,1,3,5,5-pentamethyl-1,3,5-triphenyl $C_6H_5Si(CH_3)_2OSi(CH_3)(C_6H_5)OSi(CH_3)_2C_6H_5$	422.75		$169^{0.7}$		$1.0227^{20/4}$	1.5280^{20}	bz, lig	Am70, 1116
14581	Tristearin or Glycerol tristearate $C_{17}H_{35}CO_2CH(CH_2O_2CC_{17}H_{35})_2$	891.50	cr (eth, peth)		73	$0.8559^{90/4}$	1.4399^{80}	ace, chl, CS_2	B2⁴, 1233
14582	Tristearylamine or Trioctadecylamine $(C_{18}H_{37})_3N$	774.48			54.6			eth, bz, chl	B4⁴, 829
14583	Tritetracontane $C_{43}H_{88}$ $(CH_3(CH_2)_{41}CH_3$	605.17		332^3	85.5	$0.7812^{90/4}$	1.4340^{90}		B1³, 593
14584	Tris-(4-tolyl) amine $(4\text{-}CH_3C_6H_4)_3N$	287.40	cr (aa)		117			eth, ace, bz, chl	B12³, 2034
14585	3,3′,4″-Tritolylamine $(3\text{-}CH_3C_6H_4)_2N(C_6H_4CH_3\text{-}4)$	287.40	nd (al)		89.90			al	B12¹, 415

PHYSICAL CONSTANTS OF ORGANIC COMPOUNDS (Continued)

No.	Name, Synonyms, and Formula	Mol. wt.	Color, crystalline form, specific rotation and λ_{max} (log ε)	b.p. °C	m.p. °C	Density	n_D	Solubility	Ref.
14586	Bis-(tribromomethyl) trisulfide $(CBr_3)_2S_3$	599.63	rh pr (eth)	d	125d			bz, chl, peth	$B3^{2}$, 107
14587	1,3,5-Trithiane or Trithioformaldehyde $C_3H_6S_3$	138.26	hex (bz), pr (w), nd (al)	sub	220	$1.6374^{24/4}$		bz	$B19^{4}$, 4711
14588	1,3,5-Trithiane, 2,4,6-trimethyl-(α-form) or Trithioacetaldehyde $2,4,6\text{-}(CH_3)_3(C_3H_3S_3)$	180.34	mcl (al, ace)	245-8	101			al, eth, ace, bz, chl	$B19^{4}$, 4719
14589	1,3,5-Trithiane, 2,4,6-trimethyl-(β-form) $1,3,5\text{-}(CH_3)_3(C_3H_3S_3)$	180.34	rh nd (ace)	245-8	126-7			al, eth, ace, bz, chl	$B19^{4}$, 4718
14590	1,3,5-Trithiane-2,4,6-trimethyl-2,4-6-triphenyl $C_{24}H_{24}S_3$	408.63	nd (al)		122			eth, ace, chl	$B19^{4}$, 4775
14591	1,3,5-Trithiane, 2,4,6-triphenyl (α-form) $2,4,6(C_6H_5)_3C_3H_3S_3$	366.55	nd (bz-al)	d	167			bz, chl	$B19^{4}$, 4774
14592	1,3,5-Trithiane, 2,4,6-triphenyl-(β-form) $2,4,6\text{-}(C_6H_5)_3C_3H_3S_3$	366.55	cr (bz)		229-30			bz	$B19^{4}$, 4773
14593	α-Tritisterol $C_{30}H_{50}O$	426.73	nd (MeOH-ace), $[\alpha]^{20}{}_{D}$ +54.3 (al)		114-5			al, peth, chl	$B6^{3}$, 2911
14594	β-Tritisterol $C_{30}H_{50}O$	426.73	nd (MeOH), $[\alpha]_D$ +49.2 (al)		97			al	H20, 424
14595	Tropacocaine or Benzoyl-ψ-tropeine $C_{15}H_{19}NO_2$	245.32	pl or tab	d	49	$1.0426^{100/4}$	1.5080^{100}	al, eth, bz, chl, peth	$B21^{4}$, 174
14596	Tropacocaine hydrochloride $C_{15}H_{19}NO_2 \cdot HCl$	281.78	pl (al)		283d			w	$B21^{4}$, 174
14597	Tropane or 2,3-Dihydro-8-methylnortropidine $C_8H_{15}N$	125.21		167		$0.9259^{15/15}$	1.4732^{20}		$B20^{4}$, 1963
14598	3-Tropanol $C_8H_{15}NO$	141.21		229	64			w, al, eth, chl	$B21^{4}$, 168
14599	3-Tropanone $C_8H_{13}NO$	139.20		$224\text{-}5^{714}$, 113^{25}	42-4	$1.9872^{100/4}$	1.4598^{100}	al, eth, ace, bz	$B21^{4}$, 3299
14600	Tropeine, benzoyl $C_{15}H_{19}NO_2$	245.32	cr (eth)	175-70	41-2			w, al, eth	$B21^{4}$, 173
14601	Tropeine, benzoyl, dihydrate $C_{15}H_{19}NO_2 \cdot 2H_2O$	281.35	pl (w)		58			w, al, eth	B21, 19
14602	Tropine or 3-Tropanol $C_8H_{15}NO$	141.21	hyg pl (eth)	229	64			w, al, eth, chl	$B21^{2}$, 17
14603	Tropinic acid [(d)] $C_8H_{13}NO_4$	187.20	cr (w, dil al), $[\alpha]^{t}{}_{D}$ +14.8 (w)		253d			w	B22, 123
14604	Tropinic acid [(dl)] $C_8H_{13}NO_4$	187.20	nd (dil al)		251d			w	B22, 124
14605	Tropinic acid [(l)] $C_8H_{13}NO_4$	187.20	cr (a), $[\alpha]^{20}{}_{D}$ −14.8 (w)		243			w	B22, 124
14606	Tropinone or 3-Tropanone $C_8H_{13}NO$	139.20	lo nd (peth)	$224\text{-}5^{714}$, 113^{25}	42-4	$1.9872^{100/4}$	1.4598^{100}	al, eth, ace, bz	$B21^{2}$, 225
14607	Truxane or Di-indene $C_{18}H_{16}$	232.33	pl (peth)		116			al, eth, aa	$B5^{3}$, 2253
14608	Truxane, tetraphenyl or Triphenyl allene dimer $(C_6H_5)_4C_{18}H_{12}$	536.72	lf (bz-al)		210			chl	$B5^{3}$, 2790
14609	α-Truxilline or Cocamine $C_{38}H_{46}N_2O_8$	602.77	amor pw		[ca] 80			al, eth, bz, chl	B22, 202
14610	Tryptophan [(D)] or α-Amino-β-indolyl propionic acid $C_{11}H_{12}N_2O_2$	204.23	pl (dil al) $[\alpha]^{20}{}_{D}$ +33 (w)		281-2			al	$B22^{4}$, 6765
14611	Tryptophan [(DL)] $C_{11}H_{12}N_2O_2$	204.23	pl (50% al)		282, (293)				$B22^{4}$, 6768
14612	Tryptophan [(L)] $C_{11}H_{12}N_2O_2$	204.23	lf or pl (dil al), $[\alpha]^{20}{}_{D}$ +6.1 (1N NaOH, p=11)		290-2d			w	$B22^{4}$, 6765
14613	Tryptophan, [N]-acetyl-[(DL)] $C_{13}H_{14}N_2O_3$	246.27	pl (dil al)		206-7			al, eth	$B22^{4}$, 6782

No.	Name, Synonyms, and Formula	Mol. wt.	Color, crystalline form, specific rotation and λ_{max} (log ε)	b.p. °C	m.p. °C	Density	n_D	Solubility	Ref.
14614	Tryptophan, [N] -acetyl-[(L)] $C_{13}H_{14}N_2O_3$	246.27	nd (dil MeOH), $[\alpha]^{15}/_D$ +25 (95% al, c=1)		189-90			w, al	B22[4], 6781
14615	Tryptophan, [N] -methyl [(DL)] $C_{12}H_{14}N_2O_2$	218.26	nd (dil al)		297d			al	B22[4], 6776
14616	Tryptophan, [N] -methyl [(L)] or Abrin $C_{12}H_{14}N_2O_2$	218.26	pr (w) $[\alpha]^{21}/_D$ +44.4 (dil HCl)		295d			al	B22[4], 6776
14617	Tuduranine $C_{18}H_{19}NO_3$	297.35	nd (eth), $[\alpha]^{20}/_D$ −127.5 (al)		204			al, eth, ace	B21[4], 2643
14618	Turanose or 3-*O*-α-D-glucoside-D-fructose $C_{12}H_{22}O_{11}$	342.30	pr (w-al, MeOH) $[\alpha]^{20}/_D$ +27.2→ +75.8 (w,c=4,mut)		168			w, al	B17[4], 3092
14619	Tyramine or [p] -(β -aminoethyl)phenol 4-$(H_2NCH_2CH_2)C_6H_4OH$	137.18	pl or nd (bz), cr (al) nd (w)	205-7[25], 165-7[2]	164-5			al, xyl	B13[3], 1637
14620	Tyrosine [(D)] or 2-Amino-3-(4-hydroxyphenyl)propionic acid $(4\text{-}HOC_6H_4)CH_2CH(NH_2)CO_2H$	181.19	cr (w), $[\alpha]^{25}/_D$ +10.3 (1N HCl,c=4)		310-4d			dil HCl	B14[3], 1504
14621	Tyrosine [(DL)] $(4\text{-}HOC_6H_4)CH_2CH(NH_2)CO_2H$	181.19	nd		340d				B14[3], 1506
14622	Tyrosine [(L)] $(4\text{-}HOC_6H_4)CH_2CH(NH_2)CO_2H$	181.19	nd (w), $[\alpha]^{22}/_D$ −10.6 (1N HCl,c=4)	sub	342-4d (sealed tube)				B14[3], 1504
14623	Tyrosine amide [(L)] $(4\text{-}HOC_6H_4)CH_2CH(NH_2)CONH_2$	180.21	pl or pr (al), $[\alpha]^{20}/_D$ +19.5 (w)		153-4			w, al, MeOH	B14[3], 1511
14624	Tyrosine-3-bromo [(L)] $(3\text{-}Br\text{-}4\text{-}HOC_6H_3)CH_2CH(NH_2)CO_2H$	260.09	cr (w+1), nd (w+2)		246-9d				B14[2], 377
14625	Tyrosine-3,5-dibromo [(L)] $(3,5\text{-}Br_2\text{-}4\text{-}HOC_6H_2)CH_2CH(NH_2)CO_2H$	338.98	nd or pl (w+2), $[\alpha]^{20}/_D$ −5.5 (1N HCl, c=5)		245 (anh)				B14[3], 1561
14626	Tyrosine, ethyl ester [(L)] $(4\text{-}HOC_6H_4)CH_2CH(NH_2)CO_2C_2H_5$	209.25	pr (AcOEt), $[\alpha]^{20}/_D$ +20.4 (MeOH)		108-9			al, bz, AcOEt, MeOH	B14[3], 1510
14627	Tyrosine, [N] -methyl-[(L)] $(4\text{-}HOC_6H_4)CH_2CH(NHCH_3)CO_2H$	195.22	nd $[\alpha]^{21}/_D$ +19.8 (dil HCl)		257, (d 280)				B14[3], 1513
14628	Tyrosine, methyl ester $(4\text{-}HOC_6H_4)CH_2CH(NH_2)CO_2CH_3$	195.22	pr (AcOEt), $[\alpha]^{20}/_D$ +25.7 (MeOH)		136-7			al, MeOH	B14[3], 1510

PHYSICAL CONSTANTS OF ORGANIC COMPOUNDS (Continued)

No.	Name, Synonyms, and Formula	Mol. wt.	Color, crystalline form, specific rotation and λ_{max} (log ε)	b.p. °C	m.p. °C	Density	n_D	Solubility	Ref.
14629	Umbellatine $C_{21}H_{21}NO_8$	415.40	ye nd	......	206-7	......	......	al	C48, 10034
14630	1,10-Undecadiyne $HC{\equiv}C(CH_2)_7C{\equiv}CH$	148.25	......	83^{12}	−17	$0.8182^{21/4}$	1.453^{21}	ace, bz	$B1^{2}$, 248
14631	Undecanal or Undecylaldehyde $CH_3(CH_2)_9CHO$	170.30	......	117^{18}	−4	$0.8251^{23/4}$	1.4520^{20}	al, eth	$B1^{4}$, 3374
14632	Undecanal oxime $CH_3(CH_2)_9CH{=}NOH$	185.31	wh nd (dil MeOH)	......	72	......	......	w, al, eth	$B1^{4}$, 3374
14633	Undecanal, 2-methyl $CH_3(CH_2)_8CH(CH_3)CHO$	184.32	......	114^{10}	......	$0.830^{15/4}$	1.4321^{20}	al, eth	$B1^{4}$, 3383
14634	Undecane $C_{11}H_{24}$	156.31	......	196, 75^{10}	−25.6	$0.7402^{20/4}$	1.4398^{20}	al	$B1^{4}$, 487
14635	Undecane, 1-amino or Undecylamine $CH_3(CH_2)_9CH_2NH_2$	171.33	cr (eth, al)	242, 112.3^{10}	17	$0.7979^{20/4}$	1.4398^{20}	al	$B4^{4}$, 792
14636	Undecane, 1-bromo-11-fluoro $F(CH_2)_{11}Br$	253.20	......	$95^{0.6}$	......	......	1.4518^{25}	al, eth	$B1^{4}$, 490
14637	Undecanoic acid or Undecylic acid $CH_3(CH_2)_9CO_2H$	186.29	cr (ace)	280, 164^{15}	28.6	$0.8907^{20/4}$	1.4294^{45}	al, eth, ace, bz, chl	$B2^{4}$, 1068
14638	Undecanoic acid, 2-bromo $CH_3(CH_2)_8CHBrCO_2H$	265.19	......	$178\text{-}83^{14}$	10	......	......	......	$B2^{4}$, 1073
14639	Undecanoic acid, 10-bromo $CH_3CHBr(CH_2)_8CO_2H$	265.19	cr (peth-eth, bz, ace)	......	35.7	......	......	al	$B2^{3}$, 861
14640	Undecanoic acid, 11-bromo $BrCH_2(CH_2)CO_2H$	265.19	nd (lig)	188^{18}	57	......	......	al, eth, ace, bz	$B2^{4}$, 1073
14641	Undecanoic acid, 10,11-dibromo $BrCH_2CH(Br)(CH_2)_8CO_2H$	344.09	cr	......	38	......	......	al	$B2^{4}$, 1075
14642	Undecanoic acid, ethyl ester or Ethyl undecanoate $CH_3(CH_2)_9CO_2C_2H_5$	214.35	......	131^{14}	−15	$0.8633^{20/4}$	1.4285^{20}	al, eth, ace, bz	$B2^{3}$, 858
14643	Undecanoic acid, 11-fluoro $F(CH_2)_{10}CO_2H$	204.28	......	$113\text{-}5^{0.25}$	36	......	......	al, eth, lig	$B2^{4}$, 1071
14644	Undecanoic acid, 4-hydroxy-8-lactone $C_7H_{15}CHCH_2CH_2CO$ (ring via O)	184.28	......	286, 162^{13}	......	$0.9494^{20/4}$	1.4512^{20}	al	$B17^{4}$, 4260
14645	Undecanoic acid -α-lactone or 2-Undecalactone $C_{11}H_{20}O_2$	184.28	......	286, 162^{13}	......	$0.9494^{20/4}$	1.4512^{20}	al	$B17^{2}$, 284
14646	1-Undecanol or 1-Undecylalcohol $CH_3(CH_2)_9CH_2OH$	172.31	......	243, 131^{15}	19	$0.8298^{20/4}$	1.4392^{20}	al, eth	$B1^{4}$, 1835
14647	1-Undecanol, 2-methyl $CH_3(CH_2)_8CH(CH_3)CH_2OH$	186.34	......	$129\text{-}31^{12}$	......	$0.8300^{15/4}$	1.4382^{20}	al, eth	$B1^{4}$, 1855
14648	2-Undecanol [(d)] $CH_3(CH_2)_8CHOHCH_3$	172.31	(bz), $[\alpha]^{20}_{D}$ + 10.3 (bz)	128^{20}	12	$0.8270^{20/4}$	1.4369^{20}	al, eth	$B1^{3}$, 1774
14649	2-Undecanol [(dl)] $CH_3(CH_2)_8CHOHCH_3$	172.31	......	228, 119^{12}	0	0.8268^{19}	1.4369^{20}	al, eth	$B1^{4}$, 1838
14650	2-Undecanol -[(l)] $CH_3(CH_2)_8CHOHCH_3$	172.31	$[\alpha]_D$ −0.02	231-3	......	0.8302^{20}	1.4381^{20}	al, eth	$B1^{3}$, 1774
14651	3-Undecanol *(l)* $CH_3(CH_2)_7CHOHCH_2CH_3$	172.31	$[\alpha]^{20}_{D}$ −6.2 (al)	229	17	$0.8295^{20/4}$	1.4367^{20}	al, eth, bz	$B1^{2}$, 462
14652	5-Undecanol $C_6H_{13}CHOH(C_4H_9)$	172.31	......	229, 107^{12}	−3.5	$0.8292^{20/4}$	1.4354^{24}	al, eth	$B1^{4}$, 1839
14653	6-Undecanol $C_5H_{11}CH(OH(C_5H_{11})$	172.31	cr (ace)	228, $117\text{-}8^{16}$	25	$0.8334^{20/4}$	1.4374^{20}	al, ace	$B1^{4}$, 1839
14654	1-Undecanethiol or Undecyl mercaptan $CH_3(CH_2)_9CH_2SH$	188.37	......	257.4	−3	0.8448^{20}	1.4585^{20}	......	$B1^{4}$, 1838 2(1.1000)
14655	2-Undecanone $CH_3(CH_2)_8COCH_3$	170.30	......	231-2, 106^{12}	15	$0.8250^{20/4}$	1.4291^{20}	al, eth, ace, bz, chl	$B1^{4}$, 3374
14656	3-Undecanone $CH_3(CH_2)_7COCH_2CH_3$	170.30	......	227, $104\text{-}6^{11}$	12	$0.8272^{20/4}$	1.4296^{20}	al, eth	$B1^{4}$, 3375
14657	4-Undecanone $CH_3(CH_2)_6CO(CH_2)_2CH_3$	170.30	......	106^{13}	4-5	$0.8274^{25/4}$	1.4248^{24}	al, eth	$B1^{4}$, 3375
14658	4-Undecanone, 7-ethyl,2-methyl $C_4H_9CH(C_2H_5)CH_2CH_2COCH_2CH(CH_3)_2$	212.38	......	252-3, $101\text{-}3^{4}$	......	$0.8362^{20/4}$	1.4370^{20}	al, eth, ace, bz	$B1^{3}$, 2920
14659	5-Undecanone $CH_3(CH_2)_5CO(CH_2)_3CH_3$	170.30	......	227, $105\text{-}6^{12}$	2	$0.8278^{19/4}$	1.4275^{18}	al, eth	$B1^{4}$, 3375
14660	6-Undecanone or *n*-Caprone $(C_5H_{11})_2CO$	170.30	......	228, 145.7^{29}	14-5	$0.8308^{20/4}$	1.4270^{20}	al, eth	$B1^{4}$, 3376
14661	Undecanonitrile $CH_3(CH_2)_9CN$	167.29	......	253, 143^{22}	......	$0.8254^{30/4}$	1.4293^{30}	al, eth	$B3^{2}$, 860

No.	Name, Synonyms, and Formula	Mol. wt.	Color, crystalline form, specific rotation and λ_{max} (log ε)	b.p. °C	m.p. °C	Density	n_D	Solubility	Ref.
14662	Undecasiloxane, tetra cosamethyl $(CH_3)_3SiO[Si(CH_3)_2O]_9Si(CH_3)_3$	829.77		322.8, $202^{4.7}$		0.930^{20}	1.3994^{20}	bz	$B4^3$, 1881
14663	1-Undecene $CH_3(CH_2)_8CH{=}CH_2$	154.30		192.7, 72^{10}	−49.2	$0.7503^{20/4}$	1.4261^{20}	eth, chl, lig	$B1^4$, 910
14664	2-Undecene *(trans)* $CH_3(CH_2)_7CH{=}CHCH_3$	154.30		192-3, 75^{10}	−48.3	$0.7528^{20/4}$	1.4292^{20}		$B1^4$, 911
14665	5-Undecene *(trans)* $CH_3(CH_2)_4CH{=}CH(CH_2)_3CH_3$	154.30		192, 73^{10}	−61.1	$0.7497^{20/4}$	1.4285^{20}	eth, chl, lig	$B1^4$, 911
14666	9-Undecenoic acid *(trans)* $CH_3CH{=}CH(CH_2)_7CO_2H$	184.28		273, $121\text{-}3^{0.7}$	19	$0.9119^{25/0}$	1.4519^{20}	eth	$B2^4$, 1616
14667	10-Undecenoic acid $CH_2{=}CH(CH_2)_8CO_2H$	184.28	cr	275, 165^{13}	24-5	$0.9072^{24/4}$	1.4486^{24}	al, eth	$B2^4$, 1612
14668	10- Undecenoic acid, ethyl ester or Ethyl-10-undecenoate $CH_2{=}CH(CH_2)_8CO_2C_2H_5$	212.33		264-5, 131.5^{16}	−38	0.8827^{15}	1.4449^{23}	al, eth	$B2^4$, 1614
14669	10- Undecenoic acid, methyl ester $H_2C{=}CH(CH_2)_8CO_2CH_3$	198.31		248, 124^{10}	−27.5	0.889^{15}	1.4393^{20}	al, eth, aa	$B2^4$, 1613
14670	10 -Undecenoyl chloride $CH_2{=}CH(CH_2)_8COCl$	202.72		$120\text{-}2^{10}$		$0.944^{20/4}$	1.454^{20}		$B2^4$, 1615
14671	10-Undecen-1-ol or Undecenyl alchohol $CH_2{=}CH(CH_2)_8CH_2OH$	170.30		250, $132\text{-}3^{15}$	−2	$0.8495^{15/4}$	1.4500^{20}	al, eth	$B1^4$, 2194
14672	1-Undecene-3-yne $CH_3(CH_2)_6C{\equiv}CCH{=}CH_2$	150.26		74^9		$0.7962^{20/4}$	1.4606^{20}		$B1^3$, 1054
14673	1- Undecyne $CH_3(CH_2)_8C{\equiv}CH$	152.28		196, 73.2^{10}	−25	$0.7728^{20/4}$	1.4306^{20}	al, eth, ace, bz	$B1^4$, 1064
14674	5-Undecyne $CH_3(CH_2)_4C{\equiv}C(CH_2)_3CH_3$	152.28		198, 78.3^{10}	−74.1	$0.7753^{20/4}$	1.4369^{20}		$B1^4$, 1064
14675	9-Undecynoic acid $CH_3C{\equiv}C(CH_2)_8CO_2H$	182.26	pl (dil al)	170^{15}	61			w, al, eth	$B2^3$, 1471
14676	Uneicosane $CH_3(CH_2)_{19}CH_3$	296.58	cr (w)	358.5, 203^{10}	40.5	$0.7917^{20/4}$	1.4441^{20}	peth	$B1^4$, 569
14677	Uneicosane, 1-cyclopentyl $C_5H_9C_{21}H_{43}$	364.70		420	42	0.8286^{20}	1.4602^{20}		$B5^4$, 200 (3.1031)
14678	Uneicosane, 11-decyl $(C_{10}H_{21})_3CH$	436.85		282^{10}	10.0	$0.8116^{20/4}$	1.4540^{20}		$B1^4$, 595
14679	Uneicosane, 11-(2,2-dimethyl propyl) $(CH_3)_3CCH_2CH(C_{10}H_{21})_2$	366.71		238^{10}	−21	$0.8031^{20/4}$	1.4491^{20}		$B1^4$, 585
14680	Uneicosane, 11-phenyl $C_6H_5CH(C_{10}H_{21})_2$	372.68		$205^{1.0}$	20.8	$0.8531^{20/4}$	1.4788^{20}		$B5^4$, 1226
14681	Uneicosanoic acid $CH_3(CH_2)_{19}CO_2H$	326.56	nd (ace)		82			eth, bz, chl	$B2^4$, 1285
14682	10-Uneicosene, 11-Phenyl $CH_3(CH_2)_8CH{=}C(C_6H_5)(CH_2)_9CH_3$	370.66		203^1	48.2	$0.8638^{20/4}$	1.4922^{20}		$B5^4$, 2057
14683	Untriacontane $CH_3(CH_2)_{29}CH_3$	436.85	lf (AcOEt)	458, 302^{15}	67.9	$0.781^{68/4}$	1.4278^{90}	peth	$B1^4$, 594
14684	Untriacontanoic acid or Melissic acid $CH_3(CH_2)_{29}CO_2H$	466.83	sc or nd (al or ace)		93			bz	$B2^2$, 382
14685	16-Untriacontanone or Palmitone $(C_{15}H_{31})_2CO$	450.83	lf (al)		83	$0.7947^{91/4}$	1.4297^{94}	eth	$B1^4$, 3413
14686	Uracil $C_4H_4N_2O_2$	112.09	nd (w)		338			al, eth	$B24^2$, 168
14687	Uracil, 5-amino $C_4H_5N_3O_2$	127.10	nd (w)		d				$B24^2$, 266
14688	Uracil, 5,6 dihydro or β -Lactylurea $C_4H_6N_2O_2$	114.10	nd (w)		275-6			w, al, chl	$B24^2$, 140
14689	Uracil, 5,6-dihydro-5-methyl or Dihydrothymine $C_5H_8N_2O_2$	128.13	(w or al)		264-5			w	$B24^1$, 306
14690	Urcil, 5-hydroxy or Isobarbituric acid $C_4H_4N_2O_3$	128.09	pr (w)		d>300			w	$B24^1$, 408
14691	Uracil, 1-methyl $C_5H_6N_2O_2$	126.11	pr (w)		179			w, al	$B24^2$, 170
14692	Uracil, 3-methyl $C_5H_6N_2O_2$	126.11	pr (al) nd (w)		232			w, al	$B24^2$, 170
14693	Uracil, 5-methyl or Thymine $C_5H_6N_2O_2$	126.11	nd (al), pl (w)	sub	326				$B24^2$, 183
14694	Uracil, 6-methyl $C_5H_6N_2O_2$	126.11	oct, pr or nd (w, al)		270-80d			w, al	$B24^2$, 182

No.	Name, Synonyms, and Formula	Mol. wt.	Color, crystalline form, specific rotation and λ_{max} (log ε)	b.p. °C	m.p. °C	Density	n_D	Solubility	Ref.
14695	Uracil, 5-nitro $C_4H_3N_3O_4$	157.09	gold nd (al)		exp> 300			al	B24², 171
14696	Uracil, 2-thio or 2-Thiouracil $C_4H_4N_2OS$	128.15	pr (w, al)		>340d				B24², 171
14697	6-Uracil carboxylic acid or Orotic acid $C_5H_4N_2O_4$	156.10	pr (w + 1)		347d				B25², 249
14698	6-Uracil carboxylic acid, monohydrate $C_5H_4N_2O_4 . H_2O$	174.11	pr (w)		125-30d				B25², 249
14699	Uranin or Fluorescein, sodium salt $C_{20}H_{10}O_5Na_2$	376.28	ye pw					w, al	B19⁴, 2909
14700	Urea or Carbamide H_2NCONH_2	60.06	tetr pr (al)	d	135	$1.3230^{20/4}$	1.484	w, al, Py	B3⁴, 94
14701	Urea, 1-acetyl $CH_3CONHCONH_2$	102.09		sub 180-190	218			al	B3³, 119
14702	Urea, 1-acetyl-3-methyl $CH_3CONHCONHCH_3$	116.12	tcl (w, al), pr (w)	d	180-1			w	B4⁴, 207
14703	Urea, 1-allyl CH_2=$CHCH_2NHCONH_2$	100.12	nd (al)		85			**w, al**	B4⁴, 1070
14704	Urea, 1-allyl-3-phenyl CH_2=$CHCH_2NHCONHC_6H_5$	176.22	nd (bz)		115-6			al, bz	B12³, 765
14705	Urea, 1-(4-amino benzene sulfonyl) or 1-Sulfonyl urea $4\text{-}H_2NC_6H_4SO_2NHCONH_2$	215.23	(w)	sub 320	146-8d				B14³, 1968
14706	Urea, 1-benzoyl $C_6H_5CONHCONH_2$	164.16	fl or lf (al)		214-5				B9³, 1115
14707	Urea, 1-benzyl or Phthalamic acid $C_6H_5CH_2NHCONH_2$	150.18	nd (al)	d 200	149			al, ace	B12³, 2271
14708	Urea, 1-(2-bromo-2-ethyl-butanoyl) or Adalin. Uradal $H_2NCONH[COCBr(C_2H_5)_2]$	237.10	rh (dil al)		118	1.544^{25}		ace, bz	B3⁴, 117
14710	Urea, 1-(2-bromo-3-methyl butanoyl) or Adabine Bromural $H_2NCONH[COCHBrCH(CH_3)_2]$	223.07	nd or lf (to)	sub	154 (160)	1.56^{15}		al, eth, ace, bz	B3³, 123
14711	Urea, 1-(2-bromo phenyl) $2\text{-}BrC_6H_4NHCONH_2$	215.05	nd (al)		202			bz, chl	B12³, 1419
14712	Urea, 1-(3-bromophenyl) $3\text{-}BrC_6H_4NHCONH_2$	215.05	nd (bz, al)		164-5			al, eth	B12, 634
14713	Urea, 1-(4-bromophenyl) $4\text{-}BrC_6H_4NHCONH_2$	215.05	nd (bz, al)	d 260	225-7			al, eth, bz, aa	B12³, 1452
14714	Urea, 1-butyl $C_4H_9NHCONH_2$	116.16	ta (w), nd (bz)		96			al, eth	B4⁴, 578
14715	Urea, 1-isobutyl $H_2NCONH\text{-}i\text{-}C_4H_9$	116.16	pr (w), nd (ace)		141				B4⁴, 648
14716	Urea, 1-*sec*-butyl $C_2H_5CH(CH_3)NHCONH_2$	116.16	nd $[\alpha]^{20}{}_D$ + 24.1 (al, c=1.5)		166-(d) 169-70-(dl)			al, eth	B4⁴, 622
14717	Urea, 1-*tert*-butyl $(CH_3)_3CNHCONH_2$	116.16	nd (w dil al)	sub >100	191d			w, al	B4⁴, 665
14718	Urea, 1-(2-chlorophenyl) $(2\text{-}ClC_6H_4)NHCONH_2$	170.60	pr (w)		152			w, al, ace	B12³, 1293
14719	Urea, 3-(4-chlorophenyl)-1,1-dimethyl or CNV Weed Killer $3\text{-}(4\text{-}ClC_6H_4)NHCON(CH_3)_2$	198.65	wh pl (MeOH)		170-1				C51, 16534
14720	Urea, 1,3-diacetyl $(CH_3CONH)_2CO$	144.13	nd (aa, 50% al)	sub 179-80d	154-5			eth, ace, bz	B3³, 121
14721	Urea, 1,3-dibutyl $(C_4H_9NH)_2CO$	172.27			72-4				B4⁴, 578
14722	Urea, 1,3-bis-(2,4-dinitro phenyl) $(2,4(NO_2)_2C_6H_3NH)_2CO$	392.24	lf or pr (al)		204				B12³, 1690
14723	Urea, 1,3-bis-(2-ethoxyphenyl) $(2\text{-}C_2H_5OC_6H_4NH)_2CO$	300.36	pr (dil al)		125			al	B13³, 816
14724	Urea, 1,3-bis-(4-ethoxyphenyl) $(4\text{-}C_2H_5OC_6H_4NH)_2CO$	300.36	nd (aa), pr (abs al)		255-6			al	B13³, 1112
14725	Urea, 1,1-diethyl $H_2NCON(C_2H_5)_2$	116.16	pl nd (eth)	$94\text{-}6^{0.02}$	75			w, al, eth, bz, lig	B4⁴, 380
14726	Urea, 1,3-diethyl $(C_2H_5NH)_2CO$	116.16	ta (lig), dlq nd (al)	263	112.5	1.0415	1.4616^{40}	w, al, eth	B4⁴, 370
14727	Urea, 1,3-diethyl-1,3-diphenyl $[C_6H_5N(C_2H_5)]_2CO$	268.36	(al, w)		79			al	B12³, 882

No.	Name, Synonyms, and Formula	Mol. wt.	Color, crystalline form, specific rotation and λ_{max} (log ε)	b.p. °C	m.p. °C	Density	n_D	Solubility	Ref.
14728	Urea, 1,3-bis-(hydroxymethyl) or N,N′-Dimethylol urea $(HOCH_2NH)_2CO$	120.11	pr (abs al), pl (w-al)	d 260	126	1.49^{25}		w, al	$B3^4$, 107
14729	Urea, 1,3-bis-(1-hydroxy-2,2,2-trichloro ethyl) or Dichloral urea. Cragherbicide $[CCl_3CH(OH)NH]_2CO$	354.83			196			al, ace	$B3^3$, 116
14730	Urea, 1,1-dimethyl $H_2NCON(CH_3)_2$	88.11	mcl pr (al, chl)		182	1.255		w	$B4^4$, 224
14731	Urea, 1,3-dimethyl $(CH_3NH)_2CO$	88.11	rh bipym (chl-eth)	268-70	108	1.142		w, al	$B4^4$, 207
14732	Urea, 1,3-dimethyl-1,3-diphenyl $[C_6H_5N(CH_3)]_2CO$	240.30	pl (al)	350	121			w, al, ace, chl	$B12^3$, 875
14733	Urea, 1,1-di(β-naphthyl) $(\beta\text{-}C_{10}H_7)_2NCONH_2$	312.37	nd (al)		192-3				B12, 1297
14734	Urea, 1,3-di(α-naphthyl) $(\alpha\text{-}C_{10}H_7NH)_2CO$	312.37	nd (aa, Py)	sub	296			Py	$B12^3$, 2926
14735	Urea, 1,3-di(β-naphthyl) $(\beta\text{-}C_{10}H_7NH)_2CO$	312.37	nd (ace, aa)		310			aa	$B12^3$, 3033
14736	Urea, 1,1-diphenyl or Acardite $(C_6H_5)_2NCONH_2$	212.25	ta (al)	d	189	1.276		al, eth, chl	$B12^3$, 893
14737	Urea, 1,3-diphenyl or Carbanilide $(C_6H_5NH)_2CO$	212.25	rh bipym, pr (al)	262d	238	1.239		eth, aa	$B12^3$, 767
14738	Urea, 1,3-diphenyl-1-methyl $C_6H_5NHCON(CH_3)C_6H_5$	226.28	nd (al, xyl), cr (lig)	203-5d	106			eth, bz, chl	$B12^3$ 878
14739	Urea, 1-(3-ethoxyphenyl) $(3\text{-}C_2H_5OC_6H_4)NHCONH_2$	180.21	nd		112				$B13^3$, 960
14740	Urea, 1-(4-ethoxyphenyl) or Dulcin $(4\text{-}C_2H_5OC_6H_4)NHCONH_2$	180.21	lf (dil al), pl (w)	d	173-4			al, AcOEt	$B13^3$, 1109
14741	Urea, 1-ethyl $C_2H_5NHCONH_2$	88.11	nd (bz, al-eth)	d	92-3			w, al, eth, bz, chl	$B4^4$, 369
14742	Urea, 1-ethyl-1-phenyl $C_6H_5N(C_2H_5)CONH_2$	164.21	ta (peth), cr (aa)		62-3			w, eth, ace	$B12^3$, 882
14743	Urea, 1-ethyl-3-phenyl $C_2H_5NHCONHC_6H_5$	164.21	nd (dil al)		104			al	$B12^3$, 761
14744	Urea hydrochloride $(H_2N)_2CO.HCl$	96.52			145d				$B3^4$, 102
14745	Urea, 1-hydroxy $HONHCONH_2$	76.06	nd (al)	d	141			w	$B3^4$, 170
14746	Urea, 1-hydroxymethyl or Methylol urea $HOCH_2NHCONH_2$	90.08	pr (al)		111			w, al, aa, MeOH	$B3^4$, 105
14747	Urea, 1-(2-iodo-3-methyl butanoyl) or α-Iodo isovaleryl urea $H_2NCONHCOCHICH(CH_3)_2$	270.07	lf (al)		180-1			al	$B3^1$, 29
14748	Urea, 1-(4-methoxyphenyl) $(4\text{-}CH_3OC_6H_4)NHCONH_2$	166.18	pl (w)		168			eth	$B13^3$, 1096
14749	Urea, 1-methyl $CH_3NHCONH_2$	74.08	rh pr (w, al)	d	103			w, al	$B4^4$, 205
14750	Urea, 1-methyl-1-nitroso $H_2NCON(NO)CH_3$	103.08	col or yesh pl (eth)		123-4d			al, eth, ace, bz, chl	$B4^4$, 272
14751	Urea, 1-methyl-3-phenyl $C_6H_5NHCONHCH_3$	150.18			151			w, al, bz	$B12^3$, 761
14752	Urea, 1-(2-methyl-2-butyl) $H_2NCONHC(CH_3)_2C_2H_5$	130.19	mcl (w)		162				$B4^4$, 695
14753	Urea, 1-α-naphthyl $\alpha\text{-}C_{10}H_7NHCONH_2$	186.21	nd (al)		219-20			al, eth	B12, 1253
14754	Urea, 1-(β-naphthyl) $\beta\text{-}C_{10}H_7NHCONH_2$	186.21	nd (al)		219			al	$B12^3$, 3029
14755	Urea nitrate $H_2NCONH_2.HNO_3$	123.07	mcl lf (w)		157d	$1.690^{20/4}$		al	$B3^4$, 102
14756	Urea, 1-nitro $H_2NCONHNO_2$	105.05	nd (al-peth), lf or pr(al)	exp	158-9			al, eth, ace, aa	$B3^4$, 248
14757	Urea oxalate $2(H_2NCONH_2).HO_2CCO_2H$	210.15	mcl (w + l)		173d	1.585		w	$B3^2$, 48
14758	Urea, 1-isopentyl $H_2NCONH\text{-}(i\text{-}C_5H_{11})$	130.19	pl (dil al)		96 (150)			al	$B4^4$, 695
14759	Urea, 1-(2-phenoxy ethyl) $(C_6H_5OCH_2CH_2)NHCONH_2$	180.21	nd (w + 2), cr (50% al)		120-1			al	$B6^1$, 91

No.	Name, Synonyms, and Formula	Mol. wt.	Color, crystalline form, specific rotation and λ_{max} (log ε)	b.p. °C	m.p. °C	Density	n_D	Solubility	Ref.
14760	Urea, 1-phenyl or Phenyl carbamide $C_6H_5NHCONH_2$	136.15	nd or pl (w), tab (al)	238	147			al, AcOEt	B12^3, 760
14761	Urea, 1-phenylacetyl $C_6H_5CH_2CONHCONH_2$	178.19	(al)		209			al	B9^3, 2208
14762	Urea, propionyl or Propionyl urea $CH_3CH_2CONHCONH_2$	116.12	cr (w)		210-11			w	B3^3, 121
14763	Urea, 1-propyl $C_3H_7NHCONH_2$	102.14	pr (al)		110			al	B4^4, 482
14764	Urea, 1,1,3,3-tetraethyl $[(C_2H_5)_2N]_2CO$	172.27		209, 94-5^{12}		0.919$^{20/4}$	1.4474^{20}		B4^4, 380
14765	Urea, 1,1,3,3-tetramethyl $[(CH_3)_2N]_2CO$	116.16		166-7, 63-4^{12}	−1.2	0.9687^{20}	1.4496^{23}		B4^4, 225
14766	Urea, 1,1,3,3-tetraphenyl $[(C_6H_5)_2N]_2CO$	364.45	rh (bz)		183	1.222			B12^3, 894
14767	Urea, 1-(2-tolyl) $(2\text{-}CH_3C_6H_4)NHCONH_2$	150.18	lf (al, w)		195-6			al, eth	B12^3, 1867
14768	Urea, 1-(3-tolyl) $(3\text{-}CH_3C_6H_4)NHCONH_2$	150.18	lf (w)		142			w, al	B12^3, 1968
14769	Urea, 1-(4-tolyl) $(4\text{-}CH_3C_6H_4)NHCONH_2$	150.18	nd (w), pl (w-aa)		183			al, ace	B12^3, 2082
14770	Urea, 1,1,3-trimethyl $(CH_3)_2NCONHCH_3$	102.14	pr (eth)	232.5^{764}	75.5			w, al	B4^4, 224
14771	1-Urea carboxylic acid, ethyl ester or Allophanic acid, ethyl ester $H_2NCONHCO_2C_2H_5$	132.12	nd (w, bz)	d	195				B3^4, 127
14772	Urethane, N-amino $H_2NNHCO_2C_2H_5$	104.11	cr	198d, 93^9	46			al, eth	B3^4, 174
14773	Uric acid or 2,6,8-Purine trione $C_5H_4N_4O_3$	168.11	rh pr or pl	d	d	1.89		glycerol	B26^2, 293
14774	Uric acid, 1-methyl $C_6H_6N_4O_3$	182.14	nd		400				B26^2, 299
14775	Uric acid, 3-methyl $C_6H_6N_4O_3$	182.14	pr (w + 1)		>350	1.6104^{25}			B26^2, 299
14776	Uric acid, 7-methyl $C_6H_6N_4O_3$	182.14	pl (w)		370-80d	1.706$^{25/4}$			B26^2, 299
14777	Uric acid, 9-methyl $C_6H_6N_4O_3$	182.14			380-400d			w	B26^2, 299
14778	Uric acid, 1,3,7-trimethyl $C_8H_{10}N_4O_3$	210.19			373-5				B26^2, 301
14779	Uridine or 1-β-D-Ribofuranosyluracil $C_9H_{12}N_2O_6$	244.20	nd (aq al), $[\alpha]^{20}{}_D$ + 4 (w)		165			w, Py	J1947, 358
14780	Uridylic acid or Uridine-3′-phosphate $C_9H_{13}N_2O_9P$	324.18	pr (MeOH), $[\alpha]^{20}{}_D$ + 10.5 (w)		202d			w, MeOH	J1948, 746
14781	Urocanic acid or 4-Imidazolyl acrylic acid $C_6H_6N_2O_2$	138.13	pr(w + 2)		*cis* 175-6, *trans* 218-24			w, ace	B25^2, 121
14782	Urochloralic acid or 2,2,2-trichloroethyl-β-D-glucuronide $C_8H_{11}O_7Cl_3$	325.23	nd		142			w, al	B18^4, 5113
14783	Ursodesoxycholic acid or 3α-7β-Dihydroxy-5β-cholanic acid $C_{24}H_{40}O_4$	325.53	pl (al), $[\alpha]^{20}{}_D$ + 57 (abs al, c=2)		203			al	B10^3, 1635
14784	Ursolic acid or Malol. Urson $C_{30}H_{48}O_3$	456.71	pr (al), $[\alpha]^{21}{}_D$ + 72.4 (MeOH)		284			eth, ace, chl, MeOH	B10^3, 1038
14785	Usnic acid (d) or Usninic acid $C_{18}H_{16}O_7$	344.32	ye orh pr (ace), $[\alpha]^{20}{}_D$ + 469 (chl, c=0.7)		204			al, eth	B18^4, 3522
14786	Usnic acid (dl) $C_{18}H_{16}O_7$	344.32			193-4		1.710	al, eth, chl	B18^4, 3523

No.	Name, Synonyms, and Formula	Mol. wt.	Color, crystalline form, specific rotation and λ_{max} (log ε)	b.p. °C	m.p. °C	Density	n_D	Solubility	Ref.
14787	Usnic acid (l) $C_{18}H_{16}O_7$	344.32	$[\alpha]^{20/}{}_D$ −480		203			al, eth, chl	B18[4], 3522
14788	Uzarin $C_{35}H_{54}O_{14}$	698.81	pr, $[\alpha]_D$ −27 (Py)		268-70			Py	B18[4], 1497

PHYSICAL CONSTANTS OF ORGANIC COMPOUNDS (Continued)

No.	Name, Synonyms, and Formula	Mol. wt.	Color, crystalline form, specific rotation and λ_{max} (log ε)	b.p. °C	m.p. °C	Density	n_D	Solubility	Ref.
14789	Vacciniin or 6-*O*-Benzoyl-D-glucose $C_{13}H_{16}O_7$	284.27	amor (aq ace + 1w), $[\alpha]^{21}/_D$ +48 (al)		104-6			w, al, ace	B31, 123
14790	Valeraldehyde or Pentanal $CH_3(CH_2)_3CHO$	86.13		103	−91.5	$0.8095^{20/4}$	1.3944^{20}	al, eth	$B1^4$, 3268
14791	Valeraldehyde, diethyl acetyl or Pentanal diethyl acetal $CH_3(CH_2)_3CH(OC_2H_5)_2$	160.26		59^{12}		0.829^{22}	1.4029^{22}		$B1^4$, 3269
14792	Valeraldehyde, 3-hydroxy-2-methyl or Propionaldol $C_2H_5CH(OH)CH(CH_3)CHO$	116.16		$94\text{-}6^{23}$		$0.986^{25/4}$	1.4502^{20}	w, al, eth, ace	$B1^4$, 4022
14793	Valeraldehyde, 3-hydroxy-2,2,4-trimethyl $(CH_3)_2CHCH(OH)C(CH_3)_2CHO$	144.21		$118\text{-}20^{14}$		$0.9482^{20/4}$	1.4501^{20}		$B1^4$, 4049
14794	Valeraldehyde, 4-hydroxy (dl) $CH_3CH(OH)CH_2CH_2CHO$	102.13		$63\text{-}5^{10}$		$1.019^{20/4}$	1.4359^{17}	w, ace	$B1^4$, 4002
14795	Valeraldehyde, 4-hydroxy (l) $CH_3CH(OH)CH_2CH_2CHO$	102.13	$[\alpha]^{23}/_D$ −7.8	$43\text{-}6^1$				w, ace	$B1^2$, 872
14796	Valeraldehyde, 2-methyl $C_3H_7CH(CH_3)CHO$	100.16		116^{737}				w, eth, ace	$B1^4$, 3304
14797	Valeraldehyde, 4-methyl-2-oxo or Formyl isobutyl ketone $(CH_3)_2CHCH_2COCHO$	114.14	ye-gr	$45\text{-}6^{12}$					$B1^1$, 406
14798	Valeraldehyde oxime or Valeraldoxime $CH_3(CH_2)_3CH{=}NOH$	101.15	cr		52			eth	$B1^3$, 2798
14799	Valeraldehyde, 2-oxo or Butyryl formaldehyde C_3H_7COCHO	100.12		112, 36^{16}			1.4043^{25}	w, al, eth	$B1^4$, 3658
14800	Valeraldehyde, 4-oxo or Levalinaldehyde $CH_3COCH_2CH_2CHO$	100.12		186-8d, 70^{12}	<−21	$1.0184^{21/4}$	1.4257^{22}	w, al, eth, ace, bz	$B1^4$, 3659
14801	Valeramide $CH_3(CH_2)_3CONH_2$	101.15	micalike mcl pl (peth, al)		106 (114)	0.8735^{110}	1.4183^{110}	w, al, eth	$B2^4$, 874
14802	Valeramide, 2-bromo-4-methyl (d) $(CH_3)_2CHCH_2CHBrCONH_2$	194.07	(w), $[\alpha]^{20}/_D$ −48.3 (al, c=5.9)		118			al, eth, chl	$B2^2$, 291
14803	Valeramide, N,N-dimethyl $CH_3(CH_2)_3CON(CH_3)_2$	129.20		141^{100}	−51	$0.8962^{25/4}$	1.4419^{25}	w, al, eth	$B4^3$, 127
14804	Valeramide, 4-methyl $(CH_3)_2CHCH_2CH_2CONH_2$	115.18	nd (al)		120-1			w, al	$B2^4$, 946
14805	Valeramide, 4-methyl-N-phenyl or Isocaproanilide $(CH_3)_2CHCH_2CH_2CONHC_6H_5$	191.27	nd (bz, dil al)		112			eth	$B12^3$, 478
14806	Valeramide, N-phenyl $C_4H_9CONHC_6H_5$	177.25	mcl pr (al), cr (peth)		63			eth	$B12^3$, 476
14807	Valeramide, 2,2,4-trimethyl $(CH_3)_2CHCH_2C(CH_3)_2CONH_2$	143.23	lf (lig)		71			al	$B2^2$, 305
14808	Valeric acid or Pentanoic acid $CH_3(CH_2)_3CO_2H$	102.13		186, 82.7^{10}	−33.8	$0.9391^{20/4}$	$1.4085^{20}/_4$	w, al, eth	$B2^4$, 868
14810	Valeric acid, 2-acetyl, ethyl ester $CH_3COCH(C_3H_7)CO_2C_2H_5$	172.22		$210\text{-}2^{749}$, 90^{10}		$0.9682^{15/4}$	1.4271^{20}	al, eth	$B3^4$, 1602
14811	Valeric acid, 2-amino (D+) $C_3H_7CH(NH_2)CO_2H$	117.15	lf (w), $[\alpha]^{23}/_D$ +32 (6NHCl, c=2)	sub	ca 307			w	$B4^3$, 1331
14812	Valeric acid, 2-amino (DL) or Norvaline $C_3H_7CH(NH_2)CO_2H$	117.15	lf (al, w)	sub	303 (sealed tube)			w	$B4^3$, 1333
14813	Valeric acid, 2-amino (L-) $C_3H_7CH(NH_2)CO_2H$	117.15	cr (dil al, w), $[\alpha]^{20}/_D$ −23 (20% HCl, c=10)	sub	ca−305			w	$B4^3$, 1332
14814	Valeric acid, 4-amino (D) $CH_3CH(NH_2)CH_2CH_2CO_2H$	117.15	cr (dil al), $[\alpha]^{20}/_D$ +12 (w, p=10)		214			w	$B4^3$, 1342
14815	Valeric acid, 4-amino (DL) $CH_3CH(NH_2)_2CH_2CH_2CO_2H$	117.15	cr (w)	d	199 (214)			w	$B4^3$, 1342
14816	Valeric acid, 5-amino $H_2N(CH_2)_4CO_2H$	117.15	lf (dil al)	d	157-8d			w	$B4^3$, 1343
14817	Valeric acid anhydride $(C_4H_9CO)_2O$	186.25		218^{754}, 111^{15}	−56.1	$0.924^{20/4}$	1.4171^{26}	al, eth	$B2^4$, 874

PHYSICAL CONSTANTS OF ORGANIC COMPOUNDS (Continued)

No.	Name, Synonyms, and Formula	Mol. wt.	Color, crystalline form, specific rotation and λ_{max} (log ε)	b.p. °C	m.p. °C	Density	n_D	Solubility	Ref.
14818	Valeric acid, 2-bromo (dl) $C_3H_7CHBrCO_2H$	181.03	....	118^{12}	......	1.381^{20}	....	al, eth	B2[4], 883
14819	Valeric acid, 2-bromo, ethyl ester $C_3H_7CHBrCO_2C_2H_5$	209.08	....	190-2, $92\text{-}4^{18}$	......	$1.226^{18/4}$	1.4496^{20}	al, eth	B2[3], 681
14820	Valeric acid, 2-bromo-4-methyl (d) $(CH_3)_2CHCH_2CHBrCO_2H$	195.06	oil, $[\alpha]^{20/}{}_D$ +29.8 (eth, c=6)	$131\text{-}2^{18}$	......	....	....	al, eth	B2[3], 747
14821	Valeric acid, 2-bromo-4-methyl (l) $(CH_3)_2CHCH_2CHBrCO_2H$	195.06	oil, $[\alpha]^{20/}{}_D$ −12.1	$94^{0.3}$	......	....	....	al, eth	B2[3], 747
14822	Valeric acid, 2-bromo-4-methyl, ethyl ester $(CH_3)_2CHCH_2CHBrCO_2C_2H_5$	223.11	pa ye	202-4, $86\text{-}7^{11}$	......	....	....	....	B2[3], 747
14823	Valeric acid, butyl ester or Butyl valarate $C_4H_9CO_2C_4H_9$	158.24	....	185.8, $84\text{-}5^{8}$	−92.8	$0.8710^{15/4}$	1.4128^{20}	al, eth	B2[3], 671
14824	Valeric acid, *iso*-butyl ester or Isobutyl valarate $C_4H_9CO_2CH_2CH(CH_3)_2$	158.24	....	179	......	$0.8625^{25/4}$	1.4046^{20}	al, eth, ace	B2[3], 671
14825	Valeric acid, *sec*-butyl ester (d) or *sec*-Butyl valarate $C_4H_9CO_2CH(CH_3)C_2H_5$	158.24	$[\alpha]^{20/}{}_D$ +20.7	174.5, 67^{18}	......	$0.8605^{20/4}$	1.4070^{20}	al, eth, bz, Py	B2[3], 671
14826	Valeric acid, 2-chloro (dl) $C_3H_7CHClCO_2H$	136.58	....	222^{763}, $133\text{-}5^{39}$	−15	1.141^{13}	1.4481^{30}	al, eth	B2[4], 878
14827	Valeric acid, 2-chloro, ethyl ester $C_3H_7CHClCO_2C_2H_5$	164.63	....	185^{752}	......	1.040^{12}	1.4307^{11}	eth	B2[3], 678
14828	Valeric acid, 3-chloro $C_2H_5CHClCH_2CO_2H$	136.58	cr (bz)	112^{10}	33	$1.1484^{20/4}$	1.4462^{20}	al, eth	B2[4], 879
14829	Valeric acid, 3-chloro, ethyl ester $C_2H_5CHClCH_2CO_2C_2H_5$	164.63	....	189, $66\text{-}7^{10}$	......	$1.0330^{20/4}$	1.4278^{20}	al, eth	B2[3], 678
14830	Valeric acid, 4-chloro $CH_3CHClCH_2CH_2CO_2H$	136.58	....	116^{10}	......	1.1514^{20}	1.4458^{20}	al, eth	B2[4], 879
14831	Valeric acid, 4-chloro, ethyl ester $CH_3CHClCH_2CH_2CO_2C_2H_5$	164.63	....	196, 70.5^{9}	......	$1.0393^{20/4}$	1.4310^{20}	al, eth	B2[4], 879
14832	Valeric acid, 5-chloro $Cl(CH_2)_4CO_2H$	136.58	....	230, $141\text{-}9^{12}$	18	$1.3416^{25/4}$	1.4555^{20}	al, eth	B2[4], 880
14833	Valeric acid, 5-chloro, ethyl ester $Cl(CH_2)_4CO_2C_2H_5$	164.63	....	205-6, 93^{16}	......	1.0561^{20}	1.4355^{20}	al, eth	B2[4], 880
14834	Valeric acid, 2,4-dimethyl (dl) $(CH_3)_2CHCH_2CH(CH_3)CO_2H$	130.19	....	$111^{0.9}$	143-5	1.0818^{27}	....	al, eth	B2[4], 979
14835	Valeric acid, 2,4-dioxo or Acetyl pyruvic acid $CH_3COCH_2COCO_2H$	130.10	nd (bz-chl), pr (bz)	130^{37} sub	55-63 (+1w) 101 (anh)	....	....	w, al, eth, ace, bz	B3[3], 1331
14836	Valeric acid, 2,4-dioxo,ethyl ester or Ethyl acetone oxalate $CH_3COCH_2COCO_2C_2H_5$	158.15	....	213-5, $111\text{-}2^{16}$	18	1.1251^{20}	1.4757^{17}	al, eth	B3[4], 1777
14837	Valeric acid, ethyl ester or Ethyl valerate $C_4H_9CO_2C_2H_5$	130.19	....	144.6^{736}, 50.5^{29}	−91.2	$0.8770^{20/4}$	1.4120^{20}	al, eth	B2[4], 872
14838	Valeric acid, 2-ethyl (dl) or 3-Hexane carboxylic acid $C_3H_7CH(C_2H_5)CO_2H$	130.19	....	209.2, $105\text{-}7^{18}$	......	$0.9361^{33/33}$	....	al, eth	B2[4], 975
14839	Valeric acid, 3-ethyl $(C_2H_5)_2CHCH_2CO_2H$	130.19	....	212, $104\text{-}5^{13}$	......	....	1.4250^{20}	....	B2[4], 976
14840	Valeric acid, 5-fluoro $F(CH_2)_4CO_2H$	120.12	....	83^{2}	......	....	1.4080^{25}	al, eth	B2[4], 876
14841	Valeric acid, furfuryl ester or Furfuryl valerate $C_4H_9CO_2CH_2(C_4H_3O)$	182.22	....	$228\text{-}9^{764}$, $82\text{-}3^{1}$	......	$1.0284^{20/4}$	....	al, eth	B17[2], 115
14842	Valeric acid, heptyl ester or Heptyl valerate $C_4H_9CO_2C_7H_{15}$	200.32	....	245.2	−46.4	0.8623^{20}	$1.4254^{15/}{}_{He}$	al, eth, ace	B2[4], 872
14843	Valeric acid, hexyl ester or Hexyl valerate $C_4H_9CO_2C_6H_{13}$	186.29	....	226.3	−63.1	0.8635^{20}	1.4228^{15}	al, eth, ace	B2[3], 671
14844	Valeric acid, 2-hydroxy or Valerolactic acid $C_3H_7CH(OH)CO_2H$	118.13	hyg pl	sub	34	....	....	w, al, eth	B3[4], 807
14845	Valeric acid, 4-hydroxy, lactone (d) or γ-Valerolactone $CH_3CHCH_2CH_2CO$ (ring –O–)	100.12	$[b\alpha]^{20/}{}_D$ +13.5 (undil)	$86\text{-}90^{14}$	......	....	....	w, al, ace	B17[2], 288
14846	Valeric acid, 4-hydroxy-lactone (dl) or γ-Valerolactone $CH_3CHCH_2CH_2CO$ (ring –O–)	100.12	....	206, $83\text{-}4^{13}$	−31	1.0465^{25}	1.4328^{20}	w, al, ace	B17[4], 4176
14847	Valeric acid, 4-hydroxy-lactone (l) $CH_3CHCH_2CH_2CO$ (ring –O–)	100.12	$[\alpha]^{20/}{}_D$ +4.6 (eth, c=10)	$78\text{-}80^{8}$	......	....	1.4322^{20}	w, al, ace	B17[4], 4176
14848	Valeric acid, 5-hydroxy-lactone or γ-Valerolactone $CH_2(CH_2)_3CO$ (ring –O–)	100.12	....	218-20, $113\text{-}4^{14}$	−12.5	1.0794^{20}	1.4503^{20}	w, al, eth	B17[4], 4169
14849	Valeric acid, methyl ester or Methyl valerate $C_4H_9CO_2CH_3$	116.16	....	126.5	......	$0.8947^{20/4}$	1.4003^{20}	al, eth, ace	B2[4], 871

PHYSICAL CONSTANTS OF ORGANIC COMPOUNDS (Continued)

No.	Name, Synonyms, and Formula	Mol. wt.	Color, crystalline form, specific rotation and λ_{max} (log ε)	b.p. °C	m.p. °C	Density	n_D	Solubility	Ref.
14850	Valeric acid, 2-methyl (d) $C_3H_7CH(CH_3)CO_2H$	116.16	$[\alpha]^{25/}_D$ + 18.5 (undil)	96^{15}	......	$0.927^{16/4}$	1.4112^{25}	w, al, eth	$B2^4$, 942
14851	Valeric acid, 2-methyl (dl) $C_3H_7CH(CH_3)CO_2H$	116.16	....	195-6, $102\text{-}5^{12}$	......	$0.9230^{20/4}$	1.413^{20}	w, al, eth	$B2^4$, 942
14852	Valeric acid, 2-methyl (l) $C_3H_7CH(CH_3)CO_2H$	116.16	$[\alpha]^{25/}_D$ −7.1 (eth)	190-3, 96^{15}	......	$0.9781^{20/4}$	1.4117^{15}	w, al, eth	$B2^3$, 740
14853	Valeric acid, 3-methyl (d) $C_2H_5CH(CH_3)CH_2CO_2H$	116.16	$[\alpha]^{20/}_D$ + 8.5 (undil)	199, $92\text{-}3^{10}$	......	$0.9276^{20/4}$	1.4158^{20}	al, eth	$B2^4$, 948
14854	Valeric acid, 3-methyl (dl) $C_2H_5CH(CH_3)CH_2CO_2H$	116.16	....	197-8	−41.6	0.9262^{20}	1.4159^{20}	al, eth	$B2^4$, 948
14855	Valeric acid, 3-methyl (l) $C_2H_5CH(CH_3)CH_2CO_2H$	116.16	$[\alpha]^{20/}_D$ −8.9 (al)	196-7, 105^{30}	......	$0.923^{25/4}$	1.4152^{20}	al, eth	$B2^4$, 948
14856	Valeric acid, 3-methyl-3-phenyl $C_2H_5C(CH_3)(C_6H_5)CH_2CO_2H$	192.26	....	174^{14}	......	1.050^{25}	1.5197^{25}	....	$B9^3$, 2548
14857	Valeric acid, 4-methyl or Isocaproic acid $(CH_3)_2CHCH_2CH_2CO_2H$	116.16	....	200-1, $86\text{-}8^{11}$	−33	$0.9225^{20/4}$	1.4144^{20}	al, eth	$B2^4$, 944
14858	Valeric acid, 4-methyl-2-phenyl or α-Phenyl isocaproic acid $(CH_3)_2CHCH_2CH(C_6H_5)CO_2H$	192.24	pr (peth)	$178\text{-}80^{20}$	78-9	....	....	al	$B9^3$, 2546
14859	Valeric acid, octyl ester or Octyl valerate $C_4H_9CO_2C_8H_{17}$	214.35	....	261.6	−42.3	0.8615^{20}	1.4273^{15}	al, eth, ace	$B2^3$, 672
14860	Valeric acid, 2-oxo $C_3H_7COCO_2H$	116.12	....	179, 81^{12}	6-7	....	....	eth, bz, chl	$B3^4$, 1558
14861	Valeric acid, 3-oxo, ethyl ester or Ethyl propionyl acetate $C_2H_5COCH_2CO_2C_2H_5$	144.17	....	191, 90^{13}	......	....	1.4230^{20}	al, eth, bz	$B3^4$, 1559
14862	Valeric acid, 4-oxo or Levulinic acid $CH_3COCH_2CH_2CO_2H$	116.12	lf or pl	245-6d, $139\text{-}40^{8}$	37.2	$1.1335^{20/4}$	1.4396^{20}	w, al, eth	$B3^4$, 1560
14863	Valeric acid, pentyl ester or Pentyl valerate $C_4H_9CO_2C_5H_{11}$	172.27	....	203.7, 103^{23}	−78.8	$0.8638^{20/4}$	1.4164^{20}	**al, eth**	$B2^4$, 872
14864	Valeric acid, 2-phenyl (d) $C_3H_7CH(C_6H_5)CO_2H$	178.23	$[\alpha]_D$ + 58.8 (chl)	165^{14}	......	1.047^{20}	....	al, eth, chl	$B9^3$, 2506
14865	Valeric acid, 2-phenyl (dl) $C_3H_7CH(C_6H_5)CO_2H$	178.23	nd (lig)	280	58	....	....	al, eth	$B9^3$, 2506
14866	Valeric acid, 4-phenyl $C_6H_5CH(CH_3)CH_2CH_2CO_2H$	178.23	....	210^{85}, 165^{12}	ca 13	$1.0554^{15/4}$	1.5167^{20}	al, eth	$B9^3$, 2505
14867	Valeric acid, 5-phenyl $C_6H_5(CH_2)_4CO_2H$	178.23	pl (w), pr (peth)	$190\text{-}3^{30}$	57-8	....	....	al	$B9^3$, 2502
14868	Valeric acid, propyl ester or Propyl valerate $C_4H_9CO_2C_3H_7$	144.21	....	167.5	......	$0.8699^{20/4}$	1.4065^{20}	al, eth, chl	$B2^4$, 872
14869	Valeric acid, isopropyl ester or Isopropyl valerate $C_4H_9CO_2CH(CH_3)_2$	144.21	....	153.5	......	$0.8579^{20/4}$	1.4061^{20}	al, eth, ace	$B2^3$, 671
14870	Valeronitrile $CH_3(CH_2)_3CN$	83.13	....	141.3, 30.9^{10}	−96	$0.8008^{20/4}$	1.3971^{20}	eth, ace, bz	$B2^4$, 875
14871	Valeronitrile, 2-hydroxy-2-propyl $(C_3H_7)_2C(OH)CN$	141.21	....	$119\text{-}20^{21}$	......	0.9077^{18}	1.4337^{18}	al	$B3^4$, 880
14872	Valeronitrile, 4-methyl or Isocapronitrile $(CH_3)_2CHCH_2CH_2CN$	97.16	....	156-7, 50^{13}	−51	$0.8030^{20/4}$	1.4059^{20}	al, eth	$B2^4$, 946
14873	Valeronitrile, 4-methyl-2-phenyl or Phenyl isocapronitrile $(CH_3)_2CHCH_2CH(C_6H_5)CN$	173.26	....	$263\text{-}6^{765}$, $136\text{-}8^{15}$	......	0.942^{16}	....	al, eth, bz, aa	$B9^1$, 220
14874	Valeronitrile, 2-phenyl $C_3H_7CH(C_6H_5)CN$	159.23	....	$254\text{-}5^{750}$, $125\text{-}8^{13}$	......	0.9425	1.5000^{20}	**al, bz**	$B9^3$, 2508
14875	Valerophenone or Butyl phenyl ketone $C_4H_9COC_6H_5$	162.23	....	248.5, $131\text{-}3^{13}$	......	$0.988^{20/20}$	1.5158^{20}	al, eth	$B7^3$, 1114
14876	Valerophenone, γ-oxo or Phenacyl acetone $C_6H_5COCH_2CH_2COCH_3$	176.22	ye oil	162^{12}	......	....	1.5250^{30}	ace	$B7^3$, 3509
14877	Valeryl chloride C_4H_9COCl	120.58	....	$107\text{-}10^{756}$	−110	1.0155^{15}	1.4200^{20}	....	$B2^4$, 874
14878	Valeryl chloride, 2-chloro $C_3H_7CHClCOCl$	155.02	....	$155\text{-}7^{763}$, $61\text{-}2^{28}$	......	1.1765^{20}	1.4465^{20}	eth	$B2^4$, 879
14879	Valeryl chloride, 4-chloro $CH_3CHClCH_2CH_2COCl$	155.02	....	61^{8}	......	....	....	eth	$B2^3$, 679
14880	Valeryl chloride, 5-chloro $Cl(CH_2)_4COCl$	155.02	....	83^{12}	......	1.210^{18}	1.4639^{20}	eth	$B2^4$, 881
14881	Valeryl chloride, 4,4-dimethyl or Neopentyl acetyl chloride $(CH_3)_3CCH_2CH_2COCl$	148.63	....	150-2, 58^{2}	......	....	1.4294^{20}	eth	$B2^3$, 780

PHYSICAL CONSTANTS OF ORGANIC COMPOUNDS (Continued)

No.	Name, Synonyms, and Formula	Mol. wt.	Color, crystalline form, specific rotation and λ_{max} (log ε)	b.p. °C	m.p. °C	Density	n_D	Solubility	Ref.
14882	Valeryl chloride, 2-ethyl $C_3H_7CH(C_2H_5)COCl$	148.63		158-60, 50^{11}				eth	$B2^{4}$, 975
14883	Valeryl chloride, 2-methyl (dl) $C_3H_7CH(CH_3)COCl$	134.61		$140\text{-}1^{745}$		$0.9781^{20/4}$	1.4330^{27}	eth	$B2^{3}$, 741
14884	Valeryl chloride, 3-methyl (dl) $C_2H_5CH(CH_3)CH_2COCl$	134.61		142-3		$0.9781^{20/4}$		eth	$B2^{4}$, 949
14885	Valeryl chloride, 4-methyl $(CH_3)_2CHCH_2CH_2COCl$	134.61		$144\text{-}5^{745}$		0.9725^{20}			$B2^{4}$, 945
14886	Valine (D) or *l*-α-Amino isovaleric acid $(CH_3)_2CHCH(NH_2)CO_2H$	117.15	pl (aq al), $[\alpha]^{20/}_{D}$ −29 (20% al)		156-7 (hyd) 293d (sealed tube)			w	$B4^{3}$, 1369
14887	Valine (DL) $(CH_3)_2CHCH(NH_2)CO_2H$	117.15		sub	298d (sealed tube)	1.316		w	$B4^{3}$, 1370
14888	Valine (L) or *d*-α-Amino isovaleric acid $(CH_3)_2CHCH(NH_2)CO_2H$	117.15	lf (w-al), $[\alpha]^{23/}_{D}$ + 22.9 (20% al, c=0.8)		93-6, 315 (sealed tube)	1.230		w	$B4^{3}$, 1365
14889	Valine, β-hydroxy-dl) $(CH_3)_2C(OH)CH(NH_2)CO_2H$	133.15	pl (dil al)		240d			w	$B4^{3}$, 1660
14890	Vanillan or 4-Hydroxy-3-methoxy benzaldehyde 4-HO-3-CH_3O-C_6H_3CHO	152.15	(i) nd (w, lig), (ii) tetr (w, lig)	285, 170^{15}	(i) 77-9, (ii) 81-2	1.056		al, eth, ace, bz	$B8^{3}$, 2011
14891	Vanillan acetate or 4-Acetoxy-3-methoxy benzaldehyde 4-CH_3CO_2-3-$CH_3OC_6H_3CHO$	194.19	nd (eth)		102-3			al, eth	$B8^{3}$, 2030
14892	Vanillan, 5-bromo 5-Br-4-HO-3-$CH_3OC_6H_2CHO$	231.05	pl (aa), nd, pl (al)		164-6				$B8^{3}$, 2050
14893	Vanillan, 5-chloro 5-Cl-4-HO-3-$CH_3OC_6H_2CHO$	186.59	tetr		165				$B8^{3}$, 2043
14894	Vanillan, 2-iodo 4-HO-2-I-3-$CH_3OC_6H_2CHO$	278.05			155-6			al	$B8^{3}$, 2058
14895	Vanillan, 5-Iodo 4-HO-5-I-3-$CH_3OC_6H_2CHO$	278.05	pa ye		180				$B8^{3}$, 2059
14896	*o*-Vanillan or 2-Hydroxy-3-methoxy benzaldehyde 2-HO-3-$CH_3OC_6H_3CHO$	152.15	lt ye-lt gr nd (w, lig)	265-6, 128^{10}	44-5			al, eth, lig	$B8^{3}$, 1979
14897	Vasicine (dl) or Peganine $C_{11}H_{12}N_2O$	188.23	nd (al)		209-10			al, ace, chl	$B23^{2}$, 342
14898	Vasicine (l) $C_{11}H_{12}N_2O$	188.23	nd (al), $[\alpha]^{14/}_{D}$ −62 (al, c=2.4)		211-2			al, chl	$B23^{2}$, 342
14899	Veraridine or 3-Veratroyl veracerine $C_{36}H_{51}NO_{11}$	673.80	yesh amor, $[\alpha]^{22/}_{D}$ + 8 (al)		180				$B21^{4}$, 6824
14900	Veratramide or 3,4-Dimethoxy amino benzaldehyde 3,4-$(CH_3O)_2C_6H_3CONH_2$	181.19	cr (w)		164			w, eth, bz	$B10^{3}$, 1427
14901	Veratramine $C_{27}H_{39}NO_2$	409.61	nd, $[\alpha]^{19/}_{D}$ −70 (MeOH)		209-10 (hyd)			bz, chl	$B21^{4}$, 2385
14902	Veratric acid or 3,4-Dimethoxy benzoic acid 3,4-$(CH_3O)_2C_6H_3CO_2H$	182.18	nd (w, aa), rh (sub)	sub	181-2			al, eth	$B10^{3}$, 1404
14903	Veratrine or Cevadine $C_{32}H_{49}NO_9$	591.74	rh (+ 2 al), $[\alpha]^{17/}_{D}$ + 12.5 (al)		205d			al, chl, Py	$B21^{4}$, 6820
14904	Veratronitrile or 3,4-Dimethoxy benzonitrile 3,4-$(CH_3O)_2C_6H_3CN$	163.18	nd (w)		67-8			bz	$B10^{2}$, 264
14905	Veratrosine or Veratramine-β-D-glucoside $C_{33}H_{49}NO_7$	571.75	nd (aq MeOH), $[\alpha]^{24/}_{D}$ −55 (al-chl, c=0.94)		242-3d				$B21^{4}$, 2386
14906	Verbenone (d) or d-4-Piperone $C_{10}H_{14}O$	150.22	$[\alpha]^{18/}_{D}$ + 249.6	227-8, $103\text{-}4^{16}$	9.8	0.9978^{20}	1.4993^{18}	w, al, ace, bz	$B7^{3}$, 583
14907	Verbenone (l) $C_{10}H_{14}O$	150.22	$[\alpha]_{D}$−144	253-5, 100^{16}	6.5	0.9731^{20}	1.4961^{20}	w, al, bz	$B7^{3}$, 583

PHYSICAL CONSTANTS OF ORGANIC COMPOUNDS (Continued)

No.	Name, Synonyms, and Formula	Mol. wt.	Color, crystalline form, specific rotation and λ_{max} (log ε)	b.p. °C	m.p. °C	Density	n_D	Solubility	Ref.
14908	α-Vetivone or α-Vetiverone $C_{15}H_{22}O$	218.34	(peth), $[\alpha]_D$ +238.2	144^{2}	51.5	$1.0035^{20/4}$	1.5370^{20}	ace	B7³, 1265
14909	β-Vetivone or β-Vetiverone $C_{15}H_{22}O$	218.34	(pentane), $[\alpha]^{20}/_D$ −38.9 (al, c=10.6)	$141\text{-}2^{2}$	44.5	$1.0001^{20/4}$	1.5309^{20}	ace	B7³, 1256
14910	Vicianose or 6-*O*-(α-L-arabinosido)-β-D-glucose $C_{11}H_{20}O_{10}$	312.27	nd (dil al), $[\alpha]^{20}/_D$ +56.5→ +39.7 (w, mut)	....	ca 210d	....	....	w	B17⁴, 2447
14911	Vicine (a β-D-glucoside) $C_{10}H_{16}N_4O_7$	304.26	nd (w, dil al + 1w), $[\alpha]^{25}/_D$ −12 (w, c=10)	....	239-42d	....	....	....	B31, 163
14912	Vincamine $C_{21}H_{26}N_2O_3$	354.45	$[\alpha]^{23}/_D$ +41 (Py)	....	231-2	....	....	....	M, 1107
14913	Violaxanthin or Zeaxanthin diepoxide $C_{40}H_{56}O_4$	600.88	red pr (MeOH, al-eth), $[\alpha]^{20}/_d$ +35 (chl, c=0.08)	....	208	....	....	al, eth	B19⁴, 1139
14914	Violuric acid or Alloxan-5-oxime $C_4H_3N_3O_4$	157.09	pa ye rh	....	203-4d	....	....	al	B24², 304
14915	Visnadin $C_{21}H_{24}O_7$	388.42	nd, $[\alpha]_D$ +9 (al)	....	85-6	....	....	al, eth	B19⁴, 2787
14916	Visnagin or 5-Methoxy-2-methyl furanochromone $C_{13}H_{10}O_4$	230.22	nd (w, MeOH)	....	144-5	....	....	chl	B19⁴, 2640
14917	Vitamin A_1 or Axerophytol $C_{20}H_{30}O$	286.46	ye pr (peth)	$137\text{-}8^{10\ 6}$	63-4	....	....	al, eth, ace, bz	B6³, 2787
14918	Vitamin B_1 or Thiamine hydrochloride $C_{12}H_{18}Cl_2N_4OS$	337.27	(i) mcl pr (MeOH-al), (ii) pl (MeOH-al, w-al)	....	(i) 233-4, (ii) 250	....	....	w	Am74, 2409
14919	Vitamin B_6 or Pyroxidin $C_8H_{11}NO_3$	169.18	nd (ace)	sub	160	....	....	w, al, ace	B21⁴, 2509
14920	Vitamin B_6 hydrochloride or Pyridoxin hydrochloride $C_8H_{11}NO_3.HCl$	205.64	pl (al, ace)	sub	206-8	....	....	w	B21⁴, 2511
14921	Vitamin K_1 $C_{31}H_{46}O_2$	450.71	$[\alpha]^{20}/_D$ −0.4 (bz 57.5%)	$140\text{-}5^{0.001}$ dec>120	−20	$0.967^{25/25}$	1.5250^{25}	al, eth, ace, bz, peth	B7³, 3792
14922	Vitamin B_{12} $C_{63}H_{88}N_{14}O_{14}CoP$	1355.42	....	>300	....	....	....	....	....
14923	Vitamin B_1-hydrochloride $C_{12}H_{17}N_4OSCl.HCl$	337.27	....	....	233-4	....	....	w	Am74, 2409
14924	Vitamin D_2 $C_{28}H_{44}O$	396.66	$[\alpha]^{25}/_D$ +82.6 (ace, c=3)	....	115-8	....	....	....	B6³, 3089
14925	Vitamin D_3 $C_{27}H_{44}O$	384.65	$[\alpha]^{20}/_D$ +84.8 (ace, c=1.6)	....	84-5	....	....	....	B6³, 2811
14926	Vomicine $C_{22}H_{24}N_2O_4$	380.44	nd (80% al), pr (ace), $[\alpha]^{22}/_D$ +80.4 (al, p=0.4)	....	282	....	....	chl	B27², 795
14927	Vinyl acetate $CH_3CO_2CH{=}CH_2$	86.09	....	72.2	−93.2	$0.9317^{20/4}$	1.3959^{20}	al, eth, ace, bz, chl	B2⁴, 176
14928	Vinyl acetylene or 1-Buten-3-yne $CH_2{=}CH\text{-}C{\equiv}CH$	52.08	....	5.1	....	$0.7095^{0/0}$	1.4161^{1}	bz	B1⁴, 1083
14929	Vinyl amine $CH_2{=}CHNH_2$	43.07	....	$55\text{-}6^{750}$	....	0.8321^{24}	....	w, al, eth	B20⁴, 3
14930	Vinyl bromide $CH_2{=}CHBr$	106.95	....	15.8	−139.5	1.4933	1.4410^{20}	al, eth, ace, bz, chl	B1⁴, 718

No.	Name, Synonyms, and Formula	Mol. wt.	Color, crystalline form, specific rotation and λ_{max} (log ε)	b.p. °C	m.p. °C	Density	n_D	Solubility	Ref.
14931	Vinyl chloride or Chloro ethylene $CH_2{=}CHCl$	62.50	gas	−13.4	−1538	$0.9106^{20/4}$	1.3700^{20}	al, eth	B1[4], 700
14932	Vinyl ether or Divinyl ether $(CH_2{=}CH)_2O$	70.09	…	28	−101	$0.773^{20/4}$	1.3989^{20}	al, eth, ace	B1[4], 2058
14933	Vinyl ether, hexachloro $(Cl_2C{=}CCl)_2O$	276.76	…	210	…	1.654^{21}	…	…	B1, 725
14934	Vinyl fluoride $CH_2{=}CHF$	46.04	gas	−72.2	−160.5	…	…	al, ace	B1[4], 694
14935	Vinyl iodide $CH_2{=}CHI$	153.95	…	56	…	2.037^{20}	1.5385^{20}	al, eth	B1[4], 722
14936	Vinyl propionate $CH_3CH_2CO_2CH{=}CH_2$	100.12	…	91-2	…	…	…	…	B2[4], 711
14937	2-Vinyl pyridine $2\text{-}(CH_2{=}CH)C_5H_4N$	105.14	…	159-60, 50-5[4]	…	$0.9985^{20/0}$	1.5495^{20}	al, eth, ace, chl	B20[4], 2884
14938	3-Vinyl pyridine $3\text{-}(CH_2{=}CH)C_5H_4N$	105.14	…	162, 67-8[18]	…	…	1.5530^{20}	al, eth	B20[4], 2886
14939	4-Vinyl pyridine $4\text{-}(CH_2{=}CH)C_5H_4N$	105.14	…	65[15]	…	$0.9800^{20/4}$	1.5449^{20}	w, al, chl	B20[4], 2887
14940	Vinyl sulfide $(CH_2{=}CH)_2S$	86.15	…	84	$0.9174^{15/4}$	…	…	al, eth, ace	B1[4], 2068
14941	2-Vinyl toluene or 2-Methyl styrene $2\text{-}CH_3C_6H_4CH{=}CH_2$	118.18	…	171, 51[9]	…	$0.9106^{20/4}$	1.5450^{20}	bz, chl	B5[4], 1367
14942	*m*-Vinyl toluene or 3-Methyl styrene $3\text{-}CH_3C_6H_4CH{=}CH_2$	118.18	…	168, 61-2[18]	−70	$0.9028^{20/20}$	1.5428^{20}	bz	B5[4], 1367
14943	*p*-Vinyl toluene or 4-Methyl styrene $4\text{-}CH_3C_6H_4CH{=}CH_2$	118.18	…	169, 63[15]	−37.8	$0.8760^{20/4}$	1.5428^{20}	bz	B5[4], 1369

No.	Name, Synonyms, and Formula	Mol. wt.	Color, crystalline form, specific rotation and λ_{max} (log ε)	b.p. °C	m.p. °C	Density	n_D	Solubility	Ref.
14944	Xanthene or Dibenzo-1,4-Puran $C_{13}H_{10}O$	182.22	wh ye lf (al)	310-2	100.5			eth, bz, chl, lig	$B17^4$, 614
14945	Xanthene, 9-hydroxy or Dibenz-γ-pyranol $C_{13}H_{10}O_2$	198.22	nd (aq al)		ca 125			al, eth, chl	$B17^4$, 1602
14946	Xanthene, 9-phenyl $C_{19}H_{14}O$	258.32	(al)		145-6			bz, aa	$B17^4$, 733
14947	9-Xanthene carboxylic acid or Xanthanoic acid $C_{14}H_{10}O_3$	226.23	nd (dil al, MeOH)		223.4			eth	$B18^4$, 4351
14948	Peri-xanthenoxanthene or 1,1-Binaphthalene-2:8′,2:8-dioxide $C_{20}H_{10}O_2$	282.30	ye pr (chl)	$400^{20.25}$	242			bz	$B19^4$, 448
14949	Xanthine or 2,6-Purinedione $C_5H_4N_4O_2$	152.11	yesh pl (w)	sub	d				$B26^2$, 260
14950	Xanthine, 2-methyl or Heteroxanthine $C_6H_6N_4O_2$	166.14	nd (w)		380d				$B26^2$, 263
14951	Xanthione or 9-Xanethenethione $C_{13}H_8OS$	212.27	red nd (al)		156			eth, bz	$B17^4$, 5301
14952	Xanthogen-diethyl or Auligen $C_2H_5OC(:S)SSC(:S)OC_2H_5$	242.30	ye nd or pl (al)	$107\text{-}9^{0.05}$	31-2	$1.2604^{25/4}$		eth, ace, bz, peth, chl	$B3^3$, 349
14953	Xanthogenic acid or Ethoxy dithioformic acid $C_2H_5OCS_2H$	122.20	unst	25d	fr−53				$B3^4$, 401
14953a	Xanthogenic acid, ethyl ester $CH_3CH_2OCS_2C_2H_5$	150.25		200		1.085^{19}	$1.5370^{18.2}$		$B3^2$, 153
14954	Xanthone or Dibenzopyrone $C_{13}H_8O_2$	196.21	nd (al)	349-50, 146^3	174			al, eth, bz, chl	$B17^4$, 5292
14955	Xanthone, 1,2-dihydroxy $C_{13}H_8O_4$	228.20	pa ye nd (aq al + 3 w)		166-7			al, Py	$B18^4$, 1678
14956	Xanthone, 1,3-dihydroxy $C_{13}H_8O_4$	228.20	nd (aq al + l w)	sub	259			al	$B18^4$, 1680
14957	Xanthone, 1,6-dihydroxy or Isoeuxanthone $C_{13}H_8O_4$	228.20	ye nd (dil al)		245-6			al, eth	$B18^4$, 1682
14958	Xanthone, 1,7-dihydroxy or Euxanthone $C_{13}H_8O_4$	228.20	ye nd (to), pl (al)	sub d	240				$B18^4$, 1682
14959	Xanthone, 1,8-dihydroxy $C_{13}H_8O_4$	228.20	ye lf (bz)		187				$B18^4$, 1683
14960	Xanthone, 2,3-dihydroxy $C_{13}H_8O_4$	228.20	ye nd (al)		294			al, ace, bz, aa	$B18^4$, 1683
14961	Xanthone, 2,7-dihydroxy or β-Isoeuxanthone $C_{13}H_8O_4$	228.20	ye nd (al, eth)	sub	>330			al, eth	$B18^2$, 357
14962	Xanthone, 3,4-dihydroxy $C_{13}H_8O_4$	228.20	pa red-ye nd (dil al + 3w)		240			al	$B18^4$, 1684
14963	Xanthone, 3,6-dihydroxy $C_{13}H_8O_4$	228.20	nd (dil al), pr (sub)	sub	300-50d			al, aa	$B18^4$, 1685
14964	Xanthone, 1,7-dihydroxy-3-methoxy or Gentisin $C_{14}H_{10}O_5$	258.23	ye rh	400 sub	266-7			al, Py	$B18^4$, 2602
14966	Xanthophyll or Lutein. Lateol $C_{40}H_{56}O_2$	568.88	ye or vt pr (eth-MeOH), $[\alpha]_{Cd}$ + 160 (chl)		196			al, eth, bz, peth	$B6^3$, 5870
14967	Xanthopterin or 2-Amino-4,6-pteridenediol. Uropteim $C_6H_5N_5O_2$	179.14	hyg ye amor or og (aa)	$98\text{-}100^{18}$	>410d	1.559			$B26^2$, 313
14968	Xanthotoxin or Ammoidin $C_{12}H_8O_4$	216.19	pr (dil al), nd (peth)		148			al	$B19^4$, 2633
14969	Xanthotoxol or 8-Hydroxy-4′:5′,6:7-furocoumarin $C_{11}H_6O_4$	202.17			251-2				$B19^4$, 2633
14970	Xanthoxyletin or Alloxanthylethin $C_{15}H_{14}O_4$	258.27	pr (MeOH, peth)		133			ace, bz, chl	$B19^4$, 2645
14971	Xanthyletin or 2,2-Dimethyl chromeno coumarin $C_{14}H_{12}O_3$	228.25	pr	$140\text{-}5^{0.1}$	131.5			al	$B19^4$, 1828
14972	o-Xylene or 1,2-Dimethyl benzene $1,2\text{-}(CH_3)_2C_6H_4$	106.17		144.4, 32^{10}	−25.2	$0.8802^{10/4}$	1.5055^{20}	**al, eth, ace, bz**	$B5^4$, 917
14973	o-Xylene, 4-chloro or 4-Chloro-1,2-dimethyl benzene $1,2\text{-}(CH_3)_2C_6H_3Cl\text{-}(4)$	140.61		194^{755}	−6	$1.0692^{15/15}$		ace, bz	$B5^3$, 816
14974	o-Xylene, 3,4-dinitro or 1,2-Dimethyl-3,4-dinitro benzene $1,2\text{-}(CH_3)_2\text{-}3,4\text{-}(NO_2)_2C_6H_2$	196.16	nd (al)		82			eth, bz	$B5^3$, 822
14975	o-Xylene, 3,5-dinitro $1,2\text{-}(CH_3)_2\text{-}3,5\text{-}(NO_2)_2C_6H_2$	196.16	ye nd (al, peth)		77			al, ace, bz, chl	$B5^3$, 823

PHYSICAL CONSTANTS OF ORGANIC COMPOUNDS (Continued)

No.	Name, Synonyms, and Formula	Mol. wt.	Color, crystalline form, specific rotation and λ_{max} (log ε)	b.p. °C	m.p. °C	Density	n_D	Solubility	Ref.
14976	o-Xylene, 3,6-dinitro $1,2-(CH_3)_2-3,6-(NO_2)_2C_6H_2$	196.16	nd (al)		89-90			al, eth, ace, bz, peth	$B5^{1}$, 181
14977	o-Xylene, 4,5-dinitro $1,2-(CH_3)_2-4,5-(NO_2)_2C_6H_2$	196.16	nd (al, bz, aa)		118			eth, ace, bz	$B5^{3}$, 823
14978	o-Xylene, 4-ethyl $1,2-(CH_3)_2-4-C_2H_5C_6H_3$	134.22		189.7, 67.8^{10}	−67	$0.8745^{20/4}$	1.5031^{20}	al, eth, ace, bz, peth	$B5^{4}$, 1069
14979	o-Xylene, 3-iodo or 3-Iodo-1,2-dimethyl benzene $3-I-1,2-(CH_3)_2C_6H_3$	232.06		228-30, $125-6^{15}$		$1.6395^{20/4}$	1.6074^{20}	ace	$B5^{3}$, 820
14980	o-Xylene, 4-iodo $4-I-1,2-(CH_30_2C_6H_3$	232.06		231-2, 111^{11}		$1.6334^{18/4}$	1.6049^{18}	ace	$B5^{3}$, 821
14981	o-Xylene, 3-methoxy or 2,3-Dimethyl anisole $2,3-(CH_3)_2C_6H_3OCH_3$	136.19		199, 85^{18}	29	0.9596^{40}	1.5120^{40}	al, eth, ace, bz	$B6^{3}$, 1723
14982	o-Xylene, 4-methoxy $4-CH_3O-1,2-(CH_3)_2C_6H_3$	136.19		204-5, $96-7^{17}$		$0.9744^{14/4}$	1.5198^{14}	al, eth, bz	$B6^{3}$, 1727
14983	o-Xylene, 3-nitro $3-NO_2-1,2-(CH_3)_2C_6H_3$	151.16	nd (al)	240, 131^{20}	15	$1.1402^{/20}$	1.5441^{20}	al	$B5^{4}$, 930
14984	o-Xylene, 4-nitro $4-NO_2-1,2-(CH_3)_2C_6H_3$	151.16	ye pr (al)	254^{750}, 143^{21}	30-1	1.112^{15}	1.5202^{20}	al	$B5^{4}$, 930
14985	o-Xylene, 3,4,5,6-tetrabromo $3,4,5,6-Br_4-1,2-(CH_3)_2C_6$	421.75	nd (bz)	374-5	262			bz	$B5^{3}$, 820
14986	*m*-Xylene or 1,3-Dimethyl benzene $1,3-(CH_3)_2C_6H_4$	106.17		139.1, 28.1^{10}	−47.9	$0.8642^{20/4}$	1.4972^{10}	**al, eth, ace, bz**	$B5^{4}$, 932
14987	m-Xylene 4-chloro $1,3-(CH_3)_2C_6H_3Cl-(4)$	140.61		187-8, 89^{24}		$1.0598^{20/20}$	1.5230^{25}	ace, bz	$B5^{3}$, 834
14988	*m*-Xylene, 5-chloro $1,3-(CH_3)_2-C_6H_3-Cl-(5)$	140.61		187-8, 66^{12}				ace, bz	$B5^{3}$, 834
14989	*m*-Xylene, 2,5-dinitro $1,3-(CH_3)_2-2,5-(NO_2)_2C_6H_2$	196.16	ye (al)		101			al, eth, ace, bz, chl	$B5^{2}$, 295
14990	*m*-Xylene, 4-ethyl $1,3-(CH_3)_2-4-C_2H_5C_6H_3$	134.22		188.4, 66.5^{10}	−62.9	$0.8763^{20/4}$	1.5038^{20}	**al, eth, ace, bz,** peth	$B5^{4}$, 1070
14991	*m*-Xylene, 5-ethyl $1,3-(CH_3)_2-5-C_2H_5C_6H_3$	134.22		183.7, 63^{10}	−84.3	$0.8648^{20/4}$	1.4981^{20}	**al, eth, ace**, bz, peth	$B5^{4}$, 1071
14992	*m*-Xylene, 2-iodo $2-I-1,3-(CH_3)_2C_6H_3$	232.06	oil	229-30, $102-3^{14}$	11.2	$1.6158^{20/4}$	1.6035^{20}	ace, bz	$B5^{4}$, 947
14993	*m*-Xylene, 4-iodo $4-I-1,3-(CH_3)_2C_6H_3$	232.06		231d, 111^{14}		$1.6282^{16/4}$	1.6008^{16}	ace, bz	$B5^{4}$, 947
14994	*m*-Xylene, 5-iodo $5-I-1,3-(CH_3)_2C_6H_3$	232.06	oil	230-1, 117^{27}		$1.6085^{18.5/4}$	$1.5967^{18.5}$	ace	$B5^{3}$, 840
14995	*m*-Xylene, 2-methoxy $2-CH_3O-1,3-(CH_3)_2C_6H_3$	136.19		182-3		$0.9619^{14/4}$	1.5053^{14}	al, eth, bz	$B6^{3}$, 1737
14996	*m*-Xylene, 4-methoxy $4-CH_3O-1,3-(CH_3)_2C_6H_3$	136.19		192, $83-4^{15}$		$0.9740^{16/4}$	1.5190^{16}	al, eth, bz	$B6^{3}$, 1744
14997	*m*-Xylene, 5-methoxy $5-CH_3O-1,3-(CH_3)_2C_6H_3$	136.19		194.5, 89^{15}		$0.9627^{15/4}$	1.5110^{20}	al, eth, bz, aa	$B6^{3}$, 1756
14998	*m*-Xylene, 2-nitro $2-NO_2-1,3-(CH_3)_2C_6H_3$	151.16		222, $84.6^{0.05}$	13	1.112^{15}	1.5202^{20}	al	$B5^{4}$, 948
14999	*m*-Xylene, 4-nitro $4-NO_2-1,3-(CH_3)_2C_6H_3$	151.16		246^{744}, 122^{18}	9	1.135^{15}	1.5473^{25}	eth, ace, bz	$B5^{4}$, 948
15000	*m*-Xylene, 5-nitro $5-NO_2-1,3-(CH_3)_2C_6H_3$	151.17	nd (al)	273^{739}	75			al, eth	$B5^{4}$, 948
15001	*m*-Xylene, 2,4,5,6-tetrabromo $2,4,5,6-Br_4-1,3-(CH_3)C_6$	421.75	nd (xyl, al)		248 (252)			bz	$B5^{3}$, 839
15002	*m*-Xylene, 2,4,6-trinitro $2,4,6-(NO_2)_3-1,3-(CH_3)_2-C_6H$	241.16	pa ye pr or lf (al-bz)		184	1.604^{19}		bz, chl, Py	$B5^{4}$, 950
15003	*p*-Xylene or 1,4-Dimethyl benzene $1,4-(CH_3)_2C_6H_4$	106.17	mcl pr (al)	138.3, 27.2^{10}	13.3	$0.8611^{20/4}$	1.4958^{20}	**al, eth, ace, bz**	$B5^{4}$, 951
15004	*p*-Xylene, 2-chloro $1,4-(CH_3)_2C_6H_3Cl-(2)$	140.61		187	1.6	$1.0589^{15/4}$		ace, bz	$B5^{4}$, 965
15005	*p*-Xylene, 2,3-dinitro $1,4-(CH_3)_2-2,3-(NO_2)_2C_6H_2$	196.16	mcl pr (al)		93			al, eth, ace, bz, chl	$B5^{4}$, 972
15006	*p*-Xylene, 2,5-dinitro $1,4-(CH_3)_2-2,5-(NO_2)_2C_6H_2$	196.16	ye nd (al)		147-8			ace, bz, chl	$B5^{4}$, 973
15007	*p*-Xylene, 2,6-dinitro $1,4-(CH_3)_2-2,6(NO_2)_2C_6H_2$	196.16	nd (al)		123-4			al, ace, bz, chl	$B5^{4}$, 973
15008	*p*-Xylene, 2-ethyl $1,4-(CH_3)_2-2-C_2H_5C_6H_3$	134.22		186.9, 64.9^{10}	−53.7	$0.8772^{20/4}$	1.5043^{20}	**al, eth, ace, bz,** peth	$B5^{4}$, 1070
15009	*p*-Xylene, 2-iodo $2-I-1,4-(CH_3)_2C_6H_3$	232.06		227-8d, $106-8^{13}$		$1.6168^{17/4}$	1.5992^{17}	ace, bz	$B5^{4}$, 970

PHYSICAL CONSTANTS OF ORGANIC COMPOUNDS (Continued)

No.	Name, Synonyms, and Formula	Mol. wt.	Color, crystalline form, specific rotation and λ_{max} (log ϵ)	b.p. °C	m.p. °C	Density	n_D	Solubility	Ref.
15010	*p*-Xylene, 2-methoxy $2\text{-}CH_3O\text{-}1,4\text{-}(CH_3)_2C_6H_3$	136.19	....	194^{772}	......	$0.9693^{11/4}$	1.5182^{11}	al, eth, bz, peth	B6³, 1772
15011	*p*-Xylene, 2-nitro $2\text{-}NO_2\text{-}1,4\text{-}(CH_3)_2C_6H_3$	151.16	pa ye	240-1, $64\text{-}5^{0.35}$	......	1.132^{15}	1.5413^{20}	al	B5⁴, 971
15012	*p*-Xylene, 2,3,5-trinitro $2,4,5\text{-}(NO_2)_3\text{-}1,4\text{-}(CH_3)_2C_6H$	241.16	mcl nd (al), lf (al-bz)	410 exp	139-40	1.59^{19}	....	al	B5³, 864
15013	2,3-Xylidine or 2,3-Dimethyl aniline $2,3\text{-}(CH_3)_2C_6H_3NH_2$	121.18	....	221-2, 106^{15}	<−15	0.9931^{20}	1.5684^{20}	al, eth	B12³, 2438
15014	2,4-Xylidine or 2,4-Dimethyl aniline $2,4\text{-}(CH_3)_2C_6H_3NH_2$	121.18	....	214, 91^{10}	−14.3	$0.9723^{20/4}$	1.5569^{20}	al, eth, bz	B12³, 2469
15015	2,5-Xylidine $2,5\text{-}(CH_3)_2C_6H_3NH_2$	121.18	ye lf (lig)	214, $92\text{-}100^{10}$	15.5	$0.9790^{21/4}$	1.5591^{21}	eth	B12³, 2503
15016	2,6-Xylidine or 2,6-Dimethyl aniline $2,6\text{-}(CH_3)_2C_6H_3NH_2$	121.18	....	214^{739}	11.2	0.9842^{20}	1.5610^{20}	al, eth	B12³, 2462
15017	3,4-Xylidine or 3,4-Dimethyl aniline $3,4\text{-}(CH_3)_2C_6H_3NH_2$	121.18	pl or pr (lig)	228	51	1.076^{18}	....	eth, lig	B12³, 2443
15018	3,5-Xylidine or 3,5-Dimethyl aniline $3,5\text{-}(CH_3)_2C_6H_3NH_2$	121.18	....	220-1, $99\text{-}100^{20}$	9.8	$0.9706^{20/4}$	1.5581^{20}	eth	B12³, 2495
15019	Xylitol $C_5H_{12}O_5$, $HOCH(CHOH)_3CH_2OH$	152.15	(i) rh (al), (ii) mcl (al) (st)	(ii) 215-7	(i) 61, (ii) 93-4	....	....	w, al, Py	B1⁴, 2832
15020	D-Xylose or Wood sugar $C_5H_{10}O_5$	150.13	mcl nd, $[\alpha]^{20}{}_D$ +22.5 (chl)	....	145	$1.525^{20/4}$	....	w	B1⁴, 4223
15021	D-Xylose osazone $C_{17}H_{20}N_4O_3$ $HOCH_2(CHOH)_2C(=NNHC_6H_5)CH=NNHC_6H_5$	328.37	pa ye nd, $[\alpha]_D$ −40.9 (al)	....	159(167 d)	....	....	al, eth, ace	B31, 61

No.	Name, Synonyms, and Formula	Mol. wt.	Color, crystalline form, specific rotation and λ_{max} (log ε)	b.p. °C	m.p. °C	Density	n_D	Solubility	Ref.
15022	Yamogenin $C_{27}H_{42}O_3$	414.63	pl, $[\alpha]^{25}{}_D$ −123		201				B19[4], 865
15023	Yobyrine $C_{19}H_{16}N_2$	272.35	nd(dil al)	150[0.01]	218-9			al, chl	B23[2], 263
15024	Yohimbine or Corynine $C_{21}H_{26}N_2O_3$	354.45	nd (dil al)	159[0.01] sub	241			al, eth, chl	B25[2], 201
15025	Yohimbine hydrochloride $C_{21}H_{26}N_2O_3.HCl$	390.91	orh nd or pl (w, dil HCl)		302			w	B25[2], 204
15025	Yohimbine-hydrochloride $C_{21}H_{26}N_2O_3.HCl$	390.91	orh nd or pl (w, dil HCl)		302			w	B25[2], 201
15026	Yohimbine-nitrate $C_{21}H_{26}N_2O_3.HNO_3$	417.46	pr (w)		276			w	B25[2], 204
15027	Yuccagenin $C_{27}H_{42}O_4$	430.63	nd (al), $[\alpha]^{25}{}_D$ −113		252			al, eth, diox	B19[4], 1083

No.	Name, Synonyms, and Formula	Mol. wt.	Color, crystalline form, specific rotation and λ_{max} (log ε)	b.p. °C	m.p. °C	Density	n_D	Solubility	Ref.
15028	Zagadinine $C_{27}H_{43}NO_7$	493.64	orh (al), nd (bz), [α] −45 (chl)		201-4			al, chl	Am35, 258
15029	Zeaxanthin or Zeaxanthol $C_{40}H_{56}O_2$	568.88	ye pr (MeOH), rh (chl-eth)	$226\text{-}9^{0.06}$	215.5			eth, ace, bz, chl, Py	B6³, 5865
15030	Zymosterol or 8(14).24(25′) cholestadienol $C_{27}H_{44}O$	384.65	pl (MeOH), nd, $[\alpha]^{20}{}_D$ + 149 (chl)	$160^{0.001}$ sub	110			ace, chl	B6³, 2828
15031	Zeaxanthin diacetate $C_{44}H_{60}O_4$	652.96			154-5				B6³, 5868

STRUCTURAL FORMULAS OF ORGANIC COMPOUNDS

In Numeric as they occur in Organic Compounds Table

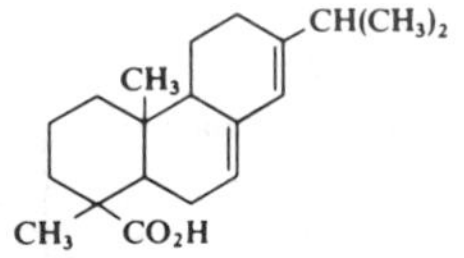

$CH(CH_3)_2$ CH_3 CH_3 CO_2H

1

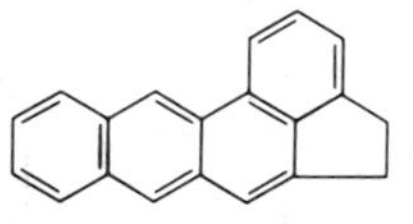

3

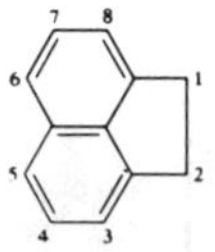

7 8 6 1 5 2 4 3

4

HO_2C

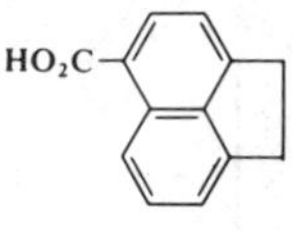

16

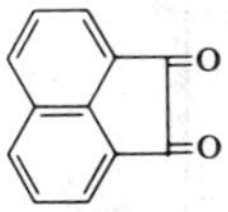

O O

17

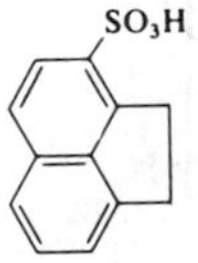

SO_3H

18

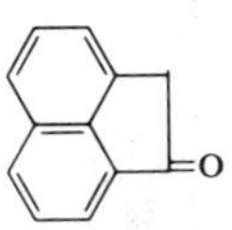

O

19

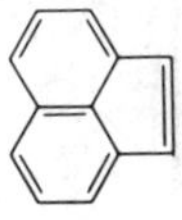

20

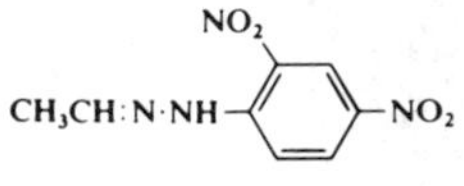

NO_2 $CH_3CH{:}N{\cdot}NH$ NO_2

22

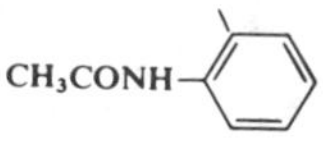

CH_3CONH

90

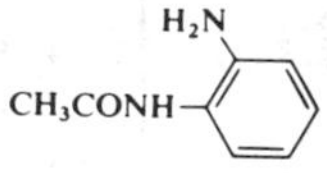

H_2N CH_3CONH

92

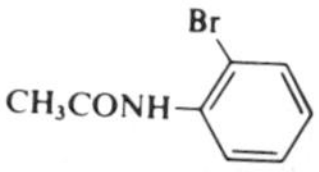

Br CH_3CONH

96

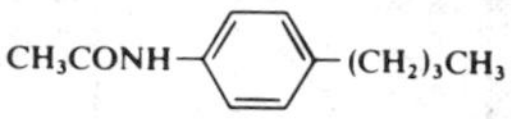

CH_3CONH $(CH_2)_3CH_3$

100

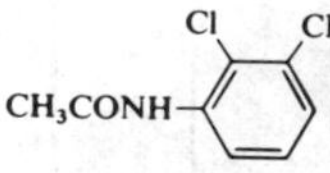

Cl CH_3 CH_3CONH

103

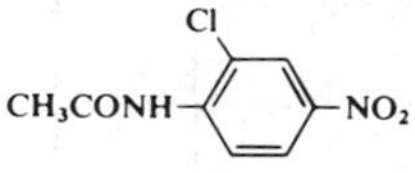

Cl CH_3CONH NO_2

104

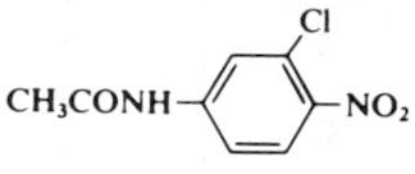

Cl CH_3CONH NO_2

111

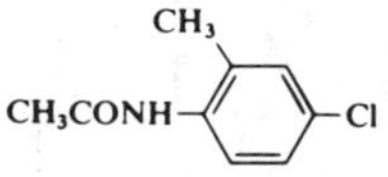

CH_3 CH_3CONH Cl

113

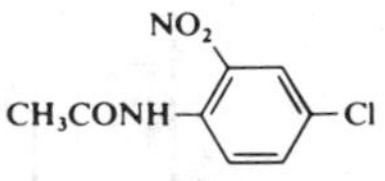

NO_2 CH_3CONH Cl

115

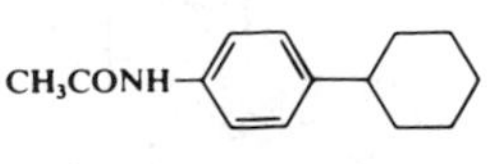

CH_3CONH

118

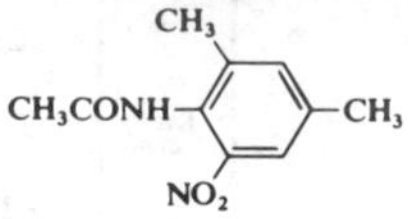

CH_3 CH_3CONH CH_3 NO_2

120

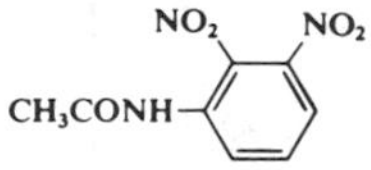

NO_2 NO_2 CH_3CONH

121

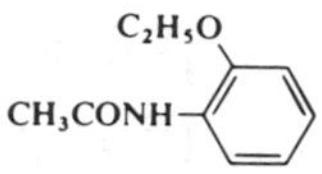

C_2H_5O CH_3CONH

128

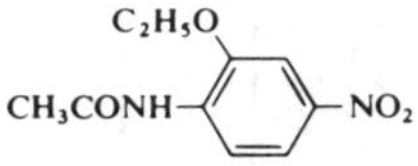

C_2H_5O CH_3CONH NO_2

129

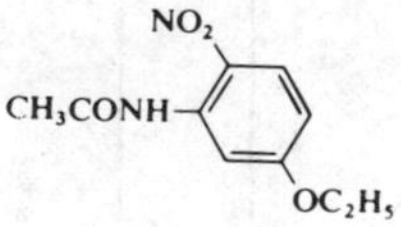

NO_2 CH_3CONH OC_2H_5

135

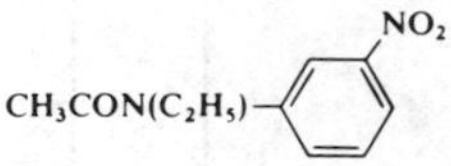

NO_2 $CH_3CON(C_2H_5)$

137

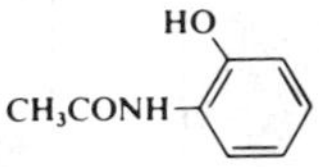

HO CH_3CONH

139

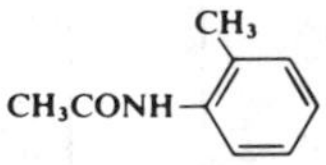

CH_3 CH_3CONH

143

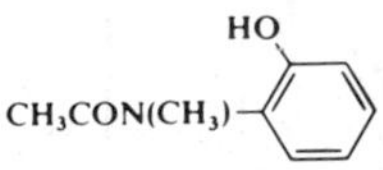

HO $CH_3CON(CH_3)$

146

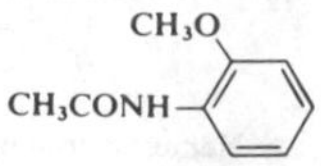

CH_3O CH_3CONH

148

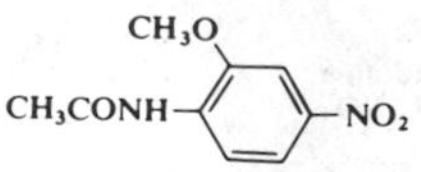

CH_3O CH_3CONH NO_2

150

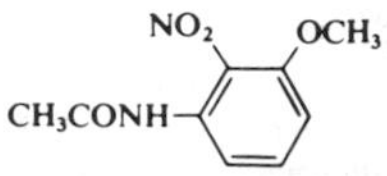

NO_2 OCH_3 CH_3CONH

154

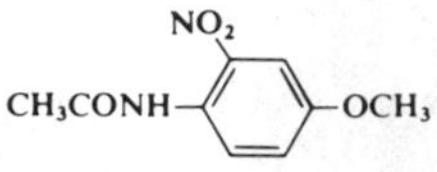

NO_2 CH_3CONH OCH_3

158

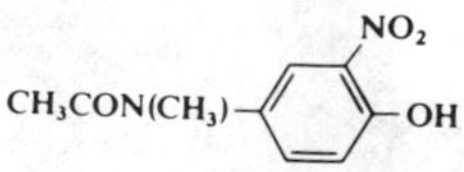

NO_2 $CH_3CON(CH_3)$ OH

162

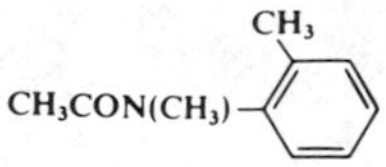

CH_3 $CH_3CON(CH_3)$

164

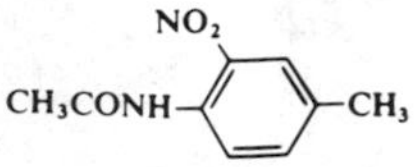

NO_2 CH_3CONH CH_3

167

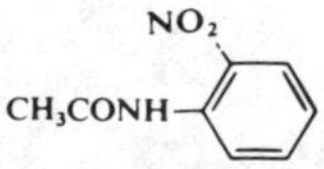

NO_2 CH_3CONH

168

CH_3CONH $(CH_2)_7CH_3$

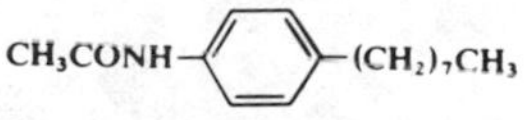

171

STRUCTURAL FORMULAS OF ORGANIC COMPOUNDS (Continued)

In Numeric as they occur in Organic Compounds Table

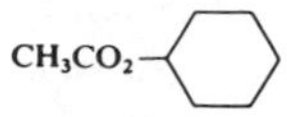

CH_3CO_2

196

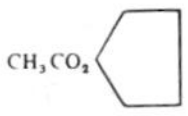

CH_3CO_2

197

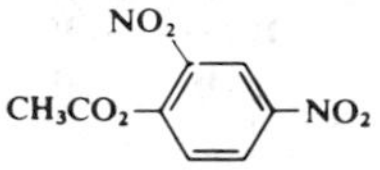

NO_2
CH_3CO_2 NO_2

201

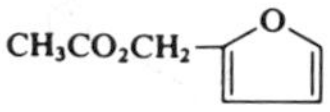

$CH_3CO_2CH_2$ O

210

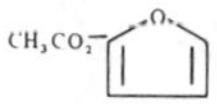

CH_3CO_2 O

211

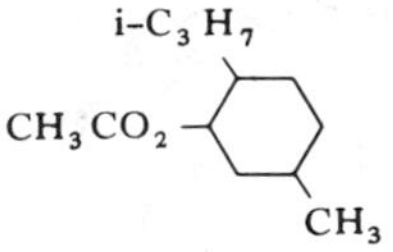

i-C_3H_7
CH_3CO_2
CH_3

221

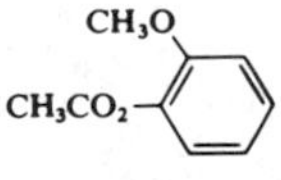

CH_3O
CH_3CO_2

224

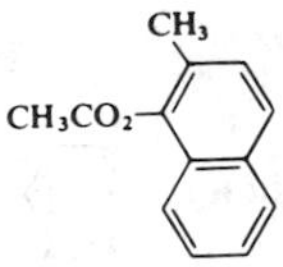

CH_3
CH_3CO_2

233

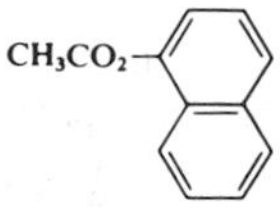

CH_3CO_2

241

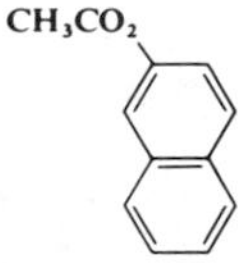

CH_3CO_2

242

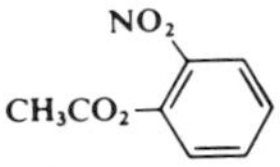

NO_2
CH_3CO_2

243

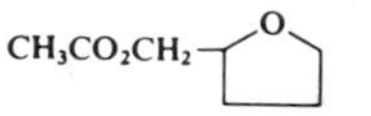

$CH_3CO_2CH_2$ O

264

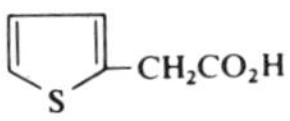

S CH_2CO_2H

265

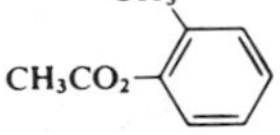

CH_3
CH_3CO_2

266

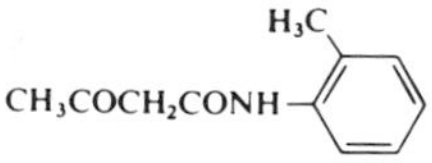

H_3C
CH_3COCH_2CONH

272

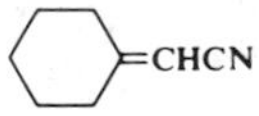

=CHCN

343

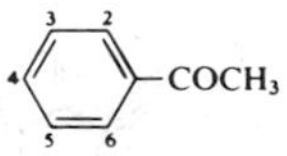

3 2
4 $COCH_3$
5 6

351

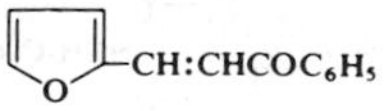

O CH:CHCOC$_6$H$_5$

395

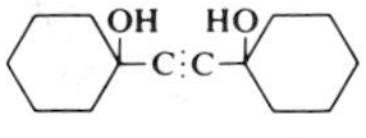

OH HO
C:C

450

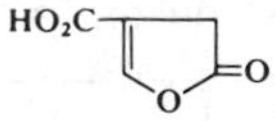

HO_2C
O O

452

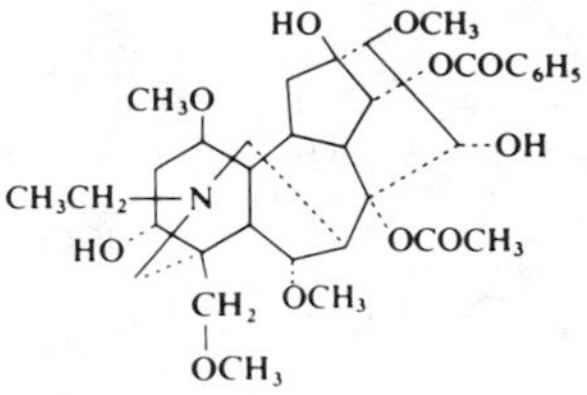

HO OCH_3
$OCOC_6H_5$
CH_3O
OH
CH_3CH_2 N
HO $OCOCH_3$
CH_2 OCH_3
OCH_3

460

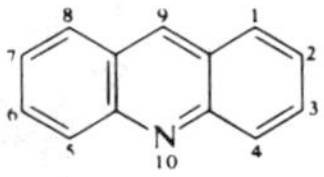

8 9 1
7 2
6 3
5 N 4
10

464

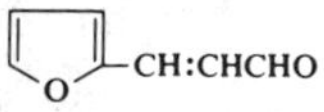

O CH:CHCHO

483

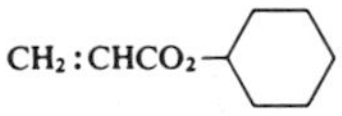

CH_2:$CHCO_2$

497

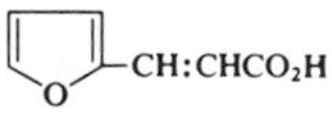

O CH:CHCO$_2$H

502

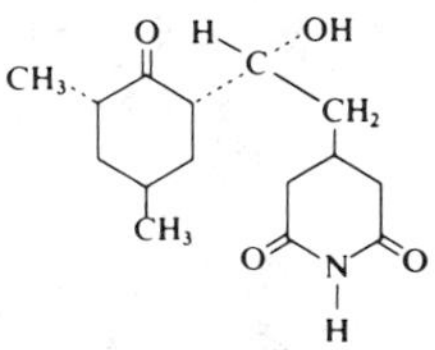

O H OH
CH_3 C
CH_2
CH_3
O N O
H

521

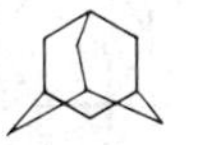

522

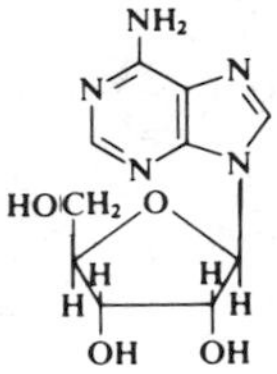

NH_2
N N
N N
HOCH$_2$ O
H H
H H
OH OH

523

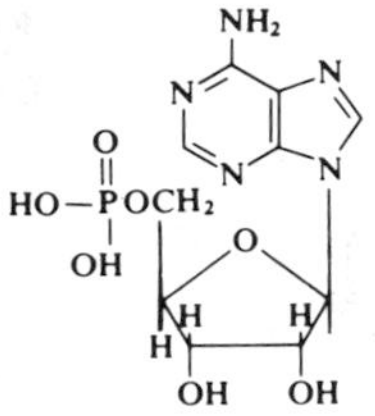

NH_2
N N
O
HO—P OCH$_2$ N N
OH O
H H
H H
OH OH

524

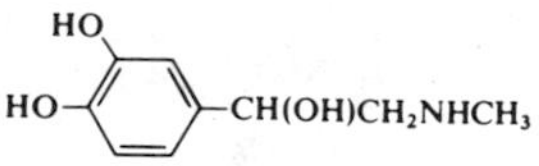

HO
HO CH(OH)CH$_2$NHCH$_3$

548

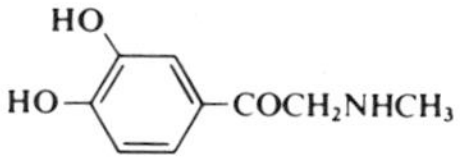

HO
HO COCH$_2$NHCH$_3$

550

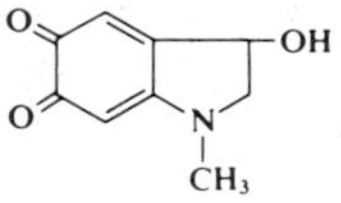

O OH
O N
CH_3

551

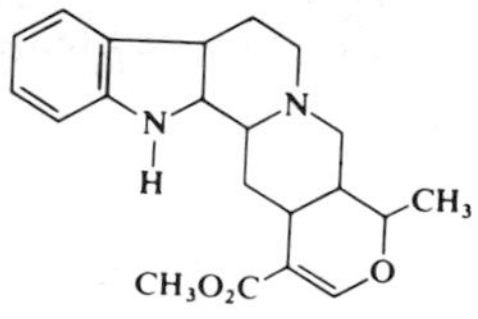

N
N
H
CH_3
CH_3O_2C O

552

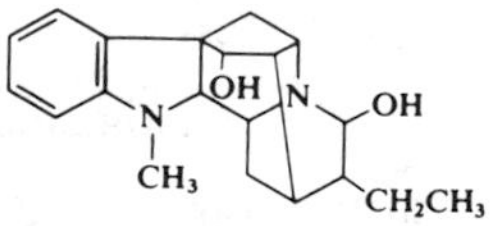

OH N OH
N
CH_3 CH_2CH_3

553

STRUCTURAL FORMULAS OF ORGANIC COMPOUNDS (Continued)

In Numeric as they occur in Organic Compounds Table

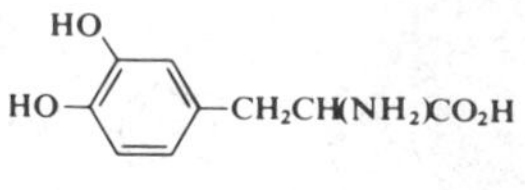

HO, HO, $CH_2CH(NH_2)CO_2H$

564

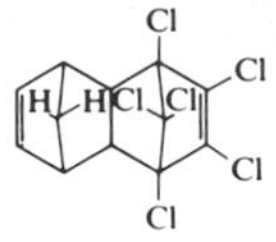

OH OH OH OH; $HOCH_2$—C—C—C—C—CHO; H H H H

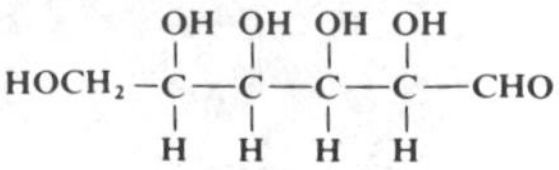

592

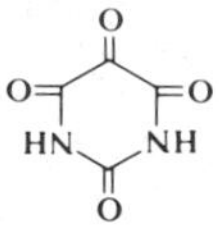

N, N^+, CH_3, CH_3O_2C, O

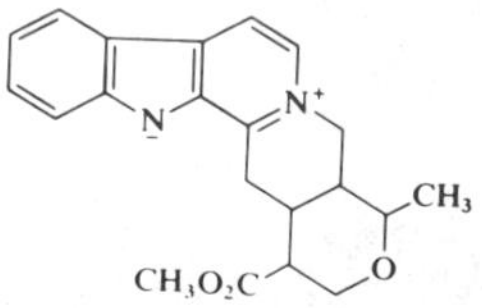

633

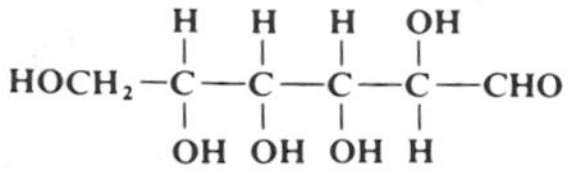

OH, H_2N, N, N, N, NH_2

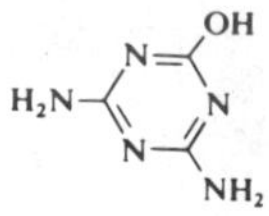

739

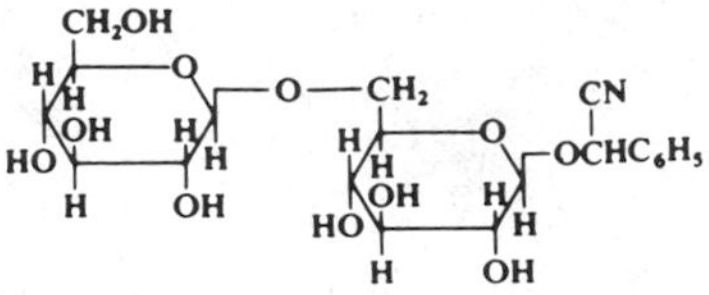

Cl, Cl, H, H, Cl, Cl, Cl, Cl

574

O, O, O, HN, NH, O

593

H H H OH; $HOCH_2$—C—C—C—C—CHO; OH OH OH H

634

CH_2OH, H, OH, HO, H, OH, O, CH_2, CN, $OCHC_6H_5$, H, OH, HO, H, OH

744

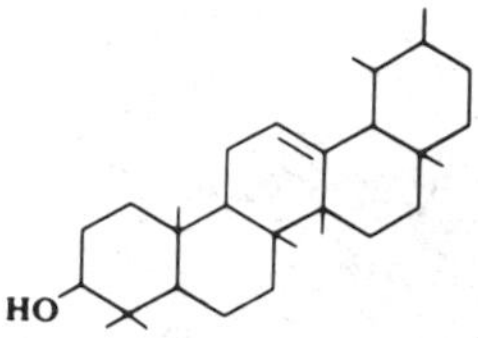

O, OH, OH, O

575

O, NH, OH, HN, CO_2H, O

594

CH_3, N, O, OH, O, N, CH_3, O, O, N, O, HO, O, N, CH_3, CH_3

636

HO

746

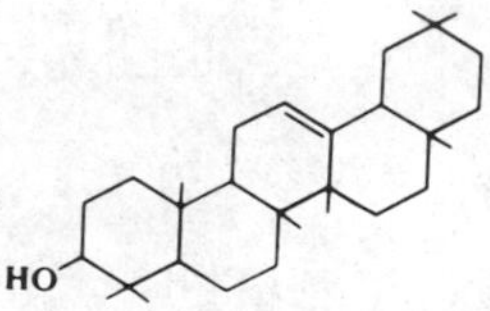

OH O, $CH(OH)CH_2CH{:}C(CH_3)_2$, OH O

576

H, N, O, OH, O, N, H, O, HO, O, N, H

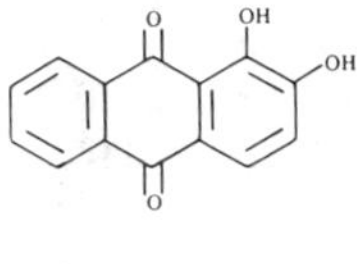

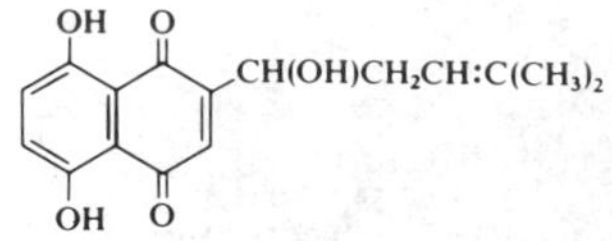

595

C_6H_5, H, N, C_6H_5, C_6H_5, N

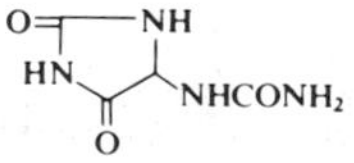

637

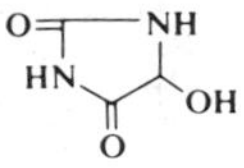

HO

747

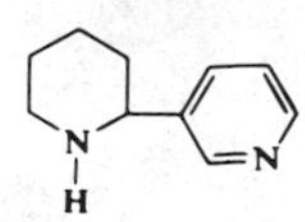

O, NH, HN, $NHCONH_2$, O

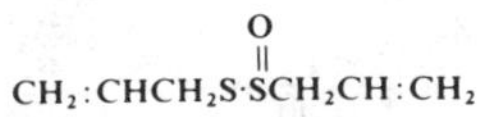

578

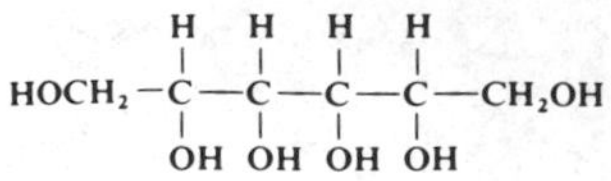

CH_3O, HO, $CH{:}CHCH_2OH$, CH_3O

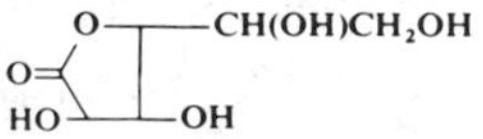

600

C_6H_5, N, C_6H_5, C_6H_5, N, C_6H_5

639

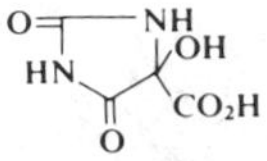

N, H, N

748

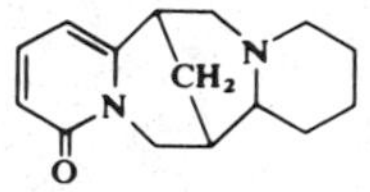

O, NH, HN, OH, O

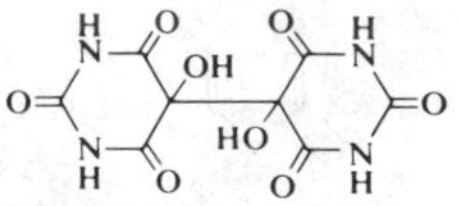

579

3, 2, 4, $NHCH_2CO_2H$, 5, 6

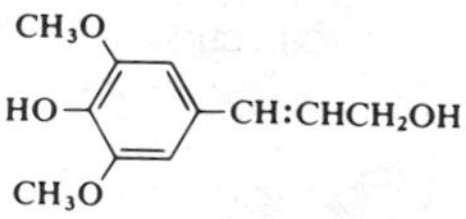

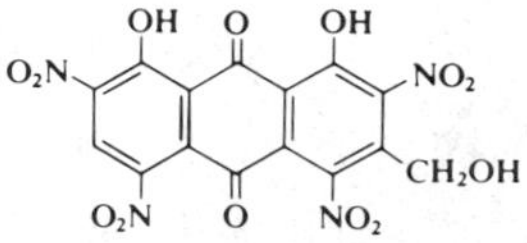

670

N, CH_2, N, O

749

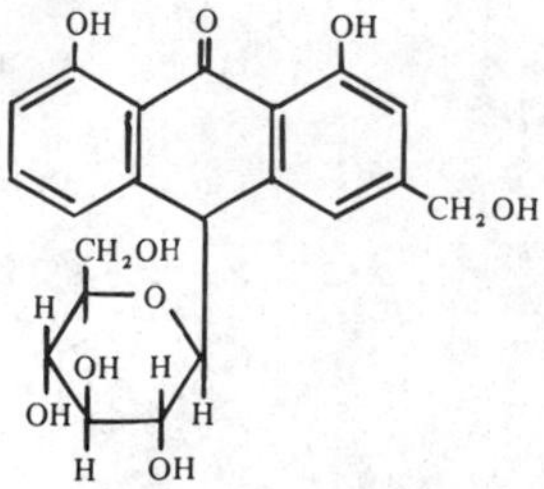

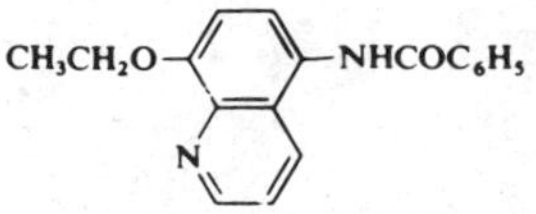

OH O OH, O_2N, NO_2, CH_2OH, O_2N, O, NO_2

631

O, NCH_2CO_2H, O

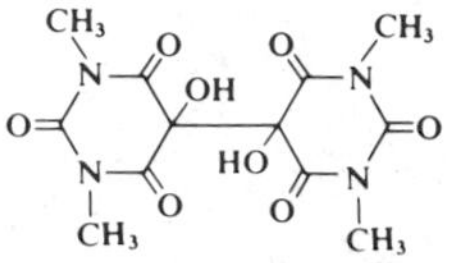

673

CH_3CH_2O, $NHCOC_6H_5$, N

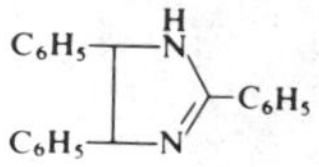

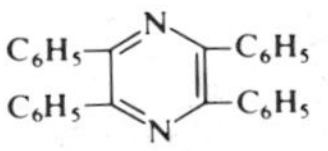

750

O

$CH_2{:}CHCH_2S{\cdot}SCH_2CH{:}CH_2$

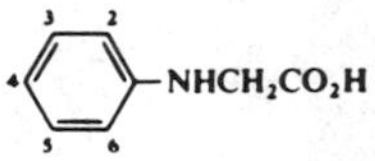

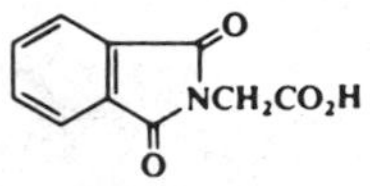

582

H H H H; $HOCH_2$—C—C—C—C—CH_2OH; OH OH OH OH

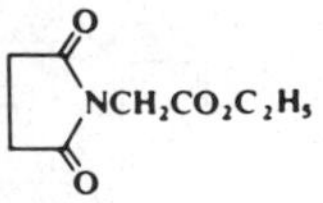

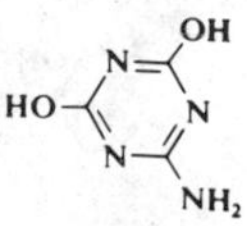

583

OH O OH, CH_2OH, CH_2OH, H, O, OH, H, OH, H, H, OH

632

O, $NCH_2CO_2C_2H_5$, O

675

O, $CH(OH)CH_2OH$, O, HO, OH

586

OH, HO, N, N, N, NH_2

738

N, H, N

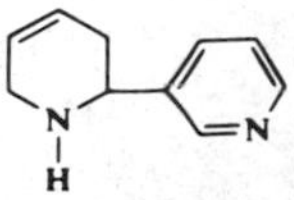

751

In Numeric as they occur in Organic Compounds Table

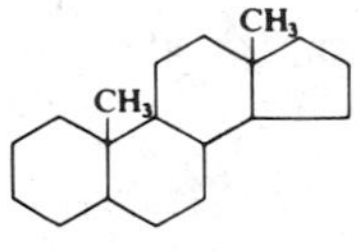

752

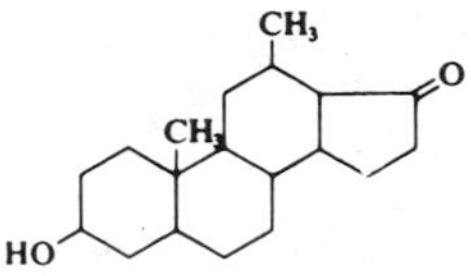

753

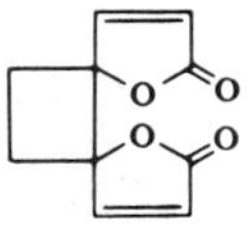

754

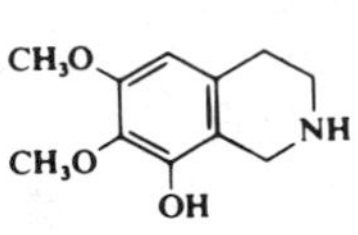

757

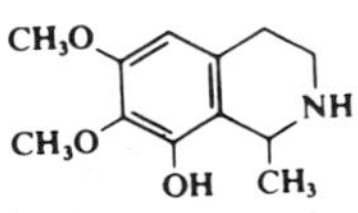

758

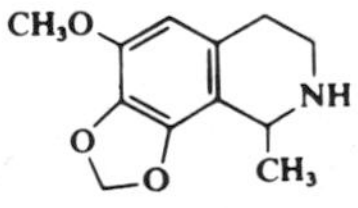

759

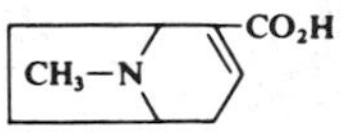

763

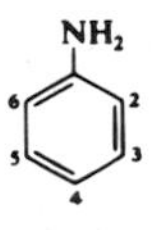

765

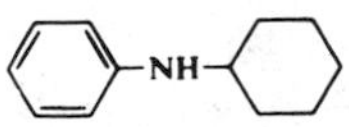

822

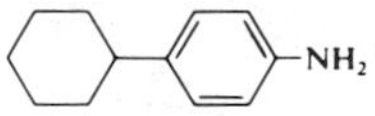

823

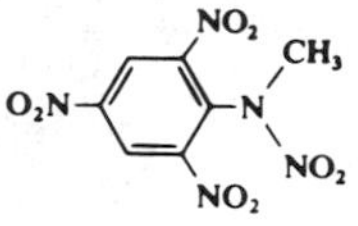

946

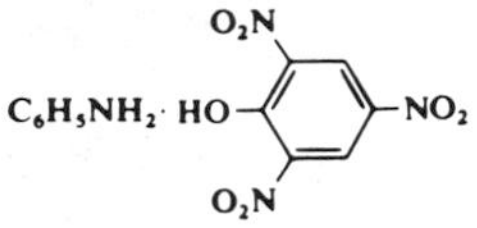

970

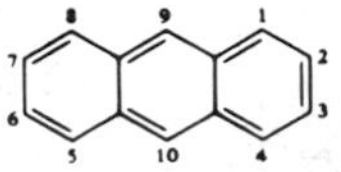

1096

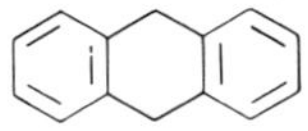

1111

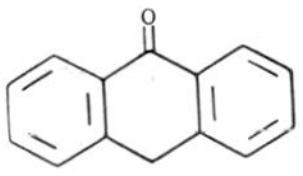

1117

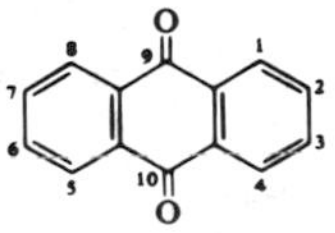

1195

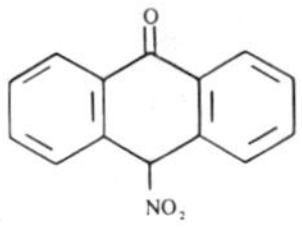

1118

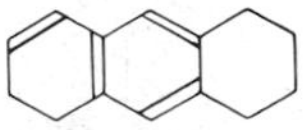

1127

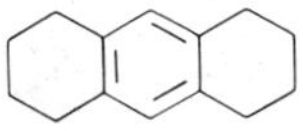

1135

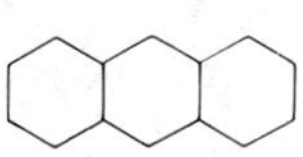

1138

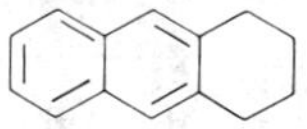

1139

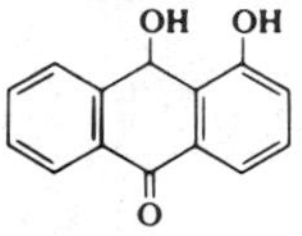

1146

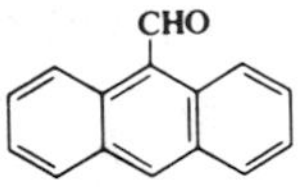

1150

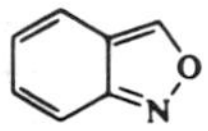

1151

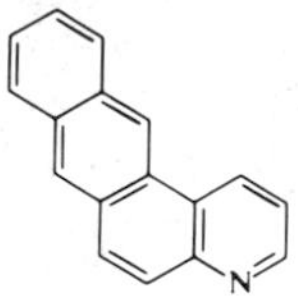

1194

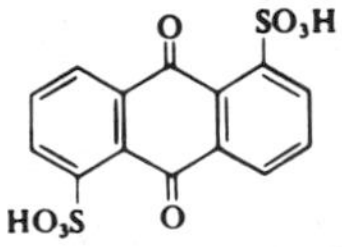

1273

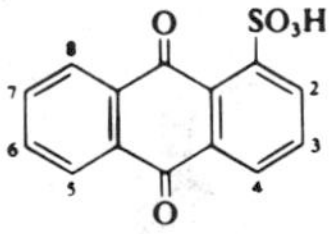

1292

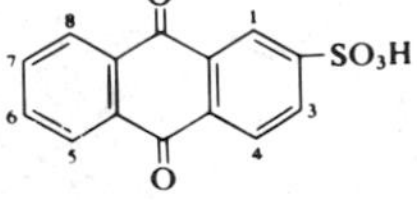

1294

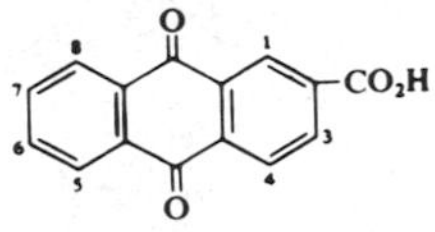

1313

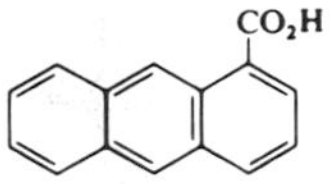

1321

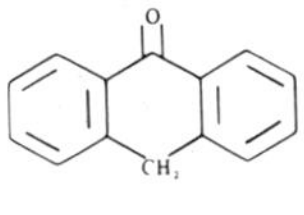

1324

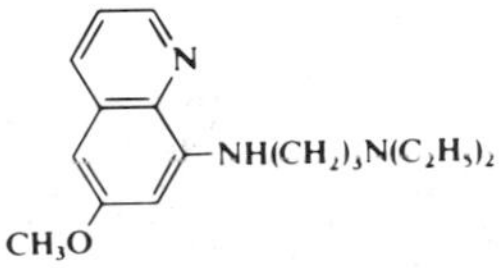

1326

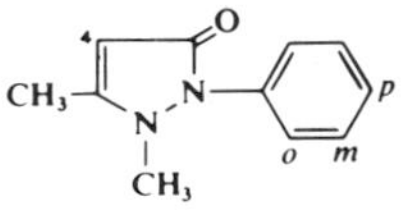

1327

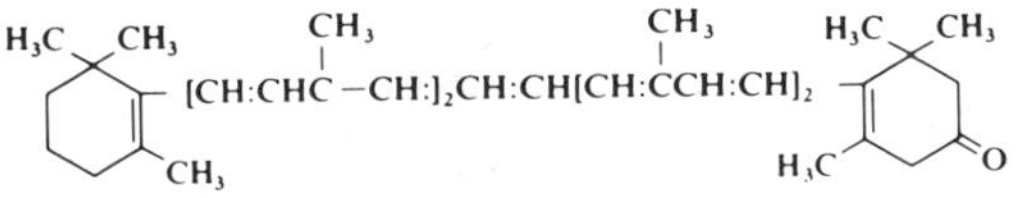

1336

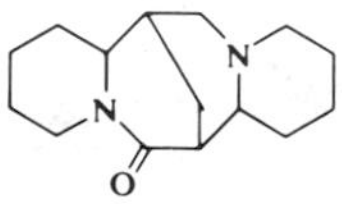

1337

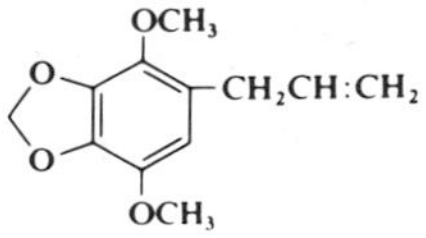

1338

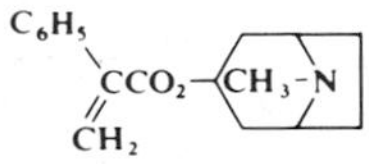

1339

STRUCTURAL FORMULAS OF ORGANIC COMPOUNDS (Continued)

In Numeric as they occur in Organic Compounds Table

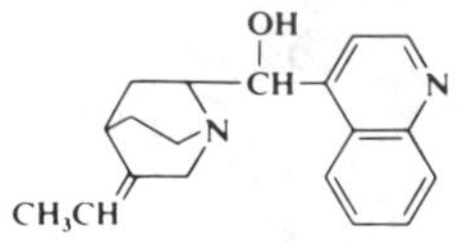

1342: OH, CH, N, N, CH_3CH

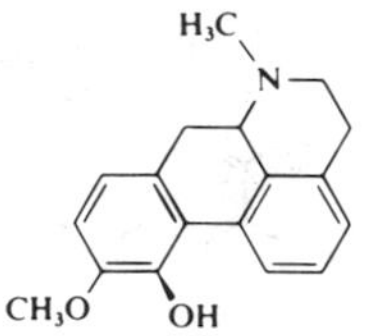

1344: H_3C, N, CH_3O, OH

1347

1350: CH_3, N, HO, OH

1353: OH, CH, N, N, CH_3CH, HO

1355: C_6H_5, N, N, O

1356: O; CH_2—C(H)(OH)—C(H)(OH)—C(OH)(H)—C(H)(OH)—

1359: O; CH_2—C(H)(OH)—C(H)(OH)—C(OH)(H)—C(OH)(H)—

1363: $HOCH_2C(OH)(H)—C(OH)(H)—C(H)(OH)—CH{:}NN(C_6H_5)_2$

1366: $HOCH_2—C(OH)(H)—C(OH)(H)—C(H)(OH)—CH_2OH$

1369: $HOCH_2—C(OH)(H)—C(OH)(H)—C(H)(OH)—CO_2H$

1375: $(CH_3)_3C$–C_6H_4–$OCH_2CH(CH_3)OSO_2CH_2CH_2Cl$

1376: $HOCH_2$—C(H)—C(H)(OH)—C(OH)(H)—C(H)(OH)—C(H)(O–C_6H_4–OH)—, ring O

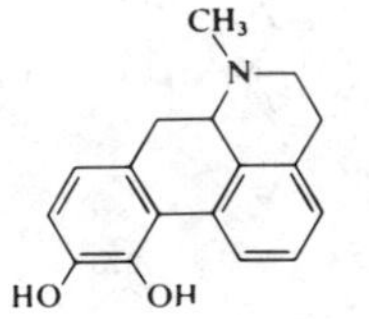

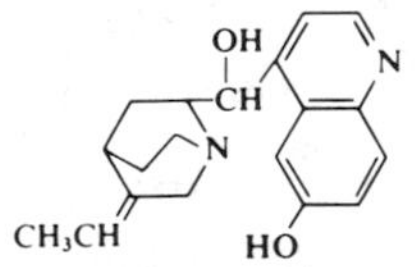

1378: HO_2C, N—CH_3

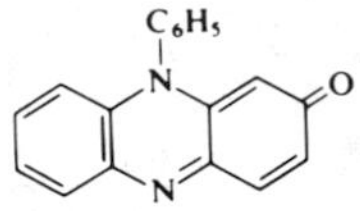

1379: CH_3O_2C, N—CH_3

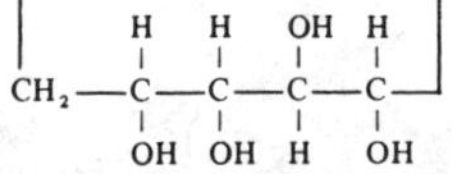

1394: As, As, As, As, As, As

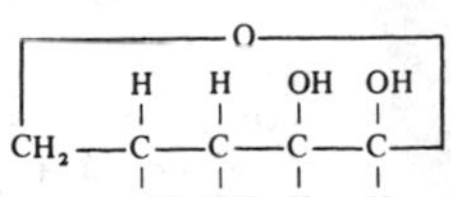

1410: CH_3O–C_6H_4–AsO

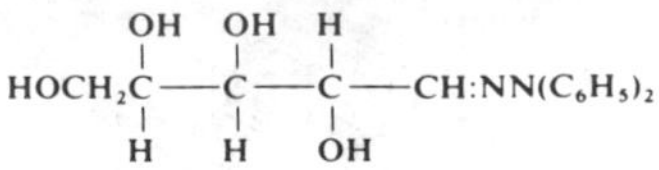

1422: CH_3, OH, CH_3, HO_2CCH, CH_3

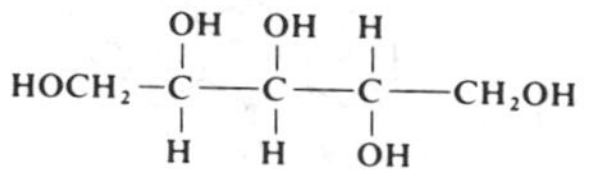

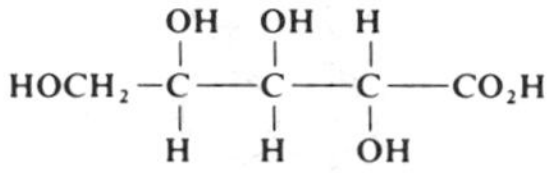

1423: H_3C, O, O, CH, CH_3, CH_3

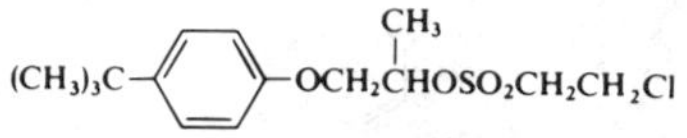

1424: O; $HOCH_2—C(H)(OH)—C(H)—C(OH)=C(OH)—C=O$

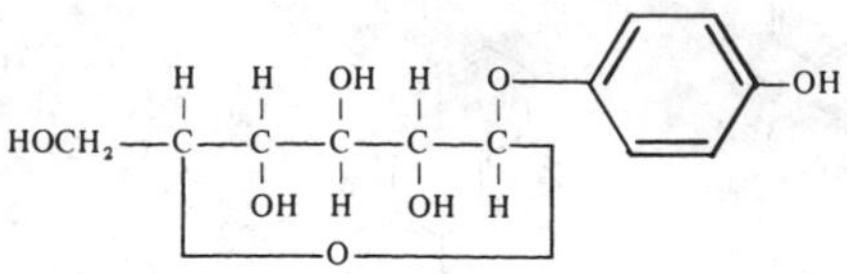

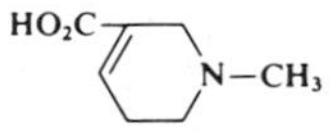

1427: $H_3C—C(OH)(H)—C(H)—C(OH)=C(OH)—C=O$, O

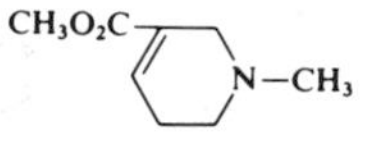

1435: N, CH_2CH_3, N, CH_3O, $COCH_3$

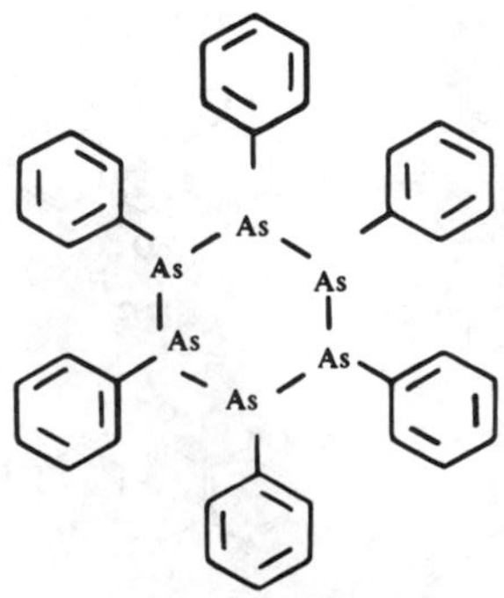

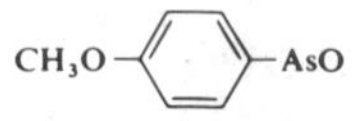

1436: N, Cl, $2HCl.2H_2O$, CH_3O, $NHCH(CH_3)\text{-}(CH_2)_3N(C_2H_5)_2$

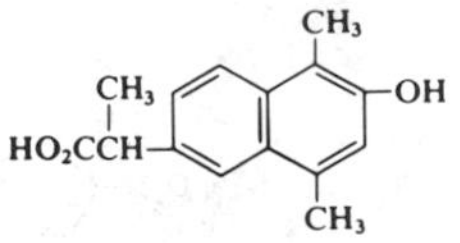

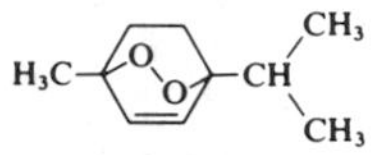

1437: O, N, CH_2, OH, CH_3

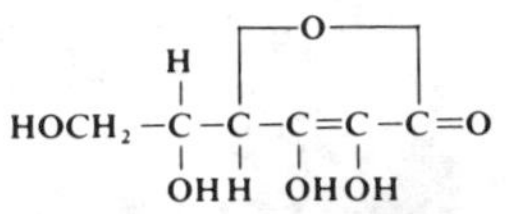

1441: $N—CH_3$, $OCOCH(C_6H_5)CH_2OH$

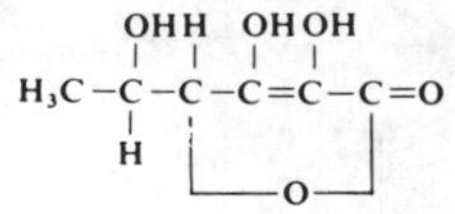

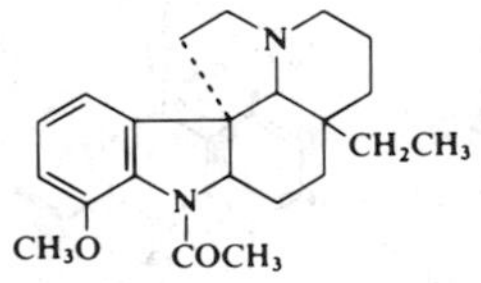

1445: $[(CH_3)_2N$–C_6H_4–$]_2C{:}NH$

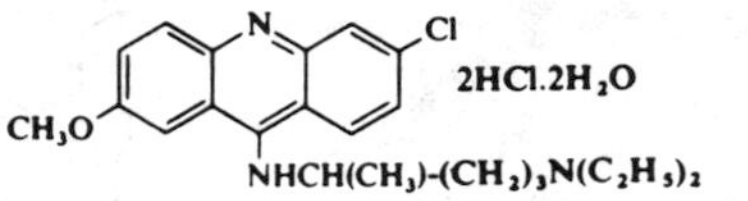

1448: O=, =C, $[$–C_6H_4–OH$]_2$

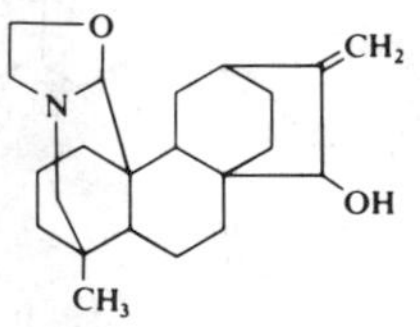

1449: Cl, CH_3, OH, $N(CH_3)_2$, OH, OH, $CONH_2$, OH, O, OH, O

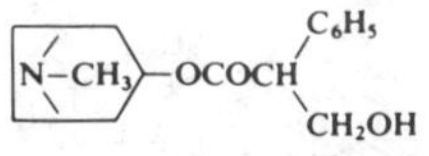

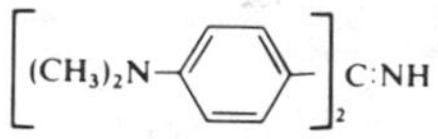

1451: OH, $CHCH_2COCH_2CO_2H$, $CH_3CH_2CH(CH_3)$, $CH(CH_3)CH_2CH_3$

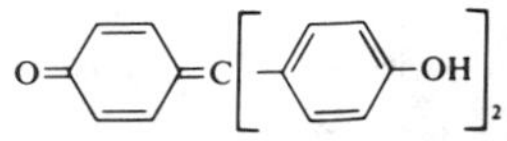

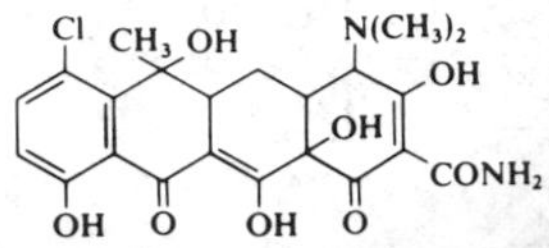

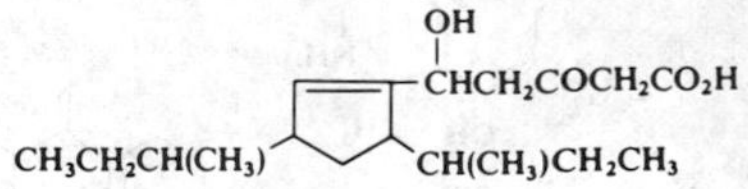

In Numeric as they occur in Organic Compounds Table

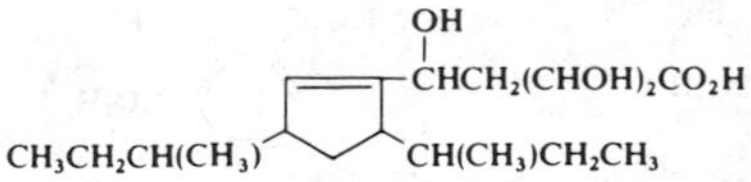

OH
CHCH$_2$(CHOH)$_2$CO$_2$H
CH$_3$CH$_2$CH(CH$_3$)
CH(CH$_3$)CH$_2$CH$_3$

1452

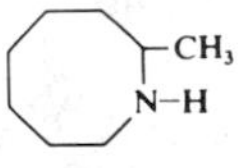

O$_2$N SO$_3$H HO N:N

1580

1603

CH$_3$ HO OCH$_3$ N CH$_2$ O CH$_2$ N CH$_3$O HO H$_3$C

1645

CH$_3$ N-H

1461

N:N SO$_3$H

1581

HN O O HN O

1608

CHO

1654

NH

1462

N:N

1583

HN O O HN O CH$_2$CH:CH$_2$

1610

NH$_2$ CH:NOH

1660

N H

1463

N:N

1586

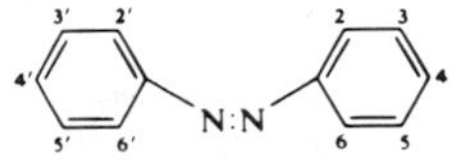

HN O O =CHC$_6$H$_5$ HN O

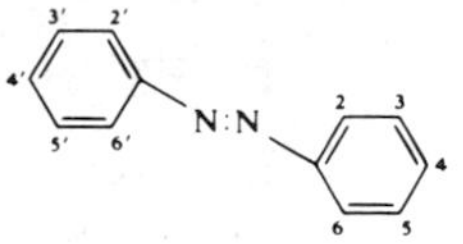

1616

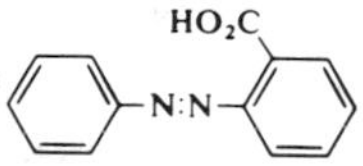

H$_2$N CH:NOH

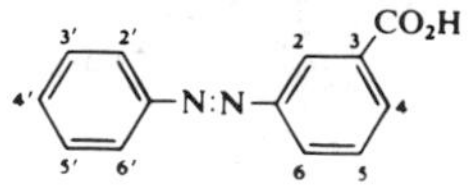

1662

N:N

1464

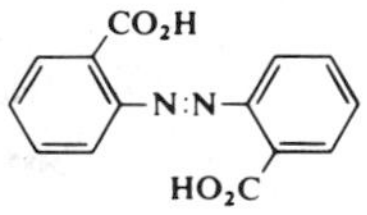

N:N

1588

HN O C$_2$H$_5$ O HN O

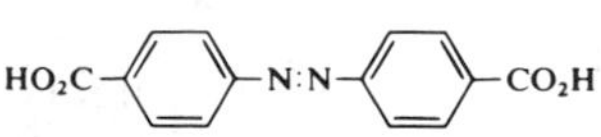

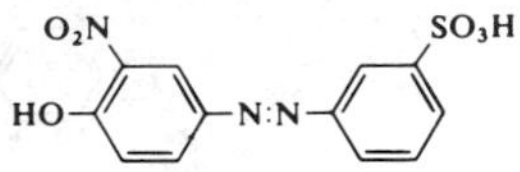

1619

Br CH(O$_2$CCH$_3$)$_2$

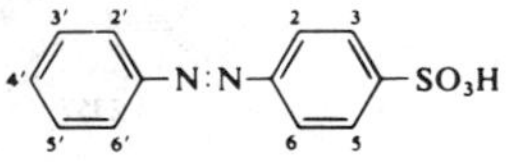

1667

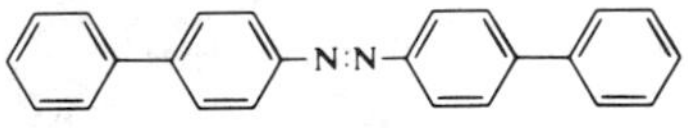

N:N

1465

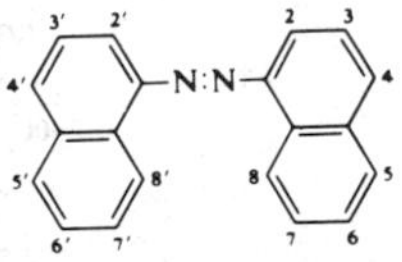

N:N

1589

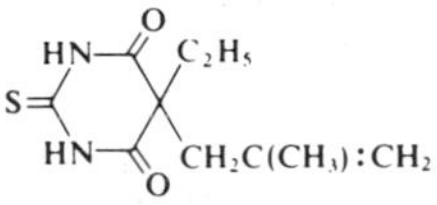

HN O C$_2$H$_5$ S HN O CH$_2$C(CH$_3$):CH$_2$

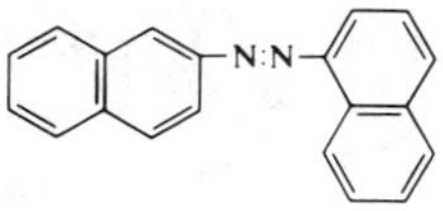

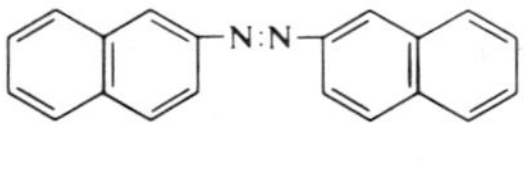

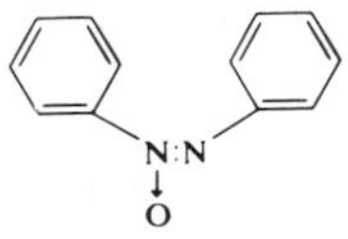

1626

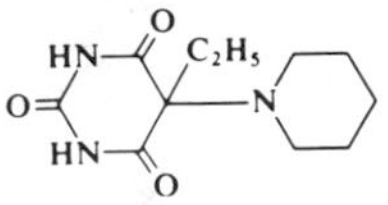

NO$_2$ CH(O$_2$CCH$_3$)$_2$

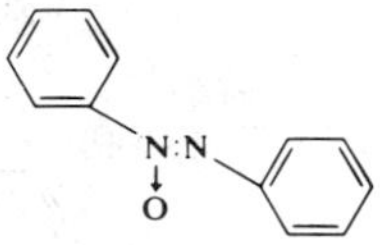

1784

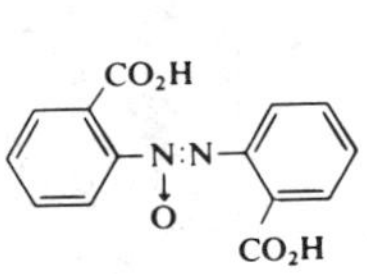

HO$_2$C N:N

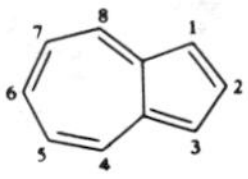

1559

N:N ↓ O

1590

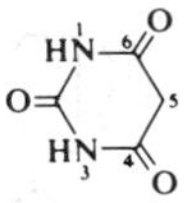

NO$_2$ CH(OCH$_3$)$_2$

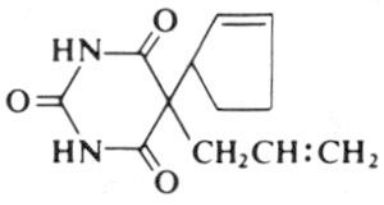

1785

CO$_2$H N:N

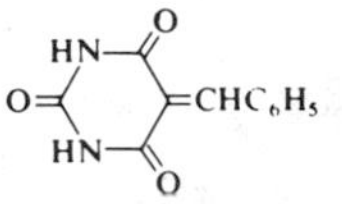

1561

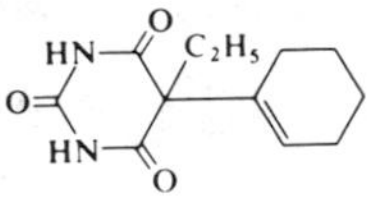

HN O C$_2$H$_5$ O N HN O

1635

O$_2$N CH(O$_2$CCH$_3$)$_2$

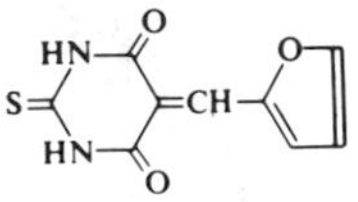

1787

CO$_2$H N:N HO$_2$C

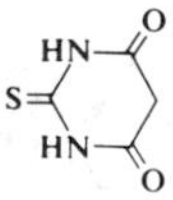

1568

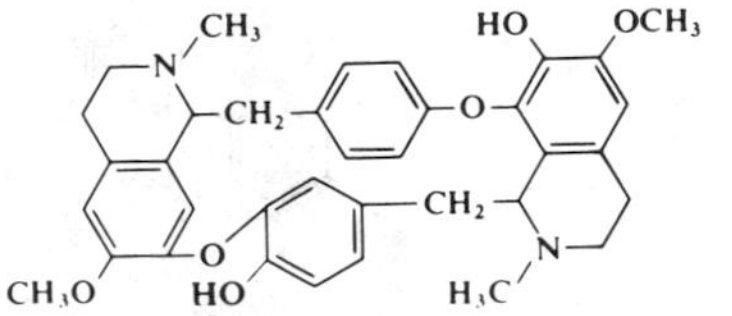

N:N ↓ O

1591

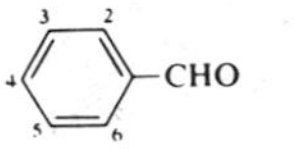

HN O S =CH HN O

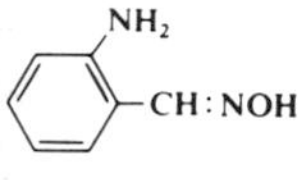

1638

O$_2$N CH(OCH$_3$)$_2$

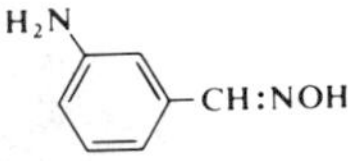

1788

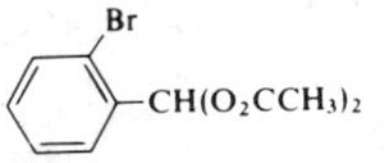

HO$_2$C N:N CO$_2$H

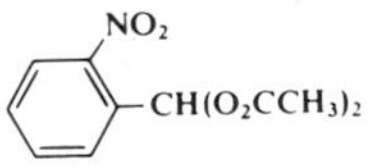

1574

CO$_2$H N:N ↓ O CO$_2$H

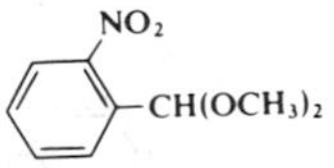

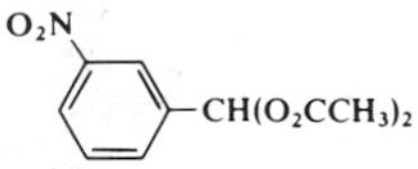

1598

HN O S HN O

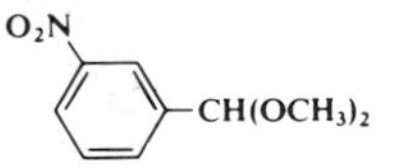

1644

CONH$_2$

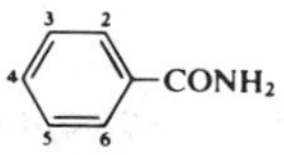

1809

STRUCTURAL FORMULAS OF ORGANIC COMPOUNDS (Continued)

In Numeric as they occur in Organic Compounds Table

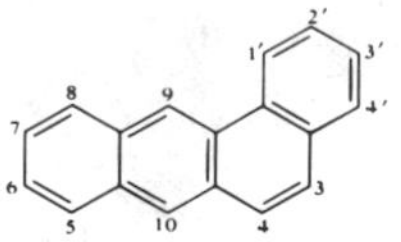

1858

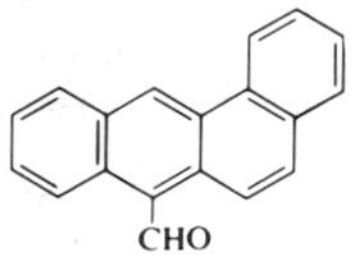

CHO

1870

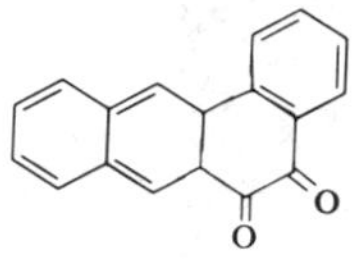

O O

1871

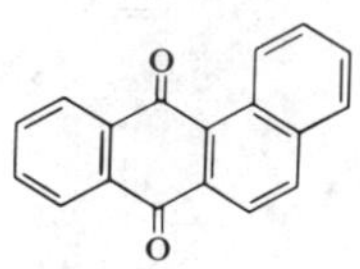

O O

1872

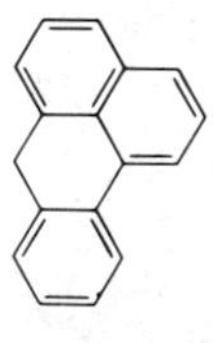

1873

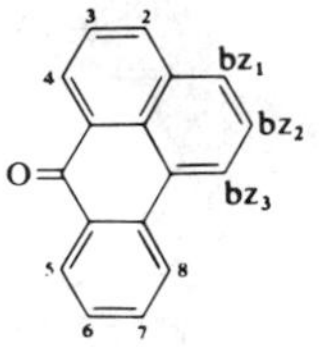

bz_1 bz_2 bz_3 O

1874

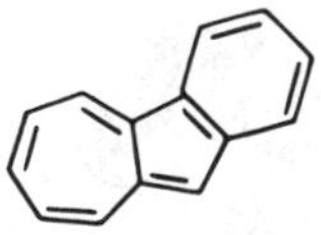

1879

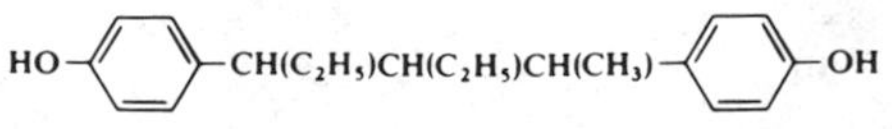

HO— —$CH(C_2H_5)CH(C_2H_5)CH(CH_3)$— —OH

1881

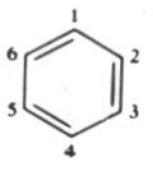

1882

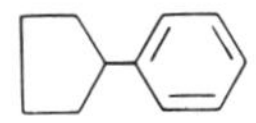

2058

2106

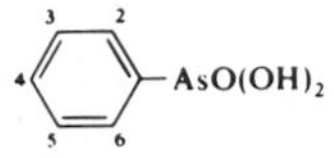

$AsO(OH)_2$

2298

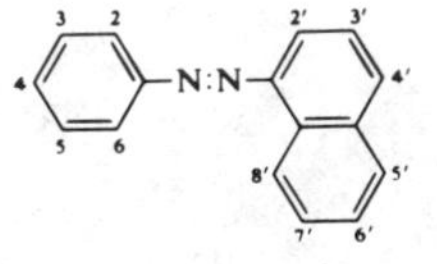

N:N

2314

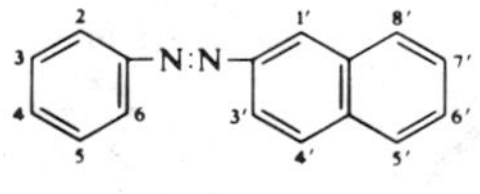

N:N

2326

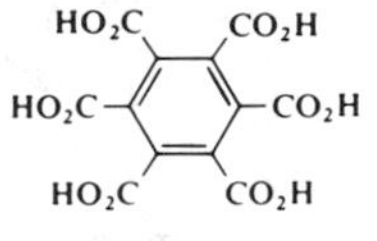

HO_2C CO_2H HO_2C CO_2H HO_2C CO_2H

2333

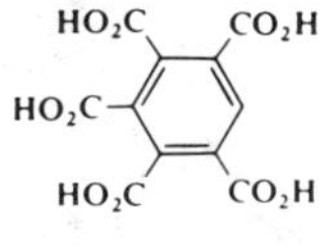

HO_2C CO_2H HO_2C HO_2C CO_2H

2334

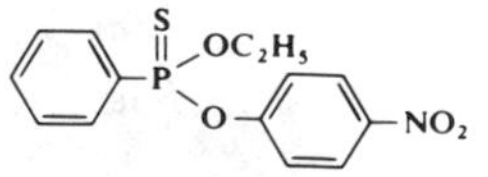

S P OC_2H_5 O NO_2

2340

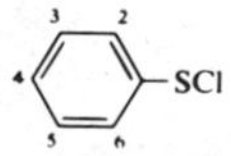

SCl

2344

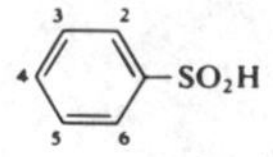

SO_2H

2348

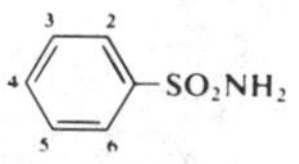

SO_2NH_2

2357

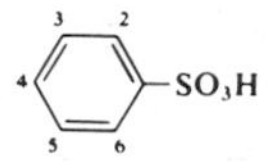

SO_3H

2378

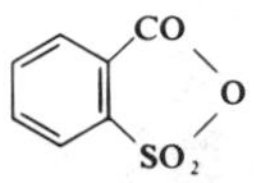

CO O SO_2

2400

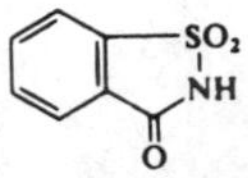

SO_2 NH O

2401

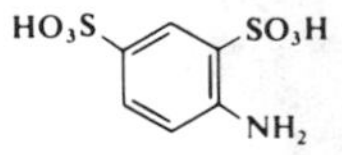

HO_3S SO_3H NH_2

2451

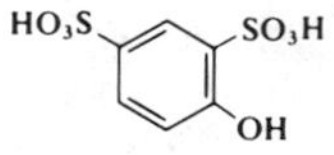

HO_3S SO_3H OH

2452

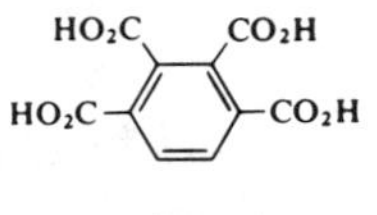

HO_2C CO_2H HO_2C CO_2H

2453

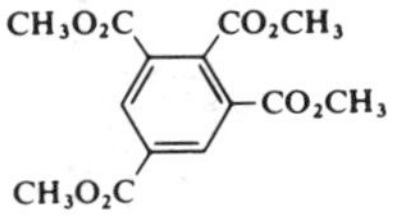

CH_3O_2C CO_2CH_3 CO_2CH_3 CH_3O_2C

2455

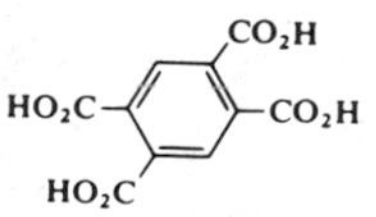

CO_2H HO_2C CO_2H HO_2C

2456

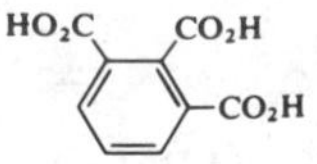

HO_2C CO_2H CO_2H

2459

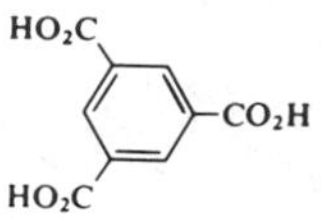

HO_2C CO_2H HO_2C

2462

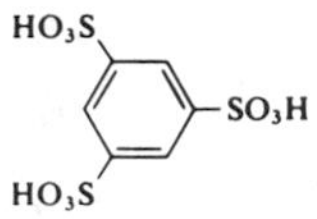

HO_3S SO_3H HO_3S

2467

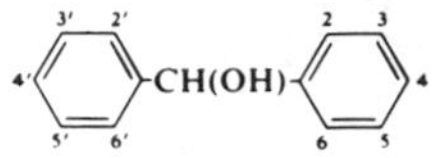

CH(OH)

2468

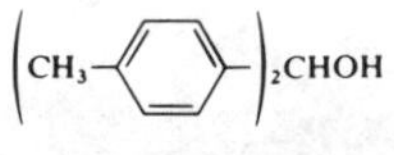

$(CH_3-C_6H_4-)_2CHOH$

2470

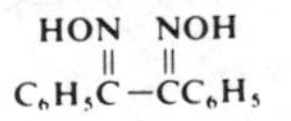

HON NOH

$C_6H_5C-CC_6H_5$

2483

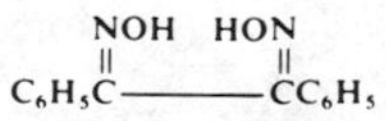

NOH HON

C_6H_5C——CC_6H_5

2484

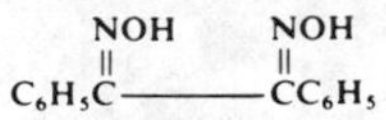

NOH NOH

C_6H_5C——CC_6H_5

2485

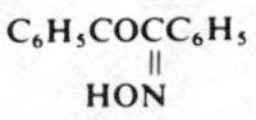

$C_6H_5COCC_6H_5$

HON

2486

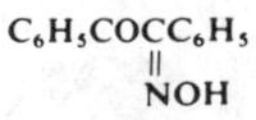

$C_6H_5COCC_6H_5$

NOH

2487

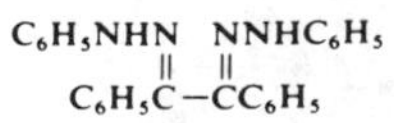

C_6H_5NHN $NNHC_6H_5$

$C_6H_5C-CC_6H_5$

2488

$NNHC_6H_5$ C_6H_5NHN

C_6H_5C——CC_6H_5

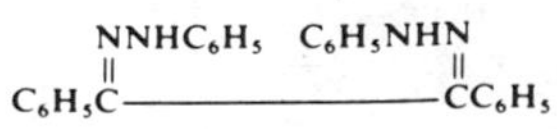

2489

In Numeric as they occur in Organic Compounds Table

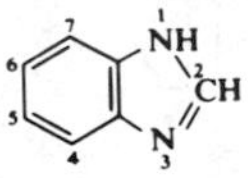

2493

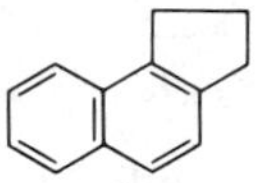

2506

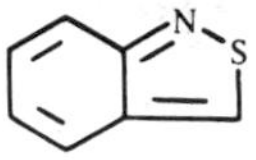

2507

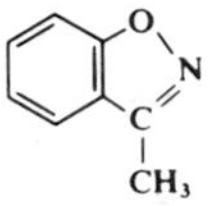

2509

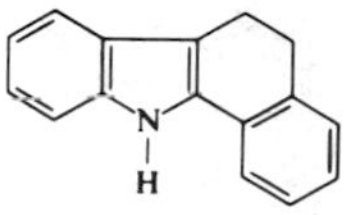

2510

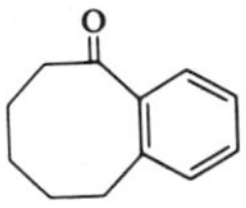

2511

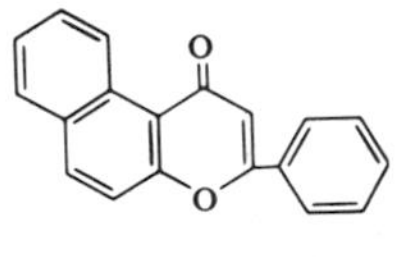

2513

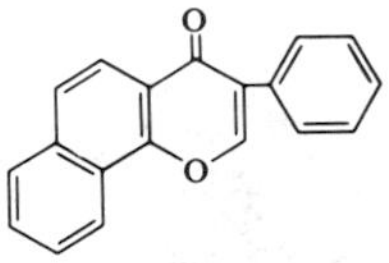

2514

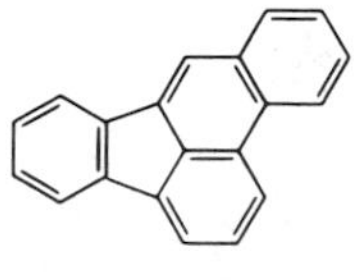

2515

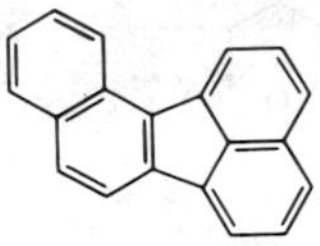

2516

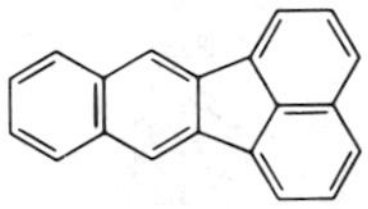

2517

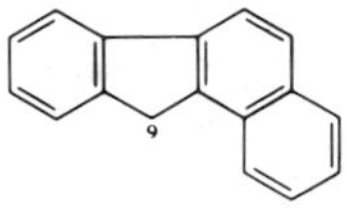

2518

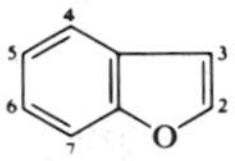

2520

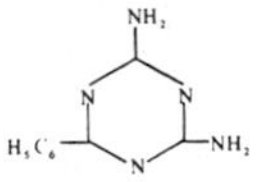

2528

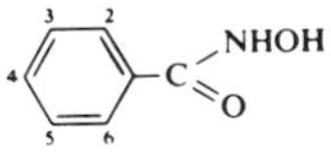

2529

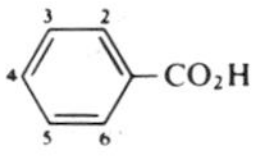

2531

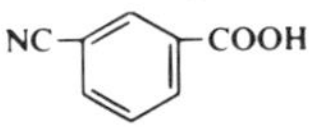

2625

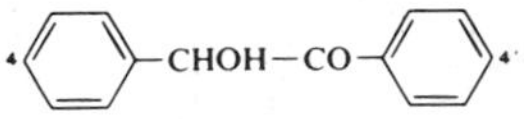

2837

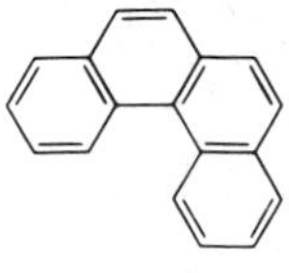

2879

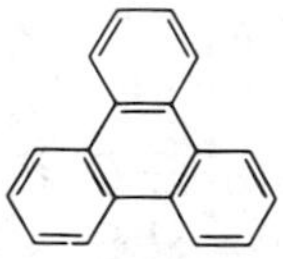

2880

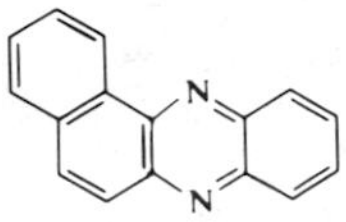

2881

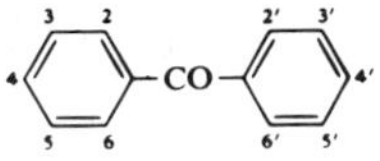

2882

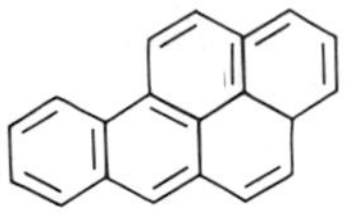

3009

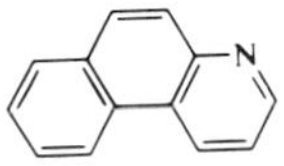

3013

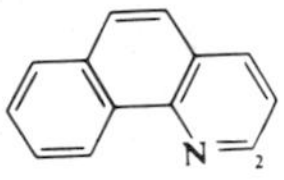

3014

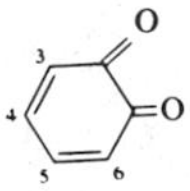

3016

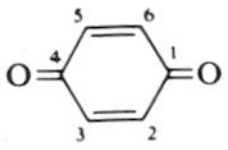

3023

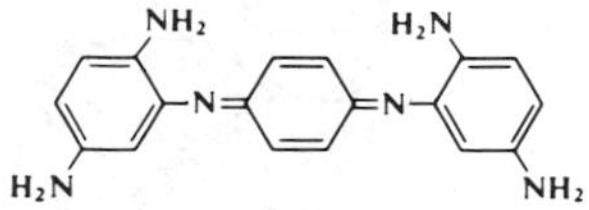

3031

3046

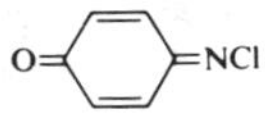

3048

3053

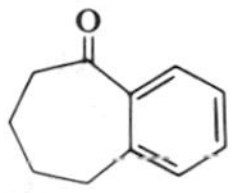

3069

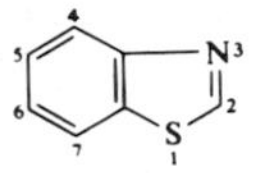

3078

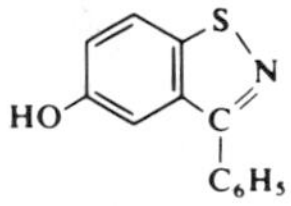

3094

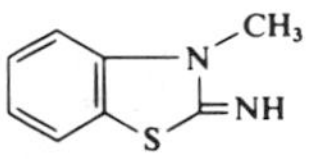

3108

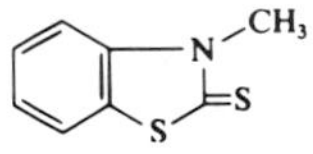

3109

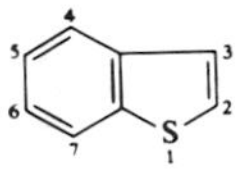

3110

STRUCTURAL FORMULAS OF ORGANIC COMPOUNDS (Continued)

In Numeric as they occur in Organic Compounds Table

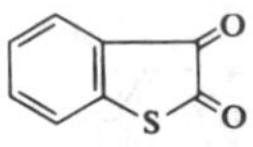

3114

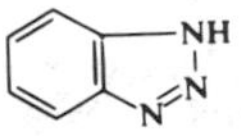

3314

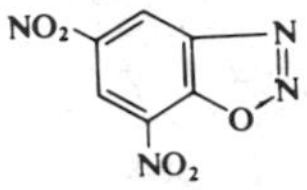

3332a

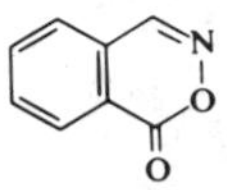

3345

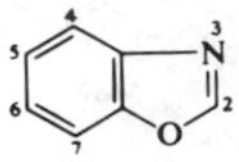

3115

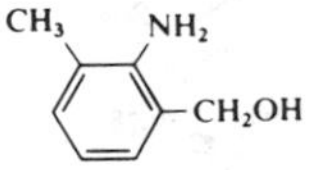

3333

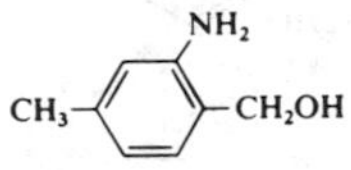

3346

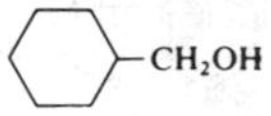

3116

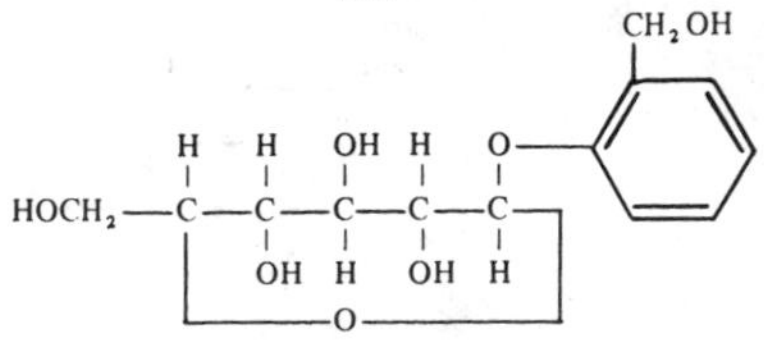

3315

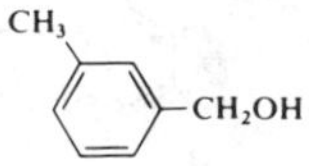

3334

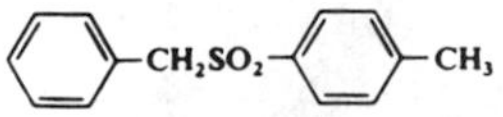

3348

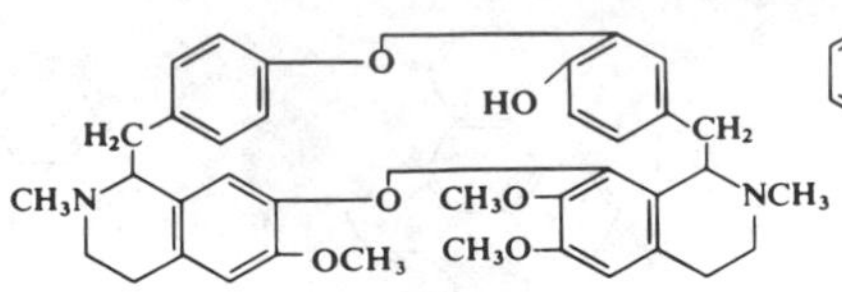

3117

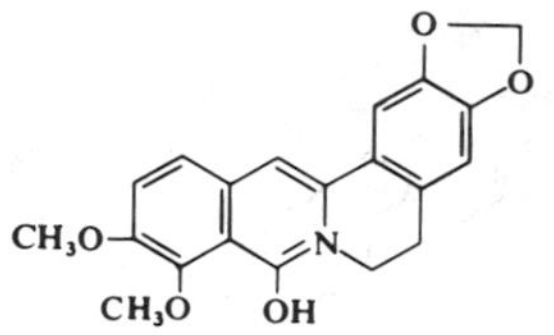

3320

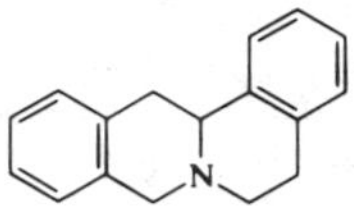

3335

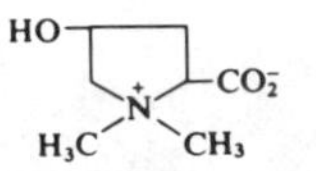

3350

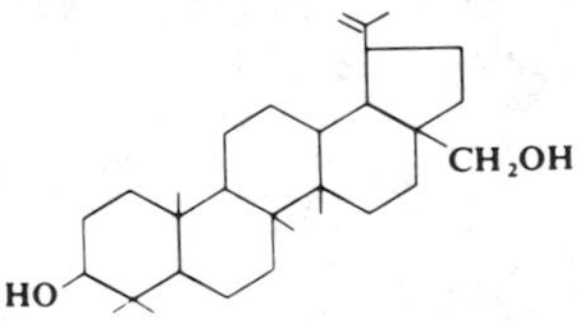

3118

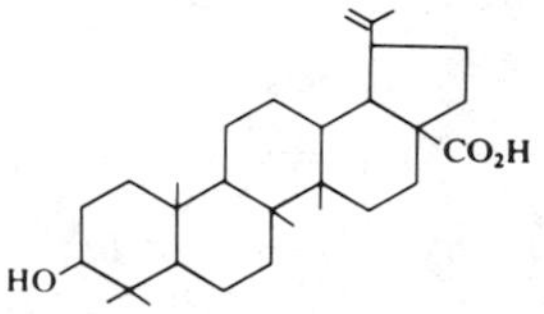

3322

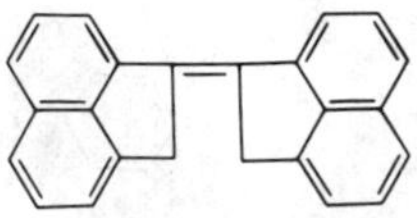

3336

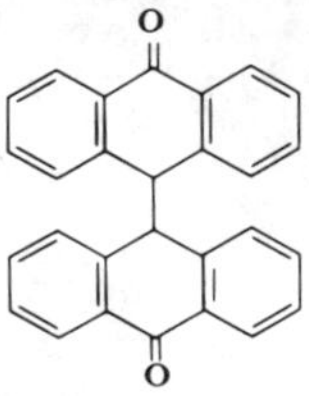

3351

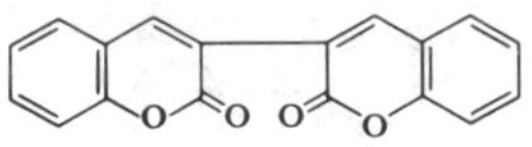

3337

3162

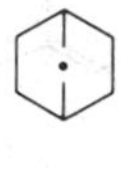

3323

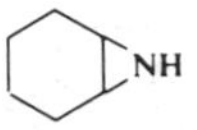

3338

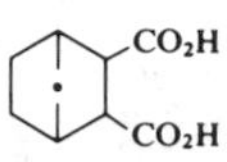

3353

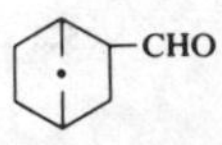

3163

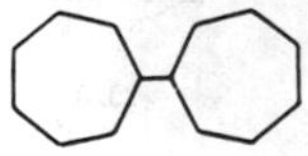

3339

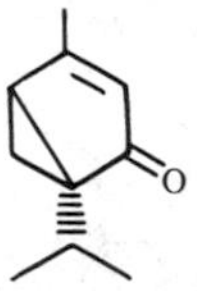

3357

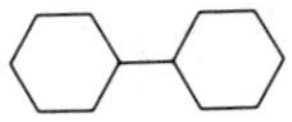

3177

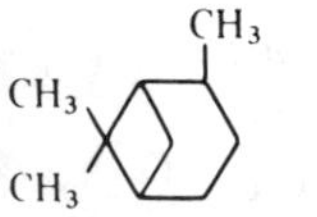

3324

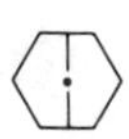

3340

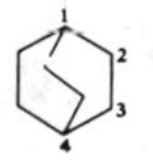

3179

3325

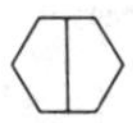

3342

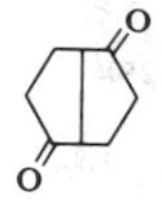

3360

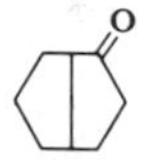

3186

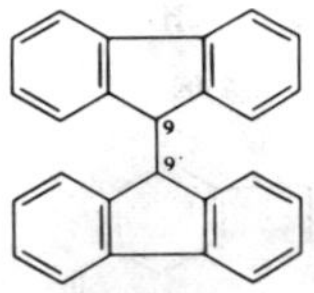

3329

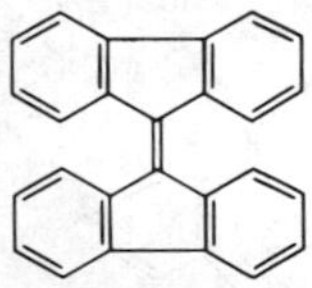

3343

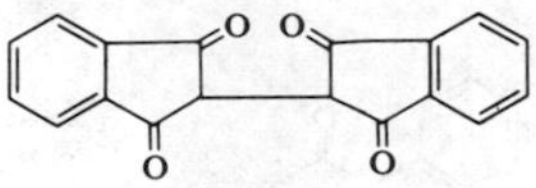

3289

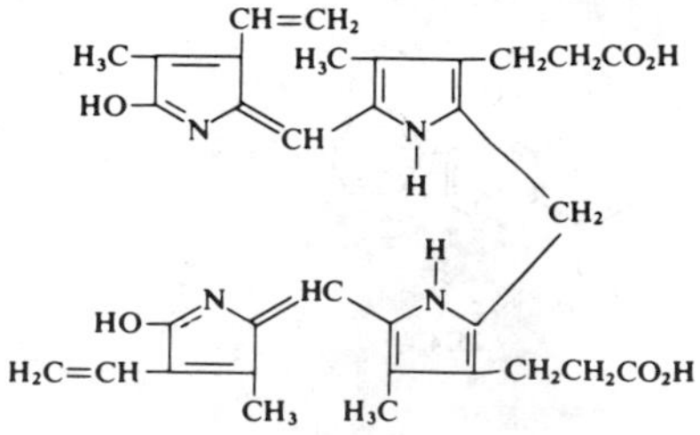

3344

In Numeric as they occur in Organic Compounds Table

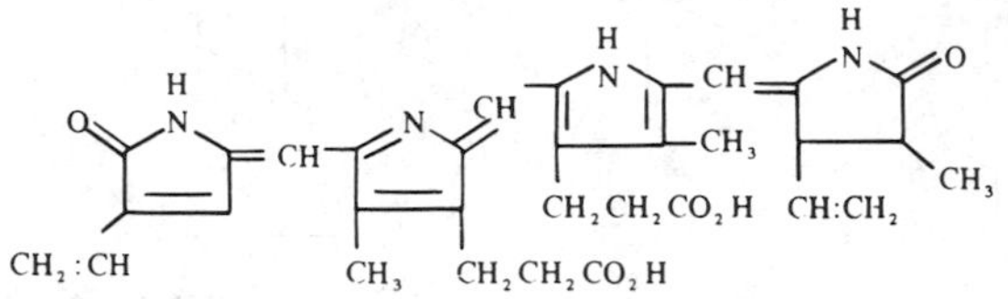

3361

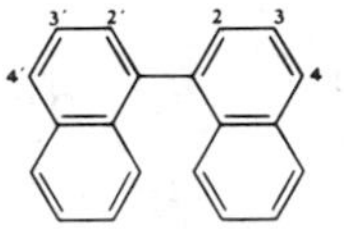

3362

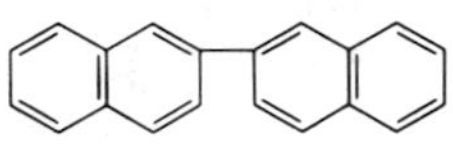

3366

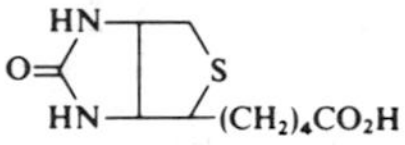

3367

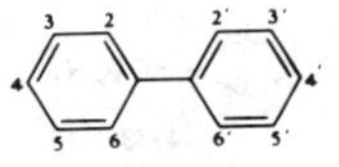

3369

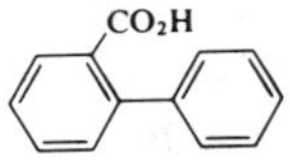

3500

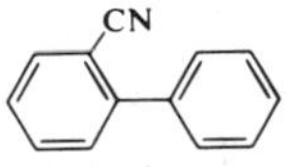

3501

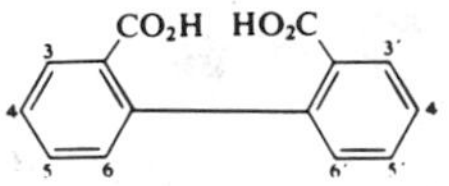

3505

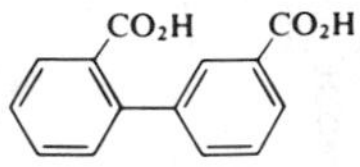

3511

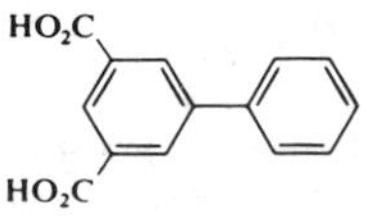

3517

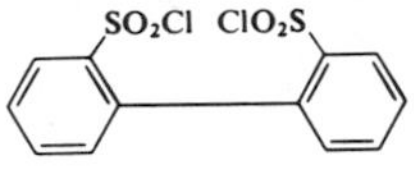

3523

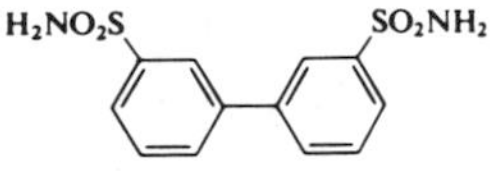

3524

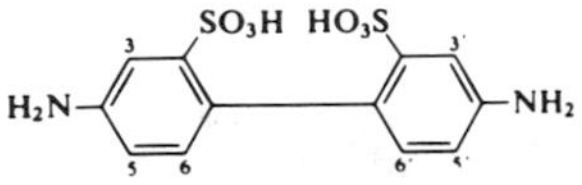

3529

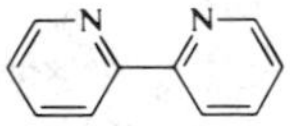

3530

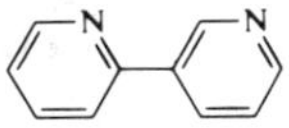

3531

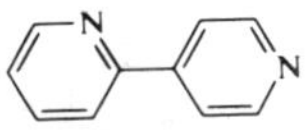

3532

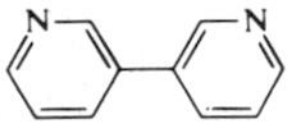

3533

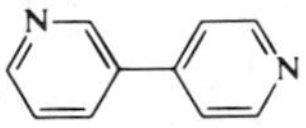

3534

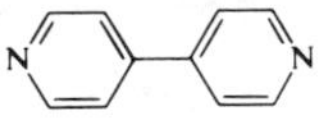

3534a

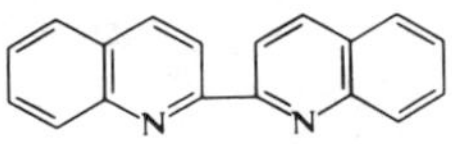

3535

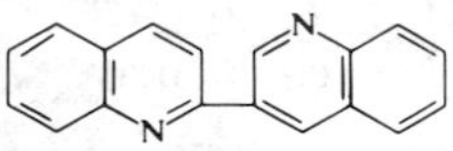

3536

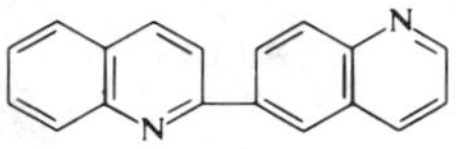

3537

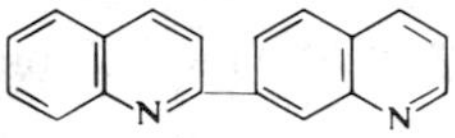

3538

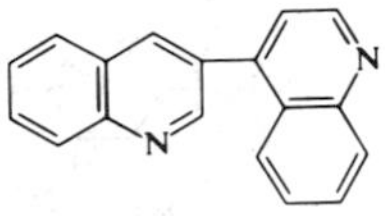

3540

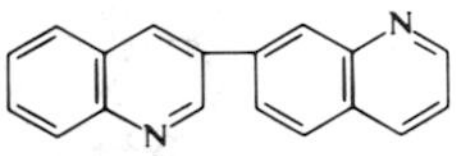

3541

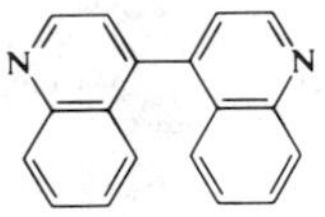

3542

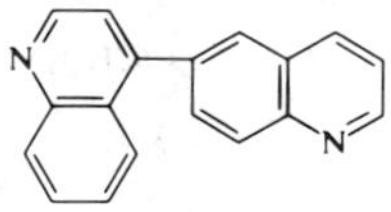

3543

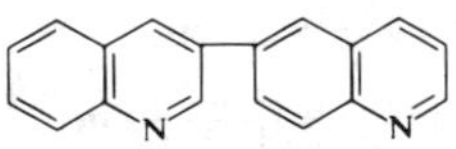

3544

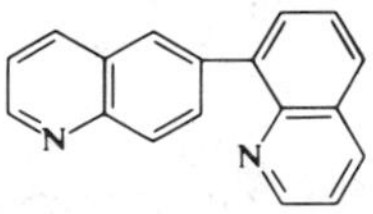

3545

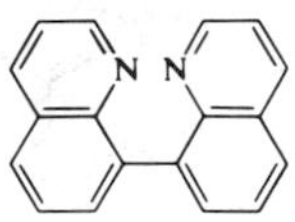

3546

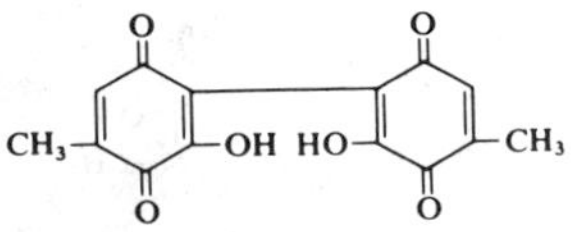

3548

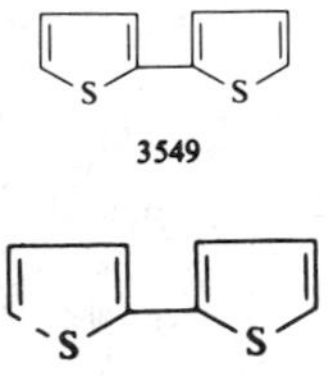

3549

3551

3570

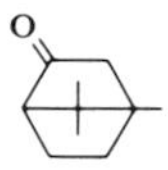

3581

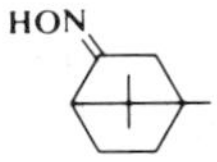

3583

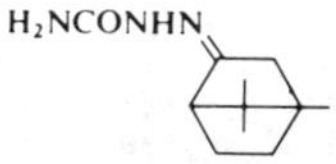

3586

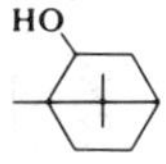

3588

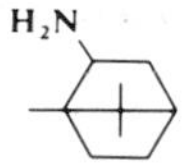

3595

3596

3597

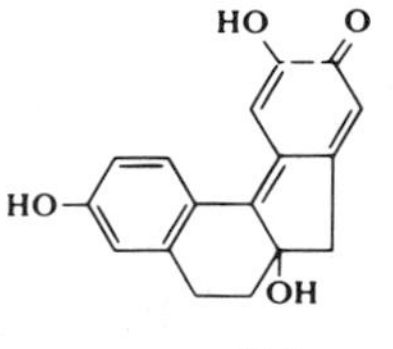

3599

STRUCTURAL FORMULAS OF ORGANIC COMPOUNDS (Continued)

In Numeric as they occur in Organic Compounds Table

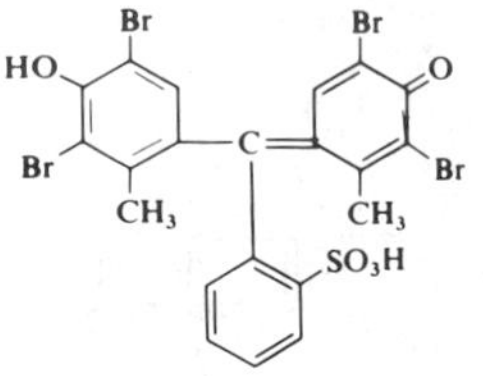

HO, Br, Br, O, Br, Br, CH_3, C, CH_3, SO_3H

3600

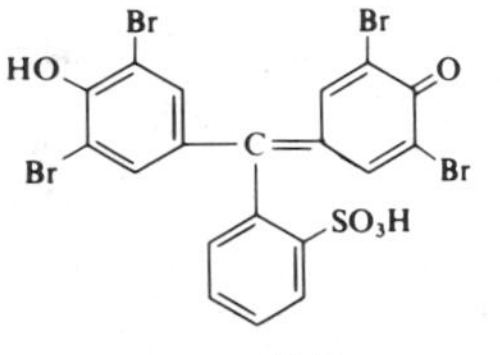

HO, Br, Br, O, Br, Br, C, SO_3H

3620

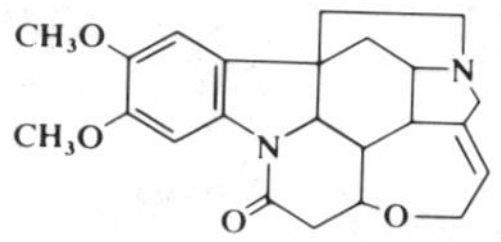

CH_3O, CH_3O, N, N, O, O

3624

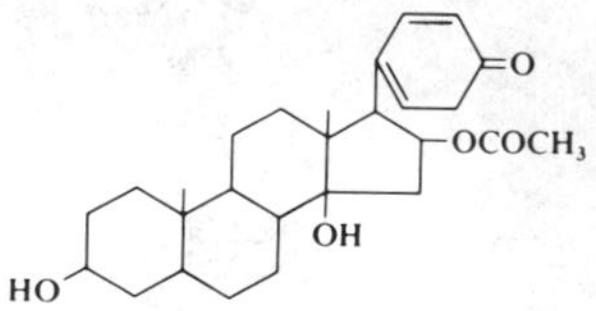

O, $OCOCH_3$, OH, HO

3628

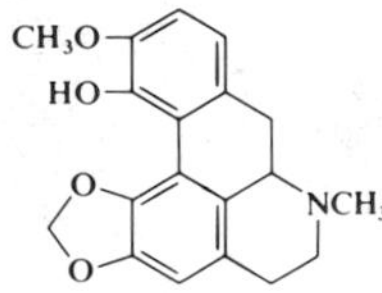

CH_3O, HO, O, O, NCH_3

3629

$CH_2CHCHCH_2$ (O, O)

3650

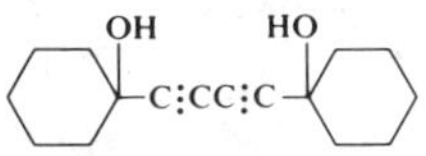

OH, HO, $C\vdots CC\vdots C$

3670

$CH_2CHCHCH_2$ (O, O)

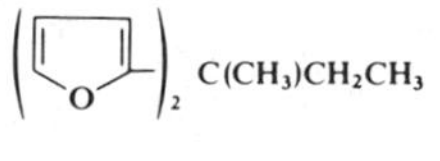

3738

$C(CH_3)CH_2CH_3$ (furyl)$_2$, O

3743

$CH_3CHCHCH_3$ (O)

3757

$CH_3CHCHCH_3$ (O)

3758

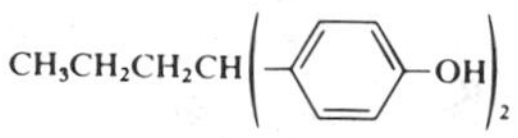

$CH_3CH_2CH_2CH$ (–C₆H₄–OH)$_2$

3765

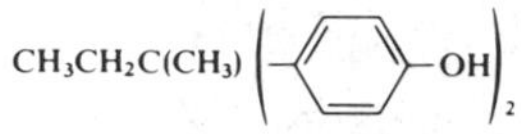

$CH_3CH_2C(CH_3)$ (–C₆H₄–OH)$_2$

3766

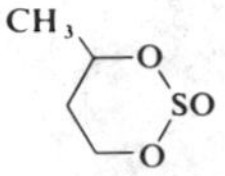

CH_3, O, SO, O

3832

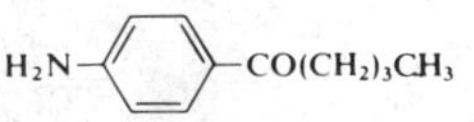

H_2N–C₆H₄–$CO(CH_2)_3CH_3$

4093

$CH_2C(CH_3)_2$ (O)

4194

$CH_3C(C_6H_5)CHCO_2C_2H_5$ (O)

4327

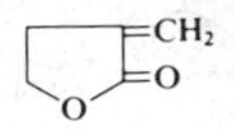

$=CH_2$, O, O

4350

CH_2CHCH_2CN (O)

4424

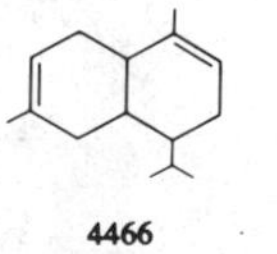

4466

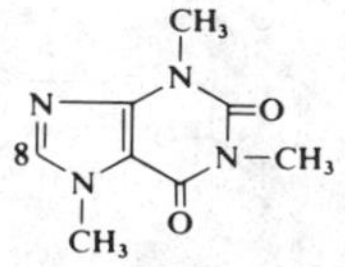

CH_3, N, N, O, N, $N-CH_3$, 8, N, CH_3, O

4467

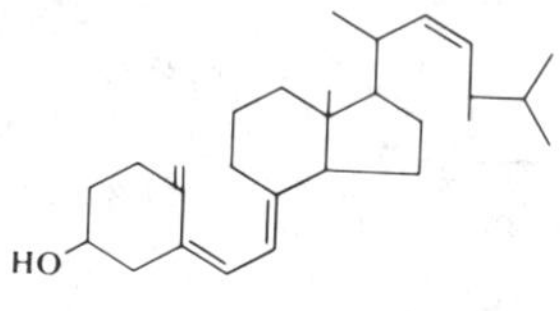

HO

4477

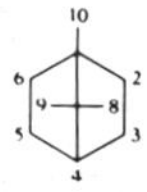

10, 6, 9, 8, 2, 5, 3, 4

4478

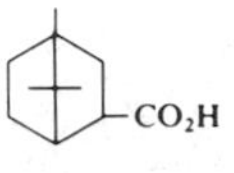

CO_2H

4479

4480

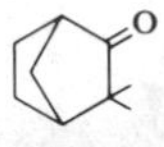

O

4483

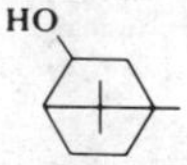

HO

4484

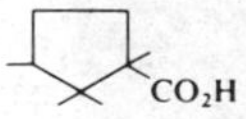

CO_2H

4486

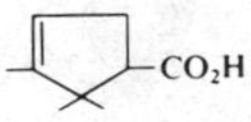

CO_2H

4489

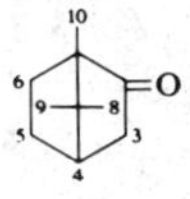

10, 6, 9, 8, O, 5, 3, 4

4490

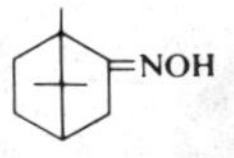

=NOH

4493

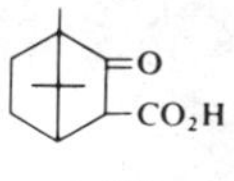

O, CO_2H

4514

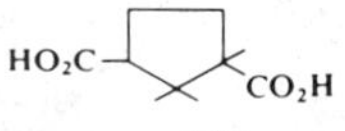

HO_2C, CO_2H

4518

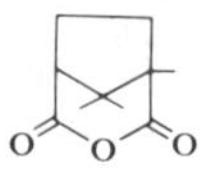

O, O, O

4520

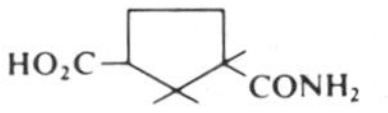

HO_2C, $CONH_2$

4525

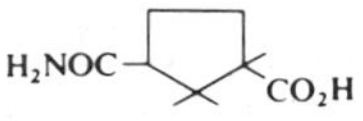

H_2NOC, CO_2H

4526

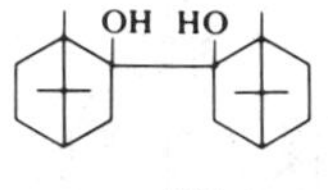

OH HO

4531

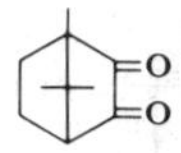

O, O

4532

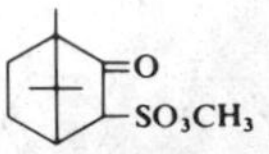

O, SO_3CH_3

4533

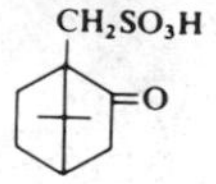

CH_2SO_3H, O

4534

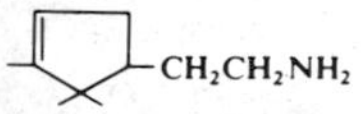

$CH_2CH_2NH_2$

4537

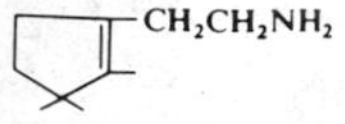

$CH_2CH_2NH_2$

4538

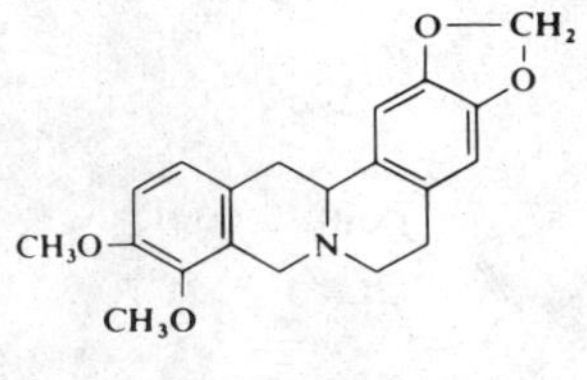

O, CH_2, O, CH_3O, N, CH_3O

4539

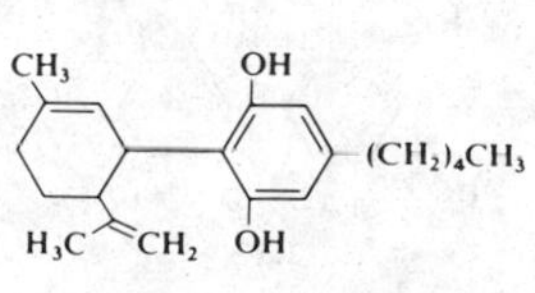

CH_3, OH, $(CH_2)_4CH_3$, H_3C, CH_2, OH

4544

STRUCTURAL FORMULAS OF ORGANIC COMPOUNDS (Continued)

In Numeric as they occur in Organic Compounds Table

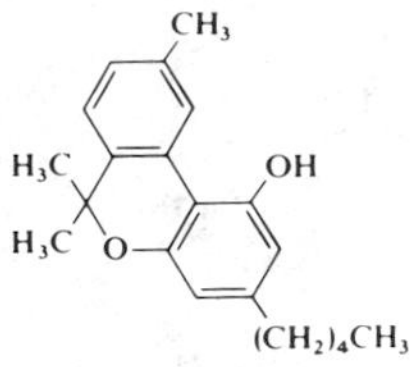

4545

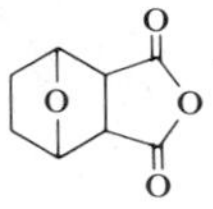

4546

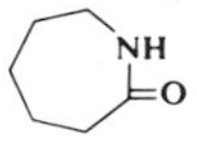

4575

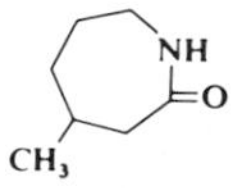

4576

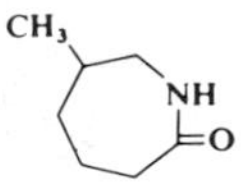

4577

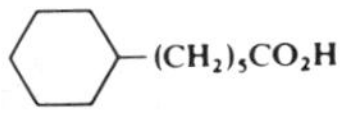

$(CH_2)_5CO_2H$

4588

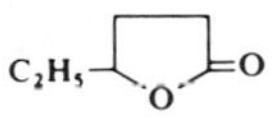

C_2H_5

4597

$CH_3(CH_2)_3$

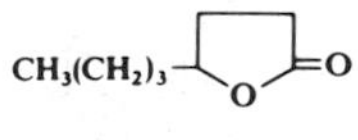

4637

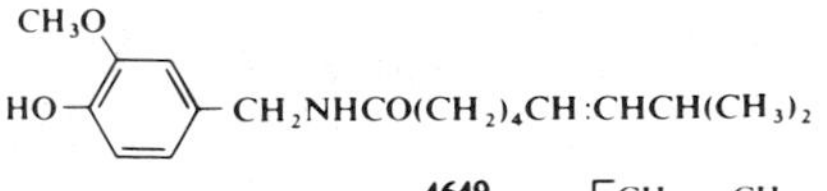

CH_3O, HO, $CH_2NHCO(CH_2)_4CH{:}CHCH(CH_3)_2$

4649

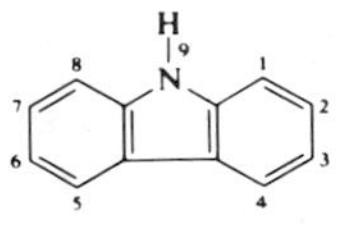

4697

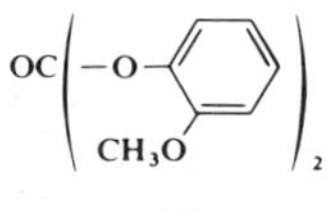

OC, CH_3O

4722

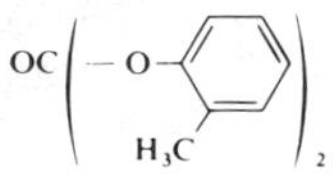

OC, H_3C

4734

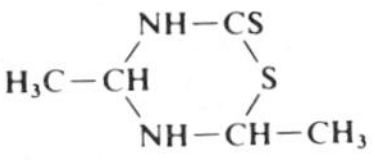

H_3C-CH, $NH-CS$, S, $NH-CH-CH_3$

4743

4744

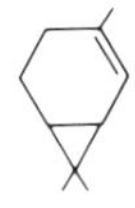

4746

HO, HO_2C, CH_3, HO, OH, HO, $CO(CHOH)_4CH_3$

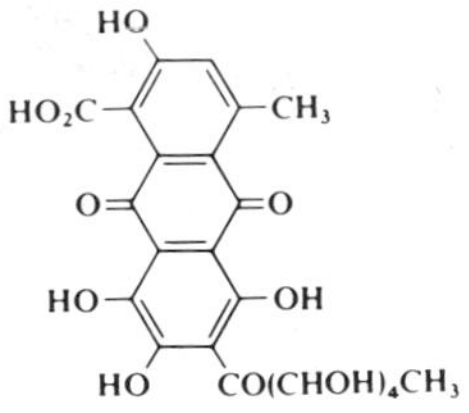

4747

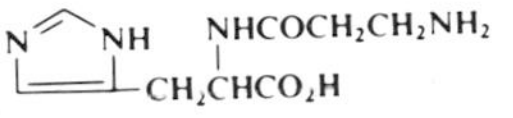

$NHCOCH_2CH_2NH_2$, CH_2CHCO_2H

4749

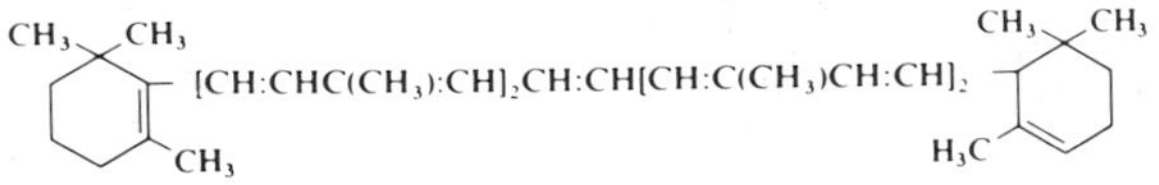

$[CH{:}CHC(CH_3){:}CH]_2CH{:}CH[CH{:}C(CH_3)CH{:}CH]_2$

4751

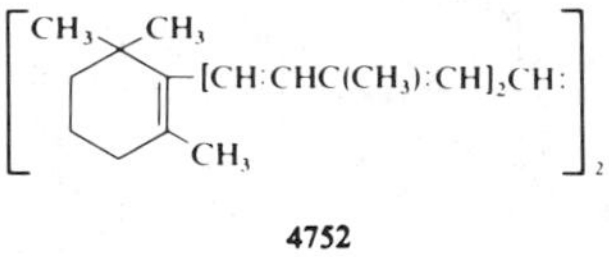

$[CH{:}CHC(CH_3){:}CH]_2CH{:}$

4752

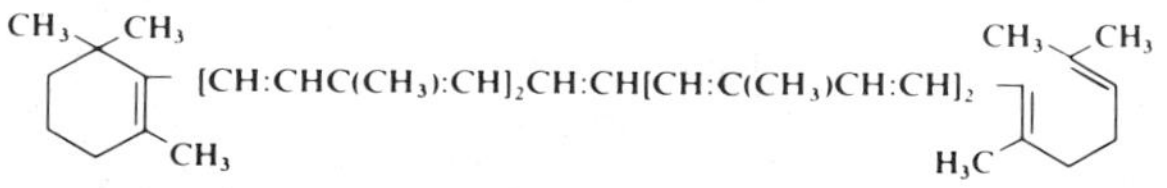

$[CH{:}CHC(CH_3){:}CH]_2CH{:}CH[CH{:}C(CH_3)CH{:}CH]_2$

4753

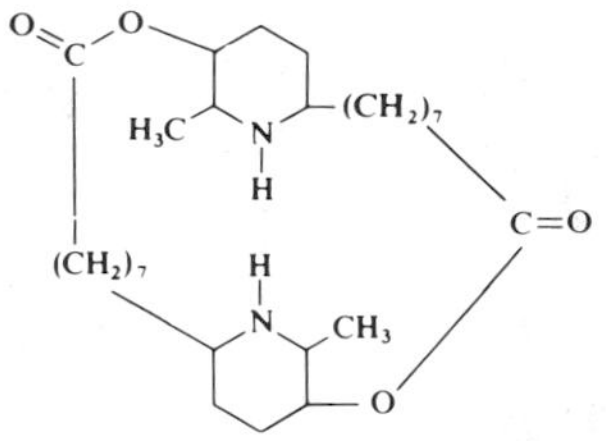

$(CH_2)_7$, $C{=}O$, $(CH_2)_7$

4754

CH(OH), CH_2, CH_3

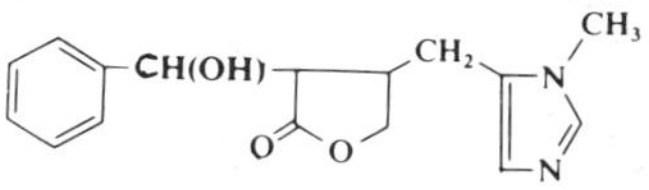

4756

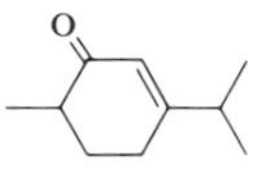

4760

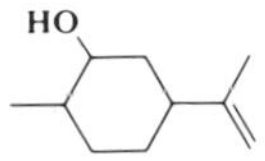

4762

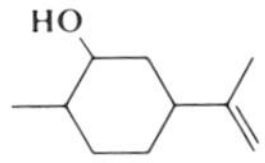

4764

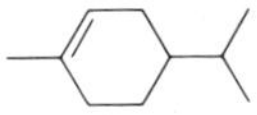

4765

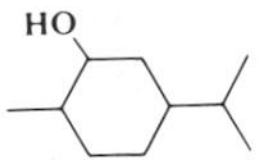

4766

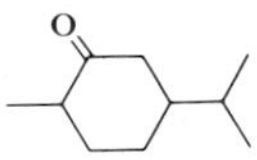

4768

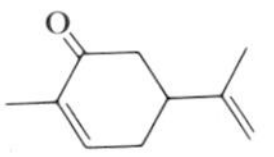

4769

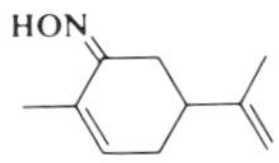

4772

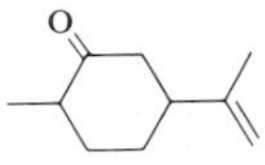

4777

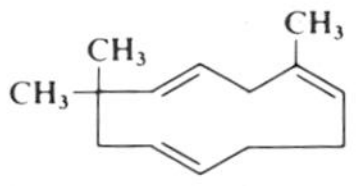

4779

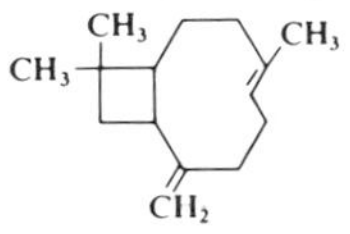

4780

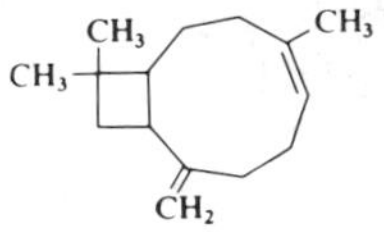

4781

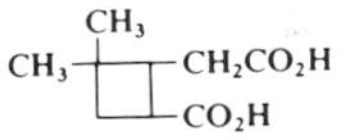

CH_2CO_2H, CO_2H

4782

CO_2H, HO

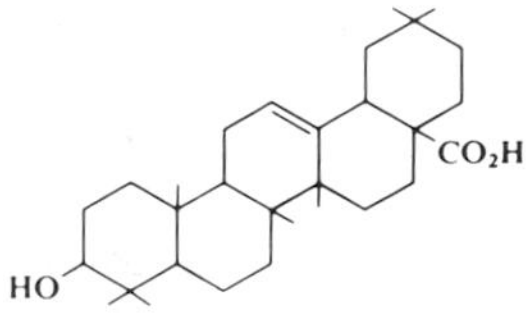

4784

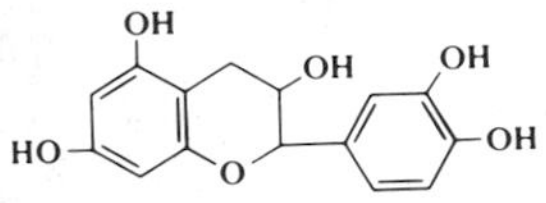

4785

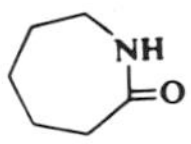

4811

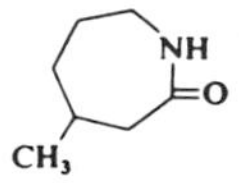

4812

STRUCTURAL FORMULAS OF ORGANIC COMPOUNDS (Continued)

In Numeric as they occur in Organic Compounds Table

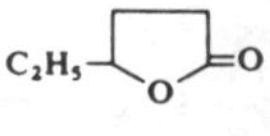

4814

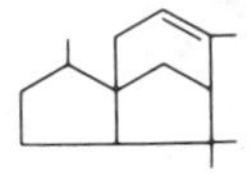

04824

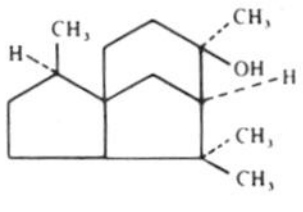

4825

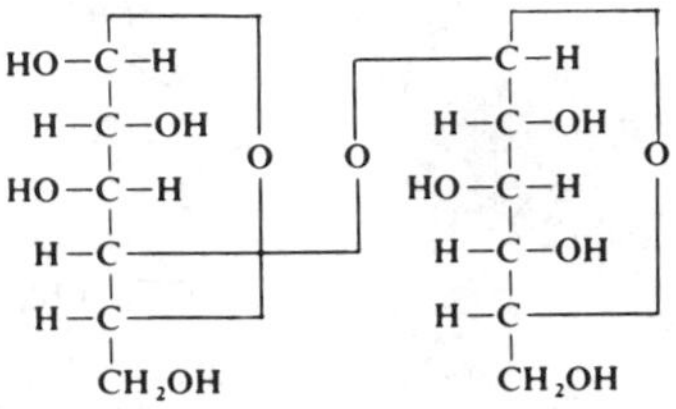

4826

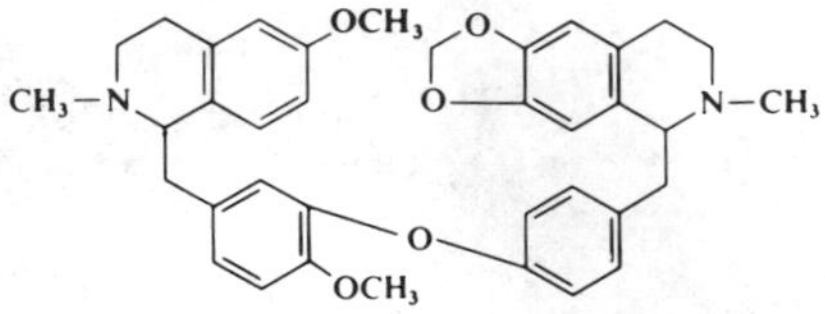

4836

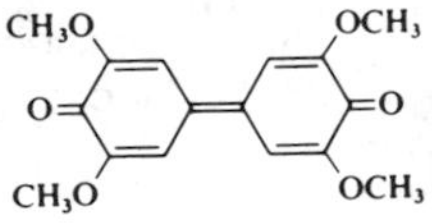

4838

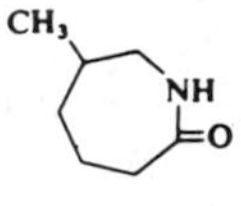

4843

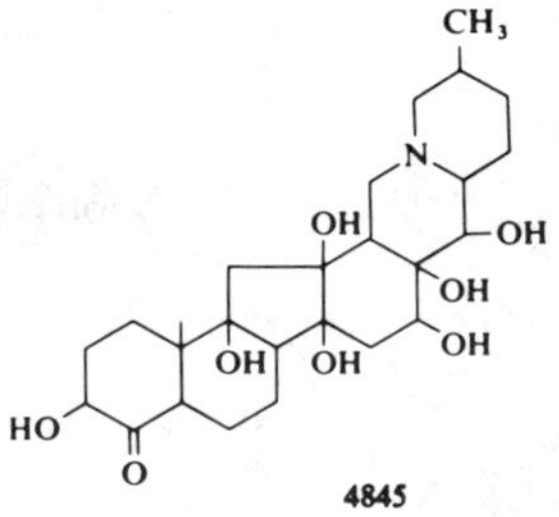

4845

4848

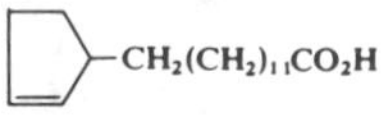

4860

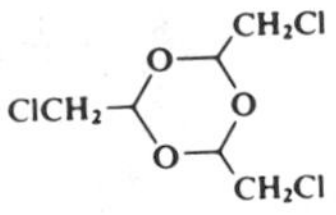

4862

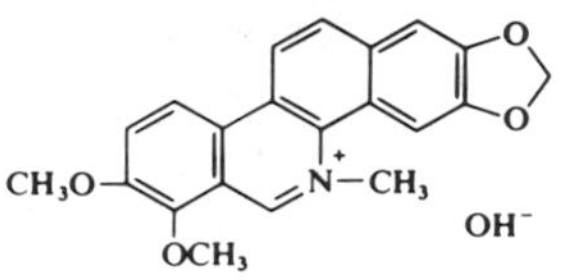

4863

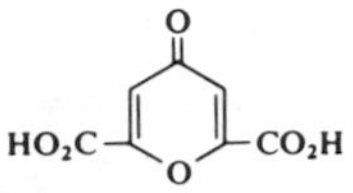

4864

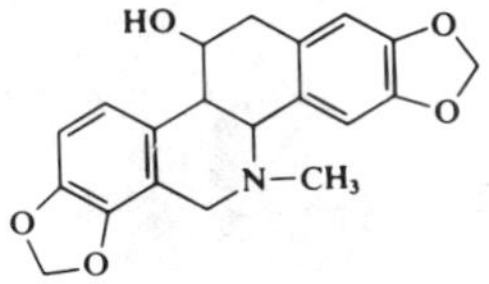

4866

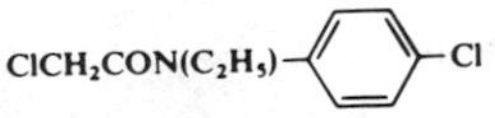

4889

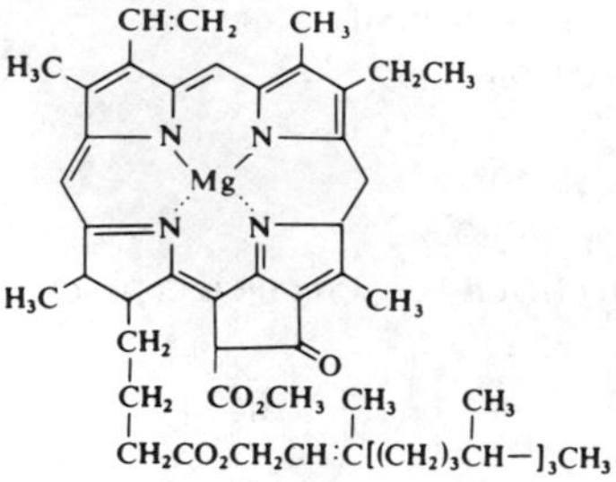

4986

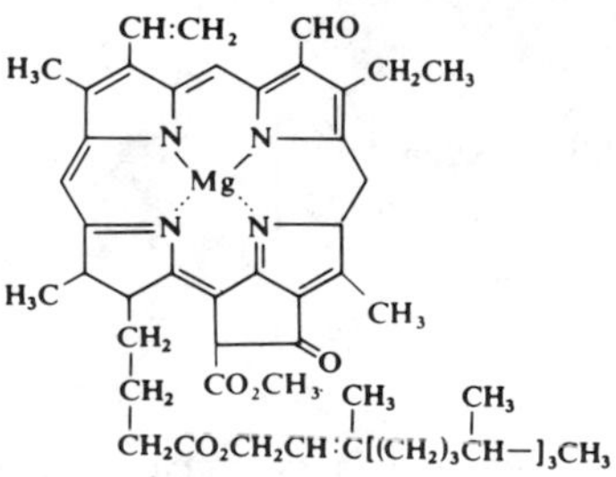

4987

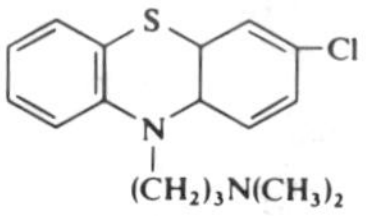

4989

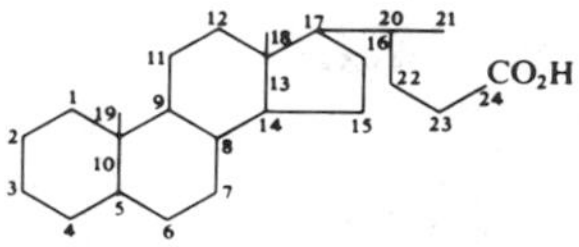

4998

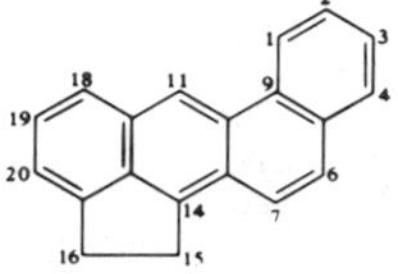

5003

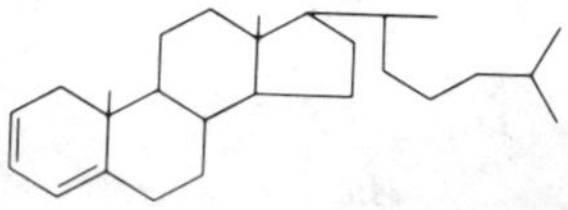

5007

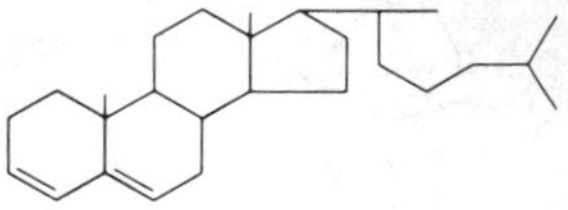

5008

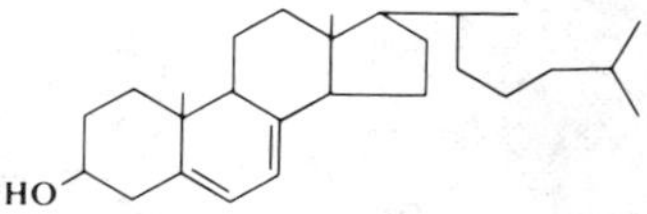

5009

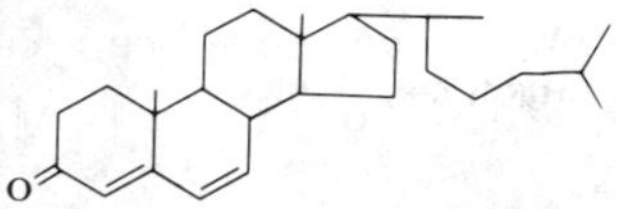

5010

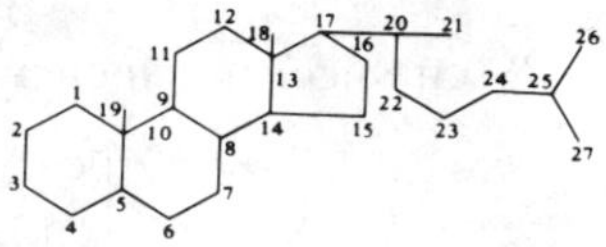

5011

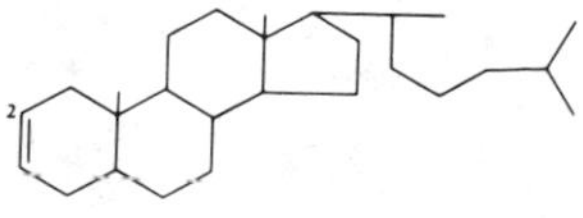

5019

In Numeric as they occur in Organic Compounds Table

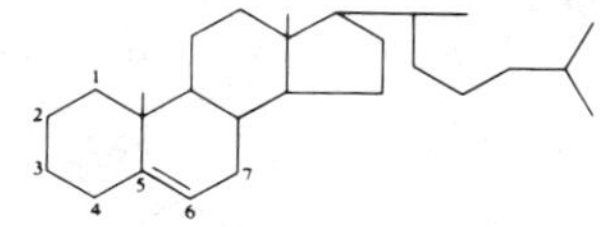

5020

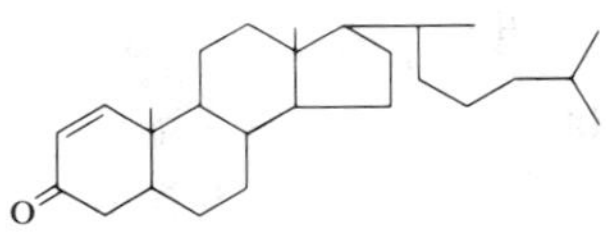

5025

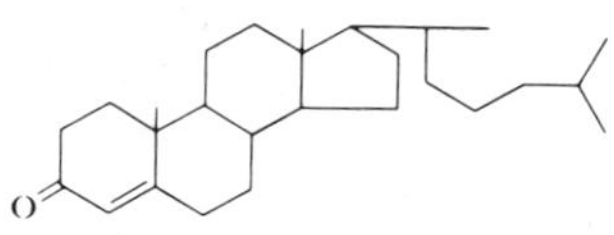

5026

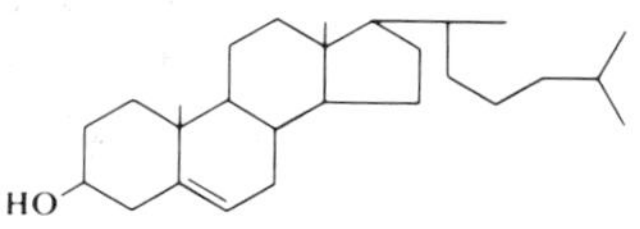

5028

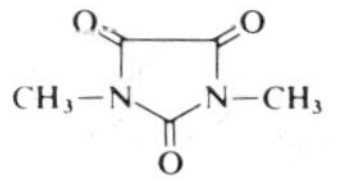

CH_3-N $N-CH_3$

5032

OH CO_2H HO OH

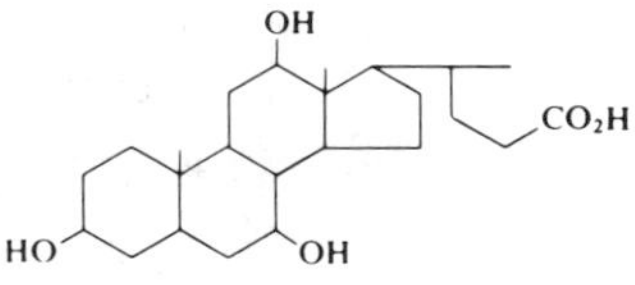

5033

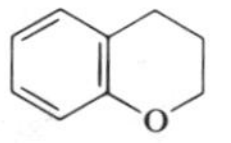

5039

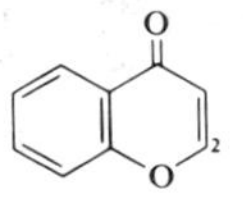

5043

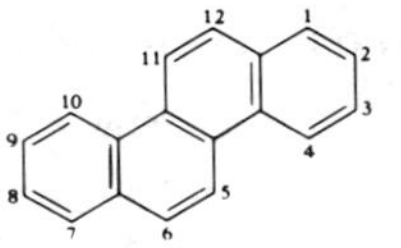

5047

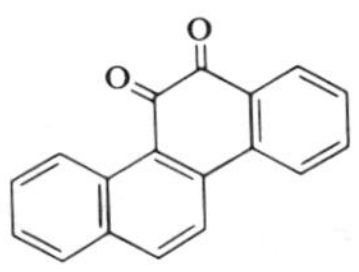

5052

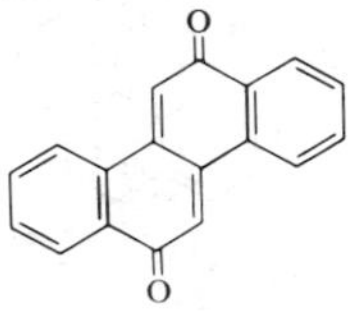

5053

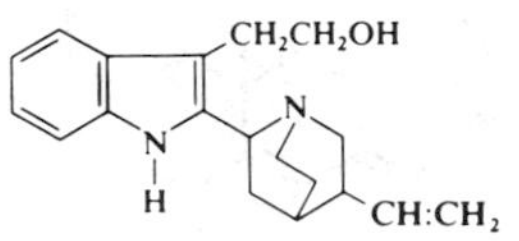

CH_2CH_2OH N H N CH:CH_2

5054

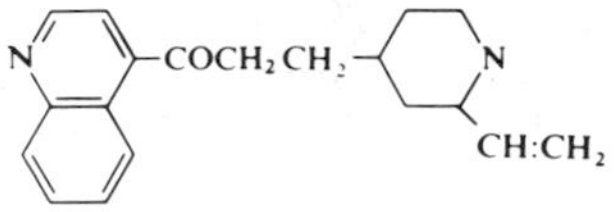

N $-COCH_2CH_2-$ N CH:CH_2

5055

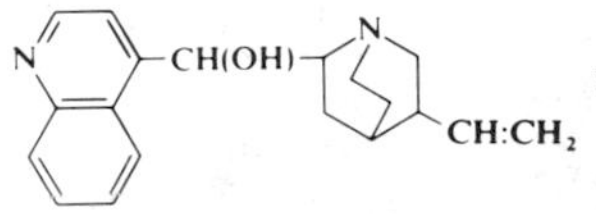

N CH(OH) N CH:CH_2

5056

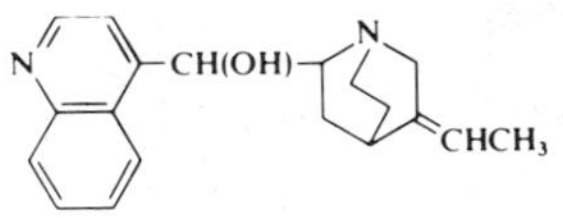

N CH(OH) N CHCH$_3$

5059

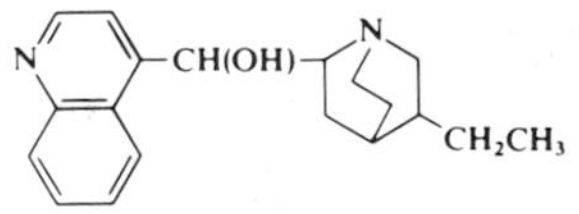

N CH(OH) N CH_2CH_3

5064

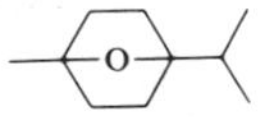

5065

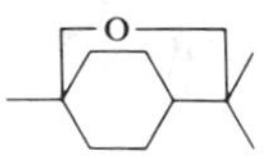

5066

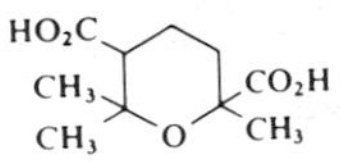

HO_2C CH_3 CH_3 O CO_2H CH_3

5067

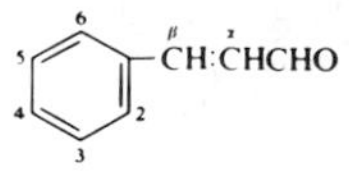

CH:CHCHO

5070

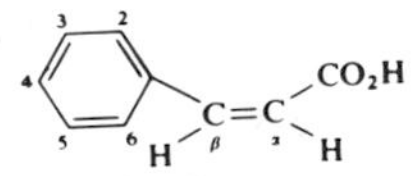

CO_2H C=C H H

5081

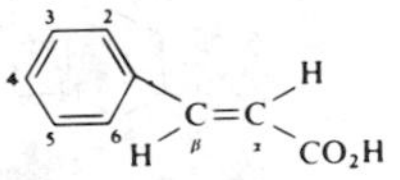

H C=C H CO_2H

5084

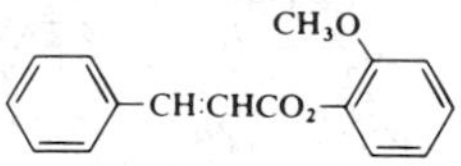

CH_3O CH:CHCO$_2$

5128

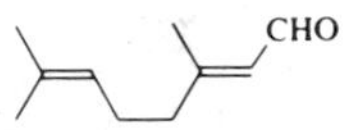

CHO

5167

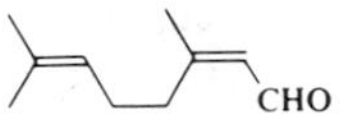

CHO

5168

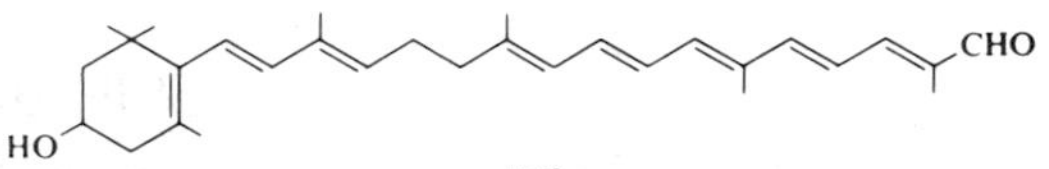

HO CHO

5169

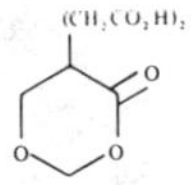

$(CH_2CO_2H)_2$

5171

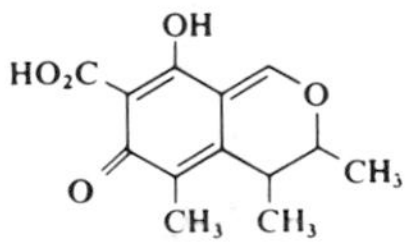

OH HO_2C O O CH_3 CH_3 CH_3

5179

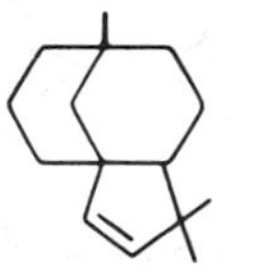

5187

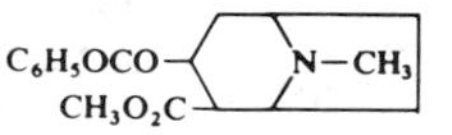

C_6H_5OCO CH_3O_2C N$-CH_3$

5188

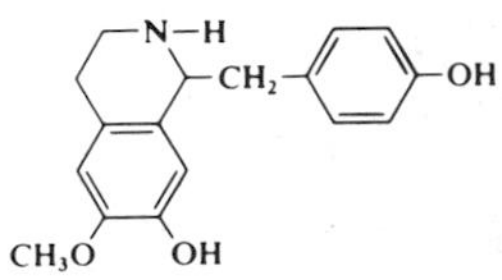

N$-$H CH_2 OH CH_3O OH

5193

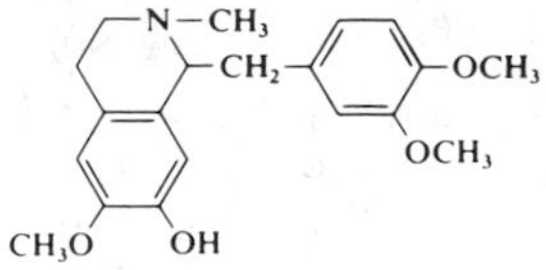

N$-CH_3$ CH_2 OCH_3 OCH_3 CH_3O OH

5194

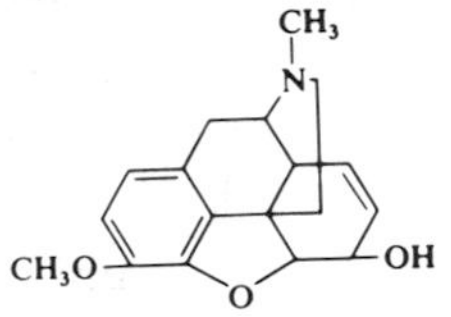

CH_3 N CH_3O O OH

5195

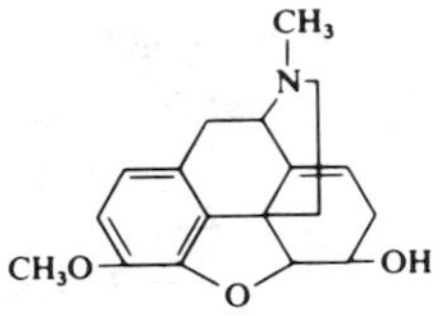

CH_3 N CH_3O O OH

5200

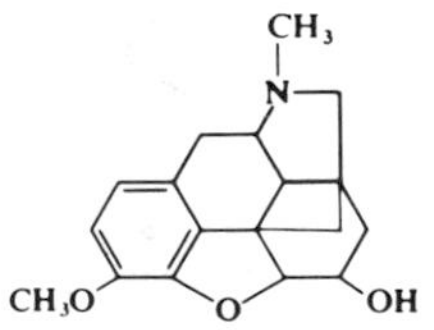

CH_3 N CH_3O O OH

5201

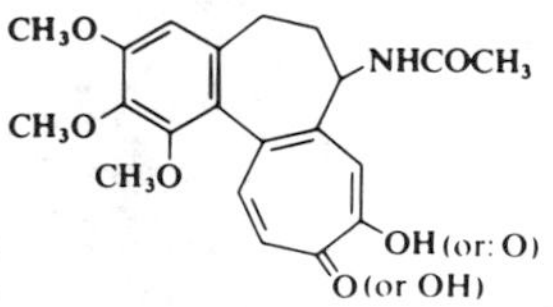

CH_3O CH_3O CH_3O NHCOCH$_3$ OH (or: O) O (or OH)

5202

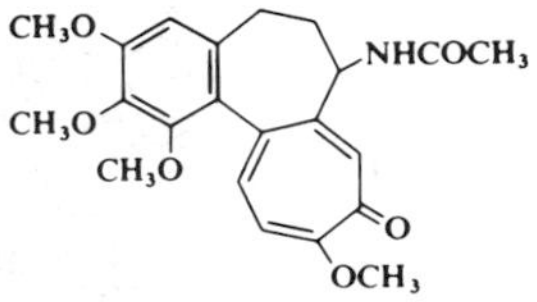

CH_3O CH_3O CH_3O NHCOCH$_3$ O OCH_3

5203

In Numeric as they occur in Organic Compounds Table

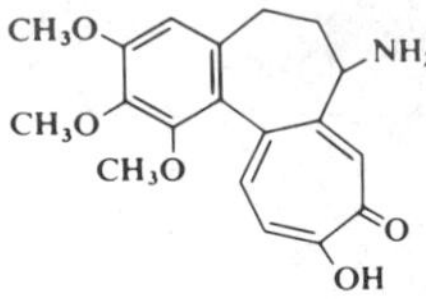

CH_3O, CH_3O, CH_3O, NH_2, OH
5204

CH_3
5221

HO, 17, $COCH_2OH$, O
5244

O, O, CO_2H
5262

5263

O, O
5264

CH_3, N, CH_3, $(CH_3)_2N$
5205

CH_2OH, H, H, OH, HO, H, OH, H, O, O, $CH:CHCH_2OH$, CH_3O
5224

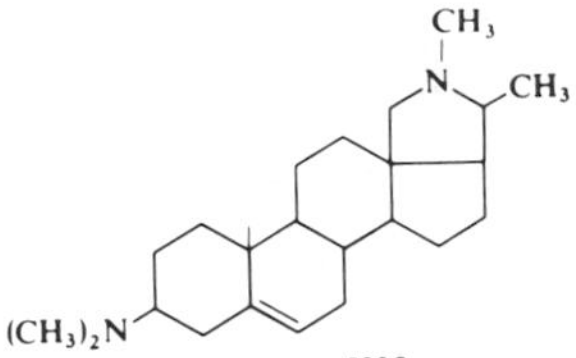

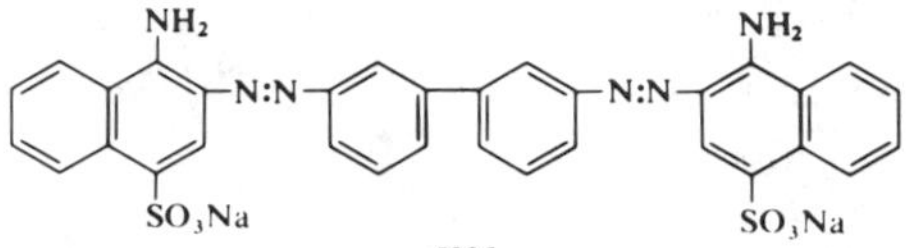

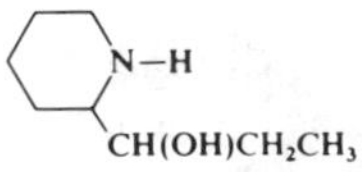

O, $COCH_2OH$, OH, O
5246

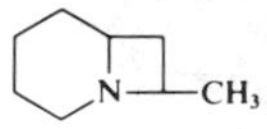

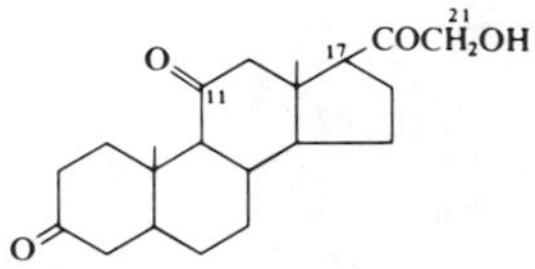

5, 4, 6, 3, 7, 8, O, O
5265

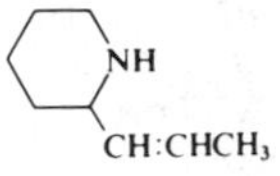

CH_3O, HO, $CH:CHCH_2OH$
5225

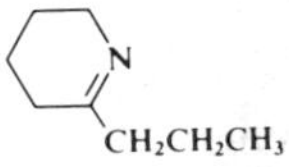

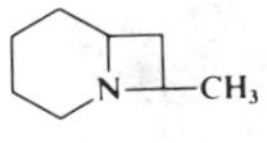

HO, N, OCH_3, CH_3O, CH_3, OCH_3
5248

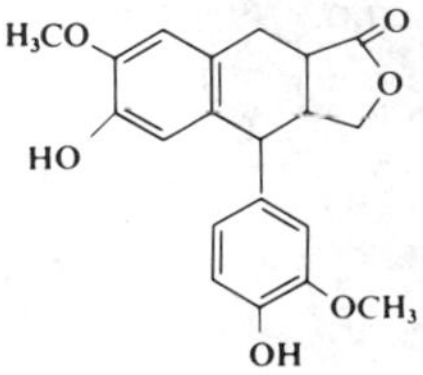

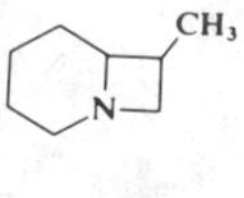

OH, CH_2COCH_3, CH, O, O
5279

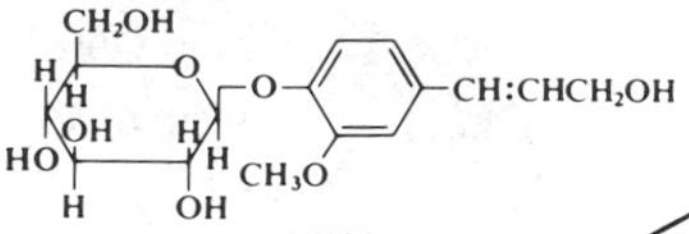

NH_2, N:N, N:N, NH_2, SO_3Na, SO_3Na
5206

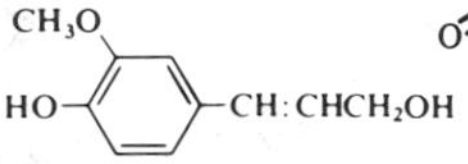

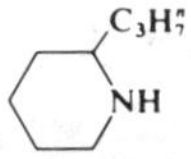

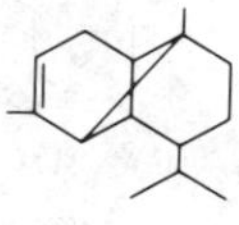

C_3H_7, NH
5226

O, O, N, CH_3, O, H_3C, O, O
5250

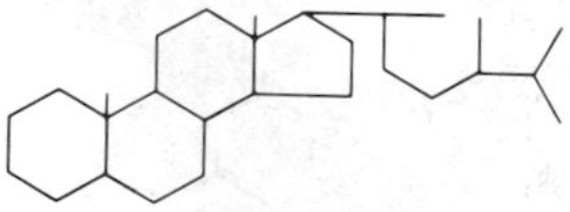

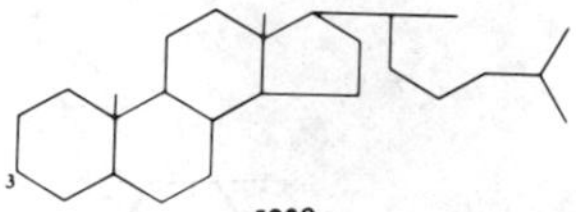

CO_2H, O, O
5295

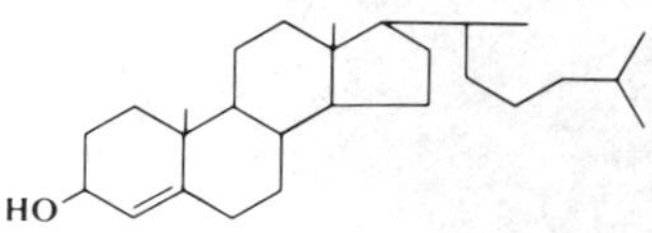

N–H, $CH(OH)CH_2CH_3$
5207

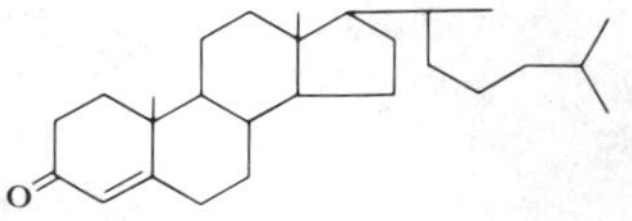

5237

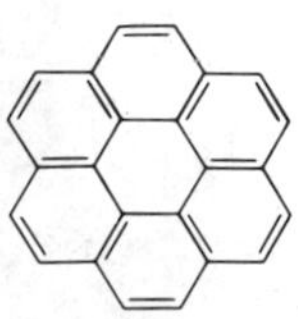

N, CH_3
5210

O, O, N, CH_3, HO, H_3C, O, O
5251

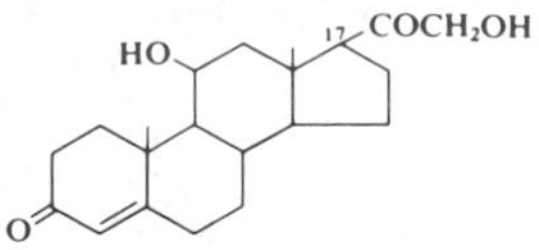

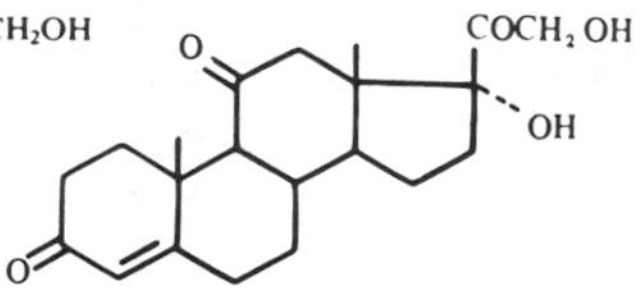

4, 5, 3, 6, 7, O, CO_2H
5296

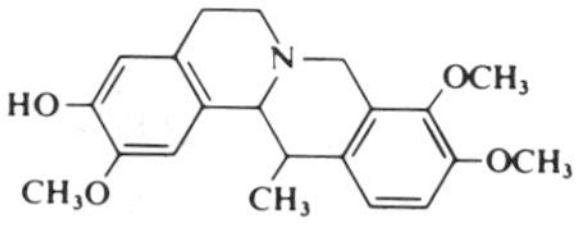

O, 11, 17, 21, $COCH_2OH$, O
5211

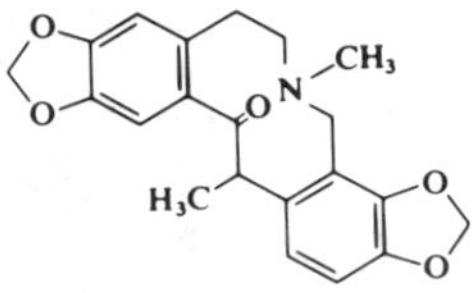

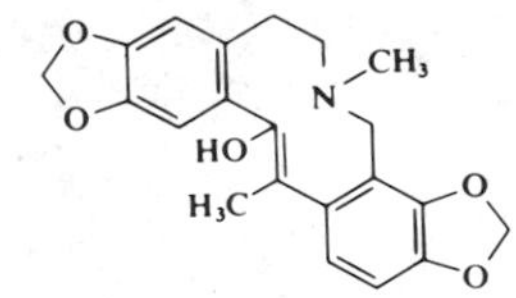

5238

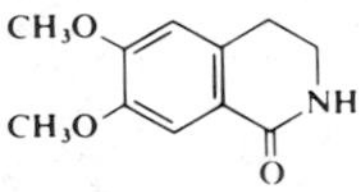

CH_3O, CH_3O, NH, O
5252

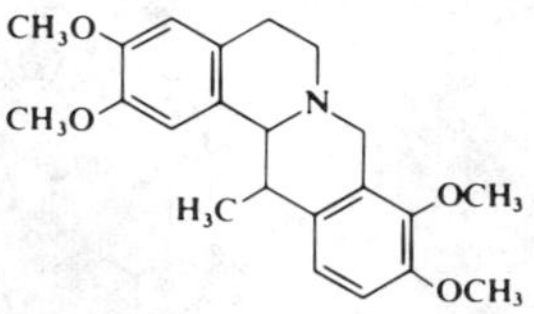

HO, O, OH, O, O
5300

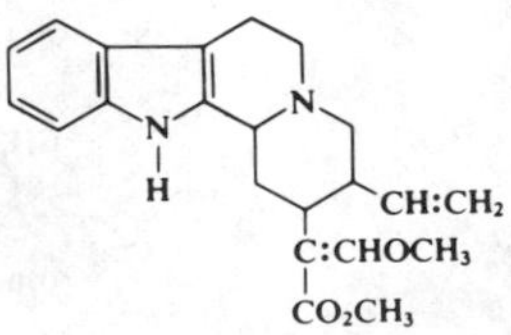

NH, $CH:CHCH_3$
5213

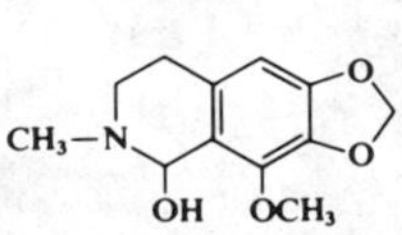

3
5239

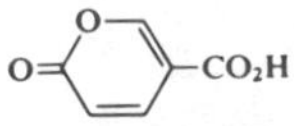

CH_3O, CH_3O, N, H_3C, OCH_3, OCH_3
5253

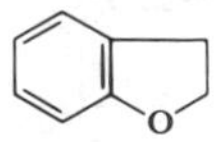

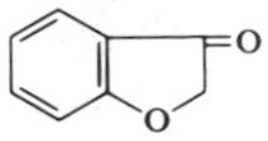

HN, NH, O, N, CH_3
5303

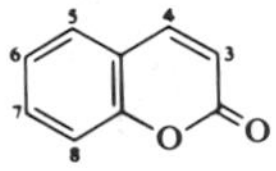

N, $CH_2CH_2CH_3$
5216

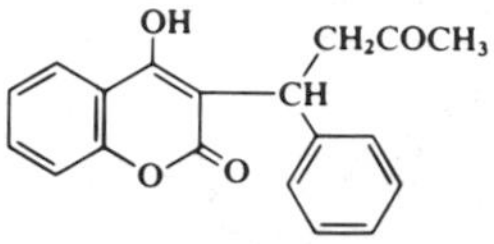

HO
5241

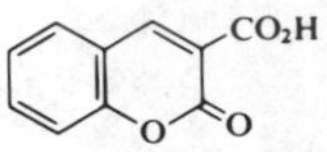

$[HO_2CC(CH_3):CHCH:CHC(CH_3):CHCH=]_2$
5379

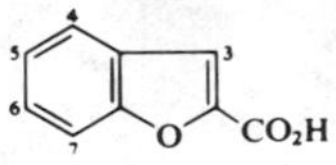

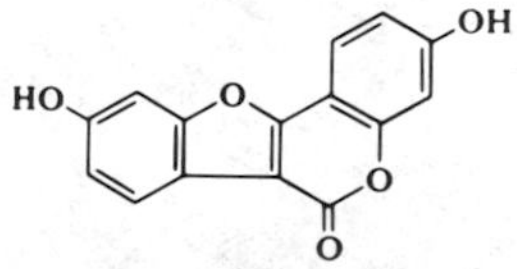

N, CH_3
5217

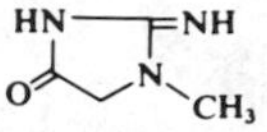

O
5242

N, N, H, $CH:CH_2$, $C:CHOCH_3$, CO_2CH_3
5258

HO, OH, O, O, O
5380

H_3CO, HO, O, O, OCH_3, OH
5220

5243

CH_3–N, O, O, OH, OCH_3
5259

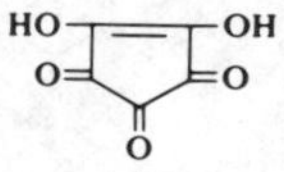

CH_3O, CH_3O, N, CH_3, O, O, O
5392

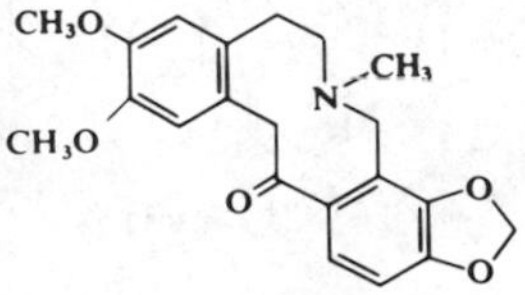

STRUCTURAL FORMULAS OF ORGANIC COMPOUNDS (Continued)

In Numeric as they occur in Organic Compounds Table

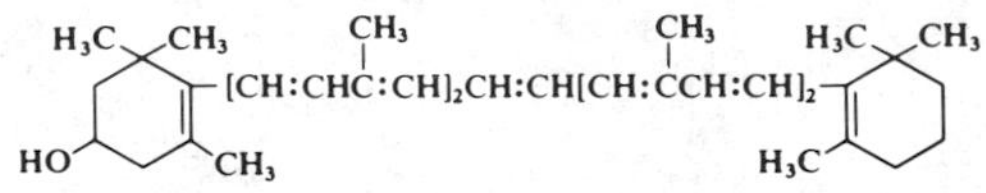

H_3C CH_3 CH_3 CH_3 H_3C CH_3
[CH:CHC:CH]$_2$CH:CH[CH:CCH:CH]$_2$
HO CH_3 H_3C

5393

$C_3H_7^i$ $CH_2CH(CH_3)CHO$

5437

NH

5466

CO$_2$H

5564

OH CH$_2$ CH$_2$

5397

5438

CH_3 CH_3 O Si O Si CH_3 CH_3 CH_3 Si CH_3 O

5473

CO$_2$H CO$_2$H

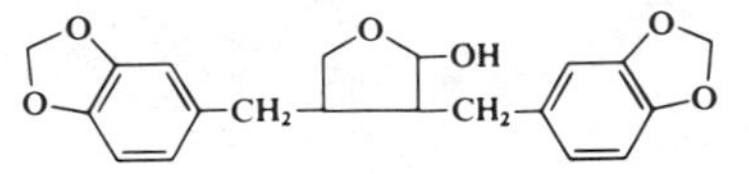

5574

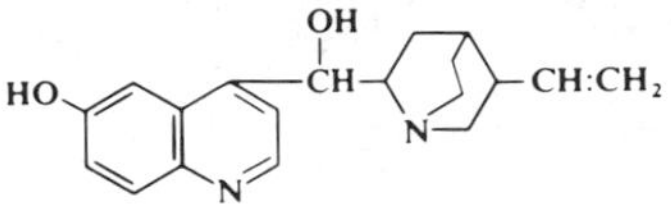

OH HO CH CH:CH$_2$ N N

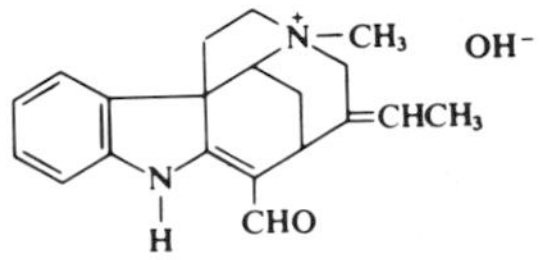

5399

CO$_2$H CO$_2$H

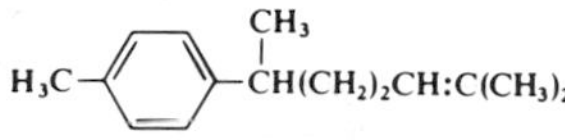

5444

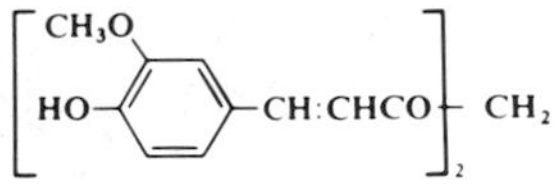

O O O O=O.8H$_2$O O O

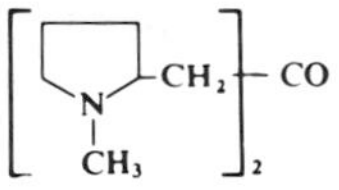

5576

5474

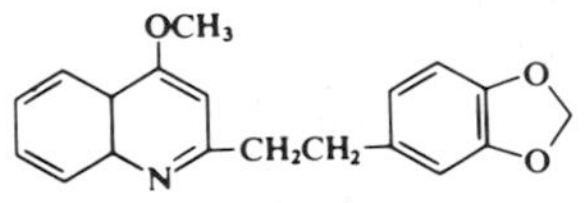

CO$_2$H CO$_2$H

5446

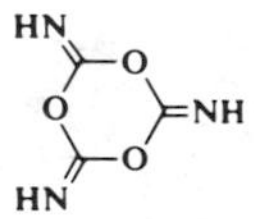

$\overset{+}{N}$—CH$_3$ OH$^-$ =CHCH$_3$ N H CHO

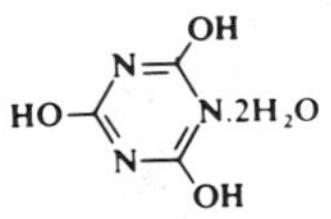

5400

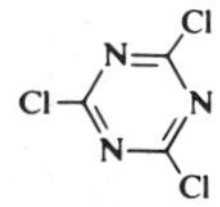

HO$_2$C CO$_2$H

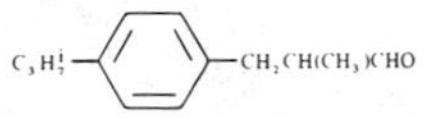

5450

O

5475

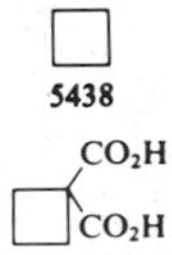

OH

5577

CH_3 H_3C CH(CH$_2$)$_2$CH:C(CH$_3$)$_2$

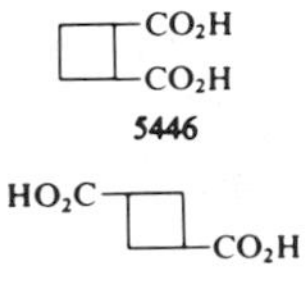

5401

O

5454

5482

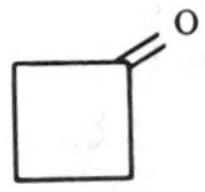

O

5621

CH$_3$O HO CH:CHCO CH$_2$]$_2$

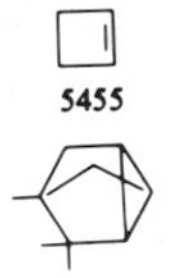

5402

5455

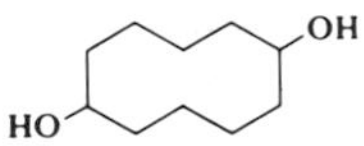

5483

SH

5647

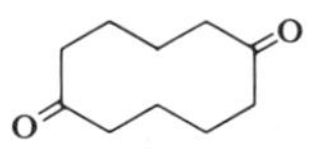

5458

CO$_2$H CO$_2$H

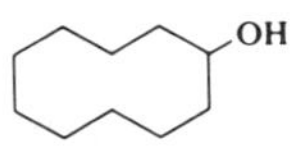

5488

OH OH

5663

CH$_2$ CO N CH$_3$]$_2$

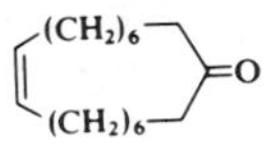

5403

HO OH

5459

CO$_2$H CO$_2$H

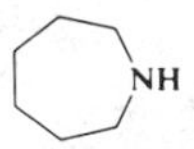

5489

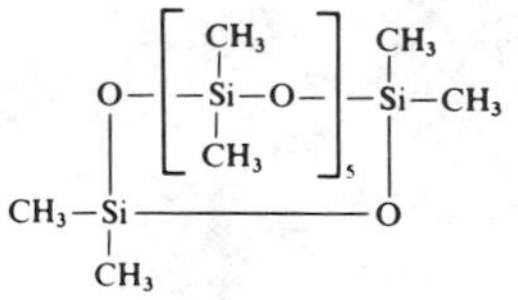

O O

5667

OCH$_3$ CH$_2$CH$_2$ N O O

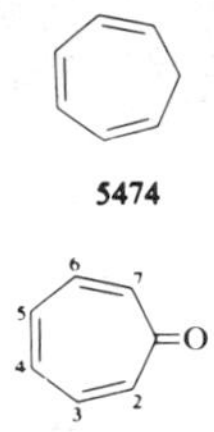

5405

O O

5460

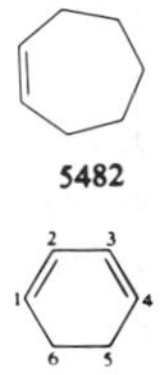

5507

CH$_3$ CH$_3$ O Si O Si CH$_3$ CH$_3$]$_4$ CH$_3$ Si CH$_3$ O

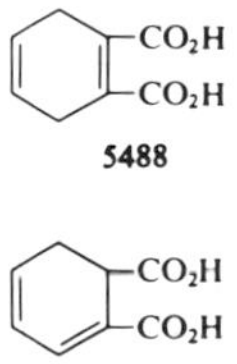

5676

OH

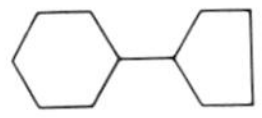

5461

CH$_2$CH$_2$OH

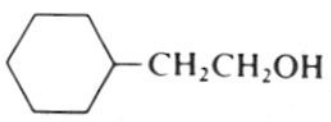

5542

HN O O =NH O HN

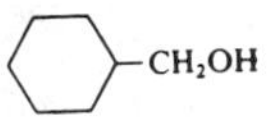

5406

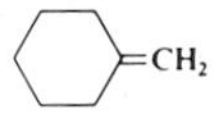

5463

CH$_2$OH

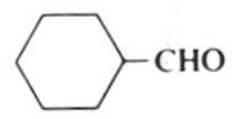

5543

CH$_3$ Si O CH$_3$ CH$_3$ Si O Si O O Si O Si CH$_3$ CH$_3$ O Si CH$_3$

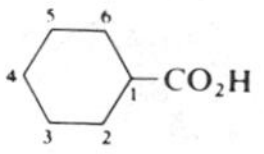

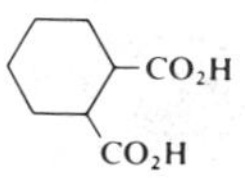

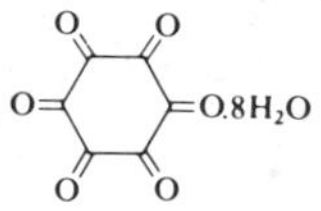

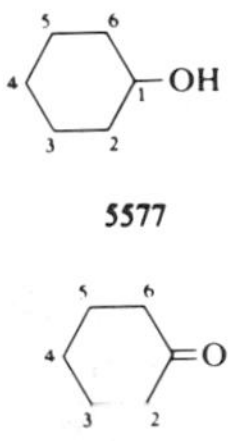

OH N N.2H$_2$O HO N OH

5434

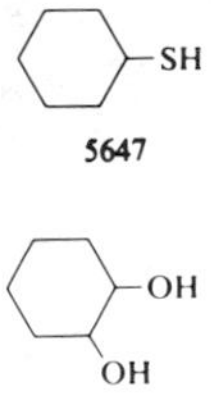

(CH$_2$)$_6$ O (CH$_2$)$_6$

5464

=CH$_2$

5551

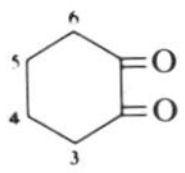

Cl N Cl N N Cl

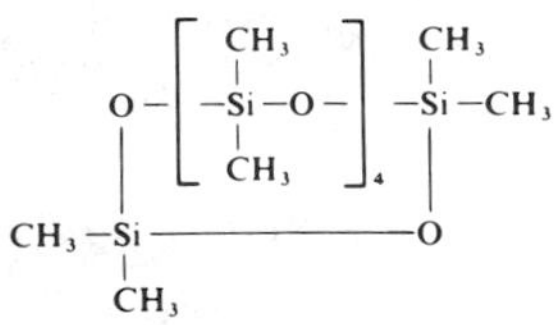

5436

5465

CHO

5563

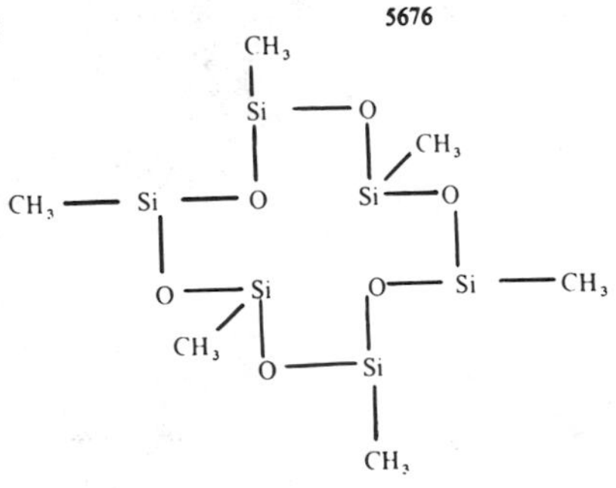

5677

STRUCTURAL FORMULAS OF ORGANIC COMPOUNDS (Continued)

In Numeric as they occur in Organic Compounds Table

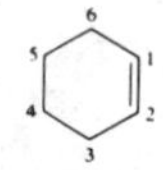

6 5 1 4 2 3
5678

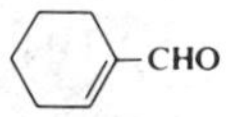

CHO
5697

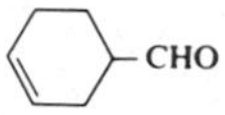

CHO
5698

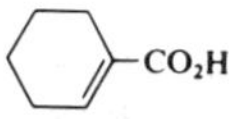

CO_2H
5699

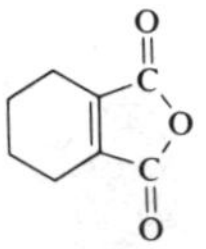

O C O C O
5702

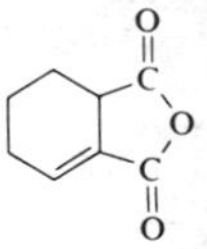

O C O C O
5703

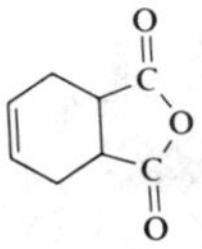

O C O C O
5704

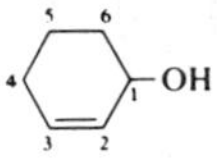

5 6 4 OH 1 3 2
5707

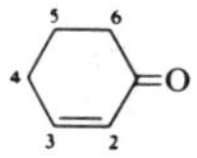

5 6 4 O 3 2
5713

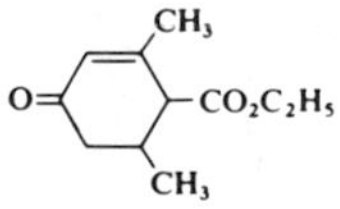

CH_3 O $CO_2C_2H_5$ CH_3
5717

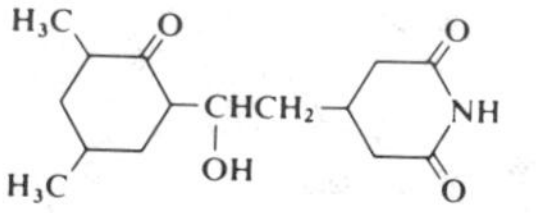

H_3C O O $CHCH_2$ NH OH H_3C O
5725

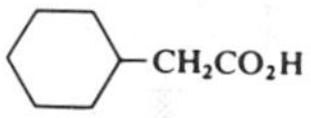

CH_2CO_2H
5726

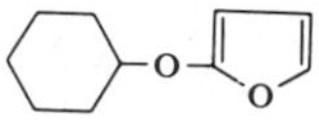

O O
5732

O
5738

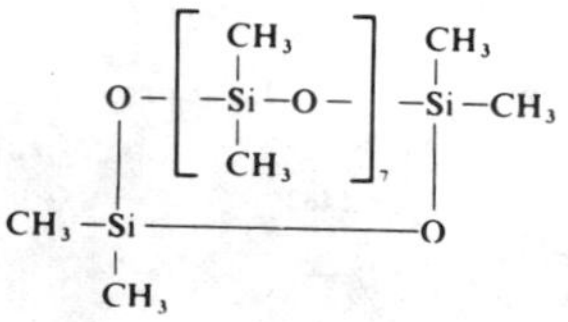

CH_3 CH_3 O —Si—O— —Si—CH_3 CH_3 7 CH_3—Si O CH_3
5739

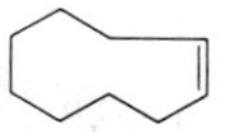

5740

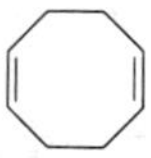

5742

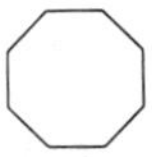

5744

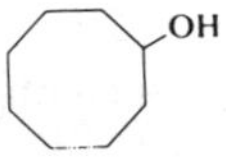

OH
5745

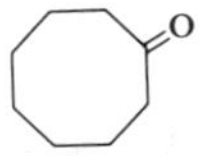

O
5746

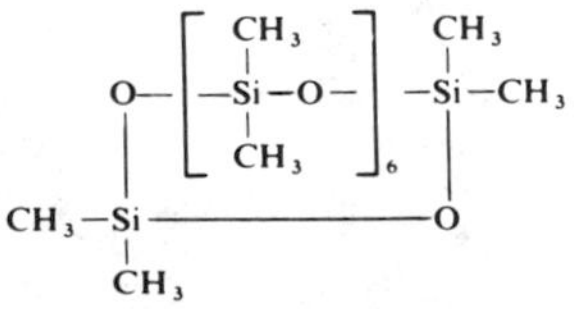

CH_3 CH_3 O —Si—O— —Si—CH_3 CH_3 6 CH_3—Si O CH_3
5748

5749

5752

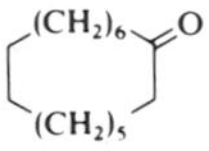

$(CH_2)_6$ O $(CH_2)_5$
5754

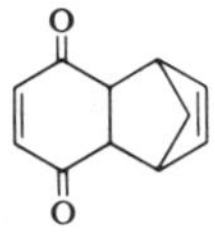

O O
5756

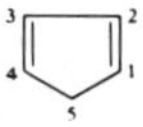

3 2 4 1 5
5756a

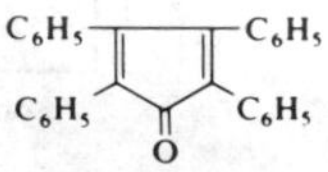

C_6H_5 C_6H_5 C_6H_5 C_6H_5 O
5760

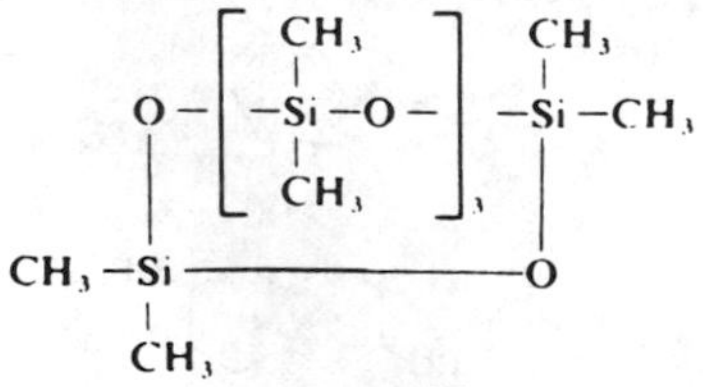

CH_3 CH_3 O —Si—O— —Si—CH_3 CH_3 3 CH_3—Si O CH_3
5761

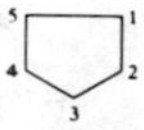

5 1 4 2 3
5762

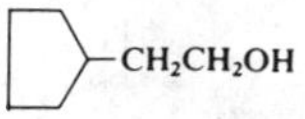

CH_2CH_2OH
5774

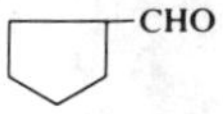

CHO
5784

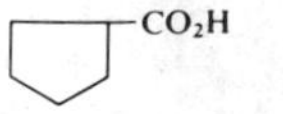

CO_2H
5785

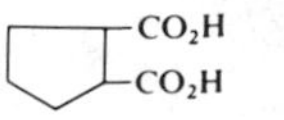

CO_2H CO_2H
5789

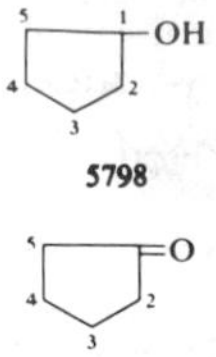

5 1 OH 4 2 3
5798

5 O 4 2 3
5801

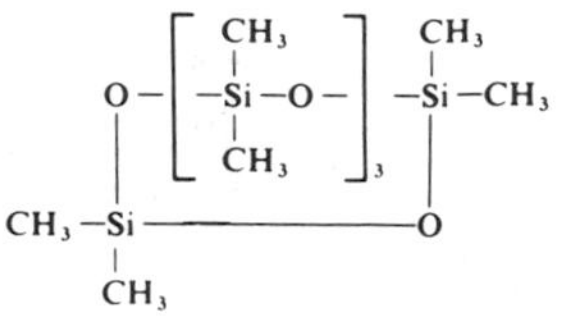

CH_3 CH_3 O —Si—O— —Si—CH_3 CH_3 3 CH_3—Si O CH_3
5805

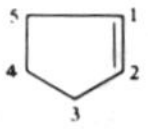

5 1 4 2 3
5806

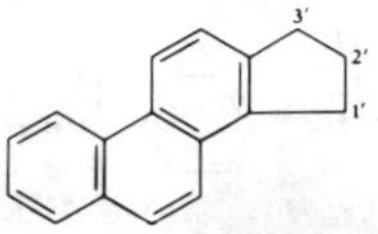

3′ 2′ 1′
5820

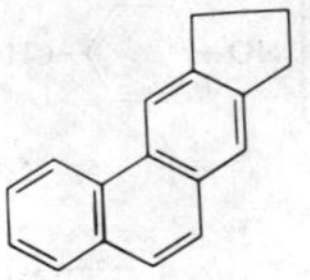

5822

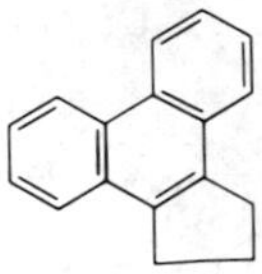

5823

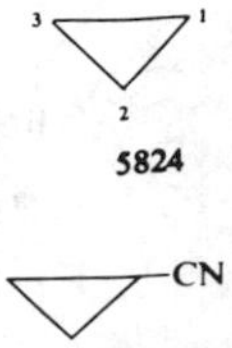

3 1 2
5824

CN
5838

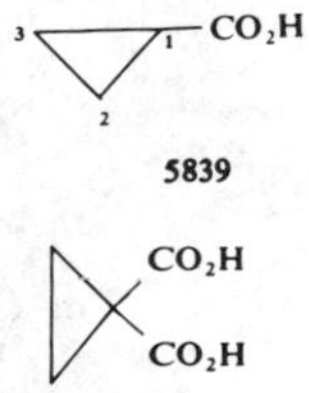

3 CO_2H 1 2
5839

CO_2H CO_2H
5847

STRUCTURAL FORMULAS OF ORGANIC COMPOUNDS (Continued)

In Numeric as they occur in Organic Compounds Table

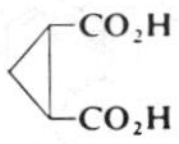

5849

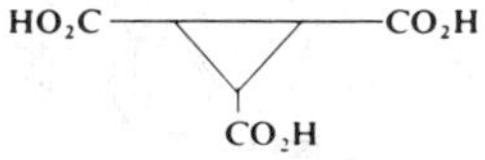

5856

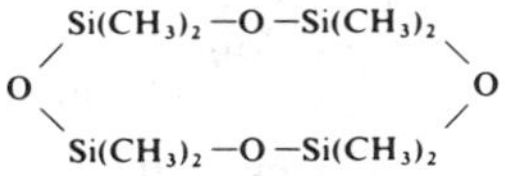

5858

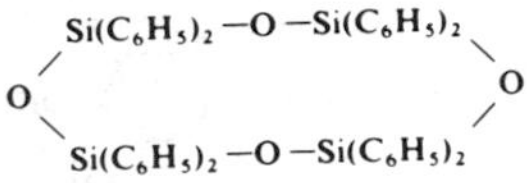

5859

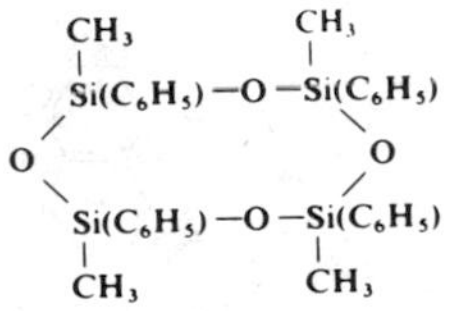

5860

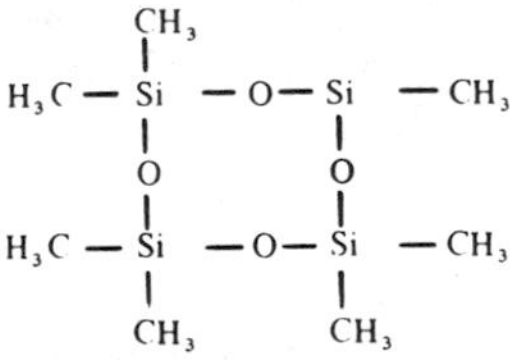

5861

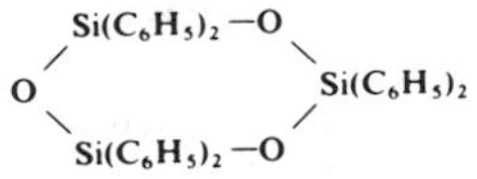

5862

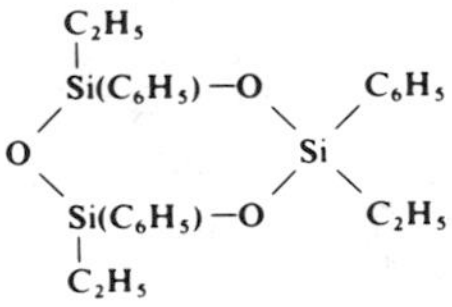

5863

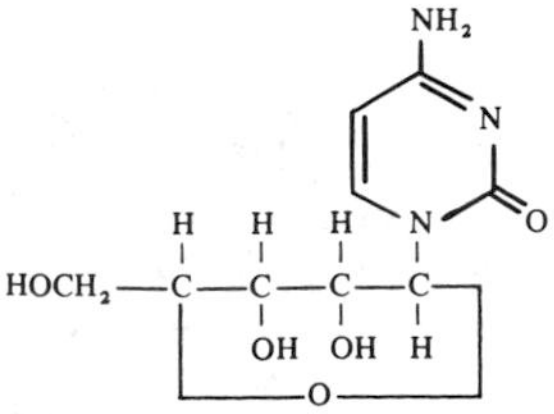

5877

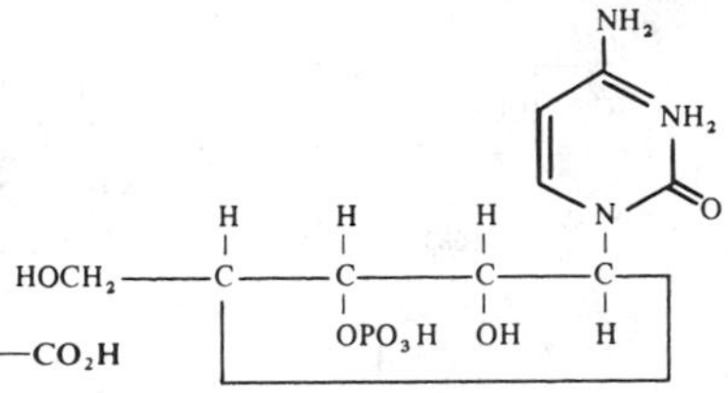

5878

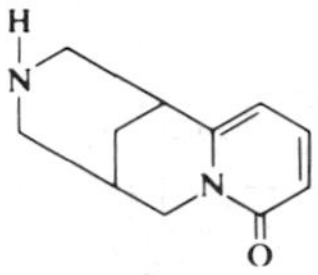

5879

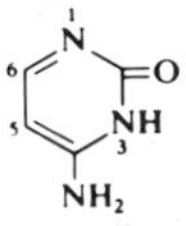

5881

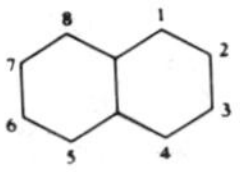

5887

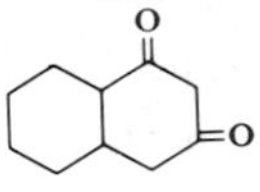

5893

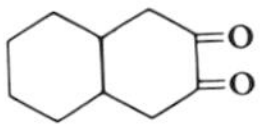

5895

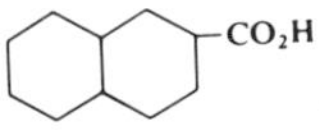

5897

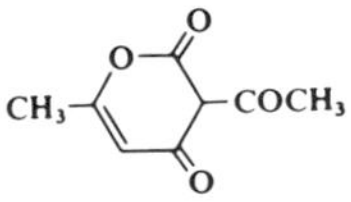

5947

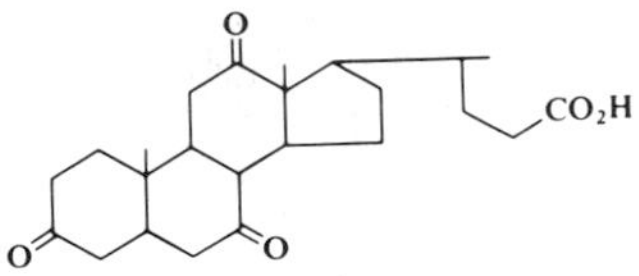

5948

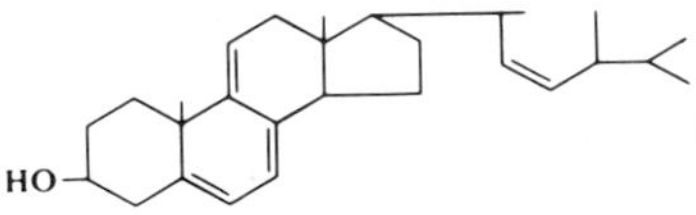

5949

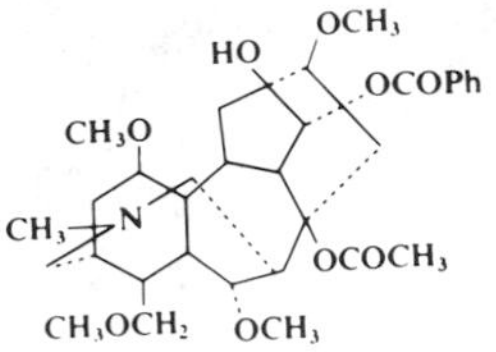

5950

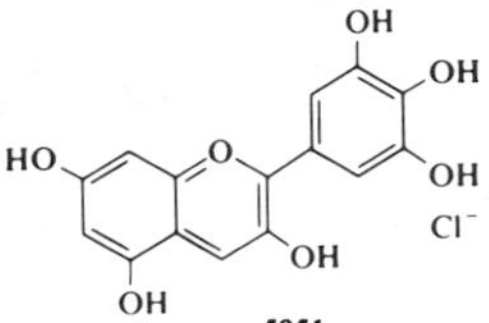

5951

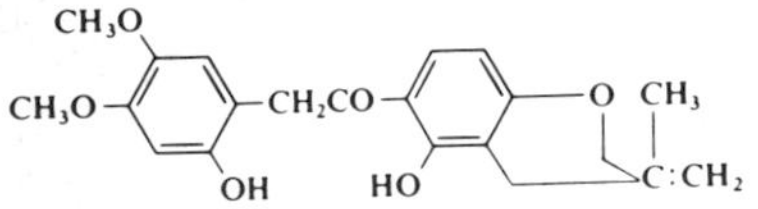

5953

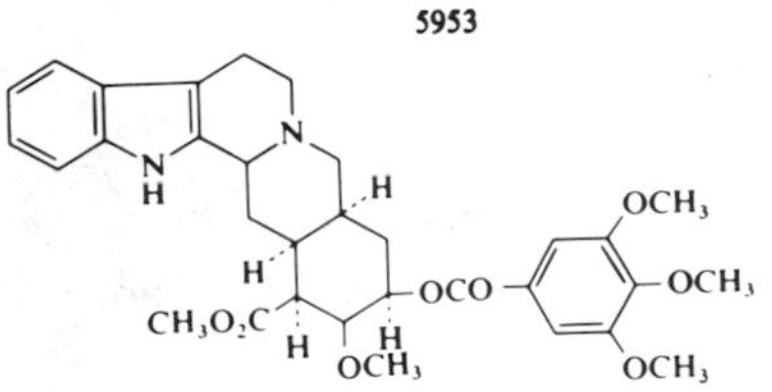

5954

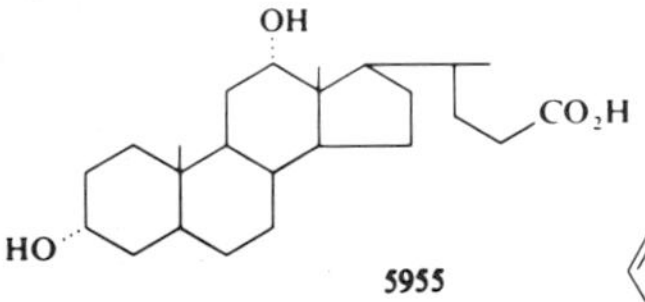

5955

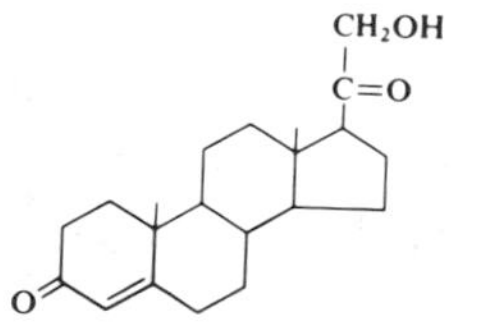

5956

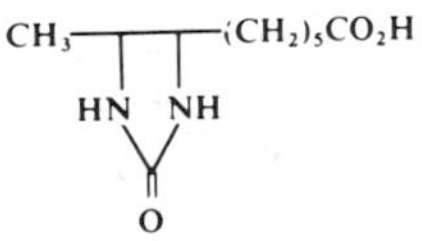

5957

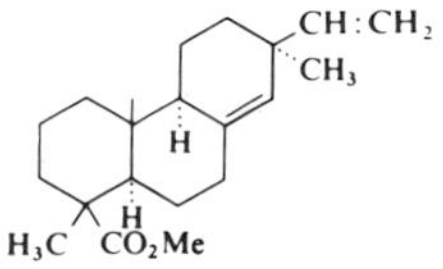

5960

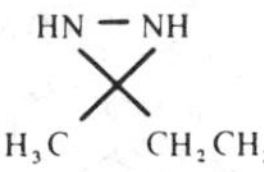

5966

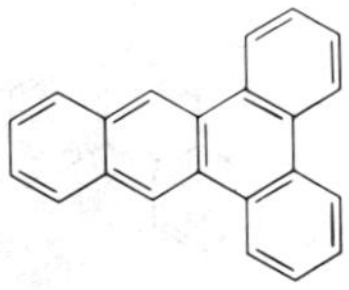

5982

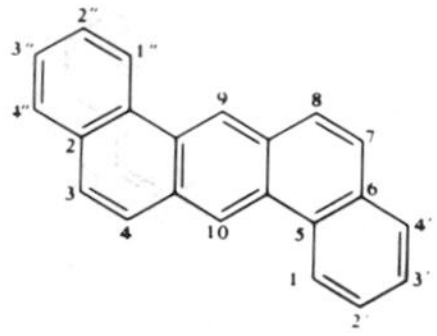

5983

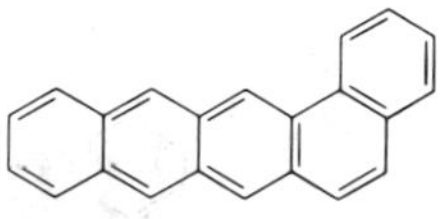

5985

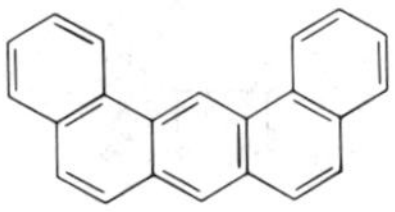

5986

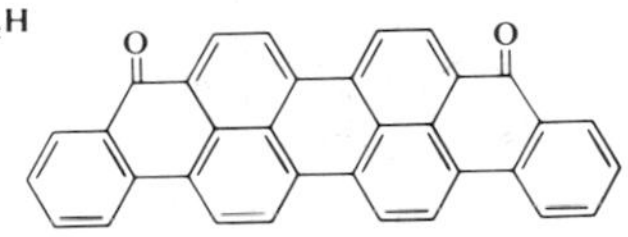

5987

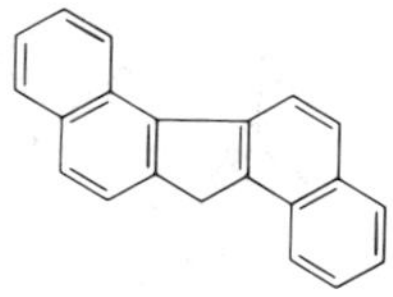

5989

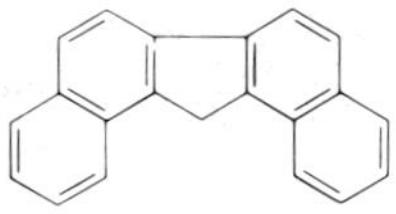

5990

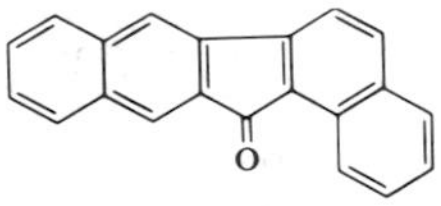

5991

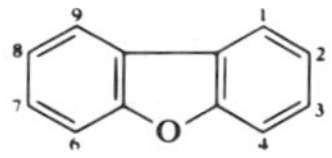

5992

In Numeric as they occur in Organic Compounds Table

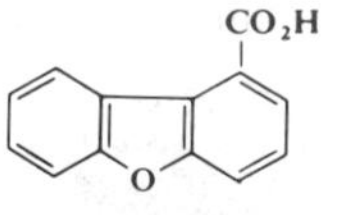

CO_2H O
6006

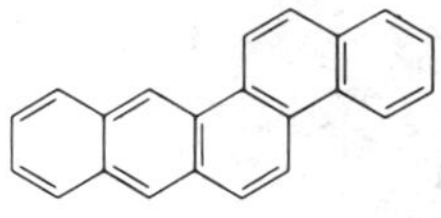

6009

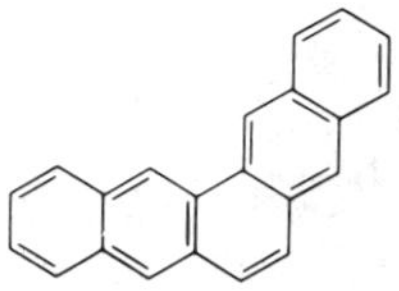

6010

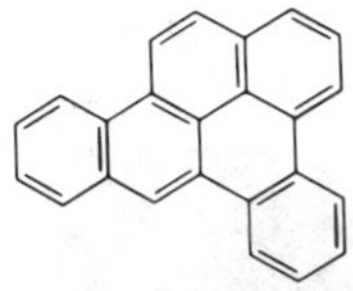

6011

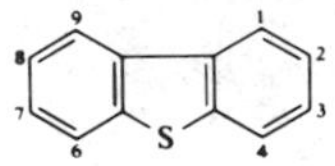

9 1 8 2 7 3 6 S 4
6012

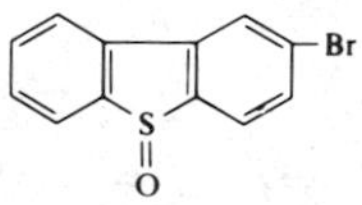

Br S O
6016

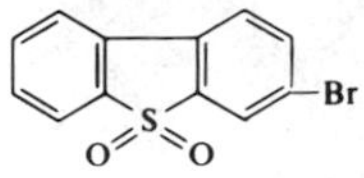

Br S O O
6017

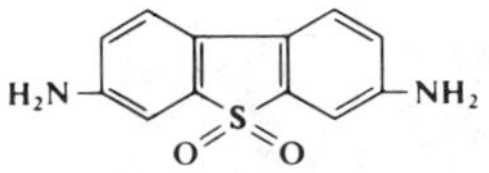

H_2N NH_2 S O O
6021

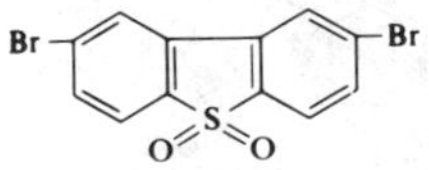

Br Br S O O
6023

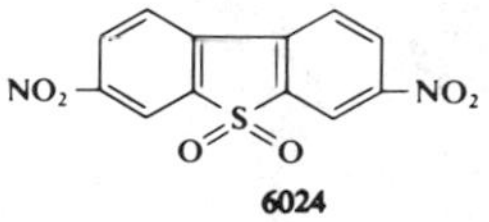

NO_2 NO_2 S O O
6024

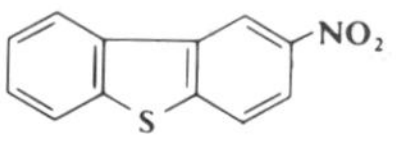

NO_2 S
6025

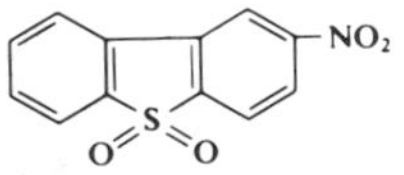

NO_2 S O O
6026

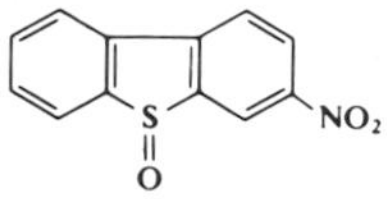

NO_2 S O
6028

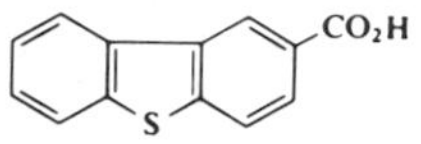

CO_2H S
6029

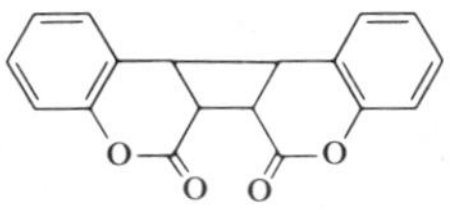

O O O O
6072

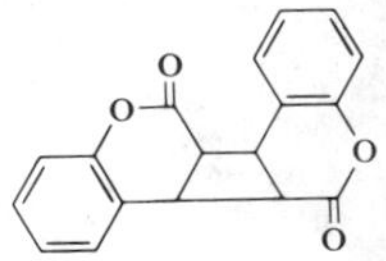

O O O O
6073

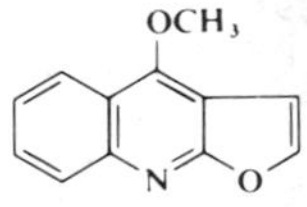

OCH_3 N O
6074

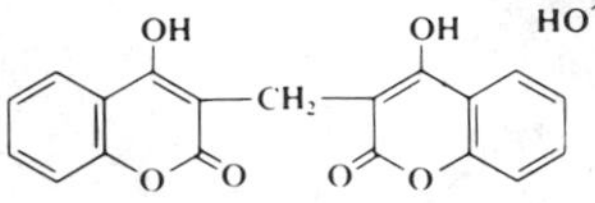

OH OH CH_2 O O O O
6075

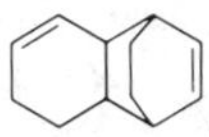

6076

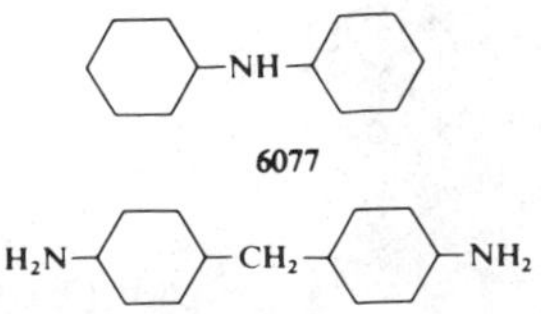

NH
6077

H_2N CH_2 NH_2
6079

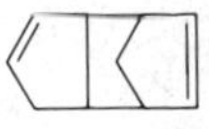

6082

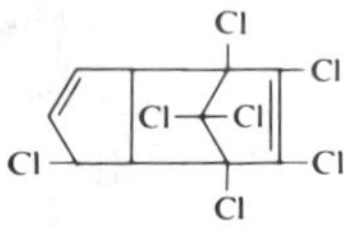

Cl Cl Cl Cl Cl Cl Cl
6083

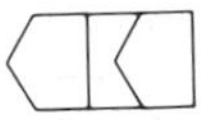

6084

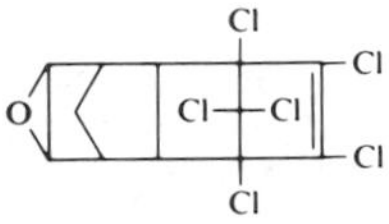

Cl Cl O Cl Cl Cl Cl
6086

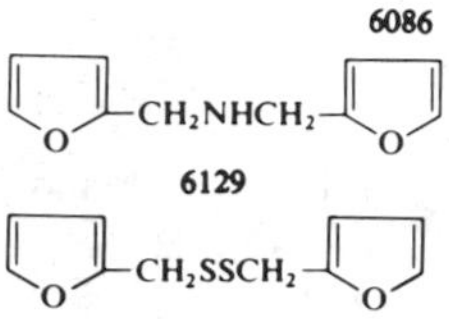

CH_2NHCH_2 O O
6129

CH_2SSCH_2 O O
6130

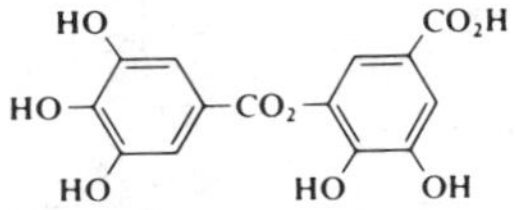

HO CO_2H HO CO_2 HO HO OH
6131

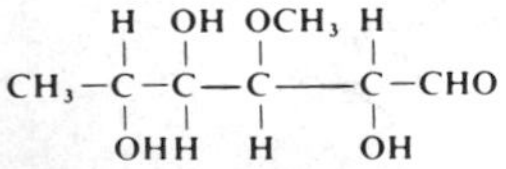

$CH_3-C(H)(OH)-C(OH)(H)-C(OCH_3)(H)-C(H)(OH)-CHO$
6132

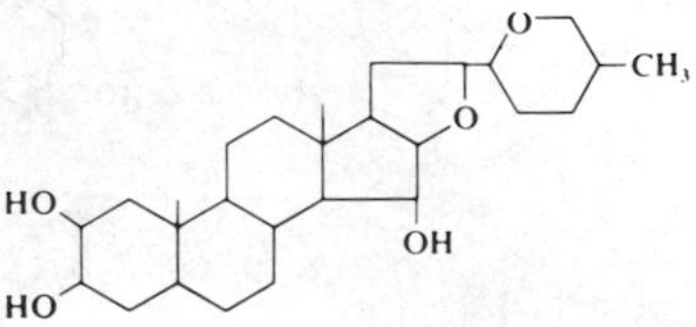

O CH_3 O HO OH HO
6133

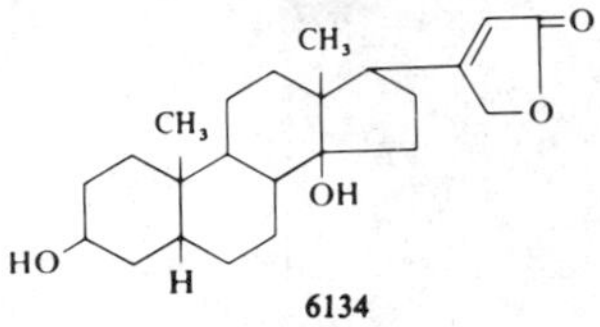

CH_3 O O CH_3 OH HO H
6134

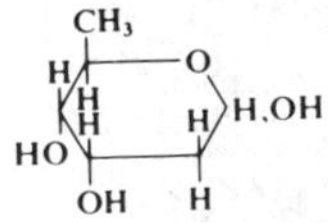

CH_3 H O H H H.OH HO H OH H
6136

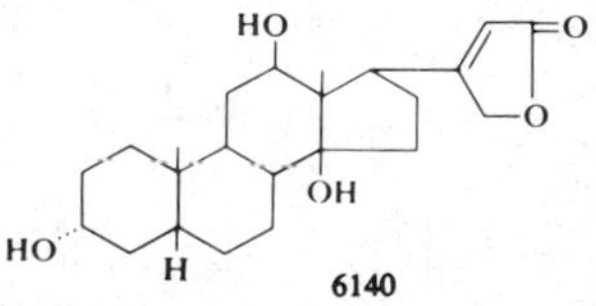

HO O O OH HO H
6140

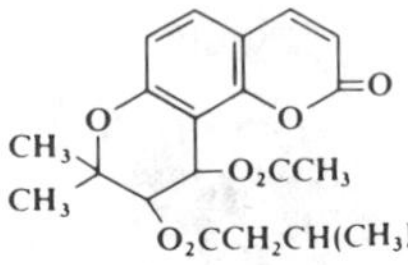

O O CH_3 CH_3 O_2CCH_3 $O_2CCH_2CH(CH_3)_2$
6145

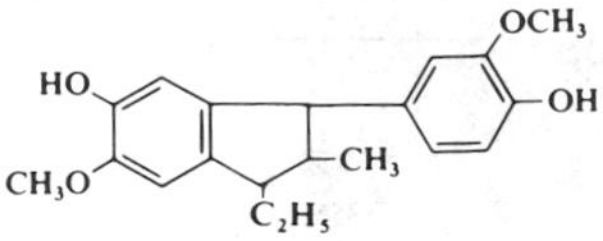

OCH_3 HO OH CH_3O CH_3 C_2H_5
6152

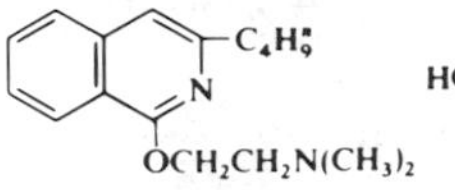

$C_4H_9^n$ N HCl $OCH_2CH_2N(CH_3)_2$
6155

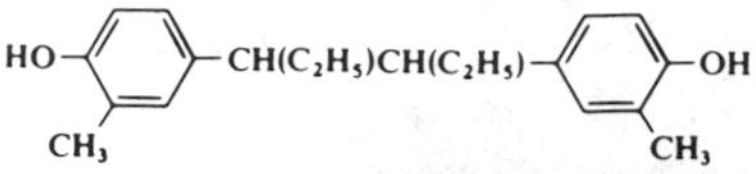

HO $CH(C_2H_5)CH(C_2H_5)$ OH CH_3 CH_3
6168

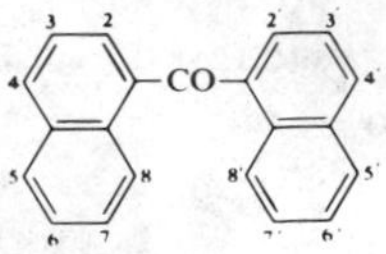

3 2 2' 3' 4 CO 4' 5 8 8' 5' 6 7 7' 6'
6181

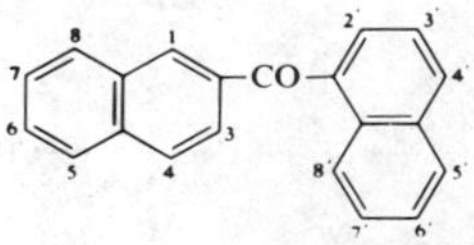

8 1 2' 3' 7 CO 4' 6 3 8' 5 4 5' 7' 6'
6182

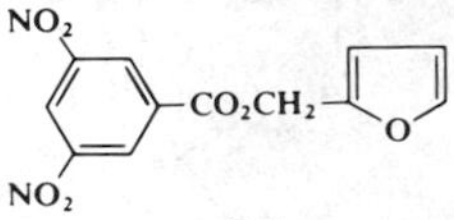

NO_2 CO_2CH_2 O NO_2
6196

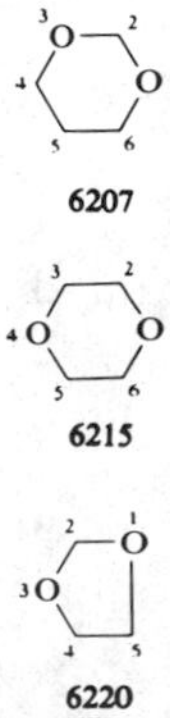

3 O 2 O 1 4 5 6
6207

3 2 4 O O 1 5 6
6215

2 O 1 3 O 4 5
6220

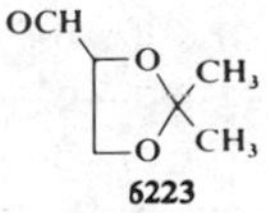

OCH O CH_3 O CH_3
6223

In Numeric as they occur in Organic Compounds Table

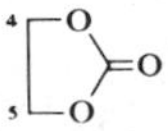

6224

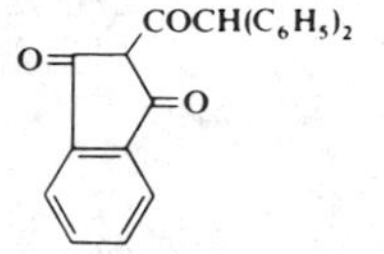

$COCH(C_6H_5)_2$

6231

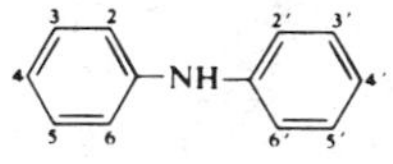

6249

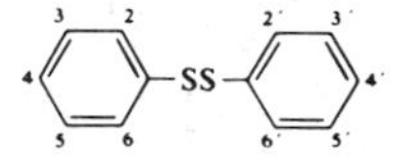

6267

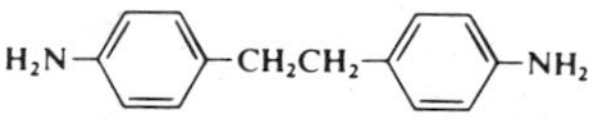

H_2N–⟨⟩–CH_2CH_2–⟨⟩–NH_2

6269

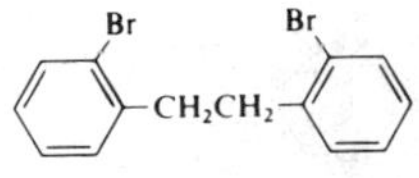

6270

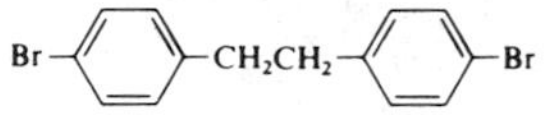

Br–⟨⟩–CH_2CH_2–⟨⟩–Br

6271

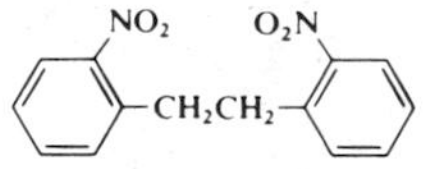

6274

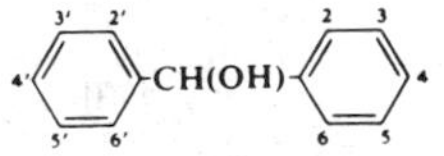

6314

$(CH_3$–⟨⟩–$)_2CHOH$

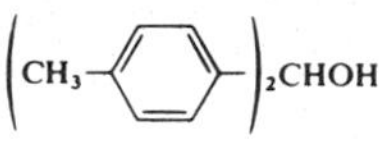

6315

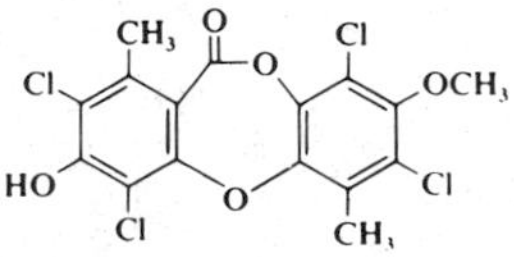

6327

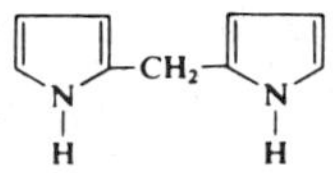

6342

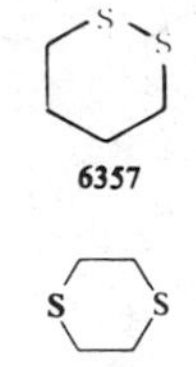

6357

6358

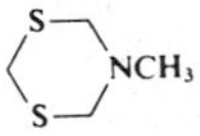

6359

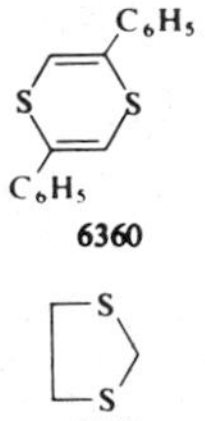

6360

6361

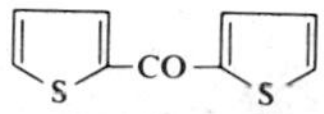

6362

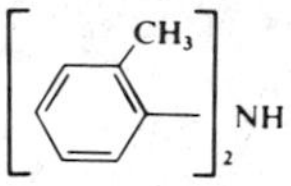

6365

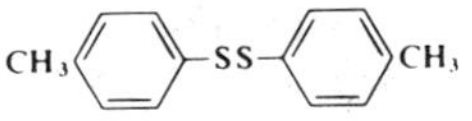

6370

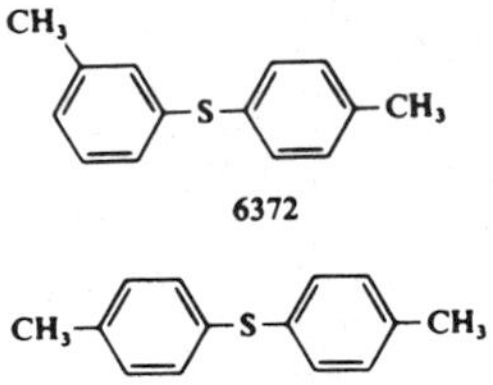

6372

6373

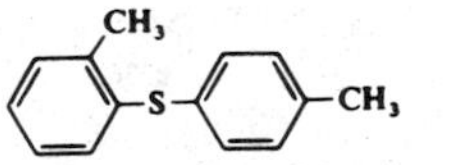

6374

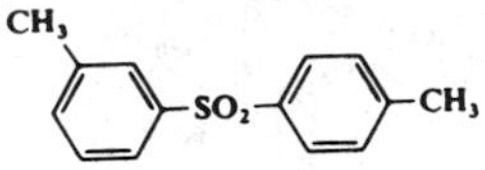

6376

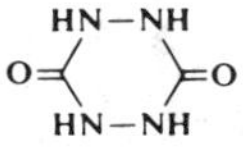

6379

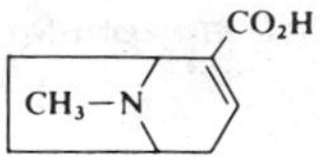

6423

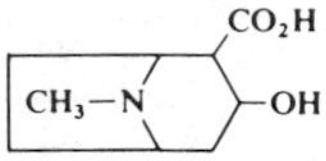

6424

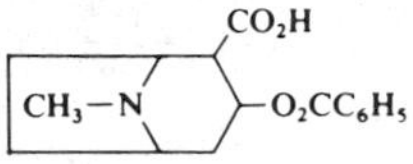

6428

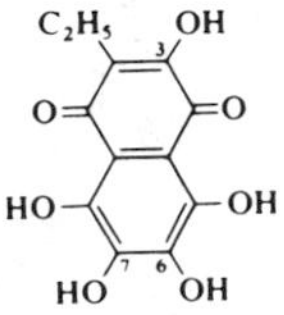

6431

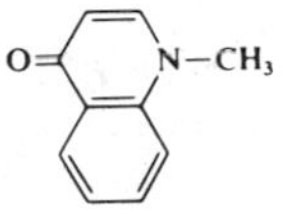

6433

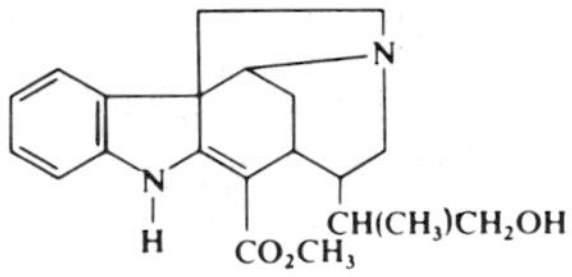

6434

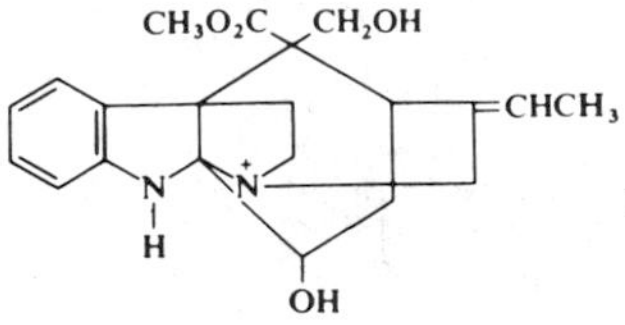

6435

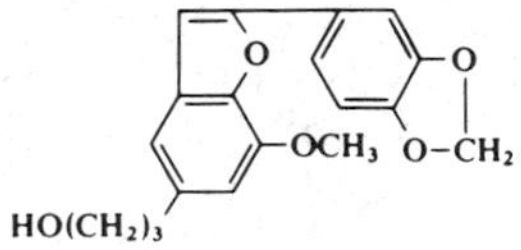

6438

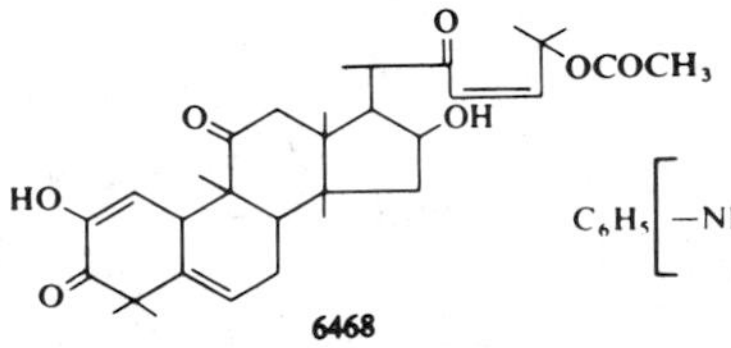

6468

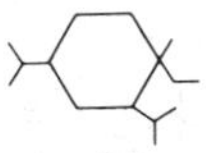

6470

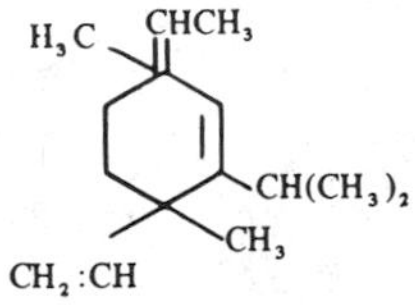

6471

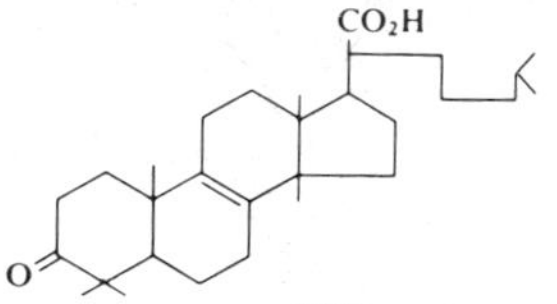

6472

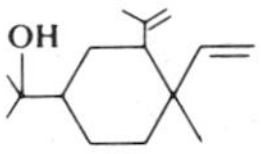

6473

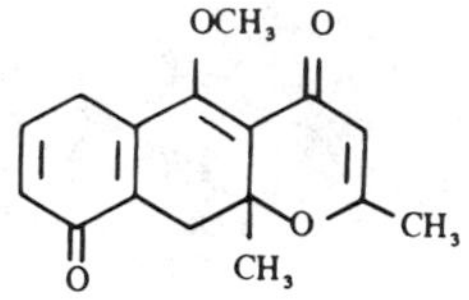

6476

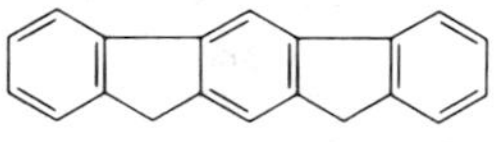

6477

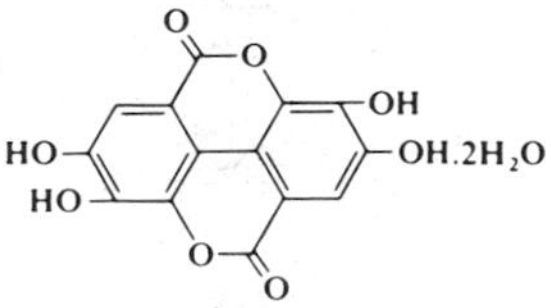

6478

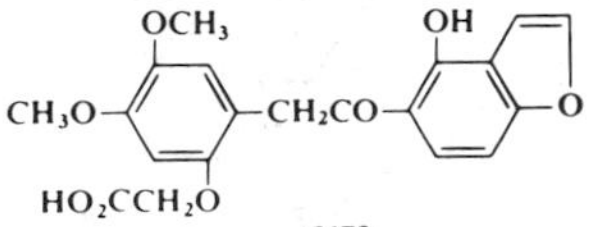

6479

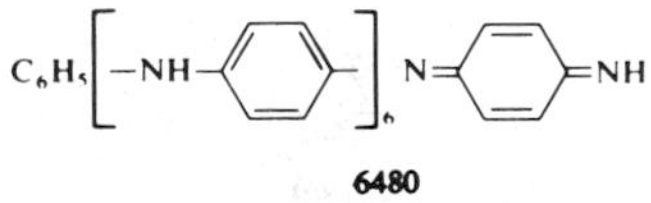

6480

STRUCTURAL FORMULAS OF ORGANIC COMPOUNDS (Continued)

In Numeric as they occur in Organic Compounds Table

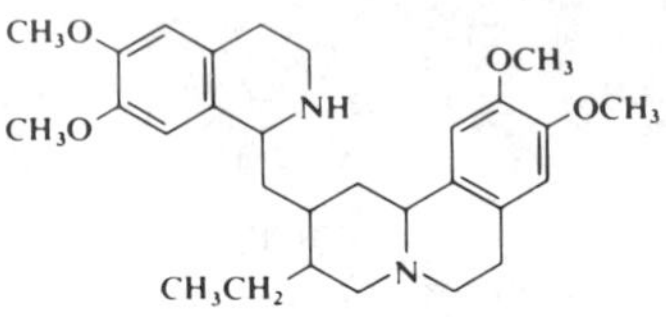

CH_3O CH_3O NH OCH_3 OCH_3 CH_3CH_2 N

6481

$H_2NCONHN$

6504

HO

6521

O OH $C_{12}H_{21}O_{11}.O$ O OH

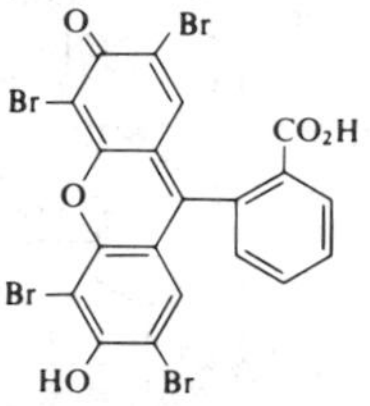

6541

O Br

6508

H H OH H
CH_3—C—C—C—C—CH_2OH
OH OH H OH

6522

O Br Br CO_2H O Br HO Br

6485

H_3C CH_3 CH_3 $CH_2CH_2NH_2$

6511

OH OH H OH
CH_3—C—C—C—C—CHO
H H OH H

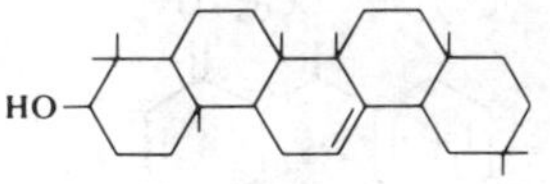

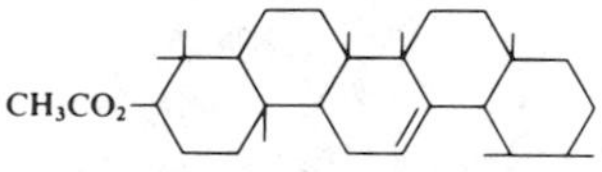

6524

HO O OH

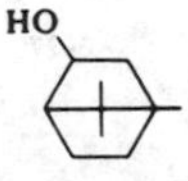

6542

OH OH OH HO O OH

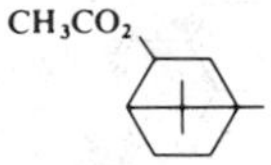

6512

HO

6525

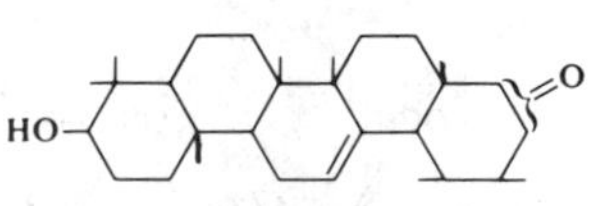

HO CH_3 CH_2 CH_3 CH_3

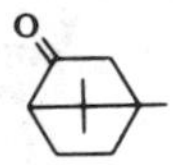

6543

H_2N—◯—CH(OH)CH(NH_2)CH_3 . 2HCl

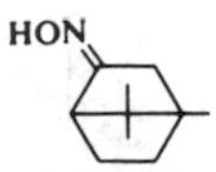

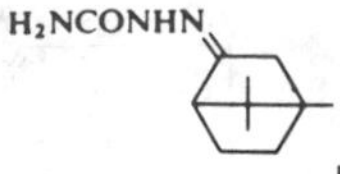

6489

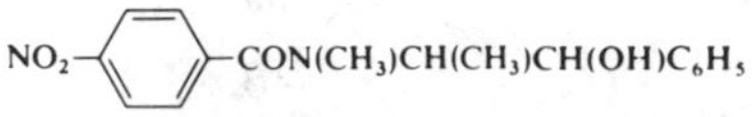

O

6526

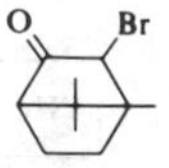

O CH_3 CH_2 CH_3 CH_3

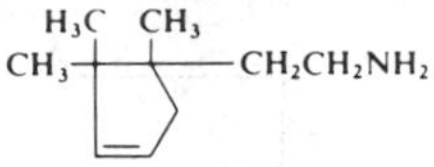

6544

NO_2—◯—CON(CH_3)CH(CH_3)CH(OH)C_6H_5

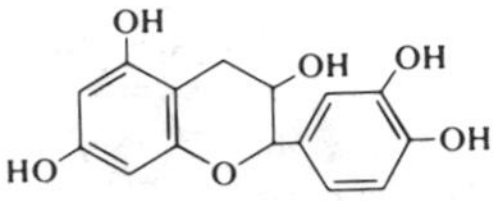

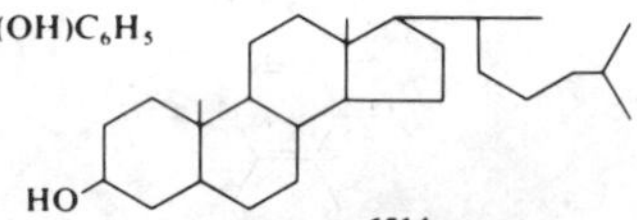

6494

HO

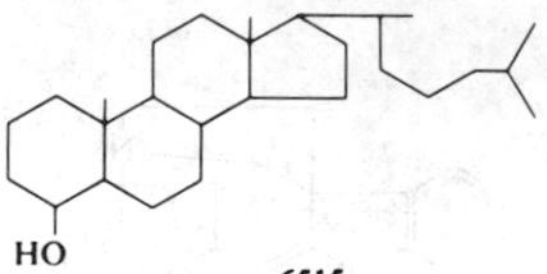

6514

N CH_2OH

6528

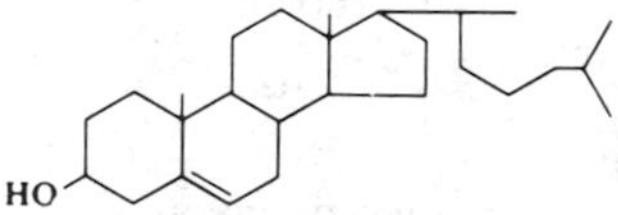

O O CH_3 CH_2 CH_3 CH_3

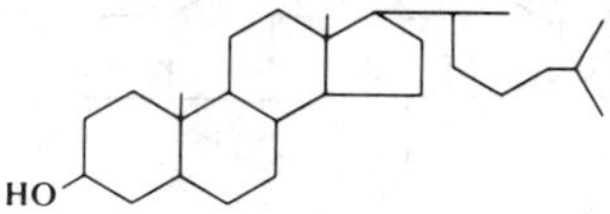

6545

HO

6496

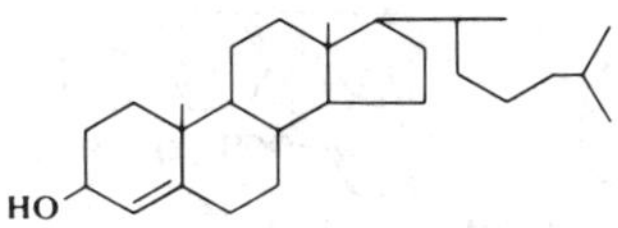

HO

6515

HO HO $CH_2CH_2NHCH_3$

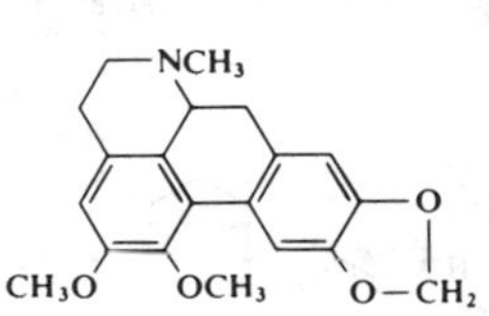

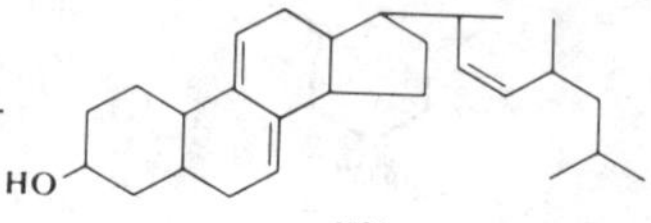

6529

CH_3CO_2

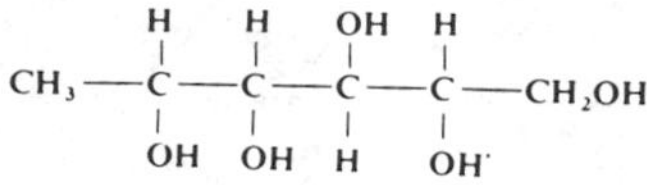

6497

HO

6516

N CH(OH) N CHCH$_3$ OCH_3

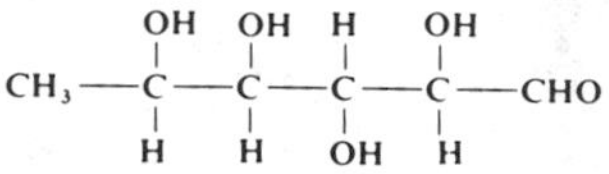

6530

HN N CH_3 $CONH_2$

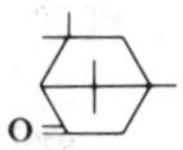

6546

HO

6499

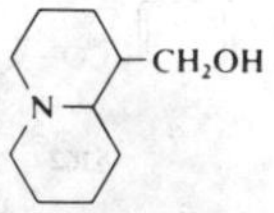

HO

6517

C_6H_5 CO_2H HO_2C C_6H_5

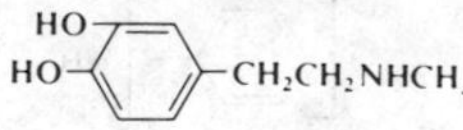

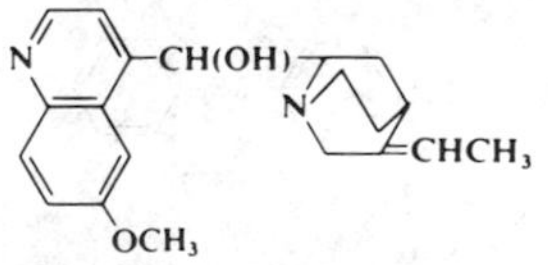

6533

CH_3CO_2

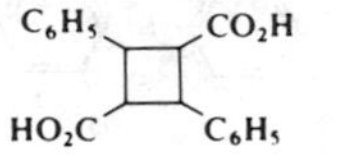

6500

O HO

6536

CH(CH_3)$_2$ OH CONH O N O N O CH_3N CH(CH_3)$_2$ N H

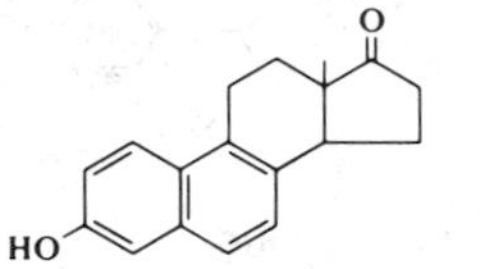

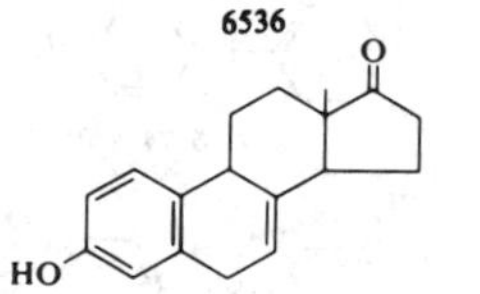

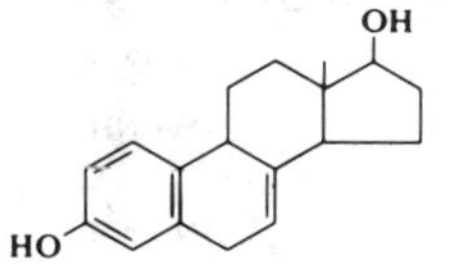

6547

HO O

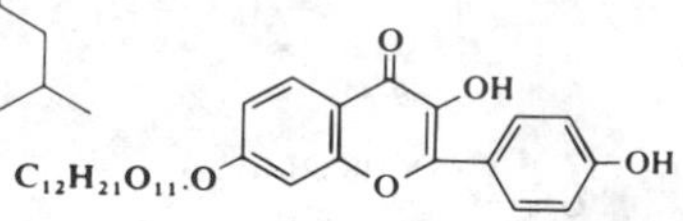

6501

HO

6518

O HO

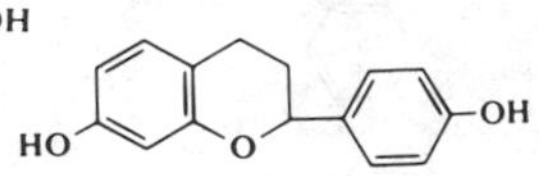

6539

O

6502

NCH_3 O CH_3O OCH_3 O—CH_2

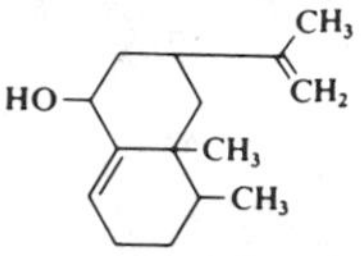

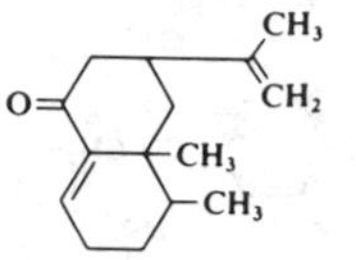

6519

OH HO

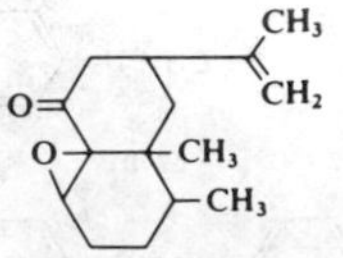

6540

CH(CH_3)$_2$ OH CONH O N O N O CH_3N $CH_2C_6H_5$ N H

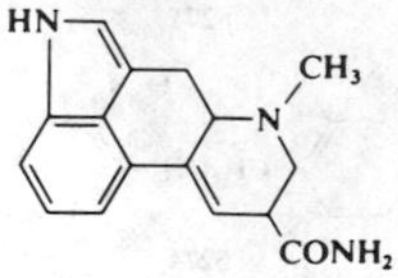

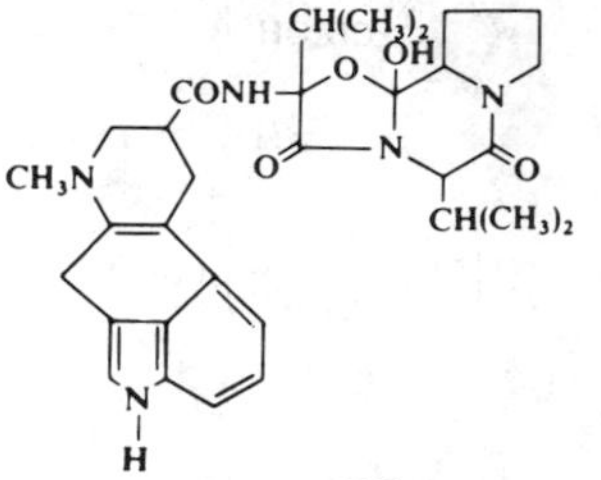

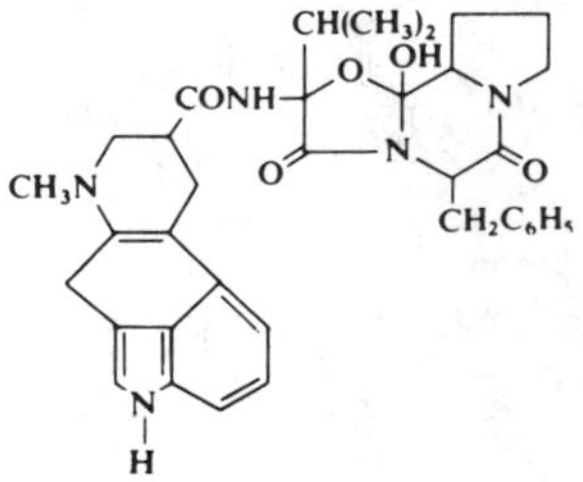

6549

HON

6503

In Numeric as they occur in Organic Compounds Table

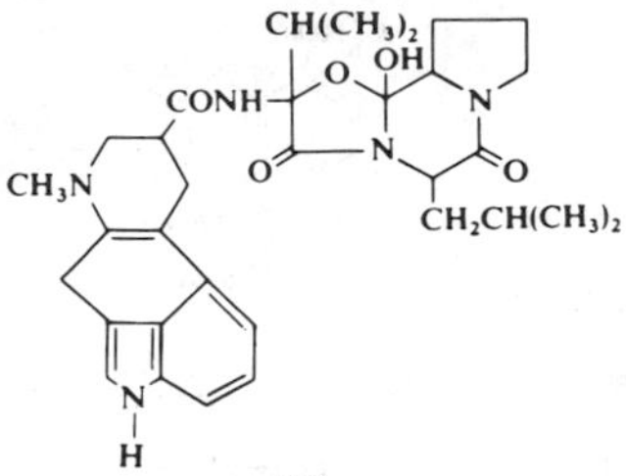

$CH(CH_3)_2$ OH CONH O N CH_3N O N O $CH_2CH(CH_3)_2$ N H

6551

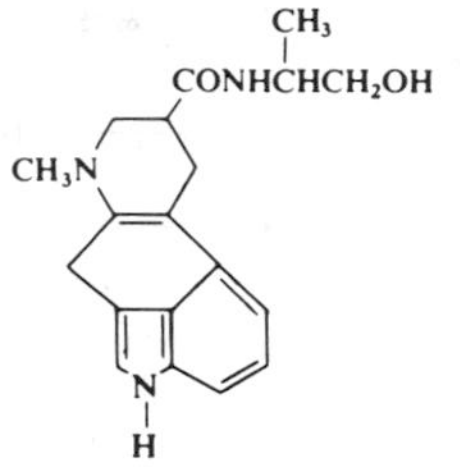

O

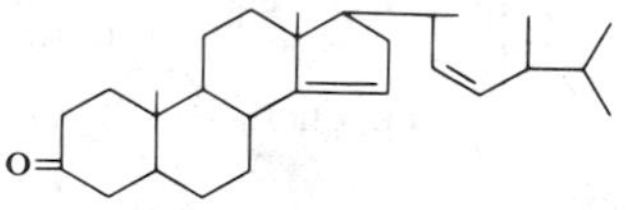

06561

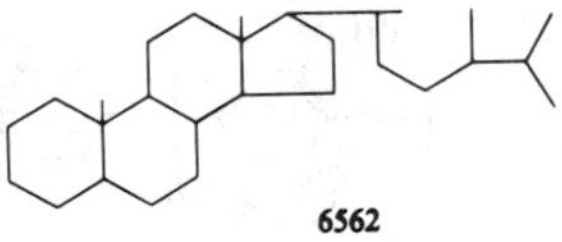

6562

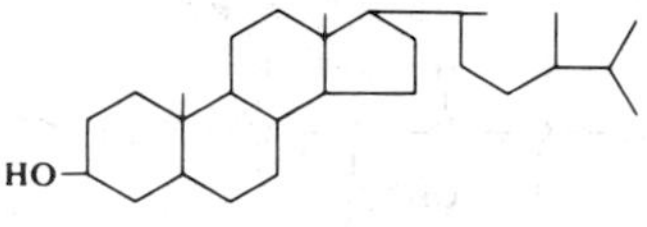

HO

6563

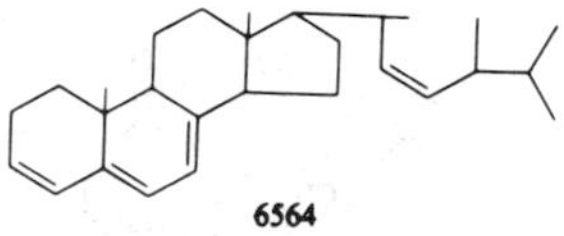

6564

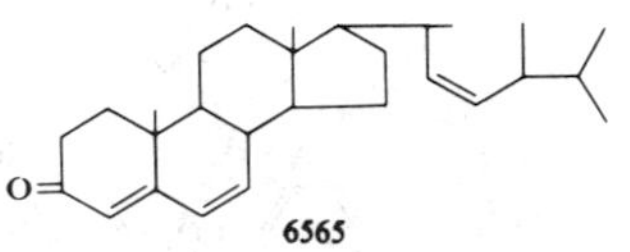

O

6565

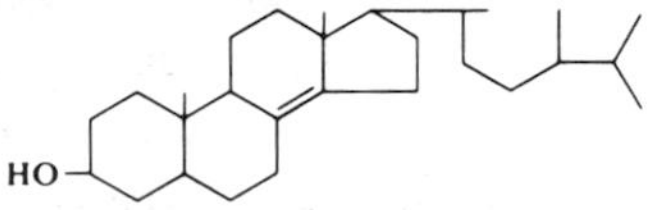

HO

6566

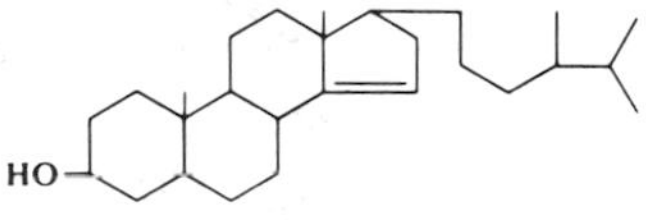

HO

6567

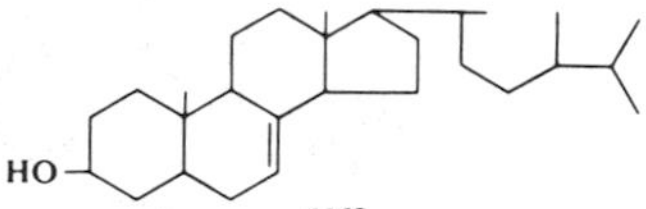

HO

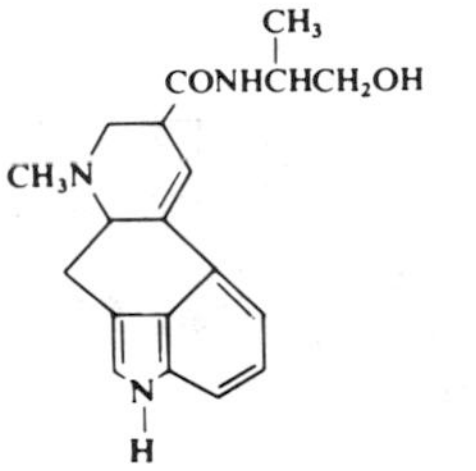

6568

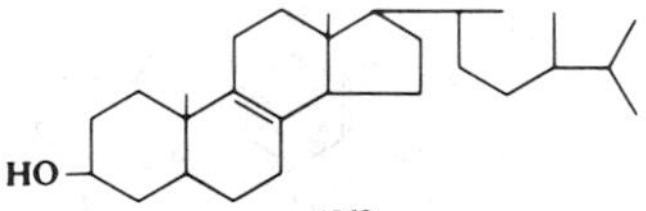

HO

6569

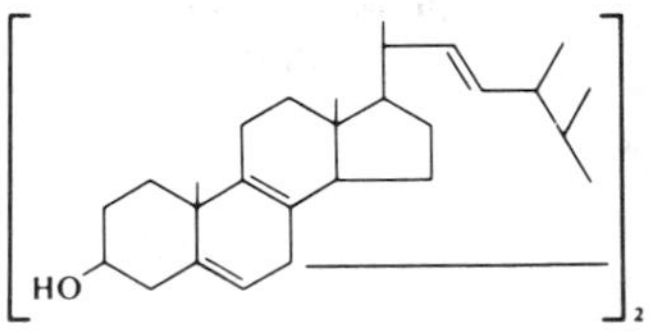

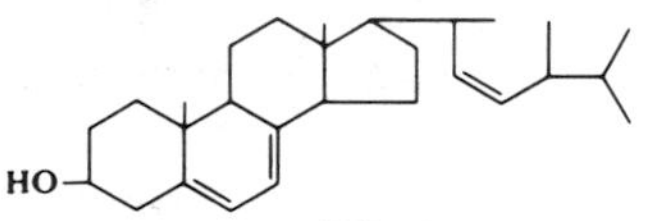

HO

6570

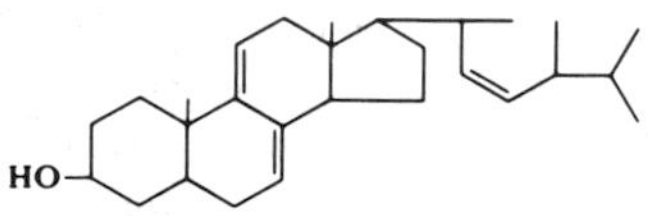

HO

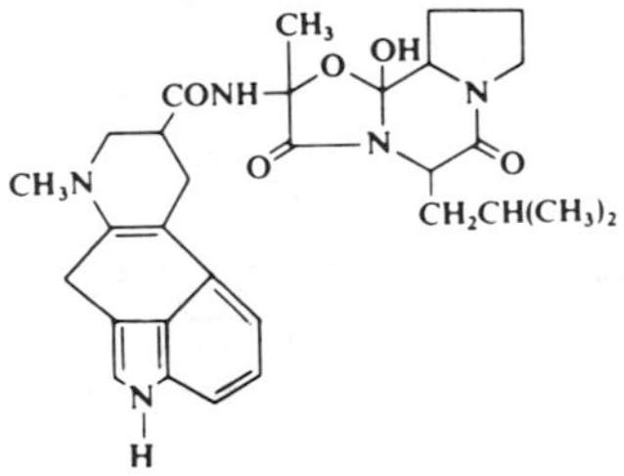

6571

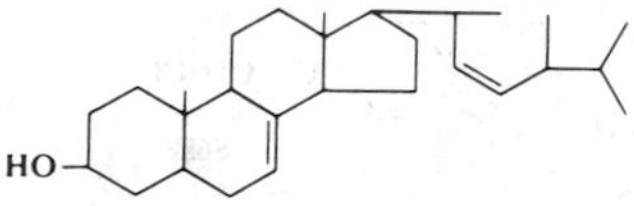

HO

6572

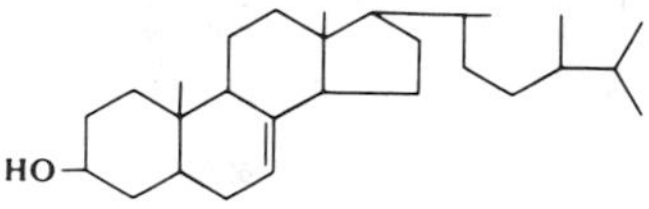

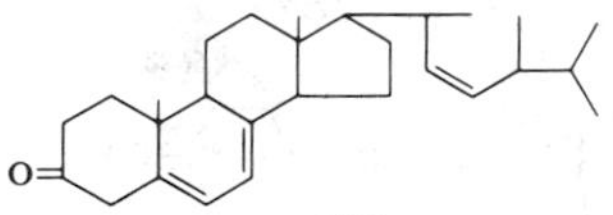

O

6573

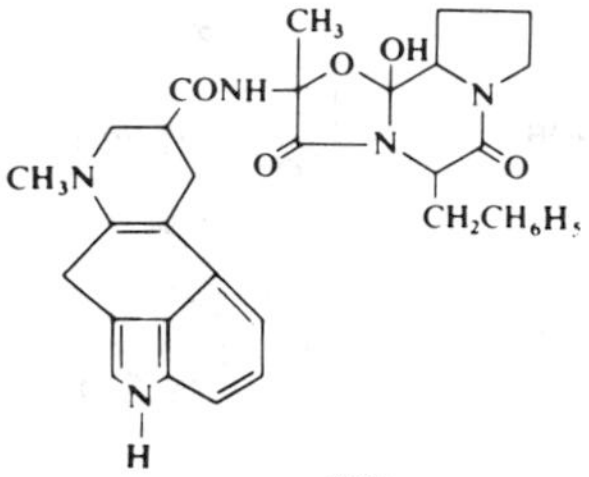

CH_3 OH CONH O N CH_3N O N O $CH_2CH_6H_5$ N H

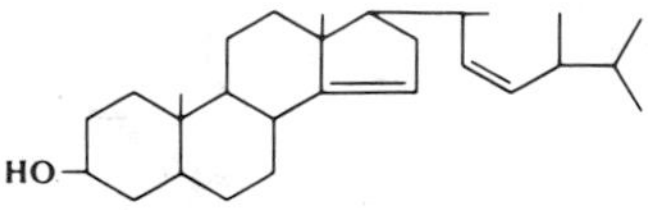

6574

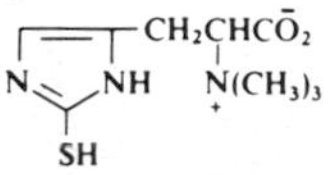

$CH_2CHC\bar{O}_2$ N NH $\overset{+}{N}(CH_3)_3$ SH

6576

R_1O OR_2 CH_3O N

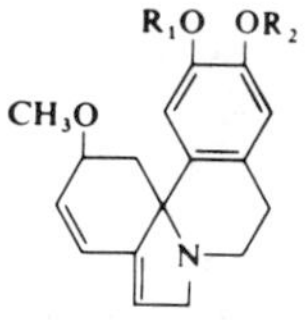

$R_1 = CH_3$ or H

$R_2 = H$ or CH_3

6579

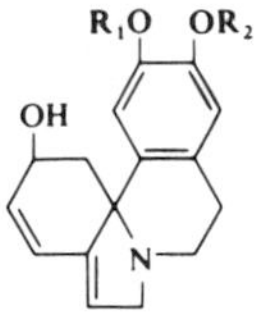

R_1O OR_2 OH N

$R_1 = CH_3$ or H

$R_2 = H$ or CH_3

6580

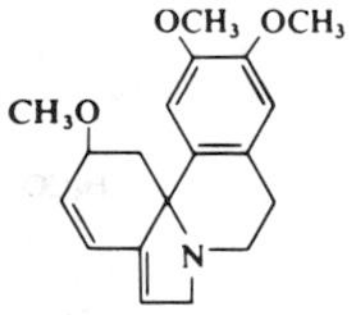

OCH_3 OCH_3 CH_3O N

6581

R_2O OR_2 CH_3O N

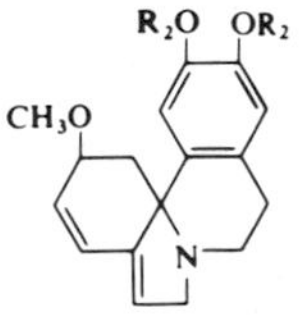

$R_1 = CH_3$ or H

$R_2 = H$ or CH_3

6582

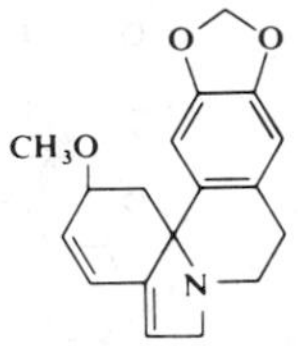

CH_3 $CONHCHCH_2OH$ CH_3N N H

6553

CH_3 $CONHCHCH_2OH$ CH_3N N H

6554

HO

6556

CH_3 OH CONH O N CH_3N O N O $CH_2CH(CH_3)_2$ N H

6557

HO

6559

HO

6560

O O CH_3O N

6584

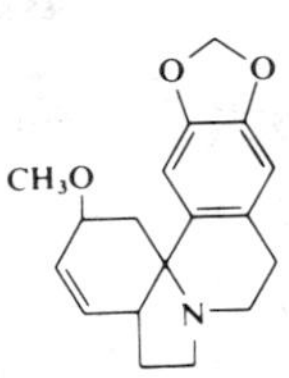

O O CH_3O N

6585

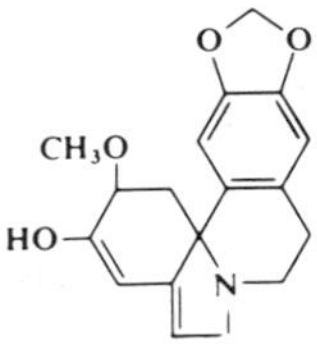

O O CH_3O HO N

6586

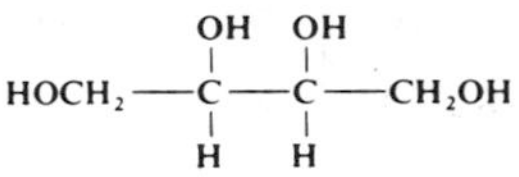

OH OH $HOCH_2$—C—C—CH_2OH H H

6589

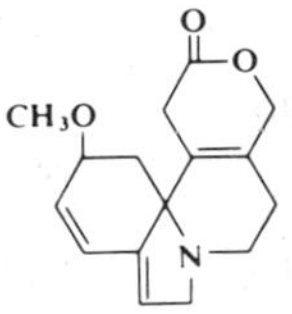

O O CH_3O N

6590

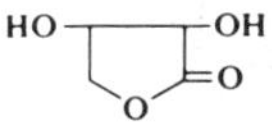

HO OH O O

6591

H H $HOCH_2$—C—C—CHO OH OH

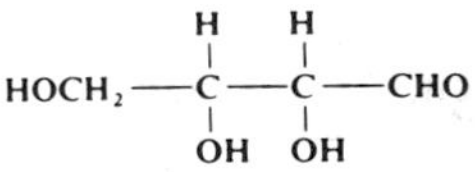

6593

In Numeric as they occur in Organic Compounds Table

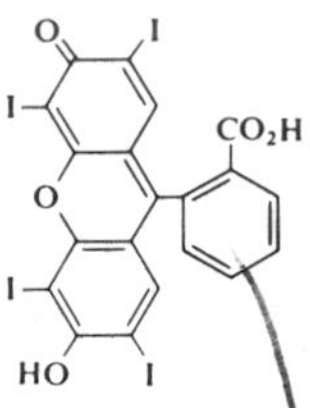

6595

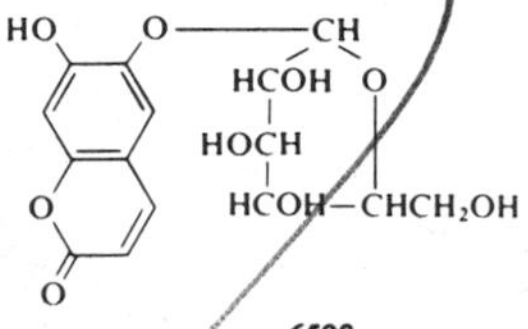

6598

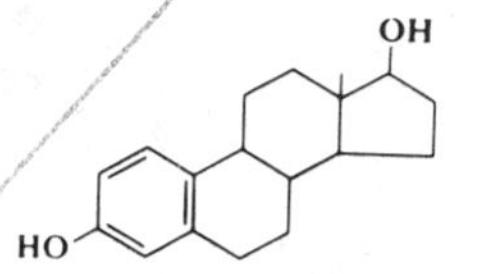

6599

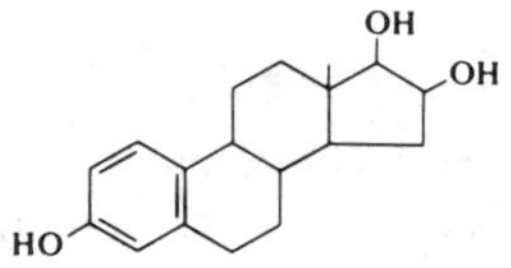

6601

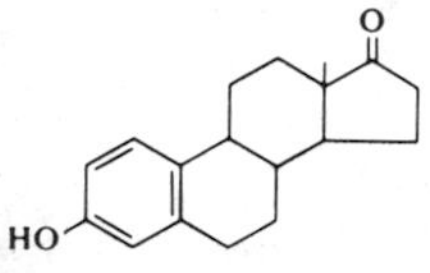

6602

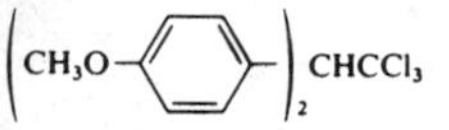

$(CH_3O-C_6H_4)_2CHCCl_3$

6605

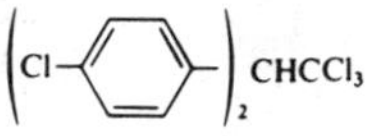

$(Cl-C_6H_4)_2CHCCl_3$

6609

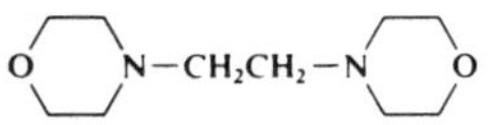

$N-CH_2CH_2-N$

6638

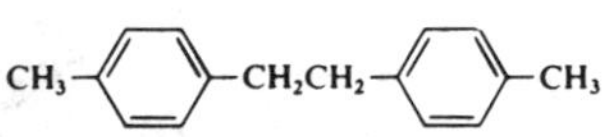

$CH_3-C_6H_4-CH_2CH_2-C_6H_4-CH_3$

6643

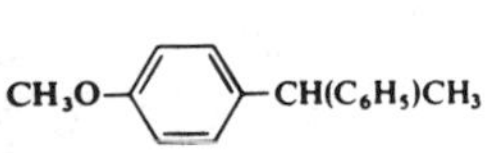

$CH_3O-C_6H_4-CH(C_6H_5)CH_3$

6655

$C_6H_{11}-CH(OH)CH(OH)-C_6H_{11}$

6681

$C_6H_5-CH(OH)CH(OH)-C_6H_5$

6682

COCO

6688

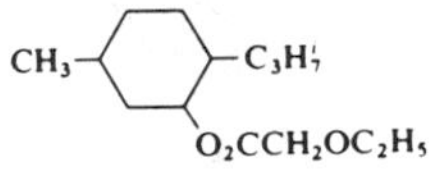

CH_3, C_3H_7, $O_2CCH_2OC_2H_5$

6742

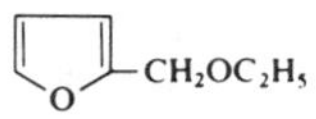

$CH_2OC_2H_5$

6794

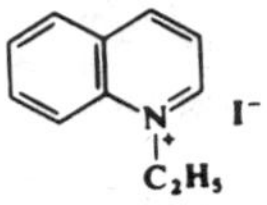

I^-, C_2H_5

6830

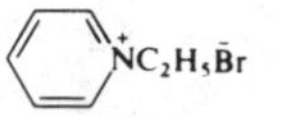

NC_2H_5Br

6831

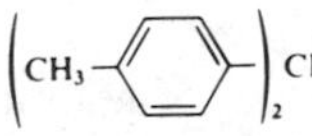

$(CH_3-C_6H_4)_2CHCH_3$

6889

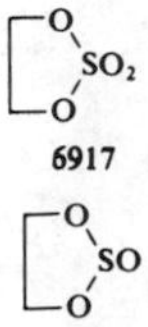

SO_2

6917

SO

6918

6919

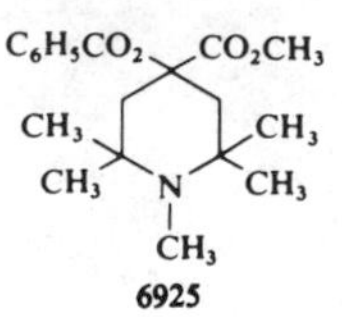

$C_6H_5CO_2$, CO_2CH_3, CH_3, CH_3, CH_3, CH_3, CH_3

6925

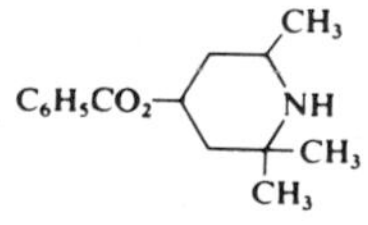

$C_6H_5CO_2$, CH_3, NH, CH_3, CH_3

6927

$CH_2:CHCH_2$, OCH_3, OH

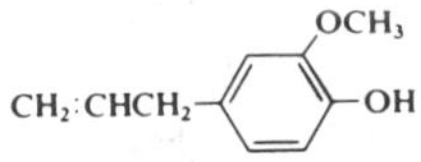

6931

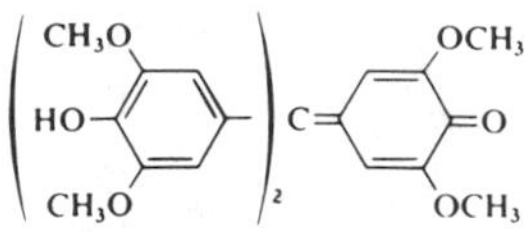

6933

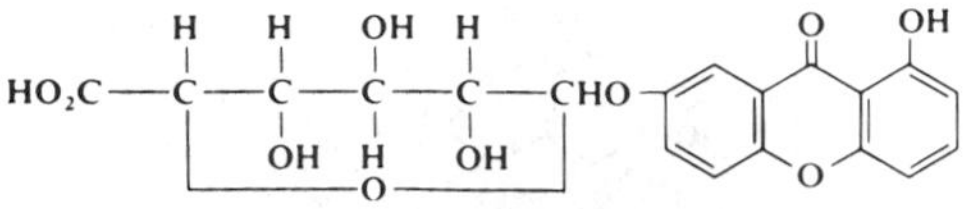

6934

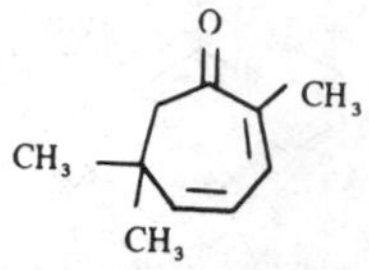

6935

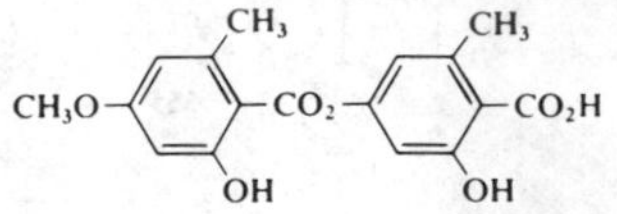

6936

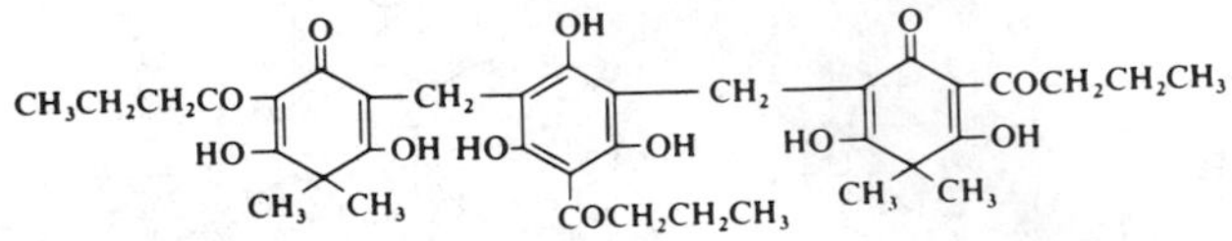

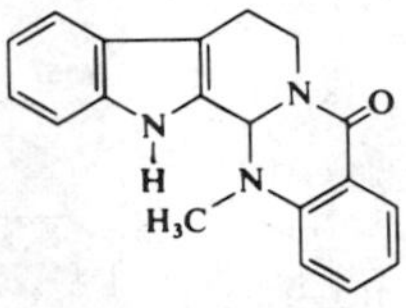

6937

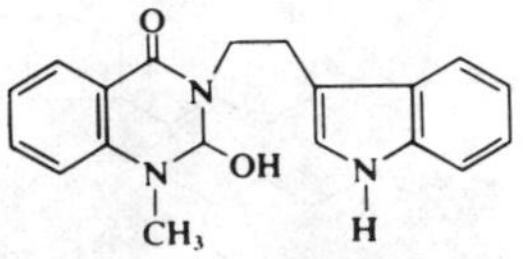

6938

$CH:CHCONHCH_2CH(CH_3)_2$

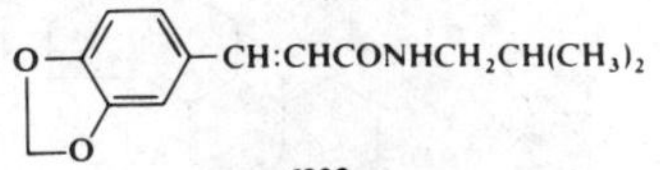

6939

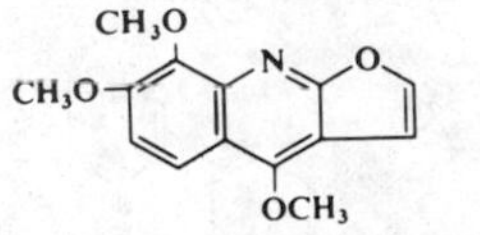

6940

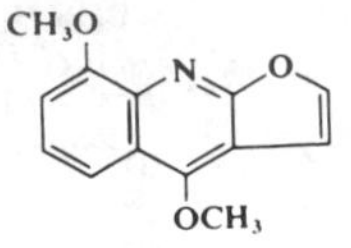

6941

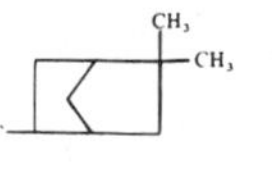

6946

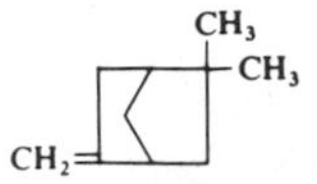

6948

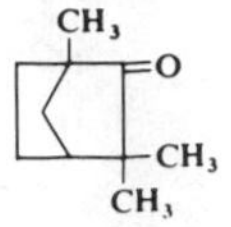

6952

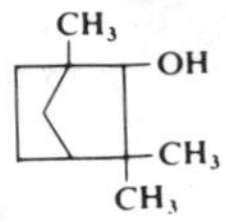

6955

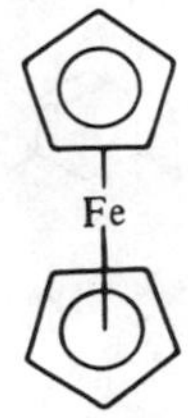

6956

Fe

6957

NH_2

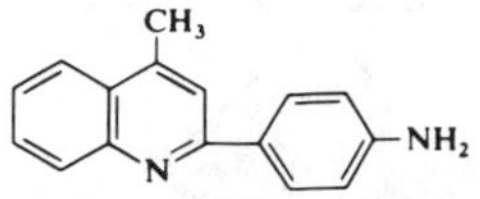

6958

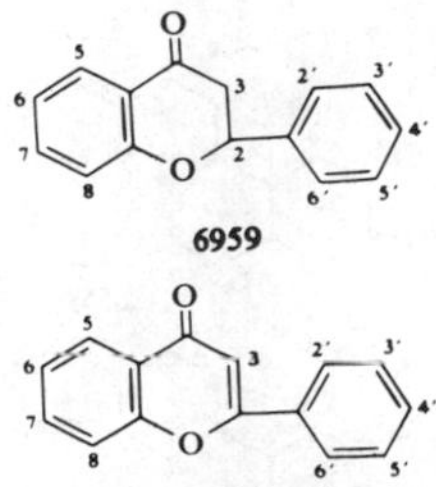

6959

6962

In Numeric as they occur in Organic Compounds Table

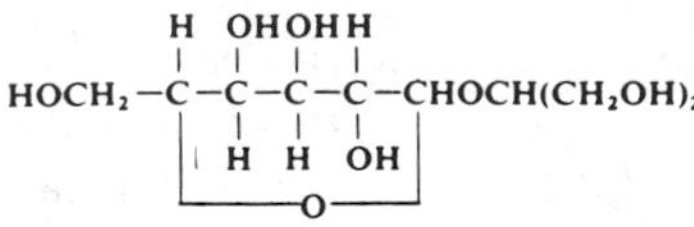

6981

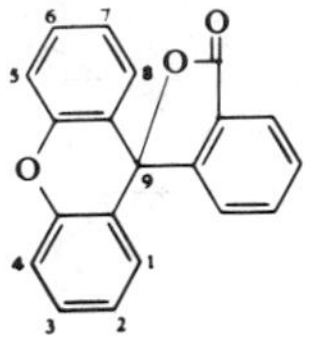

6982

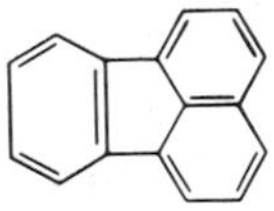

6986

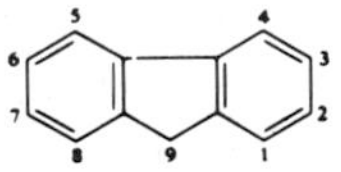

6987

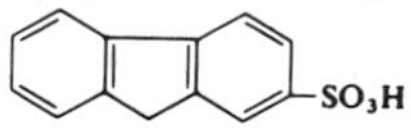

7020

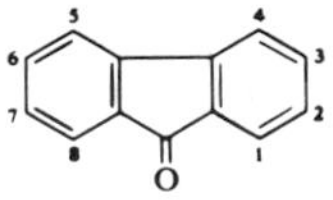

7021

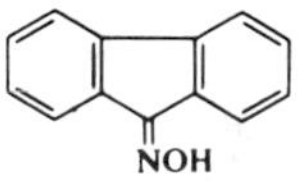

7030

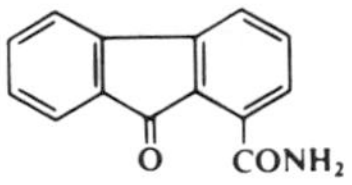

7033

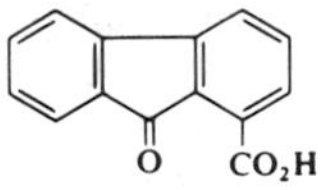

7034

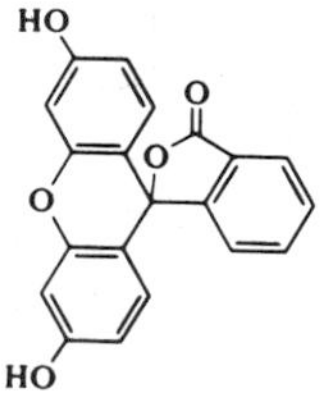

7041

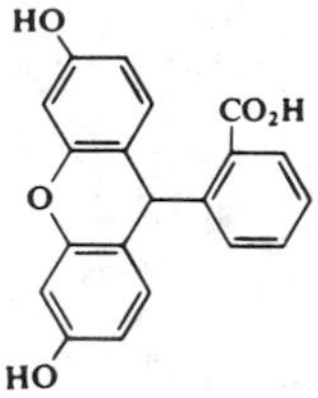

7042

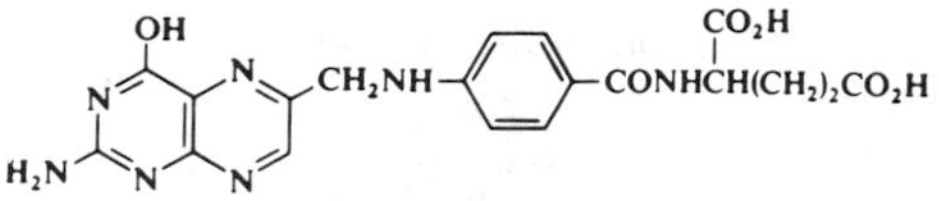

7045

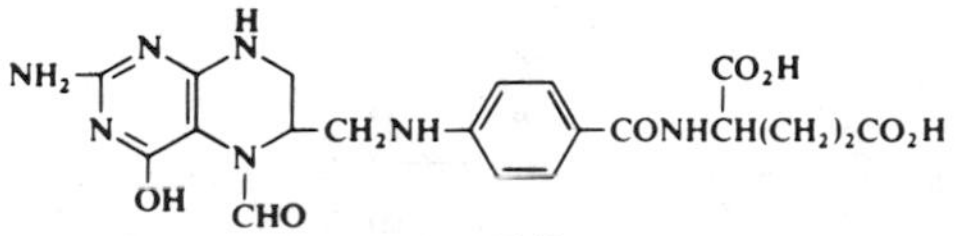

7046

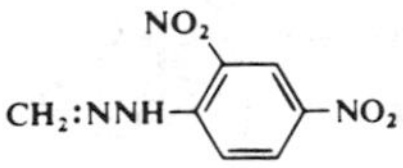

7053

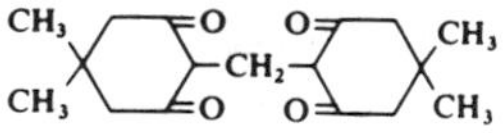

7057

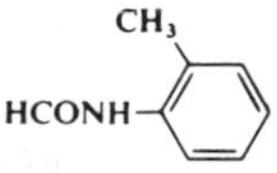

7066

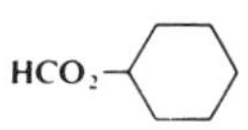

7078

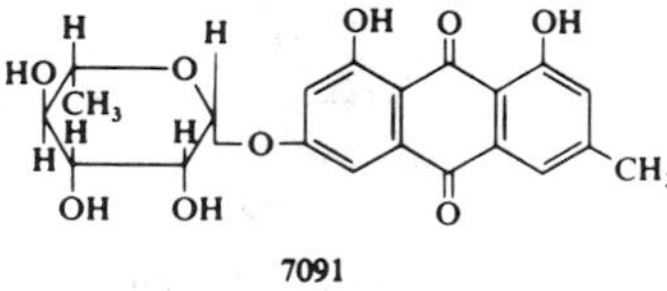

7091

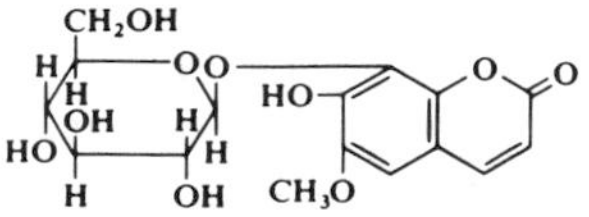

7092

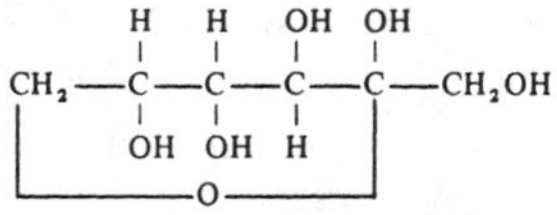

7093

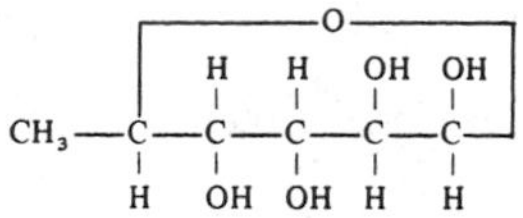

7094

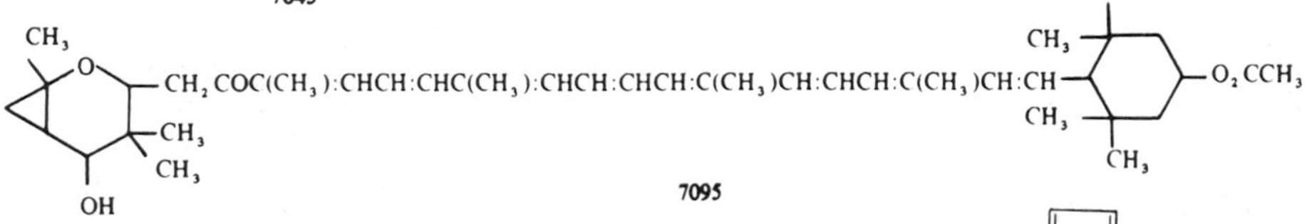

7095

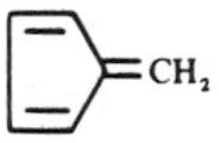

7096

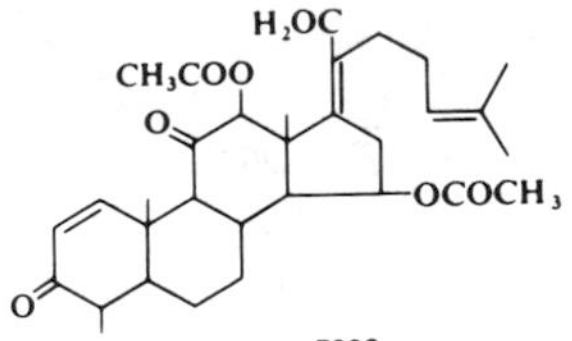

7098

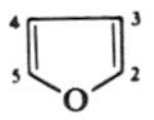

7116

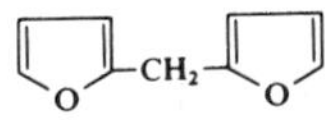

7133

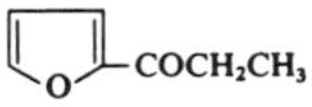

7140

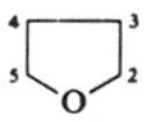

7142

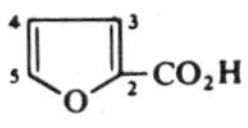

7151

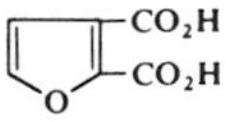

7158

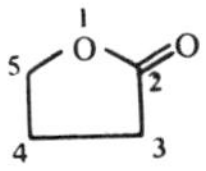

7165

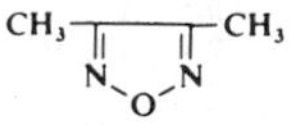

7169

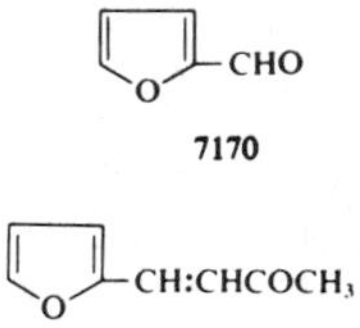

7170

7171

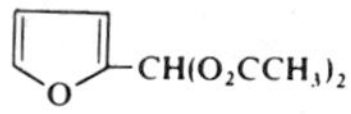

7174

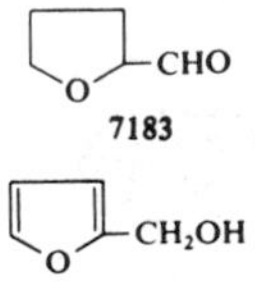

7183

7184

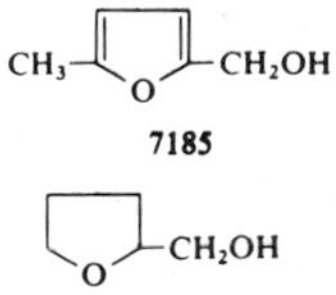

7185

7186

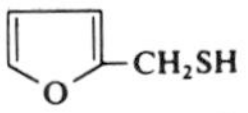

7192

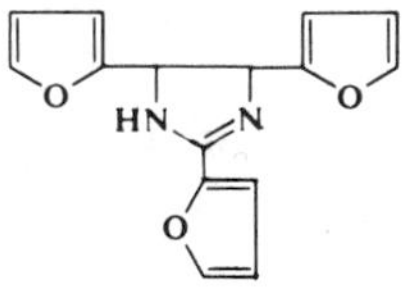

7193

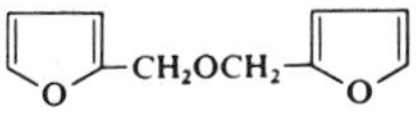

7194

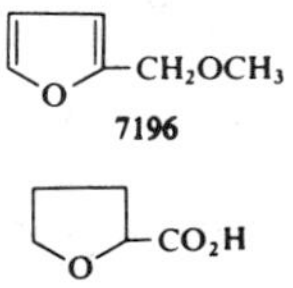

7196

7224

STRUCTURAL FORMULAS OF ORGANIC COMPOUNDS (Continued)

In Numeric as they occur in Organic Compounds Table

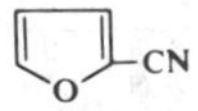

7225 — $-CN$

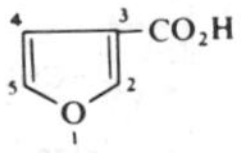

7227 — $-CO_2H$

7240 — $-COCH(OH)-$

7241 — $-CH_2CO_2H$

7242 — $-CH_2CN$

7243 — $-O(CH_2)_7CH_3$

7244 — $-O-$

7245 — $-CO-$

7246 — $-S-$

7249 — $HOCH_2-C-C-C-CCH_2OH$

7250 — $HOCH_2-C-C-C-C-CO$

7251 — $HOCH_2-C-C-C-C-CHO$

7252 — $CH_2-C-C-C-C-CHO$

7253 — $CH_3OCH_2-C-C-C-C-C$

7254 — $HCO_2-C-C-C-C-CHO$

7256 — OCH_3, CH_2CH_2, OCH_3, OCH_3

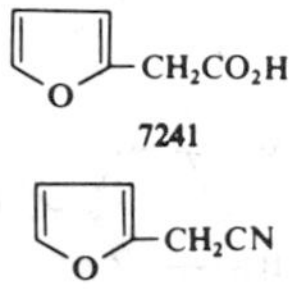

7257 — HO, CO_2H

7258 — HO, CO_2H

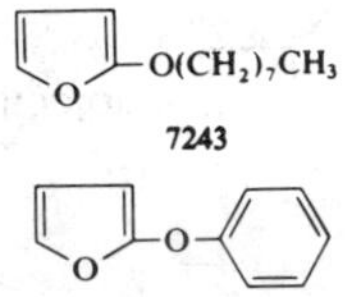

7259 — NCH_3, $CH{:}CH_2$

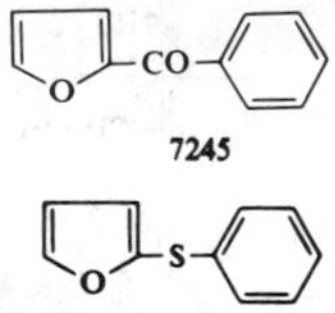

7261 — CHO

7272 — OH, HO

7275 — HO

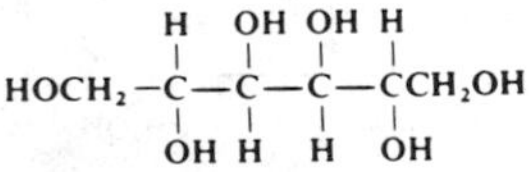

7276 — OH, HO

7278 — $HOCH_2-C-C-C-C=C-C=O$

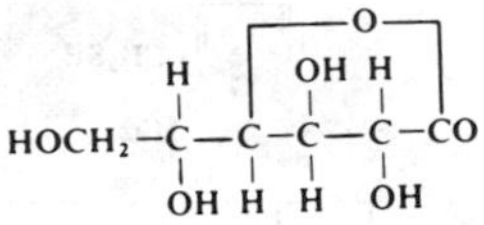

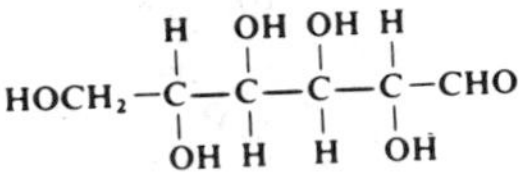

7279 — $HOCH_2-C-C-C-C-C-CHO$

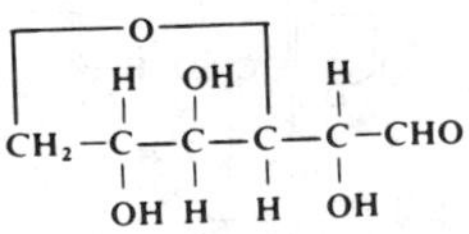

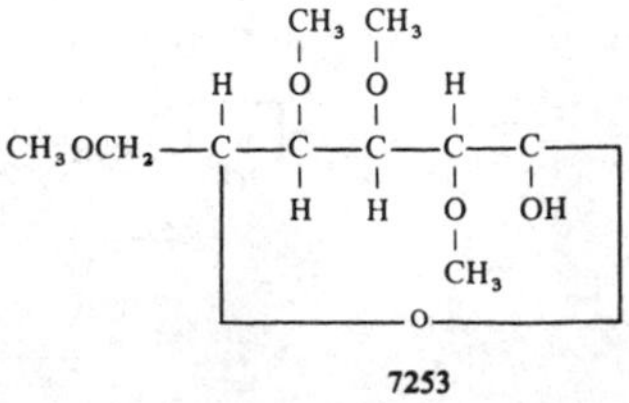

7280 — $CH_3-C-C-C-C-CHO$

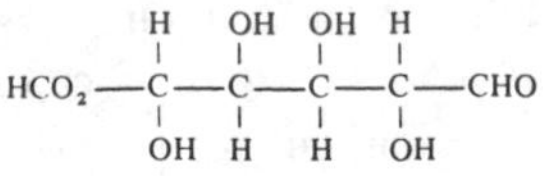

7281 — HO, CH_2OH

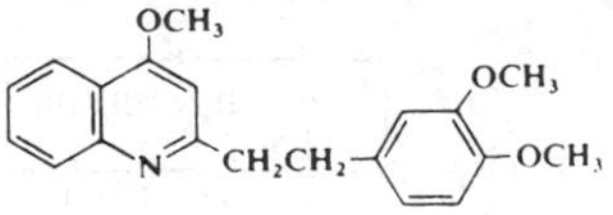

7282 — $HOCH_2-C-C-C-C-CO_2H$

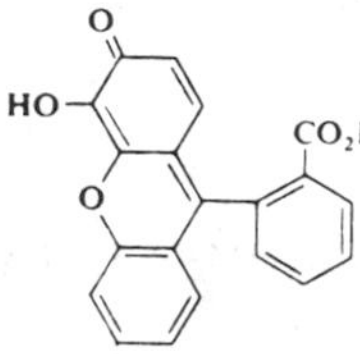

7283 — $HOCH_2-C-C-C-C-CO$

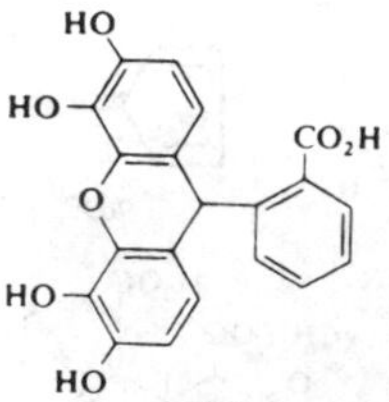

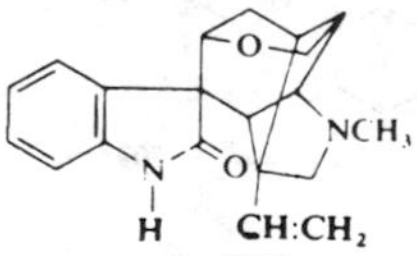

7284 — $HOCH_2-C-C-C-C-CO$

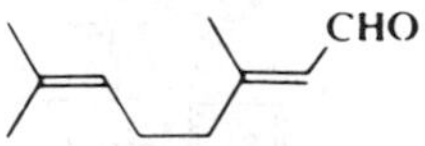

7285 — $HOCH_2-C-C-C-C-CN$

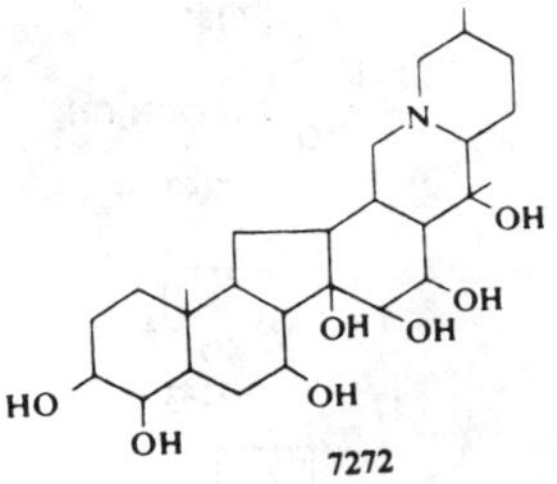

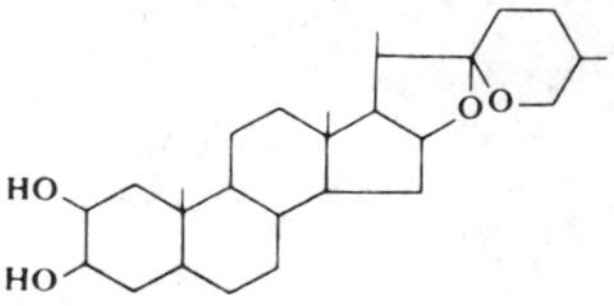

7286 — $HOCH_2-C-C-C-C-CONHNHC_6H_5$

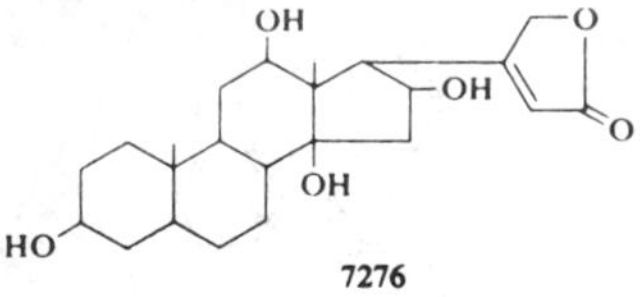

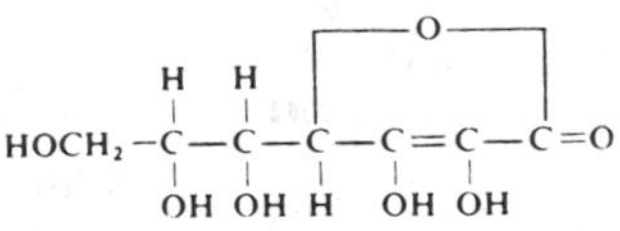

7287 — $HOCH_2-CO-C-C-C-CO_2H$

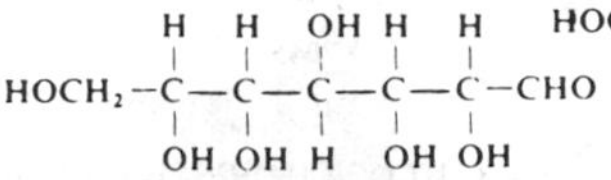

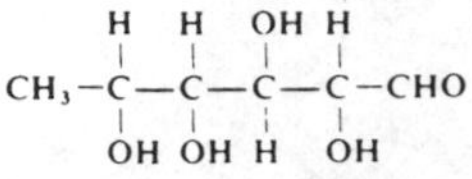

7288 — $HOCH_2-C-C-C-C-CHO$

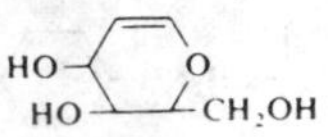

7289 — $HOCH_2-C-C-C-C-C$

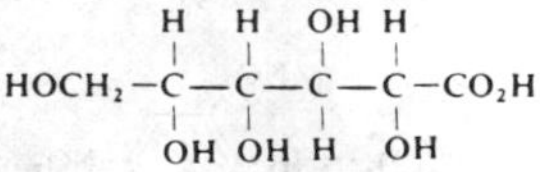

7291 — $HOCH_2-C-C-C-C-C$

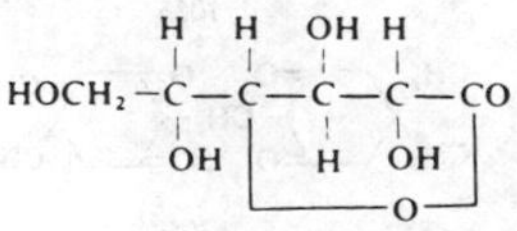

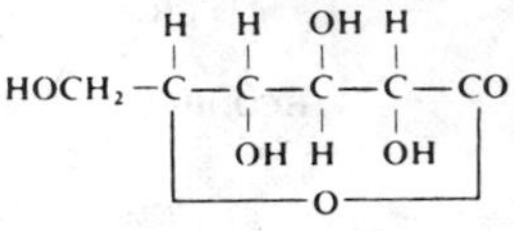

7295 — $HOCH_2-C-C-C-C-CH{:}NNHC_6H_5$

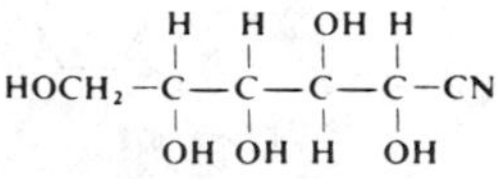

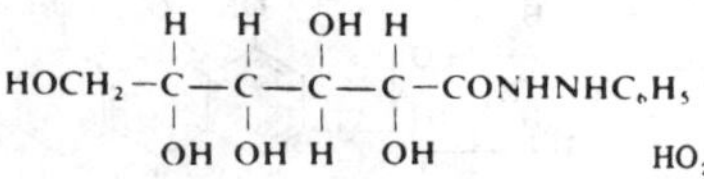

7297 — $HOCH_2-C-C-C-C(=NNHC_6H_5)-CH{:}NNHC_6H_5$

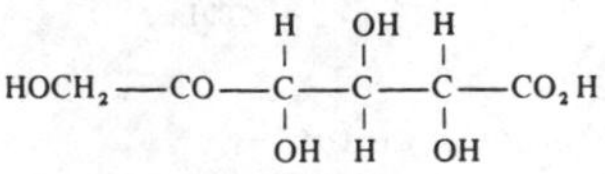

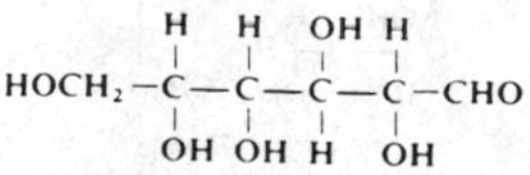

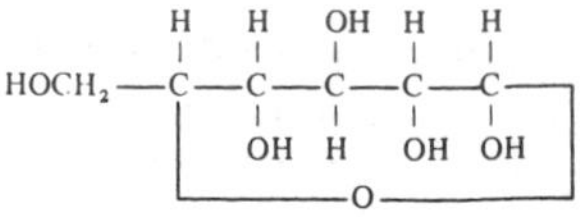

7299 — $HOCH_2-C-C-C-C-CHO$ (NH_2)

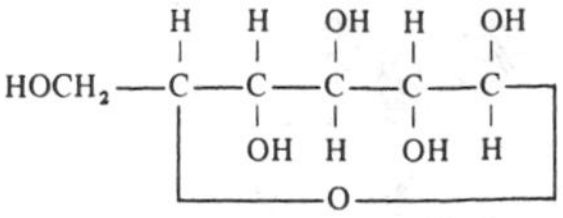

7301 — $HOCH_2-C-C-C-C-CHO$ ($NHCH_3$)

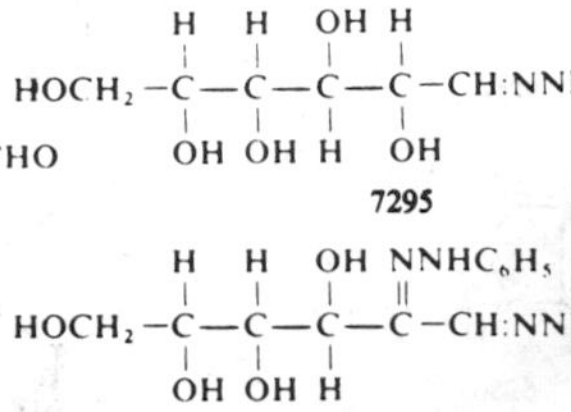

7302 — $HOCH_2-C-C-C-C-CHS$

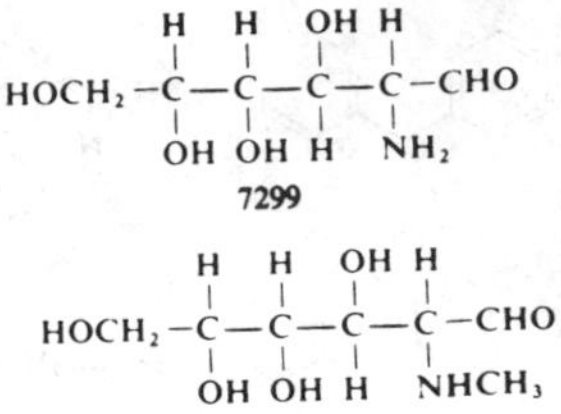

7303 — $HOCH_2-C-C-C-C-C$ (OCH_3)

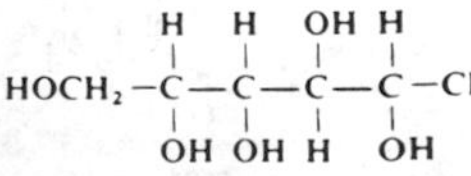

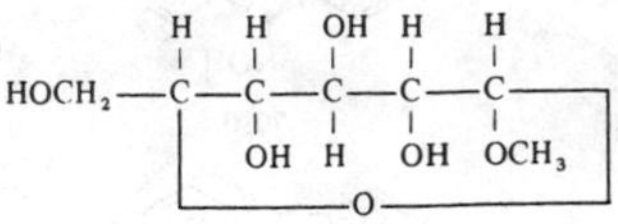

7304 — $HOCH_2-C-C-C-C-C$ (OCH_3)

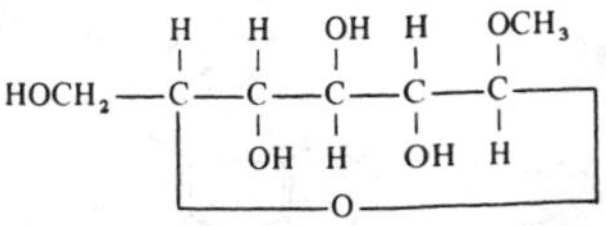

7305 — $HO_2C-C-C-C-C-C$

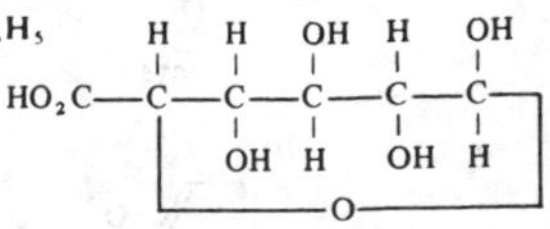

7306 — $OC-C-C-C-C-C$

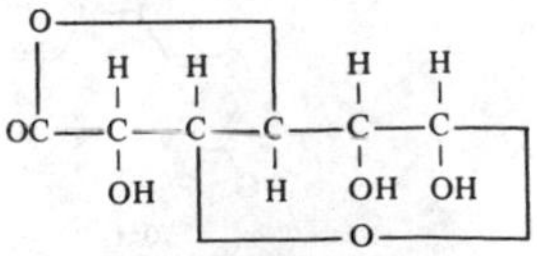

STRUCTURAL FORMULAS OF ORGANIC COMPOUNDS (Continued)

In Numeric as they occur in Organic Compounds Table

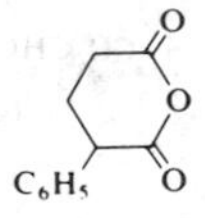

7320

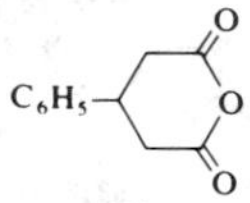

7434

7465

7477

7321

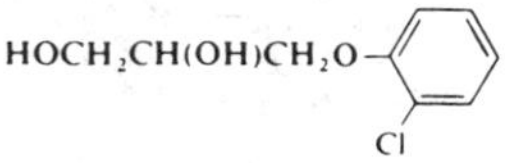

7435

7467

7478

7358

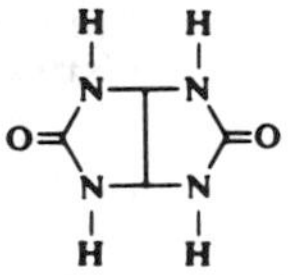

7435a

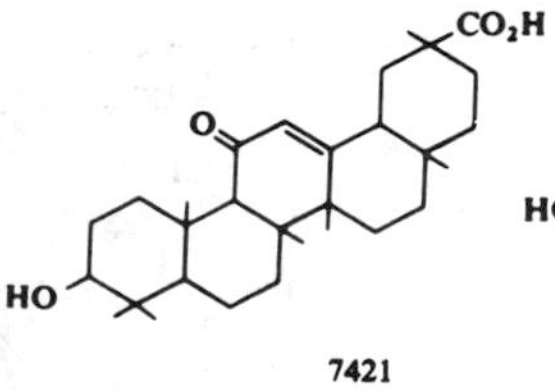

7468

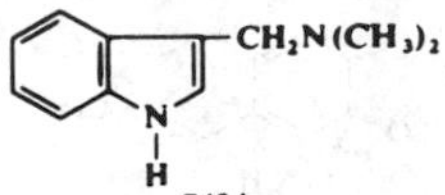

7371

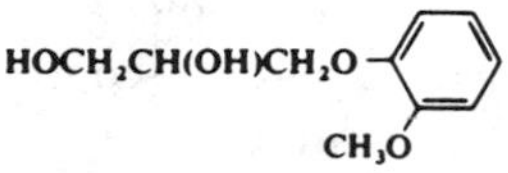

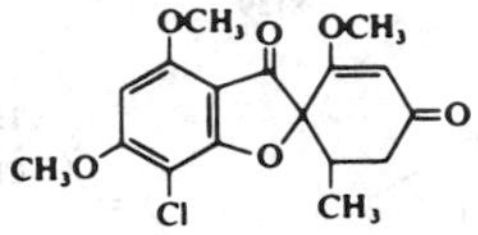

7443

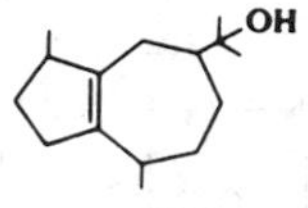

7470

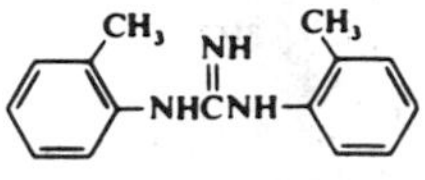

7479

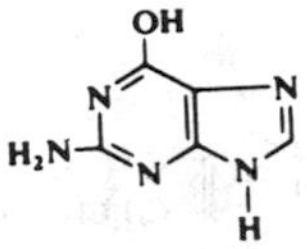

7375

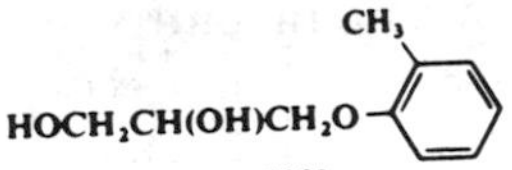

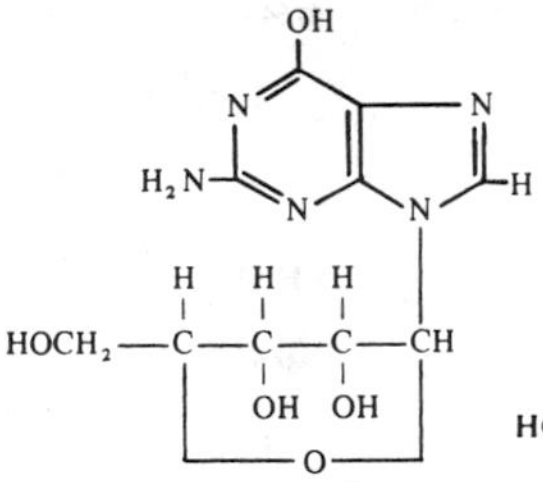

7453

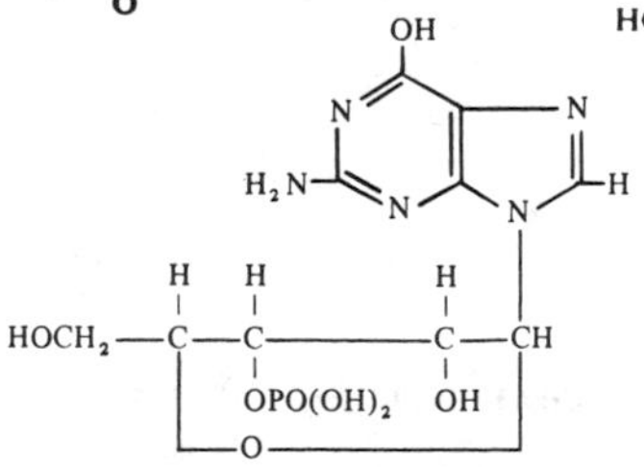

7471

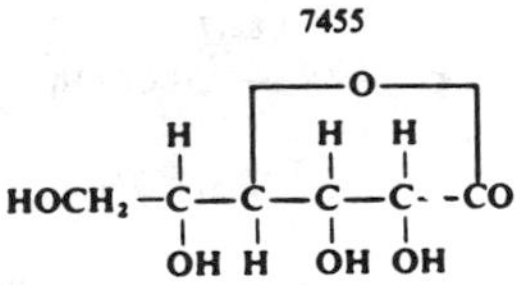

7480

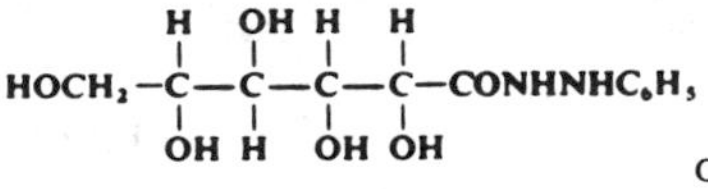

7389

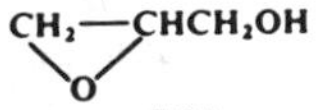

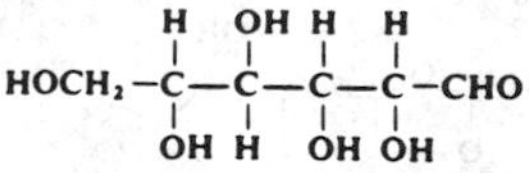

7454

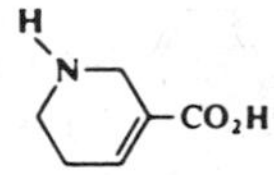

7472

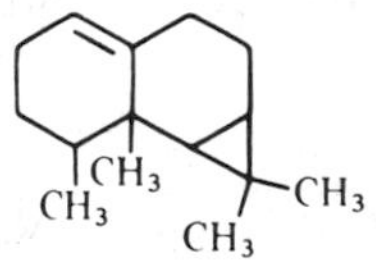

7406

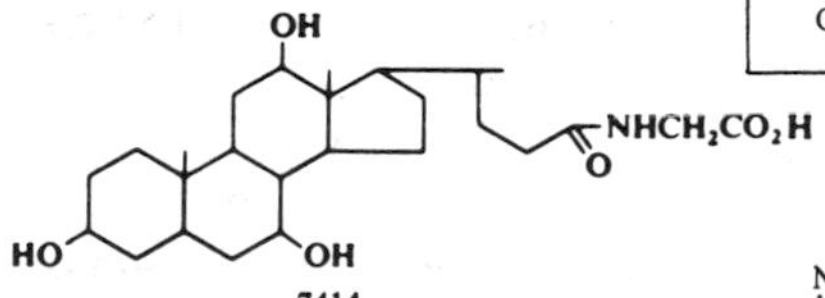

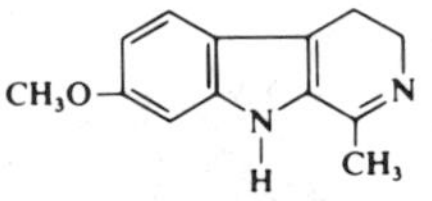

7414

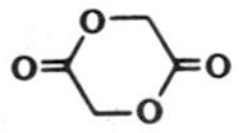

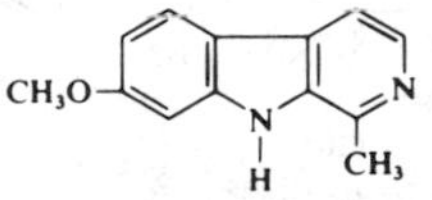

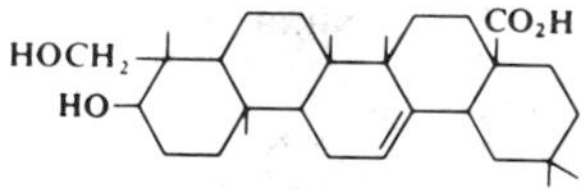

7474

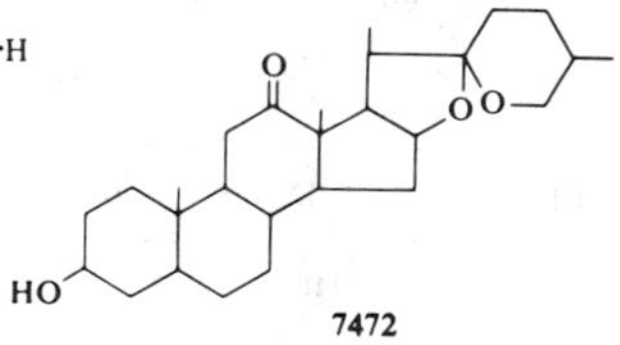

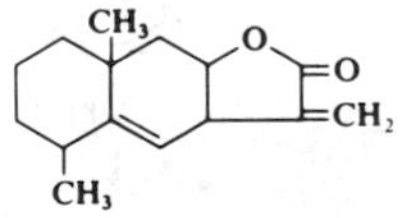

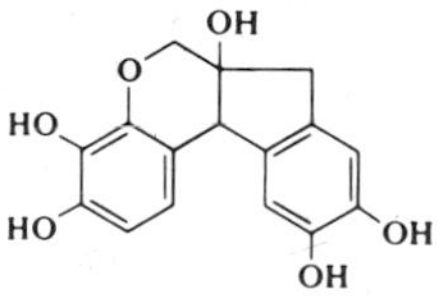

7481

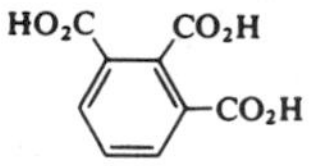

7419

7455

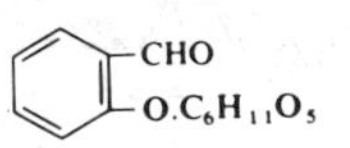

7475

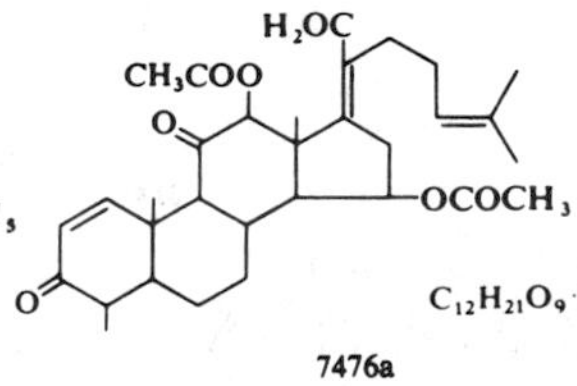

7482

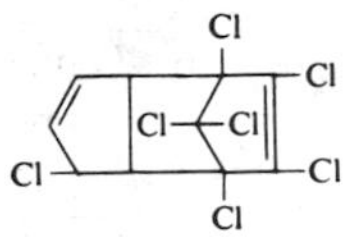

7420

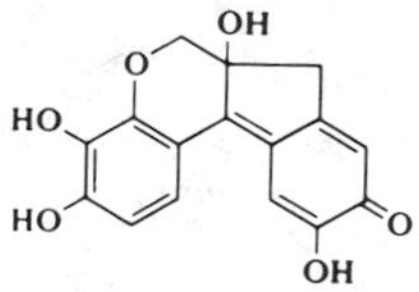

7462

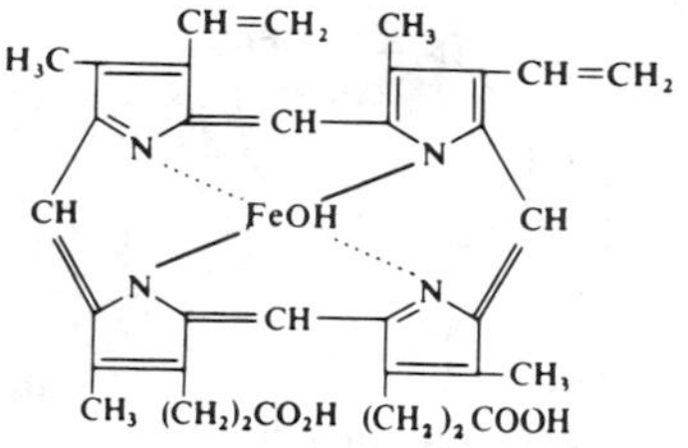

7476

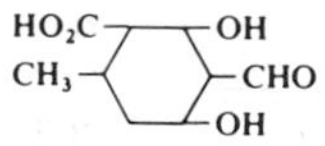

7484

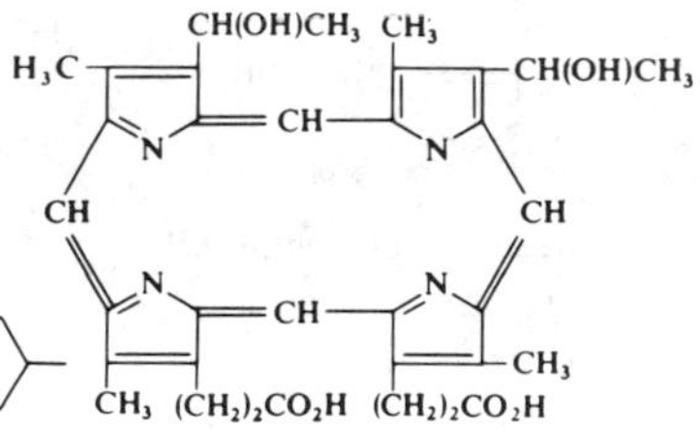

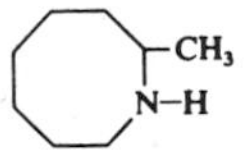

7516

7421

7464

7476a

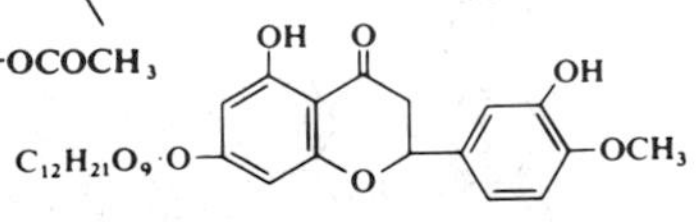

7689

STRUCTURAL FORMULAS OF ORGANIC COMPOUNDS (Continued)

In Numeric as they occur in Organic Compounds Table

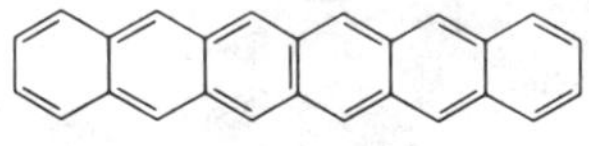

7690

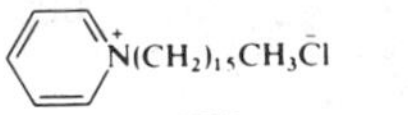

$N^{+}(CH_2)_{15}CH_3Cl^{-}$

7720

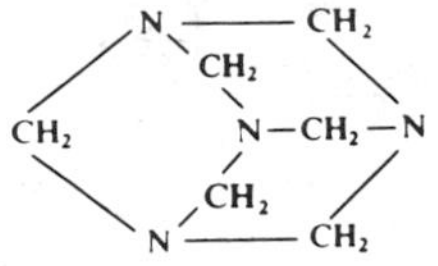

N — CH_2 — CH_2 — $N-CH_2-N$ — CH_2 — N — CH_2 — CH_2

7758

$CH_2CHC{:}CCHCH_2$ (O, O)

7997

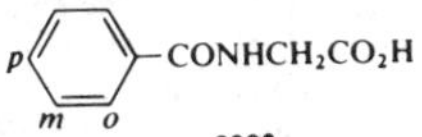

p, *m*, *o* — $CONHCH_2CO_2H$

8003

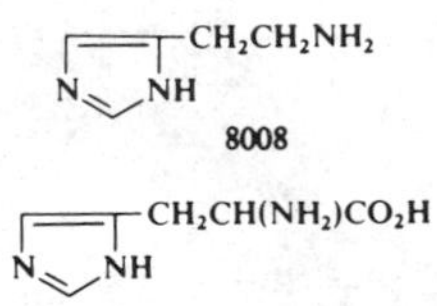

$CH_2CH_2NH_2$ (N, NH)

8008

$CH_2CH(NH_2)CO_2H$ (N, NH)

8010

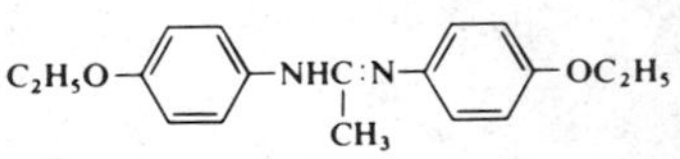

C_2H_5O—NHC:N—OC_2H_5, CH_3

8018

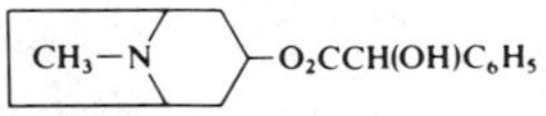

CH_3-N — $O_2CCH(OH)C_6H_5$

8020

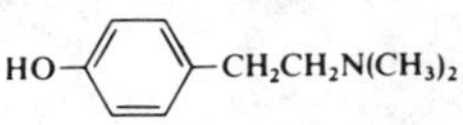

HO— —$CH_2CH_2N(CH_3)_2$

8027

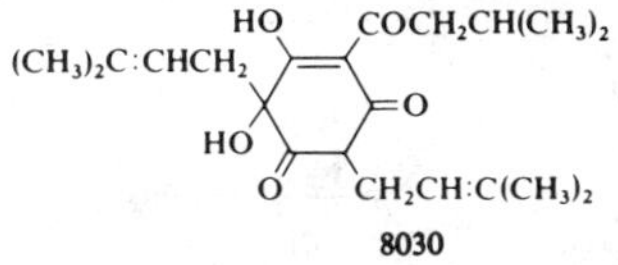

HO, $COCH_2CH(CH_3)_2$, $(CH_3)_2C{:}CHCH_2$, O, HO, O, $CH_2CH{:}C(CH_3)_2$

8030

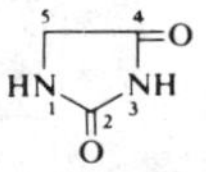

5, 4, O, HN, NH, 1, 2, 3, O

8034

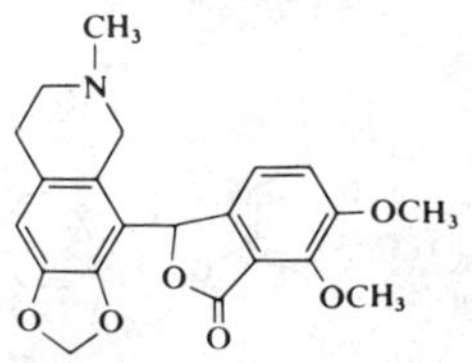

CH_3, N, OCH_3, OCH_3, O, O, O, O

8050

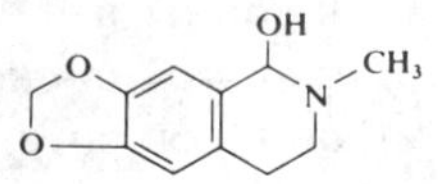

OH, O, O, N, CH_3

8052

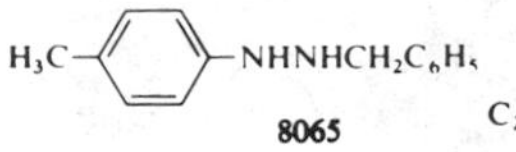

H_3C— —$NHNHCH_2C_6H_5$

8065

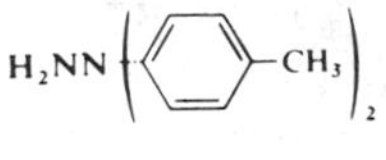

H_2NN (—CH_3)$_2$

8093

CH_3— —NHNH— —CH_3

8096

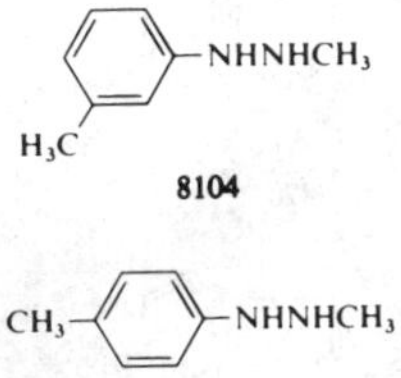

—$NHNHCH_3$, H_3C

8104

CH_3— —$NHNHCH_3$

8105

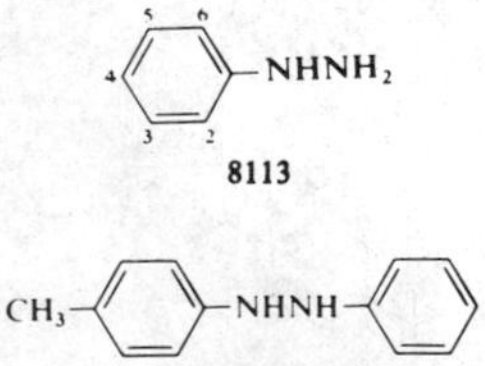

5, 6, 4, 3, 2 — $NHNH_2$

8113

CH_3— —NHNH—

8118

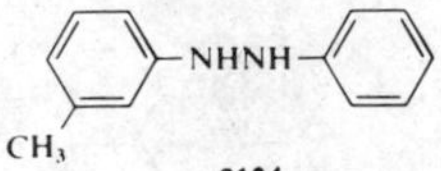

—NHNH—, CH_3

8134

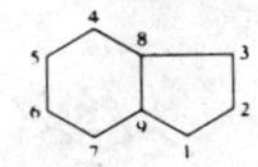

4, 8, 5, 3, 6, 2, 7, 9, 1

8138

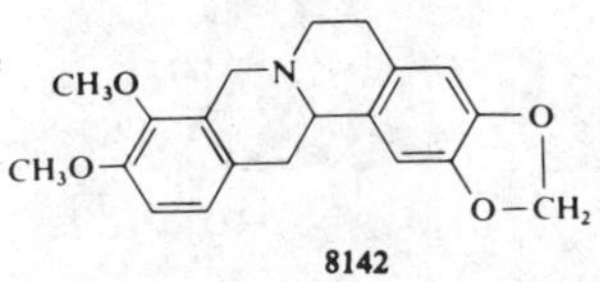

CH_3O, CH_3O, N, O, O—CH_2

8142

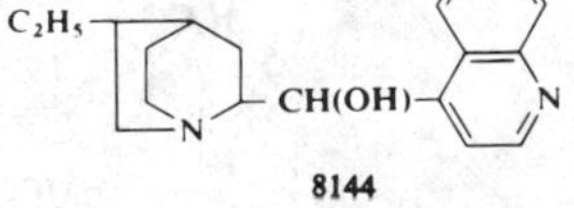

C_2H_5, N, CH(OH), N

8144

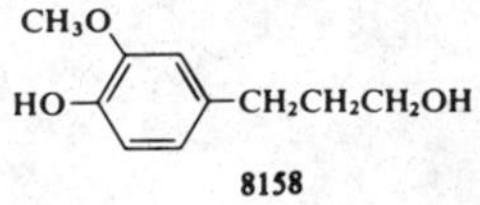

CH_3O, HO— —$CH_2CH_2CH_2OH$

8158

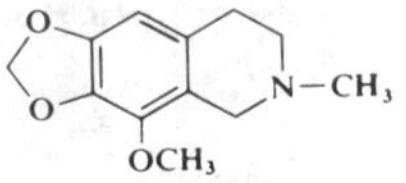

O, O, N—CH_3, OCH_3

8159

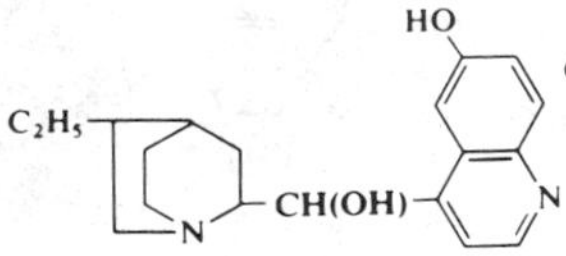

HO, C_2H_5, N, CH(OH), N

8160

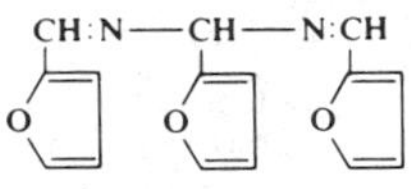

CH:N—CH—N:CH, O, O, O

8162

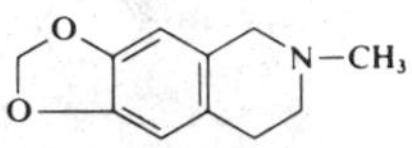

O, O, N—CH_3

8163

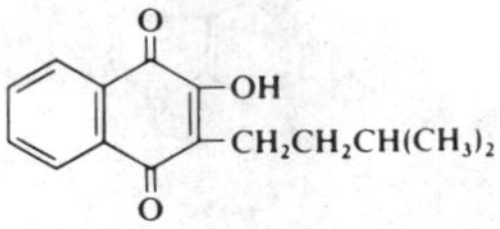

O, OH, $CH_2CH_2CH(CH_3)_2$, O

8164

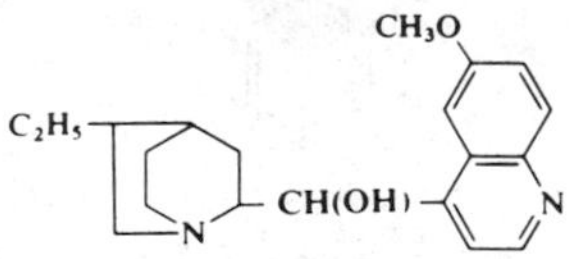

CH_3O, C_2H_5, N, CH(OH), N

8165

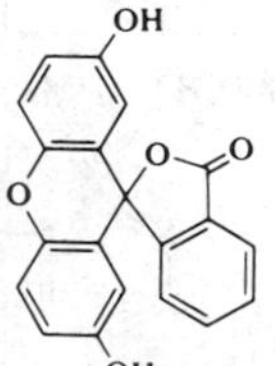

OH, O, O, O, OH

8207

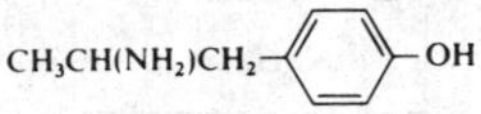

$CH_3CH(NH_2)CH_2$— —OH

8217

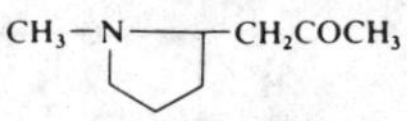

CH_3-N — CH_2COCH_3

8297

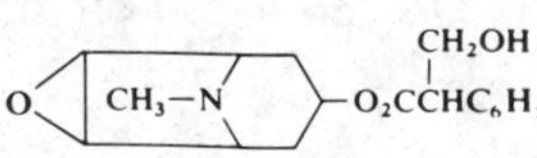

O, CH_3-N, CH_2OH, $O_2CCHC_6H_5$

8298

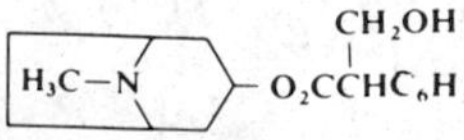

H_3C-N, CH_2OH, $O_2CCHC_6H_5$

8309

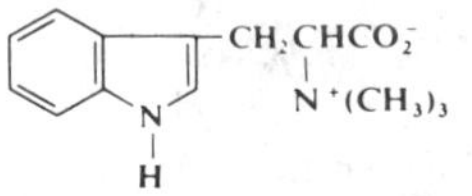

$CH_2CHCO_2^-$, $N^+(CH_3)_3$, N, H

8315

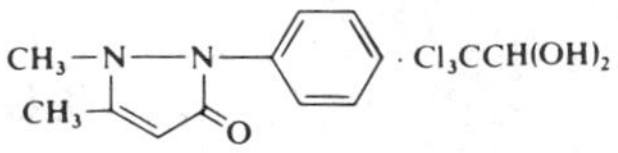

CH_3-N-N— — $\cdot\ Cl_3CCH(OH)_2$, CH_3, O

8316

O, HN, NH, N, N

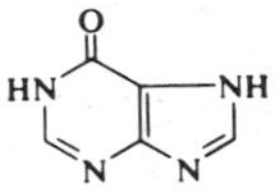

8317

H OHH OH

$HOCH_2-C-C-C-C-CH_2OH$

OHH OHH

8318

H OHH OH

$HOCH_2-C-C-C-C-CO_2H$

OHH OHH

8319

OHH OHH

$HOCH_2-C-C-C-C-CO$

H H OH

O

8320

H OHH OH

$HOCH_2C-C-C-C-CONHNHC_6H_5$

OHH OHH

8321

OHH OHH

$HOCH_2-C-C-C-C-CHO$

H OHH OH

8323

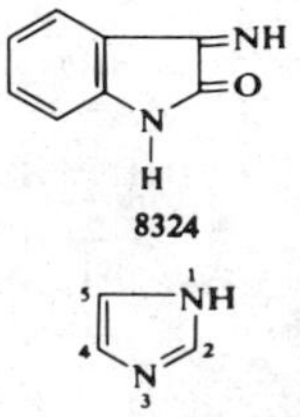

NH, O, N, H

8324

5, 1, NH, 4, N, 2, 3

8325

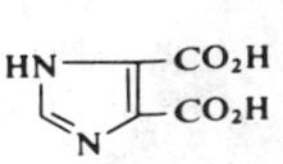

HN, CO_2H, CO_2H, N

8331

HN, NH, S

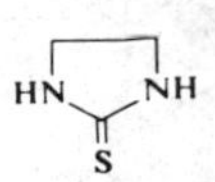

8332

In Numeric as they occur in Organic Compounds Table

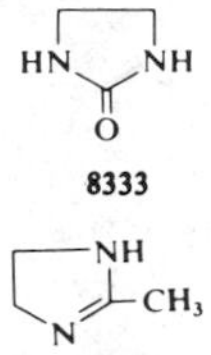

HN NH O
8333

NH CH_3 N
8334

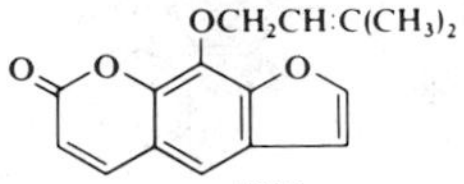

$OCH_2CH{:}C(CH_3)_2$ O O O
8336

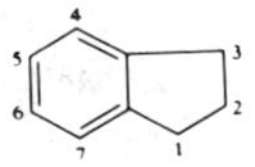

4 5 6 7 1 2 3
8338

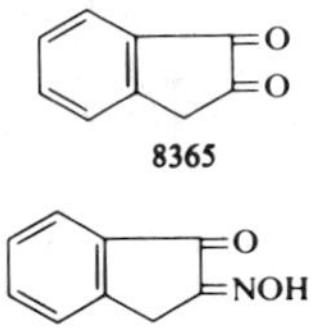

O O
8365

O NOH
8366

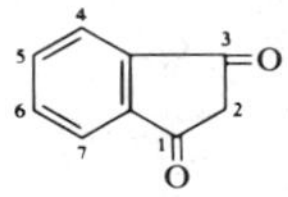

4 5 6 7 1 2 3 O O
8367

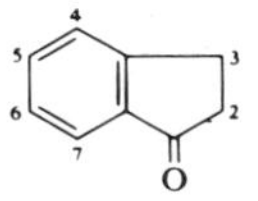

4 5 6 7 1 2 3 O
8375

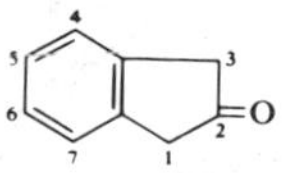

4 5 6 7 1 2 3 O
8379

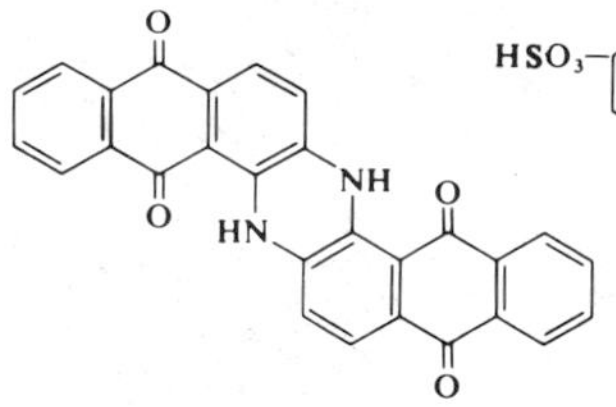

O O NH HN O O
8381

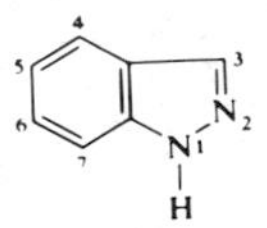

4 5 6 7 N^1 N_2 3 H
8382

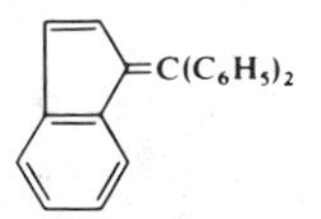

$C(C_6H_5)_2$
8383

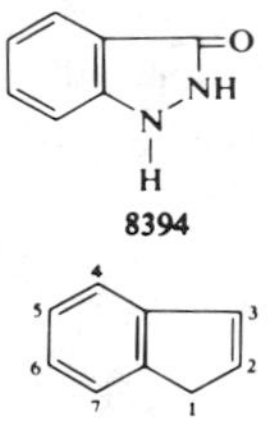

O NH N H
8394

4 5 6 7 1 2 3
8395

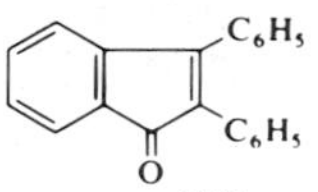

C_6H_5 C_6H_5 O
8403

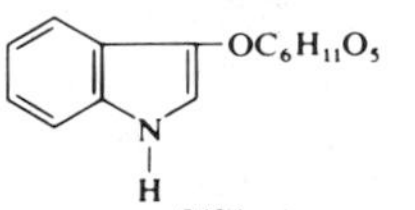

$OC_6H_{11}O_5$ N H
8404

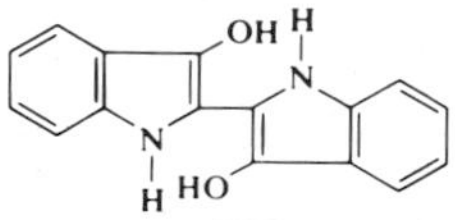

OH H N N H HO
8405

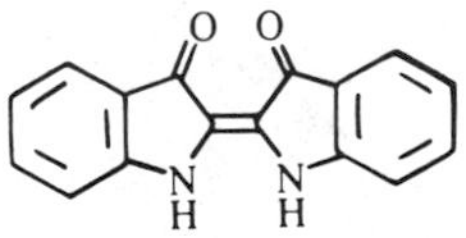

O O N H N H
8406

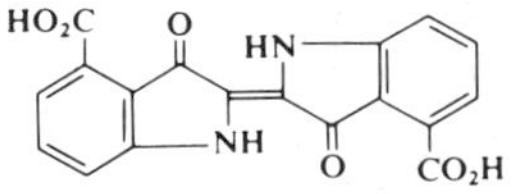

HO_2C O HN NH O CO_2H
8407

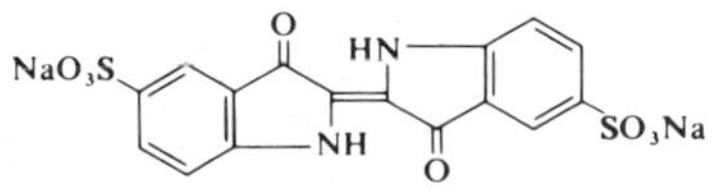

NaO_3S O HN NH O SO_3Na
8408

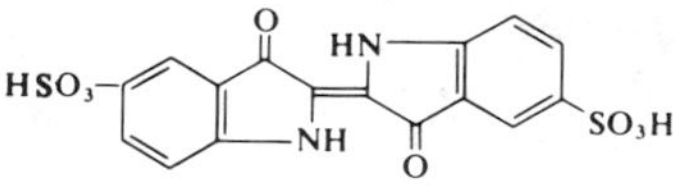

HSO_3 O HN NH O SO_3H
8409

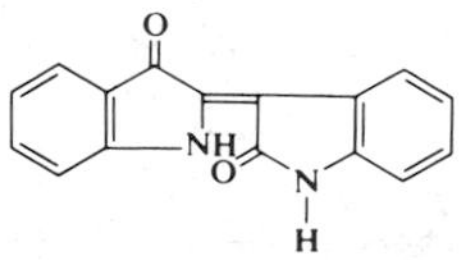

O NH O N H
8410

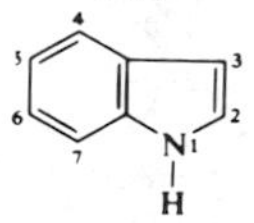

4 5 6 7 N^1 2 3 H
8411

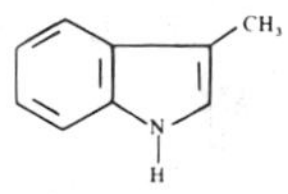

CH_3 N H
8421

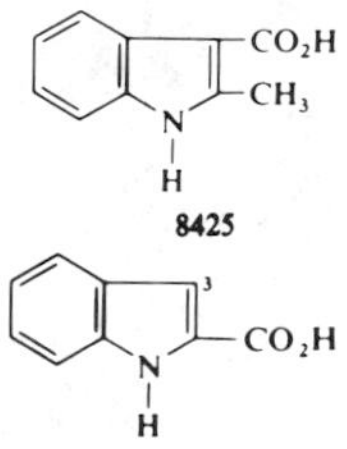

CO_2H CH_3 N H
8425

3 CO_2H N H
8426

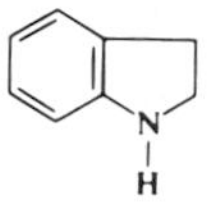

N H
8428

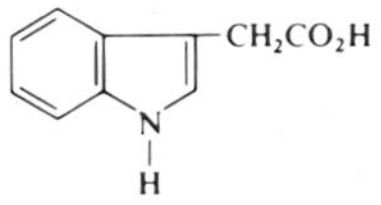

CH_2CO_2H N H
8429

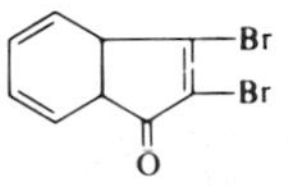

Br Br O
8430

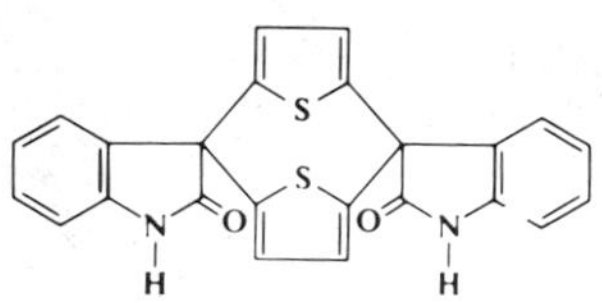

S S N H O O N H
8431

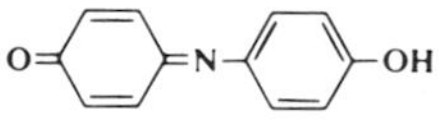

O N OH
8432

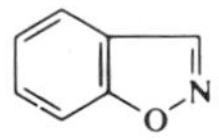

O N
8433

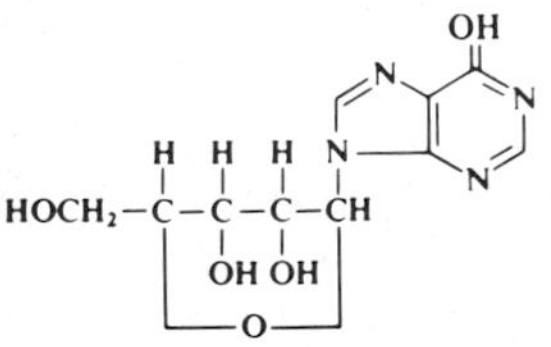

OH N N N N H H H HOCH$_2$–C–C–C–CH OH OH O
8434

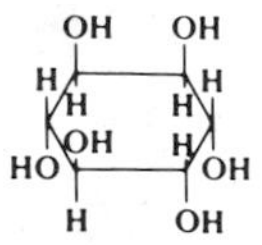

OH OH H H H H OH H HO OH H OH
8435

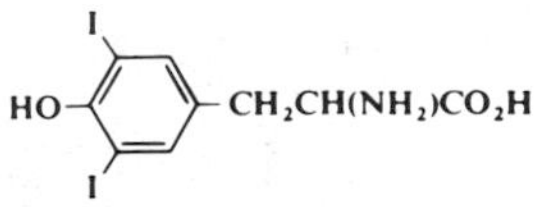

CH_3 CH_3 CH_3
8445

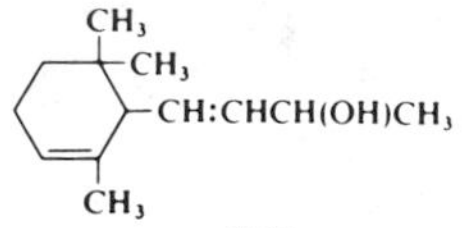

CH_3 CH_3 CH:CHCH(OH)CH_3 CH_3
8446

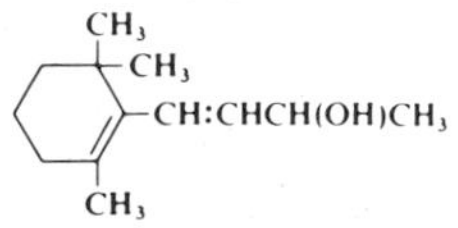

CH_3 CH_3 CH:CHCH(OH)CH_3 CH_3
8447

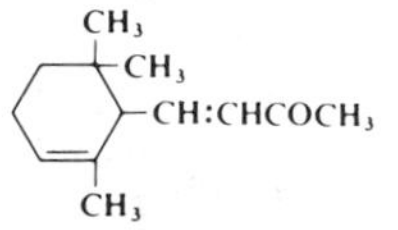

CH_3 CH_3 CH:CHCOCH_3 CH_3
8448

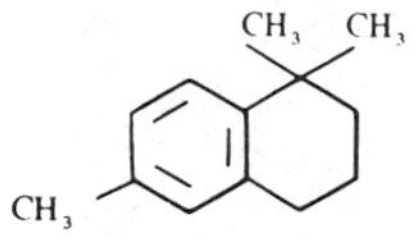

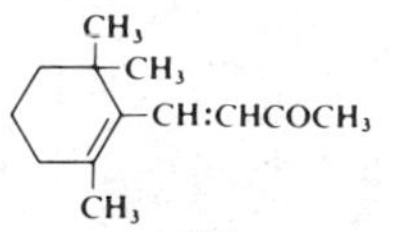

CH_3 CH_3 CH:CHCOCH_3 CH_3
8452

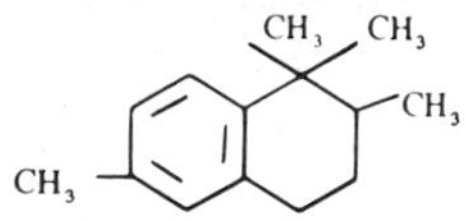

CH_3 CH_3 CH_3 CH_3
8454

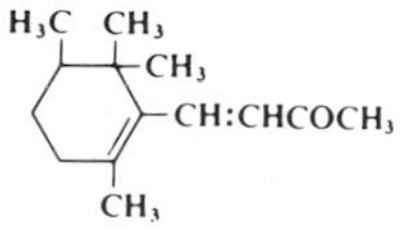

H_3C CH_3 CH_3 CH:CHCOCH_3 CH_3
8455

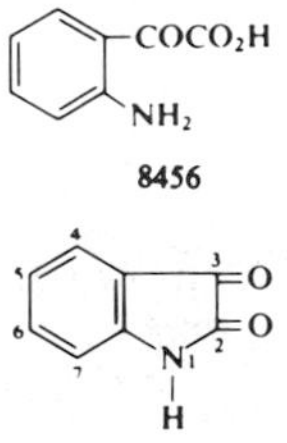

$COCO_2H$ NH_2
8456

4 5 6 7 N_1 2 3 O O H
8457

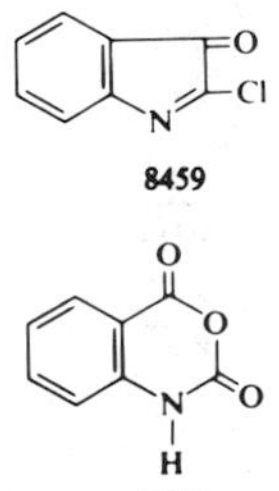

O Cl N
8459

O O N O H
8466

HO I I $CH_2CH(NH_2)CO_2H$
8442

In Numeric as they occur in Organic Compounds Table

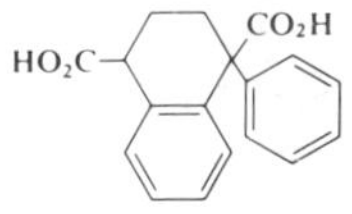

8467

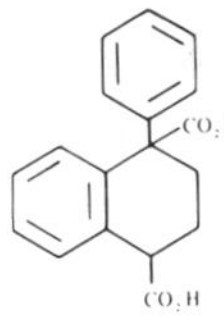

8492

8509

8527

8468

8510

8533

8471

8493

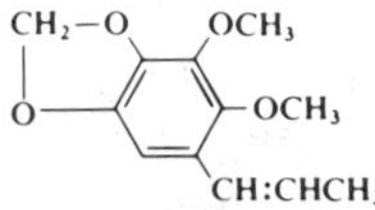

8511

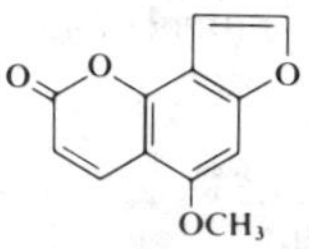

8536

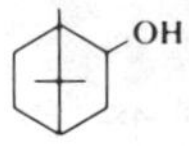

8472

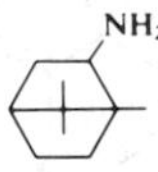

8494

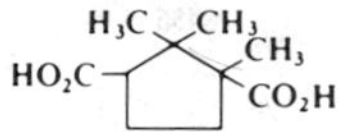

8516

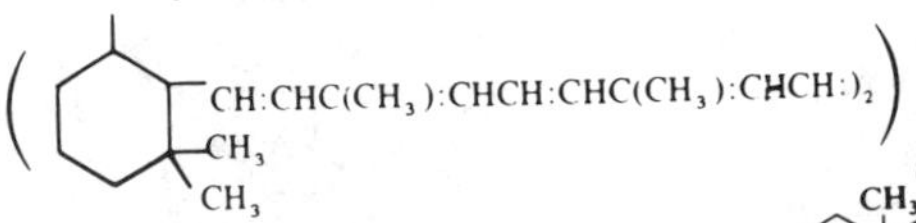

8538

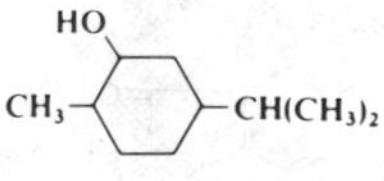

8473

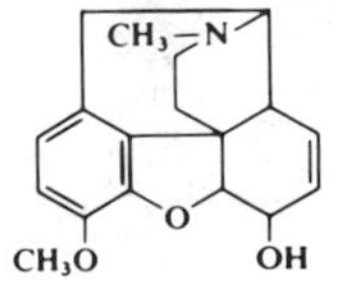

8495

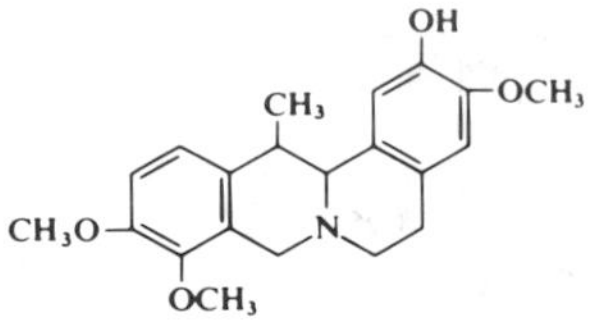

8539

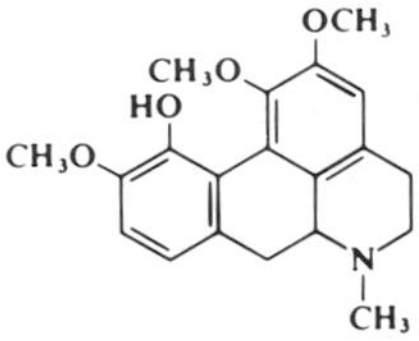

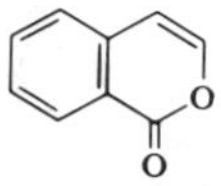

8480

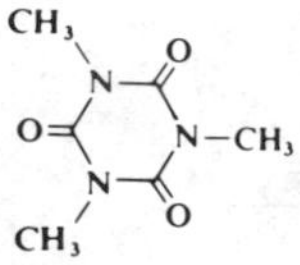

8496

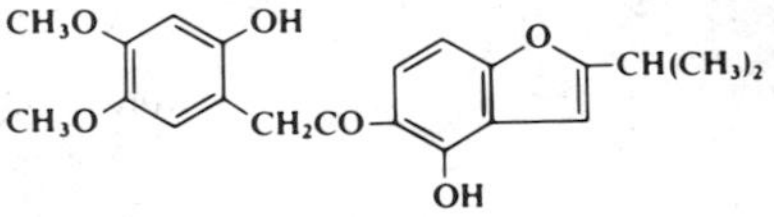

8517

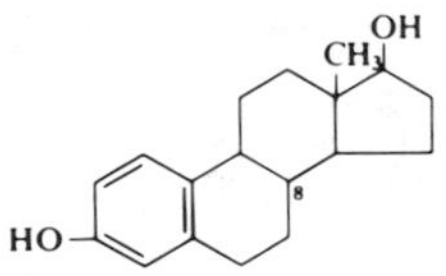

8541

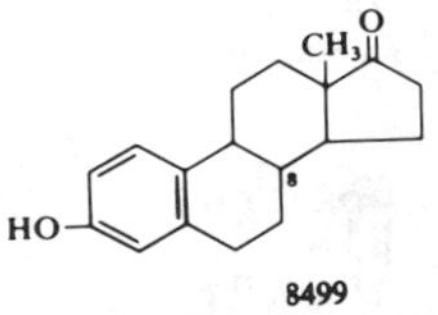

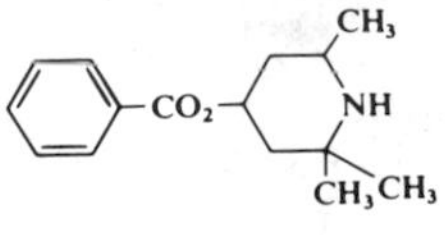

8482

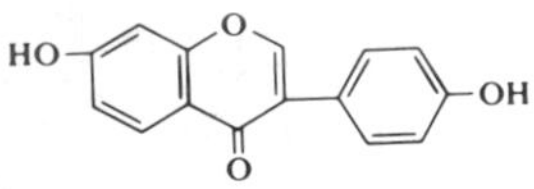

8518

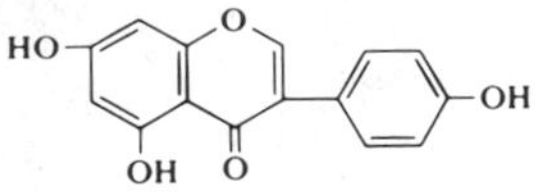

8543

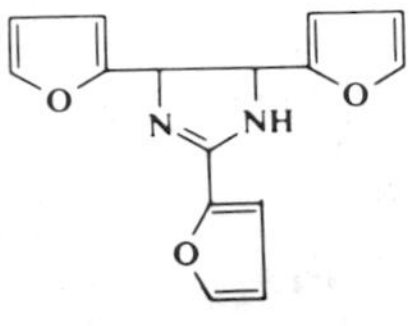

8485

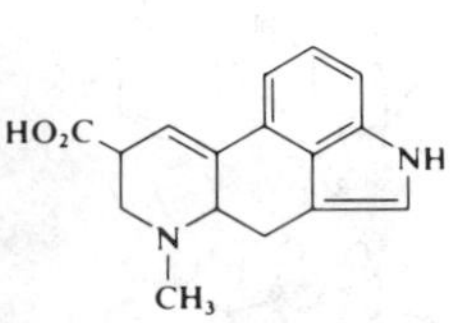

8498

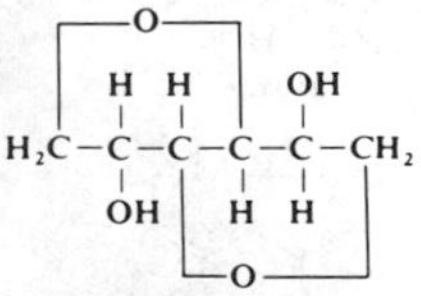

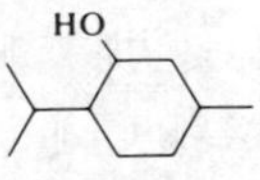

8545

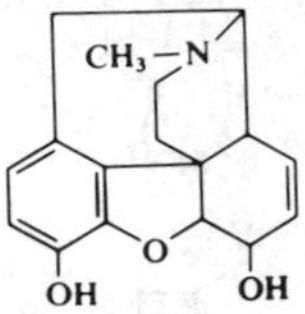

8488

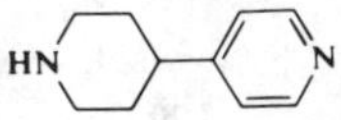

8522

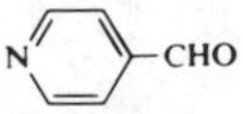

8569

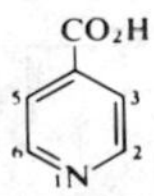

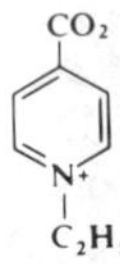

8489

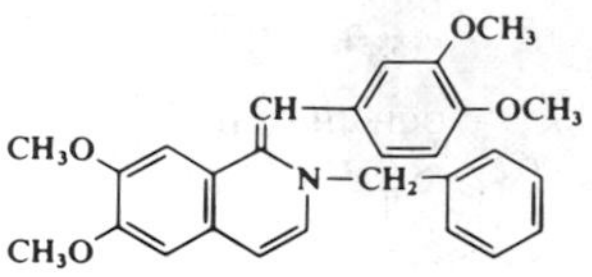

8499

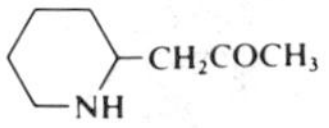

8523

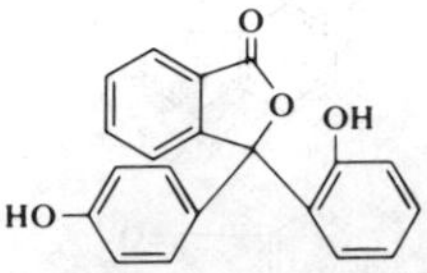

8572

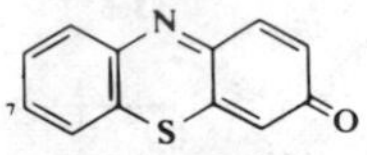

8524

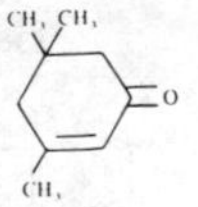

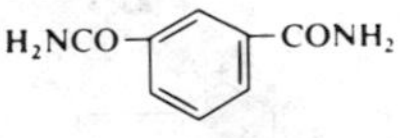

8491

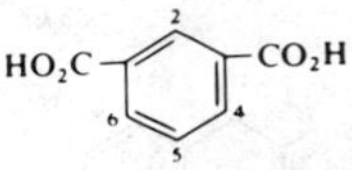

8500

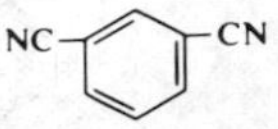

8525

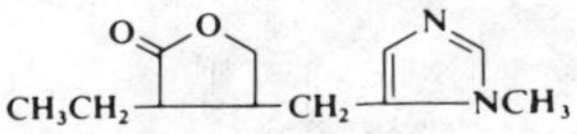

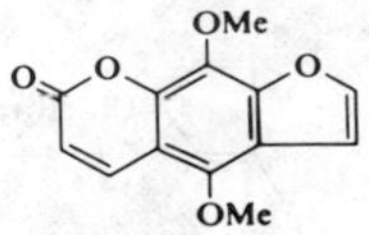

8573

In Numeric as they occur in Organic Compounds Table

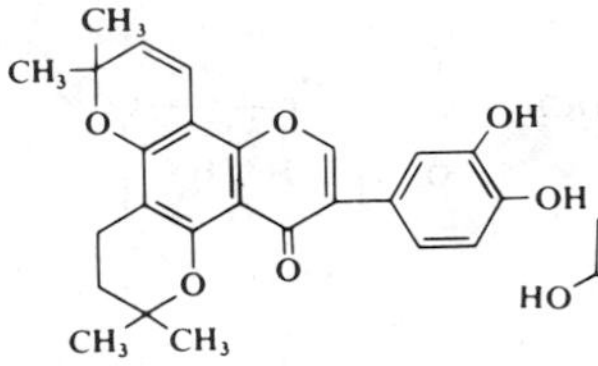

8574

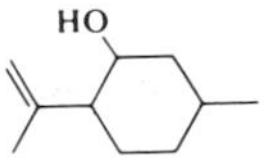

8605

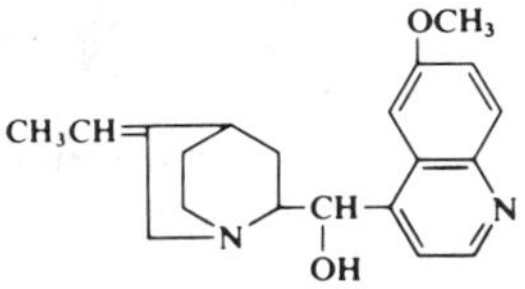

8633

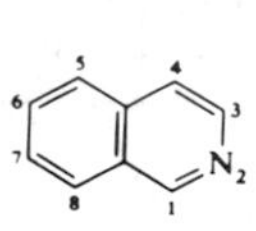

8681

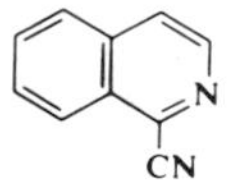

8577

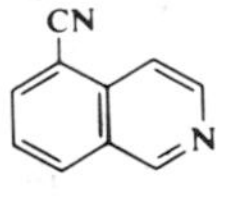

8607

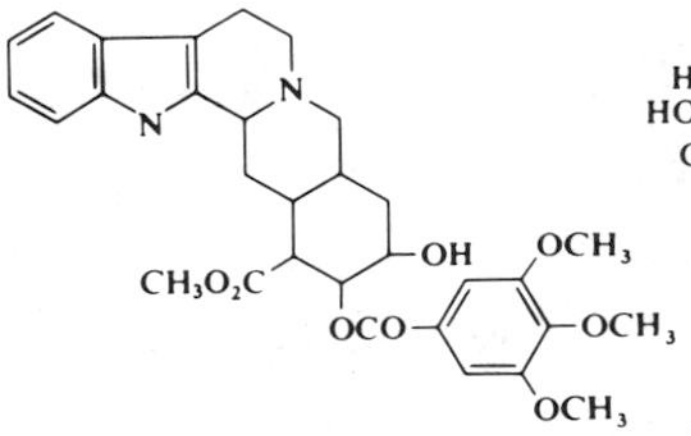

8634

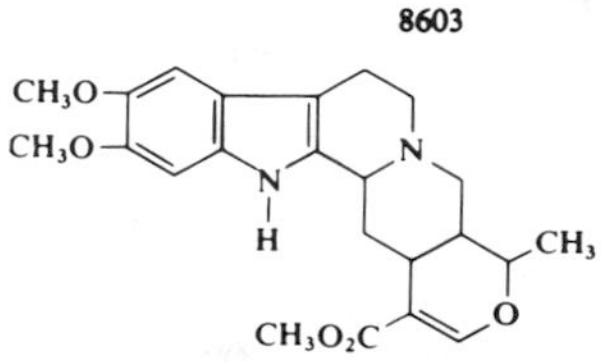

8636

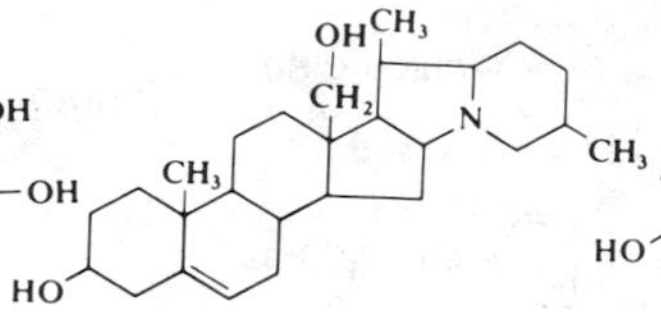

8685

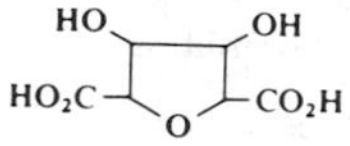

8608

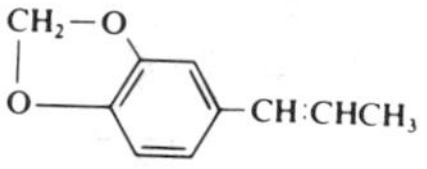

8579

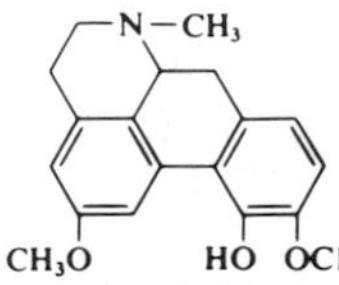

8610

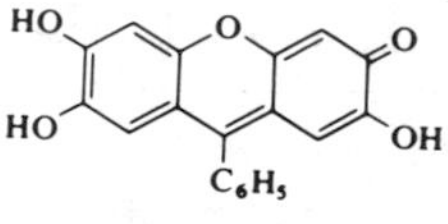

8637

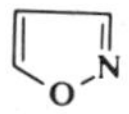

8692

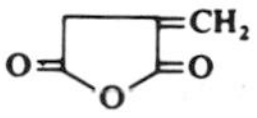

8638

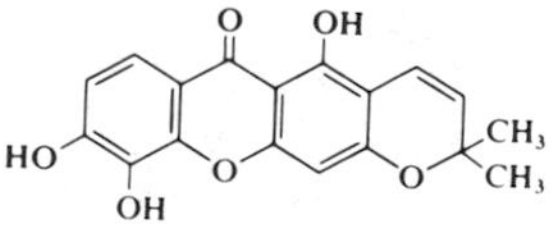

8581

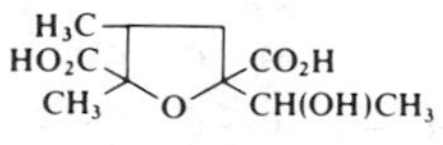

8617

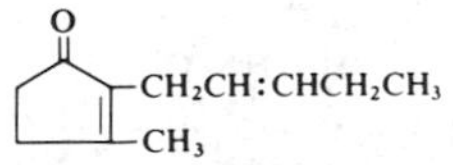

8695

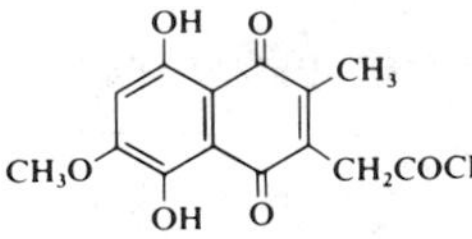

8618

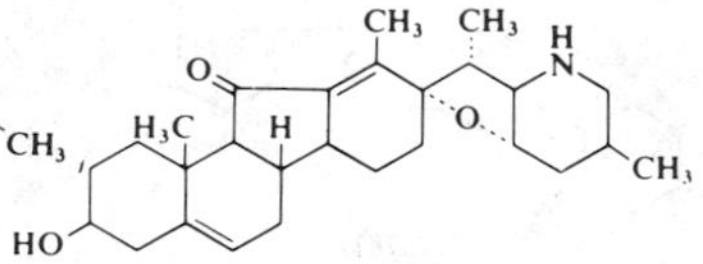

8645

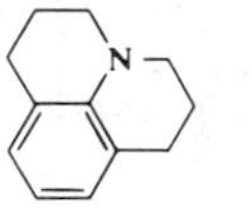

8601

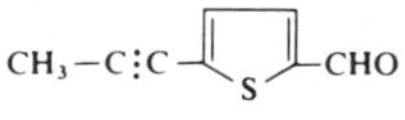

8621

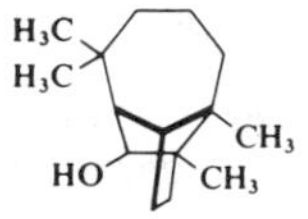

8646

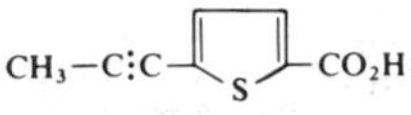

8698

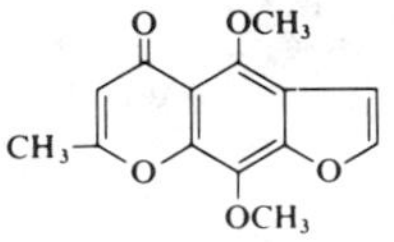

8602

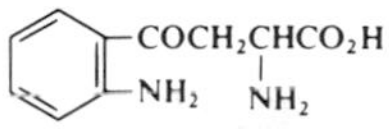

8625

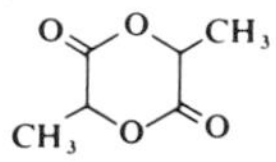

8673

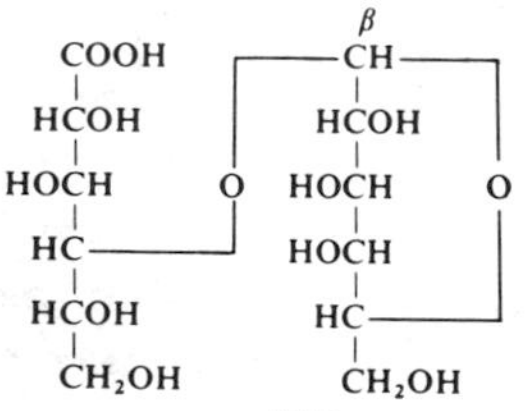

8701

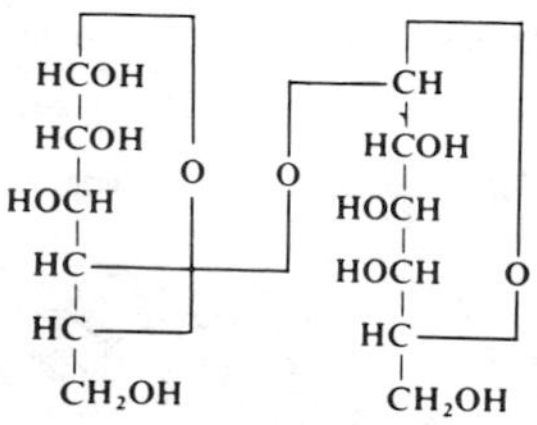

8603

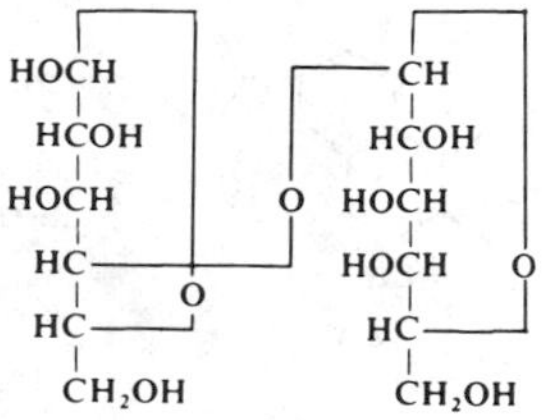

8626

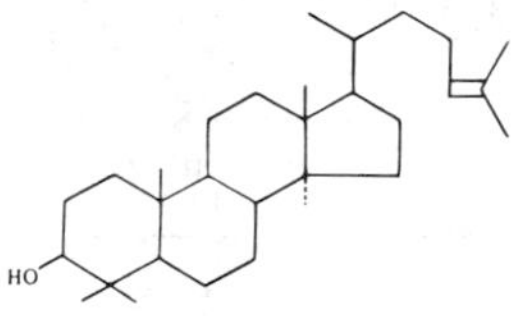

8676

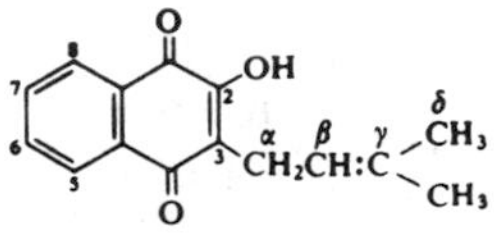

8729

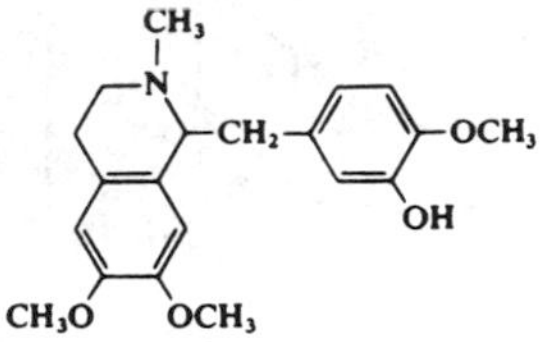

8631

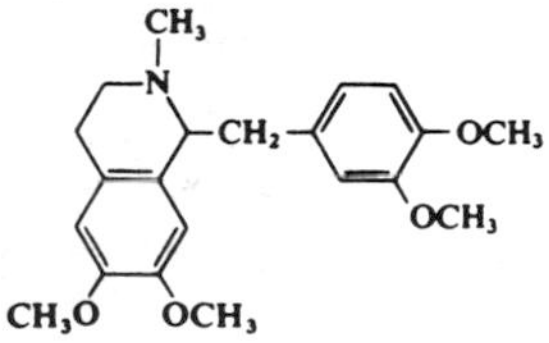

8604

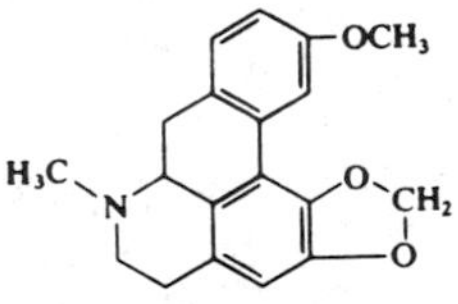

8632

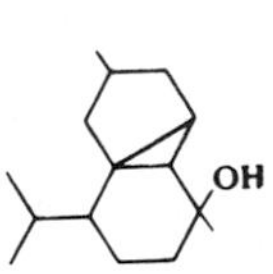

8680

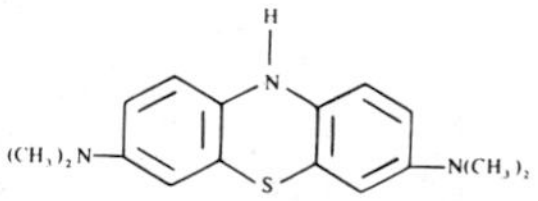

8742

In Numeric as they occur in Organic Compounds Table

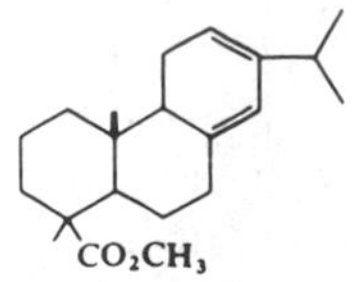

8743

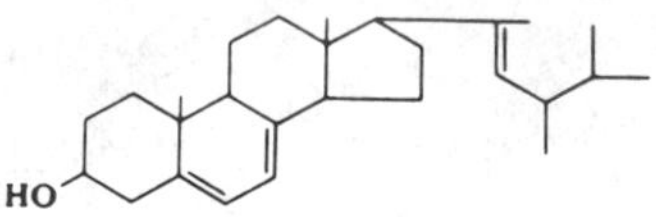

8754

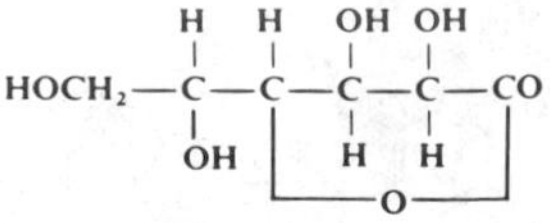

8763

8770

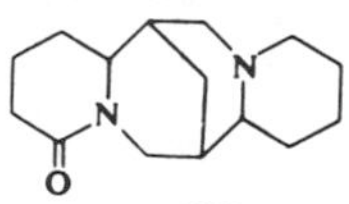

8771

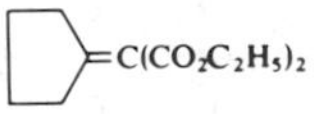

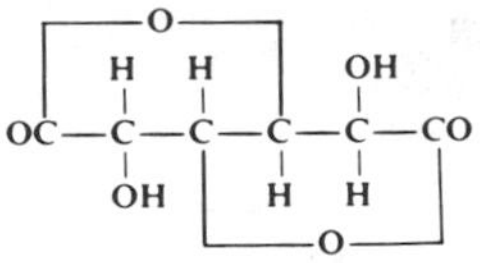

8772

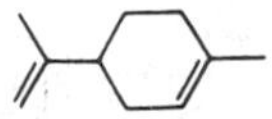

8773

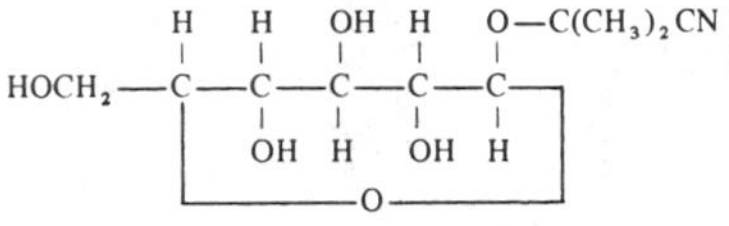

8774

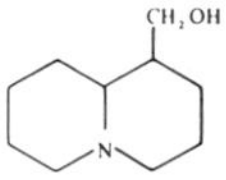

8776

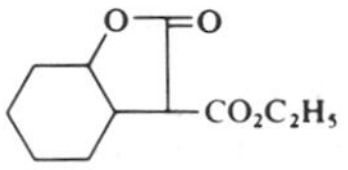

8777

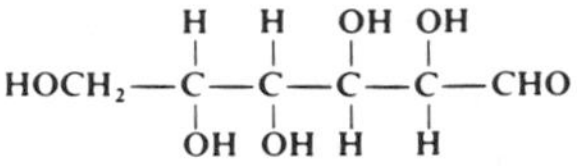

8780

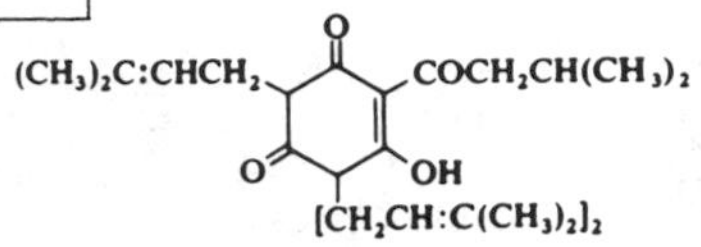

8781

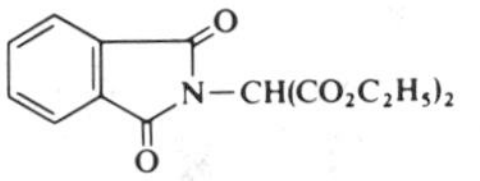

8782

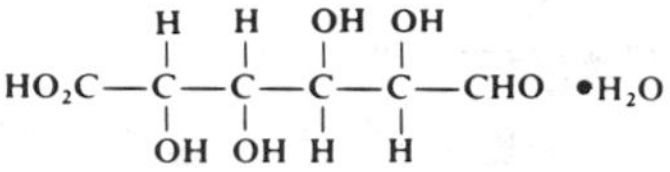

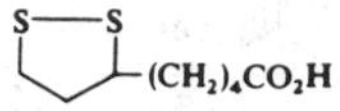

8785

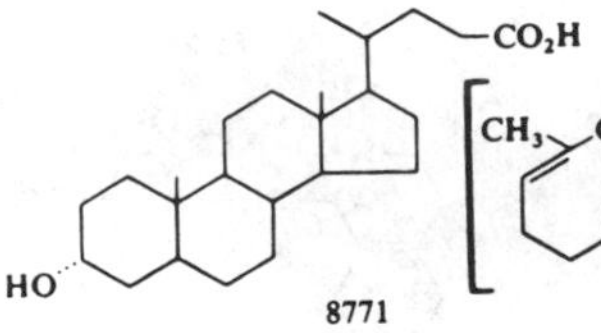

8787

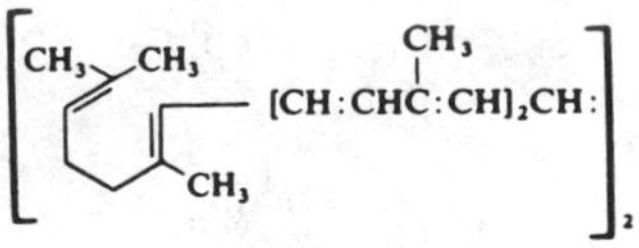

8796

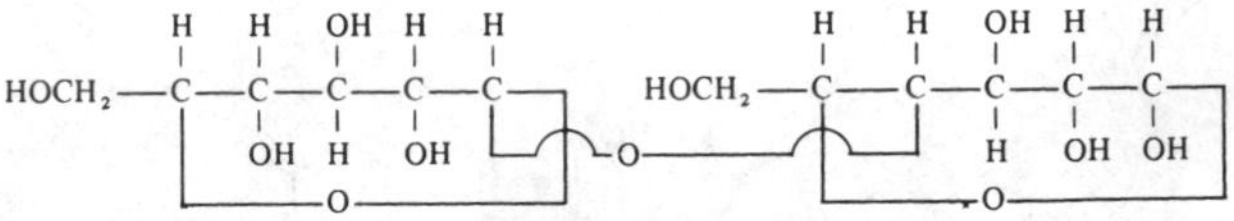

8797

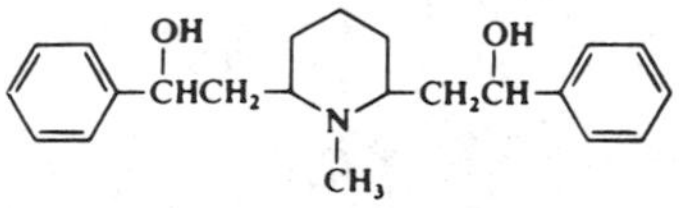

8798

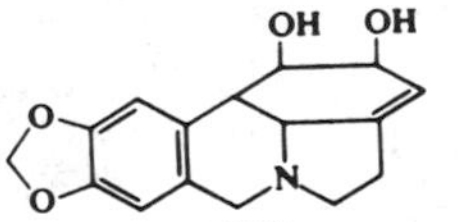

8799

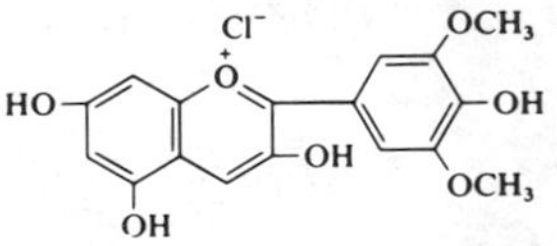

8805

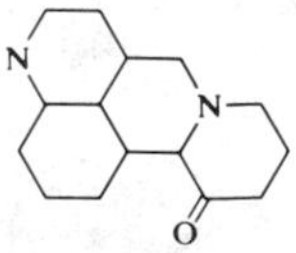

8820

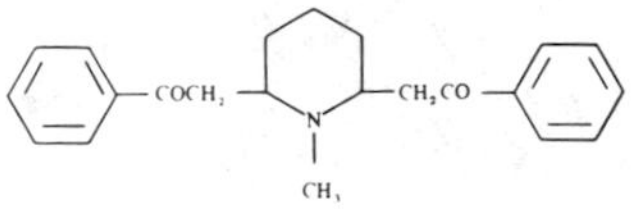

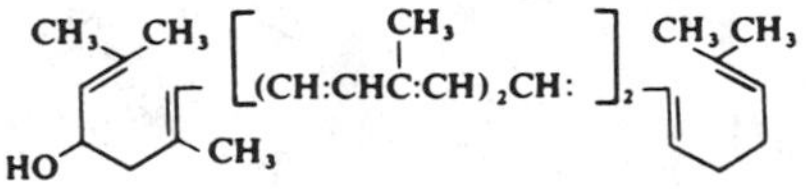

8824

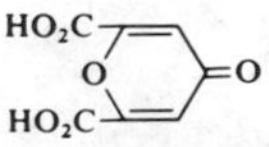

8856

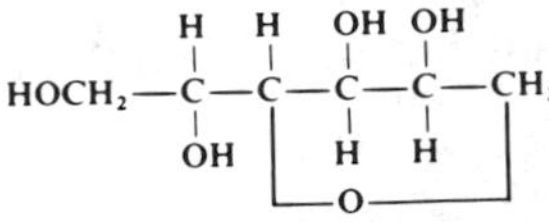

8857

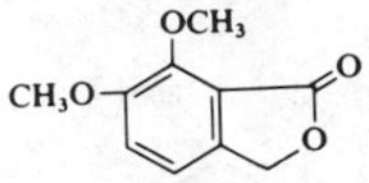

8858

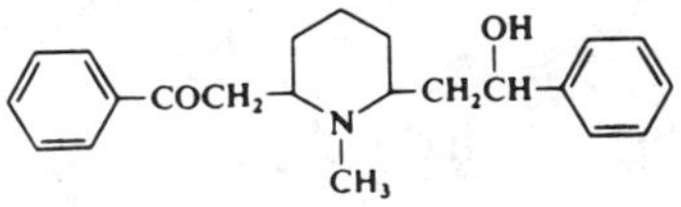

8883

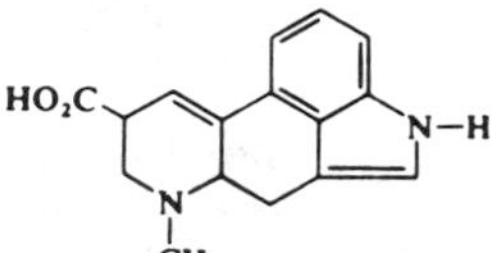

8898

8908

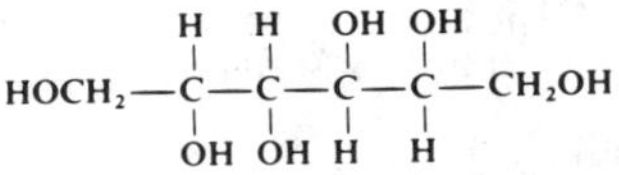

8910

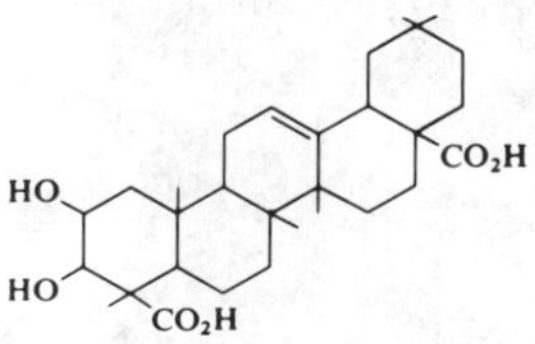

8921

8922

8926

8927

8928

8930

8932

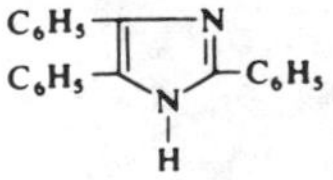

8937

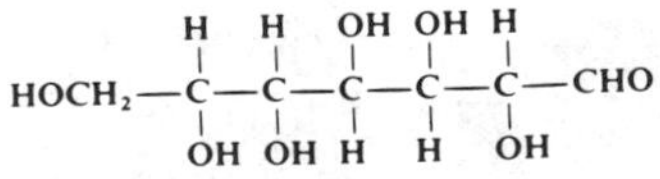

8940

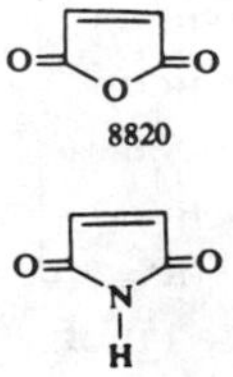

8941

8943

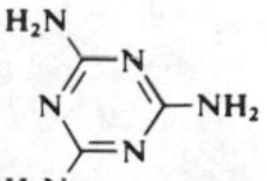

8944

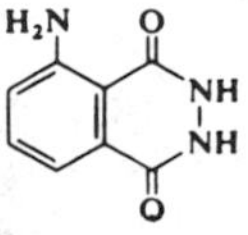

8947

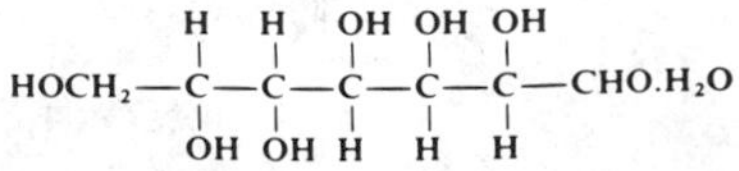

STRUCTURAL FORMULAS OF ORGANIC COMPOUNDS (Continued)

In Numeric as they occur in Organic Compounds Table

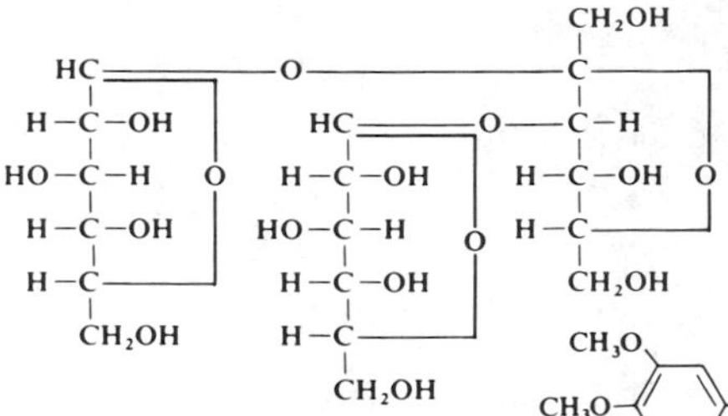

8950

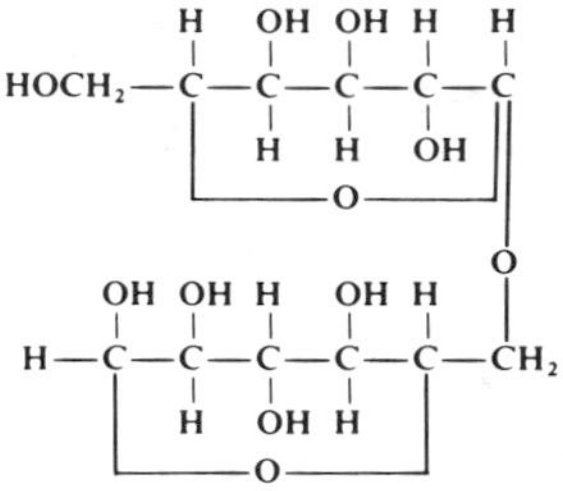

8951

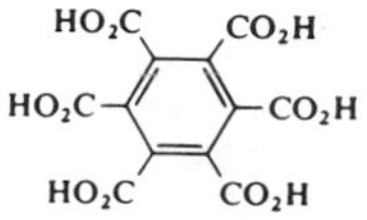

8953

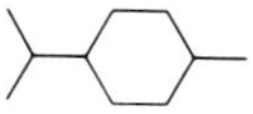

8954

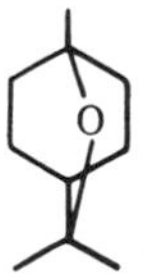

8956

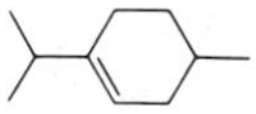

8957

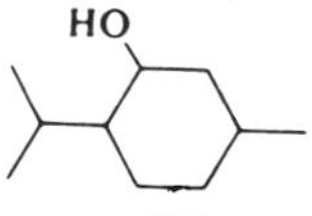

8961

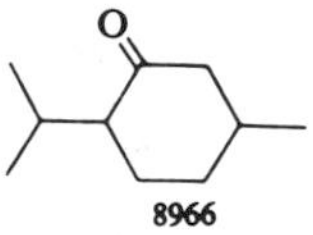

8966

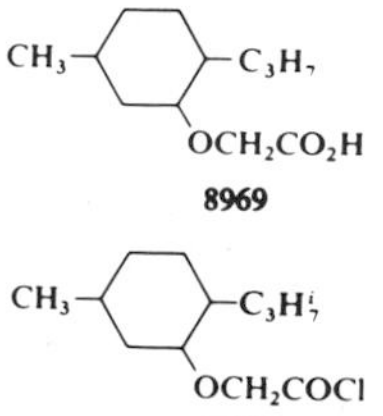

8969

8970

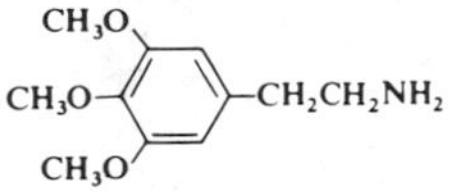

8986

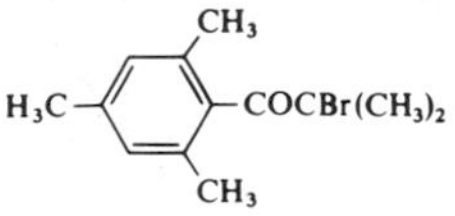

8989

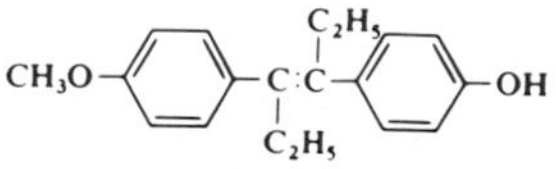

8997

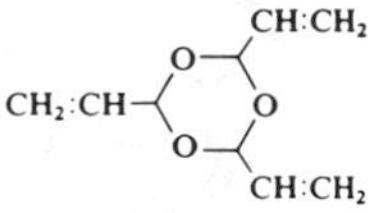

8998

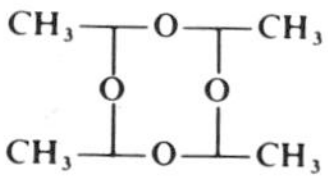

9000

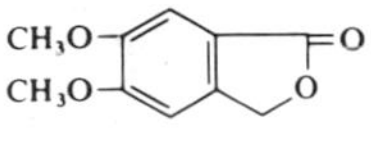

9001

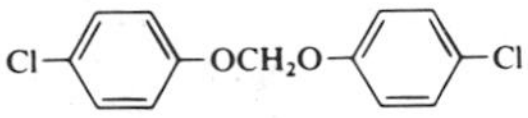

9019

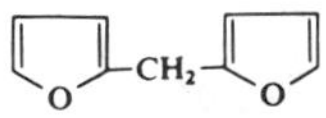

9069

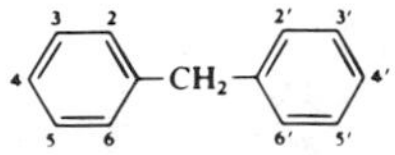

9074

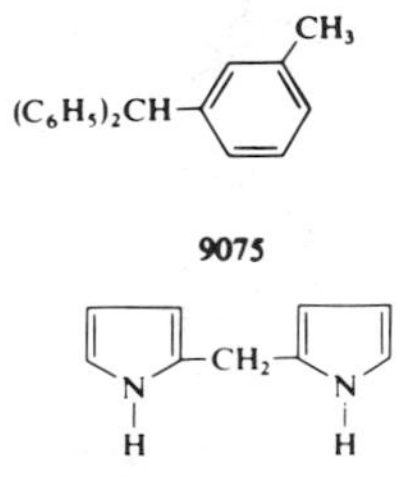

9075

9076

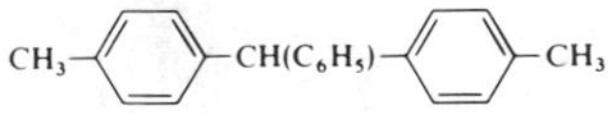

9078

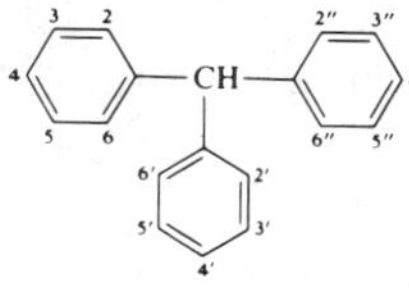

9103

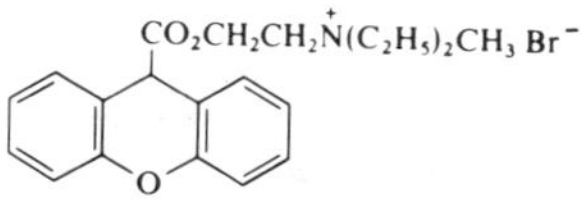

9126

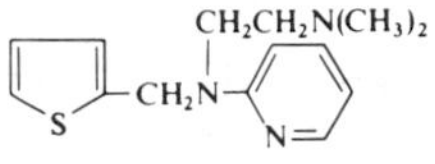

9129

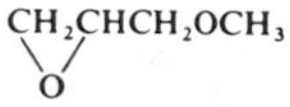

9159

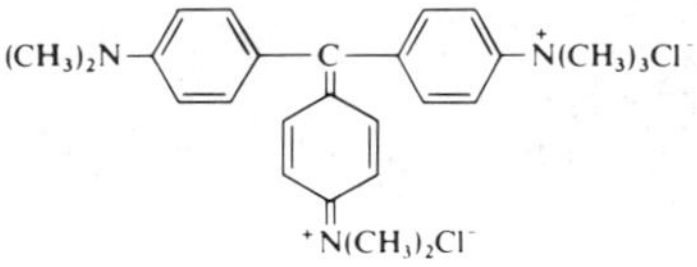

9166

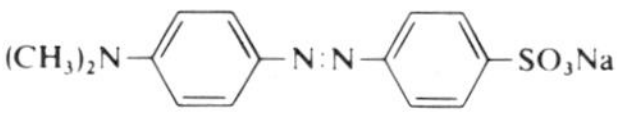

9179

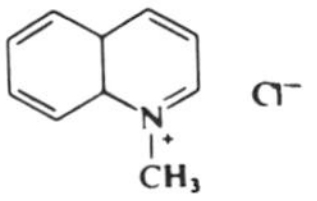

9195

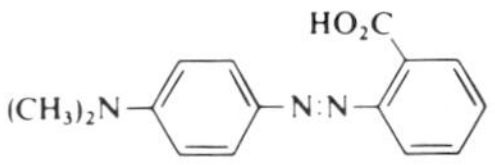

9196

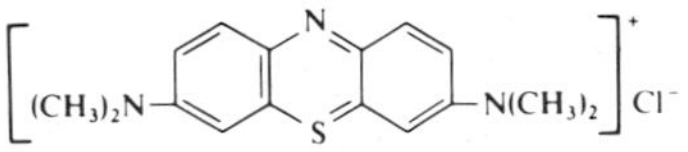

9209

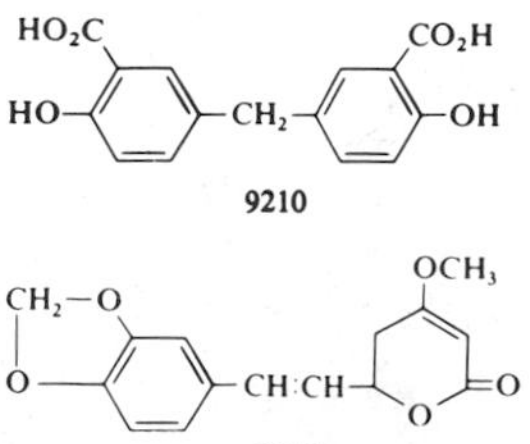

9210

9211

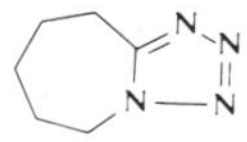

9212

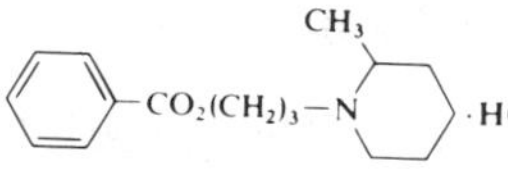

9213

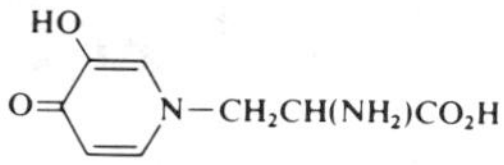

9215

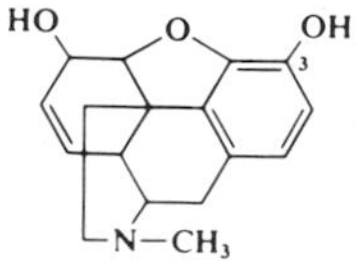

9216

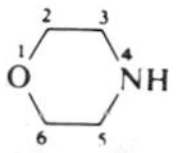

9225

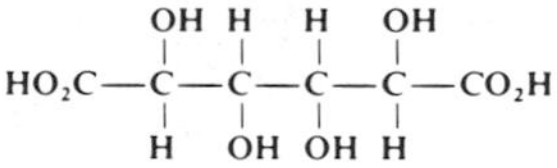

9239

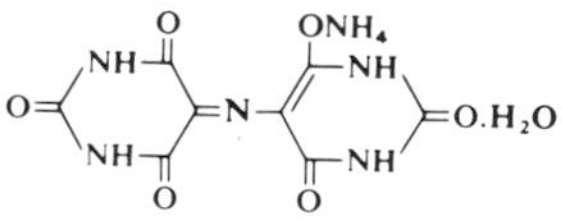

9242

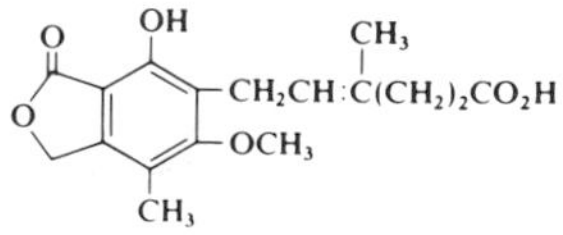

9243

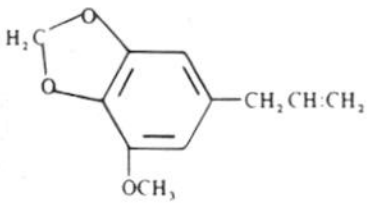

9258

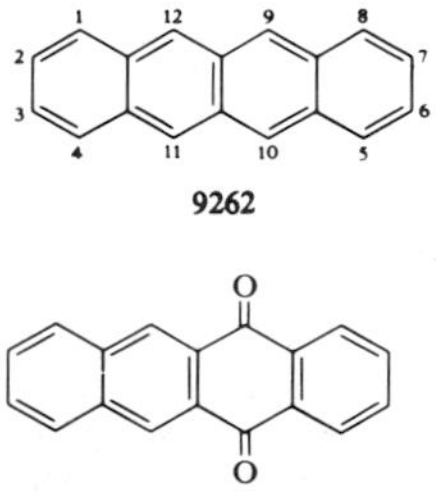

9262

9267

STRUCTURAL FORMULAS OF ORGANIC COMPOUNDS (Continued)

In Numeric as they occur in Organic Compounds Table

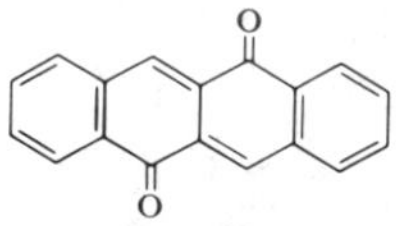

9268

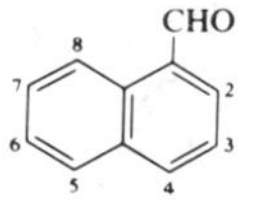

9269

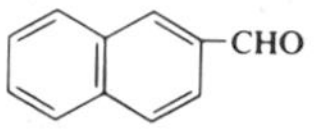

9273

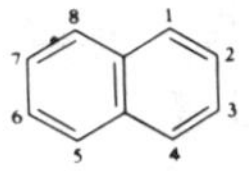

9275

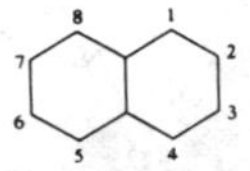

9316

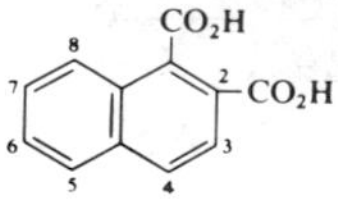

9445

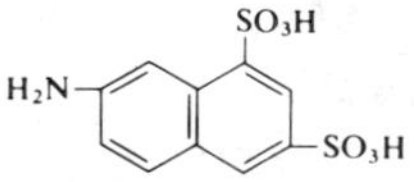

9454

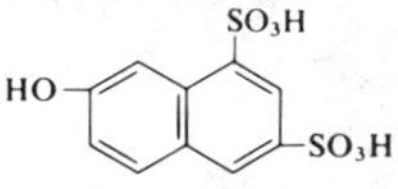

9455

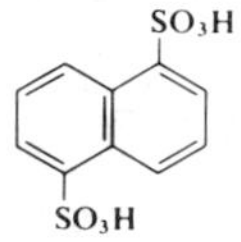

9456

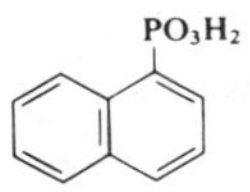

9462

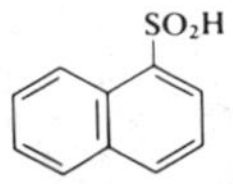

9464

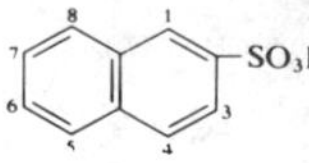

9476

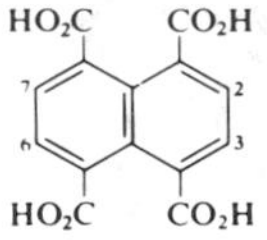

9483

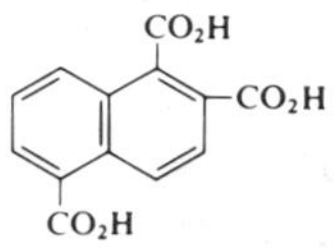

9485

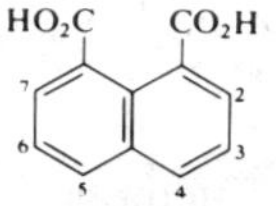

9487

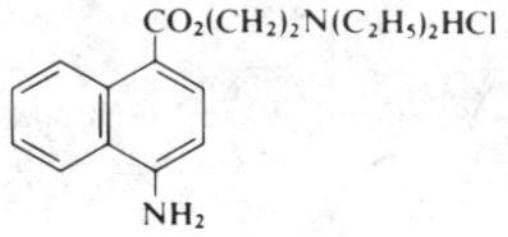

9499

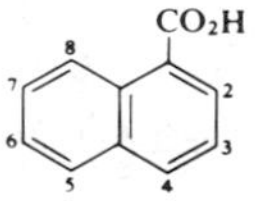

9500

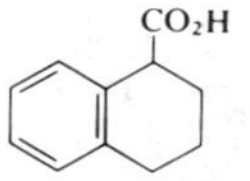

9529

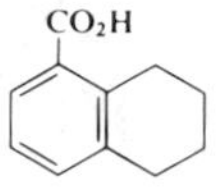

9530

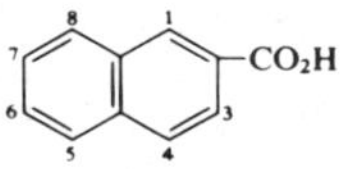

9534

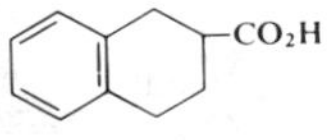

9567

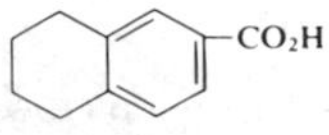

9568

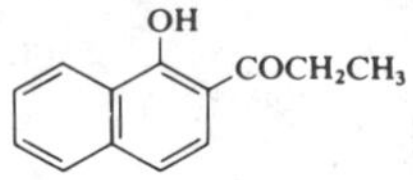

9586

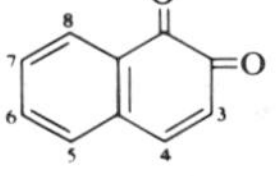

9618

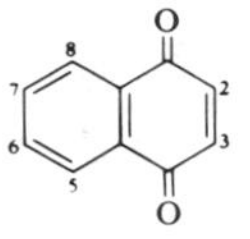

9636

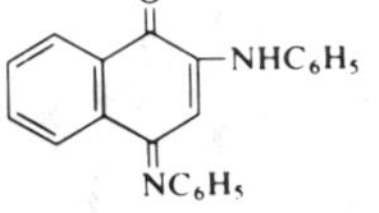

9638

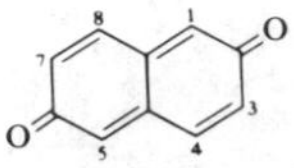

9679

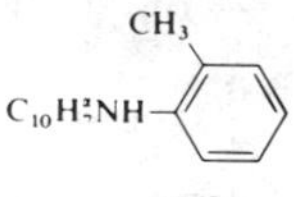

9769

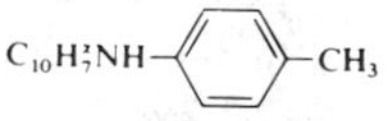

9770

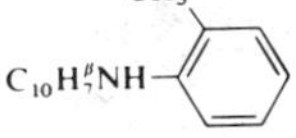

9771

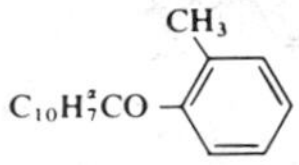

9773

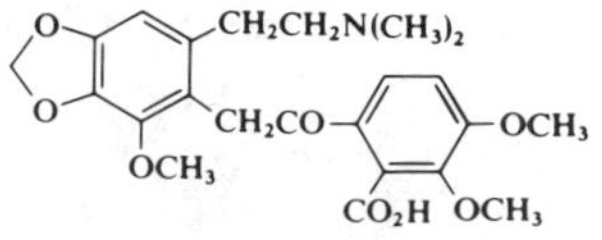

9775

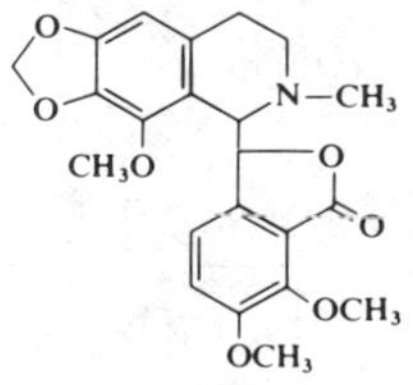

9778

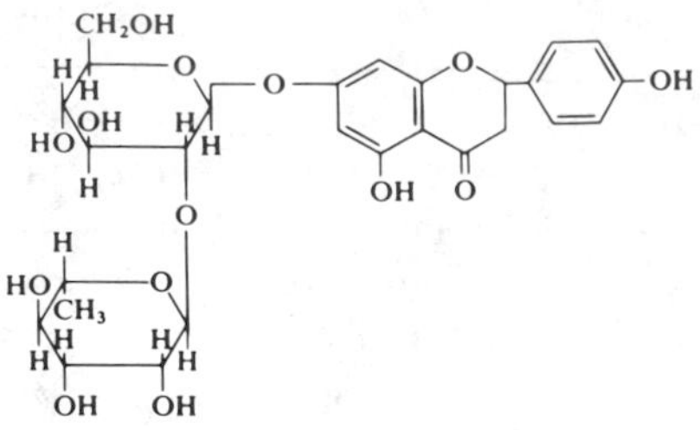

9783

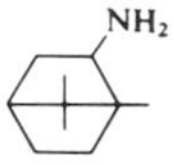

9786

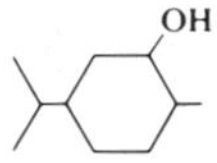

9787

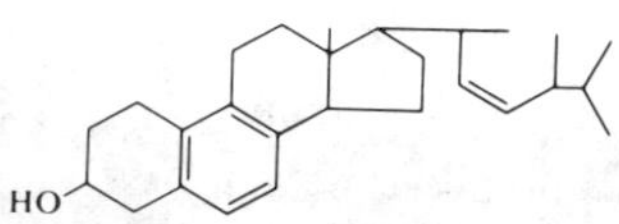

9789

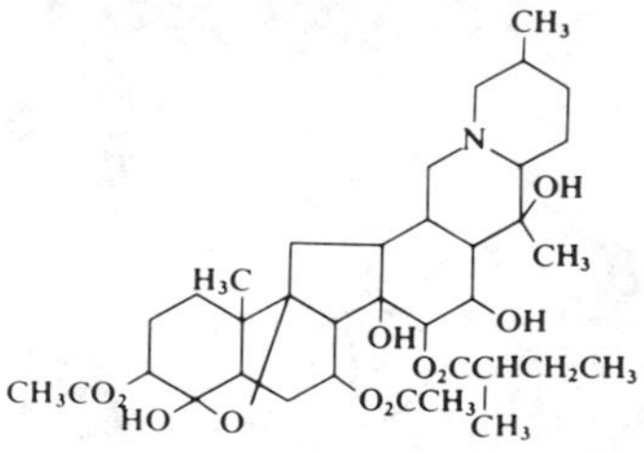

9790

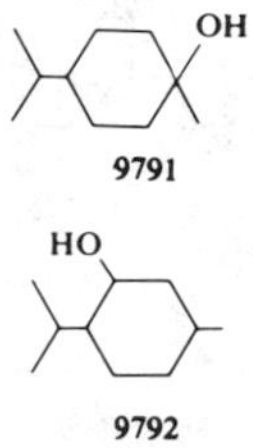

9791

9792

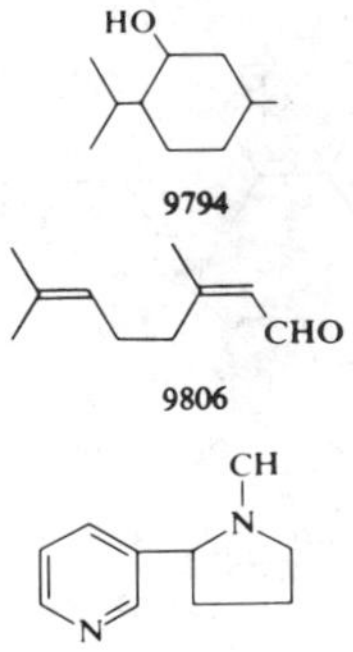

9794

9806

9812

STRUCTURAL FORMULAS OF ORGANIC COMPOUNDS (Continued)

In Numeric as they occur in Organic Compounds Table

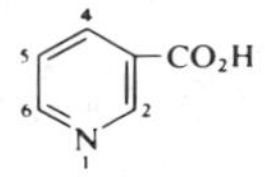

9818

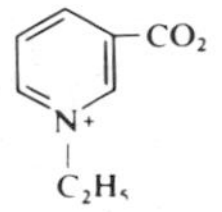

9820

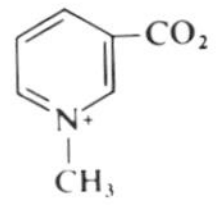

9827

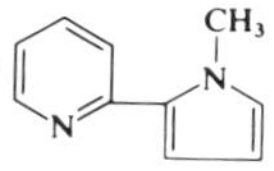

9830

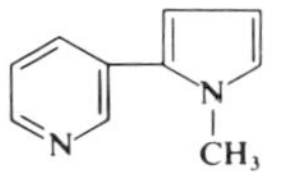

9832

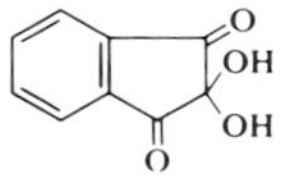

9833

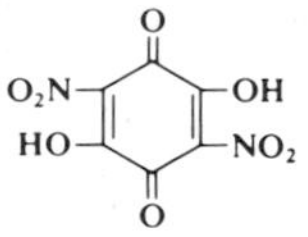

9834

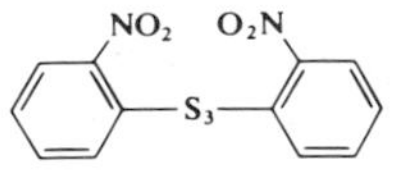

9892

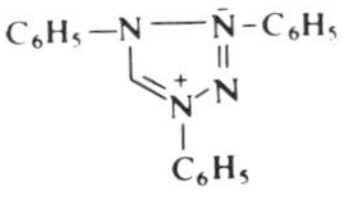

9899

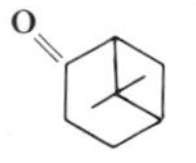

9958

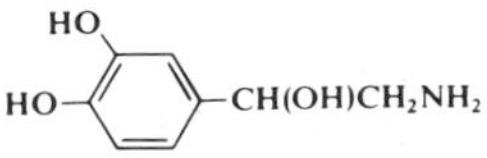

9959

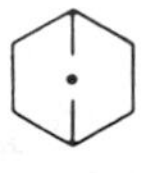

9960

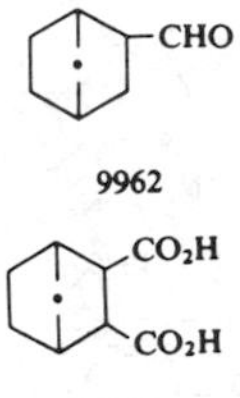

9962

9963

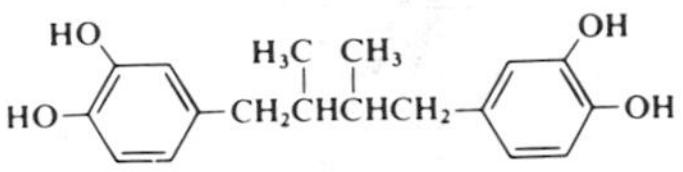

9965

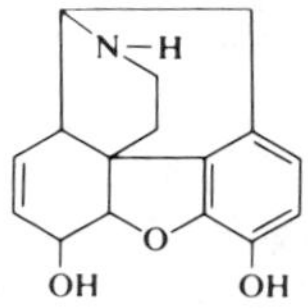

9969

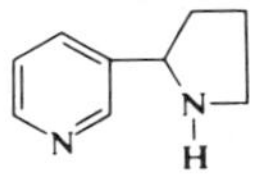

9970

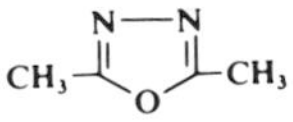

10168

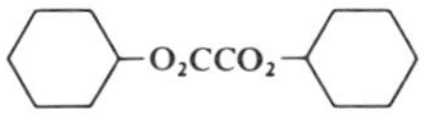

10174

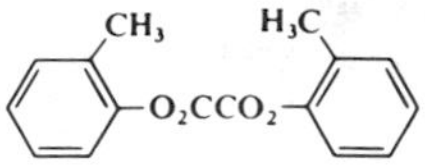

10181

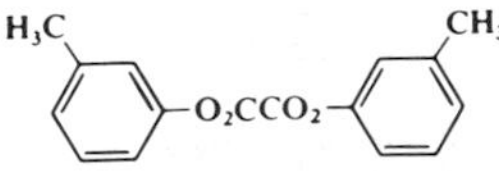

10182

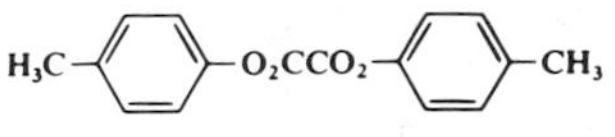

10183

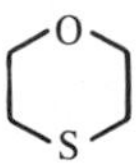

10200

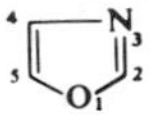

10201

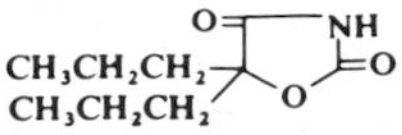

10208

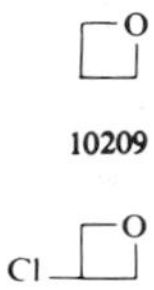

10209

10210

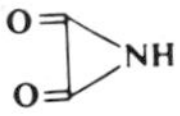

10211

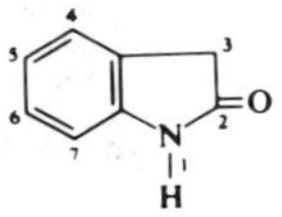

10212

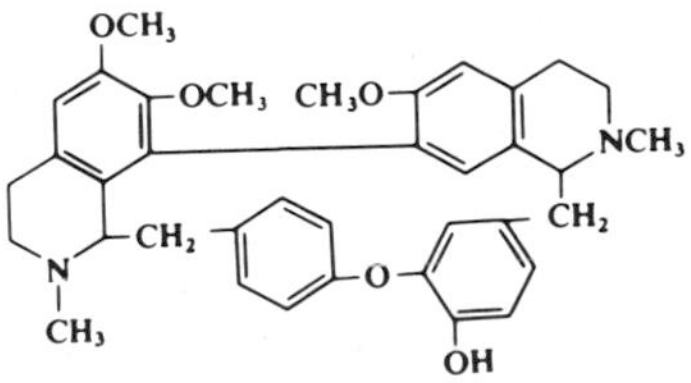

10216

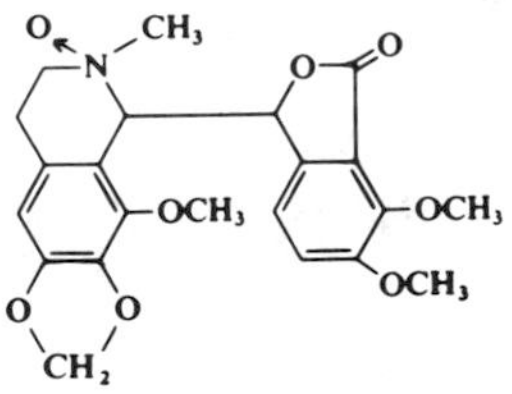

10219

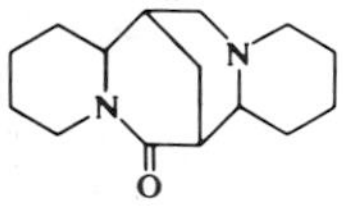

10220

10241

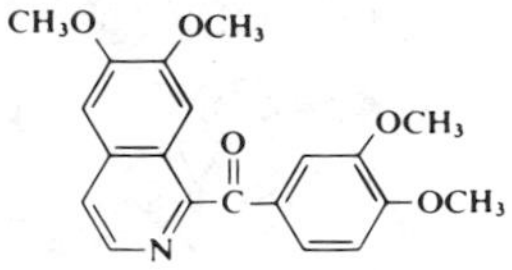

10249

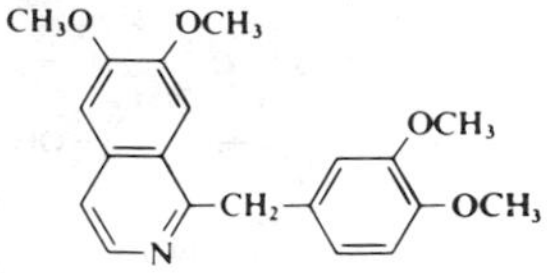

10250

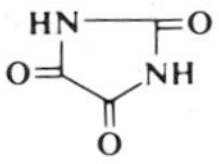

10252

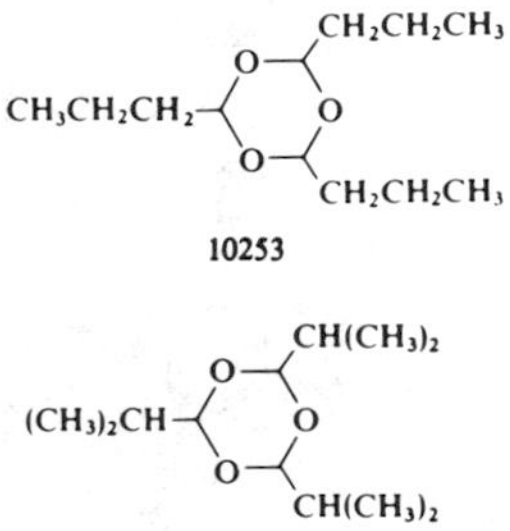

10253

10254

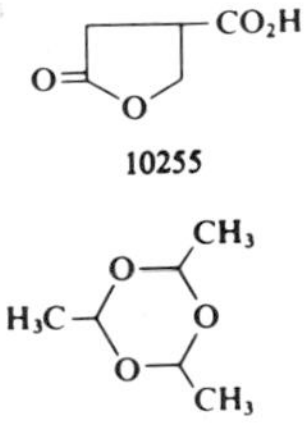

10255

10257

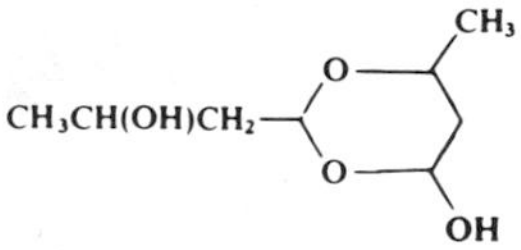

10258

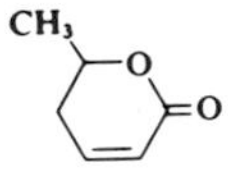

10259

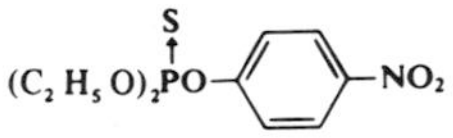

10260

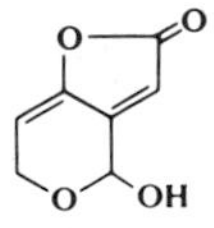

10261

In Numeric as they occur in Organic Compounds Table

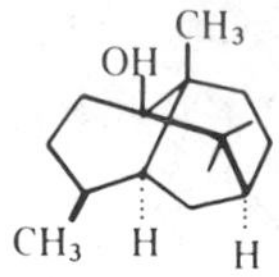

10262

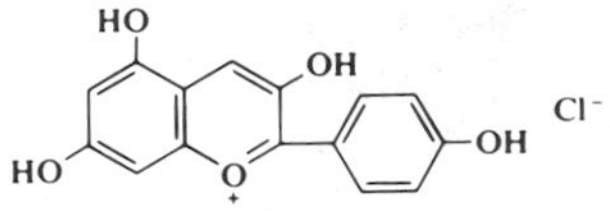

10274

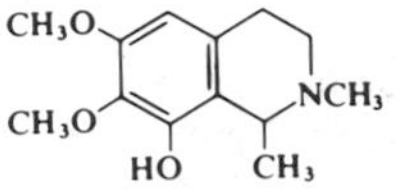

10275

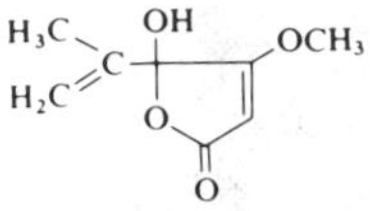

10276

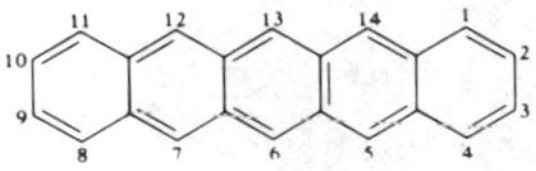

10277

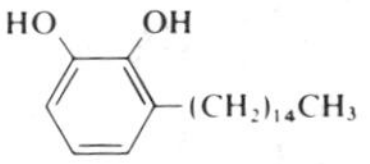

10289

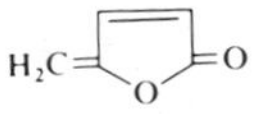

10310

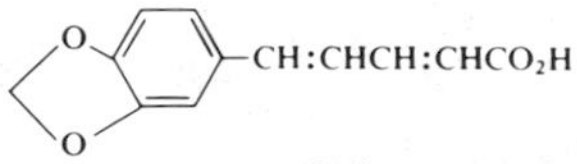

10311

10329

10330

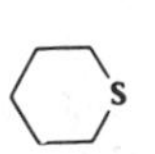

10338

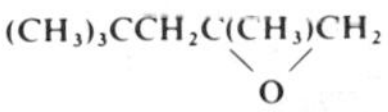

$(CH_3)_3CCH_2C(CH_3)CH_2$ O

10398

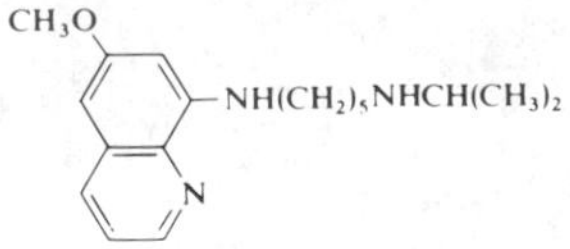

CH_3O $NH(CH_2)_5NHCH(CH_3)_2$

10511

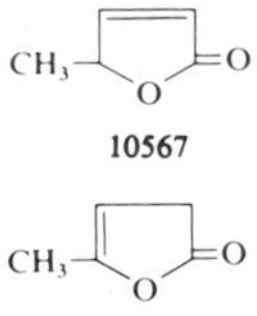

10567

10570

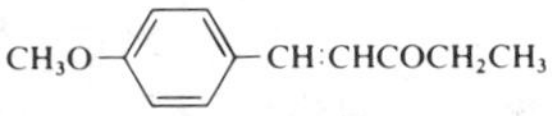

CH_3O– –$CH{:}CHCOCH_2CH_3$

10588

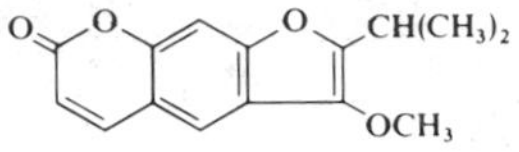

10598

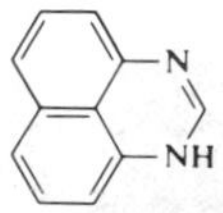

10638

$HOCH_2—C(H)(OH)—C(H)(OH)—C(OH)(H)—C(OH)(H)—C(H)(OH)—CH_2OH$

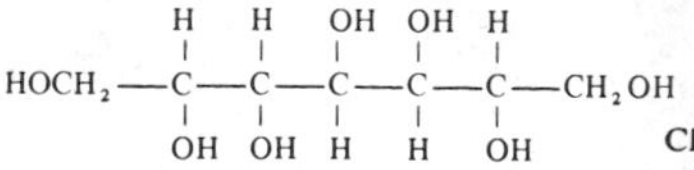

10639

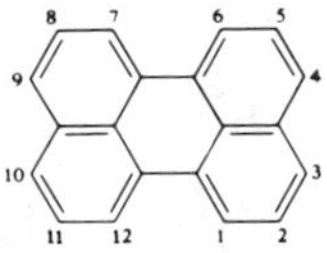

10640

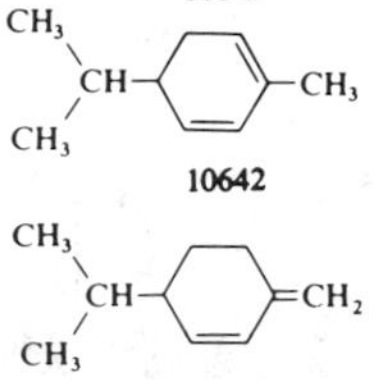

10642

10643

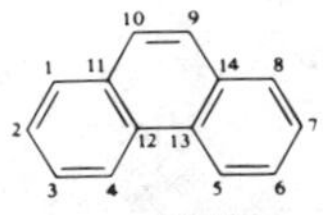

10649

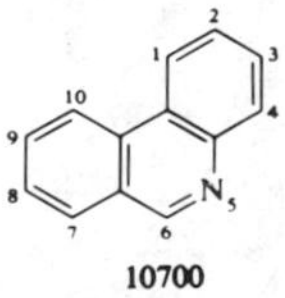

10700

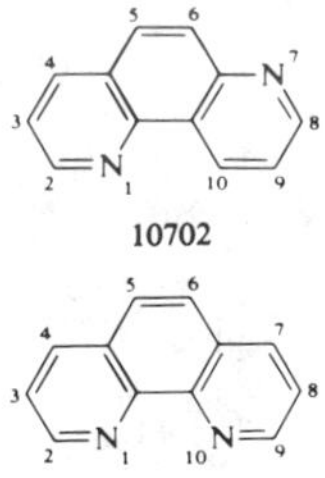

10702

10703

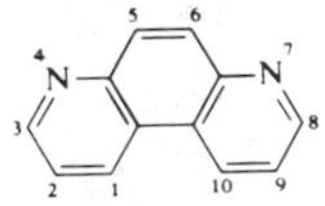

10704

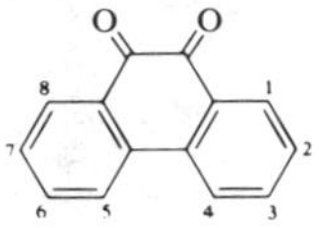

10707

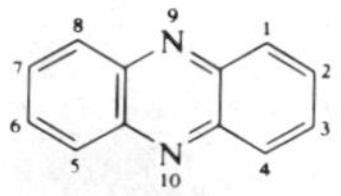

10725

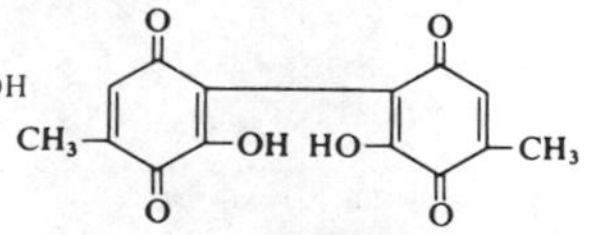

10766

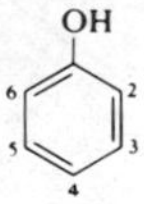

10767

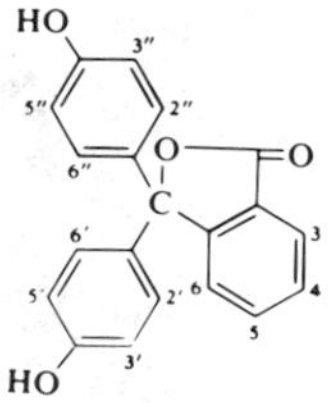

11042

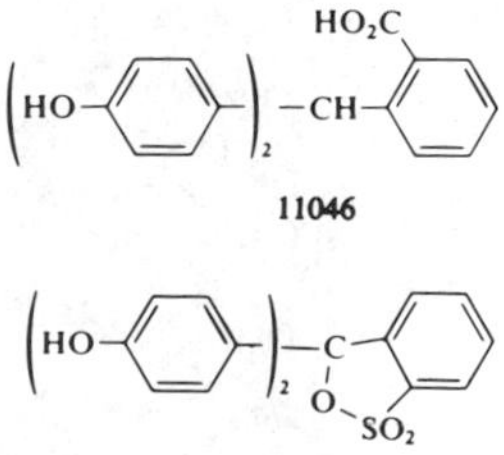

11046

11047

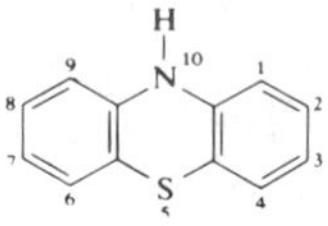

11048

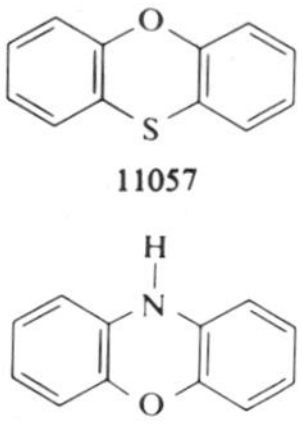

11057

11058

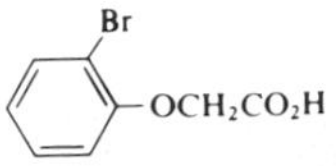

11062

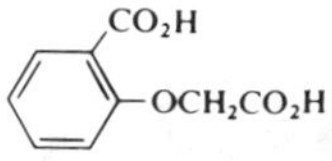

11064

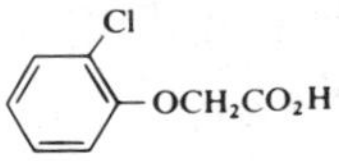

11067

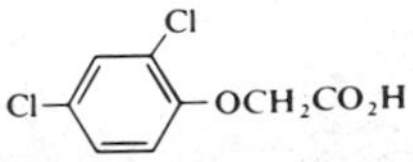

11071

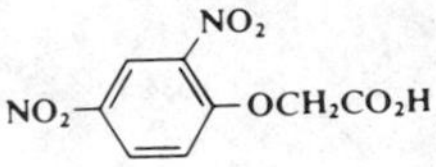

11072

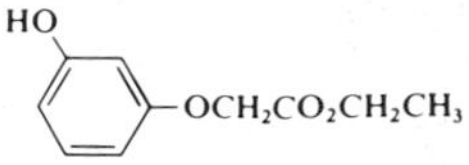

11076

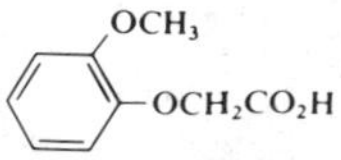

11078

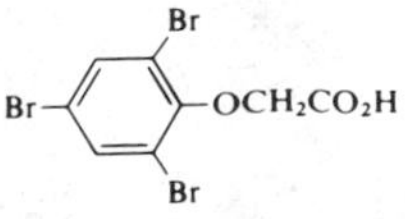

11086

In Numeric as they occur in Organic Compounds Table

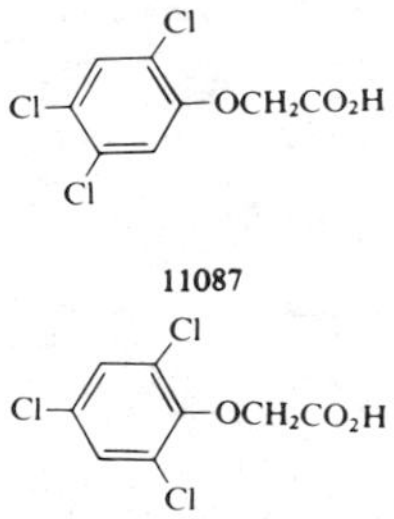

Cl, Cl, Cl, OCH_2CO_2H
11087

Cl, Cl, Cl, OCH_2CO_2H
11088

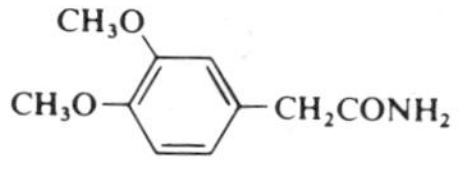

CH_3O, CH_3O, CH_2CONH_2
11094

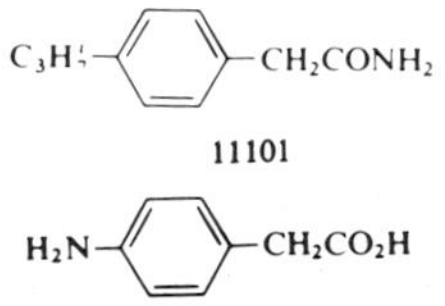

C_3H_7, CH_2CONH_2
11101

H_2N, CH_2CO_2H
11104

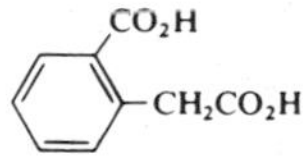

CO_2H, CH_2CO_2H
11111

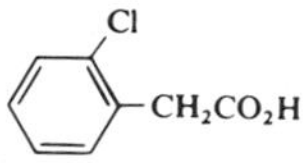

Cl, CH_2CO_2H
11117

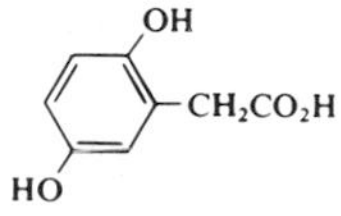

OH, CH_2CO_2H, HO
11120

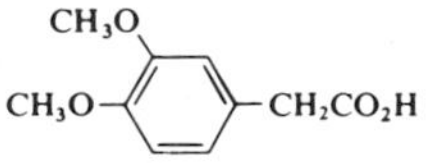

CH_3O, CH_3O, CH_2CO_2H
11122

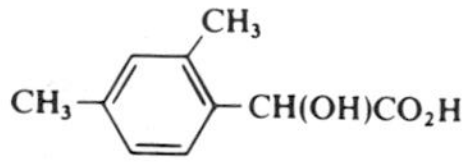

CH_3, CH_3, $CH(OH)CO_2H$
11123

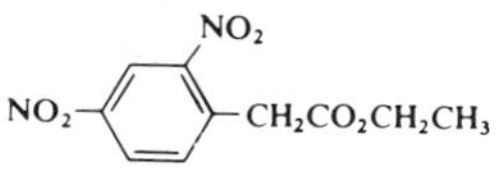

NO_2, NO_2, $CH_2CO_2CH_2CH_3$
11127

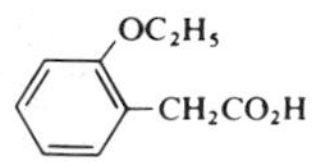

OC_2H_5, CH_2CO_2H
11129

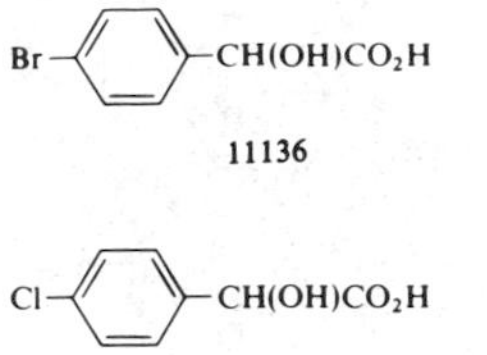

Br, $CH(OH)CO_2H$
11136

Cl, $CH(OH)CO_2H$
11137

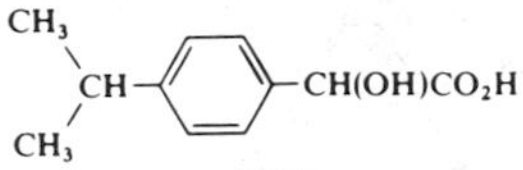

CH_3, CH, CH_3, $CH(OH)CO_2H$
11144

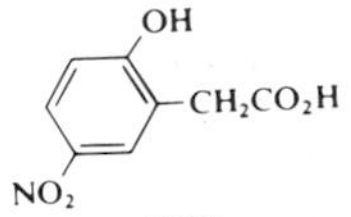

OH, CH_2CO_2H, NO_2
11148

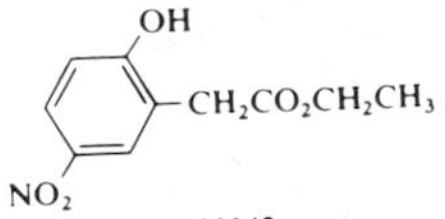

OH, $CH_2CO_2CH_2CH_3$, NO_2
11149

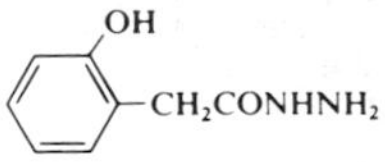

OH, $CH_2CONHNH_2$
11150

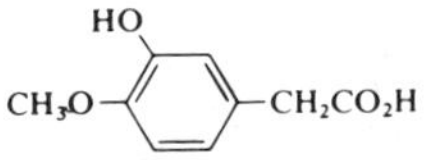

HO, CH_3O, CH_2CO_2H
11153

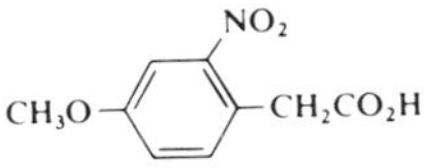

NO_2, CH_3O, CH_2CO_2H
11160

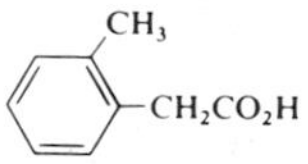

CH_3, CH_2CO_2H
11162

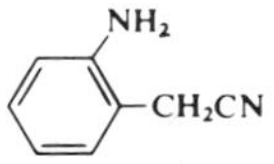

NH_2, CH_2CN
11171

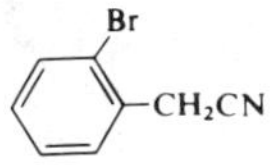

Br, CH_2CN
11174

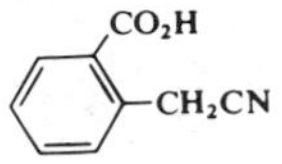

CO_2H, CH_2CN
11176

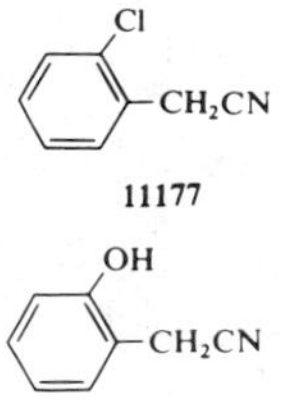

Cl, CH_2CN
11177

OH, CH_2CN
11181

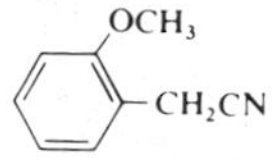

OCH_3, CH_2CN
11184

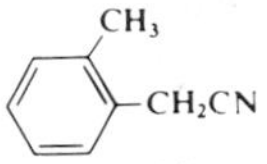

CH_3, CH_2CN
11186

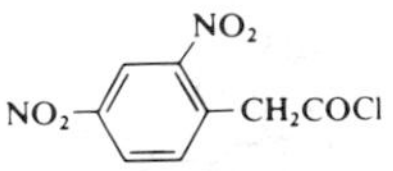

NO_2, NO_2, CH_2COCl
11195

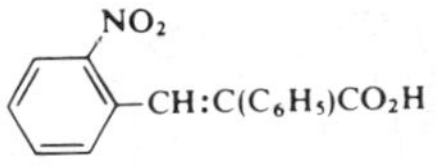

NO_2, $CH:C(C_6H_5)CO_2H$
11198

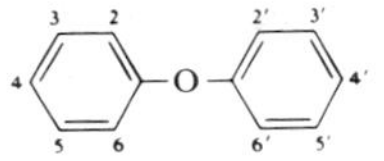

3, 2, 2', 3', 4, O, 4', 5, 6, 6', 5'
11249

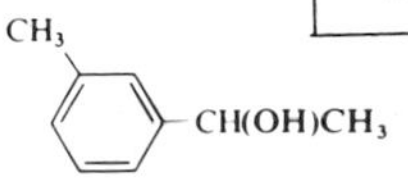

CH_3, $CH(OH)CH_3$
11254

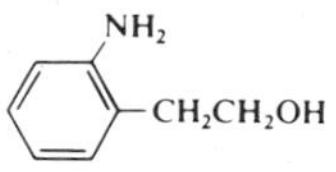

NH_2, CH_2CH_2OH
11257

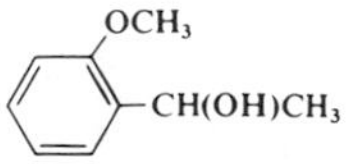

OCH_3, $CH(OH)CH_3$
11259

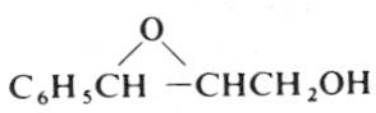

O, C_6H_5CH $-CHCH_2OH$
11266

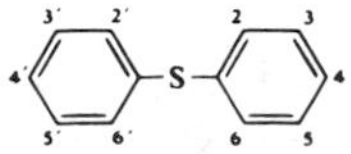

3', 2', 2, 3, 4', S, 4, 5', 6', 6, 5
11308

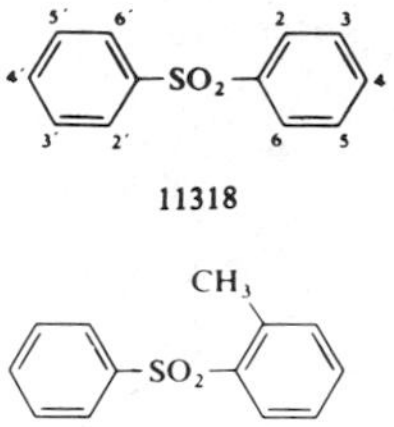

5', 6', 2, 3, 4', SO_2, 4, 3', 2', 6, 5
11318

CH_3, SO_2
11333

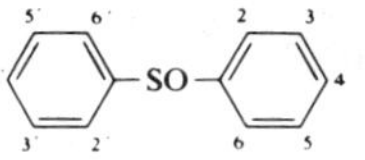

5', 6', 2, 3, 4', SO, 4, 3', 2', 6, 5
11335

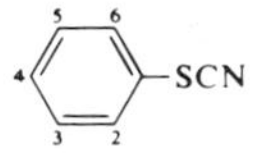

5, 6, 4, SCN, 3, 2
11341

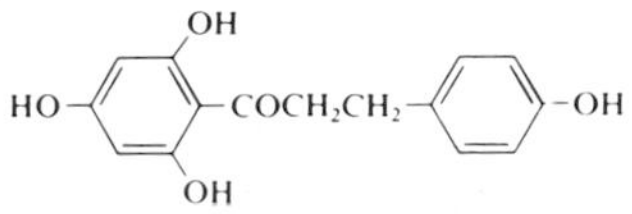

OH, HO, $COCH_2CH_2$, OH, OH
11348

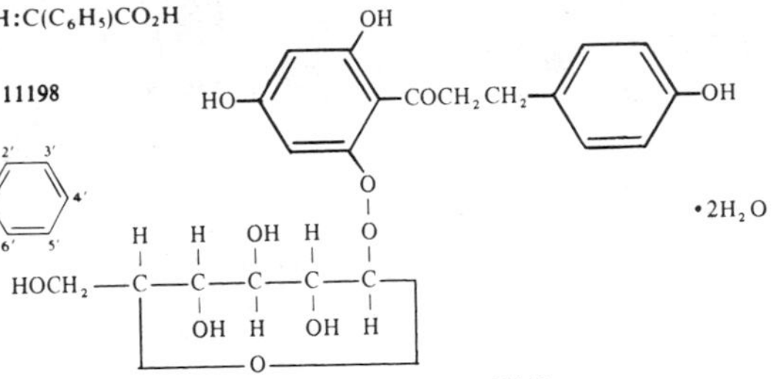

OH, HO, $COCH_2CH_2$, OH, O, H H OH H O, $HOCH_2$—C—C—C—C—C, OH H OH H, O, $\cdot 2H_2O$
11349

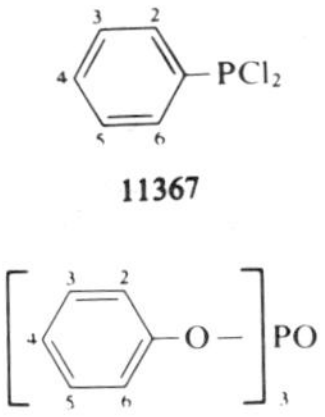

3, 2, 4, PCl_2, 5, 6
11367

[3, 2, 4, O—, 5, 6]$_3$ PO
11420

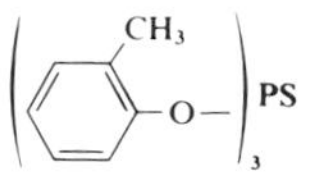

(CH_3, O—)$_3$ PS
11429

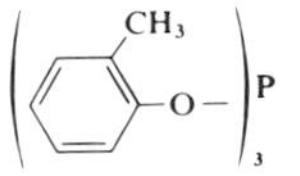

(CH_3, O—)$_3$ P
11439

In Numeric as they occur in Organic Compounds Table

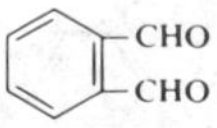

11441

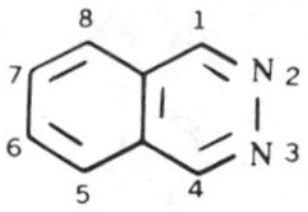

$CH_3CO-C_6H_4-OC_6H_{11}O_5$

11538

11585

11702

11446

11539

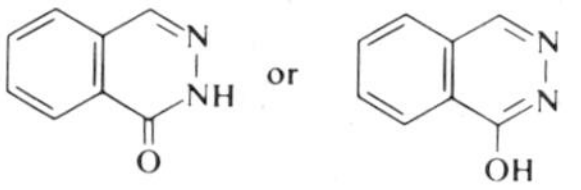

11587

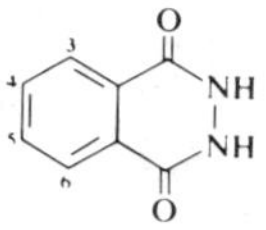

11705

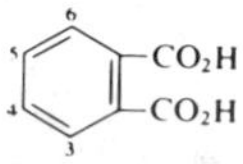

11447

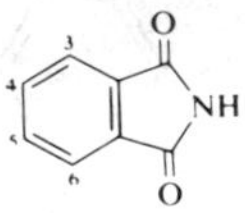

11554

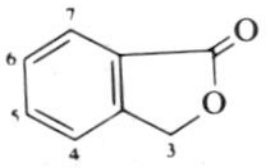

11591

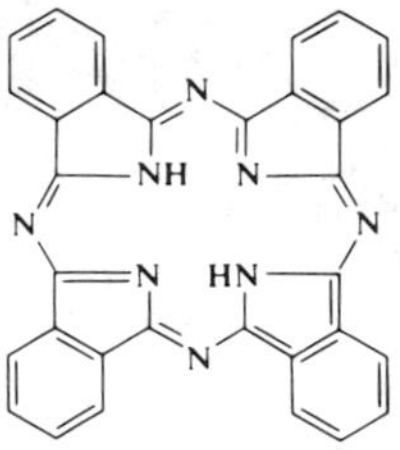

$N-COCH:CHCH:CH-$

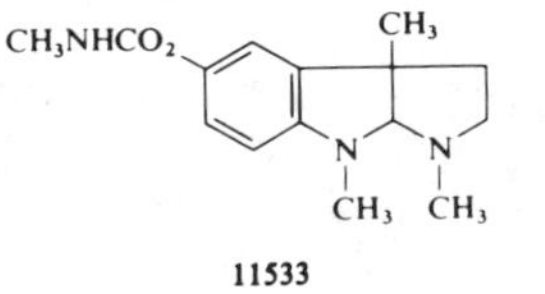

11707

11448

11556

11593

11708

11594

11712

11449

11560

11614

11714

11504

11561

11615

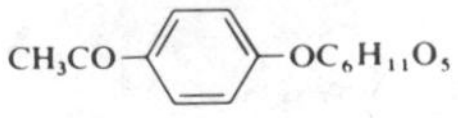

11715

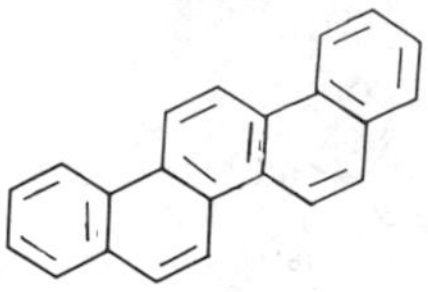

11528

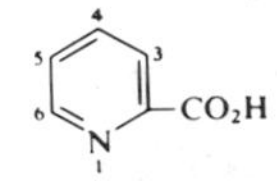

11563

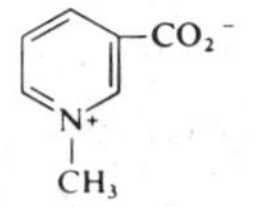

11616

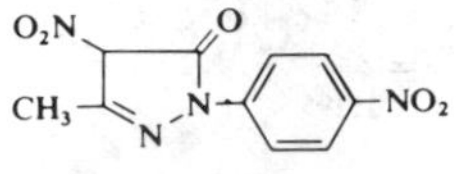

11617

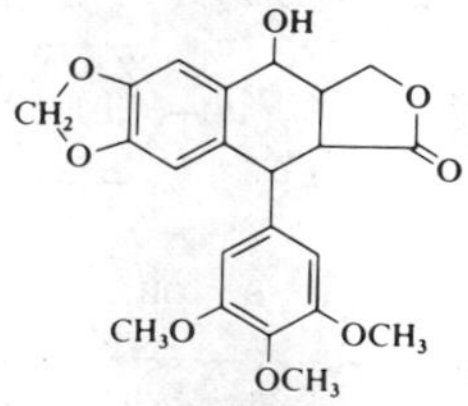

11717

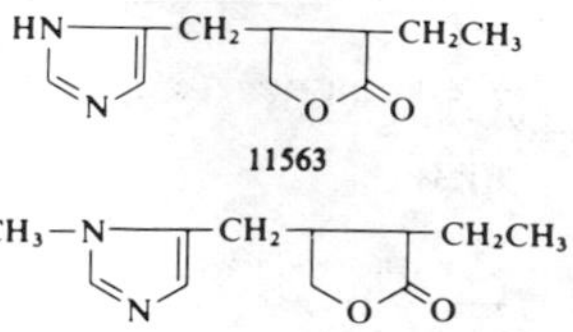

11532

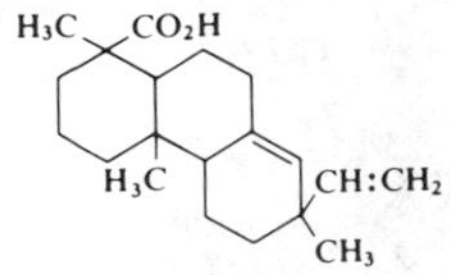

11565

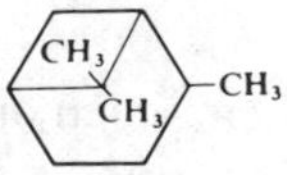

11669

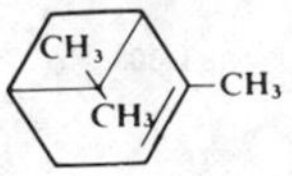

11570

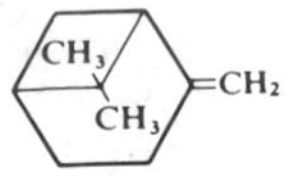

11695

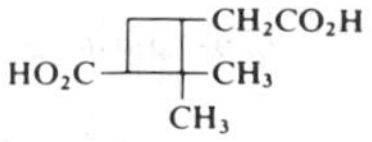

11533

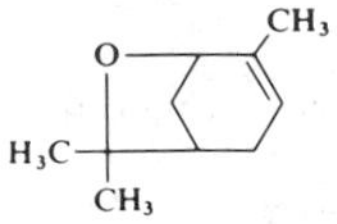

11581

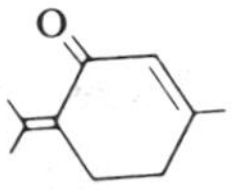

11696

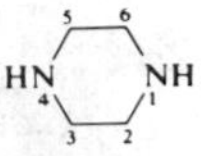

11734

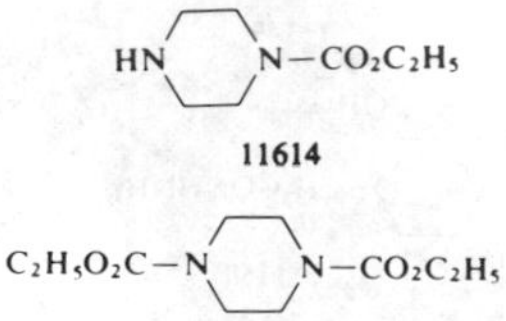

$CH_3(CH-CH_2-CH_2-CH_2)_2-C-CH_2CH_2CH:C(CH_3):CH_2$

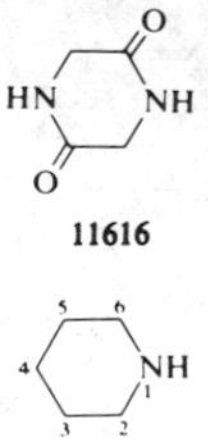

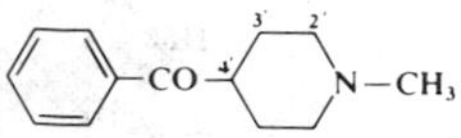

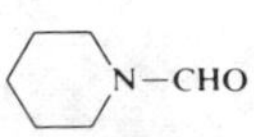

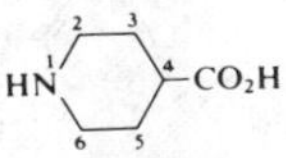

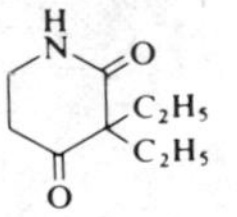

11536

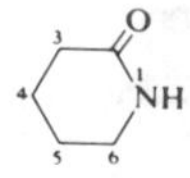

$CH_3(CH-CH_2CH_2CH_2)_3C:CHCH_2OH$

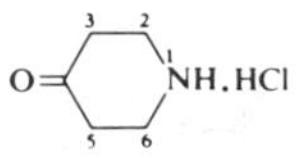

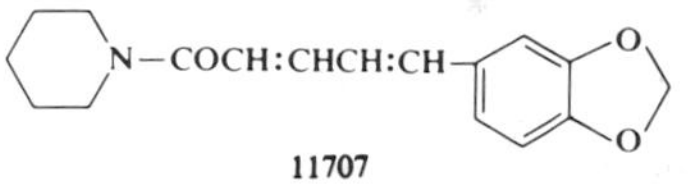

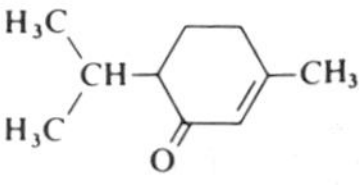

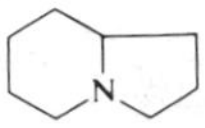

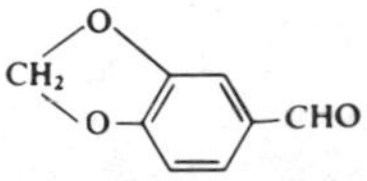

11537

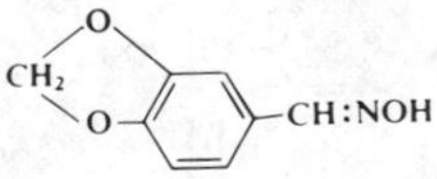

11584

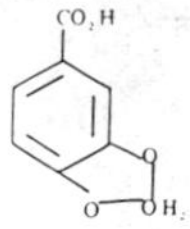

11701

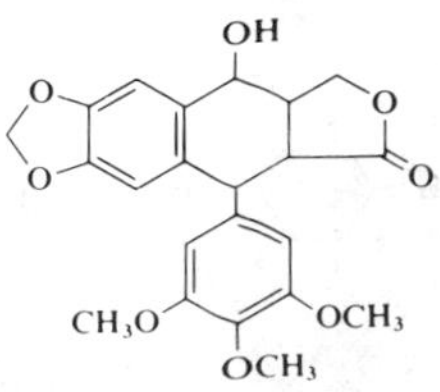

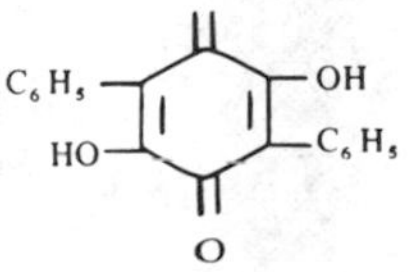

11736

STRUCTURAL FORMULAS OF ORGANIC COMPOUNDS (Continued)

In Numeric as they occur in Organic Compounds Table

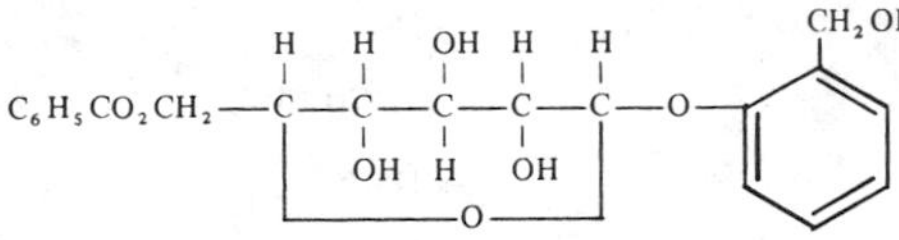

11737

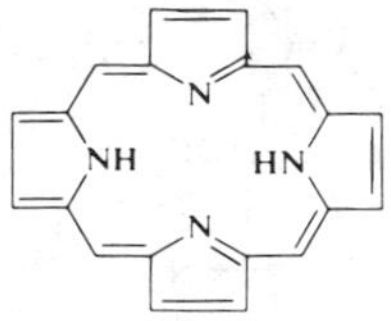

11738

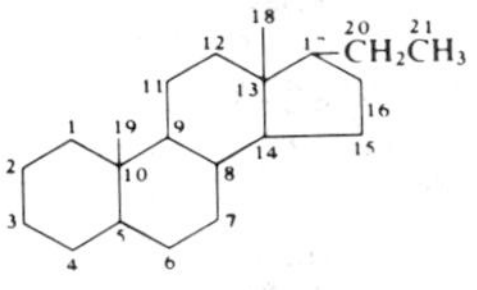

11741

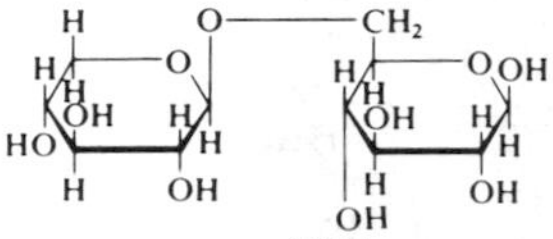

11755

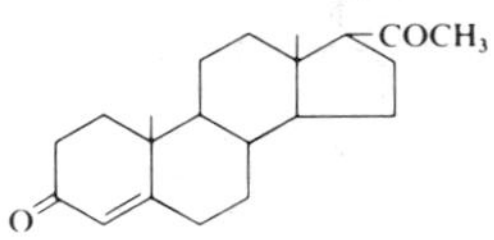

11756

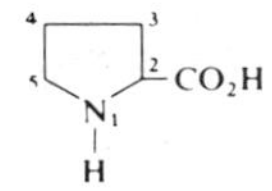

11759

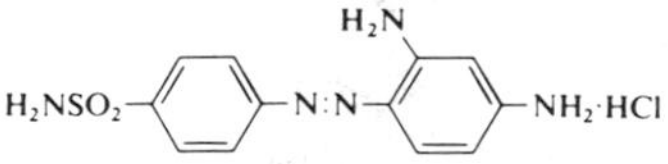

11769

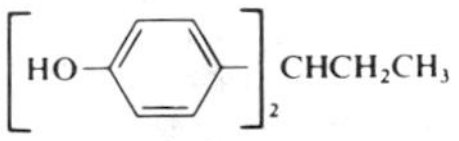

11780

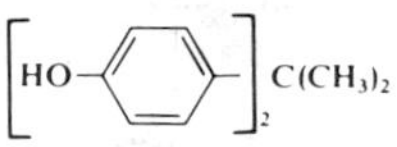

11781

CH_2CHCH_2Br (epoxide)
11791

CH_2CHCH_2Cl (epoxide)
11813

$CH_2CH(CH_3)CH_2Cl$ (epoxide)
11815

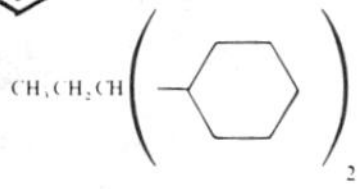

11834

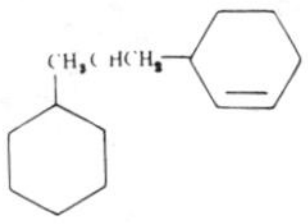

11835

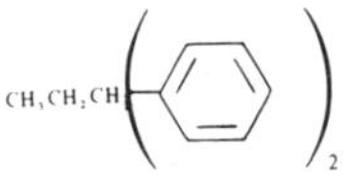

11853

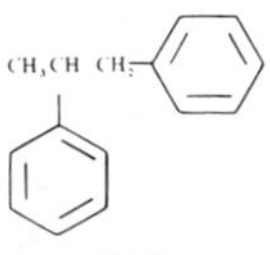

11854

CH_2CHCH_2I (epoxide)
11859

CH_2CHCCl_3 (epoxide)
11860

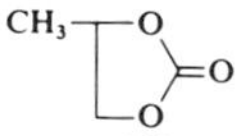

11900

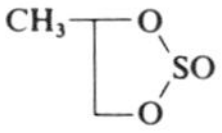

11909

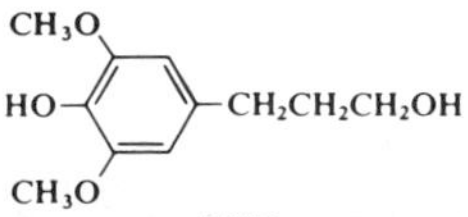

11956

CH_2CHCH_2OH (epoxide)
11958

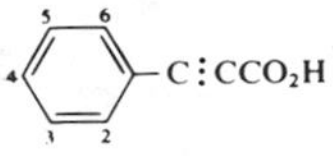

11020

12041

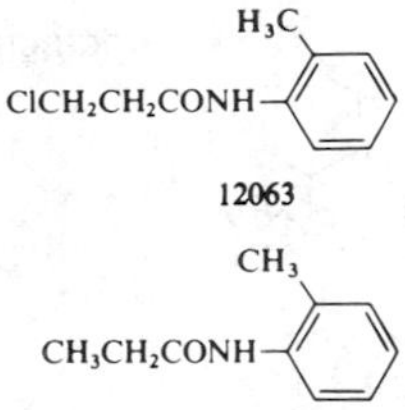

$ClCH_2CH_2CONH$—
12063

CH_3CH_2CONH—
12070

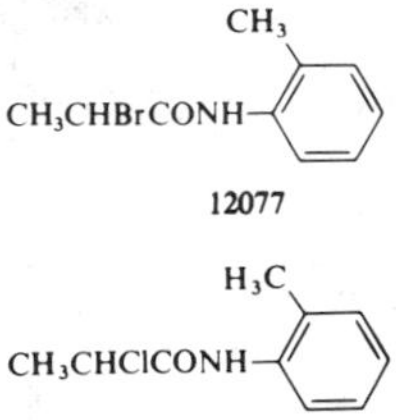

$CH_3CHBrCONH$—
12077

$CH_3CHClCONH$—
12103

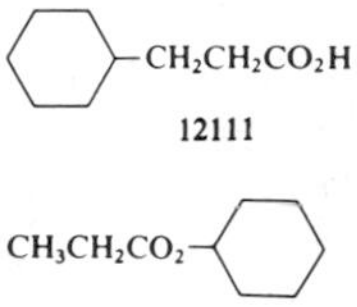

—$CH_2CH_2CO_2H$
12111

$CH_3CH_2CO_2$—
12112

CH_2CHCO_2H (epoxide)
12142

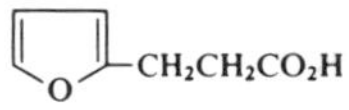

—$CH_2CH_2CO_2H$
12145

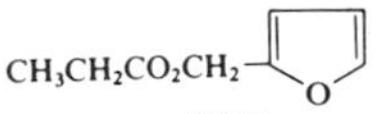

$CH_3CH_2CO_2CH_2$—
12147

12152

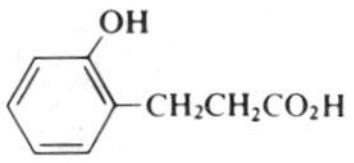

—$CH_2CH_2CO_2H$
12154

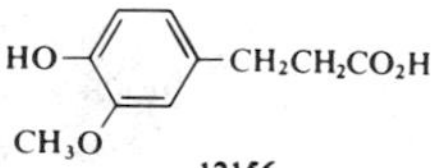

—$CH_2CH_2CO_2H$
12156

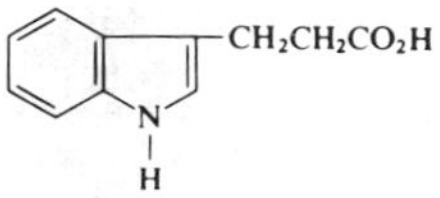

—$CH_2CH_2CO_2H$
12160

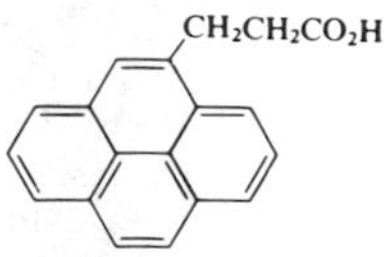

$CH_2CH_2CO_2H$
12187

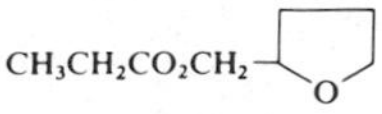

$CH_3CH_2CO_2CH_2$—
12192

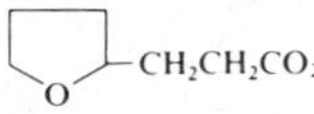

—$CH_2CH_2CO_2H$
12193

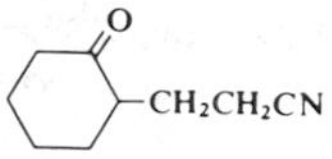

—CH_2CH_2CN
12212

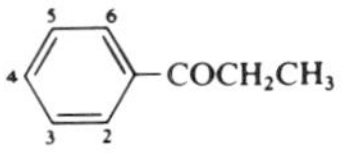

—$COCH_2CH_3$
12228

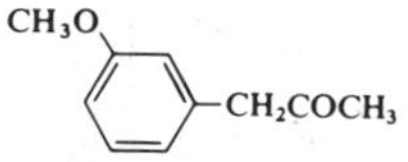

—CH_2COCH_3
12238

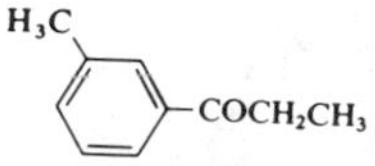

—$COCH_2CH_3$
12241

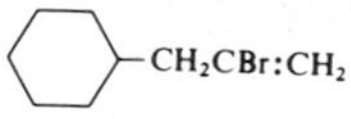

—$CH_2CBr{:}CH_2$
12261

$CH_3C{=}CH$ (epoxide)
12285

CH_2CHCH_3 (epoxide)
12300

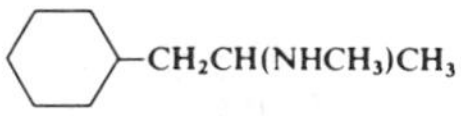

—$CH_2CH(NHCH_3)CH_3$
12305

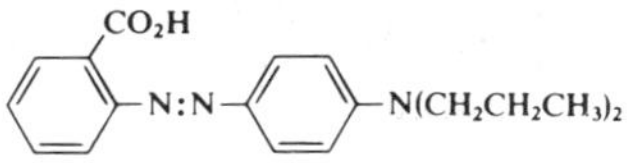

—N:N— —$N(CH_2CH_2CH_3)_2$
12320

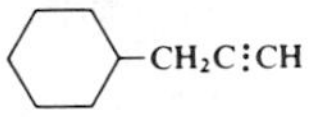

—$CH_2C{:}CH$
12336

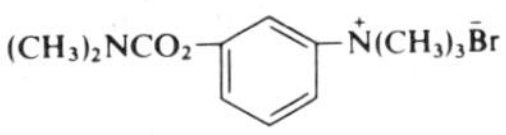

$(CH_3)_2NCO_2$— —$\overset{+}{N}(CH_3)_3\overset{-}{Br}$
12345

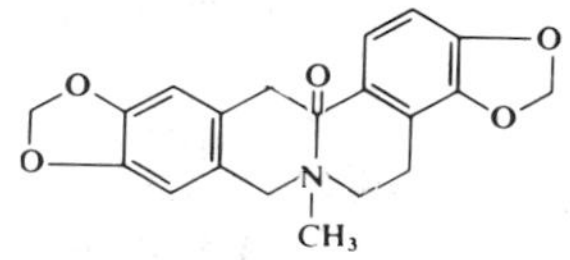

12346

In Numeric as they occur in Organic Compounds Table

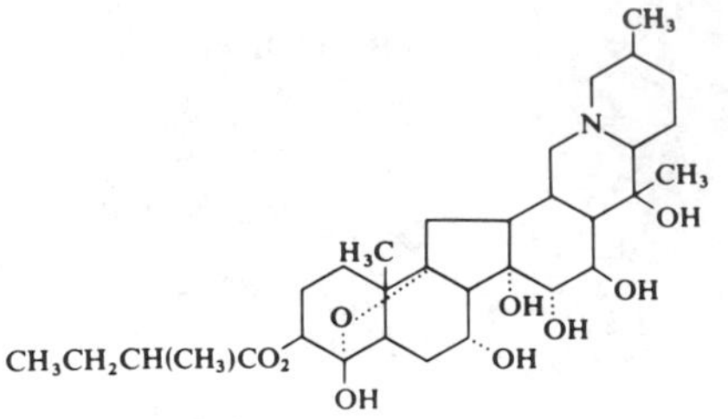

12347

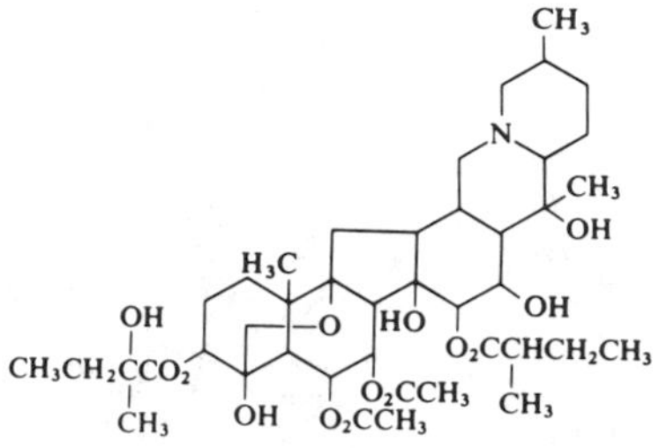

$CH_3CH_2CH(CH_3)CO_2$

12349

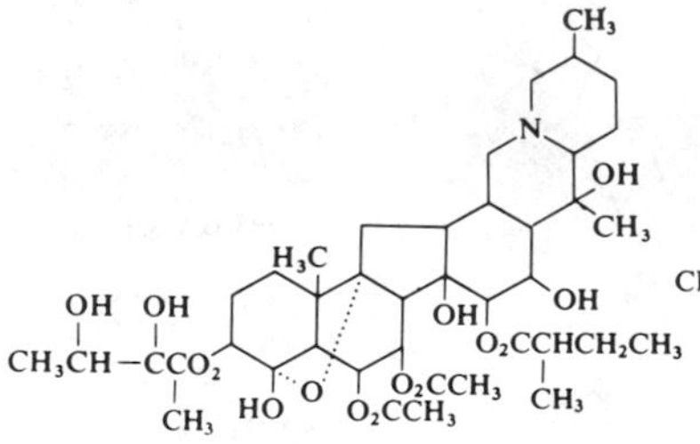

12350

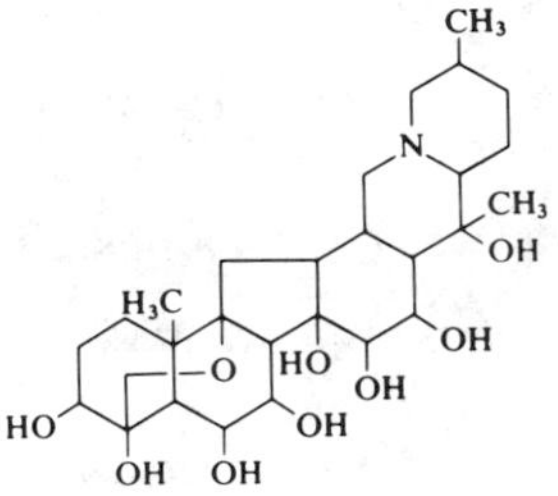

12351

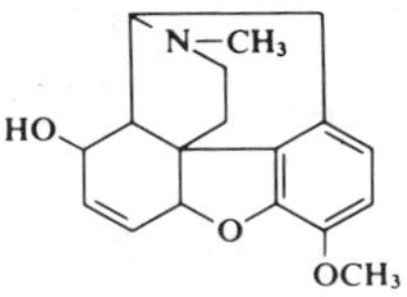

12355

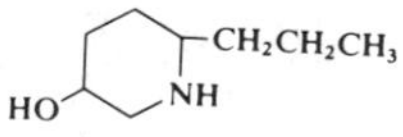

12356

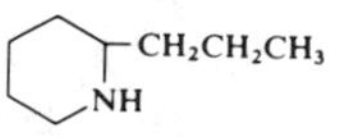

12357

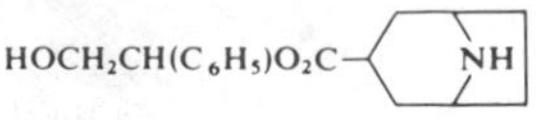

$HOCH_2CH(C_6H_5)O_2C$

12366

$C_6H_{11}O_5.O$

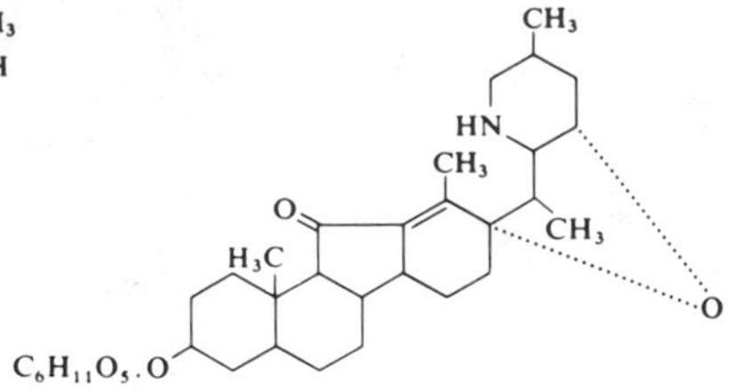

12368

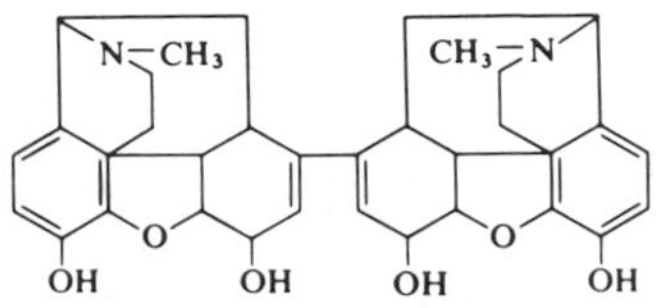

12369

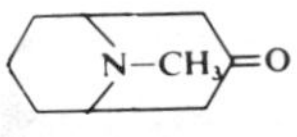

12371

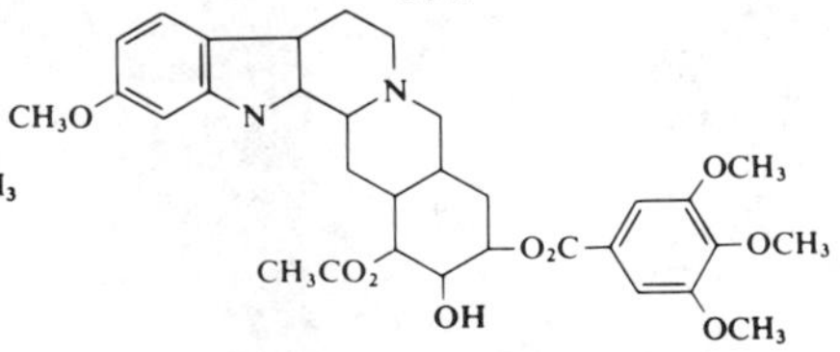

12372

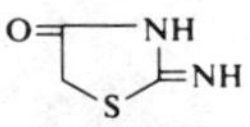

12373

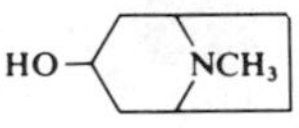

12374

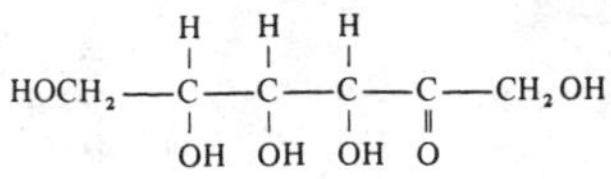

$HOCH_2-C(H)(OH)-C(H)(OH)-C(H)(OH)-C(=O)-CH_2OH$

12375

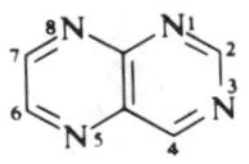

12376

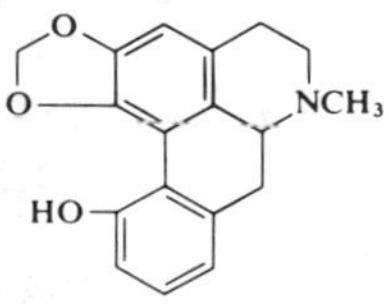

12379

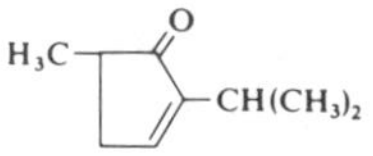

12380

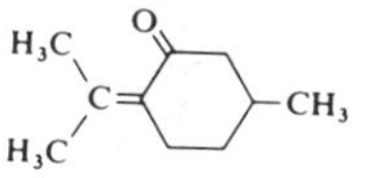

12381

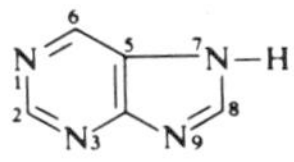

12382

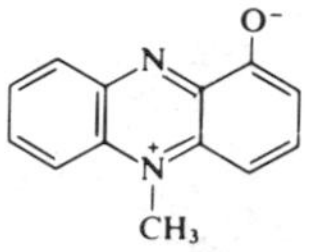

12386

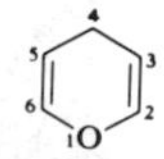

12388

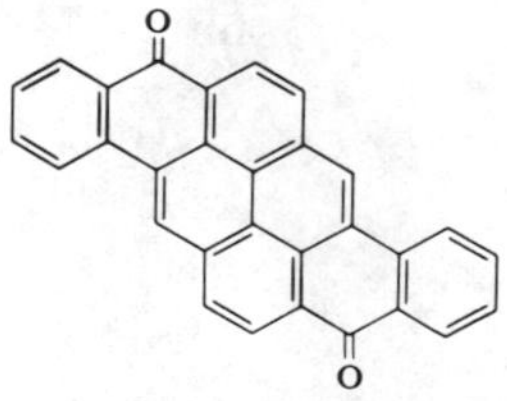

12391

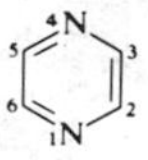

12392

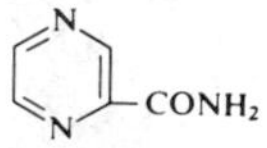

12398

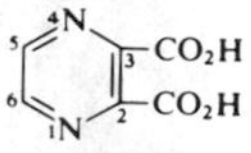

12399

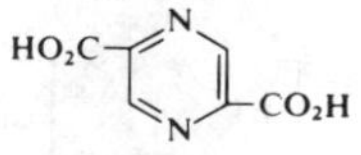

12401

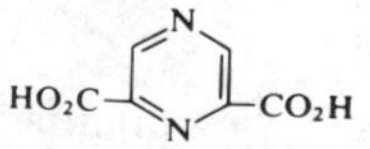

12402

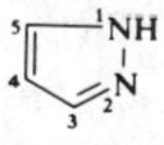

12403

STRUCTURAL FORMULAS OF ORGANIC COMPOUNDS (Continued)

In Numeric as they occur in Organic Compounds Table

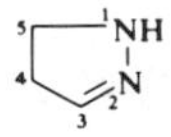

12419

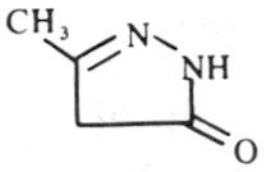

12421

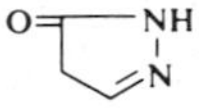

12424

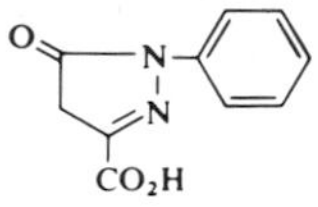

12429

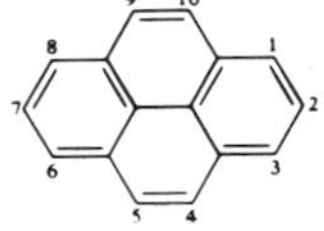

12430

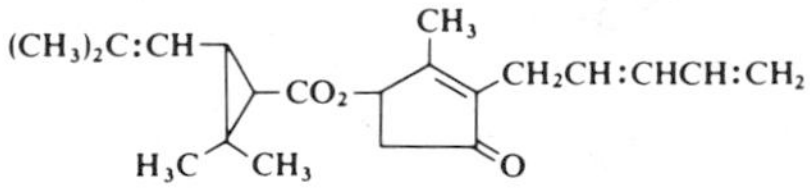

12438

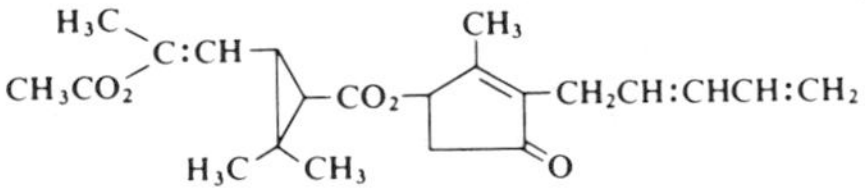

12439

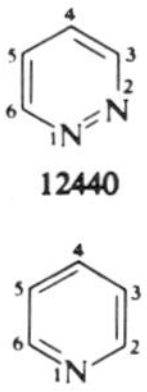

12440

12441

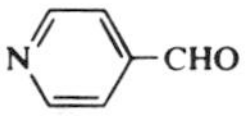

12544

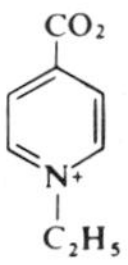

12547

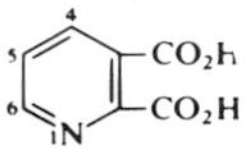

12548

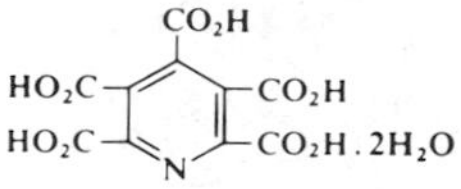

12564

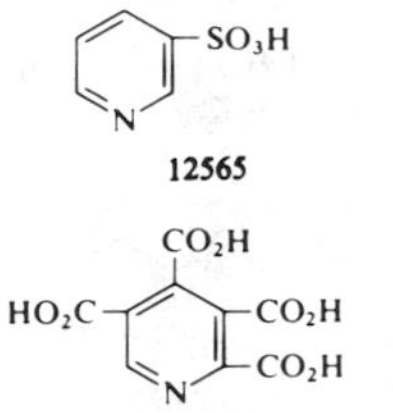

12565

12566

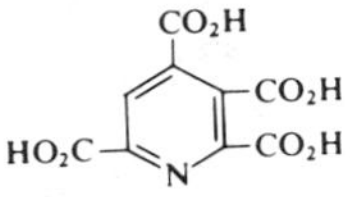

12567

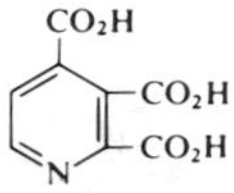

12569

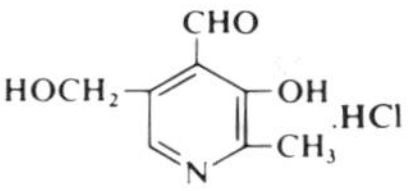

12575

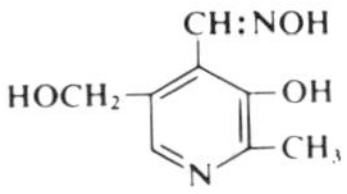

12576

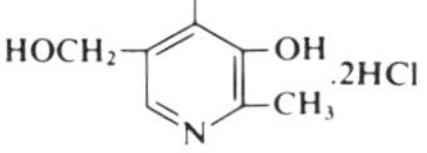

12577

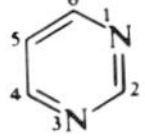

12578

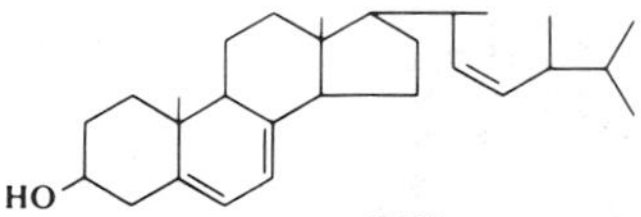

12602

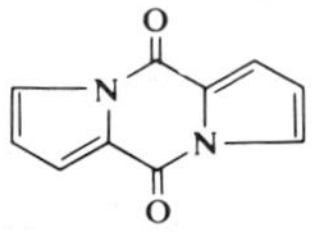

12603

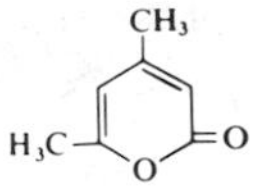

12611

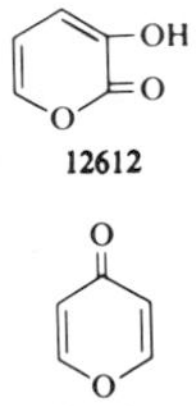

12612

12613

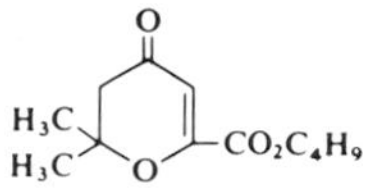

12617

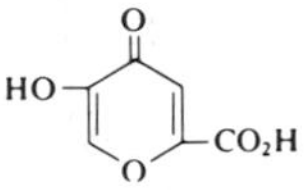

12618

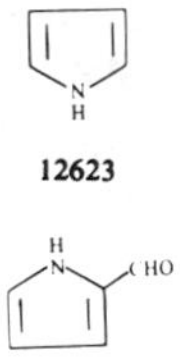

12623

12649

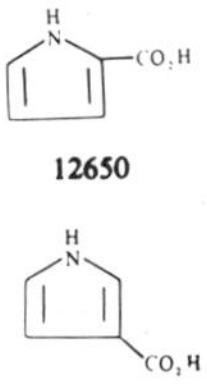

12650

12652

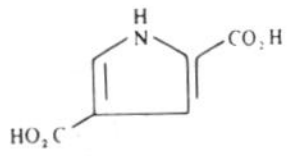

12656

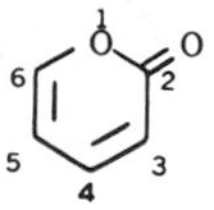

12660

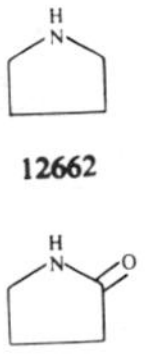

12662

12671

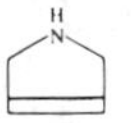

12675

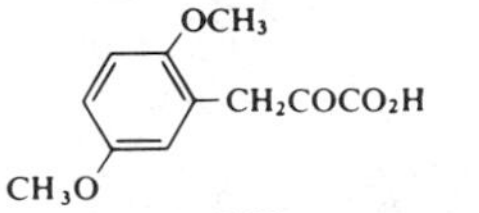

12679

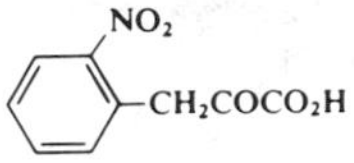

12685

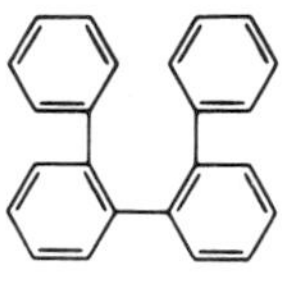

12686

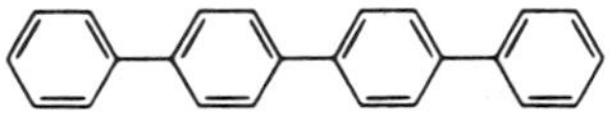

12687

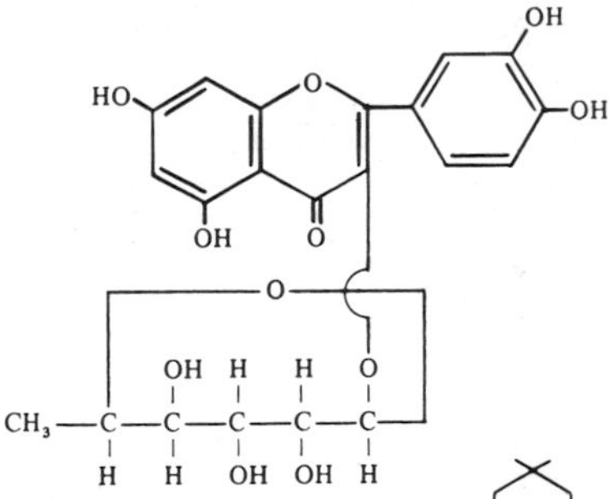

12688

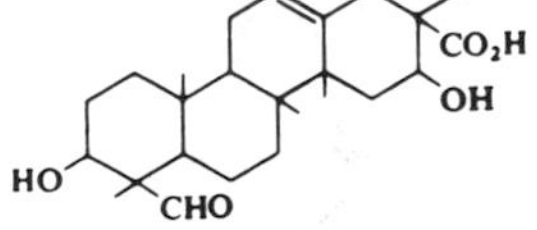

12689

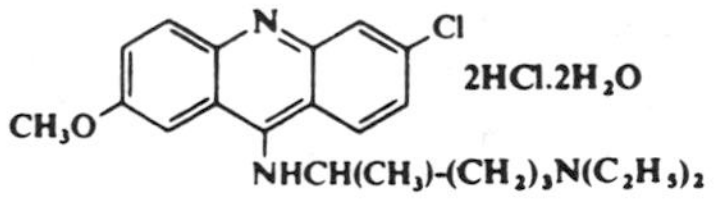

12690

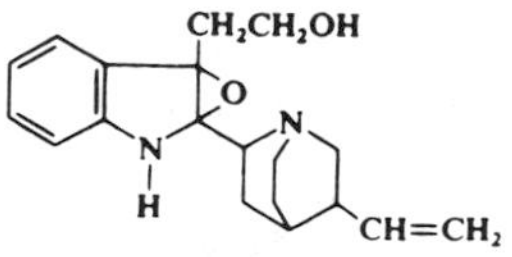

12691

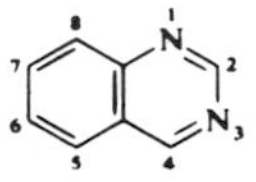

12692

In Numeric as they occur in Organic Compounds Table

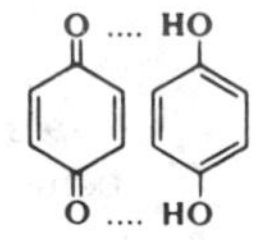

12699

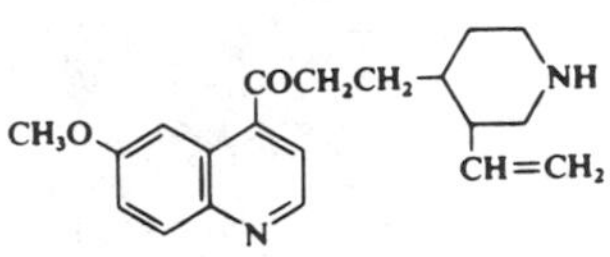

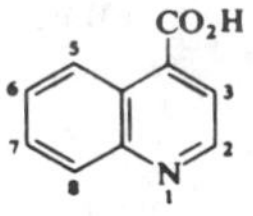

12929

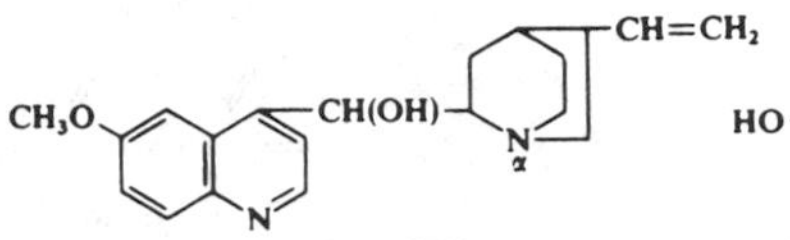

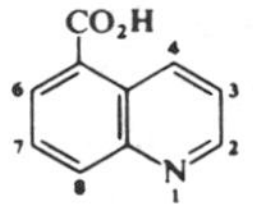

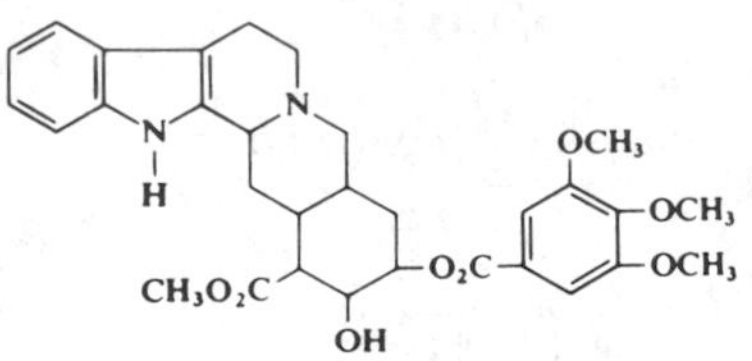

12967

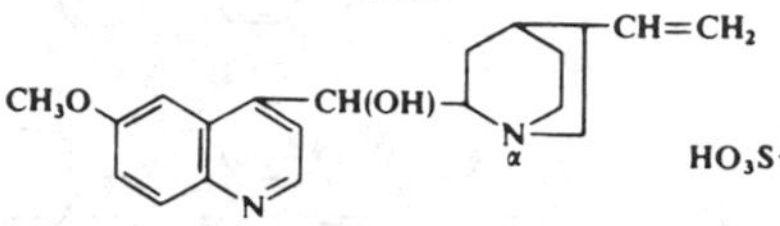

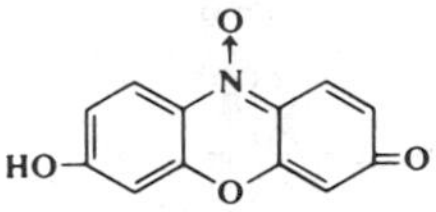

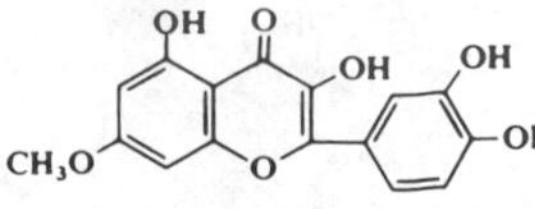

13035

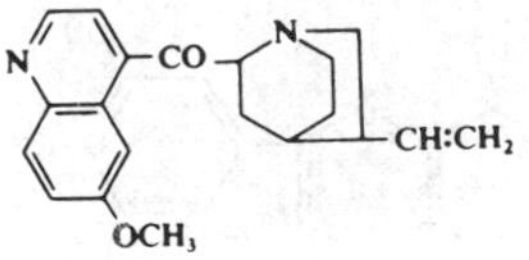

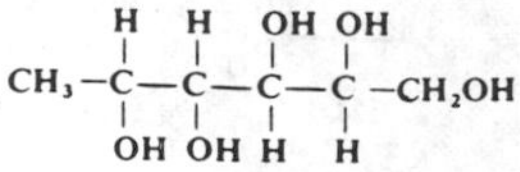

12700

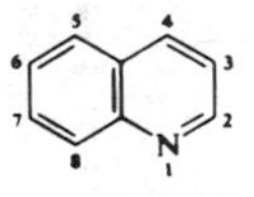

12937

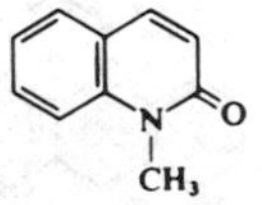

13036

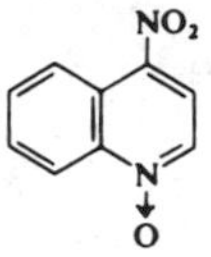

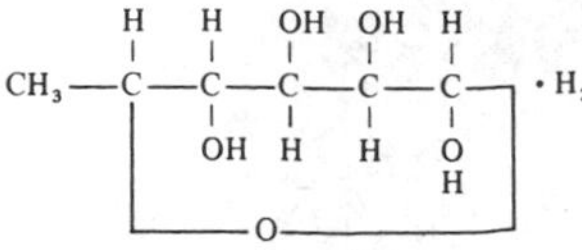

12969

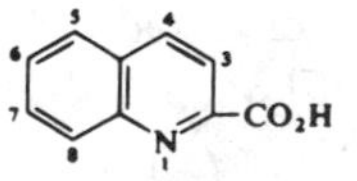

13037

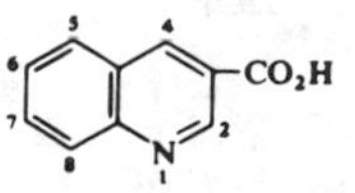

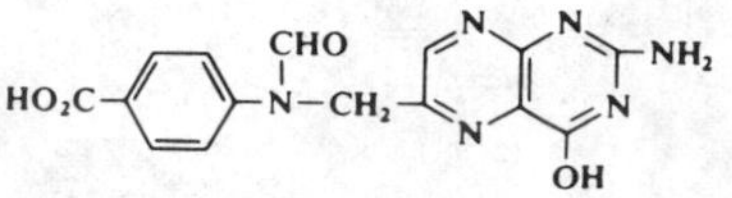

12702

12943

12970

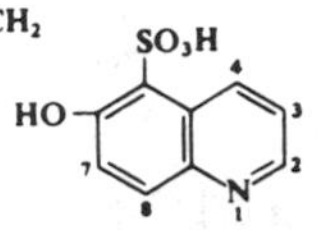

12706

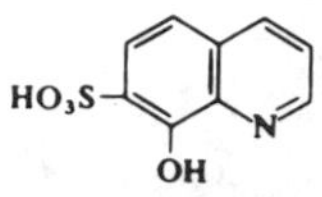

12946

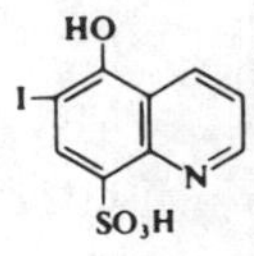

12717

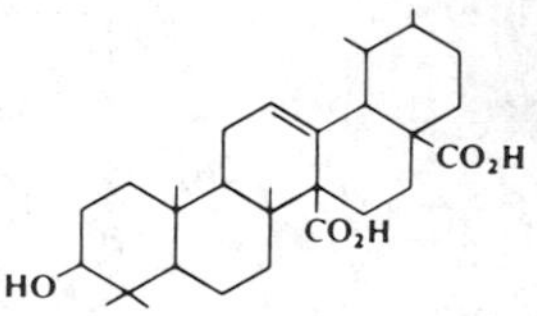

12947

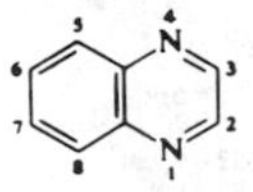

12971

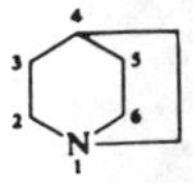

13042

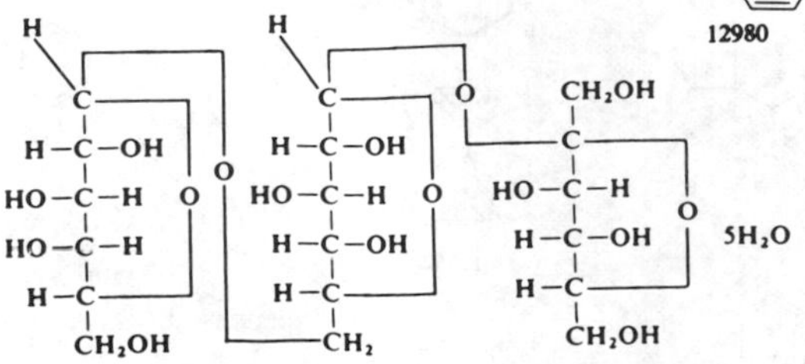

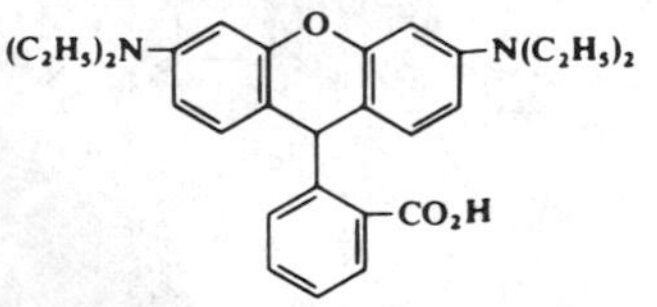

13043

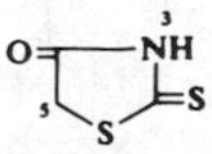

12719

12949

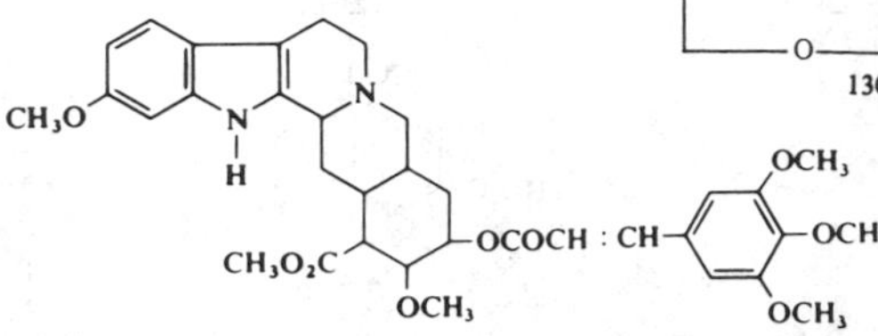

12972

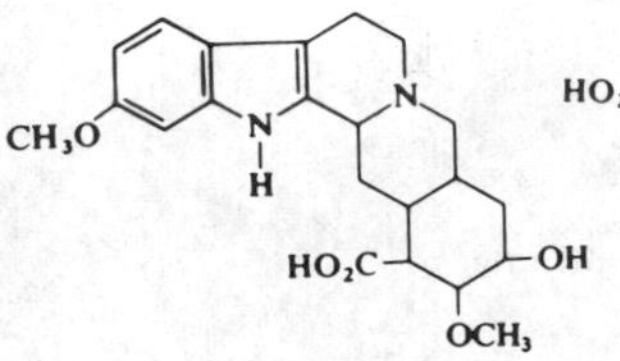

12806

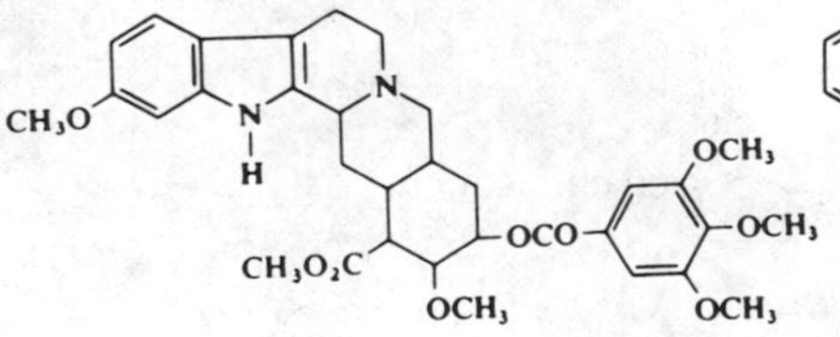

13045

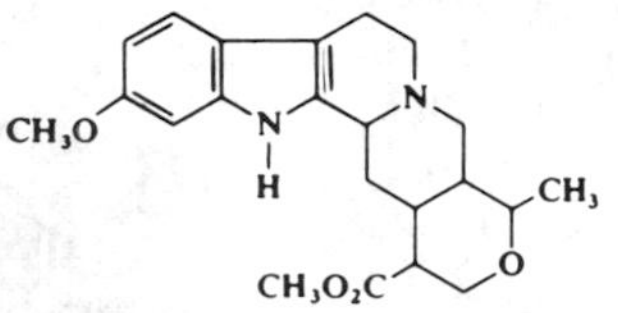

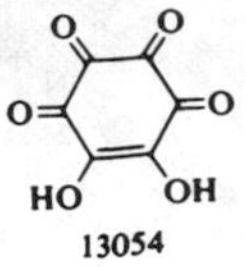

12950

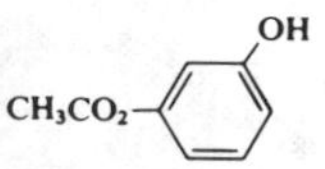

12973

13054

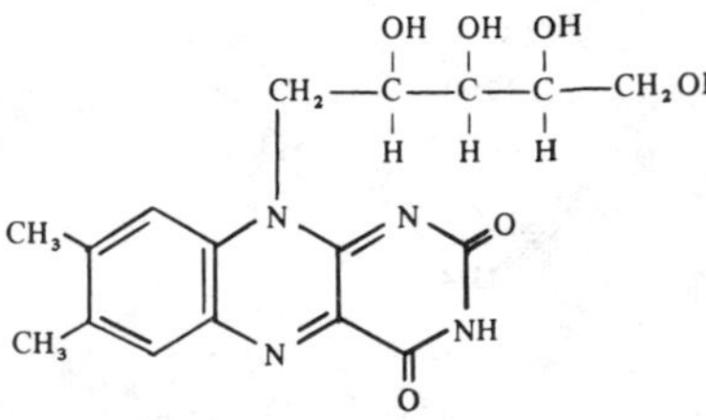

12895

12963

12980

13055

12920

12926

12966

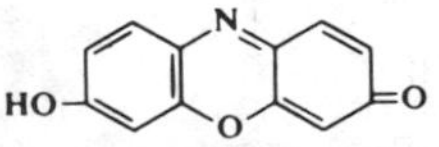

13032

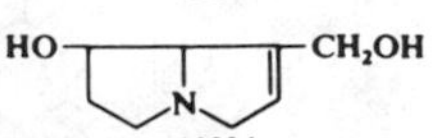

13034

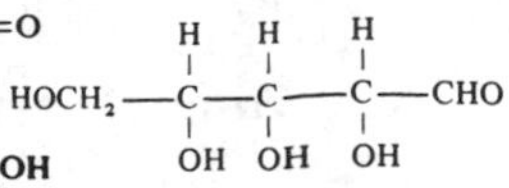

13056

In Numeric as they occur in Organic Compounds Table

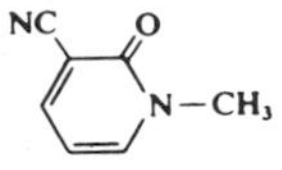

13057

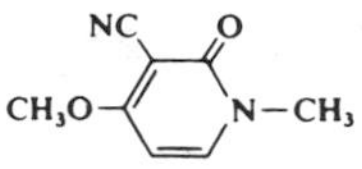

13058

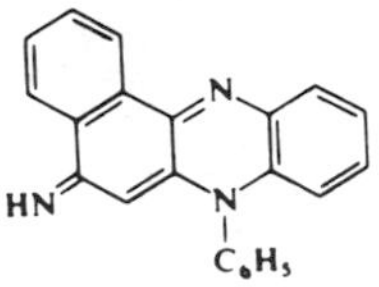

13064

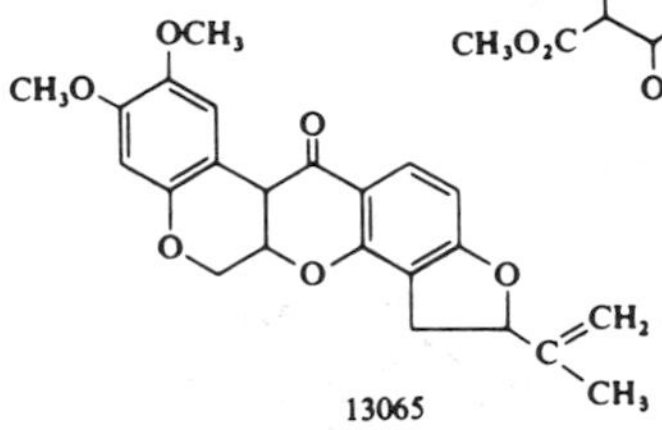

13065

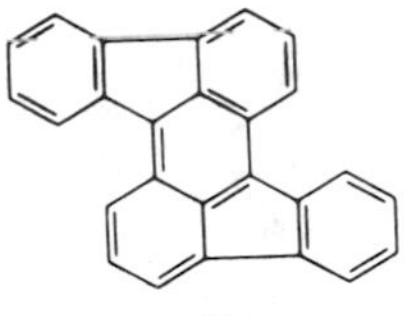

13066

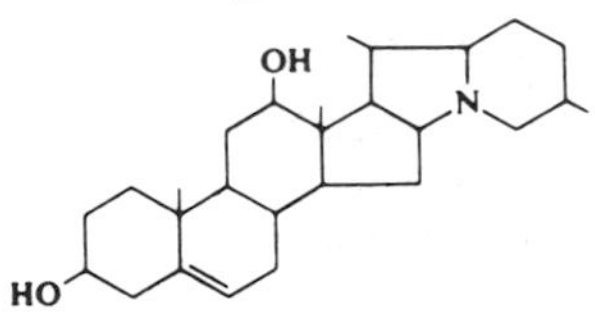

13068

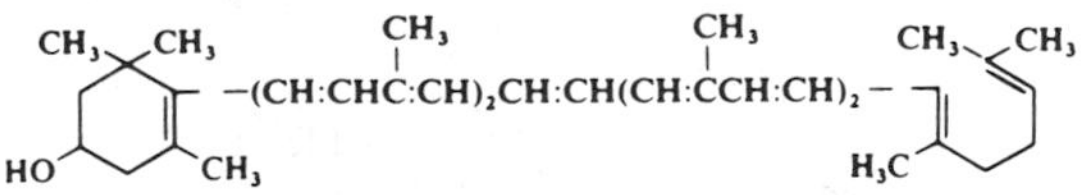

13069

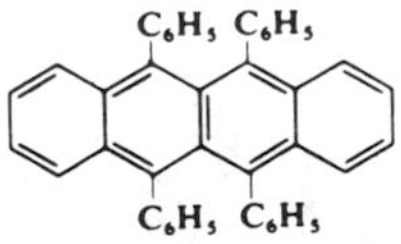

13070

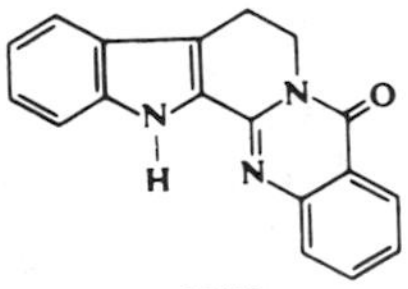

13072

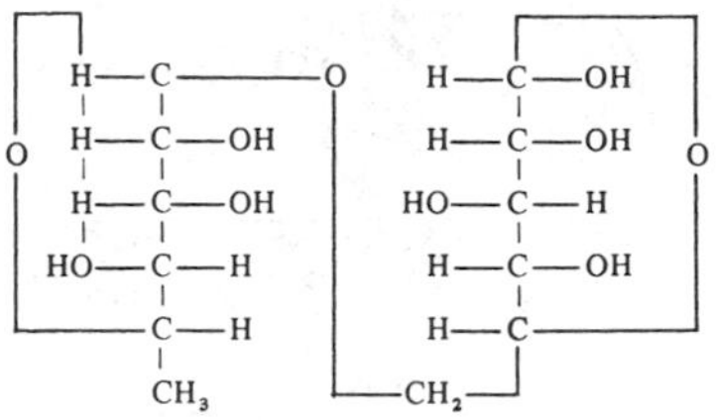

13073

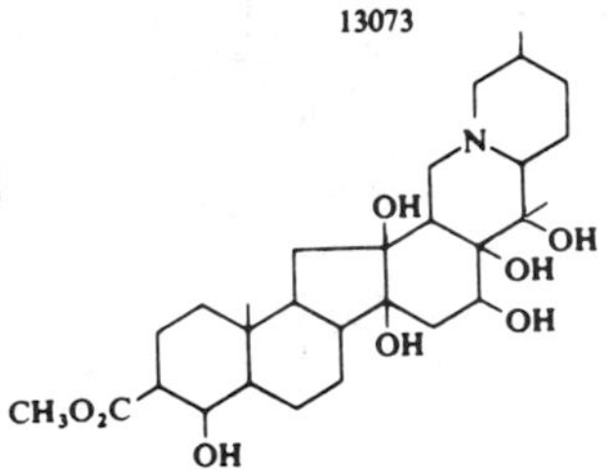

13074

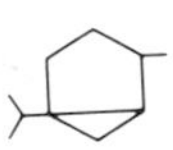

13075

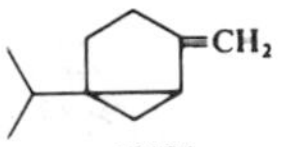

13076

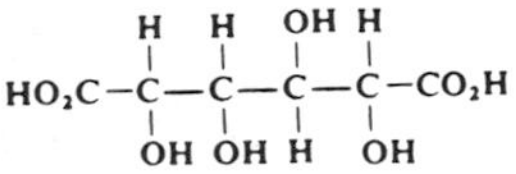

13080

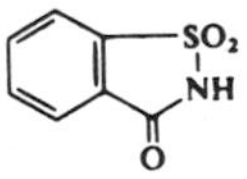

13081

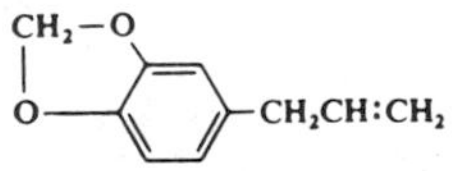

13082

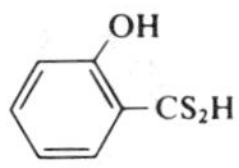

13123

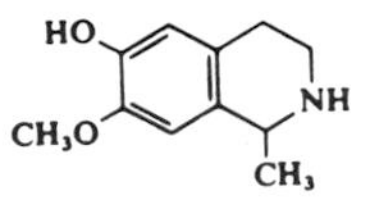

13166

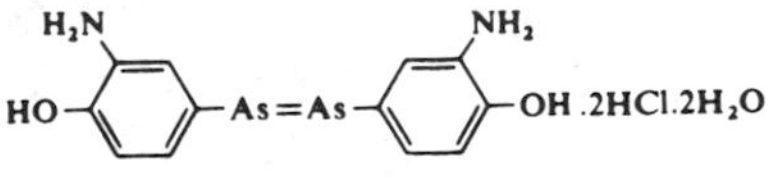

13167

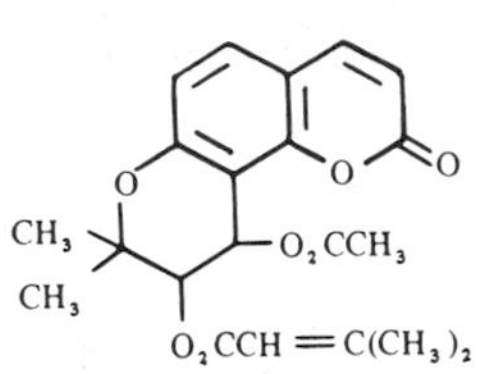

13169

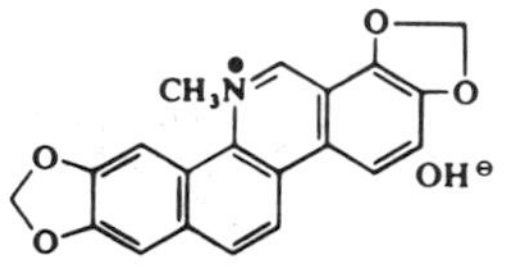

13170

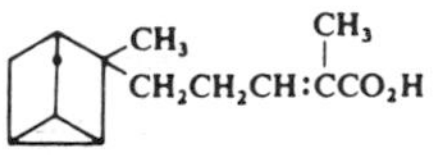

13171

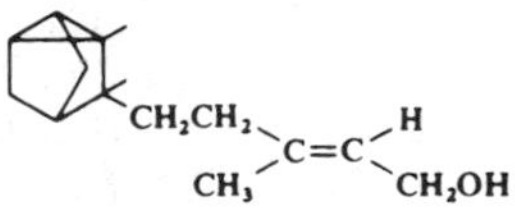

13172

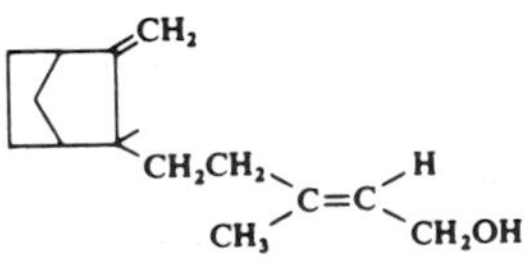

13173

13174

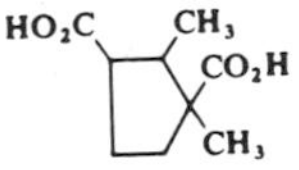

13175

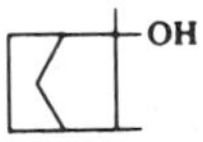

13176

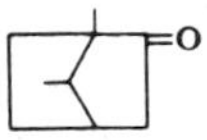

13177

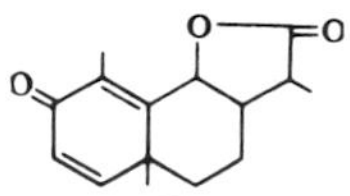

13178

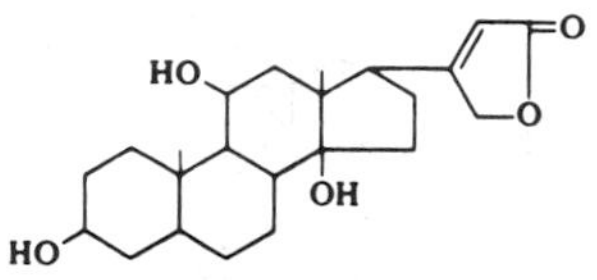

13179

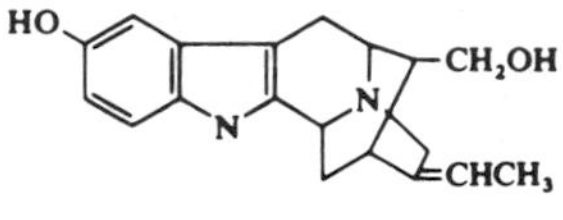

13180

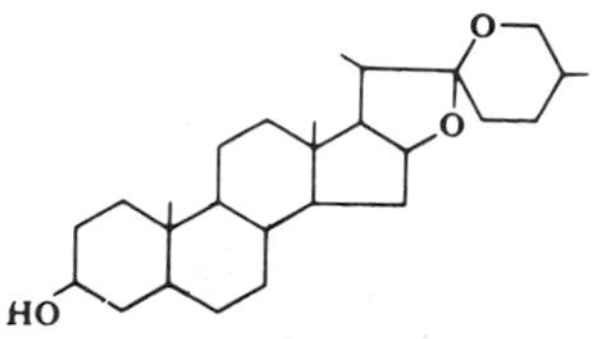

13181

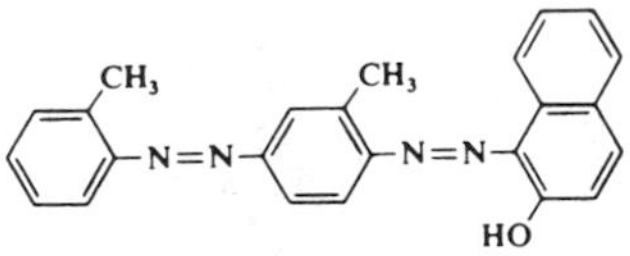

13182

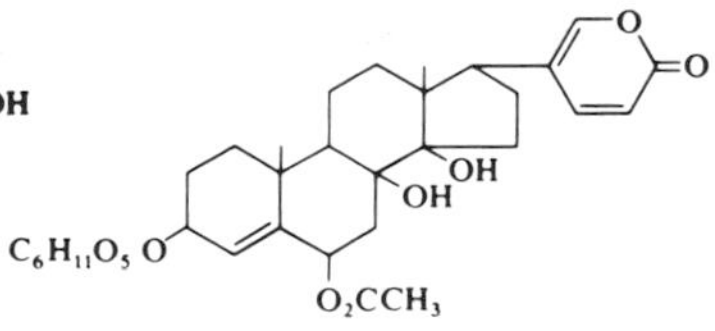

13183

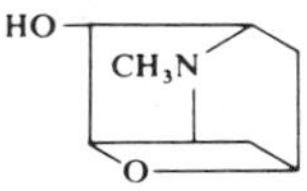

13184

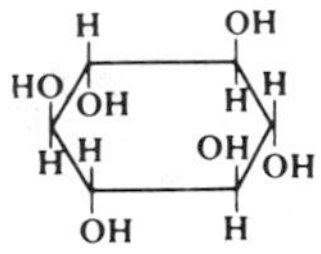

13185

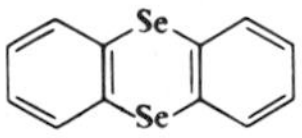

13197

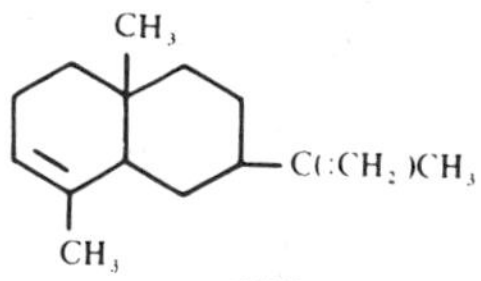

13200

STRUCTURAL FORMULAS OF ORGANIC COMPOUNDS (Continued)

In Numeric as they occur in Organic Compounds Table

CH_3
$H_2NCONHNH-$

13219

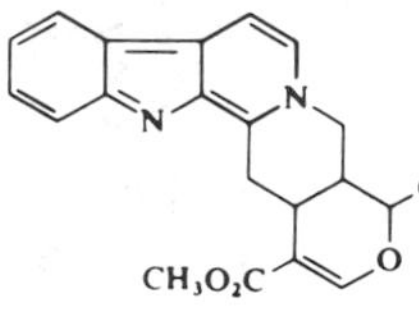

N
N
CH_3
CH_3O_2C
O

13227

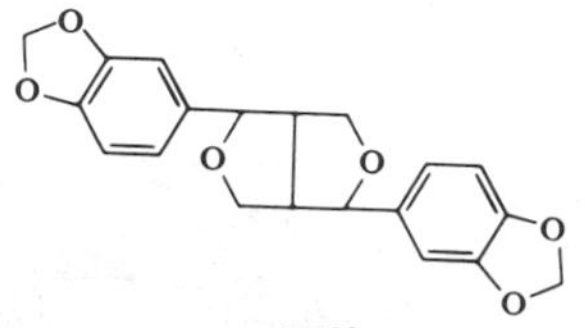

O
O
O
O
O
O

13228

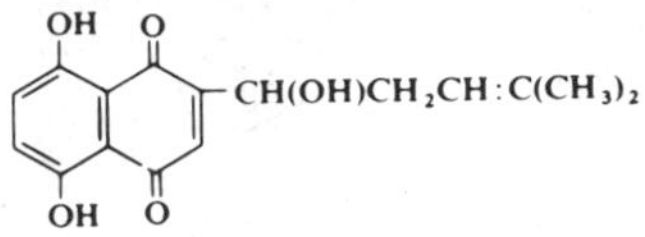

OH O
$-CH(OH)CH_2CH:C(CH_3)_2$
OH O

13229

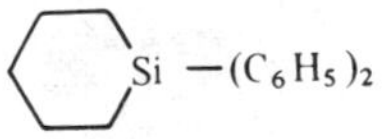

Si $-(C_6H_5)_2$

13342

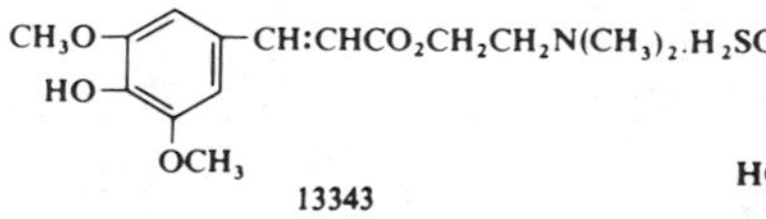

CH_3O-
HO
$-CH:CHCO_2CH_2CH_2N(CH_3)_2 \cdot H_2SO_4$
OCH_3

13343

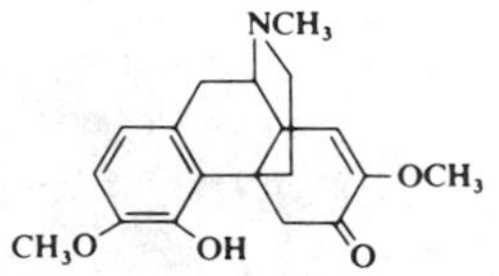

NCH_3
$-OCH_3$
CH_3O
OH
O

13345

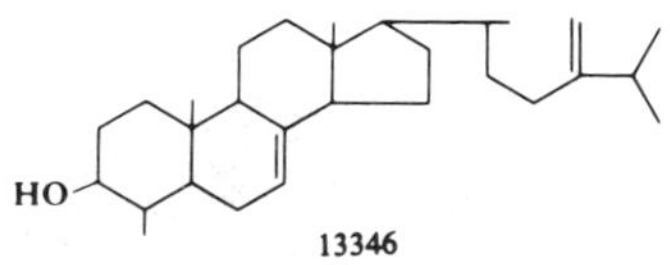

HO

13346

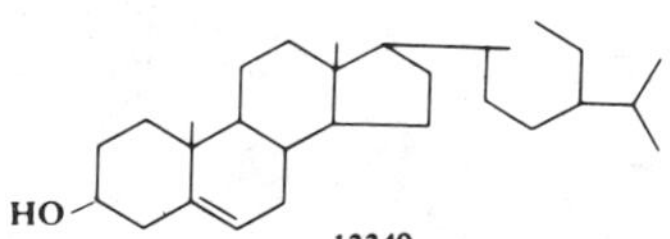

HO

13349

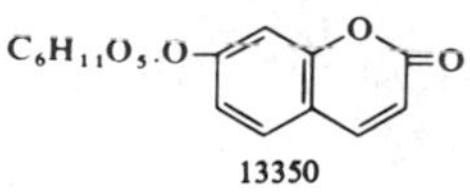

$C_6H_{11}O_5 \cdot O-$
O
=O

13350

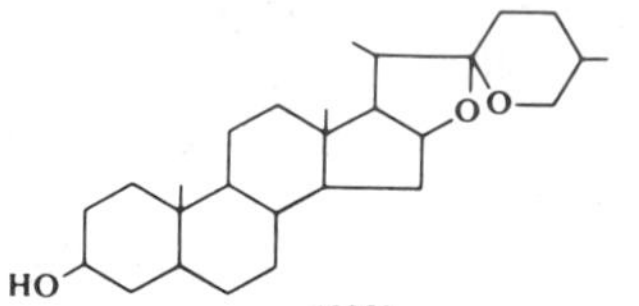

O
O
HO

13351

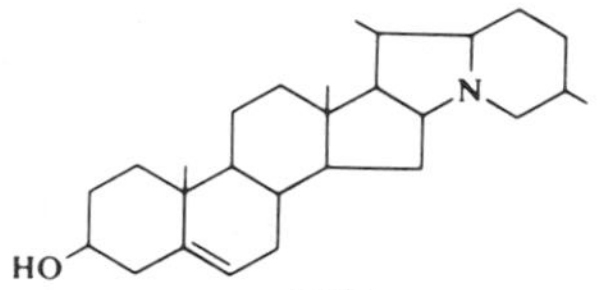

N
HO

13352

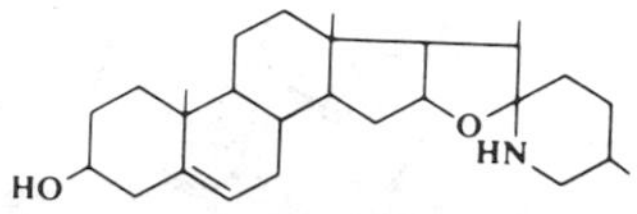

O
HN
HO

13354

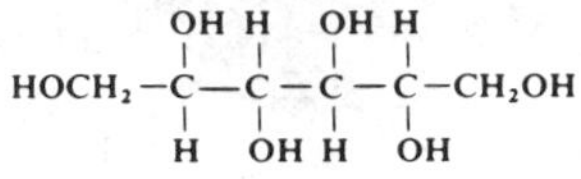

$HOCH_2-C-C-C-C-CH_2OH$ (OH H OH H / H OH H OH)

13362

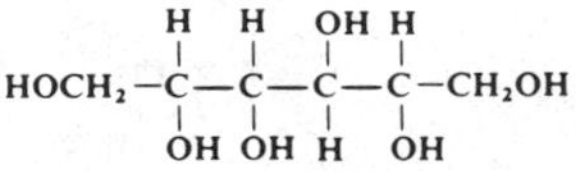

$HOCH_2-C-C-C-C-CH_2OH$ (H H OH H / OH OH H OH)

13363

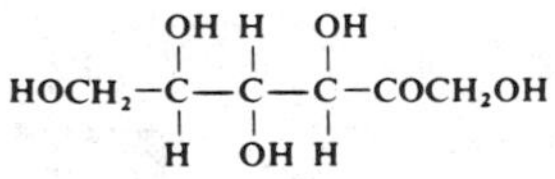

$HOCH_2-C-C-C-COCH_2OH$ (OH H OH / H OH H)

13368

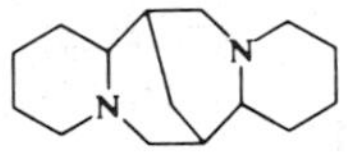

N
N

13369

O O
O O

13371

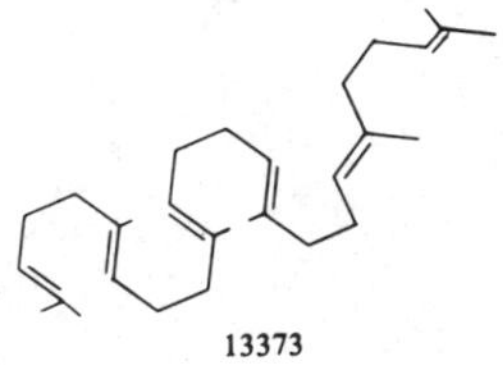

13373

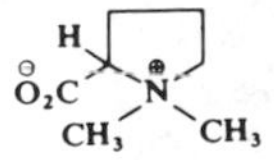

H
$\ominus O_2C$
$\oplus N$
CH_3 CH_3

13375

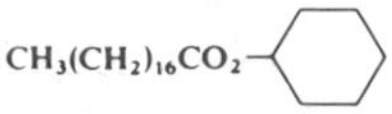

$CH_3(CH_2)_{16}CO_2-$

13386

$CH_3(CH_2)_{16}CO_2CH_2-$
O

13419

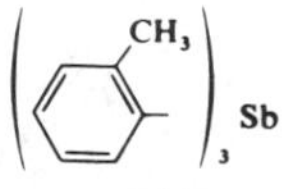

CH_3
$)_3$ Sb

13440

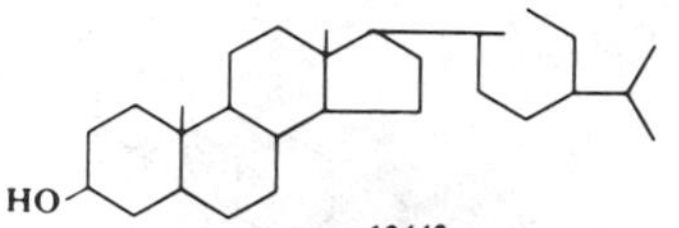

HO

13443

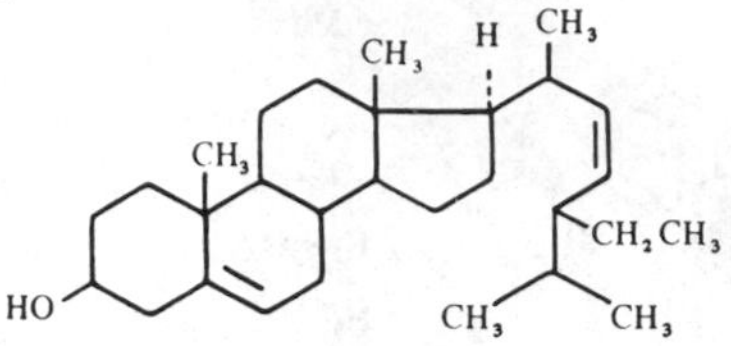

CH_3
H CH_3
CH_3
CH_2CH_3
HO
CH_3 CH_3

13444

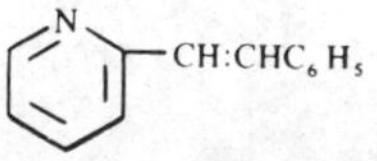

N
$-CH:CHC_6H_5$

13445

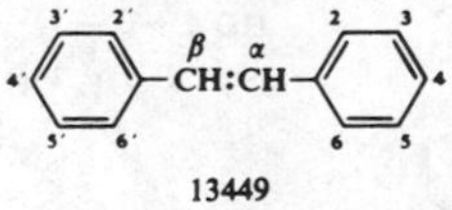

3′ 2′ 2 3
4′ β α 4
CH:CH
5′ 6′ 6 5

13449

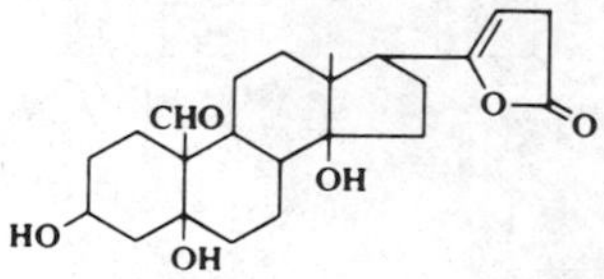

CHO
O
O
OH
HO
OH

13487

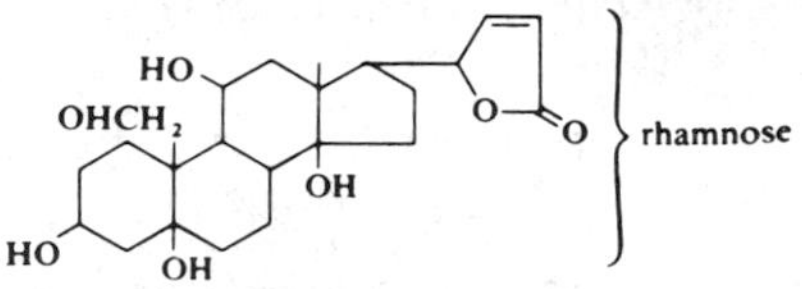

HO
$OHCH_2$
O
O
OH
HO
OH
rhamnose

13488

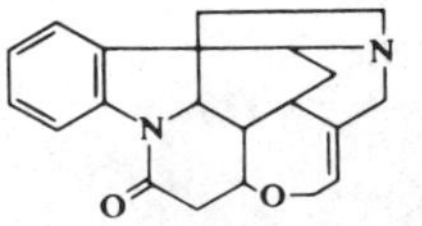

N
N
O
O

13489

$HOCH_2-C-C-C-C-CH_2$ (H H OH OH / OH H H)
O

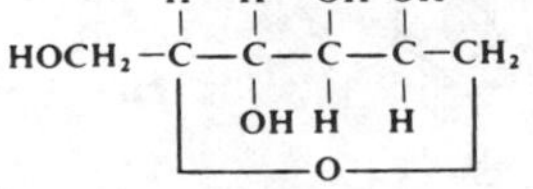

13494

In Numeric as they occur in Organic Compounds Table

13495 13619 13637 13678 13523 13620 13679 13538 13625 13638 13680 13561 13626 13639 13681 13581 13640 13588 13627 13689 13641 13628 13691 13611 13642 13692 13643 13629 13695 13613 13630 13649 13696 13631 13698 13614 13632 13675 13699 13615 13634 13700 13617 13635 13676 13701 13618 13636 13677 13702

STRUCTURAL FORMULAS OF ORGANIC COMPOUNDS (Continued)

In Numeric as they occur in Organic Compounds Table

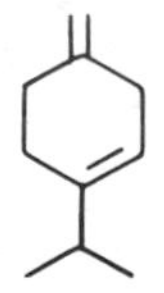

13703

13704

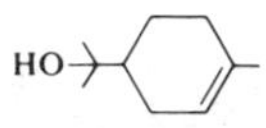

13705

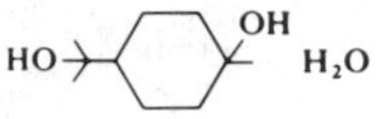

13707

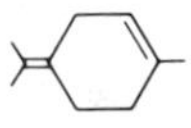

13708

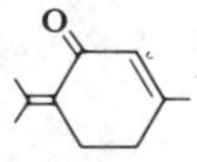

13709

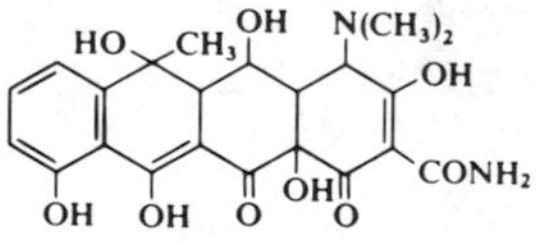

13710

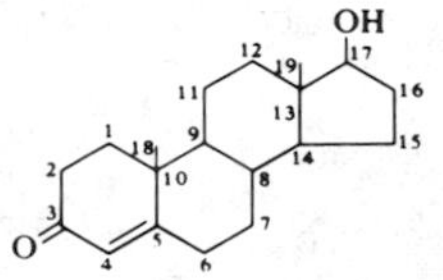

13711

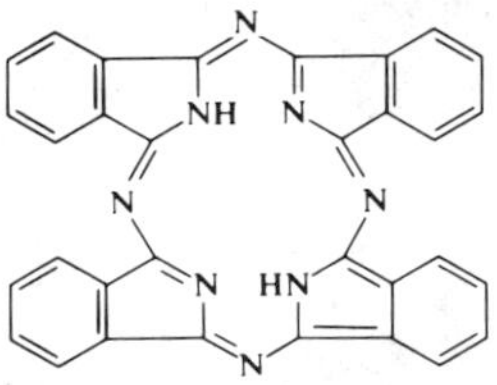

13718

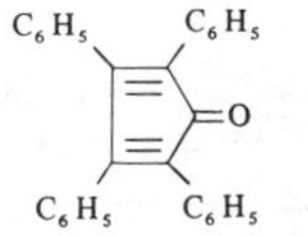

13723

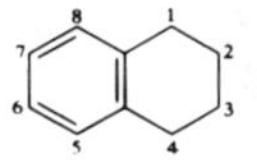

13752

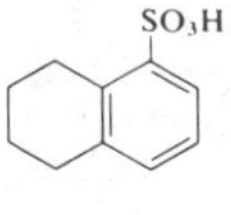

13769

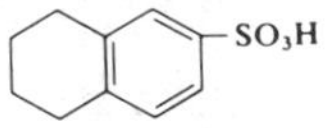

13771

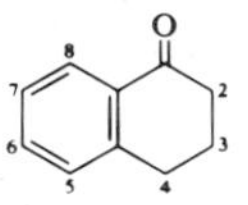

13773

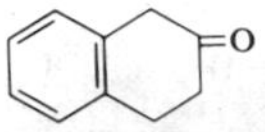

13775

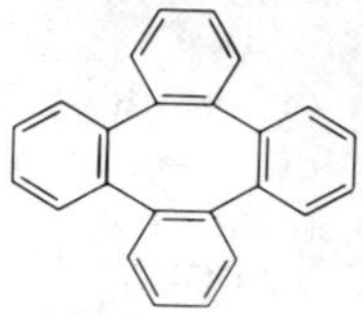

13782

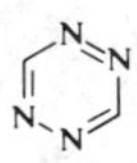

13788

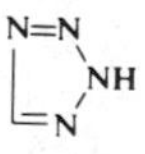

13789

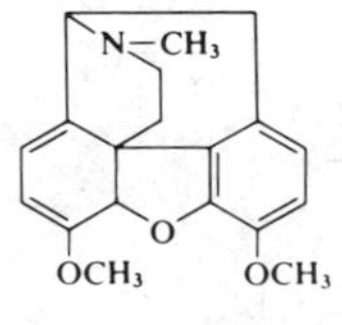

13790

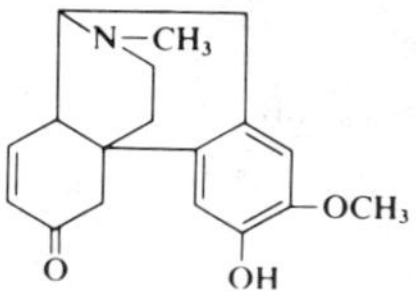

13793

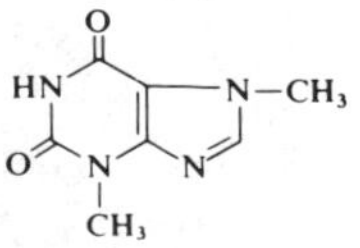

13794

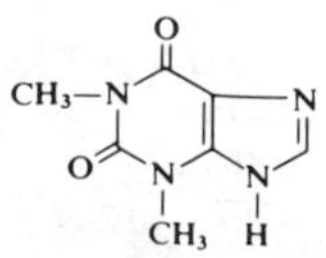

13795

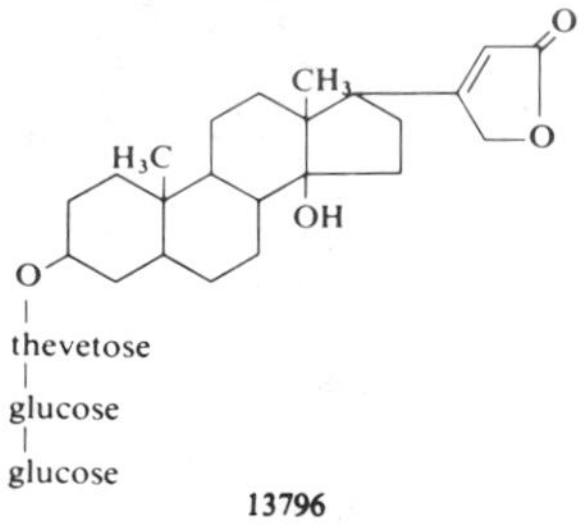

13796

13797

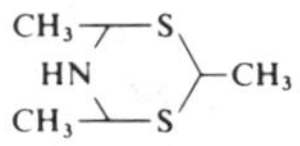

13799

13800

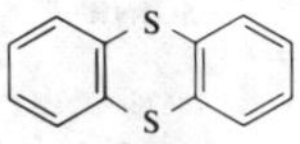

13801

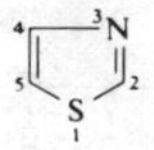

13802

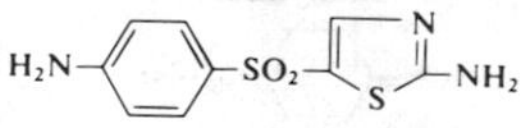

13807

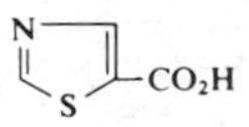

13815

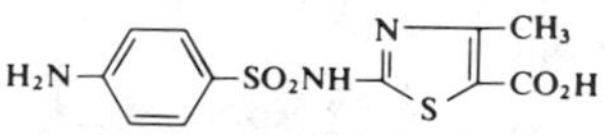

13820

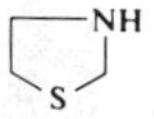

13821

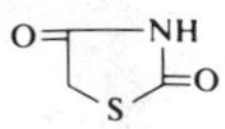

13822

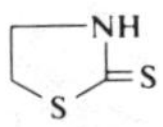

13823

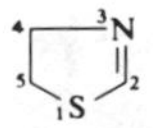

13824

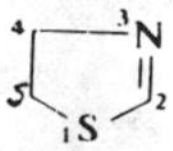

13825

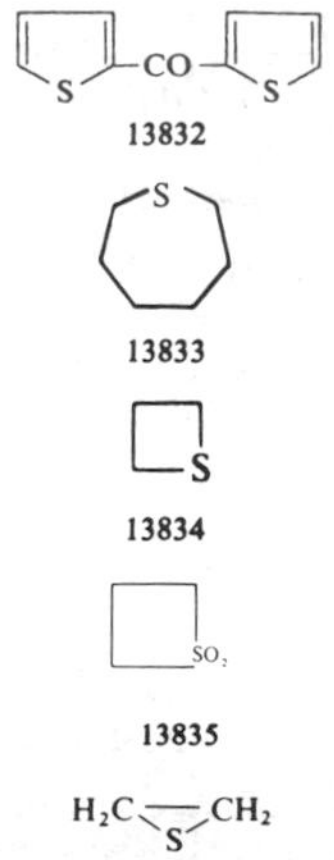

13832

13833

13834

13835

13837

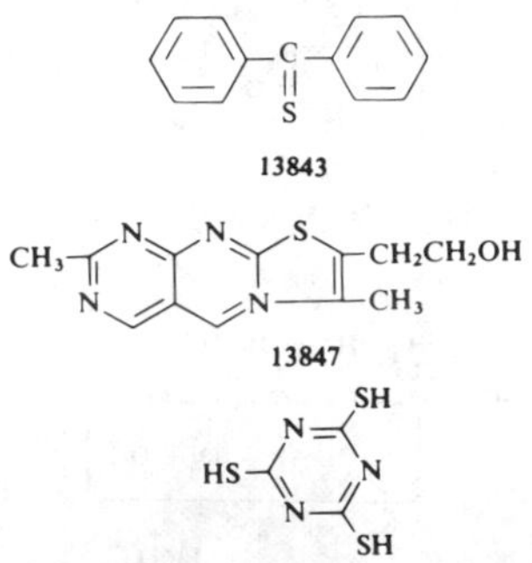

13843

13847

13848

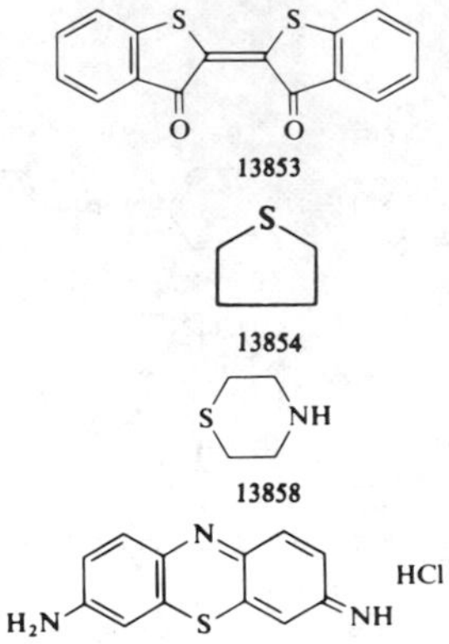

13853

13854

13858

13860

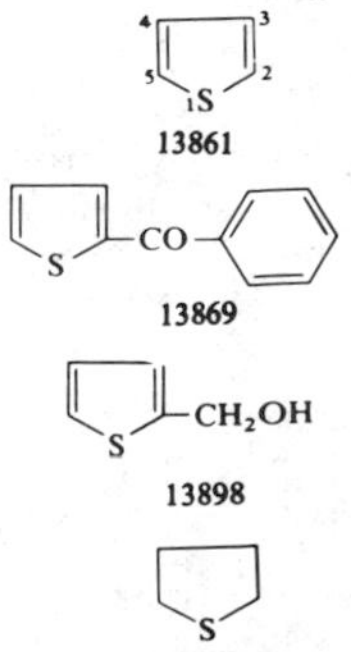

13861

13869

13898

13915

STRUCTURAL FORMULAS OF ORGANIC COMPOUNDS (Continued)

In Numeric as they occur in Organic Compounds Table

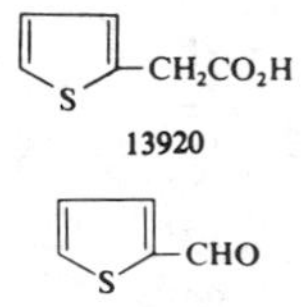

S CH_2CO_2H
13920

S CHO
13921

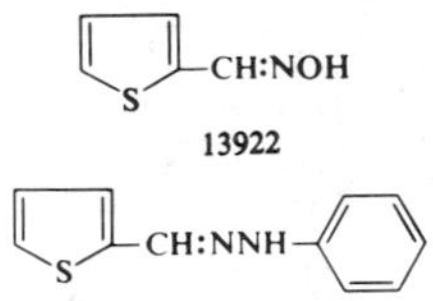

S CH:NOH
13922

S CH:NNH
13923

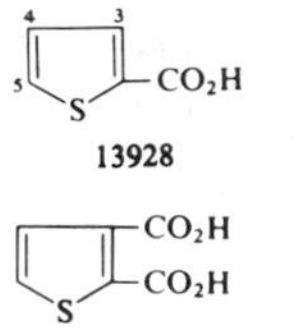

4 3 5 S CO_2H
13928

S CO_2H CO_2H
13935

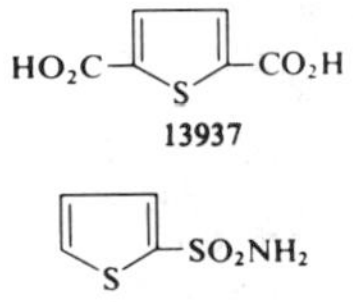

HO_2C S CO_2H
13937

S SO_2NH_2
13939

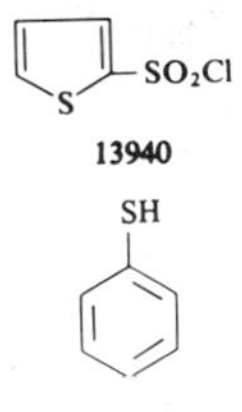

S SO_2Cl
13940

SH
13943

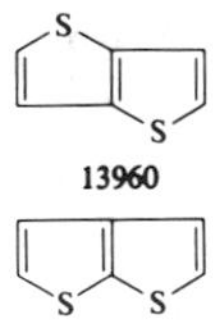

S S
13960

S S
13961

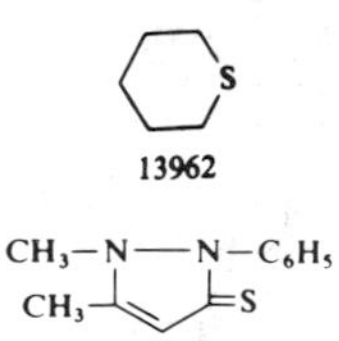

S
13962

CH_3—N——N—C_6H_5
CH_3 S
13963

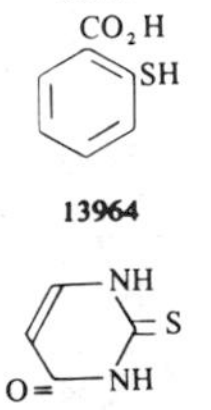

CO_2H SH
13964

NH S O= NH
13966

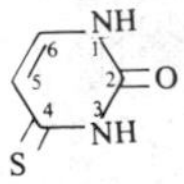

NH 6 1 5 2 =O 4 3 NH S
13969

CH_3 CH_3
NHCSNH
13994

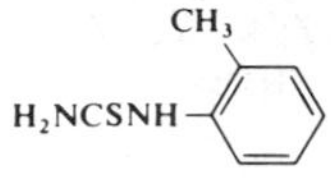

CH_3
H_2NCSNH
14013

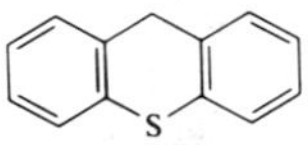

S
14018

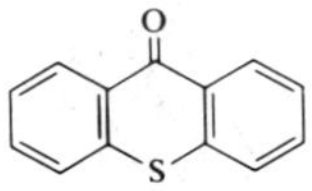

O S
14019

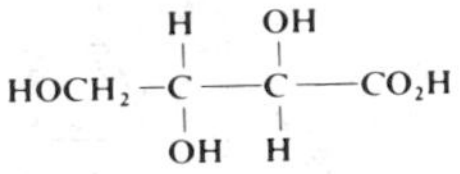

H OH
$HOCH_2$—C——C——CO_2H
OH H
14020

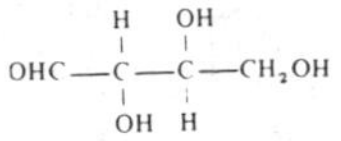

H OH
OHC—C——C—CH_2OH
OH H
14025

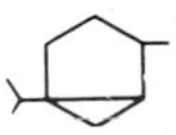

14026

14027

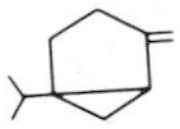

14028

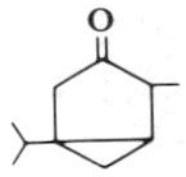

O
14030

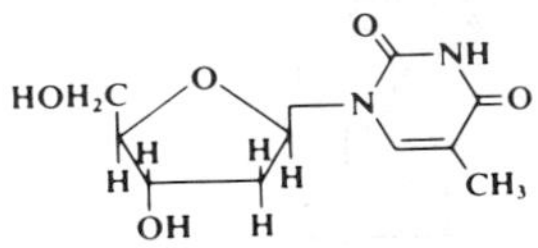

O NH
HOH_2C O N O
H H H H CH_3
OH H
14031

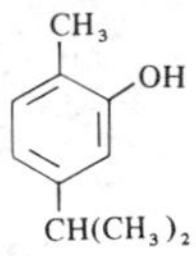

CH_3 OH
$CH(CH_3)_2$
14032

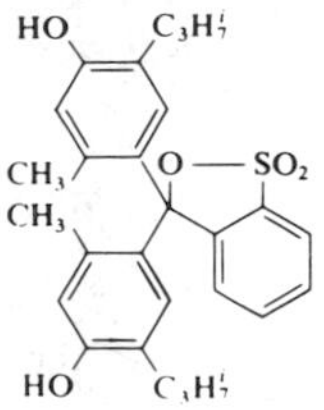

HO $C_3H_7^i$
CH_3 O — SO_2
CH_3
HO $C_3H_7^i$
14039

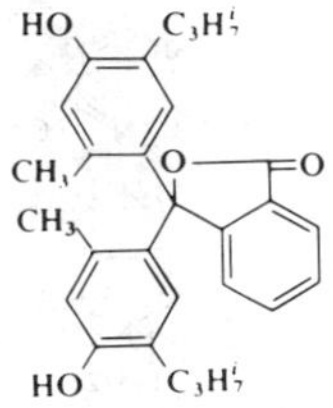

HO $C_3H_7^i$
CH_3 O O
CH_3
HO $C_3H_7^i$
14040

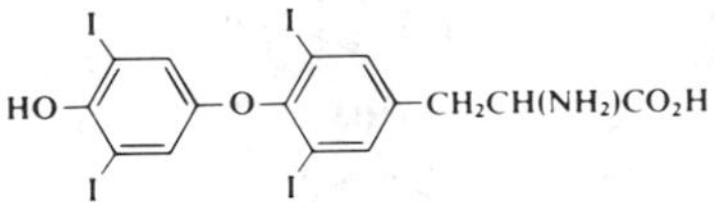

I I
HO O $CH_2CH(NH_2)CO_2H$
I I
14041

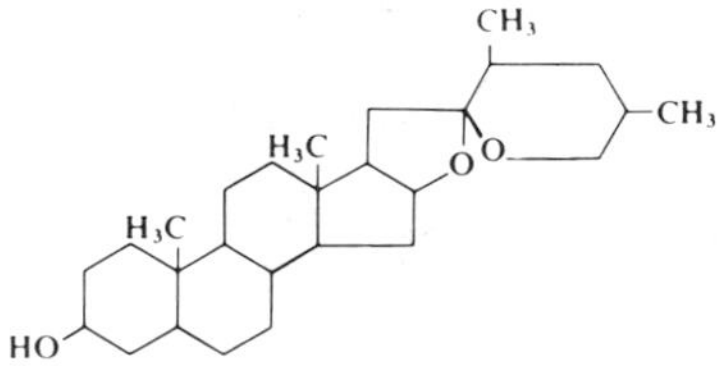

CH_3
H_3C O O CH_3
H_3C
HO
14046

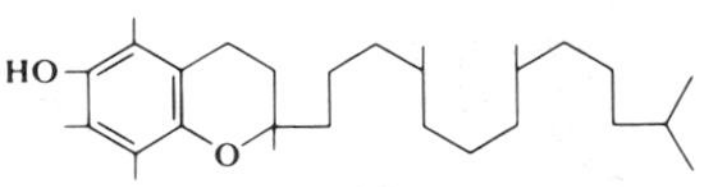

HO O
14047

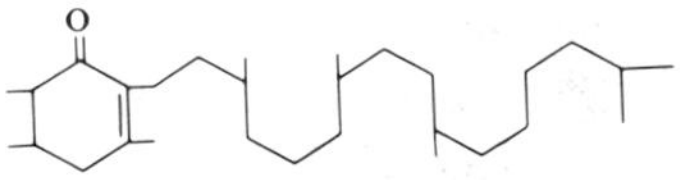

O
14049

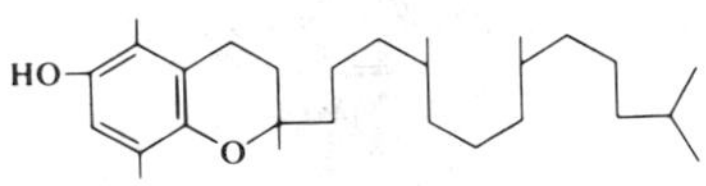

HO O
14050

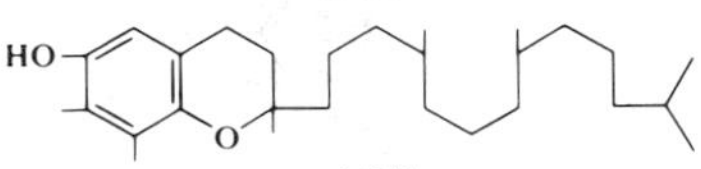

HO O
14051

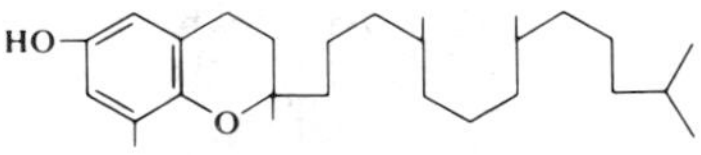

HO O
14052

$CH_3(CH_2)_3NHCONHSO_2$— —CH_3
14053

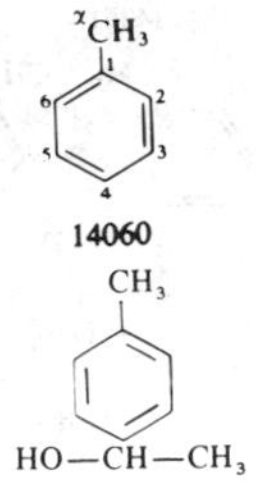

$^{\alpha}CH_3$ 1 2 3 4 5 6
14060

CH_3
HO—CH—CH_3
14131

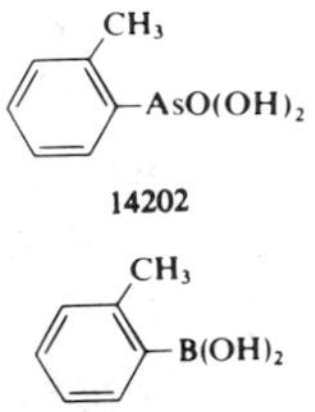

CH_3
$AsO(OH)_2$
14202

CH_3
$B(OH)_2$
14205

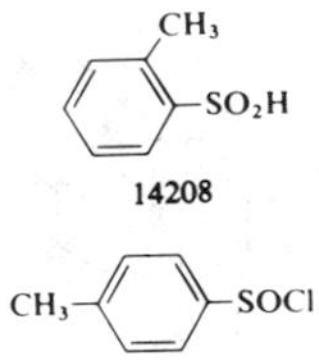

CH_3
SO_2H
14208

CH_3 SOCl
14210

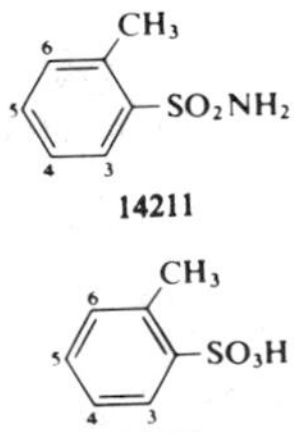

CH_3 6 5 4 3
SO_2NH_2
14211

CH_3 6 5 4 3
SO_3H
14217

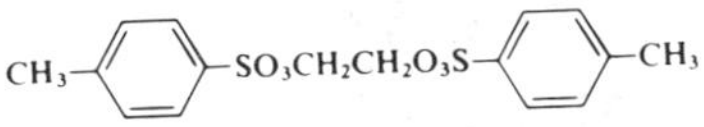

CH_3— —$SO_3CH_2CH_2O_3S$— —CH_3
14253

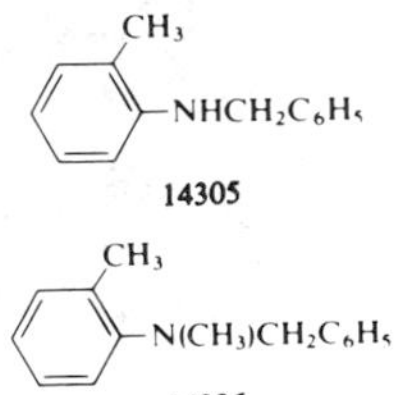

CH_3
$NHCH_2C_6H_5$
14305

CH_3
$N(CH_3)CH_2C_6H_5$
14306

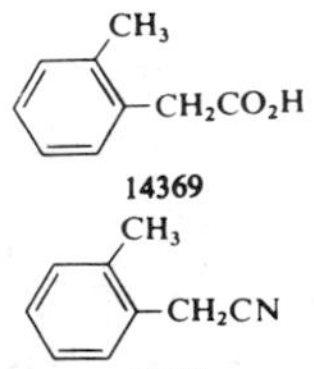

CH_3
CH_2CO_2H
14369

CH_3
CH_2CN
14370

In Numeric as they occur in Organic Compounds Table

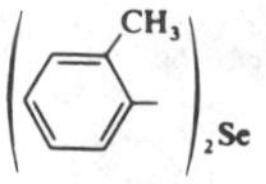

14381

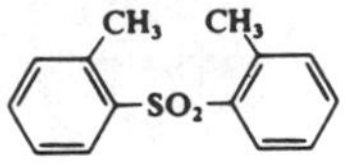

14383

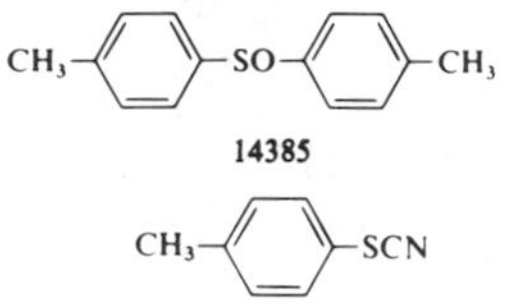

14385

14386

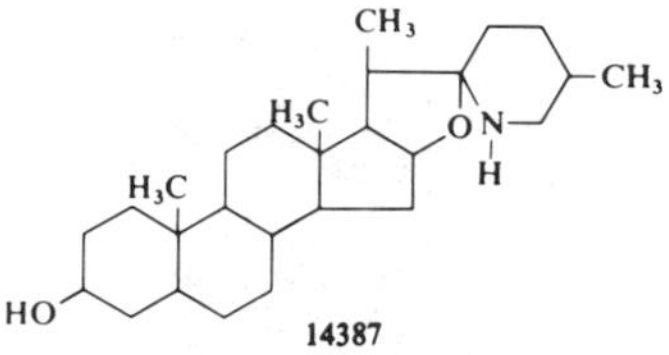

14387

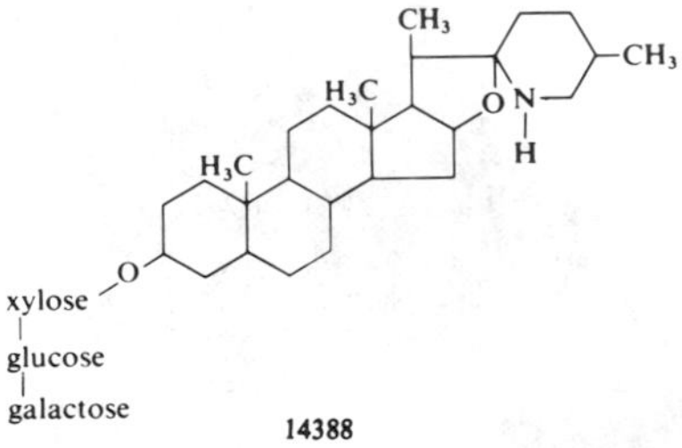

14388

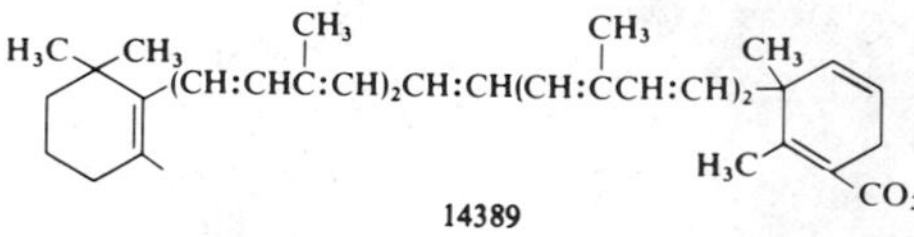

14389

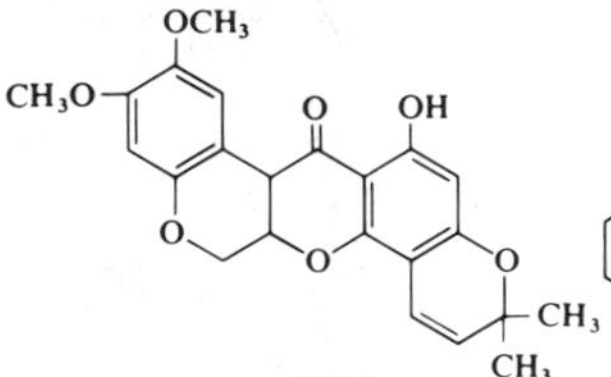

14390

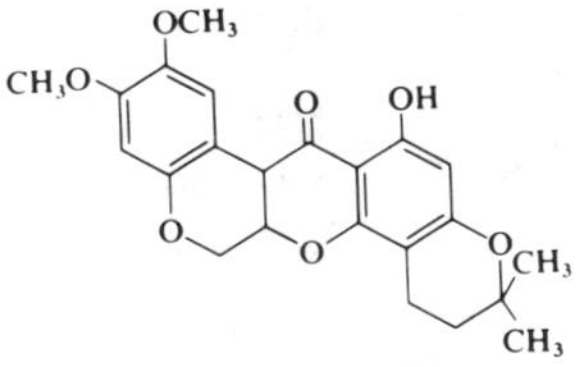

14392

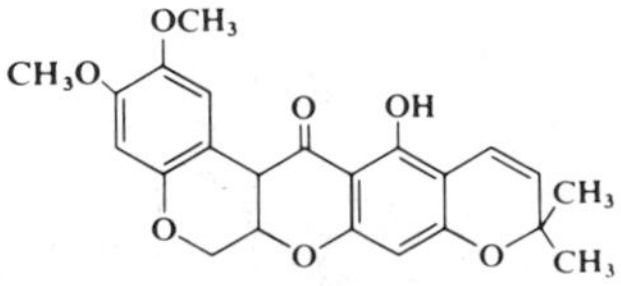

14393

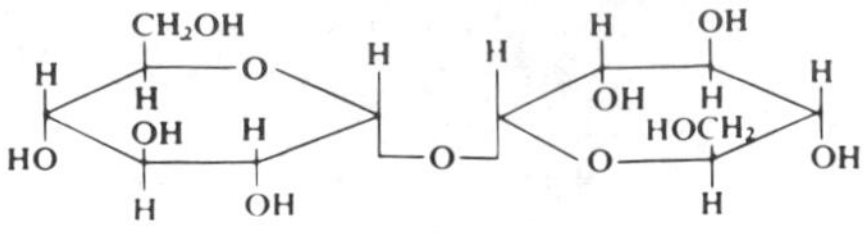

14395

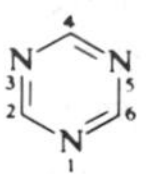

14402

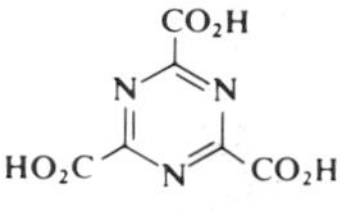

14409

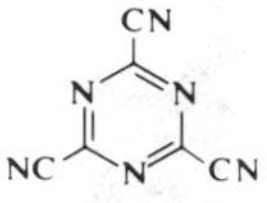

14411

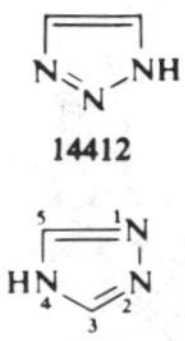

14412

14413

14456

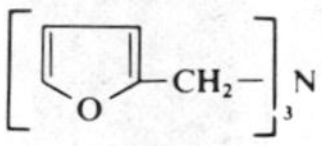

14503

14505

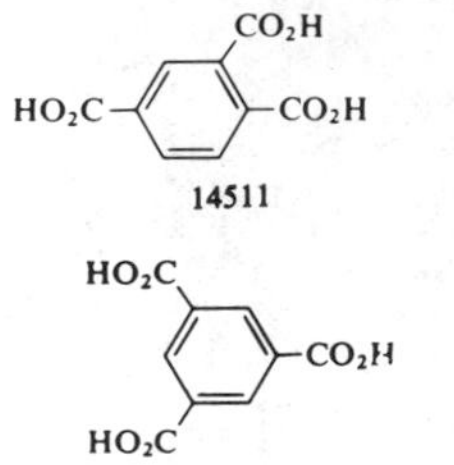

14511

14512

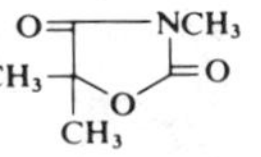

14513

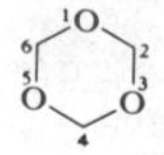

14529

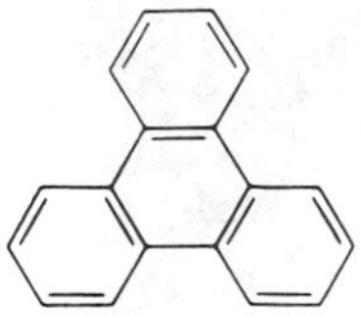

14535

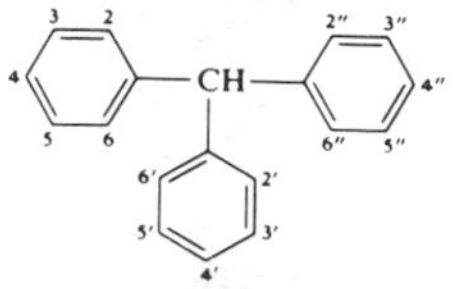

14537

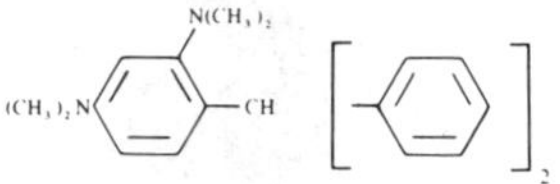

14543

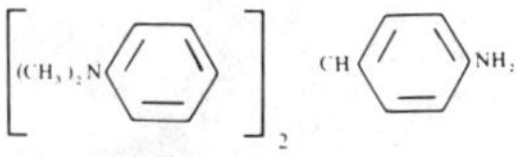

14545

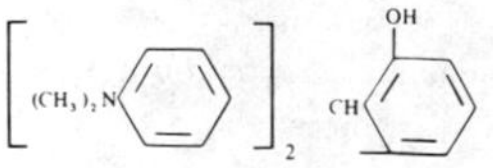

14546

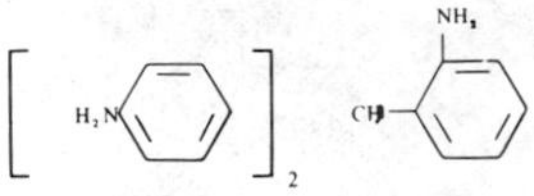

14550

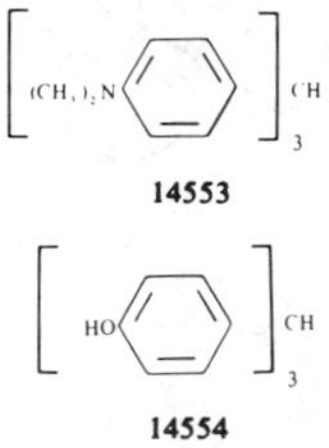

14553

14554

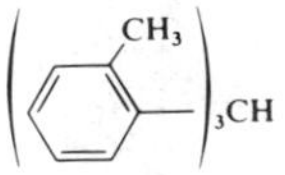

14555

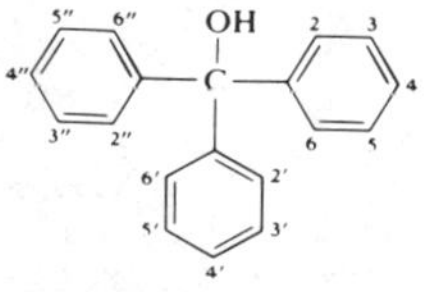

14557

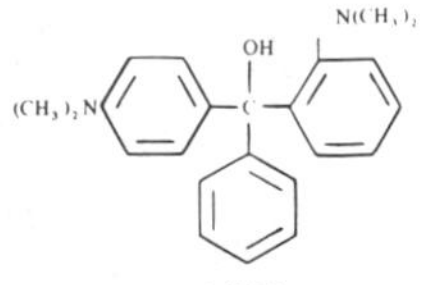

14560

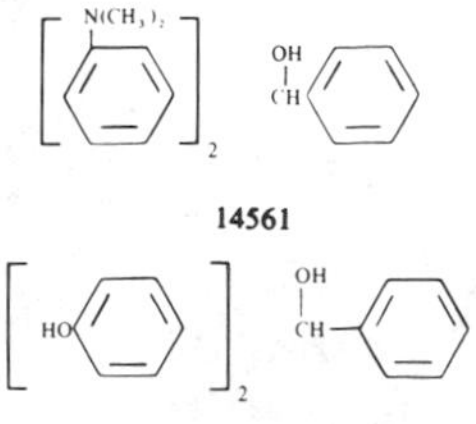

14561

14564

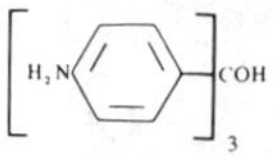

14565

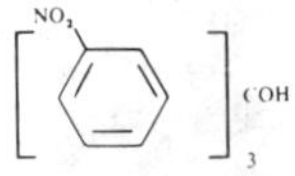

14566

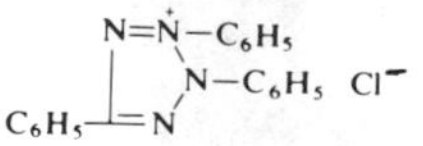

14571

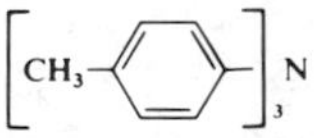

14584

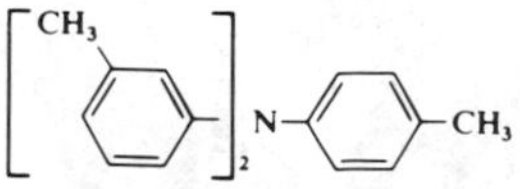

14585

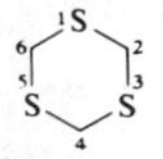

14587

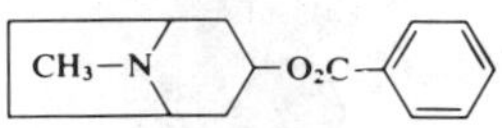

14595

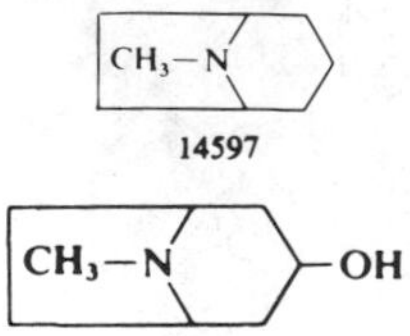

14597

14598

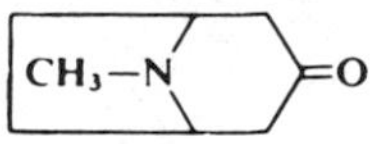

14599

STRUCTURAL FORMULAS OF ORGANIC COMPOUNDS (Continued)

In Numeric as they occur in Organic Compounds Table

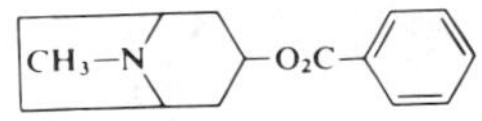

CH_3—N —O_2C—

14600

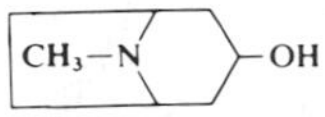

CH_3—N —OH

14602

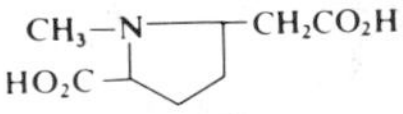

CH_3—N —CH_2CO_2H
HO_2C

14603

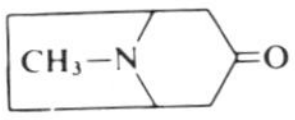

CH_3—N =O

14606

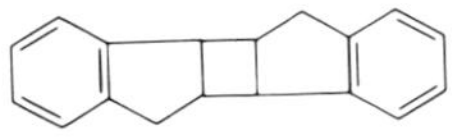

14607

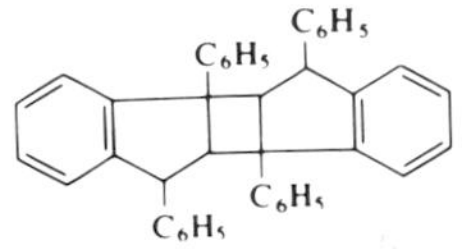

C_6H_5 C_6H_5 C_6H_5 C_6H_5

14608

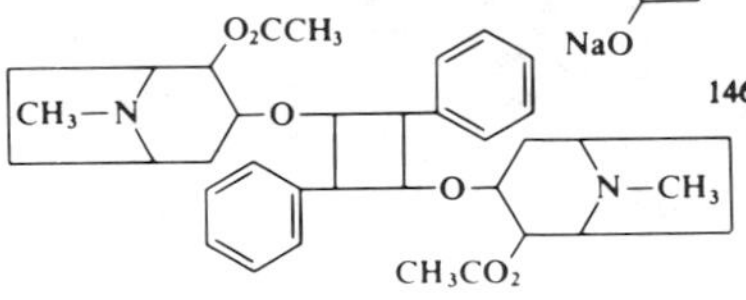

O_2CCH_3
CH_3—N —O— —O— N—CH_3
CH_3CO_2

14609

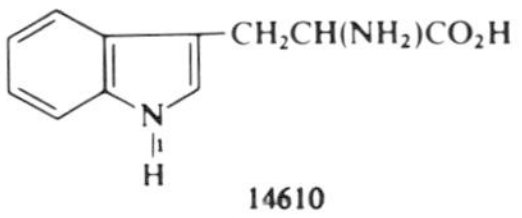

$CH_2CH(NH_2)CO_2H$
N H

14610

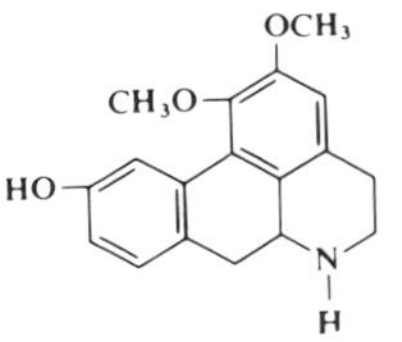

OCH_3
CH_3O
HO
N H

14617

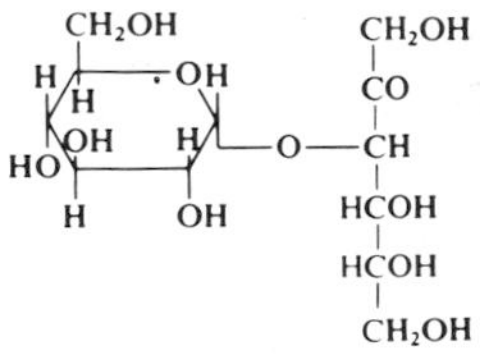

CH_2OH H OH H OH HO H H OH — O — CH_2OH CO CH HCOH HCOH CH_2OH

14618

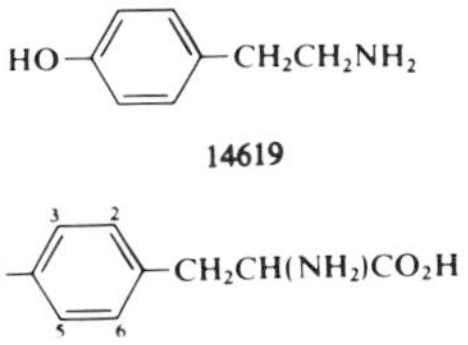

HO— —$CH_2CH_2NH_2$

14619

HO— —$CH_2CH(NH_2)CO_2H$

14620

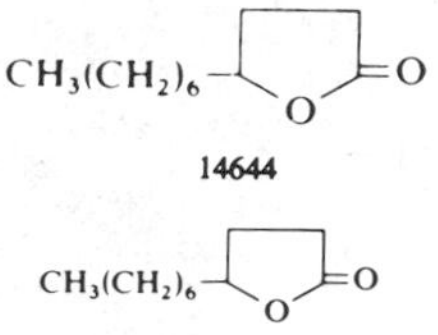

$CH_3(CH_2)_6$ O =O

14644

$CH_3(CH_2)_6$ O =O

14645

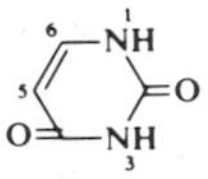

NH =O O= NH

14686

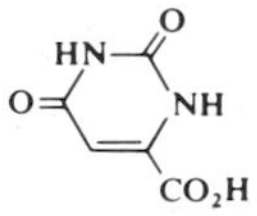

HN O O= NH CO_2H

14697

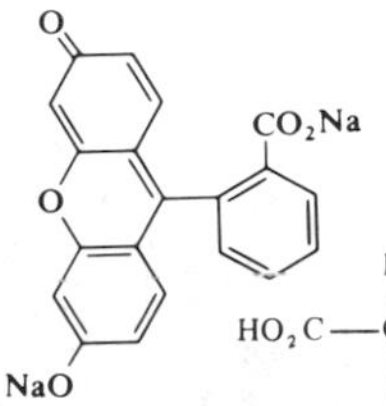

O
CO_2Na
NaO

14699

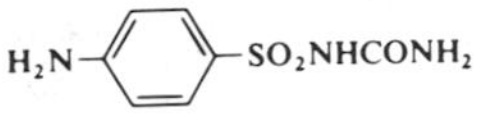

H_2N— —$SO_2NHCONH_2$

14705

Cl— —$NHCON(CH_3)_2$

14719

—NHCONH—

14737

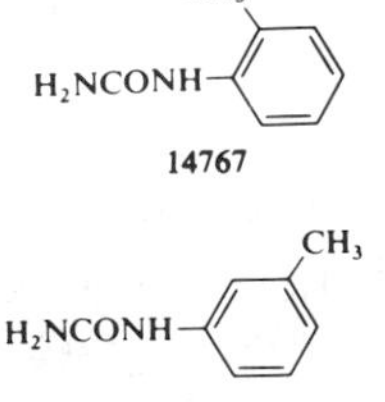

CH_3
H_2NCONH—

14767

CH_3
H_2NCONH—

14768

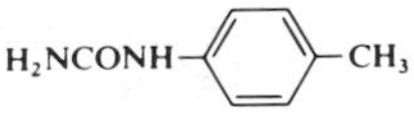

H_2NCONH— —CH_3

14769

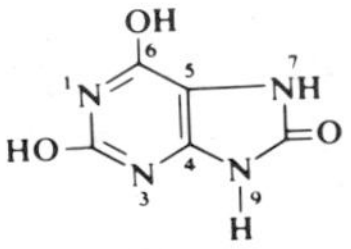

OH
N NH
HO N N =O
H

14773

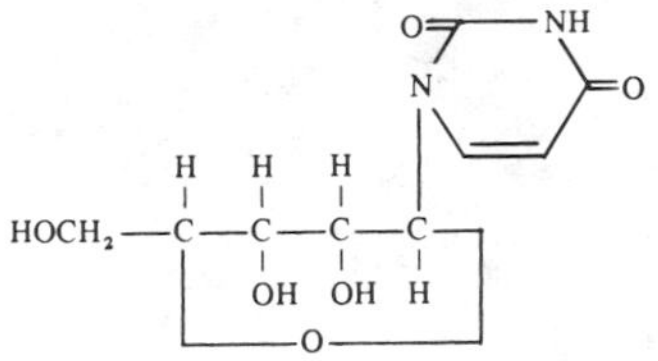

O= NH
N =O
H H H
$HOCH_2$—C—C—C—C
OH OH H
O

14779

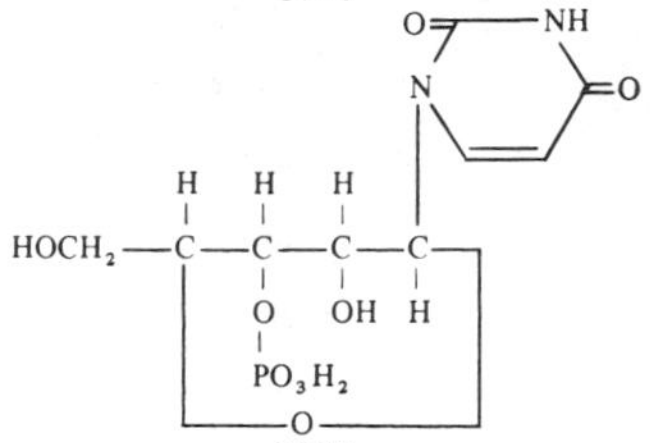

O= NH
N =O
H H H
$HOCH_2$—C—C—C—C
O OH H
PO_3H_2
O

14780

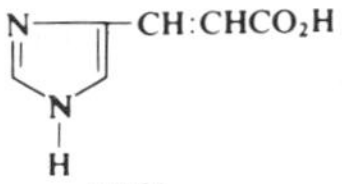

N —CH:$CHCO_2H$
N H

14781

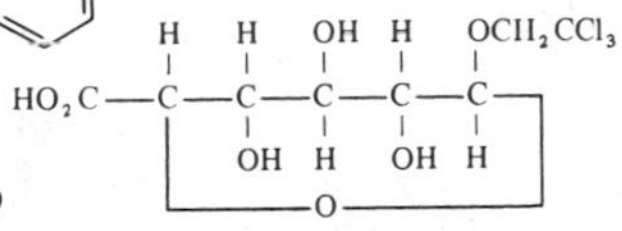

H H OH H OCH_2CCl_3
HO_2C—C—C—C—C—C
OH H OH H
O

14782

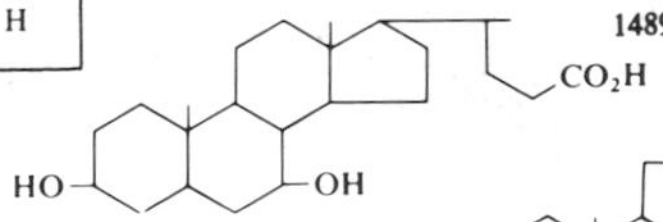

CO_2H
HO OH

14783

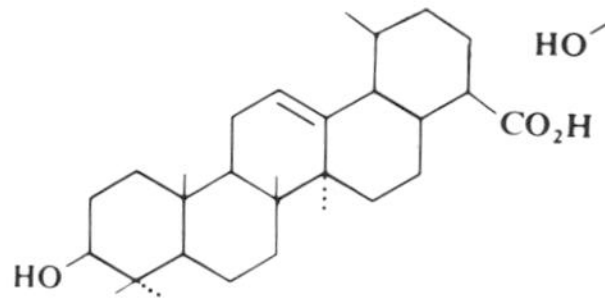

CO_2H
HO

14784

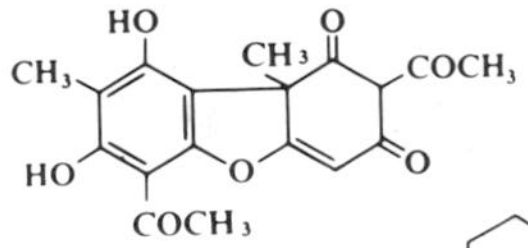

HO CH_3 O CH_3 $COCH_3$
HO O =O
$COCH_3$

14785

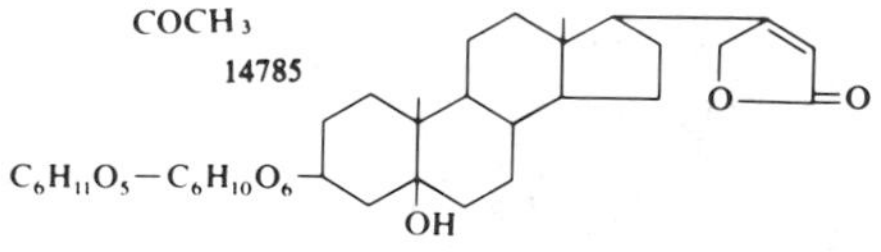

$C_6H_{11}O_5$—$C_6H_{10}O_6$
OH
O =O

14788

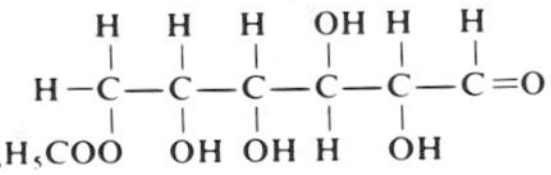

H H H OH H H
H—C—C—C—C—C—C=O
C_6H_5COO OH OH H OH

14789

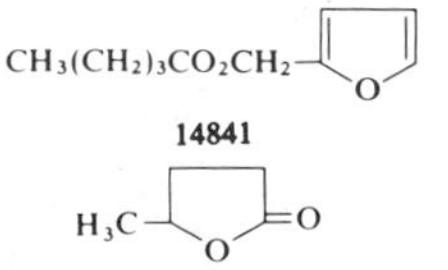

$CH_3(CH_2)_3CO_2CH_2$— O

14841

H_3C O =O

14845

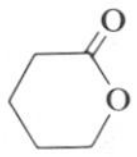

O O

14848

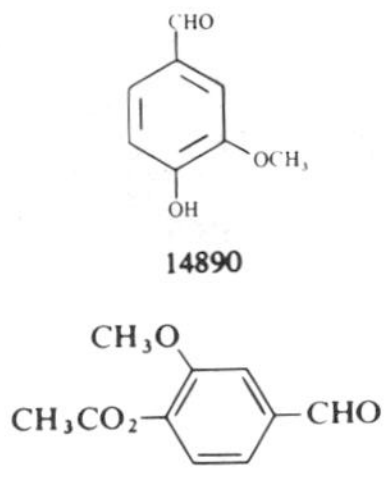

CHO
OCH_3
OH

14890

CH_3O
CH_3CO_2— —CHO

14891

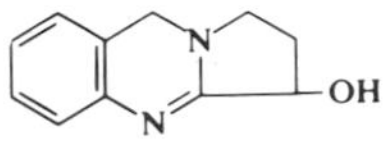

N N —OH

14897

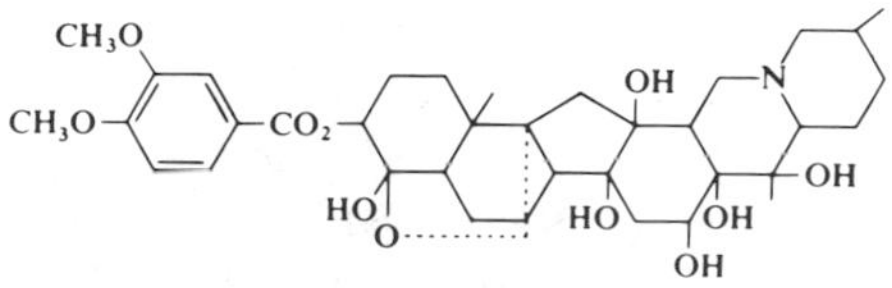

CH_3O
CH_3O— —CO_2— OH N
HO O HO OH OH OH

14899

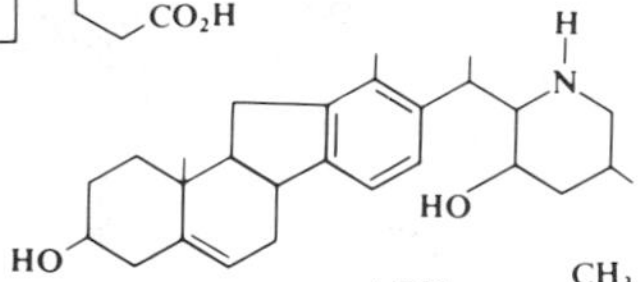

H N
HO HO

14901

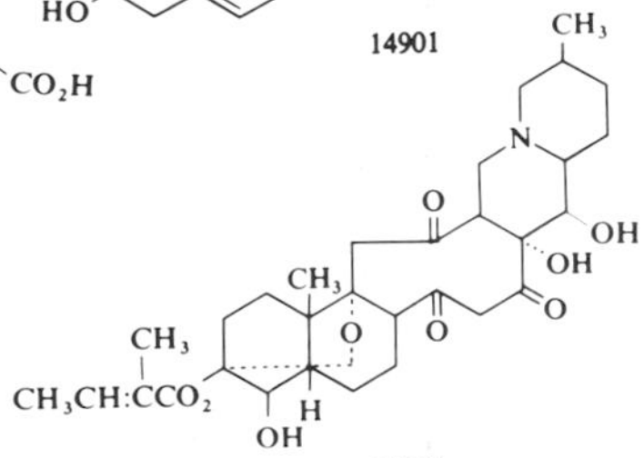

CH_3
N
O OH OH
CH_3 O O O
CH_3
$CH_3CH:CCO_2$ H OH

14903

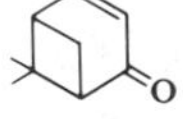

=O

14906

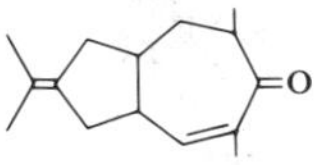

=O

14908

CH(OH) CH
H—C—OH H—C—OH
HO—C—H O O O HO—C—H
H—C—OH HO—C—H
HC CH_2
CH_2

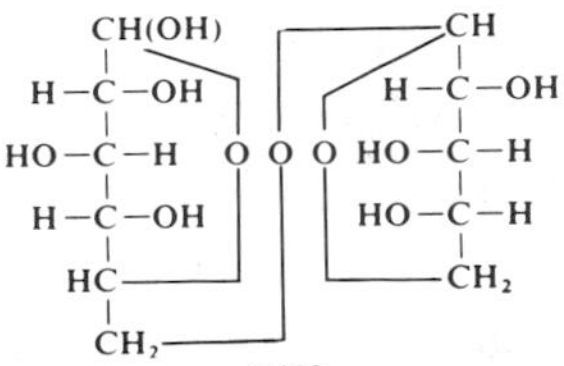

14910

STRUCTURAL FORMULAS OF ORGANIC COMPOUNDS (Continued)

In Numeric as they occur in Organic Compounds Table

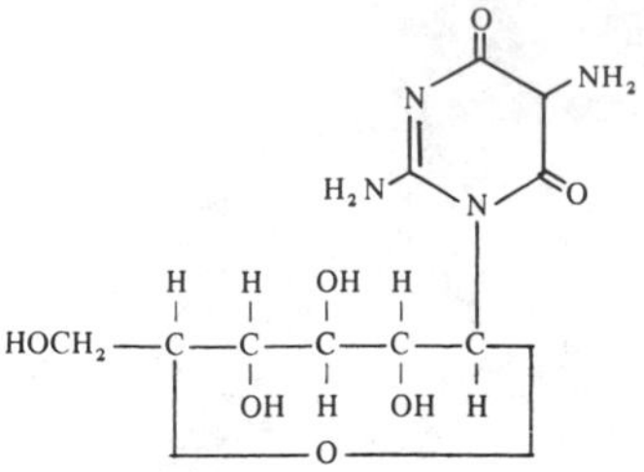

14911

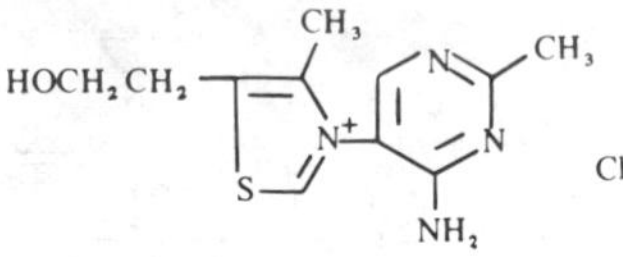

14923

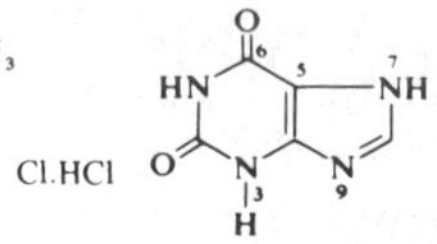

14949

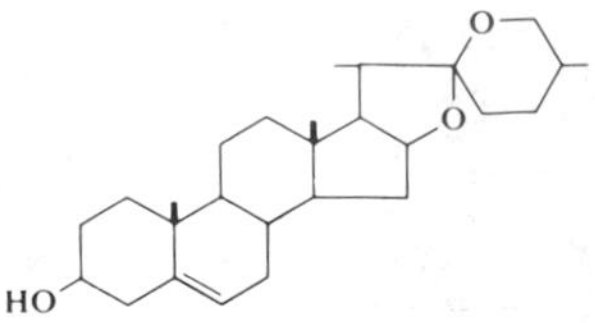

15022

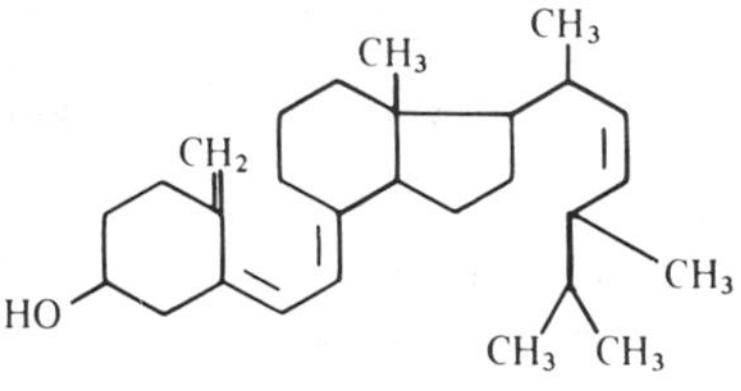

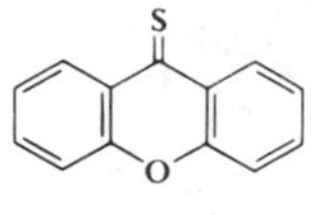

14951

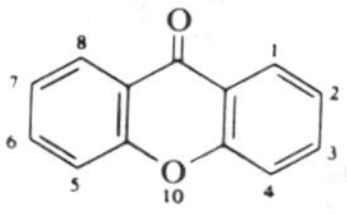

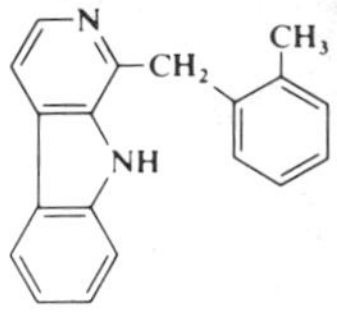

15023

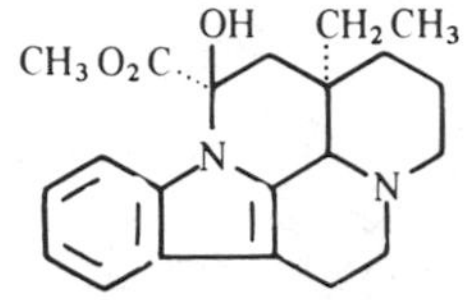

14924

14912

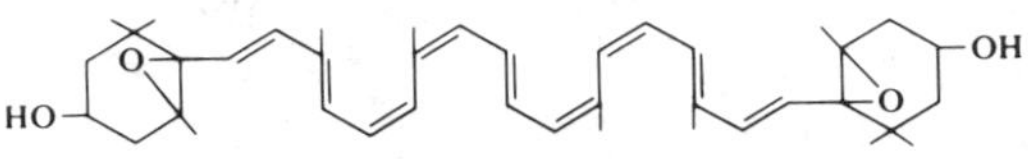

14954

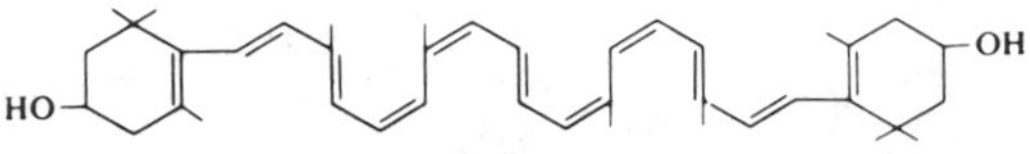

14966

14913

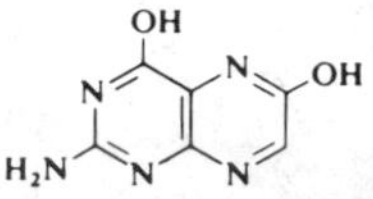

14967

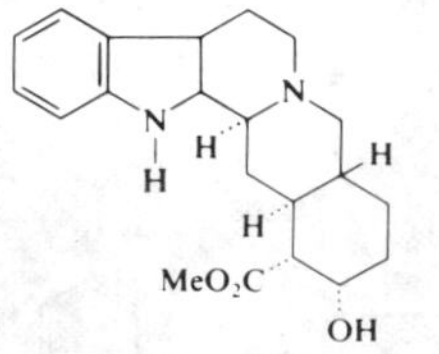

15024

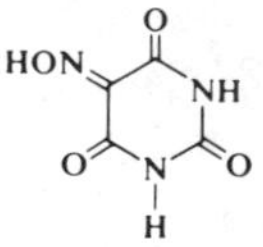

14914

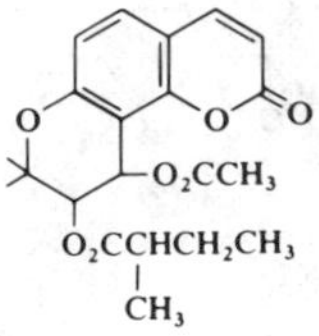

14915

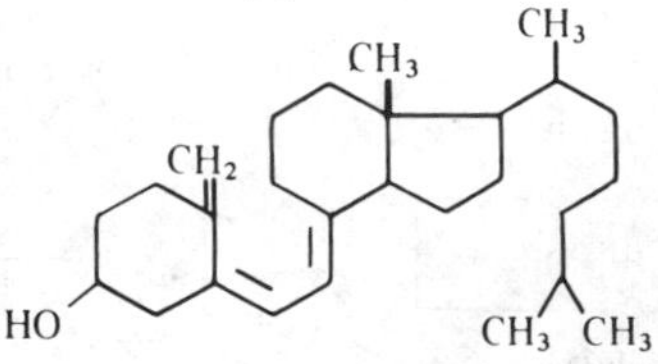

14925

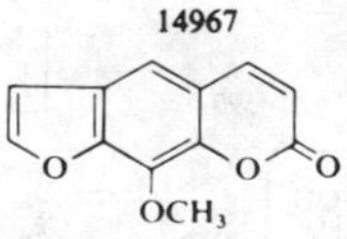

14968

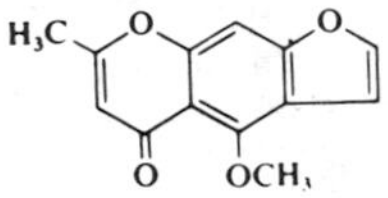

14916

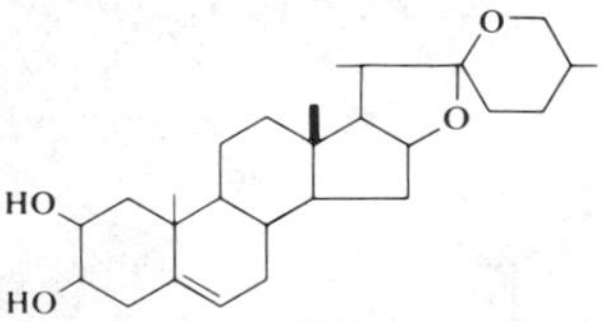

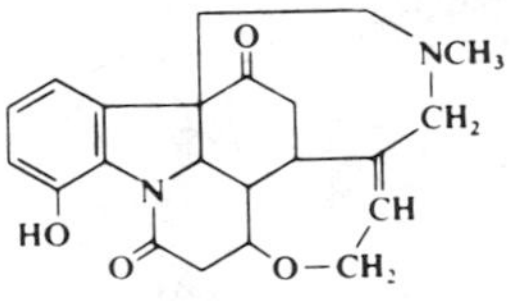

14926

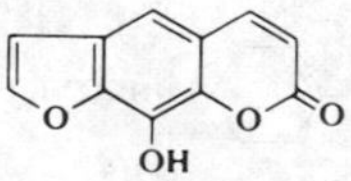

14969

15027

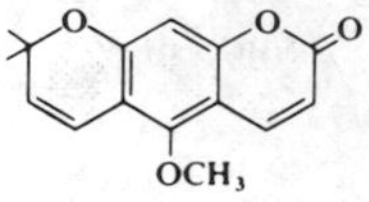

14970

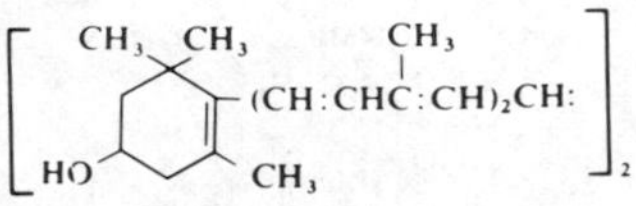

15029

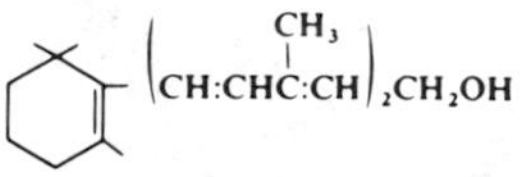

14917

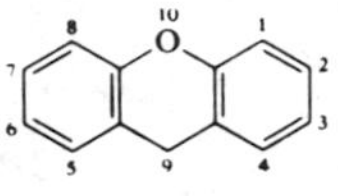

14944

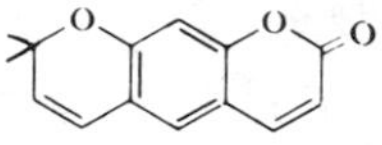

14971

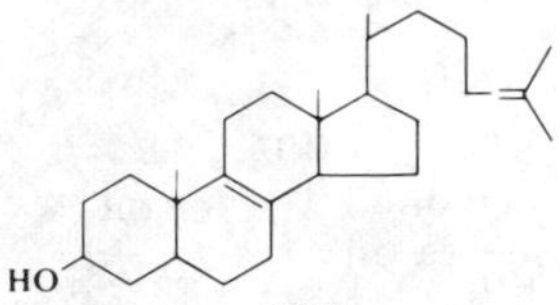

15030

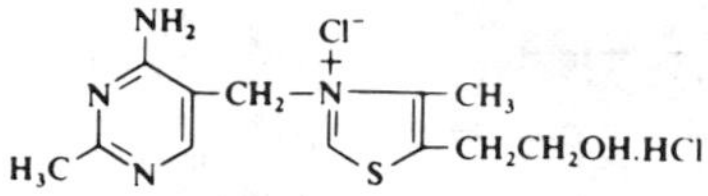

14918

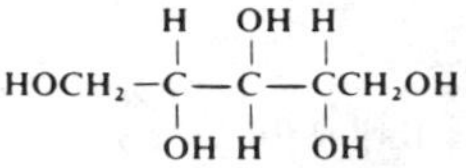

15019

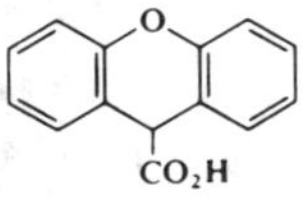

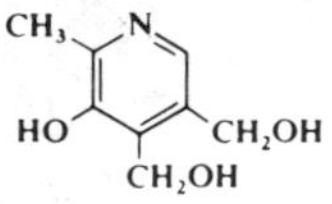

14919

14947

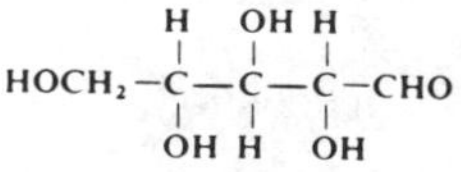

15020

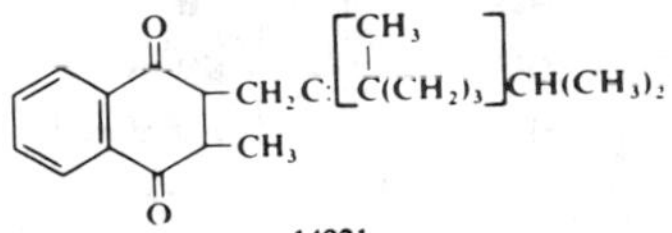

14921

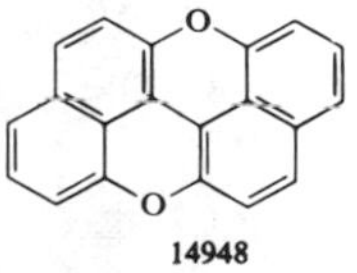

14948

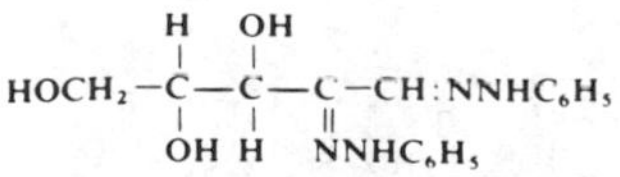

15021

MELTING POINT INDEX OF ORGANIC COMPOUNDS

Temperatures in °C; where values are not precisely known, or where there is a range of melting points, the compound is listed according to the lower temperature.

20: 8979, 9042, 9116, 9253, 9532, 9934, 10235, 11252, 12184, 12907, 13608, 14680
21: 755, 1068, 1666, 1957, 2020, 2556, 2605, 6401, 7224, 7736, 8228, 10749, 13106, 14256, 14481
22: 3240, 3469, 4801, 4942, 5159, 6802, 7488, 7576, 9059, 10007, 10839, 10905, 10924, 11006, 11179, 12328, 12578, 13419, 13651
23: 627, 983, 1791, 2090, 2765, 5626, 8528, 10382, 10684, 10951, 10997, 11928, 13427, 13498, 14412, 14475
24: 67, 99, 835, 1091, 1095, 1178, 2044, 2786, 2872, 3088, 3144, 3149, 3261, 3695, 3908, 4552, 7708, 8114, 8713, 9761, 9817, 9991,10283, 10568, 11177, 11975, 12183, 12294, 12671, 12758, 13516, 14082, 14087, 14103, 14114, 14667
25: 1886, 2128, 3184, 4010, 4071, 4076, 5269, 5359, 5577, 5592, 5745, 6282, 6635, 9074, 9227, 9343, 10313, 11424, 11794, 11994, 12076, 12594, 13110, 13412, 14436, 14653
26: 3816, 6409, 7365, 7409, 8064, 8581, 8720, 9418, 10608, 11005, 11009, 11168, 11266, 12871, 13098, 13599, 13944, 14102, 14313, 14352, 14354, 14356
27: 295, 2109, 2738, 3309, 3716, 3752, 4616, 5348, 6691, 7205, 7709, 8989, 9434, 10023, 10288, 10845, 10903, 10978, 12819, 13003, 13165, 13384, 13855, 13867, 14428, 14461
28: 79, 418, 419, 1040, 1661, 2021, 2168, 3195, 4805, 5398, 5462, 5615, 5746, 5906, 5907, 6087, 6317, 6372, 6394, 6460, 7140, 7496, 7512, 7544, 7843, 8721, 9935, 9981, 9984, 9985, 10064, 10943, 11178, 11262, 11358, 12950, 13383, 13415, 13818, 13940, 14071, 14074, 14111, 14254, 14637
29: 538, 1041, 1088, 1338, 2877, 3423, 3467, 3485, 3591, 3593, 4813, 5106, 5584, 6339, 6443, 6444, 6464, 7138, 7996, 8131, 10760, 10962, 11040, 11173, 11558, 11856, 11863, 12325, 12752, 13387, 13522, 13917, 14296, 14312, 14319, 14333, 14981
30: 442, 1772, 1904, 2759, 3071, 3192, 3378, 4912, 4980, 4982, 5297, 5304, 5988, 6141, 6216, 6290, 6400, 6446, 6932, 7497, 7746, 9052, 9247, 9760, 9847, 10008, 10233, 10635, 12600, 12935, 13360, 13369, 13411, 13722, 13725, 13914, 14476, 14984
31: 224, 1604, 1606, 2136, 2170, 3118, 3227, 3244, 4366, 4370, 4401, 5564, 5590, 5748, 6405, 6445, 7173, 7587, 8341, 8760, 10090, 10238, 10280, 12779, 13389, 13597, 13859, 14321, 14477, 14952
32: 770, 1553, 2576, 3076, 3110, 3135, 3607, 4178, 4814, 4937, 5464, 6056, 6082, 6357, 6408, 7433, 7456, 7490, 8709, 8842, 8904, 9314, 9553, 9685, 9713, 9904, 11270, 11851, 12506, 12599, 12613, 14474
33: 47, 378, 629, 849, 869, 927, 2171, 3349, 3549, 3749, 4292, 4588, 5103, 5726, 6115, 6387, 6577, 7117, 7718, 8505, 8507, 9269, 9433, 9436, 10687, 10819, 10850, 11092, 11140, 11547, 13739, 14295, 14309, 14478, 14828
34: 246, 409, 603, 768, 963, 1901, 1943, 1966, 2043, 2093, 2478, 2491, 2866, 3212, 3403, 3476, 3838, 3841, 4027, 5160, 5360, 5587, 5630, 5738, 6240, 6283, 6285, 6345, 7204, 7498, 7698, 9282, 9400, 9800, 9987, 10222, 10660, 10915, 11035, 11469, 11565, 12464, 12760, 12772, 13451, 13482, 13668, 13740, 14252, 14447, 14844
35: 320, 508, 880, 1503, 1738, 1773, 1931, 1959, 2113, 3175, 3477, 4391, 4585, 4586, 4633, 5800, 6205, 6212, 7043, 7176, 7178, 7381, 8986, 9329, 9768, 9874, 9880, 10071, 10292, 10476, 10569, 10979, 11138, 11234, 11343, 11403, 11621, 11669, 11725, 11998, 12132, 14076, 14639
36: 374, 658, 948, 1591, 1793, 2340, 2444, 3247, 3302, 3453, 4415, 4891, 5130, 5373, 5375, 5582, 6080, 6340, 6439, 6461, 6467, 7213, 7487, 9170, 9407, 10005, 10227, 10764, 11353, 11445, 12235, 12326, 12417, 12866, 13334, 14110, 14139, 14643
37: 62, 967, 998, 1774, 2960, 3185, 3221, 3246, 3490, 5162, 5619, 5941, 6812, 7807, 8340, 8588, 8746, 9376, 9531, 9705, 9840, 9914, 10286, 11127, 11139, 11714, 11911, 12090, 13049, 13130, 13404, 13731, 14072, 14080, 14330, 14453, 14862
38: 415, 938, 1028, 1066, 2356, 2439, 3611, 4558, 5041, 5213, 5318, 5344, 5667, 5699, 5919, 6369, 7491, 8095, 8623, 9055, 9175, 9279, 9732, 10456, 10486, 10590, 11135, 12048, 12405, 12764, 13107, 13192, 13410, 13754, 14085, 14148, 14186, 14261, 14641
39: 174, 236, 393, 693, 838, 1176, 1597, 1659, 2745, 2853, 2940, 3245, 5094, 5369, 6173, 6224, 7159, 7171, 7716, 9261, 10293, 10844, 10875, 11189, 11660, 11679, 11681, 11702, 12890, 13398, 13454, 13609, 13738, 14096, 14363
40: 243, 433, 1744, 1960, 2110, 2856, 2946, 3231, 3613, 3790, 4210, 4726, 4994, 5044, 5404, 5461, 5840, 6486, 6491, 6633, 6769, 7233, 7360, 7483, 7597, 8587, 8634, 8782, 10866, 11427, 12054, 12658, 13341, 13705, 13749, 14362, 14676
41: 1552, 1710, 1922, 2157, 2523, 2770, 2953, 3228, 3495, 3501, 3927, 4483, 4607, 4685, 4843, 5075, 5215, 5481, 5602, 5606, 5808, 5842, 6213, 6398, 6803, 7757, 7782, 8707, 9423, 9889, 10138, 10636, 10742, 10838, 10858, 10992, 11541, 11726, 12104, 12314, 12461, 12869, 13408, 13424, 13887, 14066, 14337, 14526, 14600

72: 14150, 14191, 14632, 14721
73: 988, 1514, 1976, 2339, 2679, 2712, 2984, 2999, 3180, 3479, 3817, 4062, 4291, 4389, 4691, 4773, 4802, 5141, 5315, 5413, 5586, 6107, 6201, 6204, 6342, 6887, 7061, 8242, 8318, 8815, 9076, 9172, 9390, 9577, 9696, 9901, 10908, 10954, 11281, 11314, 11344, 11456, 11475, 12748, 12821, 12896, 12917, 13362, 13455, 13486, 13621, 13653, 13948, 14581
74: 790, 864, 901, 1730, 2069, 2751, 3145, 3170, 3191, 3480, 3509, 4093, 5131, 5225, 5235, 5345, 5608, 5702, 5970, 6203, 6481, 6853, 8378, 8898, 9774, 10205, 10652, 10913, 11929, 12601, 12788, 12806, 12855, 12948, 13900, 14213, 14455
75: 308, 706, 734, 870, 1940, 2200, 2346, 2492, 2843, 3171, 3248, 3284, 5019, 5234, 5291, 5480, 6241, 6638, 7180, 8648, 9271, 9561, 9738, 9888, 10513, 10572, 10897, 10910, 11215, 11286, 11487, 11489, 11528, 11956, 12065, 12560, 12781, 12786, 12854, 12933, 13771, 13835, 13839, 13985, 14216, 14725, 14770, 15000
76: 265, 379, 499, 820, 894, 898, 1016, 1099, 1116, 1651, 2006, 2121, 2358, 2435, 2521, 3760, 4497, 4500, 6488, 6528, 6959, 6996, 7359, 7371, 7388, 7391, 7475, 9396, 9911, 10137, 10159, 10663, 10896, 12190, 12407, 13292, 13407, 13920, 14469
77: 780, 854, 1080, 1370, 1736, 1895, 2905, 3168, 3328, 3406, 3547, 4310, 4380, 4506, 4545, 4782, 5378, 5759, 6084, 6450, 6702, 7382, 7720, 8296, 8732, 9564, 10982, 11102, 11118, 11195, 11425, 12523, 13364, 14975
78: 990, 1135, 2013, 3018, 3112, 4215, 4235, 4318, 4498, 4501, 4505, 4820, 5010, 5144, 5174, 5674, 5703, 6033, 6196, 7375, 7920, 8083, 8601, 8705, 9671, 9872, 10702, 10828, 10861, 10892, 10999, 11115, 11412, 12063, 12653, 13010, 13906, 14011, 14067, 14068, 14121, 14238, 14339, 14368, 14858
79: 59, 69, 70, 106, 128, 332, 683, 825, 1095a, 1607, 1811, 2255, 2449, 2562, 2859, 2867, 3863, 4579, 4723, 4853, 5067, 5069, 5189, 5374, 6047, 6298, 6367, 6422, 6900, 7370, 9073, 9412, 9482, 9770, 11284, 11514, 12161, 13957, 13998, 14397, 14727
80: 421, 660, 984, 1476, 1647, 1850, 1857, 2081, 2589, 3134, 3167, 4012, 4253, 5008, 5011, 5031, 5312, 6382, 7417, 7518, 7694, 8026, 8209, 8305, 8508, 8523, 8636, 8731, 9275, 10309, 10788, 10970, 11017, 11220, 11381, 11720, 13637, 14192, 14208, 14286, 14459
81: 6, 153, 233, 245, 420, 448, 705, 830, 881, 892, 897, 1133, 1470, 1593, 2479, 2583, 2897, 3413, 3783, 3835, 4783, 5026, 5132, 5242, 5897, 6034, 6722, 7069, 7367, 7387, 7925, 8733, 9136, 9277, 9300, 9388, 9393, 9401, 9586, 9755, 10870, 11280, 11333, 12061, 12071, 13395, 14329
82: 58, 1075, 1532, 1557, 1582, 1695, 1834, 1929, 2335, 2861, 2898, 2904, 4406, 4409, 5120, 5585, 6643, 7172, 7338, 8518, 8520, 9249, 9272, 9360, 9395, 10919, 11105, 11292, 11295, 11508, 11600, 12154, 12500, 12879, 13022, 13396, 13409, 13484, 14185, 14198, 14415, 14681, 14974
83: 166, 771, 1071, 1104, 1124, 2045, 2056, 2621, 2681, 2840, 2870, 3002, 3033, 3161, 3283, 3307, 3540, 3917, 4036, 4729, 4798, 5289, 6198, 6299, 6637, 7329, 7861, 8259, 8440, 8532, 8566, 8927, 9552, 9703, 9864, 9975, 10240, 10333, 10450, 10854, 10883, 11026, 11277, 11741, 12134, 12136, 12904, 13024, 13027, 13145, 13581, 13583, 13696, 13727, 13809, 14236, 14332, 14685
84: 485, 681, 759, 817, 1079, 1510, 1656, 1667, 1683, 1873, 1995, 2073, 2348, 2785, 2945, 2948, 3297, 3306, 4257, 4883, 5073, 5357, 5822, 6018, 6270, 6307, 6518, 7021, 7035, 8952, 9429, 9495, 9601, 9602, 9820, 9902, 10813, 10885, 11190, 11739, 12451, 12940, 13394, 13633, 13734, 13825, 14073, 14285, 14539, 14925
85: 412, 760, 882, 944, 968, 1131, 1346, 1526, 2005, 2070, 2077, 2201, 2425, 2543, 3032, 3061, 3390, 3391, 3426, 3697, 4669, 4689, 5322, 5370, 5993, 5996, 6562, 7324, 8071, 8082, 8417, 8544, 8596, 9324, 9529, 9559, 9687, 10344, 10653, 10796, 10899, 12163, 12554, 12771, 12800, 12805, 13009, 13980, 14583, 14703, 14915
86: 91, 234, 271, 496, 735, 1352, 1801, 2129, 2223, 2851, 3188, 3456, 4255, 4342, 4674, 4795, 4825, 5313, 5323, 5976, 5977, 5992, 6981, 7290, 7419, 8008, 9278, 9584, 9782, 10284, 10430, 10751, 11016, 11159, 12901, 12910, 12923, 13401, 13426, 13612, 13907, 14209, 14402
87: 7, 18, 93, 97, 102, 148, 348, 498, 565, 828, 855, 871, 1590, 2033, 2067, 2075, 2261, 2672, 2928, 3086, 3151, 3178, 3189, 3334, 4046, 4313, 4791, 4803, 4862, 5002, 6194, 6493, 7330, 7334, 8292, 8517, 9302, 9960, 10888, 10895, 10928, 11022, 11222, 12141, 12179, 12593, 12984, 13014, 13405, 13640, 14016, 14492
88: 95, 137, 707, 860, 1076, 1662, 2015, 2422, 2432, 2537, 4690, 4704, 5795, 5895, 6587, 7700, 7851, 9246, 9415, 9661, 10190, 10514, 10698, 10856, 11162, 13423, 13525, 13812, 14369, 14400
89: 180, 317, 344, 1072, 1541, 1775, 2118,

89: 2209, 3256, 3320, 4073, 4722, 5268, 5287, 5308, 6038, 6040, 6421, 6725, 6820, 8542, 8698, 8700, 9335, 10258, 10317, 11053, 11217, 11497, 11964, 12070, 12156, 12431, 12936, 13393, 13700, 14092, 14585, 14976

90: 303, 396, 704, 736, 1486, 1516, 1757, 1784, 1804, 2125, 2332, 3030, 3055, 3109, 3486, 4326, 4479, 4676, 4710, 4793, 4859, 5352, 5559, 5644, 6362, 6588, 6901, 7002, 8108, 8325, 8434, 8826, 9091, 9466, 9974, 10224, 10645, 10676, 10881, 10906, 11446, 12435, 12625, 3413037, 13661, 13669, 13698, 13832, 15020

91: 101, 114, 781, 1078, 1799, 2003, 2281, 2522, 3395, 3400, 3601, 4650, 5288, 6179, 6297, 7131, 7181, 7427, 7435a, 8105, 8118, 8422, 8536, 8905, 9516, 10181, 10733, 10914, 11164, 11267, 12119, 12777, 12865, 12899, 13446, 13591, 14038, 14317, 14373, 14417

92: 20, 356, 573, 722, 886, 1335, 1539, 1956, 2048, 2342, 2550, 2937, 4504, 4678, 4806, 6195, 6289, 6429, 6644, 6999, 7844, 9835, 10829, 10831, 10857, 11390, 12909, 13008, 13039, 13140, 13810, 14190, 14323, 14741

93: 540, 1138, 1533, 1592, 1665, 2480, 2750, 3059, 3286, 3415, 3473, 4503, 4776, 4931, 5001, 5020, 5163, 5794, 5796, 6185, 6462, 8093, 8109, 8666, 8787, 8818, 8824, 8949, 10959, 10985, 11001, 11297, 11428, 11512, 12081, 12088, 12602, 12741, 12798, 12892, 13392, 13804, 14120, 14170, 14288, 14342, 14684, 14888, 15005

94: 168, 171, 541, 1050, 1114, 1190, 1652, 1674, 2443, 2763, 2857, 3013, 3019, 3433, 3507, 4033, 4507, 5995, 6378, 6605, 7970, 8075, 8164, 9103, 9305, 9308, 9769, 10977, 11021, 11378, 11492, 12729, 12918, 13380, 13864, 14385, 14537

95: 80, 81, 135, 274, 414, 728, 737, 858, 1027, 1412, 1704, 1894, 1965, 2204, 2259, 2427, 2482, 2540, 2603, 2894, 2935, 2968, 3024, 3210, 4570, 4709, 5310, 6035, 6083, 6280, 6677, 6678, 7320, 7484, 7998, 8439, 8673, 8675, 8831, 9657, 9771, 10231, 10972, 11023, 11236, 12612, 12708, 12858, 13056, 13119, 13146, 13388a, 13462, 14240, 14291, 14430

96: 4, 167, 363, 399, 989, 1997, 2718, 2860, 2885, 3193, 3232, 3399, 4785, 4787, 4897, 5022, 5325, 7113, 7582, 8287, 8821, 8878, 8899, 9556, 9571, 9721, 9885, 10734, 10752, 10769, 10973, 11038, 11117, 11134, 11155, 11306, 12239, 12695, 13155, 13202, 13564, 13565, 13806, 14009, 14169, 14229, 14714, 14758

97: 24, 42, 131, 380, 542, 1051, 1559, 1999, 2277, 3000, 3004, 3165, 3440, 4028, 4060, 4254, 4789, 5973, 6958, 7014, 7182, 8197, 8294, 8420, 8421, 9567, 9990, 10213, 10335, 10754, 10989, 11507, 12182, 12803, 12863, 12987, 13020, 13151, 14058, 14326, 14594

98: 50, 353, 1037, 1083, 1471, 1535, 1840, 1905, 1964, 2115, 2354, 2504, 2701, 2869, 2944, 3003, 3516, 3584, 4079, 4374, 4699, 5188, 5190, 5209, 5317, 5968, 6234, 6506, 6898, 7430, 8188, 8669, 8723, 8783, 9214, 9317, 9397, 9465, 9651a, 10926, 10971, 11060, 11122, 11242, 11303, 11495, 13004, 13163, 13975, 14135, 14168, 14346

99: 72, 96, 488, 1029, 1603, 1691, 1713, 2197, 2239, 2345, 2352, 2742, 2847, 3235, 3269, 3385, 5025, 5663, 5860, 6012, 6059, 6590, 6705, 6819, 6917, 7318, 7633, 8020, 8773, 9307, 9701, 10266, 10424, 10783, 10920, 11247, 11248, 11408, 12961, 13365, 13788, 14021, 14089

100: 641, 1902, 2813, 2824, 2884, 2929, 3038, 3115, 3172, 3796, 4030, 5021, 5107, 5460, 5576, 5864, 5896, 6962, 8055, 9313, 9439, 9478, 9834, 9852, 10686, 10705, 11108, 11213, 11223, 11549, 13033, 13046, 14271, 14470, 14944

101: 82, 377, 1003, 1164, 1477, 1939, 2758, 2868, 2985, 4568, 4899, 5320, 6648, 8116, 9135, 10649, 10842, 11031, 11059, 11587, 12677, 12829, 12864, 12962, 13176, 14215, 14297, 14588, 14989

102: 161, 357, 1366, 2061, 2315, 2397, 2599, 2810, 3037, 4569, 5240, 5264, 5334, 6295, 7130, 7236, 7460, 8336, 8909, 8943, 9697, 10237, 10694, 10880, 10963, 11211, 11324, 12982, 13403, 13440, 14269, 14544, 14891

103: 14, 149, 172, 502, 634, 815, 2258, 2931, 3130, 3416, 3481, 3583, 3585, 5261, 5294, 5395, 5704, 6304, 6503, 6509, 6585, 7007, 7093, 7108, 7325, 7459, 8547, 9345, 9413, 9420, 9728, 9772, 10204, 10696, 10964, 11082, 11129, 11701, 12546, 12725, 12928, 12983, 13589, 14243, 14396, 14749

104: 115, 133, 401, 547, 574, 775, 1150, 1345, 1518, 3514, 6181, 6564, 6591, 6925, 6998, 7754, 8845, 9464, 9490, 9611, 9750, 10655, 10657, 11352, 12450, 12791, 12797, 12804, 12872, 14197, 14247, 14287, 14743, 14789

105: 100, 108, 391, 843, 1052, 1077, 1677, 2007, 2103, 2279, 2471, 2967, 3029, 3281, 4576, 4612, 4788, 4898, 5447, 5449, 5664, 5671, 6522, 6523, 6592, 7321, 7452, 7458, 8415, 8729, 8833, 9250, 9312, 9365, 9372, 9406, 9549, 9582, 9603, 9754, 10125, 10739, 10965, 11107, 11119, 11354, 11940, 12068, 13023, 13050, 13376, 13769, 14105, 14173, 14272, 14558

106: 354, 514, 638, 1074, 1085, 1086, 1106, 1139, 1365, 1440, 1453, 1505, 1789,

159: 14301, 14384, 14414, 15021

160: 758, 1119, 1130, 1626, 1814, 1864, 2222, 2269, 2337, 2350, 2817, 2833, 3538, 4094, 4830, 5285, 7295, 8135, 8233, 8295, 8432, 8908, 9107, 9690, 10780, 10804, 11148, 11315, 11879, 12566, 12707, 12730, 12967, 13069, 13585, 14059, 14919

161: 162, 526, 1543, 1835, 2066, 2554, 2773, 2887, 3039, 3107, 3446, 5279, 5383, 7109, 8178, 8401, 8738, 9349, 9500, 10976, 11063, 12652, 13541, 13862

162: 594, 978, 1217, 1263, 1419, 1491, 1750, 1830, 1858, 1881, 2306, 2791, 2920, 3091, 3439, 4424, 5148, 5341, 5791, 6071, 6296, 6494, 6578, 7815, 8007, 8261, 8569, 9366, 9550, 9616, 9983, 12615, 13345, 13367, 13791, 14007, 14013, 14244, 14752

163: 48, 474, 784, 1112, 1515, 1843, 2257, 2510, 2584, 2723, 2804, 2845, 3081, 3595, 4998, 5256, 7023, 8171, 8924, 9353, 9644, 9842, 10334, 11323, 12900, 12996, 13030, 13540, 13816, 14202

164: 452, 696, 885, 1171, 1172, 1358, 1678, 1756, 1821, 2485, 2591, 2640, 2719, 2807, 3402, 3410, 3422, 3437, 3734, 4371, 4530, 4858, 5133, 5569, 5953, 8182, 8539, 9523, 10800, 11325, 11465, 11500, 11606, 12364, 12834, 13978, 14212, 14557, 14619, 14712, 14892, 14900

165: 94, 155, 559, 709, 1330, 1442, 1693, 1813, 2248, 2275, 2365, 2369, 2572, 2601, 2668, 2996, 3022, 3064, 3497, 3502, 3568, 5018, 5156, 5258, 5267, 5354, 6189, 6363, 6413, 6688, 7005, 7148, 7305, 7414, 8036, 8094, 8227, 8429, 8938, 9154, 9629, 9866, 9964, 11000, 11485, 11922, 12137, 12424, 12453, 12575, 12608, 12880, 13043, 13207, 13366, 13368, 13622, 13716, 13972, 13994, 14205, 14302, 14548, 14550, 14779, 14893

166: 518, 665, 1577, 1833, 2167, 2319, 2368, 2516, 2569, 2662, 3368, 4580, 7691, 9017, 9187, 9878, 10312, 10659, 10935, 11230, 11938, 12679, 12696, 12720, 13112, 13346, 13963, 14304, 14716, 14955

167: 1602, 2215, 2513, 2598, 2651, 2732, 2827, 3223, 5653, 6232, 6571, 7053, 8255, 8697, 9339, 9480, 9871, 10807, 12345, 12524, 12724, 13338, 14281, 14566, 14591

168: 22, 39, 98, 727, 1425, 1427, 1449, 2515, 2530, 2622, 2639, 2736, 2749, 2754, 2806, 2836, 4078, 4714, 5266, 6570, 6582, 7008, 7095, 7303, 8165, 8206, 8798, 8922, 8923, 9587, 9621, 9642, 10318, 10670, 10706, 10865, 11271, 11320, 12732, 12830, 13001, 13101, 13121, 13183, 13551, 13579, 13964, 14264, 14410, 14618, 14748

169: 475, 1019, 1576, 1851, 1872, 2700, 3068, 3344, 3558, 4707, 5393, 5672, 6020, 6251, 6967, 9628, 9730, 10601, 10785, 12740, 13117, 13141, 13620, 14022, 14393, 14560

170: 462, 730, 1194, 1200, 1215, 1218, 1267, 1283, 1493, 1561, 1827, 1828, 1874, 2307, 2493, 2500, 2663, 2924, 3386, 4693, 4696, 4708, 5659, 5747, 6437, 6586, 6936, 8080, 8230, 8268, 8902, 9309, 9470, 9672, 10324, 10864, 11101, 11228, 11619, 11912, 12458, 13045, 13175, 13444, 13596, 13607, 14277, 14719

171: 702, 1434, 2535, 2826, 3083, 3099, 3542, 5013, 5127, 5451, 6016, 6157, 7009, 8166, 8485, 8491, 8530, 8827, 9094, 9514, 9647, 10921, 12387, 12699, 12703, 13135, 13356, 13545, 13646, 13977, 14359

172: 120, 430, 431, 500, 1179, 1380, 1492, 1619, 1754, 1908, 2225, 3372, 4529, 4628, 5051, 5129, 5571, 6178, 6956, 6957, 7479, 7480, 8168, 8181, 8256, 9126, 9213, 9371, 9622, 9624, 9720, 10439, 10678, 11206, 11208, 11750, 13212, 13593, 13766, 14276

173: 1, 577, 733, 957, 2105, 2616, 2907, 2950, 3319, 3454, 5137, 7688, 8169, 8279, 8487, 9222, 9370, 9575, 9651, 9655, 9666, 10067, 10929, 12605, 13109, 13144, 13578, 13580, 14263, 14740, 14757

174: 960, 994, 1620, 1634, 2126, 2175, 2311, 2328, 2632, 2815, 4540, 5003, 5989, 6540, 6758, 8285, 8320, 10776, 10909, 10994, 11471, 11745, 12400, 12702, 13178, 14954

175: 778, 1180, 2303, 2702, 2734, 3104, 3124, 3529, 3582, 5089, 5118, 5252, 5306, 5850, 5851, 5852, 5853, 6086, 6302, 6476, 6505, 6549, 7149, 7273, 7476, 8035, 8046, 8167, 8273, 8324, 8620, 8796, 9105, 9445, 9892, 10196, 10821, 10917, 11008, 11337, 11482, 12133, 13210, 13227, 13552, 14553, 14570

176: 1162, 1631, 1836, 1879, 2294, 2376, 2499, 2647, 2753, 3008, 3009, 3034, 3536, 4476, 4526, 5309, 5955, 6572, 7032, 8495, 8551, 9625, 9779, 9837, 10725, 11237, 12457, 12589, 12721, 13206, 13457, 13550, 13677, 13995, 14235, 14265

177: 89, 723, 2666, 2696, 2794, 2910, 2997, 3007, 5023, 5863, 6187, 6293, 6498, 6940, 6984, 7306, 8265, 8284, 8396, 9503, 10704, 11088, 11453, 12468, 12470, 12706, 12861, 13142, 13205

178: 151, 787, 906, 916, 1144, 1336, 1392, 1551, 1875, 1878, 2214, 2608, 3624, 3820, 4491, 4492, 4753, 5179, 5202, 6583, 6600, 7259, 7317, 7424, 7444, 8038, 8238, 8438, 8585, 8819, 9351, 9510, 9619, 10806, 11321, 11568, 12722, 12775, 13982, 14034, 14280

179: 112, 1175, 1819, 2744, 2895, 4212, 4490, 5338, 5489, 5575, 5650, 6985, 7232, 7443, 11126, 12197, 13031, 13127, 14392, 14482, 14691

180: 312, 579, 907, 969, 1123, 1642, 1761, 1823, 2570, 3096, 3292, 5006, 5112, 5244, 5554, 5792,

205: 9967, 11760, 12588, 12597, 12715, 14046, 14407, 14903
206: 529, 645, 750, 1362, 1481, 1717, 1826, 2377, 2428, 3005, 5063, 5728, 6435, 8312, 8314, 9680, 9869, 11065, 13221, 13647, 14613, 14629, 14920
207: 568, 1136, 1137, 1361, 1496, 1849, 2484, 2809, 3011, 5109, 9264, 9431, 9637, 9660, 9749, 10789, 11073, 13108, 13987
208: 139, 1141, 1182, 1589, 2643, 2768, 2925, 3379, 3588, 3590, 4519, 5099, 5135, 5332, 7455, 8250, 8257, 8391, 8469, 8628, 8629, 8632, 9104, 10186, 10707, 12346, 12650, 13803, 14552, 14913
209: 461, 700, 1132, 1232, 2288, 2503, 2617, 6008, 6096, 8301, 8582, 8583, 8590, 10794, 14019, 14279, 14761, 14897, 14901
210: 513, 691, 1165, 1265, 2177, 2489, 2538, 2539, 2764, 2769, 2808, 2926, 2978, 3589, 4850, 4951, 5012, 5038, 5056, 5178, 6028, 7297, 8028, 8251, 8604, 8945, 9512, 10188, 11449, 11522, 11755, 12227, 12808, 12841, 12947, 13186, 13644, 13796, 13989, 14387, 14608, 14762
211: 512, 549, 1219, 2127, 3043, 5229, 6899, 7441, 9876, 14416, 14898
212: 1110, 1240, 1394, 1643, 2735, 4786, 5400, 5536, 6149, 6551, 7098, 7476a, 8054, 8473, 8474, 8786, 8855, 9263, 9491, 9499, 9521, 9785, 10199, 12140, 14556
213: 465, 467, 1161, 1236, 2300, 2565, 2638, 2731, 3363, 5278, 6192, 6574, 7308, 8179, 8280, 8442, 9562, 9669, 9870, 11202, 11345, 12422, 12556, 12849, 13701
214: 1292, 2714, 3054, 3518, 4276, 4542, 4706, 5991, 6670, 7311, 8234, 8263, 8416, 8475, 9883, 12354, 12581, 12812, 13084, 13100, 13466, 14395, 14706, 14814
215: 548, 1105, 1202, 1274, 1563, 1635, 2573, 2629, 2832, 5126, 5490, 8278, 8366, 8784, 9538, 10149, 10311, 11759, 12421, 12425, 13459, 14303, 15029
216: 170, 1096, 1450, 1498, 2361, 3085, 3125, 3511, 5231, 6256, 6477, 8053, 9537, 9547, 9959, 10216, 10787, 11519, 12382, 12693, 12932
217: 1323, 1382, 1389, 1391, 2517, 2625, 2812, 3506, 3521, 3818, 4274, 5124, 6255, 7164, 7445, 8021, 8185, 9528, 10197, 12402, 12824, 13815
218: 652, 2581, 3600, 4277, 4546, 6487, 6492, 8516, 9442, 9646, 9827, 10731, 11210, 11350, 11483, 11556, 11570, 12480, 12813, 12934, 12960, 13352, 13628, 14517, 14701, 15023
219: 1342, 2362, 3510, 8237, 9368, 13350, 13692, 13807, 13968, 14390, 14753, 14754
220: 12, 16, 527, 1641, 2423, 2709, 3364, 5193, 5198, 5232, 5245, 5246, 5249, 5760, 5856, 6431, 6558, 6599, 7272, 7422, 7435, 8022, 8034, 9508, 9542, 9663, 9689, 9693, 11761, 12351, 12564, 12823, 13114, 13477, 13723, 14587
221: 1260, 1645, 1646, 4521, 4522, 5230, 5251, 7268, 12964, 13166, 13553, 14039
222: 561, 647, 809, 3449, 5649, 6140, 6253, 7029, 8472, 8680, 9328, 9354, 9557, 10638, 11351, 11444, 12196, 12850, 13695, 14188
223: 744, 1253, 1853, 2733, 3451, 3628, 4520, 5392, 8694, 9506, 11735, 14947
224: 1639, 1797, 2494, 2563, 2821, 5280, 6017, 6512, 7310, 8184, 8800, 9726, 10251, 10792, 11346, 12844, 14156
225: 1121, 1748, 1861, 3215, 4755, 5110, 6002, 6423, 6496, 6878, 8465, 8828, 9426, 9526, 9540, 10336, 10778, 11044, 12348, 12576, 14713
226: 763, 1109, 1210, 1277, 1284, 2528, 3012, 3493, 7158, 8245, 8829, 12577, 14404
227: 2264, 2979, 3095, 6093, 6960, 7040, 8727, 8795, 9149, 11045, 12572, 13847
228: 2221, 2334, 2401, 2795, 2835, 3503, 6548, 6557, 7091, 7269, 8207, 10717, 11561, 12434, 12548, 12761, 13224, 13226, 14336
229: 566, 1242, 2560, 3450, 4827, 5050, 5092, 5954, 6022, 6995, 7033, 7438, 8049, 8144, 10710, 10793, 10993, 11046, 11717, 13081, 13149, 14592
230: 1562, 2161, 2488, 3090, 3522, 3548, 3626, 5274, 5282, 5877, 6146, 6262, 8160, 8591, 10766, 10862, 11052, 11524, 12531, 12574, 12592, 12790, 12846, 12927, 13132, 13776, 14568
231: 1159, 6966, 8149, 9223, 9558, 11450, 13458, 14912
232: 962, 1163, 1201, 1288, 1378, 2249, 2301, 3367, 6006, 6303, 6327, 8202, 10690, 12842, 13946, 14692
233: 2308, 3505, 5878, 6011, 8040, 8204, 10708, 13782, 14923
234: 469, 1428, 2636, 5186, 5301, 5990, 6468, 7276, 8148, 8150, 8243, 8402, 9374, 9555, 10150, 12836, 13617, 14024, 14505
235: 523, 550, 690, 701, 805, 1580, 1621, 1644, 2413, 3060, 3290, 5553, 6273, 6597, 8254, 8481, 8802, 9589, 9594, 12714, 13093, 13115, 13148, 13375, 13487, 14042
236: 1022, 1293, 1430, 2178, 3850, 6580, 7634, 8112, 8728, 9266, 9520, 9818, 10726, 12543, 12567, 12585, 13324, 13619
237: 1158, 2437, 2609, 3586, 3587, 5248, 5948, 6504, 6510, 6550, 7179, 8015, 8041, 8270, 9790, 9844, 10319, 11762, 12838, 12979, 14041
238: 578, 1101, 2290, 2460, 2483, 2653, 2927, 3031, 4467, 6281, 6539, 8468, 9186, 10325, 11504, 11764, 12852, 12970, 14511, 14737
239: 731, 1309, 1340, 1759, 3010, 5052, 5247, 7446, 7454, 8467, 11113, 13673, 13767, 14911
240: 764, 766, 1482, 2630, 2802, 3053, 3475, 4834, 5872, 5975, 8097, 8799, 8804, 9456, 10665, 11327, 12118, 14554, 14889, 14958, 14962
241: 470, 2453, 5059, 6581, 7107, 8070, 8527, 8603, 8605, 9527, 9833, 12547, 12971,

BOILING POINT INDEX OF ORGANIC COMPOUNDS

Temperatures in °C; where there is a range of values, the compound is listed according to the lower value.

−164: 9015
−129: 9093
−103: 6840
−88: 6603
−84: 443, 9089
−83: 4741
−82: 9100
−81: 9047
−79: 6651, 6663
−78: 4715, 9079
−77: 4187
−76: 6876
−72: 14934
−64: 14500
−59: 9033
−57: 13296
−56: 8640
−51: 9067
−50: 4742
−48: 12342
−47: 6676, 12025, 12257
−42: 6659, 11772
−40: 9037
−38: 581, 6618
−37: 6159, 6646
−36: 11880
−35: 6762
−34: 580, 11770
−32: 5824, 14501
−31: 1585, 9088
−30: 445
−29: 9061, 12293
−28: 324
−26: 6668, 6847, 11379
−25: 9155
−24: 4211, 6632, 7055, 9035
−23: 12333
−22: 9084, 14502
−21: 5429, 7047, 9123
−20: 9090, 13274
−14: 11374
−13: 14931
−12: 9026, 9174, 11866
−11: 1415, 3808, 11872
−7: 13285
−6: 3942, 4104, 4189, 6158, 9148
−4: 3638, 5442, 13290
−3: 12286
−2: 6170
0: 3672, 3981, 4014, 6657, 9113, 11843
1: 1584, 1863, 11360, 11867
2: 1411, 11491, 14514
3: 3715, 3980, 5456, 6628, 9020
4: 444, 3972, 5833
5: 3151, 4065, 14928
6: 3657, 4740, 6619, 9124
7: 6156, 7096, 11359
8: 4195
9: 9064, 9797, 11847
10: 3631, 3668, 6809
12: 3655, 3814, 5389, 5431, 5438, 6614, 6775, 9205
13: 5980, 6920
14: 9077
15: 14930
16: 6754, 6814
17: 11385
18: 9029
19: 13176
20: 21, 439, 3576, 3971, 5827
21: 3572, 6867, 9068, 11361
22: 12266
23: 4769, 5416, 11386
24: 6225, 9056
25: 5807, 8161, 9038, 10496, 11371, 11373, 11865, 14953
26: 4064, 7514, 10307, 13323
27: 3777, 4202, 9947, 10422
28: 5830, 14932
29: 1465, 4199, 5829, 10537, 14089
30: 6634, 10517, 10520, 13514
31: 3970, 7083, 7116, 9119
32: 448, 3761, 5966, 6627, 9191, 11777, 12250, 12264
33: 1421, 3667, 4225, 6173, 7189
34: 3661, 4068, 5831, 6165, 6791, 8575, 8642, 9041, 11857, 12300
35: 6704, 6839, 11805, 12254
36: 1407, 1409, 5441, 6166, 6804, 6807, 9023, 9906, 10340, 10539, 10540, 11693
37: 5828, 6863, 6864, 9199, 11368, 11380, 12265
38: 4013, 6610, 6770, 7932, 9168, 9190, 10357
39: 178, 3662, 4198, 6349, 13372, 14494
40: 481, 3636, 4450, 5756, 8661, 9060, 10553, 10618
41: 3964, 3977, 7778, 10316, 11842, 11926
42: 3639, 3646, 5733, 7940, 8981, 9082, 10301, 10595, 11721, 12022, 12263
43: 7134, 12222, 14795
44: 4091, 5806, 5832, 10298, 10619, 12225
45: 610, 6673, 7052, 7097, 7934, 7964, 12267, 12318, 14797
46: 616, 4717, 6625, 6630, 7532, 7537, 11795, 12253, 13513
47: 6675, 6866, 10209, 11773, 12246, 12277, 13866
48: 3335, 7982, 9994, 10308, 11362, 11771, 12042, 12260
49: 3745, 5758, 5762, 6343, 7191, 7968, 10315
50: 435, 4227, 5750, 5826, 6793, 6919, 6922, 7423, 7533, 7542, 7933, 9162
51: 197, 493, 8684, 10382, 10534
52: 480, 494, 1416, 3813, 4194, 4995, 5691, 5837, 6780, 8345, 8359, 11937, 12023, 12221
53: 4056, 5800, 6046, 9080, 10535, 11398, 12278
54: 5743, 7079, 7536, 9197, 10626, 14436
55: 3913, 3963, 4066, 4148, 5310, 5936, 6841, 8361, 9118, 10629, 11796, 12331, 13332, 14929
56: 306, 1463, 3759, 4658, 6092, 7406, 7408, 8654, 9034, 10012, 10562, 10628, 12005, 13269, 14935
57: 225, 2219, 4739, 6626, 7739, 7824, 7854, 7889, 7952, 10019, 10303, 10411, 10907, 12259, 12290, 13245
58: 604, 676, 3746, 3875, 3953, 4146, 4228, 4968, 6640, 8658, 8663, 8907, 9948, 10563, 12203, 12258
59: 3643, 3758, 4988, 5640, 6615, 7731, 8364, 9157, 9169, 10593, 11788, 12199, 12252, 12289, 12332, 14791
60: 3247, 3647, 3664, 3978, 4147, 4971, 5538, 5540, 6806, 6865, 7650, 10406, 10533, 10536, 11437, 13270, 14497
61: 1989, 3805, 5430, 6823, 6923, 7897, 7962, 9050, 9098, 10127, 10627, 11265, 12080, 12128, 14879
62: 326, 3994, 4687, 7044, 7528, 7661, 7665, 7848, 7910, 8292, 9081, 10023, 10119, 10302, 10402, 13507, 13827
63: 442, 1462, 3675, 3676, 3677, 3757, 3902, 3949, 4085, 4086, 4087, 4103a, 4154, 4953, 6825, 6826, 7136, 7543, 7941, 7947, 8078, 10187, 10408, 12060, 12285, 12340, 14794
64: 37, 3743, 3955, 3968, 4035, 4050, 4144, 4210, 4236, 7539, 7781, 9058, 9173, 10011
65: 605, 624, 3571, 6821, 7137, 7407, 7728, 8996, 9127, 9128, 11723, 12248, 12256, 12335, 12539, 14939
66: 613, 1673, 3219, 4052, 4662, 4680, 6100, 6364, 6810, 7960, 10597, 11402, 13299
67: 615, 3560, 3899, 4016, 4153, 4966, 6090, 7142, 7202, 7541,

96: 56, 607, 1420, 2053, 3951, 4053, 4121, 4870, 4913, 5454, 6110, 6658, 7050, 8537, 9013, 10356, 10516, 10596, 11699, 11828, 12666, 14850

97: 34, 53, 190, 596, 2770, 3569, 4011, 4029, 4054, 4089, 4868, 7077, 9053, 9895, 10363, 10525, 10554, 10636, 11942, 12198, 12244, 13288

98: 614, 2042, 3648, 3704, 3851, 3948, 4063, 4088, 4139, 4655, 5625, 6872, 6896, 7076, 7522, 7565, 7604, 7647, 7648, 7997, 8076, 8576, 10071, 10130, 10253, 11409, 11946, 12036, 13662, 13846, 13942, 14440, 14967

99: 438, 501, 2184, 2856, 3642, 3703, 3857, 3888, 3889, 3890, 4075, 4221, 4894, 5184, 5745, 5771, 5813, 5845, 6128, 6355, 6935, 7673, 7828, 9180, 10304, 10419, 10611, 12143, 12308, 13940

100: 168, 169, 170, 707, 830, 3566, 3702, 4129, 4313, 4502, 4869, 4972, 5533, 5545, 5889, 7072, 7730, 7737, 7807, 8121, 8433, 8759, 9009, 9071, 9072, 9085, 10259, 11158, 11383, 11697, 11907, 11965, 12316, 12918, 13258, 14495

101: 262, 572, 3683, 4007, 4095, 4485, 4881, 4885, 5467, 5484, 5522, 6215, 6500, 6647, 6818, 7194, 7267, 8098, 8353, 10052, 10504, 10538, 10624, 11040, 11729, 11786, 13283, 13314

102: 211, 3074, 3681, 3712, 3905, 4124, 4351, 4445, 5551, 5688, 6697, 6902, 7119, 7241, 7994, 9788,

102: 9923, 10495, 10586, 10621, 10633, 11868, 12288, 12306, 13301, 13574, 13761

103: 35, 176, 3254, 3982, 4433, 4958, 5773, 5809, 6716, 7120, 7785, 8064, 8219, 8650, 8961, 10157, 10339, 10526, 10550, 11849, 12150, 12215, 13503, 14790

104: 3678, 3731, 3935, 4203, 4220, 4418, 4976, 5381, 5686, 5687, 6108, 6834, 7678, 9032, 9956, 10341, 10565, 10599, 10610, 11394, 12105, 12279, 13894

105: 1402, 1691, 1968, 1974, 3810, 3885, 3946, 4165, 4282, 4600, 4916, 4974, 5580, 5940, 5962, 6207, 6417, 7155, 7207, 7215, 7227, 7244, 7681, 9341, 9836, 9893, 10449, 10888, 11730, 11827, 11996, 12085, 12204, 13039, 13610, 13836, 13864, 13924

106: 612, 3331, 3565, 3796, 3926, 4130, 4201, 5462, 5585, 5854, 5890, 7075, 7798, 8363, 8490, 8810, 9802, 9886, 10083, 11404, 11578, 11617, 11658, 11785, 12033, 12165, 14657

107: 343, 495, 876, 1969, 3641, 3831, 4219, 4317, 4665, 4738, 4763, 4940, 4962, 5469, 6611, 6777, 7139, 7869, 8120, 8226, 9066, 10152, 10354, 10519, 10551, 11663, 11698, 11732, 11890, 14877, 14952

108: 436, 440, 2509, 3224, 3696, 3811, 4070, 4082, 4109, 4112, 4166, 4598, 4691, 4957, 5103, 5185, 5420, 5765, 5812, 6069, 6101, 6621, 6760, 6763, 6862, 7511, 7787, 10202,

108: 10438, 11726, 11830, 11963, 12191, 13845

109: 221, 258, 1953, 2130, 3203, 3632, 3653, 3992, 4167, 4307, 4464, 4879, 4977, 5630, 6164, 6331, 6772, 7056, 7144, 7802, 7804, 8965, 9044, 10546, 11970, 12043, 12189, 13331, 14428

110: 24, 222, 1405, 1752, 1970, 2536, 3146, 3991, 4436, 4657, 4792, 4880, 4938, 4965, 5685, 6351, 6613, 7110, 7248, 7510, 7521, 7663, 7803, 7966, 8260, 8489, 9000, 9027, 10064, 10122, 10335, 10360, 10418, 10542, 10799, 11396, 12087, 12095, 12166, 12315, 12338, 13294, 13341, 13434, 14060

111: 3113, 3658, 3714, 3763, 3990, 4155, 4407, 5581, 5622, 5903, 5963, 6341, 6345, 7058, 7801, 9087, 10361, 11434, 12414, 14834

112: 187, 188, 345, 627, 630, 3708, 3900, 3909, 3910, 3914, 3960, 4150, 4622, 4887, 4954, 5964, 6068, 6130, 6346, 6402, 6664, 6817, 6861, 7172, 7515, 7677, 7805, 7958, 8220, 8476, 9342, 10413, 10448, 10457, 10617, 11254, 11886, 11995, 12051, 12280, 12504, 12612, 12665, 13496, 13497, 13901, 14130, 14799, 14828

113: 3723, 4051, 4315, 4983, 5106, 5475, 5768, 5782, 6674, 6706, 7800, 7944, 9798, 10013, 10421, 11957, 12010, 12344, 13265, 13430, 14643

114: 282, 315, 1089, 2120, 3701, 3901, 4111, 4396, 4457, 5825, 5857, 6211, 6645,

114: 7141, 7176, 7762, 7870, 8357, 9014, 9799, 10069, 10108, 10420, 10509, 10576, 11041, 11395, 11825, 11903, 12039, 12299, 12461, 12639, 13926, 14127, 14402, 14529, 14633

115: 46, 55, 194, 535, 3616, 3700, 3920, 3962, 4823, 4973, 4982, 4984, 5482, 5508, 5689, 5783, 5905, 5932, 6208, 6470, 6656, 6803, 7126, 7568, 7577, 7799, 8477, 8876, 9120, 9159, 9236, 9429, 9611, 10072, 10368, 10400, 10543, 10584, 11549, 12339, 12392, 12441, 12667, 13236, 13902, 14135, 14210, 14457, 14465

116: 189, 506, 2051, 2507, 3249, 3699, 3733, 3940, 3999, 4077, 4306, 4365, 4443, 4458, 4949, 5511, 6138, 6783, 6849, 7206, 9150, 9227, 10369, 10474, 10502, 10564, 11776, 11813, 11993, 12123, 13498, 13751, 13841, 13873, 14043, 14128, 14129, 14796, 14830

117: 175, 186, 1843, 2026, 3245, 3772, 3860, 3921, 3947, 4036, 4069, 4164, 4229, 4452, 5330, 5474, 5664, 6091, 6099, 6109, 7562, 7566, 7570, 7571, 7763, 7806, 9008, 10203, 10350, 10544, 10614, 11664, 11665, 12220, 12952, 13879, 14631

118: 347, 1398, 2523, 3775, 3836, 3852, 3853, 3903, 4401, 4421, 4944, 5465, 5732, 6827, 6854, 7063, 7646, 7808, 10015, 10032, 10352, 10384, 10401, 10466, 10501, 10549, 10579, 10589, 11728, 11783, 11789, 11992, 12032, 12167,

233: 72, 797, 1670, 1976, 2083, 2723, 3934, 6059, 6352, 6448, 7200, 7394, 8783, 10827, 10890, 11905, 13914, 14032

234: 755, 995, 1068, 1128, 1973, 2034, 2596, 2968, 4804, 5818, 7672, 8860, 9344, 10428, 11169, 11660, 11925, 12182, 12241, 12642, 13082, 13125, 13601, 13602, 13775, 13944, 14282

235: 180, 1040, 1393, 1703, 1796, 1919, 1931, 2044, 2057, 2283, 2336, 2854, 3171, 3833, 3883, 4009, 4760, 5311, 5346, 5349, 5350, 5413, 5557, 6038, 6182, 6474, 6691, 7230, 7400, 8398, 8809, 8846, 8901, 9995, 10162, 10280, 10774, 10834, 10840, 10924, 10978, 11290, 11399, 11530, 11710, 11983, 12594, 12610, 12995, 13312, 13536, 14294, 14342, 14460, 14528, 14532

236: 3218, 3241, 3271, 4387, 6386, 7307, 7322, 7326, 8284, 8613, 10223, 10566, 10819, 10915, 11969, 12356, 13599, 14095, 14471

237: 1063, 1456, 1807, 1920, 1932, 2081, 2092, 2614, 4673, 4757, 5700, 5860, 6913, 7709, 8099, 8748, 9577, 9578, 10140, 10750, 10911, 10980, 10984, 11614, 12673, 13758, 14313, 14364

238: 86, 799, 800, 976, 1975, 2295, 3103, 3140, 3340, 4208, 4281, 4431, 4805, 6174, 6856, 7381, 8153, 8445, 8851, 9201, 9489, 10761, 10762, 10826, 10866, 11439, 12242, 12719, 13952, 13953, 14080,

238: 14149, 14427, 14679, 14760

239: 877, 937, 1061, 1900, 4623, 4624, 5757, 7352, 10078, 10842, 11089, 13155, 13757

240: 414, 427, 801, 1027, 1082, 1490, 1648, 1749, 1902, 3179, 4489, 4582, 4631, 4785, 4930, 5150, 5699, 5901, 5916, 6383, 6449, 6888, 7989, 8100, 8418, 8590, 8893, 8988, 9829, 9989, 10043, 10838, 10841, 11147, 11293, 11900, 11917, 12838, 12853, 12928, 13156, 13246, 14085, 14198, 14308, 14329, 14352, 14386, 14531, 14983, 15011

241: 370, 791, 947, 3246, 4084, 4555, 4718, 4797, 4808, 5348, 6900, 8375, 9400, 12692, 14104, 14312, 14332

242: 1930, 2045, 2148, 2542, 2588, 3144, 3547, 4106, 5731, 6535, 8544, 8581, 9125, 9813, 10171, 11173, 11188, 11236, 14303, 14356, 14635

243: 392, 746, 910, 932, 998, 1021, 1094, 1774, 1928, 2095, 2243, 2595, 2604, 3148, 3263, 4191, 4562, 6665, 8113, 8194, 8859, 11377, 11555, 12181, 13057, 14646

244: 198, 1062, 2567, 3197, 4369, 4496, 4507, 5183, 5723, 7839, 8124, 8125, 9391, 11186, 11425, 11966, 13012, 13759, 14115, 14182, 14370

245: 378, 413, 422, 532, 836, 925, 1005, 1026, 1658, 1734, 1896, 2050, 2103, 2149, 2164, 3073, 3138, 3139, 3578, 4428, 4720, 4758, 4788, 5045, 5339, 5366, 5428, 5676, 5843, 6111, 6113, 6114,

245: 6214, 7598, 7786, 7930, 8542, 8597, 8746, 8849, 9138, 9812, 10759, 11035, 11080, 11187, 11420, 11902, 12232, 12243, 12611, 12869, 12959, 13003, 13295, 13543, 13633, 14011, 14033, 14311, 14372, 14419, 14588, 14589, 14842, 14862

246: 15, 389, 426, 2043, 2242, 3195, 4593, 5237, 5386, 6150, 6401, 6800, 7432, 8183, 8189, 8222, 8223, 8594, 9814, 10932, 11030, 12359, 12371, 13005, 14096, 14187, 14999

247: 368, 1702, 1733, 1745, 3114, 3442, 4852, 4860, 4861, 8152, 8340, 9600, 9705, 10647, 11018, 11110, 11863, 12627, 12877, 12878, 12891, 13090, 13678, 14103

248: 912, 3088, 3479, 4159, 4413, 5201, 6107, 6224, 8593, 10736, 10837, 10839, 10919, 11027, 12614, 12960, 13184, 14319, 14669, 14875

249: 376, 423, 1000, 1737, 1777, 2168, 2531, 3183, 4495, 4985, 5158, 6329, 6957, 10760, 10905, 11864, 11967, 13111, 13754

250: 1, 352, 1901, 2245, 2587, 2940, 3193, 3766, 4375, 4488, 5011, 5195, 5327, 6230, 6570, 7104, 7337, 7358, 7503, 7582, 7586, 7597, 7840, 8068, 8133, 8423, 8437, 8655, 8861, 9251, 9754, 10773, 10951, 10997, 11001, 11074, 11142, 11274, 11366, 11440, 11475, 11781, 12304, 12484, 12671, 13379, 13520, 13560, 13650, 13737, 14336, 14365, 14671

251: 777, 837, 850, 931, 1017, 1514, 2244, 2431, 2684, 2852, 4438, 5692, 6745, 8862, 11177, 12907, 13002, 14108

252: 835, 1008, 1080, 1955, 1956, 3181, 5450, 6213, 6410, 9141, 11314, 11421, 11572, 12455, 13271, 13743, 14658

253: 161, 243, 1522, 3480, 5070, 6354, 6931, 8608, 8844, 10775, 10889, 11139, 11365, 13726, 14309, 14320, 14661, 14907

254: 913, 935, 1093, 1400, 1885, 2241, 2566, 2593, 2737, 3147, 3378, 3433, 3561, 3564, 4578, 5472, 8411, 8776, 10740, 11550, 12994, 13537, 14354, 14874, 14984

255: 361, 983, 1091, 1177, 2076, 2085, 2090, 2218, 2284, 2746, 2872, 3072, 3369, 3487, 4486, 4568, 5318, 6077, 6409, 8372, 8674, 8720, 8848, 9932, 10267, 10874, 10985, 11294, 11357, 12482, 12768, 12819, 13722, 13773, 14088, 14157, 14330

256: 424, 1012, 1092, 1178, 1743, 1963, 2061, 2873, 3288, 3458, 6212, 6906, 7399, 7858, 8595, 9145, 9306, 10820, 10943, 11211, 11364, 11702, 12617, 12774, 13021, 13191, 13549, 14075

257: 603, 948, 1954, 2084, 3173, 3301, 5160, 5162, 6441, 7850, 8325, 8414, 8631, 10290, 10946, 11020, 11234, 11249, 11677, 12890, 13909, 14654

258: 136, 328, 415, 1028, 1066, 1722, 1961, 2093, 2338, 2743, 2858, 2865, 4374, 7390, 7667, 8598, 9299, 9377, 9378, 12238, 12411, 12887,

258: 13062, 14260, 14361, 14398
259: 71, 982, 2691, 3184, 5187, 6037, 6394, 6765, 8721, 9237, 10898, 11656, 12882, 12988, 14036
260: 164, 203, 372, 747, 749, 813, 842, 847, 965, 1007, 1608, 1912, 2256, 2683, 2720, 2864, 3464, 3549, 4642, 5008, 5359, 5439, 5457, 6397, 6609, 7330, 8057, 9140, 9229, 10194, 10431, 11412, 11414, 11424, 11426, 11565, 12021, 12903, 12905, 12916, 13104, 13107, 13420, 13917, 13928, 14079, 14086, 14413
261: 798, 1004, 2575, 3133, 4592, 5130, 7593, 8156, 8386, 8572, 9137, 9979, 12214, 12772, 12810, 13538, 14478, 14859
262: 174, 828, 990, 1095, 2582, 2741, 3149, 4824, 7551, 7576, 8154, 8712, 10977, 11413, 11914, 12607, 12776, 12886, 13597, 14411, 14737
263: 166, 875, 1189, 1735, 2849, 2859, 5159, 5319, 7545, 8173, 8329, 9261, 9358, 9363, 9439, 10042, 10871, 11438, 11714, 12193, 13738, 14110, 14275, 14476, 14507, 14726, 14873
264: 878, 2686, 3071, 4524, 5631, 6282, 6747, 9074, 9361, 9755, 10058, 10912, 11416, 12811, 12818, 12883, 13011, 14668
265: 20, 528, 1226, 1962, 2167, 2217, 3131, 3132, 4501, 4505, 4616, 4719, 5082, 6387, 6577, 7838, 8420, 8421, 8596, 8643, 9284, 9360, 10062, 10878, 10895, 10986, 11102, 11164, 11217, 11899, 12179, 12180, 12764, 12825, 12935, 13023, 13606, 14092, 14373, 14896
266: 173, 1176, 2745, 4338, 5321, 9357, 9910, 11016, 12816, 12854, 13281, 14491
267: 94, 329, 514, 1038, 1440, 2063, 2091, 2659, 3489, 4700, 7670, 8382, 10158, 10178, 10763, 10996, 11197, 11226, 12240, 12779, 12855, 12956, 13302, 13413, 14077, 14105, 14159
268: 852, 874, 1169, 2097, 2660, 5086, 5155, 7585, 8879, 9359, 9365, 9425, 10295, 10296, 11311, 11879, 12826, 13200, 13531, 14731
269: 1011, 8785, 9144, 9171, 9389, 9756, 10159, 10873, 12773, 13106, 14073
270: 257, 377, 397, 458, 849, 975, 1188, 1288, 1603, 2116, 2717, 2738, 3161, 3191, 3460, 3461, 3765, 4379, 4521, 4550, 4818, 4820, 5917, 5947, 7642, 7719, 8283, 8765, 9364, 9970, 10044, 10285, 10644, 11427, 11592, 11834, 12521, 12555, 12752, 13103, 13179, 13489, 13903
271: 2263, 4686, 5122, 6087, 6141, 7065, 7319, 11031, 12643
272: 841, 949, 2704, 2779, 3488, 3530, 4327, 4415, 4871, 5269, 6686, 7575, 7688, 7693, 7899, 8419, 9222, 10876, 11571, 11835, 13126, 14158
273: 371, 671, 771, 3253, 3462, 3469, 4083, 5001, 7361, 8710, 9831, 11719, 11839, 12420, 12522, 14090, 14318, 14666, 15000
274: 834, 1067, 1785, 2661, 3403, 3485, 4466, 4497, 5297, 8495, 8517, 8588, 9172, 9298, 9390, 9424, 9706, 11076, 12750, 13548
275: 457, 839, 1039, 1412, 1980, 2262, 2688, 2759, 3150, 3306, 4606, 5032, 5427, 6749, 7249, 7475, 7607, 9142, 9847, 9850, 9874, 11780, 12111, 13029, 13667, 13668, 13748, 14290, 14461, 14667
276: 748, 846, 1065, 2727, 3476, 3737, 6395, 7592, 8330, 8571, 9258, 9808, 10598, 10892, 12463, 12998, 13153, 13276, 13755, 14169
277: 964, 2223, 2224, 2724, 4635, 6124, 6669, 6870, 8228, 8717, 11461, 12996, 14481
278: 1709, 1899, 7067, 11025, 12730, 12758, 12991, 13060, 14487, 14490
279: 4, 822, 3620, 6115, 6309, 8147, 8870, 10990, 11622
280: 884, 927, 1911, 2174, 2220, 2713, 3308, 3465, 3532, 3736, 4551, 4629, 4690, 4784, 6308, 6798, 6811, 7360, 7587, 8248, 8326, 8388, 8422, 8587, 8634, 9375, 9753, 9817, 9832, 10215, 11263, 11853, 11854, 12283, 12501, 12523, 12754, 12792, 13373, 13655, 13657, 13660, 13661, 14637, 14865
281: 67, 99, 2927, 3467, 5148, 6206, 9291, 11526, 12500, 12814, 12911, 13656, 13805
282: 2062, 2952, 6388, 6812, 7220, 8508, 8558, 9376, 10224, 10872, 11023, 11221, 11570, 11856, 12870, 12875, 13061, 13530, 14678
283: 3389, 3478, 5317, 5808, 7641, 8424, 10765, 10859, 10922, 11473, 12874, 12906, 12975, 12980, 13313, 14093
284: 952, 3404, 3496, 4840, 4981, 7715, 7721, 10764, 14375
285: 538, 1763, 2080, 2105, 2110, 2111, 2624, 2907, 3212, 3390, 5154, 6268, 6371, 6642, 7118, 7245, 8169, 8221, 8282, 8415, 8494, 9334, 9337, 9387, 9704, 10124, 11060, 11514, 11517, 11718, 12909, 12910, 13855, 14376, 14890
286: 385, 503, 703, 2109, 2498, 3752, 4523, 5658, 6310, 6641, 6737, 7342, 8224, 8637, 9331, 10121, 10139, 10998, 11185, 11611, 12464, 12767, 12915, 13666, 14644, 14645
287: 1453, 4430, 4839, 5174, 5992, 7701, 9294, 12423, 12427, 12465, 12913, 13588
288: 853, 2173, 2208, 2850, 3468, 5131, 5537, 5657, 6463, 7066, 8331, 9388, 9571, 10000, 11284, 12782
289: 353, 2175, 3753, 5398, 6036, 6320, 7357, 7707, 9296, 9765, 11410, 12284, 12961, 13753, 14109
290: 1809, 3338, 4382, 4634, 5123, 5748, 5862, 6044, 7354, 7596, 8200, 9564, 10277, 11054, 11415, 11528, 12654, 12760, 12821, 13892, 14181
291: 1454, 1834, 2125, 3406, 3533, 8281, 9314, 9330, 9362, 9559, 10294, 11836, 12653, 13727
292: 988, 5288, 8349, 8523, 8729, 9269, 9297, 13729, 14254
293: 354, 1135, 6319, 6987, 8389, 9422, 9712, 10860, 11372, 11512, 12313, 14473
294: 689, 1338, 1791, 2506, 5173, 9287, 10293, 11495

FORMULA INDEX FOR ORGANIC COMPOUNDS

FORMULA INDEX FOR ORGANIC COMPOUNDS (Continued)

$C_6H_4N_2O_5$: 10942

$C_6H_4N_2O_6$: 13000

$C_6H_4N_2O_7S$: 02412, 02413

$C_6H_4N_4$: 12376

$C_6H_4N_4O_2$: 02198, 02199, 02200, 09819

$C_6H_4N_4O_6$: 00997

$C_6H_4O_2$: 03016, 03023

$C_6H_4O_3$: 07164

$C_6H_4O_4$: 03043, 05262

$C_6H_4O_4S$: 13935, 13936

$C_6H_4O_4S_2$: 13937

$C_6H_4O_5$: 07158, 07160, 07162, 12618

$C_6H_4O_6$: 03066, 04864

$C_6H_4S_2$: 13960, 13961

$C_6H_5AsCl_2$: 01400

$C_6H_5AsF_2$: 01405

C_6H_5AsO: 01413

$C_6H_5BF_2$: 03569

C_6H_5Br: 01921

$C_6H_5BrN_2O_2$: 00775, 00785, 00786, 00788, 00790

$C_6H_5BrN_3O_4$: 00778

C_6H_5BrO: 10814, 10819, 10826

C_6H_5BrOS: 13864

$C_6H_5BrO_2$: 04791, 08172, 12982, 12983, 12984

$C_6H_5BrO_2S$: 02351

$C_6H_5BrO_3S$: 02396, 02397

C_6H_5BrS: 13948

$C_6H_5Br_2N$: 00824, 00825, 00827, 00828, 00830, 00831

$C_6H_5Br_2NO$: 10782, 10783, 10805

$C_6H_5Br_3N_2$: 08126, 11275

$C_6H_5Br_4N$: 00986

C_6H_5Cl: 02000, 07931

$C_6H_5ClN_2$: 02330

$C_6H_5ClN_2O_2$: 00806, 00810, 00814, 00815, 00818, 00819, 00820, 00821

$C_6H_5ClN_2O_3$: 10778, 10780, 10803

C_6H_5ClO: 10843, 10850, 10852

C_6H_5ClOS: 02356, 13865

$C_6H_5ClO_2$: 04792, 04793, 08173, 12987, 12988, 12989

$C_6H_5ClO_2S$: 02352, 02431

$C_6H_5ClO_3$: 07233

$C_6H_5ClO_3S$: 02404

C_6H_5ClS: 02344, 13950, 13951

$C_6H_5Cl_2N$: 00835, 00836, 00837, 00838, 00841, 00842

$C_6H_5Cl_2NO$: 10784, 10797, 10806, 10807, 10808

$C_6H_5Cl_2NO_2S$: 02358

$C_6H_5Cl_2N_2O_3$: 10779

$C_6H_5Cl_2OP$: 02338

$C_6H_5Cl_2P$: 11367

$C_6H_5Cl_2Sb$: 13434

$C_6H_5Cl_3$: 07933

$C_6H_5Cl_3N_2$: 08127, 11276

$C_6H_5Cl_3Si$: 13311

$C_6H_5Cl_4P$: 02339

$C_6H_5Cl_5$: 02223, 05689, 05690

C_6H_5F: 02153

C_6H_5FO: 10952, 10953

$C_6H_5FO_2S$: 02450

$C_6H_5F_3Si$: 13314

$C_6H_5F_7O_2$: 04334

C_6H_5I: 02169

C_6H_5IO: 02177, 10957, 10958, 10959

$C_6H_5IO_2$: 02178, 04806, 08192, 13007, 13008

$C_6H_5I_2N$: 00858, 00860, 00861, 00864, 00865

C_6H_5NO: 02219, 07242, 08524, 12540, 12542, 12544

$C_6H_5NO_2$: 02194, 08525, 09818, 10991, 11554, 12541, 12543, 12545

$C_6H_5NO_2S$: 13956, 13957

$C_6H_5NO_3$: 09824, 09825, 09826, 10988, 10989, 10990, 13016

$C_6H_5NO_3S$: 02401

$C_6H_5NO_4$: 04817, 08195, 12656, 13014, 13015

$C_6H_5NO_4S$: 02354, 02355

$C_6H_5NO_5S$: 02425, 02426, 02427

$C_6H_5NO_6S$: 02419

$C_6H_5N_3$: 02260, 03115

$C_6H_5N_3O_3$: 02332

$C_6H_5N_3O_4$: 00906, 00907, 00908, 00909

$C_6H_5N_3O_5$: 10785

$C_6H_5N_5O$: 12378

$C_6H_5N_5O_2$: 12377, 14967

$C_6H_5N_5O_6$: 08128

C_6H_6: 01882, 07096, 07749, 07752, 07753, 07755

$C_6H_6AsClO_3$: 02302, 02303, 02304

$C_6H_6AsNO_5$: 02308, 02309, 02310

C_6H_6BrN: 00770, 00777, 00779

$C_6H_6BrNO_2S$: 02366, 02367, 02368

$C_6H_6Br_6$: 05535, 05536

C_6H_6ClN: 00803, 00804, 00807, 00811

C_6H_6ClNO: 02032, 02033, 10777, 10781, 10796, 10802, 10804

$C_6H_6ClNO_2$: 09823

$C_6H_6ClNO_2S$: 02372, 02373, 02374

$C_6H_6ClNO_3S$: 02381, 02382

$C_6H_6Cl_2$: 07934

$C_6H_6Cl_2N_2$: 08075, 11212, 11213, 11223, 11224, 11228, 11229

$C_6H_6Cl_2Si$: 13262

$C_6H_6Cl_4$: 07742

$C_6H_6Cl_6$: 05537, 05538, 05539, 05540, 07966

$C_6H_6Cl_9O_3P$: 11438

C_6H_6FN: 00922, 00923, 00924

$C_6H_6FNO_2S$: 02375

C_6H_6IN: 00926, 00927, 00928

$C_6H_6N_2$: 03046

$C_6H_6N_2O$: 00957, 09816, 11553

$C_6H_6N_2O_2$: 00951, 00952, 00954, 00955, 03053, 11278, 14781

$C_6H_6N_2O_3$: 10787, 10788, 10789, 10790, 10798, 10809, 10810

$C_6H_6N_4O_2$: 14950

$C_6H_6N_4O_3$: 14774, 14775, 14776, 14777

$C_6H_6N_4O_4$: 07179, 08088, 08089

$C_6H_6N_8O$: 09231

C_6H_6O: 10767

C_6H_6OS: 10960, 10961, 10962, 13863, 13926

$C_6H_6O_2$: 02103, 02104, 02105, 04788, 07117, 07177, 08169, 12974

$C_6H_6O_2S$: 00265, 02348, 13866, 13931

$C_6H_6O_2Se$: 02341

$C_6H_6O_2Si$: 02342

$C_6H_6O_3$: 00211, 02274, 02276, 02278, 07156, 07157, 07176, 07210, 07211, 07214, 07215, 07235, 07236, 07238, 07239, 07241, 08293, 11350, 12604, 12615

$C_6H_6O_3S$: 02378, 02379

$C_6H_6O_4$: 02248, 02249, 04214, 07154, 07209, 07743, 07744, 09240, 09241, 12616

$C_6H_6O_4S$: 02417, 02418

$C_6H_6O_6$: 00455, 00456, 02165, 05856, 08930, 08931, 12027, 12028

$C_6H_6O_7S_2$: 02452

$C_6H_6O_9S_3$: 02467

C_6H_6S: 13943

$C_6H_6S_2$: 04805, 08188, 13003

C_6H_6Se: 11011

C_6H_7: 07751

$C_6H_7AsO_3$: 02298

$C_6H_7AsO_4$: 02305, 02306, 02307

$C_6H_7BO_2$: 11210

$C_6H_7BrN_2$: 08066, 08067, 11272, 11273

$C_6H_7BrO_2$: 05672

$C_6H_7ClN_2$: 02003, 02004, 02005, 02006, 02007

C_6H_7ClO: 13361

$C_6H_7ClO_4$: 07105, 08810

C_6H_7ClSi: 13242

$C_6H_7Cl_2F_3O_2$: 12124

$C_6H_7Cl_2N$: 00805, 00809

$C_6H_7Cl_3$: 07730

$C_6H_7F_3O_3$: 00301

C_6H_7N: 00765, 11540, 11546, 11548, 12516

C_6H_7NO: 10776, 10791, 10799, 11277, 12503, 12504, 12511, 12514, 12515, 12624, 12625

C_6H_7NOS: 13862

$C_6H_7NOS_2$: 13053

$C_6H_7NO_2$: 00856, 00857, 08526

$C_6H_7NO_2S$: 02357

$C_6H_7NO_3$: 03020

$C_6H_7NO_3S$: 02359, 02376, 02380, 02387, 02391, 13624

$C_6H_7NO_4S$: 02390

$C_6H_7NO_6S_2$: 02451

C_6H_7NS: 13944, 13945, 13947

$C_6H_7N_3O$: 08530

$C_6H_7N_3O_2$: 08108, 08109, 08110, 11218, 11219

$C_6H_7O_2P$: 02335, 11295

$C_6H_7O_2S$: 13932

$C_6H_7O_3P$: 02337

$C_6H_7O_3Sb$: 02343

$C_6H_7O_4P$: 11408

C_6H_7P: 11375

C_6H_8: 05483, 05486, 07935, 07936, 07981, 07983,

C_6H_8: 10594, 10597
$C_6H_8AsNO_3$: 02299, 02300, 02301
C_6H_8BrNO: 10792
$C_6H_8Br_2O$: 05623
$C_6H_8Br_2O_3$: 00287
C_6H_8ClN: 00925
C_6H_8ClNO: 10793
C_6H_8ClNS: 13946
$C_6H_8Cl_2$: 07740
$C_6H_8Cl_2O_2$: 00546
$C_6H_8Cl_2O_3$: 00288
$C_6H_8Cl_2O_4$: 06884, 10173
$C_6H_8Cl_2O_5S$: 02408, 02409
C_6H_8INO: 10794
$C_6H_8N_2$: 00545, 02061, 02062, 02063, 07754, 08113, 08114, 11211, 11221, 11226, 11541, 11547, 11549, 11550, 12393, 12395, 12396, 12517, 12518, 97784
$C_6H_8N_2O$: 10861, 10863, 10864, 10865
$C_6H_8N_2O_2S$: 02363, 02364, 02365, 13622, 13816
$C_6H_8N_2O_3$: 00950
$C_6H_8N_2O_3S$: 02406, 02407, 02420
$C_6H_8N_6O_{18}$: 08925
$C_6H_8N_{11}$: 08946
C_6H_8O: 05635, 05713, 07128, 07129, 07132, 07724, 07985, 07986, 13355
$C_6H_8O_2$: 00487, 04026, 04216, 05667, 05671, 05674, 06795, 07185, 07196, 07745, 10259, 13357
$C_6H_8O_3$: 05788, 07108, 07127
$C_6H_8O_4$: 05444, 05446, 05447, 05448, 05449, 05450, 05451, 07107, 07861, 08673, 08674, 08675, 08814, 08833, 08902
$C_6H_8O_5$: 00543, 00544, 01427
$C_6H_8O_6$: 01424, 01425, 01426, 07306, 11938
$C_6H_8O_7$: 05170, 08607, 11940
$C_6H_8O_8$: 11939
C_6H_8S: 13888, 13889, 13890, 13891, 13894, 13895
C_6H_8Si: 13309
C_6H_9Br: 05680, 05681,
C_6H_9Br: 05766
$C_6H_9BrO_2$: 04029
$C_6H_9BrO_3$: 00281, 00282
$C_6H_9BrO_4$: 13542
C_6H_9Cl: 05682, 07727, 07982, 10300, 10302, 10303, 10621, 10629
$C_6H_9ClN_2$: 08115
C_6H_9ClO: 05626, 05627, 05628
$C_6H_9ClO_2$: 04019, 04021, 04034
$C_6H_9ClO_3$: 00285, 00286
$C_6H_9Cl_3$: 07953
$C_6H_9Cl_3O_2$: 04388, 14433, 14434, 14435, 14436
$C_6H_9Cl_3O_3$: 04862
$C_6H_9Cl_5$: 07823
$C_6H_9F_3O_2$: 14495, 14496
C_6H_9N: 12630, 12632, 12635, 12636
C_6H_9NO: 13356
C_6H_9NOS: 13811
$C_6H_9NO_2$: 07467, 08528, 11557, 13599
$C_6H_9NO_3$: 00070, 14513
$C_6H_9NO_4$: 10189
$C_6H_9NO_4S$: 00978
C_6H_9NS: 05737
$C_6H_9N_3$: 02258, 02259, 12211, 12581, 12582, 12587, 12592
$C_6H_9N_3O$: 11020
$C_6H_9N_3O_2$: 08010, 08011, 08012
$C_6H_9N_3O_3$: 00454, 08495
$C_6H_9O_2$: 04025
$C_6H_9O_3$: 00279, 06752
$C_6H_9O_4$: 00031
C_6H_{10}: 03649, 04198, 05678, 05815, 05816, 07725, 07726, 07728, 07731, 07738, 07991, 07993, 07995, 10306, 10627, 10631, 12269
$C_6H_{10}Br_2$: 05508, 05509, 05510, 05511, 05512, 05514
$C_6H_{10}Br_2O_2$: 04315, 04316
$C_6H_{10}ClNO$: 04913, 12665
$C_6H_{10}ClNO_2$: 04919
$C_6H_{10}ClN_2O_3$: 08017
$C_6H_{10}Cl_2$: 05513, 05515, 05516, 07946, 07947
$C_6H_{10}Cl_2N_2O$: 10862
$C_6H_{10}Cl_2O_2$: 06062, 12128
$C_6H_{10}Cl_3NO$: 04875, 14428
$C_6H_{10}CrO_2$: 04813
$C_6H_{10}I_2O_7S$: 02410
$C_6H_{10}NO_4P$: 00969
$C_6H_{10}N_2$: 12418
$C_6H_{10}N_2O_2$: 05670, 05675
$C_6H_{10}O$: 00611, 03653, 03654, 04124, 05452, 05528, 05621, 05707, 05712, 05784, 05802, 05803, 05804, 06902, 07746, 07747, 07939, 07940, 07976, 07980, 08991, 10515, 10589, 10592, 10635, 13360
$C_6H_{10}OS$: 00627
$C_6H_{10}OS_2$: 00582
$C_6H_{10}O_2$: 00184, 00350, 00525, 04023, 04037, 04038, 04043, 04418, 04597, 04598, 04614, 05387, 05390, 05637, 05785, 07735, 07736, 07737, 07837, 07846, 07971, 07972, 08001, 09008, 10568, 10569, 12073, 14797
$C_6H_{10}O_2S$: 04386
$C_6H_{10}O_2S_2$: 06691
$C_6H_{10}O_2S_4$: 14952
$C_6H_{10}O_3$: 00290, 00291, 00292, 00297, 04044, 04326, 04607, 06223, 06751, 06753, 07848, 07867, 08654, 08751, 12074
$C_6H_{10}O_4$: 00528, 00541, 06886, 06921, 07281, 07334, 07838, 08517, 08887, 08899, 09212, 10175, 13558, 13562, 13564, 13565
$C_6H_{10}O_5$: 04829, 05959, 06153, 07252, 07416, 07462, 07463, 08655, 08744, 13377
$C_6H_{10}O_6$: 00586, 07250, 07283, 07284, 08320, 08928, 08929, 13642, 13660, 13661, 13662, 13669
$C_6H_{10}O_7$: 07254, 07287, 07305, 08937, 08938
$C_6H_{10}O_8$: 00585, 09239, 13080
$C_6H_{10}S$: 00626, 05648
$C_6H_{10}S_3$: 00630, 05964
C_6H_{11}: 05836
$C_6H_{11}Br$: 05495, 05729
$C_6H_{11}BrN_2O_2$: 14710
$C_6H_{11}BrO_2$: 03603, 03604, 04284, 04290, 04398, 04582, 04583, 04585, 04586, 14820, 14821
$C_6H_{11}Cl$: 03990, 05506, 05730, 07943, 07944, 07945, 07956, 07962, 07965, 10524, 10552
$C_6H_{11}ClO$: 04453, 04454, 04455, 04456, 04617, 05582, 05583, 05584, 14883, 14884, 14885
$C_6H_{11}ClO_2$: 04300, 04307, 04310, 04402, 04925, 04926, 04927, 04970, 04971, 12099, 12110
$C_6H_{11}Cl_2N_3O_2$: 08015, 08016
$C_6H_{11}Cl_3O_2$: 00054, 14444
$C_6H_{11}F$: 05533
$C_6H_{11}FO_2$: 04591
$C_6H_{11}I$: 05544
$C_6H_{11}IN_2O_2$: 14747
$C_6H_{11}N$: 03335, 03776, 04425, 04615, 05963, 10604, 14872
$C_6H_{11}NO$: 04575, 04611, 05644, 11695, 12213, 12672, 12673
$C_6H_{11}NO_2$: 00060, 04027, 05552, 07845, 09226, 11696, 12069
$C_6H_{11}NO_3$: 00526, 10190
$C_6H_{11}NO_4$: 00529, 12158
$C_6H_{11}NO_5$: 07285
$C_6H_{11}NS$: 10605, 10606, 10616, 13827
$C_6H_{11}N_2O_3$: 00557
$C_6H_{11}N_3O_4$: 05178, 14504
$C_6H_{11}N_5$: 02221
$C_6H_{11}O$: 07979
C_6H_{12}: 03963, 03964, 03968, 05440, 05491, 05776, 07941, 07954, 07955, 07960, 07961, 07994, 10533, 10534, 10535, 10559, 10560, 10561, 10562, 10563
$C_6H_{12}BrF$: 07766
$C_6H_{12}BrNO$: 04242, 14802
$C_6H_{12}Br_2$: 07785, 07786, 07787
$C_6H_{12}ClF$: 07771
$C_6H_{12}ClNO$: 04880, 04882, 04883, 04902
$C_6H_{12}Cl_2$: 03732, 03734, 03735, 07788, 07789, 07790, 07791, 07792, 07793, 07795
$C_6H_{12}Cl_2O$: 04991, 04992, 04993, 12255, 12256
$C_6H_{12}Cl_2O_2$: 00030, 00033, 06058, 06885, 07050
$C_6H_{12}Cl_2S$: 12322, 12323

$C_{10}H_8Br_4$: 13765

$C_{10}H_8ClN$: 09700, 09701, 09734, 12765, 12766, 12767, 12769, 12770, 12771, 12773, 12777, 12778, 12781, 12783, 12784

$C_{10}H_8ClNO_2$: 13596

$C_{10}H_8Cl_4$: 13766

$C_{10}H_8FN$: 09709

$C_{10}H_8IN$: 09743

$C_{10}H_8N_2$: 01903, 01904, 01905, 03530, 03531, 03532, 03533, 03534

$C_{10}H_8N_2O$: 09751, 12919

$C_{10}H_8N_2OS$: 08037

$C_{10}H_8N_2OS_2$: 13047

$C_{10}H_8N_2O_2$: 09408, 09409, 09414, 09415, 09628, 09660, 09716, 09717, 09718, 09719, 09720, 09721, 09747, 09748, 09749, 09750, 12879, 12880, 12881, 12884, 12885, 12888, 12889, 12892, 12893

$C_{10}H_8N_2O_2S$: 08036, 08043

$C_{10}H_8N_2O_3$: 12429, 12867, 12868, 12872, 12873, 12876

$C_{10}H_8N_4O_5$: 11560

$C_{10}H_8O$: 09571, 09588

$C_{10}H_8OS$: 07246

$C_{10}H_8O_2$: 02521, 05288, 05289, 05290, 05291, 05292, 05293, 07244, 08401, 08402, 09345, 09346, 09348, 09350, 09351, 09352, 09353, 09354, 09355

$C_{10}H_8O_2S$: 09464, 09465

$C_{10}H_8O_3$: 00488, 05284, 05286, 05287, 05298, 13582, 13583

$C_{10}H_8O_3S$: 09466, 09476

$C_{10}H_8O_4$: 00754, 05099, 05100, 05101, 05270, 05273, 05283, 07205, 07240, 08841, 09427

$C_{10}H_8O_4S$: 09470, 09471, 09472, 09473, 09479, 09480, 09481

$C_{10}H_8O_5$: 05274

$C_{10}H_8O_6S_2$: 09456, 09457, 09458

$C_{10}H_8O_7S_2$: 09455, 09461

$C_{10}H_8O_8S_2$: 09460

$C_{10}H_8S$: 09387, 09388

$C_{10}H_8S_2$: 09356

$C_{10}H_9$: 01886

$C_{10}H_9BrO$: 04058

$C_{10}H_9ClN$: 09195

$C_{10}H_9ClO_2$: 05106

$C_{10}H_9ClO_3$: 11194

$C_{10}H_9N$: 08593, 08594, 08595, 08596, 08597, 08598, 09688, 09722, 12642, 12643, 12878, 12882, 12883, 12886, 12887, 12890, 12891

$C_{10}H_9NO$: 00303, 06433, 08412, 08592, 09593, 09594, 09595, 09596, 09597, 09598, 12806, 12836, 12837, 12838, 12840, 12842, 12845, 12846, 12847, 12849, 12850, 12851, 12853, 12855, 12856, 12857, 12858, 12859, 12869, 12870, 12871, 12875, 12948

$C_{10}H_9NO_2$: 11514, 13605

$C_{10}H_9NO_3$: 11515

$C_{10}H_9NO_3S$: 09467, 09469, 09477

$C_{10}H_9NO_4$: 05133, 05135, 05136, 05137, 05141, 05145, 05148

$C_{10}H_9NO_4S$: 09468

$C_{10}H_9NO_5$: 05134

$C_{10}H_9NO_6$: 08564, 08567, 11486

$C_{10}H_9NO_6S_2$: 09454

$C_{10}H_9NO_7S_2$: 09459

$C_{10}H_9NO_{12}$: 12564

$C_{10}H_9N_3O_3$: 12426

$C_{10}H_9O_3P$: 09462

$C_{10}H_{10}$: 02135, 02136, 03663, 03664, 04200, 08399, 08400, 09340, 09343

$C_{10}H_{10}BrNO_2$: 00275

$C_{10}H_{10}BrNO_3$: 00651

$C_{10}H_{10}Br_2O$: 03923

$C_{10}H_{10}ClN$: 09711, 09742

$C_{10}H_{10}ClNO_2$: 00276

$C_{10}H_{10}ClNO_3$: 00653

$C_{10}H_{10}ClO_2$: 05103

$C_{10}H_{10}Fe$: 06957

$C_{10}H_{10}N_2$: 08106, 08107, 09317, 09319, 09322, 09324, 09325, 09327, 09328, 09830, 09831, 09832, 12728, 12730, 12732, 12736, 12737, 12739, 12740, 12743, 12744, 12747, 12748, 12955, 12956

$C_{10}H_{10}N_2O$: 12422, 12423, 12427, 12746

$C_{10}H_{10}N_2O_4S$: 12428

$C_{10}H_{10}N_2O_6$: 02679, 02680, 06201, 06202, 11127

$C_{10}H_{10}N_2O_6S_2$: 01576

$C_{10}H_{10}N_4O_2S$: 13614, 13625

$C_{10}H_{10}O$: 04064, 05074, 05075, 07194, 13773, 13775

$C_{10}H_{10}O_2$: 00489, 00601, 01658, 02060, 02542, 03133, 04046, 04059, 04060, 04061, 05130, 05131, 05132, 08608, 09676, 13082

$C_{10}H_{10}O_2S_2$: 06130

$C_{10}H_{10}O_3$: 00257, 00397, 02540, 03132, 03159, 04045, 05073, 05129, 10644, 12075, 13090

$C_{10}H_{10}O_4$: 04794, 05127, 08174, 08558, 08559, 08838, 08943, 09001, 11134, 11135, 11473, 11477, 11718, 12991, 12992, 13578, 13579, 13580, 13690, 14891

$C_{10}H_{10}O_5$: 08840, 11352

$C_{10}H_{10}O_6$: 01898, 08556, 11471

$C_{10}H_{11}Br$: 03947, 03984, 03987

$C_{10}H_{11}BrO_2$: 11106, 12090

$C_{10}H_{11}ClN_2O_2$: 02002

$C_{10}H_{11}ClO$: 00373, 03152, 10648

$C_{10}H_{11}ClO_2S$: 13770, 13772

$C_{10}H_{11}Cl_2NO$: 04889

$C_{10}H_{11}N$: 04431, 04432, 08414, 08415, 09705

$C_{10}H_{11}NO$: 04429, 04430, 10213

$C_{10}H_{11}NO_2$: 00062, 00091, 00271, 06129, 08429

$C_{10}H_{11}NO_3$: 00558, 00559, 00560, 00646, 05085, 10194, 13534

$C_{10}H_{11}NO_4$: 08547, 08549, 08551, 11451, 13683

$C_{10}H_{11}NO_5$: 11149

$C_{10}H_{11}N_3O_2S_2$: 13619

$C_{10}H_{11}O_2$: 03036, 03037, 08504, 08505

$C_{10}H_{12}$: 02184, 02189, 03973, 03974, 03975, 05834, 06082, 08350, 08351, 08352, 08353, 13510, 13511, 13512, 13752, 14061, 14153, 14160

$C_{10}H_{12}BrNO$: 12077

$C_{10}H_{12}ClNO$: 12063, 12103

$C_{10}H_{12}N_2$: 00751

$C_{10}H_{12}N_2O$: 14704

$C_{10}H_{12}N_2O_2$: 02059, 08828

$C_{10}H_{12}N_2O_3$: 00120, 00137, 00138, 01620, 08646, 10450

$C_{10}H_{12}N_2O_4$: 00129, 00130, 00133, 00134, 00135, 02127, 04652

$C_{10}H_{12}N_2O_5$: 14037

$C_{10}H_{12}N_2S$: 13975

$C_{10}H_{12}N_4$: 09339

$C_{10}H_{12}N_4O_5$: 08434

$C_{10}H_{12}O$: 00387, 00388, 00389, 00621, 00622, 00623, 00755, 01796, 01807, 03933, 03934, 04437, 04439, 06768, 12059, 12241, 12242, 13761

$C_{10}H_{12}O_2$: 00260, 00328, 00329, 00392, 00393, 02777, 02778, 02779, 02780, 02781, 02782, 02827, 02828, 02829, 02830, 02831, 02832, 03063, 03067, 04378, 04379, 04381, 04382, 05479, 06213, 06214, 06801, 06802, 06803, 06931, 08153, 08507, 10773, 10775, 10904, 10911, 10924, 10967, 11130, 12238, 12240, 14267, 14282, 14294

$C_{10}H_{12}O_3$: 00222, 00258,

$C_{10}H_{12}O_3$: 00385, 00507, 01004, 01007, 01011, 01739, 02718, 02737, 04374, 04376, 05225, 08287, 08914, 08915, 08916, 09137, 09140, 09144, 11074, 11123, 11124, 11125, 11129, 11138, 11139, 11140, 13155, 13156

$C_{10}H_{12}O_3S$: 13769, 13771

$C_{10}H_{12}O_4$: 00386, 01805, 01806, 02648, 04546, 08026, 08281, 11076, 11122, 12156, 13136

$C_{10}H_{12}O_4S$: 13938

$C_{10}H_{12}O_5$: 02819, 02820, 02824, 02825, 02826, 07371, 13119

$C_{10}H_{13}Br$: 01923, 01973, 30197

$C_{10}H_{13}BrN_2O_3$: 01617

$C_{10}H_{13}BrO$: 01922

$C_{10}H_{13}Cl$: 02001, 05868

$C_{10}H_{13}ClO$: 14036

$C_{10}H_{13}ClO_2$: 06620

$C_{10}H_{20}N_2O_2$: 06638, 13186

$C_{10}H_{20}N_2O_4$: 03697

$C_{10}H_{20}N_2S_4$: 06109

$C_{10}H_{20}O$: 03900, 04547, 04767, 05183, 05184, 05185, 05461, 05581, 05898, 05926, 05927, 05928, 07631, 08489, 08490, 08518, 08519, 08520, 08961, 08962, 08963, 09787, 09788, 09791, 09792, 09793, 09794, 09795, 09796

$C_{10}H_{20}O_2$: 00206, 00229, 00230, 00231, 00232, 00247, 00248, 00249, 00250, 00251, 04337, 04373, 04550, 04587, 04632, 05459, 05917, 07599, 07925, 08219, 10271, 12148, 13707, 14644, 14863

$C_{10}H_{20}O_2S$: 08978, 08982

$C_{10}H_{20}O_3$: 03939

$C_{10}H_{20}O_4$: 05662, 06113

$C_{10}H_{20}O_5$: 05605

$C_{10}H_{21}$: 04548

$C_{10}H_{21}Br$: 05901, 05903

$C_{10}H_{21}Cl$: 05904

$C_{10}H_{21}ClO$: 05922

$C_{10}H_{21}F$: 05909

$C_{10}H_{21}FO$: 05923

$C_{10}H_{21}I$: 05910

$C_{10}H_{21}N$: 05505, 05517, 11647, 11675, 11676, 12305

$C_{10}H_{21}NO$: 04249, 04549

$C_{10}H_{21}NO_2$: 05913, 05940

$C_{10}H_{21}NO_3$: 05939, 08020

$C_{10}H_{21}NO_7S$: 06582

$C_{10}H_{22}$: 05899, 07573, 07574, 09926, 09927, 09928, 09929, 10045, 10046, 10047

$C_{10}H_{22}NO_4$: 10248

$C_{10}H_{22}N_2O_2$: 09236

$C_{10}H_{22}O$: 05921, 05924, 05925, 06815, 07264, 07265, 07266, 07626, 07904, 10081, 10082, 10089, 10090, 10091, 10602, 10603

$C_{10}H_{22}O_2$: 03836, 05915, 07843, 10062

$C_{10}H_{22}O_3$: 10160, 10161

$C_{10}H_{22}O_4$: 14490

$C_{10}H_{22}O_4S$: 10614, 10615

$C_{10}H_{22}O_5$: 13748

$C_{10}H_{22}S$: 04170, 05916, 10612, 10613

$C_{10}H_{22}S_2$: 06229, 06230

$C_{10}H_{23}$: 05937, 07836

$C_{10}H_{23}N$: 05900, 06226, 06228

$C_{10}H_{23}NO_2$: 06097

$C_{10}H_{30}$: 02256

$C_{10}H_{30}O_3Si_4$: 13785, 13786

$C_{10}H_{30}O_5Si_5$: 05805

$C_{11}H_6N_2O_6$: 09518

$C_{11}H_6O_4$: 14969

$C_{11}H_6O_{10}$: 02334

$C_{11}H_7BrO_2$: 09508, 09509, 09510, 09544, 09545, 09640

$C_{11}H_7ClO$: 09532, 09570

$C_{11}H_7ClO_2$: 09511, 09512, 09513, 09514, 09546, 09547, 09548

$C_{11}H_7ClO_3$: 09555, 09558

$C_{11}H_7IO_2$: 09523

$C_{11}H_7N$: 09531, 09569

$C_{11}H_7NO$: 09756, 09757

$C_{11}H_7NO_4$: 09525, 09526, 09527, 09528, 09565, 09566

$C_{11}H_7NS$: 09758, 09759, 09767, 09768

$C_{11}H_8Br_2$: 09292

$C_{11}H_8N_2$: 10638

$C_{11}H_8N_2O_3$: 01616

$C_{11}H_8N_4O_7$: 12524

$C_{11}H_8O$: 09269, 09273

$C_{11}H_8OS$: 13869, 13870

$C_{11}H_8O_2$: 07118, 07245, 09272, 09274, 09500, 09534, 09632, 09633, 09673

$C_{11}H_8O_3$: 09520, 09521, 09522, 09554, 09557, 09562, 09563, 09666, 09671

$C_{11}H_8O_4$: 07221, 09653, 09655

$C_{11}H_9Br$: 09293, 09295

$C_{11}H_9BrO$: 09605

$C_{11}H_9Cl$: 09314, 09315

$C_{11}H_9N$: 12521, 12522, 12523

$C_{11}H_9NO$: 09498, 09533, 09710, 09741

$C_{11}H_9NO_2$: 09392, 09393, 09394, 09395, 09396, 09397, 09398, 09401, 09402, 09403, 09404, 09405, 09406, 09407, 09501, 09502, 09503, 09504, 09505, 09506, 09536, 09537, 09538, 09539, 09540, 09541, 09542, 12923

$C_{11}H_9NO_3$: 12931

$C_{11}H_9N_3O$: 06510

$C_{11}H_{10}$: 09391, 09400

$C_{11}H_{10}BrNO_2$: 11510, 11511

$C_{11}H_{10}N_2$: 09496, 09497

$C_{11}H_{10}N_2O$: 07182, 12720, 12721, 12722, 12723, 12724, 12725, 14753, 14754

$C_{11}H_{10}N_2O_3$: 01641

$C_{11}H_{10}N_2S$: 13923, 14006, 14008

$C_{11}H_{10}O$: 05818, 09171, 09172, 09384, 09389, 09390, 09584, 09610, 10297

$C_{11}H_{10}O_2$: 05759, 09515, 09516, 09517, 09549, 09550, 09551, 10312, 12021

$C_{11}H_{10}O_3$: 05297, 07320, 07321

$C_{11}H_{10}O_4$: 05275

$C_{11}H_{10}S$: 13903

$C_{11}H_{11}BrN_2O$: 01333

$C_{11}H_{11}BrO_4$: 01667, 01674

$C_{11}H_{11}ClN_2O_5$: 04979

$C_{11}H_{11}Cl_2NO$: 04886

$C_{11}H_{11}N$: 09287, 09288, 09712, 09713, 09714, 09715, 09744, 09745, 09746, 12627, 12810, 12811, 12816, 12818, 12819, 12821, 12825, 12826

$C_{11}H_{11}NO$: 12812, 12813, 12814, 12815, 12817, 12820, 12822, 12823, 12824, 12827, 12828, 12830

$C_{11}H_{11}NO_2$: 05427, 08425, 12160, 13589

$C_{11}H_{11}NO_4$: 05140, 05144, 05147

$C_{11}H_{11}NO_5$: 01434

$C_{11}H_{11}NO_6$: 01784, 01787, 01790

$C_{11}H_{11}N_2O_4S_2$: 13820

$C_{11}H_{11}N_3O_2S$: 13626

$C_{11}H_{12}$: 09341, 09342

$C_{11}H_{12}Br_2O_2$: 12120, 12121

$C_{11}H_{12}ClNO$: 04873

$C_{11}H_{12}ClNO_3$: 00654

$C_{11}H_{12}IN$: 06830

$C_{11}H_{12}N_2$: 09320, 09323

$C_{11}H_{12}N_2O$: 01327, 01328, 14897, 14898

$C_{11}H_{12}N_2O_2$: 08041, 08042, 14610, 14611, 14612

$C_{11}H_{12}N_2O_6$: 02671, 02672, 06193, 06194

$C_{11}H_{12}N_2S$: 13963

$C_{11}H_{12}N_4O_2S$: 13617

$C_{11}H_{12}O$: 03069, 05072, 05439, 05457, 08376, 10590, 13774

$C_{11}H_{12}O_2$: 00334, 04062, 04281, 05120, 05121, 05122, 05161, 09529, 09530, 09567, 09568, 14876

$C_{11}H_{12}O_3$: 02784, 03131, 04280, 09258, 12175

$C_{11}H_{12}O_4$: 01694, 02543, 03310, 05112, 05113, 05114, 05115, 05116, 05118, 07338, 08896, 13126, 13539, 13540, 14107

$C_{11}H_{12}O_5$: 05119, 12679, 12680, 13637

$C_{11}H_{12}O_6$: 04865

$C_{11}H_{13}N$: 04428, 08424, 14874

$C_{11}H_{13}NO$: 06735, 10215

$C_{11}H_{13}NO_2$: 00273, 08163

$C_{11}H_{13}NO_3$: 05252, 08052, 11206, 11207, 11208

$C_{11}H_{13}NO_4$: 00685, 01155, 09880

$C_{11}H_{13}N_3O$: 01330, 01331, 01332

$C_{11}H_{13}N_3O_6$: 01997, 01999

$C_{11}H_{14}$: 02058, 05779, 05931, 08343, 08344, 08345, 08346, 08347, 08348, 08990, 13524, 13762, 13763, 13764

$C_{11}H_{14}BrNO$: 04261

$C_{11}H_{14}ClNO_3$: 08054

$C_{11}H_{14}Cl_2N_2$: 09321

$C_{11}H_{14}N_2$: 07434

$C_{11}H_{14}N_2O$: 05879

$C_{11}H_{14}N_2O_2$: 00272

$C_{11}H_{14}O$: 00426, 00427, 04159, 04438, 05040, 08613, 08988, 11240, 11731, 12034, 12359, 14875

$C_{11}H_{14}O_2$: 01885, 02116, 02587, 02588, 02589, 02590, 02591, 04368, 04380, 04396, 04415, 06162, 08152, 11035, 14864, 14865, 14867

$C_{11}H_{14}O_3$: 00505, 01708, 01709, 02683, 02684, 02686, 02688, 02715, 06745, 06747, 06749, 08266, 08288, 08918, 08919, 08920, 11144, 11145, 11146, 12181, 12183, 13103, 13104,

FORMULA INDEX FOR ORGANIC COMPOUNDS (Continued)

$C_{18}H_{14}O$: 03268, 03269, 09773, 09774
$C_{18}H_{14}O_3$: 05090
$C_{18}H_{14}O_6$: 08625
$C_{18}H_{14}O_8$: 13550
$C_{18}H_{15}As$: 01417
$C_{18}H_{15}AsO$: 01418
$C_{18}H_{15}AsS$: 01419
$C_{18}H_{15}B$: 03578
$C_{18}H_{15}Bi$: 03547
$C_{18}H_{15}ClSe$: 14568
$C_{18}H_{15}ClSi$: 13246
$C_{18}H_{15}FSe$: 14569
$C_{18}H_{15}FSi$: 13295
$C_{18}H_{15}N$: 14534
$C_{18}H_{15}NO$: 14602
$C_{18}H_{15}N_3$: 01557
$C_{18}H_{15}OP$: 11382
$C_{18}H_{15}O_3P$: 11435
$C_{18}H_{15}O_4P$: 11420
$C_{18}H_{15}P$: 11381
$C_{18}H_{15}Sb$: 13438
$C_{18}H_{16}$: 05821, 07938, 14607
$C_{18}H_{16}N_2$: 08129
$C_{18}H_{16}N_2O$: 02318, 02319, 02320, 02321, 02327
$C_{18}H_{16}N_2O_2$: 00750
$C_{18}H_{16}N_2O_6S$: 12861
$C_{18}H_{16}O_2$: 10722, 13493
$C_{18}H_{16}O_4$: 06533, 08467, 08468, 08469, 08470, 13767
$C_{18}H_{16}O_7$: 14785, 14786, 14787
$C_{18}H_{16}Si$: 13334
$C_{18}H_{17}NO_3$: 12379
$C_{18}H_{17}N_3O$: 02328
$C_{18}H_{18}$: 10686, 13033
$C_{18}H_{18}N_2O_3$: 01335
$C_{18}H_{18}N_4$: 03031
$C_{18}H_{18}O_2$: 06536, 06537, 06538
$C_{18}H_{18}O_4$: 03447, 03508, 03835, 13543, 13561
$C_{18}H_{18}O_6$: 13650
$C_{18}H_{19}NO_2$: 01344
$C_{18}H_{19}NO_3$: 06584, 14617
$C_{18}H_{19}NO_4$: 06586
$C_{18}H_{19}N_3$: 08413
$C_{18}H_{19}N_4O_4$: 07297, 07298
$C_{18}H_{20}$: 08356
$C_{18}H_{20}N_{10}O_{16}$: 01386, 01387
$C_{18}H_{20}O_2$: 06539
$C_{18}H_{21}NO_3$: 05195, 05200, 06578, 06579, 06583, 06585, 08491, 12355, 13793
$C_{18}H_{22}$: 03481
$C_{18}H_{22}BrNO_3$: 08021
$C_{18}H_{22}ClNO_3$: 08022
$C_{18}H_{22}N_2O_2$: 08018
$C_{18}H_{22}N_2O_5$: 11567
$C_{18}H_{22}N_2O_6$: 00550
$C_{18}H_{22}N_2S$: 11050
$C_{18}H_{22}O_2$: 03426, 03427, 03429, 06540, 06602, 08499
$C_{18}H_{22}O_3$: 06115
$C_{18}H_{22}O_4$: 09965
$C_{18}H_{22}O_8$: 02457
$C_{18}H_{23}ClN_2O_2$: 08019
$C_{18}H_{23}NO_3$: 05201
$C_{18}H_{23}NO_4$: 05196, 06428
$C_{18}H_{23}N_3$: 01447
$C_{18}H_{24}NO_7P$: 05198
$C_{18}H_{24}O_2$: 06599, 06600, 08498
$C_{18}H_{24}O_2Si$: 13252
$C_{18}H_{24}O_3$: 06601
$C_{18}H_{26}$: 02106
$C_{18}H_{26}ClNO_5$: 05197
$C_{18}H_{26}O_4$: 11474
$C_{18}H_{26}O_{12}$: 13365
$C_{18}H_{27}NO_3$: 04649
$C_{18}H_{27}N_3O$: 10511
$C_{18}H_{28}N_2O_4S$: 00742
$C_{18}H_{28}O$: 08718
$C_{18}H_{28}O_2$: 08713
$C_{18}H_{29}NO$: 08705
$C_{18}H_{30}$: 02137, 02162, 09978
$C_{18}H_{30}Br_6O_2$: 13390, 13402
$C_{18}H_{30}O$: 11025
$C_{18}H_{30}O_2$: 06474, 06475, 08768, 08769, 09995, 09996
$C_{18}H_{30}O_4$: 01452
$C_{18}H_{32}Br_4O_2$: 13416
$C_{18}H_{32}CaN_2O_{10}$: 10246, 10247
$C_{18}H_{32}I_2O_2$: 10001
$C_{18}H_{32}N_2O_5S$: 10244
$C_{18}H_{32}O_2$: 04860, 04861, 08764, 08767, 09976, 09977, 10010, 13420
$C_{18}H_{32}O_4$: 13401, 13426
$C_{18}H_{32}O_5$: 01451
$C_{18}H_{32}O_6$: 14426
$C_{18}H_{32}O_{16}$: 08950, 12966
$C_{18}H_{32}Si$: 13310
$C_{18}H_{33}N$: 10145
$C_{18}H_{34}$: 10007, 10008, 10009
$C_{18}H_{34}Br_2O_2$: 06464, 13387
$C_{18}H_{34}O$: 09997, 10136
$C_{18}H_{34}O_2$: 06463, 10000, 10003, 10139
$C_{18}H_{34}O_3$: 00533, 10002, 13059, 13403, 13405, 13407, 13409
$C_{18}H_{34}O_4$: 13188
$C_{18}H_{35}ClO$: 13427
$C_{18}H_{35}N$: 13424
$C_{18}H_{35}NO$: 06462, 10137
$C_{18}H_{36}$: 09998, 09999
$C_{18}H_{36}Br_2$: 09986
$C_{18}H_{36}ClNO$: 04877
$C_{18}H_{36}O$: 06467, 09979, 09994, 10004, 10005, 10146
$C_{18}H_{36}O_2$: 00214, 04558, 05919, 07497, 09992, 10229, 13381
$C_{18}H_{36}O_2S$: 08979
$C_{18}H_{36}O_3$: 10232, 13392, 13393, 13394, 13395, 13396
$C_{18}H_{36}O_4$: 13388
$C_{18}H_{37}Br$: 09984
$C_{18}H_{37}Cl$: 09985
$C_{18}H_{37}I$: 09987
$C_{18}H_{37}NO$: 13379
$C_{18}H_{38}$: 09981
$C_{18}H_{38}O$: 09993, 13428
$C_{18}H_{38}O_2$: 07699, 09990
$C_{18}H_{38}S$: 09991
$C_{18}H_{39}N$: 09982, 14507
$C_{18}H_{40}ClN$: 09983
$C_{18}H_{54}O_7Si_8$: 10099
$C_{18}H_{54}O_9Si_9$: 05739
$C_{19}H_{10}Br_4O_5S$: 03620
$C_{19}H_{11}N$: 01767
$C_{19}H_{12}O$: 01870
$C_{19}H_{12}O_2$: 02513, 02514
$C_{19}H_{12}O_5$: 08617
$C_{19}H_{12}O_6$: 06075
$C_{19}H_{12}O_7$: 05301
$C_{19}H_{13}N$: 00479
$C_{19}H_{13}NO$: 04699
$C_{19}H_{13}N_3O_6$: 14556
$C_{19}H_{13}N_3O_7$: 14566, 14567
$C_{19}H_{14}$: 01862, 01863, 01864, 01865, 01866, 01867, 01868, 01869, 05049, 05050, 05051, 07019
$C_{19}H_{14}N_2O_2$: 01533, 01539, 01542
$C_{19}H_{14}O$: 07011, 14946
$C_{19}H_{14}O_2$: 12187
$C_{19}H_{14}O_3$: 01448
$C_{19}H_{14}O_5$: 10323
$C_{19}H_{14}O_5S$: 11047
$C_{19}H_{15}Br$: 06844
$C_{19}H_{15}Cl$: 14540
$C_{19}H_{15}ClN_4$: 14571
$C_{19}H_{15}N$: 04700
$C_{19}H_{15}NO$: 01823
$C_{19}H_{15}NO_2$: 12935
$C_{19}H_{15}N_3$: 05970
$C_{19}H_{16}$: 03389, 03390, 09103, 14537
$C_{19}H_{16}N_2$: 01852, 01854, 02971, 15023
$C_{19}H_{16}N_2O$: 01483
$C_{19}H_{16}N_2S$: 14017
$C_{19}H_{16}N_4$: 09899
$C_{19}H_{16}O$: 06934, 14557
$C_{19}H_{16}O_3$: 10159, 14554, 14564
$C_{19}H_{16}O_4$: 05279
$C_{19}H_{17}$: 06679
$C_{19}H_{17}ClN_2$: 01853
$C_{19}H_{17}N$: 03210, 14538, 14539
$C_{19}H_{17}NO_2$: 05405, 12933
$C_{19}H_{17}N_3$: 07450, 07451
$C_{19}H_{17}N_3O$: 06937, 06938, 13221, 13222, 13223
$C_{19}H_{18}N_2$: 01583, 14541
$C_{19}H_{18}O$: 06725
$C_{19}H_{18}O_3$: 10320, 10321, 10322
$C_{19}H_{18}O_3Si$: 13302
$C_{19}H_{18}O_5$: 06438
$C_{19}H_{18}Si$: 13303
$C_{19}H_{19}NO_3$: 08701
$C_{19}H_{19}NO_4$: 03629, 03630
$C_{19}H_{19}N_3$: 09104, 14550, 14551, 14552
$C_{19}H_{19}N_3O$: 14565
$C_{19}H_{19}N_3O_2$: 00466
$C_{19}H_{19}N_7O_6$: 07045
$C_{19}H_{21}NO_3$: 08610, 13790
$C_{19}H_{21}NO_5$: 04863, 05204
$C_{19}H_{22}ClNO_3$: 13792
$C_{19}H_{22}N_2O$: 01341, 01342, 05055, 05056, 05059, 05060
$C_{19}H_{22}N_2O_2$: 05399, 13180
$C_{19}H_{22}O_2$: 07967
$C_{19}H_{23}NO_3$: 01346, 13791
$C_{19}H_{23}NO_4$: 09217, 13345
$C_{19}H_{23}N_3O_2$: 06553, 06554, 06555, 12320
$C_{19}H_{24}ClNO_3$: 09224
$C_{19}H_{24}Cl_2N_2O$: 05061
$C_{19}H_{24}N_2O$: 05054, 05064, 08144
$C_{19}H_{24}N_2O_2$: 05236, 08160, 12691
$C_{19}H_{24}O_2$: 08997
$C_{19}H_{25}ClN_2O_2$: 05057
$C_{19}H_{26}N_2O_4$: 01354
$C_{19}H_{27}ClN_2O_3$: 01343, 05062
$C_{19}H_{27}NO_4$: 06925
$C_{19}H_{28}ClNO_4$: 06926
$C_{19}H_{28}N_2O$: 10637
$C_{19}H_{28}O_2$: 13711
$C_{19}H_{30}O_2$: 00753, 06498, 08708

SUBLIMATION DATA FOR ORGANIC COMPOUNDS

Compiled by Mansel Davies

The tables quote the parameters from what appear to be the best data in the literature expressed in the form

$$\log_{10} p(\text{mm}) = A - B/T$$

and the temperature range for which they apply. The corresponding heats and entropies (taking the standard state of the vapor to be

ΔH (sublimation) = 2.303 R.B. cal/mol
ΔS (sublimation) = 2.303R(A − 2.881) cal/mol K.

Compound	Temp. range °C	A	B	Ref.
Acenaphthene	18 to 37	11.758	4290.5	1
Acetamide	25 to 77	11.8468	4050.1	2
Acetic acid	−35 to + 10	8.502	2177.4	3
Acetic Acid. *m*-cresyl ester	2 to 44	9.759	3170	14
Acetophenone, 1-chloro-	5 to 50	13.779	4740	14
Acetophenone, 1-chloro-*o*-nitro	23 to 54	14.24	5413	14
Acetophenone, 1-chloro-*m*-nitro	26 to 70	14.080	5700	14
Acetophenome p-methoxy	3 to 27	11.367	4056	1
Acetone, benzoyl	5 to 26	12.317	4375	1
Adipic acid	86 to 133	15.463	6757	4
Anthracene	65 to 80	12.638	5320	5
Anthracene	105 to 125	12.002	5102	6
Anthracene, 9,10 diphenyl	208 to 229	16.058	8213	22
Anthraquinone	224 to 286	12.305	5747	3
Arachidic acid	63 to 73	25.453	10,424	23
Arsine, Diphenylcyano	23 to 53	10.724	4420	14
Azobenzene (*cis*)	30 to 60	9.652	3914	7
Azobenzene (*trans*)	30 to 60	9.721	3911	7
Behenic acid	71 to 79	23.604	10,100	23
Benzanthrone	—	13.416	6030	8
Benzene	−30 to 5	9.846	2309	3
Benzene	−58 to −30	9.556	2241	3
Benzene, *p*-chloroiodo	30 to 50	9.819	3200	15
Benzene, -dichloro	10 to 50	11.985	3570	17
Benzene, *α*-hexachloro	51 to 71	11.950	4850	14
Benzene, *β*-hexachloro	95 to 117	11.790	5375	14
Benzene, *γ*-hexchloro	60 to 92	15.515	6022	14
Benzene,*λ*-hexachloro	55 to 75	12.635	5100	14
Benzene, -hexamethyl		11.070	4215	37
Benzene, 1, 2, 3,-trichloro	16 to 30	10.662	3440	6
Benzene, 1, 2, 4-trichloro	6 to 25	10.445	3254	6
Benzene, 1,3,5-trichloro	9 to 28	9.176	2956	6
Benzil	45 to 67	12.708	5140	1
Benzoic Acid	70 to 114	12.870	4776	9
Benzoic acid, *p*-hydroxy	125 to 160	13.623	6063	9
Benzoic acid, *o*-methoxy	80 to 95	11.871	4746	9
Benzophenone (stable)	16 to 42	17.46	4966	10
Benzophenone, (meta stable)	11 to 25	17.19	4818	10
Benzoquinone	—	10.00	3280	18
Benzoquinone, 2.6-dichloro	1 to 42	9.85	3670	18
Benzoquinone, trichloro	28 to 54	12.03	4630	18
Benzoquinone, tetrachloro	60 to 83	12.06	5170	18
Benzoquinone, *p*-xylo	0 to 20	11.53	4030	18
Bibenzyl	13 to 34	12.194	4386	1
Biphenyl	6 to 26	11.168	3959	1
Butyramide	25 to 68	12.739	4546	2
Butyramide	63 to 109	12.594	4513	2
Camphor	0 to 180	8.799	2797	3
Capramide	80 to 97	16.471	6577	2
Capramide, *N*-methyl	30 to 52	14.594	5371	11
Capric acid	16 to 28	17.130	6119	23
Caproamide	65 to 95	13.328	4968	2
*ε*Caprolactam	21 to 41	11.839	4339	34
Caprylamide	52 to 101	14.920	5783	2
Carbamic acid, *n*-butyl ester	19 to 43	14.582	4919	11
Carbamic acid, ethyl ester	19 to 43	14.090	4646	11
Carbamic acid, *n*-hexyl ester	18 to 41	14.748	5018	11
Carbamic acid, methyl ester	14 to 32	11.966	3883	11
Carbon tetrabromide (monoclinic)	22 to 46	9.3867	2841	12
(cubic)	48 to 56	8.5670	2579	12
Carbon tetrachloride	−64 to −48	9.089	2027	13
o-Cresol, 3, 5-dinitro	17 to 51	14.140	5400	14
Cyclohexane	−5 to + 5	8.594	1953	3
Cyclo trimethylene-trinitramine	110 to 138	11.870	5850	16
Diphenylamine	25 to 51	12.434	4654	21
Dodecanedioic acid	102 to 123	17.728	8006	4
Eicosanedioic acid	107 to 122	18.185	8644	4
Enanthamide	72 to 93	13.617	5182	2
Ethane, 1.1 p dichloro diphenyl tri-chloro	66 to 100	14.191	6160	14
Ethane, hexachloro (cubic)	13 to 174	8.731	2677	26
Ethane, hexachloro (triclinic)	13 to 174	9.890	3077	2b

Compound	Temp. range C	A	B	Ref.
Ethylene dibromide	−21 to +8	9.884	2606	13
Ethylene, *trans* di-iodo	−8 to 20	5.86	2130	19
Fluorene	33 to 49	11.325	4324	5
Formic acid	−5 to +8	12.486	3160	36
2 Fuoric acid	44 to 55	14.62	5667	25
Hendecanoic acid	20 to 28	16.432	6037	24
Heneicosanoic acid	68 to 73	22.602	9642	24
Heptadecanoic acid	48 to 58	21.836	8769	24
Hydroquinone tetrachloro	77 to 86	10.08	4650	18
Hydroquinone tetrachloro, *p*-xylo	59 to 88	12.36	5280	18
Lauramide	76 to 95	19.169	7980	2
Lauric acid	22 to 41	19.897	7322	23
Methane	−194 to −184	7.651	5169	3
Methane, triphenyl	52 to 76	12.661	5228	1
Myristamide	85 to 100	20.940	8746	2
Myristic acid	38 to 52	18.740	7291	23
Naphthalene	6 to 21	11.597	3783	5
1-Naphthol	25 to 39	13.074	4873	1
	39 to 50	11.526	4389	1
2-Naphthol	25 to 39	13.356	5109	1
	39 to 58	11.660	4579	1
Nonadecanoic acid	58 to 64	35.916	13,815	24
n-Octadecane	15 to 25	22.83	7995	27
Oxalic acid, anhyd.	60 to 105	12.223	4727	29
Oxalic acid, anhyd. (*α*)	38 to 52	13.17	5130	28
Oxalic acid, anhyd. (*β*)	38 to 50	12.57	4875	28
Oxamic acid	82 to 90	12.58	5639	30
Oxamide	80 to 96	12.57	5893	30
Palmitamide	91 to 105	22.690	9489	2
Palmitic acid	46 to 60	20.217	8069	23
Pelargonamide	80 to 97	15.249	5997	2
Pentadecanoic acid	38 to 48	23.110	8813	24
Pentaerythritol (tetrag)	106 to 135	16.17	7528	28
Pentaerythritol, tetranitrate	97 to 138	17.73	7750	16
Phenanthrene	37 to 50	11.388	4519	5
Phenol	5 to 32	11.421	3540	14
Phenol, *p*-acetyl	47 to 75	12.216	5003	31
Phenol, *p*-benzyl	40 to 62	12.600	5072	31
Phenol, *p*-*tert* butyl	8 to 30	12.332	4402	31
Phenol, 2-*tert* butyl-4- methyl	2 to 20	11.685	4036	31
Phenol, 4-*tert* butyl-2-methyl	3 to 24	11.199	3952	31
Phenol, *p*- formyl	39 to 63	11.795	4762	31
Phenol, *p*-methoxy	5 to 27	13.132	4624	31
Phenol, *o*-phenyl	19 to 40	11.754	4331	31
Phenol, *p*-phenyl	54 to 74	12.056	5068	31
Phenol, 2:4:6-tri*tert* butyl	18 to 40	11.507	4383	31
Phthalic anhydride	30 to 60	12.249	4632	32
Propionamide	45 to 73	12.041	4139	2
Pyrene	72 to 85	11.270	4904	5
Pyrrole 2-carboxylic acid	77 to 81	16.60	6633	25
Rubeanic acid	87 to 105	12.713	5515	30
Salicylic acid	95 to 134	12.859	4969	9
Sebacic acid	102 to 130	18.911	8395	4
Stearamide	94 to 106	24.449	10,230	2
Steraric acid	57 to 67	21.180	8696	23
Suberic acid	106 to 134	16.937	7472	4
Succinic acid	99 to 128	14.068	6132	4
d-Tartaric acid, dimethyl ester	35 to 44	16.610	5903	20
dl-Tartaric acid, dimenthyl ester	42 to 85	16.127	5941	20
Thapsic acid	104 to 125	17.165	7885	4
2-Thenoic acid	42 to 50	13.53	5065	25
Thymol	0 to 40	14.201	4766	14
Toluene, 2, 4, 6-trinitro	50 to 143	15.34	6180	33
Tridecanoic acid	31 to 39	20.939	7764	24
Valeramide	60 to 101	12.846	4666	2

References

1. Aihara, *Bull. Chem. Soc.*, Japan
2. Davies, Jones, and Thomas, *Trans. Faraday Soc.*, 55, 1100, 1959.
3. *CRC Handbook of Chemistry and Physics*, 41st Ed. p. 2428 et seq.
4. Davies and Thomas, *Trans. Faraday, Soc.*, 56, 185, 1960.
5. Bradley and Cleasby, *J. Chem. Soc.*, 1690, 1953.
6. Sears and Hopke, *J. Am. Chem. Soc.*, 71, 1632, 1949.
7. Bright et al., *Research*, 3, 185, 1950.
8. Inokuchi et al., *Bull. Chem. Soc.*, Japan, 25, 299, 1952.
9. Davies and Jones, *Trans. Soc.*, 50, 1042, 1954.
10. Neumann and Volker, *Z. Physik. Chem.*, 161A, 33, 1932.
11. Davies and Jones, *Trans. Soc.*, 55, 1329, 1959.
12. Bradley and Drury, ibid. , 55, 1844, 1959.
13. Nitta and Seki, *J. Chem. Soc.*, Japan, 69, 85, 1948.
14. Balson, *Trans. Faraday Soc.*, 43, 54, 1947.
15. Ewald, ibid., 49, 1401, 1953.
16. Edwards, ibid., 49, 152, 1953.
17. Darkis et al., *Ind. Eng. Chem.*, 32, 946, 1940.
18. Coolidge and Coolidge, *J. Am. Chem. Soc.* 49, 100, 1927.
19. Broadway and Fraser, *J. Chem. Soc.*, 429, 1933.
20. Crowell and Jones, *J. Phys. Chem.*, 58, 666, 1954.
21. Aihara, *J. Chem. Soc.*, Japan, 74, 437, 1953.
22. Stevens, *J. Chem. Soc.*, 2973, 1953.
23. Davies, Malpass, and Stenhagan, *Ark. Kemi.*
24. Thomas, M. Sc. thesis, Univ. of Wales, 1959.
25. Bradley and Care, *J. Chem. Soc.*, 1688, 1953.
26. Ivin and Dainton, *Trans. Soc.*, 43, 32, 1947.
27. Bradley and Shellard, *Proc. Roy. Soc.*, 198A, 239, 1949.
28. Bradley and Cotson, *J. Chem. Soc.*, 1684, 1953.
29. Noyes and Wobbe, *J. Am. Chem. Soc.*, 48, 1882, 1926.
30. Bradley and Cleasby, *J. Chem. Soc.*, 1681, 1953.
31. Aihara, *Bull Chem. Soc.*, Japan.
32. Crooks and Feetham, *J. Chem. Soc.*, 899, 1946.
33. Edwards, *Trans. Faraday Soc.*, 46, 423, 1950.
34. Aihara, *J. Chem. Soc.*, Japan, 74, 631, 1953.
35. Seki and Suzuki, *Bull. Chem. Soc.*, Japan, 70, 387, 1949.
36. Coolidge, *J. Am. Chem. Soc.*, 53, 1874, 1930.
37. Nitta et al., *J. Chem. Soc.*, Japan, 70, 387, 1949.

HEATS OF FUSION OF SOME ORGANIC COMPOUNDS

William E. Acree, Jr.

Compounds in this table are listed in order of increasing number of carbon and hydrogen atoms in the molecules. Melting point temperatures are listed in degrees Celcius and heats of fusion in calories per gram, joules per gram, and joules per gram molecular weight.

Formula	Compound	Mol. wt	M.P. (°C)	Heat of fusion (H_f) cal/g	J/g	J/mol
$CHCl_3$	Trichloromethane	119.38	−63.6	17.62	73.72	8,800
CHN	Hydrogen cyanide	27.03	−13.4	74.38	311.21	8,412
CH_2Cl_2	Dichloromethane	84.93	−95.14	16.89	70.67	6,002
CH_2N_2	Cyanamide	42.04	44.0	49.81	208.41	8,761
CH_2O_2	Formic acid	46.03	8.3	66.05	276.35	12,720
CH_3D	Monodeuteromethane	17.05	−182.7	12.76	62.97	910
CH_3Br	Bromomethane	94.94	−93.7	15.05	62.97	5,978
CH_4	Methane	16.04	−182.5	13.96	58.41	936
CD_4	Deuteromethane	20.07	−183.4	10.75	44.98	902
CH_4O	Methanol	32.04	−97.9	23.70	99.16	3,177
CH_4S	Methyl mercaptan	48.10	−121.0	29.35	122.8	5,906
CH_5N	Methylamine	31.06	−93.5	47.20	197.48	6,133
$CBrCl_3$	Bromotrichloromethane	198.27	−5.7	3.05	12.76	2,539
CCl_2O	Phosgene	98.92	−127.9	13.86	57.99	5,736
CCl_3NO_2	Chloropicrin	164.38	−64.0	48.16	201.50	33,122
CCl_4	Tetrachloromethane	153.82	−23.0	5.09	21.30	3,276
CS_2	Carbon disulfide	76.13	−111.5	13.80	57.74	4,395
$C_2HCl_3O_2$	Trichloroacetic acid	163.39	57.5	8.60	35.98	5,878
$C_2H_2Br_2Cl_2$	1,2-Dibromo-1,1-dichloromethane	256.75	−66.9	7.73	32.34	8,303
$C_2H_2Cl_2O_2$	Dichloroacetic acid	128.94	10.8	14.21	59.45	7,665
$C_2H_3Br_3$	1,1,2-Tribromomethane	266.76	−29.2	8.16	34.14	9,107
$C_2H_3ClO_2$	α-Chloroacetic acid	94.50	61.2	31.06	129.96	12,281
$C_2H_3ClO_2$	β-Chloroacetic acid	94.50	56.0	35.12	146.94	13,885
$C_2H_3Cl_3$	1,1,1-Trichloroethane	133.40	−30.4	4.90	20.50	2,734
$C_2H_3Cl_3$	1,1,2-Trichloroethane	133.40	−36.6	20.68	86.53	11,543
$C_2H_3F_3$	1,1,1-Trifluoroethane	84.04	−111.3	17.61	73.68	6,192
$C_2H_4Br_2$	1,2-Dibromoethane	187.86	9.93	13.97	57.70	10,839
$C_2H_4Cl_2$	1,2-Dichlororoethane	99.96	−35.5	21.12	88.37	8,833
$C_2H_4O_2$	Acetic acid	60.05	16.6	45.91	192.09	11,535
C_2H_5Cl	Chloroethane	64.51	−138.3	16.49	68.99	4,450
C_2H_6	Ethane	30.07	−183.3	22.73	95.10	2,859
C_2H_6O	Dimethyl ether	46.07	−141.5	25.62	107.19	4,938
C_2H_6O	Ethanol	46.07	−114.5	26.05	108.99	5,021
$C_2H_6O_2$	Ethylene glycol	62.07	−11.5	43.26	181.00	11,234
C_2H_6S	Dimethyl sulfide	62.13	−98.3	30.73	128.57	7,988
C_2H_6S	Ethanethiol	62.13	121.0	19.14	80.08	4,975
$C_2H_6S_2$	Methyl disulfide	94.19	−120.5	23.32	97.57	9,190
C_2H_7N	Dimethylamine	45.08	−92.2	31.51	131.84	5,943
$C_2H_8N_2$	Ethylene diamine	60.10	11.1	89.81	35.77	22,583
C_3H_3N	Acrylonitrile	53.06	−83.5	28.06	117.40	6,229
$C_3H_4O_2$	Acrylic acid	72.06	12.3	37.03	154.93	11,164
$C_3H_5Br_3$	1,2,3-Tribromopropane	280.78	16.19	20.24	84.68	23,776
$C_3H_5N_3O_3$	Trinitroglycerol	131.09	12.3	23.02	96.32	12,627
C_3H_6	Cyclopropane	42.08	−127.4	30.92	129.37	5,444
C_3H_6	Propene	42.08	−185.3	17.06	71.38	3,004
$C_3H_6Br_2$	1,3-Dibromopropane	201.89	−34.2	16.10	67.36	13,599
$C_3H_6Cl_2$	1,2-Dichloropropane	112.99	−100.5	13.53	56.61	6,396
C_3H_6O	Acetone	58.08	−94.8	23.42	97.99	5,691
C_3H_7Cl	2-Chloropropane	78.54	−117.2	22.48	94.06	7,387
C_3H_7N	Cyclopropylamine	57.10	−35.39	55.18	230.87	13,183
$C_3H_7NO_2$	Ethyl carbamate	89.09	48.7	40.85	170.92	15,227
C_3H_8	Propane	44.10	−181.7	19.11	79.96	3,526
C_3H_8O	Propanol	60.10	−126.1	20.66	86.44	5,195
C_3H_8O	Isopropanol	60.10	−89.5	21.37	89.41	5,373
$C_3H_8O_3$	Glycerol	92.09	18.2	47.95	200.62	8,475
C_3H_9N	Triethylamine	59.11	−117.1	26.47	110.75	6,546
$C_4H_4N_2$	Succinonitrile	80.09	54.5	11.71	48.99	3,924
$C_4H_4O_3$	Succinic anhydride	100.07	119.0	48.47	203.93	20,407
C_4H_4S	Thiophene	84.14	−39.4	14.11	59.04	4,968
C_4H_5N	Pyrrole	67.09	−23.41	28.17	117.86	7,907
C_4H_6	1,3-Butadiene	54.09	−108.9	35.28	147.61	7,984
C_4H_6	2-Butyne	54.09	−32.36	40.80	170.71	9,234
$C_4H_6O_2$	Crotonic acid	86.09	72.0	25.32	105.94	9,120
$C_4H_6O_2$	*cis*-Crotonic acid	86.09	71.2	34.90	146.02	12,571
$C_4H_6O_2$	γ-Butyrolactone	86.09	−43.37	26.57	111.17	9,571
$C_4H_6O_4$	Dimethyl oxalate	118.09	54.35	42.64	178.41	21,068
$C_4H_6O_4$	Succinic acid	118.09	183.8	66.68	278.99	32,946

Formula	Compound	Mol. wt	M.P. (°C)	Heat of fusion (H_f) cal/g	J/g	J/mol
C_4H_8	Isobutene	56.11	−140.4	25.25	105.65	5,928
C_4H_8	*cis*-2-Butene	56.11	−138.9	32.30	135.14	7,583
$C_4H_8N_2S$	Allyl thiourea	116.18	77.0	33.45	139.95	16,259
C_4H_8O	Tetrahydrofuran	72.11	−108.39	28.31	118.45	8,541
C_4H_8O	2-Butanone	72.11	−86.67	27.97	117.03	8,439
$C_4H_8O_2$	Ethyl acetate	88.11	−83.6	28.43	118.95	10,481
$C_4H_8O_2$	*n*-Butyric acid	88.11	−5.7	30.04	125.69	11,075
$C_4H_8O_2$	*p*-Dioxane	88.11	11.0	34.85	145.81	12,847
C_4H_9Br	2-Bromobutane	137.02	−112.7	12.01	50.25	6,885
C_4H_{10}	*n*-Butane	58.12	−138.3	19.18	80.25	4,664
C_4H_{10}	Isobutane	58.12	−159.42	18.96	79.33	4,611
$C_4H_{10}O$	*n*-Butanol	74.12	−89.8	29.93	125.23	9,282
$C_4H_{10}O$	*tert*-Butanol	74.12	25.4	21.88	91.55	6,786
$C_4H_{10}O$	Ethyl ether	74.12	−116.3	23.45	98.11	7,272
$C_4H_{12}Si$	Tetramethylsilane	88.19	−99.04	18.64	77.99	6,878
C_5H_8	Cyclopentane	68.12	−135.1	11.80	49.37	3,363
C_5H_8	Isoprene	68.12	−145.9	16.8	70.29	4,788
C_5H_8	Methylene cyclobutane	68.12	−134.6	20.22	84.60	5,763
C_5H_8	1,4-Pentadiene	68.12	−148.8	21.55	90.17	6,142
$C_5H_8O_2$	δ-Valerolactone	100.12	−10.33	25.14	105.19	10,532
$C_5H_8O_3$	Levulinic acid	116.12	33.0	18.97	79.37	9,216
$C_5H_8O_4$	Glutaric acid	132.12	97.8	37.81	158.20	20,901
C_5H_{10}	1-Pentene	70.13	−166.2	19.81	82.89	5,813
C_5H_{10}	*cis*-2-Pentene	70.13	−151.4	24.25	101.46	7,115
C_5H_{10}	*trans*-2-Pentene	70.13	−140.2	28.48	119.16	8,357
C_5H_{10}	Cyclopentane	70.13	−93.8	2.07	8.66	607
$C_5H_{10}O_2$	Valeric acid	102.13	−59.0	27.83	116.44	11,892
$C_5H_{11}N$	Cyclopentylamine	85.15	−82.7	23.33	97.61	8,311
C_5H_{12}	*n*-Pentane	72.15	−129.7	27.89	116.69	8,419
C_5H_{12}	Isopentane	72.15	−159.9	17.05	71.34	5,147
C_5H_{12}	2,2-Dimethylpropane	72.15	−16.6	10.79	45.15	3,258
$C_5H_{12}O$	1-Pentanol	88.15	−78.9	26.65	111.50	9,829
C_6HCl_5O	Pentachlorophenol	266.34	189.3	15.39	64.39	17,150
$C_6H_3Br_3O$	2,4,6-Tribromophenol	330.80	93.0	13.38	55.98	18,518
$C_6H_3Cl_3$	1,2,3-Trichlorobenzene	181.45	53.7	27.00	112.97	20,498
$C_6H_3Cl_3$	1,3,5-Trichlorobenzene	181.45	63.5	23.97	100.29	18,198
$C_6H_3Cl_3O$	2,4,5-Trichlorophenol	197.45	68.5	28.87	120.79	23,850
C_6H_4BrI	*o*-Bromoiodobenzene	282.91	21.0	12.18	50.96	14,417
C_6H_4BrI	*m*-Bromoiodobenzene	282.91	9.3	10.27	42.97	12,157
C_6H_4BrI	*p*-Bromoiodobenzene	282.91	90.1	16.16	67.61	19,128
$C_6H_4Br_2$	*o*-Dibromobenzene	235.91	1.8	12.78	53.47	12,614
$C_6H_4Br_2$	*m*-Dibromobenzene	235.91	−6.9	13.38	55.98	13,206
$C_6H_4Br_2$	*p*-Dibromobenzene	235.91	86.0	20.55	85.98	20,284
$C_6H_4Br_2O$	2,4-Dibromophenol	251.91	12.0	13.97	58.45	14,724
$C_6H_4ClNO_2$	*m*-Chloronitrobenzene	157.56	44.4	29.38	122.93	19,369
$C_6H_4ClNO_2$	*p*-Chloronitrobenzene	157.56	83.5	31.51	131.84	20,773
$C_6H_4Cl_2$	*o*-Dichlorobenzene	147.00	−16.7	21.02	87.95	12,929
$C_6H_4Cl_2$	*m*-Dichlorobenzene	147.00	−24.8	20.55	85.98	12,639
$C_6H_4Cl_2$	*p*-Dichlorobenzene	147.00	52.7	27.89	116.69	17,153
$C_5H_4Cl_2O$	2,3-Dichlorophenol	163.00	56.8	31.32	131.04	21,360
$C_6H_4Cl_2O$	2,4-Dichlorophenol	163.00	44.8	29.46	123.26	20,091
$C_6H_4Cl_2O$	2,5-Dichlorophenol	163.00	57.8	32.89	137.61	22,430
$C_6H_4Cl_2O$	2,6-Dichlorophenol	163.00	66.8	32.47	135.85	22,144
$C_6H_4Cl_2O$	3,4-Dichlorophenol	163.00	67.8	30.69	128.41	20.931
$C_6H_4Cl_2O$	3,5-Dichlorophenol	163.00	67.8	30.07	125.81	20,507
$C_6H_4I_2$	*o*-Diiodobenzene	329.91	23.4	10.15	42.47	14,011
$C_6H_4I_2$	*m*-Diiodobenzene	329.91	34.2	11.54	48.28	15,928
$C_6H_4I_2$	*p*-Diiodobenzene	329.91	129.0	16.20	67.78	22,361
$C_6H_4N_2O_4$	*o*-Dinitrobenzene	168.11	116.93	32.25	134.93	22,683
$C_6H_4N_2O_4$	*m*-Dinitrobenzene	168.11	89.7	24.70	103.34	17,372
$C_6H_4N_2O_4$	*p*-Dinitrobenzene	168.11	173.5	39.99	167.32	28,128
$C_6H_4N_2O_5$	2,3-Dinitrophenol	184.11	143.8	34.06	142.51	26,238
$C_6H_4N_2O_5$	2,4-Dinitrophenol	184.11	114.8	31.38	131.29	24,172
$C_6H_4N_2O_5$	2,5-Dinitrophenol	184.11	107.8	30.80	128.87	23,726
$C_6H_4N_2O_5$	2,6-Dinitrophenol	184.11	62.8	25.41	106.32	19,575
$C_6H_4N_2O_5$	3,4-Dinitrophenol	184.11	133.8	32.94	137.82	25,374
$C_6H_4O_2$	*p*-Benzoquinone	108.10	112.9	40.97	171.42	18,531
C_6H_5Br	Bromobenzene	157.01	−30.6	16.17	67.66	10,623
C_6H_5BrO	4-Bromophenol	173.01	63.5	20.50	85.77	14,839
C_6H_5Cl	Chlorobenzene	112.56	−45.2	20.40	85.35	9,607
C_6H_5ClO	2-Chlorophenol	128.56	9.8	23.28	97.40	12,522
C_6H_5ClO	3-Chlorophenol	128.56	32.6	27.71	115.94	14,905
C_6H_5ClO	4-Chlorophenol	128.56	42.7	26.15	109.41	14,066

Formula	Compound	Mol. wt	M.P. (°C)	Heat of fusion (H_f) cal/g	J/g	J/mol
C_6H_5F	Fluorobenzene	96.10	−42.21	28.12	117.65	11,306
C_6H_5I	Iodobenzene	204.01	−31.3	11.43	47.82	9,756
$C_6H_5NO_2$	Nitrobenzene	123.11	5.7	22.50	94.14	11,590
$C_6H_5NO_3$	2-Nitrophenol	139.11	44.8	29.97	125.39	17,443
$C_6H_5NO_3$	3-Nitrophenol	139.11	96.8	32.98	137.99	19,196
$C_6H_5NO_3$	4-Nitrophenol	139.11	113.8	31.36	131.21	18,253
C_6H_6	Benzene	78.11	5.53	30.45	127.40	9,951
$C_6H_6N_2O_2$	*o*-Nitroaniline	138.13	71.2	27.88	116.65	16,113
$C_6H_6N_2O_2$	*m*-Nitroaniline	138.13	147.0	36.50	152.72	21,095
$C_6H_6N_2O_2$	*p*-Nitroaniline	138.13	114.0	40.97	171.42	23,678
C_6H_6O	Phenol	94.11	40.9	28.67	119.96	11,289
$C_6H_6O_2$	1,2-Dihydroxybenzene	110.11	105.0	49.40	206.69	22,759
$C_6H_6O_2$	1,3-Dihydroxybenzene	110.11	110.0	46.22	193.38	21,293
$C_6H_6O_2$	1,4-Dihydroxybenzene	110.11	172.3	58.84	246.19	27,108
C_6H_6S	Thiophene	110.17	−14.9	24.90	104.18	11,478
C_6H_7N	Aniline	93.13	−6.3	27.09	113.34	10,555
$C_6H_8N_2$	Phenylhydrazine	108.14	19.6	36.31	151.92	16,429
$C_6H_8O_4$	Methyl fumarate	144.13	102.0	57.93	242.38	34,934
C_6H_{10}	Cyclohexene	82.15	−103.5	9.58	40.08	3,293
$C_6H_{10}O_4$	Adipic acid	146.14	153.2	57.00	238.49	34,853
C_6H_{12}	Methylcyclopentane	84.16	−142.5	19.68	82.34	6,930
C_6H_{12}	Cyclohexane	84.16	6.6	7.47	31.25	2,630
C_6H_{12}	Tetramethylethylene	84.16	−74.6	15.51	64.89	5,461
$C_6H_{12}O$	Cyclohexanol	100.16	25.46	4.19	17.53	1,756
$C_6H_{12}O_2$	ϵ-Caprolactone	116.16	−1.02	28.94	121.08	14,065
C_6H_{14}	2,2-Dimethylbutane	86.18	−99.0	1.61	6.74	581
C_6H_{14}	2,3-Dimethylbutane	86.18	−128.8	2.22	9.29	801
C_6H_{14}	*n*-Hexane	86.18	−95.3	36.27	151.75	13,078
C_6H_{14}	2-Methylpentane	86.18	−153.7	17.38	72.72	6,267
$C_6H_{14}O$	Isopropyl ether	102.18	−86.8	25.79	17.91	11,026
$C_6H_{14}O$	*n*-Propyl ether	12.18	−126.1	20.66	86.44	8,832
$C_6Cl_5NO_2$	Pentachloronitrobenzene	295.34	144.8	14.90	62.34	18,412
C_6Cl_6	Hexachlorobenzene	284.78	231.8	20.02	83.76	23,853
$C_7H_3Cl_3O_2$	2,3,6-Trichlorobenzoic acid	225.46	129.5	25.28	105.77	23,847
$C_7H_3I_3O_2$	2,3,5-Triiodobenzoic acid	499.81	230.6	15.41	64.48	32,228
$C_7H_5ClO_2$	2-Chlorobenzoic acid	156.57	140.2	39.27	164.31	25,726
$C_7H_5ClO_2$	3-Chlorobenzoic acid	156.57	154.2	36.39	152.26	23,839
$C_7H_5ClO_2$	4-Chlorobenzoic acid	156.57	239.7	49.23	205.98	32,250
$C_7H_5NO_4$	2-Nitrobenzoic acid	167.12	145.8	40.06	167.61	28,011
$C_7H_5NO_4$	3-Nitrobenzoic acid	167.12	141.1	27.59	115.44	19,292
$C_7H_5NO_4$	4-Nitrobenzoic acid	167.12	239.2	52.80	220.92	36,920
$C_7H_5N_3O_6$	2,4,6-Trinitrotoluene	227.13	80.83	22.34	93.47	21,230
$C_7H_6N_2O_4$	2,4-Dinitrotoluene	182.14	70.14	26.40	110.46	20,119
$C_7H_6O_2$	Benzoic acid	122.12	122.4	33.89	141.80	17,317
C_7H_7Br	4-Bromotoluene	171.04	28.0	20.86	87.28	14,928
$C_7H_7NO_2$	2-Aminobenzoic acid	137.14	145.0	35.98	150.54	20,645
$C_7H_7NO_2$	3-Aminobenzoic acid	137.14	179.5	38.03	159.12	21,822
$C_7H_7NO_2$	4-Aminobenzoic acid	137.14	188.5	36.46	152.55	20,921
$C_7H_7NO_3$	4-Nitro-5-methylphenol	153.14	127.8	42.77	178.95	27,404
$C_7H_7NO_3$	2-Nitro-5-methylphenol	153.14	29.6	32.45	135.77	20,792
C_7H_8	Toluene	92.14	−94.99	17.77	74.35	6,851
C_7H_8	1,3,5-Cycloheptatriene	92.14	−75.23	3.01	12.59	1,160
C_7H_8O	Benzyl alcohol	108.14	−15.2	19.83	82.97	8,972
C_7H_8O	2-Methylphenol	108.14	29.8	30.81	128.91	13,940
C_7H_8O	3-Methylphenol	108.14	11.8	20.80	87.03	9,411
C_7H_8O	4-Methylphenol	108.14	35.8	26.27	109.91	11,886
C_7H_9O	*p*-Toluidine	109.15	43.3	39.90	166.94	18,222
$C_7H_{12}O_4$	Pimelic acid	160.17	104.3	41.22	172.46	27,623
C_7H_{14}	1-Heptene	98.19	−119.7	30.82	128.95	12,662
C_7H_{14}	Methylcyclohexane	98.19	−126.6	16.43	68.74	6,750
C_7H_{14}	Cycloheptane	98.19	−8.03	4.58	19.16	1,881
C_7H_{16}	*n*-Heptane	100.20	−90.6	33.78	141.34	14,162
C_7H_{16}	2-Methylhexane	100.20	−118.2	21.16	88.53	8,871
C_7H_{16}	2,2-Dimethylpentane	100.20	−123.8	13.98	58.49	5,861
C_7H_{16}	2,4-Dimethylpentane	100.20	−119.9	15.95	66.73	6,686
C_7H_{16}	3,3-Dimethylpentane	100.20	−134.9	16.86	70.54	7,068
C_7H_{16}	3-Ethylpentane	100.20	−118.6	22.78	95.31	9,550
C_7H_{16}	2,2,3-Trimethylbutane	100.20	−25.0	5.25	21.97	2,201
$C_8H_6Cl_4$	Tetrachloro-*o*-xylene	243.95	86.0	21.02	87.95	21,455
$C_8H_6Cl_4$	Tetrachloro-*p*-xylene	243.95	95.0	22.10	92.47	22,588
$C_8H_8Br_2$	α-α′-Dibromo-*o*-xylene	263.96	95.0	24.25	101.46	26,781
$C_8H_8Br_2$	α-α′-Dibromo-*m*-xylene	263.96	77.0	21.45	89.75	23,690
$C_8H_8Cl_2$	α-α′-Dichloro-*o*-xylene	175.06	55.0	29.03	121.46	21,263

HEATS OF FUSION OF SOME ORGANIC COMPOUNDS (continued)

Formula	Compound	Mol. wt	M.P. (°C)	Heat of fusion (H_f) cal/g	J/g	J/mol
$C_8H_8Cl_2$	α-α′-Dichloro-*m*-xylene	175.06	34.0	26.64	111.46	19,512
$C_8H_8Cl_2$	α-α′-Dichloro-*p*-xylene	175.06	100.0	32.73	136.94	23,973
$C_8H_8O_2$	Phenylacetic acid	136.15	76.7	25.44	106.44	14,492
$C_8H_8O_2$	*o*-Toluic acid	136.15	103.7	35.40	148.11	20,165
$C_8H_8O_2$	*m*-Toluic acid	136.15	108.75	27.59	115.44	15,717
$C_8H_8O_2$	*p*-Toluic acid	136.15	179.6	39.90	166.94	22,729
C_8H_{10}	*o*-Xylene	106.17	−25.2	30.64	128.20	13,611
C_8H_{10}	*m*-Xylene	106.17	−47.8	26.01	108.83	11,554
C_8H_{10}	*p*-Xylene	106.17	13.2	37.83	158.28	16,805
$C_8H_{10}O$	2,3-Dimethylphenol	122.17	72.8	41.13	172.09	21,024
$C_8H_{10}O$	2,5-Dimethylphenol	122.17	74.8	45.73	191.33	23.375
$C_8H_{10}O$	2,6-Dimethylphenol	122.17	45.7	36.97	154.68	18,897
$C_8H_{10}O$	3,4-Dimethylphenol	122.17	60.8	35.46	148.36	18,125
$C_8H_{10}O$	3,5-Dimethylphenol	122.17	63.6	35.21	147.32	17,998
C_8H_{12}	1,5-Cyclooctadiene	108.18	−69.17	21.71	90.83	9,826
C_8H_{14}	Bicyclooctane	108.18	174.63	18.10	75.73	8,192
$C_8H_{14}O_4$	Suberic acid	172.18	142.13	40.01	167.40	28,823
C_8H_{16}	Cyclooctane	112.21	14.83	5.13	21.46	2,408
C_8H_{16}	Ethylcyclohexane	112.21	−33.3	17.75	74.27	8,334
C_8H_{16}	*trans*-1,1-Dimethylcyclohexane	112.21	33.3	4.38	18.33	2,057
C_8H_{16}	*cis*-1,2-Dimethylcyclohexane	112.21	−49.9	3.50	14.64	1,643
C_8H_{16}	*trans*-1,2-Dimethylcyclohexane	112.21	−88.2	22.35	93.51	10,493
C_8H_{16}	*cis*-1,3-Dimethylcyclohexane	112.21	−75.6	23.05	96.44	10,822
C_8H_{16}	*trans*-1,3-Dimethylcyclohexane	112.21	−90.1	21.01	87.91	9,864
C_8H_{16}	*cis*-1,4-Dimethylcyclohexane	112.21	−87.4	19.82	82.93	9,306
C_8H_{16}	*trans*-1,4-Dimethylcyclohexane	112.21	−36.9	26.27	109.91	12,333
$C_8H_{16}O_2$	Caprylic acid	144.21	16.3	35.40	148.11	21,359
C_8H_{18}	*n*-Octane	114.23	−56.8	43.21	180.79	20,652
C_8H_{18}	3-Methylheptane	114.23	−120.5	23.81	99.62	11,380
C_8H_{18}	4-Methylheptane	114.23	121.0	22.68	94.89	10,839
C_8H_{18}	2,2,4-Trimethylpentane	114.23	−107.3	18.92	79.16	9,042
$C_8H_{18}N_2$	1,1-Dimethylazoethane	142.24	−14.6	17.27	72.26	10,278
$C_8H_{18}N_2O$	1,1-Dimethylazoxyethane	158.24	15.2	17.40	72.84	11,526
C_9H_7N	Quinoline	129.16	−15.6	19.98	83.60	10,798
$C_9H_8O_2$	Cinnamic acid	148.16	133.0	36.50	152.72	22,627
$C_9H_8O_2$	Allocinnamic acid	148.16	68.0	27.35	114.43	16,954
$C_9H_{10}O_2$	Hydrocinnamic acid	150.18	48.0	28.14	117.74	17,682
C_9H_{12}	1,2,4-Trimethylbenzene	120.19	−43.8	7.47	31.25	3,756
C_9H_{12}	1,2,3-Trimethylbenzene	120.19	−25.4	16.65	69.66	8,372
C_9H_{12}	1,3,5-Trimethylbenzene	120.19	−44.7	18.90	79.08	9,505
C_9H_{12}	*n*-Propylbenzene	120.19	−99.56	18.43	77.11	9,268
$C_9H_{16}O_4$	Azelaic acid	188.22	106.8	41.49	173.59	32,673
C_9H_{18}	*n*-Propylcyclohexane	126.24	−94.9	19.64	82.17	10,373
$C_9H_{18}O_2$	Pelargonic acid	158.24	12.35	30.63	128.16	20,280
C_9H_{20}	*n*-Nonane	128.26	−53.5	28.83	120.62	15,471
C_9H_{20}	2,2,3,3-Tetramethylpentane	128.26	−9.8	4.35	18.20	2,334
C_9H_{20}	2,2,4,4-Tetramethylpentane	128.26	−66.54	18.16	75.98	9,745
C_9H_{20}	3,3-Diethylepentane	128.26	−33.1	18.80	78.66	10,089
$C_{10}H_7Br$	α-Bromonaphthalene	207.07	−1.8	17.50	73.22	15,162
$C_{10}H_7Br$	β-Bromonaphthalene	207.07	58.8	13.82	57.82	11,973
$C_{10}H_7Cl$	α-Chloronaphthalene	162.62	−2.5	18.96	79.33	12,901
$C_{10}H_7Cl$	β-Chloronaphthalene	162.62	58.8	21.60	90.37	14,696
$C_{10}H_7$	α-Iodonaphthalene	254.07	6.8	14.97	62.63	15,912
$C_{10}H_7I$	β-Iodonaphthalene	254.07	54.35	15.09	63.14	16,042
$C_{10}H_7NO_2$	α-Nitronaphthalene	173.17	56.7	25.44	106.44	18,432
$C_{10}H_8$	Naphthalene	128.17	78.2	35.66	149.20	19,123
$C_{10}H_8O$	α-Naphthol	144.17	94.0	38.68	161.84	23,332
$C_{10}H_8O$	β-Naphthol	144.17	123.0	29.03	121.46	17,511
$C_{10}H_9N$	α-Aminonaphthalene	143.19	50.0	24.19	101.21	14,492
$C_{10}H_9N$	β-Aminonaphthalene	143.19	113.0	39.41	164.89	23,611
$C_{10}H_{14}$	*n*-Butylbenzene	134.22	−87.9	19.98	83.60	11,221
$C_{10}H_{14}$	1,2,4,5-Tetramethylbenzene	134.22	79.3	37.40	156.48	21,003
$C_{10}H_{14}$	1,2,3,4-Tetramethylbenzene	134.22	−7.7	20.0	83.68	11,232
$C_{10}H_{14}$	1-Methyl-4-isopropylbenzene	134.22	−68.9	17.10	71.55	9,603
$C_{10}H_{14}O$	Thymol	150.22	51.5	27.47	114.93	17,265
$C_{10}H_{18}O_4$	Sebacic acid	202.25	130.8	48.22	201.75	40,804
$C_{10}H_{20}$	*n*-Butylcyclohexane	140.27	−74.73	24.12	100.92	14,156
$C_{10}H_{20}O_2$	*n*-Capric acid	172.27	31.99	38.87	162.63	28,016
$C_{10}H_{22}$	*n*-Decane	142.28	−29.7	48.34	202.25	28,776
$C_{11}H_8O_2$	α-Naphthoic acid	172.18	161.0	27.61	115.52	19,890
$C_{11}H_8O_2$	β-Naphthoic acid	172.18	185.0	32.68	136.73	23,542
$C_{11}H_{10}$	2-Methylnaphthalene	142.20	34.4	20.11	89.14	11,965
$C_{11}H_{14}$	1,1-Dimethylindan	142.20	−45.8	19.60	82.01	11,662

HEATS OF FUSION OF SOME ORGANIC COMPOUNDS (continued)

Formula	Compound	Mol. wt	M.P. (°C)	Heat of fusion (H_f) cal/g	J/g	J/mol
$C_{11}H_{14}$	4,6-Dimethylindan	142.20	−16.70	21.05	88.07	12,524
$C_{11}H_{14}$	4,7-Dimethylindan	142.20	−0.52	22.09	92.42	13,142
$C_{11}H_{20}O_4$	Undecanedioic acid	216.28	111.8	43.82	183.34	39,653
$C_{11}H_{22}O_2$	*n*-Undecilic acid	186.29	28.25	32.20	134.72	25,097
$C_{11}H_{24}$	*n*-Undecane	156.31	−25.6	34.12	142.76	22,315
$C_{12}H_8S$	Dibenzothiophene	184.26	97.8	19.85	83.05	15,303
$C_{12}H_9N$	Carbazole	167.21	243.0	42.05	175.94	29,419
$C_{12}H_{10}$	Acenaphthene	154.21	93.4	33.38	139.66	21,537
$C_{12}H_{10}$	Biphenyl	154.21	69.0	28.83	120.62	18,601
$C_{12}H_{10}N_2$	Azobenzene	182.22	67.1	28.91	120.96	22,041
$C_{12}H_{10}N_2O$	Azoxybenzene	198.22	36.0	21.62	90.46	17,931
$C_{12}H_{11}N$	Diphenylamine	168.23	52.98	25.23	105.56	17,864
$C_{12}H_{12}$	1,4-Dimethylnaphthalene	156.23	6.8	24.32	101.75	15,896
$C_{12}H_{12}$	2,3-Dimethylnaphthalene	156.23	104.8	38.40	160.67	25,101
$C_{12}H_{12}$	2,6-Dimethylnaphthalene	156.23	110.17	38.33	160.37	25,055
$C_{12}H_{12}$	2,7-Dimethylnaphthalene	156.23	95.66	35.72	149.45	23,349
$C_{12}H_{12}$	1,8-Dimethylnaphthalene	156.23	63.18	24.12	100.92	15,767
$C_{12}H_{12}N_2$	Hydrazobenzene	184.24	134.0	22.89	95.77	17,645
$C_{12}H_{16}$	Cyclohexylbenzene	160.26	7.3	22.82	95.48	15,302
$C_{12}H_{22}O_4$	Dodecanedioic acid	230.30	129.3	52.48	219.58	50,569
$C_{12}H_{24}O_2$	*n*-Lauric acid	200.32	43.22	43.72	182.92	36,643
$C_{12}H_{26}$	*n*-Dodecane	170.34	−9.6	51.33	214.76	36,582
$C_{12}Cl_{10}$	Decachlorobiphenyl	498.66	304.5	18.90	79.08	39,434
$C_{13}H_8Cl_2O$	*p,p'*-Dichlorobenzophenone	251.11	146.8	28.67	119.96	30,123
$C_{13}H_{10}$	Fluorene	166.22	114.8	28.15	117.78	19,577
$C_{13}H_{10}O$	Benzophenone	182.22	47.88	23.86	99.83	18,191
$C_{13}H_{13}N$	Benzylaniline	183.25	32.37	21.86	91.46	16,760
$C_{13}H_{18}$	1,1,4,6-Tetramethylindan	174.29	0.36	21.59	90.33	15,743
$C_{13}H_{18}$	1,1,4,7-Tetramethylindan	174.29	27.6	−15.47	64.73	11,282
$C_{13}H_{24}O_4$	Tridecanedioic acid	244.33	114.3	44.31	185.39	45,296
$C_{14}H_8O_2$	Anthroquinone	208.22	284.8	37.48	156.82	32,653
$C_{14}H_{10}$	Anthracene	178.23	219.5	38.66	161.75	28,829
$C_{14}H_{10}$	Phenanthrene	178.23	99.24	22.08	92.38	16,465
$C_{14}H_{10}O_2$	Benzil	213.23	95.2	22.15	92.68	19,762
$C_{14}H_{28}O_2$	Myristic acid	228.37	53.96	47.49	198.70	45,377
$C_{16}H_{10}$	Pyrene	202.26	151.2	20.22	84.60	17,111
$C_{16}H_{10}$	Fluoranthene	202.26	107.8	22.30	93.30	18,871
$C_{16}H_{32}$	*n*-Decylcyclohexane	224.43	−1.72	41.10	171.96	38,593
$C_{16}H_{32}O_2$	Palmitic acid	256.43	61.82	39.18	163.93	42,037
$C_{16}H_{34}O$	1-Hexanedecanol	242.44	49.27	33.80	141.42	34,286
$C_{18}H_{12}$	Chrysene	228.29	258.2	27.38	114.56	26,153
$C_{18}H_{12}$	Triphenylene	228.29	200.3	26.28	109.96	25,103
$C_{18}H_{12}$	1,2-Benzanthracene	228.29	161.1	22.38	93.64	21,377
$C_{18}H_{12}$	3,4-Benzophenanthrene	228.29	61.5	17.08	71.46	16,314
$C_{18}H_{14}$	*p*-Terphenyl	230.31	210.1	36.84	154.14	35,500
$C_{18}H_{14}O_3$	Cinnamic anhydride	278.31	48.0	28.14	117.74	32,768
$C_{18}H_{34}O_2$	Elaidic acid	282.47	44.4	52.08	217.90	61,550
$C_{18}H_{36}O_2$	Stearic acid	284.48	68.82	47.54	198.91	56,586
$C_{18}H_{38}$	*n*-Octadecane	254.5	28.2	57.65	241.21	61,388
$C_{19}H_{40}$	*n*-Nonadecane	268.53	32.1	40.78	170.62	45,817
$C_{20}H_{12}$	Perylene	252.31	280.7	30.08	125.85	31,753
$C_{20}H_{12}$	1,2-Benzopyrene	252.31	181.3	15.69	65.65	16,564
$C_{20}H_{12}$	3,4-Benzopyrene	252.31	181.0	16.41	68.66	17,324
$C_{20}H_{42}$	*n*-Eicosane	282.55	36.8	59.11	247.32	69,880
$C_{21}H_{16}$	1,2′-Dinaphthylmethane	268.36	96.4	27.21	113.85	30,553
$C_{21}H_{44}$	*n*-Heneicosane	296.58	40.5	38.44	160.83	47,700
$C_{22}H_{12}$	1,12-Benzoperylene	276.34	281.0	15.02	62.84	17,365
$C_{22}H_{12}$	*o*-Phenylenepyrene	276.34	162.0	18.60	77.82	21,505
$C_{22}H_{14}$	1,2:3,4-Dibenzanthracene	278.35	280.3	22.17	92.76	25,820
$C_{22}H_{14}$	1,2:5,6-Dibenzanthracene	278.35	271.0	26.76	119.96	31,164
$C_{22}H_{46}$	*n*-Docosane	310.61	44.4	37.67	157.61	49,955
$C_{23}H_{46}$	*n*-Tricosane	322.62	47.6	30.74	128.62	41,495
$C_{24}H_{12}$	Coronene	300.36	437.3	15.28	63.93	19,202
$C_{24}H_{14}$	3,4:9,10-Dibenzopyrene	302.37	283.6	22.03	92.17	27,869
$C_{24}H_{14}$	1,2:3,4-Dibenzopyrene	302.37	228.0	19.51	81.63	24,682
$C_{24}H_{14}$	1,2:4,5-Dibenzopyrene	302.37	247.0	24.11	100.88	30,503
$C_{24}H_{18}$	*p*-Quaterphenyl	306.41	314.0	29.48	123.34	37,793
$C_{24}H_{50}$	*n*-Tetracosane	338.66	50.9	38.74	162.09	54,893
$C_{25}H_{52}$	*n*-Pentacosane	352.69	53.7	39.13	163.72	57,742
$C_{26}H_{14}$	1,12-Phenyleneperylene	326.40	268.3	12.65	52.93	17,276
$C_{27}H_{56}$	*n*-Heptacosane	380.76	59.0	37.93	158.70	60,427
$C_{28}H_{58}$	*n*-Octacosane	394.77	61.4	39.14	163.76	64,648

HEATS OF VAPORIZATION OF ORGANIC COMPOUNDS

Formula	Name	ΔH$_v$
$CBrN$	Cyanogen bromide	10,882.8
CBr_4	Carbon tetrabromide	10,771.4
$CBrF_3$	Bromotrifluoromethane	—
CBr_2F_2	Dibromodifluoromethane	—
$CClF_3$	Chlorotrifluoromethane	3,996.3
$CClN$	Cyanogen chloride	5,243.4
CCl_2F_2	Dichlorodifluoromethane	8,363.1
CCl_2O	Phosgene	6,224.3
CCl_3F	Trichlorofluoromethane	6,424.1
CCl_3NO_2	Trichloronitromethane	9,109.7
CCl_4	Carbon tetrachloride	8,271.5
		7,628.8
CFN	Cyanogen fluoride	5,875.3
CF_4	Carbon tetrafluoride	3,016.5
$CHBr_3$	Tribromomethane	9,673.3
$CHClF_2$	Chlorodifluoromethane	5,212.9
$CHCl_2F$	Dichlorofluoromethane	6,286.8
$CHCl_3$	Chloroform	7,500.5
CHF_3	Trifluoromethane	—
CHN	Hydrogen cyanide	7,338.8
CH_2Br_2	Dibromomethane	8,722.0
CH_2Cl_2	Dichloromethane	7,572.3
CH_2O	Formaldehyde	5,917.9
CH_2O_2	Formic acid	9,896.5
CH_3ASCl_2	Dichloromethylarsine	9,636.8
CH_3BO	Borine carbonyl	4,867.6
CH_3Br	Methyl bromide	5,925.9
CH_3Cl	Methyl chloride	5,375.3
CH_3Cl_3Si	Trichloromethylsilane	7,450.0
CH_3F	Methyl fluoride	3,986.4
CH_3I	Methyl iodide	6,616.5
CH_3NO	Formamide	15,556.6
CH_3NO_2	Nitromethane	9,210.9
CH_4	Methane	2,128.8
CH_4Cl_2Si	Dichloromethylsilane	7,011.0
CH_4O	Methanol	9,377.2
		8,978.8
CH_4S	Methanethiol	6,331.9
CH_5ClSi	Chloromethylsilane	6,349.5
CH_5N	Methylamine	6,469.5
CH_6Si	Methylsilane	4,683.6
CH_9NSi_2	2-Methyldisilazane	7,185.6
CIN	Cyanogen iodide	14,065.4
CN_4O_8	Tetranitromethane	9,848.7
CO	Carbonmonoxide	1,613.3
COS	Carbonyl sulfide	4,992.2
$COSe$	Carbonyl selenide	5,366.5
CO_2	Carbon dioxide	5,539.0
$CSSe$	Carbon selenosulfide	8,003.0
CS_2	Carbon disulfide	6,786.8
C_2BrCl_2O	Trichloroacetyl bromide	9,673.9
C_2HClF_2	1-Chloro-2,2-difluoroethylene	—
C_2ClF_3	1-Chloro-1,2,2-trifluoroethylene	5,421.5
$C_2Cl_2F_2$	1,2-Dichloro-1,2-difluoroethylene	7,185.6
		—
C_2F_4	Tetrafluoroethylene	
$C_2Cl_2F_4$	1,1-Dichloro-1,2,2,2-tetrafluoroethane	—
$C_2Cl_2F_4$	1,2-Dichloro-1,1,2,2-tetrafluoroethane	6,134.6
$C_2Cl_3F_3$	1,1,2-Trichloro-1,2,2-trifluoroethane	7,115.4
C_2Cl_4	Tetrachloroethylene	9,240.5
CCl_4F_2	Tetrachlorodifluoroethane	—
$C_2Cl_4F_2$	1,1,2,2-Tetrachloro-1,2-difluoroethane	8,746.2
$C_2Cl_3F_3$	1,1,2-Trifluoro-1,2,2-trichloroethane	—
C_2Cl_6	Hexachloroethane	11,711.3
C_2ClF_5	Chloropentafluoroethane	—
C_2F_6	Hexafluoroethane	—
C_2HBr_3O	Tribromoacetaldehyde	11,057.8
C_2HCl_3	Trichloroethylene	8,314.7
C_2HCl_3O	Trichloroacetaldehyde	8,469.2
$C_2HCl_3O_2$	Trichloroacetic acid	13,817.9
C_2HCl_5	Pentachloroethane	9,800.1
C_2H_2	Acetylene	4,665.8
$C_2H_2Br_4$	1,1,1,2-Tetrabromoethane	14,517.3
$C_2H_2Br_4$	1,1,2,2-Tetrabromoethane	12,911.5
$C_2H_2O_4$	Oxalic acid	21,630.6
$C_2H_2Cl_2$	*cis*-1,2-Dichloroethylene	7,420.6
$C_2H_2Cl_2$	*trans*-1,2-Dichloroethylene	7,243.1
$C_2H_2Cl_2$	1,1-Dichloroethylene	7,211.8
$C_2H_2F_2$	1,1-Difluoroethylene	—
$C_2H_2Cl_2O_2$	Dichloroacetic acid	12,952.9
$C_2H_2Cl_4$	1,1,1,2-Tetrachloroethane	9,296.5
		8,725.6
$C_2H_2Cl_4$	1,1,2,2-Tetrachloroethane	9,917.1
C_2H_3Br	1-Bromoethylene	6,076.9
$C_2H_3BrO_2$	Bromoacetic acid	13,537.8
$C_2H_3Br_3$	1,1,2-Tribromoethane	11,874.1
C_2H_3Cl	1-Chloroethylene	6,263.0
$C_2H_3ClO_2$	Chloroacetic acid	13,134.5
$C_2H_3Cl_3$	1,1,1-Trichloroethane	8,012.7
$C_2H_3Cl_3$	1,1,2-Trichloroethane	9,163.2
$C_2H_3ClF_2$	1-Chloro-1,1-difluoroethane	—
$C_2H_3F_3$	1,1,1-Trifluoroethane	—
$C_2H_3Cl_3O_2$	Trichloroacetaldehyde-hydrate	12,141.5
C_2H_3F	1-Fluoroethylene	4,198.1
C_2H_3N	Acetonitrile	8,173.2
C_2H_3NS	Methyl thiocyanate	9,424.1
C_2H_3NS	Methyl isothiocyanate	7,990.1
C_2H_4	Ethylene	3,453.7
C_2H_4BrCl	1-Bromo-2-chloroethane	9,314.9
C_2H_4BrCl	1-Bromo-2-chloroethane	8,995.6
$C_2H_4Br_2$	1,2-Dibromoethane	9,229.4
$C_2H_4Cl_2$	1,1-Dichloroethane	7,288.0
$C_2H_4Cl_2$	1,2-Dichloroethane	7,950.7
$C_2H_4F_2$	1,1-Difluoroethane	6,068.8
C_2H_4O	Acetaldehyde	7,267.8
		6,622.1
C_2H_4O	Ethylene oxide	6,823.3
$C_2H_4O_2$	Acetic acid	9,963.9
		9,486.6
$C_2H_4O_2$	Methyl formate	7,027.8
$C_2H_4O_2S$	Mercaptoacetic acid	13,790.7
C_2H_5Br	Ethyl bromide	6,843.1
C_2H_5Cl	Ethyl chloride	6,310.6
C_2H_5ClO	2-Chloroethanol	10,740.6
$C_2H_5Cl_3Si$	Trichloroethylsilane	9,457.8
$C_2H_5Cl_3OSi$	Trichloroethoxysilane	8,811.4
C_2H_5F	Ethyl fluoride	5,519.5
$C_2H_5F_3Si$	Ethyltrifluorosilane	6,945.7
C_2H_5I	Ethyl iodide	7,851.8
C_2H_5NO	Acetamide	14,025.3
C_2H_5NO	Acetaldoxime	11,317.8
$C_2H_5NO_2$	Nitroethane	9,531.1
$C_2H_5N_3O_2$	Di(nitrosomethyl)amine	10,326.7
C_2H_6	Ethane	3,739.5
$C_2H_6Cl_2Si$	Dichlorodimethylsilane	7,995.7
C_2H_6O	Ethyl alcohol	9,673.9
C_2H_6O	Dimethyl ether	5,409.8
$C_2H_6O_2$	1,2-Ethanediol	14,032.4
C_2H_6S	Dimethyl sulfide	6,742.3
C_2H_6S	Ethane thiol	6,728.7
C_2H_6Sb	Dimethylantimony	12,075.7
C_2H_7N	Ethylamine	6,845.1
C_2H_7N	Dimethylamine	6,660.0
$C_2H_8N_2$	1,2-Ethanediamine	10,510.5
C_2H_8Si	Dimethylsilane	5,497.8
$C_2H_{10}B_2$	Dimethyldiborane	5,696.7
$C_2H_{11}NSi_2$	2-Ethyldisilazane	7,348.3
C_2N_2	Cyanogen	6,597.3
C_3H_3N	Acrylonitrile	7,941.4
C_3H_4	Propadiene	5,141.2
C_3H_4	Propyne	5,632.4

HEATS OF VAPORIZATION OF ORGANIC COMPOUNDS (continued)

Formula	Name	ΔH$_v$
$C_3H_4Br_2$	2,3-Dibromopropene	9,886.2
$C_3H_4Cl_2O_2$	Methyl dichloroacetate	10,820.5
C_3H_4O	2-Propenal	7,628.8
C_3H_4O	Acrylic acid	10,955.1
$C_3H_4O_3$	Pyruvic acid	11,815.7
$C_3H_5Br_3$	1,2,3-Tribromopropane	12,047.1
C_3H_5Cl	1-Chloropropene	6,594.3
C_3H_5Cl	Allyl chloride	7,386.8
C_3H_5ClO	Epichlorohydrine	9,815.4
$C_3H_5ClO_2$	Methyl chloroacetate	10,815.0
$C_3H_5Cl_3$	1,1,1-Trichloropropane	8,933.9
$C_3H_5Cl_3$	1,2,3-Trichloropropane	10,714.3
$C_3H_5Cl_3Si$	Allyltrichlorosilane	9,386.1
C_3H_5N	Propionitrile	8,769.0
C_3H_5NO	3-Hydroxypropionitrile	13,287.2
C_3H_5NS	Ethylisothiocyanate	9,574.7
$C_3H_5N_3O_9$	Nitroglycerine	13,753.1
C_3H_6	Propene	4,697.4
C_3H_6	Cyclopropane	5,897.7
C_3H_6BrNO	2-Bromo-2-nitrosopropane	9,619.6
$C_3H_6Br_2$	1,2-Dibromopropane	9,801.9
$C_3H_6Br_2$	1,3-Dibromopropane	10,374.4
$C_3H_6Br_2O$	2,3-Dibromo-1-propanol	13,190.0
$C_3H_6Cl_2$	1,2-Dichloropropane	8,428.5
$C_3H_6Cl_2O$	1,3-Dichloro-2-propanol	12,067.6
C_3H_6O	Acetone	7,641.5
C_3H_6O	Allyl alcohol	10,577.7
C_3H_6O	Propylene oxide	7,295.8
$C_3H_6O_2$	Propanoic acid	12,454.4
$C_3H_6O_2$	Methyl acetate	7,732.8
$C_3H_6O_2$	Ethyl formate	7,511.7
$C_3H_6O_3$	Methyl glycolate	11,105.0
$C_3H_6O_3$	Methoxyacetic acid	13,451.0
C_3H_7Br	n-Propyl bromide	8,029.8
C_3H_7Br	2-Bromopropane	7,591.7
C_3H_7Cl	n-Propyl chloride	7,485.7
		6,905.8
C_3H_7Cl	2-Chloropropane	6,855.2
$C_3H_7Cl_3Si$	Trichloroisopropylsilane	8,973.3
C_3H_7I	n-Propyl iodide	8,467.1
C_3H_7I	2-Iodopropane	8,243.4
	Propionamide	14,554.0
C_3H_7NO	1-Nitropropane	9,949.9
$C_3H_7NO_2$	2-Nitropropane	9,476.9
$C_3H_7NO_2$	Ethyl carbamate	13,078.6
$C_3H_7NO_2$	Propane	4,550.0
C_3H_8		4,811.8
C_3H_8O	*n*-Propanol	11,298.8
		10,421.1
C_3H_8O	Isopropanol	10,063.5
C_3H_8O	Ethyl methyl ether	6,388.3
$C_3H_8O_2$	1,2-Propanediol	13,575.2
$C_3H_8O_2$	1,3-Propanediol	13,782.3
$C_3H_8O_2$	2-Methoxyethanol	9,893.8
$C_3H_8O_2$	Glycerol	18,188.9
C_3H_8S	Methyl ethyl sulfide	—
C_3H_8S	Propanethiol	7,855.3
C_3H_9B	Trimethylborine	5,375.4
C_3H_9ClSi	Chlorotrimethylsilane	7,589.1
C_3H_9Ga	Trimethylgallium	7,758.8
C_3H_9N	n-Propylamine	7,408.0
C_3H_9N	Trimethylamine	6,361.7
$C_3H_9O_4P$	Trimethyl phosphate	11,019.7
$C_3H_{12}B_2$	Trimethyldiborane	6,981.8
C_3O_2	Carbon suboxide	6,446.3
C_3S_2	Carbon subsulfide	10,466.0
$C_4Cl_6O_3$	Trichloroacetic anhydride	12,929.0
C_3F_8	Octafluoropropane	
C_4H_4O	Furan	—
C_4H_2	1,3-Butadiyne	7,761.0
$C_4H_2Br_2O_3$	α,β-Dibromomaleic anhydride	12,579.2
$C_4H_2Cl_2O_2$	*trans*-Fumaryl chloride	11,251.0
$C_4H_2O_3$	Maleic anhydride	12,122.3
$C_4H_3NO_2S$	2-Nitrothiophene	11,926.2
C_4H_4	Butenyne	6,677.2
$C_4H_4Cl_2O_2$	Succinyl chloride	12,466.1
$C_4H_4Cl_2O_3$	Chloroacetic anhydride	14,645.1
$C_4H_4O_3$	Succinic anhydride	14,726.0
$C_4H_4O_4$	1,4-Dioxane-2,6-dione	14,013.6
C_4H_4S	Thiophene	8,748.3
C_4H_4Se	Selenophene	7,766.1
$C_4H_5ClO_2$	α-Chlorocrotonic acid	15,440.1
$C_4H_5ClO_3$	Ethyl chloroglyoxylate	10,268.4
$C_4H_5Cl_3O_2$	Ethyl trichloroacetate	11,625.1
C_4H_5N	3-Butenenitrile	9,447.8
C_4H_5N	Methacrylonitrile	8,083.8
C_4H_5N	*cis*-Crotononitrile	8,905.4
C_4H_5N	*trans*-Crotononitrile	9,227.1
$C_4H_5NO_2$	Succinimide	16,422.0
C_4H_5NS	Allylisothiocyanate	9,967.8
C_4H_6	1,2-Butadiene	6,539.1
C_4H_6	1,3-Butadiene	5,688.2
C_4H_6	Cyclobutene	6,167.5
C_4H_6	1-Butyne	6,596.9
C_4H_6	2-Butyne	7,868.5
$C_4H_6Cl_2O_2$	Ethyl dichloroacetate	10,842.8
$C_4H_6Cl_2O_2$	2-Chloroethyl chloroacetate	12,588.7
$C_4H_6O_2$	*cis*-Crotonic acid	12,964.7
$C_4H_6O_2$	*trans*-Crotonic acid	13,252.2
$C_4H_6O_2$	Methyl acrylate	8,598.0
$C_4H_6O_2$	Methacrylic acid	12,526.6
$C_4H_6O_2$	Vinyl acetate	8,470.4
$C_4H_6O_3$	Acetic anhydride	10,930.4
$C_4H_6O_4$	Dimethyl oxalate	11,519.4
C_4H_7Br	*cis*-1-Bromo-1-butene	8,300.2
C_4H_7Br	*trans*-1-Bromo-1-butene	8,515.7
C_4H_7Br	2-Bromo-1-butene	8,389.7
C_4H_7Br	*cis*-2-Bromo-2-butene	8,486.3
C_4H_7Br	*trans*-2-Bromo-2-butene	8,238.1
C_4H_7BrO	1-Bromo-2-butanone	10,980.7
C_4H_7BrO	2-Methylpropionyl bromide	10,974.6
$C_4H_7Br_3$	1,1,2-Tribromobutane	11,936.5
$C_4H_7Br_3$	1,2,2-Tribromobutane	11,622.3
$C_4H_7Br_3$	2,2,3-Tribromobutane	11,664.2
$C_4H_7ClO_2$	Ethyl chloroacetate	10,522.6
$C_4H_8Cl_2$	1,1-Dichloro-2-methylpropane	9,111.1
$C_4H_7Cl_3$	1,2,3-Trichlorobutane	9,447.0
C_4H_7N	Butyronitrile	9,462.9
$C_4H_7NO_2$	Diacetamide	14,508.1
C_4H_8	1-Butene	5,996.7
C_4H_8	*cis*-2-Butene	6,401.0
C_4H_8	*trans*-2-Butene	6,221.6
C_4H_8	2-Methylpropene	5,742.9
C_4H_8	Cyclobutane	6,464.8
C_4H_8BrClO	2-Bromoethyl-2-chloroethyl ether	12,010.5
$C_4H_8Br_2$	1,2-Dibromobutane	10,182.1
$C_4H_8Br_2$	*dl*-2,3-Dibromobutane	10,136.1
$C_4H_8Br_2$	*meso*-2,3-Dibromobutane	9,966.9
$C_4H_8Br_2$	1,4-Dibromobutane	11,369.3
$C_4H_8Br_2O$	Di(2-bromoethyl)ether	12,454.4
$C_4H_8Cl_2$	1,2-Dichlorobutane	8,850.6
$C_4H_8Cl_2$	2,3-Dichlorobutane	8,975.3
$C_4H_8Cl_2$	1,1-Dichloro-2-methylpropane	8,795.6
$C_4H_8Cl_2$	1,2-Dichloro-2-methylpropane	9,260.1
$C_4H_8Cl_2$	1,3-Dichloro-2-methylpropane	10,519.7
$C_4H_8Cl_2O$	Di(chloroethyl)ether	11,376.8
C_4H_8O	1,2-Epoxy-2-methylpropane	7,066.6
C_4H_8O	Methyl ethyl ketone	8,149.5
$C_4H_8O_2$	Dioxane	8,546.2
$C_4H_8O_2$	*n*-Butyric acid	11,881.2
$C_4H_8O_2$	Isobutyric acid	11,182.8
$C_4H_8O_2$	Ethyl acetate	8,301.1
$C_4H_8O_2$	Methyl propanoate	8,356.2
$C_4H_8O_2$	*n*-Propyl formate	8,208.1
$C_4H_8O_2$	Isopropyl formate	8,230.2
$C_4H_8O_3$	α-Hydroxyisobutyric acid	15,967.0
$C_4H_8O_3$	Ethyl glycolate	11,318.1

Formula	Name	ΔH$_v$
C_4H_9Br	*n*-Butyl bromide	8,789.1
C_4H_9BrO	1-Bromo-2-butanol	13,473.7
C_4H_9Cl	*n*-Butyl chloride	8,144.8
C_4H_9Cl	*sec*-Butyl chloride	7,407.9
C_4H_9Cl	Isobutyl chloride	8,045.1
C_4H_9Cl	*tert*-Butyl chloride	6,876.0
$C_4H_9ClO_2$	2-(2-Chloroethoxy)ethanol	14,082.1
C_4H_9I	*n*-Butyl iodide	—
C_4H_9I	1-Iodo-2-methylpropane	9,650.7
$C_4H_9NO_2$	Ethyl methylcarbamate	12,161.2
$C_4H_9NO_2$	Propyl carbamate	14,071.8
$C_4H_9N_2O_2$	Di(nitrosoethyl)amine	10,894.8
C_4H_{10}	*n*-Butane	5,801.2
C_4H_{10}	2-Methylpropane	5,084.4
		5,416.2
$C_4H_{10}Cl_2Si$	Dichlorodiethylsilane	10,038.6
$C_4H_{10}F_2Si$	Diethyldifluorosilane	8,214.9
$C_4H_{10}O$	*n*-Butyl alcohol	10,970.5
$C_4H_{10}O$	*sec*-Butyl alcohol	10,712.3
$C_4H_{10}O$	Isobutyl alcohol	10,936.0
$C_4H_{10}O$	*tert*-Butyl alcohol	10,413.2
$C_4H_{10}O$	Diethyl ether	6,946.2
$C_4H_{10}O$	Methyl Propyl ether	7,409.7
$C_4H_{10}O_2$	1,3-Butanediol	10,479.1
$C_4H_{10}O_2$	2,3-Butanediol	13,708.6
$C_4H_{10}O_2$	1,2-Dimethoxyethane	7,681.0
$C_4H_{10}O_2S$	2,2-Thiodiethanol	6,597.0
$C_4H_{10}O_3$	Diethylene glycol	16,146.7
$C_4H_{10}O_3$	1,2,3-Butanetriol	16,345.8
$C_4H_{10}O_3S$	Diethyl sulfite	10,783.0
$C_4H_{10}O_4S$	Diethyl sulfate	12,518.2
$C_4H_{10}S$	*n*-Butanethiol	—
$C_4H_{10}S$	Diethyl sulfide	8,210.8
$C_4H_{10}Se$	Diethyl selenide	9,274.7
$C_4H_{10}Zn$	Diethyl zinc	9,162.3
$C_4H_{11}N$	Diethyl amine	7,307.5
$C_4H_{11}N$	Isobutylamine	478.3
$C_4H_{12}Cl_2Si_2$	1,3-Diethoxytetramethyl-disiloxane	9,881.6
$C_4H_{12}Pb$	Tetramethyllead	8,843.8
$C_4H_{12}Si$	Tetramethylsilane	6,439.2
$C_4H_{12}Sn$	Tetramethyl tin	7,897.8
$C_4H_{14}B_2$	Tetramethyldiborane	7,517.1
C_4F_8	Octafluorocyclobutane	—
C_4F_{10}	Perfluoro-*n*-butane	—
C_5H_4BrN	3-Bromopyridine	10,863.7
C_5H_4Cln	2-Chloropyridine	10,614.5
$C_5H_4O_2$	2-Furaldehyde	11,614.6
$C_5H_4O_3$	Citraconic anhydride	12,307.8
C_5H_5N	Pyridine	9,649.4
$C_5H_6Cl_2O_2$	Glutaryl chloride	13,192.1
$C_5H_6N_2$	Glutaronitrile	13,767.5
$C_5H_6O_2$	Furfuryl alcohol	12,815.8
$C_5H_6O_3$	Glutaric anhydride	14,814.1
$C_5H_6O_3$	Pyrotartaric anhydride	13,251.2
C_5H_6S	2-Methylthiophene	8,884.2
C_5H_6S	3-Methylthiophene	9,084.1
$C_5H_7ClO_3$	Propyl chloroglyoxylate	11,430.0
C_5H_7N	Tiglonitrile	8,704.6
C_5H_7N	Angelonitrile	9,707.5
C_5H_7N	α-Ethylacrylonitrile	8,679.1
$C_5H_7NO_2$	Ethyl cyanoacetate	15,615.6
C_5H_8	Cyclopentene	—
C_5H_8	Isoprene	6,901.8
C_5H_8	1,3-Pentadiene	7,313.9
C_5H_8	1,4-Pentadiene	6,826.6
C_5H_8O	Tiglaldehyde	9,009.2
$C_5H_8O_2$	Levulinaldehyde	11,483.8
$C_5H_8O_2$	Tiglic acid	13,756.5
$C_5H_8O_2$	α-Valerolactone	11,537.0
$C_5H_8O_2$	α-Ethylacrylic acid	14,417.8
$C_5H_8O_2$	Ethyl acrylate	9,259.4
$C_5H_8O_2$	Methyl methacrylate	8,974.9
$C_5H_8O_3$	Levulinic acid	17,795.0
$C_5H_8O_4$	Glutaric acid	22,085.2
$C_5H_8O_4$	Dimethyl malonate	12,608.1
$C_5H_9ClO_2$	Ethyl α-chloropropionate	11,032.8
$C_5H_9ClO_2$	Isopropyl chloroacetate	10,575.7
C_5H_9N	Valeronitrile	9,931.3
C_5H_9NO	α-Hydroxybutyronitrile	13,577.0
C_5H_{10}	1-Pentene	6,931.2
C_5H_{10}	2-Pentene	—
C_5H_{10}	3-Methyl-2-butene	7,112.8
C_5H_{10}	2-Methyl-1-butene	6,474.6
C_5H_{10}	3-Methyl-1-butene	—
C_5H_{10}	Cyclopentane	7,411.1
C_5H_{10}	Methylcyclobutane	6,413.2
$C_5H_{10}Br_2$	1,2-Dibromopentane	11,130.0
$C_5H_{10}Br_2$	1,2-Dibromo-2-methylbutane	7,616.9
$C_5H_{10}Br_2$	1,3-Dibromo-3-methylbutane	10,639.6
$C_5H_{10}Cl_2Si$	Allyldichloroethylsilane	9,833.9
$C_5H_{10}O$	Diethyl ketone	11,183.0
$C_5H_{10}O$	Methyl n-propyl ketone	11,240.6
$C_5H_{10}O$	Methyl isopropyl ketone	11,073.2
$C_5H_{10}Cl_2O$	2-Chloroethyl 2-chloroisopropyl ether	11,420.8
$C_5H_{10}Cl_2O$	2-Chloroethyl 2-chloropropyl ether	11,316.9
	Di(2-chloroethoxy)methane	12,908.0
$C_5H_{10}Cl_2O_2$	4-Hydroxy-3-methyl-2-butanone	13,639.4
$C_5H_{10}O_2$	Valeric acid	13,370.3
$C_5H_{10}O_2$	Isovaleric acid	12,951.1
$C_5H_{10}O_2$	Ethyl propanoate	8,877.8
$C_5H_{10}O_2$	*n*-Propyl acetate	8,921.1
$C_5H_{10}O_2$	Isopropyl acetate	8,794.8
$C_5H_{10}O_2$	Methyl butyrate	8,886.0
$C_5H_{10}O_2$	Methyl isobutyrate	8,593.3
$C_5H_{10}O_2$	*n*-Butyl formate	9,285.9
$C_5H_{10}O_2$	Isobutyl formate	8,678.8
$C_5H_{10}O_2$	*sec*-Butyl formate	8,975.7
$C_5H_{10}O_2$	*tert*-Butyl formate	8,955.3
$C_5H_{10}O_2$	Diethyl carbonate	10,159.0
$C_5H_{10}O_3$	1-Bromopentane	—
$C_5H_{11}Br$	1-Bromo-3-methylbutane	9,282.7
$C_5H_{11}Br$	1-Chloropentane	—
$C_5H_{11}Br$	1-Iodopentane	—
$C_5H_{11}I$	1-Iodo-3-methylbutane	9,951.6
$C_5H_{11}I$	Piperidine	8,911.8
$C_5H_{11}N$	Pentanoic acid	—
$C_5H_{11}NO_2$	Isobutyl carbamate	13,897.1
$C_5H_{11}NO_3$	Isoamyl nitrate	10,817.2
C_5H_{12}	*n*-Pentane	6,595.1
C_5H_{12}	2-Methylbutane	6,470.8
C_5H_{12}	2,2-Dimethylpropane	5,648.6
$C_5H_{12}O$	Amyl alcohol	12,495.5
$C_5H_{12}O$	Isoamyl alcohol	12,497.9
$C_5H_{12}O$	2-Pentanol	12,086.2
$C_5H_{12}O$	*tert*-Amyl alcohol	11,239.2
$C_5H_{12}O$	Ethyl propyl ether	7,092.7
$C_5H_{12}O$	Methyl n-butyl ether	—
$C_5H_{12}O_3$	2,3,4-Pentanetriol	19,694.4
$C_5H_{12}S$	1-Pentanethiol	—
$C_5H_{14}OSi$	Ethoxytrimethylsilane	8,030.6
$C_5H_{14}Si$	Ethyltrimethylsilane	7,633.4
$C_5H_{14}Sn$	Ethyltrimethyltin	8,820.9
$C_6Cl_4O_2$	Chloranil	21,514.3
C_6Cl_6	Hexachlorobenzene	15,199.1
C_6HCl_5	Pentachlorobenzene	15,124.2
C_6HCl_5O	Pentachlorophenol	16,742.6
$C_6H_2BrCl_3O$	3-Bromo-2,4,6-trichlorophenol	15,231.9
$C_6H_2Cl_4$	1,2,3,4-Tetrachlorobenzene	12,872.5
$C_6H_2Cl_4$	1,2,3,5-Tetrachlorobenzene	11,982.1
$C_6H_2Cl_4$	1,2,4,5-Tetrachlorobenzene	12,828.8
$C_6H_2Cl_4O$	2,3,4,6-Tetrachlorophenol	15,362.7
$C_6H_3BrCl_2O$	2-Bromo-4,6-dichlorophenol	13,829.1
$C_6H_3Cl_3$	1,2,3-Trichlorobenzene	11,349.5
$C_6H_3Cl_3$	1,2,4-Trichlorobenzene	11,425.1
$C_6H_3Cl_3$	1,3,5-Trichlorobenzene	11,211.0

Formula	Name	ΔH$_v$
$C_6H_3Cl_3O$	2,4,5-Trichlorophenol	13,237.0
$C_6H_3Cl_3O$	2,4,6-Trichlorophenol	14,092.8
$C_6H_4Br_2$	1,4-Dibromobenzene	13,047.8
C_6H_4BrCl	1,4-Bromochlorobenzene	16,671.8
		11,451.1
$C_6H_4Cl_2$	1,2-Dichlorobenzene	10,943.0
$C_6H_4Cl_2$	1,3-Dichlorobenzene	10,446.8
$C_6H_4Cl_2$	1,4-Dichlorobenzene	17,260.5
		10,611.0
$C_6H_4Cl_2O$	2,4-Dichlorophenol	13,230.4
$C_6H_4Cl_2O$	2,6-Dichlorophenol	13,472.0
$C_6H_4Cl_3N$	2,4,6-Trichloroaniline	22,297.3
$C_6H_5AsCl_2$	Dichlorophenylarsine	12,229.5
C_6H_5Br	Bromobenzene	10,157.7
C_6H_5Cl	Chlorobenzene	10,098.0
		9,067.3
C_6H_5ClO	2-Chlorophenol	10,341.1
C_6H_5ClO	3-Chlorophenol	11,979.7
C_6H_5ClO	4-Chlorophenol	12,281.6
$C_6H_5ClO_2S$	Benzenesulfonylchloride	12,621.0
$C_6H_5Cl_2O_2P$	Phenyl dichlorophosphate	13,319.6
$C_6H_5Cl_3Si$	Trichlorophenylsilane	11,385.9
C_6H_5F	Fluorobenzene	7,980.4
$C_6H_5F_3Si$	Trifluorophenylsilane	9,171.6
C_6H_5I	Iodobenzene	10,277.2
		10,377.8
$C_6H_5NO_2$	Nitrobenzene	12,168.2
$C_6H_5NO_3$	2-Nitrophenol	12,497.3
C_6H_6	1,5-Hexadiene-3-yne	8,288.0
C_6H_6	Benzene	10,254.3
		8,146.5
C_6H_6ClN	2-Chloroaniline	12,441.0
C_6H_6ClN	3-Chloroaniline	13,385.6
C_6H_6ClN	4-Chloroaniline	12,832.8
C_6H_6ClO	4-Chlorophenol	12,964.7
$C_6H_6N_2O_2$	2-Nitroaniline	15,284.0
$C_6H_6N_2O_2$	3-Nitroaniline	15,996.3
$C_6H_6N_2O_2$	4-Nitroaniline	17,220.2
C_6H_6O	Phenol	11,891.5
$C_6H_6O_2$	Pyrocatechol	13,779.7
$C_6H_6O_2$	Resorcinol	16,400.8
$C_6H_6O_2$	Hydroquinone	18,734.0
$C_6H_6O_3$	Pyrogallol	15,731.8
C_6H_6S	Benzenethiol	11,320.1
C_6H_7N	Aniline	11,307.6
C_6H_7N	2-Picoline	9,933.2
C_6H_7N	3-Methylpyridine (β-picoline)	—
C_6H_8	1,3-Cyclohexadiene	—
$C_6H_8Cl_2O_4$	Ethylene-bis-chloroacetate	16,499.1
$C_6H_8N_2$	1,3-Phenylenediamine	14,761.1
$C_6H_8N_2$	Phenylhydrazine	13,711.9
$C_6H_8O_3$	α-Methylglutaric anhydride	14,204.9
$C_6H_8O_3$	α,α-Dimethylsuccinic anhydride	13,683.1
$C_6H_8O_4$	Dimethyl maleate	12,615.7
C_6H_{10}	Cyclohexene	—
C_6H_{10}	1,5-Hexadiene	—
$C_6H_{10}Cl_2O_2$	Isobutyl dichloroacetate	11,733.1
$C_6H_{10}Cl_2Si$	Diallyldichlorosilane	10,462.8
$C_6H_{10}O$	Cyclohexanone	10,037.6
$C_6H_{10}O$	Mesityl oxide	10,109.4
$C_6H_{10}O_2$	Isocaprolactone	11,685.0
$C_6H_{10}O_3$	Propionic anhydride	11,572.6
$C_6H_{10}O_3$	Ethyl acetoacetate	11,842.0
$C_6H_{10}O_3$	Methyl levulinate	12,249.8
$C_6H_{10}O_4$	Adipic acid	19,570.2
$C_6H_{10}O_4$	Diethyl oxalate	14,016.9
$C_6H_{10}O_4$	Glycol diacetate	12,496.1
$C_6H_{10}O_5$	Dimethyl-*l*-malate	14,127.6
$C_6H_{10}O_6$	Dimethyl-*d*-tartrate	15,372.6
$C_6H_{10}O_6$	Dimethyl-*dl*-tartrate	14,999.1
$C_6H_{10}S$	Diallyl sulfide	9,652.6
$C_6H_{11}BrO_2$	Ethyl α-bromoisobutyrate	10,635.8
$C_6H_{11}ClO_2$	*sec*-Butylchloroacetate	11,152.0
$C_6H_{11}N$	Capronitrile	10,492.3
C_6H_{12}	1-Hexene	7,787.6
C_6H_{12}	2-Hexene	—
C_6H_{12}	Cyclohexane	7,830.9
C_6H_{12}	Methylcyclopentane	7,940.0
$C_6H_{12}Cl_2O$	Dichlorodiisopropyl ether	11,881.1
$C_6H_{12}Cl_2O_2$	bis(2-Chloroethyl)acetal	13,497.1
$C_6H_{12}O$	2-Hexanone	12,358.3
$C_6H_{12}O$	4-Methyl-2-pentanone	11,669.6
$C_6H_{12}O$	Allyl propyl ether	8,621.5
$C_6H_{12}O$	Allyl isopropyl ether	8,637.5
$C_6H_{12}O$	Cyclohexanol	11,935.8
$C_6H_{12}O_2$	Caproic acid	16,189.4
$C_6H_{12}O_2$	Isocaproic acid	14,874.8
$C_6H_{12}O_2$	4-Hydroxy-4-methyl-2-pentanone	11,718.8
$C_6H_{11}O_2$	Methyl pentanoate	—
$C_6H_{12}O_2$	Methyl isovalerate	9,567.5
$C_6H_{12}O_2$	Ethyl *n*-butyrate	9,468.5
$C_6H_{12}O_2$	Ethyl isobutyrate	8,945.7
$C_6H_{12}O_2$	*n*-Propyl propanoate	9,857.2
$C_6H_{12}O_2$	*n*-Butyl acetate	—
$C_6H_{12}O_2$	Isobutyl acetate	9,300.8
$C_6H_{12}O_2$	*n*-Amyl formate	—
$C_6H_{12}O_2$	Isoamyl formate	9,438.2
$C_6H_{12}O_3$	Paraformaldehyde	10,348.2
C_6H_{14}	Hexane	7,627.2
C_6H_{14}	2-Methylpentane	7,676.6
C_6H_{14}	3-Methylpentane	7,743.9
C_6H_{14}	2,2-Dimethylbutane	7,271.0
C_6H_{14}	2,3-Dimethylbutane	7,120.0
$C_6H_{14}O$	1-Hexanol	12,708.5
$C_6H_{14}O$	2-Hexanol	12,386.5
$C_6H_{14}O$	3-Hexanol	11,157.9
$C_6H_{14}O$	2-Methyl-1-pentanol	12,036.6
$C_6H_{14}O$	2-Methyl-2-pentanol	11,132.0
$C_6H_{14}O$	2-Methyl-4-pentanol	10,985.5
$C_6H_{14}O$	Ethyl butyl ether	—
$C_6H_{14}O$	Di-*n*-propyl ether	8,229.6
$C_6H_{14}O$	Diisopropyl ether	7,777.3
$C_6H_{14}O_2$	Acetal	9,853.9
$C_6H_{14}O_2$	1,2-Diethoxyethane	8,102.6
$C_6H_{14}O_3$	Di(2-methoxyethyl)ether	11,105.2
$C_6H_{14}O_3$	Diethyleneglycol-diethyl ether	12,669.0
$C_6H_{14}O_3$	Dipropyleneglycol	14,610.4
$C_6H_{14}O_4$	Triethyleneglycol	17,097.1
$C_6H_{15}N$	Dipropyl sulfide	—
$C_6H_{15}N$	Di-*n*-Propylamine	—
$C_6H_{15}N$	Triethylamine	—
$C_6H_{15}B$	Triethylboron	2,535.0
$C_6H_{15}ClSi$	Chlorotriethylsilane	9,806.9
$C_6H_{15}O_4P$	Triethyl phosphate	11,549.9
$C_6H_{15}Tl$	Triethylthallium	9,458.6
$C_6H_{16}O_2Si$	Diethoxydimethylsilane	9,758.2
$C_6H_{16}Si$	Trimethylpropylsilane	7,964.6
$C_6H_{16}Sn$	Trimethylpropyltin	9,659.6
$C_6H_{18}Cl_2O_2Si_3$	1,5-Dichlorohexamethyltrisiloxane	11,391.5
$C_6H_{18}O_3Si_3$	Hexamethylcyclotrisiloxane	10,503.3
$C_7H_3Cl_2F_3$	3,4-Dichloro-α,α,α-trifluorotoluene	10,253.5
$C_7H_4ClF_3$	2-Chloro-α,α,α-trifluorotoluene	10,016.9
$C_7H_4Cl_4$	2-α,α,α-Tetrachlorotoluene	12,501.3
C_7H_5BrO	Benzoyl bromide	12,070.8
C_7H_5ClO	Benzoyl chloride	11,438.0
$C_7H_5Cl_2$	α,α,α-Trichlorotoluene	12,168.6
$C_7H_5F_3$	α,α,α-Trifluorotoluene	8,869.7
C_7H_5N	Benzonitrile	11,341.0
C_7H_5N	Phenyl isocyanide	10,736.7
C_7H_5NO	Phenyl isocyanate	10,556.7
$C_7H_5NO_3$	2-Nitrobenzaldehyde	13,773.6
$C_7H_5NO_3$	3-Nitrobenzaldehyde	14,726.9
C_7H_5NS	Phenyl isothiocyanate	12,132.7

Formula	Name	ΔH$_v$
$C_7H_6Cl_2$	α,α-Dichlorotoluene	11,075.9
C_7H_6O	Benzaldehyde	11,657.8
$C_7H_6O_2$	Benzoic acid	15,253.3
		16,295.1
C_7H_6O	Salicylaldehyde	11,536.5
$C_7H_6O_2$	4-Hydroxybenzaldehyde	16,043.4
$C_7H_6O_3$	Salicylic acid	18,920.7
C_7H_7Br	α-Bromotoluene	11,360.4
C_7H_7Br	2-Bromotoluene	11,365.0
C_7H_7Br	3-Bromotoluene	10,537.1
C_7H_7Br	4-Bromotoluene	10,076.2
C_7H_7BrO	4-Bromoanisole	12,075.4
C_7H_7Cl	α-Chlorotoluene	11,158.7
C_7H_7Cl	2-Chlorotoluene	10,279.3
C_7H_7Cl	3-Chlorotoluene	10,081.1
C_7H_7Cl	4-Chlorotoluene	10,151.7
C_7H_7F	2-Fluorotoluene	9,164.8
C_7H_7F	3-Fluorotoluene	9,251.8
C_7H_7F	4-Fluorotoluene	9,281.0
C_7H_7I	2-Iodotoluene	11,380.7
$C_7H_7NO_2$	2-Nitrotoluene	12,239.1
$C_7H_7NO_2$	3-Nitrotoluene	11,831.1
$C_7H_7NO_2$	4-Nitrotoluene	11,915.0
C_7H_8	Toluene	9,368.5
		8,580.5
$C_7H_8Cl_2Si$	Benzyldichlorosilane	13,128.7
$C_7H_8Cl_2Si$	Dichloromethylphenylsilane	11,464.7
$C_7H_8Cl_2Si$	Dichloro-4-tolysilane	13,125.7
C_7H_8O	Anisole	10,440.9
C_7H_8O	Benzyl alcohol	14,093.2
C_7H_8O	*o*-Cresol	12,487.3
C_7H_8O	*m*-Cresol	13,483.8
C_7H_8O	*p*-Cresol	13,611.7
$C_7H_8O_2$	3,5-Dimethyl-1,2-pyrone	14,470.6
$C_7H_8O_2$	2-Methoxyphenol	13,425.8
$C_7H_8O_3$	Ethyl 2-furoate	12,144.0
C_7H_9N	2,6-Dimethylpyridine	—
C_7H_9N	Benzylamine	11,703.2
C_7H_9N	*N*-Methylaniline	11,982.3
C_7H_9N	2-Toluidine	12,663.4
C_7H_9N	3-Toluidine	12,104.1
$C_7H_{10}O_4$	Dimethyl citraconate	12,917.3
$C_7H_{10}O_4$	Dimethyl itaconate	15,613.7
$C_7H_{10}O_4$	*trans*-Dimethyl mesaconate	12,688.1
$C_7H_{11}NO_2$	2-Cyano-2-butyl acetate	12,720.8
$C_7H_{12}O_2$	Butyl acrylate	10,194.0
C_7H_9N	4-Toluidine	12,428.6
C_7H_9NO	2-Methoxyaniline	13,684.6
$C_7H_{10}N_2$	Toluene-2,4-diamine	15,928.1
$C_7H_{10}N_2$	4-Tolylhydrazine	15,063.1
$C_7H_{10}O_3$	Trimethylsuccinic anhydride	12,196.7
$C_7H_{12}O_3$	Ethyl levulinate	12,733.6
$C_7H_{12}O_4$	Pimelic acid	19,840.8
$C_7H_{12}O_4$	Diethyl malonate	12,227.7
$C_7H_{13}ClO$	Enanthyl chloride	15,242.7
$C_7H_{13}N$	Heptanonitrile	10,830.5
C_7H_{14}	Ethylcyclopentane	8,797.7
C_7H_{14}	2-Heptene	8,643.2
C_7H_{14}	Methylcyclohexane	8,549.2
$C_7H_{14}O$	Emanthaldehyde	11,413.4
$C_7H_{14}O$	2-Heptanone	12,478.9
$C_7H_{14}O$	4-Heptanone	13,451.9
$C_7H_{14}O$	2,5-Dimethyl-3-pentanone	12,266.9
$C_7H_{14}O_2$	Enanthic acid	15,893.8
$C_7H_{14}O_2$	Methyl caproate	10,676.8
$C_7H_{14}O_2$	Ethyl isovalerate	10,183.9
$C_7H_{14}O_2$	Propyl butyrate	10,283.7
$C_7H_{14}O_2$	Propyl isobutyrate	10,259.7
$C_7H_{14}O_2$	Isopropyl isobutyrate	9,717.6
$C_7H_{14}O_2$	Isobutyl propionate	10,495.8
$C_7H_{14}O_2$	Isoamyl acetate	10,494.9
C_7H_{16}	Perfluoro-*n*-Heptane	—
C_7H_{16}	*n*-Heptane	8,928.8
		8,409.6
C_7H_{16}	2-Methylhexane	8,538.7
C_7H_{16}	3-Methylhexane	8,596.3
C_7H_{16}	3-Ethylpentane	8,642.8
C_7H_{16}	2,2-Dimethylpentane	8,106.7
C_7H_{16}	2,3-Dimethylpentane	8,390.9
C_7H_{16}	2,4-Dimethylpentane	8,167.1
C_7H_{16}	3,3-Dimethylpentane	8,145.4
C_7H_{16}	2,2,3-Trimethylbutane	7,767.1
$C_7H_{16}O$	*n*-Heptanol	13,920.9
$C_7H_{16}O_3$	Triethyl orthoformate	10,935.0
$C_7H_{18}O_3Si$	Triethoxymethylsilane	10,306.7
$C_7H_{18}Si$	Butyltrimethylsilane	9,206.0
$C_7H_{18}Si$	Triethylmethylsilane	9,232.5
$C_8H_4Cl_2O_2$	Phthaloyl chloride	13,716.0
$C_8H_4O_3$	Phthalic anhydride	13,919.0
$C_8H_5Cl_2N$	α,α-Dichlorophenylacetonitrile	12,829.9
$C_8H_5Cl_5$	Pentachloroethylbenzene	13,728.7
$C_8H_6Cl_2$	2,3-Dichlorostyrene	12,827.2
$C_8H_6Cl_2$	2,4-Dichlorostyrene	12,511.7
$C_8H_6Cl_2$	2,5-Dichlorostyrene	12,592.5
$C_8H_6Cl_2$	2,6-Dichlorostyrene	12,186.0
$C_8H_6Cl_2$	3,4-Dichlorostyrene	12,626.5
$C_8H_6Cl_2$	3,5-Dichlorostyrene	12,511.7
$C_8H_6Cl_4$	3,4,5,6-Tetrachloro-1,2-xylene	14,763.1
$C_8H_6Cl_4$	1,2,3,5-Tetrachloro-4-ethylbenzene	12,980.3
$C_8H_6O_2$	Phenylglyoxal	13,731.6
$C_8H_6O_2$	Phthalide	14,021.6
$C_8H_6O_3$	Piperonal	14,425.5
C_8H_7Cl	3-Chlorostyrene	10,990.2
C_8H_7ClO	Phenylacetyl chloride	12,627.1
C_8H_7N	2-Tolunitrile	11,557.7
C_8H_7N	4-Tolunitrile	11,562.8
C_8H_7N	Phenylacetonitrile	12,796.2
C_8H_7N	2-Tolyl isocyanide	11,303.3
$C_8H_7NO_4$	2-Nitrophenyl acetate	16,875.3
C_8H_7NS	2-Methylbenzothiazole	14,492.3
C_8H_8	Styrene	9,634.7
$C_8H_8Br_2$	(1,2-Dibromoethyl)benzene	14,874.7
$C_8H_8Cl_2$	1,2-Dichloro-3-ethylbenzene	11,784.3
$C_8H_8Cl_2$	1,2-Dichloro-4-ethylbenzene	11,711.5
$C_8H_8Cl_2$	1,4-Dichloro-2-ethylbenzene	11,262.7
C_8H_8O	Acetophenone	11,731.5
$C_8H_8O_2$	Phenylacetate	12,174.9
$C_8H_8O_2$	Phenylacetic acid	15,568.7
$C_8H_8O_2$	Anisaldehyde	13,581.8
$C_8H_8O_2$	Methyl benzoate	12,077.2
$C_8H_8O_2$	Methyl salicylate	12,658.8
$C_8H_8O_3$	Vanillin	15,703.2
$C_8H_8O_4$	Dihydroacetic acid	14,663.8
C_8H_9Br	2-Bromo-1,4-xylene	11,603.7
C_8H_9Br	1-Bromo-4-ethylbenzene	10,170.0
C_8H_9Br	(2-Bromoethyl)benzene	12,152.5
C_8H_9Cl	1-Chloro-2-ethylbenzene	10,749.7
C_8H_9Cl	1-Chloro-3-ethylbenzene	10,724.1
C_8H_9Cl	1-Chloro-4-ethylbenzene	10,659.9
C_8H_9ClO	1-Chloro-2-ethoxybenzene	12,411.1
C_8H_9ClO	4-Chlorophenylethyl alcohol	14,298.5
$C_8H_{10}Cl_2Si$	Dichlorophenylethylsilane	11,895.1
C_8H_9NO	Acetanilide	15,474.1
$C_8H_9NO_2$	Methyl anthranilate	13,186.3
$C_8H_9NO_2$	4-Nitro-1,3-xylene	12,948.0
C_8H_{10}	Ethylbenzene	9,301.3
C_8H_{10}	*o*-Xylene	9,998.5
C_8H_{10}	*m*-Xylene	9,904.2
C_8H_{10}	*p*-Xylene	9,809.9
$C_8H_{10}Cl_2OSi$	Dichloroethoxyphenylsilane	12,516.5
$C_8H_{10}Cl_2Si$	Dichloroethylphenylsilane	11,721.2
$C_8H_{10}O$	2-Ethylphenol	12,516.7
$C_8H_{10}O$	3-Ethylphenol	13,856.4
$C_8H_{10}O$	4-Ethylphenol	13,437.9
$C_8H_{10}O$	Xylenol	—

Formula	Name	ΔH_v
C_8H_{10}	2,3-Xylenol	13,106.9
$C_8H_{10}O$	2,4-Xylenol	13,130.2
$C_8H_{10}O$	2,5-Xylenol	13,130.2
$C_8H_{10}O$	3,4-Xylenol	13,991.0
$C_8H_{10}O$	3,5-Xylenol	13,767.7
$C_8H_{10}O$	Phenetole	11,075.8
$C_8H_{10}O$	α-Methyl benzyl alcohol	13,087.4
$C_8H_{10}O$	Phenylethylalcohol	13,307.4
$C_8H_{10}O_2$	4,6-Dimethylresorcinol	12,433.1
$C_8H_{10}O_2$	2-Phenoxyethanol	14,368.3
$C_8H_{10}O_6$	Diethyl dioxosuccinate	13,973.3
$C_8H_{11}ClSi$	Chlorodimethylphenylsilane	11,382.2
$C_8H_{11}N$	*N*-Ethylaniline	11,817.0
$C_8H_{11}N$	*N,N*-Dimethylaniline	11,320.4
$C_8H_{11}N$	4-Ethylaniline	12,679.9
C;8$H_{11}N$	2,4-Xylidine	13,099.2
$C_8H_{11}N$	2,6-Xylidine	11,742.6
$C_8H_{11}NO$	2-Phenetidine	13,877.8
$C_8H_{11}NO$	2-Anilinoethanol	15,643.2
$C_8H_{12}AsNO_2$	Dimethyl arsanilate	11,277.7
$C_8H_{12}Cl_2O_5$	Diethyleneglycol-bis-chloroacetate	19,830.5
$C_8H_{12}O_4$	Diethyl maleate	12,908.0
$C_8H_{12}O_4$	Diethyl fumarate	12,747.4
$C_8H_{12}Si$	Dimethylphenylsilane	10,274.2
$C_8H_{14}O_3$	Ethyl-α-ethylacetoacetate	12,344.2
$C_8H_{14}O_4$	Propyl levulinate	13,354.4
$C_8H_{14}O_4$	Isopropyl levulinate	12,689.6
$C_8H_{14}O_4$	Dipropyl oxalate	13,056.4
$C_8H_{14}O_4$	Diisopropyl oxalate	12,949.3
$C_8H_{14}O_4$	Diethyl succinate	13,076.1
$C_8H_{14}O_4$	Diethyl isosuccinate	12,087.6
$C_8H_{14}O_4$	Suberic acid	21,089.8
$C_8H_{14}O_5$	Diethyl malate	14,202.9
$C_8H_{14}O_6$	Diethyl-*dl*-tartrate	15,150.4
$C_8H_{14}O_6$	Diethyl-*d*-tartrate	15,517.8
$C_8H_{15}Br$	(2-Bromoethyl)cyclohexane	11,462.7
$C_8H_{15}N$	*n*-Caprylonitrile	12,221.8
C_8H_{15}NO:i3	Ethyl-*N,N*-diethyloxamate	13,758.4
C_8H_{16}	1-Octene	—
C_8H_{16}	2-Octene	—
C_8H_{16}	2-Methyl-2-heptene	9,643.8
C_8H_{16}	1,1-Dimethylcyclohexane	8,949.1
C_8H_{16}	*cis*-1,2-Dimethylcyclohexane	9,364.9
C_8H_{16}	*trans*-1,2-Dimethylcyclohexane	9,097.1
C_8H_{16}	*cis*-1,3-Dimethylcyclohexane	9,232.6
C_8H_{16}	*trans*-1,3-Dimethylcyclohexane	9,080.3
C_8H_{16}	*cis*-1,4-Dimethylcyclohexane	9,188.9
C_8H_{16}	*trans*-1,4-Dimethylcyclohexane	8,951.2
C_8H_{16}	Ethylcyclohexane	9,441.2
$C_8H_{16}O$	Caprylaldehyde	21,201.0
$C_8H_{16}O$	Cyclohexaneethanol	13,152.4
$C_8H_{16}O$	6-Methyl-3-hepten-2-ol	13,864.1
$C_8H_{15}O$	6-Methyl-5-hepten-2-ol	13,999.1
$C_8H_{16}O$	2-Octanone	11,649.2
$C_8H_{16}O$	2,2,4-Trimethyl-3-pentanone	12,854.6
$C_8H_{16}O_2$	Caprylic acid	16,745.7
$C_8H_{16}O_2$	Ethyl isocaproate	10,826.7
$C_8H_{16}O_2$	Propyl isovalerate	10,715.7
$C_8H_{16}O$	Isobutyl butyrate	10,283.9
$C_8H_{16}O_2$	Isobutyl isobutyrate	10,706.3
$C_8H_{16}O_2$	Amylisopropionate	10,567.2
$C_8H_{16}ClO_4$	Tetraethyleneglycol-chlorohydrin	16,371.2
$C_8H_{17}I$	1-Iodooctane	11,625.1
$C_8H_{17}NO_2$	Ethyl-*l*-leucinate	11,383.5
C_8H_{18}	Octane	9,221.0
C_8H_{18}	2-Methylheptane	9,362.0
C_8H_{18}	3-Methylheptane	9,432.0
C_8H_{18}	4-Methylheptane	9,404.8
C_8H_{18}	2,2-Dimethylhexane	8,927.8
C_8H_{18}	2,3-Dimethylhexane	9,224.9
C_8H_{18}	2,4-Dimethylhexane	9,086.6
C_8H_{18}	2,5-Dimethylhexane	9,110.2
C_8H_{18}	3,3-Dimethylhexane	9,065.2
C_8H_{18}	3,4-Dimethylhexane	9,239.4
C_8H_{18}	3-Ethylhexane	9,416.3
C_8H_{18}	2,2,3-Trimethylpentane	8,861.1
C_8H_{18}	2,2,4-Trimethylpentane	8,548.0
C_8H_{18}	2,3,3-Trimethylpentane	8,960.9
C_8H_{18}	2,3,4-Trimethylpentane	8,988.2
C_8H_{18}	2-Methyl-3-3ethylpentane	9,134.3
C_8H_{18}	3-Methyl-3-ethylpentane	9,028.7
C_8H_{18}	2,2,3,3-Tetramethylbutane	10,351.5
$C_8H_{18}N_2$	Tetramethylpiperazine	11,187.5
C_8H_{18}	*n*-Octanol	14,262.4
C_8H_{18}	2-Octanol	12,468.4
$C_8H_{18}O$	Di*n*-butyl ether	—
$C_8H_{18}O$	Methyl heptyl ether	—
$C_8H_{18}O_2$	1,2-Dipropoxy ethane	6,370.7
$C_8H_{18}O_3$	Diethylene glycolbutyl ether	14,127.0
$C_8H_{18}O_5$	Tetraethylene glycol	21,296.6
$C_8H_{18}S$	Di *n*-butyl sulfide	11,183.6
$C_8H_{18}S_2$	Dibutyl disulfide	8,254.1
$C_8H_{19}N$	Diisobutylamine	10,058.3
$C_8H_{20}O_4Si$	Tetraethoxysilane	10,968.6
$C_8H_{20}Pb$	Tetraethyllead	12,959.7
$C_8H_{20}Si$	Amyltrimethylsilane	9,659.6
$C_8H_{20}Si$	Tetraethylsilane	9,893.0
$C_8H_{20}Sb_2$	Tetraethylbistibine	12,975.4
$C_8H_{22}O_3Si_2$	1,3-Dichlorotetramethyl-disiloxane	11,261.9
$C_8H_{24}Cl_2O_3Si_4$	1,7-Dichlorooctamethyl-tetrasiloxane	12,602.9
$C_8H_{24}O_2Si_3$	Octamethyltrisiloxane	10,956.0
$C_8H_{24}O_4Si_4$	Octamethylcyclotetra siloxane	11,515.0
$C_9H_6O_2$	Coumarin	15,202.7
C_9H_7N	Quinoline	12,575.4
C_9H_7N	Isoquinoline	12,847.6
C_9H_8	Indene	10,496.7
C_9H_8O	Cinnamylaldehyde	14,048.4
$C_9H_8O_2$	*trans*-Cinnamic acid	17,492.9
C_9H_9N	Skatole	15,232.7
$C_9H_9NO_4$	Ethyl 3-nitrobenzoate	15,056.1
C_9H_{10}	α-Methyl styrene	10,214.6
C_9H_{10}	β-Methyl styrene	10,701.3
C_9H_{10}	2-Methyl styrene	—
C_9H_{10}	3-Methyl styrene	—
C_9H_{10}	4-Methyl styrene	10,724.2
$C_9H_{10}O$	2,4-Xylaldehyde	13,618.4
$C_9H_{10}O$	Cinnamyl alcohol	13,421.6
$C_9H_{10}O$	Propiophenone	12,407.6
$C_9H_{10}O$	3-Vinylanisole	12,756.4
$C_9H_{10}O$	3-Vinylanisole	12,735.8
$C_9H_{10}O$	4-Vinylanisole	12,554.7
$C_9H_{10}O_2$	Benzylacetate	12,107.2
$C_9H_{10}O_2$	Ethyl benzoate	11,981.5
$C_9H_{10}O_2$	Hydrocinnamic acid	15,411.9
$C_9H_{10}O_2$	Ethyl salicylate	13,030.1
$C_9H_{11}NO$	*N*-Methylacetanilide	13,235.2
$C_9H_{11}NO_2$	Ethyl carbanilate	19,791.8
C_9H_{12}	1,2,3-Trimethylbenzene	10,781.9
C_9H_{12}	1,2,4-Trimethylbenzene	10,710.2
C_9H_{12}	1,3,5-Trimethylbenzene	10,516.8
C_9H_{12}	*o*-Ethyl toluene	10,488.8
C_9H_{12}	*m*-Ethyl toluene	10,416.6
C_9H_{12}	*p*-Ethyl toluene	10,461.1
C_9H_{12}	Isopropylbenzene	10,335.3
C_9H_{12}	*N*-Propylbenzene	10,424.1
$C_9H_{12}O$	2-Ethylanisole	11,642.8
$C_9H_{12}O$	3-Ethylanisole	11,616.7
$C_9H_{12}O$	4-Ethylanisole	11,625.7
$C_9H_{12}O$	3-Phenyl-1-propanol	14,493.9
$C_9H_{12}O$	2-Isopropylphenol	13,402.3
$C_9H_{12}O$	3-Isopropylphenol	13,292.2
$C_9H_{12}O$	4-Isopropylphenol	13,878.7
$C_9H_{12}O$	Benzyl ethyl ether	11,315.5

HEATS OF VAPORIZATION OF ORGANIC COMPOUNDS (continued)

Formula	Name	ΔH$_v$
$C_9H_{13}ClOSi$	Chloroethoxymethyl-phenylsilane	12,270.3
$C_9H_{13}N$	2,4,5-Trimethylaniline	13,975.0
$C_9H_{13}N$	*N,N*-Dimethyl-*O*-toluidine	11,648.3
$C_9H_{13}N$	*N,N*-Dimethyl-4-toluidine	12,738.4
$C_9H_{13}N$	4-Cumidine	12,137.9
$C_9H_{14}O$	Phorone	12,557.2
$C_9H_{14}O$	Isophorone	11,277.6
$C_9H_{14}O_4$	*cis*-Diethyl citraconate	12,913.2
$C_9H_{14}O_4$	Diethyl itaconate	12,075.8
$C_9H_{14}O_4$	Diethyl mesaconate	13,326.1
$C_9H_{14}O_7$	Trimethyl citrate	15,807.7
$C_9H_{16}O_3$	Isobutyl levulinate	13,571.2
$C_9H_{16}O$	Azelaic acid	20,944.2
$C_9H_{16}O_4$	Diethyl ethylmalonate	12,842.0
$C_9H_{16}O_4$	Diethyl glutarate	13,261.5
$C_9H_{18}O$	2-Nonanone	11,529.5
$C_9H_{18}O$	Di-isobutyl ketone	—
$C_9H_{18}O$	Azelaldehyde	12,143.4
$C_9H_{18}O_2$	Pelargonic acid	17,807.8
$C_9H_{18}O_2$	Methyl caprylate	11,914.9
$C_9H_{18}O_2$	Isobutyl isovalerate	10,999.7
$C_9H_{18}O_2$	Isoamyl butyrate	11,104.5
$C_9H_{18}O_2$	Isoamyl isobutyrate	10,870.6
$C_9H_{19}I$	1-Iodononane	14,853.0
C_9H_{20}	*n*-Nonane	10,456.9
C_9H_{20}	2,6-Dimethylheptane	—
C_9H_{20}	2-Methyloctane	—
C_9H_{20}	3-Methyloctane	—
$C_9H_{20}O$	1-Nonanol	13,849.2
$C_9H_{20}O$	Diisobutyl carbinol	—
$C_9H_{20}O_3$	Dipropyleneglycol isopropyl ether	12,583.8
$C_9H_{20}O_4$	Tripropyleneglycol	15,291.4
$C_9H_{22}Si$	Hexyltrimethylsilane	10,264.9
$C_9H_{22}Si$	Triethylpropylsilane	10,709.3
$C_{10}H_7Br$	1-Bromonaphthalene	13,274.9
$C_{10}H_7Cl$	1-Chloronaphthalene	13,570.5
$C_{10}H_{12}$	Dicyclopentadiene	10,165.9
$C_{10}H_8$	Naphthalene	17,065.2
		12,311.6
$C_{10}H_8Cl_2Si$	Dichloro-1-naphthysilane	16,325.3
$C_{10}H_8O$	1-Naphthol	14,205.6
$C_{10}H_8O$	2-Naphthol	14,138.5
$C_{10}H_9N$	1-Naphthylamine	14,529.5
$C_{10}H_9N$	2-Naphthylamine	14,679.6
$C_{10}H_9N$	2-Methylquinoline	14,154.0
$C_{10}H_{10}$	1,3-Divinylbenzene	11,384.7
$C_{10}H_{10}O$	4-Phenyl-3-buten-2-one	13,913.9
$C_{10}H_{10}O_2$	α-Methylcinnamic acid	18,149.4
$C_{10}H_{10}O_2$	Methyl cinnamate	13,325.5
$C_{10}H_{10}O_2$	Safrole	13,255.8
$C_{10}H_{10}O_4$	1,2-Phenylene diacetate	14,986.0
$C_{10}H_{10}O_4$	Dimethylphthalate	14,922.2
$C_{10}H_{12}$	2,4-Dimethylstyrene	11,454.0
$C_{10}H_{12}$	2,5-Dimethylstyrene	11,283.5
$C_{10}H_{12}$	3-Ethylstyrene	11,285.7
$C_{10}H_{12}$	4-Ethylstyrene	11,146.6
$C_{10}H_{12}$	Tetralin	11,613.0
$C_{10}H_{12}O$	Anethole	13,006.8
$C_{10}H_{12}O$	4-Methylpropiophenone	12,505.0
$C_{10}H_{12}O$	Estragole	12,879.3
$C_{10}H_{12}O$	Cuminal	12,668.0
$C_{10}H_{12}O$	4-Vinylphenetole	13,728.7
$C_{10}H_{12}O_2$	Eugenol	13,907.8
$C_{10}H_{12}O_2$	Isoeugenol	14,084.2
$C_{10}H_{12}O_2$	Chavibetol	14,527.7
$C_{10}H_{12}O_2$	Propyl benzoate	12,318.7
$C_{10}H_{12}O_3$	2-Phenoxyethyl acetate	14,070.3
$C_{10}H_{13}ClO$	2-Chloroethyl-α-methylbenzyl ether	12,969.2
$C_{10}H_{13}Cl_2O_2P$	4-*tert*-Butylphenyl dichlorophosphate	13,711.0
$C_{10}H_{14}$	1,2,3,4-Tetramethylbenzene	12,258.0
$C_{10}H_{14}$	1,2,3,5-Tetramethylbenzene	12,358.4
$C_{10}H_{14}$	1,2,4,5-Tetramethylbenzene	12,583.6
$C_{10}H_{14}$	4-Ethyl-1,3-xylene	11,070.4
$C_{10}H_{14}$	5-Ethyl-1,3-xylene	11,045.5
$C_{10}H_{14}$	2-Ethyl-1,4-xylene	11,144.6
$C_{10}H_{14}$	1,2-Diethylbenzene	11,695.5
$C_{10}H_{14}$	1,3-Diethylbenzene	10,993.9
$C_{10}H_{14}$	1,4-Diethylbenzene	10,746.3
$C_{10}H_{14}$	1-Methyl-2-isopropylbenzene	—
$C_{10}H_{14}$	1-Methyl-4-isopropylbenzene	11,038.7
$C_{10}H_{14}$	*N*-Butylbenzene	11,052.1
$C_{10}H_{14}$	Isobutylbenzene	8,567.8
$C_{10}H_{14}$	*sec*-Butylbenzene	11,069.3
$C_{10}H_{14}$	*tert*-Butylbenzene	10,705.5
$C_{10}H_{14}O$	Carvacrol	13,765.7
$C_{10}H_{14}O$	Carbone	12,796.2
$C_{10}H_{14}O$	Cuminyl alcohol	13,799.2
$C_{10}H_{14}O$	4-Ethylphenetole	13,766.4
$C_{10}H_{14}O$	2-Isopropyl-5-methylphenol	13,352.8
$C_{10}H_{14}O$	4-Isobutylphenol	14,053.5
$C_{10}H_{14}O$	4-*sec*-Butylphenol	13,690.2
$C_{10}H_{14}O$	2-*sec*-Butylphenol	12,781.3
$C_{10}H_{14}O$	2-*tert*-Butylphenol	13,112.3
$C_{10}H_{14}O$	4-*tert*-Butylphenol	13,787.7
$C_{10}H_{14}N_2$	Nicotine	12,337.1
$C_{10}H_{15}N$	*N*-Diethylaniline	12,539.2
$C_{10}H_{15}NO_2$	*N*-Phenyliminodiethanol	17,482.1
$C_{10}H_{16}$	Camphene	10,505.4
$C_{10}H_{16}$	Dipentene	10,538.3
$C_{10}H_{16}$	*d*-Limonene	10,508.4
$C_{10}H_{16}$	Myrcene	10,704.8
$C_{10}H_{16}$	α-Phellandrene	11,139.5
$C_{10}H_{16}$	α-Pinene	9,813.6
$C_{10}H_{16}$	β-Pinene	10,235.8
$C_{10}H_{16}$	Terpenoline	12,030.8
$C_{10}H_{16}AsNO_3$	Diethyl arsanilinate	12,973.9
$C_{10}H_{16}O$	*d*-Camphor	12,800.9
		11,978.0
$C_{10}H_{16}O$	*l*-dihydrocarvone	11,825.9
$C_{10}H_{16}O$	α-Citral	13,255.5
$C_{10}H_{16}O$	*d*-Fenchone	11,273.4
$C_{10}H_{16}O$	Pulegone	13,395.4
$C_{10}H_{16}O$	α-Thujone	11,950.8
$C_{10}H_{16}OSi$	Ethoxydimethylphenylsilane	11,718.6
$C_{10}H_{16}O_2$	Campholenic acid	16,324.1
$C_{10}H_{16}O_2$	Diosphenol	13,644.0
$C_{10}H_{16}O_2$	Fencholic acid	16,442.8
$C_{10}H_{18}$	*cis*-Decalin	10,515.4
$C_{10}H_{18}$	*trans*-Decalin	8,749.1
$C_{10}H_{18}O$	*d*-Citronellal	12,305.1
$C_{10}H_{18}O$	Cineol	10,570.8
$C_{10}H_{18}O$	Dihydrocarveol	13,698.5
$C_{10}H_{18}O$	*dl*-Fenchyl alcohol	12,955.9
$C_{10}H_{18}O$	Geraniol	14,060.7
$C_{10}H_{18}O$	*d*-Linalool	12,269.7
$C_{10}H_{18}O$	Nerol	13,366.1
$C_{10}H_{18}O$	α-Terpineol	12,754.5
$C_{10}H_{18}O_2$	Citronellic acid	16,455.4
$C_{10}H_{18}O_3$	Amyl levulinate	14,321.7
$C_{10}H_{18}O_3$	Isoamyl levulinate	13,867.9
$C_{10}H_{18}O_4$	Diethyl ethylmethylmalonate	12,345.6
$C_{10}H_{18}O_4$	Diethyl adipate	14,240.6
$C_{10}H_{18}O_4$	Diisobutyl oxalate	13,343.1
$C_{10}H_{18}O_4$	Dipropyl succinate	13,975.7
$C_{10}H_{18}O_4$	Sebacic acid	21,978.3
$C_{10}H_{18}O_6$	Dipropyl-*d*-tartrate	15,754.0
$C_{10}H_{18}O_6$	Diisopropyl-*d*-tartrate	15,836.6
$C_{10}H_{19}N$	Camphylamine	13,224.1
$C_{10}H_{20}$	Methane	10,293.1
$C_{10}H_{20}$	1-Decene	10,233.3
$C_{10}H_{20}$	*n*-Decane	10,912.0
$C_{10}H_{20}Br_2$	1,2-Dibromodecane	16,407.7
$C_{10}H_{20}O$	Decanol	14,065.1
$C_{10}H_{20}O$	Citronellol	14,214.1

Formula	Name	ΔH_v
$C_{10}H_{20}O$	Capraldehyde	13,154.9
$C_{10}H_{20}O$	*l*-Menthol	13,475.3
$C_{10}H_{20}O$	Decan-2-one	12,114.7
$C_{10}H_{20}O_2$	Capric acid	19,372.6
$C_{10}H_{20}O_2$	Isoamyl isovalerate	11,040.8
$C_{10}H_{22}$	2,7-Dimethyloctane	10,339.3
$C_{10}H_{22}O$	Diisoamyl ether	11,072.2
$C_{10}H_{22}O_2$	2-Butyl-2-ethylbutane-1,3-diol	15,833.7
$C_{10}H_{22}O$	Dihydrocitronellol	16,769.8
$C_{10}H_{22}O_3$	Dipropylene glycol monobutyl ether	13,721.1
$C_{10}H_{22}S$	Diisoamyl sulfide	11,829.9
$C_{10}H_{24}Si$	Heptyltrimethylsilane	10,987.3
$C_{10}H_{24}Si$	Butyltriethylsilane	11,124.0
$C_{10}H_{28}O_4Si_3$	1,5-Diethoxyhexamethyl trisiloxane	12,586.4
$C_{10}H_{30}O_3Si_4$	Decamethyltetrasiloxane	11,981.2
$C_{10}H_{30}O_5Si_5$	Decamethylcyclopenta siloxane	12,272.1
$C_{11}H_8O_2$	1-Naphthoic acid	22,581.4
$C_{11}H_8O_2$	2-Naphthoic acid	22,630.8
$C_{11}H_{10}$	1-Methylnaphthalene	—
$C_{11}H_{12}O_2$	Ethyl-*trans*-cinnamate	13,639.9
$C_{11}H_{12}O_2$	1 Phenyl-1,3-pentanedione	15,033.9
$C_{11}H_{12}O_3$	Ethyl benzoylacetate	17,115.4
$C_{11}H_{12}O_3$	Myristicine	14,471.4
$C_{11}H_{14}$	1-Phenylpentane	—
$C_{11}H_{14}$	2,4,5-Trimethylstyrene	12,076.1
$C_{11}H_{14}$	2,4,6-Trimethylstyrene	11,588.8
$C_{11}H_{14}$	4-Isopropylstryene	11,471.0
$C_{11}H_{14}O$	Isobutyrophenone	12,878.8
$C_{11}H_{14}O$	Pivalophenone	13,221.3
$C_{11}H_{14}O$	2,3,5-Trimethylacetophenone	14,283.6
$C_{11}H_{14}O_2$	Isobutyl benzoate	13,105.8
$C_{11}H_{14}O_2$	4-Allylveratrole	15,027.1
$C_{11}H_{16}$	Pentamethylbenzene	—
$C_{11}H_{16}$	3,5-Diethyltoluene	11,167.4
$C_{11}H_{16}$	1,2,4-Trimethyl-5-ethylbenzene	12,145.3
$C_{11}H_{16}$	1,3,5-Trimethyl-2-ethylbenzene	11,677.3
$C_{11}H_{16}$	3-Ethylcumene	11,233.5
$C_{11}H_{16}$	4-Ethylcumene	11,425.6
$C_{11}H_{16}$	*sec*-Amylbenzene	11,886.0
$C_{11}H_{16}O$	4-*tert*-Butyl-2-cresol	13,798.1
$C_{11}H_{16}O$	2-*tert*-Butyl-4-cresol	14,037.9
$C_{11}H_{16}O$	4-*tert*-Amylphenol	13,154.3
$C_{11}H_{16}O_5$	Ethylcamphoronic anhydride	16,373.6
$C_{11}H_{18}O_2$	Bornyl formate	12,276.0
$C_{11}H_{18}O_2$	Geranyl formate	13,189.7
$C_{11}H_{18}O_2$	Neryl formate	12,959.3
$C_{11}H_{18}O_2Si$	Diethoxymethylphenyl silane	13,267.3
$C_{11}H_{18}O_5$	Diethyl-gamma-oxoazelate	17,543.6
$C_{11}H_{20}O_2$	10-Hendecenoic acid	17,247.5
$C_{11}H_{20}O_2$	Menthyl formate	12,077.7
$C_{11}H_{20}O_2$	2-Ethylhexyl acrylate	12,522.5
$C_{11}H_{20}O_2$	Octyl acrylate	12,957.5
$C_{11}H_{20}O_3$	Hexyl levulinate	14,626.2
$C_{11}H_{22}O$	Hendecan-2-one	14,353.5
$C_{11}H_{22}O_2$	Methyl caprate	13,831.7
$C_{11}H_{22}O_2$	Hendecanoic acid	14,689.9
$C_{11}H_{24}$	Undecane	11,481.7
$C_{11}H_{24}O$	Hendecan-2-ol	14,216.2
$C_{11}H_{26}Si$	Trimethyloctylsilane	12,285.8
$C_{11}H_{26}Si$	Amyltriethylsilane	11,859.7
$C_{12}H_9Br$	4-Bromobiphenyl	13,493.4
$C_{12}H_9BrO$	2-Bromo-4-phenylphenol	13,589.9
$C_{12}H_9Cl$	2-Chlorobiphenyl	13,925.7
$C_{12}H_9Cl$	4-Chlorobiphenyl	14,017.4
$C_{12}H_9ClO$	2-Chloro-3-phenylphenol	15,258.0
$C_{12}H_9ClO$	2-Chloro-6-phenylphenol	15,508.4
$C_{12}H_9Cl_2PO$	2-Xenyl dichlorophosphate	17,127.6
$C_{12}H_9N$	Carbazole	15,421.6
$C_{12}H_{10}$	Acenaphthene	13,078.5
$C_{12}H_{10}$	Diphenyl	12,910.0
$C_{12}H_{10}ClPO_3$	Diphenyl chlorophosphate	13,191.2
$C_{12}H_{10}Cl_2Si$	Dichlorodiphenylsilane	14,968.5
$C_{12}H_{10}F_2Si$	Difluorodiphenylsilane	12,913.3
$C_{12}H_{10}N_2$	Azobenzene	14,786.7
$C_{12}H_{10}O$	1-Acetonaphthone	16,095.5
$C_{12}H_{10}O$	2-Acetonaphthone	16,496.7
$C_{12}H_{10}O$	Diphenyl ether	12,325.5
$C_{12}H_{10}O$	2-Phenylphenol	15,397.8
$C_{12}H_{10}O$	4-Phenylphenol	16,974.3
$C_{12}H_{10}S$	Diphenyl sulfide	13,974.8
$C_{12}H_{10}S$	Diphenyl disulfide	17,452.0
$C_{12}H_{10}Se$	Diphenyl selenide	14,603.4
$C_{12}H_{11}N$	Diphenylamine	14,920.3
$C_{12}H_{12}$	1-Ethylnaphthalene	12,751.3
$C_{12}H_{12}N_2$	1.1-Diphenylhydrazine	15,940.4
$C_{12}H_{14}N_2O_5$	2-Cyclohexyl-4,6-dinitrophenol	19,100.0
$C_{12}H_{14}O_3$	Eugenyl acetate	15,120.7
$C_{12}H_{14}O_4$	Apiole	16,881.7
$C_{12}H_{14}O_4$	Diethyl phthalate	15,383.0
$C_{12}H_{16}$	2,5-Diethylstyrene	12,150.3
$C_{12}H_{16}$	Phenylcyclohexane	13,345.6
$C_{12}H_{16}O_2$	Isoamyl benzoate	12,782.9
$C_{12}H_{18}$	Hexamethylbenzene	—
$C_{12}H_{18}$	1,2,4-Triethylbenzene	11,957.9
$C_{12}H_{18}$	1,3,4-Triethylbenzene	12,215.0
$C_{12}H_{18}$	1,3,5-Triethylbenzene	—
$C_{12}H_{18}$	1,2-Diisopropylbenzene	11,751.4
$C_{12}H_{18}$	1,3-Diisopropylbenzene	11,498.9
$C_{12}H_{18}O$	2-*tert*-Butyl-4-ethylphenol	13,994.0
$C_{12}H_{18}O$	4-*tert*-Butyl-2,5-xylenol	14,477.9
$C_{12}H_{18}O$	4-*tert*-Butyl-2,6-xylenol	14,142.5
$C_{12}H_{18}O$	6-*tert*-Butyl-2,4-xylenol	13,882.4
$C_{12}H_{18}O$	6-*tert*-Butyl-3,4-xylenol	14,848.3
$C_{12}H_{20}O_2$	*d*-Bornyl acetate	11,838.7
$C_{12}H_{20}O_2$	Geranyl acetate	13,879.9
$C_{12}H_{20}O_2$	Linalyl acetate	12,910.6
$C_{12}H_{20}O_3Si$	Triethoxyphenylsilane	14,117.9
$C_{12}H_{20}O_7$	Triethyl citrate	14,818.4
$C_{12}H_{21}PO_4$	Trimethallyl phosphate	12,566.1
$C_{12}H_{22}O_2$	Citronellyl acetate	15,781.3
$C_{12}H_{22}O_2$	Menthyl acetate	12,819.2
$C_{12}H_{22}O_4$	Dimethyl sebacate	14,861.3
$C_{12}H_{22}O_4$	Diisoamyl oxalate	14,123.7
$C_{12}H_{22}O_4$	Diisobutyl-*d*-tartrate	14,874.9
$C_{12}H_{24}$	1-Dodecene	12,587.8
$C_{12}H_{24}$	Triisobutylene	10,790.4
$C_{12}H_{24}O$	Dodecan-2-one	14,138.7
$C_{12}H_{24}O$	Lauraldehyde	13,644.2
$C_{12}H_{24}O_2$	Lauric acid	16,585.3
$C_{12}H_{26}$	*n*-Dodecane	11,857.7
$C_{12}H_{26}O$	Dodecyl alcohol	15,166.6
$C_{12}H_{26}O_4$	Tripropylene glycol monoisopropyl ether	14,171.5
$C_{12}H_{27}N$	Triisobutylamine	12,390.6
$C_{12}H_{27}N$	Dodecylamine	14,836.4
$C_{12}H_{28}Si$	Triethylhexylsilane	12,119.4
$C_{12}H_{34}O_5Si_4$	1,7-Diethoxyoctamethyl-tetrasiloxane	14,095.9
$C_{12}H_{36}O_4Si_5$	Dodecamethylpentasiloxane	12,942.6
$C_{12}H_{36}O_6Si_6$	Dodecamethyl-cyclohexasiloxane	13,760.6
$C_{13}H_9N$	Acridine	15,174.6
$C_{13}H_{10}$	Fluorene	13,682.8
$C_{13}H_{10}O$	Benzophenone	14,725.4
$C_{13}H_{10}O_2$	Phenyl benzoate	14,181.7
$C_{13}H_{10}O_2$	Salol	15,441.6
$C_{13}H_{12}$	Diphenylmethane	13,089.4
$C_{13}H_{12}O$	Benzhydrol	15,220.2
$C_{13}H_{12}O$	Benzyl phenyl ether	14,156.7
$C_{13}H_{12}O$	1-Proprionaphthone	16,630.8

Formula	Name	ΔH$_v$
$C_{13}H_{13}ClSi$	Chloromethyldiphenylsilane	14,924.6
$C_{13}H_{14}Si$	Methyldiphenylsilane	15,396.8
$C_{13}H_{14}$	2-Isopropylnaphthalene	13,036.9
$C_{13}H_{18}O$	Enanthophenone	15,597.7
$C_{13}H_{20}$	Heptylbenzene	13,535.4
$C_{13}H_{20}O$	α-Ionone	14,253.4
$C_{13}H_{22}O_2$	Bornyl propionate	13,245.0
$C_{13}H_{26}O$	2-Tridecanone	14,416.1
$C_{13}H_{26}O_2$	Methyl laurate	14,853.5
$C_{13}H_{26}O_2$	Tridecanoic acid	19,214.8
$C_{13}H_{28}$	Tridecane	12,991.3
$C_{13}H_{28}O_4$	Tripropyleneglycol monobutyl ether	15,937.6
$C_{13}H_{30}Si$	Decyltrimethylsilane	13,311.1
$C_{13}H_{30}Si$	Triethylheptylsilane	13,298.3
$C_{14}H_8O_2$	Anthraquinone	21,163.1
$C_{14}H_8O_4$	1,4-Dihydroxyanthraquinone	17,677.9
$C_{14}H_{10}$	Anthracene	16,823.6
$C_{14}H_{10}$	Phenanthrene	14,184.0
$C_{14}H_{10}O_2$	Benzil	15,046.4
$C_{14}H_{10}O_3$	Benzoic anhydride	16,060.9
$C_{14}H_{12}$	1,1-Diphenylethylene	13,778.1
$C_{14}H_{12}$	*trans*-Diphenylethylene	15,010.1
$C_{14}H_{12}O$	Desoxybenzoin	15,642.1
$C_{14}H_{12}O$	Benzoin	15,952.5
$C_{14}H_{14}$	Dibenzyl	13,387.6
$C_{14}H_{14}O$	2-Isobutyronaphthone	17,133.8
$C_{14}H_{15}O$	Dibenzylamine	16,260.1
$C_{14}H_{15}N$	Ethyldiphenylamine	14,569.4
$C_{14}H_{20}Cl_2$	1,2-Dichlorotetraethylbenzene	14,629.0
$C_{14}H_{20}Cl_2$	1,4-Dichlorotetraethylbenzene	13,397.5
$C_{14}H_{20}O_3$	2-(4-*tert*-Butylphenoxy) ethyl acetate	16,017.6
$C_{14}H_{22}$	1,2,3,4-Tetraethylbenzene	12,763.5
$C_{14}H_{22}O$	2,4-Di-*tert*-butylphenol	14,237.7
$C_{14}H_{24}O_2$	Bornyl butyrate	13,746.1
$C_{14}H_{24}O_2$	Bornyl isobutyrate	13,501.8
$C_{14}H_{24}O_2$	Geranyl butyrate	16,086.4
$C_{14}H_{24}O_2$	Geranyl isobutyrate	15,699.5
$C_{14}H_{26}O_4$	Diethyl sebacate	16,819.6
$C_{14}H_{28}O$	2-Tetradecanone	15,102.7
$C_{14}H_{28}O$	Myristaldehyde	14,088.9
$C_{14}H_{28}O_2$	Myristic acid	18,380.1
$C_{14}H_{29}Cl$	1-Chlorotetradecane	14,083.5
$C_{14}H_{30}$	Tetradecane	13,750.0
$C_{14}H_{31}N$	Tetradecylamine	14,840.8
$C_{14}H_{32}Si$	Triethyloctysilane	12,954.8
$C_{14}H_{40}O_6Si_5$	1,9-Diethoxydeca methylpentasiloxane	15,296.9
$C_{14}H_{42}O_5Si_6$	Tetradecamethylhexasiloxane	13,800.0
$C_{14}H_{42}O_7Si_7$	Tetradecamethylcyclo heptasiloxane	14,263.8
$C_{15}H_{14}O$	1,3-Diphenyl-2-propanone	15,429.8
$C_{15}H_{14}O_2$	1-Biphenyloxy-2,3-epoxypropane	16,160.6
$C_{15}H_{16}O_2$	4,4-Isopropylidenebisphenol	23,254.0
$C_{15}H_{18}O$	Isocapronaphthone	17,360.3
$C_{15}H_{18}OSi$	Ethoxymethyldiphenylsilane	16,106.4
$C_{15}H_{20}O_2$	Helenin	26,532.7
$C_{15}H_{24}$	Cadinene	15,518.3
$C_{15}H_{24}O$	2,6-Di-*tert*-butyl-4-cresol	14,338.6
$C_{15}H_{24}O$	4,6-Di-*tert*-butyl-2-cresol	14,006.9
$C_{15}H_{24}O$	4,6-Di-*tert*-butyl-3-cresol	15,464.6
$C_{15}H_{26}O$	Champacol	14,655.9
$C_{15}H_{26}O_6$	Triethyl camphoronate	16,112.2
$C_{15}H_{30}O_2$	Methyl myristate	16,051.0
$C_{15}H_{32}$	Pentadecane	14,635.9
$C_{15}H_{32}O_5$	Tetrapropylene glycol monoisopropyl ether	16,494.6
$C_{15}H_{34}Si$	Dodecyltrimethylsilane	14,374.6
$C_{16}H_{14}O_2$	Benzyl cinnamate	20,840.6
$C_{16}H_{18}O$	Di(α-methylbenzyl)ether	14,628.1
$C_{16}H_{20}O_2Si$	Diethoxydiphenylsilane	15,828.8
$C_{16}H_{22}O_4$	Dibutyl phthalate	17,747.0
$C_{16}H_{25}Cl$	Pentaethylchlorobenzene	13,707.3
$C_{16}H_{26}$	Pentaethylbenzene	13,670.1
$C_{16}H_{26}O$	2,6-Di-*tert*-butyl-4-ethylphenol	14,438.0
$C_{16}H_{26}O$	4,6-Di-*tert*-butyl-3-ethylphenol	15,954.8
$C_{16}H_{30}O$	Muscone	14,722.5
$C_{16}H_{31}N$	Palmitonitrile	16,433.7
$C_{16}H_{32}$	1-Hexadecene	15,634.7
$C_{16}H_{32}$	Tetraisobutylene	12,937.2
$C_{16}H_{32}O$	2-Hexadecanone	15,194.4
$C_{16}H_{32}O$	Palmitaldehyde	15,454.2
$C_{16}H_{32}O_2$	Palmitic acid	17,603.6
$C_{16}H_{34}$	Hexadecane	15,405.5
$C_{16}H_{34}O$	Cetyl alcohol	14,483.4
$C_{16}H_{34}N$	Cetylamine	15,238.0
$C_{16}H_{34}Si$	Decyltriethylsilane	15,393.7
$C_{16}H_{46}O_7Si_6$	1,1,1-Diethoxydodeca methylhexasiloxane	15,945.3
$C_{16}H_{48}O_6Si_7$	Hexadecamethylheptasiloxane	14,841.5
$C_{16}H_{48}O_8Si_8$	Hexadecamethylcyclooctasiloxane	14,986.3
$C_{17}H_{10}O$	Benzanthrone	18,309.6
$C_{17}H_{18}O_3$	4-*tert*-Butylphenyl salicylate	16,455.6
$C_{17}H_{24}O_2$	Menthyl benzoate	16,804.5
$C_{17}H_{34}O$	2-Heptadecanone	16,559.8
$C_{17}H_{34}O_2$	Methyl palmitate	17,003.5
$C_{17}H_{36}$	Heptadecane	15,608.5
$C_{17}H_{38}Si$	Tetradecyltrimethylsilane	16,439.7
$C_{18}H_{12}Cl_3O_3PS$	Tri-2-chlorophenylthio phosphate	24,386.1
$C_{18}H_{15}O_4P$	Triphenyl phosphate	19,272.3
$C_{18}H_{30}$	Hexaethylbenzene	14,184.9
$C_{18}H_{30}O$	2,4,6-Tri-*tert*-butylphenol	14,703.7
$C_{18}H_{34}O_2$	Oleic acid	20,326.7
$C_{18}H_{34}O_2$	Elaidic acid	19,538.0
$C_{18}H_{36}O$	Stearaldehyde	16,555.6
$C_{18}H_{36}O_2$	Stearic acid	19,306.6
$C_{18}H_{38}$	Octadecane	15,447.0
$C_{18}H_{38}$	2-Methylheptadecane	16,095.9
$C_{18}H_{38}O$	1-Octadecanol	17,508.0
$C_{18}H_{39}N$	Ethylcetylamine	15,718.3
$C_{18}H_{52}O_8Si_7$	1,1,3-Diethoxytetradeca methylheptasiloxane	16,765.8
$C_{18}H_{54}O_7Si_8$	Octadecamethyloctasiloxane	15,270.3
$C_{19}H_{16}$	Triphenylmethane	34,470.8
$C_{19}H_{40}$	Nonadecane	16,497.3
$C_{20}H_{20}OSi$	Ethoxytriphenylsilane	20,214.2
$C_{20}H_{43}N$	Diethylhexadecylamine	15,871.3
$C_{20}H_{58}O_9Si_8$	1,1,5-Diethoxyhexadeca methyloctasiloxane	17,626.6
$C_{20}H_{60}O_8Si_9$	Eicosamethylnonasiloxane	19,522.9
$C_{21}H_{21}O_4P$	Tritolyl phosphate	20,835.9
$C_{21}H_{44}$	Heneicosane	17,702.2
$C_{22}H_{42}O_2$	Erucic acid	23,655.2
$C_{22}H_{42}O_2$	Brassidic acid	24,085.7
$C_{22}H_{46}$	Docosane	16,941.1
$C_{22}H_{56}O_9Si_{10}$	Docosamethyldecasiloxane	21,878.6
$C_{23}H_{48}$	Tricosane	19,082.1
$C_{24}H_{50}$	Tetracosane	19,642.5
$C_{24}H_{72}O_{10}Si_{11}$	Tetracosamethylhendecasiloxane	23,941.2
$C_{25}H_{52}$	Pentacosane	20,815.9
$C_{26}H_{54}$	Hexacosane	21,605.7
$C_{27}H_{33}O_4P$	Dicarvacryl-2-tolyl phosphate	24,233.3
$C_{27}H_{56}$	Heptacosane	21,958.1
$C_{28}H_{58}$	Octacosane	24,144.2
$C_{29}H_{60}$	Nonacosane	24,816.8
$C_{32}H_{34}ClO_4P$	Dicarvacryl-mono-(6-chloro-2-xenyl)-phosphate	25,299.5

SOLUBILITY PARAMETERS OF ORGANIC COMPOUNDS

While the solubility of organic compounds in selected solvents may be estimated from the "like dissolves like" rule of thumb, more useful information is available under the "Solubility" heading of the table "Physical Constants of Organic Compounds" in Section C of this book. This table provides qualitative data on solubility in water (w), ethanol (al), ethyl ether (eth), acetone (ace), and benzene (bz).

More quantitative solubility data for nonpolar organic compounds may be calculated from the Hildebrand expression for the square root of the cohesive energy density which is defined as the solubility parameter (δ). As shown by the following expression, δ values may be calculated if information for ΔH_v, Kelvin temperature (T), molecular weight (M), and density (D) is available:

$$\delta = \left(\frac{\Delta E_v}{V}\right)^{1/2} = \left(\frac{D(\Delta H_v - RT)}{M}\right)^{1/2}$$

The value for ΔH_v may be found in the table "Heats of Vaporization of Organic Compounds" immediately following this table. Values for D and M are listed under physical constants in Section C. Thus, the δ value for heptane may be calculated as follows:

$$\delta = \left[\frac{0.684\ (8670 - 1.99(298))}{100}\right]^{1/2} = 7.4\ \mathrm{H}$$

The dimensions for δ are $(\mathrm{cal\ cm^{-3}})^{1/2}$ but the Hildebrand unit (H) is used for convenience. It is important to note that the law of mixtures applies for δ values of mixed nonpolar solvents. Thus, the δ value for an equimolar mixture of heptane and carbon disulfide (δ = 10.0 H) is 8.7 H.

When ΔH_v values are not available, Small's molar attraction constants shown in Table I may be used.[a] As illustrated in the following expression which is solved for heptane, the summation of these constants (ΣG) may be used to estimate δ values at 298 K:

$$\delta = \frac{D\Sigma G}{M} = \frac{(2 \times 214) + 5(133)0.684}{100} = 7.5\ \mathrm{H}$$

Solvents such as acetone (δ = 9.9 H) and water (δ = 23.4 H) are completely miscible when a large difference in δ values exists. The critical δ range for solubility of solid solutes and liquid solutes is less than that for liquids, and the critical δ range for nonpolar polymers in nonpolar liquids is less than 2 H at temperatures below 50°C.

These δ values are most useful for nonpolar solvents that are listed in Section A of Table II. Some consideration must be given to the dipole-dipole interactions in more polar solvents that are listed in Section B of Table II. The values shown for hydrogen-bonded solvents that are shown in Section C of Table II must be used with more discretion because of the stronger intermolecular forces present. However, these values are much more useful than the "like dissolves like" rule of thumb and can be used to predict the solubility of most solutes in most solvents.

[a] Small, P. A., *J. Appl. Chem.*, 3, 71, 1953.

Table I
MOLAR-ATTRACTION CONSTANTS AT 298 K

Group	G	Group	G
$-CH_3$	214	CO ketones	275
$-CH_2-$ single bonded	133	COO esters	310
$-CH<$	28	CN	410
$>C<$	93	Cl (mean)	260
$CH_2=$	190	Cl single	270
$-CH=$ double bonded	111	Cl twinned as in $>CCl_2$	260
$>C=$	19	Cl triple as in $-CCl_3$	250
$CH{\equiv}C-$	285	Br single	340
$-C{\equiv}C-$	222	I single	425
Phenyl	735	CF_2 n-fluorocarbons only	150
Phenylene (*o,m,p*)	658	CF_3 n-fluorocarbons only	274
Naphthyl	1146	S sulfides	225
Ring, 5 membered	105–115	SH thiols	315
Ring, 6 membered	95–105	ONO_2 nitrates	~440
Conjugation	20–30	NO_2 (aliphatic nitro-compounds)	~440
H (variable)	80–100	PO_4 (organic phosphates)	~500
O ether	70	Si (in silicones)	–38

Table compiled by R. B. Seymour.

Table II
SOLUBILITY PARAMETER VALUES

A. Nonpolar Solvents

Name	δ (H)	Name	δ (H)
Acetic acid nitrile (acetonitrile)	11.9	Benzene, isopropyl (cumene)	8.5
Anthracene	9.9	Benzene, 1-isopropyl-4-methyl (*p*-cymene)	8.2
Benzene	9.2	Benzene, nitro	10.0
Benzene, chloro	9.5	Benzene, propyl	8.6
Benzene, 1,2-dichloro	10.0	Benzene, 1,3,5-trimethyl (mesitylene)	8.8
Benzene, ethyl	8.8	Benzoic acid nitrile (benzonitrile)	8.4

Table II (continued)
SOLUBILITY PARAMETER VALUES

A. Nonpolar Solvents

Name	δ (H)	Name	δ (H)
Biphenyl, perchloro	8.8	Hexene-1	7.4
1,3-Butadiene	7.1	Malonic acid dinitrile (malononitrile)	15.1
1,3-Butadiene, 2-methyl (isoprene)	7.4	Methane	5.4
Butane	6.8	Methane, bromo	9.6
Butanoic acid nitrile	10.5	Methane, dichloro (methylene chloride)	9.7
Carbon disulfide	10.0	Methane, dichloro-difluoro (Freon 12®)	5.5
Carbon tetrachloride	8.6	Methane, dichloro, manofluoro (Freon 21®)	8.3
Chloroform	9.3	Methane, nitro	12.7
Cyclohexane	8.2	Methane, tetrachloro-difluoro (Freon 112®)	7.8
Cyclohexane, methyl	7.8	Methane, trichloro-monofluoro (Freon 11®)	7.6
Cyclohexane, perfluoro	6.0	Naphthalene	9.9
Cyclopentane	8.7	Nonane	7.8
Decalin	8.8	Octane	7.6
Decane	8.0	Pentane	7.0
Dimethyl sulfide	9.4	Pentane, 1-bromo	7.6
Ethane	6.0	Pentane, 1-chloro	8.3
Ethane, bromo (ethyl bromide)	9.6	Pentanoic acid, nitrile (valeronitrile)	9.6
Ethane, chloro (ethyl chloride)	9.2	Pentene-1	6.9
Ethane, 1,2-dibromo	10.4	Phenanthrene	9.8
Ethane, 1,1-dichloro (ethylidene chloride)	8.9	Propane	6.4
Ethane, difluoro-tetrachloro (Freon 112®)	7.8	Propane, 1-bromo	8.9
Ethane, nitro	11.1	Propane, 2,2-dimethyl (neopentane)	6.3
Ethane, pentachloro	9.4	Propane, 1-nitro	16.3
Ethane, 1,1,2,2-tetrachloro	9.7	Propane-2-nitro	9.9
Ethanethiol (ethyl mercaptan)	9.2	Propene (propylene)	6.5
Ethane, 1,1,2-trichloro	9.6	Propene, 2-methyl (isobutylene)	6.7
Ethane trichloro-trifluoro (Freon 113®)	7.3	Propenoic acid nitrile (acrylonitrile)	10.5
Ethene, (ethylene)	6.1	Propionic acid nitrile	10.8
Ethene, tetrachloro (perchloroethylene)	9.3	Styrene	9.3
Ethene, trichloro	9.2	Terphenyl, hydrogenated	9.0
Heptane	7.4	Tetralin	9.5
Heptane, perfluoro	5.8	Toluene	8.9
Hexane	7.3	Xylene, *m*-	8.8

B. Moderately Polar Solvents

Name	δ (H)	Name	δ (H)
Acetic acid, butyl ester	8.5	Ethylene glycol, monomethyl ether (methyl Cellosolve)	11.4
Acetic acid, ethyl ester	9.1	Formic acid amide, *N,N*-diethyl	10.6
Acetic acid, methyl ester	9.6	Formic acid amide, *N,N*-dimethyl	12.1
Acetic acid, pentyl ester	8.0	Formic acid, ethyl ester	9.4
Acetic acid, propyl ester	8.8	Formic acid, methyl ester	10.2
Acetic acid amide, *N,N*-diethyl	9.9	Formic acid, 2-methylbutyl ester	8.0
Acetic acid amide, *N,N*-dimethyl	10.8	Formic acid, propyl ester	9.2
Acrylic acid, butyl ester	8.4	Furan	9.4
Acrylic acid, ethyl ester	8.6	Furan, tetrahydro	9.1
Acrylic acid, methyl ester	8.9	Furfural	11.2
Adipic acid, dioctyl ester	8.7	2-Heptanone	8.5
Aniline, *N,N*-dimethyl	9.7	Hexanoic acid, 6-aminolactam (ε-caprolactam)	12.7
Benzene, 1-methoxy-4-propenyl (anethole)	8.4	Hexanoic acid, 6-hydroxylactone (caprolactone)	10.1
Benzoic acid, ethyl ester	8.2	Isophorone	9.1
Benzoic acid, methyl ester	10.5	Lactic acid, butyl ester	9.4
Butanal	9.0	Lactic acid, ethyl ester	10.0
Butane, 1-iodo	8.6	Methacrylic acid, butyl ester	8.3
Butanoic acid, 4-hydroxylactone (butyrolactone)	12.6	Methacryllc acid, ethyl ester	8.5
2-Butanone	9.3	Methacrylic acid, methyl ester	8.8
Carbonic acid, diethyl ester	8.8	Oxalic acid, diethyl ester	8.6
Carbonic acid, dimethyl ester	9.9	Oxalic acid, dimethyl ester	11.0
Cyclohexanone	9.9	Oxirane (ethylene oxide)	11.1
Cyclopentanone	10.4	Pentane, 1-iodo	8.4
2-Decanone	7.8	2-Pentanone	8.7
Diethylene glycol, monobutyl ether (butyl carbitol)	9.5	Pentanone-2,4-hydroxy,4-methyl (diacetone alcohol)	9.2
Diethylene glycol, monoethyl ether (ethyl carbitol)	10.2	Pentanone-2,4-methyl (mesityl oxide)	9.0
Dimethyl sulfoxide	12.0	Phosphoric acid, triphenyl ester	8.6
1,4-Dioxane	10.0	Phosphoric acid, tri-2-tolyl ester	8.4
Ethene, chloro (vinyl chloride)	7.8	Phthalic acid, dibutyl ester	9.3
Ether, 1,1-dichloroethyl	10.0	Phthalic acid, diethyl ester	10.0
Ether, diethyl	7.4	Phthalic acid, dihexyl ester	8.9
Ether, dimethyl	8.8	Phthalic acid, dimethyl ester	10.7
Ether, dipropyl	7.8	Phthalic acid, di-2-methylnonyl ester	7.2
Ethylene glycol, monobutyl ether (butyl Cellosolve®)	9.5	Phthalic acid dioctyl ester	7.9
		Phthalic acid, dipentyl ester	9.1
Ethylene glycol, monoethyl ether (ethyl Cellosolve)	10.5	Phthalic acid, dipropyl ester	9.7
		Propane, 1,2-epoxy (propylene oxide)	9.2

Table II (continued)
SOLUBILITY PARAMETER VALUES

B. Moderately Polar Solvents

Name	δ (H)	Name	δ (H)
Sebacic acid, dioctyl ester	8.6	Propionic acid, ethyl ester	8.4
Stearic acid, butyl ester	7.5	Propionic acid, methyl ester	8.9
Sulfone, diethyl	12.4	4-Pyrone	13.4
Sulfone, dimethyl	14.5	2-Pyrrolidone, 1-methyl	11.3
Sulfone, dipropyl	11.3	Sebacic acid, dibutyl ester	9.2

C. Hydrogen-bonded Solvents

Name	δ (H)	Name	δ (H)
Acetic acid	10.1	1-Hexanol	10.7
Acetic acid amide, *N*-ethyl	12.3	1-Hexanol-2-ethyl	9.5
Acetic acid, dichloro	11.0	Maleic acid anhydride	13.6
Acetic acid, anhydride	10.3	Methacrylic acid	11.2
Acrylic acid	12.0	Methacrylic acid amide, *N*-Methyl	14.6
Amine, diethyl	8.0	Methanol	14.5
Amine, ethyl	10.0	Methanol, 2-furil (furfuryl alcohol)	12.5
Amine, methyl	11.2	1-Nonanol	8.4
Ammonia	16.3	Pentane, 1-amino	8.7
Aniline	10.3	1,3-Pentanediol, 2-methyl	10.3
1,3-Butanediol	10.9	1-Pentanol	11.6
1,4-Butanediol	10.0	2-Pentanol	12.1
2,3-Butanediol	8.7	Piperidine	11.1
1-Butanol	13.6	2-Piperidone	11.4
2-Butanol	12.6	1,2-Propanediol	10.8
1-Butanol, 2-ethyl	11.9	1-Propanol	10.5
1-Butanol, 2-methyl	11.5	2-Propanol	10.0
Butyric acid	10.5	1-Propanol, 2-methyl	10.5
Cyclohexanol	10.6	2-Propanol, 2-methyl	11.4
Diethylene glycol	11.8	2-Propenol (allyl alcohol)	12.1
1-Dodecanol	9.9	Propionic acid	8.1
Ethanol	10.0	Propionic acid anhydride	12.7
Ethanol, 2-chloro (ethylene chlorohydrin)	12.6	1,2-Propanediol	12.2
Ethylene glycol	10.7	Pyridine	14.6
Formic acid	14.7	2-Pyrrolidone	12.1
Formic acid amide, *N*-ethyl	10.8	Quinoline	13.9
Formic acid amide, *N*-methyl	15.4	Succinic acid anhydride	16.1
Glycerol	9.9	Tetraethylene glycol	16.5
2,3-Hexanediol	10.2	Toluene, 3-hydroxy (meta cresol)	10.3
1,3-Hexanediol-2-ethyl	23.4	Water	9.4

Table compiled by R. B. Seymour.

MISCIBILITY OF ORGANIC SOLVENT PAIRS

Table A

Doctor J. S. Drury

Industrial and Engineering Chemistry Vol. 44, No. 11, Nov. 1952

(Reprinted by permission)

The classifications were made by shaking together 5 ml. of each of the solvents listed in a test tube for 1 minute, then allowing the mixture to settle. If no interfacial meniscus was observed, the solvent pair was considered miscible. If such a meniscus was present, the solvent pair was regarded as immiscible. The classification of immiscible is a qualitative one since solvent pairs may exhibit some degree of partial miscibility while existing as separate phases. Solvent pairs possessing a pronounced degree of partial miscibility are designated by the symbol Is.

Compound number	Compounds	Acetone	Acetyl acetone	2-Amino-2-methyl-1-propanol	Aniline	Benzaldehyde	Benzene	Benzin	Benzyl alcohol	Butyl acetate	Butyl alcohol	*n*-Butyl ether	Capryl alcohol	Carbon tetrachloride	Diacetone alcohol	Diethanolamine	Diethyl cellosolve	Diethyl ether	Dimethylaniline	Ethyl alcohol	Ethyl benzoate	Ethylene glycol	2-Ethylhexanol	Formamide	Furfuryl alcohol	Glycerol	Hydroxyethyl-ethylenediamine	Isoamyl alcohol	Methyl isobutyl ketone	Nitromethane	Dibutoxytetra-ethylene glycol	Pyridine	Triethanolamine	Trimethylene glycol
1	Acetone	..	M	M	..	M	M	M	M	M	M	M	M	M	M	M	M	M	M	M	M	M	M	M	M	I	M	M	M	M	M	M	..	M
2	Acetyl acetone	M	..	R	..	M	M	M	M	M	M	M	M	M	M	R	M	M	M	M	M	M	M	M	M	I	R	M	M	M	M	M	..	M
3	Adiponitrile	M	M	M	M	..	M	..	M	M	M	I	..	I	M	..	M	I	M	M	M	I	I	M	M	I	M	I	M	..	M	M	M	M
4	2-Amino-2-methyl-1-propanol	M	R	..	..	M	M	I	M	M	M	Is	M	M	R	M	M	M	M	M	M	M	M	M	M	M	M	M	M	M	M	M	..	M
5	Benzaldehyde	M	M	M	..	..	M	M	M	M	M	M	M	M	M	I	M	M	M	M	M	Is	M	M	M	Is	R	M	M	M	M	M	..	M
6	Benzene	M	M	M	..	M	..	M	M	M	M	M	M	M	Is	I	M	M	M	M	M	I	M	I	M	I	I	M	M	I	M	M	..	I
7	Benzin	M	M	I	..	M	M	..	I	M	M	M	M	M	I	I	M	M	M	M	M	M	M	I	I	I	Is	M	M	M	M	M	..	I
8	Benzonitrile	M	M	M	M	..	M	..	M	M	M	M	..	M	M	..	M	M	M	M	M	I	M	I	M	I	M	M	M	..	M	M	M	I
9	Benzothiazole	M	M	M	M	..	M	..	M	M	M	M	..	M	M	..	M	M	M	M	M	M	M	I	M	I	M	M	M	..	M	M	M	M
10	Benzyl alcohol	M	M	M	..	M	M	I	..	M	M	M	M	M	M	M	M	M	M	M	M	M	M	M	M	M	M	M	M	M	M	M	..	M
11	Benzyl mercaptan	M	M	I	M	..	M	..	M	M	M	M	..	M	M	..	M	M	M	M	M	I	M	I	M	I	M	M	M	..	M	M	R	I
12	Butyl acetate	M	M	M	..	M	M	M	M	..	M	M	M	M	M	I	M	M	M	M	M	Is	M	I	M	I	I	M	M	M	M	M	..	Is
13	Butyl alcohol	M	M	M	..	M	M	M	M	M	..	M	M	M	M	M	M	M	M	M	M	M	M	M	M	M	M	M	M	M	M	M	..	M
14	*n*-Butyl ether	M	M	Is	..	M	M	M	M	M	M	..	M	M	M	I	M	M	M	M	M	I	M	I	M	I	I	M	M	I	M	M	..	I
15	Capryl alcohol	M	M	M	..	M	M	M	M	M	M	M	..	M	M	M	M	M	M	M	M	M	M	I	M	I	M	M	M	Is	M	M	..	M
16	Carbon tetrachloride	M	M	M	..	M	M	M	M	M	M	M	M	..	Is	I	M	M	M	M	M	I	M	I	M	I	I	M	M	M	M	M	..	I
17	Diacetone alcohol	M	M	R	..	M	Is	I	M	M	M	M	M	Is	..	M	M	M	M	M	M	M	M	M	M	I	R	M	M	M	M	M	..	M
18	Diethanolamine	M	R	M	..	I	I	I	M	I	M	I	M	I	M	..	I	I	Is	M	M	M	M	M	M	M	M	M	I	I	I	M	..	M
19	Diethyl Cellosolve	M	M	M	..	M	M	M	M	M	M	M	M	M	M	I	..	M	M	M	M	M	M	M	M	I	I	M	M	M	M	M	..	M
20	Diethyl ether	M	M	M	..	M	M	M	M	M	M	M	M	M	M	I	M	..	M	M	M	I	M	I	M	I	I	M	M	M	M	M	..	I
21	Dimethylaniline	M	M	M	..	M	M	M	M	M	M	M	M	M	M	Is	M	M	..	M	M	I	M	I	M	I	I	M	M	M	M	M	..	I
22	Di-*N*-propylaniline	M	M	I	M	..	M	..	M	M	M	M	..	M	I	..	M	M	M	M	M	I	M	I	M	I	I	M	M	..	M	M	I	I
23	Ethyl alcohol	M	M	M	..	M	M	M	M	M	M	M	M	M	M	M	M	M	M	..	M	M	M	M	M	M	M	M	M	M	M	M	..	M
24	Ethyl benzoate	M	M	M	..	M	M	M	M	M	M	M	M	M	M	M	M	M	M	M	..	I	M	I	M	I	M	M	M	M	M	M	..	Is
25	Ethyl isothiocyanate	M	M	R	M	..	M	..	M	M	M	M	..	M	M	..	M	M	M	M	M	I	M	I	M	I	R	M	M	..	M	M	M	I
26	Ethyl thiocyanate	M	M	M	M	..	M	..	M	M	M	M	..	M	M	..	M	M	M	M	M	I	M	I	M	I	M	M	M	..	M	M	M	I
27	Ethylene glycol	M	M	M	..	Is	I	M	M	Is	M	I	M	I	M	M	M	I	I	M	I	..	M	M	M	M	M	M	I	I	M	M	..	M
28	2-Ethylhexanol	M	M	M	..	M	M	M	M	M	M	M	M	M	M	M	M	M	M	M	M	M	..	I	M	I	M	M	M	I	M	M	..	M
29	Formamide	M	M	M	..	M	I	I	M	I	M	I	i	I	M	M	M	I	I	M	I	M	I	..	M	M	M	M	Is	M	M	M	..	M
30	Furfuryl alcohol	M	M	M	..	M	M	I	M	M	M	M	M	M	M	M	M	M	M	M	M	M	M	M	..	M	M	M	M	M	M	M	..	M
31	Glycerol	I	I	M	..	I	I	I	M	I	M	I	I	I	I	M	I	I	I	M	I	M	I	M	M	..	M	I	I	I	I	M	..	M
32	Hydroxyethyl-ethylenediamine	M	R	M	..	R	I	Is	M	I	M	I	M	I	R	M	I	I	I	M	M	M	M	M	M	M	..	M	M	M	M	M	..	M
33	Isoamyl alcohol	M	M	M	..	M	M	M	M	M	M	M	M	M	M	M	M	M	M	M	M	M	M	M	M	I	M	..	M	M	M	M	..	M
34	Isoamyl sulfide	M	M	I	M	..	M	..	M	M	M	M	..	M	I	..	M	M	M	M	M	I	M	I	I	I	I	M	M	..	M	M	I	I
35	Isobutyl mercaptan	M	M	M	M	..	M	..	M	M	M	M	..	M	M	..	M	M	M	M	M	I	M	I	M	I	I	M	M	..	M	M	R	R
36	Methyl disulfide	M	M	M	M	..	M	..	M	M	M	M	..	M	M	..	M	M	M	M	M	I	M	I	M	I	I	M	M	..	M	M	I	R
37	Methyl isobutyl ketone	M	M	M	..	M	M	M	M	M	M	M	M	M	M	I	M	M	M	M	M	I	M	Is	M	I	M	M	..	M	M	M	..	I
38	Nitromethane	M	M	M	..	M	I	M	M	M	M	I	Is	M	M	I	M	M	M	M	M	I	I	M	M	I	M	M	M	..	M	M	..	I
39	Dibutoxytetra-ethylene glycol	M	M	M	..	M	M	M	M	M	M	M	M	M	M	I	M	M	M	M	M	M	M	M	M	I	M	M	M	M	..	M	..	M
40	Pyridine	M	M	M	..	M	M	M	M	M	M	M	M	M	M	M	M	M	M	M	M	M	M	M	M	M	M	M	M	M	M	..	..	M
41	Tri-*n*-butylamine	M	M	I	I	..	M	..	M	M	M	M	..	M	I	..	M	M	M	M	M	I	M	I	M	I	I	M	M	..	M	M	I	I
42	Trimethylene glycol	M	M	M	..	M	I	I	M	Is	M	I	M	I	M	M	M	I	I	M	Is	M	M	M	M	M	M	M	I	I	M	M	..	..

MISCIBILITY OF ORGANIC SOLVENT PAIRS (Continued)

Tables B and C

W. M. Jackson and J. S. Drury

The classifications were made at 20°C in the following manner. One-milliliter portions of each solvent comprising a pair were shaken together for approximately a minute. If no interfacial meniscus was observed after the contents of the tube were allowed to settle, the solvent pair was considered to be miscible, M. If a meniscus was observed without apparent change in the volume of either solvent, the pair was regarded as immiscible, I. This classification is a qualitative one, since solvent pairs may exhibit various degrees of partial miscibility while existing as separate phases. If an obvious change occurred in the volume of each solvent, but a meniscus was present, the pair was classified as partially miscible, S. The designation R indicates that the two solvents reacted.

Table B

Compound number	Compounds	Acetone	Isoamyl acetate	*n*-Amyl cyanide	Benzene	Benzyl ether	2-Bromoethyl acetate	Chloroform	Cinnamaldehyde	Di-*n*-amylamine	Di-*n*-butyl carbonate	Diethylacetic acid	Diethylenetriamine	Diethyl formamide	Diisobutyl ketone	Diisopropylamine	Di-*n*-propyl aniline	Ethyl alcohol	Ethyl benzoate	Ethyl ether	Ethyl phenylacetate	Heptadecanol[a]	3-Heptanol	*n*-Heptyl acetate	*n*-Hexyl ether	Methyl isopropyl ketone	4-Methyl-*n*-valeric acid	*o*-Phenetidine	Sulfuric acid (concd.)	Tetradecanol[a]	Tri-*n*-butyl phosphate	Triethylene glycol	Triethylenetetramine	2,6,8-Trimethyl 4-nonanone	Compound number
1	Acetone	..	M	M	M	M	M	M	M	M	M	M	M	M	M	M	M	M	M	M	M	M	M	M	M	M	M	M	R	M	M	M	M	M	1
2	Isoamyl acetate	M	..	M	M	M	M	M	M	M	M	M	M	M	M	M	M	M	M	M	M	M	M	M	M	M	M	M	R	M	M	I	M	M	2
3	*n*-Amyl cyanide	M	M	..	M	M	M	M	M	M	M	M	M	M	M	M	M	M	M	M	M	M	M	M	M	M	M	M	R	M	M	M	M	M	3
4	Benzene	M	M	M	..	M	S	M	M	M	M	M	M	M	M	M	M	M	M	M	M	M	M	M	M	M	M	M	I	M	M	S	M	M	4
5	Benzyl ether	M	M	M	M	..	M	M	M	M	M	M	M	M	M	M	M	M	M	M	M	M	M	M	M	M	M	M	R	M	M	I	M	M	5
6	2-Bromoethyl acetate	M	M	M	S	M	..	M	M	R	M	M	R	R	M	R	M	M	M	M	M	M	M	M	S	M	M	M	R	M	M	M	R	M	6
7	Chloroform	M	M	M	M	M	M	..	M	M	M	M	M	M	M	M	M	M	M	M	M	M	M	M	M	M	M	M	I	M	M	M	M	M	7
8	Cinnamaldehyde	M	M	M	M	M	M	M	..	M	M	M	R	M	M	M	M	M	M	M	M	M	M	M	M	M	M	R	R	M	M	M	R	M	8
9	Di-*n*-amylamine	M	M	M	M	M	R	M	M	..	M	R	I	R	M	M	M	M	M	M	M	M	M	M	M	M	M	M	R	M	M	S	I	M	9
10	Di-*n*-butyl carbonate	M	M	M	M	M	M	M	M	M	..	M	M	M	M	M	M	M	M	M	M	M	M	M	M	M	M	M	R	M	M	I	I	M	10
11	Diethylacetic acid	M	M	M	M	M	M	M	M	R	M	..	R	M	M	R	M	M	M	M	M	M	M	M	M	M	M	M	R	M	M	M	R	M	11
12	Diethylenetriamine	M	M	M	M	M	R	M	R	I	M	R	..	R	M	M	M	M	M	M	M	M	M	R	I	R	R	M	R	M	M	M	M	I	12
13	Diethyl formamide	M	M	M	M	M	R	M	M	R	M	M	R	..	M	R	M	M	M	M	M	M	M	M	I	M	M	M	R	M	M	M	R	M	13
14	Diisobutyl ketone	M	M	M	M	M	M	M	M	M	M	M	M	M	..	M	M	M	M	M	M	M	M	M	M	M	M	M	R	M	M	I	M	M	14
15	Diisopropylamine	M	M	M	M	M	R	M	M	M	M	R	M	R	M	..	M	M	M	M	M	M	M	M	M	M	R	M	R	M	M	M	M	M	15
16	Di-*n*-propylaniline	M	M	M	M	M	M	M	M	M	M	M	M	M	M	M	..	M	M	M	M	M	M	M	M	M	M	M	R	M	M	I	M	M	16
17	Ethyl alcohol	M	M	M	M	M	M	M	M	M	M	M	M	M	M	M	M	..	M	M	M	M	M	M	M	M	M	M	M	M	M	M	M	M	17
18	Ethyl benzoate	M	M	M	M	M	M	M	M	M	M	M	M	M	M	M	M	M	..	M	M	M	M	M	M	M	M	M	M	M	M	M	M	M	18
19	Ethyl ether	M	M	M	M	M	M	M	M	M	M	M	M	M	M	M	M	M	M	..	M	M	M	M	M	M	M	M	R	M	M	I	M	M	19
20	Ethyl phenylacetate	M	M	M	M	M	M	M	M	M	M	M	M	M	M	M	M	M	M	M	..	M	M	M	M	M	M	M	R	M	M	M	M	M	20
21	Heptadecanol[a]	M	M	M	M	M	M	M	M	M	M	M	M	M	M	M	M	M	M	M	M	..	M	M	M	M	M	M	R	M	M	I	M	M	21
22	3-Heptanol	M	M	M	M	M	M	M	M	M	M	M	M	M	M	M	M	M	M	M	M	M	..	M	M	M	M	M	R	M	M	M	M	M	22
23	*n*-Heptyl acetate	M	M	M	M	M	M	M	M	M	M	M	R	M	M	M	M	M	M	M	M	M	M	..	M	M	M	M	R	M	M	I	R	M	23
24	*n*-Hexyl ether	M	M	M	M	M	S	M	M	M	M	M	I	I	M	M	M	M	M	M	M	M	M	M	..	M	M	M	R	M	M	I	I	M	24
25	Methyl isopropyl ketone	M	M	M	M	M	M	M	M	M	M	M	R	M	M	M	M	M	M	M	M	M	M	M	M	..	M	M	R	M	M	M	R	M	25
26	4-Methyl-*n*-valeric acid	M	M	M	M	M	M	M	M	M	M	M	R	M	M	R	M	M	M	M	M	M	M	M	M	M	..	M	R	M	M	M	R	M	26
27	*o*-Phenetidine	M	M	M	M	M	M	M	R	M	M	M	M	M	M	M	M	M	M	M	M	M	M	M	M	M	M	..	R	M	M	M	M	M	27
28	Sulfuric acid (concd.)	R	R	R	I	R	R	I	R	R	R	R	R	R	R	R	R	M	M	R	R	R	R	R	R	R	R	R	..	R	R	R	R	R	28
29	Tetradecanol[a]	M	M	M	M	M	M	M	M	M	M	M	M	M	M	M	M	M	M	M	M	M	M	M	M	M	M	M	R	..	M	I	M	M	29
30	Tri-*n*-butyl phosphate	M	M	M	M	M	M	M	M	M	M	M	M	M	M	M	M	M	M	M	M	M	M	M	M	M	M	M	R	M	..	M	M	M	30
31	Triethylene glycol	M	I	M	S	I	M	M	M	S	I	M	M	M	I	M	I	M	M	I	M	I	M	I	I	M	M	M	R	I	M	..	M	I	31
32	Triethylenetetramine	M	M	M	M	M	R	M	R	I	I	R	M	R	M	M	M	M	M	M	M	M	M	R	I	R	R	M	R	M	M	M	.	I	32
33	2,6,8-Trimethyl 4-nonanone	M	M	M	M	M	M	M	M	M	M	M	I	M	M	M	M	M	M	M	M	M	M	M	M	M	M	M	R	M	M	I	I	..	33

[a] Union Carbide name.

MISCIBILITY OF ORGANIC SOLVENT PAIRS (Continued)

Table C

Compound number	Compounds	Acetone	Isoamyl acetate	*n*-Amyl cyanide	Anisaldehyde	Benzene	Benzyl ether	Chloroform	*o*-Cresol	Diisobutyl ketone	Diethylacetic acid	Diethyl formamide	Di-*n*-propyl aniline	Ethyl alcohol	Ethyl ether	3-Heptanol	*n*-Heptyl acetate	*n*-Hexyl ether	α-Methylbenzylamine	α-Methylbenzyldiethanolamine	α-Methylbenzyldimethylamine	α-Methylbenzylethanolamine	2-Methyl-5-ethylpyridine	Methyl isopropyl ketone	4-Methyl-*n*-valeric acid	*o*-Phenetidine	2-Phenylethylamine	Isopropanolamine	Pyridine	Salicylaldehyde	Tetradecanol[a]	Tri-*n*-butyl phosphate	Triethylenetetramine	2,6,8-Trimethyl 4-nonanone
1	1,3-Butylene glycol	M	I	M	I	I	I	M	M	I	M	M	I	M	S	M	I	I	M	M	M	M	M	M	M	M	M	M	M	M	M	M	M	I
2	2,3-Butylene glycol	M	M	M	M	S	I	M	M	M	M	M	I	M	M	M	M	I	M	M	M	M	M	M	M	M	M	M	M	M	M	M	M	I
3	2-Chloroethanol	M	M	M	M	M	M	M	M	M	M	M	M	M	M	M	M	M	R	M	M	M	R	M	M	M	R	R	M	M	M	M	M	M
4	3-Chloro-1,2-propanediol	M	M	M	M	I	M	M	M	M	M	M	I	M	M	M	M	I	R	M	M	M	M	M	M	M	R	R	M	M	S	M	R	S
5	Dibutyl hydrogen phosphite	M	M	M	M	M	M	M	M	M	M	M	M	M	M	M	M	M	M	M	M	M	M	M	M	M	M	M	M	M	M	M	M	M
6	Diethylene glycol dibutyl ether	M	M	M	M	M	M	M	M	M	M	M	M	M	M	M	M	M	R	M	S	M	M	M	M	M	R	R	M	M	M	M	R	M
7	Diethylene glycol diethyl ether	M	M	M	M	M	M	M	M	M	M	M	M	M	M	M	M	M	M	M	M	M	M	M	M	M	M	M	M	M	M	M	M	M
8	Diethylene glycol monobutyl ether	M	M	M	M	M	M	M	M	M	M	M	M	M	M	M	M	M	M	M	M	M	M	M	M	M	M	M	M	M	M	M	M	M
9	Diethylene glycol monoethyl ether	M	M	M	M	M	M	M	M	M	M	M	M	M	M	M	M	I	M	M	M	M	M	M	M	M	M	M	M	M	M	M	M	M
10	Diethylene glycol monomethyl ether	M	M	M	M	M	M	M	M	M	M	M	M	M	M	M	M	I	M	M	M	M	M	M	M	M	M	M	M	M	M	M	M	M
11	Dipropylene glycol	M	M	M	M	M	M	M	M	M	M	M	M	M	M	M	M	I	M	M	M	M	M	M	M	M	M	M	M	M	M	M	M	M
12	Ethylene diacetate	M	M	M	M	M	M	M	M	M	M	M	M	M	M	M	M	I	M	M	M	M	M	M	M	M	M	M	M	M	M	M	M	M
13	Ethylene glycol	M	I	I	I	I	I	S	M	I	M	M	I	M	I	M	I	I	M	M	M	M	M	I	M	M	M	M	M	I	I	S	M	I
14	Ethyl glycol ethylbutyl ether	M	M	M	M	M	M	M	M	M	M	M	M	M	M	M	M	M	M	M	M	M	M	M	M	M	M	M	M	M	M	M	M	M
15	Ethylene glycol monobutyl ether	M	M	M	M	M	M	M	M	M	M	M	M	M	M	M	M	M	M	M	M	M	M	M	M	M	M	M	M	M	M	M	M	M
16	Ethylene glycol monoethyl ether	M	M	M	M	M	M	M	M	M	M	M	M	M	M	M	M	M	M	M	M	M	M	M	M	M	M	M	M	M	M	M	M	M
17	Ethylene glycol monomethyl ether	M	M	M	M	M	M	M	M	M	M	M	M	M	M	M	M	M	M	M	M	M	M	M	M	M	M	M	M	M	M	M	M	M
18	Ethylene glycol monophenyl ether	M	M	M	M	M	M	M	M	M	M	M	M	M	M	M	M	M	M	M	M	M	M	M	M	M	M	M	M	M	M	M	M	M
19	Glycerol	I	I	I	I	I	I	I	M	I	I	M	I	M	I	I	I	I	M	M	I	M	M	I	I	I	M	M	M	I	I	I	M	I
20	1,2-Propanediol	M	M	M	M	I	I	M	M	I	M	M	I	M	S	M	I	I	M	M	M	M	M	M	M	M	M	M	M	M	M	M	M	I
21	1,3-Propanediol	M	I	I	I	I	I	M	M	I	M	M	I	M	I	M	I	I	M	M	M	M	M	M	M	M	M	M	M	I	S	M	M	I
22	Triethylene glycol	M	I	M	M	S	I	M	M	I	M	M	I	M	I	M	I	I	M	M	M	M	M	M	M	M	M	M	M	M	I	M	M	I
23	Triethyl phosphate	M	M	M	M	M	M	M	M	M	M	M	M	M	M	M	M	M	M	M	M	M	M	M	M	M	M	M	M	M	M	M	M	M
24	Trimethylene chlorohydrin	M	M	M	M	M	M	M	M	M	M	M	M	M	M	M	M	M	R	M	M	M	R	M	M	M	R	R	M	M	M	M	R	M

[a] Union Carbide name.

Compiled by Erwin Di Cyan, Ph.D.

The field of steroids has expanded considerably and rapidly in degree and in kind, because synthetic steroids have been synthesized which though resembling the hormones in the body have no natural counterpart, but exert an effect comparable to those of the natural hormones.

In fact, the term *steroid hormone* thus becomes a misnomer when applied to the newer synthetically prepared steroids which do not have a counterpart in the body of man or other animals—as prednisone. (A hormone, by definition, is a material with certain functions and characteristics, *secreted by the ductless glands*. That part of the definition cannot be met by prednisone or by similar steroids as these are not secreted by the ductless, or endocrine glands.)

All the hormones as well as the synthetic analogues have in common the cyclopentanophenanthrene nucleus. Although chemically very similar, a comparatively slight structural change is in many instances productive of substances which have physiologically dissimilar effects, often acting upon different physiologic systems. But in many cases a small change in structure will result merely in an accentuation of certain effects.

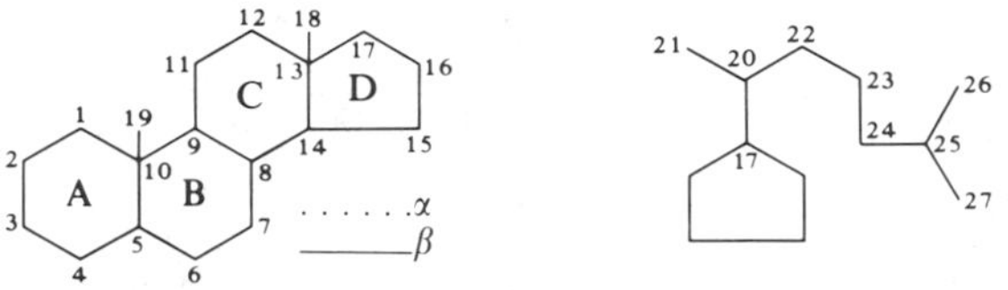

The Cyclopentanophenanthrene Nucleus

Classification. Classification becomes a bizarre problem by reason of the (a) overlapping uses to which these substances are put, and (b) the multiple purposes for which the hormones or synthetic substances are used. Indeed, the steroids may be classified by structure; that however would be uninformative to the student as to their use. Classification by origin, as adrenal, would also be unsuitable because, for example, a number of the adrenal corticosteroids are not found in the adrenal cortex at all, but merely resemble the natural hormones found in the adrenal cortex.

For those reasons the hormonal or hormonelike entries in the tables are classified by-and-large, by their predominant pharmacologic effects. Even that classification has its disparities as for example, the use of male sex hormones, i.e. the androgens, is neither limited to men, nor to uses which entail their effect upon male sex characteristics.

Uses. Originally, the use of steroid hormones was largely based upon one or more of the following predicates:

(a) To supplement the progressively declining secretion of a specific hormone due to natural biologic aging of the organism; in the menopause as an example of such declining secretion, a female sex hormone is used for such supplementation;

(b) To make available to the body a specific hormone, the natural secretion of which is inhibited because of a congenital or developmental anomaly; the underdevelopment of male secondary sex characteristics is an example of such an inhibited secretion, in which a male sex hormone is used—and correspondingly, female sex hormones in underdevelopment in females;

(c) To cause a reversal of hormonal balance in the treatment of diseases peculiar to a sex; for example, in the case of cancer of the female breast, a male sex hormone is administered, and in cancer of the prostate, a female sex hormone is used;

(d) To mimic a natural function, as menstruation, by the administration of estrogens—on withdrawal of which bleeding occurs; or by the alternate use of estrogenic and progestational—both female sex hormones.

(e) To delay a function, as ovulation, as in oral contraceptives, or *birth control pills*.

Since the finding that cortisone ameliorates the symptoms of rheumatoid arthritis (1949) the adrenal corticosteroid hormones and especially the synthetically prepared steroid analogues which have no natural counterpart in the body, have been successfully employed in the treatment of diseases not related to sex or sex function.

Androgens and Anabolic Agents. The agents listed in the tables under this classification have the effect of male sex hormones (androgens) i.e., to stimulate sexual maturation, in the "male climacteric," etc. But all androgens have in greater or lesser degree the ability to stimulate muscle development, i.e., an anabolic effect. Among the synthetically prepared agents which have no counterpart in the body (Methandrostenolone or Oxymetholone) are those which have a lessened androgenic, but a heightened anabolic effect. These qualities are determined by biological tests on animals but principally confirmed by clinical use in man. The anabolic effect includes remineralization of bone, which may be partially demineralized (osteoporosis) by age, or by certain drugs, as the adrenal corticosteroids (q.v.).

Anabolic agents are used for muscle and bone nutrition in men as well as women. The reason for the high interest in synthetic steroidal substances for anabolic use, is based on the need for materials, which within a given effective dose have a greater anabolic-to-androgenic ratio than such androgens as methyl testosterone. Otherwise, the administration of androgens to women produces manifestations of virilism, such as growth of hair on the face, a deepening of the voice, etc. Androgens are also used in the female in the suppression of excessive bleeding and in the treatment of cancer of the breast and cervix. (For other androgen-like agents, see also Progestogens and Progestins.)

Estrogens. Estrogenic agents hasten sexual maturation in the female. Therefore, they are used in underdevelopment in the female. The widest use of estrogens is in the treatment of the menopause, in which they supplement from without, the secretion of natural estrogens by the ovary, which begins to decline at about the 40th year. The menopause is usually a slow process, and the declining secretion gives rise to various symptoms during the time that the secretion declines, until adjustment to the new status takes place. The menopause, a period of physical and psychological stress, is made less precipitous by estrogens.

Frequently, a menopause must be quickly induced, as in cancer of the ovary or in uterine hemorrhage. This is done by radiation or by the removal of the uterus. Severe vasomotor symptoms occur when the menopause is thus suddenly induced. Estrogens —among other drugs—are used in the amelioration of these symptoms.

Estrogens (especially diethylstilbestrol which though not a hormone has an estrogenic effect) are also used in the control of cancer of the prostate in the male. Note the inverse correspondence to the use of male sex hormones in cancer of the breast in the female.

Progestogens and Progestins (Including 19-Norsteroid Compounds). The agents under that listing include progesterone, a female sex hormone, as well as progestins, i.e., synthetic progesterone-like compounds which have no natural counterpart in the body. Their use includes a variety of conditions: functional uterine bleeding, absence of menstruation (amenorrhea) used at times with estrogens, painful menstruation (dysmenorrhea), infertility, habitual abortion in order to maintain pregnancy, and in fact, to suppress ovulation hence their use as antifertility drugs. Certain progestins—as norethindrone combined with an estrogen, are the principal components of birth control pills—suppressing ovulation, there is no egg to fertilize, hence conception does not take place.

Adrenal Corticosteroids, Including Antiinflammatory, Antiallergic and Antirheumatic Agents. The adrenal cortex secretes a large number of hormones. They usually differ from each other in the accentuation of some phases of their properties. Virtually all of the cortical hormones are catabolic, thus having an effect in this respect, diametrically opposed to the androgens which are anabolic. Nearly all the cortical hormones—differing in degree from each other—cause retention of sodium and water by the body and hasten the excretion of potassium. These effects are utilized in the treatment of adrenal insufficiency or Addison's disease, in which conversely, there is an undue excretion of sodium and a strong retention of potassium. Desoxycorticosterone is used in Addison's disease because it has a particularly strong sodium retaining and potassium excreting effect.

Since the finding in 1949 of the usefulness of cortisone in profoundly reducing the symptoms of rheumatoid arthritis, the adrenal corticosteroids, including hydrocortisone, a natural hormone secreted by the adrenal cortex, and particularly the synthetic analogues not found in the body, as prednisone, have been used in the treatment of a wide variety of inflammatory diseases—especially diseases of collagen tissue. The same antiinflammatory effect is also brought into use in the reduction of inflammations associated with diseases of the skin, allergy, asthma, and in such systematic diseases as disseminated lupus erythematosus, also a collagen disease.

The drawbacks of cortisone, also shared in lesser measure by hydrocortisone, gave the impetus to the synthesis of steroidal substances not native to the body but differing somewhat from cortisone and hydrocortisone, in order to reduce the drawbacks attendant to the use of the latter. The sideeffects—especially those of cortisone—are retention of water and sodium, excretion of potassium, loss of mineral from bone leading to osteoporosis and fractures, hypertension, at times diabetes, personality changes or gastric ulcer. Prednisone and prednisolone among others (see tables) are two such steroidal synthetics which have the effects of cortisone, but fewer or less severe sideeffects. Whereas the synthetic steroidal substances are superior to cortisone with respect to lessened sideeffects, it cannot be said that the sideeffects are absent—they vary in degree from substance to substance.

Diuretic, Antidiuretic and Local Anesthetic Agents. Aldosterone, a natural hormone of the adrenal cortex promotes retention in the body of sodium and water, and facilitates excretion of potassium. Hence its effect is almost diametrically opposed to diuretics—especially the thiazide diuretics. Aldosterone is much more active in this respect than desoxycorticosterone, and is used in the treatment of Addison's disease, a hypofunction of the adrenal glands.

Spironolactone is an antagonist to aldosterone—the latter when elaborated in the body in excessive amounts gives rise to a syndrome called aldosteronism. Spironolactone, a synthetically produced steroid does not have a natural counterpart in the body, is diuretic when mercurial or thiazide diuretics are ineffective; it prevents sodium retention and potassium excretion—effects opposite to aldosterone. Hence spironolactone is used in aldosteronism, against edema, in the treatment of congestive heart failure and in other conditions in which an accumulation of water, and water-retaining salt, is to be corrected.

Doses. The amount of substance which comprises a dose of steroid hormones, or of the steroidal synthetics varies from substance to substance—from 0.1 mg for an estradiol ester, to 50 mg for a 19-norsteroid compound. The dose is conditioned upon the order of activity of the substance, the purpose for which it is administered, as well as the patient's response. However, as additional steroids for hormonal use are synthesized—especially those with adrenocortical activity, their average dose is usually smaller than the previously available steroid. The smaller effective dose of the more recent steroid is cited as an advantage over the previously available steroid.

However, a smaller dose cannot be claimed as an inherent advantage of a new steroid in comparison with an existing one, unless the lower dosage exhibits either greater or more prolonged activity or lesser sideeffects. One cannot meaningfully compare a dose, milligram for milligram, without taking into consideration if a heightened effect of the smaller dose produces fewer sideeffects. For example, it does not make any difference if a given effect and the same accompanying sideeffects are produced by a 50 mg or a 5 mg dose.

ADRENAL CORTICOSTEROIDS, INCLUDING ANTIINFLAMMATORY, ANTIALLERGIC AND ANTIRHEUMATIC AGENTS

Names & synonyms:	BETAMETHASONE; 9α-fluoro-16β-methylprednisolone; 16β-methyl-11β,17α,21-trihydroxy-9α-fluoro-1,4-pregnadiene-3,20-dione.	BETAMETHASONE ACETATE; 9α-fluoro-16β-methylprednisolone-21-acetate.	BETAMETHASONE DISODIUM PHOSPHATE; 9α-fluoro-16β-methylprednisolone-21-disodium phosphate.
Formulae:	CH_2OH, CO, HO, H_3C, OH, CH_3, H_3C, F, O $C_{22}H_{29}O_5F$	$H_2COCOCH_3$, C=O, HO, H_3C, OH, CH_3, H_3C, F, O $C_{24}H_{31}O_6F$	$H_2COPO_3Na_2$, C=O, HO, H_3C, OH, CH_3, H_3C, F, O $C_{22}H_{28}O_8FNa_2P$
Molecular weight	392.5	434.5	516.4
Melting point (°C)	240 (dec.)	200 to 220 (dec.)	decomposes
Specific rotation	$(\alpha)\frac{25}{D}$ +112 to +120 (100 mg. in 10 ml. dioxane)	$(\alpha)\frac{25}{D}$ +120 to +128 (100 mg. in 10 ml. dioxane)	$(\alpha)\frac{25}{D}$ +99 to +105 (100 mg. in 10 ml. water)
Absorption max.	239 mμ, E(1 %, 1 cm) 390, methanol	239 mμ, methanol	241 mμ, water

Names & synonyms:	CHLOROPREDNISONE ACETATE; 6α-chloroprednisone acetate; 6α-chloro-$\Delta^{1,4}$-pregnadien-17β,21-diol-3,11,20-trione 21-acetate.	CORTICOSTERONE; 11,21-dihydroxyprogesterone; Δ^4-pregnene-11β,21-diol-3,20-dione; 11β,21-dihydroxy-4-pregnene-3,20-dione; Kendall compound B; Reichstein substance H.	CORTISONE; 17-hydroxy-11-dehydrocorticosterone; 17α,21-dihydroxy-4-pregnene-3,11,20-trione; Δ^4-pregnene-17α,21-diol-3,11,20-trione; Kendall compound E; Wintersteiner compound F.
Formulae:	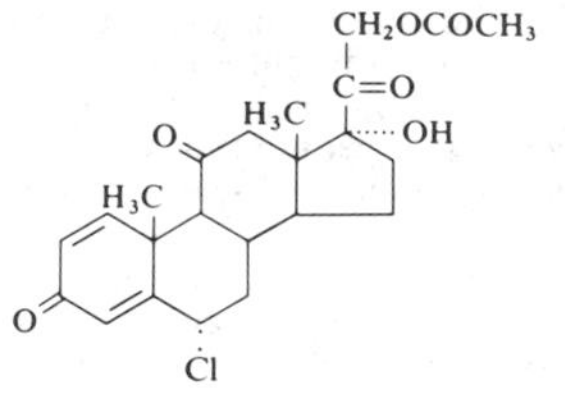	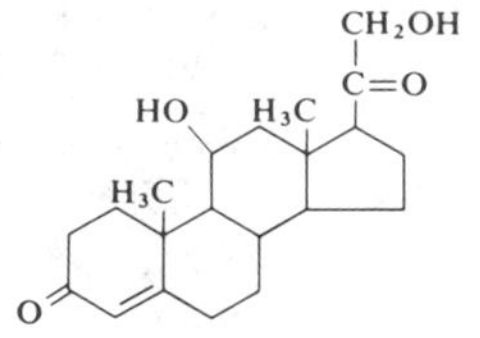 $C_{21}H_{30}O_4$	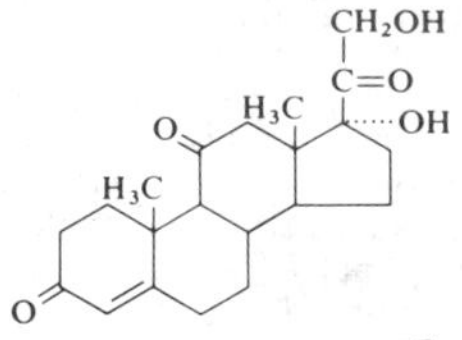 $C_{21}H_{28}O_5$
Molecular weight	436.6	346.40	360.4
Melting point (°C)	207–213	180–182	220–224
Specific rotation	$(\alpha)\frac{25}{D}$ +137 to +142 (100 mg. in 10 ml. chloroform)	$(\alpha)\frac{15}{D}$ +222 (110 mg. in 10 ml. alcohol)	$(\alpha)\frac{25}{D}$ +209 (120 mg. in 10 ml. alcohol)
Absorption max.		240 mμ	237 mμ

Names & synonyms:	DESOXYCORTICOSTERONE; deoxycorticosterone; 11-desoxycorticosterone; 21-hydroxyprogesterone; 4-pregnen-21-ol-3,20-dione; Kendall desoxy compound B; Reichstein substance Q.	DESOXYCORTICOSTERONE ACETATE; DCA; 11-desoxycorticosterone acetate.	DESOXYCORTICOSTERONE PIVALATE; desoxycorticosterone trimethylacetate; 21-hydroxy-4-pregnene-3,20-dione pivalate.
Formulae:	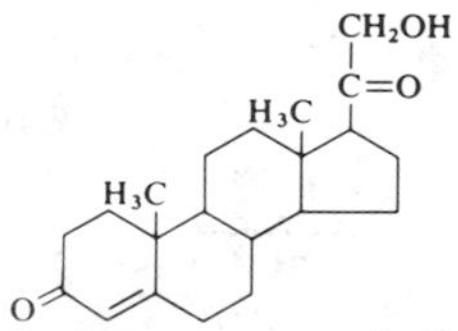 $C_{21}H_{30}O_3$	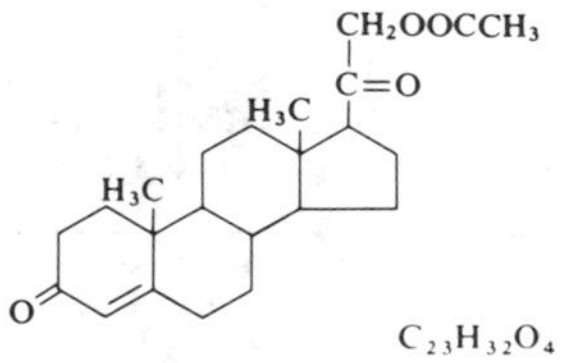 $C_{23}H_{32}O_4$	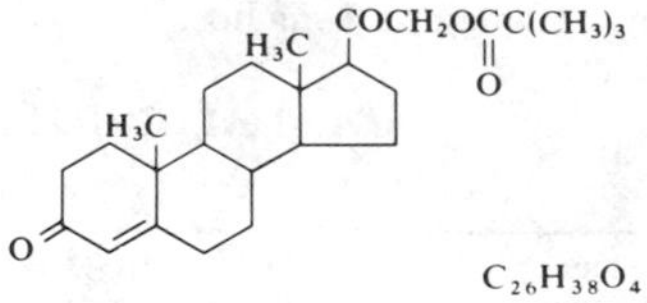 $C_{26}H_{38}O_4$
Molecular weight	330.2	372.4	414.6
Melting point (°C)	140–142	154–160	198–204
Specific rotation	$(\alpha)\frac{22}{D}$ +176 – +178 (100 mg. in 10 ml. alcohol)	$(\alpha)\frac{20}{D}$ +168 – +178 (100 mg. in 10 ml. dioxane)	$(\alpha)\frac{25}{D}$ +157 ±4 (1 % in dioxane)
Absorption max.	240 mμ		240 mμ (in ethanol)

Names & synonyms:	DEXAMETHASONE; hexadecadrol; 9α-fluoro-16α-methyl prednisolone; 9α-fluoro-11β,17α-21-trihydroxy-16α-methyl-1,4-pregnadiene-3,20-dione; 16α-methyl-9α-fluoro-1,4-pregnadiene-11β,17α-21-triol-3,20-dione; 16α-methyl-9α-fluoro-Δ^1-hydrocortisone; 1-dehydro-16α-methyl-9α-fluorohydrocortisone.	DICHLORISONE ACETATE; 9α-11β-dichloro-1,4-pregnadiene-17α,21-diol-3,20-dione-21-acetate	FLUOCINOLONE ACETONIDE; 6α,9α-difluoro-16α hydroxyprednisolone-16,17-acetonide; 6α,9α-difluoro-16α,17α-isopropylidenediosy-1,4-pregnadiene-3,20-dione.
Formulae:	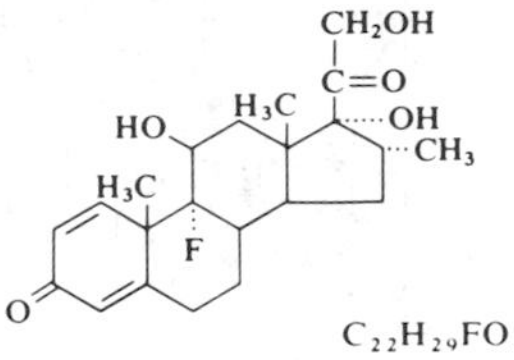 $C_{22}H_{29}FO_5$	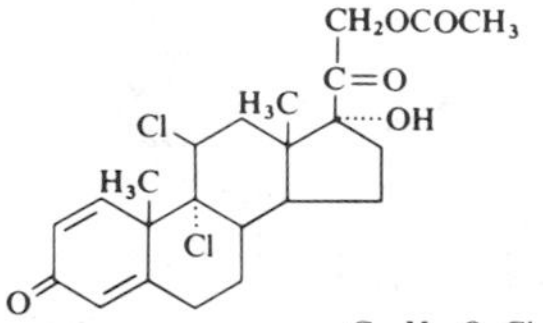 $C_{23}H_{28}O_5Cl$	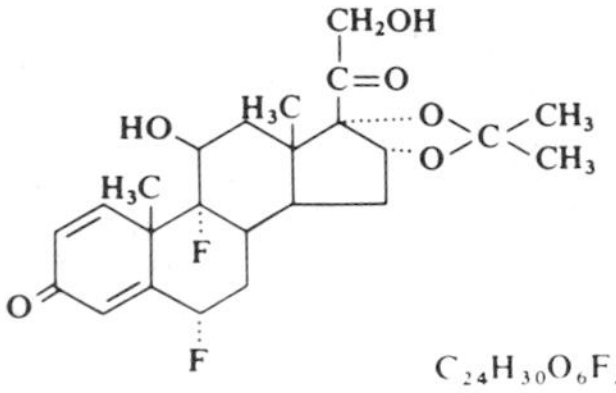 $C_{24}H_{30}O_6F_2$
Molecular weight	392.4	455.3	452.50
Melting point (°C)	262–264	235 (dec.)	255–266
Specific rotation	$(\alpha)^{25}_{D}$ +78 (100 mg. in 10 ml. dioxane)	$(\alpha)^{25}_{D}$ +160 – 168 (100 mg. in 10 ml. dioxane)	not less than +95° and not more than +105°C at 25°C.
Absorption max.		237 mμ – 316 – 337 (ε^1_1)	237 mμ ± 1 mμ

Names & synonyms:	FLUOROHYDROCORTISONE; fludrocortisone; 9α-fluorohydrocortisone; 9α-fluorocortisol; fluohydrisone; 9α-fluoro-11β,17α,21-trihydroxy-4-pregnene-3,20-dione; 9α-fluoro-17-hydroxycorticosterone.	FLUOROMETHOLONE; 9α-fluoro-11β,17α-dihydroxy-6α-methyl-1,4-pregnadiene-3,20-dione; 21-desoxy-9α-fluoro-6α-methyl-prednisolone.	FLUPREDNISOLONE; 6α-fluoroprednisolone; 6α-fluoro-1-dehydrohydrocortisone; 6α-fluoro-11β,17α,21-trihydroxy-1,4-pregnadiene-3,20-dione.
Formulae:	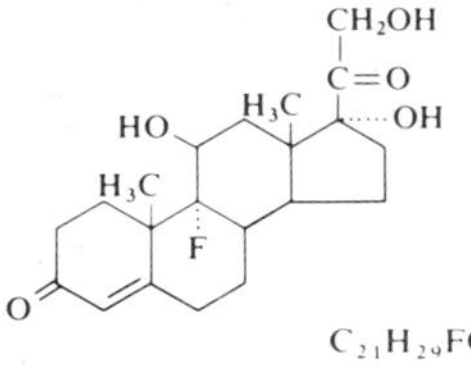 $C_{21}H_{29}FO_5$	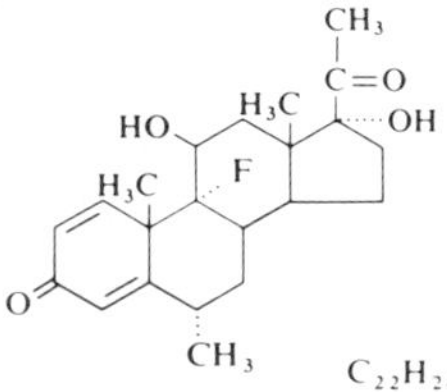 $C_{22}H_{24}FO_4$	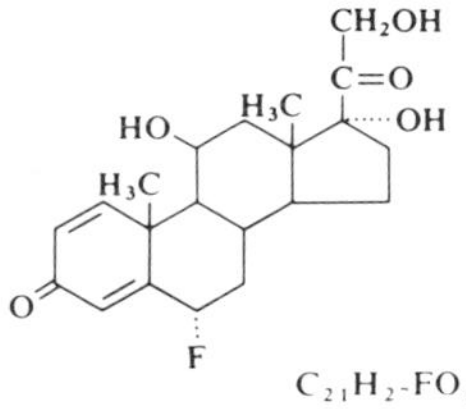 $C_{21}H_{27}FO_5$
Molecular weight	380.4	376.4	378.4
Melting point (°C)	260–262 (dec.)	290 (dec.)	205–210
Specific rotation	$(\alpha)^{23}_{D}$ +139 (55 mg. in 10 ml. alcohol)	$(\alpha)^{25}_{D}$ +56 (pyridine)	$(\alpha)_D$ +88 (dioxane)
Absorption max.		239 mu (a_M = 15,050) methanol	λ_{max} 241.5 mu (ε 16,000)

Names & synonyms:	FLURANDRENOLONE; 6-fluoro-16α-hydroxyhydrocortisone-16,17-acetonide; 6α-fluoro-11β,21-dehydroxy-16α,17α-isopropylidenedioxy-pregna-4-ene-3,20-dione.	HYDROCORTISONE; cortisol; 17-hydroxycorticosterone; hydrocortisone free alcohol; 11β,17α,21-trihydroxy-4-pregnene-3,20-dione; 4-pregnene-11β,17α,21-triol-3,20-dione; Kendall compound F; Reichstein substance M.	HYDROCORTISONE ACETATE; cortisol acetate; hydrocortisone-21-acetate; 17-hydroxycorticosterone-21-acetate.
Formulae:	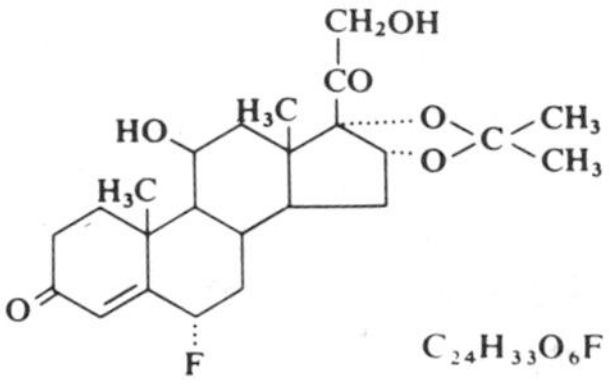 $C_{24}H_{33}O_6F$	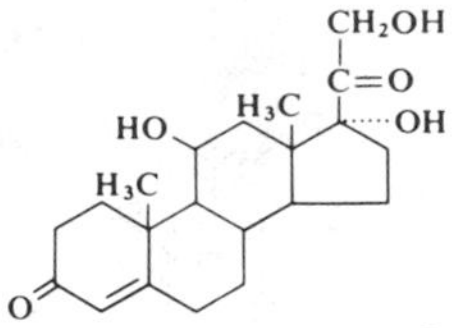 $C_{21}H_{30}O_5$	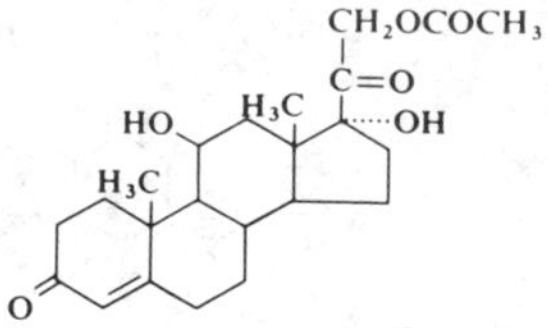 $C_{23}H_{32}O_6$
Molecular weight	436.5	362.5	404.5
Melting point (°C)	240–250	215–220 (dec.)	223 (dec.)
Specific rotation	$(\alpha)\frac{25}{D} = +145$ (1% in $CHCl_3$)	$(\alpha)\frac{25}{D}$ +150 – +156 (100 mg. in 10 ml.) dioxane)	$(\alpha)\frac{25}{D}$ +158 – +165 (100 mg. in 10 ml. dioxane)
Absorption max.	236 mμ (methanol)	242 mμ	242 mμ (methanol)

Names & synonyms:	HYDROCORTISONE SODIUM SUCCINATE; 11β, 17α, 21-trihydroxy-4-pregnene-3,20-dione, 21 hydrogen succinate, sodium salt; hydrocortisone, 21 hydrogen succinate, sodium salt.	METHYLPREDNISOLONE; 6α-methylprednisolone; Δ^1-6α-methylhydrocortisone; 1-dehydro-6α-methylhydrocortisone; 11β, 17α,21-trihydroxy-6α-methyl-1,4-pregnadiene-3,20-dione.	METHYLPREDNISOLONE SODIUM SUCCINATE; 1-dehydro-6α-methylhydrocortisone, 21-hydrogen succinate, sodium salt; 6α-methylprednisolone 21-hydrogen succinate, sodium salt; 11β, 17α, 21-trihydroxy-6α-methyl-1,4-pregnadiene-3, 20-dione, 21-hydrogen succinate, sodium salt.
Formulae:	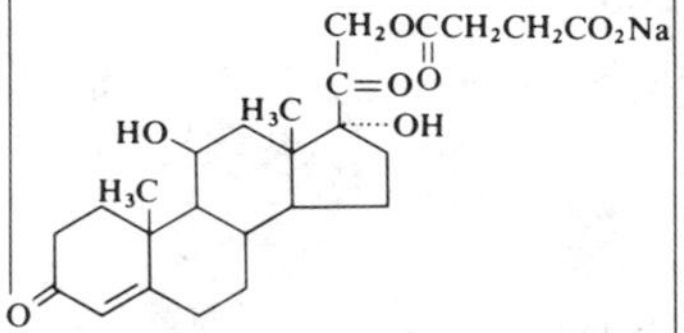 $C_{25}H_{33}O_8Na$	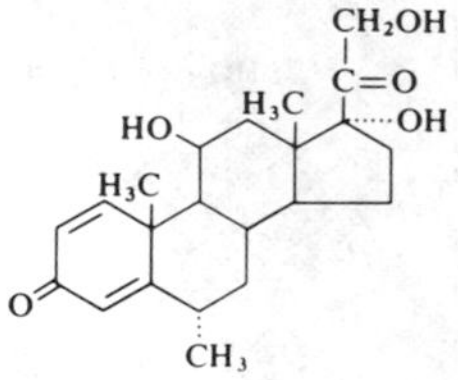 $C_{22}H_{30}O_5$	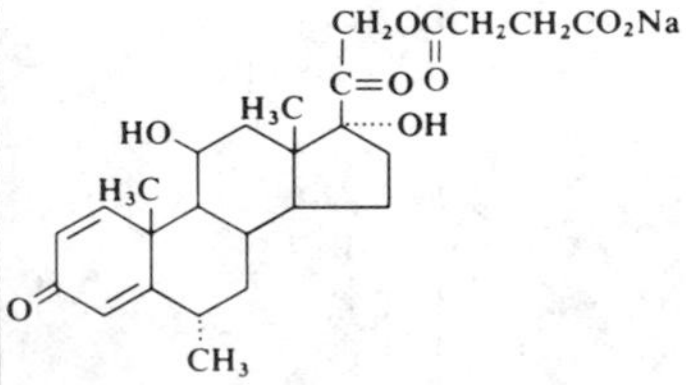 $C_{26}H_{33}O_8Na$
Molecular weight	484.5	374.5	496.5
Melting point (°C)	decomposes	230–240 (dec.)	decomposes
Specific rotation	$(\alpha)_D$ +140 ± 5 (alcohol)	$(\alpha)\frac{25}{D}$ +85 (dioxane)	$(\alpha)_D$ +100 ± 4 (alcohol)
Absorption max.	λ 242 mμ (ε 15,700)	243 mμ	λ_{max} 242 mμ (ε 14,500)

Names & synonyms:	PARAMETHASONE; 6α-fluoro-16α-methylprednisolone; 6α-fluoro-11β-17α,21-trihydroxy-16α-methyl-1,4-pregnadiene-3,20-dione.	PARAMETHASONE ACETATE; 6α-fluoro-16α-methylprednisolone-21-acetate; 6α-fluoro-16α-methylpregna-1,4-diene-11β,21-diol-3,20-dione-21-acetate; 6α-fluoro-17β,17α,21-trihydroxy-16α-methyl-1,4-pregnadiene-3,20-dione-21-acetate.	PREDNISOLONE; metacortandralone; Δ[1]-dehydrocortisol; delta F; Δ[1]-hydrocortisone; Δ[1]-dehydrohydrocortisone; 1,4-pregnadiene-3,20-dione-11β,17α,21-triol; 11β,17α,21-trihydroxy-1,4-pregnadiene-3,20-dione.
Formulae:	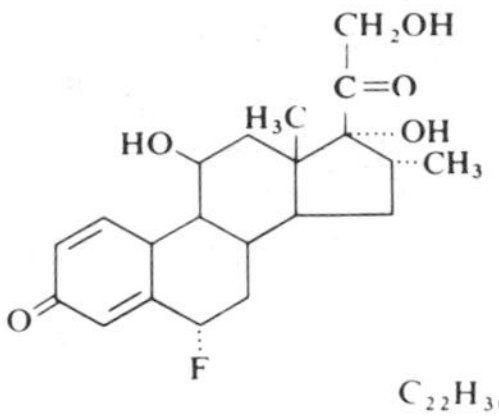 $C_{22}H_{30}O_5$	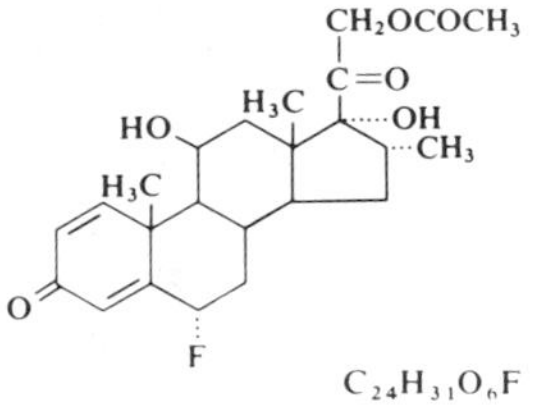 $C_{24}H_{31}O_6F$	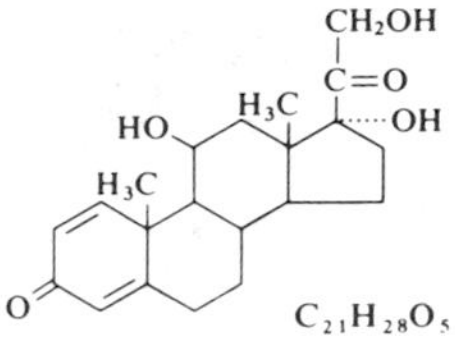 $C_{21}H_{28}O_5$
Molecular weight	392.45	434.5	360.4
Melting point (°C)	228–241	233–246	240 (dec.)
Specific rotation	+59 to +69 at 25°C	$(\alpha)\frac{25}{D}$ +72 (1 % in $CHCl_3$)	$(\alpha)\frac{25}{D}$ +97 – +103 (100 mg. in 10 ml. dioxane)
Absorption max.	242 mμ	242 mμ (methanol)	242 mμ (ε = 15,000) methanol

Names & synonyms:	PREDNISOLONE PHOSPHATE SODIUM; disodium prednisolone 21-phosphate.	PREDNISOLONE PIVALATE; prednisolone trimethylacetate; 11β,17α,21-trihydroxy-1,4-pregnadiene-3,20-dione 21-pivalate.
Formulae:	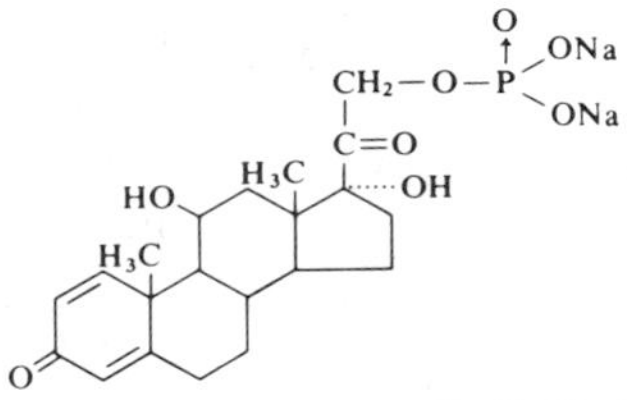 $C_{21}H_{27}Na_2O_8P$	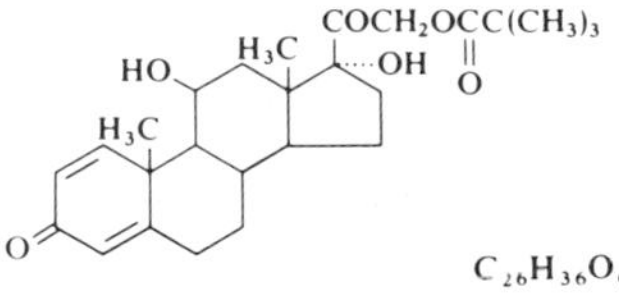 $C_{26}H_{36}O_6$
Molecular weight	484.4	444.6
Melting point (°C)		229
Specific rotation	$(\alpha)\frac{25}{D}$ +102.5 (100 mg. in 10 ml. H_2O)	+108 ±4 (1 % in dioxane)
Absorption max.	243 mμ	240 and 263 mμ (in absolute ethanol)

Names & synonyms:	PREDNISONE; metacortandricin; Δ^1-dehydrocortisone; delta E; Δ^1-cortisone; 1,4-pregnadiene-17α,21-diol-3,11,20-trione; 17α,21-dihydroxy-1,4-pregnadiene-3,11,20-trione.	TRIAMCINOLONE; 9α-fluoro-16α-hydroxyprednisolone; 9α-fluoro-11β,16α,17α,21-tetrahydroxy-1,4-pregnadiene-3,20-dione.
Formulae:	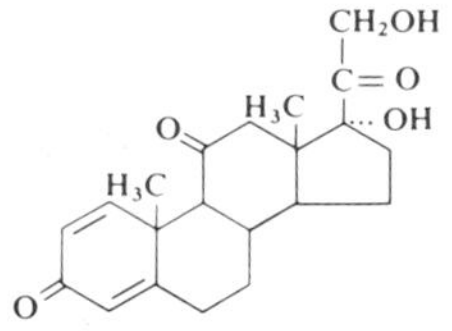 $C_{21}H_{26}O_5$	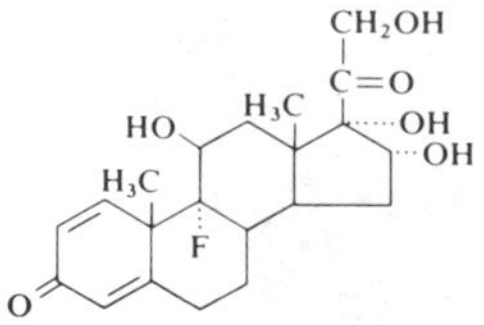 $C_{21}H_{27}FO_6$
Molecular weight	358.4	394.4
Melting point (°C)	225 (dec.)	260–262.5 (dec.)
Specific rotation	$(\alpha)^{25}_{D}$ +167 – +175 (100 mg. in 10 ml. dioxane)	$(\alpha)^{25}_{D}$ +75 (200 mg. in 100 ml. acetone)
Absorption max.	239 mμ (ε = 15,500) methanol	238 mμ (ε = 15,800)

Names & synonyms:	TRIAMCINOLONE ACETONIDE; 9α-fluoro-11β,21-dihydroxy-16α,17α-isopropylidenedioxy-1,4-pregnadiene-3,20-dione; 9α-fluoro-16α-hydroxyprednisolone 16,17-acetonide.	TRIAMCINOLONE DIACETATE; 16α,21-diacetoxy-9α-fluoro-11β,17α-dihydroxy-1,4-pregnadiene-3,20-dione; 9α-fluoro-16α-hydroxyprednisolone 16,21-diacetate.
Formulae:	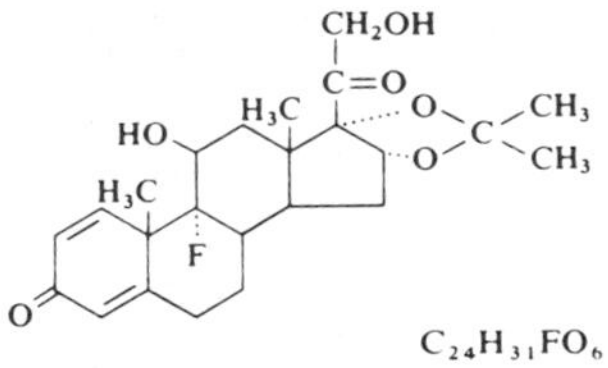 $C_{24}H_{31}FO_6$	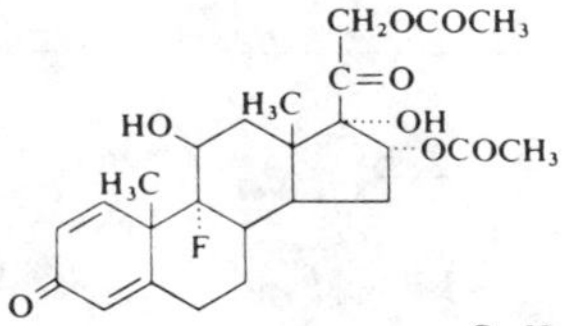 $C_{25}H_{31}FO_8$
Molecular weight	434.4	478.49
Melting point (°C)	274–278 (dec.); 292–294	variable: 158–235
Specific rotation	$(\alpha)^{25}_{D}$ +109 – +112 (53.7 mg. in 10 ml. chloroform)	$(\alpha)^{25}_{D}$ +22 (78.8 mg. in 10 ml. chloroform)
Absorption max.	238–239 mμ (ε = 14,600)	239 mμ (ε = 15,200)

ANDROGENS AND ANABOLIC AGENTS

Names & synonyms:	ANDROSTERONE; cis-androsterone; 3α-hydroxy-17-androstanone; androstane-3α-ol-17-one.	FLUOXYMESTERONE; 9α-fluoro-11β-hydroxy-17α-methyltestosterone 9α-fluoro-11β,17β-dihydroxy-17α-methyl-4-androsten-3-one.	ALDOSTERONE; electrocortin; 18-oxocorticosterone; 18-formyl-11β,21-dihydroxy-4-pregnene-3,20-dione.
Formulae:	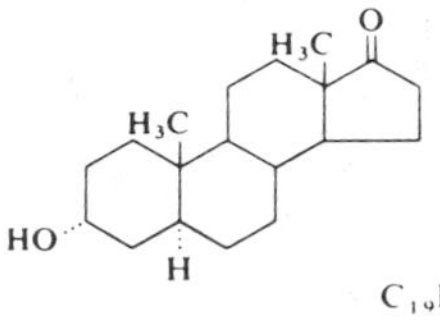 $C_{19}H_{30}O_2$	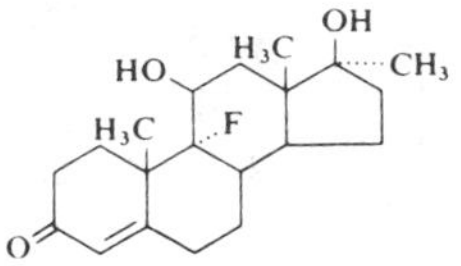 $C_{20}H_{29}FO_3$	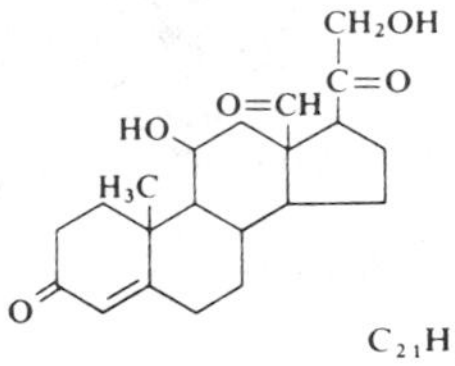 $C_{21}H_{28}O_5$
Molecular weight	290.4	336.4	360.4
Melting point (°C)	185–185.5	270 (dec.)	108–112 (hydrate); 164 (anhydrous)
Specific rotation	$(\alpha)\frac{15}{D}$ +85 – +90 (150 mg. in 10 ml. dioxane)	$(\alpha)\frac{25}{D}$ +107 – +109 (alcohol)	$(\alpha)\frac{25}{D}$ +161 (10 mg. in 10 ml. chloroform)
Absorption max.		240 mμ (ε = 16,700) alcohol	240 mμ (log ε = 4.20 monohydr.; ε mol. 15,000 anhydr.)

Names & synonyms:	HYDROXYDIONE SODIUM; 21-hydroxypregnane-3,20-dione-21-sodium hemisuccinate.	SPIRONOLACTONE; 3-(3-oxo-7α-acetylthio-17β-hydroxy-4-androsten-17α-yl)-propionic acid γ lactone.
Formulae:	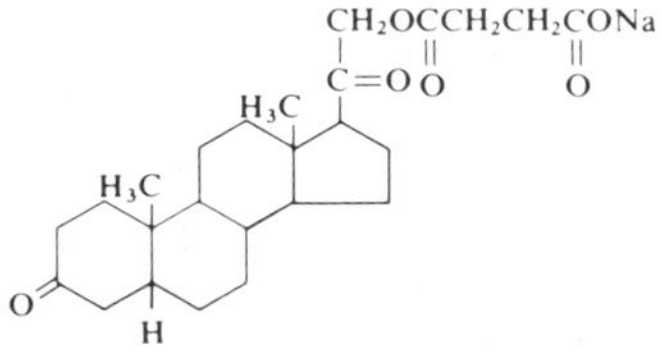 $C_{25}H_{35}O_6Na$	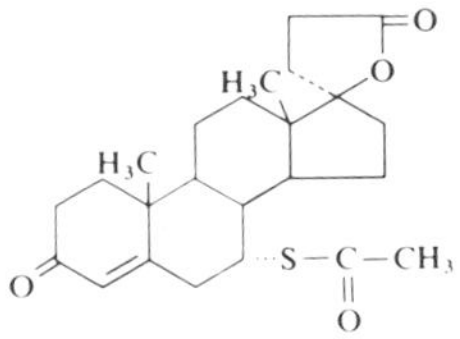 $C_{24}H_{32}O_4S$
Molecular weight	454.5	416.5
Melting point (°C)	193–203 (dec.)	135 (preliminary) 202 (dec.)
Specific rotation	$(\alpha)\frac{25}{D}$ +95 (chloroform) for free acid.	$(\alpha)\frac{25}{D}$ −34 (chloroform)
Absorption max.	280 mμ (ε = 93.2)	ε^{238} = 20,200

Names & synonyms:	METHANDROSTENOLONE; 17α-methyl-17β-hydroxy-1,4-androstadien-3-one.	METHYLANDROSTENEDIOL; MAD; methandriol; 17α-methyl-5-androsten-3β,17β-diol.	METHYL TESTOSTERONE; 17-methyl testosterone; 17α-methyl-Δ^4-androsten-17-β-ol-3-one; 17(β)-hydroxy-17(α-methyl-4-androsten-3-one.
Formulae:	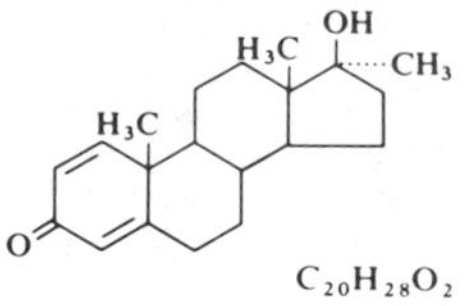 $C_{20}H_{28}O_2$	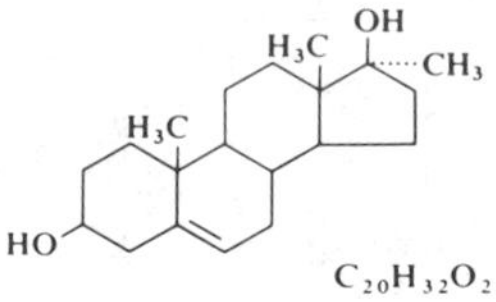 $C_{20}H_{32}O_2$	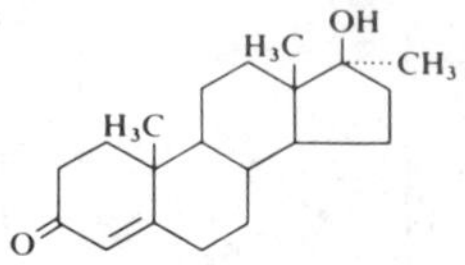 $C_{20}H_{30}O_2$
Molecular weight	300.4	304.4	302.4
Melting point (°C)	166–167	205–207	161–166
Specific rotation	$(\alpha)\frac{20}{D}$ +9 – +17 (100 mg. in 10 ml. alcohol)	$(\alpha)\frac{20}{D}$ – 73 (100 mg. in 10 ml. alcohol)	$(\alpha)\frac{25}{D}$ +69 – +75 (100 mg. in 10 ml. dioxane)
Absorption max.			

Names & synonyms:	NORETHANDROLONE; 17α-ethyl-19-nortestosterone; 17α-ethyl-17-hydroxy-4-norandrosten-3-one; 17α-ethyl-17-hydroxy-19-norandrost-4-en-3-one.	OXANDROLONE; 17β-hydroxy-17α-methyl-2-oxa-5α-androstane-3-one.
Formulae:	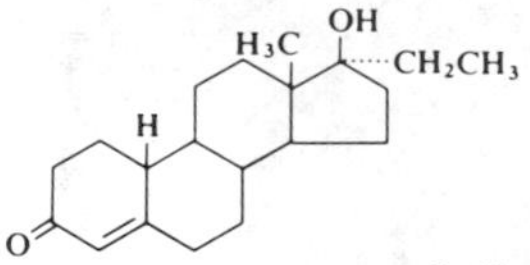 $C_{20}H_{30}O_2$	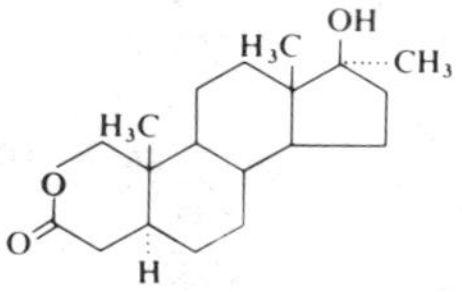 $C_{19}H_{30}O_3$
Molecular weight	302.4	306.4
Melting point (°C)	130–136	230–233
Specific rotation	$(\alpha)\frac{25}{D}$ +21 (dioxane)	$(\alpha)\frac{25}{D}$ – 21 (1 % in chloroform)
Absorption max.	240 mμ (ε = 16,500)	None

Names & synonyms:	OXYMETHOLONE: 17β-hydroxy-2-hydroxymethylene-17α-methyl-3-androstanone; 2-hydroxymethylene-17-α-methyl dihydrotestosterone.	PROMETHOLONE: 2α-methyl-dihydro-testosterone propionate; 2α-methyl-5α-androstane-17β-ol-3-one-propionate.
Formulae:	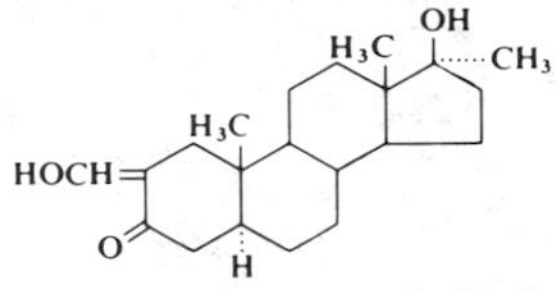 $C_{21}H_{32}O_3$	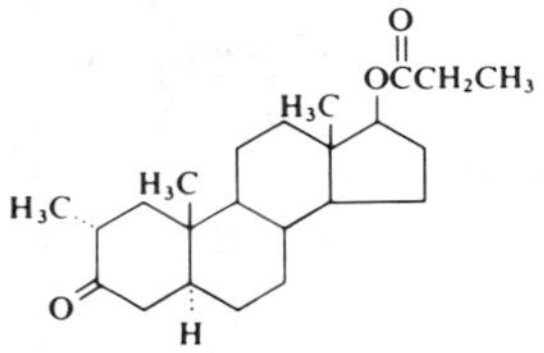
Molecular weight	332.4	360.5
Melting point (°C)	182	124–130
Specific rotation	$(\alpha)\frac{25}{D}$ = +36 (200 mg. in 10 ml. dioxane)	$(\alpha)\frac{25}{D}$ +22 – +29 (200 mg. in 10 ml. chloroform)
Absorption max.	E_1^1 = 547 at 315 mμ (in alkaline methanol made 0.01 N with NaOH)	without significant absorption from 220–300 mμ (methanol)

Names & synonyms:	TESTOSTERONE: trans-testosterone; Δ^4-androsten-17-β-ol-3-one; 17β-hydroxy-4-androsten-3-one.	TESTOSTERONE CYPIONATE: testosterone cyclopentylpropionate; 17β-hydroxy-4-androsten-3-one, cyclopentanepropionate.
Formulae:	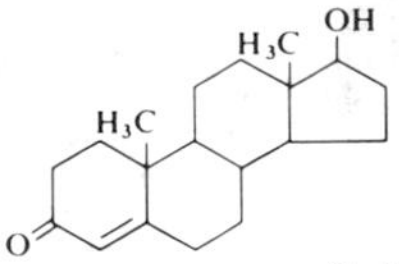 $C_{19}H_{28}O_2$	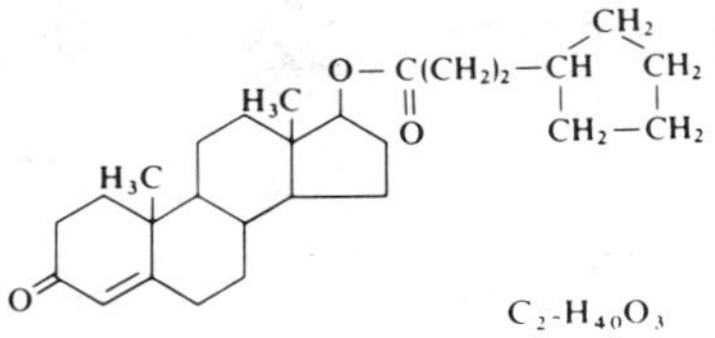 $C_{27}H_{40}O_3$
Molecular weight	288.4	412.6
Melting point (°C)	151–156	100–102
Specific rotation	$(\alpha)\frac{24}{D}$ + 109 (400 mg. in 10 ml. alcohol)	$(\alpha)_D$ +88.5 ± 3.5 ($CHCl_3$)
Absorption max.	238 mμ	λ_{max} 241 mμ (ε 16,125)

Names & synonyms:	TESTOSTERONE ENANTHATE; testosterone heptanoate; 17β-hydroxyandrost-4-en-3-one-17-enanthate.	TESTOSTERONE PHENYLACETATE; 17β-hydroxy-4-androsten-3-one phenylacetate; testosterone α-toluate.	TESTOSTERONE PROPIONATE; Δ^4-androstene-17-β-propionate-3-one.
Formulae:	$C_{26}H_{40}O_3$	$C_{27}H_{34}O_3$	$C_{22}H_{32}O_3$
Molecular weight	400.6	406.5	344.4
Melting point (°C)	34–39	129–131	118–122
Specific rotation	$(\alpha)\frac{25}{D}$ +77 – +82 (2% in dioxane)	$(\alpha)\frac{25}{D}$ +101 ±3 (1% in chloroform)	$(\alpha)\frac{25}{D}$ +83 – +90 (100 mg. in 10 ml. dioxane)
Absorption max.	241 mμ (in ethanol)	241 mμ (in ethanol)	

ESTROGENS

Names & synonyms:	EQUILENIN; 3-hydroxy-17-keto-$\Delta^{1,3,5-10,6,8}$ estrapentaene; 1,3,5–10,6,8-estrapentaen-3-ol-17-one.	EQUILIN; 3-hydroxy-17-keto-$\Delta^{1,3,5\ 10,7}$ estratetraene; 1,3,5,7-estratetraen-3-ol-17-one.	ESTRADIOL (formerly called α-estradiol); β-estradiol; dihydrofolliculin; dihydroxyestrin; 1,3,5-estratriene-3,17β-diol; 3,17-dihydroxy-$\Delta^{1,3,5\ 10}$-estratriene; 3,17-epidihydroxyestratriene.
Formulae:	$C_{18}H_{18}O_2$	$C_{18}H_{20}O_2$	$C_{18}H_{24}O_2$
Molecular weight	266.3	268.3	272.3
Melting point (°C)	258–259	236–240	173–179
Specific rotation	$(\alpha)\frac{25}{D}$ +89 (dioxane)	$(\alpha)\frac{25}{D}$ +308 (200 mg. in 10 ml. dioxane); +325 (200 mg. in 10 ml. alcohol).	$(\alpha)\frac{25}{D}$ +76 – +83 (100 mg. in 10 ml. dioxane)
Absorption max.	231, 270, 282, 292, 325, 340 mμ	283–285 mμ	225, 280 mμ

Names & synonyms:	ESTRADIOL BENZOATE; β-estradiol-3-benzoate; estradiol monobenzoate.	ESTRADIOL CYPIONATE; estradiol cyclopentylpropionate; β-estradiol 17-cyclopentanepropionate; 1,3,5(10)-estratriene-3,17β-diol,17-cyclopentanepropionate.
Formulae:	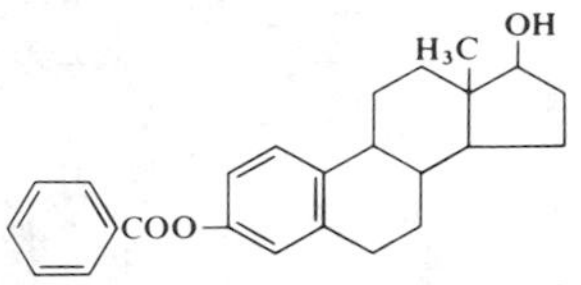 $C_{25}H_{28}O_3$	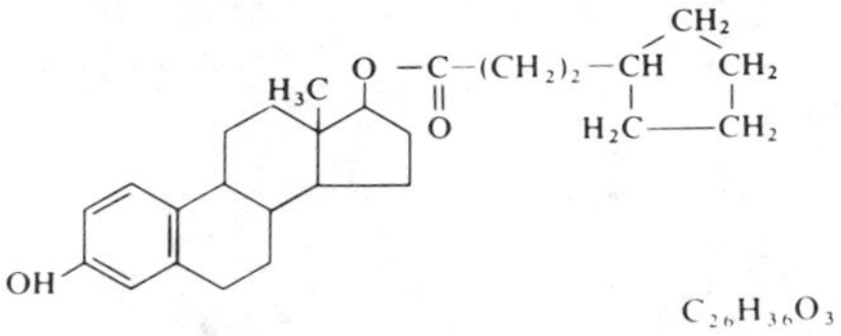 $C_{26}H_{36}O_3$
Molecular weight	376.4	396.6
Melting point (°C)	191–196	151–154
Specific rotation	$(\alpha)\frac{25}{D}$ +58 – +63 (200 mg. in 10 ml. dioxane)	$(\alpha)_D$ +41.5 ±3.5 (dioxane)
Absorption max.		223 mμ

Names & synonyms:	ESTRADIOL DIPROPIONATE; α-estradiol dipropionate; 17β-estradiol dipropionate.	ESTRIOL; trihydroxyestrin; $\Delta^{1,3,5-10}$-estratriene-3-16-cis-17-trans-diol; 1,3,5-estratriene-3,16α,17β-triol.	ESTRONE; folliculin; ketohydroxyestrin; 1,3,5-estratrien-3-ol-17-one.
Formulae:	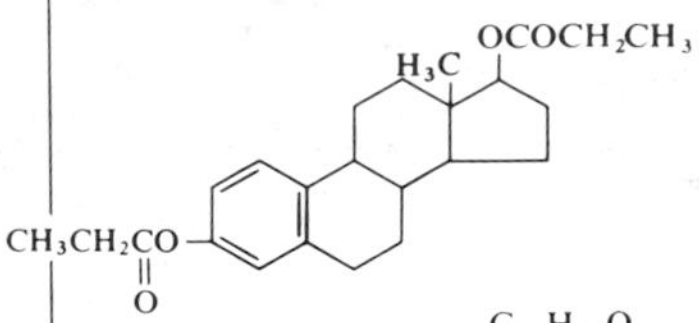 $C_{24}H_{32}O_4$	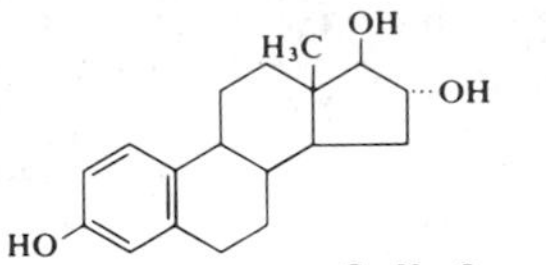 $C_{18}H_{24}O_3$	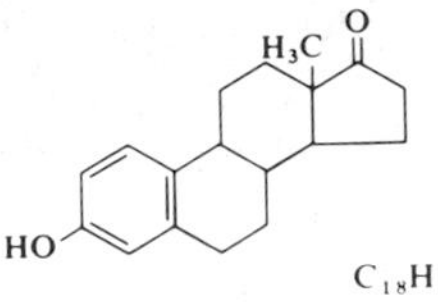 $C_{18}H_{22}O_2$
Molecular weight	384.5	288.3	270.3
Melting point (°C)	104–109	282	258–262
Specific rotation	$(\alpha)\frac{25}{D}$ +39 ±2 (1 % in dioxane)	$(\alpha)\frac{25}{D}$ +53 – +63 (40 mg. in 1 ml. dioxane)	$(\alpha)\frac{25}{D}$ +158 – +168 (100 mg. in 10 ml. dioxane)
Absorption max.	268 mμ	280 mμ	283–285 mμ

STEROID HORMONES AND OTHER STEROIDAL SYNTHETICS (Continued)

Names & synonyms:	ESTRONE BENZOATE	ETHYNYL ESTRADIOL; 17-ethinyl estradiol; 17α-ethynyl-1,3,5-estratriene-3,17β-diol.	MESTRANOL; ethynylestradiol 3-methyl ether; 3-methoxy-17α-ethynyl-1,3,5(10)-estratriene-17β-ol; 17α-ethynyl-estradiol-3-methyl ether; 3-methoxy-19-nor-17α-pregna-1,3,5,trien-20-yn-17-ol.
Formulae:	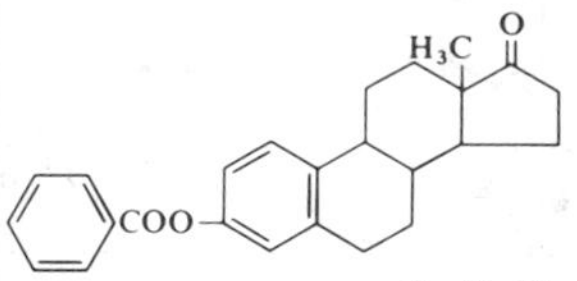 $C_{25}H_{26}O_3$	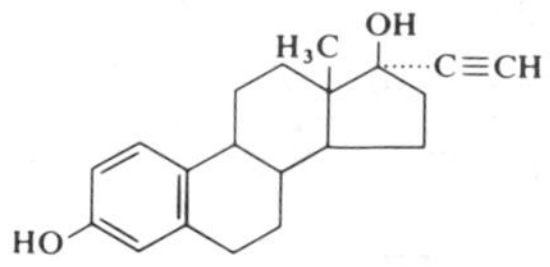 $C_{20}H_{24}O_2$	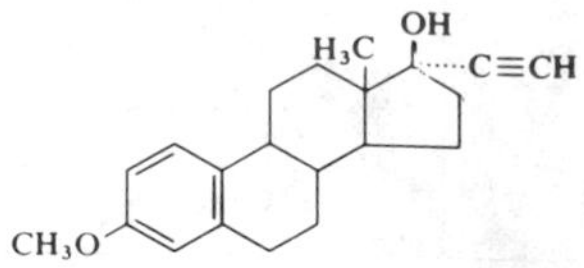 $C_{21}H_{26}O_2$
Molecular weight	374.4	296.4	310.4
Melting point (°C)	220	141–146	148–154
Specific rotation	$(\alpha)\frac{25}{D}$ +120 (dioxane)	$(\alpha)\frac{25}{D}$ +1 – +10 (100 mg. in 10 ml. dioxane)	$(\alpha)\frac{25}{D}$ +2 to +8 (200 mg. in 10 ml. dioxane)
Absorption max.		248 mμ	278 to 287 mμ (methanol)

PROGESTOGENS AND PROGESTINS (INCLUDING 19-NORSTEROID COMPOUNDS)

Names & synonyms:	ACETOXYPREGNENOLONE; 21-acetoxypregnenolone; prebediolone acetate; Δ^5-pregnene-3β,21-diol-20-one-21-monoacetate; 21-acetoxy-5-pregnene-3-ol-20-one; 3-hydroxy-21-acetoxy-5-pregnen-20-one.	ANAGESTONE ACETATE; 6α-methyl-4-pregnen-17α-ol-20-one acetate; 17α-acetoxy-6α-methylpregn-4-en-20-one; 17α-acetoxy-6α-methyl-4-pregnen-20-one.	CHLORMADINONE ACETATE; 6-chloro-Δ^6-dehydro-17α-acetoxyprogesterone; 6-chloro-$\Delta^{4,6}$-pregnadiene-17α-ol-3,20-dioneacetate.
Formulae:	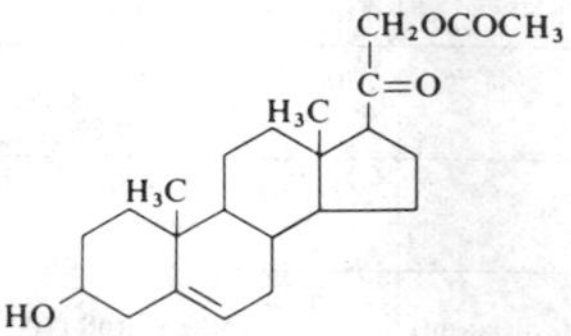 $C_{23}H_{34}O_4$	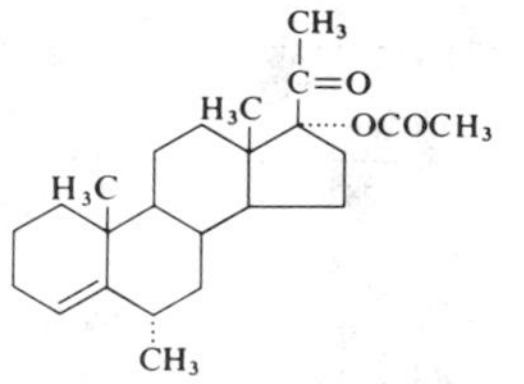 $C_{24}H_{36}O_3$	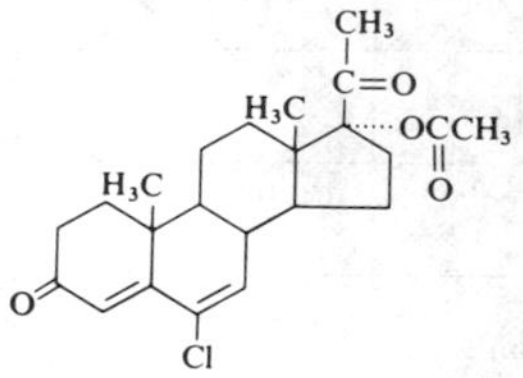 $C_{23}H_{29}ClO_4$
Molecular weight	374.5	372.6	404.9
Melting point (°C)	184–185	172–178	204–212
Specific rotation	$(\alpha)\frac{20}{D}$ +37 – +43 (dioxane)	$(\alpha)\frac{25}{D}$ +40 to +45 (10 mg. in 10 ml. chloroform)	$(\alpha)\frac{25}{D}$ 0 to −6 (200 mg. in 10 ml. chloroform)
Absorption max.			284 mμ (methanol) Log ε = 4.34 ±0.02

STEROID HORMONES AND OTHER STEROIDAL SYNTHETICS (Continued)

Names & synonyms:	DIMETHISTERONE; 6α,21-dimethylethisterone; 6α,21-dimethyl-17β-hydroxy-17α-pregn-4-en-20-yn-3-one; 6α-methyl-17α-propynylandrost-4-en-17β-ol-3-one; 17β-hydroxy-6α-methyl-17α-(prop-1-ynyl)-androst-4-ene-3-one.	ETHISTERONE; anhydrohydroxyprogesterone; ethinyl testosterone; pregneninolone; 17α ethynyl testosterone; 17α-ethynyl-17β-hydroxy-4-androsten-3-one.	ETHYNODIOL DIACETATE; 17α-ethynyl-4-estrene-3β,17β-diol-17-diacetate; 19-nor-17α-pregn-4-en-20-yne-3β,17-diol diacetate.
Formulae:	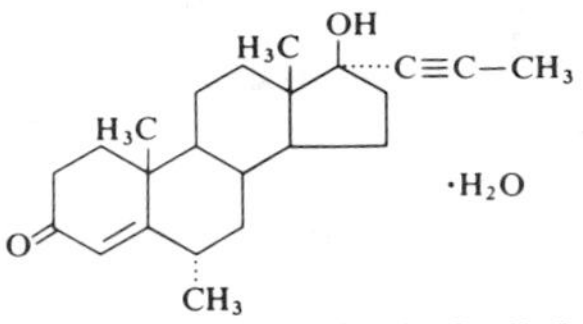 $C_{23}H_{32}O_2 \cdot H_2O$	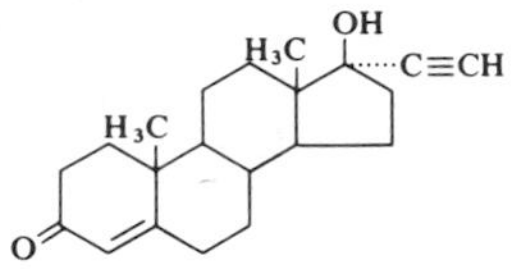 $C_{21}H_{28}O_2$	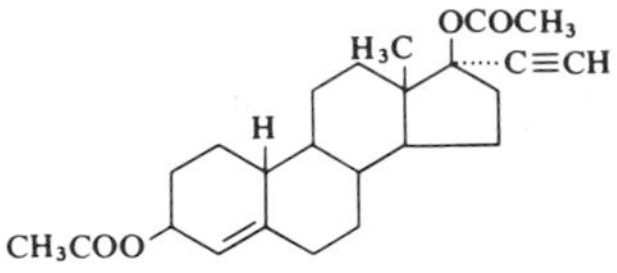 $C_{24}H_{32}O_4$
Molecular weight	358.5	312.4	384.5
Melting point (°C)	App. 100 (dec.)	266–273	126–132
Specific rotation	$(\alpha)^{20}_{D}$ +16.5 to +18.5 (2 % solution in chloroform) (calculated to the anhydrous basis)	$(\alpha)^{25}_{D}$ −32° (100 mg. in 10 ml. pyridine)	$(\alpha)^{25}_{D}$ −74 (1 % in chloroform)
Absorption max.	App. 240 mμ (anhydrous ethanol) $E^{1\%}_{1}$ cm = 443	241 mμ (methanol)	None

Names & synonyms:	FLUROGESTONE ACETATE; 17α-acetoxy-9α-fluoro-11β-hydroxy-4-pregnene-3,20-dione.	HYDROXYMETHYLPROGESTERONE; medroxyprogesterone; 17α-hydroxy-6α-methylprogesterone; 17α-hydroxy-6α-methyl-4-pregnene-3,20-dione.	HYDROXYMETHYLPROGESTERONE ACETATE; medroxyprogesterone acetate; 17α-hydroxy-6α-methylprogesterone acetate; 17α-hydroxy-6α-methyl-4-pregnene-3,20-dione acetate.
Formulae:	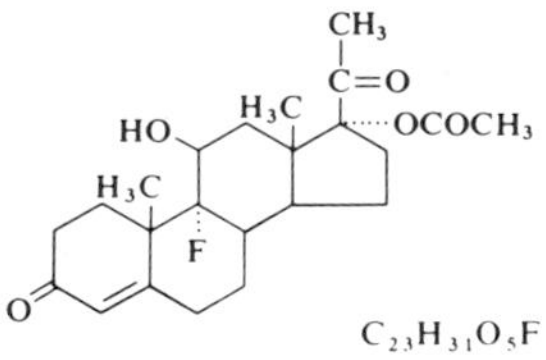 $C_{23}H_{31}O_5F$	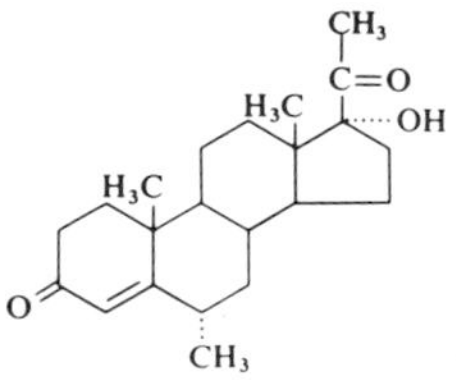 $C_{22}H_{32}O_3$	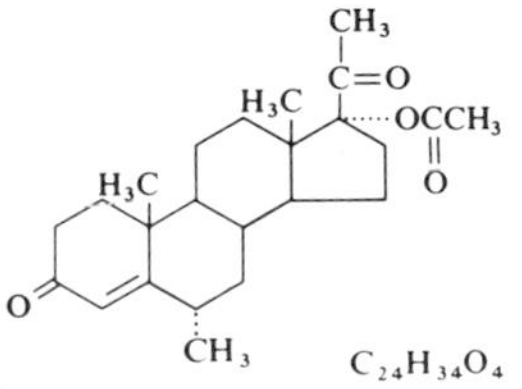 $C_{24}H_{34}O_4$
Molecular weight	406.5	344.5	386.5
Melting point (°C)	250–251	220–223.5	202–207
Specific rotation	$(\alpha)^{25}_{D}$ +78	$(\alpha)^{25}_{D}$ +75	$(\alpha)^{25}_{D}$ +51 (dioxane)
Absorption max.	238 mμ (ε = 17,100)	241 mμ (ε = 16,150)	241 mμ (α_M = 16,500) ethanol

Names & synonyms:	HYDROXYPROGESTERONE; 17α-hydroxyprogesterone; 17α-hydroxy-4-pregene-3,20 dione; 4-pregnen-17α-ol-3,20-dione.	HYDROXYPROGESTERONE ACETATE; 17α-acetoxyprogesterone; 17α-hydroxyprogesterone acetate; 17α-hydroxy-4-pregnene-3,20 dione acetate.
Formulae:	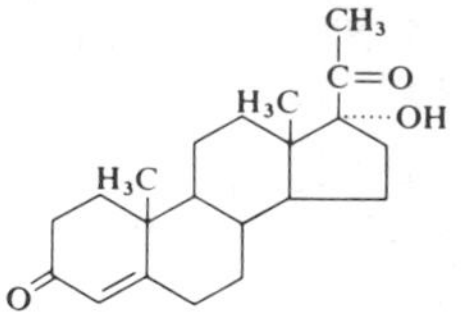 $C_{21}H_{30}O_3$	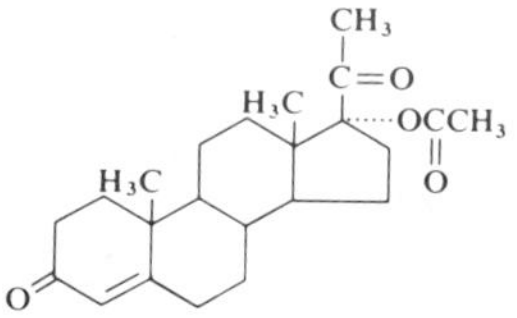 $C_{23}H_{32}O_4$
Molecular weight	330.4	372.5
Melting point (°C)	276	249–250
Specific rotation	$(\alpha)^{17}_{D}$ +105 (104 mg. in 10 ml. chloroform)	$(\alpha)^{25}_{D}$ +72 (chloroform)
Absorption max.		240 mμ (a_M = 16,875) ethanol

Names & synonyms:	HYDROXYPROGESTERONE CAPROATE; 17α-hydroxyprogesterone caproate; 17α-hydroxy-4-pregnene-3,20-dione caproate.	MELENGESTROL ACETATE; MGA; 17α-hydroxy-6-methyl-16-methylene-4,6-pregnadiene-3, 20-dione acetate; 6-dehydro-17-hydroxy-6-methyl-16-methylene-progesterone acetate.
Formulae:	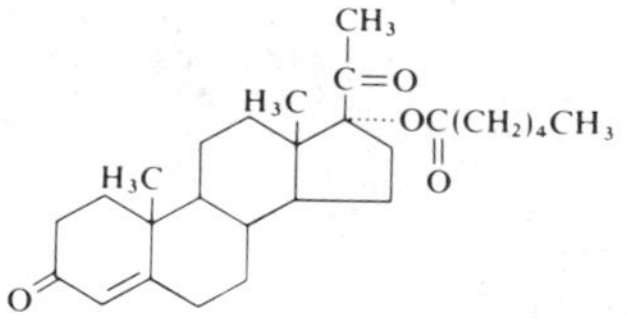 $C_{27}H_{40}O_4$	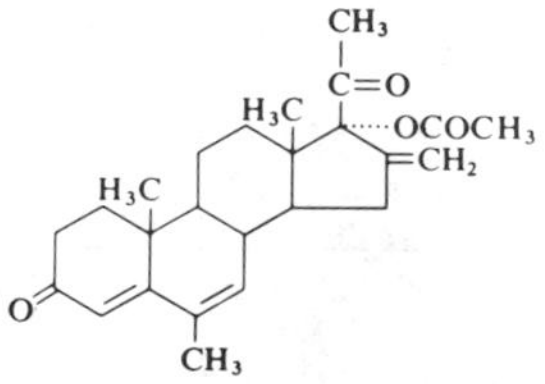 $C_{25}H_{32}O_4$
Molecular weight	428.6	396.51
Melting point (°C)	121–123	215–227
Specific rotation	$(\alpha)^{25}_{D}$ +57 (chloroform)	$(\alpha)_D$ −127 to −135 (in $CHCl_3$)
Absorption max.		288 mμ (ε^1_1 = 24,000) (ethanol)

Names & synonyms:	NORETHINDRONE; Norethisterone; 17α-ethynyl-19-nortestosterone; 17α-ethynyl-17-hydroxy-19-nor-17α-4-en-20-yn-3-one.	NORETHINDRONE ACETATE; 17α-ethinyl-19-nortestosterone acetate.
Formulae:	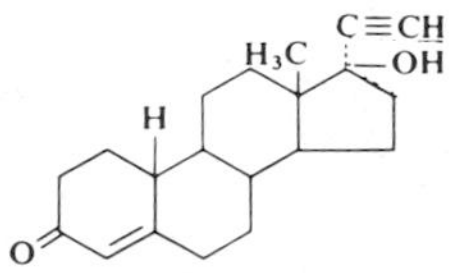 $C_{20}H_{26}O_2$	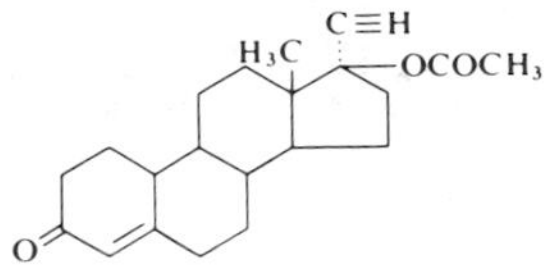 $C_{22}H_{28}O_3$
Molecular weight	298.4	340.4
Melting point (°C)	202 and 208	157–163
Specific rotation	$(\alpha)\frac{25}{D}$ −30 – −35 (200 mg. in 10 ml. dioxane)	$(\alpha)\frac{25}{D}$ −32 – −35 (200 mg. in 10 ml. dioxane)
Absorption max.	α (1 %, 1 cm) λ 240 = 535 ± 15	α (1 %, 1 cm.) λ 240 = 490 to 520 (505 ± 15) (ethanol)

Names & synonyms:	NORETHISTERONE; norethindrone; 19-norethisterone; 17α-ethynyl-19-nor-Δ^4-androstan-17β-ol-3-one; 17α-ethynyl-19-nor-testosterone; 17-hydroxy-3-oxo-19-nor-17α-pregn-4-ene-20-yne; 17-hydroxy-19-nor-17α-pregn-4-en-20-yn-3-one.	NORETHYNODREL; 17α-ethynyl-17β-hydroxy-5(10)-estren-3-one.
Formulae:	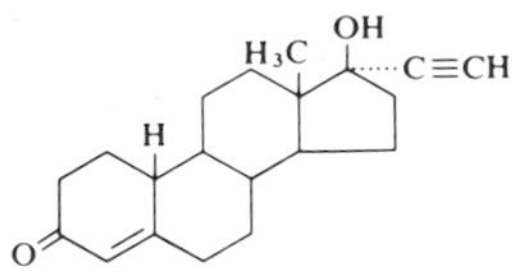 $C_{20}H_{26}O_2$	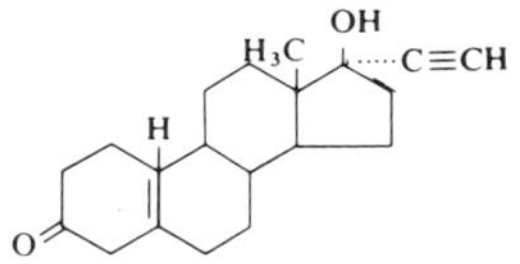 $C_{20}H_{26}O_2$
Molecular weight	298.4	298.4
Melting point (°C)	200–207	174–184
Specific rotation	$(\alpha)\frac{25}{D}$ −30 – −38 (200 mg. in 10 ml. dioxane)	$(\alpha)\frac{25}{D}$ +125 (dioxane)
Absorption max.	240 mμ (ε_1^1 = 576)	

Names & synonyms:	NORMETHISTERONE; 19-normethisterone; normethandrolone; metalutin; normetandrone; 17α-methyl-19-nor-Δ^4-androsten-17β-ol-3-one; 17α-methyl-19-nor-testosterone; 17β-hydroxy-3-oxo-17α-methyl-estra-4-ene; 17β-hydroxy-17-methyl-estr-4-en-3-one.	PREGNENOLONE; Δ^5-pregnenolone; Δ^5-pregnen-3β-ol-20-one; 17β(1-ketoethyl)-Δ^5-androstene-3β-ol.	PROGESTERONE; progestin; progestone; pregnendione; Δ^4-pregnene-3,20-dione.
Formulae:	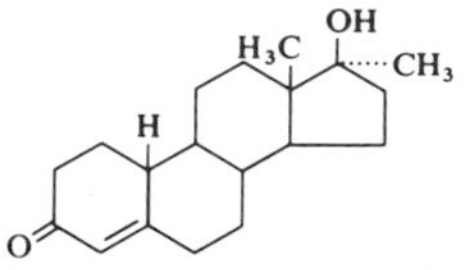 $C_{19}H_{28}O_2$	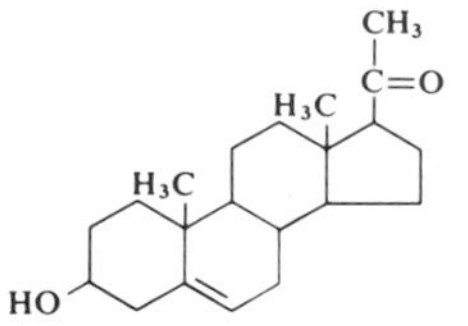 $C_{21}H_{32}O_2$	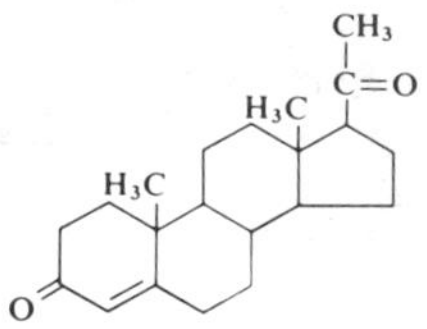 $C_{21}H_{30}O_2$
Molecular weight	288.4	308.4	314.4
Melting point (°C)	153–158	193	(β) isomer 121; (α) isomer 127–131
Specific rotation	$(\alpha)\frac{25}{D}$ +25 to +29 (200 mg. in 10 ml. chloroform)	$(\alpha)\frac{20}{D}$ +28 – +30 (alcohol)	$(\alpha)\frac{20}{D}$ +172 – +182 (200 mg. in 10 ml. dioxane)
Absorption max.	241 mμ – 565 ±15		240 mμ

DIURETIC, ANTIDIURETIC AND LOCAL ANESTHETIC AGENTS

Names & synonyms:	ALDOSTERONE; electrocortin; 18-oxocorticosterone; 18-formyl-11β,21-dihydroxy-4-pregnene-3,20-dione.	HYDROXYDIONE SODIUM; 21-hydroxypregnane-3,20-dione-21-sodium hemisuccinate.	SPIRONOLACTONE; 3-(3-oxo-7α-acetylthio-17β-hydroxy-4-androsten-17α-yl)-propionic acid γ lactone.
Formulae:	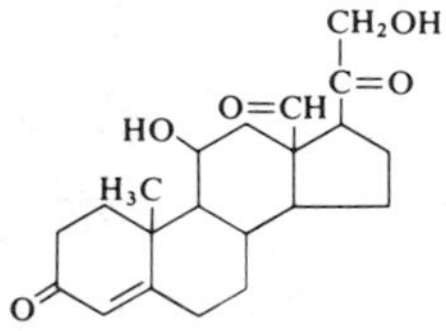 $C_{21}H_{28}O_5$	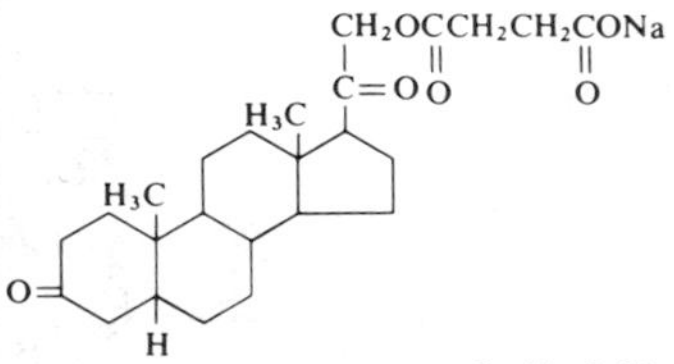 $C_{25}H_{35}O_6Na$	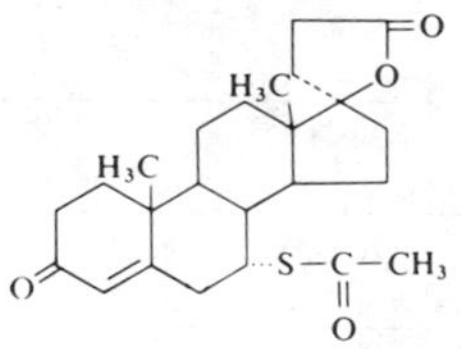 $C_{24}H_{32}O_4S$
Molecular weight	360.4	454.5	416.5
Melting point (°C)	108–112 (hydrate); 164 (anhydrous)	193–203 (dec.)	135; 202 (dec.)
Specific rotation	$(\alpha)\frac{25}{D}$ +161 (10 mg. in 10 ml. chloroform)	$(\alpha)\frac{25}{D}$ +95 (chloroform) for free acid.	$(\alpha)\frac{25}{D}$ – 34 (chloroform)
Absorption max.	240 mμ (log ε = 4.20 monohydr.; ε mol. 15,000 anhydr.)	280 mμ (ε = 93.2)	ε^{238} = 20,200

PROPERTIES OF THE AMINO ACIDS

IONIZATION CONSTANTS AND pH VALUES AT THE ISOELECTRIC POINTS OF THE AMINO ACIDS IN WATER AT 25°C

The majority of the recorded values are true thermodynamic constants calculated from electrometric force measurements of cells without liquid junctions. The values for the constants given in the table were derived from the classical, the zwitterionic (Bjerrum), and the acidic (Bronsted) formulations of ionization and the corresponding mass law expressions. pH values at the isoelectric points were calculated from the expression, $pI = \frac{1}{2}(pk_{a'} + pk_w - pk_{b1})$. The error is approximately 0.5% when this expression is used to calculate pI values for cystine, tyrosine, and diiodotyrosine.

	Classical				Zwitterionic				Acidic					
Amino acid	pk_{a1}	pk_{a2}	pk_{b1}	pk_{b2}	pK_{A1}	pK_{A2}	pK_{B1}	pP_{B2}	pK_1	pK_2	pK_3	pK_4	pI	Ref.
DL-Alanine	9.866	—	11.649	—	2.348	—	4.131	—	2.348	9.866	—	—	6.107	1
L-Arginine	12.48	—	4.96	11.99	2.01	—	1.52	4.96	2.01	9.04	12.48	—	10.76	2
L-Aspartic acid	3.86	9.82	11.93	—	2.10	3.86	4.18	—	2.10	3.86	9.82	—	2.98	3
L-Cystine	8.00	10.25	11.95	12.96	1.04	2.05	3.75	6.00	1.04	2.05	8.00	10.25	5.02	4
Diiodo-L-tyrosine	6.48	7.82	11.88	—	2.12	6.48	6.18	—	2.12	6.48	7.82	—	4.29	5,6
L-Glutamic acid	4.07	9.47	11.90	—	2.10	4.07	4.53	—	2.10	4.07	9.47	—	3.08	7
Glycine	9.778	—	11.647	—	2.350	—	4.219	—	2.350	9.778	—	—	6.064	8
L-Histidine	9.18	—	7.90	12.23	1.77	—	4.82	7.90	1.77	6.10	9.18	—	7.64	3
Hydroxy-L-proline	9.73	—	12.08	—	1.92	—	4.27	—	1.92	9.73	—	—	5.82	9
DL-Isoleucine	9.758	—	11.679	—	2.318	—	4.239	—	2.318	9.758	—	—	6.038	1
DL-Leucine	9.744	—	11.669	—	2.328	—	4.253	—	2.328	9.744	—	—	6.036	1
L-Lysine	10.53	—	5.05	11.82	2.18	—	3.47	5.05	2.18	8.95	10.53	—	9.47	2
DL-Methionine	9.21	—	11.72	—	2.28	—	4.79	—	2.28	9.21	—	—	5.74	10
DL-Phenylalanine	9.24	—	11.42	—	2.58	—	4.76	—	2.58	9.24	—	—	5.91	11
L-Proline	10.60	—	12.0	—	2.00	—	3.40	—	2.00	10.60	—	—	6.3	12
DL-Serine	9.15	—	11.79	—	2.21	—	4.85	—	2.21	9.15	—	—	5.68	9
L-Tryptophan	9.39	—	11.62	—	2.38	—	4.61	—	2.38	9.39	—	—	5.88	13
L-Tyrosine	9.11	10.07	11.80	—	2.20	9.11	3.93	—	2.20	9.11	10.07	—	5.63	6
DL-Valine	9.719	—	11.711	—	2.286	—	4.278	—	2.286	9.719	—	—	6.002	1

REFERENCES

1. Smith, P. K., Taylor, A. C., and Smith, E. R. B., *J. Biol. Chem.*, 122, 109, 1937—1938.
2. Schmidt, C. L. A., Kirk, P. L., and Appleman, W. K., *J. Biol. Chem.*, 88, 285, 1930.
3. Greenstein, J. P., *J. Biol. Chem.*, 93, 479, 1931.
4. Borsook, H., Ellis, E. L., and Huffman, H. M., *J. Biol. Chem.*, 117, 281, 1937.
5. Dalton, J. B., Kirk, P. L., and Schmidt, C. L. A., *J. Biol. Chem.*, 88, 589, 1930.
6. Winnek, P. S. and Schmidt, C. L. A., *J. Gen. Physiol.*, 18, 889, 1935.
7. Simms, H. S., *J. Gen. Physiol.*, 11, 629, 1928; 12, 231, 1928.
8. Owen, B. B., *J. Am. Chem. Soc.*, 56, 24, 1934.
9. Kirk, P. L. and Schmidt, C. L. A., *J. Biol. Chem.*, 81, 237, 1929.
10. Emerson, O. H., Kirk, P. L., and Schmidt, C. L. A., *J. Biol. Chem.*, 92, 449, 1931.
11. Miyamoto, S. and Schmidt, C. L. A., *J. Biol. Chem.*, 90, 165, 1931.
12. McCay, C. M. and Schmidt, C. L. A., *J. Gen. Physiol.*, 9, 333, 1926.
13. Schmidt, C. L. A., Appleman, W. K., and Kirk, P. L., *J. Biol. Chem.*, 85, 137, 1929—1930.

IONIZATION CONSTANTS OF THE AMINO ACIDS IN AQUEOUS ETHANOL SOLUTIONS

Amino acid	pK_1	pK_2	pK_3	Volume % ethanol	Temperature (°C)	Ref.
Alanine	3.55	10.02	—	72	25	1
Arginine	3.34	9.40	14.1	72	25	1
Aspartic acid	2.85	5.20	10.51	72	25	1
Glutamic acid	3.16	5.63	10.75	72	25	2
Glycine	2.66	9.82	—	10	19.5	2
	2.96	9.76	—	40	19.5	2
	3.46	9.82	—	72	25	1

Amino acid	pK_1	pK_2	pK_3	Volume % ethanol	Temperature (°C)	Ref.
	3.79	9.99	—	90	19.5	2
Histidine	3.00	5.85	9.45	72	25	1
Isoleucine	3.69	9.81	—	72	25	1
Lysine	2.75	8.95	10.53	48	25	1
	3.56	8.95	10.49	84	25	1
Proline	3.04	10.55	—	72	25	1
Valine	3.60	9.73	—	72	25	1

REFERENCES

1. Jukes, T. H. and Schmidt, C. L. A., *J. Biol. Chem.*, 105, 359, 1934.
2. Michaelis, L. and Mizutani, M., *Z. Physik. Chem.*, 116, 135, 1925.

IONIZATION CONSTANTS OF THE AMINO ACIDS IN AQUEOUS FORMALDEHYDE SOLUTION[a]

Amino acid	Mole % formaldehyde				
	0.99	3.95	5.60	10.0	17.9
DL-Alanine	8.36	7.42	6.96[b]	6.56	6.10
L-Arginine	—	3.45[c]	3.40[d]	—	—
L-Aspartic acid	—	—	7.21[d]	≤3.8[e] 6.85[f]	—
L-Glutamic acid	—	—	6.91[d]	≤4.2[e] 6.8[f]	—
Glycine	7.16	6.08	5.92[b]	5.34	5.04
L-Histidine	—	7.90[c]	7.90[d]	—	—
Hydroxy-L-proline	—	—	7.19[d]	—	—
L-Leucine	8.44	7.50	6.92[d]	6.62	6.20

Amino acid	Mole % formaldehyde				
	0.99	3.95	5.60	10.0	17.9
DL-Leucine	8.44	7.48	—	6.60	6.20
L-Lysine	—	7.35[c]	7.15[d]	—	—
L-Phenylalanine	—	—	6.62[d]	5.9[e]	—
DL-Phenylalanine	8.09	7.16	6.80[b]	6.35	6.13
L-Proline	—	—	7.78[d]	—	—
DL-Serine	6.66	5.74	5.63[b]	—	4.94
L-Tryptophan	—	—	6.88[d]	—	—
L-Tyrosine	—	—	7.50[d]	6.2[e] >9[f]	—
DL-Valine	8.52	7.65	7.47[b]	—	6.52

[a] pK_2 at 22°.[1]
[b] pK_2 at 22°.[2]
[c] pK_2 at 30° for arginine and pK_3 at 30° for histidine and lysine.[3]
[d] pK_2 at 30°, pK_3 at 30° for histidine and lysine.[4]
[e] pK_2 at 25° for aspartic acid, glutamic acid, phenylalanine, and tyrosine.[5]
[f] pK_3 at 30° for aspartic acid, glutamic acid, and tyrosine.[5]

REFERENCES

1. Dunn, M. S. and Weiner, J. G., *J. Biol. Chem.*, 117, 381, 1937.
2. Dunn, M. S. and Loshakoff, A., *J. Biol. Chem.*, 113, 691, 1936.
3. Levy, M., *J. Biol. Chem.*, 109, 365, 1935.
4. Levy, M. and Silberman, D. E., *J. Biol. Chem.*, 118, 723, 1937.
5. Harris, L. J., *Proc. R. Soc. London, Series B*, 95, 440, 1923—1924.

SPECIFIC ROTATIONS OF THE AMINO ACIDS USING SODIUM LIGHT (5893 Å)

Abbreviations

c —grams of solute per 100 mℓ of solution.
d —density of the solution.
p —grams of solute per 100 g of solution.
l —length of the tube in decimeters.
α —observed rotation in angular degrees.
$[\alpha]$ —specific rotation in angular degrees calculated from
$[\alpha]_\lambda^t = \alpha \times 100/c \times l = \alpha \times 100/p \times d \times l$ where t is temperature in °C and λ is wave length of the incident light in Ångstroms.

A —prepared from a protein or other naturally occurring material.
B—prepared by resolution of the inactive synthetic form.
C—prepared by resolution of the inactive racemized form.
D—prepared from the inactive synthetic form by a biological method.
E—prepared from the inactive racemized form by a biological method.
?—source not given.

Source	c	Solvent	d	p	Moles acid or base per mole amino acid	l	Temp. (°C)	α	$[\alpha]$	Ref.
L-Alanine										
A	5.790	0.97 *N*HCl	1.033	5.605	1.5	2	15	+1.70	+14.7	1
A	10.3	Water	1.03	1.00	0	2	22	+0.55	+2.7	2
A	1.781	3 *N*NaOH	—	—	15	2	20	—	+3.0	3
D-Alanine										
B	71.344	6 *N*HCl	—	—	39.4	2	30.4	−0.392	−14.6	4
L-Arginine										
A	1.653	6.0 *N*HCl	—	—	63	4.001	23.4	+1.777	+26.9	5
A	3.48	Water	—	—	0	2	20	—	+12.5	6
A	0.87	0.50 *N*NaOH	—	—	10	2	20	—	+11.8	6
L-Aspartic acid										
A	2.002	6.0 *N*HCl	—	—	39	4.001	24.0	+1.972	+24.6	7
A	1.3300	Water	—	—	0	3	18	—	4.7	3
A	1.3300	3 *N*NaOH	—	—	30	3	18	—	−1.7	3
D-Aspartic acid										
C	4.289	0.97 *N*HCl	1.032	4.156	3	1	20	−1.09	−25.5	8
L-Cystine										
A	0.9974	1.02 *N*HCl	1.0181	0.9797	24.6	2	24.35	−4.277	−214.40	9
A	0.400	0.20 *N*NaOH	—	—	12	2	18.5	—	−70.0	3
D-Cystine										
C	—	1 *N*HCl	—	1	24	—	20	—	+223	10
Diiodo-D-tyrosine										
A	5.08	1.1 *N*HCl	1.05	4.84	9.4	1	20	+0.15	+2.89	11
A	4.41	13.4 *N*NH_4OH	0.9779	4.51	132	1	20	+0.10	+2.27	11
L-Glutamic acid										
A	1.002	6.0 *N*HCl	—	—	87	4.001	22.4	+1.25	+31.2	12
A	1.471	Water	—	—	0	2	18	—	+11.5	3
A	1.471	1 *N*NaOH	—	—	10	2	18	—	+10.96	3
D-Glutamic acid										
C	5.425	0.37 *N*HCl	1.0233	5.3011	1	1	20	−1.63	−30.05	8
L-Histidine										
A	1.480	6.0 *N*HCl	—	—	63	4.001	22.7	+0.766	+13.0	7
A	1.128	Water	1.0012	1.127	0	4	25.00	−1.714	−39.01	13
A	0.775	0.50 *N*NaOH	—	—	10	2	20	—	−10.9	6
D-Histidine										
?	4.000	1.0 *N*HCl	—	—	4	1	20	−0.407	−10.2	14
B	2.66	Water	—	—	0	2	23	+2.11	+39.8	14
Hydroxy-L-proline										
A	1.31	1.0 *N*HCl	—	—	10	2	20	—	−47.3	6
A	1.001	Water	—	—	0	4.001	22.5	−3.009	−75.2	7
A	0.655	0.50 *N*NaOH	—	—	10	2	20	—	−70.6	6
Hydroxy-D-proline										
B	4.48	Water	1.03	4.35	0	1	21	+3.37	+75.2	16

Source	c	Solvent	d	p	Moles acid or base per mole amino acid	l	Temp. (°C)	α	[α]	Ref.
Allo-Hydroxy-L-proline										
B	2.617	Water	1.014	2.581	0	1	18	−1.52	−58.1	16
Allo-Hydroxy-D-proline										
B	2.530	Water	1.013	2.998	0	1	17	+1.48	+58.5	16
L-Isoleucine										
B	5.09	6.1 *N*HCl	1.098	4.64	15	1	20	+2.07	+40.61	17
B	3.10	Water	1.008	3.08	0	2	20	+0.70	+11.29	17
A	3.34	0.33 *N*NaOH	1.017	3.28	1.3	2	20	+0.74	+11.09	18
D-Isoleucine										
B	4.53	6.1 *N*HCl	1.083	4.18	17	1	20	−1.85	−40.86	17
B	3.12	Water	1.006	3.10	0	2	20	−0.66	−10.55	17
D-allo-Isoleucine										
D	5.14	6.0 *N*HCl	1.094	4.70	15.0	2	20	−3.80	−36.95	19
B	2.00	Water	—	—	0	1	20	−0.285	−14.2	20
L-allo-Isoleucine										
B	3.97	6.0 *N*HCl	—	—	20	1	20	+1.50	−38.1	20
B	2.00	Water	—	—	0	1	20	+0.28	+14.0	20
L-Leucine										
A	1.999	6.0 *N*HCl	—	—	38	4.001	25.9	+1.212	+15.1	5
A	2.001	Water	—	—	0	4.001	24.7	−0.863	−10.8	5
A	1.31	3.00 *N*NaOH	—	—	30	2	20	—	+7.6	3
D-Leucine										
?	4.0	6.0 *N*HCl	1.1	3.664	19	2	20	+1.26	−15.6	21
?	—	Water	—	2.08	0	2	20	+0.43	+10.34	38
L-Lysine										
A	2.00	6.0 *N*HCl	—	—	43	4	22.9	+1.652	+25.9	5
A	6.496	Water	—	—	0	2	20	+1.90	+14.6	22
D-Lysine										
B	2.00	0.27 *N*HCl	—	—	2	2	20	−0.939	−23.48	23
L-Methionine										
B	0.80	Water	—	—	0	2	25	−0.13	−8.11	24
D-Methionine										
B	0.80	0.2001 *N*HCl	—	—	4	2	25	−0.34	−21.18	24
B	0.80	Water	—	—	0	2	25	+0.13	+8.12	24
B	0.80	0.6 *N*$NaHCO_3$	—	—	11	2	25	−0.12	−7.47	24
L-Phenylalanine										
B	1.936	Water	1.0040	1.928	0	2	20	−1.36	−35.14	27
D-Phenylalanine										
B	3.814	5.4 *N*HCl	1.0895	3.501	23	2	20	+0.54	+7.07	28
B	2.043	Water	1.0045	2.034	0	2	20	+1.43	+35.0	27
L-Proline										
A	0.575	0.50 *N*HCl	—	—	10	2	20	—	−52.6	6
A	1.001	Water	—	—	0	4.001	23.4	−3.402	−85.0	7
B	2.42	0.6 *N*KOH	1.031	2.35	3	1	20	−2.25	−93.0	29
D-Proline										
B	3.90	Water	1.01	3.865	0	1	20	+3.18	+81.5	29
L-Serine										
B	9.344	1 *N*HCl	1.0465	8.929	1	1	25	+1.35	+14.45	30
B	10.414	Water	1.0414	9.997	0	2	20	−1.42	−6.83	30
D-Serine										
B	9.359	1 *N*HCl	1.0465	8.943	1	1	25	−1.34	−14.32	30
B	10.412	Water	1.0414	9.998	0	2	20	+1.43	+6.87	30
D-Threonine										
B	—	Water	—	1.092	0	2	26	−0.625	−28.3	31
L-Threonine										
B	—	Water	—	1.331	0	2	26	+0.780	+28.4	31
D-allo-Threonine[a]										
B	—	Water	—	1.634	0	2	26	−0.302	−9.1	31
L-allo-Threonine										
B	—	Water	—	1.643	0	2	26	+0.320	+9.6	31
L-Thyroxine										
A	—	0.13 *N*NaOH in 70% EtOH by weight	—	3	3	1	—	−0.147	−4.4	32
L-Tryptophan										
A	1.02	0.50 *N*HCl	—	—	10	2	20	—	+2.4	6
A	1.004	Water	—	—	0	4.001	22.7	−1.266	−31.5	7
A	2.426	0.5 *N*NaOH	1.0243	2.368	4.2	1	20	+0.15	+6.17	33
D-Tryptophan										
C	0.5024	Water	—	—	0	2	25	+0.326	+32.45	34
L-Tyrosine										
B	4.40	6.3 *N*HCl	1.116	3.94	28	2	20	−0.76	−8.64	35
A	0.906	3.0 *N*NaOH	—	—	60	3	18	—	−13.2	3
D-Tyrosine										
B	5.1484	6.3 *N*HCl	1.1175	4.6071	24	2	20	+0.89	+8.64	35
L-Valine										
B	3.4	6.0 *N*HCl	1.1	3.05	20	2	20	+1.93	+28.8	36
B	3.58	Water	1.007	3.56	0	2	20	+0.46	+6.42	36
D-Valine										
B	3.2	6.0 *N*HCl	1.1	2.91	21	2	20	−1.86	−29.04	36
E	6.24	Water	1.00	6.24	0	1	20	−0.37	−6.06	37

[a] The levorotatory allothreonine probably belongs to the D family and its enantiomorph to the L family.

REFERENCES

1. Clough, G. W., *J. Chem. Soc.*, 113, 526, 1918.
2. Fischer, E. and Raske, K., *Ber*, 40, 3717, 1907.
3. Lutz, O. and Jirgensons, B., *Ber.*, 63, 448, 1930.
4. Dunn, M. S., Butler, A. W., and Naiditch, M. J., unpublished data.
5. Dunn, M. S. and Courtney, G., unpublished data.
6. Lutz, O. and Jirgensons, B., *Ber.*, 64, 1221, 1931.
7. Dunn, M. S. and Stoddard, M. P., unpublished data.
8. Fischer, E., *Ber.*, 32, 2451, 1899.
9. Toennies, G. and Lavine, T. F., *J. Biol. Chem.*, 89, 153, 1930.
10. Loring, H. S. and du Vigneaud, V., *J. Biol. Chem.*, 107, 267, 1934.
11. Abderhalden, E. and Guggenheim, M., *Ber.*, 41, 1237, 1908.
12. Dunn, M. S. and Sexton, E. L., unpublished data.
13. Dunn, M. S. and Frieden, E. H., unpublished data.
14. Cox, G. J. and Berg, C. P., *J. Biol. Chem.*, 107, 497, 1934.
15. Dakin, H. D., *Biochem. J.*, 13, 398, 1919.
16. Leuchs, H. and Bormann, K., *Ber.*, 52, 2086, 1919.
17. Locquin, R., *Bull. Soc. Chim.*, (4)1, 601, 1907.
18. Ehrlich, F., *Ber.*, 37, 1809, 1904.
19. Ehrlich, F., *Ber.*, 40, 2538, 1907.
20. Abderhalden, E. and Zeisset, W., *Z. Physiol. Chem.*, 196, 121, 1931.
21. Fischer, E., and Warburg, O., *Ber.*, 38, 3997, 1905.
22. Vickery, H. B., private communication, April, 1940.
23. Berg, C. P., *J. Biol. Chem.*, 115, 9, 1936; private communication, June, 1940.
24. Windus, W. and Marvel, C. S., *J. Am. Chem. Soc.*, 53, 3490, 1931.
27. Fischer, E. and Schoeller, W., *Ann.*, 357, 1, 1907.
28. Fischer, E. and Mouneyrat, A., *Ber.*, 33, 2383, 1900.
29. Fischer, E. and Zemplén, G., *Ber.*, 42, 2989, 1909.
30. Fischer, E. and Jacobs, W. A., *Ber.*, 39, 2942, 1906.
31. West, H. D. and Carter, H. E., *J. Biol. Chem.*, 119, 109, 1937; private communication from H. E. Carter, July, 1940.
32. Foster, G. L., Palmer, W. W., and Leland, J. P., *J. Biol. Chem.*, 115, 467, 1936.
33. Abderhalden, E. and Baumann, L., *Z. Physiol. Chem.*, 55, 412, 1908.
34. Berg, C. P., J. Biol. Chem., 100, 79 1933; private communication, July, 1940.
35. Fischer, E., *Ber.*, 32, 3638, 1899.
36. Fischer, E., *Ber.*, 39, 2320, 1906.
37. Ehrlich, F. and Wendel, A., *Biochem. Z.*, 8, 399, 1908.
38. Ehrlich, F., *Biochem. Z.*, 1, 8, 1906.

SOLUBILITIES OF THE AMINO ACIDS IN GRAMS PER 100 GRAMS OF WATER

Amino acid	Temperature (°C)					Ref.
	0°	25°	50°	75°	100°	
DL-Alanine	12.11	16.72	23.09	31.89	44.04	1
L-Alanine	12.73	16.65	21.79	28.51	37.30	1
DL-Aspartic acid	0.262	0.778	2.000	4.456	8.594	1
L-Aspartic acid	0.209	0.500	1.199	2.875	6.893	1
L-Cystine[c] $\times 10^2$	0.502	1.096	2.394	5.229	11.42	2
Diiodo-DL-tyrosine × 10	0.149	0.340	0.773	—	—	3
Diiodo-L-tyrosine × 10	0.204	0.617	1.862	5.62	17.00	1
DL-Glutamic acid	0.855	2.054	4.934	11.86	28.49	1
L-Glutamic acid	0.341	0.864	2.186	5.532	14.00	1
Glycine	14.18	24.99	39.10	54.39	67.17	1
L-Histidine	—	4.19	—	—	—	4
Hydroxy-L-proline	28.86	36.11	45.18	51.67[a]	—	5
DL-Isoleucine	1.826	2.229	3.034	4.607	7.802	1
L-Isoleucine	3.791	4.117	4.818	6.076	8.255	2
DL-Leucine	0.797	0.991	1.406	2.276	4.206	1
L-Leucine	2.270	2.426[c]	2.887[b]	3.823	5.638	1
DL-Methionine	1.818	3.381	6.070	10.52	17.60	2
DL-Phenylalanine	0.997	1.411	2.187	3.708	6.886	1
L-Phenylalanine	1.983	2.965	4.431	6.624	9.900	2
L-Proline $\times 10^{-1}$	12.74	16.23	20.67	23.90[a]	—	3
DL-Serine	2.204	5.023	10.34	19.21	32.24	2
L-Tryptophan	0.823	1.136	1.706	2.795	4.987	2
DL-Tyrosine × 10	0.147	0.351	0.836	—	—	3
L-Tyrosine × 10	0.196	0.453	1.052	2.438	5.650	1
D-Tyrosine × 10	0.196	0.453	1.052	—	—	3
DL-Valine	5.98	7.09	9.11	12.61	18.81	1
L-Valine	8.34	8.85	9.62	10.24[a]	—	6

[a] Value at 65°.

[b] Dunn and Stoddard[7] report 2.19 g at 25° for L-leucine rendered methionine-free by repeated recrystallization from 6 *N* HCl. Hlynka[8] found 2.20 g at 25° and 2.66 g at 50° for L-leucine rendered methionine-free (by S. W. Fox[9]) by fractional crystallization of the formyl derivative and identical values for D-leucine obtained by resolution of the DL form.

[c] The following values were found by Loring and du Vigneaud[10]: DL-cystine (0.0049 g), D-cystine (0.0108 g), and *meso*-cystine (0.0056 g) at 25°.

REFERENCES

1. Dalton, J. B. and Schmidt, C. L. A., *J. Biol. Chem.*, 103, 549, 1933.
2. Dalton, J. B. and Schmidt, C. L. A., *J. Biol. Chem.*, 109, 241, 1935.
3. Winnek, P. S. and Schmidt, C. L. A., *J. Gen. Physiol.*, 18, 889, 1934—1935.
4. Dunn, M. S., Frieden, E. H., and Brown, H. V., unpublished data.
5. Tomiyama, T. and Schmidt, C. L. A., *J. Gen. Physiol.*, 19, 379, 1935—1936.
6. Dalton, J. B. and Schmidt, C. L. A., *J. Gen. Physiol.*, 19, 767, 1935—1936.
7. Dunn, M. S. and Stoddard, M. P., unpublished data.
8. Hlynka, I., Thesis (1939), California Institute of Technology, Pasadena, California.
9. Fox, S. W., *Science*, 84, 163, 1936.
10. Loring, H. S. and du Vigneaud, V., *J. Biol. Chem.*, 107, 270, 1934.

SOLUBILITIES OF THE AMINO ACIDS IN GRAMS PER 100 GRAMS OF WATER-ETHANOL MIXTURES

% ethanol by volume	Temp. °C	Grams amino acid per 100 grams solvent	Ref.
DL-Alanine			
24.93	0.00	3.84	1
50.10	0.00	1.16	1
74.50	0.00	0.305	1
95.14	0.00	0.0167	1
10	25	12.25	2
24.93	24.97	7.09	1
50.10	24.97	2.52	1
74.20	24.97	0.573	1
95.14	25.09	0.0329	1
25.28	45.16	10.6	1
50.10	44.96	4.25	1
74.20	44.98	0.949	1
95.14	45.19	0.0545	1
24.93	64.96	15.9	1
50.10	64.94	6.68	1
74.20	64.94	1.48	1
95.09	65.15	0.0851	1
DL-Aspartic acid			
24.93	0.03	0.0703	1
50.10	0.03	0.0267	1
74.20	0.02	0.0111	1
24.55	25.06	0.266	1
50.25	25.06	0.0992	1
74.28	25.14	0.0317	1
95.14	25.07	0.0020	1
24.74	45.25	0.680	1
50.18	45.25	0.255	1
74.28	45.27	0.0608	1
95.14	45.21	0.0042	1
24.93	64.91	1.53	1
50.10	64.91	0.588	1
74.20	65.07	0.132	1
95.14	65.00	0.0129	1
L-Aspartic acid			
20	25	0.204	3
50	25	0.0633	3
70	25	0.0224	3
90	25	0.0034	3
L-Glutamic acid			
24.74	0.01	0.0855	1
50.18	0.01	0.0371	1
74.28	0.03	0.0163	1
24.56	25.05	0.292	1
50.25	25.08	0.131	1
74.35	25.07	0.0370	1
95.14	25.04	0.0044	1
24.55	45.01	0.811	1
50.18	45.27	0.378	1
74.35	44.93	0.0885	1
95.14	45.20	0.0127	1
Glycine			
24.93	0.02	3.95	1
50.10	0.02	1.03	1
74.50	0.02	0.200	1
95.09	0.01	0.0080	1
10	25	17.13	2
24.93	24.97	8.72	1
50.10	24.97	2.47	1
74.20	24.97	0.448	1
95.14	25.09	0.0172	1
24.93	44.98	15.0	1
50.10	44.98	4.62	1
74.20	44.97	0.756	1
95.14	45.19	0.0294	1
24.93	65.11	24.5	1
50.10	65.10	8.03	1
74.20	65.07	1.23	1
95.14	65.00	0.0488	1
L-Isoleucine			
80	20	0.46	4
80	78—80	1.16	4
L-allo-Isoleucine			
80	20	0.81	4
80	78—80	1.97	4
DL-Leucine			
24.93	0.00	0.251	1
50.10	0.00	0.118	1
74.50	0.00	0.0693	1
95.14	0.00	0.0116	1
10	25	0.771	2
24.93	24.97	0.493	1
50.10	24.97	0.318	1
74.20	24.97	0.175	1
95.14	25.09	0.0258	1
24.93	45.24	0.853	1
50.10	45.24	0.633	1
74.50	45.18	0.323	1
95.14	45.18	0.0471	1
24.93	65.16	1.45	1
50.10	65.20	1.16	1
74.20	65.15	0.584	1
95.09	65.07	0.0844	1
L-Leucine			
20	25	1.33	2
60	25	0.641	2
90	25	0.123	2
L-Proline			
100	19	1.5	5
DL-Serine			
24.93	0.00	0.1530	1
50.10	0.00	0.146	1
74.50	0.00	0.0304	1
95.14	0.00	0.0008	1
24.93	25.14	1.54	1
50.10	25.14	0.461	1
74.50	25.10	0.0840	1
95.14	25.09	0.0028	1
24.93	45.15	3.14	1
50.10	45.04	0.985	1
74.20	45.04	0.185	1
95.14	45.18	0.0058	1
24.93	65.26	5.99	1
50.10	65.25	1.88	1
74.50	65.24	0.318	1
95.14	65.01	0.0152	1
DL-Threonine			
95	25	0.07[a]	6
DL-allo-Threonine			
95	25	0.03[a]	6
L-Tyrosine			
95	17	0.10	7
DL-Tyrosine			
95.09	0.00	0.0031	8
25.28	24.85	0.0285	8
50.99	24.75	0.0226	8
74.63	24.75	0.0117	8
95.09	25.24	0.0032	8
25.28	45.15	0.0630	8
50.99	45.16	0.0513	8
74.63	44.93	0.0230	8
95.09	44.98	0.0035	8
95.09	65.06	0.0067	8
DL-Valine			
24.93	0.02	2.10	1
50.10	0.02	0.769	1
74.20	0.02	0.269	1
95.14	0.01	0.0277	1
10	25	5.50	2
25.28	24.85	3.30	1
50.99	24.85	1.53	1
74.35	24.93	0.570	1
95.14	25.04	0.0569	1
24.55	44.91	5.10	1
50.25	44.92	2.74	1
74.35	44.92	0.999	1
95.14	45.21	0.0979	1
24.55	65.07	7.44	1
50.10	64.94	4.49	1
74.20	64.34	1.62	1
95.09	65.15	0.167	1
L-Valine			
20	25	5.11	2
40	25	2.93	2
60	25	1.61	2
80	25	0.52	2

[a] Grams per 100 m*l* of solution.

REFERENCES

1. Dunn, M. S. and Ross, F. J., *J. Biol. Chem.*, 125, 309, 1938.
2. Cohn, E. J., McMeekin, T. L., Edsall, J. T., and Weare, J. H., *J. Am. Chem. Soc.*, 56, 2270, 1934.
3. McMeekin, T. L., Cohn, E. J., and Weare, J. H., *J. Am. Chem. Soc.*, 57, 626, 1935.
4. Abderhalden, E. and Zeisset, W., *Z. Physiol. Chem.*, 196, 121, 1931.
5. Kapfhammer, J. and Eck, R., *Z. Physiol. Chem.*, 170, 294, 1927.
6. West, H. D. and Carter, H. E., *J. Biol. Chem.*, 119, 109, 1937.
7. Stutzer, A., *Z. Anal. Chem.*, 31, 501, 1892.
8. Dunn, M. S. and Ross, F. J., unpublished data.

SOLUBILITIES OF THE AMINO ACIDS IN GRAMS PER 100 GRAMS OF ORGANIC SOLVENT

Solvent	Grams amino acid per 100 grams solvent	Temp. (°C)	Ref.
DL-Alanine			
Ethanol	0.0087	25	1
L-Aspartic acid			
Ethanol	0.000196	25	2
L-Glutamic acid			
Ethanol	0.000347	25	2
Ethanol	0.0056	44.93	3
Glycine			
Acetone	0.000291	25	4
Butanol	0.000892	25	4
Ethanol	0.0037	25	1
Formamide	0.558	25	4
Methanol	0.0407	25	4
L-Isoleucine			
Ethanol	0.09	20	5
Ethanol	0.13	78—80	5
L-allo-Isoleucine			
Ethanol	0.13	20	5
Ethanol	0.19	78—80	5
L-Leucine			
Ethanol	0.0217	25	1
L-Proline			
Ethanol	1.5	19	6
DL-Valine			
Ethanol	0.0136	0.03	3
Ethanol	0.019	25	1

REFERENCES

1. Cohn, E. J., McMeekin, T. L., Edsall, J. T., and Weare, J. H., *J. Am. Chem. Soc.*, 56, 2270, 1934.
2. McMeekin, T. L., Cohn, E. J., and Weare, J. H., *J. Am. Chem. Soc.*, 57, 626, 1935.
3. Dunn, M. S. and Ross, F. J., *J. Biol. Chem.*, 125, 309, 1938.
4. McMeekin, T. L., Cohn, E. J., and Weare, J. H., *J. Am. Chem. Soc.*, 58, 2173, 1936.
5. Abderhalden, E. and Zeisset, W., *Z. Physiol. Chem.*, 196, 121, 1931.
6. Kapfhammer, J. and Eck, R., *Z. Physiol. Chem.*, 170, 294, 1927.

DENSITIES OF CRYSTALLINE AMINO ACIDS

Amino acid	Density	Ref.
DL-Alanine	1.424	1
L-Alanine	1.401	2
β-Alanine	1.404	1
DL-α-Amino-*n*-butyric acid	1.231	1
α-Aminoisobutyric acid	1.278	1
L-Arginine	1.1	3
L-Aspartic acid	1.66	3
DL-Glutamic acid	1.460	4
L-Glutamic acid	1.538	4
Glycine[a]	1.601	3
	1.607	1
DL-Leucine	1.191	1
L-Leucine	1.165	1
DL-Methionine	1.340	5
DL-Serine	1.537	5
L-Tyrosine	1.456	1
DL-Valine	1.316	1
L-Valine	1.230	1

[a] The density of glycine at 50° is 1.5753 according to Houck[6] who concluded that the figure 1.1607, reported by Curtius[7] and reproduced in chemical handbooks, is a typographical error.

REFERENCES

1. Cohn, E. J., McMeekin, T. L., Edsall, J. T., and Weare, J. H., *J. Am. Chem. Soc.*, 56, 2270, 1934.
2. Dalton, J. B. and Schmidt, C. L. A., *J. Biol. Chem.*, 103, 549, 1933.
3. Huffman, H. M., Ellis, E. L., and Fox, S. W., *J. Am. Chem. Soc.*, 58, 1728, 1936; Huffman, H. M., Fox, S. W., and Ellis, E. L., *J. Am. Chem. Soc.*, 59, 2144, 1937.
4. Schmidt, C. L. A., *Chemistry of the Amino Acids and Proteins*, Charles C Thomas, Springfield, 1938, p. 900.
5. Albrecht, G. and Dunn, M. S., unpublished data.
6. Houck, R. C., *J. Am. Chem. Soc.*, 52, 2420, 1930.
7. Curtius, T., *J. Prakt. Chem.*, 26, 145, 1882.

CARBOHYDRATES

These data for carbohydrates were compiled originally for the Biology Data Book by M. L. Wolfram, G. G. Maher and R. G. Pagnucco (1964). Data are reproduced here by permission of the copyright owners of the above publication, the Federation of American Societies for Experimental Biology, Washington, D.C. pp. 351–359.

All data are for crystalline substances, unless otherwise specified. Selection of substances was restricted to natural carbohydrates found free (or in chemical combination and released on hydrolysis) and to biological oxidation products of the natural carbohydrates. The nomenclature conforms with that of the British-American report as published in the *Journal of Organic Chemistry*, 28:281 (1963). Substances have been arranged alphabetically under the name of the parent sugar within groups formulated according to increasing carbon content (excluding carbon in substituents), with synonymous common names in parentheses. **Melting Point:** b.p. = boiling point; d. = decomposes; s. = sinters. **Specific Rotation** was determined in water at concentrations of 1–5 g per 100 ml. of solution and at 20°–25°C, unless otherwise specified; other temperatures or wavelengths are shown in brackets; c = grams solute per 100 ml of solution.

Part I. NATURAL MONOSACCHARIDES: ALDOSES AND KETOSES

	Substance (Synonym)	Chemical Formula	Melting Point °C	Specific Rotation $[\alpha]_D$
	(A)	(B)	(C)	(D)
	Aldoses			
1	D-Glyceraldehyde	$C_3H_6O_3$	....	+13.5 ± 0.5 (syrup)
2	D-Glyceraldehyde, 3-deoxy-3,3-*C*-bis-(hydroxymethyl)- (Cordycepose)	$C_5H_{10}O_4$	....	−26 (c 0.6, C_2H_5OH
3	D-Glyceraldehyde, 3,3-bis(*C*-hydroxymethyl)- (Apiose)	$C_5H_{10}O_5$	....	+5.6 (c 10) [15°] syrup
4	β-D-Arabinose	$C_5H_{10}O_5$	155	−175 → −103
5	D-Arabinose, 2-*O*-methyl-	$C_6H_{12}O_5$	Syrup	−102
6	α-L-Arabinose	$C_5H_{10}O_5$	158 amorphous	+55.4 → +105
7	β-L-Arabinose	$C_5H_{10}O_5$	160	+190.6 → +104.5
8	DL-Arabinose	$C_5H_{10}O_5$	163.5–164.5	None
9	α-L-Lyxose	$C_5H_{10}O_5$	105	+5.8 → +13.5
10	L-Lyxose, 5-deoxy-3-*C*-formyl- (Streptose)	$C_6H_{10}O_5$	....	....
11	L-Lyxose, 3-*C*-formyl- (Hydroxystreptose)	$C_6H_{10}O_6$	....	....
12	Pentose, 4,5-anhydro-5-deoxy-D-*erythro*-	$C_5H_8O_3$	....	....
13	Pentose, 2-deoxy-D-*erythro*-	$C_5H_{10}O_4$	96–98	−91 → −58
14	D-Ribose	$C_5H_{10}O_5$	87	−23.1 → −23.7
15	D-Ribose, 2-*C*-hydroxymethyl- (Hamamelose)	$C_6H_{12}O_6$	....	−7.1 [λ578]
16	α-D-Xylose	$C_5H_{10}O_5$	145	+93.6 → +18.8
17	D-Xylose, 5-deoxy-	$C_5H_{10}O_4$	....	+16
18	β-D-Xylose, 2-*O*-methyl-	$C_6H_{12}O_5$	137–138	−21 → +34
19	α-D-Xylose, 3-*O*-methyl-	$C_6H_{12}O_5$	95	+45 → +19
20	D-Allose, 6-deoxy-	$C_6H_{12}O_5$	140–143 146–148	+1.6 [18°] (c 0.6) −4.7 → 0
21	D-Allose, 6-deoxy-2,3-di-*O*-methyl- (Mycinose)	$C_8H_{16}O_5$	102–106	−46 → −29
22	Amicetose (a trideoxy hexose)	$C_6H_{12}O_3$	Oil, b.p. 65–70	+28.6 ($CHCl_3$)
23	Antiarose	$C_6H_{12}O_5$	....	Levo
24	α-D-Galactose	$C_6H_{12}O_6$	167	+150.7 → +80.2
25	β-D-Galactose	$C_6H_{12}O_6$	143–145	+52.8 → +80.2
26	D-Galactose, 3,6-anhydro-	$C_6H_{10}O_5$	....	+21.3 [10°]
27	α-D-Galactose, 6-deoxy- (D-Fucose; Rhodeose)	$C_6H_{12}O_5$	140–145	+127 → +76.3 (c 10)
28	D-Galactose, 6-deoxy-3-*O*-methyl- (Digitalose)	$C_7H_{14}O_5$	106[1], 119[2]	+106
29	D-Galactose, 6-deoxy-4-*O*-methyl-	$C_7H_{14}O_5$	131–132	+82
30	D-Galactose, 6-deoxy-2,3-di-*O*-methyl-	$C_8H_{16}O_5$	....	+73
31	α-D-Galactose, 3-*O*-methyl-	$C_7H_{14}O_6$	144–147	+150.6 → +108.6
32	α-D-Galactose, 6-*O*-methyl-	$C_7H_{14}O_6$	122–123	+117 → +77.3
33	L-Galactose	$C_6H_{12}O_6$	....	*See* D-Galactose
34	α-L-Galactose, 3,6-anhydro-	$C_6H_{10}O_5$	....	−39.4 → −25.2
35	α-L-Galactose, 6-deoxy- (L-Fucose)	$C_6H_{12}O_5$	145	−124.1 → −76.4
36	L-Galactose, 6-deoxy-2-*O*-methyl-	$C_7H_{14}O_5$	149–150	−75 ± 4 (c 0.5)
37	L-Galactose, 6-sulfate	$C_6H_{12}O_9S$	....	−47 (c 0.2) (Na salt)
38	DL-Galactose	$C_6H_{12}O_6$	143–144, 163	None (racemic)
39	α-D-Glucose	$C_6H_{12}O_6$	146, 83 (H_2O)	+112 → +52.7
40	β-D-Glucose	$C_6H_{12}O_6$	148–150	+18.7 → +52.7
41	D-Glucose, 6-acetate	$C_7H_{14}O_7$	135	+48
42	D-Glucose, 2,3-di-*O*-methyl-	$C_8H_{16}O_6$	85–86, 121	+50
43	D-Glucose, 6-*O*-benzoyl- (Vaccinin)	$C_{13}H_{16}O_7$	Amorphous	+48 (C_2H_5OH)
44	α-D-Glucose, 6-deoxy- (Chinovose; Epirhamnose; Glucomethylose; Isorhamnose; Isorhodeose; Quinovose)	$C_6H_{12}O_5$	139–140	+73.3 → +29.7 (c 8)
45	α-D-Glucose, 6-deoxy-3-*O*-methyl- (D-Thevetose)	$C_7H_{14}O_5$	116	+84 → +33

	Substance (Synonym)	Chemical Formula	Melting Point °C	Specific Rotation $[\alpha]_D$
	(A)	(B)	(C)	(D)
	Aldoses (Con't)			
46	D-Glucose, 6-sulfonic acid, 6-deoxy- (6-Sulfoquinovose)	$C_6H_{12}O_8S$	173–174	+87[3]
47	D-Glucose, 3-*O*-methyl-	$C_7H_{14}O_6$	162–167	+98 → +59.5
48	α-L-Glucose	$C_6H_{12}O_6$	141–143	−95.5 → −51.4
49	L-Glucose, 6-deoxy-3-*O*-methyl- (L-Thevetose)	$C_7H_{14}O_5$	126–129	−36.9 ± 2
50	D-Gulose, 6-deoxy-	$C_6H_{12}O_5$	....	....
51	Hexose, 2-deoxy-D-*arabino*-[4]	$C_6H_{12}O_5$	148	+46.6 [18°]
52	Hexose, 2,6-dideoxy-3-*O*-methyl-D-*arabino*- (D-Oleandrose)	$C_7H_{14}O_4$	....	−11
53	Hexose, 3,6-dideoxy-D-*arabino*- (Tyvelose)	$C_6H_{12}O_4$	....	+24 ± 2
54	Hexose, 2,6-dideoxy-3-*O*-methyl-L-*arabino*- (L-Oleandrose)	$C_7H_{14}O_4$	62–63	+11.9 ± 2.5
55	Hexose, 3,6-dideoxy-L-*arabino*- (Ascarylose)	$C_6H_{12}O_4$	....	−24 ± 2
56	Hexose, 2,6-dideoxy-3-*O*-methyl-D-*lyxo*- (Diginose)	$C_7H_{14}O_4$	90–92	+56 ± 4
57	Hexose, 2,6-dideoxy-L-*lyxo*- (L-Fucose, 2-deoxy-)	$C_6H_{12}O_4$	103–106	−61.6
58	Hexose, 2,6-dideoxy-3-*O*-methyl-L-*lyxo*-	$C_7H_{14}O_4$	78–85	−65
59	Hexose, 2,6-dideoxy-D-*ribo*- (Digitoxose; D-Altrose, 2,6-dideoxy-)	$C_6H_{12}O_4$	110	+46.4
60	Hexose, 2,6-dideoxy-3-*O*-methyl-D-*ribo*- (Cymarose)	$C_7H_{14}O_4$	93	+52
61	Hexose, 3,6-dideoxy-D-*ribo*- (Paratose)	$C_6H_{12}O_4$	....	+10 ± 2 (*c* 0.9)
62	Hexose, 4,6-dideoxy-3-*O*-methyl-D-*ribo*- (D-Gulose, 4,6-dideoxy-3-*O*-methyl-; Chalcose)	$C_7H_{14}O_4$	96–99	+120 → +76
63	Hexose, 2,6-dideoxy-D-*xylo*- (Boivinose)	$C_6H_{12}O_4$	96–98	−3.9 → +3.9
64	Hexose, 2,6-dideoxy-3-*O*-methyl-D-*xylo*- (Sarmentose)	$C_7H_{14}O_4$	78–79	+12 → +15.8
65	Hexose, 3,6-dideoxy-D-*xylo*- (Abequose)	$C_6H_{12}O_4$	....	−3.2 ± 0.6
66	Hexose, 2,6-dideoxy-3-*C*-methyl-L-*xylo*- (Mycarose)	$C_7H_{14}O_4$	129–129	−31.1
67	Hexose, 2,6-dideoxy-3-*C*-methyl-3-*O*-methyl-L-*xylo*-(Cladinose)	$C_8H_{16}O_4$	oil, b.p. 120–132 (0.25 mm)	−23.1
68	Hexose, 3,6-dideoxy-L-*xylo*- (Colitose)	$C_6H_{12}O_4$	....	+4 (H_2O); −51 ± 2 (CH_3OH)
69	D-Idose[5]	$C_6H_{12}O_6$	....	....
70	L-Idose, 1,6-anhydro-	$C_6H_{10}O_5$	....	....
71	α-D-Mannose	$C_6H_{12}O_6$	133	+29.3 → +14.5
72	β-D-Mannose	$C_6H_{12}O_6$	132	−16.3 → +14.5
73	D-Mannose, 6-deoxy- (D-Rhamnose)	$C_6H_{12}O_5$	86–90	−7.0
74	α-L-Mannose, 6-deoxy-monohydrate (L-Rhamnose)	$C_6H_{14}O_6$	93–94	−8.6 → +8.2
75	β-L-Mannose, 6-deoxy-	$C_6H_{12}O_5$	123–125	+38.4 → +8.9
76	L-Mannose, 6-deoxy-2-*O*-methyl-	$C_7H_{14}O_5$	....	....
77	L-Mannose, 6-deoxy-3-*O*-methyl- (L-Acofriose)	$C_7H_{14}O_5$	114–115	+30 [18°]
78	L-Mannose, 6-deoxy-2,4-di-*O*-methyl-	$C_8H_{16}O_5$	82	−19 [16°]
79	L-Mannose, 6-deoxy-5-*C*-methyl-4-*O*-methyl-(Noviose)	$C_8H_{16}O_5$	128–130	+19.9 (50% C_2H_5OH)
80	Rhodinose (a 2,3,6-trideoxyhexose)	$C_6H_{12}O_3$	....	−11 ± 1.6
81	D-Talose	$C_6H_{12}O_6$	128–132	+16.9
82	D-Talose, 6-deoxy- (D-Talomethylose)	$C_6H_{12}O_5$	129–131	+20.6
83	L-Talose, 6-deoxy- (L-Talomethylose)	$C_6H_{12}O_5$	116–118	−19.5 ± 2 [18°]
84	L-Talose, 6-deoxy-2-*O*-methyl- (L-Acovenose)	$C_7H_{14}O_5$	....	−19.4
85	Heptose, D-*glycero*-D-*galacto*-	$C_7H_{14}O_7$	139–140	+47 → +64 (*c* 0.5)
86	Heptose, D-*glycero*-D-*manno*-	$C_7H_{14}O_7$	....	....
87	Heptose, D-*glycero*-L-*manno*-	$C_7H_{14}O_7$	....	....
	Ketoses			
88	Dihydroxyacetone	$C_3H_6O_3$	80 (dimer)	None
89	Tetrulose, L-*glycero*-[8] (L-Erythrulose; Ketoerythritol; L-Threulose)	$C_4H_8O_4$	Syrup	+12
90	Pentulose, D-*erythro*- (Adonose; D-Ribulose)	$C_5H_{10}O_5$	Syrup	+16.6 [27°]
91	Pentulose, L-*erythro*- (L-Ribulose)	$C_5H_{10}O_5$	....	−16.6

Part I. NATURAL MONOSACCHARIDES: ALDOSES AND KETOSES (Continued)

	Substance (Synonym)	Chemical Formula	Melting Point °C	Specific Rotation $[\alpha]_D$
	(A)	(B)	(C)	(D)
	Ketoses (Con't)			
92	Pentulose, D-*threo*- (D-Xylulose)	$C_5H_{10}O_5$	....	−33
93	Pentulose, 5-deoxy-D-*threo*-	$C_5H_{10}O_4$	....	−5 ± 1 (CH_3OH)
94	Pentulose, L-*threo*- (L-Xylulose; L-Lyxulose; Xyloketose)	$C_5H_{10}O_5$	Syrup	+33.1
95	Hexulose, β-D-*arabino*-(β-D-Fructose; Levulose)	$C_6H_{12}O_6$	102–104[7]	−133.5 → −92
96	Hexulose, 6-deoxy-D-*arabino*- (D-Rhamnulose)	$C_6H_{12}O_5$	....	−13 ± 2
97	Hexulose, D-*lyxo*- (D-Tagatose)	$C_6H_{12}O_6$	131–132	+2.7 → −4, −5
98	5-Hexulose, D-*lyxo*	$C_6H_{12}O_6$	158	−86.6
99	Hexulose, 6-deoxy-L-*lyxo*- (L-Fuculose)	$C_6H_{12}O_5$	....	....
100	Hexulose, D-*ribo*- (D-Psicose)	$C_6H_{12}O_6$	Amorphous	+4.7
101	Hexulose, L-*xylo*- (L-Sorbose)	$C_6H_{12}O_6$	159–161	−43.1
102	Hexulose, 6-deoxy-L-*xylo*-	$C_6H_{12}O_5$	88	−25 ± 2 (*c* 0.7)
103	Heptulose, D-*altro*- (Sedoheptulose; Sedoheptose)	$C_7H_{14}O_7$	Amorphous	+2.5 (*c* 10)
104	Heptulose·hemihydrate, L-*galacto*- (Perseulose)	$C_7H_{14}O_7 \cdot \frac{1}{2}H_2O$	110–115	−90 → −80
105	Heptulose, L-*gulo*-	$C_7H_{14}O_7$	....	−28
106	Heptulose, D-*ido*-	$C_7H_{14}O_7$	172	−34 ± 8 (*c* 0.3)
107	Heptulose, D-*manno*- (Mannoketoheptose; D-Mannotagatoheptose)	$C_7H_{14}O_7$	152	+29.4
108	Heptulose, D-*talo*-	$C_7H_{14}O_7$	....	....
109	Octulose, D-*glycero*-L-*galacto*-	$C_8H_{16}O_8$	....	−57, −43.4 → −13.4
110	Octulose, D-*glycero*-D-*manno*-	$C_8H_{16}O_8$	....	+20 (CH_3OH)

[1] Original melting point. [2] Melting point after four-months' storage. [3] As a methyl glycoside cyclohexylamine salt. [4] Included because of speculations concerning it in biological processes. [5] Either D-idose or L-altrose is in the polysaccharide varianose. [6] Early literature refers to this as D-erythrose. [7] The $\cdot\frac{1}{2}H_2O$ and $\cdot 2H_2O$ forms also exist.

Part II. NATURAL MONOSACCHARIDES: AMINO SUGARS

	Substance (Synonym)	Chemical Formula	Melting Point °C	Specific Rotation $[\alpha]_D$
	(A)	(B)	(C)	(D)
	Aldosamines			
1	D-Ribose, 3-amino-3-deoxy-	$C_5H_{11}NO_4$	158–158.5 d.	−24.6 (hydrochloride)
2	D-Galactose, 2-amino-2-deoxy- (Galactosamine; Chondrosamine)	$C_6H_{13}NO_5$	185	+121 → +80 (hydrochloride)
3	α-L-Galactose, 2-amino-2,6 dideoxy- (L-Fucosamine)	$C_6H_{13}NO_4$	192–193 d.	−119 → −92 [27°] (hydrochloride)
4	α-D-Glucose, 2-amino-2-deoxy- (Glucosamine; Chitosamine)	$C_6H_{13}NO_5$	88	+100 → +47.5
5	β-D-Glucose, 2-amino-2-deoxy-	$C_6H_{13}NO_5$	110–111	+28 → +47.5
6	D-Glucose, 3-amino-3-deoxy- (Kanosamine)	$C_6H_{13}NO_5$	128 d.	+19 [14°]
7	D-Glucose, 6-amino-6-deoxy-	$C_6H_{13}NO_5$	161–162 d.	+23 → +50.1 (hydrochloride)
8	D-Glucose, 2,6-diamino-2,6-dideoxy- (Neosamine C)	$C_6H_{14}N_2O_4$	>230	+61.5 (dihydrochloride)
9	D-Glucose, 3,6-dideoxy-3-dimethylamino- (Mycaminose)	$C_8H_{17}NO_4$	115–116	+31 (hydrochloride)
10	D-Glucose, 4,6-dideoxy-4-dimethylamino-	$C_8H_{17}NO_4$	192–193	+45.5 (hydrochloride)
11	L-Glucose, 2-deoxy-2-methylamino-	$C_7H_{15}NO_5$	130–132	−64
12	D-Gulose, 2-amino-1,6-anhydro-2-deoxy-	$C_6H_{11}NO_4$	250–260 d.	+41 ± 2 (hydrochloride)
13	D-Gulose, 2-amino-2-deoxy-	$C_6H_{13}NO_5$	152–162 d.	+5.6 → −18.7 (hydrochloride)
14	Hexose, 3,4,6-trideoxy-3-dimethylamino-D-*xylo*- (Desosamine; Picrocine)	$C_8H_{17}NO_3$	189–191 d.	+49.5 (*c* 10) (hydrochloride)
15	Hexose, a 4-acetamido-2-amino-2,4,6-trideoxy-	$C_8H_{16}N_2O_4$	216–219	+115 → +94 [26°] (*c* 0.05)
16	Hexose, an amino-deoxy-3-*O*-carboxyethyl-	$C_9H_{17}NO_7$	....	....
17	Hexose, a 2,6-diamino-2,6-dideoxy- (Neosamine B; Paramose)	$C_6H_{14}N_2O_4$	135–150 d.	+17.5 (*c* 0.9 (hydrochloride)

Part II. NATURAL MONOSACCHARIDES: AMINO SUGARS (Continued)

	Substance (Synonym)	Chemical Formula	Melting Point °C	Specific Rotation $[\alpha]_D$
	(A)	(B)	(C)	(D)
	Aldosamines (Con't)			
18	Hexose, a 3-dimethylamino-2,3,6-trideoxy- (Rhodosamine)	$C_8H_{17}NO_3$	....	....
19	D-Mannose, 2-amino-2-deoxy- (Mannosamine)	$C_6H_{13}NO_5$	142 d.	−4.3 (*c* 9) (hydrochloride)
20	D-Mannose, 3-amino-3,6-dideoxy- (Mycosamine)	$C_6H_{13}NO_4$	162	−11.5 (hydrochloride)
21	D-Talose, 2-amino-2-deoxy- (Talosamine)	$C_6H_{13}NO_5$	151–153	+3.4 → −5.7 (*c* 0.9) (hydrochloride)
22	L-Talose, 2-amino-2,6-dideoxy- (Pneumosamine)	$C_6H_{13}NO_4$	162–163	+6.9 → +10.4 (hydrochloride)
	Ketosamines			
23	Pentulose, 1-(*o*-carboxyanilino)-1-deoxy-D-*erythro*-	$C_{12}H_{14}NO_6$	....	....
24	Hexulose, 1-(*o*-carboxyanilino)-1-deoxy-D-*arabino*-	$C_{13}H_{16}NO_7$	....	....
25	Hexulose, 5-amino-5-deoxy-L-*xylo*-	$C_6H_{13}NO_5$	174–176	−62
26	Hexulose, 6-deoxy-6-(*N*-methylacetamido)-L-*xylo*-	$C_9H_{17}NO_6$	....	....

Part III. NATURAL ALDITOLS AND INOSITOLS (with Inososes and Inosamines)

	Substance (Synonym)	Chemical Formula	Melting Point °C	Specific Rotation $[\alpha]_D$
	(A)	(B)	(C)	(D)
	Alditols			
1	Glycerol	$C_3H_8O_3$	20	None
2	Glycerol, 1-deoxy- (1,2-Propane-diol)[1]	$C_3H_8O_2$	Oil, b.p. 188–189	None (racemic)
3	Erythritol	$C_4H_{10}O_4$	118–120	None (meso)
4	Erythritol, 1,4-dideoxy- (2,3-Butyleneglycol)	$C_4H_{10}O_2$	25, 34	None (meso)
5	D-Threitol, 1,4-dideoxy-	$C_4H_{10}O_2$	19	−13.0
6	L-Threitol, 1,4-dideoxy-	$C_4H_{10}O_2$	....	+10.2
7	DL-Threitol, 1,4-dideoxy-	$C_4H_{10}O_2$	7.6	None (racemic)
8	D-Arabinitol	$C_5H_{12}O_5$	103	+7.82 (*c* 8, borax solution)
9	L-Arabinitol	$C_5H_{12}O_5$	101–102	−32 (*c* 0.4, 5% molybdate)
10	Ribitol (Adomitol)	$C_5H_{12}O_5$	102	None (meso)
11	Galactitol (Dulcitol)	$C_6H_{14}O_6$	186–188	None (meso)
12	D-Glucitol (Sorbitol)	$C_6H_{14}O_6$	112	−1.8 [15°]
13	D-Glucitol, 1,5-anhydro- (Polygalitol)	$C_6H_{12}O_5$	140–141	+42.4
14	L-Iditol	$C_6H_{14}O_6$	73.5	−3.5 (*c* 10)
15	D-Mannitol	$C_6H_{14}O_6$	166	−0.21
16	D-Mannitol, 1,5-anhydro- (Styracitol)	$C_6H_{12}O_5$	157	−49.9
17	Heptitol, D-*glycero*-D-*galacto*- (Heptitol, L-*glycero*-D-*manno*-; Perseitol)	$C_7H_{16}O_7$	183–185, 188	−1.1
18	Heptitol, D-*glycero*-D-*gluco*- (Heptitol, L-*glycero*-D-*talo*-; β-Sedoheptitol)	$C_7H_{16}O_7$	131–132	+46 (5% NH_4 molybdate)
19	Heptitol, D-*glycero*-D-*manno*- (Heptitol, D-*glycero*-D-*talo*-; Volemitol)	$C_7H_{16}O_7$	153	+2.65
20	Octitol, D-*erythro*-D-*galacto*-	$C_8H_{18}O_8 \cdot H_2O$	169–170	−11 (5% NH_4 molybdate)
	Inositols			
21	Betitol (a dideoxy inositol)	$C_6H_{12}O_4$	224	....
22	Bioinosose (*scyllo*-Inosose; *myo*-Inosose-2; a deoxy keto inositol)	$C_6H_{10}O_6$	198–200	None (meso)
23	*h*-Bornesitol (a *myo*-inositol monomethyl ether)	$C_7H_{14}O_6$	200	+31.6
24	*l*-Bornesitol (a *myo*-inositol monomethyl ether)	$C_7H_{14}O_6$	205–206	−32.1
25	Conduritol (a 2,3-dehydro-2,3-dideoxyinositol)	$C_6H_{10}O_4$	142–143	None (meso)
26	Cordycepic acid (a tetrahydroxycyclo hexanecarboxylic acid)[2]	$C_7H_{12}O_6$	....	....
27	Dambonitol (a *myo*-inositol dimethyl ether)	$C_8H_{16}O_6$	206	None (meso)
28	DL-Inositol	$C_6H_{12}O_6$	253	None (racemic)

Part III. NATURAL ALDITOLS AND INOSITOLS (with Inososes and Inosamines) (Continued)

	Substance (Synonym)	Chemical Formula	Melting Point °C	Specific Rotation $[\alpha]_D$
	(A)	(B)	(C)	(D)
	Inositols (Con't)			
29	*d*-Inositol	$C_6H_{12}O_6$	....	+60
30	*l*-Inositol	$C_6H_{12}O_6$	240	−65
31	Laminitol (a *C*-methyl *myo*-inositol)	$C_7H_{14}O_6$	266–269	−3
32	Liriodendritol (a *myo*-inositol dimethyl ether)	$C_8H_{16}O_6$	224	−25
33	*muco*-Inositol monomethyl ether	$C_7H_{14}O_6$	322–325	....
34	*myo*-Inositol (*meso*-Inositol)	$C_6H_{12}O_6$	217–218	None (meso)
35	*d*-*myo*-Inosose-1 (a deoxy keto inositol)	$C_6H_{10}O_6$	138–139	+19.6
36	Mytilitol (a *C*-methyl *scyllo*-inositol)	$C_7H_{14}O_6$	259	None (meso)
37	*neo*-Inosamine-2 (a deoxy amino inositol)	$C_6H_{13}O_5N$	239–241 d.	None (meso)
38	*d*-Ononitol (a *myo*-inositol monomethyl ether)	$C_7H_{14}O_6$	172	+6.6
39	*h*-Pinitol (a *dextro*-inositol monomethyl ether)	$C_7H_{14}O_6$	186	+65.5
40	*l*-Pinitol (a *levo*-inositol monomethyl ether)	$C_7H_{14}O_6$	186	−65
41	*l*-Quebrachitol (a *levo*-inositol monomethyl ether)	$C_7H_{14}O_6$	190–191	−80.2 [28°]
42	*d*-Quercitol (a deoxy *dextro*-inositol)	$C_6H_{12}O_5$	235	+24.2
43	*d*-Quinic acid (a trideoxy carboxy *dextro*-inositol)	$C_7H_{12}O_6$	164	+44 (*c* 10)
44	*l*-Quinic acid (a trideoxy carboxy *levo*-inositol)	$C_7H_{12}O_6$	162	−42.1
45	Quinic acid, 5-dehydro-	$C_7H_{10}O_6$	140–142 (138 s.)	−82.4 [28°]
46	Scyllitol (*scyllo*-Inositol; Cocositol)	$C_6H_{12}O_6$	352–353	None (meso)
47	Sequoyitol (a *myo*-inositol monomethyl ether)	$C_7H_{14}O_6$	234–235	None (meso)
48	Shikimic acid (a 3,4-anhydro-quinic acid)	$C_7H_{10}O_5$	183–184	−200 [16°]
49	Shikimic acid, 5-dehydro-	$C_7H_8O_5$	150–152	−57.5 [28°] (EtOH)
50	Streptamine (2,4-diaminodideoxy-scyllitol)	$C_6H_{14}O_4N_2$	88, 210–250 d.	None (meso)
51	Streptamine, 2-deoxy-	$C_6H_{14}O_3N_2$	....	None (meso)
52	Streptadine (1,3-Dideoxy-1,3-diguanidino-scyllitol)	$C_8H_{18}N_6O_4$	....	None (meso)
53	Viburnitol (a deoxy *levo*-inositol)[3]	$C_6H_{12}O_5$	174	−73.9

[1] The 1-phosphate ester of this diol is said to occur in brain tissue and sea-urchin eggs. [2] Strong evidence that cordycepic acid is really D-mannitol. [3] Not an enantiomorph of *d*-quercitol; other isomeric relationship is involved.

Part IV. NATURAL ALDONIC, URONIC, AND ALDARIC ACIDS

	Substance (Synonym)	Chemical Formula	Melting Point °C	Specific Rotation $[\alpha]_D$
	(A)	(B)	(C)	(D)
	Aldonic Acids			
1	D-Glyceric acid	$C_3H_6O_4$	Gum	Dextro
2	L-Glyceric acid	$C_3H_6O_4$	Gum	Levo
3	D-Arabinonic acid	$C_5H_{10}O_6$	114–116	+10.5 (*c* 6)
4	L-Arabinonic acid	$C_5H_{10}O_6$	118–119	−9.6 → −41.7[1]
5	L-Arabinonic-1,4-lactone	$C_5H_8O_5$	97–99	−72
6	D-Ribonic acid	$C_5H_{10}O_6$	112–113	−17.0
7	D-Xylonic acid	$C_5H_{10}O_6$	....	−2.9 → +20.1[1]
8	L-Xylonic acid	$C_5H_{10}O_6$	....	−91.8[1]
9	D-Altronic acid	$C_6H_{12}O_7$	....	+11.5 → +24.8[1] (Ca salt, *N* HCl)
10	D-Galactonic acid	$C_6H_{12}O_7$	122	−11.2 → +57.6[1]
11	D-Gluconic acid	$C_6H_{12}O_7$	130–132 (110–112 s.)	−6.7 → +11.9[1]
12	L-Gulonic acid	$C_6H_{12}O_7$	Exists only in soln.	[ca. 0°]
13	Hexsonic acid, 2-deoxy-D-*arabino*-	$C_6H_{12}O_6$	93–95	+68 (lactone)
14	2-Hexulosonic acid, D-*arabino*-	$C_6H_{10}O_7$	....	−81.7 (Na salt)
15	2-Hexulosonic acid, 3-deoxy-D-*erythro*-	$C_6H_{10}O_6$	....	−29.2 (*c* 6, Ca salt)
16	2-Hexulosonic acid, D-*lyxo*-	$C_6H_{10}O_7$	169	−5
17	5-Hexulosonic acid, D-*arabino*-	$C_6H_{10}O_7$	108–109	....
18	5-Hexulosonic acid, D-*xylo*-	$C_6H_{10}O_7$	....	−14.5
19	D-Mannonic acid	$C_6H_{12}O_7$	....	−15.6
20	D-Gluconic acid, *O*-β-D-galactopyranosyl- (1 → 4)- (Lactobionic acid)	$C_{12}H_{22}O_{12}$	....	+25.1 (Ca salt)

Part IV. NATURAL ALDONIC, URONIC, AND ALDARIC ACIDS (Continued)

	Substance (Synonym)	Chemical Formula	Melting Point °C	Specific Rotation $[\alpha]_D$
	(A)	(B)	(C)	(D)
	Uronic Acids			
21	L-Lyxuronic acid	$C_5H_8O_6$	...	...
22	β-D-Galacturonic acid	$C_6H_{10}O_7$	160	+27 → +55.6
23	α-D-Galacturonic acid·monohydrate	$C_6H_{12}O_8$	159–160 (110–115 s.)	+97.9 → +50.9
24	D-Galacturonic acid, 2-amino-2-deoxy-	$C_6H_{11}O_6N$	160 d.	+84.5 (pH 2 HCl)
25	β-D-Glucuronic acid	$C_6H_{10}O_7$	156	+11.7 → +36.3
26	D-Glucuronic acid, 2-amino-2-deoxy-	$C_6H_{11}O_6N$	120–172 d.	+55
27	D-Glucuronic acid, 3-*O*-methyl-	$C_7H_{12}O_7$	Syrup	+6
28	L-Guluronic acid	$C_6H_{10}O_7$	...	...
29	L-Iduronic acid	$C_6H_{10}O_7$	...	+30
30	β-D-Mannuronic acid	$C_6H_{10}O_7$	165–167	−47.9 → −23.9
31	α-D-Mannuronic acid·monohydrate	$C_6H_{12}O_8$	110 s., 120–130 d.	+16 → −6.1 (*c* 6.8)
	Aldaric Acids			
32	D-Tartaric acid	$C_4H_6O_6$	170	−15
33	L-Tartaric acid	$C_4H_6O_6$	170	+15 [15°]
34	L-Malic acid	$C_4H_6O_5$	100	−2.3 (*c* 8.4)

[1] Equilibrates with the lactone.

WAXES

These data for waxes were compiled originally for the Biology Data Book by A. H. Warth. Data are reproduced here by permission of the copyright owners of the above publication, the American Societies for Experimental Biology, Washington, D.C. p. 382.

Specific Gravity (column C) was calculated at the specified temperature, degrees centigrade, and referred to water at the same temperature. **Density**, shown in parentheses (column C), and **Refractive Index** (column D) were measured at the specified temperature, degrees centigrade.

	Wax	Melting Point °C	Specific Gravity or (Density)	Refractive Index $n_D^{°C}$	Iodine Value	Acid Value	Saponification Value
	(A)	(B)	(C)	(D)	(E)	(F)	(G)
1	Bamboo leaf	79–80	$(0.961^{25°})$	...	7.8[1]	14.5	43.4
2	Bayberry (myrtle)	46.7–48.8	$(0.985^{15°})$	$1.436^{80°}$	2.9[2]–3.9[3]	3.5	20.5–21.7
3	Beeswax, crude	62–66	$(0.927–0.970^{15°})$	$1.439–1.483^{40°}$	6.8–16.4[2]	16.8–35.8	89.3–149.0
4	Beeswax, white, U.S.P.	61–69	$(0.959–0.975^{15°})$	$1.447–1.465^{65°}$	7–11[3]	17–24	90–96
5	Beeswax, yellow	62–65	$(0.960–0.964^{15°})$	$1.443–1.449^{65°}$	6–11	18–24	90–97
6	Candelilla, refined	67–69	$(0.982–0.986^{15°})$	$1.454–1.463^{85°}$	14.4–20.4	12.7–18.1	35–86
7	Cape berry[4]	40.5–45.0	$(1.004–1.007^{15°})$	$1.450^{45°}$	0.6–2.4	2.5–3.7	211–215
8	Carandá	79.7–84.5	$(0.990^{25°})$	...	8.0–8.9	5.0–9.5	64.5–78.5
9	Carnauba	83–86	$0.990–1.001^{15°}$	$1.467–1.472^{40°}$	7.2–13.5	2.9–9.7	78–95
10	Castor oil, hydrogenated	83–88	$(0.980–0.990^{20°})$	...	2.5–8.5	1.0–5.0	177–181
11	Chinese insect	81.5–84.0	$0.950–0.970^{15°}$	$1.457^{40°}$	1.4	0.2–1.5	73–93
12	Cotton	68–71	$0.959^{15°}$	...	24.5	32	70.6
13	Cranberry	207–218	$(0.970–0.975^{15°})$	...	44.2–53.2[2]	42.2–59.1	131–134
14	Douglas-fir bark	59.0–72.8	$(1.030^{25°})$	$1.468^{80°}$	25.8–62.5	58.6–80.7	112–200
15	Esparto	67.5–78.1	$0.988^{15°}$	...	22–23	22.7–23.9	69.8–79.3
16	Flax	61.5–69.8	$0.908–0.985^{15°}$	...	21.6–28.8	17.5–48.3	77.5–101.5
17	Ghedda, E. Indian beeswax	60.5–66.4	$0.956–0.973^{15°}$	$1.440^{50°}$	5.6–12.6	5.8–7.9	84.5–118.3
18	Indian corn	80–81	...	...	4.2[2]	1.9	120.3
19	Japan wax	48–53	$0.975–0.993^{15°}$	...	4.5–12.5	6–20	206.5–237.5
20	Jojoba	11.2–11.8	$0.864–0.899^{25°}$	$1.465^{25°}$	81.7–88.4[2]	0.2–0.6	92.2–95.0
21	Madagascar	88	...	...	3.2–5.3	17.7–28.0	140.0–159.6
22	Microcrystalline, amber	64–91	$0.913–0.943^{15°}$	$1.424–1.452^{80°}$	0	0	0
23	Microcrystalline, white	71–89	$0.928–0.941^{15°}$	$1.441^{80°}$	0	0	0
24	Montan, crude	76–86	$(1.010–1.020^{25°})$	...	13.9–17.6	22.7–31.0	59.4–92.0
25	Montan, refined	77–84	$(1.010–1.030^{25°})$	...	10–14	24–43	72–103
26	Orange peel	44.0–46.5	$0.985^{15°}$	$1.502^{20°}$	115.7[2]	48.3	120.9
27	Ouricury, refined	79.0–83.8	$1.053^{15°}$	...	6.9–7.8[2]	3.4–21.1	61.8–85.8
28	Ozocerite, refined	74.4–75.0	$0.907–0.920^{15°}$	...	0	0	0
29	Palm	74–86	$(0.991–1.045^{15°})$	...	8.9–16.9[2]	5.0–10.6	64.5–104.0
30	Paraffin, American	49–63	$0.896–0.925^{15°}$	$1.442–1.448^{80°}$	0	0	0
31	Peat wax, natural	73–76	$0.980^{15°}$	...	16–40	60.0–73.3	73.9–136.0
32	Rice bran, refined	75.3–79.9	...	$1.469^{30°}$	11.1–19.4	15–17	56.9–104.4

WAXES (Continued)

	Wax	Melting Point °C	Specific Gravity or (Density)	Refractive Index $n\frac{°C}{D}$	Iodine Value	Acid Value	Saponification Value
	(A)	(B)	(C)	(D)	(E)	(F)	(G)
33	Shellac wax	79–82	$0.971–0.980^{15°}$	....	6.0–8.8[3]	12.1–24.3	63.8–83.0
34	Sisal hemp	74–81	$1.007–1.010^{15°}$	....	28–29[2]	16–19[2]	56–58
35	Sorghum grain	77–82	....	....	15.7–20.9	10.1–16.2	16–44
36	Spanish moss	79–80	....	....	33.0	25.0	120.4
37	Spermaceti	42–50	$0.905–0.945^{15°}$	$1.440^{70°}$	4.8–5.9	2.0–5.2	108–134
38	Sugarcane, crude	52–67	$0.988–0.998^{25°}$	....	32–84	24–57	128–177
39	Sugarcane, double-refined	77–82	$0.961–0.979^{25°}$	$1.510^{25°}$	13–29	8–23	55–95
40	Wool wax, refined	36–43	$0.932–0.945^{15°}$	$1.478–1.482^{40°}$	15.0–46.9	5.6–22.0	80–127

[1] Wijs test. [2] Hanus test. [3] Hubl test. [4] *Myrica cordifolia.*

TRADE NAMES OF DYESTUFF INTERMEDIATES

Trade name	Chemical name
A acid	1,7-Hydroxynaphthalene-3,6-disulfonic acid
Acetyl H acid	N-Acetyl-1-amino-8-naphthol-3,6-disulfonic acid
Alen's acid	1-Naphthylamine-3,6-disulfonic acid (also Freund's ac.)
Alizarin	1,2-Dihydroxyanthraquinone
Amido acid	2-Amino-7-hydroxynaphthalene-5-sulfonic acid
Amido J acid	2-Naphthylamine-5,7-disulfonic acid
Amino G acid	7-Amino-1,3-naphthalene disulfonic acid 2-Naphthylamine-6,8-disulfonic acid
Amino R acid	3-Amino-2,7-naphthalene disulfonic acid 2-Naphthylamine-3,6-disulfonic acid
Aminophenolic acid V	1-Amino-3-oxybenzene-5-sulfonic acid
Aminophenol sulfonic acid III	1-Amino-3-oxybenzene-6-sulfonic acid
Andresen's acid	1-Naphthol-3,8-disulfonic acid
Anisidine	o-Aminophenol methylether
Anthrachrysone	1,3,5,7-Tetrahydroanthraquinone
Anthraflavic acid	2,6-Dihydroxyanthraquinone
Anthranilic acid	o-Aminobenzoic acid
Anthrarufin	1,5-Dihydroxyanthraquinone
Anthranol	9-Hydroxyanthracene
α Anthrol	1-Hydroxyanthracene
Armstrong's acid	Naphthalene-1,5-disulfonic acid
Armstrong & Wynne acid	1-Naphthol-3-sulfonic acid
Armstrong & Wynne acid II	2-Naphthylamino-5,7-disulfonic acid
B acid	8-Amino-1-naphthol-4,6-disulfonic acid
Badische acid	2-Naphthylamino-8-sulfonic acid
Bayer's acid	2-Naphthol-8-sulfonic acid
Benzidine	p,p′-Diaminodiphenyl
Bronner's acid	2-Naphthylamino-6-sulfonic acid
β acid	Anthraquinone-2-sulfonic acid
C acid CLT acid	6-Chloro-m-toluidine-4-sulfonic acid
Casella's acid	2-Naphthol-7-sulfonic acid (F acid)
Chicago acid	1-Amino-8-naphthyl-2,4-disulfonic acid
Chloro H acid	8-Chloro-1-naphthol-3,6-disulfonic acid
Chromogene I	4,5-Dihydroxy-2,7-naphthalene disulfonic acid
Chromotrope acid	1,8-Dihydronaphthalene-3,6-disulfonic acid
Chromotropic acid	4,5-Dihydroxy-2,7-naphthalene disulfonic acid
Chrysazine	1,8-Dihydroxyanthraquinone
Cleve's acid	1-Naphthylamine-3-sulfonic acid
Cleve's acid	1-Naphthylamine-5-sulfonic acid
Cleve's acid	1-Naphthylamine-6-sulfonic acid
Cleve's acid	1-Naphthylamine-7-sulfonic acid
α Coccinic acid	1-Methyl-5-oxybenzyl-2,4-dicarbonic acid
Cresidine	3-Amino-4-methoxytoluene
Cresotic acid	Cresol carboxylic acid
Croceine acid	2-Naphthol-1-sulfonic acid
DS	4,4′-Diamino-2,2′-stilbene disulfuric acid
DTS	Dihydrothio-p-toluidine sulfonic acid
Dahl's acid	2-Naphthylamine-5-sulfonic acid
Dahl's acid II	1-Naphthylamine-4,6-disulfonic acid
Dahl's acid III	1-Naphthylamine-4,7-disulfonic acid
Dimethyl-γ-acid	7-Dimethylamino-1-naphthol sulfonic acid
Dioxy G acid	1,7-Dihydroxynaphthalene-3-sulfonic acid
Dioxy J acid	1,6-Dihydroxynaphthalene-3-sulfonic acid
Dioxy S acid	1,8-Dihydroxynaphthalene-4-sulfonic acid
Diphenylblack base	p-Aminodiphenylamine

Trade name	Chemical name
Disulfo acid S	1-Naphthylamine-4,8-disulfonic acid
δ acid	1-Naphthol-4,8-disulfonic acid 1-Naphthylamine-4,8-disulfonic acid
Ebert & Merz acid	Naphthalene-2,7-disulfonic acid Naphthalene-2,6-disulfonic acid
Ethyl-γ-acid	7-Ethylamino-1-naphthol-3-sulfonic acid
Ethyl F acid	7-Ethylamino-2-naphthalene sulfonic acid
Ewer & Pick's acid	Naphthalene-1,6-disulfonic acid
ϵ acid	1-Naphthol-3,8-disulfonic acid 1-Naphthylamine-3,8-disulfonic acid
F acid	2-Naphthol-7-sulfonic acid (Casella's acid)
Fast Black B base	4,4′-Diamino diphenylamine
Fast Blue base	Dianisidine
Fast Blue Red O base	3-Nitro-p-phenetidine
Fast Bordeaux GP	3-Nitro-p-anisidine
Fast Orange GR	o-Nitroaniline
Fast Orange R	m-Nitroaniline
Fast Red base AL	α-Aminoanthraquinone
Fast Red B base	5-Nitro-o-anisidine
Fast Red GG base	p-Nitroaniline
Fast Red GL base	3-Nitro-p-toluidine
Fast Red 3 GL base	4-Chloro-2-nitroaniline
Fast Red RL base	5-Nitro-o-toluidine
Fast Scarlet G base	4-Nitro-α-toluidine
Fast Scarlet R base	4-Nitro-O-anisidine
Forsling's acid I	2-Naphthylamine-8-sulfonic acid
Forsling's acid II	2-Naphthylamine-5-sulfonic acid
Freund's acid	1-Naphthylamine-3,6-disulfonic acid
G acid	2-Naphthol-6,8-disulfonic acid
GR acid	α-Naphthol-3,6-disulfonic acid
Gallic acid	3,4,5-Trihydroxybenzoic acid
γ acid	2-Amino-8-naphthol-6-sulfonic acid
H acid	1-Amino-8-naphthol-3,6-disulfonic acid
Histazarin	2,3-Dihydroxyanthraquinone
Isoanthraflavic acid	2,7-Dihydroxyanthraquinone
J acid	2-Amino-5-naphthol-7-sulfonic acid
K acid	1-Amino-8-naphthol-4,6-disulfonic acid
Kalle's acid	1-Naphthylamine-2,7-disulfonic acid
Ketone base	Tetramethyl aminobenzophenone
Koch's acid	1-Naphthylamine-3,6,8-trisulfonic acid
L acid	1-Naphthol-5-sulfonic acid
Laurent's acid	1-Naphthylamine-5-sulfonic acid
Lepidine	4-Methylquinoline
Leucotrop	Phenyldimethyl benzylammonium chloride
M acid	1-Amino-5-naphthol-7-sulfonic acid
Mesidine	2,4,6-Thimethylaniline
Metanilic acid	Aniline-m-sulfonic acid
Methyl-γ-acid	7-Methyl-8-naphthol disulfonic acid
Michler's hydrol	Tetramethyl diaminobenzohydrol
Michler's ketone	Tetramethyl diaminobenzophenone
Myrbane oil	Nitrobenzene
Naphthacetol	1-Acetylamino-4-naphthol
Naphthazarin	5,8-Dihydroxy-1,4-naphthoquinone
Naphthionic acid	1-Naphthylamine-4-sulfonic acid
o-Naphthionic acid	1-Naphthylamine-2-sulfonic acid
Naphthol AS	Anilide of hydronaphthoic acid
Naphthoresorcine	1,3-Dihydroxynaphthalene
Nekal BX	Na-salt of 1,4-bis,sec-butylnaphthalene-6-sulfonic acid
Nevile and Winther's acid	1-Naphthol-4-sulfonic acid
Nigrotic acid	1,3,6,7-Dihydroxysulfonaphthoic acid
Nitron 1,2,4-acid	1-Amino-8-nitro-7-naphthol-4-sulfonic acid
Nitroso base	p-Nitrodimethyl aniline
NW acid	Nevile and Winther s acid
Oxy L acid	1-Naphthol-5-sulfonic acid
Oxy Tobias acid	β-Naphthol-1-sulfonic acid
Peri acid	1-Naphthylamine-8-sulfonic acid
p-Phenetidine	p-Aminophenol ethylether
Phenyl gamma acid	2-Phenylamine-8-naphthol-6-sulfonic acid
Phenyl Peri acid	Phenyl-1-naphthylamine-8-sulfonic acid
Phosxgene	Carbonyl chloride
Phthalic acid	O-Benzenedicarbolic acid
Picramic acid	2-Amino-4,6-dinitrophenol
Picric acid	2,4,6-Trinitrophenol
Pirio's acid	4-Amino-1-naphthalene sulfonic acid
Primuline base	p-Toluidine heated with sulfur
Purpurine	1,2,4-Trihydroxyanthraquinone
Pyrogallol	1,2,3-Trihydroxybenzene
Quinaldine	2-Methylquinoline
Quinazarin	1,4-Dihydroxyanthraquinone

Trade name	Chemical name
R acid	2-Naphthol-3,6-disulfonic acid
2 R acid	2-Amino-8-naphthol-3,6-disulfonic acid
Red acid	1,5-Dihydroxynaphthalene-3,7-disulfonic acid
RG acid	1-Naphthol-3,6-disulfonic acid
Resorcinol	1,3-Dihydroxybenzene
Rumpff acid	2-Naphthol-8-sulfonic acid (Croceine acid)
S acid	1-Amino-8-naphthol-4-sulfonic acid
2 S acid	1-Amino-8-naphthol-2,4-disulfonic acid
Salicylic acid	o-Hydroxybenzoic acid
Schäffer's acid	2-Naphthol-6-sulfonic acid
Schäffer and Baum acid	α-Naphthol-2-sulfonic acid
Schollkopf's acid	1-Naphthol-4,8-disulfonic acid 1-Naphthylamine-4,8-disulfonic acid 1-Naphthylamine-8-sulfonic acid
Sulfanilic acid	Aniline-p-sulfonic acid
Thiocarbanilide	Diphenylthiourea
Tobias acid	2-Naphthylamine-1-sulfonic acid
Tolidine	Di-p-aminoditolyl
Toluidine	Aminotoluene
Violet acid	α-Naphthol-3,6-disulfuric acid
Xylidine	Aminoxylene
Y acid	2-Naphthol-6,8-disulfuric acid
Yellow acid	1,3-Dihydroxynaphthalene-5,7-disulfuric acid
1:2:4 acid	1-Amino-2-naphthol-4-sulfonic acid

NOMENCLATURE OF SOME MONOMERS AND POLYMERS

Calvin E. Schildknecht

The small molecules from which industrial polymers are formed are of two general types. The first are the unsaturated or ethylenic compounds. They are often called vinyl-type monomers. Table 1 gives formulas, common names and newer indexing names used by *Chemical Abstracts*. The second type of monomer useful in polymer synthesis are better called intermediates. These are shown in Table 2. Recent names as 2,5-Furandione for Maleic anhydride and 1,3-isobenzofurandione for Phthalic anhydride are employed in *Chemical Abstracts* for indexing, but older names are generally retained in the abstracts. Recent books[1-3] have used capital letters for abbreviations of polymer types, e.g., PVA for polyvinyl alcohols and VC-VAC for vinyl chloride-vinyl acetate copolymers. Preceding small letters may show proved structures, for example, i-PP for substantially isotactic or stereoregular polypropylenes. Experimentally determined physical properties may be added using small numbers and letters following for example, x-i-PP-mp 150-170C indicates a crosslinked isotactic polypropylene of crystal melting range 150-170C. Abbreviations used to indicate the type of polymerization processes are presented in Table 3.

Table 1
NAMES OF SOME ETHYLENIC MONOMERS

Formula	Common Name	New Name
$CH_2=CH_2$	Ethylene	Ethene
$CH_2=CH$	Vinyl (group)	Ethenyl
$CH_2=CHC_6H_5$	Styrene	Ethylbenzene
$CH_2=CHOOCCH_3$	Vinyl acetate	Ethenyl acetate
$CH_2=CHCl$	Vinyl chloride	Chloroethene
$CH_2=CHOCH_3$	Methyl vinyl ether	Methoxyethene
$CH_2=CHOCH=CH_2$	Divinyl ether	1,1-Oxybis (ethene)
$CH_2=CCl_2$	Vinylidene chloride	1,1-Dichloroethene
$CH_2=CHCOOH$	Acrylic acid	2-Propenoic acid
$CH_2=CHCN$	Acrylonitrile	2-Propenenitrile
$CH_2=CHCONH_2$	Acrylamide	2-Propenamide
$CH_2=CHCOOC_2H_5$	Ethyl acrylate	Ethyl 2-Propenoate
$CH_2=C(CH_3)COOH$	Methacrylic acid	2-Methyl-2-propenoic acid
$CH_2=C(CH_3)COOCH_3$	Methyl methacrylate	Ethyl 2-propenoate
$CH_2=CHCH_3$	Propylene	Propene
$CH_2=CHCH_2$	Allyl (group)	2-Propenyl
$CH_2=CHCH_2OH$	Allyl alcohol	2-Propen-1-ol
$CH_2=CHCH_2OOCCH_3$	Allyl acetate	2-Propenyl ethanoate

Table 2

SOME INTERMEDIATES FOR POLYMERS

Formula	Common name	New name
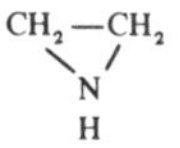 CH_2-CH_2 \\ N / H (ring)	Ethylene imine	Aziridine
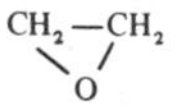 CH_2-CH_2 \\ O / (ring)	Ethylene oxide	Oxirane
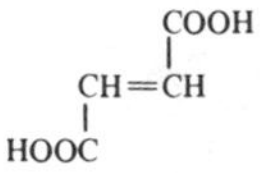 HOOC–CH=CH–COOH	Fumaric acid	2-Butenedioic acid (Z)
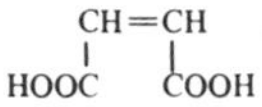 CH=CH (HOOC, COOH)	Maleic acid	2-Butenedioic acid (E)
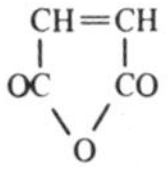 CH=CH, OC CO, O (ring)	Maleic anhydride	2,5-Furandione
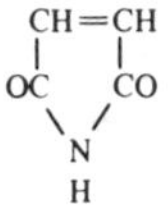 CH=CH, OC CO, N–H (ring)	Maleimide	1 H-Pyrrole-2,5-dione
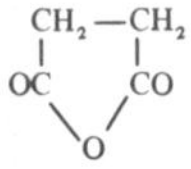 CH_2-CH_2, OC CO, O (ring)	Succinic anhydride	Dihydro-2,5-furandione
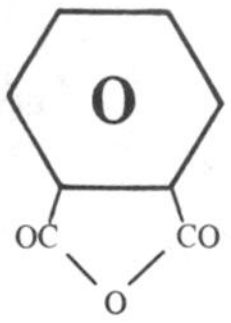 (O benzene ring) OC CO, O	Phthalic anhydride	1,3-Isobenzofurandione
$HOOC(CH_2)_4COOH$	Adipic acid	Hexanedioic acid
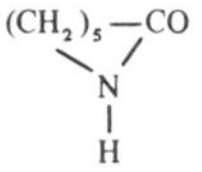 $(CH_2)_5-CO$, N–H (ring)	Caprolactam	Hexahydro-2H-azepin-2-one

Table 3
ABBREVIATIONS USED IN POLYMERIZATION PROCESSES

PE	An ethylene high polymer	TMA	Trimellitic anhydride
hd-PE	High density polyethylene	TDI	Toluene diisocyanate
S-AN	Copolymer of styrene and acrylonitrile	PP	A propylene high polymer
		i-PP	Substantially isotactic PP
PIB	An isobutene high polymer	syn-PP	Substantially syndiotactic PP (if it exists)
PMP	Poly-4-methyl-1-pentene		
HD	1,4-Hexadiene	PB	Poly-1-butene
EP	Copolymer of ethylene + propylene	AA	Acrylamide
		DAA	Diacetone acrylamide
EPDM	Ethylene-propylene-difunctional monomer copolymer	PVAC	Vinyl acetate polymer
		PVC	Vinyl chloride polymer
VAC	Vinyl acetate	PVA	Polyvinyl alcohol
VC	Vinyl chloride	PMMA	Methyl methacrylate polymer
AN	Acrylonitrile	PS	A polystyrene
MMA	Methyl methacrylate	DGEBA	Diglycidyl ether of bisphenol A
S	Styrene	UF	Ureo-formaldehyde polymer
ECH	Epichlorohydrin	THF	Tetrahydrofuran
PF	Phenol-formaldehyde polymer	PET	Polyethylene terephthalate
MF	Melamine-formaldehyde polymer	PBT	Polybutylene terephthalate
		PO	1,2-Propylene oxide
DMP	2,6-dimethylphenol	PBI	Polybenzimidazole
DEB	m-Diethynylbenzene	MDI	Methylene bis (4-phenyl isocyanate)
DMT	Dimethyl terephthalate		
BHET	bis(2-Hydroxyethyl)terephtholate	MOCA	Methylene bis (o-chloroaniline)
		PABM	Polyaminobismaleimide
BCMO	3,3-bis (Chloromethyl)oxelane	TAHT	1,3,5-Triacryloyl hexahydrotriazine
PMA	Pyromellitic dionhydride		
BPDA	Benzophenone tetracarboxylate dianhydride	HMM	Hexamethylalmelamine
		HMMM	Hexamethoxymethylolmelamine

REFERENCES

1. **Schildknecht, C. E.**, *Mod. Paint Coat.*, 69 (6), 41—45, 1979.
2. **Schildknecht, C. E.**, *Vinyl and Related Polymers*, (1952); *Polymer Processes*, (editor, 1956); *Allyl Compounds and Their Polymers*, (1973); *Polymerization Processes*, (editor, 1977), Wiley-Interscience, New York.
3. **Billmeyer, F. W.**, *Textbook of Polymer Science*, 2nd ed., Wiley-Interscience, New York, 1971.

IONIC EXCHANGE RESINS

ANION EXCHANGE RESINS

The following table is divided into two parts; the first lists properties of some anionic resins and the second, properties of some cationic resins.

Character S = strong W = weak	Trade name	Manu-facturer*	Active group	Matrix	Effective pH	Selectivity	Order of selectivity	Total exchange capacity; meq/ml	Total exchange capacity; meq/gm	Maximum thermal stability; °C	Physical form; s = sphere b = beads	Standard mesh range	Ionic form as shipped	Shipping density; lb./cu. ft.
S	Dowex 1	1	Trimethyl benzyl ammonium	Polystyrene	0–14	Cl/H approx. 25	I, NO_3, Br, Cl, Acetate, OH, F	1.33	3.5	OH^- 50 Cl^- 150	s	20–50 (wet)	Cl^-	44
S	Dowex 21 K	1	Trimethyl benzyl ammonium	Polystyrene	0–14	Cl/H approx. 15	I, NO_3, Br, Cl, Acetate, OH, F	1.25	4.5	OH^- 50 Cl^- 150	s	20–50 (wet)	Cl^-	43
S	Duolite A-101 D	2	Quaternary ammonium	Polystyrene	0–14			1.4	4.2	OH 60 Cl^- 100	b	16–50	Cl^-	
S	Ionac A-540	3	Quaternary ammonium	Polystyrene	0–14			1.0	3.6	salt 100 OH^- 60	b	16–50	salt	43–66
S	Dowex 2	1	Dimethyl ethanol benzyl ammonium	Polystyrene	0–14	Cl/H approx. 1.5	I, NO_3, Br, Cl, Acetate, OH, F	1.33	3.5	OH^- 30 Cl^- 150	s	20–50 (wet)	Cl^-	44
S	Duolite A-102 D	2	Quaternary ammonium	Polystyrene	0–14			1.4	4.2	OH^- 40 Cl^- 100	b	16–50	Cl^-	
S	Ionac A-550	3	Dimethyl ethanol benzyl ammonium	Polystyrene	0–14			1.3	3.5	salt 100 OH^- 40	b	16–50	salt	43–46
W	Duolite A-30 B	2	Tertiary amine; Quaternary ammonium	Epoxy polyamines	0–9			2.6	8.7	80	b	16–50	salt	
W	Ionac A-300	3	Tertiary amine; Quaternary ammonium	Epoxy amine	0–12			1.8	5.5	40	g	16–50	salt	19–21
W	Duolite A-6	2	Tertiary amine	Phenolic	0–5			2.4	7.6	60	g	16–50	salt	
W	Duolite A-7	2	Secondary amine	Phenolic	0–4			2.4	9.1	40	g	16–50	salt	

CATION EXCHANGE RESINS

Character S = strong W = weak	Trade name	Manu-facturer*	Active group	Matrix	Effective pH	Selectivity	Order of selectivity	Total exchange capacity; meq/ml	Total exchange capacity; meq/mg	Maximum thermal stability; °C	Physical form; s = sphere b = beads	Standard mesh range	Ionic form as shipped	Shipping density; lb./cu. ft.
S	Dowex 50	1	Nuclear sulfonic acid	Polystyrene	0–14	Na/H approx. 1.2	Ag, Cs, Rb, K, NH_4, Na, H, Li, Ba, Sr, Ca, Mg, Be	Na^+ 1.9 H^+ 1.7	Na^+ 4.8 H^+ 5.0	150	s	20–50 (wet)	H^+ or Na^+	H^+ 50 Na^+ 53
S	Dowex MPC-1	4	Nuclear sulfonic acid	Polystyrene	0–14			1.6–1.8 H^+ form	4.5–4.9 H^+ form	150	b	20–40 (wet)	Na^+	50
S	Duolite C-20	2	Nuclear sulfonic acid	Polystyrene	0–14			2.2	5.1	150	b	16–50	Na^+	
S	Ionac 240	3	Nuclear sulfonic acid	Polystyrene	0–14			1.9	4.6	140 (Na^+) 130 (H^+)	b	16–50	Na^+	50–55
S	Duolite C-3	2	Methylene sulfonic	Phenolic	0–9			1.1	2.9	60	g	16–50	H^+	
W	Dowex CCR-1	4	Carboxylic	Phenolic	0–9					38	g	20–50 (wet)	H^+ (dry)	21
W	Duolite ES-63	2	Phosphonic	Polystyrene	4–14			3.3	6.5	100	b	16–50	H^+	
W	Duolite ES-80	2	Aliphatic	Acrylic	6–14			3.5	10.2	100	b	16–50	H^+	

* 1. Dow
2. Diamond Shamrock
3. Ionac
4. Nalco

LIMITS OF SUPERHEAT OF PURE LIQUIDS

From the *Journal of Physical and Chemical Reference Data*, Volume 14, No. 3, 695, 1985. Reproduced by permission of the copyright owners, the American Chemical Society and the American Institute of Physics, and the author, C. T. Avedisian. One should refer to the original publication for discussions of the significance of the limit of superheat, limiting liquid superheat as a physical property, experimental methods for measuring limits of superheat of liquids, nucleation rates commensurate with experimental conditions, and criteria for selection of data in the table. In addition to the data in the table below, the original table contains data on limits of superheat for 27 binary mixtures and one ternary (Ethane, n-Propane, and n-Butane).

The homogeneous nucleation limit, or limit of superheat, represents the deepest penetration of a liquid into the domain on metastable states. At constant pressure and composition, it is the highest temperature below the critical point a liquid can sustain without undergoing a phase transition; at constant temperature, it is the lowest pressure. The practical significance of this limit resides in the consequences of the phase transition that eventually occurs when this limit is reached.

NOTE: In the column, *J [nuclei/(cm³·s)]*, 1E+2 means 1×10^2, 1E−2 means 1×10^{-2}, etc.

Substance	P [MPa]	T [K]	J [nuclei/(cm³-s)]
Ar Argon	-1.220	85.0	1E+02
	0.101	130.8	1E+02
	0.190	131.2	1E+02
	0.260	131.5	1E+05
	0.360	131.8	1E+01
	0.410	131.9	1E+05
	0.600	132.8	1E+05
	0.810	133.5	1E+03
	1.100	134.3	1E+01
	1.150	135.1	1E+05
	1.400	135.3	1E+01
	1.420	136.0	1E+05
	1.720	137.1	1E+05
	2.140	138.6	1E+05
	2.450	139.5	1E+05
	2.710	141.3	1E+05
H_2 Hydrogen	.076	27.8	1E-02
	.149	27.9	1E-02
	.381	29.4	1E-02
	.751	30.6	1E-02
	.834	30.8	1E-02
H_2O Water	-27.70	283.2	1E+03
	0.101	553.0	1E+06
	0.101	575.2	1E+15
	1.293	580.4	1E+21
	2.519	584.9	1E+21
	2.710	588.3	1E+21
	5.000	593.6	1E+21
	6.808	600.4	1E+21
	8.500	606.5	1E+21
	9.731	607.2	1E+21
	10.746	610.3	1E+21
	11.978	615.6	1E+21
	12.873	616.7	1E+21
	13.731	620.2	1E+21
	15.789	627.0	1E+21
	17.556	632.3	1E+21
	20.113	642.2	1E+21
He I Helium I	0.012	4.05	1E+07
	0.017	4.12	1E+07
	0.037	4.22	1E+07
	0.054	4.31	1E+07
	0.066	4.37	1E+07
	0.081	4.45	1E+07
	0.100	4.55	1E+07
	0.112	4.62	1E+07
	0.129	4.70	1E+07
	0.143	4.76	1E+07
He II Helium II	-0.06	2.09	1E+05
Kr Krypton	0.400	182.5	1E+05
	0.820	184.3	1E+05
	1.200	187.0	1E+05
	1.410	187.6	1E+05
	1.630	189.1	1E+05
	1.900	189.9	1E+05
	2.200	192.1	1E+05
	2.430	192.9	1E+05
	2.800	194.8	1E+05
	3.140	196.6	1E+05
	3.460	198.0	1E+05
	3.800	199.4	1E+05
N_2 Nitrogen	-1.010	75.0	1E+02
	0.101	110.0	1E+00
	0.410	111.4	1E+05
	0.520	112.0	1E+05
	0.610	112.1	1E+05
	0.700	112.7	1E+05
	0.820	113.2	1E+05
	0.940	113.8	1E+05
	1.060	114.2	1E+05
	1.210	114.8	1E+05
	1.240	115.2	1E+05
	1.330	115.5	1E+05
	1.360	115.6	1E+05
	1.460	116.2	1E+05
	1.590	116.8	1E+05
	1.620	117.0	1E+05
	1.730	117.6	1E+05
	1.770	117.7	1E+05
	1.870	118.3	1E+05
	1.920	118.4	1E+05
	2.070	119.1	1E+05
N_2O_3 Nitrogen-tetroxide	0.154	395.6	1E+02
	0.554	396.2	1E+02
	0.980	398.2	1E+01
	2.000	401.5	1E+01
	3.040	405.2	1E+01
	3.920	408.1	1E+01
	4.500	410.2	1E+01
	5.000	412.5	1E+01
	5.500	414.5	1E+01
	6.000	416.4	1E+01
O_2 Oxygen	-1.520	75.0	1E+02
	0.101	134.1	1E+00
	0.400	135.4	1E+05
	0.500	136.2	1E+05
	0.680	136.5	1E+05
	0.920	137.4	1E+05
	1.060	137.5	1E+05
	1.180	138.3	1E+05
	1.350	138.9	1E+05
	1.480	139.3	1E+05
	1.740	140.7	1E+05
	2.030	141.9	1E+05
	2.260	142.8	1E+05
	2.500	143.6	1E+05
	2.700	144.5	1E+05
	2.970	145.9	1E+05
SO_2 Sulphur-dioxide	0.101	323.2	1E+02
Xe Xenon	0.500	254.1	1E+05
	0.830	256.3	1E+05
	1.070	257.2	1E+05
	1.260	258.2	1E+05
	1.470	259.6	1E+05
	1.550	260.3	1E+05
	1.680	261.0	1E+05
	1.750	261.6	1E+05
	1.860	261.9	1E+05
	1.970	262.8	1E+05
	2.070	263.4	1E+05
	2.170	263.8	1E+05
	2.370	265.2	1E+05
	2.480	266.1	1E+05
	2.630	266.9	1E+05
	2.750	267.5	1E+05
	2.850	267.8	1E+05
	2.970	269.1	1E+05
	3.050	269.7	1E+05
	3.130	270.0	1E+05
	3.450	272.0	1E+05
	3.630	273.0	1E+05
$ClCHF_2$ Chloro-difluoro-methane	0.101	327.8	1E+04
	0.236	328.2	1E+04
	0.280	329.4	1E+04
	0.510	330.8	1E+04
	0.560	331.5	1E+04
	0.710	332.4	1E+04
	0.810	332.9	1E+04
	0.910	334.2	1E+04
CCl_4 Carbontetra-chloride	-27.60	268.2	1E+03
$CHCl_3$ Chloroform	-31.70	258.2	1E+03
	0.101	466.2	1E+02
CH_2Cl_2 Methylene-chloride	0.101	394.8	1E+01
CH_3Cl Chloro-methane	0.101	366.2	1E+05
CH_4 Methane	0.400	167.6	1E+05
	0.620	168.3	1E+05
	0.820	169.3	1E+05
	1.030	170.5	1E+05
	1.230	171.4	1E+05
	1.430	172.1	1E+05
	1.630	173.1	1E+05
	1.830	174.0	1E+05
	2.030	175.2	1E+05
	2.220	176.4	1E+05
	2.430	177.6	1E+05
	2.630	178.6	1E+05
	2.820	180.0	1E+05
CH_4O Methanol	0.101	458.4	1E+01
	0.101	461.2	1E+05
	0.101	466.2	1E+18
	0.600	469.2	1E+19
	1.050	471.2	1E+20
	2.030	476.7	1E+16
	2.030	478.2	1E+20
	3.000	482.2	1E+21
	4.000	488.7	1E+22
	4.980	494.7	1E+22
	5.970	501.2	1E+23
	6.960	507.7	1E+23
$C_2Cl_2H_2F_2$ Dichloro-difluoro-ethane	0.221	342.5	1E+06
	0.427	344.3	1E+06
	0.462	344.7	1E+06
	0.655	346.6	1E+06
	0.896	348.8	1E+06
	0.931	349.0	1E+06
	1.227	351.7	1E+06
	1.489	354.4	1E+06
	1.917	358.8	1E+06
	2.399	363.7	1E+06
	2.910	369.0	1E+06
	3.289	373.0	1E+06
	3.323	373.4	1E+06
	3.585	376.2	1E+06
	3.634	376.9	1E+06
C_2H_3Cl Chloroethane	0.101	374.1	1E+05
C_2H_3F Fluoroethene	0.101	290.1	1E+05
C_2H_3N Acetonitrile	0.101	497.0	1E+06
$C_2H_4F_2$ 1,1-Difluoro-ethane	0.101	343.6	1E+05
$C_2H_4O_2$ Acetic acid	-28.80	292.7	1E+03
	0.101	526.2	1E+06
$C_2H_4O_2$ Methyl-formate	0.101	423.2	1E+01
C_2H_5Br Ethylbromide	0.101	422.2	1E+01
C_2H_5Cl Ethylchloride	0.101	399.2	1E+01
C_2H_6 Ethane	0.101	269.2	1E+05

Substance	P [MPa]	T [K]	J [nuclei/(cm^3-s)]
C_2H_6O Ethanol	0.101	464.1	1E+01
	0.101	466.0	1E+04
	0.101	471.5	1E+17
	0.580	474.2	1E+19
	0.980	471.0	1E+02
	1.070	477.2	1E+20
	1.540	481.7	1E+20
	2.030	484.2	1E+21
	2.520	486.7	1E+21
	3.010	490.2	1E+21
	3.500	494.2	1E+22
C_3H_3N Acrylonitrile	0.101	489.0	1E+05
C_3H_4 Propadiene	0.101	346.2	1E+05
C_3H_4 Propyne	0.101	356.8	1E+05
C_3H_6 Cyclopropane	0.101	350.7	1E+05
C_3H_6 Propene	0.101	325.6	1E+05
C_3H_6O Acetone	0.101	454.5	1E+01
	0.101	456.4	1E+02
	0.101	458.7	1E+13
	0.101	462.7	1E+18
	0.980	462.6	1E+01
$C_3H_6O_2$ Ethylformate	0.101	428.5	1E+01
$C_3H_6O_2$ Methylacetate	0.101	416.6	1E+01
C_3H_8 N-Propane	0.101	326.4	1E+06
	0.302	332.8	1E+04
	0.491	336.8	1E+04
	0.715	339.1	1E+04
	0.907	343.2	1E+04
C_3H_8O N-Propanol	0.101	487.4	1E+04
	0.101	493.0	1E+15
	0.101	495.7	1E+18
C_3H_8O Isopropanol	0.101	473.0	1E+06
C_4H_6 1,3-Butadene	0.101	377.3	1E+05
C_4H_8 1-Butene	0.101	371.0	1E+05
C_4H_8 Butylene	0.101	510.2	1E+16
C_4H_8 Cis-2-Butene	0.101	385.4	1E+05
C_4H_8 Trans-2-Butene	0.101	379.7	1E+05
C_4H_8 2-Methylpropene	0.101	369.6	1E+05
C_4H_{10} N-Butane	0.101	377.6	1E+05
C_4H_{10} 2-Methylpropane	0.101	361.0	1E+05
$C_4H_{10}O$ N-Butanol	0.101	509.6	1E+02
	0.101	511.9	1E+04
	0.101	513.2	1E+13
	0.101	516.2	1E+16
	0.101	518.2	1E+18
	0.980	519.4	1E+02
$C_4H_{10}O$ Ether	-1.75	293.	1E+02
	-1.52	402.7	1E+04
	-1.22	407.6	1E+04
	-1.12	409.2	1E+04
	-1.00	410.2	1E+04
	-0.74	413.4	1E+04
	0.101	417.5	1E+02
	0.101	425.7	1E+19
	0.211	419.4	1E+01
	0.415	420.1	1E+01
	0.480	427.7	1E+18
	0.500	421.1	1E+02
	0.641	424.3	1E+02
	0.777	426.3	1E+01
	0.880	432.7	1E+18
	1.000	428.4	1E+01
	1.280	436.7	1E+01
	1.366	433.6	1E+01
	1.442	435.1	1E+02
	1.575	437.2	1E+01
	1.660	440.7	1E+19
	1.865	441.2	1E+01
	2.089	443.3	1E+01
	2.450	450.7	1E+21
	2.850	455.7	1E+20
$C_4H_{10}O$ Isobutanol	0.101	437.2	1E+01
$C_4H_{11}N$ Diethylamine	0.101	408.5	1E+01
C_5F_{12} Perfluoro-pentane	0.101	381.5	1E+06
	0.300	385.4	1E+06
	0.500	388.7	1E+06
	0.700	392.2	1E+06
	0.890	396.4	1E+06
	1.090	399.0	1E+06
	1.280	403.1	1E+06
	1.480	407.4	1E+06
C_5H_8 Cyclopentene	0.101	451.4	1E+06
C_5H_{10} Cyclopentane	0.101	455.1	1E+06
C_5H_{10} 1-Pentene	0.101	417.2	1E+06
C_5H_{12} 2,2-Dimethyl-propane	0.101	386.1	1E+06
C_5H_{12} Isopentane	0.101	409.2	1E+01
	0.101	411.7	1E+07
C_5H_{12} N-Pentane	0.101	418.8	1E+04
	0.101	426.2	1E+18
	0.490	423.7	1E+02
	0.880	429.1	1E+02
	1.280	435.3	1E+02
	2.600	451.2	1E+06
$C_5H_{12}O$ N-Pentanol	0.101	532.2	1E+17
C_6F_6 Hexafluoro-benzene	0.101	464.8	1E+01
	0.101	467.9	1E+06
	0.500	469.9	1E+02
	0.570	474.1	1E+06
	1.000	477.1	1E+02
	1.050	480.4	1E+06
	1.540	486.0	1E+06
	2.030	494.2	1E+06
C_6F_{14} Perfluoro-hexane	0.101	409.8	1E+06
	0.300	414.4	1E+06
	0.500	418.5	1E+06
	0.700	422.3	1E+06
	0.880	425.6	1E+06
	1.050	430.3	1E+06
	1.240	434.6	1E+06
C_6H_5Br Bromobenzene	0.101	534.2	1E+02
C_6H_5Cl Chlorobenzene	0.101	523.2	1E+02
C_6H_6 Benzene	-15.0	291.2	1E+03
	0.101	498.9	1E+02
	0.101	510.2	1E+18
	0.490	502.2	1E+02
	0.580	514.2	1E+19
	0.980	509.2	1E+02
	1.070	516.7	1E+18
	1.470	513.8	1E+02
	1.540	520.7	1E+19
	2.030	525.7	1E+19
	2.520	532.2	1E+20
	3.010	537.7	1E+17
	3.500	544.7	1E+18
C_6H_7N Aniline	-30.0	272.2	1E+03
	0.101	535.2	1E+02
C_6H_{12} Cyclohexane	0.101	490.8	1E+02
	0.300	493.1	1E+02
	0.420	495.2	1E+06
	0.720	499.7	1E+06
	0.950	501.7	1E+06
	0.980	502.1	1E+02
	1.110	504.2	1E+06
	1.350	506.2	1E+06
	1.700	512.2	1E+06
	2.160	518.2	1E+06
	2.370	519.2	1E+06
	2.550	523.2	1E+06
C_6H_{12} 1-Hexyne	0.101	465.2	1E+06
C_6H_{12} Methylcyclo-pentane	0.101	476.1	1E+06
C_6H_{14} 2,3-Dimethyl-butane	0.101	446.4	1E+06
C_6H_{14} N-Hexane	0.101	453.5	1E+02
	0.101	454.9	1E+05
	0.101	459.2	1E+13
	0.101	463.7	1E+20
	0.290	465.2	1E+15
	0.420	461.7	1E+06
	0.490	459.3	1E+02
	0.490	468.2	1E+22
	0.760	466.7	1E+06
	0.980	467.0	1E+02
	0.980	475.2	1E+23
	1.080	471.7	1E+06
	1.120	478.2	1E+15
	1.280	474.7	1E+06
	1.420	475.7	1E+06
	1.590	479.7	1E+06
	1.600	486.2	1E+16
	1.720	481.7	1E+06
	1.960	487.7	1E+17
	2.060	493.2	1E+16
	2.390	496.7	1E+06
	2.570	501.2	1E+16
$C_6H_{14}O$ N-Hexanol	0.101	555.7	1E+04
C_7F_8 Octafluoro-toluene	0.101	485.3	1E+01
	0.490	489.7	1E+01
	0.980	499.8	1E+01
C_7F_{16} Perfluoro-heptane	0.101	434.8	1E+06
	0.230	436.9	1E+06
	0.400	440.5	1E+06
	0.570	444.4	1E+06
	0.770	448.3	1E+06
	0.920	452.7	1E+06
	1.070	456.1	1E+06
	1.150	459.0	1E+06
	1.280	461.3	1E+06
C_7H_8 Toluene	0.101	526.7	1E+02
C_7H_{14} Methylcyclo-hexane	0.101	510.4	1E+06
C_7H_{16} N-Heptane	0.101	486.9	1E+06
	0.101	493.7	1E+18
	0.294	489.2	1E+06
	0.392	490.7	1E+06
	0.490	493.7	1E+06
	0.589	494.2	1E+06
	0.736	498.7	1E+06
	0.952	500.7	1E+06
	1.275	505.2	1E+06
	1.373	509.7	1E+06
	1.570	512.7	1E+06
	1.736	515.2	1E+06
	1.805	516.7	1E+06
	2.001	519.7	1E+06
$C_7H_{16}O$ N-Heptanol	0.101	566.3	1E+04
C_8F_{18} Perfluoro-octane	0.101	457.0	1E+06
	0.300	461.1	1E+06
	0.500	467.1	1E+06

Substance	P [MPa]	T [K]	J [nuclei/(cm³-s)]
	0.700	471.2	1E+06
	0.890	476.9	1E+06
	1.090	482.8	1E+06
	1.190	484.1	1E+06
C_8H_{10} Cyclo-octane	0.101	560.7	1E+06
C_8H_{10} 2,3-dimethyl-benzene	0.101	508.2	1E+02
C_8H_{16} 1-Octene	0.101	510.3	1E+06
C_8H_{18} N-Octane	0.101	513.8	1E+06
	0.377	519.3	1E+04

Substance	P [MPa]	T [K]	J [nuclei/(cm³-s)]
	0.653	525.2	1E+04
	0.929	528.6	1E+04
	1.204	532.4	1E+04
C_8H_{18} 2,2,4-Trimethyl-pentane	0.101	488.5	1E+06
$C_8H_{18}O$ N-Octanol	0.101	586.0	1E+04
C_9F_{20} Perfluoro-nonane	0.101	478.5	1E+06
	0.300	484.4	1E+06
	0.500	489.3	1E+06
	0.700	493.3	1E+06
	0.890	499.7	1E+06

Substance	P [MPa]	T [K]	J [nuclei/(cm³-s)]
	1.090	505.7	1E+06
C_9H_{20} N-Nonane	0.101	538.5	1E+06
$C_{10}F_{22}$ Perfluoro-decane	0.101	497.1	1E+06
	0.300	503.2	1E+06
	0.500	508.6	1E+06
	0.700	515.6	1E+06
	0.890	521.2	1E+06
	1.090	527.7	1E+06
$C_{10}H_{22}$ N-Decane	0.101	558.3	1E+06
$C_{12}H_{10}O$ Diphenylether	0.101	703.2	1E+17

AZEOTROPES

Zdzislaw M. Kurtyka

GENERAL CLASSIFICATION OF AZEOTROPES AND THEIR SYMBOLISM

Although the first reported observations of the appearance of minimum vapor pressure in binary mixtures were in 1802 by Dalton,[1] the term "azeotrope" was not introduced until 1911 by Wade and Merriman.[2]

Azeotrope from the Greek "not to boil with change" or "to boil unchanged" means literally the same as the English "constant-boiling", i.e., the vapor boiling from a liquid has the same composition as the liquid.

A liquid mixture of two or more components which can be separated by distillation was termed a "zeotrope" by Swietoslawski.[3] Such a mixture is also called a "nonazeotrope" or a "non-azeotropic" system in many countries. However, some objection may be placed against the use of the terms "nonazeotrope" and "non-azeotropic" system due to the use of two negatives in the same word. In Greek "a" means "non". Therefore, accuracy indicates that the word nonazeotrope should be replaced by non-non-azeotrope. It is obvious the latter is awkward and that it is reasonable to replace it with Swietoslawski's suggestion, zeotrope.

Azeotropic systems may be classified broadly in relation to the character of the extremum (maximum or minimum), the number of components in the system and whether they form one or more liquid phases.

Positive and Negative Azeotropes

In 1926 Lecat[4] proposed to divide azeotropes into positive and negative. Positive azeotropes are characterized by a minimum boiling temperature at constant preassure, i.e. a maximum in the vapor pressure at constant temperature. Negative azeotropes, on the other hand, have a maximum boiling temperature and a minimum vapor pressure.

In Anglo-Saxon literature a positive azeotrope is equivalent to a pressure-maximum azeotrope, while a negative one — to a pressure-minimum azeotrope. This description of azeotropes cannot be extended to systems exhibiting neither a minimum nor a maximum in either boiling temperature or vapor pressure (saddle or positive-negative azeotropes).

One may say at this point that neither azeotrope nor zeotrope, thus described, gives any indication whether the vapor phase consists of one, two or more liquid phases. To make this perfectly clear, the terms homo- and heteroazeotrope, and homo- and heterozeotrope were introduced. For practical purposes negative azeotropes are divided into three groups.[5]

In view of the fact that the number of different types of azeotropes continue to increase, symbols are given below for positive and negative azeotropes and zeotropes.

Type of homoazeotrope	Type of zeotrope
1. (A,B) binary positive	$(A,B)_z$ binary positive
2. A,B,C) ternary positive	$(A,B,C)_z$ ternary positive
3. (A,B,C,D)quaternary positive	$(A,B,C,D)_z$quaternary positive
4. [(−)A,B]binary negative	$[(-)A,B]_z$ binary negative

Symbols are also used for heteroazeotropes. For example, there is a combination of letters, dots, and dashes. There is one letter for each of the components in the system, one dash for each of the phases in the azeotrope and one less dot than there are dashes. The system may be illustrated by the following two examples. The system benzene-ethanol-water forms a two-phase heteroazeotrope. Symbolically this is written as (B,E,W, -·-). The system nitromethane-water-n-paraffin forms three liquid phases at the boiling temperature. This applies for n-paraffins from heptane to tridecane. If the symbol H_i is assigned to the n-parafffin, the azeotrope will be designated by the symbols $(N,W,H_i$,-·-·-).

Saddle or Positive-Negative Azeotropes

In spite of the fact that a ternary saddle azeotrope was predicted by Ostwald at the end of the last century, the first azeotrope of this type was found in 1945 by Ewell and Welch[7] in the system acetone-chloroform-methanol.

Saddle azeotropic systems, also called positive-negative systems, exhibit a hyperbolic point which is neither a minimum nor a maximum in either boiling temperature or vapor pressure, and are characterized by the presence of a "top-ridge" line. They also exhibit some peculiar properties called distillation anomalies.[7-10]

In general, ternary saddle azeotropes are classified according to the number of binary negative systems forming such an azeotrope. From this point of view, ternary saddle azeotropes may be divided into bipositive-negative and binegative-positive azeotropes. All possible types of ternary bipositive-negative azeotropes were found.[11-14] The bipositive-negative azeotropes may be designated by the symbols [(−)A,B(±)H], where, A, B, and H are the components forming these azeotropes. For the ternary binegative-positive azeotropes with B, E, and C as the components, the symbols are [(+)B,E(−)C] (two binary negative azeotropes [(−)C,B] and [(−)C,E] occur in this system). This type of a ternary saddle azeotrope is rather a rare phenomenon.[15,16]

Although the terms "saddle azeotrope" and "positive-negative" azeotrope are equivalent in the context of ternary systems, the latter is more preferable when discussing multicomponent systems.

THE COMMONNESS OF THE PHENOMENON OF AZEOTROPY

Azeotropes occur in organic and inorganic systems, although most of the known azeotropes are formed by organic compounds. This is because the very large number of organic compounds boil without decomposition in the easily accessible ranges of temperature and pressure.

In the last 2 decades of the 19th century the appearance of an extremum vapor pressure and boiling temperature was believed to be a rare phenomenon. For this reason Ostwald[17] used the term "ausgezeichnete Lösungen" to emphasize the phenomenon was not often encountered. This view survived in certain circles for many years despite convincing evidence the phenomenon of azeotropy is definitely a common one.

The first investigator who provided evidence of the commonness of azeotropy was Ryland.[18] His word dealt with 80 systems, 45 of which were azeotropic; 80 further new azeotropes were described in Lecats doctoral dissertation, published in 1908 to 1909. Lecat was able to list 1000 azeotropes in his monograph 10 years later.[19] Among them were numerous binary and ternary heteroazeotropes. This monograph showed that the appearance of maximum or miminum vapor pressures of binary and ternary mixtures should not be regarded as a rare phenomenon. From then on many physical chemists and technologists began to take an interest in the theoretical and practical application of azeotropy.

In 1949 Lecat[20] published the azeotropic data for 13,290 binary systems. The number of azeotropes reached 6287 or 47% of the systems examined.

In 1973 Horsley[21] published his Azeotropic Data-III. In this volume 15,823 binary, 725 ternary, 21 quaternary, and 2 quinary systems are reported. The number of the azeotropic systems is as follows: binary 7945 (52%), ternary 371 (51%), quaternary 9 (43%), and quinary 1.

It is interesting to see that 119 binary azeotropes (47% of the systems examined), including 32 negative, occur in the systems composed either of two inorganic compounds (elements) or an inorganic-organic compound system; 768 binary systems contain water as one component (among them 665 [86%] are azeotropic). The ternary positive-negative (saddle) azeotropes occur in 40 systems; 267 (72%) ternary azeotropic systems contain water as one component. There are also 4 ternary negative azeotropes. As far as the quaternary systems are concerned, 8 systems form positive azeotropes and one, a positive-negative.

In the last decade, studies of vapor-liquid equilibria were developed rapidly, mostly at a constant temperature, with the aim of correlating and predicting the equilibrium parameters. Unfortunately, a large number of investigators did not report whether an azeotrope or a zeotrope is formed in a particular system. In view of the fact that the number of the systems examined in that period of time is small, compared to that of the systems examined previously, it becomes clear that the current situation regarding the commonness of the phenomenon of azeotropy remains essentially unchanged.

From the existing experimental work one might conclude that the frequency of occurence of azeotropes diminishes as the number of components in the system increases, and that multicomponent azeotropes should be expected to be very rare. Such a point of view, however, would be wrong. Although the conditions of the formation of multicomponent azeotropes are complex, they are not difficult to fulfil in practice.[22-24]

THE AZEOTROPIC RANGE

The idea of relating a certain "azeotropic ability" to the chemical character of a substance is already apparent in the first works of Lecat,[19] where he arranged the experimental data on azeotropes according to the chemical character of the components.

The term "azeotropic range" is due to Swietoslawski.[25] There is also another closely related term known, namely, the "relative azeotropic effect",[26] but its use is very limited. Malesinski[27] developed the concept of Swietoslawski's azeotropic range on a general assumption that the components of the system form a regular solution.

The symmetrical azeotropic range, Z, is given by the formula

$$(\delta_1^{1/2} + \delta_2^{1/2}) = Z_{12} = 0.5Z \qquad (1)$$

where z_{12} is the half-value of the symmetrical azeotropic range and the quantities δ_1 and δ_2, known otherwise as azeotropic deviations, for positive azeotropes are defined by $\delta_1 = T_1 - T^{Az}$ and $\delta_2 = T_2 - T^{Az}$; T_1, T_2 are the boiling temperatures of the pure components and T^{Az} is the boiling temperature of the azeotrope. For negative azeotropes we have $\delta_1 = T^{Az} - T_1$ and $\delta_2 = T^{Az} - T_2$. The symmetrical azeotropic range means that both parts of the range (the upper and the lower part) are equal.

Equation 1 which is also valid at high pressures[28] enables us to compute z_{12} of any binary azeotropic system, provided that the boiing temperature of the pure components and of the azeotrope are known. It should be remembered that by convention the z_{12} values are positive for positive azeotropic systems, and negative for negative ones.

The knowledge of the azeotropic ranges usually leads to a better understanding of the distillation course of complex azeotropic mixtures, e.g. high- and low-temperature coal tars, petroleum and synthetic gasoline. In addition, the azeotropic ranges appear in the equations for calculating the boiling temperatures and compositions of ternary homoazeotropes. These equations were found useful in predicting the existence of a large number of ternary azeotropes.[24]

THE PREDICTION OF AZEOTROPIC DATA IN BINARY SYSTEMS

Empirical Correlations

Lecat[19] first observed that the composition of a binary azeotrope is related to the difference between the boiling temperatures of the components. He used a power series to relate the above quantities for the systems formed by a common substance with members of a homologous series.

His relation may be written in the form

$$x_1 = A_0 + A_1\Delta + A_2\Delta^2 + A_3\Delta^3 + \cdots \tag{2}$$

where x_1 is the weight fraction of component 1 in the azeotrope and Δ is the absolute difference between the boiling temperatures of the components.

Lecat[29] proposed also a relation between Δ and the azeotropic deviation, δ, i.e. the absolute difference between the boiling temperatures of the azeotrope and that of the more volatile component, for positive azeotropes, and the less volatile component, for negative azeotropes. The Δ—δ relation, otherwise known as Lecat's rule is

$$\delta = C_0 + C_1\Delta + C_2\Delta^2 + C_3\Delta^3 + \cdots \tag{3}$$

For most cases the terms with Δ higher than Δ^2 may be neglected to give

$$\delta = C_0 + C_1\Delta + C_2\Delta^2 \tag{4}$$

The existence of an approximate relation of this kind was expected for a series of binary regular solutions.[30] For instance, the constants of Equation 4 for the azeotropes of ethanol with aliphatic halogen derivatives are $C_o = 12$, $C_1 = -0.5$ and $C_2 = 0.00526$. The general agreement with Equation 4 for that series is satisfactory, despite that these alcohol solutions show large deviations from regularity.

In the 1940s several graphical correlations of empirical nature for the composition of azeotropes within a series of organic compounds were reported, notably by Mair et al.,[31] Horsley,[32] Meissner and Greenfield,[33] Skolnik,[26] and by Seymour.[34] A decade later Johnson and Madonis[35] described another correlation that included a number of other series of compounds.

The empirical treatment suffered mainly due to many variations in the form of the equations and the number of constants required for their evaluation.

A further attempt in improving the situation in this field was made by Seymour et al.[36] They proposed a correlation between x and Δt in the form of a master equation to correlate, among other things, the data from other series already reported. This equation is

$$\log(10\,\frac{x_1}{x_2} = mf(\tau)\,\Delta t + b \tag{5}$$

where x_1 and x_2 are the mole fractions of component 1 and 2 in the azeotrope, respectively, Δt is the difference between the boiling temperatures of the two components, m and b are constants and $f(\tau) = \tau_1/\tau_2$. The quantity τ is defined as the ratio of the boiling temperatures (°K) of a compound and a hypothetical n-paraffin of the same molecular weight. The differences between the calculated and observed azeotropic compositions are reported for 15 series of organic compounds.

An approximate linear relation between the logarithm of the mole fraction, $\log x_i$, of the main and the secondary azeotropic agent, and the average condensation temperature, and was found for several ternary homo- and heteropolyazeotropic mixtures.[37] Empirical correlations for the azeotropic composition apply to both homo- and heteroazeotropes.

Composition of a Binary Homoazeotrope

The theory of regular solutions offers methods for calculating the composition of a binary (positive or negative) azeotrope from certain properties of the pure components and of the azeotrope.

Generally speaking there are three known methods serving this purpose. Two of them involve the boiling temperatures of the pure components, T_1, T_2, and that of the azeotrope, T^{Az}, and were developed by Prigogine[30] and Malesinski.[23] The third method is based on the activity coefficients of the components at the azeotropic point, and is due to Kireev.[38]

The Prigogine equation for equal molar vaporization entropies of the components, $\Delta S_1^o = \Delta S_2^o$, may be written in the form

$$x_2 = \alpha(1 + \alpha)^{-1} \tag{6}$$

were α is the square root of the ratio of azeotropic deviations, i.e., $(\delta_1/\delta_2)^{1/2}$, and x_2 is themole fraction of components 2 in the azeotrope.

When $\Delta S_1^o \neq \Delta S_2^o$, Equation 6 takes the form

$$x_2 = \alpha'(1 + \alpha')^{-1} \tag{7}$$

In this case $\alpha' = c\alpha$ and $c = (\Delta S_1^o/\Delta S_2^o)^{1/2}$.

The Malesinski equation for the components having equal molar vaporization entropies is given by

$$x_2 = 0.5 + \frac{T_1 - T_2}{2z_{12}} \qquad (8)$$

where z_{12} is the half-value of the symmetrical azeotropic range. For systems with unequal molar vaporization entropies, Malesinski's equation requires evaluation of z_u (the upper part of the azeotropic range), the quantity which is not easily available.

Recently, Equations 6, 7, and 8 were evaluated with regard to their usefulness for the calculation of the composition of binary azeotropes.[39]

The simplest and most suitable form of Kireev's equation for the case in which $\Delta S_1^o = \Delta S_2^o$, is the expression

$$x_2 = (1 + b)^{-1} \qquad (9)$$

where b is the square root of the ratio of the logarithms of the activity coefficients of the components γ_2 and γ_1, i.e. $(\ln \gamma_2/\ln\gamma_2)^{1/2}$. At the azeotropic point the composition of the liquid and the vapor are equal, $x_2 = y_2$, and in the case of an ideal vapor phase the expressions for γ_1 and γ_2 are given by

$$\gamma_1 = P/p_1^0 \text{ and } \alpha_2 = P/p_2^0$$

In these relations p_1^o and p_2^o are the vapor pressures of the pure components at the boiling temperature of the aceotrope, and P is the total pressure of the mixture at equilibrium.

Equation 9 for $\Delta S_1^o \neq \Delta S_2^o$ becomes

$$x_2 = (1 + b')^{-1} \qquad (10)$$

where $b' = cb$ and $c = (\Delta S_2^o/\Delta S_1^o)^{1/2}$.

The possible error in the composition of the binary azeotrope under isobaric conditions is due not only to the deviations of the system from regularity but also to the change of the regular solutions constant, A_{12}, with temperature, and to the differences in the vaporization entropies of the pure components. One of the causes of the deviations of the system from regularity is the nonideality of the vapor phase.

The effect of the nonideality of the vapor phase and the differences in the vaporization entropies of the components on the azeotropic composition were studied by Kurtyka and Kurtyka[40] on the systems of acetic acid with n-paraffins. Acetic acid is associated (dimerized) in the vapor phase and its vaporization entropy is 14.85 cal/g-mole, while those of the hydrocarbons are ~20 cal/g-mole.

To get a general idea about those effects on the azeotropic composition, the results obtained for some systems of acetic acid with n-paraffins are reproduced in part from this paper. The system, $\Delta x_2 = x_{2(calcd.)} - x_{2(obsd.)}$, with x_2 computed by Equation 9; Δx_2, with x_2 computed with the corrections for the dimerization of the acid in the vapor phase, and Δx_2, with x_2 computed by Equation 10 are: acetic acid — n-heptane, 12.3, 1.8, 9.3; acetic acid — n-octane, 12.8, 3.8, 9.2, and acetic acid — n-nonane, 9.8 mole %, 4.3 mole %, and 7.0 mole %, respectively.

These results show that the Δx_2 values are reduced to a reasonable magnitude when the corrections for the dimerization of the acid are taken into account. This simply means that the dimerization of acetic acid in the vapor phase is the dominant factor contributing to the large differences in Δx_2. The effect of the differences in the vaporization entropies of the components on the azeotropic composition as exemplified by the systems containing acetic acid and n-paraffins is small compared to the deviations from regularity caused, among other things, by the dimerization of acetic acid in the vapor phase.

Equation 9 has a built-in unfavorable factor because it involves the vapor pressure, p_i^o, of each pure component at the boiling temperature of the azeotrope. p_i^o values are almost exclusively computed by means of the Antoine equation. Therefore, the accuracy of the computed p_i^o is related to the constants of that equation. However, the effect of the differences in p_i^o values on the azeotropic composition is usually small, except in cases where the boiling temperature of the azeotrope is close to the boiling temperature of one component of the system.

The Azeotropic Data of a Binary Heteroazeotrope

Many systems of two or more components exhibit a limited solubility and most of them form positive heteroazeotropes. In binary systems it is easy to predict from the data on critical solution temperatures (CST) whether an azeotrope that occurs at a certain temperature is a homoazeotrope or heteroazeotrope.[41]

To describe the vapor-liquid equilibrium in a heterogeneous system, it is necessary to introduce certain simplifying assumptions.

The simplest case is obtained upon the assumption that liquids are completely immiscible in each other and that the vapor phase is ideal.

Then the composition of a binary heteroazeotrope, y_1, is given by the formula

$$x_1 = y_1 = \frac{p_1^0\,(T^{Az})}{p_1^0\,(T^{Az}) + p_2^0\,(T^{Az})} \qquad (11)$$

where p_1^o and p_2^o are the vapor pressures of pure components 1 and 2, respectively, expressed as functions of the boiling temperature of the azeotrope, T^{Az}, or simply, at the boiling temperature of the azeotrope.

And for the boiling temperature of the azeotrope, T^{Az} we have the relation

$$P = p_1^0 (T^{Az}) + p_2^0 (T^{Az}) \tag{12}$$

where P is the total pressure at equilibrium.

In the cases in which the condition of complete immiscibility is not satisfied, we introduce the correction factors in terms of the activities of the components, α_1 and α_2 in the liquid phase.

Accordingly, the expression for the composition of an azeotrope takes the form

$$y_1 = \frac{p_1^0 (T^{Az}) \alpha_1}{p_1^0 (T^{Az}) \alpha_1 + p_2^0 (T^{Az}) \alpha_2} \tag{13}$$

And the boiling temperature of the azeotrope may be calculated from

$$P = p_1^0 (T^{Az}) \alpha_1 + p_2^0 (T^{Az}) \alpha_2 \tag{14}$$

There is also another possibility, namely to assume that the activities of both components in the liquid phase are equal, i.e., $\alpha_1 = \alpha_2 = \alpha$. The systems behavior of which is well described by a common activity, are those of nitromethane with n-paraffins.[42] The computation of the activities, α_1 and α_2 of the components is a straightforward procedure involving the solution of Equations 13 and 14.

Equations 11 and 12 were found to be satisfied for the systems of aromatic and n-paraffin hydrocarbons with water. Examples of the systems in which considerable miscibility of the components occurs are: n-butanol-water, aniline-water, and acetonitrile-n-paraffins. For instance, the solubility of aniline in water at 90°C is 6.4 g/100 mℓ.

For calculating p_1^0 and p_2^0 at the boiling temperatures of the respective azeotropes, the Antoine equation is recommended.[43] The constants of the Antoine equation, viz. $\log p^\circ = A - B/(C + t)$, where p° is expressed in millimeters Hg and t is in this case the boiling temperature of the azeotrope (°C), are compiled for quite a large number of organic compounds.[44,45]

For example, the activities of the components, α_1 and α_2 in the systems aniline-water[2] and n-butanol-water were found to be: 0.315, 1.00, and 0.649 and 0.9824, respectively.

The relations, thus described, can be easily extended to ternary and multicomponent heteroazeotropes.

THE PREDICTION OF AZEOTROPIC DATA IN TERNARY SYSTEMS

To separate nonideal liquid mixtures by fractional distillation, it is important to establish, among other things, whether the mixtures to be separated form azeotropes. In view that the experimental methods for determining the azeotropic composition are rather difficult and time-consuming operations, expecially in the case of azeotropes containing three and more components, several computational methods were developed to predict the composition of ternary and multicomponent azeotropes.

In general, two approaches in this area may be distinguished. The first one, originated by Haase[46,47] and developed by Malesinski,[48] is based on the theory of regular solutions and is restricted to homoazeotropes. The Malesinski method makes it possible to predict the appearance of ternary azeotropes of various types and to calculate their composition and boiling temperatures. The second approach, which is not limited only to the azeotropic points, is based on the use of various equations of empirical and semiempirical nautre, that relate the liquid-phase activity coefficients to the composition of the liquid phase.

The Azeotropic Data of a Ternary Homoazeotrope from the Binary Azeotropic Data

The composition of a ternary homoazeotrope (1,2,3) under isobaric conditions, when the vaporization entropies of the components are equal, is related to the two pairs of binary azeotropes by the equations.

Pairs (1,2) and (2,3):

$$x_1 = \frac{x_1^{(1,2)} + a x_3^{(2,3)}}{1 - ab} \tag{15}$$

$$x_3 = \frac{x_3^{(2,3)} + b x_1^{(1,2)}}{1 - ab} \tag{16}$$

where

$$a = \frac{z_{13} - z_{23} - z_{12}}{2z_{12}} \quad \text{and} \quad b = \frac{z_{13} - z_{23} - z_{12}}{2z_{23}}$$

$x_1^{(1,2)}$ and $x_3^{(2,3)}$ are the mole fractions of components 1 and 3 in the binary azeotropes (1,2) and (2,3), respectively; x_1 and x_3 are the mole fractions of the above components in the ternary azeotrope and z_{12}, z_{13}, and z_{23} are the half-values of the symmetrical azeotropic range.

When the vaporization entropies of the components are not equal, the regular solution constants, A_{12}, A_{13}, and A_{23} take the place of z_{12}, $z_{13}z$, and z_{23}, respectively.

Pairs (1,2), and (2,3):

$$x_1 = \frac{x_1^{(1,3)} + cx_2^{(2,3)}}{1 - cd} \tag{17}$$

$$x_2 = \frac{x_2^{(2,3)} + dx_1^{(1,3)}}{1 - cd} \tag{18}$$

where

$$c = \frac{z_{12} - z_{23} - z_{13}}{2z_{13}} \quad \text{and} \quad d = \frac{z_{12} - z_{23} - z_{13}}{2z_{23}}$$

Pairs (1,2) and (1,3):

$$x_2 = \frac{x_2^{(1,2)} + ex_3^{(1,3)}}{1 - ef} \tag{19}$$

$$x_3 = \frac{x_3^{(1,3)} + fx_2^{(1,2)}}{1 - ef} \tag{20}$$

where

$$e = \frac{z_{23} - z_{13} - z_{12}}{2z_{12}} \quad \text{and} \quad f = \frac{z_{23} - z_{13} - z_{12}}{2z_{13}}$$

The values of z_{12}, z_{13}, and z_{23} should be computed from Equation 1.

The ternary system is zeotropic if, for example, the composition of one component of the system is zero or takes a negative value.

In the case when one binary system constituting the ternary system is zeotropic, the respective z_{ij} value may be estimated from that of any close member of a homologous series, its isomers or closely related substances.

The sources of error in the calculated composition of the ternary azeotrope are similar to those of the binary azeotrope.

The boiling temperature of a ternary homoazeotrope, T, with component 2 as the reference component, can be computed from the equation

$$T = T_2 - \frac{\delta_2^{(1,2)} + \left(\frac{\delta_2^{(1,2)}}{z_{12}} \cdot \frac{\delta_2^{(2,3)}}{z_{23}}\right)^{1/2} (z_{13} - z_{23} - z_{12}) + \delta_2^{(2,3)}}{1 - \frac{(z_{13} - z_{23} - z_{12})^2}{4z_{12}z_{23}}} \tag{21}$$

where $\delta_2^{(1,2)}$ and $\delta_2^{(2,3)}$ are the azeotropic depressions or elevations in the azeotropes (1,2) and (2,3) in relation to component 2, e.g. $\delta_2^{(1,2)} = T_2 - T^{(1,2)}$: $T^{(1,2)}$ is the boiling temperature of the azeotrope (1,2).

By convention, the $\delta_i^{(i,j)}$ and z_{ij} are positive for a positive azeotrope and negative for a negative one.

By interchanging the components, the boiling temperature of the ternary homoazeotrope may be computed from the two remaining sets of the pairs of the components.

For mixtures which exactly fulfil the requirements for regular solutions, the result is independent of the choice of the reference component. If there are deviations from regularity, then the values of z_{ij} or $\delta_i^{(i,j)}$ depend on which experimentally determined quantity was used in the calculations.

For positive azeotropes Equations 15 to 21 usually give good results for the calculated azeotropic data.[23] In the case of positive-negative (saddle) azeotropes the agreement between the calculated and observed azeotropic data is less satisfactory. But this is understandable in view that these systems are complex mixtures, which often contain polar and associated components, e.g. alcohols and low-molecular fatty acids.

For the series of ternary saddle systems acetic acid-pyridine(2-picoline)-n-paraffins, the results obtained for the calculated azeotropic compositions improved considerably, when the corrections for the association (dimerization) of acetic acid in the vapor phase were taken into account.[49]

Equations 15 to 21 made it possible to predict the existence of a large number of ternary saddle (positive-negative) azeotropes that may appear in the course of fractional distillation of certain fractions of coal tar.[24]

It has been found, for instance, that ternary saddle azeotropes are formed in the series of ternary systems aniline-phenol-n-paraffins (ranging from nonane to tetradecane). Only one saddle azeotrope of this type was examined in the system aniline(1)-phenol(2)-tridecane(3).[50] For this system the differences between the calculated and observed azeotropic compositions, Δx_1, Δx_2, and Δx_3, and that of the boiling temperatures of the azeotrope, ΔT, are: 1.0, −2.1, and 1.0 mole %, and −0.11°C, respectively.

In general, good results for ternary homoazeotropes are obtained for systems which show moderate deviations from regularity, any of the binary azeotropes is close to what is called a tangent azeotrope,[23] and the differences between the boiling temperature of the ternary azeotrope and those of the components are not large.[24]

It should be also mentioned that two empirical correlations for the azeotropic composition in the series of ternary saddle azeotropes were proposed by Zeiborak.[51]

Composition of a Ternary Homoazeotrope from Vapor-Liquid Equilibrium Data of the Binary Systems

A general method for predicting the vapor-liquid equilibrium data in ternary and multicomponent systems can be restricted to the azeotropic point by the use of the relative volatility, α_{ij}, defined as

$$\alpha_{ij} = \frac{y_i x_j}{x_i y_j} = \frac{p_i^0 \gamma_i}{p_j^0 \gamma_j} = 1 \qquad (22)$$

In relation (Equation 22) p_i^0 and p_j^0 are the vapor pressures of components i and j at the boiling temperature of the azeotrope, and γ_i and γ_j are the liquid-phase activity coefficients of these components.

The method involves minimization of the function, f, which for a ternary azeotrope becomes

$$f = |\alpha_{13} - 1| + |\alpha_{23} - 1| \qquad (23)$$

α_{13} and α_{23} are the relative volatilities of the pairs of components (1,3) and (2,3), respectively.

The value of the function, f, sufficiently close to zero, corresponds to the azeotropic composition.

This method was used by Aristovicz and Stepanova[52] for the calculation of the azeotropic composition in 19 ternary systems and 1 quaternary system. The results obtained were good. In this procedure the liquid-phase activity coefficients were correlated by the Wilson equation.[53,54]

For ternary and multicomponent heteroazeotropes the above procedure remains essentially unchanged but requires the use of an equation that is applicable to partially miscible systems, e.g. the NRTL (Non-Random, Two-Liquid) equation.[55]

The Azeotropic Data of a Ternary Heteroazeotrope

On the basis of the arguments similar to those described previously, which are applicable to heteroazeotropic systems of any number of components, we can obtain the expressions for the composition and the boiling temperature of a ternary heteroazeotrope.

For the case in which the components are immiscible in each other, the expressions for y_1, y_2, and T^{Az} of a ternary heteroazeotrope are

$$y_1 = \frac{p_1^0}{p_1^0 + p_2^0 + p_3^0} \qquad (24)$$

$$y_2 = \frac{p_2^0}{p_1^0 + p_2^0 + p_3^0} \qquad (25)$$

and

$$P = p_1^0 + p_2^0 + p_3^0 \qquad (26)$$

where p_1^0, p_2^0 and p_3^0 are the vapor pressures of the pure components at the boiling temperature of the azeotrope, P is the total pressure of the mixture at equilibrium, and y_1 and y_2 are the mole fractions of component 1 and 2 in the vapor of the azeotrope, respectively.

Other cases, in which the condition of complete immiscibility of the components is not fulfilled, may be described by introducing the activities of the components, α_1, α_2, and α_3.

The procedure, which is straightforward, and involves the multiplication of each p^0_i by α_i in expressions (24-26), will not be reproduced here.

The activities, α_1, α_2, and α_3 are related to P, p_1^0, p_2^0, and p_3^0 by

$$\alpha_1 = \frac{P - p_2^0 \alpha_2 - p_3^0 \alpha_3}{p_1^0} \qquad (27)$$

$$\alpha_2 = \frac{Py_2}{p_2^0} \qquad (28)$$

and

$$\alpha_3 = \frac{Py_3}{p_3^0} \qquad (29)$$

Only one series of ternary heteroazeotropes with three liquid phases has been investigated to date.[6] The heteroazeotropes occur in the systems nitromethane-water-n-paraffins. For this series satisfactory agreement was obtained between the calculated (using Equations 24 and 25) and observed azeotropic compositions, but the calculated boiling temperatures of the azeotropes by Equation 26 were found to be much lower than those observed. Low calculated boiling temperatures are due to the relatively high miscibility of nitromethane in water at the respective boiling temperatures of the azeotropes. The difference is lower for the systems with higher-boiling hydrocarbons and tends to the difference between the calculated and observed boiling temperatures for the nitromethane-water system.

It is interesting to note that in some cases the behavior of a series of ternary heteroazeotropes, in which two components remain unchanged for the series, may be described by the activities α_1, α_2, and α_3, which are common for each component within the whole series. This case is exemplified by the series of ternary heteroazeotropes water-pyridine-n-paraffins.[56]

Studies of ternary heteroazeotropes based on the theory of regular solutions were made by Malesinska and Malesinski[57] and by Stecki.[58]

All the examined heteroazeotropes were found to be positive. The existence of negative heteroazeotropes is rather doubtful.

REFERENCES

1. **Dalton, J.,** *Mem. Manchester Phil. Soc.*, 5, 585, 1802; *Ann. Phil.*, 9, 186, 1817.
2. **Wade, J. and Merriman, R. W.,** *J. Chem. Soc. Trans.*, 99, 997, 1911.
3. **Swietoslawski, W.,** *Ebulliometric Measurements*, Reinhold, New York, 1945.
4. **Lecat, M.,** *Compt. Rend.*, 183, 880, 1926.
5. **Swietoslawski, W.,** *Rocz. Chem.* 26, 632, 1952; *Bull. Acad. Polon. Sci. Cl. III*, 1, 63, 1953.
6. **Malesinska, B. and Malesinski, W.,** *Bull. Acad. Polon. Sci. Ser. Sci. Chim.*, 11, 475, 1963.
7. **Ewell, R. L. and Welch, L. M.,** *Ind. Eng. Chem.*, 37, 1244, 1945.
8. **Lang, H.,** *Z. Physik. Chem.*, 196, 278, 1950.
9. **Swietoslawski, W. and Trabczynski, W.,** *Bull. Acad. Polon. Sci. Cl. III*, 3, 333, 1955.
10. **Galska-Krajewska, A.,** *Bull. Acad. Polon. Sci. Ser. Sci. Chim.*, 10, 45, 51, 1962.
11. **Zieborak, K.,** *Bull. Acad. Polon. Sci. Cl. III*, 3, 53, 1955.
12. **Kurtyka, Z. M.,** *J. Chem. Eng. Data*, 16, 310, 1971.
13. **Kurtyka, Z.,** *Bull. Acad. Polon. Sci. Ser. Sci. Chim.*, 9, 741, 1961.
14. **Zieborak, K. and Wyrzykowska-Stankiewicz, D.,** *Bull. Acad. Polon. Sci. Ser. Sci. Chim.*, 8, 137, 1960.
15. **Orszagh, A., Lelakowska, J., and Beldowicz, M.,** *Bull. Acad. Polon. Sci. Ser. Sci. Chim. Geol. et Geogr.*, 6, 419, 1958.
16. **Orszagh, A., Lelakowska, J., and Radecki, A.,** *Bull. Acad. Polon. Sci. Ser. Sci. Chim. Geol. et Geogr.*, 6, 605, 1958.
17. **Ostwald, W.,** *Lehrbuch der Allgemeinen Chimie*, Vol. 2, Engelman, Leipzig, 1899.
18. **Ryland, G.,** *Am. Chem. J.*, 22, 384, 1899; *Chem. News*, 81, 15, 42, 50, 1900.
19. **Lecat, M.,** *L'Azeotropisme*, Lamartin, Bruxelles, 1918.
20. **Lecat, M.,** *Tables Azeotropiques*, Vol. 1, L'Auteur, Bruxelles, 1949.
21. **Horsley, L. H.,** *Azeotropic Data-III*, American Chemical Society, Washington, D.C., 1973.
22. **Swietoslawski, W.,** *Azeotropy and Polyazeotrophy*, Pergamon Press, Oxford, London, 1963.
23. **Malesinski, W.,** *Azeotropy and Other Theoretical Problems of Vapour-Liquid Equilibrium*, Wiley-Interscience, New York, 1965.
24. **Kurtyka, Z. M.,** *Azeotropy and Its Applications*, to be published.
25. **Swietoslawski, W.,** *Bull. Acad. Sci. Polon. Ser. A*, 19, 29, 1950, *Przem. Chem.*, 7, 363, 1951.
26. **Skolnik, H.,** *Ind. Eng. Chem.*, 40, 442, 1948.
27. **Malesinski, W.,** *Bull. Acad. Polon. Sci. Cl. III*, 3, 601, 1955; 4, 295, 1956.
28. **Zawisza, A.,** *Bull. Acad. Polon. Sci. Ser. Sci. Chim.*, 9, 141, 1961.
29. **Lecat, M.,** *Azeotropisme et Distillation, Traite de Chimie Organique, Vol. I*, V. Grignard, Ed., Mason et Cie., Paris, 1935.
30. **Prigogine, I., and Defay, R.,** *Chemical Thermodynamics*, translated by D. H. Everett, Longmans, Green, London, 1954.
31. **Mair, B. J., Glasgow, A. R., and Rossini, F. D.,** *J. Res. Bur. Std.*, 27, 39, 1941.
32. **Horsley, L. H.,** *Anal. Chem.*, 19, 508, 1947.
33. **Meissner, H. P. and Greenfield, S. H.,** *Ind. Eng. Chem.*, 40, 438, 1948.
34. **Seymour, K. M.,** Abstracts 50-I, 110th National Meeting of the American Chemical Sciety, Chicago, September 1946.
35. **Johnson, A. I. and Madonis, J. A.,** *Can. J. Chem. Eng.*, 37, 71, 1959.
36. **Seymour, K. M., Carmichael, R. H., Carter, J., Ely, J., Isaacs, E., King, J., Taylor, R., and Northern, T.,** *Ind. Eng. Chem. Fundam.*, 16, 200, 1977.
37. **Orszagh, A.,** *Rocz. Chem.*, 29, 623, 636, 1955.
38. **Kireev, V. A.** *Acta Physicochim. URSS*, 14, 371, 1941.
39. **Kurtyka, Z. M. and Kurtyka, A.,** *Ind. Eng. Chem. Fundam.*, 19, 225, 1980.
40. **Kurtyka, Z. M. and Kurtyka, Z.,** *Ind. Eng. Chem. Fundam.*, in press.
41. **Francis, A. W.,** *Critical Solution Temperatures*, American Chemical Society, Washington, D.C. 1961.

42. **Malesınska, B . and Malesinski, W.,** *Bull. Acad. Polon. Sci. Ser. Sci. Chim.*, 11, 469, 1963.
43. **Antoine, C.,** *Compt. Rend.*, 107, 681, 836, 1143, 1888.
44. **Hala, E., Wichterle, I., Polak J., and Boublik, T.,** *Vapour-Liquid Equilibrium Data at Normal Pressures*, Pergamon Press, Oxford, London, 1968.
45. **Hirata, M., Ohe, S., and Nagahama, K.,** *Computer-Aided Data Book of Vapour-Liquid Equilibria*, Kodansha-Elsevier, Tokyo, Amsterdam, 1975.
46. **Haase, R.,** *Z. Physik. Chem.*, 195, 362, 1950.
47. **Haase, R.,** *Termodynamik des Mischphasen*, Springer-Verlag, Berlin, 1956.
48. **Malesinski, W.,** *Bull. Acad. Polon. Sci. Cl. III*, 4, 701, 709, 1956; 5, 177, 183, 1957.
49. **Zeiborak, K. and Wyrzykowska-Stankiewicz, D.,** *Bull. Acad. Polon. Sci. Ser. Sci. Chim. Geol. et Geogr.*, 6, 755, 1958.
50. **Stadnicki, J. S.,** *Bull. Acad. Polon. Sci. Ser. Sci. Chim.*, 10, 357, 1962.
51. **Zieborak, K.,** *Bull. Acad. Polon. Sci. Cl. III*, 3, 531, 1955.
52. **Aristovicz, V. Y. and Stepanova, E. I.,** *Zhur. Prikl. Khim.*, 43, 2192, 1970.
53. **Wilson, G. M.,** *J. Am. Chem. Soc.*, 86, 127, 1964.
54. **Prausnitz, J. M.,** *Molecular Thermodynamics of Fluid-Phase Equilibria*, Prentice Hall, Englewood Cliffs, New Jersey, 1969.
55. **Renon, H. and Prausnitz, J. M.,** *A. I. Ch. E. Journal*, 14, 135, 1968.
56. **Trabczynski, W.,** *Bull. Acad. Polon. Sci. Ser. Sci. Chim. Geol. et Geogr.*, 6, 269, 1958.
57. **Malesinska, B. and Malesinski, W.,** *Bull. Acad. Polon. Sci. Ser. Sci. Chim.*, 12, 861, 867, 1964.
58. **Stecki, J.,** *Bull. Acad. Polon. Sci. Cl. III*, 5, 421, 1957; *Ser. Sci. Chim. Geol. et Geogr.*, 6, 47, 1958.

TABLES OF AZEOTROPES AND ZEOTROPES

In Tables 1, 2, and 3 the different types of azeotropes are identified as:

1. Homoazeotrope, positive; no marking
2. Homoazeotrope, negative; N
3. Homoazeotrope, saddle or positive-negative; S
4. Heteroazeotrope, positive, with two phases; H
5. Heteroazeotrope, positive, with three phases; H-3

Throughout Tables 1,2, and 3 the azeotropic composition is expressed as weight percent (wt %). In Table 1 only the weight percent of component 2, x_2, the variable component, is listed. However, in Tables 2, and 3 the weight percents of all of the components are listed.

Compounds are listed in the following tables according to the empirical formula convention employed by Chemical Abstracts. As further assistance in locating particular systems the following index has been arranged in alphabetical order with respect to component X_1 of each system. The entry number for the particular system is in the second column of the list.

Table 1
BINARY SYSTEMS

Table 1 (Continued)
BINARY SYSTEMS

Component X_1	Entry No.
Butyronotile	764
Camphene	1715
Capric acid	1722
Caproic acid	1338
Capronitrile	1325
Caprylic acid	1622
Carbon disulfide	204
Carbon tetrachloride	185
Carvacol	1696
Carvone	1703
Chloroacetic acid	354
Chloroform	232
m-Cresol	1473
o-Cresol	1462
p-Cresol	1485
Cumene	1655
Cyclohexanol	1327
Cyclohexanone	1297
Diethylene glycol	1008
Doxane	811
Dipropylene glycol	1386
Enanthic acid	1514
Ethanol	460
2-Ethoxyethanol	994
Ethyl	
acetate	821
acetoacetate	1301
alcohol	460
aniline	1592
benzene	1551
bromoacetate	750
carbamate	631
chloroacetate	756
formate	563
fumarate	1613
lactate	1136
maleate	1616
methyl sulfide	712
nitrate	455
oxalate	1319
phenol, para	1554
pyruvate	1071
salicylate	1651
succinate	1619
Ethylene glycol	476
Ethylene sulfide	411
Ethylenediamine	522
Ethylidine diacetate	1313
Formic acid	242
2-Furaldehyde	1044
Glycol diacetate	1324
Glycol monoacetate	865
Glycerol	687
Guaiacol	1497
1-Heptanol	1521
n-Heptyl alcohol	1521
Hexachloroethane	337
1-Hexanol	1357
n-Hexyl alcohol	1357
Hydrogen chloride	14
Hydrogen cyanide	11
Hydrogen fluoride	15
Indole	1527
Iodobenzene	1168
Iodoethane	420
Iodomethane	272
Isoamyl alcohol	1148
Isoamyl benzoate	1735
Isoamyl formate	1345
Isoamyl oxalate	1739
Isobutyl nitrate	902
Isobutyronitrile	767
Isobutryic acid	829
Isopropyl alcohol	663
Isopropyl lactate	1354
Isopropyl methyl sulfide	1038
Isovaleric acid	1105
Levulinic acid	1074
2,4-Lutidine	1501
Menthol	301
Mesitol	1662
Mesitylene	1656
Methanol	301
Methyl	
acetate	567
acetoacetate	1085
acetophenone, para	1643
alcohol	301
aniline	1503
anisole, para	1562
butyrate	1120
chloroacetate	534
disulfide	520
formate	414
fumarate	1284
lactate	876
maleate	1290
1-Methylnaphthalene	1724
2-Methylnaphthalene	1728
2-Methyl-2-propanol	949
2-Methylthiophene	1063
3-Methylthiophene	1067
2-Octanol	1630
sec-Octyl alcohol	1630
Naphthalene	1679
Nitrobenzene	1159
Nitroethane	453
Nitromethane	276
1-Nitropropane	654
o-Nitrotoluene	1426
p-Nitrotoluene	1437
Pelargonic acid	1669
2-Pentanone	1096
Perfluorobutyric acid	724
Phenethyl alcohol	1568
o-Phenetidine	1600
p-Phenetidine	1607
Phenol	1190

Table 1 (Continued)
BINARY SYSTEMS

Component X_1	Entry No.	Component X_1	Entry No.
Phenyl acetate	1538	Pyridine	1056
Phenyl benzoate	1741	Pyrogallol	1252
Phenyl ether	1732	Pyrrol	734
o-Phenylenediamine	1278	Pyruvic acid	527
3-Phenylpropanol	1664	Qunialdine	1683
Phosphorus oxychloride	16	Quinoline	1635
Phosphorus trichloride	17	Resorcinol	1238
2-Picoline	1273	Silicon tetrachloride	20
3-Picoline	1276	Tetrachlorothylene	327
Pinacol	1377	Tetrahydrothiophene	888
1-Propanethiol	716	Thioacetic acid	380
1-Propanol	673	Thiophene	727
Propioamide	593	Thymol	1707
Propionic acid	574	Tin chloride	26
Propionitrile	541	Toluene	1442
Propiophenone	1648	o-Toluidine	1505
Propyl		Trichloroacetic acid	346
acetate	1123	Trichloroethylene	342
alcohol	673	Trichlorofluoromethane	173
benzene	1659	Trinitromethane	176
benzoate	1690	Triethylene glycol	1391
formate	858	Valeric acid	1128
isovalerate	1626	Water	28
lactate	1350	2,4-Xylenol	1576
nitrite	661	3,4-Xylenol	1582
propionate	1348	2,4-Xylidine	1595
Pseudocumene	1660		

Table 2
TERNARY SYSTEMS

Component X_1	Entry No.	Component X_1	Entry No.
Acetic acid	108	Hydrogen chloride	5
Acetone	141	Hydrogen cyanide	3
Acetonitrile	106	Hydrogen fluoride	7
Aniline	167	Isobutyl alcohol	161
Argon	1	Isobutyl lactate	176
1-Butanol	158	Isopropyl alcohol	151
2-Butanone	154	Methanol	101
Butyric acid	156	Methyl formate	
Carbon tetrachloride	92	Nitromethane	129
Chlorine trifluoride	4	Phenol	163
Chloroform	93	1-Propanol	152
m-Cresol	175	1-Propanol, 2-methyl	161
m,p-Cresol (mixture)	172	2-Propanol	151
p-Dioxane	157	Propionic acid	145
Ethanol	130	Propyl lactate	171
Ethyl benzene	166, 177	Pyridine	162
Ethylene glycol	135	Silicon tetrafluoride	8
Hydrogen bromide	2	Trichloroethylene	107

Table 3
QUATERNARY AND QUINARY SYSTEMS

Component X_1	Entry No.	Component X_1	Entry No.
Acetic acid	18	Hydrogen cyanide	1
Acetone	20	Isopropyl alcohol	21
Chloroform	16	Water	3

BINARY SYSTEMS

No.	System	B.P. (°C)	Azeotropic data: Compn. X_2 (wt. %)	Azeotropic data: B.P. (°C)
	Aluminum chloride	183		
1	-tantalum chloride (2)	242	90.4	235
	Boron fluoride	−100		
2	-boron hydride	−92	22.8	−106
3	-acetonitrile, N	81.6	38	101
4	-methyl formate, N	31.9	47	91
5	-methyl ether, N	−21	40	127
6	-ethyl formate, N	54.1	52	102
7	-methyl acetate, N	57.1	52	110
8	-ethyl ether, N	34.5	52	125
9	-ethyl propionate, N	99.15	60	116
	Boron hydride	−92.5		
10	-hydrogen chloride	−85	36	−94
	Hydrogen cyanide	26		
11	-methyl alcohol	64.7	Zeotropic	
12	-methyl formate	31.7	48	24
13	-ethyl nitrite	17.4	85	16.5
	Hydrogen chloride	−85		
14	-methyl ether, N	−22	62	−2
	Hydrogen fluoride	19.4		
15	-ethyl ether, N	34.5	60	74
	Phosphorus oxychloride	107.2		
16	-titanium tetrachloride, N	136.5	53.4	143.2
	Phosphorus trichloride	76		
17	-cyclohexane	80.75	Zeotropic	
18	-2,3-dimethylpentane	80.5	27	74.2
19	-2,2,3-trimethylbutane	80.9	23	74.5
	Silicon tetrachloride	56.9		
20	-carbon tetrachloride	76.75	Zeotropic	
21	-chloroform	61.0	30	55.6
22	-nitromethane	101	6	53.8
23	-acetonitrile	82	9.4	49.0
24	-acrylonitrile	79	11	51.2
25	-propionitrile	97	8	55.6
	Tin chloride	113.85		
26	-toluene	110.7	48	109.15
27	-2,5-dimethylhexane	109.4	60	107.5
	Water	100		
28	-hydrogen chloride, N	−85	20.22	108.58
29	-hydrogen bromide, N	−73	47.5	126
30	-hydrogen iodide, N	−34	57	127
31	-hydrogen fluoride, N	19.4	35.6	111.35
32	-nitric acid, N	86	67.4	120.7
33	-hydrogen peroxide	152.1	Zeotropic	
34	-hydrazine, N	113.8	67.7	120
35	-carbon tetrachloride, H	76.75	95.9	66
36	-carbon disulfide, H	46.25	97.2	42.6
37	-chloroform, H	61	97.2	56.1
38	-formic acid, N	100.75	77.4	107.2
39	-nitromethane, H	101.2	76.4	83.59
40	-tetrachloroethylene, H	121	82.8	88.5
41	-trichloroethylene	86.2	83	73.4
42	-acetonitrile, H	80.1	83.7	76.5
43	-acetic acid	118.1	Zeotropic	
44	-acetamide, H	221.2	Zeotropic	
45	-nitroethane, H	114.07	71.5	87.22
46	-ethyl nitrate	87.68	78	74.35
47	-ethyl alcohol	78.32	96	78.17
48	-methyl sulfate	189.1	27	98.6
49	-acrylonitrile, H	77.2	85.7	70.6
50	-acrolein, H	52.8	97.4	52.4
51	-acetone	56.1	Zeotropic	
52	-allyl alcohol	96.9	72.3	88.9
53	-propionaldehyde, H	47.9	98	47.5
54	-ethyl formate	54.2	95	52.6
55	-propionic acid	141.1	17.7	99.9
56	-trioxane	114.5	70	91.4
57	-l-chloropropane	46.6	97.8	44
58	-isopropyl alcohol	82.3	87.4	80.3
59	-propyl alcohol, 740 mm	97.3	71.7	87
60	-perfluorobutyric acid	122.0	29	97
61	-crotonic acid	189	2.2	99.9
62	-methyl acrylate	80	92.8	71
63	-ethyl chloroacetate	143.5	54.9	95.2
64	-butyronitrile, H	117.6	67.5	88.7
65	-isobutyronitrile, H	103	77	82.5
66	-ethyl vinyl ether, H	35.5	98.5	34.6
67	-butyric acid	163.5	3	99.4
68	-ethyl acetate	77.15	91.53	70.38
69	-isopropyl formate	68.8	97	65.0
70	-propyl formate	80.9	97.7	71.6
71	-methyl lactate	144.8	20	99
72	-butyl alcohol, H	117.4	57.5	92.7
73	-*sec*-butyl alcohol	99.5	73.2	87.0
74	-pyridine	115.5	58.7	93.6
75	-furfuryl alcohol	169.35	20	98.5
76	-furfurylamine	144	26	99
77	-isoprene	34.1	99.86	32.4
78	-cyclopentanone	130.8	57.6	94.6
79	-allyl acetate	104.1	85.3	83
80	-cyclopentanol	140.85	42	96.25
81	-valeraldehyde, H	103.3	81	83
82	-butyl formate	106.6	85.5	83.8
83	-isopropyl acetate	88.6	89.4	76.6
84	-isovaleric acid	176.5	18.4	99.5
85	-methyl butyrate	102.65	88.5	82.7
86	-methyl isobutyrate	92.3	93.2	77.7
87	-valeric acid	188.5	11	99.8
88	-piperidine	105.8	65	92.8
89	-n-pentane, H	36.1	98.6	34.6
90	-n-amyl alcohol, H	137.8	45.6	95.8
91	-*tert*-amyl alcohol	102.25	72.5	87.35
92	-2-pentanol	119.3	63.5	91.7
93	-N-methylbutylamine	91.1	85	82.7
94	-chlorobenzene	131.8	71.6	90.2
95	-nitrobenzene, H	210.85	—	98.6
96	-benzene, H	80.1	91.17	69.25
97	-phenol	182	9.2	99.52
98	-aniline, H, 742 mm	184.3	19.2	98.6
99	-2-picoline	129.5	52	93.5
100	-3-picoline	144.1	40	97
101	-4-picoline	144.3	37.2	97.35
102	-2,5-dimethylfuran	93.3	88.3	77.0
103	-cyclohexene, H	82.75	91.07	70.8
104	-ethyl crotonate	137.8	62	93.5
105	-ethylene glycol diacetate	190.8	15.4	99.7
106	-butyl chloroacetate	181.9	24.5	98.12
107	-cyclohexane, H	80.8	91.6	69.5
108	-amyl formate	132	71.6	91.6
109	-butyl acetate	126.2	71.3	90.2
110	-ethyl butyrate	120.1	78.5	87.9
111	-isoamyl formate	124.2	79	90.2
112	-isobutyl acetate	117.2	83.5	87.4
113	-isopropyl propionate	110.3	80.1	85.2
114	-propyl propionate	122.1	77	88.9
115	-paraldehyde, H	124	71.5	90
116	-n-hexane, H	68.7	94.4	61.6
117	-butyl ethyl ether, H	92.2	88.1	76.6
118	-n-hexyl alcohol, H	157.1	32.8	97.8
119	-acetal	103.6	85.5	82.6
120	-pinacol	174.35	Zeotropic	
121	-toluene, H	110.7	86.5	84.1
122	-anisole	153.85	59.5	95.5
123	-benzyl alcohol, H	205.2	9	99.9
124	-guaiacol	205.0	12.5	99.5
125	-2,6-lutidine, H	144.0	48.2	96.02
126	-o-toluidine	199.7	84.6	—
127	-p-toluidine, H	200.4	86.2	—
128	-2-heptanone	149	52	95
129	-3-heptanone	147.6	57.8	94.6
130	-4-heptanone	143.7	59.5	94.3
131	-ethyl valerate	145.45	60	94.5
132	-isoamyl acetate	142	63.7	93.8
133	-isobutyl propionate	136.85	47.8	92.75
134	-n-heptane, H	98.4	87.1	79.2
135	-benzyl formate	202.3	20	99.2
136	-methyl benzoate	199.45	20.8	99.08
137	-phenyl acetate	195.7	24.9	98.9
138	-ethylbenzene,H	136.2	67.0	92.0
139	-m-xylene, H	139	64.2	92
140	-N-ethylaniline	204.8	16.1	99.2
141	-1-octene, H	121.28	71.3	88.0
142	-hexyl acetate	171.0	39	97.4
143	-isoamyl propionate	160.3	51.5	96.55
144	-isobutyl butyrate	156.8	54	96.3
145	-isobutyl isobutyrate	147.3	60.6	95.5
146	-propyl isovalerate	155.8	54.8	96.2
147	-n-octane, H	125.7	75.5	89.6
148	-isooctane, H	99.3	88.9	78.8
149	-butyl ether, H	142.6	67	92.9
150	-n-octyl alcohol, H	195.15	10	99.4
151	-dibutylamine, H	159.6	49.5	97

No.	System	B.P. (°C)	Azeotropic data Compn. X_2 (wt. %)	Azeotropic data B.P. (°C)
152	-quinoline, H	237.3	3.4	—
153	-ethyl benzoate	212.4	16.0	99.4
154	-cumene, H	152.4	56.2	95
155	-mesitylene, H	164.6	—	96.5
156	-triallylamine, H	151.1	62	95
157	-isoamyl butyrate	178.5	36.5	98.05
158	-isobutyl carbonate	190.3	26	98.6
159	-n-nonane, H	150.8	18	94.8
160	-naphthalene, H	218.0	16	98.8
161	-methyl phthalate, H	283.2	2.5	99.95
162	-nicotine, H	—	2.5	99.85
163	-camphene, H	159.6	—	96.0
164	-n-decane, H	173.3	—	97.2
165	-n-undecane, H	194.5	4.0	98.85
166	-o-phenyl phenol	—	1.25	99.95
167	-phenyl ether, H	259.3	3.25	99.33
168	-ethyl phthalate, H	298.5	2.0	99.98
169	-isoamyl benzoate, H	262.3	4.4	99.9
170	-n-dodecane, H	214.5	2	99.45
171	-dihexylamine, H	239.8	7.2	99.8
172	-tributylamine, H	213.9	20.3	99.65
	Trichlorofluoromethane	24.9		
173	-acetaldehyde	20.2	45	15.6
174	-methyl formate	32	18	20.0
175	-2-methylbutane	27	8	23.16
	Trichloronitromethane	111.9		
176	-acetic acid	118.1	19.5	107.65
177	-ethyl alcohol	78.3	66	77.5
178	-isopropyl alcohol	82.4	65	81.95
179	-propyl alcohol	97.2	41.5	94.05
180	-isoamyl alcohol	131.9	7	111.15
181	-n pentanol	119.8	17	108.0
182	-toluene	110.75	Zeotropic	
183	-methylcyclohexane	101.15	73	100.8
184	-n-heptane	98.4	93	98.32
	Carbon tetrachloride	76.75		
185	-carbon disulfide	46.25	Zeotropic	
186	-chloroform	62.1	Zeotropic	
187	-formic acid	100.7	18.5	66.65
188	-nitromethane	101.2	17	71.3
189	-methyl alcohol	64.7	20.56	55.7
190	-acetonitrile	81.6	17	65.1
191	-acetic acid	118.1	1.54	76
192	-ethyl alcohol	78.3	15.8	65.04
193	-acrylonitrile	77.3	21	66.2
194	-acetone	56.15	88.5	56.08
195	-propyl alcohol	97.25	7.9	73.4
196	-thiophene	84	Zeotropic	
197	-butyl nitrite	78.2	30	75.3
198	-butyl alcohol	117.75	2.4	76.55
199	-ethyl ether	34.6	Zeotropic	
200	-pyridine	115.5	Zeotropic	
201	-benzene	80.1	Zeotropic	
202	-n-heptane	98.4	Zeotropic	
203	-o-xylene	143.6	Zeotropic	
	Carbon disulfide	46.25		
204	-chloroform	61.2	Zeotropic	
205	-formic acid	100.75	17	42.55
206	-nitromethane	101.2	81.4	41.2
207	-methyl alcohol	64.7	29	39.8
208	-acetic acid	118.1	Zeotropic	
209	-propyl nitrite	47.75	38	40.15
210	-ethyl alcohol	78.3	9	42.6
211	-acetone	56.15	33	39.25
212	-propyl alcohol	97.1	5.5	45.65
213	-ethyl acetate	76.7	3	46.1
214	-n-pentane	36.15	89	35.7
215	-n-hexane	68.95	Zeotropic	
216	-toluene	110.7	Zeotropic	
	Bromodichloromethane	90.2		
217	-nitromethane	101.2	25	87.3
218	-methyl alchol	64.7	40	63.8
219	-ethyl alcohol	78.3	28	75.5
220	-ethyl acetate	77.1	12	90.55
221	-benzene	80.2	Zeotropic	
222	-cyclohexane	80.75	Zeotropic	
223	-n-hexane	68.8	Zeotropic	
	Bromoform	149.5		
224	-formic acid	100.75	48	97.4
225	-acetamide	221.15	Zeotropic	
226	-butyric acid	162.45	6.8	146.8

No.	System	B.P. (°C)	Azeotropic data Compn. X_2 (wt. %)	Azeotropic data B.P. (°C)
227	-phenol	182.2	Zeotropic	
228	-aniline	184.35	Zeotropic	
229	-toluene	110.65	Zeotropic	
230	-o-cresol	191.1	Zeotropic	
231	-α-pinene	155.8	25	146.5
	Chloroform	61.2		
232	-formic acid	100.75	15	59.15
233	-methyl alcohol	64.7	12.6	53.43
234	-ethyl alcohol	78.3	7	59.35
235	-acetone, N	56.5	21.9	64.4
236	-propyl alcohol	97.2	Zeotropic	
	Chloroform	61.2		
237	-p-dioxane	101	Zeotropic	
238	-cyclohexane	80.75	Zeotropic	
239	-methylcyclopentane	72.0	20	60.5
240	-n-hexane	68.7	16.5	60.4
241	-toluene	110.65	Zeotropic	
	Formic acid	100.75		
242	-nitromethane	101.22	54.5	97.07
243	-trichloroethylene	86.95	75	74.1
244	-tetrachloroethylene	121.1	50	88.15
245	-acetic acid	118.1	Zeotropic	
246	-nitroethane	114.2	Zeotropic	
247	-ethyl ether	34.6	Zeotropic	
248	-ethyl sulfide	92.2	65	82.2
249	-pyridine, N	115.5	38.6	127.43
250	-2-methylbutane	27.95	96	27.2
251	-n-pentane	36.15	80	34.2
252	-bromobenzene	156.1	32	98.1
253	-chlorobenzene	131.75	41	93.7
254	-fluorobenzene	84.9	73	73.0
255	-benzene	80.2	69	71.05
256	-aniline	184.35	Zeotropic	
257	-2-picoline, N	129	75	158.0
258	-cyclohexane	80,75	30	70.7
259	-methylcyclopentane	72.0	71	63.3
260	-n-hexane	68.95	72	60.6
261	-propyl sulfide	141.5	17	98.0
262	-isopropyl sulfide	120.5	38	93.5
263	-toluene	110.7	50	85.8
264	-o-chlorotoluene	159.3	17	100.2
265	-methylcyclohexane	101.1	53.5	80.2
266	-n-heptane	98.45	43.5	78.2
267	-styrene	145.8	27	97.75
268	-o-xylene	143.6	26	95.5
269	-m-xylene	139.0	28.2	92.8
270	-n-octane	125.8	37	90.5
	Bromomethane	3.65		
271	-methyl alcohol	64.7	0.45	3.55
	Iodomethane	42.5		
272	-methyl alcohol	64.7	4.5	37.8
273	-ethyl alcohol	78.3	3.2	41.2
274	-acetone	56.15	5	42.4
275	-n-hexane	68.85	Zeotropic	
	Nitromethane	101.2		
276	-methyl alcohol	64.7	90.9	64.4
277	-acetic acid	118.1	4	101.2
278	-ethyl alcohol	78.3	71	76.05
279	-propyl alcohol	97.15	51.6	89.09
280	-p-dioxane	101.35	43.5	100.55
281	-n-butyl alcohol	117.73	28.6	98.0
282	-n-pentane, H	36.07	99	35
283	-cyclohexane	80.75	73.5	69.5
284	-methylcyclopentane	72.0	77	64.2
285	-n-hexane, H	68.74	81.5	61.7
286	-toluene	110.75	45	96.5
287	-n-heptane, 748 mm Hg, H	98.4	64.4	79.7
288	-styrene	145.8	Zeotropic	
289	-o-xylene	144.3	Zeotropic	
290	-n-octane, 748 mm; H	125.75	44.8	90.23
291	-cumene	152.8	Zeotropic	
292	-mesitylene	164.6	Zeotropic	
293	-n-nonane, 748 mm; H	150.85	28.4	96.14
294	-n-decane, 748 mm; H	174.12	16.1	98.81
295	-n-undecane, 748 mm; H	194.5	9.3	100.01
296	-n-dodecane, 748 mm; H	216.0	4.2	100.60
	Methyl nitrate	64.8		
297	-methyl alcohol	64.65	27	52.5
298	-cyclohexane	80.75	23	61.0
299	-n-hexane	68.8	44	56.0
300	-n-heptane	98.4	Zeotropic	

BINARY SYSTEMS (Continued)

No.	System	B.P. (°C)	Azeotropic data: Compn. X_2 (wt. %)	Azeotropic data: B.P. (°C)
	Methyl alcohol	64.7		
301	-trichloroethylene	87	62	59.3
302	-bromoethane	38	94.7	34.9
303	-acetic acid	118.1	Zeotropic	
304	-acetone	56.15	88	55.5
305	-methyl acetate	57.1	81	53.5
306	-thiophene	84	83.6	59.71
307	-methyl acrylate	80	46	62.5
308	-p-dioxane	101.05	Zeotropic	
309	-ethyl sulfide	92.2	38	61.2
310	-pyridine	115.4	Zeotropic	
311	-cyclopentane	49.4	86	38.8
312	-isobutyl formate	97.9	5	64.6
313	-piperidine	106.4	Zeotropic	
314	-n-pentane	36.15	93	30.85
315	-chlorobenzene	132.0	Zeotropic	
316	-fluorobenzene	85.15	68	59.7
317	-benzene	80.1	60.9	57.5
318	-cyclohexane	80.7	63.6	53.9
319	-toluene	110.6	27.5	63.5
320	-methylcyclohexane	100.8	46	59.2
321	-n-heptane, H	98.45	48.5	59.1
322	-o-xylene	143.6	Zeotropic	
323	-n-octane	125.75	32.5	62.75
324	-n-nonane	150.7	16.6	64.1
325	-n-decane	173.8	Zeotropic	
326	-methyl *tert*-butyl ether	55.06	85.7	51.27
	Tetrachloroethylene	121.1		
327	-acetic acid	118.1	38.5	107.35
328	-acetamide	221.2	2.6	120.45
329	-ethylene glycol	197.4	6	119.1
330	-acetone	56.1	Zeotropic	
331	-propionic acid	140.9	8.5	119.1
332	-propyl alcohol	97.25	48	94.05
333	-n-butyl alcohol	117.7	32	110.0
334	-pyridine	115.4	48.5	112.85
335	-n-amyl alcohol	138.2	15	117.0
336	-toluene	110.75	Zeotropic	
	Hexachloroethane	185		
337	-trichloroacetic acid	196	15	181
338	-phenol	182.2	30	173.7
339	-aniline	184.35	34	176.75
340	-benzyl alcohol	205.15	12	182.0
341	-p-cresol	201.7	10	183.0
	Trichloroethylene	86.9		
342	-acetic acid	118.1	3.8	86.5
343	-benzene	80.2	Zeotropic	
344	-cyclohexane	80.7	83.4	80.5
345	-n-heptane	98.45	Zeotropic	
	Trichloroacetic acid	197.55		
346	-pentachloroethane	161.95	96.5	161.8
347	-naphthalene	218.05	Zeotropic	
348	-butylbenzene	183.1	80	181.3
	Bromoacetic acid	205.1		
349	-o-dichlorobenzene	179.5	84	177.0
350	-o-bromotoluene	181.5	82	179.0
351	-acetophenone	202.0	30	206.5
352	-butylbenzene	183.1	75	179.5
353	-cymene	176.7	85	174.7
	Chloroacetic acid	189.35		
354	-bromobenzene	156.1	89	154.3
355	-phenol	181.5	Zeotropic	
356	-m-bromotoluene	183.8	70	174
357	-p-bromotoluene	185.0	66	174.1
358	-styrene	145.8	86	144.8
359	-o-xylene	144.3	88	143.5
360	-m-xylene	139.2	93	139.05
361	-n-octane	125.75	Zeotropic	
362	-cumene	152.8	79	150.8
363	-mesitylene	164.6	83	162.0
364	-pseudocumene	168.2	66	162.8
365	-naphthalene	218.05	22	187.1
366	-cymene	176.7	58	169.0
367	-n-decane	173.3	58	165.2
368	-1,3,5-triethylbenzene	215.5	25	185.5
	Acetonitrile	81.6		
369	-acetic acid	118.1	Zeotropic	
370	-ethyl alcohol	78.3	56	72.5
371	-pyridine	115.5	Zeotropic	
372	-isoprene	34.1	97.6	33.7
373	-isopropyl acetate	89.5	40.0	79.5
374	-toluene	110.7	20	81.4
375	-ethylbenzene	136.2	Zeotropic	
376	-n-undecane	195.4	Zeotropic	
	Acetaldehyde	20.4		
377	-acetone	56.15	Zeotropic	
378	-ethyl ether	34.5	23.5	18.9
379	-benzene	80.1	Zeotropic	
	Thioacetic acid	89.5		
380	-benzene	80.15	Zeotropic	
381	-cyclohexane	80.75	Zeotropic	
382	-methylcyclopentane	72.0	Zeotropic	
	Acetic acid	118.1		
383	-nitroethane	114.2	70	112.4
384	-dioxane	101.35	23	119.5
385	-acetone	56.1	Zeotropic	
386	-pyridine, N	115.5	48.9	138.1
387	-2-picoline, N	129.3	59.6	144.12
388	-3-picoline, N	144	69.6	152.5
389	-4-picoline, N	144.3	69.7	154.3
390	-benzene	80.2	98.0	80.05
391	-cyclohexane	80.75	90.4	78.8
392	-n-hexane	68.6	94.0	68.25
393	-isopropyl sulfide	120	52	111.5
394	-toluene	110.7	71.9	100.6
395	-triethylamine, N	89	33	163
396	-2,6-lutidine, N	144.0	77.1	148.1
397	-methylcyclohexane	101.1	69	96.3
398	-n-heptane	98.25	67	91.72
399	-styrene	145.2	14.3	116.8
400	-ethylbenzene	136.15	34	114.65
401	-o-xylene	143.6	22	116.6
402	-m-xylene	139.0	27.5	115.35
403	-p-xylene	138.4	28	115.25
404	-ethylcyclohexane	131.8	—	107.9
405	-n-octane	125.75	46.3	105.7
406	-cumene	152.8	16	116.0
407	-mesitylene	164.6	Zeotropic	
408	-n-nonane	150.8	31	112.9
409	-n-decane	173.3	20.5	116.75
410	-n-undecane	194.5	5	117.87
	Ethylene sulfide	55.7		
411	-acetone	56.15	43	51.5
412	-n-hexane	68.8	Zeotropic	
413	-2,3-dimethylbutane	58.0	35	54.0
	Methyl formate	31.7		
414	-ethyl ether	34.6	45	28.4
415	-isoprene	34.1	50	22.5
416	-2-methylbutane	27.95	53	17.05
417	-n-pentane	36.15	47	21.8
418	-n-hexane	69.0	Zeotropic	
419	-2,3-dimethylbutane	58.0	15	30.5
	Iodoethane	72.3		
420	-ethyl alcohol	78.3	14	63
421	-propyl alcohol	97.2	7	70
422	-n-hexane	68.85	24	68.0
	Acetamide	221.2		
423	-benzaldehyde	179.2	93.5	178.6
424	-methylaniline	196.25	86	193.8
425	-m-cresol	202.1	Zeotropic	
426	-styrene	145.8	88	144
427	-o-xylene	144.3	89	142.6
428	-m-xylene	139.0	90	138.4
429	-p-xylene	138.2	92	137.75
430	-2,4-xylenol	210.5	Zeotropic	
431	-3,4-xylenol	226.8	4	221.1
432	-ethylaniline	205.5	82	199.0
433	-quinoline	237.3	Zeotropic	
434	-indene	183.0	82.5	177.2
435	-naphthalene	218.05	73	199.55
436	-safrol	235.9	68	208.8
437	-eugenol	255.0	12	220.8
438	-p-cymene	176.7	81	170.5
439	-diethylaniline	217.05	76	198.05
440	-camphene	159.6	88	155.5
441	-dipentene	177.7	82	169.15
442	-camphor	209.1	77	199.8
443	-isoamyl valerate	192.7	84	184.85
444	-isoamyl sulfide	214.8	83	199.5
445	-1-methylnaphthalene	245.1	56.2	209.8
446	-2-methylnaphthalene	241.15	60	208.25
447	-acenaphthene	277.9	35.8	217.1
448	-biphenyl	255.9	49.5	212.95
449	-phenyl ether	259.3	48	214.55

No.	System	B.P. (°C)	Azeotropic data Compn. X_2 (wt. %)	B.P. (°C)
450	-diphenylmethane	265.6	43.5	215.15
451	-1,2-diphenylethane	284	32	218.2
452	-benzyl ether	297	Zeotropic	
	Nitroethane	114.2		
453	-n-hexane, H	68.74	89.4	59.4
454	-toluene	110.75	75	106.2
	Ethyl nitrate	87.68		
455	-thiophene	84.7	Zeotropic	
456	-benzene	80.15	88	80.03
457	-cyclohexane	80.75	64	74.5
458	-n-hexane	68.8	76	66.25
459	-n-heptane	98.4	37	82.6
	Ethyl alcohol	78.3		
460	-acrylonitrile	77.3	59	70.8
461	-acetone	56.1	Zeotropic	
462	-ethyl sulfide	92.2	44	72.6
463	-pyridine	115.4	Zeotropic	
464	-cyclopentane	49.4	92.5	44.7
465	-n-pentane	36.15	95	34.3
466	-fluorobenzene	85.15	25	70.0
467	-benzene	80.1	68.3	67.9
468	-cyclohexane	80.8	70.8	64.8
469	-n-hexane	68.95	79	58.68
470	-propyl ether	90.4	56	74.4
471	-toluene	110.7	32	76.7
472	-ethylbenzene	136.15	Zeotropic	
473	-p-xylene	138.3	Zeotropic	
474	-n-octane	125.6	22	77.0
	Ethylene glycol	197.4		
475	-pyridine	115.5	Zeotropic	
476	-benzene	80.2	Zeotropic	
477	-phenol	182.2	Zeotropic	
478	-aniline	184.35	76	180.55
479	-o-bromotoluene	181.75	75	166.8
480	-o-nitrotoluene	221.75	51.5	188.55
481	-toluene	110.6	97.7	110.1
482	-m-toluidine	200.3	58	188.55
483	-o-cresol	191.1	73	189.6
484	-m-cresol	202.1	40	195.2
485	-2,6-lutidine	144.0	Zeotropic	
486	-n-heptane	98.45	97	97.9
487	-styrene	145.8	83.5	139.5
488	-m-xylene	139.1	93.45	135.1
489	-p-xylene	138.4	93.6	134.5
490	-3,4-xylenol	226.8	11	197.2
491	-2,4,6-collidine	171.3	90.3	170.5
492	-2,4-xylidine	214.0	53	188.6
493	-butyl ether	142.1	93.6	139.5
494	-quinoline	237.3	20.5	196.35
495	-indene	183.0	74	168.4
496	-cumene	152.8	82	147.0
497	-mesitylene	164.6	87	156
498	-propylbenzene	158.8	81	152
499	-cymene	176.7	74.5	163.2
500	-camphene	159.5	80	152.5
501	-camphor	209.1	60	186.15
502	-menthol	216.3	48.5	188.55
503	-n-decane	173.3	77	161.0
504	-naphthalene	218.05	49	183.9
505	-1-methylnaphthalene	245.1	40.0	190.25
506	-2-methylnaphthalene	241.15	42.8	189.1
507	-acenaphthene	277.9	25.8	194.65
508	-biphenyl	256.1	33.5	192.25
509	-fluorene	296.4	18	196.0
510	-diphenylmethane	265.6	31.5	193.3
511	-benzyl phenyl ether	286.5	13	195.5
512	-n-tridecane	234.0	45	188.0
513	-anthracene	340	1.7	197
514	-stilbene	306.4	13	196.8
	Methyl sulfide	37.3		
515	-acetone	56.15	Zeotropic	
516	-isoprene	34.3	65	32.5
517	-cyclopentane	49.35	12.5	37.1
518	-n-pentane	36.15	53.4	31.8
519	-2,2-dimethylbutane	49.7	20.2	36.5
	Methyl disulfide	109.44		
520	-n-heptane	98.4	73.7	96.44
521	-2,3-dimethylhexane	109.15	51.8	102.84
	Ethylenediamine	116.5		
522	-n-butyl alcohol	117.7	64.3	124.7
523	-benzene	80.1	Zeotropic	

No.	System	B.P. (°C)	Azeotropic data Compn. X_2 (wt. %)	B.P. (°C)
524	-toluene	110.7	69.2	104
	Acrylonitrile	77.3		
525	-isopropyl alcohol	82.55	44	71.7
526	-benzene	80.2	53	73.3
	Pyruvic acid	166.8		
527	-propionic acid	141.3	Zeotropic	
528	-benzene	80.15	Zeotropic	
529	-toluene	110.75	92.5	110.05
530	-o-xylene	144.3	72	137.0
531	-ethylbenzene	136.15	78	130.5
532	-mesitylene	164.6	60	151.2
533	-propylbenzene	159.3	63	147.6
	Methyl chloroacetate	129.95		
534	-isobutyl alcohol	107.85	88	107.55
535	-cyclopentanol	140.85	23	127.5
536	-amyl alcohol	138.2	30	126.8
537	-isoamyl alcohol	131.3	39.5	124.9
538	-ethylbenzene	136.15	37.5	127.2
539	-m-xylene	139.2	10	128.25
540	-p-xylene	138.45	15	128.3
	Propionitrile	97.2		
541	-propyl alcohol	97.2	50	90.5
542	-n-hexane	68.8	91	63.5
543	-ethylbenzene	136.15	Zeotropic	
	Acetone	56.15		
544	-methyl acetate	57	51.7	55.8
545	-diethylamine	55.5	61.8	51.4
546	-pyridine	115.4	Zeotropic	
547	-cyclopentane	49.3	64	41.0
548	-n-pentane	36.15	80	32.5
549	-benzene	80.1	Zeotropic	
550	-cyclohexane	80.75	32.5	53.0
551	-n-hexane	68.95	41	49.8
552	-isopropyl ether	69.0	39	54.2
553	-n-heptane	98.4	10.5	55.85
	Allyl alcohol	96.95		
554	-ethyl sulfide	92.1	55	85.1
555	-pyridine	115.4	Zeotropic	
556	-benzene	80.2	82.64	76.75
557	-cyclohexane	80.8	42	74.0
558	-n-hexane	68.95	95.5	65.5
559	-methylcyclohexane	101.1	58	85.0
560	-m-xylene	139.0	Zeotropic	
561	-2,5-dimethylhexane	109.4	50	89.3
562	-n-octane	125.75	32	93.4
	Ethyl formate	54.1		
563	-n-pentane	36.2	70	32.5
564	-benzene	80.2	Zeotropic	
565	-methylcyclopentane	72.0	25	51.2
566	-n-hexane	68.95	33	49.0
	Methyl acetate	56.95		
567	-cyclopentane	49.3	62.1	43.2
568	-benzene	80.2	0.3	56.7
569	-cyclohexane	80.7	22.0	55.5
570	-n-hexane	68.95	39.3	51.75
571	-n-heptane	98.45	3.55	56.65
572	-2-methylhexane	90.0	11.4	56.0
573	-2,2,3-trimethylbutane	80.9	25.8	55.1
	Propionic acid	141.0		
574	-pyridine, N	115.5	32.8	148.6
575	-2-picoline, N	129.3	45.0	154.5
576	-chlorobenzene	132.0	82	128.9
577	-benzene	80.15	Zeotropic	
578	-o-xylene	143.6	57	135.4
579	-p-xylene	138.2	66	132.5
580	-n-hexane	68.85	Zeotropic	
581	-n-heptane	98.15	98	97.82
582	-n-octane	125.12	78.5	120.89
583	-n-nonane	150.67	46.0	134.27
584	-n-decane	174.06	19.5	139.76
585	-n-undecane	193.85	Zeotropic	
586	-propyl sulfide	141.5	55	136.5
587	-quinoline	237.5	Zeotropic	
588	-cumene	152.8	35	139.0
589	-mesitylene	164.0	23	139.3
590	-propylbenzene	158.0	25	139.5
591	-camphene	159.6	35	138.0
592	-α-pinene	155.8	41.5	136.4
	Propionamide	222.2		
593	-p-bromochlorobenzene	196.4	84	189.5
594	-p-dibromobenzene	220.25	78	204.9

No.	System	B.P. (°C)	Azeotropic data: Compn. X_2 (wt. %)	Azeotropic data: B.P. (°C)
595	-iodobenzene	188.45	90	183.5
596	-nitrobenzene	210.75	76	205.4
597	-o-nitrophenol	217.25	75.2	211.15
598	-phenol	182.2	Zeotropic	
599	-p-bromotoluene	185.0	90	181.0
600	-m-nitrotoluene	230.8	56	214.5
601	-toluene	110.75	Zeotropic	
602	-o-cresol	191.1	Zeotropic	
603	-m-cresol	202.2	Zeotropic	
604	-o-toluidine	200.35	97.5	200.25
605	-m-toluidine	203.1	Zeotropic	
606	-acetophenone	202	85	200.35
607	-methyl salicylate	222.35	66	210.55
608	-o-xylene	144.3	98	144.0
609	-dimethylaniline	194.15	84.5	190.5
610	-3,4-xylidine	225.5	72	217.2
611	-quinoline	237.3	Zeotropic	
612	-indene	182.6	88	179.5
613	-ethyl benzoate	212.6	75	205.0
614	-cumene	152.8	96	151.8
615	-mesitylene	164.6	90	162.3
616	-naphthalene	218.05	68.5	204.65
617	-cymene	176.7	85	172.8
618	-carvone	231.0	52	214.5
619	-camphene	159.6	87	156.5
620	-camphor	209.1	83	203.5
621	-borneol	213.4	78	209.2
622	-n-decane	173.3	88.2	168
623	-1-methylnaphthalene	245.1	48	213.8
624	-2-methylnaphthalene	241.15	50	213.0
625	-n-undecane	194.5	79	183
626	-acenaphthene	277.9	25	220.8
627	-biphenyl	256.1	45	216.0
628	-n-dodecane	216.0	68.4	193.0
629	-fluorene	295	10	221.5
630	-diphenylmethane	265.6	40	218.2
	Ethyl carbamate	185.25		
631	-bromobenzene	156.1	90.2	153.95
632	-iodobenzene	188.45	67	174.5
633	-nitrobenzene	210.75	12	184.95
634	-phenol	182.2	46.5	190.75
635	-benzonitrile	191.1	43	182.1
636	-anisole	153.85	95	153.5
637	-2,4-xylenol	210.5	Zeotropic	
638	-n-octyl alcohol	195.2	27.5	183.5
639	-isobutyl sulfide	172.0	77	166.5
640	-indene	182.6	65	172.65
641	-cumene	152.8	94	151.5
642	-mesitylene	164.6	78	159.0
643	-propylbenzene	159.3	85	157.0
644	-pseudocumene	168.2	75	161.4
645	-naphthalene	218.0	23	184.05
646	-butylbenzene	183.1	63	172.0
645	-camphene	159.6	85	157.0
648	-limonene	177.6	68	168.07
649	-camphor	209.1	16	184.85
650	-2-methylnaphthalene	241.15	Zeotropic	
651	-amyl ether	187.4	63	171.0
652	-isoamyl ether	173.35	73	163.15
653	-methyl pelargonate	213.8	15	184.3
	1-Nitropropane	131		
654	-propyl alcohol	97.15	91.2	96.95
655	-n-butyl alcohol	117.73	67.8	115.3
656	-isobutyl alcohol	107.89	84.8	105.28
657	-n-heptane	98.43	86.5	96.6
658	-ethylbenzene	136.19	44.0	129.0
659	-n-octane	125.66	65.8	115.8
660	-n-nonane	150.8	38.4	126.6
	Propyl nitrite	47.75		
661	-n-pentane	36.15	91	35.8
662	-cyclopentane	49.3	46	45.5
	Isopropyl alcohol	82.45		
663	-butylamine	77.8	40	74.7
664	-n-pentane	36.15	94	35.5
665	-fluorobenzene	85.15	70	74.5
666	-benzene	80.2	66.3	71.74
667	-cyclohexane	80.7	68	69.4
668	-n-hexane	68.85	77	62.7
669	-toluene	110.6	31	80.6
670	-n-heptane	98.45	49.5	76.4
671	-o-xylene	144.3	Zeotropic	
672	-n-octane	124.75	16	81.6
	Propyl alcohol	97.2		
673	-dioxane	101.35	45	95.3
674	-butyl formate	106.8	36	95.5
675	-chlorobenzene	132	20	96.5
676	-fluorobenzene	85.15	82	80.2
677	-benzene	80.2	83.1	77.12
678	-cyclohexane	80.75	81.5	74.69
679	-toluene	110.6	48.8	92.5
680	-methylcyclohexane	100.8	65.2	87.0
681	-n-heptane	98.4	65.3	84.6
682	-styrene	145.8	92	97.0
683	-o-xylene	143.6	Zeotropic	
684	-m-xylene	139.2	6	97.08
685	-p-xylene	138.4	7.8	96.88
686	-n-octane	125.6	30	93.9
	Glycerol	290.5		
687	-p-chloronitrobenzene	239.1	87	235.6
688	-triethylene glycol	288.7	63	285.1
689	-m-nitrotoluene	230.8	87	228.8
690	-p-cresol	201.7	Zeotropic	
691	-methyl salicylate	222.35	92.5	221.4
269	-3,4-xylenol	226.8	Zeotropic	
693	-o-xylene	143.6	Zeotropic	
694	-quinoline	237.3	Zeotropic	
695	-ethyl salicylate	233.7	89.7	230.5
696	-naphthalene	218.05	90	215.2
697	-safrol	235.9	85.5	231.3
698	-methyl phthalate	283.2	69	271.5
699	-estragol	215.6	92.5	213.5
700	-eugenol	254.5	96	251.3
701	-propyl benzoate	230.85	92	228.8
702	-carvone	231.0	97	230.85
703	-2-methylnaphthalene	241.15	83.5	233.7
704	-acenaphthene	277.9	71	259.1
705	-biphenyl	254.9	75	246.1
706	-phenyl ether	259.3	78	247.9
707	-1,3,5-triethylbenzene	215.5	92	212.9
708	-bornyl acetate	227.7	90	226.0
709	-diphenylmethane	265.6	73	250.8
710	-benzyl phenyl ether	286.5	70	264.5
711	-benzyl ether	297.0	64	269.5
	Ethyl methyl sulfide	66.61		
712	-cyclohexane	80,75	Zeotropic	
713	-methylcyclopentane	71.85	35.9	65.6
714	-n-hexane	68.75	43.4	63.94
715	-2,2-dimethylpentane	79.2	11.8	66.37
	1-Propanethiol	67.3		
716	-thiophene	84.7	Zeotropic	
717	-cyclohexane	80.75	2.4	67.77
718	-2,3-dimethylbutane	58.0	83.7	57.54
719	-n-hexane	68.75	47.4	64.35
720	-2-methylpentane	60.27	76.1	59.2
721	-isopropyl ether	68.3	35	66.0
722	-2,2-dimethylpentane	79.2	18.7	67.2
723	-2,2,3-trimethylbutane	80.97	12.6	67.57
	Perfluorobutyric acid	122.0		
724	-ethyl-benzene	136.15	20	115.4
725	-m-xylene	139.0	17	117.5
726	-p-xylene	138.4	18	117.6
	Thiophene	84.7		
727	-benzene	80.15	Zeotropic	
728	-cyclohexane	80.8	58.8	77.9
729	-methylcyclopentane	71.85	86	71.47
730	-n-hexane	68.75	88.8	68.46
731	-2,3-dimethylpentane	89.9	36	80.9
732	-2,4-dimethylpentane	80.55	57.3	76.58
733	-n-heptane	98.4	16.8	83.09
	Pyrrol	129.2		
734	-chlorobenzene	131.75	57	124.5
735	-isopropyl sulfide	120.5	80	117.5
736	-propyl sulfide	140.8	35	127.5
737	-toluene	110.75	Zeotropic	
	Methyl pyruvate	137.5		
738	-isoamyl acetate	142.1	35	135.0
739	-m-xylene	139.2	50	130.0
	Methyl oxalate	163.3		
740	-p-dichlorobenzene	174.35	35	162.05
741	-pinacol	174.35	19	163.15
742	-o-bromotoluene	181.5	2	164.1
743	-butyl butyrate	166.4	42	160.5
744	-ethyl caproate	167.7	40	161.0
745	-indene	182.6	17	163.6

BINARY SYSTEMS (Continued)

No.	System	B.P. (°C)	Azeotropic data: Compn. X_2 (wt. %)	Azeotropic data: B.P. (°C)
746	-mesitylene	164.0	50.2	154.8
747	-naphthalene	218.0	Zeotropic	
748	-2,7-dimethyloctane	160.6	55	147.0
749	-1,3,5-triethylbenzene	215.5	Zeotropic	
	Ethyl bromoacetate	158.8		
750	-butyric acid	164.0	16	157.4
751	-isobutyric acid	154.6	60	153.0
752	-bromobenzene	156.1	72	155.3
753	-cyclohexanol	160.8	35	155.5
754	-o-chlorotoluene	159.3	48	156.2
755	-propylbenzene	159.3	50	155.8
	Ethyl chloroacetate	143.55		
756	-isoamyl acetate	142.1	60	141.7
757	-isoamyl alcohol	131.3	77	131.0
758	-allyl sulfide	139.35	78	138.5
759	-propyl butyrate	142.8	53	141.7
760	-ethylbenzene	136.15	82	135.3
761	-o-xylene	144.3	42	140.2
762	-m-xylene	139.0	68	137.45
763	-butyl ether	142.4	55	139.8
	Butyronitrile	117.9		
764	-n-butyl alcohol	117.8	50	113.0
765	-toluene	110.75	73	107.0
766	-methylcyclohexane	101.15	80	90.5
	Isobutyronitrile	103.85		
767	-benzene	80.15	Zeotropic	
768	-methylcyclohexane	101.15	60	85.5
769	-n-heptane	98.4	62	80.5
	2-Butanone	79.6		
770	-methyl propionate	79.85	40	79.0
771	-ethyl acetate	77.1	88.2	77.05
772	-1-chlorobutane	78.5	62	77.0
773	-butyl nitrite	78.2	70	76.7
774	-*tert*-butyl alcohol	82.45	31	78.7
775	-butylamine	77.8	65	74.0
776	-fluorobenzene	84.9	25	79.3
777	-benzene	80.1	56	78.33
778	-cyclohexane	80.75	60	71.8
779	-n-hexane	68.8	71.4	64.2
780	-n-heptane	98.5	30	77.0
781	-2,5-dimethylhexane	109.4	5	79.0
	Butyraldehyde	74.8		
782	-benzene	80.1	Zeotropic	
783	-n-hexane	68.7	74	60.0
	Butyric acid	164.0		
784	-iodobutane	130.4	97.5	129.8
785	-2-furaldehyde	161.45	57.5	159.4
786	-pyridine	115.5	8.0	163.2
787	-propyl chloroacetate	162.5	60	160.5
788	-isoamyl nitrate	149.75	88	147.85
789	-p-dichlorobenzene	174.4	43	162.0
790	-chlorobenzene	132.0	97.2	131.75
791	-o-bromotoluene	181.5	28	163.0
792	-m-bromotoluene	184.3	20.5	163.62
793	-p-bromotoluene	185.0	25	161.5
794	-anisole	153.85	88	152.85
795	-n-heptane	98.4	Zeotropic	
796	-styrene	145.8	85	143.5
797	-ethylbenzene	136.15	96	135.8
798	-o-xylene	144.3	90	143.0
799	-m-xylene	139.0	94	138.5
800	-p-xylene	138.45	94.5	137.8
801	-indene	182.6	16	163.65
802	-cumene	152.8	80	149.5
803	-mesitylene	164.8	62	158.0
804	-propylbenzene	158.9	72	154.5
805	-pseudocumene	169	55	159.5
806	-naphthalene	218.1	Zeotropic	
807	-butylbenzene	183.1	25	162.5
808	-cymene	176.7	40	161.0
809	-camphene	159.6	97.2	152.3
810	-n-undecane	194.5	15.5	162.4
	Dioxane	101.35		
811	-ethyl acetate	77.1	Zeotropic	
812	-1-bromobutane	101.5	53	98.0
813	-pyridine	115.5	Zeotropic	
814	-piperidine	106.4	Zeotropic	
815	-*tert*-amyl alcohol	102.35	20	100.65
816	-benzene	80.15	Zeotropic	
817	-cyclohexane	80.75	75.4	79.5
818	-ethyl borate	118.6	8	100.7
819	-toluene	110.75	Zeotropic	
820	-n-heptane	98.4	56	91.85
	Ethyl acetate	77.1		
821	-butyl nitrite	78.2	29	76.3
822	-isobutyl nitrite	67.1	Zeotropic	
823	-*tert*-butyl alcohol	82.45	27	76.0
824	-benzene	80.15	Zeotropic	
825	-cyclohexane	80.75	44	71.6
826	-methylcyclopentane	72.0	62	67.2
827	-n-hexane	68.7	60.1	65.15
828	-methylcyclohexane	101.1	Zeotropic	
	Isobutyric acid	154.6		
829	-iodobutane	130.4	93	128.8
830	-ethyl pyruvate	155.5	40	153.0
831	-bromobenzene	156.15	65	148.6
832	-chlorobenzene	132.0	92	131.2
833	-phenol	182.2	Zeotropic	
834	-1-bromohexane	156.5	65	148.0
835	-o-bromotoluene	181.5	15	153.9
836	-toluene	110.75	Zeotropic	
837	-anisole	153.85	58	149.0
838	-styrene	145.8	73	142.0
839	-o-xylene	144.3	78	141.0
840	-m-xylene	139.0	85	136.9
841	-p-xylene	138.4	87	136.4
842	-cumene	152.8	65	146.8
843	-propylbenzene	158.9	51	149.3
844	-pseudocumene	168.2	37	152.3
845	-cymene	176.7	20	153.4
846	-camphene	159.6	55	148.1
847	-d-limonene	177.8	22	152.5
848	-2,7-dimethyloctane	160.2	52	148.55
849	-isoamyl ether	173.2	7	154.2
	Methyl propionate	79.85		
850	-1-chlorobutane	78.05	62	76.8
851	-butyl nitrite	78.2	88	77.7
852	-n-butyl alcohol	117.8	Zeotropic	
853	-benzene	80.2	48	79.45
854	-cyclohexane	80.75	48	75.0
855	-methylcyclopentane	72.0	72	69.5
856	-propyl ether	90.5	Zeotropic	
857	-methylcyclohexane	101.1	11.5	79.3
	Propyl formate	80.85		
858	-1-chlorobutane	78.5	62	76.1
859	-butyl nitrite	78.2	65	76.8
860	-*tert*-butyl alcohol	82.6	60	78.0
861	-benzene	80.2	53	78.5
862	-cyclohexane	80.75	52	75.0
863	-n-hexane	68.95	70.5	63.6
864	-n-heptane	98.5	29	78.2
	Glycol monoacetate	190.9		
865	-phenol, N	182.2	35	197.5
866	-m-bromotoluene	184.3	68	182.0
867	-o-cresol, N	191.1	49	199.45
868	-m-cresol, N	202.2	69	206.5
869	-p-cresol, N	201.7	67	206.0
870	-n-octyl alcohol	195.2	29	189.5
871	-indene	182.6	80	180.0
872	-naphthalene	218.0	Zeotropic	
873	-amyl ether	187.5	58	180.8
874	-isoamyl ether	173.2	72	170.2
875	-1,3,5-triethylbenzene	215.5	Zeotropic	
	Methyl lactate	143.8		
876	-phenol	182.2	Zeotropic	
877	-anisole	153.85	18	142.8
878	-4-heptanone	143.55	53	142.7
879	-ethyl valerate	145.45	42	140.0
880	-methyl caproate	149.8	30	141.7
881	-m-xylene	139.0	57.5	131.2
882	-p-xylene	138.2	60	130.8
883	-n-octane	125.8	70	120.3
884	-butyl ether	142.8	58	137.0
885	-cumene	152.8	38	137.8
886	-camphene	159.6	15	140.0
887	-2,7-dimethyloctane	160.1	32	137.8
	Tetrahydrothiophene	118.8		
888	-pyridine	115.4	55	113.5
889	-1-methylpyrrol	112.8	82	111.5
890	-ethylcyclohexane	131.85	19.3	117.46
891	-2-methylheptane	117.70	61.8	113.96
892	-n-octane	125.7	39.7	117.79
	Butyl nitrite	78.2		
893	-benzene	80.15	25	77.95
894	-cyclohexane	80.75	37	76.5

No.	System	B.P. (°C)	Azeotropic data Compn. X_2 (wt. %)	Azeotropic data B.P. (°C)
895	-n-hexane	68.8	82	68.5
896	-methylcyclohexane	101.15	Zeotropic	
897	-n-heptane	98.4	Zeotropic	
	Isobutyl nitrite	67.1		
898	-benzene	80.15	Zeotropic	
899	-cyclohexane	80.75	Zeotropic	
900	-methylcyclopentane	72.0	32	65.9
901	-n-hexane	68.8	46	65.0
	Isobutyl nitrate	123.5		
902	-n-butyl alcohol	117.8	55	112.8
903	-isobutyl alcohol	107.85	64	105.6
904	-chlorobenzene	131.75	Zeotropic	
905	-propyl sulfide	141.5	Zeotropic	
906	-toluene	110.75	Zeotropic	
907	-ethylbenzene	136.15	Zeotropic	
	n-Butyl alcohol	117.75		
908	-pyridine	115.5	31	118.6
909	-butyl formate	106.6	76.4	105.8
910	-ethyl carbonate	125.9	37	116.5
911	-chlorobenzene	132.0	44	115.3
912	-fluorobenzene	84.9	Zeotropic	
913	-benzene	80.1	Zeotropic	
914	-2-picoline	129.4	Zeotropic	
915	-cyclohexene	82.7	95	82.0
916	-cyclohexane	80.75	90.5	79.8
917	-hexaldehyde	128.3	22.9	116.8
918	-ethyl isobutyrate	110.1	83	109.2
919	-isoamyl formate	123.8	31	115.9
920	-isobutyl acetate	117.2	50	114.5
921	-methyl isoyalerate	116.3	60	113.5
922	-paraldehyde	123.9	48	115.75
923	-n-hexane	68.95	96.8	68.2
924	-acetal	103.55	87	101.0
925	-isopropyl sulfide	120.5	55	112.0
926	-ethyl borate	118.6	48	113.0
927	-toluene	110.7	72.2	105.5
928	-methylcyclohexane	100.8	80	95.3
929	-n-heptane	98.4	82	93.85
930	-ethylbenzene	136.15	34.9	115.85
931	-o-xylene	143.6	25	116.8
932	-m-xylene	139.0	28.5	116.5
933	-p-xylene	138.3	32	115.7
934	-n-octane	125.75	54.8	108.45
935	-butyl ether	142.1	17.5	117.65
936	-isobutyl ether	122.3	52	113.5
937	-n-nonane	150.7	28.5	115.9
938	-2,7-dimethyloctane	160.2	Zeotropic	
	***sec*-Butyl alcohol**	99.5		
939	-butyl formate	106.8	32	98.0
940	-ethyl propionate	99.15	53	95.7
941	-benzene	80.15	84.6	78.5
942	-cyclohexane	80.75	82	76.0
943	-methylcyclopentane	72.0	88.5	69.7
944	-propyl ether	90.4	78	87.0
945	-toluene	110.7	45	95.3
946	-methylcyclohexane	101.5	61.8	89.7
947	-n-heptane	98.4	63.3	88.1
948	-isooctane	99.3	66.2	88.0
	***tert*-Butyl alcohol**	82.9		
949	-ethyl sulfide	92.1	30	79.8
950	-flurobenzene	85.15	69	76.0
951	-benzene	80.2	63.4	73.95
952	-cyclohexane	80.7	65.8	71.2
953	-methylcyclopentane	72.0	74	66.6
954	-n-hexane	68.85	78	63.7
955	-isopropyl ether	68.3	92.1	67.3
956	-propyl ether	90.4	48	79.0
957	-toluene	110.7	Zeotropic	
958	-methylcyclohexane	100.8	34	78.8
959	-n-heptane	98.45	38	78.0
960	-p-xylene	138.45	Zeotropic	
961	-2,5-dimethylhexane	109.2	23	81.5
	Ethyl ether	34.6		
962	-isoprene	34.3	52	33.2
963	-2-methyl-2-butene	37.1	15	34.2
964	-n-pentane	36.16	44	33.7
965	-benzene	80.2	Zeotropic	
966	-n-hexane	68.85	Zeotropic	
	Isobutyl alcohol	108.0		
967	-2-pentanone	102.35	81	101.8
968	-3-pentanone	102.05	80	101.7
969	-butyl formate	106.8	60	103.0
970	-methyl butyrate	102.65	75	101.3
971	-propyl acetate	101.6	83	101.0
972	-n-pentane	36.15	Zeotropic	
973	-chlorobenzene	132.0	37	107.1
974	-fluorobenzene	84.9	91	84.0
975	-benzene	80.1	92.6	79.3
976	-cyclohexene	82.7	85.8	80.5
977	-cyclohexane	80.75	86	78.3
978	-methylcyclopentane	72.0	95	71.0
979	-isobutyl vinyl ether	83.0	93.8	82.7
980	-ethyl isobutyrate	110.1	48	105.5
981	-n-hexane	68.9	97.5	68.3
982	-propyl ether	90.55	90	89.5
983	-acetal	103.55	80	98.2
984	-isopropyl sulfide	100.5	27	105.8
985	-toluene	110.7	55	101.2
986	-methylcyclohexane	100.8	68	92.6
987	-n-heptane	98.45	73	90.8
988	-ethylbenzene	136.15	20	107.2
989	-p-xylene	138.4	11.4	107.1
990	-1,3-dimethylcyclohexane	120.7	44	102.2
991	-2,5-dimethylhexane	109.2	58	98.7
992	-2,2,4-trimethylpentane	99.3	73	92.0
993	-butyl ether	142.4	Zeotropic	
	2-Ethoxyethanol	135.3		
994	-toluene	110.75	89.2	110.15
995	-methylcyclohexane	101.15	85	98.6
996	-propyl butyrate	143.7	28	133.5
997	-n-heptane	98.4	86	96.5
998	-styrene	145.8	45	130.0
999	-ethylbenzene	136.15	52	127.8
1000	-p-xylene	138.45	50	128.6
1001	-n-octane	125.75	62	116.0
1002	-cumene	152.8	33	133.2
1003	-propylbenzene	159.3	20	134.6
1004	-camphene	159.6	35	131.0
	Methyl propyl ether	38.95		
1005	-2-methyl-2-butene	37.15	75	36.3
1006	-n-pentane	36.2	78	35.6
1007	-isoprene	34.3	Zeotropic	
	Diethylene glycol	245.5		
1008	-p-dibromobenzene	220.25	87	212.85
1009	-nitrobenzene	210.75	90	210.0
1010	-o-nitrophenol	217.2	89.5	216.0
1011	-pyrocatechol, N	245.9	54	259.5
1012	-m-nitrotoluene	230.8	75	224.2
1013	-methyl salicylate	222.95	85	220.55
1014	-p-cresol	202.0	Zeotropic	
1015	-ethyl fumarate	217.85	90	217.1
1016	-quinoline	237.3	71	233.6
1017	-benzyl acetate	215.0	93	214.85
1018	-naphthalane	218.0	78	212.6
1019	-isosafrol	252.0	54	233.5
1020	-safrol	235.9	67	225.5
1021	-methyl phthalate	283.7	3.7	245.4
1022	-thymol	232.9	87	232.25
1023	-1-methylnaphthalene	244.6	55	277.0
1024	-2-methylnaphthalene	241.15	61	225.45
1025	-biphenyl	256.1	52	232.65
1026	-acenaphthene	277.9	38	239.6
1027	-1,3,5-triethylbenzene	215.5	78	210.0
1028	-bornyl acetate	227.6	82	223.0
1029	-fluorene	295.0	20	243.0
1030	-diphenylmethane	265.4	48	236.0
1031	-benzyl phenyl ether	286.5	20	241.5
	1-Butanethiol	97.8		
1032	-benzene	80.15	Zeotropic	
1033	-pyridine	115.4	Zeotropic	
1034	-n-heptane	98.42	50.6	95.45
1035	-2-methylhexane	90.05	84.6	89.74
1036	-3-methylhexane	91.95	77.2	91.2
1037	-2,5-dimethylhexane	109.1	12	98.22
	Isopropyl methyl sulfide	84.76		
1038	-cyclohexane	80.85	70	79.76
1039	-3-methylhexane	91.6	17.6	84.38
1040	-2,4-dimethylpentane	80.55	70.3	79.39
	Methyl propyl sulfide	95.47		
1041	-ethylcyclopentane	103.45	9.3	95.41
1042	-methylcyclohexane	101.05	22.0	95.06
1043	-3-methylhexane	91.6	67.05	90.53
	2-Furaldehyde	161.45		
1044	-n-heptane	98.4	94.7	98.3

No.	System	B.P. (°C)	Azeotropic data Compn. X_2 (wt. %)	B.P. (°C)
1045	-ethylbenzene	136.15	Zeotropic	
1046	-o-xylene	143.6	87	140.5
1047	-m-xylene	139.0	88	138.4
1048	-p-xylene	138.4	80	138.0
1049	-cumene	152.8	73	148.5
1050	-mesitylene	164.6	40	155.2
1051	-pseudocumene	168.2	33	157.0
1052	-propylbenzene	159.2	58	151.4
1053	-cymene	176.7	32	157.8
1054	-camphene	159.5	60	146.75
1055	-cineol	176.35	41	157.25
	Pyridine	115.4		
1056	-piperidine	105.8	92	106.1
1057	-phenol, N	181.4	86.9	183.1
1058	-toluene	110.75	77.8	110.1
1059	-n-heptane	98.4	74.7	95.6
1060	-n-octane	125.75	43.9	109.5
1061	-n-nonane	150.7	10.1	115.1
1062	-n-decane	173.3	Zeotropic	
	2-Methylthiophene	111.92		
1063	-n-heptane	98.4	97.8	97.77
1064	-2-methylheptane	117.7	32.2	109.77
1065	-2,2-dimethylhexane	106.85	66.8	104.62
1066	-2,5-dimethylhexane	109.15	60.4	106.12
	3-Methylthiophene	114.96		
1067	-ethylcyclopentane	103.45	96.1	102.82
1068	-n-octane	125.75	18.0	114.15
1069	-2-methylheptane	117.7	41.2	111.86
1070	-2,5-dimethylhexane	109.15	68.3	107.12
	Ethyl pyruvate	155.1		
1071	-bromobenzene	156.1	52	149.5
1072	-m-xylene	139.2	70	137.2
1073	-cumene	152.8	55	146.2
	Levulinic acid	252.0		
1074	-m-nitrotoluene	230.8	85	229.5
1075	-p-nitrotoluene	238.9	78	236.4
1076	-methyl salicylate	222.95	94	222.75
1077	-3,4-xylenol	226.8	Zeotropic	
1078	-ethyl salicylate	233.8	82	230.5
1079	-naphthalene	218.0	89	216.7
1080	-safrol	235.9	83	232.5
1081	-1-methylnaphthalene	244.6	64	237.0
1082	-2-methylnaphthalene	241.15	71	234.55
1083	-isobutyl benzoate	241.9	75	238.6
1084	-1,3,5-triethylbenzene	215.5	89	214.0
	Methyl acetoacetate	169.5		
1085	-siobutyl sulfide	172.0	42	166.0
1086	-mesitylene	164.6	57	159.5
1086	-cymene	176.7	44	165.0
1088	-camphene	159.6	60	152.8
1089	-isoamyl ether	173.2	40	160.5
	Methyl malonate	181.4		
1090	-acetophenone	202.0	61	201.0
1091	-naphthalene	218.0	Zeotropic	
1092	-butylbenzene	183.2	48	173.0
1093	-cymene	176.7	60	169.0
1094	-camphene	159.6	74	154.6
1095	-d-limonene	177.8	52	167.3
	2-Pentanone	102.25		
1096	-methyl butyrate	102.65	50	101.9
1097	-toluene	110.7	Zeotropic	
1098	-methylcyclohexane	101.15	60	95.2
1099	-n-heptane	98.4	66	93.2
	Butyl formate	106.8		
1100	-*tert*-amyl alcohol	102.35	65	101.0
1101	-benzene	80.15	Zeotropic	
1102	-pinacolone	106.2	62	106.0
1103	-methylcyclohexane	101.15	65	96.0
1104	-n-heptane	98.45	60	90.7
	Isovaleric acid	176.5		
1105	-ethyl acetoacetate	180.4	23	176.1
1106	-ethyl oxalate	185.65	16	176.3
1107	-o-xylene	144.3	95	143.8
1108	-butyl sulfide	185.0	27	175.0
1109	-indene	183.0	40	173.0
1110	-cumene	152.8	88	152.0
1111	-mesitylene	164.6	81	162.5
1112	-pseudocumene	168.2	77	165.7
1113	-naphthalene	218.05	Zeotropic	
1114	-butylbenzene	183.1	50	173.0
1115	-cymene	175.3	62	170.8
1116	-camphene	159.6	83	156.5
1117	-cineol	176.3	57.5	175.0
1118	-n-decane	173.3	67	167.0
1119	-n-tridecane	234.0	Zeotropic	
	Methyl butyrate	102.65		
1120	-methylcyclohexane	101.1	55	97.0
1121	-n-heptane	98.45	65	95.1
1122	-n-octane	125.8	Zeotropic	
	Propyl acetate	101.6		
1123	-*tert*-amyl alcohol	102.0	42	99.5
1124	-benzene	80.2	Zeotropic	
1125	-cyclohexane	80.75	Zeotropic	
1126	-n-hexane	69.0	Zeotropic	
1127	-acetal	103.55	32	101.25
	Valeric acid	186.35		
1128	-phenol	182.2	Zeotropic	
1129	-indene	182.6	70	178.5
1130	-mesitylene	164.6	90	164.0
1131	-naphthalene	218.0	4	186.0
1132	-cymene	176.7	78	176.5
1133	-camphene	159.6	92	158.5
1134	-amyl ether	187.5	55	181.5
1135	-isoamyl ether	173.2	87.5	171.8
	Ethyl lactate	154.1		
1136	-toluene	110.75	Zeotropic	
1137	-o-xylene	144.3	70	140.2
1138	-p-xylene	138.45	83	136.6
1139	-cumene	152.8	52	143.5
1140	-mesitylene	164.9	27	150.05
1141	-pseudocumene	168.2	27	152.4
1142	-camphene	159.5	45	144.95
	n-Amyl alcohol	138.2		
1143	-benzene	80.2	Zeotropic	
1144	-phenol	182.2	Zeotropic	
1145	-amyl formate	132.0	57	131.4
1146	-ethylbenzene	136.15	60	129.8
1147	-p-xylene	138.45	58.1	130.9
	Isoamyl alcohol	131.9		
1148	-bromobenzene	156.15	15	131.65
1149	-butyl acetate	126.0	83.5	125.85
1150	-paraldehyde	124.0	78.0	123.5
1151	-o-fluorotoluene	114.0	86.0	112.1
1152	-toluene	110.7	90	109.7
1153	-n-heptane	98.45	93	97.7
1154	-ethylbenzene	136.15	51	125.7
1155	-n-octane	125.8	70	117.0
1156	-butyl ether	142.1	35	129.8
1157	-cumene	152.8	6	131.6
1158	-camphene	159.6	76	130.9
	Nitrobenzene	210.75		
1159	-aniline	184.35	Zeotropic	
1160	-methyl maleate	204.05	93	203.9
1161	-benzyl alcohol	205.25	62	204.2
1162	-3,4-xylenol	226.8	Zeotropic	
1163	-ethyl benzoate	212.5	19	210.6
1164	-camphor	208.9	65	208.4
1165	-borneol	215.0	42	207.8
1166	-1,3,5-triethylbenzene	215.5	Zeotropic	
1167	-ethyl bornyl ether	204.9	70	203.0
	Iodobenzene	188.55		
1168	-nitrobenzene	210.75	Zeotropic	
1169	-phenol	181.5	47	177.7
1170	-ethyl oxalate	185.65	52	181.0
1171	-caproic acid	205.15	12	186.8
1172	-isocaproic acid	199.5	15	185.5
1173	-benzyl alcohol	205.2	12	187.75
1174	-p-cresol	201.7	10	188.1
1175	-o-toluidine	200.35	Zeotropic	
1176	-isobutyl lactate	182.15	70	180.5
1177	-indene	182.6	Zeotropic	
1178	-isoamyl butyrate	178.5	Zeotropic	
1179	-butylbenzene	183.1	Zeotropic	
	Benzene	80.15		
1180	-aniline	184.35	Zeotropic	
1181	-cyclohexene	82.1	35.3	78.9
1182	-cyclohexane	80.75	48.1	77.56
1183	-methylcyclopentane	71.85	84	71.7
1184	-n-hexane	69.0	95.3	68.5
1185	-2,2-dimethylpentane	79.1	53.7	75.85
1186	-2,3-dimethylpentane	89.79	21.2	79.4
1187	-2,4-dimethylpentane	80.8	51.7	75.2
1188	-n-heptane	98.4	0.7	80.1
1189	-2,2,4-trimethylpentane	99.2	2.3	80.1

No.	System	B.P. (°C)	Azeotropic data Compn. X_2 (wt. %)	B.P. (°C)
	Phenol	182.2		
1190	-aniline, N	183.91	58.1	185.84
1191	-2-picoline, N	129.2	24.6	185.5
1192	-3-picoline, N	143.5	29.8	188.93
1194	-4-picoline, N	144.8	32.5	190.0
1195	-ethylene diacetate, N	189.86	60.8	195.53
1196	-benzaldehyde	179.2	49.0	175.6
1197	-o-cresol	191.1	Zeotropic	
1198	-2,4-lutidine, N	159.0	43.0	193.4
1199	-2,6-lutidine, N	144.0	27.5	185.5
1200	-o-toluidine	200.35	Zeotorpic	
1201	-2,4,6-collidine, N	171.0	47.7	195.23
1202	-n-octyl alcohol	195.15	87	195.4
1203	-*sec*-octyl alcohol	179.0	50	184.5
1204	-indene	182.2	53	177.8
1205	-mesitylene	164.5	79	163.5
1206	-pseudocumene	168.2	75	166.0
1207	-naphthalene	218.1	Zeotropic	
1208	-butylbenzene	183.1	54	175.0
1209	-camphene	159.6	78	156.1
1210	-n-decane	173.3	65	168.0
1211	-2,7-dimethyloctane	160.25	94	159.5
1212	-amyl ether	187.5	22	180.2
1213	-isoamyl ether	173.2	85	172.2
1214	-isoamyl sulfide	214.8	Zeotropic	
1215	-1,3,5-triethylbenzene	215.5	Zeotropic	
1216	-n-tridecane	235.42	16.9	180.56
	Pyrocatechol	245.9		
1217	-indole	253.5	85.0	255.0
1218	-o-phenetidine, N	232.5	8	246.0
1219	-p-phenetidine, N	249.9	66	253.8
1220	-quinoline, N	237.4	39	257.9
1221	-naphthalene	218.05	88.5	217.45
1222	-quinaldine, N	246.5	52.0	252.5
1223	-safrole	235.9	77.0	233.55
1224	-isosafrole	252.0	30	243.0
1225	-eugenol	254.8	1.5	245.85
1226	-carvone	231.0	29	248.3
1227	-thymol	232.9	83	232.2
1228	-1-methylnapthalene	244.9	60	235.1
1229	-2-methylnaphthalene	241.15	63	233.25
1230	-acenaphthene	277.9	16	245.25
1231	-biphenyl	255.9	43.5	239.85
1232	-phenyl ether	259.3	40.7	242.0
1233	-1,3,5-triethylbenzene	215.5	91.1	214.7
1234	-fluorene	295.0	Zeotropic	
1235	-diphenyl methane	265.6	35.0	243.05
1236	-n-tridecane	234.0	70.0	229.7
1237	-1,2-diphenylethane	284.9	Zeotropic	
	Resorcinol	281.4		
1238	-naphthalene	218.05	Zeotropic	
1239	-1-naphthol	288.0	30	280.2
1240	-2-naphthol	295.0	15	280.8
1241	-methyl phthalate	283.7	62	287.5
1242	-1-methylnaphthalene	244.6	85.5	243.1
1243	-2-methylnaphthalene	241.15	89.5	240.05
	Resorcinol	281.4		
1244	-p-*tert*-amylphenol	266.5	85	265.8
1245	-acenaphthene	277.9	59	266.2
1246	-biphenyl	255.9	79	252.15
1247	-phenyl ether	259.3	77	255.65
1248	-fluorene	295.0	52	274.0
1249	-n-tridecane	234.0	88	233.25
1250	-stilbene	306.5	44	277.5
1251	-1,2-diphenylethane	284.9	53	269.7
	Pyrogallol	309.0		
1252	-2-naphthol	295.0	22	293.5
1253	-acenaphthene	277.9	80	272.8
1254	-biphenyl	256.1	90	253.5
	Aniline	184.35		
1255	-o-cresol, N	191.1	92	191.25
1256	-n-octyl alcohol	195.2	17	183.95
1257	-o-xylene	144.3	Zeotropic	
1258	-indene	182.6	58.5	179.75
1259	-mesitylene	164.7	88.0	164.35
1260	-pseudocumene	169.35	86.5	168.64
1261	-naphthalene	218.0	Zeotropic	
1262	-butylbenzene	183.1	54	177.8
1263	-n-nonane	150.7	86.5	149.2
1264	-n-decane	174.6	64	167.28
1265	-amyl ether	187.5	45	177.5
1266	-isoamyl ether	173.2	72	169.35

No.	System	B.P. (°C)	Azeotropic data Compn. X_2 (wt. %)	B.P. (°C)
1267	-2-methylnaphthalene	241.15	Zeotropic	
1268	-n-undecane	194.5	42.5	175.31
1269	-1,3,5-triethylbenzene	215.5	Zeotropic	
1270	-n-dodecane	216.5	28.5	180.37
1271	-n-tridecane	235.4	13.8	182.94
1272	-n-tetradecane	252.5	4.8	183.90
	2-Picoline	129.3		
1273	-n-octane	125.75	58.0	121.12
1274	-n-nonane	150.7	15.9	129.2
1275	-n-decane	174.6	Zeotropic	
	3-Picoline	144.0		
1276	-allyl sulfide	139.35	70	135.5
1277	-2,6-lutidine	144.06	27.3	143.5
	o-Phenylenediamine	258.6		
1278	-isosafrole	252.0	70	249.2
1279	-isafrole	235.9	Zeotropic	
1280	-biphenyl	256.1	63	249.7
1281	-phenyl ether	259.0	54	251.2
1282	-diphenylmethane	265.4	30	254.0
1283	-1,2-diphenylethane	284.5	Zeotropic	
	Methyl fumarate	193.25		
1284	-m-bromotoluene	184.3	84	183.65
1285	-o-cresol, N	191.1	40	197.8
1286	-m-cresol, N	202.2	28	204.3
1287	-benzyl ethyl ether	185.0	68	183.5
1288	-naphthalene	218.0	Zeotropic	
1289	-dipentene	177.7	30	172.5
	Methyl maleate	204.05		
1290	-caproic acid	205.15	37	201.5
1291	-o-cresol, N	191.1	22	204.65
1292	-m-cresol, N	202.2	45	208.75
1293	-p-cresol, N	201.7	44	208.6
1294	-naphthalene	218.0	13	203.7
1295	-borneol	215.0	22	202.95
1296	-isoamyl sulfide	214.8	18	203.0
	Cyclohexanone	155.7		
1297	-n-hexyl alcohol	157.85	6	155.65
1298	-cumene	152.8	35	152.0
1299	-camphene	159.6	42.5	150.55
1300	-2,7-dimethyloctane	160.1	45	151.5
	Ethyl acetoacetate	180.4		
1301	-phenetole	170.45	76	169.8
1302	-isobutyl sulfide	172.0	90	171.0
1303	-indene	182.6	32	177.15
1304	-propylbenzene	159.3	76	158.3
1305	-pseudocumene	168.2	63	165.2
1306	-butylbenzene	183.1	48	172.0
1307	-cymene	176.7	59	170.5
1308	-camphene	159.6	70	156.15
1309	-dipentene	177.7	57	169.05
1310	-d-limonene	177.8	57	169.05
1311	-2,7-dimethyloctane	160.1	76	156.0
1312	-amyl ether	187.5	30	174.5
	Ethylidene diacetate	168.5		
1313	-phenetole	170.45	44	164.5
1314	-butyl butyrate	166.4	63	163.5
1315	-ethyl caproate	167.7	55	164.0
1316	-*sec*-octyl alcohol	180.4	6.5	168.3
1317	-cineole	176.35	34	164.95
1318	-isoamyl ether	173.2	43	161.5
	Ethyl oxalate	185.65		
1319	-o-cresol, N	191.1	64	194.1
1320	-camphene	159.6	84	158.5
1321	-2,7-dimethyloctane	160.1	78	159.5
1322	-amyl ether	187.5	46	177.7
1323	-isoamyl ether	173.2	71	170.15
	Glycol diacetate	186.3		
1324	-o-cresol, N	191.1	65	194.5
	Capronitrile	163.9		
1325	-cumene	152.8	82	150.8
1326	-camphene	159.6	65	143.0
	Cyclohexanol	160.8		
1327	-o-xylene	143.6	86	143.0
1328	-m-xylene	139.0	95	138.9
1329	-indene	181.7	25	160.0
1330	-propylbenzene	158.8	60	153.8
1331	-naphthalene	218.05	Zeotropic	
1332	-cymene	176.7	28	159.5
1333	-camphene	159.5	59	151.9
1334	-cineole	176.35	8	160.55
	Butyl acetate	126.0		
1335	-paraldehyde	124.35	91	124.25

No.	System	B.P. (°C)	Azeotropic data: Compn. X_2 (wt. %)	Azeotropic data: B.P. (°C)
1336	-n-octane	125.8	48	119.0
1337	-butyl ether	142.1	5	125.9
	Caproic acid	205.3		
1338	-m-cresol	202.2	87	201.9
1339	-guaiacol	205.05	58	200.8
1340	-acetophenone	202.0	68	200.5
1341	-naphthalene	218.05	39	203.75
1342	-1-methylnaphthalene	244.6	Zeotropic	
1343	-2-methylnaphthalene	241.15	Zeotropic	
1344	-1,3,5-triethylbenzene	215.5	37	202.0
	Isoamyl formate	123.8		
1345	-paraldehyde	124.1	44	123.0
1346	-ethylbenzene	136.15	Zeotropic	
1347	-isobutyl ether	122.3	35	121.5
	Propyl propionate	122.5		
1348	-toluene	110.75	Zeotropic	
1349	-n-octane	125.8	40	118.2
	Propyl lactate	171.7		
1350	-o-cresol	191.1	Zeotropic	
1351	-isobutyl sulfide	172.0	52	169.0
1352	-mesitylene	164.6	72	160.5
1353	-isoamyl ether	173.2	47	167.5
	Isopropyl lactate	166.8		
1354	-o-cresol	191.1	Zeotropic	
1355	-mesitylene	164.6	40	159.5
1356	-camphene	159.6	70	154.2
	n-Hexyl alcohol	157.8		
1357	-o-cresol	191.1	zeotropic	
1358	-anisole	153.85	63.5	151.0
1359	-m-xylene	139.0	85	138.3
1360	-cumene	152.8	65	149.5
1361	-mesitylene	164.6	45	153.5
1362	-pesudocumene	168.2	32	156.3
1363	-propylbenzene	158.8	45	152.5
	Acetal	103.55		
1364	-methylcyclohexane	101.15	60	99.65
1365	-n-heptane	98.45	72	97.75
1366	-2,5-dimethylhexane	109.3	25	103.0
1367	-n-octane	125.75	Zeotropic	
	2-Butoxyethanol	171.15		
1368	-benzaldehyde	179.2	9	170.95
1369	-o-cresol, N	191.1	85	191.55
1370	-phenetole	170.45	48	167.1
1371	-isobutyl sulfide	172.0	58	163.8
1372	-mesitylene	164.6	68	162.0
1373	-butylbenzene	183.4	26.6	169.6
1374	-camphene	159.6	70	154.5
1375	-dipentene	177.7	47	164.0
1376	-cineole	176.35	41.5	168.9
	Pinacol	174.35		
1377	-o-cresol, N	191.1	92	191.5
1378	-p-cresol	201.7	Zeotropic	
1379	-n-octane	125.75	Zeotropic	
1380	-pseudocumene	168.2	62	162.9
1381	-propylbenzene	159.3	72	156.3
1382	-naphthalene	218.05	Zeotropic	
1383	-p-cymene	176.7	50	167.7
1384	-cineole	176.35	55	168.5
1385	-isoamyl ether	173.4	60	167.2
	Dipropylene glycol	229.2		
1386	-p-cresol	201.7	Zeotropic	
1387	-methyl salicylate	222.95	65	213.0
1388	-isosafrole	252.0	40	225.5
1389	-safrole	235.9	50	222.0
1390	-2-methylnaphthalene	241.1	Zeotropic	
	Triethylene glycol	288.7		
1391	-methyl phthalate	283.2	67	277.0
1392	-1-methylnaphthalene	244.6	Zeotropic	
1393	-acenaphthene	277.9	65	271.5
1394	-biphenyl	256.1	90	255.3
1395	-fluorene	294.0	Zeotropic	
1396	-phenyl benzoate	315.0	20	286.0
1397	-diphenylmethane	265.4	80	263.0
1398	-stilbene	306.5	40	284.5
1399	-1,2-diphenylethane	284.5	58	275.5
	Benzonitrile	191.1		
1400	-o-cresol, N	191.1	51	195.95
1401	-m-cresol, N	202.2	89	202.5
1402	-p-cresol, N	201.7	86	202.1
1403	-o-toluidine	200.35	Zeotropic	
1404	-isoamyl butyrate	181.05	92	180.85
1405	-cineole	176.35	86	175.6
1406	-amyl ether	187.5	58	180.5
1407	-isoamyl ether	173.2	84	171.4
	Benzaldehyde	179.2		
1408	-o-cresol, N	191.1	77	192.0
1409	-p-cresol	2017	Zeotropic	
1410	-naphthalene	218.0	Zeotropic	
1411	-p-cymene	175.3	72	171.0
1412	-d-limonene	177.8	57	171.2
1413	-camphene	159.6	84.5	158.45
1414	-isoamyl ether	173.2	62.5	168.6
	Benzoic acid	250.8		
1415	-p-nitrotoluene	238.9	89	237.4
1416	-3,4-xylenol	226.8	Zeotropic	
1417	-propyl succinate	250.5	57	248.0
1418	-naphthalene	218.05	95	217.7
1419	-1-methylnaphthalene	244.6	73	239.6
1420	-2-methylnaphthalene	241.15	75	237.25
1421	-biphenyl	277.9	49.5	246.05
1422	-phenyl ether	259.3	41	247.3
1423	-fluorene	295.0	Zeotropic	
1424	-diphenylmethane	265.6	18	248.95
1425	-1,2-diphenylethane	284.0	Zeotropic	
	o-Nitrotoluene	221.75		
1426	-benzyl alcohol	205.2	91	204.75
1427	-methyl salicylate	222.95	14	221.65
1428	-3,4-xylenol	226.8	Zeotropic	
1429	-2,4-xylidine	214.0	Zeotropic	
1430	-naphthalene	218.0	Zeotropic	
1431	-diethylaniline	217.05	88	216.85
1432	-geraniol	229.6	19	220.7
1433	-menthol	216.3	66	214.65
1434	-n-decyl alcohol	232.8	15	221.0
1435	-2-methylnaphthalene	241.15	Zeotropic	
1436	-bornyl acetate	227.6	27	221.15
	p-Nitrotoluene	238.9		
1437	-quinoline	237.3	92	237.2
1438	-safrole	235.9	82	234.5
1439	-geraniol	229.6	75	228.8
1440	-n-decyl alcohol	232.8	67	231.5
1441	-bornyl acetate	227.6	90	227.45
	Toluene	110.7		
1442	-2,6-lutidine	144.0	Zeotropic	
1443	-ethylcyclopentane	103.5	93	103.0
1444	-n-heptane	98.4	Zeotropic	
1445	-2,5-dimethylhexane	109.4	65	107.0
1446	-2-methylheptane	117.6	18	110.3
1447	-2,3,4-trimethylpentane	113.5	40	109.5
	Benzyl alcohol	205.2		
1448	-o-cresol	191.1	Zeotropic	
1449	-m-cresol, N	202.2	39	207.1
1450	-p-cresol, N	201.7	38	206.8
1451	-methylaniline	196.25	70	195.8
1452	-o-toluidine	200.35	Zeotropic	
1453	-3,4-xylenol	226.8	Zeotropic	
1454	-dimethylaniline	194.05	93.5	193.9
1455	-ethylaniline	205.5	50.0	202.8
1456	-2,4-xylidine	214.0	Zeotropic	
1457	-naphthalene	218.05	40	204.1
1458	-diethylaniline	217.05	28	20.42
1459	-d-limonene	177.8	89	176.4
1460	-borneol	215.0	14.2	205.07
1461	-1,3,5-triethylbenzene	215.5	43	203.2
	o-Cresol	191.1		
1462	-benzylamine, N	185.0	33	201.45
1463	-phenyl acetate, N	195.7	64	198.5
1464	-2,4,6-collidine, N	171.3	37	197.2
1465	-n-octyl alcohol	195.15	62	196.9
1466	-butyl sulfide	185.0	75	183.8
1467	-indene	183.0	91	182.9
1468	-naphthalene	218.05	Zeotropic	
1469	-terpinene	181.5	72	177.8
1470	-terpinolene	184.6	66	179.5
1471	-thymene	179.7	27	176.6
1472	-camphor, N	209.1	85	209.85
	m-Cresol	202.2		
1473	-o-toluidine, N	200.35	38.5	203.65
1474	-m-toluidine, N	203.1	47	205.5
1475	-p-toluidine, N	200.55	38	204.3
1476	-phenyl acetate, N	195.7	30	204.4
1477	-2,4,6-collidine, N	171.3	27	206.2
1478	-isoamyl lactate, N	202.4	50	207.6
1479	-n-octyl alcohol, N	195.15	38	203.3

No.	System	B.P. (°C)	Azeotropic data Compn. X_2 (wt. %)	B.P. (°C)
1480	-propiophenone, N	217.7	83	218.6
1481	-phorone, N	197.8	45	206.5
1482	-naphthalene	218.05	Zeotropic	
1483	-camphor, N	209.1	63.5	213.35
1484	-1,3,5-triethylbenzene	215.5	Zeotropic	
	p-Cresol	201.6		
1485	-o-toluidine, N	200.35	43	203.5
1486	-m-toluidine, N	203.1	53	204.9
1487	-p-toluidine, N	200.55	43	204.05
1488	-o-anisidine	219.0	Zeotropic	
1489	-acetophenone, N	202.0	53.5	208.4
1490	-benzyl formate, N	202.4	58.0	207.0
1491	-methyl benzoate, N	199.4	60.0	204.35
1492	-phenyl acetate, N	195.7	32	204.3
1493	-isoamyl lactate, N	202.4	52	207.25
1494	-n-octyl alcohol, N	195.2	30	202.25
1495	-camphor, N	209.1	69.5	213.5
1496	-ethyl caprylate, N	208.35	75	209.5
	Guaiacol	205.05		
1497	-acetophenone	202.0	32.5	205.25
1498	-m-toluidine	203.1	Zeotropic	
1499	-ethylaniline	205.5	45	204.4
1500	-ethyl caprylate, N	208.35	85	208.9
	2,4-Lutidine	159.0		
1501	-n-nonane	150.7	67.75	148.3
1502	-n-undecane	195.4	Zeotropic	
	Methylaniline	196.25		
1503	-n-octyl alcohol	195.2	55	193.0
1504	-d-limonene	177.8	97	174.5
	o-Toluidine	200.35		
1505	-acetophenone, N	202.0	68	203.65
1506	-n-octyl alcohol	195.2	77	194.7
1507	-n-decane	174.6	87	173.76
1508	-n-undecane	195.5	60.3	188.25
1509	-n-dodecane	216.5	37	195.75
1510	-n-tridecane	234.6	14.5	199.45
	o-Anisidine	219.0		
1511	-naphthalene	218.0	50	217.0
1512	-2-methylnaphthalene	241.15	Zeotropic	
1513	-1,3,5-triethylbenzene	215.5	65	214.5
	Enanthic acid	222.0		
1514	-ethyl fumarate	217.85	78	216.4
1515	-ethyl maleate	223.3	50	220.0
1516	-ethyl succinate	217.85	80	216.0
1517	-propiophenone	217.7	80	216.5
1518	-naphthalene	218.0	70	214.2
1519	-biphenyl	256.1	Zeotropic	
1520	1,3,5-triethylbenzene	215.5	73	211.0
	n-Heptyl alcohol	176.15		
1521	-benzyl methyl ether	167.8	80	167.0
1522	-p-methylanisole	177.05	48	173.3
1523	-phenetole	170.45	72	169.0
1524	-p-cymene	176.0	52	173.0
1526	-isoamyl ether	173.35	63	170.35
	Indole	253.5		
1527	-carvacrol	237.85	12	254.5
1528	-p-*tert*-amylphenol	266.5	88	268.0
	Acetophenone	202.		
1529	-p-ethylphenol N	218.8	85	219.5
1530	-2,4-xylenol, N	210.5	70	213.0
1531	-3,4-xylenol	226.8	Zeotropic	
1532	-dimethylaniline	194.15	Zeotropic	
1533	-ethylaniline	205.5	Zeotropic	
1534	-2,4-xylidine	214.0	Zeotropic	
1535	-n-octyl alcohol	195.2	87.5	194.95
1536	-naphthalene	218.0	Zeotropic	
1537	-1,3,5-triethylbenzene	215.5	Zeotropic	
	Phenyl acetate	195.7		
1538	-2,4-xylenol	210.5	Zeotropic	
1539	-n-octyl alcohol	195.15	47	192.4
1540	-indene	182.6	Zeotropic	
1541	-naphthalene	218.05	Zeotropic	
1542	-thymene	179.7	82	179.3
1543	-linalool	198.6	39	193.5
	Methyl salicylate	222.95		
1544	-phenethyl alcohol	219.4	57	218.0
1545	-3,4-xylenol	226.8	Zeotropic	
1546	-ethyl maleate	223.3	40	221.95
1547	-quinoline	237.3	Zeotropic	
1548	-geraniol	229.7	3	222.2
1549	-menthol	216.4	85	216.25
1550	-n-tridecane	234.0	Zeotropic	

No.	System	B.P. (°C)	Azeotropic data Compn. X_2 (wt. %)	B.P. (°C)
	Ethylbenzene	136.15		
1551	-ethylcyclohexane	131.8	85	131.2
1552	-n-octane	125.75	Zeotropic	
1553	-n-nonane	150.7	Zeotropic	
	p-Ethylphenol	218.8		
1554	-ethyl fumarate, N	217.85	52	223.0
1555	-ethyl maleate, N	223.3	62	226.3
1556	-p-methylacetophenone, N	226.35	70	229.5
1557	-benzyl acetate, N	215.0	40	221.0
1558	-ethyl benzoate, N	212.5	20	219.8
1559	-naphthalene	218.0	55	215.0
1560	-diethylaniline	217.05	40	214.0
1561	-1,3,5-triethylbenzene	215.5	60	212.0
	p-Methylanisole	177.05		
1562	-*sec*-octyl alcohol	180.4	21	176.3
1563	-pseudocumene	169.0	Zeotropic	
1564	-butyl isovalerate	177.6	42	176.4
1565	-butylbenzene	183.2	Zeotropic	
1566	-cineole	176.35	65	175.35
1567	-isoamyl ether	173.2	70.5	172.5
	Phenethyl alcohol	219.4		
1568	-3,4-xylenol	226.8	Zeotropic	
1569	-2,4-xylidine	214.0	Zeotropic	
1570	-naphthalene	218.05	56	214.2
1571	-diethylaniline	217.05	60	213.95
1572	-borneol	213.4	80	213.0
1573	-menthol	216.3	70	215.05
1574	-1-methylnaphthalene	244.9	Zeotropic	
1575	-biphenyl	256.1	Zeotropic	
	2,4-Xylenol	210.5		
1576	-ethyl fumarate, N	217.85	68	219.65
1577	-quinoline, N	237.3	92	239.0
1578	-p-methylacetophenone, N	226.35	15	227.0
1579	-propiophenone, N	217.7	35	221.0
1580	-benzyl acetate, N	215.0	64	216.8
1581	-camphor, N	209.1	50	217.0
	3,4-Xylenol	226.8		
1582	-o-phenetidine, N	232.5	92	232.65
1583	-ethyl fumarate, N	217.85	35	228.2
1584	-ethyl maleate, N	223.3	45	230.0
1585	-quinoline, N	237.3	65	241.95
1586	-p-methylacetophenone, N	226.35	49	231.35
1587	-propiophenone, N	217.7	33	228.5
1588	-naphthalene	218.0	84	217.6
1589	-diethylaniline	217.05	92	217.0
1590	-camphor, N	209.1	27	227.55
1591	-n-tridecane	234.0	42	223.5
	Ethylaniline	205.5		
1592	-n-octyl alcohol	195.2	85	194.9
1593	-naphthalene	218.0	Zeotropic	
1594	-camphor	209.1	Zeotropic	
	2,4-Xylidine	217.4		
1595	-menthol	216.3	30	213.5
1596	-n-undecane	195.5	88	194.98
1597	-n-dodecane	216.5	63	209.8
1598	-n-tridecane	234.6	29	215.28
1599	-n-tetradecane	252.0	2.5	217.38
	o-Phenetidine	232.5		
1600	-ethyl salicylate	233.8	18	232.2
1601	-naphthalene	218.0	Zeotropic	
1602	-safrole	235.9	14	232.38
1603	-anethole	235.7	25	232.25
1604	-carvacrol	237.85	87	238.0
1605	-thymol, N	232.9	54.9	234.3
1606	-2-methylnaphthalene	241.15	Zeotropic	
	p-Phenetidine	249.9		
1607	-safrole	235.9	Zeotropic	
1608	-isosafrole	252.0	36	248.8
1609	-1-methylnaphthalene	244.6	73	243.95
1610	-2-methylnaphthalene	241.15	85	240.85
1611	-biphenyl	256.1	10	249.5
1612	-phenyl ether	259.0	15	249.75
	Ethyl fumarate	217.85		
1613	-naphthalene	218.0	42	216.7
1614	-thymol, N	232.9	87.5	233.35
1615	-menthol	216.3	70	216.0
	Ethyl maleate	223.3		
1616	-p-methylacetophenone	226.35	12	223.15
1617	-naphthalene	218.0	77	217.65
1618	-thymol, N	232.9	73	234.9
	Ethyl succinate	217.25		

No.	System	B.P. (°C)	Azeotropic data: Compn. X_2 (wt. %)	Azeotropic data: B.P. (°C)
1619	-propiophenone	217.7	33	216.7
1620	-naphthalene	218.05	38.5	216.3
1621	-2-methylnaphthalene	241.15	Zeotropic	
	Caprylic acid	238.5		
1622	-naphthalene	218.05	94	216.2
1623	-carvacrol	237.85	75	237.6
1624	-1-methylnaphthalene	244.6	48	233.5
1625	-2-methylnaphthalene	241.15	52	235.0
	Propyl isovalerate	155.7		
1626	-cumene	152.8	Zeotropic	
1627	-propylbenzene	158.9	Zeotropic	
1628	-camphene	159.6	35	145.0
1629	-nopinene	163.8	25	155.0
	***sec*-Octyl alcohol**	179.0		
1630	butylbenzene	183.1	50	178.2
1631	-p-cymene	176.7	56	174.0
1632	-thymene	179.7	48	176.0
1633	-cineole	176.35	73.5	175.85
1634	-amyl ether	187.5	14	178.8
	Quinoline	237.3		
1635	-mesitol, N	220.5	15	240.4
1636	-safrole	235.9	73	235.15
1637	-carvacrol, N	237.85	52	244.3
1638	-thymol, N	232.9	45	243.1
1639	-1-methylnaphthalene	244.6	Zeotropic	
1640	2-methylnaphththalene	241.15	7	237.25
1641	-p-*tert*-amylphenol, N	266.5	94	267.5
1642	-biphenyl	256.1	Zeotropic	
	p-Methylacetophenone	226.35		
1643	-thymol, N	232.9	68	234.9
1644	-geraniol	229.6	5	226.25
1645	-citronellol	224.4	68	223.7
1646	-2-methylnaphthalene	241.15	Zeotropic	
1647	-bornyl acetate	227.6	40	225.8
	Propiophenone	217.7		
1648	-benzyl acetate	215.0	Zeotropic	
1649	-borneol	215.0	Zeotropic	
1650	-1,3,5-triethylbenzene	215.5	75	215.4
	Ethylsalicylate	233.8		
1651	-safrole	235.9	12	233.65
1652	-geraniol	229.7	60	228.5
1653	-n-decyl alcohol	232.9	52	230.5
1654	-2-methylnaphthalene	241.15	Zeotropic	
	Cumene	152.8		
1655	-n-nonane	150.75	77	148.0
	Mesitylene	164.6		
1656	-propylbenzene	159.3	Zeotropic	
1657	-camphene	159.6		
1658	2,7-dimethyloctane	160.1	72	158.6
	Propylbenzene	159.3		
1659	-camphene	159.6	53	158.0
	Pseudocumene	168.2		
1660	-p-cymene	176.7	Zeotropic	
1661	-n-decane	173.3	25	166.5
	Mesitol	230.5		
1662	-naphthalene	218.0	63	215.5
1663	-1,3,5-triethylbenzene	215.5	70	213.0
	3-Phenylpropanol	235.6		
1664	-naphthalene	218.05	~80	217.8
1665	-safrole	235.9	53	233.8
1666	-anethole	235.7	52	234.0
1667	-thymol, N	232.9	38	237.5
1668	-biphenyl	254.9	—	235.4
	Pelargonic acid	254.0		
1669	-naphthalene	218.0	Zeotropic	
1670	-isosafrole	252.0	65	249.5
1671	-eugenol	254.8	48	250.5
1672	-thymol	232.9	Zeotropic	
1673	-1-methylnaphthalene	244.6	82	243.0
1674	-2-methylnaphthalene	241.15	90	240.2
1675	-biphenyl	256.1	55	250.0
1676	-phenyl ether	259.0	45	250.5
1677	-1,3,5-triethylbenzene	215.5	Zeotropic	
1678	-diphenylmethane	265.4	25	252.7
	Naphthalene	218.0		
1679	-borneol	213.4	65	213.0
1680	-citronellol	224.5	30	217.8
1681	-menthol	216.4	74.5	215.5
1682	-n-tridecane	234.0	Zeotropic	
	Quinaldine	246.5		

No.	System	B.P. (°C)	Azeotropic data: Compn. X_2 (wt. %)	Azeotropic data: B.P. (°C)
1683	-safrole	235.9	Zeotropic	
1684	-carvacrol, N	237.85	33	250.8
1685	-thymol, N	232.9	20	250.0
	Methyl phthalate	283.2		
1686	-acenaphthene	277.9	66.5	276.35
1687	-biphenyl	255.9	Zeotropic	
1688	-diphenylmethane	265.6	Zeotropic	
1689	-1,2-diphenylethane	284.0	47	280.5
	Propyl benzoate	230.85		
1690	-carvacrol, N	237.85	82	238.85
1691	-thymol, N	232.8	55	235.5
1692	-2-methylnaphthalene	241.15	Zeotropic	
	p-Cymene	176.7		
1693	-dipentene	177.7	40	175.8
1694	-d-limonene	177.8	25	174.5
1695	-cineoloe	176.35	55	176.2
	Carvacrol	237.85		
1696	-carvenone, N	234.5	45	243.0
1697	-menthenone, N	222.5	25	239.5
1698	-propyl succinate, N	250.5	75	251.5
1699	-n-decyl alcohol	232.8	Zeotropic	
1700	-isobutyl benzoate, N	241.9	67	243.85
1701	-biphenyl	256.1	Zeotropic	
1702	-bornyl acetate, N	227.6	25	238.2
	Carvone	230.95		
1703	-thymol, N	232.9	52	238.65
1704	-geraniol	229.6	60	229.2
1705	-n-decyl alcohol	232.8	19	230.85
1706	-2-methylnaphthalene	241.15	Zeotropic	
	Thymol	232.9		
1707	-carvenone, N	234.5	50	241.0
1708	-pulegone, N	223.8	35	235.3
1709	-geraniol	229.6	42.5	225.6
1710	-menthone, N	209.5	8	233.2
1711	-2-methylnaphthalene	241.15	Zeotropic	
1712	-isobutyl benzoate, N	242.15	80	243.2
1713	-1,3,5-triethylbenzene	215.5	Zeotropic	
1714	-bornyl acetate, N	227.7	40	235.6
	Camphene	159.6		
1715	-dipentene	177.7	Zeotropic	
1716	-2,7-dimethyloctane	160.25	38	158.0
	Borneol	211.8		
1717	-methol	216.4	Zeotropic	
1718	-1,3,5-triethylbenzene	215.5	38	212.2
	Menthol	216.3		
1719	-2-methylnaphthalene	241.15	Zeotropic	
1720	-terpineol methyl ether	216.2	50	215.3
1721	-1,3,5-triethylbenzene	215.5	~45	214.0
	Capric acid	268.8		
1722	-1-methylnaphthalene	244.6	Zeotropic	
1723	-diphenylmethane	265.4	72	262.5
	1-Methylnaphthalene	244.6		
1724	-2-methylnaphthalene	241.15	Zeotropic	
1725	-biphenyl	256.1	Zeotropic	
1726	-phenyl ether	259.0	Zeotropic	
1727	-diphenylmethane	265.4	Zeotropic	
	2-Methylnaphthalene	241.15		
1728	-isobutyl benzoate	241.9	40	240.8
	p-*tert*-Amylphenol	266.5		
1729	-acenaphthene	277.9	Zeotropic	
1730	-fluorene	295.0	Zeotropic	
1731	-diphenylmethane	265.4	60	263.0
	Phenyl ether	259.0		
1732	-isoamyl benzoate	262.05	10	258.9
1733	-isoamyl oxalate	268.0	Zeotropic	
1734	-diphenylmethane	265.6	Zeotropic	
	Isoamyl benzoate	262.0		
1735	-isoamyl oxalate	268.0	Zeotropic	
1736	-diphenylmethane	265.6		
	1,3,5-Triethylbenzene	215.5		
1737	-bornyl acetate	227.2	Zeotropic	
1738	-bornyl ethyl ether	204.9	Zeotropic	
	Isoamyl oxalate	268.0		
1739	-diphenylmethane	265.4	86	265.25
1740	-1,2-diphenylethane	284.5	Zeotropic	
	Phenyl benzoate	315.0		
1741	-stilbene	306.5	Zeotropic	
1742	-benzyl ether	297.0	Zeotropic	
	Benzyl phenyl ether	286.5		
1743	-1,2-diphenylethane	284.5	Zeotropic	

TERNARY SYSTEMS

No.	System	B.P. (°C)	Azeotropic data: Compn. (wt. %)	Azeotropic data: B.P. (°C)	Type of azeotrope	Ref.
1	Argon	−186			—	1
	Nitrogen	−195	Zeotropic			
	Oxygen	−183	90—120°K			
2	Hydrogen bromide	−67	10.4		H	2
	Water	100	11.0	105		
	Chlorobenzene	131.8	78.6			
3	Hydrogen cyanide	26			—	3
	Acetonitrile	81.6	Zeotropic			
	Acrolein	52.45				
4	Chlorine trifluoride	—			—	4
	Hydrogen fluoride	19.4	Zeotropic			
	Uranium hexafluoride	56				
5	Hydrogen chloride	−80	15.8		H	5
	Water	100	64.8	107.33		
	Phenol	182	19.4			
6	Hydrogen chloride	−80	5.3		H	5
	Water	100	20.2	96.9		
	Chlorobenzene	131.8	74.5			
7	Hydrogen fluoride	19.4	10		N	6
	Fluosilicic acid	—	36	116.1		
	Water	100	54			
8	Silicon tetrafluoride	—	24.6		—	7
	Hexafluoroethane	−78	32.7	−104		
	Ethane	−88	42.7			
9	Water	100	4.5		H	8
	Carbon tetrachloride	76.75	85.5	62		
	Ethyl alcohol	78.3	10			
10	Water	100	5		H	8
	Carbon tetrachloride	76.75	84	65.15		
	Allyl alcohol	96.95	11			
11	Water	100	4.05		H	9
	Carbon tetrachloride	76.7	91.0	65		
	sec-Butyl alcohol	99	4.95			
12	Water	100	1.6		H	10
	Carbon disulfide	46.25	93.4	41.3		
	Ethyl alcohol	78.3	5.0			
13	Water	100			—	11
	Carbon disulfide	46.25	Zeotropic			
	Dioxane	101.4				
14	Water	100			—	12
	Chloroform	61	Zeotropic			
	Formic acid	100.75				
15	Water	100	1.3		—	13
	Chloroform	61	90.5	52.3		
	Methyl alcohol	64.7	8.2			
16	Water	100	2.3		—	14
	Chloroform	61	94.2	55.3		
	Ethyl alcohol	78.3	3.5			
17	Water	100	18.6		S	15
	Formic acid	100.8	71.9	107.2		
	Propionic acid	140.7	9.5			
18	Water	100	19.5		S	16
	Formic acid	100.8	75.9	107.62		
	Butyric acid	162.4	4.6			
19	Water	100	15.5		S	16
	Formic acid	100.8	66.8	107.02		
	Isobutyric acid	154	17.7			
20	Water	100	21.3		S	16
	Formic acid	100.8	76.3	107.64		
	Isovaleric acid	176.5	2.4			
21	Water	100			—	16
	Formic acid	100.8	Zeotropic			
	Valeric acid	186				
22	Water	100	2.1		H-3	17
	Nitromethane	101.2	6.5	33.1		
	n-Pentane	36.07	91.4			
23	Water	100	7.88		H-3; 748 mm Hg	18
	Nitromethane	101.2	29.73	71.43		
	n-Heptane	98.43	62.39			
24	Water	100	12.4		H-3; 748 mm Hg	18
	Nitromethane	101.2	44.25	77.35		
	n-Octane	125.7	43.35			
25	Water	100	17.4		H-3; 748 mm Hg	18
	Nitromethane	101.2	58.3	80.72		
	n-Nonane	150.8	24.3			
26	Water	100	19.1		H-3; 748 mm Hg	18
	Nitromethane	101.2	68.1	82.35		
	n-Decane	174.12	12.8			

TERNARY SYSTEMS (Continued)

No.	System	B.P. (°C)	Azeotropic data Compn. (wt. %)	B.P. (°C)	Type of azeotrope	Ref.
27	Water	100	20.6		H-3; 748 mm Hg	18
	Nitromethane	101.2	73.3	82.82		
	n-Undecane	194.5	6.1			
28	Water	100	21.5		H-3; 748 mm Hg	18
	Nitromethane	101.2	75.3	83.13		
	n-Dodecane	214.5	3.2			
29	Water	100	22.8		H-3; 748 mm Hg	18
	Nitromethane	101.2	75.4	83.21		
	n-Tridecane	234	1.8			
30	Water	100			—	19
	Methyl alcohol	64.7	Zeotropic			
	Ethyl alcohol	78.3				
31	Water	100	5.26		—	20
	Methyl alcohol	64.7	81.20	67.85		
	Methyl chloroacetate	131.4	13.54			
32	Water	100			—	21
	Methyl alcohol	64.7	Zeotropic			
	Methyl acetate	57.1				
33	Water	100	0.6		—	22
	Methyl alcohol	64.7	5.4	30.2		
	Isoprene	34.0	94.0			
34	Water	100	12.45		—	23
	Tetrachloroethylene	120.8	66.75	81.18		
	n-Propyl alcohol	97.2	20.8			
35	Water	100	6.4		—	24
	Trichloroethylene	86.95	73.1	67		
	Acetonitrile	81.6	20.5			
36	Water	100	5.5			25
	Trichloroethylene	86.95	78.4	67.0		
	Ethyl alcohol	78.3	16.1			
37	Water	100	7		—	26
	Trichloroethylene	86.95	73	69.4		
	Isopropyl alcohol	82.45	20			
38	Water	100			—	24
	Acetonitrile	81.6	Zeotropic			
	Acetone	56.4				
39	Water	100	1		—	26
	Acetonitrile	81.6	44	72.9		
	Ethyl alcohol	78.3	55			
40	Water	100			—	27
	Acetonitrile	81.6	Zeotropic			
	Diethylamine	55.5				
41	Water	100	8.2		—	24
	Acetonitrile	81.6	23.3	66		
	Benzene	80.2	68.5			
42	Water	100	3.5		—	13
	Acetonitrile	81.6	9.6	68.6		
	Triethylamine	89.7	86.9			
43	Water	100			—	8
	Acetic acid	118.1	Zeotropic			
	Toluene	110.7				
44	Water	100	8.4		H	17
	Nitroethane	114.07	9.3	59.5		
	n-Hexane	68.74				
45	Water	100	11.5		H	17
	Nitroethane	114.07	24.5	75.1		
	n-Heptane	98.43	64.0			
46	Water	100	8.7		—	26
	Ethyl alcohol	78.3	20.3	69.5		
	Acrylonitrile	77.2	71.0			
47	Water	100	4.8		—	26
	Ethyl alcohol	78.3	87.9	78.0		
	Crotonaldehyde	102.4	7.3			
48	Water	100	9.0		—	28
	Ethyl alcohol	78.3	8.4	70.23		
	Ethyl acetate	77.05	82.6			
49	Water	100	7.5		—	26
	Ethyl alcohol	78.3	42.5	81.8		
	Butylamine	77.8	50.0			
50	Water	100	6.3		—	26
	Ethyl alcohol	78.3	8.6	62		
	Butyl methyl ether	70.3	85.1			
51	Water	100	7.4		H	29
	Ethyl alcohol	78.3	18.5	64.86		
	Benzene	80.2	74.1			
52	Water	100	4.8		H	30
	Ethyl alcohol	78.3	19.7	62.60		
	Cyclohexane	80.75	75.5			

TERNARY SYSTEMS (Continued)

No.	System	B.P. (°C)	Azeotropic data Compn. (wt. %)	 B.P. (°C)	Type of azeotrope	Ref.
53	Water	100	9		—	31
	Ethyl alcohol	78.3	13	74.7		
	Triethylamine	89.4	78			
54	Water	100	12		H	26
	Ethyl alcohol	78.3	37	74.4		
	Toluene	110.6	51			
55	Water	100	3		H	26
	Ethyl alcohol	78.3	12	56.0		
	n-Hexane	68.7	85			
56	Water	100	6.1		H	26
	Ethyl alcohol	78.3	33.0	68.8		
	n-Heptane	98.45	60.9			
57	Water	100	0.4		—	32
	Acetone	56.7	7.6	32.5		
	Isoprene	34.7	92.0			
58	Water	100	8.5		H	33
	Allyl alcohol	96.95	5.1	59.7		
	n-Hexane	68.95	86.4			
59	Water	100	12.5		H	26
	Isopropyl alcohol	82.3	40.5	83		
	Butylamine	77.8	47.0			
60	Water	100	7.5		H	34
	Isopropyl alcohol	82.45	19.0	66.3		
	Benzene	80.2				
61	Water	100	7.5		H	8
	Isopropyl alcohol	82.45	18.5	64.3		
	Cyclohexane	80.75	74.0			
62	Water	100	13.1		H	26
	Isopropyl alcohol	82.3	38.2	76.3		
	Toluene	110.6	48.7			
63	Water	100	17.0		—	35
	Propyl alcohol	97.3	10.0	82.45		
	Propyl acetate	101.6	73.0			
64	Water	100	7.6		740mm Hg; H	26
	Propyl alcohol	97.2	10.1	67		
	Benzene	80.1	82.3			
65	Water	100	8.5		H	8
	Propyl alcohol	97.2	10.0	66.55		
	Cyclohexane	80.75	81.5			
66	Water	100	19.2		H	36
	n-Butyl alcohol	117.75	2.9	61.5		
	n-Hexane	68.95	77.9			
67	Water	100	41.4		H	36
	n-Butyl alcohol	117.75	7.6	78.1		
	n-Heptane	98.4	51.0			
68	Water	100	60.0		H	36
	n-Butyl alcohol	117.75	14.6	86.1		
	n-Octane	125.75	25.4			
69	Water	100	69.9		H	36
	n-Butyl alcohol	117.75	18.3	90.0		
	n-Nonane	150.7	11.8			
70	Water	100	29.9		H	26
	n-Butyl alcohol	117.75	34.6	90.6		
	Butyl ether	142.1	35.5			
71	Water	100	5		H	37
	2-Butanone	79.6	35	63.6		
	Cyclohexane	80.7	60			
72	Water	100	4		H	37
	Butyraldehyde	74.8	21	55.0		
	n-Hexane	68.7	75			
73	Water	100	8.9		H	14
	sec-Butyl alcohol	99.6	10.8	69.7		
	Cyclohexane	80.75	80.3			
74	Water	100	9		H	13
	sec-Butyl alcohol	99.4	19	76.3		
	Isooctane	99.0	72			
75	Water	100	8.1		H	38
	tert-Butyl alcohol	82.55	21.4	67.3		
	Benzene	80.2	70.5			
76	Water	100			—	22
	tert-Butyl alcohol	82.55	Zeotropic			
	Isoprene	34.0				
77	Water	100	8		H	8
	tert-Butyl alcohol	82.55	21	65.0		
	Cyclohexane	80.75	71			
78	Water	100			H	38
	Isobutyl alcohol	108	Zeotropic			
	Benzene	80.2				
79	Water	100	17.9		H	39
	Isobutyl alcohol	108	16.4	81.3		
	Toluene	110.7	65.7			

TERNARY SYSTEMS (Continued)

No.	System	B.P. (°C)	Azeotropic data Compn. (wt. %)	B.P. (°C)	Type of azeotrope	Ref.
80	Water	100			—	37
	Ethyl acrylate	99.3	Zeotropic			
	Isopropyl ether	68.3				
81	Water	100			—	40
	Toluene	110.7	Zeotropic			
	Benzyl alcohol	204.7				
82	Water	100			—	2
	Pyridine	115.5	Zeotropic			
	Benzene	80.1				
83	Water	100	14.0		H	41
	Pyridine	115.5	15.5	78.6		
	n-Heptane	98.4	70.5			
84	Water	100	22.5		H	41
	Pyridine	115.5	25.5	86.7		
	n-Octane	125.75	52.0			
85	Water	100	30.5		H	41
	Pyridine	115.5	37.0	90.5		
	n-Nonane	150.7	32.5			
86	Water	100	35.5		H	41
	Pyridine	115.5	45.5	92.3		
	N- Decane	173.3	19.0			
87	Water	100	38.5		H	41
	Pyridine	115.5	51.0	93.1		
	n-Undecane	194.5	10.5			
88	Water	100	40.5		H	41
	Pyridine	115.5	54.5	93.5		
	n-Dodecane	216.0	5.0			
89	Water	100	32.4		H	42
	Isoamyl alcohol	131.5	19.6	89.8		
	Isoamyl formate	124.2	48.0			
90	Water	100	44.8		—	43
	Isoamyl alcohol	131.5	31.2	93.6		
	Isoamyl acetate	142.0	24.0			
91	Water	100			—	37
	2-Picoline	129.2	Zeotropic			
	Paraldehyde	124.5				
92	Carbon tetrachloride	76.8			—	44
	Methyl alcohol	64.7	Zeotropic			
	Benzene	80.1				
93	Chloroform	61			—	45
	Formic Acid	100.75	Zeotropic			
	Acetic acid	118.1				
94	Chloroform	61	47		S	46
	Methyl alcohol	64.7	23	57.5		
	Acetone	56.1	30			
95	Chloroform	61	65.3		S	47
	Ethyl alcohol	78.3	10.4	63.2		
	Acetone	56.1	24.3			
96	Chloroform	61.0	56.1		—	48
	Ethyl alcohol	78.3	9.5	57.3		
	n-Hexane	68.7	34.4			
97	Chloroform	61.2	68.8		—	48
	Acetone	56.5	3.6	60.79		
	n-Hexane	68.7	27.6			
98	Chloroform	61.0			—	49
	Acetone	56.4	Zeotropic			
	Toluene	110.7				
99	Chloroform	61.2	79.70		S	50
	Ethyl formate	54.1	5.3	61.97		
	Isopropyl bromide	59.4	15.7			
100	Chloroform	61.2			—	51
	2-Bromopropane	59.4	Zeotropic			
	Isopropyl formate	68.8				
101	Methyl alcohol	64.7	17.4		—	26
	Acetone	56.1	5.8	53.7		
	Methyl acetate	56.3	76.8			
102	Methyl alcohol	64.7	14.6		—	52
	Acetone	56.25	30.8	47		
	n-Hexane	68.95	59.6			
103	Methyl alcohol	64.7	17.8		—	53
	Methyl acetate	57	48.6	50.8		
	Cyclohexane	80.75	33.6			
104	Methyl alcohol	64.7	14.6		—	14
	Methyl acetate	56.3	36.8	47.4		
	n-Hexane	68.7	48.6			
105	Methyl alcohol	64.7			—	8
	Benzene	80.1	Zeotropic			
	Cyclohexane	80.75				
106	Acetonitrile	81.6	34		—	26
	Ethyl alcohol	78.3	8	70.1		
	Triethylamine	89.7	58			

TERNARY SYSTEMS (Continued)

No.	System	B.P. (°C)	Azeotropic data Compn. (wt. %)	B.P. (°C)	Type of azeotrope	Ref.
107	Trichloroethylene	87.2			—	54
	Benzene	80.1	Zeotropic			
	Cyclohexane	80.7				
108	Acetic acid	118.1	23		S	55
	Acetic anhydride	139.6	55	134.4		
	Pyridine	115.5	22			
109	Acetic acid	118.1	3.4		S	56
	Pyridine	115.5	10.6	98.5		
	n-Heptane	98.4	86.0			
110	Acetic acid	118.1	10.4		S	56
	Pyridine	115.5	20.1	115.7		
	n-Octane	125.75	69.5			
111	Acetic acid	118.1	20.7		S	57
	Pyridine	115.5	29.4	128.0		
	n-Nonane	150.7	49.9			
112	Acetic acid	118.1	31.4		S	56
	Pyridine	115.5	38.2	134.1		
	n-Decane	173.3	30.4			
113	Acetic acid	118.1	37.5		S	58
	Pyridine	115.5	43.5	137.1		
	n-Undecane	194.5	19.0			
114	Acetic acid	118.1	13.5		S	59
	Pyridine	115.5	25.2	129.08		
	Ethylbenzene	136.5	61.3			
115	Acetic acid	118.1	17.7		S	57
	Pyridine	115.5	30.5	132.2		
	o-Xylene	143.6	51.8			
116	Acetic acid	118.1	10.2		S	60
	Pyridine	115.5	22.5	129.22		
	p-Xylene	138.4	67.3			
117	Acetic acid	118.1	15		—	61
	Isoamyl alcohol	132	54	132		
	Isoamyl acetate	142	31			
118	Acetic acid	118.1	7.6		—	62
	Benzene	80.1	34.4	77.2		
	Cyclohexane	80.75	58.0			
119	Acetic acid	118.1	3.6		S	63
	2-Picoline	129.45	24.8	121.3		
	n-Octane	125.75	71.6			
120	Acetic acid	118.1	12.8		S	63
	2-Picoline	129.45	38.4	135.0		
	n-Nonane	150.7	48.8			
121	Acetic acid	118.1	19.9		—	63
	2-Picoline	129.45	46.8	141.3		
	n-Decane	173.3	33.3			
122	Acetic acid	118.1	30.5		—	63
	2-Picoline	129.45	55.2	143.4		
	n-Undecane	194.5	14.3			
123	Acetic acid	118.1			—	64
	2,6-Lutidine	144.0	Zeotropic			
	n-Octane	125.75				
124	Acetic acid	118.1	12.6		S	64
	2,6-Lutidine	144.0	74.3	147.0		
	n-Decane	173.3	13.1			
125	Acetic acid	118.1	75.0		S	58
	2,6-Lutidine	144.0	13.8	163.0		
	n-Undecane	194.5	11.2			
126	Acetic acid	118.1			—	65
	Ethylbenzene	136.15	Zeotropic			
	n-Nonane	150.7				
127	Acetic acid	118.1			—	66
	Acetic anhydride	139.6	Zeotropic			
	Methylene diacetate	164.0				
128	Methyl formate	31.9			—	14
	Ethyl ether	34.6	Zeotropic			
	n-Pentane	36.15				
129	Nitroethane	114.2	31.7		S	67
	p-Dioxane	101.3	17.7	102.87		
	Isobutyl alcohol	108	50.6			
130	Ethyl alcohol	78.3	29.6		—	68
	Benzene	80.1	12.8	64.7		
	Cyclohexane	80.75	57.6			
131	Ethyl alcohol	78.3			—	69
	Benzene	80.1	Zeotropic			
	n-Hexane	68.7				
132	Ethyl alcohol	78.3			—	70
	Aniline	184.35	Zeotropic			
	Toluene	110.7				
133	Ethyl alcohol	78.3			—	70
	Aniline	184.35	Zeotropic			
	n-Heptane	98.4				

TERNARY SYSTEMS (Continued)

No.	System	B.P. (°C)	Azeotropic data Compn. (wt. %)	B.P. (°C)	Type of azeotrope	Ref.
134	Ethyl alcohol	78.3			—	70
	Toluene	110.7	Zeotropic			
	n-Heptane	98.4				
135	Ethylene glycol	197.4			—	71
	Pyridine	115.5	Zeotropic			
	Phenol	181.4				
136	Ethylene glycol	197.4	5.9		S	71
	Phenol	181.4	79.1	185.01		
	2-Picoline	128.8	15.0			
137	Ethylene glycol	197.4	15.9		S	71
	Phenol	181.4	67.7	186.41		
	3-Picoline	143.5	16.4			
138	Ethylene glycol	197.4	8.7		S	71
	Phenol	181.4	74.6	185.04		
	2,6-Lutidine	144.0	16.7			
139	Ethylene glycol	197.4	29.5		S	71
	Phenol	181.4	54.8	188.55		
	2,4,6-Collidine	171.0	15.7			
140	Ethylene glycol	197.45	33.6		S	72
	o-Cresol	191.0	62.4	189.65		
	2,4,6-Collidine	171.3	4.0			
141	Acetone	56.1	51.1		—	14
	Methyl acetate	56.3	5.6	49.7		
	n-Hexane	68.7	43.3			
142	Acetone	56.1			—	73
	2-Butanone	79.6	Zeotropic			
	Ethyl acetate	77.0				
143	Acetone	56.4			—	74
	Benzene	80.1	Zeotropic			
	Cyclohexane	80.75				
144	Acetone	56.1				75
	Benzene	80.1	Zeotropic			
	Toluene	110.7				
145	Propionic acid	140.7	55.5		S	58
	Pyridine	115.5	26.4	147.1		
	n-Undecane	194.5	18.1			
146	Propionic acid	141.05	4.5		S	76
	2-Picoline	129.3	10.5	123.7		
	n-Octane	125.4	85.0			
147	Propionic acid	141.05	16.5		S	76
	2-Picoline	129.3	21.5	140.1		
	n-Nonane	150.6				
148	Propionic acid	141.05	29.5		S	76
	2-Picoline	129.3	32.0	149.33		
	n-Decane	174.0	38.5			
149	Propionic acid	141.05	43.0		S	76
	2-Picoline	129.3	40.0	153.4		
	n-Undecane	194.8	17.0			
151	Propionic acid	141.05			S	76
	2-Picoline	129.3	Zeotropic			
	n-Dodecane	216.1				
151	Isopropyl alcohol	82.3	31.1		—	76
	Benzene	80.1	15.0	69.1		
	Cyclohexane	80.75	53.9			
152	Propyl alcohol	97.2	15.5		—	77
	Benzene	80.1	30.4	73.81		
	Cyclohexane	80.75	54.2			
153	Propyl alcohol	97.2			—	78
	Benzene	80.1	Zeotropic			
	n-Heptane	98.4				
154	2-Butanone	79.6			—	79
	Ethyl acetate	77.1	Zeotropic			
	n-Hexane	68.7				
155	2-Butanone	79.6			—	80
	Benzene	80.1	Zeotropic			
	Cyclohexane	80.75				
156	Butyric acid	162.45			—	58
	Pyridine	115.5	Zeotropic			
	n-Undecane	194.5				
157	p-Dioxane	101.1	44.3		S	81
	Isobutyl alcohol	107.0	26.7	101.8		
	Toluene	110.7				
158	n-Butyl alcohol	117.75	11.9		—	82
	Pyridine	115.5	20.7	108.7		
	Toluene	110.7	67.4			
159	n-Butyl alcohol	117.75	4		—	83
	Benzene	80.1	48	77.42		
	Cyclohexane	80.75	48			
160	n-Butyl alcohol	117.7			—	84
	Benzene	80.1	Zeotropic			
	n-Heptane	98.4				

TERNARY SYSTEMS (Continued)

No.	System	B.P. (°C)	Azeotropic data Compn. (wt. %)	B.P. (°C)	Type of azeotrope	Ref.
161	Isobutyl alcohol	107.0	43.2		—	85
	Benzene	80.1	47.0	77.2		
	Cyclohexane	80.75				
162	Pyridine	115.5	8.6		S	86
	Isoamyl alcohol	131.0	4.1	110.79		
	Toluene	110.7				
163	Phenol	181.4	33.5		—	87
	Aniline	183.95	48.5	184.45		
	n-Tridecane	234.0	18.0			
164	Phenol	182.0	26.4		S	88
	Ethylene diacetate	186.0	34.4	194.45		
	Phenyl acetate	195.7	39.2			
165	Phenol	181.4	19.88		S	89
	2,4-Lutidine	159.0	21.52	181.78		
	n-Undecane	194.5	58.60			
166	Ethylbenzene	136.15			—	59
	Pyridine	115.5	Zeotropic			
	n-Nonane	150.7				
167	Aniline	184.35			—	70
	Toluene	110.7	Zeotropic			
	n-Heptane	98.4				
168	Aniline	184.35			—	8
	Benzyl alcohol	205.5	Zeotropic			
	d-Limonene	177.8				
169	Aniline	184.35			—	8
	sec-Octyl alcohol	178.7	Zeotropic			
	d-Limonene	177.8				
170	Aniline	184.35			—	8
	o-Bromotoluene	181.75	Zeotropic			
	sec-Octyl alcohol	178.7				
171	Propyl lactate	171.7	31		—	8
	Phenetole	171.5	33	163.0		
	Menthene	170.8	36			
172	m-,p-Cresol (mixt.)	202	81		S	90
	Pyridine bases (mixt.)	143	9	202.81		
	Naphthalene	218.1	10			
173	m-, p-Cresol (mixt.)	202	65.5		S	90
	Pyridine bases (mixt.)	157	16.5	202.03		
	Naphthalene	218.1				
174	m-, p-Cresol (mixt.)	202	62		S	90
	Pyridine bases (mixt.)	163	17	202.39		
	Naphthalene	218.1	21			
175	m-Cresol	202.8	61.5		S	91
	2,4,6-Collidine	171.3	20.8	205.82		
	Naphthalene	217.9	17.7			
176	Isobutyl lactate	182.15			—	8
	sec-Octyl alcohol	178.7	Zeotropic			
	Terpinene	180.5				
177	Ethylbenzene	136.1			—	92
	Isopropylbenzene	152.8	Zeotropic			
	Butylbenzene	183.1				

REFERENCES

1. **Narinskii,** *Tr. Vses. Nauch. Issled. Inst. Kisloror. Mashinostr.,* 11, 3, 1967.
2. Dow Chemical Co., unpublished data.
3. **Sokolov, Sevryogova, Zhavoronskov,** *Rev. Chim. (Bucharest),* 20, 169, 1969.
4. **Ellis, Johnson,** *J. Inorg. Nucl. Chem.,* 6, 194, 199, 1958.
5. **Prahl, Mathes,** *Angew. Chem.,* 47, 11, 1934.
6. **Munter, Aepli, Kossatz,** *Ind. Eng. Chem.,* 39, 427, 1947.
7. **Calfee, Fukuhara, Bigelow,** *J. Amer. Chem. Soc.,* 61, 3552, 1939.
8. **Lecat,** *L'Azeotropisme,* Lamartin, Bruxelles, 1918.
9. **Marinichev, Susarev,** *Zhur. Fiz. Khim.,* 43, 1132, 1969.
10. **Ghysels,** *Bull. Soc. Chim. Belges,* 33, 57, 1924.
11. **De Mol,** *Ingr. Chim.,* 22, 262, 1938.
12. **Conti, Othmer, Gilmont,** *J. Chem. Eng. Data,* 5, 301, 1960.
13. **Kudryavtseva, Susarev, Eisen,** *Zhur. Fiz. Khim.,* 43, 437, 1969.
14. **Kudryavtseva, Eisen, Susarev,** *Zhur. Fiz. Khim.* 40, 1285, 1652, 1966.
15. **Kushner, Tatsievskaya, Serafimov,** *Zhur. Fiz. Khim.,* 41, 237, 1967.
16. **Kushner, Tatsievskaya, Serafimov,** *Zhur. Fiz. Khim.,* 42, 2248, 1968.
17. **Riddick,** Commercial Solvents Corp., unpublished data.
18. **Malesinska, Malesinski,** *Bull. Acad. Polon. Sci. Ser. Sci. Chim.,* 11, 475, 1963.

19. **Delzenne,** *J. Chem. Eng. Data,* 3, 224, 1958.
20. **Calices, Hannotte,** *Ingr. Chim.,* 20, 1, 1936.
21. **Balashov, Serafimov, Bessonova,** *Zhur. Fiz. Khim.,* 40, 2294, 1966.
22. **Lesteva, Kachalova, Morozova, Ogorodnikov, Trenke,** *Zhur. Prikl. Khim.,* 40, 1808, 1967.
23. **Malesinska,** *Bull. Acad. Polon. Sci. Ser. Sci. Chim.,* 12, 853, 1964.
24. **Pratt, Preprint,** Trans. Inst. Chem. Engrs. (London), March 1947.
25. **Licht, Denzler,** *Chem. Eng. Progr.,* 44, 627, 1948.
26. Union Carbide Chemicals, *Alcohols,* 1961.
27. Union Carbide Chemicals, unpublished data.
28. **Merriman,** *J. Chem. Soc. Trans.,* 103, 1790, 1801, 1913.
29. **Young, Fortey,** *J. Chem. Soc. Trans.,* 81, 717, 1902.
30. **Zieborak, Galska,** *Bull. Acad. Polon. Sci. Cl. III,* 3, 383, 1955.
31. **Tyerman,** Br. Pat. 590.713, 1947.
32. **Patterson,** U.S. Pat. 2,407.997, 1946.
33. **Kogan, Tolstova,** *Zhur. Fiz. Khim.,* 33, 276, 1959.
34. Yorizane, Yoshimura, *Hiroshima Daigaku Kogakuba Kenkya Hokoku,* 13, 41, 1965.
35. Smirnova, Moraczevskii, Storonkin, *Vest. Leningrad Univ.,* 14, 70, 1959.
36. **Kogan, Fridman, Deizenrot,** *Zhur. Prikl. Khim.* 30, 1339, 1957.
37. Union Carbide Chemicals Co., unpublished data.
38. **Young, Fortey,** *J. Chem. Soc. Trans.,* 81, 739, 1902.
39. **Frolov, Loginova, Nazarova,** *Zhur. Fiz. Khim.* 43, 2632, 1969.
40. **Susarev, Gorbunov,** *Zhur. Prikl. Khim.,* 36, 459, 1963.
41. **Trabczynski,** *Bull. Acad. Polon. Sci. Ser. Sci. Chim. Geol. Geogr.,* 6, 269, 1958.
42. **Hannotte,** *Bull. Soc. Chim. Belges,* 35, 85, 1926.
43. **Hyatt,** U.S. Patent 2,176.500, 1939.
44. **Hirata, Hirose,** *Mem. Fac. Technol. Tokyo Metrop. Univ.,* No. 11, 876, 1961.
45. **Conti, Othmer, Gilmont,** *J. Chem. Eng. Data,* 5, 301, 1960.
46. **Ewell, Welch,** *Ind. Eng. Chem.,* 37, 1224, 1945.
47. **Morachevskii, Leontev,** *Zhur. Fiz. Khim.,* 34, 2347, 1960.
48. **Kidryavtseva, Susarev,** *Zhur. Prikl. Khim.,* 36, 1231, 1471, 1710, 2025, 1963.
49. **Satapathy et al.,** *J. Appl. Chem. (London),* 6, 261, 1956.
50. **Orszagh, Lelakowska, Beldowicz,** *Bull. Acad. Polon. Sci. Ser. Sci. Chim. Geol. Geogr.,* 6, 419, 1958.
51. **Lelakowska,** *Bull. Acad. Polon. Sci. Ser. Sci. Chim. Geol. Geogr.,* 6, 645, 1958.
52. **Forman,** U.S. Patent 2,581.789, 1952.
53. **Fisher,** U.S. Patent 2,341.433, 1944.
54. **Rao, Dakshinamurty, Rao,** *J. Sci. Ind. Res. (India),* 20B, 218, 1961.
55. **Jones,** *J. Chem. Eng. Data,* 7, 13, 1962.
56. **Zieborak,** *Bull. Acad. Polon. Sci. Cl. III,* 3, 531, 1955.
57. **Zieborak, Wyrzykowska-Stankiewicz,** *Bull. Acad. Polon. Sci. Ser. Sci. Chim. Geol. Geogr.,* 7, 247, 1959.
58. **Zieborak, Wyrzykowska-Stankiewicz,** *Bull. Acad. Polon. Sci. Ser. Sci. Chim. Geol. Geogr.,* 6, 517, 1958.
59. **Galska-Krajewska, Zieborak,** *Rocz. Chem.,* 36, 119, 1962.
60. **Galska-Krajewska,** *Bull. Acad. Polon. Sci. Ser. Sci. Chim.,* 9, 455, 1961.
61. **Krokhin,** *Zhur. Fiz. Khim.,* 43, 442, 1969.
62. **Baradarajan, Satyanarayana,** *J. Chem. Eng. Data,* 13, 148, 1968.
63. **Zieborak, Wyrzykowska-Stankiewicz,** *Bull. Acad. Polon. Sci. Ser. Sci. Chim. Geol. Geogr.,* 6, 377, 1958.
64. **Zieborak, Kaczorowana-Badyoczek, Maczynska,** *Rocz. Chem.,* 29, 783, 1955.
65. **Zieborak, Galska-Krajewska,** *Bull. Acad. Polon. Sci. Ser. Sci. Chim. Geol. Geogr.,* 7, 253, 1959.
66. **Tatscheff et al.,** *Z. Phys. Chem. (Leipzig),* 237, 52, 1968.
67. **Malesinska, Malinsinski,** *Bull. Acad. Polon. Sci. Ser. Sci. Chim.,* 8, 191, 1960.
68. **Morachevskii, Zharov,** *Zhur. Prikl. Khim.,* 36, 2771, 1963.
69. **Yuan, Ho, Keshpande, Lu,** *J. Chem. Eng. Data,* 8, 549, 1963.
70. **Hollo, Ember, Lengyel, Weig,** *Acta Chim. Acad. Sci. Hung.,* 13, 307, 1957.
71. **Razniewska,** *Rocz. Cham.,* 38, 851, 1964.
72. **Kurtyka,** *Bull. Acad. Polon. Sci. Cl. III,* 4, 49, 1956.
73. **Babicz, Ivanchikova, Serafimov,** *Zhur. Prikl. Khim.,* 42, 1354, 1969.
74. **Kurmanadharao, Krishnamurty, Rao,** *Rec. Trav. Chim.,* 76, 769, 1957.
75. **Vitman, Zharov,** *Zhur. Prikl. Khim.,* 42, 2858, 1969.
76. **Trabczynski,** *Bull. Acad. Polon. Sci. Ser. Sci. Chim.,* 12, 335, 1965.
76. **Nagata,** *Can. J. Chem. Eng.,* 42, 82, 1964.
77. **Moraczevskii, Cheng,** *Zhur. Fiz. Khim.,* 35, 2535, 1961.
78. **Fu, Lu,** *J. Chem. Eng. Data,* 13, 6, 1968.
79. **Gorbunova, Lutigina, Malenko,** *Zhur. Prikl. Khim.,* 38, 374, 622, 1965.
80. **Donald, Ridgway,** *J. Appl. Chem. (London),* 8, 403, 408, 1958.

81. **Wyrzykowska-Stankiewicz, Zieborak,** *Bull. Acad. Polon. Sci. Ser. Sci. Chim.*, 8, 655, 1960.
82. **Hollo, Lengyel,** *Ind. Eng. Chem.*, 51, 957, 1959.
83. **Zieborak, Galska-Krajewska,** *Bull. Acad. Polon. Sci. Ser. Sci. Chim. Geol. Geogr.*, 6, 763, 1958.
84. **Vijayaraghavan, Deshpande, Kuloor,** *J. Chem. Eng. Data*, 12, 13, 1967.
85. **Nataraj, Rao,** *Trans. Indian Inst. Chem. Eng.*, 95, 1968.
86. **Zieborak, Wyrzykowska-Stankiewicz,** *Bull. Acad. Polon. Sci. Ser. Sci. Chim.*, 8, 137, 1960.
87. **Stadnicki,** *Bull. Acad. Polon. Sci. Ser. Sci. Chim.*, 10, 357, 1962.
88. **Orszagh, Lelakowska, Radecki,** *Bull. Acad. Polon. Sci. Ser. Sci. Chim. Geol. Geogr.*, 6, 605, 1958.
89. **Fahmy, Assal,** *Bull. Acad. Polon. Sci. Ser. Sci. Chim.*, 14, 773, 1966.
90. **Zieborak, Markowska-Majewska,** *Bull. Acad. Polon. Sci. Cl. III*, 2, 341, 1954.
91. **Kurtyka,** *Bull. Acad. Polon. Sci. Ser. Sci. Chim.*, 9, 741, 1961.
92. **Linek, Fried, Pick,** *Coll. Czech. Chem. Commun.*, 30, 1358, 1965.

QUATERNARY AND QUINARY SYSTEMS

No.	System	B.P. (°C)	Azeotropic data Compn. (wt. %)	B.P. (°C)	Type	Ref.
1	Hydrocyanic acid	26			—	1
	Water	100.0	Zeotropic			
	Acrylonitrile	77.3				
	Acrolein	52.4				
2	Hydrocyanic acid	26			—	1
	Acetonitrile	81.6	Zeotropic			
	Acrylonitrile	77.3				
	Acrolein	52.4				
3	Water	100.0			—	2
	Formic acid	100.8	Zeotropic			
	Acetic acid	118.1				
	Butyric acid	162.4				
4	Water	100.0	7.38		H	3
	Nitromethane	101.2	20.65	76.88		
	Tetrachloroethylene	120.8	59.45			
	n-Propyl alcohol	97.2	12.52			
5	Water	100.0	9.86		H	3
	Nitromethane	101.2	34.40	77.06		
	Tetrachloroethylene	120.8	32.60			
	n-Octane	125.75	23.14			
6	Water	100.0	—		H	3
	Tetrachloroethylene	120.8	—	80.98		
	n-Propyl alcohol	97.2	—			
	n-Octane	125.75	—			
7	Water	100.0	9.98		H	3
	Nitromethane	101.2	41.00	76.34		
	n-Propyl alcohol	97.2	12.42			
	n-Octane	125.75	36.60			
8	Water	100.0			—	4
	Acetonitrile	81.6	Zeotropic			
	Ethyl alcohol	78.3				
	Triethylamine	89.7				
9	Water	100.0	8.7		—	5
	Ethyl alcohol	78.3	11.1	70		
	Crotonaldehyde	102.2	0.1			
	Ethyl acetate	77.1	80.1			
10	Water	100.0	6.1		H	6
	Ethyl alcohol	78.3	19.2	62.14		
	Benzene	80.1	20.4			
	Cyclohexane	80.75	54.3			
11	Water	100.0			—	7
	Ethyl alcohol	78.3	Zeotropic			
	Benzene	80.1				
	n-Hexane	68.95				
12	Water	100.0			—	7
	Ethyl alcohol	78.3	Zeotropic			
	Benzene	80.1				
	Methylcyclohexane	100.88				
13	Water	100.0	6.8		H	8
	Ethyl alcohol	78.3	18.7	64.97		
	Benzene	80.1	62.4			
	n-Heptane	98.4	12.1			
14	Water	100.0	6.7		H	8
	Ethyl alcohol	78.3	17.7	64.69		
	Benzene	80.1	61.4			
	Isooctane	99.3	14.1			
15	Water	100.0			—	9
	1-Chlorobutane	78.44	Zeotropic			
	n-Butyl alcohol	117.73				
	Butyl ether	—				

No.	System	B.P. (°C)	Azeotropic data Compn. (wt. %)	B.P. (°C)	Type	Ref.
16	Chloroform	61.2			—	10
	Methyl alcohol	64.6	Zeotropic			
	Methyl acetate	56.9				
	Benzene	80.1				
17	Acetic acid	118.1	17		S	11
	Pyridine	115.4	27	127.9		
	Ethylbenzene	136.4	18			
	n-Nonane	150.8	38			
18	Acetic acid	118.1			—	12
	Pyridine	115.4	Zeotropic			
	p-Xylene	138.4				
	n-Nonane	150.8				
19	Acetone	56.1			—	13
	Isopropyl alcohol	82.3	Zeotropic			
	Benzene	80.1				
	Toluene	110.7				
20	Acetone	56.1			—	14
	Benzene	80.1	Zeotropic			
	Cyclohexane	80.9				
	Toluene	110.7				
21	Isopropyl alcohol	82.3			—	15
	2-Butanone	79.6	Zeotropic			
	Benzene	80.1				
	Cyclohexane	80.9				
22	Water	100.0	9.45		H	3
	Nitromethane	101.2	37.30			
	Tetrachloroethylene	120.8	21.15	76.5		
	n-Propyl alcohol	97.2	10.58			
	n-Octane	125.75	21.52			
23	Chloroform	61.2			—	10
	Methyl alcohol	64.6				
	Acetone	56.15	Zeotropic			
	Methyl acetate	56.9				
	Benzene	80.1				

REFERENCES

1. **Sokolov, Sevryoguva, Zhavoronkov,** *Teor. Osn. Khim. Tekhnol.*, 3, 288, 1969.
2. **Kushner, Lebedeva, Tatsievskaya, Serafimov,** *Shur. Prikl. Khim.*, 42, 1104, 1968.
3. **Malesinska,** *Bull. Acad. Polon. Sci. Ser. Sci. Chim.*, 12, 853, 1964.
4. Union Carbide Chemicals, unpublished data.
5. Eastman Chemical Products, unpublished data.
6. **Zieborak, Galska,** *Bull. Acad. Polon. Sci. Cl. III*, 3, 383, 1955.
7. **Swietoslawski, Zieborak, Galska-Krajewska,** *Bull. Acad. Polon. Sci. Ser. Sci. Chim. Geol. Geogr.*, 7, 43, 1959.
8. **Swietoslawski, Zieborak,** *Bull. Acad. Polon. Sci. Ser. A*, 9, 13, 1950.
9. **Riddick,** Commercial Solvents, unpublished data.
10. **Hudson, van Winkle,** *J. Chem. Eng. Data*, 14, 310, 1969.
11. **Zieborak, Galska-Krajewska,** *Bull. Acad. Polon. Sci. Ser. Sci. Chim. Geol. Geogr.*, 7, 253, 1959.
12. **Galska-Krajewska,** *Bull. Acad. Polon. Sci. Ser. Sci. Chim.*, 9, 455, 1961.
13. **Vitman, Zharov,** *Zhur. Prikl. Khim.*, 42, 2858, 1969.
14. **Vitman, Markova,** *Zhur. Prikl. Khim.*, 42, 2360, 1969.
15. **Lutugina, Kolbina,** *Zhur. Prikl. Khim.*, 41, 2766, 1968.

SELECTED VALUES OF CHEMICAL THERMODYNAMIC PROPERTIES

(From National Bureau of Standards Technical Notes 270-3, 270-4, 270-5, 270-6, 270-7 and 270-8

D. D. Wagman, W. H. Evans, V. B. Parker, R. H. Schumm, S. M. Bailey, I. Halow, K. L. Churney, and R. L. Nuttall

The compounds listed in the table represent only a small fraction of those for which data are given in the six Technical Notes referenced above. Copies of these Technical Notes may be purchased from the Superintendent of Documents, U.S. Government Printing Office, Washington, D.C. 20402. Conversion of units used in the table to SI units or other units may be accomplished by employing the following factors.

CONVERSION FACTORS FOR UNITS OF MOLECULAR ENERGY

	J mol^{-1}	cal mol^{-1}	cm^3 atm mol^{-1}	kWh mol^{-1}	Btu lb^{-1} mol^{-1}	cm^{-1} $molecule^{-1}$	eV $molecule^{-1}$
J mol^{-1} =	1	2.390057×10^{-1}	9.86923	2.77778×10^{-7}	0.429923	8.35940×10^{-2}	1.036409×10^{-5}
cal mol^{-1} =	4.18400^a	1	41.2929	1.162222×10^{-6}	1.798796	3.49757×10^{-1}	4.33634×10^{-5}
cm^3 atm mol^{-1} =	0.1013250^a	2.42173×10^{-2}	1	2.81458×10^{-8}	4.35619×10^{-2}	8.47016×10^{-3}	1.050141×10^{-6}
kWh mol^{-1} =	3,600,000	860,421	3.55292×10^{7}	1	1,547,721	300,938	37.3107
Btu lb^{-1} mol^{-1} =	2.32600^a	5.55927×10^{-1}	22.9558	6.46111×10^{-7}	1	1.944396×10^{-1}	2.41069×10^{-5}
cm^{-1} $molecule^{-1}$ =	11.96258	2.85912	118.0614	3.32294×10^{-6}	5.14299	1	1.239812×10^{-4}
eV $molecule^{-1}$ =	96487.0^a	23060.9	952,252	2.68019×10^{-2}	41482.0	8065.73	1

^{a} These numbers represent the fundamental values used in deriving the data in the table. The remaining factors were obtained by applying the relationships:

$$n_{ij} = n_{ik} \cdot n_{kj} \qquad n_{ii} = n_{ik} \cdot n_{ki} = 1$$

INTRODUCTION

Substances and Properties Included in the Tables

The tables contain values of the enthalpy and Gibbs energy of formation, enthalpy, entropy and heat capacity at 298.15 K (25°C), and the enthalpy of formation at 0 K, for inorganic substances and organic molecules contianing not more than two carbon atoms.

No values are given in these tables for metal alloys or other solid solutions, fused salts, or for substances of undefined chemical composition.

Physical States

The physical state of each substance is indicated in the column headed "State" as crystalline solid (c), liquid (l), glassy or amorphous (amorp), or gaseous (g). For solutions, the physical state is that normal for the indicated solvent at 298.15 K. Isomeric substances or various crystalline modifications of a given substance are designated by a number following the letter designation, as c2, g2, etc.

Definition of Symbols

The symbols used in these tables are defined as follows: P = pressure; V = volume; T = absolute temperature; S = entropy; H = enthalpy (heat content); G = H − TS = Gibbs energy (formerly the free energy); $C_p = (dH/dT)_p$ = heat capacity at constant pressure.

Conventions Regarding Pure Substances

The values of the thermodynamic properties of the pure substances given in these tables are for the substances in their standard states (indicated by the superscript ° on the thermodynamic symbol). These standard states are defined as follows:

1. For a pure solid or liquid, the standard state is the substance in the condensed phase under a pressure of one atmosphere.*
2. For a gas the standard state is the hypothetical ideal gas at unit fugacity, in which state the enthalpy is that of the real gas at the same temperature and zero pressure.

The values of ΔHf° and ΔGf° given in the tables represent the change in the appropriate thermodynamic quantity when one gram-formula weight of the substance in its standard state is formed isothermally at the indicated temperature from the elements, each in its appropriate standard reference state. The standard reference state at 298.15 K for each element except phosphorus has been chosen to be the standard state that is thermodynamically stable at that temperature and one atmosphere pressure. For phosphorus the standard reference state is the crystalline white form; the more stable forms have not been well characterized thermochemically. The same reference states have been maintained for the elements at 0 K except for the liquid elements bromine and mercury, for which the reference states have been chosen as the stable crystalline forms.

The value of $H°_{298} - H°_0$ represents the enthalpy difference for the given substance between 298.15 K and 0 K. If the indicated standard state at 298.15 K is the gas, the corresponding state at 0 K is the hypothetical ideal gas; if the state at 298.15 K is solid or liquid, the corresponding state at 0 K is the thermodynamically stable crystalline solid, unless otherwise specifically indicated.

The values of S° represent the virtual or "thermal" entropy of the substance in the standard state at 298.15 K, omitting contributions from nuclear spins. Isotope mixing effects, etc., are also excluded except in the case of the hydrogen-deuterium system. Where data have been available only for a particular isotope, they have been corrected when possible to the normal isotopic composition.

The values of the enthalpies of formation of gaseous ionic species are computed on the convention that the value of ΔHf° for the electron is zero. Conversions between 0 and 298.15 K are claculated using the value of $H°_{298} - H°_0$ = 1.481 kcal per mole of electrons, and assuming that the values of $H°_{298} - H°_0$ for the ionized and nonionized molecules are the same.

Conventions Regarding Solutions

For all dissolved substances the composition of the solvent is indicated following the chemical formula of the solute. In most instances the number of moles of solvent associated with 1 mole of solute is stated explicitly. In some cases the concentration of the solute cannot be specified. For aqueous solutions this is indicated in the State column by "au" (aqueous, unspecified). Such solutions may be assumed to be "dilute".

* One standard atmosphere equals 101325 pascal.

The standard state for a strong electrolyte in aqueous solution is the ideal solution at unit mean molality (unit activity). For a non-dissociating solute in aqueous solution the standard state is the ideal solution at unit molality.

The value of ΔHf° for a solute in its standard state is equal to the apparent molal enthalpy of formation of the substance in the infinitely dilute solution, since the enthalpy of dilution of an ideal solution is zero. At this dilution the partial molal enthalpy is equal to the apparent molal quantity. At concentrations other than the standard state, the value of ΔHf° represents the apparent enthalpy of the reaction of formation of the solution from the elements comprising the solute, each in its standard reference state, and the appropriate total number of moles of solvent. In this representation the value of ΔHf° for the solvent is not required. The experimental value for an enthalpy of dilution is obtained directly as the difference between the two values of ΔHf° at the corresponding concentrations. At finite concentrations the partial molal enthalpy of formation differs from the apparent enthalpy. In some instances the partial molal enthalpy of formation is given in the Tables. In this case the concentration designation is preceded by "D:".

The values for the thermodynamic properties for an individual ion in aqueous solution are for that undissociated ion in the standard state and are based on the convention that ΔHf°, ΔGf°, S°, and Cp° for H^+(a) are zero. The properties of the neutral strong electrolyte in aqueous solution in the standard state are equal to the sum of these values for the appropriate number of ions assumed to constitute the molecule of the given electrolyte. By adopting the above convention with respect to H^+(a), it follows that for an individual ionic species the G−H−S relation becomes

$$\Delta Gf^\circ = \Delta Hf^\circ - T\{\Delta Sf^\circ + (n/2)S^\circ[H_2(gs)]\}$$

with n = the algebraic value of the ionic charge. For neutral electrolytes and gaseous ions the normal consistency relation holds (see below).

Unit of Energy and Fundamental Constants

All of the energy values given in these tables are expressed in terms of the thermochemical calorie. This unit, defined as equal to 4.1840 joules, has been generally accepted for presentation of chemical thermodyanmic data. Values reported in other units have been converted to calories by means of the conversion factors for molecular energy given in the table which precedes this discussion.

The following values of the fundamental physical constants have been used in these calculations:

- R = gas constant = 8.3143 ± 0.0012 J/deg mol = 1.98717 ± 0.00029 cal/deg mol
- F = Faraday constant = 96487.0 ± 1.6 coulombs/mol = 23060.9 ± 0.4 cal/V equivalent
- Z = Nhc = 11.96258 ± 0.00107 J/cm^{-1} mol = 2.85912 ± 0.00026 cal/cm^{-1} mol
- c_2 = second radiation constant = hc/k = 1.43879 ± 0.00015 cm deg 0°C = 273.15 K

These constants are consistent with those given in the Table of Physical Constants, recommended by the National Academy of Sciences — National Research Council.* The formula weights listed in the tables have been calculated for the empirical molecular formula given in the Formula and Description column.

Internal Consistency of the Tables

The various aspects of internal consistency are specified below:

1. Subsidiary and auxiliary quantities used. All of the values given in these tables have been calculated from the original articles, using consistent values for all subsidiary and auxiliary quantities. The original data were corrected where possible for differences in energy units, molecular weights, temperature scales, etc. Thus we have sought to maintain a uniform scale of energies for all substances in the tables.

2. Physical and thermodynamic relationships for the tabulated properties of a substance. The tabulated values of the properties of a substance satisfy all the known physical and thermodynamic relationships among these properties. The quantities ΔHf°, ΔGf°, and S° at 298.15 K satisfy the relation (within the assumed uncertainty)

$$\Delta Gf^\circ = \Delta Hf^\circ - T\,\Delta Sf^\circ$$

Substance			0 K	298.15 K (25°C)				
Formula and Description	State	Formula weight	ΔHf°_0 kcal/mol	ΔHf°	ΔGf° kcal/mol	$H^\circ_{298} - H^\circ_0$	S° cal/deg mol	C°_p
ACTINIUM								
Ac	c	227.0280	0	0	0	—	13.5	6.5
	g	227.0280	—	97.0	87.6	1.481	44.92	4.98
ALUMINUM								
Al	c	26.9815	0	0	0	1.094	6.77	5.82
	g	—	77.44	78.0	68.3	1.654	39.30	5.11
Al^+	g	—	215.476	217.517	—	—	—	—
Al^{2+}	g	—	649.663	653.185	—	—	—	—
Al^{3+}	g	—	1305.70	1310.70	—	—	—	—
std. state, m = 1	aq	—	—	−127.	−116.	—	−76.9	—
Al_2	g	53.9630	116.	116.14	103.57	2.33	55.7	8.7
AlO_2^-								
std. state, m = 1	aq	58.9803	—	−219.6	−196.7	—	−5.	—
Al_2O_3								
α, corundum	c	101.9612	−397.59	−400.5	−378.2	2.394	12.17	18.89
δ	c	—	—	−398.	—	—	—	—
ϱ	c	—	—	−391.	—	—	—	—
ϰ	c	—	—	−397.	—	—	—	—
γ	c	—	—	−395.	—	—	—	—
	amorp	—	—	−390.	—	—	—	—
$Al_2O_3 \cdot H_2O$								
boehmite	c	119.9765	—	−472.0	−436.3	—	23.15	31.37
diaspore	c	—	—	−478.	−440.	—	16.86	25.22
$Al_2O_3 \cdot 3H_2O$								
gibbsite	c	156.0072	—	−612.5	−546.7	—	33.51	44.49
bayerite	c	—	—	−610.1	—	—	—	—
AlH_3	c	30.0054	—	−11.	—	—	—	—
$AlOH^{2+}$								
std. state, m = 1	aq	43.9889	—	—	−165.9	—	—	—
$Al(OH)_3$	amorp	78.0036	—	−305.	—	—	—	—
AlF_3	c	83.9767	−358.02	−359.5	−340.6	2.778	15.88	17.95
AlF_3	g	—	−287.01	−287.9	−284.0	3.37	66.2	14.97
un-ionized; std. state, m = 1	aq	—	—	−363.	−338.	—	−6.	—
ionized; std. state, m = 1	aq	—	—	−366.	−316.	—	−86.8	—

* NBS Technical News Bulletin, October 1963. See also Report of the CODATA Task Group on Fundamental Constants, CODATA Bulletin 11, December 1973.

Substance			0 K	298.15 K (25°C)				
Formula and Description	State	Formula weight	ΔHf°_0 kcal/mol	ΔHf°	ΔGf° kcal/mol	$H^\circ_{298} - H^\circ_0$	S° cal/deg mol	C°_p
$AlCl_3$	c	133.3405	−168.02	−168.3	−150.3	4.104	26.45	21.95
	g	—	—	−139.4	—	—	—	—
std. state, m = 1	aq	—	—	−247.	−210.	—	−36.4	—
$AlCl_3 \cdot 6H_2O$	c	241.4325	—	−643.3	—	—	—	—
Al_2Cl_6	g	266.6810	—	−308.5	−291.7	—	117.	—
$AlBr_3$	c	266.7085	—	−126.0	—	—	—	24.3
	g	—	—	−101.6	—	—	—	—
std. state, m = 1	aq	—	—	−214.	−191.	—	−17.8	—
Al_2Br_6	g	533.4170	—	−232.0	—	—	—	—
AlI_3	c	407.6947	—	−75.0	−71.9	—	38.	23.6
	g	—	—	−49.6	—	—	—	—
	aq	—	—	−165.8	—	—	—	—
Al_2I_6	g	815.3894	—	−123.5	—	—	—	—
Al_2S_3	c	150.1550	—	−173.	—	—	—	—
$Al_2(SO_4)_3$	c	342.1478	—	−822.38	−740.95	—	57.2	62.00
std. state, m = 1	aq	—	—	−906.	−766.	—	−139.4	—
$Al_2(SO_4)_3 \cdot 6H_2O$	c	450.2398	—	−1269.53	−1104.82	—	112.1	117.8
Al_2Se_3	c	290.843	—	−135.	—	—	—	—
Al_2Te_3	c	436.763	—	−78.	—	—	—	—
AlN	c	40.9882	−74.80	−76.0	−68.6	0.925	4.82	7.20
$Al(NO_3)_3$								
std. state, m = 1	aq	212.9962	—	−276.	−196.	—	28.1	—
$Al(NO_3)_3 \cdot 6H_2O$	c	321.0882	—	−681.28	−526.74	—	111.8	103.5
$AlCl_3 \cdot NH_3$	c	150.3711	—	−212.6	—	—	—	—
$AlCl_3 \cdot 3NH_3$	c	184.4323	—	−283.0	—	—	—	—
$AlCl_3 \cdot 6NH_3$	c	235.5242	—	−363.4	—	—	—	94.
$AlBr_3 \cdot NH_3$	c	283.7391	—	−177.6	—	—	—	—
$AlBr_3 \cdot 3NH_3$	c	317.8003	—	−252.7	—	—	—	—
$AlBr_3 \cdot 6NH_3$	c	368.8922	—	−343.0	—	—	—	—
$AlI_3 \cdot NH_3$	c	424.7253	—	−119.0	—	—	—	—
$AlI_3 \cdot 3NH_3$	c	458.7865	—	−205.9	—	—	—	—
$AlI_3 \cdot 6NH_3$	c	509.8784	—	−312.9	—	—	—	—
$NH_4Al(SO_4)_2$	c	237.1433	—	−562.2	−487.2	—	51.7	54.12
std. state, m = 1	aq	—	—	−593.	−491.	—	−40.2	—
$NH_4Al(SO_4)_2 \cdot 12H_2O$	c	453.3274	—	−1420.26	−1180.21	—	166.6	163.3
AlP	c	57.9553	—	−39.8	—	—	—	—
$AlPO_4$								
berlinite	c	121.9529	−411.40	−414.4	−386.7	3.528	21.70	22.27
$AlAs$	c	101.9031	—	−27.8	—	—	—	—
Al_4C_3	c	143.9594	−48.71	−49.9	−46.9	3.936	21.26	27.91
$Al(CH_3)_3$	c	72.0867	−29.76	—	—	—	—	—
	liq	—	—	−32.6	−2.4	8.114	50.05	37.19
	g	—	—	−17.7	—	—	—	—
$Al_2(CH_3)_6$	g	144.1734	—	−55.19	−2.34	—	125.4	—
$Al_2Si_2O_7 \cdot 2H_2O$								
kaolinite	c	258.1615	—	−979.6	−903.0	—	48.5	58.62
halloysite	c	—	—	−975.0	−898.5	—	48.6	58.86
$Al_6Si_2O_{13}$								
mullite	c	426.0532	—	−1632.8	−1541.2	—	61.	77.94
ANTIMONY								
Sb								
III	c	121.75	0	0	0	1.410	10.92	6.03
IV, explosive	amorp	—	—	2.54	—	—	—	—
	g	—	62.63	62.7	53.1	1.481	43.06	4.97
Sb^+	g	—	261.91	263.46	—	—	—	—
Sb^{2+}	g	—	643.1	646.1	—	—	—	—
Sb^{3+}	g	—	1227.1	1231.6	—	—	—	—
Sb^{4+}	g	—	2245.4	2251.4	—	—	—	—
SbO	g	137.749	48.	47.67	—	2.122	—	—
SbO^+								
std. state, m = 1	aq	—	—	—	−42.33	—	—	—
SbO_2^-								
std. state, m = 1	aq	153.749	—	—	−81.32	—	—	—
Sb_2O_3	aq	291.498	—	−164.9	—	—	—	—
Sb_2O_4	c	307.498	—	−216.9	−190.2	—	30.4	27.39
Sb_2O_5	c	323.497	—	−232.3	−198.2	—	29.9	—
SbH_3	g	124.774	36.625	34.681	35.31	2.502	55.61	9.81
$HSbO_2$								
undissoc.; std. state, m = 1	aq	154.757	—	−116.6	−97.4	—	11.1	—
$Sb(OH)_3$	c	172.772	—	—	−163.8	—	—	—
undissoc.; std. state, m = 1	aq	—	—	−184.9	−154.1	—	27.8	—
H_3SbO_4	aq	188.772	—	−216.8	—	—	—	—
SbF_3	c	178.745	—	−218.8	—	—	—	—
$SbCl_3$	c	228.109	—	−91.34	−77.37	—	44.0	25.8
$SbCl_3$	g	—	−74.57	−75.0	−72.0	4.269	80.71	18.33
$SbCl_5$	liq	299.015	—	−105.2	−83.7	—	72.	—
$SbOCl$	c	173.202	—	−89.4	—	—	—	—
	g	—	−25.	−25.5	—	—	—	—
$SbBr_3$	c	361.477	—	−62.0	−57.2	—	49.5	—
	g	—	−41.03	−46.5	−53.5	4.727	89.09	19.17
in CS_2	—	—	—	−58.4	—	—	—	—
SbI_3	c	502.463	—	−24.0	—	—	—	—
	aq	—	—	−23.6	—	—	—	—
Sb_2S_3								
black	c	339.692	—	−41.8	−41.5	—	43.5	28.65
orange	amorp	—	—	−35.2	—	—	—	—
$Sb_2(SO_4)_3$	c	531.685	—	−574.2	—	—	—	—
ARGON								
Ar	g	39.948	0	0	0	1.481	36.9822	4.9679
std. state, m = 1	aq	—	—	−2.9	3.9	—	14.2	—
Ar^+	g	—	363.42	364.90	—	—	—	—
Ar^{2+}	g	—	1000.5	1003.5	—	—	—	—
Ar^{3+}	g	—	1943.9	1948.3	—	—	—	—
ARSENIC								
As								
α, gray, metallic	c	74.9216	0	0	0	1.226	8.4	5.89
γ, yellow, cubic	c	—	—	3.5	—	—	—	—
β	amorp	—	—	1.0	—	—	—	—
	g	—	72.04	72.3	62.4	1.481	41.61	4.968
As^+	g	—	298.38	300.12	—	—	—	—
As^{2+}	g	—	764.42	767.64	—	—	—	—

Substance			0 K	298.15 K (25°C)				
Formula and Description	State	Formula weight	ΔHf°_0 kcal/mol	ΔHf°	ΔGf° kcal/mol	$H^\circ_{298} - H^\circ_0$	S° cal/deg mol	C°_p
As^{3+}	g	—	1417.44	1422.14	—	—	—	—
As^{4+}	g	—	2573.58	2579.76	—	—	—	—
As_2	g	149.8432	53.30	53.1	41.1	2.251	57.2	8.366
AsO	g	90.9210	16.88	16.72	—	2.101	—	—
AsO^+								
undissoc.; std. state, m = 1	aq	—	—	—	−39.15	—	—	—
$AsO_2{}^-{}_{3-}$								
std. state, m = 1	aq	106.9204	—	−102.54	−83.66	—	9.7	—
AsO_4								
std. state, m = 1	aq	138.9192	—	−212.27	−155.00	—	−38.9	—
As_2O_5	c	229.8402	—	−221.05	−187.0	—	25.2	27.85
$As_2O_5 \cdot 4H_2O$	c	301.9016	—	−503.0	—	—	—	—
AsH_3	g	77.9455	17.70	15.88	16.47	2.438	53.22	9.10
$HAsO_2$								
undissoc.; std. state, m = 1	aq	107.9284	—	−109.1	−96.25	—	30.1	—
$HAsO_4{}^{2-}$								
undissoc.; std. state, m = 1	aq	139.9272	—	−216.62	−170.82	—	−0.4	--
$H_2AsO_3{}^-$								
undissoc.; std. state, m = 1	aq	124.9357	—	−170.84	−140.35	—	26.4	—
$H_2AsO_4{}^-$								
undissoc.; std. state, m = 1	aq	140.9351	—	−217.39	−180.04	—	28.	—
H_3AsO_4	c	141.9431	—	−216.6	—	—	—	—
	aq	—	—	−216.2	—	—	—	—
AsF_3	liq	131.9168	—	−196.3	−185.04	—	43.31	30.25
	g	—	−186.82	−187.80	−184.22	3.413	69.07	15.68
$AsCl_3$	liq	181.2806	—	−72.9	−62.0	—	51.7	—
	g	—	−62.12	−62.5	−59.5	4.137	78.17	18.10
$AsBr_3$	c	314.6486	—	−47.2	—	—	—	—
	g	—	−25.55	−31.	−38.	4.569	86.94	18.92
AsI_3	c	455.6348	−13.91	−13.9	−14.2	5.964	50.92	25.28
	g	—	—	—	—	4.834	92.79	19.27
As_2S_2	c	213.9712	—	−34.1	—	—	—	—
As_2S_3	c	246.0352	—	−40.4	−40.3	—	39.1	27.8
$NH_4H_2AsO_3$								
std. state, m = 1, ($NH_4{}^+$ + $H_2AsO_3{}^-$)	aq	143.9743	—	−202.51	−159.32	—	53.5	—
$(NH_4)_2HAsO_4$	c	176.0043	—	−282.4	—	—	—	—
std. state, m = 1, ($2NH_4{}^+$ + $HAsO_4{}^{2-}$)	aq	—	—	−279.96	−208.76	—	53.8	—
BARIUM								
Ba	c	137.34	0	0	0	1.65	15.0	6.71
	g	—	43.2	43.	35.	1.481	40.663	4.968
Ba^+	g	—	163.39	164.67	—	—	—	—
Ba^{2+}	g	—	394.09	396.86	—	—	—	—
std. state, m = 1	aq	—	—	−128.50	−134.02	—	+2.3	—
BaO_2	c	169.339	—	−151.6	—	—	—	16.0
$BaO_2 \cdot H_2O$	c	187.354	—	−222.3	—	—	—	—
$BaO_2 \cdot 8H_2O$	c	313.462	—	−718.6	—	—	—	—
BaH	g	138.348	53.6	53.	47.	2.082	52.29	7.19
BaH_2	c	139.356	—	−42.7	—	—	—	—
$Ba(OH)_2$	c	171.355	—	−225.8	—	—	—	—
$Ba(OH)_2 \cdot H_2O$	c	189.370	—	−298.4	—	—	—	—
$Ba(OH)_2 \cdot 3H_2O$	c	225.401	—	−442.0	—	—	—	—
$Ba(OH)_2 \cdot 8H_2O$	c	315.477	—	−798.8	−667.6	—	102.	—
BaF_2	c	175.337	−288.19	−288.5	−276.5	3.452	23.03	17.02
	g	—	−195.0	−195.5	−198.0	3.262	71.91	12.85
std. state, m = 1	aq	—	—	−287.50	−267.30	—	−4.3	—
$BaCl_2$	c	208.246	−205.35	−205.2	−193.7	3.993	29.56	17.96
	g	—	−125.37	−125.7	−128.5	3.511	77.72	13.43
std. state, m = 1	aq	—	—	−208.40	−196.76	—	29.3	—
$BaCl_2 \cdot H_2O$	c	226.261	—	−277.4	−252.32	—	39.9	—
$BaCl_2 \cdot 2H_2O$	c	244.277	—	−348.98	−309.86	—	48.5	38.71
$Ba(ClO_2)_2$	c	272.244	−162.6	−127.0	—	—	47.	—
$Ba(ClO_3)_2$	c	304.242	—	−184.4	—	—	—	—
$Ba(ClO_4)_2$	c	336.241	—	−191.2	—	—	—	—
$Ba(ClO_4)_2 \cdot 3H_2O$	c	390.287	—	−404.3	−303.7	—	94.	—
$BaBr_2$	c	297.158	—	−181.0	−176.1	—	35.	—
	g	—	−101.4	−105.	−113.	3.9	79.	14.7
$BaBr_2$								
std. state, m = 1	aq	—	—	−186.60	−183.72	—	41.7	—
∞ H_2O	aq	—	—	−186.60	—	—	—	—
$BaBr_2 \cdot H_2O$	c	315.173	—	−255.3	—	—	—	—
$BaBr_2 \cdot 2H_2O$	c	333.189	—	−326.5	−294.1	—	54.	—
$Ba(BrO_3)_2$	c	393.154	—	−171.65	−130.1	—	59.	—
$Ba(BrO_3)_2$								
std. state, m = 1	aq	—	—	−160.56	−125.16	—	79.6	—
$Ba(BrO_3)_2 \cdot H_2O$	c	411.170	−236.99	−243.84	−188.6	9.94	68.9	53.5
BaI_2	c	391.149	—	−143.9	—	—	—	—
std. state, m = 1	aq	—	—	−154.88	−158.68	—	55.5	—
$BaI_2 \cdot H_2O$	c	409.164	—	−219.7	—	—	—	—
$BaI_2 \cdot 2H_2O$	c	427.179	—	−290.8	—	—	—	—
$BaI_2 \cdot 7H_2O$	c	517.256	—	−639.7	—	—	—	—
$Ba(IO_3)_2$	c	487.145	−243.31	−245.5	−206.7	8.84	59.6	44.8
std. state, m = 1	aq	—	—	−234.3	−195.2	—	58.9	—
$Ba(IO_3)_2 \cdot H_2O$	c	505.161	—	−316.0	−263.9	—	71.	—
BaS	c	169.404	—	−110.	−109.	—	18.7	11.80
Ba_2S_2	g	338.808	—	−90.	—	—	—	—
$BaSO_3$	c	217.402	—	−281.9	—	—	—	—
$BaSO_4$	c	233.402	—	−352.1	−325.6	—	31.6	24.32
std. state, m = 1	aq	—	—	−345.82	−311.99	—	7.1	—
BaS_2O_3	c	249.466	—	—	—	—	—	40.7
$BaSl_2O_6$	aq	297.464	—	−415.4	—	—	—	—
$Ba(HSO_3)_2$	aq	299.480	—	−430.0	—	—	—	—
$BaI_2 \cdot 4SO_2$	c	647.400	—	−470.2	—	—	—	—
$BaSe$	c	216.30	—	−89.	—	—	—	—
$BaSeO_3$	c	264.298	—	−248.7	−231.4	—	40.	—
$BaSeO_4$	c	280.298	—	−274.0	−249.7	—	42.	—
BaN_2	c	165.353	—	−41.	—	—	—	—
$Ba(N_3)_2 \cdot H_2O$	c	239.396	—	−73.7	−25.1	—	45.	—
$Ba(NO_2)_2$	c	229.351	—	−183.6	—	—	—	—

SELECTED VALUES OF CHEMICAL THERMODYNAMIC PROPERTIES (Continued)

Substance			0 K	298.15 K (25°C)				
Formula and Description	State	Formula weight	ΔHf°_0 kcal/mol	ΔHf°	ΔGf° kcal/mol	$H^\circ_{298} - H^\circ_0$	S° cal/deg mol	C°_p
	aq	—	—	−178.5	—	—	—	—
$Ba(NO_3)_2$	c	261.350	—	−237.11	−190.42	—	51.1	36.18
std. state, m = 1	aq	—	—	−227.62	−187.24	—	72.3	—
$BaHPO_4$	c	233.319	—	−433.7	—	—	—	—
$Ba(H_2PO_2)_2$	c	267.317	—	−421.2	—	—	—	—
$Ba(H_2PO_2)_2 \cdot H_2O$	c	285.332	—	−490.5	—	—	—	—
$Ba(H_2PO_4)_2$	c	331.315	—	−747.	—	—	—	—
$BaHAsO_4 \cdot H_2O$	c	295.283	—	−412.6	—	—	—	—
$Ba(H_2AsO_4)_2 \cdot 2H_2O$	c	455.241	—	−696.9	—	—	—	—
BaC_2	c	161.362	—	−18.				
$BaCO_3$								
witherite	c	197.349	—	−290.7	−271.9	—	26.8	20.40
std. state, m = 1	aq	—	—	−290.34	−260.19	—	−11.3	—
BaC_2O_4	c	225.360	—	−327.1	—	—	—	—
$BaC_2O_4 \cdot 2H_2O$	c	261.391	—	−471.1	—	—	—	—
$Ba(HCO_3)_2$								
std. state, m = 1	aq	259.375	—	−459.28	−414.54	—	45.9	—
$Ba(C_2H_3O_2)_2$	c	255.430	—	−354.8	—	—	—	—
std. state, m = 1	aq	—	—	−360.82	−310.60	—	43.7	—
$BaCN_2$	c	177.365	—	−63.6	—	—	—	—
$Ba(CN)_2$	c	189.376	—	−52.2	—	—	—	—
	aq	—	—	−55.0	—	—	—	—
$BaO \cdot SiO_2$	c	213.424	—	−388.05	−368.13	—	26.2	21.51
glassy	amorp	—	—	−376.	—	—	—	—
$BaO \cdot 2SiO_2$	c	273.509	—	−609.0	−576.2	—	36.6	32.05
$2BaO \cdot SiO_2$	c	366.764	—	−546.8	−519.8	—	42.1	32.24
$2BaO \cdot 3SiO_2$	c	486.933	—	−1000.2	−947.2	—	61.7	53.68
$Ba_2Fe(CN)_6$	aq	486.634	—	−145.7	—	—	—	—
$Ba_2Fe(CN)_6 \cdot 6H_2O$	c	594.726	—	−567.0	—	—	—	—
$Ba_3[Fe(CN)_6]_2$								
std. state, m = 1	aq	835.928	—	−116.9	−53.5	—	136.1	—
$BaMnO_4$								
barium manganate	c	256.276	—	—	−267.5	—	—	—
$BaCrO_4$	c	253.334	—	−345.6	−321.53	—	379	—
$BaMoO_4$	c	297.278	—	−370.	−344.1	—	33.	33.6
$BaWO_4$	c	385.188	—	−407.	—	—	—	—
$BaTiO_3$	c	233.238	—	−396.7	−375.8	—	25.8	24.49
$BaZrO_3$	c	276.558	—	−425.3	−405.0	—	29.8	24.31
BERYLLIUM								
Be	c	9.0122	0	0	0	0.466	2.27	3.93
	g	—	76.49	77.5	68.5	1.481	32.543	4.968
Be^+	g	—	291.474	293.965	—	—	—	—
Be^{2+}	g	—	711.426	715.398	—	—	—	—
Std. state, m = 1	aq	—	—	−91.5	−90.75	—	−31.0	—
BeO	c	25.0116	−144.86	−145.7	−138.7	0.669	3.38	6.10
	g	—	—	28.	—	—	—	—
BeO_2^{2-}								
std. state, m = 1	aq	41.0110	—	−189.0	−153.0	—	−38.	—
BeH	g	10.0202	75.	75.6	68.3	2.062	42.21	6.95
BeH_2	c	11.0282	—	−4.60	—	—	—	—
$Be(OH)_2$								
α	c	43.0269	—	−215.7	−194.8	—	12.4	—
β	c	—	—	−216.5	−195.4	—	12.	—
fresh precipitated	amorp	—	—	−214.6	—	—	—	—
$Be_3(OH)_3^3+$								
std. state, m = 1	aq	78.0587	—	—	−430.6	—	—	—
BeF_2								
α, quartz	c	47.0090	−244.85	−245.4	−234.1	2.024	12.75	12.39
β, cristobalite	c	—	—	−244.7	—	—	—	—
glassy	amorp	—	—	−244.3	—	—	—	—
$BeCl_2$								
α	c	79.9182	−117.40	−117.2	−106.5	2.863	19.76	15.50
β	c	—	−118.57	−118.5	−107.3	2.729	18.12	14.92
$BeCl_2 \cdot 4H_2O$	c	151.9796	—	−432.2	—	—	—	—
Be_2Cl_4	g	159.8364	—	−202.	—	—	—	—
$BeBr_2$	c	168.8302	—	−84.5	—	—	—	—
BeI_2	c	262.8210	—	−46.0	—	—	—	—
BeS	c	41.0762	—	−56.0	—	—	—	—
$BeSO_4$								
α, tetragonal	c	105.0738	−285.505	−288.05	−261.44	3.125	18.62	20.48
std. state, m = 1	aq	—	—	−308.8	−268.72	—	−26.2	—
$BeSO_4 \cdot H_2O$	c	123.0891	—	−364.2	—	—	—	—
$BeSO_4 \cdot 2H_2O$	c	141.1045	−429.814	−435.74	−381.99	5.865	39.01	36.63
$BeSO_4 \cdot 4H_2O$								
tetragonal	c	177.1352	−569.682	−579.29	−497.29	8.306	55.68	51.77
$BeSeO_4$	c	151.970	—	−212.8	—	—	—	—
std. state, m = 1	aq	—	—	−234.7	−196.2	—	−18.1	—
$BeSeO_4 \cdot 2H_2O$	c	188.000	—	−360.3	—	—	—	—
$BeSeO_4 \cdot 4H_2O$	c	224.031	—	−505.0	—	—	—	—
Be_3N_2								
α, cubic	c	55.0500	—	−140.6	—	—	—	—
β, hexagonal	c	—	—	−136.5	—	—	—	—
$Be(NO_3)_2$	aq	133.0220	—	−191.0	—	—	—	—
Be_2C	c	30.0356	—	−28.0	—	—	—	—
$BeCO_3$	c	69.0216	—	−245.	—	—	—	—
Be_2SiO_4	c	110.1080	−510.77	−513.7	−485.8	2.922	15.37	22.84
$Be(BO_2)_2$	g	94.6318	—	−325.	—	—	—	—
$BeO \cdot Al_2O_3$								
chrysoberyl	c	126.9728	−546.32	−549.9	−520.7	3.128	15.84	25.19
$BeO \cdot 3Al_2O_3$	c	330.8952	−1335.65	−1344.9	−1271.6	8.153	42.0	63.38
$BeMoO_4$	c	168.950	—	−328.	—	—	—	—
$PuBe_{13}$	c	356.21	—	−36.	—	—	—	—
BISMUTH								
Bi	c	208.980	0	0	0	1.536	13.56	6.10
	g	—	49.56	49.5	40.2	1.481	44.669	4.968
Bi^+	g	—	217.65	219.07	—	—	—	—
Bi^{2+}	g	—	602.5	605.4	—	—	—	—
Bi^{3+}	g	—	1192.0	1196.4	—	—	—	—
std. state, m = 1	aq	—	—	—	19.8	—	—	—
Bi_2	g	417.960	53.12	52.5	—	2.454	—	8.83

Substance			0 K	298.15 K (25°C)				
Formula and Description	State	Formula weight	ΔHf°_0 kcal/mol	ΔHf°	ΔGf° kcal/mol	$H^\circ_{298} - H^\circ_0$	S° cal/deg mol	C°_p
BiO^+								
std. state, m = 1	aq	—	—	—	−35.0	—	—	—
Bi_2O_3	c	465.9582	—	−137.16	−118.0	—	36.2	27.13
$BiO \cdot OH$	c	241.9868	—	—	−88.0	—	—	—
$Bi(OH)_3$	c	260.0021	—	−170.0	—	—	—	—
$BiCl_3$	c	315.339	—	−90.6	−75.3	—	42.3	25.
$BiCl_4^-$								
std. state, m = 1	aq	350.792	—	—	−115.1	—	—	—
$BiCl_6^{3-}$								
std. state, m = 1	aq	421.698	—	—	−178.51	—	—	—
$BiOCl$	c	260.4324	—	−87.7	−77.0	—	28.8	—
$Bi(OH)_2Cl$	c	278.4477	—	—	−128.71	—	—	—
$BiOBr$	c	304.8884	—	—	−71.0	—	—	—
BiI_3	c	589.6932	—	—	−41.9	—	—	—
BiI_4								
std. state, m = 1	aq	716.5976	—	—	−49.9	—	—	—
BiS	g	241.044	—	43.	29.	—	68.	—
Bi_2S_3	c	514.152	—	−34.2	−33.6	—	47.9	29.2
$Bi_2(SO_4)_3$	c	706.1448	—	−608.1	—	—	—	—
Bi_2Te_3	c	800.760	−18.43	−18.5	−18.4	7.387	62.36	28.8
$BiAsO_4$	c	347.8992	—	—	−148.	—	—	—
BORON								
B								
β	c	10.811	0	0	0	0.290	1.40	2.65
	amorp	—	—	0.9	—	0.315	1.56 + X*	2.86
	g	—	133.28	134.5	124.0	1.511	36.65	4.971
B^+	g	—	324.64	327.34	—	—	—	—
B^{++}	g	—	904.74	908.92	—	—	—	—
B^{+++}	g	—	1779.43	1785.09	—	—	—	—
B^{+++++}	g	—	15606.2	15614.9	—	—	—	—
BO_2^-								
std. state, m = 1	aq	—	—	−184.60	−162.27	—	−8.9	—
B_2O_3	c	69.6202	−302.731	−304.20	−285.30	2.223	12.90	15.04
	amorp	—	—	−299.84	−282.6	—	18.6	14.6
	g	—	−201.4	−201.67	−198.85	3.426	66.85	15.98
$B_4O_7^{--}$								
std. state, m = 1	aq	155.2398	—	—	−622.6	—	—	—
BH	g	11.8190	106.7	107.46	100.29	2.065	41.05	6.97
BH_3	g	13.8349	—	24.	—	—	—	—
BH_4^-								
std. state, m = 1	aq	14.8429	—	11.51	27.31	—	26.4	—
B_2H_6	g	27.6698	12.29	8.5	20.7	2.857	55.45	13.60
		—	—	6.0	23.0	—	39.4	—
B_4H_{10}	g	53.3237	—	15.8	—	—	—	—
H_3BO_3	c	61.8331	−258.312	−261.55	−231.60	—	21.23	19.45
	g	—	—	−237.6	—	—	—	—
un-ionized; std. state, m = 1	aq	—	—	−256.29	−231.56	—	38.8	—
$B(OH)_4^-$								
std. state, m = 1	aq	78.8405	—	−321.23	−275.65	—	24.5	—
$H_2B_4O_7$								
std. state, m = 1, (undissoc.)	aq	157.2557	—	—	−650.1	—	—	—
BF_3	g	67.8062	−271.082	−271.75	−267.77	2.784	60.71	12.06
BF_4^-								
std. state, m = 1	aq	86.8046	—	−376.4	−355.4	—	43.	—
B_2F_4	g	97.6156	−343.48	−344.2	−337.1	4.08	75.8	18.90
BCl_3	c	117.170	−105.40	—	—	—	—	—
BCl_3	liq	117.170	—	−102.1	−92.6	6.88	49.3	25.5
	g	—	−96.28	−96.50	−92.91	3.362	69.31	14.99
$BClF_2$	g	84.2608	—	−212.8	−209.4	—	65.	—
BCl_2F	g	100.7154	—	−154.2	−150.9	—	68.	—
BBr_3	liq	250.538	—	−57.3	−57.0	—	54.9	—
	g	—	−43.83	−49.15	−55.56	3.755	77.47	16.20
$BBrF_2$	g	128.7168	—	—	—	3.054	68.42	13.49
$BFBr_2$	g	189.6274	—	—	—	3.386	74.06	14.89
BCl_2Br	g	161.626	—	—	—	3.477	74.16	15.39
$BClBr_2$	g	206.082	—	—	—	3.609	76.90	15.78
BI_3	g	391.5242	18.	17.00	4.96	4.024	83.43	16.92
BS	g	42.875	81.	81.74	69.02	2.085	51.65	7.18
BN	c	24.8177	−60.10	−60.8	−54.6	0.628	3.54	4.71
NH_4BO_2								
std. state, m = 1	aq	60.8484	—	−216.27	−181.24	—	18.2	—
NH_4BO_3	aq	76.8478	—	−195.4	—	—	—	—
B_2C	g	33.6332	181.	—	—	—	—	—
B_4C	c	55.2552	−16.93	−17.	−17.	1.343	6.48	12.62
$B(CH_3)_3$	liq	55.9162	—	−34.2	−7.7	—	57.1	—
	g	—	−23.38	−29.7	−8.6	3.831	75.2	21.15
$B(C_2H_5)_3$	liq	97.9974	−42.54	−46.5	2.2	13.02	80.47	57.65
$BH(OCH_3)_2$								
dimethoxyborane	liq	73.8879	—	−144.7	−113.2	—	57.	—
$B(OCH_3)_3$								
trimethoxyborane	liq	103.9144	—	−223.2	−178.0	11.45	67.8	45.89
$B(OC_2H_5)_3$								
triethoxyborane	liq	145.9958	—	−250.8	—	—	—	—
BSi_2	g	66.983	175.	—	—	—	—	—
$BSiC$	g	50.9082	164.	—	—	—	—	—
BROMINE								
Br	g	79.909	28.189	26.741	19.701	1.481	41.805	4.968
Br^+	g	—	301.38	301.41	—	—	—	—
Br^{2+}	g	—	799.2	800.7	—	—	—	—
Br^{3+}	g	—	1627.0	1630.0	—	—	—	—
Br^-	g	—	−53.0	−55.9	—	—	—	—
std. state, m = 1	aq	—	—	−29.05	−24.85	—	19.7	−33.9
Br_2	c	159.818	0	—	—	—	—	—
	liq	—	—	0	0	5.859	36.384	18.090
	g	—	10.923	7.387	0.751	2.323	58.641	8.61
std. state, m = 1	aq	—	—	−0.62	0.94	—	31.2	—
Br_3^-								
std. state, m = 1	aq	239.727	—	−31.17	−25.59	—	51.5	—
Br_5^-								
std. state, m = 1	aq	399.545	—	−34.0	−24.8	—	75.7	—

* X = undetermined residual entropy.

Substance			0 K	298.15 K (25°C)				
Formula and Description	State	Formula weight	ΔHf°_0 kcal/mol	ΔHf°	ΔGf° kcal/mol	$H^\circ_{298} - H^\circ_0$	S° cal/deg mol	C°_p
BrO^-								
std. state, m = 1	aq	95.9084	—	−22.5	−8.0	—	10.	—
BrO_2	c	111.9078	—	11.6	—	—	—	—
BrO_3^-								
std. state, m = 1	aq	127.9072	—	−16.03	4.43	—	38.65	—
BrO_4^-								
std. state, m = 1	aq	—	—	3.1	28.2	—	47.7	—
HBr	g	80.9170	−6.826	−8.70	−12.77	2.067	47.463	6.965
std. state, m = 1	aq	—	—	−29.05	−24.85	—	19.7	−33.9
$HBrO_3$								
std. state, m = 1	aq	128.9152	—	−16.03	4.43	—	38.65	—
BrF_3	liq	136.9042	—	−71.9	−57.5	—	42.6	29.78
	g	—	−58.41	−61.09	−54.84	3.416	69.89	15.92
BrCl	g	115.362	5.28	3.50	−0.23	2.245	57.36	8.36
CADMIUM								
Cd								
γ	c	112.40	0	0	0	1.491	12.37	6.21
α	c	—	—	−0.14	−0.14	—	12.37	—
	g	—	26.78	26.77	18.51	1.481	40.066	4.968
in Hg; two-phase amalgam	—	—	—	−5.078	−2.328	—	3.145	—
Cd^+	g	—	234.18	235.65	—	—	—	—
Cd^{2+}	g	—	624.09	627.04	—	—	—	—
std. state, m = 1	aq	—	—	−18.14	−18.542	—	−17.5	—
CdO	c	128.399	—	−61.7	−54.6	—	13.1	10.38
$Cd(OH)_2$								
precipitated	c	146.415	—	−134.0	−113.2	—	23.	—
std. state, m = 1	aq	—	—	−128.08	−93.73	—	−22.6	—
undissoc.;	aq	—	—	—	−105.8	—	—	—
CdF_2	c	150.397	—	−167.4	−154.8	—	18.5	—
std. state, m = 1	aq	—	—	−177.14	−151.82	—	−24.1	—
$CdCl_2$	c	183.306	−93.677	−93.57	−82.21	3.791	27.55	17.85
std. state, m = 1	aq	—	—	−98.04	−81.286	—	9.5	—
undissoc.; std. state, m = 1	aq	—	—	−96.8	−85.88	—	29.1	—
$CdCl_2 \cdot H_2O$	c	201.321	—	−164.54	−140.310	—	40.1	—
$CdCl_2 \cdot 5/2H_2O$	c	228.344	—	−270.54	−225.644	—	54.3	—
$Cd(ClO_4)_2$								
std. state, m = 1	aq	311.301	—	−79.96	−22.66	—	69.5	—
$Cd(ClO_4)_2 \cdot 6H_2O$	c	419.393	—	−490.6	—	—	—	—
$CdCl_2 \cdot 2HCl \cdot 7H_2O$	c	382.335	—	−654.8	—	—	—	—
$CdBr_2$	c	272.218	−72.455	−75.57	−70.82	4.235	32.8	18.32
std. state, m = 1 1	aq	—	—	−76.24	−68.24	—	21.9	—
$CdBr_2 \cdot 4H_2O$	c	344.279	—	−356.73	−298.287	—	75.6	—
CdI_2	c	366.209	−48.52	−48.6	−48.13	4.565	38.5	19.11
std. state m = 1	aq	—	—	−44.52	−43.20	—	35.7	—
CdS	c	144.464	—	−38.7	−37.4	—	15.5	—
$CdSO_4$	c	208.462	−220.720	−223.06	−196.65	4.354	29.407	23.80
std. state, m = 1	aq	—	—	−235.46	−196.51	—	−12.7	—
$CdSO_4 \cdot H_2O$	c	226.477	−292.087	−296.26	−255.46	5.582	36.814	32.16
$CdSO_4 \cdot 8/3H_2O$	c	256.502	−406.960	−413.33	−350.224	8.497	54.883	50.97
$CdSO_4 \cdot 2\frac{1}{2}H_2SO_4$	c	453.655	—	−769.6	—	—	—	—
CdSe	c	191.36	—	—	—	—	—	—
$CdSeO_3$	c	239.358	—	−137.5	−119.0	—	34.0	—
std. state, m = 1	aq	—	—	−139.8	−106.9	—	−14.4	—
$CdSeO_4$	c	255.358	—	−151.3	−127.1	—	39.3	—
std. state, m = 1	aq	—	—	−161.3	−124.0	—	−4.6	—
$CdSeO_4 \cdot H_2O$	c	273.373	—	−225.2	—	—	—	—
CdTe	c	240.00	—	−22.1	−22.0	—	24.	—
$Cd(NO_3)_2$	c	236.410	—	−109.06	—	—	—	—
std. state, m = 1	aq	—	—	−117.26	−71.76	—	52.5	—
$Cd(NO_3)_2 \cdot 2H_2O$	c	272.440	—	−252.30	—	—	—	—
$Cd(NO_3)_2 \cdot 4H_2O$	c	308.471	—	−394.11	—	—	—	—
$Cd(NH_3)^{2+}$								
std. state, m = 1	aq	129.431	—	—	−28.4	—	—	—
$Cd(NH_3)_2^{2+}$								
std. state, m = 1	aq	146.461	—	−63.6	−38.0	—	34.6	—
$Cd(NH_3)_4^{2+}$								
std. state, m = 1	aq	180.522	—	−107.6	−54.1	—	80.4	—
$CdCl_2 \cdot 2NH_3$	c	217.367	—	−152.0	−106.1	—	51.	—
$CdCl_2 \cdot 4NH_3$	c	251.428	—	−195.4	−116.2	—	79.	—
$CdCl_2 \cdot 6NH_3$	c	285.490	—	−237.9	−126.2	—	109.2	—
$CdBr_2 \cdot NH_3$	c	289.249	—	−103.3	—	—	—	—
$CdBr_2 \cdot 2NH_3$	c	306.279	—	−131.5	—	—	—	—
$CdBr_2 \cdot 6NH_3$	c	374.402	—	−218.4	—	—	—	—
$CdI_2 \cdot 2NH_3$	c	400.270	—	−104.0	—	—	—	—
$CdI_2 \cdot 6NH_3$	c	468.392	—	−173.0	—	—	—	—
$Cd_3(PO_4)_2$	c	527.143	—	—	−587.1	—	—	—
$CdCO_3$	c	172.409	—	−179.4	−160.0	—	22.1	—
CdC_2O_4	c	200.420	—	−218.1	—	—	—	—
std. state, m = 1	aq	—	—	−215.3	−179.6	—	−6.6	—
undissoc.; std. state, m = 1	aq	200.420	—	—	−185.1	—	—	—
$CdC_2O_4 \cdot 3H_2O$	c	254.466	—	—	−360.4	—	—	—
$CdCN^+$								
std. state, m = 1	aq	138.418	—	—	15.4	—	—	—
$Cd(CN)_2$	c	164.436	—	38.8	—	—	—	—
std. state, m = 1	aq	—	—	53.9	63.9	—	27.5	—
undissoc.;	aq	—	—	—	49.7	—	—	—
$Cd(CN)_3^-$								
std. state, m = 1	aq	190.454	—	—	84.8	—	—	—
$Cd(CN)_4^{2-}$								
std. state, m = 1	aq	216.471	—	102.3	121.3	—	77.	—
$Cd(CNS)_2$	c	228.564	—	12.43	—	—	—	—
std. state, m = 1	aq	—	—	18.40	25.76	—	51.4	—
undissoc.; std. state, m = 1	aq	—	—	—	23.2	—	—	—
$CdSiO_3$	c	188.484	—	−284.20	−264.0	—	23.3	21.17
$Cd(BO_2)_2$	c	198.020	—	—	−354.87	—	—	—
CALCIUM								
Ca	c	40.08	0	0	0	1.364	9.90	6.05
	g	—	42.48	42.6	34.5	1.481	36.992	4.968
Ca^+	g	—	183.45	185.05	—	—	—	—
Ca^{2+}	g	—	457.21	460.29	—	—	—	—

SELECTED VALUES OF CHEMICAL THERMODYNAMIC PROPERTIES (Continued)

Substance			0 K	298.15 K (25°C)				
Formula and Description	State	Formula weight	ΔHf_0° kcal/mol	ΔHf°	ΔGf° kcal/mol	$H_{298}^\circ - H_0^\circ$	S° cal/deg mol	C_p°
std. state, m = 1	aq	—	—	−129.74	−132.30	—	−12.7	—
CaO	c	56.079	—	−151.79	−144.37	—	9.50	10.23
	g	—	+11.	—	—	—	—	—
CaO_2	c	72.079	—	−156.0	—	—	—	—
CaH	g	41.088	55.	54.7	47.9	2.076	48.19	7.11
CaH_2	c	42.096	—	−44.5	−35.2	—	10.	—
$CaOH^+$								
std. state, m = 1	aq	—	—	—	−171.7	—	—	—
$Ca(OH)_2$	c	74.095	—	−235.68	−214.76	—	19.93	20.91
	g	—	—	−130.	—	—	—	—
std. state, m = 1	aq	—	—	−239.68	−207.49	—	−17.8	—
CaF	g	59.078	−64.76	−65.0	−71.2	2.181	54.8	8.03
CaF_2	c	78.077	—	−291.5	−279.0	—	16.46	16.02
	g	—	−186.35	−186.8	−188.9	3.025	65.55	12.25
std. state, m = 1	aq	—	—	−288.74	−265.58	—	−19.3	—
$CaCl$	g	75.533	−23.23	−23.4	−29.7	2.292	57.70	8.58
$CaCl_2$	c	110.986	—	−190.2	−178.8	—	25.0	17.35
	g	—	−112.76	−112.7	−114.54	3.613	69.35	14.18
std. state, m = 1	aq	—	—	−209.64	−195.04	—	14.3	—
$CaCl_2 \cdot H_2O$	c	129.001	—	−265.1	—	—	—	—
$CaCl_2 \cdot 2H_2O$	c	147.017	—	−335.3	—	—	—	—
$CaCl_2 \cdot 4H_2O$	c	183.047	—	−480.3	—	—	—	—
$CaCl_2 \cdot 6H_2O$	c	219.078	—	−623.3	—	—	—	—
$CaOCl_2$	c	126.985	—	−178.4	—	—	—	—
$CaOCl_2 \cdot H_2O$	c	145.001	—	−249.1	—	—	—	—
$Ca(OCl)_2$	aq	142.985	—	−180.3	—	—	—	—
$Ca(ClO_2)_2$	c	174.984	—	−162.1	—	—	—	—
$Ca(ClO_4)_2$	c	238.981	—	−176.09	—	—	—	—
std. state, m = 1	aq	—	—	−191.56	−136.42	—	74.3	—
8 H_2O	aq	—	—	−188.36	—	—	—	—
10 H_2O	aq	—	—	−189.16	—	—	—	—
$Ca(ClO_4)_2 \cdot 4H_2O$	c	311.043	—	−465.8	−352.97	—	103.6	—
$CaBr_2$	c	199.898	—	−163.2	−158.6	—	31.	—
	g	—	—	−95.2	—	—	—	—
std. state, m = 1	aq	—	—	−187.84	−182.00	—	26.7	—
$CaBr_2 \cdot 6H_2O$	c	307.990	—	−599.0	−514.6	—	98.	—
$Ca(BrO_3)_2$	c	295.894	—	−163.9	—	—	—	—
CaI_2	c	293.889	—	−127.5	−126.4	—	34.	—
	g	—	—	−65.	—	—	—	—
std. state, m = 1	aq	—	—	−156.12	−156.96	—	40.5	
$CaI_2 \cdot 8H_2O$	c	438.012	—	−700.2	—	—	—	—
$Ca(IO_3)_2$	c	389.885	—	−239.6	−200.6	—	55.	—
$Ca(IO_3)_2 \cdot H_2O$	c	407.901	—	−309.1	—	—	—	—
$Ca(IO_3)_2 \cdot 6H_2O$	c	497.977	—	−664.6	−542.0	—	108.	—
CaS	c	72.144	—	−115.3	−114.1	—	13.5	11.33
	g	—	+32.	—	—	—	—	—
$CaSO_3$	c	120.142	—	—	—	—	24.23	21.92
$CaSO_3 \cdot 2H_2O$	c	156.173	—	−418.9	−371.7	—	44.	42.7
$CaSO_4$								
insol., anhydrite	c	136.142	—	−342.76	−315.93	—	25.5	23.82
sol., α	c	—	—	−340.64	−313.93	—	25.9	23.95
sol., β	c	—	—	−339.58	−312.87	—	25.9	23.67
std. state, m = 1	aq	—	—	−347.06	−310.27	—	−7.9	—
$CaSO_4 \cdot 1/2H_2O$	c	145.149	—	−376.85	−343.41	—	31.2	28.54
macro; α								
micro; β	c	—	—	−376.35	−343.18	—	32.1	29.69
$CaSO_4 \cdot 2H_2O$								
selenite	c	172.172	—	−483.42	−429.60	—	46.4	44.46
$CaSe$	c	119.04	—	−88.0	−86.8	—	16.	—
$CaSeO_4$	c	183.038	—	−265.25	—	—	—	—
$CaSeO_4 \cdot 2H_2O$	c	219.068	—	−407.9	−355.4	—	53.	—
$Ca(N_3)_2$	c	124.120		+3.5	—	—	—	—
Ca_3N_2	c	148.253	—	−103.	—	—	—	—
$Ca(NO_2)_2$	c	132.091	—	−177.2	—	—	—	—
in 800 H_2O	aq	—	—	−179.5	—	—	—	—
$Ca(NO_2)_2 \cdot 4H_2O$	c	204.152	—	−450.7	—	—	—	—
$Ca(NO_3)_2$	c	164.090	—	−224.28	−177.63	—	46.2	35.70
std. state, m = 1	aq	—	—	−228.86	−185.52	—	57.3	—
∞ H_2O	aq	—	—	−228.86	—	—	—	—
$Ca(NO_3)_2 \cdot 2H_2O$	c	200.120	—	−368.25	−293.82	—	64.4	—
$Ca(NO_3)_2 \cdot 3H_2O$	c	218.136	—	−439.3	−351.8	—	76.3	—
$Ca(NO_3)_2 \cdot 4H_2O$	c	236.151	—	−509.64	−409.53	—	89.7	—
$Ca(NH_2)_2$								
Ca_3P_2	c	182.188	—	−121.	—	—	—	—
$Ca(PO_3)_2$								
β	c	198.024	—	—	—	5.715	35.05	34.68
glassy	amorp	—	—	−587.0	—	—	—	—
$Ca_2P_2O_7$								
β	c	254.103	−792.88	−798.0	−748.6	7.430	45.23	44.89
$Ca_3(PO_4)_2$								
β, low temp. form	c	310.183	—	−984.9	−928.5	—	56.4	54.45
α, high temp form	c	—	—	−982.3	−926.3	—	57.58	55.35
std. state, m = 1	aq	—	—	−999.8	−883.9	—	−144.	—
$CaHPO_4$	c	136.059	−430.299	−433.65	−401.83	4.455	26.62	26.30
std. state, m = 1	aq	—	—	−438.57	−392.64	—	−20.7	—
$CaHPO_4 \cdot 2H_2O$	c	172.090	−568.032	−574.47	−515.00	7.490	45.28	47.10
$Ca(H_2PO_2)_2$	c	170.057	—	−418.9	—	—	—	—
std. state, m = 1	aq	—	—	−423.1	—	—	—	—
$Ca(H_2PO_4)_2$	c	234.055	—	−742.04	—	—	—	—
std. state, m = 1	aq	—	—	−749.38	−672.64	—	30.5	
$Ca(H_2PO_4)_2 \cdot H_2O$	c	252.070	−805.547	−814.93	−730.98	9.950	62.1	61.86
$Ca_3(AsO_4)_2$	c	398.078	—	−788.4	−732.1	—	54.	—
hydrated precipitate	—	—	—	−799.	—	—	—	—
$CaHAsO_4$	aq	180.007	—	−345.6	—	—	—	—
$Ca(H_2AsO_4)_2$	aq	321.950	—	−563.6	—	—	—	—
CaC_2	c	64.102	−15.14	−14.3	−15.5	2.711	16.72	14.99
$CaCO_3$								
calcite	c	100.089	—	−288.46	−269.80	—	22.2	19.57
aragonite	c	—	—	−288.51	−269.55	—	21.2	19.42
std. state, m = 1	aq	—	—	−291.58	−258.47	—	−26.3	—

SELECTED VALUES OF CHEMICAL THERMODYNAMIC PROPERTIES (Continued)

Substance			0 K	298.15 K (25°C)				
Formula and Description	State	Formula weight	ΔHf_0° kcal/mol	ΔHf°	ΔGf° kcal/mol	$H_{298}^\circ - H_0^\circ$	S° cal/deg mol	C_p°
CaC_2O_4	c	128.100	—	-325.2	—	—	—	—
std. state, m = 1	aq	—	—	-326.9	-293.37	—	-1.8	—
$CaC_2O_4 \cdot H_2O$	c	146.115	—	-400.30	-361.85	—	37.4	36.52
$CaCN_2$	c	80.105	—	-83.8	—	—	—	—
$Ca(CN)_2$	c	92.116	—	-44.1	—	—	—	—
	aq	—	—	-56.9	—	—	—	—
$CaO \cdot SiO_2$								
wollastonite	c	116.164	—	-390.76	-370.39	—	19.58	20.38
pseudowollastonite	c	—	—	-389.2	-369.2	—	20.88	20.67
glassy	amorp	—	—	-382.65	—	—	—	—
$2CaO \cdot SiO_2$								
β	c	172.244	-548.95	-551.5	-524.1	5.098	30.53	30.78
γ	c	—	-551.25	-554.0	-526.1	4.898	28.87	30.27
$3CaO \cdot SiO_2$	c	228.323	—	-700.1	-665.4	—	40.3	41.08
$3CaO \cdot 2SiO_2$								
rankinite	c	288.408	-942.23	-946.7	-899.0	8.424	50.38	51.24
$CaO \cdot Al_2O_3$	c	158.041	-552.87	-556.0	-527.9	4.569	27.30	28.87
$CaO \cdot 2Al_2O_3$	c	260.002	-944.98	-950.7	-901.2	7.286	42.50	48.00
$2CaO \cdot Al_2O_3$	c	214.120	—	-707	—	—	—	—
$3CaO \cdot Al_2O_3$	c	270.199	-853.21	-857.5	-815.4	8.216	49.2	50.16
$CaO \cdot Al_2O_3 \cdot 2SiO_2$								
anorthite, triclinic	c	278.210	—	-1009.2	-955.5	—	48.4	50.46
anorthite, hexagonal	c	—	—	-1004.3	-949.8	—	45.8	49.76
glassy	amorp	—	—	-991.8	—	—	—	—
$CaO \cdot Fe_2O_3$	c	215.772	-361.787	-363.37	-337.67	6.076	34.74	36.71
$2CaO \cdot Fe_2O_3$	c	271.851	-508.804	-511.30	-478.44	7.564	45.12	46.19
$CaCrO_4$	aq	156.074	—	-340.9	—	—	—	—
$CaMoO_3$	c	184.018	—	-296.	—	—	—	—
$CaMoO_4$	c	200.018	—	-368.4	-342.9	—	29.3	27.32
	g	—	-197.	—	—	—	—	—
std. state, m = 1	aq	—	—	-368.2	-332.2	—	-6.2	—
$CaWO_4$	c	287.928	-391.272	-393.20	-367.71	4.775	30.21	27.28
std. state, m = 1	aq	—	—	-386.8	—	—	—	—
$CaZrO_3$	c	179.298	—	-422.3	-401.8	—	23.92	23.88
$CaHfO_3$	c	266.568	—	-433.0	—	—	—	—
$CaMgC_2O_6$								
dolomite	c	184.411	-552.93	-556.0	-517.1	6.210	37.09	37.65
$2CaO \cdot 5MgO \cdot 8SiO_2 \cdot H_2O$								
tremolite	c	812.410	—	-2954.	-2780.	23.34	131.2	156.7
$CaUO_4$	c	342.107	—	-478.4	—	—	—	—
CARBON								
C								
graphite, Acheson spectroscopic	c	12.0112	0	0	0	0.251	1.372	2.038
diamond	c	—	0.5797	0.4533	0.6930	0.125[a]	0.568	1.4615
	g	—	169.98	171.291	160.442	1.562	37.7597	4.9805
C^+	g	—	429.628	432.420	—	—	—	—
C^{2+}	g	—	991.900	996.173	—	—	—	—
C^{3+}	g	—	2095.98	2101.73	—	—	—	—
CO	g	28.0106	-27.199	-26.416	-32.780	2.0716	47.219	6.959
std. state, m = 1	aq	—	—	-28.91	-28.66	—	25.0	—
in CH_3COOH	—	—	—	-26.416	-28.15	—	31.7	—
CO^+	g	—	295.9	298.16	—	—	—	—
CO^{2+}	g	—	942.	945.5	—	—	—	—
CO_2	g	44.0100	-93.963	-94.051	-94.254	2.2378	51.06	8.87
undissoc.; std. state, m = 1	aq	—	—	-98.90	-92.26	—	28.1	—
CO_2^+	g	—	223.8	225.23	—	—	—	—
CO_3^{2-}								
std. state, m = 1	aq	60.0094	—	-161.84	-126.17	—	-13.6	—
CH	g	13.0191	141.6	142.4	—	—	—	—
CH^+	g	—	398.1	400.4	—	—	—	—
CH_2	g	14.0271	93.9	93.7	—	—	—	—
CH_2^+	g	—	333.6	334.9	—	—	—	—
CH_3	g	15.0351	34.0	33.2	—	—	—	—
CH_3^+	g	—	261.0	261.7	—	—	—	—
CH_4	g	16.0430	-15.970	-17.88	-12.13	2.388	44.492	8.439
std. state, m = 1	aq	—	—	-21.28	-8.22	—	20.0	—
CH_4^+	g	—	277.1	276.7	—	—	—	—
HCO	g	29.0185	-4.2	-4.12	-7.76	2.386	53.68	8.26
$HCOO^-$								
std. state, m = 1	aq	45.0180	—	-101.71	-83.9	—	22.	-21.0
HCO_3^-								
std. state, m = 1	aq	61.0174	—	-165.39	-140.26	—	21.8	—
HCHO	g	30.0265	-27.1	-28.	-27.	2.394	52.26	8.46
unhydrolyzed	aq	—	—	-35.9	—	—	—	—
HCOOH	liq	—	—	-101.51	-86.38	—	30.82	23.67
	g	—	—	-90.48	—	—	—	—
un-ionized; std. state, m = 1	aq	—	—	-101.68	-89.0	—	39.	—
ionized; std. state, m = 1	aq	—	—	-101.71	-83.9	—	22.	-21.0
in ∞ H_2O	aq	—	—	-101.71	—	—	—	—
H_2CO_3								
std. state, m = 1, (undissoc.)	aq	62.0253	—	-167.22	-148.94	—	44.8	—
CH_3OH	liq	32.0424	—	-57.04	-39.76	—	30.3	19.5
	g	-45.355	-47.96	-38.72	2.731	57.29	10.49	
std. state, m = 1	aq	—	—	-58.779	-41.92	—	31.8	—
CF_3	g	69.0064	-113.	-114.	—	—	—	—
CF_3^+	g	—	119.9	120.4	—	—	—	—
CF_4	g	88.0048	-219.6	-221.	-210.	3.043	62.50	14.60
COF_2	g	66.0074	-150.95	-151.7	-148.0	2.642	61.78	11.19
CH_3F	g	34.0335	—	—	—	2.422	53.25	8.96
CH_2F_2	g	52.0239	-104.97	-106.8	-100.2	2.555	58.94	10.25
CHF_3	g	70.0143	-162.84	-164.5	-156.3	2.764	62.04	12.20
CCl_3	g	118.3702	14.	14.	—	—	—	—
CCl_4	liq	153.8232	—	-32.37	-15.60	—	51.72	31.49
	g	—	-24.08	-24.6	-14.49	4.117	74.03	19.91
$COCl_2$	g	98.9166	-51.89	-52.3	-48.9	3.067	67.74	13.78
CH_3Cl	g	50.4881	-17.426	-19.32	-13.72	2.489	56.04	9.74
std. state, m = 1	aq	—	—	-24.3	-12.3	—	34.6	—

[a] Relative to C, diamond.

SELECTED VALUES OF CHEMICAL THERMODYNAMIC PROPERTIES (Continued)

Substance			0 K	298.15 K (25°C)				
Formula and Description	State	Formula weight	ΔHf°_0 kcal/mol	ΔHf°	ΔGf° kcal/mol	$H^\circ_{298} - H^\circ_0$	S° cal/deg mol	C°_p
CH_2Cl_2	liq	84.9331	—	−29.03	−16.09	—	42.5	23.9
	g	—	−20.462	−22.10	−15.75	2.830	64.56	12.18
CF_3Cl	g	104.4594	−164.8	−166.	−156.	3.293	68.16	15.98
CF_2Cl_2	g	120.9140	−113.0	−114.	−105.	3.543	71.86	17.27
$CFCl_3$	liq	137.3686	—	−72.02	−56.61	—	53.86	29.05
	g	—	−65.2	−66.	−57.	3.843	74.05	18.66
COFCl	g	86.4620	—	—	—	2.845	66.11	12.52
CH_2ClF	g	68.4785	—	—	—	2.689	63.17	11.24
$CHClF_2$	g	86.4689	—	—	—	2.955	67.11	13.35
$CHCl_2F$	g	102.9235	—	—	—	3.170	70.02	14.56
CBr_4								
monoclinic	c	331.6472	—	4.5	11.4	—	50.8	34.5
	g	26.10	19.	16.	4.873	85.55	21.79	
$COBr_2$	liq	187.8286	—	−30.4	—	—	—	—
	g	—	−19.19	−23.0	−26.5	3.340	73.85	14.78
CH_3Br	g	94.9441	−4.72	−8.4	−6.2	2.536	58.86	10.14
in C_2H_5OH	—	—	—	−13.08	—	—	—	—
CH_2Br_2	g	173.8451	—	—	—	3.020	70.06	13.07
$CHBr_3$	liq	252.7461	—	−6.8	−1.2	—	52.8	31.
	g	—	10.24	4.	2.	3.811	79.07	17.02
CF_3Br	g	148.9154	−150.72	−153.6	−147.3	3.457	71.14	16.57
CCl_3Br	g	198.2792	−8.81	−11.0	−5.1	4.285	79.55	20.38
CH_3I	liq	141.9395	—	−3.7	3.2	—	39.0	30.
	g	—	5.38	3.1	3.5	2.585	60.71	10.54
CHI_3	c	393.7323	—	33.7	—	—	—	—
	g	—	—	—	—	4.106	85.1	17.92
CF_3I	g	195.9108	—	—	—	3.579	73.44	16.94
CH_2ClI	g	176.3845	—	—	—	3.002	70.7	13.02
CH_2IBr	g	220.8405	—	—	—	3.102	73.5	13.46
CS_2	liq	76.1392	—	21.44	15.60	—	36.17	18.1
	g	—	27.86	28.05	16.05	2.547	56.82	10.85
	aq	—	—	21.3	—	—	—	—
COS	g	60.0746	−33.991	−33.96	−40.47	2.373	55.32	9.92
CH_3SH	liq	48.1070	—	−11.08	−1.85	—	40.44	21.64
	g	—	−2.885	−5.34	−2.23	2.898	60.96	12.01
CN	g	26.0178	108.	109.	102.	2.07	48.4	6.97
CNO^-								
std. state, m = 1	aq	42.0172	—	−34.9	−23.3	—	25.5	—
HCN	liq	27.0258	—	26.02	29.86	—	26.97	16.88
	g	—	32.39	32.3	29.8	2.208	48.20	8.57
std. state, m = 1	aq	—	—	36.0	41.2	—	22.5	—
HCN^+								
CH_3NH_2								
methylamine	liq	31.0577	—	−11.3	8.5	—	35.90	—
	g	—	—	−5.49	7.67	—	58.15	12.7
std. state, m = 1	aq	—	—	−16.77	4.94	—	29.5	—
CH_2N_2								
diazomethane	g	—	—	—	—	2.887	58.02	12.55
NH_2CN								
cyanamide	c	—	—	14.1	—	—	—	—
NH_4CN	c	44.0564	—	0.10	—	—	—	32.
std. state, m = 1	aq	—	—	4.3	22.2	—	49.6	—
	aq	—	—	7.7	—	—	—	—
$C{=}NH(NH_2)_2$								
guanidine	c	59.0711	—	−18.1	—	—	—	—
HNCO								
isocyanic acid	g	43.0252	—	—	—	2.615	56.85	10.72
HCNO								
cyanic acid, std. state, m = 1, ionized	aq	—	—	−34.9	−23.3	—	25.5	—
un-ionized, std. state, m = 1	aq	—	—	−36.90	−28.0	—	34.6	—
CH_3NO_2								
nitromethane	liq	61.0406	—	−27.03	−3.47	—	41.05	25.33
	g	—	−14.546	−17.86	−1.65	3.083	65.69	13.70
CH_3ONO								
methyl nitrite	g	61.0406	—	−16.5	—	—	—	—
CH_3NO_3	liq	77.0400	−38.82	−38.0	−10.4	8.26	51.9	37.6
	g	—	—	−29.8	−9.4	—	76.1	—
NH_4HCO_3	c	79.0559	—	−203.0	−159.2	—	28.9	
std. state, m = 1	aq	—	—	−197.06	−159.23	—	48.9	—
NH_4CNO	c	60.0558	—	−72.75	—	—	—	—
std. state, m = 1	aq	—	—	−66.6	−42.3	—	52.6	—
$CO(NH_2)_2$								
urea	c	60.0558	—	−79.71	−47.19	—	25.00	22.26
NH_2COONH_4								
ammonium carbamate	c	78.0712	—	−154.17	−107.09	—	31.9	—
$(NH_4)_2CO_3$								
std. state, m = 1	aq	96.0865	—	−225.18	−164.11	—	40.6	—
CNBr	c	105.9268	—	33.58	—	—	—	—
	g	—	46.07	44.5	39.5	2.648	59.32	11.22
CNI	c	152.9222	—	39.71	44.22	—	23.0	—
	g	—	54.04	53.9	47.0	2.724	61.35	11.54
std. state, m = 1	aq	—	—	42.5	44.95	—	29.9	—
CNS^-								
thiocyanate ion std. state, m = 1	aq	58.0818	—	18.27	22.15	—	34.5	−9.6
HCNS								
thiocyanic acid, undissoc. std. state, m = 1	aq	—	—	—	23.31	—	—	—
ionized; std. state, m = 1	aq	—	—	18.27	22.15	—	34.5	−9.6
NH_4CNS								
ammonium thiocyanate	c	76.1204	—	−18.8	—	—	—	—
std. state, m = 1	aq	—	—	−13.40	3.18	—	61.6	9.5
in 200 H_2O	aq	—	—	−13.4	—	—	—	—
$CS(NH_2)_2$								
thiourea	c	76.1204	—	−21.1	—	—	—	—
$C_2O_4^{2-}$								
std. state, m = 1	aq	88.0199	—	−197.2	−161.1	—	10.9	—
C_2H_4	g	28.0542	14.515	12.49	16.28	2.525	52.45	10.41
std. state, m = 1	aq	—	—	8.69	19.43	—	29.2	—

Substance			0 K	298.15 K (25°C)				
Formula and Description	State	Formula weight	ΔHf_0° kcal/mol	ΔHf°	ΔGf° kcal/mol	$H_{298}^\circ - H_0^\circ$	S° cal/deg mol	C_p°
C_2H_5								
ethyl radical	g	29.0622	28.	25.	31.	—	59.2	—
$C_2H_5^+$	g	—	222.	220.	—	—	—	—
C_2H_6	g	30.0701	−16.523	−20.24	−7.86	2.856	54.85	12.58
std. state, m = 1	aq	—	—	−24.40	−4.09	—	28.3	—
$C_2H_6^+$	g	—	252.2	250.0	—	—	—	—
$HC_2O_4^-$								
std. state, m = 1	aq	89.0279	—	−195.6	−166.93	—	35.7	—
CH_2CO								
ketene	g	42.0376	−13.86	−14.6	−14.8	2.819	59.16	12.37
$(COOH)_2$								
oxalic acid	c	90.0358	—	−197.7	—	—	—	28.
std. state, m = 1	aq	—	—	−197.2	−161.1	—	10.9	—
CH_3COO^-								
std. state, m = 1	aq	59.0450	—	−116.16	−88.29	—	20.7	−1.5
$C_2H_3O_3^-$								
glycolate ion, std. state m = 1	aq	75.0444	—	−155.9	—	—	—	—
C_2H_4O								
ethylene oxide	liq	44.0536	—	−18.60	−2.83	—	36.77	21.02
	g	—	−9.589	−12.58	−3.12	2.596	57.94	11.45
CH_3CHO								
acetaldehyde	liq	—	—	−45.96	−30.64	—	38.3	—
$1/3\,(CH_3CHO)_3$								
paraldehyde	liq	—	—	−54.73	—	—	—	—
$1/4\,(CH_3CHO)_4$								
metaldehyde	c	—	—	−56.2	—	—	—	—
$HCOOCH_3$								
methyl formate	liq	60.0530	—	−90.60	—	—	—	29.
	g	—	—	−83.7	—	—	—	—
CH_3COOH								
acetic acid	liq	60.0530	—	−115.8	−93.2	—	38.2	29.7
	g	—	−99.972	−103.31	−89.4	3.286	67.5	15.9
ionized; std. state, m = 1	aq	—	—	−116.16	−88.29	—	20.7	−1.5
un-ionized; std. state, m = 1	aq	—	—	−116.10	−94.8	—	42.7	—
$CH_2OHCOOH$								
hydroxyacetic acid (glycolic acid)	c	76.0524	—	−158.7	—	—	—	—
$CH(OH)_2COOH$								
dihydroxyacetic acid (glyoxylic acid)	c	92.0518	—	−199.7	—	—	—	—
$CH_3CH_2O^-$								
std. state, m = 1	aq	45.0616	—	—	−24.5	—	—	—
C_2H_5OH								
ethanol	liq	46.0695	—	−66.37	−41.80	—	38.4	26.64
	g	—	−51.969	−56.19	−40.29	3.390	67.54	15.64
std. state, m = 1	aq	—	—	−68.9	−43.44	—	35.5	—
C_2H_5OH								
in ∞ H_2O	aq	—	—	−68.9	—	—	—	—
$(CH_2OH)_2$								
ethylene glycol	liq	—	—	−108.70	−77.25	—	39.9	35.8
C_2F_4								
tetrafluoroethylene	g	100.0159	−154.68	−155.5	−147.2	3.903	71.69	19.23
C_2F_6								
hexafluoroethane	g	138.0127	−308.0	−310.	−290.	4.87	79.4	25.5
$CH{\equiv}CF$	g	44.0287	—	—	—	2.739	55.34	12.52
CH_3CH_2F								
ethyl fluoride	g	48.0606	—	—	—	3.06	63.2	14.0
$CH_2{=}CF_2$	g	—	−76.95	−78.6	−73.0	2.980	63.6	14.36
CH_3CHF_2	g	66.0510	−110.98	−114.3	−100.6	3.34	67.5	16.2
CH_3CF_3								
1,1,1-trifluoroethane	g	84.0414	−172.93	−176.0	−159.5	3.631	66.87	18.69
CF_3CH_2F	g	102.0318	—	—	—	4.079	75.85	20.82
C_2Cl_4								
tetrachloroethylene	liq	165.8343	—	−12.5	1.1	—	63.8	33.7
	g	—	−2.70	−2.9	5.4	4.686	81.5	22.69
C_2Cl_6								
I, cubic	c	236.7403	—	−46.0	—	—	—	—
II, monoclinic	c	—	—	−47.9	—	—	—	—
III, triclinic	c	—	—	−48.5	—	—	—	—
$CH_2{=}CHCl$								
vinyl chloride	liq	62.4992	—	3.5	—	—	—	—
	g	—	10.31	8.5	12.4	2.825	63.07	12.84
$1/n(CH_2{=}CHCl)_n$								
polyvinyl chloride	c	—	—	−22.5	—	—	—	14.2
C_2H_5Cl								
ethyl chloride	liq	64.5152	—	−32.63	−14.20	—	45.60	24.94
	g	—	−23.331	−26.81	−14.45	3.179	65.94	15.01
$CHCl{=}CHCl$								
cis-1,2-dichloroethylene	liq	—	—	−6.6	5.27	—	47.42	27.
trans-1,2-dichloroethylene	liq	—	—	−5.53	6.52	—	46.81	27.
CH_2ClCH_2Cl								
1,2-dichloroethane	liq	—	—	−39.49	−19.03	—	49.84	30.9
	g	—	−28.357	−31.02	−17.67	4.08	73.68	18.8
$CHCl{=}CCl_2$								
trichloroethylene	liq	131.3893	—	−10.1	2.9	—	54.6	28.8
	g	—	−1.032	−1.86	4.31	3.975	77.6	19.18
CH_3COCl								
acetyl chloride	liq	78.4986	—	−65.44	−49.73	—	48.0	28.
	g	—	−56.054	−58.20	−49.20	3.53	70.5	16.2
CH_2ClCH_2OH								
ethylene chlorohydrin	liq	80.5146	—	−70.6	—	—	—	—
$CHCl_2COOH$								
dichloracetic acid	liq	128.9430	—	−119.0	—	—	—	44.
ionized	aq	—	—	−122.4	—	—	—	—
un-ionized	aq	—	—	−120.4	—	—	—	—
CCl_3CHO								
chloral (trichloroacetaldehyde)	liq	147.3887	—	−56.45	—	—	—	36.
CCl_3COOH								
trichloroacetic acid	c	163.3881	—	−120.7	—	—	—	—
ionized	aq	—	—	−123.4	—	—	—	—

Substance			0 K	298.15 K (25°C)				
Formula and Description	State	Formula weight	ΔHf°_0 kcal/mol	ΔHf°	ΔGf° kcal/mol	$H^\circ_{298} - H^\circ_0$	S°	C°_p cal/deg mol
$CCl_3CH(OH)_2$								
chloral hydrate	c	165.4040	—	−137.7	—	—	—	34.
	g	—	—	−107.2	—	—	—	—
in 150 $CHCl_3$	—	—	—	−131.2	—	—	—	—
$CF_2{=}CFCl$	g	116.4705	−132.04	−132.7	−125.2	4.096	76.96	20.06
CF_3CCl_3	g	187.3765	—	—	—	5.61	88.6	28.8
$CF_2ClCFCl_2$	liq	—	—	−188.37	—	—	—	41.5
	g	—	—	−181.5	—	—	—	—
CCl_3CF_2Cl	g	203.8311	−115.75	−117.1	−97.3	5.65	91.5	29.5
$CF_2{=}CHCl$	g	98.4801	−74.25	−75.4	−69.1	3.570	72.39	17.23
CH_3CF_2Cl	liq	100.4960	—	—	—	—	—	31.4
CBr_3CBr_3								
hexabromoethane	g	503.4763	—	—	—	7.108	105.6	33.30
$CH_2{=}CHBr$								
vinyl bromide	g	106.9552	22.26	18.7	19.3	2.905	65.90	13.27
C_2H_5Br								
ethyl bromide	liq	108.9712	—	−21.99	−6.64	—	47.5	24.1
	g	—	−10.188	−15.42	−6.34	3.259	68.50	15.42
in 2000 CH_3OH	—	—	−21.71	—	—	—	—	
$CHBr{=}CHBr$								
cis 1,2-dibromoethylene	g	185.8562	—	—	—	3.491	74.38	16.44
trans, 1,2-dibromoethylene	g	—	—	—	—	3.682	74.90	16.79
CH_3CHBr_2	g	187.8722	—	—	—	3.94	78.3	19.3
CH_2BrCH_2Br								
ethylene bromide	liq	—	—	−19.4	−5.0	—	53.37	32.51
	g	—	—	−9.16	−2.47	—	79.1	20.7
CH_3COBr								
acetyl bromide	liq	122.9546	—	−53.39	—	—	—	—
$CF_2 = CFBr$	g	160.9265	—	—	—	4.238	80.0	20.51
CF_3CF_2Br	g	198.9233	—	—	—	4.98	85.7	26.1
$CF_2 = CBr_2$	g	221.8371	—	—	—	4.537	83.5	21.58
CF_2BrCF_2Br	g	259.8339	—	−186.5	—	—	—	—
$CF_2 = CHBr$	g	142.9361	—	—	—	3.682	75.1	17.63
CF_3CH_2Br	g	162.9424	—	—	—	4.299	80.59	21.67
CHF_2CF_2Br	g	180.9329	—	−197.0	—	—	—	—
CH_2ClCH_2Br	liq	143.4162	—	—	—	—	—	31.1
$CHClBrCHClBr$	g	256.7622	—	−8.8	—	—	—	—
$CF_2 = CBrCl$	g	177.3811	—	—	—	4.385	82.0	21.16
$CF_2BrCHCl_2$	g	213.8421	—	−107.9	—	—	—	—
$CI{\equiv}CI$	g	277.8311	—	—	—	3.901	74.80	16.81
C_2I_4	c	531.6399	—	73.	—	—	—	—
$CH_2 = CHI$								
vinyl iodide	g	153.9506	—	—	—	3.027	68.1	13.84
C_2H_5I								
ethyl iodide	liq	155.9666	—	−9.6	3.5	—	50.6	27.5
CH_2ICH_2I								
ethylene iodide	c	281.8630	—	0.1	13.8	—	47.	—
	g	—	—	15.9	18.8	—	83.2	19.2
C_2H_5SH								
ethanethiol	liq	—	—	−17.53	−1.28	—	49.48	28.17
	g	—	−6.940	−10.95	−1.05	3.617	70.77	17.37
$(CH_3)_2SO$								
dimethyl sulfoxide	liq	78.1335	—	−48.6	−23.7	—	45.0	35.2
	g	—	−31.427	−35.96	−19.48	4.132	73.20	21.26
$(CH_3)_2SO_2$								
dimethyl sulfone	c	94.1329	—	−107.8	−72.3	—	34.	—
	g	—	−83.3	−88.7	−65.2	4.3	74.2	23.9
$C_2H_5HSO_4$								
ethyl sulfuric acid	aq	126.1317	—	−209.3	—	—	—	—
$N{\equiv}C{-}C{\equiv}N$								
cyanogen	g	52.0357	73.386	73.84	71.07	3.028	57.79	13.58
CH_3CN								
acetonitrile	liq	—	—	12.8	23.7	—	35.76	21.86
	g	—	22.58	20.9	25.0	2.892	58.67	12.48
$C_2H_5NH_2$								
ethylamine	liq	45.0848	—	−17.7	—	—	—	31.
	g	—	—	−11.27	—	—	—	16.7
$(CH_3)_2NH$								
dimethylamine	liq	—	—	−10.5	16.7	—	43.58	32.9
	g	—	—	−4.41	16.35	—	65.24	16.9
std. state, m = 1	aq	—	—	−16.88	13.85	—	31.8	—
$NH_2CH_2CH_2NH_2$								
ethylenediamine	liq	—	—	−5.82	—	—	—	50.
CH_3CONH_2								
acetamide	c	—	—	−76.0	—	—	—	16.
$C_2H_5NO_2$								
nitroethane	liq	75.0676	—	−33.5	—	—	—	33.
	g	—	—	−23.56	—	—	—	—
CH_3CH_2ONO								
ethyl nitrite	liq	—	—	−30.8	—	—	—	—
NH_2CH_2COOH								
glycine	c	—	−121.415	−126.22	−88.09	3.867	24.74	23.71
ionized; std. state, m = 1	aq	—	—	−112.280	−75.278	—	28.54	—
un-ionized, std. state, m = 1	aq	—	—	−122.846	−88.618	—	37.84	—
$CH_3CH_2ONO_2$								
ethyl nitrate	c	—	−45.023	—	—	—	—	—
	liq	—	—	−45.49	−10.29	9.242	59.08	40.7
CH_3COONH_4								
ammonium acetate	c	77.0836	—	−147.26	—	—	—	—
std. state, m = 1	aq	—	—	−147.83	−107.26	—	47.8	17.6
$CH_2OHCOONH_4$								
ammonium glycolate	c	93.0830	—	−190.6	—	—	—	—
$(CH = NOH)_2$								
glyoxime	c	—	—	−21.2	—	—	—	—
$(CH_3)_2NH_2NO_3$								
dimethylammonium nitrate	c	108.0977	—	−83.7	—	—	—	—
std. state, m = 1	aq	—	—	−78.30	−27.41	—	76.2	—
$(NH_4)_2C_2O_4$								
ammonium oxalate	c	—	—	−268.4	—	—	—	54.
std. state, m = 1	aq	—	—	−260.5	−199.0	—	65.1	—
$(NH_4)_2C_2O_4 \cdot H_2O$	c	142.1124	—	−340.7	—	—	—	—

Substance			0 K	298.15 K (25°C)				
Formula and Description	State	Formula weight	ΔHf_0° kcal/mol	ΔHf°	ΔGf° kcal/mol	$H_{298}^\circ - H_0^\circ$	S° cal/deg mol	C_p°
CH_3NCS								
methyl isothiocyanate	c	73.1169	—	19.0	—	—	—	—
	g	—	33.46	31.3	34.5	3.464	69.29	15.65
CH_3SCN								
methyl thiocyanate	liq	—	—	28.4	—	—	—	—
	g	—	—	38.3	—	—	—	—
CH_3COOCH_3	liq	74.0801	—	−106.42	—	—	—	—
$(CH_3)_3N$								
trimethylamine	liq	59.1119	—	−11.0	24.1	—	49.82	32.31
	g	—	—	−5.81	23.65	—	68.6	—
std. state, m = 1	aq	—	—	−18.17	22.22	—	31.9	—
$(C_2H_5)_2O$								
diethyl ether	liq	74.1237	—	−66.82	—	—	—	—
	g	—	—	−60.26	—	—	—	—
$(C_2H_5)_2S$								
diethyl sulfide	liq	90.1883	—	−28.43	2.81	—	64.36	40.97
	g	—	−13.15	−19.86	4.34	5.467	87.96	27.97
$(C_2H,)_2NH$								
diethylamine	liq	73.1390	—	−24.7	—	—	—	—
	g	—	—	−17.07	—	—	—	—
CERIUM								
Ce								
γ	c	140.12	0	0	0	1.8	17.2	6.44
	g	—	101.2	101.	92	1.594	45.807	5.515
Ce^+	g	—	227.3	228.5	—	—	—	—
Ce^{2+}	g	—	478.	480.	—	—	—	—
Ce^{3+}	g	—	942.7	947.4	—	—	—	—
std. state, m = 1	aq	—	—	−166.4	−160.6	—	−49.	—
CeO_2	c	172.119	−256.69	−258.80	−244.9	2.478	14.89	14.73
Ce_2O_3	c	328.238	−427.64	−429.3	−407.8	5.13	36.0	27.4
CeH_2	c	142.136	−47.83	−49.	−39.	1.776	13.3	9.78
CeF_3	c	197.115	—	—	—	4.237	27.5	22.3
$CeF_3 \cdot H_2O$	c	215.131	—	−472.4	—	—	—	—
$CeCl_3$	c	246.479	—	−251.8	−233.7	—	36.	20.9
	g	—	—	−174.	—	—	—	—
std. state, m = 1	aq	—	—	−286.2	−254.7	—	−9.	—
$CeCl_3 \cdot 7H_2O$	c	372.587	—	−757.5	—	—	—	—
CeOCl	c	191.572	—	−239.	—	—	—	—
$CeClO_4^{2+}$								
std. state, m = 1	aq	239.571	—	−209.1	−165.2	—	−37.	—
$CeBr^{2+}$								
std. state, m = 1	aq	220.029	—	—	−186.3	—	—	—
CeI_3	c	520.833	—	−155.3	—	—	—	—
$Ce(I_3)_3$	c	664.828	—	−332.	—	—	—	—
$Ce(IO_3)_3 \cdot 2H_2O$	c	700.858	—	—	−378.3	—	—	—
CeS_2	c	204.248	—	−146.3	—	—	—	—
Ce_2S	g	312.304	73.	71.6	60.0	3.25	81.	12.9
Ce_2S_3	c	376.432	—	−284.	—	—	—	—
$CeSO_4^+$								
std. state	aq	236.182	—	−380.2	−343.3	—	−17.	—
$Ce(SO_4)_2^-$								
std. state	aq	332.243	—	−595.9	−523.6	—	2.	—
$Ce_2(SO_4)_3$	c	568.425	—	−945.1	—	—	—	—
$Ce_2(SO_4)_3 \cdot 5H_2O$	c	658.502	—	—	—	—	—	132.
$Ce_2(SO_4)_3 \cdot 8H_2O$	c	712.548	—	—	−1320.6	—	—	—
$Ce(NO_3)_3$	c	326.135	—	−293.0	—	—	—	—
$Ce(NO_3)_3 \cdot 3H_2O$	c	380.181	—	−516.	—	—	—	—
$Ce(NO_3)_3 \cdot 4H_2O$	c	398.196	—	−588.9	—	—	—	—
$Ce(NO_3)_3 \cdot 6H_2O$	c	434.227	—	−729.14	—	—	—	—
CeC_2	c	164.142	—	−15.	15.2	—	20.	—
	g	—	136.	136.2	122.9	2.47	64.	10.5
CeC_4	g	188.165	167.1	168.	152.	3.68	73.	17.3
$CeCrO_3$	c	240.114	—	−368.	−347.	—	25.	—
CESIUM								
Cs	c	132.9054	0	0	0	1.843	20.37	7.69
	g	132.9054	18.542	18.180	11.748	1.481	41.942	4.968
Cs^+	g	132.9054	108.337	109.456	—	—	—	—
Cs^{2+}	g	132.9054	686.63	689.23	—	—	—	—
Cs^+	a	132.9054	—	−61.73	−69.79	—	31.80	−2.5
CsO_2	c	164.9042	—	−68.4	—	—	—	—
Cs_2O	c	281.8102	−82.142	−82.64	−73.65	4.225	35.10	18.16
	g	281.8102	—	−37.	—	—	—	—
CsH	c	133.9134	—	−12.950	—	—	—	—
	g	133.9134	28.4	27.7	23.1	2.114	51.40	7.54
CsOH	c	149.9128	—	−99.72	—	—	—	—
	g	149.9128	−57.9	−59.	−59.1	2.828	60.88	11.88
	a	149.9128	—	−116.70	−107.38	—	29.23	—
$CsOH \cdot H_2O$	c	167.9282	—	−180.22	—	—	—	—
CsF	c	151.9038	−132.205	−132.3	−125.6	2.802	22.18	12.21
	g	151.9038	−85.21	−85.8	−89.8	2.306	58.11	8.57
CsCl	c	168.3584	−105.926	−105.89	−99.08	2.976	24.18	12.54
	g	168.3584	−56.89	−57.41	−61.62	2.42	61.15	8.83
CsClO	a	184.3578	—	−87.3	−78.6	—	42.	—
$CsClO_2$	a	200.3572	—	−77.6	−65.7	—	56.0	—
$CsClO_3$	c	216.3566	—	−98.4	−73.6	—	37.3	—
	a	216.3566	—	−86.58	−71.71	—	70.6	—
$CsClO_4$	c	232.3560	−104.136	−105.90	−75.13	5.325	41.84	25.88
	a	232.3560	—	−92.64	−71.85	—	75.3	—
	c	212.8144	−95.358	−96.99	−93.55	3.140	27.02	12.65
	g	212.8144	−47.69	−50.0	−57.6	2.46	63.89	8.86
CsBrO	a	228.8138	—	−84.2	−77.8	—	42.	—
$CsBrO_3$	c	260.8126	—	−89.82	−68.11	—	39.1	—
	a	260.8126	—	−77.76	−65.36	—	70.45	—
CsI	c	259.8098	−82.652	−82.84	−81.40	3.232	29.41	12.62
	g	259.8098	−35.40	−36.3	−45.7	2.52	65.77	8.95
CsIO	a	275.8092	—	−87.4	−79.0	—	30.5	—
$CsIO_3$	c	307.8080	—	—	−103.7	—	—	—
$CsIO_4$	a	307.8080	—	−114.6	−100.4	—	60.1	—
$CsIO_4$	c	323.8074	—	—	−91.0	—	—	—
Cs_2S	c	297.8748	—	−86.0	—	—	—	—
	a	297.8748	—	−115.6	−119.1	—	60.1	—

SELECTED VALUES OF CHEMICAL THERMODYNAMIC PROPERTIES (Continued)

Substance			0 K	298.15 K (25°C)				
Formula and Description	State	Formula weight	ΔHf°_0 kcal/mol	ΔHf°	ΔGf° kcal/mol	$H^\circ_{298} - H^\circ_0$	S°	C°_p cal/deg mol
Cs_2SO_3	c	345.8730	—	−271.2	—	—	—	—
	a	345.8730	—	−275.4	−255.9	—	56.6	—
Cs_2SO_4	c	361.8724	−342.63	−344.89	−316.36	6.63	50.65	32.24
	g	361.8724	—	−263.6	—	—	—	—
	a	361.8724	—	−340.78	−317.55	—	68.4	—
$CsHSO_3$								
from HSO_3	a	213.9756	—	−211.40	−195.94	—	65.2	—
$CsHSO_4$	c	229.9750	—	−276.8	—	—	—	—
from HSO_4^-	a	229.9750	—	−273.81	−250.48	—	63.3	—
Cs_2Se	a	344.7708	—	—	−108.7	—	—	—
Cs_2SeO_3	a	392.7690	—	−245.2	−228.0	—	67.	—
Cs_2SeO_4	c	408.7684	—	−272.34	—	—	—	—
	a	408.7684	—	−266.7	−245.1	—	76.5	—
CsN_3	c	174.9255	—	−4.7	—	—	—	—
$CsNO_2$	a	178.9109	—	−86.7	−77.5	—	61.2	—
$CsNO_3$	c	194.9103	—	−120.93	−97.18	—	37.1	—
	g	194.9103	—	−89.4	—	—	—	—
	a	194.9103	—	−111.29	−96.40	—	66.8	−23.7
$CsPO_3$	c	211.8774	—	−296.7	—	—	—	—
Cs_3PO_4	a	493.6876	—	−490.5	−452.9	—	42.	—
$Cs_4P_2O_7$	a	705.5650	—	−789.7	−737.9	—	99.2	—
CsH_2PO_4	c	229.8928	—	−374.0	—	—	—	—
$CsAsO_2$	a	239.8258	—	−164.27	−153.45	—	41.5	—
Cs_3AsO_4	a	537.6354	—	−397.46	−364.37	—	56.5	—
$Cs_2C_2O_4$								
oxalate	c	353.8308	—	—	—	7.45	56.92	—
	a	353.8308	—	−320.7	−300.7	—	74.5	—
$CsHCO_3$	c	193.9228	—	−230.9	—	—	—	—
from HCO_3^-	a	193.9228	—	−227.12	−210.05	—	53.6	—
$CsHC_2O_4$								
from $HC_2O_4^-$	a	221.9334	—	−257.3	−236.72	—	67.5	—
CH_3COOCs								
acetate	a	191.9506	—	−177.89	−158.08	—	52.5	—
$CsCN$	c	158.9233	—	—	—	4.330	33.40	15.70
	a	158.9233	—	−25.7	−28.6	—	54.3	—
$CsCNS$								
thiocyanate	a	190.9873	—	−43.46	−47.64	—	66.3	—
$CsMnO_4$	a	251.8410	—	−191.1	−176.7	—	77.5	—
	a	664.7288	—	−305.7	−280.5	—	125.0	—
Cs_2CrO_4	c	381.8044	—	−341.57	—	—	—	—
$Cs_2Cr_2O_7$	c	481.7986	−497.023	−499.24	−456.09	10.671	78.89	55.34
	a	481.7986	—	−479.7	−450.6	—	126.2	—
Cs_2WO_4	a	513.6584	—	−380.6	—	—	—	—
$CsVO_3$	a	231.8456	—	−274.0	−257.1	—	44.	—
Cs_3VO_4	a	513.6558	—	—	−424.3	—	—	—
Cs_3UO_4	c	567.8374	−459.01	−461.0	−431.7	7.365	52.50	36.51
CHLORINE								
Cl	g	35.453	28.68	29.082	25.262	1.499	39.457	5.220
Cl^+	g	—	328.86	330.74	—	—	—	—
Cl^{2+}	g	—	877.81	881.17	—	—	—	—
Cl^{3+}	g	—	1798.26	1803.10	—	—	—	—
Cl^-	g	—	−57.7	−58.8	—	—	—	—
std. state, m = 1	aq	—	—	−39.952	−31.372	—	13.5	−32.6
Cl_2	g	70.906	0	0	0	2.193	53.288	8.104
std. state, m = 1	aq	—	—	−5.6	1.65	—	29.	—
Cl_3^-								
std. state, m = 1	aq	106.359	—	—	−28.8	—	—	—
ClO	g	51.4524	24.36	24.34	23.45	2.114	54.14	7.52
ClO^-								
std. state, m = 1	aq	—	—	−25.6	−8.8	—	10.	—
ClO_2	g	67.4518	25.09	24.5	28.8	2.580	61.36	10.03
std. state, m = 1	aq	—	—	17.9	28.7	—	39.4	—
ClO_2^-								
std. state, m = 1	aq	—	—	−15.9	4.1	—	24.2	—
ClO_3	g	83.4512	—	37.	—	—	—	—
ClO_3^-								
std. state, m = 1	aq	—	—	−23.7	−0.8	—	38.8	—
ClO_4								
std. state, m = 1	aq	99.4506	—	−30.91	−2.06	—	43.5	—
Cl_2O	g	86.9054	19.71	19.2	23.4	2.719	63.60	10.85
HCl	g	36.4610	−22.020	−22.062	−22.777	2.066	44.646	6.96
std state, m = 1	aq	—	—	−39.952	−31.372	—	13.5	−32.6
HCl								
∞ H_2O	aq	—	—	−39.952	—	—	—	—
$HClO$	g	52.4604	—	—	—	2.440	56.54	8.88
undissoc.; std. state, m = 1	aq	—	—	−28.9	−19.1	—	34.	—
$HClO_2$								
undissoc.; std. state, m = 1	aq	68.4598	—	−12.4	1.4	—	45.0	—
$HClO_3$								
std. state, m = 1	aq	84.4592	—	24.85	−1.92	—	38.8	—
$HClO_4$	liq	100.4586	—	−9.70	—	—	—	—
std. state, m = 1	aq	—	—	−30.91	−2.06	—	43.5	—
$HClO_4$								
∞ H_2O	aq	—	—	−30.91	—	—	—	—
ClF	g	54.4514	−13.0	−13.02	−13.37	2.127	52.05	7.66
ClF_3	liq	92.4482	—	−45.3	—	—	—	—
CHROMIUM								
Cr	c	51.996	0	0	0	0.970	5.68	5.58
	g	—	94.29	94.8	84.1	1.481	41.68	4.97
Cr^+	g	—	250.30	252.29	—	—	—	—
Cr^{2+}	g	—	630.7	634.2	—	—	—	—
std. state, m = 1	aq	—	—	−34.3	—	—	—	—
Cr^{3+}	g	—	1345.	1350.	—	—	—	—
CrO_3	c	99.9942	—	−140.9	—	—	—	—
	g	—	—	−92.2	—	—	—	—
CrO_4^{2-}								
std. state, m = 1	aq	115.9936	—	−210.60	−173.96	—	12.00	—
Cr_2O_3	c	151.9902	—	−272.4	−252.9	—	19.4	28.38
$Cr_2O_7^{2-}$								
std. state, m = 1	aq	215.9878	—	−356.2	−311.0	—	62.6	

Substance			0 K	298.15 K (25°C)				
Formula and Description	State	Formula weight	ΔHf°_0 kcal/mol	ΔHf°	ΔGf° kcal/mol	$H^\circ_{298} - H^\circ_0$	S° cal/deg mol	C°_p
Cr_3O_4	c	219.9856	—	−366.	—	—	—	—
$HCrO_4^-$								
std. state, m = 1	aq	117.0016	—	−209.9	−182.8	—	44.0	—
$Cr(OH)_3$								
precipitated	c	103.0181	—	−254.3	—	—	—	—
CrF_2	c	89.9928	—	−186.	—	—	—	—
	g	—	—	−99.	—	—	—	—
CrF_3	c	108.9912	−276.22	−277.	−260.	3.357	22.44	18.82
$CrCl_2$	c	122.902	−94.93	−94.5	−85.1	3.593	27.56	17.01
	g	—	—	−30.7	—	—	—	—
	aq	—	—	−114.2	—	—	—	—
$CrCl_2 \cdot 2H_2O$								
light green	c	158.9327	—	−237.1	—	—	—	—
$CrCl_2 \cdot 3H_2O$								
pale blue	c	176.9480	—	−308.9	—	—	—	—
$CrCl_2 \cdot 4H_2O$								
dark blue	c	194.9634	—	−384.4	—	—	—	—
$CrCl_3$	c	158.355	−132.96	−133.0	−116.2	4.22	29.4	21.94
CrO_2Cl_2	liq	154.9008	—	−138.5	−122.1	—	53.0	—
	g	—	−127.68	−128.6	−119.9	4.32	78.8	20.2
$CrBr_2$	c	211.814	—	−72.2	—	—	—	—
	g	—	—	−17.	—	—	—	—
$(CrBr_2)_2$	g	423.628	—	−84.	—	—	—	—
CrI_2	c	305.8048	—	−37.5	—	—	—	—
	g	—	—	24.	—	—	—	—
	aq	—	—	−60.1	—	—	—	—
CrI_3	c	432.7092	—	−49.0	—	—	—	—
$CrICl_2$	c	249.8064	—	−100.	—	—	—	—
$CrIBr_2$	c	338.7184	—	−79.	—	—	—	—
$Cr_2(SO_4)_3$	c	392.1768	—	—	—	—	—	67.5
CrN	c	66.0027	—	−29.8	—	—	—	11.0
Cr_2N	c	117.9987	—	−30.5	—	—	—	—
NH_4HCrO_4								
std. state, m = 1	aq	135.0402	—	−241.6	−201.8	—	71.1	—
$(NH_4)_2CrO_4$	c	152.0708	—	−279.0	—	—	—	—
std. state, m = 1	aq	—	—	−273.94	−211.90	—	66.2	—
in 300 H_2O	aq	—	—	−273.5	—	—	—	—
$(NH_4)_2Cr_2O_7$	c	252.0650	—	−431.8	—	—	—	—
std. state, m = 1	aq	—	—	−419.5	−348.9	—	116.8	—
Cr_3C_2	c	180.0103	−19.51	−19.3	−19.5	3.621	20.42	23.53
Cr_7C_3	c	400.0054	—	−38.7	−39.9	—	48.0	49.92
$Cr_{23}C_6$	c	1267.9749	—	−87.2	−89.3	—	145.8	149.2
$Cr(CO)_6$	c	220.0593	—	−257.4	—	—	—	—
	g	—	—	−240.4	—	—	—	—
$PbCrO_4$	c	323.184	—	−222.5	—	—	—	—
Tl_2CrO_4	c	524.734	—	−225.8	−205.9	—	67.5	—
Ag_2CrO_4	c	331.7336	—	−174.89	−153.40	—	52.0	34.00
std. state, m = 1	aq	—	—	−160.13	−137.09	—	46.8	—
$FeCr_2O_4$	c	223.8366	—	−345.3	−321.2	—	34.9	31.94
COBALT								
Co								
α, hexagonal	c	58.9332	0	0	0	1.139	7.18	5.93
β, f. c. cubic	c	—	—	0.11	0.06	—	7.34	—
	g	—	101.119	101.5	90.9	1.520	42.879	5.502
Co^+	g	—	282.486	284.348	—	—	—	—
Co^{2+}	g	—	675.82	679.17	—	—	—	—
std. state, m = 1	aq	—	—	−13.9	−13.0	—	−27.	—
Co^{3+}	g	—	1448.35	1453.18	—	—	—	—
std. state, m = 1	aq	—	—	22.	32.	—	−73.	—
Co_2CoO	c	74.9326	—	−56.87	−51.20	—	12.66	13.20
Co_3O_4	c	240.7972	—	−213.	−185.	—	24.5	29.5
$Co(OH)_2$								
blue, precipitated	c	92.9479	—	—	−107.6	—	—	—
pink, precipitated	c	—	—	−129.0	−108.6	—	19.	—
pink, precipitated, aged	c	—	—	—	−109.5	—	—	—
std. state, m = 1	aq	—	—	−123.8	−88.2	—	−32.	—
undissoc.; std. state, m = 1	aq	—	—	—	−100.8	—	—	—
$Co(OH)_3$								
precipitated	c	109.9553	—	−171.3	—	—	—	—
CoF_3	c	115.9284	—	−193.8	—	—	—	—
$CoCl_2$	c	129.8392	−74.74	−74.7	−64.5	3.375	26.09	18.76
std. state, m = 1	aq	—	—	−93.8	−75.7	—	0.	—
$CoCl_2$								
in ∞ H_2O	aq	—	—	−93.8	—	—	—	—
$CoCl_2 \cdot H_2O$	c	147.8545	—	−147.	—	—	—	—
$CoCl_2 \cdot 2H_2O$	c	165.8699	—	−220.6	−182.8	—	45.	—
$CoCl_2 \cdot 6H_2O$	c	237.9312	—	−505.6	−412.4	—	82.	—
$Co(ClO_4)_2$								
std. state, m = 1	aq	257.8344	—	−75.7	−17.1	—	60.	—
in ∞ H_2O	aq	—	—	−75.7	—	—	—	—
$CoBr_2$	c	218.7512	—	−52.8	—	—	—	19.0
std. state, m = 1	aq	—	—	−72.0	−62.7	—	12.	—
$CoBr_2 \cdot 6H_2O$	c	326.8432	—	−482.8	—	—	—	—
CoI_2	c	312.7420	—	−21.2	—	—	—	—
std. state, m = 1	aq	—	—	−40.3	−37.7	—	26.	—
$Co(IO_3)_2$								
std. state, m = 1	aq	408.7384	—	−119.7	−74.2	—	30.	—
$Co(IO_3)_2 \cdot 2H_2O$	c	444.7691	—	−258.6	−190.2	—	64.	—
CoS	c	90.9972	—	−19.8	—	—	—	—
Co_2S_3								
precipitated	c	214.0584	—	−35.2	—	—	—	—
$CoSO_4$	c	154.9948	—	−212.3	−187.0	—	28.2	—
std. state, m = 1	aq	—	—	−231.2	−191.0	—	−22.	—
$CoSO_4 \cdot 6H_2O$	c	263.0868	−630.257	−641.4	−534.35	13.525	87.86	84.46
$CoSO_4 \cdot 7H_2O$	c	281.1022	−699.547	−712.22	−591.26	15.097	97.05	93.33
$CoSe$	c	137.893	—	−14.6	—	—	—	—
$CoTe_2$	c	314.133	—	−31.	—	—	—	—
$Co(NO_3)_2$	c	182.9430	—	−100.5	—	—	—	—
std. state, m = 1	aq	—	—	−113.0	−66.2	—	43.	—
$[Co(NH_3)]^{2+}$								
std. state, m = 1	aq	75.9638	—	−34.7	−22.1	—	3.	—

Substance			0 K	298.15 K (25°C)				
Formula and Description	State	Formula weight	ΔHf°_0 kcal/mol	ΔHf°	ΔGf° kcal/mol	$H^\circ_{298} - H^\circ_0$	S° cal/deg mol	C°_p
$[Co(NH_3)_2]^{2+}$								
std. state, m = 1	aq	92.9944	—	—	−30.5	—	—	—
$[Co(NH_3)_3]^{2+}$								
std. state, m = 1	aq	110.0250	—	—	−38.1	—	—	—
$[Co(NH_3)_4]^{2+}$								
std. state, m = 1	aq	127.0556	—	—	−45.3	—	—	—
$[Co(NH_3)_6]^{3+}$								
std. state, m = 1	aq	161.1169	—	−139.8	−38.9	—	40.0	—
$[Co(NH_3)_6]N_3^{3+}$								
std. state, m = 1	aq	203.1370	—	—	42.9	—	60	—
$[Co(NH_3)_6](NO_3)_3$	c	347.1316	—	−306.4	−125.5	—	107.	—
std. state, m = 1	aq	—	—	−288.5	−117.4	—	140.	—
$CoBr_2 \cdot NH_3$	c	235.7818	—	−85.1	—	—	—	—
$CoBr_2 \cdot 2NH_3$								
rose	c	252.8124	—	−116.0	—	—	—	—
$[Co(NH_3)_6]Br^{2+}$								
std. state, m = 1	aq	241.0259	—	−171.7	−60.5	—	39.	—
$[Co(NH_3)_6]Br_2$	c	320.9349	—	−216.4	—	—	—	—
$[Co(NH_3)_6]Br_3$	c	400.8439	−217.52	−239.7	−119.8	12.18	77.7	78.1
std. state. m = 1	aq	—	—	−227.0	−112.2	—	94.	—
Co_2P	c	148.8402	—	−45.	—	—	—	—
$Co_3(PO_4)_2$	c	366.7424	—	—	−573.3	—	—	—
$CoHPO_4$	c	154.9126	—	—	−282.5	—	—	—
$CoAs$	c	133.8548	—	−9.7	—	—	—	—
$CoAs_2$	c	208.7764	—	−14.7	—	—	—	—
Co_2As	c	192.7880	—	−9.5	—	—	—	—
Co_2As_3	c	342.6312	—	−23.3	—	—	—	—
Co_3As_2	c	326.6428	—	−19.4	—	—	—	—
Co_5As_2	c	444.5092	—	−19.0	—	—	—	—
$Co_3(AsO_4)_2$	c	454.6380	—	—	−387.4	—	—	—
$CoSb$	c	180.683	—	−10.	—	—	—	—
$CoSb_2$	c	302.433	—	−13.	—	—	—	—
$CoSb_3$	c	424.183	—	−16.	—	—	—	—
Co_2C	c	129.8776	—	−10.	—	—	—	—
$CoCo_3$	c	118.9426	—	−170.4	—	—	—	—
CoC_2O_4	c	146.9531	—	−203.5	—	—	—	—
std. state, m = 1	aq	—	—	−211.1	−174.1	—	−16.	—
$Co(C_2O_4)_2^{2-}$								
std. state, m = 1	aq	234.9730	—	−408.5	−344.8	—	26.	—
$CoSi$	c	87.0192	−23.87	−24.0	−23.6	1.78	10.3	10.6
$CoSi_2$	c	115.1052	—	−24.6	—	—	—	—
$CoSi_3$	c	143.1912	—	−25.6	—	—	—	—
Co_2SiO_4	c	209.9500	—	−353.	—	—	—	—
COPPER								
Cu	c	63.54	0	0	0	1.196	7.923	5.840
	g	—	80.58	80.86	71.37	1.481	39.74	4.968
Cu^+	g	—	258.752	260.513	—	—	—	—
std. state, m = 1	aq	—	—	17.13	11.95	—	9.7	—
Cu^{2+}	g	—	726.69	729.93	—	—	—	—
std. state, m = 1	aq	—	—	15.48	15.66	—	−23.8	—
Cu^{3+}	g	—	1576.1	1580.8	—	—	—	—
Cu_2	g	127.08	115.7	115.72	103.24	2.370	57.71	8.75
CuO	c	79.539	—	−37.6	−31.0	—	10.19	10.11
Cu_2O	c	143.079	—	−40.3	−34.9	—	22.26	15.21
CuH	c	64.548	—	5.1	—	—	—	—
$Cu(OH)_2$	c	97.555	—	−107.5	—	—	—	—
std. state, m = 1	aq	—	—	−94.46	−59.53	—	−28.9	—
CuF^+								
std. state, m = 1	aq	82.538	—	−62.4	−52.7	—	−16.	—
CuF_2	c	101.537	—	−129.7	—	—	—	—
$CuF_2 \cdot 2H_2O$	c	137.567	—	—	−234.6	—	—	—
$CuCl$	c	98.993	—	−32.8	−28.65	—	20.6	11.6
$CuCl^+$								
std. state, m = 1	aq	—	—	—	−16.3	—	—	—
$CuCl_2$	c	134.446	−52.79	−52.6	−42.0	3.581	25.83	17.18
$CuCl_2 \cdot 2\,H_2O$	c	170.477	—	−196.3	−156.8	—	40.	—
$CuCl_2$								
std. state. m = 1	aq	134.446	—	—	−57.4	—	—	—
$Cu(ClO_4)_2$								
std. state, m = 1	aq	262.441	—	−46.34	11.54	—	63.2	—
$Cu(ClO_4)_2 \cdot 6H_2O$	c	370.533	—	−460.9	—	—	—	—
Cu_2OCl_2	c	213.985	—	−90.	—	—	—	—
$CuBr$	c	143.449	−23.76	−25.0	−24.1	2.893	22.97	13.08
$CuBr^+$								
std. state, m = 1	aq	—	—	—	−11.9	—	—	—
$CuBr_2$	c	223.358	—	−33.9	—	—	—	—
$CuBr_2 \cdot 4H_2O$	c	295.419	—	−317.0	—	—	—	—
$CuBr_2 \cdot 3Cu(OH)_2$	c	516.022	—	−378.1	−306.2	—	68.	—
$Cu(BrO_3)_2 \cdot 3Cu(OH)_2$	c	612.019	—	—	−244.2	—	—	—
CuI	c	190.444	—	−16.2	−16.6	—	23.1	12.92
$Cu(IO_3)_2$								
std. state, m = 1	aq	413.345	—	−90.3	−45.5	—	32.8	—
$Cu(IO_3)_2 \cdot H_2O$	c	431.361	—	−165.4	−112.0	—	59.1	—
$Cu(IO_3)_2 \cdot 3Cu(OH)_2$	c	706.009	—	—	−160.0	—	—	—
CuS	c	95.604	—	−12.7	−12.8	—	15.9	11.43
Cu_2S								
α	c	159.144	—	−19.0	−20.6	—	28.9	18.24
$CuSO_4$	c	159.602	—	−184.36	−158.2	—	26.	23.9
std. state, m = 1	aq	—	—	−201.84	−162.31	—	−19.0	—
$CuSO_4$								
in ∞ H_2O	aq	—	—	−201.84	—	—	—	—
$CuSO_4 \cdot H_2O$	c	177.617	—	−259.52	−219.46	—	34.9	32.
$CuSO_4 \cdot 3H_2O$	c	213.648	—	−402.56	−334.65	—	52.9	49.
$CuSO_4 \cdot 5H_2O$	c	249.678	—	−544.85	−449.344	—	71.8	67.
Cu_2SO_4	c	223.142	—	−179.6	—	—	—	—
$CuO \cdot CuSO_4$	c	239.141	—	−223.8	—	—	—	—
$CuSO_4 \cdot 2Cu(OH)_2$								
antlerite	c	354.711	—	—	−345.8	—	—	—
$CuSO_4 \cdot 3Cu(OH)_2$								
brochantite	c	452.266	—	—	−434.5	—	—	—

Substance			0 K	298.15 K (25°C)				
Formula and Description	State	Formula weight	ΔHf°_0 kcal/mol	ΔHf°	ΔGf° kcal/mol	$H^\circ_{298} - H^\circ_0$	S° cal/deg mol	C°_p
$CuSO_4 \cdot 3Cu(OH)_2 \cdot H_2O$								
langite	c	470.281	—	−594.	−488.6	—	80.	—
CuSe	c	142.50	—	−9.45	—	—	—	—
$CuSe_2$	c	221.46	—	−10.3	—	—	—	—
Cu_2Se	c	206.04	—	−14.2	—	—	—	—
Cu_2Te	c	254.68	—	5.	—	—	—	—
CuN_3	c	105.560	—	66.7	82.4	—	24.	—
$Cu(N_3)_2$	c	147.580	—	143.0	—	—	—	—
Cu_3N	c	204.627	—	17.8	—	—	—	22.
$Cu(NO_3)_2$	c	187.550	—	−72.4	—	—	—	—
std. state, m = 1	aq	—	—	−83.64	−37.56	—	46.2	—
$Cu(NO_3)_2$								
in ∞ H_2O	aq	—	—	−83.64	—	—	—	—
$Cu(NO_3)_2 \cdot 3H_2O$	c	241.596	—	−290.9	—	—	—	—
$Cu(NO_3)_2 \cdot 6H_2O$	c	295.642	—	−504.5	—	—	—	—
$Cu(NH_3)^{2+}$								
std. state, m = 1	aq	80.571	—	−9.3	3.72	—	2.9	—
$Cu(NH_3)_2^{2+}$								
std. state, m = 1	aq	97.601	—	−34.0	−7.28	—	26.6	—
$Cu(NH_3)_3^{2+}$								
std. state, m = 1	aq	114.632	—	−58.7	−17.48	—	47.7	—
$Cu(NH_3)_4^{2+}$								
std. state, m = 1	aq	131.662	—	−83.3	−26.60	—	65.4	—
$Cu(NH_3)_5^{2+}$								
std. state, m = 1	aq	148.693	—	—	−32.13	—	—	—
$CuSO_4 \cdot NH_3$	c	176.632	—	−218.0	—	—	—	—
$CuSO_4 \cdot 2NH_3$	c	193.663	—	−248.2	—	—	—	—
$CuSO_4 \cdot 5NH_3$	c	244.755	—	−328.1	—	—	—	—
CuP_2	c	125.488	—	−29.	—	—	—	—
$Cu_3(PO_4)_2$	c	380.563	—	—	−490.3	—	—	—
Cu_3As	c	265.542	—	−2.8	—	—	—	—
$Cu_3(AsO_4)_2$	c	468.458	—	—	−310.9	—	—	—
std. state, m = 1	aq	—	—	−378.10	−263.02	—	−149.2	—
CuC_2O_4	c	151.560	—	—	−158.2	—	—	—
std. state, m = 1	aq	—	—	−181.7	−145.5	—	−12.9	—
undissoc.; std. state, m = 1	aq	—	—	—	−153.1	—	—	—
$CuCO_3 \cdot Cu(OH)_2$								
malachite	c	221.104	—	−251.3	−213.6	—	44.5	—
$2CuCO_3 \cdot Cu(OH)_2$								
azurite	c	344.653	—	−390.1	—	—	—	—
CuCN	c	89.558	—	23.0	26.6	—	20.2	—
$Cu(CN)_2^-$								
std. state, m = 1	aq	115.576	—	—	61.6	—	—	—
$Cu(CN)_3^{2-}$								
std. state, m = 1	aq	141.594	—	—	96.5	—	—	—
$Cu(CN)_4^{3-}$								
std. state, m = 1	aq	167.611	—	—	135.4	—	—	—
CuONC								
cuprous fulminate	c	105.557	—	26.3	—	—	—	—
CuCNS	c	121.622	—	—	16.7	—	—	—
std. state, m = 1	aq	—	—	35.40	34.10	—	44.2	—
$(Cu(CNS)_2$								
std. state, m = 1	aq	179.704	—	52.02	59.96	—	45.2	—
CuAl	c	90.522	—	−9.8	—	—	—	—
$CuAl_2$	c	117.503	—	−9.75	—	—	—	—
Cu_2Al	c	154.062	—	−16.5	—	—	—	—
Cu_3Al	c	217.602	—	−16.8	—	—	—	—
Cu_3Al_2	c	244.583	—	−26.2	—	—	—	—
DYSPROSIUM								
Dy	c	162.50	0	0	0	2.116	17.87	6.73
	g	—	70.04	69.4	60.8	1.48	46.97	4.97
Dy^+	g	—	206.8	207.8	—	—	—	—
Dy^{2+}	g	—	476.	477.0	—	—	—	—
Dy^{3+}								
std. state, m = 1	aq	—	—	−167.	−159.	—	−55.2	5.
DyO	g	178.499	−19.	—	—	—	—	—
Dy_2O_3	c	372.998	−443.02	−445.3	−423.4	5.04	35.8	27.79
DyF	g	181.498	—	−43.	—	—	—	—
DyF^{2+}								
std. state, m = 1	aq	—	—	—	−231.7	—	—	—
$DyCl_3$								
β	c	268.859	—	−239.	—	—	—	—
γ	c	—	—	−236.	—	—	—	—
std. state, m = 1	aq	—	—	−286.	−253.	—	−14.8	−93.
$DyCl_3$								
in ∞ H_2O	aq	—	—	−286.	—	—	—	—
$DyCl_3 \cdot 6H_2O$	c	376.951	−676.83	−686.	−586.	14.42	96.00	82.7
DyI_3	c	543.213	—	−145.	—	—	—	—
$Dy(IO_3)_3$	c	687.208	—	−329.	—	—	—	—
DyC_2	g	186.522	206.	206.1	193.2	2.47	64.	10.5
ERBIUM								
Er	c	167.26	0	0	0	1.765	17.49	6.72
	g	—	76.08	75.8	67.1	1.48	46.72	4.97
Er^+	g	—	216.8	218.0	—	—	—	—
Er^{2+}	g	—	492.	495.	—	—	—	—
Er^{3+}								
std. state, m = 1	aq	—	—	−168.6	−159.9	—	−58.4	5.
ErO	g	183.259	−13.	—	—	—	—	—
Er_2O_3	c	382.518	−451.69	−453.6	−432.3	4.78	37.2	25.93
ErH_2	c	169.276	—	−49.0	—	—	—	—
Er^2H_2	c	171.288	—	−49.3	—	—	—	—
ErH_3	c	170.284	—	−58.	—	—	—	—
ErF	g	186.258	—	−45.	—	—	—	—
ErF_2	g	205.257	—	−164.	—	—	—	—
ErF_3	c	224.255	—	−409.	—	—	—	—
	g	—	—	−294.	—	—	—	—
$ErCl_3$	c	273.619	—	−238.7	—	—	—	24.
$ErCl_3$	g	—	—	−167.	—	—	—	—
$ErCl_3 \cdot 6H_2O$	c	381.711	−677.92	−687.0	−586.6	14.39	95.3	82.0
$Er(BrO_3)_3 \cdot 9H_2O)$	c	713.120	—	−844.9	—	—	—	—

Substance			0 K	298.15 K (25°C)				
Formula and Description	State	Formula weight	ΔHf°_0 kcal/mol	ΔHf°	ΔGf° kcal/mol	$H^\circ_{298} - H^\circ_0$	S° cal/deg mol	C°_p
ErI_3	c	547.973	—	−146.5	—	—	—	—
$Er(IO_3)_3$	c	691.968	—	−330.	—	—	—	—
ErC_2	g	191.282	138.	138.2	125.4	2.47	63.	10.5
EUROPIUM								
Eu	c	151.96	0	0	0	1.913	18.59	6.61
	g	—	42.232	41.9	34.0	1.481	45.097	4.968
Eu^+	g	—	172.93	174.08	—	—	—	—
Eu^{2+}	g	—	432.	435.	—	—	—	—
std. state, m = 1	aq	—	—	−126.	−129.1	—	−2.	—
Eu^{3+}								
std. state, m = 1	aq	—	—	−144.6	−137.2	—	−53.	2.
EuO	c	167.959	—	−141.5	−133.1	—	15.	—
Eu_2O_3								
cubic	c	351.918	—	−397.4	—	—	—	29.6
monoclinic	c	—	—	−394.7	−372.1	—	35.	29.2
Eu_3O_4	c	519.878	—	−543.	−512.	—	49.	—
$Eu(OH)_3$	c	202.982	—	—	−285.5	—	—	—
EuF	g	170.958	—	−70.	—	—	—	—
EuF_3	c	208.955	—	—	—	—	—	—
$EuCl_2$	c	222.866	—	−197.	—	—	—	—
	g	—	—	−110.	—	—	—	—
$EuCl_3$	c	258.319	—	−223.7	—	—	—	—
std. state, m = 1	aq	—	—	−264.4	−231.3	—	−13.	−96.
$EuCl_3 \cdot 6H_2O$	c	366.411	—	−665.6	−565.5	—	97.3	87.7
$Eu(BrO_3)_3 \cdot 9H_2O$	c	697.820	—	−823.4	—	—	—	—
$Eu(IO_3)_3$	c	676.668	—	−308.4	—	—	—	—
EuS	g	184.024	27.	—	—	—	—	—
$Eu_2(SO_4)_3 \cdot 8H_2O$	c	736.228	—	—	—	—	160.6	146.0
EuC_2	c	175.982	—	−15.	−16.	—	24.	—
FLUORINE								
F	g	18.9984	18.38	18.88	14.80	1.558	37.917	5.436
F^+	g	—	420.16	422.14	—	—	—	—
F^{2+}	g	—	1226.98	1230.44	—	—	—	—
F^{3+}	g	—	2672.0	2676.9	—	—	—	—
F^{4+}	g	—	4684.2	4690.6	—	—	—	—
F^-	g	—	−63.7	−64.7	—	—	—	—
std. state, m = 1	aq	—	—	−79.50	−66.64	—	−3.3	−25.5
F_2	g	37.9968	0	0	0	2.108	48.44	7.48
F_2^+	g	—	365.1	366.6	—	—	—	—
FO	g	34.9978	41.	41.	—	—	—	—
F_2O	g	53.9962	−4.7	−5.2	−1.1	2.604	59.11	10.35
HF	liq	20.0064	—	−71.65	—	—	18.02 + x[a]	12.35
	g	—	−64.789	−64.8	−65.3	2.055	41.508	6.963
ionized; std. state, m = 1	aq	—	—	−79.50	−66.64	—	−3.3	−25.5
HF								
∞ H_2O	aq	—	—	−79.50	—	—	—	—
HF_2^-								
std. state, m = 1	aq	39.0048	—	−155.34	−138.18	—	22.1	—
XeF_4	c	207.294	—	−62.5	—	—	—	—
FRANCIUM								
Fr	cs	223.0000	0	0	0	—	22.8	—
FrF	c	241.9984	—	—	—	2.80	26.0	12.6
FrCl	c	258.4530	—	—	—	3.10	27.0	12.80
FrBr	c	302.9040	—	—	—	3.30	31.0	12.90
FrI	c	349.9045	—	—	—	3.40	33.0	12.90
GADOLINIUM								
Gd	c	157.25	0	0	0	2.178	16.27	8.85
	g	—	95.353	95.0	86.0	1.825	46.416	6.584
Gd^+	g	—	237.0	238.1	—	—	—	—
Gd^{2+}	g	—	517.	519.	—	—	—	—
Gd^{3+}								
std. state, m = 1	aq	—	—	−164.	−158.	—	−49.2	0.
GdO	g	173.249	−17.	—	—	—	—	—
Gd_2O_3								
monoclinic	c	362.498	—	−434.9	—	—	—	25.5
cubic	c	—	—	—	—	4.45	36.0	25.22
GdH_2	c	159.266	—	−45.5	—	—	—	—
GdF	g	176.248	—	−41.	—	—	—	—
GdF_2	g	195.247	—	−169.	—	—	—	—
GdF_3	g	214.245	—	−310.	—	—	—	—
$GdCl_3$	c	263.609	—	−241.	—	—	—	21.
std. state, m = 1	aq	—	—	−284.	−253.	—	−8.8	−98.
$GdCl_3 \cdot 6H_2O$	c	371.701	−675.66	−685.	−586.	14.43	97.56	83.0
$Gd(BrO_3)_3 \cdot 9H_2O$	c	703.110	—	−842.9	—	—	—	—
GdI_3	c	537.963	—	−142.	—	—	—	—
$Gd(IO_3)_3$	c	681.958	—	−327.	—	—	—	—
GdS	g	189.314	38.	—	—	—	—	—
$Gd_2(SO_4)_3 \cdot 8H_2O$	c	746.808	—	−1513.	−1322.	—	155.8	140.5
$Gd(NO_3)_3 \cdot 6H_2O$	c	451.357	—	—	—	19.16	133.2	106.2
$GdPO_4 \cdot H_2O$	c	270.237	—	—	−490.	—	—	—
GdC_2	c	181.272	—	−25.	—	—	—	—
	g	—	128.2	128.2	—	2.5	—	—
GALLIUM								
Ga	c	69.72	—	—	—	1.331	9.77	6.18
	liq	—	—	1.33	—	—	—	—
	g	—	66.	66.2	57.1	1.566	40.38	6.06
Ga^+	g	—	204.32	206.00	—	—	—	—
Ga^{2+}	g	—	677.38	680.54	—	—	—	—
std. state, m = 1	aq	—	—	—	−21.	—	—	—
Ga^{3+}	g	—	1385.	1390.	—	—	—	—
std. state, m = 1	aq	—	—	−50.6	−38.0	—	−79.	—
Ga^{4+}	g	—	2865.	2871.	—	—	—	—
GaO	g	85.719	67.	66.8	60.6	2.127	55.2	7.66
Ga_2O	c	155.439	—	−85.	—	—	—	—
Ga_2O_3								
β, rhombic	c	187.438	—	−260.3	−238.6	—	20.31	22.00
GaH	g	70.728	53.	52.7	46.3	2.07	46.69	7.00
$Ga(OH)_3$	c	120.742	—	−230.5	−198.7	—	24.	—
GaF	g	88.718	−60.	−60.2	—	2.167	—	7.95
GaCl	g	105.173	−19.	−19.1	−25.4	2.29	57.4	8.50

[a] x = Undetermined residual entropy

Substance			0 K	298.15 K (25°C)				
Formula and Description	State	Formula weight	ΔHf°_0 kcal/mol	ΔHf°	ΔGf° kcal/mol	$H^\circ_{298} - H^\circ_0$	S° cal/deg mol	C°_p
$GaCl_3$	c	176.079	—	−125.4	−108.7	—	34.	—
	g	—	—	−107.0	—	—	—	—
GaBr	g	149.629	−10.	−11.9	−21.5	2.37	60.2	8.70
$GaBr_3$	c	309.447	—	−92.4	−86.0	—	43.	—
GaI	g	196.624	7.4	6.9	—	2.41	—	8.76
GaI_3	c	450.433	—	−57.1	—	—	—	—
	g	—	—	−34.0	—	—	—	—
$Ga_2(SO_4)_3$	c	427.625	—	—	—	—	—	62.4
GaN	c	83.727	—	−26.4	—	—	—	—
GaP	c	100.694	—	−21.	—	—	—	—
$GaPO_4$	c	164.691	—	—	−310.1	—	—	—
GaAs	c	144.642	—	−17.	−16.2	—	15.34	11.05
GaSb	c	191.47	—	−10.0	−9.3	—	18.18	11.60
Ga_2C_2	g	163.462	—	134.	—	—	—	—
GERMANIUM								
Ge	c	72.59	0	0	0	1.105	7.43	5.580
	g	—	89.34	90.0	80.3	1.768	40.103	7.345
Ge^+	g	—	271.18	273.32	—	—	—	—
Ge^{2+}	g	—	638.63	642.25	—	—	—	—
Ge^{3+}	g	—	1427.85	1432.95	—	—	—	—
GeO								
brown	c	88.589	—	−62.6	−56.7	—	12.	—
yellow	c	—	—	—	−49.5	—	—	—
	g	—	−11.	−11.04	−17.49	2.102	53.58	7.39
GeO_2								
hexagonal	c	104.589	—	−131.7	−118.8	—	13.21	12.45
	amorp	—	—	−128.4	—	—	—	—
Ge_2O_2	g	177.179	—	−112.	—	—	—	—
Ge_3O_3	g	265.768	—	−212.	—	—	—	—
GeH_4	g	76.622	24.29	21.7	27.1	2.567	51.87	10.76
Ge_2H_6	liq	151.228	—	32.82	—	—	—	—
	g	—	—	38.8	—	—	—	—
Ge_3H_8	liq	225.834	—	46.3	—	—	—	—
	g	—	—	54.2	—	—	—	—
H_2GeO_3	aq	122.604	—	−195.73	—	—	—	—
GeF	g	91.588	−8.	−7.97	—	2.185	—	8.30
GeF_2	g	110.587	−121.	—	—	—	—	—
GeF_4	g	148.584	—	—	—	4.163	72.36	19.56
GeCl	g	108.043	37.	37.09	29.7	2.290	59.	8.81
$GeCl_4$	liq	214.402	—	−127.1	−110.6	—	58.7	—
GeH_3Cl	g	111.067	—	—	—	2.865	63.00	13.08
$GeHCl_3$	g	—	—	—	—	4.192	79.06	19.40
GeBr	g	152.499	58.	56.32	—	2.355	—	8.87
$GeBr_2$	g	232.408	—	−15.0	−25.5	—	79.1	—
$GeBr_4$	liq	392.226	—	−83.1	−79.2	—	67.1	—
	g	—	−64.61	−71.7	−76.0	5.736	94.66	24.34
GeI_2	c	326.399	—	−21.	−20.	—	32.	—
	g	—	—	11.2	−1.0	—	76.	—
GeI_4	c	580.208	—	−33.9	−34.5	—	64.8	—
	g	—	−12.31	−13.6	−25.4	6.12	102.49	24.89
GeS	c	104.654	—	−16.5	−17.1	—	17.	—
	g	—	22.0	22.	10.	2.185	56.	8.05
GeS_2	c	136.718	—	−45.3	—	—	—	—
GeSe	c	151.55	—	−22.0	—	—	—	—
GeTe	c	200.19	—	−6.	—	—	—	—
Ge_3N_4	c	273.797	—	−15.1	—	—	—	—
GeP	c	103.564	—	−5.	−4.	—	15.	—
GeC	g	84.601	150.	151.	—	—	—	—
GeC_2	g	96.612	142.	143.	—	—	—	—
Ge_2C	g	157.191	130.	131.	—	—	—	—
GeSi	g	100.676	126.	127.	—	—	—	—
Ge_2Si	g	173.266	123.	124.	—	—	—	—
Ge_3Si	g	245.856	121.	122.	—	—	—	—
GOLD								
Au	c	196.967	—	—	—	1.436	11.33	6.075
	g	—	87.46	87.5	78.0	1.481	43.115	4.968
Au^+	g	—	300.20	301.73	—	—	—	—
Au^{2+}	g	—	773.0	776.0	—	—	—	—
AuO_3^{3-}								
std. state, m = 1	aq	244.9652	—	—	−12.4	—	—	—
AuH	g	197.9750	70.9	70.5	63.5	2.068	50.441	6.968
$HAuO_3^{2-}$								
std. state, m = 1	aq	245.9732	—	—	−34.0	—	—	—
$H_2AuO_3^-$								
std. state, m = 1	aq	246.9811	—	—	−52.2	—	—	—
$Au(OH)_3$								
precipitated	c	247.9891	—	−101.5	−75.77	—	45.3	—
AuF_3	c	253.9622	—	−86.9	—	—	—	—
AuCl	c	232.420	—	−8.3	—	—	—	—
$AuCl_2^-$								
std. state, m = 1	aq	267.873	—	—	−36.13	—	—	—
$AuCl_3$	c	303.326	—	−28.1	—	—	—	—
$AuCl_3 \cdot 2H_2O$	c	339.357	—	−170.9	—	—	—	—
$AuCl_4^-$								
std. state, m = 1	aq	338.779	—	−77.0	−56.22	—	63.8	—
$HAuCl_4$								
std. state, m = 1	aq	339.787	—	−77.0	−56.22	—	63.8	—
$HAuCl_4 \cdot 3H_2O$	c	393.833	—	−284.9	—	—	—	—
$HAuCl_4 \cdot 4H_2O$	c	411.848	—	−355.9	—	—	—	—
AuBr	c	276.876	—	−3.34	—	—	—	—
$AuBr_2^-$								
std. state, m = 1	aq	356.785	—	−30.7	−27.49	—	52.5	—
$AuBr_3$	c	436.695	—	−12.73	—	—	—	—
$AuBr_4^-$								
std. state, m = 1	aq	516.603	—	−45.8	−40.0	—	80.3	—
$HAuBr_4 \cdot 5H_2O$	c	607.688	—	−398.7	—	—	—	—
AuI	c	323.8714	—	0.	—	—	—	—
$AuSb_2$	c	440.467	—	−2.6	—	—	—	—
$Au(CN)_2^-$								
std. state, m = 1	aq	249.003	—	57.9	68.3	—	41.	—

Substance			0 K	298.15 K (25°C)				
Formula and Description	State	Formula weight	ΔHf_0° kcal/mol	ΔHf°	ΔGf° kcal/mol	$H_{298}^\circ - H_0^\circ$	S°	C_p° cal/deg mol
$Au(SCN)_2^-$								
std. state, m = 1	aq	313.131	—	—	60.2	—	—	—
$Au(SCN)_4^-$								
std. state, m = 1	aq	429.294	—	—	134.2	—	—	—
$Au(SCN)_5^{2-}$								
std. state, m = 1	aq	487.376	—	—	156.4	—	—	—
$Au(SCN)_6^{3-}$								
std. state, m = 1	aq	545.458	—	—	178.5	—	—	—
AuSn	c	315.657	—	−7.28	−7.15	—	23.22	12.06
$AuSn_2$	c	434.347	—	−10.14	−9.07	—	32.4	—
$AuSn_4$	c	671.727	—	−9.25	−9.04	—	59.9	—
$AuPb_2$	c	611.347	—	−1.5	—	—	—	—
AuIn	c	311.787	—	−10.8	—	—	—	—
$AuIn_2$	c	426.607	—	−18.0	—	—	—	—
AuCd	c	309.367	—	−9.28	−9.34	—	23.9	—
AuCu	c	260.507	—	−4.46	−4.40	—	19.1	11.9
$AuCu_3$	c	387.587	—	−6.84	−6.97	—	35.7	24.0
HAFNIUM								
Hf								
α, hexagonal	c	178.49	0	0	0	1.397	10.41	6.15
	g	—	147.92	148.0	137.8	1.481	44.642	4.972
Hf^+	g	—	309.	311.	—	—	—	—
Hf^{2+}	g	—	653.	656.	—	—	—	—
Hf^{3+}	g	—	1190.	1195.	—	—	—	—
HfO	g	194.489	—	12.	—	—	—	—
HfO_2	c	210.489	—	−273.6	−260.1	—	14.18	14.40
HfF_4								
monoclinic	c	254.484	—	−461.4	−437.5	—	27.	—
$HfCl_4$	c	320.302	—	−236.70	−215.42	—	45.6	28.80
HfN	c	192.497	—	−88.3	—	—	—	—
HfC	c	190,501	—	−60.1	—	—	—	—
HfB	c	189.301	—	−47.	—	—	—	—
HfB_2	c	200.112	−80.09	−80.3	−79.4	1.77	10.2	11.89
HELIUM								
He	g	4.0026	0	0	0	1.481	30.1244	4.9679
std. state, m = 1	aq	—	—	−0.4	4.6	—	13.3	—
He^+	g	—	566.978	568.459	—	—	—	—
HOLMIUM								
Ho	c	164.930	0	0	0	1.91	18.0	6.49
	g	—	72.33	71.9	63.3	1.48	46.72	4.97
Ho^+	g	—	211.1	212.2	—	—	—	—
Ho^{2+}	g	—	483.	486.	—	—	—	—
Ho^{3+}								
std. state, m = 1	aq	—	—	−168.5	−161.0	—	−54.2	4.
HoO	g	180.929	−22.	—	—	—	—	—
Ho_2O_3	c	377.858	−447.59	−449.5	−428.1	5.02	37.8	27.48
HoH_2	c	166.946	—	−51.7	—	—	—	—
HoF	g	183.928	—	−43.	—	—	—	—
HoF_2	g	202.927	—	−163.	—	—	—	—
HoF_3	c	221.925	—	−408.	—	—	—	—
	g	—	—	−294.	—	—	—	—
$HoCl_3$	c	271.289	—	−240.3	—	—	—	21.
	g	—	—	−168.	—	—	—	—
$HoCl_3$								
∞ H_2O	aq	—	—	−288.4	—	—	—	—
$HoCl_3 \cdot 6H_2O$	c	379.381	−678.80	−687.9	−588.0	14.49	97.08	83.0
HoOCl	c	216.382	—	−239.2	—	—	—	—
HoI_3	c	545.643	—	−149.0	—	—	—	—
HoS	g	196.994	43.	—	—	—	—	—
HoAs	c	239.852	—	−72	—	—	—	—
HoC_2	c	188.952	—	−26.	−26.7	—	23.	—
	g	—	135.	135.1	—	2.47	—	10.5
HoC_4	g	212.975	165.	—	—	—	—	—
Ho_2C_3	c	365.893	—	−56.	—	—	—	—
HoAu	g	361.897	100.	99.06	—	2.41	—	8.77
HYDROGEN								
H	g	1.0080	51.626	52.095	48.581	1.481	27.391	4.9679
^1H	g	1.0078	51.626	52.095	48.581	1.481	27.391	4.9679
^2H	g	2.0141	52.524	52.981	49.360	1.481	29.455	4.9679
H^+	g	1.0080	365.211	367.161	—	—	—	—
std. state, m = 1	aq	—	—	0	0	—	0	0
H^-	g	—	34.40	33.39	—	—	—	—
H_2	g	2.0159	0	0	0	2.0238	31.208	6.889
$^1H_2$	g	2.0156	0	0	0	2.0238	31.208	6.889
$^2H_2$	g	4.0282	0	0	0	2.0481	34.620	6.978
$^1H^2H$	g	3.0219	0.079	0.076	−0.350	2.0328	34.343	6.978
H_2								
std. state, m = 1	aq	2.0159	—	−1.0	4.2	—	13.8	—
H_2^+	g	—	355.74	357.23	—	—	—	—
OH	g	17.0074	9.25	9.31	8.18	2.1070	43.890	7.143
OH^-	g	—	−32.3	−33.67	—	—	—	—
std. state, m = 1	aq	—	—	−54.970	−37.594	—	−2.57	−35.5
HO_2	g	33.0068	6.	5.	—	—	—	—
H_2O	liq	18.0153	—	−68.315	−56.687	—	16.71	17.995
^2H_2O	liq	20.0276	—	−70.411	−58.195	—	18.15	20.16
$^1H^2HO$	liq	19.0213	—	−69.285	−57.817	—	18.95	—
H_2O	g	18.0153	−57.102	−57.796	−54.634	2.3667	45.104	8.025
^1H_2O	g	18.0150	−57.102	−57.796	−54.634	2.3667	45.103	8.025
^2H_2O	g	20.0276	−58.855	−59.560	−56.059	2.3801	47.378	8.19
$^1H^2HO$	g	19.0213	−57.927	−58.628	−55.719	2.3721	47.658	8.08
H_2O_2	liq	34.0147	—	−44.88	−28.78	−26.2	—	21.3
	g	—	−31.08	−32.58	−25.24	2.594	55.6	10.3
undissoc.; std. state, m = 1	aq	—	—	−45.69	−32.05	—	34.4	—
H_2O_2								
∞ H_2O	aq	—	—	−45.69	—	—	—	—
INDIUM								
In	c	114.82	0	0	0	1.578	13.82	6.39
	g	—	58.25	58.15	49.89	1.48	41.51	4.98
In^+	g	—	191.68	193.06	—	—	—	—
std. state, m = 1	aq	—	—	—	−2.9	—	—	—

Substance			0 K	298.15 K (25°C)				
Formula and Description	State	Formula weight	ΔHf_0° kcal/mol	ΔHf°	ΔGf° kcal/mol	$H_{298}^\circ - H_0^\circ$	S° cal/deg mol	C_p°
In^{2+}	g	—	626.82	629.68	—	—	—	—
std. state, m = 1	aq	—	—	—	−12.1	—	—	—
In^{3+}	g	—	1273.2	1277.56	—	—	—	—
std. state, m = 1	aq	—	—	−25.	−23.4	—	−36.	—
InO	g	130.819	92.	93.	87.1	2.14	56.5	7.78
In_2O_3	c	277.638	—	−221.27	−198.55	—	24.9	22.
InH	g	115.828	52.	51.5	45.49	2.075	49.60	7.07
$InOH$	g	131.827	−18.	−19.	—	—	—	—
InF	g	133.818	−48.2	−48.61	—	2.198	—	—
$InCl_3$	c	221.179	—	−128.4	—	—	—	—
In_2Cl_3	g	335.999	—	−103.6	—	—	—	—
$InBr$	c	194.729	—	−41.9	−40.4	—	27.	—
	g	—	−11.5	−13.6	−22.54	2.406	61.99	8.76
$InBr_3$	c	354.547	—	−102.5	—	—	—	—
	g	—	—	−67.4	—	—	—	—
InI	c	241.724	—	−27.8	−28.8	—	31.	—
	g	—	2.52	1.8	−9.0	2.437	63.87	8.80
InI_3	c	495.533	—	−57.	—	—	—	—
	g	—	—	−28.8	—	—	—	—
InS	c	146.884	—	−33.0	−31.5	—	16.	—
	g	—	—	57.	—	—	—	—
In_2S	g	261.704	15.	2.9	—	76.	—	—
In_2S_3	c	325.832	—	−102.	−98.6	—	39.1	28.20
$In_2(SO_4)_3$	c	517.825	—	−666.	−583.	—	65.	67.
InN	c	128.827	—	−4.2	—	—	—	—
InP	c	145.794	—	−21.2	−18.4	—	14.3	10.86
$InAs$	c	189.742	—	−14.0	−12.8	—	18.1	11.42
$InSb$	c	236.57	—	−7.3	−6.1	—	20.6	11.82
$InSb_2$	g	358.32	—	75.	—	—	—	—
IODINE								
I	g	126.9044	25.631	25.535	16.798	1.481	43.184	4.968
I^+	g	—	266.77	268.16	—	—	—	—
I^{2+}	g	—	707.22	710.09	—	—	—	—
I^-	g	—	−45.4	−47.0	—	—	—	—
std. state, m = 1	aq	—	—	−13.19	−12.33	—	26.6	−34.0
I_2	c	253.8088	0	0	0	3.154	27.757	13.011
	g	—	15.659	14.923	4.627	2.418	62.28	8.82
std. state, m = 1	aq	—	—	5.4	3.92	—	32.8	—
I_3^-								
std. state, m = 1	aq	380.7132	—	−12.3	−12.3	—	57.2	—
IO^-								
std. state, m = 1	aq	—	—	−25.7	−9.2	—	−1.3	—
IO_3^-								
std. state, m = 1	aq	174.9026	—	−52.9	−30.6	—	28.3	—
IO_4^-								
std. state, m = 1	aq	190.9020	—	−36.2	−14.0	—	−53.	—
I_2O_5	c	333.8058	—	−37.78	—	—	—	—
HI	g	127.9124	6.850	6.33	0.41	2.069	49.351	6.969
std. state, m = 1	aq	—	—	−13.19	−12.33	—	26.6	−34.0
HI								
in ∞ H_2O	aq	—	—	−13.19	—	—	—	—
HIO								
undissoc.; std. state, m = 1	aq	143.9118	—	−33.0	−23.7	—	22.8	—
HIO_3	c	175.9106	—	−55.0	—	—	—	—
undissoc.; std. state, m = 1	aq	—	—	−50.5	−31.7	—	39.9	—
IF	g	145.9028	−22.40	−22.86	−28.32	2.174	56.42	7.99
IF_5	liq	221.8964	—	−206.7	—	—	—	—
ICl								
α	c	162.3574	—	−8.4	—	—	—	—
	liq	—	—	−5.71	−3.25	—	32.3	—
	g	—	4.64	4.25	−1.30	2.282	59.140	8.50
std. state, m = 1	aq	—	—	—	−4.1	—	—	—
ICl_3	c	233.2634	—	−21.4	−5.34	—	40.0	—
IBr	c	206.8134	—	−2.5	—	—	—	—
	g	—	11.90	9.76	0.89	2.367	61.822	8.71
std. state, m = 1	aq	—	—	—	−1.0	—	—	—
IRIDIUM								
Ir	c	192.2	0	0	0	1.260	8.48	6.00
	g	—	158.78	159.0	147.7	1.481	46.240	4.968
Ir^+	g	—	373.	375.	—	—	—	—
IrO_2	c	224.20	—	−65.5	—	—	—	13.7
IrO_3	g	240.20	—	1.9	—	—	—	—
IrF_6	c	306.19	—	−138.54	−110.34	—	59.2	—
$IrCl$	c	227.65	—	−19.5	—	—	—	—
$IrCl_3$	c	298.56	—	−58.7	—	—	—	—
IrS_2	c	256.33	—	−33.	—	—	—	—
Ir_2S_3	c	480.59	—	−56.	—	—	—	—
IRON								
Fe								
α	c	55.847	0	0	0	1.073	6.52	6.00
	g	—	98.94	99.5	88.6	1.6374	43.112	6.137
Fe^+	g	—	281.12	283.16	—	—	—	—
Fe^{2+}	g	—	654.2	657.8	—	—	—	—
std. state, m = 1	aq	—	—	−21.3	−18.85	—	−32.9	—
Fe^{3+}	g	—	1360.8	1365.9	—	—	—	—
std. state, m = 1	aq	—	—	−11.6	−1.1	—	−75.5	—
$Fe_{0.947}O$								
wustite	c	68.8865	−63.85	−63.64	−58.89	2.26	13.74	11.50
FeO	c	71.8464	—	−65.0	—	—	—	—
Fe_2O_3								
hematite	c	159.6922	−195.46	−197.0	−177.4	3.719	20.89	24.82
Fe_3O_4								
magnetite	c	231.5386	−265.80	−267.3	−242.7	5.87	35.0	34.28
$FeO(OH)$								
goethite	c	88.8538	—	−133.6	—	—	—	—
$Fe(OH)_2$								
precipitated	c	89.8617	—	−136.0	−116.3	—	21.	—
$Fe(OH)_3$								
precipitated	c	106.8691	—	−196.7	−166.5	—	25.5	—
undissoc.; std. state, m = 1	aq	—	—	—	−157.6	—	—	—

Substance			0 K	298.15 K (25°C)				
Formula and Description	State	Formula weight	ΔHf_0° kcal/mol	ΔHf°	ΔGf° kcal/mol	$H_{298}^\circ - H_0^\circ$	S° cal/deg mol	C_p°
FeF_2	c	93.8438	—	—	—	3.049	20.79	16.28
FeF_2^+								
std. state, m = 1	aq	—	—	−166.4	−150.2	—	−15.	—
FeF_3	aq	112.8422	—	−242.9	—	—	—	—
$FeCl_2$	c	126.753	−82.313	−81.69	−72.26	3.889	28.19	18.32
std. state, m = 1	aq	—	—	−101.2	−81.59	—	−5.9	—
$FeCl_2 \cdot 2H_2O$	c	162.7837	—	−227.8	—	—	—	—
$FeCl_2 \cdot 4H_2O$	c	198.8143	—	−370.3	—	—	—	—
$FeCl_3$	c	162.206	−95.828	−95.48	−79.84	4.710	34.0	23.10
std. state, m = 1	aq	—	—	−131.5	−95.2	—	−35.0	—
$FeCl_3 \cdot 6H_2O$	c	270.2980	—	−531.5	—	—	—	—
$FeOCl$	c	107.2994	—	−90.1	—	—	—	—
$Fe(ClO_4)_2$								
std. state, m = 1	aq	254.7482	—	−83.1	−22.97	—	54.1	—
$FeBr_2$	c	215.665	—	−59.7	—	—	—	—
std. state, m = 1	aq	–	—	−79.4	−68.55	—	6.5	—
$FeBr_3$	c	295.574	—	−64.1	—	—	—	—
std. state, m = 1	aq	—	—	−98.8	−75.7	—	−16.4	—
FeI_2	c	309.6558	—	−27.0	—	—	—	—
std. state, m = 1	aq	—	—	−47.7	−43.51	—	20.3	—
FeI_3	g	436.5602	—	17.	—	—	—	—
std. state, m = 1	aq	—	—	−51.2	−38.1	—	4.3	—
$Fe_{1.000}S$								
Iron-rich pyrrhotite, α	c	87.911	−24.01	−23.9	−24.0	2.235	14.41	12.08
FeS_2								
pyrite	c	119.975	−41.72	−42.6	−39.9	2.302	12.65	14.86
markasite	c	—	—	−37.0	—	—	—	—
$FeSO_4$	c	151.9086	—	−221.9	−196.2	—	25.7	24.04
std. state, m = 1	aq	—	—	−238.6	−196.82	—	−28.1	—
$FeSO_4 \cdot H_2O$	c	169.9239	—	−297.25	—	—	—	—
$FeSO_4 \cdot 4H_2O$	c	223.9700	—	−508.9	—	—	—	—
$FeSO_4 \cdot 7H_2O$	c	278.0160	—	−720.50	−599.97	—	97.8	94.28
$FeSe$	c	134.807	—	−18.0	—	—	—	—
$FeTe$	c	183.447	—	−15.0	—	—	—	—
Fe_4N	c	237.3947	—	−2.5	0.9	—	37.	—
$Fe(NO_3)_3$	aq	241.8617	—	−161.3	—	—	—	—
std. state, m = 1	aq	—	—	−160.3	−80.9	—	29.5	—
FeP	c	86.8208	—	−30.	—	—	—	—
FeP_2	c	117.7946	—	−46.	—	—	—	—
Fe_2P	c	142.6678	—	−39.	—	—	—	—
Fe_3P	c	198.5148	—	−39.	—	—	—	—
$FePO_4$	c	150.8184	—	−310.1	—	—	—	—
$FePO_4 \cdot 2H_2O$								
strengite	c	186.8491	−445.28	−451.3	−396.2	6.607	40.93	43.15
Fe_3C								
α, cementite	c	179.5522	—	6.0	4.8	—	25.0	25.3
$FeCO_3$								
siderite	c	115.8564	—	−177.00	−159.35	—	22.2	19.63
$Fe(CO)_5$	liq	195.8998	—	−185.0	−168.6	—	80.8	57.5
	g	—	—	−175.4	−166.65	—	106.4	—
$HFe(CN)_6^{3-}$								
std. state, m = 1	aq	212.9621	—	108.9	160.40	—	42.	—
$H_2Fe(CN)_6^{2-}$								
std.state, m = 1	aq	213.9700	—	108.9	157.37	—	52.	—
$H_3Fe(CN)_6^-$								
std. state, m = 1	aq	—	—	108.9	—	—	—	—
$Fe(SCN)^{2+}$								
thiocyanate; std. state, m = 1	aq	113.9288	—	5.6	17.0	—	−31.	—
$FeSi$	c	83.933	−17.67	−17.6	−17.6	1.91	11.0	11.4
$FeSi_2$								
β-lebeanite	c	112.019	−19.19	−19.4	−18.7	2.40	13.3	15.79
$FeSi_{2.33}$								
α-lebeanite	c	121.287	−14.03	−14.	−14.	2.89	16.6	17.62
Fe_3Si	c	195.627	−22.55	−22.4	−22.6	4.14	24.8	23.50
Fe_2SiO_4								
fayalite	c	203.7776	—	−353.7	−329.6	—	34.7	31.76
$FeAl_2O_4$	c	173.8076	—	−470.	−442.	—	25.4	29.53
KRYPTON								
Kr	g	83.80	0	0	0	1.481	39.1905	4.9679
std. state, m = 1	aq	—	—	−3.7	3.6	—	14.7	—
Kr^+	g	—	322.84	324.32	—	—	—	—
Kr^{2+}	g	—	889.47	892.43	—	—	—	—
Kr^{3+}	g	—	1741.6	1746.0	—	—	—	—
LANTHANUM								
La								
α	c	138.91	0	0	0	1.593	13.6	6.48
	g	—	103.084	103.0	94.07	1.509	43.563	5.438
La^+	g	—	231.690	233.087	—	—	—	—
La^{2+}	g	—	487.	490.	—	—	—	—
La^{3+}	g	—	929.2	933.2	—	—	—	—
std. state, m = 1	aq	—	—	−169.0	−163.4	—	−52.0	−3.
LaO	g	154.909	−28.5	−29.01	−34.72	2.121	57.27	7.59
La_2O	g	293.819	−2.	−3.2	—	3.00	—	12.0
La_2O_2	g	309.819	−145.	−146.6	—	3.67	—	16.2
La_2O_3	c	325.818	−427.19	−428.7	−407.7	4.742	30.43	26.00
LaH_2	c	140.926	—	−48.3	—	—	—	—
$La(OH)_3$	c	189.932	—	−337.0	—	—	—	—
LaF	g	157.908	—	—	—	2.173	56.9	7.98
LaF_3	g	195.905	—	—	—	4.14	78.7	17.4
$LaCl_3$	c	245.269	—	−256.0	—	—	—	26.0
std. state, m = 1	aq	—	—	−288.9	−257.5	—	−12.	−101.
$LaCl_3$								
in ∞ H_2O	aq	—	—	−288.9	—	—	—	—
$LaCl_3 \cdot 7H_2O$	c	371.376	—	−759.7	−648.5	17.12	110.6	103.0
$La(BrO_3)_3 \cdot 9H_2O$	c	684.770	—	−846.0	—	—	—	—
LaI_3	c	519.623	—	−159.4	—	—	—	—
$La(IO_3)_3$	c	663.618	—	−334.	−270.4	—	62.	—
LaS	c	170.974	−108.9	−109.	−107.9	2.6	17.5	14.
La_2S_3	c	374.012	—	−289	—	—	—	—
$La_2(SO_4)_3$	c	566.005	—	−942.0	—	—	—	67.
$La_2(SO_4)_3 \cdot 9H_2O$	c	728.143	—	−1589.	—	—	—	152.

Substance			0 K	298.15 K (25°C)				
Formula and Description	State	Formula weight	ΔHf°_0 kcal/mol	ΔHf°	ΔGf° kcal/mol	$H^\circ_{298} - H^\circ_0$	S° cal/deg mol	C°_p
LaN	c	152.917	—	−72.5	—	—	—	—
$La(NO_3)_3$	c	324.925	—	−299.8	—	—	—	—
$La(NO_3)_3$								
in ∞ H_2O	aq	—	—	−317.7	—	—	—	—
$La(NO_3)_3 \cdot 3H_2O$	c	378.971	—	−520.0	—	—	—	—
$La(NO_3)_3 \cdot 4H_2O$	c	396.986	—	−592.3	—	—	—	—
$La(NO_3)_3 \cdot 6H_2O$	c	433.017	—	−732.23	—	—	—	—
LaC_2	c	162.932	—	−17.	−17.3	—	17.	—
$La_2(CO_3)_3$	c	457.848	—	—	−750.9	—	—	—
LaB_6	c	203.776	—	−31.	—	—	—	—
LaAu	g	335.877	111.4	110.8	98.2	2.46	67.	8.8
LEAD								
Pb	c	207.19	0	0	0	1.644	15.49	6.32
	g	—	46.76	46.6	38.7	1.481	41.889	4.968
Pb^+	g	—	217.795	219.116	—	—	—	—
Pb^{2+}	g	—	564.44	567.25	—	—	—	—
std. state, m = 1	aq	—	—	−0.4	−5.83	—	2.5	—
Pb^{3+}	g	—	1300.93	1305.22	—	—	—	—
Pb^{4+}	g	—	2276.9	2282.7	—	—	—	—
PbO								
yellow	c	223.189	−51.766	−51.466	−44.91	2.207	16.42	10.94
red	c	—	—	−52.34	−45.16	—	15.9	10.95
PbO_2	c	239.189	—	−66.3	−51.95	—	16.4	15.45
Pb_2O_3	c	462.378	—	—	—	—	36.3	25.74
Pb_3O_4	c	685.568	—	−171.7	−143.7	—	50.5	35.1
$Pb(OH)_2$	c	241.205	—	—	−108.1	—	—	—
precipitated	c	—	—	−123.3	—	—	—	—
PbF_2	c	245.187	—	−158.7	−147.5	—	26.4	—
std. state, m = 1	aq	—	—	−159.4	−139.11	—	−4.1	—
PbF_4	c	283.184	—	−225.1	—	—	—	—
$PbCl_2$	c	278.096	—	−85.90	−75.98	—	32.5	—
ionized; std. state, m = 1	aq	—	—	−80.3	−68.57	—	29.5	—
$PbCl_4$	liq	349.002	—	−78.7	—	—	—	—
$PbBr_2$	c	367.008	—	−66.6	−62.60	—	38.6	19.15
ionized; std. state, m = 1	aq	—	—	−58.5	−55.53	—	41.9	—
PbI_2	c	460.999	−41.81	−41.94	−41.50	4.666	41.79	18.49
ionized; std. state, m = 1	aq	—	—	−26.8	−30.49	—	55.7	—
PbS	c	239.254	—	−24.0	−23.6	—	21.8	11.83
$PbSO_3$	c	287.252	—	−160.1	—	—	—	—
$PbSO_4$	c	303.252	−217.82	−219.87	−194.36	4.795	35.51	24.667
PbS_2O_3	c	319.316	—	−161.1	—	—	—	—
PbSe	c	286.15	—	−24.6	−24.3	—	24.5	12.0
$PbSeO_4$	c	350.148	—	−145.6	−120.7	—	40.1	—
$PbTeO_3 \cdot 0.667H_2O$ (amorp)	—	394.804	—	−185.5	—	—	—	—
$Pb_3(PO_4)_2$	c	811.513	—	—	—	—	84.4	61.25
$PbCO_3$	c	267.199	—	−167.1	−149.5	—	31.3	20.89
PbC_2O_4								
std. state, m = 1	c	295.210	—	−197.6	−166.9	—	34.9	25.2
$Pb(CH_3)_4$	liq	267.330	—	23.4	—	—	—	—
$Pb(C_2H_5)_4$	liq	323.439	—	12.6	—	—	—	—
	g	—	—	26.19	—	—	—	—
$PSiO_3$	c	283.274	—	−273.83	−253.86	—	26.2	21.52
Pb_2SiO_4	c	506.464	—	−325.8	−299.4	—	44.6	32.78
LITHIUM								
Li	c	6.941	0	0	0	1.106	6.96	5.92
	g	6.941	37.715	38.09	30.28	1.481	33.14	4.968
Li^+	g	6.941	162.045	162.42	—	—	—	—
Li^{2+}	g	6.941	—	1936.7	—	—	—	—
Li^{3+}	g	6.941	—	4760.5	—	—	—	—
Li^+	a	6.941	—	−66.56	−70.10	—	3.2	16.4
Li								
in D:99Hg		6.941	—	−19.60	—	—	—	—
LI								
Hg:x		6.941	—	—	−19.5	—	—	—
LiO	g	22.9404	18.10	18.1	12.5	2.14	50.40	7.75
LiO_2	g	38.9398	—	—	—	2.60	58.27	10.34
Li_2O	c	29.8814	−141.393	−142.91	−134.13	1.732	8.98	12.93
	g	29.8814	−38.18	−38.4	−43.4	3.03	55.30	11.91
Li_2O_2	c	45.8808	—	−151.6	—	—	—	—
	a	45.8808	—	−158.8	—	—	—	—
LiH	c	7.9490	−20.425	−21.64	−16.34	0.903	4.782	6.66
LiD	c	8.9551	−20.684	−21.73	−16.18	1.085	5.640	8.20
LiOH	c	23.9484	−114.517	−115.90	−104.92	1.772	10.23	11.87
LiOD	c	24.9545	—	−116.8	—	—	—	—
LiF	c	25.9394	−146.657	−147.22	−140.47	1.547	8.52	9.94
Li_2F_2	g	51.8788	−223.67	−224.8	−224.7	3.19	61.79	15.09
Li_3F_3	g	77.8182	−360.30	−361.9	−357.	4.88	76.	24.5
$LiHF_2$	c	45.9458	−223.66	−225.22	−209.11	2.826	17.0	16.77
LiCl	c	42.394	−97.68	−97.66	−91.87	2.224	14.18	11.47
	g	42.394	−46.663	−46.7	−51.8	2.165	50.840	7.94
	a	42.394	—	−106.51	−101.48	—	16.7	−16.2
$LiCl \cdot H_2O$	c	60.4094	—	−170.31	−151.01	—	24.58	—
$LiCl \cdot 2H_2O$	c	78.4248	—	−242.03	—	—	—	—
$LiCl \cdot 3H_2O$	c	96.4402	—	−313.4	—	—	—	—
Li_2Cl_2	g	84.788	−141.20	−141.9	−142.5	3.70	69.0	17.26
$LiClO_3$	c	90.3922	—	−88.2	—	—	—	—
$LiClO_4$	c	106.3916	—	−91.06	—	—	—	—
	a	106.3916	—	−97.472	−72.2	—	46.7	−1.8
$LiClO_4 \cdot H_2O$	c	124.4070	—	−166.6	−121.8	—	37.1	—
$LiClO_4 \cdot 3H_2O$	c	160.4378	—	−310.22	−239.30	—	60.9	—
LiBr	c	86.850	—	−83.942	−81.74	—	17.75	—
	a	86.850	—	−95.612	−94.95	—	22.9	−17.5
$LiBr \cdot H_2O$	c	104.8654	—	−158.36	−142.05	—	26.2	—
$LiBr \cdot 2H_2O$	c	122.8808	—	−230.1	−200.9	—	38.8	—
$LiBrO_3$	c	134.8482	—	−82.93	—	—	—	—
LiI	c	133.8454	−64.663	−64.63	−64.60	2.716	20.74	12.20
	a	133.8454	—	−79.75	−82.4	—	29.8	−17.6
$LiI \cdot H_2O$	c	151.8608	—	−141.09	−127.0	—	29.4	—
$LiI \cdot 2H_2O$	c	169.8762	—	−212.81	−186.5	—	44.	—
$LiI \cdot 3H_2O$	c	187.8916	—	−284.93	—	—	—	—
$LiIO_3$	c	181.8436	—	−120.31	—	—	—	—

Substance			0 K	298.15 K (25°C)				
Formula and Description	State	Formula weight	ΔHf°_0 kcal/mol	ΔHf°	ΔGf° kcal/mol	$H^\circ_{298} - H^\circ_0$	S° cal/deg mol	C°_p
	a	181.8436	—	−119.46	−100.70	—	31.4	−13.2
Li_2S	c	45.946	—	−105.5	—	—	—	—
Li_2S_2	c	78.010	—	−104.7	—	—	—	—
Li_2SO_3	c	93.9442	—	−281.3	—	—	—	—
Li_2SO_4	c	109.9436	−340.367	−343.33	−315.91	4.452	27.5	28.10
	a	109.9436	—	−350.44	−318.18	—	11.3	−37.2
$LiSO_4 \cdot H_2O$	c	127.9590	−410.02	−414.8	−374.2	5.697	39.1	36.11
$Li_2SO_4 \cdot D_2O$	c	129.9712	—	−416.9	−375.6	—	40.	—
$LiHS$	c	40.0130	—	−60.1	—	—	—	—
$LiNO_2$	c	52.9465	—	−89.0	−72.2	—	23.	—
	a	52.9465	—	−91.56	−77.8	—	32.7	−6.9
$LiNO_3$	c	68.9459	—	−115.47	−91.1	—	21.5	—
	a	68.9459	—	−116.12	−96.7	—	38.3	−4.3
$LiNO_3 \cdot 3H_2O$	c	122.9921	—	−328.5	−263.8	—	53.4	—
Li_3PO_4	c	115.7944	—	−500.9	—	—	—	—
Li_2C_2	c	37.9044	—	−14.2	—	—	—	—
Li_2CO_3	c	73.8914	−288.652	−290.6	−270.58	3.627	21.60	23.69
Li_2SiO_3	c	89.9662	−391.26	−393.9	−372.2	3.453	19.08	23.68
Li_2SiF_6	c	155.9584	—	−704.4	—	—	—	—
$LiBO_2$	c	49.7508	−245.37	−246.7	−233.3	2.144	12.3	14.3
$LiBH_4$	c	21.7840	−43.20	−45.6	−29.9	3.049	18.13	19.73
	a	21.7840	—	−55.05	−42.8	—	29.7	—
$LiAlSi_2O_6$								
α spodumene	c	186.0909	—	−730.1	−688.71	—	30.90	38.0
β spodumene	c	186.0909	—	−723.4	−683.79	—	36.90	38.9
Li_2CrO_4	c	129.8756	—	−331.9	—	—	—	—
	a	129.8756	—	−343.7	−314.2	—	18.5	—
Li_2MoO_4	c	173.8196	—	−363.36	−336.9	—	30.	—
Li_2WO_4	c	261.7296	—	−335.	—	—	—	—
Li_2TiO_3	c	109.7802	−396.780	−399.3	−377.6	3.953	21.93	26.54
Li_2ZrO_3	c	153.1002	—	−420.7	—	—	—	—
Li_4ZrO_4	c	182.9816	—	−567.	—	—	—	—
Li_4PuF_8	c	418.8012	—	−1033.	—	—	—	—
LUTETIUM								
Lu	c	174.97	0	0	0	1.524	12.18	6.42
	g	—	102.242	102.2	96.7	1.482	44.142	4.986
Lu^+	g	—	227.36	228.80	—	—	—	—
Lu^{2+}	g	—	547.	549.9	—	—	—	—
Lu^{3+}								
std. state, m = 1	aq	—	—	−159.	150.	—	−63.	6.
Lu_2O_3	c	397.938	−446.93	−448.9	−427.6	4.192	26.28	24.32
$LuCl_3$	c	281.329	—	−226.0	—	—	—	—
std. state, m = 1	aq	—	—	−279.	−244.	—	−23.	−92.
$LuCl_3 \cdot 6H_2O$	c	389.421	−667.41	−676.6	−576.3	13.96	89.9	82.0
$LuOCl$	c	226.422	—	−227	—	—	—	—
$Lu(BrO_3)_3 \cdot 9H_2O$	c	720.830	—	−833.8	—	—	—	—
LuI_3	c	555.683	—	−131.	—	—	—	—
$Lu(IO_3)_3$	c	699.678	—	320.	—	—	—	—
LuS	g	207.034	48.	—	—	—	—	—
MAGNESIUM								
Mg	c	24.312	0	0	0	1.195	7.81	5.95
	g	—	35.014	35.30	27.04	1.481	35.502	4.968
Mg^+	g	—	211.333	213.100	—	—	—	—
Mg^{2+}	g	—	558.052	561.299	—	—	—	—
std. state, m = 1	aq	—	—	−111.58	−108.7	—	−33.0	—
Mg^{3+}	g	—	2406.08	2410.81	—	—	—	—
Mg^{4+}	g	—	4927.14	4933.35	—	—	—	—
MgO								
macrocrystal (periclase)	c	40.3114	−142.813	−143.81	−136.10	1.235	6.44	8.88
microcrystal	c	—	−141.954	142.92	−135.27	1.266	6.67	9.00
MgH_2	c	26.3279	−16.05	−18.0	−8.6	1.270	7.43	8.45
$Mg(OH)_2$	c	58.3267	−218.402	−220.97	−199.23	2.725	15.10	18.41
precipitate	amorp	—	—	−220.0	—	—	—	—
std. state, m = 1	aq	—	—	−221.52	−183.9	—	−38.1	—
MgF	g	43.3104	−52.89	−53.0	−59.2	2.143	52.79	7.78
MgF_2	c	62.3088	−267.57	−268.5	−255.8	2.370	13.68	14.72
$MgCl_2$	c	95.218	−153.180	−153.28	−141.45	3.288	21.42	17.06
$MgCl_2$								
in ∞ H_2O	aq	—	—	−191.48	—	—	—	—
$MgCl_2 \cdot H_2O$	c	113.2333	—	−231.03	−205.98	—	32.8	27.55
$MgCl_2 \cdot 2H_2O$	c	131.2487	—	−305.86	−267.24	—	43.0	38.05
$MgCl_2 \cdot 4H_2O$	c	167.2793	—	−453.87	−388.03	—	63.1	57.70
$MgCl_2 \cdot 6H_2O$	c	203.3100	—	−597.28	−505.49	—	87.5	75.30
$Mg(ClO_4)_2$	c	223.2132	—	−135.97	—	—	—	—
std. state, m = 1	aq	—	—	−173.40	−112.8	—	54.0	—
$Mg(ClO_4)_2$								
in ∞ H_2O	aq	—	—	−173.40	—	—	—	—
$Mg(ClO_4)_2 \cdot 2H_2O$	c	259.2439	—	−291.3	—	—	—	—
$Mg(ClO_4)_2 \cdot 4H_2O$	c	295.2746	—	−439.1	—	—	—	—
$Mg(ClO_4)_2 \cdot 6H_2O$	c	331.3052	—	−584.5	−445.3	—	124.5	—
$MgBr_2$	c	184.130	—	−125.3	−120.4	—	28.0	—
	g	—	—	−74.0	—	—	—	—
std. state, m = 1	aq	—	—	−169.68	−158.4	—	6.4	—
$MgBr_2$								
in ∞ H_2O	aq	—	—	−169.68	—	—	—	—
MgI_2	c	278.1208	—	−87.0	−85.6	—	31.0	—
	g	—	—	−41.	—	—	—	—
std. state, m = 1	aq	—	—	−137.96	−133.4	—	20.2	—
in ∞ H_2O	aq	—	—	−137.96	—	—	—	—
MgS	c	56.376	−82.44	−82.7	−81.7	1.992	12.03	10.89
$MgSO_3$	c	104.3742	—	−241.0	—	—	—	—
$MgSO_4$	c	120.3736	—	−307.1	−279.8	—	21.9	23.06
undissociated; std. state,	aq	—	—	−324.1	−289.74	—	−1.7	—
m = 1	aq	—	—	−328.90	−286.7	—	−28.2	—
$MgSO_4$								
in ∞ H_2O	aq	—	—	−328.90	—	—	—	—
$MgSO_4 \cdot 2H_2O$	c	156.4043	—	−453.2	—	—	—	—
$MgSO_4 \cdot 4H_2O$	c	192.4350	—	−596.7	—	—	—	—
$MgSO_4 \cdot 6H_2O$	c	228.4656	—	−737.8	−629.1	—	83.2	83.20
$MgSO_4 \cdot 7H_2O$	c	246.4810	—	−809.92	−686.4	—	89.	—

SELECTED VALUES OF CHEMICAL THERMODYNAMIC PROPERTIES (Continued)

Substance			0 K	298.15 K (25°C)				
Formula and Description	State	Formula weight	ΔHf_0° kcal/mol	ΔHf°	ΔGf° kcal/mol	$H_{298}^\circ - H_0^\circ$	S° cal/deg mol	C_p°
$Mg(NO_3)_2$	c	148.3218	—	−188.97	−140.9	—	39.2	33.92
std. state, m = 1	aq	—	—	−210.70	−161.9	—	37.0	—
$Mg(NO_3)_2$								
in ∞ H_2O	aq	—	—	−210.70	—	—	—	—
$Mg(NO_3)_2 \cdot 2H_2O$	c	184.3525	—	−336.8	—	—	—	—
$Mg(NO_3)_2 \cdot 6H_2O$	c	256.4138	—	−624.59	−497.3	—	108.	—
$Mg_2P_2O_7$	c	222.5674	—	—	—	6.467	37.02	42.53
$Mg_3(PO_4)_2$	c	262.8788	−896.98	−903.6	−845.8	7.825	45.22	51.02
MgC_2	c	48.3343	—	+20.	—	—	—	—
Mg_2C_3	c	84.6574	—	+17	—	—	—	—
$MgCO_3$								
magnesite	c	84.3214	—	−261.9	−241.9	—	15.7	18.05
$MgCO_3 \cdot 3H_2O$								
nesquehonite	c	138.3674	—	—	−412.6	—	—	—
$MgCO_3 \cdot 5H_2O$								
lansfordite	c	174.3980	—	—	−525.7	—	—	—
MgC_2O_4	c	112.3319	—	−303.3	—	—	—	—
std. state, m = 1	aq	—	—	−308.8	−269.8	—	−22.1	—
Mg_2Si	c	76.710	—	−18.6	−18.0	—	18.	17.6
$MgSiO_3$								
clinoenstatite	c	100.3962	−368.039	−370.22	−349.46	2.895	16.19	19.45
Mg_2SiO_4								
forsterite	c	140.7076	−516.42	−519.6	−491.2	4.129	22.74	28.32
$Mg_3Si_2O_5(OH)_4$								
chrysotile	c	277.1345	—	−1043.4	−965.1	—	52.9	65.41
antigorite	c	—	—	—	—	—	53.2	65.47
$Mg_3Si_4O_{10}(OH)_2$								
talc	c	379.2887	−1405.57	−1415.5	−1324.8	11.20	62.3	76.9
$Mg_2Al_4Si_5O_{18}$								
cordierite	c	584.9692	—	−2177.	−2055.	—	97.3	108.1
$MgCrO_4$	c	140.3056	—	−321.1	—	—	—	—
$MgCr_2O_4$	c	192.3016	—	−426.3	−398.9	—	25.34	30.30
$MgMoO_4$	c	184.250	—	−334.81	−309.69	—	28.4	26.57
$MgWO_4$	c	272.160	−363.88	−366.3	−339.6	4.115	24.18	26.14
MgV_2O_6								
metavanadate	c	222.1924	—	−526.19	−487.43	—	38.4	39.47
$Mg_2V_2O_7$								
pyrovanadate	c	262.5038	—	−677.80	−632.24	—	47.9	48.63
$MgTiO_3$								
metatitanate	c	120.210	−373.69	−375.9	−354.8	3.240	17.82	21.96
Mg_2TiO_4								
orthotitanate	c	160.522	−514.32	−517.5	−489.2	4.502	26.13	30.75
$MgUO_4$	c	326.339	—	−443.9	−418.2	—	31.5	30.6
MgU_3O_{10}	c	898.393	—	—	—	—	80.9	73.
MANGANESE								
Mn								
α	c	54.9380	0	0	0	1.194	7.65	6.29
β	c	—	—	—	—	1.234	8.22	6.34
γ	c	—	.343	.37	.34	1.221	7.75	6.59
	g	—	66.77	67.1	57.0	1.481	41.49	4.97
Mn^+	g	—	238.20	240.0	—	—	—	—
Mn^{++}	g	—	598.87	602.1	—	—	—	—
std. state, m = 1	aq	—	—	−52.76	−54.5	—	−17.6	12.
MnO	c	70.9374	—	−92.07	−86.74	—	14.27	10.86
MnO_2	c	86.9368	—	−124.29	−111.18	—	12.68	12.94
precipitated	amorp	—	—	−120.1	—	—	—	—
Mn_2O_3	c	157.8742	—	−229.2	−210.6	—	26.4	25.73
Mn_3O_4	c	228.8116	—	−331.7	−306.7	—	37.2	33.38
MnH	g	55.9460	—	—	—	2.077	51.03	7.035
$Mn(OH)_2$								
precipitated	amorp	88.9527	—	−166.2	−147.0	—	23.7	—
MnF	g	73.9364	−5.	−5.2	—	—	—	—
MnF_2	c	92.9348	—	—	—	3.11	22.05	15.96
$MnCl_2$	c	125.8440	−115.245	−115.03	−105.29	3.602	28.26	17.43
std. state, m = 1	aq	—	—	−132.66	−117.3	—	9.3	−53.
$MnCl_2 \cdot H_2O$	c	143.8593	—	−188.8	−166.4	—	41.6	
$MnCl_2 \cdot 2H_2O$	c	161.8747	—	−261.0	−225.2	—	52.3	—
$MnCl_2 \cdot 4H_2O$	c	197.9054	—	−403.3	−340.3	—	72.5	
MnBr	g	134.8470	20.8	19.1	—	—	—	—
$MnBr_2$	c	214.7560	—	−92.0	—	—	—	—
std. state, m = 1	aq	—	—	−110.9	—	—	—	—
$MnBr_2 \cdot H_2O$	c	232.7713	—	−168.5	—	—	—	—
$MnBr_2 \cdot 4H_2O$	c	286.8174	—	−380.1	—	—	—	—
MnI	g	181.8424	25.7	25.5	—	—	—	—
MnI_2								
std. state, m = 1	aq	308.7468	—	−79.1	—	—	—	—
$MnI_2 \cdot 2H_2O$	c	344.7748	—	−201.4	—	—	—	—
$MnI_2 \cdot 4H_2O$	c	380.8082	—	−343.9	—	—	—	—
$Mn(IO_3)_2$	c	404.7432	—	−160.	−124.4	—	63.	—
MnS								
green	c	87.0020	—	−51.2	−52.2	—	18.7	11.94
precipitated, pink	amorp	—	—	−51.1	—	—	—	—
$MnSO_4$	c	150.9996	—	−254.60	−228.83	—	26.8	24.02
std. state, m = 1	aq	—	—	−270.1	−232.5	—	−12.8	−58.
$MnSO_4$								
in ∞ H_2O	aq	—	—	−270.1	—	—	—	—
$MnSO_4 \cdot H_2O$								
α	c	169.0149	—	−329.0	—	—	—	—
β	c	—	—	−322.2	—	—	—	—
$MnSO_4 \cdot 4H_2O$	c	223.0610	—	−539.7	—	—	—	—
$MnSO_4 \cdot 5H_2O$	c	241.0763	—	−610.2	—	—	—	78.
$MnSO_4 \cdot 7H_2O$	c	277.1070	—	−750.3	—	—	—	—
$Mn(N_3)_2$								
manganese azide	c	138.9782	—	92.2	—	—	—	—
Mn_5N_2	c	302.7034	—	−48.8	—	—	—	—
$Mn(NO_3)_2$	c	178.9478	—	−137.73	—	—	—	—
std. state, m = 1	aq	—	—	−151.9	−107.8	—	52.	−29.
MnP	c	85.9118	—	−27.	—	—	—	—
MnP_3	c	147.8594	—	−51.	—	—	—	—
$Mn_3(PO_4)_2$	c	354.7568	—	−744.9	—	—	—	—
$MnHPO_4$	c	150.9174	—	—	−332.5	—	—	—

Substance			0 K	298.15 K (25°C)				
Formula and Description	State	Formula weight	ΔHf°_0 kcal/mol	ΔHf°	ΔGf° kcal/mol	$H^\circ_{298} - H^\circ_0$	S° cal/deg mol	C°_p
Mn_3C	c	176.8252	—	1.1	1.3	—	23.6	22.33
Mn_7C_3	c	420.5995	—	−10.	—	—	—	—
$MnCO_3$								
natural	c	114.9474	—	−213.7	−195.2	—	20.5	19.48
precipitated	c	—	—	−211.1	—	—	—	—
MnC_2O_4	c	142.9579	—	−245.9	—	—	—	—
undissoc.; std. state, m = 1	aq	—	—	−248.5	−221.0	—	16.1	—
$MnC_2O_4 \cdot 2H_2O$	c	178.9886	—	−389.2	−338.2	—	48.	—
$MnC_2O_4 \cdot 3H_2O$	c	197.0039	—	−459.1	—	—	—	—
$MnSiO_3$	c	131.0222	—	−315.7	−296.5	—	21.3	20.66
glassy	amorp	—	—	−307.2	—	—	—	—
Mn_2SiO_4	c	201.9596	—	−413.6	−390.1	—	39.0	31.04
MERCURY								
Hg	c	200.59	0	—	—	—	—	—
	liq	—	—	0	0	2.233	18.17	6.688
	g	—	15.407	14.655	7.613	1.481	41.79	4.968
std. state, m = 1	aq	—	—	9.0	9.4	—	17.	—
Hg^{+}	g	—	256.10	256.82	—	—	—	—
Hg^{2+}	g	—	688.63	690.83	—	—	—	—
std. state, m = 1	aq	—	—	40.9	39.30	—	−7.7	—
Hg^{3+}	g	—	1478.	1480.	—	—	—	—
Hg_2^{2+}								
std. state, m = 1	aq	—	—	41.2	36.70	—	20.2	—
HgO								
red, orthorhombic	c	216.589	—	−21.71	−13.995	—	16.80	10.53
yellow	c	—	—	−21.62	−13.964	—	17.0	—
hexagonal	c	—	—	−21.4	−13.92	—	17.6	—
HgH	g	201.598	58.36	57.20	51.63	2.078	52.46	7.16
$Hg(OH)_2$								
undissoc.; std. state, m = 1	aq	234.605	—	−84.9	−65.7	—	34.	—
Hg_2F_2	c	439.177	—	—	−104.1	—	—	—
$HgCl^{+}$								
std. state, m = 1	aq	—	—	−4.5	−1.3	—	18.	—
$HgCl_2$	c	271.496	—	−53.6	−42.7	—	34.9	—
undissoc.; std. state, m = 1	aq	—	—	−51.7	−41.4	—	37.	—
Hg_2Cl_2	c	472.086	—	−63.39	−50.377	—	46.0	—
$HgBr_2$	c	360.408	—	−40.8	−36.6	—	41.	
undissoc.; std. state, m = 1	aq	—	—	−38.4	−34.2	—	41.	—
Hg_2Br_2	c	560.998	—	−49.45	−43.278	—	52.	—
HgI	g	327.494	32.9	31.64	21.14	2.546	67.26	8.99
HgI_2								
red	c	454.399	—	−25.2	−24.3	—	43.	—
yellow	c	—	—	−24.6	—	—	—	—
Hg_2I_2	c	654.989	—	−29.00	−26.53	—	55.8	—
HgS								
red	c	232.654	—	−13.9	−12.1	—	19.7	11.57
black	c	—	—	−12.8	−11.4	—	21.1	—
Hg_2SO_4	c	497.242	—	−177.61	−149.589	—	47.96	31.54
$Hg_2(N_3)_2$	c	485.220	—	142.0	178.4	—	49.	—
HgC_2O_4	c	288.610	—	−162.1	—	—	—	—
Hg_2CO_3	c	461.189	—	−132.3	−111.9	—	43.	—
$Hg_2C_2O_4$	c	489.200	—	—	−141.8	—	—	—
$HgCH_3$	g	215.625	—	40.	—	—	—	—
$Hg(CH_3)_2$	liq	230.660	—	14.3	33.5	—	50.	—
$Hg(C_2H_5)_2$	liq	258.715	—	7.2	—	—	—	—
	g	—	—	18.0	—	—	—	—
$Hg(CN)_2$	c	252.626	—	63.0	—	—	—	—
undissoc.; std. state, m = 1	aq	—	—	66.5	74.6	—	39.5	—
$Hg(CN)_3^{-}$								
std. state, m = 1	aq	278.644	—	94.9	110.7	—	53.8	—
$Hg(CN)_4^{2-}$								
std. state, m = 1	aq	304.662	—	125.8	147.8	—	73.	—
$Hg(ONC)_2$								
mercuric fulminate	c	284.624	—	64.	—	—	—	—
MOLYBDENUM								
Mo	c	95.94	0	0	0	1.098	6.85	5.75
	g	—	156.92	157.3	146.4	1.4812	43.461	4.968
Mo^{+}	g	—	320.6	322.5	—	—	—	—
Mo^{2+}	g	—	693.2	696.5	—	—	—	—
Mo^{3+}	g	—	1319.0	1323.8	—	—	—	—
Mo^{4+}	g	—	2388.9	2395.2	—	—	—	—
Mo^{5+}	g	—	3799.4	3807.2	—	—	—	—
MoO	g	111.939	—	101.	—	—	—	—
MoO_2	c	127.939	—	−140.76	−127.40	—	11.06	13.38
	g	—	—	3.	—	—	—	—
MoO_3	c	143.939	—	−178.08	−159.66	—	18.58	17.92
	g	—	—	−78.	—	—	—	—
	aq	—	—	−172.5	—	—	—	—
MoO_4	aq	159.938	—	−158.0	—	—	—	—
MoO_4^{2-}								
std. state, m = 1	aq	—	—	−238.5	−199.9	—	6.5	—
MoO_5	aq	175.937	—	−139.6	—	—	—	—
H_2MoO_4								
white	c	161.954	—	−250.0	—	—	—	—
$H_2MoO_4 \cdot H_2O$								
yellow	c	179.969	—	−325.	—	—	—	—
MoF_6	liq	209.930	−381.733	−378.95	−352.08	10.205	62.06	40.58
	g	—	−370.608	−372.29	−351.88	5.74	83.75	28.82
$MoCl_2$	c	166.846	—	−67.4	—	—	—	—
$MoCl_3$	c	202.299	—	−92.5	—	—	—	—
$MoCl_4$	c	237.752	—	−114.8	—	—	—	—
$MoCl_5$	c	273.205	—	−126.0	—	—	—	—
MoO_2Cl_2	c	198.845	—	−171.4	—	—	—	—
$MoO_2Cl_2 \cdot H_2O$	c	216.860	—	−245.4	—	—	—	—
$MoOCl_4$	c	253.751	—	−153.0	—	—	—	—
$MoBr_2$	c	255.758	—	−62.4	—	—	—	18.3
$MoBr_4$	c	415.576	—	−76.8	—	—	—	—
MoO_2Br_2	c	287.757	—	−150.4	—	—	—	—
MoS_2	c	160.068	−55.52	−56.2	−54.0	2.528	14.96	15.19
Mo_2S_3	c	288.072	—	−87.	—	—	—	—

Substance			0 K	298.15 K (25°C)				
Formula and Description	State	Formula weight	ΔHf°_0 kcal/mol	ΔHf°	ΔGf° kcal/mol	$H^\circ_{298} - H^\circ_0$	S° cal/deg mol	C°_p
Mo_2N	c	205.887	—	−19.50	—	—	—	—
MoC	c	107.951	—	−2.4	—	—	—	—
Mo_2C	c	203.891	—	−10.9	—	—	—	—
$Mo(CO)_6$	c	264.003	—	−234.9	−209.8	—	77.9	57.90
$FeMoO_4$	c	215.785	—	−257.	−233.	—	30.9	28.31
NEODYMIUM								
Nd	c	144.24	0	0	0	1.73	17.1	6.56
	g	—	78.53	78.3	69.9	1.498	45.243	5.280
Nd^+	g	—	205.1	206.3	—	—	—	—
Nd^{2+}	g	—	452.	455.	—	—	—	—
Nd^{3+}								
std. state, m = 1	aq	—	—	−166.4	−160.5	—	−49.4	−5.
NdO	g	160.239	−30.2	—	—	—	—	—
Nd_2O_3								
hexagonal	c	336.478	−430.50	−432.1	−411.3	5.00	37.9	26.60
NdH_2	c	146.256	—	−46.	—	—	—	—
NdF	g	163.238	—	−38.	—	—	—	—
NdF_2	g	182.237	—	−165.	—	—	—	—
NdF_3	c	201.235	—	−396.	—	—	—	—
$NdCl_2$	c	215.146	—	−163.	—	—	—	—
$NdCl_3$	c	250.599	—	−248.8	—	—	—	27.
$NdCl_3$								
$\infty\ H_2O$	aq	—	—	−286.3	—	—	—	—
$NdCl_3 \cdot 6H_2O$	c	358.691	−678.72	−687.0	−588.1	15.14	99.7	86.25
NdI_3	c	524.953	—	−152.8	—	—	—	—
$Nd(IO_3)_3$	c	668.948	—	−332.	—	—	—	—
NdS	g	176.304	33.	—	—	—	—	—
Nd_2S_3	c	384.672	—	−284.	−280.2	6.160	44.28	29.28
$Nd_2(SO_4)_3 \cdot 8H_2O$	c	720.788	—	−1513.1	—	—	160.9	144.9
Nd_2Se_3	c	525.36	—	—	—	7.13	53.6	31.1
$Nd_2(SeO_3)_3 \cdot 8H_2O$								
amorp.	c	813.477	—	−1230.3	—	—	—	—
$Nd_2(SeO_4)_3 \cdot 5H_2O$	c	807.430	—	−1100.4	—	—	—	—
$Nd(NO_3)_3$	c	330.255	—	−294.2	—	—	—	—
$Nd(NO_3)_3 \cdot 3H_2O$	c	384.301	—	−515.	—	—	—	—
$Nd(NO_3)_3 \cdot 4H_2O$	c	402.316	—	−588.6	—	—	—	—
$Nd(NO_3)_3 \cdot 6H_2O$	c	438.347	—	−728.39	—	—	—	—
NdC_2	g	168.262	130.5	130.75	117.9	2.48	6.3	10.6
$Nd_2(CO_3)_3$	c	468.508	—	—	−744.5	—	—	—
NEON								
Ne	g	20.183	0	0	0	1.481	34.9471	4.9679
std. state, m = 1	aq	—	—	−1.1	4.6	—	15.8	—
Ne^+	g	—	497.29	498.77	—	—	—	—
Ne^{2+}	g	1444.7	1447.6	—	—	—	—	—
Ne^{3+}	g	—	2915.	2919.	—	—	—	—
Ne^{4+}	g	—	5156	5162	—	—	—	—
Ne^{5+}	g	—	8072.	8079.	—	—	—	—
NICKEL								
Ni	c	58.71	0	0	0	1.144	7.14	6.23
	g	—	102.213	102.7	91.9	1.631	43.519	5.583
Ni^+	g	—	278.275	280.243	—	—	—	—
Ni^{2+}	g	—	696.87	700.32	—	—	—	—
std. state, m = 1	aq	—	—	−12.9	−10.9	—	−30.8	—
Ni^{3+}	g	—	1508.0	1512.9	—	—	—	—
NiO	c	74.709	−56.7	−57.3	−50.6	1.6	9.08	10.59
Ni_2O_3	c	165.418	—	−117.0	—	—	—	—
$Ni(OH)_2$	c	92.725	—	−126.6	−106.9	—	21.	—
std. state, m = 1	aq	—	—	−122.8	−86.1	—	−35.9	—
$Ni(OH)_3$								
precipitated	c	109.732	—	−160.	—	—	—	—
NiF_2	c	96.707	−155.18	−155.7	−144.4	2.729	17.59	15.31
$NiCl_2$	c	129.616	−73.077	−72.976	−61.918	3.438	23.34	17.13
$NiCl_2$								
in ∞H_2O	aq	—	—	−92.8	—	—	—	—
$NiCl_2 \cdot 2H_2O$	c	165.647	—	−220.4	−181.7	—	42.	—
$NiCl_2 \cdot 4H_2O$	c	201.677	—	−362.5	−295.2	—	58.	—
$NiCl_2 \cdot 6H_2O$	c	237.708	—	−502.67	−409.54	—	82.3	—
$Ni(ClO_4)_2$								
std. state, m = 1	aq	257.611	—	−74.7	−15.0	—	56.2	—
$Ni(ClO_4)_2 \cdot 6H_2O$	c	365.703	—	−486.6	—	—	—	—
$NiBr_2$	c	218.528	—	−50.7	—	—	—	—
std. state, m = 1	aq	—	—	−71.0	−60.6	—	8.6	—
$NiBr_2 \cdot 3H_2O$	c	272.574	—	−274.0	—	—	—	—
NiI_2	c	312.519	—	−18.7	—	—	—	—
std. state, m = 1	aq	—	—	−39.3	−35.6	—	22.4	—
$Ni(IO_3)_2$	c	408.515	—	−116.9	−78.0	—	51.	—
NiS	c	90.774	—	−19.6	−19.0	—	12.66	11.26
precipitated	c	—	—	−18.5	—	—	—	—
Ni_3S_2	c	240.258	—	−48.5	−47.1	—	32.0	28.12
$NiSO_4$	c	154.772	—	−208.63	−181.6	—	22.	33.
std. state, m = 1	aq	—	—	−230.2	−188.9	—	−26.0	—
$NiSO_4$								
$\infty\ H_2O$	aq	—	—	−230.2	—	—	—	—
$NiSO_4 \cdot 4H_2O$	c	226.833	—	−502.9	—	—	—	—
$NiSO_4 \cdot 6H_2O$								
α, tetragonal, green	c	262.864	−628.887	−641.21	−531.78	12.391	79.94	78.36
β, monoclinic, blue	c	—	—	−638.7	—	—	—	—
$NiSO_4 \cdot 7H_2O$	c	280.879	−697.670	−711.36	−588.49	14.085	90.57	87.14
Ni_3N	c	190.137	—	0.2	—	—	—	—
$Ni(NO_3)_2$	c	182.720	—	−99.2		—	—	—
std. state, m = 1	aq	—	—	−112.0	−64.2	—	39.2	—
$Ni(NO_3)_2 \cdot 3H_2O$	c	236.766	—	−317.0	—	—	—	—
$Ni(NO_3)_2 \cdot 6H_2O$	c	290.812	—	−528.6	—	—	—	111.
Ni_3C	c	188.141	—	16.1	—	—	—	—
$NiCO_3$	c	118.719	—	—	−146.4	—	—	—
$Ni(CO)_4$	liq	170.752	—	−151.3	−140.6	—	74.9	48.9
	g	—	−144.877	−144.10	−140.36	7.074	98.1	34.70
$Ni(HCO_2)_2$	c	148.746	—	−208.4	—	—	—	—
$Ni(CN)_2$								
precipitated	c	110.746	—	30.5	—	—	—	—
std. state, m = 1	aq	—	—	59.1	71.5	—	14.2	—

Substance			0 K	298.15 K (25°C)				
Formula and Description	State	Formula weight	ΔHf°_0 kcal/mol	ΔHf°	ΔGf° kcal/mol	$H^\circ_{298} - H^\circ_0$	S° cal/deg mol	C°_p
$Ni(CNS)_2$	c	174.874	—	22.8	—	—	—	—
NIOBIUM								
Nb	c	92.906	0	0	0	1.255	8.70	5.88
	g	—	172.758	173.5	162.8	1.997	44.490	7.208
Nb^+	g	—	330.32	332.54	—	—	—	—
Nb^{2+}	g	—	660.55	664.25	—	—	—	—
Nb^{3+}	g	—	1238.0	1243.1	—	—	—	—
Nb^{4+}	g	—	2121.3	2127.9	—	—	—	—
NbO	c	108.9054	—	−97.0	−90.5	—	11.5	9.86
	g	—	51.2	51.	44.	2.099	57.09	7.36
NbO_2	c	124.9048	−189.19	−190.3	−177.0	2.222	13.03	13.74
	g	—	—	−51.3	−52.3	—	61.0	—
Nb_2O_5								
(high temp. form)	c	265.809	−451.63	−454.0	−422.1	5.325	32.80	31.57
NbF_5	c	187.898	−432.68	−433.5	−406.1	5.707	38.3	32.2
$NbCl_5$	c	270.171	—	−190.6	−163.3	—	50.3	35.4
$NbOCl_2$	c	179.8114	—	−185.1	—	—	—	—
$NbOCl_3$	c	215.2644	—	−210.2	−187.	—	34.	—
$NbBr_5$	c	492.451	—	−132.9	—	—	—	—
$NbOBr_3$	c	348.6174	—	−179.3	—	—	—	—
NbI_5	c	727.428	—	−64.2	—	—	—	—
NbN	c	106.9127	−55.35	−56.2	−49.2	1.439	8.25	9.32
Nb_2N	c	199.8187	—	−59.9	—	—	—	—
NbC	c	104.9172	−33.12	−33.2	−32.7	1.422	8.46	8.81
Nb_2C	c	197.8232	—	−45.4	−44.4	—	15.3	14.48
NITROGEN								
N	g	14.0067	112.534	112.979	108.886	1.481	36.613	4.968
N^+	g	—	447.663	449.589	—	—	—	—
N^{2+}	g	—	1130.55	1133.96	—	—	—	—
N^{3+}	g	—	2224.52	2229.41	—	—	—	—
N^{4+}	g	—	4011.04	4017.41	—	—	—	—
N^{5+}	g	—	6268.41	6276.26	—	—	—	—
N_2	g	28.0134	0	0	0	2.072	45.77	6.961
N_3	g	—	45.	43.2	—	—	—	—
std. state, m = 1	aq	—	—	65.76	83.2	—	25.8	—
NO	g	30.0061	21.45	21.57	20.69	—	50.347	7.133
NO_2	g	46.0055	8.60	7.93	12.26	2.438	57.35	8.89
NO_2^-								
std. state, m = 1	aq	—	—	−25.0	−7.7	—	29.4	−23.3
NO_3^-								
nitrate; std. state, m = 1	aq	62.0049	—	−49.56	−26.61	—	35.0	−20.7
N_2O	g	44.0128	20.435	19.61	24.90	2.284	52.52	9.19
N_2O_3	liq	76.0116	—	12.02	—	—	—	—
	g	—	21.628	20.01	33.32	3.566	74.61	15.68
N_2O_4	liq	92.0110	—	−4.66	23.29	—	50.0	34.1
	g	—	4.49	2.19	23.38	3.918	72.70	18.47
N_2O_5	c	108.0104	—	−10.3	27.2	—	42.6	34.2
	g	—	5.7	2.7	27.5	4.237	85.0	20.2
NH_3	g	17.0306	−9.34	−11.02	−3.94	2.388	45.97	8.38
undissoc.; std. state, m = 1	aq	—	—	−19.19	−6.35	—	26.6	—
NH_4^+								
std. state, m = 1	aq	18.0386	—	−31.67	−18.97	—	27.1	19.1
N_2H_4	liq	32.0453	—	12.10	35.67	—	28.97	23.63
	g	—	26.18	22.80	38.07	2.743	56.97	11.85
undissoc.; std. state, m = 1	aq	—	—	8.20	30.6	—	33.	—
HN_3	liq	43.0281	—	63.1	78.2	—	33.6	—
	g	—	71.82	70.3	78.4	2.599	57.09	10.44
undissoc.; std. state, m = 1	aq	—	—	62.16	76.9	—	34.9	—
HNO_2								
cis	g	47.0135	−17.12	−18.64	−10.27	2.608	59.43	10.70
trans	g	—	−17.68	−19.15	−10.82	2.652	59.54	11.01
cis-trans mixture, equil.	g	—	—	−19.0	−11.0	—	60.7	10.9
undissoc.; std. state, m = 1	aq	—	—	−28.5	−12.1	—	32.4	—
HNO_3	liq	63.0129	—	−41.61	−19.31	—	37.19	26.26
	g	—	−29.94	−32.28	−17.87	2.815	63.64	12.75
std. state, m = 1	aq	—	—	−49.56	−26.61	—	35.0	−20.7
HNO_3								
∞ H_2O	aq	—	—	−49.56	—	—	—	—
NH_4OH	liq	35.0460	—	−86.33	−60.74	—	39.57	37.02
undissoc.; std. state, m = 1	aq	—	—	−87.505	−63.04	—	43.3	—
ionized; std. state, m = 1	aq	—	—	−86.64	−56.56	—	24.5	−16.4
NH_4OH								
∞ H_2O	aq	—	—	−86.64	—	—	—	—
NH_4NO_2	c	64.0441	—	−61.3	—	—	—	—
std. state, m = 1	aq	—	—	−56.7	−26.7	—	56.5	−4.2
NH_4NO_3	c	80.0435	—	−87.37	−43.98	—	36.11	33.3
std. state, m = 1	aq	—	—	−81.23	−45.58	—	62.1	−1.6
NH_4NO_3								
∞ H_2O	aq	—	—	−81.23	—	—	—	—
$(NH_4)_2O$	liq	52.0766	—	−102.94	−63.84	—	63.94	59.08
NF_3	g	71.0019	−28.43	−29.8	−19.9	2.827	62.29	12.7
NH_4F	c	37.0370	−107.41	−110.89	−83.36	2.655	17.20	15.60
std. state, m = 1	aq	—	—	−111.17	−85.61	—	23.8	−6.4
NCl_3	liq	120.3657	—	55.	—	—	—	—
NOCl	g	65.4591	12.81	12.36	15.79	2.716	62.52	10.68
NH_4Cl	c	53.4916	—	−75.15	−48.51	—	22.6	20.1
std. state, m = 1	aq	—	—	−71.62	−50.34	—	40.6	−13.5
NH_4Cl								
∞ H_2O	aq	—	—	−71.62	—	—	—	—
$NH_2OH \cdot HCl$	c	69.4910	—	−75.9	—	—	—	22.2
NH_4ClO								
std. state, m = 1	aq	69.4910	—	−57.3	−27.8	—	37.	—
NH_4ClO_2								
std. state, m = 1	aq	85.4904	—	−47.6	−14.9	—	51.3	—
std. state, m = 1	aq	101.4898	—	−56.52	−20.89	—	65.9	—
NH_4ClO_4	c	117.4892	—	−70.58	−21.25	—	44.5	—
std. state, m = 1	aq	—	—	−62.58	−21.03	—	70.6	—
NH_4Br	c	94.9477	—	−64.73	−41.9	—	27.	23.
std. state, m = 1	aq	—	—	−60.72	−43.82	—	46.8	−14.8
NH_4Br								
∞ H_2O	aq	—	—	−60.72	—	—	—	—

SELECTED VALUES OF CHEMICAL THERMODYNAMIC PROPERTIES (Continued)

Substance: Formula and Description	State	Formula weight	0 K: ΔHf°_0 kcal/mol	298.15 K (25°C): ΔHf° kcal/mol	ΔGf° kcal/mol	$H^\circ_{298} - H^\circ_0$ kcal/mol	S° cal/deg mol	C°_p cal/deg mol
NH_4BrO								
std. state, m = 1	aq	113.9470	—	−54.2	−27.0	—	37.	—
NH_4BrO_3								
std. state, m = 1	aq	145.9458	—	−47.70	−14.54	—	65.75	—
NH_4Br_2Cl								
std. state, m = −72.4	aq	213.3096	—	−72.4	−49.7	—	72.2	—
NH_4I	c	144.9430	—	−48.14	−26.9	—	28.	—
std. state, m = 1	aq	—	—	−44.86	−31.30	—	53.7	−14.9
NH_4I								
∞ H_2O	aq	—	—	−44.86	—	—	—	—
NH_4IO								
std. state, m = 1	aq	160.9424	—	−57.4	−28.2	—	25.8	—
NH_4IO_3	c	192.9412	—	−92.2	—	—	—	—
std. state, m = 1	aq	—	—	−84.6	−49.6	—	55.4	—
NH_4IO_4	aq	208.9406	—	−67.9	—	—	—	—
NH_4HS	c	51.1106	—	−37.5	−12.1	—	23.3	—
std. state, m = 1 (NH_4^+ + HS^-)	aq	—	—	−35.9	−16.09	—	42.1	—
$(NH_4)_2S$								
std. state, m = 1	aq	68.1412	—	−55.4	−17.4	—	50.7	—
$(NH_4)_2S_2$								
std. state, m = 1	aq	100.2052	—	−56.1	−18.9	—	61.0	—
$(NH_4)_2S_3$								
std. state, m = 1	aq	132.2692	—	−57.1	−20.3	—	70.0	—
H_2NSO_3H								
sulfamic acid	c	97.0928	—	−161.3	—	—	—	—
	aq	—	—	−156.3	—	—	—	—
NH_4HSO_3	c	99.1088	—	−183.7	—	—	—	—
NH_4HSO_4	c	115.1082	—	−245.45	—	—	—	—
$SO_2(NH_2)_2$								
sulfamide	c	96.1081	—	−129.3	—	—	—	—
$(NH_4)_2SO_3$	c	116.1394	—	−211.6	—	—	—	—
std. state, m = 1	aq	—	—	−215.2	−154.2	—	47.2	—
$(NH_4)_2SO_4$	c	132.1388	−282.23	−215.56	—	52.6	44.81	
std. state, m = 1, ($2NH_4^+$ + SO_4^{2-})	aq	—	—	−280.66	−215.91	—	59.0	−31.8
OSMIUM								
Os	c	190.2	0	0	0	—	7.8	5.9
	g	—	—	189.	178.	1.481	46.000	4.968
Os^+	g	—	—	391.91	—	—	—	—
Os^{2+}	g	—	—	785.	—	—	—	—
OsO_3	g	238.20	—	−67.8	—	—	—	—
OsO_4								
yellow	c	254.20	—	−94.2	−72.9	—	34.4	—
white	c	—	—	−92.2	−72.6	—	40.1	—
$Os(OH)_4$	amorp	258.23	—	—	−161.0	—	—	—
$OsCl_3$	c	296.56	—	−45.5	—	—	—	—
$OsCl_4$	c	332.01	—	−60.9	—	—	—	—
OsS_2	c	254.33	—	−34.9	—	—	—	—
O	g	15.9994	58.983	59.553	55.389	1.607	38.467	5.237
O^+	g	—	373.019	375.070	—	—	—	—
O^{2+}	g	—	1183.73	1187.26	—	—	—	—
O^{3+}	g	—	2450.87	2455.88	—	—	—	—
O^{4+}	g	—	4236.1	4242.6	—	—	—	—
O^{5+}	g	—	6862.8	6870.8	—	—	—	—
O_2	g	31.9988	0	0	0	2.0746	49.003	7.016
std. state, m = 1	aq	—	—	−2.8	3.9	—	26.5	—
O_3	g	47.9982	34.74	34.1	39.0	2.4736	57.08	9.37
	aq	—	—	30.1	—	—	—	—
PALLADIUM								
Pd	c	106.4	0	0	0	1.299	8.98	6.21
	g	—	90.2	90.4	81.2	1.481	39.90	4.968
Pd^+	g	—	282.436	284.117	—	—	—	—
Pd^{2+}	g	—	730.5	733.6	—	—	—	—
std. state, m = 1	aq	—	—	35.6	42.2	—	−44.	—
Pd^{3+}	g	—	1490.	1494.	—	—	—	—
PdO	c	122.40	—	−20.4	—	—	—	7.5
Pd_2H	c	213.81	—	−4.7	—	—	—	—
$Pd(OH)_2$								
precipitated	c	140.41	—	−94.4	—	—	—	—
$Pd(OH)_4$								
precipitated	c	174.43	—	−171.1	—	—	—	—
$PdCl_2$	c	177.31	—	−47.5	—	—	—	—
$PdBr_2$	c	266.22	—	−24.9	—	—	—	—
PdI_2	c	360.21	—	−15.1	−17.1	—	43.	—
PdS	c	138.46	—	−18.	−16.	—	11.	—
PdS_2	c	170.53	—	−19.4	−17.8	—	19.	—
Pd_4S	c	457.66	—	−16.	−16.	—	43.	—
$Pd(CN)_2$	c	158.44	—	49.1	—	—	—	—
$Pd(CNS)_2$	c	222.56	—	—	56.0	—	—	—
PHOSPHORUS								
P								
α, white	c	30.9738	0	0	0	1.281*	9.82	5.698
red, triclinic	c	—	−3.78	−4.2	−2.9	0.862	5.45	5.07
black	c	—	—	−9.4	—	—	—	—
red	amorp	—	—	−1.8	—	—	—	—
	g	—	75.	75.20	66.51	1.481	38.978	4.968
in CS_2	—	—	—	0.5	—	—	—	—
P^+	g	—	328.20	329.88	—	—	—	—
P^{2+}	g	—	781.51	784.67	—	—	—	—
P^{3+}	g	—	1477.11	1481.75	—	—	—	—
P^{4+}	g	—	2661.67	2667.79	—	—	—	—
P^{5+}	g	—	4161.2	4168.8	—	—	—	—
P_2	g	61.9476	34.94	34.5	24.8	2.126	52.108	7.66
P_4	g	123.8952	15.83	14.08	5.85	3.378	66.89	16.05
PO	g	46.9732	−6.7	−6.8	−12.4	2.245	53.22	7.59
P_4O_6	c	219.8916	—	−392.0	—	—	—	—
P_4O_{10}								
hexagonal	c	283.8892	−705.82	−713.2	−644.8	8.117	54.70	50.60
	amorp	—	—	−727.	—	—	—	—
PH_3	g	33.9977	3.20	1.3	3.2	2.420	50.22	8.87

Substance			0 K	298.15 K (25°C)				
Formula and Description	State	Formula weight	ΔHf°_0 kcal/mol	ΔHf°	ΔGf° kcal/mol	$H^\circ_{298} - H^\circ_0$	S° cal/deg mol	C°_p
std. state, m = 1	aq	—	—	−2.27	6.05	—	28.7	—
HPO_3	c	79.9800	—	−226.7	—	—	—	—
H_3PO_3	c	81.9959	—	−230.5	—	—	—	—
H_3PO_4	c	97.9953	−301.29	−305.7	−267.5	4.059	26.41	25.35
ionized, std. state, m = 1	aq	—	—	−305.3	−243.5	—	−53.	—
PF	g	49.9722	—	—	—	2.117	53.74	7.56
PF_3	g	87.9690	−218.25	−219.6	−214.5	3.092	65.28	14.03
PF_5	g	125.9658	—	−381.4	—	—	—	—
PCl_3	liq	137.3328	—	−76.4	−65.1	—	51.9	—
	g	—	−67.85	−68.6	−64.0	3.817	74.49	17.17
PCl_5	c	208.2388	—	−106.0	—	—	—	—
$POCl_3$	c	153.3322	−145.81	—	—	—	—	—
PBr_3	liq	270.7008	—	−44.1	−42.0	—	57.4	—
	g	—	−27.47	−33.3	−38.9	4.240	83.17	18.16
PBr_5	c	430.5188	—	−64.5	—	—	—	—
$POBr_3$	c	286.7002	—	−109.6	—	—	—	—
PI_3	c	411.6870	—	−10.9	—	—	—	—
	g	—	—	—	—	4.542	89.45	18.73
PH_4I	c	161.9101	—	−16.7	0.2	—	29.4	26.2
P_2S_3	c	158.1396	—	−19.2	—	—	—	—
PN	g	44.9805	26.5	26.26	20.97	2.080	50.45	7.10
P_3N_5	c	162.9549	—	−71.4	—	—	—	36.
$NH_4H_2PO_4$	c	115.0259	—	−345.38	−289.33	—	36.32	34.00
$(NH_4)_2HPO_3$	aq	116.0571	—	−294.9	—	—	—	—
$(NH_4)_2HPO_4$	c	132.0565	—	−374.50	—	—	—	45.
$(NH_4)_3PO_4$	c	149.0871	—	−399.6	—	—	—	—
std. state, m = 1	aq	—	—	−400.3	−300.4	—	28.	—
PLATINUM								
Pt	c	195.09	0	0	0	1.372	9.95	6.18
	g	—	134.90	135.1	124.4	1.572	45.960	6.102
Pt^+	g	—	341.6	343.3	—	—	—	—
std. state, m = 1	aq	—	—	—	60.9	—	—	—
PtO_2	g	227.089	—	41.0	40.1	62.	—	—
Pt_3O_4	c	649.268	—	−39.	—	—	—	—
$Pt(OH)_2$	c	229.105	—	−84.1	—	—	—	—
PtCl	c	230.543	—	−13.5	—	—	—	—
$PtCl_2$	c	265.996	—	−29.5	—	—	—	—
std. state, m = 1	aq	—	—	—	—	—	—	—
$PtCl_3$	c	301.449	—	−43.5	—	—	—	—
$PtCl_4$	c	336.902	—	−55.4	—	—	—	—
$HPtCl_5 \cdot 2H_2O$	c	409.394	—	−242.0	—	—	—	—
$H_2PtCl_6 \cdot 6H_2O$	c	517.916	—	566.7	—	—	—	—
PtBr	c	274.999	—	−9.2	—	—	—	—
$PtBr_2$	c	354.908	—	−19.6	—	—	—	—
$PtBr_3$	c	434.817	—	−28.9	—	—	—	—
$PtBr_4$	c	514.726	—	−37.4	—	—	—	—
PtI_4	c	702.708	—	−17.4	—	—	—	—
PtS	c	227.154	−19.020	−19.5	−18.2	1.946	13.16	10.37
PtS_2	c	259.218	−25.333	−26.0	−23.8	2.813	17.85	15.75
POLONIUM								
Po	c	210.	0	0	0	—	—	—
Po^{2+}								
std. state, m = 1	aq	—	—	—	17.	—	—	—
Po^{4+}								
std. state, m = 1	aq	—	—	—	70.	—	—	—
$Po(OH)_4$	c	278.0	—	—	−130.	—	—	—
PoS	c	242.1	—	—	−1.	—	—	—
POTASSIUM								
K	c	39.1020	0	0	0	1.695	15.34	7.07
	g	39.1020	21.544	21.33	14.49	1.481	38.295	4.968
K^+	g	39.1020	—	122.92	—	—	—	—
K^{2+}	g	39.1020	—	853.70	—	—	—	—
K^{3+}	g	39.1020	—	1909.5	—	—	—	—
K^{4+}	g	39.1020	—	3315.6	—	—	—	—
K^+	a	39.1020	—	−60.32	−67.70	—	24.5	5.2
K								
in 88.81 Hg		39.1020	—	—	−24.41	—	—	—
KO_2	c	71.1008	—	−68.10	−57.23	—	27.9	18.53
KO_3	c	87.1002	—	−62.2	—	—	—	—
K_2O	c	94.2034	—	−86.4	—	—	—	—
	g	94.2034	—	−15.	—	—	—	—
K_2O_2	c	110.2028	—	−118.1	−101.6	—	24.4	—
	g	110.2028	—	−38.	—	—	—	—
KH	c	40.1100	—	−13.80	—	—	—	—
KD	c	41.1160	—	−13.21	—	—	—	—
KOH	c	56.1094	−100.681	−101.521	−90.61	−2.904	18.85	15.51
	a	56.1094	—	−115.29	−105.29	—	21.9	−30.3
in ∞ H_2O	a	56.1094	—	−115.29	—	—	—	—
KF	c	58.1004	−135.223	−135.58	−128.53	2.392	15.91	11.72
KF								
in ∞ H_2O	a	58.1004	—	−139.82	—	—	—	—
$KF \cdot 2H_2O$	c	94.1312	—	−278.112	−244.17	—	37.1	—
KHF_2	cα	78.1068	−220.560	−221.72	−205.48	3.655	24.92	18.39
KCl	c	74.5550	−104.310	−104.385	−97.79	2.717	19.74	12.26
	a	74.5550	—	−100.27	−99.07	—	38.0	−27.4
KClO	a	90.5544	—	−85.9	−76.5	—	35.	—
$KClO_2$	a	106.5538	—	−76.2	−63.6	—	48.7	—
$KClO_3$	c	122.5532	—	−95.06	−70.82	—	34.2	23.96
	a	122.5532	—	−85.17	−69.62	—	63.3	—
$KClO_4$	c	138.5526	−101.525	−103.43	−72.46	5.036	36.1	26.86
	a	138.5526	—	−91.23	−69.76	—	68.0	—
KBr	c	119.0110	−92.414	−94.120	−90.98	2.919	22.92	12.50
	g	119.0110	−40.832	−43.04	−50.89	2.416	59.85	8.824
	a	119.0110	—	−89.37	−92.55	—	44.2	−28.7
KBr_3	a	278.8290	—	−91.49	−93.29	—	76.0	—
KBrO	a	135.0104	—	−82.8	−75.7	—	34.	—
$KBrO_3$	c	167.0092	−83.957	−86.10	−64.82	5.593	35.65	28.72
	a	167.0092	—	−76.35	−63.27	—	63.15	—
$KBrO_4$	c	183.0086	−65.619	−68.80	−41.70	5.593	40.65	28.72
	a	183.0086	—	−57.2	−39.5	—	72.2	—
KI	c	166.0065	−78.137	−78.370	−77.651	3.039	25.41	12.65

SELECTED VALUES OF CHEMICAL THERMODYNAMIC PROPERTIES (Continued)

Substance			0 K	298.15 K (25°C)				
Formula and Description	State	Formula weight	ΔHf°_0 kcal/mol	ΔHf°	ΔGf° kcal/mol	$H^\circ_{298} - H^\circ_0$	S° cal/deg mol	C°_p
KI_3	c	419.8152	—	−78.4	—	—	—	—
	a	419.8152	—	−72.6	−80.0	—	81.7	—
KIO	a	182.0059	—	−86.0	−76.9	—	23.2	—
KIO_3	c	214.0047	−118.54	−119.83	−100.00	5.09	36.20	25.45
	a	214.0047	—	−113.2	−98.3	—	52.8	—
KIO_4	c	230.0041	—	−111.67	−86.38	—	42.	—
	a	230.0041	—	−96.5	−81.7	—	77.	—
K_2S	c	110.2680	—	−91.0	−87.0	—	25.0	
	a	110.2680	—	−112.7	−114.9	—	45.5	
K_2SO_3	c	158.2662	—	−269.0	—	—	—	—
K_2SO_3	a	158.2662	—	−272.5	−251.7	—	42.	—
K_2SO_4	c	174.2656	−341.126	−343.64	−315.83	6.079	41.96	31.42
	a	174.2656	—	−337.96	−313.37	—	53.8	−60.
$K_2S_2O_3$	c	190.3302	—	−280.5	—	—	—	—
	a	190.3302	—	−276.5	−260.3	—	65.	—
$K_2S_2O_7$	c	254.3278	—	−474.8	−428.2	—	61.	—
	a	254.3278	—	−455.5	—	—	—	—
$KHSO_4$	c	136.1716	—	−277.4	−246.5	—	33.0	—
	a	136.1716	—	272.40	−248.39	—	56.0	−15.
KNO_2								
rhombic	c	85.1075	−88.45	−88.39	−73.28	4.871	36.35	25.67
	a	85.1075	—	−85.3	−75.4	—	53.9	—
KNO_3	c	101.1069	−116.86	−118.22	−94.39	4.488	31.80	23.04
KNO_3	a	101.1069	—	−109.88	−94.31	—	59.5	−15.5
KPO_3	c	118.0740	—	—	—	3.886	25.93	21.56
	a	118.0740	—	−293.8	—	—	—	—
K_3PO_4	c	212.2774	—	−466.1	—	—	—	—
	a	212.2774	—	−486.3	−446.6	—	21.0	—
K_3AsO_4	a	256.2252	—	−393.23	−358.10	—	34.6	—
KH_2AsO_4	c	180.0372	−278.60	−282.2	−247.6	5.490	37.05	30.29
K_2CO_3	c	138.2134	−273.76	−275.1	−254.2	5.417	37.17	27.35
$K_2C_2O_4$								
oxalate	c	166.2240	—	−321.9	—	—	—	—
HCOOK								
formate	c	84.1200	—	−162.46	—	—	—	—
$KHCO_3$	c	100.1194	—	−230.2	−206.4	—	27.6	—
CH_3COOK								
acetate	c	98.1472	—	−172.8	—	—	—	—
	a	98.1472	—	−176.48	−155.99	—	45.2	3.7
KCN	c	65.1199	−28.174	−27.0	−24.35	4.157	30.71	15.84
	g	65.1199	21.494	21.7	15.34	3.188	62.57	12.51
KCNO								
cyanate	c	81.1193	—	−100.06	—	—	—	—
	a	81.1193	—	−95.2	−91.0	—	50.0	—
KCNS	c	97.1839	−47.980	−47.84	−42.62	4.176	29.70	21.16
	a	97.1839	—	−42.05	−45.55	—	59.0	−4.4
K_2SiO_3	c	154.2882	—	—	—	5.230	34.9	28.3
K_2SiF_6	c	220.2804	—	−706.5	−668.9	—	54.0	—
$KAl(SO_4)_2$	c	258.2067	—	−590.4	−535.4	—	48.9	46.12
$KAl(SO_4)_2 \cdot 3H_2O$	c	312.2529	—	−808.1	−711.0	—	75.0	—
$KAl(SO_4)_2 \cdot 12H_2O$	c	474.3915	—	−1448.8	−1228.9	—	164.3	155.6
$KAl_3Si_3O_{10}(OH)_2$								
muscovite	c	398.3133	—	−1430.3	−1340.5	—	73.2	—
$K_3Fe(CN)_6$	c	329.2604	—	−59.7	−31.0	—	101.83	—
	a	329.2604	—	−46.7	−28.8	—	138.1	
$K_4Fe(CN)_6$	c	368.3624	−141.30	−142.0	−108.3	14.871	100.1	79.40
$K_4Fe(CN)_6$	a	368.3624	—	−132.4	−104.71	—	120.7	—
$KMnO_4$	c	158.0376	—	−200.1	−176.3	—	41.04	28.10
	a	158.0376	—	—	—	—	—	−14.4
K_2CrO_4	c	194.1976	−333.796	−335.5	−309.7	6.805	47.83	34.89
	a	194.1976	—	−331.24	−309.36	—	61.0	—
$C_2Cr_2O_7$	c	294.1918	—	−492.7	−449.8	—	69	52.4
	a	294.1918	—	−476.8	−446.4	—	111.6	—
NaK	l	62.0918	—	1.5	—	—	—	—
Na_2K	l	85.0816	—	2.0	—	—	—	—
NaK_2	l	101.1938	—	2.5	—	—	—	—
PRASEODYMIUM								
Pr	c	140.907	0	0	0	1.74	17.5	6.50
	g	—	85.25	85.0	76.7	1.487	45.339	5.105
Pr^+	g	—	210.3	211.5	—	—	—	—
Pr^{2+}	g	—	453.6	456.3	—	—	—	—
Pr^{3+}	g	—	952.3	956.5	—	—	—	—
std. state, m = 1	aq	—	—	−168.4	−162.3	—	−50.	−7.
Pr^{4+}	g	—	1851.	1856.	—	—	—	—
Pr^{5+}	g	—	3176.	3183.	—	—	—	—
PrO	g	156.906	−38.	—	—	—	—	—
PrO_2	c	172.906	—	−226.9	—	—	—	28.06
Pr_2O_3								
hexagonal	c	329.812	—	−432.5	—	—	—	28.06
cubic	c	—	—	−432.5	—	—	—	—
PrH_2	c	142.923	−45.47	−47.4	−36.9	1.83	13.6	9.8
$PrOH^{2+}$								
std. state, m = 1	aq	157.914	—	—	−206.	—	—	—
$Pr(OH)_2^+$								
std. state, m = 1	aq	174.922	—	—	−257.	—	—	—
$Pr(OH)_3$	c	191.929	—	—	−307.1	—	—	—
$PrCl_3$	c	247.266	—	−252.6	—	—	—	24.
	g	—	—	−187.	—	—	—	—
std. state, m = 1	aq	—	—	−288.3	−256.4	—	−10.	−105.
PrOCl	c	192.359	—	−242.	—	—	—	—
PrI_3	c	521.620	—	−156.4	—	—	—	—
$Pr(IO_3)_3$	c	665.615	—	−333.8	—	—	—	—
$PrSO_4^+$								
std. state, m = 1	aq	236.969	—	−382.3	−345.1	—	−17.	—
$Pr(SO_4)_2^-$								
std. state, m = 1	aq	333.030	—	−598.4	−525.6	—	0.	—
$Pr(NO_3)_3$	c	326.922	—	−293.8	—	—	—	—
in HNO_3(aq)	aq	—	—	−316.7	—	—	—	—

Substance			0 K	298.15 K (25°C)				
Formula and Description	State	Formula weight	ΔHf°_0 kcal/mol	ΔHf°	ΔGf° kcal/mol	$H^\circ_{298} - H^\circ_0$	S° cal/deg mol	C°_p
$Pr(NO_3)_3 \cdot 6H_2O$	c	435.014	—	−731.05	—	—	—	—
PrC	c	152.918	—	−13.0	—	—	—	—
PrC_2	g	164.929	131.	131.3	118.7	2.48	62.6	10.6
$Pr_2(CO_3)_3$	c	461.842	—	−768.	—	—	—	—
PROMETHIUM								
147_{Pm}	g	146.915	—	—	—	1.545	44.692	5.797
PROTACTINIUM								
Pa	c	231.0359	0	0	0	—	12.4	—
	g	231.0359	—	145.	134.6	1.518	47.31	5.48
Pa^{4+}	a	231.0359	—	−148.0	—	—	—	—
in HCl + 3.43 H_2O:A		231.0359	—	−159.6	—	—	—	—
Pa^{5+}								
in HCl + 3.43 H_2O:A		231.0359	—	−161.8	—	—	—	—
Pa^{4+}								
in HCl + 8.16 H_2O:A		231.0359	—	−144.8	—	—	—	—
in HCl + 54.4 H_2O:A		231.0359	—	−147.7	—	—	—	—
PaO_2	c	263.0347	—	—	—	—	17.8	—
$PaCl_4$	c	372.8479	—	−249.3	−227.7	—	46.0	—
in HCl + 3.43 H_2O:Au		372.8479	—	−290.1	—	—	—	—
in HCl + 8.16 H_2O:Au		372.8479	—	−291.6	—	—	—	—
in HCl + 54.4 H_2O:Au		372.8479	—	−304.9	—	—	—	—
$PaCl_5$	c	408.3009	—	−273.6	−247.2	—	57.	—
	g	408.3009	—	−251.	−236.	—	94.	—
in HCl + 3.43 H_2O:Au		408.3009	—	−322.2	—	—	—	—
$PaBr_4$	c	550.6719	—	−197.0	−188.3	—	56.0	—
in HCl + 8.16 H_2O:Au		550.6719	—	−248.70	—	—	—	—
in HCl + 54.4 H_2O:Au		550.6719	—	−262.44	—	—	—	—
$PaBr_5$	c	630.5809	—	−206.	−196.	—	69.	—
	g	630.5809	—	−180.	−182.	—	111.	—
in HCl + 3.43 H_2O:Au		630.5809	—	−264.8	—	—	—	—
$PaOBr_2$	c	406.8533	—	−239.	—	—	—	—
PaI_4	c	738.6535	—	−123.2	—	—	—	—
in HCl + 54.4 H_2O:Au		738.6535	—	−199.1	—	—	—	—
RADIUM								
Ra	c	226.025	0	0	0	—	17.	—
	g	—	—	38.	31.	1.481	42.15	4.97
Ra^+	g	—	—	161.22	—	—	—	—
Ra^{2+}	g	—	—	396.70	—	—	—	—
std. state, m = 1	aq	—	—	−126.1	−134.2	—	13.	—
RaO	c	242.0244	—	−125.	—	—	—	—
$RaCl_2$	c	296.931	—	—	—	—	32.	—
std. state, m = 1	aq	—	—	−206.0	−196.9	—	40.	—
$RaCl_2 \cdot 2H_2O$	c	332.9617	—	−350.	−311.4	—	51.	—
$Ra(IO_3)_2$	c	575.8302	—	−245.4	−207.6	—	65.	—
$RaSO_4$	c	322.0866	—	−351.6	−326.4	—	33.	—
std. state, m = 1	aq	—	—	−343.4	−312.2	—	18.	—
$Ra(NO_3)_2$	c	350.0348	—	−237.	−190.3	—	53.	—
std. state, m = 1	aq	—	—	−225.2	−187.4	—	83.	—
RADON								
Rn	g	222.	0	0	0	1.481	42.09	4.968
Rn^+	g	—	247.86	249.34	—	—	—	—
RHENIUM								
Re	c	186.2	0	0	0	1.296	8.81	6.09
	g	—	183.8	184.0	173.2	1.481	45.131	4.968
Re^+	g	—	365.5	367.1	—	—	—	—
std. state, m = 1	aq	—	—	—	−8.	—	—	—
ReO_2	c	218.20	—	—	−88.	—	—	—
ReO_3	c	234.20	—	−144.6	—	—	—	—
$HReO_4$	c	251.21	—	−182.2	−156.9	—	37.8	—
std. state, m = 1	aq	—	—	−188.2	−166.0	—	48.1	−3.2
$HReO_4$								
∞ H_2O	aq	—	—	−188.2	—	—	—	—
$ReCl_3$	c	292.56	−62.73	−63.	−45.	4.319	29.6	22.08
$ReCl_5$	c	363.47	—	−89.	—	—	—	—
H_2ReCl_4	c	330.03	—	−152.	—	—	—	—
$ReBr_3$	c	425.93	—	−40.	—	—	—	—
ReS_2	c	250.33	—	−43.	—	—	—	—
RHODIUM								
Rh	c	102.905	0	0	0	1.174	7.53	5.97
	g	—	132.79	133.1	122.1	1.483	44.383	5.022
Rh^+	g	—	304.90	306.69	—	—	—	—
Rh^{2+}	g	—	721.8	725.0	—	—	—	—
Rh^{3+}	g	—	1438.	1443.	—	—	—	—
RhO	g	118.9044	—	92.	—	—	—	—
RhO^+	g	—	—	308.	—	—	—	—
RhO_2	g	134.9038	—	44.	—	—	—	—
Rh_2O_3	c	253.8082	—	−82.	—	—	—	24.8
$RhCl_2$	g	173.811	—	30.3	—	—	—	—
$RhCl_3$	c	209.264	—	−71.5	—	—	—	—
$RhCl_6^{3-}$	aq	315.623	—	−202.8	—	—	—	—
$RhCl_3 \cdot 3(C_2H_5)_2S$	c	479.8289	—	−166.	—	—	—	—
RUBIDIUM								
Rb	c	85.4678	0	0	0	1.790	18.35	7.424
	g	85.4678	19.639	19.330	12.690	1.481	40.626	4.968
Rb^+	g	85.4678	115.965	117.137	—	—	—	—
Rb^{2+}	g	85.4678	745.10	747.76	—	—	—	—
Rb^{3+}	g	85.4678	1660.0	1664.1	—	—	—	—
Rb^+	a	85.4678	—	−60.03	−67.87	—	29.04	—
Rb								
in 185 Hg		85.4678	—	—	−25.25	—	—	—
RbO_2	c	117.4666	—	−66.6	—	—	—	—
Rb_2O	c	186.9350	—	−81.	—	—	—	—
Rb_2O_2	c	202.9344	—	−112.8	—	—	—	—
RbH	c	86.4758	—	−12.5	—	—	—	—
RbOH	c	102.4752	—	−99.95	—	—	—	—
$RbOH \cdot H_2O$	c	120.4906	—	−178.98	—	—	—	—
$RbOH \cdot 2H_2O$	c	138.5060	—	−251.73	—	—	—	—
RbF	c	104.4662	—	−133.3	—	—	—	—
	a	104.4662	—	−139.53	−134.510	—	25.70	—
$RbHF_2$	c	124.4726	−219.52	−220.5	−204.5	3.932	28.70	18.97

SELECTED VALUES OF CHEMICAL THERMODYNAMIC PROPERTIES (Continued)

Substance			0 K	298.15 K (25°C)				
Formula and Description	State	Formula weight	ΔHf°_0 kcal/mol	ΔHf°	ΔGf° kcal/mol	$H^\circ_{298} - H^\circ_0$	S°	C°_p cal/deg mol
$RbCl$	c	120.9208	−104.080	−104.05	−97.47	2.917	22.92	12.52
$RbCl_3$	a	191.8268	—	—	−96.7	—	—	—
$RbClO$	a	136.9202	—	−85.6	−76.7	—	39.	—
$RbClO_2$	a	152.9196	—	−75.9	−63.8	—	53.2	—
$RbClO_3$	c	168.9190	—	−96.3	−71.8	—	36.3	24.66
	a	168.9190	—	−84.88	−69.77	—	67.80	—
$RbClO_4$	c	184.9184	—	−104.50	−73.54	—	39.2	—
	a	184.9184	—	−90.94	−69.93	—	72.5	—
$RbBr$	c	165.3768	−92.714	−94.31	−91.25	3.124	26.28	12.63
	a	165.3768	—	−89.08	−92.72	—	48.74	—
$RbBr_3$	c	325.1948	—	−100.0	—	—	—	—
	a	325.1948	—	−91.20	−93.46	—	80.5	—
$RbBr_5$	a	485.0128	—	−94.0	−92.7	—	104.7	—
$RbBrO_3$	c	213.3750	—	−87.78	−66.47	—	38.5	—
$RbBrO_3$	a	213.3750	—	−76.06	−63.44	—	67.69	—
$RbBrO_4$	a	229.3744	—	−56.9	−39.7	—	76.7	—
$RbBrCl_2$	c	236.2828	—	−116.2	—	—	—	—
RbI	c	212.3722	−79.603	−79.78	−78.60	3.190	28.30	12.71
	a	212.3722	—	−73.22	−80.20	—	55.6	—
RbI_3	c	466.1810	—	−82.8	−81.0	—	53.9	—
	a	466.1810	—	−72.3	−80.2	—	86.2	—
$RbIO_3$	c	260.3704	—	—	−101.90	—	—	—
	a	260.3704	—	−112.90	−98.50	—	57.30	—
Rb_2S	c	202.9996	—	−86.2	—	—	—	—
	a	202.9996	—	−112.2	−115.2	—	54.6	—
Rb_2S_2	a	235.0636	—	−112.9	−116.7	—	64.9	—
Rb_2SO_3	a	250.9978	—	−272.0	−252.0	—	51.	
Rb_2SO_4	c	266.9972	−340.795	−343.12	−314.76	6.458	47.19	32.04
Rb_2SO_4	a	266.9972	—	−337.38	−313.71	—	62.9	—
$RbNO_3$	c	147.4727	—	−118.32	−94.61	—	35.2	24.4
	a	147.4727	—	−109.59	−94.48	—	64.0	—
Rb_3PO_4	a	351.3748	—	−485.4	−447.1	—	34.	—
$Rb_4P_2O_7$	a	515.8146	—	−782.9	−730.2	—	88.	—
$RbAsO_2$	a	192.3882	—	−162.57	−151.53	—	38.7	—
Rb_3AsO_4	a	395.3226	—	−392.36	−358.61	—	48.2	—
Rb_2CO_3	c	230.9450	−270.41	−271.5	−251.2	5.851	43.34	28.11
	a	230.9450	—	−281.90	−261.91	—	44.5	—
$RbHCO_3$	c	146.4852	—	−230.2	−206.4	—	29.0	—
	a	146.4852	—	−225.42	−208.13	—	50.84	—
$RbCN$	c	111.4857	—	—	—	4.159	33.67	16.20
	a	111.4857	—	−24.0	−26.7	—	51.5	—
$RbCNO$								
cyanate	a	127.4851	—	−94.9	−91.2	—	54.5	—
$RbCNS$								
thiocyanate	a	143.5497	—	−41.76	−45.72	—	63.5	—
$RbBO_2$	c	128.2776	−231.0	−232.0	−218.2	3.181	22.54	17.7
$RbBH_4$	a	100.3108	—	−48.52	−40.56	—	55.4	—
$RbMnO_4$	a	204.4034	—	−189.4	−174.8	—	74.7	—
Rb_2MnO_4	a	289.8712	—	−276.	−255.4	—	72.	—
Rb_2CrO_4	c	286.9292	—	−338.0	—	—	—	—
	a	286.9292	—	−330.66	−309.70	—	70.08	—
$Rb_2Cr_2O_7$	a	386.9234	—	−476.3	−446.7	—	120.7	—
RUTHENIUM								
Ru	c	101.07	0	0	0	1.100	6.82	5.75
	g	—	153.210	153.6	142.4	1.490	44.550	5.144
Ru^+	g	—	323.07	324.94	—	—	—	—
Ru^{2+}	g	—	709.6	713.0	—	—	—	—
Ru^{3+}	g	—	1366.0	1370.9	—	—	—	—
RuO_4	c	133.069	—	−72.9	—	—	—	—
$RuCl_3$								
black	c	207.429	—	−49.	—	—	—	—
$RuCl_4$	g	242.882	—	−12.4	—	—	—	—
$RuBr_3$	c	340.797	—	−33.	—	—	—	—
RuI_3	c	481.783	—	−15.7	—	—	—	—
RuS_2	c	165.198	—	−47.	—	—	—	—
SAMARIUM								
Sm	c	150.35	0	0	0	1.81	16.63	7.06
	g	—	49.26	49.4	41.3	1.953	43.722	7.255
Sm^+	g	—	179.1	180.7	—	—	—	—
Sm^{2+}	g	—	434.	438.	—	—	—	—
std. state, m = 1	aq	—	—	—	−118.9	—	—	—
Sm^{3+}								
std. state, m = 1	aq	—	—	−165.3	−159.3	—	−50.6	−5.
SmO	g	166.349	−31.	—	—	—	—	—
Sm_2O_3								
monoclinic	c	348.698	−433.89	−435.7	−414.6	5.02	36.1	27.37
cubic	c	—	—	—	—	5.0	—	26.86
SmF	g	169.348	—	−63.	—	—	—	—
SmF_3	c	207.345	—	−425	—	—	—	—
$SmF_3 \cdot H_2O$	c	216.353	—	−436.2	—	—	—	—
$SmCl_3$	c	256.709	—	−245.2	—	—	—	—
∞ H_2O	aq	—	—	−285.2	—	—	—	—
$SmCl_3 \cdot 6H_2O$	c	364.801	—	−686.0	−587.1	—	99.	86.4
SmI_3	c	531.063	—	−148.2	—	—	—	—
$Sm(IO_3)_3$	c	675.058	—	−330.	—	—	—	—
$Sm_2(SO_3)_3$	c	540.887	—	—	−697.4	—	—	—
$Sm_2(SO_4)_3$	c	588.885	—	−931.9	—	—	—	—
$Sm(NO_3)_3$	c	336.365	—	−289.7	—	—	—	—
SmC_2	c	174.372	—	−17.	−18.1	—	23.	—
$Sm_2(CO_3)_3$	c	480.728	—	—	−741.4	—	—	—
SCANDIUM								
Sc	c	44.956	0	0	0	1.247	8.28	6.10
	g	—	89.87	90.3	80.32	1.674	41.75	5.28
Sc^+	g	—	240.69	—	—	—	—	—
Sc^{2+}	g	—	535.87	—	—	—	—	—
Sc^{3+}	g	—	1106.86	—	—	—	—	—
std. state, m = 1	aq	—	—	−146.8	−140.2	—	−61.	—
Sc^{4+}	g	—	2801.1	—	—	—	—	—

Substance			0 K	298.15 K (25°C)				
Formula and Description	State	Formula weight	ΔHf°_0 kcal/mol	ΔHf°	ΔGf° kcal/mol	$H^\circ_{298} - H^\circ_0$	S° cal/deg mol	C°_p
Sc^{5+}	g	—	4914.9	—	—	—	—	—
ScO	g	60.955	−13.5	−13.68	−19.90	2.100	53.65	7.38
Sc_2O	g	105.911	−6.	−6.9	—	2.7	—	11.2
Sc_2O_3	c	137.910	−453.95	−456.22	−434.85	3.34	18.4	22.52
$Sc(OH)_3$	c	95.978	—	−325.9	−294.8	—	24.	—
ScF	g	63.9544	−33.	−33.2	−39.3	2.138	53.11	7.74
ScF_2	g	82.9528	−153.	−153.5	−156.6	2.84	67.0	11.5
ScF_3	c	101.9512	—	−389.4	−371.8	—	22.	—
$ScCl_2$	g	115.862	—	—	—	3.1	72.5	12.6
$ScCl_3$	c	151.315	—	−221.1	—	—	—	—
in 5500 H_2O	aq	—	—	−268.8	—	—	—	—
$ScCl_3 \cdot 6H_2O$	c	259.411	—	−666.6	—	—	—	—
$Sc(OH)_2Cl$	c	114.424	—	−303.	−276.3	—	26.	—
$ScBr_3$	c	284.683	—	−177.6	—	—	—	—
ScI_2	g	298.765	—	—	—	3.4	81.6	13.3
ScS	g	77.020	41.9	41.8	29.7	2.2	56.3	8.0
$Sc_2(SO_4)_3$	c	378.097	—	—	—	—	—	62.0
ScC_2	g	68.978	143.	143.6	—	—	—	—
$Sc(HCO_2)_3$	c	180.009	—	—	—	—	—	54.
$Sc_2(C_2O_4)_3$	c	353.971	—	—	—	—	—	124.
SELENIUM								
Se								
hexagonal, black	c	78.96	0	0	0	1.319	10.144	6.062
monoclinic, red	c	—	—	1.6	—	—	—	—
	g	—	54.11	54.27	44.71	1.4815	42.21	4.978
glassy	amorp	—	—	1.2	—	—	—	—
Se^{2-}								
std. state, m = 1	aq	—	—	—	30.9	—	—	—
Se_2	g	157.92	35.26	34.9	23.0	2.275	60.2	8.46
Se_6	g	473.76	—	39.2	—	—	—	—
SeO	g	94.959	13.0	12.75	6.41	2.108	55.9	7.47
SeO_2	c	110.959	—	−53.86	—	—	—	—
	aq	—	—	−52.97	—	—	—	—
SeO_3	c	126.958	—	−39.9	—	—	—	—
Se_2O_5	c	237.917	—	−97.6	—	—	—	—
H_2SeO_3	c	128.974	—	−125.35	—	—	—	—
undissoc.; std. state, m = 1	aq	—	—	−121.29	−101.87	—	49.7	—
	aq	—	—	−121.24	—	—	—	—
H_2SeO_4	c	144.974	—	−126.7	—	—	—	—
$H_2SeO_4 \cdot H_2O$	c	162.989	—	−200.9	—	—	—	—
	liq	—	—	−196.1	—	—	—	—
SeF_6	g	192.950	−264.1	−267.	−243.	4.740	74.99	26.4
$SeCl_2$	g	149.866	—	−7.6	—	—	—	—
$SeCl_4$	c	220.772	—	−43.8	—	—	—	—
SILICON								
Si	c	28.086	0	0	0	0.769	4.50	4.78
	amorp	—	—	1.0	—	—	—	—
	g	—	107.86	108.9	98.3	1.805	40.12	5.318
Si^{+}	g	—	295.83	298.35	—	—	—	—
Si^{2+}	g	—	672.71	676.71	—	—	—	—
Si^{3+}	g	—	1444.50	1449.98	—	—	—	—
Si^{4+}	g	—	2485.50	2492.46	—	—	—	—
Si^{5+}	g	—	6331.3	6339.8	—	—	—	—
Si^{6+}	g	—	11062.6	11072.6	—	—	—	—
Si_2	g	56.172	141.32	142.	128.	2.22	54.92	8.22
Si_3	g	84.258	146.4	147.	—	2.9	—	12.9
SiO	g	44.0854	−24.08	−23.8	−30.2	2.082	50.55	7.15
SiO_2								
α, quartz	c	60.0848	—	−217.72	−204.75	1.657	10.00	10.62
α, cristobalite	c	—	—	−217.37	−204.46	1.671	10.20	10.56
SiO_2								
α, tridymite	c	—		−217.27	−204.42	1.693	10.4	10.66
	amorp	—	—	−215.94	−203.33	—	11.2	10.6
	g	—	—	−77.	—	—	—	—
	aq	—	—	−214.4	—	—	—	—
SiH	g	29.0940	86.	86.28	—	2.069	—	—
SiH_4	g	32.1179	10.30	8.2	13.6	2.517	48.88	10.24
Si_2H_6	g	62.2198	23.04	19.2	30.4	3.768	65.14	19.31
Si_3H_8	liq	92.3218	—	22.1	—	—	—	—
	g	—	—	28.9	—	—	—	—
H_2SiO_3	c	78.1001	—	−284.1	−261.1	—	32.	—
undissoc.; std. state, m = 1	aq	—	−282.7	−258.0	—	26.	—	
H_4SiO_4	c	96.1155	—	−354.0	−318.6	—	46.	—
SiF	g	47.0844	1.	1.7	−5.8	2.260	53.94	7.80
SiF_2	g	66.0828	−147.75	−148.	−150.	2.630	60.38	10.49
SiF_4	g	104.0796	−384.66	−385.98	−375.88	3.663	67.49	17.60
std. state, m = 1	aq	—	—	—	−384.2	—	—	—
SiF_6^{2-}								
std. state, m = 1	aq	142.0764	—	−571.0	−525.7	—	29.2	—
H_2SiF_6	aq	144.0923	—	—	—	—	—	—
SiCl	g	63.539	45.	45.39	—	2.267	—	8.81
$SiCl_2$	g	98.992	−39.61	−39.59	−42.35	2.98	67.0	12.16
$SiCl_4$	liq	169.898	—	−164.2	−148.16	—	57.3	34.73
	g	—	−156.508	−157.03	−147.47	4.633	79.02	21.57
SiBr	g	107.995	51.30	50.	—	2.40	—	9.23
$SiBr_4$	liq	347.722	—	−109.3	−106.1	—	66.4	—
	g	—	—	−99.3	−103.2	5.317	90.29	23.21
SiI_4	c	535.7036	—	−45.3	—	—	—	—
SiH_3I	g	158.0143	—	—	—	2.886	64.73	13.00
SiS	g	60.150	26.6	26.88	14.56	2.135	53.43	7.71
SiS_2	c	92.214	—	−49.5	—	—	—	—
Si_3N_4								
α	c	140.2848	—	−177.7	−153.6	—	24.2	—
SiC								
β, cubic	c	40.0972	−15.36	−15.6	−15.0	0.781	3.97	6.42
α, hexagonal	c	—	—	−15.0	−14.4	—	3.94	6.38
	g	—	175.6	177.	—	2.4	—	—
SiC_2	g	52.1083	145.7	147.	—	2.5	—	—
Si_2C	g	68.1832	131.1	132.	—	2.7	—	—
Si_2C_2	g	80.1943	167.	—	—	—	—	—

Substance			0 K	298.15 K (25°C)				
Formula and Description	State	Formula weight	ΔHf_0° kcal/mol	ΔHf°	ΔGf° kcal/mol	$H_{298}^\circ - H_0^\circ$	S° cal/deg mol	C_p°
Si_2C_3	g	92.2054	176.	—	—	—	—	—
Si_3C	g	96.2692	161.	—	—	—	—	—
SILVER								
Ag	c	107.870	0	0	0	1.373	10.17	6.059
	g	—	67.90	68.01	58.72	1.481	41.321	4.9679
Ag^+	g	—	242.00	243.59	—	—	—	—
std. state, m = 1	aq	—	—	25.234	18.433	—	17.37	5.2
Ag_2O	c	231.7394	−7.034	−7.42	−2.68	3.397	29.0	15.74
Ag_2O_2	c	247.7388	—	−5.8	6.6	—	28.	21.
Ag_2O_3	c	263.7382	—	8.1	29.0	—	24.	—
AgF	c	126.8684	—	−48.9	—	—	—	—
std. state, m = 1	aq	—	—	−54.27	−48.21	—	14.1	−20.3
in ∞ H_2O	aq	—	—	−54.27	—	—	—	—
AgF_2	c	145.8668	—	−86.	—	—	—	—
AgCl	c	143.323	—	−30.370	−26.244	—	23.0	12.14
std. state, m = 1	aq	—	—	−14.718	−12.939	—	30.9	−27.4
$AgClO_2$	c	175.3218	—	2.10	18.1	—	32.16	20.87
$AgClO_3$	c	191.3212	—	−7.24	15.4	—	34	—
std. state, m = 1	aq	—	—	0.38	16.51	—	56.2	—
$AgClO_4$	c	207.3206	—	−7.44	—	—	—	—
std. state, m = 1	aq	—	−5.68	16.37	—	60.9	—	
AgBr	c	187.779	—	−23.99	−23.16	—	25.6	12.52
std. state, m = 1	aq	—	—	−3.82	−6.42	—	37.1	−28.7
$AgBrO_3$	c	235.7772	—	−2.5	−17.04	—	36.3	—
std. state, m = 1	aq	—	—	9.20	22.86	—	56.02	—
AgI	c	234.7744	—	−14.78	−15.82	—	27.6	13.58
std. state, m = 1	aq	—	—	12.04	6.10	—	44.0	−28.8
AgI_2^-								
std. state, m = 1	aq	361.6788	—	—	−20.8	—	—	—
AgI_3^{2-}								
std. state, m = 1	aq	488.5832	—	−43.5	−36.8	—	60.5	—
AgI_4^{3-}								
std. state, m = 1	aq	615.4876	—	—	−50.1	—	—	—
$AgIO_3$	c	282.7726	—	−40.9	−22.4	—	35.7	24.60
std. state, m = 1	aq	—	—	−27.7	−12.2	—	45.7	—
Ag_2S								
α, orthorhombic	c	247.804	−8.126	−7.79	−9.72	4.136	34.42	18.29
β	c	—	—	−7.03	−9.43	—	36.0	—
Ag_2SO_3	c	295.8022	—	−117.3	−98.3	—	37.8	—
std. state, m = 1	aq	—	—	−101.4	−79.4	—	27.8	—
Ag_2SO_4	c	311.8016	—	−171.10	−147.82	—	47.9	31.40
std. state, m = 1	aq	—	—	−166.85	−141.10	—	39.6	−60.
Ag_2Se	c	294.70	−9.42	−9.	−10.6	4.48	36.02	19.54
AgN_3	c	149.8901	—	73.8	89.9	—	24.9	—
std. state, m = 1	aq	—	—	90.99	101.6	—	43.2	
Ag_3N	c	337.6167	—	47.6	—	—	—	—
$AgNO_2$	c	153.8755	—	−10.77	4.56	—	30.64	19.17
std. state, m = 1	aq	—	—	10.7	9.5	—	46.8	−18.1
$AgNO_3$	c	169.8749	—	−29.73	−8.00	—	33.68	22.24
nitrate; std. state, m = 1	aq	—	—	−24.33	−8.18	—	52.4	−15.5
$Ag(NH_3)_2^+$								
std. state, m = 1	aq	141.9312	—	−26.60	−4.12	—	58.6	—
$Ag(NH_3)_2NO_3$	c	203.9361	—	−85.0	—	—	—	—
std. state, m = 1	aq	—	—	−76.16	−30.73	—	93.6	—
$Ag(NH_3)_2Cl$								
std. state, m = 1	aq	177.3842	—	−66.55	−35.49	—	72.1	—
undissoc.; std. state, m = 1	aq	—	—	—	−35.01	—	—	—
$Ag(NH_3)_2Br$								
std. state, m = 1	aq	221.8402	—	−55.65	−28.97	—	78.3	—
Ag_3PO_4	c	418.5814	—	—	−210.	—	—	—
AgCN	c	133.8878	34.354	34.9	37.5	3.206	25.62	15.95
std. state, m = 1	aq	—	—	61.2	59.6	—	39.9	—
$Ag(CN)_2$								
std. state, m = 1	aq	159.9057	—	64.6	73.0	—	46.	—
AgONC								
silver fulminate	c	149.8872	—	43.	—	—	—	—
AgSCN	c	165.9518	—	21.0	24.23	—	31.3	15.
std. state, m = 1	aq	—	—	43.50	40.58	—	51.9	−4.4
SODIUM								
Na	c	22.9898	0	0	0	1.54	12.24	6.75
	g	22.9898	25.709	25.65	18.354	1.481	36.712	4.968
Na^+	g	22.9898	—	145.55	—	—	—	—
Na^{2+}	g	22.9898	—	1237.48	—	—	—	—
Na^{3+}	g	22.9898	—	2891.0	—	—	—	—
Na^{4+}	g	22.9898	—	5173.5	—	—	—	—
Na^{5+}	g	22.9898	—	8366.	—	—	—	—
Na^+	a	22.9898	—	−57.39	−62.593	—	14.1	11.1
NaO	g	38.9892	25.4	25.	19.7	2.22	54.6	8.3
NaO_2	c	54.9886	−62.96	−62.2	−52.2	4.37	27.7	17.24
Na_2O	c	61.9790	−97.85	−99.00	−89.74	2.964	17.94	16.52
Na_2O_2	c	77.9784	−120.70	−122.10	−107.00	3.75	22.70	21.33
NaOH	c	39.9972	−100.641	−101.723	−90.709	2.507	15.405	14.23
	g	39.9972	−48.6	−49.5	−50.2	2.72	54.57	11.56
in ∞ H_2O	a	39.9972	—	−112.36	—	—	—	—
NaF	c	41.9882	−136.542	−137.105	−129.902	2.031	12.30	11.20
	a	41.9882	—	−136.89	−129.23	—	10.8	−14.4
NaCl	c	58.4428	−98.168	−98.268	−91.815	2.536	17.24	12.07
	g	58.4428	−41.9	−42.22	−47.00	2.298	54.90	8.55
	a	58.4428	—	−97.34	−93.965	—	27.6	−21.5
in ∞ H_2O	a	58.4428	—	−97.34	—	—	—	—
NaClO	a	74.4422	—	−83.0	−71.4	—	24.	—
in 400 H_2O		74.4422	—	−82.8	—	—	—	—
$NaClO_2$	c	90.4416	—	−73.38	—	—	—	—
	a	90.4416	—	−73.3	−58.5	—	38.3	—
$NaClO_3$	c	106.4410	—	−87.422	−62.697	—	29.5	—
	a	106.4410	—	−82.24	−64.51	—	52.9	—
$NaClO_4$	c	122.4404	—	−91.61	−60.93	—	34.0	—
	a	122.4404	—	−88.30	−64.65	—	57.6	—

Substance			0 K	298.15 K (25°C)				
Formula and Description	State	Formula weight	ΔHf_0° kcal/mol	ΔHf° kcal/mol	ΔGf° kcal/mol	$H_{298}^\circ - H_0^\circ$ kcal/mol	S° cal/deg mol	C_p° cal/deg mol
NaBr	c	102.8988	−84.596	−86.296	−83.409	2.770	20.75	12.28
	g	102.8988	−32.08	−34.2	−42.31	2.346	57.62	8.68
	a	102.8988	—	−86.440	−87.440	—	33.8	−22.8
$NaBrO$	a	118.8982	—	−79.9	−70.6	—	24.	—
$NaBrO_3$	c	150.8970	—	−79.85	−58.04	—	30.8	—
	a	150.8970	—	−73.42	−58.16	—	52.8	—
$NaBrO_4$	a	166.8964	—	−54.30	−34.40	—	61.80	—
NaI	c	149.8942	−68.593	−68.78	−68.37	2.93	23.55	12.45
	a	149.8942	—	−70.58	−74.92	—	40.7	−22.9
NaI_3	c	403.7030	—	−56.2	—	—	—	—
NaI_3	a	407.7030	—	−69.7	−74.9	—	71.3	—
$NaIO_3$	c	197.8924	—	−115.150	—	—	—	22.0
	a	197.8924	—	−110.30	−93.20	—	42.4	—
$NaIO_3 \cdot H_2O$	c	215.9078	—	−186.30	−151.56	—	38.8	—
$NaIO_3 \cdot 5H_2O$	c	287.9694	—	−466.60	—	—	—	—
$NaIO_4$	c	213.8918	—	−102.60	−77.22	—	39.0	—
	a	213.8918	—	−93.60	−76.60	—	67.	—
in 2,000 H_2O		213.8918	—	−93.808	—	—	—	—
$NaIO_3 \cdot 3H_2O$	c	267.9380	—	—	−247.8	—	—	—
NaH_4IO_6								
from $H_4IO_6^-$	au	249.9226	—	−237.8	—	—	—	—
$Na_2H_3IO_6$								
from $H_3IO_6^{2-}$	au	271.9044	—	−294.4	—	—	—	—
$NaIO_2F_2$	c	219.8898	—	−202.50	—	—	—	—
$NaICl_2$	c	220.8002	—	−96.0	—	—	—	—
	a	220.8002	—	—	−101.1	—	—	—
$NaICl_4$	c	291.7062	—	−112.0	—	—	—	—
$NaIBr_2$	c	309.7122	—	−83.0	—	—	—	—
	a	309.7122	—	—	−92.0	—	—	—
$NaBrI_2$	a	356.7076	—	−88.0	−88.9	—	61.3	—
Na_2S	c	78.0436	—	−87.2	−83.6	—	20.0	—
	a	78.0436	—	−106.9	−104.7	—	24.7	—
Na_2S_2	c	110.1076	—	−94.9	−90.5	—	25.	—
Na_2S_2	a	110.1076	—	−107.6	−106.2	—	35.0	—
Na_2SO_3	c	126.0418	−261.21	−263.1	−242.0	5.36	34.88	28.74
	a	126.0418	—	−266.70	−241.50	—	21.0	—
Na_2SO_4	c	142.0412	−328.789	−331.52	−303.59	5.551	35.75	30.64
	a	142.0412	—	−332.10	−303.16	—	33.0	−48.
$Na_2SO_4 \cdot 10H_2O$	c	322.1952	—	−1034.24	−871.75	—	141.5	—
$Na_2S_2O_3$	c	158.1058	—	−268.4	−245.7	—	37.	—
	a	158.1058	—	−270.65	−250.0	—	44.0	—
$Na_2S_2O_3 \cdot 5H_2O$	c	248.1828	—	−623.31	−533.0	—	89.0	—
$Na_2S_2O_4$	c	174.1052	—	−294.5	—	—	—	—
$NaHSO_4$	c	120.0594	—	−269.0	−237.3	—	27.0	—
from HSO_4	a	120.0594	—	−269.47	−243.28	—	45.6	−9.
Na_2SeO_4	c	188.9372	—	255.5	—	—	—	—
	a	188.9372	—	258.0	−230.7	—	41.1	—
NaN_3	c	65.0099	6.316	5.19	22.41	3.522	23.15	18.31
	a	65.0099	—	8.37	20.6	—	39.9	—
$NaNO_2$	c	68.9953	—	−85.72	−68.02	—	24.8	—
	a	68.9953	—	−82.4	−70.3	—	43.5	−12.2
$NaNO_3$	c	84.9947	−110.248	−111.82	−87.73	4.115	27.85	22.20
	a	84.9947	—	−106.95	−89.20	—	49.1	−9.6
$NaNH_2$	c	39.0125	−27.84	−29.6	−15.3	2.842	18.38	15.81
$NaNH_3$	c	40.0205	—	−16.27	−2.94	—	37.2	—
Na_3PO_4	c	163.9408	−454.79	−458.27	−427.55	6.566	41.54	36.68
	a	163.9408	—	−477.5	−431.3	—	−11.	—
$(NaPO_3)_3$	c	305.8854	−865.9	−873.	−808.	10.720	68.47	62.00
$Na_4P_2O_7$	c	265.9026	−756.2	−762.	−709.7	10.180	64.60	57.63
$Na_5P_3O_{10}$								
form I, quenched	c	367.8644	−1043.57	−1051.4	−978.5	14.070	91.25	78.16
form II	c2	367.8644	−1045.76	−1054.0	−980.0	13.67	87.37	77.72
in 5,200 H_2O		367.8644	—	−1068.1	—	—	—	—
NaH_2PO_2	c	87.9784	—	−200.5	—	—	—	—
NaH_2PO_3	c	103.9778	—	−288.0	—	—	—	—
NaH_2PO_4	c	119.9772	−363.06	−367.3	−331.3	4.75	30.47	27.93
$NaH_2PO_4 \cdot H_2O$	c	137.9926	—	−438.1	—	—	—	—
Na_2HPO_3	c	125.9596	—	−336.8	—	—	—	—
Na_2HPO_4	c	141.9590	−413.92	−417.8	−384.4	5.646	35.97	32.34
$Na_2HPO_4 \cdot 2H_2O$	c	177.9898	—	−560.7	−499.2	—	52.9	—
$Na_2HPO_4 \cdot 7H_2O$	c	268.0668	—	−913.4	−784.0	—	103.87	—
$Na_2HPO_4 \cdot 12H_2O$	c	358.1438	—	−1266.2	−1068.0	—	151.49	—
$Na_2H_2P_2O_7$	c	221.9390	−654.06	−660.8	−602.9	8.184	52.63	47.36
$NaAsO_2$	c	129.9102	—	−157.87	—	—	—	
	a	129.9102	—	−159.93	−146.25	—	23.8	—
Na_3AsO_4	c	207.8886	—	−368.	—	—	—	—
	a	207.8886	—	−384.44	−342.78	—	3.4	—
Na_3BiO_4	c	341.9470	—	−291.	—	—	—	—
Na_2CO_3	c	105.9890	−268.76	−270.24	−249.64	4.959	32.26	26.84
	a	105.9890	—	−276.62	−251.36	—	14.6	—
$Na_2CO_3 \cdot H_2O$	c	124.0044	−338.87	−342.08	−307.22	6.296	40.18	34.80
$Na_2CO_3 \cdot 7H_2O$	c	232.0968	—	−764.81	−648.8	—	102.	—
$Na_2CO_3 \cdot 10H_2O$	c	286.1430	−959.62	−975.46	−819.36	21.21	134.8	131.53
$Na_2C_2O_4$	c	133.9996	—	−315.0	—	—	—	34.
HCOONa	c	68.0078	−158.19	−159.3	−143.4	3.767	24.80	19.76
$NaHCO_3$	c	84.0072	−225.15	−227.25	−203.4	3.81	24.3	20.94
$NaOCH_3$	c	54.0244	−85.41	−87.9	−70.46	3.374	26.43	16.60
$NaC_2H_3O_2$	c	82.0350	—	−169.41	−145.14	—	29.4	19.1
	a	82.0350	—	−173.55	−150.88	—	34.8	9.6
$NaC_2H_3O_2 \cdot 3H_2O$	c	136.0812	—	−383.2	−317.6	—	58.	—
$NaOC_2H_5$	c	68.0516	—	−98.90	—	—	—	—
	a	68.0516	—	−113.	—	—	—	—
CCl_3COONa								
sodium trichloroacetate	c	185.3700	—	−178.9	—	—	—	—
NaCN								
(C,I,cubic)	c	49.0077	−22.56	−20.91	−18.27	4.480	27.63	16.82
(c,II,orthorhombic)	c2	49.0077	—	−21.69	—	—	—	—

Substance			0 K	298.15 K (25°C)				
Formula and Description	State	Formula weight	ΔHf_0° kcal/mol	ΔHf°	ΔGf° kcal/mol	$H_{298}^\circ - H_0^\circ$	S°	C_p° cal/deg mol
$NaCN \cdot 1/2H_2O$	c	58.0154	—	−56.35	—	—	—	—
$NaCN \cdot 2H_2O$	c	85.0385	—	−162.47	—	—	—	—
$NaCNO$	c	65.0071	—	−96.89	−85.6	—	23.1	20.7
	a	65.0071	—	−92.3	−85.9	—	39.6	—
$NaCNS$	c	81.0717	—	−40.75	—	—	—	—
	a	81.0717	—	−39.12	−40.44	—	48.6	1.5
Na_2SiO_3								
sodium metasilicate	c	122.0638	—	−371.63	−349.19	—	27.21	—
	gl	122.0638	—	−368.1	—	—	—	—
$Na_2SiO_3 \cdot 5H_2O$	c	212.1408	—	−728.6	—	—	—	—
$Na_2SiO_3 \cdot 9H_2O$	c	284.2024	—	−1010.7	—	—	—	—
$Na_2Si_2O_5$	c	182.1486	—	−589.8	−555.0	—	39.21	—
sodium disilicate, stable up to 951K [formerly β]								
	c2	182.1486	—	−589.22	—	—	—	—
stable 951K to m.pt.(1147K)[formerly α]								
unstable	c3	182.1486	—	−587.28	—	—	—	—
	gl	122.0638	—	−368.1	—	—	—	—
Na_4SiO_4	c	184.0428	—	—	—	—	46.76	—
$NaBO_2$	c	65.7996	−232.38	−233.5	−220.06	2.780	17.57	15.76
$NaBO_2 \cdot 2H_2O$	c	101.8304	—	−378.	−337.0	—	37.	—
$NaBO_2 \cdot 4H_2O$	c	137.8612	—	−520.	−451.3	—	55.	—
$NaBO_3 \cdot 4H_2O$	c	153.8606	—	−505.3	—	—	—	—
$Na_2B_4O_7$	c	201.2194	−782.36	−786.6	−740.0	7.262	45.30	44.64
	am	201.2194	—	−781.8	−735.4	—	46.1	—
	a	201.2194	—	—	−747.8	—	—	—
$Na_2B_4O_7 \cdot 4H_2O$	c	273.2810	—	−1077.3	—	—	—	—
$Na_2B_4O_7 \cdot 5H_2O$	c	291.2964	—	−1147.8	—	—	—	—
$Na_2B_4O_7 \cdot 10H_2O$								
borax	c	381.3734	—	−1503.0	−1318.5	—	140.	147.
$NaBH_4$	c	37.8328	−43.100	−45.08	−29.62	3.890	24.21	20.74
	a	37.8328	—	−45.88	−35.28	—	40.5	—
$NaBF_4$	c	109.7944	−440.04	−440.9	−418.30	5.191	34.73	28.74
$NaBF_4$	a	109.7944	—	−433.8	−418.0	—	58.	—
$NaAlO_2$	c	81.9701	—	−271.30	−256.06	—	16.90	17.52
	a	81.9701	—	−277.0	−259.4	—	9.	—
$NaAl(SO_4)_2 \cdot 12H_2O$ alum	c	458.2793	—	−1434.69	—	—	—	—
$NaAlSiO_4$								
nepheline, nephelite	c	142.0549	—	−500.2	−472.8	—	29.7	
$NaAlSi_2O_6$								
jadeite	c	202.1397	—	−724.4	−681.7	—	31.9	—
dehydrated analcite	c2	202.1397	—	−713.5	−673.8	—	41.9	39.30
$NaHg$	c	223.5798	—	−11.3	−9.76	—	25.	—
$NaHg_2$	c	424.1698	—	−18.2	−16.2	—	42.	—
$NaHg_4$	c	825.3498	—	−21.2	−17.7	—	73.	—
Na_3Hg	c	269.5594	—	−11.0	−10.84	—	54.	—
Na_3Hg_2	c	470.1494	—	−21.9	−20.7	—	69.	—
Na_5Hg_2	c	516.1290	—	—	−21.7	—	—	—
Na_7Hg_8	c	1765.6486	—	−84.4	−76.0	—	203.	—
$Na_3Fe(CN)_6$	a	280.9238	—	−37.9	−13.5	—	106.9	—
$Na_4Fe(CN)_6$	a	303.9136	—	−120.7	−84.28	—	79.1	—
$NaMnO_4$	a	141.9254	—	−186.8	−169.5	—	59.8	—
Na_2MnO_4	c	164.9152	—	−276.3	—	—	—	—
	a	164.9152	—	−271.	−244.9	—	42.	—
Na_2CrO_4	c	161.9732	−318.2	−320.8	−295.17	6.323	42.21	33.97
	a	161.9732	—	−325.38	−299.15	—	40.2	—
$Na_2Cr_2O_7$	c	261.9674	—	−472.9	—	—	—	—
	a	216.9674	—	−471.0	−436.2	—	90.8	—
Na_2MoO_4	c	205.9172	−348.63	−350.89	−323.71	6.070	38.17	33.87
$Na_2MoO_4 \cdot 2H_2O$	c	241.9480	—	−492.1	−437.46	—	57.5	—
Na_2WO_4	c	293.8272	−367.83	−370.2	−342.86	6.05	38.6	33.41
$Na_2W_2O_7$	c	525.6754	−571.44	−574.8	−529.84	9.36	60.8	51.36
$NaVO_3$	c	121.9300	−272.31	−273.85	−254.33	4.217	27.17	23.32
Na_3VO_4	c	183.9090	−417.36	−420.14	−391.45	7.093	45.4	39.40
$Na_4V_2O_7$	c	305.8390	−693.73	−697.62	−650.46	11.75	76.1	64.47
$NaNbO_3$	c	163.8940	—	−314.5	−294.7	—	28.	—
Na_2TiO_3	c	141.8778	−377.88	−380.3	−357.6	4.917	29.08	30.03
$Na_2Ti_2O_5$	c	221.7766	—	—	—	6.910	41.56	41.68
$NaUO_3$	c	309.0170	—	−360.	—	—	—	—
Na_2UO_4								
α form	c	348.0062	−450.02	−452.5	−424.90	6.268	39.68	35.05
β form	c2	348.0062	—	−450.2	—	—	—	—
Na_3UO_4	c	370.9960	−481.15	−484.0	−454.4	7.435	47.37	41.35
STRONTIUM								
Sr	c	87.62	0	0	0	—	12.5	6.3
	g	—	—	39.3	31.3	1.481	39.32	4.968
Sr^{+}	g	—	—	172.11	—	—	—	—
Sr^{2+}	g	—	—	427.96	—	—	—	—
std. state, m = 1	aq	—	—	−130.45	−133.71	—	−7.8	—
Sr^{3+}	g	—	—	1435.	—	—	—	—
Sr^{4+}	g	—	—	2751.	—	—	—	—
SrO	c	103.619	—	−141.5	−134.3	—	13.0	10.76
	g	—	—	−2.	—	—	—	—
SrO_2	c	119.619	—	−151.4	—	—	—	—
$SrO_2 \cdot 8H_2O$	c	263.742	—	−722.6	—	—	—	—
Sr_2O	c	191.239	—	−154.7	—	—	—	—
SrH	g	88.628	—	+52.	45.	2.080	50.80	7.17
SrH_2	c	89.636	—	−43.1	—	—	—	—
$SrOH$	g	104.627	—	−41.	—	—	—	—
$Sr(OH)_2$	c	121.635	—	−229.2	—	—	—	—
	g	—	—	−135.	—	—	—	—
in 800 H_2O	aq	—	—	−240.1	—	—	—	—
SrF	g	106.618	—	−69.0	−75.1	2.219	57.31	8.27
SrF_2	c	125.617	—	−290.7	−278.4	3.125	19.63	16.73
	g	—	—	−182.7	−185.3	3.192	69.54	12.66
$SrCl$	g	123.073	—	−30.2	−36.5	2.338	60.2	8.73
$SrCl_2$								
α	c	158.526	—	−198.1	−186.7	3.880	27.45	18.07
	g	—	—	−116.1	−118.6	3.454	74.26	13.33
$SrCl_2 \cdot H_2O$	c	176.541	—	−271.7	−247.7	—	41.	28.7
$SrCl_2 \cdot 2H_2O$	c	194.557	—	−343.7	−306.4	—	52.	38.3

SELECTED VALUES OF CHEMICAL THERMODYNAMIC PROPERTIES (Continued)

Substance			0 K	298.15 K (25°C)				
Formula and Description	State	Formula weight	ΔHf_0° kcal/mol	ΔHf°	ΔGf° kcal/mol	$H_{298}^\circ - H_0^\circ$	S° cal/deg mol	C_p°
$SrCl_2 \cdot 6H_2O$	c	266.618	—	−627.1	−535.67	—	93.4	—
$Sr(ClO_4)_2$	c	286.521	—	−182.31	—	—	—	—
std. state, m = 1	aq	—	—	−192.27	−137.83	—	79.2	—
∞ H_2O	aq	—	—	−192.27	—	—	—	—
$SrBr_2$	c	247.438	—	−171.5	−166.6	4.261	32.29	18.01
	g	—	—	−98.	−106.	3.8	77.3	14.5
std. state, m = 1	aq	—	—	−188.55	−183.41	—	31.6	—
∞ H_2O	aq	—	—	−188.55	—	—	—	—
$SrBr_2$	c	247.438	—	−171.5	−166.6	4.261	32.29	18.01
	g	—	—	−98.	−106	3.8	77.3	14.5
std. state, m = 1	aq	—	—	−188.55	−183.41	—	31.6	—
$Sr(IO_3)_2$	c	437.425	—	−243.6	−204.4	—	56.	—
$Sr(IO_3)_2 \cdot H_2O$	c	455.441	—	−313.2	−260.4	—	66.	—
$Sr(IO_3)_2 \cdot 6H_2O$	c	545.517	—	−666.8	−543.7	—	109.	—
SrS	c	119.684	—	−112.9	−111.8	—	16.3	11.64
$SrSO_3$	c	167.682	—	−281.3	—	—	—	—
$SrSO_4$	c	183.682	—	−347.3	−320.5	—	28.	—
precipitate	c	—	—	−346.5	—	—	—	—
std. state, m = 1	aq	—	—	−347.77	−311.68	—	−3.0	—
$SrSeO_3$	c	214.578	—	−250.4	—	—	—	—
$SrSeO_4$	c	230.578	—	−273.1	—	—	—	—
$Sr(N_3)_2$	c	171.660	—	+2.1	—	—	—	—
Sr_3N_2	c	290.873	—	−93.5	—	—	—	—
$Sr(NO_2)_2$	c	179.631	—	−182.3	—	—	—	—
In 800 H_2O	aq	—	—	−180.5	—	—	—	—
$Sr(NO_3)_2$	c	211.630	—	−233.80	−186.46	6.854	46.50	35.83
std. state, m = 1	aq	—	—	−229.57	−186.93	—	62.2	—
$Sr(NO_3)_2$								
in ∞ H_2O	aq	—	—	−229.57	—	—	—	—
Sr_3P_2	c	324.808	—	−152.	—	—	—	—
$Sr_3(PO_4)_2$	c	452.803	—	−985.4	—	—	—	—
$SrHPO_4$	c	183.599	—	−435.4	−403.6	—	29.	—
$Sr(H_2PO_4)_2$	c	281.595	–	−749.2	—	—	—	—
$Sr_3(AsO_4)_2$	c	540.698	—	−792.8	−736.2	—	61.	—
hydrated precipitate	—	—	—	−803.	—	—	—	—
$SrHAsO_4$	aq	227.547	—	−345.7	—	—	—	—
$Sr(H_2AsO_4)_2$	aq	369.490	—	−563.2	—	—	—	—
SrC_2	c	111.642	—	−18.	—	—	—	—
$SrCO_3$								
strontianite	c	147.629	—	−291.6	−272.5	—	23.2	19.46
std. state, m − 1	aq	—	—	−292.29	−259.88	—	−21.4	—
SrC_2O_4	c	175.640	—	−327.6	—	—	—	—
std. state, m = 1	aq	—	—	−327.6	−294.8	—	+2.1	—
$SrCN_2$	c	127.645	—	−72.5	—	—	—	—
$Sr(CN)_2$	aq	139.656	—	−57.0	—	—	—	—
$Sr(CN)_2 \cdot 4H_2O$	c	211.717	—	−333.7	—	—	—	—
$SrSiO_3$	c	163.704	—	−390.5	−370.4	—	23.1	21.16
Sr_2SiO_4	c	267.324	—	−550.8	−523.7	—	36.6	32.09
$SrMoO_3$	c	231.558	—	−306.	—	—	—	—
$SrMoO_4$	c	247.558	—	−370.	—	—	—	—
$SrWO_4$	c	335.468	—	−391.9	−366.	—	33.	—
$SrTiO_3$	c	183.518	—	−399.71	−379.64	—	26.0	23.51
Sr_2TiO_4	c	287.138	—	−546.7	−520.7	—	38.0	34.34
$SrZrO_3$	c	226.838	—	−422.4	−402.2	—	27.5	24.71
SULFUR								
S								
rhombic	c	32.064	0	0	0	1.054	7.60	5.41
monoclinic	c	—	—	0.08	—	—	—	—
	g	—	66.1	66.636	56.951	1.591	40.084	5.658
S^+	g	—	304.95	306.97	—	—	—	—
S^{2+}	g	—	844.8	848.2	—	—	—	—
S^{3+}	g	—	1653.0	1657.9	—	—	—	—
S^{4+}	g	—	2743.6	2750.0	—	—	—	—
S^{5+}	g	—	4415.	4422.	—	—	—	—
S^{6+}	g	—	6445.	6654.	—	—	—	—
S_2	g	64.128	30.647	30.68	18.96	2.141	54.51	7.76
S_2^{2-}								
std. state, m = 1	aq	—	—	7.2	19.0	—	6.8	—
S_3	g	96.192	—	31.7	—	—	—	—
S_4	g	128.256	—	32.7	—	—	—	—
S_5	g	160.320	—	29.6	—	—	—	—
S_6	g	192.384	—	24.5	—	—	—	—
SO	g	48.0634	1.5	1.496	−4.741	2.087	53.02	7.21
SO_2	liq	64.0628	—	−76.6	—	—	—	—
	g	—	−70.336	−70.944	−71.748	2.521	59.30	9.53
undissoc.; std. state, m = 1	aq	—	—	−77.194	−71.871	—	38.7	—
SO_2								
in 10,000 H_2O	aq	—	—	−80.584	—	—	—	—
SO_3								
I, β	c	80.0622	—	−108.63	−88.19	—	12.5	—
	liq	—	—	−105.41	−88.04	—	22.85	—
	g	—	−93.21	−94.58	−88.69	2.796	61.34	12.11
SO_3^{2-}								
std. state, m = 1	aq	—	—	−151.9	−116.3	—	−7.	—
SO_4^{2-}								
std. state, m = 1	aq	96.0616	—	−217.32	−177.97	—	4.8	−70.
$S_2O_3^{2-}$	aq	112.1262	—	−155.9	—	—	—	—
$S_2O_4^{2-}$								
std. state, m = 1	aq	128.1256	—	−180.1	−143.5	—	22.	—
$S_2O_7^{2-}$	aq	176.1238	—	−334.9	—	—	—	—
$S_2O_8^{2-}$								
std. state, m = 1	aq	192.1232	—	−320.0	−265.4	—	59.3	—
H_2S	g	34.0799	−4.232	−4.93	−8.02	2.379	49.16	8.18
std. state, m = 1	aq	—	—	−9.5	−6.66	—	29.	—
HSO_3								
std. state, m = 1	aq	81.0702	—	−149.67	−126.15	—	33.4	—
HSO_4^-								
std. state, m = 1	aq	97.0696	—	−212.08	−180.69	—	31.5	−20.
H_2SO_3								
undissoc.; std. state, m = 1	aq	82.0781	—	−145.51	−128.56	—	55.5	—
H_2SO_4	c	98.0775	−194.069	—	—	—	—	—

Substance			0 K	298.15 K (25°C)				
Formula and Description	State	Formula weight	ΔHf_0° kcal/mol	ΔHf°	ΔGf° kcal/mol	$H_{298}^\circ - H_0^\circ$	S° cal/deg mol	C_p°
	liq	—	—	−194.548	−164.938	6.748	37.501	33.20
std. state, m = 1	aq	—	—	−217.32	−177.97	—	4.8	−70.
H_2SO_4								
in ∞ H_2O	aq	—	—	−217.32	—	—	—	—
$H_2SO_4 \cdot 1H_2O$	liq	116.0929	—	−269.508	−227.182	—	50.56	51.35
$H_2SO_4 \cdot 2H_2O$	liq	134.1082	—	−341.085	−286.770	—	66.06	62.34
$H_2SO_4 \cdot 3H_2O$	liq	152.1236	—	−411.186	−345.178	—	82.55	76.23
$H_2SO_4 \cdot 4H_2O$	liq	170.1389	—	−480.688	−403.001	—	99.09	91.35
$H_2SO_4 \cdot 6.5\ H_2O$	liq	215.1772	—	−653.264	−546.403	—	140.51	136.30
$H_2S_2O_4$								
std. state, m = 1	aq	130.1415	—	—	−147.4	—	—	—
undissoc.								
$H_2S_2O_6$	aq	162.1403	—	−286.4	—	—	—	—
$H_2S_2O_7$	c	178.1397	—	−304.4	—	—	—	27.
$H_2S_2O_8$								
std. state, m = 1	aq	194.1391	—	−320.0	−265.4	—	59.3	—
SF_4	g	108.0576	−183.4	−185.2	−174.8	3.482	69.77	17.45
SF_6	g	146.0544	−285.7	−289.	−264.2	4.056	69.72	23.25
std. state, m = 1	aq	—	—	−293.0	−259.3	—	39.8	—
$SOCl_2$	liq	118.9694	—	−58.7	—	—	—	29.
	g	—	−50.07	−50.8	−47.4	3.559	74.01	15.9
in C_6H_6	—	—	—	−59.7	—	—	—	—
SO_2Cl_2	liq	134.9688	—	−94.2	—	—	—	32.
	g	—	−85.50	−87.0	−76.5	3.825	74.53	18.4
in C_6H_6	—	—	—	−94.2	—	—	—	—
TANTALUM								
Ta	c	180.948	0	0	0	1.347	9.92	6.06
	g	—	186.765	186.9	176.7	1.482	44.241	4.985
Ta^+	g	—	368.72	370.33	—	—	—	—
TaO	g	196.9474	60.3	60.	53.	2.10	57.6	7.31
TaO_2	g	212.9468	−40.7	−41.	−43.	3.1	64.	12.5
Ta_2O_5								
β	c	441.8930	—	−489.0	−456.8	—	34.2	32.30
Ta_2H	c	362.9040	−6.93	−7.8	−16.5	2.84	18.9	21.7
TaF_5	c	275.9400	—	−454.97	—	—	—	—
$TaCl_3$	c	287.307	—	−132.2	—	—	—	—
$TaCl_4$	c	322.760	—	−167.7	—	—	—	—
$TaCl_5$	c	358.213	—	−205.3	—	—	—	—
$TaBr_5$	c	580.493	—	−143.0	—	—	—	—
TaS_2	c	245.076	—	−111.	—	—	—	—
TaN	c	194.9547	—	−60.1	—	—	—	9.7
Ta_2N	c	375.9027	—	−65.	—	—	—	—
TaC	c	192.9592	−34.96	−35.0	−34.6	1.56	10.11	8.79
Ta_2C	c	373.9072	—	−51.0	−50.8	—	20.7	—
$TaSi_2$	c	237.120	—	−28.	—	—	—	—
Ta_5Si_3	c	988.998	—	−76.	—	—	—	—
TaB_2	c	202.570	—	−46.	—	—	—	—
TECHNETIUM								
Tc	c	98.906	0	0	0	—	—	—
	g	—	—	162.	—	1.481	43.25	4.97
Tc^+	g	—	—	331.	—	—	—	—
Tc_2O_7	c	309.8078	—	−266.	—	—	—	—
$HTcO_4$	c	163.9115	—	−167.	—	—	—	—
std. state, m = 1	aq	—	—	−173.	—	—	—	—
TELLURIUM								
Te	c	127.60	0	0	0	1.463	11.88	6.15
	g	—	47.	47.02	37.55	1.481	43.65	4.968
	amorp	—	—	2.7	—	—	—	—
Te^+	g	—	254.76	256.26	—	—	—	—
Te^{2+}	g	—	684.	687.	—	—	—	—
Te^{3+}	g	—	1390.	1394.	—	—	—	—
Te^{4+}	g	—	2262.	2268.	—	—	—	—
Te^{5+}	g	—	3652.	3659.	—	—	—	—
TeO	g	143.599	16.	15.6	9.2	2.093	57.7	7.19
TeO_2	c	159.599	—	−77.1	−64.6	—	19.0	—
H_2Te	g	129.616	—	23.8	—	—	—	—
H_2TeO_3								
std. state, m = 1	aq	177.614	—	—	−76.2	—	—	—
H_6TeO_6	c	229.644	—	−310.4	—	—	—	—
TeF_6	g	241.590	—	−315.	—	—	—	—
$TeCl_4$	c	269.412	—	−78.0	—	—	—	33.1
$TeBr_4$	c	447.236	—	−45.5	—	—	—	—
TERBIUM								
Tb	c	158.924	0	0	0	2.250	17.50	6.91
	g	—	93.36	92.9	83.6	1.79	48.63	5.87
Tb^+	g	—	228.3	229.3	—	—	—	—
Tb^{2+}	g	—	494.	496.	—	—	—	—
Tb^{3+}								
std. state, m = 1	aq	—	—	−163.2	−155.8	—	−54.	4.
TbO	g	174.923	−19.	—	—	—	—	—
TbO_2	c	190.923	—	−232.2	—	—	—	—
Tb_2O_3	c	365.846	—	−445.8	—	—	—	27.7
$TbCl_3$	c	265.283	—	−238.3	—	—	—	—
std. state, m = 1	aq	—	—	−283.0	−249.9	—	−14.	−94.
$TbCl_3$								
in ∞ H_2O	aq	—	—	−283.0	—	—	—	—
$TbCl_3 \cdot 6H_2O$	c	373.375	—	−683.4	−583.4	—	96.4	—
$Tb(BrO_3)_3$								
in 39.1 H_2O (satd)	aq	542.646	—	−214.1	—	—	—	—
$Tb(IO_3)_3$	c	683.632	—	−326.	—	—	—	—
TbC_2	g	182.946	212.	211.7	198.7	2.47	64.	10.5
$Tb_2(CO_3)_3$	c	497.876	—	−795.7	—	—	—	—
THALLIUM								
Tl	c	204.37	0	0	0	1.632	15.45	6.29
	g	—	43.701	43.55	35.24	1.481	43.225	4.968
in Hg	liq	—	—	0.076	−0.062	—	15.80	—
Tl^+	g	—	184.553	185.883	—	—	—	—
std. state, m = 1	aq	—	—	1.28	−7.74	—	30.0	—
Tl^{2+}	g	—	655.636	658.447	—	—	—	—
Tl^{3+}	g	—	1343.5	1347.8	—	—	—	—
std. state, m = 1	aq	—	—	47.0	51.3	—	−46.	—

Substance			0 K	298.15 K (25°C)				
Formula and Description	State	Formula weight	ΔHf°_0 kcal/mol	ΔHf°	ΔGf° kcal/mol	$H^\circ_{298} - H^\circ_0$	S°	C°_p cal/deg mol
Tl_2O	c	424.7394	—	−42.7	−35.2	—	30.	—
Tl_2O_3	c	456.7382	—	—	−74.5	—	—	—
Tl_2O_4	c	472.7376	—	—	−83.0	—	—	—
$TlOH$	c	221.3774	—	−57.1	−46.8	—	21.	—
$TlOH$								
std. state, m = 1	aq	—	—	−53.69	−45.33	—	27.4	—
$Tl(OH)_3$	c	255.3922	—	—	−121.2	—	—	—
TlF	c	223.3684	—	−77.6	—	—	—	—
std. state, m = 1	aq	—	—	−78.22	−74.38	—	26.7	—
$TlCl$	c	239.8230	−49.091	−48.79	−44.20	3.030	26.59	12.17
std. state, m = 1	aq	—	—	−38.67	−39.11	—	43.5	—
$TlCl_3$	c	310.7290	—	−75.3	—	—	—	—
$TlCl_3$								
std. state, m = 1,								
$Tl^{3+} + 3Cl^-$	aq	—	—	−72.9	−42.8	—	−5.5	—
$TlClO_3$								
std. state, m = 1	aq	287.8212	—	−22.4	−8.5	—	68.8	
$TlBr$	c	284.2790	—	−41.4	−40.00	—	28.8	—
std. state, m = 1	aq	—	—	−27.77	−32.59	—	49.7	—
$TlBr_3$	aq	444.0970	—	−59.7	—	—	—	—
std. state, m = 1	aq	—	—	−40.2	−23.2	—	13.	—
$TlBrO_3$	c	332.2772	—	−32.6	−12.70	—	40.3	—
std. state, m = 1	aq	—	—	−18.7	−7.3	—	69.0	—
TlI	c	331.2744	—	−29.6	−29.97	—	30.5	—
std. state, m = 1	aq	—	—	−11.91	−20.07	—	56.6	—
$TlIO_3$	c	379.2726	—	−63.9	−45.86	—	42.2	—
std. state,m = 1	aq	—	—	−51.6	−38.3	—	58.3	—
Tl_2S	c	440.8040	—	−23.2	−22.4	—	36.	—
Tl_2SO_4	c	504.8016	—	−222.7	−198.49	—	55.1	—
std. state, m = 1	aq	—	—	−214.76	−193.45	—	64.8	—
TlN_3	c	246.3901	—	55.8	70.38	—	35.1	—
$TlNO_3$	c	266.3749	—	−58.30	−36.44	—	38.4	23.78
undissoc.; std. state, m = 1	aq	—	—	−48.93	−34.85	—	64.5	—
Tl_2CO_3	c	468.7493	—	−167.3	−146.9	—	37.1	—
$TlCNS$	c	262.4518	—	6.8	9.21	—	39.	—
undissoc.; std. state, m = 1	aq	—	—	16.59	13.32	—	58.2	—
THORIUM								
Th	c	232.0381	0	0	0	1.556	12.76	6.53
	g	232.0381	143.08	143.0	133.26	1.481	45.420	4.97
Th^+	g	232.0381	—	283.1	—	—	—	—
Th^{2+}	g	232.0381	—	550.6	—	—	—	—
Th^{3+}	g	232.0381	—	1012.	—	—	—	—
Th^{4+}	g	232.0381	—	1677.	—	—	—	—
	a	232.0381	—	−183.8	−168.5	—	−101.0	—
ThO	g	248.0375	−5.5	−6.0	−12.0	2.109	57.350	7.47
ThO_2	c	264.0369	−292.01	−293.12	−279.35	2.523	15.590	14.76
ThH_2	c	234.0541	−31.4	−33.4	−23.9	1.616	12.120	8.77
ThF	g	251.0365	—	—	—	2.23	61.50	8.28
ThF^{3+}	a	251.0365	—	−264.5	−246.1	—	−71.7	—
ThF_4	c	308.0317	−499.24	−499.90	−477.30	5.114	33.950	26.420
	g	308.0317	−417.1	−418.0	−409.7	4.90	81.7	22.2
$ThCl$	g	267.4911	—	—	—	2.35	64.3	8.71
$ThCl_2$	g	302.9441	—	—	—	3.36	75.8	13.2
$ThCl_4$	c	373.8501	—	−283.6	−261.6	—	45.5	—
$ThCl_4 \cdot 2H_2O$	c	409.8809	—	−436.0	—	—	—	—
$ThCl_4 \cdot 4H_2O$	c	445.9117	—	−587.8	—	—	—	—
$ThCl_4 \cdot 7H_2O$	c	499.9579	—	−804.4	—	—	—	—
$ThOCl_2$	c	318.9435	—	−294.5	−276.3	—	29.5	—
$ThBr_4$	c	551.6741	—	−230.7	−221.6	—	55.	—
	g	551.6741	−175.26	−182.3	−187.5	6.23	103.	25.1
$ThBr_4 \cdot 7H_2O$	c	677.7819	—	−757.2	—	—	—	—
$ThBr_2 \cdot 10H_2O$	c	731.8281	—	−975.1	—	—	—	—
$ThOBr_2$	c	407.8555	—	−283.	—	—	—	—
ThI	g	358.9425	—	—	—	2.49	68.9	8.98
ThI_2	g	485.8469	—	—	—	3.69	85.0	13.7
ThI_3	g	612.7513	—	—	—	5.12	102.7	19.6
ThI_4	c	739.6557	—	−158.9	−156.6	—	61.	—
	g	739.6557	−108.89	−110.1	−123.1	6.65	112.	25.4
ThS	c	264.1021	—	−94.5	−93.4	—	16.68	—
ThS_2	c	296.1661	—	−149.7	−148.2	—	23.0	—
Th_2S_3	c	560.2682	—	−259.0	−257.4	—	43.	—
ThN	c	246.0448	−92.93	−93.5	−86.9	2.020	13.40	10.8
	g	246.0448	—	118.	—	—	—	—
Th_3N_4	c	752.1411	—	−314.3	−289.9	—	48.	—
$Th(NO_3)_4$	c	480.0577	—	−344.5	—	—	—	—
ThP	c	263.0119	—	−83.2	−81.54	—	17.0	—
	g	263.0119	128.5	128.	116.	2.29	63.9	8.5
Th_3P_4	c	820.0095	—	−273.0	−266.0	—	53.0	—
ThC	c	244.0493	—	−29.60	—	—	—	—
$ThC_{1.94}$	c	255.3398	−35.4	−35.	−35.3	2.447	16.37	13.55
ThC_2	g	256.0605	—	173.	—	—	—	—
THULIUM								
Tm	c	168.934	0	0	0	1.767	17.69	6.46
	g	—	55.786	55.5	47.2	1.481	45.412	4.968
Tm^+	g	—	198.3	199.5	—	—	—	—
Tm^{2+}	g	—	476.	479.	—	—	—	—
Tm^{3+}	g	—	1023.	1027.	—	—	—	—
std. state, m = 1	aq	—	—	−166.8	−158.2	—	−58.	6.
TmO	g	184.933	−19.	—	—	—	—	—
Tm_2O_3	c	385.866	−449.75	−451.4	−428.9	4.99	33.4	27.9
$TmCl_3$	c	275.293	—	−235.8	—	—	—	—
	g	—	—	−166.	—	—	—	—
std. state, m = 1	aq	—	—	−286.6	−252.3	—	−18.	−92.
$TmCl_3$								
in ∞ H_2O	aq	—	—	−286.6	—	—	—	—
$TmCl_3 \cdot 6H_2O$	c	383.385	—	—	—	—	95.5	—
$TmOCl$	c	220.386	—	−236.0	—	—	—	—
TmI_3	c	549.647	—	−143.8	—	—	—	—
$Tm(IO_3)_3$	c	693.642	—	−328.	—	—	—	—
TmC_2	c	192.956	—	−22.	—	—	—	—

Substance			0 K	298.15 K (25°C)				
Formula and Description	State	Formula weight	ΔHf°_0 kcal/mol	ΔHf°	ΔGf° kcal/mol	$H^\circ_{298} - H^\circ_0$	S° cal/deg mol	C°_p
TIN								
Sn								
I, white	c	118.69	0	0	0	1.505	12.32	6.45
II, gray	c	—	−0.371	−0.50	0.03	1.376	10.55	6.16
	g	—	72.18	72.2	63.9	1.485	40.243	5.081
Sn^{+}	g	—	241.54	243.04	—	—	—	—
Sn^{2+}	g	—	578.97	581.95	—	—	—	—
in aq HCl, std. state, m = 1	aq	—	—	−2.1	−6.5	—	−4.	—
Sn^{3+}	g	—	1282.9	1287.4	—	—	—	—
SnO	c	134.689	—	−68.3	−61.4	—	13.5	10.59
	g	—	—	—	—	2.117	55.45	7.55
SnO_2	c	150.689	—	−138.8	−124.2	—	12.5	12.57
SnH_4	g	122.722	41.78	38.9	45.0	2.669	54.39	11.70
$Sn(OH)_2$								
precipitated	c	152.705	—	−134.1	−117.5	—	37.	—
$Sn(OH)_4$								
precipitated	c	186.719	—	−265.3	—	—	—	—
$SnCl_2$	c	189.596	—	−77.7	—	—	—	—
un-ionized, in aq HCl; std. state, m = 1	aq	—	—	−78.8	−71.6	—	41.	—
$SnCl_2 \cdot 2H_2O$	c	225.627	—	−220.2	—	—	—	—
$SnCl_4$	liq	260.502	—	−122.2	−105.2	—	61.8	39.5
in aq HCl; std. state, m = 1	aq	—	—	−152.5	−124.9	—	26.	—
$SnBr_2$	c	278.508	—	−58.2	—	—	—	—
	aq	—	—	−56.8	—	—	—	—
$SnBr_4$	c	438.326	—	−90.2	−83.7	—	63.2	—
SnI_2	c	372.499	—	−34.3	—	—	—	—
SnI_4	c	626.308	—	—	—	—	—	20.3
SnS	c	150.754	—	−24.	−23.5	—	18.4	11.77
SnS_2	c	182.818	—	—	—	—	20.9	16.76
TITANIUM								
Ti	c	47.90	0	0	0	1.149	7.32	5.98
	g	—	111.65	112.3	101.6	1.802	43.066	5.839
Ti^{+}	g	—	268.9	271.1	—	—	—	—
Ti^{2+}	g	—	582.1	585.7	—	—	—	—
Ti^{3+}	g	—	1216.0	1221.1	—	—	—	—
Ti^{4+}	g	—	2213.8	2220.4	—	—	—	—
TiO								
α	c	63.899	−123.49	−124.2	−118.3	1.48	8.31	12.
TiO_2								
anatase	c	79.899	−223.44	−224.6	−211.4	2.062	11.93	13.26
Brookite	c	—	—	−225.1	—	—	—	—
rutile	c	—	−224.6	−225.8	−212.6	2.065	12.03	13.15
	amorp	—	—	−210.	—	—	—	—
Ti_2O_3	c	143.798	−361.52	−363.5	−342.8	3.431	18.83	23.27
TiH_2	c	49.916	−26.61	−28.6	−19.2	1.18	7.1	7.2
$TiCl_2$	c	118.806	−122.64	−122.8	−111.0	3.18	20.9	16.69
$TiCl_3$	c	154.259	−172.86	−172.3	−156.2	5.00	33.4	23.22
$TiCl_4$	c	189.712	−195.75	—	—	—	—	—
	liq	—	—	−192.2	−176.2	—	60.31	34.70
$TiBr_2$	c	207.718	—	−96.	—	—	—	—
$TiBr_3$	c	287.627	−126.65	−131.1	−125.2	5.49	42.2	24.31
$TiBr_4$	c	367.536	−141.36	−147.4	−140.9	6.825	58.2	31.43
	g	—	−124.14	−131.3	−135.8	5.71	95.2	24.1
TiI_2	c	301.709	—	−63.	—	—	—	—
TiI_4	c	555.518	—	−89.8	−88.8	—	59.6	30.03
	g	—	—	−66.4	—	—	—	—
TiS	c	79.964	—	−57.	—	—	—	—
TiS_2	c	112.028	—	—	—	—	18.73	16.23
TiN	c	61.907	−79.93	−80.8	−74.0	1.311	7.23	8.86
TiP	c	78.874	—	−67.6	—	—	—	—
TiC	c	59.911	−43.80	−44.1	−43.2	1.101	5.79	8.04
TiB_2	c	69.522	−77.0	−77.4	−76.4	1.333	6.81	10.58
TUNGSTEN								
W	c	183.85	0	0	0	1.190	7.80	5.80
	g	—	202.70	203.0	192.9	1.486	41.549	5.093
W^{+}	g	—	386.8	388.6	—	—	—	—
WO	g	199.849	—	108.	—	—	—	—
WO_2	c	215.849	−139.762	−140.94	−127.61	2.087	12.08	13.41
WO_3	c	231.848	−200.100	−201.45	−182.62	2.952	18.14	17.63
WO_4^{2-}								
std. state, m = 1	aq	247.848	—	−257.1	—	—	—	—
H_2WO_4	c	249.864	—	−270.5	—	—	—	—
WS_2	c	247.978	—	−50.	—	—	—	—
WC	c	195.861	—	−9.69	—	—	—	—
W_2C	c	379.711	—	−6.3	—	—	—	—
$W(CO)_6$	c	351.913	—	−227.9	—	—	—	—
Fe_7W_6	c	1494.029	—	—	—	—	—	77.7
$FeWO_4$	c	303.695	—	−276.	−252.	—	31.5	27.39
$MnWO_4$	c	302.786	—	−311.9	—	—	—	29.7
URANIUM								
U	c	238.0290	0	0	0	1.521	12.00	6.612
	g	238.0290	127.97	128.0	117.4	1.553	47.72	5.663
U^{+}	g	238.0290	270.81	272.32	—	—	—	—
U^{2+}	g	238.0290	520.	—	—	—	—	—
U^{3+}	g	238.0290	960.	—	—	—	—	—
	g	238.0290	—	−116.9	−113.6	—	−46.	—
U^{4+}	a	238.0290	—	−141.3	−126.9	—	−98.	—
UO	g	254.0284	—	5.	—	—	—	—
UO_2	c	270.0278	−258.40	−259.3	−246.6	2.696	18.41	15.20
	g	270.0278	−110.85	−111.3	−112.7	3.15	65.6	12.28
UO_2	a	270.0278	—	—	−230.1	—	—	—
UO_2^{2+}	a	270.0278	—	−243.7	−227.9	—	−23.3	—
UO_3								
γ, orthorhombic	c	286.0272	−291.35	−292.5	−273.9	3.486	22.97	19.52
ε form, triclinic, red	c2	286.0272	—	−291.0	—	—	—	—
	c3	286.0272	−289.96	−291.0	−272.6	3.596	23.76	19.57
α, orthorhombic, prev. described as hexagonal	c4	286.0272	−290.526	−291.65	−273.02	3.509	23.02	19.44

SELECTED VALUES OF CHEMICAL THERMODYNAMIC PROPERTIES (Continued)

Substance			0 K	298.15 K (25°C)				
Formula and Description	State	Formula weight	ΔHf_0° kcal/mol	ΔHf°	ΔGf° kcal/mol	$H_{298}^\circ - H_0^\circ$	S° cal/deg mol	C_p°
β, orthorhombic, orange-red;								
δ, cubic, dark red	c5	286.0272	—	−291.5	—	—	—	—
amorphous, orange	am	286.0272	—	−288.8	—	—	—	—
	g	286.0272	−195.	—	—	—	—	—
$UO_3 \cdot H_2O$								
β, orthorhombic	c	304.0426	—	−366.6	−333.4	—	30.	—
ε form, monoclinic	c2	304.0426	—	−366.0	—	—	—	—
α, transition to β 278.3k	c3	304.0426	—	−365.2	—	—	—	—
U_3O_7								
β, tetragonal	c	826.0828	−816.38	−819.1	−775.1	9.108	59.88	51.51
α, tetragonal	c2	826.0828	—	—	—	9.009	59.19	51.09
U_3O_8								
α, orthorhombic	c	842.0822	−851.75	−854.4	−805.4	10.216	67.54	56.97
UF	g	257.0274	−5.7	−6.	−13.	2.28	60	9.04
UF_2	g	276.0258	−134.6	−135.	−138.	3.28	71.	16.0
UF_3	c	295.0242	−360.3	−360.6	−344.2	4.392	29.50	22.73
UF_4								
monoclinic	c	314.0226	−458.75	−459.1	−437.4	5.390	36.25	27.73
UF_4	g	314.0226	−382.72	−383.7	−377.5	4.76	88.	21.8
	g	333.0210	−462.8	−464.	−452.	5.6	93.	26.2
UF_6	c	352.0194	−524.80	−525.1	−494.4	7.545	54.4	39.86
UOF_2	c	292.0252	—	−358.3	−341.5	—	28.5	—
UCl_3	c	344.3880	−207.61	−207.1	−191.	5.318	38.0	24.5
UCl_4	c	379.8410	−243.97	−243.6	−222.3	6.28	47.1	29.16
UCl_6	c	450.7470	−261.8	−261.	−230.	8.90	68.3	42.0
$UOCl_2$	c	324.9344	−254.83	−255.0	−238.1	4.586	33.06	22.72
UO_2Cl_2	c	340.9338	−296.67	−297.3	−274.0	5.157	35.98	25.78
UBr_3	c	477.7560	—	−167.1	−161.0	—	46.	26.0
UBr_4	c	557.6650	—	−191.8	−183.5	—	57.0	30.6
$UOBr_2$	c	413.8464	−229.27	−232.7	−222.2	4.989	37.66	23.42
UO_2Br_2	c	429.8458	—	−271.9	−254.9	—	40.5	—
UI_3	c	618.7422	—	−110.1	−109.9	—	53.	26.8
UI_4	c	745.6466	—	−122.4	−121.1	—	63.	32.1
US_2								
β	c	302.1570	−126.1	−126.	−125.8	3.698	26.39	17.84
UO_2SO_4								
β	c	366.0894	—	−441.0	−402.4	—	37.0	34.7
$U(SO_4)_2$	c	430.1522	—	−554.	—	—	—	—
UN	c	252.0357	−69.12	−69.5	−63.5	2.173	14.92	11.37
$UO_2(NO_3)_2$	c	394.0376	—	−322.5	−264.1	—	58.	—
UP	c	269.0028	−63.8	−64.	−63.	2.58	18.7	11.9
UP_2	c	299.9766	−72.6	−73.	−71.	3.679	24.3	19.12
U_3P_4	c	837.9822	−199.2	−200.	−196.	8.87	61.8	41.8
UC	c	250.0402	−23.90	−23.5	−23.7	2.176	14.15	11.98
U_2C_3	c	512.0916	−44.43	−43.4	−44.8	4.829	32.93	25.66
UO_2CO_3	c	330.0372	—	−404.2	−373.5	—	33.	—
USi	c	266.1150	—	−19.2	—	—	—	—
USi_2	c	294.2010	—	−31.2	—	—	—	—
USi_3	c	322.2870	—	−31.6	—	—	—	—
VANADIUM								
V	c	50.942	0	0	0	1.109	6.91	5.95
	g	—	122.12	122.90	180.32	1.8898	43.544	6.217
V^+	g	—	277.550	279.811	—	—	—	—
V^{2+}	g	—	615.39	619.13	—	—	—	—
V^{3+}	g	—	1292.69	1297.91	—	—	—	—
V^{4+}	g	—	2369.8	2376.5	—	—	—	—
VO	c	66.9414	—	−103.2	−96.6	—	9.3	10.86
VO^{2+}								
std. state	aq	—	—	−116.3	−106.7	—	−32.0	—
VO_2	g	82.9408	—	−57.	—	—	—	—
V_2O_3	c	149.8822	—	−291.3	−272.3	—	23.5	24.67
V_2O_4								
α	c	165.8816	—	−341.1	−315.1	—	24.5	27.96
V_2O_5	c	181.8810	—	−370.6	−339.3	—	31.3	30.51
V_3O_5	c	232.8230	—	−462.	−434.	—	39.	—
V_4O_7	c	315.7638	—	−631.	−587.	—	52.	—
V_6O_{13}	c	513.6442	—	−1062.	—	—	—	—
HVO_4^{2-}								
std. state	aq	115.9476	—	−277.0	−233.0	—	4.	—
$H_2VO_4^-$								
std. state	aq	116.9555	—	−280.6	−244.0	—	29.	—
$HV_2O_7^{3-}$								
std. state	aq	214.8878	—	—	−428.4	—	—	—
$H_3V_2O_7^-$								
std. state	aq	216.9037	—	—	−445.5	—	—	—
VF_3	c	107.9372	—	—	—	—	23.18	21.62
VF_4	c	126.9356	—	−335.4	—	—	—	—
VF_5	liq	145.9340	—	−353.8	−328.2	—	42.0	—
VCl_2	c	121.848	—	−108.	−97.	—	23.2	17.26
VCl_3	c	157.301	—	−138.8	−122.2	—	31.3	22.27
VCl_4	liq	192.754	—	−136.1	−120.4	—	61.	—
VOCl	c	102.3944	—	−145.	−133.	—	18.	—
VBr_2	c	210.760	—	−87.3	—	—	—	—
VBr_3	c	290.669	—	−103.6	—	—	—	—
VI_2	c	304.7508	—	−60.1	—	—	—	—
VI_3	c	431.6552	—	−64.7	—	—	—	—
VI_4	g	558.5596	—	−29.3	—	—	—	—
V_2S_3	c	198.076	—	−227.	—	—	—	—
VN	c	64.9487	—	−51.9	−45.7	—	8.91	9.08
XENON								
Xeg	g	131.30	0	0	0	1.481	40.5290	4.9679
std. state, m = 1	aq	—	—	−4.2	3.2	—	15.7	—
Xe^+	g	—	279.72	281.20	—	—	—	—
Xe^{2+}	g	—	768.8	771.8	—	—	—	—
Xe^{3+}	g	—	1509.6	1514.1	—	—	—	—
YTTERBIUM								
Yb	c	173.04	0	0	0	1.604	14.31	6.39
	g	—	36.479	36.4	28.3	1.481	41.352	4.968
Yb^+	g	—	180.70	182.05	—	—	—	—
Yb^{2+}	g	—	462.	464.8	—	—	—	—
std. state, m = 1	aq	—	—	—	−126.	—	—	—

Substance			0 K	298.15 K (25°C)				
Formula and Description	State	Formula weight	ΔHf°_0 kcal/mol	ΔHf°	ΔGf° kcal/mol	$H^\circ_{298} - H^\circ_0$	S° cal/deg mol	C°_p
Yb^{3+}	g	—	1043.	1047.3	—	—	—	—
std. state, m = 1	aq	—	—	−161.2	−153.9	—	−57.	6.
Yb_2O_3	c	394.078	−432.08	−433.7	−412.7	4.69	31.8	27.57
YbH	g	174.048	—	—	—	2.083	52.66	7.178
$YbCl_2$	c	243.946	—	−191.1	—	—	—	—
$YbCl_3$	c	279.399	—	−229.4	—	—	—	—
std. state, m = 1	aq	—	—	−281.1	−248.0	—	−17.	−92.
$YbCl_3$								
in ∞ H_2O	aq	—	—	−281.1	—	—	—	—
$YbCl_3 \cdot 6H_2O$	c	387.491	—	−680.2	−580.6	—	94.6	81.6
YbOCl	c	224.492	—	−229.9	—	—	—	—
$Yb(IO_3)_3$	c	697.748	—	−322.	—	—	—	—
$Yb(NO_3)_3$								
in 200 H_2O	aq	359.055	—	−308.997	—	—	—	—
YbC_2	c	197.062	—	−17.9	−18.5	—	19.	—
YTTRIUM								
Y	c	88.905	0	0	0	1.426	10.62	6.34
	g	—	100.49	100.7	91.1	1.639	42.87	6.18
Y^+	g	—	247.62	—	—	—	—	—
Y^{2+}	g	—	529.88	—	—	—	—	—
Y^{3+}	g	—	1003.1	—	—	—	—	—
std. state, m = 1	aq	—	—	−172.9	−165.8	—	−60.	—
Y^{4+}	g	—	2428.	—	—	—	—	—
YO	g	104.9044	−9.	−9.3	−15.5	2.115	55.88	7.53
Y_2O	g	193.8094	2.	1.0	—	2.9	—	11.7
Y_2O_2	g	209.8088	−126.	−127.4	—	3.5	—	15.8
Y_2O_3	c	225.8082	−453.40	−455.38	−434.19	3.983	23.68	24.50
YH_2	c	90.9209	−51.95	−54.0	−44.3	1.403	9.18	8.24
YH_3	c	91.9289	−61.15	−64.0	−49.9	1.613	10.02	10.36
YF	g	107.9034	−32.7	−33.	−39.	2.163	55.38	7.92
YF_2	g	126.9018	—	—	—	2.9	69.3	11.7
YF_3	c	145.900	—	−410.8	−393.1	—	24.	—
YCl	g	124.358	48.	47.8	41.5	2.286	58.33	8.56
YCl^{2+}								
std. state, m = 1	aq	—	—	−214.0	−198.7	—	−46.	—
YCl_3	c	195.264	—	−239.0	—	—	—	—
$YCl_3 \cdot 6H_2O$	c	303.356	—	−691.3	−592.1	—	92.	—
YBr_2	g	248.723	—	—	—	3.4	80.0	13.2
YI_2	g	342.7138	—	—	—	3.5	84.0	13.5
YI_3	c	469.6182	—	−147.4	—	—	—	—
$Y(IO_3)_3$	c	613.6128	—	—	−271.2	—	—	—
YS	g	120.969	42.	41.7	29.7	2.2	58.	8.2
$Y_2(SO_4)_3$	c	465.995	—	—	−866.8	—	—	138.
YC_2	c	112.9274	—	−26.	−26.	—	13.	—
$Y_2(CO_3)_3$	c	357.8379	—	—	−752.4	—	—	—
ZINC								
Zn	c	65.37	0	0	0	1.350	9.95	6.07
	g	—	31.114	31.245	22.748	1.481	38.450	4.968
Zn^+	g	—	247.740	249.352	—	—	—	—
Zn^{2+}	g	—	662.00	665.09	—	—	—	—
std. state, m = 1	aq	—	—	−36.78	−35.14	—	−26.8	11.
ZnO	c	81.369	—	−83.24	−76.08	—	10.43	9.62
ZnO_2^{2-}								
std. state, m = 1	aq	97.369	—	—	−91.85	—	—	—
$HZnO_2^-$								
std. state, m = 1	aq	98.377	—	—	−109.26	—	—	—
$Zn(OH)_2$								
γ	c	99.385	—	—	−132.38	—	—	—
β	c	—	—	−153.42	−132.31	—	19.4	—
e	c	—	—	−153.74	−132.68	—	19.5	17.3
precipitated	—	—	—	−153.5	—	—	—	—
$Zn(OH)_2$								
std. state, m = 1	aq	—	—	−146.72	−110.33	—	−31.9	−60.
$Zn(OH)_4^{2-}$								
std. state, m = 1	aq	133.399	—	—	−205.23	—	—	—
ZnF_2	c	103.367	−182.06	−182.7	−170.5	2.821	17.61	15.69
$ZnCl_2$	c	136.276	−99.255	−99.20	−88.296	3.598	26.64	17.05
std. state, m = 1	aq	—	—	−116.68	−97.88	—	0.2	−54.
in ∞ H_2O	aq	—	—	−116.68	—	—	—	—
$Zn(ClO_4)_2$								
std. state, m = 1	aq	264.271	—	−98.60	−39.26	—	60.2	—
$Zn(ClO_4)_2 \cdot 6H_2O$	c	372.363	—	−509.89	—	—	—	—
$ZnBr_2$	c	225.188	—	−78.55	−74.60	—	33.1	—
std. state, m = 1	aq	—	—	−94.88	−84.84	—	12.6	−57.
$ZnBr_2 \cdot 2H_2O$	c	261.219	—	−224.0	−191.1	—	47.5	—
ZnI_2	c	319.179	—	−49.72	−49.94	—	38.5	—
std. state, m = 1	aq	—	—	−63.16	−59.80	—	25.2	−57.
$Zn(IO_3)_2$	c	415.175	—	—	−103.68	—	—	—
std. state, m = 1	aq	—	—	−142.6	−96.3	—	29.8	—
ZnS								
wurtzite	c	97.434	—	−46.04	—	—	—	—
sphalerite	c	—	—	−49.23	−48.11	—	13.8	11.0
$ZnSO_4$	c	161.432	—	−234.9	−209.0	—	28.6	—
$ZnSO_4$								
std. state, m = 1	aq	—	—	−254.10	−213.11	—	−22.0	−59.
in ∞ H_2O	aq	—	—	−254.10	—	—	—	—
$ZnSO_4 \cdot H_2O$	c	179.447	—	−311.78	−270.58	—	33.1	—
$ZnSO_4 \cdot 6H_2O$	c	269.524	—	−663.83	−555.64	—	86.9	85.49
$ZnSO_4 \cdot 7H_2O$	c	287.539	—	−735.60	−612.59	—	92.9	91.64
$Zn(N_3)_2$	c	149.410	—	52.	—	—	—	—
undissoc.; std. state, m = 1	aq	—	—	—	129.6	—	—	—
Zn_3N_2	c	224.123	—	−5.4	—	—	—	26.
$Zn(NO_3)_2$	c	189.380	—	−115.6	—	—	—	—
ionized; std. state, m = 1	aq	—	—	−135.90	−88.36	—	43.2	−30.
$Zn(NO_3)_2$								
in ∞ H_2O	aq	—	—	−135.90	—	—	—	—
$Zn(NO_3)_2 \cdot H_2O$	c	207.395	—	−192.4	—	—	—	—
$Zn(NO_3)_2 \cdot 2H_2O$	c	225.410	—	−265.36	—	—	—	—
$Zn(NO_3)_2 \cdot 4H_2O$	c	261.441	—	−406.10	—	—	—	—
$Zn(NO_3)_2 \cdot 6H_2O$	c	279.472	—	−551.30	−423.79	—	109.2	77.2

Substance			0 K	298.15 K (25°C)				
Formula and Description	State	Formula weight	ΔHf°_0 kcal/mol	ΔHf°	ΔGf° kcal/mol	$H^\circ_{298} - H^\circ_0$	S° cal/deg mol	C°_p
Zn_3P_2	c	258.058	—	−113.	—	—	—	—
$Zn(PO_3)_2$	c	223.314	—	−497.9	—	—	—	—
$Zn_2(P_2O_7)$	c	304.683	—	−600.0	—	—	—	—
$Z_3(PO_4)_2$	c	386.053	—	−691.3	—	—	—	—
$ZnCO_3$	c	125.379	—	−194.26	−174.85	—	19.7	19.05
$ZnCO_3 \cdot H_2O$	c	143.395	—	—	−232.0	—	—	—
ZnC_2O_4								
std. state, m = 1	aq	153.390	—	−234.0	−196.2	—	−15.9	—
$Zn(CH_3)_2$	liq	95.440	—	5.6	—	—	—	—
$Zn(C_2H_5)_2$	liq	123.494	—	2.5	—	—	—	—
$Zn(CN)_2$	c	117.406	—	22.9	—	—	—	—
$ZnSiO_3$	c	141.454	—	−301.2	—	—	—	—
Zn_2SiO_4	c	222.824	—	−391.19	−364.06	—	31.4	29.48
ZIRCONIUM								
Zr								
α, hexagonal	c	91.22	0	0	0	1.322	9.32	6.06
	g	—	145.19	145.5	135.4	1.629	43.32	6.37
Zr^+	g	—	302.9	304.7	—	—	—	—
Zr^{2+}	g	—	605.7	609.0	—	—	—	—
Zr^{3+}	g	—	1135.9	1140.6	—	—	—	—
Zr^{4+}	g	—	1927.8	1934.0	—	—	—	—
ZrO	g	107.219	—	15.	—	—	—	—
ZrO_2								
α, monoclinic	c	123.219	−261.734	−263.04	−249.24	2.091	12.04	13.43
ZrO_2								
hydrated ppt	—	—	—	−260.4	—	—	—	—
ZrO_3								
ppt	c	139.218	—	−241.	—	—	—	—
ZrH_2								
zirconium hydride	c	93.236	−38.34	−40.4	−30.8	1.284	8.37	7.40
ZrF_4								
β, monoclinic	c	167.214	−455.44	−456.8	−432.6	4.183	25.00	24.79
ZrCl	c	126.673	—	−63.	—	—	—	—
$ZrCl_2$	c	162.126	—	−120.	—	—	—	—
$ZrCl_3$	c	197.579	—	−179.	—	—	—	—
$ZrCl_4$	c	233.032	−234.60	−234.35	−212.7	5.957	43.4	28.63
$ZrOCl_2$	aq	178.125	—	−280.3	—	—	—	—
$ZrBr_4$	c	410.856	—	−181.8	—	—	—	—
$ZrOBr_2$	aq	267.037	—	−259.9	—	—	—	—
ZrI_4	c	598.838	—	−115.1	—	—	—	—
ZrS_2	c	155.348	—	−135.3	—	—	—	—
$Zr(SO_4)_2$	c	283.343	—	−529.9	—	—	—	41.
$Zr(SO_4)_2 \cdot H_2O$	c	301.359	—	−610.4	—	—	—	—
$Zr(SO_4)_2 \cdot 4H_2O$	c	355.405	—	−825.6	—	—	—	—
ZrN	c	105.227	−86.42	−87.2	−80.4	1.575	9.29	9.66
ZrC	c	103.231	−48.33	−48.5	−47.7	1.401	7.96	9.06
ZrSi	c	119.306	—	−37.	—	—	—	—
$ZrSi_2$	c	147.392	—	−38.	—	—	—	—
Zr_2Si	c	210.526	—	−50.	—	—	—	—
Zr_3Si	c	301.746	—	−52.	—	—	—	—
Zr_3Si_2	c	329.832	—	−92.	—	—	—	—
$ZrSiO_4$	c	183.304	−483.32	−486.0	−458.7	3.562	20.1	23.58
ZrB_2	c	112.842	−77.69	−78.0	−77.0	1.590	8.59	11.53

VALUES OF CHEMICAL THERMODYNAMIC PROPERTIES OF HYDROCARBONS

The values in this table are for the ideal gas state at 298.15 K. The units for ΔHf°, ΔFf°, and Log_{10} Kf are Kcal/g mol. The units for absolute entropy, S°, are cal/°K g mol.

It is frequently possible to calculate values for compounds not listed since the following increments are known for an addition of a methylene group, CH_2, to the following types of compounds:

Normal alkyl cyclohexanes
Normal alkyl benzenes
Normal alkyl cyclopentanes
Normal monoolefins (1-alkenes)
Normal acetylenes (1-alkynes)

For each of the above types of compounds the increments per CH_2 group are

ΔHf°:	-4.926 kcal/g mol
ΔFf°:	-2.048 kcal/g mol
Log_{10} Kf:	-1.5012 kcal/g mol
S°:	-9.183 cal/deg g mol

Relationships to SI units – The symbols cal $mole^{-1}$ deg^{-1} and gibbs/mol are identical and refer to units of calories per degree-mole. These units can be converted to SI units of joules per degree-mole by multiplying the tabulated values by 4.184. Similarly, values in kilocalories per mole can be converted to joules per mole by multiplying with the factor 4184. For further discussions of the SI system and for conversions from other units, see *Pure and Applied Chemistry*, 21, 1, 1970.

Formula	Compound	ΔHf°	ΔFf°	Log_{10} Kf	S°
CH_4	Methane	-17.889	-12.140	8.8985	44.50
C_2H_2	Ethyne (acetylene)	54.194	50.000	-36.6490	47.997
C_2H_4	Ethene (ethylene)	12.496	16.282	-11.9345	52.54
C_2H_6	Ethane	-20.236	- 7.860	5.7613	54.85
C_3H_4	Propadiene (allene)	45.92	48.37	-35.4519	58.30
C_3H_4	Propyne (methyl-acetylene)	44.319	46.313	-33.9469	59.30
C_3H_6	Propene (propylene)	4.879	14.990	-10.9875	63.80
C_3H_8	Propane	-24.820	- 5.614	4.1150	64.51
C_4H_6	1,2-Butadiene	39.55	48.21	-35.3377	70.03
C_4H_6	1,3-Butadiene	26.75	36.43	-26.7004	66.62
C_4H_6	1-Butyne (ethyl acetylene)	39.70	48.52	-35.5616	69.51
C_4H_6	2-Butyne (dimethylacetylene)	35.374	44.725	32.7823	67.71
C_4H_8	1-Butene	0.280	17.217	-12.6199	73.48
C_4H_8	cis-2-Butene	-1.362	16.046	-11.7618	71.90
C_4H_8	trans-2-Butene	-2.405	15.315	-11.2255	70.86
C_4H_8	2-Methylpropane (isobutene)	- 3.343	14.582	10.6888	70.17
C_4H_{10}	*n*-Butane	-29.812	- 3.754	2.7516	74.10
C_4H_{10}	2-Methylpropane (isobutane)	-31.452	- 4.296	3.1489	70.42
C_5H_8	1-Pentyne	34.50	50.17	-36.7712	79.10
C_5H_8	2-Pentyne	30.80	46.41	34.0177	79.30
C_5H_8	3-Methyl-1-butyne	32.60	49.12	-36.0061	76.23
C_5H_8	1,2-Pentadiene	34.80	50.29	-36.861	79.7
C_5H_8	cis-1,3-Pentadiene (cis-piperylene)	18.70	34.88	25.563	77.5
C_5H_8	trans-1,3-Pentadiene (trans-piperylene)	18.60	35.07	-25.707	76.4
C_5H_8	1,4-Pentadiene	25.20	40.69	-29.824	79.7
C_5H_8	2,3-Pentadiene	33.10	49.22	36.074	77.6
C_5H_8	3-Methyl-1,2-butadiene	31.00	47.47	34.657	76.4
C_5H_8	2-Methyl-1,3-butadiene (isoprene)	18.10	34.87	25.560	75.44
C_5H_{10}	1-Pentene	- 5.000	18.787	13.7704	83.08
C_5H_{10}	cis-2-Pentene	- 6.710	17.173	12.5874	82.76
C_5H_{10}	trans-2-Pentene	- 7.590	16.575	-12.1495	81.81
C_5H_{10}	2-Methyl-1-butene	- 8.680	15.509	-11.3680	81.73
C_5H_{10}	3-Methyl-1-butene	- 6.920	17.874	-13.1017	79.70
C_5H_{10}	2-Methyl-2-butene	-10.170	14.267	-10.4572	80.90
C_5H_{10}	Cyclopentane	-18.46	9.23	- 6.7643	70.00
C_5H_{12}	*n*-Pentane	-35.00	- 1.96	1.4366	83.27
C_6H_6	Benzene	19.820	30.989	-22.7143	64.34
C_6H_{10}	1-Hexyne	29.55	52.19	-38.258	88.27
C_6H_{12}	1-Hexene	- 9.96	20.80	-15.2491	92.25
C_6H_{12}	cis-2-Hexene	-11.56	19.18	-14.0549	92.35
C_6H_{12}	trans-2-Hexene	-12.56	18.46	-13.5291	91.40
C_6H_{12}	cis-3-Hexene	-11.56	19.66	-14.4094	90.73
C_6H_{12}	trans-3-Hexene	-12.56	18.86	-13.8262	90.04
C_6H_{12}	2-Methyl-1-pentene	-13.56	17.48	-12.8135	91.32
C_6H_{12}	3-Methyl-1-pentene	-11.02	20.28	-14.8655	90.45
C_6H_{12}	4-Methyl-1-pentene	-11.66	19.90	-14.5865	89.58
C_6H_{12}	2-Methyl-2-pentene	-14.96	16.34	-11.9780	90.45
C_6H_{12}	cis-3-Methyl-2-pentene	-14.32	16.98	-12.4471	90.45
C_6H_{12}	trans-3-Methyl-2-pentene	-14.32	16.74	-12.2697	91.26
C_6H_{12}	cis-4-Methyl-2-pentene	-13.26	18.40	-13.4903	89.23
C_6H_{12}	trans-4-Methyl-2-pentene	-14 26	17.77	-13.0216	88.02
C_6H_{12}	2-Ethyl-1-butene	-12.92	18.51	-13.5690	90.01
C_6H_{12}	2,3-Dimethyl-1-butene	-14.78	17.43	-12.7782	89.39
C_6H_{12}	3,3-Dimethyl-1-butene	-14.25	19.04	-13.9578	83.79
C_6H_{12}	2,3-Dimethyl-2-butene	-15.91	16.52	-12.1073	86.67
C_6H_{12}	Methylcyclopentane	-25.50	8.55	- 6.2649	81.24
C_6H_{12}	Cyclohexane	-29.43	7.59	- 5.5605	71.28
C_6H_{14}	*n*-Hexane	-39.96	0.05	0.037	92.45

Formula	Compound	ΔHf°	ΔFf°	$Log_{10}Kf$	S°
C_7H_8	Methylbenzene (toluene)	11.950	29.228	-21.4236	76.42
C_7H_{12}	1-Heptyne	24.62	54.24	-39.759	97.25
C_7H_{12}	1-Heptene	-14.85	22.84	-16.742	101.43
C_7H_{14}	Ethylcyclopentane	-30.37	10.59	- 7.7632	90.62
C_7H_{14}	1,1-Dimethylcyclopentane	-33.05	9.33	- 6.8372	85.87
C_7H_{14}	1,cis-2-Dimethylcyclopentane	-30.96	10.93	- 8.0107	87.51
C_7H_{14}	1,trans-2-Dimethylcyclopentane	-32.67	9.17	- 6.7224	87.67
C_7H_{14}	1,cis-3-Dimethylcyclopentane	-31.93	9.91	- 7.2648	87.67
C_7H_{14}	1,trans-3-Dimethylcyclopentane	-32.47	9.37	- 6.8690	87.67
C_7H_{14}	Methylcyclohexane	-36.99	6.52	- 4.7819	82.06
C_7H_{16}	*n*-Heptane	-44.89	2.09	- 1.532	101.64
C_8H_8	Ethenylbenzene (styrene)	35.32	51.10	-37.4532	82.48
C_8H_{10}	Ethylbenzene	7.120	31.208	-22.8750	86.15
C_8H_{10}	1,2-Dimethylbenzene (*o*-xylene)	4.540	29.177	-21.3860	84.31
C_8H_{10}	1,3-Dimethylbenzene (*m*-xylene)	4.120	28.405	-20.8202	85.49
C_8H_{10}	1,4-Dimethylbenzene (*p*-xylene)	4.290	28.952	-21.2214	84.23
C_8H_{14}	1-Octyne	19.70	56.29	-41.260	106.63
C_8H_{16}	1-Octene	-19.82	24.89	-18.244	110.61
C_8H_{16}	*n*-Propylcyclopentane	-35.39	12.54	- 9.195	99.80
C_8H_{16}	1,1-Dimethylcyclohexane	-43.26	8.42	- 6.174	87.24
C_8H_{16}	cis-1,2-Dimethylcyclohexane	-41.15	9.85	- 7.225	89.51
C_8H_{16}	trans-1,2-Dimethylcyclohexane	-43.02	8.24	- 6.038	88.65
C_8H_{16}	cis-1,3-Dimethylcyclohexane	-44.16	7.13	- 5.228	88.54
C_8H_{16}	trans-1,3-Dimethylcyclohexane	-42.20	8.68	- 6.363	89.92
C_8H_{16}	cis-1,4-Dimethylcyclohexane	-42.22	9.07	- 6.650	88.54
C_8H_{16}	trans-1,4-Dimethylcyclohexane	-44.12	7.58	- 5.552	87.19
C_8H_{18}	*n*-Octane	-49.82	4.14	- 3.035	110.82
C_8H_{18}	2-Methylheptane	-51.50	3.06	- 2.243	108.81
C_8H_{18}	3-Methylheptane	-50.82	3.29	- 2.412	110.32
C_8H_{18}	4-Methylheptane	-50.69	4.00	- 2.932	108.35
C_8H_{18}	3-Ethylhexane	-50.40	3.95	- 2.895	109.51
C_8H_{18}	2,2-Dimethylhexane	-53.71	2.56	- 1.876	103.06
C_8H_{18}	2,3-Dimethylhexane	-51.13	4.23	- 3.101	106.11
C_8H_{18}	2,4-Dimethylhexane	-52.44	2.80	- 2.052	106.51
C_8H_{18}	2,5-Dimethylhexane	-53.21	2.50	- 1.832	104.93
C_8H_{18}	3,3-Dimethylhexane	-52.61	3.17	- 2.324	104.70
C_8H_{18}	3,4-Dimethylhexane	-50.91	4.14	- 3.035	107.15
C_8H_{18}	2-Methyl-3-ethylpentane	-50.48	5.08	- 3.724	105.43
C_8H_{18}	3-Methyl-3-ethylpentane	-51.38	4.76	- 3.489	103.48
C_8H_{18}	2,2,3-Trimethylpentane	-52.61	4.09	- 2.998	101.62
C_8H_{18}	2,2,4-Trimethylpentane	-53.57	3.13	- 2.294	101.62
C_8H_{18}	2,3,3-Trimethylpentane	-51.73	4.52	- 3.313	103.14
C_8H_{18}	2,3,4-Trimethylpentane	-51.97	4.32	- 3.167	102.99
C_8H_{18}	2,2,3,3-Tetramethylbutane	-53.99	4.88	- 3.577	94.34
C_9H_{10}	Isopropenylbenzene (α-methylstyrene; 2-phenyl-1-propene)	27.00	49.84	-36.531	91.70
C_9H_{10}	1-Methyl-2-ethenylbenzene (*o*-Methylstyrene)	28.30	51.14	-37.484	91.70
C_9H_{10}	1-Methyl-3-ethenylbenzene (*m*-methylstyrene)	27.60	50.02	-36.665	93.1
C_9H_{10}	1-Methyl-4-ethenylbenzene (*p*-methylstyrene)	27.40	50.24	-36.825	91.7
C_9H_{12}	*n*-Propylbenzene	1.870	32.810	-24.049	95.74
C_9H_{12}	Isopropylbenzene (Cumene)	0.940	32.738	-23.996	92.87
C_9H_{12}	1,3,5-Trimethylbenzene (Mesitylene)	- 3.840	28.172	-20.6497	92.15
C_9H_{16}	1-Nonyne	14.77	58.34	-42.761	115.82
C_9H_{18}	1-Nonene	-24.74	26.94	-19.747	119.80
C_9H_{18}	*n*-Butylcyclopentane	-40.22	14.69	-10.768	108.99
C_9H_{20}	*n*-Nonane	-54.74	6.18	- 4.536	120.00
$C_{10}H_{14}$	*n*-Butylbenzene	- 3.30	34.62	-25.374	104.91
$C_{10}H_{18}$	1-Decyne	9.85	60.39	-44.262	125.00
$C_{10}H_{10}$	1-Decene	-29.67	28.99	-21.249	128.98
$C_{10}H_{22}$	*n*-Decane	-59.67	8.23	6.037	129.19
$C_{11}H_{22}$	1-Undecene	-34.60	31.03	-22.745	138.16
$C_{11}H_{24}$	*n*-Undecane	-64.60	10.28	- 7.539	138.37
$C_{12}H_{24}$	1-Dodecene	-39.52	33.08	-24.297	147.34

From Rossini, F. D., Pitzer, K. S., Arnett, R. L., Braun, R. M., and Pimentel, G. C., *Selected Values of Physical and Thermodynamic Properties of Hydrocarbons and Related Compounds,* Carnegie Press, Pittsburgh, 1953.

KEY VALUES FOR THERMODYNAMICS

The following table was prepared from data published in CODATA Bulletin 28 (April, 1978) which is a report of the CODATA Task Group on Key Values of Thermodynamics. A bibliography, references citing methods for calculations and measurements, bases for selection of data and limitations of the data are presented in Bulletin 28. One may contact CODATA by writing to CODATA Secretariat, 51 Boulevard de Montmorency, 75016 Paris, France.

The recommended values were derived from a reconsideration of all the tentative data published in CODATA Special Report 4, March 1977 ("Tentative set of key values for thermodynamics. Part VI"), which took place at a meeting of the Task Group held in Lund, Sweden, in August 1977. It was decided that values of $\Delta_f H°$ (298.15 K) for NO^-_3 (aq) and all values for Cd(g), Ge(g) and U(g) should remain tentative for the present, that P_4O_{10}(cr) and PbO(cr, yellow) should be dropped from the program, and that the remaining values given in CODATA Special Report 4 should advance to recommended status, albeit with some minor numerical adjustments.

In the following tables, the species are listed in the "Standard Order of Arrangement", as used in most modern compendia of thermodynamic data. The reference state for each element at 298.15 K is the thermodynamically stable standard state except for phosphorus, for which the "white" crystal modification has been selected, as it is the most reproducible; there is a subtlety concerning the reference state for tin, where the "white" form (thermodynamcally stable at 298.15 K) is taken as the reference state at all temperatures down to zero, even though that form is known to be metastable below 286 K.

The usual definitions of standard states have been adopted. For crystalline solids (cr) and liquids (l) the standard state is that of the pure substance (in a stated crystallographic modification, where appropriate) under a pressure of 101 325 Pa. For gases (g), the standard state is that of the ideal gas at a pressure of 101 325 Pa. For species in aqueous solution (aq), the standard state is the hypothetical ideal solution at unit activity (molality scale); the properties of ideal aqueous ionic solutions are taken equal to the sum of the properties of the individual ions. It should be noted that values of H° (298.15 K) − H° (0) for gases relate to the hypothetical ideal-gas state at zero temperature, while values of H° (298.15 K) − H° (0) for both liquids and crystalline solids relate to crystalline solids at zero temperature.

Relative atomic masses were taken from the recommendations of the IUPAC Commission on Atomic Weights, 1970. Data in the tables relate to the natural mixture of isotopic species; nuclear spin contributions have been ignored. The joule table is the primary table; values in the calorie table were derived from the corresponding values in joules by dividing by 4.184, with retention of sufficient decimal places to assure accurate reconversion.

For consistency with earlier work, the following values of the fundamental constants were employed in the calculations: gas constant, R = (8.314 33 ± 0.000 80) $J \cdot K^{-1} \cdot mol^{-1}$; Faraday constant, F = (96 487.0 ± 1.0) $J \cdot V^{-1} \cdot mol^{-1}$; constant relating wave number and energy, $N_A hc$ = (0.119 625 6 ± 0.000 002 6) $J \cdot m \cdot mol^{-1}$. These values differ very slightly from those recommended by CODATA in 1973 (CODATA Bulletin 11, December 1973). However, adoption of the 1973 values of the fundamental constants would change the values in the present tables by far less than their assigned uncertainties, so the thermodynamic data where reported may be said to be consistent with the 1973 set of fundamental constants.

CODATA RECOMMENDED KEY VALUES FOR THERMODYNAMICS, 1977
(To convert joules to calories multiply by 0.239006)

Substance	State	$\Delta_f H°$ (298.15 K) $kJ \cdot mol^{-1}$	S° (298.15 K) $J \cdot K^{-1} \cdot mol^{-1}$	H° (298.15 K) − H° (0) $kJ \cdot mol^{-1}$
O	g	249.17 ± 0.10	160.946 ± 0.020	6.728 ± 0.003
O_2	g	0	205.037 ± 0.033	8.682 ± 0.004
H	g	217.997 ± 0.006	114.604 ± 0.015	6.197 ± 0.002
H^+	aq	0	0	—
H_2	g	0	130.570 ± 0.033	8.468 ± 0.003
OH^-	aq	−230.025 ± 0.045	−10.71 ± 0.20	—
H_2O	l	−285.830 ± 0.042	69.950 ± 0.080	13.293 ± 0.020
H_2O	g	−241.814 ± 0.042	188.724 ± 0.040	9.908 ± 0.008
He	g	0	126.039 ± 0.012	6.197 ± 0.002
Ne	g	0	146.214 ± 0.016	6.197 ± 0.002
Ar	g	0	154.732 ± 0.020	6.197 ± 0.002
Kr	g	0	163.971 ± 0.020	6.197 ± 0.002
Xe	g	0	169.573 ± 0.020	6.197 ± 0.002
F	g	79.39 ± 0.30	158.640 ± 0.020	6.518 ± 0.004
F^-	aq	−335.35 ± 0.65	−13.18 ± 0.54	—
F_2	g	0	202.685 ± 0.040	8.825 ± 0.004
HF	g	−273.30 ± 0.70	173.655 ± 0.035	8.599 ± 0.004
Cl	g	121.302 ± 0.008	165.076 ± 0.020	6.272 ± 0.003
Cl^-	aq	−167.080 ± 0.088	56.73 ± 0.16	—
Cl_2	g	0	222.965 ± 0.040	9.180 ± 0.008
HCl	g	−92.31 ± 0.13	186.786 ± 0.033	8.640 ± 0.004
Br	g	111.86 ± 0.12	174.904 ± 0.020	6.197 ± 0.020
Br^-	aq	−121.50 ± 0.15	82.84 ± 0.20	—
Br_2	l	0	152.210 ± 0.040	24.52 ± 0.13
Br_2	g	30.91 ± 0.11	245.350 ± 0.054	9.724 ± 0.012
HBr	g	−36.38 ± 0.17	198.585 ± 0.033	8.648 ± 0.004
I	g	106.762 ± 0.040	180.673 ± 0.020	6.197 ± 0.002
I^-	aq	−56.90 ± 0.84	106.70 ± 0.20	—
I_2	cr	0	116.139 ± 0.080	13.196 ± 0.040
I_2	g	62.421 ± 0.080	260.567 ± 0.063	10.117 ± 0.021
HI	g	26.36 ± 0.80	206.480 ± 0.040	8.657 ± 0.006
S	cr, rhombic	0	32.054 ± 0.050	4.412 ± 0.060
S	g	276.98 ± 0.25	167.715 ± 0.035	6.657 ± 0.004
S_2	g	128.49 ± 0.30	228.055 ± 0.050	9.131 ± 0.008
SO_2	g	−296.81 ± 0	248.11 ± 0.06	10.548 ± 0.013
SO_4^{-2}	aq	−909.60 ± 0.40	18.83 ± 0.50	—
N	g	472.68 ± 0.40	153.189 ± 0.020	6.197 ± 0.002
N_2	g	0	191.502 ± 0.025	8.669 ± 0.003
NO_3^-	aq	—	146.94 ± 0.85	—
NH_3	g	−45.94 ± 0.35	192.67 ± 0.08	10.046 ± 0.008

CODATA RECOMMENDED KEY VALUES FOR THERMODYNAMICS, 1977
(To convert joules to calories multiply by 0.239006) (continued)

Substance	State	$\Delta_f H°$ (298.15 K) kJ·mol⁻¹	S° (298.15 K) J·K⁻¹·mol⁻¹	H° (298.15 K) − H° (0) kJ·mol⁻¹
NH_4^+	aq	−133.26 ± 0.25	111.17 ± 0.75	—
P	cr, white	0	41.09 ± 0.25	5.360 ± 0.015
P	g	316.5 ± 1.0	163.085 ± 0.020	6.197 ± 0.002
P_2	g	144.0 ± 2.0	218.01 ± 0.04	8.903 ± 0.0008
P_4	g	58.9 ± 0.3	279.9 ± 0.5	14.10 ± 0.24
C	cr	0	5.74 ± 0.12	1.050 ± 0.020
C	g	716.67 ± 0.44	157.988 ± 0.020	6.535 ± 0.006
CO	g	−110.53 ± 0.17	197.556 ± 0.032	8.673 ± 0.008
CO_2	g	−393.51 ± 0.13	213.677 ± 0.040	9.364 ± 0.008
Si	cr	0	18.81 ± 0.08	3.217 ± 0.008
Si	g	450 ± 8	167.870 ± 0.035	7.550 ± 0.004
SiO_2	cr, α-quartz	−910.7 ± 1.0	41.46 ± 0.20	6.916 ± 0.020
SiF_4	g	−1614.95 ± 0.85	282.65 ± 0.40	15.36 ± 0.05
Ge	Cr, cubic	0	31.09 ± 0.13	4.636 ± 0.015
GeO_2	cr, tetrag	−580.2 ± 1.2	39.71 ± 0.15	7.230 ± 0.020
GeF_4	g	−1190.15 ± 0.50	301.8 ± 1.0	17.30 ± 0.080
Sn	cr, white	0	51.18 ± 0.08	6.323 ± 0.008
Sn	g	301.2 ± 1.7	168.380 ± 0.020	6.215 ± 0.002
Sn^{+2}	aq	−8.9 ± 0.8	−15.8 ± 4.0	—
SnO	cr	−285.93 ± 0.70	57.17 ± 0.30	8.736 ± 0.022
SnO_2	cr	−580.78 ± 0.40	52.3 ± 1.2	8.76 ± 0.08
Pb	cr	0	64.80 ± 0.30	6.870 ± 0.020
Pb	g	195.20 ± 0.80	175.270 ± 0.020	6.197 ± 0.002
Pb^{+2}	aq	0.92 ± 0.25	17.7 ± 0.8	—
$PbSO_4$	cr	−919.94 ± 0.90	148.49 ± 0.40	20.050 ± 0.040
B	cr	0	5.90 ± 0.08	1.222 ± 0.008
B	g	560 ± 12	153.325 ± 0.035	6.315 ± 0.004
B_2O_3	cr	−1273.5 ± 1.4	53.97 ± 0.30	9.301 ± 0.040
BF_3	g	−1135.95 ± 0.80	254.31 ± 0.10	11.650 ± 0.020
Al	cr	0	28.35 ± 0.08	4.565 ± 0.010
Al	g	329.7 ± 4.0	164.440 ± 0.030	6.919 ± 0.004
Al_2O_3	cr, α-corundum	−1675.7 ± 1.3	50.92 ± 0.10	10.016 ± 0.020
AlF_3	cr	−1510.4 ± 1.3	66.5 ± 0.4	11.62 ± 0.04
Zn	cr	0	41.63 ± 0.13	5.657 ± 0.020
Zn	g	130.42 ± 0.20	160.875 ± 0.025	6.197 ± 0.002
Zn^{+2}	aq	−153.39 ± 0.20	−109.6 ± 0.7	—
ZnO	cr	−350.46 ± 0.27	43.64 ± 0.40	6.933 ± 0.040
Cd	cr	0	51.80 ± 0.15	6.247 ± 0.004
Cd^{+2}	aq	−75.88 ± 0.60	−72.8 ± 1.2	—
CdO	cr	−258.1 ± 0.8	54.8 ± 1.7	8.41 ± 0.08
$CdSO_4 \cdot 8/3H_2O$	cr	−1729.55 ± 0.80	229.66 ± 0.40	35.56 ± 0.04
Hg	l	0	75.90 ± 0.12	9.342 ± 0.008
Hg	g	61.38 ± 0.04	174.860 ± 0.020	6.197 ± 0.002
Hg^{+2}	aq	170.16 ± 0.20	−36.32 ± 0.80	—
Hg_2^{+2}	aq	166.82 ± 0.20	65.52 ± 0.80	—
HgO	cr, red	−90.83 ± 0.12	70.25 ± 0.30	9.117 ± 0.025
Hg_2Cl_2	cr	−265.45 ± 0.30	191.6 ± 1.5	23.25 ± 0.20
Hg_2SO_4	cr	−743.41 ± 0.50	200.71 ± 0.20	26.070 ± 0.030
Cu	cr	0	33.15 ± 0.08	5.004 ± 0.008
Cu	g	337.6 ± 1.2	166.285 ± 0.025	6.197 ± 0.002
Cu^{+2}	aq	65.69 ± 0.80	−97.1 ± 1.2	—
$CuSO_4$	cr	−771.1 ± 1.2	109.2 ± 0.4	16.86 ± 0.08
Ag	cr	0	42.55 ± 0.21	5.745 ± 0.020
Ag	g	284.9 ± 0.8	172.883 ± 0.025	6.197 ± 0.002
Ag^+	aq	105.750 ± 0.085	73.38 ± 0.40	—
AgCl	cr	−127.070 ± 0.085	96.23 ± 0.20	12.033 ± 0.040
Th	cr	0	53.39 ± 0.40	6.510 ± 0.020
Th	g	598 ± 6	190.06 ± 0.40	6.197 ± 0.002
ThO_2	cr	−1226.4 ± 3.5	65.23 ± 0.20	10.560 ± 0.020
U	cr	0	50.20 ± 0.20	6.364 ± 0.020
UO_2	cr	−1085.0 ± 1.0	77.03 ± 0.20	11.280 ± 0.020
UO_2^{+2}	aq	−1019.2 ± 2.5	−98.3 ± 4.0	—
UO_3	cr, gamma	−1223.8 ± 2.0	96.11 ± 0.40	14.585 ± 0.080
U_3O_8	cr	−3574.8 ± 2.5	282.55 ± 0.50	42.74 ± 0.10
Be	cr	0	9.50 ± 0.08	1.950 ± 0.020
Be	g	324 ± 5	136.165 ± 0.020	6.197 ± 0.002
BeO	cr	−609.4 ± 2.5	13.77 ± 0.04	2.837 ± 0.008
Mg	cr	0	32.68 ± 0.10	5.000 ± 0.030
Mg	g	147.10 ± 0.80	148.535 ± 0.020	6.197 ± 0.002
MgO	cr	−601.5 ± 0.3	26.95 ± 0.15	5.160 ± 0.020
MgF_2	cr	−1124.2 ± 1.2	57.2 ± 0.4	9.92 ± 0.06

CODATA RECOMMENDED KEY VALUES FOR THERMODYNAMICS, 1977
(To convert joules to calories multiply by 0.239006) (continued)

Substance	State	$\Delta_f H°$ (298.15 K) kJ·mol^{-1}	S° (298.15 K) J·K^{-1}·mol^{-1}	H° (298.15 K) − H° (0) kJ·mol^{-1}
Ca	cr	0	41.6 ± 0.4	5.73 ± 0.04
Ca	g	177.8 ± 0.8	154.775 ± 0.020	6.197 ± 0.002
Ca^{+2}	aq	−543.10 ± 0.80	−56.4 ± 0.4	—
CaO	cr	−635.09 ± 0.90	38.1 ± 0.4	6.75 ± 0.06
Li	cr	0	29.12 ± 0.20	4.632 ± 0.040
Li^{+}	aq	−278.455 ± 0.090	11.30 ± 0.35	—
Na	cr	0	51.30 ± 0.20	6.460 ± 0.020
Na^{+}	aq	−240.300 ± 0.065	58.41 ± 0.20	—
K	cr	0	64.68 ± 0.20	7.088 ± 0.020
K^{+}	aq	−252.17 ± 0.10	101.04 ± 0.25	—
Rb	cr	0	76.78 ± 0.30	7.489 ± 0.020
Rb^{+}	aq	−251.12 ± 0.13	120.46 ± 0.40	—
Cs	cr	0	85.23 ± 0.40	7.711 ± 0.020
Cs^{+}	aq	−258.04 ± 0.13	132.84 ± 0.40	—

HEAT OF DILUTION OF ACIDS

From National Standards Reference Data Systems NSRDS-NBS 2

Vivian B. Parker

ΔH_{diln}, the integral heat of dilution, is the change in enthalpy, per mole of solute, when a solution of concentration m_1 is diluted to a final finite concentration m_2. When the dilution is carried out by addition of an infinite amount of solvent, so the final solution is infinitely dilute, the enthalpy change is the integral heat of dilution to infinite dilution. Since Φ_L, the relative apparent molal enthalpy, is equal to and opposite in sign to this, only Φ_L is referred to here.

Φ_L, cal/mole, at 25°C

n	m	HF	HCl	$HClO_4$	HBr	HI	HNO_3	CH_2O_2	$C_2H_4O_2$
∞	0.00	0	0	0	0	0	0	0	0
500,000	.000111	300	5	5	5	5	5	9	40
100,000	.000555	900	10	10	9	9	11	13	50
50,000	.00111	1,300	16	14	13	12	15	20	53
20,000	.00278	1,800	25	22	22	20	23	23	55
10,000	.00555	2,130	34	30	31	29	31	25	58
7,000	.00793	2,250	40	35	37	34	36	26	59
5,000	.01110	2,360	47	40	44	41	42	26	61
4,000	.01388	2,450	54	43	49	46	46	27	62
3,000	.01850	2,550	60	47	56	52	51	28	62
2,000	.02775	2,700	74	54	68	63	59	28	63
1,500	.03700	2,812	85	58	77	71	65	29	64
1,110	.05000	2,927	97	62	89	81	73	29	65
1,000	.05551	2,969	102	62	92	84	76	29	65
900	.0617	2,989	107	63	97	88	78	30	66
800	.0694	3,015	113	64	102	92	81	31	67
700	.0793	3,037	120	65	108	96	84	32	68
600	.0925	3,057	129	65	115	102	88	32	68
555.1	.1000	3,060	133	65	119	105	89	32	69
500	.1110	3,077	140	65	124	108	92	32	70
400	.1388	3,097	156	64	135	116	97	33	72
300	.1850	3,126	176	61	150	125	103	34	76
277.5	.2000	3,129	182	59	155	128	105	35	79
200	.2775	3,142	212	50	176	140	117	36	82
150	.3700	3,148	242	36	197	154	118	39	88
111.0	.5000	3,156	280	18	225	170	119	42	97
100	.5551	3,160	295	+12	235	176	120	44	101
75	.7401	3,167	343	−14	270	194	121	49	113
55.51	1.0000	3,179	405	−48	314	223	121	54	130
50	1.1101	3,184	431	−61	331	234	121	56	147
40	1.3877	3,192	493	−91	379	260	121	60	155
37.00	1.5000	3,194	518	−103	398	269	121	62	162
30	1.8502	3,200	595	−138	455	301	124	65	183
27.75	2.0000	3,203	627	−149	477	315	126	66	192
25	2.2202	3,208	674	−162	510	336	130	67	204
22.20	2.5000	3,211	732	−173	550	365	139	68	218
20	2.7753	3,214	792	−182	590	396	149	69	233
18.50	3.0000	3,216	838	−187	624	427	159	69	245
15.86	3.500	3,221	946	−196	709	503	189	69	268
15	3.7004	3,227	988	−195	743	536	203	69	277
13.88	4.0000	3,234	1,052	−188	796	588	229	69	291
12.33	4.5000	3,246	1,171	−175	887	676	265	69	313
12	4.6255	3,249	1,190	−170	911	700	277	69	318
11.10	5.0000	3,256	1,271	−150	983	764	313	69	333
10	5.5506	3,265	1,396	−117	1,097	855	368	68	353
9.5	5.8427	3,269	1,462	−97	1,156	920	400	68	363
9.251	6.0000	3,272	1,498	−84	1,196	950	418	67	368
9.0	6.1674	3,274	1,535	−72	1,230	980	437	67	373
8.5	6.5301	3,278	1,618	−40	1,313	1,050	480	66	383
8.0	6.9383	3,282	1,710	+4	1,401	1,115	530	65	392
7.929	7.0000	3,283	1,725	11	1,416	1,130	538	65	394
7.5	7.4008	3,286	1,820	61	1,497	1,210	595	63	402
7.0	7.9295	3.290	1,942	135	1,608	1,325	661	61	411
6.938	8.0000	3,291	1,960	146	1,622	1,340	667	61	412
6.5	8.5394	3,296	2,090	229	1,738	1,450	745	58	420
6.167	9.0000	3,302	2,202	306	1,845	1,570	805	55	426
6.0	9.2510	3,305	2,265	348	1,903	1,630	840	53	429
5.551	10.0000	3,316	2,447	481	2,078	1,820	940	49	436
5.5	10.0920	3,317	2,472	499	2,102	1,850	950	49	437
5.0	11.1012	3,335	2,721	730	2,344	2,100	1,098	43	445
4.5	12.3346	3,362	3,025	1,144	2,655	2,460	1,270	37	453
4.0	13.8765	3,400	3,404	1,574	3,089	2,960	1,495	29	462
3.700	15.0000	3,428	3,680	1,893	3,415	3,350	1,645	26	469
3.5	15.8589	3,450	3,882	2,150	3,668	3,660	1,770	21	473
3.25	17.0788	3,483	4,160	2,460	4,005	4,110	1,920	17	481
3.0	18.5020	3,520	4,460	2,880	4,370	4,630	2,101	13	488
2.775	20.0000	3,557	4,750	3,300	4,760	5,190	2,270	9	496
2.5	22.2024	3,607	5,180	4,000	5,300	6,000	2,520	+4	506
2.0	27.7530	3,712	6,260	5,500	6,650	.....	3,060	−5	528
1.5	37.0040	.....	8,240	.....	8,530	.....	3,770	−13	532
1.0	55.506	.....	10,900	.....	11,670	.....	4,715	+11	518
0.5	111.012	.....	......	.....	......	.....	.....	77	495
0.25	222.02	.....	......	.....	......	.....	.....	129	......

HEATS OF SOLUTION

From National Standards Reference Data Systems NSRDS-NBS 2

Vivian B. Parker

ΔH°_∞ 25°C for uni-univalent electrolytes in H_2O

Substance	State	ΔH°_∞	Substance	State	ΔH°_∞	Substance	State	ΔH°_∞
		cal/mole			*cal/mole*			*cal/mole*
HF	g	−14,700	$LiBr\cdot 2H_2O$	c	−2,250	KCl	c	4,115
HCl	g	−17,888	$LiBrO_3$	c	340	$KClO_3$	c	9,890
$HClO_4$	l	−21,215	LiI	c	−15,130	$KClO_4$	c	12,200
$HClO_4\cdot H_2O$	c	−7,875	$LiI\cdot H_2O$	c	−7,090	KBr	c	4,750
HBr	g	−20,350	$LiI\cdot 2H_2O$	c	−3,530	$KBrO_3$	c	9,830
HI	g	−19,520	$LiI\cdot 3H_2O$	c	140	KI	c	4,860
HIO_3	c	2,100	$LiNO_2$	c	−2,630	KIO_3	c	6,630
HNO_3	l	−7,954	$LiNO_2\cdot H_2O$	c	1,680	KNO_2	c	3,190
HCOOH	l	−205	$LiNO_3$	c	−600	KNO_3	c	8,340
CH_3COOH	l	−360				$KC_2H_3O_2$	c	−3,665
			NaOH	c	−10,637	KCN	c	2,800
NH_3	g	−7,290	$NaOH\cdot H_2O$	c	−5,118	KCNO	c	4,840
NH_4Cl	c	3,533	NaF	c	218	KCNS	c	5,790
NH_4ClO_4	c	8,000	NaCl	c	928	$KMnO_4$	c	10,410
NH_4Br	c	4,010	$NaClO_2$	c	80			
NH_4I	c	3,280	$NaClO_2\cdot 3H_2O$	c	6,830	RbOH	c	−14,900
NH_4IO_3	c	7,600	$NaClO_3$	c	5,191	$RbOH\cdot H_2O$	c	−4,310
NH_4NO_2	c	4,600	$NaClO_4$	c	3,317	$RbOH\cdot 2H_2O$	c	210
NH_4NO_3	c	6,140	$NaClO_4\cdot H_2O$	c	5,380	RbF	c	−6,240
$NH_4C_2H_3O_2$	c	−570	NaBr	c	−144	$RbF\cdot H_2O$	c	−100
NH_4CN	c	4,200	$NaBr\cdot 2H_2O$	c	4,454	$RbF\cdot 1\frac{1}{2}H_2O$	c	320
NH_4CNS	c	5,400	$NaBrO_3$	c	6,430	RbCl	c	4,130
CH_3NH_3Cl	c	1,378	NaI	c	−1,800	$RbClO_3$	c	11,410
$(CH_3)_3NHCl$	c	350	$NaI\cdot 2H_2O$	c	3,855	$RbClO_4$	c	13,560
$N(CH_3)_4Cl$	c	975	$NaIO_3$	c	4,850	RbBr	c	5,230
$N(CH_3)_4Br$	c	5,800	$NaNO_2$	c	3,320	$RbBrO_3$	c	11,700
$N(CH_3)_4I$	c	10,055	$NaNO_3$	c	4,900	RbI	c	6,000
			$NaC_2H_3O_2$	c	−4,140	$RbNO_3$	c	8,720
$AgClO_4$	c	1,760	$NaC_2H_3O_2\cdot 3H_2O$	c	4,700			
$AgNO_2$	c	8,830	NaCN	c	290	CsOH	c	−17,100
$AgNO_3$	c	5,400	$NaCN\cdot \frac{1}{2}H_2O$	c	790	$CsOH\cdot H_2O$	c	−4,900
			$NaCN\cdot 2H_2O$	c	4,440	CsF	c	−8,810
LiOH	c	−5,632	NaCNO	c	4,590	$CsF\cdot H_2O$	c	−2,500
$LiOH\cdot H_2O$	c	−1,600	NaCNS	c	1,632	$CsF\cdot 1\frac{1}{2}H_2O$	c	−1,300
LiF	c	1,130				CsCl	c	4,250
LiCl	c	−8,850	KOH	c	−13,769	$CsClO_4$	c	13,250
$LiCl\cdot H_2O$	c	−4,560	$KOH\cdot H_2O$	c	−3,500	CsBr	c	6,210
$LiClO_4$	c	−6,345	$KOH\cdot 1\frac{1}{2}H_2O$	c	−2,500	$CsBrO_3$	c	12,060
$LiClO_4\cdot 3H_2O$	c	7,795	KF	c	−4,238	CsI	c	7,970
LiBr	c	−11,670	$KF\cdot 2H_2O$	c	1,666	$CsNO_3$	c	9,560
$LiBr\cdot H_2O$	c	−5,560						

HEAT CAPACITY OF AQUEOUS SOLUTIONS OF VARIOUS ACIDS

From National Standards Reference Data Systems NSRDS-NBS 2

Vivian B. Parker

Φ_C is the apparent molal heat capacity of the solute, equal to $[(1000 + mM_2)C - 1000C^\circ]/m$ where C and C° are the specific heats (per unit mass) of the solution and pure solvent, respectively, m is the molality, and M_2 is the molecular weight of the solute.

Φ_C, cal/deg mole, at 25°C

n	*m*	HF	HCl	HBr	HI	HIO_3	HNO_3	CH_2O_2	$C_2H_4O_2$	$C_3H_6O_2$
∞	0.00	−25.5	−32.6	−33.9	−34.0	−29.6	−20.7	−21.0	−1.5	+26.7
500,000	.000111	−23.0	.....	.....	.....	.....	.....	−9.8	+25.8	38.0
100,000	.000555	−18.8	−32.4	−33.8	−33.9	−29.4	−20.6	−1.2	32.6	45.1
50,000	.00111	−16.6	−32.4	−33.7	−33.8	−29.3	−20.5	+3.1	34.0	48.3
20,000	.00278	−12.4	−32.3	−33.6	−33.7	−29.1	−20.4	10.2	35.7	52.2
10,000	.00555	−8.6	−32.2	−33.4	−33.5	−28.5	−20.3	12.7	36.9	54.1
7,000	.00793	−6.7	−32.1	−33.4	−33.4	−28.4	−20.2	13.7	37.4	54.8
5,000	.01110	−4.9	−32.0	−33.3	−33.3	−28.1	−20.1	14.3	37.8	55.3
4,000	.01388	−3.6	−31.9	−33.2	−33.2	−27.7	−20.1	14.7	37.9	55.6
3,000	.01850	−2.2	−31.8	−33.1	−33.1	−27.2	−20.0	15.1	38.1	56.1
2,000	.02775	−0.5	−31.6	−32.9	−32.9	−25.8	−19.8	15.8	38.5	56.7
1,500	.03700	+0.4	−31.4	−32.7	−32.7	−24.9	−19.6	16.4	38.7	57.1
1,000	.05551	1.6	−31.2	−32.5	−32.5	−23.0	−19.3	16.8	39.0	57.7
900	.0617	1.8	−31.2	−32.4	−32.4	−22.3	−19.2	17.0	39.0	57.8
800	.0694	2.0	−31.1	−32.3	−32.3	−21.5	−19.1	17.2	39.1	57.9
700	.0793	2.2	−30.9	−32.2	−32.1	−20.2	−19.0	17.3	39.2	58.0
600	.0925	2.4	−30.8	−32.1	−32.0	−18.7	−18.9	17.5	39.3	58.3
500	.1110	2.7	−30.6	−31.9	−31.7	−16.7	−18.7	17.7	39.4	58.4
400	.1388	2.8	−30.3	−31.6	−31.4	−13.9	−18.4	18.0	39.4	58.6
300	.1850	3.2	−30.0	−31.2	−30.9	−10.1	−17.9	18.3	39.4	58.8
200	.2775	3.3	−29.3	−30.6	−30.2	−4.7	−17.2	18.7	39.3	58.9
150	.3700	3.5	−28.8	−30.1	−29.6	−0.7	−16.4	18.9	39.2	58.8
100	.5551	3.8	−27.8	−29.1	−28.4	+4.8	−15.0	19.0	39.1	58.7
75	.7401	4.1	−27.0	−28.5	−27.5	9.2	−13.7	19.2	38.9	58.6
50	1.1101	4.5	−25.6	−26.8	−25.9	15.7	−11.5	19.8	38.6	58.3
40	1.3877	4.9	−24.8	−25.9	−24.8	19.0	−9.9	19.8	38.4	58.0
30	1.8502	5.3	−23.6	−24.5	−23.3	23.1	−7.3	20.0	38.0	57.1
25	2.2202	5.6	−22.7	−23.6	−22.2	25.7	−5.4	20.1	37.7	56.2
20	2.7753	5.7	−21.5	−22.2	−20.8	.....	−2.7	20.2	37.1	55.0
15	3.7004	6.0	−19.8	−20.3	.....	.....	+1.4	20.4	36.3	53.1
12	4.6255	6.2	−18.2	−18.5	.....	.....	5.0	20.4	35.4	51.3
10	5.5506	6.3	−16.8	−16.8	.....	.....	8.5	20.6	34.8	49.7
9.5	5.8427	6.4	−16.4	−16.3	.....	.....	9.2	20.6	34.7	.....
9.0	6.1674	6.4	−15.8	−15.7	.....	.....	10.3	20.6	34.5	.....
8.5	6.5301	6.5	−15.4	−15.1	.....	.....	11.4	20.7	34.4	.....
8.0	6.9383	6.6	−14.8	−14.4	.....	.....	12.5	20.7	34.2	.....
7.5	7.4008	6.7	−14.2	−13.2	.....	.....	13.7	20.8	34.0	.....
7.0	7.9295	6.9	−13.5	−12.7	.....	.....	14.9	20.8	33.8	.....
6.5	8.5394	7.0	−12.7	−11.9	.....	.....	16.1	20.8	33.5	.....
6.0	9.2510	7.1	−11.8	−10.8	.....	.....	17.1	20.9	33.3	.....
5.5	10.0920	7.2	−10.8	−9.6	.....	.....	18.3	20.9	33.0	.....
5.0	11.1012	7.3	−9.6	−8.2	.....	.....	19.3	21.0	32.8	.....
4.5	12.3346	7.4	−8.7	−6.8	.....	.....	20.4	21.1	32.5	.....
4.0	13.8765	7.5	−6.6	−5.5	.....	.....	21.3	21.2	32.2	.....
3.5	15.8589	7.6	−4.7	−4.0	.....	.....	22.1	21.3	31.8	.....
3.25	17.0788	7.6	.....	−3.2	.....	.....	22.6	21.4	31.7	.....
3.0	18.5020	7.7	.....	−2.3	.....	.....	23.0	21.5	31.5	.....
2.5	22.2024	7.8	.....	−0.4	.....	.....	23.8	21.6	31.2	.....
2.0	27.7530	7.9	.....	.....	.....	.....	24.6	21.7	30.8	.....
1.5	37.0040	.....	.....	.....	.....	.....	25.2	21.8	30.4	.....
1.0	55.506	.....	.....	.....	.....	.....	25.7	22.0	30.1	.....

THERMODYNAMIC FORMULAS

Compiled by Doctor E. A. Coomes

Legend:

- p = Pressure
- V = Volume
- T = Temperature
- n = Number of mols
- S = Entropy
- U = Internal energy (some books use E)
- Cp = Molal specific heat at constant pressure
- β = Coefficient volume expansion
- K = Compressibility
- H = U + pV = Total heat or enthalpy
- A = U − TS = Helmholtz free energy
- G = H − TS = Gibbs' free energy (some books use F).

Use of Table — Partial derivatives of the first order for the eight fundamental thermodynamic variables, namely, p, V, T, U, S, H, A, G, may be obtained in terms of $(\partial V/\partial T)_p$, $(\partial V/\partial p)_T$, and $(\partial H/\partial T)_p$; the latter three are connected to measurable quantities as follows:

$$\frac{1}{V}\left(\frac{\partial V}{\partial T}\right)_p = \beta; \quad -\frac{1}{V}\left(\frac{\partial V}{\partial p}\right)_T = K; \quad \left(\frac{\partial H}{\partial T}\right)_p = nC_p$$

Computation by the Table — The method of using the table will become apparent in working several examples. Suppose it is desired to know $(\partial H/\partial p)_s$ in terms of p, V, and T. Under caption "Constant" move horizontally to column marked "S"; across from "H" beside caption "differential" find "$-VnCp/T$." Across from "p" in column "S" find "$-nCp/T$"; $(\partial H/\partial p)_s$ is found by taking the ratio of the two:

$$(\partial H/\partial p)_S = (-VnC_p/T)/(-nC_p/T) = V$$

To find $(\partial S/\partial V)_T$ in terms of p, V, T, move to column "T" under "constant." Opposite "S" beside "differential" find "$(\partial V/\partial T)_p$"; opposite "V" find "$-(\partial V/\partial p)_T$".

Taking the ratio:

$$(\partial S/\partial V)_T = (\partial V/\partial T)_p / -(\partial V/\partial p)_T = (\partial p/\partial T)_V$$

Differential \ Constant	T	p	V	S
T	0	1	$\left(\frac{\partial V}{\partial p}\right)_T$	$-\left(\frac{\partial V}{\partial T}\right)_p$
p	−1	0	$-\left(\frac{\partial V}{\partial T}\right)_p$	$-\frac{nC_p}{T}$
V	$-\left(\frac{\partial V}{\partial p}\right)_T$	$\left(\frac{\partial V}{\partial T}\right)_p$	0	$\left(-\frac{1}{T}\right)\left[nC_p\left(\frac{\partial V}{\partial p}\right)_T + T\left(\frac{\partial V}{\partial T}\right)_p^2\right]$
S	$\left(\frac{\partial V}{\partial T}\right)_p$	$\frac{nC_p}{T}$	$\left(\frac{1}{T}\right)\left[nC_p\left(\frac{\partial V}{\partial p}\right)_T + T\left(\frac{\partial V}{\partial T}\right)_p^2\right]$	0
U	$T\left(\frac{\partial V}{\partial T}\right)_p + p\left(\frac{\partial V}{\partial p}\right)_T$	$nC_p - p\left(\frac{\partial V}{\partial T}\right)_p$	$nC_p\left(\frac{\partial V}{\partial p}\right)_T + T\left(\frac{\partial V}{\partial T}\right)_p^2$	$\left(\frac{p}{T}\right)\left[nC_p\left(\frac{\partial V}{\partial p}\right)_T + T\left(\frac{\partial V}{\partial T}\right)_p^2\right]$
H	$-V + T\left(\frac{\partial V}{\partial T}\right)_p$	nC_p	$nC_p\left(\frac{\partial V}{\partial p}\right)_T + T\left(\frac{\partial V}{\partial T}\right)_p^2 - V\left(\frac{\partial V}{\partial T}\right)$	$-\frac{VnC_p}{T}$
A	$p\left(\frac{\partial V}{\partial p}\right)_T$	$-S - p\left(\frac{\partial V}{\partial T}\right)_p$	$-S\left(\frac{\partial V}{\partial p}\right)_T$	$\left(\frac{1}{T}\right)\left[pnC_p\left(\frac{\partial V}{\partial p}\right)_T + pT\left(\frac{\partial V}{\partial T}\right)_p^2 + TS\left(\frac{\partial V}{\partial T}\right)_p\right]$
G	−V	−S	$-V\left(\frac{\partial V}{\partial T}\right)_p - S\left(\frac{\partial V}{\partial p}\right)_T$	$\left(-\frac{1}{T}\right)\left[nC_pV - TS\left(\frac{\partial V}{\partial T}\right)_p\right]$

Differential \ Constant	U	H	A	G
T	$-T\left(\frac{\partial V}{\partial T}\right)_p - p\left(\frac{\partial V}{\partial p}\right)_T$	$V - T\left(\frac{\partial V}{\partial T}\right)_p$	$-p\left(\frac{\partial V}{\partial p}\right)_T$	V
p	$-nC_p + p\left(\frac{\partial V}{\partial T}\right)_p$	$-nC_p$	$S + p\left(\frac{\partial V}{\partial T}\right)_p$	S
V	$-nC_p\left(\frac{\partial V}{\partial p}\right)_T - T\left(\frac{\partial V}{\partial T}\right)_p^2$	$-nC_p\left(\frac{\partial V}{\partial p}\right)_T - T\left(\frac{\partial V}{\partial T}\right)_p^2 + V\left(\frac{\partial V}{\partial T}\right)_p$	$S\left(\frac{\partial V}{\partial p}\right)_T$	$V\left(\frac{\partial V}{\partial T}\right)_p + S\left(\frac{\partial V}{\partial p}\right)_T$
S	$\left(-\frac{p}{T}\right)\left[nC_p\left(\frac{\partial V}{\partial p}\right)_T + T\left(\frac{\partial V}{\partial T}\right)_p^2\right]$	$\frac{VnC_p}{T}$	$\left(-\frac{1}{T}\right)\left[pnC_p\left(\frac{\partial V}{\partial p}\right)_T + pT\left(\frac{\partial V}{\partial T}\right)_p^2 + TS\left(\frac{\partial V}{\partial T}\right)_p\right]$	$\left(\frac{1}{T}\right)\left[nC_pV - TS\left(\frac{\partial V}{\partial T}\right)_p\right]$
U	0	$V\left[nC_p - p\left(\frac{\partial V}{\partial T}\right)_p\right] + p\left[nC_p\left(\frac{\partial V}{\partial p}\right)_T + T\left(\frac{\partial V}{\partial T}\right)_p^2\right]$	$-p\left[nC_p\left(\frac{\partial V}{\partial p}\right)_T + T\left(\frac{\partial V}{\partial T}\right)_p^2\right] - S\left[T\left(\frac{\partial V}{\partial T}\right)_p + p\left(\frac{\partial V}{\partial p}\right)_T\right]$	$V\left[nC_p - p\left(\frac{\partial V}{\partial T}\right)_p\right] - S\left[T\left(\frac{\partial V}{\partial T}\right)_p + p\left(\frac{\partial V}{\partial p}\right)_T\right]$
H	$-V\left[nC_p - p\left(\frac{\partial V}{\partial T}\right)_p\right] - p\left[nC_p\left(\frac{\partial V}{\partial p}\right)_T + T\left(\frac{\partial V}{\partial T}\right)_p^2\right]$	0	$-\left[S + p\left(\frac{\partial V}{\partial T}\right)_p\right] \times \left[V - T\left(\frac{\partial V}{\partial T}\right)_p - pnC_p\left(\frac{\partial V}{\partial p}\right)_T\right]$	$VnC_p + VS - TS\left(\frac{\partial V}{\partial T}\right)_p$
A	$p\left[nC_p\left(\frac{\partial V}{\partial p}\right)_T + T\left(\frac{\partial V}{\partial T}\right)_p^2\right] + S\left[T\left(\frac{\partial V}{\partial T}\right)_p + p\left(\frac{\partial V}{\partial p}\right)_T\right]$	$-\left[S + p\left(\frac{\partial V}{\partial T}\right)_p\right] \times \left[V - T\left(\frac{\partial V}{\partial T}\right)_p - pnC_p\left(\frac{\partial V}{\partial p}\right)_T\right]$	0	$-S\left[V + p\left(\frac{\partial V}{\partial p}\right)_T\right] + pV\left(\frac{\partial V}{\partial T}\right)_p$
G	$-V\left[nC_p - p\left(\frac{\partial V}{\partial T}\right)_p\right] + S\left[T\left(\frac{\partial V}{\partial T}\right)_p + p\left(\frac{\partial V}{\partial p}\right)_T\right]$	$-VnC_p - VS + TS\left(\frac{\partial V}{\partial T}\right)_p$	$S\left[V + p\left(\frac{\partial V}{\partial p}\right)_T\right] + pV\left(\frac{\partial V}{\partial T}\right)_p$	0

BUFFER SOLUTIONS

OPERATIONAL DEFINITIONS OF pH

Prepared by R. A. Robinson

The operational definition of pH is:

$$pH = pH(s) + E/k$$

where E is the e.m.f. of the cell:

$$H_2|\text{Solution, pH}|\text{Saturated KCl}|\text{Solution, pH(s)}|H_2$$

the half cell on the left containing the solution whose pH is being measured and that on the right a standard buffer mixture of known pH; $k = 2.303RT/F$, where R is the gas constant, T the temperature in degrees Kelvin and F the value of the faraday.

Alternatively, the cell:

Glass electrode|Solution, pH|Saturated calomel electrode

can be used, the glass electrode being calibrated using a standard buffer mixture or, if possible, two standard buffer mixtures whose pH values lie on either side of that of the solution which is being measured. Suitable standard buffer mixtures are:

0.05 M potassium hydrogen phthalate (pH = 4.008 at 25°C)
0.025 M potassium dihydrogen phosphate
0.025 M disodium hydrogen phosphate (pH = 6.865 at 25°C)
0.01 M borax (pH = 9.180 at 25°C)

For most purposes pH can be equated to $-\log_{10} \gamma_{H^+} m_{H^+}$, i.e., to the negative logarithm of the hydrogen ion activity. There is a small difference between those two quantities if pH > 9.2 or pH < 4.0, given by:

$$-\log \gamma_{H^+} m_{H^+} = pH + 0.014(pH - 9.2) \text{ for } pH > 9.2$$
$$= pH + 0.009(4.0 - pH) \text{ for } pH < 4.0$$

It should be noted that in the table titled "Solutions giving Round Values of pH at 25°C" it is $-\log \gamma_{H^+} m_{H^+}$ and not pH which is quoted when there is a difference between them.

References:

R. G. Bates, "Electrometric pH Determinations: Theory and Practice" Wiley, New York, 1954.

R. A. Robinson and R. H. Stokes, "Electrolyte Solutions," 2nd edition, Butterworths, London; Academic Press, Inc. New York, 1959. R. C. Bates, J. Res. of N.B.S. 66 A, 179 (1962).

National Bureau of Standards

R. G. Bates and S. F. Acree, Res. **34,** 373 (1945); W. J. Hammer, C. D. Pinching and S. F. Acree, ibid. **36,** 47 (1946); G. G. Manor, N. J. DeLollis, P. W. Lindwall and S. F. Acree, ibid., **36,** 543 (1946); R. G. Bates, ibid., **39,** 411 (1947); R. G. Bates, V. E. Bower, R. G. Miller and E. R. Smith, ibid., **47,** 433 (1951); V. E. Bower, R. G. Bates and E. R. Smith, ibid., **51,** 189 (1953); V. E. Bower and R. G. Bates, ibid., **55,** 197 (1955); R. G. Bates, V. E. Bower and E. R. Smith, ibid., **56,** 305 (1956); V. E. Bower and R. G. Bates, ibid., **59,** 261; R. G. Bates and V. E. Bower, Anal. Chem., **28,** 1322 (1956).

PROPERTIES OF STANDARD AQUEOUS BUFFER SOLUTIONS AT 25°C

Solution	Buffer substance	Molality m	Weight of salt in air per liter solution	Density g/ml	Molarity M	Dilution value $\Delta pH_{\frac{1}{2}}$	ΔpH_s^a	Buffer value, equiv. per pH	Temp coeff., dpH_s/dt. Units per °C
Tetroxalate	$KH_3(C_2O_4)_2 \cdot 2H_2O$	0.05	12.61	1.0032	0.04962	+0.186	−0.0028	0.070	+0.001
Tartrate	$KHC_4H_4O_6$, sat. sol'n. at 25°C	0.0341	......	1.0036	0.034	+0.049	−0.0003	0.027	−0.0014
Phthalate	$KHC_8O_4H_4$	0.05	10.12	1.0017	0.04958	+0.052	−0.0009	0.016	+0.0012
Phosphate	KH_2PO_4 + Na_2HPO_4	0.025[b]	3.39 3.53	1.0028	0.0249[b]	+0.080	−0.0006	0.029	−0.0028
Phosphate	KH_2PO_4 + Na_2HPO_4	0.008695[c] 0.03043[d]	1.179 4.30	1.0020	0.008665[c] 0.03032[d]	+0.07[e]	−0.0005	0.016	−0.0028
Borax	$Na_2B_4O_7 \cdot 10H_2O$	0.01	3.80	0.9996	0.009971	+0.01	−0.0001	0.020	−0.0082
Calcium hydroxide	$Ca(OH)_2$, sat. sol'n. at 25°C	0.0203	......	0.9991	0.02025	−0.28	+0.0014	0.09	−0.033

[a] ΔpH_s = pH_s (M Molar solution) − pH_s (m molal solution).
[b] Concentration of each phosphate salt.
[c] KH_2PO_4.
[d] Na_2HPO_4.
[e] Calculated value.

BUFFER SOLUTIONS (Continued)

SOLUTIONS GIVING ROUND VALUES OF pH AT 25°C

Reproduced from "Electrolyte Solutions" by permission from Robinson and Stokes, authors, and Butterworth's Scientific Publications.

A*		B*		C*		D*		E*	
pH	x	pH	x	pH	x	pH	x	pH	x
1.00	67.0	2.20	49.5	4.10	1.3	5.80	3.6	7.00	46.6
1.10	52.8	2.30	45.8	4.20	3.0	5.90	4.6	7.10	45.7
1.20	42.5	2.40	42.2	4.30	4.7	6.00	5.6	7.20	44.7
1.30	33.6	2.50	38.8	4.40	6.6	6.10	6.8	7.30	43.4
1.40	26.6	2.60	35.4	4.50	8.7	6.20	8.1	7.40	42.0
1.50	20.7	2.70	32.1	4.60	11.1	6.30	9.7	7.50	40.3
1.60	16.2	2.80	28.9	4.70	13.6	6.40	11.6	7.60	38.5
1.70	13.0	2.90	25.7	4.80	16.5	6.50	13.9	7.70	36.6
1.80	10.2	3.00	22.3	4.90	19.4	6.60	16.4	7.80	34.5
1.90	8.1	3.10	18.8	5.00	22.6	6.70	19.3	7.90	32.0
2.00	6.5	3.20	15.7	5.10	25.5	6.80	22.4	8.00	29.2
2.10	5.1	3.30	12.9	5.20	28.8	6.90	25.9	8.10	26.2
2.20	3.9	3.40	10.4	5.30	31.6	7.00	29.1	8.20	22.9
		3.50	8.2	5.40	34.1	7.10	32.1	8.30	19.9
		3.60	6.3	5.50	36.6	7.20	34.7	8.40	17.2
		3.70	4.5	5.60	38.8	7.30	37.0	8.50	14.7
		3.80	2.9	5.70	40.6	7.40	39.1	8.60	12.2
		3.90	1.4	5.80	42.3	7.50	40.9	8.70	10.3
		4.00	0.1	5.90	43.7	7.60	42.4	8.80	8.5
						7.70	43.5	8.90	7.0
						7.80	44.5	9.00	5.7
						7.90	45.3		
						8.00	46.1		

F*		G*		H*		I*		J*	
pH	x	pH	x	pH	x	pH	x	pH	x
8.00	20.5	9.20	0.9	9.60	5.0	10.90	3.3	12.00	6.0
8.10	19.7	9.30	3.6	9.70	6.2	11.00	4.1	12.10	8.0
8.20	18.8	9.40	6.2	9.80	7.6	11.10	5.1	12.20	10.2
8.30	17.7	9.50	8.8	9.90	9.1	11.20	6.3	12.30	12.8
8.40	16.6	9.60	11.1	10.00	10.7	11.30	7.6	12.40	16.2
8.50	15.2	9.70	13.1	10.10	12.2	11.40	9.1	12.50	20.4
8.60	13.5	9.80	15.0	10.20	13.8	11.50	11.1	12.60	25.6
8.70	11.6	9.90	16.7	10.30	15.2	11.60	13.5	12.70	32.2
8.80	9.6	10.00	18.3	10.40	16.5	11.70	16.2	12.80	41.2
8.90	7.1	10.10	19.5	10.50	17.8	11.80	19.4	12.90	53.0
9.00	4.6	10.20	20.5	10.60	19.1	11.90	23.0	13.00	66.0
9.10	2.0	10.30	21.3	10.70	20.2	12.00	26.9		
		10.40	22.1	10.80	21.2				
		10.50	22.7	10.90	22.0				
		10.60	23.3	11.00	22.7				
		10.70	23.8						
		10.80	24.25						

*A. 25 ml of 0.2 molar KCl + x ml of 0.2 molar HCl.
*B. 50 ml of 0.1 molar potassium hydrogen phthalate + x ml of 0.1 molar HCl.
*C. 50 ml of 0.1 molar potassium hydrogen phthalate + x ml of 0.1 molar NaOH.
*D. 50 ml of 0.1 molar potassium dihydrogen phosphate + x ml 0.1 molar NaOH.
*E. 50 ml of 0.1 molar tris(hydroxymethyl) aminomethane + x ml of 0.1 M HCl.
*F. 50 ml of 0.025 molar borax + x ml of 0.1 molar HCl.
*G. 50 ml of 0.025 molar borax + x ml of 0.1 molar NaOH.
*H. 50 ml of 0.05 molar sodium bicarbonate + x ml of 0.1 molar NaOH.
*I. 50 ml of 0.05 molar disodium hydrogen phosphate + x ml of 0.1 molar NaOH.
*J. 25 ml of 0.2 molar KCl + x ml of 0.2 molar NaOH.
Final Volume of Mixtures = 100 ml

BUFFER SOLUTIONS (Continued)

STANDARD VALUES OF pH AT TEMPERATURE 0–95°C

Temperature	Tetroxalate 0.05 molal	Tartrate 0.0341 molal (sat'd at 25°C)	Phthalate 0.05 molal	Phosphate[a]	Phosphate[b]	Borax 0.01 molal	Calcium hydroxide (sat'd at 25°C)
0	1.666	.....	4.003	6.984	7.534	9.464	13.423
5	1.668	.....	3.999	6.951	7.500	9.395	13.207
10	1.670	.....	3.998	6.923	7.472	9.332	13.003
15	1.672	.....	3.999	6.900	7.448	9.276	12.810
20	1.675	.....	4.002	6.881	7.429	9.225	12.627
25	1.679	3.557	4.008	6.865	7.413	9.180	12.454
30	1.683	3.552	4.015	6.853	7.400	9.139	12.289
35	1.688	3.549	4.024	6.844	7.389	9.102	12.133
38	1.691	3.548	4.030	6.840	7.384	9.081	12.043
40	1.694	3.547	4.035	6.838	7.380	9.068	11.984
45	1.700	3.547	4.047	6.834	7.373	9.038	11.841
50	1.707	3.549	4.060	6.833	7.367	9.011	11.705
55	1.715	3.554	4.075	6.834	.....	8.985	11.574
60	1.723	3.560	4.091	6.836	.....	8.962	11.449
70	1.743	3.580	4.126	6.845	.....	8.921	......
80	1.766	3.609	4.164	6.859	.....	8.885	......
90	1.792	3.650	4.205	6.877	.....	8.850	......
95	1.806	3.674	4.227	6.886	.....	8.833	......

[a] Solution 0.025 m KH_2PO_4 and 0.025 m Na_2HPO_4.
[b] Solution 0.008695 m KH_2PO_4 and 0.03043 m Na_2HPO_4.

APPROXIMATE pH VALUES

The following tables give approximate pH values for a number of substances such as acids, bases, foods, biological fluids, etc. All values are rounded off to the nearest tenth and are based on measurements made at 25° C. A few buffer systems with their pH values are also given.

From Modern pH and Chlorine Control, W. A. Taylor & Co., by permission

ACIDS

Hydrochloric, N	0.1
Hydrochloric, 0.1N	1.1
Hydrochloric, 0.01N	2.0
Sulfuric, N	0.3
Sulfuric, 0.1N	1.2
Sulfuric, 0.01N	2.1
Orthophosphoric, 0.1N	1.5
Sulfurous, 0.1N	1.5
Oxalic, 0.1N	1.6
Tartaric, 0.1N	2.2
Malic, 0.1N	2.2
Citric, 0.1N	2.2
Formic, 0.1N	2.3
Lactic, 0.1N	2.4
Acetic, N	2.4
Acetic, 0.1N	2.9
Acetic, 0.01N	3.4
Benzoic, 0.01N	3.1
Alum, 0.1N	3.2
Carbonic (saturated)	3.8
Hydrogen sulfide, 0.1N	4.1
Arsenious (saturated)	5.0
Hydrocyanic, 0.1N	5.1
Boric, 0.1N	5.2

BASES

Sodium hydroxide, N	14.0
Sodium hydroxide, 0.1N	13.0
Sodium hydroxide, 0.01N	12.0
Potassium hydroxide, N	14.0
Potassium hydroxide, 0.1N	13.0
Potassium hydroxide, 0.01N	12.0
Sodium metasilicate, 0.1N	12.6
Lime (saturated)	12.4
Trisodium phosphate, 0.1N	12.0
Sodium carbonate, 0.1N	11.6
Ammonia, N	11.6
Ammonia, 0.1N	11.1
Ammonia, 0.01N	10.6
Potassium cyanide, 0.1N	11.0
Magnesia (saturated)	10.5
Sodium sesquicarbonate, 0.1M	10.1
Ferrous hydroxide (saturated)	9.5
Calcium carbonate (saturated)	9.4
Borax, 0.1N	9.2
Sodium bicarbonate, 0.1N	8.4

BIOLOGIC MATERIALS

Blood, plasma, human	7.3–7.5
Spinal fluid, human	7.3–7.5
Blood, whole, dog	6.9–7.2
Saliva, human	6.5–7.5
Gastric contents, human	1.0–3.0
Duodenal contents, human	4.8–8.2
Feces, human	4.6–8.4
Urine, human	4.8–8.4
Milk, human	6.6–7.6
Bile, human	6.8–7.0

FOODS

Apples	2.9–3.3
Apricots	3.6–4.0
Asparagus	5.4–5.8
Bananas	4.5–4.7
Beans	5.0–6.0
Beers	4.0–5.0
Beets	4.9–5.5
Blackberries	3.2–3.6
Bread, white	5.0–6.0
Butter	6.1–6.4
Cabbage	5.2–5.4
Carrots	4.9–5.3
Cheese	4.8–6.4
Cherries	3.2–4.0
Cider	2.9–3.3
Corn	6.0–6.5
Crackers	6.5–8.5
Dates	6.2–6.4
Eggs, fresh white	7.6–8.0
Flour, wheat	5.5–6.5
Gooseberries	2.8–3.0
Grapefruit	3.0–3.3
Grapes	3.5–4.5
Hominy (lye)	6.8–8.0
Jams, fruit	3.5–4.0
Jellies, fruit	2.8–3.4
Lemons	2.2–2.4
Limes	1.8–2.0
Maple syrup	6.5–7.0
Milk, cows	6.3–6.6
Olives	3.6–3.8
Oranges	3.0–4.0
Oysters	6.1–6.6
Peaches	3.4–3.6
Pears	3.6–4.0
Peas	5.8–6.4
Pickles, dill	3.2–3.6
Pickles, sour	3.0–3.4
Pimento	4.6–5.2
Plums	2.8–3.0
Potatoes	5.6–6.0
Pumpkin	4.8–5.2
Raspberries	3.2–3.6
Rhubarb	3.1–3.2
Salmon	6.1–6.3
Sauerkraut	3.4–3.6
Shrimp	6.8–7.0
Soft drinks	2.0–4.0
Spinach	5.1–5.7
Squash	5.0–5.4
Strawberries	3.0–3.5
Sweet potatoes	5.3–5.6
Tomatoes	4.0–4.4
Tuna	5.9–6.1
Turnips	5.2–5.6
Vinegar	2.4–3.4
Water, drinking	6.5–8.0
Wines	2.8–3.8

Indicator	Approximate pH range	Color-change	Preparation
Methyl Violet	0.0–1.6	yel to bl	0.01–0.05% in water
Crystal Violet	0.0–1.8	yel to bl	0.02% in water
Ethyl Violet	0.0–2.4	yel to bl	0.1 g in 50 ml of MeOH + 50 ml of water
Malachite Green	0.2–1.8	yel to bl grn	water
Methyl Green	0.2–1.8	yel to bl	0.1% in water
2-(p-dimethylaminophenylazo)pyridine	0.2–1.8 4.4–5.6	yel to bl red to yel	0.1% in EtOH
o-Cresolsulfonephthalein (Cresol Red)	0.4–1.8 7.0–8.8	yel to red yel to red	0.1 g in 26.2 ml 0.01N NaOH + 223.8 ml water
Quinaldine Red	1.0–2.2	col to red	1% in EtOH
p-(p-dimethylaminophenylazo)-benzoic acid, Na-salt (Paramethyl Red)	1.0–3.0	red to yel	EtOH
m-(p-anilnophenylazo)benzene sulfonic acid, Na-salt (Metanil Yellow)	1.2–2.4	red to yel	0.01% in water
4-Phenylazodiphenylamine	1.2–2.6	red to yel	0.01 g in 1 ml 1N HCl + 50 ml EtOH + 49 ml water
Thymolsulfonephthalein (Thymol Blue)	1.2–2.8 8.0–9.6	red to yel yel to bl	0.1 g in 21.5 ml 0.01N NaOH + 229.5 ml water
m-Cresolsulfonephthalein (Metacresol Purple)	1.2–2.8 7.4–9.0	red to yel yel to purp	0.1 g in 26.2 ml 0.01N NaOH + 223.8 ml water
p-(p-anilinophenylazo)benzenesulfonic acid, Na-salt (Orange IV)	1.4–2.8	red to yel	0.01% in water
4-o-Tolylazo-o-toluidine	1.4–2.8	or to yel	water
Erythrosine, disodium salt	2.2–3.6	or to red	0.1% in water
Benzopurpurine 48	2.2–4.2	vt to red	0.1% in water
N,N-dimethyl-p-(m-tolylazo)aniline	2.6–4.8	red to yel	0.1% in water
4,4′-Bix(2-amino-1-naphthylazo)2,2′-stilbenedisulfonic acid	3.0–4.0	purp to red	0.1 g in 5.9 ml 0.05N NaOH + 94.1 ml water
Tetrabromophenolphthaleinethyl ester, K-salt	3.0–4.2	yel to bl	0.1% in EtOH
3′,3″,5′,5″-tetrabromophenolsulfonephthalein (Bromophenol Blue)	3.0–4.6	yel to bl	0.1 g in 14.9 ml 0.01N NaOH + 235.1 ml water
2,4-Dinitrophenol	2.8–4.0	col to yel	saturated water solution
N,N-Dimethyl-p-phenylazoaniline (p-Dimethylaminoazobenzene)	2.8–4.4	red to yel	0.1 g in 90 ml in EtOH + 10 ml water
Congo Red	3.0–5.0	blue to red	0.1% in water
Methyl Orange-Xylene Cyanole solution	3.2–4.2	purp to grn	ready solution
Methyl Orange	3.2–4.4	red to yel	0.01% in water
Ethyl Orange	3.4–4.8	red to yel	0.05–0.2% in water or aqueous EtOH
4-(4-Dimethylamino-1-naphthylazo)-3-methoxybenzenesulfonic acid	3.5–4.8	vt to yel	0.1% in 60% EtOH
3′,3″,5′,5″-Tetrabromo-m-cresolsulfonephthalein (Bromocresol Green)	3.8–5.4	yel to blue	0.1 g in 14.3 ml 0.01N NaOH + 235.7 ml water
Resazurin	3.8–6.4	or to vt	water
4-Phenylazo-1-naphthylamine	4.0–5.6	red to yel	0.1% in EtOH
Ethyl Red	4.0–5.8	col to red	0.1 g in 50 ml MeOH + 50 ml water
2-(p-Dimethylaminophenylazo)-pyridine	0.2–1.8 4.4–5.6	yel to red red to yel	0.1% in EtOH
4-(p-ethoxyphenylazo)-m-phenylenediamine monohydrochloride	4.4–5.8	or to yel	0.1% in water
Lacmoid	4.4–6.2	red to bl	0.2% in EtOH
Alizarin Red S	4.6–6.0	yel to red	dilute solution in water
Methyl Red	4.8–6.0	red to yel	0.02 g in 60 ml EtOH + 40 ml water

Indicator	Approximate pH range	Color-change	Preparation
Propyl Red	4.8–6.6	red to yel	EtOH
5′,5″-Dibromo-o-cresolsulfone-phthalein (Bromocresol Purple)	5.2–6.8	yel to purp	0.1 g in 18.5 ml 0.01N NaOH + 231.5 ml water
3′,3″-Dichlorophenolsulfonephthalein (Chlorophenol Red)	5.2–6.8	yel to red	0 1 g in 23.6 ml 0.01N NaOH + 226.4 ml water
p-Nitrophenol	5.4–6.6	col to yel	0.1% in water
Alizarin	5.6–7.2	yel to red	0.1% in MeOH
	11.0–12.4	red to purp	
2-(2,4-Dinitrophenylazo)-1-naphthol-3, 6-disulfonic acid, di-Na salt	6.0–7.0	yel to bl	0.1% in water
3′,3″-Dibromothymolsulfonephthalein (Bromothymol Blue)	6.0–7.6	yel to bl	0.1 g in 16 ml 0.01N NaOH + 234 ml water
6,8-Dinitro-2,4-(1H)quinazolinedione (m-Dinitrobenzoylene urea)	6.4–8.0	col to yel	25 g in 115 ml M NaOH + 50 ml boiling water 0.292 g of NaCl in 100 ml water
Brilliant Yellow	6.6–7.8	yel to or	1% in water
Phenolsulfonephthalein (Phenol Red)	6.6–8.0	yel to red	0.1 g in 28.2 ml 0.01N NaOH + 221.8 ml water
Neutral Red	6.8–8.0	red to amb	0.01 g in 50 ml EtOH + 50 ml water
m-Nitrophenol	6.8–8.6	col to yel	0.3% in water
o-Cresolsulfonephthalein (Cresol Red)	0.0–1.0	red to yel	0.1 g in 26.2 ml 0.01N
	7.0–8.8	yel to red	NaOH + 223.8 ml water
Curcumin	7.4–8.6	yel to red	EtOH
	10.2–11.8		
m-Cresolsulfonephthalein (Metacresol	1.2–2.8	red to yel	0.1 g in 26.2 ml 0.01N
Purple)	7.4–9.0	yel to purp	NaOH + 223.8 ml water
4,4′-Bis(4-amino-1-naphthylazo) 2,2′stilbene disulfonic acid	8.0–9.0	bl to red	0.1 g in 5.9 ml 0.05N NaOH + 94.1 ml water
Thymolsulfonephthalein (Thymol Blue)	1.2–2.8	red to yel	0.1 g in 21.5 ml 0.01N
	8.0–9.6		NaOH + 228.5 ml water
o-Cresolphthalein	8.2–9.8	col to red	0.04% in EtOH
p-Naphtholbenzene	8.2–10.0	or to bl	1% in dil. alkali
Phenolphthalein	8.2–10.0	col to pink	0.05 g in 50 ml EtOH + 50 ml water
Ethyl-bis(2,4-dimethylphenyl)acetate	8.4–9.6	col to bl	saturated solution in 50% acetone alcohol
Thymolphthalein	9.4–10.6	col to bl	0.04 g in 50 ml EtOH + 50 ml water
5-(p-Nitrophenylazo)salicylic acid, Na-salt (Alizarin Yellow R)	10.1–12.0	yel to red	0.01% in water
p-(2,4-Dihydroxyphenylazo)benzene-sulfonic acid, Na-salt	11.4–12.6	yel to or	0.1% in water
5,5′-Indigodisulfonic acid, di-Na-salt	11.4–13.0	bl to yel	water
2,4,6-Trinitrotoluene	11.5–13.0	col to or	0.1–0.5% in EtOH
1,3,5-Trinitrobenzene	12.0–14.0	col to or	0.1–0.5% in EtOH
Clayton Yellow	12.2–13.2	yel to amb	0.1% in water

CONVERSION FORMULAE FOR SOLUTIONS HAVING CONCENTRATIONS EXPRESSED IN VARIOUS WAYS

A = Weight per cent of solute
B = Molecular weight of solvent
E = Molecular weight of solute
F = Grams of solute per liter of solution
G = Molality
M = Molarity
N = Mole fraction
R = Density of solution grams per cc

Concentration of solute—SOUGHT	Concentration of solute—GIVEN				
	A	N	G	M	F
A	—	$\frac{100N \times E}{N \times E + (1 - N)B}$	$\frac{100G \times E}{1000 + G \times E}$	$\frac{M \times E}{10R}$	$\frac{F}{10R}$
N	$\frac{\frac{A}{E}}{\frac{A}{E} + \frac{100 - A}{B}}$	—	$\frac{B \times G}{B \times G + 1000}$	$\frac{B \times M}{M(B - E) + 1000R}$	$\frac{B \times F}{F(B - E) + 1000R \times E}$
G	$\frac{1000A}{E(100 - A)}$	$\frac{1000N}{B - N \times B}$	—	$\frac{1000M}{1000R - (M \times E)}$	$\frac{1000F}{E(1000R - F)}$
M	$\frac{10R \times A}{E}$	$\frac{1000R \times N}{N \times E + (1 - N)B}$	$\frac{1000R \times G}{1000 + E \times G}$	—	$\frac{F}{E}$
F	10AR	$\frac{1000R \times N \times E}{N \times E + (1 - N)B}$	$\frac{1000R \times G \times E}{1000 + G \times E}$	M × E	—

ELECTROCHEMICAL SERIES

Petr Vanýsek

There are three tables for this Electrochemical Series. Each table lists standard reduction potentials, E° values, at 298.15 K (25°C), and at a pressure of 101.325 kPa (1 atm.). Table 1 is an alphabetical listing of the elements according to the symbols for the elements. Thus, data for Silver (Ag) precedes those for Aluminum (Al). Table 2 lists only those reduction reactions which have E° values positive to the potential of the Standard Hydrogen Electrode. In Table 2, the reactions are listed in the order of increasing positive potential and range from 0.000 V to +3.053 V. Table 3 lists only those reduction reactions which have E° values negative to the potential of the Standard Hydrogen Electrode. In Table 3, reactions are listed in the order of increasing negative potential and range from −0.017 to −4.10 V.

Table 1
ALPHABETICAL LISTING

Reaction	E°, V
$Ag^+ + e \rightleftharpoons Ag$	0.7996
$Ag^{2+} + e \rightleftharpoons Ag^+$	1.980
$Ag(ac) + e \rightleftharpoons Ag + (ac)^-$	0.643
$AgBr + e \rightleftharpoons Ag + Br^-$	0.07133
$AgBrO_3 + e \rightleftharpoons Ag + BrO_3^-$	0.546
$Ag_2C_2O_4 + 2\,e \rightleftharpoons 2Ag + C_2O_{44}^{2-}$	0.4647
$AgCl + e \rightleftharpoons Ag + Cl^-$	0.22233
$AgCN + e \rightleftharpoons Ag + CN^-$	−0.017
$Ag_2CO_3 + 2\,e \rightleftharpoons 2\,Ag + CO_3^{2-}$	0.47
$Ag_2CrO_4 + 2\,e \rightleftharpoons 2\,Ag + CrO_4^{2-}$	0.4470
$AgF + e \rightleftharpoons Ag + F^-$	0.779
$Ag_4[Fe(CN)_6] + 4\,e \rightleftharpoons 4\,Ag + [Fe(CN)_6]^{4-}$	0.1478
$AgI + e \rightleftharpoons Ag + I^-$	−0.15224
$AgIO_3 + e \rightleftharpoons Ag + IO_3^-$	0.354
$Ag_2MoO_4 + 2\,e \rightleftharpoons 2\,Ag + MoO_4^{2-}$	0.4573
$AgNO_2 + e \rightleftharpoons Ag + NO_2^-$	0.564
$Ag_2O + H_2O + 2\,e \rightleftharpoons 2\,Ag + 2\,OH^-$	0.342
$Ag_2O_3 + H_2O + 2\,e \rightleftharpoons 2\,AgO + 2\,OH^-$	0.739
$2\,AgO + H_2O + 2\,e \rightleftharpoons Ag_2O + 2\,OH^-$	0.607
$AgOCN + e \rightleftharpoons Ag + OCN^-$	0.41
$Ag_2S + 2\,e \rightleftharpoons 2\,Ag + S^{2-}$	−0.691
$Ag_2S + 2\,H^+ + 2\,e \rightleftharpoons 2\,Ag + H_2S$	−0.0366
$AgSCN + e \rightleftharpoons Ag + SCN^-$	0.08951
$Ag_2SeO_3 + 2\,e \rightleftharpoons 2\,Ag + SeO_4^{2-}$	0.3629
$Ag_2SO_4 + 2\,e \rightleftharpoons 2\,Ag + SO_4^{2-}$	0.654
$Ag_2WO_4 + 2\,e \rightleftharpoons 2\,Ag + WO_4^{2-}$	0.4660
$Al^{3+} + 3\,e \rightleftharpoons Al$	−1.662
$H_2AlO_3^- + H_2O + 3\,e \rightleftharpoons Al + 4\,OH^-$	−2.33
$AlF_6^{3-} + 3\,e \rightleftharpoons Al + 6\,F^-$	−2.069
$As + 3H^+ + 3\,e \rightleftharpoons AsH_3$	−0.608
$As_2O_3 + 6\,H^+ + 6\,e \rightleftharpoons 2\,As + 3\,H_2O$	0.234
$HAsO_2 + 3\,H^+ + 3\,e \rightleftharpoons As + 2\,H_2O$	0.248
$AsO_2^- + 2\,H_2O + 3\,e \rightleftharpoons As + 4\,OH^-$	−0.68
$H_3AsO_4 + 2\,H^+ + 2\,e^- \rightleftharpoons HAsO_2 + 2\,H_2O$	0.560
$AsO_4^{3-} + 2\,H_2O + 2\,e \rightleftharpoons AsO_2^- + 4\,OH^-$	−0.71
$Au^+ + e \rightleftharpoons Au$	1.692
$Au^{3+} + 2\,e \rightleftharpoons Au^+$	1.401
$Au^{3+} + 3\,e \rightleftharpoons Au$	1.498
$AuBr_2^- + e \rightleftharpoons Au + 2\,Br^-$	0.959
$AuBr_4^- + 3\,e \rightleftharpoons Au + 4\,Br^-$	0.854
$AuCl_4^- + 3\,e \rightleftharpoons Au + 4\,Cl^-$	1.002
$Au(OH)_3 + 3\,H^+ + 3\,e \rightleftharpoons Au + 3\,H_2O$	1.45
$H_2BO_3^- + 5\,H_2O + 8\,e \rightleftharpoons BH_4^- + 8\,OH^-$	−1.24
$H_2BO_3^- + H_2O + 3\,e \rightleftharpoons B + 4\,OH^-$	−1.79
$H_3BO_3 + 3\,H^+ + 3\,e \rightleftharpoons B + 3\,H_2O$	−0.8698
$Ba^{2+} + 2\,e \rightleftharpoons Ba$	−2.912
$Ba^{2+} + 2\,e \rightleftharpoons Ba(Hg)$	−1.570
$Ba(OH)_2 + 2\,e \rightleftharpoons Ba + 2\,OH^-$	−2.99
$Be^{2+} + 2\,e \rightleftharpoons Be$	−1.847
$Be_2O_3^{2-} + 3\,H_2O + 4\,e \rightleftharpoons 2\,Be + 6\,OH^-$	−2.63

Table 1 (continued)
ALPHABETICAL LISTING

Reaction	E°, V
p-benzoquinone + 2 H^+ + 2 e $\rightleftharpoons$ hydroquinone	0.6992
$BiCl_4^-$ + 3 e $\rightleftharpoons$ Bi + 4 Cl^-	0.16
Bi_2O_3 + 3 H_2O + 6 e $\rightleftharpoons$ 2 Bi + 6 OH^-	−0.46
Bi_2O_4 + 4 H^+ + 2 e $\rightleftharpoons$ 2 BiO^+ + 2 H_2O	1.593
BiO^+ + 2 H^+ + 3 e $\rightleftharpoons$ Bi + H_2O	0.320
BiOCl + 2 H^+ + 3 e $\rightleftharpoons$ Bi + Cl^- + H_2O	0.1583
Br_2(aq) + 2 e $\rightleftharpoons$ 2 Br^-	1.0873
Br_2(1) + 2 e $\rightleftharpoons$ 2 Br^-	1.066
HBrO + H^+ + 2 e $\rightleftharpoons$ Br^- + H_2O	1.331
HBrO + H^+ + e $\rightleftharpoons$ 1/2Br_2(aq) + H_2O	1.574
HBrO + H^+ + e $\rightleftharpoons$ 1/2Br_2(ℓ) + H_2O	1.596
BrO^- + H_2O + 2 e $\rightleftharpoons$ Br^- + 2 OH^-	0.761
BrO_3^- + 6 H^+ + 5 e $\rightleftharpoons$ 1/2Br_2 + 3 H_2O	1.482
BrO_3^- + 6 H^+ + 6 e $\rightleftharpoons$ Br^- + 3 H_2O	1.423
BrO_3^- + 3 H_2O + 6 e $\rightleftharpoons$ Br^- + 6 OH^-	0.61
Ca^+ + e $\rightleftharpoons$ Ca	−3.80
Ca^{2+} + 2 e $\rightleftharpoons$ Ca	−2.868
Calomel electrode, 1 molal KCl	0.2800
Calomel electrode, 1 mol/ℓ KCl (NCE)	0.2801
Calomel electrode, 0.1 mol/ℓ KCl	0.3337
Calomel electrode, saturated KCl (SCE)	0.2412
Calomel electrode, saturated NaCl (SSCE)	0.2360
$Ca(OH)_2$ + 2 e $\rightleftharpoons$ Ca + 2 OH^-	−3.02
Cd^{2+} + 2 e $\rightleftharpoons$ Cd	−0.4030
Cd^{2+} + 2 e $\rightleftharpoons$ Cd(Hg)	−0.3521
$Cd(OH)_2$ + 2 e $\rightleftharpoons$ Cd(Hg) + 2 OH^-	−0.809
$CdSO_4$ + 2 e $\rightleftharpoons$ Cd + SO_4^{2-}	−0.246
Ce^{3+} + 3 e $\rightleftharpoons$ Ce	−2.483
Ce^{3+} + 3 e $\rightleftharpoons$ Ce(Hg)	−1.4373
Ce^{4+} + e $\rightleftharpoons$ Ce^{3+}	1.61
$CeOH^{3+}$ + H^+ + e $\rightleftharpoons$ Ce^{3+} + H_2O	1.715
Cl_2(g) + 2 e $\rightleftharpoons$ Cl^-	1.35827
HClO + H^+ + e $\rightleftharpoons$ 1/2Cl_2 + H_2O	1.611
HClO + H^+ 2 e $\rightleftharpoons$ Cl^- + H_2O	1.482
ClO^- + H_2O + 2 e $\rightleftharpoons$ Cl^- + 2 OH^-	0.81
ClO_2 + H^+ + e $\rightleftharpoons$ $HClO_2$	1.277
$HClO_2$ + 2 H^+ + 2 e $\rightleftharpoons$ HClO + H_2O	1.645
$HClO_2$ + 3 H^+ + 3 e $\rightleftharpoons$ 1/2Cl_2 + 2 H_2O	1.628
$HClO_2$ + 3 H^+ + 4 e $\rightleftharpoons$ Cl^- + 2 H_2O	1.570
ClO_2^- + H_2O + 2 e $\rightleftharpoons$ ClO^- + 2 OH^-	0.66
ClO_2^- + 2 H_2O + 4 e $\rightleftharpoons$ Cl^- + 4 OH^-	0.76
ClO_2(aq) + e $\rightleftharpoons$ ClO_2^-	0.954
ClO_3^- + 2 H^+ + e $\rightleftharpoons$ ClO_2 + H_2O	1.152
ClO_3^- + 3 H^+ + 2 e $\rightleftharpoons$ $HClO_2$ + H_2O	1.214
ClO_3^- + 6 H^+ + 5 e $\rightleftharpoons$ 1/2Cl_2 + 3 H_2O	1.47
ClO_3^- + 6 H^+ + 6 e $\rightleftharpoons$ Cl^- + 3 H_2O	1.451
ClO_3^- + H_2O + 2 e $\rightleftharpoons$ ClO_2^- + 2 OH^-	0.33
ClO_3^- + 3 H_2O + 6 e $\rightleftharpoons$ Cl^- + 6 OH^-	0.62
ClO_4^- + 2 H^+ + 2 e $\rightleftharpoons$ ClO_3^- + H_2O	1.189
ClO_4^- + 8 H^+ + 7 e $\rightleftharpoons$ 1/2Cl_2 + 4 H_2O	1.39
ClO_4^- + 8 H^+ + 8 e $\rightleftharpoons$ Cl^- + 4 H_2O	1.389
ClO_4^- + H_2O + 2 e $\rightleftharpoons$ ClO_3^- + 2 OH^-	0.36
$(CN)_2$ + 2 H^+ + 2 e $\rightleftharpoons$ 2 HCN	0.373
2 HCNO + 2 H^+ + 2 e $\rightleftharpoons$ $(CN)_2$ + 2 H_2O	0.330
$(CNS)_2$ + 2 e $\rightleftharpoons$ 2 CNS^-	0.77
Co^{2+} + 2 e $\rightleftharpoons$ Co	−0.28
Co^{3+} + e $\rightleftharpoons$ Co^{2+} (2 mol/ℓ H_2SO_4)	1.83
$[Co(NH_3)_6]^{3+}$ + e $\rightleftharpoons$ $[Co(NH_3)_6]^{2+}$	0.108
$Co(OH)_2$ + 2 e $\rightleftharpoons$ Co + 2 OH^-	−0.73
$Co(OH)_3$ + e $\rightleftharpoons$ $Co(OH)_2$ + OH^-	0.17
CO_2 + 2 H^+ + 2 e $\rightleftharpoons$ HCOOH	−0.199
Cr^{2+} + 2 e $\rightleftharpoons$ Cr	−0.913
Cr^{3+} + e $\rightleftharpoons$ Cr^{2+}	−0.407
Cr^{3+} + 3 e $\rightleftharpoons$ Cr	−0.744
$Cr_2O_7^{2-}$ + 14 H^+ + 6 e $\rightleftharpoons$ 2 Cr^{3+} + 7 H_2O	1.232
CrO_2^- + 2 H_2O + 3 e $\rightleftharpoons$ Cr + 4 OH^-	−1.2
$HCrO_4^{2-}$ + 7 H^+ + 3 e $\rightleftharpoons$ Cr^{3+} + 4 H_2O	1.350
CrO_4^- + 4 H_2O + 3 e $\rightleftharpoons$ $Cr(OH)_3$ + 5 OH^-	−0.13
$Cr(OH)_3$ + 3 e $\rightleftharpoons$ Cr + 3 OH^-	−1.48
Cs^+ + e $\rightleftharpoons$ Cs	−2.92
Cu^+ + e $\rightleftharpoons$ Cu	0.521
Cu^{2+} + e $\rightleftharpoons$ Cu^+	0.153
Cu^{2+} + 2 e $\rightleftharpoons$ Cu	0.3419
Cu^{2+} + 2 e $\rightleftharpoons$ Cu(Hg)	0.345
Cu^{2+} + 2 CN^- + e $\rightleftharpoons$ $[Cu(CN)_2]^-$	1.103
CuI_2^- + e $\rightleftharpoons$ Cu + 2 I^-	0.00
Cu_2O + H_2O + 2 e $\rightleftharpoons$ 2 Cu + 2 OH^-	−0.360
$Cu(OH)_2$ + 2 e $\rightleftharpoons$ Cu + 2 OH^-	−0.222
2 $Cu(OH)_2$ + 2 e $\rightleftharpoons$ Cu_2O + 2 OH^- + H_2O	−0.080
D^+ + e $\rightleftharpoons$ 1/2D_2	−0.0034
2 D^+ + 2 e $\rightleftharpoons$ D_2	−0.044
Eu^{2+} + 2 e $\rightleftharpoons$ Eu	−3.395
Eu^{3+} + 3 e $\rightleftharpoons$ Eu	−2.407
Eu^{3+} + e $\rightleftharpoons$ Eu^{2+}	−0.36
F_2 + 2 H^+ + 2 e $\rightleftharpoons$ 2 HF	3.053
F_2 + 2 e $\rightleftharpoons$ 2 F^-	2.866
F_2O + 2 H^+ + 4 e $\rightleftharpoons$ H_2O + 2 F^-	2.153
Fe^{2+} + 2 e $\rightleftharpoons$ Fe	−0.447
Fe^{3+} + 3 e $\rightleftharpoons$ Fe	−0.037
Fe^{3+} + e $\rightleftharpoons$ Fe^{2+}	0.771
$[Fe(CN)_6]^{3-}$ + e $\rightleftharpoons$ $[Fe(CN)_6]^{4-}$	0.358
FeO_4^{2-} + 8 H^+ + 3 e $\rightleftharpoons$ Fe^{3+} + 4 H_2O	2.20
$Fe(OH)_3$ + e $\rightleftharpoons$ $Fe(OH)_2$ + OH^-	−0.56
$[Fe(phenanthroline)_3]^{3+}$ + e $\rightleftharpoons$ $[Fe(phen)_3]^{2+}$	1.147
$[Fe(phen)_3]^{3+}$ + e $\rightleftharpoons$ $[Fe(phen)_3]^{2+}$ (1 mol/ℓ H_2SO_4)	1.06
$[Ferricinium]^+$ + e $\rightleftharpoons$ ferrocene	0.400
Ga^{3+} + 3 e $\rightleftharpoons$ Ga	−0.560
$H_2GaO_3^-$ + H_2O + 3 e $\rightleftharpoons$ Ga + 4 OH^-	−1.219
Ge^{2+} + 2 e $\rightleftharpoons$ Ge	0.24
Ge^{4+} + 4 e $\rightleftharpoons$ Ge	0.124
Ge^{4+} + 2 e $\rightleftharpoons$ Ge^{2+}	0.00
GeO_2 + 2 H^+ + 2 e $\rightleftharpoons$ GeO + H_2O	−0.118
H_2GeO_3 + 4 H^+ + 4 e $\rightleftharpoons$ Ge + 3 H_2O	−0.182
2 H^+ + 2 e $\rightleftharpoons$ H_2	0.00000
H_2 + 2 e $\rightleftharpoons$ 2 H^-	−2.23
HO_2 + H^+ + e $\rightleftharpoons$ H_2O_2	1.495
2 H_2O + 2 e $\rightleftharpoons$ H_2 + 2 OH^-	−0.8277
H_2O_2 + 2 H^+ + 2 e $\rightleftharpoons$ 2 H_2O	1.776
HfO^{2+} + 2 H^+ + 4 e $\rightleftharpoons$ Hf + H_2O	−1.724
HfO_2 + 4 H^+ + 4 e $\rightleftharpoons$ Hf + 2 H_2O	−1.505
$HfO(OH)_2$ + H_2O + 4 e $\rightleftharpoons$ Hf + 4 OH^-	−2.50
Hg^{2+} + 2 e $\rightleftharpoons$ Hg	0.851
2 Hg^{2+} + 2 e $\rightleftharpoons$ Hg_2^{2+}	0.920
Hg_2^{2+} + 2 e $\rightleftharpoons$ 2 Hg	0.7973

Table 1 (continued)
ALPHABETICAL LISTING

Reaction	E°, V
$Hg_2(ac)_2 + 2\,e \rightleftharpoons 2\,Hg + 2\,(ac)^-$	0.51163
$Hg_2Br_2 + 2\,e \rightleftharpoons 2\,Hg + 2\,Br^-$	0.13923
$Hg_2Cl_2 + 2\,e \rightleftharpoons 2\,Hg + 2\,Cl^-$	0.26808
$Hg_2HPO_4 + 2\,e \rightleftharpoons 2\,Hg + HPO_4^{2-}$	0.6359
$Hg_2I_2 + 2\,e \rightleftharpoons 2\,Hg + 2\,I^-$	−0.0405
$Hg_2O + H_2O + 2\,e \rightleftharpoons 2\,Hg + 2\,OH^-$	0.123
$HgO + H_2O + 2\,e \rightleftharpoons Hg + 2\,OH^-$	0.0977
$Hg_2SO_4 + 2\,e \rightleftharpoons 2\,Hg + SO_4^{2-}$	0.6125
$I_2 + 2\,e \rightleftharpoons 2\,I^-$	0.5355
$I_3^- + 2\,e \rightleftharpoons 3\,I^-$	0.536
$H_3IO_6 + 2\,e \rightleftharpoons IO_3^- + 3\,OH^-$	0.7
$H_5IO_6 + H^+ + 2\,e \rightleftharpoons IO_3^- + 3\,H_2O$	1.601
$2\,HIO + 2\,H^+ + 2\,e \rightleftharpoons I_2 + 2\,H_2O$	1.439
$HIO + H^+ + 2\,e \rightleftharpoons I^- + H_2O$	0.987
$IO^- + H_2O + 2\,e \rightleftharpoons I^- + 2\,OH^-$	0.485
$2\,IO_3^- + 12\,H^+ + 10\,e \rightleftharpoons I_2 + 6\,H_2O$	1.195
$IO_3^- + 6\,H^+ + 6\,e \rightleftharpoons I^- + 3\,H_2O$	1.085
$IO_3^- + 2\,H_2O + 4\,e \rightleftharpoons IO^- + 4\,OH^-$	0.15
$IO_3^- + 3\,H_2O + 6\,e \rightleftharpoons I^- + 6\,OH^-$	0.26
$In^+ + e \rightleftharpoons In$	−0.14
$In^{2+} + e \rightleftharpoons In^+$	−0.40
$In^{3+} + e \rightleftharpoons In^{2+}$	−0.49
$In^{3+} + 2\,e \rightleftharpoons In^+$	−0.443
$In^{3+} + 3e \rightleftharpoons In$	−0.3382
$Ir^{3+} + 3\,e \rightleftharpoons Ir$	1.156
$[IrCl_6]^{2-} + e \rightleftharpoons [IrCl_6]^{3-}$	0.8665
$[IrCl_6]^{3-} + 3\,e \rightleftharpoons Ir + 6\,Cl^-$	0.77
$Ir_2O_3 + 3\,H_2O + 6\,e \rightleftharpoons 2\,Ir + 6\,OH^-$	0.098
$K^+ + e \rightleftharpoons K$	−2.931
$La^{3+} + 3\,e \rightleftharpoons La$	−2.522
$La(OH)_3 + 3\,e \rightleftharpoons La + 3\,OH^-$	−2.90
$Li^+ + e \rightleftharpoons Li$	−3.0401
$Mg^+ + e \rightleftharpoons Mg$	−2.70
$Mg^{2+} + 2\,e \rightleftharpoons Mg$	−2.372
$Mg(OH)_2 + 2\,e \rightleftharpoons Mg + 2\,OH^-$	−2.690
$Mn^{2+} + 2\,e \rightleftharpoons Mn$	−1.185
$Mn^{3+} + 3 \rightleftharpoons Mn^{2+}$	1.5415
$MnO_2 + 4\,H^+ + 2\,e \rightleftharpoons Mn^{2+} + 2\,H_2O$	1.224
$MnO_4^- + e \rightleftharpoons MnO_4^{2-}$	0.558
$MnO_4^- + 4\,H^+ + 3\,e \rightleftharpoons MnO_2 + 2\,H_2O$	1.679
$MnO_4^- + 8\,H^+ + 5\,e \rightleftharpoons Mn^{2+} + 4\,H_2O$	1.507
$MnO_4^- + 2\,H_2O + 3\,e \rightleftharpoons MnO_2 + 4\,OH^-$	0.595
$MnO_4^{2-} + 2\,H_2O + 2\,e \rightleftharpoons MnO_2 + 4\,OH^-$	0.60
$Mn(OH)_2 + 2\,e \rightleftharpoons Mn + 2\,OH^-$	−1.56
$Mn(OH)_3 + e \rightleftharpoons Mn(OH)_2 + OH^-$	0.15
$Mo^{3+} + 3\,e \rightleftharpoons Mo$	−0.200
$N_2 + 2\,H_2O + 6\,H^+ + 6\,e \rightleftharpoons 2\,NH_4OH$	0.092
$3\,N_2 + 2\,H^+ + 2\,e \rightleftharpoons 2\,NH_3$	−3.09
$N_5^+ + 3\,H^+ + 2\,e \rightleftharpoons 2\,NH_4^+$	1.275
$N_2O + 2\,H^+ + 2\,e \rightleftharpoons N_2 + H_2O$	1.766
$H_2N_2O_2 + 2\,H^+ + 2\,e \rightleftharpoons N_2 + 2\,H_2O$	2.65
$N_2O_4 + 2\,e \rightleftharpoons 2\,NO_2^-$	0.867
$N_2O_4 + 2\,H^+ + 2\,e \rightleftharpoons 2\,HNO_2$	1.065
$N_2O_4 + 4\,H^+ + 4\,e \rightleftharpoons 2\,NO + 2\,H_2O$	1.035
$2\,NH_3OH^+ + H^+ + 2\,e \rightleftharpoons N_2H_5^+ + 2\,H_2O$	1.42
$2\,NO + 2\,e \rightleftharpoons N_2O_2^{2-}$	0.10
$2\,NO + 2\,H^+ + 2\,e \rightleftharpoons N_2O + H_2O$	1.591
$2\,NO + H_2O + 2\,e \rightleftharpoons N_2O + 2\,OH^-$	0.76
$HNO_2 + H^+ + e \rightleftharpoons NO + H_2O$	0.983
$2\,HNO_2 + 4\,H^+ + 4\,e \rightleftharpoons H_2N_2O_2 + 2\,H_2O$	0.86
$2\,HNO_2 + 4\,H^+ + 4\,e \rightleftharpoons N_2O + 3\,H_2O$	1.297
$NO_2^- + H_2O + 3\,e \rightleftharpoons NO + 2\,OH^-$	−0.46
$2\,NO_2^- + 2\,H_2O + 4\,e \rightleftharpoons N_2^{2-} + 4\,OH^-$	−0.18
$2\,NO_2^- + 3\,H_2O + 4\,e \rightleftharpoons N_2O + 6\,OH^-$	0.15
$NO_3^- + 3\,H^+ + 2\,e \rightleftharpoons HNO_2 + H_2O$	0.934
$NO_3^- + 4\,H^+ + 3\,e \rightleftharpoons NO + 2\,H_2O$	0.957
$2\,NO_3^- + 4\,H^+ + 2\,e \rightleftharpoons N_2O_4 + 2\,H_2O$	0.803
$NO_3^- + H_2O + 2\,e \rightleftharpoons NO_2^- + 2\,OH^-$	0.01
$2\,NO_3^- + 2\,H_2O + 2\,e \rightleftharpoons N_2O_4 + 4\,OH^-$	−0.85
$Na^+ + e \rightleftharpoons Na$	−2.71
$Nb^{3+} + 3\,e \rightleftharpoons Nb$	−1.099
$Nb_2O_5 + 10\,H^+ + 10\,e \rightleftharpoons 2\,Nb + 5\,H_2O$	−0.644
$Nd^{3+} + 3\,e \rightleftharpoons Nd$	−2.431
$Ni^{2+} + 2\,e \rightleftharpoons Ni$	−0.257
$Ni(OH)_2 + 2\,e \rightleftharpoons Ni + 2\,OH^-$	−0.72
$NiO_2 + 4\,H^+ + 2\,e \rightleftharpoons Ni^{2+} + 2\,H_2O$	1.678
$NiO_2 + 2\,H_2O + 2\,e \rightleftharpoons Ni\,(OH)_2 + 2\,OH^-$	−0.490
$Np^{3+} + 3\,e \rightleftharpoons Np$	−1.856
$Np^{4+} + e \rightleftharpoons Np^{3+}$	0.147
$NpO_2 + H_2O + H^+ + e \rightleftharpoons Np(OH)_3$	−0.962
$O_2 + 2\,H^+ + 2\,e \rightleftharpoons H_2O_2$	0.695
$O_2 + 4\,H^+ + 4\,e \rightleftharpoons 2\,H_2O$	1.229
$O_2 + H_2O + 2\,e \rightleftharpoons HO_2^- + OH^-$	−0.076
$O_2 + 2\,H_2O + 2\,e \rightleftharpoons H_2O_2 + 2\,OH^-$	−0.146
$O_2 + 2\,H_2O + 4\,e \rightleftharpoons 4\,OH^-$	0.401
$O_3 + 2\,H^+ + 2\,e \rightleftharpoons O_2 + H_2O$	2.076
$O_3 + H_2O + 2\,e \rightleftharpoons O_2 + 2\,OH^-$	1.24
$O(g) + 2\,H^+ + 2\,e \rightleftharpoons H_2O$	2.421
$OH + e \rightleftharpoons OH^-$	2.02
$HO_2^- + H_2O + 2\,e \rightleftharpoons 3\,OH^-$	0.878
$OsO_4 + 8\,H^+ + 8\,e \rightleftharpoons Os + 4\,H_2O$	0.85
$P(red) + 3\,H^+ + 3\,e \rightleftharpoons PH_3(g)$	−0.111
$P(white) + 3\,H^+ + 3\,e \rightleftharpoons PH_3(g)$	−0.063
$P + 3\,H_2O + 3\,e \rightleftharpoons PH_3(g) + 3\,OH^-$	−0.87
$H_2PO_2^- + e \rightleftharpoons P + 2\,OH^-$	−1.82
$H_3PO_2 + H^+ + 3\,e \rightleftharpoons P + 2\,H_2O$	−0.508
$H_3PO_3 + 2\,H^+ + 2\,e \rightleftharpoons H_3PO_2 + H_2O$	−0.499
$H_3PO_3 + 3\,H^+ + 3\,e \rightleftharpoons P + 3\,H_2O$	−0.454
$HPO_3^{2-} + 2\,H_2O + 2\,e \rightleftharpoons H_2PO_2^- + 3\,OH^-$	−1.65
$HPO_3^{2-} + 2\,H_2O + 3\,e \rightleftharpoons P + 5\,OH^-$	−1.71
$H_3PO_4 + 2\,H^+ + 2\,e \rightleftharpoons H_3PO_3 + H_2O$	−0.276
$PO_4^{3-} + 2\,H_2O + 2\,e \rightleftharpoons HPO_3^{2-} + 3\,OH^-$	−1.05
$Pb^{2+} + 2\,e \rightleftharpoons Pb$	−0.1262
$Pb^{2+} + 2\,e \rightleftharpoons Pb(Hg)$	−0.1205
$PbBr_2 + 2\,e \rightleftharpoons Pb + 2\,Br^-$	−0.284
$PbCl_2 + 2\,e \rightleftharpoons Pb + 2\,Cl^-$	−0.2675
$PbF_2 + 2\,e \rightleftharpoons Pb + 2\,F^-$	−0.3444
$PbHPO_4 + 2\,e \rightleftharpoons Pb + HPO_4^{2-}$	−0.465
$PbI_2 + 2\,e \rightleftharpoons Pb + 2\,I^-$	−0.365
$PbO + H_2O + 2\,e \rightleftharpoons Pb + 2\,OH^-$	−0.580
$PbO_2 + 4\,H^+ + 2\,e \rightleftharpoons Pb^{2+} + 2\,H_2O$	1.455
$HPbO_2^- + H_2O + 2\,e \rightleftharpoons Pb + 3\,OH^-$	−0.537
$PbO_2 + H_2O + 2\,e \rightleftharpoons PbO + 2\,OH^-$	0.247
$PbO_2 + SO_4^{2-} + 4\,H^+ + 2\,e \rightleftharpoons PbSO_4 + 2\,H_2O$	1.6913
$PbSO_4 + 2\,e \rightleftharpoons Pb + SO_4^{2-}$	−0.3588

Table 1 (continued)
ALPHABETICAL LISTING

Reaction	E°, V
$PbSO_4 + 2e \rightleftharpoons Pb(Hg) + SO_4^{2-}$	−0.3505
$Pd^{2+} + 2e \rightleftharpoons Pd$	0.951
$[PdCl_4]^{2-} + 2e \rightleftharpoons Pd + 4Cl^-$	0.591
$[PdCl_6]^{2-} + 2e \rightleftharpoons [PdCl_4]^{2-} + 2Cl^-$	1.288
$Pd(OH)_2 + 2e \rightleftharpoons Pd + 2OH^-$	0.07
$Pt^{2+} + 2e \rightleftharpoons Pt$	1.118
$[PtCl_4]^{2-} + 2e \rightleftharpoons Pt + 4Cl^-$	0.755
$[PtCl_6]^{2-} + 2e \rightleftharpoons [PtCl_4]^{2-} + 2Cl^-$	0.68
$Pt(OH)_2 + 2e \rightleftharpoons Pt + 2OH^-$	0.14
$Pu^{3+} + 3e \rightleftharpoons Pu$	−2.031
$Pu^{4+} + e \rightleftharpoons Pu^{3+}$	1.006
$Pu^{5+} + e \rightleftharpoons Pu^{4+}$	1.099
$PuO_2(OH)_2 + 2H^+ + 2e \rightleftharpoons Pu(OH)_4$	1.325
$PuO_2(OH)_2 + H^+ + e \rightleftharpoons PuO_2OH + H_2O$	1.062
$Rb^+ + e \rightleftharpoons Rb$	−2.98
$Re^{3+} + 3e \rightleftharpoons Re$	0.300
$ReO_4^- + 4H^+ + 3e \rightleftharpoons ReO_2 + 2H_2O$	0.510
$ReO_2 + 4H^+ + 4e \rightleftharpoons Re + 2H_2O$	0.2513
$ReO_4^- + 2H^+ + e \rightleftharpoons ReO_3 + H_2O$	0.768
$ReO_4^- + 4H_2O + 7e \rightleftharpoons Re + 8OH^-$	−0.584
$ReO_4^- + 8H^+ + 7e \rightleftharpoons Re + 4H_2O$	0.368
$Rh^+ + e \rightleftharpoons Rh$	0.600
$Rh^{2+} + 2e \rightleftharpoons Rh$	0.600
$Rh^{3+} + 3e \rightleftharpoons Rh$	0.758
$[RhCl_6]^{3-} + 3e \rightleftharpoons Rh + 6Cl^-$	0.431
$Ru^{2+} + 2e \rightleftharpoons Ru$	0.455
$Ru^{3+} + e \rightleftharpoons Ru^{2+}$	0.2487
$RuO_2 + 4H^+ + 2e \rightleftharpoons Ru^{2+} + 2H_2O$	1.120
$RuO_4^- + e \rightleftharpoons RuO_4^{2-}$	0.59
$RuO_4 + e \rightleftharpoons RuO_4^-$	1.00
$S + 2e \rightleftharpoons S^{2-}$	−0.47627
$S + 2H^+ + 2e \rightleftharpoons H_2S(aq)$	0.142
$S + H_2O + 2e \rightleftharpoons HS^- + OH^-$	−0.478
$2S + 2e \rightleftharpoons S_2^{2-}$	−0.42836
$S_2O_6^{2-} + 4H^+ + 2e \rightleftharpoons 2H_2SO_3$	0.564
$S_2O_8^{2-} + 2e \rightleftharpoons 2SO_4^{2-}$	2.010
$S_2O_8^{2-} + 2H^+ + 2e \rightleftharpoons 2HSO_4^-$	2.123
$S_4O_6^{2-} + 2e \rightleftharpoons 2S_2O_3^{2-}$	0.08
$2H_2SO_3 + H^+ + 2e \rightleftharpoons HS_2O_4^- + 2H_2O$	−0.056
$H_2SO_3 + 4H^+ + 4e \rightleftharpoons S + 3H_2O$	0.449
$2SO_3^{2-} + 2H_2O + 2e \rightleftharpoons S_2O_4^{2-} + 4OH^-$	−1.12
$2SO_3^{2-} + 3H_2O + 4e \rightleftharpoons S_2O_3^{2-} + 6OH^-$	−0.571
$SO_4^{2-} + 4H^+ + 2e \rightleftharpoons H_2SO_3 + H_2O$	0.172
$2SO_4^{2-} + 4H^+ + 2e \rightleftharpoons S_2O_6^{2-} + H_2O$	−0.22
$SO_4^{2-} + H_2O + 2e \rightleftharpoons SO_3^{2-} + 2OH^-$	−0.93
$Sb + 3H^+ + 3e \rightleftharpoons SbH_3$	−0.510
$Sb_2O_3 + 6H^+ + 6e \rightleftharpoons 2Sb + 3H_2O$	0.152
Sb_2O_5 (senarmontite) $+ 4H^+ + 4e \rightleftharpoons Sb_2O_3 + 2H_2O$	0.671
Sb_2O_5 (valentinite) $+ 4H^+ + 4e \rightleftharpoons Sb_2O_3 + 2H_2O$	0.649
$Sb_2O_5 + 6H^+ + 4e \rightleftharpoons 2SbO^+ + 3H_2O$	0.581
$SbO^+ + 2H^+ + 3e \rightleftharpoons Sb + 2H_2O$	0.212
$SbO_2^- + 2H_2O + 3e \rightleftharpoons Sb + 4OH^-$	−0.66
$SbO_3^- + H_2O + 2e \rightleftharpoons SbO_2^- + 2OH^-$	−0.59
$Sc^{3+} + 3e \rightleftharpoons Sc$	−2.077
$Se + 2e \rightleftharpoons Se^{2-}$	−0.924
$Se + 2H^+ + 2e \rightleftharpoons H_2Se(aq)$	−0.399
$H_2SeO_3 + 4H^+ + 4e \rightleftharpoons Se + 3H_2O$	−0.74
$SeO_3^{2-} + 3H_2O + 4e \rightleftharpoons Se + 6OH^-$	−0.366
$SeO_4^{2-} + 4H^+ + 2e \rightleftharpoons H_2SeO_3 + H_2O$	1.151
$SeO_4^{2-} + H_2O + 2e \rightleftharpoons SeO_3^{2-} + 2OH^-$	0.05
$SiF_6^{2-} + 4e \rightleftharpoons Si + 6F^-$	−1.24
SiO_2 (quartz) $+ 4H^+ + 4e \rightleftharpoons Si + 2H_2O$	0.857
$SiO_3^{2-} + 3H_2O + 4e \rightleftharpoons Si + 6OH^-$	−1.697
$Sn^{2+} + 2e \rightleftharpoons Sn$	−0.1375
$Sn^{4+} + 2e \rightleftharpoons Sn^{2+}$	0.151
$HSnO_2^- + H_2O + 2e \rightleftharpoons Sn + 3OH^-$	−0.909
$Sn(OH)_6^{2-} + 2e \rightleftharpoons HSnO_2^- + 3OH^- + H_2O$	−0.93
$Sr^+ + e \rightleftharpoons Sr$	−4.10
$Sr^{2+} + 2e \rightleftharpoons Sr$	−2.89
$Sr^{2+} + 2e \rightleftharpoons Sr(Hg)$	−1.793
$Sr(OH)_2 + 2e \rightleftharpoons Sr + 2OH^-$	−2.88
$Ta_2O_5 + 10H^+ + 10e \rightleftharpoons 2Ta + 5H_2O$	−0.750
$Tc^{2+} + 2e \rightleftharpoons Tc$	0.400
$TcO_4^- + 4H^+ + 3e \rightleftharpoons TcO_2 + 2H_2O$	0.782
$Te + 2e \rightleftharpoons Te^{2-}$	−1.143
$Te + 2H^+ + 2e \rightleftharpoons H_2Te$	−0.793
$Te^{4+} + 4e \rightleftharpoons Te$	0.568
$TeO_2 + 4H^+ + 4e \rightleftharpoons Te + 2H_2O$	0.593
$TeO_3^{2-} + 3H_2O + 4e \rightleftharpoons Te + 6OH^-$	−0.57
$TeO_4^- + 8H^+ + 7e \rightleftharpoons Te + 4H_2O$	0.472
$H_6TeO_6 + 2H^+ + 2e \rightleftharpoons TeO_2 + 4H_2O$	1.02
$Th^{4+} + 4e \rightleftharpoons Th$	−1.899
$ThO_2 + 4H^+ + 4e \rightleftharpoons Th + 2H_2O$	−1.789
$Th(OH)_4 + 4e \rightleftharpoons Th + 4OH^-$	−2.48
$Ti^{2+} + 2e \rightleftharpoons Ti$	−1.630
$Ti^{3+} + e \rightleftharpoons Ti^{2+}$	−0.368
$TiO_2 + 4H^+ + 2e \rightleftharpoons Ti^{2+} + 2H_2O$	−0.502
$TiOH^{3+} + H^+ + e \rightleftharpoons Ti^{3+} + H_2O$	−0.055
$Tl^+ + e \rightleftharpoons Tl$	−0.336
$Tl^+ + e \rightleftharpoons Tl(Hg)$	−0.3338
$Tl^{3+} + 2e \rightleftharpoons Tl^+$	1.252
$TlBr + e \rightleftharpoons Tl + Br^-$	−0.658
$TlCl + e \rightleftharpoons Tl + Cl^-$	−0.5568
$TlI + e \rightleftharpoons Tl + I^-$	−0.752
$Tl_2O_3 + 3H_2O + 4e \rightleftharpoons 2Tl^+ + 6OH^-$	0.02
$TlOH + e \rightleftharpoons Tl + OH^-$	−0.34
$Tl(OH)_3 + 2e \rightleftharpoons TlOH + 2OH^-$	−0.05
$Tl_2SO_4 + 2e \rightleftharpoons Tl + SO_4^{2-}$	−0.4360
$U^{3+} + 3e \rightleftharpoons U$	−1.798
$U^{4+} + e \rightleftharpoons U^{3+}$	−0.607
$UO_2^+ + 4H^+ + e \rightleftharpoons U^{4+} + 2H_2O$	0.612
$UO_2^{2+} + e \rightleftharpoons UO^+{}_2$	0.062
$UO_2^{2+} + 4H^+ + 2e \rightleftharpoons U^{4+} + 2H_2O$	0.327
$UO_2^{2+} + 4H^+ + 6e \rightleftharpoons U + 2H_2O$	−1.444
$V^{2+} + 2e \rightleftharpoons V$	−1.175
$V^{3+} + e \rightleftharpoons V^{2+}$	−0.255
$VO^{2+} + 2H^+ + e \rightleftharpoons V^{3+} + H_2O$	0.337
$VO_2^+ + 2H^+ + e \rightleftharpoons VO^{2+} + H_2O$	0.991
$V(OH)_4^+ + 2H^+ + e \rightleftharpoons VO^{2+} + 3H_2O$	1.00
$V(OH)_4^+ + 4H^+ + 5e \rightleftharpoons V + 4H_2O$	−0.254
$W_2O_5 + 2H^+ + 2e \rightleftharpoons 2WO_2 + H_2O$	−0.031

Table 1 (continued)
ALPHABETICAL LISTING

Reaction	E°, V
$WO_2 + 4 H^+ + 4 e \rightleftharpoons W + 2 H_2O$	−0.119
$WO_3 + 6 H^+ + 6 e \rightleftharpoons W + 3 H_2O$	−0.090
$2 WO_3 + 2 H^+ + 2 e \rightleftharpoons W_2O_5 + H_2O$	−0.029
$Y^{3+} + 3 e \rightleftharpoons Y$	−2.372
$Zn^{2+} + 2 e \rightleftharpoons Zn$	−0.7618
$Zn^{2+} + 2 e \rightleftharpoons Zn(Hg)$	−0.7628
$ZnO_2^{2-} + 2 H_2O + 2 e \rightleftharpoons Zn + 4 OH^-$	−1.215
$ZnSO_4 \cdot 7H_2O + 2 e \rightleftharpoons Zn(Hg) + SO_4^{2-}$ (Sat'd $ZnSO_4$)	−0.7993
$ZrO_2 + 4 H^+ + 4 e \rightleftharpoons Zr + 2 H_2O$	−1.553
$ZrO(OH)_2 + H_2O + 4 e \rightleftharpoons Zr + 4 OH^-$	−2.36

Table 2
REDUCTION REACTIONS HAVING E° VALUES MORE POSITIVE THAN THAT OF THE STANDARD HYDROGEN ELECTRODE

Reaction	E°, V
$2 H^+ + 2 e \rightleftharpoons H_2$	0.00000
$CuI_2^- + e \rightleftharpoons Cu + 2 I^-$	0.00
$Ge^{4+} + 2 e \rightleftharpoons Ge^{2+}$	0.00
$NO_3^- + H_2O + 2 e \rightleftharpoons NO_2^- + 2 OH^-$	0.01
$Tl_2O_3 + 3 H_2O + 4 e \rightleftharpoons 2 Tl^+ + 6 OH^-$	0.02
$SeO_4^{2-} + H_2O + 2 e \rightleftharpoons SeO_3^{2-} + 2 OH^-$	0.05
$UO_2^{2+} + e \rightleftharpoons UO_2^+$	0.062
$Pd(OH)_2 + 2 e \rightleftharpoons Pd + 2 OH^-$	0.07
$AgBr + e \rightleftharpoons Ag + Br^-$	0.07133
$S_4O_6^{2-} + 2 e \rightleftharpoons 2 S_2O_3^{2-}$	0.08
$AgSCN + e \rightleftharpoons Ag + SCN^-$	0.8951
$N_2 + 2 H_2O + 6 H^+ + 6 e \rightleftharpoons 2 NH_4OH$	0.092
$HgO + H_2O + 2 e \rightleftharpoons Hg + 2 OH^-$	0.0977
$Ir_2O_3 + 3 H_2O + 6 e \rightleftharpoons 2 Ir + 6 OH^-$	0.098
$2 NO + 2 e \rightleftharpoons N_2O_2^{2-}$	0.10
$[Co(NH_3)_6]^{3+} + e \rightleftharpoons [Co(NH_3)_6]^{2+}$	0.108
$Hg_2O + H_2O + 2 e \rightleftharpoons 2 Hg + 2 OH^-$	0.123
$Ge^{4+} + 4 e \rightleftharpoons Ge$	0.124
$Hg_2Br_2 + 2 e \rightleftharpoons 2 Hg + 2 Br^-$	0.13923
$Pt(OH)_2 + 2 e \rightleftharpoons Pt + 2 OH^-$	0.14
$S + 2H^+ + 2 e \rightleftharpoons H_2S(aq)$	0.142
$Np^{4+} + e \rightleftharpoons Np^{3+}$	0.147
$Ag_4[Fe(CN)_6] + 4 e \rightleftharpoons 4 Ag + [Fe(CN)_6]^{4-}$	0.1478
$IO_3^- + 2 H_2O + 4 e \rightleftharpoons IO^- + 4 OH^-$	0.15
$Mn(OH)_3 + e \rightleftharpoons Mn(OH)_2 + OH^-$	0.15
$2 NO_2^- + 3 H_2O + 4 e \rightleftharpoons N_2O + 6 OH^-$	0.15
$Sn^{4+} + 2 e \rightleftharpoons Sn^{2+}$	0.151
$Sb_2O_3 + 6 H^+ + 6 e \rightleftharpoons 2 Sb + 3 H_2O$	0.152
$Cu^{2+} + e \rightleftharpoons Cu^+$	0.153
$BiOCl + 2 H^+ + 3 e \rightleftharpoons Bi + Cl^- + H_2O$	0.1583
$Bi(Cl)_4^- + 3 e \rightleftharpoons Bi + 4 Cl^-$	0.16
$Co(OH)_3 + e \rightleftharpoons Co(OH)_2 + OH^-$	0.17
$SO_4^{2-} + 4 H^+ + 2 e \rightleftharpoons H_2SO_3 + H_2O$	0.172
$SbO^+ + 2 H^+ + 3 e \rightleftharpoons Sb + 2 H_2O$	0.212
$AgCl + e \rightleftharpoons Ag + Cl^-$	0.22233
$As_2O_3 + 6 H^+ + 6 e \rightleftharpoons 2 As + 3 H_2O$	0.234
Calomel electrode, saturated NaCl (SSCE)	0.2360
$Ge^{2+} + 2 e \rightleftharpoons Ge$	0.24
Calomel electrode, saturated KCl	0.2412
$PbO_2 + H_2O + 2 e \rightleftharpoons PbO + 2 OH^-$	0.247
$HAsO_2 + 3 H^+ + 3_e \rightleftharpoons As + 2 H_2O$	0.248
$Ru^{3+} + e \rightleftharpoons Ru^{2+}$	0.2487
$ReO_2 + 4 H^+ + 4 e \rightleftharpoons Re + 2 H_2O$	0.2513
$IO_3^- + 3 H_2O + 6 e \rightleftharpoons I^- + OH^-$	0.26
$Hg_2Cl_2 + 2 e \rightleftharpoons 2 Hg + 2 Cl^-$	0.26808
Calomel electrode, molal KCl	0.2800
Calomel electrode, 1 mol/1 KCl (NCE)	0.2801
$Re^{3+} + 3 e \rightleftharpoons Re$	0.300
$BiO^+ + 2 H^+ + 3 e \rightleftharpoons Bi + H_2O$	0.320
$UO_2^{2+} + 4 H^+ + 2 e \rightleftharpoons U^{4+} + 2 H_2O$	0.327
$ClO_3^- + H_2O + 2 e \rightleftharpoons ClO_2^- + 2 OH^-$	0.33
$2 HCNO + 2 H^+ + 2 e \rightleftharpoons (CN)_2 + 2 H_2O$	0.330
Calomel electrode, 0.1 mol/1 KCl	0.3337
$VO^{2+} + 2 H^+ + e \rightleftharpoons V^{3+} + H_2O$	0.337
$Cu^{2+} + 2 e \rightleftharpoons Cu$	0.3419
$Ag_2O + H_2O + 2 e \rightleftharpoons 2 Ag + 2 OH^-$	0.342
$Cu^{2+} + 2 e \rightleftharpoons Cu(Hg)$	0.345
$AgIO_3 + e \rightleftharpoons Ag + IO^-3$	0.354
$[Fe(CN)_6]^{3-} + e \rightleftharpoons [Fe(CN)_6]^{4-}$	0.358
$ClO_4^- + H_2O + 2 e \rightleftharpoons ClO^-_3 + 2 OH^-$	0.36
$Ag_2SeO_3 + 2 e \rightleftharpoons 2 Ag + SeO_3^{2-}$	0.3629
$ReO_4^- + 8 H^+ + 7 e \rightleftharpoons Re + 4 H_2O$	0.368
$(CN)_2 + 2 H^+ + 2 e \rightleftharpoons 2 HCN$	0.373
$[Ferricinium]^+ + e \rightleftharpoons$ ferrocene	0.400
$Tc^{2+} + 2 e \rightleftharpoons Tc$	0.400
$O_2 + 2 H_2O + 4 e \rightleftharpoons 4 OH^-$	0.401
$AgOCN + e \rightleftharpoons Ag + OCN^-$	0.41
$[RhCl_6]^{3-} + 3 e \rightleftharpoons Rh + 6 Cl^-$	0.431
$Ag_2CrO_4 + 2 e \rightleftharpoons 2 Ag + CrO_4^{2-}$	0.4470
$H_2SO_3 + 4 H^+ + 4 e \rightleftharpoons S + 3 H_2O$	0.449
$Ru^{2+} + 2 e \rightleftharpoons Ru$	0.455
$Ag_2MoO_4 + 2 e \rightleftharpoons 2 Ag + MoO_4^{2-}$	0.4573
$Ag_2C_2O_4 + 2 e \rightleftharpoons 2 Ag + C_2O_4^{2-}$	0.4647
$Ag_2WO_4 + 2 e \rightleftharpoons 2 Ag + WO_4^{2-}$	0.4660
$Ag_2CO_3 + 2 e \rightleftharpoons 2 Ag + CO_3^{2-}$	0.47
$TeO_4^- + 8 H^+ + 7 e \rightleftharpoons Te + 4 H_2O$	0.472
$IO^- + H_2O + 2 e \rightleftharpoons I^- + 2 OH^-$	0.485
$NiO_2 + 2 H_2O + 2 e \rightleftharpoons Ni(OH)_2 + 2 OH^-$	0.490
$ReO_4^- + 4 H^+ + 3 e \rightleftharpoons ReO_2 + 2 H_2O$	0.510
$Hg_2(ac)_2 + 2 e \rightleftharpoons 2 Hg + 2 (ac)^-$	0.51163
$Cu^+ + e \rightleftharpoons Cu$	0.521
$I_2 + 2 e \rightleftharpoons 2 I^-$	0.5355
$I_3^- + 2 e \rightleftharpoons 3 I^-$	0.536
$AgBrO_3 + e \rightleftharpoons Ag + BrO_3^-$	0.546
$MnO_4^- + e \rightleftharpoons MnO_4^{2-}$	0.558
$H_3AsO_4 + 2 H^+ + 2 e^- \rightleftharpoons HAsO_2 + 2 H_2O$	0.560
$S_2O_6^{2-} + 4 H^+ + 2 e \rightleftharpoons 2 H_2SO_3$	0.564
$AgNO_2 + e \rightleftharpoons Ag + NO_2^-$	0.564
$Te^{4+} + 4 e \rightleftharpoons Te$	0.568
$Sb_2O_5 + 6 H^+ + 4 e \rightleftharpoons 2 SbO^+ + 3 H_2O$	0.581
$RuO_4^- + e \rightleftharpoons RuO_4^{2+}$	0.59
$[PdCl_4]^{2-} + 2 e \rightleftharpoons Pd + 4 Cl^-$	0.591

Table 2 (continued)
REDUCTION REACTIONS HAVING E° VALUES MORE POSITIVE THAN THAT OF THE STANDARD HYDROGEN ELECTRODE

Reaction	E°, V
$TeO_2 + 4 H^+ + 4 e \rightleftharpoons Te + 2 H_2O$	0.593
$MnO_4^- + 2 H_2O + 3 e \rightleftharpoons MnO_2 + 4 OH^-$	0.595
$Rh^{2+} + 2 e \rightleftharpoons Rh$	0.600
$Rh^+ + e \rightleftharpoons Rh$	0.600
$MnO_4^{2-} + 2 H_2O + 2 e \rightleftharpoons MnO_2 + 4 OH^-$	0.60
$2 AgO + H_2O + 2 e \rightleftharpoons Ag_2O + 2 OH^-$	0.607
$BrO_3^- + 3 H_2O + 6 e \rightleftharpoons Br^- + 6 OH^-$	0.61
$UO_2^+ + 4 H^+ + e \rightleftharpoons U^{4+} + 2 H_2O$	0.612
$Hg_2SO_4 + 2 e \rightleftharpoons 2 Hg + SO_4^{2-}$	0.6125
$ClO_3^- + 3 H_2O + 6 e \rightleftharpoons Cl^- + 6 OH^-$	0.62
$Hg_2HPO_4 + 2 e \rightleftharpoons 2 Hg + HPO_4^{2-}$	0.6359
$Ag(ac) + e \rightleftharpoons Ag + (ac)^-$	0.643
Sb_2O_5(valentinite) $+ 4 H^+ + 4 e \rightleftharpoons Sb_2O_3 + 2 H_2O$	0.649
$Ag_2SO_4 + 2 e \rightleftharpoons 2 Ag + SO_4^{2-}$	0.654
$ClO_2^- + H_2O + 2 e \rightleftharpoons ClO^- + 2 OH^-$	0.66
Sb_2O_5(senarmontite) $+ 4 H^+ + 4 e \rightleftharpoons Sb_2O_3 + 2 H_2O$	0.671
$[PtCl_6]^{2-} + 2 e \rightleftharpoons [PtCl_4]^{2-} + 2 Cl^-$	0.68
$O_2 + 2 H^+ + 2 e \rightleftharpoons H_2O_2$	0.695
p-benzoquinone $+ 2 H^+ + 2 e \rightleftharpoons$ hydroquinone	0.6992
$H_3IO_6 + 2 e \rightleftharpoons IO_3^- + 3 OH^-$	0.7
$Ag_2O_3 + H_2O + 2 e \rightleftharpoons 2 AgO + 2 OH^-$	0.739
$[PtCl_4]^{2-} + 2 e \rightleftharpoons Pt + 4 Cl^-$	0.755
$Rh^{3+} + 3 e \rightleftharpoons Rh$	0.758
$ClO_2^- + 2 H_2O + 4 e \rightleftharpoons Cl^- + 4 OH^-$	0.76
$2 NO + H_2O + 2 e \rightleftharpoons N_2O + 2 OH^-$	0.76
$BrO^- + H_2O + 2 e \rightleftharpoons Br^- + 2 OH^-$	0.761
$ReO_4^- + 2 H^+ + e \rightleftharpoons ReO_3 + H_2O$	0.768
$(CNS)_2 + 2 e \rightleftharpoons 2 CNS^-$	0.77
$[IrCl_6]^{3-} + 3 e \rightleftharpoons Ir + 6 Cl^-$	0.77
$Fe^{3+} + e \rightleftharpoons Fe^{2+}$	0.771
$Ag(F) + e \rightleftharpoons Ag + F^-$	0.779
$TcO_4^- + 4 H^+ + 3 e \rightleftharpoons TcO_2 + 2 H_2O$	0.782
$Hg_2^{2+} + 2 e \rightleftharpoons 2 Hg$	0.7973
$Ag^+ + e \rightleftharpoons Ag$	0.7996
$2 NO_3^- + 4 H^+ + 2 e \rightleftharpoons N_2O_4 + 2 H_2O$	0.803
$ClO^- + H_2O + 2 e \rightleftharpoons Cl^- + 2 OH^-$	0.841
$OsO_4 + 8 H^+ + 8 e \rightleftharpoons Os + 4 H_2O$	0.85
$Hg^{2+} + 2 e \rightleftharpoons Hg$	0.851
$AuBr_4^- + 3 e \rightleftharpoons Au + 4 Br^-$	0.854
SiO_2(quartz) $+ 4 H^+ + 4 e \rightleftharpoons Si + 2 H_2O$	0.857
$2 HNO_2 + 4 H^+ + 4 e \rightleftharpoons H_2N_2O_2 + H_2O$	0.86
$[IrCl_6]^{2-} + e \rightleftharpoons [IrCl_6]^{3-}$	0.8665
$N_2O_4 + 2 e \rightleftharpoons 2 NO_2^-$	0.867
$HO_2^- + H_2O + 2 e \rightleftharpoons 3 OH^-$	0.878
$2 Hg^{2+} + 2 e \rightleftharpoons Hg_2^{2+}$	0.920
$NO_3^- + 3 H^+ + 2 e \rightleftharpoons HNO_2 + H_2O$	0.934
$Pd^{2+} + 2 e \rightleftharpoons Pd$	0.951
$ClO_2(aq) + e \rightleftharpoons ClO_2^-$	0.954
$NO_3^- + 4 H^+ + 3 e \rightleftharpoons NO + 2 H_2O$	0.957
$AuBr_2^- + e \rightleftharpoons Au + 2 Br^-$	0.959
$HNO_2 + H^+ + e \rightleftharpoons NO + H_2O$	0.983
$HIO + H^+ + 2 e \rightleftharpoons I^- + H_2O$	0.987
$VO_2^+ + 2 H^+ + e \rightleftharpoons VO^{2+} + H_2O$	0.991
$RuO_4 + e \rightleftharpoons RuO_4^-$	1.00
$V(OH)_4^+ + 2 H^+ + e \rightleftharpoons VO^{2+} + 3 H_2O$	1.00
$AuCl_4^- + 3 e \rightleftharpoons Au + 4 Cl^-$	1.002
$Pu^{4+} + e \rightleftharpoons Pu^{3+}$	1.006
$H_6TeO_6 + 2 H^+ + 2 e \rightleftharpoons TeO_2 + 4 H_2O$	1.02
$N_2O_4 + 4 H^+ + 4 e \rightleftharpoons 2 NO + 2 H_2O$	1.035
$[Fe(phen)_3]^{3+} + e \rightleftharpoons [Fe(phen)_3]^{2+}$ (1 (mol/ℓ H_2SO_4)	1.06
$PuO_2(OH)_2 + H^+ + e \rightleftharpoons PuO_2OH + H_2O$	1.062
$N_2O_4 + 2 H^+ + 2 e \rightleftharpoons 2 HNO_2$	1.065
$Br_2(1) + 2 e \rightleftharpoons 2 Br^-$	1.066
$IO_3^- + 6 H^+ + 6 e \rightleftharpoons I^- + 3 H_2O$	1.085
$Br_2(aq) + 2 e \rightleftharpoons 2 Br^-$	1.0873
$Pu^{5+} + e \rightleftharpoons Pu^{4+}$	1.099
$Cu^{2+} + 2 CN^- + e \rightleftharpoons [Cu(CN)_2]^-$	
$Pt^{2+} + 2 e \rightleftharpoons Pt$	1.118
$RuO_2 + 4 H^+ + 2 e \rightleftharpoons Ru^{2+} + 2 H_2O$	1.120
$[Fe(phenanthroline)_3]^{3+} + e \rightleftharpoons [Fe(phen)_3]^{2+}$	1.147
$SeO_4^{2-} + 4 H^+ + 2 e \rightleftharpoons H_2SeO_3 + H_2O$	1.151
$ClO_3^- + 2 H^+ + e \rightleftharpoons ClO_2 + H_2O$	1.152
$Ir^{3+} + 3 e \rightleftharpoons Ir$	1.156
$ClO_4^- + 2 H^+ + 2 e \rightleftharpoons ClO_3^- + H_2O$	1.189
$2 IO_3^- + 12 H^+ + 10 e \rightleftharpoons I_2\ 6 H_2O$	1.195
$ClO^-_3 + 3 H^+ + 2 e \rightleftharpoons HClO_2 + H_2O$	1.214
$MnO_2 + 4 H^+ + 2 e \rightleftharpoons Mn^{2+} + 2 H_2O$	1.224
$O_2 + 4 H^+ + 4 e \rightleftharpoons 2 H_2O$	1.229
$Cr_2O_7^{2-} + 14 H^+ + 6 e \rightleftharpoons 2 Cr^{3+} + 7 H_2O$	1.232
$O_3 + H_2O + 2 e \rightleftharpoons O_2 + 2OH^-$	1.24
$Tl^{3+} + 2 e \rightleftharpoons Tl^+$	1.252
$N_2H_5^+ + 3 H^+ + 2 e \rightleftharpoons 2 NH_4^+$	1.275
$ClO_2 + H^+ + e \rightleftharpoons HClO_2$	1.277
$[PdCl_6]^{2-} + 2 e \rightleftharpoons [PdCl_4]^{2-} + 2 Cl^-$	1.288
$2 HNO_2 + 4 H^+ + 4 e \rightleftharpoons N_2O + 3 H_2O$	1.297
$PuO_2(OH)_2 + 2 H^+ + 2 e \rightleftharpoons Pu(OH)_4$	1.325
$HBrO + H^+ + 2 e \rightleftharpoons Br^- + H_2O$	1.331
$HCrO_4^- + 7 H^+ + 3 e \rightleftharpoons Cr^{3+} + 4 H_2O$	1.350
$Cl_2(g) + 2 e \rightleftharpoons Cl^-$	1.35827
$ClO_4^- + 8 H^+ + 8 e \rightleftharpoons Cl^- + 4 H_2O$	1.389
$ClO_4^- + 8 H^+ + 7 e \rightleftharpoons {}^1/_2Cl_2 + 4 H_2O$	1.39
$Au^{3+} + 2 e \rightleftharpoons Au^+$	1.401
$2 NH_3OH^+ + H^+ + 2 e \rightleftharpoons N_2H_5^+ + 2 H_2O$	1.42
$BrO_3^- + 6 H^+ + 6 e \rightleftharpoons Br^- + 3 H_2O$	1.423
$2 HIO + 2 H^+ + 2 e \rightleftharpoons I_2 + 2 H_2O$	1.439
$Au(OH)_3 + 3 H^+ + 3 e \rightleftharpoons Au^- + 3 H_2O$	1.45
$310_3^- + 6 H^+ + 6 e \rightleftharpoons Cl^- + 3 H_2O$	1.451
$PbO_2 + 4 H^+ + 2 e \rightleftharpoons Pb^{2+} + 2 H_2O$	1.455
$ClO_3^- + 6 H^+ + 5 e \rightleftharpoons {}^1/_2Cl_2 + 3 H_2O$	1.47
$BrO_3^- + 6 H^+ + 5 e \rightleftharpoons {}^1/_2Br_2 + 3 H_2O$	1.482
$HClO + H^+ + 2 e \rightleftharpoons Cl^- + H_2O$	1.482
$HO_2 + H^+ + e \rightleftharpoons H_2O_2$	1.495
$Au^{3+} + 3 e \rightleftharpoons Au$	1.498
$MnO_4^- + 8 H^+ + 5 e \rightleftharpoons Mn^{2+} + 4 H_2O$	1.507
$Mn^{3+} + e \rightleftharpoons Mn^{2+}$	1.5415
$HClO_2 + 3 H^+ + 4 e \rightleftharpoons Cl^- + 2 H_2O$	1.570
$HBrO + H^+ + e \rightleftharpoons {}^1/_2Br_2(aq) + H_2O$	1.574

Table 2 (continued)
REDUCTION REACTIONS HAVING E° VALUES MORE POSITIVE THAN THAT OF THE STANDARD HYDROGEN ELECTRODE

Reaction	E°, V
$2\ NO + 2\ H^+ + 2\ e \rightleftharpoons N_2O + H_2O$	1.591
$Bi_2O_4 + 4\ H^+ + 2\ e \rightleftharpoons 2\ BiO^+ + 2\ H_2O$	1.593
$HBrO + H^+ + e \rightleftharpoons 1/2\ Br_2(\ell) + H_2O$	1.596
$H_5IO_6 + H^+ + 2\ e \rightleftharpoons IO_3^- + 3\ H_2O$	1.601
$Ce^{4+} + e \rightleftharpoons Ce^{3+}$	1.61
$HClO + H^+ + e \rightleftharpoons 1/2Cl_2 + H_2O$	1.611
$HClO_2 + 3\ H^+ + 3\ e \rightleftharpoons 1/2Cl_2 + 2\ H_2O$	1.628
$HClO_2 + 2\ H^+ + 2\ e \rightleftharpoons HClO + H_2O$	1.645
$NiO_2 + 4\ H^+ + 2\ e \rightleftharpoons Ni^{2+} + 2\ H_2O$	1.678
$MnO_4^- + 4\ H^+ + 3\ e \rightleftharpoons MnO_2 + 2\ H_2O$	1.679
$PbO_2 + SO_4^{2-} + 4\ H^+ + 2\ e \rightleftharpoons PbSO_4 + 2\ H_2O$	1.6913
$Au^+ + e \rightleftharpoons Au$	1.692
$CeOH^{3+} + H^+ + e \rightleftharpoons Ce^{3+} + H_2O$	1.715
$N_2O + 2\ H^+ + 2\ e \rightleftharpoons N_2 + H_2O$	1.766
$H_2O_2 + 2\ H^+ + 2\ e \rightleftharpoons 2\ H_2O$	1.776
$Co^{3+} + e \rightleftharpoons Co^{2+}$ (2 mol/ℓ H_2SO_4)	1.83
$Ag^{2+} + e \rightleftharpoons Ag^+$	1.980
$S_2O_8^{2-} + 2\ e \rightleftharpoons 2\ SO_4^{2-}$	2.010
$OH + e \rightleftharpoons OH^-$	2.02
$O_3 + 2\ H^+ + 2\ e \rightleftharpoons O_2 + H_2O$	2.076
$S_2O_8^{2-} + 2\ H^+ + 2\ e \rightleftharpoons 2\ HSO_4^-$	2.123
$F_2O + 2\ H^+ + 4\ e \rightleftharpoons H_2O + 2\ F^-$	2.153
$FeO_4^{2-} + 8\ H^+ + 3\ e \rightleftharpoons Fe^{3+} + 4\ H_2O$	2.20
$O(g) + 2\ H^+ + 2\ e \rightleftharpoons H_2O$	2.421
$H_2N_2O_2 + 2\ H^+ + 2\ e \rightleftharpoons N_2 + 2\ H_2O$	2.65
$F_2 + 2\ e \rightleftharpoons 2\ F^-$	2.866
$F_2 + 2\ H^+ + 2\ e \rightleftharpoons 2\ HF$	3.053

Table 3
REDUCTION REACTIONS HAVING E° VALUES MORE NEGATIVE THAN THAT OF THE STANDARD HYDROGEN ELECTRODE

Reaction	E°, V
$2\ H^+ + 2\ e \rightleftharpoons H_2$	0.00000
$AgCN + e \rightleftharpoons Ag + CN^-$	−0.017
$2\ WO_3 + 2\ H^+ + 2\ e \rightleftharpoons W_2O_5 + H_2O$	−0.029
$W_2O_5 + 2\ H^+ + 2\ e \rightleftharpoons 2\ WO_2 + H_2O$	−0.031
$D^+ + e \rightleftharpoons 1/2D_2$	−0.0034
$Ag_2S + 2\ H^+ + 2\ e \rightleftharpoons 2\ Ag + H_2S$	−0.0366
$Fe^{3+} + 3\ e \rightleftharpoons Fe$	−0.037
$Hg_2I_2 + 2\ e \rightleftharpoons 2\ Hg + 2\ I^-$	−0.0405
$2\ D^+ + 2\ e \rightleftharpoons D_2$	−0.044
$Tl(OH)_3 + 2\ e \rightleftharpoons TlOH + 2\ OH^-$	−0.05
$TiOH^{3+} + H^+ + e \rightleftharpoons Ti^{3+} + H_2O$	−0.055
$2\ H_2SO_3 + H^+ + 2\ e \rightleftharpoons HS_2O_4^- + 2\ H_2O$	−0.056
$P(white) + 3\ H^+ + 3\ e \rightleftharpoons PH_3(g)$	−0.063
$O_2^- + H_2O + 2\ e \rightleftharpoons HO_2^- + OH^-$	−0.076
$2\ Cu(OH)_2 + 2\ e \rightleftharpoons Cu_2O + 2\ OH^- + H_2O$	−0.080
$WO_3 + 6\ H^+ + 6\ e \rightleftharpoons W + 3\ H_2O$	−0.090
$P(red) + 3\ H^+ + 3\ e \rightleftharpoons PH_3(g)$	−0.111
$GeO_2 + 2\ H^+ + 2\ e \rightleftharpoons GeO + H_2O$	−0.118
$WO_2 + 4\ H^+ + 4\ e \rightleftharpoons W + 2\ H_2O$	−0.119
$Pb^{2+} + 2\ e \rightleftharpoons Pb(Hg)$	−0.1205
$Pb^{2+} + 2\ e \rightleftharpoons Pb$	−0.1262
$CrO_4^{2-} + 4\ H_2O + 3\ e \rightleftharpoons Cr(OH)_3 + 5\ OH^-$	−0.13
$Sn^{2-} + 2\ e \rightleftharpoons Sn$	−0.1375
$In^+ + e \rightleftharpoons In$	−0.14
$O_2 + 2\ H_2O + 2\ e \rightleftharpoons H_2O_2 + 2\ OH^-$	−0.146
$AgI + e \rightleftharpoons Ag + I^-$	−0.15224
$2\ NO_2^- + 2\ H_2O + 4\ e \rightleftharpoons N_2O_2^{2-} + 4\ OH^-$	−0.18
$H_2GeO_3 + 4\ H^+ + 4\ e \rightleftharpoons Ge + 3\ H_2O$	−0.182
$CO_2 + 2\ H^+ + 2\ e \rightleftharpoons HCOOH$	−0.199
$Mo^{3+} + 3\ e \rightleftharpoons Mo$	−0.200
$2\ SO_3^{2-} + 4\ H^+ + 2\ e \rightleftharpoons S_2O_6^{2-} + H_2O$	−0.22
$Cu(OH)_2 + 2\ e \rightleftharpoons Cu + 2\ OH^-$	−0.222
$CdSO_4 + 2\ e \rightleftharpoons Cd + SO_4^{2-}$	−0.246
$V(OH)_4^+ + 4\ H^+ + 5\ e \rightleftharpoons V + 4\ H_2O$	−0.254
$V^{3+} + e \rightleftharpoons V^{2+}$	−0.255
$Ni^{2+} + 2\ e \rightleftharpoons Ni$	−0.257
$PbCl_2 + 2\ e \rightleftharpoons Pb + 2\ Cl^-$	−0.2675
$H_3PO_4 + 2\ H^+ + 2\ e \rightleftharpoons H_3PO_3 + H_2O$	−0.276
$Co^{2+} + 2\ e \rightleftharpoons Co$	−0.28
$PbBr_2 + 2\ e \rightleftharpoons Pb + 2\ Br^-$	−0.284
$Tl^+ + e \rightleftharpoons Tl(Hg)$	−0.3338
$Tl^+ + e \rightleftharpoons Tl$	−0.336
$In^{3+} + 3\ e \rightleftharpoons In$	−0.3382
$TlOH + e \rightleftharpoons Tl + OH^-$	−0.34
$PbF_2 + 2\ e \rightleftharpoons Pb + 2\ F^-$	−0.3444
$PbSO_4 + 2\ e \rightleftharpoons Pb(Hg) + SO_4^{2-}$	−0.3505
$Cd^{2+} + 2\ e \rightleftharpoons Cd(Hg)$	−0.3521
$PbSO_4 + 2\ e \rightleftharpoons Pb + SO_4^{2-}$	−0.3588
$Cu_2O + H_2O + 2\ e \rightleftharpoons 2\ Cu + 2\ OH^-$	−0.360
$Eu^{3+} + e \rightleftharpoons Eu^{2+}$	−0.36
$PbI_2 + 2\ e \rightleftharpoons Pb + 2\ I^-$	−0.365
$SeO_3^{2-} + 3\ H_2O + 4\ e \rightleftharpoons Se + 6\ OH^-$	−0.366
$Ti^{3+} + e \rightleftharpoons Ti^{2+}$	−0.368
$Se + 2\ H^+ + 2\ e \rightleftharpoons H_2Se(aq)$	−0.399
$In^{2+} + e \rightleftharpoons In^+$	−0.40
$Cd^{2+} + 2\ e \rightleftharpoons Cd$	−0.4030
$Cr^{3+} + e \rightleftharpoons Cr^{2+}$	−0.407
$2\ S + 2\ e \rightleftharpoons S_2^{2-}$	−0.42836
$Tl_2SO_4 + 2\ e \rightleftharpoons Tl + SO_4^{2-}$	−0.4360
$In^{3+} + 2\ e \rightleftharpoons In^+$	−0.443
$Fe^{2+} + 2\ e \rightleftharpoons Fe$	−0.447
$H_3PO_3 + 3\ H^+ + 3\ e \rightleftharpoons P + 3\ H_2O$	−0.454
$Bi_2O_3 + 3\ H_2O + 6\ e \rightleftharpoons 2\ Bi + 6\ OH^-$	−0.46
$NO_2^- + H_2O + e \rightleftharpoons NO + 2\ OH$	−0.46
$PbHPO_4 + 2\ e \rightleftharpoons Pb + HPO_4^{2-}$	−0.465
$S + 2\ e \rightleftharpoons S^{2-}$	−0.47627
$S + H_2O + 2\ e \rightleftharpoons HS^- + OH^-$	−0.478
$In^{3+} + e \rightleftharpoons In^{2+}$	−0.49
$H_3PO_3 + 2\ H^+ + 2\ e \rightleftharpoons H_3PO_2 + H_2O$	−0.499
$TiO_2 + 4\ H^+ + 2\ e \rightleftharpoons Ti^{2+} + 2\ H_2O$	−0.502
$H_3PO_2 + H^+ + e \rightleftharpoons P + 2\ H_2O$	−0.508

Table 3 (continued)
REDUCTION REACTIONS HAVING E° VALUES MORE NEGATIVE THAN THAT OF THE STANDARD HYDROGEN ELECTRODE

Reaction	E°, V
$Sb + 3\ H^+ + 3\ e \rightleftharpoons SbH_3$	−0.510
$HPbO_2^- + H_2O + 2\ e \rightleftharpoons Pb + 3\ OH^-$	−0.537
$TlCl + e \rightleftharpoons Tl + Cl^-$	−0.5568
$Ga^{3+} + 3\ e \rightleftharpoons Ga$	−0.560
$Fe(OH)_3 + e \rightleftharpoons Fe(OH)_2 + OH^-$	−0.56
$TeO_3^{2-} + 3\ H_2O + 4\ e \rightleftharpoons Te + 6\ OH^-$	−0.57
$2\ SO_3^{2-} + 3\ H_2O + 4\ e \rightleftharpoons S_2O_3^{2-} + 6\ OH^-$	−0.571
$PbO + H_2O + 2\ e \rightleftharpoons Pb + 2\ OH^-$	−0.580
$ReO_2^- + 4\ H_2O + 7\ e \rightleftharpoons Re + 8\ OH^-$	−0.584
$SbO_3^- + H_2O + 2\ e \rightleftharpoons SbO_2^- + 2\ OH^-$	−0.59
$U^{4+} + e \rightleftharpoons U^{3+}$	−0.607
$As + 3\ H^+ + 3\ e \rightleftharpoons AsH_3$	−0.608
$Nb_2O_5 + 10\ H^+ + 10\ e \rightleftharpoons 2\ Nb + 5\ H_2O$	−0.644
$TlBr + e \rightleftharpoons Tl + Br^-$	−0.658
$SbO_2^- + 2\ H_2O + 3\ e \rightleftharpoons Sb + 4\ OH^-$	−0.66
$AsO_2^- + 2\ H_2O + 3\ e \rightleftharpoons As + 4\ OH^-$	−0.68
$Ag_2S + 2\ e \rightleftharpoons 2\ Ag + S^{2-}$	−0.691
$AsO_4^{3-} + 2\ H_2O + 2\ e \rightleftharpoons AsO_2^- + 4\ OH^-$	−0.71
$Ni(OH)_2 + 2\ e \rightleftharpoons Ni + 2\ OH^-$	−0.72
$Co(OH)_2 + 2\ e \rightleftharpoons Co + 2\ OH^-$	−0.73
$H_2SeO_3 + 4\ H^+ + 4\ e \rightleftharpoons Se + 3\ H_2O$	−0.74
$Cr^{3+} + 3\ e \rightleftharpoons Cr$	−0.744
$Ta_2O_5 + 10\ H^+ + 10\ e \rightleftharpoons 2\ Ta + 5\ H_2O$	−0.750
$TlI + e \rightleftharpoons Tl + I^-$	−0.752
$Zn^{2+} + 2\ e \rightleftharpoons Zn$	−0.7618
$Zn^{2+} + 2\ e \rightleftharpoons Zn(Hg)$	−0.7628
$Te + 2\ H^+ + 2\ e \rightleftharpoons H_2Te$	−0.793
$ZnSO_4{\cdot}7H_2O + 2\ e \rightleftharpoons Zn(Hg) + SO_4^{2-}$ (Sat'd $ZnSO_4$)	−0.7993
$Cd(OH)_2 + 2\ e \rightleftharpoons Cd(Hg) + 2\ OH^-$	−0.809
$2\ H_2O + 2\ e \rightleftharpoons H_2 + 2\ OH^-$	−0.8277
$2\ NO_3^- + 2\ H_2O + 2\ e \rightleftharpoons N_2O_4 + 4\ OH^-$	−0.85
$H_3BO_3 + 3\ H^+ + 3\ e \rightleftharpoons B + 3\ H_2O$	−0.8698
$P + 3\ H_2O + 3\ e \rightleftharpoons PH_3(g) + 3\ OH^-$	−0.87
$HSnO_2^- + H_2O + 2\ e \rightleftharpoons Sn + 3\ OH^-$	−0.909
$Cr^{2+} + 2\ e \rightleftharpoons Cr$	−0.913
$Se + 2\ e \rightleftharpoons Se^{2-}$	−0.924
$SO_4^{2-} + H_2O + 2\ e \rightleftharpoons SO_3^{2-} + 2\ OH^-$	−0.93
$Sn(OH)_6^{2-} + 2\ e \rightleftharpoons HSnO_2^- + 3\ OH^- + H_2O$	−0.93
$NpO_2 + H_2O + H^+ + e \rightleftharpoons Np(OH)_3$	−0.962
$PO_4^{3-} + 2\ H_2O + 2\ e \rightleftharpoons HPO_3^{2-} + 3\ OH^-$	−1.05
$Nb^{3+} + 3\ e \rightleftharpoons Nb$	−1.099
$2\ SO_3^{2-} + 2\ H_2O + 2\ e \rightleftharpoons S_2O_4^{2-} + 4\ OH^-$	−1.12
$Te + 2\ e \rightleftharpoons Te^{2-}$	−1.143
$V^{2+} + 2\ e \rightleftharpoons V$	−1.175
$Mn^{2+} + 2\ e \rightleftharpoons Mn$	−1.185
$CrO_2^- + 2\ H_2O + 3\ e \rightleftharpoons Cr + 4\ OH^-$	−1.2
$ZnO_2^- + 2\ H_2O + 2\ e \rightleftharpoons Zn + 4\ OH^-$	−1.215
$H_2GaO_3^- + H_2O + 3\ e \rightleftharpoons Ga + 4\ OH^-$	−1.219
$H_2BO_3^- + 5\ H_2O + 8\ e \rightleftharpoons BH_4^- + 8\ OH^-$	−1.24
$SiF_6^{2-} + 4\ e \rightleftharpoons Si + 6\ F^-$	−1.24
$Ce^{3+} + 3\ e \rightleftharpoons Ce(Hg)$	−1.4373
$UO_2^{2+} + 4\ H^+ + 6\ e \rightleftharpoons U + 2\ H_2O$	−1.444
$Cr(OH)_3 + 3\ e \rightleftharpoons Cr + 3\ OH^-$	−1.48
$HfO_2 + 4\ H^+ + 4\ e \rightleftharpoons Hf + 2\ H_2O$	−1.505
$ZrO_2 + 4\ H^+ + 4\ e \rightleftharpoons Zr + 2\ H_2O$	−1.553
$Mn(OH)_2 + 2\ e \rightleftharpoons Mn + 2\ OH^-$	−1.56
$Ba^{2+} + 2\ e \rightleftharpoons Ba(Hg)$	−1.570
$Ti^{2+} + 2\ e \rightleftharpoons Ti$	−1.630
$HPO_3^{2-} + 2\ H_2O + 2\ e \rightleftharpoons H_2PO_2^- + 3\ OH^-$	−1.65
$Al^{3+} + 3\ e \rightleftharpoons Al$	−1.662
$SiO_3^{2-}\ H_2O + 4\ e \rightleftharpoons Si + 6\ OH^-$	−1.697
$HPO_3^{2-} + 2\ H_2O + 3\ e \rightleftharpoons P + 5\ OH^-$	−1.71
$HfO^{2+} + 2\ H^+ + 4\ e \rightleftharpoons Hf + H_2O$	−1.724
$ThO_2 + 4\ H^+ + 4\ e \rightleftharpoons Th + 2\ H_2O$	−1.789
$H_2BO_3^- + H_2O + 3\ e \rightleftharpoons B + 4\ OH^-$	−1.79
$Sr^{2+} + 2\ e \rightleftharpoons Sr(Hg)$	−1.793
$U^{3+} + 3\ e \rightleftharpoons U$	−1.798
$H_2PO_2^- + e \rightleftharpoons P + 2\ OH^-$	−1.82
$Be^{2+} + 2\ e \rightleftharpoons Be$	−1.847
$Np^{3+} + 3\ e \rightleftharpoons Np$	−1.856
$Th^{4+} + 4\ e \rightleftharpoons Th$	−1.899
$Pu^{3+} + 3\ e \rightleftharpoons Pu$	−2.031
$AlF_6^{3-} + 3\ e \rightleftharpoons Al + 6\ F^-$	−2.069
$Sc^{3+} + 3\ e \rightleftharpoons Sc$	−2.077
$H_2 + 2\ e \rightleftharpoons 2\ H^-$	−2.23
$H_2AlO_3^- + H_2O + 3\ e \rightleftharpoons Al + 4\ OH^-$	−2.33
$ZrO(OH)_2 + H_2O + 4\ e \rightleftharpoons Zr + 4\ OH^-$	−2.36
$Mg^{2+} + 2\ e \rightleftharpoons Mg$	−2.372
$Y^{3+} + 3\ e \rightleftharpoons Y$	−2.372
$Eu^{3+} + 3\ e \rightleftharpoons Eu$	−2.407
$Nd^{3+} + 3\ e \rightleftharpoons Nd$	−2.431
$Th(OH)_4 + 4\ e \rightleftharpoons Th + 4\ OH^-$	−2.48
$Ce^{3+} + 3\ e \rightleftharpoons Ce$	−2.483
$HfO(OH)_2 + H_2O + 4\ e \rightleftharpoons Hf + 4\ OH^-$	−2.50
$La^{3+} + 3\ e \rightleftharpoons La$	−2.522
$Be_2O_3^{2-} + 3\ H_2O + 4\ e \rightleftharpoons 2\ Be + 6\ OH^-$	−2.63
$Mg(OH)_2 + 2\ e \rightleftharpoons Mg + 2\ OH^-$	−2.690
$Mg^+ + e \rightleftharpoons Mg$	−2.70
$Na^+ + e \rightleftharpoons Na$	−2.71
$Ca^{2+} + 2\ e \rightleftharpoons Ca$	−2.868
$Sr(OH)_2 + 2\ e \rightleftharpoons Sr + 2\ OH^-$	−2.88
$Sr^{2+} + 2\ e \rightleftharpoons Sr$	−2.89
$La(OH)_3 + 3\ e \rightleftharpoons La + 3\ OH^-$	−2.90
$Ba^{2+} + 2\ e \rightleftharpoons Ba$	−2.912
$Cs^+ + e \rightleftharpoons Cs$	−2.92
$K^+ + e \rightleftharpoons K$	−2.931
$Rb^+ + e \rightleftharpoons Rb$	−2.98
$Ba(OH)_2 + 2\ e \rightleftharpoons Ba + 2\ OH^-$	−2.99
$Ca(OH)_2 + 2\ e \rightleftharpoons Ca + 2\ OH^-$	−3.02
$Li^+ + e \rightleftharpoons Li$	−3.0401
$3\ N_2 + 2\ H^+ + 2\ e \rightleftharpoons 2\ NH_3$	−3.09
$Eu^{2+} + 2\ e \rightleftharpoons Eu$	−3.395
$Ca^+ + e \rightleftharpoons Ca$	−3.80
$Sr^+ + e \rightleftharpoons Sr$	−4.10

DISSOCIATION CONSTANTS OF ORGANIC BASES IN AQUEOUS SOLUTION

No.	Compound	Temp. °C	Step	pK_a	K_a
a1	Acetamide	25		0.63	2.34×10^{-1}
a2	Acridine	20		5.58	2.63×10^{-6}
a3	α-Alanine	25		2.345	4.52×10^{-3}
a4	Alanine, glycyl-	25		3.153	7.03×10^{-4}
a5	Alanine, methoxy- (DL)	25		2.037	9.18×10^{-3}
a6	Alanine, phenyl	25	2	9.18	6.61×10^{-10}
a7	Allothreonine	25	1	2.108	7.80×10^{-3}
	Allothreonine	25	2	9.096	8.02×10^{-10}
a8	*n*-Amylamine	25		10.63	2.34×10^{-11}
a9	Aniline	25		4.63	2.34×10^{-5}
a10	Aniline, *n*-allyl	25		4.17	6.76×10^{-5}
a11	Aniline, 4-(p-aminobenzoyl)	25	1	2.932	1.17×10^{-3}
a12	Aniline, 4-benzyl	25		2.17	6.76×10^{-3}
a13	Aniline, 2-bromo	25		2.53	2.95×10^{-3}
a14	Aniline, 3-bromo	25		3.58	2.63×10^{-4}
a15	Aniline, 4-bromo	25		3.86	1.38×10^{-4}
a16	Aniline, 4-bromo-N,N-dimethyl	25		4.232	5.86×10^{-5}
a17	Aniline, o-chloro	25		2.65	2.24×10^{-3}
a18	Aniline, m-chloro	25		3.46	3.47×10^{-4}
a19	Aniline, p-chloro	25		4.15	7.08×10^{-5}
a20	Aniline, 3-chloro-N,N-dimethyl	20		3.837	1.46×10^{-4}
a21	Aniline, 4-chloro-N,N-dimethyl	20		4.395	4.03×10^{-5}
a22	Aniline, 3,5-dibromo	25		2.34	4.57×10^{-3}
a23	Aniline, 2,4-dichloro	22		2.05	8.91×10^{-3}
a24	Aniline, N,N-diethyl	22		6.61	2.46×10^{-7}
a25	Aniline, N,N-dimethyl	25		5.15	7.08×10^{-6}
a26	Aniline, N,N-dimethyl-3-nitro	25		2.626	2.37×10^{-3}
a27	Aniline, N-ethyl	24		5.12	7.59×10^{-6}
a28	Aniline, 2-fluoro	25		3.20	6.31×10^{-4}
a29	Aniline, 3-fluoro	25		3.50	3.16×10^{-4}
a30	Aniline, 4-fluoro	25		4.65	2.24×10^{-5}
a31	Aniline, 2-iodo	25		2.60	2.51×10^{-3}
a32	Aniline, N-methyl	25		4.848	1.41×10^{-5}
a33	Aniline, 4-methylthio	25		4.35	4.46×10^{-5}
a34	Aniline, 3-nitro	25		2.466	3.42×10^{-3}
a35	Aniline, 4-nitro	25		1.0	1.00×10^{-1}
a36	Aniline, 2-sulfonic acid	25	2	2.459	3.47×10^{-3}
a37	Aniline, 3-sulfonic acid	25	2	3.738	1.82×10^{-4}
a38	Aniline, 4-sulfonic acid	25	2	3.227	5.92×10^{-4}
a39	*o*-Anisidine	25		4.52	3.02×10^{-5}
a40	*m*-Anisidine	25		4.23	5.89×10^{-5}
a41	*p*-Anisidine	25		5.34	4.57×10^{-6}
a42	Arginine	25	1	1.8217	1.51×10^{-2}
		25	2	8.9936	1.01×10^{-9}
a43	Asparagine	20	1	2.213	6.12×10^{-3}
		20	2	8.85	1.41×10^{-9}
a44	Asparagine, glycyl	25	1	2.942	1.14×10^{-3}
		18	2	8.44	3.63×10^{-9}
a45	DL-Aspartic acid	1	1	2.122	7.55×10^{-3}
		1	2	4.006	1.00×10^{-4}
a46	Azetidine (Trimethylimidine)	25		11.29	5.12×10^{-12}
a47	Aziridine	25		8.01	9.77×10^{-9}
b1	Benzene, 4-aminoazo	25		2.82	1.51×10^{-3}
b2	Benzene, 2-aminoethyl (β-Phenylamine)	25		9.84	1.45×10^{-10}
b3	Benzene, 4-dimethylaminoazo	25		3.226	5.94×10^{-4}
b4	Benzidine	30	1	4.66	2.19×10^{-5}
		30	2	3.57	2.69×10^{-4}
b5	Benzimidazole	25		5.532	2.94×10^{-6}
b6	Benzimidazole, 2-ethyl	25		6.18	6.61×10^{-7}
b7	Benzimidazole, 2-methyl	25		6.19	6.46×10^{-7}
b8	Benzimidazole, 2-phenyl	25	1	5.23	5.89×10^{-6}
		25	2	11.91	1.23×10^{-12}
b9	Benzoic acid, 2-amino (Anthranilic acid)	25	1	2.108	7.80×10^{-3}
		25	2	4.946	1.13×10^{-5}
b10	Benzoic acid, 4-amino	25	1	2.501	3.15×10^{-3}
		25	2	4.874	1.33×10^{-5}
b11	Benzylamine	25		9.33	4.67×10^{-10}
b12	Betaine	0		1.83	1.48×10^{-2}
b13	Biphenyl, 2-amino	22		3.82	1.51×10^{-4}
b14	Bornylamine(trans-)	25		10.17	6.76×10^{-11}
b15	Brucine	25	1	8.28	5.24×10^{-9}
b16	Butane, 1-amino-3-methyl	25		10.60	2.51×10^{-11}
b17	Butane, 2-amino-2-methyl	19		10.85	1.41×10^{-11}
b18	Butane, 1,4-diamino (Putrescine)	10	1	11.15	7.08×10^{-12}
		10	2	9.71	1.95×10^{-10}
b19	*n*-Butylamine	20		10.77	1.69×10^{-11}
b20	*t*-Butylamine	18		10.83	1.48×10^{-11}
b21	Butyric acid, 4-amino	25	1	4.0312	9.31×10^{-5}
		25	2	10.5557	2.78×10^{-11}
b22	*n*-Butyric acid, glycyl-2-amino	25	1	3.1546	7.01×10^{-4}
c1	Cacodylic acid	25	1	1.57	2.69×10^{-2}
			2	6.27	5.37×10^{-7}
c2	β-Chlortriethyl-ammonium	25		8.80	1.59×10^{-9}
c3	Cinnoline	20		2.37	4.27×10^{-3}
c4	Codeine	25		8.21	6.15×10^{-9}
c5	Cyclohexaneamine, *n*-butyl	25		11.23	5.89×10^{-12}
c6	Cyclohexylamine	24		10.66	2.19×10^{-11}
c7	Cystine	30	1	1.90	1.25×10^{-2}
		30	2	8.24	5.76×10^{-9}
d1	*n*-Decylamine	25		10.64	2.29×10^{-11}
d2	Diethylamine	40		10.489	3.24×10^{-11}
d3	Diisobutylamine	21		10.91	1.23×10^{-11}
d4	Diisopropylamine	28.5		10.96	1.09×10^{-11}
d5	Dimethylamine	25		10.732	1.85×10^{-11}
d6	*n*-Diphenylamine	25		0.79	1.62×10^{-1}
d7	*n*-Dodecaneamine (Laurylamine)	25		10.63	2.35×10^{-11}
e1	*d*-Ephedrine	10		10.139	7.26×10^{-11}
e2	*l*-Ephedrine	10		9.958	1.10×10^{-10}
e3	Ethane, 1-amino-3-methoxy	10		9.89	1.29×10^{-10}
e4	Ethane, 1,2-bismethylamino	25	1	10.40	3.98×10^{-11}
		25	2	8.26	5.50×10^{-9}
e5	Ethanol, 2-amino	25		9.50	3.16×10^{-10}
e6	Ethylamine	20		10.807	1.56×10^{-11}
e7	Ethylenediamine	0	1	10.712	1.94×10^{-11}
		0	2	7.564	2.73×10^{-8}
g1	*l*-Glutamic acid	25	1	2.13	7.41×10^{-3}
		25	2	4.31	4.90×10^{-5}
e2	Glutamic acid, α-monoethyl	25	1	3.846	1.42×10^{-4}
		25	2	7.838	1.45×10^{-8}
e3	*l*-Glutamine	—		9.28	5.25×10^{-10}
e4	*l*-Glutathione	25	2	3.59	2.57×10^{-4}
e5	Glycine	25	1	2.3503	4.46×10^{-3}
		25	2	9.7796	1.68×10^{-10}
e6	Glycine, *n*-acetyl	25		3.6698	2.14×10^{-4}
e7	Glycine, dimethyl	5		10.3371	4.60×10^{-11}
e8	Glycine, glycyl	25		3.1397	7.25×10^{-4}
e9	Glycine, glycylglycyl	25	1	3.225	5.96×10^{-4}
		25	2	8.090	8.13×10^{-9}
e10	Glycine, leucyl	25	1	3.25	5.62×10^{-4}
		25	2	8.28	5.25×10^{-9}
e11	Glycine, methyl (Sarcosine)	25	1	2.21	6.16×10^{-3}
		25	2	10.12	7.58×10^{-11}
g12	Glycine, phenyl	25	1	1.83	1.48×10^{-2}
		25	2	4.39	4.07×10^{-5}
g13	Glycine, N,*n*-propyl	25	1	2.35	4.46×10^{-3}
		25	2	10.19	6.46×10^{-11}
g14	Glycine, tetraglycyl	20	1	3.10	7.94×10^{-4}
		20	2	8.02	9.55×10^{-9}
g15	Glycylserine	25	1	2.9808	1.04×10^{-3}
		25	2	8.38	4.17×10^{-9}
h1	Hexadecaneamine	25		10.63	2.35×10^{-11}
h2	Heptane, 1-amino	25		10.66	2.19×10^{-11}
h3	Heptane, 2-amino	19		10.88	1.58×10^{-11}
h4	Heptane, 2-methylamino	17		10.99	1.02×10^{-11}
h5	Hexadecaneamine	25		10.61	2.46×10^{-11}
h6	Hexamethylene-diamine	0	1	11.857	1.39×10^{-12}
		0	2	10.762	1.73×10^{-11}
h7	Hexanoic acid, 6-amino	25	1	4.373	4.23×10^{-5}
		25	2	10.804	1.57×10^{-11}
h8	*n*-Hexylamine	25		10.56	2.75×10^{-11}
h9	*dl*-Histidine	25	1	1.80	1.58×10^{-2}
		25	2	6.04	9.12×10^{-7}
		25	3	9.33	4.67×10^{-10}
h10	Histidine, β-alanyl (Carnosine)	20	1	2.73	1.86×10^{-3}
		20	2	6.87	1.35×10^{-7}
		20	3	9.73	1.48×10^{-10}
i1	Imidazol	25		6.953	1.11×10^{-7}
i2	Imidazol, 2,4-dimethyl	25		8.359	5.50×10^{-9}
i3	Imidazol, 1-methyl (Oxalmethyline)	25		6.95	1.12×10^{-7}
i4	Indane, 1-amino (*d*-1-Hydrindamine	22.5		9.21	6.17×10^{-10}
i5	Isobutyric acid, 2-amino	25	1	2.357	4.30×10^{-3}
		25	2	10.205	6.23×10^{-11}
i6	Isoleucine	25	1	2.318	4.81×10^{-3}

No.	Compound	Temp. °C	Step	pK_a	K_a
		25	2	9.758	1.74×10^{-10}
i7	Isoquinoline (Leucoline)	20		5.42	3.80×10^{-6}
i8	Isoquinoline, 1-amino	20		7.59	2.57×10^{-8}
i9	Isoquinoline, 7-hydroxy	20	1	5.68	2.09×10^{-6}
		20	2	8.90	1.26×10^{-9}
l1	*L*-Leucine	25	1	2.328	4.70×10^{-3}
		25	2	9.744	1.80×10^{-10}
l2	Leucine, glycyl	25		3.18	6.61×10^{-4}
m1	Methionine	25	1	2.22	6.02×10^{-3}
		25	2	9.27	5.37×10^{-10}
m2	Methylamine	25		10.657	2.70×10^{-11}
m3	Morphine	25		8.21	6.16×10^{-9}
m4	Morpholine	25		8.33	4.67×10^{-9}
n1	Naphthalene, 1-amino-6-hydroxy	25		3.97	1.07×10^{-4}
n2	Naphthalene, dimethylamino	25		4.566	2.72×10^{-5}
n3	α-Naphthylamine	25		3.92	1.20×10^{-4}
n4	β-Naphthylamine	25		4.16	6.92×10^{-5}
n5	α-Naphthylamine, *n*-methyl	27		3.67	2.13×10^{-4}
n6	Neobornylamine(cis-)	25		10.01	9.77×10^{-11}
n7	Nicotine	25	1	8.02	9.55×10^{-9}
		25	2	3.12	7.59×10^{-4}
n8	*n*-Nonylamine	25		10.64	2.29×10^{-11}
n9	Norleucine	25		2.335	4.62×10^{-3}
o1	Octadecaneamine	25		10.60	2.51×10^{-11}
o2	Octylamine	25		10.65	2.24×10^{-11}
o3	Ornithine	25	1	1.705	1.97×10^{-2}
		25	2	8.690	2.04×10^{-9}
p1	Papaverine	25		6.40	3.98×10^{-7}
p2	Pentane, 3-amino	17		10.59	2.57×10^{-11}
p3	Pentane, 3-amino-3-methyl	16		11.01	9.77×10^{-12}
p4	*n*-Pentadecylamine	25		10.61	2.46×10^{-11}
p5	Pentanoic acid, 5-amino(Valeric acid)	25	1	4.270	5.37×10^{-5}
		25	2	10.766	1.71×10^{-11}
p6	Perimidine	20		6.35	4.47×10^{-7}
p7	Phenanthridine	20		5.58	2.63×10^{-6}
p8	1,10-Phenanthroline	25		4.84	1.44×10^{-5}
p9	*o*-Phenetidine (2-Ethoxyaniline)	28		4.43	3.72×10^{-5}
p10	*m*-Phenetidine (3-Ethoxyaniline)	25		4.18	6.60×10^{-5}
p11	*p*-Phenetidine (4-Ethoxyaniline)	28		5.20	6.31×10^{-6}
p12	α-Picoline	20		5.97	1.07×10^{-6}
p13	β-Picoline	20		5.68	2.09×10^{-6}
p14	γ-Picoline	20		6.02	9.55×10^{-7}
p15	Pilocarpine	30		6.87	1.35×10^{-7}
p16	Piperazine	23.5	1	9.83	1.48×10^{-10}
		23.5	2	5.56	2.76×10^{-6}
p17	Piperazine, 2,5-dimethyl(trans-)	25	1	9.66	2.19×10^{-10}
		25	2	5.20	6.31×10^{-6}
p18	Piperidine	25		11.123	7.53×10^{-12}
p19	Piperidine, 3-acetyl	25		3.18	6.61×10^{-4}
p20	Piperidine, 1-*n*-butyl	23		10.47	3.39×10^{-11}
p21	Piperidine, 1,2-dimethyl	25		10.22	6.03×10^{-11}
p22	Piperidine, 1-ethyl	23		10.45	3.55×10^{-11}
p23	Piperidine, 1-methyl	25		10.08	8.32×10^{-11}
p24	Piperidine, 2,2,6,6-tetramethyl	25		11.07	8.51×10^{-12}
p25	Piperidine, 2,2,4-trimethyl	30		11.04	9.12×10^{-12}
p26	Proline	25	1	1.952	1.11×10^{-2}
		25	2	10.640	2.29×10^{-11}
p27	Proline, hydroxy	25	1	1.818	1.52×10^{-2}
		25	2	9.662	2.18×10^{-10}
p28	Propane, 1-amino-2,2-dimethyl	25		10.15	7.08×10^{-11}
p29	Propane, 1,2-diamino	25	1	9.82	1.52×10^{-10}
		25	2	6.61	2.46×10^{-7}
p30	Propane, 1,3-diamino	10	1	10.94	1.15×10^{-11}
		10	2	9.03	9.33×10^{-10}
p31	Propane, 1,2,3-triamino	20	1	9.59	2.57×10^{-10}
		20	2	7.95	1.12×10^{-8}
p32	Propanoic acid, 3-amino (β-Alanine)	25	1	3.551	2.81×10^{-4}
		25	2	10.238	5.78×10^{-11}
p33	Propylamine	20		10.708	1.96×10^{-11}
p34	Pteridine	20		4.05	8.91×10^{-5}
p35	Pteridine, 2-amino-4,6-dihydroxy	20	2	6.59	2.57×10^{-7}
		20	3	9.31	4.90×10^{-10}
p36	Pteridine, 2-amino-4-hydroxy	20	1	2.27	5.37×10^{-3}
		20	2	7.96	1.10×10^{-8}
p37	Pteridine, 6-chloro	20		3.68	2.09×10^{-4}
p38	Pteridine, 6-hydroxy-4-methyl	20	1	4.08	8.32×10^{-5}
			2	6.41	3.89×10^{-7}
p39	Purine	20	1	2.30	5.01×10^{-3}
		20	2	8.96	1.10×10^{-9}
p40	Purine, 6-amino (Adenine)	25	1	4.12	7.59×10^{-5}
		25	2	9.83	1.48×10^{-10}
p41	Purine, 2-dimethylamino	20	1	4.00	1.00×10^{-4}
		20	2	10.24	5.75×10^{-11}
p42	Purine, 8-hydroxy	20	1	2.56	2.75×10^{-3}
		20	2	8.26	9.49×10^{-9}
p43	Pyrazine	27		0.65	2.24×10^{-1}
p44	Pyrazine, 2-methyl	27		1.45	3.54×10^{-2}
p45	Pyrazine, methylamino	25		3.39	4.07×10^{-4}
p46	Pyridazine	20		2.24	5.76×10^{-3}
p47	Pyrimidine, 2-amino	20		3.45	3.54×10^{-4}
p48	Pyrimidine, 2-amino-4,6-dimethyl	20		4.82	1.51×10^{-5}
p49	Pyrimidine, 2-amino-5-nitro	20		0.35	4.46×10^{-1}
p50	Pyridine	25		5.25	5.62×10^{-6}
p51	Pyridine, 2-aldoxime	20	1	3.59	2.57×10^{-4}
		20	2	10.18	6.61×10^{-11}
p52	Pyridine, 2-amino	20		6.82	1.51×10^{-7}
p53	Pyridine, 4-amino	25		9.1141	7.69×10^{-10}
p54	Pyridine, 2-benzyl	25		5.13	7.41×10^{-6}
p55	Pyridine, 3-bromo	25		2.84	1.45×10^{-3}
p56	Pyridine, 3-chloro	25		2.84	1.45×10^{-3}
p57	Pyridine, 2,5-diamino	20		6.48	3.31×10^{-7}
p58	Pyridine, 2,3-dimethyl (2,3-Lutidine)	25		6.57	2.69×10^{-7}
p59	Pyridine, 2,4-dimethyl (2,4-Lutidine)	25		6.99	1.02×10^{-7}
p60	Pyridine, 3,5-dimethyl (3,5-Lutidine)	25		6.15	7.08×10^{-7}
p61	Pyridine, 2-ethyl	25		5.89	1.28×10^{-6}
p62	Pyridine, 2-formyl	20		3.80	1.59×10^{-4}
p63	Pyridine, 2-hydroxy (2-Pyridol)	20	1	0.75	9.82×10^{-1}
		20	2	11.65	2.24×10^{-12}
p64	Pyridine, 4-hydroxy	20	1	3.20	6.31×10^{-4}
		20	2	11.12	7.59×10^{-12}
p65	Pyridine, methoxy	25		6.47	3.30×10^{-7}
p66	Pyridine, 4-methylamino	20		9.65	2.24×10^{-10}
p67	Pyridine, 2,4,6-trimethyl	25		7.43	3.72×10^{-8}
p68	Pyrrolidine	25		11.27	5.37×10^{-12}
p69	Pyrrolidine, 1,2-dimethyl	26		10.20	6.31×10^{-11}
p70	Pyrrolidine, *n*-methyl	25		10.32	4.79×10^{-11}
q1	Quinazoline	20		3.43	3.72×10^{-4}
q2	Quinazoline, 5-hydroxy	20	1	3.62	2.40×10^{-4}
		20	2	7.41	3.89×10^{-8}
q3	Quinine	25	1	8.52	3.02×10^{-9}
		25	2	4.13	7.41×10^{-5}
q4	Quinoline	20		4.90	1.25×10^{-5}
q5	Quinoline, 3-amino	20		4.91	1.23×10^{-5}
q6	Quinoline, 3-bromo	25		2.69	2.04×10^{-3}
q7	Quinoline, 8-carboxy	25		1.82	1.51×10^{-2}
q8	Quinoline, 3-hydroxy (3-Quinolinol)	20	1	4.28	5.25×10^{-5}
		20	2	8.08	8.32×10^{-9}
q9	Quinoline, 8-hydroxy (8-Quinolinol)	20	1	5.017	1.21×10^{-6}
		25	2	9.812	1.54×10^{-10}
q10	Quinoline, 8-hydroxy-5-sulfo	25	1	4.112	7.73×10^{-5}
		25	2	8.757	1.75×10^{-9}
q11	Quinoline, 6-methoxy	20		5.03	9.33×10^{-6}
q12	Quinoline, 2-methyl (Quinaldine)	20		5.83	1.48×10^{-6}
q13	Quinoline, 4-methyl (Lepidine)	20		5.67	2.14×10^{-6}
q14	Quinoline, 5-methyl	20		5.20	6.31×10^{-6}
q15	Quinoxaline (Quinazine)	20		0.56	3.63×10^{-1}
s1	Serine (2-amino-3-hydroxypropanoic acid)	25	1	2.186	5.49×10^{-3}
		25	2	9.208	6.19×10^{-10}
s2	Strychnine	25		8.26	5.49×10^{-9}
t1	Taurine (2-Aminoethane sulfonic acid)	25	2	9.0614	8.69×10^{-10}
t2	Tetradecaneamine (Myristilamine)	25		10.62	2.40×10^{-11}
t3	Thiazole	20		2.44	3.63×10^{-3}
t4	Thiazole, 2-amino	20		5.36	4.36×10^{-5}
t5	Threonine	25	1	2.088	8.16×10^{-3}
		25	2	9.10	7.94×10^{-10}
t6	*o*-Toluidine	25		4.44	3.63×10^{-5}
t7	*m*-Toluidine	25		4.73	1.86×10^{-5}
t8	*p*-Toluidine	25		5.08	8.32×10^{-6}

No.	Compound	Temp. °C	Step	pK_a	K_a
t9	1,3,5-Triazine, 2,4,6-triamino......	25		5.00	1.00×10^{-5}
t10	Tridecaneamine.......	25		10.63	2.35×10^{-11}
t11	Triethylamine........	18		11.01	9.77×10^{-12}
t12	Trimethylamine.......	25		9.81	1.55×10^{-10}
t13	Tryptophan..........	25	1	2.43	3.72×10^{-3}
		25	2	9.44	3.63×10^{-10}
t14	Tyrosine.............	25	2	9.11	7.76×10^{-10}
		25	3	10.13	7.41×10^{-11}
t15	Tyrosineamide........	25		7.33	4.68×10^{-8}
u1	Urea................	21		0.10	7.94×10^{-1}
v1	Valine...............	25	1	2.286	5.17×10^{-3}
		25	2	9.719	1.91×10^{-10}

DISSOCIATION CONSTANTS OF INORGANIC BASES IN AQUEOUS SOLUTIONS AT 298K

There is some arbitrariness about the designation of bases, rather than their conjugate acids, for tabulation. There are, nonetheless, a number of substances which are usually thought of as bases: some are listed here. Bases with pK less than zero are shown as "strong". These values describe the thermodynamic quotient of the first ionization of the base dissolved in aqueous solution at "infinite dilution". Concentration quotients may be profoundly affected by concentration, and by the nature and concentrations of other solutes. Many hydroxo-complexes are susceptible to polymerization. Consequently, the pK of $M(OH)_n$ must be used with some caution, especially if the cation has a large effective charge:radius radio. For more specific information, consult current specialist monographs.

NOTE: A = may react with excess strong base. B = Bronsted base: proton acceptor. I = basic properties may be obscured by low solubility. P = equilibria may be obscured by formation of polynuclear complexes. S = very sensitive to ionic medium. T = approximate value derived from kinetic results.

Base	pK_b	Notes
NH_3	4.75	B
N_2H_4	6.05	B
$N_2H_5^+$	14	B
NH_2OH	8.04	B
PH_3	28	B,T
LiOH	0.2	S
MOH	Strong	(M = Na, K, Rb, Cs)
$Be(OH)_2$	5.75	A,P
$BeOH^+$	8.6	A,P
$Mg(OH)_2$	I	I,P
$MgOH^+$	2.6	S,P
$M(OH)_2$	Strong	I (M = Ca, Sr, Ba)
$CaOH^+$	1.2	S
$SrOH^+$	0.7	S
$BaOH^+$	0.6	S
$Al(OH)_3$	8.3	A,P,S
$Al(OH)_2^+$	9.7	A,P,S
$Zn(OH)_2$	6.1	A,I,P,S
$Cd(OH)_2$	10.3	A,I,P,S
$Hg(OH)_2$	11.2	I,S
AgOH	2	A,I

DISSOCIATION CONSTANTS OF ORGANIC ACIDS IN AQUEOUS SOLUTIONS

Compound	T°C	Step	K	pK
Acetic	25		1.76×10^{-5}	4.75
Acetoacetic	18		2.62×10^{-4}	3.58
Acrylic	25		5.6×10^{-5}	4.25
Adipamic	25		2.35×10^{-5}	4.63
Adipic	25	1	3.71×10^{-5}	4.43
Adipic	25	2	3.87×10^{-6}	5.41
d-Alanine	25		1.35×10^{-10}	9.87
Allantoin	25		1.10×10^{-9}	8.96
Alloxanic	25		2.3×1^{-7}	6.64
α-Aminoacetic (glycine)	25		1.67×10^{-10}	9.78
o-Aminobenzoic	25		1.07×10^{-7}	6.97
m-Aminobenzoic	25		1.67×10^{-5}	4.78
p-Aminobenzoic	25		1.2×10^{-5}	4.92
o-Aminobenzosulfonic	25		3.3×10^{-3}	2.48
m-Aminobenzosulfonic	25		1.85×10^{-4}	3.73
p-Aminobenzosulfonic	25		5.81×10^{-4}	3.24
Anisic	25		3.38×10^{-5}	4.47
o-*β*-Anisylpropionic	25		1.59×10^{-5}	4.80
m-*β*-Anisylpropionic	25		2.24×10^{-5}	4.65
p-*β*-Anisylpropionic	25		2.04×10^{-5}	4.69
Ascorbic	24	1	7.94×10^{-5}	4.10
Ascorbic	16	2	1.62×10^{-12}	11.79
DL-Aspartic	25	1	1.38×10^{-4}	3.86
DL-Aspartic	25	2	1.51×10^{-10}	9.82
Barbituric	25		9.8×10^{-5}	4.01
Benzoic	25		6.46×10^{-5}	4.19
Benzosulfonic	25		2×10^{-1}	0.70
Bromoacetic	25		2.05×10^{-3}	2.69
o-Bromobenzoic	25		1.45×10^{-3}	2.84
m-Bromobenzoic	25		1.37×10^{-4}	3.86
n-Butyric	20		1.54×10^{-5}	4.81
iso-Butyric	18		1.44×10^{-5}	4.84
Cacodylic	25		6.4×10^{-7}	6.19
n-Caproic	18		1.43×10^{-5}	4.83
iso-Caproic	18		1.46×10^{-5}	4.84
Chloroacetic	25		1.40×10^{-3}	2.85
o-Chlorobenzoic	25		1.20×10^{-3}	2.92
m-Chlorobenzoic	25		1.51×10^{-4}	3.82
p-Chlorobenzoic	25		1.04×10^{-4}	3.98
α-Chlorobutyric	R.T.		1.39×10^{-3}	2.86
β-Chlorobutyric	R.T.		8.9×10^{-5}	4.05
γ-Chlorobutyric	R.T.		3.0×10^{-5}	4.52
o-Chlorocinnamic	25		5.89×10^{-5}	4.23
m-Chlorocinnamic	25		5.13×10^{-5}	4.29
p-Chlorocinnamic	25		3.89×10^{-5}	4.41
o-Chlorophenoxyacetic	25		8.91×10^{-4}	3.05
m-Chlorophenoxyacetic	25		7.94×10^{-4}	3.10
o-Chlorophenylacetic	25		1.18×10^{-5}	4.07
m-Chlorophenylacetic	25		7.25×10^{-5}	4.14
p-Chlorophenylacetic	25		6.46×10^{-5}	4.19
β-(*o*-Chlorophenyl) propionic	25		2.63×10^{-4}	4.58
β-(*m*-Chlorophenyl) propionic	25		2.57×10^{-5}	4.59
β-(*p*-Chlorophenyl) propionic	25		2.46×10^{-5}	4.61
α-Chloropropinic	25		1.47×10^{-3}	2.83
β-Chloropropionic	25		1.04×10^{-4}	3.98
cis-Cinnamic	25		1.3×10^{-4}	3.89
trans-Cinnamic	25		3.65×10^{-5}	4.44
Citric	20	1	7.10×10^{-4}	3.14
Citric	20	2	1.68×10^{-5}	4.77
Citric	20	3	4.1×10^{-7}	6.39
o-Cresol	25		6.3×10^{-11}	10.20
m-Cresol	25		9.8×10^{-11}	10.01
p-Cresol	25		6.7×10^{-11}	10.17
Crotonic (*trans-*)	25		2.03×10^{-5}	4.69
Cyanoacetic	25		3.65×10^{-3}	2.45
γ-Cyanobutyric	25		3.80×10^{-3}	2.42
o-Cyanophenoxyacetic	25		1.05×10^{-3}	2.98
m-Cyanophenoxyacetic	25		9.33×10^{-4}	3.03
p-Cyanophenoxyacetic	25		1.18×10^{-3}	2.93
Cyanopropionic	25		3.6×10^{-3}	2.44
Cyclohexane-1:1-dicarboxylic	25	1	3.55×10^{-4}	3.45
Cyclohexane-1:1-dicarboxylic	25	2	7.76×10^{-7}	6.11
Cyclopropane-1:1-dicarboyxlic	25	1	1.51×10^{-2}	1.82
Cyclopropane-1:1-dicarboxylic	25	2	3.72×10^{-8}	7.43
DL-Cysteine	30	1	7.25×10^{-9}	8.14
DL-Cysteine	30	2	4.6×10^{-11}	10.34
L-Cystine	25	1	1.4×10^{-8}	7.85
LCystine	25	2	1.4×10^{-10}	9.85
Deuteroacetic (in D_2O)	25		5.5×10^{-4}	5.25
Dichloroacetic	25		3.32×0^{-2}	1.48

Compound	T°C	Step	K	pK
Dichloroacetylacetic	?		7.8×10^{-3}	2.11
Dichlorophenol (2,3-)	25		3.6×10^{-8}	7.44
Dihydroxybenzoic (2,2-)	25		1.14×10^{-3}	2.94
Dihydroxybenzoic (2,5-)	25		1.08×10^{-3}	2.97
Dihydroxybenzoic (3,4-)	25		3.3×10^{-5}	4.48
Dihydroxybenzoic (3,5-)	25		9.1×10^{-5}	4.04
Dihydroxymalic	25		1.12×10^{-2}	1.92
Dihydroxytartaric	25		1.2×10^{-2}	1.92
Dimethylglycine	25		1.3×10^{-10}	9.89
Dimethylmalic	25	1	6.83×10^{-4}	3.17
Dimethylmalic	25	2	8.72×10^{-7}	6.06
Dimethylmalonic	25		7.08×10^{-4}	3.15
Dinicotinic	25		1.6×10^{-3}	2.80
Dinitrophenol (2,4-)	15		1.1×10^{-4}	3.96
Dinitrophenol (3,6-)	15		7.1×10^{-6}	5.15
Diphenylacetic	25		1.15×10^{-4}	3.94
Ethylbenzoic	25		4.47×10^{-5}	4.35
Ethylphenylacetic	25		4.27×10^{-5}	4.37
Fluorobenzoic	17		1.25×10^{-3}	2.90
Formic	20		1.77×10^{-4}	3.75
Fumaric *(trans-)*	18	1	9.30×10^{-4}	3.03
Fumaric *(trans-)*	18	2	3.62×10^{-5}	4.44
Furancarboxylic	25		7.1×10^{-4}	3.15
Furoic	25		6.76×10^{-4}	3.17
Gallic	25		3.9×10^{-5}	4.41
Glutaramic	25		3.98×10^{-5}	4.60
Glutaric	25	1	4.58×10^{-5}	4.31
Glutaric	25	2	3.89×10^{-6}	5.41
Glycerol	25		7×10^{-15}	14.15
Glycine	25		1.67×10^{-10}	9.78
Glycol	25		6×10^{-15}	14.22
Glycolic	25		1.48×10^{-4}	3.83
Heptanoic	25		1.28×10^{-5}	4.89
Hexahydrobenzoic	25		1.26×10^{-5}	4.90
Hexanoic	25		1.31×10^{-5}	4.88
Hippuric	25		$1.57 \times 10\text{um}^{4}$	3.80
Histidine	25		6.7×10^{-10}	9.17
Hydroquinone	20		4.5×10^{-11}	10.35
o-Hydroxybenzoic	19	1	1.07×10^{-3}	2.97
o-Hydroxybenzoic	18	2	4×10^{-14}	13.40
m-Hydroxybenzoic	19	1	8.7×10^{-5}	4.06
m-Hydroxybenzoic	19	2	1.2×10^{-10}	9.92
p-Hydroxybenzoic	19	1	3.3×10^{-5}	4.48
p-Hydroxybenzoic	19	2	4.8×10^{-10}	9.32
β-Hydroxybutyric	25		2×10^{-5}	4.70
γ-Hydroxybutyric	25		1.9×10^{-5}	4.72
β-Hydroxypropionic	25		3.1×10^{-5}	4.51
γ-Hydroxyquinoline	20		3.1×10^{-10}	9.51
Iodoacetic	25		7.5×10^{-4}	3.12
o-Iodobenzoic	25		1.4×10^{-3}	2.85
m-Iodobenzoic	25		1.6×10^{-4}	3.80
Itaconic	25	1	1.40×10^{-4}	3.85
Itaconic	25	2	3.56×10^{-6}	5.45
Lactic	100		8.4×10^{-4}	3.08
Lutidinic	25		7.0×10^{-3}	2.15
Lysine	25		2.95×10^{-11}	10.53
Maleic	25	1	1.42×10^{-2}	1.83
Maleic	25	2	8.57×10^{-7}	6.07
Malic	25	1	3.9×10^{-4}	3.40
Malic	25	2	7.8×10^{-6}	5.11
Malonic	25	1	1.49×10^{-2}	2.83
Malonic	25	2	2.03×10^{-6}	5.69
DL-Mandelic	25		1.4×10^{-4}	3.85
Mesaconic	25	1	8.22×10^{-4}	3.09
Mesaconic	25	2	1.78×10^{-5}	4.75
Mesitylenic	25		4.8×10^{-5}	4.32
Methyl-*o*-aminobenzoic	25		4.6×10^{-5}	5.34
Methyl-*m*-aminobenzoic	25		8×10^{-6}	5.10
Methyl-*p*-aminobenzoic	25		9.2×10^{-6}	5.04
o-Methylcinnamic	25		3.16×10^{-5}	4.50
m-Methylcinnamic	25		3.63×10^{-5}	4.44
p-Methylcinnamic	25		2.76×10^{-5}	4.56
β-Methylglutaric	25		5.75×10^{-5}	4.24
n-Methylglycine	18		1.2×10^{-10}	9.92
Methylmalonic	25		1.17×10^{-4}	3.07
Methylsuccinic	25	1	7.4×10^{-5}	4.13
Methylsuccinic	25	2	2.3×10^{-4}	5.64
o-Monochlorophenol	25		3.2×10^{-9}	8.49
m-Monochlorophenol	25		1.4×10^{-9}	8.85
p-Monochlorophenol	25		6.6×10^{-10}	9.18
Naphthalenesulfonic	25		2.7×10^{-1}	0.57
α-Naphthoic	25		2×10^{-4}	3.70
β-Naphthoic	25		6.8×10^{-5}	4.17
α-Naphthol	25		4.6×10^{-10}	9.34
β-Naphthol	25		3.1×10^{-10}	9.51
Nitrobenzene	0		1.05×10^{-4}	3.98
o-Nitrobenzoic	18		6.95×10^{-3}	2.16
m-Nitrobenzoic	25		3.4×10^{-4}	3.47
p-Nitrobenzoic	25		3.93×10^{-4}	3.41
o-Nitrophenol	25		6.8×10^{-8}	7.17
m-Nitrophenol	25		5.3×10^{-9}	8.28
p-Nitrophenol	25		7×10^{-8}	7.15
o-Nitrophenylacetic	25		1.00×10^{-4}	4.00
m-Nitrophenylacetic	25		1.07×10^{-4}	3.97
p-Nitrophenylacetic	25		1.41×10^{-4}	3.85
o-β-Nitrophenylpropionic	25		3.16×10^{-5}	4.50
p-β-Nitrophenylopropionic	25		3.39×10^{-5}	4.47
Nonanic	25		1.09×10^{-5}	4.96
Octanoic	25		1.28×10^{-5}	4.89
Oxalic	25	1	5.90×10^{-2}	1.23
Oxalic	25	2	6.40×10^{-5}	4.19
Phenol	20		1.28×10^{-10}	9.89
Phenylacetic	18		5.2×10^{-5}	4.28
o-Phenylbenzoic	25		3.47×10^{-4}	3.46
γ-Phenylbutyric	25		1.74×10^{-5}	4.76
α-Phenylpropionic	25		2.27×10^{-5}	4.64
β-Phenylpropionic	25		4.25×10^{-5}	4.37
o-Phthalic	25	1	1.3×10^{-3}	2.89
o-Phthalic	25	2	3.9×10^{-8}	5.51
m-Phthalic	25	1	2.9×10^{-4}	3.54
m-Phthalic	18	2	2.5×10^{-5}	4.60
p-Phthalic	25	1	3.1×10^{-4}	3.51
p-Phthalic	16	2	1.5×10^{-5}	4.82
Picric	25		4.2×10^{-1}	0.38
Pimelic	25		3.09×10^{-5}	4.71
Propionic	25		1.34×10^{-5}	4.87
iso-Propylbenzoic	25		3.98×10^{-5}	4.40
2-Pyridinecarboxylic	25		3×10^{-6}	5.52
3-Pyridinecarboxylic	25		1.4×10^{-5}	4.85
4-Pyridinecariboxylic	25		1.1×10^{-5}	4.96
Pyrocatechol	20		1.4×10^{-10}	9.85
Quinolinic	25		3×10^{-3}	2.52
Resorcinol	25		1.55×10^{-10}	9.81
Saccharin	18		2.1×10^{-12}	11.68
Suberic	25		2.99×10^{-5}	4.52
Succinic	25	1	6.89×10^{-5}	4.16
Succinic	25	2	2.47×10^{-6}	5.61
Sulfanilic	25		5.9×10^{-4}	3.23
α-Tataric	25	1	1.04×10^{-3}	2.98
α-Tartaric	25	2	4.55×10^{-5}	4.34
meso-Tartaric	25	1	6×10^{-4}	3.22
meso-Tartaric	25	2	1.53×10^{-5}	4.82
Theobromine	18		1.3×10^{-8}	7.89
Terephthalic	25		3.1×10^{-4}	3.51
Thioacetic	25		4.7×10^{-4}	3.33
Thiophenecarboxylic	25		3.3×10^{-4}	3.48
o-Toluic	25		1.22×10^{-4}	3.91
m-Toluic	25		5.32×10^{-5}	4.27
p-Toluic	25		4.33×10^{-5}	4.36
Trichloroacetic	25		2×10^{-1}	0.70
Trichlorophenol	25		1×10^{-6}	6.00
Trihydroxybenzoic (2,4,6-)	25		2.1×10^{-2}	1.68
Trimethylacetic	18		9.4×10^{-6}	5.03
Trinitrophenol (2,4,6-)	25		4.2×10^{-1}	0.38
Tryptophan	25		4.2×10^{-10}	9.38
Tyrosine	17		3.98×10^{-9}	8.40
Uric	12		1.3×10^{-4}	3.89
n-Valeric	18		1.51×10^{-5}	4.82
iso-Valeric	25		1.7×10^{-5}	4.77
Veronal	25		3.7×10^{-8}	7.43
Vinylacetic	25		4.57×10^{-5}	4.34
Xanthine	40		1.24×10^{-10}	9.91

DISSOCIATION CONSTANTS OF ACIDS IN WATER AT VARIOUS TEMPERATURES

Acids		0°	5°	10°	15°	20°	25°	30°	35°	40°	45°	50°
Formic	$K_A \cdot 10^4$	1.638	1.691	1.728	1.749	1.765	1.772	1.768	1.747	1.716	1.685	1.650
Acetic	$K_A \cdot 10^5$	1.657	1.700	1.729	1.745	1.753	1.754	1.750	1.728	1.703	1.670	1.633
Propionic	$K_A \cdot 10^5$	1.274	1.305	1.326	1.336	1.338	1.336	1.326	1.310	1.280	1.257	1.229
n-Butyric	$K_A \cdot 10^5$	1.563	1.574	1.576	1.569	1.542	1.515	1.484	1.439	1.395	1.347	1.302
Chloracetic	$K_A \cdot 10^3$	1.528	—	1.488	—	—	1.379	—	—	1.230	—	—
Lactic	$K_A \cdot 10^4$	1.287	—	—	—	—	1.374	—	—	—	—	1.270
Glycollic	$K_A \cdot 10^4$	1.334	—	—	—	—	1.475	—	—	—	—	1.415
Oxalic	$K_{2A} \cdot 10^5$	5.91	5.82	5.70	5.55	5.40	5.18	4.92	4.67	4.41	4.09	3.83
Malonic	$K_{2A} \cdot 10^6$	2.140	2.165	2.152	2.124	2.076	2.014	1.948	1.863	1.768	1.670	1.575
Phosphoric	$K_A \cdot 10^3$	8.968	—	—	—	—	7.516	—	—	—	—	5.495
Phosphoric	$K_{2A} \cdot 10^8$	4.85	5.24	5.57	5.89	6.12	6.34	6.46	6.53	6.58	6.59	6.55
Boric	$K_A \cdot 10^{10}$	—	3.63	4.17	4.72	5.26	5.79	6.34	6.86	7.38	—	8.32
Carbonic	$K_{1A} \cdot 10^7$	2.64	3.04	3.44	3.81	4.16	4.45	4.71	4.90	5.04	5.13	5.19
Phenol-sulfonic	$K_{2A} \cdot 10^{10}$	4.45	5.20	6.03	6.92	7.85	8.85	9.89	10.94	12.00	13.09	14.16
Glycine	$K_{1A} \cdot 10^7$	—	3.82	3.99	4.17	4.32	4.46	4.57	4.66	4.73	4.77	4.79
Citric	$K_{1A} \cdot 10^4$	6.03	6.31	6.69	6.92	7.21	7.45	7.66	7.78	7.96	7.99	8.04
	$K_{2A} \cdot 10^5$	1.45	1.54	1.60	1.65	1.70	1.73	1.76	1.77	1.78	1.76	1.75
	$K_{3A} \cdot 10^7$	4.05	4.11	4.14	4.13	4.09	4.02	3.99	3.78	3.69	3.45	3.28

Reproducibility between various workers is about $\pm$ (0.01—0.02) $\cdot 10^5$.
All values are on the m-scale.

DISSOCIATION CONSTANTS OF INORGANIC ACIDS IN AQUEOUS SOLUTIONS

(Approximately 0.1—0.01 N)

Compound	T°C	Step	K	pK
Arsenic	18	1	5.62×10^{-3}	2.25
Arsenic	18	2	1.70×10^{-7}	6.77
Arsenic	18	3	3.95×10^{-12}	11.60
Arsenious	25		6×10^{-10}	9.23
o-Boric	20	1	7.3×10^{-10}	9.14
o-Boric	20	2	1.8×10^{-13}	12.74
o-Boric	20	3	1.6×10^{-14}	13.80
Carbonic	25	1	4.30×10^{-7}	6.37
Carbonic	25	2	5.61×10^{-11}	10.25
Chromic	25	1	1.8×10^{-1}	0.74
Chromic	25	2	3.20×10^{-7}	6.49
Germanic	25	1	2.6×10^{-9}	8.59
Germanic	25	2	1.9×10^{-13}	12.72
Hyrocyanic	25		4.93×10^{-10}	9.31
Hydrofluoric	25		3.53×10^{-4}	3.45
Hydrogen sulfide	18	1	9.1×10^{-8}	7.04
Hydrogen sulfide	18	2	1.1×10^{-12}	11.96
Hydrogen peroide	25		2.4×10^{-12}	11.62
Hypobromous	25		2.06×10^{-9}	8.69
Hypochlorous	18		2.95×10^{-5}	4.53
Hypoiodous	25		2.3×10^{-11}	10.64
Iodic	25		1.69×10^{-1}	0.77
Nitrous	12.5		4.6×10^{-4}	3.37
Periodic	25		2.3×10^{-2}	1.64
o-Phosphoric	25	1	7.52×10^{-3}	2.12
o-Phosphoric	25	2	6.23×10^{-8}	7.21
o-Phosphoric	18	3	2.2×10^{-13}	12.67
Phosphorous	18	1	1.0×10^{-2}	2.00
Phosphorous	18	2	2.6×10^{-7}	6.59
Pyrophosphoric	18	1	1.4×10^{-1}	0.85
Pyrophosphoric	18	2	3.2×10^{-2}	1.49
Pyrophosphoric	18	3	1.7×10^{-6}	5.77
Pyrophosphoric	18	4	6×10^{-9}	8.22
Selenic	25	2	1.2×10^{-2}	1.92
Selenious	25	1	3.5×10^{-2}	2.46
Selenious	25	2	5×10^{-3}	7.31
m-Silicic	R.T.	1	2×10^{-10}	9.70
m-Silicic	R.T.	2	1×10^{-12}	12.00
o-Silicic	30	1	2.2×10^{-10}	9.66
o-Silicic	30	2	2×10^{-12}	11.70
o-Silicic	30	3	1×10^{-12}	12.00
o-Silicic	30	4	1×10^{-12}	12.00
Sulfuric	25	2	1.20×10^{-2}	1.92
Sulfurous	18	1	1.54×10^{-2}	1.81
Sulfurous	18	2	1.02×10^{-7}	6.91
Telluric	18	1	2.09×10^{-8}	7.68
Telluric	18	2	6.46×10^{-12}	11.29
Tellurous	25	1	3×10^{-8}	2.48
Tellurous	25	2	2×10^{-8}	7.70
Tetraboric	25	1	$\sim 10^{-4}$	$\sim$ 4.00
Tetraboric	25	2	$\sim 10^{-9}$	$\sim$ 9.00

DISSOCIATION CONSTANTS (K_b) OF AQUEOUS AMMONIA FROM 0 TO 50°C

Temperature (°C)	pK_b	K_b
0	4.862	1.374×10^{-5}
5	4.830	1.479×10^{-5}
10	4.804	1.570×10^{-5}
15	4.782	1.652×10^{-5}
20	4.767	1.710×10^{-5}
25	4.751	1.774×10^{-5}
30	4.740	1.820×10^{-5}
35	4.733	1.849×10^{-5}
40	4.730	1.862×10^{-5}
45	4.726	1.879×10^{-5}
50	4.723	1.892×10^{-5}

Values of K_b accurate to ±0.005; determined by e.m.f. method by: Bates, R. G. and Pinching, G. D., *J. Am. Chem. Soc.*, 72, 1393, 1950.

ION PRODUCT OF WATER SUBSTANCE

William L. Marshall and E. U. Franck

Pressure (bars)	Temperature (°C)																	
	0	25	50	75	100	150	200	250	300	350	400	450	500	600	700	800	900	1000
Sat'd Vapor	14.938	13.995	13.275	12.712	12.265	11.638	11.289	11.191	11.406	12.30	—	—	—	—	—	—	—	—
250	14.83	13.90	13.19	12.63	12.18	11.54	11.16	11.01	11.14	11.77	19.43	21.59	22.40	23.27	23.81	24.23	24.59	24.93
500	14.72	13.82	13.11	12.55	12.10	11.45	11.05	10.85	10.86	11.14	11.88	13.74	16.13	18.30	19.29	19.92	20.39	20.80
750	14.62	13.73	13.04	12.48	12.03	11.36	10.95	10.72	10.66	10.79	11.17	11.89	13.01	15.25	16.55	17.35	17.93	18.39
1,000	14.53	13.66	12.96	12.41	11.96	11.29	10.86	10.60	10.50	10.54	10.77	11.19	11.81	13.40	14.70	15.58	16.22	16.72
1,500	14.34	13.53	12.85	12.29	11.84	11.16	10.71	10.43	10.26	10.22	10.29	10.48	10.77	11.59	12.50	13.30	13.97	14.50
2,000	14.21	13.40	12.73	12.18	11.72	11.04	10.57	10.27	10.08	9.98	9.98	10.07	10.23	10.73	11.36	11.98	12.54	12.97
2,500	14.08	13.28	12.62	12.07	11.61	10.92	10.45	10.12	9.91	9.79	9.74	9.77	9.86	10.18	10.63	11.11	11.59	12.02
3,000	13.97	13.18	12.53	11.98	11.53	10.83	10.34	9.99	9.76	9.61	9.54	9.53	9.57	9.78	10.11	10.49	10.89	11.24
3,500	13.87	13.09	12.44	11.90	11.44	10.74	10.24	9.88	9.63	9.47	9.37	9.33	9.34	9.48	9.71	10.02	10.35	10.62
4,000	13.77	13.00	12.35	11.82	11.37	10.66	10.16	9.79	9.52	9.34	9.22	9.16	9.15	9.23	9.41	9.65	9.93	10.13
5,000	13.60	12.83	12.19	11.66	11.22	10.52	10.00	9.62	9.34	9.13	8.99	8.90	8.85	8.85	8.95	9.11	9.30	9.42
6,000	13.44	12.68	12.05	11.53	11.09	10.39	9.87	9.48	9.18	8.96	8.80	8.69	8.62	8.57	8.61	8.72	8.86	8.97
7,000	13.31	12.55	11.93	11.41	10.97	10.27	9.75	9.35	9.04	8.81	8.64	8.51	8.42	8.34	8.34	8.40	8.51	8.64
8,000	13.18	12.43	11.82	11.30	10.86	10.17	9.64	9.24	8.93	8.68	8.50	8.36	8.25	8.13	8.10	8.13	8.21	8.38
9,000	13.04	12.31	11.71	11.20	10.77	10.07	9.54	9.13	8.82	8.57	8.37	8.22	8.10	7.95	7.89	7.89	7.95	8.12
10,000	12.91	12.21	11.62	11.11	10.68	9.98	9.45	9.04	8.71	8.46	8.25	8.09	7.96	7.78	7.70	7.68	7.70	7.85

Note: Data in the following table were calculated from the equation, $\log_{10} K^*_w = A + B/T + C/T^2 + D/T^3 + (E + F/T\ G/T^2) \log_{10} \varrho^*_w$, where, $K^*_w = K_w/(\text{mol kg}^{-1})$, and $\varrho^*_w = \varrho_w/(\text{g cm}^{-3})$. The parameters are:

$A = -4.098$ $\quad$ $C = +2.2362 \times 10^5 K^2$ $\quad$ $E = +13.957$ $\quad$ $G = +8.5641 \times 10^5 K^2$

$B = -3245.2K$ $\quad$ $D = -3.984 \times 10^7 K^3$ $\quad$ $F = -1262.3K$

Users of this table may wish to refer to the reference cited for the background for this international formulation of the ion product of water substance.

From *J. Phys. Chem. Ref. Data*, 10, 295, 1981. With permission.

IONIZATION CONSTANT FOR WATER (K_w)

$-\log_{10} K_w$	Temperature °C.	$-\log_{10} K_w$	Temperature °C.
14.9435	0	13.8330	30
14.7338	5	13.6801	35
14.5346	10	13.5348	40
14.3463	15	13.3960	45
14.1669	20	13.2617	50
14.0000	24	13.1369	55
13.9965	25	13.0171	60

IONIZATION CONSTANTS FOR DEUTERIUM OXIDE FROM 10 TO 50°C

From NBS Technical Note 400
The subscript *m* indicates values on the molal scale, whereas the subscript *c* indicates values on the molar scale.

t(°C)	pK_m	pK_c
10	15.526	15.439
20	15.136	15.049
25	14.955	14.869
30	14.784	14.699
40	14.468	14.385
50	14.182	14.103

STANDARD SOLUTIONS FOR CALIBRATING CONDUCTIVITY CELLS

Grams of KCl per 1 kg of solution in vacuum	κ [$10^{-3} \times S^{-1} m^{-1}$] at		
	0°C	18°C	25°C
71.135 2	0.651 44	0.977 90	1.112 87
7.419 13	0.071 344	0.111 612	0.128 496
0.745 263	0.007 732 6	0.012 199 2	0.014 080 7

From Jones, G. and Bradshaw, B. C., *J. Am. Chem. Soc.*, 55, 1780, 1933, converted from (int. ohm)$^{-1}$ cm^{-1}.

EQUIVALENT CONDUCTIVITIES, λ, OF SOME ELECTROLYTES IN AQUEOUS SOUTION AT 25°C

Petr Vanýsek

The values of λ are given in 10^{-4} m^2 S mol l^{-1}

Compound	Concentration, mol^{-1}							
	Infinite dilution	0.0005	0.001	0.005	0.01	0.02	0.05	0.1
			λ					
$AgNO_3$	133.29	131.29	130.45	127.14	124.70	121.35	115.18	109.09
$1/2BaCl_2$	139.91	135.89	134.27	127.96	123.88	119.03	111.42	105.14
$1/2CaCl_2$	135.77	131.86	130.30	124.19	120.30	115.59	108.42	102.41
$1/2Ca(OH)_2$	258	—	—	233	226	214	—	—
$1/2CuSO_4$	133.6	121.6	115.20	94.02	83.08	72.16	59.02	50.55
HCl	425.95	422.53	421.15	415.59	411.80	407.04	398.89	391.13
KBr	151.9	—	—	146.02	143.36	140.41	135.61	131.32
KCl	149.79	147.74	146.88	143.48	141.20	138.27	133.30	128.90
$KClO_4$	139.97	138.69	137.80	134.09	131.39	127.86	121.56	115.14
$1/3K_3Fe(CN)_6$	174.5	166.4	163.1	150.7	—	—	—	—
$1/4K_4Fe(CN)_6$	184		167.16	146.02	134.76	122.76	107.65	97.82
$KHCO_3$	117.94	116.04	115.28	112.18	110.03	107.17	—	—
KI	150.31	—	—	144.30	142.11	139.38	134.90	131.05
KIO_4	127.86	125.74	124.88	121.18	118.45	114.08	106.67	98.2
KNO_3	144.89	142.70	141.77	138.41	132.75	132.34	126.25	120.34
$KMnO_4$	134.8	—	133.3	—	126.5	—	—	113
KOH	271.5	—	234	230	228	—	219	213
$KReO_4$	128.20	126.03	125.12	121.31	118.49	114.49	106.40	97.40
$1/3LaCl_3$	145.9	139.6	137.0	127.5	121.8	115.3	106.2	99.1
LiCl	114.97	113.09	112.34	109.35	107.27	104.60	100.06	95.81
$LiClO_4$	105.93	104.13	103.39	100.52	98.56	96.13	92.15	88.52
$1/2MgCl_2$	129.34	125.55	124.15	118.25	114.49	109.99	103.03	97.05
NH_4Cl	149.6	—	146.7	134.4	141.21	138.25	133.22	128.69
NaCl	126.39	124.44	123.68	120.59	118.45	115.70	111.01	106.69
$NaClO_4$	117.42	115.58	114.82	111.70	109.54	106.91	102.35	98.38
NaI	126.88	125.30	124.19	121.19	119.18	116.64	112.73	108.73
$NaOOCCH_3$	91.0	89.2	88.5	85.68	83.72	81.20	76.88	72.76
$NaOOC_2H_5$	85.88	84.20	83.50	80.86	79.01	76.59	—	—
$NaOOC_3H_7$	82.66	81.00	80.27	77.54	75.72	73.35	69.29	65.24
NaOH	247.7	245.5	244.6	240.7	237.9	—	—	—
Na picrate	80.45	—	78.6	757	73.7	—	66.3	61.8
$1/2Na_2SO_4$	129.8	125.68	124.09	117.09	112.38	106.73	97.70	89.94
$1/2SrCl_2$	135.73	131.84	130.27	124.18	120.23	115.48	108.20	102.14
$1/2ZnSO_4$	132.7	121.3	114.47	95.44	84.87	74.20	61.17	52.61

EQUIVALENT IONIC CONDUCTIVITIES EXTRAPOLATED TO INFINITE DILUTION IN AQUEOUS SOLUTIONS AT 25°C

Petr Vanýsek

Ion	Λ_o (10^{-4} m^2 S mol^{-1})	Ion	Λ_o (10^{-4} m^2 S mol^{-1})	Ion	Λ_o (10^{-4} m^2 S mol^{-1})
Inorganic cations		**Inorganic anions**		**Inorganic anions**	
Ag^+	61.9	$1/3Dy^{3+}$	65.6	NH_4^+	73.5
$1/3Al^{3+}$	61	$1/3Er^{3+}$	65.9	$N_2H_5^+$	59
$1/2Ba^{2+}$	63.6	$1/3Eu^{3+}$	67.8	Na^+	50.08
$1/2Be^{2+}$	45	$1/2Fe^{2+}$	54	$1/3Nd^{3+}$	69.4
$1/2Ca^{2+}$	59.47	$1/3Fe^{3+}$	68	$1/2Ni^{2+}$	50
$1/2Cd^{2+}$	54	$1/3Gd^{3+}$	67.3	$1/4[Ni_2(trien)_3]^{4+}$	52
$1/3Ce^{3+}$	69.8	H^+	349.65	$1/2Pb^{2+}$	71
$1/2Co^{2+}$	55	$1/2Hg_2^{2+}$	68.6	$1/3Pr^{3+}$	69.5
$1/3[Co(NH_3)_6]^{3+}$	101.9	$1/2Hg^{2+}$	63.6	$1/2Ra^{2+}$	66.8
$1/3[Co(en)_3]^{3+}$	74.7	$1/3Ho^{3+}$	66.3	Rb^+	77.8
$1/6[Co_2(trien)_3]^{6+}$	69	K^+	73.48	$1/3Sc^{3+}$	64.7
$1/3Cr^{3+}$	67	$1/3La^{3+}$	69.7	$1/3Sm^{3+}$	68.5
Cs^+	77.2	Li^+	38.66	$1/2Sr^{2+}$	59.4
$1/2Cu^{2+}$	53.6	$1/2Mg^{2+}$	53.0	Tl^+	74.7
D^+ (deuterium)	213.7 (18°)	$1/2Mn^{2+}$	53.5	$1/3Tm^{3+}$	65.4

EQUIVALENT IONIC CONDUCTIVITIES EXTRAPOLATED TO INFINITE DILUTION IN AQUEOUS SOLUTIONS AT 25°C (continued)

Ion	Λ_o (10^{-4} m^2 S mol^{-1})
Inorganic cations	
$1/2UO_2^{2+}$	32
$1/3Y^{3+}$	62
$1/3Yb^{3+}$	65.6
$1/2Zn^{2+}$	52.8
$Au(CN)_2^-$	50
$Au(CN)_4^-$	36
$B(C_6H_5)_4^-$	21
Br^-	78.1
Br_3^-	43
BrO_3^-	55.7
CN^-	78
CNO^-	64.6
$1/2CO_3^{2-}$	69.3
Cl^-	76.31
ClO_2^-	52
ClO_3^-	64.6
ClO_4^-	67.3
$1/3[Co(CN)_6]^{3-}$	98.9
$1/2CrO_4^{2-}$	85
F^-	55.4
$1/4[Fe(CN)_6]^{4-}$	110.4

Ion	Λ_o (10^{-4} m^2 S mol^{-1})
Inorganic cations	
$1/3[Fe(CN)_6]^{3-}$	100.9
$H_2AsO_4^-$	34
HCO_3^-	44.5
HF_2^-	75
$1/2HPO_4^{2-}$	33
$H_2PO_4^-$	33
$H_2PO_2^-$	46
HS^-	65
HSO_3^-	50
HSO_4^-	50
$H_2SbO_4^-$	31
I^-	76.8
IO_3^-	40.5
IO_4^-	54.5
MnO_4^-	61.3
MoO_4^-	74.5
$N(CN)_2^-$	54.5
NO_2^-	71.8
NO_3^-	71.42
$NH_2SO_3^-$	48.6
N_3^-	69

Ion	Λ_o (10^{-4} m^2 S mol^{-1})
Inorganic cations	
OCN^-	64.6
OH^-	198
PF_6^-	56.9
$1/2PO_3F^{2-}$	63.3
$1/2PO_4^{3-}$	69.0
$1/4P_2O_7^{4-}$	96
$1/3P_3O_9^{3-}$	83.6
$1/5P_3O_{10}^{5-}$	109
$1/3P_4O_{13}^{2-}$	94
ReO_4^-	54.9
SCN^-	66
$1/2SO_3^{2-}$	79.9
$1/2SO_4^{2-}$	80.0
$1/2S_2O_3^{2-}$	85.0
$1/2S_2O_4^{2-}$	66.5
$1/2S_2O_6^{2-}$	93
$1/2S_2O_8^{2-}$	86
$Sb(OH)_6^-$	31.9
$SeCN^-$	64.7
$1/2SeO_4^{2-}$	75.7
$1/2WO_4^{2-}$	69

Organic cations

Ion	Λ_o
Benzyltrimethylammonium	34.6
i-Butylammonium	38
Butyltrimethylammonium	33.6
n-Decylpyridinium	29.5
Decyltrimethylammonium	24.4
Diethylammonium	42.0
Dimethylammonium	51.8
Dipropylammonium	30.1
n-Dodecylammonium	23.8
Dodecyltrimethylammonium	22.6
Ethanolammonium	42.2
Ethylammonium	47.2
Ethyltrimethylammonium	40.5
Hexadecyltrimethylammonium	20.9
Hexyltrimethylammonium	29.6
Histidyl	23.0
Hydroxyethyltrimethylarsonium	39.4
Methylammonium	58.7
Octadecylpyridinium	20
Octadecyltributylammoniuim	16.6
Octadecyltriethylammonium	17.9
Octadecyltrimethylammonium	19.9
Octadecyltripropylammonium	17.2
Octyltrimethylammonium	26.5
Pentylammonium	37
Piperidinium	37.2
Propylammonium	40.8
Pyrilammonium	24.3
Tetra-*n*-butylammonium	19.5
Tetradecyltrimethylammonium	21.5
Tetraethylammonium	32.6
Tetramethylammonium	44.9
Tetra-*i*-pentylammonium	17.9
Tetra-*n*-pentylammonium	17.5
Tetra-*n*-propylammonium	23.4
Triethylammonium	34.3
Triethylsulfonium	36.1
Trimethylammonium	47.23
Trimethylhexylammonium	34.6
Trimethylsulfonium	51.4
Tripropylammonium	26.1

Organic anions

Ion	Λ_o
Acetate	40.9
p-Anisate	29.0
$1/2$Azelate^{2-}	40.6
Benzoate	32.4
Bromoacetate	39.2
Bromobenzoate	30
n-Butyrate	32.6
Chloroacetate	42.2
m-Chlorobenzoate	31
o-Chlorobenzoate	30.2
$1/3$Citrate^{3-}	70.2
Crotonate	33.2
Cyanoacetate	43.4
Cyclohexane carboxylate	28.7
1/2 1,1-Cyclopropane-dicarboxylate^{2-}	53.4
Decylsulfate	26
Dichloroacetate	38.3
$1/2$Diethylbarbiturate^{2-}	26.3
Dihydrogencitrate	30
$1/2$Dimethylmalonate^{2-}	49.4
3,5-Dinitrobenzoate	28.3
Dodecylsulfate	24
Ethylmalonate	49.3
Ethylsulfate	39.6
Fluoroacetate	44.4
Fluorobenzoate	33
Formate	54.6
$1/2$Fumarate^{2-}	61.8
$1/2$Glutarate^{2-}	52.6
Hydrogenoxalate	40.2
Isovalerate	32.7
Iodoacetate	40.6
Lactate	38.8
$1/2$Malate^{2-}	58.8

EQUIVALENT IONIC CONDUCTIVITIES EXTRAPOLATED TO INFINITE DILUTION IN AQUEOUS SOLUTIONS AT 25°C (continued)

Organic anions		Organic anions	
1/2Maleate^{2-}	61.9	Picrate	30.37
1/2Malonate^{2-}	63.5	Pivalate	31.9
Methylsulfate	48.8	Propionate	35.8
Naphtylacetate	28.4	Propylsulfate	37.1
1/2Oxalate^{2-}	74.11	Salicylate	36
Octylsulfate	29	1/2Suberate^{2-}	36
Phenylacetate	30.6	1/2Succinate^{2-}	58.8
1/2*o*-Phtalate^{2-}	52.3	1/2Tartarate^{2-}	59.6
1/2*m*-Phtalate^{2-}	54.7	Trichloroacetate	36.6

ACTIVITY COEFFICIENTS OF ACIDS, BASES AND SALTS

Petr Vanýsek

The following coefficients are valid at 25°C. The concentrations are expressed as molalities.

	0.1	0.2	0.3	0.4	0.5	0.6	0.7	0.8	0.9	1.0
$AgNO_3$	0.734	0.657	0.606	0.567	0.536	0.509	0.485	0.464	0.446	0.429
$AlCl_3$	0.337	0.305	0.302	0.313	0.331	0.356	0.388	0.429	0.479	0.539
$Al_2(SO_4)_3$	0.035	0.0225	0.0176	0.0153	0.0143	0.014	0.0142	0.0149	0.0159	0.0175
$BaCl_2$	0.500	0.444	0.419	0.405	0.397	0.391	0.391	0.391	0.392	0.395
$BeSO_4$	0.150	0.109	0.0885	0.0769	0.0692	0.0639	0.0600	0.0570	0.0546	0.0530
$CaCl_2$	0.518	0.472	0.455	0.448	0.448	0.453	0.460	0.470	0.484	0.500
$CdCl_2$	0.2280	0.1638	0.1329	0.1139	0.1006	0.0905	0.0827	0.0765	0.0713	0.0669
$Cd(NO_3)_2$	0.513	0.464	0.442	0.430	0.425	0.423	0.423	0.425	0.428	0.433
$CdSO_4$	0.150	0.103	0.0822	0.0699	0.0615	0.0553	0.0505	0.0468	0.0438	0.0415
$CoCl_2$	0.522	0.479	0.463	0.459	0.462	0.470	0.479	0.492	0.511	0.531
$CrCl_3$	0.331	0.298	0.294	0.300	0.314	0.335	0.362	0.397	0.436	0.481
$Cr(NO_3)_3$	0.319	0.285	0.279	0.281	0.291	0.304	0.322	0.344	0.371	0.401
$Cr_2(SO_4)_3$	0.0458	0.0300	0.0238	0.0207	0.0190	0.0182	0.0181	0.0185	0.0194	0.0208
$CsBr$	0.754	0.694	0.654	0.626	0.603	0.586	0.571	0.558	0.547	0.538
$CsCl$	0.756	0.694	0.656	0.628	0.606	0.589	0.575	0.563	0.553	0.544
CsI	0.754	0.692	0.651	0.621	0.599	0.581	0.567	0.554	0.543	0.533
$CsNO_3$	0.733	0.655	0.602	0.561	0.528	0.501	0.478	0.458	0.439	0.422
$CsOH$	0.795	0.761	0.744	0.739	0.739	0.742	0.748	0.754	0.762	0.771
$CsOAc$	0.799	0.771	0.761	0.759	0.762	0.768	0.776	0.783	0.792	0.802
Cs_2SO_4	0.456	0.382	0.338	0.311	0.291	0.274	0.262	0.251	0.242	0.235
$CuCl_2$	0.508	0.455	0.429	0.417	0.411	0.409	0.409	0.410	0.413	0.417
$Cu(NO_3)_2$	0.511	0.460	0.439	0.429	0.426	0.427	0.431	0.437	0.445	0.455
$CuSO_4$	0.150	0.104	0.0829	0.0704	0.0620	0.0559	0.0512	0.0475	0.0446	0.0423
$FeCl_2$	0.5185	0.473	0.454	0.448	0.450	0.454	0.463	0.473	0.488	0.506
HBr	0.805	0.782	0.777	0.781	0.789	0.801	0.815	0.832	0.850	0.871
HCl	0.796	0.767	0.756	0.755	0.757	0.763	0.772	0.783	0.795	0.809
$HClO_4$	0.803	0.778	0.768	0.766	0.769	0.776	0.785	0.795	0.808	0.823
HI	0.818	0.807	0.811	0.823	0.839	0.860	0.883	0.908	0.935	0.963
HNO_3	0.791	0.754	0.735	0.725	0.720	0.717	0.717	0.718	0.721	0.724
H_2SO_4	0.2655	0.2090	0.1826	—	0.1557	—	0.1417	—	—	0.1316
KBr	0.772	0.722	0.693	0.673	0.657	0.646	0.636	0.629	0.622	0.617
KCl	0.770	0.718	0.688	0.666	0.649	0.637	0.626	0.618	0.610	0.604
$KClO_3$	0.749	0.681	0.635	0.599	0.568	0.541	0.518	—	—	—
K_2CrO_4	0.456	0.382	0.340	0.313	0.292	0.276	0.263	0.253	0.243	0.235
KF	0.775	0.727	0.700	0.682	0.670	0.661	0.654	0.650	0.646	0.645
$K_3Fe(CN)_6$	0.268	0.212	0.184	0.167	0.155	0.146	0.140	0.135	0.131	0.128
$K_4Fe(CN)_6$	0.139	0.0993	0.0808	0.0693	0.0614	0.0556	0.0512	0.0479	0.0454	—
KH_2PO_4	0.731	0.653	0.602	0.561	0.529	0.501	0.477	0.456	0.438	0.421
KI	0.778	0.733	0.707	0.689	0.676	0.667	0.660	0.654	0.649	0.645
KNO_3	0.739	0.663	0.614	0.576	0.545	0.519	0.496	0.476	0.459	0.443
$KOAc$	0.796	0.766	0.754	0.750	0.751	0.754	0.759	0.766	0.774	0.783
KOH	0.798	0.760	0.742	0.734	0.732	0.733	0.736	0.742	0.749	0.756
$KSCN$	0.769	0.716	0.685	0.663	0.646	0.633	0.623	0.614	0.606	0.599
K_2SO_4	0.441	0.360	0.316	0.286	0.264	0.246	0.232	—	—	—
$LiBr$	0.796	0.766	0.756	0.752	0.753	0.758	0.767	0.777	0.789	0.803
$LiCl$	0.790	0.757	0.744	0.740	0.739	0.743	0.748	0.755	0.764	0.774
$LiClO_4$	0.812	0.794	0.792	0.798	0.808	0.820	0.834	0.852	0.869	0.887
LiI	0.815	0.802	0.804	0.813	0.824	0.838	0.852	0.870	0.888	0.910
$LiNO_3$	0.788	0.752	0.736	0.728	0.726	0.727	0.729	0.733	0.737	0.743
$LiOH$	0.760	0.702	0.665	0.638	0.617	0.599	0.585	0.573	0.563	0.554
$LiOAc$	0.784	0.742	0.721	0.709	0.700	0.691	0.689	0.688	0.688	0.689
Li_2SO_4	0.468	0.398	0.361	0.337	0.319	0.307	0.297	0.289	0.282	0.277
$MgCl_2$	0.529	0.489	0.477	0.475	0.481	0.491	0.506	0.522	0.544	0.570

ACTIVITY COEFFICIENTS OF ACIDS, BASES AND SALTS (continued)

	0.1	**0.2**	**0.3**	**0.4**	**0.5**	**0.6**	**0.7**	**0.8**	**0.9**	**1.0**
$MgSO_4$	0.150	0.107	0.0874	0.0756	0.0675	0.0616	0.0571	0.0536	0.0508	0.0485
$MnCl_2$	0.516	0.469	0.450	0.442	0.440	0.443	0.448	0.455	0.466	0.479
$MnSO_4$	0.150	0.105	0.0848	0.0725	0.0640	0.0578	0.0530	0.0493	0.0463	0.0439
NH_4Cl	0.770	0.718	0.687	0.665	0.649	0.636	0.625	0.617	0.609	0.603
NH_4NO_3	0.740	0.677	0.636	0.606	0.582	0.562	0.545	0.530	0.516	0.504
$(NH_4)_2SO_4$	0.439	0.356	0.311	0.280	0.257	0.240	0.226	0.214	0.205	0.196
$NaBr$	0.782	0.741	0.719	0.704	0.697	0.692	0.689	0.687	0.687	0.687
$NaCl$	0.778	0.735	0.710	0.693	0.681	0.673	0.667	0.662	0.659	0.657
$NaClO_3$	0.772	0.720	0.688	0.664	0.645	0.630	0.617	0.606	0.597	0.589
$NaClO_4$	0.775	0.729	0.701	0.683	0.668	0.656	0.648	0.641	0.635	0.629
Na_2CrO_4	0.464	0.394	0.353	0.327	0.307	0.292	0.280	0.269	0.261	0.253
NaF	0.765	0.710	0.676	0.651	0.632	0.616	0.603	0.592	0.582	0.573
NaH_2PO_4	0.744	0.675	0.629	0.593	0.563	0.539	0.517	0.499	0.483	0.468
NaI	0.787	0.751	0.735	0.727	0.723	0.723	0.724	0.727	0.731	0.736
$NaNO_3$	0.762	0.703	0.666	0.638	0.617	0.599	0.583	0.570	0.558	0.548
NaOAc	0.791	0.757	0.744	0.737	0.735	0.736	0.740	0.745	0.752	0.757
NaOH	0.766	0.727	0.708	0.697	0.690	0.685	0.681	0.679	0.678	0.678
NaSCN	0.787	0.750		0.720	0.715	0.712	0.710	0.710	0.711	0.712
Na_2SO_4	0.445	0.365	0.320	0.289	0.266	0.248	0.233	0.221	0.210	0.201
$NiCl_2$	0.522	0.479	0.463	0.460	0.464	0.471	0.482	0.496	0.515	0.563
$NiSO_4$	0.150	0.105	0.0841	0.0713	0.0627	0.0562	0.0515	0.0478	0.0448	0.0425
$Pb(NO_3)_3$	0.395	0.308	0.260	0.228	0.205	0.187	0.172	0.160	0.150	0.141
RbBr	0.763	0.706	0.673	0.650	0.632	0.617	0.605	0.595	0.586	0.578
RbCl	0.764	0.709	0.675	0.652	0.634	0.620	0.608	0.599	0.590	0.583
RbI	0.762	0.705	0.671	0.647	0.629	0.614	0.602	0.591	0.583	0.575
$RbNO_3$	0.734	0.658	0.606	0.565	0.534	0.508	0.485	0.465	0.446	0.430
RbOAc	0.796	0.767	0.756	0.753	0.755	0.759	0.766	0.773	0.782	0.792
Rb_2SO_4	0.451	0.374	0.331	0.301	0.279	0.263	0.249	0.238	0.228	0.219
$SrCl_2$	0.511	0.462	0.442	0.433	0.430	0.431	0.434	0.441	0.449	0.461
$TlClO_4$	0.730	0.652	0.599	0.559	0.527	—	—	—	—	—
$TlNO_3$	0.702	0.606	0.545	0.500	—	—	—	—	—	—
UO_2Cl_2	0.544	0.510	0.520	0.505	0.517	0.532	0.549	0.571	0.595	0.620
UO_2SO_4	0.150	0.102	0.0807	0.0689	0.0611	0.0566	0.0515	0.0483	0.0458	0.0439
$ZnCl_2$	0.515	0.462	0.432	0.411	0.394	0.380	0.369	0.357	0.348	0.339
$Zn(NO_3)_2$	0.531	0.489	0.474	0.469	0.473	0.480	0.489	0.501	0.518	0.535
$ZnSO_4$	0.150	0.140	0.0835	0.0714	0.0630	0.0569	0.0523	0.0487	0.0458	0.0435

MOLECULAR ELEVATION OF THE BOILING POINT*

Molecular elevation of the boiling point showing the elevation of the boiling point in degrees C due to the addition of one gram molecular weight of the dissolved substance to 1000 g of any one of the solvents below. The correction in the last column gives the number of degrees to be subtracted for each mm. of difference between the barometric reading and 760 mm.

Solvent	K_B	Barometric correction per mm.	Solvent	K_B	Barometric correction per mm.
Acetic acid	3.07	0.0008	Ethyl acetate	2.77	0.0007
Acetone	1.71	0.0004	Ethyl ether	2.02	0.0005
Aniline	3.52	0.0009	*n*-Hexane	2.75	0.0007
Benzene	2.53	0.0007	Methanol (methyl alcohol)	0.83	0.0002
Bromobenzene	6.26	0.0016	Methyl acetate	2.15	0.0005
Carbon bisulfide	2.34	0.0006	Nitrobenzene	5.24	0.0013
Carbon tetrachloride	5.03	0.0013	*n*-Octane	4.02	0.0010
Chloroform	3.63	0.0009	Phenol	3.56	0.0009
Cyclohexane	2.79	0.0007	Toluene	3.33	0.0008
Ethanol (ethyl alcohol)	1.22	0.0003	Water	0.512	0.0001

* Most values from Hoyt, C. S. and Fink, C. K., Journal of Physical Chemistry, Vol. 41, No. 3., March, 1937.

MOLECULAR DEPRESSION OF THE FREEZING POINT

Showing the depression of the freezing point due to the addition of one gram molecular weight of dissolved substance, for various solvents.

Solvent	Depression for one gram molecular weight dissolved in 100 g °C	Solvent	Depression for one gram molecular weight dissolved in 100 g °C
Acetic acid	39.0	Naphthalene	68—69
Benzene	49.0	Nitrobenzene	70.0
Benzophenone	98.0	Phenol	74.0
Diphenyl	80.0	Stearic acid	45.0
Diphenylamine	86.0	Triphenyl methane	124.5
Ethylene dibromide	118.0	Urethane	51.4
Formic acid	27.7	Water	18.5—18.7

CORRECTION OF BOILING POINTS TO STANDARD PRESSURE

By H. B. Hass and R. F. Newton

This correction may be made by using the equation:

$$\Delta t = \frac{(273.1 + t)(2.8808 - \log p)}{\phi + .15(2.8808 - \log p)} \quad (1)$$

where Δt = degrees C to be added to the observed boiling point.
t = the observed boiling point.
$\log p$ = the logarithm of the observed pressure in millimeters of mercury.

$$\phi = \frac{\Delta H_{vap}}{2.303RT_b} = \frac{\Delta S_{vap}}{2.303R}$$

The value of ϕ may be estimated from the graph and the table. Substances not included in the table may be classified by grouping them with compounds which bear a close physical or structural resemblance to them.

Example 1. Benzene boils at 20°C. at 75 mm pressure. What is its normal boiling point? We do not find benzene in the table but we find hydrocarbons in group 2, and a group 2 compound with a boiling point of 20° has a ϕ of 4.6.

Substituting in the equation

$$\Delta t = \frac{(273.1 + 20)(2.8808 - 1.8751)}{4.60 + .15(2.8808 - 1.8751)} = 62°$$

Adding this to 20° gives 82° as a first approximation.

The graph shows that the ϕ for a compound of group 2 boiling at 82° is 4.72 instead of 4.60 which we originally used. Since ϕ is in the denominator, this increase will lower our Δt by the ratio, 4.60/4.72, or the corrected Δt is $62 \times 4.60/4.72 = 60.4$. Adding Δt to t, gives 80.4° as a second approximation.

The formula can best be used in a slightly different form when the reverse calculation is desired, *i.e.*, when one calculates the vapor pressure at a given temperature, lower than the normal boiling point.

$$2.8808 - \log p = \frac{\phi \Delta t}{273.1 + t - .15\Delta t} \quad (2)$$

Example 2. Alcohol boils at 78.4°C. What is its vapor pressure at 20°C.? Substituting in equation 2:

CORRECTION OF BOILING POINTS TO STANDARD PRESSURE (Continued)

$$2.8808 - \log p = \frac{6.06 \times 58.4}{293.1 - (.15 \times 58.4)} = 1.245$$

$$\log p = 2.8808 - 1.245 = 1.6358$$

$$p = 43.2 \text{ mm.}$$

Here no second approximation is necessary, since the correct value of ϕ was taken immediately, the normal boiling point having been known.

Compound	Group	Compound	Group
Acetaldehyde	3	Amines	3
Acetic acid	4	*n*-Amyl alcohol	8
Acetic anhydride	6	Anthracene	1
Acetone	3	Anthraquinone	1
Acetophenone	4	Benzaldehyde	2
Benzoic acid	5	Hydrogen cyanide	3
Benzonitrile	2	Isoamyl alcohol	7
Benzophenone	2	Isobutyl alcohol	8
Benzyl alcohol	5	Isobutyric acid	6
Butylethylene	1	Isocaproic acid	7
Butyric acid	7	Methane	1
Camphor	2	Methanol	7
Carbon monoxide	1	Methyl amine	5
Carbon oxysulfide	2	Methyl benzoate	3
Carbon suboxide	2	Methyl ether	3
Carbon sulfoselenide	2	Methyl ethyl ether	3
m.p. Chloroanilines	3	Methyl ethyl ketone	2
Chlorinated derivatives	Same group as though Cl were H	Methyl fluoride	3
		Methyl formate	4
o.m.p. Cresols	4	Methyl salicylate	2
Cyanogen	4	Methyl silicane	1
Cyanogen chloride	3	α, β Naphthols	3
Dibenzyl ketone	2	Nitrobenzene	3
Dimethyl amine	4	Nitromethane	3
Dimethyl oxalate	4	o.m.p. Nitrotoluenes	2
Dimethyl silicane	2	o.m.p. Nitrotoluidines	2
Esters	3	Phenanthrene	1
Ethanol	8	Phenol	5
Ethers	2	Phosgene	2
Ethylamine	4	Phthalic anhydride	2
Ethylene glycol	7	Propionic acid	5
Ethylene oxide	3	*n*-Propyl alcohol	8
Formic acid	3	Quinoline	2
Glycol diacetate	4	Sulfides	2
Halogen derivatives	Same group as though halogen were hydrogen.	Tetranitromethane	3
		Trichloroethylene	1
		Valeric acid	7
Heptylic acid	7	Water	6
Hydrocarbons	2		

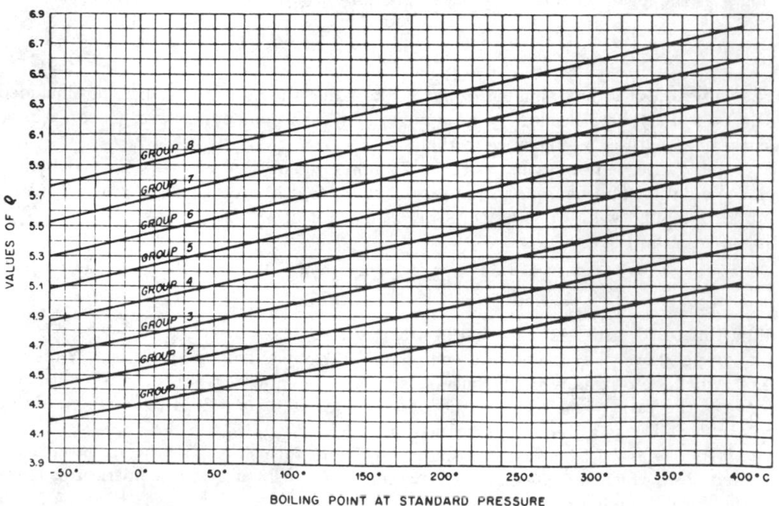

VAN DER WAALS' CONSTANTS FOR GASES

(Calculated from Amagat units in Landolt-Bornstein Physical Chemical Tables)

Van der Waals' equation is an equation of state for real gases. It may be written

$$\left(P + \frac{a}{V^2}\right)(V - b) = RT \text{ for one mole.} \qquad \text{or} \qquad \left(P + \frac{n^2a}{V^2}\right)(V - nb) = nRT \text{ for } n \text{ moles.}$$

The term a is a measure of the attractive force between the molecules. The term b is due to the finite volume of the molecules and to their general incompressibility. It is known that a and b vary to some extent with temperature.

The values for a and b in the following table are those to be used when the pressure is in atmospheres and the volume is in liters. Thus R in the above equation will be 0.08206 liter atmospheres per mole per degree. T is degrees Kelvin.

Name	Formula	a $\frac{(\text{liters})^2 \times \text{atm.}}{(\text{mole})^2}$	b $\frac{\text{liters}}{\text{mole}}$
Acetic acid	CH_3CO_2H	17.59	0.1068
Acetic anhydride	$(CH_3CO)_2O$	19.90	0.1263
Acetone	$(CH_3)_2CO$	13.91	0.0994
Acetonitrile	CH_3CN	17.58	0.1168
Acetylene	C_2H_2	4.390	0.05136
Ammonia	NH_3	4.170	0.03707
Amyl formate	$HCO_2C_5H_{11}$	27.58	0.1730
Amylene	C_5H_{10}	15.90	0.1207
*Iso*amylene	C_5H_{10}	18.08	0.1405
Aniline	$C_6H_5NH_2$	26.50	0.1369
Argon	A	1.345	0.03219
Benzene	C_6H_6	18.00	0.1154
Benzonitrile	C_6H_5CN	33.39	0.1724
Bromobenzene	C_6H_5Br	28.56	0.1539
n-Butane	C_4H_{10}	14.47	0.1226
iso-Butane	C_4H_{10}	12.87	0.1142
iso-Butyl acetate	$CH_3CO_2C_4H_9$	28.50	0.1833
iso-Butyl alcohol	C_4H_9OH	17.03	0.1143
iso-Butyl benzene	$C_6H_5C_4H_9$	38.59	0.2144
iso-Butyl formate	$HCO_2C_4H_9$	22.54	0.1476
Butyronitrile	C_3H_7CN	25.72	0.1596
Capronitrile	$C_5H_{11}CN$	34.16	0.1984
Carbon dioxide	CO_2	3.592	0.04267
Carbon disulfide	CS_2	11.62	0.07685
Carbon monoxide	CO	1.485	0.03985
Carbon oxysulfide	COS	3.933	0.05817
Carbon tetrachloride	CCl_4	20.39	0.1383
Chlorine	Cl_2	6.493	0.05622
Chlorobenzene	C_6H_5Cl	25.43	0.1453
Chloroform	$CHCl_3$	15.17	0.1022
m-Cresol	$CH_3C_6H_4OH$	31.38	0.1607
Cyanogen	C_2N_2	7.667	0.06901
Cyclohexane	C_6H_{12}	22.81	0.1424
Cymene	$C_{10}H_{14}$	42.16	0.2336
Decane	$C_{10}H_{12}$	48.55	0.2905
Di-isobutyl	C_8H_{18}	34.97	0.2296
Diethylamine	$(C_2H_5)_2NH$	19.15	0.1392
Dimethylamine	$(CH_3)_2NH$	10.38	0.08570
Dimethylaniline	$C_6H_5N(CH_3)_2$	37.49	0.1970
Diphenyl	$(C_6H_5)_2$	52.79	0.2480
Diphenyl methane	$(C_6H_5)_2CH_2$	38.20	0.2240
Dipropylamine	$(C_3H_7)_2NH$	27.72	0.1820
Di-isopropyl	$(C_3H_7)_2$	23.13	0.1669
Durene	$C_{10}H_{14}$	45.32	0.2424
Ethane	C_2H_6	5.489	0.06380
Ethyl acetate	$CH_3CO_2C_2H_5$	20.45	0.1412
Ethyl alcohol	C_2H_5OH	12.02	0.08407
Ethylamine	$C_2H_5NH_2$	10.60	0.08409
Ethyl benzene	$C_2H_5C_6H_5$	28.60	0.1667
Ethyl butyrate	$C_3H_7CO_2C_2H_5$	30.07	0.1919
Ethyl isobutyrate	$C_3H_7CO_2C_2H_5$	28.87	0.1994
Ethyl chloride	C_2H_5Cl	10.91	0.08651
Ethyl ether	$(C_2H_5)_2O$	17.38	0.1344
Ethyl formate	$HCO_2C_2H_5$	14.80	0.1056
Ethyl mercaptan	C_2H_5SH	11.24	0.08098
Ethyl propionate	$C_2H_5CO_2C_2H_5$	24.39	0.1615
Ethyl sulfide	$(C_2H_5)_2S$	18.75	0.1214
Ethylene	C_2H_4	4.471	0.05714
Ethylene bromide	$(CH_2Br)_2$	13.98	0.08664
Ethylene chloride	$(CH_2Cl)_2$	16.91	0.1086
Ethylidene chloride	CH_3CHCl_2	15.50	0.1073
Fluorobenzene	C_6H_5F	19.93	0.1286
Germanium tetrachloride	$GeCl_4$	22.60	0.1485
Helium	He	0.03412	0.02370
n-Heptane	C_7H_{16}	31.51	0.2065
n-Hexane	C_6H_{14}	24.39	0.1735
Hydrogen	H_2	0.2444	0.02661
Hydrogen bromide	HBr	4.451	0.04431
Hydrogen chloride	HCl	3.667	0.04081
Hydrogen selenide	H_2Se	5.268	0.04637
Hydrogen sulfide	H_2S	4.431	0.04287
Iodobenzene	C_6H_5I	33.08	0.1656
Krypton	Kr	2.318	0.03978
Mercury	Hg	8.093	0.01696
Mesitylene	$(CH_3)_3C_6H_3$	34.32	0.1979
Methane	CH_4	2.253	0.04278
Methyl acetate	$CH_3CO_2CH_3$	15.29	0.1091
Methyl alcohol	CH_3OH	9.523	0.06702
Methylamine	CH_3NH_2	7.130	0.05992
Methyl butyrate	$C_3H_7CO_2CH_3$	23.94	0.1569
Methyl isobutyrate	$C_3H_7CO_2CH_3$	24.50	0.1637
Methyl chloride	CH_3Cl	7.471	0.06483
Methyl ether	$(CH_3)_2O$	8.073	0.07246
Methyl ethyl ether	$CH_3OC_2H_5$	11.95	0.09775
Methyl ethyl sulfide	$CH_3SC_2H_5$	19.23	0.1304
Methyl fluoride	CH_3F	4.631	0.05264
Methyl formate	HCO_2CH_3	10.84	0.08068
Methyl propionate	$C_2H_5CO_2CH_3$	19.91	0.1360
Methyl sulfide	$(CH_3)_2S$	12.87	0.09213
Methyl valerate	$C_4H_9CO_2CH_3$	28.96	0.1845
Naphthalene	$C_{10}H_8$	39.74	0.1937
Neon	Ne	0.2107	0.01709
Nitric oxide	NO	1.340	0.02789
Nitrogen	N_2	1.390	0.03913
Nitrogen dioxide	NO_2	5.284	0.04424
Nitrous oxide	N_2O	3.782	0.04415
n-Octane	C_8H_{18}	37.32	0.2368
Oxygen	O_2	1.360	0.03183
n-Pentane	C_5H_{12}	19.01	0.1460
iso-Pentane	C_5H_{12}	18.05	0.1417
Phenetole	$C_6H_5OC_2H_5$	35.16	0.1963
Phosphine	PH_3	4.631	0.05156
Phosphonium chloride	PH_4Cl	4.054	0.04545
Phosphorus	P	52.94	0.1566
Propane	C_3H_8	8.664	0.08445
Propionic acid	$C_2H_5CO_2H$	20.11	0.1187
Propionitrile	C_2H_5CN	16.44	0.1064
Propyl acetate	$CH_3CO_2C_3H_7$	24.63	0.1619
Propyl alcohol	C_3H_7OH	14.92	0.1019
Propylamine	$C_3H_7NH_2$	14.99	0.1090
Propyl benzene	$C_6H_5C_3H_7$	35.85	0.2028
iso-Propyl benzene	$C_6H_5C_3H_7$	35.64	0.2025
Propyl chloride	C_3H_7Cl	15.91	0.1141
Propyl formate	$HCO_2C_3H_7$	18.95	0.1280
Propylene	C_2H_6	8.379	0.08272
Pseudo-cumene	$C_6H_3(CH_3)_3$	36.61	0.2021
Silicon fluoride	SiF_4	4.195	0.05571
Silicon tetrahydride	SiH_4	4.320	0.05786
Stannic chloride	$SnCl_4$	26.91	0.1642
Sulfur dioxide	SO_2	6.714	0.05636
Thiophene	C_4H_4S	20.72	0.1270
Toluene	$C_6H_5CH_3$	24.06	0.1463
Triethylamine	$(C_2H_5)_3N$	27.17	0.1831
Trimethylamine	$(CH_3)_3N$	13.02	0.1084
Xenon	Xe	4.194	0.05105
m-Xylene	$C_6H_4(CH_3)_2$	30.36	0.1772
o-Xylene	$C_6H_4(CH_3)_2$	29.98	0.1755
p-Xylene	$C_6H_4(CH_3)_2$	30.93	0.1809
Water	H_2O	5.464	0.03049

VAN DER WAAL'S RADII IN Å

N	1.5	Te	2.20
P	1.9	H	1.2
As	2.0	F	1.35
Sb	2.2	Cl	1.80
O	1.40	Br	1.95
S	1.85	I	2.15
Se	2.00		

EMERGENT STEM CORRECTION FOR LIQUID-IN-GLASS THERMOMETERS

Accurate thermometers are calibrated with the entire stem immersed in the bath which determines the temperature of the thermometer bulb. However, for reasons of convenience it is common practice when using a thermometer to permit its stem to extend out of the apparatus. Under these conditions both the stem and the mercury in the exposed stem are at a temperature different from that of the bulb. This introduces an error into the observed temperature. Since the coefficient of thermal expansion of glass is less than that of mercury, the observed temperature will be less than the true temperature if the bulb is hotter than the stem and greater than the true temperature, providing the thermal gradient is reversed. For exact work the magnitude of this error can only be determined by experiment. However, for most purposes it is sufficiently accurate to apply the following equation which takes into account the difference of the thermal expansion of glass and mercury:

$$T_c = T_o + F \times L(T_o - T_m)$$

Where

T_c = corrected temperature
T_o = observed temperature
T_m = mean temperature of exposed stem. The mean temperature of the exposed stem may be determined by fastening the bulb of a second thermometer against the midpoint of the exposed liquid column.
L = the length of the exposed column in degrees above the surface of the substance whose temperature is being determined.
F = correction factor. For approximate work and when the liquid in the thermometer is mercury a value for F of 0.00016 is generally used. For more accurate work with mercury filled thermometers values as given in the following table are used. For thermometers filled with organic liquids it is customary to use 0.001 for the value of F.

Values of F for various glasses

Tm°C.	Corning 0041	Corning 8800	Corning 8810	Jena 16 III	Jena 59 III
50	0.000157	0.000166	0.000156	0.000158	0.000164
150	0.000159	0.000167	0.000157	0.000158	0.000165
250	0.000163	0.000168	0.000161	0.000161	0.000170
350	0.000168	0.000173	0.000166	—	0.000177

PRESSURE OF AQUEOUS VAPOR

Vapor Pressure of Ice

Pressure of aqueous vapor over ice in mm of Hg for temperatures from −98 to 0°C.

Temp. °C	0.0	0.2	0.4	0.6	0.8
−90	0.000070	0.000048	0.000033	0.000022	0.000015
−80	0.00040	0.00029	0.00020	0.00014	0.00010
−70	0.00194	0.00143	0.00105	0.00077	0.00056
−60	0.00808	0.00614	0.00464	0.00349	0.00261
−50	0.02955	0.0230	0.0178	0.0138	0.0106
−40	0.0966	0.0768	0.0609	0.0481	0.0378
−30	0.2859	0.2318	0.1873	0.1507	0.1209
−29	0.317	0.311	0.304	0.298	0.292
−28	0.351	0.344	0.337	0.330	0.324
−27	0.389	0.381	0.374	0.366	0.359
−26	0.430	0.422	0.414	0.405	0.397
−25	0.476	0.467	0.457	0.448	0.439
−24	0.526	0.515	0.505	0.495	0.486
−23	0.580	0.569	0.558	0.547	0.536
−22	0.640	0.627	0.615	0.603	0.592
−21	0.705	0.691	0.678	0.665	0.652
−20	0.776	0.761	0.747	0.733	0.719
−19	0.854	0.838	0.822	0.806	0.791
−18	0.939	0.921	0.904	0.887	0.870
−17	1.031	1.012	0.993	0.975	0.956
−16	1.132	1.111	1.091	1.070	1.051
−15	1.241	1.219	1.196	1.175	1.153
−14	1.361	1.336	1.312	1.288	1.264
−13	1.490	1.464	1.437	1.411	1.386
−12	1.632	1.602	1.574	1.546	1.518
−11	1.785	1.753	1.722	1.691	1.661
−10	1.950	1.916	1.883	1.849	1.817
−9	2.131	2.093	2.057	2.021	1.985
−8	2.326	2.285	2.246	2.207	2.168
−7	2.537	2.493	2.450	2.408	2.367
−6	2.765	2.718	2.672	2.626	2.581
−5	3.013	2.962	2.912	2.862	2.813
−4	3.280	3.225	3.171	3.117	3.065
−3	3.568	3.509	3.451	3.393	3.336
−2	3.880	3.816	3.753	3.691	3.630
−1	4.217	4.147	4.079	4.012	3.946
−0	4.579	4.504	4.431	4.359	4.287

VAPOR PRESSURE OF WATER BELOW 100°C

Pressure of aqueous vapor over water in mm of Hg for temperatures from −15.8 to 100°C. Values for fractional degrees between 50 and 89 were obtained by interpolation.

Temp. °C	0.0	0.2	0.4	0.6	0.8
−15	1.436	1.414	1.390	1.368	1.345
−14	1.560	1.534	1.511	1.485	1.460
−13	1.691	1.665	1.637	1.611	1.585
−12	1.834	1.804	1.776	1.748	1.720
−11	1.987	1.955	1.924	1.893	1.863
−10	2.149	2.116	2.084	2.050	2.018
−9	2.326	2.289	2.254	2.219	2.184
−8	2.514	2.475	2.437	2.399	2.362
−7	2.715	2.674	2.633	2.593	2.553
−6	2.931	2.887	2.843	2.800	2.757
−5	3.163	3.115	3.069	3.022	2.976
−4	3.410	3.359	3.309	3.259	3.211
−3	3.673	3.620	3.567	3.514	3.461
−2	3.956	3.898	3.841	3.785	3.730
−1	4.258	4.196	4.135	4.075	4.016
−0	4.579	4.513	4.448	4.385	4.320
0	4.579	4.647	4.715	4.785	4.855
1	4.926	4.998	5.070	5.144	5.219
2	5.294	5.370	5.447	5.525	5.605
3	5.685	5.766	5.848	5.931	6.015

VAPOR PRESSURE OF WATER BELOW 100°C (continued)

Temp. °C	0.0	0.2	0.4	0.6	0.8
4	6.101	6.187	6.274	6.363	6.453
5	6.543	6.635	6.728	6.822	6.917
6	7.013	7.111	7.209	7.309	7.411
7	7.513	7.617	7.722	7.828	7.936
8	8.045	8.155	8.267	8.380	8.494
9	8.609	8.727	8.845	8.965	9.086
10	9.209	9.333	9.458	9.585	9.714
11	9.844	9.976	10.109	10.244	10.380
12	10.518	10.658	10.799	10.941	11.085
13	11.231	11.379	11.528	11.680	11.833
14	11.987	12.144	12.302	12.462	12.624
15	12.788	12.953	13.121	13.290	13.461
16	13.634	13.809	13.987	14.166	14.347
17	14.530	14.715	14.903	15.092	15.284
18	15.477	15.673	15.871	16.071	16.272
19	16.477	16.685	16.894	17.105	17.319
20	17.535	17.753	17.974	18.197	18.422
21	18.650	18.880	19.113	19.349	19.587
22	19.827	20.070	20.316	20.565	20.815
23	21.068	21.324	21.583	21.845	22.110
24	22.377	22.648	22.922	23.198	23.476
25	23.756	24.039	24.326	24.617	24.912
26	25.209	25.509	25.812	26.117	26.426
27	26.739	27.055	27.374	27.696	28.021
28	28.349	28.680	29.015	29.354	29.697
29	30.043	30.392	30.745	31.102	31.461
30	31.824	32.191	32.561	32.934	33.312
31	33.695	34.082	34.471	34.864	35.261
32	35.663	36.068	36.477	36.891	37.308
33	37.729	38.155	38.584	39.018	39.457
34	39.898	40.344	40.796	41.251	41.710
35	42.175	42.644	43.117	43.595	44.078
36	44.563	45.054	45.549	46.050	46.556
37	47.067	47.582	48.102	48.627	49.157
38	49.692	50.231	50.774	51.323	51.879
39	52.442	53.009	53.580	54.156	54.737
40	55.324	55.91	56.51	57.11	57.72
41	58.34	58.96	59.58	60.22	60.86
42	61.50	62.14	62.80	63.46	64.12
43	64.80	65.48	66.16	66.86	67.56
44	68.26	68.97	69.69	70.41	71.14
45	71.88	72.62	73.36	74.12	74.88
46	75.65	76.43	77.21	78.00	78.80
47	79.60	80.41	81.23	82.05	82.87
48	83.71	84.56	85.42	86.28	87.14
49	88.02	88.90	89.79	90.69	91.59
50	92.51	93.5	94.4	95.3	96.3
51	97.20	98.2	99.1	100.1	101.1
52	102.09	103.1	104.1	105.1	106.2
53	107.20	108.2	109.3	110.4	111.4
54	112.51	113.6	114.7	115.8	116.9
55	118.04	119.1	120.3	121.5	122.6
56	123.80	125.0	126.2	127.4	128.6
57	129.82	131.0	132.3	133.5	134.7
58	136.08	137.3	138.5	139.9	141.2
59	142.60	143.9	145.2	146.6	148.0
60	149.38	150.7	152.1	153.5	155.0
61	156.43	157.8	159.3	160.8	162.3
62	163.77	165.2	166.8	168.3	169.8
63	171.38	172.9	174.5	176.1	177.7
64	179.31	180.9	182.5	184.2	185.8
65	187.54	189.2	190.9	192.6	194.3
66	196.09	197.8	199.5	201.3	203.1
67	204.96	206.8	208.6	210.5	212.3
68	214.17	216.0	218.0	219.9	221.8
69	223.73	225.7	227.7	229.7	231.7
70	233.7	235.7	237.7	239.7	241.8
71	243.9	246.0	248.2	250.3	252.4
72	254.6	256.8	259.0	261.2	263.4
73	265.7	268.0	270.2	272.6	274.8
74	277.2	279.4	281.8	284.2	286.6
75	289.1	291.5	294.0	296.4	298.8
76	301.4	303.8	306.4	308.9	311.4
77	314.1	316.6	319.2	322.0	324.6
78	327.3	330.0	332.8	335.6	338.2
79	341.0	343.8	346.6	349.4	352.2
80	355.1	358.0	361.0	363.8	366.8
81	369.7	372.6	375.6	378.8	381.8
82	384.9	388.0	391.2	394.4	397.4
83	400.6	403.8	407.0	410.2	413.6
84	416.8	420.2	423.6	426.8	430.2
85	433.6	437.0	440.4	444.0	447.5
86	450.9	454.4	458.0	461.6	465.2
87	468.7	472.4	476.0	479.8	483.4
88	487.1	491.0	494.7	498.5	502.2
89	506.1	510.0	513.9	517.8	521.8
90	525.76	529.77	533.80	537.86	541.95
91	546.05	550.18	554.35	558.53	562.75
92	566.99	571.26	575.55	579.87	584.22
93	588.60	593.00	597.43	601.89	606.38
94	610.90	615.44	620.01	624.61	629.24
95	633.90	638.59	643.30	648.05	652.82
96	657.62	662.45	667.31	672.20	677.12
97	682.07	687.04	692.05	697.10	702.17
98	707.27	712.40	717.56	722.75	727.98
99	733.24	738.53	743.85	749.20	754.58
100	760.00	765.45	770.93	776.44	782.00
101	787.57	793.18	798.82	804.50	810.21

VAPOR PRESSURE OF WATER ABOVE 100°C.

Based on values given by Keyes in the International Critical Tables.

Temp. °C	Pressure mm	Pressure Pounds per sq. in.	Temp. °F	Temp. °C	Pressure mm	Pressure Pounds per sq. in.	Temp. °F	Temp. °C	Pressure mm	Pressure Pounds per sq. in.	Temp. °F	Temp. °C	Pressure mm	Pressure Pounds per sq. in.	Temp. °F
100	760.	14.696	212.0	127	1850.83	35.789	260.6	154	3970.24	76.772	309.2	181	7694.24	148.782	357.8
101	787.51	15.228	213.8	128	1907.83	36.891	262.4	155	4075.88	78.815	311.0	182	7872.08	152.221	359.6
102	815.86	15.776	215.6	129	1966.35	38.023	264.2	156	4183.80	80.901	312.8	183	8052.96	155.719	361.4
103	845.12	16.342	217.4	130	2026.10	39.180	266.0	157	4293.24	83.018	314.6	184	8236.88	159.275	363.2
104	875.06	16.921	219.2	131	2087.42	40.364	267.8	158	4404.96	85.178	316.4	185	8423.84	162.890	365.0
105	906.07	17.521	221.0	132	2150.42	41.582	269.6	159	4519.72	87.397	318.2	186	8616.12	166.609	366.8
106	937.92	18.136	222.8	133	2214.64	42.824	271.4	160	4636.00	89.646	320.0	187	8809.92	170.356	368.6
107	970.60	18.768	224.6	134	2280.76	44.103	273.2	161	4755.32	91.953	321.8	188	9007.52	174.177	370.4
108	1004.42	19.422	226.4	135	2347.26	45.389	275.0	162	4876.92	94.304	323.6	189	9208.16	178.057	372.2
109	1038.92	20.089	228.2	136	2416.34	46.724	276.8	163	5000.04	96.685	325.4	190	9413.36	182.025	374.0
110	1074.56	20.779	230.0	137	2488.16	48.113	278.6	164	5126.96	99.139	327.2	191	9620.08	186.022	375.8
111	1111.20	21.487	231.8	138	2560.67	49.515	280.4	165	5256.16	101.638	329.0	192	9831.36	190.107	377.6
112	1148.74	22.213	233.6	139	2634.84	50.950	282.2	166	5386.88	104.165	330.8	193	10047.20	194.281	379.4
113	1187.42	22.961	235.4	140	2710.92	52.421	284.0	167	5521.40	106.766	332.6	194	10265.32	198.499	381.2
114	1227.25	23.731	237.2	141	2788.44	53.920	285.8	168	5658.20	109.412	334.4	195	10488.76	202.819	383.0
115	1267.98	24.519	239.0	142	2867.48	55.448	287.6	169	5798.04	112.116	336.2	196	10715.24	207.199	384.8
116	1309.94	25.330	240.8	143	2948.80	57.020	289.4	170	5940.92	114.879	338.0	197	10944.76	211.637	386.6
117	1352.95	26.162	242.6	144	3031.64	58.622	291.2	171	6085.32	117.671	339.8	198	11179.60	216.178	388.4
118	1397.18	27.017	244.4	145	3116.76	60.268	293.0	172	6233.52	120.537	341.6	199	11417.48	220.778	390.2
119	1442.63	27.896	246.2	146	3203.40	61.944	294.8	173	6383.24	123.432	343.4	200	11659.16	225.451	392.0
120	1489.14	28.795	248.0	147	3292.32	63.663	296.6	174	6538.28	126.430	345.2	201	11905.40	230.213	393.8
121	1536.80	29.717	249.8	148	3382.76	65.412	298.4	175	6694.08	129.442	347.0	202	12155.44	235.048	395.6
122	1586.04	30.669	251.6	149	3476.24	67.220	300.2	176	6852.92	132.514	348.8	203	12408.52	239.942	397.4
123	1636.36	31.642	253.4	150	3570.48	69.042	302.0	177	7015.56	135.659	350.6	204	12666.16	244.924	399.2
124	1687.81	32.637	255.2	151	3667.00	70.908	303.8	178	7180.48	138.848	352.4	205	12929.12	250.008	401.0
125	1740.93	33.664	257.0	152	3766.56	72.833	305.6	179	7349.20	142.110	354.2	206	13197.40	255.196	402.8
126	1795.12	34.712	258.8	153	3866.88	74.773	307.4	180	7520.20	145.417	356.0	207	13467.96	260.428	404.6

VAPOR PRESSURE OF WATER ABOVE 100°C (continued)

Temp. °C	Pressure mm	Pressure Pounds per sq. in.	Temp. °F
208	13742.32	265.733	406.4
209	14022.76	271.156	408.2
210	14305.48	276.623	410.0
211	14595.04	282.222	411.8
212	14888.40	287.895	413.6
213	15184.80	293.626	415.4
214	15488.04	299.490	417.2
215	15792.80	305.383	419.0
216	16104.40	311.408	420.8
217	16420.56	317.522	422.6
218	16742.04	323.738	424.4
219	17067.32	330.028	426.2
220	17395.64	336.377	428.0
221	17731.56	342.872	429.8
222	18072.80	349.471	431.6
223	18417.84	356.143	433.4
224	18766.68	362.888	435.2
225	19123.12	369.781	437.0
226	19482.60	376.732	438.8
227	19848.92	383.815	440.6
228	20219.80	390.987	442.4
229	20596.76	398.276	444.2
230	20978.28	405.654	446.0
231	21365.12	413.134	447.8
232	21757.28	420.717	449.6
233	22154.00	428.388	451.4
234	22558.32	436.207	453.2
235	22967.96	444.128	455.0
236	23382.92	452.152	456.8
237	23802.44	460.264	458.6
238	24229.56	468.523	460.4
239	24661.24	476.871	462.2
240	25100.52	485.365	464.0
241	25543.60	493.933	465.8
242	25994.28	502.647	467.6
243	26449.52	511.450	469.4
244	26912.36	520.400	471.2
245	27381.28	529.467	473.0
246	27855.52	538.638	474.8
247	28335.84	547.926	476.6
248	28823.76	557.360	478.4
249	29317.00	566.898	480.2
250	29817.84	576.583	482.0
251	30324.00	586.370	483.8
252	30837.76	596.305	485.6
253	31356.84	606.342	487.4
254	31885.04	616.556	489.2
255	32417.80	626.858	491.0
256	32957.40	637.292	492.8
257	33505.36	647.888	494.6
258	34059.40	658.601	496.4
259	34618.76	669.417	498.2
260	35188.00	680.425	500.0
261	35761.80	691.520	501.8
262	36343.20	702.763	503.6
263	36932.20	714.152	505.4
264	37529.56	725.703	507.2
265	38133.00	737.372	509.0
266	38742.52	749.158	510.8
267	39361.92	761.135	512.6
268	39986.64	773.215	514.4
269	40619.72	785.457	516.2
270	41261.16	797.861	518.0
271	41910.20	810.411	519.8
272	42566.08	823.094	521.6
273	43229.56	835.923	523.4
274	43902.16	848.929	525.2
275	44580.84	862.053	527.0
276	45269.40	875.367	528.8
277	45964.04	888.799	530.6
278	46669.32	902.437	532.4
279	47382.20	916.222	534.2
280	48104.20	930.183	536.0
281	48833.80	944.291	537.8
282	49570.24	958.532	539.6
283	50316.56	972.963	541.4
284	51072.76	987.586	543.2
285	51838.08	1002.385	545.0
286	52611.76	1017.345	546.8
287	53395.32	1032.497	548.6
288	54187.24	1047.810	550.4
289	54989.04	1063.314	552.2
290	55799.20	1078.980	554.0
291	56612.40	1094.705	555.8
292	57448.40	1110.871	557.6
293	58284.40	1127.036	559.4
294	59135.60	1143.496	561.2
295	59994.40	1160.102	563.0
296	60860.80	1176.856	564.8
297	61742.40	1193.903	566.6
298	62624.00	1210.950	568.4
299	63528.40	1228.439	570.2
300	64432.80	1245.927	572.0
301	65352.40	1263.709	573.8
302	66279.60	1281.638	575.6
303	67214.40	1299.714	577.4
304	68156.80	1317.937	579.2
305	69114.40	1336.454	581.0
306	70072.00	1354.971	582.8
307	71052.40	1373.929	584.6
308	72048.00	1393.181	586.4
309	73028.40	1412.139	588.2
310	74024.00	1431.390	590.0
311	75042.40	1451.083	591.8
312	76076.00	1471.070	593.6
313	77117.20	1491.203	595.4
314	78166.00	1511.484	597.2
315	79230.00	1532.058	599.0
316	80294.00	1552.632	600.8
317	81373.20	1573.501	602.6
318	82467.60	1594.663	604.4
319	83569.60	1615.972	606.2
320	84686.80	1637.575	608.0
321	85819.20	1659.472	609.8
322	86959.20	1681.516	611.6
323	88114.40	1703.854	613.4
324	89277.20	1726.339	615.2
325	90447.60	1748.971	617.0
326	91633.20	1771.897	618.8
327	92826.40	1794.969	620.6
328	94042.40	1818.483	622.4
329	95273.60	1842.291	624.2
330	96512.40	1866.245	626.0
331	97758.80	1890.346	627.8
332	99020.40	1914.742	629.6
333	100297.20	1939.431	631.4
334	101581.60	1964.267	633.2
335	102881.20	1989.398	635.0
336	104196.00	2014.822	636.8
337	105526.00	2040.540	638.6
338	106871.20	2066.552	640.4
339	108224.00	2092.710	642.2
340	109592.00	2119.163	644.0
341	110967.60	2145.763	645.8
342	112358.40	2172.657	647.6
343	113749.20	2199.550	649.4
344	115178.00	2227.179	651.2
345	116614.40	2254.954	653.0
346	118073.60	2283.171	654.8
347	119532.80	2311.387	656.6
348	121014.80	2340.044	658.4
349	122504.40	2368.848	660.2
350	124001.60	2397.799	662.0
351	125521.60	2427.191	663.8
352	127049.20	2456.730	665.6
353	128599.60	2486.710	667.4
354	130157.60	2516.837	669.2
355	131730.80	2547.258	671.0
356	133326.80	2578.119	672.8
357	134945.60	2609.422	674.6
358	136579.60	2641.018	676.4
359	138228.80	2672.908	678.2
360	139893.20	2705.093	680.0
361	141572.80	2737.571	681.8
362	143275.20	2770.490	683.6
363	144992.80	2803.703	685.4
364	146733.20	2837.357	687.2
365	148519.20	2871.892	689.0
366	150320.40	2906.722	690.8
367	152129.20	2941.698	692.6
368	153960.80	2977.116	694.4
369	155815.20	3012.974	696.2
370	157692.40	3049.273	698.0
371	159584.80	3085.866	699.8
372	161507.60	3123.047	701.6
373	163468.40	3160.963	703.4
374	165467.20	3199.613	705.2

OXYGEN SOLUBILITY IN AQUEOUS ELECTROLYTE SOLUTIONS

Pressure: 1 Atmosphere (0.101 325 MPa)
Temperature: 310.2K

The Bunsen coefficients, α, and the standard deviation among a number of tests performed under identical conditions are presented. The units are mℓ/(mℓ atm). The Bunsen coefficient α is defined as the gas volume at STP (0.101 325 MPa and 273.15K) absorbed per unit volume of pure liquid at the temperature of the measurement.

The following data are from *Industrial & Engineering Chemistry Fundamentals*, 25, 778—779, 1986, Werner Lang and Rolf Zander. They are reproduced with permission of the copyright owner, the American Chemical Society.

Electrolyte	Concn of solution (mol/dm^3)	$10^4\alpha$
HCl	2.005	211 ± 4
	3.050	198 ± 7
	3.910	191 ± 4
	4.000	185 ± 2
$AlCl_3$	0.501	173 ± 1
	0.745	144 ± 3
	1.006	119 ± 2
	1.081	113 ± 3
	2.009	63 ± 3
	2.503	41 ± 2
$BaCl_2$	0.507	178 ± 3
	0.997	131 ± 2
	1.509	96 ± 2
$CaCl_2$	0.494	184 ± 2
	0.747	162 ± 2
	0.987	144 ± 1
	1.021	141 ± 1
	1.421	110 ± 2
	1.490	107 ± 1
	2.985	51 ± 2
	3.557	37 ± 1
	3.894	34 ± 1
	4.477	23 ± 1
$CdCl_2$	0.260	213 ± 4
	0.479	196 ± 1
	0.505	194 ± 3
	0.523	193 ± 3
	0.747	179 ± 3
	0.760	178 ± 2
	0.966	164 ± 1
	0.997	163 ± 2
	1.025	161 ± 3
	1.046	164 ± 3
	1.264	149 ± 3
	1.457	141 ± 3
	1.491	141 ± 3
	1.503	138 ± 3
	1.922	120 ± 1
	1.996	116 ± 2
	2.041	116 ± 2
	2.029	115 ± 1
$CeCl_3$	0.498	159 ± 7
	0.979	111 ± 2
	1.974	51 ± 1
	2.462	33 ± 4
$CoCl_2$	0.501	184 ± 2
	0.749	162 ± 2
	0.993	144 ± 3
	1.494	114 ± 2
CsCl	0.515	212 ± 5
	1.001	192 ± 2
	1.028	187 ± 4
	1.505	170 ± 3
	1.993	152 ± 3
	2.203	138 ± 2
	2.517	132 ± 3
	3.002	119 ± 2
	3.382	105 ± 1
	4.003	97 ± 1
	4.028	98 ± 1
	5.003	78 ± 2
	6.003	64 ± 2
	6.122	60 ± 2
$CuCl_2$	0.497	192 ± 2
	0.519	189 ± 1
	0.763	169 ± 1
	0.994	155 ± 1
	1.002	156 ± 1
	1.499	122 ± 1
	1.517	124 ± 1
$FeCl_3$	0.500	186 ± 3
	1.016	148 ± 2
	2.064	84 ± 7
KCl	0.503	208 ± 1
	1.002	178 ± 2
	1.502	158 ± 5
	1.992	134 ± 3
	2.976	105 ± 1
$LaCl_3$	0.514	162 ± 2
	0.992	112 ± 1
	1.993	50 ± 1
	2.478	37 ± 2
LiCl	0.985	194 ± 4
	1.069	189 ± 8
	1.482	174 ± 6
	1.993	159 ± 7
	2.336	150 ± 3
	3.978	109 ± 3
$MgCl_2$	0.503	190 ± 3
	0.523	187 ± 4
	0.982	153 ± 1
	0.998	149 ± 3
	1.463	120 ± 2
	1.592	113 ± 2
	1.745	106 ± 3
	2.126	91 ± 1
	2.879	64 ± 2
$MnCl_2$	0.840	159 ± 2
	1.230	134 ± 3
	1.756	106 ± 2
	2.127	88 ± 1
NaCl	1.001	177 ± 4
	1.265	164 ± 8
	1.503	152 ± 2
	2.016	130 ± 3
	2.989	100 ± 2
	3.030	98 ± 3
	4.017	71 ± 2
NH_4Cl	1.002	200 ± 3
	2.008	172 ± 1
	2.215	160 ± 1
	3.001	144 ± 2
$NiCl_2$	0.744	167 ± 2
	0.991	145 ± 1
	1.490	117 ± 2
RbCl	0.990	187 ± 3
	1.014	184 ± 7
	1.359	166 ± 2
	1.984	143 ± 1
	2.957	112 ± 2
$ZnCl_2$	0.250	213 ± 7
	0.493	190 ± 3
	0.762	176 ± 3
	0.816	174 ± 2
	1.017	164 ± 4
	1.069	158 ± 3
	1.250	154 ± 1
	1.460	149 ± 1
	1.939	132 ± 3
	2.005	132 ± 2
	2.959	108 ± 2
HNO_3	1.000	236 ± 4
	2.000	230 ± 2
	4.000	221 ± 4
$Al(NO_3)_3$	0.308	202 ± 5
	0.602	171 ± 3
	1.095	127 ± 7
$Ba(NO_3)_2$	0.158	222 ± 3
	0.298	207 ± 4
$Ca(NO_3)_2$	0.497	195 ± 4
	0.990	155 ± 6
	1.961	101 ± 3
$Cd(NO_3)_2$	0.551	194 ± 5
	0.740	181 ± 3
	1.548	133 ± 2
	2.046	110 ± 3
$Ce(NO_3)_3$	0.997	137 ± 3
	1.862	85 ± 2
	2.952	50 ± 3
$Co(NO_3)_2$	0.502	194 ± 4
	0.753	177 ± 4
	1.002	157 ± 2
	1.487	126 ± 2
$CsNO_3$	0.314	226 ± 2
	0.325	228 ± 2
	0.615	215 ± 2
	0.624	209 ± 1
	0.930	198 ± 2
$Cu(NO_3)_2$	0.298	213 ± 5
	0.497	198 ± 1
	0.613	188 ± 3
	0.987	160 ± 1
	1.267	144 ± 1
$Fe(NO_3)_3$	0.498	189 ± 5
	0.739	169 ± 1
	0.999	147 ± 3
	1.037	147 ± 1
KNO_3	0.746	201 ± 1
	1.013	191 ± 3
	1.510	169 ± 2
	1.852	161 ± 6
$La(NO_3)_3$	0.493	184 ± 2
	0.966	141 ± 2
	2.227	69 ± 2
$LiNO_3$	0.955	201 ± 4
	1.916	169 ± 4
	2.070	164 ± 3
	2.869	140 ± 3
$Mg(NO_3)_2$	0.861	175 ± 6
	0.992	162 ± 2
	1.663	124 ± 2
	2.238	99 ± 3
$Mn(NO_3)_2$	0.506	198 ± 2
	1.011	161 ± 3
	1.482	132 ± 5
	1.936	112 ± 1

Electrolyte	Concn of solution (mol/dm^3)	$10^4\alpha$	Electrolyte	Concn of solution (mol/dm^3)	$10^4\alpha$	Electrolyte	Concn of solution (mol/dm^3)	$10^4\alpha$
$NaNO_3$	0.762	200 ± 3		0.741	155 ± 3		1.499	95 ± 3
	1.017	189 ± 4		0.959	139 ± 1		1.783	82 ± 1
	1.533	166 ± 3		1.000	133 ± 1		2.151	66 ± 1
	2.078	151 ± 4	$Fe_2(SO_4)_3$	0.487	148 ± 2	$KHSO_4$	0.565	192 ± 2
NH_4NO_3	0.998	210 ± 4		0.723	115 ± 2		0.771	176 ± 1
	1.991	185 ± 4		0.940	94 ± 2		1.011	162 ± 3
	2.050	181 ± 4	K_2SO_4	0.247	200 ± 3		1.484	141 ± 2
	3.120	158 ± 2		0.297	190 ± 2		1.522	137 ± 3
$Ni(NO_3)_2$	0.752	173 ± 1		0.397	177 ± 2		1.638	133 ± 2
	0.852	167 ± 2		0.405	174 ± 3		1.798	128 ± 3
	1.385	135 ± 1		0.500	166 ± 3		1.805	127 ± 2
	2.062	99 ± 1		0.582	157 ± 2		2.017	115 ± 2
$RbNO_3$	0.487	219 ± 3	Li_2SO_4	0.510	177 ± 2	$NaHSO_4$	0.489	194 ± 2
	0.522	216 ± 5		1.007	127 ± 2		0.519	193 ± 1
	0.958	193 ± 3		1.903	71 ± 1		0.776	175 ± 3
	1.006	192 ± 3	$MgSO_4$	0.497	174 ± 3		0.995	159 ± 2
	1.058	192 ± 5		0.750	148 ± 4		1.018	159 ± 2
	1.360	177 ± 4		0.996	128 ± 1		1.503	135 ± 1
	1.989	158 ± 1		1.196	112 ± 2		1.574	132 ± 2
	2.017	155 ± 7		1.202	113 ± 2		1.939	118 ± 3
$Th(NO_3)_4$	0.488	174 ± 1		1.500	89 ± 2		2.628	92 ± 2
	0.741	145 ± 2		2.003	66 ± 3	NH_4HSO_4	1.031	176 ± 3
	0.974	121 ± 3		2.013	66 ± 2		1.503	153 ± 2
	1.385	92 ± 1	$MnSO_4$	0.503	179 ± 2		1.982	135 ± 2
$Zn(NO_3)_2$	0.499	196 ± 3		0.691	161 ± 4		2.643	116 ± 1
	0.756	178 ± 3		0.998	132 ± 3		3.010	108 ± 2
	0.995	159 ± 5		1.386	108 ± 2	$CsOH$	0.890	174 ± 1
	1.239	146 ± 3		1.407	108 ± 1		2.142	108 ± 3
	1.506	128 ± 4		1.740	88 ± 1		2.325	100 ± 1
	1.651	119 ± 1	Na_2SO_4	0.498	163 ± 2	KOH	0.938	169 ± 1
H_2SO_4	1.000	209 ± 7		0.748	136 ± 2		1.030	164 ± 2
	1.500	189 ± 5		0.883	120 ± 2		1.844	122 ± 1
	1.600	190 ± 1		0.993	113 ± 2		2.135	108 ± 2
	2.000	177 ± 5		1.325	87 ± 1		2.311	102 ± 1
	2.500	163 ± 4		1.487	76 ± 2		2.329	100 ± 1
$Al_2(SO_4)_3$	0.197	175 ± 2		1.762	63 ± 1		3.077	79 ± 1
	0.298	151 ± 3	$(NH_4)_2SO_4$	1.016	140 ± 2		3.502	67 ± 1
	0.396	129 ± 2		1.047	137 ± 2		4.848	42 ± 1
$CdSO_4$	1.012	132 ± 2		1.988	86 ± 2		4.871	37 ± 1
	1.583	96 ± 3		2.007	82 ± 2	$LiOH$	1.015	175 ± 1
	1.989	70 ± 2		2.498	66 ± 2		1.856	135 ± 1
	2.510	50 ± 2		2.984	51 ± 1		3.075	95 ± 2
	2.735	45 ± 1		3.499	44 ± 1		4.059	65 ± 4
	3.026	36 ± 1	$NiSO_4$	0.499	176 ± 2	$NaOH$	1.000	164 ± 1
$CoSO_4$	0.497	174 ± 2		0.749	150 ± 1		1.139	155 ± 2
	0.767	150 ± 1		1.181	116 ± 3		2.000	114 ± 1
	1.001	128 ± 2		1.489	95 ± 3		2.105	104 ± 1
	1.494	96 ± 1	Rb_2SO_4	0.401	181 ± 1		2.122	106 ± 1
Cs_2SO_4	0.500	171 ± 1		0.803	138 ± 2		3.035	74 ± 1
	1.015	120 ± 1		1.199	104 ± 1		4.071	49 ± 1
	1.525	86 ± 1	$ZnSO_4$	0.501	170 ± 2	$RbOH$	1.112	157 ± 2
	1.909	66 ± 1		0.719	148 ± 5		2.070	109 ± 3
$CuSO_4$	0.496	177 ± 1		1.006	122 ± 4		3.187	76 ± 3
	0.591	168 ± 4		1.481	98 ± 2			

HEAT OF COMBUSTION
For Organic Compounds

The heat of combustion is given in kilogram calores per gram molecular weight of the substance when combustion takes place at atmospheric pressure and at either 20°C or 25°C. If the data are for 20°C there is no asterisk for the numerical value of the heat of combustion. If the numerical value is for 25°C there is an asterisk marking the numerical value of the heat of combustion. The final products of combustion are gaseous carbon dioxide, liquid water and nitrogen gas for C, H, N compounds. For method of computing heats of formation see statement following this table.

Name	Formula	Physical state	Heat of combustion, kg. calories
Acetaldehyde	CH_3CHO	liquid	278.77*
Acetamide	CH_3CONH_2	solid	282.6
Acetanilide	$CH_3CONHC_6H_5$	solid	1010.4
Acetic acid	CH_3CO_2H	liquid	209.02*
Acetic anhydride	$(CH_3CO)_2O$	liquid	431.70*
Acetone	$(CH_3)_2CO$	liquid	427.92*
Acetonitrile	CH_3CN	liquid	302.4
Acetophenone	$C_6H_5COCH_3$	liquid	991.60*
Acetylacetone	$CH_3COCH_2COCH_3$	liquid	615.9
Acetylene	$(CH)_2$	gas	310.61*
Acrolein	CH_2CHO	liquid	389.6
Acrylic acid	CH_2CO_2H	liquid	327.0*
Adipic acid	$(CH_2)_4(CH_2H)_2$	solid	668.29*
Alanine	$CH_3CH(NH_2)CO_2H$	solid	387.7
Aldol, see β-hydroxybutyr-aldehyde			
Alizarin, see Dihydroxyanthraquinone			
Allyl alcohol	CH_2CH_2OH	liquid	442.4
Allylene	CH_3C	gas	465.1
p-Aminoazobenzene	$H_2NC_6H_4N_2C_6H_5$	solid	1574.0
p-Aminophenol	$HOC_6H_4NH_2$	solid	760.0
Amygdalin	$C_{20}H_{27}O_{11}N$	solid	2348.4
Amyl acetate	$C_4H_9CO_2C_2H_5$	liquid	1042.5
Amyl aocohol	$(CH_3)_2CHCH_2CH_2OH$	liquid	793.7
Amylene	C_5H_{10}	liquid	803.4
Anethole	$C_{10}H_{12}O$	solid	1324.4
Aniline	$C_6H_5NH_2$	liquid	811.7
p-Anisidine	$CH_3OC_6H_4NH_2$	solid	924.0
Anisole	$C_6H_5OCH_3$	liquid	905.1
Anthracene	$C_{14}H_{10}$	solid	1712.0*
Anthraquinone	$C_{14}H_8O_2$	solid	1544.5
Arabinose	$C_5H_{10}O_5$	solid	559.9
Arabitol	$C_5H_{12}O_5$	solid	661.2
Arachidic acid	$C_{20}H_{40}O_2$	solid	3025.9
Azelaic acid	$(CH_2)_7(CO_2H)_2$	solid	1141.7
Azobenzene	$(C_6H_5N)_2$	solid	1545.9
Azoxybenzene	$(C_6H_5N)_2O$	solid	1534.5
Behenic acid	$C_{22}H_{44}O_2$	solid	3338.4
Benzalacetone	$C_6H_5CHCOCH_3$	solid	1257.4
Benzaldehyde	C_6H_5CHO	liquid	843.2*
Benzamide	$C_6H_5CONH_2$	solid	847.6
Benzanilide	$C_6H_5CONHC_6H_5$	solid	1575.5
Benzene	C_6H_6	liquid	780.96*
Benzenediazonium nitrate	$C_6H_5N_2NO_3$	solid	782.6
Benzidine	$(C_6H_4NH_2)_2$	solid	1560.9
Benzil	$(C_6H_5CO)_2$	solid	1624.6
Benzoic acid*	$C_6H_5CO_2H$	solid	771.24*
Benzoic anhydride	$(C_6H_5CO)_2O$	solid	1555.1
Benzoin	$C_6H_5.CHOH.COC_6H_5$	solid	1671.4
Benzonitrile	C_6H_5CN	liquid	865.5
Benzophenone	$(C_6H_5)_2CO$	solid	1556.5
Benzoyl chloride	C_6H_5COCl	liquid	782.8
Benzoyl peroxide	$(C_6H_5CO)_2O_2$	solid	1551.7
Benzyl alcohol	$C_6H_5CH_2OH$	liquid	894.3
Benzylamine	$C_6H_5CH_2NH_2$	liquid	969.4
Benzyl carbylamine	$C_6H_5CH_2NC$	liquid	1046.5
Benzyl chloride	$C_6H_5CH_2Cl$	liquid	886.4
Benzyl cyanide	$C_6H_5CH_2CN$	liquid	1023.5
Borneol	$C_{10}H_{18}O$	liquid	1469.6
Brucine	$C_{23}H_{26}O_4N_2$	gas	687.68*
n-Butyl alcohol	C_4H_9OH	liquid	639.53*
tert-Butyl alcohol, see Trimethyl carbinol			
n-Butylamine	$C_4H_9NH_2$	liquid	710.6
sec-Butylamine	$(CH_3)(C_2H_5)NH_2$	liquid	713.0
tert-Butylamine	$(CH_3)_3CNH_2$	liquid	716.0
tert-Butylbenzene	$C_6H_5C(CH_3)_3$	liquid	1400.4
n-Butyramide	$C_3H_7CONH_2$	solid	596.0
n-Butyric acid	$C_3H_7CO_2H$	liquid	521.87*
n-Butyronitrile	C_3H_7CN	liquid	613.3
Caffeine	$C_8H_{10}O_2N_4$	solid	1014.2
Camphene	$C_{10}H_{16}$	solid	1468.8
Camphor	$C_{10}H_{16}O$	solid	1411.0
Cane sugar, see Sucrose			
Capric acid	$C_9H_{18}O_2$	solid	1453.07*
Caproic acid	$C_5H_{11}CO_2H$	liquid	834.49*
Carbon disulfide	CS_2	liquid	246.6
Carbon subnitride	$(C.CN)_2$	solid	514.8

* 25°C

HEAT OF COMBUSTION
For Organic Compounds (Continued)

Name	Formula	Physical state	Heat of combustion. kg. calories
Carbon tetrachloride	CCl_4	liquid	37.3
Carbonyl sulfide	COS	gas	130.5
Carvacrol	$C_{10}H_{14}O$	liquid	1354.5
Cetyl alcohol	$C_{16}H_{34}O$	solid	2504.4
Cetyl palmitate	$C_{32}H_{64}O$	solid	4872.8
Chloracetic acid	$ClCH_2CO_2H$	solid	171.0
o-Chlorobenzoic acid	$ClC_6H_4CO_2H$	solid	734.5
Chloroform	$CHCl_3$	liquid	89.2
Chrysene	$C_{18}H_{12}$	solid	2139.1
Cinnamic acid *(trans)*	$C_6H_5CH{:}CHCO_2H$	solid	1040.2
Cinnamic aldehyde	$C_6H_5CH{:}CHCHO$	liquid	1112.3
Cinnamic anhydride	$C_{18}H_{14}O_3$	solid	2091.3
d-Citrene	$C_{10}H_{16}$	liquid	1483.0
Citric acid (anhydr)	$C_6H_8O_7$	solid	468.6*
Codeine	$C_{18}H_{21}O_3N.H_2O$	solid	2327.6
Coniine	$C_8H_{17}N$	liquid	1275.5
Creatine (anhydr)	$C_4H_9O_2N_3$	solid	559.8
Creatinine	$C_4H_7ON_3$	solid	563.4
o-Cresol	$CH_3C_6H_4OH$	liquid	882.6
o-Cresol	$CH_3C_6H_4OH$	solid	882.72*
m-Cresol	$CH_3C_6H_4OH$	liquid	880.5
p-Cresol	$CH_3C_6H_4OH$	liquid	882.5
p-Cresol	$CH_3C_6H_4OH$	solid	883.99*
m-Cresolmethyl ether	$CH_3C_6H_4OCH_3$	liquid	1057.0
Crotonaldehyde	C_3H_5CHO	liquid	542.1
Cyanoacetic acid	$NCCH_2CO_2H$	solid	298.8
Cyanogen	$(CN)_2$	gas	258.3
Cyclobutane	C_4H_8	liquid	650.22*
Cycloheptane	$(CH_2)_7$	liquid	1087.3
Cycloheptanol	$CH_2(CH_2)_5CHOH$	liquid	1050.2
Cycloheptene	C_7H_{12}	liquid	1099.09*
Cyclohexane	$(CH_2)_6$	liquid	936.87*
Cyclohe xanol	$CH_2(CH_2)_4CHOH$	liquid	890.7
Cyclohexene, see Tetrahydrobenzene			
Cyclopentane	$(CH_2)_5$	liquid	786.55*
Cyclopropane, see Trimethylene			
Cymene	$C_6H_4(CH_3)(CH_3COCO_3)$-(1,4)	liquid	1402.8
Decahydronaphthalene *(cis)*	$C_{10}H_{18}$	liquid	1502.5
Decahydronaphthalene *(trons)*	$C_{10}H_{18}$	liquid	1499.5
Decane	$C_{10}H_{22}$	liquid	1610.2
Dextrose, see Glucose			
Diallyl	$(CH_2CHCH_2)_2$	vapor	903.4
Diamyl ether	$(C_5H_{11})_2O$	liquid	1609.3
Diamylene	$C_{10}H_{20}$	liquid	1582.2
Dibenzyl	$(C_6H_5CH_2)_2$	solid	1810.6
Dibenzyl amine	$(C_6H_5CH_2)_2NH$	solid	1853.0
o-Dichlorobenzene	$C_6H_4Cl_2$	liquid	671.8
Diethylacetic acid	$(C_2H_5)_2CHCO_2H$	liquid	830.8
Diethyl amine	$(C_2H_5)_2NH$	liquid	716.9
Diethylaniline	$C_6H_5(C_2H_5)_2$	liquid	1451.6
Diethyl carbonate	$CO(OC_2H_5)_2$	liquid	647.9
Diethyl ether	$(C_2H_5)_2O$	liquid	657.52*
Diethyl ketone	$(C_2H_5)_2CO$	liquid	735.6
Diethyl malonate	$CH_2(CO_2C_2H_5)_2$	liquid	860.4
Diethyl oxalate	$(CO_2C_2H_5)_2$	liquid	716.0
Diethyl succintae	$(CH_2CO_2C_2H_5)$	liquid	1007.3
Dihydrobenzene	C_6H_8	liquid	847.8
δ_1-Dihydronaphthalene	$C_{10}H_{10}$	liquid	1296.3
δ_1-Dihydronaphthalene	$C_{10}H_{10}$	solid	1298.3
Dihydroxyanthraquinone	$C_{14}H_6O_2(OH)_2$-(1,2)	solid	1448.9
Diisoamyl	$[(CH_3)_2CHCH_2CH_2]_2$	liquid	1615.8
Diisobutylene	$[(CH_3)_2CHCH_2]_2$	liquid	1252.4
Diisopropyl	$[(CH_3)_2CH]_2$	vapor	993.9
Diisopropyl ketone	$[(CH_3)_2CH]_2CO$	liquid	1045.5
Dimethyl amine	$(CH_3)_2NH$	liquid	416.7
Dimethylaniline	$C_6H_5N(CH_3)_2$	liquid	1142.7
Dimethyl carbonate	$CO(OCH_3)_2$	liquid	340.8
Dimethyl ether	$(CH_3)_2O$	gas	
Dimethylethyl carbinol	$C_2H_5(CH_3)_2CHOH$	liquid	784.6
Dimethyl fumarate	$(CHCO_2CH_3)_2$	solid	663.3*
2,5-Dimethylhexane	$(CH_3)_2CH.C_2H_4CH(CH_3)_2$	liquid	1303.3
3,4-Dimethylhexane	$[(C_2H_5)(CH_3)CH]_2$	liquid	1303.7
Dimethyl maleate	$(CHCO_2CH_3)_2$	liquid	669.4*
Dimethyl oxalate	$(CO_2CH_3)_2$	liquid	400.2*
2,2-Dimethylpentane	$(CH_3)_3C.C_3H_7$	liquid	1148.9
2,3-Dimethylpentane	$(CH_3)_7CHCH(CH_3)C_2H_5$	liquid	1148.9
2,4-Dimethylpentane	$(CH_3)_2CHCH_2CH(CH_3)_2$	liquid	1148.9
3,3-Dimethylpentane	$(CH_3)_7C(C_2H_5)_2$	liquid	1147.9
Dimethyl phthalate	$C_6H_4(CO_7CH_3)_2$	liquid	1119.7
Dimethyl succinate	$(CH_2CO_7CH_3)_2$	solid	706.3*
m-Dinitrobenzene	$C_6H_4(NO_2)_2$	solid	696.8

HEAT OF COMBUSTION
For Organic Compounds (Continued)

Name	Formula	Physical state	Heat of combustion. kg. calories
Dinitrophenol	$C_6H_3(OH)(NO_2)_2$—(1, 2,4)	solid	648.0
Dinitrotoluene	$C_6H_3)(NO_2)_2$—(1,2,4)	solid	852.8
Diphenyl	$(C_6H_5)_2$	solid	1493.6
Diphenyl amine	$(C_6H_5)_2NH$	solid	1536.2
Diphenyl carbinol	$(C_6H_5)_2CHOH$	solid	1615.4
Diphenylmethane	$(C_6H_5)_2CH_2$	solid	1655.0
Diphenylnitrosamine	$(C_6H_5)_2N.NO$	solid	1532.6
Dipropargyl	$(CH{:}\ C.CH_2)_2$	vapor	882.9
Dipropyl ketone	$(C_3H_7)_2CO$	liquid	1050.5
Dulcitol	$C_6H_{14}O_6$	solid	729.1
Durene	$C_6H_2(CH_3)_4$—(1,2,4,5)	solid	1393.6
Eicosane	$C_{20}H_{42}$	solid	3183.1
Erythritol	$C_4H_{10}O_4$	solid	504.1
Ethane	C_2H_6	gas	372.81*
Ethine, see Acetylene			
Ethyl acetate	$CH_3CO_7C_2H_5$	liquid	536.9
Ethyl acetoacetate	$CH_3COCH_2CO_2C_2H_5$	liquid	690.8
Ethyl alcohol	C_2H_6OH	liquid	326.68*
Ethyl amine	$C_2H_5NH_2$	liquid	409.5*
Ethylaniline	$C_6H_5NHC_2H_5$	liquid	1121.5
Ethylbenzene	$C_2H_5C_6H_5$	liquid	1091.2
Ethyl benzoate	$C_6H_5CO_2C_2H_5$	liquid	1098.7
Ethyl bromide	C_2H_5Br	vapor	340.5
Ethyl n-butyrate	$C_3H_7CO_2C_2H_5$	liquid	851.2
Ethyl carbylamine	C_2H_5NC	liquid	477.1
Ethyl chloride	C_2H_5Cl	vapor	316.7
Ethylcycloheptane	$C_2H_5C_7H_{13}$	liquid	1406.8
Ethyl formate	$HCO_2C_2H_5$	liquid	391.7
3-Ethylhexane	$(C_2H_5)_2CH.C_3H_7$	liquid	1302.3
Ethyl iodide	C_2H_5I	liquid	356.0
Ethyl isobutyrate	$(CH_3)_2CHCH_2CO_2C_2H_5$	liquid	845.7
Ethyl isocyanate	C_2H_5NCO	liquid	424.5
Ethyl nitrate	$C_2H_5ONO_2$	vapor	322.4
Ethyl nitrite	C_2H_5ONO	vapor	332.6
3-Ethylpentane	$(C_2H_5)_3CH$	liquid	1149.9
Ethyl propionate	$C_2H_5CO_2C_2H_5$	liquid	690.8
Ethyl salicylate	$HOC_6H_4CO_2C_2H_5$	liquid	1051.2
Ethyl valerate	$C_4H_9CO_2C_2H_5$	liquid	1017.5
Ethylene	CH_2CH_2	gas	337.23*
Ethylene chloride	$(CH_7Cl)_2$	vapor	271.0
Ethylene diamine	$(CH_2NH_2)_3$	liquid	452.6
Ethylene glycol	$(CH_2OH)_2$	liquid	281.9
Ethylene iodide	$(CH_2I)_2$	solid	324.8
Ethylene oxide	CH_2CH_2O	liquid	302.1
Ethylidene chloride	CH_3CHCl_2	liquid	267.1
Eugenol	$C_{10}H_{12}O_2$	liquid	1286.6
Fenchane	$C_{10}H_{18}$	liquid	1502.6
Fluorene	$(C_6H_4)_2{:}\ CH_2$	solid	1584.9
Fluorobenzene	C_6H_5F	liquid	747.2
Formaldehyde	CH_2O	gas	136.42*
Formamide	$HCONH_2$	solid	134.9
Formic acid	HCO_2H	liquid	60.86*
β-D-Fructose	$C_6H_{12}O_6$	solid	672.0*
Fumaric acid *(trans)*	$(CHCO_2H)_2$	solid	318.99*
Furfural	C_4H_3OCHO	liquid	559.5
a-D-Galactose	$C_6H_{12}O_6$	solid	670.1*
Gallic acid	$C_6H_2(OH)_3CO_2H$—(1,3,5,6)	solid	633.7
a-D-Glucose	$C_6H_{12}O_6$	solid	669.94*
Glutaric acid	$(CH_2)_3(CO_2H)_2$	solid	514.08*
Glycerol	$(CH_2OH)_2CHOH$	liquid	397.0
Glyceryl tributyrate	$C_{15}H_{26}O_6$	liquid	1941.1
Glycine	$H_2NCH_2CO_2H$	solid	232.67*
Glycogen	$(C_6H_{10}O_5)x$ per kg	solid	4186.8
Glycollic acid	CH_7OHCO_2H	solid	166.1*
Glycylglycine	$C_4H_8O_3N_2$	solid	470.7
n-Heptaldehyde	$CH_3(CH_2)_5CHO$	liquid	1062.2*
n-Heptane	C_7H_{16}	liquid	1149.9
Heptine-l	$CH{:}\ C(CH_2)_4CH_3$	liquid	1091.2
n-Heptyl alcohol	$CH_3(CH_2)_5CH_2OH$	liquid	1104.9
Heptyl amine	$C_7H_{15}NH_2$	liquid	1178.9
Heptylic acid	$C_7H_{14}O_2$	liquid	986.1
n-Hexane	C_6H_{14}	liquid	995.01*
Hexachlorbenzene	C_6Cl_6	solid	509.0
Hexachlorethane	C_2Cl_6	solid	110.0
Hexadecane	$C_{16}II_{34}$	solid	2559.1
Hexahydronaphthalene	$C_{10}H_{14}$	liquid	1419.3
Hexamethylbenzene	$C_6(CH_3)_6$	solid	1711.9
Hexamethylenetetramine	$(CH_2)_6N_4$	solid	1006.7
Hexamethylethane	$(CH_3)_3Cl_2$	solid	1301.8
Hexyl amine	$C_6H_{13}NH_2$	liquid	1022.2
Hexylene	C_6H_{12}	liquid	952.6

HEAT OF COMBUSTION
For Organic Compounds (Continued)

Name	Formula	Physical state	Heat of combustion. kg. calories
Hippuric acid	$C_6H_5CONHCH_2CO_2H$	solid	1012.4
Hydantoic acid	$C_3H_6O_3N_2$	solid	308.6
Hydrazobenzene	$(C_6H_5NH)_2$	solid	1597.3
Hydroquinol	$C_6H_4(OH)_2$	solid	681.78*
Hydroquinoldimethyl ether	$(CH_3O)_2C_6H_4$	solid	1014.7
p-Hydroxyazobenzene	$HOC_6H_4N_2C_6H_5$	solid	1502.0
o-Hydroxybenzaldehyde	$C_6H_4(OH)CHO$	liquid	796.4*
m-Hydroxybenzaldehyde	$C_6H_4(OH)CHO$	solid	788.7
p-Hydroxybenzaldehyde	$C_6H_4(OH)CHO$	solid	792.7
m-Hydroxybenzoic acid	$HOC_6H_4CO_2H$	solid	726.1
p-Hydroxybenzoic acid	$HOC_6H_4CO_2H$	solid	725.4
β-Hydroxybutyraldehyde	$CH_3CHOHCH_2CHO$	liquid	546.6
Indigo	$C_{16}H_{10}O_2N_2$	solid	1815.0
Indole	C_8H_7N	solid	1022.2
Inositol	$C_6H_{12}O_6$	solid	662.1
Iodoform	CHI_3	solid	161.9
Isoamyl amine	$(CH_3)_2CHC_2H_4NH_2$	liquid	866.8
Isobutane	$(CH_3)_2CH$	gas	683.4
Isobutyl alcohol	$(CH_3)_2CHCH_2OH$	liquid	638.2
Isobutyl amine	$C_4H_9NII_2$	liquid	713.6
Isobutylene	$(CH_3)_2C:CH_2$	gas	647.2
Isobutyraldehyde	$(CH_3)_2CHCHO$	vapor	596.8
Isobutyramide	$(CH_3)_2CHCONH_2$	solid	595.9
Isobutyric acid	$(CH_3)_2CHCO_2H$	liquid	517.4
Isoeugenol	$C_{10}H_{12}O_2$	liquid	1277.6
Isopentane	C_5H_{12}	gas	843.5(?)
Isopentane	C_5H_{12}	liquid	838.3(?)
Isophthalic acid	$C_6H_4(CO_2H)_2$	solid	768.3
Isopropyl alcohol	$(CH_3)_2CHOH$	liquid	474.8
Isopropylbenzene	$(CH_3)_2CHC_6H_5$	liquid	1247.3
Isopropyltoluene	$C_6H_4(CH_3)(CH_3CHC_3)$—(1,3)	liquid	1409.5
Isopropyltoluene, see Cymene			
Isosafrole	$C_{10}H_{10}O_2$	liquid	1233.9
Lactic acid, DL	$CH_3CHOHCO_2H$	liquid	326.8*
Lactose (anhydr.)	$C_{12}H_{22}O_{11}$	solid	1350.0*
Lauric acid	$C_{12}H_{24}O_2$	solid	1763.25*
Leucine	$C_6H_{13}O_2N$	solid	855.6
d-Limonene	$C_{10}H_{16}$	liquid	1471.2
Maleic acid *(cis)*	$(CHCO_2H)_2$	solid	323.89*
Maleic anhydride	$(CHCO)_2O$	solid	332.10*
l-Malic acid	$(CHOHCH_2):(CO_2H)_2$	solid	317.37*
Malonic acid	$CH_2(CO_2H)_2$	solid	205.82*
Maltose	$C_{12}H_{22}O_{11}$	solid	1349.3*
Mandelic acid	$C_6H_5CHOHCO_2H$	solid	890.3
d-Mannitol	$C_6H_{14}O_6$	solid	727.6
Menthene	$C_{10}H_{18}$	liquid	1523.2
Menthol	$C_{10}H_{20}O$	solid	1508.8
Mesitylene	$(CH_3)_3C_6H_3$—(1,3,5)	liquid	1243.6
Mesityl oxide	$(CH_3)_2C:CHCOCH_3$	liquid	846.7
Mesotartaric acid	$(CHOH)_2(CO_2H)_2$	solid	276.0
Methane	CH_4	gas	212.79*
Methyl acetate	$CH_3CO_2CH_3$	liquid	381.2
Methyl alcohol	CH_3OH	liquid	173.64*
Methyl amine	CH_3NH_2	liquid	253.5*
Methylaniline	$C_6H_5NHCH_3$	liquid	973.5
Methyl benzoate	$C_6H_5CO_2CH_3$	liquid	945.9*
Methyl bromide	CH_3Br	vapor	184.0
Methyl butyl ketone	$CH_3COC_4H_9$	liquid	895.2
Methyl *tert*-butyl ketone, see Pinacoline			
Methyl butyrate	$C_3H_7COOCH_3$	liquid	692.8
Methyl carbylamine	CH_3NC	liquid	320.1
Methyl chloride	CH_3Cl	gas	164.2
Methyl cinnamate	$C_{10}H_{10}O_2$	solid	1213.0
Methylcyclobutane	$CH_3CHCH_2CH_2CH_2$	liquid	784.2
Methylcycloheptane	$CH_3C_7H_{13}$	liquid	1244.5
Methylcyclohexane	$CH_3C_6H_{11}$	liquid	1091.8
Methylcyclopentane	$CH_3CH.C_3H_6CH_2$	liquid	937.9
Methyldiethyl carbinol	$CH_3(C_2H_5)_2CHOH$	liquid	927.0
Methylene chloride	CH_2Cl_2	vapor	106.8
Methylene iodide	CH_2I_2	liquid	178.4
Methylethyl ether	$CH_3OC_2H_5$	vapor	503.69*
Methylethyl ketone	$CH_3COC_2H_5$	liquid	584.17*
Methyl formate	HCO_2CH_3	liquid	234.1*
2-Methylheptane	$(CH_3)_2CH.C_5H_{11}$	liquid	1306.1
2-Methylhexane	$(CH_3)_2CHC_4H_9$	liquid	1148.9
3-Methylhexane	$(C_2H_5)(CH_3)CHC_3H_7$	liquid	1148.9
Methylhexyl ketone	$CH_3COC_6H_{13}$	liquid	1205.1
Methyl iodide	CH_3I	liquid	194.7
Methyl isobutyrate	$(CH_3)_2CHCO_2CH_3$	liquid	694.2
Methyl isocyanate	CH_3NCO	liquid	269.4
Methylisopropyl ketone	$CH_3COCH(CH_3)_2$	liquid	733.9
Methyl lactate	$CH_3CHOHCO_2CH_3$	liquid	497.2

HEAT OF COMBUSTION
For Organic Compounds (Continued)

Name	Formula	Physical state	Heat of combustion. kg. calories
Methyl propionate	$C_2H_5CO_2CH_3$	vapor	552.3
Methylpropyl ketone	$CH_3COC_3H_7$	liquid	740.78*
Methyl salicylate	$HOC_6H_4CO_2CH_3$	liquid	898.6*
Milk sugar, see Lactose			
Morphine	$C_{17}H_{19}O_3N.H_2O$	solid	2146.3
Mucic acid	$C_6H_{10}O_8$	solid	483.0*
Myristic acid	$C_{14}H_{28}O_2$	solid	2073.91*
Naphthalene	$C_{10}H_8$	solid	1231.8*
α-Naphthoic acid	$C_{10}H_7CO_2H$	solid	1231.8
β-Naphthoic acid	$C_{10}H_7CO_2H$	solid	1227.6
α-Naphthol	$C_{10}H_7OH$	solid	1185.4
β-Naphthol	$C_{10}H_7OH$	solid	1187.2
α-Naphthonitrile	$C_{10}H_7CN$	solid	1326.2
β-Naphthonitrile	$C_{10}H_7CN$	solid	1321.0
α-Naphthoquinone	$C_{10}H_6O_2$	solid	1100.8
β-Naphthoquinone	$C_{10}H_6O_2$	solid	1106.4
α-Naphthyl amine	$C_{10}H_7NH_2$	solid	1263.5
β-Naphthyl amine	$C_{10}H_7NH_2$	solid	1261.0
Narceine	$C_{23}H_{27}O_8N.2H_2O$	solid	2802.9
Narcotine	$C_{22}H_{23}O_7N$	solid	2644.5
Nicotine	$C_{10}H_{14}N_2$	liquid	1427.7
o-Nitraniline	$C_6H_4(NH_2)(NO_2)$	solid	765.8
m-Nitraniline	$C_6H_4(NH_2)(NO_2)$	solid	765.2
p-Nitraniline	$C_6H_4(NH_2)(NO_2)$	solid	761.0
m-Nitrobenzaldehyde	$O_2NC_6H_4CHO$	solid	800.4
Nitrobenzene	$C_6H_5NO_2$	liquid	739.2
m-Nitrobenzoic acid	$O_2NC_6H_4CO_2H$	solid	729.1
Nitroethane	$C_2H_5NO_2$	liquid	322.2
Nitroglycerine, see Trinitroglycerol			
Nitromethane	CH_3NO_2	liquid	169.4
o-Nitrophenol	$HOC_6H_6NO_2$	solid	689.1
m-Nitrophenol	$HOC_5H_4NO_2$	solid	684.4
p-Nitrophenol	$HOC_6H_4NO_2$	solid	688.8
Nitropropane	$C_3H_7NO_2$	liquid	477.9
p-Nitrotoluene	$CH_3C_6H_4NO_2$	liquid	897.0
p-Nitrotoluene	$CH_3C_6H_6NO_2$	solid	888.6
Octahydronaphthalene	$C_{10}H_{16}$	liquid	1461.7
n-Octane	C_8H_{18}	liquid	1302.7
Octyl alcohol	$C_8H_{18}O$	liquid	1262.0
Oleic acid	$C_{18}H_{34}O_2$	liquid	2657.4*
Oxalic acid, *a*	$(CO_2H)_2$	solid	58.7*
Oxamide	$(CONH_2)_2$	solid	203.2
Palmitic acid	$C_{16}H_{32}O_2$	solid	2384.76*
Papaverine	$C_{20}H_{21}O_4N$	solid	2478.1
Pentamethylbenzene	$C_6H(CH_3)_5$	solid	1554.0
n-Pentane	C_5H_{12}	gas	845.16*
n-Pentane	C_8H_{12}	liquid	838.78*
Phenacetin	$C_{10}H_{13}O_2N$	solid	1285.2
Phenanthraquinone	$C_{14}H_8O_2$	solid	1544.0
Phenanthrene	$C_{14}H_{10}$	solid	1685.6*
Phenetole	$C_6H_5OC_2H_5$	liquid	1060.3
Phenol	C_6H_5OH	solid	729.80*
Phenylacetic acid	$C_6H_5CH_2CO_2H$	solid	930.4*
Phenylacetylene	C_6H_5C	liquid	1024.2
Phenylalanine	$C_9H_{11}O_2N$	solid	1111.3
p-Phenylenediamine	$C_6H_4(NH_2)_2$	solid	843.4
Phenylethylene, see Styrene			
Phenylglycine	$C_2H_5NHCH_2CO_2H$	solid	955.1
Phenylhydrazine	$C_6H_5N_2H_3$	solid	875.4
Phenylhydroxylamine	C_6H_5NHOH	liquid	803.7
Phenyl iodide	C_6H_5I	liquid	770.7
Phloroglucinol	$C_6H_3(OH)_3$	solid	635.7
o-Phthalic acid	$C_6H_4(CO_2H)_2$	solid	770.44*
Phthalic anhydride	$C_6H_4(CO_2O$	solid	783.4
Phthalimide	$C_8H_5O_2N$	solid	849.5
Picric acid	$C_6H_2(OH)(NO_2)_3$—(1,2,4,6)	solid	611.8
Pinacoline	$CH_3COC(CH_3)_3$	solid	891.8
Piperidine	$C_5H_{11}N$	liquid	826.6
Piperonal	$C_8H_6O_3$	solid	870.7
Propane	C_3H_8	gas	530.57*
Propine, see Allylene			
Propionaodehyde	C_2H_5CHO	liquid	434.1*
Propionamide	$C_2H_5CONH_2$	solid	439.9
Propionic acid	$C_2H_5CO_2H$	liquid	365.03*
Propionic anhydride	$(C_2H_5CO)_2O$	liquid	746.6
Propionitrile	C_2H_5CN	liquid	456.4
n-Propyl alcohol	C_3H_7OH	liquid	482.75*
Propyl amine	$C_3H_7NH_2$	liquid	565.3*
n-Propylbenzene	$C_3H_7C_6H_5$	liquid	1246.4
Propyl bromide	C_3H_7Br	vapor	497.3
Propyl carbylamine	C_3H_7NC	liquid	639.6

HEAT OF COMBUSTION
For Organic Compounds (Continued)

Name	Formula	Physical state	Heat of combustion. kg. calories
Propyl chloride	C_3H_7Cl	vapor	478.3
Propylene	CH_3CH_2	gas	490.2
Propylene glycol	$CH_3CHOHCH_2OH$	liquid	431.0
n-Propyl iodide	C_3H_7I	liquid	514.3
n-Propyltoluene	$C_6H_4(CH_3)(C_3H_7)$—(1,3)	liquid	1405.4
Pseudocumene	$C_6H_3(CH_3)_3$—(1,2,4)	liquid	1241.7
Pyridine	C_5H_5N	liquid	665.0*
Pyrocatechol	$C_6H_4(OH)_2$	solid	683.0*
Pyrogallol	$C_6H_3(OH)_3$	solid	638.7
Pyrrole	C_4H_5N	liquid	567.7
Quercitol	$C_6H_{12}O_5$	solid	704.2
Quinoline	C_9H_7N	liquid	1123.5
Quinone	$O:C_6H_4:O$	solid	656.6
Raffinose	$C_{18}H_{32}O_{16}$	solid	2025.5
Retene	$C_{18}H_{18}$	solid	2306.8
Resorcinol	$C_6H_4(OH)_2$	solid	681.30*
Resorcinoldimethyl ether	$(CH_3O)_2C_6H_4$	liquid	1022.6
Rhamnose	$C_6H_{12}O_6$	solid	718.3
Safrole	$C_{10}H_{10}O_2$	liquid	1244.1
Salicylaldehyde, see o-Hydroxybenzaldehyde			
Salicylic acid*	$HOC_6H_4CO_2H$—(1,2)	solid	722.4**
Sarcosine	$CH_3NHCH_2CO_2H$	solid	401.1
Sebacic acid	$(CH_2)_8(CO_2H)_2$	solid	1297.3
Skatole	C_9H_9N	liquid	1170.5
d-Sorbose	$C_6H_{12}O_6$	solid	668.3
Starch	$(C_6H_{10}O_5)x$ per kg	solid	4178.8
Stearic acid	$C_{18}H_{36}O_2$	solid	2696.12*
Strychnine	$C_{21}H_{22}O_2N_2$	solid	2685.7
Styrene	$C_6H_5CH_2$	liquid	1047.1
Suberic acid	$(CH_2)_6(CO_2H)_2$	solid	985.2
Succinic acid	$(CH_2CO_2H)_2$	solid	356.36*
Succinic acid nitrile	$(CH_2CN)_2$	liquid	545.7
Succinic anhydride	$(CH_2CO)_2O$	solid	369.0*
Succinimide	$C_4H_5O_2N$	solid	437.9
Sucrose	$C_{12}H_{22}O_{11}$	solid	1348.2*
Sylvestrene	$C_{10}H_{15}$	liquid	1464.7
l-Tartaric acid	$(CHOH)_2(CO_2H)_2$	solid	274.7*
d,l-Tartaric acid (anhydr)	$(CHOH)_2(CO_2H)_2$	solid	272.6*
Terephthalic acid	$C_6H_4(CO_2H)_2$	solid	770.4
Terpin hydrate	$C_{10}H_{22}O_3$	solid	1451.0
Terpineol	$C_{10}H_{18}O$	solid	1469.5
Tetrahydrobenzene	C_6H_{10}	liquid	891.9
Tetrahydronaphthalene	$C_{10}H_{12}$	liquid	1352.4
Tetramethylmethane	$(CH_3)_4C$	gas	842.6
Tetraphenylmethane	$(C_6H_5)_4C$	solid	3102.4
Tetryl	$C_6H_5N_5O_8$	solid	842.3
Thebaine	$C_{19}H_{21}O_3N$	solid	2441.3
Thiophene	C_4H_4S	liquid	670.5
Thujane	$C_{10}H_{18}$	liquid	1506.4
Thymol	$C_{10}H_{14}O$	liquid	1353.4
Thymol	$C_{10}H_{14}O$	solid	1349.7
Thymoquinone	$C_{10}H_{12}O_2$	solid	1271.3
Toluene	$CH_2C_6H_5$	liquid	934.2
o-Toluic acid	$CH_3C_6H_4CO_2H$	solid	928.9
m-Toluic acid	$CH_3C_6H_4CO_2H$	solid	928.6
p-Toluic acid	$CH_3C_6H_4CO_2H$	solid	926.9
o-Toluidine	$CH_3C_6H_4NH_2$	liquid	964.3
m-Toluidine	$CH_3C_6H_4NH_2$	liquid	965.3
p-Toluidine	$Ch_3C_6H_4NH_2$	solid	958.4
o-Tolunitrile	$CH_3C_6H_4CN$	liquid	1030.3
Toluquinone	$C_7H_6O_3$	solid	803.2
Triaminotriphenyl carbinol	$(C_6H_4NH_2)_2COH$	solid	2483.5
Tribenzyl amine	$(C_6H_5CH_2)_3N$	solid	2762.1
Trichloracetic acid	$Cl_2C.CO_2H$	solid	92.8
Triethyl amine	$(C_2H_5)_2N$	liquid	1036.8
Triethyl carbinol	$(C_2H_5)_3CHOH$	liquid	1080.0
Triisoamyl amine	$[(CH_3)_2CHCH_2CH_2]_3N$	liquid	2459.3
Triisobutyl amine	$[(CH_3)_2CHCH_2]_3N$	liquid	1973.6
Trimethyl amine	$(CH_3)_3N$	liquid	578.6
2,2,3-Trimethylbutane	$(CH_3)_2C.CH(CH_3)_2$	liquid	1147.9
Trimethyl carbinol	$(CH_3)_2COH$	liquid	629.3
Trimethylene	$CH_2CH_2CH_2$	gas	499.89*
Trimethylethylene	$(CH_3)_2CCH_3$	liquid	796.0
Trimethylethylene	$(CH_3)_2CCH_3$	vapor	803.6
2,2,4-Trimethylpentane	$(CH_3)_3C.CH_2CH(CH_3)_2$	liquid	1303.9
Trinitrobenzene	$C_6H_2(NO_2)_3$—(1,3,5)	solid	663.7
Trinitroglycerol	$C_3H_5(NO_3)_3$	liquid	368.4
Trinitrotoluene	$C_5H_2(CH_3)(NO_2)_3$—(1,2,4,6)	solid	820.7
Triphenyl amine	$(C_6H_5)_3N$	solid	2267.8
Triphenylbenzene	$C_6H_3(C_6H_5)_3$—(1,3,5)	solid	2936.7
Triphenyl carbinol	$(C_6H_5)_3CHOH$	sed	2340.8

** Recommended as a secondary thermochemical standard.

HEAT OF COMBUSTION
For Organic Compounds (Continued)

Name	Formula	Physical state	Heat of combustion. kg. calories
Triphenylmethane	$(C_6H_5)_2CH$	solid	2388.7
Triphenyl methyl	$(C_6H_5)_3C$	solid	2378.5
Tyrosine	$C_9H_{11}O_3N$	solid	1070.2
Undecyclic acid	$C_{11}H_{22}O_2$	solid	1615.9
Urea	$(NH_2)_2CO$	solid	150.97*
Urethane	$NH_2CO_2C_2H_5$	solid	397.2
Uric acid	$C_5H_4O_3N_4$	solid	460.2
n-Valeric acid	$C_4H_9CO_2H$	liquid	678.12*
Vanillin	$C_6H_3(OH)(OCH_3)CHO$ (1,2,4)	solid	914.1
o-Xylene	$(CH_3)_2C_6H_4$	liquid	1091.7
m-Xylene	$(CH_3)_7C_6H_4$	liquid	1088.4
p-Xylene	$(CH_3)_2C_6H_4$	liquid	1089.1
Xylose	$C_5H_{10}O_5$	solid	559.0*

HEAT OF FORMATION

The thermal change involved in the formation of 1 mol of substance from its elements is the heat of formation, ΔH, of the substance. If all of the substances involved in the reaction are each in their standard states and each substance is at unit activity, the thermal change is the standard heat of formation, ΔH°. By definition, all elements in their standard state have a heat of formation of zero. By further definition, the sign of ΔH is negative if heat is evolved and is positive if heat is absorbed. Thus, for the thermochemical reaction:

$$7C\ (s) + 3H_2(g) + O_2(g) = C_6H_5COOH\ (s) \qquad \Delta H^\circ = -771.24\ \text{kcal}$$

(benzoic acid)

Heats of formation may be calculated from heats of combustion. The heat of formation of compound "A" is equal to the sum of the heats of formation of the products of combustion of compound "A" minus the heat of combustion of compound "A". Some heats of formation of some products of combustion of organic compounds are:

Substance	ΔH, heat of formation (kcal/g mole)
Free Elements	0
CO	− 26.416
CO_2	− 93.963
½ H_2O (1), from 1 H	− 34.158
H_2O (1)	− 68.317
HF (Dilute aqueous solution)	− 76.531
HCl (Dilute aqueous solution)	− 39.850
HBr (Dilute aqueous solution)	− 28.958
HI (Dilute aqueous solution)	− 13.106
HNO_3 (Dilute aqueous solution)	− 48.484
H_2SO_4 (Dilute aqueous solution)	−213.552
SO_2 (g)	− 70.336

Two examples of calculations are:

Example I

Calculate the heat of formation of methane (CH_4) where

Heat of combustion of CH_4 = −210.8 kcal/g mol
Heat of formation of CO_2 = −93.963 kcal/g mol
Heat of formation of H_2O = −68.317 kcal/g mol

and where the combustion reaction occurs according to the following equation:

$$CH_4(gas) + 2O_2(gas) = CO_2(gas) + 2H_2O\ (liquid)$$

Heat of formation of CH_4 equals:

	Heat of formation of CO_2	=	−93.963 kcal
+	Two times heat of formation of H_2O	=	2(−68.317)kcal
−	Heat of combustion of CH_4	=	−(−210.8) kcal
			−19.8 kcal/g mol

Example II

Calculate the heat of formation of ethylene (C_2H_4) where

Heat of combustion of C_2H_4 = −337.23 kcal/g mol

and where the combustion reaction is as follows:

$$C_2H_4(gas) + 3O_2(gas) = 2CO_2(gas) + 2H_2O\ (liquid)$$

	Two times heat of formation of CO_2	=	2(−93.963)kcal
+	Two times heat of formation of H_2O	=	2(−68.317)kcal
−	Heat of combustion of C_2H_4	=	−(−337.23) kcal
			+12.67 kcal/g mol

LOWERING OF VAPOR PRESSURE BY SALTS IN AQUEOUS SOLUTIONS

The table gives the reduction of the vapor pressure in millimeters due to the presence of the number of grammolecules of salt per liter of water given at the head of the columns, at the temperature 100° C, at which temperature the vapor pressure of pure water is 760 millimeters.

(From Smithsonian Tables.)

Substance	0.5	1.0	2.0	3.0	4.0	5.0	6.0	8.0	10.0
$Al_2(SO_4)_3$	12.8	36.5							
$AlCl_3$	22.5	61.0	179.0	318.0					
BaS_2O_6	6.6	15.4	34.4						
$Ba(OH)_2$	12.3	22.5	39.0						
$Ba(NO_3)_2$	13.5	27.0							
$Ba(ClO_3)_2$	15.8	33.3	70.5	108.2					
$BaCl_2$	16.4	36.7	77.6						
$BaBr_2$	16.8	38.8	91.4	150.0	204.7				
CaS_2O_8	9.9	23.0	56.0	106.0					
$Ca(NO_3)_2$	16.4	34.8	74.6	139.3	161.7	205.4			
$CaCl_2$	17.0	39.8	95.3	166.6	241.5	319.5			
$CaBr_2$	17.7	44.2	105.8	191.0	283.3	368.5			
$CdSO_4$	4.1	8.9	18.1						
CdI_2	7.6	14.8	33.5	52.7					
$CdBr_2$	8.6	17.8	36.7	55.7	80.0				
$CdCl_2$	9.6	18.8	36.7	57.0	77.3	99.0			
$Cd(NO_3)_2$	15.9	36.1	78.0	122.2					
$Cd(ClO_3)_2$	17.5								
$CoSO_4$	5.5	10.7	22.9	45.5					
$CoCl_2$	15.0	34.8	83.0	136.0	186.4				
$Co(NO_3)_2$	17.3	39.2	89.0	152.0	218.7	282.0	332.0		
$FeSO_4$	5.8	10.7	24.0	42.4					
H_3BO_3	6.0	12.3	25.1	38.0	51.0				
H_3PO_4	6.6	14.0	28.6	45.2	62.0	81.5	103.0	146.9	189.5
H_3AsO_4	7.3	15.0	30.2	46.4	64.9				
H_2SO_4	12.9	26.5	62.8	104.0	148.0	198.4	247.0	343.2	
KH_2PO_4	10.2	19.5	33.3	47.8	60.5	73.1	85.2		
KNO_3	10.3	21.1	40.1	57.6	74.5	88.2	102.1	126.3	148.0
$KClO_3$	10.6	21.6	42.8	62.1	80.0				
$KBrO_3$	10.9	22.4	45.0						
$KHSO_4$	10.9	21.9	43.3	65.3	85.5	107.8	129.9	170.0	
KNO_2	11.1	22.8	44.8	67.0	90.0	110.5	130.7	167.0	198.8
$KClO_4$	11.5	22.3							
KCl	12.2	24.4	48.8	74.1	100.9	128.5	152.2		
$KHCO_3$	11.6	23.6	59.0	77.6	104.2	132.0	160.0	210.0	255.0
KI	12.5	25.3	52.2	82.6	112.2	141.5	171.8	225.5	278.5
$K_2C_2O_4$	13.9	28.3	59.8	94.2	131.0				
K_2WO_4	13.9	33.0	75.0	123.8	175.4	226.4			
K_2CO_3	14.4	31.0	68.3	105.5	152.0	209.0	258.5	350.0	
KOH	15.0	29.5	64.0	99.2	140.0	181.8	223.0	309.5	387.8
K_2CrO_4	16.2	29.5	60.0						
$LiNO_3$	12.2	25.9	55.7	88.9	122.2	155.1	188.0	253.4	309.2
$LiCl$	12.1	25.5	57.1	95.0	132.5	175.5	219.5	311.5	393.5
$LiBr$	12.2	26.2	60.0	97.0	140.0	186.3	241.5	341.5	438.0
Li_2SO_4	13.3	28.1	56.8	89.0					
$LiHSO_4$	12.8	27.0	57.0	93.0	130.0	168.0			
LiI	13.6	28.6	64.7	105.2	154.5	206.0	264.0	357.0	445.0
Li_2SiFl_6	15.4	34.0	70.0	106.0					

Substance	0.5	1.0	2.0	3.0	4.0	5.0	6.0	8.0	10.0
$LiOH$	15.9	37.4	78.1						
Li_2CrO_4	16.4	32.6	74.0	120.0	171.0				
$MgSO_4$	6.5	12.0	24.5	47.5					
$MgCl_2$	16.8	39.0	100.5	183.3	277.0	377.0			
$Mg(NO_3)_2$	17.6	42.0	101.0	174.8					
$MgBr_2$	17.9	44.0	115.8	205.3	298.5				
$MgH_2(SO_4)_2$	18.3	46.0	116.0						
$MnSO_4$	6.0	10.5	21.0						
$MnCl_2$	15.0	34.0	76.0	122.3	167.0	209.0			
NaH_2PO_4	10.5	20.0	36.5	51.7	66.8	82.0	96.5	126.7	157.1
$NaHSO_4$	10.9	22.1	47.3	75.0	100.2	126.1	148.5	189.7	231.4
$NaNO_3$	10.6	22.5	46.2	68.1	90.3	111.5	131.7	167.8	198.8
$NaClO_3$	10.5	23.0	48.4	73.5	98.5	123.3	147.5	196.5	223.5
$(NaPO_3)_6$	11.6								
$NaOH$	11.8	22.8	48.2	77.3	107.5	139.1	172.5	243.3	314.0
$NaNO_2$	11.6	24.4	50.0	75.0	98.2	122.5	146.5	189.0	226.2
Na_2HPO_4	12.1	23.5	43.0	60.0	78.7	99.8	122.1		
$NaHCO_3$	12.9	24.1	48.2	77.6	102.2	127.8	152.0	198.0	239.4
Na_2SO_4	12.6	25.0	48.9	74.2					
$NaCl$	12.3	25.2	52.1	80.0	111.0	143.0	176.5		
$NaBrO_3$	12.1	25.0	54.1	81.3	108.8	136.0			
$NaBr$	12.6	25.9	57.0	89.2	124.2	159.5	197.5	268.0	
NaI	12.1	25.6	60.2	99.5	136.7	177.5	221.0	301.5	370.0
$Na_4P_2O_7$	13.2	22.0							
Na_2CO_3	14.3	27.3	53.5	80.2	111.0				
$Na_2C_2O_4$	14.5	30.0	65.8	105.8	146.0				
Na_2WO_4	14.8	33.6	71.6	115.7	162.6				
Na_3PO_4	16.5	30.0	52.5						
$(NaPO_3)_3$	17.1	36.5							
NH_4NO_3	12.8	22.0	42.1	62.7	82.9	103.8	121.0	152.2	180.0
$(NH_4)_2SiFl_6$	11.5	25.0	44.5						
NH_4Cl	12.0	23.7	45.1	69.3	94.2	118.5	138.2	179.0	213.8
NH_4HSO_4	11.5	22.0	46.8	71.0	94.5	118.	139.0	181.2	218.0
$(NH_4)_2SO_4$	11.0	24.0	46.5	69.5	93.0	117.0	141.8		
NH_4Br	11.9	23.9	48.8	74.1	99.4	121.5	145.5	190.2	228.5
NH_4I	12.9	25.1	49.8	78.5	104.5	132.3	156.0	200.0	243.5
$NiSO_4$	5.0	10.2	21.5						
$NiCl_2$	16.1	37.0	86.7	147.0	212.8				
$Ni(NO_3)_2$	16.1	37.3	91.3	156.2	235.0				
$Pb(NO_3)_2$	12.3	23.5	45.0	63.0					
$Sr(SO_3)_2$	7.2	20.3	47.0						
$Sr(NO_3)_2$	15.8	31.0	64.0	97.4	131.4				
$SrCl_2$	16.8	38.8	91.4	156.8	223.3	281.5			
$SrBr_2$	17.8	42.0	101.1	179.0	267.0				
$ZnSO_4$	4.9	10.4	21.5	42.1	66.2				
$ZnCl_2$	9.2	18.7	46.2	75.0	107.0	153.0	195.0		
$Zn(NO_3)_2$	16.6	39.0	93.5	157.5	223.8				

THERMAL CONDUCTIVITY OF GASES

The values in this table are given as cal/(sec)(cm²)(°C/cm) × 10^{-6}. To convert these values to Btu/(hr) (ft²) (°F/ft) × 10^{-6} multiply by 241.909.

Gas °F °C	−400 −240	−300 −184.4	−200 −128.9	−100 −73.3	−40 −40	−20 −28.9	0 −17.8	20 −6.7	40 4.4	60 15.6	80 26.7	100 37.8	120 48.9	200 93.3
Acetylene	.....	.....	.....	28.10	34.71	37.19	39.67	42.15	45.04	47.94	50.83	53.72	56.62	69.43
Air	.....	.....	.....	.....	50.09	52.15	54.22	56.24	58.31	60.34	62.20	64.22	66.04	.....
Ammonia	.....	.....	.....	.....	43.39	45.87	48.35	50.83	53.31	55.79	58.68	61.58	64.47	.....
Argon	.....	.....	.....	.....	34.30	35.95	37.19	38.85	40.09	41.33	42.57	44.22	45.46	.....
Bromine	.....	.....	.....	.....	.....	.....	9.09	.....	.....	.....	.....	11.57	.....	.....
n-Butane	.....	.....	.....	.....	.....	.....	.....	30.99	33.06	35.54	38.02	40.91	43.39	54.14
i-Butane	.....	.....	.....	.....	.....	.....	.....	32.65	33.89	36.37	38.85	41.74	44.22	55.79
Carbon dioxide	.....	.....	.....	.....	27.90	29.75	31.70	33.68	35.62	37.61	39.67	41.74	43.81	.....
Carbon disulfide	.....	.....	.....	.....	.....	.....	14.05	15.29	16.53	17.77	19.01	19.84	.....	.....
Carbon monoxide	.....	.....	.....	.....	47.94	50.00	51.95	53.85	55.87	57.86	59.92	61.99	63.89	.....
Chlorine	.....	.....	.....	.....	15.29	16.53	17.36	18.18	19.01	20.25	21.08	21.90	23.14	.....
Deuterium	.....	.....	.....	.....	274.82	285.15	295.07	305.81	309.95	322.34	334.74	343.01	355.40	.....
Ethane	.....	.....	.....	23.97	32.65	35.54	38.43	41.33	44.63	47.94	51.24	54.55	58.27	74.39
Ethanol	.....	.....	.....	.....	.....	.....	.....	29.34	30.99	32.65	34.71	36.78	.....	.....
Ethylamine	.....	.....	.....	.....	.....	.....	.....	31.41	33.47	35.54	37.61	39.67	42.15	.....
Ethylene	.....	.....	.....	26.86	33.06	35.54	38.02	40.50	43.39	46.29	49.18	52.07	54.96	68.19
Fluorine	.....	18.18	30.58	43.39	50.83	52.90	55.38	57.86	59.92	61.99	64.06	66.12	68.19	76.04
Helium	84.31	163.24	221.51	274.8	304.99	314.49	324.00	333.50	343.42	352.10	360.36	368.63	376.07	.....
Hydrogen	59.92	142.57	227.29	308.7	357.47	371.93	388.46	405.00	417.39	433.92	446.32	458.72	471.11	.....
Hydrogen bromide	.....	.....	.....	.....	15.29	16.11	16.49	17.77	18.60	19.84	20.66	21.49	.....	.....
Hydrogen chloride	.....	.....	.....	.....	25.62	26.86	28.51	29.75	30.99	32.23	33.89	35.12	.....	.....
Hydrogen cyanide	.....	.....	.....	.....	.....	.....	23.97	25.62	26.86	28.10	29.75	30.99	32.65	.....
Hydrogen sulfide	.....	.....	.....	.....	.....	.....	28.10	29.75	31.41	33.47	.....	36.78	.....	.....
Krypton	.....	.....	.....	.....	.....	.....	19.84	.....	.....	.....	.....	23.56	.....	.....
Methane	.....	22.32	36.86	52.07	61.37	64.55	67.86	71.08	74.39	78.11	81.83	85.54	89.26	106.62
Neon	.....	.....	.....	.....	97.94	100.84	104.14	107.03	109.93	112.82	115.71	118.19	121.09	.....
Nitric oxide	.....	.....	30.91	42.40	49.01	51.24	53.39	55.54	57.65	59.76	61.99	64.06	66.12	74.39
Nitrogen	.....	20.25	33.06	44.22	50.42	52.48	54.55	56.20	58.27	60.34	62.40	64.06	65.71	.....
Nitrous oxide	.....	.....	.....	.....	28.93	30.91	32.90	35.04	37.15	39.30	41.45	43.81	46.08	.....
Oxygen	.....	18.84	31.66	43.72	50.54	52.81	54.96	57.24	59.43	61.58	63.64	65.91	68.19	76.87
n-Propane	.....	.....	.....	.....	27.69	29.75	32.23	34.71	37.19	39.67	42.47	45.46	48.35	60.75
R-11(CCl_3F)	.....	.....	.....	.....	.....	12.81	13.64	14.88	15.70	16.53	17.77	18.60	.....	.....
R-12(CCl_2F_2)	.....	.....	.....	.....	.....	17.36	18.60	19.42	20.66	21.49	22.73	23.56	.....	.....
R-21($CHCl_2F$)	.....	.....	.....	.....	.....	.....	21.90	22.32	22.73	23.14	23.56	23.97	.....	.....
R-22($CHClF_2$)	.....	.....	.....	.....	.....	.....	24.80	25.62	26.45	27.28	28.10	28.93	.....	.....
Water	.....	.....	.....	.....	.....	.....	34.71	36.78	38.85	40.50	42.57	44.63	46.70	54.96

THERMAL CONDUCTIVITY OF GASEOUS HELIUM, NITROGEN AND WATER

From NSRDS-NBS 8

R. W. Powell, C. Y. Ho, and P. E. Liley

The thermal conductivity, k, is given in the units Milliwatt cm^{-1} $°K^{-1}$. To convert to Cal(*gm*) hr^{-1} cm^{-1} $°K^{-1}$ multiply the values listed in the table by 0.860421

T (K)	k He	k N_2
0.08	0.00044	
0.09	0.00053	
0.10	0.00064	
0.15	0.00130	
0.20	0.00231	
0.25	0.0039	
0.30	0.0062	
0.35	0.0089	
0.40	0.0120	
0.45	0.0154	
0.5	0.0187	
0.6	0.0231	
0.7	0.0252	
0.8	0.0262	
0.9	0.0266	
1.0	0.0269	
1.25	0.0281	
1.5	0.0306	
2.0	0.0393	
2.5	0.0502	
3.0	0.0607	
3.5	0.0732	
4.0	0.0803	
4.5	0.0879	
5.0	0.0962	
6	0.1113	
7	0.1247	
8	0.1393	
9	0.1523	
10	0.1640	
12	0.1866	
14	0.2067	
16	0.2259	
18	0.2435	
20	0.2582	
25	0.2962	
30	0.3330	
35	0.3669	
40	0.4000	
45	0.4314	
50	0.4623	(0.0485)*
60	0.521	(0.0578)*
70	0.578	(0.0670)*
80	0.631	0.0762
90	0.679	0.0852
100	0.730	0.0941
110	0.776	0.1030
120	0.819	0.1119
130	0.863	0.1208
140	0.907	0.1296

T (K)	k He	k N_2	k H_2O
150	0.950	0.1385	
160	0.992	0.1474	
170	1.033	0.1562	
180	1.072	0.1651	
190	1.112	0.1739	
200	1.151	0.1826	
210	1.190	0.1908	
220	1.228	0.1989	
230	1.266	0.2067	
240	1.304	0.2145	
250	1.338	0.2222	(0.140)*
260	1.372	0.2298	(0.148)*
270	1.405	0.2374	(0.156)*
280	1.437	0.2449	0.164
290	1.468	0.2524	0.172
300	1.499	0.2598	0.181
310	1.530	0.2671	0.189
320	1.560	0.2741	0.197
330	1.590	0.2808	0.205
340	1.619	0.2874	0.214
350	1.649	0.2939	0.222
360	1.678	0.3002	0.231
370	1.708	0.3065	0.239
380	1.737	0.3127	0.248
390	1.766	0.3189	0.256
400	1.795	0.3252	0.264
410	1.824	0.3314	0.273
420	1.853	0.3376	0.282
430	1.882	0.3438	0.291
440	1.914	0.3501	0.300
450	1.947	0.3564	0.307
460	1.980	0.3626	0.317
470	2.013	0.3688	0.327
480	2.046	0.3749	0.337
490	2.080	0.3808	0.347
500	2.114	0.3864	0.357
510	2.15	0.392	0.368
520	2.18	0.398	0.378
530	2.22	0.403	0.389
540	2.25	0.408	0.400
550	2.29	0.414	0.411
560	2.33	0.420	0.422
570	2.36	0.425	0.432
580	2.40	0.431	0.443
590	2.43	0.436	0.454
600	2.47	0.441	0.464
610	2.51	0.446	0.475
620	2.54	0.452	0.486
630	2.58	0.457	0.497
640	2.61	0.462	0.508

T (K)	k He	k N_2	k H_2O
650	2.64	0.467	0.518
660	2.67	0.472	0.529
670	2.69	0.478	0.540
680	2.72	0.483	0.551
690	2.75	0.488	0.562
700	2.78	0.493	0.572
710	2.81	0.498	0.58
720	2.84	0.503	0.59
730	2.87	0.508	0.60
740	2.90	0.513	0.62
750	2.92	0.517	0.63
760	2.95	0.522	0.64
770	2.98	0.526	0.65
780	3.01	0.531	0.66
790	3.04	0.536	0.67
800	3.07	0.541	0.68
810	3.09	0.546	0.69
820	3.12	0.551	0.70
830	3.15	0.555	0.71
840	3.18	0.559	0.72
850	3.21	0.564	0.73
860	3.23	0.569	0.74
870	3.26	0.574	0.75
880	3.29	0.578	0.76
890	3.32	0.583	0.77
900	3.35	0.587	0.78
910	3.37	0.592	
920	3.40	0.596	
930	3.43	0.600	
940	3.46	0.605	
950	3.49	0.609	
960	3.52	0.613	
970	3.54	0.618	
980	3.57	0.622	
990	3.60	0.626	
1000	3.63	0.631	
1050	3.76	0.651	
1100	3.89	0.672	
1150	4.03	0.693	
1200	4.16	0.713	
1250	4.29	0.733	
1300	4.43	0.754	
1350	4.55	0.775	
1400	4.69	0.797	
1450	4.82	0.819	
1500	4.94	0.842	
1550	5.07	0.867	
1600	5.21	0.893	
1650	5.33	0.921	
1700	5.45	0.950	

T (K)	k He	k N_2
1750	5.57	0.981
1800	5.70	1.013
1850	5.83	1.046
1900	5.96	1.080
1950	6.08	1.113
2000	6.20	1.146
2100	6.44	1.207
2200	6.69	1.263
2300	6.93	1.314
2400	7.16	1.361
2500	7.39	1.406
2600	7.62	1.449
2700	7.85	1.494
2800	8.07	1.542
2900	8.29	1.590
3000	8.51	1.640
3100	8.72	1.691
3200	8.95	1.743
3300	9.16	1.795
3400	9.37	1.853
3500	9.58	1.915
3600	9.79	
3700	10.00	
3800	10.22	
3900	10.43	
4000	10.64	
4100	10.85	
4200	11.06	
4300	11.27	
4400	11.48	
4500	11.69	
4600	11.90	
4700	12.11	
4800	12.31	
4900	12.51	
5000	12.71	

THERMAL CONDUCTIVITY OF DIELECTRIC CRYSTALS

Name	Remarks	Conductivity nw/cm deg K 83°K	273°K
Marble	Small crystals, 99.9% $CaCO_3$	42	33
Do	99.99% $CaCo_3$	54	38
Do	Large crystals	50	33
Calcite	Main crystal axis perpendicular to rod axis	180	46
Do	Main crystal axis parallel to rod axis	293	54
Sylvite	Natural crystal	159	75
KCl	Pressed at 8000 atm	314	88
KCl	From a melt	402	92
NaCl	Do	343	92
NaCl	Pressed at 8000 atm	251	71
Rock salt	Do	180	63
Sylvite	Do	343	84
KCl	Pressed at 1250 atm	243	75
KCl	Pressed at 2500 atm	368	92
KCl	Pressed at 8900 atm	402	96
KBr	Pressed at 8000 atm	92	38
NaBr	Do	50	25
KI	Do	121	29
KF	Do	234	71
NaF	Do	519	105
RbI	Do	59	33
RbCl	Do	29	21
90% KBr, 10% KCl	Do	50	29
75% KBr, 25% KCl	Do	29	21
50% KBr, 50% KCl	Do	25	25
25% KBr, 75% KCl	Pressed at 8000 atm	46	33
10% KBr, 90% KCl	Do	80	50
50% KCl, 50% NaCl	Do	188	71
KNO_2	Do	17	21
Mercuric chloride	Do	17	13
NH_4Cl	Do	109	25
NH_2Br	Do	67	25
$Ba(NO_3)_2$	Do	33	13
Copper sulfate		29	21
Magnesium sulfate		25	25
$K_4Fe(CN_6)$		17	17
Chrom alum		13	21
Potassium alum		13	21
Potassium bichromate	Main crystal axis perpendicular to rod axis	17	21
Do	Main crystal axis parallel to rod axis	17	17
Topaz	Mineral		234
Zincblend	Do	63	264
Beryll	Do	88	84
Tourmaline	Do	38	46

THERMAL CONDUCTIVITY OF ORGANIC COMPOUNDS[a]

Substance	k	t, °C	t, °F
Acetaldehyde	0.0004089	21	69.8
Acetic acid	0.0004109	20	68
Acetic anhydride	0.0005286	21	69.8
Acetone	0.0004750	−80	−112
	0.0004543	16	61
	0.0004031	75	167
Allyl alcohol	0.0004295	30	86
Amyl acetate(n)	0.0003085	20	68
(iso)	0.000310	20	68
Amyl alcohol(n)	0.0003874	30—100	86—212
(iso)	0.0003531	30	86
Amyl bromide (n)	0.0002350	18.64.4	
Aniline	0.0004237	16.5	61.5
Benzene	0.0003780	22.5	72.5
	0.0003275	50	122
	0.0003630	60	140
	0.0002870	140	284
Bromobenzene	0.0002664	20	68
Butyl acetate (n)	0.000327	20	68
Butyl alcohol (n)	0.0003663	20	68
Carbon tetrachloride	0.0002470	20	68
	0.0002333	50	122
Chlorobenzene	0.0003457	30—100	86—212
Chlorotoluene (p)	0.000310	20	68
Chloroform	0.0002891	16	61
	0.000246	20	68
Cresol(m)	0.0003581	20	68
(p)	0.000345	20.1	68.2
Cumene	0.000298	20	68
Cymene (p)	0.0003217	30	86
Decane	0.0003349	30	86
Diethyl ether	0.0003283	30	86
Dichloroethane, 1—2	0.000302	20	68
Di-isopropyl ether	0.000262	20	68
Ethyl acetate	0.0003560	16	60.8
Ethyl alcohol	0.0003995	20	68
Ethyl benzene	0.0003160	20	68
Ethyl bromide	0.0002862	30	86
Ethyl ether	0.0003283	30	86
Ethyl iodide	0.0002651	30	86
Ethylene glycol	0.0006236	20	68
	0.0006323	15	122
	0.0006443	80	176
Freon-12 (CCl_2F_2)	0.0002310	0—75	32—167
Freon-21 ($CHCl_2F$)	0.0003180	0—75	32—167
Freon-22 ($CHClF_2$)	0.0002309	40	104
Freon-113 (CCl_2FCCCl_2F)	0.0002379	0—80	32—176
Freon-114 ($C_2H_2F_4$)	0.0002127	0—75	32—167
Glycerol	0.000703	20	68
Heptane (n)	0.0003354	30	86
Heptyl alcohol	0.0003882	70—100	86—212
Hexane (n)	0.0003287	30—100	86—212
Hexyl alcohol (n)	0.0003857	30—100	86—212
Iodobenzene	0.0002874	30—100	86—212
Mesitylene	0.0003246	20	68
Methyl alcohol	0.0004832	20	68
Methyl aniline	0.0004419	21.5	70.5
Methyl chloride	0.0004597	−15(−) + 30	5—86
Methyl cyclohexane	0.0003052	30	86
Methylene chloride	0.0002908	0	32
Nitrobenzene	0.0003907	30—100	86—212
Nitromethane	0.0005142	30	86
Nonane (n)	0.0003374	30—100	86—212
Nonyl alcohol (n)	0.0004014	30—100	86-212
Octane (n)	0.0003469	30	86
Octyl alcohol (n)	0.0003973	30—100	86—212
Oleic acid	0.0005514	26.5	79.7
Palmitic acid	0.0004097	72.5	162.5
Pentachloroethane	0.0002994	20	68
Pentane (n)	0.0003221	30	86
Phenetole	0.0003577	−20	−4
Phenyl hydrazine	0.0004121	25	69.8
Propyl acetate (sio)	0.000321	20	68
Propyl alcohol (iso)	0.0003362	20	68
Propylene chloride	0.0002994	20—50	68—122
Propylene glycol (1—2)	0.0004799	20—80	68—176
Stearic acid	0.0003824	72.5	162.5
Tetrachloroethane (sym)	0.000272	20	68
Tetrachloroethylene	0.0003866	20	68
Toluene	0.0003804	−80	−112
	0.0003221	20	68
	0.0002808	80	176
Trichloroethylene	0.0003246	−60	−76
	0.0002775	20	68
Triethylamine	0.0003498	−80	−112
	0.0002891	20	68
	0.0002664	44.4	112
Xylene (o)	0.0003411	−20(−) + 80	(−4)−176
Xylene (m)	0.0003767	25	77

[a]The values in this table are given as cal/(sec)(cm^2)(°C/cm). To convert these values to Btu/(hr)(ft$_2$)(°F/ft) multiply by 242.08 .

THERMAL CONDUCTIVITY OF INORGANIC COMPOUNDS

Ammonia	0.0001198	−15(−)+30	5—86
Argon	0.0002895	−183	−297
	0.0001677	−133	−207
	0.000553	−105	−157.5
	0.0000409	−75	−102.5
Carbon dioxide	0.0002040	−50	−58
	0.0002412	−40	−40
	0.0002664	−30	−22
	0.0002746	−20	−4
	0.0002495	0	32
	0.0001677	30	86
Nitrogen	0.0003400	−196	−321.5
	0.0002028	−158	−253
	0.0000640	−105	−155
Oxygen	0.0000500	(−207)—(−191)	(−340)—(−312)
	0.0000504	(−178)—(−182)	(−288)—(−295)
Water	0.001326	−3	27
	0.001372	+7	45
	0.001456	27	81
	0.001522	47	117
	0.001575	67	153
	0.001625	97	207
	0.001635	107	225
	0.001628	157	315
	0.001580	197	387
	0.001463	247	441
	0.001288	297	567
	0.001004	347	657

THERMAL CONDUCTIVITY OF MISCELLANEOUS SUBSTANCES

Chlorinated diphenyl 1242	0.0002936	30—100	86—212
Chlorinated diphenyl 1248	0.0002808	30—100	86—212
Kerosene	0.0003572	30	86
Light heat transfer oil	0.0003159	30—100	86—212
Petroleum ether	0.0003118	30	86
Red oil	0.0003366	30	86
Transformer oil	0.0004242	70—100	86—212

THERMAL CONDUCTIVITY OF MATERIALS
(Bureau of Standards Letter Circular No. 227)

D = Density in pound per cubic foot.
K = Thermal conductivity in B.T.U. per hour, square foot, and temperature gradient of 1 degree Fahrenheit per inch thickness. The lower the conductivity, the greater the insulating values.

Soft Flexible Materials in Sheet Form

		D	K
Dry zero	Kapok between burlap or paper	1.0	0.24
		2.0	0.25
Cabots quilt	Eel grass between kraft paper	3.4	0.25
		4.6	0.26
Hair felt	Felted cattle hair	11.0	0.26
		13.0	0.26
Balsam wool	Chemically treated wood fiber	2.2	0.27
Hairinsul	75% hair 25% jute	6.3	0.27
	50% hair 50% jute	6.1	0.26
Linofelt	Flax fibers between paper	4.9	0.28
Thermofelt	Jute and asbestos fibers, felted	10.0	0.37
	Hair and asbestos fibers, felted	7.8	0.28

Loose Materials

		D	K
Rock wool	Fibrous material made from rock also made in sheet form, felted and confined with wire netting	6.0	0.26
		10.0	0.27
		14.0	0.28
		18.0	0.29
Glass wool	Pyrex glass, curled	4.0	0.29
		10.0	0.29
Sil-O-Cel	Powdered diatomaceous earth	10.6	0.31
Regranulated cork	Fine particles	9.4	0.30
	about $^3/_{16}$ inch particles	8.1	0.31
Thermofill	Gypsum in powdered form	26	0.52
		34	0.60
Sawdust	Various	12.0	0.41
	redwood	10.9	0.42
Savings	Various, from planer	8.8	0.41
Charcoal	From maple, beech and birch, coarse	13.2	0.36
	6 mesh	15.2	0.37
	20 mesh	19.2	0.39

Semiflexible Materials in Sheet Form

		D	K
Flaxlinum	Flax fiber	13.0	0.31
Fibrofelt	Flax and rye fiber	13.6	0.32

Semiflexible Materials in Sheet Form

		D	K
Flaxlinum	Flax fiber	13.0	0.31
Fibrofelt	Flax and rye fiber	13.6	0.32

Semirigid Materials in Board Form

		D	K
Corkboard	No added binder; very low density	5.4	0.25
Corkboard	No added binder; low density	7.0	0.27
Corkboard	No added binder; medium density	10.6	0.30
Corkboard	No added binder; High density	14.0	0.34
Eureka	Corkboard with asphaltic binder	14.5	0.32
Rock Cork	Rock wood block with binder Also called "Tucork"	14.5	0.326
Lith	Board containing rock wool, flax and straw pulp	14.3	0.40

Stiff Fibrous Materials in Sheet Form

		D	K
Insulite	Wood pulp	16.2	0.34
		16.9	0.34
Celotex	Sugar cane fiber	13.2	0.34
		14.8	0.34

	K =
*Masonite	0.33
*Inso-board	0.33
*Malzewood	0.33 to 0.39
*Cornstalk Pith Board	0.24 to 0.30
*Maftex	0.34

Cellular Gypsum

	D	K
Insulex or Pyrocell	8	0.35
	12	0.44
	18	0.59
	24	0.77
	30	1.00

Woods (Across Grain)

	D	K
Balsa	7.3	0.33
	8.8	0.38
	20	0.58
Cypress	29	0.67
White pine	32	0.78
Mahogany	34	0.90
Virginia pine	34	0.98
Oak	38	1.02
Maple	44	1.10

Miscellaneous Bulding Materials
(Data taken from various sources)

	K		K
Cinder concrete	2 to 3	Limestone	4 to 9
Building gypsum	About 3	Concrete	6 to 9
Plaster	2 to 5	Sandstone	8 to 16
Building brick	3 to 6	Marble	14 to 20
Glass	5 to 6	Granite	13 to 28

* From various commercial laboratories and the work of O. R. Sweeney at Iowa State College.

THERMAL CONDUCTIVITY DATA ON CERAMIC MATERIALS

Description[a]	Class[b]	Water Abs. %	Bulk Density g/cc	Thermal[c] Conductivity 100°F	200°F	300°F
Single Crystals						
Silicon carbide	5	—	—	52.0	50.0	49.0
Periclase	5	—	—	26.7	22.5	19.5
Sapphire, c-axis	5	—	—	20.2	16.0	14.0
Sapphire, a-axis	5	—	—	18.7	15.0	12.9
Topaz, a-axis	5	—	—	10.8	9.4	7.9
Kyanite, c-axis	5	—	—	10.00	8.6	7.4
Kyanite, b-axis	5	—	—	9.6	8.3	7.1
Spinel, $MgO \cdot Al_2O_3$	5	—	—	6.80	6.20	5.50
Quartz, c-axis	4	—	—	6.40	5.40	5.20
Quartz, A-axis	4	—	—	3.40	3.00	2.60
Rutile, c-axis	5	—	—	5.60	4.80	4.40
Rutile, a-axis	5	—	—	3.20	3.20	3.20
Fluorite	5	—	—	5.30	4.37	3.45
Beryl, aquamarine, c-axis	4	—	—	3.18	3.15	3.12
Beryl, aquamarine, a-axis	4	—	—	2.52	2.52	2.52
Zircon, a-axis	4	—	—	2.45	2.45	2.45
Zircon, c-axis	4	—	—	2.34	2.34	2.35
Polycrystalline Single Oxide Ceramics						
Pure BeO, hot pressed	2	0.03	2.97	125.0	104.0	92.0
MgO (spec. pure)	1	0.83	3.21	21.2	18.4	16.0
$SnO_2$98%	1	0.03	6.62	17.5	15.0	12.7
ZnO (yellow)	1	0.00	5.28	16.8	14.6	12.5
ZnO gray)	1	0.03	5.20	13.6	11.8	10.2
CuO (100%)	1	0.04	6.76	10.2	9.00	7.80
ThO_2, hot pressed	2	—	9.58	8.00	7.02	6.50
CeO_2	1	0.00	6.20	6.63	6.29	5.20
Mn_2O_4	1	0.02	4.21	4.18	3.80	3.41
PbO (100%)	1	0.38	7.98	1.6	1.25	0.98

[a] Composition: 90%, MgO, 10%, Al_2O_2 designates weight percent. $Li_2O{:}4B_2O_3$ designates mole composition, does not indicate compound formation.
[b] Classification: 1 = research body; 2 = industrial research body; 3 = commercial body; 4 = natural mineral; 5 = synthetic mineral.
[c] Thermal conductivity: Units in Btu/(hr) (sq ft) (°F/ft); to convert to cal/(sec) (sq cm)(°C/cm) multiply by 0.00413. (I) = determination made with high vacuum apparatus, inconel thermodes. No letters following value, determination made with high vacuum thermal conductivity apparatus, co-per thermodes.

By permission from Engineering Research Bulletin No. 40 Rutgers University, 1958.

THERMAL CONDUCTIVITIES OF GLASSES BETWEEN −150 AND +100°C

E. H. Ratcliffe

Type of glass	Approximate silica contents (wt. %)	Approximate contents other oxides normally present in quantity (wt. %)		Estimated approximate thermal conductivity at various temperatures: Temperature (°C)	Thermal conductivity $\left(\frac{\text{cal cm}}{\text{cm}^2\text{ s deg C}}\right) \times 10^4$
(a) Vitreous silica	100	....		−150	20.0
				−100	25.0
				−50	28.8
				0	31.5
				50	33.7
				100	35.4
(b) 'Vycor' glass	96	B_2O_3	3	−100	24
				0	30
				100	34
(c) General information					
'Crown' glasses	50–75	Various		−100	12–20.5
				30	19–26
				100	21–29
'Flint' glasses	20–55	Various		−100	9–15
				30	13–21
				100	15–23
(d) Pyrex type chemically-resistant borosilicate glasses	80–81	B_2O_3	12–13	−100	21
		Na_2O	4	0	26
		Al	2	100	30
(e) Borosilicate crown glasses	60–65	B_2O_3	15–20	−100	16–17.5
				0	21–22.5
				100	24–25.5
	65–70	B_2O_3	10–15	−100	17.5–19
				0	22.5–24
				100	25.5–27
	70–75	B_2O_3	5–10	−100	19–20.5
				0	24.5–26
				100	27.5–29
(f) (i) Zinc crown glasses	55–65	ZnO	5–15	−100	21–22
		Remainder		0	26–27
		B_2O_3, Al_2O_3		100	28–30
		ZnO	5–15	−100	14–17
		Remainder		0	17–21
		Na_2O, K_2O		100	20–23
		ZnO	15–25	−100	21–22
		Remainder		0	26–27
		B_2O_3, Al_2O_3		100	27–29
		ZnO	15–25	−100	16–19
		Remainder		0	20–23
		Na_2O, K_2O		100	22–25

THERMAL CONDUCTIVITIES OF GLASSES BETWEEN −150 AND +100°C (Continued)

Type of glass	Approximate silica contents (wt. %)	Approximate contents other oxides normally present in quantity (wt. %)		Estimated approximate thermal conductivity at various temperatures: Temperature (°C)	Thermal conductivity $\left(\frac{\text{cal cm}}{\text{cm}^2\text{ s deg C}}\right) \times 10^4$
(f) (ii) Zinc crown glasses	65–75	ZnO	5–15	−100	21–22
		Remainder		0	27–28
		B_2O_3, Al_2O_3		100	29–31
		ZnO	5–15	−100	17–20
		Remainder		0	21–25
		Na_2O, K_2O		100	24–27
		ZnO	15–25	−100	21–23
		Remainder		0	27–28
		B_2O_3, Al_2O_3		100	29–30
		ZnO	15–25	−100	16–20
		Remainder		0	20–24
		Na_2O, K_2O		100	25–29
(g) Barium crown glasses	31	B_2O_3	12	−100	13
		Al_2O_3	8	0	17
		BaO	48	100	19
	41	B_2O_3	6	−100	14
		Al_2O_3	2	0	18
		ZnO	8	100	20
		BaO	43		
	47	B_2O_3	4	−100	15
		Na_2O	1	0	18
		K_2O	7	100	21
		ZnO	8		
		BaO	32		
	65	B_2O_3	2	−100	17
		Na_2O	5	0	21
		K_2O	15	100	24
		ZnO	2		
		BaO	10		
(h) Borate glasses					
Borate flint glass	9	B_2O_3	36	−100	13
		Na_2O	1	0	16
		K_2O	2	100	19
		PbO	36		
		Al_2O_3	10		
		ZnO	6		
Borate flint glass	.....	B_2O_3	56	−100	12
		Al_2O_3	12	0	16
		PbO	32	100	20
Borate flint glass	.....	B_2O_3	43	−100	9
		Al_2O_3	5	0	13
		PbO	52	100	17
Borate glass	4	B_2O_3	55	−100	15
		Al_2O_3	14	0	19
		PbO	11	100	21
		K_2O	4		
		ZnO	12		
Borate crown glass	.....	B_2O_3	64	−100	12
		Na_2O	8	0	16
		K_2O	3	100	20
		BaO	4		
		PbO	3		
		Al_2O_3	18		
Light borate crown glass	.....	B_2O_3	69	−100	13
		Na_2O	8	0	17
		BaO	5	100	21
		Al_2O_3	18		
Zinc borate glass	.....	B_2O_3	40	−100	16
		ZnO	60	0	18
				100	20
(i) Phosphate crown glasses					
Potash phosphate glass	.....	P_2O_5	70	0	18
		B_2O_3	3	100	20
		K_2O	12		
	.....	Al_2O_3	10		
		MgO	4		
Baryta phosphate glass	.....	P_2O_5	60	45	18
		B_2O_3	3		
		Al_2O_3	8		
		BaO	28		

THERMAL CONDUCTIVITIES OF GLASSES BETWEEN −150 AND +100°C (Continued)

Type of glass	Approximate silica contents (wt. %)	Approximate contents other oxides normally present in quantity (wt. %)		Estimated approximate thermal conductivity at various temperatures	
				Temperature (°C)	Thermal conductivity $\left(\frac{\text{cal cm}}{\text{cm}^2\text{ s deg C}}\right) \times 10^4$
(j) Soda-lime glasses	75	Na_2O	17	−100	18
		CaO	8	0	23
				100	26
	75	Na_2O	12	−100	21
		CaO	13	0	26
				100	28
	72	Na_2O	15	−100	19
		CaO	11	0	24
		Al_2O_3	2	100	27
	65	Na_2O	25	−100	16
		CaO	10	0	20
				100	23
	65	Na_2O	15	−100	20
		CaO	20	0	24
				100	26
	60	Na_2O	20	−100	18
		CaO	20	0	22
				100	24
(k) Other crown glasses					
Crown glass	75	Na_2O	9	−100	19
		K_2O	11	0	24
		CaO	5	100	26
High dispersion crown glass	68	Na_2O	16	−100	16
		ZnO	3	0	20
		PbO	13	100	24
(l) Miscellaneous flint glasses					
(i) Silicate flint glasses					
Light flint glasses	65	PbO	25	−100	16–17
		Others	10	0	21–22
				100	24–25
	55	PbO	35	−100	14–16
		Others	10	0	18–20
				100	21–22
Ordinary flint glass	45	PbO	45	−100	12–14
		Others	10	0	16–18
				100	19–20
Heavy flint glass	35	PbO	60	−100	11–12
		Others	5	0	14–15
				100	17–18
Very heavy flint glasses	25	PbO	73	−100	10–11
		Others	2	0	13–14
				100	15–16
	20	PbO	80	−100	10
				0	12
				100	14
(ii) Borosilicate flint glass	33	B_2O_3	31	−100	15
		PbO	25	0	20
		Al_2O_3	7	100	23
		K_2O	3		
		Na_2O	1		
(iii) Barium flint glass	50	BaO	24	−100	14
		PbO	6	0	17
		K_2O	8	100	20
		Na_2O	3		
		ZnO	8		
		Sb_2O_3	1		
(m) Other glasses					
(i) Potassium glass	59	K_2O	33	50	21–22
		CaO	8		
(ii) Iron glasses	63	Fe_2O_3	10	−100	19
		Na_2O	17	0	23
		MgO	4	100	25
		CaO	3		
		Al_2O_3	2		
	67	Fe_2O_3	15	0	21–22
		Na_2O_3	18	100	24–25
	62	Fe_2O_3	20	0	20.5–21.5
		Na_2O	18	100	23–24
(ii) Rock glasses					
Obsidian				0	32
				100	35
Artificial diabase				0	27
				100	30

THERMAL CONDUCTIVITY OF CERTAIN METALS

From NSRDS-NBS 8

R. W. Powell, C. Y. Ho, and P. E. Liley

The thermal conductivity, k, is given in the units Watt $cm^{-1}\ °K^{-1}$.
To convert to Cal(*gm*) $hr^{-1}\ cm^{-1}\ °C^{-1}$ multiply the values listed in the tables by 860.421
To convert to Btu $hr^{-1}\ ft^{-1}\ °F^{-1}$ multiply the values listed in the tables by 57.818.
ρ_0 is the residual electrical resistivity and the value of ρ at 4.2°K is used approximately as ρ_0.

T,K	Aluminum 99.996⁺% ρ_0 = 0.00315 μohm cm	Copper 99.999⁺% ρ_0 = 0.000851 μohm cm	Gold 99.999⁺% ρ_0 = 0.0055 μohm cm	Iron 99.998⁺% ρ_0 = 0.0327 μohm cm	Manganin	Platinum 99.999% ρ_0 = 0.0106 μohm cm	Silver 99.999⁺% ρ_0 = 0.00062 μohm cm	Tungsten 99.99⁺% ρ_0 = 0.0017 μohm cm
0	0	0	0	0	0	0	0	0
1	7.8	28.7	4.4	0.75	0.0007	2.31	39.4	14.4
2	15.5	57.3	8.9	1.49	0.0018	4.60	78.3	28.7
3	23.2	85.5	13.1	2.24	0.0031	6.79	115	42.6
4	30.8	113	17.1	2.97	0.0046	8.8	147	55.6
5	38.1	138	20.7	3.71	0.0062	10.5	172	67.1
6	45.1	159	23.7	4.42	0.0078	11.8	187	76.2
7	51.5	177	26.0	5.13	0.0095	12.6	193	82.4
8	57.3	189	27.5	5.80	0.0111	12.9	190	85.3
9	62.2	195	28.2	6.45	0.0128	12.8	181	85.1
10	66.1	196	28.2	7.05	0.0145	12.3	168	82.4
11	69.0	193	27.7	7.62	0.0162	11.7	154	77.9
12	70.8	185	26.7	8.13	0.0180	10.9	139	72.4
13	71.5	176	25.5	8.58	0.0197	10.1	124	66.4
14	71.3	166	24.1	8.97	0.0215	9.3	109	60.4
15	70.2	156	22.6	9.30	0.0232	8.4	96	54.8
16	68.4	145	20.9	9.56	0.0250	7.6	85	49.3
18	63.5	124	17.7	9.88	0.0285	6.1	66	40.0
20	56.5	105	15.0	9.97	0.0322	4.9	51	32.6
25	40.0	68	10.2	9.36	0.0410	3.15	29.5	20.4
30	28.5	43	7.6	8.14	0.0497	2.28	19.3	13.1
35	21.0	29	6.1	6.81	0.0583	1.80	13.7	8.9
40	16.0	20.5	5.2	5.55	0.067	1.51	10.5	6.5
45	12.5	15.3	4.6	4.50	0.075	1.32	8.4	5.07
50	10.0	12.2	4.2	3.72	0.082	1.18	7.0	4.17
60	6.7	8.5	3.8	2.65	0.097	1.01	5.5	3.18
70	5.0	6.7	3.58	2.04	0.110	0.90	4.97	2.76
80	4.0	5.7	3.52	1.68	0.120	0.84	4.71	2.56
90	3.4	5.14	3.48	1.46	0.127	0.81	4.60	2.44
100	3.0	4.83	3.45	1.32	0.133	0.79	4.50	2.35
150	2.47	4.28	3.35	1.04	0.156	0.762	4.32	2.10
200	2.37	4.13	3.27	0.94	0.172	0.748	4.30	1.97
250	2.35	4.04	3.20	0.865	0.193	0.737	4.28	1.86
273	2.36	4.01	3.18	0.835	0.206	0.734	4.28	1.82
300	2.37	3.98	3.15	0.803	0.222	0.730	4.27	1.78
350	2.40	3.94	3.13	0.744	0.250	0.726	4.24	1.70
400	2.40	3.92	3.12	0.694	(0.279)	0.722	4.20	1.62
500	2.37	3.88	3.09	0.613	(0.338)	0.719	4.13	1.49
600	2.32	3.83	3.04	0.547	(0.397)	0.720	4.05	1.39
700	2.26	3.77	2.98	0.487		0.723	3.97	1.33
800	2.20	3.71	2.92	0.433		0.729	3.89	1.28
900	2.13	3.64	2.85	0.380		0.737	3.82	1.24
1000	[0.93]**	3.57	(2.78)	0.326		0.748	(3.74)	1.21
1100	[0.96]	3.50	(2.71)	0.297		0.760	(3.66)	1.18
1200	[0.99]	3.42	(2.62)	0.282		0.775	(3.58)	1.15
1300	[1.02]	(3.34)†	(2.51)	0.299		0.791		1.13
1400				0.309		0.807		1.11
1500				0.318		0.824		1.09
1600				(0.327)		0.842		1.07
						0.860		1.05
						0.877		1.03
						(0.895)		1.02
						(0.913)		1.00
								0.98
								0.96
								0.94
								0.925
								0.915
								0.905
								0.900
								(0.895)

* In the table the third significant figure is given only for the purpose of comparison and for smoothness and is not indicative of the degree of accuracy.

** Values in square brackets are for liquid state.

† Values in parentheses are extrapolated.

‡ Estimated.

THERMAL CONDUCTIVITY OF CERTAIN LIQUIDS

From NSRDS-NBS 8
R. W. Powell, C. Y. Ho, and P. E. Liley

The thermal conductivity, k, is given in the units Milliwatt $cm^{-1}\,°K^{-1}$. To convert to Cal(*gm*) $hr^{-1}\,cm^{-1}\,°K^{-1}$ multiply the values listed in the table by 0.860421

T (K)	Helium	Nitrogen	Argon	Carbon tetra-chloride	Diphenyl	m-Terphenyl	Toluene	Water
2.4	0.192							
2.6	0.193							
2.8	0.197							
3.0	0.204							
3.2	0.214							
3.4	0.227							
3.6	0.241							
3.8	0.260							
4.0	0.282							
4.2	0.307							
4.4	(0.335)‡							
4.6	(0.366)‡							
4.8	(0.400)‡							
5.0	(0.437)‡							
5.2	(0.477)‡							
60		1.692†						
65		1.598						
70		1.504						
75		1.411						
80		1.320‡	1.315†					
85		1.229‡	1.258					
90		1.140‡	1.200‡					
95		1.051‡	1.141‡					
100		0.965‡	1.082‡					
105		0.879‡	1.023‡					
110		0.794‡	0.963‡					
115		0.710‡	0.903‡					
120		0.627‡	0.842‡					
125		0.544‡	0.780‡					
130			0.717‡					
135			0.654‡					
140			0.591‡					
145			0.527‡					
150			0.463‡				(1.719)†	
160							(1.694)†	
170							(1.669)†	
180							1.644	
190							1.619	
200							1.594	
210							1.569	
220							1.543	
230				(1.169)†			1.518	
240				(1.150)†			1.492	
250				1.131			1.467	5.22†
260				1.112			1.442	5.39†
270				1.093			1.416	5.55†
280				1.074			1.391	5.74
290				1.055			1.365	5.92
300				1.036			1.340	6.09
310				1.017			1.315	6.23
320				0.997			1.289	6.37
330				0.978	(1.402)†		1.264	6.48
340				0.959	(1.387)†		1.238	6.59
350				0.940	1.373	(1.361)†	1.213	6.68
360				(0.921)	1.359	(1.356)†	1.188	6.75
370				(0.902)	1.345	1.351	1.162	6.80
380				(0.882)	1.331	1.346	1.137	6.84‡
390				(0.863)	1.316	1.341	(1.112)‡	6.86‡
400				(0.844)	1.302	1.335	(1.086)‡	6.86‡
410				(0.825)	1.288	1.329	(1.061)‡	6.86‡
420				(0.806)	1.274	1.323	(1.036)‡	6.84‡
430				(0.787)	1.259	1.317	(1.013)‡	6.81‡
440				(0.768)	1.245	1.310	(0.985)‡	6.78‡
450				(0.749)	1.231	1.304	(0.959)‡	6.73‡
460					1.217	1.297	(0.933)‡	6.67‡
470					1.202	1.290	(0.908)‡	6.61‡
480					1.188	1.283	(0.885)‡	6.53‡
490					1.174	1.276	(0.862)‡	6.45‡
500					1.160	1.268	(0.839)‡	6.35‡
510					1.146	1.261		6.24‡
520					1.131	1.254		6.12‡
530					1.117‡	1.246		5.99‡
540					1.103‡	1.238		5.86‡
550					1.089‡	1.230		5.71‡
560					1.074‡	1.222		5.55‡
570					1.060‡	1.213		5.39‡
580					1.046‡	1.205		5.20‡
590					1.032‡	1.197		5.01‡
600					1.018‡	1.188		4.81‡
610						1.180		4.60‡
620						1.172		4.40‡
630						1.163		(4.20)‡
640						1.155‡		(4.01)‡
650						1.146‡		

† Extrapolated for the supercooled liquid. [Approximate n.m.p. in K: N_2, 63; A, 84; CCl_4, 250; $C_{12}H_{10}$, 342; m-$C_{18}H_{14}$, 361; p-$C_{18}H_{14}$, 486; C_7H_{10}, 178; H_2O, 273.1].

‡Under saturation vapor pressure [Approximate n.b.p. in K: He, 4.3; N_2, 77; A, 88; CCl_4, 350; $C_{12}H_{10}$, 528; m-$C_{18}H_{14}$, 637; p-$C_{18}H_{14}$, 658; C_7H_{10}, 384; H_2O, 373].

THERMAL CONDUCTIVITY OF THE ELEMENTS

Data contained in the following table were extracted from the extensive compilation prepared by C. Y. Ho, R. W. Powell, and P. E. Liley under the National Standard Reference Data System (NSRDS) of the National Bureau of Standards project at the Thermophysical Properties Research Center (TPRC) at Purdue University and published in the Journal of Physical and Chemical Reference Data, *1*, 279-421 (1972). The data in the table below are used with the permission of the authors and the copyright owners, the American Institute of Physics and the American Chemical Society. Users are referred to their more extensive compilation for conductivities at temperatures other than those listed in the table below, and also to obtain an understanding of the basis of selection of recommended and provisional values. Temperatures are in kelvins (K) and conductivities, k, in watt per centimeter kelvin (W cm^{-1} K^{-1}), except as noted. If the numerical value of k has a superscript m, the units of k are milliwatt per centimeter kelvin (mW cm^{-1} K^{-1}). To convert the listed conductivities to units other than those in the tables, one should make use of the conversion factors listed in the table "Conversion Factors for Units of Thermal Conductivity", which follows this table. Conductivity values listed with an asterisk*, are provisional values.

Element	State or Condition	Conductivity at 273.2K	298.2K	373.2K
Aluminum	Solid	2.36	2.37	2.40
Antimony	Polycrystalline	0.255	0.244	0.219
Argon	Gas at 1 atm.	0.1619^{m} (270K)	0.1772^{m} (300K)	0.2103^{m} (370K)
Arsenic	Solid, Gray, Polycrystalline	0.539*	0.502*	0.427*
Barium	Solid	0.185*	0.184* (295K)	
Beryllium	Polycrystalline	2.18*	2.01	1.68
Bismuth	Solid			
	// to triagonal axis	0.0554	0.0530	0.0481
	⊥ to triagonal axis	0.0953	0.0919	0.0844
	Polycrystalline	0.0822	0.0792	0.0722
Boron	Solid	0.318	0.274	0.188
Bromine	Saturated liquid	1.30^{m}* (270K)	1.22^{m}* (300K)	1.06^{m}* (370K)
	Saturated vapor		0.048^{m}* (300K)	
	Gas	0.042^{m}* (270K)	0.048^{m}* (300K)	0.057^{m}* (350K)
Cadmium	Solid			
	// to c-axis	0.835	0.830	0.816
	⊥ to c-axis	1.04	1.04	1.02
	Polycrystalline	0.975	0.969	0.953
Calcium	Solid	2.06*	2.01*	1.92*
Carbon	Solid, Amorphous	0.0150	0.0159	0.0182
	Solid, Type I (Diamond)	9.94	9.90	7.03*
	Solid, Type IIa (Diamond)	26.2	23.2	17.0*
	Solid, Type IIb (Diamond)	15.2	13.6	10.2*
	Solid, Acheson graphite			
	// to axis of extrusion	1.69	1.65	1.50
	⊥ to axis of extrusion	1.21	1.19	1.11
	Solid, AGOT graphite			
	// to axis of extrusion	2.28	2.21	1.95
	⊥ to axis of extrusion	1.41	1.38	1.22
	Solid, ATJ graphite			
	// to molding pressure	0.984	0.982	0.933
	⊥ to molding pressure	1.31	1.29	1.21
	Solid, AWG graphite			
	// to molding pressure	0.807	0.796	0.733
	⊥ to molding pressure	1.32	1.28	1.16
	Solid, Pyrolytic graphite			
	// to layer planes	21.3	19.6	15.1
	⊥ to layer planes	0.0636	0.0573	0.0442
	Solid, 875S graphite			
	// to axis of extrusion	1.97*	1.92*	1.75*
	⊥ to axis of extrusion	1.49*	1.46*	1.34*
	Solid, 890S graphite			
	// to axis of extrusion	1.87*	1.83*	1.66*
	⊥ to axis of extrusion	1.51*	1.48*	1.36*
Cerium	Solid, Polycrystalline	0.108*	0.113	0.128*
Cesium	Solid	0.361*	0.359 (301.9K)	
	Liquid		0.197 (301.9K)	0.201
Chlorine	Saturated liquid	1.49^{m}* (270K)	1.34^{m}* (300K)	0.95^{m}* (370K)
	Saturated vapor	0.082^{m}* (270K)	0.097^{m}* (300K)	0.155^{m}* (370K)

THERMAL CONDUCTIVITY OF THE ELEMENTS (*Continued*)

Element	State or Condition	Conductivity at 273.2K	298.2K	373.2K
	Gas, 1 atm.	0.078[m]	0.089[m]	0.114[m]
		(270K)	(300K)	(370K)
Chromium	Solid, Polycrystalline	0.965	0.939	0.921
Cobalt	Solid, Polycrystalline	1.05	1.00	0.890
Copper	Solid	4.03	4.01	3.95
Dysprosium	Solid			
	∥ to c-axis	0.114*	0.117*	
	⊥ to c-axis	0.101*	0.103*	
	Polycrystalline	0.105*	0.107*	0.108*
Erbium	Solid			
	∥ to c-axis	0.187*	0.184*	
	⊥ to c-axis	0.127*	0.126*	
	Polycrystalline	0.147*	0.145*	0.140*
Europium	Solid	0.140*	0.139*	
Fluorine	Gas, 1 atm.	0.251[m]	0.279[m]	0.344[m]
		(270K)	(300K)	(370K)
Gadolinium	Solid			
	∥ to c-axis	0.104*	0.108*	
	⊥ to c-axis	0.103*	0.103*	
	Polycrystalline	0.103*	0.105*	
Gallium	Solid			
	∥ to a-axis	0.410	0.408	
	∥ to b-axis	0.884	0.883	
	∥ to c-axis	0.160	0.159	
	Liquid		0.281	0.328
			(302.93K)	
Germanium	Solid	0.667	0.602	0.465
Gold	Solid	3.19	3.18	3.13
Hafnium	Solid, Polycrystalline	0.233*	0.230	0.224
Helium	Solid, ^{3}He	0.033	0.020	0.0021
		(0.9K)	(1K)	(2K)
	Solid, ^{4}He	0.650	0.245	0.0018
		(0.9K)	(1K)	(2K)
	Liquid, saturated; He-I	0.191[m]	0.232[m]	0.434[m]
		(2.5K)	(3.5K)	(5K)
	Gas, 1 atm.	1.411[m]	1.520[m]	1.766[m]
Holmium	Solid			
	∥ to c-axis	0.215*	0.222*	
	⊥ to c-axis	0.136*	0.138*	
	Polycrystalline	0.159*	0.162*	0.170*
Hydrogen	Solid, Normal Hydrogen	2.30	0.0158	0.0090
		(4K)	(10K)	(15K)
	Liquid, saturated;	1.022[m]	1.269[m]	0.60[m]*
	Normal Hydrogen	(15K)	(25K)	(33K)
	Gas, 1 atm.	1.665[m]	1.815[m]	2.106[m]
	Normal Hydrogen			
	Liquid, saturated;	0.824[m]	0.998[m]	0.58[m]
	para-Hydrogen	(14K)	(25K)	(32K)
	Vapor, saturated;	0.081[m]*	0.242[m]*	0.58[m]*
	para-Hydrogen	(10K)	(25K)	(32K)
	Gas, 1 atm.;	1.768[m]*	1.880[m]*	2.126[m]*
	para-Hydrogen			
	Deuterium:			
	Liquid, saturated	1.26[m]	1.37[m]	0.83[m]*
		(20K)	(30K)	(38K)
	Vapor, saturated	0.084[m]*	0.26[m]*	0.83[m]*
		(20K)	(30K)	(38K)
	Gas, 1 atm.	1.294[m]*	1.406[m]*	1.66[m]*
		(270K)	(300K)	(370K)
	Tritium:			
	Liquid, saturated	1.25	1.34	0.68
		(21K)	(30K)	(44K)
Indium	Solid, Polycrystalline	0.837	0.818	0.762
Iodine	Solid	4.81[m]*	4.49[m]*	3.75[m]*
			(300K)	(386.8K)
	Liquid, saturated			1.16[m]*
				(386.8K)
Iridium	Solid	1.48	1.47	1.45
Iron	Solid	0.865	0.804	0.720
	Armco Iron	0.747	0.728	0.676
Krypton	Solid	0.4[m]*	17[m]	2.5[m]
		(1K)	(10K)	(116K)
	Liquid, saturated			0.931[m]
				(116K)
	Vapor, saturated	0.0406[m]*	0.0554[m]*	0.21[m]*
		(120K)	(150K)	(210K)
	Gas	0.0860[m]	0.0949[m]	0.1145[m]
		(270K)	(300K)	(370K)

Element	State or Condition	Conductivity at 273.2K	298.2K	373.2K
Lanthanum	Solid, Polycrystalline	0.131	0.134	0.145
Lead	Solid	0.356	0.353	0.344
Lithium	Solid	0.859	0.848	0.818
Lutetium	Solid			
	∥ to c-axis	0.236*	0.232*	
	⊥ to c-axis	0.140*	0.138*	
	Polycrystalline	0.167*	0.164*	
Magnesium	Solid, Polycrystalline	1.57	1.56	1.54
Manganese	Solid	0.0768*	0.0781*	
Mercury	Liquid	0.0782	0.0830	0.0947
Molybdenum	Solid	1.39*	1.38*	1.35*
Neodymium	Solid, Polycrystalline	0.165*	0.165*	0.167*
Neon	Gas	0.461^{m}* (270K)	0.493^{m}* (300K)	0.563^{m}* (370K)
Neptunium	Solid		0.063* (300K)	
Nickel	Solid	0.941	0.909	0.827
Niobium	Solid	0.533	0.537	0.548
Nitrogen	Solid	56^{m} (4K)	17^{m} (10K)	3.2^{m} (25K)
	Liquid, saturated	1.60^{m} (65K)	0.966^{m} (100K)	0.37^{m}* (126K)
	Vapor, saturated	0.061^{m}* (65K)	0.111^{m}* (100K)	0.37^{m}* (126K)
	Gas, 1 atm.	0.2374^{m} (270K)	0.2598^{m} (300K)	0.3065^{m} (370K)
Osmium	Solid			
	∥ to c-axis	2.93 (2K)	14.3 (10K)	15.4 (30K)
	⊥ to c-axis	1.76 (2K)	8.65 (10K)	11.1 (30K)
	Polycrystalline	2.09 (2K)	10.2 (10K)	12.4 (30K)
	Polycrystalline	0.880*	0.876*	0.870*
Oxygen	Liquid, saturated	1.501^{m} (90K)	1.023^{m} (125K)	0.41^{m} (155K)
	Vapor, saturated	0.081^{m}* (90K)	0.135^{m}* (125K)	0.41^{m}* (155K)
	Gas, 1 atm.	0.2424 (270K)	0.2674 (300K)	0.3204 (370K)
Palladium	Solid	0.716*	0.718	0.730
Phosphorus	Solid			
	Black (Polycrystalline)	0.132	0.121	
	White	0.00250*	0.00236*	
	Liquid, White			0.00181
Platinum	Solid	0.717	0.716	0.717
Plutonium	Solid, polycrystalline	0.0616*	0.0670*	0.0790* (350K)
Potassium	Solid	1.036*	1.025	
	Liquid			0.532
Praeseodymium	Solid, polycrystalline	0.120	0.125	0.134
Promethium	Solid, polycrystalline		0.179*	0.184*
Radium	Solid		0.186 (293.2K)	
Radon	Gas, 1 atm.	0.0327^{m}* (270K)	0.0364^{m}* (300K)	0.0445^{m}* (370K)
Rhenium	Solid, polycrystalline	0.486	0.480	0.466
Rhodium	Solid	1.51	1.50	1.47
Rubidium	Solid	0.583*	0.582	0.581 (312.04K)
	Liquid			0.333 (312.04K)
Ruthenium	Solid, polycrystalline	1.17	1.17	1.15
Samarium	Solid, polycrystalline	0.133*	0.133*	0.133*
Scandium	Solid, polycrystalline	0.157	0.158	
Selenium	Solid			
	∥ to c-axis	0.0481	0.0452	0.0483
	⊥ to c-axis	0.0137	0.0131	0.0139
	Amorphous	0.00428	0.00519	0.00818 (323.2K)
Silicon	Solid	1.68	1.49	1.08
Silver	Solid	4.29	4.29	4.26
Sodium	Solid	1.42	1.42	1.32 (371K)
Strontium	Solid, polycrystalline	0.364*	0.354*	0.325*

THERMAL CONDUCTIVITY OF THE ELEMENTS (*Continued*)

Element	State or Condition	Conductivity at 273.2K	298.2K	373.2K
Sulfur	Solid, polycrystalline	0.00287	0.00270	0.00154
	Solid, amorphous	0.00200	0.00205	0.00216* (350K)
	Liquid			0.00129 (392.2K)
Tantalum	Solid	0.574	0.575	0.577
Technetium	Solid, polycrystalline	0.509*	0.506	0.501
Tellurium	Solid			
	∥ to c-axis	0.0360	0.0338	0.0292
	⊥ to c-axis	0.0208	0.197	0.173
Terbium	Solid			
	∥ to c-axis	0.138*	0.147*	
	⊥ to c-axis	0.0900*	0.0956*	
	Polycrystalline	0.104*	0.111*	
Thallium	Solid, polycrystalline	0.469	0.461	0.443
Thorium	Solid	0.540*	0.540*	0.543*
Thulium	Solid			
	∥ to c-axis	0.242*	0.242*	
	⊥ to c-axis	0.140*	0.141*	
	Polycrystalline	0.168*	0.169*	
Tin	Solid			
	∥ to c-axis	0.527	0.516	0.489
	⊥ to c-axis	0.759	0.743	0.704
	Polycrystalline	0.682	0.668	0.632
Titanium	Solid, polycrystalline	0.224	0.219	0.207
Tungsten	Solid	1.77	1.73	1.63
Uranium	Solid, polycrystalline	0.270	0.275	0.291
Vanadium	Solid	0.307*	0.307	0.310
Xenon	Liquid, saturated	0.31[m] (270K)	0.16[m]* (290K)	
	Vapor, saturated	0.084[m]*	0.16[m]*	
	Gas, 1 atm.	0.0514[m] (270K)	0.0569[m] (300K)	0.0695[m] (370K)
Ytterbium	Solid	0.354*	0.349*	0.343*
Yttrium	Solid, polycrystalline	0.170*	0.172*	0.177*
Zinc	Solid, polycrystalline	1.17	1.16	1.12
Zirconium	Solid, polycrystalline	0.232*	0.227 (300K)	0.218

THERMAL CONDUCTIVITY OF ROCKS

Rock	Temperature, °C	Conductivity, Kcal m^{-1} hr^{-1} deg^{-1}	Heat Capacity, cal g^{-1} deg^{-1}
Granite	0	3.02	0.192
	50	2.81	–
	100	2.59	–
	200	2.34	0.228
	300	2.12	–
	400	–	0.258
Marble	118	1.44	0.21
	196	1.29	0.24
	245	1.19	–
	360	0.95	0.271
Dolomitic	130	1.41	–
limestone	181	1.37	–
	268	1.29	–
	377	1.15	–
	0	1.65	0.17
Shale	100	1.51	–
	120	1.33	–
	188	1.41	0.24
	304	1.26	0.245
	0	4.9	–
Sandstone	100	3.82	0.26
(quartzitic)	200	3.24	–

CONVERSION FACTORS FOR UNITS OF THERMAL CONDUCTIVITY

MULTIPLY

by appropriate factor to OBTAIN→	$Btu_{IT}\ h^{-1}\ ft^{-1}\ F^{-1}$	Btu_{IT} in. $h^{-1}\ ft^{-2}\ F^{-1}$	$Btu_{th}\ h^{-1}\ ft^{-1}\ F^{-1}$	Btu_{th} in. $h^{-1}\ ft^{-2}\ F^{-1}$	$cal_{IT}\ s^{-1}\ cm^{-1}\ C^{-1}$	$cal_{th}\ s^{-1}\ cm^{-1}\ C^{-1}$
$Btu_{IT}\ h^{-1}\ ft^{-1}\ F^{-1}$	1	12	1.00067	12.0080	4.13379 x 10^{-3}	4.13656 x 10^{-3}
Btu_{IT} in. $h^{-1}\ ft^{-2}\ F^{-1}$	8.33333 x 10^{-2}	1	8.33891 x 10^{-2}	1.00067	3.44482 x 10^{-4}	3.44713 x 10^{-4}
$Btu_{th}\ h^{-1}\ ft^{-1}\ F^{-1}$	0.999331	11.9920	1	12	4.13102 x 10^{-3}	4.13379 x 10^{-3}
Btu_{th} in. $h^{-1}\ ft^{-2}\ F^{-1}$	8.32776 x 10^{-2}	0.999331	8.33333 x 10^{-2}	1	3.44252 x 10^{-4}	3.44482 x 10^{-4}
$cal_{IT}\ s^{-1}\ cm^{-1}\ C^{-1}$	2.41909 x 10^{2}	2.90291 x 10^{3}	2.42071 x 10^{2}	2.90485 x 10^{3}	1	1.00067
$cal_{th}\ s^{-1}\ cm^{-1}\ C^{-1}$	2.41747 x 10^{2}	2.90096 x 10^{3}	2.41909 x 10^{2}	2.90291 x 10^{3}	0.999331	1
$kcal_{th}\ h^{-1}\ m^{-1}\ C^{-1}$	0.671520	8.05824	0.671969	8.06363	2.77592 x 10^{-3}	2.77778 x 10^{-3}
$J\ s^{-1}\ cm^{-1}\ K^{-1}$	57.7789	6.93347 x 10^{2}	57.8176	6.93811 x 10^{2}	0.238846	0.239006
$W\ cm^{-1}\ K^{-1}$	57.7789	6.93347 x 10^{2}	57.8176	6.93811 x 10^{2}	0.238846	0.239006
$W\ m^{-1}\ K^{-1}$	0.577789	6.93347	0.578176	6.93811	2.38846 x 10^{-3}	2.39006 x 10^{-3}
$mW\ cm^{-1}\ K^{-1}$	5.77789 x 10^{-2}	0.693347	5.78176 x 10^{-2}	0.693811	2.38846 x 10^{-4}	2.39006 x 10^{-4}

MULTIPLY

by appropriate factor to OBTAIN→	$kcal_{th}\ h^{-1}\ m^{-1}\ C^{-1}$	$J\ s^{-1}\ cm^{-1}\ K^{-1}$	$W\ cm^{-1}\ K^{-1}$	$W\ m^{-1}\ K^{-1}$	$mW\ cm^{-1}\ K^{-1}$
$Btu_{IT}\ h^{-1}\ ft^{-1}\ F^{-1}$	1.48916	1.73073 x 10^{-2}	1.73073 x 10^{-2}	1.73073	17.3073
Btu_{IT} in. $h^{-1}\ ft^{-2}\ F^{-1}$	0.124097	1.44228 x 10^{-3}	1.44228 x 10^{-3}	0.144228	1.44228
$Btu_{th}\ h^{-1}\ ft^{-1}\ F^{-1}$	1.48816	1.72958 x 10^{-2}	1.72958 x 10^{-2}	1.72958	17.2958
Btu_{th} in. $h^{-1}\ ft^{-2}\ F^{-1}$	0.124014	1.44131 x 10^{-3}	1.44131 x 10^{-3}	0.144131	1.44131
$cal_{IT}\ s^{-1}\ cm^{-1}\ C^{-1}$	3.60241 x 10^{2}	4.1868	4.1868	4.1868 x 10^{2}	4.1868 x 10^{3}
$cal_{th}\ s^{-1}\ cm^{-1}\ C^{-1}$	3.6 x 10^{2}	4.184	4.184	4.184 x 10^{2}	4.184 x 10^{3}
$kcal_{th}\ h^{-1}\ m^{-1}\ C^{-1}$	1	1.16222 x 10^{-2}	1.16222 x 10^{-2}	1.16222	11.6222
$J\ s^{-1}\ cm^{-1}\ K^{-1}$	86.0421	1	1	1 x 10^{2}	1 x 10^{3}
$W\ cm^{-1}\ K^{-1}$	86.0421	1	1	1 x 10^{2}	1 x 10^{3}
$W\ m^{-1}\ K^{-1}$	0.860421	1 x 10^{-2}	1 x 10^{-2}	1	10
$mW\ cm^{-1}\ K^{-1}$	8.60421 x 10^{-2}	1 x 10^{-3}	1 x 10^{-3}	0.1	1

Reproduced by permission of the publishers and copyright owners of the 1967 ASME Steam Tables. Further data and information on the thermodynamic and transport properties of steam and water are contained in the above ASME publication. It is obtainable from The American Society of Mechanical Engineers, United Engineering Center, 345 East 47th Street, New York, New York 10017.

Properties of Saturated Steam and Saturated Water

Temp. F	Press. psia	Volume, ft³/lbm			Enthalpy, Btu/lbm			Entropy, Btu/lbm ×F			Temp. F
		Water v_f	Evap. v_{fg}	Steam v_g	Water h_f	Evap. h_{fg}	Steam h_g	Water s_f	Evap. s_{fg}	Steam s_g	
705.47	3208.2	0.05078	0.00000	0.05078	906.0	0.0	906.0	1.0612	0.0000	1.0612	705.47
705.0	3198.3	0.04427	0.01304	0.05730	873.0	61.4	934.4	1.0329	0.0527	1.0856	705.0
704.5	3187.8	0.04233	0.01822	0.06055	861.9	85.3	947.2	1.0234	0.0732	1.0967	704.5
704.0	3177.2	0.04108	0.02192	0.06300	854.2	102.0	956.2	1.0169	0.0876	1.1046	704.0
703.5	3166.8	0.04015	0.02489	0.06504	848.2	115.2	963.5	1.0118	0.0991	1.1109	703.5
703.0	3156.3	0.03940	0.02744	0.06684	843.2	126.4	969.6	1.0076	0.1087	1.1163	703.0
702.5	3145.9	0.03878	0.02969	0.06847	838.9	136.1	974.9	1.0039	0.1171	1.1210	702.5
702.0	3135.5	0.03824	0.03173	0.06997	835.0	144.7	979.7	1.0006	0.1246	1.1252	702.0
701.5	3125.2	0.03777	0.03361	0.07138	831.5	152.6	984.0	0.9977	0.1314	1.1291	701.5
701.0	3114.9	0.03735	0.03536	0.07271	828.2	159.8	988.0	0.9949	0.1377	1.1326	701.0
700.5	3104.6	0.03697	0.03701	0.07397	825.2	166.5	991.7	0.9924	0.1435	1.1359	700.5
700.0	3094.3	0.03662	0.03857	0.07519	822.4	172.7	995.2	0.9901	0.1490	1.1390	700.0
699.0	3073.9	0.03600	0.04149	0.07749	817.3	184.2	1001.5	0.9858	0.1590	1.1447	699.0
698.0	3053.6	0.03546	0.04420	0.07966	812.6	194.6	1007.2	0.9818	0.1681	1.1499	698.0
697.0	3033.5	0.03498	0.04674	0.08172	808.4	204.0	1012.4	0.9783	0.1764	1.1547	697.0
696.0	3013.4	0.03455	0.04916	0.08371	804.4	212.8	1017.2	0.9749	0.1841	1.1591	696.0
695.0	2993.5	0.03415	0.05147	0.08563	800.6	221.0	1021.7	0.9718	0.1914	1.1632	695.0
694.0	2973.7	0.03379	0.05370	0.08749	797.1	228.8	1025.9	0.9689	0.1983	1.1671	694.0
693.0	2954.0	0.03345	0.05587	0.08931	793.8	236.1	1029.9	0.9660	0.2048	1.1708	693.0
692.0	2934.5	0.03313	0.05797	0.09110	790.5	243.1	1033.6	0.9634	0.2110	1.1744	692.0
690.0	2895.7	0.03256	0.06203	0.09459	784.5	256.1	1040.6	0.9583	0.2227	1.1810	690.0
688.0	2857.4	0.03204	0.06595	0.09799	778.8	268.2	1047.0	0.9535	0.2337	1.1872	688.0
686.0	2819.5	0.03157	0.06976	0.10133	773.4	279.5	1052.9	0.9490	0.2439	1.1930	686.0
684.0	2782.1	0.03114	0.07349	0.10463	768.2	290.2	1058.4	0.9447	0.2537	1.1984	684.0
682.0	2745.1	0.03074	0.07716	0.10790	763.3	300.4	1063.6	0.9406	0.2631	1.2036	682.0
680.0	2708.6	0.03037	0.08080	0.11117	758.5	310.1	1068.5	0.9365	0.2720	1.2086	680.0
678.0	2672.5	0.03002	0.08440	0.11442	753.8	319.4	1073.2	0.9326	0.2807	1.2133	678.0
676.0	2636.8	0.02970	0.08799	0.11769	749.2	328.5	1077.6	0.9287	0.2892	1.2179	676.0
674.0	2601.5	0.02939	0.09156	0.12096	744.7	337.2	1081.9	0.9249	0.2974	1.2223	674.0
672.0	2566.6	0.02911	0.09514	0.12424	740.2	345.7	1085.9	0.9212	0.3054	1.2266	672.0
670.0	2532.2	0.02884	0.09871	0.12755	735.8	354.0	1089.8	0.9174	0.3133	1.2307	670.0
668.0	2498.1	0.02858	0.10229	0.13087	731.5	362.1	1093.5	0.9137	0.3210	1.2347	668.0
666.0	2464.4	0.02834	0.10588	0.13421	727.1	370.0	1097.1	0.9100	0.3286	1.2387	666.0
664.0	2431.1	0.02811	0.10947	0.13757	722.9	377.7	1100.6	0.9064	0.3361	1.2425	664.0
662.0	2398.2	0.02789	0.11306	0.14095	718.8	385.1	1103.9	0.9028	0.3434	1.2462	662.0
660.0	2365.7	0.02768	0.11663	0.14431	714.9	392.1	1107.0	0.8995	0.3502	1.2498	660.0
658.0	2333.5	0.02748	0.12023	0.14771	711.1	399.0	1110.1	0.8963	0.3570	1.2533	658.0
656.0	2301.7	0.02728	0.12387	0.15115	707.4	405.7	1113.1	0.8931	0.3637	1.2567	656.0
654.0	2270.3	0.02709	0.12754	0.15463	703.7	412.2	1115.9	0.8899	0.3702	1.2601	654.0
652.0	2239.2	0.02691	0.13124	0.15816	700.0	418.7	1118.7	0.8868	0.3767	1.2634	652.0
650.0	2208.4	0.02674	0.13499	0.16173	696.4	425.0	1121.4	0.8837	0.3830	1.2667	650.0
648.0	2178.1	0.02657	0.13876	0.16534	692.9	431.1	1124.0	0.8806	0.3893	1.2699	648.0
646.0	2148.0	0.02641	0.14258	0.16899	689.4	437.2	1126.6	0.8776	0.3954	1.2730	646.0
644.0	2118.3	0.02625	0.14644	0.17269	685.9	443.1	1129.0	0.8746	0.4015	1.2761	644.0
642.0	2088.9	0.02610	0.15033	0.17643	682.5	448.9	1131.4	0.8716	0.4075	1.2791	642.0
640.0	2059.9	0.02595	0.15427	0.18021	679.1	454.6	1133.7	0.8686	0.4134	1.2821	640.0
638.0	2031.2	0.02580	0.15824	0.18405	675.8	460.2	1136.0	0.8657	0.4193	1.2850	638.0
636.0	2002.8	0.02566	0.16226	0.18792	672.4	465.7	1138.1	0.8628	0.4251	1.2879	636.0
634.0	1974.7	0.02553	0.16633	0.19185	669.1	471.1	1140.2	0.8599	0.4307	1.2907	634.0
632.0	1947.0	0.02539	0.17044	0.19583	665.9	476.4	1142.2	0.8571	0.4364	1.2934	632.0
630.0	1919.5	0.02526	0.17459	0.19986	662.7	481.6	1144.2	0.8542	0.4419	1.2962	630.0
628.0	1892.4	0.02514	0.17880	0.20394	659.5	486.7	1146.1	0.8514	0.4474	1.2988	628.0
626.0	1865.6	0.02501	0.18306	0.20807	656.3	491.7	1148.0	0.8486	0.4529	1.3015	626.0
624.0	1839.0	0.02489	0.18737	0.21226	653.1	496.6	1149.8	0.8458	0.4583	1.3041	624.0
622.0	1812.8	0.02477	0.19173	0.21650	650.0	501.5	1151.5	0.8430	0.4636	1.3066	622.0
620.0	1786.9	0.02466	0.19615	0.22081	646.9	506.3	1153.2	0.8403	0.4689	1.3092	620.0
618.0	1761.2	0.02455	0.20063	0.22517	643.8	511.0	1154.8	0.8375	0.4742	1.3117	618.0
616.0	1735.9	0.02444	0.20516	0.22960	640.8	515.6	1156.4	0.8348	0.4794	1.3141	616.0
614.0	1710.8	0.02433	0.20976	0.23409	637.8	520.2	1158.0	0.8321	0.4845	1.3166	614.0
612.0	1686.1	0.02422	0.21442	0.23865	634.8	524.7	1159.5	0.8294	0.4896	1.3190	612.0
610.0	1661.6	0.02412	0.21915	0.24327	631.8	529.2	1160.9	0.8267	0.4947	1.3214	610.0
608.0	1637.3	0.02402	0.22394	0.24796	628.8	533.6	1162.4	0.8240	0.4997	1.3238	608.0
606.0	1613.4	0.02392	0.22881	0.25273	625.9	537.9	1163.8	0.8214	0.5048	1.3261	606.0
604.0	1589.7	0.02382	0.23374	0.25757	622.9	542.2	1165.1	0.8187	0.5097	1.3284	604.0
602.0	1566.3	0.02373	0.23875	0.26248	620.0	546.4	1166.4	0.8161	0.5147	1.3307	602.0
600.0	1543.2	0.02364	0.24384	0.26747	617.1	550.6	1167.7	0.8134	0.5196	1.3330	600.0
598.0	1520.4	0.02354	0.24900	0.27255	614.3	554.7	1169.0	0.8108	0.5245	1.3353	598.0
596.0	1497.8	0.02345	0.25425	0.27770	611.4	558.8	1170.2	0.8082	0.5293	1.3375	596.0
594.0	1475.4	0.02337	0.25958	0.28294	608.6	562.8	1171.4	0.8056	0.5342	1.3398	594.0
592.0	1453.3	0.02328	0.26499	0.28827	605.7	566.8	1172.6	0.8030	0.5390	1.3420	592.0
590.0	1431.5	0.02319	0.27049	0.29368	602.9	570.8	1173.7	0.8004	0.5437	1.3442	590.0
588.0	1410.0	0.02311	0.27608	0.29919	600.1	574.7	1174.8	0.7978	0.5485	1.3464	588.0
586.0	1388.6	0.02303	0.28176	0.30478	597.3	578.5	1175.9	0.7953	0.5532	1.3485	586.0
584.0	1367.6	0.02295	0.28753	0.31048	594.6	582.4	1176.9	0.7927	0.5580	1.3507	584.0
582.0	1346.7	0.02287	0.29340	0.31627	591.8	586.1	1178.0	0.7902	0.5627	1.3528	582.0
580.0	1326.2	0.02279	0.29937	0.32216	589.1	589.9	1179.0	0.7876	0.5673	1.3550	580.0

Quantities for saturated liquid v_f h_f s_f

Quantities for saturated vapor v_g h_g s_g

Increment for evaporation v_{fg} h_{fg} s_{fg}

STEAM TABLES (Continued)

Properties of Saturated Steam and Saturated Water

Temp. F	Press. psia	Volume, ft³/lbm Water v_f	Evap. v_{fg}	Steam v_g	Enthalpy, Btu/lbm Water h_f	Evap. h_{fg}	Steam h_g	Entropy, Btu/lbm ×F Water s_f	Evap. s_{fg}	Steam s_g	Temp. F
580.0	1326.17	0.02279	0.29937	0.32216	589.1	589.9	1179.0	0.7876	0.5673	1.3550	580.0
578.0	1305.84	0.02271	0.30544	0.32816	586.4	593.6	1179.9	0.7851	0.5720	1.3571	578.0
576.0	1285.74	0.02264	0.31162	0.33426	583.7	597.2	1180.9	0.7825	0.5766	1.3592	576.0
574.0	1265.89	0.02256	0.31790	0.34046	581.0	600.9	1181.8	0.7800	0.5813	1.3613	574.0
572.0	1246.26	0.02249	0.32429	0.34678	578.3	604.5	1182.7	0.7775	0.5859	1.3634	572.0
570.0	1226.88	0.02242	0.33079	0.35321	575.6	608.0	1183.6	0.7750	0.5905	1.3654	570.0
568.0	1207.72	0.02235	0.33741	0.35975	572.9	611.5	1184.5	0.7725	0.5950	1.3675	568.0
566.0	1188.80	0.02228	0.34414	0.36642	570.3	615.0	1185.3	0.7699	0.5996	1.3696	566.0
564.0	1170.10	0.02221	0.35099	0.37320	567.6	618.5	1186.1	0.7674	0.6041	1.3716	564.0
562.0	1151.63	0.02214	0.35797	0.38011	565.0	621.9	1186.9	0.7650	0.6087	1.3736	562.0
560.0	1133.38	0.02207	0.36507	0.38714	562.4	625.3	1187.7	0.7625	0.6132	1.3757	560.0
558.0	1115.36	0.02201	0.37230	0.39431	559.8	628.6	1188.4	0.7600	0.6177	1.3777	558.0
556.0	1097.55	0.02194	0.37966	0.40160	557.2	632.0	1189.2	0.7575	0.6222	1.3797	556.0
554.0	1079.96	0.02188	0.38715	0.40903	554.6	635.3	1189.9	0.7550	0.6267	1.3817	554.0
552.0	1062.59	0.02182	0.39479	0.41660	552.0	638.5	1190.6	0.7525	0.6311	1.3837	552.0
550.0	1045.43	0.02176	0.40256	0.42432	549.5	641.8	1191.2	0.7501	0.6356	1.3856	550.0
548.0	1028.49	0.02169	0.41048	0.43217	546.9	645.0	1191.9	0.7476	0.6400	1.3876	548.0
546.0	1011.75	0.02163	0.41855	0.44018	544.4	648.1	1192.5	0.7451	0.6445	1.3896	546.0
544.0	995.22	0.02157	0.42677	0.44834	541.8	651.3	1193.1	0.7427	0.6489	1.3915	544.0
542.0	978.90	0.02151	0.43514	0.45665	539.3	654.4	1193.7	0.7402	0.6533	1.3935	542.0
540.0	962.79	0.02146	0.44367	0.46513	536.8	657.5	1194.3	0.7378	0.6577	1.3954	540.0
538.0	946.88	0.02140	0.45237	0.47377	534.2	660.6	1194.8	0.7353	0.6621	1.3974	538.0
536.0	931.17	0.02134	0.46123	0.48257	531.7	663.6	1195.4	0.7329	0.6665	1.3993	536.0
534.0	915.66	0.02129	0.47026	0.49155	529.2	666.6	1195.9	0.7304	0.6708	1.4012	534.0
532.0	900.34	0.02123	0.47947	0.50070	526.8	669.6	1196.4	0.7280	0.6752	1.4032	532.0
530.0	885.23	0.02118	0.48886	0.51004	524.3	672.6	1196.9	0.7255	0.6796	1.4051	530.0
528.0	870.31	0.02112	0.49843	0.51955	521.8	675.5	1197.3	0.7231	0.6839	1.4070	528.0
526.0	855.58	0.02107	0.50819	0.52926	519.3	678.4	1197.8	0.7206	0.6883	1.4089	526.0
524.0	841.04	0.02102	0.51814	0.53916	516.9	681.3	1198.2	0.7182	0.6926	1.4108	524.0
522.0	826.69	0.02097	0.52829	0.54926	514.4	684.2	1198.6	0.7158	0.6969	1.4127	522.0
520.0	812.53	0.02091	0.53864	0.55956	512.0	687.0	1199.0	0.7133	0.7013	1.4146	520.0
518.0	798.55	0.02086	0.54920	0.57006	509.6	689.9	1199.4	0.7109	0.7056	1.4165	518.0
516.0	784.76	0.02081	0.55997	0.58079	507.1	692.7	1199.8	0.7085	0.7099	1.4183	516.0
514.0	771.15	0.02076	0.57096	0.59173	504.7	695.4	1200.2	0.7060	0.7142	1.4202	514.0
512.0	757.72	0.02072	0.58218	0.60289	502.3	698.2	1200.5	0.7036	0.7185	1.4221	512.0
510.0	744.47	0.02067	0.59362	0.61429	499.9	700.9	1200.8	0.7012	0.7228	1.4240	510.0
508.0	731.40	0.02062	0.60530	0.62592	497.5	703.7	1201.1	0.6987	0.7271	1.4258	508.0
506.0	718.50	0.02057	0.61722	0.63779	495.1	706.3	1201.4	0.6963	0.7314	1.4277	506.0
504.0	705.78	0.02053	0.62938	0.64991	492.7	709.0	1201.7	0.6939	0.7357	1.4296	504.0
502.0	693.23	0.02048	0.64180	0.66228	490.3	711.7	1202.0	0.6915	0.7400	1.4314	502.0
500.0	680.86	0.02043	0.65448	0.67492	487.9	714.3	1202.2	0.6890	0.7443	1.4333	500.0
498.0	668.65	0.02039	0.66743	0.68782	485.6	716.9	1202.5	0.6866	0.7486	1.4352	498.0
496.0	656.61	0.02034	0.68065	0.70100	483.2	719.5	1202.7	0.6842	0.7528	1.4370	496.0
494.0	644.73	0.02030	0.69415	0.71445	480.8	722.1	1202.9	0.6818	0.7571	1.4389	494.0
492.0	633.03	0.02026	0.70794	0.72820	478.5	724.6	1203.1	0.6793	0.7614	1.4407	492.0
490.0	621.48	0.02021	0.72203	0.74224	476.1	727.2	1203.3	0.6769	0.7657	1.4426	490.0
488.0	610.10	0.02017	0.73641	0.75658	473.8	729.7	1203.5	0.6745	0.7700	1.4444	488.0
486.0	598.87	0.02013	0.75111	0.77124	471.5	732.2	1203.7	0.6721	0.7742	1.4463	486.0
484.0	587.81	0.02009	0.76613	0.78622	469.1	734.7	1203.8	0.6696	0.7785	1.4481	484.0
482.0	576.90	0.02004	0.78148	0.80152	466.8	737.2	1204.0	0.6672	0.7828	1.4500	482.0
480.0	566.15	0.02000	0.79716	0.81717	464.5	739.6	1204.1	0.6648	0.7871	1.4518	480.0
478.0	555.55	0.01996	0.81319	0.83315	462.2	742.1	1204.2	0.6624	0.7913	1.4537	478.0
476.0	545.11	0.01992	0.82958	0.84950	459.9	744.5	1204.3	0.6599	0.7956	1.4555	476.0
474.0	534.81	0.01988	0.84632	0.86621	457.5	746.9	1204.4	0.6575	0.7999	1.4574	474.0
472.0	524.67	0.01984	0.86345	0.88329	455.2	749.3	1204.5	0.6551	0.8042	1.4592	472.0
470.0	514.67	0.01980	0.88095	0.90076	452.9	751.6	1204.6	0.6527	0.8084	1.4611	470.0
468.0	504.83	0.01976	0.89885	0.91862	450.7	754.0	1204.6	0.6502	0.8127	1.4629	468.0
466.0	495.12	0.01973	0.91716	0.93689	448.4	756.3	1204.7	0.6478	0.8170	1.4648	466.0
464.0	485.56	0.01969	0.93588	0.95557	446.1	758.6	1204.7	0.6454	0.8213	1.4667	464.0
462.0	476.14	0.01965	0.95504	0.97469	443.8	761.0	1204.8	0.6429	0.8256	1.4685	462.0
460.0	466.87	0.01961	0.97463	0.99424	441.5	763.2	1204.8	0.6405	0.8299	1.4704	460.0
458.0	457.73	0.01958	0.99467	1.01425	439.3	765.5	1204.8	0.6381	0.8342	1.4722	458.0
456.0	448.73	0.01954	1.01518	1.03472	437.0	767.8	1204.8	0.6356	0.8385	1.4741	456.0
454.0	439.87	0.01950	1.03616	1.05567	434.7	770.0	1204.8	0.6332	0.8428	1.4759	454.0
452.0	431.14	0.01947	1.05764	1.07711	432.5	772.3	1204.8	0.6308	0.8471	1.4778	452.0
450.0	422.55	0.01943	1.07962	1.09905	430.2	774.5	1204.7	0.6283	0.8514	1.4797	450.0
448.0	414.09	0.01940	1.10212	1.12152	428.0	776.7	1204.7	0.6259	0.8557	1.4815	448.0
446.0	405.76	0.01936	1.12515	1.14452	425.7	778.9	1204.6	0.6234	0.8600	1.4834	446.0
444.0	397.56	0.01933	1.14874	1.16806	423.5	781.1	1204.6	0.6210	0.8643	1.4853	444.0
442.0	389.49	0.01929	1.17288	1.19217	421.3	783.2	1204.5	0.6185	0.8686	1.4872	442.0
440.0	381.54	0.01926	1.19761	1.21687	419.0	785.4	1204.4	0.6161	0.8729	1.4890	440.0
438.0	373.72	0.01923	1.22293	1.24216	416.8	787.5	1204.3	0.6136	0.8773	1.4909	438.0
436.0	366.03	0.01919	1.24887	1.26806	414.6	789.7	1204.2	0.6112	0.8816	1.4928	436.0
434.0	358.46	0.01916	1.27544	1.29460	412.4	791.8	1204.1	0.6087	0.8859	1.4947	434.0
432.0	351.00	0.01913	1.30266	1.32179	410.1	793.9	1204.0	0.6063	0.8903	1.4966	432.0
430.0	343.67	0.01909	1.33055	1.34965	407.9	796.0	1203.9	0.6038	0.8946	1.4985	430.0

STEAM TABLES (Continued)

Properties of Saturated Steam and Saturated Water

Temp. F	Press. psia	Volume, ft³/lbm Water v_f	Evap. v_{fg}	Steam v_g	Enthalpy, Btu/lbm Water h_f	Evap. h_{fg}	Steam h_g	Entropy, Btu/lbm ×F Water s_f	Evap. s_{fg}	Steam s_g	Temp. F
430.0	343.674	0.01909	1.3306	1.3496	407.9	796.0	1203.9	0.6038	0.8946	1.4985	430.0
428.0	336.463	0.01906	1.3591	1.3782	405.7	798.0	1203.7	0.6014	0.8990	1.5004	428.0
426.0	329.369	0.01903	1.3884	1.4075	403.5	800.1	1203.6	0.5989	0.9034	1.5023	426.0
424.0	322.391	0.01900	1.4184	1.4374	401.3	802.2	1203.5	0.5964	0.9077	1.5042	424.0
422.0	315.529	0.01897	1.4492	1.4682	399.1	804.2	1203.3	0.5940	0.9121	1.5061	422.0
420.0	308.780	0.01894	1.4808	1.4997	396.9	806.2	1203.1	0.5915	0.9165	1.5080	420.0
418.0	302.143	0.01890	1.5131	1.5320	394.7	808.2	1202.9	0.5890	0.9209	1.5099	418.0
416.0	295.617	0.01887	1.5463	1.5651	392.5	810.2	1202.8	0.5866	0.9253	1.5118	416.0
414.0	289.201	0.01884	1.5803	1.5991	390.3	812.2	1202.6	0.5841	0.9297	1.5137	414.0
412.0	282.894	0.01881	1.6152	1.6340	388.1	814.2	1202.4	0.5816	0.9341	1.5157	412.0
410.0	276.694	0.01878	1.6510	1.6697	386.0	816.2	1202.1	0.5791	0.9385	1.5176	410.0
408.0	270.600	0.01875	1.6877	1.7064	383.8	818.2	1201.9	0.5766	0.9429	1.5195	408.0
406.0	264.611	0.01872	1.7253	1.7441	381.6	820.1	1201.7	0.5742	0.9473	1.5215	406.0
404.0	258.725	0.01870	1.7640	1.7827	379.4	822.0	1201.5	0.5717	0.9518	1.5234	404.0
402.0	252.942	0.01867	1.8037	1.8223	377.3	824.0	1201.2	0.5692	0.9562	1.5254	402.0
400.0	247.259	0.01864	1.8444	1.8630	375.1	825.9	1201.0	0.5667	0.9607	1.5274	400.0
398.0	241.677	0.01861	1.8862	1.9048	372.9	827.8	1200.7	0.5642	0.9651	1.5293	398.0
396.0	236.193	0.01858	1.9291	1.9477	370.8	829.7	1200.4	0.5617	0.9696	1.5313	396.0
394.0	230.807	0.01855	1.9731	1.9917	368.6	831.6	1200.2	0.5592	0.9741	1.5333	394.0
392.0	225.516	0.01853	2.0184	2.0369	366.5	833.4	1199.9	0.5567	0.9786	1.5352	392.0
390.0	220.321	0.01850	2.0649	2.0833	364.3	835.3	1199.6	0.5542	0.9831	1.5372	390.0
388.0	215.220	0.01847	2.1126	2.1311	362.2	837.2	1199.3	0.5516	0.9876	1.5392	388.0
386.0	210.211	0.01844	2.1616	2.1801	360.0	839.0	1199.0	0.5491	0.9921	1.5412	386.0
384.0	205.294	0.01842	2.2120	2.2304	357.9	840.8	1198.7	0.5466	0.9966	1.5432	384.0
382.0	200.467	0.01839	2.2638	2.2821	355.7	842.7	1198.4	0.5441	1.0012	1.5452	382.0
380.0	195.729	0.01836	2.3170	2.3353	353.6	844.5	1198.0	0.5416	1.0057	1.5473	380.0
378.0	191.080	0.01834	2.3716	2.3900	351.4	846.3	1197.7	0.5390	1.0103	1.5493	378.0
376.0	186.517	0.01831	2.4279	2.4462	349.3	848.1	1197.4	0.5365	1.0148	1.5513	376.0
374.0	182.040	0.01829	2.4857	2.5039	347.2	849.8	1197.0	0.5340	1.0194	1.5534	374.0
372.0	177.648	0.01826	2.5451	2.5633	345.0	851.6	1196.7	0.5314	1.0240	1.5554	372.0
370.0	173.339	0.01823	2.6062	2.6244	342.9	853.4	1196.3	0.5289	1.0286	1.5575	370.0
368.0	169.113	0.01821	2.6691	2.6873	340.8	855.1	1195.9	0.5263	1.0332	1.5595	368.0
366.0	164.968	0.01818	2.7337	2.7519	338.7	856.9	1195.6	0.5238	1.0378	1.5616	366.0
364.0	160.903	0.01816	2.8002	2.8184	336.5	858.6	1195.2	0.5212	1.0424	1.5637	364.0
362.0	156.917	0.01813	2.8687	2.8868	334.4	860.4	1194.8	0.5187	1.0471	1.5658	362.0
360.0	153.010	0.01811	2.9392	2.9573	332.3	862.1	1194.4	0.5161	1.0517	1.5678	360.0
358.0	149.179	0.01809	3.0117	3.0298	330.2	863.8	1194.0	0.5135	1.0564	1.5699	358.0
356.0	145.424	0.01806	3.0863	3.1044	328.1	865.5	1193.6	0.5110	1.0611	1.5721	356.0
354.0	141.744	0.01804	3.1632	3.1812	326.0	867.2	1193.2	0.5084	1.0658	1.5742	354.0
352.0	138.138	0.01801	3.2423	3.2603	323.9	868.9	1192.7	0.5058	1.0705	1.5763	352.0
350.0	134.604	0.01799	3.3238	3.3418	321.8	870.6	1192.3	0.5032	1.0752	1.5784	350.0
348.0	131.142	0.01797	3.4078	3.4258	319.7	872.2	1191.9	0.5006	1.0799	1.5806	348.0
346.0	127.751	0.01794	3.4943	3.5122	317.6	873.9	1191.4	0.4980	1.0847	1.5827	346.0
344.0	124.430	0.01792	3.5834	3.6013	315.5	875.5	1191.0	0.4954	1.0894	1.5849	344.0
342.0	121.177	0.01790	3.6752	3.6931	313.4	877.2	1190.5	0.4928	1.0942	1.5871	342.0
340.0	117.992	0.01787	3.7699	3.7878	311.3	878.8	1190.1	0.4902	1.0990	1.5892	340.0
338.0	114.873	0.01785	3.8675	3.8853	309.2	880.5	1189.6	0.4876	1.1038	1.5914	338.0
336.0	111.820	0.01783	3.9681	3.9859	307.1	882.1	1189.1	0.4850	1.1086	1.5936	336.0
334.0	108.832	0.01781	4.0718	4.0896	305.0	883.7	1188.7	0.4824	1.1134	1.5958	334.0
332.0	105.907	0.01779	4.1788	4.1966	302.9	885.3	1188.2	0.4798	1.1183	1.5981	332.0
330.0	103.045	0.01776	4.2892	4.3069	300.8	886.9	1187.7	0.4772	1.1231	1.6003	330.0
328.0	100.245	0.01774	4.4030	4.4208	298.7	888.5	1187.2	0.4745	1.1280	1.6025	328.0
326.0	97.506	0.01772	4.5205	4.5382	296.6	890.1	1186.7	0.4719	1.1329	1.6048	326.0
324.0	94.826	0.01770	4.6418	4.6595	294.6	891.6	1186.2	0.4692	1.1378	1.6071	324.0
322.0	92.205	0.01768	4.7669	4.7846	292.5	893.2	1185.7	0.4666	1.1427	1.6093	322.0
320.0	89.643	0.01766	4.8961	4.9138	290.4	894.8	1185.2	0.4640	1.1477	1.6116	320.0
318.0	87.137	0.01764	5.0295	5.0471	288.3	896.3	1184.7	0.4613	1.1526	1.6139	318.0
316.0	84.688	0.01761	5.1673	5.1849	286.3	897.9	1184.1	0.4586	1.1576	1.6162	316.0
314.0	82.293	0.01759	5.3096	5.3272	284.2	899.4	1183.6	0.4560	1.1626	1.6185	314.0
312.0	79.953	0.01757	5.4566	5.4742	282.1	901.0	1183.1	0.4533	1.1676	1.6209	312.0
310.0	77.667	0.01755	5.6085	5.6260	280.0	902.5	1182.5	0.4506	1.1726	1.6232	310.0
308.0	75.433	0.01753	5.7655	5.7830	278.0	904.0	1182.0	0.4479	1.1776	1.6256	308.0
306.0	73.251	0.01751	5.9277	5.9452	275.9	905.5	1181.4	0.4453	1.1827	1.6279	306.0
304.0	71.119	0.01749	6.0955	6.1130	273.8	907.0	1180.9	0.4426	1.1877	1.6303	304.0
302.0	69.038	0.01747	6.2689	6.2864	271.8	908.5	1180.3	0.4399	1.1928	1.6327	302.0
300.0	67.005	0.01745	6.4483	6.4658	269.7	910.0	1179.7	0.4372	1.1979	1.6351	300.0
298.0	65.021	0.01743	6.6339	6.6513	267.7	911.5	1179.2	0.4345	1.2031	1.6375	298.0
296.0	63.084	0.01741	6.8259	6.8433	265.6	913.0	1178.6	0.4317	1.2082	1.6400	296.0
294.0	61.194	0.01739	7.0245	7.0419	263.5	914.5	1178.0	0.4290	1.2134	1.6424	294.0
292.0	59.350	0.01738	7.2301	7.2475	261.5	915.9	1177.4	0.4263	1.2186	1.6449	292.0
290.0	57.550	0.01736	7.4430	7.4603	259.4	917.4	1176.8	0.4236	1.2238	1.6473	290.0
288.0	55.795	0.01734	7.6634	7.6807	257.4	918.8	1176.2	0.4208	1.2290	1.6498	288.0
286.0	54.083	0.01732	7.8916	7.9089	255.3	920.3	1175.6	0.4181	1.2342	1.6523	286.0
284.0	52.414	0.01730	8.1280	8.1453	253.3	921.7	1175.0	0.4154	1.2395	1.6548	284.0
282.0	50.786	0.01728	8.3729	8.3902	251.2	923.2	1174.4	0.4126	1.2448	1.6574	282.0
280.0	49.200	0.01726	8.6267	8.6439	249.2	924.6	1173.8	0.4098	1.2501	1.6599	280.0

STEAM TABLES (Continued)

Properties of Saturated Steam and Saturated Water

Temp. F	Press. psia	Volume, ft³/lbm Water v_f	Evap. v_{fg}	Steam v_g	Enthalpy, Btu/lbm Water h_f	Evap. h_{fg}	Steam h_g	Entropy, Btu/lbm ×F Water s_f	Evap. s_{fg}	Steam s_g	Temp. F
280.0	49.200	0.017264	8.627	8.644	249.17	924.6	1173.8	0.4098	1.2501	1.6599	280.0
278.0	47.653	0.017246	8.890	8.907	247.13	926.0	1173.2	0.4071	1.2554	1.6625	278.0
276.0	46.147	0.017228	9.162	9.180	245.08	927.5	1172.5	0.4043	1.2607	1.6650	276.0
274.0	44.678	0.017210	9.445	9.462	243.03	928.9	1171.9	0.4015	1.2661	1.6676	274.0
272.0	43.249	0.017193	9.738	9.755	240.99	930.3	1171.3	0.3987	1.2715	1.6702	272.0
270.0	41.856	0.017175	10.042	10.060	238.95	931.7	1170.6	0.3960	1.2769	1.6729	270.0
268.0	40.500	0.017157	10.358	10.375	236.91	933.1	1170.0	0.3932	1.2823	1.6755	268.0
266.0	39.179	0.017140	10.685	10.703	234.87	934.5	1169.3	0.3904	1.2878	1.6781	266.0
264.0	37.894	0.017123	11.025	11.042	232.83	935.9	1168.7	0.3876	1.2933	1.6808	264.0
262.0	36.644	0.017106	11.378	11.395	230.79	937.3	1168.0	0.3847	1.2988	1.6835	262.0
260.0	35.427	0.017089	11.745	11.762	228.76	938.6	1167.4	0.3819	1.3043	1.6862	260.0
258.0	34.243	0.017072	12.125	12.142	226.72	940.0	1166.7	0.3791	1.3098	1.6889	258.0
256.0	33.091	0.017055	12.520	12.538	224.69	941.4	1166.1	0.3763	1.3154	1.6917	256.0
254.0	31.972	0.017039	12.931	12.948	222.65	942.7	1165.4	0.3734	1.3210	1.6944	254.0
252.0	30.883	0.017022	13.358	13.375	220.62	944.1	1164.7	0.3706	1.3266	1.6972	252.0
250.0	29.825	0.017006	13.802	13.819	218.59	945.4	1164.0	0.3677	1.3323	1.7000	250.0
248.0	28.796	0.016990	14.264	14.281	216.56	946.8	1163.4	0.3649	1.3379	1.7028	248.0
246.0	27.797	0.016974	14.744	14.761	214.53	948.1	1162.7	0.3620	1.3436	1.7056	246.0
244.0	26.826	0.016958	15.243	15.260	212.50	949.5	1162.0	0.3591	1.3494	1.7085	244.0
242.0	25.883	0.016942	15.763	15.780	210.48	950.8	1161.3	0.3562	1.3551	1.7113	242.0
240.0	24.968	0.016926	16.304	16.321	208.45	952.1	1160.6	0.3533	1.3609	1.7142	240.0
238.0	24.079	0.016910	16.867	16.884	206.42	953.5	1159.9	0.3505	1.3667	1.7171	238.0
236.0	23.216	0.016895	17.454	17.471	204.40	954.8	1159.2	0.3476	1.3725	1.7201	236.0
234.0	22.379	0.016880	18.065	18.082	202.38	956.1	1158.5	0.3446	1.3784	1.7230	234.0
232.0	21.567	0.016864	18.701	18.718	200.35	957.4	1157.8	0.3417	1.3842	1.7260	232.0
230.0	20.779	0.016849	19.364	19.381	198.33	958.7	1157.1	0.3388	1.3902	1.7290	230.0
229.0	20.394	0.016842	19.707	19.723	197.32	959.4	1156.7	0.3373	1.3931	1.7305	229.0
228.0	20.015	0.016834	20.056	20.073	196.31	960.0	1156.3	0.3359	1.3961	1.7320	228.0
227.0	19.642	0.016827	20.413	20.429	195.30	960.7	1156.0	0.3344	1.3991	1.7335	227.0
226.0	19.274	0.016819	20.777	20.794	194.29	961.3	1155.6	0.3329	1.4021	1.7350	226.0
225.0	18.912	0.016812	21.149	21.166	193.28	962.0	1155.3	0.3315	1.4051	1.7365	225.0
224.0	18.556	0.016805	21.529	21.545	192.27	962.6	1154.9	0.3300	1.4081	1.7380	224.0
223.0	18.206	0.016797	21.917	21.933	191.26	963.3	1154.5	0.3285	1.4111	1.7396	223.0
222.0	17.860	0.016790	22.313	22.330	190.25	963.9	1154.2	0.3270	1.4141	1.7411	222.0
221.0	17.521	0.016783	22.718	22.735	189.24	964.6	1153.8	0.3255	1.4171	1.7427	221.0
220.0	17.186	0.016775	23.131	23.148	188.23	965.2	1153.4	0.3241	1.4201	1.7442	220.0
219.0	16.857	0.016768	23.554	23.571	187.22	965.8	1153.1	0.3226	1.4232	1.7458	219.0
218.0	16.533	0.016761	23.986	24.002	186.21	966.5	1152.7	0.3211	1.4262	1.7473	218.0
217.0	16.214	0.016754	24.427	24.444	185.21	967.1	1152.3	0.3196	1.4293	1.7489	217.0
216.0	15.901	0.016747	24.878	24.894	184.20	967.8	1152.0	0.3181	1.4323	1.7505	216.0
215.0	15.592	0.016740	25.338	25.355	183.19	968.4	1151.6	0.3166	1.4354	1.7520	215.0
214.0	15.289	0.016733	25.809	25.826	182.18	969.0	1151.2	0.3151	1.4385	1.7536	214.0
213.0	14.990	0.016726	26.290	26.307	181.17	969.7	1150.8	0.3136	1.4416	1.7552	213.0
212.0	14.696	0.016719	26.782	26.799	180.17	970.3	1150.5	0.3121	1.4447	1.7568	212.0
211.0	14.407	0.016712	27.285	27.302	179.16	970.9	1150.1	0.3106	1.4478	1.7584	211.0
210.0	14.123	0.016705	27.799	27.816	178.15	971.6	1149.7	0.3091	1.4509	1.7600	210.0
209.0	13.843	0.016698	28.324	28.341	177.14	972.2	1149.4	0.3076	1.4540	1.7616	209.0
208.0	13.568	0.016691	28.862	28.878	176.14	972.8	1149.0	0.3061	1.4571	1.7632	208.0
207.0	13.297	0.016684	29.411	29.428	175.13	973.5	1148.6	0.3046	1.4602	1.7649	207.0
206.0	13.031	0.016677	29.973	29.989	174.12	974.1	1148.2	0.3031	1.4634	1.7665	206.0
205.0	12.770	0.016670	30.547	30.564	173.12	974.7	1147.9	0.3016	1.4665	1.7681	205.0
204.0	12.512	0.016664	31.135	31.151	172.11	975.4	1147.5	0.3001	1.4697	1.7698	204.0
203.0	12.259	0.016657	31.736	31.752	171.10	976.0	1147.1	0.2986	1.4728	1.7714	203.0
202.0	12.011	0.016650	32.350	32.367	170.10	976.6	1146.7	0.2971	1.4760	1.7731	202.0
201.0	11.766	0.016643	32.979	32.996	169.09	977.2	1146.3	0.2955	1.4792	1.7747	201.0
200.0	11.526	0.016637	33.622	33.639	168.09	977.9	1146.0	0.2940	1.4824	1.7764	200.0
199.0	11.290	0.016630	34.280	34.297	167.08	978.5	1145.6	0.2925	1.4856	1.7781	199.0
198.0	11.058	0.016624	34.954	34.970	166.08	979.1	1145.2	0.2910	1.4888	1.7798	198.0
197.0	10.830	0.016617	35.643	35.659	165.07	979.7	1144.8	0.2894	1.4920	1.7814	197.0
196.0	10.605	0.016611	36.348	36.364	164.06	980.4	1144.4	0.2879	1.4952	1.7831	196.0
195.0	10.385	0.016604	37.069	37.086	163.06	981.0	1144.0	0.2864	1.4985	1.7848	195.0
194.0	10.168	0.016598	37.808	37.824	162.05	981.6	1143.7	0.2848	1.5017	1.7865	194.0
193.0	9.956	0.016591	38.564	38.580	161.05	982.2	1143.3	0.2833	1.5050	1.7882	193.0
192.0	9.747	0.016585	39.337	39.354	160.05	982.8	1142.9	0.2818	1.5082	1.7900	192.0
191.0	9.541	0.016578	40.130	40.146	159.04	983.5	1142.5	0.2802	1.5115	1.7917	191.0
190.0	9.340	0.016572	40.941	40.957	158.04	984.1	1142.1	0.2787	1.5148	1.7934	190.0
189.0	9.141	0.016566	41.771	41.787	157.03	984.7	1141.7	0.2771	1.5180	1.7952	189.0
188.0	8.947	0.016559	42.621	42.638	156.03	985.3	1141.3	0.2756	1.5213	1.7969	188.0
187.0	8.756	0.016553	43.492	43.508	155.02	985.9	1140.9	0.2740	1.5246	1.7987	187.0
186.0	8.568	0.016547	44.383	44.400	154.02	986.5	1140.5	0.2725	1.5279	1.8004	186.0
185.0	8.384	0.016541	45.297	45.313	153.02	987.1	1140.2	0.2709	1.5313	1.8022	185.0
184.0	8.203	0.016534	46.232	46.249	152.01	987.8	1139.8	0.2694	1.5346	1.8040	184.0
183.0	8.025	0.016528	47.190	47.207	151.01	988.4	1139.4	0.2678	1.5379	1.8057	183.0
182.0	7.850	0.016522	48.172	48.189	150.01	989.0	1139.0	0.2662	1.5413	1.8075	182.0
181.0	7.679	0.016516	49.178	49.194	149.00	989.6	1138.6	0.2647	1.5446	1.8093	181.0
180.0	7.511	0.016510	50.208	50.225	148.00	990.2	1138.2	0.2631	1.5480	1.8111	180.0

STEAM TABLES (Continued)

Properties of Saturated Steam and Saturated Water

Temp. F	Press. psia	Volume, ft³/lbm Water v_f	Evap. v_{fg}	Steam v_g	Enthalpy, Btu/lbm Water h_f	Evap. h_{fg}	Steam h_g	Entropy, Btu/lbm×F Water s_f	Evap. s_{fg}	Steam s_g	Temp. F
180.0	7.5110	0.016510	50.21	50.22	148.00	990.2	1138.2	0.2631	1.5480	1.8111	180.0
179.0	7.3460	0.016504	51.26	51.28	147.00	990.8	1137.8	0.2615	1.5514	1.8129	179.0
178.0	7.1840	0.016498	52.35	52.36	145.99	991.4	1137.4	0.2600	1.5548	1.8147	178.0
177.0	7.0250	0.016492	53.46	53.47	144.99	992.0	1137.0	0.2584	1.5582	1.8166	177.0
176.0	6.8690	0.016486	54.59	54.61	143.99	992.6	1136.6	0.2568	1.5616	1.8184	176.0
175.0	6.7159	0.016480	55.76	55.77	142.99	993.2	1136.2	0.2552	1.5650	1.8202	175.0
174.0	6.5656	0.016474	56.95	56.97	141.98	993.8	1135.8	0.2537	1.5684	1.8221	174.0
173.0	6.4182	0.016468	58.18	58.19	140.98	994.4	1135.4	0.2521	1.5718	1.8239	173.0
172.0	6.2736	0.016463	59.43	59.45	139.98	995.0	1135.0	0.2505	1.5753	1.8258	172.0
171.0	6.1318	0.016457	60.72	60.74	138.98	995.6	1134.6	0.2489	1.5787	1.8276	171.0
170.0	5.9926	0.016451	62.04	62.06	137.97	996.2	1134.2	0.2473	1.5822	1.8295	170.0
169.0	5.8562	0.016445	63.39	63.41	136.97	996.8	1133.8	0.2457	1.5857	1.8314	169.0
168.0	5.7223	0.016440	64.78	64.80	135.97	997.4	1133.4	0.2441	1.5892	1.8333	168.0
167.0	5.5911	0.016434	66.21	66.22	134.97	998.0	1133.0	0.2425	1.5926	1.8352	167.0
166.0	5.4623	0.016428	67.67	67.68	133.97	998.6	1132.6	0.2409	1.5961	1.8371	166.0
165.0	5.3361	0.016423	69.17	69.18	132.96	999.2	1132.2	0.2393	1.5997	1.8390	165.0
164.0	5.2124	0.016417	70.70	70.72	131.96	999.8	1131.8	0.2377	1.6032	1.8409	164.0
163.0	5.0911	0.016412	72.28	72.30	130.96	1000.4	1131.4	0.2361	1.6067	1.8428	163.0
162.0	4.9722	0.016406	73.90	73.92	129.96	1001.0	1131.0	0.2345	1.6103	1.8448	162.0
161.0	4.8556	0.016401	75.56	75.58	128.96	1001.6	1130.6	0.2329	1.6138	1.8467	161.0
160.0	4.7414	0.016395	77.27	77.29	127.96	1002.2	1130.2	0.2313	1.6174	1.8487	160.0
159.0	4.6294	0.016390	79.02	79.04	126.96	1002.8	1129.8	0.2297	1.6210	1.8506	159.0
158.0	4.5197	0.016384	80.82	80.83	125.96	1003.4	1129.4	0.2281	1.6245	1.8526	158.0
157.0	4.4122	0.016379	82.66	82.68	124.95	1004.0	1129.0	0.2264	1.6281	1.8546	157.0
156.0	4.3068	0.016374	84.56	84.57	123.95	1004.6	1128.6	0.2248	1.6318	1.8566	156.0
155.0	4.2036	0.016369	86.50	86.52	122.95	1005.2	1128.2	0.2232	1.6354	1.8586	155.0
154.0	4.1025	0.016363	88.50	88.52	121.95	1005.8	1127.7	0.2216	1.6390	1.8606	154.0
153.0	4.0035	0.016358	90.55	90.57	120.95	1006.4	1127.3	0.2199	1.6426	1.8626	153.0
152.0	3.9065	0.016353	92.66	92.68	119.95	1007.0	1126.9	0.2183	1.6463	1.8646	152.0
151.0	3.8114	0.016348	94.83	94.84	118.95	1007.6	1126.5	0.2167	1.6500	1.8666	151.0
150.0	3.7184	0.016343	97.05	97.07	117.95	1008.2	1126.1	0.2150	1.6536	1.8686	150.0
149.0	3.6273	0.016337	99.33	99.35	116.95	1008.7	1125.7	0.2134	1.6573	1.8707	149.0
148.0	3.5381	0.016332	101.68	101.70	115.95	1009.3	1125.3	0.2117	1.6610	1.8727	148.0
147.0	3.4508	0.016327	104.10	104.11	114.95	1009.9	1124.9	0.2101	1.6647	1.8748	147.0
146.0	3.3653	0.016322	106.58	106.59	113.95	1010.5	1124.5	0.2084	1.6684	1.8769	146.0
145.0	3.2816	0.016317	109.12	109.14	112.95	1011.1	1124.0	0.2068	1.6722	1.8789	145.0
144.0	3.1997	0.016312	111.74	111.76	111.95	1011.7	1123.6	0.2051	1.6759	1.8810	144.0
143.0	3.1195	0.016308	114.44	114.45	110.95	1012.3	1123.2	0.2035	1.6797	1.8831	143.0
142.0	3.0411	0.016303	117.21	117.22	109.95	1012.9	1122.8	0.2018	1.6834	1.8852	142.0
141.0	2.9643	0.016298	120.05	120.07	108.95	1013.4	1122.4	0.2001	1.6872	1.8873	141.0
140.0	2.8892	0.016293	122.98	123.00	107.95	1014.0	1122.0	0.1985	1.6910	1.8895	140.0
139.0	2.8157	0.016288	125.99	126.01	106.95	1014.6	1121.6	0.1968	1.6948	1.8916	139.0
138.0	2.7438	0.016284	129.09	129.11	105.95	1015.2	1121.1	0.1951	1.6986	1.8937	138.0
137.0	2.6735	0.016279	132.28	132.29	104.95	1015.8	1120.7	0.1935	1.7024	1.8959	137.0
136.0	2.6047	0.016274	135.55	135.57	103.95	1016.4	1120.3	0.1918	1.7063	1.8980	136.0
135.0	2.5375	0.016270	138.93	138.94	102.95	1016.9	1119.9	0.1901	1.7101	1.9002	135.0
134.0	2.4717	0.016265	142.40	142.41	101.95	1017.5	1119.5	0.1884	1.7140	1.9024	134.0
133.0	2.4074	0.016260	145.97	145.98	100.95	1018.1	1119.1	0.1867	1.7178	1.9046	133.0
132.0	2.3445	0.016256	149.64	149.66	99.95	1018.7	1118.6	0.1851	1.7217	1.9068	132.0
131.0	2.2830	0.016251	153.42	153.44	98.95	1019.3	1118.2	0.1834	1.7256	1.9090	131.0
130.0	2.2230	0.016247	157.32	157.33	97.96	1019.8	1117.8	0.1817	1.7295	1.9112	130.0
129.0	2.1642	0.016243	161.32	161.34	96.96	1020.4	1117.4	0.1800	1.7335	1.9134	129.0
128.0	2.1068	0.016238	165.45	165.47	95.96	1021.0	1117.0	0.1783	1.7374	1.9157	128.0
127.0	2.0507	0.016234	169.70	169.72	94.96	1021.6	1116.5	0.1766	1.7413	1.9179	127.0
126.0	1.9959	0.016229	174.08	174.09	93.96	1022.2	1116.1	0.1749	1.7453	1.9202	126.0
125.0	1.9424	0.016225	178.58	178.60	92.96	1022.7	1115.7	0.1732	1.7493	1.9224	125.0
124.0	1.8901	0.016221	183.23	183.24	91.96	1023.3	1115.3	0.1715	1.7533	1.9247	124.0
123.0	1.8390	0.016217	188.01	188.03	90.96	1023.9	1114.9	0.1697	1.7573	1.9270	123.0
122.0	1.7891	0.016213	192.94	192.95	89.96	1024.5	1114.4	0.1680	1.7613	1.9293	122.0
121.0	1.7403	0.016208	198.01	198.03	88.96	1025.0	1114.0	0.1663	1.7653	1.9316	121.0
120.0	1.6927	0.016204	203.25	203.26	87.97	1025.6	1113.6	0.1646	1.7693	1.9339	120.0
119.0	1.6463	0.016200	208.64	208.66	86.97	1026.2	1113.2	0.1629	1.7734	1.9362	119.0
118.0	1.6009	0.016196	214.20	214.21	85.97	1026.8	1112.7	0.1611	1.7774	1.9386	118.0
117.0	1.5566	0.016192	219.93	219.94	84.97	1027.3	1112.3	0.1594	1.7815	1.9409	117.0
116.0	1.5133	0.016188	225.84	225.85	83.97	1027.9	1111.9	0.1577	1.7856	1.9433	116.0
115.0	1.4711	0.016184	231.93	231.94	82.97	1028.5	1111.5	0.1559	1.7897	1.9457	115.0
114.0	1.4299	0.016180	238.21	238.22	81.97	1029.1	1111.0	0.1542	1.7938	1.9480	114.0
113.0	1.3898	0.016177	244.69	244.70	80.98	1029.6	1110.6	0.1525	1.7980	1.9504	113.0
112.0	1.3505	0.016173	251.37	251.38	79.98	1030.2	1110.2	0.1507	1.8021	1.9528	112.0
111.0	1.3123	0.016169	258.26	258.28	78.98	1030.8	1109.8	0.1490	1.8063	1.9552	111.0
110.0	1.2750	0.016165	265.37	265.39	77.98	1031.4	1109.3	0.1472	1.8105	1.9577	110.0
109.0	1.2385	0.016162	272.71	272.72	76.98	1031.9	1108.9	0.1455	1.8146	1.9601	109.0
108.0	1.2030	0.016158	280.28	280.30	75.98	1032.5	1108.5	0.1437	1.8188	1.9626	108.0
107.0	1.1684	0.016154	288.09	288.11	74.99	1033.1	1108.1	0.1419	1.8231	1.9650	107.0
106.0	1.1347	0.016151	296.16	296.18	73.99	1033.6	1107.6	0.1402	1.8273	1.9675	106.0
105.0	1.1017	0.016147	304.49	304.50	72.99	1034.2	1107.2	0.1384	1.8315	1.9700	105.0

STEAM TABLES (Continued)

Properties of Saturated Steam and Saturated Water

Temp. F	Press. psia	Volume, ft³/lbm Water v_f	Evap. v_{fg}	Steam v_g	Enthalpy, Btu/lbm Water h_f	Evap. h_{fg}	Steam h_g	Entropy, Btu/lbm×F Water s_f	Evap. s_{fg}	Steam s_g	Temp. F
105.0	1.10174	0.016147	304.5	304.5	72.990	1034.2	1107.2	0.1384	1.8315	1.9700	105.0
104.0	1.06965	0.016144	313.1	313.1	71.992	1034.8	1106.8	0.1366	1.8358	1.9725	104.0
103.0	1.03838	0.016140	322.0	322.0	70.993	1035.4	1106.3	0.1349	1.8401	1.9750	103.0
102.0	1.00789	0.016137	331.1	331.1	69.995	1035.9	1105.9	0.1331	1.8444	1.9775	102.0
101.0	0.97818	0.016133	340.6	340.6	68.997	1036.5	1105.5	0.1313	1.8487	1.9800	101.0
100.0	0.94924	0.016130	350.4	350.4	67.999	1037.1	1105.1	0.1295	1.8530	1.9825	100.0
99.0	0.92103	0.016127	360.5	360.5	67.001	1037.6	1104.6	0.1278	1.8573	1.9851	99.0
98.0	0.89356	0.016123	370.9	370.9	66.003	1038.2	1104.2	0.1260	1.8617	1.9876	98.0
97.0	0.86679	0.016120	381.7	381.7	65.005	1038.8	1103.8	0.1242	1.8660	1.9902	97.0
96.0	0.84072	0.016117	392.8	392.9	64.006	1039.3	1103.3	0.1224	1.8704	1.9928	96.0
95.0	0.81534	0.016114	404.4	404.4	63.008	1039.9	1102.9	0.1206	1.8748	1.9954	95.0
94.0	0.79062	0.016111	416.3	416.3	62.010	1040.5	1102.5	0.1188	1.8792	1.9980	94.0
93.0	0.76655	0.016108	428.6	428.6	61.012	1041.0	1102.1	0.1170	1.8837	2.0006	93.0
92.0	0.74313	0.016105	441.3	441.3	60.014	1041.6	1101.6	0.1152	1.8881	2.0033	92.0
91.0	0.72032	0.016102	454.5	454.5	59.016	1042.2	1101.2	0.1134	1.8926	2.0059	91.0
90.0	0.69813	0.016099	468.1	468.1	58.018	1042.7	1100.8	0.1115	1.8970	2.0086	90.0
89.0	0.67653	0.016096	482.2	482.2	57.020	1043.3	1100.3	0.1097	1.9015	2.0112	89.0
88.0	0.65551	0.016093	496.8	496.8	56.022	1043.9	1099.9	0.1079	1.9060	2.0139	88.0
87.0	0.63507	0.016090	511.9	511.9	55.024	1044.4	1099.5	0.1061	1.9105	2.0166	87.0
86.0	0.61518	0.016087	527.5	527.5	54.026	1045.0	1099.0	0.1043	1.9151	2.0193	86.0
85.0	0.59583	0.016085	543.6	543.6	53.027	1045.6	1098.6	0.1024	1.9196	2.0221	85.0
84.0	0.57702	0.016082	560.3	560.3	52.029	1046.1	1098.2	0.1006	1.9242	2.0248	84.0
83.0	0.55872	0.016079	577.6	577.6	51.031	1046.7	1097.7	0.0988	1.9288	2.0275	83.0
82.0	0.54093	0.016077	595.5	595.6	50.033	1047.3	1097.3	0.0969	1.9334	2.0303	82.0
81.0	0.52364	0.016074	614.1	614.1	49.035	1047.8	1096.9	0.0951	1.9380	2.0331	81.0
80.0	0.50683	0.016072	633.3	633.3	48.037	1048.4	1096.4	0.0932	1.9426	2.0359	80.0
79.0	0.49049	0.016070	653.2	653.2	47.038	1049.0	1096.0	0.0914	1.9473	2.0387	79.0
78.0	0.47461	0.016067	673.8	673.9	46.040	1049.5	1095.6	0.0895	1.9520	2.0415	78.0
77.0	0.45919	0.016065	695.2	695.2	45.042	1050.1	1095.1	0.0877	1.9567	2.0443	77.0
76.0	0.44420	0.016063	717.4	717.4	44.043	1050.7	1094.7	0.0858	1.9614	2.0472	76.0
75.0	0.42964	0.016060	740.3	740.3	43.045	1051.2	1094.3	0.0839	1.9661	2.0500	75.0
74.0	0.41550	0.016058	764.1	764.1	42.046	1051.8	1093.8	0.0821	1.9708	2.0529	74.0
73.0	0.40177	0.016056	788.8	788.8	41.048	1052.4	1093.4	0.0802	1.9756	2.0558	73.0
72.0	0.38844	0.016054	814.3	814.3	40.049	1052.9	1093.0	0.0783	1.9804	2.0587	72.0
71.0	0.37549	0.016052	840.8	840.9	39.050	1053.5	1092.5	0.0764	1.9852	2.0616	71.0
70.0	0.36292	0.016050	868.3	868.4	38.052	1054.0	1092.1	0.0745	1.9900	2.0645	70.0
69.0	0.35073	0.016048	896.9	896.9	37.053	1054.6	1091.7	0.0727	1.9948	2.0675	69.0
68.0	0.33889	0.016046	926.5	926.5	36.054	1055.2	1091.2	0.0708	1.9996	2.0704	68.0
67.0	0.32740	0.016044	957.2	957.2	35.055	1055.7	1090.8	0.0689	2.0045	2.0734	67.0
66.0	0.31626	0.016043	989.0	989.1	34.056	1056.3	1090.4	0.0670	2.0094	2.0764	66.0
65.0	0.30545	0.016041	1022.1	1022.1	33.057	1056.9	1089.9	0.0651	2.0143	2.0794	65.0
64.0	0.29497	0.016039	1056.5	1056.5	32.058	1057.4	1089.5	0.0632	2.0192	2.0824	64.0
63.0	0.28480	0.016038	1092.1	1092.1	31.058	1058.0	1089.0	0.0613	2.0242	2.0854	63.0
62.0	0.27494	0.016036	1129.2	1129.2	30.059	1058.5	1088.6	0.0593	2.0291	2.0885	62.0
61.0	0.26538	0.016035	1167.6	1167.6	29.059	1059.1	1088.2	0.0574	2.0341	2.0915	61.0
60.0	0.25611	0.016033	1207.6	1207.6	28.060	1059.7	1087.7	0.0555	2.0391	2.0946	60.0
59.0	0.24713	0.016032	1249.1	1249.1	27.060	1060.2	1087.3	0.0536	2.0441	2.0977	59.0
58.0	0.23843	0.016031	1292.2	1292.2	26.060	1060.8	1086.9	0.0516	2.0491	2.1008	58.0
57.0	0.23000	0.016029	1337.0	1337.0	25.060	1061.4	1086.4	0.0497	2.0542	2.1039	57.0
56.0	0.22183	0.016028	1383.6	1383.6	24.059	1061.9	1086.0	0.0478	2.0593	2.1070	56.0
55.0	0.21392	0.016027	1432.0	1432.0	23.059	1062.5	1085.6	0.0458	2.0644	2.1102	55.0
54.0	0.20625	0.016026	1482.4	1482.4	22.058	1063.1	1085.1	0.0439	2.0695	2.1134	54.0
53.0	0.19883	0.016025	1534.7	1534.8	21.058	1063.6	1084.7	0.0419	2.0746	2.1165	53.0
52.0	0.19165	0.016024	1589.2	1589.2	20.057	1064.2	1084.2	0.0400	2.0798	2.1197	52.0
51.0	0.18469	0.016023	1645.9	1645.9	19.056	1064.7	1083.8	0.0380	2.0849	2.1230	51.0
50.0	0.17796	0.016023	1704.8	1704.8	18.054	1065.3	1083.4	0.0361	2.0901	2.1262	50.0
49.0	0.17144	0.016022	1766.2	1766.2	17.053	1065.9	1082.9	0.0341	2.0953	2.1294	49.0
48.0	0.16514	0.016021	1830.0	1830.0	16.051	1066.4	1082.5	0.0321	2.1006	2.1327	48.0
47.0	0.15904	0.016021	1896.5	1896.5	15.049	1067.0	1082.1	0.0301	2.1058	2.1360	47.0
46.0	0.15314	0.016020	1965.7	1965.7	14.047	1067.6	1081.6	0.0282	2.1111	2.1393	46.0
45.0	0.14744	0.016020	2037.7	2037.8	13.044	1068.1	1081.2	0.0262	2.1164	2.1426	45.0
44.0	0.14192	0.016019	2112.8	2112.8	12.041	1068.7	1080.7	0.0242	2.1217	2.1459	44.0
43.0	0.13659	0.016019	2191.0	2191.0	11.038	1069.3	1080.3	0.0222	2.1271	2.1493	43.0
42.0	0.13143	0.016019	2272.4	2272.4	10.035	1069.8	1079.9	0.0202	2.1325	2.1527	42.0
41.0	0.12645	0.016019	2357.3	2357.3	9.031	1070.4	1079.4	0.0182	2.1378	2.1560	41.0
40.0	0.12163	0.016019	2445.8	2445.8	8.027	1071.0	1079.0	0.0162	2.1432	2.1594	40.0
39.0	0.11698	0.016019	2538.0	2538.0	7.023	1071.5	1078.5	0.0142	2.1487	2.1629	39.0
38.0	0.11249	0.016019	2634.1	2634.2	6.018	1072.1	1078.1	0.0122	2.1541	2.1663	38.0
37.0	0.10815	0.016019	2734.4	2734.4	5.013	1072.7	1077.7	0.0101	2.1596	2.1697	37.0
36.0	0.10395	0.016020	2839.0	2839.0	4.008	1073.2	1077.2	0.0081	2.1651	2.1732	36.0
35.0	0.09991	0.016020	2948.1	2948.1	3.002	1073.8	1076.8	0.0061	2.1706	2.1767	35.0
34.0	0.09600	0.016021	3061.9	3061.9	1.996	1074.4	1076.4	0.0041	2.1762	2.1802	34.0
33.0	0.09223	0.016021	3180.7	3180.7	0.989	1074.9	1075.9	0.0020	2.1817	2.1837	33.0
32.018	0.08865	0.016022	3302.4	3302.4	0.0003	1075.5	1075.5	0.0000	2.1872	2.1872	32.018
*32.0	0.08859	0.016022	3304.7	3304.7	−0.0179	1075.5	1075.5	−0.0000	2.1873	2.1873	32.0

*The states here shown are metastable

STEAM TABLES (Continued)

Specific Heat at constant pressure of Steam and of Water

Temp. F	c_p, Btu/lbm × F															Temp. F
Press., psia	1	1.5	2	3	4	6	8	10	15	20	30	40	60	80	100	Press., psia
Sat. Water	0.998	0.998	0.999	1.000	1.000	1.002	1.003	1.004	1.007	1.010	1.014	1.019	1.026	1.033	1.039	Sat. Water
Sat. Steam	0.450	0.452	0.454	0.458	0.461	0.466	0.471	0.475	0.485	0.493	0.508	0.521	0.543	0.564	0.582	Sat. Steam
1500	0.559	0.559	0.559	0.559	0.559	0.559	0.559	0.559	0.559	0.559	0.560	0.560	0.560	0.561	0.561	1500
1480	0.557	0.557	0.557	0.557	0.557	0.557	0.557	0.558	0.558	0.558	0.558	0.558	0.559	0.559	0.559	1480
1460	0.556	0.556	0.556	0.556	0.556	0.556	0.556	0.556	0.556	0.556	0.556	0.557	0.557	0.557	0.558	1460
1440	0.554	0.554	0.554	0.554	0.554	0.554	0.554	0.554	0.554	0.554	0.555	0.555	0.555	0.556	0.556	1440
1420	0.552	0.552	0.552	0.552	0.552	0.552	0.553	0.553	0.553	0.553	0.553	0.553	0.554	0.554	0.555	1420
1400	0.551	0.551	0.551	0.551	0.551	0.551	0.551	0.551	0.551	0.551	0.551	0.552	0.552	0.553	0.553	1400
1380	0.549	0.549	0.549	0.549	0.549	0.549	0.549	0.549	0.549	0.549	0.550	0.550	0.550	0.551	0.551	1380
1360	0.547	0.547	0.547	0.547	0.547	0.547	0.547	0.547	0.548	0.548	0.548	0.548	0.549	0.549	0.550	1360
1340	0.546	0.546	0.546	0.546	0.546	0.546	0.546	0.546	0.546	0.546	0.546	0.547	0.547	0.548	0.548	1340
1320	0.544	0.544	0.544	0.544	0.544	0.544	0.544	0.544	0.544	0.544	0.545	0.545	0.545	0.546	0.546	1320
1300	0.542	0.542	0.542	0.542	0.542	0.542	0.542	0.542	0.542	0.543	0.543	0.543	0.544	0.544	0.545	1300
1280	0.540	0.540	0.540	0.540	0.540	0.540	0.540	0.541	0.541	0.541	0.541	0.541	0.542	0.543	0.543	1280
1260	0.538	0.539	0.539	0.539	0.539	0.539	0.539	0.539	0.539	0.539	0.539	0.540	0.540	0.541	0.541	1260
1240	0.537	0.537	0.537	0.537	0.537	0.537	0.537	0.537	0.537	0.537	0.538	0.538	0.539	0.539	0.540	1240
1220	0.535	0.535	0.535	0.535	0.535	0.535	0.535	0.535	0.535	0.536	0.536	0.536	0.537	0.537	0.538	1220
1200	0.533	0.533	0.533	0.533	0.533	0.533	0.533	0.533	0.534	0.534	0.534	0.534	0.535	0.536	0.536	1200
1180	0.531	0.531	0.531	0.531	0.531	0.531	0.532	0.532	0.532	0.532	0.532	0.533	0.533	0.534	0.535	1180
1160	0.529	0.529	0.530	0.530	0.530	0.530	0.530	0.530	0.530	0.530	0.530	0.531	0.532	0.532	0.533	1160
1140	0.528	0.528	0.528	0.528	0.528	0.528	0.528	0.528	0.528	0.528	0.529	0.529	0.530	0.531	0.531	1140
1120	0.526	0.526	0.526	0.526	0.526	0.526	0.526	0.526	0.526	0.527	0.527	0.527	0.528	0.529	0.530	1120
1100	0.524	0.524	0.524	0.524	0.524	0.524	0.524	0.524	0.525	0.525	0.525	0.526	0.526	0.527	0.528	1100
1080	0.522	0.522	0.522	0.522	0.522	0.522	0.522	0.523	0.523	0.523	0.523	0.524	0.525	0.525	0.526	1080
1060	0.520	0.520	0.520	0.520	0.520	0.521	0.521	0.521	0.521	0.521	0.522	0.522	0.523	0.524	0.524	1060
1040	0.518	0.519	0.519	0.519	0.519	0.519	0.519	0.519	0.519	0.519	0.520	0.520	0.521	0.522	0.523	1040
1020	0.517	0.517	0.517	0.517	0.517	0.517	0.517	0.517	0.517	0.518	0.518	0.518	0.519	0.520	0.521	1020
1000	0.515	0.515	0.515	0.515	0.515	0.515	0.515	0.515	0.515	0.516	0.516	0.517	0.518	0.519	0.519	1000
980	0.513	0.513	0.513	0.513	0.513	0.513	0.513	0.513	0.514	0.514	0.514	0.515	0.516	0.517	0.518	980
960	0.511	0.511	0.511	0.511	0.511	0.511	0.512	0.512	0.512	0.512	0.513	0.513	0.514	0.515	0.516	960
940	0.509	0.509	0.509	0.509	0.509	0.510	0.510	0.510	0.510	0.510	0.511	0.511	0.512	0.514	0.515	940
920	0.507	0.508	0.508	0.508	0.508	0.508	0.508	0.508	0.508	0.509	0.509	0.510	0.511	0.512	0.513	920
900	0.506	0.506	0.506	0.506	0.506	0.506	0.506	0.506	0.506	0.507	0.507	0.508	0.509	0.510	0.512	900
880	0.504	0.504	0.504	0.504	0.504	0.504	0.504	0.504	0.505	0.505	0.506	0.506	0.508	0.509	0.510	880
860	0.502	0.502	0.502	0.502	0.502	0.502	0.503	0.503	0.503	0.503	0.504	0.505	0.506	0.507	0.509	860
840	0.500	0.500	0.500	0.500	0.500	0.501	0.501	0.501	0.501	0.502	0.502	0.503	0.504	0.506	0.507	840
820	0.498	0.498	0.499	0.499	0.499	0.499	0.499	0.499	0.499	0.500	0.501	0.501	0.503	0.504	0.506	820
800	0.497	0.497	0.497	0.497	0.497	0.497	0.497	0.497	0.498	0.498	0.499	0.500	0.501	0.503	0.505	800
780	0.495	0.495	0.495	0.495	0.495	0.495	0.495	0.496	0.496	0.496	0.497	0.498	0.500	0.502	0.503	780
760	0.493	0.493	0.493	0.493	0.493	0.494	0.494	0.494	0.494	0.495	0.496	0.497	0.499	0.500	0.502	760
740	0.491	0.491	0.491	0.492	0.492	0.492	0.492	0.492	0.493	0.493	0.494	0.495	0.497	0.499	0.501	740
720	0.490	0.490	0.490	0.490	0.490	0.490	0.490	0.491	0.491	0.492	0.493	0.494	0.496	0.498	0.500	720
700	0.488	0.488	0.488	0.488	0.488	0.488	0.489	0.489	0.490	0.490	0.491	0.492	0.495	0.497	0.500	700
680	0.486	0.486	0.486	0.486	0.487	0.487	0.487	0.487	0.488	0.489	0.490	0.491	0.494	0.496	0.499	680
660	0.484	0.485	0.485	0.485	0.485	0.485	0.485	0.486	0.486	0.487	0.489	0.490	0.493	0.496	0.499	660
640	0.483	0.483	0.483	0.483	0.483	0.484	0.484	0.484	0.485	0.486	0.487	0.489	0.492	0.495	0.499	640
620	0.481	0.481	0.481	0.481	0.482	0.482	0.482	0.483	0.483	0.484	0.486	0.488	0.491	0.495	0.499	620
600	0.479	0.480	0.480	0.480	0.480	0.480	0.481	0.481	0.482	0.483	0.485	0.487	0.491	0.495	0.499	600
580	0.478	0.478	0.478	0.478	0.478	0.479	0.479	0.480	0.481	0.482	0.484	0.486	0.491	0.495	0.500	580
560	0.476	0.476	0.476	0.477	0.477	0.477	0.478	0.478	0.479	0.481	0.483	0.485	0.490	0.496	0.501	560
540	0.475	0.475	0.475	0.475	0.475	0.476	0.476	0.477	0.478	0.480	0.482	0.485	0.491	0.497	0.503	540
520	0.473	0.473	0.473	0.474	0.474	0.475	0.475	0.476	0.477	0.479	0.482	0.485	0.491	0.498	0.505	520
500	0.472	0.472	0.472	0.472	0.473	0.473	0.474	0.475	0.476	0.478	0.481	0.485	0.492	0.500	0.508	500
480	0.470	0.470	0.470	0.471	0.471	0.472	0.473	0.473	0.475	0.477	0.481	0.485	0.493	0.502	0.511	480
460	0.469	0.469	0.469	0.469	0.470	0.471	0.472	0.472	0.475	0.477	0.481	0.486	0.495	0.505	0.516	460
440	0.467	0.467	0.468	0.468	0.469	0.470	0.470	0.471	0.474	0.476	0.481	0.487	0.498	0.509	0.522	440
420	0.466	0.466	0.466	0.467	0.467	0.468	0.470	0.471	0.473	0.476	0.482	0.488	0.501	0.514	0.528	420
400	0.464	0.465	0.465	0.466	0.466	0.467	0.469	0.470	0.473	0.476	0.483	0.490	0.504	0.520	0.536	400
380	0.463	0.463	0.464	0.464	0.465	0.466	0.468	0.469	0.473	0.477	0.484	0.492	0.509	0.527	0.546	380
360	0.462	0.462	0.462	0.463	0.464	0.466	0.467	0.469	0.473	0.477	0.486	0.495	0.515	0.536	0.558	360
340	0.460	0.461	0.461	0.462	0.463	0.465	0.467	0.469	0.473	0.478	0.488	0.499	0.521	0.546	0.572	340
320	0.459	0.460	0.460	0.461	0.462	0.464	0.467	0.469	0.474	0.480	0.491	0.504	0.530	0.558	1.036	320
300	0.458	0.459	0.459	0.460	0.462	0.464	0.466	0.469	0.475	0.482	0.495	0.509	0.539	1.029	1.029	300
280	0.457	0.458	0.458	0.460	0.461	0.464	0.467	0.469	0.477	0.484	0.500	0.516	1.022	1.022	1.022	280
260	0.456	0.457	0.457	0.459	0.461	0.464	0.467	0.470	0.478	0.487	0.505	1.017	1.017	1.017	1.016	260
240	0.455	0.456	0.457	0.458	0.460	0.464	0.468	0.471	0.481	0.491	1.012	1.012	1.012	1.012	1.012	240
220	0.454	0.455	0.456	0.458	0.460	0.464	0.468	0.473	0.484	1.008	1.008	1.008	1.008	1.008	1.008	220
200	0.453	0.454	0.455	0.458	0.460	0.465	0.470	0.475	1.005	1.005	1.005	1.005	1.005	1.005	1.005	200
180	0.452	0.454	0.455	0.458	0.460	0.466	1.003	1.003	1.003	1.003	1.003	1.003	1.003	1.002	1.002	180
160	0.451	0.453	0.455	0.458	0.461	1.001	1.001	1.001	1.001	1.001	1.001	1.001	1.001	1.001	1.001	160
140	0.451	0.453	0.454	1.000	1.000	1.000	1.000	1.000	0.999	0.999	0.999	0.999	0.999	0.999	0.999	140
120	0.450	0.452	0.999	0.999	0.999	0.999	0.999	0.999	0.999	0.999	0.998	0.998	0.998	0.998	0.998	120
100	0.998	0.998	0.998	0.998	0.998	0.998	0.998	0.998	0.998	0.998	0.998	0.998	0.998	0.998	0.998	100
80	0.998	0.998	0.998	0.998	0.998	0.998	0.998	0.998	0.998	0.998	0.998	0.998	0.998	0.998	0.998	80
60	1.000	1.000	1.000	1.000	1.000	1.000	1.000	1.000	1.000	1.000	1.000	1.000	0.999	0.999	0.999	60
40	1.004	1.004	1.004	1.004	1.004	1.004	1.004	1.004	1.004	1.004	1.004	1.004	1.004	1.004	1.003	40
32	1.007	1.007	1.007	1.007	1.007	1.007	1.007	1.007	1.007	1.007	1.007	1.007	1.007	1.007	1.006	32

STEAM TABLES (Continued)

Specific Heat at constant pressure of Steam and of Water

Temp. F / Press., psia	150	200	300	400	600	800	1000	1500	2000	3000	4000	6000	8000	10000	15000	Temp. F / Press., psia
	c_p, Btu/lbm × F															
Sat. Water	1.054	1.067	1.093	1.118	1.168	1.224	1.286	1.492	1.841	7.646	—	—	—	—	—	Sat. Water
Sat. Steam	0.624	0.661	0.729	0.792	0.915	1.046	1.191	1.667	2.557	13.66	—	—	—	—	—	Sat. Steam
1500	0.562	0.563	0.565	0.567	0.571	0.576	0.580	0.590	0.601	0.623	0.645	0.691	0.737	0.780	0.868	1500
1480	0.561	0.562	0.564	0.566	0.570	0.575	0.579	0.590	0.601	0.623	0.647	0.694	0.742	0.786	0.878	1480
1460	0.559	0.560	0.562	0.565	0.569	0.573	0.578	0.589	0.601	0.624	0.648	0.698	0.747	0.793	0.888	1460
1440	0.557	0.559	0.561	0.563	0.568	0.572	0.577	0.589	0.600	0.625	0.650	0.701	0.753	0.800	0.900	1440
1420	0.556	0.557	0.559	0.562	0.566	0.571	0.576	0.588	0.600	0.625	0.651	0.705	0.759	0.808	0.909	1420
1400	0.554	0.555	0.558	0.560	0.565	0.570	0.575	0.587	0.600	0.626	0.653	0.709	0.765	0.817	0.926	1400
1380	0.553	0.554	0.556	0.559	0.564	0.569	0.574	0.587	0.600	0.627	0.655	0.714	0.773	0.827	0.939	1380
1360	0.551	0.552	0.555	0.558	0.563	0.568	0.573	0.586	0.600	0.628	0.657	0.719	0.781	0.838	0.953	1360
1340	0.549	0.551	0.553	0.556	0.561	0.567	0.572	0.586	0.600	0.629	0.660	0.725	0.790	0.850	0.968	1340
1320	0.548	0.549	0.552	0.555	0.560	0.566	0.571	0.585	0.600	0.630	0.663	0.731	0.800	0.864	0.983	1320
1300	0.546	0.548	0.550	0.553	0.559	0.565	0.570	0.585	0.600	0.632	0.666	0.738	0.811	0.879	0.998	1300
1280	0.545	0.546	0.549	0.552	0.558	0.564	0.570	0.585	0.600	0.634	0.669	0.746	0.824	0.897	1.014	1280
1260	0.543	0.544	0.547	0.550	0.556	0.563	0.569	0.585	0.601	0.636	0.673	0.755	0.838	0.918	1.033	1260
1240	0.541	0.543	0.546	0.549	0.555	0.562	0.568	0.584	0.601	0.638	0.678	0.765	0.855	0.942	1.053	1240
1220	0.540	0.541	0.544	0.548	0.554	0.561	0.567	0.584	0.602	0.641	0.683	0.777	0.875	0.969	1.072	1220
1200	0.538	0.540	0.543	0.546	0.553	0.560	0.567	0.584	0.603	0.644	0.689	0.790	0.897	1.000	1.095	1200
1180	0.536	0.538	0.541	0.545	0.552	0.559	0.566	0.584	0.604	0.647	0.696	0.805	0.922	1.033	1.117	1180
1160	0.535	0.536	0.540	0.544	0.551	0.558	0.565	0.585	0.606	0.652	0.704	0.823	0.952	1.070	1.143	1160
1140	0.533	0.535	0.539	0.542	0.550	0.557	0.565	0.585	0.607	0.656	0.713	0.843	0.986	1.107	1.167	1140
1120	0.531	0.533	0.537	0.541	0.549	0.557	0.565	0.586	0.609	0.662	0.723	0.866	1.025	1.149	1.190	1120
1100	0.530	0.532	0.536	0.540	0.548	0.556	0.564	0.587	0.612	0.668	0.735	0.893	1.070	1.193	1.220	1100
1080	0.528	0.530	0.534	0.538	0.547	0.555	0.564	0.588	0.615	0.676	0.749	0.924	1.120	1.242	1.240	1080
1060	0.527	0.529	0.533	0.537	0.546	0.555	0.564	0.590	0.618	0.685	0.765	0.960	1.176	1.295	1.260	1060
1040	0.525	0.527	0.532	0.536	0.545	0.555	0.565	0.592	0.622	0.695	0.783	1.002	1.238	1.351	1.282	1040
1020	0.523	0.526	0.530	0.535	0.545	0.555	0.565	0.594	0.627	0.707	0.804	1.051	1.306	1.399	1.298	1020
1000	0.522	0.524	0.529	0.534	0.544	0.555	0.566	0.597	0.633	0.721	0.829	1.110	1.382	1.471	1.306	1000
980	0.520	0.523	0.528	0.533	0.544	0.555	0.567	0.601	0.640	0.737	0.858	1.180	1.475	1.531	1.312	980
960	0.519	0.521	0.527	0.532	0.543	0.556	0.568	0.605	0.648	0.756	0.893	1.267	1.598	1.595	1.310	960
940	0.517	0.520	0.526	0.531	0.543	0.556	0.570	0.610	0.658	0.778	0.934	1.376	1.708	1.639	1.299	940
920	0.516	0.519	0.525	0.531	0.544	0.558	0.573	0.617	0.669	0.803	0.984	1.520	1.819	1.667	1.281	920
900	0.515	0.518	0.524	0.530	0.544	0.559	0.576	0.624	0.683	0.834	1.048	1.716	1.932	1.660	1.259	900
880	0.513	0.516	0.523	0.530	0.545	0.561	0.580	0.633	0.699	0.872	1.130	1.993	2.000	1.633	1.232	880
860	0.512	0.515	0.523	0.530	0.546	0.564	0.584	0.644	0.718	0.918	1.240	2.316	2.019	1.593	1.212	860
840	0.511	0.514	0.522	0.530	0.548	0.568	0.590	0.657	0.740	0.977	1.395	2.653	1.978	1.547	1.192	840
820	0.510	0.514	0.522	0.531	0.550	0.572	0.597	0.672	0.767	1.054	1.620	2.886	1.888	1.503	1.175	820
800	0.509	0.513	0.522	0.532	0.553	0.577	0.605	0.690	0.800	1.160	1.967	2.872	1.768	1.459	1.157	800
780	0.508	0.513	0.522	0.533	0.557	0.584	0.615	0.712	0.840	1.312	2.550	2.547	1.670	1.416	1.142	780
760	0.507	0.512	0.523	0.535	0.561	0.592	0.628	0.738	0.892	1.542	4.462	2.156	1.576	1.370	1.126	760
740	0.507	0.512	0.524	0.537	0.567	0.602	0.642	0.770	0.960	1.913	8.119	1.886	1.493	1.332	1.114	740
720	0.506	0.512	0.525	0.540	0.574	0.613	0.660	0.811	1.052	2.584	3.458	1.696	1.421	1.290	1.100	720
700	0.506	0.513	0.528	0.544	0.582	0.627	0.681	0.861	1.181	6.145°	2.237	1.557	1.358	1.250	1.089	700
680	0.506	0.514	0.530	0.549	0.592	0.644	0.707	0.927	1.365	2.469	1.789	1.450	1.303	1.217	1.079	680
660	0.507	0.515	0.534	0.555	0.604	0.665	0.738	1.015	1.639	1.851	1.587	1.369	1.256	1.187	1.071	660
640	0.507	0.517	0.538	0.562	0.619	0.690	0.777	1.135	2.219	1.601	1.454	1.303	1.216	1.157	1.063	640
620	0.509	0.519	0.543	0.571	0.637	0.720	0.826	1.308	1.614	1.455	1.362	1.252	1.184	1.136	1.056	620
600	0.510	0.522	0.550	0.582	0.659	0.757	0.888	1.586	1.453	1.358	1.295	1.211	1.157	1.118	1.052	600
580	0.513	0.526	0.558	0.595	0.685	0.804	0.969	1.393	1.351	1.289	1.243	1.178	1.134	1.102	1.046	580
560	0.516	0.531	0.568	0.611	0.717	0.862	1.079	1.309	1.281	1.237	1.202	1.151	1.115	1.087	1.039	560
540	0.519	0.538	0.580	0.630	0.756	0.937	1.272	1.249	1.229	1.196	1.169	1.128	1.098	1.074	1.031	540
520	0.524	0.545	0.594	0.653	0.804	1.035	1.221	1.204	1.189	1.164	1.142	1.109	1.083	1.062	1.024	520
500	0.530	0.554	0.611	0.680	0.865	1.187	1.181	1.169	1.157	1.137	1.120	1.092	1.069	1.051	1.017	500
480	0.537	0.565	0.632	0.714	1.159	1.154	1.150	1.140	1.131	1.115	1.101	1.077	1.057	1.041	1.010	480
460	0.545	0.578	0.657	0.755	1.132	1.128	1.125	1.117	1.110	1.096	1.084	1.064	1.047	1.033	1.004	460
440	0.556	0.594	0.687	1.113	1.110	1.107	1.104	1.098	1.092	1.080	1.070	1.052	1.038	1.025	0.999	440
420	0.568	0.614	0.724	1.094	1.091	1.089	1.087	1.081	1.076	1.067	1.058	1.042	1.029	1.018	0.994	420
400	0.583	0.636	1.079	1.078	1.076	1.074	1.072	1.067	1.063	1.055	1.047	1.034	1.022	1.011	0.990	400
380	0.601	1.066	1.065	1.065	1.063	1.061	1.059	1.056	1.052	1.044	1.038	1.026	1.015	1.006	0.986	380
360	0.622	1.054	1.054	1.053	1.052	1.050	1.049	1.045	1.042	1.036	1.030	1.019	1.009	1.001	0.982	360
340	1.045	1.044	1.044	1.043	1.042	1.040	1.039	1.036	1.033	1.028	1.022	1.013	1.004	0.996	0.979	340
320	1.036	1.036	1.035	1.034	1.033	1.032	1.031	1.028	1.026	1.021	1.016	1.007	0.999	0.992	0.976	320
300	1.028	1.028	1.028	1.027	1.026	1.025	1.024	1.022	1.019	1.015	1.010	1.002	0.995	0.988	0.973	300
280	1.022	1.022	1.021	1.021	1.020	1.019	1.018	1.016	1.014	1.009	1.005	0.998	0.991	0.985	0.971	280
260	1.016	1.016	1.016	1.015	1.014	1.013	1.013	1.011	1.009	1.005	1.001	0.994	0.988	0.982	0.968	260
240	1.012	1.011	1.011	1.011	1.010	1.009	1.008	1.006	1.004	1.001	0.997	0.991	0.985	0.979	0.966	240
220	1.008	1.008	1.007	1.007	1.006	1.005	1.005	1.003	1.001	0.998	0.994	0.988	0.982	0.977	0.964	220
200	1.005	1.004	1.004	1.004	1.003	1.002	1.002	1.000	0.998	0.995	0.992	0.986	0.980	0.975	0.963	200
180	1.002	1.002	1.002	1.001	1.001	1.000	0.999	0.998	0.996	0.993	0.989	0.983	0.978	0.973	0.961	180
160	1.000	1.000	1.000	0.999	0.999	0.998	0.997	0.996	0.994	0.991	0.987	0.981	0.976	0.971	0.959	160
140	0.999	0.999	0.998	0.998	0.997	0.997	0.996	0.994	0.992	0.989	0.986	0.980	0.974	0.969	0.958	140
120	0.998	0.998	0.997	0.997	0.996	0.996	0.995	0.993	0.991	0.988	0.984	0.978	0.972	0.967	0.957	120
100	0.997	0.997	0.997	0.996	0.996	0.995	0.994	0.992	0.990	0.986	0.983	0.976	0.970	0.965	0.955	100
80	0.998	0.997	0.997	0.996	0.995	0.994	0.994	0.991	0.989	0.985	0.981	0.974	0.968	0.962	0.951	80
60	0.999	0.999	0.998	0.997	0.996	0.995	0.994	0.991	0.989	0.984	0.979	0.970	0.963	0.956	0.942	60
40	1.003	1.003	1.002	1.001	1.000	0.998	0.997	0.993	0.989	0.983	0.976	0.965	0.954	0.945	0.920	40
32	1.006	1.006	1.005	1.004	1.002	1.000	0.999	0.994	0.990	0.983	0.975	0.962	0.949	0.937	0.904	32

°Critical point.

STEAM TABLES (Continued)

Thermal Conductivity of Steam and Water

Temp. F	k, (Btu/hr × ft. × F) × 10³												
Press., psia	1	2	5	10	20	50	100	200	500	1000	2000	5000	7500
Sat. Water	364.0	373.1	383.8	390.4	395.2	397.4	394.7	386.2	361.7	327.6	271.8	—	—
Sat. Steam	11.6	12.2	13.0	13.8	14.8	16.6	18.4	21.1	27.2	36.5	61.3	—	—
1500	63.7	63.7	63.7	63.7	63.7	63.8	64.0	64.3	65.4	67.1	70.7	82.0	92.2
1450	61.4	61.4	61.5	61.5	61.5	61.6	61.8	62.1	63.2	64.9	68.5	80.1	90.6
1400	59.2	59.2	59.2	59.2	59.3	59.4	59.6	59.9	60.9	62.7	66.3	78.2	89.2
1350	57.0	57.0	57.0	57.0	57.1	57.2	57.3	57.7	58.7	60.5	64.2	76.3	87.9
1300	54.8	54.8	54.8	54.8	54.8	54.9	55.1	55.5	56.5	58.3	62.0	74.6	86.9
1250	52.6	52.6	52.6	52.6	52.6	52.7	52.9	53.2	54.3	56.1	59.9	73.0	86.3
1200	50.4	50.4	50.4	50.4	50.4	50.5	50.7	51.0	52.1	53.9	57.8	71.6	86.2
1150	48.2	48.2	48.2	48.2	48.2	48.3	48.5	48.9	49.9	51.8	55.7	70.5	87.0
1100	46.0	46.0	46.0	46.0	46.1	46.2	46.3	46.7	47.8	49.6	53.7	69.8	89.0
1050	43.9	43.9	43.9	43.9	43.9	44.0	44.2	44.6	45.6	47.5	51.8	69.7	93.4
1000	41.7	41.7	41.8	41.8	41.8	41.9	42.1	42.4	43.5	45.5	50.0	70.7	102.9
950	39.6	39.6	39.7	39.7	39.7	39.8	40.0	40.3	41.4	43.5	48.3	73.5	115.5
900	37.6	37.6	37.6	37.6	37.6	37.7	37.9	38.3	39.4	41.5	46.8	80.2	138.7
850	35.5	35.6	35.6	35.6	35.6	35.7	35.9	36.3	37.4	39.7	45.6	96.7	178.8
800	33.6	33.6	33.6	33.6	33.6	33.7	33.9	34.3	35.5	37.9	44.9	129.6	223.2
750	31.6	31.6	31.6	31.6	31.7	31.8	32.0	32.3	33.6	36.3	45.2	202.5	258.3
700	29.7	29.7	29.7	29.7	29.8	29.9	30.1	30.4	31.8	35.0	47.5[⊙]	262.8	295.1
650	27.8	27.8	27.9	27.9	27.9	28.0	28.2	28.6	30.1	34.1	55.7	304.3	326.7
600	26.0	26.0	26.1	26.1	26.1	26.2	26.4	26.9	28.7	34.1	301.9	333.7	349.3
550	24.3	24.3	24.3	24.3	24.4	24.5	24.7	25.2	27.5	36.1	333.7	356.1	368.0
500	22.6	22.6	22.6	22.6	22.7	22.8	23.0	23.6	26.9	350.8	357.4	373.8	383.6
450	21.0	21.0	21.0	21.0	21.0	21.2	21.4	22.3	368.1	370.6	375.3	387.9	396.5
400	19.4	19.4	19.4	19.4	19.5	19.6	20.0	21.3	383.0	384.9	388.5	398.6	406.4
350	17.9	17.9	17.9	17.9	18.0	18.2	18.8	392.0	392.9	394.4	397.4	406.1	413.2
300	16.5	16.5	16.5	16.5	16.6	16.9	396.9	397.2	398.0	399.3	402.0	409.9	416.4
250	15.1	15.1	15.1	15.2	15.3	396.9	397.0	397.3	398.1	399.4	402.1	409.7	415.8
200	13.8	13.8	13.9	14.0	391.6	391.6	391.8	392.1	393.0	394.4	397.2	404.9	410.6
150	12.7	12.7	380.5	380.5	380.6	380.7	380.8	381.1	382.1	383.7	386.7	394.7	400.3
100	363.3	363.3	363.3	363.3	363.3	363.4	363.6	363.9	365.0	366.6	369.8	378.3	384.1
50	339.1	339.1	339.1	339.1	339.2	339.3	339.4	339.8	340.8	342.5	345.7	354.6	361.0
32	328.6	328.6	328.6	328.6	328.6	328.7	328.9	329.2	330.3	331.9	335.1	344.1	350.8

[⊙] Critical point.

PHYSICAL PROPERTIES OF FLUOROCARBON REFRIGERANTS

Property No.	Refrigerant name		11	12	13	13B1	14	21
1	Formula		CCl_3F	CCl_2F_2	$CClF_3$	$CBrF_3$	CF_4	$CHCl_2F$
2	Molecular weight		137.37	120.91	104.46	148.92	88.01	102.92
3	Normal boiling point; °C		23.82	−29.79	−81.4	−57.75	−127.96	8.92
4	Normal freezing point; °C		−111	−158	−181	−168	−184	−135
5	Critical temperature; °C		198	112	28.9	67	−45.67	178.5
6	Critical pressure; atm		43.5	40.6	38.2	39.1	36.96	51
7	Critical volume; cm/mol		247	217	181	200	141	197
8	Critical density; g/cm		0.554	0.558	0.578	0.745	0.626	0.522
9	Density of liquid at 25°C; g/cm		1.467	1.311	$1.298^{-30^{\circ}C}$	1.538	$1.317^{-80^{\circ}C}$	1.366
10	Density of saturated vapor at B.P.; g/liter		5.86	6.33	7.01	8.71	7.62	4.57
11	Specific heat of liquid at 25°C; cal/g		0.208	0.232	$0.247^{-30^{\circ}C}$	0.208	$0.294^{-80^{\circ}C}$	0.256
12	Specific heat of vapor at 25°C and 1 atm; cal/g		$0.142^{38^{\circ}C}$	0.145	0.158	0.112	0.169	0.140
13	Heat of vaporization at B.P.; cal/g		43.10	39.47	35.47	28.38	32.49	57.86
14	Thermal conductivity at 25°C; Btu/(hr)(ft)(°F)	liquid	0.050	0.041	0.020	0.025	$0.040^{-100^{\circ}F}$	0.063
		vapor; 1 atm	0.00484	0.00557	—	—	—	0.0569
15	Viscosity at 25°C; centipoise	liquid	0.42	0.26	0.016	0.15	0.020	0.34
		vapor; 1 atm	0.011	0.013	—	0.016	—	0.011
16	Surface tension at 25°C; dyne cm		18	9	$14^{-73.3^{\circ}C}$	4	$14^{-73.3^{\circ}C}$	18
17	Refractive index of liquid at 25°C		1.374	1.287	$1.199^{-73.3^{\circ}C}$	1.238	$1.151^{-73.3^{\circ}C}$	1.354
18	Dielectric constant	liquid	$2.28^{29^{\circ}C}$	$2.13^{29^{\circ}C}$	—	—	—	$5.34^{28^{\circ}C}$
		vapor at 0.5 atm	$1.10019^{29^{\circ}C}$	$1.0016^{29^{\circ}C}$	$1.0013^{29^{\circ}C}$	—	$1.0006^{24.5^{\circ}C}$	$1.0035^{30^{\circ}C}$
19	Solubility in water at 25°C and 1 atm; wt %		0.011	0.028	0.009	0.03	0.0015	0.95
20	Solubility of water in compound at 25°C; wt %		0.011	0.009	—	$0.0095^{21.1^{\circ}C}$	—	0.13
21	Toxicity; Group number (See separate table for definition of group number.)		5a	6	probably 6	6	probably 6	<4 >5

PHYSICAL PROPERTIES OF FLUOROCARBON REFRIGERANTS (Continued)

Property No.	22	23	112	113	114	114B2	115	116	500	502
1	$CHClF_2$	CHF_3	$C_2Cl_4F_2$	$C_2Cl_3F_3$	$C_2Cl_2F_4$	$C_2Br_2F_4$	C_2ClF_5	C_2F_6	*	**
2	80.47	70.01	203.83	187.38	170.91	259.83	154.47	138.01	105.5	120.7
3	−40.75	−82.03	92.8	47.57	3.77	47.26	−38.7	−78.2	−33.5	−45.6
4	−160	−155.2	26	−35	−94	−110.5	−106	−100.6	−158.9	—
5	96	25.9	278	214.1	145.7	214.5	80	19.7	105.4	179.89
6	49.12	47.7	34	33.7	32.2	34	30.8	29.4	43.7	590.3
7	165	133	370	325	293	329	259	225	—	290
8	0.525	0.525	0.55	0.576	0.582	0.790	0.596	0.612	0.497	0.56
9	1.194	0.670	$1.634^{30°C}$	1.565	1.456	2.163	1.291	$1.587^{-73.3°C}$	$1.138^{30°C}$	1.242
10	4.72	4.66	7.02	7.38	7.83	—	8.37	9.01	—	6.05
11	0.300	1.553	—	0.218	0.243	0.166	0.285	$0.232^{-73.3°C}$	$0.161^{30°C}$	—
12	0.157	0.176	—	$0.161^{60°C}$	0.170	—	0.164	$0.182^{0 mm}$	—	—
13	55.81	57.23	37 (est)	35.07	32.51	25 (est)	30.11	27.97	—	42.48
14	0.052	0.008	0.040	0.038	0.034	0.027	0.026	$0.045^{-100°F}$	—	0.038
	0.00678	—	—	$0.0045^{0.5 atm}$	0.00646	—	0.00803	0.0098 (est)	—	—
15	0.23	0.016 (est)	1.21	0.68	0.38	0.72	0.26	—	$0.292^{-15°C}$	0.25
	0.013	—	—	$0.01^{0.1 atm}$	0.011	—	0.013	—	—	—
16	8	$15^{-73.3°C}$	$23^{30°C}$	19	12	18	5	$16^{-73.3°C}$	—	8
17	1.256	$1.251^{-73.3°C}$	1.413	1.354	1.288	1.367	1.214	$1.206^{-73.3°C}$	—	1.235
18	$6.11^{24°C}$	—	$2.52^{25°C}$	$2.41^{25°C}$	$2.26^{25°C}$	$2.34^{25°C}$	—	—	—	—
	$1.0035^{25.4°C}$	—	—	—	$1.0021^{26.5°C}$	—	$1.0018^{27.4°C}$	—	—	—
19	0.30	0.10	0.12 (sat. pr)	0.017 (sat. pr)	0.013	—	0.006	—	—	—
20	0.13	—	—	0.011	0.009	—	—	—	—	0.560
21	5a	probably 6	<4 >5	<4 >5	6	5a	6	probably 6	5a	5a

* Azeotrope of CCl_2F_2 (73.8 wt %) and $C_2H_4F_2$ (26.2 wt %) ** Azeotrope of $CHClF_2$ (48.8 wt %) and C_2ClF_5 (51.2 wt %).

UNDERWRITERS' LABORATORIES' CLASSIFICATION OF COMPARATIVE LIFEHAZARD OF GASES AND VAPORS

(Group number definition)

Group	Definition	Examples
1	Gases or vapors which in concentrations of the order of 1/2 to 1% for durations of exposure of the order of 5 min are lethal or produce serious injury.	Sulfur dioxide
2	Gases or vapors which in concentrations of the order of 1/2 to 1 % for duration of exposure of the order of 1/2 hr are lethal or produce serious injury	Ammonia, methyl bromide
3	Gases or vapors which in concentrations of the order of 2 to 2 1/2 % for durations of exposure of the order of 1 hr are lethal or produce serious injury.	Bromochloromethane carbon tetrachloride, chloroform, methyl formate
4	Gases or vapors which in concentrations of the order of 2 to 2 1/2 % for durations of exposure of the order of 2 hr are lethal or produce serious injury.	Dichloroethylene methyl chloride, ethylbromide
Between 4 and 5	Appear to classify as somewhat less toxic than Group 4.	Methylene chloride, ethyl chloride.
		Refrigerant 112[a]
	Much less toxic than group 4 but somewhat more toxic than Group 5.	Refrigerant 113
		Refrigerant 21
5a	Gases or vapors much less toxic than Group 4 but more toxic than Group 6.	Refrigerant 11
		Refrigerant 22
		Refrigerant 114B2
		Refrigerant 502
		Carbon dioxide
5b	Gases or vapors which available data indicate would classify as either Group 5a or Group 6.	Ethane, propane, butane
6	Gases or vapors which in concentrations up to at least about 20% by volume for duration of exposure of the order of 2 hr do not appear to produce injury.	Refrigerant 13B1
		Refrigerant 12
		Refrigerant 114
		Refrigerant 115
		Refrigerant 13[a]
		Refrigerant 14[a]
		Refrigerant 23[a]
		Refrigerant 116[a]
		Refrigerant C318[a]

[a] Not tested by U.L. but estimated to belong in group indicated.

THERMAL CONDUCTIVITY OF LIQUID FLUOROCARBONS

To convert from W/(m)(K) to Btu/(hr)(ft)(°F) divided by 1.7296.
To convert from W/(m)(K) to cal/(sec)(cm)(°C) divide by 418.4

Fluorocarbon	Formula	Temperature, K	Conductivity, W/(m)(K)
12	CCl_2F_2	277.2	94.14
		298.1	97.49
		303.8	103.76
		317.9	110.04
		329.9	117.15
		346.8	126.36
22	$CHClF_3$	289.5	107.11
		302.8	114.64
		327.8	126.76
		346.8	140.16
114	$C_2Cl_2F_4$	303.6	113.39
		316.8	122.59
		328.7	130.96
		343.0	140.58
13B1	$CBrF_3$	277.3	91.21
		282.3	93.30
		303.5	103.34
		318.7	111.71
		331.9	118.41
		342.8	123.01
		346.9	128.03
C-318	C_4F_8	280.1	112.97
		287.3	117.57
		298.0	130.96
		310.7	141.42
		323.2	148.11
		332.2	156.48
		342.9	158.99
		348.4	165.69
		350.8	166.94

MISCELLANEOUS PROPERTIES OF COMMON REFRIGERANTS

Refrigerant	Formula	Flash Point °F.	Ignition Temp. °F.	Explosive Limits % by Volume: Lower	Explosive Limits % by Volume: Upper	Vapor Density (Air = 1)	Boiling Point °F	Threshold Limit Value* Parts per Million in Air	Water Soluble
Ammonia	NH_3	—	1204	16	25	0.59	−28	100	yes
Bromotrifluoromethane (Kulene-131)	CF_3Br		nonflammable			5.25	−73.6	—	no
Butane	C_4H_{10}	−76	806	1.8	8.4	2.04	33	—	no
Carbon dioxide	CO_2		nonflammable			1.53	−108	5000	yes
Carbon tetrachloride	CCL_4		nonflammable			5.32	170	25	no
Dichlorodifluoromethane (Freon-12)	CCl_2F_2		nonflammable			4.17	−21.6	—	no
Dichlorodifluoromethane, 73.8% Ethylidene fluoride, 26.2% (Carrene-7)	CCl_2F_2 CH_3CHF_2		nonflammable			3.24	−28.0	—	no
Dichloromonofluoromethane (Freon-21)	$CHCl_2F$		practically nonflammable			3.55	48	—	no
Dichlorotetrafluoroethane (Freon-114)	$C_2Cl_2F_4$		practically nonflammable			5.89	38	—	no
Ethane	C_2H_6	<20	950	3.0	12.5	1.04	−128	—	no
Ethylene	C_2H_4	<20	842	3.1	32	0.972	−155	—	yes
Isobutane	$(CH_3)_3CH$	<20	1010	1.8	8.4	2.01	14	—	no
Methyl chloride	CH_3Cl	632	1170	10.7	11.4	1.78	−11	100	yes
Monochlorodifluoromethane (Freon-22)	$CHClF_2$		practically nonflammable			2.9	−41	—	yes
Monochlorotrifluoromethane (Freon-13)	$CClF_3$		nonflammable			3.6	−112	—	—
Propane	C_3H_8	<20	871	2.2	9.5	1.56	−45	—	no
Propylene	C_3H_6	<20	927	2.4	10.3	1.49	−53	—	yes
Sulfur dioxide	SO_2		nonflammable			2.2	14	10	yes
Tetrafluoromethane (Freon-14)	CF_4		nonflammable			3.0	−198	—	no
Trichloroethylene	C_2HCl_3		nonflammable at normal temperature			4.53	189	200	no
Trichloromonofluoromethane (Freon-11)(Carrene-2)	CCL_3F		nonflammable			4.7	75.3	—	no
Trichlorotrifluoroethane (Freon-113)	$C_2Cl_3F_3$		practically nonflammable			6.4	118	—	no

* Maximum average atmospheric concentration of contaminants to which workers may be exposed for an eight-hour work day without injury to health. (American Conference of Governmental Industrial Hygienists: "Threshold Limit Values for 1954.) Where blanks appear in this column no published information was available on threshold limit values.

HYGROMETRIC AND BAROMETRIC TABLES

CONVERSION TABLE FOR BAROMETRIC READINGS

U.S. inches to cm.

Inches	.00	.01	.02	.03	.04	.05	.06	.07	.08	.09
27.0	68.580	.606	.631	.656	.682	.707	.733	.758	.783	.809
27.1	.834	.860	.885	.910	.936	.961	.987	*.012	*.037	*.063
27.2	69.088	.114	.139	.164	.190	.215	.241	.266	.291	.317
27.3	.342	.368	.393	.418	.444	.469	.495	.520	.545	.571
27.4	.596	.622	.647	.672	.698	.723	.749	.774	.799	.825
27.5	.850	.876	.901	.926	.952	.977	*.002	*.028	*.053	*.079
27.6	70.104	.130	.155	.180	.206	.231	.257	.282	.307	.333
27.7	.358	.384	.409	.434	.460	.485	.511	.536	.561	.587
27.8	.612	.638	.663	.688	.714	.739	.765	.790	.815	.841
27.9	.866	.892	.917	.942	.968	.993	*.018	*.044	*.069	*.095
28.0	71.120	.146	.171	.196	.222	.247	.273	.298	.323	.349
28.1	.374	.400	.425	.450	.476	.501	.527	.552	.577	.603
28.2	.628	.654	.679	.704	.730	.755	.781	.806	.831	.857
28.3	.882	.908	.933	.958	.984	*.009	*.035	*.060	*.085	*.111
28.4	72.136	.162	.187	.212	.238	.263	.289	.314	.339	.365
28.5	.390	.416	.441	.466	.492	.517	.543	.568	.593	.619
28.6	.644	.670	.695	.720	.746	.771	.797	.822	.847	.873
28.7	.898	.924	.949	.974	*.000	*.025	*.051	*.076	*.101	*.127
28.8	73.152	.178	.203	.228	.254	.279	.305	.330	.355	.381
28.9	.406	.432	.457	.482	.508	.533	.559	.584	.609	.635

Inches	.00	.01	.02	.03	.04	.05	.06	.07	.08	.09
29.0	.660	.686	.711	.736	.762	.787	.813	.838	.863	.889
29.1	.914	.940	.965	.990	*.016	*.041	*.067	*.092	*.117	*.143
29.2	74.168	.194	.219	.244	.270	.295	.321	.346	.371	.397
29.3	.422	.448	.473	.498	.524	.549	.575	.600	.625	.651
29.4	.676	.702	.727	.752	.778	.803	.829	.854	.879	.905
29.5	.930	.956	.981	*.006	*.032	*.057	*.083	*.108	*.133	*.159
29.6	75.184	.210	.235	.260	.286	.311	.337	.362	.387	.413
29.7	.438	.464	.489	.514	.540	.565	.591	.616	.641	.667
29.8	.692	.718	.743	.768	.794	.819	.845	.870	.895	.921
29.9	.946	.972	.997	*.022	*.048	*.073	*.099	*.124	*.149	*.175
30.0	76.200	.226	.251	.277	.302	.327	.353	.378	.404	.429
30.1	.454	.480	.505	.531	.556	.581	.607	.632	.658	.683
30.2	.708	.734	.759	.785	.810	.835	.861	.886	.912	.937
30.3	.962	.988	*.013	*.039	*.064	*.089	*.115	*.140	*.166	*.191
30.4	77.216	.242	.267	.293	.318	.343	.369	.394	.420	.445
30.5	.470	.496	.521	.547	.572	.597	.623	.648	.674	.699
30.6	.724	.750	.775	.801	.826	.851	.877	.902	.928	.953
30.7	.978	*.004	*.029	*.055	*.080	*.105	*.131	*.156	*.182	*.207
30.8	78.232	.258	.283	.309	.334	.359	.385	.410	.436	.461
30.9	.486	.512	.537	.563	.588	.613	.639	.664	.690	.715

U.S. Inches to Millibars

Based on the relation 1 inch of mercury at 32°F represents a pressure of 33.8639 millibars.
Note: Figures in last nine columns to be preceded by 7, 8, 9 or 10 as indicated in column 2.

Inches	.00	.01	.02	.03	.04	.05	.06	.07	.08	.09
23.0	7 78.87	79.21	79.55	79.89	80.22	80.56	80.90	81.24	81.58	81.92
23.1	7 82.26	82.59	82.93	83.27	83.61	83.95	84.29	84.63	84.97	85.30
23.2	7 85.64	85.98	86.32	86.66	87.00	87.34	87.67	88.01	88.35	88.69
23.3	7 89.03	89.37	89.71	90.04	90.38	90.72	91.06	91.40	91.74	92.08
23.4	7 92.42	92.75	93.09	93.43	93.77	94.11	94.45	94.79	95.12	95.46
23.5	7 95.80	96.14	96.48	96.82	97.16	97.49	97.83	98.17	98.51	98.85
23.6	7 99.19	99.53	99.87	*00.20	*00.54	*00.88	*01.22	*01.56	*01.90	*02.24

Inches	.00	.01	.02	.03	.04	.05	.06	.07	.08	.09
23.7	8 02.57	02.91	03.25	03.59	03.93	04.27	04.61	04.94	05.28	05.62
23.8	8 05.96	06.30	06.64	06.98	07.32	07.65	07.99	08.33	08.67	09.01
23.9	8 09.35	09.69	10.02	10.36	10.70	11.04	11.38	11.72	12.06	12.39
24.0	8 12.73	13.07	13.41	13.75	14.09	14.43	14.77	15.10	15.44	15.78
24.1	8 16.12	16.46	16.80	17.14	17.47	17.81	18.15	18.49	18.83	19.17
24.2	8 19.51	19.85	20.18	20.52	20.86	21.20	21.54	21.88	22.22	22.55
24.3	8 22.89	23.23	23.57	23.91	24.25	24.59	24.92	25.26	25.60	25.94

U.S. Inches to Millibars (Continued)

Inches	.00	.01	.02	.03	.04	.05	.06	.07	.08	.09
24.4	8 26.28	26.62	26.96	27.30	27.63	27.97	28.31	28.65	28.99	29.33
24.5	8 29.67	30.00	30.34	30.68	31.02	31.36	31.70	32.04	32.37	32.71
24.6	8 33.05	33.39	33.73	34.07	34.41	34.75	35.08	35.42	35.76	36.10
24.7	8 36.44	36.78	37.12	37.45	37.79	38.13	38.47	38.81	39.15	39.49
24.8	8 39.82	40.16	40.50	40.84	41.18	41.52	41.86	42.20	42.53	42.87
24.9	8 43.21	43.55	43.89	44.23	44.57	44.90	45.24	45.58	45.92	46.26
25.0	8 46.60	46.94	47.27	47.61	47.95	48.29	48.63	48.97	49.31	49.65
25.1	8 49.98	50.32	50.66	51.00	51.34	51.68	52.02	52.35	52.69	53.03
25.2	8 53.37	53.71	54.05	54.39	54.72	55.06	55.40	55.74	56.08	56.42
25.3	8 56.76	57.10	57.43	57.77	58.11	58.45	58.79	59.13	59.47	59.80
25.4	8 60.14	60.48	60.82	61.16	61.50	61.84	62.17	62.51	62.85	63.19
25.5	8 63.53	63.87	64.21	64.55	64.88	65.22	65.56	65.90	66.24	66.58
25.6	8 66.92	67.25	67.59	67.93	68.27	68.61	68.95	69.29	69.62	69.96
25.7	8 70.30	70.64	70.98	71.32	71.66	72.00	72.33	72.67	73.01	73.35
25.8	8 73.69	74.03	74.37	74.70	75.04	75.38	75.72	76.06	76.40	76.74
25.9	8 77.08	77.41	77.75	78.09	78.43	78.77	79.11	79.45	79.78	80.12
26.0	8 80.46	80.80	81.14	81.48	81.82	82.15	82.49	82.83	83.17	83.51
26.1	8 83.85	84.19	84.53	84.86	85.20	85.54	85.88	86.22	86.56	86.90
26.2	8 87.23	87.57	87.91	88.25	88.59	88.93	89.27	89.60	89.94	90.28
26.3	8 90.62	90.96	91.30	91.64	91.98	92.31	92.65	92.99	93.33	93.67
26.4	8 94.01	94.35	94.68	95.02	95.36	95.70	96.04	96.38	96.72	97.05
26.5	8 97.39	97.73	98.07	98.41	98.75	99.09	99.43	99.76	*00.10	*00.44
26.6	9 00.78	01.12	01.46	01.80	02.13	02.47	02.81	03.15	03.49	03.83
26.7	9 04.17	04.50	04.84	05.18	05.52	05.86	06.20	06.54	06.88	07.21
26.8	9 07.55	07.89	08.23	08.57	08.91	09.25	09.58	09.92	10.26	10.60
26.9	9 10.94	11.28	11.62	11.95	12.29	12.63	12.97	13.31	13.65	13.99
27.0	9 14.33	14.66	15.00	15.34	15.68	16.02	16.36	16.70	17.03	17.37
27.1	9 17.71	18.05	18.39	18.73	19.07	19.40	19.74	20.08	20.42	20.76
27.2	9 21.10	21.44	21.78	22.11	22.45	22.79	23.13	23.47	23.81	24.15
27.3	9 24.48	24.82	25.16	25.50	25.84	26.18	26.52	26.85	27.19	27.53
27.4	9 27.87	28.21	28.55	28.89	29.23	29.56	29.90	30.24	30.58	30.03
27.5	9 31.26	31.60	31.93	32.27	32.61	32.95	33.29	33.63	33.97	34.31
27.6	9 34.64	34.98	35.32	35.66	36.00	36.34	36.68	37.01	37.35	37.69
27.7	9 38.03	38.37	38.71	39.05	39.38	39.72	40.06	40.40	40.74	41.08
27.8	9 41.42	41.76	42.09	42.43	42.77	43.11	43.45	43.79	44.13	44.46
27.9	9 44.80	45.14	45.48	45.82	46.16	46.50	46.83	47.17	47.51	47.85
28.0	9 48.19	48.53	48.87	49.21	49.54	49.88	50.22	50.56	50.90	51.24
28.1	9 51.58	51.91	52.25	52.59	52.93	53.27	53.61	53.95	54.28	54.62
28.2	9 54.96	55.30	55.64	55.98	56.32	56.66	56.99	57.33	57.67	58.01
28.3	9 58.35	58.69	59.03	59.36	59.70	60.04	60.38	60.72	61.06	61.40
28.4	9 61.73	62.07	62.41	62.75	63.09	63.43	63.77	64.11	64.44	64.78
28.5	9 65.12	65.46	65.80	66.14	66.48	66.81	67.15	67.49	67.83	68.17
28.6	9 68.51	68.85	69.18	69.52	69.86	70.20	70.54	70.88	71.22	71.56
28.7	9 71.89	72.23	72.57	72.91	73.25	73.59	73.93	74.26	74.60	74.94
28.8	9 75.28	75.62	75.96	76.30	76.63	76.97	77.31	77.65	77.99	78.33
28.9	9 78.67	79.01	79.34	79.68	80.02	80.36	80.70	81.04	81.38	81.71
29.0	9 82.05	82.39	82.73	83.07	83.41	83.75	84.08	84.42	84.76	85.10
29.1	9 85.44	85.78	86.12	86.46	86.79	87.13	87.47	87.81	88.15	88.49
29.2	9 88.83	89.16	89.50	89.84	90.18	90.52	90.86	91.20	91.53	91.87
29.3	9 92.21	92.55	92.89	93.23	93.57	93.91	94.24	94.58	94.92	95.26
29.4	9 95.60	95.94	96.28	96.61	96.95	97.29	97.63	97.97	98.31	98.65
29.5	9 98.99	99.32	99.66	00.00	*00.34	*00.68	*01.02	*01.36	*01.69	*02.03
29.6	10 02.37	02.71	03.05	03.39	03.73	04.06	04.40	04.74	05.08	05.42
29.7	10 05.76	06.10	06.44	06.77	07.11	07.45	07.79	08.13	08.47	08.81
29.8	10 09.14	09.48	09.82	10.16	10.50	10.84	11.18	11.51	11.85	12.19
29.9	10 12.53	12.87	13.21	13.55	13.89	14.22	14.56	14.90	15.24	15.58
30.0	10 15.92	16.26	16.59	16.93	17.27	17.61	17.95	18.29	18.63	18.96
30.1	10 19.30	19.64	19.98	20.32	20.66	21.00	21.34	21.67	22.01	22.35
30.2	10 22.69	23.03	23.37	23.71	24.04	24.38	24.72	25.06	25.40	25.74
30.3	10 26.08	26.41	26.75	27.09	27.43	27.77	28.11	28.45	28.79	29.12
30.4	10 29.46	29.80	30.14	30.48	30.82	31.16	31.49	31.83	32.17	32.51
30.5	10 32.85	33.19	33.53	33.86	34.20	34.54	34.88	35.22	35.56	35.90
30.6	10 36.24	36.57	36.91	37.25	37.59	37.93	38.27	38.61	38.94	39.28
30.7	10 39.62	39.96	40.30	40.64	40.98	41.31	41.65	41.99	42.33	42.67
30.8	10 43.01	43.35	43.69	44.02	44.36	44.70	45.04	45.38	45.72	46.06
30.9	10 46.39	46.73	47.07	47.41	47.75	48.09	48.43	48.76	49.10	49.44
31.0	10 49.78	50.12	50.46	50.80	51.14	51.47	51.81	52.15	52.49	52.83
31.1	10 53.17	53.51	53.84	54.18	54.52	54.86	55.20	55.54	55.88	56.22
31.2	10 56.55	56.89	57.23	57.57	57.91	58.25	58.59	58.92	59.26	59.60
31.3	10 59.94	60.28	60.62	60.96	61.29	61.63	61.97	62.31	62.65	62.99
31.4	10 63.33	63.67	64.00	64.34	64.68	65.02	65.36	65.70	66.04	66.37
31.5	10 66.71	67.05	67.39	67.73	68.07	68.41	68.74	69.08	69.42	69.76
31.6	10 70.10	70.44	70.78	71.12	71.45	71.79	72.13	72.47	72.81	73.15
31.7	10 73.49	73.82	74.16	74.50	74.84	75.18	75.52	75.86	76.19	76.53
31.8	10 76.87	77.21	77.55	77.89	78.23	78.57	78.90	79.24	79.58	79.92
31.9	10 80.26	80.60	80.94	81.27	81.61	81.95	82.29	82.63	82.97	83.31

Centimeters to Millibars

Based on the relation 1 centimeter of mercury at 0°C represents a pressure of 13.3322 millibars.

Note: Figures in last nine columns to be preceded by 9.

Centi-meters	.00	.01	.02	.03	.04	.05	.06	.07	.08	.09
68.0	9 06.59	06.72	06.86	06.99	07.12	07.26	07.39	07.52	07.66	07.79
68.1	9 07.92	08.06	08.19	08.32	08.46	08.59	08.72	08.86	08.99	09.12
68.2	9 09.26	09.39	09.52	09.66	09.79	09.92	10.06	10.19	10.32	10.46
68.3	9 10.59	10.72	10.86	10.99	11.12	11.26	11.39	11.52	11.66	11.79
68.4	9 11.92	12.06	12.19	12.32	12.46	12.59	12.72	12.86	12.99	13.12
68.5	9 13.26	13.39	13.52	13.66	13.79	13.92	14.06	14.19	14.32	14.46
68.6	9 14.59	14.72	14.86	14.99	15.12	15.26	15.39	15.52	15.66	15.79
68.7	9 15.92	16.06	16.19	16.32	16.46	16.59	16.72	16.86	16.99	17.12
68.8	9 17.26	17.39	17.52	17.66	17.79	17.92	18.06	18.19	18.32	18.46
68.9	9 18.59	18.72	18.86	18.99	19.12	19.26	19.39	19.52	19.66	19.79
69.0	9 19.92	20.06	20.19	20.32	20.46	20.59	20.72	20.86	20.99	21.12
69.1	9 21.26	21.39	21.52	21.65	21.79	21.92	22.05	22.19	22.32	22.45
69.2	9 22.59	22.72	22.85	22.99	23.12	23.25	23.39	23.52	23.65	23.79
69.3	9 23.92	24.05	24.19	24.32	24.45	24.59	24.72	24.85	24.99	25.12
69.4	9 25.25	25.39	25.52	25.65	25.79	25.92	26.05	26.19	26.32	26.45
69.5	9 26.59	26.72	26.85	26.99	27.12	27.25	27.39	27.52	27.65	27.79
69.6	9 27.92	28.05	28.19	28.32	28.45	28.59	28.72	28.85	28.99	29.12
69.7	9 29.25	29.39	29.52	29.65	29.79	29.92	30.05	30.19	30.32	30.45
69.8	9 30.59	30.72	30.85	30.99	31.12	31.25	31.39	31.52	31.65	31.79
69.9	9 31.92	32.05	32.19	32.32	32.45	32.59	32.72	32.85	32.99	33.12
70.0	9 33.25	33.39	33.52	33.65	33.79	33.92	34.05	34.19	34.32	34.45
70.1	9 34.59	34.72	34.85	34.99	35.12	35.25	35.39	35.52	35.65	35.79
70.2	9 35.92	36.05	36.19	36.32	36.45	36.59	36.72	36.85	36.99	37.12
70.3	9 37.25	37.39	37.52	37.65	37.79	37.92	38.05	38.19	38.32	38.45
70.4	9 38.59	38.72	38.85	38.99	39.12	39.25	39.39	39.52	39.65	39.79
70.5	9 39.92	40.05	40.19	40.32	40.45	40.59	40.72	40.85	40.99	41.12
70.6	9 41.25	41.39	41.52	41.65	41.79	41.92	42.05	42.19	42.32	42.45
70.7	9 42.59	42.72	42.85	42.99	43.12	43.25	43.39	43.52	43.65	43.79
70.8	9 43.92	44.05	44.19	44.32	44.45	44.59	44.72	44.85	44.99	45.12
70.9	9 45.25	45.39	45.52	45.65	45.79	45.92	46.05	46.19	46.32	46.45
71.0	9 46.59	46.72	46.85	46.99	47.12	47.25	47.39	47.52	47.65	47.79
71.1	9 47.92	48.05	48.19	48.32	48.45	48.59	48.72	48.85	48.99	49.12
71.2	9 49.25	49.39	49.52	49.65	49.79	49.92	50.05	50.19	50.32	50.45
71.3	9 50.59	50.72	50.85	50.99	51.12	51.25	51.39	51.52	51.65	51.79
71.4	9 51.92	52.05	52.19	52.32	52.45	52.59	52.72	52.85	52.99	53.12
71.5	9 53.25	53.39	53.52	53.65	53.79	53.92	54.05	54.19	54.32	54.45
71.6	9 54.59	54.72	54.85	54.99	55.12	55.25	55.39	55.52	55.65	55.79
71.7	9 55.92	56.05	56.19	56.32	56.45	56.59	56.72	56.85	56.99	57.12
71.8	9 57.25	57.39	57.52	57.65	57.79	57.92	58.05	58.19	58.32	58.45
71.9	9 58.59	58.72	58.85	58.99	59.12	59.25	59.39	59.52	59.65	59.79
72.0	9 59.92	60.05	60.19	60.32	60.45	60.59	60.72	60.85	60.98	61.12
72.1	9 61.25	61.38	61.52	61.65	61.78	61.92	62.05	62.18	62.32	62.45
72.2	9 62.58	62.72	62.85	62.98	63.12	63.25	63.38	63.52	63.65	63.78
72.3	9 63.92	64.05	64.18	64.32	64.45	64.58	64.72	64.85	64.98	65.12
72.4	9 65.25	65.38	65.52	65.65	65.78	65.92	66.05	66.18	66.32	66.45
72.5	9 66.58	66.72	66.85	66.98	67.12	67.25	67.38	67.52	67.65	67.78
72.6	9 67.92	68.05	68.18	68.32	68.45	68.58	68.72	68.85	68.98	69.12
72.7	9 69.25	69.38	69.52	69.65	69.78	69.92	70.05	70.18	70.32	70.45
72.8	9 70.58	70.72	70.85	70.98	71.12	71.25	71.38	71.52	71.65	71.78
72.9	9 71.92	72.05	72.18	72.32	72.45	72.58	72.72	72.85	72.98	73.12
73.0	9 73.25	73.38	73.52	73.65	73.78	73.92	74.05	74.18	74.32	74.45
73.1	9 74.58	74.72	74.85	74.98	75.12	75.25	75.38	75.52	75.65	75.78
73.2	9 75.92	76.05	76.18	76.32	76.45	76.58	76.72	76.85	76.98	77.12
73.3	9 77.25	77.38	77.52	77.65	77.78	77.92	78.05	78.18	78.32	78.45
73.4	9 78.58	78.72	78.85	78.98	79.12	79.25	79.38	79.52	79.65	79.78
73.5	9 79.92	80.05	80.18	80.32	80.45	80.58	80.72	80.85	80.98	81.12
73.6	9 81.25	81.38	81.52	81.65	81.78	81.92	82.05	82.18	82.32	82.45
73.7	9 82.58	82.72	82.85	82.98	83.12	83.25	83.38	83.52	83.65	83.78
73.8	9 83.92	84.05	84.18	84.32	84.45	84.58	84.72	84.85	84.98	85.12
73.9	9 85.25	85.38	85.52	85.65	85.78	85.92	86.05	86.18	86.32	86.45
74.0	9 86.58	86.72	86.85	86.98	87.12	87.25	87.38	87.52	87.65	87.78
74.1	9 87.92	88.05	88.18	88.32	88.45	88.58	88.72	88.85	88.98	89.12
74.2	9 89.25	89.38	89.52	89.65	89.78	89.92	90.05	90.18	90.32	90.45
74.3	9 90.58	90.72	90.85	90.98	91.12	91.25	91.38	91.52	91.65	91.78
74.4	9 91.92	92.05	92.18	92.32	92.45	92.58	92.72	92.85	92.98	93.12
74.5	9 93.25	93.38	93.52	93.65	93.78	93.92	94.05	94.18	94.32	94.45
74.6	9 94.58	94.72	94.85	94.98	95.12	95.25	95.38	95.52	95.65	95.78
74.7	9 95.92	96.05	96.18	96.32	96.45	96.58	96.72	96.85	96.98	97.12
74.8	9 97.25	97.38	97.52	97.65	97.78	97.92	98.05	98.18	98.32	98.45
74.9	9 98.58	98.72	98.85	98.98	99.12	99.25	99.38	99.52	99.65	99.78
75.0	9 99.92	*00.05	*00.18	*00.31	*00.45	*00.58	*00.71	*00.85	*00.98	*01.11
75.1	10 01.25	01.38	01.51	01.65	01.78	01.91	02.05	02.18	02.31	02.45
75.2	10 02.58	02.71	02.85	02.98	03.11	03.25	03.38	03.51	03.65	03.78
75.3	10 03.91	04.05	04.18	04.31	04.45	04.58	04.71	04.85	04.98	05.11
75.4	10 05.25	05.38	05.51	05.65	05.78	05.91	06.05	06.18	06.31	06.45
75.5	10 06.58	06.71	06.85	06.98	07.11	07.25	07.38	07.51	07.65	07.78
75.6	10 07.91	08.05	08.18	08.31	08.45	08.58	08.71	08.85	08.98	09.11
75.7	10 09.25	09.38	09.51	09.65	09.78	09.91	10.05	10.18	10.31	10.45
75.8	10 10.58	10.71	10.85	10.98	11.11	11.25	11.38	11.51	11.65	11.78
75.9	10 11.91	12.05	12.18	12.31	12.45	12.58	12.71	12.85	12.98	13.11
76.0	10 13.25	13.38	13.51	13.65	13.78	13.91	14.05	14.18	14.31	14.45
76.1	10 14.58	14.71	14.85	14.98	15.11	15.25	15.38	15.51	15.65	15.78
76.2	10 15.91	16.05	16.18	16.31	16.45	16.58	16.71	16.85	16.98	17.11
76.3	10 17.25	17.38	17.51	17.65	17.78	17.91	18.05	18.18	18.31	18.45
76.4	10 18.58	18.71	18.85	18.98	19.11	19.25	19.38	19.51	19.65	19.78
76.5	10 19.91	20.05	20.18	20.31	20.45	20.58	20.71	20.85	20.98	21.11
76.6	10 21.25	21.38	21.51	21.65	21.78	21.91	22.05	22.18	22.31	22.45
76.7	10 22.58	22.71	22.85	22.98	23.11	23.25	23.38	23.51	23.65	23.78
76.8	10 23.91	24.05	24.18	24.31	24.45	24.58	24.71	24.85	24.98	25.11
76.9	10 25.25	25.38	25.51	25.65	25.78	25.91	26.05	26.18	26.31	26.45
77.0	10 26.58	26.71	26.85	26.98	27.11	27.25	27.38	27.51	27.65	27.78
77.1	10 27.91	28.05	28.18	28.31	28.45	28.58	28.71	28.85	28.98	29.11
77.2	10 29.25	29.38	29.51	29.65	29.78	29.91	30.05	30.18	30.31	30.45
77.3	10 30.58	30.71	30.85	30.98	31.11	31.25	31.38	31.51	31.65	31.78
77.4	10 31.91	32.05	32.18	32.31	32.45	32.58	32.71	32.85	32.98	33.11
77.5	10 33.25	33.38	33.51	33.65	33.78	33.91	34.05	34.18	34.31	34.45
77.6	10 34.58	34.71	34.85	34.98	35.11	35.25	35.38	35.51	35.65	35.78
77.7	10 35.91	36.05	36.18	36.31	36.45	36.58	36.71	36.85	36.98	37.11
77.8	10 37.25	37.38	37.51	37.65	37.78	37.91	38.05	38.18	38.31	38.45
77.9	10 38.58	38.71	38.85	38.98	39.11	39.24	39.38	39.51	39.64	39.78

TEMPERATURE CORRECTION FOR BAROMETER READINGS

Brass Scale — Metric Units

To reduce readings of a mercurial barometer with a brass scale to 0°C subtract the appropriate quantity as found in the table. These values are based on the coefficient of expansion of mercury $(181792 + 0.175t + 0.035116t^2) \times 10^{-9}$, and of brass 0.0000184 per °C. Corrections are in millimeters.

Temp. °C	Observed height in millimeters																	
	620	630	640	650	660	670	680	690	700	710	720	730	740	750	760	770	780	790
0	0.00	0.00	0.00	0.00	0.00	0.00	0.00	0.00	0.00	0.00	0.00	0.00	0.00	0.00	0.00	0.00	0.00	0.00
1	.10	.10	.10	.11	.11	.11	.11	.11	.11	.12	.12	.12	.12	.12	.12	.13	.13	.13
2	.20	.21	.21	.21	.22	.22	.22	.23	.23	.23	.24	.24	.24	.25	.25	.25	.25	.26
3	.30	.31	.31	.32	.32	.33	.33	.34	.34	.35	.35	.36	.36	.37	.37	.38	.38	.39
4	.40	.41	.42	.42	.43	.44	.44	.45	.46	.46	.47	.48	.48	.49	.50	.50	.51	.52
5	0.51	0.51	0.52	0.53	0.54	0.55	0.56	0.56	0.57	0.58	0.59	0.60	0.60	0.61	0.62	0.63	0.64	0.64
6	.61	.62	.63	.64	.65	.66	.67	.68	.69	.70	.71	.71	.72	.73	.74	.75	.76	.77
7	.71	.72	.73	.74	.75	.77	.78	.79	.80	.81	.82	.83	.85	.86	.87	.88	.89	.90
8	.81	.82	.84	.85	.86	.87	.89	.90	.91	.93	.94	.95	.97	.98	.99	1.01	1.02	1.03
9	.91	.92	.94	.95	.97	.98	1.00	1.01	1.03	1.04	1.06	1.07	1.09	1.10	1.12	1.13	1.15	1.16
10	1.01	1.03	1.04	1.06	1.08	1.09	1.11	1.13	1.14	1.16	1.17	1.19	1.21	1.22	1.24	1.26	1.27	1.29
11	1.11	1.13	1.15	1.17	1.18	1.20	1.22	1.24	1.26	1.27	1.29	1.31	1.33	1.35	1.36	1.38	1.40	1.42
12	1.21	1.23	1.25	1.27	1.29	1.31	1.33	1.35	1.37	1.39	1.41	1.43	1.45	1.47	1.49	1.51	1.53	1.55
13	1.31	1.34	1.36	1.38	1.40	1.42	1.44	1.46	1.48	1.50	1.53	1.55	1.57	1.59	1.61	1.63	1.65	1.67
14	1.41	1.44	1.46	1.48	1.51	1.53	1.55	1.57	1.60	1.62	1.64	1.67	1.69	1.71	1.73	1.76	1.78	1.80
15	1.52	1.54	1.56	1.59	1.61	1.64	1.66	1.69	1.71	1.74	1.76	1.78	1.81	1.83	1.86	1.88	1.91	1.93
16	1.62	1.64	1.67	1.69	1.72	1.75	1.77	1.80	1.82	1.85	1.88	1.90	1.93	1.96	1.98	2.01	2.03	2.06
17	1.72	1.74	1.77	1.80	1.83	1.86	1.88	1.91	1.94	1.97	1.99	2.02	2.05	2.08	2.10	2.13	2.16	2.19
18	1.82	1.85	1.88	1.91	1.93	1.96	1.99	2.02	2.05	2.08	2.11	2.14	2.17	2.20	2.23	2.26	2.29	2.32
19	1.92	1.95	1.98	2.01	2.04	2.07	2.10	2.13	2.17	2.20	2.23	2.26	2.29	2.32	2.35	2.38	2.41	2.44
20	2.02	2.05	2.08	2.12	2.15	2.18	2.21	2.25	2.28	2.31	2.34	2.38	2.41	2.44	2.47	2.51	2.54	2.57
21	2.12	2.15	2.19	2.22	2.26	2.29	2.32	2.36	2.39	2.43	2.46	2.50	2.53	2.56	2.60	2.63	2.67	2.70
22	2.22	2.26	2.29	2.33	2.36	2.40	2.43	2.47	2.51	2.54	2.58	2.61	2.65	2.69	2.72	2.76	2.79	2.83
23	2.32	2.36	2.40	2.43	2.47	2.51	2.54	2.58	2.62	2.66	2.69	2.73	2.77	2.81	2.84	2.88	2.92	2.96
24	2.42	2.46	2.50	2.54	2.58	2.62	2.66	2.69	2.73	2.77	2.81	2.85	2.89	2.93	2.97	3.01	3.05	3.08
25	2.52	2.56	2.60	2.64	2.68	2.72	2.77	2.81	2.85	2.89	2.93	2.97	3.01	3.05	3.09	3.13	3.17	3.21
26	2.62	2.66	2.71	2.75	2.79	2.83	2.88	2.92	2.96	3.00	3.04	3.09	3.13	3.17	3.21	3.26	3.30	3.34
27	2.72	2.77	2.81	2.85	2.90	2.94	2.99	3.03	3.07	3.12	3.16	3.20	3.25	3.29	3.34	3.38	3.42	3.47
28	2.82	2.87	2.91	2.96	3.00	3.05	3.10	3.14	3.19	3.23	3.28	3.32	3.37	3.41	3.46	3.51	3.55	3.60
29	2.92	2.97	3.02	3.06	3.11	3.16	3.21	3.25	3.30	3.35	3.39	3.44	3.49	3.54	3.58	3.63	3.68	3.72
30	3.02	3.07	3.12	3.17	3.22	3.27	3.32	3.36	3.41	3.46	3.51	3.56	3.61	3.66	3.71	3.75	3.80	3.85
31	3.12	3.17	3.22	3.27	3.32	3.37	3.43	3.48	3.53	3.58	3.63	3.68	3.73	3.78	3.83	3.88	3.93	3.98
32	3.22	3.28	3.33	3.38	3.43	3.48	3.54	3.59	3.64	3.69	3.74	3.79	3.85	3.90	3.95	4.00	4.05	4.11
33	3.32	3.38	3.43	3.48	3.54	3.59	3.64	3.70	3.75	3.81	3.86	3.91	3.97	4.02	4.07	4.13	4.18	4.23
34	3.42	3.48	3.53	3.59	3.64	3.70	3.75	3.81	3.87	3.92	3.98	4.03	4.09	4.14	4.20	4.25	4.31	4.36
35	3.52	3.58	3.64	3.69	3.75	3.81	3.86	3.92	3.98	4.03	4.09	4.15	4.21	4.26	4.32	4.38	4.43	4.49

Brass Scale — English Units

Standard Temperature of scale 62°F; of mercury, 32°F. Zero correction at 28.5°F; subtract corrections above, add below. Owing to the difference in the standard temperature of English and metric scales, readings taken in inches to be reduced to centimeters should first be corrected for temperature.

Temp. °F	Observed height in inches																	
	23.0 in.	23.5 in.	24.0 in.	24.5 in.	25.0 in.	25.5 in.	26.0 in.	26.5 in.	27.0 in.	27.5 in.	28.0 in.	28.5 in.	29.0 in.	29.5 in.	30.0 in.	30.5 in.	31.0 in.	31.5 in.
0	+.060	+.061	+.063	+.064	+.065	+.067	+.068	+.069	+.070	.072	.073	.075	.076	.077	.078	.080	.081	.082
2	.056	.057	.058	.060	.061	.062	.063	.065	.065	.067	.068	.069	.070	.072	.073	.074	.075	.077
4	.052	.053	.054	.055	.056	.057	.058	.060	.061	.062	.063	.064	.065	.066	.067	.069	.070	.071
6	.047	.048	.049	.051	.052	.053	.054	.055	.056	.057	.058	.059	.060	.061	.062	.063	.064	.065
8	.043	.044	.045	.046	.047	.048	.049	.050	.051	.052	.053	.054	.054	.056	.056	.057	.058	.059
10	.039	.040	.041	.042	.042	.043	.044	.045	.046	.047	.047	.048	.049	.050	.051	.052	.053	.054
12	.035	.036	.036	.037	.038	.039	.039	.040	.041	.042	.042	.043	.044	.045	.045	.046	.047	.048
14	.031	.031	.032	.033	.033	.034	.035	.035	.036	.037	.037	.038	.039	.039	.040	.041	.041	.042
16	.026	.027	.028	.028	.029	.029	.030	.031	.031	.032	.032	.033	.033	.034	.034	.035	.036	.036
18	.022	.023	.023	.024	.024	.025	.025	.026	.026	.027	.027	.028	.028	.029	.029	.030	.030	.031
20	.018	.018	.019	.019	.020	.020	.020	.021	.021	.022	.022	.022	.023	.023	.024	.024	.024	.025
22	.014	.014	.014	.015	.015	.015	.016	.016	.016	.017	.017	.017	.017	.018	.018	.018	.019	.019
24	.010	.010	.010	.010	.011	.011	.011	.011	.011	.012	.012	.012	.012	.012	.013	.013	.013	.013
26	.005	.006	.006	.006	.006	.006	.006	.006	.006	.007	.007	.007	.007	.007	.007	.007	.007	.003
28	+.001	+.001	+.001	+.001	+.001	+.001	+.001	+.002	+.002	+.002	+.002	+.002	+.002	+.002	+.002	+.002	+.002	+.002
30	−.003	−.003	−.003	−.003	−.003	−.003	−.003	−.003	−.003	−.003	−.003	−.004	−.004	−.004	−.004	−.004	−.004	−.004
32	.007	.007	.007	.008	.008	.008	.008	.008	.008	.008	.009	.009	.009	.009	.009	.009	.009	.010
34	.011	.011	.012	.012	.012	.012	.013	.013	.013	.013	.014	.014	.014	.014	.015	.015	.015	.015
36	.015	.016	.016	.016	.017	.017	.017	.018	.018	.018	.019	.019	.019	.020	.020	.020	.021	.021
38	.020	.020	.020	.021	.021	.022	.022	.023	.023	.023	.024	.024	.025	.025	.026	.026	.026	.027
40	.024	.024	.025	.025	.026	.026	.027	.027	.028	.028	.029	.030	.030	.031	.031	.032	.032	.033
42	.028	.029	.029	.030	.030	.031	.032	.032	.033	.033	.034	.035	.035	.036	.036	.037	.038	.038
44	.032	.033	.033	.034	.035	.036	.036	.037	.038	.038	.039	.040	.040	.041	.042	.043	.043	.044
46	.036	.037	.038	.039	.039	.040	.041	.042	.043	.043	.044	.045	.046	.047	.047	.048	.049	.050
48	.040	.041	.042	.043	.044	.045	.046	.047	.047	.048	.049	.050	.051	.052	.053	.054	.054	.055
50	.045	.046	.046	.048	.048	.050	.050	.052	.052	.053	.054	.055	.056	.057	.058	.059	.060	.061
52	.049	.050	.051	.052	.053	.054	.055	.056	.057	.058	.059	.061	.061	.063	.064	.065	.066	.067
54	.053	.054	.055	.057	.057	.059	.060	.061	.062	.063	.064	.066	.067	.068	.069	.070	.071	.073
56	.057	.058	.060	.061	.062	.063	.064	.066	.067	.068	.069	.071	.072	.073	.074	.076	.077	.078
58	.061	.063	.064	.065	.066	.068	.069	.071	.072	.073	.074	.076	.077	.079	.080	.081	.082	.084
60	.065	.067	.068	.070	.071	.073	.074	.076	.077	.078	.080	.081	.082	.084	.085	.087	.088	.090
62	.069	.071	.073	.074	.076	.077	.079	.080	.082	.083	.085	.086	.088	.089	.091	.092	.094	.095
64	.074	.075	.077	.079	.080	.082	.083	.085	.086	.088	.090	.092	.093	.095	.096	.098	099	.101
66	.078	.079	.081	.083	.085	.087	.088	.090	.091	.093	.095	.097	.098	.100	.101	.103	.105	.107
68	.082	.084	.085	.088	.089	.091	.093	.095	.096	.098	.100	.102	.103	.105	.107	.109	.110	.113
70	.086	.088	.090	.092	.094	.096	.097	.100	.101	.103	.105	.107	.109	.111	.112	.115	.116	.118

TEMPERATURE CORRECTION FOR BAROMETER READINGS (continued)

Temp. °F	Observed height in inches 23.0 in.	23.5 in.	24.0 in.	24.5 in.	25.0 in.	25.5 in.	26.0 in.	26.5 in.	27.0 in.	27.5 in.	28.0 in.	28.5 in.	29.0 in.	29.5 in.	30.0 in.	30.5 in.	31.0 in.	31.5 in.
72	.090	.092	.094	.096	.098	.100	.102	.104	.106	.108	.110	.112	.114	.116	.118	.120	.122	.124
74	.094	.096	.098	.101	.103	.105	.107	.109	.111	.113	.115	.117	.119	.121	.123	.126	.127	.130
76	.098	.101	.103	.105	.107	.110	.111	.114	.116	.118	.120	.122	.124	.127	.128	.131	.133	.135
78	.103	.105	.107	.110	.112	.114	.116	.119	.120	.123	.125	.128	.129	.132	.134	.137	.138	.141
80	.107	.109	.111	.114	.116	.119	.121	.123	.125	.128	.130	.133	.135	.137	.139	.142	.144	.147
82	.111	.113	.116	.119	.121	.123	.125	.128	.130	.133	.135	.138	.140	.143	.145	.148	.149	.152
84	.115	.118	.120	.123	.125	.128	.130	.133	.135	.138	.140	.143	.145	.148	.150	.153	.155	.158
86	.119	.122	.124	.127	.130	.133	.135	.138	.140	.143	.145	.148	.150	153	.155	.159	.161	.164
88	.123	.126	.129	.132	.134	.137	.139	.143	.145	.148	.150	.153	.155	.159	.161	.164	.166	.169
90	.127	.130	.133	.136	.138	.142	.144	.147	.150	.153	.155	.158	.161	.164	.166	.170	.172	.175
92	.132	.134	.137	.141	.143	.146	.149	.152	.154	.158	.160	.163	.166	.169	.172	.175	.177	.181
94	.136	.139	.142	.145	.147	.151	.153	.157	.159	.163	.165	.169	.171	.175	.177	.180	.183	.186
96	.140	.143	.146	.150	.152	.155	.158	.161	.164	.168	.170	.174	.176	.180	.182	.186	.188	.192
98	.144	.147	.150	.154	.156	.160	.163	.166	.169	.172	.175	.179	.181	.185	.188	.191	.194	.197
100	.148	.151	.154	.158	.161	.164	.167	.171	.174	.177	.180	.184	.187	.190	.193	.197	.200	.203

TEMPERATURE CORRECTION, GLASS SCALE

Metric

To reduce readings of a mercurial barometer with a glass scale to 0°C. subtract the appropriate quantity as found in table.

Temp °C.	Observed height in centimeters. 70 cm.	71 cm.	72 cm.	73 cm.	74 cm.	75 cm.	76 cm.	77 cm.	78 cm.	Temp. °C.	Observed height in centimeters. 70 cm.	71 cm.	72 cm.	73 cm.	74 cm.	75 cm.	76 cm.	77 cm.	78 cm.
0	0.000	0.000	0.000	0.000	0.000	0.000	0.000	0.000	0.000	15	0.181	0.184	0.186	0.189	0.191	0.193	0.196	0.198	0.201
1	.012	.012	.013	.013	.013	.013	.013	.013	.014	16	.194	.196	.199	.201	.204	.207	.209	.212	.214
2	.025	.025	.025	.026	.026	.026	.026	.027	.027	17	.205	.208	.210	.213	.216	.219	.221	.224	.227
3	.036	.036	.037	.037	.038	.038	.039	.039	.040	18	.217	.220	.223	.226	.229	.232	.235	.238	.241
4	.048	.049	.049	.050	.051	.051	.052	.053	.053	19	.230	.233	.236	.239	.242	.245	.248	.251	.254
5	.060	.061	.062	.063	.064	.064	.065	.066	.067	20	.242	.245	.248	.252	.255	.258	.261	.264	.268
6	.073	.074	.074	.076	.077	.077	.078	.079	.080	21	.254	.258	.261	.264	.268	.271	.275	.278	.281
7	.085	.086	.087	.088	.089	.091	.092	.093	.094	22	.266	.269	.273	.276	.280	.283	.287	.290	.294
8	.096	.098	.099	.100	.101	.103	.104	.105	.107	23	.278	.282	.285	.289	.293	.296	.300	.304	.308
9	.109	.110	.111	.113	.114	.116	.117	.119	.120	24	.290	.294	.298	.302	.306	.310	.313	.317	.321
10	.121	.122	.124	.126	.127	.129	.130	.132	.134	25	.303	.307	.311	.315	.319	.323	.327	.331	.335
11	.133	.135	.137	.138	.140	.142	.144	.146	.147	26	.315	.319	.323	.327	.332	.336	.340	.344	.348
12	.144	.146	.148	.150	.152	.154	.156	.158	.160	27	.326	.331	.335	.339	.344	.348	.352	.357	.361
13	.157	.159	.161	.163	.165	.167	.169	.171	.174	28	.339	.343	.348	.352	.357	.361	.366	.370	.375
14	.169	.171	.174	.176	.178	.180	.183	.185	.187	29	.351	.356	.360	.365	.370	.374	.379	.384	.388
										30	.363	.368	.373	.378	.383	.387	.392	.397	.402

WEIGHT IN GRAMS OF A CUBIC METER OF SATURATED AQUEOUS VAPOR

(From Smithsonian Tables)

Mass in grams per cubic meter.

Temp. °C	0.0	1.0	2.0	3.0	4.0	5.0	6.0	7.0	8.0	9.0
−20	1.074	.988	.909	.836	.768	.705	.646	.592	.542	.496
−10	2.358	2.186	2.026	1.876	1.736	1.605	1.483	1.369	1.264	1.165
− 0	4.847	4.523	4.217	3.930	3.660	3.407	3.169	2.946	2.737	2.541
+ 0	4.847	5.192	5.559	5.947	6.360	6.797	7.260	7.750	8.270	8.819
+10	9.399	10.01	10.66	11.35	12.07	12.83	13.63	14.48	15.37	16.31
+20	17.30	18.34	19.43	20.58	21.78	23.05	24.38	25.78	27.24	28.78
+30	30.38	32.07	33.83	35.68	37.61	39.63	41.75	43.96	46.26	28.67

EFFICIENCY OF DRYING AGENTS

Compiled by John H. Yoe

A. Drying agents depending upon chemical action (absorption) for their efficiency:*

Substance	Residual water, mg per liter of dry air**	Reference
P_2O_5	<1 mg in 40,000 1,	Morley, Am. J. Sci., 34, 199 (1887); J.A.C.S., 26, 1171 (1904).
$Mg(ClO_4)_2$ anhyd.	"Unweighable" in 2101,	Willard and Smith, J.A.C.S., 44, 2255 (1922).
BaO	0.00065	Bower, Bur. Std. J. Res., 12, 241 (1934).
KOH fused	0.002	Baxter and Starkweather, J.A.C.S., 38, 2038 (1916).
CaO	0.003	Bower, loc. cit.
H_2SO_4	0.003	Baxter and Starkweather, loc. cit.
$CaSO_4$ anhyd.	0.005	Bower loc. cit.
Al_2O_3	0.005	Ibid.
KOH sticks	0.014	Ibid.
NaOH fused	0.16	Baxter and Starkweather, loc. cit.
$CaBr_2$	0.18	Baxter and Warren, J.A.C.S., 33, 340 (1911).
$CaCl_2$ fused	0.34	Baxter and Starkweather, loc. cit.
NaOH sticks	0.80	Bower loc. cit.
$Ba(ClO_4)_2$	0.82	Ibid.
$ZnCl_2$	0.85	Baxter and Warren, loc. cit.
$ZnBr_2$	1.16	Ibid.
$CaCl_2$ granular	1.5	Bower, loc. cit.
$CuSO_4$ anhyd.	2.8	Ibid.

B. Drying agents depending upon physical action (adsorption) for their efficiency:* Alumina (low temperature fired), asbestos, charcoal, clay and porcelain (low temperature fired), glass wool, kieselguhr, silica gel, refrigeration.

* It should be noted that the efficiency of some drying agents (e.g. Al_2O_3 and anhydrous $CaCl_2$, and probably also BaO, anhydrous $Mg(ClO_4)_2$, $Mg(ClO_4)_2 \cdot 3H_2O$, anhydrous $Ba(ClO_4)_2$, and $CaSO_4$) depends upon both adsorption and absorption.

** 30°C. for Bower's values; others 25°C. or room temp.

REDUCTION OF BAROMETER TO SEA LEVEL

The correction to be added to reduce barometric readings to "sea level" values depends principally on three factors: The temperature of the air column (assumed) from the station to sea level, the altitude of the station, and the value of the reading itself. Two tables are provided. Table I is entered with the altitude and assumed temperature and is a fator "2000 m" taken out. Table II is entered with the above factor and the approximate barometer reading and the final correction taken out.

The correction is to be added. If B_0 is the corrected or sea level value; B the barometer reading at the station; C the correction,—

$$C = B_0 - B = B\,(10^m - 1)$$

The actual barometer reading at the station should be corrected for temperature of the mercury column by the usual methods before entering the tables or applying the sea level correction.

A complete explanation of the theory of the corrections and a more extended set of tables will be found in the Smithsonian Meterological Tables.

Latitude Factor

The influence of the latitude on the value of the correction is usually negligible, being overshadowed by uncertainties in the assumed temperature of the air column. For cases where this correction is desirable the table below is provided. The value of the temperature-altitude factor "2000 m" obtained in Table I is corrected for latitude by subtracting for latitudes 0-45° and adding for latitudes from 45-90° the values found. With this corrected value of "2000 m" Table II is entered for the value of the correction.

LATITUDE FACTOR

To be used in connection with Tables I and II, either English or metric units, to obtain latitude corrections to temperature-altitude factor. For latitudes 0-45° subtract the correction. For latitudes 45-90° add the correction.

Temp.-Alt. from Table I	Latitude			
	0°	15°	30°	45°
100	0.3	0.2	0.1	0.0
200	0.5	0.5	0.3	0.0
300	0.8	0.7	0.4	0.0
	90°	75°	60°	45°

METRIC UNITS—TABLE I

Values of the temperature-altitude factor (2000 m.) for entering table II.

Altitude in meters	Assumed temperature of air column °C									
	−16°	−8°	0°	+4°	+8°	+12°	+16°	+20°	+24°	+28°
10	1.2	1.1	1.1	1.1	1.0	1.0	1.0	1.0	1.0	1.0
50	5.8	5.6	5.4	5.3	5.2	5.2	5.1	5.0	4.9	4.9
100	11.5	11.2	10.8	10.7	10.5	10.3	10.2	10.0	9.9	9.7
150	17.3	16.7	16.2	16.0	15.7	15.5	15.3	15.0	14.8	14.6
200	23.0	22.3	21.6	21.3	21.0	20.7	20.3	20.0	19.7	19.5
250	28.8	27.9	27.0	26.6	26.2	25.8	25.4	25.0	24.7	24.3
300	34.5	33.5	32.5	32.0	31.5	31.0	30.5	30.1	29.6	29.2
350	40.3	39.0	37.9	37.3	36.7	36.2	35.6	35.1	34.6	34.0
400	46.0	44.6	43.3	42.6	42.0	41.3	40.7	40.1	39.5	38.9
450	51.8	50.2	48.7	47.9	47.2	46.5	45.8	45.1	44.4	43.8
500	57.5	55.8	54.1	53.3	52.4	51.6	50.9	50.1	49.4	48.6
550	63.3	61.4	59.5	58.6	57.7	56.8	55.9	55.1	54.3	53.5
600	69.0	66.9	64.9	63.9	62.9	62.0	61.0	60.1	59.2	58.3
650	74.8	72.5	70.3	69.2	68.2	67.1	66.1	65.1	64.2	63.2
700	80.6	78.1	75.7	74.6	73.4	72.3	71.2	70.1	69.1	68.1
750	86.3	83.7	81.1	79.9	78.7	77.5	76.3	75.1	74.0	72.9
800	92.1	89.2	86.5	85.2	83.9	82.6	81.4	80.1	79.0	77.8
850	97.8	94.8	92.0	90.5	89.2	87.8	86.4	85.2	83.9	82.7
900	103.6	100.4	97.4	95.9	94.4	93.0	91.5	90.2	88.8	87.5
950	109.3	106.0	102.8	101.2	99.6	98.1	96.6	95.2	93.8	92.4
1000	115.1	111.5	108.2	106.5	104.9	103.3	101.7	100.2	98.7	97.3
1050	120.8	117.1	113.6	111.8	110.1	108.4	106.8	105.2	103.6	102.1
1100	126.6	122.7	119.0	117.2	115.4	113.6	111.9	110.2	108.6	107.0
1150	132.3	128.3	124.4	122.5	120.6	118.8	117.0	115.2	113.5	111.8
1200	138.1	133.8	129.8	127.8	125.9	123.9	122.0	120.2	118.4	116.7
1250	143.8	139.4	135.2	133.1	131.1	129.1	127.1	125.2	123.4	121.6
1300	149.6	145.0	140.6	138.5	136.3	134.3	132.2	130.2	128.3	126.4
1350	155.3	150.6	146.0	143.8	141.6	139.4	137.3	135.2	133.2	131.3
1400	161.1	156.2	151.4	149.1	146.8	144.6	142.4	140.2	138.2	136.2
1450	166.8	161.7	156.8	154.5	152.1	149.7	147.5	145.3	143.1	141.0

Altitude in meters	Assumed temperature of air column °C									
	−16°	−8°	0°	+4°	+8°	+12°	+16°	+20°	+24°	+28°
1500	172.6	167.3	162.3	159.8	157.3	154.9	152.5	150.3	148.0	145.9
1550	178.3	172.9	167.7	165.1	162.6	160.1	157.6	155.3	153.0	150.7
1600	184.1	178.5	173.1	170.4	167.8	165.2	162.7	160.3	157.9	155.6
1650	189.8	184.0	178.5	175.7	173.0	170.4	167.8	165.3	162.8	160.5
1700	195.6	189.6	183.9	181.1	178.3	175.6	172.9	170.3	167.8	165.3
1750	201.4	195.2	189.3	186.4	183.5	180.7	178.0	175.3	172.7	170.2
1800	207.1	200.8	194.7	191.7	188.8	185.9	183.1	180.3	177.6	175.0
1850	212.9	206.3	200.1	197.0	194.0	191.0	188.1	185.3	182.6	179.9
1900	218.6	211.9	205.5	202.4	199.3	196.2	193.2	190.3	187.5	184.8
1950	224.4	217.5	210.9	207.7	204.5	201.4	198.3	195.3	192.4	189.6
2000	230.1	223.0	216.3	213.0	209.7	206.5	203.4	200.3	197.4	194.5
2050	235.9	228.6	221.7	218.3	215.0	211.7	208.5	205.3	202.3	199.3
2100	241.6	234.2	227.1	223.7	220.2	216.8	213.5	210.4	207.2	204.2
2150	247.4	239.8	232.5	229.0	225.5	222.0	218.6	215.4	212.2	209.1
2200	253.1	245.4	237.9	234.3	230.7	227.2	223.7	220.4	217.1	213.9
2250	258.9	250.9	243.4	239.6	235.9	232.3	228.8	225.4	222.0	218.8
2300	264.6	256.5	248.8	245.0	241.2	237.5	233.9	230.4	227.0	223.6
2350	270.4	262.1	254.2	250.3	246.4	242.7	239.0	235.4	231.9	228.5
2400	276.1	267.7	259.6	255.6	251.7	247.8	244.0	240.4	236.8	233.4
2450	281.9	273.2	265.0	260.9	256.9	253.0	249.1	245.4	241.8	238.2
2500	287.6	278.8	270.4	266.2	262.2	258.1	254.2	250.4	246.7	243.1
2550	293.4	284.4	275.8	271.6	267.4	263.3	259.3	255.4	251.6	247.9
2600	299.1	290.0	281.2	276.9	272.6	268.5	264.4	260.4	256.6	252.8
2650	304.9	295.5	286.6	282.2	277.9	273.6	269.5	265.4	261.5	257.7
2700	310.6	301.1	292.0	287.5	283.1	278.8	274.5	270.4	266.4	262.5
2750	316.4	306.7	297.4	292.9	288.4	283.9	279.6	275.4	271.4	267.4
2800	322.1	312.3	302.8	298.2	293.6	289.1	284.7	280.4	276.3	272.2
2850	327.9	317.8	308.2	303.5	298.8	294.3	289.8	285.4	281.2	277.1
2900	333.6	323.4	313.6	308.8	304.1	299.4	294.9	290.4	286.2	282.0
2950	339.4	329.0	319.0	314.2	309.3	304.6	299.9	295.5	291.1	286.8
3000	345.1	334.5	324.4	319.5	314.6	309.7	305.0	300.5	296.0	291.7

METRIC UNITS — TABLE II

Values of Correction to be Added

Temp.-alt.-factor	Barometer reading 780 mm	760 mm	740 mm	720 mm	700 mm
1	0.9	0.9	0.9	0.8	0.8
5	4.5	4.4	4.3	4.2	4.0
10	9.0	8.8	8.6	8.3	8.1
15	13.6	13.2	12.9	12.5	12.2
20	18.2	17.7	17.2	16.8	16.3
25	22.8	22.2	21.6	21.0	20.4
30	27.4	26.7	26.0	25.3	24.6
35	—	31.2	30.4	29.6	28.8

Temp.-alt.-factor	760 mm	740 mm	720 mm	700 mm	680 mm	660 mm
40	35.8	34.9	33.9	33.0	32.0	31.1
45	40.4	39.3	38.3	37.2	36.2	35.1
50	45.0	43.8	42.7	41.5	40.3	39.1
55	49.7	48.4	47.1	45.8	44.5	43.1
60	—	52.9	51.5	50.1	48.6	47.2
65	—	57.5	55.9	54.4	52.8	51.3
70	—	62.1	60.4	58.7	57.1	55.4
75	—	66.7	64.9	63.1	61.3	59.5

Temp.-alt.-factor	720 mm	700 mm	680 mm	660 mm	640 mm
80	69.5	67.5	65.6	63.7	61.7
85	74.0	72.0	69.9	67.9	65.8
90	78.6	76.4	74.2	72.1	69.9
95	83.2	80.9	78.6	76.3	74.0
100	87.9	85.4	83.0	80.5	78.1
105	—	89.9	87.4	84.8	82.2
110	—	94.5	91.8	89.1	86.4
115	—	99.1	96.3	93.4	90.6
120	—	103.7	100.7	97.8	94.8
125	—	108.3	105.3	102.2	99.1

Temp.-alt.-factor	680 mm	660 mm	6[illegible]	620 mm	600 mm
125	105.3	102.2	99.1	96.0	92.9
130	109.8	106.6	103.3	100.1	96.9
135	114.3	111.0	107.6	104.3	100.9
140	118.9	115.4	111.9	108.4	104.9
145	123.5	119.9	116.3	112.6	109.0
150	128.2	124.4	120.6	116.9	113.1
155	—	128.9	125.0	121.1	117.2
160	—	133.5	129.4	125.4	121.4
165	—	138.1	133.9	129.7	125.5
170	—	142.7	138.4	134.0	129.7

Temp.-alt.-factor	Barometer reading 640 mm	620 mm	600 mm	580 mm	560 mm
170	138.4	134.0	129.7	125.4	121.1
175	142.9	138.4	133.9	129.5	125.0
180	147.4	142.8	138.2	133.6	129.0
185	151.9	147.2	142.4	137.7	132.9
190	156.5	151.6	146.7	141.8	136.9
195	161.1	156.1	151.0	146.0	141.0
200	165.7	160.5	155.4	150.2	145.0
205	170.4	165.0	159.7	154.4	149.1
210	—	169.6	164.1	158.6	153.2
215	—	174.1	168.5	162.9	157.3

Temp.-alt.-factor	620 mm	600 mm	580 mm	560 mm	540 mm
215	174.1	168.5	162.9	157.3	151.7
220	178.7	172.9	167.2	161.4	155.7
225	183.3	177.4	171.5	165.6	159.7
230	188.0	181.9	175.8	169.8	163.7
235	192.6	186.4	180.2	174.0	167.8
240	—	191.0	184.6	178.2	171.9
245	—	195.5	189.0	182.5	176.0
250	—	200.1	193.4	186.8	180.1
255	—	204.7	197.9	191.1	184.3
260	—	209.4	202.4	195.4	188.4

Temp.-alt.-factor	580 mm	560 mm	540 mm	520 mm
260	202.4	195.4	188.4	181.5
265	206.9	199.8	192.6	185.5
270	211.5	204.2	196.9	189.6
275	216.0	208.6	201.1	193.7
280	220.6	213.0	205.4	197.8
285	225.2	217.5	209.7	201.9
290	229.9	222.0	214.0	206.1
295	—	226.5	218.4	210.3
300	—	231.0	222.8	214.5

Temp.-alt.-factor	560 mm	540 mm	520 mm	500 mm	480 mm
305	235.6	227.2	218.8	210.3	201.9
310	240.2	231.6	223.0	214.4	205.9
315	244.8	236.0	227.3	218.6	209.8
320	249.4	240.5	231.6	222.7	213.8
325	254.1	245.0	236.0	226.9	217.8
330	—	249.6	240.3	231.1	221.8
335	—	254.1	244.7	235.3	225.9
340	—	258.7	249.1	239.6	230.0
345	—	263.3	253.6	243.8	234.1

ENGLISH UNITS — TABLE I

Values of the temperature-altitude factor (2000 m.) for entering table II

Altitude feet	Assumed temperature of air column °F −20	0	+10	+20	+30	+40	+50	+60	+70	+80
200	7.4	7.1	6.9	6.8	6.6	6.5	6.3	6.2	6.1	6.0
400	14.8	14.1	13.8	13.5	13.2	13.0	12.7	12.4	12.2	11.9
600	22.2	21.2	20.7	20.3	19.9	19.5	19.0	18.6	18.2	17.9
800	29.6	28.3	27.7	27.1	26.5	25.9	25.4	24.8	24.3	23.8
1000	37.0	35.3	34.6	33.8	33.1	32.4	31.7	31.1	30.4	29.8
1200	44.3	42.4	41.5	40.6	39.7	38.9	38.1	37.3	36.5	35.8
1400	51.7	49.5	48.4	47.4	46.4	45.4	44.4	43.5	42.6	41.7
1600	59.1	56.5	55.3	54.1	53.0	51.9	50.8	49.7	48.7	47.7
1800	66.5	63.6	62.2	60.9	59.6	58.4	57.1	55.9	54.7	53.6
2000	73.9	70.6	69.1	67.7	66.2	64.8	63.4	62.1	60.8	59.6
2200	81.3	77.7	76.0	74.4	72.9	71.3	69.8	68.3	66.9	65.5
2400	88.7	84.8	82.9	81.2	79.5	77.8	76.1	74.5	73.0	71.5
2600	96.1	91.8	89.9	87.9	86.1	84.3	82.5	80.7	79.1	77.5
2800	103.5	98.9	96.8	94.7	92.7	90.8	88.8	87.0	85.1	83.4
3000	110.9	106.0	103.7	101.5	99.3	97.2	95.2	93.2	91.2	89.4
3200	118.2	113.0	110.6	108.2	106.0	103.7	101.5	99.4	97.3	95.3
3400	125.6	120.1	117.5	115.0	112.6	110.2	107.9	105.6	103.4	101.3
3600	133.0	127.2	124.4	121.8	119.2	116.7	114.2	111.8	109.5	107.2
3800	140.4	134.2	131.3	128.5	125.8	123.2	120.5	118.0	115.5	113.2
4000	147.8	141.3	138.2	135.3	132.4	129.6	126.9	124.2	121.6	119.2
4200	155.2	148.3	145.1	142.1	139.1	136.1	133.2	130.4	127.7	125.1
4400	162.6	155.4	152.0	148.8	145.7	142.6	139.6	136.6	133.8	131.1
4600	170.0	162.5	159.0	155.6	152.3	149.1	145.9	142.8	139.9	137.0

Altitude feet	Assumed temperature of air column °F −20	0	+10	+20	+30	+40	+50	+0	+70	+80
4800	177.3	169.5	165.9	162.3	158.9	155.6	152.2	149.0	145.9	143.0
5000	184.7	176.6	172.8	169.1	165.6	162.0	158.6	155.2	152.0	148.9
5200	192.1	183.7	179.7	175.9	172.2	168.5	164.9	161.5	158.1	154.9
5400	199.5	190.7	186.6	182.6	178.8	175.0	171.3	167.7	164.2	160.8
5600	206.9	197.8	193.5	189.4	185.4	181.5	177.6	173.9	170.3	166.8
5800	214.3	204.8	200.4	196.2	192.0	188.0	184.0	180.1	176.3	172.8
6000	221.7	211.9	207.3	202.9	198.7	194.4	190.3	186.3	182.4	178.7
6200	229.1	219.0	214.2	209.7	205.3	200.9	196.6	192.5	188.5	184.7
6400	236.4	226.0	221.1	216.4	211.9	207.4	203.0	198.7	194.6	190.6
6600	243.8	233.1	228.0	223.2	218.5	213.0	209.3	204.9	200.7	196.6
6800	251.2	240.1	235.0	230.0	225.1	220.4	215.7	211.1	206.7	202.5
7000	258.6	247.2	241.9	236.7	231.8	226.8	222.0	217.3	212.8	208.5
7200	266.0	254.3	248.8	243.5	238.4	233.3	228.4	223.5	218.9	214.4
7400	273.4	261.3	255.7	250.2	245.0	239.8	234.7	229.7	225.0	220.4
7600	280.8	268.4	262.6	257.0	251.6	246.3	241.0	235.9	231.1	226.4
7800	288.1	275.4	269.5	263.8	258.2	252.8	247.4	242.2	237.1	232.3
8000	295.5	282.5	276.4	270.5	264.8	259.2	253.7	248.4	243.2	238.3
8200	302.9	289.6	283.3	277.3	271.5	265.7	260.1	254.6	249.3	244.2
8400	310.3	296.6	290.2	284.0	278.1	272.2	266.4	260.8	255.4	250.2
8600	317.7	303.7	297.1	290.8	284.7	278.7	272.7	267.0	261.4	256.1
8800	325.1	310.7	304.0	297.6	291.3	285.2	279.1	273.2	267.5	262.1
9000	332.5	317.8	310.9	304.3	297.9	291.6	285.4	279.4	273.6	268.0

ENGLISH UNITS — TABLE II

Value of Correction to be Added.

Temp alt. factor	31	30	29	28	27
	in.	in.	in.	in.	in.
1	0.04	0.03	0.03	—	—
5	0.18	0.17	0.17	—	—
10	0.36	0.35	0.34	0.32	—
15	0.54	0.52	0.51	0.49	—
20	0.72	0.70	0.68	0.65	—
25	—	0.88	0.85	0.82	—
30	—	1.05	1.02	0.98	—
35	—	1.23	1.19	1.15	—
40	—	1.41	1.37	1.32	1.27
45	—	1.60	1.54	1.49	1.44
50	—	—	1.72	1.66	1.60
55	—	—	1.90	1.83	1.76
60	—	—	2.07	2.00	1.93
65	—	—	2.25	2.18	2.10
70	—	—	2.43	2.35	2.27
75	—	—	—	2.53	2.43
80	—	—	—	2.70	2.60

Temp. alt. factor	26	25	24	23	22
	in.	in.	in.	in.	in.
165	5.44	5.23	5.02	—	—
170	5.62	5.40	5.19	—	—
175	—	5.58	5.36	—	—
180	—	5.76	5.53	5.30	—
185	—	5.93	5.70	5.46	—
190	—	6.11	5.87	5.62	—
195	—	6.29	6.04	5.79	—
200	—	—	6.21	5.96	—
205	—	—	6.39	6.12	—
210	—	—	6.56	6.29	—
215	—	—	6.74	6.46	—
220	—	—	6.92	6.63	6.34
225	—	—	7.10	6.80	6.51
230	—	—	7.28	6.97	6.67
235	—	—	7.46	7.15	6.84
240	—	—	—	7.32	7.00
245	—	—	—	7.49	7.17

Temp alt. factor	28	27	26	25	24
	in.	in.	in.	in.	in.
75	2.53	2.43	2.34	—	—
80	2.70	2.60	2.51	—	—
85	2.88	2.78	2.67	—	—
90	3.06	2.95	2.84	—	—
95	3.24	3.12	3.01	—	—
100	3.42	3.29	3.17	—	—
105	3.60	3.47	3.34	3.21	—
110	—	3.85	3.51	3.38	—
115	—	3.82	3.68	3.54	—
120	—	4.00	3.85	3.70	—
125	—	4.18	4.02	3.87	—
130	—	4.36	4.20	4.04	—
135	—	4.54	4.37	4.20	—
140	—	—	4.55	4.37	4.20
145	—	—	4.72	4.54	4.36
150	—	—	4.90	4.71	4.52
155	—	—	5.08	4.88	4.69
160	—	—	5.26	5.06	4.85

Temp. alt. factor	23	22	21	20
	in.	in.	in.	in.
250	7.67	7.34	—	—
255	7.85	7.51	—	—
260	8.03	7.68	7.33	—
265	8.21	7.85	7.49	—
270	8.39	8.02	7.66	—
275	8.57	8.19	7.82	—
280	—	8.37	7.99	—
285	—	8.54	8.16	—
290	—	8.72	8.32	—
295	—	8.90	8.49	8.09
300	—	9.08	8.66	8.25
305	—	9.26	8.83	8.41
310	—	9.44	9.01	8.58
315	—	9.62	9.18	8.74
320	—	9.80	9.35	8.91
325	—	—	9.53	9.08
330	—	—	9.71	9.24
—	—	—	—	—

REDUCTION OF BAROMETER TO GRAVITY AT SEA LEVEL

Metric Units
Correction to be subtracted given in millimeters
(From Smithsonian Physical Tables)

Height above sea level in meters	Observed Height of Barometer in Millimeters						
	500	550	600	650	700	750	800
100	—	—	—	—	.02	.02	.02
200	—	—	—	—	.04	.05	.05
300	—	—	—	—	.07	.07	.07
400	—	—	—	—	.09	.10	.10
500	—	—	—	—	.11	.12	.13
600	—	—	—	.12	.13	.14	—
700	—	—	—	.14	.15	.16	—
800	—	—	—	.16	.18	.19	—
900	—	—	—	.18	.20	.22	—
1000	—	.18	.19	.20	.22	.24	—
1100	—	.19	.21	.22	.24	—	—
1200	—	.21	.23	.24	.26	—	—
1300	—	.22	.24	.26	.29	—	—
1400	—	.24	.26	.28	.31	—	—
1500	.24	.26	.28	.30	.33	—	—
1600	.25	.28	.30	.32	—	—	—
1700	.27	.30	.32	.34	—	—	—
1800	.28	.31	.34	.36	—	—	—
1900	.30	.33	.36	.39	—	—	—
2000	.31	.34	.38	.41	—	—	—
2100	.33	.36	.40	—	—	—	—
2200	.35	.38	.41	—	—	—	—
2300	.36	.40	.43	—	—	—	—
2400	.38	.42	.45	—	—	—	—
2500	.39	.43	.47	—	—	—	—

English Units

Height above sea level in feet	Observed Height in Inches						
	18	20	22	24	26	28	30
1000	—	—	—	—	.003	.003	.003
2000	—	—	—	.004	.005	.005	.006
3000	—	—	.007	.007	.008	.008	—
4000	—	—	.009	.009	.010	—	—
4500	—	—	0.10	0.10	0.11	—	—
5000	—	0.10	0.11	0.11	0.12	—	—
5500	—	0.11	0.12	0.13	—	—	—
6000	—	0.11	0.13	0.14	—	—	—
6500	0.11	0.12	0.14	0.15	—	—	—
7000	0.12	0.13	0.15	0.16	—	—	—
7500	0.13	0.14	0.16	0.17	—	—	—
8000	0.14	0.15	0.17	—	—	—	—
8500	0.15	0.16	0.18	—	—	—	—
9000	0.16	0.17	0.19	—	—	—	—
9500	0.16	0.18	0.20	—	—	—	—

REDUCTION OF BAROMETER TO LATITUDE 45°

Metric Scale

For latitudes below 45°, subtract the correction; for latitudes greater than 45° it is to be added. Corrections in cm.
(From Smithsonian Meterological Tables.)

Latitude		Observed Height of Barometer in Centimeters					
		68	70	72	74	76	78
25°	65°	0.116	0.120	0.123	0.127	0.130	0.133
26	64	.111	.115	.118	.121	.125	.128
27	63	.106	.110	.113	.116	.119	.122
28	62	.101	.104	.107	.110	.113	.116
29	61	.096	.099	.102	.104	.107	.110
30	60	.091	.094	.096	.098	.101	.104
31	59	.085	.087	.090	.092	.095	.097
32	58	.079	.082	.084	.086	.089	.091
33	57	.074	.076	.078	.080	.082	.084
34	56	.068	.070	.072	.074	.076	.078
35	55	.062	.064	.066	.067	.069	.071
36	54	.056	.058	.059	.061	.063	.064
37	53	.050	.051	.053	.054	.056	.057
38	52	.044	.045	.046	.048	.049	.050
39	51	.038	.039	.040	.041	.042	.043
40	50	.031	.032	.033	.034	.035	.036
41	49	.025	.026	.027	.027	.028	.029
42	48	.019	.019	.020	.021	.021	.022
43	47	.013	.013	.013	14	.014	.014
44	46	.006	.007	.007	.007	.007	.007

English Scale (Corrections in inches)

Latitude		Observed Height in Inches					
		25	26	27	28	29	30
25°	65°	0.043	0.044	0.046	0.048	0.050	0.051
26	64	.041	.043	.044	.046	.048	.049
27	63	.039	.041	.042	.044	.045	.047
28	62	.037	.039	.040	.042	.043	.045
29	61	.035	.037	.038	.039	.041	.042
30	60	.033	.035	.036	.037	.039	.040
31	59	.031	.032	.034	.035	.036	.037
32	58	.029	.030	.032	.033	.034	.035
33	57	.027	.028	.029	.030	.031	.032
34	56	.025	.026	.027	.028	.029	.030
35	55	.023	.024	.025	.025	.026	.027
36	54	.021	.021	.022	.023	.024	.025
37	53	.018	.019	.020	.021	.021	.022
38	52	.016	.017	.017	.018	.019	.019
39	51	.014	.014	.015	.015	.016	.017
40	50	.012	.012	.012	.013	.013	.014
41	49	.009	.010	.010	.010	.011	.011
42	48	.007	.007	.008	.008	.008	.008
43	47	.005	.005	.005	.005	.005	.006
44	46	.002	.002	.003	.003	.003	.003

RELATIVE HUMIDITY — DEW-POINT

The table gives the relative humidity of the air for temperature t and dewpoint d.
(From Smithsonian Meterological Tables.)

Depression of dew-point t-d, °C	Dew-Point (d) −10	0	+10	+20	+30	Depression of dew-point t-d, °C	Dew-Point (d) −10	0	+10	+20	+30	Depression of dew-point t-d, °C	Dew-Point (d) −10	0	+10	+20	+30
0.0	100%	100%	100%	100%	100%	5.4	66	68	70	72	74	12.0	41	44	47	49	—
0.2	98	99	99	99	99	5.6	65	67	69	71	73	12.5	39	42	45	48	—
0.4	97	97	97	98	98	5.8	64	66	69	70	72	13.0	38	41	44	46	—
0.6	95	96	96	96	97	6.0	63	66	68	70	71	13.5	37	40	43	45	—
0.8	94	94	95	95	96	6.2	62	65	67	69	71	14.0	35	38	41	44	—
1.0	92	93	94	94	94	6.4	61	64	66	68	70	14.5	34	37	40	43	—
1.2	91	92	92	93	93	6.6	60	63	65	67	69	15.0	33	36	39	42	—
1.4	90	90	91	92	92	6.8	60	62	64	66	68	15.5	32	35	38	40	—
1.6	88	89	90	91	91	7.0	59	61	63	66	68	16.0	31	34	37	39	—
1.8	87	88	89	90	90	7.2	58	60	63	65	67	16.5	30	33	36	38	—
2.0	86	87	88	88	89	7.4	57	60	62	64	66	17.0	29	32	35	37	—
2.2	84	85	86	87	88	7.6	56	59	61	63	65	17.5	28	31	34	36	—
2.4	83	84	85	86	87	7.8	55	58	60	63	65	18.0	27	30	33	35	—
2.6	82	83	84	85	86	8.0	54	57	60	62	64	18.5	26	29	32	34	—
2.8	80	82	83	84	85	8.2	54	56	59	61	63	19.0	25	28	31	33	—
3.0	79	81	82	83	84	8.4	53	56	58	60	63	19.5	24	27	30	33	—
3.2	78	80	81	82	83	8.6	52	55	57	60	62	20.0	24	26	29	32	—
3.4	77	79	80	81	82	8.8	51	54	57	59	61	21.0	22	25	27	—	—
3.6	76	77	79	80	82	9.0	51	53	56	58	61	22.0	21	23	26	—	—
3.8	75	76	78	79	81	9.2	50	53	55	58	60	23.0	19	22	24	—	—
4.0	73	75	77	78	80	9.4	49	52	55	57	59	24.0	18	21	23	—	—
4.2	72	74	76	77	79	9.6	48	51	54	56	59	25.0	17	19	22	—	—
4.4	71	73	75	77	78	9.8	48	51	53	56	58	26.0	16	18	21	—	—
4.6	70	72	74	76	77	10.0	47	50	53	55	57	27.0	15	17	20	—	—
4.8	69	71	73	75	76	10.5	45	48	51	54	—	28.0	14	16	19	—	—
5.0	68	70	72	74	75	11.0	44	47	49	52	—	29.0	13	15	18	—	—
5.2	67	69	71	73	75	11.5	42	45	48	51	—	30.0	12	14	17	—	—

RELATIVE HUMIDITY FROM WET AND DRY BULB THERMOMETER (CENT. SCALE)

This table gives the approximate relative humidity directly from the reading of the air temperature (dry bulb) (t′°C) and the wet bulb (t′°C). It is computed for a barometric pressure of 74.27 cm Hg. Errors resulting from the use of this table for air temperatures above −10°C and between 77.5 and 71 cm Hg will usually be within the errors of observation.

Condensed from Bulletin of the U.S. Weather Bureau No. 1071

t-t′t	0.2	0.4	0.6	0.8	1.0	1.2	1.4	1.6	1.8	2.0	2.2	2.4	2.6	2.8	3.0	3.2	3.4	3.6	3.8	4.0	4.5	5.0	5.5	6.0	6.5	7.0	7.5	8.0		9.0	9.5	10.0	10.5	11.0
−10	93	87	80	74	67	61	54	48	41	35	28	22	16	9																				
−9	94	88	81	75	69	63	57	51	45	39	33	27	21	15	9																			
−8	94	88	83	77	71	65	60	54	48	43	37	32	26	20	15	10																		
−7	95	89	84	78	73	67	62	57	52	46	41	36	31	25	20	15	10	5																
−6	95	90	85	79	74	69	64	59	54	49	45	40	35	30	25	20	15	11	6															
−5	95	90	86	81	76	71	66	62	57	52	48	43	39	34	29	25	20	16	11	7														
−4	95	91	86	82	77	73	68	64	59	55	51	46	42	38	33	29	25	21	17	12														
−3	96	91	87	82	78	74	70	66	62	57	53	49	45	41	37	33	29	25	21	17	8													
−2	96	92	88	84	79	75	71	68	64	60	56	52	48	44	40	37	33	29	25	22	12													
−1	96	92	88	84	81	77	73	69	66	62	58	54	51	47	43	40	36	33	29	26	17	8												
0	96	93	89	85	81	78	74	71	67	64	60	57	53	50	46	43	40	36	33	29	21	13	5											
1	97	93	90	86	83	80	76	73	70	66	63	59	56	53	49	46	43	40	36	33	25	17	10											
2	97	93	90	87	84	81	78	74	71	68	65	62	59	55	52	49	46	43	40	37	29	22	14	7										
3	97	94	91	88	84	82	78	76	72	70	67	64	61	58	55	52	49	46	43	40	33	26	19	12	5									
4	97	94	91	88	85	82	79	77	74	71	68	65	62	60	57	54	51	48	46	43	36	29	22	16	9									
5	97	94	91	88	86	83	80	77	75	72	69	67	64	61	58	56	53	51	48	45	39	33	26	20	13	7								
6	97	94	92	89	86	84	81	78	76	73	70	68	65	63	60	58	55	53	50	48	41	35	29	24	17	11	5							
7	97	95	92	89	87	84	82	79	77	74	72	69	67	64	62	59	57	54	52	50	44	38	32	26	21	15	10							
8	97	95	92	90	87	85	82	80	77	75	73	70	68	65	63	61	58	56	54	51	46	40	35	29	24	19	14	8						
9	98	95	93	90	88	85	83	81	78	76	74	71	69	67	64	62	60	58	55	53	48	42	37	32	27	22	17	12	7					
10	98	95	93	90	88	86	83	81	79	77	74	72	70	68	66	63	61	59	57	55	50	44	39	34	29	24	20	15	10	6				
11	98	95	93	91	89	86	84	82	80	78	75	73	71	69	67	65	62	60	58	56	51	46	41	36	32	27	22	18	13	9	5			
12	98	96	93	91	89	87	85	82	80	78	76	74	72	70	68	66	64	62	60	58	53	48	43	39	34	29	25	21	16	12	8			
13	98	96	93	91	89	87	85	83	81	79	77	75	73	71	69	67	65	63	61	59	54	50	45	41	36	32	28	23	19	15	11	7		
14	98	96	94	92	90	88	86	84	82	79	78	76	74	72	70	68	66	64	62	60	56	51	47	42	38	34	30	26	22	18	14	10	6	
15	98	96	94	92	90	88	86	84	82	80	78	76	74	73	71	69	67	65	63	61	57	53	48	44	40	36	32	27	24	20	16	13	9	6

t-t′t	0.5	1.0	1.5	2.0	2.5	3.0	3.5	4.0	4.5	5.0	5.5	6.0	6.5	7.0	7.5	8.0	8.5	9.0	9.5	10.0	10.5	11.0	11.5	12.0	12.5	13.0	13.5	14.0	14.5	15.0	16.0	17.0	18.0	19.0	20.0
16	95	90	85	81	76	71	67	63	58	54	50	46	42	38	34	30	26	23	19	15	12	8	5												
17	95	90	86	81	76	72	68	64	60	55	51	47	43	40	36	32	28	25	21	18	14	11	8												
18	95	91	86	82	77	73	69	65	61	57	53	49	45	41	38	34	30	27	23	20	17	14	10	7											
19	95	91	87	82	78	74	70	65	62	58	54	50	46	43	39	36	32	29	26	22	19	16	13	10	7										
20	96	91	87	83	78	74	70	66	63	59	55	51	48	44	41	37	34	31	28	24	21	18	15	12	9	6									
21	96	91	87	83	79	75	71	67	64	60	56	53	49	46	42	39	36	32	29	26	23	20	17	14	12	9	6								
22	96	92	87	83	80	76	72	68	64	61	57	54	50	47	44	40	37	34	31	28	25	22	19	17	14	11	8	6							
23	96	92	88	84	80	76	72	69	65	62	58	55	52	48	45	42	39	36	33	30	27	24	21	19	16	13	11	8	6						
24	96	92	88	84	80	77	73	69	66	62	59	56	53	49	46	43	40	37	34	31	29	26	23	20	18	15	13	10	8	5					
25	96	92	88	84	81	77	74	70	67	63	60	57	54	50	47	44	41	39	36	33	30	28	25	22	20	17	15	12	10	8					
26	96	92	88	85	81	78	74	71	67	64	61	58	54	51	49	46	43	40	37	34	32	29	26	24	21	19	17	14	12	10	5				
27	96	92	89	85	82	78	75	71	68	65	62	58	56	52	50	47	44	41	38	36	33	31	28	26	23	21	18	16	14	12	7				
28	96	93	89	85	82	78	75	72	69	65	62	59	56	53	51	48	45	42	40	37	34	32	29	27	25	22	20	18	16	13	9	5			
29	96	93	89	86	82	79	76	72	69	66	63	60	57	54	52	49	46	43	41	38	36	33	31	28	26	24	22	19	17	15	11	7			
30	96	93	89	86	83	79	76	73	70	67	64	61	58	55	52	50	47	44	42	39	37	35	32	30	28	25	23	21	19	17	13	9	5		
31	96	93	90	86	83	80	77	73	70	67	64	61	59	56	53	51	48	45	43	40	38	36	33	31	29	27	25	22	20	18	14	11	7		
32	96	93	90	86	83	80	77	74	71	68	65	62	60	57	54	51	49	46	44	41	39	37	35	32	30	28	26	24	22	20	16	12	9	5	
33	97	93	90	87	83	80	77	74	71	68	66	63	60	57	55	52	50	47	45	42	40	38	36	33	31	29	27	25	23	21	17	14	10	7	
34	97	93	90	87	84	81	78	75	72	69	66	63	61	58	56	53	51	48	46	43	41	39	37	35	32	30	28	26	24	23	19	15	12	8	5
35	97	94	90	87	84	81	78	75	72	69	67	64	61	59	56	54	51	49	47	44	42	40	38	36	34	32	30	28	26	24	20	17	13	10	7
36	97	94	90	87	84	81	78	75	73	70	67	64	62	59	57	54	52	50	48	45	43	41	39	37	35	33	31	29	27	25	21	18	15	11	8
37	97	94	91	87	84	82	79	76	73	70	68	65	63	60	58	55	53	51	48	46	44	42	40	38	36	34	32	30	28	26	23	19	16	13	10
38	97	94	91	88	84	82	79	76	74	71	68	66	63	61	58	56	54	51	49	47	45	43	41	39	37	35	33	31	29	27	24	20	17	14	11
39	97	94	91	88	85	82	79	77	74	71	69	66	64	61	59	57	54	52	50	48	46	43	42	39	38	36	34	32	30	28	25	22	18	15	12
40	97	94	91	88	85	82	80	77	74	72	69	67	64	62	59	57	54	53	51	48	46	44	42	40	38	36	35	33	31	29	26	23	20	16	14

REDUCTION OF PSYCHROMETRIC OBSERVATION

For the reduction of observations with the wet and dry bulb thermometer. Assuming the relative velocity of the air to the thermometer bulbs is at least three meters per second; if t is the temperature of the air as indicated by the dry bulb, t_w, the temperature of the wet bulb, B, the barometric pressure, and E_w, the vapor tension of water corresponding to t_2, then the actual vapor tension is

$$E = E_w - 0.00066B(t - t_w)\,[1 + 0.00115(t - t_w)] \text{ millimeters}$$

The value of the term

$$0.00066B(t - t_w)\,[1 + 0.00115\,(t - t_w)]$$

is given in the following table.

(From Miller's Laboratory Physics, Ginn & Co., Publishers, by permission.)

Barometric Pressure B in Millimeters

$t - t_w$	700	710	720	730	740	750	760	770
°C	mm	mm	mm	mm	mm	mm	mm	mm
1	0.463	0.469	0.476	0.482	0.489	0.496	0.502	0.509
2	0.926	0.939	0.953	0.966	0.979	0.992	1.006	1.019
3	1.391	1.411	1.431	1.450	1.470	1.490	1.510	1.530
4	1.857	1.883	1.910	1.936	1.963	1.989	2.016	2.042
5	2.323	2.356	2.390	2.423	2.456	2.489	2.522	2.556
6	2.791	2.831	2.871	2.911	2.951	2.990	3.030	3.070
7	3.260	3.307	3.353	3.400	3.446	3.493	3.539	3.586
8	3.730	3.783	3.837	3.890	3.943	3.996	4.050	4.103
9	4.201	4.261	4.321	4.381	4.441	4.501	4.561	4.621
10	4.673	4.740	4.807	4.873	4.940	5.007	5.074	5.140

$t - t_2$	700	710	720	730	740	750	760	770
°C	mm	mm	mm	mm	mm	mm	mm	mm
11	5.146	5.220	5.293	5.367	5.440	5.514	5.587	5.661
12	5.621	5.701	5.781	5.861	5.942	6.022	6.102	6.183
13	6.096	6.183	6.270	6.357	6.444	6.531	6.618	6.705
14	6.572	6.666	6.760	6.854	6.948	7.042	7.135	7.229
15	7.050	7.150	7.251	7.352	7.452	7.553	7.654	7.754
16	7.528	7.636	7.743	7.851	7.958	8.066	8.173	8.281
17	8.008	8.122	8.236	8.351	8.465	8.580	8.694	8.808
18	8.488	8.609	8.731	8.852	8.973	9.094	9.216	9.337
19	8.970	9.098	9.226	9.354	9.482	9.610	9.739	9.867
20	9.453	9.588	9.723	9.858	9.993	10.128	10.263	10.398

CONSTANT HUMIDITY

The following table shows the % humidity and the aqueous tension at the given temperature within a closed space when an excess of the substance indicated is in contact with a saturated aqueous solution of the given solid phase.

Solid phase	t°C.	% humidity	Aq. tension mm Hg	Solid phase	t°C.	% humidity	Aq. tension mm Hg
$H_3PO_4.\frac{1}{2}H_2O$	24	9	1.99	$NaClO_3$	20	75	13.0
$KC_2H_3O_2$	168	13	738	$(NH_4)_2SO_4$	108	75	754
$LiCl.H_2O$	20	15	2.60	$NaC_2H_3O_2.3H_2O$	20	76	13.2
$KC_2H_3O_2$	20	20	3.47	$H_2C_2O_4.2H_2O$	20	76	13.2
KF	100	22.9	174	$Na_2S_2O_3.5H_2O$	20	78	13.5
NaBr	100	22.9	174	NH_4Cl	20	79.5	13.8
NaCl,KNO_3 and $NaNO_3$	16.39	30.49	4.23	NH_4Cl	25	79.3	18.6
$CaCl_2.6H_2O$	24.5	31	7.08	NH_4Cl	30	77.5	24.4
$CaCl_2.6H_2O$	20	32.3	5.61	$(NH_4)_2SO_4$	20	81	14.1
$CaCl_2.6H_2O$	18.5	35	5.54	$(NH_4)_2SO_4$	25	81.1	19.1
CrO_3	20	35	6.08	$(NH_4)_2SO_4$	30	81.1	25.6
$CaCl_2.6H_2O$	10	38	3.47	KBr	20	84	14.6
$CaCl_26H_2O$	5	39.8	2.59	Tl_2SO_4	104.7	84.8	768
$Zn(NO_3)_2.6H_2O$	20	42	7.29	$KHSO_4$	20	86	14.9
$K_2CO_3.2H_2O$	24.5	43	9.82	$Na_2CO_3.10H_2O$	24.5	87	20.9
$K_2CO_3.2H_2O$	18.5	44	6.96	$BaCl_2.2H_2O$	24.5	88	20.1
KNO_2	20	45	7.81	K_2CrO_4	20	88	15.3
KCNS	20	47	8.16	$Pb(NO_3)_2$	103.5	88.4	760
NaI	100	50.4	383	$ZnSO_4.7H_2O$	20	90	15.6
$Ca(NO_3)_2.4H_2O$	24.5	51	11.6	$Na_2CO_3.10H_2O$	18.5	92	14.6
$NaHSO_4.H_2O$	20	52	9.03	$NaBrO_3$	20	92	16.0
$Na_2Cr_2O_7.2H_2O$	20	52	9.03	K_2HPO_4	20	92	16.0
$Mg(NO_3)_2.6H_2O$	24.5	52	11.9	$NH_4H_2PO_4$	30	92.9	29.3
$NaClO_3$	100	54	410	$NH_4H_2PO_4$	25	93	21.9
$Ca(NO_3)_2.4H_2O$	18.5	56	8.86	$Na_2SO_4.10H_2O$	20	93	16.1
$Mg(NO_3)_2.6H_2O$	18.5	56	8.86	$NH_4H_2PO_4$	20	93.1	16.2
KI	100	56.2	427	$ZnSO_4.7H_2O$	5	94.7	6.10
$NaBr.2H_2O$	20	58	10.1	$Na_2SO_3.7H_2O$	20	95	16.5
$Mg(C_2H_3O_2)_2.4H_2O$	20	65	11.3	$Na_2HPO_4.12H_2O$	20	95	16.5
$NaNO_2$	20	66	11.5	NaF	100	96.6	734
NH_4Cl and KNO_3	30	68.6	21.6	$Pb(NO_3)_2$	20	98	17.0
KBr	100	69.2	526	$CuSO_4.5H_2O$	20	98	17.0
NH_4Cl and KNO_3	25	71.2	16.7	$TlNO_3$	100.3	98.7	759
NH_4Cl and KNO_3	20	72.6	12.6	TlCl	100.1	99.7	761

CONSTANT HUMIDITY WITH SULFURIC ACID SOLUTIONS

The relative humidity and pressure of aqueous vapor of air in equilibrium conditions above aqueous solutions of sulfuric acid are given below.

Density of acid solution	Relative humidity	Vapor pressure at 20°C	Density of acid solution	Relative humidity	Vapor pressure at 20°C
1.00	100.0	17.4	1.30	58.3	10.1
1.05	97.5	17.0	1.35	47.2	8.3
1.10	93.9	16.3	1.40	37.1	6.5
1.15	88.8	15.4	1.50	18.8	3.3
1.20	80.5	14.0	1.60	8.5	1.5
1.25	70.4	12.2	1.70	3.2	0.6

For concentration of sulfuric acid solution refer to tables relating density to percent composition.

VELOCITY OF SOUND

Compiled by Gordon E. Becker, Bell Telephone Laboratories

The data for the Velocity of Sound is Various Materials were compiled from a variety of sources. For more extensive tables one is referred to the following books:
AIP Handbook, Smithsonian Tables.
Mason: Physical Acoustics and the Properties of Solids (1958).
Chalmers and Quarrell: Physical Examination of Metals (1960).
Mason: Piezoelectric Crystals and their Application to Ultrasonics (1950).
Bergmann: Der Ultraschall (Hirzel, 1954).

Definition of Terms: V_l = Velocity of plane longitudinal wave in bulk material
V_s = Velocity of plane transverse (shear) wave
V_{ext} = Velocity of longitudinal wave (extensional wave) in thin rods.

SOLIDS

Substance	Density g/cc	V_l m/sec	V_s m/sec	V_{ext} m/sec
Metals				
Aluminum, rolled	2.7	6420	3040	5000
Berylium	1.87	12890	8880	12870
Brass (70 Cu, 30 Zn)	8.6	4700	2110	3480
Copper, annealed	8.93	4760	2325	3810
Copper, rolled	8.93	5010	2270	3750
Duralumin 17S	2.79	6320	3130	5150
Gold, hard-drawn	19.7	3240	1200	2030
Iron, electrolytic	7.9	5950	3240	5120
Iron, Armco	7.85	5960	3240	5200
Lead, annealed	11.4	2160	700	1190
Lead, rolled	11.4	1960	690	1210
Magnesium, drawn, annealed	1.74	5770	3050	4940
Molybdenum	10.1	6250	3350	5400
Monel metal	8.90	5350	2720	4400
Nickel (unmagnetized)	8.85	5480	2990	4800
Nickel	8.9	6040	3000	4900
Platinum	21.4	3260	1730	2800
Silver	10.4	3650	1610	2680
Steel, mild	7.85	5960	3235	5200
Steel, 347 Stainless	7.9	5790	3100	5000
Steel (1%C)	7.84	5940	3220	5180
Steel (1%C, hardened)	7.84	5854	3150	5070
Tin, rolled	7.3	3320	1670	2730
Titanium	4.5	6070	3125	5080
Tungsten, annealed	19.3	5220	2890	4620
Tungsten, drawn	19.3	5410	2640	4320
Tungsten Carbide	13.8	6655	3980	6220
Zinc, rolled	7.1	4210	2440	3850
Various				
Fused silica	2.2	5968	3764	5760
Glass, pyrex	2.32	5640	3280	5170
Glass, heavy silicate flint	3.88	3980	2380	3720
Glass, light borate crown	2.24	5100	2840	4540
Lucite	1.18	2680	1100	1840
Nylon 6-6	1.11	2620	1070	1800
Polyethylene	0.90	1950	540	920
Polystyrene	1.06	2350	1120	2240
Rubber, butyl	1.07	1830		
Rubber, gum	0.95	1550		
Rubber neoprene	1.33	1600		
Brick	1.8			3650
Clay rock	2.2			3480
Cork	0.25			500
Marble	2.6			3810
Paraffin	0.9			1300
Tallow				390
Woods				
Ash, along the fiber				4670
Ash, across the rings				1390
Ash, along the rings				1260
Beech, along the fiber				3340
Elm, along the fiber				4120
Maple, along the fiber				4110
Oak, along the fiber				3850

LIQUIDS

Substance	Formula	Density g/cc	Velocity at 25°C m/sec	$-\partial v/\partial t$ m/sec °C
Acetone	C_3H_6O	0.79	1174	4.5
Benzene	C_6H_6	0.870	1295	4.65
Carbon disulphide	CS_2	1.26	1149	—
Carbon tetrachloride	CCl_4	1.595	926	2.7
Castor oil	$C_{11}H_{10}O_{10}$	0.969	1477	3.6
Chloroform	$CHCl_3$	1.49	987	3.4
Ethanoi	C_2H_6O	0.79	1207	4.0
Ethanol amide	C_2H_7NO	1.018	1724	3.4
Ethyl ether	$C_4H_{10}O$	0.713	985	4.87
Ethylene glycol	$C_2H_6O_2$	1.113	1658	2.1
Glycerol	$C_3H_8O_3$	1.26	1904	2.2
Kerosene	—	0.81	1324	3.6
Mercury	Hg	13.5	1450	—
Methanol	CH_4O	0.791	1103	3.2
Nitrobenzene	$C_6H_5NO_2$	1.20	1463	3.6
Turpentine	—	0.88	1255	—
Water (distilled)	H_2O	0.998	1496.7+ 2	−2.4
Water (sea)	—	1.025	1531	−2.4
Xylene hexafluoride	$C_8H_4F_6$	1.37	879	—

GASES AND VAPORS

Substance	Formula	Density g/ℓ	Velocity m/sec	$\partial v/\partial t$ m/sec °C
Gases (0°C)				
Air, dry		1.293	331.45	0.59
Ammonia	NH_3	0.771	415	
Argon	Ar	1.783	319	0.56
Carbon monoxide	CO	1.25	338	0.6
Carbon dioxide	CO_2	1.977	259	0.4
Chlorine	CL_2	3.214	206	
Deuterium	D_2		890	1.6
Ethane (10°C)	C_2H_6	1.356	308	
Ethylene	C_2H_4	1.260	317	
Helium	He	0.178	965	0.8
Hydrogen	H_2	0.0899	1284	2.2
Hydrogen bromide	HBr	3.50	200	
Hydrogen chloride	HCl	1.639	296	
Hydrogen iodide	HI	5.66	157	
Hydrogen sulfide	H_2S	1.539	289	
Illuminating (Coal gas)			453	
Methane	CH_4	0.7168	430	
Neon	Ne	0.900	435	0.8
Nitric oxide (10°C)	NO	1.34	324	
Nitrogen	N_2	1.251	334	0.6
Nitrous oxide	N_2O	1.977	263	0.5
Oxygen	O_2	1.429	316	0.56
Sulfur dioxide	SO_2	2.927	213	0.47
Vapors (97.1°C)				
Acetone	C_3H_6O		239	0.32
Benzene	C_6H_6		202	0.3
Carbon tetrachloride	CCl_4		145	
Chloroform	$CHCl_3$		171	0.24
Ethanol	C_2H_6O		269	0.4
Ethyl ether	$C_4H_{10}O$		206	0.3
Methanol	CH_4O		335	0.46
Water vapor (134°C)	H_2O		494	

SOUND VELOCITY IN WATER ABOVE 212°F

By permission from the Acoustical Society of America, Volume 31 (1959) and J. C. McDade, D. R. Pardue, A. L. Gedrich and F. Vrataric.

Temperature°F	Velocity m/sec	Velocity ft/sec	Temperature °F	Velocity m/sec	Velocity ft/sec
186.8	1552	5092	370	1368	4488
200	1548	5079	380	1353	4439
210	1544	5066	390	1337	4386
220	1538	5046	400	1320	4331
230	1532	5026	410	1302	4272
240	1524	5000	420	1283	4209
250	1516	4974	430	1264	4147
260	1507	4944	440	1244	4081
270	1497	4911	450	1220	4010
280	1487	4879	460	1200	3940
290	1476	4843	470	1180	3880
300	1465	4806	480	1160	3800
310	1453	4767	490	1140	3730
320	1440	4724	500	1110	3650
330	1426	4678	510	1090	3570
340	1412	4633	520	1070	3500
350	1398	4587	530	1040	3410
360	1383	4537	540	1010	3320
			550	980	3230

MUSICAL SCALES

EQUAL TEMPERED CHROMATIC SCALE

A_4 = 440

American Standard pitch. Adopted by the American Standards Association in 1936

Note	Frequency	Note	Frequency	Note	Frequency	Note	Frequency
C_0	16.35	C_2	65.41	C_4	261.63	C_6	1046.50
$C\#_0$	17.32	$C\#_2$	69.30	$C\#_4$	277.18	$C\#_6$	1108.73
D_0	18.35	D_2	73.42	D_4	293.66	D_6	1174.66
$D\#_0$	19.45	$D\#_2$	77.78	$D\#_4$	311.13	$D\#_6$	1244.51
E_0	20.60	E_2	82.41	E_4	329.63	E_6	1318.51
F_0	21.83	F_2	87.31	F_4	349.23	F_6	1396.91
$F\#_0$	23.12	$F\#_2$	92.50	$F\#_4$	369.99	$F\#_6$	1479.98
G_0	24.50	G_2	98.00	G_4	392.00	G_6	1567.98
$G\#_0$	25.96	$G\#_2$	103.83	$G\#_4$	415.30	$G\#_6$	1661.22
A_0	27.50	A_2	110.00	A_4	**440.00**	A_6	1760.00
$A\#_0$	29.14	$A\#_2$	116.54	$A\#_4$	466.16	$A\#_6$	1864.66
B_0	30.87	B_2	123.47	B_4	493.88	B_6	1975.53
C_1	32.70	C_3	130.81	C_5	523.25	C_7	2093.00
$C\#_1$	34.65	$C\#_3$	138.59	$C\#_5$	554.37	$C\#_7$	2217.46
D_1	36.71	D_3	146.83	D_5	587.33	D_7	2349.32
$D\#_1$	38.89	$D\#_3$	155.56	$D\#_5$	622.25	$D\#_7$	2489.02
E_1	41.20	E_3	164.81	E_5	659.26	E_7	2637.02
F_1	43.65	F_3	174.61	F_5	698.46	F_7	2793.83
F#1	46.25	$F\#_3$	185.00	$F\#_5$	739.99	$F\#_7$	2959.96
G_1	49.00	G_3	196.00	G_5	783.99	G_7	3135.96
$G\#_1$	51.91	$G\#_3$	207.65	$G\#_5$	830.61	$G\#_7$	3322.44
A_1	55.00	A_3	220.00	A_5	880.00	A_7	3520.00
$A\#_1$	58.27	$A\#_3$	233.08	$A\#_5$	932.33	$A\#_7$	3729.31
B_1	61.74	B_3	246.94	B_5	987.77	B_7	3951.07
						C_8	4186.01

EQUAL TEMPERED CHROMATIC SCALE

A_4 = 435

International Pitch, adopted 1891

Note	Frequency	Note	Frequency	Note	Frequency	Note	Frequency
C_0	16.17	C_2	64.66	C_4	258.65	C_6	1034.61
$C\#_0$	17.13	$C\#_2$	68.51	$C\#_4$	274.03	$C\#_6$	1096.13
D_0	18.15	D_2	72.58	D_4	290.33	D_6	1161.31
$D\#_0$	19.22	$D\#_2$	76.90	$D\#_4$	307.59	$D\#_6$	1230.37
E_0	20.37	E_2	81.47	E_4	325.88	E_6	1303.53
F_0	21.58	F_2	86.31	F_4	345.26	F_6	1381.04
$F\#_0$	22.86	$F\#_2$	91.45	$F\#_4$	365.79	$F\#_6$	1463.16
G_0	24.22	G_2	96.89	G_4	387.54	G_6	1550.16
$G\#_0$	25.66	$G\#_2$	102.65	$G\#_4$	410.59	$G\#_6$	1642.34
A_0	27.19	A_2	108.75	A_4	**435.00**	A_6	1740.00
$A\#_0$	28.80	$A\#_2$	115.22	$A\#_4$	460.87	$A\#_6$	1843.47
B_0	30.52	B_2	122.07	B_4	488.27	B_6	1953.08
C_1	32.33	C_3	129.33	C_5	517.31	C_7	2069.22
$C\#_1$	34.25	$C\#_3$	137.02	$C\#_5$	548.07	$C\#_7$	2192.26
D_1	36.29	D_3	145.16	D_5	580.66	D_7	2322.62
$D\#_1$	38.45	$D\#_3$	153.80	$D\#_5$	615.18	$D\#_7$	2460.73
E_1	40.74	E_3	162.94	E_5	651.76	E_7	2607.05
F_1	43.16	F_3	172.63	F_5	690.52	F_7	2762.08
$F\#_1$	45.72	$F\#_3$	182.89	$F\#_5$	731.58	$F\#_7$	2926.32
G_1	48.44	G_3	193.77	G_5	775.08	G_7	3100.33
$G\#_1$	51.32	$G\#_3$	205.29	$G\#_5$	821.17	$G\#_7$	3284.68
A_1	54.38	A_3	217.50	A_5	870.00	A_7	3480.00
$A\#_1$	57.61	$A\#_3$	230.43	$A\#_5$	921.73	$A\#_7$	3686.93
B_1	61.03	B_3	244.14	B_5	976.54	B_7	3906.17
						C_8	4138.44

SCIENTIFIC OR JUST SCALE

C_4 = 256

Note	Frequency	Note	Frequency	Note	Frequency	Note	Frequency
C_0	16	C_2	64	C_4	256	C_6	1024
D_0	18	D_2	72	D_4	288	D_6	1152
E_0	20	E_2	80	E_4	320	E_6	1280
F_0	21.33	F_2	85.33	F_4	341.33	F_6	1365.33
G_0	24	G_2	96	G_4	384	G_6	1536
A_0	26.67	A_2	106.67	A_4	426.67	A_6	1706.67
B_0	30	B_2	120	B_4	480	B_6	1920
C_1	32	C_3	128	C_5	512	C_7	2048
D_1	36	D_3	144	D_5	576	D_7	2304
E_1	40	E_3	160	E_5	640	E_7	2560
F_1	42.67	F_3	170.67	F_5	682.67	F_7	2730.67
G_1	48	G_3	192	G_5	768	G_7	3072
A_1	53.33	A_3	213.33	A_5	853.33	A_7	3413.33
B_1	60	B_3	240	B_5	960	B_7	3840
						C_8	4096

ABSORPTION AND VELOCITY OF SOUND IN STILL AIR

The following data refer only to the temperature 20C (68F). They were abstracted from an extensive compilation prepared by L. B. Evans and H. E. Bass. The entire report, Tables of Absorption and Velocity of Sound in Still Air at 68F (20C), AD-738 576 is available from National Technical Information Service, U. S. Department of Commerce, 5285 Port Royal Road, Springfield, Va. 22151.

Frequency (Hz)	Absorption (dB/1000 ft)	Absorption (dB/Km)	Absorption (dB/sec)	Velocity (1000 ft/sec)
Relative Humidity = 0%				
20.	0.154	0.51	0.174	1.126892
40.	0.327	1.07	0.368	1.127013
50.	0.384	1.26	0.433	1.127050
63.	0.436	1.43	0.491	1.127085
100.	0.509	1.67	0.573	1.127131
200.	0.560	1.84	0.631	1.127161
400.	0.596	1.96	0.672	1.127169
630.	0.645	2.11	0.727	1.127171
800.	0.692	2.27	0.780	1.127172
1250.	0.861	2.82	0.970	1.127172
2000.	1.262	4.14	1.423	1.127173
4000.	2.696	8.84	3.039	1.127178
6300.	4.541	14.89	5.118	1.127182
10000.	8.013	26.28	9.032	1.127184
12500.	10.918	35.81	12.306	1.127186
16000.	15.901	52.15	17.923	1.127187
20000.	22.978	75.37	25.901	1.127187
40000.	81.405	267.01	91.759	1.127188
63000.	196.544	644.66	221.542	1.127188
80000.	314.677	1032.14	354.700	1.127188
Relative Humidity = 5%				
20.	0.031	0.10	0.034	1.126973
40.	0.074	0.24	0.083	1.126996
50.	0.092	0.30	0.104	1.127004
63.	0.114	0.37	0.129	1.127009
100.	0.179	0.59	0.202	1.127019
200.	0.449	1.47	0.506	1.127028
400.	1.451	4.76	1.635	1.127043
630.	3.211	10.53	3.619	1.127067
800.	4.774	15.66	5.380	1.127088
1250.	9.164	30.06	10.329	1.127147
2000.	15.175	49.77	17.106	1.127228
4000.	22.685	74.41	25.573	1.127321
6300.	26.245	86.08	29.587	1.127352
10000.	30.781	100.96	34.701	1.127365
12500.	34.306	112.52	38.676	1.127369
16000.	40.263	132.06	45.391	1.127372
20000.	48.653	159.58	54.850	1.127373
40000.	115.903	380.16	130.666	1.127378
63000.	242.070	793.99	272.905	1.127381
80000.	367.063	1203.97	413.821	1.127383
Relative Humidity = 10%				
20.	0.021	0.07	0.024	1.127167
40.	0.064	0.21	0.072	1.127183
50.	0.084	0.28	0.095	1.127191
63.	0.108	0.35	0.122	1.127199
100.	0.161	0.53	0.181	1.127213
200.	0.289	0.95	0.326	1.127225
400.	0.706	2.32	0.796	1.127230
630.	1.501	4.92	1.692	1.127234
800.	2.297	7.54	2.590	1.127238
1250.	5.155	16.91	5.811	1.127254
2000.	11.658	38.24	13.141	1.127287
4000.	31.023	101.76	34.975	1.127386
6300.	47.085	154.44	53.087	1.127462
10000.	61.578	201.98	69.431	1.127522
12500.	68.146	223.52	76.837	1.127540
16000.	76.231	250.04	85.955	1.127555
20000.	85.605	280.78	96.525	1.127563
40000.	151.938	498.36	171.321	1.127576
63000.	277.191	909.19	312.555	1.127581
80000.	403.662	1324.01	455.162	1.127583

ABSORPTION AND VELOCITY OF SOUND IN STILL AIR (*Continued*)

Frequency (Hz)	Absorption (dB/1000 ft)	Absorption (dB/Km)	Absorption (dB/sec)	Velocity (1000 ft/sec)
Relative Humidity = 20%				
20.	0.013	0.04	0.014	1.127568
40.	0.045	0.15	0.051	1.127577
50.	0.066	0.22	0.074	1.127582
63.	0.093	0.30	0.105	1.127587
100.	0.164	0.54	0.185	1.127603
200.	0.285	0.93	0.321	1.127624
400.	0.476	1.56	0.537	1.127633
630.	0.789	2.59	0.890	1.127636
800.	1.103	3.62	1.244	1.127637
1250.	2.277	7.47	2.568	1.127640
2000.	5.310	17.42	5.987	1.127645
4000.	18.991	62.29	21.416	1.127670
6300.	41.151	134.98	46.406	1.127710
10000.	79.657	261.28	89.836	1.127778
12500.	103.004	337.85	116.170	1.127817
16000.	130.465	427.93	147.146	1.127859
20000.	155.809	511.05	175.736	1.127893
40000.	251.952	826.40	284.192	1.127960
63000.	382.062	1253.16	430.957	1.127977
80000.	508.369	1667.45	573.431	1.127982
Relative Humidity = 30%				
20.	0.009	0.03	0.010	1.127976
40.	0.034	0.11	0.038	1.127980
50.	0.051	0.17	0.057	1.127984
63.	0.075	0.25	0.085	1.127987
100.	0.151	0.50	0.170	1.127999
200.	0.309	1.01	0.349	1.128023
400.	0.484	1.59	0.546	1.128037
630.	0.682	2.24	0.770	1.128041
800.	0.868	2.85	0.979	1.128044
1250.	1.552	5.09	1.751	1.128045
2000.	3.333	10.93	3.760	1.128047
4000.	11.856	38.89	13.374	1.128056
6300.	27.626	90.61	31.164	1.128072
10000.	62.493	204.98	70.498	1.128105
12500.	89.659	294.08	101.148	1.128131
16000.	128.814	422.51	145.324	1.128166
20000.	171.847	563.66	193.879	1.128204
40000.	338.710	1110.97	382.172	1.128316
63000.	499.838	1639.47	563.998	1.128361
80000.	635.085	2083.08	716.614	1.128375
Relative Humidity = 40%				
20.	0.007	0.02	0.008	1.128386
40.	0.027	0.09	0.030	1.128388
50.	0.041	0.13	0.046	1.128390
63.	0.062	0.20	0.070	1.128392
100.	0.134	0.44	0.151	1.128402
200.	0.318	1.04	0.359	1.128424
400.	0.524	1.72	0.592	1.128443
630.	0.692	2.27	0.781	1.128449
800.	0.829	2.72	0.935	1.128451
1250.	1.309	4.29	1.477	1.128453
2000.	2.544	8.34	2.870	1.128456
4000.	8.523	27.96	9.618	1.128460
6300.	19.995	65.58	22.564	1.128467
10000.	47.390	155.44	53.478	1.128484
12500.	70.857	232.41	79.962	1.128499
16000.	108.308	355.25	122.228	1.128521
20000.	154.838	507.87	174.742	1.128549
40000.	380.371	1247.62	429.310	1.128663
63000.	596.091	1955.18	672.827	1.128733
80000.	753.514	2471.53	850.536	1.128758
Relative Humidity = 50%				
20.	0.006	0.02	0.006	1.128795
40.	0.022	0.07	0.025	1.128797
50.	0.034	0.11	0.038	1.128798
63.	0.052	0.17	0.058	1.128800
100.	0.117	0.38	0.132	1.128807
200.	0.313	1.03	0.353	1.128826
400.	0.563	1.85	0.636	1.128849

ABSORPTION AND VELOCITY OF SOUND IN STILL AIR (*Continued*)

Frequency (Hz)	Absorption (dB/1000 ft)	Absorption (dB/Km)	Absorption (dB/sec)	Velocity (1000 ft/sec)
630.	0.734	2.41	0.828	1.128858
800.	0.851	2.79	0.961	1.128860
1250.	1.231	4.04	1.390	1.128862
2000.	2.176	7.14	2.457	1.128865
4000.	6.752	22.15	7.622	1.128867
6300.	15.643	51.31	17.659	1.128871
10000.	37.521	123.07	42.357	1.128881
12500.	57.009	186.99	64.357	1.128890
16000.	89.594	293.87	101.143	1.128903
20000.	132.719	435.32	149.830	1.128922
40000.	382.722	1255.33	432.101	1.129020
63000.	654.606	2147.11	739.115	1.129099
80000.	844.024	2768.40	953.016	1.129133
Relative Humidity = 60%				
20.	0.005	0.02	0.005	1.129207
40.	0.018	0.06	0.021	1.129208
50.	0.029	0.09	0.032	1.129209
63.	0.044	0.15	0.050	1.129210
100.	0.103	0.34	0.117	1.129215
200.	0.301	0.99	0.339	1.129232
400.	0.593	1.94	0.669	1.129254
630.	0.782	2.57	0.884	1.129266
800.	0.896	2.94	1.012	1.129269
1250.	1.223	4.01	1.381	1.129273
2000.	1.997	6.55	2.256	1.129275
4000.	5.711	18.73	6.449	1.129277
6300.	12.962	42.51	14.638	1.129279
10000.	31.050	101.84	35.064	1.129286
12500.	47.462	155.67	53.598	1.129292
16000.	75.544	247.78	85.312	1.129300
20000.	113.958	373.78	128.694	1.129313
40000.	364.443	1195.37	411.598	1.129389
63000.	677.023	2220.64	764.675	1.129467
80000.	899.912	2951.71	1016.458	1.129508
Relative Humidity = 70%				
20.	0.004	0.01	0.005	1.129618
40.	0.016	0.05	0.018	1.129619
50.	0.025	0.08	0.028	1.129619
63.	0.039	0.13	0.044	1.129620
100.	0.092	0.30	0.104	1.129623
200.	0.284	0.93	0.321	1.129638
400.	0.611	2.01	0.691	1.129662
630.	0.829	2.72	0.937	1.129673
800.	0.947	3.11	1.070	1.129678
1250.	1.250	4.10	1.412	1.129683
2000.	1.915	6.28	2.163	1.129685
4000.	5.056	16.58	5.712	1.129687
6300.	11.197	36.73	12.649	1.129689
10000.	26.624	87.33	30.077	1.129693
12500.	40.758	133.69	46.044	1.129697
16000.	65.246	214.01	73.708	1.129704
20000.	99.365	325.92	112.254	1.129712
40000.	339.153	1112.42	383.165	1.129770
63000.	673.667	2209.63	761.137	1.129842
80000.	924.481	3032.30	1044.556	1.129884
Relative Humidity = 80%				
20.	0.004	0.01	0.004	1.130030
40.	0.014	0.05	0.016	1.130030
50.	0.022	0.07	0.025	1.130031
63.	0.034	0.11	0.039	1.130032
100.	0.082	0.27	0.093	1.130034
200.	0.267	0.88	0.302	1.130047
400.	0.620	2.03	0.701	1.130069
630.	0.870	2.85	0.983	1.130082
800.	0.998	3.27	1.128	1.130088
1250.	1.293	4.24	1.461	1.130094
2000.	1.887	6.19	2.132	1.130096
4000.	4.626	15.17	5.228	1.130098
6300.	9.976	32.72	11.274	1.130100
10000.	23.466	76.97	26.519	1.130102
12500.	35.896	117.74	40.566	1.130106
16000.	57.588	188.89	65.081	1.130110
20000.	88.144	289.11	99.612	1.130116

DIELECTRIC CONSTANTS (Continued)

Dielectric Constants of Pure Liquids

Inorganic Liquids

	Substance	ε	°C	a(or α) × 10^2	Range t_1, t_2
A	Argon	1.53_8	−191	0.34	−191.−184
$AlBr_3$	Aluminum bromide	3.38	100	0.33	100,240
AsH_3	Arsine	2.50	−100	0.43	−116,−72
BBr_3	Boron bromide	2.58	0	0.28	−70,80
Br_2	Bromine	3.09	20	0.7	0,50
CO_2	Carbon dioxide	1.60[c]	20	—	—
Cl_2	Chlorine	2.10_1	−50	0.31	−65,−33
		1.91	14	0.32	−22,14
		1.7_2	77		
		1.5_4	142		
D_2	Deuterium	1.277	20°K	0.4	18.8,21.2° K
D_2O	Deuterium oxide	78.25	25	[d]	0.4,98
F_2	Fluorine	1.54	−202	0.19	−216,−190
$GeCl_4$	Germanium tetrachloride	2.43°	25	0.240	0.55
HBr	Hydrogen bromide	7.00	−85	0.26(α)	−85,−70
HCl	Hydrogen chloride	6.35	−15	0.288(α)	−85,−15
		12.	−113	—	—
		4.6	28	—	—
HF	Hydrogen fluoride	17_5	−73	—	—
		13_4	−42		
		11_1	−27		
		84	0		
HI	Hydrogen iodide	3.39	−50	0.8	−51,−37
H_2	Hydrogen	1.228	20.4°K	0.34	14,21°K
H_2O	Water	78.54	25	[e]	0,100
		34.5_9	200	[f]	100,370
H_2O_2	Hydrogen peroxide	84.2	0	[f]	−30,20
H_2S	Hydrogen sulfide	9.26	−85.5	—	—
		9.05	−78.5	—	—
He	Helium	1.055_4	2.06°K	—	—
		1.055_9	2.30[f]		
		1.055_3	2.63		
		1.053_9	3.09		
		1.051_8	3.58		
		1.048	4.19		
I_2	Iodine	6.8	400	—	ε_r = 13.3—0.0 0.016 (±0.002) T
		$11._7$	140		
		$13._0$	168		
NH_3	Ammonia	25.	−77.7	—	—
		22.4	−33.4	—	—
		18.9	5	—	—
		17.8	15		
		16.9	25		
		16.3	35		
NOBr	Nitrosyl bromide	$13._4$	15	—	—
NOCl	Nitrosyl chloride	$18._2$	12	—	—
N_2	Nitrogen	1.454	−203	0.29	−210,−195
N_2H_4	Hydrazine	$52._9$	20	0.21(α)	0,25
N_2O	Dinitrogen oxide	1.97	−90	—	—
		1.61	0	0.6	−6,14
N_2O_4	Dinitrogen tetroxide	2.5[b]	15	—	—
O_2	Oxygen	1.507	−193	0.24	−218,−183
P	Phosphorus	4.10	34	—	—
		4.06	46		
		3.86	85		
PCl_3	Phosphorus trichloride	3.43	25	0.84	17,60
PCl_4	Phosphorus pentachloride	2.8_4	160	—	—
$POCl_3$	Phosphoryl chloride	$13._3$	22	—	—
$PSCl_3$	Thophosphoryl chloride	5.8	22	—	—
$PbCl_4$	Lead tetrachloride	2.78	20	—	—
S	Sulfur	3.52	118	[e]	—
		3.48	231		
$SOBr_2$	Thionyl bromide	9.06	20	3.0	at 20
$SOCl_2$	Thionyl chloride	9.25	20	3.9	at 20
SO_2	Sulfur dioxide	17.6	−20	0.287(α)	−65,−15
		15.0_8	0	—	—
		$14._1$	20	7.7	14,140
		2.1_0	154[h]		
SO_2	Sulfur trioxide	3.11	18	—	—
S_2Cl_2	Sulfur monochloride	4.79	15	0.146(α)	−41,15
SO_2Cl_2	Sulfuryl chloride	$10._0$	22	—	—
$SbCl_5$	Antimony pentachloride	3.22	20	0.46	2,47
Se	Selenium	5.40	250	0.25	237,301
$SiCl_4$	Silicon tetrachloride	2.4[a]	16	—	—
$SnCl_4$	Tin tetrachloride	2.87	20	0.30	−30,20
$TiCl_4$	Titanium tetrachloride	2.80	20	0.20	−20,20

Organic Liquids

	Substance	ε	°C	a(or α) × 10^2	Range t_1, t_2
CCl_4	Carbon tetrachloride	2.238	20	0.200	−10,60
CN_2O_8	Tetranitromethane	2.52_1	25	—	—
CO_2	Carbon dioxide	1.60_4[c]	0	—	—
CS_2	Carbon disulfide	2.641	20	0.268	−90,130
		3.001	−110		
		2.19	180		
$CHBr_2$	Bromoform	4.39	20	0.105(α)	10,70
$CHCl_3$	Chloroform	4.806	20	0.160(α)	0,50
		6.76	−60	—	—
		6.12	−40		
		5.61	−20		
		3.7_1	100	—	—
		3.3_2	140		
		2.9_2	180		
CHN	Hydrocyanic acid	$158._1$	0	[i]	−13,18
		$114._9$	20	0.63(α)	18,26
CH_2Br_2	Dibromomethane	7.77	10	—	—
		6.68	40		
CH_2Cl_2	Dichloromethane	9.08	20	[j]	−80,25
CH_2I_2	Diiodomethane	5.32	25	—	—
CH_1O_2	Formic acid	$58._5$[a]	16	—	—
CH_2Br	Bromomethane	9.82	0	[k]	−80,0
CH_2Cl	Chloromethane	12.6	−20	[l]	−70,−20
CH_3I	Iodomethane	7.00	20	[m]	−70,40
CH_3NO	Formamide	109.	20	72.	18,25
CH_4	Methane	1.70	−173	0.2	−181,−159
CH_4O	Methanol	32.63	25	0.264(α)	5,55
		64.	−113	—	—
		54.	−80		
		40.	−20		
CH_5N	Methylamine	11.4	−10	0.26(α)	−30,−10
		9.4	25	—	—
C_2					
C_2HCl_3	Trichloroethylene	3.4_2	ca 16	—	—
C_2HCl_3O	Chloral	4.9_4	20	0.17(α)	15,45
		7.6	−40		
		4.2	62		
$C_2HCl_3O_2$	Trichloroacetic acid	4.6	60	—	—
$C_2H_2Br_2$	*cis*-1,2Dibromoethylene	7.7_2	0	—	—
		7.0_5	25		
	trans-1,2-Dibromoethylene	2.9_7	0	—	—
		2.8_8	25		
$C_2H_2Br_4$	1,1,2,3-Tetrabromoethane	8.6	3	—	—
		7.0	22		
$C_2H_2Cl_2$	1,1-Dichloroethylene	4.6_7	16	—	—
	cis-1,2-Dichloroethylene	9.20	25	—	—
	trans-1,2-Dichloroethylene	2.14	25	—	—
$C_2H_2Cl_2O_2$	Dichloroacetic acid	8.2	22	—	—
		7.8	61		
C_2H_3ClO	Acetyl chloride	$16._9$	2	—	—
		$15._8$	22		
$C_2H_3ClO_2$	Chloroacetic acid	12.3	60	2.	60,80
C_2H_4O	Ethylene oxide	$13._9$	−1	—	—
	Acetaldehyde	$21._8$[a]	10	—	—
$C_2H_4O_2$	Acetic acid	6.15	20	—	—
		6.29	40		
		6.62	70		
	Methyl formate	8.5	20	5.	0,20
C_2H_4ClO	2-Chloroethanol (ethylene chlorohydrin)	$25._8$	25	—	—
		$13._2$	132		
C_3H_5NO	Acetamide	50.[a]	83	—	—
$C_2H_5NO_2$	Nitroethane	28.0_6	30	11.4	30,35
C_2H_6O	Ethanol	24.30	25	—	—
		24.3[5]	25	0.270(α)	−5,70
		41.0[5]	−60	0.297(α)	−110,−20
	Methyl ether	5.02	25	2.38	25,100
		2.97	110		
		2.64	120		
		2.37	125		
		2.26	126.1		
		1.90	127.6		
$(C_2H_6OSi)_n$ $n = 4$	Octamethylcyclotetrasilox-ane	2.39	20	—	—

DIELECTRIC CONSTANTS (Continued)

Dielectric Constants of Pure Liquids

Organic Liquids

	Substance	ε	°C	a(or α) × 10^2	Range t_1, t_2
$n = 5$	Decamethylcyclopentasiloxane	2.50	20	—	—
$n = 6$	Dodecamethylcyclohexasiloxane	2.59	20	—	—
$n = 7$	Tetradecamethylcycloheptasiloxane	2.68	20	—	—
$n = 8$	Hexadecamethylcyclooctasiloxane	2.74	20	—	—
$C_2H_6O_2$	Glycol	37.$_7$	25	0.224(α)	20,100
C_2H_7N	Dimethylamine	6.32	0	—	—
		5.26	25		
	C_3				
C_3H_6	Propene	1.87$_5$	20	—	—
		1.79^5	45		
		1.69	65		
		1.53$_0$	85		
		1.44$_1$	90		
		1.33$_1$	91.9[h]		
C_3H_6O	2-Propen-1-ol(Allylalcohol)	21.$_6$	15	—	—
	Acetone	20.7$_0$	25	0.205(α)	−60,40
		17.7	56	—	—
	Propionaldehyde	18.$_5$[a]	17	—	—
$C_3H_6O_2$	Propionic acid	3.30	10	—	—
		3.44	40		
	Ethyl formate	7.1$_6$	25	—	—
	Methyl acetate	6.68	25	2.2	25,40
$C_3H_6O_3$	*dl*-Lactic acid	22.	17	—	—
$C_3H_6NO_2$	Ethyl carbamate (Urethan)	14.2	50	5.2	50,70
C_3H_8	Propane	1.61	0	0.20	−90,15
C_3H_8O	1-Propanol	20.1	25	0.293(α)	20,90
		38.	−80	—	—
		29.	−34		
	2-Propanol	18.3	25	0.310(α)	20,70
$C_3H_8O_2$	1,2-Propanediol	32.$_0$	20	0.27(α)	at 20
	1,3-Propanediol	35.$_0$	20	0.23(α)	at 20
$C_3H_8O_3$	Glycerol	42.5	25	0.208(α)	0,100
C_3H_9N	Isopropylamine	5.5[b]	20	—	—
	Trimethylamine	2.44	25	0.52	0,25
	C_4				
$C_4H_2O_3$	Maleic anhydride	50[a]	60	—	—
C_4H_4O	Furan	2.95	25	—	—
C_4H_4S	Thiophene	2.76	16	—	—
C_4H_5N	Pyrrole	7.48	18	—	—
C_4H_5NS	Allyl isothiocyanate	17.$_2$[b]	18	—	—
C_4H_6O	Vinyl ether	3.94	20	—	—
$C_4H_6O_3$	Acetic anhydride	22.^{4}	1	—	—
		20.$_7$	19		
C_4H_8O	2-Butanone	18.5$_1$	20	0.207(α)	−60,60
	Butyraldehyde	13.4	26	—	—
		10.8	77		
$C_4H_8O_2$	Butyric acid	2.97	20	−0.23	10,70
	Isobutyric acid	2.71	10	—	—
		2.73	40		
	Propyl formate	7.7$_2$[a]	19	—	—
	Ethylacetate	6.02	25	1.5	at 25
		5.3$_0$	77	—	—
	Methyl propionate	5.5[n]	19	—	—
	1,4-Dioxane	2.209	25	0.170	20,50
C_4H_9NO	Morpho line	7.33	25	—	—
$C_4H_{10}O$	1-Butanol	17.8	20	0.300(α)	−40,20
		17.1	25	0.335(α)	25,70
		8.2	118	—	—
	2-Methyl-1-propanol	17.7	25	0.377(α)	20,90
		34.	−80	—	—
		26.	−34		
	2-Butanol	15.8	25	—	—
	2-Methyl-2-propanol	10.9	30	—	—
		8.49	50		
		6.89	70		
	Ethyl ether	4.335	20	2.0	at 20
		4.34^5	20	0.217(α)	−40,30
		10.4	−116	—	—
		3.97	40	0.170(α)	40,140
		2.1$_2$	180		
		1.8^9	190		
		1.5$_3$	193.3[h]		
$C_4H_{10}Zn$	Diethyl zinc	2.5$_5$	20	—	—
	C_5				
$C_5H_4O_2$	Furfural	46.^{9}	1	—	—
		41.$_9$	20		
		34.$_9$	50		
C_5H_5N	Pyridine	12.3	25	—	—
		9.4	116	—	—
C_5H_8	1,3-Pentadiene[c]	2.32	25	—	—
	2-Methyl-1,3-butadiene (Isoprene)	2.10	25	0.24	−75,25
$C_5H_8O_2$	2,4-Pentanedione (Acetylacetone)	25.$_7$[a]	20	—	—
C_5H_{10}	1-Pentene	2.100	20	—	—
	2-Methyl-1-butene	2.197	20	—	—
	Cyclopentane	1.965	20	—	—
	Ethylcyclopropane	1.933	20	—	—
$C_5H_{10}O$	2-Pentanone	15.4$_5$	20	0.195(α)	−40.80
		22.0	−60		
	3-Pentanone	17.0$_0$	20	0.225(α)	0,80
		19.4	−20		
		19.8	−40		
$C_5H_{10}O_2$	Valeric acid	2.6$_6$	20	—	—
	Isovaleric acid	2.6$_4$	20	—	—
	Methyl butyrate	5.6[n]	20	—	—
$C_5H_{11}N$	Piperidine	5.8[b]	22	—	—
C_5H_{12}	*n*-Pentane	1.844	20	0.160	−50,30
		2.011	−90		
		1.984	−70		
	2-Methylbutane	1.843	20	—	—
$C_5H_{12}O$	1-Pentanol	13.9	25	0.23(α)	15,35
	3-Methyl-1-butanol	14.7	25	—	—
		5.8^2	132		
	2-Methyl-2-butanol	5.82	25	—	—
	C_6				
C_6H_4Cl	*o*-Dichlorobenzene	9.93	25	0.194(α)	0.50
	m-Dichlorobenzene	5.04	25	0.120(α)	0,50
	p-Dichlorobenzene	2.41	50	0.18	50,80
C_6H_5Br	Bromobenzene	5.40	25	0.115(α)	0,70
C_6H_5Cl	Chlorobenzene	5.708	20	—	—
		5.621	25		
		5.71	20	0.130(α)	0,80
		7.28	−50		
		6.30	−20		
		4.21	130		
C_6H_5Cl	*o*-Chlorophenol	6.31	25	2.7	25,58
	p-Chlorophenol	9.47	55	3.7	55,65
C_6H_5I	Iodobenzene	4.63	20	—	—
$C_6H_5NO_2$	Nitrobenzene	34.82	25	0.225(α)	10,80
		20.8	130	0.164(α)	130,211
		24.9	90		
		22.7	110		
$C_6H_5NO_3$	*o*-Nitrophenol	17.$_3$	50	6.4	50,60
C_6H_6	Benzene	2.284	20	0.200	10,60
		2.073	129	—	—
		1.966	182		
C_6H_6BrN	*m*-Bromoaniline	13.$_0$[n]	19	—	—
C_6H_6ClN	*m*-Chloroaniline	13.^{4}[n]	19	—	—
$C_6H_6N_2O_2$	*o*-Nitroaniline	34.$_5$	90	3.	90,110
	p-Nitroaniline	56.$_3$	160	6.	160,180
C_6H_6O	Phenol	9.78	60	0.32(α)	40,70
C_6H_7N	Aniline	6.89	20	0.148(α)	0,50
		5.93	70	—	—
		4.54	184.6	—	—
	2-Methylpyridine (α-Picoline)	9.8[b]	20	—	—
$C_6H_8N_2$	Phenylhydrazine	7.2	23	—	—
C_6H_{10}	Cyclohexene	2.220	25	—	—
		2.6$_0$	−105	—	—
$C_6H_{10}O$	Cyclohexanone	18.3	20	—	—
		19.$_9$	−40	—	—
$C_6H_{10}O$	Ethyl acetoacetate	15.$_7$[a]	22	—	—
C_6H_{12}	Cyclohexane	2.023	20	0.160	10,60
	Methylcyclopentane	1.985	20	—	—
	Ethylcyclobutane	1.965	20	—	—
$C_6H_{12}O$	Cyclohexanol	15.0	25	0.437(α)	20,66
		7.2$_4$	100		
		4.8$_8$	150		
$C_6H_{12}O_2$	Butyl acetate	5.01	20	1.4	20,40
		6.8$_4$	−73		
	Ethyl butyrate	5.10	18	1.0	at 20
$C_6H_{12}O_3$	Paraldehyde	13.9	25	—	—
		6.29	128		
C_6H_{14}	*n*-Hexane	1.890	20	0.155	−10,50
		2.044	−90		
		1.990	−50		

Dielectric Constants of Pure Liquids

Organic Liquids

	Substance	ε	°C	a (or α) × 10^2	Range t_1, t_2
$C_6H_{14}O$	1-Hexanol	13.3	25	0.35(α)	15,35
		8.5_6	75		
	Propyl ether	3.3_9	26	—	—
	Isopropyl ether	3.88	25	1.8	0,25
$C_6H_{15}Al$	Triethyl aluminum	2.9	20	—	—
$C_6H_{15}N$	Dipropylamine	2.9[b]	21	—	—
	Triethylamine	2.42	25	—	—
$C_6H_{15}N$	Dipropylamine	2.9[b]	21	—	—
	Triethylamine	2.42	25	—	—
$C_6H_{18}OSi_2$	$(CH_3)_3Si[OSi(CH_3)_2]\,n\,CH_3$				
$n = 1$	Hexamethyldisiloxane	2.17	20	—	—
$n = 2$	Octamethyltrisiloxane	2.30	20	—	—
$n = 3$	Decamethyltetrasiloxane	2.39	20	—	—
$n = 4$	Dodecamethylpentasiloxane	2.46	20	—	—
$n = 5$	Tetradecamethylhexasiloxane	2.50	20	—	—
$n = 66$[w]		2.72	20	—	—
	C_7				
C_7H_5ClO	Benzoyl chloride	29.	0	—	—
		23.	20		
C_7H_5NO	Phenyl isocyanate	8.8[b]	20	—	—
C_7H_5NS	Phenyl isothiocyanate	$10._4$[a]	20	—	—
C_7H_6O	Benzaldehyde	$19._7$	0	—	—
		$17._8$	20		
$C_7H_6O_2$	Salicylaldehyde	$17._1$	30	7.	30,40
C_7H_7Br	*o*-Bromotoluene	4.23	58	—	—
	m-Bromotoluene	5.36	58	—	—
	p-Bromotoluene	5.49	58	—	—
C_7H_7Cl	*o*-Chlorotoluene	4.45	20	—	—
		4.16	58	—	—
	m-Chlorotoluene	5.55	20	—	—
		5.04	58	—	—
	p-Chlorotoluene	6.08	20	—	—
		5.55	58	—	—
	α-Chlorotoluene	7.0	13	—	—
$C_7H_7NO_2$	*o*-Nitrotoluene	27.4	20	15.	at 20
		$21._6$	58	—	—
		11.8	222	—	—
	m-Nitrotoluene	$23._8$	20	—	—
		$21._9$	58	—	—
	p-Nitrotoluene	$22._2$	58	—	—
C_7H_8	Toluene	2.438	0	0.0455(α)	−90,0
		2.379	25	0.243	0,90
		2.15_7	127		
		2.04_2	181		
C_7H_8O	Benzyl alcohol	13.1	20	—	—
		9.47	70	—	—
		6.6	132	—	—
	o-Cresol	11.5	25	11	25,30
	m-Cresol	11.8	25	0.41(α)	15,50
	p-Cresol	9.9_1	58	—	—
C_7H_9N	*o*-Toluidine	6.34	18	—	—
		5.71	58	—	—
		4.00	200	—	—
	m-Toluidine	5.95	18	—	—
		5.45	58	—	—
	p-Toluidine	4.98	54	—	—
	N-Methylaniline	5.97	22	—	—
	1-Heptene	2.05	20	—	—
	C_8				
C_8H_8	Styrene (Phenylethylene)	2.43	25	—	—
		2.32	75		
	Acetophenone	17.39	25	4.	at 25
		8.64	202	—	—
$C_8H_8O_3$	Phenyl acetate	5.23	20	0.7	at 20
	Methyl benzoate	6.59	20	0.14(α)	20,50
$C_8H_8O_3$	Methyl salicylate	9.41	30	3.1	30,40
C_8H_{10}	Ethylbenzene	2.412	20	—	—
	o-Xylene	2.568	20	0.266	−20,130
	m-Xylene	2.374	20	0.195	−40,180
	p-Xylene	2.270	20	0.160	20,130
C_8H_{18}	*n*-Octane	1.948	20	0.130	−50,50
		1.879	70		
		1.817	110		
	2,2,3-Trimethylpentane	1.96	20	—	—
	2,2,4-Trimethylpentane	1.940	20	0.142	−100,100
$C_8H_{18}O$	1-Octanol	10.3_4	20	0.410(α)	20,60
$C_8H_{20}O_4Si$	Tetraethyl silicate	4.1[b]	ca 20	—	—
	C_9				
C_9H_7N	Quinoline	9.00	25	—	—
		5.05	238		
	Isoquinoline	10.7	25	—	—
C_9H_8O	Cinnamaldehyde	16.9	24	—	—
$C_9H_{10}O_2$	Benzyl acetate	5.1[n]	21	—	—
	Ethyl benzoate	6.02	20	2.1	20,40
$C_9H_{10}O_3$	Ethyl salicylate	7.99	30	2.	30,40
C_9H_{12}	Isopropylbenzene (Cumene)	2.38_0	20	—	—
	1,3,5-Trimethylbenzene (Mesitylene)	2.27_9	20	—	—
C_9H_{20}	*n*-Nonane	1.972	20	0.135	−10,90
		2.059	−50		
		1.847	110		
		1.787	150		
	C_{10}				
$C_{10}H_8$	Naphthalene	2.54	85	—	—
$C_{10}H_{10}O_4$	Dimethyl phthalate	8.5	24	—	—
$C_{10}H_{16}$	*d*-Camphene	2.33	ca 40	—	—
	d-Pinene	2.64	25	—	—
	l-Pinene	2.76	20	—	—
	Terpinene	2.7[b]	21	—	—
	d-Limonene	2.3_6	20	—	—
	d-Limonene (Dipentene)	2.3[o]	20	—	—
$C_{10}H_{22}$	*n*-Decane	1.991	20	0.130	10,110
		2.050	−30		
		1.844	130		
		1.783	170		
$C_{10}H_{22}$	1-Decanol	8.1	20	—	—
$C_{11}H_{24}$	*n*-Undecane	2.005	20	0.125	10,130
		2.039	−10		
		1.838	150		
		1.781	190		
	C_{12}				
$C_{12}H_{10}$	Diphenyl	2.53	75	0.18	75,155
$C_{12}H_{10}O$	Phenyl ether	3.65	30	0.7	30,50
$C_{12}H_{11}N$	Diphenylamine	3.3	52	—	—
$C_{12}H_{26}$	*n*-Dodecane	2.015	20	0.120	10,150
		2.047	−10		
		1.776	210		
$C_{13}H_{10}$	Benzophenone	11.4	50	—	—
	C_{14}				
$C_{14}H_{15}N$	Dibenzylamine	3.6[b]	20	—	—
	C_{16}				
$C_{16}H_{32}O_2$	Palmitic acid	2.30	71	—	—
	C_{18}				
$C_{18}H_{32}O_2$	Linoleic acid	2.61	0	—	—
		2.71	20		
		2.70	70		
		2.60	120		
$C_{18}H_{34}O_2$	Oleic acid	2.46	20	—	—
		2.45	60		
		2.41	100		
$C_{18}H_{36}O_2$	Stearic acid	2.29	70	—	—
		2.26	100	—	—
	Ethyl palmitate	3.20	20	0.4	20,40
		2.71	104	—	—
		2.46	182		
	C_{19}				
$C_{19}H_{18}$	Triphenylmethane	2.45	100	0.14	94,175
$C_{19}H_{38}O_4$	Monopalmitin	5.34	67	—	—
		5.09	80		
	C_{20}				
$C_{20}H_{38}O_2$	Ethyl Oleate	3.17	28	0.48	28,122
$C_{20}H_{40}O_2$	Ethyl Stearate	2.98	40	0.6	32,50
		2.69	100		
		2.48	167		
	C_{21}				
$C_{21}H_{21}O_4P$	Tricresyl phosphate	6.9	40	—	—
	C_{22}				
$C_{22}H_{42}O_3$	Butyl oleate	4.0	25	—	—
$C_{22}H_{44}O_2$	Butyl stearate	3.11_1	30	0.53	30,35

DIELECTRIC CONSTANTS (Continued)

Dielectric Constants of Pure Liquids

[a] $f = 4 \times 10^8$ cycles/sec.

[b] $f = 3.6 \times 10^8$ cycles/sec.

[c] At pressure of 50 atmospheres.

[d] $\varepsilon = 78.25\,[1-4.617\,(10^{-3})\,t-25) + 1.22\,(10^{-5})\,(t-25)^2 - 1.7\,(10^{-8})\,(t-25)^3]$; average deviation ± 0.04%.

[e] $\varepsilon = 78.54\,[1 - 4.579\,(10^{-3})\,(t-25) + 1.19\,(10^{-5})\,(t-25)^2 - 2.8\,(10^{-8})\,(t-25)^3]$; average deviation ± 0.03%.

[f] Liquid transition and discontinuity in variation of dielectric constant with temperature at 2.295°K.

[g] Graphical data in the range 118—350°C show a miniumum near 160° and a broad maximum near 200°.

[h] Critical temperature.

[i] $\mathrm{Log}_{10}\,\varepsilon = 2.199 - 0.0079\,t + 0.00005\,t^2$.

[j] $\varepsilon = (3320/T) - 2.24$.

[k] $\varepsilon = (3320/T) - 2.34$.

[l] $\varepsilon = 12.6 - 0.061\,(t + 20) + 0.0005\,(t + 20)^2$.

[m] $\varepsilon = (2160/T) - 0.39$.

[n] $f = 5 \times 10^8$ cycles/sec.

[p] Critical temperature = 126.9°C.

[s] $\varepsilon = 84.2 - 0.62\,t + 0.0032\,t^2$.

[t] Mixture of *cis-trans* isomers.

[w] Silicone oil of average molecular weight corresponding to this formula.

[x] Value chosen to conform with the remainder of the tabulated data for this substance.

[y] $\varepsilon = 5321/T + 233.76 - 0.9297\,T + 0.001417\,T^2 - 0.000008292\,T^3$.

Dielectric Constants of Solids

Compiled by Earle C. Gregg, Jr.

Solids[a] (17 to 22°C)

Material	Frequency	Dielectric constant
Acetamide	4×10^8	4.0
Acetanilide	—	2.9
Acetic acid (2°C)	4×10^8	4.1
Aluminum oleate	4×10^8	2.40
Ammonium bromide	10^8	7.1
Ammonium chloride	10^8	7.0
Antimony trichloride	10^8	5.34
Apatite ⊥ optic axis	3×10^8	9.50
Apatite ∥ optic axis	3×10^8	7.41
Asphalt	$<3 \times 10^6$	2.68
Barium chloride (anhyd.)	6×10^7	11.4
Barium chloride ($2H_2O$)	6×10^7	9.4
Barium nitrate	6×10^7	5.9
Barium sulfate (15°C)	10^8	11.4
Beryl ⊥ optic axis	10^4	7.02
Beryl ∥ optic axis	10^4	6.08
Calcite ⊥ optic axis	10^4	8.5
Calcite ∥ optic axis	10^4	8.0
Calcium carbonate	10^6	6.14
Calcium fluoride	10^4	7.36
Calcium sulfate ($2H_2O$)	10^4	5.66
Cassiterite ⊥ optic axis	10^{12}	23.4
Cassiterite ∥ optic axis	10^{12}	24
d-Cocaine	5×10^8	3.10
Cupric oleate	4×10^8	2.80
Cupric oxide (15°C)	10^8	18.1
Cupric sulfate (anhyd.)	6×10^7	10.3
Cupric sulfate ($5H_2O$)	6×10^7	7.8
Diamond	10^8	5.5
Diphenylmethane	4×10^8	2.7
Dolomite ⊥ optic axis	10^8	8.0
Dolomite ∥	10^8	6.8
Ferrous oxide (15°C)	10^8	14.2
Iodine	10^8	4
Lead acetate	10^6	2.6
Lead carbonate (15°C)	10^8	18.6
Lead chloride	10^6	4.2
Lead monoxide (15°C)	10^8	25.9
Lead nitrate	6×10^7	37.7
Lead oleate	4×10^8	3.27
Lead sulfate	10^6	14.3
Lead sulfide (15°)	16^6	17.9
Malachite (mean)	10^{12}	7.2
Mercuric chloride	10^6	3.2
Mercurous chloride	10^6	9.4
Naphthalene	4×10^8	2.52

Material	Frequency	Dielectric constant
Phenanthrene	4×10^8	2.80
Phenol (10°C)	4×10^8	4.3
Phosphorus, red	10^8	4.1
Phosphorus, yellow	10^8	3.6
Potassium aluminum sulfate	10^6	3.8
Potassium carbonate (15°C)	10^8	5.6
Potassium chlorate	6×10^7	5.1
Potassium chloride	10^4	5.03
Potassium chromate	6×10^7	7.3
Potassium iodide	6×10^7	5.6
Potassium nitrate	6×10^7	5.0
Potassium sulfate	6×10^7	5.9
Quartz ⊥ optic axis	3×10^7	4.34
Quartz ∥ optic axis	3×10^7	4.27
Resorcinol	4×10^8	3.2
Ruby ⊥ optic axis	10^4	13.27
Ruby ∥ optic axis	10^4	11.28
Rutile ⊥ optic axis	10^8	86
Rutile ∥ optic axis	10^8	170
Selenium	10^8	6.6
Silver bromide	10^6	12.2
Silver chloride	10^6	11.2
Silver cyanide	10^6	5.6
Smithsonite ⊥ optic axis	10^{12}	9.3
Smithsonite ∥ optic axis	10^{10}	9.4
Sodium carbonate (anhyd.)	6×10^7	8.4
Sodium carbonate ($10H_2O$)	6×10^7	5.3
Sodium chloride	10^4	6.12
Sodium nitrate	—	5.2
Sodium oleate	4×10^8	2.75
Sodium perchlorate	6×10^7	5.4
Sucrose (mean)	3×10^8	3.32
Sulfur (mean)	—	4.0
Thallium chloride	10^6	46.9
p-Toluidine	4×10^8	3.0
Tourmaline ⊥ optic axis	10^4	7.10
Tourmaline ∥ optic axis	10^4	6.3
Urea	4×10^8	3.5
Zircon ⊥, ∥	10^8	12

[a] For plastics and other insulating materials, refer to table on Properties of Dielectrics.

DIELECTRIC CONSTANTS (Continued)

Table of Dielectric Constants of Reference Gases at 20°C and 1 Atmosphere

The listed values ($\varepsilon - 1$) refer to the gas at a temperature of 20°C and pressure of 1 atmosphere. The values can be adjusted to slightly different conditions without introducing more than 0.1% error by use of the following equation:

$$\frac{(\varepsilon - 1)_{t,p}}{(\varepsilon - 1)_{20°,1\ atm}} = \frac{p}{760[1 + 0.003411(t - 20)]}$$

where p = pressure in mm Hg
t = degrees C
t should be between 10 and 30°C and p between 700 and 800 mm Hg. From National Bureau of Standards Circular 537

Substance	$(\varepsilon - 1) \cdot 10^6$	Ref.
Helium	Radio frequency	
	67.8	Watson
	63.7	Hector
	64.5	Jelatis
	Microwave	
	65.6	Birnbaum
	65.2	Essen
	Optical	
	64.6	Koch
	64.5	Cuthbertson
Hydrogen	Radio frequency	
	254.0	Watson
	Microwave	
	253.4	Essen
	Optical	
	254.1	Cuthbertson
	253.6	Koch
	253.7	Kirn
	254.3	Tausz
Oxygen	Radio frequency	
	494.$_3$	Watson
	496.2	Jelatis
	Microwave*	
	494.9	Birnbaum
	495.0	Essen
	494.9	Essen
	Optical	
	494.5	Cuthbertson
	493.5	Lowery
	494.7	Tausz
	494.4	Ladenberg
Argon	Radio frequency	
	513.0	Watson
	516.4	Jelatis
	Microwave	
	517.7	Essen
	Optical	
	516.8	Cuthbertson
	517.8	Quarder
	517.0	Tausz
	516.7	Damkohler
Air (dry, CO_2 Free)	Radio frequency	
	537.$_0$	Watson
	Microwave*	
	536.6	Birnbaum
	536.6	Essen
	536.6	Essen
	Optical	
	536.9	Koch
	535.8	Meggers
	536.0	Traub
	536.7	Quarder
	536.4	Lowery
	536.5	Tausz
	536.1	Koster
	536.3	Perard
	535.8	Barrell
Nitrogen	Radio frequency	
	547.$_2$	Watson
	Microwave	
	547.3	Birnbaum
	548.0	Essen
	548.0	Essen
	Optical	
	548.9	Cuthbertson
	548.7	Koch
	547.2	Tausz
Carbon dioxide	Radio frequency	
	921.$_5$	Watson
	Microwave	
	922.4	Birnbaum
	920.6	Essen

* These values were derived from measurements of the refractive index after making allowance for the magnetic permeability of oxygen.

DIELECTRIC CONSTANTS OF GASES AT 760 MM PRESSURE

Compiled by Earl C. Gregg, Jr.

Material	Temperature °C	Frequency cycles/sec	Dielectric Constant
Acetaldehyde	100	$<3 \times 10^6$	1.0213
	0	2×10^6	
Acetone			1.0159
Acetyl chloride	0	$<3 \times 10^6$	1.0217
Acetylene	0	$<3 \times 10^6$	1.00134
Air			1.000590
Ammonia	0	$<3 \times 10^6$	1.0072
β-Amylene			1.0028
Argon	23	10^{10}	1.000545
Benzene	400	3×10^8	1.0028
Bromine	0	$<3 \times 10^6$	1.0128
Butylene			1.00319
Carbon dioxide	100	$<3 \times 10^6$	1.000985
Carbon disulfide	23	10^{10}	1.0029
Carbon monoxide	23	10^{10}	1.00070
Carbon tetrachloride	100	$<3 \times 10^6$	1.0030
Chloroform	100	$<3 \times 10^6$	1.0042
Dichlorodifluoromethane	100	$<3 \times 10^6$	1.00029
	0	$<3 \times 10^6$	
	0	$<3 \times 10^6$	
Dichlorofluoromethane	100	$<3 \times 10^6$	1.00049
Dimethylamine	0	2×10^6	1.00040
Ethane	0	$<3 \times 10^6$	1.00150
Ethylalcohol	100	$<3 \times 10^6$	1.0061
Ethylamine			1.00053
Ethyl bromide	20	$<3 \times 10^6$	1.0139
Ethyl chloride	20	$<3 \times 10^6$	1.0132
Ethyl ether	23	10^{10}	1.0049
Ethyl formate	126	$<3 \times 10^6$	1.0083
Ethyl iodide	20	$<3 \times 10^6$	1.0140
			1.0089
Ethylene	110	$<3 \times 10^6$	1.00144
Helium	140	$<3 \times 10^6$	1.0000684
n-Heptane	20	$<3 \times 10^6$	1.0035
Hydrogen	100	$<3 \times 10^6$	1.000264
Hydrogen bromide	20	$<3 \times 10^6$	1.00313
Hydrogen chloride	0	$<3 \times 10^6$	1.0046
	0	$<3 \times 10^6$	
Hydrogen iodide	0	$<10^6$	1.00234
	100	$<3 \times 10^6$	
Hydrogen sulfide	0	$<10^6$	1.00030
	100	$<3 \times 10^6$	
Mercury	180	$<3 \times 10^6$	1.00074
Methane	0	$<3 \times 10^6$	1.000944
Methyl alcohol	0	$<10^6$	1.0057
Methyl bromide	0	$<3 \times 10^6$	1.00095
Methyl chloride	0	$<3 \times 10^6$	1.00094
Methyl iodide	110	$<3 \times 10^6$	1.0063
Methylamine	120	$<3 \times 10^6$	1.0038
Methylene chloride	23	10^{10}	1.0065
Neon			1.000127
Nitrogen			1.000580
Nitromethane	23	10^{10}	1.0247
Nitrous oxide (N_2O)	23	2.5×10^{10}	1.00113
	0	$<3 \times 10^6$	
Oxygen	100	$<3 \times 10^6$	1.000523
n-Pentane	23	10^{10}	1.0025
n-Propyl chloride	20	$<3 \times 10^6$	1.0143
iso-Propyl chloride	20	$<3 \times 10^6$	1.0152
Sulfur dioxide	100	$<3 \times 10^6$	1.00075
Toluene	100	$<3 \times 10^6$	1.0043
Vinyl bromide	20	$<3 \times 10^6$	1.0081
	100	$<3 \times 10^6$	
Water (steam)	0	$<3 \times 10^6$	1.0126
	0	$<3 \times 10^6$	1.00785

PROPERTIES OF DIELECTRICS

In most cases properties have been determined by A.S.T.M. (American Society for Testing Materials) test methods at room temperature under standard conditions. Values will in general change considerably with temperature.

DIELECTRIC CONSTANTS OF SOME PLASTICS AND RUBBERS

Name	°C	Frequency (hertz) 1×10^3	1×10^6	1×10^8
Plastics				
Phenol-formaldehyde	25—27	5.15—8.61	4.45—5.05	4.1—4.5
	57	6.35	4.90	4.5
	88	8.5	5.2	4.7
Phenol-aniline-formaldehyde	25	4.50	4.31	4.11
	79	4.75	4.51	4.35
Melamine-formaldehyde	24—28	6.0—6.90	5.82—6.20	5.5—5.55
	57	6.95	5.40	4.90
	88	11.8	6.0	5.5
Urea-formaldehyde	24	6.7	6.0	5.2
	80	7.8	6.8	—
Polyamide resins				
Nylon 66	25	3.75	3.33	3.16
Nylon 610	25	3.50	3.14	3.0
	84	11.2	4.4	3.4
Cellulose acetate	26	3.50—4.48	3.28—3.90	3.05—3.40
Cellulose nitrate	27	8.4	6.6	5.2
	78	7.5	6.2	5.2
Methyl cellulose	22	6.8	5.7	4.3
Ethyl cellulose	25	3.09	3.01	2.90
Silicone resins	25	3.79—3.91	3.79—3.82	3.82
Polyethylene	−12	2.37	2.35	2.33
	23	2.26	2.26	2.26
Polyisobutylene	25	2.23	2.23	2.23
Vinylite QYNA	20	3.10	2.88	2.85
	76	3.83	3.0	2.8
	110	8.6		
Vinylite 5544	25	7.20	4.13	3.05
Vinylite 5901	25	5.5	3.4	3.0
Vinylite VU	24	5.65	3.30	2.80
	79	8.15	5.5	3.4
Vinylite VYHW	20	3.12	2.91	2.83
Vinylite VYNW	20	3.15	2.90	2.8
Polyvinyl chloride	25	4.55 (1×10^4)	3.3	—

Name	°C	Frequency (hertz) 1×10^3	1×10^6	1×10^8
Polyvinylidene and vinyl	23	4.65	3.18	2.82
chloride	84	4.94	4.40	3.2
Polychlorotrifluoroethylene	25	2.76	2.48	2.36
Polytetrafluoroethylene	22	2.1	2.1	2.1
(Teflon)	100	2.04	2.04	—
Polyvinylalcohol acetate	25	7.8	5.2	—
	85	100	10	—
Polyvinylacetals	26—27	3.02—3.12	2.86—2.92	2.67
	88	3.5	3.1	2.85
Polyacrylates				
Lucite	−12	2.9	2.63	2.50
	23	2.84	2.63	2.58
	81	3.45	2.72	2.59
Plexiglas	27	3.12	2.76	—
Polystyrene	25	2.54—2.56	2.54—2.56	2.55
	80	2.54	2.54	2.54
Styrene copolymers	25	2.55—2.95	2.55—2.80	2.55—2.77
Polyesters	25	3.22—4.3	3.12—4.0	2.94—2.98
Alkyd resins				
Alkyd isocyanate foam	25	1.223	1.218	1.20
Plaskon, clay filled	25	5.26	4.92	4.77
Plaskon, glass filled	25	5.04	4.73	4.50
Epoxy resins	25	3.63—3.67	3.52—3.62	3.32—3.35
Rubbers				
Hevea, vulcanized	27	2.94	2.74	2.42
Hevea compound	27	36	9	6.8
Gutta percha	25	2.60	2.53	2.47
Balata	25	2.50	2.50	2.42
Buna S	20	2.66	2.56	2.52
Butyl rubber compound	25	2.42	2.40	2.39
Neoprene	24	6.60	6.26	4.5
Silicon rubber	25	3.12—3.30	3.10—3.20	3.06—3.18

DIELECTRIC CONSTANTS OF CERAMICS

Material	Dielectric constant 10^6 cycles	Dielectric strength volts mil	Volume resistivity Ohms-cm (23°C)	Loss factor[a]
Alumina	4.5—8.4	40—160	10^{11}—10^{14}	0.0002—0.01
Corderite	4.5—5.4	40—250	10^{12}—10^{14}	0.004—0.012
Forsterite	6.2	240	10^{14}	0.0004
Porcelain (dry process)	6.0—8.0	40—240	10^{12}—10^{14}	0.0003—0.02
Porcelain (wet process)	6.0—7.0	90—400	10^{12}—10^{14}	0.006—0.01
Porcelain, zircon	7.1—10.5	250—400	10^{13}—10^{15}	0.0002—0.008
Steatite	5.5—7.5	200—400	10^{13}—10^{15}	0.0002—0.004
Titanates (Ba, Sr, Ca, Mg, and Pb)	15—12.000	50—300	10^{8}—10^{15}	0.0001—0.02
Titanium dioxide	14—110	100—210	10^{13}—10^{18}	0.0002—0.005

DIELECTRIC CONSTANTS OF WAXES

Material	Dielectric constant 10^6 Cycles	Dielectric strength volts mil	Volume resistivity Ohms-cm (23°C)	Loss factor[a]
Acrawax C	2.4	—	—	0.005
Beeswax, white	2.75—3.0	—	5×10^{14}	0.025
Beeswax, yellow	2.9	—	8×10^{14}	0.029
Candelilla	2.25—2.50			
Carnauba	2.75—3.0			
Cerese, brown G	2.27	—	—	0.0025
Ceresine	2.25—2.50	—	$>5 \times 10^{18}$	0.0011
Halowax 1001	~4.10	—	2×10^{13}	0.014
Halowax 1013	~4.75	—	—	0.036
Halowax 1014	~4.40	—	—	0.035
Halowax 11-314	2.94	—	—	0.00094
Microcrystalline waxes	2.2—2.5			
Opalwax	3.1	—	—	0.34

[a] Power factor × dielectric constant equals loss factor.

DIELECTRIC CONSTANTS OF WAXES (continued)

Material	Dielectric constant 10^6 Cycles	Dielectric strength volts mil	Volume resistivity Ohms-cm (23°C)	Loss factor[a]
Ozokerite wax	2.3	100—150	5×10^{14}	0.0018
Paraffin	2.0—2.5	250	10^{15}—10^{19}	0.003 (900 cps)
Parawax	2.25	—	10^{16}	0.00045
135 A.M.P. wax	2.25	—	—	0.00023

DIELECTRIC CONSTANTS OF GLASSES

Type	Dielectric constant at 100 mc (20°C)	Volume resistivity (350°C megohm-cm)	Loss factor[a]
Corning 0010	6.32	10	0.015
Corning 0080	6.75	0.13	0.058
Corning 0120	6.65	100	0.012
Pyrex 1710	6.00	2,500	0.025
Pyrex 3320	4.71	—	0.019
Pyrex 7040	4.65	80	0.013
Pyrex 7050	4.77	16	0.017
Pyrex 7052	5.07	25	0.019
Pyrex 7060	4.70	13	0.018
Pyrex 7070	4.00	1,300	0.0048
Vycor 7230	3.83	—	0.0061
Pyrex 7720	4.50	16	0.014
Pyrex 7740	5.00	4	0.040
Pyrex 7750	4.28	50	0.011
Pyrex 7760	4.50	50	0.0081
Vycor 7900	3.9	130	0.0023
Vycor 7910	3.8	1,600	0.00091
Vycor 7911	3.8	4,000	0.00072
Corning 8870	9.5	5,000	0.0085
G. E. Clear (silica glass)	3.81	4,000—30,000	0.00038
Quartz (fused)	3.75 4.1 (1 mc)	—	0.0002 (1 mc)

[a] Power factor × dielectric constant equals loss factor.

STATIC DIELECTRIC CONSTANT OF WATER SUBSTANCE[a]

The temperatures are in degrees kelvin and the pressures in megapascals. Some conversion factors for pressure which may be useful are:

Megapascals × 9.8692 = atmospheres (760 mm Hg)
Megapascals × 14.504 = pounds per square inch
Megapascals × 10^{-6} = newtons per square meter (pascals)
Megapascals × 10.1972 = kilograms per square centimeter
Megapascals × 10 = bars
Megapascals × 7.501 × 10^3 = mm Hg at 0°C
Megapascals × 4.014 × 10^3 = inches of H_2O at 4°C

P/T	273.15	298.15	323.15	348.15	373.15	398.15	423.15	448.15	473.15	498.15	523.15	548.15	573.15	623.15	673.15	723.15	773.15	823.15
10	88.28	78.85	70.27	62.59	55.76	49.70	44.30	39.47	35.11	31.13	27.43	23.90	20.39	1.23	1.17	1.14	1.11	1.10
20	88.75	79.24	70.63	62.94	56.11	50.05	44.66	39.85	35.52	31.58	27.95	24.54	21.24	14.07	1.64	1.42	1.32	1.26
30	89.20	79.63	70.98	63.28	56.44	50.39	45.01	40.22	35.91	32.01	28.43	25.11	21.95	15.66	5.91	2.07	1.68	1.51
40	89.64	80.00	71.32	63.61	56.77	50.72	45.34	40.56	36.28	32.40	28.87	25.61	22.56	16.72	10.46	3.84	2.34	1.90
50	90.07	80.36	71.66	63.93	57.08	51.03	45.67	40.89	36.63	32.78	29.28	26.08	23.10	17.55	12.16	6.57	3.45	2.48
60	90.49	80.72	71.98	64.24	57.39	51.34	45.98	41.21	36.96	33.13	29.67	26.50	23.58	18.24	13.28	8.53	4.90	3.26
70	90.90	81.07	72.30	64.54	57.69	51.64	46.28	41.52	37.28	33.47	30.03	26.90	24.02	18.84	14.16	9.87	6.31	4.20
80	91.29	81.42	72.62	64.84	57.98	51.93	46.57	41.82	37.59	33.79	30.37	27.27	24.43	19.37	14.88	10.88	7.50	5.16
90	91.67	81.75	72.92	65.13	58.27	52.21	46.86	42.11	37.89	34.10	30.70	27.62	24.81	19.85	15.50	11.70	8.47	6.06
100	92.04	82.08	73.22	65.42	58.55	52.49	47.14	42.39	38.17	34.40	31.01	27.95	25.17	20.29	16.05	12.39	9.29	6.88
125	92.89	82.84	73.93	66.09	59.19	53.12	47.78	43.05	38.86	35.13	31.78	28.76	26.03	21.26	17.21	13.77	10.88	8.53
150	93.71	83.57	74.62	66.74	59.82	53.75	48.40	43.68	39.50	35.78	32.46	29.47	26.77	22.09	18.16	14.85	12.07	9.80
175	94.48	84.28	75.27	67.36	60.42	54.34	48.98	44.27	40.10	36.39	33.09	30.12	27.45	22.83	18.98	15.74	13.04	10.81
200	95.20	84.94	75.89	67.95	61.00	54.90	49.54	44.83	40.66	36.97	33.67	30.72	28.07	23.49	19.69	16.51	13.86	11.65
225	95.87	85.58	76.50	68.53	61.55	55.44	50.08	45.36	41.20	37.51	34.22	31.28	28.64	24.09	20.33	17.19	14.56	12.38
250	96.51	86.20	77.08	69.08	62.08	55.96	50.59	45.87	41.70	38.02	34.74	31.81	29.17	24.65	20.91	17.80	15.19	13.01
300	97.69	87.34	78.17	70.14	63.10	56.94	51.55	46.82	42.65	38.97	35.69	32.77	30.15	25.65	21.94	18.85	16.25	14.07
350	98.75	88.40	79.19	71.12	64.05	57.86	52.45	47.70	43.52	39.83	36.56	33.64	31.02	26.53	22.83	19.74	17.14	14.93
400	99.72	89.39	80.13	72.03	64.94	58.74	53.30	48.53	44.33	40.64	37.36	34.43	31.81	27.32	23.62	20.52	17.89	15.66
450	100.60	90.30	81.02	72.89	65.78	59.56	54.10	49.31	45.10	41.38	38.09	35.16	32.54	28.04	24.32	21.20	18.55	16.28
500	101.42	91.16	81.84	73.69	66.57	60.33	54.85	50.05	45.82	42.09	38.78	35.84	33.21	28.70	24.96	21.82	19.14	16.83

[a] Prepared by International Association for the Properties of Steam.

DIELECTRIC CONSTANT OF DEUTERIUM OXIDE

t	ε	$-\frac{d\varepsilon}{dt}$	$-\frac{1}{\varepsilon}\frac{d\varepsilon}{dt}$
°C			
4	85.877	0.3974	4.627×10^{-3}
5	85.480	.3956	4.628
10	83.526	.3862	4.624
15	81.618	.3771	4.620
20	79.755	.3681	4.615
25	77.936	.3594	4.611
30	76.161	.3509	4.607
35	74.427	.3425	4.602
40	72.735	.3344	4.597
45	71.083	.3265	4.593
50	69.470	.3187	4.587
55	67.896	.3112	4.583
60	66.358	.3038	4.578
65	64.857	.2967	4.575
70	63.391	.2898	4.571
75	61.959	.2830	4.567
80	60.561	.2765	4.565
85	59.194	.2701	4.563
90	57.859	.2640	4.563
95	56.554	.2581	4.564
100	55.278	.2523	4.564

DIELECTRIC CONSTANTS (Continued)

Dielectric Constant of Liquid Parahydrogen vs. Temperature (°K) and Pressure (atm)

R. J. Corruccini

P atm \ T, °K	20	21	22	23	24	25	26	27	28	29	30	31	32
1	1.2297												
2	1.2302	1.2260	1.2216										
3	1.2306	1.2265	1.2221	1.2174	1.2122								
4	1.2311	1.2270	1.2227	1.2180	1.2129	1.2073	1.2010						
5	1.2315	1.2275	1.2233	1.2186	1.2136	1.2081	1.2020	1.1950					
6	1.2320	1.2280	1.2238	1.2192	1.2143	1.2089	1.2029	1.1962	1.1883				
7	1.2324	1.2285	1.2243	1.2198	1.2150	1.2097	1.2039	1.1973	1.1897	1.1805			
8	1.2329	1.2290	1.2249	1.2204	1.2157	1.2105	1.2048	1.1984	1.1911	1.1824			
9	1.2333	1.2295	1.2254	1.2210	1.2163	1.2112	1.2056	1.1994	1.1924	1.1842	1.1734		
10	1.2337	1.2300	1.2259	1.2216	1.2169	1.2119	1.2065	1.2004	1.1936	1.1857	1.1758	1.1621	
15	1.2358	1.2322	1.2284	1.2243	1.2200	1.2153	1.2103	1.2049	1.1990	1.1924	1.1847	1.1758	1.1645
20	1.2378	1.2343	1.2307	1.2268	1.2227	1.2184	1.2137	1.2088	1.2034	1.1976	1.1913	1.1839	1.1757
25	1.2396	1.2363	1.2328	1.2291	1.2253	1.2211	1.2168	1.2122	1.2073	1.2021	1.1964	1.1903	1.1832
30	1.2414	1.2382	1.2349	1.2313	1.2276	1.2237	1.2196	1.2153	1.2107	1.2059	1.2008	1.1952	1.1891
35	1.2431	1.2400	1.2368	1.2334	1.2298	1.2261	1.2222	1.2181	1.2138	1.2093	1.2046	1.1995	1.1942
40	1.2448	1.2418	1.2386	1.2354	1.2319	1.2284	1.2246	1.2208	1.2167	1.2124	1.2080	1.2033	1.1984
45	1.2464	1.2434	1.2404	1.2372	1.2339	1.2305	1.2269	1.2232	1.2193	1.2153	1.2111	1.2067	1.2021
50	1.2479	1.2450	1.2421	1.2390	1.2358	1.2325	1.2291	1.2255	1.2218	1.2179	1.2139	1.2098	1.2055
60	1.2508	1.2481	1.2453	1.2424	1.2394	1.2363	1.2331	1.2297	1.2263	1.2227	1.2191	1.2153	1.2114
70	1.2535	1.2510	1.2483	1.2455	1.2427	1.2397	1.2367	1.2336	1.2303	1.2270	1.2236	1.2201	1.2165
80	1.2561	1.2536	1.2511	1.2484	1.2457	1.2429	1.2400	1.2371	1.2340	1.2309	1.2277	1.2244	1.2211
90	1.2585	1.2561	1.2537	1.2512	1.2486	1.2459	1.2431	1.2403	1.2374	1.2345	1.2315	1.2284	1.2252
100	1.2608	1.2586	1.2562	1.2538	1.2513	1.2487	1.2461	1.2434	1.2406	1.2378	1.2349	1.2320	1.2290
120	1.2652	1.2631	1.2609	1.2586	1.2563	1.2539	1.2514	1.2489	1.2464	1.2438	1.2412	1.2385	1.2357
140	1.2693	1.2672	1.2651	1.2630	1.2608	1.2586	1.2563	1.2540	1.2516	1.2492	1.2467	1.2442	1.2417
160	1.2730	1.2711	1.2691	1.2671	1.2650	1.2629	1.2607	1.2585	1.2563	1.2540	1.2517	1.2494	1.2470
180	1.2766	1.2747	1.2728	1.2709	1.2689	1.2669	1.2649	1.2628	1.2606	1.2585	1.2563	1.2541	1.2518
200	1.2799	1.2781	1.2763	1.2745	1.2726	1.2707	1.2687	1.2667	1.2647	1.2626	1.2605	1.2584	1.2563
220	1.2831	1.2814	1.2796	1.2779	1.2760	1.2742	1.2723	1.2704	1.2685	1.2665	1.2645	1.2625	1.2605
240		1.2845	1.2828	1.2811	1.2793	1.2775	1.2757	1.2739	1.2720	1.2701	1.2682	1.2663	1.2643
260		1.2874	1.2858	1.2841	1.2824	1.2807	1.2790	1.2772	1.2754	1.2736	1.2717	1.2699	1.2680
280			1.2886	1.2870	1.2853	1.2837	1.2821	1.2803	1.2786	1.2768	1.2751	1.2733	1.2714
300			1.2914	1.2898	1.2882	1.2866	1.2850	1.2833	1.2817	1.2800	1.2782	1.2765	1.2747
320				1.2925	1.2910	1.2894	1.2878	1.2862	1.2846	1.2829	1.2813	1.2796	1.2779
340				1.2951	1.2936	1.2921	1.2905	1.2890	1.2874	1.2858	1.2842	1.2825	1.2809

Note: Values below the stepped line represent an extrapolation of p with density.

Selected Values of Electric Dipole Moments for Molecules in the Gas Phase

Ralph D. Nelson, Jr., David R. Lide, Jr., and Arthur A. Maryott

The following table was abstracted from the publication, "Selected Values of Electric Dipole Moments for Molecules in the Gas Phase" compiled by Nelson, Lide and Maryott and published as part of the National Reference Data Series—National Bureau of Standards (NSRDS—NBS 10). The publication is available from the Superintendent of Documents, U.S. Government Printing Office, Washington, D.C., 20402. Those desiring a complete listing of all compounds in the NSRDS—NBS 10, discussion of the bibliographic procedure and the principal methods of dipole moment measurement should obtain the publication.

Values of the dipole moment, μ, are expressed in the cgs system of units, since this is the system universally used by workers in the field. The numerical values are in debye units, D, ($1\ D = 10^{-18}$ electrostatic units of charge$\times$centimeters). The conversion factor to the Système International is $1\ D = 3.33564 \times 10^{-30}$ coulomb-meter.

Code symbol	Estimated accuracy of value
A	$\pm 1\%$ or, for $\mu < 1.0$ D, ± 0.01 D
B	$\pm 2\%$ or, for $\mu < 1.0$ D, ± 0.02 D
C	$\pm 5\%$ or, for $\mu < 1.0$ D, ± 0.05 D
S	$\mu \equiv 0$ on grounds of molecular symmetry
i	The significance of these values may involve some ambiguity because of the possibility of different conformations or spatial isomers.

Compounds not containing carbon

Formula	Compound name	Selected moment (debyes)	
AgCl	Silver chloride	5.73	C
$AsCl_3$	Arsenic trichloride	1.59	C
AsF_3	Arsenic trifluoride	2.59	B
AsH_3	Arsine	0.20	C
BCl_3	Boron trichloride	0	S
BF_3	Boron trifluoride	0	S
B_2H_6	Diborane	0	S
$B_3H_6N_3$	Triborotriazine (Borazine)	0	S
B_5H_9	Pentaborane	2.13	B
BaO	Barium oxide	7.95	A
BrH	Hydrogen bromide	0.82	B
BrH_3Si	Bromosilane	1.33	B
BrK	Potassium bromide	10.41	B
BrLi	Lithium bromide	7.27	A
Br_2Hg	Mercury dibromide	0	S
Br_4Sn	Tin tetrabromide	0	S
ClCs	Cesium chloride	10.42	A
ClF	Chlorine fluoride	0.88	C
$ClFO_3$	Perchloryl fluoride	0.023	A
$ClGeH_3$	Chlorogermane	2.13	A
ClH	Hydrogen chloride	1.08	B
ClH_3Si	Chlorosilane	1.31	A
ClK	Potassium chloride	10.27	A
ClLi	Lithium chloride	7.13	A
ClNa	Sodium chloride	9.00	A
$ClNO_2$	Nitryl chloride	0.53	A
ClTl	Thallium chloride	4.44	B
Cl_2F_3P	Dichlorotrifluorophosphorus	0.68	C
Cl_2H_2Si	Dichlorosilane	1.17	B
Cl_2Hg	Mercury dichloride	0	S
Cl_2OS	Thionyl chloride	1.45	B
Cl_2O_2S	Sulfuryl chloride	1.81	B
Cl_3F_2P	Trichlorodifluorophosphorus	0	S
Cl_3HSi	Trichlorosilane	0.86	B
Cl_3P	Phosphorus trichloride	0.78	C
Cl_4FP	Tetrachlorofluorophosphorus	0.21	B
Cl_4Ge	Germanium tetrachloride	0	S
Cl_4Si	Silicon tetrachloride	0	S
Cl_4Sn	Tin tetrachloride	0	S
Cl_4Ti	Titanium tetrachloride	0	S
CsF	Cesium fluoride	7.88	A
FH	Hydrogen fluoride	1.82	A
FH_3Si	Fluorosilane	1.27	B
FH_5Si_2	Fluorodisilane	1.26	A
FK	Potassium fluoride	8.60	A
FLi	Lithium fluoride	6.33	A
FNO	Nitrosyl fluoride	1.81	B
FNa	Sodium fluoride	8.16	A
FRb	Rubidium fluoride	8.55	A
FTl	Thallium fluoride	4.23	A
F_2HN	Difluoramine	1.92	A
F_2H_2Si	Difluorosilane	1.55	A
F_2N_2	*cis*-Difluorodiazine	0.16	A
F_2O	Oxygen difluoride	0.297	A
F_2OS	Thionyl fluoride	1.63	A
F_2O_2	Dioxygen difluoride	1.44	C
F_2O_2S	Sulfuryl fluoride	1.12	B
F_2S_2	Sulfur monofluoride ($S = SF_2$ isomer)	1.03	C
F_2S_2	Sulfur monofluoride (FSSF isomer)	1.45	B
F_2Si	Silicon difluoride	1.23	B
F_3HSi	Trifluorosilane	1.27	B
F_3N	Nitrogen trifluoride	0.235	A
F_3NS	Nitridotrifluorosulfur	1.91	B
F_3OP	Phosphoryl fluoride	1.76	B
F_3P	Phosphorus trifluoride	1.03	A
F_3PS	Thiophosphoryl fluoride	0.64	B
F_4N_2	Tetrafluorohydrazine, *gauche* conformation	0.26	B
F_4S	Sulfur tetrafluoride	0.632	A
F_4Si	Silicon tetrafluoride	0	S
F_5P	Phosphorus pentafluoride	0	S
F_5I	Iodine pentafluoride	2.18	C
F_6S	Sulfur hexafluoride	0	S
F_6Se	Selenium hexafluoride	0	S
F_6Te	Tellurium hexafluoride	0	S
F_6U	Uranium hexafluoride	0	S
HI	Hydrogen iodide	0.44	B
HLi	Lithium hydride	5.88	A
HN	Imidyl radical		
HNO_3	Nitric acid	2.17	A
HO	Hydroxyl radical	1.66	A
H_2O	Water	1.85	A
H_2O_2	Hydrogen peroxide	2.2	D
H_2S	Hydrogen sulfide	0.97	A
H_3N	Ammonia	1.47	A
H_3P	Phosphine	0.58	A
H_3Sb	Stibine	0.12	C
H_4N_2	Hydrazine	1.75	C
H_4Si	Silane	0	S
H_6OSi_2	Disilyl ether (disiloxane)	0.24	B
H_6Si_2	Disilane	0	S
HgI_2	Mercury diiodide	0	S
ILi	Lithium iodide	7.43	A
I_4Sn	Tin tetraiodide	0	S
NO	Nitrogen monoxide (nitric oxide)	0.153	A
NO_2	Nitrogen dioxide	0.316	A
N_2O	Dinitrogen oxide (nitrous oxide)	0.167	A
OS	Sulfur monoxide	1.55	A
OS_2	Disulfur monoxide	1.47	B
OSr	Strontium oxide	8.90	A
O_2S	Sulfur dioxide	1.63	A
O_3	Ozone	0.53	B
O_3S	Sulfur trioxide	0	S
O_4Os	Osmium tetroxide	0	S

Compounds containing carbon

Formula	Compound name	Selected moment (debyes)	
$CBrF_3$	Bromotrifluoromethane	0.65	C
CBr_2F_2	Dibromodifluoromethane	0.66	C
$CClF_3$	Chlorotrifluoromethane	0.50	A
CClN	Cyanogen chloride	2.82	B
CCl_2F_2	Dichlorodifluoromethane	0.51	C
CCl_2O	Carbonyl chloride (phosgene)	1.17	A
CCl_2S	Thiocarbonyl chloride	0.29	C
CCl_3F	Trichlorofluoromethane	0.45	C
CCl_3NO_2	Trichloronitromethane	1.89	C
CCl_4	Carbon tetrachloride	0	S
CFN	Cyanogen fluoride	2.17	C
CF_2	Carbon difluoride	0.46	B
CF_2O	Carbonyl fluoride	0.95	A
CF_3I	Iodotrifluoromethane	0.92	C
CF_3NO_2	Trifluoronitromethane	1.44	C
CF_4	Carbon tetrafluoride	0	S
CN_4O_8	Tetranitromethane	0	S
CO	Carbon monoxide	0.112	A
COS	Carbonyl sulfide	0.712	A
COSe	Carbonyl selenide	0.73	B
CO_2	Carbon dioxide	0	S
CS	Carbon monosulfide	1.98	A
CSTe	Thiocarbonyl telluride	0.17	A
CS_2	Carbon disulfide	0	S
$CHBr_3$	Tribromomethane	0.99	B
$CHClF_2$	Chlorodifluoromethane	1.42	B

Selected Values of Electric Dipole Moments for Molecules in the Gas Phase (Continued)

Compounds containing carbon—Continued

Formula	Compound name	Selected moment (debyes)	
$CHCl_2F$	Dichlorofluoromethane	1.29	B
$CHCl_3$	Trichloromethane (chloroform)	1.01	B
$CHFO$	Formyl fluoride	2.02	A
CHF_3	Trifluoromethane	1.65	A
CHN	Hydrogen cyanide	2.98	A
CHP	Methylidyne phosphide (methinophosphide)	0.390	A
CH_2Br_2	Dibromomethane	1.43	B
CH_2ClF	Chlorofluoromethane	1.82	B
CH_2ClNO_2	Chloronitromethane	2.91	B
CH_2Cl_2	Dichloromethane	1.60	B
CH_2F_2	Difluoromethane	1.97	A
CH_2N_2	Cyanogen amide (cyanamide)	4.27	C
CH_2N_2	Diazomethane	1.50	A
CH_2N_2	Diazirine	1.59	C
CH_2O	Methanal (formaldehyde)	2.33	A
CH_2O_2	Methanoic acid (formic acid)	1.41	A
CH_3BF_2	Methyl difluoroborane	1.66	B
CH_3BO	Carbonyl borane	1.80	B
CH_3Br	Bromomethane	1.81	A
CH_3Cl	Chloromethane	1.87	A
CH_3F	Fluoromethane	1.85	A
CH_3I	Iodomethane	1.62	B
CH_3NO	Hydroxyliminomethane (formaldoxime)	0.44	A
CH_3NO	Formyl amide (formamide)	3.73	B
CH_3NOS	Methyl sulfinylamine	1.70	B
CH_3NO_2	Nitromethane	3.46	A
CH_3NO_3	Methyl nitrate	3.12	B
CH_3N_3	Methyl azide	2.17	B
CH_4	Methane	0	S
CH_4F_2Si	Methyl difluorosilane	2.11	A
CH_4O	Methanol	1.70	A
CH_4S	Methanethiol (methyl mercaptan)	1.52	C
CH_5FSi	Methyl monofluorosilane	1.71	A
CH_5N	Methyl amine	1.31	B
CH_5P	Methyl phosphine	1.10	A
CH_6Ge	Methyl germane	0.643	A
CH_6OSi	Methoxysilane	1.17	B
CH_6Si	Methyl silane	0.735	A
CH_6Sn	Methyl stannane	0.68	C
C_2ClF_5	Chloropentafluorethane	0.52	C
C_2F_6	Hexafluorethane	0	S
C_2N_2	Dicyanogen (cyanogen)	0	S
C_2N_2S	Dicyano sulfide	3.02	A
C_2HCl	Chloroacetylene	0.44	A
C_2HCl_5	Pentachloroethane	0.92	C
C_2HF	Fluoroacetylene	0.73	C
C_2HF_3	Trifluoroethylene	1.40	C
C_2HF_5	Pentafluoroethane	1.54	C
C_2H_2	Acetylene	0	S
$C_2H_2Cl_2$	1, 1-Dichloroethylene	1.34	A
$C_2H_2Cl_2$	*cis*-1, 2-Dichloroethylene	1.90	B
$C_2H_2Cl_2O$	Chloroacetyl chloride	2.23	Ci
$C_2H_2Cl_4$	1,1,2,2-Tetrachloroethane	1.32	Ci
C_2H_2FN	Fluorocyanomethane	3.43	C
$C_2H_2F_2$	1,1-Difluoroethylene	1.38	A
$C_2H_2F_2$	*cis*-1,2-Difluoroethylene	2.42	A
$C_2H_2N_2O$	1,2,5-Oxadiazole	3.38	A
$C_2H_2N_2O$	1,3,4-Oxadiazole	3.04	B
$C_2H_2N_2S$	1,2,5-Thiadiazole	1.56	A
$C_2H_2N_2S$	1.3.4-Thiadiazole	3.29	B
C_2H_2O	Methylene carbonyl (ketene)	1.42	B
C_2H_3Br	Bromoethylene	1.42	B
C_2H_3Cl	Chloroethylene	1.45	B
$C_2H_3ClF_2$	1-Chloro-1,1-difluoroethane	2.14	B
C_2H_3ClO	Acetyl chloride	2.72	C
$C_2H_3Cl_3$	1,1,1-Trichloroethane	1.78	B
C_2H_3F	Fluoroethylene	1.43	A
C_2H_3FO	Acetyl fluoride	2.96	A
$C_2H_3F_3$	1,1,1-Trifluoroethane	2.32	B
$C_2H_3F_3$	1,1,2-Trifluoroethane	1.58	B
C_2H_3N	Cyanomethane (acetonitrile)	3.92	A
C_2H_3N	Isocyanomethane	3.85	B
C_2H_4	Ethylene	0	S
C_2H_4ClF	1-Chloro-2-fluoroethane, *gauche* conformation	2.72	C
$C_2H_4ClNO_2$	1-Chloro-1-nitroethane	3.27	B
$C_2H_4Cl_2$	1,1-Dichloroethane	2.06	B
$C_2H_4F_2$	1,1-Difluoroethane	2.27	B
C_2H_4Ge	Germyl acetylene	0.136	A
C_2H_4O	Oxirane (ethylene oxide)	1.89	A
C_2H_4O	Ethanal (acetaldehyde)	2.69	B
$C_2H_4O_2$	Ethanoic acid (acetic acid)	1.74	C
$C_2H_4O_2$	Methyl methanoate (methly formate)	1.77	B
C_2H_4S	Thiirane (ethylene sulfide)	1.85	A
C_2H_4Si	Silyl acetylene	0.316	A
C_2H_5Br	Bromoethane	2.03	A
C_2H_5BrO	Bromomethoxymethane	2.05	Ci
C_2H_5Cl	Chloroethane	2.05	A
C_2H_5ClO	2-Chloroethanol	1.78	Ci
C_2H_5F	Fluoroethane	1.94	B
C_2H_5I	Iodoethane	1.91	B
C_2H_5N	Iminoethane (ethyleneimine)	1.90	A
C_2H_5N	Methyliminomethane ($CH_3N = CH_2$)	1.53	B

Compounds containing carbon—Continued

Formula	Compound name	Selected moment (debyes)	
C_2H_5NO	Acetyl amine (acetamide)	3.76	Bi
C_2H_5NO	Methylaminomethanal (*N*-methylformamide)	3.83	Bi
$C_2H_5NO_2$	Nitritoethane (ethyl nitrite)	2.40	Ci
$C_2H_5NO_2$	Nitroethane	3.65	B
C_2H_6	Ethane	0	S
C_2H_6BF	Dimethyl fluoroborane	1.32	C
C_2H_6O	Ethanol	1.69	Bi
C_2H_6O	Dimethyl ether	1.30	A
C_2H_6OS	Dimethylsulfoxide	3.96	A
$C_2H_6O_2$	1,2-Ethanediol (ethylene glycol)	2.28	Ci
$C_2H_6O_2S$	Dimethyl sulfoxylate (dimethyl sulfone)	4.49	B
C_2H_6S	Ethanethiol	1.58	Bi
C_2H_6S	Dimethyl sulfide	1.50	A
C_2H_6Si	Silyl ethylene	0.66	A
$C_2H_7B_5$	2,4-Dicarbaheptaborane	1.32	B
C_2H_7N	Aminoethane (ethyl amine)	1.22	Ci
C_2H_7N	Dimethyl amine	1.03	B
C_2H_7P	Ethyl phosphine	1.17	Bi
C_2H_7P	Dimethyl phosphine	1.23	A
$C_2H_8N_2$	1,2-Diaminoethane	1.99	Ci
C_2H_8Si	Dimethyl silane	0.75	A
C_2H_8Si	Ethyl silane	0.81	B
C_3O_2	Dicarbonyl carbon (carbon suboxide)	0	S
C_3HF_3	3,3,3-Trifluoropropyne	2.36	B
C_3HN	Cyanoacetylene	3.72	A
$C_3H_2N_2$	Dicyanomethane	3.73	A
C_3H_2O	Propynal	2.47	B
$C_3H_2O_3$	Vinylene carbonate	4.55	A
C_3H_3Br	3-Bromopropyne	1.54	C
C_3H_3Cl	3-Chloropropyne	1.68	C
$C_3H_3F_3$	3,3,3-Trifluoropropene	2.45	B
C_3H_3N	Cyanoethylene	3.87	B
C_3H_3NO	Acetyl cyanide	3.45	B
C_3H_3NS	Thiazole	1.62	B
C_3H_4	Cyclopropene	0.45	A
C_3H_4	Propyne	0.781	A
C_3H_4	Propadiene (allene)	0	S
$C_3H_4Cl_2$	1,1-Dichlorocyclopropane	1.58	B
C_3H_4O	Ethylidene carbonyl (methyl ketene)	1.79	B
C_3H_4O	Propenal, *trans* conformation (acrolein)	3.12	B
$C_3H_4O_2$	2-Oxoöxetane (β-propiolactone)	4.18	A
$C_3H_4O_2$	Vinyl formate	1.49	A
C_3H_5Cl	2-Chloropropene	1.66	B
C_3H_5Cl	*cis*-1-Chloropropene	1.67	C
C_3H_5Cl	*trans*-1-Chloropropene	1.97	C
C_3H_5Cl	3-Chloropropene	1.94	Ci
C_3H_5F	*cis*-1-Fluoropropene	1.46	B
C_3H_5F	2-Fluoropropene	1.61	B
C_3H_5F	3-Fluoropropene, *cis* conformation	1.76	A
C_3H_5F	3-Fluoropropene, *gauche* conformation.	1.94	A
C_3H_5N	Cyanoethane (propionitrile)	4.02	A
C_3H_6	Cyclopropane	0	S
C_3H_6	Propene	0.366	A
$C_3H_6ClNO_2$	1-Chloro-1-nitropropane	3.48	Bi
$C_3H_6Cl_2$	1,2-Dichloropropane		i
$C_3H_6Cl_2$	1,3-Dichloropropane	2.08	Bi
$C_3H_6Cl_2$	2,2-Dichloropropane	2.27	C
C_3H_6O	Oxetane (trimethylene oxide)	1.94	A
C_3H_6O	Methyl oxirane (propylene oxide)	2.01	A
C_3H_6O	Propanone (acetone)	2.88	A
C_3H_6O	2-Propen-1-ol (allyl alcohol)	1.60	C
C_3H_6O	Propanal, *cis* conformation (propionaldehyde)	2.52	B
$C_3H_6O_2$	Propanoic acid	1.75	Ci
$C_3H_6O_2$	Methyl acetate	1.72	Ci
$C_3H_6O_2$	Ethyl formate	1.93	Ci
$C_3H_6O_3$	1,3,5-Trioxane	2.08	A
C_3H_6S	Thietane (trimethylene sulfide)	1.85	C
C_3H_6S	Methyl thiirane (propylene sulfide)	1.95	A
C_3H_7Br	1-Bromopropane	2.18	Ci
C_3H_7Br	2-Bromopropane	2.21	C
C_3H_7Cl	1-Chloropropane	2.05	Bi
C_3H_7Cl	2-Chloropropane	2.17	C
C_3H_7F	1-Fluoropropane, *gauche* conformation	1.90	C
C_3H_7F	1-Fluoropropane, *trans* conformation	2.05	B
C_3H_7I	1-Iodopropane	2.04	Ci
C_3H_7NO	N,N-Dimethylformamide	3.82	Bi
C_3H_7NO	Acetyl methylamine (*N*-Methylacetamide)	3.73	Bi
$C_3H_7NO_2$	1-Nitropropane	3.66	Bi
$C_3H_7NO_2$	2-Nitropropane	3.73	B
C_3H_8	Propane	0.084	A
C_3H_8O	1-Propanol	1.68	Bi
C_3H_8O	2-Propanol	1.66	Bi
C_3H_8O	Methoxyethane (methyl ethyl ether)	1.23	Ci
C_3H_9As	Trimethyl arsine	0.86	B
C_3H_9N	Trimethyl amine	0.612	A
C_3H_9N	1-Aminopropane (*n*-propylamine)	1.17	Ci
C_3H_9P	Trimethyl phosphine	1.19	A

Selected Values of Electric Dipole Moments for Molecules in the Gas Phase (Continued)

Compounds containing carbon—Continued

Formula	Compound name	Selected moment (debyes)	
$C_3H_{10}Si$	Trimethyl silane	0.525	A
C_4F_8	Perfluorocyclobutane	0	S
$C_4H_2N_2$	*trans*-1,2-Dicyanoethylene	0	S
$C_4H_4Cl_2$	1,4-Dichloro-2-butyne	2.10	Bi
$C_4H_4F_2$	1,1-Difluoro-1,3-butadiene (*trans* conformation)	1.29	A
C_4H_4O	Furan	0.66	A
$C_4H_4O_2$	Diketene	3.53	B
C_4H_4S	Thiophene	0.55	C
C_4H_5Cl	4-Chloro-1,2-butadiene	2.02	Ci
C_4H_5Cl	1-Chloro-2-butyne	2.19	C
C_4H_5F	2-Fluoro-1,3-butadiene (*trans* conformation)	1.42	A
C_4H_5N	Pyrrole	1.84	C
C_4H_5N	*cis*-1-Cyanopropene	4.08	B
C_4H_5N	*trans*-1-Cyanopropene	4.50	B
C_4H_5N	2-Cyanopropene (methacrylonitrile)	3.69	C
C_4H_6	Cyclobutene	0.132	A
C_4H_6	1-Butyne	0.80	C
C_4H_6	1,2-Butadiene	0.403	A
C_4H_6	1,3-Butadiene	0	S
C_4H_6O	Cyclobutanone	2.99	B
C_4H_6O	*trans*-2-Butenal (crotonaldehyde)	3.67	Bi
C_4H_6O	2-Methylpropenal (methacrolein)	2.68	Ci
C_4H_6O	3-Butene-2-one	3.16	B
C_4H_7Cl	1-Chloro-2-methylpropene	1.95	Bi
$C_4H_7Cl_3$	1,1,2-Trichloro-2-methylpropane	1.86	Ci
C_4H_7F	Fluorocyclobutane	1.94	A
C_4H_7N	1-Cyanopropane	4.07	Bi
C_4H_8	1-Butene	0.34	Ci
C_4H_8	*trans*-2-Butene	0	S
C_4H_8	2-Methylpropene	0.50	A
$C_4H_8Cl_2$	1,4-Dichlorobutane	2.22	Ci
C_4H_8O	Tetrahydrofuran	1.63	C
C_4H_8O	*cis*-2,3-Dimethyloxirane	2.03	A
C_4H_8O	Butanal	2.72	Bi
$C_4H_8O_2$	1,4-Dioxane	0	S
$C_4H_8O_2$	Ethyl acetate	1.78	Ci
C_4H_9Br	1-Bromobutane	2.08	Ci
C_4H_9Br	2-Bromobutane	2.23	Ci
C_4H_9Cl	1-Chlorobutane	2.05	Bi
C_4H_9Cl	2-Chlorobutane	2.04	Ci
C_4H_9Cl	1-Chloro-2-methylpropane	2.00	Ci
C_4H_9Cl	2-Chloro-2-methylpropane	2.13	B
C_4H_9F	2-Fluoro-2-methylpropane	1.96	A
C_4H_9I	1-Iodobutane	2.12	Ci
C_4H_9NO	Propanoyl methylamine (N-methylpropionamide)	3.61	Bi
C_4H_9NO	Acetyl dimethylamine (N,N-dimethylacetamide)	3.81	Bi
$C_4H_9NO_2$	2-Nitrito-2-methylpropane (t-butyl nitrite)	2.74	Ci
$C_4H_9NO_2$	1-Nitrobutane	3.59	Bi
$C_4H_9NO_2$	2-Nitro-2-methylpropane	3.71	B
C_4H_{10}	Butane	$\leqslant$0.05	Ci
C_4H_{10}	2-Methylpropane	0.132	A
$C_4H_{10}O$	1-Butanol	1.66	Bi
$C_4H_{10}O$	2-Methylpropan-1-ol (isobutanol)	1.64	C
$C_4H_{10}O$	Diethyl ether	1.15	Bi
$C_4H_{10}S$	Diethyl sulfide	1.54	Ci
$C_4H_{11}N$	Diethyl amine	0.92	Ci
C_5H_5N	Pyridine	2.19	B
C_5H_5N	1-Cyano-1,3-butadiene	3.90	Ci
C_5H_6	1,3-Cyclopentadiene	0.419	A
C_5H_8	Cyclopentene	0.20	B
C_5H_8	1-Pentyne	0.81	Ci
C_5H_8	2-Methyl-1,3-butadiene (*trans* conformation)	0.25	A
$C_5H_8O_2$	Acetylacetone		Ci
C_5H_9N	1-Cyanobutane	4.12	Bi
C_5H_9N	2-Cyano-2-methylpropane	3.95	A
$C_5H_{10}O_3$	Diethyl carbonate	1.10	Ci
$C_5H_{11}Br$	1-Bromopentane	2.20	Ci
$C_5H_{11}Cl$	1-Chloropentane	2.16	Ci
C_5H_{12}	2-Methylbutane	0.13	C
C_5H_{12}	2,2-Dimethylpropane	0	S
$C_6H_2Cl_2O_2$	2,5-Dichloro-1,4-cyclohexadienedione	0	S
$C_6H_4ClNO_2$	*o*-Chloronitrobenzene	4.64	B
$C_6H_4ClNO_2$	*m*-Chloronitrobenzene	3.73	B
$C_6H_4ClNO_2$	*p*-Chloronitrobenzene	2.83	B
$C_6H_4Cl_2$	*o*-Dichlorobenzene	2.50	B
$C_6H_4Cl_2$	*m*-Dichlorobenzene	1.72	C
$C_6H_4Cl_2$	*p*-Dichlorobenzene	0	S
$C_6H_4FNO_2$	*p*-Fluoronitrobenzene	2.87	B
$C_6H_4F_2$	*m*-Difluorobenzene	1.58	B
$C_6H_4N_2O_4$	*p*-Dinitrobenzene	0	S
$C_6H_4O_2$	1,4-Cyclohexadienedione (*p*-benzoquinone)	0	S
C_6H_5Br	Bromobenzene	1.70	B
C_6H_5Cl	Chlorobenzene	1.69	B
C_6H_5ClO	*p*-Chlorophenol	2.11	C
C_6H_5F	Fluorobenzene	1.60	C
C_6H_5I	Iodobenzene	1.70	C
$C_6H_5NO_2$	Nitrobenzene	4.22	B
C_6H_6	Benzene	0	S

Compounds containing carbon—Continued

Formula	Compound name	Selected moment (debyes)	
C_6H_6O	Phenol	1.45	C
C_6H_7N	Aminobenzene (aniline)	1.53	C
C_6H_8	1,3-Cyclohexadiene	0.44	B
C_6H_{10}	1-Hexyne	0.83	Ci
C_6H_{10}	3, 3-Dimethyl-1-butyne	0.66	A
$C_6H_{10}Cl_2$	*cis*-le,2a-Dichlorocyclohexane	3.11	C
$C_6H_{12}N_2$	Diisopropylidene hydrazine (dimethyl ketazine)	1.53	Bi
$C_6H_{12}O_2$	Pentyl formate (n-amyl formate)	1.90	Ci
$C_6H_{12}O_3$	2,4,6-Trimethyl-1,3,5-trioxane (paraldehyde)	1.43	C
$C_6H_{14}O$	Dipropyl ether	1.21	Ci
$C_6H_{14}O_2$	1,1-Diethoxyethane		i
$C_6H_{15}N$	Triethyl amine	0.66	Ci
$C_7H_4ClF_3$	*o*-Chloro(trifluoromethyl)benzene	3.46	B
$C_7H_4ClF_3$	*p*-Chloro(trifluoromethyl)benzene	1.58	C
$C_7H_5F_3$	(Trifluoromethyl)benzene	2.86	B
C_7H_5N	Cyanobenzene (benzonitrile)	4.18	B
C_7H_7Cl	*o*-Chlorotoluene	1.56	C
C_7H_7Cl	*p*-Chlorotoluene	2.21	B
C_7H_7F	*o*-Fluorotoluene	1.37	C
C_7H_7F	*m*-Fluorotoluene	1.86	C
C_7H_7F	*p*-Fluorotoluene	2.00	C
$C_7H_7NO_3$	*o*-Nitro(methoxy)benzene	4.83	Bi
$C_7H_7NO_3$	*m*-Nitro(methoxy)benzene	4.55	Bi
$C_7H_7NO_3$	*p*-Nitro(methoxy)benzene	5.26	B
C_7H_8	1,3,5-Cycloheptatriene	0.25	C
C_7H_8	Toluene	0.36	C
C_7H_8O	Phenylmethanol (benzyl alcohol)	1.71	C
C_7H_8O	Methoxybenzene (anisole)	1.38	C
C_7H_9NO	*o*-Amino(methoxy)benzene	1.61	Ci
$C_7H_{14}O_2$	Pentyl acetate (*n*-amyl acetate)	1.75	Ci
$C_7H_{15}Br$	1-Bromoheptane	2.16	Ci
$C_8H_4N_2$	*p*-Dicyanobenzene	0	S
C_8H_8O	Acetylbenzene (acetophenone)	3.02	B
$C_8H_8O_2$	2,5-Dimethyl-1,-4-cyclohexadienedione	0	S
C_8H_{10}	Ethylbenzene	0.59	C
C_8H_{10}	*o*-Xylene	0.62	C
C_8H_{10}	*p*-Xylene	0	S
$C_8H_{12}O_2$	Tetramethylcyclobutane-1,3-dione	0	S
$C_8H_{18}O$	Dibutyl ether	1.17	Ci
C_9H_7N	Quinoline	2.29	C
C_9H_7N	Isoquinoline	2.73	C
$C_9H_{10}O_2$	Ethyl phenylformate (ethyl benzoate)	2.00	Ci
$C_{10}H_8$	Azulene	0.80	B
$C_{10}H_{14}BeO_4$	Bis(2,4-pentanedionato) beryllium	0	S
$C_{12}H_9BrO$	*p*-Bromophenoxybenzene	1.98	C
$C_{12}H_9NO_3$	*p*-Nitrophenoxybenzene	4.54	B
$C_{12}H_{10}$	Phenylbenzene (diphenyl)	0	S
$C_{13}H_{11}BrO$	*p*-Bromophenoxy-*p*-toluene	2.45	C
$C_{14}H_{14}O$	Bis(*p*-tolyl) ether	1.54	C
$C_{15}H_{21}AlO_6$	Tris(2,4-pentanedionato) aluminum	0	S
$C_{15}H_{21}CrO_6$	Tris(2,4-pentanedionato) chromium (III)	0	S
$C_{15}H_{21}FeO_6$	Tris(2,4-pentanedionato) iron (III)	0	S
$C_{20}H_{28}O_8Th$	Tetrakis(2,4-pentanedionato) thorium	0	S

DIPOLE MOMENTS

The method of measurement of the dipole moments is indicated in the following **two tables** by the symbols:

B benzene solution
C carbon tetrachloride solution
D 1,4-dioxane solution
H n-heptane solution
St measurement of Stark effect in microwave spectrum of gas

Dipole Moments for Some Inorganic Compounds

Compound	Dipole Moment $\times 10^{-18}$ e. s. u.	Method
$AlBr_3$	5.14	B
AlI_3	2.48	B
CsCl	10.42	St
CsF	7.875	St
HF	1.92 ± 0.02	..
HCl	1.084 ± 0.003–0.007	..
NBr	0.78	..
HDSe	0.62	St
HI	0.38	..
DCl	1.084 ± 0.003–0.007	..
HNO_3	2.16	St
$HgBr_2$	0.95	B
$HgCl_2$	1.23	B
H_2O	1.87	..
H_2O_2	2.13 ± 0.05	..
H_2S	1.10	..
SO_2	1.60	..
SO_3	0.00	..
SO_2F_2	1.110	St
NH_3	1.3	..
N_2H_4	1.84	..
NO	0.16	..
NO_2	0.29	..
N_2O_4	0.37	..
NOCl	1.83	..
NOBr	1.87	..
PCl_3	0.90–1.16	..
PCl_5	0.0	..
CO	0.10	..
CO_2	0.0	..
SiD_2F_2	1.53	St
SiH_2F_2	1.54	St
$SnCl_4$	0.95	B
SnI_4	0	B
$TiCl_4$	0	C

Dipole Moments for Some Organo-metallic Compounds

Compound	Dipole Moment $\times 10^{-18}$ e. s. u.	Method
Beryllium diethyl	1.0	H
Cadmium diethyl	0.3	H
Mercury diethyl	0.0	H
Magnesium diethyl	4.8	D
Zinc diethyl	0.0	H
Beryllium diphenyl	1.6	B
Cadmium diphenyl	0.6	B
Mercury diphenyl	0.2	B
Magnesium diphenyl	4.9	D
Zinc diphenyl	0.8	B
Chromium (0), diphenyl	0	B
Chromium, ditolyl	0	B
Cobalt, mononitrosyl tricarbonyl	0.72	B
Cyclopentadienyl, chromium dicarbonyl mono nitrosyl	3.23	B
Cyclopentadienyl, manganese tri-carbonyl	3.30	B
Cyclopentadienyl, cobalt ducarbonyl	2.87	B
Cyclopentadienyl, vanadium tetra-carbonyl	3.17	B
Penta cyclopentadienyl, dicobalt	0	B
Dicyclopentadienyl, iron (II)	0	B
Dicyclopentadienyl, lead (II)	1.63	B
Dicyclopentadienyl, tin (II)	1.02	B
Ethyl lithium	0.87	B
Iron, dinitrosyl dicarbonyl	0.95	B
Iron, tetracarbonyl-diiodide	3.68	B
Iron, tetracarbonyl mono-(methyl isonitrile)	5.07	B
Iron, pentacarbonyl	0.63	B
Iron, bis(p-chlorophenyl cyclo-pentadienyl	3.12	B
Ruthenium (II), di-indenyl	0	B

Dipole Moments

Dipole Moments of Amino Acid Esters

Substance	$\mu \cdot 10^{18}$ e. s. u.
Glycine ethyl ester	2.11
α-Alanine ethyl ester	2.09
α-Aminobutyric acid ethyl ester	2.13
α-Aminovaleric acid ethyl ester	2.13
Valine ethyl ester	2.11
α-Aminocaproic acid ethyl ester	2.13
β-Alanine ethyl ester	2.14
β-Aminobutyric acid ethyl ester	2.11

Accurate to ±0.01 . 10^{-18} e. s. u.
J. Wyman, Chem. Rev., 1936, **19**, 213.

Dipole Moments of Amides

Substance	
Urea	4.56
Thiourea	4.89
Symm.-dimethylurea	4.8
Tetraethylurea	3.3
Propylurea	4.1
Acetamide	3.6
Sulfamide	3.9
Benzamide	3.6
Valeramide	3.7
Caproamide	3.9

For comprehensive list of dipole moments see Trans. Faraday Soc., 1934, **30**, General Discussion.

Dipole Moments of Some Hormones and Related Compounds in Dioxan

Compound	
Cholestane-3(β) : 7(α)-diol	2.31
Cholestane-3(β) : 7(β)-diol	2.55
Cholestane	2.98
Δ^5-Cholestane-3(β)ol-7 one	3.79
Androsterone	3.70
β-Androsterone	2.95
Δ^5-Androstene-3(β) : 17(α)-diol	2.89
Δ^5-Androstene-3(β) : 17(β)-diol	2.69
Δ^5-Androstene-3(β)ol-17 one	2.46
Testosterone	4.32
cis-Testosterone	5.17
Δ^4-Androstene-3 : 17 dione	3.32
Isophorone	3.96

Ethylenic >C=C< in a six membered ring and conjugated with >C=0 increases the dipole moment approximately by 1 Debye. Non-conjugated >C=C< in sterols decreases the dipole moment by approximately 0.49. Biological activity is not correlated with dipole moment. W. D. Kumler and G. M. Fohlen, J. Am. Chem. Soc., 1945, **67**, 437.

FINE-STRUCTURE SEPARATIONS IN ATOMIC NEGATIVE IONS

From the *Journal of Physical and Chemical Reference Data*, Volume 14, No. 3, 731, 1985. Reproduced by permission of the copyright owners, the American Chemical Society and the American Institute of Physics, and the authors, H. Hotop and W. C. Lineberger. Users of data in this table are referred to the original publication for references for the data.

Z	Negative ion	Fine-structure interval[a]	Separation (cm^{-1})	Method[b]	Z	Negative ion	Fine-structure interval[a]	Separation (cm^{-1})	Method[b]
2	$He^-(1s2s2p\ ^4P^o)$	$5/2 \rightarrow 3/2$	0.027508(27)	rf	33	$As^-(^3P)$	$2 \rightarrow 1$	1100(200)	LIE; QIE
		$5/2 \rightarrow 1/2$	0.2888(18)	rf			$2 \rightarrow 0$	1500(200)	LIE; QIE
5	$B^-(^3P)$	$0 \rightarrow 1$	4(1)	RIE			$2 \rightarrow 0$	≈1370	PT
		$0 \rightarrow 2$	9(1)	RIE	34	$Se^-(^2P)$	$3/2 \rightarrow 1/2$	2279(2)	LPT
6	$C^-(^2D)$	$3/2 \rightarrow 5/2$	3(1)	LIE	40	$Zr^-(^4F)$	$3/2 \rightarrow 5/2$	250(50)	RIE
8	$O^-(^2P)$	$3/2 \rightarrow 1/2$	177.08(5)	LPT			$5/2 \rightarrow 7/2$	330(70)	RIE
13	$Al^-(^3P)$	$0 \rightarrow 1$	26(3)	RIE			$7/2 \rightarrow 9/2$	370(70)	RIE
		$0 \rightarrow 2$	76(7)	RIE			$3/2 \rightarrow 9/2$	950(100)	RIE
14	$Si^-(^2D)$	$3/2 \rightarrow 5/2$	7(2)	LIE	41	$Nb^-(^5D)$	$0 \rightarrow 1$	110(20)	RIE
15	$P^-(^3P)$	$2 \rightarrow 1$	181(2)	LPT			$1 \rightarrow 2$	200(40)	RIE
		$2 \rightarrow 0$	263(2)	LPT			$2 \rightarrow 3$	250(40)	RIE
16	$S^-(^2P)$	$3/2 \rightarrow 1/2$	483.54(1)	LPT			$3 \rightarrow 4$	310(60)	RIE
22	$Ti^-(^4F)$	$3/2 \rightarrow 5/2$	72(7)	LIE			$0 \rightarrow 4$	860(90)	RIE
		$5/2 \rightarrow 7/2$	99(10)	LIE	45	$Rh^-(^3F)$	$4 \rightarrow 3$	2370(65)	LPES
		$7/2 \rightarrow 9/2$	124(12)	LIE			$3 \rightarrow 2$	1000(65)	LPES
		$3/2 \rightarrow 9/2$	295(15)	LIE			$4 \rightarrow 2$	3370(65)	LPES
23	$V^-(^5D)$	$0 \rightarrow 1$	35(4)	RIE	46	$Pd^-(^2D)$	$5/2 \rightarrow 3/2$[c]	3450(350)	RIE
		$1 \rightarrow 2$	70(7)	RIE	49	$In^-(^3P)$	$0 \rightarrow 1$	680(70)	RIE; QIE
		$2 \rightarrow 3$	100(10)	RIE			$0 \rightarrow 2$	1550(150)	RIE; QIE
		$3 \rightarrow 4$	125(13)	RIE	50	$Sn^-(^2D)$	$3/2 \rightarrow 5/2$	800(200)	LIE
		$0 \rightarrow 4$	330(17)	RIE	51	$Sb^-(^3P)$	$2 \rightarrow 1$	2700(500)	LIE; QIE
26	$Fe^-(^4F)$	$9/2 \rightarrow 7/2$	540(50)	RIE			$2 \rightarrow 0$	3000(500)	LIE; QIE
		$7/2 \rightarrow 5/2$	390(40)	RIE			$2 \rightarrow (1,0)$	≈2740	PT
		$5/2 \rightarrow 3/2$	270(30)	RIE	52	$Te^-(^2P)$	$3/2 \rightarrow 1/2$	5008(5)	LPT
		$9/2 \rightarrow 3/2$	1200(60)	RIE	73	$Ta^-(^5D)$	$0 \rightarrow 1$	1070(110)	LPES
27	$Co^-(^3F)$	$4 \rightarrow 3$	910(50)	LPES			$1 \rightarrow 2$	1170(120)	LPES
		$3 \rightarrow 2$	650(50)	LPES			$2 \rightarrow 3$[d]	980(200)	RIE
		$4 \rightarrow 2$	1560(50)	LPES	77	$Ir^-(^3F)$	$4 \rightarrow 3$	7600(1500)	RIE
28	$Ni^-(^2D)$	$5/2 \rightarrow 3/2$	1470(100)	LPES			$3 \rightarrow 2$	4400(900)	RIE
31	$Ga^-(^3P)$	$0 \rightarrow 1$	220(20)	RIE; QIE			$4 \rightarrow 2$	12000(1200)	RIE
		$0 \rightarrow 2$	580(50)	RIE; QIE	78	$Pt^-(^2D)$	$5/2 \rightarrow 3/2$	10000(1000)	RIE; QIE
32	$Ge^-(^2D)$	$3/2 \rightarrow 5/2$	160(30)	LIE					

[a] Total angular momentum of lower (left) and upper fine structure levels are listed.

[b] Abbreviations used: rf — radio frequency resonance technique; RIE — isoelectronic extrapolation of ratios of fine structure separations; LIE — isoelectronic extrapolation from logarithmic plot; QIE — quadratic isoelectronic extrapolation; LPT — tunable laser photodetachment threshold; LPES — laser photodetachment electron spectrometry; and PT — photodetachment threshold using conventional light sources.

[c] $J = 3/2$ not bound.

[d] J = 3 not bound.

ELECTRON AFFINITIES

Thomas M. Miller

Electron affinity is defined as the energy difference between the lowest (ground) state of the neutral and the lowest state of the corresponding negative ion. The accuracy of electron affinity measurements has been greatly improved since the advent of laser photodetachment experiments with negative ions. Electron affinities can be determined with optical precision, though a detailed understanding of atomic and molecular states and splittings is required to specify the photodetachment threshold corresponding to the electron affinity.

Atomic and molecular electron affinities are discussed in two excellent articles reviewing photodetachment studies which appear in *Gas Phase Ion Chemistry*, Vol. 3, M. T. Bowers, Ed., Academic Press, Orlando, 1984, Chapter 21 by P. S. Drzaic, J. Marks, and J. I. Brauman, "Electron Photodetachment from Gas Phase Negative Ions," p. 167; and Chapter 22 by R. D. Mead, A. E. Stevens, and W. C. Lineberger, "Photodetachment in Negative Ion Beams," p. 213. Persons interested in photodetachment details should consult these articles and the critical review of H. Hotop and W. C. Lineberger, *J. Phys. Chem. Ref. Data*, 14, 731, 1985. For simplicity in the tables below, any electron affinity which was discussed in the articles by Drzaic et al. or Hotop and Lineberger is referenced to these sources, where original references are given. A great many additional electron affinities have been provided here by G. B. Ellison, W. C. Lineberger, H. Hotop, D. G. Leopold, and K. H. Bowen. Electron affinities for the lanthanides and actinides have not been measured, but theoretical estimates have been made by S. G. Bratch, *Chem. Phys. Lett.*, 98, 113, 1983, and S. G. Bratch and J. J. Lagowski, *Chem. Phys. Lett.*, 107, 136, 1984.

For the present tabulation the value e/hc = 8065.4786 ± 0.0208 cm^{-1}/eV (Taylor, B. N., *Rev. Mod. Phys.*, 56, 531, 1984) has been used to convert electron affinities from cm^{-1} into eV. Only in the cases of O, S, and C_2H_3O is the 2.6 ppm uncertainty in e/hc significant. However, the uncertainties in the electron affinities given in the tables below do *not* include the uncertainty in e/hc.

Electron affinities for the rare earths have not been measured, but theoretical estimates have been made by S. G. Bratsch and J. J. Lagowski, *Chem. Phys. Lett.*, 107, 136, 1984. Abbreviations used in the tables: calc = calculated value; PT = photodetachment threshold using a lamp as a light source; LPT = laser photodetachment threshold; LPES = laser photoelectron spectroscopy; DA = dissociative attachment; LOGS = laser optogalvanic spectroscopy; e-scat = electron scattering; plasma = absorption or emission in a plasma; CT = charge transfer; and CD = collisional detachment.

Table 1
ATOMIC ELECTRON AFFINITIES

Atomic number	Atom	Electron affinity in eV	Uncertainty in eV	Method	Ref.	Atomic number	Atom	Electron affinity in eV	Uncertainty in eV	Method	Ref.
1	H	0.754209	0.000003	calc	1	4	Be	Not stable	—	calc	1
2	He	Not stable[a]	—	calc	1	5	B	0.277	0.010	LPES	1
3	Li	0.6180	0.0005	LPT	1	6	C	1.2629	0.0003	LPT	1

Table 1 (continued)
ATOMIC ELECTRON AFFINITIES

Atomic number	Atom	Electron affinity in eV	Uncertainty in eV	Method	Ref.	Atomic number	Atom	Electron affinity in eV	Uncertainty in eV	Method	Ref.
7	N	Not stable	—	DA	1	40	Zr	0.426	0.014	LPES	1
8	O	1.46112516	0.0000008	LPT	4	41	Nb	0.893	0.025	LPES	1
9	F	3.399	0.003	plasma	1	42	Mo	0.746	0.010	LPES	1
10	Ne	Not stable	—	calc	1	43	Tc	0.55	0.20	calc	1
11	Na	0.547930	0.000025	LPT	1	44	Ru	1.05	0.15	calc	1
12	Mg	Not stable	—	e-scat	1	45	Rh	1.137	0.008	LPES	1
13	Al	0.441	0.010	LPES	1	46	Pd	0.557	0.0008	LPES	1
14	Si	1.385	0.005	LPES	1	47	Ag	1.302	0.007	LPES	1
15	P	0.7465	0.0003	LPT	1	48	Cd	Not stable	—	e-scat	1
16	S	2.077120	0.000001	LPT	1	49	In	0.30	0.2	PT	1
17	Cl	3.617	0.003	plasma	1	50	Sn	1.112	0.004	LPES	28
18	Ar	Not stable	—	calc	1	51	Sb	1.07	0.05	PT	1
19	K	0.50147	0.00010	LPT	1	52	Te	1.9708	0.0004	LPT	1
20	Ca	Not stable	—	calc	1	53	I	3.0591	0.001	LOGS	1
21	Sc	0.188	0.020	LPES	1	54	Xe	Not stable	—	calc	1
22	Ti	0.079	0.014	LPES	1	55	Cs	0.471630	0.000025	LPT	1
23	V	0.525	0.012	LPES	1	56	Ba	Not stable	—	calc	1
24	Cr	0.666	0.012	LPES	1	57	La	0.5	0.3	calc	1
25	Mn	Not stable	—	calc	1	72	Hf	≈0	—	calc	1
26	Fe	0.151	0.003	LPES	27	73	Ta	0.322	0.012	LPES	1
27	Co	0.662	0.003	LPES	27	74	W	0.815	0.002	LPES	37
28	Ni	1.156	0.010	LPES	1	75	Re	0.15	0.15	calc	1
29	Cu	1.235	0.05	LPES	1	76	Os	1.10	0.2	calc	1
30	Zn	Not stable	—	e-scat	1	77	Ir	1.565	0.008	LPES	1
31	Ga	0.3	0.15	PT	1	78	Pt	2.128	0.002	LPT	1
32	Ge	1.233	0.003	LPES	28	79	Au	2.30863	0.00003	LPT	1
33	As	0.81	0.03	PT	1	80	Hg	Not stable	—	e-scat	1
34	Se	2.02069	0.00003	LPT	1	81	Tl	0.2	0.2	PT	1
35	Br	3.365	0.003	plasma	1	82	Pb	0.364	0.008	LPES	1
36	Kr	Not stable	—	calc	1	83	Bi	0.946	0.010	LPES	1
37	Rb	0.48592	0.00002	LPT	1	84	Po	1.9	0.3	calc	1
38	Sr	Not stable	—	calc	1	85	At	2.8	0.2	calc	1
39	Y	0.307	0.012	LPES	1	86	Rn	Not stable	—	calc	1

[a] Not stable refers to the negative ion.

Table 2
ELECTRON AFFINITIES FOR DIATOMIC MOLECULES

Molecule	Electron affinity in eV	Uncertainty in eV	Method	Ref.	Molecule	Electron affinity in eV	Uncertainty in eV	Method	Ref.
As_2	0	—	PT	2	MgI	1.899	0.018	LPES	31
AsH	1.0	0.1	PT	2	MnD	0.866	0.010	LPES	9
BO	3.12	0.09	PT	6	MnH	0.869	0.010	LPES	9
BeH	0.7	0.1	PT	2	NH	0.370	0.004	LPT	32
Br_2	2.55	0.10	CT	2	NO	0.024	0.010	LPES	2
C_2	3.39	0.15	LPT	2	NS	1.194	0.011	LPES	2
CH	1.238	0.008	LPES	2	NaBr	0.788	0.010	LPES	30
CN	3.821	0.004	LOGS	7	NaCl	0.727	0.010	LPES	30
CS	0.205	0.021	LPES	2	NaF	0.520	0.010	LPES	30
CaH	0.93	0.05	PT	2	NaI	0.865	0.010	LPES	30
Cl_2	2.38	0.10	CT	2	NiD	0.477	0.007	LPES	29
ClO	2.17	0.10	LPT	8	NiH	0.481	0.007	LPES	29
Co_2	1.110	0.008	LPES	27	O_2	0.440	0.008	LPES	2
CoD	0.680	0.010	LPES	29	OD	1.825548	0.000037	LPT	3
CoH	0.671	0.010	LPES	29	OH	1.827670	0.000021	LPT	3
CrD	0.568	0.010	LPES	29	P_2	0.589	0.025	LPES	42
CrH	0.563	0.010	LPES	29	PH	1.028	0.010	LPES	2
CrO	1.222	0.010	LPES	5	PO	1.092	0.010	LPES	2
CsCl	0.455	0.010	LPES	30	RbCl	0.544	0.009	LPES	30
Cu_2	0.842	0.010	LPES	37	Re_2	1.571	0.008	LPES	33
F_2	3.08	0.10	CT	2	S_2	1.663	0.040	LPES	2
Fe_2	0.902	0.008	LPES	27	SD	2.315	0.002	LPES	10
FeD	0.932	0.015	LPES	9	SH	2.317	0.002	LPES	10
FeH	0.934	0.011	LPES	9	SO	1.126	0.013	LPES	2
FeO	1.499	0.010	LPES	5	Se_2	1.96	0.05	LPES	38
I_2	2.55	0.05	CT	2	SeH	2.21	0.03	PT	2
IBr	2.55	0.10	CT	2	SeO	1.456	0.020	LPES	41
KBr	0.642	0.010	LPES	30	SiH	1.277	0.009	LPES	2
KCl	0.582	0.010	LPES	30	Te_2	1.92	0.07	LPES	38
KI	0.728	0.010	LPES	30	TeH	2.102	0.015	LPES	39
LiCl	0.593	0.010	LPES	30	TeO	1.697	0.022	LPES	40
MgCl	1.589	0.011	LPES	31	ZnH	≤0.95	—	PT	2
MgH	1.05	0.06	PT	2					

Table 3
ELECTRON AFFINITIES FOR TRIATOMIC MOLECULES

Molecule	Electron affinity in eV	Uncertainty in eV	Method	Ref.
AsH_2	1.27	0.03	PT	2
BO_2	3.57	0.13	CT	6
C_3	1.981	0.019	LPES	11
CD_2	0.645	0.006	LPES	12
CH_2	0.652	0.006	LPES	12
C_2H	2.94	0.10	PT	2
C_2O	1.848	0.027	LPES	13
COS	0.46	0.20	CD	2
CS_2	1.0	0.2	CD	2
CoD_2	1.465	0.013	LPES	34
CoH_2	1.450	0.014	LPES	34
Cu_3	2.4	0.1	LPES	37
DCO	0.301	0.005	LPES	35
DNO	0.330	0.015	LPES	14
DO_2	1.089	0.017	LPES	15
FeCO	1.26	0.02	LPES	2
FeO_2	1.038	0.013	LPES	34
FeH_2	1.049	0.014	LPES	34
GeH_2	1.097	0.015	LPES	28
HCO	0.313	0.005	LPES	35

Molecule	Electron affinity in eV	Uncertainty in eV	Method	Ref.
HNO	0.338	0.015	LPES	14
HO_2	1.078	0.017	LPES	15
MnD_2	0.465	0.014	LPES	34
MnH_2	0.444	0.016	LPES	34
N_3	2.70	0.12	PT	2
NH_2	0.744	0.022	PT	2
N_2O	≤1.465	—	CT	2
NO_2	2.275	0.025	LPT	2
NiCO	0.804	0.012	LPES	2
NiD_2	1.926	0.007	LPES	34
NiH_2	1.934	0.008	LPES	34
O_3	2.1028	0.0025	LPT	2
PH_2	1.271	0.010	LPES	2
S_3	2.093	0.025	LPES	16
SO_2	1.095	0.008	LPES	16
S_2O	1.878	0.008	LPES	16
SeD_2	1.038	0.013	LPES	34
SeH_2	1.049	0.014	LPES	34
SeO_2	1.83	0.05	LPES	38
SiH_2	1.124	0.020	LPES	2

Table 4
ELECTRON AFFINITIES FOR LARGER POLYATOMIC MOLECULES

Molecule	Electron affinity in eV	Uncertainty in eV	Method	Ref.
Cu_4	1.40	0.05	LPES	37
Cu_5	1.92	0.05	LPES	37
Cu_6	1.92	0.05	LPES	37
Cu_7	2.10	0.05	LPES	37
Cu_8	1.53	0.05	LPES	37
Cu_9	2.45	0.15	LPES	37
Cu_{10}	1.99	0.10	LPES	37
$Fe(CO)_2$	1.22	0.02	LPES	2
$Fe(CO)_3$	1.8	0.2	LPES	2
$Fe(CO)_4$	2.4	0.3	LPES	2
GeH_3	≤1.74	0.04	PT	2
HNO_3	0.57	0.15	CD	2
$Ni(CO)_2$	0.643	0.014	LPES	2
$Ni(CO)_3$	1.077	0.013	LPES	2
$OH(H_2O)$	<2.95	0.15	PT	2
PBr_3	1.59	0.15	CD	2
PBr_2Cl	1.63	0.20	CD	2
PCl_2Br	1.52	0.20	CD	2
PCl_3	0.82	0.10	CD	2
$POCl_2$	3.83	0.25	CD	2
$POCl_3$	1.41	0.20	CD	2
SF_4	2.35	0.10	CT	2
SF_6	0.46	0.20	CD	2
SO_3	≥1.70	0.15	CD	2
SeF_6	2.9	0.2	CD	2
SiO_3	1.386	0.022	LPES	43
SiF_3	≤2.95	0.10	PT	17
SiH_3	1.406	0.014	LPES	43
TeF_6	3.34	0.17	CD	2
UF_5	4.4	0.4	CD	2
UF_6	≥5.1	—	CD	2
WF_5	1.25	0.3	CD	18
WF_6	3.36	0.10	CT	19
	3.72	0.20	CD	18
CH_3	0.08	0.03	LPES	2
CO_3	2.69	0.14	LPES	2
CF_3Br	0.91	0.2	CD	2
CF_3I	1.57	0.2	CD	2
CH_3I	0.2	0.1	CT	2
$CO_3(H_2O)$	2.1	0.2	PT	2
CH_3NO_2	0.44	0.2	CD	2
CD_3O	1.552	0.022	LPES	2
CH_3O	1.570	0.022	LPES	2

Molecule	Electron affinity in eV	Uncertainty in eV	Method	Ref.
CD_3S	1.856	0.006	LPT	2
CH_3S	1.861	0.004	LPT	2
C_2DO	2.350	0.020	LPES	13
C_2HO	2.350	0.020	LPES	13
C_2D_2 (vinylidene-d_2)	0.49	0.02	LPES	20
C_2H_2 (vinylidene)	0.47	0.02	LPES	20
C_2H_2FO (acetyl fluoride enolate)	2.22	0.09	PT	2
C_2D_2N (cyanomethyl-d_2 radical)	1.532	0.014	LPES	21
C_2H_2N (cyanomethyl radical)	1.534	0.014	LPES	21
C_2H_2N (isocyanomethyl radical)	1.067	0.024	LPES	231
C_2D_3O (acetaldehyde-d_3 enolate)	1.81899	0.00012	LPT	22
C_2H_3O (acetaldehyde enolate)	1.82478	0.00012	LPT	22
C_2H_5N (ethyl nitrene)	0.56	0.01	PT	2
C_2D_5O (ethoxide-d_5)	1.702	0.033	LPES	23
C_2H_5O (ethoxide)	1.726	0.033	LPES	23
C_2H_5S (ethyl sulfide)	1.953	0.006	LPT	2
C_3H	1.858	0.023	LPES	11
C_3H_2	1.794	0.021	LPES	11
$C_3H_2F_3O$ (1,1,1-trifluoroacetone enolate)	2.58	0.12	PT	2
C_3H_3 (propargyl radical, $CH_2C{\equiv}CH$)	0.893	0.025	LPES	24
C_3H_2D ($CH_2C{\equiv}CD$)	0.88	0.15	LPES	24
C_3D_2H ($CD_2C{\equiv}CH$)	0.907	0.023	LPES	24
C_3H_3N (CH_3CH-CN)	1.247	0.012	LPES	21
C_3D_5 (allyl-d_5)	0.381	0.025	LPES	25
C_3H_5 (allyl)	0.362	0.019	LPES	25

Table 4 (continued)
ELECTRON AFFINITIES FOR LARGER POLYATOMIC MOLECULES

Molecule	Electron affinity in eV	Uncertainty in eV	Method	Ref.
C_3H_4D ($CH_2=CD\text{-}CH_2$)	0.373	0.019	LPES	25
C_3H_5O (acetone enolate)	1.757	0.033	LPES	23
C_3H_5O (propionaldehyde enolate)	1.611	0.023	LPES	23
$C_3H_5O_2$ (methyl acetate enolate)	1.80	0.06	PT	2
C_3H_7O (*n*-propyl oxide)	1.789	0.033	LPES	23
C_3H_7O (isopropyl oxide)	1.839	0.029	LPES	23
C_3H_7S (*n*-propyl sulfide)	2.00	0.02	PT	2
C_3H_7S (*i*-propyl sulfide)	2.02	0.02	PT	2
C_3O	1.34	0.15	LPES	11
C_3O_2	0.85	0.15	LPES	11
$C_4F_4O_3$ (tetrafluorosuccinic anhydride)	0.5	0.2	CD	2
$C_4H_2O_3$ (maleic anhydride)	1.4	0.2	CD	2
C_4H_4N (pyrrolate)	2.39	0.13	PT	2
C_4H_5O (cyclobutane enolate)	1.84	0.06	PT	2
C_4H_7O (butyraldehyde enolate)	1.67	0.05	PT	2
C_4H_6DO (2-butanone-3-d_1 enolate)	1.67	0.05	PT	2
$C_4H_5D_2O$ (2-butanone-3,3-d_2 enolate)	1.75	0.06	PT	2
C_4H_9O (*t*-butoxyl)	1.912	0.054	LPES	23
C_4H_9S (*n*-butyl sulfide)	2.03	0.02	PT	2
C_4H_9S (*t*-butyl sulfide)	2.07	0.02	PT	2
C_4O	2.05	0.15	LPES	11
C_4O_2	2.0	0.2	LPES	11
$C_5F_6O_3$ (hexafluoroglutaric anhydride)	1.5	0.2	CD	2
C_5D_5 (cyclopentadienyl-d_5)	1.790	0.008	LPES	11
C_5H_5 (cyclopentadienyl)	1.804	0.007	LPES	11
C_5H_7 (pentadienyl)	0.91	0.03	PT	2
C_5H_7O (cyclopentanone enolate)	1.62	0.06	PT	2
C_5H_9O (3-pentanone enolate)	1.69	0.05	PT	2
$C_5H_{11}O$ (neopentoxyl)	1.93	0.05	LPT	2
$C_5H_{11}S$ (*n*-pentyl sulfide)	2.09	0.02	PT	2
C_5O_2	1.2	0.2	LPES	11
$C_6Br_4O_2$ (*p*-bromanil)	2.44	0.20	CT	2
$C_6Cl_4O_2$ (*p*-chloranil)	2.76	0.20	CT	2
$C_6F_4O_2$ (*p*-fluoranil)	2.92	0.20	CT	2
C_6F_6 (hexafluorobenzene)	>1.8	0.3	CT	2
C_6F_{10} (perfluorocyclohexene)	>1.4	0.3	CT	2
C_6D_4 (*o*-benzyne-d_4)	0.551	0.010	LPES	36
C_6H_4 (*o*-benzyne)	0.560	0.010	LPES	36
C_6H_4ClO (*o*-chloroperoxide)	≤2.58	0.08	PT	2
$C_6H_4O_2$ (*p*-benzoquinone)	1.89	0.3	CD	2
C_6D_5 (phenyl-d_5)	1.092	0.020	LPES	26
C_6H_5 (phenyl)	1.102	0.020	LPES	26
C_6H_5N (phenyl nitrene)	1.46	0.02	PT	2
$C_6H_5NO_2$ (nitrobenzene)	>0.7	0.2	CT	2
C_6H_5O (phenoxide)	≤2.36	0.06	PT	2
C_6H_5S (thiophenoxide)	≤2.47	0.06	PT	2
C_6H_5NH (anilide)	1.70	0.03	PT	2
C_6H_7 (methylcyclopentadienyl)	<1.67	0.04	PT	2
C_6H_9O (cyclohexanone enolate)	1.55	0.05	PT	2
$C_6H_{11}O$ (pinacolone enolate)	1.88	0.06	PT	2
$C_6H_{11}O$ (3,3-dimethylbutananal enolate)	1.82	0.06	PT	2
C_6N_4 (TCNE)	2.3	0.3	PT	2
C_7F_8 (octafluorotoluene)	>1.7	0.3	CT	2
C_7H_6FO (*m*-fluoroacetophenone enolate)	2.218	0.010	LPT	2
C_7H_6FO (*p*-fluoroacetophenone enolate)	2.176	0.010	LPT	2
C_7H_7O (*o*-methyl phenoxide)	≤2.36	0.06	PT	2
C_7H_9 (heptatrienyl)	1.27	0.03	PT	2
C_7H_9O (2-norbornanone enolate)	1.61	0.05	PT	2
$C_7H_{11}O$ (cycloheptanone enolate)	1.48	0.04	PT	2
$C_7H_{11}O$ (2,5-dimethylcyclopentanone enolate)	1.49	0.04	PT	2
$C_7H_{13}O$ (4-heptanone enolate)	1.72	0.06	PT	2
$C_7H_{13}O$ (di-isopropyl ketone enolate)	1.46	0.04	PT	2
C_8H_7O (acetophenone enolate)	2.057	0.010	LPT	2
C_8H_7O (phenylacetaldehyde enolate)	2.10	0.08	PT	2
$C_8H_{13}O$ (cyclooctanone enolate)	1.63	0.06	PT	2
C_9H_9O (*m*-methylacetophenone enolate)	2.030	0.010	LPT	2
C_9H_9SiN (trimethylsilylnitrene)	1.43	0.10	PT	2
$C_9H_{15}O$ (cyclononanone enolate)	1.69	0.06	PT	2
$C_{10}H_{17}O$ (cyclodecanone enolate)	1.83	0.06	PT	2
$C_{12}H_4N_4$ (TCNQ)	2.8	0.3	CD	2
$C_{12}H_9$ (perinaphthenyl)	1.07	0.10	PT	2
$C_{12}H_{15}O$ (*t*-butylacetophenone enolalte)	2.032	0.010	LPT	2
$C_{12}H_{21}O$ (cyclododecanone enolate)	1.90	0.07	PT	2

REFERENCES

1. **Hotop, H. and Lineberger, W. C.,** *J. Phys. Chem. Ref. Data,* 14, 731, 1985.
2. **Drzaic, P. S., Marks, J., and Brauman, J. I.,** in *Gas Phase Ion Chemistry,* Vol. 3, Bowers, M. T., Ed., Academic Press, Orlando, 1984, p. 167.
3. **Schulz, P. A., Mead, R. D., Jones, P. L., and Lineberger, W. C.,** *J. Chem. Phys.,* 77, 1153, 1982.
4. **Neumark, D. M., Lykke, K. R., Andersen, T., and Lineberger, W. C.,** *Phys. Rev. A,* 32, 1890, 1985.
5. **Leopold, D. G., Murray, K. K., Miller, T. M., and Lineberger, W. C.,** unpublished data.
6. **Srivastava, R. D., Uy, O. M., and Farber, M.,** *Trans. Faraday Soc.,* 67, 2941, 1971.
7. **Klein, R., McGinnis, R. P., and Leone, S. R.,** *Chem. Phys. Lett.,* 100, 475, 1983.
8. **Lee, L. C., Smith, G. P., Moseley, J. T., Cosby, P. C., and Guest, J. A.,** *J. Chem. Phys.,* 70, 3237, 1979.
9. **Stevens, A. E., Fiegerle, C. S., and Lineberger, W. C.,** *J. Chem. Phys.,* 78, 5420, 1983.
10. **Breyer, F., Frey, P., and Hotop, H.,** *Z. Phys.,* A 300, 7, 1981.
11. **Oakes, J. M. and Ellison, G. B.,** to be published.
12. **Leopold, D. G., Murray, K. K., Miller, A. E. S., and Lineberger, W. C.,** *J. Chem. Phys.,* 83, 4849, 1985.
13. **Oakes, J. M., Jones, M. E., Bierbaum, V. M., and Ellison, G. B.,** *J. Phys. Chem.,* 87, 4810, 1983.
14. **Ellis, H. B., Jr. and Ellison, G. B.,** *J. Chem. Phys.,* 78, 6541, 1983.
15. **Oakes, J. M., Harding, L. B., and Ellison, G. B.,** *J. Chem. Phys.,* 83, 5400, 1985.
16. **Nimlos, M. E. and Ellison, G. B.,** to be published.
17. **Richardson, L. M., Stephenson, L. M., and Brauman, J. I.,** *Chem. Phys. Lett.,* 30, 17, 1975.
18. **Dispert, H. and Lacmann, K.,** *Chem. Phys. Lett.,* 45, 311, 1977.
19. **Viggiano, A. A. and Paulson, J. F.,** 37th Gaseous Electronics Conference, Boulder, CO, 1984. *Bull. Am. Phys. Soc.,* 30, 134, 1985.
20. **Burnett, S. M., Stevens, A. E., Fiegerle, C. S., and Lineberger, W. C.,** *Chem. Phys. Lett.,* 100, 124, 1983.
21. **Moran, S., Ellis, H. B., Jr., and Ellison, G. B.,** *J. Am. Chem. Soc.,* submitted, 1985.
22. **Mead, R. D., Lykke, K. R., Lineberger, W. C., Marks, J., and Brauman, J. I.,** *J. Chem. Phys.,* 81, 4883, 1984. See Lykke, K. R., Mead, R. D., and Lineberger, W. C., *Phys. Rev. Lett.,* 52, 2221, 1984. The electron affinities are (14717.7 ± 1.0) cm^{-1} for acetaldehyde enolate and (14671.0 ± 1.0) cm^{-1} for acetaldehyde-d_3 enolate.
23. **Ellison, G. B., Engelking, P. C., and Lineberger, W. C.,** *J. Phys. Chem.,* 86, 4873, 1982.
24. **Oakes, J. M. and Ellison, G. B.,** *J. Am. Chem. Soc.,* 105, 2969, 1983.
25. **Ellison, G. B. and Oakes, J. M.,** *J. Am. Chem. Soc.,* 106, 7734, 1984.
26. **Miller, A. E. S. and Lineberger, W. C.,** to be published.
27. **Leopold, D.G. and Lineberger, W. C.,** *J. Chem. Phys.,* 85, 51, 1986.
28. **Miller, T. M., Miller, A. E. S., and Lineberger, W. C.,** *Phys. Rev. A,* 33, 3558, 1986.
29. **Miller, A. E. S., Fiegerle, C. S., and Lineberger, W. C.,** *J. Chem. Phys.,* submitted, 1987.
30. **Miller, T. M., Leopold, D. G., Murray, K. K., and Lineberger, W. C.,** *J. Chem. Phys.,* 85, 2368, 1986.
31. **Miller, T. M. and Lineberger, W. C.,** to be published.
32. **Neumark, D. M., Lykke, K. R., Andersen, T., and Lineberger, W. C.,** *J. Chem. Phys.,* 83, 4364, 1985.
33. **Leopold, D. G., Miller, T. M., and Lineberger, W. C.,** *J. Am. Chem. Soc.,* 108, 178, 1986.
34. **Miller, A. E. S., Fiegerle, C. S., and Lineberger, W. C.,** *J. Chem. Phys.,* 84, 4127, 1986.
35. **Murray, K. K., Miller, T. M., Leopold, D. G., and Lineberger, W. C.,** *J. Chem. Phys.,* 84, 2520, 1986.
36. **Leopold, D. G., Miller, T. M., Miller, A. E. S., and Lineberger, W. C.,** *J. Am. Chem. Soc.,* 108, 1379, 1986.
37. **Leopold, D. G., Ho, J., and Lineberger, W. C.,** *J. Chem. Phys.,* submitted, 1986.
38. **Snodgrass, J. T., Coe, J. V., McHugh, K. M., Friedhoff, C. B., and Bowen, K. H.,** *J. Phys. Chem.,* to be submitted.
39. **Friedhoff, C. B., Snodgrass, J. T., Coe, J. V., McHugh, K. M., and Bowen, K. H.,** *J. Chem. Phys.,* 84, 1051, 1986.
40. **Friedhoff, C. B., Coe, J. V., Snodgrass, J. T., McHugh, K. M., and Bowen, K. H.,** *Chem. Phys. Lett.,* 124, 268, 1986.
41. **Coe, J. V., Snodgrass, J. T., Friedhoff, C. B., McHugh, K. M., and Bowen, K. H.,** *J. Chem. Phys.,* 84, 619, 1986.
42. **Snodgrass, J. T., Coe, J. V., Friedhoff, C. B., McHugh, K. M., and Bowen, K. H.,** *Chem. Phys. Lett.,* 122, 352, 1985.
43. **Nimlos, M. R. and Ellison, G. B.,** *J. Am. Chem. Soc.,* 108, 6522, 1986.

ATOMIC AND MOLECULAR POLARIZABILITIES

Thomas M. Miller

The *polarizability* of an atom or molecule describes the response of the electron cloud to an external electric field. The atomic or molecular energy shift ΔW due to an external electric field E is proportional to the electric field squared, for external fields which are weak compared to the internal electric fields between the nucleus and electron cloud. The *electric dipole polarizability* α is the constant of proportionality defined by $\Delta W = -{}^1/_2\,\alpha E^2$. The induced electric dipole moment is αE. *Hyperpolarizabilities*, coefficients of higher powers of E, are less often required. Technically, the polarizability is a tensor quantity but for spherically symmetric charge distributions reduces to a single number. In any case, an *average polarizability* is usually adequate in calculations. Frequency-dependent or *dynamic polarizabilities* are needed for electric fields which vary in time, except for frequencies which are much lower than electron orbital frequencies, where *static polarizabilities* suffice.

Polarizabilities for atoms and molecules in excited states are found to be larger than for ground states and may be positive or negative. Molecular polarizabilities are very slightly temperature dependent since the size of the molecule depends on its vibrational and rotational quantum numbers. Only in the case of hydrogen has this effect been studied enough to warrant consideration in Table 3.

Tabulated here are static average electric dipole polarizabilities for ground state atoms and molecules.

Polarizabilities are normally expressed in cgs units of cm^3. Ground state polarizabilities are in the range of $10^{-24}\,cm^3 = 1\,Å^3$ and hence are often given in $Å^3$ units. Theorists tend to use atomic units of a_0^3 where a_0 is the Bohr radius. The conversion is $\alpha(cm^3) = 0.148184 \times \alpha(a_0^3)$. Polarizabilities are rarely encountered in SI units, $C{\cdot}m^2/V = J/(V/m)^2$. The conversion from cgs units to SI units is $\alpha(C{\cdot}m^2/V) = 4\pi\epsilon_0 \times 10^{-6} \times \alpha(cm^3)$, where ϵ_0 is the permittivity of free space in SI units and the factor 10^{-6} simply converts cm^3 into m^3. Thus, $\alpha(C{\cdot}m^2/V) = 1.11265 \times 10^{-16} \times \alpha(cm^3)$. Persons measuring excited state polarizabilities by optical methods tend to use units of $MHz/(V/cm)^2$, where the energy shift, ΔW, is expressed in frequency units with a factor of h understood. The polarizability is $-2\,\Delta W/E^2$. The conversion into cgs units is $\alpha(cm^3) = 5.95531 \times 10^{-16} \times \alpha[MHz/(V/cm)^2]$.

The polarizability appears in many formulas for low-energy processes involving the valence electrons of atoms or molecules. These formulas are given below in cgs units: the polarizability α is in cm^3; masses m or μ are in grams; energies are in ergs; and electric charges are in esu, where $e = 4.8032 \times 10^{-10}$ esu. The symbol α(ν) denotes a frequency (ν) dependent polarizability, where α(ν) reduces to the static polarizability α for ν = 0. For further information and references, see T. M. Miller and B. Bederson, *Advances in Atomic and Molecular Physics*, Vol. 13, pp. 1—55, 1977. Details on polarizability-related interactions, especially in regard to hyperpolarizabilities and nonlinear optical phenomena, are given by M. P. Bogaard and B. J. Orr, in *Physical Chemistry*, Series Two, Vol. 2, Molecular Structure and Properties, A. D. Buckingham, Ed., Butterworths, London, 1975, pp. 149—194. A tabulation of tensor and hyperpolarizabilities is included. The number density in Table 1 is usually taken to be that of 1 atmosphere at STP in reporting experimental data.

Table 1
FORMULA INVOLVING POLARIZABILITY

Description	Formula	Remarks
Lorentz-Lorenz relation	$\alpha(\nu) = \frac{3}{4\pi n}\left[\frac{\eta^2(\nu) - 1}{\eta^2(\nu) + 2}\right]$	For a gas of atoms or nonpolar molecules; the gas number density is n, and the index of refraction is $\eta(\nu)$
Refraction by polar molecules	$\alpha(\nu) + \frac{d^2}{3kT} = \frac{3}{4\pi n}\left[\frac{\eta^2(\nu)-1}{\eta^2(\nu)+2}\right]$	The dipole moment is d, in esu·cm ($= 10^{-18}$ D)
Dielectric constant (dimensionless)	$K(\nu) = 1 + 4\pi n\ \alpha(\nu)$	From the Lorentz-Lorenz relation for the usual case of $K(\nu) \approx 1$
Index of refraction (dimensionless)	$\eta(\nu) = 1 + 2\pi n\ \alpha(\nu)$	From $\eta^2(\nu) = K(\nu)$
Diamagnetic susceptibility	$\chi_m \simeq e^2(a_oN_e\alpha)^{1/2}/4m_ec^2$	From the approximation that the static polarizability is given by the variational formula $\alpha = (4/9a_o)\sum_i (N_i r_i^2)^2$; N_e is the number of electrons, m_e is the electron mass, and a_o is the Bohr radius, 5.292×10^{-9} cm; a crude approximation is $\chi_m = (E_i/4m_ec^2)\alpha$ where E_i is the ionization energy
Long-range electron- or ion-molecule interaction energy	$V(r) = -e^2\alpha/2r^4$	The target molecule polarizability is α
Ion mobility in a gas	$K = 13.87/(\alpha\mu)^{1/2}$ $cm^2/(volt \cdot sec)$	This one formula is not in cgs units. It gives mobility in usual units of $cm^2/(volt \cdot sec)$ when the gas molecule α is in $Å^3$ or 10^{-24} cm^3 units and the reduced mass μ of the ion-molecule pair is in amu; classical limit; pure polarization potential
Langevin capture cross section	$\sigma(v_o) = \frac{2\pi e}{v_o}\left(\frac{\alpha}{\mu}\right)^{1/2}$	The relative velocity of approach for an ion-molecular pair is v_o; the target molecular polarizability is α and the reduced mass of the ion-molecular pair is μ
Langevin reaction rate coefficient	$k = 2\pi e\left(\frac{\alpha}{\mu}\right)^{1/2}$	"Gas kinetic" rate coefficient for an ion-molecule reaction
Rate coefficient for polar molecules	$k_d = 2\pi e\left[\left(\frac{\alpha}{\mu}\right)^{1/2} + cd\left(\frac{2}{\mu\pi kT}\right)^{1/2}\right]$	The dipole moment of the neutral is d is esu·cm; the number c is a "locking factor" that depends on α and d, and is between 0 and 1
Modified effective range cross section for electron-neutral scattering	$\sigma(k) = 4\pi A^2 + \frac{8\pi^2\mu}{3\hbar^2}e^2\alpha Ak + \ldots$	Here, k is the wavenumber of the electron, equal to its momentum divided by $\hbar$, and $\hbar$ is Planck's constant divided by 2π; A is called the "scattering length;" The reduced mass is μ
van der Waals constant between two systems A,B	$C_6 = \frac{3}{2}\left(\frac{\alpha^A\alpha^B E^A E^B}{E^A + E^B}\right)$	For the interaction potential term $V_6(r) = -C_6r^{-6}$; the dipole polarizability is α and E is an average dipole transition energy
Dipole-quadrupole constant between two systems A,B	$C_8 = \frac{15}{4}\left(\frac{\alpha^A\alpha_q^B E^A E_q^B}{E^A + E_q^B}\right) + \frac{15}{4}\left(\frac{\alpha_q^A\alpha^B E_q^A E^B}{E_q^A + E^B}\right)$	For the interaction potential term $V_8(r) = -C_8r^{-8}$; the dipolar polarizability is α and the quadrupole polarizability (the next higher order polarizability beyond dipole) is α_q; E and E_q are average dipole and quadrupole transition energies, respectively
Van der Waals constant between an atom and a surface	$C_3 = \frac{\alpha g\ E^AE^S}{8(E^A + E^S)}$	For an interaction potential $V_3(r) = -C_3r^{-3}$; E^A and E^S are characteristic energies of the atom and surface; g = 1 for a free-electron metal and $g = (\epsilon_\infty - 1)/(\epsilon_\infty + 1)$ for an ionic crystal
Relation between $\alpha(\nu)$ and oscillator strengths	$\alpha(\nu) = \frac{e^2\hbar^2}{m_e}\sum_k \frac{f_k}{(E_k)^2 - (h\nu)^2}$	Here, f_k is the oscillator strength from the ground state to an excited state k; E_k is the excitation energy of state k; this formula is especially useful for static polarizabilities ($\nu = 0$)
Dynamic polarizability	$\alpha(\nu) = \frac{\alpha\ E_r^2}{E_r^2 - (h\nu)^2}$	Approximate variation of the frequency-dependent polarizability $\alpha(\nu)$ from zero frequency up to the first dipole-allowed electronic transition, of energy E_r; the static dipole polarizability is α, i.e., for $\nu = 0$; ignoring infrared contributions

Table 1 (continued)
FORMULA INVOLVING POLARIZABILITY

Description	Formula	Remarks
Rayleigh scattering cross section	$\sigma(\nu) = \left(\frac{8\pi}{9c^4}\right)(2\pi\nu)^4 \times \left[3\alpha^2(\nu) + \frac{2}{3}\gamma^2(\nu)\right]$	The photon frequency is ν; the average polarizability is $\alpha(\nu)$ and the polarizability anisotropy (the difference between polarizabilities parallel and perpendicular to the applied field) is $\gamma(\nu)$
Verdet constant	$V(\nu) = \frac{\nu n}{2mc^2}\left(\frac{d\alpha(\nu)}{d\nu}\right)$	Defined from $\theta = V(\nu)B$, where θ is the angle of rotation of linearly polarized light through a medium of number density n, per unit length, for a longitudinal magnetic field strength B (Faraday effect)

Table 2
STATIC AVERAGE ELECTRIC DIPOLE POLARIZABILITIES FOR GROUND STATE ATOMS

Atomic Number	Atom	Polarizability (units of 10^{-24} cm^3)	Estimated accuracy (%)	Method	Ref.
1	H	0.666793	"exact"	Calc	MB77
2	He	0.204956	"exact"	Calc	MB77
		0.2050	0.1	Index/diel	NB65/OC67
3	Li	24.3	2	Beam	MB77
4	Be	5.60	2	Calc	MB77
5	B	3.03	2	Calc	MB77
6	C	1.76	2	Calc	MB77
7	N	1.10	2	Calc/index	MB77
8	O	0.802	2	Calc/index	MB77
9	F	0.557	2	Calc	MB77
10	Ne	0.3956	0.1	Diel	OC67
11	Na	23.6	2	Beam	MB77
12	Mg	10.6	2	Calc	MB77
13	Al	8.34	2	Calc	MB77
14	Si	5.38	2	Calc	MB77
15	P	3.63	2	Calc	MB77
16	S	2.90	2	Calc	MB77
17	Cl	2.18	2	Calc	MB77
18	Ar	1.6411	0.05	Index/diel	NB65/OC67
19	K	43.4	2	Beam	MB77
20	Ca	22.8	2	Calc	MB77
		25.0	8	Beam	MB77
21	Sc	17.8	25	Calc	D84
22	Ti	14.6	25	Calc	D84
23	V	12.4	25	Calc	D84
24	Cr	11.6	25	Calc	D84
25	Mn	9.4	25	Calc	D84
26	Fe	8.4	25	Calc	D84
27	Co	7.5	25	Calc	D84
28	Ni	6.8	25	Calc	D84
29	Cu	6.1	25	Calc	D84
		7.31	25	Calc	G84
30	Zn	7.1	25	Calc	MB77
		5.6	25	Calc	D84
31	Ga	8.12	2	Calc	MB77
32	Ge	6.07	2	Calc	MB77
33	As	4.31	2	Calc	MB77
34	Se	3.77	2	Calc	MB77
35	Br	3.05	2	Calc	MB77
36	Kr	2.4844	0.05	Diel	OC67
37	Rb	47.3	2	Beam	MB77
38	Sr	27.6	8	Beam	MB77
39	Y	22.7	25	Calc	D84
40	Zr	17.9	25	Calc	D84
41	Nb	15.7	25	Calc	D84
42	Mo	12.8	25	Calc	D84
43	Tc	11.4	25	Calc	D84
44	Ru	9.6	25	Calc	D84
45	Rh	8.6	25	Calc	D84
46	Pd	4.8	25	Calc	D84
47	Ag	7.2	25	Calc	D84
		8.56	25	Calc	G84
48	Cd	7.2	25	Calc	D84

Table 2 (continued)
STATIC AVERAGE ELECTRIC DIPOLE POLARIZABILITIES FOR GROUND STATE ATOMS

Atomic Number	Atom	Polarizability (units of 10^{-24} cm^3)	Estimated accuracy (%)	Method	Ref.
49	In	10.2	12	Beam	GMBSJ84
		9.1	25	Calc	D84
50	Sn	7.7	25	Calc	D84
51	Sb	6.6	25	Calc	D84
52	Te	5.5	25	Calc	D84
53	I	5.35	25	Index	A56
		4.7	25	Calc	D84
54	Xe	4.044	0.5	Diel	MB77
55	Cs	59.6	2	Beam	MB77
56	Ba	39.7	8	Beam	MB77
57	La	31.1	25	Calc	D84
58	Ce	29.6	25	Calc	D84
59	Pr	28.2	25	Calc	D84
60	Nd	31.4	25	Calc	D84
61	Pm	30.1	25	Calc	D84
62	Sm	28.8	25	Calc	D84
63	Eu	27.7	25	Calc	D84
64	Gd	23.5	25	Calc	D84
65	Tb	25.5	25	Calc	D84
66	Dy	24.5	25	Calc	D84
67	Ho	23.6	25	Calc	D84
68	Er	22.7	25	Calc	D84
69	Tm	21.8	25	Calc	D84
70	Yb	21.0	25	Calc	D84
71	Lu	21.9	25	Calc	D84
72	Hf	16.2	25	Calc	D84
73	Ta	13.1	25	Calc	D84
74	W	11.1	25	Calc	D84
75	Re	9.7	25	Calc	D84
76	Os	8.5	25	Calc	D84
77	Ir	7.6	25	Calc	D84
78	Pt	6.5	25	Calc	D84
79	Au	5.8	25	Calc	D84
		6.48	25	Calc	G84
80	Hg	5.7	25	Calc	D84
		5.1	15	Diel	MB77
81	Tl	7.6	15	Beam	NYU84
		7.5	25	Calc	D84
82	Pb	6.8	25	Calc	D84
83	Bi	7.4	25	Calc	D84
84	Po	6.8	25	Calc	D84
85	At	6.0	25	Calc	D84
86	Rn	5.3	25	Calc	D84
87	Fr	48.7	25	Calc	D84
88	Ra	38.3	25	Calc	D84
89	Ac	32.1	25	Calc	D84
90	Th	32.1	25	Calc	D84
91	Pa	25.4	25	Calc	D84
92	U	27.4	25	Calc	D84
93	Np	24.8	25	Calc	D84
94	Pu	24.5	25	Calc	D84
95	Am	23.3	25	Calc	D84
96	Cm	23.0	25	Calc	D84
97	Bk	22.7	25	Calc	D84
98	Cf	20.5	25	Calc	D84
99	Es	19.7	25	Calc	D84
100	Fm	23.8	25	Calc	D84
101	Md	18.2	25	Calc	D84
102	No	17.5	25	Calc	D84

Note: Calc = calculated value; Beam = atomic beam deflection technique; Index = determination based on the measured index of refraction; Diel = determination based on the measured dielectric constant.

REFERENCES

A56. **Atoji, M.,** *J. Chem. Phys.*, 25, 174, 1956. Semiempirical method based on molecular polarizabilities and atomic radii.

D84. **Doolen, G. D.,** Los Alamos National Laboratory, unpublished. A relativistic linear response method was used. The method is that described by A. Zangwill and P. Soven, *Phys. Rev. A*, 21, 1561, 1980. Adjustments of less than 10% have been made to these results to bring them into agreement with accurate experimental values where available, for the purpose of presenting "recommended" polarizabilities in Table 2. (T. M. Miller.)

G84 **Gollisch, H.,** *J. Phys. B*, 17, 1463, 1984. Other results and useful references are contained in this paper.

GMBSJ84. **Guella, T. P., Miller, T. M., Bederson, B., Stockdale, J. A. D., and Jaduszliwer, B.,** *Phys. Rev. A*, 29, 2977, 1984.

MB77. **Miller, T. M. and Bederson, B.,** *Adv. At. Mol. Phys.*, 13, 1, 1977. For simplicity, any value in Table 2 which has not changed since this 1977 review is referenced as MB77. Persons interested in original references and further details should consult MB77.
NB65. **Newell, A. C. and Baird, R. D.,** *J. Appl. Phys.*, 36, 3751, 1965.
NYU84. Preliminary value from the New York University group. See GMBSJ84.
OC67. **Orcutt, R. H. and Cole, R. H.,** *J. Chem. Phys.*, 46, 697, 1967. See also the later references from this group, given following Table 2.

Table 3

Average Electric Dipole Polarizabilities for Ground State Diatomic Molecules (in Units of 10^{-24} cm^3)

Molecule	Polarizability	Ref.
BH	3.32*	1
Br_2	7.02	2
CO	1.95	3
Cl_2	4.61	3
Cs_2	91.	4
D_2		
(v=0, J=0)	0.7921*	5
(293°K)	0.7954	6
DCl	2.84	2
F_2	1.38*	7
H_2		
(v=0, J=0)	0.8023*	5
(293°K)	0.8045*	5
(293°K)	0.8042	6
(322°K)	0.8059	8
HBr	3.61	3
HCl	2.63	3
	2.77	2
HD		
(v=0, J=0)	0.7976*	5
HF	2.46	3
HI	5.44	3
	5.35	2
HgCl	7.4*	9
ICl	12.3	2
K_2	61.	4
Li_2	34.	4
LiCl	3.46*	10
LiF	10.8*	11
LiH	3.84*	12
	3.68*	13
	3.88*	14
N_2	1.7403	6, 8
NO	1.70	2
Na_2	30.	4
NaLi	40.	4
O_2	1.5812	6
Rb_2	68.	4

Average Electric Dipole Polarizabilities for Ground State Triatomic Molecules (in Units of 10^{-24} cm^3)

Molecule	Polarizability	Ref.
BeH_2	4.34*	14
CO_2	2.911	8
CS_2	8.74	3
	8.86	2
D_2O	1.26	2
H_2O	1.45	2
H_2S	3.78	3
	3.95	2
HCN	2.59	3
	2.46	2
$HgBr_2$	14.5	2
$HgCl_2$	11.6	2
HgI_2	19.1	2
N_2O	3.03	8
NO_2	3.02	2†
O_3	3.21	2
OCS	5.71	2
	5.2	15
SO_2	3.72	3
	4.28	2

Average Electric Dipole Polarizabilities for Ground State Inorganic Polyatomic Molecules (Larger Than Triatomic) (in Units of 10^{-24} cm^3)

Molecule	Polarizability	Ref.
$AsCl_3$	14.9	2
AsN_3	5.75	2
BCl_3	9.47	2
	9.38	20
BF_3	3.31	2
$(BN_3)_2$	5.73	2
$(BH_2N)_3$	8.	2†
ClF_3	6.32	2
$(CsBr)_2$	54.	16
$(CsCl)_2$	42.	16
$(CsF)_2$	28.	16
$(CsI)_2$	52.	16
$GeCl_4$	15.1	2
GeH_3Cl	6.7	2†
$(HgCl)_2$	14.7	9
$(KBr)_2$	42.	16
$(KCl)_2$	32.	16
$(KF)_2$	25.	16
$(KI)_2$	36.	16
(LiBr)	19.	16
$(LiCl)_2$	13.	16
$(LiF)_2$	7.	16
$(LiI)_2$	23.	16
ND_3	1.70	2
NF_3	3.62	2
NH_3	2.26	3
	2.10	2
	2.81	20
$(NO_2)_2$	6.69	2
$(NaBr)_2$	27.	16
$(NaCl)_2$	23.	16
$(NaF)_2$	21.	16
$(NaI)_2$	27.	16
OsO_4	8.17	2
PCl_3	12.8	2
PF_5	6.10	2
PH_3	4.84	2
$(RbBr)_2$	48.	16
$(RbCl)_2$	43.	16
$(RbF)_2$	41.	16
$(RbI)_2$	46.	16
SF_6	6.54	8
$(SF_5)_2$	13.2	2
SO_3	4.84	2
SO_2Cl_2	10.5	2
SeF_6	7.33	2
SiF_4	5.45	2
SiH_4	5.44	2
$(SiH_3)_2$	11.1	2
$SiHCl_3$	10.7	2
SiH_2Cl_2	8.92	2
SiH_3Cl	7.02	2
$SnBr_4$	22.0	2
$SnCl_4$	18.0	2
	13.8	15
SnI_4	32.3	2
TeF_6	9.00	2
$TiCl_4$	16.4	2
UF_6	12.5	2

Average Electric Dipole Polarizabilities for Ground State Hydrocarbon Molecules (in Units of 10^{-24} cm^3)

Molecule	Polarizability	Ref.
CH_4		
Methane	2.593	8
C_2H_2		
Acetylene	3.33	3
	3.93	2
C_2H_4		
Ethylene	4.252	8
C_2H_6		
Ethane	4.47	3
	4.43	2
C_3H_4		
Propyne	6.18	2
C_3H_6		
Propene	6.26	2
Cyclopropane	5.66	2
C_3H_8		3
Propane	6.29	2
	6.37	
C_4H_6		
1-butyne	7.41	2†
1,3-butadiene	8.64	2
C_4H_8		
1-butene	7.97	2
	8.52	2
trans-2-butene	8.49	2
trans-2,3-epoxy		
butane	8.22*	17
2-methylpropene	8.29	2
C_4H_{10}		
n-butane	8.20	2
C_5H_6		
1,3-cyclopentadiene	8.64	2
C_5H_8		
1-pentyne	9.12	2
trans-1,3-pentadiene	10.0	2
isoprene	9.99	2
C_5H_{10}		
Cyclopentane	9.15	18
C_5H_{12}		
Pentane	9.99	2
Neopentane	10.20	18
C_6H_6		
Benzene	10.32	3
	10.74	2
C_6H_{10}		
1-hexyne	10.9	2†
2-ethyl-1,3-		
butadiene	11.8	2†

Molecule	Polarizability	Ref.
3-methyl-1,3-		
pentadiene	11.8	2†
2-methyl-1,3-		
pentadiene	12.1	2†
2,3-dimethyl-1,3-		
butadiene	11.8	2†
Cyclohexene	10.7	2†
C_6H_{12}		
Cyclohexane	11.0	18
	10.87	15
C_6H_{14}		
n-hexane	11.9	2
C_7H_8		
Toluene	12.3	2
	12.26	15
C_7H_{12}		
1-heptyne	12.8	2†
C_7H_{14}	13.1	
Methylcyclohexane		2
C_7H_{16}	13.7	2
n-heptane		
C_8H_8	15.	2
Styrene		
C_8H_{10}		
Ethylbenzene	14.2	2
o-xylene	14.9	2
p-xylene	14.1	15
m-xylene	14.9	2
	14.2	15
	14.18	15
C_8H_{16}		
Ethylcyclohexane	15.9	2
C_8H_{18}		
n-octane	15.9	2
C_9H_{12}		
Isopropylbenzene	16.0	2†
C_9H_{18}		
Isopropylcyclo-	17.2	2
hexane		
$C_{10}H_8$		
Napthalene	16.5	17
$C_{10}H_{14}$		
t-butylbenzene	17.8	2†
$C_{10}H_{20}$		
t-butylcyclohexane	19.8	2
$C_{14}H_{10}$		
Anthracene	25.4	17
Phenanthracene	38.8*	17

Average Electric Dipole Polarizabilities for Ground State Organic Halides (in Units of 10^{-24} cm^3)

Molecule	Polarizability	Ref.
CBr_2F_2		
Dibromodifluoro-	9.	2†
methane		
$CClF_3$		
Chlorotrifluoro-	5.59	8
methane	5.72	20
CCl_2F_2		
Dichlorodifluoro-	7.81	2
methane	7.93	20
CCl_2O		
Phosgene	7.29	2
CCl_2S		
Thiophosgene	10.2	2
CCl_3F		
Trichlorofluoro-	9.47	2
methane		
CCl_3NO_2		
Trichloronitro-	10.8	2†
methane		
CCl_4		
Carbon tetrachloride	11.2	2
	10.5	3
CF_4		
Carbon tetrafluoride	3.838	8
CF_2O		
Carbonylfluoride	1.88*	17

Molecule	Polarizability	Ref.
$CHBr_3$		
Bromoform	11.8	17
$CHBrF_2$		
Bromodifluoromethane	5.7	2†
$CHClF_2$		
Chlorodifluoro-	5.91	2
methane	6.38	20
$CHCl_2F$		
Dichlorofluoro-	6.82	2
methane		
$CHCl_3$		
Chloroform	9.5	8
CHF_3		
Fluoroform	3.57	8
	3.52	20
$CHFO$		
Fluoroformaldehyde	1.76*	17
CHI_3		
Iodoform	18.0	17
CH_2Br_2		
Dibromomethane	9.32	2
CH_2ClNO_2		
Chloronitromethane	6.9	2†
CH_2Cl_2		
Dichloromethane	6.48	3
	7.93	2

Molecule	Polarizability	Ref.
CH_3Br		
Bromomethane	6.03	2
	5.55	15
	5.87	20
CH_3Cl		
Chloromethane	4.72	8
	5.35	20
CH_3F		
Fluoromethane	2.97	8
CH_3I		
Iodomethane	7.97	2
C_2ClF_5		
Chloropentafluoroethane	6.3	2†
$C_2Cl_2F_4$		
1,2-dichlorotetra-fluorethane	8.5	2†
C_2Cl_3N		
Trichloroacetonitrile	6.10	18
C_2F_6		
Hexafluoroethane	6.82	2
C_2HBr		
Bromoacetylene	7.39	2
C_2HCl		
Chloroacetylene	6.07	2
C_2HCl_5		
Pentachloroethane	14.0	2
$C_2H_2Cl_2$		
cis-dichloroethylene	7.89	2
$C_2H_2Cl_2F_2$		
1,1-dichloro-2,2-di-fluoroethane	8.4	2†
$C_2H_2Cl_2O$		
Chloroacetyl chloride	8.92	2
$C_2H_2Cl_3F$		
1,2,2-trichloro-1-fluoroethane	10.2	2†
$C_2H_2Cl_4$		
1,1,2,2-tetrachloro-ethane	12.1	2†
C_2H_2ClN		
Chloroacetonitrile	6.10	18
$C_2H_2F_2$		
1,1-Difluoroethylene	**5.01**	**20**
C_2H_3Br		
Bromoethylene	7.59	2
C_2H_3Cl	6.41	
Chloroethylene		2
$C_2H_3ClF_2$		
1-chloro-1,1-difluoro-ethane	8.05	2
C_2H_3ClO		
Acetyl chloride	6.62	2
$C_2H_3ClO_2$		
ethyl chloroformate	7.1	2†
$C_2H_3Cl_3$		
1,1,1-trichloroethane	10.7	2
$C_2H_3F_3$		
1,1,1-trifluoroethane	4.4	2†
C_2H_3I		
Iodoethylene	9.3	2†
C_2H_4BrCl		
1-boromo-2-chloroethane	9.5	2†
$C_2H_4Br_2$		
1,2-dibromoethane	10.7	2†
C_2H_4ClF		
1-chloro-2-fluoroethane	6.5	2†
$C_2H_4ClNO_2$		
1-chloro-1-nitroethane	10.9	2
$C_2H_4Cl_2$		
1,1-dichloroethane	8.64	2
1,2-dichloroethane	8.	2†
C_2H_5Br		
Bromoethane	8.05	2
C_2H_5Cl	8.29	2
Chloroethane	6.4	15
	7.27	20
C_2H_5ClO		
2-chloroethanol	7.1	2†
Chloromethoxymethane	7.1	2†
C_2H_5F		
Fluoroethane	4.96	2
C_2H_5I		
Iodoethane	10.0	2
$C_3H_4Cl_2$		
Dichloropropene	10.1	2†
C_3H_5Cl		
Chloropropene	8.3	2
C_3H_5ClO		
Chloroacetone	8.4	2†
$C_3H_5ClO_2$		
Ethyl chloroformate	9.0	2†
$C_3H_6ClNO_2$		
1-chloro-1-nitropropane	10.4	2†
$C_3H_6Cl_2$		
Dichloropropane	10.9	2†
C_3H_7Br		
1-bromopropane	9.4	2†
2-bromopropane	9.6	2†
C_3H_7Cl		
Chloropropane	10.0	2
C_3H_7I		
1-iodopropane	11.5	2†
C_4H_5Cl		
4-chloro-1,2-butadiene	10.0	2†
C_4H_7Cl		
1-chloro-2-methylpropene	10.8	2
$C_4H_8Cl_2$		
1,4-dichlorobutane	12.0	2†
C_4H_9Br		
Bromobutane	13.9	2
C_4H_9Cl		
1-chlorobutane	11.3	2
1-chloro-2-methylpropane	11.1	2
2-chloro-2-methylpropane	12.5	2†
2-chlorobutane	12.4	2
C_4H_9I		
1-iodobutane	13.3	2†
$C_5H_{11}Br$		
1-bromopentane	13.1	2†
$C_5H_{11}Cl$		
1-chloropentane	12.0	2†
$C_6H_2Cl_2O_2$		
2,5-dichloro-1,4-benzoquinone	18.4	2
C_6H_4BrF		
p-bromofluorobenzene	13.4	2†
$C_6H_4ClNO_2$		
Chloronitrobenzene	14.6	2†
$C_6H_4Cl_2$	14.0	2
Dichlorobenzene	14.3	15
C_6H_4FI		
p-fluoroiodobenzene	15.5	2†
$C_6H_4FNO_2$		
p-fluoronitrobenzene	12.8	2†
$C_6H_4F_2$		
m-difluorobenzene	10.3	2†
C_6H_5Br		
Bromobenzene	14.7	2
C_6H_5Cl	14.1	2
Chlorobenzene	12.3	15
C_6H_5ClO		
Chlorophenol	13.0	2†
C_6H_5F		2
Fluorobenzene	10.3	
C_6H_5I		
Iodobenzene	15.5	2†
C_7H_7F		
Fluorotoluene	12.3	2†
$C_7H_{15}Br$		
1-bromoheptane	16.8	2†
$C_{12}H_8Br_2O$		
4,4′-dibromodiphenyl ether	27.8	2†
$C_{12}H_9BrO$		
4-bromodiphenyl ether	24.2	2†
$C_{13}H_{11}BrO$		
p-bromophenyl-p-tolyl ether	26.6	2†

Average Electric Dipole Polarizabilities for Other Ground State Organic Molecules (in Units of 10^{-24} cm^3)

Molecule	Polarizability	Ref.
CN_4O_8		
Tetranitromethane	15.3	2
CH_2O		
Formaldehyde	2.8	2†
	2.45	18
CH_2O_2		
Formic acid	3.4	2†
CH_3NO		
Foramide	4.2	2†
	4.08	18
CH_3NO_2		
Nitromethane	7.37	2
CH_4O		
Methanol	3.29	2
	3.23	15
	3.32	18
CH_5N		
ethyl amine	4.7	2
	4.01	19
C_2N_2		
Cyanogen	7.99	2
C_2H_2O		
Ketene	4.4	2†
C_2H_3N		
Acetonitrile	4.40	2†
	4.48	18
C_2H_4O	4.6	2†
Acetaldehyde	4.59	18
$C_2H_4O_2$		
Acetic acid	5.1	2†
$C_2H_4O_4$		
Formic acid dimer	12.7	2
C_2H_5NO		
Acetamide	5.67	18
-methyl foramide	5.91	18
$C_2H_5NO_2$		
Nitroethane	9.63	2
Ethyl nitrite	7.0	15
C_2H_5O		
Ethylene oxide	4.43	18
C_2H_6O	5.41	2
Ethanol	5.11	18
	5.84	2
Methyl ether	5.16	15
	5.29	20
$C_2H_6O_2$		
Ethylene glycol	5.7	2†
$C_2H_6O_2S$		
Dimethyl sulfone	7.3	2†
C_2H_6S		
Ethanethiol	7.41	2
C_2H_7N		
Ethyl amine	7.10	2
Dimethyl amine	6.37	2
$C_2H_8N_2$		
Ethylene diamine	7.2	2†
$C_3H_2N_2$		
alononitrile	5.79	18
C_3H_3N		
Acrylonitrile	8.05	2
C_3H_4O		
Propenal	6.38	2†
C_3H_5N	6.70	2
Propionitrile	6.24	18
C_3H_6O		
Acetone	6.33	15
	6.4	2†
	6.39	18
Allyl alcohol	7.65	2
Propionaldehyde	6.50	2
$C_3H_6O_2$		
Propionic acid	6.9	2†
Ethyl formate	8.01	2
Methyl acetate	6.94	2
$C_3H_6O_3$		
Dimethyl carbonate	7.7	2†
C_3H_7NO		
-methyl acetamide	7.82	18
,N-dimethyl formamide	7.81	18
$C_3H_7NO_2$		
Nitropropane	8.5	2†
C_3H_8O		
2-propanol	7.61	2
	6.97	18
1-propanol	6.74	2
Methoxyethane	7.93	2
$C_3H_8O_2$		
Dimethoxymethane	7.7	2†
C_3H_9N		
n-propylamine	9.20	2
trimethylamine	8.15	2
$C_4H_2N_2$		
Fumaronitrile	11.8	2
$C_4H_4N_2$		
Succinonitrile	8.1	2†
Pyrimidene	8.53*	17
Pyridazine	9.27*	17
$C_4H_4O_2$		
Diketene	8.0	2†
C_4H_4S		
Thiophene	9.67	2
C_4H_5N		
Methacrylonitrile	8.0	2†
Trans-crotononitrile	8.2	2†
C_4H_6O		
Crotonaldehyde	8.5	2†
Methacrylaldehyde	8.3	2†
$C_4H_6O_2$		
Biacetyl	8.2	2†
$C_4H_6O_3$		
Acetic anhydride	8.9	2†
C_4H_6S		
Divinyl sulfide	10.9	2†
C_4H_7N		
Butyronitrile	8.4	2†
Isobutyronitrile	8.05	18
C_4H_8O		
Butyraldehyde	8.2	2†
Methyl ethyl ketone	8.13	15
$C_4H_8O_2$		
Ethyl acetate	9.7	2
1,4-dioxane	10.0	2
$C_4H_8O_2$		
p-dioxane	8.60	18
2-methyl-1,3-dioxolane	9.44	15
$C_4H_9NO_2$		
1-nitrobutane	10.4	2†
2-methyl-2-nitropropane	10.3	2†
$C_4H_{10}O$		
Ethyl ether	10.2	2
	8.73	15
1-butanol	8.88	2
2-methylpropanol	8.92	2
$C_4H_{10}S$		
Ethyl sulfide	10.8	2
$C_4H_{11}N$		
n-butylamine	13.5	2
Diethylamine	10.2	2
C_5H_5N		
Pyridine	9.5	15
4-cyano-1,3-butadiene	10.5	2†
$C_5H_8O_2$		
Acetyl acetone	10.5	2†
C_5H_9N		
Valeronitrile	10.4	2
22-DMPN	9.59	18
$C_5H_{10}O$		
Diethyl ketone	9.93	15
Methyl propyl ketone	9.93	15
$C_5H_{10}O_3$		
diethyl carbonate	11.3	2
$C_5H_{12}O_2$		
Diethoxyethane	11.3	2†

Molecule	Polarizability	Ref.
$C_5H_{12}O_4$		
Tetramethyl orthocarbonate	13.	2†
$C_6H_4N_2O_4$		
p-dinitrobenzene	18.4	2
$C_6H_4O_2$		
p-benzoquinone	14.5	2
$C_6H_5NO_2$		
nitrobenzene	14.7	2
	12.92	15
C_6H_6O		
Phenol	11.1	2†
	9.94*	17
C_6H_7N		
Aniline	12.1	2†
$C_6H_8N_2$		
Phenylenediamine	13.8	2†
$C_6H_{10}O_3$		
Ethyl acetoacetate	12.9	2†
$C_6H_{12}N_2$		
Dimethylketazine	15.6	2
$C_6H_{12}O$		
Cyclohexanol	11.56	18
$C_6H_{12}O_2$		
Amyl formate	14.2	2
$C_6H_{12}O_3$		
Paraldehyde	17.9	2
$C_6H_{14}O$	12.8	2
Propyl ether	12.5	15
$C_6H_{14}O_2$		
1,1-diethoxyethane	13.2	2†
$C_6H_{15}N$		
Triethylamine	13.1	2
$C_7H_4N_2O_2$		
p-cyanonitrobenzene	19.	2
C_7H_5N		
Benzonitrile	12.5	2†
$C_7H_7NO_3$		
Nitroanisole	15.7	2†
C_7H_7O		
Anisole	13.1	2†
C_7H_9NO		
o-anisidine	14.2	2†
$C_7H_{14}O$		
Cyclohexyl methyl ether	13.4	2†
di-isopropyl ketone	13.5	15
$C_7H_{14}O_2$		
Amyl acetate	14.9	2
$C_8H_4N_2$		
p-dicyanobenzene	19.2	2
C_8H_8O		
Acetophenone	15.0	2
$C_8H_8O_2$		
2,5-dimethyl-1,4-benzoquinone	18.8	2
$C_8H_{10}O$		
Phenetole	14.9	2
$C_8H_{11}N$		
N-dimethylaniline	16.2	2†
$C_8H_{12}O_2$		
Ethyl sorbate	17.2	2†
Tetramethylcyclobutane-1,3-dione	18.6	2
$C_8H_{14}O_4$		
Diethyl succinate	16.8	2†
$C_8H_{18}O$		
n-butyl ether	17.2	2
$C_9H_{10}O_2$		
Ethyl benzoate	16.9	2†
$C_{10}H_{14}BeO_4$		
Beryllium acetylacetonate	34.1	2
$C_{12}H_9NO_3$		
4-nitrodiphenyl ether	24.7	2†
$C_{14}H_{14}O$		
di-p-tolyl ether	24.9	2†
$C_{15}H_{21}AlO_6$		
Aluminum acetylacetonate	51.9	2
$C_{15}H_{21}CrO_6$	53.7	2
Chromium acetylacetonate		
$C_{15}H_{21}FeO_6$		
Ferric acetylacetonate	58.1	2
$C_{20}H_{18}O_8Th$		
Thorium acetylacetonate	79.	2

Note: All polarizabilities in Table 3 are experimental values except those values marked by an asterisk (*), which indicates a calculated result. The experimental polarizabilities are mostly determined by measurements of a dielectric constant or refractive index which are quite accurate (0.5% or better). However, one should treat many of the results with several percent of caution because of the age of the data and because some of the results refer to optical frequencies rather than static. Comments given with the references are intended to allow one to judge the degree of caution required. Interested persons should consult these references. In many cases, the reference given is to a theoretical paper in which the experimental results are quoted. These papers, noted in the References, contain valuable information on polarizability calculations and experimental data which often includes the tensor components of the polarizability.

REFERENCES

1. **McCullough, E. A., Jr.,** *J. Chem. Phys.*, 63, 5050, 1975. This calculation is for α_{zz}, not $\bar{\alpha}$.
2. **Maryott, A. A. and Buckley, F.,** U. S. National Bureau of Standards Circular No. 537, 1953. A tabulation of dipole moments, dielectric constants, and molar refractions measured between 1910 and 1952, and used here to determine polarizabilities if no more recent result exists. The polarizability is $3/(4\pi N_0)$ times the molar polarization or molar refraction, where N_0 is Avogadro's number. The value $3/(4\pi N_0) = 0.3964308 \times 10^{-24}$ cm^3 was used for this conversation. A dagger (†) following the reference number in the present table indicates that the polarizability was derived from the molar refraction and hence may not include some low-frequency contributions to the static polarizability; these "static" polarizabilities are therefore low by 1 to 30%.
3. **Hirschfelder, J. O., Curtis, C. F., and Bird, R. B.,** *Molecular Theory of Gases and Liquids*, John Wiley & Sons, New York, 1954, 950. Fundamental information on molecular polarizabilities.
4. **Miller, T. M. and Bederson, B.,** *Adv. At. Mol. Phys.*, 13, 1, 1977. Review emphasizing atomic polarizabilities and measurement techniques. The data quoted in Table 3 are accurate to 8 to 12%.
5. **Kolos, W. and Wolniewicz, L.,** *J. Chem. Phys.*, 46, 1426, 1967. Highly accurate molecular hydrogen calculations.
6. **Newell, A. C. and Baird, R. C.,** J. Appl. Phys., 36, 3751, 1965. Highly accurate refractive index measurements at 47.7 GHz (essentially static).
7. **Jao, T. C., Beebe, N. H. F., Person, W. B., and Sabin, J. R.,** *Chem. Phys. Lett.*, 26, 47, 1974. Tensor polarizabilities, derivatives, and other results are reported.
8. **Orcutt, R. H. and Cole, R. H.,** *J. Chem. Phys.*, 46, 697, 1967; Sutter, H. and Cole, R. H., *J. Chem. Phys.*, 52, 132, 1970; Bose, T. K. and Cole, R. H., *J. Chem. Phys.*, 52, 140, 1970 and 54, 3829, 1971; Nelson, R. D. and Cole, R. H., *J. Chem. Phys.*, 54, 1971; Bose, T. K., Sochanski, J. S., and Cole, R. H., *J. Chem. Phys.*, 57, 3592, 1972; Kirouac, S. and Bose, T. K., *J. Chem. Phys.*, 59, 3043, 1973 and 64, 1580, 1976. Highly accurate dielectric constant measurements. These modern data give the most accurate polarizabilities available. A criticism of these data in the case of *polar* molecules is given in Ref. 20, p. 2905.
9. **Huestis, D. L.,** Technical Report #MP 78—25, SRI International (project PYU 6158), Menlo Park, CA 94025. Molar refractions for mercury-chlorine compounds are analyzed.
10. **Bounds, D. G., Clarke, J. H. R., and Hinchliffe, A.,** *Chem. Phys. Lett.*, 45, 367, 1977. Theoretical tensor polarizability for LiCl.
11. **Kolker, H. J. and Karplus, M.,** *J. Chem. Phys.*, 39, 2011, 1963. Theoretical.
12. **Gutschick, V. P. and McKoy, V.,** *J. Chem. Phys.*, 58, 2397, 1973. Theoretical tensor polarizabilities.
13. **Gready, J. E., Bacskay, G. B., and Hush, N. S.,** *Chem. Phys.*, 22, 141, 1977 and 23, 9, 1977. Theoretical.
14. **Amos, A. T. and Yoffe, J. A.,** *J. Chem. Phys.*, 63, 4723, 1975. Theoretical.

15. **Stuart, H. A.,** *Landolt-Börnstein Zahlenwerte und Funktionen*, Vol. 1, Part 3, Eucken, A. and Hellwege, K. H., Eds., Springer-Verlag, Berlin, 1951, p. 511. Tabulation of molecular polarizabilities. Two misprints in the chemical symbols have been corrected.
16. **Kremens, R., Bederson, B., Jaduszliwer, B., Stockdale, J., and Tino, A.,** *J. Chem. Phys.*, 81, 1676, 1984. Guella, T. P., Miller, T. M., Bederson, B., and Jaduszliwer, B., *Bull. Am. Phys. Soc.*, 29, 797, 1984; also, to be published. Average polarizability measurements and semi-empirical calculations. The data quoted in Table 3 are accurate to about 14%.
17. **Marchese, F. T. and Jaffé,** *Theoret. Chim. Acta (Berlin)*, 45, 241, 1977. Theoretical and experimental tensor polarizabilities are tabulated in this paper.
18. **Applequist, J., Carl, J. R., and Fung, K.-K.,** *J. Am. Chem. Soc.*, 94, 2952, 1972. Excellent reference on the calculation of molecular polarizabilities, including extensive tables of tensor polarizabilities, both theoretical and experimental, at 589.3 nm wavelength.
19. **Bridge, N. J. and Buckingham, A. D.,** *Proc. R. Soc. (London)*, A295, 334, 1966. Measured tensor polarizabilities at 633 nm wavelength.
20. **Barnes, A. N. M., Turner, D. J., and Sutton, L. E.,** Dielectric constants yielding polarizabilities accurate from 0.3—8%, *Trans. Faraday Soc.*, 67, 2902, 1971.

IONIZATION POTENTIALS[a]

Z	Element	Spectrum I	II	III	IV	V	VI	VII	VIII	IX	X	XI	XII	XIII	XIV	XV	XVI	XVII	XVIII	XIX	XX	XXI
1	H	13.598																				
2	He	24.587	54.416																			
3	Li	5.392	75.638	122.451																		
4	Be	9.322	18.211	153.893	217.713																	
5	B	8.298	25.154	37.930	259.368	340.217																
6	C	11.260	24.383	47.887	64.492	392.077	489.981															
7	N	14.534	29.601	47.448	77.472	97.888	552.057	667.029														
8	O	13.618	35.116	54.934	77.412	113.896	138.116	739.315	871.387													
9	F	17.422	34.970	62.707	87.138	114.240	157.161	185.182	953.886	1103.089												
10	Ne	21.564	40.962	63.45	97.11	126.21	157.93	207.27	239.09	1195.797	1362.164											
11	Na	5.139	47.286	71.64	98.91	138.39	172.15	208.47	264.18	299.87	1465.091	1648.659										
12	Mg	7.646	15.035	80.143	109.24	141.26	186.50	224.94	265.90	327.95	367.53	1761.802	1962.613									
13	Al	5.986	18.828	28.447	119.99	153.71	190.47	241.43	284.59	330.21	398.57	442.07	2085.983	2304.080								
14	Si	8.151	16.345	33.492	45.141	166.77	205.05	246.52	303.17	351.10	401.43	476.06	523.50	2437.676	2673.108							
15	P	10.486	19.725	30.18	51.37	65.023	230.43	263.22	309.41	371.73	424.50	479.57	560.41	611.85	2816.943	3069.762						
16	S	10.360	23.33	34.83	47.30	72.68	88.049	280.93	328.23	379.10	447.09	504.78	564.65	651.63	707.14	3223.836	3494.099					
17	Cl	12.967	23.81	39.61	53.46	67.8	98.03	114.193	348.28	400.05	455.62	529.26	591.97	656.69	749.74	809.39	3658.425	3946.193				
18	Ar	15.759	27.629	40.74	59.81	75.02	91.007	124.319	143.456	422.44	478.68	538.95	618.24	686.09	755.73	854.75	918	4120.778	4426.114			
19	K	4.341	31.625	45.72	60.91	82.66	100.0	117.56	154.86	175.814	503.44	564.13	629.09	714.02	787.13	861.77	968	1034	4610.955	4933.931		
20	Ca	6.113	11.871	50.908	67.10	84.41	108.78	127.7	147.24	188.54	211.270	591.25	656.39	726.03	816.61	895.12	974	1087	1157	5129.045	5469.738	
21	Sc	6.54	12.80	24.76	73.47	91.66	111.1	138.0	158.7	180.02	225.32	249.832	685.89	755.47	829.79	926.00						
22	Ti	6.82	13.58	27.491	43.266	99.22	119.36	140.8	168.5	193.2	215.91	265.23	291.497	787.33	861.33	940.36						
23	V	6.74	14.65	29.310	46.707	65.23	128.12	150.17	173.7	205.8	230.5	255.04	308.25	336.267	895.58	974.02						
24	Cr	6.766	16.50	30.96	49.1	69.3	90.56	161.1	184.7	209.3	244.4	270.8	298.0	355	384.30	1010.64						
25	Mn	7.435	15.640	33.667	51.2	72.4	95	119.27	196.46	221.8	248.3	286.0	314.4	343.6	404	435.3	1136.2					
26	Fe	7.870	16.18	30.651	54.8	75.0	99	125	151.06	235.04	262.1	290.4	330.8	361.0	392.2	457	489.5	1266.1				
27	Co	7.86	17.06	33.50	51.3	79.5	102	129	157	186.13	276	305	336	379	411	444	512	546.8	1403.0			
28	Ni	7.635	18.168	35.17	54.9	75.5	108	133	162	193	224.5	321.2	352	384	430	464	499	571	607.2	1547		
29	Cu	7.726	20.292	36.83	55.2	79.9	103	139	166	199	232	266	368.8	401	435	484	520	557	633	671	1698	
30	Zn	9.394	17.964	39.722	59.4	82.6	108	134	174	203	238	274	310.8	419.7	454	490	542	579	619	698	738	1856
31	Ga	5.999	20.51	30.71	64																	
32	Ge	7.899	15.934	34.22	45.71	93.5																
33	As	9.81	18.633	28.351	50.13	62.63	127.6															
34	Se	9.752	21.19	30.820	42.944	68.3	81.70	155.4														
35	Br	11.814	21.8	36	47.3	59.7	88.6	103.0	192.8													
36	Kr	13.999	24.359	36.95	52.5	64.7	78.5	111.0	126	230.39												
37	Rb	4.177	27.28	40	52.6	71.0	84.4	99.2	136	150	277.1											
38	Sr	5.695	11.030	43.6	57	71.6	90.8	106	122.3	162	177	324.1										
39	Y	6.38	12.24	20.52	61.8	77.0	93.0	116	129	146.52	191	206	374.0									
40	Zr	6.84	13.13	22.99	34.34	81.5																
41	Nb	6.88	14.32	25.04	38.3	50.55	102.6	125														
42	Mo	7.099	16.15	27.16	46.4	61.2	68	126.8	153													
43	Te	7.28	15.26	29.54																		
44	Ru	7.37	16.76	28.47																		
45	Rh	7.46	18.08	31.06																		
46	Pd	8.34	19.43	32.93																		
47	Ag	7.576	21.49	34.83																		

Z Element	Spectrum																				
	I	II	III	IV	V	VI	VII	VIII	IX	X	XI	XII	XIII	XIV	XV	XVI	XVII	XVIII	XIX	XX	XXI
48 Cd	8.993	16.908	37.48																		
49 In	5.786	18.869	28.03	54																	
50 Sn	7.344	14.632	30.502	40.734	72.28																
51 Sb	8.641	16.53	25.3	44.2	56	108															
52 Te	9.009	18.6	27.96	37.41	58.75	70.7	137														
53 I	10.451	19.131	33																		
54 Xe	12.130	21.21	32.1																		
55 Cs	3.894	25.1																			
56 Ba	5.212	10.004																			
57 La	5.577	11.06	19.175																		
58 Ce	5.47	10.85	20.20	36.72																	
59 Pr	5.42	10.55	21.62	38.95	57.45																
60 Nd	5.49	10.72																			
61 Pm	5.55	10.90																			
62 Sm	5.63	11.07																			
63 Eu	5.67	11.25																			
64 Gd	6.14	12.1																			
65 Tb	5.85	11.52																			
66 Dy	5.93	11.67																			
67 Ho	6.02	11.80																			
68 Er	6.10	11.93																			
69 Tm	6.18	12.05	23.71																		
70 Yb	6.254	12.17	25.2																		
71 Lu	5.426	13.9																			
72 Hf	7.0	14.9	23.3	33.3																	
73 Ta	7.89																				
74 W	7.98																				
75 Re	7.88																				
76 Os	8.7																				
77 Ir	9.1																				
78 Pt	9.0	18.563																			
79 Au	9.225	20.5																			
80 Hg	10.437	18.756	34.2																		
81 Tl	6.108	20.428	29.83																		
82 Pb	7.416	15.032	31.937	42.32	68.8																
83 Bi	7.289	16.69	25.56	45.3	56.0	88.3															
84 Po	8.42																				
85 At																					
86 Rn	10.748																				
87 Fr																					
88 Ra	5.279	10.147																			
89 Ac	6.9	12.1																			
90 Th		11.5	20.0	28.8																	
91 Pa																					
92 U																					
93 Np																					
94 Pu	5.8																				
95 Am	6.0																				

[a] Numerical values in this table are expressed in electron volts. The conversion factor used in converting the spectral data to electron volts was 1 eV = 8065.73 cm^{-1}.

From Moore, C. E., *Analyses of Optical Spectra,* NSRDS-NBS 34, Office of Standard Reference Data, National Bureau of Standards, Washington, D.C.

IONIZATION POTENTIALS OF MOLECULES

From data published up to July, 1966
Condensed by J. L. Franklin and Pat Haug from a compilation entitled
"Ionization Potentials, Appearance Potentials and Heats of Formation of Positive Ions"
by
J. L. Franklin, J. G. Dillard, H. M. Rosenstock, J. T. Herron, K. Draxl and F. H. Field
Published by National Standard Reference Data System

The following symbols are employed for the principal important methods:

CTS = Charge Transfer Spectra, EI = Electron Impact, PE = Photoelectron Spectroscopy, PI = Photoionization, S = Optical Spectroscopy.

Molecule	Ionization Potential ev	Method	ΔH_f of Ion Kcal/mole	Reference
H_2	15.427 ±0.002	S	356	*1*
D_2	15.46 ±0.01	PI	356	*2*
BH	9.77 ±0.05	S	333	*3*
BH_3	11.4 ±0.2	EI	279	*
B_2H_6	12.0	EI	286	*4, 5*
B_5H_9	10.5	EI	262	*
B_6H_{10}	9.3 ±0.1	EI	237	*6*
C_2	12.0 ±0.6	EI	475	*7*
C_3	12.6	EI	480	*7*
CH	11.13 ±0.22	S	399	*8*
CH_2	10.396 ±0.003	EI, S, PI	333	*9, 10, 11*
CH_3	9.83	S, PI	259	*12, 13*
CD_3	9.832 ±0.002	S	259	*12*
CH_4	12.6	PI	274	*
CD_4	12.888	PI	280	*11, 14*
C_2H_2	11.4	PI, PE	317	*
C_2D_2	11.416 ±0.006	PI	317	*15*
C_2H_3	9.4	EI	269	*
C_2H_4	10.5	S, PE	253	*16, 17*
C_2H_5	8.4	PI	219	*13*
C_2H_6	11.5	PI, PE	245	*14, 17*
$HC\equiv C-CH_2$	8.25	PI, EI		*18*
$CH_2=C=CH_2$	10.16 ±0.02	EI	280	*19*
$CH_3C\equiv CH$	10.36 ±0.01	PI	283	*20*
cyclo-C_3H_4	9.95	EI	296	*21*
C_3H_5 (allyl)	8.15	EI	216	*22, 8*
cyclo-C_3H_5	8.05	EI	239	*23*
C_3H_6	9.73	S, PI	229	*
cyclo-C_3H_6	10.09 ±0.02	PI	245	*24*
n-C_3H_7	8.1	PI	209	*13*
iso-C_3H_7	7.5	PI	190	*13*
C_3H_8	11.1	PI, PE	231	*
C_4H_2	10.2 ±0.1	EI		*25*
C_4H_4	9.87	EI	294	*25, 26*
$CH_3CH=C=CH_2$	9.57 ±0.02	EI	259	*19*
$CH_2=CHCH=CH_2$	9.07	PI, PE	236	*
$C_2H_5C\equiv CH$	10.18 ±0.01	PI	274	*20*
$CH_3C\equiv CCH_3$	9.9 ±0.1	EI	263	*25*
cyclo-C_4H_7	7.88 ±0.05	EI	213	*23*
$CH_3CH=CHCH_2$	7.71 ±0.05	EI	203	*27, 8*
$CH_2=C(CH_3)CH_2$	8.03 ±0.05	EI	203	*27, 8*
1-C_4H_8	9.6	PI	221	*
cis-2-C_4H_8	9.13	PI	209	*24, 28*
trans-2-C_4H_8	9.13	PI	208	*24, 28*
iso-C_4H_8	9.23 ±0.02	PI	209	*24, 28*
cyclo-C_4H_8	10.58	EI	250	*23*
n-C_4H_9	8.64 ±0.05	EI	218	*29*
sec-C_4H_9	7.93 ±0.05	EI	192	*29*
iso-C_4H_9	8.35 ±0.05	EI	205	*29*
tert-C_4H_9	7.42 ±0.07	EI	176	*29*
n-C_4H_{10}	10.63 ±0.03	PI	215	*30*
iso-C_4H_{10}	10.57	PI	212	*24*
cyclo-C_5H_6	8.97	EI	239	*31, 32*
$C_2H_5CH=C=CH_2$	9.42	EI	252	*21*
$CH_3CH=CHCH=CH_2$	8.68	EI	219	*21*
$CH_3CH=C=CHCH_3$	9.26	EI	247	*21*
$CH_2=CHCH_2CH=CH_2$	9.58	EI	246	*21*
$CH_2=CHC(CH_3)=CH_2$	8.845 ±0.005	PI	235	*24*
cyclo-C_5H_8	9.01 ±0.01	PI	216	*24*
cyclo-C_5H_9	7.79 ±0.02	EI	194	*23*
1-C_5H_{10}	9.50 ±0.02	PI	214	*24*
cis-2-C_5H_{10}	9.11	EI	203	*21*
trans-2-C_5H_{10}	9.06	EI	201	*21*
$(CH_3)_2CHCH=CH_2$	9.51 ±0.03	PI	212	*24*
$C_2H_5C(CH_3)=CH_2$	9.12 ±0.02	PI	202	*24*
$(CH_3)_2C=CHCH_3$	8.67 ±0.02	PI	189	*24*
cyclo-$C_3H_4(CH_3)_2$	9.77 ±0.02	EI	225	*33*
cyclo-C_5H_{10}	10.53 ±0.05	PI	224	*24*
tert-C_5H_{11}	7.12 ±0.1	EI	164	*34, 35*
neo-C_5H_{11}	8.33 ±0.1	EI	196	*34*
n-C_5H_{12}	10.35	PI	204	*24*
iso-C_5H_{12}	10.32	PI	201	*24*
neo-C_5H_{12}	10.35	PI	199	*24*
C_6H_4 (benzyne)	9.6	EI		*
cyclo-C_6H_5	9.2	PI, PE	284	*36, 37*
cyclo-C_6H_6	9.24	S, PI	233	*
$CH\equiv CCH=CHCH=CH_2$	9.50	EI	307	*38*
$C_2H_5C\equiv CC\equiv CH$	9.25	EI	307	*38*
$CH_3C\equiv CCH_2C\equiv CH$	9.75	EI	319	*38*
$CH_3C\equiv CC\equiv CCH_3$	9.20	EI	301	*38*
$CH\equiv CCH_2CH_2C\equiv CH$	10.35	EI	338	*38*
C_6H_8 (1-methylcyclopentadiene)	8.43 ±0.05	EI	218	*31*
C_6H_8 (2-methylcyclopentadiene)	8.46 ±0.05	EI	219	*31*
cyclo-C_6H_{10}	8.72	PE	199	*17*
cyclo-C_6H_{11}	7.7	EI	185	*23*
1-C_6H_{12}	9.45 ±0.02	PI	208	*35, 24*
$(CH_3)_2C=C(CH_3)_2$	8.30	PI	175	*28*
cyclo-C_6H_{12}	9.8	PI, PE	197	*17, 24*
n-C_6H_{14}	10.18	PI	195	*24*
iso-C_6H_{14}	10.12	PI	192	*24*
$(C_2H_5)_2CHCH_3$	10.08	PI	191	*24*
$C_2H_5C(CH_3)_3$	10.06	PI	188	*24*
$(CH_3)_2CHCH(CH_3)_2$	10.02	PI	189	*24*
cyclo-$C_6H_5CH_2$	7.76 ±0.08	EI	216	*39*
cyclo-C_7H_7	6.240 ±0.01	S	209	*40*
cyclo-$C_6H_5CH_3$	8.82 ±0.01	PI	215	*
cyclo-C_7H_8	8.5	EI	240	*32, 41, 42*
bicyclo-(2.2.1)C_7H_8	8.67	EI	267	*42*
bicyclo-(3.2.0)C_7H_8	9.37	EI	246	*41*
C_7H_{10} (1,2-dimethylcyclopentadiene)	8.1 ±0.1	EI	204	*31*
C_7H_{10} (5,5-dimethylcyclopentadiene)	8.22 ±0.05	EI	206	*31*
C_7H_{10} (1,3-cycloheptadiene)	8.55	EI	219	*41*
C_7H_{10} (norbornene)	8.95 ±0.15	EI	237	*43*
C_7H_{12} (4-methylcyclohexane)	8.91 ±0.01	PI	198	*24*
cyclo-$C_6H_{11}CH_3$	9.85 ±0.03	PI	190	*24*
n-C_7H_{16}	9.90 ±0.05	PI	183	*37*
cyclo-$C_6H_5C\equiv CH$	8.815 ±0.005	PI	279	*24*
C_8H_8 (styrene)	8.47 ±0.02	PI	232	*24*
C_8H_8 (cyclotatetraene)	8.0	PI, PE	255	*17, 24*
C_8H_8 (cubane)	8.74 ±0.15	EI	350	*44*
m-$C_6H_4CH_3CH_2$	7.65 ±0.03	EI	206	*39*
p-$C_6H_4CH_3CH_2$	7.46 ±0.03	EI	202	*39*
cyclo-$C_6H_5C_2H_5$	8.76 ±0.01	PI	209	*24*
o-$C_6H_4(CH_3)_2$	8.56	PI	202	*
m-$C_6H_4(CH_3)_2$	8.58	PI, PE	202	*
p-$C_6H_4(CH_3)_2$	8.44	PI	199	*
C_8H_{10} (7-methylcycloheptatriene)	8.39 ±0.1	EI	231	*45*
C_8H_{10} (1-methylspiroheptadiene)	8.02 ±0.1	EI	229	*45*
C_8H_{10} (2-methylspiroheptadiene)	8.07 ±0.1	EI	230	*45*
C_8H_{10} (6-methylspiroheptadiene)	8.40 ±0.1	EI	239	*45*
C_8H_{12} (1,2,3-trimethylcyclopentadiene)	7.96 ±0.05	EI	194	*31*
C_8H_{12} (1,5,5-trimethylcyclopentadiene)	8.00 ±0.1	EI	193	*31*

* Average of several values.

Molecule	Ionization Potential ev	Method	ΔH_f of Ion Kcal/mole	Reference
C_8H_{12} (4-vinylcyclohexene)	8.93 ±0.02	PI	224	*24*
cis-1,2-cyclo-$C_6H_{10}(CH_3)_2$	10.08 ±0.02	EI	191	*46*
trans-1,2-cyclo-$C_6H_{10}(CH_3)_2$	10.08 ±0.03	EI	189	*46*
C_8H_{18} (2,2,4-trimethylpentane)	9.86	PI	174	*24*
C_8H_{18} (2,2,3,3-tetramethylbutane)	9.79	EI	184	*8*
C_9H_8 (indene)	8.81	EI	246	*47*
$C_6H_5C(CH_3){=}CH_2$	8.35 ±0.01	PI	220	*24*
cyclo-C_6H_5(n-C_3H_7)	8.72 ±0.01	PI	203	*24*
cyclo-C_6H_5(iso-C_3H_7)	8.69 ±0.01	PI	201	*24*
1,2,3 cyclo-$C_6H_3(CH_3)_3$	8.48	PI	193	*28*
1,2,4 cyclo-$C_6H_3(CH_3)_3$	8.27	PI	187	*28*
1,3,5 cyclo-$C_6H_3(CH_3)_3$	8.4	PI	190	*
$C_{10}H_8$ (naphthalene)	8.12	PI	220	*24, 48*
$C_{10}H_8$ (azulene)	7.42	S	243	*49, 50*
$C_6H_4C_3H_7CH_2$	7.42	EI	188	*39*
cyclo-C_6H_5 (n-C_4H_9)	8.69 ±0.01	PI	197	*24*
$C_{10}H_{14}$ (sec-butylbenzene)	8.68 ±0.01	PI	196	*24*
$C_{10}H_{14}$ (tert-butylbenzene)	8.68 ±0.01	PI	193	*24*
$C_{10}H_{14}$ (1,2,3,5-tetramethylbenzene)	8.47 ±0.05	EI	185	*24*
$C_{10}H_{14}$ (1,2,4,5-tetramethylbenzene)	8.03	PI, PE	174	*
$C_{10}H_{18}$ (cis-decaline)	9.61 ±0.02	EI	181	*51*
$C_{10}H_{18}$ (trans-decaline)	9.61 ±0.02	EI	178	*51*
$C_{11}H_9$ (1-naphthyl methyl)	7.35	EI	208	*52*
$C_{11}H_9$ (2-naphthyl methyl)	7.56 ±0.05	EI	217	*52*
$C_{11}H_{10}$ (methylnaphthalene)	7.96 ±0.01	PI	209	*24*
$C_{11}H_{10}$ (2-methylnaphthalene)	7.955 ±0.01	PI	209	*24*
$C_6H(CH_3)_5$	7.92 ±0.02	PI	155	*53*
$C_{11}H_{18}$ (hexamethylcyclopentadiene)	7.74 ±0.05	EI	165	*31*
$C_{12}H_{10}$ (biphenyl)	8.27 ±0.01	PI	230	*24*
cyclo-$C_6(CH_3)_6$	7.85 ±0.02	PI	152	*28*
$C_{13}H_{10}$ (fluorene)	8.63	EI	243	*47*
$C_{14}H_{10}$ (diphenylacetylene)	8.85 ±0.05	EI	303	*54*
$C_{14}H_{10}$ (anthracene)	7.55	EI	228	*55*
$C_{14}H_{10}$ (phenanthrene)	8.1	EI	233	*54, 55*
$C_{18}H_{12}$ (1,2-benzanthracene)	8.01	EI	251	*56*
$C_{18}H_{30}$ (1-phenyldodecane)	9.05 ±0.1	EI	165	*57*
$C_{18}H_{30}$ (3-phenyldodecane)	8.95 ±0.1	EI	163	*57*
$C_{19}H_{32}$ (7-phenyltridecanc)	8.91 ±0.1	EI	157	*57*
$C_{26}H_{46}$ (1-phenylicosane)	9.34 ±0.1	EI	132	*57*
$C_{26}H_{46}$ (2-phenylicosane)	9.22 ±0.1	EI	129	*57*
$C_{26}H_{46}$ (3-phenylicosane)	8.95 ±0.1	EI	123	*57*
$C_{26}H_{46}$ (4-phenylicosane)	9.01 ±0.1	EI	125	*57*
$C_{26}H_{46}$ (5-phenylicosane)	9.04 ±0.1	EI	125	*57*
$C_{26}H_{46}$ (7-phenylicosane)	8.97 ±0.1	EI	124	*57*
$C_{26}H_{46}$ (9-phenylicosane)	9.06 ±0.1	EI	126	*57*
$(CH_3)_3B$	8.8 ±0.2	EI	173	*58*
$(C_2H_5)_3B$	9.0 ±0.2	EI	170	*58*
N_2	15.576	S	359	*59*
NH	13.10 ±0.05	EI	382	*60*
NH_2	11.4 ±0.1	EI	304	*61*
NH_3	10.2	S, PI, PE	223	*
N_2H_2	9.85 ±0.1	EI		*62*
N_2H_4	8.74 ±0.06	PI	224	*63*
CN	14.3	EI	430	*64, 65*
HCN	13.8	EI	351	*26, 66*
C_2N_2	13.6	EI	387	*8, 64*
CH_5N	8.97	PI	201	*67*
C_2H_2N	10.9	EI	298	*68*
CH_3CN	12.2	PI	302	*14, 24*
cyclo-C_2H_5N	9.94 ±0.15	EI	255	*69*
$C_2H_5NH_2$	8.86 ±0.02	PI	193	*24*
$(CH_3)_2NH$	8.24 ±0.02	PI	186	*24*
$CH_2{=}CHCN$	10.91 ±0.01	PI	296	*24*
C_2H_5CN	11.84 ±0.02	PI	289	*24*
C_3H_7N	9.1 ±0.15	EI	225	*70*
$(CH_2)_3NH$	9.1 ±0.15	EI	225	*70*
n-$C_3H_7NH_2$	8.78 ±0.02	PI	185	*24*
iso-$C_3H_7NH_2$	8.72 ±0.03	PI	183	*24*
$(CH_3)_3N$	7.82 ±0.02	PI	175	*24*
$CH_2{=}CHCH_2CN$	10.39 ±0.01	PI	281	*24*
C_4H_5N (pyrrole)	8.20 ±0.01	PI	215	*24*
$(CH_3)_2CCN$	9.15 ±0.1	EI	239	*68*
n-C_3H_7CN	11.67 ±0.05	PI	280	*24*
C_4H_9N (pyrrolidine)	8.41	PE	192	*17*
n-$C_4H_9NH_2$	8.71 ±0.03	PI	179	*24*
sec-$C_4H_9NH_2$	8.70	PI	177	*24*
iso-$C_4H_9NH_2$	8.70	PI	177	*24*
tert-$C_4H_9NH_2$	8.64	PI	173	*24*
$(C_2H_5)_2NH$	8.01 ±0.01	PI	163	*24*
C_5H_5N (pyridine)	9.3	S, PI	247	*24, 71*
C_6H_7N (aniline)	7.7	PI	202	*
C_6H_7N (2-picoline)	9.02 ±0.03	PI	232	*24*
C_6H_7N (3-picoline)	9.04 ±0.03	PI	236	*24*
C_6H_7N (4-picoline)	9.04 ±0.03	PI	233	*24*
$C_6H_{13}N$ (cyclohexylamine)	8.86	PE	181	*17*
(n-$C_3H_7)_2NH$	7.84 ±0.02	PI	153	*24*
(iso-$C_3H_7)_2NH$	7.73 ±0.03	PI	148	*24*
$(C_2H_5)_3N$	7.50 ±0.02	PI	147	*24*
cyclo-C_6H_5CN	9.705 ±0.01	PI	277	*24*
C_7H_9N (n-methylaniline)	7.32	PI	192	*72, 73*
C_7H_9N (m-toluidine)	7.50 ±0.02	PI	189	*48*
C_7H_9N (2,3-lutidine)	8.85 ±0.02	PI	218	*24*
C_7H_9N (2,4-lutidine)	8.85 ±0.03	PI	218	*24*
C_7H_9N (2,6-lutidine)	8.85 ±0.02	PI	218	*24*
$C_6H_5CH_2CN$	9.40 ±0.5	EI	259	*74*
m-$C_6H_4CH_3CN$	9.66 ±0.05	EI	271	*74*
p-$C_6H_4CH_3CN$	9.76	EI	273	*75*
cyclo-$C_6H_5NHC_2H_5$	7.56	CTS	193	*76*
cyclo-$C_6H_5N(CH_3)_2$	7.12	PI	185	*72, 73*
(n-$C_4H_9)_2NH$	7.69 ±0.03	PI	140	*24*
$C_9H_{13}N$ (N-n-propylaniline)	7.54	CTS	188	*76*
$C_9H_{13}N$ (N-ethyl-N-methylaniline)	7.37	CTS	185	*76*
$C_9H_{13}N$ (N,N-dimethyl-o-toluidine)	7.37	CTS	184	*76*
$C_9H_{13}N$ (N,N-dimethyl-m-toluidine)	7.35	CTS	181	*76*
$C_9H_{13}N$ (N,N-dimethyl-p-toluidine)	7.33	CTS	181	*76*
(n-$C_3H_7)_3N$	7.23	PI	207	*24*
$C_{10}H_{15}N$ (N-n-butylaniline)	7.53	CTS	183	*76*
$C_{10}H_{15}N$ (N,N-diethylaniline)	6.99	CTS	172	*76*
$C_{10}H_{15}N$ (N,N-dimethyl-p-ethylaniline)	7.38	CTS	177	*76*
$C_{10}H_{15}N$ (N,N-2,4-tetramethylaniline)	7.17	CTS	171	*76*
$C_{10}H_{15}N$ (N,N,2,6-tetramethylaniline)	7.22	CTS	173	*76*
$C_{10}H_{15}N$ (N,N,3,5-tetramethylaniline)	7.25	CTS	172	*76*
$C_{11}H_{17}N$ (N,N-diethyl-p-toluidene)	6.93	CTS	164	*76*
$C_{11}H_{17}N$ (N,N-dimethyl-p-isopropylaniline)	7.41	CTS	174	*76*
$(C_6H_5)_2NH$	7.25 ±0.03	PI	223	*77*
$C_{12}H_{19}N$ (N,N-di-n-propylaniline)	6.96	CTS	163	*76*
$C_{12}H_{19}N$ (N,N-dimethyl-p-tert-butylaniline)	7.43	CTS	165	*76*
$C_{14}H_{23}N$ (N,N-di-n-butylaniline)	6.95	CTS	153	*76*

* Average of several values.

Molecule	Ionization Potential ev	Method	ΔH_f of Ion Kcal/mole	Reference
$(C_6H_5)_3N$	6.86 ±0.03	PI	243	77
CH_2N_2 (diazirine)	10.18 ±0.05	EI	314	78
CH_2N_2 (diazomethane)	8.999 ±0.001	S	257	79
CH_6N_2 (methylhydrazine)	8.00 ±0.06	PI	207	63
$CH_3N{=}NCH_3$	8.65 ±0.1	EI	243	80
$C_2H_8N_2$ (1,1-dimethylhydrazine)	7.67 ±0.05	PI	197	81
$C_2H_8N_2$ (1,2-dimethylhydrazine)	7.75 ±0.1	EI	200	82
$(CH_3)_3N_2H$	7.93 ±0.1	EI	202	82
o-$C_4H_4N_2$ (o-diazine)	9.9	EI	275	83
m-$C_4H_4N_2$ (m-diazine)	9.9	EI	277	83
p-$C_4H_4N_2$ (p-diazine)	9.8	EI	274	83
$1,1(C_2H_5)_2N_2H_2$	7.59 ±0.05	PI	184	63
$(CH_3)_4N_2$	7.76 ±0.05	EI	196	82
$C_5H_4NNH_2$	8.97 ±0.05	EI	244	84
$C_5H_{14}N_2$ (1-methyl-1-n-butylhydrazine)	7.62 ±0.05	PI	180	63
$(CH_3)_2NC_6H_4N(CH_3)_2$ [p-bis(dimethylamino)-benzene]	6.9	CTS	180	85
CH_3N_3	9.5 ±0.1	EI	276	86
O_2	12.063 ±0.001	PI	278	*
O_3	12.3 ±0.1	PE	318	87
OH	13.17 ±0.1	EI	313	88
H_2O	12.6	PI	233	*
D_2O	12.6	PI	232	37
HO_2	11.53 ±0.02	EI	271	89
H_2O_2	11.0	EI	233	90, 91
Li_2O	6.8	EI	120	92, 93
CO	14.013 ±0.004	S	297	94
CO_2	13.769 ±0.03	S	223	96
NO	9.25	PI, S	235	*
N_2O	12.894	S	317	95
NO_2	9.79	PI	233	97, 98
CHO	9.8	EI	221	*
CH_2O	10.88	PI	223	24, 48
CH_3OH	10.84	PI, PE	202	17, 24
CH_3CO	10.3	PI	152	*
CH_3CHO	10.2	PI	196	*
C_2H_4O (ethylene oxide)	10.6	PI, S	231	*
C_2H_5OH	10.49	PI	185	24, 30
CH_3OCH_3	9.98	S, PI	186	24, 99
$CH_2{=}CHCHO$	10.10 ±0.01	PI	210	24
C_2H_5CHO	9.98	PI	181	24
CH_3COCH_3	9.69	PI	171	*
$CH_2{=}CHCH_2OH$	9.67 ±0.05	PI	191	24
$CH_2{=}CHOCH_3$	8.93 ±0.02	PI	178	24
C_3H_6O (propyleneoxide)	10.22 ±0.02	PI	214	24
C_3H_6O (trimethylene-oxide)	9.667 ±0.005	S	199	100
n-C_3H_7OH	10.1	PI	172	24, 101
iso-C_3H_7OH	10.15	PI	169	24
C_4H_4O (furan)	8.89	S, PI	197	24, 102
$CH_3CH{=}CHCHO$	9.73 ±0.01	PI	194	24
n-C_3H_7CHO	9.86 ±0.02	PI	174	24
iso-C_3H_7CHO	9.74 ±0.03	PI	169	24
$C_2H_5COCH_3$	9.5	PI	161	*
cyclo-C_4H_8O	9.42	S	174	100
n-C_4H_9OH	10.04	PI	165	24
$C_2H_5OC_2H_5$	9.6	PI	161	24
C_5H_8O (cyclo-pentanone)	9.26 ±0.01	PI	163	24
C_5H_8O (dihydropyran)	8.34 ±0.01	PI	164	24
n-C_4H_9CHO	9.82 ±0.05	PI	168	24
iso-C_4H_9CHO	9.71 ±0.05	PI	164	24
n-$C_3H_7COCH_3$	9.37 ±0.02	PI	154	103
iso-$C_3H_7COCH_3$	9.30 ±0.02	PI	151	24, 103
$(C_2H_5)_2CO$	9.32 ±0.01	PI	153	24
cyclo-$C_5H_{10}O$	9.25 ±0.01	S	161	100
C_6H_5O	8.84	EI	226	36
C_6H_6O (phenol)	8.51	PI	173	24, 48
$(CH_3)_2C{=}CHCOCH_3$	9.08 ±0.03	PI	168	24
$C_6H_{10}O$ (cyclo-hexanone)	9.14 ±0.01	PI	152	24
n-$C_4H_9COCH_3$	9.35	PI	149	24, 103
iso-$C_4H_9COCH_3$	9.30	PI	147	24, 103
tert-$C_4H_9COCH_3$	9.17 ±0.03	PI	140	24
$(n\text{-}C_3H_7)_2O$	9.27 ±0.05	PI	147	24
$(iso\text{-}C_3H_7)_2O$	9.20 ±0.05	PI	142	24
C_6H_5CHO	9.52	PI	209	24
C_7H_6O (tropone)	9.68 ±0.02	EI	240	104
cyclo-$C_6H_5CH_2OH$	9.14 ±0.05	EI	186	74
cyclo-$C_6H_5OCH_3$	8.21 ±0.02	PI	173	24
C_7H_8O (m-cresol)	8.52 ±0.05	EI	165	74
n-$C_5H_{11}COCH_3$	9.33 ±0.03	PI	143	24
C_8H_8O (acetophenone)	9.27 ±0.03	PI	191	24
C_8H_8O (p-methyl-benzaldehyde)	9.33 ±0.05	EI	194	105
$C_8H_{10}O$ (benzyl methyl ether)	8.85 ±0.03	PI	184	30
$C_8H_{10}O$ (phenyl ethyl ether)	8.13 ±0.02	PI	167	24
$C_8H_{10}O$ (m-methylanisole)	8.31 ±0.05	EI	169	74
$C_9H_{10}O$ (phenyl ethyl ketone)	9.27 ±0.05	EI	189	74
$C_9H_{10}O$ (m-methyl-acetophenone)	9.15 ±0.05	EI	182	74
$C_{12}H_{10}O$ (phenyl ether)	8.82 ±0.05	EI	220	106
$C_{13}H_{10}O$ (benzophenone)	9.4	EI	229	106, 105
$C_{14}H_{12}O$ (p-methyl-benzophenone)	9.13 ±0.05	EI	214	105
HCOOH	11.05 ±0.01	PI	164	24
CH_3COOH	10.69±0.03	PI	135	24
$HCOOCH_3$	10.815 ±0.005	PE	166	165
C_2H_5COOH	10.24 ±0.03	PI	127	24
$HCOOC_2H_5$	10.61 ±0.01	PI	156	24
CH_3COOCH_3	10.27 ±0.02	PI	138	24
$(CH_3O)_2CH_2$	10.00 ±0.05	PI	145	24
$CH_3COOCH{=}CH_2$	9.19 ±0.05	PI	137	24
$CH_3COCOCH_3$	9.24 ±0.03	PI	135	24
n-C_3H_7COOH	10.16 ±0.05	PI	121	24
iso-C_3H_7COOH	10.02 ±0.05	PI	115	24
$HCOOCH_2CH_2CH_3$	10.54 ±0.01	PI	149	24
$CH_3COOC_2H_5$	10.11 ±0.02	PI	126	24
$C_2H_5COOCH_3$	10.15 ±0.03	PI	127	24
$C_4H_8O_2$ (p-dioxane)	9.13 ±0.03	PI	126	24
$(CH_3O)_2CHCH_3$	9.65 ±0.03	PI	129	24
$C_4H_4O_2$ (2-furfur-aldehyde)	9.21 ±0.01	PI	187	24
$CH_3COCH_2COCH_3$	8.87 ±0.03	PI	122	24
$HCOO(CH_2)_3CH_3$	10.50 ±0.02	PI	144	24
$HCOOCH_2CH(CH_3)_2$	10.46 ±0.02	PI	139	24
$CH_3COOCH_2CH_2CH_3$	10.04 ±0.03	PI	121	24
$CH_3COOCH(CH_3)_2$	9.99 ±0.03	PI	116	24
$C_2H_5COOC_2H_5$	10.00 ±0.02	PI	119	24
n-$C_3H_7COOCH_3$	10.07 ±0.03	PI	125	24
iso-$C_3H_7COOCH_3$	9.98 ±0.02	PI	121	24
$(C_2H_5O)_2CH_2$	9.70 ±0.05	PI	134	24
p-$C_6H_4O_2$	9.67 ±0.02	PI	198	48
$CH_3COOC_4H_9$	9.56 ±0.03	PI	104	48
$CH_3COO(CH_2)_3CH_3$	10.01	PI	114	24
CH_3COOCH_2-$CH(CH_3)_2$	9.97	PI	111	24
CH_3COOCH-$(CH_3)C_2H_5$	9.91 ±0.03	PI	110	24
C_6H_5COOH	9.73 ±0.09	EI	152	105
p-HOC_6H_4CHO	9.32 ±0.02	EI	157	105
$C_8H_8O_2$ (α-hydroxy-acetophenone)	9.33 ±0.05	EI	159	74
$C_6H_5COOCH_3$	9.35 ±0.06	EI	144	105
$C_8H_8O_2$ (p-methoxy-benzaldehyde)	8.60 ±0.03	EI	150	105
$C_8H_8O_2$ (m-hydroxy-acetophenone)	8.67 ±0.05	EI	134	74
$C_8H_8O_2$ (p-hydroxy-acetophenone)	8.70 ±0.03	EI	135	105
$C_9H_{10}O_2$ (α-methoxy-acetophenone)	8.60 ±0.05	EI	142	74
$C_9H_{10}O_2$ (m-methoxy-acetophenone)	8.53 ±0.05	EI	140	74
$C_9H_{10}O_2$ (p-methoxy-acetophenone)	8.62 ±0.05	EI	142	105
$C_9H_{10}O_2$ (methyl p-toluate)	8.94 ±0.04	EI	130	105
$C_{13}H_{10}O_2$ (p-hydroxy-benzophenone)	8.59 ±0.05	EI	165	105
$C_{13}H_{10}O_2$ (phenyl benzoate)	8.98 ±0.05	EI	177	106
$C_{14}H_{10}O_2$ (benzil)	8.78 ±0.05	EI	181	106

* Average of several values.

Molecule	Ionization Potential ev	Method	ΔH_f of Ion Kcal/mole	Reference
$CH_3OCH_2COOCH_3$	9.56 ±0.05	EI	88	74
$C_9H_{10}O_3$ (methyl p-methoxybenzoate)	8.43 ±0.04	EI	90	105
$(C_6H_5)_2CO$	9.01 ±0.05	EI	122	106
CH_3CONH_2	9.77 ±0.02	PI	171	24
$HCON(CH_3)_2$	9.12 ±0.02	PI	160	24
$CH_3CONHCH_3$	8.90 ±0.02	PI	150	24
$CH_3CON(CH_3)_2$	8.81 ±0.03	PI	145	24
C_5H_4NOH	9.70 ±0.05	EI	209	84
$HCON(C_2H_5)_2$	8.89 ±0.02	PI	145	24
C_6H_5NO (2-pyridine-carboxaldehyde)	9.75 ±0.05	EI	227	84
C_6H_5NO (4-pyridine-carboxaldehyde)	10.12 ±0.05	EI	235	84
$CH_3CON(C_2H_5)_2$	8.60 ±0.02	PI	130	24
C_6H_5NCO	8.77 ±0.02	PI	222	24
C_7H_7NO (benzamide)	9.4 ±0.2	EI	197	107, 108
C_7H_7NO (p-amino-benzaldehyde)	8.25 ±0.02	EI	182	105
C_7H_9NO (p-methoxy-analine)	7.82	EI	169	75
C_8H_9NO (acetanilide)	8.39 ±0.10	EI	171	108
C_8H_9NO (m-amino-acetophenone)	8.09 ±0.05	EI	171	74
C_8H_9NO (p-amino-acetophenone)	8.17 ±0.02	EI	172	105
C_9H_7NO (α-cyano-acetophenone)	9.56 ±0.05	EI	235	74
CH_3NO_2	11.1	PI	238	14, 24
$C_2H_5NO_2$	10.88 ±0.05	PI	226	24
n-$C_3H_7NO_2$	10.81 ±0.03	PI	221	24
iso-$C_3H_7NO_2$	10.71 ±0.05	PI	217	24
$C_6H_5NO_2$	9.92	PI	246	24
$C_7H_6NO_2$ (m-nitro-benzyl radical)	8.56 ±0.1	EI	227	39
$C_7H_7NO_2$ (m-nitro-toluene)	9.65 ±0.05	EI	233	74
$C_7H_7NO_2$ (p-nitro-toluene)	9.82	EI	237	75
$C_8H_9NO_2$	8.08 ±0.01	EI	122	90
$C_6H_6N_2O_2$ (o-nitro-aniline)	8.66	EI	215	75
$C_6H_6N_2O_2$ (m-nitro-aniline)	8.7	EI	216	75
$C_6H_6N_2O_2$ (p-nitro-aniline)	8.85	EI	219	75
$C_2H_5NO_3$	11.22	PI	222	24
n-$C_3H_7ONO_2$	11.07 ±0.02	PI	213	24
$C_6H_5NO_3$ (p-nitro-phenol)	9.52	EI	187	75
$C_7H_5NO_3$ (p-nitro-benzaldehyde)	10.27 ±0.01	EI	217	105
$C_8H_7NO_3$ (m-nitro-acetophenone)	9.89 ±0.05	EI	201	74
$C_8H_7NO_3$ (p-nitro-acetophenone)	10.07 ±0.02	EI	205	105
$C_8H_7NO_4$ (methyl-p-nitrobenzoate)	10.20 ±0.03	EI	160	105
F_2	15.7	S	362	109
HF	15.77 ±0.02	EI	299	110
BF	11.3	EI	233	111, 112
BF_3	15.5	EI	87	*
CF_2	11.8	EI	237	113, 114
C_2F_4	10.12	PI	78	28
C_6F_6	9.97	PI		28
NF_2	11.9	EI	284	115, 116
trans-N_2F_2	13.1 ±0.1	EI	322	115
NF_3	13.2 ±0.2	EI	275	115, 123
N_2F_4	12.04 ±0.10	EI	276	124
OF_2	13.6	EI	309	117, 118
XeF_2	11.5 ±0.2	S	239	119
CH_2F	9.35	EI	207	29
CH_3F	12.85 ±0.01	EI	229	120
C_2H_3F	10.37	PI	211	28, 121
cyclo-C_6H_5F	9.2	S, PI	186	*
C_7H_7F	8.9	PI	172	24
CHF_2	9.45	EI	143	29, 122
$C_2H_2F_2$	10.30	PI	159	28, 121
o-$C_6H_4F_2$	9.31	PI	147	121
p-$C_6H_4F_2$	9.15	PI	140	121
C_2HF_3	10.14	PI	122	28
$CH_2=CHCF_3$	10.9	PI	93	28
$C_7H_5F_3$	9.68	S, PI	84	24, 125
$C_7H_{11}F_3$ (trifluoro-methylcyclohexane)	10.46 ±0.02	PI	37	24
C_6H_5OF (o-fluoro-phenol)	8.66 ±0.01	PI	132	24
Na_2	4.90 ±0.01	PI	147	126
AlF	9.8	EI	166	127, 128
$(CH_3)_4Si$	9.9	EI	171	129, 130
PH_3	9.98	PI	231	131
PF_3	9.71	PI	4	131
CH_3PH_2	9.72 ±0.15	EI	217	132
$C_2H_5PH_2$	9.47 ±0.5	EI	206	132
$(CH_3)_3P$	8.60 ±0.2	EI	175	132
$(C_6H_5)_3P$	7.36 ±0.05	PI	242	77
S_6	9.7	EI	248	133
S_7	9.2 ±0.3	EI		133
HS	10.50 ±0.1	EI	276	134
H_2S	10.4	S	235	135
CS_2	10.080	S	261	95
SO_2	12.34 ±0.02	PI	214	30
CH_3S	8.06 ±0.1	EI	218	134
CH_3SH	9.440 ±0.005	PI	212	24
C_2H_4S (ethylene sulfide)	8.87 ±0.15	EI	224	69
C_2H_5SH	9.285 ±0.005	PI	203	24
CH_3SCH_3	8.685 ±0.005	PI	191	24
C_3H_6S (propylene sulfide)	8.6 ±0.2	EI	218	136
$(CH_2)_3S$	8.9 ±0.15	EI	220	70
n-C_3H_7SH	9.195 ±0.005	PI	197	24
$C_2H_5SCH_3$	8.55 ±0.01	PI	183	24
C_4H_4S (thiophene)	8.860 ±0.005	PI	229	24
$CH_3SCH_2-CH=CH_2$	8.70 ±0.2	EI	211	137
$(CH_2)_4S$	8.57 ±0.75	EI	190	70
n-C_4H_9SH	9.14 ±0.02	PI	191	24
$C_2H_5SC_2H_5$	8.430 ±0.005	PI	175	24
n-$C_3H_7SCH_3$	8.8 ±0.15	EI	183	138
iso-$C_3H_7SCH_3$	8.7 ±0.2	EI	179	137
C_6H_5S	8.63 ±0.1	EI	250	134
C_6H_5SH	8.32 ±0.01	PI	217	138
C_6H_8S (2-ethyl-thiophene)	8.8 ±0.2	EI	215	139
$(n\text{-}C_3H_7)_2S$	8.5	PI, EI	170	24, 143
$C_6H_5S_2CH_3$	8.9	EI	229	137
$C_7H_{10}S$ (2-propyl-thiophene)	8.6 ±0.2	EI	205	139
$C_8H_{12}S$	8.8	EI	221	137
$C_8H_{12}S$ (2-butyl-thiophene)	8.5 ±0.2	EI	198	139
CH_3SSCH_3	8.46 ±0.03	PI	189	24
$C_2H_5SSC_2H_5$	8.27 ±0.03	PI	173	24
CH_3SSSCH_3	8.80 ±0.15	EI	203	140
COS	11.17 ±0.01	PI	224	138
SO_2F_2	13.3 ±0.1	EI	102	141
CH_3NCS	9.25 ±0.03	PI	245	24
CH_3SCN	10.065 ±0.01	PI	270	24
C_2H_5NCS	9.14 ±0.03	PI	237	24
C_2H_5SCN	9.89 ±0.01	PI	261	24
C_7H_5NS (phenyl-isothiocyanate)	8.520 ±0.005	PI	260	24
$C_6H_5CH_2SCN$	9.06 ±0.05	EI	274	74
NH_2CSNH_2	8.50 ±0.05	EI	194	142
$NH_2CSNHCH_3$	8.29 ±0.05	EI	188	142
$NH_2CSNHCH=CH_2$	8.29 ±0.05	EI	213	142
$NH_2CSN(CH_3)_2$	8.34 ±0.05	EI	186	142
$CH_3NHCSNHCH_3$	8.17 ±0.05	EI	184	142
$CH_3NHCSN(CH_3)_2$	7.93 ±0.05	EI	176	142
$C_5H_{12}N_2S$	7.98 ±0.05	EI	170	142
$(CH_3)_2NCSN(CH_3)_2$	7.95 ±0.05	EI	173	142
CH_3COSH	10.00 ±0.02	PI	179	24
Cl_2	11.48 ±0.01	PI	265	24
HCl	12.74	PI	272	*
LiCl	10.1	EI	186	144
CCl_3	8.78 ±0.05	EI	214	*
CCl_4	11.47 ±0.01	PI	240	24
C_2Cl_4	9.32 ±0.01	PI	212	24, 28
PCl_3	9.91	PI	160	131
CH_2Cl	9.32	EI	244	29
CH_3Cl	11.3	S, PI	239	14, 145
C_2H_3Cl	9.996	S, PI	239	*
C_2H_5Cl	10.97	PI	226	24
$CH_3C\equiv CCl$	9.9 ±0.1	EI	267	25
n-C_3H_7Cl	10.82 ±0.03	PI	219	24

* Average of several values.

Molecule	Ionization Potential ev	Method	ΔH_f of Ion Kcal/mole	Reference
iso-C_3H_7Cl	10.78 ±0.02	PI	211	*24*
n-C_4H_9Cl	10.67 ±0.03	PI	210	*24*
sec-C_4H_9Cl	10.65 ±0.03	PI	210	*24*
iso-C_4H_9Cl	10.66 ±0.03	PI	208	*24*
tert-C_4H_9Cl	10.61 ±0.03	PI	202	*24*
C_6H_5Cl	9.07	PI	222	*24*
$C_6H_5CH_2Cl$	9.19 ±0.05	EI	219	*74*
o-$C_6H_4ClCH_3$	8.83 ±0.02	PI	208	*24*
m-$C_6H_4ClCH_3$	8.83 ±0.02	PI	207	*24*
p-$C_6H_4ClCH_3$	8.69 ±0.02	PI	204	*24*
C_7H_9Cl (endo-5-chloro-2-norbornene)	9.1 ±0.15	EI	233	*43*
C_7H_9Cl (exo-5-chloro-2-norbornene)	9.15 ±0.15	EI	234	*43*
C_7H_9Cl (3-chloro-nortricyclene)	9.51 ±0.15	EI	234	*43*
$CHCl_2$	9.30	EI	245	*29*
CH_2Cl_2	11.35 ±0.02	PI	240	*24*
cis-$C_2H_2Cl_2$	9.65	PI	223	*
trans-$C_2H_2Cl_2$	9.64	PI	224	*
CH_2ClCH_2Cl	11.12 ±0.05	PI	225	*24*
$CH_2{=}CClCH_2Cl$	9.82 ±0.03	PI	218	*24*
1,2-$C_3H_6Cl_2$	10.87 ±0.05	PI	215	*24*
1,3-$C_3H_6Cl_2$	10.85 ±0.05	PI	215	*24*
o-$C_6H_4Cl_2$	9.06	PI	217	*24, 28*
m-$C_6H_4Cl_2$	9.12 ±0.01	PI	217	*24*
p-$C_6H_4Cl_2$	8.95	PI	212	*28*
$CHCl_3$	11.42 ±0.03	PI	239	*24*
C_2HCl_3	9.45	PI	216	*
$CHCl_2CHCl_2$	11.10 ±0.05	EI	220	*146*
CNCl	12.49 ±0.04	EI	321	*147*
CF_3Cl	12.91 ±0.03	PI	132	*24*
C_2F_3Cl	10.4 ±0.2	EI	107	*148*
C_6F_5Cl	10.4 ±0.1	EI		*149*
CF_2Cl_2	12.31 ±0.05	PI	170	*24*
$CF_3CCl{=}CClCF_3$	10.36 ±0.1	PI		*24*
$CFCl_3$	11.77 ±0.02	PI	205	*24*
CF_3CCl_3	11.78 ±0.03	PI		*24*
$CFCl_2CF_2Cl$	11.99 ±0.02	PI	95	*24*
ClO_3F	13.6 ±0.2	EI	308	*150*
C_5H_4NCl (2-chloro-pyridene)	9.91 ±0.05	EI	255	*84*
C_5H_4NCl (4-chloro-pyridene)	10.15 ±0.05	EI	260	*84*
CH_3COCl	11.02 ±0.05	PI	196	*24*
CH_3COCH_2Cl	9.99	EI	173	*105, 151*
o-$C_6H_4(OH)Cl$	9.28	EI	181	*75*
p-$C_6H_4(OH)Cl$	9.07	EI	175	*75*
C_6H_5COCl	9.70 ±0.01	EI	195	*105*
p-C_6H_4ClCHO	9.61 ±0.01	EI	201	*105*
C_8H_7OCl (α-chloro-acetophenone)	9.5	EI	195	*74, 105*
C_8H_7OCl (p-chloro-acetophenone)	9.47 ±0.05	EI	190	*105*
C_8H_7OCl (p-methyl-benzylchloride)	9.37 ±0.01	EI	187	*105*
$C_6H_4ClCOC_6H_5$	9.68 ±0.01	EI	227	*105*
$CH_2ClCOOCH_3$	10.53 ±0.05	EI	138	*74*
p-$CH_3OC_6H_4COCl$	8.87 ±0.05	EI	149	*105*
p-ClC_6H_4COCl	9.58 ±0.03	EI	192	*105*
cis-C_2H_2FCl	9.86	PI	191	*24, 121*
trans-C_2H_2FCl	9.87	PI	191	*24, 121*
o-C_6H_4FCl	9.155 ±0.01	PI	180	*24*
m-C_6H_4FCl	9.21 ±0.01	PI	180	*24*
p-C_6H_4FCl	9.43 ±0.02	EI	185	*152*
CHF_2Cl	12.45 ±0.05	PI	174	*24*
cis-C_2HF_2Cl	9.86 ±0.02	PI	147	*121*
trans-C_2HF_2Cl	9.83 ±0.02	PI	147	*121*
CH_3CF_2Cl	11.98 ±0.01	PI		*24*
n-$C_3F_7CH_2Cl$	11.84 ±0.02	PI		*24*
$CHFCl_2$	12.39 ±0.20	EI	217	*153*
$(CH_3)_3SiCl$	10.58 ±0.04	EI	160	*130*
CH_3SiCl_3	11.36 ±0.03	EI	136	*153*
$CH_2{=}CHSiCl_3$	10.79 ±0.02	PI	148	*24*
$C_2H_5SiCl_3$	10.74 ±0.04	EI	117	*153*
iso-$C_3H_7SiCl_3$	10.28 ±0.1	EI	100	*153*
C_4H_3ClS	8.68 ±0.01	PI	217	*24*
$C_6H_4NO_2COCl$	10.66 ±0.01	EI	219	*105*
CaF	5.8	EI	75	*
$C_5H_5V(CO)_4$	8.2 ±0.3	EI	9	*154*
$Cr(CO)_6$	8.03 ±0.03	PI	−55	*155*
$C_5H_5Mn(CO)_3$	8.3 ±0.4	EI	83	*154*
$Fe(CO)_5$	7.95 ±0.03	PI	8	*155*
$C_5H_5Co(CO)_2$	8.3 ±0.2	EI	136	*154*
$Ni(CO)_4$	8.28 ±0.03	PI	47	*155*
$(CH_3)_4Ge$	9.2 ±0.2	EI	177	*156*
As_4	9.07 ±0.07	EI	244	*157*
AsH_3	10.03	PI	247	*131*
$AsCl_3$	11.7 ±0.1	EI	208	*158*
$(CH_3)_3As$	8.3 ±0.1	EI	202	*158*
$(C_6H_5)_3As$	7.34 ±0.07	PI	250	*77*
Br_2	10.54 ±0.03	PI	250	*24, 159*
HBr	11.62 ±0.03	PI	259	*24*
$MgBr_2$	10.65 ±0.15	EI	172	*160*
BrCl	11.1 ±0.2	EI	259	*161*
CH_3Br	10.53	S, PI	234	*
C_2H_3Br	9.80	PI	243	*
C_2H_5Br	10.29	S	222	*24, 162*
$CH_3C{\equiv}CBr$	10.1 ±0.1	EI	283	*25*
$CH_3CH{=}CHBr$	9.30 ±0.05	PI	224	*24*
n-C_3H_7Br	10.18 ±0.01	PI	216	*24*
iso-C_3H_7Br	10.075 ±0.01	PI	208	*24*
n-C_4H_9Br	10.125 ±0.01	PI	208	*24*
sec-C_4H_9Br	9.98 ±0.01	PI	206	*24*
iso-C_4H_9Br	10.09 ±0.02	PI	208	*24*
tert-C_4H_9Br	9.89 ±0.03	PI	201	*24*
n-$C_5H_{11}Br$	10.10 ±0.02	PI	205	*24*
C_6H_5Br	8.98 ±0.02	PI	231	*24*
o-$C_6H_4BrCH_3$	8.78 ±0.01	PI	218	*24*
m-$C_6H_4BrCH_3$	8.81 ±0.02	PI	218	*24*
p-$C_6H_4BrCH_3$	8.67 ±0.02	PI	215	*24*
CH_2Br_2	10.49 ±0.02	PI	241	*24*
cis-$C_2H_2Br_2$	9.45	PI	241	*
trans-$C_2H_2Br_2$	9.46	PI	240	*
CH_3CHBr_2	10.19 ±0.03	PI	240	*24*
1,3-$C_3H_6Br_2$	10.07 ±0.02	PI	221	*24*
$CHBr_3$	10.51 ±0.02	PI	246	*24*
C_2HBr_3	9.27	PI	240	*24, 28*
CNBr	11.95 ±0.08	EI	320	*147*
CF_3Br	11.89	EI	121	*
C_5H_4NBr (2-bromo-pyridine)	9.65 ±0.05	EI	261	*84*
C_5H_4NBr (4-bromo-pyridine)	9.94 ±0.05	EI	267	*84*
CH_3COBr	10.55 ±0.05	PI	197	*24*
C_6H_4BrOH	9.04	EI	187	*75*
$CH_2BrCOOCH_3$	10.37 ±0.05	EI	146	*74*
C_6H_4FBr	8.99 ±0.03	PI	187	*24*
CF_2BrCH_2Br	10.83 ±0.01	PI	160	*24*
C_4H_3BrS	8.63 ±0.01	PI	228	*24*
CH_2ClBr	10.77 ±0.01	PI	236	*24*
CH_2BrCH_2Cl	10.63 ±0.03	PI	227	*24*
$CHCl_2Br$	10.88 ±0.05	EI	237	*146*
$(CH_3)_3SiBr$	10.24 ±0.02	EI	171	*130*
$Mo(CO)_6$	8.12 ±0.03	PI	−31	*155*
RuO_4	12.33 ±0.23	EI	240	*163*
$(CH_3)_4Sn$	8.25 ±0.15	EI	186	*129*
SbH_3	9.58	PI	256	*131*
$(C_6H_5)_3Sb$	7.3 ±0.1	PI	255	*77*
I_2	9.28 ±0.02	PI	229	*24*
HI	10.39	PI	246	*24, 159*
ICl	10.31 ±0.02	EI	242	*164*
IBr	9.98 ±0.03	EI	240	*164*
CH_3I	9.54	S, PI	223	*24, 145*
C_2H_5I	9.33	S, PI	213	*24, 162*
n-C_3H_7I	9.26 ±0.01	PI	208	*24*
iso-C_3H_7I	9.17 ±0.02	PI	201	*24*
n-C_4H_9I	9.21 ±0.01	PI	202	*24*
sec-C_4H_9I	9.09 ±0.02	PI	198	*24*
iso-C_4H_9I	9.18 ±0.02	PI	200	*24*
tert-C_4H_9I	9.02 ±0.03	PI	193	*24*
n-$C_5H_{11}I$	9.19 ±0.01	PI	197	*24*
C_6H_5I	8.73 ±0.03	PI	238	*24*
o-$C_6H_4ICH_3$	8.62 ±0.01	PI	228	*24*
m-$C_6H_4ICH_3$	8.61 ±0.03	PI	226	*24*
p-$C_6H_4ICH_3$	8.50 ±0.01	PI	224	*24*
$W(CO)_6$	8.18 ±0.03	PI	−20	*155*
OsO_4	12.97 ±0.12	EI	219	*163*
$(CH_3)_2Hg$	9.0	EI	233	*143, 156*
$(C_2H_5)_2Hg$	8.5 ±0.1	EI	221	*143*
(iso-$C_3H_7)_2Hg$	7.6 ±0.1	EI	188	*143*
CH_3HgCl	11.5 ±0.2	EI	253	*143*
$(CH_3)_4Pb$	8.0 ±0.4	EI	217	*129*
$(C_6H_5)_3Bi$	7.3 ±0.1	PI	288	*77*

* Average of several values.

REFERENCES

1. Beutler, H. and Junger, H. O. *Zeit. f. Physik* **100**, 80 (1936).
2. Dibeler, V. H., Reese, R. M. and Krauss, M. *Adv. Mass Spectry.* **3**, 471 (1966).
3. Bauer, S. H., Herzberg, G. and Johns, J. W. C. *J. Mol. Spectr.* **13**, 256 (1964).
4. Margrave, J. L. *J. Phys. Chem.* **61**, 38 (1957).
5. Koski, W. S., Kaufman, J. J., Pachucki, C. F. and Shipko, F. J. *J. Am. Chem. Soc.* **80**, 3202 (1958).
6. Fehlner, T. P. and Koski, W. S. *J. Am. Chem. Soc.* **86**, 581 (1964).
7. Drowart, J., DeMaria, G. and Inghram, M. G. *J. Chem. Phys.* **29**, 1015 (1958).
8. Field, F. H. and Franklin, J. L. *Electron Impact Phenomena and the Properties of Gaseous Ions*, Academic Press, Inc., New York, N.Y. (1957).
9. Herzberg, G. *Can. J. Phys.* **39**, 1511 (1961).
10. Waldron, J. D., *Metropolitan Vickers Gazette* **27**, 66 (1956).
11. Dibeler, V. H., Krauss, M., Reese, R. M. and Harllee, F. N. *J. Chem. Phys.* **42**, 3791 (1965).
12. Herzberg, G. and Shoosmith, J., *Can J. Phys.* **34**, 523 (1956).
13. Elder, F. A., Giese, C., Steiner, B. and Inghram, M. *J. Chem. Phys.* **36**, 3292 (1962).
14. Nicholson, A. J. C., *J. Chem. Phys.* **43**, 1171 (1965).
15. Dibeler, V. H. and Reese, R. M. *J. Res. Natl. Bur. Std.* **A68**, 409 (1964).
16. Zelikoff, M. and Watanabe, K., *J. Opt. Soc. Am.* **43**, 756 (1953).
17. Al-Joboury, M. I. and Turner, D. W., *J. Chem. Soc. London* **4434** (1964).
18. Farmer, J. B. and Lossing, F. P., *Can. J. Chem.* **33**, 861 (1955).
19. Collin, J. and Lossing, F. P., *J. Am. Chem. Soc.* **79**, 5848 (1957).
20. Watanabe, K. and Namioka, T., *J. Chem. Phys.* **24**, 915 (1956).
21. Collin, J. and Lossing, F. P., *J. Am. Chem. Soc.* **81**, 2064 (1959).
22. Dorman, F. H. *J. Chem. Phys.* **43**, 3507 (1965).
23. Pottie, R. F., Harrison, A. G. and Lossing, F. P. *J. Am. Chem. Soc.* **83**, 3204 (1961).
24. Watanabe, K., Nakayama, T. and Mottl, J. *J. Quant. Spectrosc. Radiat. Transfer* **2**, 369 (1962).
25. Coats, F. H. and Anderson, R. C. *J. Am. Chem. Soc.* **79**, 1340 (1957).
26. Varsel, C. J., Morrell, F. A., Resnik, F. E. and Powell, W. A. *Anal. Chem.* **32**, 182 (1960).
27. McDowell, C. A., Lossing, F. P., Henderson, I. H. S. and Farmer, J. B. *Can. J. Chem.* **34**, 345 (1956).
28. Bralsford, R., Harris, P. V. and Price, W. C. *Proc. Roy. Soc.* **A258**, 459 (1960).
29. Lossing, F. P., Kebarle, P. and DeSousa, J. B. *Adv. Mass Spectrometry* **431** (1959).
30. Watanabe, K. *J. Chem. Phys.* **26**, 542 (1957).
31. Meyer, F. and Harrison, A. G. *Can. J. Chem.* **42**, 2256 (1964).
32. Harrison, A. G., Honnen, L. R., Dauben, H. J. and Lossing, F. P. *J. Am. Chem. Soc.* **82**, 5593 (1960).
33. Natalis, P. and Laune, *J. Bull. Soc. Chim. Belg.* **73**, 944 (1964).
34. Taubert, R. and Lossing, F. P. *J. Am. Chem. Soc.* **84**, 1523 (1962).
35. Steiner, B., Giese, C. F. and Inghram, M. G. *J. Chem. Phys.* **34**, 189 (1961).
36. Fisher, I. P., Palmer, T. F. and Lossing, F. P. *J. Am. Chem. Soc.* **86**, 2741 (1964).
37. Brehm, B. *Z. Naturforschg.* **21a**, 196 (1966).
38. Momigny, J., Brakier, L. and D Or, L. *Bull. Classe Sci. Acad. Roy. Belg.* **48**, 1002 (1962).
39. Harrison, A. G., Kebarle, P. and Lossing, F. P. *J. Am. Chem. Soc.* **83**, 777 (1961).
40. Thrush, B. A. and Zwolenik, J. J. *Disc. Faraday Soc.* **35**, 196 (1963).
41. Lifshitz, C. and Bauer, S. H. *J. Phys. Chem.* **67**, 1629 (1963).
42. Meyerson, S., McCollum, J. D. and Rylander, P. N. *J. Am. Chem. Soc.* **83**, 1401 (1961).
43. Steele, W. C., Jennings, B. H., Botyos, G. L. and Dudek, G. O. *J. Org. Chem.* **30**, 2886 (1965).
44. Kybett, B. D., Carroll, S., Natalis, P., Bonnell, D. W., Margrave, J. L. and Franklin, J. L. *J. Am. Chem. Soc.* **88**, 626 (1966).
45. Meyer, F., Haynes, P., McLean, S. and Harrison, A. G. *Can. J. Chem.* **43**, 211 (1965).
46. Natalis, P. *Bull. Soc. Chim. Belg.* **73**, 961 (1964).
47. Pottie, R. F. and Lossing, F. P. *J. Am. Chem. Soc.* **85**, 269 (1963).
48. Vilesov, F. I. and Terenin, A. N. *Dokl. Phys. Chem., Proc. Acad. Sci.* **USSR 115**, 539 (1957).
49. Clark, L. B. *J. Chem. Phys.* **43**, 2566 (1965).
50. Kitagawa, T., Harada, Y., Inokuchi, H. and Kodera, K. *J. Mol. Spectr.* **19**, 1 (1966).
51. Natalis, P. *Bull. Soc. Roy. Sci. Liege* **31**, 803 (1962).
52. Harrison, A. G. and Lossing, F. P. *J. Am. Chem. Soc.* **82**, 1052 (1960).
53. Vilesov, F. I. *J. Phys. Chem.* **USSR 35**, 986 (1961).
54. Natalis, P. and Franklin, J. L. *J. Phys. Chem.* **69**, 2935 (1965).
55. Wacks, M. E. and Dibeler, V. H. *J. Chem. Phys.* **31**, 1557 (1959).
56. Wacks, M. E. *J. Chem. Phys.* **41**, 1661 (1964).
57. King, A. B. *J. Chem. Phys.* **42**, 3526 (1965).
58. Law, R. W. and Margrave, J. L. *J. Chem. Phys.* **25**, 1086 (1956).
59. Lofthus, A. *The Molecular Spectrum of Nitrogen* Department of Physics, University of Oslo, Blindern, **Norway, Spectroscopic Report No. 2**, 1 (1960).
60. Reed, R. I. and Snedden, W. *J. Chem. Soc.* **4132** (1959).
61. Foner, S. N. and Hudson, R. L. *J. Chem. Phys.* **29**, 442 (1958).
62. Foner, S. N. and Hudson, R. L. *J. Chem. Phys.* **28**, 719 (1958).
63. Akopyan, M. E. and Vilesov, F. I. *Kinetics and Catalysis*, **4**, 32 (1963).
64. Dibeler, V. H., Reese, R. M. and Franklin, J. L. *J. Am. Chem. Soc.* **83**, 1813 (1961).
65. Berkowitz, J. *J. Chem. Phys.* **36**, 2533 (1962).
66. Morrison, J. D. and Nicholson, A. J. C. *J. Chem. Phys.* **20**, 1021 (1952).
67. Watanabe, K. and Mottl, J. R. *J. Chem. Phys.* **26**, 1773 (1957).
68. Pottie, R. F. and Lossing, F. P. *J. Am. Chem. Soc.* **83**, 4737 (1961)
69. Gallegos, E. and Kiser, R. W. *J. Phys. Chem.* **65**, 1177 (1961).
70. Gallegos, E. J. and Kiser, R. W. *J. Phys. Chem.* **66**, 136 (1962).
71. Amr El-Sayed, M. F., Kasha, M. and Tanaka, Y. *J. Chem. Phys.* **34**, 334 (1961).
72. Akopyan, M. E. and Vilesov, F. I. *Dokl. Phys. Chem., Proc. Acad. Sci.* **USSR 158**, 965 (1964).
73. Kurbatov, B. L., Vilesov, F. I. and Terenin, A. N. *Soviet Physics-Doklady* **6**, 883 (1962).
74. Pignataro, S., Foffani, A., Innorta, G. and Distefano, G. *Z. Phys. Chem. Neue Folge* **49**, 291 (1966).
75. Crable, G. F. and Kearns, G. L. *J. Phys. Chem.* **66**, 436 (1962).
76. Farrell, P. G. and Newton, J. *J. Phys. Chem.* **69**, 3506 (1965).
77. Vilesov, F. I. and Zaitsev, V. M. *Dokl. Phys. Chem., Proc. Acad. Sci.* **USSR 154**, 117 (1964).
78. Paulett, G. S. and Ettinger, R. *J. Chem. Phys.* **39**, 825 (1963).
79. Merer, A. J., *Can. J. Phys.* **42**, 1242 (1964).
80. Gowenlock, B. G., Majer, J. R. and Snelling, D. R. *Trans. Faraday Soc.* **58**, 670 (1962).
81. Akopyan, M. E., Vilesov, F. I. and Terenin, A. N. *Izv. Akad. Nauk USSR, Ser. Fiz.* **27**, 1083 (1963).
82. Dibeler, V. H., Franklin, J. L. and Reese, R. M. *J. Am. Chem. Soc.* **81**, 68 (1959).
83. Momigny, J., Urbain, J. and Wankenne, H. *Bull. Soc. Roy. Sci. Liege* **34**, 337 (1965).
84. Basila, M. R. and Clancy, D. J. *J. Phys. Chem.* **67**, 1551 (1963).
85. Finch, A. C. M. *J. Chem. Soc.* **2272** (1964).
86. Franklin, J. L., Dibeler, V. H., Reese, R. M. and Krauss, M. *J. Am. Chem. Soc.* **80**, 298 (1958).
87. Radwan, T. N. and Turner, D. W. *J. Chem. Soc.* **Sect. A85** (1966).
88. Foner, S. N. and Hudson, R. L. *Advances in Chem.* **Ser. 34** (1962).
89. Foner, S. N. and Hudson, R. L. *J. Chem. Phys.* **23**, 1364 (1955).
90. Lindeman, L. P. and Guffy, J. C. *J. Chem. Phys.* **29**, 247 (1958).
91. Foner, S. N. and Hudson, R. L. *J. Chem. Phys.* **36**, 2676 (1962).
92. Berkowitz, J., Chupka, W. A., Blue, G. D. and Margrave, J. L. *J. Phys. Chem.*, **63**, 644 (1959).
93. White, D., Seshadri, K. S., Dever, D. F., Mann, D. E. and Linevski, M. J. *J. Chem. Phys.* **39**, 2463 (1963).
94. Krupenie, P. H. *The Band Spectrum of Carbon Monoxide* Institute for Basic Standards, National Bureau of Standards, Washington, D.C., **NSRDS-NBS 5**, 1 (1966).
95. Tanaka, Y., Jursa, A. S. and LeBlanc, F. J. *J. Chem. Phys.* **32**, 1205 (1960).
96. Tanaka, Y., Jursa, A. S. and LeBlanc, F. J. *J, Chem. Phys.* **32**, 1199 (1960).
97. Nakayama, T., Kitamura, M. Y. and Watanabe, K. *J. Chem. Phys.* **30**, 1180 (1959).
98. Frost, D. C., Mak, D. and McDowell, C. A. *Can. J. Chem.* **40**, 1064 (1962).
99. Hernandez, G. J. *J. Chem. Phys.* **38**, 1644 (1963).
100. Hernandez, G. J. *J. Chem. Phys.* **38**, 2233 (1963).
101. Chupka, W. A. *J. Chem. Phys.* **30**, 191 (1959).
102. Watanabe, K. and Nakayama, T. *J. Chem. Phys.* **29**, 48 (1958).
103. Murad, E. and Inghram, M. G. *J. Chem. Phys*, **40**, 3263 (1964).
104. Higasi, K., Nozoe, T. and Omura, I. *Bull. Chem. Soc. Japan* **30**, 408 (1957).
105. Foffani, A., Pignataro, S., Cantone, B. and Grasso, F. *Z. Phys. Chem. Neue Folge* **42**, 221 (1964).
106. Natalis, P. and Franklin, J. L. *J. Phys. Chem.* **69**, 2943 (1965).
107. Cotter, J. L. *J. Chem. Soc.* **5742** (1965).
108. Cotter, J. L. *J. Chem. Soc.* **5477** (1964).
109. Iczkowski, R. P. and Margrave, J. L. *J. Chem. Phys.* **30**, 403 (1959).
110. Frost, D. C. and McDowell, C. A. *Can. J. Chem.* **36**, 39 (1958).
111. Hildenbrand, D. L. and Muran, E. *J. Chem. Phys.* **43**, 1400 (1965).
112. Marriott, J. and Craggs, J. D. *J. Electronics and Control* **3**, 194 (1957).
113. Fisher, I. P., Homer, J. B. and Lossing, F. P. *J. Am. Chem. Soc.* **87**, 957 (1965).
114. Pottie, R. F. *J. Chem. Phys.* **42**, 2607 (1965).
115. Herron, J. T. and Dibeler, V. H. *J. Res. Natl. Bur. Std.* **65**, 405 (1961).
116. Loughran, E. D. and Mader, C. J. *Chem. Phys.* **32**, 1578 (1960).
117. Frost, D. C. and McDowell, C. A. *The Determination of Ionization and Dissociation Potentials of Molecules by Radiation with Electrons* Department of Chemistry, University of British Columbia, Vancouver 8, B. C., Canada, **AFCRL-TR-60-423 1** (1960).
118. Dibeler, V. H., Reese, R. M. and Franklin, J. L. *J. Chem. Phys.* **27**, 1296 (1957).
119. Wilson, E. G. Jortner, J. and Rice, S. A. *J. Am. Chem. Soc.* **85**, 813 (1963).
120. Frost, D. C. and McDowell, C. A. *Proc. Roy. Soc.* **A241**, 194 (1957).
121. Momigny, J. *Nature* **199**, 1179 (1963).
122. Martin, R. H., Lampe, F. W. and Taft, R. W. *J. Am. Chem. Soc.* **88**, 1353 (1966).
123. Reese, R. M. and Dibeler, V. H. *J. Chem. Phys.* **24**, 1175 (1956).
124. Herron, J. T. and Dibeler, V. H. *J. Chem. Phys.* **33**, 1595 (1960).
125. Hammond, V. J., Price, W. C., Teegan, J. P. and Walsh, A. D. *Disc. Faraday Soc.* **9**, 53 (1950).
126. Hudson, R. D. *J. Chem. Phys.* **43**, 1790 (1965).
127. Margrave, J. L. *J. Chem. Phys.* **41**, 2250 (1964).

REFERENCES (Continued)

128. Ehlert, T. C. and Margrave, J. L. *J. Am. Chem. Soc.* **86**, 3901 (1964).
129. Hobrock, B. G. and Kiser, R. W. *J. Phys. Chem.* **65**, 2186 (1961).
130. Hess, G. G., Lampe, F. W. and Sommer, L. H. *J. Am. Chem. Soc.* **87**, 5327 (1965).
131. Price, W. C. and Passmore, T. R. *Disc. Faraday Soc.* **35**, 232 (1963).
132. Wada, Y. and Kiser, R. W. *J. Phys. Chem.* **68**, 2290 (1964).
133. Berkowitz, J. and Chupka, W. A. *J. Chem. Phys.* **40**, 287 (1964).
134. Palmer, T. F. and Lossing, F. P. *J. Am. Chem. Soc.* **84**, 4661 (1962).
135. Price, W. C. *Bull. Am. Phys. Soc.* **10**, 9 (1935).
136. Hobrock, B. G. and Kiser, R. W. *J. Phys. Chem.* **66**, 1551 (1962).
137. Hobrock, B. G. and Kiser, R. W. *J. Phys. Chem.* **67**, 648 (1963).
138. Hobrock, B. G. and Kiser, R. W. *J. Phys. Chem.* **66**, 1648 (1962).
139. Khvostenko, V. I. *Russ. J. Phys. Chem.* **36**, 197 (1962).
140. Hobrock, B. G. and Kiser, R. W. *J. Phys. Chem.* **67**, 1283 (1963).
141. Reese, R. M., Dibeler, V. H. and Franklin, J. L. *J. Chem. Phys.* **29**, 880 (1958).
142. Baldwin, M., Maccoll, A., Kirkien-Konasiewicz, A. and Saville, B. *Chem. Ind.* **286** (1966).
143. Gowenlock, B. G., Kay, J. and Majer, J. R. *Trans. Faraday Soc.* **59**, 2463 (1963).
144. Berkowitz, J., Tasman, H. A. and Chupka, W. A. *J. Chem. Phys.* **36**, 2170 (1962).
145. Price, W. C. *J. Chem. Phys.* **4**, 539 (1936).
146. Harrison, A. G. and Shannon, T. W. *Can. J. Chem.* **40**, 1730 (1962).
147. Herron, J. T. and Dibeler, V. H. *J. Am. Chem. Soc.* **82**, 1555 (1960).
148. Margrave, J. L. *J. Chem. Phys.* **31**, 1432 (1959).
149. Majer, J. R. and Patrick, C. R. *Trans. Faraday Soc.* **58**, 17 (1962).
150. Dibeler, V. H., Reese, R. M. and Mann, D. E. *J. Chem. Phys.* **27**, 176 (1957).
151. Foffani, A., Pignataro, S., Cantone, B. and Grasso, F. *Nuovo Cimento* **29**, 918 (1963).
152. Momigny, J. and Wirtz-Cordier, A. M. *Ann. Soc. Sci. Bruxelles* **76**, 164 (1962).
153. Hobrock, D. L. and Kiser, R. W. *J. Phys. Chem.* **68**, 575 (1964).
154. Winters, R. E. and Kiser, R. W. *J. Organometal. Chem.* **4**, 190 (1965).
155. Vilesov, F. I. and Kurbatov, B. L. *Dokl. Phys. Chem., Proc. Acad. Sci.* **USSR 140**, 792 (1961).
156. Hobrock, B. G. and Kiser, R. W. *J. Phys. Chem.* **66**, 155 (1962).
157. Westmore, J. B., Mann, K. H. and Tickner, A. W. *J. Phys. Chem.* **68**, 606 (1964).
158. Cullen, W. R. and Frost, D. C. *Can. J. Chem.* **40**, 390 (1962).
159. Morrison, J. D., Hurzeler, H., Inghram, M. G. and Stanton, H. E. *J. Chem. Phys.* **33**, 821 (1960).
160. Berkowitz, J. and Marquart, J. R. *J. Chem. Phys.* **37**, 1853 (1962).
161. Irsa, A. P. and Friedman, L. *J. Inorg. Nucl. Chem.* **6**, 77 (1958).
162. Price, W. C. *J. Chem. Phys.* **4**, 547 (1936).
163. Dillard, J. G. and Kiser, R. W. *J. Phys. Chem.* **69**, 3893 (1965).
164. Frost, D. C. and McDowell, C. A. *Can. J. Chem.* **38**, 407 (1960).
165. Thomas, R. K., *Proc. R. Soc. London Ser. A,* 331, 249, 1972.

ELECTRON WORK FUNCTIONS OF THE ELEMENTS

Compiled by Herbert B. Michaelson, 1977

The measured values cited for polycrystalline and single-crystal specimens are selected as being the best available data at this time. The selection is based on (1) The validity of the experimental technique (e.g., vacua of 10^{-9} or 10^{-10} Torr, clean surfaces, and identification of crystal-face distribution and other surface conditions), and (2) Best agreement with preferred values and theoretical values of the true work function (given variously by Fomenko,[1] Rivière,[2] Trasatti,[3] and Lang and Kohn[4]). Experimental data that are not well substantiated according to these criteria are listed in *italics*. Crystallographic directions for single-crystal data are indicated by parentheses.

Abbreviations apply to the experimental method: T, thermionic; P, photoelectric; CPD, contact potential difference; F, field emission. Important distinctions among such measurements are discussed in the Rivière[2] paper, pp. 180 to 198.

Element	Experimental value, ϕ (eV)	Experimental method	Ref.
Ag	*4.26*	P	5
	4.64 (100)	P	5
	4.52 (110)	P	5
	4.74 (111)	P	6
Al	4.28	P	7
	4.41 (100)	P	8
	4.06 (110)	P	7
	4.24 (111)	P	8
As	*3.75*	P	9
Au	5.1	P	10
	5.47 (100)	P	11
	5.37 (110)		11
	5.31 (111)		11
B	*4.45*	T	12
Ba	*2.7*	T	13
Be	4.98	P	14
Bi	*4.22*	P	15
C	*5.0*	CPD	16
Ca	2.87	P	17
Cd	*4.22*	CPD	18
Ce	2.9	P	10
Co	5.0	P	10
Cr	4.5	P	10
Cs	2.14	P	19
Cu	4.65	P	10
	4.59 (100)	P	20
	4.48 (110)	P	20
	4.94 (111)	P	20
	4.53 (112)	P	20
Eu	2.5	P	10
Fe	4.5	P	10
	4.67 (100)	P	21
	4.81α (111)	P	22
	4.70α	P	23
	4.62β	P	23
	4.68γ	P	23
Ga	4.2	CPD	24
Ge	5.0	CPD	25
	4.80 (111)	P	26
Gd	3.1	P	10
Hf	3.9	P	10
Hg	4.49	P	27
In	4.12	P	28
Ir	*5.27*	T	29
	5.42 (110)	F	30
	5.76 (111)	F	30
	5.67 (100)	F	31
	5.00 (210)	F	31
K	2.30	P	32
La	3.5	P	10
Li	*2.9*	F	33
Lu	*3.3*	CPD	34
Mg	*3.66*	P	35
Mn	4.1	P	10
Mo	4.6	P	10
	4.53 (100)	P	36
	4.95 (110)	P	36
	4.55 (111)	P	36
	4.36 (112)	P	36
	4.50 (114)	P	36
	4.55 (332)	P	36
Na	2.75	P	37
Nb	4.3	P	10
	4.02 (001)	T	38
	4.87 (110)	T	38
	4.36 (111)	T	38
	4.63 (112)	T	38
	4.29 (113)	T	38
	3.95 (116)	T	38
	4.18 (310)	T	38
Nd	3.2	P	10
Ni	5.15	P	10
	5.22 (100)	P	39
	5.04 (110)	P	39
	5.35 (111)	P	39
Os	*4.83*	T	29
Pb	4.25	P	40
Pd	5.12	P	31
	5.6 (111)	P	41
Pt	5.65	P	10
	5.7 (111)	P	41
Rb	*2.16*	P	27

Element	Experimental value, ϕ (eV)	Experimental method	Ref.
Re	*4.96*	T	29
	5.75 (1011)	F	33
Rh	4.98	P	31
Ru	4.71	P	31
Sb	4.55 (amorph.)	–	42
	4.7 (100)	–	43
Sc	3.5	P	10
Se	5.9	P	44
Si	4.85n	CPD	40
	*4.91*p (100)	CPD	45
	4.60p (111)	P	46
Sm	2.7	P	10
Sn	*4.42*	CPD	47
Sr	*2.59*	T	48
Ta	4.25	T	29
	4.15 (100)	T	49
	4.80 (110)	T	49
	4.00 (111)	T	49
Tb	3.0	P	50
Te	4.95	P	44
Th	3.4	T	51
Ti	4.33	P	10
Tl	*3.84*	CPD	52
U	3.63	P & CPD	53
	3.73 (100)	P & CPD	54
	3.90 (110)	P & CPD	54
	3.67 (113)	P & CPD	54
V	4.3	P	10
W	4.55	CPD	55
	4.63 (100)	F	30
	5.25 (110)	F	30
	4.47 (111)	F	30
	4.18 (113)	CPD	56
	4.30 (116)	T	57
Y	3.1	P	10
Zn	*4.33*	P	15
	4.9 (0001)	CPD	58
Zr	4.05	P	10

REFERENCES

1. **Fomenko, V. S.,** *Emission Properties of Materials,* 3rd ed., Naukova Dumka, Kiev, 1970 (in Russian).
2. **Rivière, J. C.,** *Solid State Surface Science,* Green, M., Ed., Vol. 1, Marcel Dekker, 1969, chap. 4.
3. **Trasatti, S.,** *Chim. Ind.* (Milan), 53(6), 559, 1971.
4. **Lang, N. D. and Kohn, W.,** *Phys. Rev. B,* 3(4), 1215, 1971.
5. **Dweydari, A. W. and Mee, C. H. B.,** *Phys. Status Solidi A,* 27, 223, 1975.
6. **Dweydari, A. W. and Mee, C. H. B.,** *Phys. Status Solidi A,* 17, 247, 1973.
7. **Eastment, R. M. and Mee, C. H. B.,** *J. Phys. F,* 3, 1738, 1973.
8. **Grepstad, J. K., Gartland, P. O., and Slagsvold, B. J.,** *Surf. Sci.,* 57, 348, 1976.
9. **Raisin, C. and Pinchaux, R.,** *Solid State Commun.,* 16, 941, 1975.
10. **Eastman, D. E.,** *Phys. Rev. Sect. B,* 2, 1, 1970.
11. **Potter, H. C. and Blakeley, J. M.,** *J. Vac. Sci. Technol.,* 12, 635, 1975 and Potter, H. C., Ph.D. thesis Cornell University, Materials Science Center Rep. No. 1353, 1970.
12. **Adirovich, E. I. and Gol'dshtein, L. M.,** *Fiz. Tverdogo Tela* (Leningrad), 9, 1258, 1967.
13. **Bondarenko, B. V. and Makhov, V. I.,** *Sov. Phys. Solid State,* 12(7), 1522, 1971.
14. **Gustafsson, Broden, and Nilsson,** *J. Phys. F,* 4, 2351, 1974.
15. **Suhrmann, R. and Wedler, G.,** *Z. Angew. Phys.,* 14, 70, 1962.
16. **Robrieux, B., Faure, R., and Dussaulcy, J. P.,** *C. R. Acad. Sci. Ser. B,* 278(14), 659, 1974.
17. **Gaudart, L. and Riviora, R.,** *Appl. Opt.,* 10, 2336, 1971.
18. **Anderson, P. A.,** *Phys. Rev.,* 98, 1739, 1955.
19. **Boutry, G. A. and Dormont, H.,** *Philips Tech. Rev.,* 30, 225, 1969.
20. **Gartland, P. O.,** *Phys. Norv.,* 6(3,4), 201, 1972.
21. **Ueda, K. and Shimizu, R.,** *Jp. J. Appl. Phys.,* 11(6), 916, 1972.
22. **Kobayashi, H. and Kato, S.,** *Surf. Sci.,* 18(2), 341, 1969.
23. **Cardwell, A.,** *Phys. Rev.,* 92, 554, 1953.
24. **Osipova, E. V., Shurmovskaya, N. A., and Burshtein, R. Kh.,** *Elektrokhimiya,* 5(10), 1139, 1969 (in Russian).
25. **Boiko, B. A., Gorodetskii, D. A., and Yas'ko, A. A.,** *Sov. Phys. Solid State,* 15(11), 2101, 1974.
26. **Gobeli, G. W. and Allen, F. G.,** *Surf. Sci.,* 2, 402, 1964.
27. **Lazarev, V. B. and Malov, Y. I.,** *Fiz. Met. Metalloved.,* 24(3), 565, 1967.
28. **Peisner, J., Roboz, P., and Barna, P. B.,** *Phys. Stat. A,* 4, K187, 1971.
29. **Wilson, R. G.,** *J. Appl. Phys.,* 37, 3170, 1966.
30. **Strayer, R. W., Mackie, W., and Swanson, L. W.,** *Surf. Sci.,* 34, 225, 1973.
31. **Nieuwenhuys, Bouwman, and Sachtler,** *Thin Solid Films,* 21, 51, 1974.
32. **Van Oirschot, Th. G. J., van den Brink, M., and Sachtler, W. H. M.,** *Surf. Sci.,* 29, 189, 1972.
33. **Ovchinnikov, A. P. and Tsarev, B. M.,** *Sov. Phys. Solid State,* 9(12), 2766, 1968.
34. **Bondarenko, B. V. and Makhov, V. I.,** *Sov. Phys. Solid State,* 12, 2986, 1971.
35. **Garron, R.,** *C. R. Acad. Sci.,* 258, 1458, 1964.
36. **Berge, Gartland, and Slagsvold,** *Surf. Sci.,* 43, 275, 1974.
37. **Whitefield, R. J. and Brady, J. J.,** *Phys. Rev. Lett.,* 26(7), 380, 1971.
38. **Leblanc, R. P., Vanbrugghe, B. C., and Girouard, F. E.,** *Can. J. Phys.,* 52, 1589, 1974.
39. **Baker, B. G., Johnson, E. B., and Maire, G. I. C.,** *Surf. Sci.,* 24, 572, 1971.
40. **Thanailakis, A.,** *Inst. Phys. Conf. Ser.,* p. 59, 1974.
41. **Demuth, J. E.,** *Chem. Phys. Lett.,* 45, 12, 1977.
42. **Gorodetskii, D. A. and Yas'ko, A. A.,** *Sov. Phys. Solid State,* 13(11), 2928, 1972.
43. **Gorodetskii, D. A. and Yas'ko, A. A.,** *Sov. Phys. Solid State,* 13(5), 1085, 1971.
44. **Williams, R. H. and Polanco, J. I.,** *J. Phys. C,* 7, 2745, 1974.
45. **Allen, F. G.,** *J. Phys. Chem. Solids,* 8, 119, 1959.
46. **Allen, F. G. and Gobeli, G. W.,** *J. Appl. Phys.,* 35, 597, 1964.
47. **Simmons, J. G.,** *Phys. Rev. Lett.,* 10, 10, 1963.
48. **Alleau, T.,** *Surface Phenomena in Thermionic Emitters, Round Table Conf.,* Inst. Tech. Phys. Julich Nucl. Res. Establ., Julich, Germany, 1969, p. 54 (in English).
49. **Protopopov, Mikheeva, Shreinberg, and Shuppe,** *Fiz. Tverdogo Tela,* 8(4), 1140, 1966.
50. **Nemchenok, R. L., Strakovskaya, S. E., and Titenskii, A. I.,** *Fiz. Tverdogo Tela* 11(9), 2692, 1969.
51. **Estrup, P. J., Anderson, J. R., and Danforth, W. E.,** *Surf. Sci.,* 4, 286, 1966.
52. **Klein, O. and Lange, E.,** *Z. Elektrochem.,* 44, 542, 1938.
53. **Hopkins, B. J. and Sargood, A. J.,** *Nuovo Cimento,* 5, 459, 1967.
54. **Lea, C. and Mee, C. H. B.,** *J. Appl. Phys.,* 39, 5890, 1968.
55. **Hopkins, B. J. and Rivière, J. C.,** *Proc. Phys. Soc.* (London), 81, 590, 1963.
56. **Love, H. M. and Dyer, G. L.,** *Can. J. Phys.,* 40, 1837, 1962.
57. **Sultanov, V. M.,** *Radio Eng. Electron.,* 9, 252, 1964 (English translation).
58. **Baker, J. M. and Blakeley, J. M.,** *Surf. Sci.,* 32, 45, 1972.

TABLES OF PROPERTIES OF SEMICONDUCTORS

Compiled by Dr. Brian Randall Pamplin

The term "semiconductor" is applied to a material in which electric current is carried by electrons or holes and whose electrical conductivity when extremely pure rises exponentially with temperature and may be increased from this low "intrinsic" value by many orders of magnitude by "doping" with electrically active impurities.

Semiconductors are characterised by an energy gap in the allowed energies of electrons in the material which separates the normally filled energy levels of the *valence band* (where "missing" electrons behave like positively charged current carriers "holes") and the *conduction band* (where electrons behave rather like a gas of free negatively charged carriers with an effective mass dependent on the material and the direction of the electrons' motion). This energy gap depends on the nature of the material and varies with direction in anisotropic crystals. It is slightly dependent on temperature and pressure, and this dependence is usually almost linear at normal temperatures and pressures.

The data is presented in three tables. Table I "General Properties of Semiconductors" lists the main crystallographic and semiconducting properties of a large number of semiconducting materials in three main categories; "Tetrahedral Semiconductors" in which every atom is tetrahedrally co-ordinated to four nearest neighbour atoms (or atomic sites) as for example in the diamond structure; "Octahedral Semiconductors" in which every atom is octahedrally co-ordinated to six nearest neighbour atoms—as for example in the halite structure; and "Other Semiconductors".

Table II gives more detailed information about some better known semiconductors, while Table III gives some information about the electronic energy band structure parameters of the best known materials.

TABLE I

GENERAL PROPERTIES OF SEMICONDUCTORS

(listed by Crystal Structure)

Substance	Lattice Parameters (A° Room temperature)	Density (gm/cc)	Melting Point (°K)	Minimum Room Temperature Energy Gap (eV)	Thermal Conductivity (mW cm^{-1} K^{-1})	Heat of Formation k cal/mole	Mobility (Room Temperature) (cm²/V.s) Electrons	Holes	Remarks
PART A ADAMANTINE SEMICONDUCTORS									
§A1 Diamond Structure Elements (Strukturbericht symbol A4, Space Group Fd3m - O_h^7)									
C	3.5597	3.51	4300	5.4	2000	161	1800	1400	
Si	5.43072	2.3283	1685	1.107	1240	77.5	1900	500	
Ge	5.65754	5.3234	1231	0.67	640	69.5	3800	1820	
α - Sn	6.4912	5.765	503	0.08		64	2500	2400	
§A2 Sphalerite (Zinc Blende) Structure Compounds (Strukturbericht symbol B3 Space Group F $\bar{4}$ 3m - T_d^2)									
I VII Compounds									
CuF	4.255								
CuCl	5.4057	3.53	695						
CuBr	5.6905	4.72	770	2.94		115			
CuI	6.0427	5.63	878			105			
AgBr						102	4000		
AgI	6.473	5.67				93	30		
II VI Compounds									
BeS	4.865	2.36							
BeSe	5.139	4.315							
BeTe	5.626	5.090							
BePo	8.38	7.3							
Zn0	4.63								see § A3
ZnS	5.4093	4.079	2100	3.54	140	114	180	5(400°C)	see also § A3
ZnSe	5.6676	5.42	1790	2.58	140	101	540	28	
ZnTe	6.101	5.72	1568	2.26	140	90	340	100	
ZnPo									
CdS	5.5818		1750						see § A3
CdSe	6.05		1512						see § A3
CdTe	6.477	5.86	1365	1.44	55	81	1200	50	
CdPo									
HgS	5.8517	7.73	~2020						usually cinnabar
HgSe	6.084	8.25	1070	-0.10	10	59	20000		
HgTe	6.460	8.17	943	-0.15	20	58	25000	350	
III V Compounds									
BN	3.615	3.49	3000	~4	200	195			
BP(L.T.)	4.538	2.9		~6			500	70	
BAs	4.777								
AlP	5.451	2.85	1770	2.5					
AlAs	5.6622	3.81	1870	2.16		150	1200	420	
AlSb	6.1355	4.218	1330	1.60	600	140	200–400	550	
GaP	5.4505	4.13	1750	2.24	1100	152	300	100	
GaAs	5.65315	5.316	1510	1.35	370	128	8800	400	
GaSb	6.0954	5.619	980	0.67	270	118	4000	1400	
InP	5.86875	4.787	1330	1.27	800	134	4600	150	
InAs	6.05838	5.66	1215	0.36	290	114	33000	460	
InSb	6.47877	5.775	798	0.165	160	107	78000	750	
Other Sphalerite Structure Compounds									
MnS	5.60								see also § A3
MnSe	5.82								see also § A3
β - SiC	4.348	3.21	3070	2.3			4000		
Ga_2Te_3	5.899	5.75	1063	~1.0	~14	65			
In_2Te_3(H.T.)	6.150	5.8	940	~1.0	~8	47.4	~10		
$MgGeP_2$	5.652								
$ZnSnP_2$	5.65			2.1					
$ZnSnAs_2$(H.T.)	5.851	5.53	1050	~0.7	70				

TABLE 1

GENERAL PROPERTIES OF SEMICONDUCTORS (Continued)

Substance	Lattice Parameters (A° Room temperature)		Density (gm/cc)	Melting Point (°K)	Minimum Room Temperature Energy Gap (eV)	Thermal Conductivity (mW cm⁻¹ K⁻¹)	Heat of Formation k cal/mole	Mobility (Room Temperature) (cm²/V.s) Electrons	Holes	Remarks
§ A3 Wurtzite (Zincite) Structure Compounds (Strukturbericht symbol B4, Space Group P $6_3mc - C^4_{6v}$)										
I VII Compounds										
CuCl	3.91	6.42		T_c 680°K						
CuBr	4.06	6.66		T_c 658°K						
CuI	4.31	7.09								
AgI	4.580	7.494			2.63					
II VI Compounds										
BeO	2.698	4.380		2800						
MgTe	4.54	7.39	3.85	~2800						
ZnO	3.24950	5.2069	5.66	2250	3.2	6	154	180		
ZnS	3.8140	6.2576	4.1	2100	3.67		110			
ZnSe	3.996	6.626		1793						
ZnTe	4.27	6.99		1568						
CdS	4.1348	6.7490	4.82	1748	2.42		96	400		
CdSe	4.299	7.010	5.66	1512	1.74		90	650		
CdTe	4.57	7.47			1.50					
III V Compounds										
BP(H.T.)	3.562	5.900								
AlN	3.111	4.978	3.26	~2500	6.02		197			
GaN	3.190	5.189	6.10	1500	3.34		157			
InN	3.533	5.693	6.88	1200	2.0		133			
Other Wurtzite Structure Compounds										
MnS	3.985	6.45								
MnSe	4.12	6.72								
SiC	3.076	5.048								
MnTe	4.078	6.701			~1.0					
Al_2S_3	3.579	5.829	2.55		4.1		426			
Al_2Se_3	3.890	6.30	3.91		3.1		367			
§ A$\bar{4}$ Chalcopyrite Structure Compounds (Strukturbericht symbol $E1_1$, Space Group I $\bar{4}$ 2d $- D^{12}_{2d}$)										
I III VI_2 Compounds										
$CuAlS_2$	5.323	10.44	3.47		2.5					
$CuAlSe_2$	5.617	10.92	4.70	1270	1.1					
$CuAlTe_2$	5.976	11.80	5.50	1160	0.88					
$CuGaS_2$	5.360	10.49	4.35							
$CuGaSe_2$	5.618	11.01	5.56	1310	0.96, 1.63					
$CuGaTe_2$	6.013	11.93	5.99	1150	0.82, 1.0					
$CuInS_2$	5.528	11.08	4.75		1.2					
$CuInSe_2$	5.785	11.57	5.77	1250	0.86, 0.92	37				
$CuInTe_2$	6.179	12.365	6.10	970	0.95	49				
$CuTlS_2$	5.580	11.17	6.32							
$CuTlSe_2$(L.T.)	5.844	11.65	7.11	680	1.07					
$CuFeS_2$	5.25	10.32		1150	0.53					
$CuFeSe_2$				850	0.16					
$CuLaS_2$	5.65	10.86								
$AgAlS_2$	5.707	10.28	3.94							
$AgAlSe_2$	5.968	10.77	5.07	1220	0.7					
$AgAlTe_2$	6.309	11.85	6.18	1000	0.56					
$AgGaS_2$	5.755	10.28	4.72							
$AgGaSe_2$	5.985	10.90	5.84	1120	1.66					
$AgGaTe_2$	6.301	11.96	6.05	990	1.1	10				
$AgInS_2$(L.T.)	5.828	11.19	5.00		1.9					
$AgInSe_2$	6.102	11.69	5.81	1053	1.18	30				
$AgInTe_2$	6.42	12.59	6.12	965	0.96, 0.52					
$AgFeS_2$	5.66	10.30	4.53							
II IV V_2 Compounds										
$ZnSiP_2$	5.400	10.441	3.39	1640	2.3			1000		
$ZnGeP_2$	5.465	10.771	4.17	1295	2.2					
$CdSiP_2$	5.678	10.431	4.00	~1470	2.2			1000		
$CdGeP_2$	5.741	10.775	4.48	~1060	1.8					
$CdSnP_2$	5.900	11.518			1.5					
$ZnSiAs_2$	5.61	10.88	4.70	~1350	1.7				50	
$ZnGeAs_2$	5.672	11.153	5.32	~1150	0.85	110				
$ZnSnAs_2$	5.8515	11.704	5.53	~ 910	0.65	150			300	disorders at 910°K
$CdSiAs_2$	5.884	10.882								
$CdGeAs_2$	5.9427	11.2172	5.60	~ 903	0.53	40		70	25	disorders at 903°K
$CdSnAs_2$	6.0944	11.9182	5.72	880	0.26	70		22000	250	
§ A5 "Defect Chalcopyrite" Structure Compounds (Strukturbericht symbol E3, Space Group I $\bar{4}$ $- S^2_4$)										
$ZnAl_2Se_4$	5.503	10.90	4.37							
$ZnAl_2Te_4$(?)	5.904	12.05	4.95							
$ZnGa_2S_4$(?)	5.274	10.44	3.80							
$ZnGa_2Se_4$(?)	5.496	10.99	5.21							
$ZnGa_2Te_4$(?)	5.937	11.87	5.67		1.35					
$ZnIn_2Se_4$	5.711	11.42	5.44	1250	1.82					
$ZnIn_2Te_4$	6.122	12.24	5.83	1075	1.2					
$CdAl_2S_4$	5.564	10.32	3.06							
$CdAl_2Se_4$	5.747	10.68	4.54							
$CdAl_2Te_4$(?)	6.011	12.21	5.10							
$CdGa_2S_4$	5.577	10.08	4.03							
$CdGa_2Se_4$	5.743	10.73	5.32							
$CdGa_2Te_4$	6.093	11.81	5.77							
$CdIn_2Te_4$	6.205	12.41	5.9	1060	(1.26 or 0.9)					

STANDARD CALIBRATION TABLES FOR THERMOCOUPLES

The following tables which represent the Temperature-E.M.F. functions of various thermocouples should be used with appropriate correction curves if precise results are desired. These curves must be determined for each individual couple by plotting ΔE, the difference between the observed and the standard E.M.F., against the standard E.M.F. at three or more fixed temperature points. The value ΔE as shown by such a correction curve is then subtracted algebraically from the observed E.M.F. to give the true E.M.F. reading.

In the following tables the fixed or "cold junction" is at 0°C.; when the cold junction is not maintained at 0°C. the readings of the E.M.F. must be corrected as follows: $Et = E_{(t-tc)} + Etc$ where $E_{(t-tc)}$ is the observed reading, *Etc* is the E.M.F. for the temperature corresponding to the cold junction temperature as read from the standard table and *Et* is the E.M.F. produced by the jot junction corrected to the value which would be obtained with the cold junction at 0°C. The temperature corresponding to *Et* is then obtained by reference to the standard table.

Since the E.M.F.-temperature function is not linear the cold junction should be maintained at a temperature very close to that at whch the thermocouple was calibrated. Otherwise considerable error will result despite the above correction.

PLATINUM VERSUS PLATINUM-10-PERCENT RHODIUM THERMOCOUPLES

(Electromotive Force in Absolute Millivolts. Temperatures in Degrees C (Int. 1948). Reference Junctions at 0°C.)

°C	0	10	20	30	40	50	60	70	80	90
					Millivolts					
0	0	0.06	0.11	0.17	0.24	0.30	0.36	0.43	0.50	0.57
100	0.64	0.72	0.79	0.87	0.95	1.03	1.11	1.19	1.27	1.35
200	1.44	1.52	1.61	1.69	1.78	1.87	1.96	2.05	2.14	2.23
300	2.32	2.41	2.50	2.59	2.69	2.78	2.87	2.97	3.06	3.16
400	3.25	3.35	3.44	3.54	3.64	3.73	3.83	3.93	4.02	4.12
500	4.22	4.32	4.42	4.52	4.62	4.72	4.82	4.92	5.02	5.12
600	5.22	5.33	5.43	5.53	5.64	5.74	5.84	5.95	6.05	6.16
700	6.26	6.37	6.47	6.58	6.68	6.79	6.90	7.01	7.11	7.22
800	7.33	7.44	7.55	7.66	7.77	7.88	7.99	8.10	8.21	8.32
900	8.43	8.55	8.66	8.77	8.88	9.00	9.11	9.23	9.34	9.46
1000	9.57	9.6	9.80	9.92	10.04	10.15	10.27	10.39	10.51	10.62
1100	10.74	10.86	10.98	11.10	11.22	11.34	11.46	11.58	11.70	11.82
1200	11.94	12.06	12.18	12.30	12.42	12.54	12.66	12.78	12.90	13.02
1300	13.14	13.26	13.38	13.50	13.62	13.74	13.86	13.98	14.10	14.22
1400	14.34	14.46	14.58	14.70	14.82	14.94	15.05	15.17	15.29	15.41
1500	15.53	15.65	15.77	15.89	16.01	16.12	16.24	16.36	16.48	16.60
1600	16.72	16.83	16.95	17.07	17.19	17.31	17.42	17.54	17.66	17.77
1700	17.89	18.01	18.12	18.24	18.36	18.47	18.59	—	—	—

(Electromotive Force in Absolute Millivolts. Temperatures in Degrees F.* Reference Junctions at 32°F.)

°F	0	10	20	30	40	50	60	70	80	90
					Millivolts					
0	—	—	—	—	0.02	0.06	0.09	0.12	0.15	0.19
100	0.22	0.26	0.29	0.33	0.36	0.40	0.44	0.48	0.52	0.56
200	0.60	0.64	0.68	0.72	0.76	0.80	0.84	0.89	0.93	0.97
300	1.02	1.06	1.11	1.15	1.20	1.24	1.29	1.33	1.38	1.43
400	1.47	1.52	1.57	1.62	1.66	1.71	1.76	1.81	1.86	1.91
500	1.96	2.01	2.06	2.11	2.16	2.21	2.26	2.31	2.36	2.41
600	2.46	2.51	2.56	2.61	2.66	2.72	2.77	2.82	2.87	2.92
700	2.98	3.03	3.08	3.14	3.19	3.24	3.29	3.35	3.40	3.45
800	3.51	3.56	3.61	3.67	3.72	3.78	3.83	3.88	3.94	3.99
900	4.05	4.10	4.16	4.21	4.26	4.32	4.37	4.43	4.49	4.54
1000	4.60	4.65	4.71	4.76	4.82	4.87	4.93	4.99	5.04	5.10
1100	5.16	5.21	5.27	5.33	5.38	5.44	5.50	5.56	5.61	5.67
1200	5.73	5.78	5.84	5.90	5.96	6.02	6.07	6.13	6.19	6.25
1300	6.31	6.37	6.42	6.48	6.54	6.60	6.66	6.72	6.78	6.84
1400	6.90	6.96	7.02	7.08	7.14	7.20	7.26	7.32	7.38	7.44
1500	7.50	7.56	7.62	7.68	7.74	7.80	7.86	7.93	7.99	8.05
1600	8.11	8.17	8.23	8.30	8.36	8.42	8.48	8.55	8.61	8.67
1700	8.73	8.80	8.86	8.92	8.98	9.05	9.11	9.17	9.24	9.30
1800	9.37	9.43	9.49	9.56	9.62	9.69	9.75	9.82	9.88	9.94
1900	10.01	10.07	10.14	1.20	10.27	10.33	10.40	10.47	10.53	10.60
2000	10.66	10.73	10.79	10.86	10.93	10.99	11.06	11.12	11.19	11.26
2100	11.32	11.39	11.46	11.52	11.59	11.66	11.72	11.79	11.86	11.92
2200	11.99	12.06	12.12	12.19	12.26	12.32	12.39	12.46	12.52	12.59
2300	12.66	12.72	12.79	12.86	12.92	12.99	13.06	13.12	13.19	13.26
2400	13.33	13.39	13.46	13.53	13.59	13.66	13.73	13.79	13.86	13.92
2500	13.99	14.06	14.12	14.19	14.26	14.32	14.39	14.46	14.52	14.59
2600	14.66	14.72	14.79	14.86	14.92	14.99	15.05	15.12	15.19	15.25
2700	15.32	15.39	15.45	15.52	15.58	15.65	15.72	15.78	15.85	15.91
2800	15.98	16.05	16.11	16.18	16.24	16.31	16.37	16.44	16.51	16.57
2900	16.64	16.70	16.77	16.83	16.90	16.97	17.03	17.10	17.16	17.23
3000	17.29	17.36	17.42	17.49	17.55	17.62	17.68	17.75	17.81	17.88
3100	17.94	18.01	18.07	18.14	18.20	18.27	18.33	18.40	18.46	18.53
3200	18.59	18.66	—	—	—	—	—	—	—	—

* Based on the International Temperature Scale of 1948.

PLATINUM VERSUS PLATINUM-13-PERCENT RHODIUM THERMOCOUPLES

(Electromotive Force in Absolute Millivolts. Temperatures in Degrees C (Int. 1948) Reference Junctions at 0°C.)

°C	0	10	20	30	40	50	60	70	80	90
					Millivolts					
0	0.00	0.06	0.11	0.17	0.23	0.30	0.36	0.43	0.50	0.57
100	0.65	0.72	0.80	0.88	0.96	1.04	1.12	1.21	1.29	1.38
200	1.47	1.55	1.64	1.73	1.83	1.92	2.01	2.11	2.20	2.30
300	2.40	2.49	2.59	2.69	2.79	2.89	2.99	3.09	3.19	3.30
400	3.40	3.50	3.61	3.71	3.82	3.92	4.03	4.13	4.24	4.35
500	4.46	4.56	4.67	4.78	4.89	5.00	5.12	5.23	5.34	5.45
600	5.56	5.68	5.79	5.91	6.02	6.14	6.25	6.37	6.49	6.60
700	6.72	6.84	6.96	7.08	7.20	7.32	7.44	7.56	7.68	7.80
800	7.92	8.05	8.17	8.29	8.42	8.54	8.67	8.80	8.92	9.05
900	9.18	9.30	9.43	9.56	9.69	9.82	9.95	10.08	10.21	10.34
1000	10.47	10.60	10.74	10.87	11.00	11.14	11.27	11.41	11.54	11.68
1100	11.82	11.94	12.09	12.23	12.37	12.50	12.64	12.78	12.92	13.06
1200	13.19	13.33	13.47	13.61	13.75	13.89	14.03	14.17	14.30	14.44
1300	14.58	14.72	14.86	15.00	15.14	15.28	15.42	15.55	15.69	15.83
1400	15.97	16.11	16.25	16.39	16.52	16.66	16.80	16.94	17.08	17.22
1500	17.36	17.49	17.63	17.77	17.91	18.04	18.18	18.32	18.45	18.59
1600	18.73	18.86	19.00	19.14	19.27	19.41	19.55	19.68	19.82	19.95
1700	20.09	—	—	—	—	—	—	—	—	—

(Electromotive Force in Absolute Millivolts. Temperatures in Degrees F* Reference Junctions at 32°F.)

°F	0	10	20	30	40	50	60	70	80	90
					Millivolts					
0	—	—	—	—	0.02	0.06	0.09	0.12	0.15	0.19
100	0.22	0.26	0.29	0.33	0.36	0.40	0.44	0.48	0.52	0.56
200	0.60	0.64	0.68	0.72	0.76	0.81	0.85	0.89	0.94	0.98
300	1.03	1.08	1.12	1.17	1.21	1.26	1.31	1.36	1.41	1.46
400	1.50	1.55	1.60	1.65	1.70	1.75	1.81	1.86	1.91	1.96
500	2.01	2.07	2.12	2.17	2.22	2.28	2.33	2.38	2.44	2.49
600	2.55	2.60	2.66	2.71	2.77	2.82	2.88	2.94	2.99	3.05
700	3.10	3.16	3.22	3.27	3.33	3.39	3.45	3.50	3.56	3.62
800	3.68	3.74	3.79	3.85	3.91	3.97	4.03	4.09	4.14	4.21
900	4.26	4.32	4.38	4.44	4.50	4.56	4.62	4.69	4.75	4.81
1000	4.87	4.93	4.99	5.05	5.12	5.18	5.24	5.30	5.36	5.43
1100	5.49	5.55	5.61	5.68	5.74	5.81	5.87	5.93	6.00	6.06
1200	6.13	6.19	6.25	6.32	6.38	6.45	6.51	6.58	6.64	6.71
1300	6.77	6.84	6.90	6.97	7.04	7.10	7.17	7.24	7.30	7.37
1400	7.44	7.50	7.57	7.64	7.71	7.77	7.84	7.91	7.98	8.05
1500	8.12	8.18	8.25	8.32	8.39	8.46	8.53	8.60	8.67	8.74
1600	8.81	8.88	8.95	9.02	9.09	9.16	9.23	9.30	9.37	9.45
1700	9.52	9.59	9.66	9.73	9.80	9.87	9.95	10.02	10.09	10.16
1800	10.24	10.31	10.38	10.46	10.53	10.60	10.68	10.75	10.82	10.90
1900	10.97	11.05	11.12	11.20	11.27	11.35	11.42	11.50	11.58	11.65
2000	11.73	11.80	11.88	11.95	12.03	12.11	12.18	12.26	12.34	12.41
2100	12.49	12.56	12.64	12.72	12.80	12.87	12.95	13.03	13.10	13.18
2200	13.26	13.33	13.41	13.49	13.56	13.64	13.72	13.80	13.87	13.95
2300	14.03	14.10	14.18	14.26	14.34	14.41	14.49	14.57	14.64	14.72
2400	14.80	14.88	15.95	15.03	15.11	15.18	15.26	15.34	15.42	15.49
2500	15.57	15.65	15.72	15.80	15.88	15.95	16.03	16.11	16.19	16.26
2600	16.34	16.42	16.49	16.57	16.65	16.73	16.80	16.88	16.96	17.03
2700	17.11	17.19	17.26	17.34	17.42	17.49	17.57	17.65	17.72	17.80
2800	17.88	17.95	18.03	18.10	18.18	18.26	18.33	18.41	18.48	18.56
2900	18.64	18.71	18.79	18.86	18.94	19.02	19.09	19.17	19.24	19.32
3000	19.39	19.47	19.55	19.62	19.70	19.77	19.85	19.92	20.00	20.08
3100	20.15	—	—	—	—	—	—	—	—	—

CHROMEL-ALUMEL THERMOCOUPLES

(Electromotive Force in Absolute Millivolts. Temperatures in Degrees C (Int. 1948) Reference Junctions at 0°C)

°C	0	1	2	3	4	5	6	7	8	9
	Millivolts									
−190	−5.60	−5.62	−5.63	−5.65	−5.67	−5.68	−5.70	−5.71	−5.73	−5.74
−180	−5.43	−5.45	−5.46	−5.48	−5.50	−5.52	−5.53	−5.55	−5.57	−5.58
−170	−5.24	−5.26	−5.28	−5.30	−5.32	−5.34	−5.35	−5.37	−5.39	−5.41
−160	−5.03	−5.05	−5.08	−5.10	−5.12	−5.14	−5.16	−5.18	−5.20	−5.22
−150	−4.81	−4.84	−4.86	−4.88	−4.90	−4.92	−4.95	−4.97	−4.99	−5.01
−140	−4.58	−4.60	−4.62	−4.65	−4.67	−4.70	−4.72	−4.74	−4.77	−4.79
−130	−4.32	−4.35	−4.37	−4.40	−4.42	−4.45	−4.48	−4.50	−4.52	−4.55
−120	−4.06	−4.08	−4.11	−4.14	−4.16	−4.19	−4.22	−4.24	−4.27	−4.30
−110	−3.78	−3.81	−3.84	−3.86	−3.89	−3.92	−3.95	−3.98	−4.00	−4.03
−100	−3.49	−3.52	−3.55	−3.58	−3.61	−3.64	−3.66	−3.69	−3.72	−3.75
−90	−3.19	−3.22	−3.25	−3.28	−3.31	−3.34	−3.37	−3.40	−3.43	−3.46
−80	−2.87	−2.90	−2.93	−2.96	−3.00	−3.03	−3.06	−3.09	−3.12	−3.16
−70	−2.54	−2.57	−2.61	−2.64	−2.67	−2.71	−2.74	−2.77	−2.80	−2.84
−60	−2.29	−2.24	−2.27	−2.30	−2.34	−2.37	−2.41	−2.33	−2.47	−2.51
−50	−1.86	−1.89	−1.93	−1.96	−2.00	−2.03	−2.07	−2.10	−2.13	−2.17
−40	−1.50	−1.54	−1.57	−1.61	−1.64	−1.68	−1.72	−1.75	−1.79	−1.82
−30	−1.14	−1.17	−1.21	−1.25	−1.28	−1.32	−1.36	−1.39	−1.43	−1.37
−20	−0.77	−0.80	−0.84	−0.88	−0.92	−0.95	−0.99	−1.03	−1.06	−1.10
−10	−0.39	−0.42	−0.46	−0.50	−0.54	−0.58	−0.62	−0.66	−0.69	−0.73
(−)0	−0.00	−0.04	−0.08	−0.12	−0.16	−0.19	−0.23	−0.27	−0.31	−0.35
(+)0	0.00	0.04	0.08	0.12	0.16	0.20	0.24	0.28	0.32	0.36
10	0.40	0.44	0.48	0.52	0.56	0.60	0.64	0.68	0.72	0.76
20	0.80	0.84	0.88	0.92	0.96	1.00	1.04	1.08	1.12	1.16
30	1.20	1.24	1.28	1.32	1.36	1.40	1.44	1.49	1.53	1.57
40	1.61	1.65	1.69	1.73	1.77	1.81	1.85	1.90	1.94	1.98
50	2.02	2.06	2.10	2.14	2.18	2.23	2.27	2.31	2.35	2.39
60	2.43	2.47	2.51	2.56	2.60	2.64	2.68	2.72	2.76	2.80
70	2.85	2.89	2.93	2.97	3.01	3.05	3.10	3.14	3.18	3.22
80	3.26	3.30	3.35	3.39	3.43	3.47	3.51	3.56	3.60	3.63
90	3.68	3.72	3.76	3.81	3.85	3.89	3.93	3.97	4.01	4.06
100	4.10	4.14	4.18	4.22	4.26	4.31	4.35	4.39	4.43	4.47
110	4.51	4.55	4.60	4.64	4.68	4.72	4.76	4.80	4.84	4.88
120	4.92	4.96	5.01	5.05	5.09	5.13	5.17	5.21	5.25	5.29
130	5.33	5.37	5.41	5.45	5.49	5.53	5.57	5.61	5.65	5.69
140	5.73	5.77	5.81	5.85	5.89	5.93	5.97	6.01	6.05	6.09
150	6.13	6.17	6.21	6.25	6.29	6.33	6.37	6.41	6.45	6.49
160	6.53	6.57	6.61	6.65	6.69	6.73	6.77	6.81	6.85	6.89
170	6.93	6.97	7.01	7.05	7.09	7.13	7.17	7.21	7.25	7.29
180	7.33	7.37	7.41	7.45	7.49	7.53	7.57	7.61	7.65	7.69
190	7.73	7.77	7.81	7.85	7.89	7.93	7.97	8.01	8.05	8.09
200	8.13	8.17	8.21	8.25	8.29	8.33	8.37	8.41	−8.46	8.50
210	8.54	8.58	8.62	8.66	8.70	8.74	8.78	8.82	8.86	8.90
220	8.94	8.98	9.02	9.06	9.10	9.14	9.18	9.22	9.26	9.30
230	9.34	9.38	9.42	9.46	9.50	9.54	9.59	9.63	9.67	9.71
240	9.75	9.79	9.83	9.87	9.91	9.95	9.99	10.03	10.07	10.11
250	10.16	10.20	10.24	10.28	10.32	10.36	10.40	10.44	10.48	10.52
260	10.57	10.61	10.65	10.69	10.73	10.77	10.81	10.85	10.89	10.93
270	10.98	11.02	11.06	11.10	11.14	11.18	11.22	11.26	11.30	11.34
280	11.39	11.43	11.47	11.51	11.55	11.59	11.63	11.67	11.72	11.76
290	11.80	11.84	11.88	11.92	11.96	12.01	12.05	12.09	12.13	12.17
300	12.21	12.25	12.29	12.34	12.38	12.42	12.46	12.50	12.54	12.58
310	12.63	12.67	12.71	12.75	12.79	12.83	12.88	12.92	12.96	13.00
320	13.04	13.08	13.12	13.17	13.21	13.25	13.29	13.33	13.37	13.42
330	13.46	13.50	13.54	13.58	13.62	13.67	13.71	13.75	13.79	13.83
340	13.88	13.92	13.96	14.00	14.04	14.09	14.13	14.17	14.21	14.25
350	14.29	14.34	14.38	14.42	14.46	14.50	14.55	14.59	14.63	14.67
360	14.71	14.76	14.80	14.84	14.88	14.92	14.97	15.01	15.05	15.09
370	15.13	15.18	15.22	15.26	15.30	15.34	15.39	15.43	15.47	15.51
380	15.55	15.60	15.64	15.68	15.72	15.76	15.81	15.85	15.89	15.93
390	15.98	16.02	16.06	16.10	16.14	16.19	16.23	16.27	16.31	16.36
400	16.40	16.44	16.48	16.52	16.57	16.61	16.65	16.69	16.74	16.78
410	16.82	16.86	16.91	16.95	16.99	17.03	17.07	17.12	17.16	17.20
420	17.24	17.29	17.33	17.37	17.41	17.46	17.50	17.54	17.58	17.62
430	17.67	17.71	17.75	17.79	17.84	17.88	17.92	17.96	18.01	18.05
440	18.09	18.13	18.17	18.22	18.26	18.30	18.34	18.39	18.43	18.47
450	18.51	18.56	18.60	18.64	18.68	18.73	18.77	18.81	18.85	18.90
460	18.94	18.98	19.02	19.07	19.11	19.15	19.19	19.24	19.28	19.32
470	19.36	19.41	19.45	19.49	19.54	19.58	19.62	19.66	19.71	19.75
480	19.79	19.84	19.88	19.92	19.96	20.01	20.03	20.09	20.13	20.18
490	20.22	20.26	20.31	20.35	20.39	20.43	20.48	20.53	20.56	20.60
500	20.65	20.69	20.73	20.77	20.82	20.86	20.90	20.94	20.99	21.03
510	21.07	21.11	21.16	21.20	21.24	21.28	21.32	21.37	21.41	21.45
520	21.50	21.54	21.58	21.63	21.67	21.71	21.75	21.80	21.84	21.88
530	21.92	21.97	22.01	22.05	22.09	22.14	22.18	22.22	22.26	22.31
540	22.35	22.39	22.43	22.48	22.52	22.56	22.61	22.65	22.69	22.73
550	22.78	22.82	22.86	22.90	22.95	22.99	23.03	23.07	23.12	23.16
560	23.20	23.25	23.29	23.33	23.38	23.42	23.46	23.50	23.54	23.59
570	23.63	23.67	23.72	23.76	23.80	23.84	23.89	23.93	23.97	24.01
580	24.06	24.10	24.14	24.18	24.23	24.27	24.31	24.36	24.40	24.44
590	24.49	24.53	24.57	24.61	25.65	24.70	24.74	24.78	24.83	24.87
600	24.91	24.95	25.00	25.04	25.08	25.12	25.17	25.21	25.25	25.29
610	25.34	25.38	25.42	25.47	25.51	25.55	25.59	25.64	25.68	25.72
620	25.76	25.81	25.85	25.89	25.93	25.98	26.02	26.06	26.10	26.15
630	26.19	26.23	26.27	26.32	26.36	26.40	26.44	26.48	26.53	26.57
640	26.61	26.65	26.70	26.74	26.78	26.82	26.86	26.91	26.95	26.99
650	27.03	27.07	27.12	27.16	27.20	27.24	27.28	27.33	27.37	27.41
660	27.45	27.49	27.54	27.58	27.62	27.66	27.71	27.75	27.79	27.83
670	27.87	27.92	27.96	28.00	28.04	28.08	28.13	28.17	28.21	28.25
680	28.29	28.34	28.38	28.42	28.46	28.50	28.55	28.59	28.63	28.67
690	28.72	28.76	28.80	28.84	28.88	28.93	28.97	29.01	29.05	29.10
700	29.14	29.18	29.22	29.26	29.30	29.35	29.39	29.43	29.47	29.52
710	29.56	29.60	29.64	29.68	29.72	29.77	29.81	29.85	29.89	29.93
720	29.97	30.02	30.06	30.10	30.14	30.18	30.23	30.27	30.31	30.35
730	30.39	30.44	30.48	30.52	30.56	30.60	30.65	30.69	30.73	30.77
740	30.81	30.85	30.90	30.94	30.98	31.02	31.06	31.10	31.15	31.19
750	31.23	31.27	31.31	31.35	31.40	31.44	31.48	31.52	31.56	31.60
760	31.65	31.69	31.73	31.77	31.81	31.85	31.90	31.94	31.98	32.02
770	32.06	32.10	32.15	32.19	32.23	32.27	32.31	32.35	32.39	32.43
780	32.48	32.52	32.56	32.60	32.64	32.68	32.72	32.76	32.81	32.85
790	32.89	32.93	32.97	33.01	33.05	33.09	33.13	33.18	33.22	33.26
800	33.30	33.34	33.38	33.42	33.46	33.50	33.54	33.59	33.63	33.67
810	33.71	33.75	33.79	33.83	33.87	33.91	33.95	33.99	34.04	34.08
820	34.12	34.16	34.20	34.24	34.28	34.32	34.36	34.40	34.44	34.48
830	34.53	34.57	34.61	34.65	34.69	34.73	34.77	34.81	34.85	34.89
840	34.93	34.97	35.02	35.06	35.10	35.14	35.18	35.22	35.26	35.30
850	35.34	35.38	35.42	35.46	35.50	35.54	35.58	35.63	35.67	35.71
860	35.75	35.79	35.83	35.87	35.91	35.95	35.99	36.03	36.07	36.11
870	36.15	36.19	36.23	36.27	36.31	36.35	36.39	36.43	36.47	36.51
880	36.55	36.59	36.63	36.67	36.72	36.76	36.80	36.84	36.88	36.92
890	36.96	37.00	37.04	37.08	37.12	37.16	37.20	37.24	37.28	37.32
900	37.36	37.40	37.44	37.48	37.52	37.56	37.60	37.64	37.68	37.72
910	37.76	37.80	37.84	37.88	37.92	37.96	38.00	38.04	38.08	38.12
920	38.16	38.20	38.24	38.28	38.32	38.36	38.40	38.44	38.48	38.52
930	38.56	38.60	38.64	38.68	38.72	38.76	38.80	38.84	38.88	38.92
940	38.95	38.99	39.03	39.07	39.11	39.15	39.19	39.23	39.27	39.31
950	39.35	39.39	39.43	39.47	39.51	39.55	39.59	39.63	39.67	39.71
960	39.75	39.79	39.83	39.86	39.90	39.94	39.98	40.02	40.06	40.10
970	40.14	40.18	40.22	40.26	40.30	40.34	40.38	40.41	40.45	40.49
980	40.53	40.57	40.61	40.65	40.69	40.73	40.77	40.81	40.85	40.89
990	30.92	40.96	41.00	41.04	41.08	41.12	41.16	41.20	41.24	41.28
1000	41.31	41.35	41.39	41.43	41.47	41.51	41.55	41.59	41.63	41.67
1010	41.70	41.74	41.78	41.82	41.86	41.90	41.94	41.98	42.02	42.05
1020	42.09	42.13	42.17	42.21	42.25	42.29	42.33	42.36	42.40	42.44
1030	42.48	42.52	42.56	42.60	42.63	42.67	42.71	42.75	42.79	42.83
1040	42.87	42.90	42.94	42.98	43.02	43.06	43.10	43.14	43.17	43.21
1050	43.25	43.29	43.33	43.37	43.41	43.44	43.48	43.52	43.56	43.60
1060	43.63	43.67	43.71	43.75	43.79	43.83	43.87	43.90	43.94	43.98
1070	44.02	44.06	44.10	44.13	44.17	44.21	44.25	44.29	44.33	44.36
1080	44.40	44.44	44.48	44.52	44.55	44.59	44.63	44.67	44.71	44.74
1090	44.78	44.82	44.86	44.90	44.93	44.97	45.01	45.05	45.09	45.12
1100	45.16	45.20	45.24	45.27	45.31	45.35	45.39	45.43	45.46	45.50
1110	45.54	45.58	45.62	45.65	45.69	45.73	45.77	45.80	45.84	45.88
1120	45.92	45.96	45.99	46.03	46.07	46.11	46.14	46.18	46.22	46.26
1130	46.29	46.33	46.37	46.41	46.44	46.48	46.52	46.56	46.59	46.63
1140	46.67	46.70	46.74	46.78	46.82	46.85	46.89	46.93	46.97	47.00
1150	47.04	47.08	47.12	47.15	47.19	47.23	47.26	47.30	47.34	47.38
1160	47.41	47.45	47.49	47.52	47.56	47.60	47.63	47.67	47.71	47.75
1170	47.78	47.82	47.86	47.89	47.93	47.97	48.00	48.04	48.08	48.12
1180	48.15	48.19	48.23	48.26	48.30	48.34	48.37	48.41	48.45	48.48
1190	48.52	48.56	48.59	48.63	48.67	48.70	48.74	48.78	48.81	48.85
1200	48.89	48.92	48.96	49.00	49.03	49.07	49.11	49.14	49.18	49.22
1210	49.25	49.29	49.32	49.36	49.40	49.43	49.47	49.51	49.54	49.58
1220	49.62	49.65	49.69	49.72	49.76	49.80	49.83	49.87	49.90	49.94
1230	49.98	50.01	50.05	50.08	50.12	50.16	50.19	50.23	50.26	50.30
1240	50.34	50.37	50.41	50.44	50.48	50.52	50.55	50.59	50.62	50.66
1250	50.69	50.73	50.77	50.80	50.84	50.87	50.91	50.94	50.98	51.02
1260	51.05	51.09	51.12	51.16	51.19	51.23	51.27	51.30	51.34	51.37
1270	51.41	51.44	51.48	51.51	51.55	51.58	51.62	51.66	51.69	51.73
1280	51.76	51.80	51.83	51.87	51.90	51.94	51.97	52.01	52.04	52.08
1290	52.11	52.15	52.18	52.22	52.25	52.29	52.32	52.36	52.39	52.43
1300	52.46	52.50	52.53	52.57	52.60	52.64	52.67	52.71	52.74	52.78
1310	52.81	52.85	52.88	52.92	52.95	52.99	53.02	53.06	53.09	53.13
1320	53.16	53.20	53.23	53.27	53.30	53.34	53.37	53.41	53.44	53.47
1330	53.51	53.54	53.58	53.61	53.65	53.68	53.72	53.75	53.79	53.82
1340	53.85	53.89	53.92	53.96	53.99	54.03	54.06	54.10	54.13	54.16
1350	54.20	54.23	54.27	54.30	54.34	54.37	54.40	54.44	54.47	54.51
1360	54.54	54.57	54.61	54.64	54.68	54.71	54.74	54.78	54.81	54.85
1370	54.88	54.91	—	—	—	—	—	—	—	—

* Based on the International Temperature Scale of 1948.

CHROMEL-ALUMEL THERMOCOUPLES (Continued)

(Electromotive Force in Absolute Millivolts. Temperatures in Degrees F* Reference Junctions at 32°F.)

°F	0	1	2	3	4	5	6	7	8	9
	Millivolts									
−300	−5.51	−5.52	−5.53	−5.54	−5.54	−5.55	−5.56	−5.57	−5.58	−5.59
−290	−5.41	−5.42	−5.43	−5.44	−5.45	−5.46	−5.47	−5.48	−5.49	−5.50
−280	−5.30	−5.31	−5.32	−5.34	−5.35	−5.36	−5.37	−5.38	−5.39	−5.40
−270	−5.20	−5.21	−5.22	−5.23	−5.24	−5.25	−5.26	−5.27	−5.28	−5.29
−260	−5.08	−5.09	−5.10	−5.12	−5.13	−5.14	−5.15	−5.16	−5.17	−5.18
−250	−4.96	−4.97	−4.99	−5.00	−5.01	−5.02	−5.03	−5.04	−5.06	−5.07
−240	−4.84	−4.85	−4.86	−4.88	−4.89	−4.90	−4.91	−4.92	−4.94	−4.95
−230	−4.71	−4.72	−4.74	−4.75	−4.76	−4.77	−4.79	−4.80	−4.81	−4.82
−220	−4.58	−4.59	−4.60	−4.62	−4.63	−4.64	−4.66	−4.67	−4.68	−4.70
−210	−4.44	−4.45	−4.46	−4.48	−4.49	−4.51	−4.52	−4.53	−4.55	−4.56
−200	−4.29	−4.31	−4.32	−4.34	−4.35	−4.36	−4.38	−4.39	−4.41	−4.42
−190	−4.15	−4.16	−4.18	−4.19	−4.21	−4.22	−4.24	−4.25	−4.26	−4.28
−180	−4.00	−4.01	−4.03	−4.04	−4.06	−4.07	−4.09	−4.10	−4.12	−4.13
−170	−3.84	−3.86	−3.88	−3.89	−3.91	−3.92	−3.94	−3.95	−3.97	−3.98
−160	−3.69	−3.70	−3.72	−3.73	−3.75	−3.76	−3.78	−3.80	−3.81	−3.83
−150	−3.52	−3.54	−3.56	−3.57	−3.59	−3.60	−3.62	−3.64	−3.65	−3.67
−140	−3.36	−3.38	−3.39	−3.41	−3.42	−3.44	−3.46	−3.47	−3.49	−3.51
−130	−3.19	−3.20	−3.22	−3.24	−3.25	−3.27	−3.29	−3.31	−3.32	−3.34
−120	−3.01	−3.03	−3.05	−3.06	−3.08	−3.10	−3.12	−3.13	−3.15	−3.17
−110	−2.83	−2.85	−2.87	−2.89	−2.90	−2.92	−2.94	−2.96	−2.98	−2.99
−100	−2.65	−2.67	−2.69	−2.71	−2.72	−2.74	−2.76	−2.78	−2.80	−2.82
−90	−2.47	−2.49	−2.50	−2.52	−2.54	−2.56	−2.58	−2.60	−2.62	−2.63
−80	−2.28	−2.30	−2.32	−2.34	−2.36	−2.37	−2.39	−2.41	−2.43	−2.45
−70	−2.09	−2.11	−2.13	−2.15	−2.17	−2.18	−2.20	−2.22	−2.24	−2.26
−60	−1.90	−1.92	−1.94	−1.96	−1.97	−1.99	−2.01	−2.03	−2.05	−2.07
−50	−1.70	−1.72	−1.74	−1.76	−1.78	−1.80	−1.82	−1.84	−1.86	−1.88
−40	−1.50	−1.52	−1.54	−1.56	−1.58	−1.60	−1.62	−1.64	−1.66	−1.68
−30	−1.30	−1.32	−1.34	−1.36	−1.38	−1.40	−1.42	−1.44	−1.46	−1.48
−20	−1.10	−1.12	−1.14	−1.16	−1.18	−1.20	−1.22	−1.24	−1.26	−1.28
−10	−0.89	−0.91	−0.93	−0.95	−0.97	−0.99	−1.01	−1.03	−1.06	−1.08
(−)0	−0.68	−0.70	−0.72	−0.75	−0.77	−0.79	−0.81	−0.83	−0.85	−0.87
(+)0	−0.68	−0.66	−0.64	−0.62	−0.60	−0.58	−0.56	−0.54	−0.52	−0.49
10	−0.47	−0.45	−0.43	−0.41	−0.39	−0.37	−0.34	−0.32	−0.30	−0.28
20	−0.26	−0.24	−0.22	−0.19	−0.17	−0.15	−0.13	−0.11	−0.09	−0.07
30	−0.04	−0.02	0.00	0.02	0.04	0.07	0.09	0.11	0.13	0.15
40	0.18	0.20	0.22	0.24	0.26	0.29	0.31	0.33	0.35	0.37
50	0.40	0.42	0.44	0.46	0.48	0.51	0.53	0.55	0.57	0.60
60	0.62	0.64	0.65	0.68	0.71	0.73	0.75	0.77	0.80	0.82
70	0.84	0.86	0.88	0.91	0.93	0.95	0.97	1.00	1.02	1.04
80	1.06	1.09	1.11	1.13	1.15	1.18	1.20	1.22	1.24	1.27
90	1.29	1.31	1.33	1.36	1.38	1.40	1.43	1.45	1.47	1.49
100	1.52	1.54	1.56	1.58	1.61	1.63	1.65	1.68	1.70	1.72
110	1.74	1.77	1.79	1.81	1.84	1.86	1.88	1.90	1.93	1.95
120	1.97	2.00	2.02	2.04	2.06	2.09	2.11	2.13	2.16	2.18
130	2.20	2.23	2.25	2.27	2.29	2.32	2.34	2.36	2.39	2.41
140	2.43	2.46	2.48	2.50	2.52	2.55	2.57	2.59	2.62	2.64
150	2.66	2.69	2.71	2.73	2.75	2.78	2.80	2.82	2.85	2.87
160	2.89	2.92	2.94	2.96	2.98	3.01	3.03	3.05	3.08	3.10
170	3.12	3.15	3.17	3.19	3.22	3.24	3.26	3.29	3.31	3.33
180	3.36	3.38	3.40	3.43	3.45	3.47	3.49	2.52	3.54	3.56
190	3.59	3.61	3.63	3.66	3.68	3.70	3.73	3.75	3.77	3.80
200	3.82	3.84	3.87	3.89	3.91	3.94	3.96	3.98	4.01	4.03
210	4.05	4.08	4.10	4.12	4.15	4.17	4.19	4.21	4.24	4.26
220	4.28	4.31	4.33	4.35	4.38	4.40	4.42	4.44	4.47	4.49
230	4.51	4.54	4.56	4.58	4.61	4.63	4.65	4.67	4.70	4.72
240	4.74	4.77	4.79	4.81	4.83	4.86	4.88	4.90	4.92	4.95
250	4.97	4.99	5.02	5.04	5.06	5.08	5.11	5.13	5.15	5.17
260	5.20	5.22	5.24	5.26	5.29	5.31	5.33	5.35	5.38	5.40
270	5.42	5.44	5.47	5.49	5.51	5.53	5.56	5.58	5.60	5.62
280	5.65	5.67	5.69	5.71	5.73	5.76	5.78	5.80	5.82	5.85
290	5.87	5.89	5.91	5.93	5.96	5.98	6.00	6.02	6.05	6.07
300	6.09	6.11	6.13	6.16	6.18	6.20	6.22	6.25	6.27	6.29
310	6.31	6.33	6.36	6.38	6.40	6.42	6.45	6.47	6.49	6.51
320	6.53	6.56	6.58	6.60	6.62	6.65	6.67	6.69	6.71	6.73
330	6.76	6.78	6.80	6.82	6.84	6.87	6.89	6.91	6.93	6.96
340	6.98	7.00	7.02	7.04	7.07	7.09	7.11	7.13	7.15	7.18
350	7.20	7.22	7.24	7.26	7.29	7.31	7.33	7.35	7.38	7.40
360	7.42	7.44	7.46	7.49	7.51	7.53	7.55	7.58	7.60	7.62
370	7.64	7.66	7.69	7.71	7.73	7.75	7.78	7.80	7.82	7.84
380	7.87	7.89	7.91	7.93	7.95	7.98	8.00	8.02	8.04	8.07
390	8.09	8.11	8.13	8.16	8.18	8.20	8.22	8.24	8.27	8.29
400	8.31	8.33	8.36	8.38	8.40	8.42	8.45	8.47	8.49	8.51
410	8.54	8.56	8.58	8.60	8.62	8.65	8.67	8.69	8.71	8.74
420	8.76	8.78	8.80	8.82	8.85	8.87	8.89	8.91	8.94	8.96
430	8.98	9.00	9.03	9.05	9.07	9.09	9.12	9.14	9.16	9.18
440	9.21	9.23	9.25	9.27	9.30	9.32	9.34	9.36	9.39	9.41
450	9.43	9.45	9.48	9.50	9.52	9.54	9.57	9.59	9.61	9.63
460	9.66	9.68	9.70	9.73	9.75	9.77	9.79	9.82	9.84	9.86
470	9.88	9.91	9.93	9.95	9.97	10.00	10.02	10.04	10.06	10.09
480	10.11	10.13	10.16	10.18	10.20	10.22	10.25	10.27	10.29	10.31
490	10.34	10.36	10.38	10.40	10.43	10.45	10.47	10.50	10.52	10.54
500	10.57	10.59	10.61	10.63	10.66	10.68	10.70	10.72	10.75	10.77

* Based on on the International Temperature Scale of 1948.

°F	0	1	2	3	4	5	6	7	8	9
	Millivolts									
510	10.79	10.82	10.84	10.86	10.88	10.91	10.93	10.95	10.98	11.00
520	11.02	11.04	11.07	11.09	11.11	11.13	11.16	11.18	11.20	11.23
530	11.25	11.27	11.29	11.32	11.34	11.36	11.39	11.41	11.43	11.45
540	11.48	11.50	11.52	11.55	11.57	11.59	11.61	11.64	11.66	11.68
550	11.71	11.73	11.75	11.78	11.80	11.82	11.84	11.87	11.89	11.91
560	11.94	11.98	11.98	12.01	12.03	12.05	12.07	12.10	12.12	12.14
570	12.17	12.19	12.21	12.24	12.26	12.28	12.30	12.33	12.35	12.37
580	12.40	12.42	12.44	12.47	12.49	12.51	12.53	12.56	12.58	12.60
590	12.63	12.65	12.67	12.70	12.72	12.74	12.76	12.79	12.81	12.83
600	12.86	12.88	12.90	12.93	12.95	12.97	13.00	13.02	13.04	13.06
610	13.09	13.11	13.13	13.16	13.18	13.20	13.23	13.25	13.27	13.30
620	13.32	13.34	13.36	13.39	13.41	13.44	13.46	13.48	13.50	13.53
630	13.55	13.57	13.60	13.62	13.63	13.67	13.69	13.71	13.74	13.76
640	13.78	13.81	13.83	13.85	13.88	13.90	13.92	13.95	13.97	13.99
650	14.02	14.04	14.06	14.09	14.11	14.13	14.15	14.18	14.20	14.22
660	14.25	14.27	14.29	14.32	14.34	14.36	14.39	14.41	14.43	14.46
670	14.48	14.50	14.53	14.55	14.57	14.60	14.62	14.64	14.67	14.69
680	14.71	14.74	14.76	14.78	14.81	14.83	14.85	14.88	14.90	14.92
690	14.95	14.97	14.99	15.02	15.04	15.06	15.09	15.11	15.13	15.16
700	15.18	15.20	15.23	15.25	15.27	15.30	15.32	15.34	15.37	15.39
710	15.41	15.44	15.46	15.48	15.51	15.53	15.55	15.58	15.60	15.62
720	15.65	15.67	15.69	15.72	15.74	15.76	15.79	15.81	15.83	15.86
730	15.88	15.90	15.93	15.95	15.98	16.00	16.02	16.05	16.07	16.09
740	16.12	16.14	16.16	16.19	16.21	16.23	16.26	16.28	16.30	16.33
750	16.35	16.37	16.40	16.42	16.45	16.47	16.49	16.52	16.54	16.56
760	16.59	16.61	16.63	16.66	16.68	16.70	16.73	16.75	16.77	16.80
770	16.82	16.84	16.87	16.89	16.92	16.94	16.96	16.99	17.01	17.03
780	17.06	17.08	17.10	17.13	17.15	17.17	17.20	17.22	17.24	17.27
790	17.29	17.31	17.34	17.36	17.39	17.41	17.43	17.46	17.48	17.50
800	17.53	17.55	17.57	17.60	17.62	17.64	17.67	17.69	17.71	17.74
810	17.76	17.78	17.81	17.83	17.86	17.88	17.90	17.93	17.95	17.97
820	18.00	18.02	18.04	18.07	18.09	18.11	18.14	18.16	18.18	18.21
830	18.23	18.25	18.28	18.30	18.33	18.35	18.37	18.40	18.42	18.44
840	18.47	18.49	18.51	18.54	18.56	18.58	18.61	18.63	18.65	18.68
850	18.70	18.73	18.75	18.77	18.80	18.82	18.84	18.87	18.89	18.91
860	18.94	18.96	18.90	19.01	19.03	19.06	19.08	19.10	19.13	19.15
870	19.18	19.20	19.22	19.25	19.27	19.29	19.32	19.34	19.36	19.39
880	19.41	19.44	19.46	19.48	19.51	19.53	19.55	19.58	19.60	19.63
890	19.65	19.67	19.70	19.72	19.75	19.77	19.79	19.82	19.84	19.86
900	19.89	19.91	19.94	19.96	19.98	20.01	20.03	20.05	20.08	20.10
910	20.13	20.15	20.17	20.20	20.22	20.24	20.27	20.29	20.32	20.24
920	20.36	20.39	20.41	20.43	20.46	20.48	20.50	20.53	20.55	20.58
930	20.60	20.62	20.65	20.67	20.69	20.72	20.74	20.76	20.79	20.81
940	20.84	20.86	20.88	20.91	20.93	20.95	20.98	21.00	21.03	21.05
950	21.07	21.10	21.12	21.14	21.17	21.19	21.21	21.24	21.26	21.28
960	21.31	21.33	21.36	21.38	21.40	21.43	21.45	21.47	21.50	21.52
970	21.54	21.57	21.59	21.62	21.64	21.66	21.69	21.71	21.73	21.76
980	21.78	21.81	21.83	21.85	21.88	21.90	21.92	21.95	21.97	21.99
990	22.02	22.04	22.07	22.09	22.11	22.14	22.16	22.18	22.21	22.23
1000	22.26	22.28	22.30	22.33	22.35	22.37	22.40	22.42	22.44	22.47
1010	22.49	22.52	22.54	22.56	22.59	22.61	22.63	22.66	22.68	22.71
1020	22.73	22.75	22.78	22.80	22.82	22.85	22.87	22.90	22.92	22.94
1030	22.97	22.99	23.01	23.04	23.06	23.08	23.11	23.13	23.16	23.18
1040	23.20	23.23	23.25	23.27	23.30	23.32	23.35	23.37	23.39	23.42
1050	23.44	23.46	23.49	23.51	23.54	23.56	23.58	23.61	23.63	23.65
1060	23.68	23.70	23.72	23.75	23.77	23.80	23.82	23.84	23.87	23.89
1070	23.91	23.94	23.96	23.99	24.01	24.03	24.06	24.08	24.10	24.13
1080	24.15	24.18	24.20	24.22	24.25	24.27	24.29	24.32	24.34	24.36
1090	24.39	24.41	24.44	24.46	24.49	24.51	24.53	24.55	24.58	24.60
1100	24.63	24.65	24.67	24.70	24.72	24.74	24.77	24.79	24.82	24.84
1110	24.86	24.89	24.91	24.93	24.96	24.98	25.01	25.03	25.05	25.08
1120	25.10	25.12	25.15	25.17	25.20	25.22	25.24	25.27	25.29	25.31
1130	25.34	25.36	25.38	25.41	25.43	25.46	25.48	25.50	25.53	25.55
1140	25.57	25.60	25.62	25.65	25.67	25.69	25.72	25.74	25.76	25.79
1150	25.81	25.83	25.86	25.88	25.91	25.93	25.95	25.98	26.00	26.02
1160	26.05	26.07	26.09	26.12	26.14	26.16	26.19	26.21	26.24	26.26
1170	26.28	26.31	26.33	26.35	26.38	26.40	26.42	26.45	26.47	26.49
1180	26.52	26.54	26.56	26.59	26.61	26.63	26.66	26.68	26.70	26.73
1190	26.75	26.77	26.80	26.82	26.85	26.87	26.89	26.91	26.94	26.96
1200	26.98	27.01	27.03	27.06	27.08	27.10	27.12	27.15	27.17	27.20
1210	27.22	27.24	27.27	27.29	27.31	27.34	27.36	27.38	27.40	27.43
1220	27.45	27.48	27.50	27.52	27.55	27.57	27.59	27.62	27.64	27.66
1230	27.69	27.71	27.73	27.76	27.78	27.80	27.83	27.85	27.87	27.90
1240	27.92	27.94	27.97	27.99	28.01	28.04	28.06	28.08	28.11	28.13
1250	28.15	28.18	28.20	28.22	28.25	28.27	28.29	28.32	28.34	28.37
1260	28.39	28.41	28.44	28.46	28.48	28.50	28.53	28.55	28.58	28.60
1270	28.62	28.65	28.67	28.69	28.72	28.74	28.76	28.79	28.81	28.83
1280	28.86	28.88	28.90	28.93	28.95	28.97	29.00	29.02	29.04	29.07
1290	29.09	29.11	29.14	29.16	29.18	29.21	29.23	29.25	29.28	29.30
1300	29.32	29.35	29.37	29.39	29.42	29.44	29.46	29.49	29.51	29.53
1310	29.56	29.58	29.60	29.63	29.65	29.67	29.70	29.72	29.74	29.77
1320	29.79	29.81	29.84	29.86	29.88	29.91	29.93	29.95	29.97	30.00

CHROMEL-ALUMEL THERMOCOUPLES (Continued)

°F	0	1	2	3	4	5	6	7	8	9
	Millivolts									
1330	30.02	30.05	30.07	30.09	30.11	30.14	30.16	30.18	30.21	30.23
1340	30.25	30.28	30.30	30.32	30.35	30.37	30.39	30.42	30.44	30.46
1350	30.49	30.51	30.53	30.56	30.58	30.60	30.63	30.65	30.67	30.70
1360	30.72	30.74	30.77	30.79	30.81	30.83	30.86	30.88	30.90	30.93
1370	30.95	30.97	31.00	31.02	31.04	31.07	31.09	31.11	31.14	31.16
1380	31.18	31.21	31.23	31.25	31.28	31.30	31.32	31.34	31.37	31.39
1390	31.42	31.44	31.46	31.48	31.51	31.53	31.55	31.58	31.60	31.62
1400	31.65	31.67	31.69	31.72	31.74	31.76	31.78	31.81	31.83	31.85
1410	31.88	31.90	31.92	31.95	31.97	31.99	32.02	32.04	32.06	32.08
1420	32.11	32.13	32.15	32.18	32.20	32.22	32.25	32.27	32.29	32.31
1430	32.34	32.36	32.38	32.41	32.43	32.45	32.48	32.50	32.52	32.54
1440	32.57	32.59	32.61	32.64	32.66	32.68	32.70	32.73	32.75	32.77
1450	32.80	32.82	32.84	32.86	32.89	32.91	32.93	32.96	32.98	33.00
1460	33.02	33.05	33.07	33.09	33.12	33.14	33.16	33.18	33.21	33.23
1470	33.25	33.28	33.30	33.32	33.34	33.37	33.39	33.41	33.43	33.46
1480	33.48	33.50	33.53	33.55	33.57	33.59	33.62	33.64	33.66	33.69
1490	33.71	33.73	33.75	33.78	33.80	33.82	33.84	33.87	33.89	33.91
1500	33.93	33.96	33.98	34.00	34.03	34.05	34.07	34.09	34.12	34.14
1510	34.16	34.18	34.21	34.23	34.25	34.28	34.30	34.32	34.34	34.37
1520	34.39	34.41	34.43	34.46	34.48	34.50	34.53	34.55	34.57	34.59
1530	34.62	34.64	34.66	34.68	34.71	34.73	34.75	34.77	34.80	34.82
1540	34.84	34.87	34.89	34.91	34.93	34.96	34.98	35.00	35.02	35.05
1550	35.07	35.09	35.11	35.14	35.16	35.18	35.21	35.23	35.25	35.27
1560	35.29	35.32	35.34	35.36	35.39	35.41	35.43	35.45	35.48	35.50
1570	35.52	35.54	35.57	35.59	35.61	35.63	35.66	35.68	35.70	35.72
1580	35.75	35.77	35.79	35.81	35.84	35.86	35.88	35.90	35.93	35.95
1590	35.97	35.99	36.02	36.04	36.06	36.08	36.11	36.13	36.15	36.17
1600	36.19	36.19	36.22	36.24	36.29	36.31	36.33	36.35	36.37	36.40
1610	36.42	36.44	36.46	36.49	36.51	36.53	36.55	36.58	36.60	36.62
1620	36.64	36.67	36.69	36.71	36.73	36.76	36.78	36.80	36.82	36.84
1630	36.87	36.89	36.91	36.93	36.96	36.98	37.00	37.02	37.05	37.07
1640	37.09	37.11	37.14	37.16	37.18	37.20	37.23	37.25	37.27	37.29
1650	37.31	37.34	37.36	37.38	37.40	37.43	37.45	37.47	37.49	37.52
1660	37.54	37.56	37.58	37.60	37.63	37.65	37.67	37.69	37.72	37.74
1670	37.76	37.78	37.81	37.83	37.85	37.87	37.89	37.92	37.94	37.96
1680	37.98	38.01	38.03	38.05	38.07	38.09	38.12	38.14	38.16	38.18
1690	38.20	38.23	38.25	38.27	38.29	38.32	38.34	38.36	38.38	38.40
1700	38.43	38.45	38.47	38.49	38.51	38.54	38.56	38.58	38.60	38.62
1710	38.65	38.67	38.69	38.71	38.73	38.76	38.78	38.80	38.82	38.84
1720	38.87	38.89	38.91	38.93	38.95	38.98	39.00	39.02	39.04	39.06
1730	39.09	39.11	39.13	39.15	39.17	39.20	39.22	39.24	39.26	39.28
1740	39.31	39.33	39.35	39.37	39.39	39.42	39.44	39.46	39.48	39.50
1750	39.53	39.55	39.57	39.59	39.61	39.64	39.66	39.68	39.70	39.72
1760	39.75	39.77	39.79	39.81	39.83	39.86	39.88	39.69	39.92	39.94
1770	39.96	39.99	40.01	40.03	40.05	40.07	40.10	40.12	40.14	40.16
1780	40.18	40.20	40.23	40.25	40.27	40.29	40.31	40.34	40.36	40.38
1790	40.40	40.42	40.44	40.47	40.49	40.51	40.53	40.55	40.58	40.60
1800	40.62	40.64	40.66	40.68	40.71	40.73	40.75	40.77	40.79	40.82
1810	40.84	40.86	40.88	40.90	40.92	40.95	40.97	40.99	41.01	41.03
1820	41.05	41.08	41.10	41.12	41.14	41.16	41.18	41.21	41.23	41.25
1830	41.27	41.29	41.31	41.34	41.36	41.38	41.40	41.42	41.45	41.47
1840	41.49	41.51	41.53	41.55	41.57	41.60	41.62	41.64	41.66	41.68
1850	41.70	41.73	41.75	41.77	41.79	41.81	41.83	41.85	41.88	41.90
1860	41.92	41.94	41.96	41.99	42.01	42.03	42.05	42.07	42.09	42.11
1870	42.14	42.16	42.18	42.20	42.22	42.24	42.26	42.29	42.31	42.33
1880	42.35	42.37	42.39	42.42	42.44	42.46	42.48	42.50	42.52	42.55
1890	42.57	42.59	42.61	42.63	42.65	42.67	42.69	42.72	42.74	42.76
1900	42.78	42.80	42.82	42.84	42.87	42.89	42.91	42.93	42.95	42.97
1910	42.99	43.01	43.04	43.06	43.08	43.10	43.12	43.14	43.17	43.19

°F	0	1	2	3	4	5	6	7	8	9
	Millivolts									
1920	43.21	43.23	43.25	43.27	43.29	43.21	43.34	43.36	43.38	43.40
1930	43.42	43.44	43.47	43.49	43.51	43.53	43.55	43.57	43.59	43.61
1940	43.63	43.66	43.68	43.70	43.72	43.75	43.76	43.78	43.81	43.83
1950	43.85	43.87	43.89	43.91	43.93	43.95	43.98	44.00	44.02	44.04
1960	44.06	44.08	44.10	44.13	44.15	44.17	44.19	44.21	44.23	44.25
1970	44.27	44.30	44.32	44.34	44.36	44.38	44.40	44.42	44.44	44.47
1980	44.49	44.51	44.53	44.55	44.57	44.59	44.61	44.63	44.66	44.68
1990	44.70	44.72	44.74	44.76	44.78	44.80	44.82	44.85	44.87	44.89
2000	44.91	44.93	44.95	44.97	44.99	45.01	45.03	45.06	45.08	45.10
2010	45.12	45.14	45.16	45.18	45.20	45.22	45.24	45.27	45.29	45.31
2020	45.33	45.35	45.87	45.39	45.41	45.43	45.45	45.48	45.50	45.52
2030	45.54	45.56	45.58	45.60	45.62	45.64	45.66	45.69	45.71	45.73
2040	45.75	45.77	45.79	45.81	45.83	45.85	45.87	45.90	45.92	45.94
2050	45.96	45.98	46.00	46.02	46.04	46.06	46.08	46.11	46.13	46.15
2060	46.17	46.19	46.21	46.23	46.25	46.27	46.29	46.31	46.33	46.36
2070	46.38	46.40	46.42	46.44	46.46	46.48	46.50	46.52	46.54	46.56
2080	46.58	46.60	46.63	46.65	46.67	46.69	46.71	46.73	46.75	46.77
2090	46.79	46.81	46.83	46.85	46.87	46.90	46.92	46.94	46.96	46.98
2100	47.00	47.02	47.04	47.06	47.08	47.10	47.12	47.14	47.17	47.19
2110	47.21	47.23	47.25	47.27	47.29	47.31	47.33	47.35	47.37	47.39
2120	47.41	47.43	47.45	47.47	47.49	47.52	47.54	47.56	47.58	47.60
2130	47.62	47.64	47.66	47.68	47.70	47.72	47.74	47.76	47.78	47.80
2140	47.82	47.84	47.86	47.89	47.91	47.93	47.95	47.97	47.99	48.01
2150	48.03	48.05	48.07	48.09	48.11	48.13	48.15	48.17	48.19	48.21
2160	48.23	48.25	48.27	48.29	48.32	48.34	48.36	48.38	48.40	48.42
2170	48.44	48.46	48.48	48.50	48.52	48.54	48.56	48.58	48.60	48.62
2180	48.64	48.66	48.68	48.70	48.72	48.74	48.76	48.79	48.81	48.83
2190	48.85	48.87	48.89	48.91	48.93	48.95	48.97	48.99	49.01	49.03
2200	49.05	49.07	49.09	49.11	49.13	49.15	49.17	49.19	49.21	49.23
2210	49.25	49.27	49.29	49.31	49.33	49.35	49.37	49.39	49.41	49.43
2220	49.45	49.47	49.49	49.51	49.53	49.55	49.57	49.59	49.61	49.63
2230	49.65	49.67	49.69	49.71	49.73	49.76	49.78	49.80	49.82	49.84
2240	49.86	49.88	49.90	49.92	49.94	49.96	49.98	50.00	50.02	50.04
2250	50.06	50.08	50.10	50.12	50.14	50.16	50.18	50.20	50.22	50.24
2260	50.26	50.28	50.30	50.32	50.34	50.36	50.38	50.40	50.42	50.44
2270	50.46	50.48	50.50	50.52	50.54	50.56	50.57	50.59	50.61	50.63
2280	50.65	50.67	50.69	50.71	50.73	50.75	50.77	50.79	50.81	50.83
2290	50.85	50.87	50.89	50.91	50.93	50.95	50.97	50.99	51.01	51.03
2300	51.05	51.07	51.09	51.11	51.13	51.15	51.17	51.19	51.21	51.23
2310	51.25	51.27	51.29	51.31	51.33	51.35	51.37	51.39	51.41	51.43
2320	51.45	51.47	51.48	51.50	51.52	51.54	51.56	51.58	51.60	51.62
2330	51.64	51.66	51.68	51.70	51.72	51.74	51.76	51.78	51.80	51.82
2340	51.84	51.86	51.88	51.90	51.92	51.94	51.96	51.98	52.00	52.01
2350	52.03	52.05	52.07	52.09	52.11	52.13	52.15	52.17	52.19	52.21
2360	52.23	52.25	52.27	52.29	52.31	52.33	52.35	52.37	52.39	52.41
2370	52.42	52.44	52.46	52.48	52.50	52.52	52.54	52.56	52.58	52.60
2380	52.62	52.64	52.66	52.68	52.70	52.72	52.74	52.76	52.77	52.79
2390	52.81	52.83	52.85	52.87	52.89	52.91	52.93	52.95	52.97	52.99
2400	53.01	53.03	53.05	53.07	53.08	53.10	53.12	53.14	53.16	53.18
2410	53.20	53.22	53.24	53.26	53.28	53.30	53.32	53.34	53.35	53.37
2420	53.39	53.41	53.43	53.45	53.47	53.49	53.51	53.53	53.55	53.57
2430	53.59	53.60	53.62	53.64	53.66	53.68	53.70	53.72	53.74	53.76
2440	53.78	53.80	53.82	53.83	53.85	53.87	53.89	53.91	53.93	53.95
2450	53.97	53.99	54.01	54.03	54.04	54.06	54.08	54.10	54.12	54.14
2460	54.16	54.18	54.20	54.22	54.24	54.25	54.27	54.29	54.31	54.33
2470	54.35	54.37	54.39	54.41	54.43	54.44	54.46	54.48	54.50	54.52
2480	54.54	54.56	54.58	54.60	54.62	54.63	54.65	54.67	54.69	54.71
2490	54.73	54.75	54.77	54.79	54.81	54.82	54.84	54.86	54.88	54.90
2500	54.92	—	—	—	—	—	—	—	—	—

IRON-CONSTANTAN THERMOCOUPLES (MODIFIED 1913)

(Electromotive Force in Absolute Millivolts. Temperatures in Degrees C (Int. 1948). Reference Junctions at 0°C.)

°C	0	1	2	3	4	5	6	7	8	9
	Millivolts									
−190	−7.66	−7.69	−7.71	−7.73	−7.76	−7.78				
−180	−7.40	−7.43	−7.46	−7.49	−7.51	−7.54	−7.56	−7.59	−7.61	−7.64
−170	−7.12	−7.15	−7.18	−7.21	−7.24	−7.27	−7.30	−7.32	−7.35	−7.38
−160	−6.82	−6.85	−6.88	−6.91	−6.94	−6.97	−7.00	−7.03	−7.06	−7.09
−150	−6.50	−6.53	−5.56	−6.60	−6.63	−6.66	−6.69	−6.72	−6.76	−6.79
−140	−6.16	−6.19	−6.22	−6.26	−6.29	−6.33	−6.36	−6.40	−6.43	−6.46
−130	−5.80	−5.84	−5.87	−5.91	−5.94	−5.98	−6.01	−6.05	−6.08	−6.12
−120	−5.42	−5.46	−5.50	−5.54	−5.58	−5.61	−5.65	−5.69	−5.72	−5.76
−110	−5.03	−5.07	−5.11	−5.15	−5.19	−5.23	−5.27	−5.31	−5.35	−5.38
−100	−4.63	−4.67	−4.71	−4.75	−4.79	−4.83	−4.87	−4.91	−4.95	−4.99
−90	−4.21	−4.25	−4.30	−4.34	−4.38	−4.42	−4.46	−4.50	−4.55	−4.59
−80	−3.78	−3.82	−3.87	−3.91	−3.96	−4.00	−4.04	−4.08	−4.13	−4.17
−70	−3.34	−3.38	−3.43	−3.47	−3.52	−3.56	−3.60	−3.65	−3.69	−3.74
−60	−2.89	−2.94	−2.98	−3.03	−3.07	3.12	−3.16	−3.21	−3.25	−3.30
−50	−2.43	−2.48	−2.52	−2.57	−2.62	−2.66	−2.71	−2.75	−2.80	−2.84

°C	0	1	2	3	4	5	6	7	8	9
	Millivolts									
−40	−1.96	−2.01	−2.06	−2.10	−2.15	−2.20	−2.24	−2.29	−2.34	−2.38
−30	−1.48	−1.53	−1.58	−1.63	−1.67	−1.72	−1.77	−1.82	−1.87	−1.91
−20	−1.00	−1.04	−1.09	−1.14	−1.19	−1.24	−1.29	−1.34	−1.39	−1.43
−10	−0.50	−0.55	−0.60	−0.65	−0.70	−0.75	−0.80	−0.85	−0.90	−0.95
(−)0	0.00	−0.05	−0.10	−0.15	−0.20	−0.25	−0.30	−0.35	−0.40	−0.45
(+)0	0.00	0.05	0.10	0.15	0.20	0.25	0.30	0.35	0.40	0.45
10	0.50	0.56	0.61	0.66	0.71	0.76	0.81	0.86	0.91	0.97
20	1.02	1.07	1.12	1.17	1.22	1.28	1.33	1.38	1.43	1.48
30	1.54	1.59	1.64	1.69	1.74	1.80	1.85	1.90	1.95	2.00
40	2.06	2.11	2.16	2.22	2.27	2.32	2.37	2.42	2.48	2.53
50	2.58	2.64	2.69	2.74	2.80	2.85	2.90	2.96	3.01	3.06
60	3.11	3.17	3.22	3.27	3.33	3.38	3.43	3.49	3.54	3.60
70	3.65	3.70	3.76	3.81	3.86	3.92	3.97	4.02	4.08	4.13
80	4.19	4.24	4.29	4.35	4.40	4.46	4.51	4.56	4.62	4.67
90	4.73	4.78	4.83	4.89	4.94	5.00	5.05	5.10	5.16	5.21

IRON-CONSTANTAN THERMOCOUPLES (MODIFIED 1913) (Continued)

°C	0	1	2	3	4	5	6	7	8	9
	Millivolts									
100	5.27	5.32	5.38	5.43	5.48	5.54	5.59	5.65	5.70	5.76
110	5.81	5.86	5.92	5.97	6.03	6.08	6.14	6.19	6.25	6.30
120	6.36	6.41	6.47	6.52	6.58	6.63	6.68	6.74	6.79	6.85
130	6.90	6.96	7.01	7.07	7.12	7.18	7.23	7.29	7.34	7.40
140	7.45	7.51	7.56	7.62	7.67	7.73	7.78	7.84	7.89	7.95
150	8.00	8.06	8.12	8.17	8.23	8.28	8.34	8.39	8.45	8.50
160	8.56	8.61	8.67	8.72	8.78	8.84	8.89	8.95	9.00	9.06
170	9.11	9.17	9.22	9.28	9.33	9.39	9.44	9.50	9.56	9.61
180	9.67	9.72	9.78	9.83	9.89	9.95	10.00	10.06	10.11	10.17
190	10.22	10.28	10.34	10.39	10.45	10.50	10.56	10.61	10.67	10.72
200	10.78	10.84	10.89	10.95	11.00	11.06	11.12	11.17	11.23	11.28
210	11.34	11.39	11.45	11.50	11.56	11.62	11.67	11.73	11.78	11.84
220	11.89	11.95	12.00	12.06	12.12	12.17	12.23	12.28	12.34	12.39
230	12.45	12.50	12.56	12.62	12.67	12.73	12.78	12.84	12.89	12.95
240	13.01	13.06	13.12	13.17	13.23	13.28	13.34	13.40	13.45	13.51
250	13.56	13.62	13.67	13.73	13.78	13.84	13.89	13.95	14.00	14.06
260	14.12	14.17	14.23	14.28	14.34	14.39	14.45	14.50	14.56	14.61
270	14.67	14.72	14.78	14.83	14.89	14.94	15.00	15.06	15.11	15.17
280	15.22	15.28	15.33	15.39	15.44	15.50	15.55	15.61	15.66	15.72
290	15.77	15.83	15.88	15.94	16.00	16.05	16.11	16.16	16.22	16.27
300	16.33	16.38	16.44	16.49	16.55	16.60	16.66	16.71	16.77	16.82
310	16.88	16.93	16.99	17.04	17.10	17.15	17.21	17.26	17.32	17.37
320	17.43	17.48	17.54	17.60	17.65	17.71	17.76	17.82	17.87	17.93
330	17.98	18.04	18.09	18.15	18.20	18.26	18.32	18.37	18.43	18.48
340	18.54	18.59	18.65	18.70	18.76	18.81	18.87	18.92	18.98	19.03
350	19.09	19.14	19.20	19.26	19.31	19.37	19.42	19.48	19.53	19.59
360	19.64	19.70	19.75	19.81	19.86	19.92	19.97	20.03	20.08	20.14
370	20.20	20.25	20.31	20.36	20.42	20.47	20.53	20.58	20.64	20.69
380	20.75	20.80	20.86	20.91	20.97	21.02	21.08	21.13	21.19	21.24
390	21.30	21.35	21.41	21.46	21.52	21.57	21.63	21.68	21.74	21.79
400	21.85	21.90	21.96	22.02	22.07	22.13	22.18	22.24	22.29	22.35
410	22.40	22.46	22.51	22.57	22.62	22.68	22.73	22.79	22.84	22.90
420	22.95	23.01	23.06	23.12	23.17	23.23	23.28	23.34	23.39	23.45
430	23.50	23.56	23.61	23.67	23.72	23.78	23.83	23.89	23.94	24.00

°C	0	1	2	3	4	5	6	7	8	9
	Millivolts									
440	24.06	24.11	24.17	24.22	24.28	24.33	24.39	24.44	24.50	24.55
450	24.61	24.66	24.72	24.77	24.83	24.88	24.94	25.00	25.05	25.11
460	25.16	25.22	25.27	25.33	25.38	25.44	25.49	25.55	25.60	25.66
470	25.72	25.77	25.83	25.88	25.94	25.99	26.05	26.10	26.16	26.22
480	26.27	26.33	26.38	26.44	26.49	26.55	26.61	26.66	26.72	26.77
490	26.83	26.89	26.94	27.00	27.05	27.11	27.17	27.22	27.28	27.33
500	27.39	27.45	27.50	27.56	27.61	27.67	27.73	27.78	27.84	27.90
510	27.95	28.01	28.07	28.12	28.18	28.23	28.29	28.35	28.40	28.46
520	28.52	28.57	28.63	28.69	28.74	28.80	28.86	28.91	28.97	29.02
530	29.08	29.14	29.20	29.25	29.31	29.37	29.42	29.48	29.54	29.59
540	29.65	29.71	29.76	29.82	29.88	29.94	29.99	30.05	30.11	30.16
550	30.22	30.28	30.34	30.39	30.45	30.51	30.57	30.62	30.68	30.74
560	30.80	30.85	30.91	30.97	31.02	31.08	31.14	31.20	31.26	31.31
570	31.37	31.43	31.49	31.54	31.60	31.66	31.72	31.78	31.83	31.89
580	31.95	32.01	32.06	32.12	32.18	32.24	32.30	32.36	32.41	32.47
590	32.53	32.59	32.65	32.71	32.76	32.82	32.88	32.94	33.00	33.06
600	33.11	33.17	33.23	33.29	33.35	33.41	33.46	33.52	33.58	33.64
610	33.70	33.76	33.82	33.88	33.94	33.99	34.05	34.11	34.17	34.23
620	34.29	34.35	34.41	34.47	34.53	34.58	34.64	34.70	34.76	34.82
630	34.88	34.94	35.00	35.06	35.12	35.18	35.24	35.30	35.36	35.42
640	35.48	35.54	35.60	36.66	35.72	35.78	35.84	35.90	35.96	36.02
650	36.08	36.14	36.20	36.26	36.32	36.38	36.44	36.50	36.56	36.62
660	36.69	36.75	36.81	36.87	36.93	36.99	37.05	37.11	37.18	37.24
670	37.30	37.36	37.42	37.48	37.54	37.60	37.66	37.73	37.79	37.85
680	37.91	37.97	38.04	38.10	38.16	38.22	38.28	38.34	38.41	38.47
690	38.53	38.59	38.66	38.72	38.78	38.84	38.90	38.97	39.03	39.09
700	39.15	39.22	39.28	39.34	39.40	39.47	39.53	39.59	39.65	39.72
710	39.78	39.84	39.91	39.97	40.03	40.10	40.16	40.22	40.28	40.35
720	40.41	40.48	40.54	40.60	40.66	40.73	40.79	40.86	40.92	40.98
730	41.05	41.11	41.17	41.24	41.30	41.36	41.43	41.49	41.56	41.62
740	41.68	41.75	41.81	41.87	41.94	42.00	42.07	42.13	42.19	42.26
750	42.32	42.38	42.45	42.51	42.58	42.64	42.70	42.77	42.83	42.90
760	42.96	—	—	—	—	—	—	—	—	—

(Electromotive Force in Absolute Millivolts. Temperatures in Degrees F* Reference Junctions at 32°F.)

°F	0	1	2	3	4	5	6	7	8	9
	Millivolts									
−310	−7.66	−7.68	−7.69	−7.70	−7.71	−7.73	−7.74	−7.75	−7.76	−7.78
−300	−7.52	−7.54	−7.55	−7.57	−7.58	−7.59	−7.61	−7.62	−7.64	−7.65
−290	−7.38	−7.39	−7.40	−7.42	−7.44	−7.45	−7.46	−7.48	−7.49	−7.51
−280	−7.22	−7.24	−7.25	−7.27	−7.28	−7.30	−7.31	−7.33	−7.34	−7.36
−270	−7.06	−7.07	−7.09	−7.11	−7.12	−7.14	−7.15	−7.17	−7.19	−7.20
−260	−6.89	−6.90	−6.92	−6.94	−6.96	−6.97	−6.99	−7.01	−7.02	−7.04
−250	−6.71	−6.73	−6.75	−6.77	−6.78	−6.80	−6.82	−6.84	−6.85	−6.87
−240	−6.53	−6.55	−6.57	−6.59	−6.61	−6.62	−6.64	−6.66	−6.68	−6.70
−230	−6.35	−6.37	−6.38	−6.40	−6.42	−6.44	−6.46	−6.48	−6.50	−6.52
−220	−6.16	−6.18	−6.19	−6.21	−6.23	−6.25	−6.27	−6.29	−6.31	−6.33
−210	−5.96	−5.98	−6.00	−6.02	−6.04	−6.06	−6.08	−6.10	−6.12	−6.14
−200	−5.76	−5.78	−5.80	−5.82	−5.84	−5.86	−5.88	−5.90	−5.92	−5.94
−190	−5.55	−5.57	−5.59	−5.61	−5.63	−5.65	−5.67	−5.70	−5.72	−5.74
−180	−5.34	−5.36	−5.38	−5.40	−5.42	−5.44	−5.46	−5.49	−5.51	−5.53
−170	−5.12	−5.14	−5.16	−5.19	−5.21	−5.23	−5.25	−5.27	−5.30	−5.32
−160	−4.90	−4.92	−4.94	−4.97	−4.99	−5.01	−5.03	−5.06	−5.08	−5.10
−150	−4.68	−4.70	−4.72	−4.74	−4.76	−4.79	−4.81	−4.83	−4.86	−4.88
−140	−4.44	−4.47	−4.49	−4.51	−4.54	−4.56	−4.58	−4.61	−4.63	−4.65
−130	−4.21	−4.23	−4.26	−4.28	−4.30	−4.33	−4.35	−4.38	−4.40	−4.42
−120	−3.97	−4.00	−4.02	−4.04	−4.07	−4.09	−4.12	−4.14	−4.16	−4.19
−110	−3.73	3.76	−3.78	−3.81	−3.83	−3.85	−3.88	−3.90	−3.93	−3.95
−100	−3.49	−3.51	−3.54	−3.56	−3.59	−3.61	−3.64	−3.66	−3.68	−3.71
−90	−3.24	−3.27	−3.29	−3.32	−3.34	−3.36	−3.39	−3.41	−3.44	−3.46
−80	−2.99	−3.02	−3.04	−3.07	−3.09	−3.12	−3.14	−3.17	−3.19	−3.22
−70	−2.74	−2.76	−2.79	−2.81	−2.84	−2.86	−2.89	−2.92	−2.94	−2.97
−60	−2.48	−2.51	−2.53	−2.56	−2.58	−2.61	−2.64	−2.66	−2.69	−2.71
−50	−2.22	−2.25	−2.27	−2.30	−2.33	−2.35	−2.38	−2.40	−2.43	−2.46
−40	−1.96	−1.99	−2.01	−2.04	−2.06	−2.09	−2.12	−2.14	−2.17	−2.20
−30	−1.70	−1.72	−1.75	−1.78	−1.80	−1.83	−1.86	−1.88	−1.91	−1.94
−20	−1.43	−1.46	−1.48	−1.51	−1.54	−1.56	−1.59	−1.62	−1.64	−1.67
−10	−1.16	−1.19	−1.21	−1.24	−1.27	−1.29	−1.32	−1.35	−1.38	−1.40
(−)0	−0.89	−0.91	−0.94	−0.97	−1.00	−1.02	−1.05	−1.08	−1.10	−1.13
+(0)	−0.89	−0.86	−0.83	−0.80	−0.78	−0.75	−0.72	−0.70	−0.67	−0.64
10	−0.61	−0.58	−0.56	−0.53	−0.50	−0.48	−0.45	−0.42	−0.39	−0.36
20	−0.34	−0.31	−0.28	−0.25	−0.22	−0.20	−0.17	−0.14	−0.11	−0.09
30	−0.06	−0.03	0.00	0.03	0.05	0.08	0.11	0.14	0.17	0.19
40	0.22	0.25	0.28	0.31	0.34	0.36	0.39	0.42	0.45	0.48
50	0.50	0.53	0.56	0.59	0.62	0.65	0.67	0.70	0.73	0.76
60	0.79	0.82	0.84	0.87	0.90	0.93	0.96	0.99	1.02	1.04
70	1.07	1.10	1.13	1.16	1.19	1.22	1.25	1.28	1.30	1.33
80	1.36	1.39	1.42	1.45	1.48	1.51	1.54	1.56	1.59	1.62

°F	0	1	2	3	4	5	6	7	8	9
	Millivolts									
90	1.65	1.68	1.71	1.74	1.77	1.80	1.83	1.85	1.88	1.91
100	1.94	1.97	2.00	2.03	2.06	2.09	2.12	2.14	2.17	2.20
110	2.23	2.26	2.29	2.32	2.35	2.38	2.41	2.44	2.47	2.50
120	2.52	2.55	2.58	2.61	2.64	2.67	2.70	2.73	2.76	2.79
130	2.82	2.85	2.88	2.91	2.94	2.97	3.00	3.03	3.06	3.08
140	3.11	3.14	3.17	3.20	3.23	3.26	3.29	3.32	3.35	3.38
150	3.41	3.44	3.47	3.50	3.53	3.56	3.59	3.62	3.65	3.68
160	3.71	3.74	3.77	3.80	3.83	3.86	3.89	3.92	3.95	3.98
170	4.01	4.04	4.07	4.10	4.13	4.16	4.19	4.22	4.25	4.28
180	4.31	4.34	4.37	4.40	4.43	4.46	4.49	4.52	4.55	4.58
190	4.61	4.64	4.67	4.70	4.73	4.76	4.79	4.82	4.85	4.88
200	4.91	4.94	4.97	5.00	5.03	5.06	5.09	5.12	5.15	5.18
210	5.21	5.24	5.27	5.30	5.33	5.36	5.39	5.42	5.45	5.48
220	5.51	5.54	5.57	5.60	5.63	5.66	5.69	5.72	5.75	5.78
230	5.81	5.84	5.87	5.90	5.93	5.96	5.99	6.02	6.05	6.08
240	6.11	6.14	6.17	6.20	6.24	6.27	6.30	6.33	6.36	6.39
250	6.42	6.45	6.48	6.51	6.54	6.57	6.60	6.63	6.66	6.69
260	6.72	6.75	6.78	6.81	6.84	6.87	6.90	6.93	6.96	7.00
270	7.03	7.06	7.09	7.12	7.15	7.18	7.21	7.24	7.27	7.30
280	7.33	7.36	7.39	7.42	7.45	7.48	7.51	7.54	7.58	7.61
290	7.64	7.67	7.70	7.73	7.76	7.79	7.82	7.85	7.88	7.91
300	7.94	7.97	8.00	8.04	8.07	8.10	8.13	8.16	8.19	8.22
310	8.25	8.28	8.31	8.34	8.37	8.40	8.44	8.47	8.50	8.53
320	8.56	8.59	8.62	8.65	8.68	8.71	8.74	8.77	8.80	8.84
330	8.87	8.90	8.93	8.96	8.99	9.02	9.05	9.08	9.11	9.14
340	9.17	9.20	9.24	9.27	9.30	9.33	9.36	9.39	9.42	9.45
350	9.48	9.51	9.54	9.58	9.61	9.64	9.67	9.70	9.73	9.76
360	9.79	9.82	9.85	9.88	9.92	9.95	9.98	10.01	10.04	10.07
370	10.10	10.13	10.16	10.19	10.22	10.25	10.28	10.32	10.35	10.38
380	10.41	10.44	10.47	10.50	10.53	10.56	10.60	10.63	10.66	10.69
390	10.72	10.75	10.78	10.81	10.84	10.87	10.90	10.94	10.97	11.00
400	11.03	11.06	11.00	11.12	11.15	11.18	11.21	11.24	11.28	11.31
410	11.34	11.37	11.40	11.43	11.46	11.49	11.52	11.55	11.58	11.62
420	11.65	11.68	11.71	11.74	11.77	11.80	11.83	11.86	11.89	11.92
430	11.96	11.99	12.02	12.05	12.08	12.11	12.14	12.17	12.20	12.23
440	12.26	12.30	12.33	12.36	12.39	12.42	12.45	12.48	12.51	12.54
450	12.57	12.60	12.64	12.67	12.70	12.73	12.76	12.79	12.82	12.85
460	12.88	12.91	12.94	12.98	13.01	13.04	13.07	13.10	13.13	13.16
470	13.19	13.22	13.25	13.28	13.31	13.34	13.38	13.41	13.44	13.47
480	13.50	13.53	13.56	13.59	13.62	13.65	13.68	13.72	13.75	13.78
490	13.81	13.84	13.87	13.90	13.93	13.96	13.99	14.02	14.05	14.08

* Based on the International Temperature Scale of 1948.

IRON-CONSTANTAN THERMOCOUPLES (MODIFIED 1913) (Continued)

Millivolts

°F	0	1	2	3	4	5	6	7	8	9
500	14.12	14.15	14.18	14.21	14.24	14.27	14.30	14.33	14.36	14.39
510	14.42	14.45	14.48	14.52	14.55	14.58	14.61	14.64	14.67	14.70
520	14.73	14.76	14.79	14.82	14.85	14.88	14.91	14.94	14.98	15.01
530	15.04	15.07	15.10	15.13	15.16	15.19	15.22	15.25	15.28	15.31
540	15.34	15.37	15.40	15.44	15.47	15.50	15.53	15.56	15.59	15.62
550	15.65	15.68	15.71	15.74	15.77	15.80	15.84	15.87	15.90	15.93
560	15.96	15.99	16.02	16.05	16.08	16.11	16.14	16.17	16.20	16.23
570	16.26	16.30	16.33	16.36	16.39	16.42	16.45	16.48	16.51	16.54
580	16.57	16.60	16.63	16.66	16.69	16.72	16.75	16.78	16.82	16.85
590	16.88	16.91	16.94	16.97	17.00	17.03	17.06	17.09	17.12	17.15
600	17.18	17.21	17.24	17.28	17.31	17.34	17.37	17.40	17.43	17.46
610	17.49	17.52	17.55	17.58	17.61	17.64	17.68	17.71	17.74	17.77
620	17.80	17.83	17.86	17.89	17.92	17.95	17.98	18.01	18.04	18.08
630	18.11	18.14	18.17	18.20	18.23	18.26	18.29	18.32	18.35	18.38
640	18.41	18.44	18.47	18.50	18.54	18.57	18.60	18.63	18.66	18.69
650	18.72	18.75	18.78	18.81	18.84	18.87	18.90	18.94	18.97	19.00
660	19.03	19.06	19.09	19.12	19.15	19.18	19.21	19.24	19.27	19.30
670	19.34	19.37	19.40	19.43	19.46	19.49	19.52	19.55	19.58	19.61
680	19.64	19.67	19.70	19.74	19.77	19.80	19.83	19.86	19.89	19.92
690	19.95	19.98	20.01	20.04	20.07	20.10	20.13	20.16	20.20	20.23
700	20.26	20.29	20.32	20.35	20.38	20.41	20.44	20.47	20.50	20.53
710	20.56	20.59	20.62	20.66	20.69	20.72	20.75	20.78	20.81	20.84
720	20.87	20.90	20.93	20.96	20.99	21.02	21.05	21.08	21.11	21.14
730	21.18	21.21	21.24	21.27	21.30	21.33	21.36	21.39	21.42	21.45
740	21.48	21.51	21.54	21.57	21.60	21.64	21.67	21.70	21.73	21.76
750	21.79	21.82	21.85	21.88	21.91	21.94	21.97	22.00	22.03	22.06
760	22.10	22.13	22.16	22.19	22.22	22.25	22.28	22.31	22.34	22.37
770	22.40	22.43	22.46	22.49	22.52	22.55	22.58	22.62	22.65	22.68
780	22.71	22.74	22.77	22.80	22.83	22.86	22.89	22.92	22.95	22.98
790	23.01	23.04	23.08	23.11	23.14	23.17	23.20	23.23	23.26	23.29
800	23.32	23.35	23.38	23.41	23.44	23.47	23.50	23.53	23.56	23.60
810	23.63	23.66	23.69	23.72	23.75	23.78	23.81	23.84	23.87	23.90
820	23.93	23.96	23.99	24.02	24.06	24.09	24.12	24.15	24.18	24.21
830	24.24	24.27	24.30	24.33	24.36	24.39	24.42	24.45	24.48	24.52
840	24.55	24.58	24.61	24.64	24.67	24.70	24.73	24.76	24.79	24.82
850	24.85	24.88	24.91	24.94	24.98	25.01	25.04	25.07	25.10	25.13
860	25.16	25.19	25.22	25.25	25.28	25.32	25.35	25.38	25.41	25.44
870	25.47	25.50	25.53	25.56	25.59	25.62	25.65	25.68	25.72	25.75
880	25.78	25.81	25.84	25.87	25.90	25.93	25.96	25.99	26.02	26.06
890	26.09	26.12	26.15	26.18	26.21	26.24	26.27	26.30	26.33	26.36
900	26.40	26.43	26.46	26.49	26.52	26.55	26.58	26.61	26.64	26.67
910	26.70	26.74	26.77	26.80	26.83	26.86	26.89	26.92	26.95	26.98
920	27.02	27.05	27.08	27.11	27.14	27.17	27.20	27.23	27.26	27.30
930	27.33	27.36	27.39	27.42	27.45	27.48	27.51	27.54	27.58	27.61
940	27.64	27.67	27.70	27.73	27.76	27.80	27.83	27.86	27.89	27.92
950	27.95	27.98	28.02	28.05	28.08	28.11	28.14	28.17	28.20	28.23

Millivolts

°F	0	1	2	3	4	5	6	7	8	9
960	28.26	28.30	28.33	28.36	28.39	28.42	28.45	28.48	28.52	28.55
970	28.58	28.61	28.64	28.67	28.70	28.74	28.77	28.80	28.83	28.86
980	28.89	28.92	28.96	28.99	29.02	29.05	29.08	29.11	29.14	29.18
990	29.21	29.24	29.27	29.30	29.33	29.37	29.40	29.43	29.46	29.49
1000	29.52	29.56	29.59	29.62	29.65	29.68	29.71	29.75	29.78	29.81
1010	29.84	29.87	29.90	29.94	29.97	30.00	30.03	30.06	30.10	30.13
1020	30.16	30.19	30.22	30.25	30.28	30.32	30.35	30.38	30.41	30.44
1030	30.48	30.51	30.54	30.57	30.60	30.64	30.67	30.70	30.73	30.76
1040	30.80	30.83	30.86	30.89	30.92	30.96	30.99	31.02	31.05	31.08
1050	31.12	31.15	31.18	31.21	31.24	31.28	31.31	31.34	31.37	31.40
1060	31.44	31.47	31.50	31.53	31.56	31.60	31.63	31.66	31.69	31.72
1070	31.76	31.79	31.82	31.85	31.88	31.92	31.95	31.98	32.01	32.05
1080	32.08	32.11	32.14	32.18	32.21	32.24	32.27	32.30	32.34	32.37
1090	32.40	32.43	32.47	32.50	32.53	32.56	32.60	32.63	32.66	32.69
1100	32.72	32.76	32.79	32.82	32.86	32.89	32.92	32.95	32.98	33.02
1110	33.05	33.08	33.11	33.15	33.18	33.21	33.24	33.28	33.31	33.34
1120	33.37	33.41	33.44	33.47	33.50	33.54	33.57	33.60	33.64	33.67
1130	33.70	33.73	33.76	33.80	33.83	33.86	33.90	33.93	33.96	33.99
1140	34.03	34.06	34.09	34.12	34.16	34.19	34.22	34.26	34.29	34.32
1150	34.36	34.39	34.42	34.45	34.49	34.52	34.55	34.58	34.62	34.65
1160	34.68	34.72	34.75	34.78	34.82	34.85	34.88	34.92	34.95	34.98
1170	35.01	35.05	35.08	35.11	35.15	35.18	35.21	35.25	35.28	35.31
1180	35.35	35.38	35.41	35.45	35.48	35.51	35.54	35.58	35.61	35.64
1190	35.68	35.71	35.74	35.78	35.81	35.84	35.88	35.91	35.94	35.98
1200	36.01	36.05	36.08	36.11	36.15	36.18	36.21	36.25	36.28	36.31
1210	36.35	36.38	36.42	36.45	36.48	36.52	36.55	36.58	36.62	36.65
1220	36.69	36.72	36.75	36.79	36.82	36.86	36.89	36.92	36.96	36.99
1230	37.02	37.06	37.09	37.13	37.16	37.20	37.23	37.26	37.30	37.33
1240	37.36	37.40	37.43	37.47	37.50	37.54	37.57	37.60	37.64	37.67
1250	37.71	37.74	37.78	37.81	37.84	37.88	37.91	37.95	37.98	38.02
1260	38.05	38.08	38.12	38.15	38.19	38.22	38.26	38.29	38.32	38.36
1270	38.39	38.43	38.46	38.50	38.53	38.57	38.60	38.64	38.67	38.70
1280	38.74	38.77	38.81	38.84	38.88	38.91	38.95	38.98	39.02	39.05
1290	39.08	39.12	39.15	39.19	39.22	39.26	39.29	39.33	39.36	39.40
1300	39.43	39.47	39.50	39.54	39.57	39.61	39.64	39.68	39.71	39.75
1310	39.78	39.82	39.85	39.89	39.92	39.96	39.99	40.03	40.06	40.10
1320	40.13	40.17	40.20	40.24	40.27	40.31	40.34	40.38	40.41	40.45
1330	40.48	40.52	40.55	40.59	40.62	40.66	40.69	40.73	40.76	40.80
1340	40.83	40.87	40.90	40.94	40.98	41.01	41.05	41.08	41.12	41.15
1350	41.19	41.22	41.26	41.29	41.33	41.36	41.40	41.43	41.47	41.50
1360	41.54	41.58	41.61	41.65	41.68	41.72	41.75	41.79	41.82	41.86
1370	41.90	41.93	41.97	42.00	42.04	42.07	42.11	42.14	42.18	42.22
1380	42.25	42.29	42.32	42.36	42.39	42.43	42.46	42.50	42.53	42.57
1390	42.61	42.64	42.68	42.71	42.75	42.78	42.82	42.85	42.89	42.92
1400	42.96	—	—	—	—	—	—	—	—	—

TEMPERATURE-E. M. F. VALUES FOR COPPER-CONSTANTAN

E. M. F. values are in millivolts; reference junctions at 0°C.;
temperatures are in degrees C.
Roeser and Wensel, National Bureau of Standards

Millivolts

°C	0	10	20	30	40	50	60	70	80	90
−200	−5.54									
−100	−3.35	−3.62	−3.89	−4.14	−4.38	−4.60	−4.82	−5.02	−5.20	−5.38
0	0	−0.38	−0.75	−1.11	−1.47	−1.81	−2.14	−2.46	−2.77	−3.06
0	0	0.39	0.79	1.19	1.61	2.03	2.47	2.91	3.36	3.81
100	4.28	4.75	5.23	5.71	6.20	6.70	7.21	7.72	8.23	8.76
200	9.29	9.82	10.36	10.91	11.46	12.01	12.57	13.14	13.71	14.28
300	14.86	15.44	16.03	16.62	17.22	17.82	18.42	19.03	19.64	20.25
400	20.87									

E. M. F. values are in millivolts; reference junctions at 32°F.;
temperatures are in degrees F.
Roeser and Wensel, National Bureau of Standards

Millivolts

°F	0	10	20	30	40	50	60	70	80	90
−300	−5.28									
−200	−4.11	−4.25	−4.38	−4.50	−4.63	−4.75	−4.86	−4.97	−5.08	−5.18
−100	−2.56	−2.73	−2.90	−3.06	−3.22	−3.38	−3.53	−3.68	−3.83	−3.97
0	−0.67	−0.87	−1.07	−1.27	−1.47	−1.66	−1.84	−2.03	−2.21	−2.39
0	−0.67	−0.46	−0.26	−0.04	+0.17	0.39	0.61	0.83	1.06	1.29
100	1.52	1.75	1.99	2.23	2.47	2.71	2.96	3.21	3.46	3.71
200	3.97	4.22	4.48	4.75	5.01	5.28	5.55	5.82	6.09	6.37
300	6.64	6.92	7.21	7.49	7.77	8.06	8.35	8.64	8.93	9.23
400	9.52	9.82	10.12	10.42	10.72	11.03	11.33	11.64	11.95	12.26
500	12.57	12.89	13.20	13.52	13.83	14.15	14.47	14.79	15.12	15.44
600	15.77	16.10	16.42	16.75	17.08	17.42	17.75	18.08	18.42	18.75
700	19.09	19.43	19.77	20.11	20.45	20.80				

REFERENCE TABLE FOR Pt TO Pt—10 PER CENT Rh THERMOCOUPLE

Emfs are expressed in microvolts and temperatures in °C. Cold junctions at 0°C. Roeser and Wensel, National Bureau of Standards

E(μv)	0	1,000	2,000	3,000	4,000	5,000	6,000	7,000	8,000	9,000	10,000	11,000	12,000	13,000	14,000	15,000	16,000	17,000
0	0	146.9	265.0	373.7	477.7	578.1	675.3	769.5	861.0	950.2	1,037.2	1,122.3	1,206.4	1,290.0	1,373.8	1,458.0	1,542.6	1,627.8
	17.7	12.5	11.2	10.5	10.2	9.9	9.5	9.3	9.0	8.8	8.6	8.5	8.3	8.3	8.4	8.4	8.5	8.6
100	17.7	159.4	276.2	384.2	487.9	588.0	684.8	778.8	870.0	959.0	1,045.8	1,130.8	1,214.7	1,298.3	1,382.2	1,466.4	1,551.1	1,636.4
	16.7	12.3	11.1	10.5	10.2	9.8	9.5	9.2	9.0	8.8	8.6	8.4	8.4	8.4	8.4	8.4	8.5	8.5
200	34.4	171.7	287.3	394.7	498.1	597.8	694.3	788.0	879.0	967.8	1,054.4	1,139.2	1,223.1	1,306.7	1,390.6	1,474.8	1,559.6	1,644.9
	15.8	12.1	11.0	10.5	10.1	9.8	9.5	9.2	9.0	8.7	8.5	8.4	8.3	8.4	8.4	8.5	8.5	8.6
300	50.2	183.8	298.3	405.2	508.2	607.6	703.8	797.2	888.0	976.5	1,062.9	1,147.6	1,231.4	1,315.1	1,399.0	1,483.3	1,568.1	1,653.5
	15.2	12.0	11.0	10.5	10.1	9.8	9.5	9.2	9.0	8.8	8.6	8.4	8.4	8.4	8.4	8.5	8.5	8.6
400	65.4	195.8	309.3	415.7	518.3	617.4	713.3	806.4	897.0	985.3	1,071.5	1,156.0	1,239.8	1,323.5	1,407.4	1,491.8	1,576.6	1,662.1
	14.6	11.8	10.9	10.4	10.1	9.7	9.4	9.2	8.9	8.7	8.5	8.4	8.4	8.3	8.4	8.4	8.5	8.6
500	80.0	207.6	320.2	426.1	528.4	627.1	722.7	815.6	905.9	994.0	1,080.0	1,164.4	1,248.2	1,331.8	1,415.8	1,500.2	1,585.1	1,670.7
	14.1	11.7	10.8	10.4	10.0	9.7	9.4	9.1	8.9	8.7	8.5	8.4	8.3	8.4	8.4	8.5	8.6	8.6
600	94.1	219.3	331.0	436.5	538.4	636.8	732.1	824.7	914.8	1,002.7	1,088.5	1,172.8	1,256.5	1,340.2	1,424.2	1,508.7	1,593.7	1,679.3
	13.7	11.6	10.7	10.3	10.0	9.7	9.4	9.1	8.9	8.6	8.5	8.4	8.5	8.4	8.5	8.5	8.5	8.6
700	107.8	230.9	341.7	446.8	548.4	646.5	741.5	833.8	923.7	1,011.3	1,097.0	1,181.2	1,264.9	1,348.6	1,432.7	1,517.2	1,602.2	1,687.9
	13.3	11.5	10.7	10.3	9.9	9.6	9.4	9.1	8.9	8.7	8.4	8.4	8.3	8.4	8.4	8.4	8.5	8.6
800	121.1	242.4	352.4	457.1	558.3	656.1	750.9	842.9	932.6	1,020.0	1,105.4	1,189.6	1,273.2	1,357.0	1,441.1	1,525.6	1,610.7	1,696.5
	13.0	11.3	10.7	10.3	9.9	9.6	9.3	9.1	8.8	8.6	8.5	8.4	8.4	8.4	8.4	8.5	8.6	8.6
900	134.1	253.7	363.1	467.4	568.2	665.7	760.2	852.0	941.4	1,028.6	1,113.9	1,198.0	1,281.6	1,365.4	1,449.5	1,534.1	1,619.3	1,705,1
	12.8	11.3	10.6	10.3	9.9	9.6	9.3	9.0	8.8	8.6	8.4	8.4	8.4	8.4	8.5	8.5	8.5	8.6
1,000	146.9	265.0	373.7	477.7	578.1	675.3	769.5	861.0	950.2	1,037.2	1,122.3	1,206.4	1,290.0	1,373.8	1,458.0	1,542.6	1,627.8	1,713.7

Emfs are expressed in microvolts and temperatures in °C. Cold junctions at 0°C. Roeser and Wensel, National Bureau of Standards

E(μv)	0	1,000	2,000	3,000	4,000	5,000	6,000	7,000	8,000	9,000	10,000	11,000	12,000	13,000	14,000	15,000	16,000	17,000
0	32.0	296.4	509.0	704.7	891.9	1,072.6	1,247.5	1,417.1	1,581.8	1,742.4	1,899.0	2,052.1	2,203.5	2,354.0	2,504.8	2,656.4	2,808.7	2,962.0
	31.9	22.5	20.1	19.0	18.4	17.7	17.1	16.7	16.2	15.8	15.5	15.2	15.0	15.0	15.2	15.1	15.3	15.4
100	63.9	318.9	529.1	723.7	910.3	1,090.3	1,264.6	1,433.8	1,598.0	1,758.2	1,914.5	2,067.3	2,218.5	2,369.0	2,520.0	2,671.5	2,824.0	2,977.4
	30.0	22.1	20.0	18.9	18.3	17.7	17.1	16.6	16.2	15.8	15.4	15.2	15.0	15.1	15.1	15.2	15.3	15.4
200	93.9	341.0	549.1	742.6	928.6	1,108.0	1,281.7	1,450.4	1.614.2	1,774.0	1,929.9	2,082.5	2,233.5	2,384.1	2,535.1	2,686.7	2,839.3	2,992.8
	28.5	21.8	19.9	18.8	18.2	17.7	17.1	16.6	16.2	15.7	15.4	15.2	15.0	15.1	15.1	15.2	15.3	15.5
300	122.4	362.8	569.0	761.4	946.8	1,125.7	1,298.8	1,467.0	1,630.4	1,789.7	1,945.3	2,097.7	2,248.5	2,399.2	2,550.2	2,701.9	2,854.6	3,008.3
	27.3	21.6	19.8	18.8	18.1	17.6	17.1	16.5	16.1	15.8	15.4	15.1	15.1	15.1	15.1	15.3	15.3	15.5
400	149.7	384.4	588.8	780.2	964.9	1,143.3	1,315.9	1,483.5	1,646.5	1,805.5	1,960.7	2,112.8	2,263.6	2,414.3	2,565.3	2,717.2	2,869.9	3,023.8
	26.3	21.3	19.6	18.8	18.1	17.5	17.0	16.5	16.1	15.7	15.3	15.1	15.1	15.0	15.1	15.2	15.3	15.5
500	176.0	405.7	608.4	799.0	983.0	1,160.8	1,332.9	1,500.0	1,662.6	1,821.2	1,976.0	2,127.9	2,278.7	2,429.3	2,580.4	2,732.4	2.885.2	3,039.3
	25.4	21.0	19.4	18.7	18.1	17.5	16.9	16.5	16.0	15.6	15.3	15.1	15.0	15.1	15.2	15.3	15.4	15.4
600	201.4	426.7	627.8	817.7	1,001.1	1,178.3	1,349.8	1,516.5	1,678.6	1,836.8	1,991.3	2,143.0	2,293.7	2,444.4	2,595.6	2,747.7	2,900.6	3,054.7
	24.6	20.9	19.3	18.6	18.0	17.4	16.9	16.4	16.0	15.6	15.2	15.2	15.1	15.1	15.2	15.3	15.4	15.5
700	226.0	447.6	647.1	836.3	1,019.1	1,195.7	1,366.7	1,532.9	1,694.6	1,852.4	2,006.5	2,158.2	2,308.8	2,459.5	2,610.8	2,763.0	2,916.0	3,070.2
	24.0	20.7	19.3	18.5	17.9	17.3	16.9	16.3	16.0	15.6	15.2	15.1	15.0	15.1	15.2	15.2	15.3	15.5
800	250.0	468.3	666.4	854.8	1,037.0	1,213.0	1,383.6	1,549.2	1,710.6	1,868.0	2,021.7	2,173.3	2,323.8	2,474.6	2,626.0	2,778.2	2,931.3	3,085.7
	23.4	20.4	19.2	18.6	17.8	17.3	16.8	16.3	15.9	15.5	15.2	15.1	15.1	15.1	15.2	15.2	15.4	15.5
900	273.4	488.7	685.6	873.4	1,054.8	1,230.3	1,400.4	1,565.5	1,726.5	1,883.5	2,036.9	2,188.4	2,338.9	2,489.7	2,641.2	2,793.4	2,946.7	3,101.2
	23.0	20.3	19.1	18.5	17.8	17.2	16.7	16.3	15.9	15.5	15.2	15.1	15.1	15.1	15.2	15.3	15.3	15.5
1,000	296.4	509.0	704.7	891.9	1,072.6	1,247.5	1,417.1	1,581.8	1,742.4	1,899.0	2,052.1	2,203.5	2,354.0	2,504.8	2,656.4	2,808.7	2,962.0	3,116.7

REFERENCE TABLE FOR Pt TO Pt—13 PER CENT Rh THERMOCOUPLE

Emfs are expressed in microvolts and temperatures in °C. Cold junctions at 0°C. Roeser and Wensel, National Bureau of Standards

E(μv)	0	1,000	2,000	3,000	4,000	5,000	6,000	7,000	8,000	9,000	10,000	11,000	12,000	13,000	14,000	15,000	16,000	17,000	18,000	19,000
0	0	145.3	258.8	361.0	457.4	549.8	638.3	723.5	806.0	886.1	964.1	1,040.0	1,113.9	1,186.9	1,259.3	1,331.8	1,404.3	1,476.9	1,550.0	1,623.6
	17.9	12.2	10.6	9.9	9.5	9.0	8.7	8.4	8.1	7.9	7.6	7.4	7.3	7.2	7.3	7.2	7.3	7.3	7.4	7.4
100	17.9	157.5	269.4	370.9	466.9	558.8	647.0	731.9	814.1	894.0	971.7	1,047.4	1,121.2	1.194.1	1,266.6	1,339.0	1,411.6	1,484.2	1,557.4	1,631.0
	16.7	12.0	10.5	9.8	9.4	9.0	8.6	8.4	8.1	7.8	7.7	7.5	7.4	7.3	7.2	7.2	7.3	7.3	7.4	7.4
200	34.6	169.5	279.9	380.7	476.3	567.8	655.6	740.3	822.2	901.8	979.4	1,054.9	1,128.6	1,201.4	1,273.8	1,346.2	1,418.9	1,491.5	1,564.8	1,638.4
	15.8	11.7	10.4	9.8	9.4	9.0	8.6	8.3	8.1	7.9	7.6	7.4	7.3	7.3	7.3	7.3	7.2	7.3	7.3	7.4
300	50.4	181.2	290.3	390.5	485.7	576.8	644.2	748.6	830.3	909.7	987.0	1,062.3	1,135.9	1,208.7	1,281.1	1.353.5	1,426.1	1.498.8	1,572.1	1.645.8
	15.1	11.5	10.3	9.7	9.3	8.9	8.6	8.3	8.0	7.8	7.7	7.4	7.3	7.2	7.2	7.3	7.2	7.3	7.4	7.4
400	65.5	192.7	300.6	400.2	495.0	585.7	672.8	756.9	838.3	917.5	994.7	1,069.7	1.143.2	1,215.9	1,288.3	1,360.8	1,433.3	1,506.1	1,579.5	1,653.2
	14.5	11.4	10.2	9.7	9.3	8.9	8.5	8.2	8.0	7.8	7.6	7.4	7.3	7.3	7.2	7.2	7.2	7.3	7.3	7.4
500	80.0	204.1	310.8	409.9	504.3	594.6	681.3	765.1	846.3	925.3	1,002.3	1,077.1	1,150.5	1,223.2	1,295.5	1,368.0	1,440.5	1,513.4	1,586.8	1,660.6
	13.9	11.2	10.2	9.6	9.2	8.8	8.5	8.2	8.0	7.8	7.6	7.3	7.3	7.2	7.2	7.3	7.3	7.3	7.4	7.4
600	93.9	215.3	321.0	419.5	513.5	603.4	689.8	773.3	854.3	933.1	1,009.9	1,084.4	1.157.8	1,230.4	1,302.7	1,375.3	1447.8	1,520.7	1,594.2	1,668.0
	13.4	11.1	10.1	9.5	9.1	8.8	8.5	8.2	8.0	7.8	7.6	7.4	7.3	7.2	7.2	7.3	7.2	7.3	7.3	7.4
700	107.3	226.4	331.1	429.0	522.6	612.2	698.3	781.5	862.3	940.9	1,017.5	1,091.8	1,165.1	1,237.6	1,309.9	1,382.6	1,455.0	1,528.0	1,601.5	1,675.4
	13.0	10.9	10.0	9.5	9.1	8.7	8.4	8.2	8.0	7.8	7.5	7.4	7.2	7.2	7.3	7.2	7.3	7.3	7.4	7.4
800	120.3	237.3	341.1	438.5	531.7	620.9	706.7	789.7	870.3	948.7	1,025.0	1,099.2	1,172.3	1,244.8	1,317.2	1.389.8	1,462.3	1,525.3	1,608.9	1,682.8
	12.6	10.8	10.0	9.5	9.1	8.7	8.4	8.1	7.9	7.7	7.5	7.3	7.3	7.3	7.3	7.3	7.3	7.4	7.3	7.4
900	132.9	248.1	351.1	448.0	540.8	629.6	715.1	797.8	878.2	956.4	1,032.5	1,106.5	1,179.6	1,252.1	1,324.5	1,397.1	1,469.6	1,542.7	1,616,2	1,690.2
	12.4	10.7	9.9	9.4	9.0	8.7	8.4	8.2	7.9	7.7	7.5	7.4	7.3	7.2	7.3	7.2	7.3	7.3	7.4	7.4
1,000	145.3	258.8	361.0	457.4	549.8	638.3	723.5	806.0	886.1	964.1	1,040.0	1,113.9	1,186.9	1,259.3	1,331.8	1,404.3	1,476.9	1,550.0	1,623.6	1,697.6

Emfs are expressed in microvolts and temperatures in °F. Cold junctions at 32°F. Roeser and Wensel, National Bureau of Standards

E(μv)	0	1,000	2,000	3,000	4,000	5,000	6,000	7,000	8,000	9,000	10,000	11,000	12,000	13,000	14,000	15,000	16,000	17,000	18,000	19,000
0	32.0	293.5	497.8	681.8	855.3	1,021.6	1,180.9	1334.3	1,482.8	1,627.0	1,767.4	1,904.0	2,037.0	2,168.4	2,298.7	2,429.2	2,559.7	2,690.4	2,822.0	2,954.5
	32.2	22.0	19.1	17.8	17.0	16.2	15.6	15.1	14.6	14.2	13.8	13.4	13.2	13.0	13.1	13.0	13.1	13.2	13.3	13.3
100	64.2	315.5	516.9	699.6	872.3	1,037.8	1,196.5	1,349.4	1,497.4	1,641.2	1,781.2	1,917.4	2,050.2	2,181.4	2,311.8	2,442.2	2,572.8	2,703.6	2,835.3	2,967.8
	30.1	21.6	18.9	17.7	16.2	16.2	15.6	15.1	14.6	14.1	13.7	13.4	13.2	13.1	13.0	13.0	13.1	13.1	13.3	13.3
200	94.3	337.1	535.8	717.3	889.3	1,054.0	1,212.1	1,364.5	1,512.0	1,655.3	1,794.9	1,930.8	2,063.4	2,194.5	2,324.8	2,455.2	2,585.9	2,716.7	2,848.6	2,981.1
	28.4	21.1	18.7	17.6	16.9	16.2	15.5	15.0	14.5	14.1	13.7	13.3	13.2	13.1	13.1	13.1	13.0	13.1	13.2	13.3
300	122.7	358.2	554.5	734.9	906.2	1,070.2	1,227.6	1,379.5	1,526.5	1,669.4	1,808.6	1,944.1	2,076.6	2,207.6	2,337.9	2,468.3	2,598.9	2,729.8	2,861.8	2,994.4
	27.2	20.7	18.6	17.5	16.8	16.1	15.4	14.9	14.4	14.1	13.8	13.3	13.2	13.0	13.0	13.1	13.0	13.2	13.3	13.4
400	149.9	378.9	573.1	752.4	923.0	1.086.3	1,243.0	1,394.4	1,540.9	1,683.5	1,822.4	1,957.4	2.089.8	2,220.6	2,350.9	2.481.4	2,611.9	2,743.0	2,875.1	3,007.8
	26.1	20.4	18.4	17.4	16.7	16.0	15.3	14.8	14.4	14.0	13.7	13.3	13.1	13.1	13.0	13.0	13.0	13.1	13.2	13.3
500	176.0	399.3	591.5	769.8	939.7	1,102.3	1,258.3	1,409.2	1,555.3	1,697.5	1,836.1	1,970.7	2,102.9	2.233.7	2,363.9	2,494.4	2,624.9	2,756.1	2,888.3	3,021.1
	25.0	20.2	18.3	17.3	16.6	15.9	15.3	14.8	14.4	14.1	13.7	13.3	13.1	13.0	13.0	13.1	13.1	13.2	13.3	13.3
600	201.0	419.5	609.8	787.1	956.3	1,118.2	1,273.6	1,424.0	1,569.7	1,711.6	1,849.8	1,984.0	2,116.0	2,246.7	2,376.9	2,507.5	2,638.0	2,769.3	2,901.6	3,034.4
	24.1	20.0	18.2	17.2	16.4	15.8	15.3	14.8	14.4	14.0	13.6	13.3	13.1	13.0	13.0	13.1	13.0	13.1	13.2	13.3
700	225.1	439.5	628.0	804.3	972.7	1,134.0	1,288.9	1,438.8	1,584.1	1,725.6	1,863.4	1,997.3	2,129.1	2,259.7	2,389.9	2,520.6	2,651.0	2,782.4	2,914.8	3,047.7
	23.4	19.7	18.0	17.1	16.4	15.7	15.2	14.7	14.4	14.0	13.6	13.3	13.1	13.0	13.1	13.0	13.1	13.2	13.2	13.3
800	248.5	459.2	646.0	821.4	989.1	1,149.7	1,304.1	1,453.5	1,598.5	1,739.6	1,877.0	2,010.6	2,142.2	2,272.7	2,403.0	2,533.6	2,664,1	2,795.6	2,928.0	3,061.0
	22.7	19.4	18.0	17.0	16.3	15.6	15.1	14.6	14.3	13.9	13.5	13.2	13.1	13.0	13.1	13.1	13.1	13.2	13.2	13.4
900	271.2	478.6	664.0	838.4	1,005.4	1,165.3	1,319.2	1,468.1	1,612.8	1,753.5	1,890.5	2,023.8	2,155.3	2,285.7	2,416.1	2,546.7	2,677.2	2,808.8	2,941.2	3,074.4
	22.3	19.2	17.8	16.9	16.2	15.6	15.1	14.7	14.2	13.9	13.5	13.2	13.1	13.0	13.1	13.0	13.2	13.2	13.3	13.3
1,000	293.5	497.8	681.8	855.3	1,021.6	1,180.9	1,334.3	1,482.8	1,627.0	1,767.4	1,904.0	2,037.0	2,168.4	2,298.7	2,429.2	2,559.7	2,690.4	2,822.0	2,954.5	3,087.7

INTRODUCTION TO X-RAY CROSS SECTIONS

Alex F. Burr

These tables are part of an extensive report published by W. H. McMaster, et al. as UCRL 50174 and available from the National Technical Information Service, Springfield, Va. 22151. Section I of UCRL 50174 describes the data base and the treatment given it, Section II contains the total cross sections between 1 and 1000 keV for all the elements, Section III contains results used in producing Section II, and Section IV contains total cross sections for selected energies and is reproduced in part here. To obtain these values existing experimental x-ray total cross section data and theoretical cross section calculations were surveyed. The coherent (Rayleigh) scattering cross sections and the incoherent (Compton) scattering cross sections were computed. The photo-electric cross sections were obtained by least squares fitting of experimental data, theory, and interpolation of experiment and theory. The following table contains cross sections interpolated from the basic compilation at those wavelengths of most use to x-ray crystallographers. The wavelengths chosen were selected to correspond to those given in the International Tables for X-Ray Crystallography. The energy-to-wavelength conversion is given below.

Table I. Energy-to-wavelength conversion

Target radiation	Å	keV	Target radiation	Å	keV
Ag K$\bar{\alpha}$	0.5608	22.105	Ni K$\bar{\alpha}$	1.6591	7.472
Kβ_1	0.4970	24.942	Kβ_1	1.5001	8.265
Pd K$\bar{\alpha}$	0.5869	21.125	Co K$\bar{\alpha}$	1.7902	6.925
Kβ_1	0.5205	23.819	Kβ_1	1.6208	7.649
Rh K$\bar{\alpha}$	0.6147	20.169	Fe K$\bar{\alpha}$	1.9373	6.400
Kβ_1	0.5456	22.724	Kβ_1	1.7565	7.058
Mo K$\bar{\alpha}$	0.7107	17.444	Mn K$\bar{\alpha}$	2.1031	5.895
Kβ_1	0.6323	19.608	Kβ_1	1.9102	6.490
Zn K$\bar{\alpha}$	1.4364	8.631	Cr K$\bar{\alpha}$	2.2909	5.412
Kβ_1	1.2952	9.572	Kβ_1	2.0848	5.947
Cu K$\bar{\alpha}$	1.5418	8.041	Ti K$\bar{\alpha}$	2.7496	4.509
Kβ_1	1.3922	8.905	Kβ_1	2.5138	4.932

Table II. Total Cross Section in cm^2/g

Z KEV	1 H	2 He	3 Li	4 Be	5 B	6 C	7 N	8 O	9 F	10 Ne	11 Na	12 Mg	13 Al	14 Si	15 P	16 S	17 Cl	18 Ar	19 K	20 Ca
4.51	.432	.661	2.10	5.63	12.9	25.6	43.0	64.8	91.8	125	168	220	264	336	386	464	518	577	679	805
4.93	.421	.550	1.62	4.25	9.69	19.4	32.6	49.4	70.3	95.9	129	171	206	263	304	364	411	456	542	638
5.41	.412	.465	1.24	3.18	7.23	14.5	24.4	37.2	53.1	72.7	98.5	131	158	203	236	282	322	356	427	500
5.90	.405	.405	.986	2.45	5.53	11.1	18.7	28.6	41.1	56.4	76.6	102	124	160	186	223	256	282	342	398
5.95	.405	.400	.964	2.39	5.39	10.8	18.2	27.9	40.0	54.9	74.7	99.6	121	156	182	217	250	276	334	389
6.40	.400	.362	.798	1.92	4.28	8.55	14.5	22.2	32.0	44.0	59.9	80.2	97.5	126	148	177	205	225	275	319
6.49	.400	.355	.770	1.84	4.10	8.18	13.9	21.3	30.6	42.2	57.5	77	93.7	121	142	170	197	217	264	307
6.93	.397	.329	.659	1.52	3.36	6.68	11.3	17.4	25.1	34.7	47.3	63.5	77.5	100	118	141	165	181	222	257
7.06	.396	.322	.631	1.44	3.17	6.30	10.7	16.5	23.7	32.8	44.7	60.1	73.4	95.1	112	134	156	172	211	244
7.47	.394	.303	.555	1.23	2.67	5.28	8.96	13.8	19.9	27.6	37.7	50.8	62.2	80.7	95.2	114	134	147	181	209
7.65	.393	.297	.528	1.15	2.49	4.92	8.33	12.9	18.6	25.7	35.2	47.4	58.1	75.4	89.1	107	125	137	170	196
8.04	.391	.284	.477	1.01	2.14	4.22	7.14	11.0	16.0	22.1	30.3	40.9	50.2	65.3	77.3	92.5	109	120	148	171
8.27	.390	.277	.452	.936	1.98	3.88	6.56	10.1	14.7	20.4	27.9	37.6	46.3	60.2	71.3	85.5	101	111	138	159
8.63	.389	.268	.417	837	1.74	3.40	5.74	8.87	12.8	17.9	24.4	33.0	40.7	53.0	62.9	75.4	89.4	97.7	122	141
8.91	.388	.262	.394	774	1.59	3.09	5.22	8.06	11.7	16.2	22.2	30.1	37.1	48.4	57.4	68.9	81.8	89.3	112	129
9.57	.386	.250	.349	.651	1.30	2.49	4.19	6.47	9.36	13.1	17.9	24.2	30.0	39.1	46.6	55.9	66.7	72.7	91.4	106
17.44	.373	.202	.197	.245	.345	.535	.790	1.15	1.58	2.21	2.94	3.98	5.04	6.53	7.87	9.63	11.6	12.6	16.2	19
19.61	.370	.197	.187	.222	.293	.429	.605	.855	1.15	1.60	2.10	2.83	3.59	4.62	5.57	6.84	8.26	8.95	11.5	13.6
20.17	.369	.196	.185	.217	.283	.408	.570	.799	1.07	1.48	1.94	2.60	3.30	4.26	5.13	6.30	7.61	8.24	10.6	12.5
21.13	.368	.195	182	.210	.268	.379	.519	.717	.952	1.31	1.70	2.28	2.89	3.71	4.47	5.50	6.63	7.18	9.24	10.9
22.11	.366	.193	.179	.205	.256	.354	.476	.648	.851	1.16	1.50	2.00	2.54	3.25	3.91	4.82	5.80	6.28	8.08	9.57
22.72	.366	.192	177	.201	.249	.340	.452	.610	.795	1.08	1.39	1.86	2.35	3.00	3.61	4.45	5.35	5.79	7.45	8.84
23.82	.364	.191	.175	.196	.239	.319	.416	.553	.711	.963	1.23	1.63	2.06	2.62	3.15	3.88	4.66	5.04	6.49	7.71
24.94	.363	.189	.173	.192	.229	.301	.385	.504	.640	.861	1.09	1.44	1.81	2.30	2.76	3.40	4.08	4.41	5.67	6.75

Table II. Total Cross Section in cm^2/g (Continued)

Z KEV	21 Sc	22 Ti	23 V	24 Cr	25 Mn	26 Fe	27 Co	28 Ni	29 Cu	30 Zn	31 Ga	32 Ge	33 As	34 Se	35 Br	36 Kr	37 Rb	38 Sr	39 Y	40 Zr
4.51	819	111	125	143	160	188	206	240	257	280	309	329	368	403	435	464	508	552	599	648
4.93	658	86.8	97.3	111	125	147	161	188	201	220	242	258	288	317	342	366	400	436	473	511
5.41	521	571	75.1	85.7	96.1	113	125	146	155	172	187	200	224	246	266	285	312	339	369	399
5.90	420	459	513	67.4	75.6	88.9	98.4	115	123	137	148	158	178	195	211	226	248	270	294	317
5.95	411	449	501	65.8	73.8	86.8	96.1	113	120	134	144	155	173	190	206	221	242	263	287	310
6.40	339	370	411	462	59.9	70.4	78.3	91.8	97.4	110	117	126	142	156	169	181	198	215	235	254
6.49	327	357	396	445	57.6	67.7	75.3	88.3	93.6	106	113	122	136	150	162	174	191	207	227	244
6.93	276	301	333	375	405	56.3	62.9	73.8	78.1	88.7	94.2	102	114	125	136	146	160	174	190	205
7.06	262	286	316	357	385	53.3	59.6	70.0	74.1	84.3	89.3	96.8	108	119	129	138	152	165	181	195
7.47	226	246	271	307	331	367	50.9	59.8	63.2	72.4	76.2	82.9	92.5	101	110	119	130	141	155	167
7.65	212	231	255	288	311	346	47.7	56.1	59.2	68.0	71.4	77.8	86.8	95.1	104	111	122	132	145	157
8.04	186	202	223	252	273	304	339	48.8	51.5	59.5	62.1	67.9	75.7	82.9	90.3	97	106	115	127	137
8.27	173	188	206	234	253	283	315	45.2	47.7	55.3	57.5	63.0	70.1	77.8	83.7	89.9	98.5	107	118	127
8.63	153	167	183	208	225	253	281	306	42.3	49.2	50.9	56.0	62.2	68.0	74.2	79.8	87.4	94.7	105	113
8.91	141	153	168	191	207	234	259	283	38.7	45.3	46.7	51.4	57.0	62.3	68.1	73.2	80.2	86.8	96.2	103
9.57	116	126	138	157	170	194	214	236	245	37.4	38.1	42.3	46.7	51.0	55.8	60.0	65.7	71.0	79.0	84.8
17.44	21.0	23.3	25.2	29.3	31.9	37.7	41.0	47.2	49.3	55.5	56.9	60.5	66.0	68.8	74.7	79.1	83.0	88.0	97.6	16.1
19.61	15.0	16.7	18.1	21.0	22.9	27.2	29.5	34.2	35.8	40.3	41.7	44.3	48.6	51.2	55.6	58.6	62.1	65.6	72.6	75.2
20.17	13.8	15.4	16.7	19.4	21.1	25.1	27.2	31.6	33.1	37.2	38.6	41.0	45.1	47.6	51.7	54.5	57.8	61.0	67.5	70.1
21.13	12.1	13.4	14.6	17.0	18.5	22.0	23.8	27.7	29.0	32.7	34.0	36.1	39.7	42.1	45.6	48.1	51.1	54.0	59.7	62.1
22.11	10.5	11.8	12.8	14.9	16.2	19.3	20.9	24.3	25.5	28.7	29.9	31.8	35.0	37.2	40.4	42.5	45.3	47.9	53.0	55.2
22.72	9.72	10.9	11.8	13.7	15.0	17.8	19.3	22.5	23.6	26.6	27.7	29.4	32.4	34.6	37.5	39.5	42.1	44.5	49.2	51.4
23.82	8.47	9.48	10.3	12.0	13.1	15.6	16.9	19.7	20.7	23.3	24.3	25.8	28.5	30.5	33.1	34.8	37.2	39.3	43.5	45.5
24.94	7.40	8.30	9.02	10.5	11.5	13.7	14.8	17.3	18.2	20.4	21.4	22.7	25.1	26.9	29.2	30.7	32.9	34.8	38.5	40.4

Z KEV	41 Nb	42 Mo	43 Tc	44 Ru	45 Rh	46 Pd	47 Ag	48 Cd	49 In	50 Sn	51 Sb	52 Te	53 I	54 Xe	55 Cs	56 Ba	57 La	58 Ce	59 Pr	60 Nd
4.51	697	738	786	832	892	928	987	1064	1151	1128	997	753	293	300	324	334	355	383	414	433
4.93	552	585	621	660	708	739	785	842	906	899	926	843	921	683	259	266	282	304	330	344
5.41	432	457	486	518	555	581	617	659	706	709	733	769	835	755	803	587	223	240	261	271
5.90	344	365	387	414	444	466	495	526	561	569	592	617	666	701	742	660	677	521	210	218
5.95	336	357	378	404	434	455	484	514	548	557	579	603	651	685	725	645	662	509	205	213
6.40	276	293	310	333	357	375	398	422	449	460	479	497	535	565	597	615	636	592	450	464
6.49	266	282	299	320	344	361	384	407	433	443	462	479	516	545	575	593	613	571	610	447
6.93	223	237	251	269	289	304	324	342	363	374	391	404	434	459	484	499	519	559	596	532
7.06	212	225	238	256	275	289	308	325	345	356	373	385	413	437	460	475	494	532	567	506
7.47	182	193	204	220	236	249	265	279	295	307	322	331	355	376	395	408	427	459	488	506
7.65	170	181	192	207	222	234	249	262	277	289	303	312	333	353	372	384	402	432	459	476
8.04	149	158	168	181	194	205	218	229	242	253	267	273	292	310	325	336	354	379	402	418
8.27	138	147	156	168	180	190	203	213	225	236	248	254	271	288	302	312	329	352	374	389
8.63	122	130	138	149	160	169	180	189	200	210	221	226	241	256	269	278	294	314	333	347
8.91	112	120	127	137	147	156	166	174	183	193	204	208	222	236	248	256	271	290	307	320
9.57	92.0	98.2	104	113	121	128	137	143	151	159	168	172	183	195	204	211	224	240	253	265
17.44	17.0	18.4	19.8	21.3	23.1	24.4	26.4	27.7	29.1	31.2	33.0	33.9	36.3	38.3	40.4	42.4	45.3	48.6	50.8	53.3
19.61	81.2	13.3	14.3	15.4	16.7	17.6	19.1	20.1	21.2	22.6	23.9	24.7	26.5	27.9	29.5	31.0	33.1	35.5	37.1	38.9
20.17	75.6	79.3	13.2	14.2	15.4	16.3	17.7	18.6	19.6	20.9	22.1	22.8	24.6	25.8	27.3	28.7	30.7	33.0	34.4	36.0
21.13	67.1	70.3	73.0	12.5	13.5	14.3	15.5	16.4	17.3	18.4	19.4	20.1	21.7	22.7	24.1	25.4	27.0	29.1	30.3	31.7
22.11	59.6	62.5	65.0	11.0	11.9	12.6	13.7	14.5	15.3	16.2	17.1	17.7	19.2	20.0	21.3	22.5	23.9	25.7	26.8	28.0
22.72	55.5	58.2	60.6	63.8	11.0	11.6	12.7	13.4	14.1	15.0	15.8	16.4	17.8	18.6	19.8	20.9	22.1	23.9	24.9	26.0
23.82	49.1	51.5	53.7	56.6	58.3	10.2	11.1	11.8	12.4	13.2	13.8	14.4	15.7	16.3	17.4	18.4	19.5	21.0	21.9	22.9
24.94	43.6	45.7	47.7	50.3	52.0	57.0	97.8	10.4	11.0	11.6	12.1	12.7	13.9	14.4	15.4	16.3	17.2	18.6	19.3	20.2

X-RAY CROSS SECTIONS (Continued)

Table II. Total Cross Section in cm^2/g (Continued)

Z KEV	61 Pm	62 Sm	63 Eu	64 Gd	65 Tb	66 Dy	67 Ho	68 Er	69 Tm	70 Yb
4.51	455	473	503	510	546	568	589	615	644	664
4.93	361	375	398	405	432	449	465	485	508	524
5.41	285	295	313	319	339	352	363	380	397	410
5.90	229	237	251	256	271	281	290	303	317	327
5.95	224	232	245	250	265	275	283	296	310	319
6.40	186	192	203	207	219	227	234	244	255	263
6.49	476	185	195	200	211	219	225	235	246	254
6.93	401	412	165	170	179	185	190	198	207	214
7.06	536	392	420	161	170	176	181	189	197	204
7.47	535	475	361	369	147	152	156	163	170	175
7.65	503	446	477	347	367	143	146	153	160	165
8.04	441	454	418	427	322	337	128	134	140	145
8.27	410	422	450	397	420	313	333	125	131	135
8.63	366	376	401	410	375	392	296	318	117	120
8.91	336	347	369	377	400	360	272	292	289	111
9.57	278	287	305	312	330	344	364	336	239	232
17.44	55.5	58.0	61.2	62.8	66.8	68.9	72.1	75.6	79.0	80.2
19.61	40.5	42.4	44.7	46.0	48.9	50.4	52.8	55.1	57.9	59.2
20.17	37.6	39.3	41.5	42.6	45.3	46.7	48.9	51.0	53.8	55.0
21.13	33.1	34.7	36.6	37.6	40.0	41.2	43.2	45.0	47.5	48.7
22.11	29.3	30.7	32.4	33.3	35.4	36.5	38.3	39.8	42.1	43.2
22.72	27.1	28.5	30.1	30.9	32.9	33.9	35.6	37.0	39.1	40.2
23.82	23.9	25.1	26.5	27.3	29.0	29.9	31.4	32.6	34.5	35.5
24.94	21.1	22.2	23.4	24.1	25.6	26.4	27.8	28.8	30.6	31.5

Z KEV	71 Lu	72 Hf	73 Ta	74 W	75 Re	76 Os	77 Ir	78 Pt	79 Au	80 Hg
4.51	696	720	736	753	796	824	871	934	906	958
4.93	549	568	581	598	631	653	688	734	720	760
5.41	430	445	455	470	496	512	540	572	568	598
5.90	343	355	363	378	397	410	431	455	457	480
5.95	335	347	355	369	388	401	422	444	447	469
6.40	276	286	293	306	321	332	348	365	371	388
6.49	266	276	282	295	310	320	336	351	358	375
6.93	225	233	238	250	262	270	283	295	303	317
7.06	214	222	227	238	249	257	270	281	289	302
7.47	184	191	195	206	215	222	233	241	250	261
7.65	173	180	184	194	203	209	219	226	236	246
8.04	152	158	162	171	178	184	192	198	208	216
8.27	142	147	150	159	166	171	179	184	194	202
8.63	126	131	134	142	149	153	160	164	174	180
8.91	117	121	124	132	137	141	148	151	161	167
9.57	247	236	103	110	114	118	123	125	134	139
17.44	84.2	86.3	89.5	95.8	98.7	100	103	109	111	115
19.61	62.0	64.2	66.1	70.6	72.5	74.1	77.2	80.2	82.3	85.3
20.17	57.6	59.7	61.4	65.5	67.3	68.9	71.9	74.6	76.5	79.4
21.13	51.0	52.9	54.4	58.0	59.5	61.1	63.8	66.0	67.8	70.4
22.11	45.3	46.9	48.3	51.4	52.8	54.2	56.8	58.6	60.2	62.6
22.72	42.1	43.7	44.9	47.8	49.1	50.5	52.9	54.5	56.0	58.3
23.82	37.2	38.6	39.7	42.2	43.3	44.6	46.8	48.2	49.5	51.7
24.94	33.0	34.2	35.2	37.4	38.4	39.6	41.6	42.7	43.9	45.9

Z KEV	81 Tl	82 Pb	83 Bi	86 Rn	90 Th	92 U	95 Pu
4.51	991	1035	1066	1174	1098	1084	960
4.93	785	820	847	930	993	862	1023
5.41	617	645	667	731	844	774	803
5.90	494	517	536	586	678	672	731
5.95	483	505	524	573	663	657	772
6.40	400	418	435	474	549	545	638
6.49	386	403	419	458	530	526	615
6.93	326	341	355	387	449	446	520
7.06	311	325	338	369	428	425	495
7.47	268	280	293	318	370	368	427
7.65	253	264	276	300	348	347	402
8.04	222	232	243	264	307	306	353
8.27	207	216	227	246	286	285	329
8.63	185	194	203	220	256	256	294
8.91	171	179	188	203	237	236	271
9.57	142	149	156	169	197	197	225
17.44	119	123	126	117	99.5	96.7	48.8
19.61	88.3	90.6	93.5	101	73.3	72.6	79.0
20.17	82.0	84.1	87.0	93.8	95.0	67.8	73.7
21.13	72.7	74.5	77.2	83.3	97.2	84.2	65.7
22.11	64.6	66.1	68.7	74.1	86.3	86.9	58.8
22.72	60.1	61.5	64.0	69.1	80.3	81.1	76.4
23.82	53.2	54.4	56.7	61.2	71.0	72.1	78.3
24.94	47.2	48.2	50.4	54.4	62.9	64.2	69.8

X-RAY WAVELENGTHS

J. A. Bearden

These tables were originally published as the final report to the U.S. Atomic Energy Commission as Report NYO-10586 in partial fulfillment of Contract AT(30-1)-2543. The tables were later reproduced in *Review of Modern Physics*. The data may also be obtained from the Superintendent of Documents, U.S. Government Printing Office, Washington, D. C. 20402 in the publication NSRDS-NBS 14. Persons seeking discussion of the experimental work, conventions, secondary standards, etc. will find these in *Review of Modern Physics*, Vol. 39, No. 1, 78-124, January 1967.

THE W $K\alpha_1$ WAVELENGTH STANDARD

A wavelength standard should possess characteristics which permit its ready redetermination in other laboratories by different techniques. Considering all of the factors involved in the selection of a wavelength standard, the W $K\alpha_1$ line is superior to any other x-ray or γ-ray wavelength. Its advantages as the x-ray wavelength standard are discussed in *Review of Modern Physics* Vol. 39, page 82 (1967).

$$\lambda W\ K\alpha_1 = 0.2090100\ \text{Å} \pm 5\ \text{ppm}.$$

This numerical value of the wavelength of the W $K\alpha_1$ line is used to define the *x-ray wavelength standard* by the relation

$$\lambda(W\ K\alpha_1) = 0.2090100\ \text{Å}^*.$$

This is a new unit of length which may differ from the angstrom by ±5 ppm (probable error), *but as a wavelength standard it has no error*. In order to clearly indicate that this unit is not exactly an angstrom, it has been designated Å*.

Wavelengths tabulated normally refer to the pure element in its solid form. However, there are many instances in which such data are not available. For example, rare gases are of necessity almost always used in the gaseous form, while the rare-earth elements were customarily used in the form of salts.

In high precision work there is some ambiguity as to exactly what feature of a line profile should be taken to be the "true wavelength." In double-crystal work the line peak is usually employed. In crystallography the centroid is widely used; in photographic work with visual observation of the plates, there is involved some subjective criterion of the observer which it is difficult to define precisely. In this survey the peak of the line profile has been adopted as the standard criterion.

In the study of the X-ray literature, the wavelengths of a number of lines were noted which appeared inconsistent with the remaining data. A Moseley-type diagram was constructed, and if the value was clearly outside estimated probable error, it was assumed that an experimental or typographical error had occurred, and the interpolated value was listed in the table. Such cases are marked with a dagger (†) as a superscript to the wavelength. For elements of atomic number 85 through 89 and 91, there are no measured lines of the *K* series and very few of other series except for 88 radium and 91 protactinium. Likewise there are very few measurements for 43 technetium and 54 xenon. In these cases, interpolated values are listed for the more prominent lines and marked with a dagger (†).

X-RAY WAVELENGTHS

X-ray wavelengths in Å* units and in keV. The probable error (p.e.) is the error in the last digit of wavelength. Designation indicates both conventional Siegbahn notation (if applicable) and transition, e.g., $\beta_1\ L_{II}M_{IV}$ denotes a transition between the L_{II} and M_{IV} levels, which is the $L\beta_1$ line in Siegbahn notation.

Designation	Å*	p.e.	keV	Å*	p.e.	keV
	3 Lithium			**4 Beryllium**		
$\alpha\ KL$	228.	1	0.0543	114.	1	0.1085
	5 Boron			**6 Carbon**		
$\alpha\ KL$	67.6	3	0.1833	44.7	3	0.277
	7 Nitrogen			**8 Oxygen**		
$\alpha\ KL$	31.6	4	0.3924	23.62	3	0.5249
	9 Fluorine			**10 Neon**		
$\alpha_{1,2}\ KL_{II,III}$	18.32	2	0.6768	14.610	3	0.8486
$\beta\ KM$				14.452	5	0.8579
	11 Sodium			**12 Magnesium**		
$\alpha_{1,2}\ KL_{II,III}$	11.9101	9	1.0410	9.8900	2	1.25360
$\beta\ KM$	11.575	2	1.0711	9.521	2	1.3022
$L_{II,III}M$	407.1	5	0.03045	251.5	5	0.0493
$L_I L_{II,III}$	376	1	0.0330	317	1	0.0392
	13 Aluminum			**14 Silicon**		
$\alpha_2\ KL_{II}$	8.34173	9	1.48627	7.12791	9	1.73938
$\alpha_1\ KL_{III}$	8.33934	9	1.48670	7.12542	9	1.73998
$\beta\ KM$	7.960	2	1.5574	6.753	1	1.8359
$L_{II,III}$	171.4	5	0.0724	135.5	4	0.0915
$L_I L_{II,III}$	290.	1	0.0428			
	15 Phosphorus			**16 Sulfur**		
$\alpha_2\ KL_{II}$	6.160†	1	2.0127	5.37496	8	2.30664
$\alpha_1\ KL_{III}$	6.157†	1	2.0137	5.37216	7	2.30784
$\beta\ KM$	5.796	2	2.1390			
$\beta_1\ KM$				5.0316	2	2.4640
$\beta_2\ KM$				5.0233	3	2.4681
$L_{II,III}M$	103.8	4	0.1194			
$l,\eta\ L_{II,III}M_I$				83.4	3	0.1487
	17 Chlorine			**18 Argon**		
$\alpha_2\ KL_{II}$	4.7307	1	2.62078	4.19474	5	2.95563
$\alpha_1\ KL_{III}$	4.7278	1	2.62239	4.19180	5	2.95770
$\beta\ KM$	4.4034	3	2.8156			
$\beta_{1,3}\ KM_{II,III}$				3.8860	2	3.1905
$\eta\ L_{II}M_I$	67.33	9	0.1841	55.9†	1	0.2217
$l\ L_{III}M_I$	67.90	9	0.1826	56.3†	1	0.2201
	19 Potassium			**20 Calcium**		
$\alpha_2\ KL_{II}$	3.7445	2	3.3111	3.36166	3	3.68809
$\alpha_1\ KL_{III}$	3.7414	2	3.3138	3.35839	3	3.69168
$\beta_{1,3}\ KM_{II,III}$	3.4539	2	3.5896	3.0897	2	4.0127
$\beta_5\ KM_{IV,V}$	3.4413	4	3.6027	3.0746	3	4.0325

Designation	Å*	p.e.	keV	Å*	p.e.	keV
	19 Potassium *(Cont.)*			**20 Calcium** *(Cont.)*		
$\eta\ L_{II}M_I$	47.24	2	0.2625	40.46	2	0.3064
β_1				35.94	2	0.3449
$l\ L_{III}M_I$	47.74	1	0.25971	40.96	2	0.3027
$\alpha_{1,2}\ L_{III}M_{IV,V}$				36.33	2	0.3413
$M_{II,III}N_I$	692	9	0.0179	525.	9	0.0236
	21 Scandium			**22 Titanium**		
$\alpha_2\ KL_{II}$	3.0342	1	4.0861	2.75216	2	4.50486
$\alpha_1\ KL_{III}$	3.0309†	1	4.0906	2.74851	2	4.51084
$\beta_{1,3}\ KM_{II,III}$	2.7796	2	4.4605	2.51391	2	4.93181
$\beta_5\ KM_{IV,V}$	2.7634	3	4.4865	2.4985	2	4.9623
$\eta\ L_{II}M_I$	35.13	2	0.3529	30.89	3	0.4013
$\beta_1\ L_{II}M_{IV}$	31.02	2	0.3996	27.05	2	0.4584
$l\ L_{III}M_I$	35.59	3	0.3483	31.36	2	0.3953
$\alpha_{1,2}\ L_{III}M_{IV,V}$	31.35	3	0.3954	27.42	2	0.4522
	23 Vanadium			**24 Chromium**		
$\alpha_2\ KL_{II}$	2.50738	2	4.94464	2.293606	3	5.40551
$\alpha_1\ KL_{III}$	2.50356	2	4.95220	2.28970	2	5.41472
$\beta_{1,3}\ KM_{II,III}$	2.28440	2	5.42729	2.08487	2	5.94671
$\beta_5\ KM_{IV,V}$	2.26951	6	5.4629	2.07087	6	5.9869
$\beta_{3,4}\ L_I M_{II,III}$	21.19†	9	0.585	18.96	2	0.654
$\eta\ L_{II}M_I$	27.34	3	0.4535	24.30	3	0.5102
$\beta_1\ L_{II}M_{IV}$	23.88	4	0.5192	21.27	1	0.5828
$l\ L_{III}M_I$	27.77	1	0.4465	24.78	1	0.5003
$\alpha_{1,2}\ L_{III}M_{IV,V}$	24.25	3	0.5113	21.64	3	0.5728
$M_{II,III}M_{IV,V}$	337.	9	0.037	309.	9	0.040
	25 Manganese			**26 Iron**		
$\alpha_2\ KL_{II}$	2.10578	2	5.88765	1.939980	9	6.39084
$\alpha_1\ KL_{III}$	2.101820	9	5.89875	1.936042	9	6.40384
$\beta_{1,3}\ KM_{II,III}$	1.91021	2	6.49045	1.75661	2	7.05798
$\beta_5\ KM_{IV,V}$	1.8971	1	6.5352	1.7442	1	7.1081
$\beta_{3,4}\ L_I M_{II,III}$	17.19	2	0.721	15.65	2	0.792
$\eta\ L_{II}M_I$	21.85	2	0.5675	19.75	4	0.628
$\beta_1\ L_{II}M_{IV}$	19.11	2	0.6488	17.26	1	0.7185
$l\ L_{III}M_I$	22.29	1	0.5563	20.15	1	0.6152
$\alpha_{1,2}\ L_{III}M_{IV,V}$	19.45	1	0.6374	17.59	2	0.7050
$M_{II,III}M_{IV,V}$	273.	6	0.045	243.	5	0.051
	27 Cobalt			**28 Nickel**		
$\alpha_2\ KL_{II}$	1.792850	9	6.91530	1.661747	8	7.46089
$\alpha_1\ KL_{III}$	1.788965	9	6.93032	1.657910	8	7.47815
$\beta_{1,3}\ KM_{II,III}$	1.62079	2	7.64943	1.500135	8	8.26466
$\beta_5\ KM_{IV,V}$	1.60891	3	7.7059	1.48862	4	8.3286
$\beta_{3,4}\ L_I M_{II,III}$	14.31	3	0.870	13.18	1	0.941
$\eta\ L_{II}M_I$	17.87	3	0.694	16.27	3	0.762
$\beta_1\ L_{II}M_{IV}$	15.666	8	0.7914	14.271	6	0.8688
$l\ L_{III}M_I$	18.292	8	0.6778	16.693	9	0.7427
$\alpha_{1,2}\ L_{III}M_{IV,V}$	15.972	6	0.7762	14.561	3	0.8515
$M_{II,III}M_{IV,V}$	214.	6	0.058	190.	2	0.0651

Designation	Å*	p.e.	keV	Å*	p.e.	keV
	29 Copper			**30 Zinc**		
$\alpha_2\ KL_{II}$	1.544390	2	8.02783	1.439000	8	8.61578
$\alpha_1\ KL_{III}$	1.540562	2	8.04778	1.435155	7	8.63886
$\beta_3\ KM_{II}$	1.3926	1	8.9029			
$\beta_{1,3}\ KM_{II,III}$	1.392218	9	8.90529	1.29525	2	9.5720
$\beta_2\ KN_{II,III}$				1.28372	2	9.6580
$\beta_5\ KM_{IV,V}$	1.38109	3	8.9770	1.2848	1	9.6501
$\beta_{3,4}\ L_IM_{II,III}$	12.122	8	1.0228	11.200	7	1.1070
$\eta\ L_{II}M_I$	14.90	2	0.832	13.68	2	0.906
$\beta_1\ L_{II}M_{IV}$	13.053	3	0.9498	11.983	3	1.0347
$l\ L_{III}M_I$	15.286	9	0.8111	14.02	2	0.884
$\alpha_{1,2}\ L_{III}M_{IV,V}$	13.336	3	0.9297	12.254	3	1.0117
$M_{II,III}M_{V,V}$	173.	3	0.072	157.	3	0.079

Designation	Å*	p.e.	keV	Å*	p.e.	keV
	31 Gallium			**32 Germanium**		
$\alpha_2\ KL_{II}$	1.34399	1	9.22482	1.258011	9	9.85532
$\alpha_1\ KL_{III}$	1.340083	9	9.25174	1.254054	9	9.88642
$\beta_3\ KM_{II}$	1.20835	5	10.2603	1.12936	9	10.9780
$\beta_1\ KM_{III}$	1.20789	2	10.2642	1.12894	2	10.9821
$\beta_2\ KN_{II,III}$	1.19600	2	10.3663	1.11686	2	11.1008
$\beta_5\ KM_{IV,V}$	1.1981	2	10.348	1.1195	1	11.0745
$\beta_4\ L_IM_{II}$				9.640	2	1.2861
$\beta_3\ L_IM_{III}$				9.581	2	1.2941
$\beta_{3,4}\ L_IM_{II,III}$	10.359†	8	1.197			
$\eta\ L_{II}M_I$	12.597	2	0.9842	11.609	2	1.0680
$\beta_1\ L_{II}M_{IV}$	11.023	2	1.1248	10.175	1	1.2185
$l\ L_{III}M_I$	12.953	2	0.9572	11.965	4	1.0362
$\alpha_{1,2}\ L_{III}M_{IV,V}$	11.292	1	1.09792	10.4361	8	1.18800

Designation	Å*	p.e.	keV	Å*	p.e.	keV
	33 Arsenic			**34 Selenium**		
$\alpha_2\ KL_{II}$	1.17987	1	10.50799	1.10882	2	11.1814
$\alpha_1\ KL_{III}$	1.17588	1	10.54372	1.10477	2	11.2224
$\beta_3\ KM_{II}$	1.05783	5	11.7203	0.99268	5	12.4896
$\beta_1\ KM_{III}$	1.05730	2	11.7262	0.99218	3	12.4959
$\beta_2\ KN_{II,III}$	1.04500	3	11.8642	0.97992	5	12.6522
$\beta_5\ KM_{IV,V}$	1.0488	1	11.822	0.9843	1	12.595
$\beta_{3,4}\ L_IM_{II,III}$	8.929	1	1.3884	8.321†	9	1.490
$\eta\ L_{II}M_I$	10.734	1	1.1550	9.962	1	1.2446
$\beta_1\ L_{II}M_{IV}$	9.4141	8	1.3170	8.7358	5	1.41923
$l\ L_{III}M_I$	11.072	1	1.1198	10.294	1	1.2044
$\alpha_{1,2}\ L_{III}M_{IV,V}$	9.6709	8	1.2820	8.9900	5	1.37910
M_VN_{III}				230.	2	0.0538

Designation	Å*	p.e.	keV	Å*	p.e.	keV
	35 Bromine			**36 Krypton**		
$\alpha_2\ KL_{II}$	1.04382	2	11.8776	0.9841	1	12.598
$\alpha_1\ KL_{III}$	1.03974	2	11.9242	0.9801	1	12.649
$\beta_3\ KM_{II}$	0.93327	5	13.2845	0.8790	1	14.104
$\beta_1\ KM_{III}$	0.93279	2	13.2914	0.8785	1	14.112
$\beta_2\ KN_{II,III}$	0.92046	2	13.4695	0.8661	1	14.315
$\beta_5\ KM_{IV,V}$	0.9255	1	13.396	0.8708	2	14.238
$\beta_4\ KN_{IV,V}$				0.8653	2	14.328
$\beta_4\ L_IM_{II}$				7.304	5	1.697
$\beta_3\ L_IM_{III}$				7.264	5	1.707

Designation	Å*	p.e.	keV	Å*	p.e.	keV
	35 Bromine (*Cont.*)			**36 Krypton** (*Cont.*)		
$\beta_{3,4}\ L_IM_{II,III}$	7.767†	9	1.596			
$\eta\ L_{II}M_I$	9.255	1	1.3396			
$\beta_1\ L_{II}M_{IV}$	8.1251	5	1.52590	7.576†	3	1.6366
γ_5				7.279	5	1.703
$l\ L_{III}M_I$	9.585	1	1.2935			
$\alpha_{1,2}\ L_{III}M_{IV,V}$	8.3746	5	1.48043	7.817†	3	1.5860
β_6				7.510	4	1.6510
$L_{II}N_{III}$				7.250	5	1.710
M_IM_{II}	184.6	3	0.0672			
M_IM_{III}	164.7	3	0.0753			
$M_{II}M_{IV}$	109.4	3	0.1133			
$M_{II}N_I$	76.9	2	0.1613			
$M_{III}M_{IV,V}$	113.8	3	0.1089			
	79.8	3	0.1554			
$\zeta_2\ M_{IV}N_{II}$	191.1	2	0.06488			
$M_{IV}N_{III}$	189.5	3	0.0654			
$\zeta_1\ M_VN_{III}$	192.6	2	0.06437			

Designation	Å*	p.e.	keV	Å*	p.e.	keV
	37 Rubidium			**38 Strontium**		
$\alpha_2\ KL_{II}$	0.92969	1	13.3358	0.87943	1	14.0979
$\alpha_1\ KL_{III}$	0.925553	9	13.3953	0.87526	1	14.1650
$\beta_3\ KM_{II}$	0.82921	3	14.9517	0.78345	3	15.8249
$\beta_1\ KM_{III}$	0.82868	2	14.9613	0.78292	2	15.8357
$\beta_2\ KN_{II,III}$	0.81645	3	15.1854	0.77081	3	16.0846
$\beta_5\ KM_{IV,V}$	0.8219	1	15.085	0.7764	1	15.969
$\beta_4\ KN_{IV,V}$	0.8154	2	15.205	0.76989	5	16.104
$\beta_4\ L_IM_{II}$	6.8207	3	1.81771	6.4026	3	1.93643
$\beta_3\ L_IM_{III}$	6.7876	3	1.82659	6.3672	3	1.94719
$\gamma_{2,3}\ L_IN_{II,III}$	6.0458	3	2.0507	5.6445	3	2.1965
$\eta\ L_{II}M_I$	8.0415	4	1.54177	7.5171	3	1.64933
$\beta_1\ L_{II}M_{IV}$	7.0759	3	1.75217	6.6239	3	1.87172
$\gamma_5\ L_{II}N_{IV}$	6.7553	3	1.83532	6.2961	3	1.96916
$l\ L_{III}M_I$	8.3636	4	1.48238	7.8362	3	1.58215
$\alpha_2\ L_{III}M_{IV}$	7.3251	3	1.69256	6.8697	3	1.80474
$\alpha_1\ L_{III}M_V$	7.3183	2	1.69413	6.8628	2	1.80656
$\beta_6\ L_{III}N_I$	6.9842	3	1.77517	6.5191	3	1.90181
M_IM_{III}	144.4	3	0.0859			
$M_{II}M_{IV}$	91.5	2	0.1355	85.7	2	0.1447
$M_{II}N_I$	57.0	2	0.2174	51.3	1	0.2416
$M_{III}M_{IV,V}$	96.7	2	0.1282	91.4	2	0.1357
$M_{III}N_I$	59.5	2	0.2083	53.6	1	0.2313
$\zeta_2\ M_{IV}N_{II}$	127.8	2	0.0970			
$M_{IV}N_{III}$	126.8	2	0.0978			
$\zeta_2\ M_{IV}N_{II,III}$				108.0	2	0.1148
$\zeta_1\ M_VN_{III}$	128.7	2	0.0964	108.7	1	0.1140

Designation	Å*	p.e.	keV	Å*	p.e.	keV
	39 Yttrium			**40 Zirconium**		
$\alpha_2\ KL_{II}$	0.83305	1	14.8829	0.79015	1	15.6909
$\alpha_1\ KL_{III}$	0.82884	1	14.9584	0.78593	1	15.7751
$\beta_3\ KM_{II}$	0.74126	3	16.7258	0.70228	4	17.654
$\beta_1\ KM_{III}$	0.74072	2	16.7378	0.70173	3	17.6678
$\beta_2\ KN_{II,III}$	0.72864	4	17.0154	0.68993	4	17.970
$\beta_5\ KM_{IV,V}$	0.7345	1	16.879	0.6959	1	17.815

Designation	Å*	p.e.	keV	Å*	p.e.	keV
	39 Yttrium (*Cont.*)			**40 Zirconium** (*Cont.*)		
$\beta_4\ KN_{IV,V}$	0.72776	5	17.036	0.68901	5	17.994
$\beta_4\ L_IM_{II}$	6.0186	3	2.0600	5.6681	3	2.1873
$\beta_3\ L_IM_{III}$	5.9832	3	2.0722	5.6330	3	2.2010
$\gamma_{2,3}\ L_IN_{II,III}$	5.2830	3	2.3468	4.9536	3	2.5029
$\eta\ L_IM_I$	7.0406	3	1.76095	6.6069	3	1.87654
$\beta_1\ L_{III}M_{IV}$	6.2120	3	1.99584	5.8360	3	2.1244
$\gamma_5\ L_{II}N_I$	5.8754	3	2.1102	5.4977	3	2.2551
$\gamma_1\ L_{II}N_{IV}$				5.3843	3	2.3027
$l\ L_{III}M_I$	7.3563	3	1.68536	6.9185	3	1.79201
$\alpha_2\ L_{III}M_{IV}$	6.4558	3	1.92047	6.0778	3	2.0399
$\alpha_1\ L_{III}M_V$	6.4488	2	1.92256	6.0705	2	2.04236
$\beta_6\ L_{III}N_I$	6.0942	3	2.0344	5.7101	3	2.1712
$\beta_{2,15}$				5.5863	3	2.2194
$M_{II}M_{IV}$	81.5	2	0.1522	76.7	2	0.1617
$M_{II}N_I$	46.48	9	0.267			
$M_{III}M_V$				80.9	3	0.1533
$M_{III}N_I$	48.5	2	0.256			
$M_{III}M_{IV,V}$	86.5	2	0.1434			
$\zeta\ M_{IV,V}N_{II,III}$	93.4	2	0.1328	82.1	2	0.1511
$M_{IV,V}O_{II,III}$				70.0	4	0.177
	41 Niobium			**42 Molybdenum**		
$\alpha_2\ KL_{II}$	0.75044	1	16.5210	0.713590	6	17.3743
$\alpha_1\ KL_{III}$	0.74620	1	16.6151	0.709300	1	17.47934
$\beta_3\ KM_{II}$	0.66634	3	18.6063	0.632872	9	19.5903
$\beta_1\ KM_{III}$	0.66576	2	18.6225	0.632288	9	19.6083
β_2^{II}				0.62107	5	19.963
$\beta_2\ KN_{II,III}$	0.65416	4	18.953	0.62099	2	19.9652
$\beta_4\ KN_{IV,V}$	0.65318	5	18.981			
$\beta_5^{II}\ KM_{IV}$				0.62708	5	19.771
$\beta_5^{I}\ KM_V$				0.62692	5	19.776
$\beta_4\ KN_{IV,V}$				0.62001	9	19.996
$\beta_4\ L_IM_{II}$	5.3455	3	2.3194	5.0488	3	2.4557
$\beta_3\ L_IM_{III}$	5.3102	3	2.3348	5.0133	3	2.4730
$\gamma_{2,3}\ L_IN_{II,III}$	4.6542	2	2.6638	4.3800	2	2.8306
$\eta\ L_{II}M_I$	6.2109	3	1.99620	5.8475	3	2.1202
$\beta_1\ L_{II}M_{IV}$	5.4923	3	2.2574	5.17708	8	2.39481
$\gamma_5\ L_{II}N_I$	5.1517	3	2.4066	4.8369	2	2.5632
$\gamma_1\ L_{II}N_{IV}$	5.0361	3	2.4618	4.7258	2	2.6235
$l\ L_{III}M_I$	6.5176	3	1.90225	6.1508	3	2.01568
$\alpha_2\ L_{III}M_{IV}$	5.7319	3	2.1630	5.41437	8	2.28985
$\alpha_1\ L_{III}M_V$	5.7243	2	2.16589	5.40655	8	2.29316
$\beta_6\ L_{III}N_I$	5.3613	3	2.3125	5.0488	5	2.4557
$\beta_{2,15}\ L_{III}N_{IV,V}$	5.2379	3	2.3670	4.9232	2	2.5183
$M_{II}M_{IV}$	72.1	3	0.1718	68.9	2	0.1798
$M_{II}N_I$	38.4	3	0.323	35.3	3	0.351
$M_{II}N_{IV}$	33.1	2	0.375			
$M_{III}M_V$	78.4	2	0.1582	74.9	1	0.1656
$M_{III}N_I$	40.7	2	0.305	37.5	2	0.331
$\gamma\ M_{III}N_{IV,V}$	34.9	2	0.356			
$\zeta\ M_{IV,V}N_{II,III}$	72.19	9	0.1717	64.38	7	0.1926
$M_{IV,V}O_{II,III}$	61.9	2	0.2002	54.8	2	0.2262

Designation	Å*	p.e.	keV	Å*	p.e.	keV
	43 Technetium			**44 Ruthenium**		
$\alpha_2\ KL_{II}$	0.67932†	3	18.2508	0.647408	5	19.1504
$\alpha_1\ KL_{III}$	0.67502†	3	18.3671	0.643083	4	19.2792
$\beta_3\ KM_{II}$	0.60188†	4	20.599	0.573067	4	21.6346
$\beta_1\ KM_{III}$	0.60130†	4	20.619	0.572482	4	21.6568
$\beta_2\ KN_{II,III}$	0.59024†	5	21.005	0.56166	3	22.074
$\beta_5^{II}\ KM_{IV}$				0.5680	2	21.829
$\beta_5^{I}\ KM_V$				0.56785	9	21.834
β_4				0.56089	9	22.104
$\beta_4\ L_{II}M_{IV}$				4.5230	2	2.7411
$\beta_3\ L_IM_{III}$				4.4866	3	2.7634
$\gamma_{2,3}\ L_IN_{II,III}$				3.8977	2	3.1809
$\eta\ L_{II}M_I$				5.2050	2	2.38197
$\beta_1\ L_{II}M_{IV}$	4.8873†	8	2.5368	4.62058	3	2.68323
$\gamma_5\ L_{II}N_I$				4.2873	2	2.8918
$\gamma_1\ L_{II}N_{IV}$				4.1822	2	2.9645
$l\ L_{III}M_I$				5.5035	3	2.2528
$\alpha_2\ L_{III}M_{IV}$				4.85381	7	2.55431
$\alpha_1\ L_{III}M_V$	5.1148†	3	2.4240	4.84575	5	2.55855
$\beta_6\ L_{III}N_I$				4.4866	3	2.7634
$\beta_{2,15}\ L_{III}N_{IV,V}$				4.3718	2	2.8360
$M_{II}M_{IV}$				62.2	1	0.1992
$M_{II}N_I$				32.3	2	0.384
$M_{II}N_{IV}$				25.50	9	0.486
$M_{III}M_V$				68.3	1	0.1814
$\gamma\ M_{III}N_{IV,V}$				26.9	1	0.462
$\zeta\ M_{IV,V}N_{II,III}$				52.34	7	0.2369
$M_{IV,V}O_{II,III}$				44.8	1	0.2768
	45 Rhodium			**46 Palladium**		
$\alpha_2\ KL_{II}$	0.617630	4	20.0737	0.589821	3	21.0201
$\alpha_1\ KL_{III}$	0.613279	4	20.2161	0.585448	3	21.1771
$\beta_3\ KM_{II}$	0.546200	4	22.6989	0.521123	4	23.7911
$\beta_1\ KM_{III}$	0.545605	4	22.7236	0.520520	4	23.8187
$\beta_2^{II}\ KN_{II}$	0.53513	5	23.168			
$\beta_2\ KN_{II,III}$	0.53503	2	23.1728	0.510228	4	24.2991
$\beta_5^{II}\ KM_{IV}$	0.54118	9	22.909			
$\beta_5^{I}\ KM_V$	0.54101	9	22.917			
$\beta_4\ KN_{IV,V}$	0.53401	9	23.217	0.5093	2	24.346
$\beta_5\ KM_{IV,V}$				0.51670	9	23.995
$\beta_4\ L_IM_{II}$	4.2888	2	2.8908	4.0711	2	3.0454
$\beta_3\ L_IM_{III}$	4.2522	2	2.9157	4.0346	2	3.0730
$\gamma_{2,3}\ L_IN_{II,III}$	3.6855	2	3.3640	3.4892	2	3.5533
$\eta\ L_{II}M_I$	4.9217	2	2.5191	4.6605	2	2.6603
$\beta_1\ L_{II}M_{IV}$	4.37414	4	2.83441	4.14622	5	2.99022
$\gamma_5\ L_{II}N_I$	4.0451	2	3.0650	3.8222	2	3.2437
$\gamma_1\ L_{II}N_{IV}$	3.9437	2	3.1438	3.7246	2	3.3287
$l\ L_{III}M_I$	5.2169	3	2.3765	4.9525	3	2.5034
$\alpha_2\ L_{III}M_{IV}$	4.60545	9	2.69205	4.37588	7	2.83329
$\alpha_1\ L_{III}M_V$	4.59743	9	2.69674	4.36767	5	2.83861
$\beta_6\ L_{III}N_I$	4.2417	2	2.9229	4.0162	2	3.0870
$\beta_{2,15}\ L_{III}N_{IV,V}$	4.1310	2	3.0013	3.90887	4	3.17179
$\beta_{10}\ L_IM_{IV}$				3.7988	2	3.2637

TABLE A (*Continued*)

Designation	Å*	p.e.	keV	Å*	p.e.	keV
	45 Rhodium (*Cont.*)			**46 Palladium** (*Cont.*)		
$\beta_9\ L_IM_V$				3.7920	2	3.2696
$M_IN_{II,III}$				20.1	2	0.616
$M_{II}M_{IV}$	59.3	1	0.2090	56.5	1	0.2194
$M_{II}N_I$	28.1	2	0.442	26.2	2	0.474
$M_{II}N_{IV}$				22.1	1	0.560
$M_{II}\ M_V$	65.5	1	0.1892	62.9	1	0.1970
$M_{II}{}^{I}N_I$	29.8	1	0.417	27.9	1	0.445
$\gamma\ M_{III}N_{IV,V}$	25.01	9	0.496	23.3†	1	0.531
$\zeta\ M_{IV,V}N_{II,III}$	47.67	9	0.2601	43.6	1	0.2844
$M_{IV,V}O_{II,III}$	40.9	2	0.303	37.4	2	0.332
	47 Silver			**48 Cadmium**		
$\alpha_2\ KL_{II}$	0.563798	4	21.9903	0.539422	3	22.9841
$\alpha_1\ KL_{III}$	0.5594075	6	22.16292	0.535010	3	23.1736
$\beta_3\ KM_{II}$	0.497685	4	24.9115	0.475730	5	26.0612
$\beta_1\ KM_{III}$	0.497069	4	24.9424	0.475105	6	26.0955
$\beta_2\ KN_{II,III}$	0.487032	4	25.4564	0.465328	7	26.6438
$\beta_5\ KM_{IV,V}$	0.49306	2	25.145			
$\beta_4\ KN_{IV,V}$	0.48598	3	25.512			
$\beta_4\ L_IM_{II}$	3.87023	5	3.20346	3.68203	9	3.36719
$\beta_3\ L_IM_{III}$	3.83313	9	3.23446	3.64495	9	3.40145
$\gamma_2\ L_IN_{II}$	3.31216	9	3.7432	3.1377	2	3.9513
$\gamma_3\ L_IN_{III}$	3.30635	9	3.7498			
$\eta\ L_{II}M_I$	4.4183	2	2.8061	4.19315	9	2.95675
$\beta_1\ L_{II}M_{IV}$	3.93473	3	3.15094	3.73823	4	3.31657
$\gamma_5\ L_{II}N_I$	3.61638	9	3.42832	3.42551	9	3.61935
$\gamma_1\ L_{III}N_{IV}$	3.52260	4	3.51959	3.33564	6	3.71686
$l\ L_{III}M_I$	4.7076	2	2.6337	4.48014	9	2.76735
$\alpha_2\ L_{III}M_{IV}$	4.16294	5	2.97821	3.96496	6	3.12691
$\alpha_1\ L_{III}M_V$	4.15443	3	2.98431	3.95635	4	3.13373
$\beta_6\ L_{III}N_I$	3.80774	9	3.25603	3.61467	9	3.42994
$\beta_{2,15}\ L_{III}N_{IV,V}$	3.70335	3	3.34781	3.51408	4	3.52812
$\beta_{10}\ L_IM_{IV}$	3.61158	9	3.43287	3.4367	2	3.6075
$\beta_9\ L_IM_V$	3.60497	9	3.43917	3.43015	9	3.61445
$M_IN_{II,III}$	18.8	2	0.658			
$M_{II}M_{IV}$	54.0	1	0.2295	52.0	2	0.2384
$M_{II}N_I$				22.9	2	0.540
$M_{II}N_{IV}$	20.66	7	0.600	19.40	7	0.639
$M_{III}M_V$	60.5	1	0.2048	58.7	2	0.2111
$M_{III}N_I$	26.0	1	0.478	24.5	1	0.507
$\gamma\ M_{III}N_{IV,V}$	21.82	7	0.568	20.47	7	0.606
$M_{IV}O_{II,III}$				30.4	1	0.408
$\zeta\ M_{IV,V}N_{II,III}$	39.77	7	0.3117	36.8	1	0.3371
M_VN_I	24.4	2	0.509			
M_VO_{III}				30.8	1	0.403
$M_{IV,V}O_{II,III}$	33.5	3	0.370			
	49 Indium			**50 Tin**		
$\alpha_2\ KL_{II}$	0.516544	3	24.0020	0.495053	3	25.0440
$\alpha_1\ KL_{III}$	0.512113	3	24.2097	0.490599	3	25.2713
$\beta_3\ KM_{II}$	0.455181	4	27.2377	0.435877	5	28.4440

Designation	Å*	p.e.	keV	Å*	p.e.	keV
	49 Indium (*Cont.*)			**50 Tin** (*Cont.*)		
$\beta_1\ KM_{III}$	0.454545	4	27.2759	0.435236	5	28.4860
$\beta_2\ KN_{II,III}$	0.44500	1	27.8608	0.425915	8	29.1093
$KO_{II,III}$	0.44374	3	27.940	0.42467	3	29.195
$\beta_5{}^{II}\ KM_{IV}$	0.45098	2	27.491	0.43184	3	28.710
$\beta_5{}^{I}\ KM_V$	0.45086	2	27.499	0.43175	3	28.716
$\beta_4\ KN_{IV,V}$	0.44393	4	27.928	0.42495	3	29.175
$\beta_4\ L_IM_{II}$	3.50697	9	3.5353	3.34335	9	3.7083
$\beta_3\ L_IM_{III}$	3.46984	9	3.5731	3.30585	3	3.7500
$\gamma_{2,3}\ L_IN_{II,III}$	2.9800	2	4.1605	2.8327	2	4.3768
$\gamma_4\ L_IO_{II,III}$	2.9264	2	4.2367	2.7775	2	4.4638
$\eta\ L_{II}M_I$	3.98327	9	3.11254	3.78876	9	3.27234
$\beta_1\ L_{II}M_{IV}$	3.55531	4	3.48721	3.38487	3	3.66280
$\gamma_5\ L_{II}N_I$	3.24907	9	3.8159	3.08475	9	4.0192
$\gamma_1\ L_{II}N_{IV}$	3.16213	4	3.92081	3.00115	3	4.13112
$l\ L_{III}M_I$	4.26873	9	2.90440	4.07165	9	3.04499
$\alpha_2\ L_{III}M_{IV}$	3.78073	6	3.27929	3.60891	4	3.43542
$\alpha_1\ L_{III}M_V$	3.77192	4	3.28694	3.59994	3	3.44398
$\beta_6\ L_{III}N_I$	3.43606	9	3.60823	3.26901	9	3.7926
$\beta_{2,15}\ L_{III}N_{IV,V}$	3.33838	3	3.71381	3.17505	3	3.90486
$\beta_7\ L_{III}O_I$	3.324	4	3.730	3.1564	3	3.9279
$\beta_{10}\ L_IM_{IV}$	3.27404	9	3.7868	3.12170	9	3.9716
$\beta_9\ L_IM_V$	3.26763	9	3.7942	3.11513	9	3.9800
$M_{II}M_{IV}$				47.3	1	0.2621
$M_{II}N_I$				20.0	1	0.619
$M_{II}N_{IV}$				16.93	5	0.733
$M_{III}M_V$				54.2	1	0.2287
$M_{III}N_I$				21.5	1	0.575
$\gamma\ M_{III}N_{IV,V}$				17.94	5	0.691
$M_{IV}O_{II,III}$				25.3	1	0.491
$\zeta\ M_{IV,V}N_{II,III}$				31.24	9	0.397
M_VO_{III}				25.7	1	0.483
	51 Antimony			**52 Tellurium**		
$\alpha_2\ KL_{II}$	0.474827	3	26.1108	0.455784	3	27.2017
$\alpha_1\ KL_{III}$	0.470354	3	26.3591	0.451295	3	27.4723
$\beta_3\ KM_{II}$	0.417737	4	29.6792	0.400659	4	30.9443
$\beta_1\ KM_{III}$	0.417085	3	29.7256	0.399995	5	30.9957
$\beta_2\ KN_{II,III}$	0.407973	5	30.3895	0.391102	6	31.7004
$KO_{II,III}$	0.40666	1	30.4875	0.38974	1	31.8114
$\beta_5{}^{II}\ KM_{IV}$	0.41388	1	29.9560			
$\beta_5{}^{I}\ KM_V$	0.41378	1	29.9632			
$\beta_4\ KN_{IV,V}$	0.40702	1	30.4604			
$\beta_4\ L_IM_{II}$	3.19014	9	3.8864	3.04661	9	4.0695
$\beta_3\ L_IM_{III}$	3.15258	9	3.9327	3.00893	9	4.1204
$\gamma_{2,3}\ L_IN_{II,III}$	2.6953	2	4.5999	2.5674	2	4.8290
$\gamma_4\ L_IO_{II,III}$	2.6398	2	4.6967	2.5113	2	4.9369
$\eta\ L_{II}M_I$	3.60765	9	3.43661	3.43832	9	3.60586
$\beta_1\ L_{II}M_{IV}$	3.22567	4	3.84357	3.07677	6	4.02958
$\gamma_5\ L_{II}N_I$	2.93187	9	4.2287	2.79007	9	4.4437
$\gamma_1\ L_{II}N_{IV}$	2.85159	3	4.34779	2.71241	6	4.5709
$l\ L_{III}M_I$	3.88826	9	3.18860	3.71696	9	3.33555
$\alpha_2\ L_{III}M_{IV}$	3.44840	6	3.59532	3.29846	9	3.7588

Designation	Å*	p.e.	keV	Å*	p.e.	keV
	51 Antimony (*Cont.*)			**52 Tellurium** (*Cont.*)		
$\alpha_1\ L_{III}M_V$	3.43941	4	3.60472	3.28920	6	3.76933
$\beta_6\ L_{III}N_I$	3.11513	9	3.9800	2.97088	9	4.1732
$\beta_{2,15}\ L_{III}N_{IV,V}$	3.02335	3	4.10078	2.88217	8	4.3017
$\beta_7\ L_{III}O_I$	3.0052	3	4.1255	2.8634	3	4.3298
$\beta_{10}\ L_IM_{IV}$	2.97917	9	4.1616	2.84679	9	4.3551
$\beta_9\ L_IM_V$	2.97261	9	4.1708	2.83897	9	4.3671
$M_{II}M_{IV}$	45.2	1	0.2743			
$M_{II}N_I$	18.8	1	0.658	17.6	1	0.703
$M_{II}N_{IV}$	15.98	5	0.776			
$M_{III}M_V$	52.2	1	0.2375	50.3	1	0.2465
$M_{III}N_I$	20.2	1	0.612	19.1	1	0.648
$\gamma\ M_{III}N_{IV,V}$	16.92	4	0.733	15.93	4	0.778
$M_{IV}O_{II,III}$				21.34	5	0.581
$\zeta\ M_{IV,V}N_{II,III}$	28.88	8	0.429	26.72	9	0.464
M_VO_{III}				21.78	5	0.569

Designation	Å*	p.e.	keV	Å*	p.e.	keV
	53 Iodine			**54 Xenon**		
$\alpha_2\ KL_{II}$	0.437829	7	28.3172	$0.42087^{\dagger}$	2	29.458
$\alpha_1\ KL_{III}$	0.433318	5	28.6120	$0.41634^{\dagger}$	2	29.779
$\beta_3\ KM_{II}$	0.384564	4	32.2394	$0.36941^{\dagger}$	2	33.562
$\beta_1\ KM_{III}$	0.383905	4	32.2947	$0.36872^{\dagger}$	2	33.624
$\beta_2\ KN_{II,III}$	$0.37523^{\dagger}$	2	33.042	$0.36026^{\dagger}$	3	34.415
$\beta_4\ L_IM_{II}$	2.91207	9	4.2575			
$\beta_3\ L_IM_{III}$	2.87429	9	4.3134			
$\gamma_{2,3}\ L_IN_{II,III}$	2.4475	2	5.0657			
$\gamma_4\ L_IO_{II,III}$	2.3913	2	5.1848			
$\eta\ L_{II}M_I$	3.27979	9	3.7801			
$\beta_1\ L_{II}M_{IV}$	2.93744	6	4.22072			
$\gamma_5\ L_{II}N_I$	2.65710	9	4.6660			
$\gamma_1\ L_{II}N_{IV}$	2.58244	8	4.8009			
$l\ L_{III}M_I$	3.55754	9	3.48502			
$\alpha_2\ L_{III}M_{IV}$	3.15791	6	3.92604			
$\alpha_1\ L_{III}M_V$	3.14860	6	3.93765	$3.0166^{\dagger}$	2	4.1099
$\beta_6\ L_{III}N_I$	2.83672	9	4.3706			
$\beta_{2,15}\ L_{III}N_{IV,V}$	2.75053	8	4.5075			
$\beta_7\ L_{III}O_I$	2.7288	3	4.5435			
$\beta_{10}\ L_IM_{IV}$	2.72104	9	4.5564			
$\beta_9\ L_IM_V$	2.71352	9	4.5690			

Designation	Å*	p.e.	keV	Å*	p.e.	keV
	55 Cesium			**56 Barium**		
$\alpha_2\ KL_{II}$	0.404835	4	30.6251	0.389668	5	31.8171
$\alpha_1\ KL_{III}$	0.400290	4	30.9728	0.385111	4	32.1936
$\beta_3\ KM_{II}$	0.355050	4	34.9194	0.341507	4	36.3040
$\beta_1\ KM_{III}$	0.354364	7	34.9869	0.340811	3	36.3782
$\beta_2\ KN_{II,III}$	0.34611	2	35.822	0.33277	1	37.257
$KO_{II,III}$				0.33127	2	37.426
$\beta_5^{II}\ KM_{IV}$				0.33835	2	36.643
$\beta_5^{I}\ KM_V$				0.33814	2	36.666
$\beta_4\ KN_{IV,V}$				0.33229	2	37.311
$\beta_4\ L_IM_{II}$	2.6666	2	4.6494	2.5553	2	4.8519
$\beta_3\ L_IM_{III}$	2.6285	2	4.7167	2.5164	2	4.9269
$\gamma_2\ L_IN_{II}$	2.2371	2	5.5420	2.1387	2	5.7969
$\gamma_3\ L_IN_{III}$	2.2328	2	5.5527	2.1342	2	5.8092

Designation	Å*	p.e.	keV	Å*	p.e.	keV
	55 Cesium (*Cont.*)			**56 Barium** (*Cont.*)		
$\gamma_4\ L_IO_{II,III}$	2.1741	2	5.7026	2.0756	3	5.9733
$\eta\ L_{II}M_I$	2.9932	2	4.1421	2.8627	3	4.3309
$\beta_1\ L_{II}M_{IV}$	2.6837	2	4.6198	2.56821	5	4.82753
$\gamma_5\ L_{II}N_I$	2.4174	2	5.1287	2.3085	3	5.3707
$\gamma_1\ L_{II}N_{IV}$	2.3480	2	5.2804	2.2415	2	5.5311
$l\ L_{III}M_I$	3.2670	2	3.7950	3.1355	2	3.9541
$\alpha_2\ L_{III}M_{IV}$	2.9020	2	4.2722	2.78553	5	4.45090
$\alpha_1\ L_{III}M_V$	2.8924	2	4.2865	2.77595	5	4.46626
$\beta_6\ L_{III}N_I$	2.5932	2	4.7811	2.4826	2	4.9939
$\beta_{2,15}\ L_{III}N_{IV,V}$	2.5118	2	4.9359	2.40435	6	5.1565
$\beta_7\ L_{III}O_I$	2.4849	2	4.9893	2.3806	2	5.2079
$\beta_{10}\ L_IM_{IV}$	2.4920	2	4.9752	2.3869	2	5.1941
$\beta_9\ L_IM_V$	2.4783	2	5.0026	2.3764	2	5.2171
$\gamma\ M_{III}N_{IV,V}$				12.75	3	0.973
$M_{IV}O_{II}$				15.91	5	0.779
$M_{IV}O_{III}$				15.72	9	0.789
$\zeta\ M_VN_{III}$				20.64	4	0.601
M_VO_{III}				16.20	5	0.765
$N_{IV}O_{II}$	188.6	1	0.06574	163.3	2	0.07590
$N_{IV}O_{III}$	183.8	1	0.06746	159.0	2	0.07796
N_VO_{III}	190.3	1	0.06515	164.6	2	0.07530

Designation	Å*	p.e.	keV	Å*	p.e.	keV
	57 Lanthanum			**58 Cerium**		
$\alpha_2\ KL_{II}$	0.375313	2	33.0341	0.361683	2	34.2789
$\alpha_1\ KL_{III}$	0.370737	2	33.4418	0.357092	2	34.7197
$\beta_3\ KM_{II}$	0.328686	4	37.7202	0.316520	4	39.1701
$\beta_1\ KM_{III}$	0.327983	3	37.8010	0.315816	2	39.2573
$\beta_2\ KN_{II,III}$	0.320117	7	38.7299	0.30816	1	40.233
$KO_{II,III}$	0.31864	2	38.909	0.30668	2	40.427
$\beta_5^{II}\ KM_{IV}$	0.32563	2	38.074	0.31357	2	39.539
$\beta_5^{I}\ KM_V$	0.32546	2	38.094	0.31342	2	39.558
$\beta_4\ KN_{IV,V}$	0.31931	2	38.828	0.30737	2	40.337
$\beta_4\ L_IM_{II}$	2.4493	3	5.0620	2.3497	4	5.2765
$\beta_3\ L_IM_{III}$	2.4105	3	5.1434	2.3109	3	5.3651
$\gamma_2\ L_IN_{II}$	2.0460	4	6.060	1.9602	3	6.3250
$\gamma_3\ L_IN_{III}$	2.0410	4	6.074	1.9553	3	6.3409
$\gamma_4\ L_IO_{II,III}$	1.9830	4	6.252	1.8991	4	6.528
$\eta\ L_{II}M_I$	2.740	3	4.525	2.6203	4	4.7315
$\beta_1\ L_{II}M_{IV}$	2.45891	5	5.0421	2.3561	3	5.2622
$\gamma_5\ L_{II}N_I$	2.2056	4	5.621	2.1103	3	5.8751
$\gamma_1\ L_{II}N_{IV}$	2.1418	3	5.7885	2.0487	4	6.052
$\gamma_8\ L_{II}O_I$				2.0237	4	6.126
$l\ L_{III}M_I$	3.006	3	4.124	2.8917	4	4.2875
$\alpha_2\ L_{III}M_{IV}$	2.67533	5	4.63423	2.5706	3	4.8230
$\alpha_1\ L_{III}M_V$	2.66570	5	4.65097	2.5615	2	4.8402
$\beta_6\ L_{III}N_I$	2.3790	4	5.2114	2.2818	3	5.4334
$\beta_{2,15}\ L_{III}N_{IV,V}$	2.3030	3	5.3835	2.2087	2	5.6134
$\beta_7\ L_{III}O_I$	2.275	3	5.450	2.1701	2	5.7132
$\beta_{10}\ L_IM_{IV}$	2.290	3	5.415	2.1958	5	5.646
$\beta_9\ L_IM_V$	2.282	3	5.434	2.1885	3	5.6650
$\gamma\ M_{III}N_{IV,V}$	12.08	4	1.027	11.53	1	1.0749
$\beta\ M_{IV}N_{VI}$	14.51	5	0.854	13.75	4	0.902
$\zeta\ M_VN_{III}$	19.44	5	0.638	18.35	4	0.676
$\alpha\ M_VN_{VI,VII}$	14.88	5	0.833	14.04	2	0.883

Designation	Å*	p.e.	keV	Å*	p.e.	keV
	57 Lanthanum (*Cont.*)			**58 Cerium** (*Cont.*)		
$M_VO_{II,III}$				14.39	5	0.862
$N_{IV,V}O_{II,III}$	152.6	6	0.0812	144.4	6	0.0859

Designation	Å*	p.e.	keV	Å*	p.e.	keV
	59 Praseodymium			**60 Neodymium**		
$\alpha_2\ KL_{II}$	0.348749	2	35.5502	0.336472	2	36.8474
$\alpha_1\ KL_{III}$	0.344140	2	36.0263	0.331846	2	37.3610
$\beta_3\ KM_{II}$	0.304975	5	40.6529	0.294027	3	42.1665
$\beta_1\ KM_{III}$	0.304261	4	40.7482	0.293299	2	42.2713
$\beta_2\ KN_{II,III}$	0.29679	2	41.773	0.2861†	1	43.33
$\beta_4\ L_IM_{II}$	2.2550	4	5.4981	2.1669	3	5.7216
$\beta_3\ L_IM_{III}$	2.2172	3	5.5918	2.1268	2	5.8294
$\gamma_2\ L_IN_{II}$	1.8791	4	6.598	1.8013	4	6.883
$\gamma_3\ L_IN_{III}$	1.8740	4	6.616	1.7964	4	6.902
$\gamma_4\ L_IO_{II,III}$	1.8193	4	6.815	1.7445	4	7.107
$\eta\ L_{II}M_I$	2.512	3	4.935	2.4094	4	5.1457
$\beta_1\ L_{II}M_{IV}$	2.2588	3	5.4889	2.1669	2	5.7216
$\gamma_5\ L_{II}N_I$	2.0205	4	6.136	1.9355	4	6.406
$\gamma_1\ L_{II}N_{IV}$	1.9611	3	6.3221	1.8779	2	6.6021
$\gamma_8\ L_{II}O_I$	1.9362	4	6.403	1.8552	5	6.683
$l\ L_{III}M_I$	2.7841	4	4.4532	2.6760	4	4.6330
$\alpha_2\ L_{III}M_{IV}$	2.4729	3	5.0135	2.3807	3	5.2077
$\alpha_1\ L_{III}M_V$	2.4630	2	5.0337	2.3704	2	5.2304
$\beta_6\ L_{III}N_I$	2.1906	4	5.660	2.1039	3	5.8930
$\beta_{2,15}\ L_{III}N_{IV,V}$	2.1194	4	5.850	2.0360	3	6.0894
$\beta_7\ L_{III}O_I$	2.0919	4	5.927	2.0092	3	6.1708
$\beta_{10}\ L_IM_{IV}$	2.1071	4	5.884	2.0237	3	6.1265
$\beta_9\ L_IM_V$	2.1004	4	5.903	2.0165	3	6.1484
$\gamma\ M_{III}N_{IV,V}$	10.998	9	1.1273	10.505	9	1.180
$\beta\ M_{IV}N_{VI}$	13.06	2	0.950	12.44	2	0.997
$\zeta\ M_VN_{III}$	17.38	4	0.714	16.46	4	0.753
$\alpha\ M_VN_{VI,VII}$	13.343	5	0.9292	12.68	2	0.978
$N_{IV,V}N_{VI,VII}$	113.	1	0.1095	107.	1	0.116
$N_{IV,V}O_{II,III}$	136.5	4	0.0908	128.9	7	0.0962

Designation	Å*	p.e.	keV	Å*	p.e.	keV
	61 Promethium			**62 Samarium**		
$\alpha_2\ KL_{II}$	0.324803	4	38.1712	0.313698	2	39.5224
$\alpha_1\ KL_{III}$	0.320160	4	38.7247	0.309040	2	40.1181
$\beta_3\ KM_{II}$	0.28363†	4	43.713	0.27376	2	45.289
$\beta_1\ KM_{III}$	0.28290†	3	43.826	0.27301	2	45.413
$\beta_2\ KN_{II,III}$	0.2759†	1	44.94	0.2662	1	46.58
$KO_{II,III}$				0.26491	3	46.801
$\beta_5\ KM_{IV,V}$				0.27111	3	45.731
$\beta_4\ L_IM_{II}$				2.00095	6	6.1963
$\beta_3\ L_IM_{III}$	2.0421	4	6.071	1.96241	3	6.3180
$\gamma_2\ L_IN_{II}$				1.66044	6	7.4668
$\gamma_3\ L_IN_{III}$				1.65601	3	7.4867
$\gamma_4\ L_IO_{II,III}$				1.60728	3	7.7137
$\eta\ L_{II}M_I$				2.21824	3	5.5892
$\beta_1\ L_{II}M_{IV}$	2.0797	4	5.961	1.99806	3	6.2051
$\gamma_5\ L_{II}N_I$				1.77934•	3	6.9678
$\gamma_1\ L_{II}N_{IV}$	1.7989	9	6.892	1.72724	3	7.1780
$\gamma_6\ L_{II}O_{IV}$				1.6966	9	7.3076
$l\ L_{III}M_I$				2.4823	4	4.9945

Designation	Å*	p.e.	keV	Å*	p.e.	keV
	61 Promethium (*Cont.*)			**62 Samarium** (*Cont.*)		
$\alpha_2\ L_{III}M_{IV}$	2.2926	4	5.4078	2.21062	3	5.6084
$\alpha_1\ L_{III}M_V$	2.2822	3	5.4325	2.1998	2	5.6361
$\beta_6\ L_{III}N_I$				1.94643	3	6.3697
$\beta_{2,15}\ L_{III}N_{IV,V}$	1.9559	6	6.339	1.88221	3	6.5870
$\beta_7\ L_{III}O_I$				1.85626	3	6.6791
$\beta_5\ L_{III}O_{IV,V}$				1.84700	9	6.7126
$\beta_{10}\ L_IM_{IV}$				1.86990	3	6.6304
$\beta_9\ L_IM_V$				1.86166	3	6.6597
$\gamma\ M_{III}N_{IV,V}$				9.600	9	1.291
$\beta\ M_{IV}N_{VI}$				11.27	1	1.0998
$\zeta\ M_VN_{III}$				14.91	4	0.831
$\alpha\ M_VN_{VI,VII}$				11.47	3	1.081
$N_{IV,V}N_{VI,VII}$				98.	1	0.126
$N_{IV,V}O_{II,III}$				117.4	4	0.1056

Designation	Å*	p.e.	keV	Å*	p.e.	keV
	63 Europium			**64 Gadolinium**		
$\alpha_2\ KL_{II}$	0.303118	2	40.9019	0.293038	2	42.3089
$\alpha_1\ KL_{III}$	0.298446	2	41.5422	0.288353	2	42.9962
$\beta_3\ KM_{II}$	0.264332	5	46.9036	0.25534	2	48.555
$\beta_1\ KM_{III}$	0.263577	5	47.0379	0.25460	2	48.697
$\beta_2\ KN_{II,III}$	0.256923	8	48.256	0.24816	3	49.959
$KO_{II,III}$	0.255645	7	48.497	0.24687	3	50.221
$\beta_5\ KM_{IV,V}$				0.25275	3	49.052
$\beta_4\ L_IM_{II}$	1.9255	2	6.4389	1.8540	2	6.6871
$\beta_3\ L_IM_{III}$	1.8867	2	6.5713	1.8150	2	6.8311
$\gamma_2\ L_IN_{II}$	1.5961	2	7.7677	1.5331	2	8.087
$\gamma_3\ L_IN_{III}$	1.5903	2	7.7961	1.5297	2	8.105
$\gamma_4\ L_IO_{II,III}$	1.5439	1	8.0304	1.4839	2	8.355
$\eta\ L_{II}M_I$	2.1315	2	5.8166	2.0494	1	6.0495
$\beta_1\ L_{II}M_{IV}$	1.9203	2	6.4564	1.8468	2	6.7132
$\gamma_5\ L_{II}N_I$	1.7085	2	7.2566	1.6412	2	7.5543
$\gamma_1\ L_{II}N_{IV}$	1.6574	2	7.4803	1.5924	2	7.7858
$\gamma_8\ L_{II}O_I$	1.6346	2	7.5849	1.5707	2	7.894
$\gamma_6\ L_{II}O_{IV}$	1.6282	2	7.6147	1.5644	2	7.925
$l\ L_{III}M_I$	2.3948	2	5.1772	2.3122	2	5.3621
$\alpha_2\ L_{III}M_{IV}$	2.1315	2	5.8166	2.0578	2	6.0250
$\alpha_1\ L_{III}M_V$	2.1209	2	5.8457	2.0468	2	6.0572
$\beta_6\ L_{III}N_I$	1.8737	2	6.6170	1.8054	2	6.8671
$\beta_{2,15}\ L_{III}N_{IV,V}$	1.8118	2	6.8432	1.7455	2	7.1028
$\beta_7\ L_{III}O_I$	1.7851	2	6.9453	1.7203	2	7.2071
$\beta_5\ L_{III}O_{IV,V}$	1.7772	2	6.9763	1.7130	2	7.2374
$\beta_{10}\ L_IM_{IV}$	1.7993	3	6.890	1.7315	3	7.160
$\beta_9\ L_IM_V$	1.7916	3	6.920	1.7240	3	7.192
$L_IO_{IV,V}$				1.4807	3	8.373
$\gamma\ M_{III}N_{IV,V}$	9.211	9	1.346	8.844	9	1.402
$\beta\ M_{IV}N_{VI}$	10.750	7	1.1533	10.254	6	1.2091
$\zeta\ M_VN_{III}$	14.22	2	0.872	13.57	2	0.914
$\alpha\ M_VN_{VI,VII}$	10.96	3	1.131	10.46	3	1.185
$N_{IV,V}O_{II,III}$	112.0	6	0.1107			

Designation	Å*	p.e.	keV	Å*	p.e.	keV
	65 Terbium			**66 Dysprosium**		
$\alpha_2\ KL_{II}$	0.283423	2	43.7441	0.274247	2	45.2078
$\alpha_1\ KL_{III}$	0.278724	2	44.4816	0.269533	2	45.9984
$\beta\ KM_{II}$	0.24683	2	50.229	0.23862	2	51.957

Designation	Å*	p.e.	keV	Å*	p.e.	keV
	65 Terbium (*Cont.*)			**66 Dysprosium** (*Cont.*)		
$\beta_1\ KM_{III}$	0.24608	2	50.382	0.23788	2	52.119
$\beta_2\ KN_{II,III}$	0.2397†	2	51.72	0.2317†	2	53.51
$KO_{II,III}$	0.23858	3	51.965	0.23056	3	53.774
$\beta_5\ KM_{IV,V}$				0.23618	3	52.494
$\beta_4\ L_IM_{II}$	1.7864	2	6.9403	1.72103	7	7.2039
$\beta_3\ L_IM_{III}$	1.7472	2	7.0959	1.6822	2	7.3702
$\gamma_2\ L_IN_{II}$	1.4764	2	8.398	1.42278	7	8.7140
$\gamma_3\ L_IN_{III}$	1.4718	2	8.423	1.41640	7	8.7532
$\gamma_4\ L_IO_{II,III}$	1.4276	2	8.685	1.37459	7	9.0195
$\eta\ L_{II}M_I$	1.9730	2	6.2839	1.89743	7	6.5342
$\beta_1\ L_{II}M_{IV}$	1.7768	3	6.978	1.71062	7	7.2477
$\gamma_5\ L_{II}N_I$	1.5787	2	7.8535	1.51824	7	8.1661
$\gamma_1\ L_{II}N_{IV}$	1.5303	2	8.102	1.47266	7	8.4188
$\gamma_8\ L_{II}O_I$	1.5097	2	8.212			
$\gamma_6\ L_{II}O_{IV}$	1.5035	2	8.246	1.44579	7	8.5753
$l\ L_{III}M_I$	2.2352	2	5.5467	2.15877	7	5.7431
$\alpha_2\ L_{III}M_{IV}$	1.9875	2	6.2380	1.91991	3	6.4577
$\alpha_1\ L_{III}M_V$	1.9765	2	6.2728	1.90881	3	6.4952
$\beta_6\ L_{III}N_I$	1.7422	2	7.1163	1.68213	7	7.3705
$\beta_{2,15}\ L_{III}N_{IV,V}$	1.6830	2	7.3667	1.62369	7	7.6357
$\beta_7\ L_{III}O_I$	1.6585	2	7.4753	1.60447	7	7.7272
$\beta_5\ L_{III}O_{IV,V}$	1.6510	2	7.5094	1.58837	7	7.8055
$\beta_{10}\ L_IM_{IV}$	1.6673	3	7.436	1.60743	9	7.7130
$\beta_9\ L_IM_V$				1.59973	9	7.7501
$L_IO_{IV,V}$	1.4228	3	8.714			
$\gamma\ M_{III}N_{IV,V}$	8.486	9	1.461	8.144	9	1.522
$\beta\ M_{IV}N_{VI}$	9.792	6	1.2661	9.357	6	1.3250
$\zeta\ M_VN_{III}$	12.98	2	0.955	12.43	2	0.998
$\alpha\ M_VN_{VI,VII}$	10.00	2	1.240	9.59	2	1.293
$N_{IV,V}N_{VI,VII}$	86.	1	0.144	83.	1	0.149
$N_{IV,V}O_{II,III}$	102.2	4	0.1213	97.2	8	0.128
	67 Holmium			**68 Erbium**		
$\alpha_2\ KL_{II}$	0.265486	2	46.6997	0.257110	2	48.2211
$\alpha_1\ KL_{III}$	0.260756	2	47.5467	0.252365	2	49.1277
$\beta_3\ KM_{II}$	0.23083	2	53.711	0.22341	2	55.494
$\beta_1\ KM_{III}$	0.23012	2	53.877	0.22266	2	55.681
$\beta_2\ KN_{II,III}$	0.2241†	2	55.32	0.2167†	2	57.21
$KO_{II,III}$	0.22305	3	55.584	0.21581	3	57.450
$\beta_5\ KM_{IV,V}$	0.22855	3	54.246	0.22124	3	56.040
$\beta_4\ L_IM_{II}$	1.6595	2	7.4708	1.6007	1	7.7453
$\beta_3\ L_IM_{III}$	1.6203	2	7.6519	1.5616	1	7.9392
$\gamma_2\ L_IN_{II}$	1.3698	2	9.051	1.3210	2	9.385
$\gamma_3\ L_IN_{III}$	1.3643	2	9.087	1.3146	1	9.4309
$\gamma_4\ L_IO_{II,III}$	1.3225	2	9.374	1.2752	2	9.722
$\eta\ L_{II}M_I$	1.8264	2	6.7883	1.7566	1	7.0579
$\beta_1\ L_{II}M_{IV}$	1.6475	2	7.5253	1.5873	1	7.8109
$\gamma_5\ L_{II}N_I$	1.4618	2	8.481	1.4067	3	8.814
$\gamma_1\ L_{II}N_{IV}$	1.4174	2	8.747	1.3641	2	9.089
$\gamma_8\ L_{II}O_I$	1.3983	2	8.867			
$\gamma_6\ L_{II}O_{IV}$	1.3923	2	8.905	1.3397	3	9.255
$l\ L_{III}M_I$	2.0860	2	5.9434	2.015	1	6.152
$\alpha_2\ L_{III}M_{IV}$	1.8561	2	6.6795	1.7955	2	6.9050
$\alpha_1\ L_{III}M_V$	1.8450	2	6.7198	1.78425	9	6.9487
$\beta_6\ L_{III}N_I$	1.6237	2	7.6359	1.5675	2	7.909
$\beta_{2,15}\ L_{III}N_{IV,V}$	1.5671	2	7.911	1.51399	9	8.1890
$\beta_7\ L_{III}O_I$				1.4941	3	8.298
$\beta_5\ L_{III}O_{IV,V}$	1.5378	2	8.062	1.4848	3	8.350

Designation	Å*	p.e.	keV	Å*	p.e.	keV
	67 Holmium (*Cont.*)			**68 Erbium** (*Cont.*)		
$\beta_{10}\ L_IM_{IV}$	1.5486	3	8.006	1.4941	3	8.298
$L_IO_{IV,V}$	1.3208	3	9.387			
$\beta_9\ L_IM_V$				1.4855	5	8.346
$M_{II}N_{IV}$				7.60	1	1.632
$\gamma\ M_{III}N_{IV,V}$	7.865	9	1.576			
$\gamma\ M_{III}\ V$				7.546	8	1.643
$\beta\ M_{IV}N_{VI}$	8.965	4	1.3830	8.592	3	1.4430
$\zeta\ M_VN_{III}$	11.86	1	1.0450	11.37	1	1.0901
$\alpha\ M_VN_{VI,VII}$	9.20	2	1.348	8.82	1	1.406
$N_{IV}N_{VI}$				72.7	9	0.171
$N_VN_{VI,VII}$				76.3	7	0.163
	69 Thulium			**70 Ytterbium**		
$\alpha_2\ KL_{II}$	0.249095	2	49.7726	0.241424	2	51.3540
$\alpha_1\ KL_{III}$	0.244338	2	50.7416	0.236655	2	52.3889
$\beta_3\ KM_{II}$	0.21636	2	57.304	0.2096†	1	59.14
$\beta_1\ KM_{III}$	0.21556	2	57.517	0.20884	8	59.37
$\beta_2\ KN_{II,III}$	0.2098†	2	59.09	0.2033†	2	60.98
$KO_{II,III}$	0.20891	2	59.346	0.20226	2	61.298
$\beta_5\ KM_{IV,V}$	0.21404	2	57.923	0.20739	2	59.782
$\beta_4\ L_IM_{II}$	1.5448	2	8.026	1.49138	3	8.3132
$\beta_3\ L_IM_{III}$	1.5063	2	8.231	1.45233	5	8.5367
$\gamma_2\ L_IN_{II}$	1.2742	2	9.730	1.22879	7	10.0897
$\gamma_3\ L_IN_{III}$	1.2678	2	9.779	1.22232	5	10.1431
$\gamma_4\ L_IO_{II,III}$	1.2294	2	10.084	1.1853	1	10.4603
$\eta\ L_{II}M_I$	1.6963	2	7.3088	1.63560	5	7.5802
$\beta_1\ L_{II}M_{IV}$	1.5304	2	8.101	1.47565	5	8.4018
$\gamma_5\ L_{II}N_I$	1.3558	2	9.144	1.3063	1	9.4910
$\gamma_1\ L_{II}N_{IV}$	1.3153	2	9.426	1.26769	5	9.8701
$\gamma_8\ L_{II}O_I$				1.24923	5	9.9246
$\gamma_6\ L_{II}O_{IV}$	1.2905	2	9.607	1.24271	3	9.9766
$l\ L_{III}M_I$	1.9550	2	6.3419	1.89415	5	6.5455
$\alpha_2\ L_{III}M_{IV}$	1.7381	2	7.1331	1.68285	5	7.3673
$\alpha_1\ L_{III}M_V$	1.7268†	2	7.1799	1.67189	4	7.4156
$\beta_6\ L_{III}N_I$	1.5162	2	8.177	1.4661	1	8.4563
$\beta_{2,15}\ L_{III}N_{IV,V}$	1.4640	2	8.468	1.41550	5	8.7588
$\beta_7\ L_{III}O_I$				1.3948	1	8.8889
$\beta_5\ L_{III}O_{IV,V}$	1.4349	2	8.641	1.38696	7	8.9390
$\beta_{10}\ L_IM_{IV}$	1.4410	3	8.604	1.3915	1	8.9100
$\beta_9\ L_IM_V$	1.4336	3	8.648	1.3838	1	8.9597
L_IO_I				1.1886	1	10.4312
$L_IO_{IV,V}$	1.2263	3	10.110	1.1827	1	10.4833
$L_{II}M_{II}$				1.58844	9	7.8052
$L_{II}O_{II,III}$				1.2453	1	9.9561
$t\ L_{III}M_{II}$				1.83091	9	6.7715
$L_{III}O_{II,III}$				1.3898	1	8.9209
$M_{III}N_I$				8.470	9	1.464
$\gamma\ M_{III}N_V$				7.024	8	1.765
$\beta\ M_{IV}N_{VI}$	8.249	7	1.503	7.909	2	1.5675
$\zeta\ M_VN_{III}$				10.48	1	1.183
$\alpha\ M_VN_{VI,VII}$	8.48	1	1.462	8.149	5	1.5214
$N_{IV}N_{VI}$				65.1	7	0.190
$N_VN_{VI,VII}$				69.3	5	0.179
	71 Lutetium			**72 Hafnium**		
$\alpha_2\ KL_{II}$	0.234081	2	52.9650	0.227024	3	54.6114
$\alpha_1\ KL_{III}$	0.229298	2	54.0698	0.222227	3	55.7902
$\beta_3\ KM_{II}$	0.20309†	4	61.05	0.19686†	4	62.98

Designation	Å*	p.e.	keV	Å*	p.e.	keV
	71 Lutetium (*Cont.*)			**72 Hafnium** (*Cont.*)		
$\beta_1\ KM_{III}$	0.20231†	3	61.283	0.19607†	3	63.234
$\beta_2\ KN_{II,III}$	0.1969†	2	62.97	0.1908†	2	64.98
$KO_{II,III}$	0.19589	2	63.293			
$\beta_5\ KM_{IV,V}$	0.20084	2	61.732			
$\beta_4\ L_IM_{II}$	1.44056	5	8.6064	1.39220	5	8.9054
$\beta_3\ L_IM_{III}$	1.40140	5	8.8469	1.35300	5	9.1634
$\gamma_2\ L_IN_{II}$	1.1853	2	10.460	1.14442	5	10.8335
$\gamma_3\ L_IN_{III}$	1.17953	4	10.5110	1.13841	5	10.8907
$\gamma'_4\ L_IO_{II}$				1.10376	5	11.2326
$\gamma_4\ L_IO_{II,III}$	1.1435	1	10.8425	1.10303	5	11.2401
$\eta\ L_{II}M_I$	1.5779	1	7.8575	1.52325	5	8.1393
$\beta_1\ L_{II}M_{IV}$	1.42359	3	8.7090	1.37410	5	9.0227
$\gamma_5\ L_{II}N_I$	1.2596	1	9.8428	1.21537	5	10.2011
$\gamma_1\ L_{II}N_{IV}$	1.22228	4	10.1434	1.17900	5	10.5158
$\gamma_8\ L_{II}O_I$	1.2047	1	10.2915	1.16138	5	10.6754
$\gamma_6\ L_{II}O_{IV}$	1.1987	1	10.3431	1.15519	5	10.7325
$l\ L_{III}M_I$	1.8360	1	6.7528	1.78145	5	6.9596
$\alpha_2\ L_{III}M_{IV}$	1.63029	5	7.6049	1.58046	5	7.8446
$\alpha_1\ L_{III}M_{IV}$	1.61951	3	7.6555	1.56958	5	7.8990
$\beta_6\ L_{III}N_I$	1.4189	1	8.7376	1.37410	5	9.0227
$\beta_{15}\ L_{III}N_{IV}$	1.3715	1	9.0395	1.32783	5	9.3371
$\beta_2\ L_{III}N_V$	1.37012	3	9.0489	1.32639	5	9.3473
$\beta_7\ L_{III}O_I$	1.34949	5	9.1873	1.30564	5	9.4958
$\beta_5\ L_{III}O_{IV,V}$	1.34183	7	9.2397	1.29761	5	9.5546
L_IM_I				1.43025	9	8.6685
$\beta_{10}\ L_IM_{IV}$	1.3430	2	9.232	1.29819	9	9.5503
$\beta_9 L_IM_V$	1.3358	1	9.2816	1.29025	9	9.6090
L_IN_{IV}	1.16227	9	10.6672	1.12250	9	11.0451
$\gamma_{11}\ L_IN_V$	1.16107	9	10.6782	1.12146	9	11.0553
L_IO_I				1.10664	9	11.2034
L_IO_{IV}				1.10086	9	11.2622
$L_{II}M_{II}$	1.53333	9	8.0858	1.48064	9	8.3735
$\beta_{17}\ L_{II}M_{III}$				1.43643	9	8.6312
$L_{II}N_V$				1.17788	9	10.5258
$v\ L_{II}N_{VI}$				1.15830	9	10.7037
$L_{II}O_{II,III}$	1.2014	1	10.3198			
$t\ L_{III}M_{II}$	1.7760	1	6.9810	1.72305	9	7.1954
$s\ L_{III}M_{III}$				1.66346	9	7.4532
$L_{III}N_{II}$				1.35887	9	9.1239
$L_{III}N_{III}$				1.35053	9	9.1802
$u\ L_{III}N_{VI,VII}$				1.30165	9	9.5249
$L_{III}O_{II,III}$	1.34524	9	9.2163			
$M_{III}N_I$				7.887	9	1.572
$\gamma\ M_{III}N_V$	6.768	6	1.832	6.544	4	1.895
ζ_2				9.686	7	1.2800
$\beta\ M_{IV}N_{VI}$	7.601	2	1.6312	7.303	1	1.6976
ζ_1				9.686	7	1.2800
$\alpha\ M_VN_{VI,VII}$	7.840	2	1.5813	7.539	1	1.6446
$N_{IV}N_{VI}$	63.0	5	0.197			
$N_VN_{VI,VII}$	65.7	2	0.1886			
	73 Tantalum			**74 Tungsten**		
$\alpha_2\ KL_{II}$	0.220305	8	56.277	0.213828	2	57.9817
$\alpha_1\ KL_{III}$	0.215497	4	57.532	0.2090100	Std	59.31824
$\beta_3\ KM_{II}$	0.190890	2	64.9488	0.185181	2	66.9514
$\beta_1\ KM_{III}$	0.190089	4	65.223	0.184374	2	67.2443
$\beta_2^{II}\ KN_{II}$	0.185188	9	66.949	0.17960	1	69.031
$\beta_2^{I}\ KN_{III}$	0.185011	8	67.013	0.179421	7	69.101

Designation	Å*	p.e.	keV	Å*	p.e.	keV
	73 Tantalum (*Cont.*)			**74 Tungsten** (*Cont.*)		
$KO_{II,III}$	0.184031	7	67.370	0.178444	5	69.479
KL_I				0.21592	4	57.42
$\beta_5^{II}\ KM_{IV}$	0.188920	6	65.626	0.183264	5	67.652
$\beta_5^{I}\ KM_V$	0.188757	6	65.683	0.183092	7	67.715
$\beta_4\ KN_{IV,V}$	0.18451	1	67.194	0.17892	2	69.294
$\beta_4\ L_IM_{II}$	1.34581	3	9.2124	1.30162	5	9.5252
$\beta_3\ L_IM_{III}$	1.30678	3	9.4875	1.26269	5	9.8188
$\gamma_2\ L_IN_{II}$	1.1053	1	11.217	1.06806	3	11.6080
$\gamma_3\ L_IN_{III}$	1.09936	4	11.2776	1.06200	6	11.6743
$\gamma'_4\ L_IO_{II}$	1.06544	3	11.6366	1.02863	3	12.0530
$\gamma_4\ L_IO_{III}$	1.06467	3	11.6451	1.02775	3	12.0634
$\eta\ L_{II}M_I$	1.47106	5	8.4280	1.42110	3	8.7243
$\beta_1\ L_{II}M_{IV}$	1.32698	3	9.3431	1.281809	9	9.67235
$\gamma_5\ L_{II}N_I$	1.1729	1	10.5702	1.13235	3	10.9490
$\gamma_1\ L_{II}N_{IV}$	1.13794	3	10.8952	1.09855	3	11.2859
$\gamma_8\ L_{II}O_I$	1.1205	1	11.0646	1.08113	4	11.4677
$\gamma_6\ L_{II}O_{IV}$	1.11388	3	11.1306	1.07448	5	11.5387
$l\ L_{III}M_I$	1.72841	5	7.1731	1.6782	1	7.3878
$\alpha_2\ L_{III}M_{IV}$	1.53293	2	8.0879	1.48743	2	8.3352
$\alpha_1\ L_{III}M_V$	1.52197	2	8.1461	1.47639	2	8.3976
$\beta_6\ L_{III}N_I$	1.33094	8	9.3153	1.28989	7	9.6117
$\beta_{15}\ L_{III}N_{IV}$	1.28619	5	9.6394	1.24631	3	9.9478
$\beta_2\ L_{III}N_V$	1.28454	2	9.6518	1.24460	3	9.9615
$\beta_7\ L_{III}O_I$	1.26385	5	9.8098	1.22400	4	10.1292
$\beta_5\ L_{III}O_{IV,V}$	1.2555	1	9.8750	1.21545	3	10.2004
L_IM_I				1.3365	3	9.277
$\beta_{10}\ L_IM_{IV}$	1.2537	2	9.889	1.21218	3	10.2279
$\beta_9\ L_IM_V$	1.2466	2	9.946	1.20479	7	10.2907
L_IN_I	1.11521	9	11.1173			
L_IN_{IV}	1.08377	7	11.4398	1.0468	2	11.844
$\gamma_{11}\ L_IN_V$	1.08205	7	11.4580	1.0458	1	11.856
$L_IN_{VI,VII}$	1.06357	9	11.6570			
L_IO_I	1.06771	9	11.6118	1.0317	3	12.017
$L_IO_{IV,V}$	1.06192	9	11.6752	1.0250	2	12.095
$L_{II}M_{II}$	1.43048	9	8.6671			
$\beta_{17}\ L_{II}M_{III}$	1.3864	1	8.9428	1.3387	2	9.261
$L_{II}M_V$	1.31897	9	9.3998	1.2728	2	9.741
$L_{II}N_{II}$	1.1600	2	10.688	1.1218	3	11.052
$L_{II}N_{III}$	1.1553	1	10.7316	1.1149	2	11.120
$L_{II}N_V$	1.13687	9	10.9055			
$v\ L_{II}N_{VI}$	1.1158	1	11.1113	1.0771	1	11.510
$L_{II}O_{II}$	1.11789	9	11.0907			
$L_{II}O_{III}$	1.11693	9	11.1001	1.0792	2	11.488
$t\ L_{III}M_{II}$	1.67265	9	7.4123	1.6244	3	7.632
$s\ L_{III}M_{III}$	1.61264	9	7.6881	1.5642	3	7.926
$L_{III}N_{II}$	1.3167	1	9.4158	1.2765	2	9.712
$L_{III}N_{III}$	1.3086	1	9.4742	1.2672	2	9.784
$u\ L_{III}N_{VI,VII}$	1.25778	4	9.8572	1.21868	5	10.1733
$L_{III}O_{II,III}$	1.2601	3	9.839	1.2211	2	10.153
M_IN_{III}	5.40	2	2.295	5.172	9	2.397
$M_IO_{II,III}$				4.44	2	2.79
$M_{II}N_I$				6.28	2	1.973
$M_{II}N_{IV}$	5.570	4	2.226	5.357	4	2.314
$M_{III}N_I$	7.612	9	1.629	7.360	8	1.684
$M_{III}N_{IV}$	6.353	5	1.951	6.134	4	2.021
$\gamma\ M_{III}N_V$	6.312	4	1.964	6.092	3	2.035
$M_{III}O_I$	5.83	2	2.126	5.628	8	2.203
$M_{III}O_{IV,V}$	5.67	3	2.19			

Designation	Å*	p.e.	keV	Å*	p.e.	keV
	73 Tantalum (*Cont.*)			**74 Tungsten** (*Cont.*)		
ζ_2 $M_{IV}N_{II}$	9.330	5	1.3288	8.993	5	1.3787
$M_{IV}N_{III}$	8.90	2	1.393	8.573	8	1.446
β $M_{IV}N_{VI}$	7.023	1	1.7655	6.757	1	1.8349
$M_{IV}O_{II}$	7.09	2	1.748	6.806	9	1.822
ζ_1 M_VN_{III}	9.316	4	1.3308	8.962	4	1.3835
α $M_VN_{VI,VII}$	7.252	1	1.7096			
α_2 M_VN_{VI}				6.992	2	1.7731
α_1 M_VN_{VII}				6.983	1	1.7754
M_VO_{III}	7.30	2	1.700	7.005	9	1.770
$N_{II}N_{IV}$				54.0	2	0.2295
$N_{IV}N_{VI}$	58.2	1	0.2130	55.8	1	0.2221
$N_VN_{VI,VII}$	61.1	2	0.2028			
N_VN_{VI}				59.5	3	0.208
N_VN_{VII}				58.4	1	0.2122
	75 Rhenium			**76 Osmium**		
α_2 KL_{II}	0.207611	1	59.7179	0.201639	2	61.4867
α_1 KL_{III}	0.202781	2	61.1403	0.196794	2	63.0005
β_3 KM_{II}	0.179697	3	68.994	0.174431	3	71.077
β_1 KM_{III}	0.178880	3	69.310	0.173611	3	71.413
β_2^{II} KN_{II}	0.17425	1	71.151	0.16910	1	73.318
β_2^{I} KN_{III}	0.174054	6	71.232	0.168906	6	73.402
$KO_{II,III}$	0.17308	1	71.633	0.16798	1	73.808
β_5^{II} KM_{IV}	0.17783	1	69.719	0.17262	1	71.824
β_5^{I} KM_V	0.17766	1	69.786	0.17245	1	71.895
β_4 $KN_{IV,V}$	0.17362	2	71.410	0.16842	2	73.615
β_4 L_IM_{II}	1.25917	5	9.8463	1.21844	5	10.1754
β_3 L_IM_{III}	1.22031	5	10.1598	1.17955	7	10.5108
γ_2 L_IN_{II}	1.03233	5	12.0098	0.99805	5	12.4224
γ_3 L_IN_{III}	1.02613	7	12.0824	0.99186	5	12.4998
γ'_4 L_IO_{II}	0.99334	5	12.4813	0.96033	8	12.910
γ_4 L_IO_{III}	0.99249	5	12.4920	0.95938	8	12.923
η $L_{II}M_I$	1.37342	5	9.0272	1.32785	7	9.3370
β_1 $L_{II}M_{IV}$	1.23858	2	10.0100	1.19727	7	10.3553
γ_5 $L_{II}N_I$	1.09388	5	11.3341	1.05693	5	11.7303
γ_1 $L_{II}N_{IV}$	1.06099	5	11.6854	1.02503	5	12.0953
γ_8 $L_{II}O_I$	1.04398	5	11.8758	1.00788	5	12.3012
γ_6 $L_{II}O_{IV}$	1.03699	9	11.956	1.00107	5	12.3848
l $L_{III}M_I$	1.63056	5	7.6036	1.58498	7	7.8222
α_2 $L_{III}M_{IV}$	1.44396	5	8.5862	1.40234	5	8.8410
α_1 $L_{III}M_V$	1.43290	4	8.6525	1.39121	5	8.9117
β_6 $L_{III}N_I$	1.25100	5	9.9105	1.21349	5	10.2169
β_{15} $L_{III}N_{IV}$	1.20819	5	10.2617	1.17167	5	10.5816
β_2 $L_{III}N_V$	1.20660	4	10.2752	1.16979	8	10.5985
β_7 $L_{III}O_I$	1.18610	5	10.4529	1.14933	8	10.7872
β_5 $L_{III}O_{IV,V}$	1.17721	5	10.5318	1.1405	1	10.8711
β_{10} L_IM_{IV}	1.17218	5	10.5770	1.13353	5	10.9376
β_9 L_IM_V	1.16487	4	10.6433	1.12637	6	11.0071
L_IN_I	1.0420	1	11.899			
L_IN_{IV}	1.0119	1	12.252	0.9772	3	12.687
γ_{11} L_IN_V	1.0108	1	12.266	0.9765	3	12.696
L_IO_I	0.9965	1	12.442	0.96318	7	12.8721
$L_IO_{IV,V}$	0.9900	1	12.524	0.95603	5	12.9683
$L_{II}M_{II}$	1.3366	1	9.2761	1.2934	2	9.586
β_{17} $L_{II}M_{III}$	1.2927	1	9.5910	1.2480	2	9.934

Designation	Å*	p.e.	keV	Å*	p.e.	keV
	75 Rhenium (*Cont.*)			**76 Osmium** (*Cont.*)		
$L_{II}M_V$	1.2305	1	10.0753	1.18977	7	10.4205
$L_{II}N_{II}$	1.0839	1	11.438			
$L_{II}N_{III}$	1.0767	1	11.515	1.03973	5	11.9243
v $L_{II}N_{VI}$	1.0404	1	11.917	1.0050	2	12.337
$L_{II}O_{III}$	1.0397	1	11.925	1.0047	2	12.340
t $L_{III}M_{II}$	1.5789	1	7.8525	1.5347	2	8.079
s $L_{III}M_{III}$	1.5178	1	8.1682	1.4735	2	8.414
$L_{III}N_I$				1.20086	7	10.3244
$L_{III}N_{III}$	1.2283	1	10.0933			
u $L_{III}N_{VI,VII}$	1.1815	1	10.4931	1.14537	7	10.8245
M_IN_{III}				4.79	2	2.59
$M_{II}N_I$				5.81	2	2.133
$M_{II}N_{IV}$				4.955	4	2.502
$M_{III}N_I$				6.89	2	1.798
$M_{III}N_{IV}$	5.931	5	2.090	5.724	5	2.166
γ $M_{III}N_V$	5.885	2	2.1067	5.682	4	2.182
ζ_2 $M_{IV}N_{II}$	8.664	5	1.4310	8.359	5	1.4831
$M_{IV}N_{III}$	8.239	8	1.505			
β $M_{IV}N_{VI}$	6.504	1	1.9061	6.267	1	1.9783
ζ_1 M_VN_{III}	8.629	4	1.4368	8.310	4	1.4919
α $M_VN_{VI,VII}$	6.729	1	1.8425	6.490	1	1.9102
$N_{IV}N_{VI}$				51.9	1	0.2388
$N_VN_{VI,VII}$				54.7	2	0.2266
	77 Iridium			**78 Platinum**		
α_2 KL_{II}	0.195904	2	63.2867	0.190381	4	65.122
α_1 KL_{III}	0.191047	2	64.8956	0.185511	4	66.832
β_3 KM_{II}	0.169367	2	73.2027	0.164501	3	75.368
β_1 KM_{III}	0.168542	2	73.5608	0.163675	3	75.748
β_2^{II} KN_{II}	0.16415	1	75.529	0.15939	1	77.785
β_2^{I} KN_{III}	0.163956	7	75.619	0.15920	1	77.878
$KO_{II,III}$	0.163019	5	76.053	0.15826	1	78.341
β_5^{II} KM_{IV}	0.16759	2	73.980	0.16271	2	76.199
β_5^{I} KM_V	0.167373	9	74.075	0.16255	3	76.27
β_4 $KN_{IV,V}$	0.16352	2	75.821	0.15881	2	78.069
β_4 L_IM_{II}	1.17958	3	10.5106	1.14223	5	10.8543
β_3 L_IM_{III}	1.14085	3	10.8674	1.10394	5	11.2308
γ_2 L_IN_{II}	0.96545	3	12.8418	0.93427	5	13.2704
γ_3 L_IN_{III}	0.95931	5	12.9240	0.92791	5	13.3613
γ'_4 L_IO_{II}	0.92831	3	13.3555	0.89747	4	13.8145
γ_4 L_IO_{III}	0.92744	3	13.3681	0.89659	4	13.8281
η $L_{II}M_I$	1.28448	3	9.6522	1.2429	2	9.975
β_1 $L_{II}M_{IV}$	1.15781	3	10.7083	1.11990	2	11.0707
γ_5 $L_{II}N_I$	1.02175	5	12.1342	0.9877	2	12.552
γ_1 $L_{II}N_{IV}$	0.99085	3	12.5126	0.95797	3	12.9420
γ_8 $L_{II}O_I$	0.97409	3	12.7279	0.9411	1	13.173
γ_6 $L_{II}O_{IV}$	0.96708	4	12.8201	0.9342	2	13.271
l $L_{III}M_I$	1.54094	3	8.0458	1.4995	2	8.268
α_2 $L_{III}M_{IV}$	1.36250	5	9.0995	1.32432	2	9.3618
α_1 $L_{III}M_V$	1.35128	3	9.1751	1.31304	3	9.4423
β_6 $L_{III}N_I$	1.17796	3	10.5251	1.14355	5	10.8418
β_{15} $L_{III}N_{IV}$	1.13707	3	10.9036			
β_2 $L_{III}N_V$	1.13532	3	10.9203	1.10200	3	11.2505
β_7 $L_{III}O_I$	1.11489	3	11.1205	1.08168	3	11.4619

Designation	Å*	p.e.	keV	Å*	p.e.	keV
	77 Iridium (*Cont.*)			**78 Platinum** (*Cont.*)		
$\beta_5\ L_{III}O_{IV,V}$	1.10585	3	11.2114	1.0724	2	11.561
L_IM_I	1.2102	2	10.245	1.16962	9	10.6001
$\beta_{10}\ L_IM_{IV}$	1.09702	4	11.3016	1.06183	7	11.6762
$\beta_9\ L_IM_V$	1.08975	5	11.3770	1.05446	5	11.7577
L_IN_I	0.9766	2	12.695	0.9455	2	13.113
L_IN_{IV}	0.9459	2	13.108			
$\gamma_{11}\ L_IN_V$	0.9446	2	13.126	0.9143	2	13.560
$L_IO_{IV,V}$	0.9243	3	13.413			
L_IO_I				0.8995	2	13.784
L_IO_{IV}				0.8943	1	13.864
L_IO_V				0.8934	1	13.878
$L_{II}M_{II}$	1.2502	3	9.917	1.213	1	10.225
$\beta_{17}\ L_{II}M_{III}$	1.2069	2	10.273	1.1667	1	10.6265
$L_{II}M_V$	1.1489	2	10.791	1.1129	2	11.140
$L_{II}N_{II}$	1.0120	2	12.251	0.9792	2	12.661
$L_{II}N_{III}$	1.0054	3	12.332	0.97173	4	12.7588
$v\ L_{II}N_{VI}$	0.97161	6	12.7603	0.93931	5	13.1992
$L_{II}O_{III}$	0.96979	5	12.7843			
$t\ L_{III}M_{II}$	1.4930	3	8.304	1.4530	2	8.533
$s\ L_{III}M_{III}$	1.4318	2	8.659	1.3895	2	8.923
$L_{III}N_{II}$	1.16545	5	10.6380	1.1310	2	10.962
$L_{III}N_{III}$	1.1560	3	10.725	1.1226	2	11.044
$u\ L_{III}N_{VI,VII}$	1.11145	4	11.1549	1.07896	5	11.4908
$L_{III}O_{II,III}$	1.10923	6	11.1772	1.0761	3	11.521
M_IN_{III}	4.631†	9	2.677	4.460	9	2.780
$M_{II}N_{IV}$	4.780	4	2.594	4.601	4	2.695
$M_{III}N_I$	6.669	9	1.859	6.455	9	1.921
$M_{III}N_{IV}$	5.540	5	2.238	5.357	5	2.314
$\gamma\ M_{III}N_V$	5.500	4	2.254	5.319	4	2.331
$M_{III}O_I$				4.876	9	2.543
$M_{III}O_{IV,V}$	4.869	9	2.546	4.694	8	2.641
$\zeta_2\ M_{IV}N_{II}$	8.065	5	1.5373	7.790	5	1.592
$M_{IV}N_{III}$	7.645	8	1.622	7.371	8	1.682
$\beta\ M_{IV}N_{VI}$	6.038	1	2.0535	5.828	1	2.1273
$\zeta_1\ M_VN_{III}$	8.021	4	1.5458	7.738	4	1.6022
$\alpha_2\ M_VN_{VI}$	6.275	3	1.9758	6.058	3	2.047
$\alpha_1\ M_VN_{VII}$	6.262	1	1.9799	6.047	1	2.0505
M_VO_{III}				5.987	9	2.071
$N_{IV}N_{VI}$	50.2	1	0.2470	48.1	2	0.258
$N_VN_{VI,VII}$	52.8	1	0.2348	50.9	1	0.2436
	79 Gold			**80 Mercury**		
$\alpha_2\ KL_{II}$	0.185075	2	66.9895	0.179958	3	68.895
$\alpha_1\ KL_{III}$	0.180195	2	68.8037	0.175068	3	70.819
$\beta_3\ KM_{II}$	0.159810	2	77.580	0.155321	3	79.822
$\beta_1\ KM_{III}$	0.158982	3	77.984	0.154487	3	80.253
$\beta_2^{II}\ KN_{II}$	0.15483	2	80.08	0.15040	2	82.43
$\beta_2^{I}\ KN_{III}$	0.154618	9	80.185	0.15020	2	82.54
$KO_{II,III}$	0.153694	7	80.667	0.14931	2	83.04
KL_I	0.18672	4	66.40			
$\beta_5^{II}\ KM_{IV}$	0.158062	7	78.438			
$\beta_5^{I}\ KM_V$	0.157880	5	78.529			
$\beta_5\ KM_{IV,V}$				0.15353	2	80.75
$\beta_4\ KN_{IV,V}$	0.154224	5	80.391	0.14978	2	82.78
$\beta_4\ L_IM_{II}$	1.10651	3	11.2047	1.07222	7	11.5630
$\beta_3\ L_IM_{III}$	1.06785	9	11.6103	1.03358	7	11.9953

Designation	Å*	p.e.	keV	Å*	p.e.	keV
	79 Gold (*Cont.*)			**80 Mercury** (*Cont.*)		
$\gamma_2\ L_IN_{II}$	0.90434	3	13.7095	0.87544	7	14.162
$\gamma_3\ L_IN_{III}$	0.89783	5	13.8090	0.86915	7	14.265
$\gamma'_4\ L_IO_{II}$	0.86816	4	14.2809	0.84013	7	14.757
$\gamma_4\ L_IO_{III}$	0.86703	4	14.2996	0.83894	7	14.778
$\eta\ L_{II}M_I$	1.20273	3	10.3083	1.1640	1	10.6512
$\beta_1\ L_{II}M_{IV}$	1.08353	3	11.4423	1.04868	5	11.8226
$\gamma_5\ L_{II}N_I$	0.95559	3	12.9743	0.92453	7	13.410
$\gamma_1\ L_{II}N_{IV}$	0.92650	3	13.3817	0.89646	5	13.8301
$\gamma_8\ L_{II}O_I$	0.90989	5	13.6260	0.87995	7	14.090
$\gamma_6\ L_{II}O_{IV}$	0.90297	3	13.7304	0.87319	7	14.199
$l\ L_{III}M_I$	1.45964	9	8.4939	1.4216	1	8.7210
$\alpha_2\ L_{III}M_{IV}$	1.28772	3	9.6280	1.25264	7	9.8976
$\alpha_1\ L_{III}M_V$	1.27640	3	9.7133	1.24120	5	9.9888
$\beta_6\ L_{III}N_I$	1.11092	3	11.1602	1.07975	7	11.4824
$\beta_{15}\ L_{III}N_{IV}$	1.07188	5	11.5667	1.04151	7	11.9040
$\beta_2\ L_{III}N_V$	1.07022	3	11.5847	1.03975	7	11.9241
$\beta_7\ L_{III}O_I$	1.04974	8	11.8106	1.01937	7	12.1625
$\beta_5\ L_{III}O_{IV,V}$	1.04044	3	11.9163	1.00987	7	12.2769
L_IM_I	1.13525	5	10.9210	1.0999	2	11.272
$\beta_{10}\ L_IM_{IV}$	1.02789	7	12.0617	0.9962	2	12.446
$\beta_9\ L_IM_V$	1.02063	7	12.1474	0.9871	2	12.560
L_IN_I	0.9131	1	13.578	0.8827	2	14.045
L_IN_{IV}	0.88563	7	13.999			
$\gamma_{11}\ L_IN_V$	0.88433	7	14.020	0.85657	7	14.474
L_IO_I	0.87074	5	14.2385	0.8452	2	14.670
$L_IO_{IV,V}$	0.86400	5	14.3497	0.8350	2	14.847
$L_{II}M_{II}$	1.1708	1	10.5892	1.1387	5	10.888
$\beta_{17}\ L_{II}M_{III}$	1.12798	5	10.9915	1.0916	5	11.358
$L_{II}M_V$	1.0756	2	11.526			
$L_{II}N_{III}$	0.9402	2	13.186	0.90894	7	13.640
$v\ L_{II}N_{VI}$	0.90837	5	13.6487	0.87885	7	14.107
$L_{II}O_{II}$	0.90746	7	13.662	0.8784	1	14.114
$L_{II}O_{III}$	0.90638	7	13.679	0.8758	1	14.156
$t\ L_{III}M_{II}$	1.41366	7	8.7702	1.3746	2	9.019
$s\ L_{III}M_{III}$	1.35131	7	9.1749	1.3112	2	9.455
$L_{III}N_{II}$	1.09968	7	11.2743	1.0649	2	11.642
$L_{III}N_{III}$	1.09026	7	11.3717	1.0585	1	11.713
$u\ L_{III}N_{VI,VII}$	1.04752	5	11.8357			
$u'\ L_{III}N_{VI}$				1.01769	7	12.1826
$u\ L_{III}N_{VII}$				1.01674	7	12.1940
$L_{III}O_{II,III}$	1.0450	2	11.865			
$L_{III}O_{II}$				1.01558	7	12.2079
$L_{III}O_{III}$				1.01404	7	12.2264
$L_{III}P_{II,III}$	1.03876	7	11.9355			
M_IN_{III}	4.300	9	2.883			
$M_{II}N_{IV}$	4.432	4	2.797			
$M_{III}N_I$	6.259	9	1.981	6.09	2	2.036
$M_{III}N_{IV}$	5.186	5	2.391			
$\gamma\ M_{III}N_V$	5.145	4	2.410	4.984†	2	2.4875
$M_{III}O_I$	4.703	9	2.636			
$M_{III}O_{IV,V}$	4.522	6	2.742			
$\zeta_2\ M_{IV}N_{II}$	7.523	5	1.648			
$M_{IV}N_{III}$	7.101	8	1.746	6.87	2	1.805
$\beta\ M_{IV}N_{VI}$	5.624	1	2.2046	5.4318†	9	2.2825
$\zeta_1\ M_VN_{III}$	7.466	4	1.6605			
$\alpha_2\ M_VN_{VI}$	5.854	3	2.118			

Designation	Å*	p.e.	keV	Å*	p.e.	keV
	79 Gold (*Cont.*)			**80 Mercury** (*Cont.*)		
$\alpha_1\ M_VN_{VII}$	5.840	1	2.1229	5.6476†	9	2.1953
M_VO_{III}	5.767	9	2.150			
$N_{IV}N_{VI}$	46.8	2	0.265	45.2†	3	0.274
$N_VN_{VI,VII}$	49.4	1	0.2510	47.9†	3	0.259
	81 Thallium			**82 Lead**		
$\alpha_2\ KL_{II}$	0.175036	2	70.8319	0.170294	2	72.8042
$\alpha_1\ KL_{III}$	0.170136	2	72.8715	0.165376	2	74.9694
$\beta_3\ KM_{II}$	0.150980	6	82.118	0.146810	4	84.450
$\beta_1\ KM_{III}$	0.150142	5	82.576	0.145970	6	84.936
$\beta_2^{II}\ KN_{II}$	0.14614	1	84.836	0.14212	2	87.23
$\beta_2^{I}\ KN_{III}$	0.14595	1	84.946	0.14191	1	87.364
$KO_{II,III}$	0.14509	1	85.451	0.141012	8	87.922
KP				0.1408	1	88.06
$\beta_5\ KM_{IV,V}$	0.14917	1	83.114			
$\beta_5^{II}\ KM_{IV}$				0.14512	2	85.43
$\beta_5^{I}\ KM_V$				0.14495	3	85.53
$\beta_4\ KN_{IV,V}$	0.14553	2	85.19	0.14155	3	87.59
$\beta_4\ L_IM_{II}$	1.03918	3	11.9306	1.0075	1	12.306
$\beta_3\ L_IM_{III}$	1.00062	3	12.3904	0.96911	7	12.7933
$\gamma_2\ L_IN_{II}$	0.84773	5	14.6251	0.8210	2	15.101
$\gamma_3\ L_IN_{III}$	0.84130	4	14.7368	0.8147	1	15.218
$\gamma'_4\ L_IO_{II}$	0.81308	5	15.2482	0.78706	7	15.752
$\gamma_4\ L_IO_{III}$	0.81184	5	15.2716	0.7858	1	15.777
$\eta\ L_{II}M_I$	1.12769	3	10.9943	1.09241	7	11.3493
$\beta_1\ L_{II}M_{IV}$	1.01513	4	12.2133	0.98291	3	12.6137
$\gamma_5\ L_{II}N_I$	0.89500	4	13.8526	0.86655	5	14.3075
$\gamma_1\ L_{II}N_{IV}$	0.86752	3	14.2915	0.83973	3	14.7644
$\gamma_8\ L_{II}O_I$	0.8513	2	14.564	0.82365	5	15.0527
$\gamma_6\ L_{II}O_{IV}$	0.8442	2	14.685	0.81683	5	15.1783
$L_{II}P_I$				0.81583	5	15.1969
$l\ L_{III}M_I$	1.38477	3	8.9532	1.34990	7	9.1845
$\alpha_2\ L_{III}M_{IV}$	1.21875	3	10.1728	1.18648	5	10.4495
$\alpha_1\ L_{III}M_V$	1.20739	4	10.2685	1.17501	2	10.5515
$\beta_6\ L_{III}N_I$	1.04963	5	11.8118	1.0210	1	12.143
$\beta_{15}\ L_{III}N_{IV}$	1.01201	3	12.2510	0.98389	7	12.6011
$\beta_2\ L_{III}N_V$	1.01031	3	12.2715	0.98221	7	12.6226
$\beta_7\ L_{III}O_I$	0.99017	5	12.5212	0.9620	1	12.888
$\beta_5\ L_{III}O_{IV,V}$	0.98058	3	12.6436	0.9526	1	13.015
L_IM_I	1.0644	2	11.648	1.0323	2	12.010
$\beta_{10}\ L_IM_{IV}$	0.96389	7	12.8626	0.9339	2	13.275
$\beta_9\ L_IM_V$	0.95675	7	12.9585	0.9268	1	13.377
L_IN_I	0.8549	1	14.503	0.82859	7	14.963
L_IN_{IV}	0.83001	7	14.937	0.80364	7	15.427
$\gamma_{11}\ L_IN_V$	0.82879	5	14.9593	0.80233	9	15.453
$L_IN_{VI,VII}$				0.7884	1	15.725
L_IO_I	0.8158	1	15.198	0.7897	1	15.699
$L_IO_{IV,V}$	0.80861	5	15.3327	0.78257	7	15.843
$L_{II}M_{II}$	1.0997	1	11.274	1.0644	2	11.648
$\beta_{17}\ L_{II}M_{III}$	1.05609	7	11.7397	1.0223	1	12.127
$L_{II}M_V$	1.00722	5	12.3093	0.9747	1	12.720
$L_{II}N_{II}$	0.882	2	14.057	0.8585	3	14.442

Designation	Å*	p.e.	keV	Å*	p.e.	keV
	81 Thallium (*Cont.*)			**82 Lead** (*Cont.*)		
$L_{II}N_{III}$	0.87996	5	14.0893	0.85192	7	14.553
$L_{II}N_V$				0.8382	2	14.791
$v\ L_{II}N_{VI}$	0.85048	5	14.5777	0.82327	7	15.060
$L_{II}O_{II}$	0.8490	1	14.604			
$L_{II}O_{III}$				0.8200	1	15.120
$t\ L_{III}M_{II}$	1.34154	5	9.2417	1.30767	7	9.4811
$s\ L_{III}M_{III}$	1.27807	5	9.7007	1.24385	7	9.9675
$L_{III}N_{II}$				1.01040	7	12.2705
$L_{III}N_{III}$	1.0286	1	12.053	1.0005	1	12.392
$u\ L_{III}N_{VI,VII}$	0.9888	1	12.538	0.96133	7	12.8968
$L_{III}O_{II}$	0.98738	5	12.5566	0.9586	1	12.934
$L_{III}O_{III}$	0.98538	5	12.5820	0.9578	1	12.945
$L_{III}P_{II,III}$	0.97926	5	12.6607	0.95118	7	13.0344
M_IN_{III}	4.013	9	3.089	3.872	9	3.202
$M_{II}N_I$				4.655	8	2.664
$M_{II}N_{IV}$	4.116	4	3.013	3.968	5	3.124
$M_{III}N_I$	5.884	8	2.107	5.704	8	2.174
$M_{III}N_{IV}$	4.865	5	2.548	4.715	3	2.630
$\gamma\ M_{III}N_V$	4.823	4	2.571	4.674	1	2.6527
$M_{III}O_I$				4.244	9	2.921
$M_{II}O_{IV,V}$	4.216	6	2.941	4.069	6	3.047
$\zeta_2\ M_{IV}N_{II}$	7.032	5	1.763	6.802	5	1.823
$M_{IV}N_{III}$				6.384	7	1.942
$\beta\ M_{IV}N_{VI}$	5.249	1	2.3621	5.076	1	2.4427
$M_{IV}O_{II}$	5.196	9	2.386	5.004	9	2.477
$\zeta_1\ M_VN_{III}$	6.974	4	1.778	6.740	3	1.8395
$\alpha_2\ M_VN_{VI}$	5.472	2	2.2656	5.299	2	2.3397
$\alpha_1\ M_VN_{VII}$	5.460	1	2.2706	5.286	1	2.3455
M_VO_{III}				5.168	9	2.399
$N_{IV}N_{VI}$				42.3	2	0.293
$N_VN_{VI,VII}$	46.5	2	0.267	45.0	1	0.2756
$N_{VI}O_{IV}$	115.3	2	0.1075	102.4	1	0.1211
$N_{VI}O_V$	113.0	1	0.10968	100.2	2	0.1237
$N_{VII}O_V$	117.7	1	0.10530	104.3	1	0.1189
	83 Bismuth			**84 Polonium**		
$\alpha_2\ KL_{II}$	0.165717	2	74.8148	0.16130†	1	76.862
$\alpha_1\ KL_{III}$	0.160789	2	77.1079	0.15636†	1	79.290
$\beta_3\ KM_{II}$	0.142779	7	86.834	0.13892†	2	89.25
$\beta_1\ KM_{III}$	0.141948	3	87.343	0.13807†	2	89.80
$\beta_2^{II}\ KN_{II}$	0.13817	1	89.733	0.13438†	2	92.26
$\beta_2^{I}\ KN_{III}$	0.13797	1	89.864	0.13418†	2	92.40
$KO_{II,III}$	0.13709	1	90.435			
$\beta_5\ KM_{IV,V}$	0.14111	1	87.860			
$\beta_4\ KN_{IV,V}$	0.13759	2	90.11			
$\beta_4\ L_IM_{II}$	0.97690	4	12.6912	0.9475	3	13.086
$\beta_3\ L_IM_{III}$	0.93855	3	13.2098	0.9091	3	13.638
$\gamma_2\ L_IN_{II}$	0.79565	3	15.5824	0.772	1	16.07
$\gamma_3\ L_IN_{III}$	0.78917	5	15.7102			
$\gamma'_4\ L_IO_{II}$	0.76198	3	16.2709			
$\gamma_4\ L_IO_{III}$	0.76087	3	16.2947			
$\gamma_{13}\ L_IP_{II,III}$	0.75690	3	16.3802			
$\eta\ L_{II}M_I$	1.05856	3	11.7122			
$\beta_1\ L_{II}M_{IV}$	0.951978	9	13.0235	0.9220	2	13.447
$\gamma_5\ L_{II}N_I$	0.83923	5	14.7732			

TABLE A (*Continued*)

Designation	Å*	p.e.	keV	Å*	p.e.	keV
	83 Bismuth (*Cont.*)			84 Polonium (*Cont.*)		
$\gamma_1\ L_{II}N_{IV}$	0.81311	2	15.2477	0.78748	9	15.744
$\gamma_8\ L_{II}O_I$	0.7973	1	15.551			
$\gamma_6\ L_{II}O_{IV}$	0.79043	3	15.6853	0.7645	2	16.218
$l\ L_{III}M_I$	1.31610	7	9.4204	1.2829	5	9.664
$\alpha_2\ L_{III}M_{IV}$	1.15536	1	10.73091	1.12548†	5	11.0158
$\alpha_1\ L_{III}M_V$	1.14386	2	10.8388	1.11386	4	11.1308
$\beta_6\ L_{III}N_I$	0.99331	3	12.4816	0.9672	2	12.819
$\beta_{15}\ L_{III}N_{IV}$	0.95702	5	12.9549	0.9312	2	13.314
$\beta_2\ L_{III}N_V$	0.95518	4	12.9799	0.92937	5	13.3404
$\beta_7\ L_{III}O_I$	0.93505	5	13.2593			
$\beta_5\ L_{III}O_{IV,V}$	0.92556	3	13.3953	0.8996	2	13.782
L_IM_I	1.0005	9	12.39			
$\beta_{10}\ L_IM_{IV}$	0.90495	4	13.7002			
$\beta_9\ L_IM_V$	0.89791	3	13.8077			
L_IN_I	0.8022	1	15.456			
L_IN_{IV}	0.7795	5	15.904			
$\gamma_{11}\ L_IN_V$	0.77728	5	15.951			
$L_IN_{VI,VII}$	0.7641	5	16.23			
$L_IO_{IV,V}$	0.75791	5	16.358			
$L_{II}M_{II}$	1.0346	9	11.98			
$\beta_{17}\ L_{II}M_{III}$	0.98913	5	12.5344			
$L_{II}M_V$	0.94419	5	13.1310			
$L_{II}N_{II}$	0.8344	9	14.86			
$L_{II}N_{III}$	0.8248	1	15.031			
$v\ L_{II}N_{VI}$	0.79721	9	15.552			
$L_{II}O_{III}$	0.79384	5	15.6178			
$t\ L_{III}M_{II}$	1.2748	1	9.7252			
$s\ L_{III}M_{III}$	1.2105	1	10.2421			
$L_{III}N_{II}$	0.98280	5	12.6151			
$L_{III}N_{III}$	0.97321	5	12.7394			
$u\ L_{III}N_{VI,VII}$	0.93505	5	13.2593			
$L_{III}O_{II}$	0.9323	2	13.298			
$L_{III}O_{III}$	0.9302	2	13.328			
$L_{III}P_{II,III}$	0.92413	4	13.4159			
M_IN_{II}	3.892	9	3.185			
M_IN_{III}	3.740	9	3.315			
$M_{II}N_{IV}$	3.834	4	3.234			
$M_{III}N_I$	5.537	8	2.239			
$M_{III}N_{IV}$	4.571	5	2.712			
$\gamma\ M_{III}N_V$	4.532	2	2.735			
$M_{III}O_I$	4.105	9	3.021			
$M_{III}O_{IV,V}$	3.932	6	3.153			
$\zeta_2\ M_{IV}N_{II}$	6.585	5	1.883			
$M_{IV}N_{III}$	6.162	8	2.012			
$\beta\ M_{IV}N_{VI}$	4.909	1	2.5255			
$M_{IV}O_{II}$	4.823	3	2.571			
$M_{IV}P_{II,III}$	4.59	2	2.70			
$\zeta_1\ M_VN_{III}$	6.521	4	1.901			
$\alpha_2\ M_VN_{VI}$	5.130	2	2.4170			
$\alpha_1\ M_VN_{VII}$	5.118	1	2.4226			
$N_IP_{II,III}$	13.30	6	0.932			
$N_{VI}O_{IV}$	91.6	1	0.1354			
$N_{VII}O_V$	93.2	1	0.1330			

Designation	Å*	p.e.	keV	Å*	p.e.	keV
	85 Astatine			**86 Radon**		
$\alpha_2\ KL_{II}$	0.15705†	2	78.95	0.15294†	3	81.07
$\alpha_1\ KL_{III}$	0.15210†	2	81.52	0.14798†	3	83.78
$\beta_3\ KM_{II}$	0.13517†	4	91.72	0.13155†	5	94.24
$\beta_1\ KM_{III}$	0.13432†	4	92.30	0.13069†	5	94.87
$\beta_2^{II}\ KN_{II}$	0.13072†	4	94.84	0.12719†	5	97.47
$\beta_2^{I}\ KN_{III}$	0.13052†	4	94.99	0.12698†	5	97.64
$\beta_3\ L_IM_{III}$	0.88135†	9	14.067	0.85436†	9	14.512
$\beta_1\ L_{II}M_{IV}$	0.89349†	9	13.876	0.86605†	9	14.316
$\gamma_1\ L_{II}N_{IV}$	0.76289†	9	16.251	0.73928†	9	16.770
$\alpha_2\ L_{III}M_{IV}$	1.09671†	5	11.3048	1.06899†	5	11.5979
$\alpha_1\ L_{III}M_V$	1.08500†	5	11.4268	1.05723†	5	11.7270
	87 Francium			**88 Radium**		
$\alpha_2\ KL_{II}$	0.14896†	3	83.23	0.14512†	2	85.43
$\alpha_1\ KL_{III}$	0.14399†	3	86.10	0.14014†	2	88.47
$\beta_3\ KM_{II}$	0.12807†	5	96.81	0.12469†	3	99.43
$\beta_1\ KM_{III}$	0.12719†	5	97.47	0.12382†	3	100.13
$\beta_2^{II}\ KN_{II}$	0.12379†	5	100.16	0.12050†	3	102.89
$\beta_2^{I}\ KN_{III}$	0.12358†	5	100.33	0.12029†	3	103.07
$\beta_4\ L_IM_{II}$				0.84071	5	14.7472
$\beta_3\ L_IM_{III}$	0.82789†	9	14.976	0.80273	5	15.4449
$\gamma_2\ L_IN_{II}$				0.68199	5	18.179
$\gamma_3\ L_IN_{III}$				0.67538	5	18.357
$\gamma'_4\ L_IO_{II}$				0.65131	5	19.036
$\gamma_4\ L_IO_{III}$				0.64965	5	19.084
$\gamma_{13}\ L_IP_{II,III}$				0.64513	5	19.218
$\eta\ L_{II}M_I$				0.90742	5	13.6630
$\beta_1\ L_{II}M_{IV}$	0.83940†	9	14.770	0.81375	5	15.2358
$\gamma_5\ L_{II}N_I$				0.71774	5	17.274
$\gamma_1\ L_{II}N_{IV}$	0.71652†	9	17.303	0.69463	5	17.849
γ_8				0.6801	1	18.230
$\gamma_6\ L_{II}O_{IV}$				0.67328	5	18.414
$L_{II}P_I$				0.6724	1	18.439
$l\ L_{III}M_I$				1.16719	5	10.6222
$\alpha_2\ L_{III}M_{IV}$	1.04230	5	11.8950	1.01656	5	12.1962
$\alpha_1\ L_{III}M_V$	1.03049	5	12.0313	1.00473	5	12.3397
$\beta_6\ L_{III}N_I$				0.87088	5	14.2362
$\beta_{15}\ L_{III}N_{IV}$				0.83722	5	14.8086
$\beta_2\ L_{III}N_V$	0.858	2	14.45	0.83537	5	14.8414
$\beta_7\ L_{III}O_I$				0.8162	1	15.190
$\beta_5\ L_{III}O_{IV,V}$				0.80627	5	15.3771
$L_{III}P_I$				0.8050	1	15.402
$\beta_{10}\ L_IM_{IV}$				0.77546	5	15.988
$\beta_9\ L_IM_V$				0.76857	5	16.131
L_IN_I				0.6874	1	18.036
L_IN_{IV}				0.6666	1	18.600
$\gamma_{11}\ L_IN_V$				0.6654	1	18.633
$L_IO_{IV,V}$				0.6468	1	19.167
$\beta_{17}\ L_{II}M_{III}$				0.8438	1	14.692
$L_{II}N_{III}$				0.7043	1	17.604
$L_{II}N_V$				0.6932	1	17.884
$L_{II}O_{II}$				0.6780	1	18.286

Designation	Å*	p.e.	keV	Å*	p.e.	keV
	87 Francium (*Cont.*)			**88 Radium** (*Cont.*)		
$L_{II}O_{III}$				0.6764	1	18.330
$L_{II}P_{II,III}$				0.6714	1	18.466
$L_{III}N_{II}$				0.8618	1	14.387
$L_{III}N_{III}$				0.8512	1	14.566
u $L_{III}N_{VI,VII}$				0.8186	1	15.146
$L_{III}P_{II,III}$				0.8038	1	15.425
	89 Actinium			**90 Thorium**		
α_2 KL_{II}	0.14141†	2	87.67	0.137829	2	89.953
α_1 KL_{III}	0.136417†	8	90.884	0.132813	2	93.350
β_3 KM_{II}	0.12143†	2	102.10	0.118268	3	104.831
β_1 KM_{III}	0.12055†	2	102.85	0.117396	9	105.609
β_2^{II} KN_{II}	0.11732†	2	105.67	0.11426	1	108.511
β_2^{I} KN_{III}	0.11711†	2	105.86	0.114040	9	108.717
$KO_{II,III}$				0.11322	1	109.500
β_5 $KM_{IV,V}$				0.116667	9	106.269
β_4 $KN_{IV,V}$				0.11366	2	109.08
β_4 L_IM_{II}				0.79257	4	15.6429
β_3 L_IM_{III}	0.77822†	9	15.931	0.75479	3	16.4258
γ_2 L_IN_{II}				0.64221	4	19.305
γ_3 L_IN_{III}				0.63559	4	19.507
γ'_4 L_IO_{II}				0.61251	4	20.242
γ_4 L_IO_{III}				0.61098	4	20.292
γ_{13} $L_IP_{II,III}$				0.60705	8	20.424
η $L_{II}M_I$				0.85446	4	14.5099
β_1 $L_{II}M_{IV}$	0.78903†	9	15.713	0.765210	9	16.2022
γ_5 $L_{II}N_I$				0.67491	4	18.370
γ_1 $L_{II}N_{IV}$	0.67351†	9	18.408	0.65313	3	18.9825
γ_8 $L_{II}O_I$				0.63898	5	19.403
γ_6 $L_{II}O_{IV}$				0.63258	4	19.599
$L_{II}P_I$				0.6316	1	19.629
$L_{II}P_{IV}$				0.62991	9	19.682
l $L_{III}M_I$				1.11508	4	11.1186
α_2 $L_{III}M_{IV}$	0.99178†	5	12.5008	0.96788	2	12.8096
α_1 $L_{III}M_V$	0.97993†	5	12.6520	0.95600	3	12.9687
β_6 $L_{III}N_I$				0.82790	8	14.975
β_{15} $L_{III}N_{IV}$				0.79539	5	15.5875
β_2 $L_{III}N_V$				0.79354	3	15.6237
β_7 $L_{III}O_I$				0.77437	4	16.0105
β_5 $L_{III}O_{IV,V}$				0.76468	5	16.213
$L_{III}P_I$				0.76338	5	16.241
$L_{III}P_{IV,V}$				0.76087	9	16.295
β_{10} L_IM_{IV}				0.7301	1	16.981
β_9 L_IM_V				0.7234	1	17.139
L_IN_I				0.64755	5	19.146
L_IN_{IV}				0.6276	1	19.755
γ_{11} L_IN_V				0.62636	9	19.794
$L_IN_{VI,VII}$				0.6160	1	20.128
L_IO_I				0.6146	1	20.174
$L_IO_{IV,V}$				0.6083	1	20.383
$L_{II}M_{II}$				0.8338	1	14.869
β_{17} $L_{II}M_{III}$				0.79257	4	15.6429
$L_{II}M_V$				0.7579	1	16.359
$L_{II}N_{III}$				0.6620	1	18.729
$L_{II}N_V$				0.6521	1	19.014

Designation	Å*	p.e.	keV	Å*	p.e.	keV
	89 Actinium (*Cont.*)			**90 Thorium** (*Cont.*)		
v $L_{II}N_{VI}$				0.64064	9	19.353
$L_{II}O_{II}$				0.6369	1	19.466
$L_{II}O_{III}$				0.6356	1	19.506
$L_{II}P_{II,III}$				0.6312	1	19.642
t $L_{III}M_I$				1.08009	9	11.4788
s $L_{III}M_{II}$				1.0112	1	12.261
$L_{III}N_{II}$				0.8190	2	15.138
$L_{III}N_{III}$				0.8082	1	15.341
u $L_{III}N_{VI,VII}$				0.77661	5	15.964
$L_{III}O_{II}$				0.7713	1	16.074
$L_{III}O_{III}$				0.7690	1	16.123
$L_{III}P_{II,III}$				0.7625	2	16.260
M_IN_{III}				2.934	8	4.23
M_IO_{III}				2.442	9	5.08
$M_{II}N_I$				3.537	9	3.505
$M_{II}N_{IV}$				3.011	2	4.117
$M_{II}O_{IV}$				2.618	5	4.735
$M_{III}N_I$				4.568	5	2.714
$M_{III}N_{IV}$				3.718	3	3.335
γ $M_{II}N_V$				3.679	2	3.370
$M_{III}O_I$				3.283	9	3.78
$M_{III}O_{IV,V}$				3.131	3	3.959
ζ_2 M_VN_{II}				5.340	5	2.322
$M_{IV}N_{III}$				4.911	5	2.524
β $M_{IV}N_{VI}$				3.941	1	3.1458
$M_{IV}O_{II}$				3.808	4	3.256
ζ_1 M_VN_{III}				5.245	5	2.364
α_2 M_VN_{VI}				4.151	2	2.987
α_1 M_VN_{VII}				4.1381	9	2.9961
M_VP_{III}				3.760	9	3.298
N P_{II}				9.44	7	1.313
N P_{III}				9.40	7	1.1319
$N_{II}O_{IV}$				11.56	5	1.072
N_IP_I				11.07	7	1.120
$N_{III}O_V$				13.8	1	0.897
$N_{IV}N_{VI}$				33.57	9	0.3693
$N_VN_{VI,VII}$				36.32	9	0.3414
$N_{VI}O_{IV}$				49.5	1	0.2505
$N_{VI}O_V$				48.2	1	0.2572
$N_{VII}O_V$				50.0	1	0.2479
$O_{III}P_{IV,V}$				68.2	3	0.1817
$O_{IV,V}Q_{II,III}$				181.	5	0.068
	91 Protactinium			**92 Uranium**		
α_2 KL_{II}	0.134343†	9	92.287	0.130968	4	94.665
α_1 KL_{III}	0.129325†	3	95.868	0.125947	3	98.439
β_3 KM_{II}	0.11523†	2	107.60	0.112296	4	110.406
β_1 KM_{III}	0.114345†	8	108.427	0.111394	5	111.300
β_2^{II} KN_{II}	0.11129†	2	111.40	0.10837	1	114.40
β_2^{I} KN_{III}	0.11107†	2	111.62	0.10818	1	114.60
$KO_{II,III}$				0.10744	1	115.39
β_5 $KM_{IV,V}$				0.11069	1	112.01
β_4 $KN_{IV,V}$				0.10780	2	115.01
β_4 L_IM_I	0.7699	1	16.104	0.747985	9	16.5753
β_3 L_IM_{III}	0.73230	5	16.930	0.71029	2	17.4550

Designation	Å*	p.e.	keV	Å*	p.e.	keV
	91 Protactinium (*Cont.*)			**92 Uranium** (*Cont.*)		
$\gamma_2\ L_IN_{II}$	0.6239	1	19.872	0.605237	9	20.4847
$\gamma_3\ L_IN_{III}$	0.6169	1	20.098	0.598574	9	20.7127
$\gamma'_4\ L_IO_{II}$				0.576700	9	21.4984
$\gamma_4\ L_IO_{II,III}$	0.5937	1	20.882	0.57499	9	21.562
γ_{13}				0.5706	1	21.729
$\eta\ L_{II}M_I$	0.8295	1	14.946	0.80509	2	15.3997
$\beta_1\ L_{II}M_{IV}$	0.74232	5	16.702	0.719984	8	17.2200
$\gamma_5\ L_{II}N_I$	0.6550	1	18.930	0.63557	2	19.5072
$\gamma_1\ L_{II}N_{IV}$	0.63358†	9	19.568	0.614770	9	20.1671
$\gamma_8\ L_{II}O_I$				0.60125	5	20.621
$\gamma_6\ L_{II}O_{IV}$	0.6133	1	20.216	0.594845	9	20.8426
$L_{II}P_{IV}$				0.59203	5	20.942
$l\ L_{III}M_I$	1.0908	1	11.366	1.06712	2	11.6183
$\alpha_2\ L_{III}M_{IV}$	0.94482†	5	13.1222	0.922558	9	13.4388
$\alpha_1\ L_{III}M_V$	0.93284	5	13.2907	0.910639	9	13.6147
$\beta_6\ L_{III}N_I$	0.8079	1	15.347	0.78838	2	15.7260
$\beta_{15}\ L_{III}N_{IV}$				0.756642	9	16.3857
$\beta_2\ L_{III}N_V$	0.7737	1	16.024	0.754681	9	16.4283
$\beta_7\ L_{III}O_I$	0.7546	2	16.431	0.73602	6	16.845
$\beta_5\ L_{III}O_{IV,V}$	0.7452	2	16.636	0.726305	9	17.0701
$L_{III}P_I$				0.72521	5	17.096
$L_{III}P_{IV,V}$				0.72240	5	17.162
$\beta_{10}\ L_IM\ _V$	0.7088	2	17.492	0.68760	5	18.031
$\beta_9\ L_IM_V$	0.7018	1	17.667	0.681014	8	18.2054
L_IN_{IV}				0.59096	5	20.979
$\gamma_{11}\ L_IN_V$				0.58986	5	21.019
$L_IO_{IV,V}$				0.5725	1	21.657
$\beta_{17}\ L_{II}M_{III}$				0.74503	5	16.641
$L_{II}N_{III}$				0.6228	1	19.907
$v\ L_{II}N_{VI}$				0.6031	1	20.556
$L_{III}O_{III}$				0.59728	5	20.758
$L_{II}P_{II,III}$				0.5930	2	20.906
$t\ L_{III}M_{II}$				1.0347	1	11.982
$s\ L_{III}M_{III}$				0.9636	1	12.866
$L_{III}N_{II}$				0.78017	9	15.892
$L_{III}N_{III}$				0.7691	1	16.120
$u\ L_{III}N_{VI,VII}$				0.738603	9	16.7859
$L_{III}O_{II}$				0.7333	1	16.907
$L_{III}O_{III}$				0.7309	1	16.962
$L_{III}P_{II,III}$				0.72426	5	17.118
M_IN_{II}				2.92	2	4.25
M_IN_{III}				2.753	8	4.50
M_IO_{III}				2.304	7	5.38
M_IP_{III}				2.253	6	5.50
$M_{II}N_I$	3.441	5	3.603	3.329	4	3.724
$M_{II}N_{IV}$	2.910	2	4.260	2.817	2	4.401
$M_{II}O_{IV}$	2.527	4	4.906	2.443	4	5.075
$M_{III}N_I$	4.450	4	2.786	4.330	2	2.863
$M_{III}N_{IV}$	3.614	2	3.430	3.521	2	3.521
$\gamma\ M_{III}N_V$	3.577	1	3.4657	3.479	1	3.563
$M_{III}O_I$	3.245	9	3.82	3.115	7	3.980
$M_{III}O_{IV,V}$	3.038	2	4.081	2.948	2	4.205
$\zeta_2\ M_{IV}N_{II}$	5.193	2	2.3876	5.050	2	2.4548
$M_{IV}N_{III}$				4.625	5	2.681
$\beta\ M_{IV}N_{VI}$	3.827	1	3.2397	3.716	1	3.3367

Designation	Å*	p.e.	keV	Å*	p.e.	keV
	91 Protactinium (*Cont.*)			**92 Uranium** (*Cont.*)		
$M_{IV}O_{II}$	3.691	2	3.359	3.576	1	3.4666
$\zeta_1\ M_VN_{III}$	5.092	2	2.4350	4.946	2	2.507
$\alpha_2\ M_VN_{VI}$	4.035	3	3.072	3.924	1	3.1595
$\alpha_1\ M_VN_{VII}$	4.022	1	3.0823	3.910	1	3.1708
N_IO_{III}				10.09	7	1.229
N_IP_{II}				8.81	7	1.41
N_IP_{III}				8.76	7	1.42
$N_{II}P_I$				10.40	7	1.192
$N_{III}O_V$				12.90	9	0.961
$N_{IV}N_{VI}$				31.8	1	0.390
$N_VN_{VI,VII}$				34.8	1	0.357
$N_{IV}O_{IV}$				43.3	2	0.286
$N_{VI}O_V$				42.1	2	0.295
$N_IP_{IV,V}$				8.60	7	1.44
	93 Neptunium			**94 Plutonium**		
$\beta_4\ L_IM_{II}$	0.72671	2	17.0607	0.70620	2	17.5560
$\beta_3\ L_IM_{III}$	0.68920†	9	17.989	0.66871	2	18.5405
$\gamma_2\ L_IN_{II}$	0.5873	5	21.11	0.57068	2	21.7251
$\gamma_3\ L_IN_{III}$	0.5810	5	21.34	0.564001	9	21.9824
$\gamma'_4\ L_IO_{II}$				0.5432	1	22.823
$\gamma_4\ L_IO_{II,III}$	0.5585	5	22.20	0.5416	1	22.891
$\eta\ L_{II}M_I$	0.7809	2	15.876	0.7591	1	16.333
$\beta_1\ L_{II}M_{IV}$	0.698478	9	17.7502	0.67772	2	18.2937
$\gamma_5\ L_{II}N_I$	0.616	1	20.12	0.5988	1	20.704
$\gamma_1\ L_{II}N_{IV}$	0.596498	9	20.7848	0.578882	9	21.4173
γ_8				0.5658	1	21.914
$\gamma_6\ L_{II}O_{IV}$	0.57699	5	21.488	0.55973	2	22.1502
$l\ L_{III}M_I$	1.0428	6	11.890	1.0226	1	12.124
$\alpha_2\ L_{III}M_{IV}$	0.901045	9	13.7597	0.88028	2	14.0842
$\alpha_1\ L_{III}M_V$	0.889128	9	13.9441	0.86830	2	14.2786
$\beta_6\ L_{III}N_I$	0.769	1	16.13	0.75148	2	16.4983
$\beta_{15}\ L_{III}N_{IV}$				0.7205	1	17.208
$\beta_2\ L_{III}N_V$	0.736230	9	16.8400	0.71851	2	17.2553
$\beta_7\ L_{III}O_I$				0.7003	1	17.705
$\beta_5\ L_{III}O_{IV,V}$	0.70814	2	17.5081	0.69068	2	17.9506
$\beta_{10}\ L_IM_{IV}$				0.6482	1	19.126
$\beta_9\ L_IM_V$				0.6416	1	19.323
$u\ L_{III}N_{VI,VII}$				0.7031	1	17.635
	95 Americium					
$\beta_4\ L_IM_{II}$	0.68639	2	18.0627			
$\beta_3\ L_IM_{III}$	0.64891	2	19.1059			
$\gamma_2\ L_IN_{II}$	0.5544	2	22.361			
$\beta_1\ L_{II}M_{IV}$	0.657655	9	18.8520			
$\gamma_1\ L_{II}N_{IV}$	0.561886	9	22.0652			
$\gamma_6\ L_{II}O_{IV}$	0.54311	2	22.8282			
$l\ L_{III}M_I$	1.0012	6	12.384			
$\alpha_2\ L_{III}M_{IV}$	0.860266	9	14.4119			
$\alpha_1\ L_{III}M_V$	0.848187	9	14.6172			
$\beta_6\ L_{III}N_I$	0.73418	2	16.8870			
$\beta_{15}\ L_{III}N_{IV}$	0.70341	2	17.6258			
$\beta_2\ L_{III}N_V$	0.701390	9	17.6765			
$\beta_5\ L_{III}O_{IV,V}$	0.67383	2	18.3996			

TABLE B. Wavelengths in numerical order of the emission lines and absorption edges.

Wavelength Å*	p.e.	Element	Designation		keV
0.10723	1	92 U	K	Abs. Edge	115.62
0.10744	1	92 U		$KO_{II,III}$	115.39
0.10780	2	92 U	$K\beta_4$	$KN_{IV,V}$	115.01
0.10818	1	92 U	$K\beta_2^{I}$	KN_{III}	114.60
0.10837	1	92 U	$K\beta_2^{II}$	KN_{II}	114.40
0.11069	1	92 U	$K\beta_5$	$KM_{IV,V}$	112.01
0.11107	2	91 Pa	$K\beta_2^{I}$	KN_{III}	111.62
0.11129	2	91 Pa	$K\beta_2^{II}$	KN_{II}	111.40
0.111394	5	92 U	$K\beta_1$	KM_{III}	111.300
0.112296	4	92 U	$K\beta_3$	KM_{II}	110.406
0.11307	1	90 Th	K	Abs. Edge	109.646
0.11322	1	90 Th		$KO_{II,III}$	109.500
0.11366	2	90 Th	$K\beta_4$	$KN_{IV,V}$	109.08
0.114040	9	90 Th	$K\beta_2^{I}$	KN_{III}	108.717
0.11426	1	90 Th	$K\beta_2^{II}$	KN_{II}	108.511
0.114345	8	91 Pa	$K\beta_1$	KM_{III}	108.427
0.11523	2	91 Pa	$K\beta_3$	KM_{II}	107.60
0.116667	9	90 Th	$K\beta_5$	$KM_{IV,V}$	106.269
0.11711	2	89 Ac	$K\beta_2^{I}$	KN_{III}	105.86
0.11732	2	89 Ac	$K\beta_2^{II}$	KN_{II}	105.67
0.117396	9	90 Th	$K\beta_1$	KM_{III}	105.609
0.118268	3	90 Th	$K\beta_3$	KM_{II}	104.831
0.12029	3	88 Ra	$K\beta_2^{I}$	KN_{III}	103.07
0.12050	3	88 Ra	$K\beta_2^{II}$	KN_{II}	102.89
0.12055	2	89 Ac	$K\beta_1$	KM_{III}	102.85
0.12143	2	89 Ac	$K\beta_3$	KM_{II}	102.10
0.12358	5	87 Fr	$K\beta_2^{I}$	KN_{III}	100.33
0.12379	5	87 Fr	$K\beta_2^{II}$	KN_{II}	100.16
0.12382	3	88 Ra	$K\beta_1$	KM_{III}	100.13
0.12469	3	88 Ra	$K\beta_3$	KM_{II}	99.43
0.125947	3	92 U	$K\alpha_1$	KL_{III}	98.439
0.12698	5	86 Rn	$K\beta_2^{I}$	KN_{III}	97.64
0.12719	5	86 Rn	$K\beta_2^{II}$	KN_{II}	97.47
0.12719	5	87 Fr	$K\beta_1$	KM_{III}	97.47
0.12807	5	87 Fr	$K\beta_3$	KM_{II}	96.81
0.129325	3	91 Pa	$K\alpha_1$	KL_{III}	95.868
0.13052	4	85 At	$K\beta_2^{I}$	KN_{III}	94.99
0.13069	5	86 Rn	$K\beta_1$	KM_{III}	94.87
0.13072	4	85 At	$K\beta_2^{II}$	KN_{II}	94.84
0.130968	4	92 U	$K\alpha_2$	KL_{II}	94.665
0.13155	5	86 Rn	$K\beta_3$	KM_{II}	94.24
0.132813	2	90 Th	$K\alpha_1$	KL_{III}	93.350
0.13418	2	84 Po	$K\beta_2^{I}$	KN_{III}	92.40
0.13432	4	85 At	$K\beta_1$	KM_{III}	92.30
0.134343	9	91 Pa	$K\alpha_2$	KL_{II}	92.287
0.13438	2	84 Po	$K\beta_2^{II}$	KN_{II}	92.26
0.13517	4	85 At	$K\beta_3$	KM_{II}	91.72
0.136417	8	89 Ac	$K\alpha_1$	KL_{III}	90.884
0.13694	1	83 Bi	K	Abs. Edge	90.534
0.13709	1	83 Bi		$KO_{II,III}$	90.435
0.13759	2	83 Bi	$K\beta_4$	$KN_{IV,V}$	90.11
0.137829	2	90 Th	$K\alpha_2$	KL_{II}	89.953
0.13797	1	83 Bi	$K\beta_2^{I}$	KN_{III}	89.864
0.13807	2	84 Po	$K\beta_1$	KM_{III}	89.80
0.13817	1	83 Bi	$K\beta_2^{II}$	KN_{II}	89.733
0.13892	2	84 Po	$K\beta_3$	KM_{II}	89.25
0.14014	2	88 Ra	$K\alpha_1$	KL_{III}	88.47
0.1408	1	82 Pb		KP	88.06
0.140880	5	82 Pb	K	Abs. Edge	88.005
0.141012	8	82 Pb		$KO_{II,III}$	87.922
0.14111	1	83 Bi	$K\beta_5$	$KM_{IV,V}$	87.860
0.14141	2	89 Ac	$K\alpha_2$	KL_{II}	87.67
0.14155	3	82 Pb	$K\beta_4$	$KN_{IV,V}$	87.59
0.14191	1	82 Pb	$K\beta_2^{I}$	KN_{III}	87.364
0.141948	3	83 Bi	$K\beta_1$	KM_{III}	87.343
0.14212	2	82 Pb	$K\beta_2^{II}$	KN_{II}	87.23
0.142779	7	83 Bi	$K\beta_3$	KM_{II}	86.834
0.14399	3	87 Fr	$K\alpha_1$	KL_{III}	86.10
0.14495	1	81 Tl	K	Abs. Edge	85.533
0.14495	3	82 Pb	$K\beta_5^{I}$	KM_{V}	85.53
0.14509	1	81 Tl		$KO_{II,III}$	85.451
0.14512	2	82 Pb	$K\beta_5^{II}$	KM_{IV}	85.43
0.14512	2	88 Ra	$K\alpha_2$	KL_{II}	85.43
0.14553	2	81 Tl	$K\beta_4$	$KN_{IV,V}$	85.19
0.14595	1	81 Tl	$K\beta_2^{I}$	KN_{III}	84.946
0.145970	6	82 Pb	$K\beta_1$	KM_{III}	84.936
0.14614	1	81 Tl	$K\beta_2^{II}$	KN_{II}	84.836
0.146810	4	82 Pb	$K\beta_3$	KM_{II}	84.450
0.14798	3	86 Rn	$K\alpha_1$	KL_{III}	83.78
0.14896	3	87 Fr	$K\alpha_2$	KL_{II}	83.23
0.14917	1	81 Tl	$K\beta_5$	$KM_{IV,V}$	83.114
0.14918	1	80 Hg	K	Abs. Edge	83.109
0.14931	2	80 Hg		$KO_{II,III}$	83.04
0.14978	2	80 Hg	$K\beta_4$	$KN_{IV,V}$	82.78
0.150142	5	81 Tl	$K\beta_1$	KM_{III}	82.576
0.15020	2	80 Hg	$K\beta_2^{I}$	KN_{III}	82.54
0.15040	2	80 Hg	$K\beta_2^{II}$	KN_{II}	82.43
0.150980	6	81 Tl	$K\beta_3$	KM_{II}	82.118
0.15210	2	85 At	$K\alpha_1$	KL_{III}	81.52
0.15294	3	86 Rn	$K\alpha_2$	KL_{II}	81.07
0.15353	2	80 Hg	$K\beta_5$	$KM_{IV,V}$	80.75
0.153593	5	79 Au	K	Abs. Edge	80.720
0.153694	7	79 Au		$KO_{II,III}$	80.667
0.154224	5	79 Au	$K\beta_4$	$KN_{IV,V}$	80.391
0.154487	3	80 Hg	$K\beta_1$	KM_{III}	80.253
0.154618	9	79 Au	$K\beta_2^{I}$	KN_{III}	80.185
0.15483	2	79 Au	$K\beta_2^{II}$	KN_{II}	80.08
0.155321	3	80 Hg	$K\beta_3$	KM_{II}	79.822
0.15636	1	84 Po	$K\alpha_1$	KL_{III}	79.290
0.15705	2	85 At	$K\alpha_2$	KL_{II}	78.95
0.157880	5	79 Au	$K\beta_5^{I}$	KM_{V}	78.529
0.158062	7	79 Au	$K\beta_5^{II}$	KM_{IV}	78.438
0.15818	1	78 Pt	K	Abs. Edge	78.381
0.15826	1	78 Pt		$KO_{II,III}$	78.341
0.15881	2	78 Pt	$K\beta_4$	$KN_{IV,V}$	78.069
0.158982	3	79 Au	$K\beta_1$	KM_{III}	77.984
0.15920	1	78 Pt	$K\beta_2^{I}$	KN_{III}	77.878
0.15939	1	78 Pt	$K\beta_2^{II}$	KN_{II}	77.785
0.159810	2	79 Au	$K\beta_3$	KM_{II}	77.580
0.160789	2	83 Bi	$K\alpha_1$	KL_{III}	77.1079
0.16130	1	84 Po	$K\alpha_2$	KL_{II}	76.862
0.16255	3	78 Pt	$K\beta_5^{I}$	KM_{V}	76.27
0.16271	2	78 Pt	$K\beta_5^{II}$	KM_{IV}	76.199
0.16292	1	77 Ir	K	Abs. Edge	76.101

Wavelength Å*	p.e.	Element	Designation		keV
0.163019	5	77 Ir		$KO_{II,III}$	76.053
0.16352	2	77 Ir	$K\beta_4$	$KN_{IV,V}$	75.821
0.163675	3	78 Pt	$K\beta_1$	KM_{III}	75.748
0.163956	7	77 Ir	$K\beta_2{}^{I}$	KN_{III}	75.619
0.16415	1	77 Ir	$K\beta_2{}^{II}$	KN_{II}	75.529
0.164501	3	78 Pt	$K\beta_3$	KM_{II}	75.368
0.165376	2	82 Pb	$K\alpha_1$	KL_{III}	74.9694
0.165717	2	83 Bi	$K\alpha_2$	KL_{II}	74.8148
0.167373	9	77 Ir	$K\beta_5{}^{I}$	KM_{V}	74.075
0.16759	2	77 Ir	$K\beta_5{}^{II}$	KM_{IV}	73.980
0.16787	1	76 Os	K	Abs. Edge	73.856
0.16798	1	76 Os		$KO_{II,III}$	73.808
0.16842	2	76 Os	$K\beta_4$	$KN_{IV,V}$	73.615
0.168542	2	77 Ir	$K\beta_1$	KM_{III}	73.5608
0.168906	6	76 Os	$K\beta_2{}^{I}$	KN_{III}	73.402
0.16910	1	76 Os	$K\beta_2{}^{II}$	KN_{II}	73.318
0.169367	2	77 Ir	$K\beta_3$	KM_{II}	73.2027
0.170136	2	81 Tl	$K\alpha_1$	KL_{III}	72.8715
0.170294	2	82 Pb	$K\alpha_2$	KL_{II}	72.8042
0.17245	1	76 Os	$K\beta_5{}^{I}$	KM_{V}	71.895
0.17262	1	76 Os	$K\beta_5{}^{II}$	KM_{IV}	71.824
0.17302	1	75 Re	K	Abs. Edge	71.658
0.17308	1	75 Re		$KO_{II,III}$	71.633
0.173611	3	76 Os	$K\beta_1$	KM_{III}	71.413
0.17362	2	75 Re	$K\beta_4$	$KN_{IV,V}$	71.410
0.174054	6	75 Re	$K\beta_2{}^{I}$	KN_{III}	71.232
0.17425	1	75 Re	$K\beta_2{}^{II}$	KN_{II}	71.151
0.174431	3	76 Os	$K\beta_3$	KM_{II}	71.077
0.175036	2	81 Tl	$K\alpha_2$	KL_{II}	70.8319
0.175068	3	80 Hg	$K\alpha_1$	KL_{III}	70.819
0.17766	1	75 Re	$K\beta_5{}^{I}$	KM_{V}	69.786
0.17783	1	75 Re	$K\beta_5{}^{II}$	KM_{IV}	69.719
0.17837	1	74 W	K	Abs. Edge	69.508
0.178444	5	74 W		$KO_{II,III}$	69.479
0.178880	3	75 Re	$K\beta_1$	KM_{III}	69.310
0.17892	2	74 W	$K\beta_4$	$KN_{IV,V}$	69.294
0.179421	7	74 W	$K\beta_2{}^{I}$	KN_{III}	69.101
0.17960	1	74 W	$K\beta_2{}^{II}$	KN_{II}	69.031
0.179697	3	75 Re	$K\beta_3$	KM_{II}	68.994
0.179958	3	80 Hg	$K\alpha_2$	KL_{II}	68.895
0.180195	2	79 Au	$K\alpha_1$	KL_{III}	68.8037
0.183092	7	74 W	$K\beta_5{}^{I}$	KM_{V}	67.715
0.183264	5	74 W	$K\beta_5{}^{II}$	KM_{IV}	67.652
0.18394	1	73 Ta	K	Abs. Edge	67.403
0.184031	7	73 Ta		$KO_{II,III}$	67.370
0.184374	2	74 W	$K\beta_1$	KM_{III}	67.2443
0.18451	1	73 Ta	$K\beta_4$	$KN_{IV,V}$	67.194
0.185011	8	73 Ta	$K\beta_2{}^{I}$	KN_{III}	67.013
0.185075	2	79 Au	$K\alpha_2$	KL_{II}	66.9895
0.185181	2	74 W	$K\beta_3$	KM_{II}	66.9514
0.185188	9	73 Ta	$K\beta_2{}^{II}$	KN_{II}	66.949
0.185511	4	78 Pt	$K\alpha_1$	KL_{III}	66.832
0.18672	4	79 Au		KL_{I}	66.40
0.188757	6	73 Ta	$K\beta_5{}^{I}$	KM_{V}	65.683
0.188920	6	73 Ta	$K\beta_5{}^{II}$	KM_{IV}	65.626
0.18982	5	72 Hf	K	Abs. Edge	65.31
0.190089	4	73 Ta	$K\beta_1$	KM_{III}	65.223
0.190381	4	78 Pt	$K\alpha_2$	KL_{II}	65.122
0.1908	2	72 Hf	$K\beta_2$	$KN_{II,III}$	64.98
0.190890	2	73 Ta	$K\beta_3$	KM_{II}	64.9488
0.191047	2	77 Ir	$K\alpha_1$	KL_{III}	64.8956
0.19585	5	71 Lu	K	Abs. Edge	63.31
0.19589	2	71 Lu		$KO_{II,III}$	63.293
0.195904	2	77 Ir	$K\alpha_2$	KL_{II}	63.2867
0.19607	3	72 Hf	$K\beta_1$	KM_{III}	63.234
0.196794	2	76 Os	$K\alpha_1$	KL_{III}	63.0005
0.19686	4	72 Hf	$K\beta_3$	KM_{II}	62.98
0.1969	2	71 Lu	$K\beta_2$	$KN_{II,III}$	62.97
0.20084	2	71 Lu	$K\beta_5$	$KM_{IV,V}$	61.732
0.201639	2	76 Os	$K\alpha_2$	KL_{II}	61.4867
0.20224	5	70 Yb	K	Abs. Edge	61.30
0.20226	2	70 Yb		$KO_{II,III}$	61.298
0.20231	3	71 Lu	$K\beta_1$	KM_{III}	61.283
0.202781	2	75 Re	$K\alpha_1$	KL_{III}	61.1403
0.20309	4	71 Lu	$K\beta_3$	KM_{II}	61.05
0.2033	2	70 Yb	$K\beta_2$	$KN_{II,III}$	60.89
0.20739	2	70 Yb	$K\beta_5$	$KM_{IV,V}$	59.782
0.207611	1	75 Re	$K\alpha_2$	KL_{II}	59.7179
0.20880	5	69 Tm	K	Abs. Edge	59.38
0.20884	8	70 Yb	$K\beta_1$	KM_{III}	59.37
0.20891	2	69 Tm		$KO_{II,III}$	59.346
0.2090100	Std.	74 W	$K\alpha_1$	KL_{III}	59.31824
0.2096	1	70 Yb	$K\beta_3$	KM_{II}	59.14
0.2098	2	69 Tm	$K\beta_2$	$KN_{II,III}$	59.09
0.213828	2	74 W	$K\alpha_2$	KL_{II}	57.9817
0.21404	2	69 Tm	$K\beta_5$	$KM_{IV,V}$	57.923
0.215497	4	73 Ta	$K\alpha_1$	KL_{III}	57.532
0.21556	2	69 Tm	$K\beta_1$	KM_{III}	57.517
0.21567	1	68 Er	K	Abs. Edge	57.487
0.21581	3	68 Er		$KO_{II,III}$	57.450
0.21592	4	74 W		KL_{I}	57.42
0.21636	2	69 Tm	$K\beta_3$	KM_{II}	57.304
0.2167	2	68 Er	$K\beta_2$	$KN_{II,III}$	57.21
0.220305	8	73 Ta	$K\alpha_2$	KL_{II}	56.277
0.22124	3	68 Er	$K\beta_5$	$KM_{IV,V}$	56.040
0.222227	3	72 Hf	$K\alpha_1$	KL_{III}	55.7902
0.22266	2	68 Er	$K\beta_1$	KM_{III}	55.681
0.22291	1	67 Ho	K	Abs. Edge	55.619
0.22305	3	67 Ho		$KO_{II,III}$	55.584
0.22341	2	68 Er	$K\beta_3$	KM_{II}	55.494
0.2241	2	67 Ho	$K\beta_2$	$KN_{II,III}$	55.32
0.227024	3	72 Hf	$K\alpha_2$	KL_{II}	54.6114
0.22855	3	67 Ho	$K\beta_5$	$KM_{IV,V}$	54.246
0.229298	2	71 Lu	$K\alpha_1$	KL_{III}	54.0698
0.23012	2	67 Ho	$K\beta_1$	KM_{III}	53.877
0.23048	1	66 Dy	K	Abs. Edge	53.793
0.23056	3	66 Dy		$KO_{II,III}$	53.774
0.23083	2	67 Ho	$K\beta_3$	KM_{II}	53.711
0.2317	2	66 Dy	$K\beta_2$	$KN_{II,III}$	53.47
0.234081	2	71 Lu	$K\alpha_2$	KL_{II}	52.9650
0.23618	3	66 Dy	$K\beta_5$	$KM_{IV,V}$	52.494
0.236655	2	70 Yb	$K\alpha_1$	KL_{III}	52.3889
0.23788	2	66 Dy	$K\beta_1$	KM_{III}	52.119
0.23841	1	65 Tb	K	Abs. Edge	52.002

Wavelength Å*	p.e.	Element		Designation	keV
0.23858	3	65 Tb		$KO_{II,III}$	51.965
0.23862	2	66 Dy	$K\beta_3$	KM_{II}	51.957
0.2397	2	65 Tb	$K\beta_2$	$KN_{II,III}$	51.68
0.241424	2	70 Yb	$K\alpha_2$	KL_{II}	51.3540
0.244338	2	69 Tm	$K\alpha_1$	KL_{III}	50.7416
0.24608	2	65 Tb	$K\beta_1$	KM_{III}	50.382
0.24681	1	64 Gd	K	Abs. Edge	50.233
0.24683	2	65 Tb	$K\beta_3$	KM_{II}	50.229
0.24687	3	64 Gd		$KO_{II,III}$	50.221
0.24816	3	64 Gd	$K\beta_2$	$KN_{II,III}$	49.959
0.249095	2	69 Tm	$K\alpha_2$	KL_{II}	49.7726
0.252365	2	68 Er	$K\alpha_1$	KL_{III}	49.1277
0.25275	3	64 Gd	$K\beta_5$	$KM_{IV,V}$	49.052
0.25460	2	64 Gd	$K\beta_1$	KM_{III}	48.697
0.25534	2	64 Gd	$K\beta_3$	KM_{II}	48.555
0.25553	1	63 Eu	K	Abs. Edge	48.519
0.255645	7	63 Eu		$KO_{II,III}$	48.497
0.256923	8	63 Eu	$K\beta_2^{I}$	$KN_{II,III}$	48.256
0.257110	2	68 Er	$K\alpha_2$	KL_{II}	48.2211
0.260756	2	67 Ho	$K\alpha_1$	KL_{III}	47.5467
0.263577	5	63 Eu	$K\beta_1$	KM_{III}	47.0379
0.264332	5	63 Eu	$K\beta_3$	KM_{II}	46.9036
0.26464	5	62 Sm	K	Abs. Edge	46.849
0.26491	3	62 Sm		$KO_{II,III}$	46.801
0.265486	2	67 Ho	$K\alpha_2$	KL_{II}	46.6997
0.2662	1	62 Sm	$K\beta_2$	$KN_{II,III}$	46.57
0.269533	2	66 Dy	$K\alpha_1$	KL_{III}	45.9984
0.27111	3	62 Sm	$K\beta_5$	$KM_{IV,V}$	45.731
0.27301	2	62 Sm	$K\beta_1$	KM_{III}	45.413
0.27376	2	62 Sm	$K\beta_3$	KM_{II}	45.289
0.274247	2	66 Dy	$K\alpha_2$	KL_{II}	45.2078
0.27431	5	61 Pm	K	Abs. Edge	45.198
0.2759	1	61 Pm	$K\beta_2$	$KN_{II,III}$	44.93
0.278724	2	65 Tb	$K\alpha_1$	KL_{III}	44.4816
0.28290	3	61 Pm	$K\beta_1$	KM_{III}	43.826
0.283423	2	65 Tb	$K\alpha_2$	KL_{II}	43.7441
0.28363	4	61 Pm	$K\beta_3$	KM_{II}	43.713
0.28453	5	60 Nd	K	Abs. Edge	43.574
0.2861	1	60 Nd	$K\beta_2$	$KN_{II,III}$	43.32
0.288353	2	64 Gd	$K\alpha_1$	KL_{III}	42.9962
0.293038	2	64 Gd	$K\alpha_2$	KL_{II}	42.3089
0.293299	2	60 Nd	$K\beta_1$	KM_{III}	42.2713
0.294027	3	60 Nd	$K\beta_3$	KM_{II}	42.1665
0.29518	5	59 Pr	K	Abs. Edge	42.002
0.29679	2	59 Pr	$K\beta_2$	$KN_{II,III}$	41.773
0.298446	2	63 Eu	$K\alpha_1$	KL_{III}	41.5422
0.303118	2	63 Eu	$K\alpha_2$	KL_{II}	40.9019
0.304261	4	59 Pr	$K\beta_1$	KM_{III}	40.7482
0.304975	5	59 Pr	$K\beta_3$	KM_{II}	40.6529
0.30648	5	58 Ce	K	Abs. Edge	40.453
0.30668	2	58 Ce		$KO_{II,III}$	40.427
0.30737	2	58 Ce	$K\beta_4^{I}$	$KN_{IV,V}$	40.337
0.30816	1	58 Ce	$K\beta_2$	$KN_{II,III}$	40.233
0.309040	2	62 Sm	$K\alpha_1$	KL_{III}	40.1181
0.31342	2	58 Ce	$K\beta_5^{I}$	KM_{V}	39.558
0.31357	2	58 Ce	$K\beta_5^{II}$	KM_{IV}	39.539
0.313698	2	62 Sm	$K\alpha_2$	KL_{II}	39.5224
0.315816	2	58 Ce	$K\beta_1$	KM_{III}	39.2573
0.316520	4	58 Ce	$K\beta_3$	KM_{II}	39.1701
0.31844	5	57 La	K	Abs. Edge	38.934
0.31864	2	57 La		$KO_{II,III}$	38.909
0.31931	2	57 La	$K\beta_4^{I}$	$KN_{IV,V}$	38.828
0.320117	7	57 La	$K\beta_2$	$KN_{II,III}$	38.7299
0.320160	4	61 Pm	$K\alpha_1$	KL_{III}	38.7247
0.324803	4	61 Pm	$K\alpha_2$	KL_{II}	38.1712
0.32546	2	57 La	$K\beta_5^{I}$	KM_{V}	38.094
0.32563	2	57 La	$K\beta_5^{II}$	KM_{IV}	38.074
0.327983	3	57 La	$K\beta_1$	KM_{III}	37.8010
0.328686	4	57 La	$K\beta_3$	KM_{II}	37.7202
0.33104	1	56 Ba	K	Abs. Edge	37.452
0.33127	2	56 Ba		$KO_{II,III}$	37.426
0.331846	2	60 Nd	$K\alpha_1$	KL_{III}	37.3610
0.33229	2	56 Ba	$K\beta_4^{II}$	KN_{IV}	37.311
0.33277	1	56 Ba	$K\beta_2$	$KN_{II,III}$	37.257
0.336472	2	60 Nd	$K\alpha_2$	KL_{II}	36.8474
0.33814	2	56 Ba	$K\beta_5^{I}$	KM_{V}	36.666
0.33835	2	56 Ba	$K\beta_5^{II}$	KM_{IV}	36.643
0.340811	3	56 Ba	$K\beta_1$	KM_{III}	36.3782
0.341507	4	56 Ba	$K\beta_3$	KM_{II}	36.3040
0.344140	2	59 Pr	$K\alpha_1$	KL_{III}	36.0263
0.34451	1	55 Cs	K	Abs. Edge	35.987
0.34611	2	55 Cs	$K\beta_2$	$KN_{II,III}$	35.822
0.348749	2	59 Pr	$K\alpha_2$	KL_{II}	35.5502
0.354364	7	55 Cs	$K\beta_1$	KM_{III}	34.9869
0.355050	4	55 Cs	$K\beta_3$	KM_{II}	34.9194
0.357092	2	58 Ce	$K\alpha_1$	KL_{III}	34.7197
0.3584	5	54 Xe	K	Abs. Edge	34.59
0.36026	3	54 Xe	$K\beta_2$	$KN_{II,III}$	34.415
0.361683	2	58 Ce	$K\alpha_2$	KL_{II}	34.2789
0.36872	2	54 Xe	$K\beta_1$	KM_{III}	33.624
0.36941	2	54 Xe	$K\beta_3$	KM_{II}	33.562
0.370737	2	57 La	$K\alpha_1$	KL_{III}	33.4418
0.37381	1	53 I	K	Abs. Edge	33.1665
0.37523	2	53 I	$K\beta_2$	$KN_{II,III}$	33.042
0.375313	2	57 La	$K\alpha_2$	KL_{II}	33.0341
0.383905	4	53 I	$K\beta_1$	KM_{III}	32.2947
0.384564	4	53 I	$K\beta_3$	KM_{II}	32.2394
0.385111	4	56 Ba	$K\alpha_1$	KL_{III}	32.1936
0.389668	5	56 Ba	$K\alpha_2$	KL_{II}	31.8171
0.38974	1	52 Te		$KO_{II,III}$	31.8114
0.38974	1	52 Te	K	Abs. Edge	31.8114
0.391102	6	52 Te	$K\beta_2$	$KN_{II,III}$	31.7004
0.399995	5	52 Te	$K\beta_1$	KM_{III}	30.9957
0.400290	4	55 Cs	$K\alpha_1$	KL_{III}	30.9728
0.400659	4	52 Te	$K\beta_3$	KM_{II}	30.9443
0.404835	4	55 Cs	$K\alpha_2$	KL_{II}	30.6251
0.40666	1	51 Sb		$KO_{II,III}$	30.4875
0.40668	1	51 Sb	K	Abs. Edge	30.4860
0.40702	1	51 Sb	$K\beta_4^{I}$	$KN_{IV,V}$	30.4604
0.407973	5	51 Sb	$K\beta_2$	$KN_{II,III}$	30.3895
0.41378	1	51 Sb	$K\beta_5^{I}$	KM_{V}	29.9632
0.41388	1	51 Sb	$K\beta_5^{II}$	KM_{IV}	29.9560
0.41634	2	54 Xe	$K\alpha_1$	KL_{III}	29.779
0.417085	3	51 Sb	$K\beta_1$	KM_{III}	29.7256

Wavelength Å*	p.e.	Element	Designation		keV
0.417737	4	51 Sb	$K\beta_3$	KM_{II}	29.6792
0.42087	2	54 Xe	$K\alpha_2$	KL_{II}	29.458
0.42467	3	50 Sn		$KO_{II,III}$	29.195
0.42467	1	50 Sn	K	Abs. Edge	29.1947
0.42495	3	50 Sn	$K\beta_4^{I}$	$KN_{IV,V}$	29.175
0.425915	8	50 Sn	$K\beta_2$	$KN_{II,III}$	29.1093
0.43175	3	50 Sn	$K\beta_5^{I}$	KM_{V}	28.716
0.43184	3	50 Sn	$K\beta_5^{II}$	KM_{IV}	28.710
0.433318	5	53 I	$K\alpha_1$	KL_{III}	28.6120
0.435236	5	50 Sn	$K\beta_1$	KM_{III}	28.4860
0.435877	5	50 Sn	$K\beta_3$	KM_{II}	28.4440
0.437829	7	53 I	$K\alpha_2$	KL_{II}	28.3172
0.44371	1	49 In	K	Abs. Edge	27.9420
0.44374	3	49 In		$KO_{II,III}$	27.940
0.44393	4	49 In	$K\beta_4^{I}$	$KN_{IV,V}$	27.928
0.44500	1	49 In	$K\beta_2$	$KN_{II,III}$	27.8608
0.45086	2	49 In	$K\beta_5^{I}$	KM_{V}	27.499
0.45098	2	49 In	$K\beta_5^{II}$	KM_{IV}	27.491
0.451295	3	52 Te	$K\alpha_1$	KL_{III}	27.4723
0.454545	4	49 In	$K\beta_1$	KM_{III}	27.2759
0.455181	4	49 In	$K\beta_3$	KM_{II}	27.2377
0.455784	3	52 Te	$K\alpha_2$	KL_{II}	27.2017
0.46407	1	48 Cd	K	Abs. Edge	26.7159
0.465328	7	48 Cd	$K\beta_2$	$KN_{II,III}$	26.6438
0.470354	3	51 Sb	$K\alpha_1$	KL_{III}	26.3591
0.474827	3	51 Sb	$K\alpha_2$	KL_{II}	26.1108
0.475105	6	48 Cd	$K\beta_1$	KM_{III}	26.0955
0.475730	5	48 Cd	$K\beta_3$	KM_{II}	26.0612
0.48589	1	47 Ag	K	Abs. Edge	25.5165
0.4859	9	47 Ag	$K\beta_4$	$KN_{IV,V}$	25.512
0.487032	4	47 Ag	$K\beta_2$	$KN_{II,III}$	25.4564
0.490599	3	50 Sn	$K\alpha_1$	KL_{III}	25.2713
0.49306	2	47 Ag	$K\beta_5$	$KM_{IV,V}$	25.145
0.495053	3	50 Sn	$K\alpha_2$	KL_{II}	25.0440
0.497069	4	47 Ag	$K\beta_1$	KM_{III}	24.9424
0.497685	4	47 Ag	$K\beta_3$	KM_{II}	24.9115
0.5092	1	46 Pd	K	Abs. Edge	24.348
0.5093	2	46 Pd	$K\beta_4$	$KN_{IV,V}$	24.346
0.510228	4	46 Pd	$K\beta_2$	$KN_{II,III}$	24.2991
0.512113	3	49 In	$K\alpha_1$	KL_{III}	24.2097
0.516544	3	49 In	$K\alpha_2$	KL_{II}	24.0020
0.51670	9	46 Pd	$K\beta_5$	$KM_{IV,V}$	23.995
0.520520	4	46 Pd	$K\beta_1$	KM_{III}	23.8187
0.521123	4	46 Pd	$K\beta_3$	KM_{II}	23.7911
0.53395	1	45 Rh	K	Abs. Edge	23.2198
0.53401	9	45 Rh	$K\beta_4^{I}$	$KN_{IV,V}$	23.217
0.535010	3	48 Cd	$K\alpha_1$	KL_{III}	23.1736
0.53503	2	45 Rh	$K\beta_2$	$KN_{II,III}$	23.1728
0.53513	5	45 Rh	$K\beta_2^{II}$	KN_{II}	23.168
0.5365	1	94 Pu	L_I	Abs. Edge	23.109
0.539422	3	48 Cd	$K\alpha_2$	KL_{II}	22.9841
0.54101	9	45 Rh	$K\beta_5^{I}$	KM_{V}	22.917
0.54118	9	45 Rh	$K\beta_5^{II}$	KM_{IV}	22.909
0.5416	1	94 Pu	$L\gamma_4$	L_IO_{III}	22.891
0.54311	2	95 Am	$L\gamma_6$	$L_{II}O_{IV}$	22.8282
0.5432	1	94 Pu	$L\gamma_4'$	L_IO_{II}	22.823
0.545605	4	45 Rh	$K\beta_1$	KM_{III}	22.7236
0.546200	4	45 Rh	$K\beta_3$	KM_{II}	22.6989
0.5544	2	95 Am	$L\gamma_2$	L_IN_{II}	22.361
0.5572	1	94 Pu	L_{II}	Abs. Edge	22.253
0.5585	5	93 Np	$L\gamma_4$	$L_IO_{II,III}$	22.20
0.5594075	6	47 Ag	$K\alpha_1$	KL_{III}	22.16292
0.55973	2	94 Pu	$L\gamma_6$	$L_{II}O_{IV}$	22.1502
0.56051	1	44 Ru	K	Abs. Edge	22.1193
0.56089	9	44 Ru	$K\beta_4$	$KN_{IV,V}$	22.104
0.56166	3	44 Ru	$K\beta_2$	$KN_{II,III}$	22.074
0.561886	9	95 Am	$L\gamma_1$	$L_{II}N_{IV}$	22.0652
0.563798	4	47 Ag	$K\alpha_2$	KL_{II}	21.9903
0.564001	9	94 Pu	$L\gamma_3$	L_IN_{III}	21.9824
0.5658	1	94 Pu	$L\gamma_8$	$L_{II}O_I$	21.914
0.56785	9	44 Ru	$K\beta_5^{I}$	KM_{V}	21.834
0.5680	2	44 Ru	$K\beta_5^{II}$	KM_{IV}	21.829
0.5695	1	92 U	L_I	Abs. Edge	21.771
0.5706	1	92 U	$L\gamma_{13}$	$L_IP_{II,III}$	21.729
0.57068	2	94 Pu	$L\gamma_2$	L_IN_{II}	21.1251
0.572482	4	44 Ru	$K\beta_1$	KM_{III}	21.6568
0.5725	1	92 U		$L_IO_{IV,V}$	21.657
0.573067	4	44 Ru	$K\beta_3$	KM_{II}	21.6346
0.57499	9	92 U	$L\gamma_4$	L_IO_{III}	21.562
0.576700	9	92 U	$L\gamma_4'$	L_IO_{II}	21.4984
0.57699	5	93 Np	$L\gamma_6$	$L_{II}O_{IV}$	21.488
0.578882	9	94 Pu	$L\gamma_1$	$L_{II}N_{IV}$	21.4173
0.5810	5	93 Np	$L\gamma_3$	L_IN_{III}	21.34
0.585448	3	46 Pd	$K\alpha_1$	KL_{III}	21.1771
0.5873	5	93 Np	$L\gamma_2$	L_IN_{II}	21.11
0.58906	1	43 Tc	K	Abs. Edge	21.0473
0.589821	3	46 Pd	$K\alpha_2$	KL_{II}	21.0201
0.58986	5	92 U	$L\gamma_{11}$	L_IN_V	21.019
0.59024	5	43 Tc	$K\beta_2$	$KN_{II,III}$	21.005
0.59096	5	92 U		L_IN_{IV}	20.979
0.5919	1	92 U	L_{II}	Abs. Edge	20.945
0.59203	5	92 U		$L_{II}P_{IV}$	20.942
0.5930	2	92 U		$L_{II}P_{II,III}$	20.906
0.5937	1	91 Pa	$L\gamma_4$	$L_IO_{II,III}$	20.882
0.594845	9	92 U	$L\gamma_6$	$L_{II}O_{IV}$	20.8426
0.596498	9	93 Np	$L\gamma_1$	$L_{II}N_{IV}$	20.7848
0.59728	5	92 U		$L_{II}O_{III}$	20.758
0.598574	9	92 U	$L\gamma_3$	L_IN_{III}	20.7127
0.5988	1	94 Pu	$L\gamma_5$	$L_{II}N_I$	20.704
0.60125	5	92 U	$L\gamma_8$	$L_{II}O_I$	20.621
0.60130	4	43 Tc	$K\beta_1$	KM_{III}	20.619
0.60188	4	43 Tc	$K\beta_3$	KM_{II}	20.599
0.6031	1	92 U	Lv	$L_{II}N_{VI}$	20.556
0.605237	9	92 U	$L\gamma_2$	L_IN_{II}	20.4847
0.6059	1	90 Th	L_I	Abs. Edge	20.464
0.60705	8	90 Th	$L\gamma_{13}$	$L_IP_{II,III}$	20.424
0.6083	1	90 Th		$L_IO_{IV,V}$	20.383
0.61098	4	90 Th	$L\gamma_4$	L_IO_{III}	20.292
0.61251	4	90 Th	$L\gamma_4'$	L_IO_{II}	20.242
0.6133	1	91 Pa	$L\gamma_6$	$L_{II}O_{IV}$	20.216
0.613279	4	45 Rh	$K\alpha_1$	KL_{III}	20.2161
0.6146	1	90 Th		L_IO_I	20.174
0.614770	9	92 U	$L\gamma_1$	$L_{II}N_{IV}$	20.1671
0.6160	1	90 Th		$L_IN_{VI,VII}$	20.128

Wavelength Å*	p.e.	Element		Designation	keV
0.616	1	93 Np	$L\gamma_5$	$L_{II}N_I$	20.12
0.6169	1	91 Pa	$L\gamma_3$	L_IN_{III}	20.098
0.617630	4	45 Rh	$K\alpha_2$	KL_{II}	20.0737
0.61978	1	42 Mo	K	Abs. Edge	20.0039
0.62001	9	42 Mo	$K\beta_4^I$	$KN_{IV,V}$	19.996
0.62099	2	42 Mo	$K\beta_2$	$KN_{II,III}$	19.9652
0.62107	5	42 Mo	$K\beta_2^{II}$	KN_{II}	19.963
0.6228	1	92 U		$L_{II}N_{III}$	19.907
0.6239	1	91 Pa	$L\gamma_2$	L_IN_{II}	19.872
0.62636	9	90 Th	$L\gamma_{11}$	L_IN_V	19.794
0.62692	5	42 No	$K\beta_5^I$	KM_V	19.776
0.62708	5	42 Mo	$K\beta_5^{II}$	KM_{IV}	19.771
0.6276	1	90 Th		L_IN_{IV}	19.755
0.6299	1	90 Th	L_{II}	Abs. Edge	19.683
0.62991	9	90 Th		$L_{II}P_{IV}$	19.682
0.6312	1	90 Th		$L_{II}P_{II,III}$	19.642
0.6316	1	90 Th		$L_{II}P_I$	19.629
0.632288	9	42 Mo	$K\beta_1$	KM_{III}	19.6083
0.63258	4	90 Th	$L\gamma_6$	$L_{II}O_{IV}$	19.599
0.632872	2	42 Mo	$K\beta_3$	KM_{II}	19.5903
0.63358	9	91 Pa	$L\gamma_1$	$L_{II}N_{IV}$	19.568
0.63557	2	92 U	$L\gamma_5$	$L_{II}N_I$	19.5072
0.63559	4	90 Th	$L\gamma_3$	L_IN_{III}	19.507
0.6356	1	90 Th		$L_{II}O_{III}$	19.506
0.6369	1	90 Th		$L_{II}O_{II}$	19.466
0.63898	5	90 Th	$L\gamma_8$	$L_{II}O_I$	19.403
0.64064	9	90 Th	Lv	$L_{II}N_{VI}$	19.353
0.6416	1	94 Pu	$L\beta_9$	L_IM_V	19.323
0.64221	4	90 Th	$L\gamma_2$	L_IN_{II}	19.305
0.643083	4	44 Ru	$K\alpha_1$	KL_{III}	19.2792
0.6445	1	88 Ra	L_I	Abs. Edge	19.236
0.64513	5	88 Ra	$L\gamma_{13}$	$L_IP_{II,III}$	19.218
0.6468	1	88 Ra		$L_IO_{IV,V}$	19.167
0.647408	5	44 Ru	$K\alpha_2$	KL_{II}	19.1504
0.64755	5	90 Th		L_IN_I	19.146
0.6482	1	94 Pu	$L\beta_{10}$	L_IM_{IV}	19.126
0.64891	2	95 Am	$L\beta_3$	L_IM_{III}	19.1059
0.64965	5	88 Ra	$L\gamma_4$	L_IO_{III}	19.084
0.65131	5	88 Ra	$L\gamma_4'$	L_IO_{II}	19.036
0.6521	1	90 Th		$L_{II}N_V$	19.014
0.65298	1	41 Nb	K	Abs. Edge	18.9869
0.65313	3	90 Th	$L\gamma_1$	$L_{II}N_{IV}$	18.9825
0.65318	5	41 Nb	$K\beta_4$	$KN_{IV,V}$	18.981
0.65416	4	41 Nb	$K\beta_2$	$KN_{II,III}$	18.953
0.6550	1	91 Pa	$L\gamma_5$	$L_{II}N_I$	18.930
0.657655	9	95 Am	$L\beta_1$	$L_{II}M_{IV}$	18.8520
0.6620	1	90 Th		$L_{II}N_{III}$	18.729
0.6654	1	88 Ra	$L\gamma_{11}$	L_IN_V	18.633
0.66576	2	41 Nb	$K\beta_1$	KM_{III}	18.6225
0.66634	3	41 Nb	$K\beta_3$	KM_{II}	18.6063
0.6666	1	88 Ra		L_IN_{IV}	18.600
0.66871	2	94 Pu	$L\beta_3$	L_IM_{III}	18.5405
0.6707	1	88 Ra	L_{II}	Abs. Edge	18.486
0.6714	1	88 Ra		$L_{II}P_{II,III}$	18.466
0.6724	1	88 Ra		$L_{II}P_I$	18.439
0.67328	5	88 Ra	$L\gamma_6$	$L_{II}O_{IV}$	18.414
0.67351	9	89 Ac	$L\gamma_1$	$L_{II}N_{IV}$	18.408
0.67383	2	95 Am	$L\beta_5$	$L_{III}O_{IV,V}$	18.3996
0.67491	4	90 Th	$L\gamma_5$	$L_{II}N_I$	18.370
0.67502	3	43 Tc	$K\alpha_1$	KL_{III}	18.3671
0.67538	5	88 Ra	$L\gamma_3$	L_IN_{III}	18.357
0.6764	1	88 Ra		$L_{II}O_{III}$	18.330
0.67772	2	94 Pu	$L\beta_1$	$L_{II}M_{IV}$	18.2937
0.6780	1	88 Ra		$L_{II}O_{II}$	18.286
0.67932	3	43 Tc	$K\alpha_2$	KL_{II}	18.2508
0.6801	1	88 Ra	$L\gamma_8$	$L_{II}O_I$	18.230
0.681014	8	92 U	$L\beta_9$	L_IM_V	18.2054
0.68199	5	88 Ra	$L\gamma_2$	L_IN_{II}	18.179
0.68639	2	95 Am	$L\beta_4$	L_IM_{II}	18.0627
0.6867	1	94 Pu	L_{III}	Abs. Edge	18.054
0.6874	1	88 Ra		L_IN_I	18.036
0.68760	5	92 U	$L\beta_{10}$	L_IM_{IV}	18.031
0.68883	1	40 Zr	K	Abs. Edge	17.9989
0.68901	5	40 Zr	$K\beta_4$	$KN_{IV,V}$	17.994
0.68920	9	93 Np	$L\beta_3$	L_IM_{III}	17.989
0.68993	4	40 Zr	$K\beta_2$	$KN_{II,III}$	17.970
0.69068	2	94 Pu	$L\beta_5$	$L_{III}O_{IV,V}$	17.9506
0.6932	1	88 Ra		$L_{II}N_V$	17.884
0.69463	5	88 Ra	$L\gamma_1$	$L_{II}N_{IV}$	17.849
0.6959	1	40 Zr	$K\beta_5$	$KM_{IV,V}$	17.815
0.698478	9	93 Np	$L\beta_1$	$L_{II}M_{IV}$	17.7502
0.7003	1	94 Pu	$L\beta_7$	$L_{III}O_I$	17.705
0.701390	9	95 Am	$L\beta_2$	$L_{III}N_V$	17.6765
0.70173	3	40 Zr	$K\beta_1$	KM_{III}	17.6678
0.7018	1	91 Pa	$L\beta_9$	L_IM_V	17.667
0.70228	4	40 Zr	$K\beta_3$	KM_{II}	17.654
0.7031	1	94 Pu	Lu	$L_{III}N_{VI,VII}$	17.635
0.70341	2	95 Am	$L\beta_{15}$	$L_{III}N_{IV}$	17.6258
0.7043	1	88 Ra		$L_{II}N_{III}$	17.604
0.70620	2	94 Pu	$L\beta_4$	L_IM_{II}	17.5560
0.70814	2	93 Np	$L\beta_5$	$L_{III}O_{IV,V}$	17.5081
0.7088	2	91 Pa	$L\beta_{10}$	L_IM_{IV}	17.492
0.709300	1	42 Mo	$K\alpha_1$	KL_{III}	17.47934
0.71029	2	92 U	$L\beta_3$	L_IM_{III}	17.4550
0.713590	6	42 Mo	$K\alpha_2$	KL_{II}	17.3743
0.71652	9	87 Fr	$L\gamma_1$	$L_{II}N_{IV}$	17.303
0.71774	5	88 Ra	$L\gamma_5$	$L_{II}N_I$	17.274
0.71851	2	94 Pu	$L\beta_2$	$L_{III}N_V$	17.2553
0.719984	8	92 U	$L\beta_1$	$L_{II}M_{IV}$	17.2200
0.7205	1	94 Pu	$L\beta_{15}$	$L_{III}N_{IV}$	17.208
0.7223	1	92 U	L_{III}	Abs. Edge	17.165
0.72240	5	92 U		$L_{III}P_{IV,V}$	17.162
0.7234	1	90 Th	$L\beta_9$	L_IM_V	17.139
0.72426	5	92 U		$L_{III}P_{II,III}$	17.118
0.72521	5	92 U		$L_{III}P_I$	17.096
0.726305	9	92 U	$L\beta_5$	$L_{III}O_{IV,V}$	17.0701
0.72671	2	93 Np	$L\beta_4$	L_IM_{II}	17.0607
0.72766	5	39 Y	K	Abs. Edge	17.038
0.72776	5	39 Y	$K\beta_4$	$KN_{IV,V}$	17.036
0.72864	4	39 Y	$K\beta_2$	$KN_{II,III}$	17.0154
0.7301	1	90 Th	$L\beta_{10}$	L_IM_{IV}	16.981
0.7309	1	92 U		$L_{III}O_{III}$	16.962
0.73230	5	91 Pa	$L\beta_3$	L_IM_{III}	16.930
0.7333	1	92 U		$L_{III}O_{II}$	16.907

Wavelength Å*	p.e.	Element		Designation	keV
0.73418	2	95 Am	$L\beta_6$	$L_{III}N_I$	16.8870
0.7345	1	39 Y	$K\beta_5$	$KM_{IV,V}$	16.879
0.73602	6	92 U	$L\beta_7$	$L_{III}O_I$	16.845
0.736230	9	93 Np	$L\beta_2$	$L_{III}N_V$	16.8400
0.738603	9	92 U	Lu	$L_{III}N_{VI,VII}$	16.7859
0.73928	9	86 Rn	$L\gamma_1$	$L_{II}N_{IV}$	16.770
0.74072	2	39 Y	$K\beta_1$	KM_{III}	16.7378
0.74126	3	39 Y	$K\beta_3$	KM_{II}	16.7258
0.74232	5	91 Pa	$L\beta_1$	$L_{II}M_{IV}$	16.702
0.74503	5	92 U	$L\beta_{17}$	$L_{II}M_{III}$	16.641
0.7452	2	91 Pa	$L\beta_5$	$L_{III}O_{IV,V}$	16.636
0.74620	1	41 Nb	$K\alpha_1$	KL_{III}	16.6151
0.747985	9	92 U	$L\beta_4$	L_IM_{II}	16.5753
0.75044	1	41 Nb	$K\alpha_2$	KL_{II}	16.5210
0.75148	2	94 Pu	$L\beta_6$	$L_{III}N_I$	16.4983
0.7546	2	91 Pa	$L\beta_7$	$L_{III}O_I$	16.431
0.754681	9	92 U	$L\beta_2$	$L_{III}N_V$	16.4283
0.75479	3	90 Th	$L\beta_3$	L_IM_{III}	16.4258
0.756642	9	92 U	$L\beta_{15}$	$L_{III}N_{IV}$	16.3857
0.75690	3	83 Bi	$L\gamma_{13}$	$L_IP_{II,III}$	16.3802
0.7571	1	83 Bi	L_I	Abs. Edge	16.376
0.7579	1	90 Th		$L_{II}M_V$	16.359
0.75791	5	83 Bi		$L_IO_{IV,V}$	16.358
0.7591	1	94 Pu	$L\eta$	$L_{II}M_I$	16.333
0.7607	1	90 Th	L_{III}	Abs. Edge	16.299
0.76087	9	90 Th		$L_{III}P_{IV,V}$	16.295
0.76087	3	83 Bi	$L\gamma_4$	L_IO_{III}	16.2947
0.76198	3	83 Bi	$L\gamma_4'$	L_IO_{II}	16.2709
0.7625	2	90 Th		$L_{III}P_{II,III}$	16.260
0.76289	9	85 At	$L\gamma_1$	$L_{II}N_{IV}$	16.251
0.76338	5	90 Th		$L_{III}P_I$	16.241
0.7641	5	83 Bi		$L_IN_{VI,VII}$	16.23
0.7645	2	84 Po	$L\gamma_6$	$L_{II}O_{IV}$	16.218
0.76468	5	90 Th	$L\beta_5$	$L_{III}O_{IV,V}$	16.213
0.765210	9	90 Th	$L\beta_1$	$L_{II}M_{IV}$	16.2022
0.76857	5	88 Ra	$L\beta_9$	L_IM_V	16.131
0.769	1	93 Np	$L\beta_6$	$L_{III}N_I$	16.13
0.7690	1	90 Th		$L_{III}O_{III}$	16.123
0.7691	1	92 U		$L_{III}N_{III}$	16.120
0.76973	5	38 Sr	K	Abs. Edge	16.107
0.7699	1	91 Pa	$L\beta_4$	L_IM_{II}	16.104
0.76989	5	38 Sr	$K\beta_4$	$KN_{IV,V}$	16.104
0.77081	3	38 Sr	$K\beta_2$	$KN_{II,III}$	16.0846
0.7713	1	90 Th		$L_{III}O_{II}$	16.074
0.772	1	84 Po	$L\gamma_2$	L_IN_{II}	16.07
0.7737	1	91 Pa	$L\beta_2$	$L_{III}N_V$	16.024
0.77437	4	90 Th	$L\beta_7$	$L_{III}O_I$	16.0105
0.77546	5	88 Ra	$L\beta_{10}$	L_IM_{IV}	15.988
0.7764	1	38 Sr	$K\beta_5$	$KM_{IV,V}$	15.969
0.77661	5	90 Th	Lu	$L_{III}N_{VI,VII}$	15.964
0.77728	5	83 Bi	$L\gamma_{11}$	L_IN_V	15.951
0.77822	9	89 Ac	$L\beta_3$	L_IM_{III}	15.931
0.77954	5	83 Bi		L_IN_{IV}	15.904
0.78017	9	92 U		$L_{III}N_{II}$	15.892
0.7809	2	93 Np	$L\eta$	$L_{II}M_I$	15.876
0.78196	5	82 Pb	L_I	Abs. Edge	15.855
0.78257	7	82 Pb		$L_IO_{IV,V}$	15.843

Wavelength Å*	p.e.	Element		Designation	keV
0.78292	2	38 Sr	$K\beta_1$	KM_{III}	15.8357
0.78345	3	38 Sr	$K\beta_3$	KM_{II}	15.8249
0.7858	1	82 Pb	$L\gamma_4$	L_IO_{III}	15.777
0.78593	1	40 Zr	$K\alpha_1$	KL_{III}	15.7751
0.78706	7	82 Pb	$L\gamma_4'$	L_IO_{II}	15.752
0.78748	9	84 Po	$L\gamma_1$	$L_{II}N_{IV}$	15.744
0.78838	2	92 U	$L\beta_6$	$L_{III}N_I$	15.7260
0.7884	1	82 Pb		$L_IN_{VI,VII}$	15.725
0.7887	1	83 Bi	L_{II}	Abs. Edge	15.719
0.78903	9	89 Ac	$L\beta_1$	$L_{II}M_{IV}$	15.713
0.78917	5	83 Bi	$L\gamma_3$	L_IN_{III}	15.7102
0.7897	1	82 Pb		L_IO_I	15.699
0.79015	1	40 Zr	$K\alpha_2$	KL_{II}	15.6909
0.79043	3	83 Bi	$L\gamma_6$	$L_{II}O_{IV}$	15.6853
0.79257	4	90 Th	$L\beta_4$	L_IM_{II}	15.6429
0.79257	4	90 Th	$L\beta_{17}$	$L_{II}M_{III}$	15.6429
0.79354	3	90 Th	$L\beta_2$	$L_{III}M_V$	15.6237
0.79384	5	83 Bi		$L_{II}O_{III}$	15.6178
0.79539	5	90 Th	$L\beta_{15}$	$L_{III}N_{IV}$	15.5875
0.79565	3	83 Bi	$L\gamma_2$	L_IN_{II}	15.5824
0.79721	9	83 Bi	Lv	$L_{II}N_{VI}$	15.552
0.7973	1	83 Bi	$L\gamma_8$	$L_{II}O_I$	15.551
0.8022	1	83 Bi		L_IN_I	15.456
0.80233	9	82 Pb	$L\gamma_{11}$	L_IN_V	15.453
0.80273	5	88 Ra	$L\beta_3$	L_IM_{III}	15.4449
0.8028	1	88 Ra	L_{III}	Abs. Edge	15.444
0.80364	7	82 Pb		L_IN_{IV}	15.427
0.8038	1	88 Ra		$L_{III}P_{II,III}$	15.425
0.8050	1	88 Ra		$L_{III}P_I$	15.402
0.80509	2	92 U	$L\eta$	$L_{II}M_I$	15.3997
0.80627	5	88 Ra	$L\beta_5$	$L_{III}O_{IV,V}$	15.3771
0.8079	1	91 Pa	$L\beta_6$	$L_{III}N_I$	15.347
0.8081	1	81 Tl	L_I	Abs. Edge	15.343
0.8082	1	90 Th		$L_{III}N_{III}$	15.341
0.80861	5	81 Tl		$L_IO_{IV,V}$	15.3327
0.81163	9	90 Th		L_IM_I	15.276
0.81184	5	81 Tl	$L\gamma_4$	L_IO_{III}	15.2716
0.81308	5	81 Tl	$L\gamma_4'$	L_IO_{II}	15.2482
0.81311	2	83 Bi	$L\gamma_1$	$L_{II}N_{IV}$	15.2477
0.81375	5	88 Ra	$L\beta_1$	$L_{II}M_{IV}$	15.2358
0.8147	1	82 Pb	$L\gamma_3$	L_IN_{III}	15.218
0.81538	5	82 Pb	L_{II}	Abs. Edge	15.2053
0.8154	2	37 Rb	$K\beta_4$	$KN_{IV,V}$	15.205
0.81554	5	37 Rb	K	Abs. Edge	15.2023
0.8158	1	81 Tl		L_IO_I	15.198
0.81583	5	82 Pb		$L_{II}P_I$	15.1969
0.8162	1	88 Ra	$L\beta_7$	$L_{III}O_I$	15.190
0.81645	3	37 Rb	$K\beta_2$	$KN_{II,III}$	15.1854
0.81683	5	82 Pb	$L\gamma_6$	$L_{II}O_{IV}$	15.1783
0.8186	1	88 Ra	Lu	$L_{III}N_{VI,VII}$	15.146
0.8190	2	90 Th		$L_{III}N_{II}$	15.138
0.8200	1	82 Pb		$L_{II}O_{III}$	15.120
0.8210	2	82 Pb	$L\gamma_2$	L_IN_{II}	15.101
0.8219	1	37 Rb	$K\beta_5$	$KM_{IV,V}$	15.085
0.82327	7	82 Pb	Lv	$L_{II}N_{VI}$	15.060
0.82365	5	82 Pb	$L\gamma_8$	$L_{II}O_I$	15.0527
0.8248	1	83 Bi		$L_{II}N_{III}$	15.031

TABLE B (*Continued*)

Wavelength Å*	p.e.	Element	Designation		keV
0.82789	9	87 Fr	$L\beta_3$	L_IM_{III}	14.976
0.82790	8	90 Th	$L\beta_6$	$L_{III}N_I$	14.975
0.82859	7	82 Pb		L_IN_I	14.963
0.82868	2	37 Rb	$K\beta_1$	KM_{III}	14.9613
0.82879	5	81 Tl	$L\gamma_{11}$	L_IN_V	14.9593
0.82884	1	39 Y	$K\alpha_1$	KL_{III}	14.9584
0.82921	3	37 Rb	$K\beta_3$	KM_{II}	14.9517
0.8295	1	91 Pa	$L\eta$	$L_{II}M_I$	14.946
0.83001	7	81 Tl		L_IN_{IV}	14.937
0.83305	1	39 Y	$K\alpha_2$	KL_{II}	14.8829
0.8338	1	90 Th		$L_{II}M_{II}$	14.869
0.8344	9	83 Bi		$L_{II}N_{II}$	14.86
0.8350	2	80 Hg		$L_IO_{IV,V}$	14.847
0.8353	1	80 Hg	L_I	Abs. Edge	14.842
0.83537	5	88 Ra	$L\beta_2$	$L_{III}N_V$	14.8414
0.83722	5	88 Ra	$L\beta_{15}$	$L_{III}N_{IV}$	14.8086
0.8382	2	82 Pb		$L_{II}N_V$	14.791
0.83894	7	80 Hg	$L\gamma_4$	L_IO_{III}	14.778
0.83923	5	83 Bi	$L\gamma_5$	$L_{II}N_I$	14.7732
0.83940	9	87 Fr	$L\beta_1$	$L_{II}M_{IV}$	14.770
0.83973	3	82 Pb	$L\gamma_1$	$L_{II}N_{IV}$	14.7644
0.84013	7	80 Hg	$L\gamma_4'$	L_IO_{II}	14.757
0.84071	5	88 Ra	$L\beta_4$	L_IM_{II}	14.7472
0.84130	4	81 Tl	$L\gamma_3$	L_IN_{III}	14.7368
0.8434	1	81 Tl	L_{II}	Abs. Edge	14.699
0.8438	1	88 Ra	$L\beta_{17}$	$L_{II}M_{III}$	14.692
0.8442	2	81 Tl	$L\gamma_6$	$L_{II}O_{IV}$	14.685
0.8452	2	80 Hg		L_IO_I	14.670
0.84773	5	81 Tl	$L\gamma_2$	L_IN_{II}	14.6251
0.848187	9	95 Am	$L\alpha_1$	$L_{III}M_V$	14.6172
0.8490	1	81 Tl		$L_{II}O_{II}$	14.604
0.85048	5	81 Tl	Lv	$L_{II}N_{VI}$	14.5777
0.8512	1	88 Ra		$L_{III}N_{III}$	14.566
0.8513	2	81 Tl	$L\gamma_8$	$L_{II}O_I$	14.564
0.85192	7	82 Pb		$L_{II}N_{III}$	14.553
0.85436	9	86 Rn	$L\beta_3$	L_IM_{III}	14.512
0.85446	4	90 Th	$L\eta$	$L_{II}M_I$	14.5099
0.8549	1	81 Tl		L_IN_I	14.503
0.85657	7	80 Hg	$L\gamma_{11}$	L_IN_V	14.474
0.858	2	87 Fr	$L\beta_2$	$L_{III}N_V$	14.45
0.8585	3	82 Pb		$L_{II}N_{II}$	14.442
0.860266	9	95 Am	$L\alpha_2$	$L_{III}M_{IV}$	14.4119
0.8618	1	88 Ra		$L_{III}N_{II}$	14.387
0.86376	5	79 Au	L_I	Abs. Edge	14.3537
0.86400	5	79 Au		$L_IO_{IV,V}$	14.3497
0.8653	2	36 Kr	$K\beta_4$	$KN_{IV,V}$	14.328
0.86552	1	36 Kr	K	Abs. Edge	14.3244
0.86605	9	86 Rn	$L\beta_1$	$L_{II}M_{IV}$	14.316
0.8661	1	36 Kr	$K\beta_2$	$KN_{II,III}$	14.315
0.86655	5	82 Pb	$L\gamma_5$	$L_{II}N_I$	14.3075
0.86703	4	79 Au	$L\gamma_4$	L_IO_{III}	14.2996
0.86752	3	81 Tl	$L\gamma_1$	$L_{II}N_{IV}$	14.2915
0.86816	4	79 Au	$L\gamma_4'$	L_IO_{II}	14.2809
0.86830	2	94 Pu	$L\alpha_1$	$L_{III}M_V$	14.2786
0.86915	7	80 Hg	$L\gamma_3$	L_IN_{III}	14.265
0.87074	5	79 Au		L_IO_I	14.2385
0.8708	2	36 Kr	$K\beta_5$	$KM_{IV,V}$	14.238
0.87088	5	88 Ra	$L\beta_6$	$L_{III}N_I$	14.2362
0.8722	1	80 Hg	L_{II}	Abs. Edge	14.215
0.87319	7	80 Hg	$L\gamma_6$	$L_{II}O_{IV}$	14.199
0.87526	1	38 Sr	$K\alpha_1$	KL_{III}	14.1650
0.87544	7	80 Hg	$L\gamma_2$	L_IN_{II}	14.162
0.8758	1	80 Hg		$L_{II}O_{III}$	14.156
0.8784	1	80 Hg		$L_{II}O_{II}$	14.114
0.8785	1	36 Kr	$K\beta_1$	KM_{III}	14.112
0.87885	7	80 Hg	Lv	$L_{II}N_{VI}$	14.107
0.8790	1	36 Kr	$K\beta_3$	KM_{II}	14.104
0.87943	1	38 Sr	$K\alpha_2$	KL_{II}	14.0979
0.87995	7	80 Hg	$L\gamma_8$	$L_{II}O_I$	14.090
0.87996	5	81 Tl		$L_{II}N_{III}$	14.0893
0.88028	2	94 Pu	$L\alpha_2$	$L_{III}M_{IV}$	14.0842
0.88135	9	85 At	$L\beta_3$	L_IM_{III}	14.067
0.8827	2	80 Hg		L_IN_I	14.045
0.88433	7	79 Au	$L\gamma_{11}$	L_IN_V	14.020
0.88563	7	79 Au		L_IN_{IV}	13.999
0.8882	2	81 Tl		$L_{II}M_{II}$	13.959
0.889128	9	93 Np	$L\alpha_1$	$L_{III}M_V$	13.9441
0.8931	1	78 Pt	L_I	Abs. Edge	13.883
0.8934	1	78 Pt		L_IO_V	13.878
0.89349	9	85 At	$L\beta_1$	$L_{II}M_{IV}$	13.876
0.8943	1	78 Pt		L_IO_{IV}	13.864
0.89500	4	81 Tl	$L\gamma_5$	$L_{II}N_I$	13.8526
0.89646	5	80 Hg	$L\gamma_1$	$L_{II}N_{IV}$	13.8301
0.89659	4	78 Pt	$L\gamma_4$	L_IO_{III}	13.8281
0.89747	4	78 Pt	$L\gamma_4'$	L_IO_{II}	13.8145
0.89783	5	79 Au	$L\gamma_3$	L_IN_{III}	13.8090
0.89791	3	83 Bi	$L\beta_9$	L_IM_V	13.8077
0.8995	2	78 Pt		L_IO_I	13.784
0.8996	2	84 Po	$L\beta_5$	$L_{III}O_{IV,V}$	13.782
0.901045	9	93 Np	$L\alpha_2$	$L_{III}M_{IV}$	13.7597
0.90259	5	79 Au	L_{II}	Abs. Edge	13.7361
0.90297	3	79 Au	$L\gamma_6$	$L_{II}O_{IV}$	13.7304
0.90434	3	79 Au	$L\gamma_2$	L_IN_{II}	13.7095
0.90495	4	83 Bi	$L\beta_{10}$	L_IM_{IV}	13.7002
0.90638	7	79 Au		$L_{II}O_{III}$	13.679
0.90742	5	88 Ra	$L\eta$	$L_{II}M_I$	13.6630
0.90746	7	79 Au		$L_{II}O_{II}$	13.662
0.90837	5	79 Au	Lv	$L_{II}N_{VI}$	13.6487
0.90894	7	80 Hg		$L_{II}N_{III}$	13.640
0.9091	3	84 Po	$L\beta_3$	L_IM_{III}	13.638
0.90989	5	79 Au	$L\gamma_8$	$L_{II}O_I$	13.6260
0.910639	9	92 U	$L\alpha_1$	$L_{III}M_V$	13.6147
0.9131	1	79 Au		L_IN_I	13.578
0.9143	2	78 Pt	$L\gamma_{11}$	L_IN_V	13.560
0.9204	1	35 Br	K	Abs. Edge	13.470
0.92046	2	35 Br	$K\beta_2$	$KN_{II,III}$	13.4695
0.9220	2	84 Po	$L\beta_1$	$L_{II}M_{IV}$	13.447
0.922558	9	92 U	$L\alpha_2$	$L_{III}M_{IV}$	13.4388
0.9234	1	83 Bi	L_{III}	Abs. Edge	13.426
0.9236	1	77 Ir	L_I	Abs. Edge	13.423
0.92413	4	83 Bi		$L_{III}P_{II,III}$	13.4159
0.9243	3	77 Ir		$L_IO_{IV,V}$	13.413
0.92453	7	80 Hg	$L\gamma_5$	$L_{II}N_I$	13.410
0.9255	1	35 Br	$K\beta_5$	$KM_{IV,V}$	13.396

Wavelength Å*	p.e.	Element		Designation	keV
0.925553	9	37 Rb	$K\alpha_1$	KL_{III}	13.3953
0.92556	3	83 Bi	$L\beta_5$	$L_{III}O_{IV,V}$	13.3953
0.92650	3	79 Au	$L\gamma_1$	$L_{II}N_{IV}$	13.3817
0.9268	1	82 Pb	$L\beta_9$	L_IM_V	13.377
0.92744	3	77 Ir	$L\gamma_4$	L_IO_{III}	13.3681
0.92791	5	78 Pt	$L\gamma_3$	L_IN_{III}	13.3613
0.92831	3	77 Ir	$L\gamma_4'$	L_IO_{II}	13.3555
0.92937	5	84 Po	$L\beta_2$	$L_{III}N_V$	13.3404
0.92969	1	37 Rb	$K\alpha_2$	KL_{II}	13.3358
0.9302	2	83 Bi		$L_{III}O_{III}$	13.328
0.9312	2	84 Po	$L\beta_{15}$	$L_{III}N_{IV}$	13.314
0.9323	2	83 Bi		$L_{III}O_{II}$	13.298
0.93279	2	35 Br	$K\beta_1$	KM_{III}	13.2914
0.93284	5	91 Pa	$L\alpha_1$	$L_{III}M_V$	13.2907
0.93327	5	35 Br	$K\beta_3$	KM_{II}	13.2845
0.9339	2	82 Pb	$L\beta_{10}$	L_IM_{IV}	13.275
0.93414	5	78 Pt	L_{II}	Abs. Edge	13.2723
0.9342	2	78 Pt	$L\gamma_6$	$L_{II}O_{IV}$	13.271
0.93427	5	78 Pt	$L\gamma_2$	L_IN_{II}	13.2704
0.93505	5	83 Bi	$L\beta_7$	$L_{III}O_I$	13.2593
0.93505	5	83 Bi	Lu	$L_{III}N_{VI,VII}$	13.2593
0.93855	3	83 Bi	$L\beta_3$	L_IM_{III}	13.2098
0.93931	5	78 Pt	Lv	$L_{II}N_{VI}$	13.1992
0.9402	2	79 Au		$L_{II}N_{III}$	13.186
0.9411	1	78 Pt	$L\gamma_8$	$L_{II}O_I$	13.173
0.94419	5	83 Bi		$L_{II}M_V$	13.1310
0.9446	2	77 Ir	$L\gamma_{11}$	L_IN_V	13.126
0.94482	5	91 Pa	$L\alpha_2$	$L_{III}M_{IV}$	13.1222
0.9455	2	78 Pt		L_IN_I	13.113
0.9459	2	77 Ir		L_IN_{IV}	13.108
0.9475	3	84 Po	$L\beta_4$	L_IM_{II}	13.086
0.95073	5	82 Pb	L_{III}	Abs. Edge	13.0406
0.95118	7	82 Pb		$L_{III}P_{II,III}$	13.0344
0.951978	9	83 Bi	$L\beta_1$	$L_{II}M_{IV}$	13.0235
0.9526	1	82 Pb	$L\beta_5$	$L_{III}O_{IV,V}$	13.015
0.95518	4	83 Bi	$L\beta_2$	$L_{III}N_V$	12.9799
0.95559	3	79 Au	$L\gamma_5$	$L_{II}N_I$	12.9743
0.9558	1	76 Os	L_I	Abs. Edge	12.972
0.95600	3	90 Th	$L\alpha_1$	$L_{III}M_V$	12.9687
0.95603	5	76 Os		$L_IO_{IV,V}$	12.9683
0.95675	7	81 Tl	$L\beta_9$	L_IM_V	12.9585
0.95702	5	83 Bi	$L\beta_{15}$	$L_{III}N_{IV}$	12.9549
0.9578	1	82 Pb		$L_{III}O_{III}$	12.945
0.95797	3	78 Pt	$L\gamma_1$	$L_{II}N_{IV}$	12.9420
0.9586	1	82 Pb		$L_{III}O_{II}$	12.934
0.95931	5	77 Ir	$L\gamma_3$	L_IN_{III}	12.9240
0.95938	8	76 Os	$L\gamma_4$	L_IO_{III}	12.923
0.96033	8	76 Os	$L\gamma_4'$	L_IO_{II}	12.910
0.96133	7	82 Pb	Lu	$L_{III}N_{VI,VII}$	12.8968
0.9620	1	82 Pb	$L\beta_7$	$L_{III}O_I$	12.888
0.96318	7	76 Os		L_IO_I	12.8721
0.9636	1	92 U	Ls	$L_{III}M_{III}$	12.866
0.96389	7	81 Tl	$L\beta_{10}$	L_IM_{IV}	12.8626
0.96545	3	77 Ir	$L\gamma_2$	L_IN_{II}	12.8418
0.96708	4	77 Ir	$L\gamma_6$	$L_{II}O_{IV}$	12.8201
0.9671	1	77 Ir	L_{II}	Abs. Edge	12.820
0.9672	2	84 Po	$L\beta_6$	$L_{III}N_I$	12.819
0.96788	2	90 Th	$L\alpha_2$	$L_{III}M_{IV}$	12.8096
0.96911	7	82 Pb	$L\beta_3$	L_IM_{III}	12.7933
0.96979	5	77 Ir		$L_{II}O_{III}$	12.7843
0.97161	6	77 Ir	Lv	$L_{II}N_{VI}$	12.7603
0.97173	4	78 Pt		$L_{II}N_{III}$	12.7588
0.97321	5	83 Bi		$L_{III}N_{III}$	12.7394
0.97409	3	77 Ir	$L\gamma_8$	$L_{II}O_I$	12.7279
0.9747	1	82 Pb		$L_{II}M_V$	12.720
0.9765	3	76 Os	$L\gamma_{11}$	L_IN_V	12.696
0.9766	2	77 Ir		L_IN_I	12.695
0.97690	4	83 Bi	$L\beta_4$	L_IM_{II}	12.6912
0.9772	3	76 Os		L_IN_{IV}	12.687
0.9792	2	78 Pt		$L_{II}N_{II}$	12.661
0.97926	5	81 Tl		$L_{III}P_{II,III}$	12.6607
0.9793	1	81 Tl	L_{III}	Abs. Edge	12.660
0.97974	1	34 Se	K	Abs. Edge	12.6545
0.97992	5	34 Se	$K\beta_2$	$KN_{II,III}$	12.6522
0.97993	5	89 Ac	$L\alpha_1$	$L_{III}M_V$	12.6520
0.9801	1	36 Kr	$K\alpha_1$	KL_{III}	12.649
0.98058	3	81 Tl	$L\beta_5$	$L_{III}O_{IV,V}$	12.6436
0.98221	7	82 Pb	$L\beta_2$	$L_{III}N_V$	12.6226
0.98280	5	83 Bi		$L_{III}N_{II}$	12.6151
0.98291	3	82 Pb	$L\beta_1$	$L_{II}M_{IV}$	12.6137
0.98389	7	82 Pb	$L\beta_{15}$	$L_{III}N_{IV}$	12.6011
0.9841	1	36 Kr	$K\alpha_2$	KL_{II}	12.598
0.9843	1	34 Se	$K\beta_5$	$KM_{IV,V}$	12.595
0.98538	5	81 Tl		$L_{III}O_{III}$	12.5820
0.9871	2	80 Hg	$L\beta_9$	L_IM_V	12.560
0.98738	5	81 Tl		$L_{III}O_{II}$	12.5566
0.9877	2	78 Pt	$L\gamma_5$	$L_{II}N_I$	12.552
0.9888	1	81 Tl	Lu	$L_{III}N_{VI,VII}$	12.538
0.98913	5	83 Bi	$L\beta_{17}$	$L_{II}M_{III}$	12.5344
0.9894	1	75 Re	L_I	Abs. Edge	12.530
0.9900	1	75 Re		$L_IO_{IV,V}$	12.524
0.99017	5	81 Tl	$L\beta_7$	$L_{III}O_I$	12.5212
0.99085	3	77 Ir	$L\gamma_1$	$L_{II}N_{IV}$	12.5126
0.99178	5	89 Ac	$L\alpha_2$	$L_{III}M_{IV}$	12.5008
0.99186	5	76 Os	$L\gamma_3$	L_IN_{III}	12.4998
0.99218	3	34 Se	$K\beta_1$	KM_{III}	12.4959
0.99249	5	75 Re	$L\gamma_4$	L_IO_{III}	12.4920
0.99268	5	34 Se	$K\beta_3$	KM_{II}	12.4896
0.99331	3	83 Bi	$L\beta_6$	$L_{III}N_I$	12.4816
0.99334	5	75 Re	$L\gamma_4'$	L_IO_{II}	12.4813
0.9962	2	80 Hg	$L\beta_{10}$	L_IM_{IV}	12.446
0.9965	1	75 Re		L_IO_I	12.442
0.99805	5	76 Os	$L\gamma_2$	L_IN_{II}	12.4224
1.0005	1	82 Pb		$L_{III}N_{III}$	12.392
1.0005	9	83 Bi		L_IM_I	12.39
1.00062	3	81 Tl	$L\beta_3$	L_IM_{III}	12.3904
1.00107	5	76 Os	$L\gamma_6$	$L_{II}O_{IV}$	12.3848
1.0012	6	95 Am	Ll	$L_{III}M_I$	12.384
1.0014	1	76 Os	L_{II}	Abs. Edge	12.381
1.0047	2	76 Os		$L_{II}O_{III}$	12.340
1.00473	5	88 Ra	$L\alpha_1$	$L_{III}M_V$	12.3397
1.0050	2	76 Os	Lv	$L_{II}N_{VI}$	12.337
1.0054	3	77 Ir		$L_{II}N_{III}$	12.332
1.00722	5	81 Tl		$L_{II}M_V$	12.3093

TABLE B (*Continued*)

Wavelength Å*	p.e.	Element		Designation	keV
1.0075	1	82 Pb	$L\beta_4$	L_IM_{II}	12.306
1.00788	5	76 Os	$L\gamma_8$	$L_{II}O_I$	12.3012
1.0091	1	80 Hg	L_{III}	Abs. Edge	12.286
1.00987	7	80 Hg	$L\beta_5$	$L_{III}O_{IV,V}$	12.2769
1.01031	3	81 Tl	$L\beta_2$	$L_{III}N_V$	12.2715
1.01040	7	82 Pb		$L_{III}N_{II}$	12.2705
1.0108	1	75 Re	$L\gamma_{11}$	L_IN_V	12.266
1.0112	1	90 Th	Ls	$L_{III}M_{III}$	12.261
1.0119	1	75 Re		L_IN_{IV}	12.252
1.0120	2	77 Ir		$L_{II}N_{II}$	12.251
1.01201	3	81 Tl	$L\beta_{15}$	$L_{III}N_{IV}$	12.2510
1.01404	7	80 Hg		$L_{III}O_{III}$	12.2264
1.01513	4	81 Tl	$L\beta_1$	$L_{II}M_{IV}$	12.2133
1.01558	7	80 Hg		$L_{III}O_{II}$	12.2079
1.01656	5	88 Ra	$L\alpha_2$	$L_{III}M_{IV}$	12.1962
1.01674	7	80 Hg	Lu	$L_{III}N_{VII}$	12.1940
1.01769	7	80 Hg	Lu'	$L_{III}N_{VI}$	12.1826
1.01937	7	80 Hg	$L\beta_7$	$L_{III}O_I$	12.1625
1.02063	7	79 Au	$L\beta_9$	L_IM_V	12.1474
1.0210	1	82 Pb	$L\beta_6$	$L_{III}N_I$	12.143
1.02175	5	77 Ir	$L\gamma_5$	$L_{II}N_I$	12.1342
1.0223	1	82 Pb	$L\beta_{17}$	$L_{II}M_{III}$	12.127
1.0226	1	94 Pu	Ll	$L_{III}M_I$	12.124
1.02467	5	74 W	L_I	Abs. Edge	12.0996
1.0250	2	74 W		$L_IO_{IV,V}$	12.095
1.02503	5	76 Os	$L\gamma_1$	$L_{II}N_{IV}$	12.0953
1.02613	7	75 Re	$L\gamma_3$	L_IN_{III}	12.0824
1.02775	3	74 W	$L\gamma_4$	L_IO_{III}	12.0634
1.02789	7	79 Au	$L\beta_{10}$	L_IM_{IV}	12.0617
1.0286	1	81 Tl		$L_{III}N_{III}$	12.053
1.02863	3	74 W	$L\gamma_4'$	L_IO_{II}	12.0530
1.03049	5	87 Fr	$L\alpha_1$	$L_{III}M_V$	12.0313
1.0317	3	74 W		L_IO_I	12.017
1.03233	5	75 Re	$L\gamma_2$	L_IN_{II}	12.0098
1.0323	2	82 Pb		L_IM_I	12.010
1.03358	7	80 Hg	$L\beta_3$	L_IM_{III}	11.9953
1.0346	9	83 Bi		L_IM_{II}	11.98
1.0347	1	92 U	Lt	$L_{III}M_{II}$	11.982
1.03699	9	75 Re	$L\gamma_6$	$L_{II}O_{IV}$	11.956
1.0371	1	75 Re	L_{II}	Abs. Edge	11.954
1.03876	7	79 Au		$L_{III}P_{II,III}$	11.9355
1.03918	3	81 Tl	$L\beta_4$	$L_{II}M_{II}$	11.9306
1.0397	1	75 Re		$L_{II}O_{III}$	11.925
1.03973	5	76 Os		$L_{II}N_{III}$	11.9243
1.03974	2	35 Br	$K\alpha_1$	KL_{III}	11.9242
1.03975	7	80 Hg	$L\beta_2$	$L_{III}N_V$	11.9241
1.04000	5	79 Au	L_{III}	Abs. Edge	11.9212
1.0404	1	75 Re	Lv	$L_{II}N_{VI}$	11.917
1.04044	3	79 Au	$L\beta_5$	$L_{III}O_{IV,V}$	11.9163
1.04151	7	80 Hg	$L\beta_{15}$	$L_{III}N_{IV}$	11.9040
1.0420	1	75 Re		L_IN_I	11.899
1.04230	5	87 Fr	$L\alpha_2$	$L_{III}M_{IV}$	11.8950
1.0428	6	93 Np	Ll	$L_{III}M_I$	11.890
1.04382	2	35 Br	$K\alpha_2$	KL_{II}	11.8776
1.04398	5	75 Re	$L\gamma_8$	$L_{II}O_I$	11.8758
1.0450	2	79 Au		$L_{III}O_{II,III}$	11.865
1.0450	1	33 As	K	Abs. Edge	11.865
1.04500	3	33 As	$K\beta_2$	$KN_{II,III}$	11.8642
1.0458	1	74 W	$L\gamma_{11}$	L_IN_V	11.856
1.0468	2	74 W		L_IN_{IV}	11.844
1.04752	5	79 Au	Lu	$L_{III}N_{VI,VII}$	11.8357
1.04868	5	80 Hg	$L\beta_1$	$L_{II}M_{IV}$	11.8226
1.0488	1	33 As	$K\beta_5$	$KM_{IV,V}$	11.822
1.04963	5	81 Tl	$L\beta_6$	$L_{III}N_I$	11.8118
1.04974	8	79 Au	$L\beta_7$	$L_{III}O_I$	11.8106
1.05446	5	78 Pt	$L\beta_9$	L_IM_V	11.7577
1.05609	7	81 Tl	$L\beta_{17}$	$L_{II}M_{III}$	11.7397
1.05693	5	76 Os	$L\gamma_5$	$L_{II}N_I$	11.7303
1.05723	5	86 Rn	$L\alpha_1$	$L_{III}M_V$	11.7270
1.05730	2	33 As	$K\beta_1$	KM_{III}	11.7262
1.05783	5	33 As	$K\beta_3$	KM_{II}	11.7203
1.0585	1	80 Hg		$L_{III}N_{III}$	11.713
1.05856	3	83 Bi	$L\eta$	$L_{II}M_I$	11.7122
1.06099	5	75 Re	$L\gamma_1$	$L_{II}N_{IV}$	11.6854
1.0613	1	73 Ta	L_I	Abs. Edge	11.682
1.06183	7	78 Pt	$L\beta_{10}$	L_IM_{IV}	11.6762
1.06192	9	73 Ta		$L_IO_{IV,V}$	11.6752
1.06200	6	74 W	$L\gamma_3$	L_IN_{III}	11.6743
1.06357	9	73 Ta		$L_IN_{VI,VII}$	11.6570
1.0644	2	82 Pb		$L_{II}M_{II}$	11.648
1.0644	2	81 Tl		L_IM_I	11.648
1.06467	3	73 Ta	$L\gamma_4$	L_IO_{III}	11.6451
1.0649	2	80 Hg		$L_{III}N_{II}$	11.642
1.06544	3	73 Ta	$L\gamma_4'$	L_IO_{II}	11.6366
1.06712	2	92 U	Ll	$L_{III}M_I$	11.6183
1.06771	9	73 Ta		L_IO_I	11.6118
1.06785	9	79 Au	$L\beta_3$	L_IM_{III}	11.6103
1.06806	3	74 W	$L\gamma_2$	L_IN_{II}	11.6080
1.06899	5	86 Rn	$L\alpha_2$	$L_{III}M_{IV}$	11.5979
1.07022	3	79 Au	$L\beta_2$	$L_{III}N_V$	11.5847
1.07188	5	79 Au	$L\beta_{15}$	$L_{III}N_{IV}$	11.5667
1.07222	7	80 Hg	$L\beta_4$	L_IM_{II}	11.5630
1.0723	1	78 Pt	L_{III}	Abs. Edge	11.562
1.0724	2	78 Pt	$L\beta_5$	$L_{III}O_{IV,V}$	11.561
1.07448	5	74 W	$L\gamma_6$	$L_{II}O_{IV}$	11.5387
1.0745	1	74 W	L_{II}	Abs. Edge	11.538
1.0756	2	79 Au		$L_{II}M_V$	11.526
1.0761	3	78 Pt		$L_{III}O_{II,III}$	11.521
1.0767	1	75 Re		$L_{II}N_{III}$	11.515
1.0771	1	74 W	Lv	$L_{II}N_{VI}$	11.510
1.07896	5	78 Pt	Lu	$L_{III}N_{VI,VII}$	11.4908
1.0792	2	74 W		$L_{II}O_{III}$	11.488
1.07975	7	80 Hg	$L\beta_6$	$L_{III}N_I$	11.4824
1.08009	9	90 Th	Lt	$L_{III}M_{II}$	11.4788
1.08113	4	74 W	$L\gamma_8$	$L_{II}O_I$	11.4677
1.08168	3	78 Pt	$L\beta_7$	$L_{III}O_I$	11.4619
1.08205	7	73 Ta	$L\gamma_{11}$	L_IN_V	11.4580
1.08353	3	79 Au	$L\beta_1$	$L_{II}M_{IV}$	11.4423
1.08377	7	73 Ta		L_IN_{IV}	11.4398
1.0839	1	75 Re		$L_{II}N_{II}$	11.438
1.08500	5	85 At	$L\alpha_1$	$L_{III}M_V$	11.4268
1.08975	5	77 Ir	$L\beta_9$	L_IM_V	11.3770
1.09026	7	79 Au		$L_{III}N_{III}$	11.3717
1.0908	1	91 Pa	Ll	$L_{III}M_I$	11.366

TABLE B (*Continued*)

Wavelength Å*	p.e.	Element	Designation		keV
1.0916	5	80 Hg	$L\beta_{17}$	$L_{II}M_{III}$	11.358
1.09241	7	82 Pb	$L\eta$	$L_{II}M_{I}$	11.3493
1.09388	5	75 Re	$L\gamma_5$	$L_{II}N_{I}$	11.3341
1.09671	5	85 At	$L\alpha_2$	$L_{III}M_{IV}$	11.3048
1.09702	4	77 Ir	$L\beta_{10}$	$L_{I}M_{IV}$	11.3016
1.09855	3	74 W	$L\gamma_1$	$L_{II}N_{IV}$	11.2859
1.09936	4	73 Ta	$L\gamma_3$	$L_{I}N_{III}$	11.2776
1.0997	1	81 Tl		$L_{II}M_{II}$	11.274
1.0997	1	72 Hf	L_{I}	Abs. Edge	11.274
1.09968	7	79 Au		$L_{III}N_{II}$	11.2743
1.0999	2	80 Hg		$L_{I}M_{I}$	11.272
1.10086	9	72 Hf		$L_{I}O_{IV}$	11.2622
1.10200	3	78 Pt	$L\beta_2$	$L_{III}N_{V}$	11.2505
1.10303	5	72 Hf	$L\gamma_4$	$L_{I}O_{III}$	11.2401
1.10376	5	72 Hf	$L\gamma_4'$	$L_{I}O_{II}$	11.2326
1.10394	5	78 Pt	$L\beta_3$	$L_{I}M_{III}$	11.2308
1.10477	2	34 Se	$K\alpha_1$	KL_{III}	11.2224
1.1053	1	73 Ta	$L\gamma_2$	$L_{I}N_{II}$	11.217
1.1058	1	77 Ir	L_{III}	Abs. Edge	11.212
1.10585	3	77 Ir	$L\beta_5$	$L_{III}O_{IV,V}$	11.2114
1.10651	3	79 Au	$L\beta_4$	$L_{I}M_{II}$	11.2047
1.10664	9	72 Hf		$L_{I}O_{I}$	11.2034
1.10882	2	34 Se	$K\alpha_2$	KL_{II}	11.1814
1.10923	6	77 Ir		$L_{III}O_{II,III}$	11.1772
1.11092	3	79 Au	$L\beta_6$	$L_{III}N_{I}$	11.1602
1.11145	4	77 Ir	Lu	$L_{III}N_{VI,VII}$	11.1549
1.1129	2	78 Pt		$L_{II}M_{V}$	11.140
1.1137	1	73 Ta	L_{II}	Abs. Edge	11.132
1.11386	4	84 Po	$L\alpha_1$	$L_{III}M_{V}$	11.1308
1.11388	3	73 Ta	$L\gamma_6$	$L_{II}O_{IV}$	11.1306
1.11489	3	77 Ir	$L\beta_7$	$L_{III}O_{I}$	11.1205
1.1149	2	74 W		$L_{II}N_{III}$	11.120
1.11508	4	90 Th	Ll	$L_{III}M_{I}$	11.1186
1.11521	9	73 Ta		$L_{I}N_{I}$	11.1173
1.1158	1	73 Ta	Lv	$L_{II}N_{VI}$	11.1113
1.11658	5	32 Ge	K	Abs. Edge	11.1036
1.11686	2	32 Ge	$K\beta_2$	$KN_{II,III}$	11.1008
1.11693	9	73 Ta		$L_{II}O_{III}$	11.1001
1.11789	9	73 Ta		$L_{II}O_{II}$	11.0907
1.1195	1	32 Ge	$K\beta_5$	$KM_{IV,V}$	11.0745
1.11990	2	78 Pt	$L\beta_1$	$L_{II}M_{IV}$	11.0707
1.1205	1	73 Ta	$L\gamma_8$	$L_{II}O_{I}$	11.0646
1.12146	9	72 Hf	$L\gamma_{11}$	$L_{I}N_{V}$	11.0553
1.1218	3	74 W		$L_{II}N_{II}$	11.052
1.12250	9	72 Hf		$L_{I}N_{IV}$	11.0451
1.1226	2	78 Pt		$L_{III}N_{III}$	11.044
1.12548	5	84 Po	$L\alpha_2$	$L_{III}M_{IV}$	11.0158
1.12637	6	76 Os	$L\beta_9$	$L_{I}M_{V}$	11.0071
1.12769	3	81 Tl	$L\eta$	$L_{II}M_{I}$	10.9943
1.12798	5	79 Au	$L\beta_{17}$	$L_{II}M_{III}$	10.9915
1.12894	2	32 Ge	$K\beta_1$	KM_{III}	10.9821
1.12936	9	32 Ge	$K\beta_3$	KM_{II}	10.9780
1.1310	2	78 Pt		$L_{III}N_{II}$	10.962
1.13235	3	74 W	$L\gamma_5$	$L_{II}N_{I}$	10.9490
1.13353	5	76 Os	$L\beta_{10}$	$L_{I}M_{IV}$	10.9376
1.13525	5	79 Au		$L_{I}M_{I}$	10.9210
1.13532	3	77 Ir	$L\beta_2$	$L_{III}N_{V}$	10.9203
1.13687	9	73 Ta		$L_{II}N_{V}$	10.9055
1.13707	3	77 Ir	$L\beta_{15}$	$L_{III}N_{IV}$	10.9036
1.13794	3	73 Ta	$L\gamma_1$	$L_{II}N_{IV}$	10.8952
1.13841	5	72 Hf	$L\gamma_3$	$L_{I}N_{III}$	10.8907
1.1387	5	80 Hg		$L_{II}M_{II}$	10.888
1.1402	1	71 Lu	L_{I}	Abs. Edge	10.8740
1.1405	1	76 Os	$L\beta_5$	$L_{III}O_{IV,V}$	10.8711
1.1408	1	76 Os	L_{III}	Abs. Edge	10.8683
1.14085	3	77 Ir	$L\beta_3$	$L_{I}M_{III}$	10.8674
1.14223	5	78 Pt	$L\beta_4$	$L_{I}M_{II}$	10.8543
1.1435	1	71 Lu	$L\gamma_4$	$L_{I}O_{II,III}$	10.8425
1.14355	5	78 Pt	$L\beta_6$	$L_{III}N_{I}$	10.8418
1.14386	2	83 Bi	$L\alpha_1$	$L_{III}M_{V}$	10.8388
1.14442	5	72 Hf	$L\gamma_2$	$L_{I}N_{II}$	10.8335
1.14537	7	76 Os	Lu	$L_{III}N_{VI,VII}$	10.8245
1.1489	2	77 Ir		$L_{II}M_{V}$	10.791
1.14933	8	76 Os	$L\beta_7$	$L_{III}O_{I}$	10.7872
1.1548	1	72 Hf	L_{II}	Abs. Edge	10.7362
1.15519	5	72 Hf	$L\gamma_6$	$L_{II}O_{IV}$	10.7325
1.1553	1	73 Ta		$L_{II}N_{III}$	10.7316
1.15536	1	83 Bi	$L\alpha_2$	$L_{III}M_{IV}$	10.73091
1.1560	3	77 Ir		$L_{III}N_{III}$	10.725
1.15781	3	77 Ir	$L\beta_1$	$L_{II}M_{IV}$	10.7083
1.15830	9	72 Hf	Lv	$L_{II}N_{VI}$	10.7037
1.1600	2	73 Ta		$L_{II}N_{II}$	10.688
1.16107	9	71 Lu	$L\gamma_{11}$	$L_{I}N_{V}$	10.6782
1.16138	5	72 Hf	$L\gamma_8$	$L_{II}O_{I}$	10.6754
1.16227	9	71 Lu		$L_{I}N_{IV}$	10.6672
1.1640	1	80 Hg	$L\eta$	$L_{II}M_{I}$	10.6512
1.16487	4	75 Re	$L\beta_9$	$L_{I}M_{V}$	10.6433
1.16545	5	77 Ir		$L_{III}N_{II}$	10.6380
1.1667	1	78 Pt	$L\beta_{17}$	$L_{II}M_{III}$	10.6265
1.16719	5	88 Ra	Ll	$L_{III}M_{I}$	10.6222
1.16962	9	78 Pt		$L_{I}M_{I}$	10.6001
1.16979	8	76 Os	$L\beta_2$	$L_{III}N_{V}$	10.5985
1.1708	1	79 Au		$L_{II}M_{II}$	10.5892
1.17167	5	76 Os	$L\beta_{15}$	$L_{III}N_{IV}$	10.5816
1.17218	5	75 Re	$L\beta_{10}$	$L_{I}M_{IV}$	10.5770
1.1729	1	73 Ta	$L\gamma_5$	$L_{II}N_{I}$	10.5702
1.17501	2	82 Pb	$L\alpha_1$	$L_{III}M_{V}$	10.5515
1.17588	1	33 As	$K\alpha_1$	KL_{III}	10.54372
1.17721	5	75 Re	$L\beta_5$	$L_{III}O_{IV,V}$	10.5318
1.1773	1	75 Re	L_{III}	Abs. Edge	10.5306
1.17788	9	72 Hf		$L_{II}N_{V}$	10.5258
1.17796	3	77 Ir	$L\beta_6$	$L_{III}N_{I}$	10.5251
1.17900	5	72 Hf	$L\gamma_1$	$L_{II}N_{IV}$	10.5158
1.17953	4	71 Lu	$L\gamma_3$	$L_{I}N_{III}$	10.5110
1.17955	7	76 Os	$L\beta_3$	$L_{I}M_{III}$	10.5108
1.17958	3	77 Ir	$L\beta_4$	$L_{I}M_{II}$	10.5106
1.17987	1	33 As	$K\alpha_2$	KL_{II}	10.50799
1.1815	1	75 Re	Lu	$L_{III}N_{VI,VII}$	10.4931
1.1818	1	70 Yb	L_{I}	Abs. Edge	10.4904
1.1827	1	70 Yb		$L_{I}O_{IV,V}$	10.4833
1.1853	1	70 Yb	$L\gamma_4$	$L_{I}O_{II,III}$	10.4603
1.1853	2	71 Lu	$L\gamma_2$	$L_{I}N_{II}$	10.460
1.18610	5	75 Re	$L\beta_7$	$L_{III}O_{I}$	10.4529
1.18648	5	82 Pb	$L\alpha_2$	$L_{III}M_{IV}$	10.4495

Wavelength Å*	p.e.	Element		Designation	keV
1.4941	3	68 Er	$L\beta_7$	$L_{III}O_I$	8.298
1.4941	3	68 Er	$L\beta_{10}$	L_IM_{IV}	8.298
1.4995	2	78 Pt	Ll	$L_{III}M_I$	8.268
1.500135	8	28 Ni	$K\beta_{1,3}$	$KM_{II,III}$	8.26466
1.5023	1	65 Tb	L_{II}	Abs. Edge	8.2527
1.5035	2	65 Tb	$L\gamma_6$	$L_{II}O_{IV}$	8.246
1.5063	2	69 Tm	$L\beta_3$	L_IM_{III}	8.231
1.5097	2	65 Tb	$L\gamma_8$	$L_{II}O_I$	8.212
1.51399	9	68 Er	$L\beta_{2,15}$	$L_{III}N_{IV,V}$	8.1890
1.5162	2	69 Tm	$L\beta_6$	$L_{III}N_I$	8.177
1.5178	1	75 Re	Ls	$L_{III}M_{III}$	8.1682
1.51824	7	66 Dy	$L\gamma_5$	$L_{II}N_I$	8.1661
1.52197	2	73 Ta	$L\alpha_1$	$L_{III}M_V$	8.1461
1.52325	5	72 Hf	$L\eta$	$L_{II}M_I$	8.1393
1.5297	2	64 Gd	$L\gamma_3$	L_IN_{III}	8.105
1.5303	2	65 Tb	$L\gamma_1$	$L_{II}N_{IV}$	8.102
1.5304	2	69 Tm	$L\beta_1$	$L_{II}M_{IV}$	8.101
1.53293	2	73 Ta	$L\alpha_2$	$L_{III}M_{IV}$	8.0879
1.5331	2	64 Gd	$L\gamma_2$	L_IN_{II}	8.087
1.53333	9	71 Lu		$L_{II}M_{II}$	8.0858
1.5347	2	76 Os	Lt	$L_{III}M_{II}$	8.079
1.5368	1	67 Ho	L_{III}	Abs. Edge	8.0676
1.5378	2	67 Ho	$L\beta_5$	$L_{III}O_{IV,V}$	8.062
1.5381	1	63 Eu	L_I	Abs. Edge	8.0607
1.540562	2	29 Cu	$K\alpha_1$	KL_{III}	8.04778
1.54094	3	77 Ir	Ll	$L_{III}M_I$	8.0458
1.5439	1	63 Eu	$L\gamma_4$	$L_IO_{II,III}$	8.0304
1.544390	9	29 Cu	$K\alpha_2$	KL_{II}	8.02783
1.5448	2	69 Tm	$L\beta_4$	L_IM_{II}	8.026
1.5486	3	67 Ho	$L\beta_{10}$	L_IM_{IV}	8.006
1.5616	1	68 Er	$L\beta_3$	L_IM_{III}	7.9392
1.5632	1	64 Gd	L_{II}	Abs. Edge	7.9310
1.5642	3	74 W	Ls	$L_{III}M_{III}$	7.926
1.5644	2	64 Gd	$L\gamma_6$	$L_{II}O_{IV}$	7.925
1.5671	2	67 Ho	$L\beta_{2,15}$	$L_{III}N_{IV,V}$	7.911
1.5675	2	68 Er	$L\beta_6$	$L_{III}N_I$	7.909
1.56958	5	72 Hf	$L\alpha_1$	$L_{III}M_V$	7.8990
1.5707	2	64 Gd	$L\gamma_8$	$L_{II}O_I$	7.894
1.5779	1	71 Lu	$L\eta$	$L_{II}M_I$	7.8575
1.5787	2	65 Tb	$L\gamma_5$	$L_{II}N_I$	7.8535
1.5789	1	75 Re	Lt	$L_{III}M_{II}$	7.8525
1.58046	5	72 Hf	$L\alpha_2$	$L_{III}M_{IV}$	7.8446
1.58498	7	76 Os	Ll	$L_{III}M_I$	7.8222
1.5873	1	68 Er	$L\beta_1$	$L_{II}M_{IV}$	7.8109
1.58837	7	66 Dy	$L\beta_5$	$L_{III}O_{IV,V}$	7.8055
1.58844	9	70 Yb		$L_{II}M_{II}$	7.8052
1.5903	2	63 Eu	$L\gamma_3$	L_IN_{III}	7.7961
1.5916	1	66 Dy	L_{III}	Abs. Edge	7.7897
1.5924	2	64 Gd	$L\gamma_1$	$L_{II}N_{IV}$	7.7858
1.5961	2	63 Eu	$L\gamma_2$	L_IN_{II}	7.7677
1.59973	9	66 Dy	$L\beta_9$	L_IM_V	7.7501
1.6002	1	62 Sm	L_I	Abs. Edge	7.7478
1.6007	1	68 Er	$L\beta_4$	L_IM_{II}	7.7453
1.60447	7	66 Dy	$L\beta_7$	$L_{III}O_I$	7.7272
1.60728	3	62 Sm	$L\gamma_4$	$L_IO_{II,III}$	7.714
1.60743	9	66 Dy	$L\beta_{10}$	L_IM_{IV}	7.7130
1.60815	1	27 Co	K	Abs. Edge	7.70954
1.60891	3	27 Co	$K\beta_5$	$KM_{IV,V}$	7.7059
1.61264	9	73 Ta	Ls	$L_{III}M_{III}$	7.6881
1.61951	3	71 Lu	$L\alpha_1$	$L_{III}M_V$	7.6555
1.6203	2	67 Ho	$L\beta_3$	L_IM_{III}	7.6519
1.62079	2	27 Co	$K\beta_{1,3}$	$KM_{II,III}$	7.64943
1.6237	2	67 Ho	$L\beta_6$	$L_{III}N_I$	7.6359
1.62369	7	66 Dy	$L\beta_{2,15}$	$L_{III}N_{IV,V}$	7.6357
1.6244	3	74 W	Lt	$L_{III}M_{II}$	7.6324
1.6271	1	63 Eu	L_{II}	Abs. Edge	7.6199
1.6282	2	63 Eu	$L\gamma_6$	$L_{II}O_{IV}$	7.6147
1.63029	5	71 Lu	$L\alpha_2$	$L_{III}M_{IV}$	7.6049
1.63056	5	75 Re	Ll	$L_{III}M_I$	7.6036
1.6346	2	63 Eu	$L\gamma_8$	$L_{II}O_I$	7.5849
1.63560	5	70 Yb	$L\eta$	$L_{II}M_I$	7.5802
1.6412	2	64 Gd	$L\gamma_5$	$L_{II}N_I$	7.5543
1.6475	2	67 Ho	$L\beta_1$	$L_{II}M_{IV}$	7.5253
1.6497	1	65 Tb	L_{III}	Abs. Edge	7.5153
1.6510	2	65 Tb	$L\beta_5$	$L_{III}O_{IV,V}$	7.5094
1.65601	3	62 Sm	$L\gamma_3$	L_IN_{III}	7.487
1.6574	2	63 Eu	$L\gamma_1$	$L_{II}N_{IV}$	7.4803
1.657910	8	28 Ni	$K\alpha_1$	KL_{III}	7.47815
1.6585	2	65 Tb	$L\beta_7$	$L_{III}O_I$	7.4753
1.6595	2	67 Ho	$L\beta_4$	L_IM_{II}	7.4708
1.66044	6	62 Sm	$L\gamma_2$	L_IN_{II}	7.467
1.661747	8	28 Ni	$K\alpha_2$	KL_{II}	7.46089
1.66346	9	72 Hf	Ls	$L_{III}M_{III}$	7.4532
1.6673	3	65 Tb	$L\beta_{10}$	L_IM_{IV}	7.436
1.6674	5	61 Pm	L_I	Abs. Edge	7.436
1.67189	4	70 Yb	$L\alpha_1$	$L_{III}M_V$	7.4156
1.67265	9	73 Ta	Lt	$L_{III}M_{II}$	7.4123
1.6782	1	74 W	Ll	$L_{III}M_I$	7.3878
1.68213	7	66 Dy	$L\beta_6$	$L_{III}N_I$	7.3705
1.6822	2	66 Dy	$L\beta_3$	L_IM_{III}	7.3702
1.68285	5	70 Yb	$L\alpha_2$	$L_{III}M_{IV}$	7.3673
1.6830	2	65 Tb	$L\beta_{2,15}$	$L_{III}N_{IV,V}$	7.3667
1.6953	1	62 Sm	L_{II}	Abs. Edge	7.3132
1.6963	2	69 Tm	$L\eta$	$L_{II}M_I$	7.3088
1.6966	9	62 Sm	$L\gamma_6$	$L_{II}O_{IV}$	7.308
1.7085	2	63 Eu	$L\gamma_5$	$L_{II}N_I$	7.2566
1.71062	7	66 Dy	$L\beta_1$	$L_{II}M_{IV}$	7.2477
1.7117	1	64 Gd	L_{III}	Abs. Edge	7.2430
1.7130	2	64 Gd	$L\beta_5$	$L_{III}O_{IV,V}$	7.2374
1.7203	2	64 Gd	$L\beta_7$	$L_{III}O_I$	7.2071
1.72103	7	66 Dy	$L\beta_4$	L_IM_{II}	7.2039
1.72305	9	72 Hf	Lt	$L_{III}M_{II}$	7.1954
1.7240	3	64 Gd	$L\beta_9$	L_IM_V	7.192
1.72724	3	62 Sm	$L\gamma_1$	$L_{II}N_{IV}$	7.178
1.7268	2	69 Tm	$L\alpha_1$	$L_{III}M_V$	7.1799
1.72841	5	73 Ta	Ll	$L_{III}M_I$	7.1731
1.7315	3	64 Gd	$L\beta_{10}$	L_IM_{IV}	7.160
1.7381	2	69 Tm	$L\alpha_2$	$L_{III}M_{IV}$	7.1331
1.7390	1	60 Nd	L_I	Abs. Edge	7.1294
1.7422	2	65 Tb	$L\beta_6$	$L_{III}N_I$	7.1163
1.74346	1	26 Fe	K	Abs. Edge	7.11120
1.7442	1	26 Fe	$K\beta_5$	$KM_{IV,V}$	7.1081
1.7445	4	60 Nd	$L\gamma_4$	$L_IO_{II,III}$	7.107
1.7455	2	64 Gd	$L\beta_{2,15}$	$L_{III}N_{IV,V}$	7.1028

Wavelength Å*	p.e.	Element	Designation		keV
1.1886	1	70 Yb		$L_{I}O_{I}$	10.4312
1.18977	7	76 Os		$L_{II}M_{V}$	10.4205
1.1958	1	31 Ga	K	Abs. Edge	10.3682
1.19600	2	31 Ga	$K\beta_2$	$KN_{II,III}$	10.3663
1.19727	7	76 Os	$L\beta_1$	$L_{II}M_{IV}$	10.3553
1.1981	2	31 Ga	$K\beta_5$	$KM_{IV,V}$	10.348
1.1985	1	71 Lu	L_{II}	Abs. Edge	10.3448
1.1987	1	71 Lu	$L\gamma_6$	$L_{II}O_{IV}$	10.3431
1.20086	7	76 Os		$L_{III}N_{II}$	10.3244
1.2014	1	71 Lu		$L_{II}O_{II,III}$	10.3198
1.20273	3	79 Au	$L\eta$	$L_{II}M_{I}$	10.3083
1.2047	1	71 Lu	$L\gamma_8$	$L_{II}O_{I}$	10.2915
1.20479	7	74 W	$L\beta_9$	$L_{I}M_{V}$	10.2907
1.20660	4	75 Re	$L\beta_2$	$L_{III}N_{V}$	10.2752
1.2069	2	77 Ir	$L\beta_{17}$	$L_{II}M_{III}$	10.273
1.20739	4	81 Tl	$L\alpha_1$	$L_{III}M_{V}$	10.2685
1.20789	2	31 Ga	$K\beta_1$	KM_{III}	10.2642
1.20819	5	75 Re	$L\beta_{15}$	$L_{III}N_{IV}$	10.2617
1.20835	5	31 Ga	$K\beta_3$	KM_{II}	10.2603
1.2102	2	77 Ir		$L_{I}M_{I}$	10.245
1.2105	1	83 Bi	Ls	$L_{III}M_{III}$	10.2421
1.21218	3	74 W	$L\beta_{10}$	$L_{I}M_{IV}$	10.2279
1.213	1	78 Pt		$L_{II}M_{II}$	10.225
1.21349	5	76 Os	$L\beta_6$	$L_{III}N_{I}$	10.2169
1.21537	5	72 Hf	$L\gamma_5$	$L_{II}N_{I}$	10.2011
1.21545	3	74 W	$L\beta_5$	$L_{III}O_{IV,V}$	10.2004
1.2155	1	74 W	L_{III}	Abs. Edge	10.1999
1.21844	5	76 Os	$L\beta_4$	$L_{I}M_{II}$	10.1754
1.21868	5	74 W	Lu	$L_{III}N_{VI,VII}$	10.1733
1.21875	3	81 Tl	$L\alpha_2$	$L_{III}M_{IV}$	10.1728
1.22031	5	75 Re	$L\beta_3$	$L_{I}M_{III}$	10.1598
1.2211	2	74 W		$L_{III}O_{II,III}$	10.153
1.22228	4	71 Lu	$L\gamma_1$	$L_{II}N_{IV}$	10.1434
1.22232	5	70 Yb	$L\gamma_3$	$L_{I}N_{III}$	10.1431
1.22400	4	74 W	$L\beta_7$	$L_{III}O_{I}$	10.1292
1.2250	1	69 Tm	L_{I}	Abs. Edge	10.1206
1.2263	3	69 Tm		$L_{I}O_{IV,V}$	10.110
1.2283	1	75 Re		$L_{III}N_{III}$	10.0933
1.22879	7	70 Yb	$L\gamma_2$	$L_{I}N_{II}$	10.0897
1.2294	2	69 Tm	$L\gamma_4$	$L_{I}O_{II,III}$	10.084
1.2305	1	75 Re		$L_{II}M_{V}$	10.0753
1.23858	2	75 Re	$L\beta_1$	$L_{II}M_{IV}$	10.0100
1.24120	5	80 Hg	$L\alpha_1$	$L_{III}M_{V}$	9.9888
1.24271	3	70 Yb	$L\gamma_6$	$L_{II}O_{IV}$	9.9766
1.2428	1	70 Yb	L_{II}	Abs. Edge	9.9761
1.2429	2	78 Pt	$L\eta$	$L_{II}M_{I}$	9.975
1.24385	7	82 Pb	Ls	$L_{III}M_{III}$	9.9675
1.24460	3	74 W	$L\beta_2$	$L_{III}N_{V}$	9.9615
1.2453	1	70 Yb		$L_{II}O_{II,III}$	9.9561
1.24631	3	74 W	$L\beta_{15}$	$L_{III}N_{IV}$	9.9478
1.2466	2	73 Ta	$L\beta_9$	$L_{I}M_{V}$	9.946
1.2480	2	76 Os	$L\beta_{17}$	$L_{II}M_{III}$	9.934
1.24923	5	70 Yb	$L\gamma_8$	$L_{II}O_{I}$	9.9246
1.2502	3	77 Ir		$L_{II}M_{II}$	9.917
1.25100	5	75 Re	$L\beta_6$	$L_{III}N_{I}$	9.9105
1.25264	7	80 Hg	$L\alpha_2$	$L_{III}M_{IV}$	9.8976
1.2537	2	73 Ta	$L\beta_{10}$	$L_{I}M_{IV}$	9.889

Wavelength Å*	p.e.	Element	Designation		keV
1.254054	9	32 Ge	$K\alpha_1$	KL_{III}	9.88642
1.2553	1	73 Ta	L_{III}	Abs. Edge	9.8766
1.2555	1	73 Ta	$L\beta_5$	$L_{III}O_{IV,V}$	9.8750
1.25778	4	73 Ta	Lu	$L_{III}N_{VI,VII}$	9.8572
1.258011	9	32 Ge	$K\alpha_2$	KL_{II}	9.85532
1.25917	5	75 Re	$L\beta_4$	$L_{I}M_{II}$	9.8463
1.2596	1	71 Lu	$L\gamma_5$	$L_{II}N_{I}$	9.8428
1.2601	3	73 Ta		$L_{III}O_{II,III}$	9.839
1.26269	5	74 W	$L\beta_3$	$L_{I}M_{III}$	9.8188
1.26385	5	73 Ta	$L\beta_7$	$L_{III}O_{I}$	9.8098
1.2672	2	74 W		$L_{III}N_{III}$	9.784
1.26769	5	70 Yb	$L\gamma_1$	$L_{II}N_{IV}$	9.7801
1.2678	2	69 Tm	$L\gamma_3$	$L_{I}N_{III}$	9.779
1.2706	1	68 Er	L_{I}	Abs. Edge	9.7574
1.2728	2	74 W		$L_{II}M_{V}$	9.741
1.2742	2	69 Tm	$L\gamma_2$	$L_{I}N_{II}$	9.730
1.2748	1	83 Bi	Lt	$L_{III}M_{II}$	9.7252
1.2752	2	68 Er	$L\gamma_4$	$L_{I}O_{II,III}$	9.722
1.27640	3	79 Au	$L\alpha_1$	$L_{III}M_{V}$	9.7133
1.2765	2	74 W		$L_{III}N_{II}$	9.712
1.27807	5	81 Tl	Ls	$L_{III}M_{III}$	9.7007
1.281809	9	74 W	$L\beta_1$	$L_{II}M_{IV}$	9.67235
1.2829	5	84 Po	Ll	$L_{III}M_{I}$	9.664
1.2834	1	30 Zn	K	Abs. Edge	9.6607
1.28372	2	30 Zn	$K\beta_2$	$KN_{II,III}$	9.6580
1.28448	3	77 Ir	$L\eta$	$L_{II}M_{I}$	9.6522
1.28454	2	73 Ta	$L\beta_2$	$L_{III}N_{V}$	9.6518
1.2848	1	30 Zn	$K\beta_5$	$KM_{IV,V}$	9.6501
1.28619	5	73 Ta	$L\beta_{15}$	$L_{III}N_{IV}$	9.6394
1.28772	3	79 Au	$L\alpha_2$	$L_{III}M_{IV}$	9.6280
1.2892	1	69 Tm	L_{II}	Abs. Edge	9.6171
1.28989	7	74 W	$L\beta_6$	$L_{III}N_{I}$	9.6117
1.29025	9	72 Hf	$L\beta_9$	$L_{I}M_{V}$	9.6090
1.2905	2	69 Tm	$L\gamma_6$	$L_{II}O_{IV}$	9.607
1.2927	1	75 Re	$L\beta_{17}$	$L_{II}M_{III}$	9.5910
1.2934	2	76 Os		$L_{II}M_{II}$	9.586
1.29525	2	30 Zn	$K\beta_{1,3}$	$KM_{II,III}$	9.5720
1.2972	1	72 Hf	L_{III}	Abs. Edge	9.5577
1.29761	5	72 Hf	$L\beta_5$	$L_{III}O_{IV,V}$	9.5546
1.29819	9	72 Hf	$L\beta_{10}$	$L_{I}M_{IV}$	9.5503
1.30162	5	74 W	$L\beta_4$	$L_{I}M_{II}$	9.5252
1.30165	9	72 Hf	Lu	$L_{III}N_{VI,VII}$	9.5249
1.30564	5	72 Hf	$L\beta_7$	$L_{III}O_{I}$	9.4958
1.3063	1	70 Yb	$L\gamma_5$	$L_{II}N_{I}$	9.4910
1.30678	3	73 Ta	$L\beta_3$	$L_{I}M_{III}$	9.4875
1.30767	7	82 Pb	Lt	$L_{III}M_{II}$	9.4811
1.3086	1	73 Ta		$L_{III}N_{III}$	9.4742
1.3112	2	80 Hg	Ls	$L_{III}M_{III}$	9.455
1.31304	3	78 Pt	$L\alpha_1$	$L_{III}M_{V}$	9.4423
1.3146	1	68 Er	$L\gamma_3$	$L_{I}N_{III}$	9.4309
1.3153	2	69 Tm	$L\gamma_1$	$L_{II}N_{IV}$	9.426
1.31610	7	83 Bi	Ll	$L_{III}M_{I}$	9.4204
1.3167	1	73 Ta		$L_{III}N_{II}$	9.4158
1.31897	9	73 Ta		$L_{II}M_{V}$	9.3998
1.3190	1	67 Ho	L_{I}	Abs. Edge	9.3994
1.3208	3	67 Ho		$L_{I}O_{IV,V}$	9.387
1.3210	2	68 Er	$L\gamma_2$	$L_{I}N_{II}$	9.385

TABLE B (*Continued*)

Wavelength Å*	p.e.	Element	Designation		keV
1.3225	2	67 Ho	$L\gamma_4$	$L_IO_{II,III}$	9.374
1.32432	2	78 Pt	$L\alpha_2$	$L_{III}M_{IV}$	9.3618
1.32639	5	72 Hf	$L\beta_2$	$L_{III}N_V$	9.3473
1.32698	3	73 Ta	$L\beta_1$	$L_{II}M_{IV}$	9.3431
1.32783	5	72 Hf	$L\beta_{15}$	$L_{III}N_{IV}$	9.3371
1.32785	7	76 Os	$L\eta$	$L_{II}M_I$	9.3370
1.33094	8	73 Ta	$L\beta_6$	$L_{III}N_I$	9.3153
1.3358	1	71 Lu	$L\beta_9$	L_IM_V	9.2816
1.3365	3	74 W		L_IM_I	9.277
1.3366	1	75 Re		$L_{II}M_{II}$	9.2761
1.3386	1	68 Er	L_{II}	Abs. Edge	9.2622
1.3387	2	74 W	$L\beta_{17}$	$L_{II}M_{III}$	9.261
1.3397	3	68 Er	$L\gamma_6$	$L_{II}O_{IV}$	9.255
1.340083	9	31 Ga	$K\alpha_1$	KL_{III}	9.25174
1.3405	1	71 Lu	L_{III}	Abs. Edge	9.2490
1.34154	5	81 Tl	Lt	$L_{III}M_{II}$	9.2417
1.34183	7	71 Lu	$L\beta_5$	$L_{III}O_{IV,V}$	9.2397
1.3430	2	71 Lu	$L\beta_{10}$	L_IM_{IV}	9.232
1.34399	1	31 Ga	$K\alpha_2$	KL_{II}	9.22482
1.34524	9	71 Lu		$L_{III}O_{II,III}$	9.2163
1.34581	3	73 Ta	$L\beta_4$	L_IM_{II}	9.2124
1.34949	5	71 Lu	$L\beta_7$	$L_{III}O_I$	9.1873
1.34990	7	82 Pb	Ll	$L_{III}M_I$	9.1845
1.35053	9	72 Hf		$L_{III}N_{III}$	9.1802
1.35128	3	77 Ir	$L\alpha_1$	$L_{III}M_V$	9.1751
1.35131	7	79 Au	Ls	$L_{III}M_{III}$	9.1749
1.35300	5	72 Hf	$L\beta_3$	L_IM_{III}	9.1634
1.3558	2	69 Tm	$L\gamma_5$	$L_{II}N_I$	9.144
1.35887	9	72 Hf		$L_{III}N_{II}$	9.1239
1.36250	5	77 Ir	$L\alpha_2$	$L_{III}M_{IV}$	9.0995
1.3641	2	68 Er	$L\gamma_1$	$L_{II}N_{IV}$	9.089
1.3643	2	67 Ho	$L\gamma_3$	L_IN_{III}	9.087
1.3692	1	66 Dy	L_I	Abs. Edge	9.0548
1.3698	2	67 Ho	$L\gamma_2$	L_IN_{II}	9.051
1.37012	3	71 Lu	$L\beta_2$	$L_{III}N_V$	9.0489
1.3715	1	71 Lu	$L\beta_{15}$	$L_{III}N_{IV}$	9.0395
1.37342	5	75 Re	$L\eta$	$L_{II}M_I$	9.0272
1.37410	5	72 Hf	$L\beta_1$	$L_{II}M_{IV}$	9.0227
1.37410	5	72 Hf	$L\beta_6$	$L_{III}N_I$	9.0227
1.37459	7	66 Dy	$L\gamma_4$	$L_IO_{II,III}$	9.0195
1.3746	2	80 Hg	Lt	$L_{III}M_{II}$	9.019
1.38059	5	29 Cu	K	Abs. Edge	8.9803
1.38109	3	29 Cu	$K\beta_2$	$KM_{IV,V}$	8.9770
1.3838	1	70 Yb	$L\beta_9$	L_IM_V	8.9597
1.38477	3	81 Tl	Ll	$L_{III}M_I$	8.9532
1.3862	1	70 Yb	L_{III}	Abs. Edge	8.9441
1.3864	1	73 Ta	$L\beta_{17}$	$L_{II}M_{III}$	8.9428
1.38696	7	70 Yb	$L\beta_5$	$L_{III}O_{IV,V}$	8.9390
1.3895	2	78 Pt	Ls	$L_{III}M_{III}$	8.923
1.3898	1	70 Yb		$L_{III}O_{II,III}$	8.9209
1.3905	1	67 Ho	L_{II}	Abs. Edge	8.9164
1.39121	5	76 Os	$L\alpha_1$	$L_{III}M_V$	8.9117
1.3915	1	70 Yb	$L\beta_{10}$	L_IM_{IV}	8.9100
1.39220	5	72 Hf	$L\beta_4$	L_IM_{II}	8.9054
1.392218	9	29 Cu	$K\beta_{1,3}$	$KM_{II,III}$	8.90529
1.3923	2	67 Ho	$L\gamma_6$	$L_{II}O_{IV}$	8.905
1.3926	1	29 Cu	$K\beta_3$	KM_{II}	8.9029
1.3948	1	70 Yb	$L\beta_7$	$L_{III}O_I$	8.8889
1.3983	2	67 Ho	$L\gamma_8$	$L_{II}O_I$	8.867
1.40140	5	71 Lu	$L\beta_3$	L_IM_{III}	8.8469
1.40234	5	76 Os	$L\alpha_2$	$L_{III}M_{IV}$	8.8410
1.4067	3	68 Er	$L\gamma_5$	$L_{II}N_I$	8.814
1.41366	7	79 Au	Lt	$L_{III}M_{II}$	8.7702
1.41550	5	70 Yb	$L\beta_{2,15}$	$L_{III}N_{IV,V}$	8.7588
1.41640	7	66 Dy	$L\gamma_3$	L_IN_{III}	8.7532
1.4174	2	67 Ho	$L\gamma_1$	$L_{II}N_{IV}$	8.747
1.4189	1	71 Lu	$L\beta_6$	$L_{III}N_I$	8.7376
1.42110	3	74 W	$L\eta$	$L_{II}M_I$	8.7243
1.4216	1	80 Hg	Ll	$L_{III}M_I$	8.7210
1.4223	1	65 Tb	L_I	Abs. Edge	8.7167
1.42278	7	66 Dy	$L\gamma_2$	L_IN_{II}	8.7140
1.4228	3	65 Tb		$L_IO_{IV,V}$	8.714
1.42359	3	71 Lu	$L\beta_1$	$L_{II}M_{IV}$	8.7090
1.4276	2	65 Tb	$L\gamma_4$	$L_IO_{II,III}$	8.685
1.43025	9	72 Hf		L_IM_I	8.6685
1.43048	9	73 Ta		$L_{II}M_{II}$	8.6671
1.4318	2	77 Ir	Ls	$L_{III}M_{III}$	8.659
1.43290	4	75 Re	$L\alpha_1$	$L_{III}M_V$	8.6525
1.4334	1	69 Tm	L_{III}	Abs. Edge	8.6496
1.4336	3	69 Tm	$L\beta_9$	L_IM_V	8.648
1.4349	2	69 Tm	$L\beta_5$	$L_{III}O_{IV,V}$	8.641
1.435155	7	30 Zn	$K\alpha_1$	KL_{III}	8.63886
1.43643	9	72 Hf	$L\beta_{17}$	$L_{II}M_{III}$	8.6312
1.439000	8	30 Zn	$K\alpha_2$	KL_{II}	8.61578
1.44056	5	71 Lu	$L\beta_4$	L_IM_{II}	8.6064
1.4410	3	69 Tm	$L\beta_{10}$	L_IM_{IV}	8.604
1.44396	5	75 Re	$L\alpha_2$	$L_{III}M_{IV}$	8.5862
1.4445	1	66 Dy	L_{II}	Abs. Edge	8.5830
1.44579	7	66 Dy	$L\gamma_6$	$L_{II}O_{IV}$	8.5753
1.45233	5	70 Yb	$L\beta_3$	L_IM_{III}	8.5367
1.4530	2	78 Pt	Lt	$L_{III}M_{II}$	8.533
1.45964	9	79 Au	Ll	$L_{III}M_I$	8.4939
1.4618	2	67 Ho	$L\gamma_5$	$L_{II}N_I$	8.481
1.4640	2	69 Tm	$L\beta_{2,15}$	$L_{III}N_{IV,V}$	8.468
1.4661	1	70 Yb	$L\beta_6$	$L_{III}N_I$	8.4563
1.47106	5	73 Ta	$L\eta$	$L_{II}M_I$	8.4280
1.4718	2	65 Tb	$L\gamma_3$	L_IN_{III}	8.423
1.47266	7	66 Dy	$L\gamma_1$	$L_{II}N_{IV}$	8.4188
1.4735	2	76 Os	Ls	$L_{III}M_{III}$	8.414
1.47565	5	70 Yb	$L\beta_1$	$L_{II}M_{IV}$	8.4018
1.4764	2	65 Tb	$L\gamma_2$	L_IN_{II}	8.398
1.47639	2	74 W	$L\alpha_1$	$L_{III}M_V$	8.3976
1.4784	1	64 Gd	L_I	Abs. Edge	8.3864
1.48064	9	72 Hf		$L_{II}M_{II}$	8.3735
1.4807	3	64 Gd		$L_IO_{IV,V}$	8.373
1.4835	1	68 Er	L_{III}	Abs. Edge	8.3575
1.4839	2	64 Gd	$L\gamma_4$	$L_IO_{II,III}$	8.355
1.4848	3	68 Er	$L\beta_5$	$L_{III}O_{IV,V}$	8.350
1.4855	5	68 Er	$L\beta_9$	L_IM_V	8.346
1.48743	2	74 W	$L\alpha_2$	$L_{III}M_{IV}$	8.3352
1.48807	1	28 Ni	K	Abs. Edge	8.33165
1.48862	4	28 Ni	$K\beta_5$	$KM_{IV,V}$	8.3286
1.49138	3	70 Yb	$L\beta_4$	L_IM_{II}	8.3132
1.4930	3	77 Ir	Lt	$L_{III}M_{II}$	8.304

Wavelength Å*	p.e.	Element		Designation	keV
1.7472	2	65 Tb	$L\beta_3$	L_IM_{III}	7.0959
1.75661	2	26 Fe	$K\beta_{1,3}$	$KM_{II,III}$	7.05798
1.7566	1	68 Er	$L\eta$	$L_{II}M_I$	7.0579
1.7676	5	61 Pm	L_{II}	Abs. Edge	7.014
1.7760	1	71 Lu	Lt	$L_{III}M_{II}$	6.9810
1.7761	1	63 Eu	L_{III}	Abs. Edge	6.9806
1.7768	3	65 Tb	$L\beta_1$	$L_{II}M_{IV}$	6.978
1.7772	2	63 Eu	$L\beta_5$	$L_{III}O_{IV,V}$	6.9763
1.77934	3	62 Sm	$L\gamma_5$	$L_{II}N_I$	6.968
1.78145	5	72 Hf	Ll	$L_{III}M_I$	6.9596
1.78425	9	68 Er	$L\alpha_1$	$L_{III}M_V$	6.9487
1.7851	2	63 Eu	$L\beta_7$	$L_{III}O_I$	6.9453
1.7864	2	65 Tb	$L\beta_4$	L_IM_{II}	6.9403
1.788965	9	27 Co	$K\alpha_1$	KL_{III}	6.93032
1.7916	3	63 Eu	$L\beta_9$	L_IM_V	6.920
1.792850	9	27 Co	$K\alpha_2$	KL_{II}	6.91530
1.7955	2	68 Er	$L\alpha_2$	$L_{III}M_{IV}$	6.9050
1.7964	4	60 Nd	$L\gamma_3$	L_IN_{III}	6.902
1.7989	9	61 Pm	$L\gamma_1$	$L_{II}N_{IV}$	6.892
1.7993	3	63 Eu	$L\beta_{10}$	L_IM_{IV}	6.890
1.8013	4	60 Nd	$L\gamma_2$	L_IN_{II}	6.883
1.8054	2	64 Gd	$L\beta_6$	$L_{III}N_I$	6.8671
1.8118	2	63 Eu	$L\beta_{2,15}$	$L_{III}N_{IV,V}$	6.8432
1.8141	5	59 Pr	L_I	Abs. Edge	6.834
1.8150	2	64 Gd	$L\beta_3$	L_IM_{III}	6.8311
1.8193	4	59 Pr	$L\gamma_4$	$L_IO_{II,III}$	6.815
1.8264	2	67 Ho	$L\eta$	$L_{II}M_I$	6.7883
1.83091	9	70 Yb	Lt	$L_{III}M_{II}$	6.7715
1.8360	1	71 Lu	Ll	$L_{III}M_I$	6.7528
1.8440	1	60 Nd	L_{II}	Abs. Edge	6.7234
1.8450	2	67 Ho	$L\alpha_1$	$L_{III}M_V$	6.7198
1.8457	1	62 Sm	L_{III}	Abs. Edge	6.7172
1.8468	2	64 Gd	$L\beta_1$	$L_{II}M_{IV}$	6.7132
1.84700	9	62 Sm	$L\beta_5$	$L_{III}O_{IV,V}$	6.7126
1.8540	2	64 Gd	$L\beta_4$	L_IM_{II}	6.6871
1.8552	5	60 Nd	$L\gamma_8$	$L_{II}O_I$	6.683
1.8561	2	67 Ho	$L\alpha_2$	$L_{III}M_{IV}$	6.6795
1.85626	3	62 Sm	$L\beta_7$	$L_{III}O_I$	6.679
1.86166	3	62 Sm	$L\beta_9$	L_IM_V	6.660
1.86990	3	62 Sm	$L\beta_{10}$	L_IM_{IV}	6.634
1.8737	2	63 Eu	$L\beta_6$	$L_{III}N_I$	6.6170
1.8740	4	59 Pr	$L\gamma_3$	L_IN_{III}	6.616
1.8779	2	60 Nd	$L\gamma_1$	$L_{II}N_{IV}$	6.6021
1.8791	4	59 Pr	$L\gamma_2$	L_IN_{II}	6.598
1.8821	3	62 Sm	$L\beta_{2,15}$	$L_{III}N_{IV,V}$	6.586
1.8867	2	63 Eu	$L\beta_3$	L_IM_{III}	6.5713
1.8934	5	58 Ce	L_I	Abs. Edge	6.548
1.89415	5	70 Yb	Ll	$L_{III}M_I$	6.5455
1.89643	5	25 Mn	K	Abs. Edge	6.5376
1.8971	1	25 Mn	$K\beta_5$	$KM_{IV,V}$	6.5352
1.89743	7	66 Dy	$L\eta$	$L_{II}M_I$	6.5342
1.8991	4	58 Ce	$L\gamma_4$	$L_IO_{II,III}$	6.528
1.90881	3	66 Dy	$L\alpha_1$	$L_{III}M_V$	6.4952
1.91021	2	25 Mn	$K\beta_{1,3}$	$KM_{II,III}$	6.49045
1.9191	1	61 Pm	L_{III}	Abs. Edge	6.4605
1.91991	3	66 Dy	$L\alpha_2$	$L_{III}M_{IV}$	6.4577
1.9203	2	63 Eu	$L\beta_1$	$L_{II}M_{IV}$	6.4564

Wavelength Å*	p.e.	Element		Designation	keV
1.9255	2	63 Eu	$L\beta_4$	L_IM_{II}	6.4389
1.9255	5	59 Pr	L_{II}	Abs. Edge	6.439
1.9355	4	60 Nd	$L\gamma_5$	$L_{II}N_I$	6.406
1.936042	9	26 Fe	$K\alpha_1$	KL_{III}	6.40384
1.9362	4	59 Pr	$L\gamma_8$	$L_{II}O_I$	6.403
1.939980	9	26 Fe	$K\alpha_2$	KL_{II}	6.39084
1.94643	3	62 Sm	$L\beta_6$	$L_{III}N_I$	6.3693
1.9550	2	69 Tm	Ll	$L_{III}M_I$	6.3419
1.9553	3	58 Ce	$L\gamma_3$	L_IN_{III}	6.3409
1.9559	6	61 Pm	$L\beta_{2,15}$	$L_{III}N_{IV,V}$	6.339
1.9602	3	58 Ce	$L\gamma_2$	L_IN_{II}	6.3250
1.9611	3	59 Pr	$L\gamma_1$	$L_{II}N_{IV}$	6.3221
1.96241	3	62 Sm	$L\beta_3$	L_IM_{III}	6.318
1.9730	2	65 Tb	$L\eta$	$L_{II}M_I$	6.2839
1.9765	2	65 Tb	$L\alpha_1$	$L_{III}M_V$	6.2728
1.9780	5	57 La	L_I	Abs. Edge	6.268
1.9830	4	57 La	$L\gamma_4$	$L_IO_{II,III}$	6.252
1.9875	2	65 Tb	$L\alpha_2$	$L_{III}M_{IV}$	6.2380
1.9967	1	60 Nd	L_{III}	Abs. Edge	6.2092
1.99806	3	62 Sm	$L\beta_1$	$L_{II}M_{IV}$	6.2051
2.00095	6	62 Sm	$L\beta_4$	L_IM_{II}	6.196
2.0092	3	60 Nd	$L\beta_7$	$L_{III}O_I$	6.1708
2.0124	5	58 Ce	L_{II}	Abs. Edge	6.161
2.015	1	68 Er	Ll	$L_{III}M_I$	6.152
2.0165	3	60 Nd	$L\beta_9$	L_IM_V	6.1484
2.0205	4	59 Pr	$L\gamma_5$	$L_{II}N_I$	6.136
2.0237	4	58 Ce	$L\gamma_8$	$L_{II}O_I$	6.126
2.0237	3	60 Nd	$L\beta_{10}$	L_IM_{IV}	6.1265
2.0360	3	60 Nd	$L\beta_{2,15}$	$L_{III}N_{IV,V}$	6.0894
2.0410	4	57 La	$L\gamma_3$	L_IN_{III}	6.074
2.0421	4	61 Pm	$L\beta_3$	L_IM_{III}	6.071
2.0460	4	57 La	$L\gamma_2$	L_IN_{II}	6.060
2.0468	2	64 Gd	$L\alpha_1$	$L_{III}M_V$	6.0572
2.0487	4	58 Ce	$L\gamma_1$	$L_{II}N_{IV}$	6.052
2.0494	1	64 Gd	$L\eta$	$L_{II}M_I$	6.0495
2.0578	2	64 Gd	$L\alpha_2$	$L_{III}M_{IV}$	6.0250
2.0678	5	56 Ba	L_I	Abs. Edge	5.996
2.07020	5	24 Cr	K	Abs. Edge	5.9888
2.07087	6	24 Cr	$K\beta_5$	$KM_{IV,V}$	5.9869
2.0756	3	56 Ba	$L\gamma_4$	$L_IO_{II,III}$	5.9733
2.0791	5	59 Pr	L_{III}	Abs. Edge	5.963
2.0797	4	61 Pm	$L\beta_1$	$L_{II}M_{IV}$	5.961
2.08487	2	24 Cr	$K\beta_{1,3}$	$KM_{II,III}$	5.94671
2.0860	2	67 Ho	Ll	$L_{III}M_I$	5.9434
2.0919	4	59 Pr	$L\beta_7$	$L_{III}O_I$	5.927
2.1004	4	59 Pr	$L\beta_9$	L_IM_V	5.903
2.101820	9	25 Mn	$K\alpha_1$	KL_{III}	5.89875
2.1039	3	60 Nd	$L\beta_6$	$L_{III}N_I$	5.8930
2.1053	5	57 La	L_{II}	Abs. Edge	5.889
2.10578	2	25 Mn	$K\alpha_2$	KL_{II}	5.88765
2.1071	4	59 Pr	$L\beta_{10}$	L_IM_{IV}	5.884
2.1103	3	58 Ce	$L\gamma_5$	$L_{II}N_I$	5.8751
2.1194	4	59 Pr	$L\beta_{2,15}$	$L_{III}N_{IV,V}$	5.850
2.1209	2	63 Eu	$L\alpha_1$	$L_{III}M_V$	5.8457
2.1268	2	60 Nd	$L\beta_3$	L_IM_{III}	5.8294
2.1315	2	63 Eu	$L\eta$	$L_{II}M_I$	5.8166
2.1315	2	63 Eu	$L\alpha_2$	$L_{III}M_{IV}$	5.8166

Wavelength Å*	p.e.	Element	Designation		keV
2.1342	2	56 Ba	$L\gamma_3$	L_IN_{III}	5.8092
2.1387	2	56 Ba	$L\gamma_2$	L_IN_{II}	5.7969
2.1418	3	57 La	$L\gamma_1$	$L_{II}N_{IV}$	5.7885
2.15877	7	66 Dy	Ll	$L_{III}M_I$	5.7431
2.166	1	58 Ce	L_{III}	Abs. Edge	5.723
2.1669	3	60 Nd	$L\beta_4$	L_IM_{II}	5.7216
2.1669	2	60 Nd	$L\beta_1$	$L_{II}M_{IV}$	5.7216
2.1673	5	55 Cs	L_I	Abs. Edge	5.721
2.1701	2	58 Ce	$L\beta_7$	$L_{III}O_I$	5.7132
2.1741	2	55 Cs	$L\gamma_4$	$L_IO_{II,III}$	5.7026
2.1885	3	58 Ce	$L\beta_9$	L_IM_V	5.6650
2.1906	4	59 Pr	$L\beta_6$	$L_{III}N_I$	5.660
2.1958	5	58 Ce	$L\beta_{10}$	L_IM_{IV}	5.646
2.1998	2	62 Sm	$L\alpha_1$	$L_{III}M_V$	5.6361
2.2048	1	56 Ba	L_{II}	Abs. Edge	5.6233
2.2056	4	57 La	$L\gamma_5$	$L_{II}N_I$	5.621
2.2087	2	58 Ce	$L\beta_{2,15}$	$L_{III}N_{IV,V}$	5.6134
2.21062	3	62 Sm	$L\alpha_2$	$L_{III}M_{IV}$	5.6090
2.2172	3	59 Pr	$L\beta_3$	L_IM_{III}	5.5918
2.21824	3	62 Sm	$L\eta$	$L_{II}M_I$	5.589
2.2328	2	55 Cs	$L\gamma_3$	L_IN_{III}	5.5527
2.2352	2	65 Tb	Ll	$L_{III}M_I$	5.5467
2.2371	2	55 Cs	$L\gamma_2$	L_IN_{II}	5.5420
2.2415	2	56 Ba	$L\gamma_1$	$L_{II}N_{IV}$	5.5311
2.253	6	92 U		M_IP_{III}	5.50
2.2550	4	59 Pr	$L\beta_4$	L_IM_{II}	5.4981
2.2588	3	59 Pr	$L\beta_1$	$L_{II}M_{IV}$	5.4889
2.261	1	57 La	L_{III}	Abs. Edge	5.484
2.2691	1	23 V	K	Abs. Edge	5.4639
2.26951	6	23 V	$K\beta_5$	$KM_{IV,V}$	5.4629
2.2737	1	54 Xe	L_I	Abs. Edge	5.4528
2.275	3	57 La	$L\beta_7$	$L_{III}O_I$	5.450
2.282	3	57 La	$L\beta_9$	L_IM_V	5.434
2.2818	3	58 Ce	$L\beta_6$	$L_{III}N_I$	5.4334
2.2822	3	61 Pm	$L\alpha_1$	$L_{III}M_V$	5.4325
2.28440	2	23 V	$K\beta_{1,3}$	$KM_{II,III}$	5.42729
2.28970	2	24 Cr	$K\alpha_1$	KL_{III}	5.41472
2.290	3	57 La	$L\beta_{10}$	L_IM_{IV}	5.415
2.2926	4	61 Pm	$L\alpha_2$	$L_{III}M_{IV}$	5.4078
2.293606	3	24 Cr	$K\alpha_2$	KL_{II}	5.405509
2.3030	3	57 La	$L\beta_{2,15}$	$L_{III}N_{IV,V}$	5.3835
2.304	7	92 U		M_IO_{III}	5.38
2.3085	3	56 Ba	$L\gamma_5$	$L_{II}N_I$	5.3707
2.3109	3	58 Ce	$L\beta_3$	L_IM_{III}	5.3651
2.3122	2	64 Gd	Ll	$L_{III}M_I$	5.3621
2.3139	1	55 Cs	L_{II}	Abs. Edge	5.3581
2.3480	2	55 Cs	$L\gamma_1$	$L_{II}N_{IV}$	5.2804
2.3497	4	58 Ce	$L\beta_4$	L_IM_{II}	5.2765
2.3561	3	58 Ce	$L\beta_1$	$L_{II}M_{IV}$	5.2622
2.3629	1	56 Ba	L_{III}	Abs. Edge	5.2470
2.3704	2	60 Nd	$L\alpha_1$	$L_{III}M_V$	5.2304
2.3764	2	56 Ba	$L\beta_9$	L_IM_V	5.2171
2.3790	4	57 La	$L\beta_6$	$L_{III}N_I$	5.2114
2.3806	2	56 Ba	$L\beta_7$	$L_{III}O_I$	5.2079
2.3807	3	60 Nd	$L\alpha_2$	$L_{III}M_{IV}$	5.2077
2.3869	2	56 Ba	$L\beta_{10}$	L_IM_{IV}	5.1941
2.3880	5	53 I	L_I	Abs. Edge	5.192

Wavelength Å*	p.e.	Element	Designation		keV
2.3913	2	53 I	$L\gamma_4$	$L_IO_{II,III}$	5.1848
2.3948	2	63 Eu	Ll	$L_{III}M_I$	5.1772
2.40435	6	56 Ba	$L\beta_{2,15}$	$L_{III}N_{IV,V}$	5.1565
2.4094	4	60 Nd	$L\eta$	$L_{II}M_I$	5.1457
2.4105	3	57 La	$L\beta_3$	L_IM_{III}	5.1434
2.4174	2	55 Cs	$L\gamma_5$	$L_{II}N_I$	5.1287
2.4292	1	54 Xe	L_{II}	Abs. Edge	5.1037
2.442	9	90 Th		M_IO_{III}	5.08
2.443	4	92 U		$M_{II}O_{IV}$	5.075
2.4475	2	53 I	$L\gamma_{2,3}$	$L_IN_{II,III}$	5.0657
2.4493	3	57 La	$L\beta_4$	L_IM_{II}	5.0620
2.45891	5	57 La	$L\beta_1$	$L_{II}M_{IV}$	5.0421
2.4630	2	59 Pr	$L\alpha_1$	$L_{III}M_V$	5.0337
2.4729	3	59 Pr	$L\alpha_2$	$L_{III}M_{IV}$	5.0135
2.4740	1	55 Cs	L_{III}	Abs. Edge	5.0113
2.4783	2	55 Cs	$L\beta_9$	L_IM_V	5.0026
2.4823	4	62 Sm	Ll	$L_{III}M_I$	4.9945
2.4826	2	56 Ba	$L\beta_6$	$L_{III}N_I$	4.9939
2.4849	2	55 Cs	$L\beta_7$	$L_{III}O_I$	4.9893
2.4920	2	55 Cs	$L\beta_{10}$	L_IM_{IV}	4.9752
2.49734	5	22 Ti	K	Abs. Edge	4.96452
2.4985	2	22 Ti	$K\beta_5$	$KM_{IV,V}$	4.9623
2.50356	2	23 V	$K\alpha_1$	KL_{III}	4.95220
2.50738	2	23 V	$K\alpha_2$	KL_{II}	4.94464
2.5099	1	52 Te	L_I	Abs. Edge	4.9397
2.5113	2	52 Te	$L\gamma_4$	$L_IO_{II,III}$	4.9369
2.5118	2	55 Cs	$L\beta_{2,15}$	$L_{III}N_{IV,V}$	4.9359
2.512	3	59 Pr	$L\eta$	$L_{II}M_I$	4.935
2.51391	2	22 Ti	$K\beta_{1,3}$	$KM_{II,III}$	4.93181
2.5164	2	56 Ba	$L\beta_3$	L_IM_{III}	4.9269
2.527	4	91 Pa		$M_{II}O_{IV}$	4.906
2.5542	5	53 I	L_{II}	Abs. Edge	4.8540
2.5553	2	56 Ba	$L\beta_4$	L_IM_{II}	4.8519
2.5615	2	58 Ce	$L\alpha_1$	$L_{III}M_V$	4.8402
2.5674	2	52 Te	$L\gamma_{2,3}$	$L_IN_{II,III}$	4.8290
2.56821	5	56 Ba	$L\beta_1$	$L_{II}M_{IV}$	4.82753
2.5706	3	58 Ce	$L\alpha_2$	$L_{III}M_{IV}$	4.8230
2.58244	8	53 I	$L\gamma_1$	$L_{II}N_{IV}$	4.8009
2.5926	1	54 Xe	L_{III}	Abs. Edge	4.7822
2.5932	2	55 Cs	$L\beta_6$	$L_{III}N_I$	4.7811
2.618	5	90 Th		$M_{II}O_{IV}$	4.735
2.6203	4	58 Ce	$L\eta$	$L_{II}M_I$	4.7315
2.6285	2	55 Cs	$L\beta_3$	L_IM_{III}	4.7167
2.6388	1	51 Sb	L_I	Abs. Edge	4.6984
2.6398	2	51 Sb	$L\gamma_4$	$L_IO_{II,III}$	4.6967
2.65710	9	53 I	$L\gamma_5$	$L_{II}N_I$	4.6660
2.66570	5	57 La	$L\alpha_1$	$L_{III}M_V$	4.65097
2.6666	2	55 Cs	$L\beta_4$	L_IM_{II}	4.6494
2.67533	5	57 La	$L\alpha_2$	$L_{III}M_{IV}$	4.63423
2.6760	4	60 Nd	Ll	$L_{III}M_I$	4.6330
2.6837	2	55 Cs	$L\beta_1$	$L_{II}M_{IV}$	4.6198
2.6879	1	52 Te	L_{II}	Abs. Edge	4.6126
2.6953	2	51 Sb	$L\gamma_{2,3}$	$L_IN_{II,III}$	4.5999
2.71241	6	52 Te	$L\gamma_1$	$L_{II}N_{IV}$	4.5709
2.71352	9	53 I	$L\beta_9$	L_IM_V	4.5690
2.7196	5	53 I	L_{III}	Abs. Edge	4.5587
2.72104	9	53 I	$L\beta_{10}$	L_IM_{IV}	4.5564

Wavelength Å*	p.e.	Element	Designation		keV
2.7288	3	53 I	$L\beta_7$	$L_{III}O_I$	4.5435
2.740	3	57 La	$L\eta$	$L_{II}M_I$	4.525
2.74851	2	22 Ti	$K\alpha_1$	KL_{III}	4.51084
2.75053	8	53 I	$L\beta_{2,15}$	$L_{III}N_{IV,V}$	4.5075
2.75216	2	22 Ti	$K\alpha_2$	KL_{II}	4.50486
2.753	8	92 U		M_IN_{III}	4.50
2.762	1	21 Sc	K	Abs. Edge	4.489
2.7634	3	21 Sc	$K\beta_5$	$KM_{IV,V}$	4.4865
2.77595	5	56 Ba	$L\alpha_1$	$L_{III}M_V$	4.46626
2.7769	1	50 Sn	L_I	Abs. Edge	4.4648
2.7775	2	50 Sn	$L\gamma_4$	$L_IO_{II,III}$	4.4638
2.7796	2	21 Sc	$K\beta_{1,3}$	$KM_{II,III}$	4.4605
2.7841	4	59 Pr	Ll	$L_{III}M_I$	4.4532
2.78553	5	56 Ba	$L\alpha_2$	$L_{III}M_{IV}$	4.45090
2.79007	9	52 Te	$L\gamma_5$	$L_{II}N_I$	4.4437
2.817	2	92 U		$M_{II}N_{IV}$	4.401
2.8294	5	51 Sb	L_{II}	Abs. Edge	4.3819
2.8327	2	50 Sn	$L\gamma_{2,3}$	$L_IN_{II,III}$	4.3768
2.83672	9	53 I	$L\beta_6$	$L_{III}N_I$	4.3706
2.83897	9	52 Te	$L\beta_9$	L_IM_V	4.3671
2.84679	9	52 Te	$L\beta_{10}$	L_IM_{IV}	4.3551
2.85159	3	51 Sb	$L\gamma_1$	$L_{II}N_{IV}$	4.34779
2.8555	1	52 Te	L_{III}	Abs. Edge	4.3418
2.8627	3	56 Ba	$L\eta$	$L_{II}M_I$	4.3309
2.8634	3	52 Te	$L\beta_7$	$L_{III}O_I$	4.3298
2.87429	9	53 I	$L\beta_3$	L_IM_{III}	4.3134
2.88217	8	52 Te	$L\beta_{2,15}$	$L_{III}N_{IV,V}$	4.3017
2.884	5	92 U	M_{III}	Abs. Edge	4.299
2.8917	4	58 Ce	Ll	$L_{III}M_I$	4.2875
2.8924	2	55 Cs	$L\alpha_1$	$L_{III}M_V$	4.2865
2.9020	2	55 Cs	$L\alpha_2$	$L_{III}M_{IV}$	4.2722
2.910	2	91 Pa		$M_{II}N_{IV}$	4.260
2.91207	9	53 I	$L\beta_4$	L_IM_{II}	4.2575
2.92	2	92 U		M_IN_{II}	4.25
2.9260	1	49 In	L_I	Abs. Edge	4.2373
2.9264	2	49 In	$L\gamma_4$	$L_IO_{II,III}$	4.2367
2.93187	9	51 Sb	$L\gamma_5$	$L_{II}N_I$	4.2287
2.934	8	90 Th		M_IN_{III}	4.23
2.93744	6	53 I	$L\beta_1$	$L_{II}M_{IV}$	4.22072
2.948	2	92 U		$M_{III}O_{IV,V}$	4.205
2.97088	9	52 Te	$L\beta_6$	$L_{III}N_I$	4.1732
2.97261	9	51 Sb	$L\beta_9$	L_IM_V	4.1708
2.97917	9	51 Sb	$L\beta_{10}$	L_IM_{IV}	4.1616
2.9800	2	49 In	$L\gamma_{2,3}$	$L_IN_{II,III}$	4.1605
2.9823	1	50 Sn	L_{II}	Abs. Edge	4.1573
2.9932	2	55 Cs	$L\eta$	$L_{II}M_I$	4.1421
3.0003	1	51 Sb	L_{III}	Abs. Edge	4.1323
3.00115	3	50 Sn	$L\gamma_1$	$L_{II}N_{IV}$	4.13112
3.0052	3	51 Sb	$L\beta_7$	$L_{III}O_I$	4.1255
3.006	3	57 La	Ll	$L_{III}M_I$	4.124
3.00893	9	52 Te	$L\beta_3$	L_IM_{III}	4.1204
3.011	2	90 Th		$M_{II}N_{IV}$	4.117
3.0166	2	54 Xe	$L\alpha_1$	$L_{III}M_V$	4.1099
3.02335	3	51 Sb	$L\beta_{2,15}$	$L_{III}N_{IV,V}$	4.10078
3.0309	1	21 Sc	$K\alpha_1$	KL_{III}	4.0906
3.0342	1	21 Sc	$K\alpha_2$	KL_{II}	4.0861
3.038	2	91 Pa		$M_{III}O_{IV,V}$	4.081

Wavelength Å*	p.e.	Element	Designation		keV
3.04661	9	52 Te	$L\beta_4$	L_IM_{II}	4.0695
3.068	5	90 Th	M_{III}	Abs. Edge	4.041
3.0703	1	20 Ca	K	Abs. Edge	4.0381
3.0746	3	20 Ca	$K\beta_5$	$KM_{IV,V}$	4.0325
3.07677	6	52 Te	$L\beta_1$	$L_{II}M_{IV}$	4.02958
3.08475	9	50 Sn	$L\gamma_5$	$L_{II}N_I$	4.0192
3.0849	1	48 Cd	L_I	Abs. Edge	4.0190
3.0897	2	20 Ca	$K\beta_{1,3}$	$KM_{II,III}$	4.0127
3.094	5	83 Bi	M_I	Abs. Edge	4.007
3.11513	9	50 Sn	$L\beta_9$	L_IM_V	3.9800
3.11513	9	51 Sb	$L\beta_6$	$L_{III}N_I$	3.9800
3.115	7	92 U		$M_{III}O_I$	3.980
3.12170	9	50 Sn	$L\beta_{10}$	L_IM_{IV}	3.9716
3.131	3	90 Th		$M_{III}O_{IV,V}$	3.959
3.1355	2	56 Ba	Ll	$L_{III}M_I$	3.9541
3.1377	2	48 Cd	$L\gamma_2$	L_IN_{II}	3.9513
3.1473	1	49 In	L_{II}	Abs. Edge	3.9393
3.14860	6	53 I	$L\alpha_1$	$L_{III}M_V$	3.93765
3.15258	9	51 Sb	$L\beta_3$	L_IM_{III}	3.9327
3.1557	1	50 Sn	L_{III}	Abs. Edge	3.9288
3.1564	3	50 Sn	$L\beta_7$	$L_{III}O_I$	3.9279
3.15791	6	53 I	$L\alpha_2$	$L_{III}M_{IV}$	3.92604
3.16213	4	49 In	$L\gamma_1$	$L_{II}N_{IV}$	3.92081
3.17505	3	50 Sn	$L\beta_{2,15}$	$L_{III}N_{IV,V}$	3.90486
3.19014	9	51 Sb	$L\beta_4$	L_IM_{II}	3.8364
3.217	5	82 Pb	M_I	Abs. Edge	3.854
3.22567	4	51 Sb	$L\beta_1$	$L_{II}M_{IV}$	3.84357
3.245	9	91 Pa		$M_{III}O_I$	3.82
3.24907	9	49 In	$L\gamma_5$	$L_{II}N_I$	3.8159
3.2564	1	47 Ag	L_I	Abs. Edge	3.8072
3.2670	2	55 Cs	Ll	$L_{III}M_I$	3.7950
3.26763	9	49 In	$L\beta_9$	L_IM_V	3.7942
3.26901	9	50 Sn	$L\beta_6$	$L_{III}N_I$	3.7926
3.27404	9	49 In	$L\beta_{10}$	L_IM_{IV}	3.7868
3.27979	9	53 I	$L\eta$	$L_{II}M_I$	3.7801
3.283	9	90 Th		$M_{III}O_I$	3.78
3.28920	6	52 Te	$L\alpha_1$	$L_{III}M_V$	3.76933
3.29846	9	52 Te	$L\alpha_2$	$L_{III}M_{IV}$	3.7588
3.30585	3	50 Sn	$L\beta_3$	L_IM_{III}	3.7500
3.30635	9	47 Ag	$L\gamma_3$	L_IN_{III}	3.7498
3.31216	9	47 Ag	$L\gamma_2$	L_IN_{II}	3.7432
3.3237	1	49 In	L_{III}	Abs. Edge	3.7302
3.324	4	49 In	$L\beta_7$	$L_{III}O_I$	3.730
3.3257	1	48 Cd	L_{II}	Abs. Edge	3.7280
3.329	4	92 U		$M_{II}N_I$	3.724
3.333	5	92 U	M_{IV}	Abs. Edge	3.720
3.33564	6	48 Cd	$L\gamma_1$	$L_{II}N_{IV}$	3.71686
3.33838	3	49 In	$L\beta_{2,15}$	$L_{III}N_{IV,V}$	3.71381
3.34335	9	50 Sn	$L\beta_4$	L_IM_{II}	3.7083
3.346	5	81 Tl	M_I	Abs. Edge	3.705
3.35839	3	20 Ca	$K\alpha_1$	KL_{III}	3.69168
3.359	5	83 Bi	M_{II}	Abs. Edge	3.691
3.36166	3	20 Ca	$K\alpha_2$	KL_{II}	3.68809
3.38487	3	50 Sn	$L\beta_1$	$L_{II}M_{IV}$	3.66280
3.42551	9	48 Cd	$L\gamma_5$	$L_{II}N_I$	3.61935
3.43015	9	48 Cd	$L\beta_9$	L_IM_V	3.61445
3.43606	9	49 In	$L\beta_6$	$L_{III}N_I$	3.60823

Wavelength Å*	p.e.	Element		Designation	keV
3.4365	1	19 K	K	Abs. Edge	3.6078
3.4367	2	48 Cd	$L\beta_{10}$	L_IM_{IV}	3.6075
3.437	1	46 Pd	L_I	Abs. Edge	3.607
3.43832	9	52 Te	$L\eta$	$L_{II}M_I$	3.60586
3.43941	4	51 Sb	$L\alpha_1$	$L_{III}M_V$	3.60472
3.441	5	91 Pa		$M_{II}N_I$	3.603
3.4413	4	19 K	$K\beta_5$	$KM_{IV,V}$	3.6027
3.44840	6	51 Sb	$L\alpha_2$	$L_{III}M_{IV}$	3.59532
3.4539	2	19 K	$K\beta_{1,3}$	$KM_{II,III}$	3.5896
3.46984	9	49 In	$L\beta_3$	L_IM_{III}	3.57311
3.478	5	80 Hg	M_I	Abs. Edge	3.565
3.479	1	92 U	$M\gamma$	$M_{III}N_V$	3.563
3.4892	2	46 Pd	$L\gamma_{2,3}$	$L_IN_{II,III}$	3.5533
3.492	5	82 Pb	M_{II}	Abs. Edge	3.550
3.497	5	92 U	M_V	Abs. Edge	3.545
3.5047	1	48 Cd	L_{III}	Abs. Edge	3.5376
3.50697	9	49 In	$L\beta_4$	L_IM_{II}	3.53528
3.51408	4	48 Cd	$L\beta_{2,15}$	$L_{III}N_{IV,V}$	3.52812
3.5164	1	47 Ag	L_{II}	Abs. Edge	3.5258
3.521	2	92 U		$M_{III}N_{IV}$	3.521
3.52260	4	47 Ag	$L\gamma_1$	$L_{II}N_{IV}$	3.51959
3.537	9	90 Th		$M_{II}N_I$	3.505
3.55531	4	49 In	$L\beta_1$	$L_{II}M_{IV}$	3.48721
3.557	5	90 Th	M_{IV}	Abs. Edge	3.485
3.55754	9	53 I	Ll	$L_{III}M_I$	3.48502
3.576	1	92 U		$M_{IV}O_{II}$	3.4666
3.577	1	91 Pa	$M\gamma$	$M_{III}N_V$	3.4657
3.59994	3	50 Sn	$L\alpha_1$	$L_{III}M_V$	3.44398
3.60497	9	47 Ag	$L\beta_9$	L_IM_V	3.43917
3.60765	9	51 Sb	$L\eta$	$L_{II}M_I$	3.43661
3.60891	4	50 Sn	$L\alpha_2$	$L_{III}M_{IV}$	3.43542
3.61158	9	47 Ag	$L\beta_{10}$	L_IM_{IV}	3.43287
3.614	2	91 Pa		$M_{III}N_{IV}$	3.430
3.61467	9	48 Cd	$L\beta_6$	$L_{III}N_I$	3.42994
3.61638	9	47 Ag	$L\gamma_5$	$L_{II}N_I$	3.42832
3.616	5	79 Au	M_I	Abs. Edge	3.428
3.629	5	45 Rh	L_I	Abs. Edge	3.417
3.634	5	81 Tl	M_{II}	Abs. Edge	3.412
3.64495	9	48 Cd	$L\beta_3$	L_IM_{III}	3.40145
3.679	2	90 Th	$M\gamma$	$M_{III}N_V$	3.370
3.68203	9	48 Cd	$L\beta_4$	L_IM_{II}	3.36719
3.6855	2	45 Rh	$L\gamma_{2,3}$	$L_IN_{II,III}$	3.3640
3.691	2	91 Pa		$M_{IV}O_{II}$	3.359
3.6999	1	47 Ag	L_{III}	Abs. Edge	3.35096
3.70335	3	47 Ag	$L\beta_{2,15}$	$L_{III}N_{IV,V}$	3.34781
3.716	1	92 U	$M\beta$	$M_{IV}N_{VI}$	3.3367
3.71696	9	52 Te	Ll	$L_{III}M_I$	3.33555
3.718	3	90 Th		$M_{III}N_{IV}$	3.335
3.7228	1	46 Pd	L_{II}	Abs. Edge	3.33031
3.7246	2	46 Pd	$L\gamma_1$	$L_{II}N_{IV}$	3.3287
3.729	5	90 Th	M_V	Abs. Edge	3.325
3.73823	4	48 Cd	$L\beta_1$	$L_{II}M_{IV}$	3.31657
3.740	9	83 Bi		M_IN_{III}	3.315
3.7414	2	19 K	$K\alpha_1$	KL_{III}	3.3138
3.7445	2	19 K	$K\alpha_2$	KL_{II}	3.3111
3.760	9	90 Th		M_VP_{III}	3.298
3.762	5	78 Pt	M_I	Abs. Edge	3.296
3.77192	4	49 In	$L\alpha_1$	$L_{III}M_V$	3.28694
3.78073	6	49 In	$L\alpha_2$	$L_{III}M_{IV}$	3.27929
3.783	5	80 Hg	M_{II}	Abs. Edge	3.277
3.78876	9	50 Sn	$L\eta$	$L_{II}M_I$	3.27234
3.7920	2	46 Pd	$L\beta_9$	L_IM_V	3.2696
3.7988	2	46 Pd	$L\beta_{10}$	L_IM_{IV}	3.2637
3.80774	9	47 Ag	$L\beta_6$	$L_{III}N_I$	3.25603
3.808	4	90 Th		$M_{IV}O_{II}$	3.256
3.8222	2	46 Pd	$L\gamma_5$	$L_{II}N_I$	3.2437
3.827	1	91 Pa	$M\beta$	$M_{IV}N_{VI}$	3.2397
3.83313	9	47 Ag	$L\beta_3$	L_IM_{III}	3.23446
3.834	4	83 Bi		$M_{II}N_{IV}$	3.234
3.835	5	44 Ru	L_I	Abs. Edge	3.233
3.87023	5	47 Ag	$L\beta_4$	L_IM_{II}	3.20346
3.87090	5	18 A	K	Abs. Edge	3.20290
3.872	9	82 Pb		M_IN_{III}	3.202
3.8860	2	18 A	$K\beta_{1,3}$	$KM_{II,III}$	3.1905
3.88826	9	51 Sb	Ll	$L_{III}M_I$	3.18860
3.892	9	83 Bi		M_IN_{II}	3.185
3.8977	2	44 Ru	$L\gamma_{2,3}$	$L_IN_{II,III}$	3.1809
3.904	5	83 Bi	M_{III}	Abs. Edge	3.176
3.9074	1	46 Pd	L_{III}	Abs. Edge	3.17298
3.90887	4	46 Pd	$L\beta_{2,15}$	$L_{III}N_{IV,V}$	3.17179
3.910	1	92 U	$M\alpha_1$	M_VN_{VII}	3.1708
3.915	5	77 Ir	M_I	Abs. Edge	3.167
3.924	1	92 U	$M\alpha_2$	M_VN_{VI}	3.1595
3.932	6	83 Bi		$M_{III}O_{IV,V}$	3.153
3.93473	3	47 Ag	$L\beta_1$	$L_{II}M_{IV}$	3.15094
3.936	5	79 Au	M_{II}	Abs. Edge	3.150
3.941	1	90 Th	$M\beta$	$M_{IV}N_{VI}$	3.1458
3.9425	5	45 Rh	L_{II}	Abs. Edge	3.1448
3.9437	2	45 Rh	$L\gamma_1$	$L_{II}N_{IV}$	3.1438
3.95635	4	48 Cd	$L\alpha_1$	$L_{III}M_V$	3.13373
3.96496	6	48 Cd	$L\alpha_2$	$L_{III}M_{IV}$	3.12691
3.968	5	82 Pb		$M_{II}N_{IV}$	3.124
3.98327	9	49 In	$L\eta$	$L_{II}M_I$	3.11254
4.013	9	81 Tl		M_IN_{III}	3.089
4.0162	2	46 Pd	$L\beta_6$	$L_{III}N_I$	3.0870
4.022	1	91 Pa	$M\alpha_1$	M_VN_{VII}	3.0823
4.0346	2	46 Pd	$L\beta_3$	L_IM_{III}	3.0730
4.035	3	91 Pa	$M\alpha_2$	M_VN_{VI}	3.072
4.0451	2	45 Rh	$L\gamma_5$	$L_{II}N_I$	3.0650
4.047	1	82 Pb	M_{III}	Abs. Edge	3.0632
4.058	5	43 Te	L_I	Abs. Edge	3.055
4.069	6	82 Pb		$M_{III}O_{IV,V}$	3.047
4.0711	2	46 Pd	$L\beta_4$	L_IM_{II}	3.0454
4.071	5	76 Os	M_I	Abs. Edge	3.045
4.07165	9	50 Sn	Ll	$L_{III}M_I$	3.04499
4.093	5	78 Pt	M_{II}	Abs. Edge	3.029
4.105	9	83 Bi		$M_{III}O_I$	3.021
4.116	4	81 Tl		$M_{II}N_{IV}$	3.013
4.1299	5	45 Rh	L_{III}	Abs. Edge	3.0021
4.1310	2	45 Rh	$L\beta_{2,15}$	$L_{III}N_{IV,V}$	3.0013
4.1381	9	90 Th	$M\alpha_1$	M_VN_{VII}	2.9961
4.14622	5	46 Pd	$L\beta_1$	$L_{II}M_{IV}$	2.99022
4.151	2	90 Th	$M\alpha_2$	M_VN_{VI}	2.987
4.15443	3	47 Ag	$L\alpha_1$	$L_{III}M_V$	2.98431

Wavelength Å*	p.e.	Element		Designation	keV
4.16294	5	47 Ag	$L\alpha_2$	$L_{III}M_{IV}$	2.97821
4.180	1	44 Ru	L_{II}	Abs. Edge	2.9663
4.1822	2	44 Ru	$L\gamma_1$	$L_{II}N_{IV}$	2.9645
4.19180	5	18 A	$K\alpha_1$	KL_{III}	2.95770
4.19315	9	48 Cd	$L\eta$	$L_{II}M_I$	2.95675
4.19474	5	18 A	$K\alpha_2$	KL_{II}	2.95563
4.198	1	81 Tl	M_{III}	Abs. Edge	2.9535
4.216	6	81 Tl		$M_{III}O_{IV,V}$	2.941
4.236	5	75 Re	M_I	Abs. Edge	2.927
4.2417	2	45 Rh	$L\beta_6$	$L_{III}N_I$	2.9229
4.244	9	82 Pb		$M_{III}O_I$	2.921
4.2522	2	45 Rh	$L\beta_3$	L_IM_{III}	2.9157
4.260	5	77 Ir	M_{II}	Abs. Edge	2.910
4.26873	9	49 In	Ll	$L_{III}M_I$	2.90440
4.2873	2	44 Ru	$L\gamma_5$	$L_{II}N_I$	2.8918
4.2888	2	45 Rh	$L\beta_4$	L_IM_{II}	2.8908
4.300	9	79 Au		M_IN_{III}	2.883
4.304	5	42 Mo	L_I	Abs. Edge	2.881
4.330	2	92 U		$M_{III}N_I$	2.863
4.355	1	80 Hg	M_{III}	Abs. Edge	2.8469
4.36767	5	46 Pd	$L\alpha_1$	$L_{III}M_V$	2.83861
4.369	1	44 Ru	L_{III}	Abs. Edge	2.8377
4.3718	2	44 Ru	$L\beta_{2,15}$	$L_{III}N_{IV,V}$	2.8360
4.37414	4	45 Rh	$L\beta_1$	$L_{II}M_{IV}$	2.83441
4.37588	7	46 Pd	$L\alpha_2$	$L_{III}M_{IV}$	2.83329
4.3800	2	42 Mo	$L\gamma_{2,3}$	$L_IN_{II,III}$	2.8306
4.3971	1	17 Cl	K	Abs. Edge	2.81960
4.4034	3	17 Cl	$K\beta$	KM	2.8156
4.407	5	74 W	M_I	Abs. Edge	2.813
4.4183	2	47 Ag	$L\eta$	$L_{II}M_I$	2.8061
4.432	4	79 Au		$M_{II}N_{IV}$	2.797
4.433	5	76 Os	M_{II}	Abs. Edge	2.797
4.436	1	43 Te	L_{II}	Abs. Edge	2.7948
4.44	2	74 W		$M_IO_{II,III}$	2.79
4.450	4	91 Pa		$M_{III}N_I$	2.786
4.460	9	78 Pt		M_IN_{III}	2.780
4.48014	9	48 Cd	Ll	$L_{III}M_I$	2.76735
4.4866	3	44 Ru	$L\beta_3$	L_IM_{III}	2.7634
4.4866	3	44 Ru	$L\beta_6$	$L_{III}N_I$	2.7634
4.518	1	79 Au	M_{III}	Abs. Edge	2.7439
4.522	6	79 Au		$M_{III}O_{IV,V}$	2.742
4.5230	2	44 Ru	$L\beta_4$	L_IM_{II}	2.7411
4.532	2	83 Bi	$M\gamma$	$M_{III}N_V$	2.735
4.568	5	90 Th		$M_{III}N_I$	2.714
4.571	5	83 Bi		$M_{III}N_{IV}$	2.712
4.572	5	83 Bi	M_{IV}	Abs. Edge	2.711
4.575	5	41 Nb	L_I	Abs. Edge	2.710
4.585	5	73 Ta	M_I	Abs. Edge	2.704
4.59	2	83 Bi		$M_{IV}P_{II,III}$	2.70
4.59743	9	45 Rh	$L\alpha_1$	$L_{III}M_V$	2.69674
4.601	4	78 Pt		$M_{II}N_{IV}$	2.695
4.60545	9	45 Rh	$L\alpha_2$	$L_{III}M_{IV}$	2.69205
4.620	5	75 Re	M_{II}	Abs. Edge	2.684
4.62058	3	44 Ru	$L\beta_1$	$L_{II}M_{IV}$	2.68323
4.625	5	92 U		$M_{IV}N_{III}$	2.681
4.630	1	43 Tc	L_{III}	Abs. Edge	2.6780
4.631	9	77 Ir		M_IN_{III}	2.677
4.6542	2	41 Nb	$L\gamma_{2,3}$	$L_IN_{II,III}$	2.6638
4.655	8	82 Pb		$M_{II}N_I$	2.664
4.6605	2	46 Pd	$L\eta$	$L_{II}M_I$	2.6603
4.674	1	82 Pb	$M\gamma$	$M_{III}N_V$	2.6527
4.686	1	78 Pt	M_{III}	Abs. Edge	2.6459
4.694	8	78 Pt		$M_{III}O_{IV,V}$	2.641
4.703	9	79 Au		$M_{III}O_I$	2.636
4.7076	2	47 Ag	Ll	$L_{III}M_I$	2.6337
4.715	3	82 Pb		$M_{III}N_{IV}$	2.630
4.719	1	42 Mo	L_{II}	Abs. Edge	2.6274
4.7258	2	42 Mo	$L\gamma_1$	$L_{II}N_{IV}$	2.6235
4.7278	1	17 Cl	$K\alpha_1$	KL_{III}	2.62239
4.7307	1	17 Cl	$K\alpha_2$	KL_{II}	2.62078
4.757	5	82 Pb	M_{IV}	Abs. Edge	2.606
4.764	5	83 Bi	M_V	Abs. Edge	2.603
4.780	4	77 Ir		$M_{II}N_{IV}$	2.594
4.79	2	76 Os		M_IN_{III}	2.59
4.815	5	74 W	M_{II}	Abs. Edge	2.575
4.823	3	83 Bi		$M_{IV}O_{II}$	2.571
4.823	4	81 Tl	$M\gamma$	$M_{III}N_V$	2.571
4.8369	2	42 Mo	$L\gamma_5$	$L_{II}N_I$	2.5632
4.84575	5	44 Ru	$L\alpha_1$	$L_{III}M_V$	2.55855
4.85381	7	44 Ru	$L\alpha_2$	$L_{III}M_{IV}$	2.55431
4.861	1	77 Ir	M_{III}	Abs. Edge	2.5505
4.865	5	81 Tl		$M_{III}N_{IV}$	2.548
4.869	9	77 Ir		$M_{III}O_{IV,V}$	2.546
4.876	9	78 Pt		$M_{III}O_I$	2.543
4.879	5	40 Zr	L_I	Abs. Edge	2.541
4.8873	8	43 Tc	$L\beta_1$	$L_{II}M_{IV}$	2.5368
4.909	1	83 Bi	$M\beta$	$M_{IV}N_{VI}$	2.5255
4.911	5	90 Th		$M_{IV}N_{III}$	2.524
4.913	1	42 Mo	L_{III}	Abs. Edge	2.5234
4.9217	2	45 Rh	$L\eta$	$L_{II}M_I$	2.5191
4.9232	2	42 Mo	$L\beta_{2,15}$	$L_{III}N_{IV,V}$	2.5183
4.946	2	92 U	$M\zeta_1$	M_VN_{III}	2.507
4.952	5	81 Tl	M_{IV}	Abs. Edge	2.504
4.9525	3	46 Pd	Ll	$L_{III}M_I$	2.5034
4.9536	3	40 Zr	$L\gamma_{2,3}$	$L_IN_{II,III}$	2.5029
4.955	4	76 Os		$M_{II}N_{IV}$	2.502
4.955	5	82 Pb	M_V	Abs. Edge	2.502
4.984	2	80 Hg	$M\gamma$	$M_{III}N_V$	2.4875
5.004	9	82 Pb		$M_{IV}O_{II}$	2.477
5.0133	3	42 Mo	$L\beta_3$	L_IM_{III}	2.4730
5.0185	1	16 S	K	Abs. Edge	2.47048
5.020	5	73 Ta	M_{II}	Abs. Edge	2.470
5.0233	3	16 S	$K\beta_x$	KM	2.4681
5.031	1	41 Nb	L_{II}	Abs. Edge	2.4641
5.0316	2	16 S	$K\beta_1$	KM	2.46404
5.0361	3	41 Nb	$L\gamma_1$	$L_{II}N_{IV}$	2.4618
5.043	5	76 Os	M_{III}	Abs. Edge	2.458
5.0488	3	42 Mo	$L\beta_4$	L_IM_{II}	2.4557
5.0488	5	42 Mo	$L\beta_6$	$L_{III}N_I$	2.4557
5.050	2	92 U	$M\zeta_2$	$M_{IV}N_{II}$	2.4548
5.076	1	82 Pb	$M\beta$	$M_{IV}N_{VI}$	2.4427
5.092	2	91 Pa	$M\zeta_1$	M_VN_{III}	2.4350
5.1148	3	43 Tc	$L\alpha_1$	$L_{III}M_V$	2.4240
5.118	1	83 Bi	$M\alpha_1$	M_VN_{VII}	2.4226

Wavelength Å*	p.e.	Element		Designation	keV
5.130	2	83 Bi	$M\alpha_2$	M_VN_{VI}	2.4170
5.145	4	79 Au	$M\gamma$	$M_{III}N_V$	2.410
5.1517	3	41 Nb	$L\gamma_5$	$L_{II}N_I$	2.4066
5.153	5	81 Tl	M_V	Abs. Edge	2.406
5.157	5	80 Hg	M_{IV}	Abs. Edge	2.404
5.168	9	82 Pb		M_VO_{III}	2.399
5.172	9	74 W		M_IN_{III}	2.397
5.17708	8	42 Mo	$L\beta_1$	$L_{II}M_{IV}$	2.39481
5.186	5	79 Au		$M_{III}N_{IV}$	2.391
5.193	2	91 Pa	$M\zeta_2$	$M_{IV}N_{II}$	2.3876
5.196	9	81 Tl		$M_{IV}O_{II}$	2.386
5.2050	2	44 Ru	$L\eta$	$L_{II}M_I$	2.38197
5.217	5	39 Y	L_I	Abs. Edge	2.377
5.2169	3	45 Rh	Ll	$L_{III}M_I$	2.3765
5.230	1	41 Nb	L_{III}	Abs. Edge	2.3706
5.234	5	75 Re	M_{III}	Abs. Edge	2.369
5.2379	3	41 Nb	$L\beta_{2,15}$	$L_{III}N_{IV,V}$	2.3670
5.245	5	90 Th	$M\zeta_1$	M_VN_{III}	2.364
5.249	1	81 Tl	$M\beta$	$M_{IV}N_{VI}$	2.3621
5.2830	3	39 Y	$L\gamma_{2,3}$	$L_IN_{II,III}$	2.3468
5.286	1	82 Pb	$M\alpha_1$	M_VN_{VII}	2.3455
5.299	2	82 Pb	$M\alpha_2$	M_VN_{VI}	2.3397
5.3102	3	41 Nb	$L\beta_3$	L_IM_{III}	2.3348
5.319	4	78 Pt	$M\gamma$	$M_{III}N_V$	2.331
5.340	5	90 Th	$M\zeta_2$	$M_{IV}N_{II}$	2.322
5.3455	3	41 Nb	$L\beta_4$	L_IM_{II}	2.3194
5.357	4	74 W		$M_{II}N_{IV}$	2.314
5.357	5	78 Pt		$M_{III}N_{IV}$	2.314
5.36	1	80 Hg	M_V	Abs. Edge	2.313
5.3613	3	41 Nb	$L\beta_6$	$L_{III}N_I$	2.3125
5.37216	7	16 S	$K\alpha_1$	KL_{III}	2.30784
5.374	5	79 Au	M_{IV}	Abs. Edge	2.307
5.37496	8	16 S	$K\alpha_2$	KL_{II}	2.30664
5.378	1	40 Zr	L_{II}	Abs. Edge	2.3053
5.3843	3	40 Zr	$L\gamma_1$	$L_{II}N_{IV}$	2.3027
5.40	2	73 Ta		M_IN_{III}	2.295
5.40655	8	42 Mo	$L\alpha_1$	$L_{III}M_V$	2.29316
5.41437	8	42 Mo	$L\alpha_2$	$L_{III}M_{IV}$	2.28985
5.4318	9	80 Hg	$M\beta$	$M_{IV}N_{VI}$	2.2825
5.435	1	74 W	M_{III}	Abs. Edge	2.2811
5.460	1	81 Tl	$M\alpha_1$	M_VN_{VII}	2.2706
5.472	2	81 Tl	$M\alpha_2$	M_VN_{VI}	2.2656
5.4923	3	41 Nb	$L\beta_1$	$L_{II}M_{IV}$	2.2574
5.4977	3	40 Zr	$L\gamma_5$	$L_{II}N_I$	2.2551
5.500	4	77 Ir	$M\gamma$	$M_{III}N_V$	2.254
5.5035	3	44 Ru	Ll	$L_{III}M_I$	2.2528
5.537	8	83 Bi		$M_{III}N_I$	2.239
5.540	5	77 Ir		$M_{III}N_{IV}$	2.238
5.570	4	73 Ta		$M_{II}N_{IV}$	2.226
5.579	1	40 Zr	L_{III}	Abs. Edge	2.2225
5.584	5	79 Au	M_V	Abs. Edge	2.220
5.5863	3	40 Zr	$L\beta_{2,15}$	$L_{III}N_{IV,V}$	2.2194
5.59	1	78 Pt	M_{IV}	Abs. Edge	2.217
5.592	5	38 Sr	L_I	Abs. Edge	2.217
5.624	1	79 Au	$M\beta$	$M_{IV}N_{VI}$	2.2046
5.628	8	74 W		$M_{III}O_I$	2.203
5.6330	3	40 Zr	$L\beta_3$	L_IM_{III}	2.2010
5.6445	3	38 Sr	$L\gamma_{2,3}$	$L_IN_{II,III}$	2.1965
5.6476	9	80 Hg	$M\alpha_1$	M_VN_{VII}	2.1953
5.650	5	73 Ta	M_{III}	Abs. Edge	2.194
5.6681	3	40 Zr	$L\beta_4$	L_IM_{II}	2.1873
5.67	3	73 Ta		$M_{III}O_{IV,V}$	2.19
5.682	4	76 Os	$M\gamma$	$M_{III}N_V$	2.182
5.704	8	82 Pb		$M_{III}N_I$	2.174
5.7101	3	40 Zr	$L\beta_6$	$L_{III}N_I$	2.1712
5.724	5	76 Os		$M_{III}N_{IV}$	2.166
5.7243	2	41 Nb	$L\alpha_1$	$L_{III}M_V$	2.16589
5.7319	3	41 Nb	$L\alpha_2$	$L_{III}M_{IV}$	2.1630
5.756	1	39 Y	L_{II}	Abs. Edge	2.1540
5.767	9	79 Au		M_VO_{III}	2.150
5.784	1	15 P	K	Abs. Edge	2.1435
5.796	2	15 P	$K\beta$	KM	2.1391
5.81	2	76 Os		$M_{II}N_I$	2.133
5.81	1	78 Pt	M_V	Abs. Edge	2.133
5.828	1	78 Pt	$M\beta$	$M_{IV}N_{VI}$	2.1273
5.83	2	73 Ta		$M_{III}O_I$	2.126
5.83	1	77 Ir	M_{IV}	Abs. Edge	2.126
5.8360	3	40 Zr	$L\beta_1$	$L_{II}M_{IV}$	2.1244
5.840	1	79 Au	$M\alpha_1$	M_VN_{VII}	2.1229
5.8475	3	42 Mo	$L\eta$	$L_{II}M_I$	2.1202
5.854	3	79 Au	$M\alpha_2$	M_VN_{VI}	2.118
5.8754	3	39 Y	$L\gamma_5$	$L_{II}N_I$	2.1102
5.884	8	81 Tl		$M_{III}N_I$	2.107
5.885	2	75 Re	$M\gamma$	$M_{III}N_V$	2.1067
5.931	5	75 Re		$M_{III}N_{IV}$	2.090
5.962	1	39 Y	L_{III}	Abs. Edge	2.0794
5.9832	3	39 Y	$L\beta_3$	L_IM_{III}	2.0722
5.987	9	78 Pt		M_VO_{III}	2.071
6.008	5	37 Rb	L_I	Abs. Edge	2.063
6.0186	3	39 Y	$L\beta_4$	L_IM_{II}	2.0600
6.038	1	77 Ir	$M\beta$	$M_{IV}N_{VI}$	2.0535
6.0458	3	37 Rb	$L\gamma_{2,3}$	$L_IN_{II,III}$	2.0507
6.047	1	78 Pt	$M\alpha_1$	M_VN_{VII}	2.0505
6.05	1	77 Ir	M_V	Abs. Edge	2.048
6.058	3	78 Pt	$M\alpha_2$	M_VN_{VI}	2.047
6.0705	2	40 Zr	$L\alpha_1$	$L_{III}M_V$	2.04236
6.073	5	76 Os	M_{IV}	Abs. Edge	2.042
6.0778	3	40 Zr	$L\alpha_2$	$L_{III}M_{IV}$	2.0399
6.09	2	80 Hg		$M_{III}N_I$	2.036
6.092	3	74 W	$M\gamma$	$M_{III}N_V$	2.035
6.0942	3	39 Y	$L\beta_6$	$L_{III}N_I$	2.0344
6.134	4	74 W		$M_{III}N_{IV}$	2.021
6.1508	3	42 Mo	Ll	$L_{III}M_I$	2.01568
6.157	1	15 P	$K\alpha_1$	KL_{III}	2.0137
6.160	1	15 P	$K\alpha_2$	KL_{II}	2.0127
6.162	8	83 Bi		$M_{IV}N_{III}$	2.012
6.173	1	38 Sr	L_{II}	Abs. Edge	2.0085
6.2109	3	41 Nb	$L\eta$	$L_{II}M_I$	1.99620
6.2120	3	39 Y	$L\beta_1$	$L_{II}M_{IV}$	1.99584
6.259	9	79 Au		$M_{III}N_I$	1.981
6.262	1	77 Ir	$M\alpha_1$	M_VN_{VII}	1.9799
6.267	1	76 Os	$M\beta$	$M_{IV}N_{VI}$	1.9783
6.275	3	77 Ir	$M\alpha_2$	M_VN_{VI}	1.9758
6.28	2	74 W		$M_{II}N_I$	1.973

Wavelength Å*	p.e.	Element		Designation	keV
6.2961	3	38 Sr	$L\gamma_5$	$L_{II}N_I$	1.96916
6.30	1	76 Os	M_V	Abs. Edge	1.967
6.312	4	73 Ta	$M\gamma$	$M_{III}N_V$	1.964
6.33	1	75 Re	M_{IV}	Abs. Edge	1.958
6.353	5	73 Ta		$M_{III}N_{IV}$	1.951
6.3672	3	38 Sr	$L\beta_3$	L_IM_{III}	1.94719
6.384	7	82 Pb		$M_{IV}N_{III}$	1.942
6.387	1	38 Sr	L_{III}	Abs. Edge	1.9411
6.4026	3	38 Sr	$L\beta_4$	L_IM_{II}	1.93643
6.4488	2	39 Y	$L\alpha_1$	$L_{III}M_V$	1.92256
6.455	9	78 Pt		$M_{III}N_I$	1.921
6.4558	3	39 Y	$L\alpha_2$	$L_{III}M_{IV}$	1.92047
6.47	1	36 Kr	L_I	Abs. Edge	1.915
6.490	1	76 Os	$M\alpha$	$M_VN_{VI,VII}$	1.9102
6.504	1	75 Re	$M\beta$	$M_{IV}N_{VI}$	1.9061
6.5176	3	41 Nb	Ll	$L_{III}M_I$	1.90225
6.5191	3	38 Sr	$L\beta_6$	$L_{III}N_I$	1.90181
6.521	4	83 Bi	$M\zeta_1$	M_VN_{III}	1.901
6.544	4	72 Hf	$M\gamma$	$M_{III}N_V$	1.895
6.560	5	75 Re	M_V	Abs. Edge	1.890
6.585	5	83 Bi	$M\zeta_2$	$M_{IV}N_{II}$	1.883
6.59	1	74 W	M_{IV}	Abs. Edge	1.880
6.6069	3	40 Zr	$L\eta$	$L_{II}M_I$	1.87654
6.6239	3	38 Sr	$L\beta_1$	$L_{II}M_{IV}$	1.87172
6.644	1	37 Rb	L_{II}	Abs. Edge	1.8661
6.669	9	77 Ir		$M_{III}N_I$	1.859
6.729	1	75 Re	$M\alpha$	$M_VN_{VI,VII}$	1.8425
6.738	1	14 Si	K	Abs. Edge	1.8400
6.740	3	82 Pb	$M\zeta_1$	M_VN_{III}	1.8395
6.7530	1	14 Si	$K\beta$	KM	1.83594
6.755	3	37 Rb	$L\gamma_5$	$L_{II}N_{IV}$	1.83532
6.757	1	74 W	$M\beta$	$M_{IV}N_{VI}$	1.8349
6.768	6	71 Lu	$M\gamma$	$M_{III}N_V$	1.832
6.7876	3	37 Rb	$L\beta_3$	L_IM_{III}	1.82659
6.802	5	82 Pb	$M\zeta_2$	$M_{IV}N_{II}$	1.823
6.806	9	74 W		$M_{IV}O_{II}$	1.822
6.8207	3	37 Rb	$L\beta_4$	L_IM_{II}	1.81771
6.83	1	74 W	M_V	Abs. Edge	1.814
6.862	1	37 Rb	L_{III}	Abs. Edge	1.8067
6.8628	2	38 Sr	$L\alpha_1$	$L_{III}M_V$	1.80656
6.8697	3	38 Sr	$L\alpha_2$	$L_{III}M_{IV}$	1.80474
6.87	1	73 Ta	M_{IV}	Abs. Edge	1.804
6.87	2	80 Hg	δ	$M_{IV}N_{III}$	1.805
6.89	2	76 Os		$M_{III}N_I$	1.798
6.9185	3	40 Zr	Ll	$L_{III}M_I$	1.79201
6.959	5	35 Br	L_I	Abs. Edge	1.781
6.974	4	81 Tl	$M\zeta_1$	M_VN_{III}	1.778
6.983	1	74 W	$M\alpha_1$	M_VN_{VII}	1.7754
6.9842	3	37 Rb	$L\beta_6$	$L_{III}N_I$	1.77517
6.992	2	74 W	$M\alpha_2$	M_VN_{VI}	1.7731
7.005	9	74 W		M_VO_{III}	1.770
7.023	1	73 Ta	$M\beta$	$M_{IV}N_{VI}$	1.7655
7.024	8	70 Yb	$M\gamma$	$M_{III}N_V$	1.765
7.032	5	81 Tl	$M\zeta_2$	$M_{IV}N_{II}$	1.763
7.0406	3	39 Y	$L\eta$	$L_{II}M_I$	1.76095
7.0759	3	37 Rb	$L\beta_1$	$L_{II}M_{IV}$	1.75217
7.09	2	73 Ta		$M_{IV}O_{II,III}$	1.748
7.101	8	79 Au		$M_{IV}N_{III}$	1.746
7.11	1	73 Ta	M_V	Abs. Edge	1.743
7.12542	9	14 Si	$K\alpha_1$	KL_{III}	1.73998
7.12791	9	14 Si	$K\alpha_2$	KL_{II}	1.73938
7.168	1	36 Kr	L_{II}	Abs. Edge	1.7297
7.250	5	36 Kr		$L_{II}N_{III}$	1.710
7.252	1	73 Ta	$M\alpha$	$M_VN_{VI,VII}$	1.7096
7.264	5	36 Kr	$L\beta_3$	L_IM_{III}	1.707
7.279	5	36 Kr	$L\gamma_5$	$L_{II}N_I$	1.703
7.30	2	73 Ta		M_VO_{III}	1.700
7.303	1	72 Hf	$M\beta$	$M_{IV}N_{VI}$	1.6976
7.304	5	36 Kr	$L\beta_4$	L_IM_{II}	1.697
7.3183	2	37 Rb	$L\alpha_1$	$L_{III}M_V$	1.69413
7.3251	3	37 Rb	$L\alpha_2$	$L_{III}M_{IV}$	1.69256
7.3563	3	39 Y	Ll	$L_{III}M_I$	1.68536
7.360	8	74 W		$M_{III}N_I$	1.684
7.371	8	78 Pt		$M_{IV}N_{III}$	1.682
7.392	1	36 Kr	L_{III}	Abs. Edge	1.6772
7.466	4	79 Au	$M\zeta_1$	M_VN_{III}	1.6605
7.503	1	34 Se	L_I	Abs. Edge	1.6525
7.510	4	36 Kr	$L\beta_6$	$L_{III}N_I$	1.6510
7.5171	3	38 Sr	$L\eta$	$L_{II}M_I$	1.64933
7.523	5	79 Au	$M\zeta_2$	$M_{IV}N_{II}$	1.648
7.539	1	72 Hf	$M\alpha$	$M_VN_{VI,VII}$	1.6446
7.546	8	68 Er	$M\gamma$	$M_{III}N_V$	1.643
7.576	3	36 Kr	$L\beta_1$	$L_{II}M_{IV}$	1.6366
7.60	1	68 Er		$M_{III}N_{IV}$	1.632
7.601	2	71 Lu	$M\beta$	$M_{IV}N_{VI}$	1.6312
7.612	9	73 Ta		$M_{III}N_I$	1.629
7.645	8	77 Ir		$M_{IV}N_{III}$	1.622
7.738	4	78 Pt	$M\zeta_1$	M_VN_{III}	1.6022
7.753	5	35 Br	L_{II}	Abs. Edge	1.599
7.767	9	35 Br	$L\beta_{3,4}$	$L_IM_{II,III}$	1.596
7.790	5	78 Pt	$M\zeta_2$	$M_{IV}N_{II}$	1.592
7.817	3	36 Kr	$L\alpha_{1,2}$	$L_{III}M_{IV,V}$	1.5860
7.8362	3	38 Sr	Ll	$L_{III}M_I$	1.58215
7.840	2	71 Lu	$M\alpha$	$M_VN_{VI,VII}$	1.5813
7.865	9	67 Ho	$M\gamma$	$M_{III}N_{IV,V}$	1.576
7.887	9	72 Hf		$M_{III}N_I$	1.572
7.909	2	70 Yb	$M\beta$	$M_{IV}N_{VI}$	1.5675
7.94813	5	13 Al	K	Abs. Edge	1.55988
7.960	2	13 Al	$K\beta$	KM	1.55745
7.984	5	35 Br	L_{III}	Abs. Edge	1.5530
8.021	4	77 Ir	$M\zeta_1$	M_VN_{III}	1.5458
8.0415	4	37 Rb	$L\eta$	$L_{II}M_I$	1.54177
8.065	5	77 Ir	$M\zeta_2$	$M_{IV}N_{II}$	1.5373
8.107	1	33 As	L_I	Abs. Edge	1.5293
8.1251	5	35 Br	$L\beta_1$	$L_{II}M_{IV}$	1.52590
8.144	9	66 Dy	$M\gamma$	$M_{III}N_{IV,V}$	1.522
8.149	5	70 Yb	$M\alpha$	$M_VN_{VI,VII}$	1.5214
8.239	8	75 Re		$M_{IV}N_{III}$	1.505
8.249	7	69 Tm	$M\beta$	$M_{IV}N_{VI}$	1.503
8.310	4	76 Os	$M\zeta_1$	M_VN_{III}	1.4919
8.321	9	34 Se	$L\beta_{3,4}$	$L_IM_{II,III}$	1.490
8.33934	9	13 Al	$K\alpha_1$	KL_{III}	1.48670
8.34173	9	13 Al	$K\alpha_2$	KL_{II}	1.48627
8.359	5	76 Os	$M\zeta_2$	$M_{IV}N_{II}$	1.4831

TABLE B (*Continued*)

Wavelength Å*	p.e.	Element	Designation		keV
8.3636	4	37 Rb	Ll	$L_{III}M_{I}$	1.48238
8.3746	5	35 Br	$L\alpha_{1,2}$	$L_{III}M_{IV,V}$	1.48043
8.407	1	34 Se	L_{II}	Abs. Edge	1.4747
8.470	9	70 Yb		$M_{III}N_{I}$	1.464
8.48	1	69 Tm	$M\alpha$	$M_{V}N_{VI,VII}$	1.462
8.486	9	65 Tb	$M\gamma$	$M_{III}N_{IV,V}$	1.461
8.487	5	69 Tm	M_{V}	Abs. Edge	1.4609
8.573	8	74 W		$M_{IV}N_{III}$	1.446
8.592	3	68 Er	$M\beta$	$M_{IV}N_{VI}$	1.4430
8.60	7	92 U		$N_{I}P_{IV,V}$	1.44
8.601	5	68 Er	M_{IV}	Abs. Edge	1.4415
8.629	4	75 Re	$M\zeta_1$	$M_{V}N_{III}$	1.4368
8.646	1	34 Se	L_{III}	Abs. Edge	1.4340
8.664	5	75 Re	$M\zeta_2$	$M_{IV}N_{II}$	1.4310
8.7358	5	34 Se	$L\beta_1$	$L_{II}M_{IV}$	1.41923
8.76	7	92 U		$N_{I}P_{III}$	1.42
8.773	1	32 Ge	L_{I}	Abs. Edge	1.4132
8.81	7	92 U		$N_{I}P_{II}$	1.41
8.82	1	68 Er	$M\alpha$	$M_{V}N_{VI,VII}$	1.406
8.844	9	64 Gd	$M\gamma$	$M_{III}N_{IV,V}$	1.402
8.847	5	68 Er	M_{V}	Abs. Edge	1.4013
8.90	2	73 Ta		$M_{IV}N_{III}$	1.393
8.929	1	33 As	$L\beta_{3,4}$	$L_{I}M_{II,III}$	1.3884
8.962	4	74 W	$M\zeta_1$	$M_{V}N_{III}$	1.3835
8.965	4	67 Ho	$M\beta$	$M_{IV}N_{VI}$	1.3830
8.9900	5	34 Se	$L\alpha_{1,2}$	$L_{III}M_{IV,V}$	1.37910
8.993	5	74 W	$M\zeta_2$	$M_{IV}N_{II}$	1.3787
9.125	1	33 As	L_{II}	Abs. Edge	1.3587
9.20	2	67 Ho	$M\alpha$	$M_{V}N_{VI,VII}$	1.348
9.211	9	63 Eu	$M\gamma$	$M_{III}N_{IV,V}$	1.346
9.255	1	35 Br	$L\eta$	$L_{II}M_{I}$	1.3396
9.316	4	73 Ta	$M\zeta_1$	$M_{V}N_{III}$	1.3308
9.330	5	73 Ta	$M\zeta_2$	$M_{IV}N_{II}$	1.3288
9.357	6	66 Dy	$M\beta$	$M_{IV}N_{VI}$	1.3250
9.367	1	33 As	L_{III}	Abs. Edge	1.3235
9.40	7	90 Th		$N_{I}P_{III}$	1.319
9.4141	8	33 As	$L\beta_1$	$L_{II}M_{IV}$	1.3170
9.44	7	90 Th		$N_{I}P_{II}$	1.313
9.5122	1	12 Mg	K	Abs. Edge	1.30339
9.517	5	31 Ga	L_{I}	Abs. Edge	1.3028
9.521	2	12 Mg	$K\beta$	KM	1.3022
9.581	2	32 Ge	$L\beta_3$	$L_{I}M_{III}$	1.2941
9.585	1	35 Br	Ll	$L_{III}M_{I}$	1.2935
9.59	2	66 Dy	$M\alpha$	$M_{V}N_{VI,VII}$	1.293
9.600	9	62 Sm	$M\gamma$	$M_{III}N_{IV,V}$	1.291
9.640	2	32 Ge	$L\beta_4$	$L_{I}M_{II}$	1.2861
9.6709	8	33 As	$L\alpha_{1,2}$	$L_{III}M_{IV,V}$	1.2820
9.686	7	72 Hf	$M\zeta_2$	$M_{IV}N_{II}$	1.2800
9.686	7	72 Hf	$M\zeta_1$	$M_{V}N_{III}$	1.2800
9.792	6	65 Tb	$M\beta$	$M_{IV}N_{VI}$	1.2661
9.8900	2	12 Mg	$K\alpha_{1,2}$	$KL_{II,III}$	1.25360
9.924	1	32 Ge	L_{II}	Abs. Edge	1.2494
9.962	1	34 Se	$L\eta$	$L_{II}M_{I}$	1.2446
10.00	2	65 Tb	$M\alpha$	$M_{V}N_{VI,VII}$	1.240
10.09	7	92 U		$N_{I}O_{III}$	1.229
10.175	1	32 Ge	$L\beta_1$	$L_{II}M_{IV}$	1.2185
10.187	1	32 Ge	L_{III}	Abs. Edge	1.2170
10.254	6	64 Gd	$M\beta$	$M_{IV}N_{VI}$	1.2091
10.294	1	34 Se	Ll	$L_{III}M_{I}$	1.2044
10.359	9	31 Ga	$L\beta_{3,4}$	$L_{I}M_{II,III}$	1.197
10.40	7	92 U		$N_{II}P_{I}$	1.192
10.4361	8	32 Ge	$L\alpha_{1,2}$	$L_{III}M_{IV,V}$	1.18800
10.46	3	64 Gd	$M\alpha$	$M_{V}N_{VI,VII}$	1.185
10.48	1	70 Yb	$M\zeta$	$M_{V}N_{III}$	1.183
10.505	9	60 Nd	$M\gamma$	$M_{III}N_{IV,V}$	1.180
10.711	5	63 Eu	M_{IV}	Abs. Edge	1.1575
10.734	1	33 As	$L\eta$	$L_{II}M_{I}$	1.1550
10.750	7	63 Eu	$M\beta$	$M_{IV}N_{VI}$	1.1533
10.828	5	31 Ga	L_{II}	Abs. Edge	1.1450
10.96	3	63 Eu	$M\alpha$	$M_{V}N_{VI,VII}$	1.131
10.998	9	59 Pr	$M\gamma$	$M_{III}N_{IV,V}$	1.1273
11.013	5	63 Eu	M_{V}	Abs. Edge	1.1258
11.023	2	31 Ga	$L\beta_1$	$L_{II}M_{IV}$	1.1248
11.072	1	33 As	Ll	$L_{III}M_{I}$	1.1198
11.07	7	90 Th		$N_{II}P_{I}$	1.120
11.100	1	31 Ga	L_{III}	Abs. Edge	1.1169
11.200	7	30 Zn	$L\beta_{3,4}$	$L_{I}M_{II,III}$	1.1070
11.27	1	62 Sm	$M\beta$	$M_{IV}N_{VI}$	1.0998
11.288	5	62 Sm	M_{IV}	Abs. Edge	1.0983
11.292	1	31 Ga	$L\alpha_{1,2}$	$L_{III}M_{IV,V}$	1.09792
11.37	1	68 Er	$M\zeta$	$M_{V}N_{III}$	1.0901
11.47	3	62 Sm	$M\alpha$	$M_{V}N_{VI,VII}$	1.081
11.53	1	58 Ce	$M\gamma$	$M_{III}N_{IV,V}$	1.0749
11.552	5	62 Sm	M_{V}	Abs. Edge	1.0732
11.56	5	90 Th		$N_{II}O_{IV}$	1.072
11.569	1	11 Na	K	Abs. Edge	1.07167
11.575	2	11 Na	$K\beta$	KM	1.0711
11.609	2	32 Ge	$L\eta$	$L_{II}M_{I}$	1.0680
11.862	1	30 Zn	L_{II}	Abs. Edge	1.04523
11.86	1	67 Ho	$M\zeta$	$M_{V}N_{III}$	1.0450
11.9101	9	11 Na	$K\alpha_{1,2}$	$KL_{II,III}$	1.04098
11.965	2	32 Ge	Ll	$L_{III}M_{I}$	1.0362
11.983	3	30 Zn	$L\beta_1$	$L_{II}M_{IV}$	1.0347
12.08	4	57 La	$M\gamma$	$M_{III}N_{IV,V}$	1.027
12.122	3	29 Cu	$L\beta_{3,4}$	$L_{I}M_{II,III}$	1.0228
12.131	1	30 Zn	L_{III}	Abs. Edge	1.02201
12.254	3	30 Zn	$L\alpha_{1,2}$	$L_{III}M_{IV,V}$	1.0117
12.43	2	66 Dy	$M\zeta$	$M_{V}N_{III}$	0.998
12.44	2	60 Nd	$M\beta$	$M_{IV}N_{VI}$	0.997
12.459	5	60 Nd	M_{IV}	Abs. Edge	0.9951
12.597	2	31 Ga	$L\eta$	$L_{II}M_{I}$	0.9842
12.68	2	60 Nd	$M\alpha$	$M_{V}N_{VI,VII}$	0.978
12.737	5	60 Nd	M_{V}	Abs. Edge	0.9734
12.75	3	56 Ba	$M\gamma$	$M_{III}N_{IV,V}$	0.973
12.90	9	92 U		$N_{III}O_{V}$	0.961
12.953	2	31 Ga	Ll	$L_{III}M_{I}$	0.9572
12.98	2	65 Tb	$M\zeta$	$M_{V}N_{III}$	0.955
13.014	1	29 Cu	L_{II}	Abs. Edge	0.95268
13.053	3	29 Cu	$L\beta_1$	$L_{II}M_{IV}$	0.9498
13.06	2	59 Pr	$M\beta$	$M_{IV}N_{VI}$	0.950
13.06	1	30 Zn	L_{I}	Abs. Edge	0.9495
13.122	5	59 Pr	M_{IV}	Abs. Edge	0.9448
13.18	2	28 Ni	$L\beta_{3,4}$	$L_{I}M_{II,III}$	0.941
13.288	1	29 Cu	L_{III}	Abs. Edge	0.93306

TABLE B (*Continued*)

Wavelength Å*	p.e.	Element		Designation	keV
13.30	6	83 Bi		$N_IP_{II,III}$	0.932
13.336	3	29 Cu	$L\alpha_{1,2}$	$L_{III}M_{IV,V}$	0.9297
13.343	5	59 Pr	$M\alpha$	$M_VN_{VI,VII}$	0.9292
13.394	5	59 Pr	M_V	Abs. Edge	0.9257
13.57	2	64 Gd	$M\zeta$	M_VN_{III}	0.914
13.68	2	30 Zn	$L\eta$	$L_{II}M_I$	0.906
13.75	4	58 Ce	$M\beta$	$M_{IV}N_{VI}$	0.902
13.8	1	90 Th		$N_{III}O_V$	0.897
14.02	2	30 Zn	Ll	$L_{III}M_I$	0.884
14.04	2	58 Ce	$M\alpha$	$M_VN_{VI,VII}$	0.883
14.22	2	63 Eu	$M\zeta$	M_VN_{III}	0.872
14.242	5	28 Ni	L_{II}	Abs. Edge	0.8706
14.271	6	28 Ni	$L\beta_1$	$L_{II}M_{IV}$	0.8688
14.3018	1	10 Ne	K	Abs. Edge	0.866889
14.31	3	27 Co	$L\beta_{3,4}$	$L_IM_{II,III}$	0.870
14.39	5	58 Ce		$M_VO_{II,III}$	0.862
14.452	5	10 Ne	$K\beta$	KM	0.8579
14.51	5	57 La	$M\beta$	$M_{IV}N_{VI}$	0.854
14.525	5	28 Ni	L_{III}	Abs. Edge	0.8536
14.561	3	28 Ni	$L\alpha_{1,2}$	$L_{III}M_{IV,V}$	0.8515
14.610	3	10 Ne	$K\alpha_{1,2}$	$KL_{II,III}$	0.8486
14.88	5	57 La	$M\alpha$	$M_VN_{VI,VII}$	0.833
14.90	2	29 Cu	$L\eta$	$L_{II}M_I$	0.832
14.91	4	62 Sm	$M\zeta$	M_VN_{III}	0.831
15.286	9	29 Cu	Ll	$L_{III}M_I$	0.8111
15.56	1	56 Ba	M_{IV}	Abs. Edge	0.7967
15.618	5	27 Co	L_{II}	Abs. Edge	0.7938
15.65	4	26 Fe	$L\beta_{3,4}$	$L_IM_{II,III}$	0.792
15.666	8	27 Co	$L\beta_1$	$L_{II}M_{IV}$	0.7914
15.72	9	56 Ba		$M_{IV}O_{III}$	0.789
15.89	1	56 Ba	M_V	Abs. Edge	0.7801
15.91	5	56 Ba		$M_{IV}O_{II}$	0.779
15.915	5	27 Co	L_{III}	Abs. Edge	0.7790
15.93	4	52 Te	$M\gamma$	$M_{III}N_{IV,V}$	0.778
15.972	6	27 Co	$L\alpha_{1,2}$	$L_{III}M_{IV,V}$	0.7762
15.98	5	51 Sb		$M_{II}N_{IV}$	0.776
16.20	5	56 Ba		M_VO_{III}	0.765
16.27	3	28 Ni	$L\eta$	$L_{II}M_I$	0.762
16.46	4	60 Nd	$M\zeta$	M_VN_{III}	0.753
16.693	9	28 Ni	Ll	$L_{III}M_I$	0.7427
16.7	1	24 Cr	L_I	Abs. Edge	0.741
16.92	4	51 Sb	$M\gamma$	$M_{III}N_{IV,V}$	0.733
16.93	5	50 Sn		$M_{II}N_{IV}$	0.733
17.19	4	25 Mn	$L\beta_{3,4}$	$L_IM_{II,III}$	0.721
17.202	5	26 Fe	L_{II}	Abs. Edge	0.7208
17.26	1	26 Fe	$L\beta_1$	$L_{II}M_{IV}$	0.7185
17.38	4	59 Pr	$M\zeta$	M_VN_{III}	0.714
17.525	5	26 Fe	L_{III}	Abs. Edge	0.7074
17.59	2	26 Fe	$L\alpha_{1,2}$	$L_{III}M_{IV,V}$	0.7050
17.6	1	52 Te		$M_{II}N_I$	0.703
17.87	3	27 Co	$L\eta$	$L_{II}M_I$	0.694
17.94	5	50 Sn	$M\gamma$	$M_{III}N_{IV,V}$	0.691
17.9	1	24 Cr	L_{II}	Abs. Edge	0.691
18.292	8	27 Co	Ll	$L_{III}M_I$	0.6778
18.32	2	9 F	$K\alpha$	KL	0.6768
18.35	4	58 Ce	$M\zeta$	M_VN_{III}	0.676
18.8	1	51 Sb		$M_{II}N_I$	0.658
18.8	2	47 Ag		$M_IN_{II,III}$	0.658
18.96	4	24 Cr	$L\beta_{3,4}$	$L_IM_{II,III}$	0.654
19.11	2	25 Mn	$L\beta_1$	$L_{II}M_{IV}$	0.6488
19.1	1	52 Te		$M_{III}N_I$	0.648
19.40	7	48 Cd		$M_{II}N_{IV}$	0.639
19.44	5	57 La	$M\zeta$	M_VN_{III}	0.638
19.45	1	25 Mn	$L\alpha_{1,2}$	$L_{III}M_{IV,V}$	0.6374
19.66	5	53 I	$M_{IV,V}$	Abs. Edge	0.631
19.75	4	26 Fe	$L\eta$	$L_{II}M_I$	0.628
20.0	1	50 Sn		$M_{II}N_I$	0.619
20.1	2	46 Pd		$M_IN_{II,III}$	0.616
20.15	1	26 Fe	Ll	$L_{III}M_I$	0.6152
20.2	1	51 Sb		$M_{III}N_I$	0.612
20.47	7	48 Cd	$M\gamma$	$M_{III}N_{IV,V}$	0.606
20.64	4	56 Ba	$M\zeta$	M_VN_{III}	0.601
20.66	7	47 Ag		$M_{II}N_{IV}$	0.600
20.7	1	24 Cr	L_{III}	Abs. Edge	0.598
21.19	5	23 Va	$L\beta_{3,4}$	$L_IM_{II,III}$	0.585
21.27	1	24 Cr	$L\beta_1$	$L_{II}M_{IV}$	0.5828
21.34	5	52 Te		$M_{IV}O_{II,III}$	0.581
21.5	1	50 Sn		$M_{III}N_I$	0.575
21.64	3	24 Cr	$L\alpha_{1,2}$	$L_{III}M_{IV,V}$	0.5728
21.78	5	52 Te		M_VO_{III}	0.569
21.82	7	47 Ag	$M\gamma$	$M_{III}N_{IV,V}$	0.568
21.85	2	25 Mn	$L\eta$	$L_{II}M_I$	0.5675
22.1	1	46 Pd		$M_{II}N_{IV}$	0.560
22.29	1	25 Mn	Ll	$L_{III}M_I$	0.5563
22.9	2	48 Cd		$M_{II}N_I$	0.540
23.32	1	8 O	K	Abs. Edge	0.5317
23.3	1	46 Pd	$M\gamma$	$M_{III}N_{IV,V}$	0.531
23.62	3	8 O	$K\alpha$	KL	0.5249
23.88	4	23 Va	$L\beta_1$	$L_{II}M_{IV}$	0.5192
24.25	3	23 Va	$L\alpha_{1,2}$	$L_{III}M_{IV,V}$	0.5113
24.28	5	50 Sn	$M_{IV,V}$	Abs. Edge	0.511
24.30	3	24 Cr	$L\eta$	$L_{II}M_I$	0.5102
24.4	2	47 Ag		M_VN_I	0.509
24.5	1	48 Cd		$M_{III}N_I$	0.507
24.78	1	24 Cr	Ll	$L_{III}M_I$	0.5003
25.01	9	45 Rh	$M\gamma$	$M_{III}N_{IV,V}$	0.496
25.3	1	50 Sn		$M_{IV}O_{II,III}$	0.491
25.50	9	44 Ru		$M_{II}N_{IV}$	0.486
25.7	1	50 Sn		M_VO_{III}	0.483
26.0	1	47 Ag		$M_{III}N_I$	0.478
26.2	2	46 Pd		$M_{II}N_I$	0.474
26.72	9	52 Te	$M\zeta$	$M_{IV,V}N_{II,III}$	0.464
26.9	1	44 Ru	$M\gamma$	$M_{III}N_{IV,V}$	0.462
27.05	2	22 Ti	$L\beta_1$	$L_{II}M_{IV}$	0.4584
27.29	1	22 Ti	$L_{II,III}$	Abs. Edge	0.4544
27.34	3	23 Va	$L\eta$	$L_{II}M_I$	0.4535
27.42	2	22 Ti	$L\alpha_{1,2}$	$L_{III}M_{IV,V}$	0.4522
27.77	1	23 Va	Ll	$L_{III}M_I$	0.4465
27.9	1	46 Pd		$M_{III}N_I$	0.445
28.1	2	45 Rh		$M_{II}N_I$	0.442
28.13	5	48 Cd	$M_{IV,V}$	Abs. Edge	0.4408
28.88	8	51 Sb	$M\zeta$	$M_{IV,V}N_{II,III}$	0.429
29.8	1	45 Rh		$M_{III}N_I$	0.417
30.4	1	48 Cd		$M_{IV}O_{II,III}$	0.408

Wavelength Å*	p.e.	Element		Designation	keV
30.8	1	48 Cd		M_VO_{III}	0.403
30.82	5	47 Ag	M_{IV}	Abs. Edge	0.4022
30.89	3	22 Ti	$L\eta$	$L_{II}M_I$	0.4013
30.99	1	7 N	K	Abs. Edge	0.4000
31.02	2	21 Sc	$L\beta_1$	$L_{II}M_{IV}$	0.3996
31.14	5	47 Ag	M_V	Abs. Edge	0.3981
31.24	9	50 Sn	$M\zeta$	$M_{IV,V}N_{II,III}$	0.397
31.35	3	21 Sc	$L\alpha_{1,2}$	$L_{III}M_{IV,V}$	0.3954
31.36	2	22 Ti	Ll	$L_{III}M_I$	0.3953
31.60	4	7 N	$K\alpha$	KL	0.3924
31.8	1	92 U		$N_{IV}N_{VI}$	0.390
32.3	2	44 Ru		$M_{II}N_I$	0.384
33.1	2	41 Nb		$M_{II}N_{IV}$	0.375
33.5	3	47 Ag		$M_{IV,V}O_{II,III}$	0.370
33.57	9	90 Th		$N_{IV}N_{VI}$	0.3693
34.8	1	92 U		$N_VN_{VI,VII}$	0.357
34.9	2	41 Nb	$M\gamma$	$M_{III}N_{IV,V}$	0.356
35.13	2	21 Sc	$L\eta$	$L_{II}M_I$	0.3529
35.13	1	20 Ca	L_{II}	Abs. Edge	0.3529
35.3	3	42 Mo		$M_{II}N_I$	0.351
35.49	1	20 Ca	L_{III}	Abs. Edge	0.34931
35.59	3	21 Sc	Ll	$L_{III}M_I$	0.3483
35.63	1	20 Ca	$L_{II,III}$	Abs. Edge	0.34793
35.94	2	20 Ca	$L\beta_1$	$L_{II}M_{IV}$	0.3449
36.32	9	90 Th		$N_VN_{VI,VII}$	0.3414
36.33	2	20 Ca	$L\alpha_{1,2}$	$L_{III}M_{IV,V}$	0.3413
36.8	1	48 Cd	$M\zeta$	$M_{IV,V}N_{II,III}$	0.3371
37.4	2	46 Pd		$M_{IV,V}O_{II,III}$	0.332
37.5	2	42 Mo		$M_{III}N_I$	0.331
38.4	3	41 Nb		$M_{II}N_I$	0.323
39.77	7	47 Ag	$M\zeta$	$M_{IV,V}N_{II,III}$	0.3117
40.46	2	20 Ca	$L\eta$	$L_{II}M_I$	0.3064
40.7	2	41 Nb		$M_{III}N_I$	0.305
40.9	2	45 Rh		$M_{IV,V}O_{II,III}$	0.303
40.96	2	20 Ca	Ll	$L_{III}M_I$	0.3027
42.1	2	92 U		$N_{VI}O_V$	0.295
42.1	1	19 K	$L_{II,III}$	Abs. Edge	0.2946
42.3	2	82 Pb		$N_{IV}N_{VI}$	0.293
43.3	2	92 U		$N_{VI}O_{IV}$	0.286
43.6	1	46 Pd	$M\zeta$	$M_{IV,V}N_{II,III}$	0.2844
43.68	1	6 C	K	Abs. Edge	0.28384
44.7	3	6 C	$K\alpha$	KL	0.277
44.8	1	44 Ru		$M_{IV,V}O_{II,III}$	0.2768
45.0	1	82 Pb		$N_VN_{VI,VII}$	0.2756
45.2	3	80 Hg		$N_{IV}N_{VI}$	0.274
45.2	1	51 Sb		$M_{II}M_{IV}$	0.2743
46.48	9	39 Y		$M_{II}N_I$	0.267
46.5	2	81 Tl		$N_VN_{VI,VII}$	0.267
46.8	2	79 Au		$N_{IV}N_{VI}$	0.265
47.24	2	19 K	Ll	$L_{II}M_I$	0.2625
47.3	1	50 Sn		$M_{II}M_{IV}$	0.2621
47.67	9	45 Rh	$M\zeta$	$M_{IV,V}N_{II,III}$	0.2601
47.74	1	19 K	Ll	$L_{III}M_I$	0.25971
47.9	3	80 Hg		$N_VN_{VI,VII}$	0.259
48.1	2	78 Pt		$N_{IV}N_{VI}$	0.258
48.2	1	90 Th		$N_{VI}O_V$	0.2572
48.5	2	39 Y		$M_{III}N_I$	0.256
49.4	1	79 Au		$N_VN_{VI,VII}$	0.2510
49.5	1	90 Th		$N_{VI}O_{IV}$	0.2505
50.0	1	90 Th		$N_{VII}O_V$	0.2479
50.2	1	77 Ir		$N_{IV}N_{VI}$	0.2470
50.3	1	52 Te		$M_{III}M_V$	0.2465
50.9	1	78 Pt		$N_VN_{VI,VII}$	0.2436
51.3	1	38 Sr		$M_{II}N_I$	0.2416
51.9	1	76 Os		$N_{IV}N_{VI}$	0.2388
52.0	2	48 Cd		$M_{II}M_{IV}$	0.2384
52.2	1	51 Sb		$M_{III}M_V$	0.2375
52.34	7	44 Ru	$M\zeta$	$M_{IV,V}N_{II,III}$	0.2369
52.8	1	77 Ir		$N_VN_{VI,VII}$	0.2348
53.6	1	38 Sr		$M_{III}N_I$	0.2313
54.0	2	74 W		$N_{II}N_{IV}$	0.2295
54.0	1	47 Ag		$M_{II}M_{IV}$	0.2295
54.2	1	50 Sn		$M_{III}M_V$	0.2287
54.7	2	76 Os		$N_VN_{VI,VII}$	0.2266
54.8	2	42 Mo		$M_{IV,V}O_{II,III}$	0.2262
55.8	1	74 W		$N_{IV}N_{VI}$	0.2221
55.9	1	18 A	$L\eta$	$L_{II}M_I$	0.2217
56.3	1	18 A	Ll	$L_{III}M_I$	0.2201
56.5	1	46 Pd		$M_{II}M_{IV}$	0.2194
57.0	2	37 Rb		$M_{II}N_I$	0.2174
58.2	1	73 Ta		$N_{IV}N_{VI}$	0.2130
58.4	1	74 W		N_VN_{VII}	0.2122
58.7	2	48 Cd		$M_{III}M_V$	0.2111
59.3	1	45 Rh		$M_{II}M_{IV}$	0.2090
59.5	3	74 W		N_VN_{VI}	0.208
59.5	2	37 Rb		$M_{III}N_I$	0.2083
60.5	1	47 Ag		$M_{III}M_V$	0.2048
61.1	2	73 Ta		$N_VN_{VI,VII}$	0.2028
61.9	2	41 Nb		$M_{IV,V}O_{II,III}$	0.2002
62.2	1	44 Ru		$M_{II}M_{IV}$	0.1992
62.9	1	46 Pd		$M_{III}M_V$	0.1970
63.0	5	71 Lu		$N_{IV}N_{VI}$	0.197
64.38	7	42 Mo	$M\zeta$	$M_{IV,V}N_{II,III}$	0.1926
65.1	7	70 Yb		$N_{IV}N_{VI}$	0.190
65.5	1	45 Rh		$M_{III}M_V$	0.1892
65.7	2	71 Lu		$N_VN_{VI,VII}$	0.1886
67.33	9	17 Cl	$L\eta$	$L_{II}M_I$	0.1841
67.6	3	5 B	$K\alpha$	KL	0.1833
67.90	9	17 Cl	Ll	$L_{III}M_I$	0.1826
68.2	3	90 Th		$O_{III}P_{IV,V}$	0.1817
68.3	1	44 Ru		$M_{III}M_V$	0.1814
68.9	2	42 Mo		$M_{II}M_{IV}$	0.1798
69.3	5	70 Yb		$N_VN_{VI,VII}$	0.179
70.0	4	40 Zr		$M_{IV,V}O_{II,III}$	0.177
72.1	3	41 Nb		$M_{II}M_{IV}$	0.1718
72.19	9	41 Nb	$M\zeta$	$M_{IV,V}N_{II,III}$	0.1717
72.7	9	68 Er		$N_{IV}N_{VI}$	0.171
74.9	1	42 Mo		$M_{III}M_V$	0.1656
76.3	7	68 Er		$N_VN_{VI,VII}$	0.163
76.7	2	40 Zr		$M_{II}M_{IV}$	0.1617
76.9	2	35 Br		$M_{II}N_I$	0.1613
78.4	2	41 Nb		$M_{III}M_V$	0.1582
79.8	3	35 Br		$M_{III}N_I$	0.1554
80.9	3	40 Zr		$M_{III}M_V$	0.1533

Wavelength Å*	p.e.	Element	Designation		keV
81.5	2	39 Y		$M_{II}M_{IV}$	0.1522
82.1	2	40 Zr	$M\zeta$	$M_{IV,V}N_{II,III}$	0.1511
83.	1	66 Dy		$N_{IV,V}N_{VI,VII}$	0.149
83.4	3	16 S	Ll, η	$L_{II,III}M_{I}$	0.1487
85.7	2	38 Sr		$M_{II}M_{IV}$	0.1447
86.	1	65 Tb		$N_{IV,V}N_{VI,VII}$	0.144
86.5	2	39 Y		$M_{III}M_{IV,V}$	0.1434
91.4	2	38 Sr		$M_{III}M_{IV,V}$	0.1357
91.5	2	37 Rb		$M_{II}M_{IV}$	0.1355
91.6	1	83 Bi		$N_{VI}O_{IV}$	0.1354
93.2	1	83 Bi		$N_{VII}O_{V}$	0.1330
93.4	2	39 Y	$M\zeta$	$M_{IV,V}N_{II,III}$	0.1328
94.	1	15 P	$L_{II,III}$	Abs. Edge	0.132
96.7	2	37 Rb		$M_{III}M_{IV,V}$	0.1282
97.2	8	66 Dy		$N_{IV,V}O_{II,III}$	0.128
98.	1	62 Sm		$N_{IV,V}N_{VI,VII}$	0.126
100.2	2	82 Pb		$N_{VI}O_{V}$	0.1237
102.2	4	65 Tb		$N_{IV,V}O_{II,III}$	0.1213
102.4	1	82 Pb		$N_{VI}O_{IV}$	0.1211
103.8	4	15 P		$L_{II,III}M$	0.1194
104.3	1	82 Pb		$N_{VII}O_{V}$	0.1189
107.	1	60 Nd		$N_{IV,V}N_{VI,VII}$	0.116
108.0	2	38 Sr	$M\zeta_2$	$M_{IV}N_{II,III}$	0.1148
108.7	1	38 Sr	$M\zeta_1$	$M_{V}N_{III}$	0.1140
109.4	3	35 Br		$M_{II}M_{IV}$	0.1133
110.6	5	29 Cu	M_{I}	Abs. Edge	0.1121
111.	1	4 Be	K	Abs. Edge	0.111
112.0	6	63 Eu		$N_{IV,V}O_{II,III}$	0.1107
113.0	1	81 Tl		$N_{VI}O_{V}$	0.10968
113.	1	59 Pr		$N_{IV,V}N_{VI,VII}$	0.1095
113.8	3	35 Br		$M_{III}M_{IV,V}$	0.1089
114.	1	4 Be	$K\alpha$	KL	0.1085
115.3	2	81 Tl		$N_{VI}O_{IV}$	0.1075
117.4	4	62 Sm		$N_{IV,V}O_{II,III}$	0.1056
117.7	1	81 Tl		$N_{VII}O_{V}$	0.10530
123.	1	14 Si	$L_{II,III}$	Abs. Edge	0.1006
126.8	2	37 Rb		$M_{IV}N_{III}$	0.0978
127.8	2	37 Rb	$M\zeta_2$	$M_{IV}N_{II}$	0.0970
128.7	2	37 Rb	$M\zeta_1$	$M_{V}N_{III}$	0.0964
128.9	7	60 Nd		$N_{IV,V}O_{II,III}$	0.0962
135.5	4	14 Si		$L_{II,III}M$	0.0915
136.5	4	59 Pr		$N_{IV,V}O_{II,III}$	0.0908
137.0	5	30 Zn	M_{II}	Abs. Edge	0.0905
142.5	1	13 Al	L_{I}	Abs. Edge	0.08701
143.9	5	30 Zn	M_{III}	Abs. Edge	0.0862
144.4	6	58 Ce		$N_{IV,V}O_{II,III}$	0.0859
144.4	3	37 Rb		$M_{I}M_{III}$	0.0859
152.6	6	57 La		$N_{IV,V}O_{II,III}$	0.0812
157.	3	30 Zn		$M_{II,III}M_{IV,V}$	0.079
159.0	2	56 Ba		$N_{IV}O_{III}$	0.07796
159.5	5	29 Cu	M_{II}	Abs. Edge	0.0777
163.3	2	56 Ba		$N_{IV}O_{II}$	0.07590
164.6	2	56 Ba		$N_{V}O_{III}$	0.07530
164.7	3	35 Br		$M_{I}M_{III}$	0.0753
166.0	5	29 Cu	M_{III}	Abs. Edge	0.0747
170.4	1	13 Al	$L_{II,III}$	Abs. Edge	0.07278
171.4	5	13 Al		$L_{II,III}M$	0.0724
173.	3	29 Cu		$M_{II,III}M_{IV,V}$	0.072
181.	5	90 Th		$O_{IV,V}Q_{II,III}$	0.068
183.8	1	55 Cs		$N_{IV}O_{III}$	0.06746
184.6	3	35 Br		$M_{I}M_{II}$	0.0672
188.4	1	28 Ni	M_{III}	Abs. Edge	0.06581
188.6	1	55 Cs		$N_{IV}O_{II}$	0.06574
189.5	3	35 Br		$M_{IV}N_{III}$	0.0654
190.3	1	55 Cs		$N_{V}O_{III}$	0.06515
190.	2	28 Ni		$M_{II,III}M_{IV,V}$	0.0651
191.1	2	35 Br	$M\zeta_2$	$M_{IV}N_{II}$	0.06488
192.6	2	35 Br	$M\zeta_1$	$M_{V}N_{III}$	0.06437
197.3	1	12 Mg	L_{I}	Abs. Edge	0.06284
202.	5	27 Co	$M_{II,III}$	Abs. Edge	0.061
203.	1	16 S		$L_{I}L_{II,III}$	0.061
214.	6	27 Co		$M_{II,III}M_{IV,V}$	0.058
224.	1	53 I	$N_{IV,V}$	Abs. Edge	0.0552
226.5	1	3 Li	K	Abs. Edge	0.05475
227.8	1	34 Se	M_{V}	Abs. Edge	0.05443
228.	1	3 Li	$K\alpha$	KL	0.0543
230.	2	34 Se		$M_{V}N_{III}$	0.0538
230.	1	26 Fe	$M_{II,III}$	Abs. Edge	0.0538
243.	5	26 Fe		$M_{II,III}M_{IV,V}$	0.051
249.3	1	12 Mg	L_{II}	Abs. Edge	0.04973
250.7	1	12 Mg	L_{III}	Abs. Edge	0.04945
251.5	5	12 Mg		$L_{II,III}M$	0.04929
273.	6	25 Mn		$M_{II,III}M_{IV,V}$	0.045
290.	1	13 Al		$L_{I}L_{II,III}$	0.0428
309.	9	24 Cr		$M_{II,III}M_{IV,V}$	0.040
317.	1	12 Mg		$L_{I}L_{II,III}$	0.0392
337.	9	23 V		$M_{II,III}M_{IV,V}$	0.0368
376.	1	11 Na		$L_{I}L_{II,III}$	0.03299
399.	5	35 Br	N_{I}	Abs. Edge	0.0311
405.	5	11 Na	$L_{II,III}$	Abs. Edge	0.0306
407.1	5	11 Na		$L_{II,III}M$	0.03045
417.	5	17 Cl	M_{I}	Abs. Edge	0.0297
444.	5	53 I	O_{I}	Abs. Edge	0.0279
525.	9	20 Ca		$M_{II,III}N_{I}$	0.0236
692.	9	19 K		$M_{II,III}N_{I}$	0.0179

X-RAY ATOMIC ENERGY LEVELS*

J. A. Bearden and A. F. Burr

These tables were originally published as the final report to the U.S. Atomic Energy Commission as Report NYO-2543-1 in partial fulfillment of Contract AT(30-1)-2543. The tables were later reproduced in *Review of Modern Physics.* The data may also be obtained from the Superintendent of Documents, U.S. Government Printing Office, Washington, D. C. 20402 in the publication NSRDS-NBS 14. Persons seeking discussion of the details of calculations, sources of energy level information and the problem of properly interpreting the experimental measurements should refer to the original publication or to *Review of Modern Physics,* Vol. 39, 125–142, January 1967.

All of the x-ray emission wavelengths have recently been reevaluated and placed on a consistent Å* scale. For most elements these data give a highly overdetermined set of equations for energy level differences, which have been solved by least-squares adjustment for each case. This procedure makes "best" use of all x-ray wavelength data, and also permits calculation of the probable error for each energy difference. Photoelectron measurements of absolute energy levels are more precise than x-ray absorption edge data. These have been used to establish the absolute scale for eighty-one elements and, in many cases, to provide additional energy level difference data. The x-ray absorption wavelengths were used for eight elements and ionization measurements for two; the remaining five were interpolated by a Moseley diagram involving the output values of energy levels from adjacent elements. Probable errors are listed on an absolute energy basis. In the original source of the present data, a table of energy levels in Rydberg units is given. Difference tables in volts, Rydbergs, and milli-Å* wavelength units, with the respective probable errors, are also included there.

Recommended values of the atomic energy levels, and probable errors in eV. Where available, photoelectron direct measurements are listed in brackets [ ] immediately under the recommended values. The measured values of the x-ray absorption energies are shown in parentheses ( ). Interpolated values are enclosed in angle brackets ⟨ ⟩.

Level	1 H	2 He	3 Li	4 Be	5 B	6 C	7 N	8 O
K	13.59811[a]	24.58678[b]	54.75±0.02 (54.75)	111.0±1.0 (111.0)	188.0±0.4 [188.0][c]	283.8±0.4 [283.8][c] (283.8)	401.6±0.4 [401.6][c]	532.0±0.4 [532.0][c]
L_I								23.7±0.4 [23.7][d]
$L_{II,III}$					4.7±0.9	6.4±1.9	9.2±0.6	7.1±0.8

Level	9 F	10 Ne	11 Na	12 Mg	13 Al	14 Si	15 P	16 S
K	685.4±0.4 [685.4][c]	866.9±0.3 (866.9)	1072.1±0.4 [1072.1][c] (1072.)	1305.0±0.4 [1305.0][c] (1303.)	1559.6±0.4 [1559.6][c] (1559.8)	1838.9±0.4 [1838.9][c]	2145.5±0.4 [2145.5][d]	2472.0±0.4 [2472.0][c] (2470.)
L_I	⟨31.⟩	⟨45.⟩	63.3±0.4 [63.3][d]	89.4±0.4 [89.4][d] (63.)	117.7±0.4 [117.7][d] (87.)	148.7±0.4 [148.7][d]	189.3±0.4 [189.3][d]	229.2±0.4 [229.2][d]
$L_{II,III}$	8.6±0.8	18.3±0.4	31.1±0.4 (31.)	51.4±0.5 (50.)	73.1±0.5 (72.8)	99.2±0.5 (100.6)	132.2±0.5 (132.)	164.8±0.7

Level	17 Cl	18 Ar	19 K	20 Ca	21 Sc	22 Ti	23 V	24 Cr
K	2822.4±0.3 [2822.4][c] (2020.)	3202.9±0.3 (3202.9)	3607.4±0.4 [3607.4][c] (3607.8)	4038.1±0.4 [4038.1][c] (4038.1)	4492.8±0.4 [4492.8][c]	4966.4±0.4 [4966.4][d] (4964.5)	5465.1±0.3 [5465.1][c] (5464.)	5989.2±0.3 [5989.2][c] (5989.)
L_I	270.2±0.4 [270.2][d]	320. ⟨320.⟩[d]	377.1±0.4 [377.1][d]	437.8±0.4 [437.8][d]	500.4±0.4 [500.4][d]	563.7±0.4 [563.7][d]	628.2±0.4 [628.2][d]	694.6±0.4 [694.6][d]
L_{II}	201.6±0.3	247.3±0.3	296.3±0.4	350.0±0.4	406.7±0.4	461.5±0.4	520.5±0.3	583.7±0.3
L_{III}	200.0±0.3	245.2±0.3	293.6±0.4	346.4±0.4	402.2±0.4	455.5±0.4	512.9±0.3	574.5±0.3
M_I	17.5±0.4	25.3±0.4	33.9±0.4	43.7±0.4	53.8±0.4	60.3±0.4	66.5±0.4	74.1±0.4
$M_{II,III}$	6.8±0.4	12.4±0.3	17.8±0.4	25.4±0.4	32.3±0.5	34.6±0.4	37.8±0.3	42.5±0.3
$M_{IV,V}$					6.6±0.5	3.7	2.2±0.3	2.3±0.4

Level	25 Mn	26 Fe	27 Co	28 Ni	29 Cu	30 Zn	31 Ga	32 Ge
K	6539.0±0.4 [6539.0][c] (6538.)	7112.0±0.9 [7111.3][c,f] (7111.2)	7708.9±0.3 [7708.9][c] (7709.5)	8332.8±0.4 [8332.8][c] (8331.6)	8978.9±0.4 [8978.9][c,g] (8980.3)	9658.6±0.6 [9658.6][g] (9660.7)	10367.1±0.5 [10367.1][g] (10368.2)	11103.1±0.7 [11103.8][g] (11103.6)
L_I	769.0±0.4 [769.0][d]	846.1±0.4 [846.1][d]	925.6±0.4 [925.6][d]	1008.1±0.4 [1008.1][d]	1096.1±0.4 [1096.0][d]	1193.6±0.9	1297.7±1.1	1414.3±0.7 [1413.6][g]
L_{II}	651.4±0.4	721.1±0.9 (720.8)	793.6±0.3 (793.8)	871.9±0.4 (870.6)	951.0±0.4 [950.0][h] (953.)	1042.8±0.6 (1045.)	1142.3±0.5	1247.8±0.7 (1249.)
L_{III}	640.3±0.4	708.1±0.9 (707.4)	778.6±0.3 (779.0)	854.7±0.4 (853.6)	931.1±0.4 [931.4][h] (933.)	1019.7±0.6 (1022.)	1115.4±0.5 (1117.)	1216.7±0.7 (1217.0)

* Wavelengths corresponding to these energy levels may be calculated from $\lambda\ (\text{nm}) = \dfrac{1.239852}{E(\text{keV})}$.

	25 Mn	26 Fe	27 Co	28 Ni	29 Cu	30 Zn	31 Ga	32 Ge
M_I	83.9±0.5	92.9±0.9	100.7±0.4	111.8±0.6	119.8±0.6	135.9±1.1	158.1±0.5	180.0±0.8
M_{II} M_{III}	48.6±0.4	54.0±0.9 (54.)	59.5±0.3 (61.)	68.1±0.4 (66.)	73.6±0.4 (75.)	86.6±0.6 (86.)	106.8±0.7 102.9±0.5	127.9±0.9 120.8±0.7
$M_{IV,V}$	3.3±0.5	3.6±0.9	2.9±0.3	3.6±0.4	1.6±0.4	8.1±0.6	17.4±0.5	28.7±0.7

	33 As	34 Se	35 Br	36 Kr	37 Rb	38 Sr	39 Y	40 Zr
K	11866.7±0.7 [11866.7][i] (11865.)	12657.8±0.7 [12657.8][g] (12654.5)	13473.7±0.4 (13470.)	14325.6±0.8 (14324.4)	15199.7±0.3 (15202.)	16104.6±0.3 (16107.)	17038.4±0.3 (17038.)	17997.6±0.4 (17999.)
L_I	1526.5±0.8 (1529.)	1653.9±3.5 (1652.5)	1782.0±0.4 [1782.0][j]	1921.0±0.6 [1921.2][k]	2065.1±0.3 [2065.4][j]	2216.3±0.3 [2216.2][l]	2372.5±0.3 [2372.7][l]	2531.6±0.3 [2531.6][l]
L_{II}	1358.6±0.7 (1358.7)	1476.2±0.7 (1474.7)	1596.0±0.4 [1596.2][j]	1727.2±0.5 [1727.2][k] (1730.)	1863.9±0.3 [1863.4][j]	2006.8±0.3 [2006.6][l] (2008.5)	2155.5±0.3 [2155.0][l] (2154.0)	2306.7±0.3 [2306.5][l] (2305.3)
L_{III}	1323.1±0.7 (1323.5)	1435.8±0.7 (1434.0)	1549.9±0.4 [1549.7][j]	1674.9±0.5 [1674.8][k] (1677.)	1804.4±0.3 [1804.6][j]	1939.6±0.3 [1939.9][l] (1941.)	2080.0±0.3 [2080.2][l] (2079.4)	2222.3±0.3 [2222.5][l] (2222.5)
M_I	203.5±0.7	231.5±0.7	256.5±0.4		322.1±0.3	357.5±0.3	393.6±0.3	430.3±0.3
M_{II}	146.4±1.2	168.2±1.3	189.3±0.4	222.7±1.1	247.4±0.3	279.8±0.3	312.4±0.4	344.2±0.4
M_{III}	140.5±0.8	161.9±1.0	181.5±0.4	213.8±1.1	238.5±0.3	269.1±0.3	300.3±0.4	330.5±0.4
M_{IV} M_V	41.2±0.7	56.7±0.8	70.1±0.4 69.0±0.4	88.9±0.8	111.8±0.3 110.3±0.3	135.0±0.3 133.1±0.3	159.6±0.3 157.4±0.3	182.4±0.3 180.0±0.3
N_I			27.3±0.5	24.0±0.8	29.3±0.3	37.7±0.3	45.4±0.3	51.3±0.3
N_{II} N_{III}	2.5±1.0	5.6±1.3	5.2±0.4 4.6±0.4	10.6±1.9	14.8±0.4 14.0±0.3	19.9±0.3	25.6±0.4	28.7±0.4

	41 Nb	42 Mo	43 Tc	44 Ru	45 Rh	46 Pd	47 Ag	48 Cd
K	18985.6±0.4 (18987.)	19999.5±0.3 (20004.)	21044.0±0.7	22117.2±0.3 (22119.)	23219.9±0.3 (23219.8)	24350.3±0.3 (24348.)	25514.0±0.3 (25516.)	26711.2±0.3 (26716.)
L_I	2697.7±0.3 [2697.7][l]	2865.5±0.3 [2866.0][l]	3042.5±0.4 [3042.5][l]	3224.0±0.3 [3224.3][l]	3411.9±0.3 [3412.0][l] (3417.)	3604.3±0.3 [3604.6][l] (3607.)	3805.8±0.3 [3806.2][m] (3807.)	4018.0±0.3 [4018.1][m] (4019.)
L_{II}	2464.7±0.3 [2464.7][l]	2625.1±0.3 [2624.5][l] (2627.)	2793.2±0.4 [2973.2][l]	2966.9±0.3 [2966.8][l] (2966.3)	3146.1±0.3 [3146.3][l] (3145.)	3330.3±0.3 [3330.3][l] (3330.3)	3523.7±0.3 [3523.6][g,m] (3526.)	3727.0±0.3 [3727.1][m] (3728.)
L_{III}	2370.5±0.3 [2370.6][l]	2520.2±0.3 [2520.2][l] (2523.2)	2676.9±0.4 [2676.9][l]	2837.9±0.3 [2837.7][l] (2837.7)	3003.8±0.3 [3003.5][g,l] (3002.)	3173.3±0.3 [3173.0][g,l] (3173.0)	3351.1±0.3 [3350.8][g] (3351.0)	3537.5±0.3 [3537.3][g] (3537.6)
M_I	468.4±0.3	504.6±0.3		585.0±0.3	627.1±0.3	669.9±0.3	717.5±0.3	770.2±0.3
M_{II}	378.4±0.4	409.7±0.4	444.9±1.5	482.8±0.3	521.0±0.3	559.1±0.3	602.4±0.3	650.7±0.3
M_{III}	363.0±0.4	392.3±0.3	425.0±1.5	460.6±0.3	496.2±0.3	531.5±0.3	571.4±0.3	616.5±0.3
M_{IV}	207.4±0.3	230.3±0.3	256.4±0.5	283.6±0.3	311.7±0.3	340.0±0.3	372.8±0.3	410.5±0.3
M_V	204.6±0.3	227.0±0.3	252.9±0.4	279.4±0.3	307.0±0.3	334.7±0.3	366.7±0.3	403.7±0.3
N_I	58.1±0.3	61.8±0.3		74.9±0.3	81.0±0.3	86.4±0.3	95.2±0.3	107.6±0.3
N_{II} N_{III}	33.9±0.4	34.8±0.4	38.9±1.9	43.1±0.4	47.9±0.4	51.1±0.4	62.6±0.3 55.9±0.3	66.9±0.4
$N_{IV,V}$	3.2±0.3	1.8±0.3		2.0±0.3	2.5±0.4	1.5±0.3	3.3±0.3	9.3±0.3

	49 In	50 Sn	51 Sb	52 Te	53 I	54 Xe	55 Cs	56 Ba
K	27939.9±0.3	29200.1±0.4 (29195.)	30491.2±0.3 (30486.)	31813.8±0.3 (31811.)	33169.4±0.4 (33167.)	34561.4±1.1 (34590.)	35984.6±0.4 (35987.)	37440.6±0.4 (37452.)
L_I	4237.5±0.3 [4237.7][m] (4237.3)	4464.7±0.3 [4464.5][g] (4464.8)	4698.3±0.3 [4698.3][m] (4698.4)	4939.2±0.3 [4939.3][m] (4939.7)	5188.1±0.3 [5188.1][j]	5452.8±0.4 (5452.8)	5714.3±0.4 [5712.7][j] (5721.)	5988.8±0.4 [5986.8][j] (5996.)
L_{II}	3938.0±0.3 [3937.8][m] (3939.3)	4156.1±0.3 [4156.2][g] (4157.)	4380.4±0.3 [4380.6][m] (4382.)	4612.0±0.3 [4612.0][m] (4612.6)	4852.1±0.3 [4852.0][j]	5103.7±0.4 (5103.7)	5359.4±0.3 [5359.5][j] (5358.)	5623.6±0.3 [5623.6][j] (5623.3)
L_{III}	3730.1±0.3 [3730.0][g] (3730.2)	3928.8±0.3 [3928.8][g] (3928.8)	4132.2±0.3 [4132.2][g] (4132.3)	4341.4±0.3 [4341.2][g] (4341.8)	4557.1±0.3 [4557.1][j]	4782.2±0.4 (4782.2)	5011.9±0.3 [5012.0][j] (5011.3)	5247.0±0.3 [5247.3][j] (5247.0)
M_I	825.6±0.3	883.8±0.3	943.7±0.3	1006.0±0.3	1072.1±0.3		1217.1±0.4	1292.8±0.4
M_{II}	702.2±0.3	756.4±0.4	811.9±0.3	869.7±0.3	930.5±0.3	999.0±2.1	1065.0±0.5	1136.7±0.5

	49 In	50 Sn	51 Sb	52 Te	53 I	54 Xe	55 Cs	56 Ba
M_{III}	664.3±0.3	714.4±0.3	765.6±0.3	818.7±0.3	874.6±0.3	937.0±2.1	997.6±0.5	1062.2±0.5
M_{IV}	450.8±0.3	493.3±0.3	536.9±0.3	582.5±0.3	631.3±0.3		739.5±0.4	796.1±0.3
M_V	443.1±0.3	484.8±0.3	527.5±0.3	572.1±0.3	619.4±0.3	672.3±0.5	725.5±0.5	780.7±0.3
N_I	121.9±0.3	136.5±0.4	152.0±0.3	168.3±0.3	186.4±0.3		230.8±0.4	253.0±0.5
N_{II}	77.4±0.4	88.6±0.4	98.4±0.5	110.2±0.5	122.7±0.5	146.7±3.1	172.3±0.6	191.8±0.7
N_{III}							161.6±0.6	179.7±0.6
N_{IV}	16.2±0.3	23.9±0.3	31.4±0.3	39.8±0.3	49.6±0.3		78.8±0.5	92.5±0.5
N_V							76.5±0.5	89.9±0.5
O_I	0.1±4.5	0.9±0.5	6.7±0.5	11.6±0.6	13.6±0.6		22.7±0.5	39.1±0.6
O_{II}	0.8±0.4	1.1±0.5	2.1±0.4	2.3±0.5	3.3±0.5		13.1±0.5	16.6±0.5
O_{III}							11.4±0.5	14.6±0.5

	57 La	58 Ce	59 Pr	60 Nd	61 Pm	62 Sm	63 Eu	64 Gd
K	38924.6±0.4 (38934.)	40443.0±0.4 (40453.)	41990.6±0.5 (42002.)	43568.9±0.4 (43574.)	45184.0±0.7 (45198.)	46834.2±0.5 (46849.)	48519.0±0.4 (48519.)	50239.1±0.5 (50233.)
L_I	6266.3±0.5 [6266.3][a]	6548.8±0.5 [6548.5][a]	6834.8±0.5 [6834.9][a]	7126.0±0.4 [7125.8][a] (7129.)	7427.9±0.8 [7427.9][b]	7736.8±0.5 [7736.2][a] (7748.)	8052.0±0.4 [8051.7][a] (8061.)	8375.6±0.5 [8375.4][a] (8386.)
L_{II}	5890.6±0.4 [5890.7][a]	6164.2±0.4 [6164.3][a]	6440.4±0.5 [6440.2][a]	6721.5±0.4 [6721.8][a] (6723.)	7012.8±0.6 [7012.8][b]	7311.8±0.4 [7312.0][a] (7313.)	7617.1±0.4 [7617.6][a] (7620.)	7930.3±0.4 [7930.5][a] (7931.)
L_{III}	5482.7±0.4 [5482.6][a]	5723.4±0.4 [5723.6][a]	5964.3±0.4 [5964.3][a]	6207.9±0.4 [6208.0][a] (6209.)	6459.3±0.6 [6459.4][b]	6716.2±0.5 [6716.8][a] (6717.)	6976.9±0.4 [6976.7][a] (6981.)	7242.8±0.4 [7242.8][a] (7243.)
M_I	1361.3±0.3	1434.6±0.6	1511.0±0.8	1575.3±0.7		1722.8±0.8	1800.0±0.5	1880.8±0.5
M_{II}	1204.4±0.6	1272.8±0.6	1337.4±0.7	1402.8±0.6	1471.4±6.2	1540.7±1.2	1613.9±0.7	1688.3±0.7
M_{III}	1123.4±0.5	1185.4±0.5	1242.2±0.6	1297.4±0.5	1356.9±1.4	1419.8±1.1	1480.6±0.6	1544.0±0.8
M_{IV}	848.5±0.4	901.3±0.6	951.1±0.6	999.9±0.6	1051.5±0.9	1106.0±0.8	1160.6±0.6	1217.2±0.6
M_V	831.7±0.4	883.3±0.5	931.0±0.6	977.7±0.6	1026.9±1.0	1080.2±0.6	1130.9±0.6	1185.2±0.6
N_I	270.4±0.8	289.6±0.7	304.5±0.9	315.2±0.8		345.7±0.9	360.2±0.7	375.8±0.7
N_{II}	205.8±1.2	223.3±1.1	236.3±1.5	243.3±1.6	242.±16.	265.6±1.9	283.9±1.0	288.5±1.2
N_{III}	191.4±0.9	207.2±0.9	217.6±1.1	224.6±1.3		247.4±1.5	256.6±0.8	270.9±0.9
$N_{IV,V}$	98.9±0.8	110.0±0.6	113.2±0.7	117.5±0.7	120.4±2.0	129.0±1.2	133.2±0.6	140.5±0.8
$N_{VI,VII}$		0.1±1.2	2.0±0.6	1.5±0.9		5.5±1.1	0.0±3.2	0.1±3.5
O_I	32.3±7.2	37.8±1.3	37.4±1.0	37.5±0.9		37.4±1.5	31.8±0.7	36.1±0.8
$O_{II,III}$	14.4±1.2	19.8±1.2	22.3±0.7	21.1±0.8		21.3±1.5	22.0±0.6	20.3±1.2

	65 Tb	66 Dy	67 Ho	68 Er	69 Tm	70 Yb	71 Lu	72 Hf
K	51995.7±0.5 (52002.)	53788.5±0.5 (53793.)	55617.7±0.5 (55619.)	57485.5±0.5 (57487.)	59389.6±0.5	61332.3±0.5 (61300.)	63313.8±0.5 (63310.)	65350.8±0.6 (65310.)
L_I	8708.0±0.5 [8707.6][a] (8717.)	9045.8±0.5 [9046.5][a]	9394.2±0.4 [9394.3][a] (9399.)	9751.3±0.4 [9751.5][a] (9757.)	10115.7±0.4 [10115.6][a] (10121.)	10486.4±0.4 [10487.3][a] (10490.)	10870.4±0.4 [10870.1][a] (10874.)	11270.7±0.4 [11271.6][b] (11274.)
L_{II}	8251.6±0.4 [8251.8][a] (8253.)	8580.6±0.4 [8580.4][a] (8583.)	8917.8±0.4 [8918.2][a] (8916.)	9264.3±0.4 [9264.3][a] (9262.)	9616.9±0.4 [9617.1][a] (9617.1)	9978.2±0.4 [9977.9][a] (9976.)	10348.6±0.4 [10349.0][a] (10345.)	10739.4±0.4 [10738.9][b] (10736.)
L_{III}	7514.0±0.4 [7514.2][a] (7515.)	7790.1±0.4 [7789.6][a] (7789.7)	8071.1±0.4 [8070.6][a] (8068.)	8357.9±0.4 [8357.6][a] (8357.5)	8648.0±0.4 [8647.8][a] (8649.6)	8943.6±0.4 [8942.6][a] (8944.1)	9244.1±0.4 [9243.8][a]	9560.7±0.4 [9560.4][b] (9558.)
M_I	1967.5±0.6	2046.8±0.4	2128.3±0.6	2206.5±0.6	2306.8±0.7	2398.1±0.4	2491.2±0.5	2600.9±0.4
M_{II}	1767.7±0.9	1841.8±0.5	1922.8±1.0	2005.8±0.6	2089.8±1.1	2173.0±0.4	2263.5±0.4	2365.4±0.4
M_{III}	1611.3±0.8	1675.6±0.9	1741.2±0.9	1811.8±0.6	1884.5±1.1	1949.8±0.5	2023.6±0.5	2107.6±0.4
M_{IV}	1275.0±0.6	1332.5±0.4	1391.5±0.7	1453.3±0.5	1514.6±0.7	1576.3±0.4	1639.4±0.4	1716.4±0.4
M_V	1241.2±0.7	1294.9±0.4	1351.4±0.8	1409.3±0.5	1467.7±0.9	1527.8±0.4	1588.5±0.4	1661.7±0.4
N_I	397.9±0.8	416.3±0.5	435.7±0.8	449.1±1.0	471.7±0.9	487.2±0.6	506.2±0.6	538.1±0.4
N_{II}	310.2±1.2	331.8±0.6	343.5±1.4	366.2±1.5	385.9±1.6	396.7±0.7	410.1±1.8	437.0±0.5
N_{III}	385.0±1.0	292.9±0.6	306.6±0.9	320.0±0.7	336.6±1.6	343.5±0.5	359.3±0.5	380.4±0.5
N_{IV}	147.0±0.8	154.2±0.5	161.0±1.0	176.7±1.2	179.6±1.2	198.1±0.5	204.8±0.5	223.8±0.4
N_V				167.6±1.5		184.9±1.3	195.0±0.4	213.7±0.5

	65 Tb	66 Dy	67 Ho	68 Er	69 Tm	70 Yb	71 Lu	72 Hf
$N_{VI,VII}$	2.6±1.5	4.2±1.6	3.7±3.0	4.3±1.4	5.3±1.9	6.3±1.0	6.9±0.5	17.1±0.5
O_I	39.0±0.8	62.9±0.5	51.2±1.3	59.8±1.7	53.2±3.0	54.1±0.5	56.8±0.5	64.9±0.4
O_{II} O_{III}	25.4±0.8	26.3±0.6	20.3±1.5	29.4±1.6	32.3±1.6	23.4±0.6	28.0±0.6	38.1±0.6 30.6±0.6

	73 Ta	74 W	75 Re	76 Os	77 Ir	78 Pt	79 Au	80 Hg
K	67416.4±0.6 (67403.)	69525.0±0.3 (69508.)	71676.4±0.4 (71658.)	73870.8±0.5	76111.0±0.5	78394.8±0.7 (78381.)	80724.9±0.5 (80720.)	83102.3±0.8
L_I	11681.5±0.3 [11680.2][p] (11682.)	12099.8±0.3 [12098.2][p] (12099.6)	12526.7±0.4 (12530.)	12968.0±0.4 (12972.)	13418.5±0.3 (13423.)	13879.9±0.4 (13883.)	14352.8±0.4 (14353.7)	14839.3±1.0 (14842.)
L_{II}	11136.1±0.3 [11136.1][p] (11132.)	11544.0±0.3 [11541.4][p] (11538.)	11958.7±0.3 [11956.9][p] (11954.)	12385.0±0.4 (12381.)	12824.1±0.3 [12824.0][e,p] (12820.)	13272.6±0.3 [13272.6][e,p] (13272.3)	13733.6±0.3 [13733.5][e,p] (13736.)	14208.7±0.7 (14215.)
L_{III}	9881.1±0.3 [9880.3][p] (9877.7)	10206.8±0.3 [10204.2][p] (10200.)	10535.3±0.3 [10534.2][p] (10531.)	10870.9±0.3 [10870.7][p] (10868.)	11215.2±0.3 [11215.1][e,p] (11212.)	11563.7±0.3 [11563.7][e,p] (11562.)	11918.7±0.3 [11918.2][e,p] (11921.)	12283.9±0.4 [12284.0][e,p] (12286.)
M_I	2708.0±0.4	2819.6±0.4	2931.7±0.4	3048.5±0.4	3173.7±1.7	3296.0±0.9	3424.9±0.3 [3424.8][p]	3561.6±1.1
M_{II}	2468.7±0.3 [2468.6][p]	2574.9±0.3 [2575.0][p]	2681.6±0.4	2792.2±0.3 [2791.9][p]	2908.7±0.3 [2909.1][p]	3026.5±0.4 [3026.5][p] (3029.)	3147.8±0.4 [3149.5][p]	3278.5±1.3
M_{III}	2194.0±0.3 [2194.1][p]	2281.0±0.3 [2281.0][p]	2367.3±0.3 [2367.3][p]	2457.2±0.4 [2457.4][p]	2550.7±0.3 [2550.5][p] (2550.5)	2645.4±0.4 [2645.5][p] (2645.9)	2743.0±0.3 [2743.1][p] (2744.0)	2847.1±0.4 [2847.1][p]
M_{IV}	1793.2±0.3 [1793.1][p]	1871.6±0.3 [1871.4][p]	1948.9±0.3 [1948.9][p]	2030.8±0.3 [2031.0][p]	2116.1±0.3 [2116.1][p]	2201.9±0.3 [2201.9][p]	2291.1±0.3 [2291.2][p] (2307.)	2384.9±0.3 [2384.9][p]
M_V	1735.1±0.3 [1735.2][p]	1809.2±0.3 [1809.3][p]	1882.9±0.3 [1882.9][p]	1960.1±0.3 [1960.2][p]	2040.4±0.3 [2040.5][p]	2121.6±0.3 [2121.6][p]	2205.7±0.3 [2206.1][p] (2220.)	2294.9±0.3 [2294.9][p]
N_I	565.5±0.5	595.0±0.4	625.0±0.4	654.3±0.5	690.1±0.4	722.0±0.6	758.8±0.4	800.3±1.0
N_{II}	464.8±0.5	491.6±0.4	517.9±0.5	546.5±0.5	577.1±0.4	609.2±0.6	643.7±0.5	676.9±2.4
N_{III}	404.5±0.4	425.3±0.5	444.4±0.5	468.2±0.6	494.3±0.6	519.0±0.6	545.4±0.5	571.0±1.4
N_{IV}	241.3±0.4	258.8±0.4	273.7±0.5	289.4±0.5	311.4±0.4	330.8±0.5	352.0±0.4	378.3±1.0
N_V	229.3±0.3	245.4±0.4	260.2±0.4	272.8±0.6	294.9±0.4	313.3±0.4	333.9±0.4	359.8±1.2
N_{VI} N_{VII}	25.0±0.4	36.5±0.4 33.6±0.4	40.6±0.4	46.3±0.6	63.4±0.4 60.5±0.4	74.3±0.4 71.1±0.5	86.4±0.4 82.8±0.5	102.2±0.5 98.5±0.5
O_I	71.1±0.5	77.1±0.4	82.8±0.5	83.7±0.6	95.2±0.4	101.7±0.4	107.8±0.7	120.3±1.3
O_{II}	44.9±0.4	46.8±0.5	45.6±0.7	58.0±1.1	63.0±0.6	65.3±0.7	71.7±0.7	80.5±1.3
O_{III}	36.4±0.4	35.6±0.5	34.6±0.6	45.4±1.0	50.5±0.6	51.7±0.7	53.7±0.7	57.6±1.3
$O_{IV,V}$	5.7±0.4	6.1±0.4	3.5±0.5		3.8±0.4	2.2±1.3	2.5±0.5	6.4±1.4

	81 Tl	82 Pb	83 Bi	84 Po	85 At	86 Rn	87 Fr	88 Ra
K	85530.4±0.6	88004.5±0.7 (88005.)	90525.9±0.7 (90534.)	93105.0±3.8	95729.9±7.7	98404.±12.	101137.±13.	103921.9±7.2
L_I	15346.7±0.4 (15343.)	15860.8±0.5 (15855.)	16387.5±0.4 (16376.)	16939.3±9.8	17493.±29.	18049.±38.	18639.±40.	19236.7±1.5 (19236.0)
L_{II}	14697.9±0.3 [14697.3][p] (14699.)	15200.0±0.4 (15205.)	15711.1±0.3 [15708.4][p] (15719.)	16244.3±2.4	16784.7±2.5	17337.1±3.4	17906.5±3.5	18484.3±1.5 (18486.0)
L_{III}	12657.5±0.3 [12656.3][e,p] (12660.)	13035.2±0.3 [13034.9][e,p] (13041.)	13418.6±0.3 [13418.3][e,p] (13426.)	13813.8±1.0 ⟨13813.8⟩	14213.5±2.0 ⟨14213.5⟩	14619.4±3.0 ⟨14619.4⟩	15031.2±3.0 ⟨15031.2⟩	15444.4±1.5 (15444.0)
M_I	3704.1±0.4	3850.7±0.5	3999.1±0.3 [3999.1][p]	4149.4±3.9	⟨4317.⟩	⟨4482.⟩	⟨4652.⟩	4822.0±1.5
M_{II}	3415.7±0.3 [3415.7][p]	3554.2±0.3 [3554.2][p]	3696.3±0.3 [3696.4][p]	3854.1±9.8	4008.±28.	4159.±38.	4327.±40.	4489.5±1.8
M_{III}	2956.6±0.3 [2956.5][p]	3066.4±0.4 [3066.3][p]	3176.9±0.3 [3176.8][p]	3301.9±9.9	3426.±29.	3538.±38.	3663.±40.	3791.8±1.7
M_{IV}	2485.1±0.3 [2485.2][p]	2585.6±0.3 [2585.5][p] (2606.)	2687.6±0.3 [2687.4][p]	2798.0±1.2	2908.7±2.1	3021.5±3.1	3136.2±3.1	3248.4±1.6

	81 Tl	82 Pb	83 Bi	84 Po	85 At	86 Rn	87 Fr	88 Ra
M_{V}	2389.3±0.3 [2389.4][p]	2484.0±0.3 [2484.2][p] (2502.)	2579.6±0.3 [2579.5][p]	2683.0±1.1	2786.7±2.1	2892.4±3.1	2999.9±3.1	3104.9±1.6
N_{I}	845.5±0.5	893.6±0.7	938.2±0.3 [938.7][p]	995.3±2.9	⟨1042.⟩	⟨1097.⟩	⟨1153.⟩	1208.4±1.6
N_{II}	721.3±0.8	763.9±0.8	805.3±0.3 [805.3][p]	851.±12.	886.±30.	929.±40.	980±42.	1057.6±1.8
N_{III}	609.0±0.5	644.5±0.6	678.9±0.3 [678.9][p]	705.±14.	740.±30.	768.±40.	810±43.	879.1±1.8
N_{IV}	406.6±0.4	435.2±0.5	463.6±0.3 [463.6][p]	500.2±2.4	533.2±3.2	566.6±4.0	603.3±4.1	635.9±1.6
N_{V}	386.2±0.5	412.9±0.6	440.0±0.3 [440.1][p]	473.4±1.3			577.±34.	602.7±1.7
N_{VI}	122.8±0.4	142.9±0.4	161.9±0.5					298.9±2.4 (N_{VI}, N_{VII})
N_{VII}	118.5±0.4	138.1±0.4	157.4±0.6					
O_{I}	136.3±0.7	147.3±0.8	159.3±0.7					254.4±2.1
O_{II}	99.6±0.6	104.8±1.0	116.8±0.7					200.4±2.0
O_{III}	75.4±0.6	86.0±1.0	92.8±0.6					152.8±2.0
O_{IV}	15.3±0.4	21.8±0.4	26.5±0.5	31.4±3.2 (O_{IV}, O_{V})				67.2±1.7 (O_{IV}, O_{V})
O_{V}	13.1±0.4	19.2±0.4	24.4±0.6					
P_{I}		3.1±1.0						43.5±2.2
$P_{II,III}$		0.7±1.0	2.7±0.7					18.8±1.8

	89 Ac	90 Th	91 Pa	92 U	93 Np	94 Pu	95 Am	96 Cm
K	106755.3±5.3	109650.9±0.9	112601.4±2.4	115606.1±1.6	118678.±33.	121818.±44.	125027.±55.	128220
L_{I}	19840.±18.	20472.1±0.5 (20464.)	21104.6±1.8 (21128.)	21757.4±0.3 (21771.)	22426.8±0.9	23097.2±1.6 (23109.)	23772.9±2.0 ⟨23772.9⟩	24460
L_{II}	19083.2±2.8	19693.2±0.4 (19683.)	20313.7±1.5 (20319.)	20947.6±0.3 (20945.)	21600.5±0.4	22266.2±0.7 (22253.)	22944.0±1.0	23779
L_{III}	15871.0±2.0 ⟨15871.0⟩	16300.3±0.3 [16299.6][q] (16299.)	16733.1±1.4 (16733.)	17166.3±0.3 [17168.5][r] (17165.)	17610.0±0.4 (17606.2)	18056.8±0.6 (18053.1)	18504.1±0.9 (18504.1)	18930
M_{I}	⟨5002.⟩	5182.3±0.3 [5182.3][q]	5366.9±1.6	5548.0±0.4	5723.2±3.6	5932.9±1.4	6120.5±7.5	6288
M_{II}	4656.±18.	4830.4±0.4 [4830.6][q]	5000.9±2.3	5182.2±0.4 [5180.9][r]	5366.2±0.7 [5366.4][s]	5541.2±1.7	5710.2±2.1	5895
M_{III}	3909.±18.	4046.1±0.4 [4046.1][q] (4041.)	4173.8±1.8	4303.4±0.3 [4303.6][r] (4299.)	4434.7±0.5 [4434.6][s]	4556.6±1.5	4667.0±2.1	4797
M_{IV}	3370.2±2.1	3490.8±0.3 [3490.7][q] (3485.)	3611.2±1.4 (3608.)	3727.6±0.3 [3728.1][r] (3720.)	3850.3±0.4 [3849.8][s]	3972.6±0.6 [3972.7][t]	4092.1±1.0	4227
M_{V}	3219.0±2.1	3332.0±0.3 [3332.1][q] (3325.)	3441.8±1.4 (3436.)	3551.7±0.3 [3551.7][r] (3545.)	3665.8±0.4 [3664.2][s]	3778.1±0.6 [3778.0][t]	3886.9±1.0	3971
N_{I}	⟨1269.⟩	1329.5±0.4 [1329.8][q]	1387.1±1.9	1440.8±0.4 [1441.3][r]	1500.7±0.8 [1500.7][s]	1558.6±0.8	1617.1±1.1	1643
N_{II}	1080.±19.	1168.2±0.4 [1168.3][q]	1224.3±1.6	1272.6±0.3 [1272.5][r]	1327.7±0.8 [1327.7][s]	1372.1±1.8	1411.8±8.3	1440
N_{III}	890.±19.	967.3±0.4 [967.6][q]	1006.7±1.7	1044.9±0.3 [1044.9][r]	1086.8±0.7 [1086.8][s]	1114.8±1.6	⟨1135.7⟩	1154
N_{IV}	674.9±3.7	714.1±0.4 [714.4][q]	743.4±2.1	780.4±0.3 [779.7][r]	815.9±0.5 [817.1][s]	848.9±0.6 [848.9][t]	878.7±1.0	
N_{V}		676.4±0.4 [676.4][q]	708.2±1.8	737.7±0.3 [737.6][r]	770.3±0.4 [773.2][s]	801.4±0.6 [801.4][t]	827.6±1.0	
N_{VI}		344.4±0.3 [344.2][q]	371.2±1.6	391.3±0.6	415.0±0.8 [415.0][s]	445.8±1.7		
N_{VII}		335.2±0.4 [335.0][q]	359.5±1.6	380.9±0.9	404.4±0.5 [404.4][s]	432.4±2.1		
O_{I}		290.2±0.8	309.6±4.3	323.7±1.1		351.9±2.4		385
O_{II}		229.4±1.1	222.9±3.9 (O_{II}, O_{III})	259.3±0.5	283.4±0.8 [283.4][s]	274.1±4.7		
O_{III}		181.8±0.4 [181.8][q]		195.1±1.3	206.1±0.7 [206.1][s]	206.5±4.7		

	89 Ac	90 Th	91 Pa	92 U	93 Np	94 Pu	95 Am	96 Cm
O_{IV}		94.3±0.4 [94.4][q]	94.1±2.8	105.0±0.5	109.3±0.7 [108.8][s]	116.0±1.2	115.8±1.3	
O_{V}		87.9±0.3 [88.1][q]		96.3±1.4	101.3±0.5 [101.4][s]	105.4±1.0	103.3±1.1	
P_{I}		59.5±1.1		70.7±1.2				
P_{II}		49.0±2.5		42.3±9.0				
P_{III}		43.0±2.5		32.3±9.0				

	97 Bk	98 Cf	99 Es	100 Fm	101 Md	102 No	103 Lw
K	[131590±40][u]	135960	139490	143090	146780	150540	154380
L_{I}	[25275±17][u]	26110	26900	27700	28530	29380	30240
L_{II}	[24385±17][u]	25250	26020	26810	27610	28440	29280
L_{III}	[19452±20][u]	19930	20410	20900	21390	21880	22360
M_{I}	[6556±21][u]	6754	6977	7205	7441	7675	7900
M_{II}	[6147±31][u]	6359	6574	6793	7019	7245	7460
M_{III}	[4977±31][u]	5109	5252	5397	5546	5688	5710
M_{IV}	4366	4497	4630	4766	4903	5037	5150
M_{V}	4132	4253	4374	4498	4622	4741	4860
N_{I}	[1755±22][u]	1799	1868	1937	2010	2078	2140
N_{II}	1554	1616	1680	1747	1814	1876	1930
N_{III}	1235	1279	1321	1366	1410	1448	1480
O_{I}	[398±22][u]	419	435	454	472	484	490

[a] J. E. Mack, 1949, as given in C. E. Moore, *Atomic Energy Levels* (U. S. National Bureau of Standards, Washington, D. C., 1949), Vol. 1, p. 1.
[b] G. Herzberg, 1957, as given in C. E. Moore, *Atomic Energy Levels* (U. S. National Bureau of Standards, Washington, D. C., 1958), Vol. 3, p. 238.
[c] See Ref. 18.
[d] A. Fahlman, D. Hamrin, R. Nordberg, C. Nordling, and K. Siegbahn, Phys. Rev. Letters **14**, 127 (1965). See also Ref. 26.
[e] See Ref. 15.
[f] See Ref. 11.
[g] C. Nordling, Arkiv Fysik **15**, 397 (1959).
[h] E. Sokolowski, C. Nordling, and K. Siegbahn, Arkiv Fysik **12**, 301 (1957).
[i] C. Nordling and S. Hagström, Arkiv Fysik **16**, 515 (1960).
[j] I. Andersson and S. Hagström, Arkiv Fysik **27**, 161 (1964).
[k] M. O. Krause, Phys. Rev. **140**, A1845 (1965).
[l] A. Fahlman, O. Hörnfeldt, and C. Nordling, Arkiv Fysik **23**, 75 (1962).
[m] P. Bergvall, O. Hörnfeldt, and C. Nordling, Arkiv Fysik **17**, 113 (1960).
[n] P. Bergvall and S. Hagström, Arkiv Fysik **17**, 61 (1960).
[o] S. Hagström, Z. Physik **178**, 82 (1964).
[p] A. Fahlman and S. Hagström, Arkiv Fysik **27**, 69 (1964).
[q] C. Nordling and S. Hagström, Z. Physik **178**, 418 (1964).
[r] C. Nordling and S. Hagström, Arkiv Fysik **15**, 431 (1959).
[s] S. Hagström, Bull. Am. Phys. Soc. **11**, 389 (1966).
[t] A. Fahlman, K. Hamrin, R. Nordberg, C. Nordling, K. Siegbahn, and L. W. Holm, Phys. Letters **19**, 643 (1966).
[u] J. M. Hollander, M. D. Holtz, T. Novakov, and R. L. Graham ,Arkiv Fysik **28**, 375 (1965).

LATTICE SPACING OF COMMON ANALYZING CRYSTALS

Crystal	Reflection plane	*d* Spacing, Å	Crystal	Reflection plane	*d* Spacing, Å
ADP[a]	101	5.31	Lead stearate		51
ADP[a]	110	5.325	LiF	200	2.014
ADP[a]	200	3.75	LiF	220	1.424
Beryl	$10\bar{1}0$	7.98	Mica	002	9.96
Calcite	100	3.036	NaCl	200	2.820
EDDT[b]	020	4.404	Oxalic acid	001	5.85
Germanium	111	3.265	PET[d]	002	4.371
Graphite	001	6.69	Quartz	$10\bar{1}0$	4.255
Gypsum	010	7.600	Quartz	$10\bar{1}1$	3.343
KAP[c]	001	13.32	Quartz	$11\bar{2}0$	2.456
KBr	200	3.29	Silicon	111	3.13
KCl	200	3.14	Topaz	303	1.356

While several of the above spacings have been measured to more than four significant figures, no more than four figures are given here because complications introduced by the index of refraction, anomalous dispersion, temperature coefficient of expansion, and crystal impurities must be considered before the additional figures are useful.

[a] Ammonium dihydrogen phosphate.
[b] Ethylenediamine d-tartrate.
[c] Potassium acid phthalate.
[d] Pentaerythritol.

RADIATIVE TRANSITION PROBABILITIES FOR *K* X-RAY LINES

$K\beta_1' = KM_2 + KM_3 + KM_{4,5}$ $\quad K\beta_2' = KN_{2,3} + KO_{2,3}$

$K\alpha = K\alpha_1 + K\alpha_2$ $\quad K\beta = K\beta_1' + K\beta_2'$

Element	$K\alpha_2/K\alpha_1$	$K\alpha_3/K\alpha_1$	$K\beta_1/K\alpha_1$	$K\beta'_2/K\alpha_1$	$K\beta_4/K\alpha_1$	$K\beta_5/K\alpha_1$	$K\beta_3/K\beta_1$	$K\beta/K\alpha$
$_{12}$Mg								0.013
$_{14}$Si								0.027
$_{16}$S								0.059
$_{18}$Ar								0.105
$_{20}$Ca	0.502							0.128
$_{22}$Ti	0.503							0.134
$_{24}$Cr	0.504							0.135
$_{26}$Fe	0.506							0.135
$_{28}$Ni	0.508							0.135
$_{30}$Zn	0.510							0.138
$_{32}$Ge	0.513							0.147
$_{34}$Se	0.515							0.157
$_{36}$Kr	0.517			0.019				0.172
$_{38}$Sr	0.520			0.030				0.180
$_{40}$Br	0.523			0.037				0.190
$_{42}$Mo	0.525			0.041				0.197
$_{44}$Ru	0.527			0.045				0.204
$_{46}$Pd	0.529			0.048				0.210
$_{48}$Cd	0.532			0.053			0.519	0.213
$_{50}$Sn	0.534			0.055			0.519	0.220
$_{52}$Te	0.537			0.058			0.519	0.225
$_{54}$Xe	0.539			0.064			0.518	0.232
$_{56}$Ba	0.543			0.070			0.518	0.237
$_{58}$Ce	0.546			0.076			0.518	0.242
$_{60}$Nd	0.549	0.11×10^{-3}		0.083			0.518	0.247
$_{62}$Sm	0.552	0.14×10^{-3}		0.086			0.517	0.250
$_{64}$Gd	0.556	0.17×10^{-3}	0.192	0.089	0.85×10^{-3}	3.02×10^{-3}	0.517	0.255
$_{66}$Dy	0.560	0.21×10^{-3}	0.198	0.089	0.92×10^{-3}	3.43×10^{-3}	0.517	0.257
$_{68}$Er	0.564	0.26×10^{-3}	0.202	0.088	0.96×10^{-3}	3.85×10^{-3}	0.518	0.260
$_{70}$Yb	0.567	0.30×10^{-3}	0.207	0.087	1.04×10^{-3}	4.23×10^{-3}	0.518	0.264
$_{72}$Hf	0.572	0.36×10^{-3}	0.212	0.085	1.16×10^{-3}	4.62×10^{-3}	0.518	0.267
$_{74}$W	0.576	0.43×10^{-3}	0.216	0.086	1.28×10^{-3}	5.04×10^{-3}	0.518	0.269
$_{76}$Os	0.580	0.51×10^{-3}	0.222	0.087	1.43×10^{-3}	5.44×10^{-3}	0.519	0.273
$_{78}$Pt	0.583	0.63×10^{-3}	0.226	0.091	1.61×10^{-3}	5.84×10^{-3}	0.520	0.275
$_{80}$Hg	0.588	0.76×10^{-3}	0.228	0.096	1.80×10^{-3}	6.24×10^{-3}	0.520	0.278
$_{82}$Pb	0.593	0.91×10^{-3}	0.228	0.102	2.02×10^{-3}	6.64×10^{-3}	0.521	0.280
$_{84}$Po	0.597	1.12×10^{-3}	0.228	0.108	2.26×10^{-3}	7.05×10^{-3}	0.522	0.283
$_{86}$Rn	0.602	1.32×10^{-3}	0.228	0.113	2.52×10^{-3}	7.48×10^{-3}	0.523	0.286
$_{88}$Ra	0.608	1.58×10^{-3}	0.230	0.117	2.80×10^{-3}	7.80×10^{-3}	0.524	0.287
$_{90}$Th	0.613	1.85×10^{-3}	0.232	0.120	3.13×10^{-3}	8.25×10^{-3}	0.525	0.288
$_{92}$U	0.619	2.15×10^{-3}	0.234	0.123	3.47×10^{-3}	8.65×10^{-3}	0.527	0.289
$_{94}$Pu	0.625		0.234	0.125			0.528	0.291
$_{96}$Cm	0.632		0.234	0.128			0.529	0.293
$_{98}$Cf	0.642		0.238	0.132			0.531	0.295
$_{100}$Fm	0.648		0.240	0.135			0.533	0.297

From Salem, S. I., Panossian, S. L., and Krause, R. A., *At. Data Nucl. Data Tables*, 14, 91, 1974. Reproduced by permission of the copyright owner, Academic Press.

RADIATIVE TRANSITION PROBABILITIES FOR *L* X-RAY LINES

The following three tables present data for the radiative transition probabilities for the L_1, L_2, and L_3 X-ray lines. The data are normalized respectively to $L\beta_3$ = 100, $L\beta_1$ = 100 and $L\alpha_1$ = 100.

L_1 X-RAY LINES NORMALIZED TO $L\beta_3$ = 100

Element	$L\beta_3$	$L\beta_4$	$L\gamma_2$	$L\gamma_3$	Element	$L\beta_3$	$L\beta_4$	$L\gamma_2$	$L\gamma_3$
$_{32}$Ge	100			17.3	$_{66}$Dy	100	61.8	19.5	28.0
$_{34}$Se	100			18.0	$_{68}$Er	100	63.5	19.8	29.0
$_{36}$Kr	100			18.2	$_{70}$Yb	100	65.5	20.7	29.8
$_{38}$Sr	100			18.8	$_{72}$Hf	100	67.8	21.2	30.7
$_{40}$Zr	100			19.0	$_{74}$W	100	70.5	21.8	31.8
$_{42}$Mo	100	70.6		19.6	$_{76}$Os	100	73.2	23.0	32.8
$_{44}$Ru	100	67.8		20.2	$_{78}$Pt	100	76.5	24.5	33.8
$_{46}$Pd	100	65.5		20.6	$_{80}$Hg	100	80.3	26.3	35.0
$_{48}$Cd	100	63.5		21.3	$_{82}$Pb	100	84.2	28.6	36.0
$_{50}$Sn	100	62.1		22.0	$_{84}$Po	100	88.5	31.3	37.2
$_{52}$Te	100	60.7		22.6	$_{86}$Rn	100	93.4	34.2	38.2
$_{54}$Xe	100	59.8		23.3	$_{88}$Ra	100	98.9	37.5	39.6
$_{56}$Ba	100	59.5		24.0	$_{90}$Th	100	104.5	41.2	41.0
$_{58}$Ce	100	59.2		24.6	$_{92}$U	100	110.2	45.0	42.6
$_{60}$Nd	100	59.4		25.4	$_{94}$Pu	100	116.2	49.5	44.0
$_{62}$Sm	100	60.0		26.3	$_{96}$Cm	100	123.0	55.7	45.7
$_{64}$Gd	100	60.8	19.2	27.0					

RADIATIVE TRANSITION PROBABILITIES FOR L X-RAY LINES (*Continued*)

L_2 X-RAY LINES NORMALIZED TO $L\beta_1$ = 100

Element	L_{β_1}	L_η	L_{γ_1}	L_{γ_6}
$_{28}$Ni	100	7.60		
$_{30}$Zn	100	6.80		
$_{32}$Ge	100	6.28		
$_{34}$Se	100	5.80		
$_{36}$Kr	100	5.35		
$_{38}$Sr	100	4.93		
$_{40}$Zr	100	4.60	3.30	
$_{42}$Mo	100	4.30	5.50	
$_{44}$Ru	100	4.00	7.33	
$_{46}$Pd	100	3.75	10.67	
$_{48}$Cd	100	3.55	10.60	
$_{50}$Sn	100	3.35	11.80	
$_{52}$Te	100	3.20	12.70	
$_{54}$Xe	100	3.00	14.00	
$_{56}$Ba	100	2.85	14.50	
$_{58}$Ce	100	2.70	15.30	
$_{60}$Nd	100	2.60	16.00	
$_{62}$Sn	100	2.45	16.50	
$_{64}$Gd	100	2.35	17.00	
$_{66}$Dy	100	2.25	17.40	
$_{68}$Er	100	2.16	17.80	
$_{70}$Yb	100	2.10	18.17	
$_{72}$Hf	100	2.08	18.43	
$_{74}$W	100	2.10	18.80	0.72
$_{76}$Os	100	2.12	19.34	1.65
$_{78}$Pt	100	2.18	19.73	2.40
$_{80}$Hg	100	2.25	20.35	3.10
$_{82}$Pb	100	2.30	20.93	3.65
$_{84}$Po	100	2.40	21.54	4.15
$_{86}$Rn	100	2.46	22.20	4.55
$_{88}$Ra	100	2.50	22.87	4.87
$_{90}$Th	100	2.60	23.43	5.02
$_{92}$U	100	2.65	24.10	5.12
$_{94}$Pu	100	2.70	24.40	5.16
$_{96}$Cm	100	2.75	25.07	5.20

L_3 X-RAY LINES NORMALIZED TO $L\alpha_1$ = 100

Element	L_{α_1}	$L_{\beta_{2,15}}$	L_{α_2}	L_{β_5}	L_{β_6}	L_ℓ
$_{26}$Fe	100					12.22
$_{28}$Ni	100					8.95
$_{30}$Zn	100					7.34
$_{32}$Ge	100					6.45
$_{34}$Se	100					7.76
$_{36}$Kr	100					5.28
$_{38}$Sr	100					4.92
$_{40}$Zr	100	0.70	11.10			4.67
$_{42}$Mo	100	5.17	11.10			4.45
$_{44}$Ru	100	9.30	11.12			4.28
$_{46}$Pd	100	11.80	11.12			4.11
$_{48}$Cd	100	14.33	11.12			4.07
$_{50}$Sn	100	16.00	11.13			4.00
$_{52}$Te	100	18.00	11.13			4.00
$_{54}$Xe	100	19.40	11.13			4.00
$_{56}$Ba	100	20.67	11.13			4.02
$_{58}$Ce	100	21.00	11.14			4.09
$_{60}$Nd	100	21.33	11.14		0.875	4.13
$_{62}$Sm	100	21.07	11.14		0.925	4.16
$_{64}$Gd	100	20.83	11.14		0.99	4.20
$_{66}$Dy	100	20.50	11.14		1.05	4.26
$_{68}$Er	100	20.04	11.15		1.12	4.33
$_{70}$Yb	100	19.40	11.15		1.17	4.47
$_{72}$Hf	100	21.33	11.15	0.30	1.21	4.59
$_{74}$W	100	22.74	11.16	0.50	1.25	4.76
$_{76}$Os	100	23.40	11.16	1.32	1.37	4.95
$_{78}$Pt	100	24.00	11.16	1.98	1.43	5.14
$_{80}$Hg	100	24.50	11.17	2.62	1.50	5.37
$_{82}$Pb	100	24.83	11.17	3.21	1.56	5.58
$_{84}$Po	100	25.13	11.17	3.73	1.62	5.80
$_{86}$Rn	100	25.60	11.18	4.25	1.68	6.00
$_{88}$Ra	100	25.92	11.18	4.73	1.76	6.26
$_{90}$Th	100	26.17	11.18	5.18	1.82	6.54
$_{92}$U	100	26.40	11.18	5.58	1.89	6.79
$_{94}$Pu	100	26.67	11.18	5.92	1.95	7.02
$_{96}$Cm	100	26.93	11.18	6.26	2.01	7.34

From Salem, S. I., Panossian, S. L., and Krause, R. A., *At. Data Nucl. Data Tables,* 14, 91, 1974. Reproduced by permission of the copyright owner. Academic Press.

NATURAL WIDTH, IN eV OF THE INDICATED K X-RAY LINES

Element	$K\alpha_1$	$K\alpha_2$	$K\beta_1$	$K\beta_3$
20Ca	1.00	0.98		
22Ti	1.45	2.13		
24Cr	2.05	2.64		
26Fe	2.45	3.20		
28Ni	3.00	3.70		
30Zn	3.40	3.96		
32Ge	3.75	4.18		
34Se	4.10	4.43		
36Kr	4.23	4.62		
38Sr	5.17	4.97		
40Zn	5.70	5.25		
42Mo	6.82	6.80		
44Ru	7.41	7.96		
46Pd	8.80	9.20		
48Cd	9.80	10.40		
50Sn	11.20	12.40	11.80	11.00
52Te	12.80	14.20	13.30	12.30
54Xe	14.20	15.10	15.30	13.43
56Ba	16.10	16.80	18.15	16.00
58Ce	18.60	19.50	20.60	17.95
60Nd	21.50	21.50	23.25	21.33
62Sm	26.00	24.70	25.65	24.65
64Gd	29.50	28.00	29.37	28.00
66Dy	33.90	32.20	32.73	32.00
68Er	37.40	35.50	36.20	35.70
70Yb	42.00	40.60	41.43	41.15
72Hf	45.30	44.30	46.00	46.10
74W	47.75	48.00	51.83	51.50
76Os	53.00	49.40	55.90	55.95
78Pt	60.30	54.30	59.98	62.13
80Hg	64.75	68.20	65.75	68.95
82Pb	68.30	79.00	72.20	74.90
84Po	73.20	86.30	78.60	82.85
86Rn	80.00	89.50	85.50	91.20
88Ra	87.00	91.20	94.20	98.95
90Th	94.70	97.00	99.70	105.00
92U	103.00	106.00	115.00	120.00

From Salem, S. I. and Lee, P. L., *At. Data Nucl. Data Tables,* **18**, 233, 1976.

NATURAL WIDTH, IN eV OF THE INDICATED L X-RAY LINES

Element	$L\alpha_1$	$L\alpha_2$	$L\beta_1$	$L\beta_2$	$L\beta_3$	$L\beta_4$	$L\gamma_1$
40Zn	1.68	1.52	1.87	5.13	5.50	5.60	3.34
42Mo	1.86	1.80	2.03	5.30	5.90	5.78	3.76
44Ru	2.03	1.98	2.18	5.45	6.35	5.96	4.15
46Pd	2.21	2.16	2.36	5.63	6.80	6.18	4.50
48Cd	2.43	2.40	2.54	5.82	7.23	6.28	4.83
50Sn	2.62	2.62	2.75	6.10	7.70	6.60	5.23
52Tc	2.88	2.88	2.96	6.25	8.22	6.82	5.60
54Xe	3.15	3.15	3.20	6.43	8.70	7.15	5.95
56Ba	3.39	3.45	3.45	6.70	9.20	7.42	6.35
58Ce	3.70	3.78	3.73	6.86	9.70	7.82	6.75
60Nd	3.93	4.08	4.00	7.18	10.30	8.15	7.16
62Sm	4.13	4.50	4.33	7.42	10.80	8.60	7.50
64Gd	4.46	4.90	4.63	7.70	11.20	9.08	7.83
66Dy	4.81	5.35	5.03	7.90	11.50	9.60	8.30
68Er	5.17	5.73	5.45	8.28	11.85	10.03	8.75
70Yb	5.40	6.22	5.90	8.58	12.20	11.00	9.20
72Hf	5.83	6.70	6.36	8.92	12.40	12.80	9.63
74W	6.50	7.20	6.90	9.06	13.10	14.60	10.20
76Os	7.04	7.70	7.42	9.60	14.60	16.50	10.65
78Pt	7.60	8.28	8.00	9.95	16.10	18.00	11.20
80Hg	8.10	8.80	8.70	10.40	17.40	19.70	11.80
82Pb	8.82	9.35	9.35	10.75	18.65	21.30	12.30
84Po	9.50	9.95	10.10	11.25	19.90	22.70	13.05
86Rn	10.03	10.50	10.65	11.65	21.00	24.00	13.55
88Ra	11.00	11.20	11.60	12.20	22.00	25.20	14.30
90Th	11.90	11.80	12.40	12.80	22.85	26.35	15.00
92U	12.40	12.40	13.50	13.30	23.70	27.50	15.70
94Pu	13.20	13.00	14.10	13.90	24.10	28.30	16.40
96Cm	14.80	13.60	15.70	14.60	25.00	29.40	17.10

DIFFRACTION DATA FOR CUBIC ISOMORPHS

From Volume 14, pages 689, 690, and 691 of the Analytical Edition of *Industrial and Engineering Chemistry,* with permission.

A 4

X Units	Substance
3.56	C (diamond)
5.42	Si
5.62	Ge
6.46	α-Sn

A 1

X Units	Substance
3.517	Ni
3.554	α-Co
3.60	Taenite (57.7% Fe, 40.8% Ni, 0.5% P)
3.608	Cu
3.63	γ-Fe (1370°K.)
3.797	Rh
3.831	Ir
3.880	Pd
3.88—4.04	Pd-H
3.912	Pt
4.041	Al
4.070	Au
4.077	Ag
4.30	Co-N
4.40—4.46	Ti-H
4.52	Ne (4°K.)
4.66	Zr-H
4.84	β-Tl
4.939	Pb
5.08	Th
5.14	α-Ce
5.296	β-La
5.43	A (4°K.)
5.56	Ca
5.59	Kr (20°K.)
5.70	Kr (92°K.)
6.05	Sr
6.20	X (88°K.)

A 2

X Units	Substance
2.861	α-Fe
2.875	α-Cr
2.90	β-Fe (1070°K.)
2.93	δ-Fe (1700°K.)
3.03	V
3.03—3.41	V-C
3.140	Mo
3.157	W
3.295	Cb
3.30	Ta
3.32	β-Ti (1200°K.)
3.46	Li (~80°K.)
3.50	Li
3.61	β-Zr (1120°K.)
4.24	Na (~80°K.)
4.29	Na
5.02	Ba
5.20	K (120°K.)
5.33	K
5.62	Rb (~80°K.)
6.05	Cs (~80°K.)

B 1

X Units	Substance
4.018	LiF
4.065	LiD
4.08	VO
4.09	LiH
4.12	(Li_2TiO_3)
4.12—4.20	$(Li_2TiO_3\text{-}MgO)$
4.13	VN
4.14	CrN
4.14	VC
4.142	$(63Li_2Fe_2O_4 \cdot 37Li_2TiO_3)$
4.173	NiO
4.207	MgO
4.282	MgO (1570°K.)
4.225	TiN
4.235	TiO
4.24	80 TiN-20 TiC
4.27	CoO
4.28	V-N
4.283	FeO (160°K.)
4.290	FeO (299°K.)
4.30	VC (ε-phase)
4.315	TiC
4.40	CbC
4.41	CbN
4.426	MnO (117°K.)
4.436	MnO (299°K.)
4.44	ScN
4.446	TaC
4.458	HfC
4.615	NaF
4.62	ZrN
4.69	CdO
4.69	ZrC
4.80	CaO
4.82	(Na_2CeO_3)
4.84	(Na_2PrO_3)
4.88	NaH
4.92	AgF
5.006	CaNH
5.13	SrO
5.14	LiCl
5.14	NdN
5.19	MgS
5.192	MnS (130°K.)
5.210	MnS (299°K.)
5.33	KF
5.45	MgSe
5.45	MnSe
5.45	SrNH
5.49	LiBr
5.52	BaO
5.545	AgCl
5.55—5.76	AgCl-AgBr
5.627	NaCl
5.63	RbF
5.68	CaS
5.69	SnAs
5.70	KH
5.755	AgBr
5.76—5.92	AgBr-AgI
5.83	NdP
5.83	NaCN
5.84	BaNH
5.87	SrS
5.91	CaSe
5.94	PbS
5.95	NaBr
5.957	EuS
5.96	NdAs
6.00	PrAs
6.00	LiI
6.01	CsF
6.04	RbH
6.05	β-NaSH (>360°K.)
6.06	CeAs
6.13	LaAs
6.14	PbSe
6.23	SrSe
6.278	KCl
6.285	SnTe
6.31	NdSb
6.345	CaTe
6.35	PrSb
6.36	BaS
6.38	CsH
6.40	CeSb
6.44	PbTe
6.462	NaI
6.45	PrBi
6.48	LaSb
6.49	CeBi
6.53	KCN
6.53	NH_4Cl (>457°K.)
6.56	RbCl
6.57	LaBi
6.58	KBr
6.59	BaSe
6.60	β-KSH (>440°K.)
6.65	SrTe
6.82	RbCN
6.86	RbBr
6.90	NH_4Br (>411°K.)
6.93	β-RbSH (470°K.)
6.99	BaTe
7.052	KI
7.10	β-CsCl (>730°K.)
7.24	NH_4I (>255°K.)
7.325	RbI

H 0_5

X Units	Substance
6.96 ± 0.04	$AgClO_4$ (453 ± 20°K.)

DIFFRACTION DATA FOR CUBIC ISOMORPHS (Continued)

X Units	Substance
7.16 ± 0.10	$NaClO_4$ (618 ± 35°K.)
7.49 ± 0.02	$KClO_4$ (598 ± 15°K.)
7.65 ± 0.05	$TlClO_4$(553°K.)
7.65 ± 0.02	NH_4ClO_4(528 ± 15°K.)
7.68 ± 0.03	$RbClO_4$(583 ± 10°K.)
7.97 ± 0.01	$CsClO_4$ (513 ± 10°K.)
B 3	
4.255	CuF
4.36	CSi IV
4.855	BeS
5.10	BeSe
5.304	$(Cu, Fe, Mo, Sn)_4(S, As, Te)_2$, cousite
5.41	CuCl
5.425	β-ZnS
5.43	AlP
5.44	GaP
5.58	BeTe
5.60	MnS (red)
5.63	AlAs
5.635	GaAs
5.655	ZnSe
5.68	CuBr
5.82	β-CdS
5.84	HgS
5.86	InP
6.04	CdSe
6.04	InAs
6.05	CuI
6.07	HgSe
6.08	ZnTe
6.103	α-Cu_2HgI_4
6.12	AlSb
6.12	GaSb
6.13	SnSb
6.383	α-Ag_2HgI_4
6.40	HgTe
6.43	CdTe
6.45	InSb
6.48	AgI
B 32	
6.195	LiGa
6.209	LiZn
6.36	LiAl
6.687	LiCd
6.786	LiIn
7.297	NaIn
7.373	$(CeMg_3)$
7.373	$(PrMg_3)$
7.473	NaTl
B 20	
4.437	NiSi
4.438	FeSi
4.438	CoSi
4.548	MnSi
4.620	CrSi
B 2	
2.603	NiBe
2.606	CoBe
2.69	CuBe
2.813	PdBe
2.82	AlNi
2.945	CuZn
2.989	CuPd
3.146	AuZn
3.156	AgZn
3.168	AgLi
3.259	AuMg
3.275	AgMg
3.287	HgLi
3.325	AgCd
3.34	AuCd (670°K.)
3.424	LiTl
3.442	HgMg
3.628	MgTl
3.67	PrZn
3.70	CeZn
3.73	AlNd
3.74	α-RbCl (83°K.)
3.75	LaZn
3.82	TlCn
3.82	PrCd
3.84	TlSb
3.835	TlCl
3.847	CaTl
3.86	NH_4Cl (<457°K.)
3.86	CeCd
3.88	MgPr
3.90	LaCd
3.97	TlBr
3.98	TlBi
4.024	SrTl
4.05	NH_4Br (<411°K.)
4.112	CsCl
4.20	TlI
4.20	CsCl (<720°K.)
4.25	CsCN
4.287	CsBr
4.29	CsSH
4.37	NH_4I (290°K.)
4.56	CsI
$D2_1$	
4.07	YB_6
4.07	ErB_6
4.10	NdB_6
4.12	GdB_6
4.12	PrB_6
4.13	CeB_6
4.13	YbB_6
4.14	CaB_6
4.15	LaB_6
4.15	ThB_6
4.19	SrB_6
4.33	BaB_6
C 1	
4.33	Be_2C
4.619	Li_2O
5.06	$(3ZrO_2 \cdot MgO)$
5.07	ZrO_2
5.08	$(95ZrO_2 \cdot 5CeO_2)$
5.13	$(95HiO_2 \cdot 5CeO_2)$
5.38	PrO_2
5.40	CeO_2
5.40	CdF_2
5.406	CuF_2
5.45	CaF_2
5.47	UO_2
5.526	$(66CaF_2 \cdot 33YF_3)$
5.53	$(91CaF_2 \cdot 9ThF_4)$
5.54	HgF_2
5.55	Na_2O
5.58	ThO_2
5.59	Cu_2S
5.704	Li_2S
5.749	Cu_2Se
5.782	SrF_2
5.796	EuF_2
5.838	$(66SrF_2 \cdot 33LaF_3$
5.91	$PtAl_2$
5.91	$PtGa_2$
5.935	β-PbF_2(520°K.)
5.99	Al_2Au
6.005	Li_2Se
6.06	$AuGa_2$
6.19	BaF_2
6.34	Mg_2Si
6.35	$PtIn_2$
6.368	RaF_2
6.379	Mg_2Ge
6.436	K_2O
6.50	Li_2Te
6.50	$AuIn_2$
6.526	Na_2S
6.763	Mg_2Sn
6.809	Na_2Se
6.81	Mg_2Pb
6.98	$SrCl_2$
7.314	Na_2Te
7.38	K_2S
7.65	RbS_2
7.676	K_2Se
8.152	K_2Te
C 15	
5.94	Be_2Cu
6.287	Be_2Ag
6.435	Be_2Ti
6.96	MgNiZn
7.03	Cu_2Mg
7.61	W_2Zr
7.79	Au_2Na

DIFFRACTION DATA FOR CUBIC ISOMORPHS (Continued)

X Units	Substance
7.91	Au_2Pb
7.94	Au_2Bi
8.02	Al_2Ca
8.04	Al_2Ce
8.16	Al_2La
9.50	Bi_2K
C 2	
5.41	FeS_2
5.42	(Fe, Ni)S_2 (6.5% Ni)
5.57	RbS_2
5.57	RuS_2
5.57	Bravoite (53.8% NiS_2, 39.1% FeS_2, 7.1% CoS_2)
5.62	OsS_2
5.64	CoS_2
5.65	(Cu, Ni, Co, Fe)(S, Se)$_2$
5.68	PtP_2
5.74	NiS_2
5.85	$CoSe_2$
5.92	$RuSe_2$
5.93	$OsSe_2$
5.94	$PtAs_2$
5.97	$PdAs_2$
6.02	$NiSe_2$
6.096	MnS_2
6.36	$RuTe_2$
6.37	$OsTe_2$
6.43	$PtSb_2$
6.44	$PdSb_2$
6.64	$AuSb_2$
6.94	$MnTe_2$
C 3	
4.25	Cu_2O
4.73	Ag_2O
F 1	
5.55	CoAsS
5.68	NiAsS
5.90	NiSbS
	(Ni, Fe)AsS, plessite
	Ni(As, Sb)S, corynite
	Ni(Sb, Bi)S, kallilite
	(Co, Ni)SbS, willyamite
$D5_3$	
8.13	Be_3N_2
9.37	(Mn, Fe)$_2O_3$
9.42	Mn_2O_3
9.74	Zn_3N_2
9.79	Sc_2O_3
9.94	Mg_3N_2
10.12	In_2O_3
10.15	Be_3P_2
10.37	Lu_2O_3
10.39	Yb_2O_3
10.52	Tm_2O_3
10.54	Er_2O_3
10.57	Tl_2O_3
10.58	Ho_2O_3
10.60	Y_2O_3
10.63	Dy_2O_3
10.70	Tb_2O_3
10.79	Gd_2O_3
10.79	Cd_3N_2
10.84	Eu_2O_3
10.85	Sm_2O_3
11.05	Nd_2O_3
11.40	α-Ca_3N_2
12.02	Mg_3P_2
12.33	Mg_3As_2
$D6_1$	
11.05	As_4O_6
11.14	Sb_4O_6
$D1_1$	
10.32	$ZrCl_4$
11.25	$TiBr_4$
(11.34)	(CBr_4) (>320°K.)
(11.62)	(CI_4)
11.89	GeI_4
11.99	SiI_4
12.00	TiI_4
12.23	SnI_4
$E2_1$	
3.67	$YAlO_3$
3.75	$CdTiO_3$
3.78	$LaAlO_3$
3.80	$CaTiO_3$
3.83	$NaWO_3$
3.85	(Na, Ce, Ca)(Ti, Cb)O_3 Loparite
3.88	$NaTaO_3$
3.89	$LaGaO_3$
3.89	$NaCbO_3$
3.91	$SrTiO_3$
3.92	$CaSnO_3$
3.97	$BaTiO_3$
3.98	$KTaO_3$
3.99	$CaZrO_3$
4.00	$KMgF_3$
4.005	$KNiF_3$
4.01	$KCbO_3$
4.03	$SrSnO_3$
4.05	$KZnF_3$
4.07	$KCoF_3$
4.07	$SrHfO_3$
4.09	$SrZrO_3$
4.18	$BaZrO_3$
4.35	$BaPrO_3$
4.38	$BaCeO_3$
4.46	KIO_3
4.48	$BaThO_3$
4.5	NH_4IO_3
4.52	$RbIO_3$
4.66	$CsIO_3$
5.12	$MgZrO_3$
5.20	$CsCdCl_3$
5.33	$CsCdBr_3$
5.44	$CsHgCl_3$
5.77	$CsHgBr_3$
$G0_3$	
6.57	$NaClO_3$
6.71	$NaBrO_3$
$G2_1$	
7.60	$Ca(NO_3)_2$
7.81	$Sr(NO_3)_2$
7.84	$Pb(NO_3)_2$
8.11	$Ba(NO_3)_2$
$H1_1$	
8.045	$NiAl_2O_4$
8.07	$CuAl_2O_4$
8.07	$CoCo_2O_4$
8.07	$MgAl_2O_4$
8.08	$CoAl_2O_4$
8.08	$ZnAl_2O_4$
8.10	$FeAl_2O_4$
8.11	(Ni, Co)(Co, Ni)$_2O_4$
8.11	(Zn, Co)Co_2O_4
8.11	$MgCo_2O_4$
8.27	$MnAl_2O_4$
8.27	(Mn, Co)(Co, Mn)$_2O_4$
$H1_1$	
8.28	$MgGa_2O_4$
8.30	$NiCr_2O_4$
8.30	$MgCr_2O_4$
8.31	$ZnCr_2O_4$
8.32	$CoCr_2O_4$
8.32	$ZnGa_2O_4$
8.35	$NiFe_2O_4$
8.35	$Cu_2Cr_2O_4$
8.35	$FeCr_2O_4$
8.36	$MgFe_2O_4$
8.38	$CoFe_2O_4$
8.38	$NiMn_2O_4$
8.40	$ZnFe_2O_4$
8.40	$FeFe_2O_4$
8.42	(Mn, Mg)Fe_2O_4
8.42	$TiCo_2O_4$
8.43	$MnCr_2O_4$
8.43	$TiMg_2O_4$
8.43	$TiZn_2O_4$
8.44	$CuFe_2O_4$
8.47	FeV_2O_4
8.49	$MnCr_2O_4$
8.50	$TiFe_2O_4$
8.54	$MnFe_2O_4$
8.58	$CdCr_2O_4$
8.58	$SnMg_2O_4$
8.61	$SnCo_2O_4$
8.63	$SnZn_2O_4$
8.67	$CdFe_2O_4$
8.67	$TiMn_2O_4$
8.81	$MgIn_2O_4$
9.26	Ag_2MoO_4

DIFFRACTION DATA FOR CUBIC ISOMORPHS (Continued)

X Units		Substance
9.4		$CoCo_2S_4$
9.45		$(Co, Ni)_3S_4$
9.46		$CuCo_2O_4$
9.5		NiN_2S_4
9.92		$ZnCr_2S_4$
10.05		$MnCr_2S_4$
10.19		$CdCr_2S_4$
12.54		$K_2Zn(Cn)_4$
12.76		$K_2Hg(CN)_4$
12.84		$K_2Cd(CN)_4$
	$H\ 5_8$	
10.08		$2Na_2SO_4 \cdot NaCl \cdot NaF$
	$H\ 4_{13}$	
12.11		$KCr(SO_4)_2 \cdot 12H_2O$
12.12		$KAl(SO_4)_2 \cdot 12H_2O$
12.15		$NH_4Al(SO_4)_2 \cdot 12H_2O$
12.20		$RbAl(SO_4)_2 \cdot 12H_2O$
12.21		$TlAl(SO_4)_2 \cdot 12H_2O$
12.31		$CsAl(SO_4)_2 \cdot 12H_2O$
12.44		$NH_3 \cdot CH_3Al(SO_4)_2 \cdot 12H_2O$ (β-alum)
	Langbeinite	
9.93		$K_2Mg(SO_4)_3$
10.2		$K_2(Ca_2Mg)SO_3)_3$
	$H\ 2_1$	
6.00		Ag_3PO_4
6.120		Ag_3AsO_4(90°K.)
6.130		Ag_3AsO_4 (380°K.)
	$H\ 2_4$	
5.37		Cu_3VS_4
	$J\ 1_1$	
8.17		K_2SiF_6
8.35		$(NH_4)_2SiF_6$
8.38		$Rb_2CrF_5H_2O$
8.41		$Tl_2CrF_5H_2O$
8.42		$(NH_4)_2VF_5H_2O$
8.42		$Rb_2VF_5H_2O$
8.45		$Tl_2VF_5H_2O$
8.45		Rb_2SiF_6
8.58		Tl_2SiF_6
8.87		Cs_2SiF_6
8.99		Cs_2GeF_6
9.73		K_2PtCl_6
9.73		K_2OsCl_6
9.76		Tl_2PtCl_6
9.84		$(NH_4)_2PtCl_6$
9.86		K_2ReCl_6
9.88		Rb_2PtCl_6
9.92		Rb_2TiCl_6
9.94		$(NH_4)_2SeCl_6$
9.97		K_2SnCl_6
9.97		Tl_2SnCl_6
9.98		Rb_2SeCl_6
10.02		Rb_2PdBr_6
10.04		$(NH_4)SnCl_6$
10.08		$Ni(NH_3)_6Cl_2$
10.10		Rb_2SnCl_6
10.10		$Co(NH_3)_6Cl_2$
10.11		Tl_2TeCl_6
10.14		$(NH_4)_2PBCl_6$
10.14		K_2TeCl_6
10.15		$Fe(NH_2)_6Cl_2$
10.16		$Mg(NH_3)_6Cl_2$
10.17		Cs_2PtCl_6
	$J\ 1_1$	
10.18		Rb_2ZrCl_6
10.18		$(NH_4)_2TeCl_6$
10.20		$Mn(NH_3)_6Cl_2$
10.20		Rb_2PbCl_6
10.22		Cs_2TiCl_6
10.23		Rb_2TeCl_6
10.25		$Zn(NH_2)_6(ClI_4)_3$
10.26		Cs_2SeCl_6
10.30		K_2OSBr_6
10.35		Cs_2SnCl_6
10.36		K_2SeBr_6
10.36		K_2PtBr_6
10.39		$Co(NH_3)_6Br_2$
10.4		$Ni(NH_2)_6Br_2$
10.41		Cs_2ZrCl_6
10.42		Cs_2PbCl_6
10.45		Cs_2TeCl_6
10.45		$Co(NH_2)H_2OSO_4Br$
10.46		$(NH_4)SeBr_6$
10.46		$Zn(NH_2)_6Br_2$
10.47		$Fe(NH_3)_6Br_2$

X Units	Substance
10.47	$Mg(NH_3)_6Br_2$
10.48	K_2SnBr_6
10.51	$Co(NH_3)_6SO_4Br$
10.52	$Mn(NH_3)_6Br_2$
10.54	$Sr_2Ni(NO_2)_6$
10.55	$Pb_2Ni(NO_2)_6$
10.57	$(NH_4)SnBr_6$
10.58	Rb_2SnBr_6
10.62	$Co(NH_3)_5H_2OSO_4I$
10.63	$Co(NH_3)_6SeO_4Br$
10.67	$Ba_2Ni(NO_2)_6$
10.71	$Ca(NH_2)_6Br_2$
10.71	$Co(NH_3)_6SO_4I$
10.77	Cs_2SnBr_6
10.79	$Co(NH_3)_6SeO_4I$
10.9	$Ni(NH_3)_6I_2$
10.91	$Co(NH_3)_6I_2$
10.96	$Zn(NH_3)_6I_2$
10.97	$Fe(NH_3)_6I_2$
10.98	$Mg(NH_2)_6I_2$
11.04	$Mn(NH_3)_6I_2$
11.04	$Cd(NH_3I_2$
11.24	$Ca(NH_3)_6I_2$
11.27	$Ni(NH_3)_6(BF_4)_2$
11.3	$Co(NH_3)_6(BF_4)_2$
(11.3)	$Zn(NH_3)_6(ClO_4)_2$
11.34	$Mg(NH_3)_6(BF_4)_2$
11.34	$Fe(NH_3)_6(BF_4)_2$
11.37	$Mn(NH_3)_6(BF_4)_2$
11.38	$Cd(NH_3)_6(BF_4)_2$
11.41	$Ni(NH_3)_6(ClO_4)_2$
11.43	$Co(NH_3)_6(ClO_4)_2$
11.46	$Ni(NH_3)_6(SO_3F)_2$
11.49	$Co(NH_2)_6(SO_3F)_2$
11.52	$Fe(NH_3)_6(ClO_4)_2$
11.53	$Mg(NH_3)_6(ClO_4)_2$
11.54	$Cd(NH_3)_6Br_2$
11.54	$Fe(NH_3)_6(SO_2F)_2$
11.58	$Mn(NH_3)_6(ClO_4)_2$
11.59	$Cd(NH_3)_6(ClO_4)_2$
11.59	$Mn(NH_2)_6(SO_2F)_2$
11.62	$Cd(NH_2)_6(SO_3F)_2$
11.91	$Ni(NH_3)_6(PF_6)_2$
11.94	$Co(NH_2)_6(PF_6)_2$
12.03	$Ni(NH_2 \cdot CH_3)_6I_2$
12.05	$Co(NH_2 \cdot CH_3)_6I_2$
12.19	$\{NH(CH_3)_3\}_2SnCl_6$
12.41	$\{S(CH_3)_2\}_2SnCl_6$
12.65	$\{N(CH_3)_4\}_2PtCl_6$
12.80	$\{S(CH_3)_2C_2H_5\}_2SnCl_6$
12.87	$\{N(CH_3)_4\}_2SnCl_6$
13.17	$\{N(CH_3)_2CH_5\}_2SnCl_6$
13.51	$\{N(CH_3)(C_2H_5)_3\}_2SnCl_6$
13.93	$\{P(CH_3)(C_2H_5)_3\}_2SnCl_6$
$J\ 2_1$ and Related Structures	
8.88	Li_3FeF_6
8.90	$(NH_4)_3AlF_6$
9.01	$(NH_4)_3CrF_6$
9.04	$(NH_4)_3VF_6$
9.10	$(NH_4)_3FeF_6$
9.10	$(NH_4)MoO_3F_2$
9.26	Na_2FeF_6
9.93	K_2FeF_6
9.96	$CuLi_2Fe(CN)_6$
10.0	$CuR_2Fe(CN)_6$, R = Na, K, Rb, NH_4, Tl
10.15	$K_2CdFe(NO_2)_6$
10.17	$K_2CaCo(NO_2)_6$
10.19	$K_2CaFe(NO_2)_6$
10.2	$Fe^{III}RFe^{II}(CN)_6$ R = Na, K, Rb, NH_4
10.22	$K_2HgFe(NO_2)_6$
10.23	$K_2SrCo(NO_2)_6$
10.25	$(NH_4)_2CaFe(NO_2)_6$
10.25	$NaTl_2Co(NO_2)_6$
10.28	$K_2CdNi(NO_2)_6$
10.28	$(NH_4)_2CdFe(NO_2)_6$
10.29	$K_2HgNi(NO_2)_6$
10.30	$K_2SrFe(NO_2)_6$
10.30	$Tl_2CaFe(NO_2)_6$
10.31	$K_2PbFe(NO_2)_6$
10.32	$K_2CaNi(NO_2)_6$
10.34	$(NH_4)_2SrFe(NO_2)_6$
10.37	$(NH_4)_2PbFe(NO_2)_6$
10.37	$Tl_2CdNi(NO_2)_6$
10.39	$Tl_2PbFe(NO_2)_6$

X Units	Substance
10.39	$NaRb_2Co(NO_2)_6$
10.40	$Tl_2SrFe(NO_2)_6$
10.4	$K_2PbCo(NO_2)_6$
10.41	$(NH_4)_2CdNi)NO_2)_6$
10.42	$Tl_2HgNi(NO_2)_6$
10.43	$K_2BaFe(NO_2)_6$
10.45	$K_2BaCo(NO_2)_6$
10.45	$K_2Co(NO_2)_6$
10.46	$(NH_4)_2HgNi(NO_2)_6$
10.47	$Rb_2HgNi(NO_2)_6$
10.49	$K_2SrNi(NO_2)_6$
10.49	$K_4Ni(NO_2)_6$
10.50	$(NH_4)_2BaFe(NO_2)_6$
10.54	$K_2LiBi(NO_2)_6$
10.55	$K_2PbNi(NO_2)_6$
10.55	$Tl_2BaFe(NO_2)_6$
10.58	$Rb_2CdNi, Cd(NO_2)_6$
10.58	$K_3Ir(NO_2)_6$
10.59	$Rb_2LiBi(NO_2)_6$
10.6	$K_2PbCu(NO_2)_6$
10.63	$K_2Rh(NO_2)_6$
10.63	$(NH_4)_2LiBi(NO_2)_6$
10.64	$Tl_2LiBi(NO_2)_6$
10.67	$K_2BaNi(NO_2)_6$
10.70	$NaCs_2Co(NO_2)_6$
10.70	$Ba_1\{Rh(NO_2)_6\}_2$
10.72	$Tl_2Co(NO_2)_6$
10.73	$Rb_3Co(NO_2)_6$
10.73	$(NH_4)_2Ir(NO_2)_6$
10.73	$Tl_2Ir(NO_2)_6$
10.77	$Rb_2Ir(NO_2)_6$
10.8	$(NH_4)_2Co(NO_2)_6$
10.81	$Cs_2Cd\{Ni, Cd(NO_2)_6\}$
10.82	$Co(NH_3)_5H_2OI_3$
10.83	$Rb_3Rh(NO_2)_6$
10.88	$K_2NaBi(NO_2)_6$
10.89	$Co(NH_3)_6I_3$
10.91	$Tl_3Rh(NO_2)_6$
10.91	$(NH_4)_3Rh (NO_2)_6$
10.94	$Cs_2LiBi(NO_2)_6$
10.95	$K_2AgBi(NO_2)_6$
10.98	$Rb_2NaBi(NO_2)_6$
10.99	$(NH_4)_2NaBi(NO_2)_6$
11.01	$Tl_2NaBi(NO_2)_6$
11.05	$Rb_2AgBi(NO_2)_6$
11.06	$Tl_2AgBi(NO_2)_6$
11.10	$(NH_4)_2AgBi(NO_2)_6$
11.15	$Cs_2NaBi(NO_2)_6$
11.15	$Cs_3Co(NO_2)_6$
11.17	$Cs_3Ir(NO_2)_6$
11.19	$Cs_3Bi(NO_2)_6$
11.19	$Cs_2AgBi(NO_2)_6$
11.21	$Co(NH_3)_6(BF_4)_3$
11.30	$Cs_3Rh(NO_2)_6$
11.32	$\{Co(NH_3)_6 \cdot H_2O\}(ClO_4)_3$
11.39	$Co(NH_3)_6(ClO_4)_3$
11.67	$Co(NH_3)_6(PF_6)$
$K\,6_1$	
7.46	SiP_2O_7
7.80	TiP_2O_7
7.98	SnP_2O_7
8.18	HfP_2O_7
8.20	ZrP_2O_7
8.61	UP_2O_7
$S\,1_4$	
11.51	$Al_2(Mg, Fe)_3(SiO_4)_3$, pyrope
11.51	$Al_2Fe_2(SiO_4)_3$, almandite
11.60	$Al_2Mn_3(SiO_4)_3$, spessartite
11.87	$Al_2Ca_2(SiO_4)_3$, grossularite
11.89	$(Al, Fe)_2Ca_3(SiO_4)_3$, hessonite
11.95	$Cr_2Ca_3(SiO_4)_3$, uvarovite
12.03	$Fe_2Ca_3(SiO_4)_3$, andradite
12.10	$(Na, Li)_3AlF_6$, cryolithionite
12.35—12.46	$(Mg, Mn)_2(Ca, Na)_3AsO_4)_3$, berzelite
$S\,6_1$	
13.68	$NaAlSi_2O_6H_2O$
$S\,0_8$	
13.82	$Al_{13}Si_5O_{20}(OH, F)_{18}Cl$, zunyite
$S\,6_2$	
8.87	$Na_4(AlSiO_4)_3Cl$, sodalite
Tetrahedrite	
10.19	$(Cu, Fe)_{12}As_4S_{13}$, binnite
10.2—10.6	$(Cu, Ag)_{10} (Zn, Fe)_2(Sb, As)_4S_{13}$

STANDARD UNITS, SYMBOLS, AND DEFINING EQUATIONS FOR FUNDAMENTAL PHOTOMETRIC AND RADIOMETRIC QUANTITIES

Submitted by Abraham Abramowitz
from Z-7.1-1967

Radiometric Quantities
(See Note at bottom of page)

Quantity*	Symbol*	Defining Equation**	Commonly Used Units	Symbol
Radiant energy	$Q, (Q_e)$		erg ‡joule kilowatt-hour	 J kWh
Radiant density	$w, (w_e)$	$w = dQ/dV$	‡ joule per cubic meter erg per cubic centimeter	J/m^3 erg/cm^3
Radiant flux	$\Phi, (\Phi_e)$	$\Phi = dQ/dt$	erg per second †watt	erg/s W
Radiant flux density at a surface				
Radiant exitance (Radiant emittance)†	$M, (M_e)$	$M = d\Phi/dA$	watt per square centimeter	W/cm^2
Irradiance	$E, (E_e)$	$E = d\Phi/dA$	‡watt per square meter, etc.	W/m^2
Radiant intensity	$I, (I_e)$	$I = d\Phi/d\omega$ (ω = solid angle through which flux from point source is radiated)	‡watt per steradian	W/sr
Radiance	$L, (L_e)$	$L = d^2\Phi/d\omega(dA \cos\theta)$ $= dI/(dA \cos\theta)$ (θ = angle between line of sight and normal to surface considered)	watt per steradian and square centimeter ‡watt per steradian and square meter	$W \cdot sr^{-1} cm^{-2}$ $W \cdot sr^{-1} m^{-2}$
Emissivity	ε	$\varepsilon = M/M_{blackbody}$ (M and $M_{blackbody}$ are respectively the radiant exitance of the measured specimen and that of a blackbody at the same temperature as the specimen)	one (numeric)	—

Note: The symbols for photometric quantities (see following table) are the same as those for the corresponding radiometric quantities (see above). When it is necessary to differentiate them the subscripts v and e respectively should be used, e.g., Q_v and Q_e.

*Quantities may be restricted to a narrow wavelength band by adding the word spectral and indicating the wavelength. The corresponding symbols are changed by adding a subscript λ, e.g., Q_λ for a spectral concentration or a λ in parentheses, e.g., $K(\lambda)$, for a function of wavelength.

**The equations in this column are given merely for identification.

***Φ_i = incident flux
Φ_a = absorbed flux
Φ_r = reflected flux
Φ_t = transmitted flux

†to be deprecated.

‡International System (SI) unit.

PHOTOMETRIC QUANTITIES

Quantity*	Symbol*	Defining Equation**	Commonly Used Units	Symbol
Absorptance	$\alpha, (\alpha_v, \alpha_e)$	$\alpha = \Phi_a/\Phi_i$***	one (numeric)	—
Reflectance	$\rho, (\rho_v, \rho_e)$	$\rho = \Phi_r/\Phi_i$***	one (numeric)	—
Transmittance	$\tau, (\tau_v, \tau_e)$	$\tau = \Phi_t/\Phi_i$***	one (numeric)	—
Luminous energy (quantity of light)	$Q, (Q_v)$	$Q_v = \int_{380}^{760} K(\lambda)\, Q_e\lambda\, d\lambda$	lumen-hour ‡lumen-second (talbot)	lm · h lm · s
Luminous density	$w, (w_v)$	$w = dQ/dV$	‡lumen-second per cubic meter	$\mathrm{lm \cdot s \cdot m^{-3}}$
Luminous flux	$\Phi, (\Phi_v)$	$\Phi = dQ/dt$	‡lumen	lm
Luminous flux density at a surface				
Luminous exitance (Luminous emittance)†	$M, (M_v)$	$M = d\Phi/dA$	lumen per square foot	$\mathrm{lm/ft^2}$
Illumination (illuminance)	$E, (E_v)$	$E = d\Phi/dA$	footcandle (lumen per square foot) ‡lux ($\mathrm{lm/m^2}$) phot ($\mathrm{lm/cm^2}$)	fc lx ph
Luminous intensity (candlepower)	$I, (I_v)$	$I = d\Phi/d\omega$ (ω = solid angle through which flux from point source is radiated)	‡candela (lumen per steradian)	cd
Luminance (photometric brightness)	$L, (L_v)$	$L = d^2\Phi/d\omega\,(dA\cos\theta)$ $= dI/(dA\cos\theta)$ (θ = angle between line of sight and normal to surface considered)	candela per unit area stilb ($\mathrm{cd/cm^2}$) nit (†$\mathrm{cd/m^2}$) footlambert ($\mathrm{cd/\pi ft^2}$) lambert ($\mathrm{cd/\pi cm^2}$) apostilb ($\mathrm{cd/\pi m^2}$)	$\mathrm{cd/in^2}$, etc. sb nt fL L asb
Luminous efficacy	K	$K = \Phi_v/\Phi_e$	‡lumen per watt	lm/W
Luminous efficiency	V	$V = K/K_{\mathrm{maximum}}$ (K_{maximum} = maximum value of $K(\lambda)$ function)	one (numeric)	—

ILLUMINATION CONVERSION FACTORS

1 lumen = 1/680 lightwatt	1 watt-second = 1 joule = 10^7 ergs
1 lumen-hour = 60 lumen-minutes	1 phot = 1 $\mathrm{lumen/cm^2}$
1 footcandle = 1 $\mathrm{lumen/ft^2}$	1 lux = 1 $\mathrm{lumen/m^2}$

Number of → Multiplied by ↘ Equals Number of ↓	Footcandles	*Lux	Phots	Milliphots
Footcandles	1	0.0929	929	0.929
*Lux	10.76	1	10,000	10
Phot	0.00108	0.0001	1	0.001
Milliphot	1.076	0.1	1,000	1

*The International Standard (SI) unit.

LINE SPECTRA OF THE ELEMENTS

Edited by Joseph Reader and Charles H. Corliss*
National Bureau of Standards

These tables were prepared under the auspices of the Committee on Line Spectra of the Elements of the National Academy of Sciences—National Research Council. They contain the outstanding spectral lines of neutral (I), singly ionized (II), doubly ionized (III), triply ionized (IV), and quadruply ionized (V) atoms. Listed are lines that appear in emission from the vacuum ultraviolet to the far infrared. For most atoms these lines were selected from much larger lists in such a way as to include the stronger observed lines in each spectral region. In a few cases prominent monoxide band heads are also given.

The data were compiled by the following contributors, whose initials are given in the headings of the tables that they prepared:

- J. G. Conway—Lawrence Berkeley Laboratory
- C. H. Corliss—National Bureau of Standards
- R. D. Cowan—Los Alamos Scientific Laboratory
- C. R. Cowley—University of Michigan
- Henry M. and Hannah Crosswhite—Argonne National Laboratory
- S. P. Davis—University of California, Berkeley
- V. Kaufman—National Bureau of Standards
- R. L. Kelly—Naval Postgraduate School
- J. F. Kielkopf—University of Louisville
- W. C. Martin—National Bureau of Standards
- T. K. McCubbin—Pennsylvania State University
- L. J. Radziemski—Los Alamos Scientific Laboratory
- J. Reader—National Bureau of Standards
- C. J. Sansonetti—National Bureau of Standards
- G. V. Shalimoff—Lawrence Berkeley Laboratory
- R. W. Stanley—Purdue University
- J. O. Stoner, Jr.—University of Arizona
- H. H. Stroke—New York University
- D. R. Wood—Wright State University
- E. F. Worden—Lawrence Livermore Laboratory
- J. J. Wynne—International Business Machines Corporation
- R. Zalubas—National Bureau of Standards

The literature references are collected at the end of the entire set of tables.

All wavelengths are given in Angstroms. Below 2000 Å the wavelengths are in vacuum; above 2000 Å the wavelengths are in air. Wavelengths given to three decimal places have an uncertainty of less than 0.001 Å and are therefore suitable for the calibration of most spectrographs. In the air region, the elements used most commonly for calibration purposes are Ne, Ar, Kr, Fe, Th, and Hg; in the vacuum region, the most common are C, N, O, Si, Cu.

A large number of the lines for neutral and singly ionized atoms were extracted from the National Bureau of Standards (NBS) Tables of Spectral-Line Intensities.[1] The intensities of these lines represent quantitative estimates of relative line strengths that take account of varying detection sensitivity at different wavelengths. They are on a linear scale. For nearly all of the other lines the intensities represent qualitative estimates of the relative strengths of lines not greatly separated in wavelength. Because different observers frequently use different scales for their intensity estimates, these intensities are useful only as a rough indication of the appearance of a spectrum. In some cases the intensity scale is not intended to be linear. In the tables of first and second spectra the intensities of the lines of the singly ionized atom relative to those of the neutral atom should be used with caution, inasmuch as the concentration of ions in a light source depends greatly on the excitation conditions.

Descriptive symbols used in the tables have the following meanings:

- c—complex
- d—line consists of two unresolved lines
- h—hazy
- l—shaded to longer wavelengths
- s—shaded to shorter wavelengths
- p—perturbed by a close line
- b—band head
- r—easily reversed
- w—wide

ACTINIUM (Ac)
Z = 89

Ac I and II
Ref. 193 — J.G.C.

Intensity		Wavelength	
		Air	
8	h	2100.00	II
20		2712.50	II
10		2726.23	II
10	h	2760.18	II
10	h	2781.56	II
20		2797.59	II
20		2806.76	II
8		2833.47	II
150	h	2847.16	II
8		2895.20	II
30		2896.82	II
30		2923.02	II
200		2994.17	II
500		3043.30	II
200		3069.36	II
100		3078.07	II
100		3086.04	II
100		3087.37	II
200		3112.83	II
100		3120.16	II
500	s	3153.09	II
600	s	3154.41	II
200	s	3164.81	II
300	s	3230.59	II
150	s	3237.70	II
500		3260.91	II
100	s	3318.01	II
200	s	3383.53	II
200		3413.84	II
500		3417.77	II
500	s	3481.16	II
200		3489.53	II
100		3529.24	II
100		3534.63	II
200	s	3554.99	II
1000	s	3565.59	II
100		3694.88	II
300	s	3756.67	II
200		3799.82	II
2000	s	3863.12	II
100		3914.47	II
400	s	4061.60	II
3000	s	4088.44	II
3000	s	4168.40	II
100		4179.98	I
20		4183.12	I
20		4194.40	I
300	s	4209.69	II
300		4359.13	II
20	l	4384.53	I
1000	l	4386.41	II
20		4396.71	I
20		4462.73	I
1000	l	4507.20	II
500		4605.45	II
10		4716.58	I
400	s	4720.16	II
300		4812.22	II
100		4945.18	II
100		4958.23	II
100		4960.87	II
150		5446.38	II
300	l	5732.05	II
400		5758.97	II
1000		5910.85	II
600	l	6164.75	II
200	l	6167.83	II
400		6242.83	II
20		6359.86	I
20	l	6691.27	I
6		7290.40	I
6	l	7866.10	I

Ac III
Ref. 193 — J.G.C.

Intensity		Wavelength	
		Air	
1000	h	2626.44	III
50	h	2682.90	III
2000	h	2952.55	III
2000	h	3392.78	III
3000		3487.59	III
2000	h	4413.09	III
3000	h	4569.87	III
8	h	5193.21	III

Ac IV
Ref. 193 — J.G.C.

Intensity		Wavelength	
		Air	
20	h	2062.00	IV
30	h	2502.12	IV
100	h	2558.08	IV
5	h	2790.83	IV
50	h	2793.90	IV
20	l	3224.7	IV

ALUMINUM (Al)
Z = 13

Al I and II
Ref. 81, 89, 144, 227, 228, 282 — E.F.W.

Intensity	Wavelength	
	Vacuum	
40	1177.43	II
50	1191.812	II
150	1350.18	II
800	1539.830	II
100	1569.385	II
125	1596.059	II
150	1625.627	II
100	1644.235	II
100	1644.809	II
1000	1670.787	II
100	1686.250	II
800	1719.440	II
500	1721.244	II
900	1721.271	II
500	1724.952	II
900	1724.984	II
350	1760.104	II
300	1761.975	II
290	1763.00	I
500	1763.869	II
700	1763.952	II
450	1765.64	I

* Charles H. Corliss is now retired.

Aluminum (Cont.)

Intensity		Wavelength	Spectrum
300		1765.815	II
450		1766.38	I
600		1767.731	II
450		1769.14	I
600		1828.588	II
400		1832.837	II
250		1834.808	II
300		1855.929	II
700		1858.026	II
120		1859.980	II
1000		1862.311	II
200		1929.978	II
150		1931.048	II
200		1932.377	II
400		1934.503	II
150		1934.713	II
150		1936.907	II
220		1939.261	II
700		1990.531	II

Air

Intensity		Wavelength	Spectrum
150		2016.052	II
150		2016.234	II
100		2016.368	II
200		2074.008	II
700		2094.264	II
150		2094.744	II
300		2094.791	II
100		2095.104	II
200		2095.141	II
60		2150.70	I
60		2181.00	I
400		2269.10	I
120		2269.22	I
60		2312.49	I
70		2313.53	I
90		2317.48	I
60		2319.06	I
140		2321.56	I
460		2367.05	I
110		2367.61	I
110		2368.11	I
180		2369.30	I
140		2370.22	I
70		2370.73	I
160		2372.07	I
850		2373.12	I
170		2373.35	I
110		2373.57	I
60		2378.40	I
60		2513.30	I
240		2567.98	I
480		2575.10	I
60		2575.40	I
80		2631.55	II
110		2637.70	II
150		2652.48	I
200		2660.39	I
160		2669.17	II
650		2816.19	II
90		2837.96	I
90		2840.10	I
150		3041.28	II
360		3050.07	I
60		3054.68	I
450		3057.14	I
90		3064.29	I
60		3066.14	I
150		3074.64	II
4500	rS	3082.153	I
7200	r	3092.710	I
1800	r	3092.839	I
150		3428.92	II
70		3439.35	I
150		3443.64	I
70		3444.86	I
70		3458.22	I
60		3479.81	I
60		3482.63	I
450		3586.56	II
360		3587.07	II
290		3587.45	II
220		3651.06	II
110		3651.10	II
150		3654.98	II
290		3655.00	II
450		3900.68	II
60		3932.00	I
4500	r	3944.006	I
9000	r	3961.520	I
110		3995.86	II
290		4226.81	II
150		4585.82	II
110		4588.19	II
550		4666.80	II
110		4898.76	II
110		4902.77	II
150		5280.21	II
70		5107.52	I
290		5283.77	II
150		5285.85	II
110		5312.32	II
220		5316.07	II
150		5371.84	II
180		5557.06	I
110		5557.95	I
450		5593.23	II
110		5853.62	II
220		5971.94	II
290		6001.76	II
220		6001.88	II
450		6006.42	II
150		6061.11	II
290		6068.43	II
110		6068.53	II
450		6073.23	II
110		6181.57	II
150		6181.68	II
290		6182.28	II
220		6182.45	II
450	h	6183.42	II
450		6201.52	II
360		6201.70	II
290		6226.18	II
360		6231.78	II
450		6243.36	II
450		6335.74	II
360		6696.02	I
230		6698.67	I
60		7083.97	I
70		7084.64	I
110		7361.57	I
140		7362.30	I
60		7606.16	I
90		7614.82	I
230		7835.31	I
290		7836.13	I
60		7993.05	I
90		8003.19	I
70		8065.97	I
110		8075.35	I
290		8640.70	II
360		8772.87	I
450		8773.90	I
110		8828.91	I
180		8841.28	I
90		8912.90	I
140		8923.56	I
60		9089.91	I
70		9139.95	I
150		9290.65	II
110		9290.75	II
150		10076.29	II
110		10768.36	I
140		10782.04	I
110		10872.98	I
230		10891.73	I
450		11253.19	I
570		11254.88	I
570		13123.41	I
450		13150.76	I
230		16718.96	I
300		16750.56	I
140		16763.36	I
300		21093.04	I
360		21163.75	I

Al III
Ref. 127 — E.F.W.

Intensity		Wavelength	
		Vacuum	
70		486.884	III
30		486.912	III
250		511.138	III
150		511.191	III
500		560.317	III
200		560.433	III
100		670.068	III
200		671.118	III
500		695.829	III
400		696.217	III
200		725.683	III
300		726.915	III
400		855.034	III
500		856.746	III
400		892.024	III
50		893.887	III
450		893.897	III
10		1162.59	III
5		1162.62	III
100		1352.81	III
5		1352.82	III
70		1352.86	III
600		1379.67	III
800		1384.13	III
700		1605.766	III
100		1611.814	III
800		1611.874	III
1000		1854.716	III
600		1862.790	III
300		1935.840	III
15		1935.863	III
200		1935.949	III
		Air	
110		2399.00	III
285		2762.77	III
220		2762.87	III
450		2906.93	III
360		3348.52	III
290		3350.88	III
870		3601.63	III
550		3601.93	III
750		3612.36	III
450		3702.11	III
550		3713.12	III
110	h	3980.14	III
110		4082.45	III
150		4088.61	III
110	h	4142.37	III
650		4149.92	III
650		4150.17	III
110	h	4364.64	III
650		4479.89	III
650		4479.97	III
760		4512.56	III
550		4528.94	III
870		4529.19	III
110		4701.15	III
150		4701.41	III
110	h	4904.10	III
110	h	5151.01	III
110	h	5163.89	III
1200		5696.60	III
1000		5722.73	III
110		6055.21	III
220	h	7635.37	III
150		7660.26	III
220		7681.97	III
360		7881.79	III
150		7882.52	III
290		7905.51	III
290	h	8243.59	III
360	h	8275.11	III
290		9571.52	III
360		9605.99	III

Al IV
Ref. 8, 146 — E.F.W.

Intensity	Wavelength	
	Vacuum	
400	124.03	IV
700	129.73	IV
800	160.07	IV
700	161.69	IV
500	1027.34	IV
800	1042.17	IV
700	1048.52	IV
500	1058.90	IV
500	1061.43	IV
600	1064.89	IV
500	1066.57	IV
600	1069.44	IV
400	1105.74	IV
600	1118.82	IV
500	1125.61	IV
400	1136.82	IV
400	1198.50	IV
400	1220.55	IV
900	1237.19	IV
600	1240.21	IV
700	1240.86	IV
700	1248.79	IV
900	1257.62	IV
800	1264.18	IV
1000	1272.76	IV
400	1337.90	IV
500	1376.62	IV
500	1388.79	IV
600	1431.94	IV
700	1441.82	IV
800	1447.51	IV
600	1457.96	IV
700	1486.89	IV
800	1494.79	IV
400	1519.07	IV
800	1537.54	IV
500	1550.19	IV
1000	1557.25	IV
500	1559.03	IV
700	1564.16	IV
900	1582.04	IV
800	1584.46	IV
400	1589.28	IV
400	1606.65	IV
400	1617.81	IV
600	1627.54	IV
500	1636.82	IV
800	1639.06	IV
1000	1818.56	IV
700	1881.16	IV
	Air	
400	2515.87	IV
500	3208.20	IV
500	3267.21	IV
600	3285.13	IV
400	3344.46	IV
500	3473.54	IV
900	3492.23	IV
800	3508.46	IV
500	3511.28	IV
700	3517.56	IV
400	3527.03	IV
500	3541.08	IV

Al V
Ref. 6 — E.F.W.

Intensity	Wavelength	
	Vacuum	
300	103.80	V
400	103.88	V
250	104.07	V
250	104.18	V
600	107.95	V
300	108.06	V
300	108.11	V
250	118.50	V
900	125.53	V
800	126.07	V
800	130.41	V
1000	130.85	V
900	131.00	V
900	131.44	V
500	132.63	V
1000	278.69	V
900	281.39	V
250	1068.26	V
500	1088.67	V
300	1090.14	V
300	1150.30	V
350	1165.42	V
250	1168.48	V
500	1287.70	V
400	1330.06	V
400	1350.52	V
400	1363.35	V
600	1369.20	V
300	1373.70	V
400	1445.87	V
300	1455.26	V
600	1475.64	V
300	1486.05	V
700	1508.37	V
1000	1526.14	V
500	1539.12	V
300	1577.90	V
350	1589.87	V

AMERICIUM (Am)
Z = 95

Am I and II
Ref. 92 — J.G.C.

Intensity		Wavelength	
		Air	
100	s	2706.35	II
100	s	2728.69	II
200	s	2756.55	II
100	l	2812.10	II
200	s	2812.92	II
1000	l	2815.28	II
100	s	2815.98	II
100	l	2831.24	II
5000	s	2832.26	II
100	l	2833.95	II
100	l	2861.92	II
100	s	2866.20	II
1000	l	2888.51	II
100	l	2893.29	II
200	l	2899.56	II
100	l	2909.86	II
200	l	2911.13	II
1000	s	2920.59	II
200	l	2927.53	II
200	l	2936.99	II
100	l	2939.08	II
500	l	2950.39	II
100	s	2957.05	II
100	l	2958.39	II
100	l	2963.02	II
1000	l	2966.71	II
1000	l	2969.29	II
1000	s	2987.24	II
500	l	2993.51	II
1000	s	3004.25	II

Americium (Cont.)

Intensity		Wavelength	
500	s	3027.99	II
100	l	3028.86	II
500	l	3038.36	II
200	s	3053.69	II
2000	s	3120.49	II
200	s	3161.83	II
100	s	3167.86	II
100	l	3203.26	II
500	s	3282.32	II
200	l	3286.67	II
100	l	3343.87	I
500	s	3362.55	II
200	l	3395.01	I
200	s	3419.66	II
200		3446.19	I
1000	l	3452.10	II
5000	l	3483.31	II
5000		3510.13	I
1000	l	3530.95	I
200	s	3562.68	II
5000		3569.16	I
100	s	3596.07	II
500		3603.41	I
5000		3673.12	I
100	l	3684.57	II
1000	s	3696.42	II
100	l	3707.86	II
5000	l	3777.50	II
5000	l	3926.25	II
1000	l	3952.58	II
100	s	4020.25	I
100	s	4035.81	I
500	s	4036.37	II
5000	s	4089.29	II
100	l	4089.32	II
100	s	4140.96	I
1000	s	4188.12	II
1000		4265.55	I
5000		4289.26	I
200	s	4309.65	II
2000	s	4324.57	II
2000	s	4441.36	II
5000	l	4509.45	II
5000	l	4575.59	II
1000	s	4593.31	II
100	l	4649.12	I
100	l	4653.45	I
5000	l	4662.79	I
2000	l	4681.65	I
2000	l	4699.70	II
1000	l	4706.80	I
2000	l	4872.22	II
200	l	4990.79	I
100	s	5000.21	I
1000	l	5020.96	II
200	l	5215.99	II
1000	s	5402.62	I
1000	s	5424.70	I
1000	l	5584.21	II
1000	s	5598.13	I
10000	l	6054.64	I
1000	l	6405.11	I
500	l	6544.16	I
500	s	6955.58	I

ANTIMONY (Sb)
Z = 51

Sb I and II
Ref. 167, 194 — L.J.R. and J.R.

Intensity		Wavelength	
		Vacuum	
1		691.20	II
1		764.43	II
1		814.85	II
1		849.39	II
2		855.08	II
4		876.84	II
4		921.07	II
6		983.57	II
6		1001.13	II
6		1009.43	II
6		1052.21	II
8		1056.27	II
8		1057.32	II
6		1073.81	II
6		1230.30	II
8		1274.98	II
8		1327.40	II
6		1358.04	II
8		1384.70	II
6		1407.83	II
10		1430.76	I
8		1436.49	II
10	h	1464.19	I
20	r	1486.57	I
40	h	1491.36	I
50	r	1512.57	I
120	r	1532.74	I
80	r	1535.06	I
6		1565.51	II
8		1576.11	II
7		1581.36	II
80	r	1599.96	I
10		1606.98	II
200	w	1612.8	I
100	w	1623.3	I
50	h	1651.20	I
20		1657.04	II
100	w	1662.6	I
50		1698.85	I
80	r	1716.93	I
150	r	1717.45	I
150	r	1723.43	I
100	r	1736.19	I
8		1736.43	II
50		1757.79	I
100	h	1765.76	I
100	r	1780.87	I
100	r	1788.24	I
150		1800.18	I
50	r	1810.50	I
80	r	1814.20	I
100		1829.50	I
60		1858.89	I
50	r	1868.17	I
300	r	1871.15	I
150	r	1882.56	I
70		1891.28	I
70		1899.39	I
100		1927.08	I
200	r	1950.39	I
80	h	1964.3	I
60		1986.05	I
6		1990.60	II
		Air	
50		2024.00	I
60	r	2029.49	I
70	r	2039.77	I
150	r	2049.57	I
50		2063.43	I
1000	r	2068.33	I
100		2079.56	I
50	r	2098.41	I
80	r	2118.48	I
100	r	2127.39	I
50	r	2137.05	I
100	r	2139.69	I
10		2141.80	II
50	r	2141.83	I
100	r	2144.86	I
50		2158.91	I
1500	r	2175.81	I
250	r	2179.19	I
6		2179.25	II
200	r	2201.32	I
300	r	2208.45	I
6		2208.50	II
150	r	2220.73	I
100		2221.98	I
120	r	2224.93	I
6		2225.15	II
300	r	2262.51	I
120		2288.98	I
150	r	2293.44	I
300	r	2306.46	I
2500	r	2311.47	I
150		2315.89	I
400	h	2373.67	I
300	h	2383.64	I
100		2395.22	I
150		2422.13	I
250		2426.35	I
400	r	2445.51	I
400		2478.32	I
150		2480.44	I
8		2480.46	II
100		2510.54	I
2000	r	2528.52	I
15		2528.54	II
10		2567.75	II
150		2574.06	I
1500	r	2598.05	I
500	r	2598.09	I
300	r	2612.31	I
200	r	2652.60	I
12		2656.55	II
300	r	2670.64	I
200	r	2682.76	I
120		2692.25	I
150	r	2718.90	I
400	r	2769.95	I
12		2851.09	II
100		2851.11	I
1000	r	2877.92	I
12		2966.10	II
15		2980.96	II
500	r	3029.83	I
12		3034.01	II
12		3040.67	II
600	r	3232.52	I
20		3241.28	II
700	r	3267.51	I
12		3383.09	II
100		3383.15	I
15		3498.46	II
12		3520.47	II
25		3637.80	II
250		3637.83	I
20		3722.78	II
200	r	3722.79	I
20		3850.22	II
200		4033.55	I
20		4033.56	II
20		4133.63	II
15		4140.54	II
15		4195.17	II
20		4219.07	II
20		4314.32	II
12		4344.83	II
12		4411.42	II
12		4446.48	II
12		4506.92	II
15		4514.50	II
30		4596.90	II
20		4599.09	II
15		4604.77	II
30		4647.32	II
20		4675.74	II
40		4711.26	II
12		4735.44	II
20		4757.81	II
20		4765.36	II
12		4766.91	II
30		4784.03	II
20		4802.01	II
20		4832.82	II
20		4877.24	II
15		4947.40	II
15		5044.56	II
12		5166.32	II
15		5176.55	II
20		5238.94	II
20		5354.24	II
15		5464.08	II
30	h	5490.32	I
40	h	5556.10	I
15		5568.13	II
30	l	5602.19	I
100	l	5632.02	I
30		5639.75	II
60	h	5830.34	I
15		5895.09	II
100		6005.21	II
20		6053.41	II
30		6079.80	II
50		6130.04	II
20		6154.94	II
12		6302.76	II
20		6611.49	I
30		6647.44	II
15		6688.01	II
6		6806.67	II
30	h	7648.28	I
80		7844.44	I
200		7924.65	I
40	h	7969.55	I
60		8411.69	I
150		8572.64	I
100		8619.55	I
30	h	8682.7	I
30		9132.21	I
400		9518.68	I
30		9866.78	I
400		9949.14	I
200		10078.49	I
300		10261.01	I
50		10364.33	I
50	h	10488.3	I
200		10585.60	I
1000		10677.41	I
800		10741.94	I
80		10794.11	I
600		10839.73	I
200		10868.58	I
400		10879.55	I
300		11012.79	I
40	h	11079.95	I
30	h	11084.98	I
50	h	11104.84	I
50	h	11108.52	I
30		11189.61	I
150		11266.23	I
30		11863.37	I
1		11957.7	I
5		12116.06	I
2		12276.6	I
2		12466.75	I

Sb III
Ref. 164 — L.J.R. and J.R.

Intensity	Wavelength	
	Vacuum	
10	691.18	III
10	698.69	III
15	722.86	III
8	724.81	III
15	732.33	III
15	999.62	III
40	1011.94	III
10	1056.58	III
40	1065.90	III
20	1069.93	III
20	1070.43	III
5	1073.76	III
30	1075.82	III
5	1078.10	III
20	1084.06	III
10	1098.34	III
10	1135.43	III
30	1151.49	III
40	1157.74	III
12	1166.96	III
50	1205.20	III
50	1210.64	III
20	1306.69	III
8	1379.58	III
20	1404.18	III
10	1429.57	III
15	1673.89	III
3	1710.23	III
15	1711.84	III
15	1725.33	III
12	1762.30	III
12	1839.32	III
10	1946.13	III
	Air	
3	2054.10	III
2	2091.85	III
5	2127.00	III
5	2507.71	III
15	2590.13	III
1	2614.20	III
12	2617.17	III
1	2617.63	III
20	2669.39	III
5	2785.87	III
20	2790.27	III
20	3336.61	III
50	3504.07	III
15	3519.06	III
15	3533.45	III
40	3559.18	III
40	3566.25	III
30	3738.90	III
40	4265.09	III
50	4352.16	III
30	4591.89	III
30	4692.91	III
1	5247.71	III
1	5690.8	III
1	5717.3	III
3	5845.5	III
5	6246.7	III
3	6287.6	III

Sb IV
Ref. 386 — L.J.R.

Intensity	Wavelength		
Vacuum			
861.5	SB	IV	1
873.5	SB	IV	1
888.3	SB	IV	1
891.2	SB	IV	1
1087.6	SB	IV	1
1099.3	SB	IV	1
1115.1	SB	IV	1
1120.4	SB	IV	1
1145.9	SB	IV	1
1151.5	SB	IV	1
1171.4	SB	IV	1
1192.9	SB	IV	1
1199.1	SB	IV	1
1499.2	SB	IV	1

Sb V
Ref. 406 — L.J.R.

Intensity	Wavelength	
	Vacuum	
3	699.22	V
1	746.06	V
6	831.00	V
1	898.02	V
8	1104.32	V
12	1226.00	V
12	1505.70	V
12	1524.47	V

ARGON (Ar)
Z = 18

Ar I and II
Ref. 190, 203, 204, 219, — E.F.W.

Vacuum

Intensity		Wavelength	
30		487.227	II
50		490.650	II
30		490.701	II
30		519.327	II
30		542.912	II
200		543.203	II
70		547.461	II
70		556.817	II
70		573.362	II
30		576.736	II
70		580.263	II
30		583.437	II
70		597.700	II
30		602.858	II
30		612.372	II
500		661.867	II
30		664.562	II
200		666.011	II
1000		670.946	II
3000		671.851	II
70		676.242	II
30		677.952	II
30		679.218	II
200		679.401	II
200		718.090	II
3000		723.361	II
500		725.548	II
70		730.930	II
200		740.269	II
200		744.925	II
70		745.322	II
20		802.859	I
100		806.471	I
60		806.869	I
30		807.218	I
40		807.653	I
50		809.927	I
120		816.232	I
70		816.464	I
80		820.124	I
120		825.346	I
120		826.365	I
150		834.392	I
100		835.002	I
100		842.805	I
180		866.800	I
150		869.754	I
180	r	876.058	I
180	r	879.947	I
150		894.310	I
1000		919.781	II
1000		932.054	II
1000	r	1048.220	I
500	r	1066.660	I

Air

Intensity	Wavelength	
5	2420.456	II
10	2516.789	II
10	2534.709	II
15	2562.087	II
25	2891.612	II
200	2942.893	II
100	2979.050	II
50	3033.508	II
50	3093.402	II
8	3200.37	I
20	3243.689	II
25	3293.640	II
20	3307.228	II
7	3319.34	I
25	3350.924	II
7	3373.47	I
25	3376.436	II
25	3388.531	II
7	3393.73	I
7	3461.07	I
70	3476.747	II
20	3478.232	II
50	3491.244	II
100	3491.536	II
70	3509.778	II
70	3514.388	II
70	3545.596	II
70	3545.845	II
7	3554.306	I
100	3559.508	II
100	3561.030	II
70	3576.616	II
25	3581.608	II
50	3582.355	II
70	3588.441	II
7	3606.522	I
25	3622.138	II
20	3639.833	II
35	3718.206	II
70	3729.309	II
50	3737.889	II
150	3765.270	II
50	3766.119	II
20	3770.369	I
20	3770.520	II
25	3780.840	II
25	3803.172	II
50	3809.456	II
7	3834.679	I
70	3850.581	II
35	3868.528	II
35	3925.719	II
50	3928.623	II
25	3932.547	II
70	3946.097	II
7	3947.505	I
35	3948.979	I
20	3979.356	II
35	3994.792	II
50	4013.857	II
50	4033.809	II
20	4035.460	II
150	4042.894	II
50	4044.418	I
100	4052.921	II
200	4072.005	II
70	4072.385	II
25	4076.628	II
35	4079.574	II
25	4082.387	II
150	4103.912	II
300	4131.724	II
35	4156.086	II
400	4158.590	I
50	4164.180	I
35	4179.297	II
50	4181.884	I
100	4190.713	I
50	4191.029	I
200	4198.317	I
400	4200.674	I
25	4218.665	II
25	4222.637	II
25	4226.988	II
100	4228.158	II
100	4237.220	II
25	4251.185	I
200	4259.362	I
100	4266.286	I
70	4266.527	II
150	4272.169	I
550	4277.528	II
20	4282.898	II
100	4300.101	I
25	4300.650	II
70	4309.239	II
200	4331.200	II
50	4332.030	II
100	4333.561	I
50	4335.338	I
25	4345.168	I
800	4348.064	II
50	4352.205	II
25	4362.066	II
50	4367.832	II
200	4370.753	II
70	4371.329	II
50	4375.954	II
150	4379.667	II
50	4385.057	II
70	4400.097	II
200	4400.986	II
400	4426.001	II
150	4430.189	II
50	4430.996	II
50	4433.838	II
20	4439.461	II
35	4448.879	II
100	4474.759	II
200	4481.811	II
100	4510.733	I
20	4522.323	I
20	4530.552	II
400	4545.052	II
20	4564.405	II
400	4579.350	II
400	4589.898	II
15	4596.097	I
550	4609.567	II
7	4628.441	I
35	4637.233	II
400	4657.901	II
15	4702.316	I
20	4721.591	II
550	4726.868	II
50	4732.053	II
300	4735.906	II
800	4764.865	II
550	4806.020	II
150	4847.810	II
50	4865.910	II
800	4879.864	II
70	4889.042	II
20	4904.752	II
35	4933.209	II
200	4965.080	II
50	5009.334	II
70	5017.163	II
70	5062.037	II
20	5090.495	II
100	5141.783	II
70	5145.308	II
5	5151.391	I
15	5162.285	I
25	5165.773	II
20	5187.746	I
20	5216.814	II
7	5221.271	I
5	5421.352	I
10	5451.652	I
25	5495.874	I
5	5506.113	I
25	5558.702	I
10	5572.541	I
35	5606.733	I
20	5650.704	I
10	5739.520	I
5	5834.263	I
10	5860.310	I
15	5882.624	I
25	5888.584	I
50	5912.085	I
15	5928.813	I
5	5942.669	I
7	5987.302	I
5	5998.999	I
5	6025.150	I
70	6032.127	I
35	6043.223	I
10	6052.723	I
20	6059.372	I
7	6098.803	I
10	6105.635	I
100	6114.923	II
10	6145.441	I
7	6170.174	I
150	6172.278	II
10	6173.096	I
10	6212.503	I
5	6215.938	I
25	6243.120	II
7	6296.872	I
15	6307.657	I
7	6369.575	I
20	6384.717	I
70	6416.307	I
25	6483.082	II
15	6538.112	I
15	6604.853	I
25	6638.221	II
20	6639.740	II
50	6643.698	II
5	6660.676	I
5	6664.051	I
25	6666.359	II
100	6677.282	I
35	6684.293	II
150	6752.834	I
5	6756.163	I
15	6766.612	I
20	6861.269	II
150	6871.289	I
5	6879.582	I
10	6888.174	I
50	6937.664	I
7	6951.478	I
7	6960.250	I
10000	6965.431	I
150	7030.251	I
10000	7067.218	I
100	7068.736	I
25	7107.478	I
25	7125.820	I
1000	7147.042	I
15	7158.839	I
70	7206.980	I
15	7265.172	I
7	7270.664	I
2000	7272.936	I
35	7311.716	I
25	7316.005	I
5	7350.814	I
70	7353.293	I
200	7372.118	I
20	7380.426	II
10000	7383.980	I
20	7392.980	I
15	7412.337	I
10	7425.294	I
25	7435.368	I
10	7436.297	I
20000	7503.869	I
15000	7514.652	I
25000	7635.106	I
15000	7723.761	I
10000	7724.207	I
10	7891.075	I
20000	7948.176	I
20000	8006.157	I
25000	8014.786	I
7	8053.308	I
20000	8103.693	I
35000	8115.311	I
10000	8264.522	I
20	8392.27	I
15000	8408.210	I
20000	8424.648	I
15000	8521.442	I
7	8605.776	I
4500	8667.944	I
20	8771.860	II
180	8849.91	I
20	9075.394	I
35000	9122.967	I
550	9194.638	I
15000	9224.499	I
400	9291.531	I
1600	9354.220	I
25000	9657.786	I
4500	9784.503	I
180	10052.06	I
30	10332.72	I
100	10467.177	II
1600	10470.054	I
13	10478.034	I
180	10506.50	I
200	10673.565	I
11	10681.773	I
7	10683.034	II
30	10733.87	I
30	10759.16	I
7	10812.896	II
11	11078.869	I
30	11106.46	I
12	11441.832	I
400	11488.109	I
200	11668.710	I
12	11719.488	I
200	12112.326	I
50	12139.738	I
50	12343.393	I
200	12402.827	I
200	12439.321	I
100	12456.12	I
200	12487.663	I
150	12702.281	I
30	12733.418	I
12	12746.232	I
200	12802.739	I
50	12933.195	I
500	12956.659	I
200	13008.264	I
200	13213.99	I
200	13228.107	I
100	13230.90	I
500	13272.64	I
1000	13313.210	I
1000	13367.111	I
30	13499.41	I
1000	13504.191	I
11	13573.617	I
30	13599.333	I
400	13622.659	I
200	13678.550	I
1000	13718.577	I
10	13825.715	I
10	13907.478	I
200	14093.640	I
100	15046.50	I
25	15172.69	I
10	15329.34	I
30	15989.49	I
30	16519.86	I
500	16940.58	I
12	18427.76	I
50	20616.23	I
30	20986.11	I
20	23133.20	I
20	23966.52	I

Ar III
Ref. 367, 372, 373, 375 — E.F.W.

Vacuum

Intensity	Wavelength	
12	769.15	III

Argon (Cont.)

Intensity	Wavelength	
10	871.10	III
9	875.53	III
12	878.73	III
8	879.62	III
9	883.18	III
10	887.40	III
7	1669.67	III
7	1673.42	III
7	1675.48	III
9	1914.40	III
7	1915.56	III

Air

Intensity	Wavelength	
10	2125.16	III
15	2133.87	III
10	2138.59	III
10	2148.73	III
15	2166.19	III
10	2168.26	III
20	2170.23	III
25	2177.22	III
8	2184.06	III
10	2188.22	III
15	2192.06	III
7	2248.73	III
10	2279.10	III
7	2281.22	III
7	2282.21	III
12	2293.03	III
10	2300.85	III
15	2302.17	III
9	2317.00	III
15	2317.47	III
12	2318.04	III
10	2319.13	III
10	2319.37	III
9	2345.17	III
7	2351.67	III
9	2360.26	III
10	2395.63	III
12	2399.15	III
10	2413.20	III
7	2415.61	III
10	2418.82	III
12	2423.52	III
12	2423.93	III
7	2443.69	III
8	2472.95	III
7	2476.10	III
12	2488.86	III
7	2631.90	III
10	2654.63	III
8	2674.02	III
9	2678.38	III
10	2724.84	III
7	2762.23	III
7	2842.88	III
8	2855.29	III
9	2884.12	III
10	3010.02	III
12	3024.05	III
12	3054.82	III
10	3064.77	III
10	3078.15	III
7	3110.41	III
7	3127.90	III
25	3285.85	III
20	3301.88	III
15	3311.25	III
7	3323.59	III
25	3336.13	III
20	3344.72	III
15	3358.49	III
7	3361.28	III
15	3391.85	III
7	3417.49	III
9	3424.25	III
8	3438.04	III
9	3471.32	III
20	3480.55	III
12	3499.67	III
15	3503.58	III
8	3511.12	III
20	3795.37	III
10	3858.32	III
7	3907.84	III
8	3960.53	III
6	4023.60	III
5	4146.70	III

Ar IV
Ref. 367, 368, 374 — E.F.W.

Vacuum

Intensity		Wavelength	
4		396.87	IV
4		398.55	IV
6		623.77	IV
10		683.28	IV
7		688.39	IV
12	p	689.01	IV
6		699.41	IV
8		700.28	IV
4		754.20	IV
5		761.47	IV
5		800.57	IV
10		801.09	IV
10		801.41	IV
5		801.91	IV
15		840.03	IV
20		843.77	IV
25		850.60	IV
5		900.36	IV
9		901.17	IV

Air

Intensity	Wavelength	
4	2299.72	IV
8	2447.71	IV
12	2513.28	IV
6	2518.40	IV
9	2525.69	IV
12	2562.17	IV
10	2568.07	IV
7	2569.53	IV
12	2599.47	IV
10	2608.06	IV
7	2608.44	IV
12	2615.68	IV
6	2619.98	IV
12	2621.36	IV
12	2624.92	IV
15	2640.34	IV
9	2682.63	IV
14	2757.92	IV
10	2776.26	IV
12	2784.47	IV
14	2788.96	IV
7	2797.11	IV
16	2809.44	IV
10	2830.25	IV
6	2874.40	IV
12	2913.00	IV
11	2926.33	IV
6	3037.98	IV
8	3077.40	IV

Ar V
Ref. 414, 421 — E.F.W.

Vacuum

Intensity		Wavelength	
3		336.56	V
3		337.56	V
6		338.00	V
2		338.43	V
2		339.01	V
3		339.89	V
3		350.88	V
2		436.67	V
5		446.00	V
8		446.95	V
4		447.53	V
18		449.06	V
4		449.49	V
3		458.12	V
2		458.98	V
6	p	461.23	V
3		462.42	V
7		463.94	V
3		522.09	V
5		524.19	V
6		527.69	V
2		554.50	V
5		558.48	V
3		635.12	V
3		705.35	V
5		709.20	V
4		715.60	V
3		715.65	V
2		725.11	V
4		822.16	V
5		827.05	V
3		827.35	V
4	p	834.88	V
2		836.13	V

ARSENIC (As)
Z = 33

As I and II
Ref. 168, 197 — R.L.K.

Vacuum

Intensity		Wavelength	
165		761.24	II
165		802.83	II
340		1021.96	II
340		1082.35	II
500		1139.40	II
615		1149.31	II
555		1181.51	II
555		1189.87	II
615		1196.38	II
615		1196.56	II
340		1207.44	II
800		1211.17	II
800		1218.10	II
340		1223.15	II
760		1241.31	II
965		1243.08	II
870		1245.67	II
800		1258.58	II
965		1263.77	II
800		1266.34	II
800		1267.59	II
715		1280.99	II
715		1287.54	II
715		1305.70	II
340		1307.74	II
760		1333.15	II
965		1341.55	II
760		1355.93	II
965		1369.77	II
800		1373.65	II
1000		1375.07	II
760		1375.78	II
800		1394.64	II
800		1400.31	II
500		1448.59	II
500		1558.88	II
500		1570.99	II
100	r	1593.60	I
500		1660.55	II
100		1758.60	I
170		1806.15	I
340		1860.34	II
1000	r	1890.42	I
500		1912.94	II
800	r	1937.59	I
585	r	1972.62	I
170	r	1990.35	I
100	r	1991.13	I
100	r	1995.43	I

Air

Intensity		Wavelength	
230	r	2003.34	I
100	r	2009.19	I
100		2013.32	I
100		2112.99	I
100		2144.08	I
135		2165.52	I
350	r	2288.12	I
350	r	2349.84	I
100	r	2370.77	I
135	r	2381.18	I
170	r	2456.53	I
340		2602.00	II
170	r	2780.22	I
300		2830.359	II
300		2831.164	II
100	r	2860.44	I
300		2884.406	II
615		2959.572	II
300		3003.819	II
300		3116.516	II
340		3842.60	II
715		4190.082	II
615		4197.40	II
615		4242.982	II
500		4315.657	II
500		4323.867	II
500		4336.64	II
500		4352.145	II
425		4352.864	II
375		4371.17	II
615		4427.106	II
615		4431.562	II
715		4458.469	II
340		4461.075	II
715		4466.348	II
500		4474.46	II
800		4494.230	II
850		4507.659	II
615		4539.74	II
715		4543.483	II
615		4602.427	II
340		4629.787	II
340		4707.586	II
340		4730.67	II
340		4888.557	II
100		5068.98	I
340		5105.58	II
500		5107.55	II
100		5121.34	I
100		5141.63	I
425		5231.38	II
500		5331.23	II
100		5408.13	I
135		5451.32	I
340		5497.727	II
425		5558.09	II
425		5651.32	II
425		6110.07	II
500		6170.27	II
300		6511.74	II
300		7092.27	II
300		7102.72	II
340		7990.53	II
300		8174.51	II
100		8428.91	I
100		8564.71	I
100		8654.14	I
135		8821.73	I
100		8869.66	I
135		9267.28	I
200		9300.61	I
230		9597.95	I
290		9626.70	I
230		9833.76	I
100		9886.05	I
140		9900.55	I
170		9915.71	I
290		9923.05	I
100		10010.63	I
290		10024.04	I
100		10453.09	I
100		10575.02	I
170		10614.07	I

As III
Ref. 163 — R.L.K.

Vacuum

Intensity	Wavelength	
185	849.9	III
185	866.3	III
510	871.7	III
325	889.0	III
325	927.5	III
325	953.6	III
325	937.2	III
325	963.8	III
120	1172.2	III
185	1209.3	III

Air

Intensity	Wavelength	
80	2926.3	III
185	2982.0	III
325	3922.6	III
185	4037.2	III

As IV
Ref. 244 — R.L.K.

Air

Intensity	Wavelength	
150	2253.1	IV
200	2263.2	IV
200	2301.0	IV
250	2417.5	IV
150	2446.1	IV
250	2454.0	IV
200	2461.4	IV
150	3108.8	IV

As V
Ref. 280 — R.L.K.

Vacuum

Intensity	Wavelength	
25	600.7	V
40	616.0	V
120	715.5	V
150	734.8	V
60	737.2	V
250	987.7	V
250	1029.5	V
40	1051.6	V
60	1056.6	V

ASTATINE (At)
Z = 85

At I
Ref. 188 — E.F.W.

Astatine (Cont.)

Intensity		Wavelength	
		Air	
8		2162.25	I
10		2244.01	I

BARIUM (Ba)
Z = 56

Ba I and II
Ref. 1, 252, 277, 279 — J.J.W.

Intensity		Wavelength	
		Vacuum	
200		1486.72	II
400		1504.01	II
300		1554.38	II
200		1572.73	II
		1573.92	II
		1630.40	II
100		1674.51	II
400		1694.37	II
		1697.16	II
		1761.75	II
		1771.03	II
		1786.93	II
100		1904.15	II
500		1924.70	II
		1985.60	II
300		1999.54	II
		Air	
		2009.20	II
400		2023.95	II
		2052.68	II
		2054.57	II
500		2214.7	II
800		2245.61	II
1000		2254.73	II
1400		2304.24	II
2000		2335.27	II
190		2347.58	II
60		2528.51	II
8	h	2596.64	I
100		2634.78	II
8		2702.63	I
18		2771.36	II
15		2785.28	I
100	r	3071.58	I
10	h	3108.21	I
8		3132.60	I
8	h	3135.72	I
10		3137.70	I
10		3155.34	I
10		3155.67	I
12		3158.05	I
12	h	3158.54	I
25		3165.60	I
15	h	3173.69	I
30		3183.16	I
15		3183.96	I
10		3193.91	I
25	h	3203.70	I
30		3221.63	I
40		3222.19	I
50		3261.96	I
60	r	3262.34	I
40		3281.50	I
15		3281.77	I
50		3322.80	I
80	h	3356.80	I
60	r	3377.08	I
20		3377.39	I
70	r	3420.32	I
25		3421.01	I
30	h	3421.48	I
40		3463.74	I
200	r	3501.11	I
80	h	3524.97	I
30	h	3531.35	I
80	h	3544.66	I
20	h	3547.68	I
100		3552.45	II
200		3567.73	II
100		3576.28	II
30		3577.62	I
80	h	3579.67	I
200		3596.57	II
40		3630.64	I
40	h	3636.83	I
20	h	3688.47	I
400		3735.75	II
200		3816.69	II
200		3842.80	II
100		3854.76	II
20		3889.33	I
1400	l	3891.78	II
20		3892.65	I
40		3909.91	I
500		3914.73	II
50		3935.72	I
20		3937.87	I
200		3939.67	II
500		3949.51	II
80		3993.40	I
30		3995.66	I
300		4036.26	II
200		4083.77	II
30	h	4084.86	I
1500	h	4130.66	II
20		4132.43	I
200		4166.00	II
500		4216.04	II
800		4267.95	II
100		4283.10	I
300		4287.80	II
200		4297.60	II
800		4309.32	II
20	h	4323.00	I
600		4325.73	II
200		4326.74	II
300		4329.62	II
80		4350.33	I
60		4402.54	I
400		4405.23	II
60	h	4488.98	I
40		4431.89	I
50	h	4493.64	I
40		4505.92	I
200		4509.63	II
60	h	4523.17	I
130		4524.93	II
65000		4554.03	II
40		4573.85	I
80		4579.64	I
30		4599.75	I
20	h	4619.92	I
25	h	4628.33	I
300		4644.10	II
30		4673.62	I
35		4691.62	I
20		4700.43	I
800		4708.94	II
40		4726.44	I
800		4843.46	II
300		4847.14	II
200		4850.84	II
30	h	4877.65	I
400		4899.97	II
15		4902.90	I
20000		4934.09	II
8		4947.35	I
1000		4957.15	II
300		4997.81	II
1000		5013.00	II
20	h	5159.94	I
20		5267.03	I
800		5361.35	II
1000		5391.60	II
200		5421.05	II
100		5424.55	I
200		5428.79	II
300		5480.30	II
200		5519.05	I
1000	r	5535.48	I
20	h	5620.40	I
10		5680.18	I
400		5777.62	I
800		5784.18	II
100		5800.23	I
20		5805.69	I
150		5826.28	I
2800		5853.68	II
15		5907.64	I
100		5971.70	I
800		5981.25	II
100		5997.09	I
300		5999.85	II
100		6019.47	I
200		6063.12	I
300		6110.78	I
400		6135.83	II
20000		6141.72	II
150		6341.68	I
500		6378.91	II
90		6450.85	I
150		6482.91	I
12000		6496.90	II
300		6498.76	I
150		6527.31	I
3000		6595.33	I
150		6654.10	I
1500		6675.27	I
1800		6693.84	I
1000		6769.62	II
600		6865.69	I
300	h	6867.85	I
1000		6874.09	II
6000		7059.94	I
2400	hS	7120.33	I
600		7195.24	I
600	hL	7228.84	I
3000		7280.30	I
1200		7392.41	I
300		7417.53	I
900	hL	7459.78	I
600		7488.08	I
450	hL	7636.90	I
600	hL	7642.91	I
1800		7672.09	I
1200		7780.48	I
180	h	7839.57	I
1500		7905.75	I
600		7911.34	I
900	h	8210.24	I
1800	h	8559.97	I
100		8710.74	II
100		8737.71	II
300	h	8799.76	I
300		8860.98	I
450		8914.99	I
300		9219.69	I
300		9308.08	I
300	h	9324.58	I
1500		9370.06	I
300		9455.92	I
450		9589.37	I
900		9608.88	I
300	h	9645.72	I
1500	hL	9830.37	I
900		10001.08	I
600		10032.10	I
1200	h	10233.23	I
300		10471.26	I
120	hL	10791.25	I
180	hL	11012.69	I
150	h	11114.42	I
240		11303.04	I
120	h	11697.45	I
120		13207.30	I
120		13810.50	I
120		14077.90	I
120		15000.40	I
120		20712.00	I
150		25515.70	I
150		29223.90	I

Ba III
Ref. III — J.J.W.

Intensity		Wavelength	
		Vacuum	
5		403.82	III
2		407.12	III
7		420.12	III
4		423.84	III
9		448.95	III
8		456.96	III
14		555.48	III
14		587.57	III
18		647.27	III
9		653.36	III
15		743.12	III
12		1097.41	III
15		1113.67	III
11		1116.01	III
14		1133.05	III
12		1151.76	III
12		1170.62	III
13		1207.29	III
11		1218.92	III
12		1224.55	III
12		1288.53	III
11		1299.18	III
11		1307.40	III
12		1308.87	III
12		1315.72	III
12		1334.01	III
11		1354.71	III
11		1369.53	III
11		1416.61	III
12		1478.85	III
12		1510.68	III
12		1514.22	III
12		1565.61	III
12		1566.12	III
12		1574.55	III
12		1596.80	III
12		1610.95	III
12		1615.78	III
12		1711.53	III
12		1861.74	III
12		1883.92	III
11		1974.76	III
		Air	
10		2001.30	III
15		2008.40	III
13		2022.45	III
10		2038.84	III
12		2070.43	III
12		2071.68	III
10		2076.00	III
12		2081.35	III
10		2134.87	III
16		2156.37	III
10		2160.76	III
20		2230.33	III
30		2280.68	III
35		2323.51	III
60		2331.10	III
25		2476.73	III
25		2505.07	III
40		2512.28	III
40		2523.83	III
25		2530.92	III
50		2559.54	III
25		2570.48	III
40		2681.89	III
30		2745.78	III
25		2938.95	III
25		2960.05	III
30		2962.48	III
20		3014.22	III
30		3043.42	III
40		3079.14	III
30		3103.92	III
30		3119.22	III
30		3152.70	III
25		3195.17	III
25		3235.04	III
25		3281.65	III
20		3286.79	III
50		3368.18	III
30		3369.68	III
25		3649.18	III
25		3926.85	III
25		3993.06	III
18	p	4053.71	III
15		4697.44	III
10		5049.55	III
10		5097.54	III
12	p	5102.25	III
10		5134.54	III
10		5998.00	III
13		6101.99	III
10		6377.11	III
10		6383.76	III
8		6526.17	III
8		7095.49	III
8		8308.69	III
8		9521.76	III

Ba IV
Ref. 78 — J.J.W.

Intensity		Wavelength	
		Vacuum	
40000		794.89	IV
50000		923.74	IV

Ba V

Ref. 259—J.R.

Intensity		Wavelength	
		Vacuum	
15		612.55	V
5		658.11	V
5		681.09	V
300		719.86	V
150		721.85	V
100		760.45	V
1000		766.87	V
100		783.61	V
100		816.41	V
40		875.69	V
300		877.41	V
15		892.28	V
200		946.26	V

BERKELIUM (Bk)
Z = 97

Bk I and II
Ref.53, 339—J.G.C.

Intensity		Wavelength	
		Air	
10000	s	2748.02	II
10000	s	2827.57	II
10000	l	2872.11	II
10000	s	2878.57	II
10000	l	2884.77	II
10000	s	2889.80	II
10000	s	2893.66	II
10000	l	2910.65	II
10000	s	2926.49	
10000	s	2927.91	
10000	l	2941.71	II
10000	l	2951.76	II
10000	l	2969.13	II
10000	l	2987.76	II
10000	l	3178.47	II
10000		3239.72	I
10000	s	3247.26	II
10000		3252.19	I
10000	s	3263.47	II
10000	l	3288.75	I

Berkelium (Cont.)

10000		3289.35	I
10000	s	3302.35	II
10000		3335.26	I
10000	s	3387.45	II
10000		3408.28	I
10000	l	3412.13	II
10000		3426.95	I
10000	l	3432.62	I
10000	l	3437.47	I
10000		3442.66	I
10000	s	3461.24	II
10000	l	3453.90	I
10000	s	3464.13	II
10000	s	3472.02	II
10000	s	3477.62	II
10000		3528.72	I
10000	l	3531.40	I
10000	l	3535.73	I
10000	s	3542.19	II
10000		3553.60	I
10000	s	3555.88	I
10000		3556.52	I
10000		3565.41	
10000	l	3567.25	II
10000		3590.32	I
10000		3595.88	I
10000		3601.12	I
10000	s	3603.20	II
10000		3604.78	
10000	l	3608.49	
10000		3609.61	
10000		3611.03	
10000	l	3611.93	
10000		3613.91	
10000		3616.62	
10000		3619.37	
10000	s	3621.81	
10000		3627.61	I
10000		3633.28	
10000	l	3637.05	I
10000		3640.26	I
10000	s	3640.93	II
10000	l	3675.59	I
10000	s	3681.22	II
10000	l	3684.43	I
10000	s	3685.21	I
10000	l	3686.74	I
10000		3692.73	I
10000		3695.37	I
10000		3703.28	I
10000	s	3704.02	I
10000		3705.26	I
10000	s	3711.14	II
10000		3712.93	I
10000		3725.39	I
10000		3739.92	I
10000		3743.05	I
10000		3745.40	I
10000		3750.08	I
10000		3751.91	I
10000		3757.35	I
10000		3757.85	I
10000	s	3771.06	II
10000		3780.72	I
10000		3781.17	I
10000		3785.38	I
10000		3788.21	I
10000		3791.42	I
10000		3796.21	I
10000		3797.12	
10000		3798.63	I
10000		3801.08	II
10000	s	3802.35	I
10000		3802.47	I
10000		3815.29	I
10000	s	3823.10	II
10000	s	3824.08	II
10000		3825.19	I
10000	s	3825.84	II
10000		3827.41	I
10000		3830.55	I
10000	l	3831.57	II
10000	s	3833.48	
10000	s	3835.97	II
10000		3842.19	I
10000	l	3846.62	I
10000		3847.63	I
10000		3855.03	I
10000	l	3859.89	II
10000	l	3877.94	II
10000		3880.11	I
10000	l	3882.60	I
10000	l	3894.55	II
10000	s	3906.09	II
10000	l	3912.16	II
10000	l	3916.37	II
10000		3921.42	I
10000	l	3928.05	II
10000	l	4147.13	II
10000	s	4189.69	II
10000	s	4197.44	II
10000	l	4329.58	I
10000		4351.50	I
10000		4363.64	I
10000		4423.01	I
10000		4466.46	I
10000		4685.70	
10000	l	4765.40	
10000	s	5056.73	I
10000	l	5118.24	I
10000	s	5135.53	II
10000		5170.61	I
10000		5197.55	I
10000		5212.53	I
10000		5271.95	I
10000		5392.03	I
10000		5394.24	I
10000	l	5404.62	I
10000		5449.63	I
10000	l	5467.47	I
10000	s	5484.58	I
10000		5512.22	II
10000	l	5537.93	I
10000	l	5556.80	I
10000	s	5557.09	I
10000		5581.21	I
10000	l	5656.54	I
10000		5659.03	I
10000		5702.24	I
10000		5910.71	I
10000	l	7040.85	I
10000		7107.85	I
10000	l	7176.22	I
10000		7249.26	
10000		7252.50	I
10000	l	7257.21	I
10000		7306.94	I
10000		7394.26	
10000	l	7511.26	I
10000		7551.12	I
10000		7579.77	I
10000	l	7729.93	I
10000	s	7903.90	I
10000	s	9319.30	I
10000	s	9429.13	I
10000	l	9801.18	I
10000	l	9862.39	II
10000	l	9879.29	I
10000	l	9892.38	I
10000		10126.20	I
10000	l	10186.58	I
10000		10292.44	I
10000	l	10527.71	I
10000	l	10570.53	I
10000	l	11293.14	I
10000	l	11500.30	I
10000	s	11575.34	I
10000	s	11793.09	I
10000	s	12159.05	I
10000		13061.13	I
10000		13498.36	I
10000	s	14196.93	I
10000		15136.10	I
10000		18352.31	
10000	l	19273.87	I
10000	s	19653.22	I
10000	s	23902.85	I
10000	s	24192.62	I

BERYLLIUM (Be)
Z = 4

Be I and II
Ref. 15, 44, 115, 134, 135, 198, 335 — J.O.S.

Intensity		Wavelength	
		Vacuum	
		82.58	II
		83.66	II
		89.16	I
		89.80	II
		90.04	II
		90.21	I
		90.67	I
		91.06	II
		91.36	II
		91.74	II
		92.19	I
		92.61	II
		93.14	II
		93.42	II
		93.93	II
		94.78	II
		95.76	II
		96.29	I
		97.24	I
		97.44	I
		97.86	I
		97.97	I
		98.12	I
		98.37	I
		98.66	I
		98.94	I
		99.19	I
		100.86	I
		101.20	I
		102.13	I
		102.49	II
		104.40	II
		104.67	I
		105.80	I
		107.26	I
		107.38	I
		714.0	II
		725.71	II
		743.58	II
8		775.37	II
20		842.06	II
		865.3	II
2		925.25	II
10		943.56	II
10		973.27	II
		981.4	II
		1020.1	II
8		1026.93	II
5		1036.32	II
15		1048.23	II
20		1143.03	II
		1155.9	II
60		1197.19	II
		1426.12	I
		1491.76	I
20		1512.30	II
60		1512.43	II
100		1661.49	I
15		1776.12	II
20		1776.34	II
		1907.	I
		1909.0	II
		1912.	I
		1919.	I
5		1929.67	I
10		1943.68	I
		1956.	I
50		1964.59	I
5		1985.13	I
		1997.95	I
		1997.98	I
60		1998.01	I
		Air	
		2033.25	I
		2033.28	I
		2033.38	I
50		2055.90	I
100		2056.01	I
10		2125.57	I
20		2125.68	I
25		2145.	I
55		2174.99	I
55		2175.10	I
		2273.5	II
		2324.6	II
		2337.0	I
950		2348.61	I
20		2350.66	I
60		2350.71	I
200		2350.83	I
2		2413.34	II
16		2413.46	II
20		2453.84	II
		2480.6	I
35		2494.54	I
35		2494.58	I
100		2494.73	I
16		2507.43	II
5		2617.99	II
20		2618.13	II
100		2650.45	I
60		2650.55	I
200		2650.62	I
60		2650.69	I
100		2650.76	I
5		2697.46	II
20		2697.58	II
20		2728.88	II
30		2738.05	I
		2764.2	II
20		2898.13	I
10		2898.19	I
20		2898.25	I
30		2986.06	I
10		2986.42	I
60		3019.33	I
30		3019.49	I
30		3019.53	I
20		3019.60	I
10		3046.52	II
30		3046.69	II
		3090.3	I
10		3110.81	I
10		3110.92	I
20		3110.99	I
		3120.	I
480		3130.42	II
320		3131.07	II
		3136.	I
		3150.	I
		3160.6	I
		3163.	I
		3168.	I
		3180.7	II
		3187.	I
20		3193.81	I
20		3197.10	II
30		3197.15	II
20		3208.60	I
		3220.	I
60		3229.63	I
2		3233.52	II
10		3241.62	II
30		3241.83	II
15		3269.02	I
100		3274.58	II
30		3274.67	II
30		3282.91	I
30		3321.01	I
30		3321.09	I
220		3321.34	I
20		3345.43	I
60		3367.63	I
		3405.6	II
5		3451.37	I
300		3455.18	I
20		3476.56	I
300		3515.54	I
10		3555.	I
100		3736.30	I
700		3813.45	I
40		3865.13	I
80		3865.42	I
1		3865.51	I
6		3865.72	I
100		3866.03	I
100		4253.05	I
60		4253.76	I
300		4360.66	II
500		4360.99	II
400		4407.94	I
		4526.6	I
		4548.	I
12		4572.66	I
700		4673.33	II
1000		4673.42	II
6		4709.37	I
200		4828.16	II
40		4849.16	I
2	h	4858.22	II
80		5087.75	I
8		5218.12	II
20		5218.33	II
3		5255.86	II
64		5270.28	II
500		5270.81	II
20		5403.04	II
20		5410.21	II
		5558.	I
10		6229.11	I
16		6279.43	II
30		6279.73	II
30		6473.54	I
60		6547.89	II
60		6558.36	II
30		6564.52	I
2	h	6636.44	II
1		6756.72	II
2		6757.13	II
30		6786.56	I
1	h	6884.22	I
6	h	6884.44	I
100		6982.75	I
6	h	7154.40	I
40	h	7154.65	I
100		7209.13	I
3		7401.20	II
2		7401.43	II
10		7551.90	I
10	h	7618.68	I
20	h	7618.88	I
60		8090.06	I
5	h	8158.99	I
10	h	8159.24	I
4		8254.07	I
10	h	8287.07	I
30		8547.36	I
60		8547.67	I
300		8801.37	I
6		8882.18	I
40		9190.45	I
20	h	9243.92	I
1	h	9343.89	II
40		9392.74	I
2		9476.43	II
16		9477.03	II
20		9847.32	I
10	h	9895.63	I
20	h	9895.96	I
80		9939.78	I
16		10095.52	II

Beryllium (Cont.)

Intensity		Wavelength	
20		10095.73	II
60		10119.92	II
80		10331.03	I
30		11066.46	I
		11173.	II
1		11173.73	II
120		11496.39	I
2	h	11625.16	II
		11659.	II
2		11660.25	II
100		12095.36	II
30		12098.18	II
100		14643.92	I
60		14644.75	I
200		16157.72	I
80		17855.38	I
120		17856.63	I
100		18143.54	I
160		31775.05	I
200		31778.70	I

Be III
Ref. 73, 102, 175 — J.O.S.

Intensity		Wavelength	
		Vacuum	
1	h	76.10	III
2		76.48	III
3		78.53	III
4		78.66	III
1	h	78.92	III
5		81.89	III
10		82.38	III
20		83.20	III
30		84.76	III
50		88.31	III
100		100.25	III
3		509.99	III
2		549.31	III
6		582.08	III
4		661.32	III
8		675.59	III
4		725.59	III
7		746.23	III
2		767.75	III
1		1114.69	III
2		1213.12	III
1		1214.32	III
2		1362.25	III
1		1401.52	III
10		1421.26	III
5		1422.86	III
1		1435.17	III
2		1440.77	III
2	h	1754.69	III
3		1917.03	III
60	h	1954.97	III
		Air	
75	h	2076.94	III
60	h	2080.38	III
25		2118.56	III
15	h	2122.27	III
15	h	2127.20	III
5		2137.25	III
5		2191.57	III
100		3720.36	III
		3720.92	III
		3722.98	III
90	h	4249.14	III
2		4485.52	III
100	h	4487.30	III
1		4495.09	III
140	h	4497.8	III
140	h	6142.01	III

Be IV
Ref. 171 — J.O.S.

Intensity		Wavelength	
		Vacuum	
		58.13	IV
		58.57	IV
		59.32	IV
		60.74	IV
		64.06	IV
		75.93	IV

BISMUTH (Bi)
Z = 83

Bi I and II
Ref. 1, 357-359 — C.H.C.

Intensity		Wavelength	
		Vacuum	
15		1058.88	II
20		1085.47	II
10		1099.20	II
8		1163.19	II
8		1167.06	II
10		1225.43	II
15		1232.78	II
10		1241.05	II
10		1265.35	II
15		1283.73	II
10		1306.18	II
20		1325.46	II
20		1329.47	II
20		1350.07	II
25		1372.61	II
15		1376.02	II
20		1393.92	II
45		1436.83	II
25		1447.94	II
50		1455.11	II
25		1462.14	II
35		1486.93	II
20		1502.50	II
40		1520.57	II
40		1533.17	II
30		1536.77	II
35		1538.06	II
20		1563.67	II
40		1573.70	II
60		1591.79	II
25		1601.58	II
40		1609.70	II
40		1611.38	II
20		1652.81	II
20		1749.29	II
80		1777.11	II
60		1787.47	II
70		1791.93	II
70		1823.80	II
100		1902.41	II
9000		1954.53	I
7000		1960.13	I
25		1989.35	II
		Air	
7000		2021.21	I
9000		2061.70	I
45	h	2068.9	II
4600		2110.26	I
2500		2133.63	I
15		2143.40	II
15		2143.46	II
60		2186.9	II
40	h	2214.0	II
360		2228.25	I
1700		2230.61	I
340		2276.58	I
16		2368.12	II
12		2368.25	II
190		2400.88	I
10		2501.0	II
25		2515.69	I
70		2524.49	I
20	h	2544.5	II
700		2627.91	I
12		2693.0	II
280	c	2696.76	I
20		2713.3	II
140	d	2730.50	I
360		2780.52	I
15		2803.42	II
11		2803.70	II
12		2805.3	II
140	c	2809.62	I
4000		2897.98	I
15		2936.7	II
3200		2938.30	I
20		2950.4	II
12		2963.4	II
2800		2989.03	I
700		2993.34	I
2400		3024.64	I
60		3034.87	I
9000	c	3067.72	I
140		3076.66	I
550	c	3397.21	I
10		3430.83	II
12		3431.23	II
500	c	3510.85	I
380	c	3596.11	I
12		3654.2	II
70	h	3792.5	II
12		3811.1	II
20		3815.8	II
10		3845.8	II
30		3863.9	II
40	h	4079.1	II
10		4097.2	II
140		4121.53	I
140		4121.86	I
75	h	4259.4	II
25		4272.0	II
70	h	4301.7	II
12	h	4339.8	II
25	h	4340.5	II
12	h	4379.4	II
25	h	4476.8	II
60	h	4705.3	II
600	c	4722.52	I
30		4730.3	II
20		4749.7	II
12		4908.2	II
10		4916.6	II
12		4969.7	II
20		4993.6	II
10		5091.6	II
50	h	5124.3	II
60	h	5144.3	II
20		5201.5	II
75	h	5209.2	II
40	h	5270.3	II
10		5397.8	II
10	c	5552.35	I
3		5599.41	I
20		5655.2	II
40	h	5719.2	II
6		5742.55	I
12		5818.3	II
20		5860.2	II
20		5973.0	II
15		6059.1	II
15		6128.0	II
6		6134.82	I
3		6475.73	I
3		6476.24	I
15		6497.7	II
10		6577.2	II
40	h	6600.2	II
50	h	6808.6	II
4	h	6991.12	I
12		7033.	II
2		7036.15	I
10	h	7381.	II
2		7502.33	I
10	h	7637.	II
10		7750.	II
3		7838.70	I
2		7840.33	I
20		7965.	II
12	h	8050.	II
15		8328.	II
15		8388.	II
30		8532.	II
2		8544.54	I
1		8579.74	I
25		8653.	II
2		8754.88	I
3		8761.54	I
25		8863.	II
2		8907.81	I
2000	d	9657.04	I
40		9827.78	I
20		10104.5	I
15		10138.8	I
20		10300.6	I
20		10536.19	I
50		11072.44	I
15		11551.6	I
1500	d	11710.37	I
40		11999.49	I
200		12165.08	I
10		12374.64	I
200		12690.04	I
100		12817.8	I
200		14330.5	I
50		16001.5	I
60		22551.6	I

Bi III
Ref. 359 — C.H.C.

Intensity		Wavelength	
		Vacuum	
1		590.73	III
5		670.76	III
4		775.16	III
6		803.65	III
7		920.93	III
6		925.48	III
25		1039.99	III
50	h	1045.76	III
30		1051.81	III
20		1139.01	III
15		1145.91	III
50		1224.64	III
40		1326.84	III
60		1346.12	III
35		1423.33	III
35		1423.52	III
60	h	1461.00	III
60	h	1606.40	III
20	h	1691.5	III
20		1834.32	III
10		1863.9	III
10		1912.12	III
10		1988.26	III
		Air	
20		2020.75	III
20		2021.15	III
10		2073.22	III
14		2073.37	III
15		2103.42	III
30		2213.55	III
75	h	2414.6	III
10		2437.6	III
30	h	2847.4	III
80	h	2855.6	III
35		3115.0	III
40	h	3451.0	III
40		3473.8	III
35		3485.5	III
15		3540.8	III
45		3613.4	III
50		3695.32	III
50		3695.68	III
12		4224.6	III
25		4327.8	III
30		4560.84	III
30		4561.54	III
40	h	4797.4	III
45	h	5079.3	III
12		6623.4	III
10	h	7381.	III
12		7551.	III
25		7598.	III
10	h	7637.	III
40		8008.	III
50		8070.	III
20		8100.	III
15		8671.	III
20		8934.	III

Bi IV
Ref. 360 — C.H.C.

Intensity		Wavelength	
		Vacuum	
6		420.7	IV
6		431.2	IV
6		790.5	IV
6		790.6	IV
8		792.5	IV
10		820.3	IV
9		822.9	IV
12		824.9	IV
15		872.6	IV
8		876.8	IV
9		916.7	IV
12		923.9	IV
15		943.3	IV
9		967.6	IV
8		968.8	IV
8		989.8	IV
24		1103.4	IV
7		1128.8	IV
6		1138.6	IV
6		1139.8	IV
7		1149.7	IV
60		1317.0	IV
30		1910.0	IV
		Air	
30		2093.	IV
100		2311.	IV
100		2326.	IV
100		2376.	IV
100		2629.	IV
100		2677.	IV
100		2767.	IV
100		2772.	IV
100		2786.	IV
100		2842.	IV
100		2924.	IV
100		2933.	IV
100		2936.	IV
100		3012.	IV
100		3042.	IV
100		3239.	IV
100		3643.	IV
100		3682.	IV
100		3734.	IV
100		3868.	IV
30		4342.	IV

Bismuth (Cont.)

Intensity		Wavelength	
30		5347.	IV

Bi V
Ref. 361 — C.H.C.

Intensity		Wavelength	
Vacuum			
1		355.77	V
1		369.52	V
1		429.78	V
1		435.63	V
2		488.39	V
1		492.72	V
3		563.62	V
2		678.87	V
6		686.88	V
1		706.54	V
5		730.71	V
10		738.17	V
6		849.86	V
5		855.68	V
15	d	864.45	V
6		880.17	V
6		929.81	V
15	d	1139.46	V

BORON (B)
Z = 5

B I and II
Ref. 66, 104, 171, 222 — R.L.K.

Intensity		Wavelength	
Vacuum			
70		693.95	II
40		731.36	II
40		731.44	II
110		882.54	II
110		882.68	II
40		984.67	II
110		1081.88	II
110		1082.07	II
110		1230.16	II
220		1362.46	II
70		1600.46	I
120		1600.73	I
160		1623.58	II
110		1623.77	II
220		1624.02	II
70		1624.16	II
160		1624.34	II
100		1663.04	I
150		1666.87	I
200		1667.29	I
150		1817.86	I
200		1818.37	I
300		1825.91	I
300		1826.41	I
110		1842.81	II
Air			
250		2066.38	I
250		2066.65	I
100		2066.93	I
300		2067.19	I
500		2088.91	I
500		2089.57	I
70		2220.30	II
40		2323.03	II
40		2328.67	II
40		2393.20	II
220		2395.05	II
40		2459.69	II
40		2459.90	II
1000		2496.77	I
1000		2497.73	I
160		2918.08	II
110		3032.26	II
70		3179.33	II
110		3323.18	II
110		3323.60	II
450		3451.29	II
285		4121.93	II
110		4194.79	II
110		4472.10	II
110		4472.85	II
70		4784.21	II
110		4940.38	II
110		6080.44	II
70		6285.47	II
70		7030.20	II
40		7031.90	II
70		8668.57	I
20		8667.22	I
800		11660.04	I
570		11662.47	I
125		15629.08	I
200		16240.38	I
250		16244.67	I
235		18994.33	I

B III
Ref. 69, 221 — R.L.K.

Intensity		Wavelength	
Vacuum			
150		518.24	III
75		518.27	III
40		411.80	III
20		510.77	III
40		510.85	III
110		677.00	III
160		677.14	III
40		758.48	III
70		758.67	III
20		1953.83	III
Air			
550		2065.78	III
450		2067.23	III
160		2077.09	III
40		2234.09	III
70		2234.59	III
40		4242.98	III
70		4243.61	III
220		4487.05	III
360		4497.73	III
110		7835.25	III
70		7841.41	III

B IV
Ref. 74 — R.L.K.

Intensity		Wavelength	
Vacuum			
10		52.68	IV
30		60.31	IV
160		344.0	IV
450		385.0	IV
285		418.7	IV
70		1112.2	IV
450		1168.9	IV
70		1170.9	IV
Air			
70		2524.7	IV
160		2530.3	IV
450		2821.68	IV
70		2824.57	IV
285		2825.85	IV

B V
Ref. 94 — R.L.K.

Intensity		Wavelength	
Vacuum			
30		41.00	V
		48.59	V
		194.37	V
		262.37	V
		512.53	V
		749.74	V

BROMINE (Br)
Z = 35

Br I and II
Ref. 122, 124, 240, 248, 316
G.V.S.

Intensity		Wavelength	
Vacuum			
300		711.68	II
250		815.48	II
350		856.19	II
1000		889.23	II
500		896.64	II
500		905.99	II
300		922.56	II
1000		948.97	II
500		984.93	II
500		1012.10	II
1000		1015.54	II
500		1037.02	II
1000		1049.00	II
450		1064.76	II
500		1071.87	II
250		1101.50	I
300		1134.59	I
250		1136.29	I
250		1177.23	I
400		1178.90	I
1000		1189.28	I
250		1189.38	I
1000		1189.50	I
500		1198.37	I
800		1209.76	I
1000		1210.73	I
750		1216.01	I
1000		1221.13	I
900		1221.87	I
1000		1223.24	I
1200		1224.41	I
1200		1226.90	I
750		1228.05	I
7500		1232.43	I
1200		1243.90	I
800		1249.59	I
1500		1251.66	I
1000		1255.80	I
1500		1259.20	I
1200		1261.66	I
1200		1266.20	I
1000		1279.48	I
1000		1286.26	I
3000		1309.91	I
3000		1316.74	I
1000		1317.37	I
2000		1317.70	I
12000		1384.60	I
3000		1449.90	I
50000		1488.45	I
30000		1531.74	I
25000		1540.65	I
30000		1574.84	I
20000		1576.39	I
25000		1582.31	I
75000		1633.40	I
Air			
350		2285.17	II
350		2287.60	II
500	h	2317.30	II
400		2336.93	II
350		2386.45	II
500		2386.70	II
300		2388.69	II
450		2388.96	II
500		2389.69	II
350		2392.21	II
400		2392.42	II
300		2488.50	II
300	h	2495.22	II
450		2521.70	II
400		2541.48	II
400		2556.92	II
350		2690.17	II
400		2713.77	II
350	h	2746.52	II
300	h	2807.55	II
400	h	2893.40	II
400	h	2917.18	II
400	h	2967.21	II
500	h	2972.26	II
300		2981.86	II
300	h	2985.87	II
300	h	2986.53	II
300	h	3016.48	II
350		3423.82	II
300	h	3606.80	II
350		3714.30	II
1200		3815.65	I
350		3834.69	II
300		3871.21	II
400		3891.63	II
300		3901.24	II
300		3914.20	II
500		3914.38	II
350		3919.51	II
400		3924.09	II
300		3929.55	II
350		3939.69	II
350		3950.61	II
500		3980.38	II
1500		3992.36	I
300		4024.04	II
300		4135.66	II
300		4140.20	II
400		4179.63	II
300		4193.45	II
1000		4223.89	II
300		4236.89	II
300		4291.39	II
2000		4365.14	I
1000		4365.60	II
1500		4425.14	I
10000		4441.74	I
10000		4472.61	I
20000		4477.72	I
1000		4490.42	I
3000		4513.44	I
15000		4525.59	I
300		4529.60	II
500		4542.92	II
3000		4575.74	I
300		4601.36	II
2500		4614.58	I
350		4622.70	II
300		4642.02	II
300		4651.98	II
500		4678.70	II
400		4693.17	II
500		4704.85	II
350		4719.76	II
400		4720.36	II
300		4728.20	II
300		4735.41	II
400		4742.64	II
2500		4752.28	I
350		4766.00	II
400		4779.40	II
4000		4780.31	I
1600		4785.19	I
500		4785.50	II
300		4802.33	II
500		4816.70	II
300		4818.46	II
350		4844.81	II
350		4848.75	II
400		4921.12	II
400		4928.79	II
450		4930.66	II
300		4945.51	II
4000		4979.76	I
300		5038.74	II
300		5054.64	II
400		5164.38	II
300		5180.01	II
500		5182.35	II
300		5193.90	II
500		5238.23	II
300		5272.68	II
350		5304.10	II
400		5330.57	II
500		5332.05	II
1200		5395.48	I
400		5422.78	II
350		5424.99	II
300		5435.07	II
1200		5466.22	I
350		5478.47	II
300		5488.79	II
300		5495.06	II
500		5506.69	II
350		5589.94	II
300		5718.71	II
300		5830.78	II
1800		5852.08	I
1600		5940.48	I
2400		6122.14	I
40000		6148.60	I
300		6161.74	II
2000		6177.39	I
1500		6335.48	I
60000		6350.73	I
400		6352.94	II
2500		6410.32	I
1800		6483.56	I
1000		6514.62	I
20000		6544.57	I
1500		6548.09	I
50000	c	6559.80	I
1000		6571.31	I
1800		6579.14	I
20000		6582.17	I
1500		6620.47	I
50000	c	6631.62	I
20000		6682.28	I
10000		6692.13	I
8000		6728.28	I
2000		6760.06	I
2000		6779.48	I
2200		6786.74	I
6500		6790.04	I
1600	c	6791.48	I
1800		6861.15	I
10000		7005.19	I
2000		7260.45	I
10000		7348.51	I
40000		7512.96	I
1600		7591.61	I
1800		7595.07	I
2000		7616.41	I
30000		7803.02	I
1200		7827.23	I
2500	s	7881.45	I
2500		7881.57	I
2500		7925.81	I
30000	c	7938.68	I
3000		7947.94	I
3000		7950.18	I

Bromine (Cont.)

Intensity		Wavelength	
8000		7978.44	I
10000		7978.57	I
30000		7989.94	I
2000		8026.35	I
2500		8026.54	I
30000		8131.52	I
1000	c	8152.65	I
10000		8153.75	I
25000		8154.00	I
5000		8246.86	I
15000		8264.96	I
75000	c	8272.44	I
20000		8334.70	I
10000		8343.70	I
1200		8384.04	I
40000		8446.55	I
4000		8477.45	I
1500		8513.38	I
1000		8557.73	I
1000		8566.28	I
20000		8638.66	I
4000		8698.53	I
10000	c	8793.47	I
15000		8819.96	I
25000		8825.22	I
4000		8888.98	I
30000		8897.62	I
6000		8932.40	I
1800		8949.39	I
9000		8964.00	I
350		9024.42	II
30000		9166.06	I
15000		9173.63	I
20000		9178.16	I
40000		9265.42	I
15000		9320.86	I
300		9434.04	II
6000		9793.48	I
10000		9896.40	I
3000		10140.08	I
6000		10237.74	I
1000		10299.62	I
1500		10377.65	I
30000		10457.96	I
1000		10742.14	I
3000		10755.92	I
1700		13217.17	I
1800		14354.57	I
1250		14888.70	I
1800		16731.19	I
1200		18568.31	I
3500		19733.62	I
1000		20281.73	I
1000		20624.67	I
1200		21787.24	I
4000		22865.65	I
1000		23513.15	I
500		28346.50	I
500		30380.85	I
600		31630.13	I
120		32693.90	I
150		34181.87	I
150		38345.75	I
120		39964.36	I

Br III
Ref. 246, 250 — G.V.S.

Intensity		Wavelength	
		Vacuum	
450		611.1	III
300		620.4	III
500		665.54	III
500		677.19	III
300		677.8	III
450		687.68	III
400		690.2	III
350		696.99	III
300		727.0	III
300		736.4	III
250		769.63	III
250		817.79	III
350		949.0	III
400		960.4	III
450		984.9	III
250		1313.5	III
250		1402.9	III
		Air	
400	h	2293.44	III
300		2313.29	III
300		2462.39	III
300		2482.60	III
350		2499.25	III
350		2529.49	III
350	h	2551.09	III
350	h	2570.83	III
300	h	2573.17	III
400	h	2584.99	III
500		2589.14	III
300	h	2594.48	III
400	h	2595.98	III
450	h	2606.20	III
350	h	2608.15	III
500	h	2613.13	III
350	h	2616.26	III
500	h	2626.52	III
350	h	2629.23	III
350	h	2639.60	III
350	h	2671.53	III
350	h	2735.83	III
300		2770.50	III
300	h	2785.28	III
300		2804.16	III
400		2926.96	III
300		2936.22	III
350		2969.00	III
400		2994.04	III
500		3020.76	III
300		3033.63	III
350		3036.45	III
500		3074.42	III
350		3091.94	III
350		3117.29	III
300		3147.81	III
400		3174.08	III
300		3321.08	III
450		3333.07	III
500		3349.64	III
300		3385.25	III
450		3447.36	III
400		3487.58	III
300		3506.47	III
450		3517.36	III
500		3540.16	III
300		3551.08	III
500		3562.43	III
450		3600.71	III
250		3693.53	III
450		3820.26	III
200		3903.95	III
350		4506.55	III
200		4519.74	III
150		5175.87	III
100		5446.80	III
100		7192.8	III
100		7673.1	III

Br IV
Ref. 139, 142, 243, 249
G.V.S.

Intensity		Wavelength	
		Vacuum	
700		379.73	IV
700		400.37	IV
1000		545.43	IV
1000		559.76	IV
1000		569.19	IV
1000		576.59	IV
1000		585.10	IV
1000		586.71	IV
1000		597.51	IV
1000		600.09	IV
1000		601.27	IV
1000		607.03	IV
1000		617.85	IV
1000		619.87	IV
1000		630.14	IV
1000		642.23	IV
1000		661.53	IV
1000		683.51	IV
1000		697.72	IV
1000		715.39	IV
1000		731.00	IV
1000		800.12	IV
1000		813.66	IV
900		1274.82	IV
1000		1703.51	IV
		Air	
1000		2133.79	IV
1000		2145.02	IV
1000		2257.21	IV
1000		2272.73	IV
1000		2307.40	IV
1000		2408.16	IV
1000		2411.58	IV
700		2491.14	IV
1000		2581.19	IV
600		2661.40	IV
700		2820.87	IV
1000		2842.88	IV
1100	h	2907.71	IV
500		3041.18	IV
500		3380.56	IV

Br V
Ref. 42 — G.V.S.

Intensity		Wavelength	
		Vacuum	
600		468.37	V
800		482.11	V
900		531.97	V
1000		547.90	V
700		549.77	V
800		621.03	V
800		632.22	V
700		645.44	V
400		652.64	V
800		657.54	V
800		679.62	V
700		812.95	V
1000		850.81	V
150		855.27	V
600		1041.60	V
1000		1069.15	V
500		1080.54	V
900		1112.13	V
1000		1143.56	V
150		1429.75	V
400		1442.60	V
150		1470.35	V

CADMIUM (Cd)
Z = 48

Cd I and II
Ref. 44, 285, 296 — R.D.C.

Intensity		Wavelength	
		Vacuum	
100		1256.00	II
150		1296.43	II
100		1326.50	II
150		1370.91	II
200		1514.26	II
200		1571.58	II
100		1668.60	II
50		1702.47	II
50		1724.41	II
100		1785.84	II
100		1827.70	II
300		1922.23	II
100		1943.54	II
40		1965.54	II
30		1986.89	II
200		1995.43	II
		Air	
100		2007.49	II
50		2032.45	II
75		2036.23	II
150		2096.00	II
1000	r	2144.41	II
50		2155.06	II
100		2187.79	II
1000		2194.56	II
1000		2265.02	II
1500	r	2288.022	I
1000		2312.77	II
200		2321.07	II
40		2376.82	II
50		2418.69	II
50		2469.73	II
40		2487.93	II
3		2491.00	I
40		2495.58	II
10		2508.91	I
50		2509.11	II
30		2516.22	II
15	h	2518.59	I
25	h	2525.196	I
50		2544.613	I
50		2551.98	II
25		2553.465	I
3		2565.789	I
500		2572.93	II
50		2580.106	I
3		2584.87	I
30		2592.026	I
25	h	2602.048	I
50		2628.979	I
40		2632.190	I
75		2639.420	I
40		2659.23	II
50	h	2660.325	I
25		2668.20	II
50		2672.62	II
100		2677.540	I
25		2677.748	I
50		2707.00	II
75		2712.505	I
50		2733.820	I
1000		2748.54	II
100	h	2763.894	I
50	h	2764.230	I
50		2774.958	I
30		2823.19	II
200		2836.900	I
25		2856.46	II
100		2868.180	I
200	r	2880.767	I
50	r	2881.224	I
200		2914.67	II
50		2927.87	II
200		2929.27	II
1000	r	2980.620	I
200	r	2981.362	I
50		2981.845	I
50		3030.60	II
150		3080.822	I
25		3081.48	II
30		3082.593	I
100		3092.34	II
200		3133.167	I
50		3146.79	II
150		3250.33	II
300		3252.524	I
300		3261.055	I
50		3343.21	II
50		3385.49	II
30		3388.88	II
800		3403.652	I
50		3417.49	II
50		3442.42	II
100		3464.43	II
1000		3466.200	I
800		3467.655	I
25		3483.08	II
150		3495.44	II
25		3499.952	I
100		3524.11	II
100		3535.69	II
1000		3610.508	I
800		3612.873	I
60		3614.453	I
20		3649.558	I
10		3981.926	I
100		4029.12	II
200		4134.77	II
50		4141.49	II
100		4285.08	II
8		4306.672	I
100		4412.41	II
3		4412.989	I
1000		4415.63	II
30		4440.45	II
8		4662.352	I
200		4678.149	I
30		4744.69	II
300		4799.912	I
50		4881.72	II
50		5025.50	II
1000	h	5085.822	I
6		5154.660	I
100		5268.01	II
100		5271.60	II
1000		5337.48	II
1000		5378.13	II
200		5381.89	II
40		5843.30	II
50		5880.22	II
300		6099.142	I
100		6111.49	I
100		6325.166	I
30		6330.013	I
400		6354.72	II
500		6359.98	II
2000		6438.470	I
400		6464.94	II
25		6567.65	II
500		6725.78	II
100		6759.19	II
30		6778.116	I
50		7237.01	II
100		7284.38	II
1000		7345.670	I
50		8066.99	II
5		8200.309	I
20		9292	I
15		11655	I
35		14491	I
80		15712	I
55	d	19125	I
25		24378	I
35		25455	I

Cd III
Ref. 296 — R.D.C.

Intensity		Wavelength	
		Vacuum	
8		677.39	III

Cadmium (Cont.)

Intensity	Wavelength	
15	684.58	III
10	720.70	III
5	1383.60	III
15	1392.10	III
10	1396.78	III
25	1416.28	III
5	1420.29	III
8	1420.54	III
5	1432.86	III
20	1446.08	III
25	1447.55	III
30	1455.74	III
8	1466.14	III
5	1471.97	III
15	1491.81	III
5	1511.01	III
10	1511.65	III
5	1513.13	III
10	1523.55	III
15	1528.40	III
30	1529.30	III
5	1532.10	III
50	1545.17	III
25	1547.57	III
10	1550.07	III
20	1550.45	III
15	1550.89	III
5	1552.18	III
5	1556.48	III
15	1560.66	III
5	1566.03	III
15	1568.98	III
10	1582.39	III
40	1601.59	III
20	1604.87	III
20	1606.64	III
10	1607.28	III
15	1608.91	III
10	1609.61	III
15	1612.51	III
15	1625.27	III
25	1628.54	III
20	1651.87	III
25	1655.63	III
30	1678.15	III
10	1699.70	III
40	1707.16	III
10	1721.93	III
40	1722.95	III
30	1725.66	III
25	1739.00	III
5	1745.69	III
40	1747.67	III
15	1748.15	III
30	1768.82	III
40	1773.06	III
30	1789.19	III
75	1793.40	III
15	1796.10	III
5	1800.57	III
40	1823.41	III
50	1844.66	III
40	1851.13	III
20	1851.37	III
40	1855.85	III
200	1856.67	III
150	1874.08	III
15	1886.49	III
15	1903.48	III
25	1909.98	III
15	1910.57	III
10	1939.59	III
15	1988.81	III
	Air	
20	2000.60	III
15	2004.07	III
15	2016.12	III
40	2039.83	III
50	2045.61	III
10	2061.25	III
75	2087.91	III
10	2097.45	III
5	2100.47	III
50	2111.60	III
5	2188.13	III
5	2218.43	III
5	2224.43	III
7	2418.24	III
10	2426.36	III
25	2499.81	III
15	2618.81	III
5	2630.56	III
20	2766.99	III
30	2805.59	III
10	3035.72	III

Cd IV
Ref. 353, 399 — R.D.C.

Intensity	Wavelength	
	Vacuum	
50	427.01	IV
20	437.88	IV
50	447.85	IV
60	480.90	IV
10	489.49	IV
70	493.00	IV
70	495.13	IV
70	498.14	IV
70	498.53	IV
80	504.09	IV
70	504.20	IV
70	504.50	IV
80	506.31	IV
60	508.01	IV
50	508.95	IV
70	509.55	IV
25	509.81	IV
70	511.40	IV
80	513.00	IV
70	514.50	IV
60	519.42	IV
15	520.97	IV
80	524.41	IV
70	524.47	IV
50	524.77	IV
70	525.10	IV
60	525.19	IV
70	527.07	IV
50	530.79	IV
80	531.09	IV
80	531.51	IV
70	534.29	IV
70	536.77	IV
50	537.24	IV
60	540.90	IV
70	541.74	IV
80	542.60	IV
80	546.55	IV
40	548.01	IV
20	548.33	IV
15	548.90	IV
25	551.27	IV
20	552.90	IV
60	553.06	IV
80	554.05	IV
25	560.26	IV
10	564.16	IV
60	567.01	IV
20	1062.23	IV
150	1118.16	IV
30	1126.00	IV
20	1134.08	IV
15	1139.04	IV
20	1154.64	IV
10	1155.73	IV
100	1164.65	IV
20	1165.78	IV
40	1167.30	IV
20	1179.73	IV
15	1183.07	IV
100	1183.40	IV
40	1194.13	IV
20	1195.63	IV
30	1196.47	IV
20	1198.93	IV
15	1215.38	IV
20	1223.52	IV
20	1246.06	IV
15	1246.56	IV
15	1249.94	IV
30	1266.47	IV
15	1274.41	IV
20	1285.63	IV
20	1287.58	IV
20	1299.46	IV
40	1304.36	IV
30	1306.07	IV
15	1316.89	IV
15	1321.85	IV
20	1325.55	IV
30	1340.97	IV
15	1346.15	IV
30	1354.78	IV
20	1358.11	IV
30	1362.55	IV
60	1370.48	IV
30	1380.98	IV
20	1397.65	IV
30	1403.68	IV
15	1406.58	IV
60	1418.89	IV
15	1429.83	IV
20	1447.54	IV
20	1452.63	IV
15	1465.97	IV
15	1466.67	IV
15	1482.95	IV
20	1491.79	IV
20	1570.20	IV
20	1598.73	IV
20	1600.42	IV
15	1622.87	IV

CALCIUM (Ca)
Z = 20

Ca I and II
Ref. 70, 150, 270 — J.J.W. and H.H.S.

Intensity	Wavelength	
	Vacuum	
24	1341.89	II
12	1342.54	II
20	1433.75	II
12	1432.50	II
20	1553.18	II
32	1554.64	II
4	1642.80	II
20	1643.77	II
36	1644.44	II
60	1649.86	II
32	1651.99	II
12	1673.86	II
20	1680.05	II
2	1680.13	II
8	1691.78	II
16	1698.18	II
20	1807.34	II
40	1814.50	II
4	1814.65	II
60	1840.06	II
40	1838.01	II
20	1843.09	II
40	1850.69	II
	Air	
	2103.24	II
	2112.76	II
	2113.15	II
	2128.75	II
	2131.51	II
	2132.30	II
2	2150.80	I
	2197.79	II
5	2200.73	I
	2208.61	II
6	2275.46	I
8	2398.56	I
7	2721.65	I
9	2994.96	I
8	2997.31	I
8	2999.64	I
9	3000.86	I
10	3006.86	I
9	3009.21	I
2	3024.94	I
2	3034.54	I
2	3045.74	I
3	3055.32	I
2	3071.57	I
2	3076.95	I
2	3080.79	I
2	3099.30	I
10	3125.18	II
5	3136.02	I
6	3140.79	I
7	3150.75	I
170	3158.87	II
180	3179.33	II
5	3180.52	I
150	3181.28	II
7	3209.96	I
8	3215.17	I
6	3215.34	I
9	3225.90	I
6	3226.15	I
5	3274.67	I
6	3286.07	I
10	3308.02	II
20	3316.51	II
10	3344.51	I
10	3347.04	II
11	3350.21	I
9	3350.36	I
12	3361.92	I
9	3362.14	I
10	3452.66	II
20	3461.87	II
9	3468.48	I
11	3474.76	I
10	3485.61	II
13	3487.60	I
10	3495.16	II
15	3624.11	I
17	3630.75	I
14	3630.97	I
20	3644.41	I
14	3644.77	I
8	3644.99	I
5	3675.29	I
6	3678.21	I
30	3683.70	II
40	3694.11	II
10	3694.36	II
170	3706.03	II
180	3736.90	II
10	3739.38	II
6	3748.35	I
8	3750.29	I
9	3753.34	I
20	3755.67	II
30	3758.39	II
9	3870.48	I
11	3872.54	I
11	3872.56	I
12	3875.78	I
12	3875.80	I
6	3889.10	I
6	3923.48	I
230	3933.66	II
9	3935.29	I
6	3946.04	I
15	3948.90	I
17	3957.05	I
220	3968.47	II
8	3972.57	I
18	3973.71	I
50	4097.10	II
15	4098.53	I
15	4098.57	I
60	4109.82	II
30	4110.28	II
40	4206.18	II
50	4220.07	II
50	4226.73	I
15	4240.46	I
24	4283.01	I
22	4289.36	I
22	4298.99	I
25	4302.53	I
23	4307.74	I
22	4318.65	I
20	4355.08	I
25	4425.44	I
26	4434.96	I
25	4435.69	I
30	4454.78	I
28	4455.89	I
20	4456.61	I
20	4472.04	II
10	4479.23	II
20	4489.18	II
23	4526.94	I
22	4578.55	I
23	4581.40	I
23	4581.47	I
24	4585.87	I
24	4585.96	I
20	4685.27	I
30	4716.74	II
40	4721.03	II
40	4799.97	II
25	4878.13	I
70	5001.48	II
80	5019.97	II
40	5021.14	II
23	5041.62	I
25	5188.85	I
22	5261.71	I
23	5262.24	I
22	5264.24	I
24	5265.56	I
25	5270.27	I
60	5285.27	II
70	5307.22	II
50	5339.19	II
27	5349.47	I
23	5512.98	I
25	5581.97	I
27	5588.76	I
24	5590.12	I
26	5594.47	I
25	5598.49	I
24	5601.29	I
24	5602.85	I
30	5857.45	I
10	5922.72	II
10	5923.69	II
27	6102.72	I
29	6122.22	I
22	6161.29	I
30	6162.17	I
22	6163.76	I
24	6166.44	I
26	6169.06	I
28	6169.56	I
35	6439.07	I
30	6449.81	I
22	6455.60	I
80	6456.87	II
34	6462.57	I
29	6471.66	I

Calcium (Cont.)

Intensity	Wavelength	
32	6493.78	I
28	6499.65	I
23	6572.78	I
30	6717.69	I
33	7148.15	I
31	7202.19	I
	7291.47	II
	7323.89	II
33	7326.15	I
30	7575.81	II
60	7581.11	II
80	7601.30	II
20	7602.32	II
40	7820.78	II
60	7843.38	II
20	8017.50	II
20	8020.50	II
70	8133.05	II
100	8201.72	II
110	8248.80	II
70	8254.73	II
14	8256.67	I
10	8338.04	I
12	8339.12	I
10	8352.39	I
11	8357.17	I
130	8498.02	II
170	8542.09	II
10	8633.95	I
160	8662.14	II
12	8842.61	I
15	8909.18	I
100	8912.07	II
110	8927.36	II
12	8967.47	I
16	9099.10	I
13	9105.62	I
12	9108.82	I
10	9171.14	I
110	9213.90	II
90	9312.00	II
100	9319.56	II
110	9320.65	II
25	9416.97	I
10	9456.80	I
10	9534.88	I
11	9548.38	I
100	9567.97	II
110	9599.24	II
80	9601.82	II
10	9604.28	I
12	9663.65	I
10	9664.41	I
14	9676.30	I
14	9688.67	I
13	9701.94	I
80	9854.74	II
110	9890.63	II
90	9931.39	II
100	10223.04	II
20	10343.81	I
13	10838.97	I
13	10861.58	I
13	10863.87	I
14	10869.50	I
14	10879.87	I
20	11838.99	II
10	11949.72	II
25	12816.04	I
24	12823.86	I
25	12909.10	I
30	13033.57	I
21	13086.44	I
24	13134.95	I
20	16150.77	I
22	16157.36	I
21	16197.04	I
20	18925.47	I
24	18970.14	I
30	19046.14	I
48	19309.20	I
49	19452.99	I
47	19505.72	I
50	19776.79	I
35	19853.10	I
34	19862.22	I
23	19917.19	I
24	19933.70	I
	21389.00	II
	21428.90	II
20	22607.93	I
25	22624.93	I
30	22651.23	I

Ca III
Ref. 25, 16 — J.J.W. and H.H.S.

Intensity	Wavelength	
	Vacuum	
6	296.96	III
9	403.72	III
7	409.95	III
5	439.69	III
5	633.59	III
5	685.41	III
5	697.55	III
5	699.09	III
5	699.89	III
6	701.39	III
5	727.66	III
8	740.55	III
6	746.25	III
5	747.98	III
5	779.61	III
5	800.30	III
5	809.93	III
5	817.06	III
6	821.57	III
6	840.56	III
6	1020.07	III
5	1034.65	III
5	1187.30	III
8	1188.61	III
8	1188.61	III
5	1190.86	III
10	1262.65	III
11	1278.39	III
10	1281.55	III
12	1286.52	III
12	1298.04	III
11	1317.70	III
10	1328.95	III
11	1335.13	III
10	1360.01	III
11	1385.43	III
11	1397.69	III
13	1453.16	III
12	1459.79	III
11	1461.88	III
15	1463.34	III
16	1484.87	III
12	1496.88	III
11	1506.88	III
20	1545.29	III
15	1555.53	III
18	1562.47	III
13	1571.27	III
13	1586.13	III
10	1762.26	III
10	1783.93	III
10	1794.22	III
12	1800.21	III
13	1807.89	III
14	1812.15	III
11	1813.59	III
12	1830.06	III
10	1860.43	III
14	1870.26	III
14	1872.37	III
10	1894.12	III
11	1910.10	III
12	1935.72	III
10	1939.68	III
11	1943.01	III
12	1948.26	III
10	1953.55	III
10	1958.97	III
13	1964.61	III
13	1967.94	III
12	1972.82	III
10	1977.01	III
10	1978.55	III
11	1981.19	III
	Air	
12	2033.36	III
12	2041.53	III
13	2078.92	III
13	2098.49	III
15	2114.41	III
17	2123.03	III
14	2129.19	III
14	2133.96	III
13	2140.36	III
16	2152.43	III
12	2171.57	III
12	2276.52	III
15	2312.08	III
14	2497.74	III
15	2541.50	III
13	2587.15	III
12	2590.41	III
15	2620.82	III
15	2634.14	III
12	2686.72	III
16	2687.76	III
15	2704.86	III
14	2771.28	III
15	2791.59	III
16	2813.88	III
17	2866.54	III
18	2869.95	III
19	2881.78	III
21	2899.79	III
19	2924.33	III
20	2988.63	III
18	2989.27	III
15	3028.59	III
19	3119.67	III
15	3367.79	III
19	3372.67	III
18	3537.77	III
15	4081.77	III
15	4153.57	III
15	4164.31	III
15	4184.20	III
18	4207.24	III
17	4233.74	III
16	4240.74	III
15	4284.39	III
20	4302.81	III
15	4329.19	III
16	4333.57	III
15	4358.38	III
19	4399.59	III
17	4406.29	III
17	4431.30	III
19	4499.88	III
18	4516.59	III
18	4572.12	III
11	4708.83	III
11	4716.27	III
10	4859.17	III
10	5008.95	III
10	5050.07	III
10	5231.82	III
11	5247.37	III
13	5271.98	III
10	5301.32	III
11	5321.29	III
10	5328.06	III
11	5570.58	III
10	5579.06	III
13	6069.98	III
10	6173.22	III
12	6213.98	III
11	6294.89	III
11	6370.11	III
10	6387.55	III
12	6424.51	III
12	6485.35	III
10	6538.78	III
10	6542.24	III
10	7308.69	III
10	7843.06	III
12	7898.46	III
10 l	8217.20	III

Ca IV
Ref. 150 — J.J.W. and H.H.S.

Intensity		Wavelength	
		Vacuum	
150		249.41	IV
150		250.15	IV
150		251.35	IV
250		296.55	IV
200		299.32	IV
200		318.09	IV
50		318.39	IV
120		321.59	IV
250		329.12	IV
150		329.39	IV
200		331.44	IV
250		331.99	IV
235		332.53	IV
150		332.81	IV
200		338.83	IV
150		339.79	IV
200		340.29	IV
200		341.29	IV
200		341.46	IV
250	c	342.45	IV
100		343.19	IV
200		343.44	IV
250		343.93	IV
200		344.96	IV
215		345.13	IV
250		374.74	IV
600		434.57	IV
100		437.27	IV
250		437.77	IV
200		438.93	IV
750		443.82	IV
50		445.02	IV
500		450.57	IV
50		454.55	IV
250		456.98	IV
250		461.09	IV
150		565.46	IV
750		656.00	IV
500		669.70	IV

Ca V
Ref. 150 — J.J.W. and H.H.S.

Intensity		Wavelength	
		Vacuum	
200		190.36	V
250		190.46	V
250		196.97	V
300		199.55	V
250		200.51	V
265		257.98	V
165		260.45	V
400		267.77	V
300		270.31	V
200		271.14	V
250		272.27	V
200		272.98	V
400		280.99	V
300		284.98	V
450	c	286.96	V
500		322.17	V
250		322.76	V
300		323.22	V
250		324.48	V
250		325.28	V
300		330.94	V
200		333.44	V
200		333.57	V
300		334.55	V
250	c	335.34	V
200		336.55	V
200		337.54	V
250		338.06	V
200		343.64	V
450		352.92	V
250		356.25	V
250		377.18	V
200		387.08	V
750		425.00	V
500		558.60	V
400		637.93	V
300		643.12	V
400		646.57	V
250		647.88	V
250		651.55	V
300		656.76	V

CALIFORNIUM (Cf)
Z = 98

Cf I and II
Ref. 52, 331 — J.G.C.

Intensity		Wavelength	
		Air	
10000		2739.31	
10000	s	2759.10	
10000		2774.52	
10000	l	2852.03	
10000	s	2855.24	
10000		3298.14	
10000		3352.71	
10000	l	3367.79	
10000		3392.22	I
10000		3481.07	
10000	l	3513.47	
10000		3531.49	I
10000		3540.98	I
10000		3598.77	I
10000		3605.32	I
10000		3612.11	II
10000		3617.49	I
10000	s	3626.76	II
10000		3659.46	
10000		3662.70	I
10000	s	3699.49	
10000	l	3722.11	II
10000		3739.35	I
10000		3785.61	I
10000	l	3789.04	II
10000	s	3893.23	II
10000	l	3993.57	II
10000		4035.45	
10000		4099.12	I
10000		4242.38	I
10000		4329.03	I
10000		4335.22	I
10000		5173.96	I
10000		5179.08	I
10000		5219.24	I
10000	s	5279.01	
10000	s	5320.09	
10000	s	5339.13	
10000		5408.88	I
10000		5726.05	I
10000		6622.83	I
10000		6631.26	I

Californium (Cont.)

Intensity		Wavelength	
10000		6677.90	I
10000		6894.59	I
10000		6927.10	II
10000	l	7074.52	I
10000	s	7307.90	I
10000		8141.29	I
10000		8241.77	I
10000		8333.85	II
10000		8423.49	II
10000		8568.83	II
10000	l	9228.52	
10000		9337.70	
10000		9649.51	
10000	s	10308.41	
10000	l	10568.83	
10000	s	10614.84	
10000		11300.19	
10000		11681.85	
10000		11941.33	
10000	l	12183.05	
10000	s	12352.72	
10000	s	12437.48	
10000	l	12789.41	
10000	l	13329.98	
10000	s	13362.98	
10000	l	13376.89	
10000	l	13474.44	
10000	l	14772.49	
10000	s	15281.32	
10000		15587.12	
10000		15675.92	
10000		16759.06	
10000	s	17626.25	
10000	l	18718.69	
10000	h	19068.71	
10000	l	19336.96	
10000	l	19576.84	
10000	l	20393.38	
10000	s	20869.98	

CARBON (C)
Z = 6

C I and II
Ref. 211 — R.L.K.

Intensity		Wavelength	
		Vacuum	
9		595.022	II
30		687.053	II
50		687.345	II
10		858.092	II
20		858.559	II
30		903.624	II
60		903.962	II
150		904.142	II
30		904.480	II
9		1009.86	II
10		1010.08	II
10		1010.37	II
80		1036.337	II
150		1037.018	II
150		1157.910	I
150		1158.019	I
150		1158.035	I
150		1188.992	I
150		1189.447	I
200		1189.631	I
300		1193.009	I
300		1193.031	I
300		1193.240	I
300		1193.264	I
100		1193.393	I
150		1193.649	I
150		1193.679	I
100		1194.064	I
100		1194.488	I
100		1261.552	I
250		1277.245	I
250		1277.282	I
300		1277.513	I
300		1277.550	I
200		1280.333	I
100		1311.363	I
9		1323.951	II
120		1329.578	I
120		1329.600	I
150		1334.532	II
300		1335.708	II
100		1354.288	I
150		1355.84	I
120		1364.164	I
100		1459.032	I
200		1463.336	I
120		1467.402	I
150		1481.764	I
150		1560.310	I
400		1560.683	I
400		1560.708	I
100		1561.341	I
400		1561.438	I
150		1656.266	I
120		1656.928	I
300		1657.008	I
120		1657.380	I
120		1657.907	I
150		1658.122	I
500		1751.823	I
1000		1930.905	I
		Air	
800		2478.56	I
250		2509.12	II
350		2512.06	II
250	h	2574.83	II
350	l	2741.28	II
250		2746.49	II
1000		2836.71	II
800		2837.60	II
800	h	2992.62	II
350		3876.19	II
350		3876.41	II
350		3876.66	II
570		3918.98	II
800		3920.69	II
250		4074.52	II
350	l	4075.85	II
800		4267.00	II
1000		4267.26	II
200		4771.75	I
200		4932.05	I
200		5052.17	I
350		5132.94	II
350		5133.28	II
350		5143.49	II
570		5145.16	II
400		5151.09	II
300		5380.34	I
250		5648.07	II
350		5662.47	II
570		5889.77	II
350		5891.59	II
200		6001.13	I
250		6006.03	I
110		6007.18	I
150		6010.68	I
300		6013.22	I
250		6014.84	I
800		6578.05	II
570		6582.88	II
200		6587.61	I
250		6783.90	II
250		7113.18	I
250		7115.19	I
250		7115.63	II
200		7116.99	I
350		7119.90	II
800		7231.32	II
1000		7236.42	II
200		7860.89	I
200		8058.62	I
520		8335.15	I
250		9061.43	I
200		9062.47	I
200		9078.28	I
250		9088.51	I
450		9094.83	I
300		9111.80	I
800		9405.73	I
150		9603.03	I
250		9620.80	I
300		9658.44	I
200		10683.08	I
300		10691.25	I
12		11619.29	I
23		11628.83	I
13		11658.85	I
47		11659.68	I
24		11669.63	I
85		11748.22	I
142		11753.32	I
114		11754.76	I
11		11777.54	I
17		11892.91	I
30		11895.75	I
26		12614.10	I
20		13502.27	I
38		14399.65	I
16		14403.25	I
61		14420.12	I
12		14429.03	I
13		14442.24	I
12		16559.66	I
50		16890.38	I
10		17338.56	I
11		17448.60	I
13		18139.80	I
23		19721.99	I

C III
Ref. 22, 211 — R.L.K.

Intensity		Wavelength	
		Vacuum	
250		371.69	III
250		371.75	III
150		371.78	III
500		386.203	III
200		450.734	III
400		459.46	III
500		459.52	III
570		459.63	III
250		511.522	III
250		535.288	III
300		538.080	III
350		538.149	III
400		538.312	III
350		574.281	III
800		977.03	III
370		1174.93	III
350		1175.26	III
330		1175.59	III
500		1175.71	III
350		1175.99	III
370		1176.37	III
		Air	
250		2162.94	III
800		2296.87	III
150		2697.75	III
110	l	2724.85	III
150	l	2725.30	III
150	l	2725.90	III
200		2982.11	III
150		4056.06	III
200		4067.94	III
250		4068.91	III
250		4070.26	III
150		4162.86	III
250	h	4186.90	III
200		4325.56	III
600		4647.42	III
520		4650.25	III
375		4651.47	III
200		4665.86	III
450		5695.92	III
150		5826.42	III
150		6744.38	III
150	h	7037.25	III
150		7612.65	III
300	h	8196.48	III
150		8332.99	III
300		8500.32	III

C IV
Ref. 66, 211 — R.L.K.

Intensity		Wavelength	
		Vacuum	
250		244.91	IV
200		289.14	IV
250		289.23	IV
570		312.42	IV
500		312.46	IV
650		384.03	IV
700		384.18	IV
400		419.52	IV
500		419.71	IV
1000		1548.202	IV
900		1550.774	IV
		Air	
200	l	2524.41	IV
300	s	2529.98	IV
200	w	4658.30	IV
250		5801.33	IV
200		5811.98	IV
90	w	7726.2	IV

C V
Ref. 211 — R.L.K.

Intensity		Wavelength	
		Vacuum	
110		34.973	V
450		40.268	V
110		227.19	V
160		248.66	V
160		248.74	V
		Air	
40		2270.91	V
5		2277.25	V
20		2277.92	V
5		4943.88	V
5		4944.56	V

CERIUM (Ce)
Z = 58

Ce I and II
Ref. 1 — C.H.C.

Intensity		Wavelength	
		Air	
130		2462.97	II
110		2518.51	II
200		2548.68	II
340		2651.01	II
120		2696.07	II
120		2706.88	II
120		2723.38	II
110		2741.96	II
100		2750.89	II
150		2761.42	II
120		2784.27	II
100		2785.35	II
100		2790.53	II
140		2791.42	II
270		2830.90	II
100		2833.31	II
250		2874.14	II
110		2908.42	II
100		2918.67	II
120		2955.94	II
110		2964.80	II
100		2972.58	II
400		2976.91	II
150		2977.46	II
120		2980.41	II
250		2990.87	II
110		2994.42	II
320		2995.64	II
400		3008.79	II
370		3017.20	II
210		3037.73	II
200		3051.98	II
350		3055.24	II
320		3056.78	II
680		3063.01	II
320		3083.67	II
250		3084.44	II
200		3090.37	II
370		3103.38	II
200		3107.47	II
320		3110.28	II
300		3111.17	II
220		3127.53	II
200		3130.33	II
240		3130.87	II
200		3144.60	II
290		3145.28	II
290		3146.41	II
290		3164.15	II
290		3169.18	II
290		3171.61	II
480		3183.52	II
240		3186.13	II
200		3190.34	II
710		3194.83	II
200		3199.28	II
990		3201.71	II
200		3218.38	II
710		3218.94	II
880		3221.17	II
330		3225.67	II
710		3227.11	II
240		3229.36	II
480		3231.24	II
710		3234.16	II
330		3234.89	II
390		3236.74	II
390		3243.37	II
200		3246.67	II
200		3260.98	II
200		3263.88	II
990		3272.25	II
330		3274.86	II
200		3279.84	II
330		3285.22	II
240		3295.28	II
200		3296.88	II
220		3300.15	II
240		3304.84	II
200		3312.22	II
240		3314.72	II
200		3317.80	II
200		3334.46	II
240		3341.87	II
330		3343.86	II
440		3344.76	II
200		3355.02	II

Cerium (Cont.)

240	3357.22	II
200	3360.54	II
240	3366.55	II
200	3371.18	II
200	3373.46	II
200	3373.73	II
480	3377.13	II
200	3383.68	II
200	3404.91	II
240	3405.98	II
290	3417.45	II
600	3422.71	II
390	3426.21	II
290	3441.21	II
480	3476.84	II
240	3482.35	II
710	3485.05	II
210	3507.94	II
600	3517.38	II
210	3520.52	II
330	3521.88	II
210	3526.68	II
600	3534.05	II
770	3539.08	II
210	3545.60	II
290	3546.19	II
240	3552.73	II
420	3555.00	II
1200	3560.80	II
210	3576.23	II
1000	3577.45	II
330	3590.60	II
390	3607.63	II
550	3609.69	II
420	3613.70	II
440	3622.15	II
380	3623.74	II
440	3623.84	II
200	3631.19	II
350	3646.97	II
260	3647.75	II
260	3647.95	II
420	3653.11	II
660	3653.67	II
310	3654.97	II
1800	3655.85	II
440	3659.23	II
350	3659.97	II
880	3660.64	II
880	3667.98	II
220	3672.18	I
350	3672.79	II
220	3679.42	II
300	3694.91	II
220	3704.98	II
1000	3709.29	II
1000	3709.93	II
1400	3716.37	II
420	3718.19	II
420	3718.38	II
210	3719.80	II
420	3725.68	II
490	3728.02	II
800	3728.42	II
320	3748.06	II
250	3751.45	II
200	3755.43	II
300	3762.98	II
680	3764.12	II
200	3765.04	II
300	3768.76	II
210	3770.76	II
300	3771.60	II
250	3776.61	II
620	3781.62	II
440	3782.52	II
200	3783.58	II
860	3786.63	II
520	3788.75	II
300	3792.32	II
2500	3801.52	II
800	3803.09	II
1000	3808.11	II
490	3809.21	II
250	3812.20	II
490	3815.85	II
470	3817.46	II
300	3819.02	II
470	3823.90	II
470	3830.55	II
490	3831.08	II
490	3834.55	II
270	3836.10	II
1100	3838.54	II
200	3843.76	II
220	3846.52	II
250	3848.10	II
860	3848.59	II
860	3853.15	II
1200	3854.18	II
1200	3854.31	II
620	3855.29	II
390	3857.02	II
370	3857.64	II
200	3862.46	II
200	3868.13	II
270	3874.68	II
620	3876.97	II
1100	3878.36	II
1500	3882.45	II
1000	3889.98	II
210	3890.75	II
210	3890.98	II
620	3895.11	II
590	3896.80	II
490	3898.27	II
270	3898.94	II
200	3903.34	II
250	3904.34	II
200	3906.92	II
770	3907.29	II
560	3908.41	II
390	3908.54	II
270	3909.31	II
230	3912.19	II
980	3912.44	II
390	3915.52	II
390	3916.14	II
230	3917.64	II
770	3918.28	II
480	3919.81	II
590	3921.73	II
560	3923.11	II
450	3924.64	II
770	3931.09	II
310	3931.37	II
230	3931.83	II
310	3933.73	II
560	3938.09	II
770	3940.34	II
310	3940.97	II
2000	3942.15	II
2700	3942.75	II
770	3943.89	II
310	3947.97	II
3100	3952.54	II
340	3953.66	II
310	3955.36	II
230	3956.06	II
980	3956.28	II
230	3958.27	II
230	3958.87	II
770	3960.91	II
390	3964.50	II
770	3967.05	II
450	3971.68	II
270	3972.07	II
270	3975.07	II
770	3978.65	II
560	3980.88	II
560	3982.89	II
310	3983.29	II
770	3984.68	II
370	3989.44	II
700	3992.39	II
370	3992.91	II
910	3993.82	II
2800	3999.24	II
230	4001.56	II
910	4003.77	II
370	4005.64	II
210	4007.59	II
2700	4012.39	II
910	4014.90	II
250	4015.88	II
200	4019.04	II
240	4022.27	II
840	4024.49	II
240	4025.15	II
840	4028.41	II
250	4030.34	II
840	4031.34	II
340	4037.67	II
2100	4040.76	II
910	4042.58	II
230	4045.21	II
620	4046.34	II
210	4051.43	II
210	4051.99	II
700	4053.51	II
450	4054.99	II
280	4062.22	II
230	4062.94	II
280	4067.28	II
420	4068.84	II
1100	4071.81	II
270	4072.92	II
1800	4073.48	II
210	4073.74	II
1500	4075.71	II
1500	4075.85	II
210	4076.24	II
420	4077.47	II
530	4078.32	II
270	4078.52	II
270	4080.44	II
670	4081.22	II
910	4083.23	II
450	4085.23	II
250	4087.36	II
230	4088.85	II
450	4101.77	II
250	4105.00	II
510	4107.42	II
200	4110.38	II
250	4111.39	II
420	4115.37	II
250	4117.01	II
200	4117.29	II
200	4117.59	II
770	4118.14	II
250	4119.02	II
310	4119.79	II
310	4119.88	II
450	4120.83	II
510	4123.24	II
510	4123.49	II
980	4123.87	II
510	4124.79	II
980	4127.37	II
250	4127.74	II
200	4128.07	II
530	4130.71	II
480	4131.10	II
2700	4133.80	II
270	4135.44	II
270	4137.47	II
2000	4137.65	II
270	4138.10	II
210	4138.35	II
770	4142.40	II
390	4144.49	II
670	4145.00	II
480	4146.23	II
280	4148.90	II
420	4149.79	II
980	4149.94	II
420	4150.91	II
1400	4151.97	II
230	4153.13	II
450	4159.03	II
310	4163.52	II
1300	4165.61	II
620	4166.88	II
250	4167.80	II
320	4169.77	II
320	4169.88	II
340	4176.70	II
340	4181.08	II
340	4185.33	II
3500	4186.60	II
530	4187.32	II
560	4193.09	II
370	4193.28	II
370	4193.87	II
630	4196.34	II
280	4198.00	II
280	4198.67	II
840	4198.72	II
240	4201.24	II
910	4202.94	II
270	4209.41	II
370	4214.04	II
310	4217.59	II
1500	4222.60	II
770	4227.75	II
390	4231.74	II
240	4234.21	II
200	4236.02	II
980	4239.92	II
390	4242.72	II
310	4245.89	II
310	4245.98	II
390	4246.72	II
1100	4248.68	II
390	4253.37	II
620	4255.79	II
200	4263.43	II
620	4270.19	II
390	4270.72	II
200	4278.86	II
280	4285.37	II
200	4288.66	II
200	4289.44	II
2000	4289.94	II
200	4296.07	II
1500	4296.67	II
420	4296.78	II
590	4299.36	II
770	4300.33	II
420	4305.14	II
770	4306.72	II
390	4309.74	II
560	4320.72	II
310	4330.45	II
310	4332.71	II
240	4336.23	II
980	4337.77	II
340	4339.31	II
700	4349.79	II
560	4352.71	II
910	4364.66	II
350	4373.82	II
530	4375.92	II
910	4382.17	II
700	4386.84	II
310	4388.01	II
1700	4391.66	II
200	4398.79	II
510	4399.20	II
350	4410.64	II
350	4410.76	II
310	4416.90	II
980	4418.78	II
200	4423.68	II
310	4427.07	II
480	4427.92	II
310	4428.44	II
650	4429.27	II
480	4444.39	II
450	4444.70	II
770	4449.34	II
620	4450.73	II
2400	4460.21	II
450	4461.14	II
420	4463.41	II
280	4467.54	II
1400	4471.24	II
450	4472.72	II
700	4479.36	II
700	4483.90	II
840	4486.91	II
250	4497.85	II
100	4506.41	I
110	4515.86	II
100	4519.59	II
770	4523.08	II
840	4527.35	II
840	4528.47	II
110	4532.49	II
110	4539.07	II
840	4539.75	II
210	4544.96	II
250	4551.30	II
650	4560.28	II
310	4560.96	II
2100	4562.36	II
420	4565.84	II
1100	4572.28	II
420	4582.50	II
130	4591.12	II
840	4593.93	II
420	4606.40	II
420	4624.90	II
1700	4628.16	II
170	4632.32	I
110	4650.51	I
130	4654.29	II
110	4669.50	II
150	4680.13	II
270	4684.61	II
200	4714.00	II
100	4714.81	II
110	4725.09	II
100	4733.52	II
310	4737.28	II
100	4739.53	II
160	4747.17	II
110	4757.84	II
100	4768.77	II
230	4773.94	II
110	4822.55	I
140	4847.77	I
180	4882.46	II
110	4943.44	I
130	4971.50	II
130	4994.63	I
210	5009.10	I
100	5011.77	II
120	5022.87	II
120	5037.78	II
120	5040.85	I
180	5044.02	II
120	5071.78	I
240	5075.35	II
470	5079.68	II
130	5112.70	I
160	5117.17	II
170	5129.57	I
110	5147.57	II
100	5149.99	I
280	5159.69	I
280	5161.48	I
190	5174.55	I
370	5187.46	II
210	5191.66	II
190	5211.92	I
260	5223.46	I
180	5229.75	I
140	5232.92	II
260	5245.92	I
130	5265.71	II

Cerium (Cont.)

340		5274.23	II
130		5296.56	I
130		5328.08	I
190		5330.54	II
450		5353.53	II
300		5393.40	II
150		5397.64	I
280		5409.23	II
110		5420.38	I
140		5449.24	I
140		5468.37	II
140		5472.29	II
260		5512.08	II
110		5556.25	I
170		5564.97	I
130		5565.97	I
100		5595.88	I
240		5601.28	I
190		5655.14	I
240		5669.96	I
120		5677.75	I
120		5692.94	I
300		5696.99	I
370		5699.23	I
240		5719.03	I
140		5773.12	I
120		5788.15	I
120		5812.92	I
230		5940.86	I
11		6000.18	I
55		6001.90	I
55		6005.86	I
15		6006.20	I
55		6006.82	I
19		6007.37	I
75		6013.42	I
23		6016.59	I
110		6024.20	I
15		6027.16	I
11		6031.26	I
23		6033.58	II
35		6034.20	II
23		6034.41	I
35		6035.49	II
110		6043.39	II
28		6045.42	I
55		6047.40	I
19		6051.80	II
23		6057.50	I
35		6058.00	I
23		6066.75	I
19		6069.46	I
35		6069.48	I
35		6072.00	I
35		6076.61	I
17		6077.16	I
17		6080.37	I
17		6081.28	I
19		6088.86	I
19		6088.96	I
35		6093.19	I
45		6098.34	II
11		6099.80	I
28		6108.74	II
15		6118.56	I
17		6118.90	I
45		6123.67	I
19		6132.00	II
19		6132.18	I
11		6135.45	I
23		6139.03	I
15		6142.92	I
35		6143.36	II
23		6146.43	I
19		6147.84	I
23		6151.72	I
19		6159.82	I
19		6162.14	I
19		6165.45	I
19		6175.28	I
35		6186.17	I
15		6187.97	I
15		6195.23	I
19		6195.53	I
19		6198.05	I
35		6208.98	I
11		6216.82	I
35		6228.94	I
19		6229.13	I
23		6232.45	II
28		6237.45	I
13		6238.71	I
11		6241.87	I
13		6242.91	I
15		6253.65	I
13		6257.99	I
15		6264.27	I
45		6272.05	II
15		6276.47	I
35		6295.58	I
28		6299.51	II
23		6300.21	I
13		6306.64	I
35		6310.01	I
15		6335.40	I
11		6337.21	I
13		6340.70	I
35		6343.95	II
35		6371.11	II
28		6386.84	I
23		6393.02	II
11		6395.16	I
11		6425.29	II
35		6430.07	I
19		6434.39	I
23		6436.40	I
19		6446.12	I
35		6458.03	I
19		6466.88	II
28		6467.39	I
35		6473.72	I
17		6490.97	I
11		6503.27	II
11		6507.16	II
23		6513.59	II
19		6517.31	I
19		6551.70	I
45		6555.65	I
23		6579.10	I
15		6606.35	I
15		6606.86	II
22		6612.06	I
10		6623.00	I
30		6628.93	I
13		6650.89	I
22		6652.72	II
10		6661.41	I
13		6665.59	I
10		6675.54	II
15		6686.60	I
26		6700.66	I
35		6704.27	I
13		6704.52	II
10		6706.04	II
15		6728.71	I
15		6729.57	I
15		6744.70	II
10		6746.90	I
30		6774.28	II
35		6775.59	I
10		6778.28	I
18		6807.81	I
10		6808.82	I
15		6818.23	I
10		6829.73	II
13		6847.25	I
12		6856.55	I
10		6893.66	I
10		6898.45	II
30		6924.81	I
10		6939.45	I
19		6973.50	II
10		6983.82	II
30		6986.02	I
12		7054.51	I
11		7058.68	II
11		7060.00	I
35		7061.75	II
11		7064.49	I
35		7086.35	II
11		7105.04	II
11		7115.08	II
10		7124.73	I
16		7141.42	I
19		7150.23	II
10		7151.67	I
16		7155.25	I
16		7156.99	II
16		7189.40	II
10		7191.72	I
11		7201.56	II
16		7201.89	I
10		7203.55	I
12		7210.67	I
19		7217.36	I
16		7235.71	II
22		7238.36	II
12		7241.73	I
25		7252.75	I
12		7262.64	I
11	h	7277.90	I
11		7296.17	I
19		7301.42	II
19		7313.45	II
25		7329.91	I
16		7334.68	II
12		7343.44	I
25		7397.77	I
11		7401.27	I
12		7417.94	II
11		7424.70	C
12		7433.08	I
11		7438.56	I
12		7444.44	I
10		7472.41	I
16		7486.57	II
11		7527.46	I
11		7527.68	I
10		7533.73	I
10		7551.25	I
12		7562.44	I
10		7562.86	I
10	h	7563.60	I
10		7603.10	I
25		7616.11	II
12		7646.08	I
10		7647.88	I
12		7682.47	I
25		7689.17	II
10		7732.33	I
16		7748.35	I
10		7797.70	I
12		7842.59	I
22		7844.94	II
16		7850.02	II
16		7851.18	II
22		7857.54	II
12		7864.49	I
10		7866.04	I
16		7898.96	II
11		7913.52	I
10		7927.30	C
10		7927.72	I
10		7934.50	II
30		8025.56	II
16		8070.71	I
10		8094.43	I
16		8120.36	I
10		8241.55	II
12		8261.09	I
16		8418.23	II
11		8495.82	I
12		8539.08	II
10		8612.64	I
10	h	8647.66	I
11	h	8702.38	II
25		8772.14	II
12		8810.84	I
30		8891.20	II

Ce III
Ref. 136, 305 — J.R.

Vacuum

Intensity	Wavelength	
100	840.24	III
20	844.11	III
40	845.02	III
20	847.88	III
200	851.18	III
200	852.63	III
200	853.47	III
60	853.78	III
200	855.16	III
200	858.30	III
400	860.15	III
200	862.25	III
40	868.74	III
20	869.51	III
40	869.84	III
20	871.15	III
40	871.27	III
20	880.68	III
60	881.75	III
60	884.04	III
20	885.22	III
30	888.39	III
80	892.75	III
20	899.32	III
100	912.77	III
40	937.04	III
40	999.26	III
20	1025.25	III
20	1025.29	III
20	1026.28	III
40	1029.37	III
20	1034.55	III
20	1041.14	III
100	1042.74	III
50	1051.61	III
70	1057.40	III
100	1057.66	III
100	1058.46	III
50	1062.99	III
30	1063.26	III
50	1063.51	III
100	1067.76	III
100	1068.69	III
20	1070.54	III
200	1072.79	III
40	1073.69	III
30	1079.35	III
20	1080.82	III
30	1088.70	III
30	1090.03	III
20	1092.48	III
20	1099.25	III
40	1100.71	III
20	1107.09	III
20	1111.19	III
20	1116.30	III
20	1125.58	III
20	1129.73	III
20	1132.74	III
100	1142.55	III
20	1192.41	III
50	1201.87	III
20	1204.05	III
20	1719.43	III
20	1796.89	III
20	1836.66	III
20	1836.99	III
30	1862.32	III
30	1950.36	III
20	1990.54	III

Air

Intensity	Wavelength	
100	2033.34	III
100	2057.65	III
100	2077.87	III
200	2083.32	III
100	2089.96	III
400	2109.07	III
100	2122.55	III
500	2136.95	III
1000	2151.44	III
3000	2166.88	III
2000	2169.48	III
5000	2180.64	III
1000	2183.71	III
2000	2203.15	III
3000	2218.11	III
5000	2222.01	III
5000	2225.08	III
3000	2227.84	III
3000	2228.05	III
5000	2242.29	III
2000	2249.25	III
2000	2264.85	III
2000	2268.20	III
2000	2287.82	III
2000	2298.70	III
3000	2300.65	III
4000	2302.09	III
5000	2317.34	III
10000	2318.64	III
5000	2324.31	III
2000	2337.66	III
5000	2350.10	III
2000	2362.54	III
2000	2367.77	III
10000	2372.34	III
5000	2377.07	III
5000	2377.48	III
10000	2380.12	III
3000	2382.28	III
3000	2385.06	III
5000	2395.04	III
3000	2406.15	III
4000	2408.08	III
2000	2410.26	III
5000	2415.60	III
2000	2417.01	III
2000	2423.02	III
3000	2428.64	III
5000	2430.24	III
10000	2431.45	III
15000	2439.80	III
3000	2441.55	III
2000	2444.78	III
10000	2454.32	III
10000	2469.95	III
3000	2471.66	III
5000	2477.25	III
8000	2479.44	III
3000	2479.51	III
10000	2483.82	III
10000	2497.50	III
3000	2503.56	III
2000	2504.43	III
20000	2531.99	III
3000	2539.27	III
3000	2557.49	III
4000	2577.67	III
2000	2578.30	III
2000	2584.71	III
10000	2603.59	III
2000	2607.96	III
2000	2615.79	III
2000	2649.38	III
2000	2662.81	III
3000	2719.30	III
2000	2730.04	III
3000	2743.71	III
4000	2748.90	III
4000	2754.87	III
4000	2768.28	III
3000	2849.40	III

Cerium (Cont.)

Intensity	Wavelength	Spectrum
2000	2861.39	III
4000	2907.05	III
10000	2923.81	III
5000	2925.26	III
10000	2931.54	III
2000	2948.53	III
5000	2973.72	III
10000	3022.75	III
50000	3031.58	III
95000	3055.59	III
20000	3056.56	III
40000	3057.23	III
20000	3057.58	III
40000	3085.10	III
20000	106.98	III
30000	3110.53	III
30000	3121.56	III
20000	3141.29	III
20000	3143.96	III
20000	3147.06	III
20000	3228.57	III
3000	3234.20	III
4000	3267.76	III
3000	3267.94	III
20000	3353.29	III
10000	3395.77	III
4000	3398.91	III
30000	3427.36	III
40000	3443.63	III
30000	3454.39	III
40000	3459.39	III
60000	3470.92	III
50000	3497.81	III
60000	3504.64	III
500	3514.41	III
50000	3544.07	III
3000	3784.29	III
800	3936.80	III
300	3957.10	III
500	4169.42	III
300	4191.70	III
500	4194.83	III
300	4213.26	III
300	4217.13	III
400	4284.77	III
300	4304.71	III
600	4346.35	III
400	4389.97	III
600	4448.32	III
500	4485.27	III
1000	4521.92	III
1000	4535.73	III
300	4576.90	III
500	4627.60	III
300	4766.07	III
500	4976.45	III
500	5650.97	III
1000	5664.20	III
500	5691.08	III
300	5710.59	III
500	5749.47	III
500	5949.83	III
2000	5962.22	III
500	5962.71	III
400	5979.56	III
1000	5983.40	III
3000	6002.63	III
10000	6032.54	III
10000	6060.91	III
500	6061.79	III
500	6097.35	III
500	6098.87	III
500	6135.10	III
300	6287.79	III
500	6308.16	III
300	6341.75	III
1000	6944.94	III
700	7739.04	III
300	7758.27	III
500	7826.80	III
300	7948.64	III
500	7960.31	III
500	7991.01	III
400	8030.80	III
300	8084.12	III
400	8177.33	III
300	8186.03	III
300	8222.16	III
300	9056.53	III
300	9079.58	III
400	9328.20	III
300	9367.03	III
300	9567.37	III
400	10458.37	III
400	10494.42	III
300	10534.36	III
400	10684.46	III
15	12756.96	III
12	12821.62	III
80	15847.58	III
80	15956.79	III
12	15960.59	III
87	16128.75	III
42	18579.82	III
38	19141.29	III
27	19377.15	III
26	19466.14	III
20	19498.14	III
55	19524.18	III
30	20685.63	III
12	21380.23	III

Ce IV
Ref. 166 — J.R.

Intensity	Wavelength	
	Vacuum	
2	447.58	IV
1	443.11	IV
8	558.92	IV
8	571.59	IV
40	741.79	IV
30	754.60	IV
12	755.75	IV
6	975.20	IV
5	1009.31	IV
2	1022.12	IV
9	1057.67	IV
1	1059.64	IV
50	1289.41	IV
75	1332.16	IV
75	1372.72	IV
2	1577.60	IV
1	1572.62	IV
15	1641.58	IV
20	1775.30	IV
20	1779.03	IV
35	1914.75	IV
10	1937.21	IV
	Air	
100	2000.42	IV
35	2003.11	IV
100	2009.94	IV
3	2433.50	IV
5	2445.50	IV

Ce V
Ref. 261 — J.R.

Intensity	Wavelength	
	Vacuum	
100	365.66	V
300	399.36	V
150	404.21	V
200	482.96	V
100	552.13	V

CESIUM (Cs)
Z = 55

Cs I and II
Ref. 82, 154, 155, 200, 263, 325— C.J.S.

Intensity		Wavelength	
		Vacuum	
250		591.04	II
2000		639.36	II
500		668.39	II
15000		718.14	II
15000		808.76	II
15000		813.84	II
35000		901.27	II
40000		926.66	II
3		1656.15	II
7		1689.46	II
2		1691.83	II
7		1717.64	II
7		1718.97	II
3		1727.79	II
3		1736.77	II
8		1807.83	II
8		1813.75	II
7		1815.16	II
18		1840.50	II
13		1859.16	II
7		1864.83	II
7		1876.72	II
18		1883.93	II
5		1914.61	II
29		1935.19	II
		Air	
72		2024.97	II
46		2028.32	II
180		2080.05	II
160		2087.20	II
100		2102.22	II
160		2205.52	II
190		2220.53	II
120		2245.81	II
1600		2267.65	II
750		2273.84	II
220		2285.41	II
250		2286.15	II
330		2315.69	II
1400		2332.46	II
240		2375.86	II
220		2379.27	II
1000		2392.86	II
370		2425.17	II
220		2543.93	II
630		2596.99	II
450		2609.43	II
170		2628.86	II
300		2699.18	II
180		2789.78	II
150		2793.31	II
680	w	2816.92	II
150		2829.04	II
140	c	2866.36	II
1600		2931.08	II
290		3151.19	II
520	c	3265.91	II
650	w	3267.12	II
850	c	3271.63	II
570		3368.57	II
150		3514.05	II
110	w	3615.01	II
170	w	3680.11	II
130		3732.56	II
1100	w	3785.44	II
2700		3805.12	II
520	c	3861.50	II
2100	c	3876.15	I
600	c	3888.61	I
3400		3896.99	II
4200		3959.51	II
2500		3965.20	II
2100		3974.25	II
8000		4039.85	II
2000		4068.78	II
320	w	4151.27	II
2000		4213.14	II
1900		4232.20	II
980		4234.41	II
14000		4264.70	II
18000	w	4277.13	II
5100		4288.38	II
1900		4300.65	II
7600		4363.30	II
3700		4373.04	II
960		4384.44	II
3900	w	4405.26	II
12000		4501.55	II
20000		4526.74	II
4100		4538.97	II
1000	c	4555.28	I
460	c	4593.17	I
100000		4603.79	II
4200	c	4616.17	II
2800		4646.52	II
990		4670.29	II
1500		4732.99	II
7000		4763.64	II
1900		4786.38	II
25000		4830.19	II
19000		4870.04	II
4900	c	4880.05	II
37000		4952.85	II
8200		4972.60	II
27000		5043.80	II
2800	c	5059.87	II
2900		5096.60	II
6500	c	5209.58	II
75000		5227.04	II
29000		5249.38	II
11000		5274.05	II
1300		5306.60	II
10000	c	5349.13	II
22000		5370.99	II
1200		5402.78	II
2900		5419.67	II
60	c	5465.94	I
37		5502.88	I
39000		5563.02	II
100		5635.21	I
210	c	5664.02	I
27		5745.72	I
4500		5814.16	II
24000		5831.14	II
59	c	5838.83	I
300		5845.14	I
51000		5925.63	II
1400	c	5984.39	II
640	c	6010.49	I
86		6034.09	I
760		6076.72	II
9800		6128.61	II
1000		6213.10	I
170		6217.60	I
320	c	6354.55	I
2000		6419.52	II
8300		6495.53	II
10000	w	6536.44	II
490		6586.51	I
530	c	6628.01	II
97		6628.66	I
8800		6646.57	II
3300	c	6723.28	I
9600		6724.47	II
200		6824.65	I
300		6870.45	I
37000		6955.50	II
4800		6973.30	I
16000		6979.67	II
980		6983.49	I
2300		7130.54	II
13000	w	7149.54	II
1600		7160.90	II
630		7188.37	II
1100		7206.04	II
790		7228.53	I
960		7248.88	II
130		7279.90	I
1100		7279.96	I
780		7369.36	II
550		7437.78	II
440	w	7523.39	II
2600	c	7608.90	I
910	w	7651.95	II
310	h	7746.98	II
2400		7852.52	II
3300		7943.88	I
22000		7997.44	II
2100		8012.98	II
3500		8015.73	I
8200	c	8047.13	II
1300		8078.50	II
510		8078.94	I
4500		8079.04	I
59000	cr	8521.13	I
680		8521.62	II
1700		8608.31	II
420	w	8695.60	II
15000	c	8761.41	I
840		8775.42	II
340	w	8857.39	II
61000	cr	8943.47	I
18000		9172.32	I
5200		9208.53	I
540	c	9212.36	II
3300	c	9220.75	II
310	c	9718.11	II
630	c	9932.91	II
1400		9994.79	II
19000		10024.36	I
4800		10123.41	I
26000		10123.60	I
3000	c	10176.02	II
2700		10379.66	II
1300		10480.93	II
4700		10504.51	II
1900	w	10807.88	II
610	w	11324.34	II
530	c	11496.56	II
410	c	11704.18	II
330	c	11797.93	II
710	c	11840.84	II
1100		12604.29	II
2000	c	12735.52	II
400	c	12746.34	II
850		13406.83	II
2900		13424.31	I
38000	c	13588.29	I
8400		13602.56	I
4200	c	13692.91	II
5700		13758.81	I
1400	c	13868.82	II
880	c	14482.21	II
55000	c	14694.91	I
350		14906.91	II
1900	c	15293.80	II
1600	c	15356.61	II
620		15445.47	II
940	c	15735.48	II
1100	c	16426.14	II
820		16535.63	I
1500		17012.32	I
340		18160.84	II
430		18179.35	II
710		18221.36	II
390		18222.40	II
510		18404.77	II
310		18407.23	II
590		18509.52	II
610		18509.85	II
100	c	18742.18	II
530	c	18921.76	II
150	c	18986.48	II
180		19924.85	II
390		20110.77	II
760		20138.47	I
200		20301.54	II
220	c	20443.87	II
92	w	21103.42	II
79	c	22344.98	II
78	c	22448.98	II
880		22811.86	I
1100		23037.98	I
3900		23344.47	I

Cesium (Cont.)

Intensity		Wavelength	
320	c	23408.41	II
180		24132.52	II
4400		24251.21	I
850		24374.96	I
120	c	24528.78	II
140		24810.89	II
170		25189.83	II
900		25220.37	II
240		25733.29	II
890	d	25763.51	II
500		25764.73	I
120		26448.83	II
340		26503.26	II
190		26727.33	II
17		26954.56	II
22		26978.68	II
23		27187.98	II
9		28649.38	II
680	c	29310.06	I
7		29542.08	II
2800		30103.27	I
9		30487.16	II
610	c	30953.06	I
1100		34900.13	I
190		36131.00	I
2	c	39177.28	I
2	d	39421.25	I
1		39424.11	I

Cs III

Ref. 78, 200, 201—C.J.S.

Intensity		Wavelength	
		Vacuum	
10000		614.01	III
2000		638.17	III
2500		666.25	III
5000		691.60	III
3500		703.89	III
20000		721.79	III
20000		722.20	III
5000		731.56	III
12000		740.29	III
7500		830.39	III
15000		920.35	III
25000	c	1054.79	III
17	c	1673.99	III
12	c	1705.25	III
10		1801.83	III
20	c	1822.40	III
11		1823.93	III
12		1824.70	III
12		1841.80	III
25		1915.50	III
25	c	1923.29	III
12		1961.33	III
17		1996.56	III
		Air	
710		2035.11	III
120		2056.43	III
330		2076.43	III
540		2077.30	III
410		2088.68	III
210		2101.63	III
200		2141.47	III
1000		2316.88	III
230		2325.95	III
390		2340.49	III
1600		2455.81	III
1600		2477.57	III
890		2485.45	III
410		2495.07	III
1400		2525.67	III
430		2573.05	III
16000		2596.86	III
390		2610.12	III
6200		2630.51	III
370		2700.32	III
710		2701.20	III
390		2776.44	III
270		2810.87	III
630		2845.70	III
3100		2859.32	III
200		2893.85	III
180		2921.13	III
3200		2976.86	III
210		3001.28	III
1700		3066.59	III
1100	c	3149.36	III
1400		3152.36	III
8400		3268.32	III
1300		3315.51	III
550		3340.60	III
430		3344.02	III
1200		3349.46	III
400		3463.45	III
580		3476.83	III
480		3559.82	III
7200		3597.45	III
1300		3608.31	III
2300		3618.19	III
300	c	3641.34	III
520		3651.08	III
4800		3661.40	III
640		3699.50	III
430		3837.46	III
2900		3888.37	III
2700		3925.60	III
680	c	4001.70	III
3100		4006.55	III
420		4006.78	III
520		4043.42	III
370		4403.86	III
1200		4410.22	III
940		4425.68	III
530		4471.48	III
1200		4506.72	III
590		4522.86	III
420	h	4620.61	III
210		4665.52	III
140		4851.59	III
370		5035.72	III
230		5380.79	III
140		5950.14	III
110		5979.97	III
150		6043.99	III
870		6079.86	III
330		6150.42	III
450		6242.96	III
510		6456.33	III
400		6753.12	III
1900	c	7219.60	III

Cs IV

Ref. 259 — J.R.

Intensity		Wavelength	
		Vacuum	
35		707.20	IV
60		759.57	IV
5		778.21	IV
500		824.80	IV
350		828.86	IV
400		868.18	IV
1000		874.84	IV
400		896.92	IV
400		923.02	IV
600		986.14	IV
300		995.14	IV
60		1019.13	IV
550		1068.91	IV
200		1282.66	IV

CHLORINE (Cl)
Z = 17

Cl I and II

Ref 238, 239 — L.J.R.

Intensity		Wavelength	
		Vacuum	
350		559.305	II
400		571.904	II
800		574.406	II
500		586.24	II
700		618.057	II
600		619.982	II
800		620.298	II
700		626.735	II
800		635.881	II
1000		636.626	II
1000		650.894	II
1000		659.811	II
1300		661.841	II
2000		663.074	II
1500		682.053	II
1500		687.656	II
1500		693.594	II
2000		725.271	II
2500		728.951	II
2000		777.562	II
5000		787.580	II
5000		788.740	II
5000		793.342	II
6000		839.297	II
8000		839.599	II
7000	p	841.41	II
5000		851.691	II
2000		888.026	II
2000		893.549	II
2000		961.499	II
30		969.92	I
40		978.284	I
25		998.372	I
25		998.432	I
75		1002.346	I
150		1013.664	I
90		1025.553	I
6000		1063.831	II
3000		1067.945	II
9000		1071.036	II
6000		1071.767	II
5000		1075.230	II
5000		1079.080	II
200		1084.667	I
200		1085.171	I
250		1085.304	I
400		1088.06	I
350		1090.271	I
250		1090.982	I
250		1092.437	I
400		1094.769	I
350		1095.148	I
350		1095.662	I
400		1095.797	I
250		1096.810	I
300		1097.369	I
200		1098.068	I
200		1099.523	I
500		1107.528	I
800		1139.214	II
800		1167.148	I
3000		1179.293	I
1200		1188.774	I
900		1201.353	I
3000		1335.726	I
10000		1347.240	I
5000		1351.657	I
12000		1363.447	I
2500		1373.116	I
20000		1379.528	I
25000		1389.693	I
20000		1389.957	I
12000		1396.527	I
500		1441.470	II
500		1528.569	II
500		1542.942	II
500		1558.144	II
500		1565.050	II
500		1857.488	II
450	h	1997.370	II
		Air	
450		2032.116	II
350	h	2088.583	II
350	h	2091.458	II
170		2427.79	II
360		2434.07	II
340		2498.53	II
470		2502.74	II
260		2546.96	II
500		2549.88	II
460		2564.84	II
320		2603.31	II
950		2658.72	II
750		2676.95	II
1200		2688.04	II
410		2912.05	II
950		2996.65	II
500		3006.06	II
950		3057.96	II
1300		3071.32	II
1400		3092.19	II
1200		3123.72	II
1900		3315.43	II
1200		3329.10	II
2500		3353.35	II
20		3726.54	I
1200		3749.96	II
1000		3781.17	II
1500		3798.76	II
1900		3805.18	II
1300		3809.46	II
1700		3820.20	II
2800		3827.59	II
4500		3833.35	II
2500		3843.20	II
3100		3845.37	II
3900		3845.65	II
1500		3845.80	II
10000		3850.99	II
7900		3851.37	II
1200		3851.65	II
25000		3860.83	II
4400		3860.99	II
1000		3861.37	II
1500		3913.87	II
1100		3916.63	II
20		3944.82	I
20		4104.79	I
10000	h	4132.50	II
65		4209.67	I
50		4226.42	I
60		4264.58	I
100		4363.27	I
100		4369.50	I
5000		4372.93	II
100		4379.90	I
100		4389.75	I
90		4390.40	I
90		4403.03	I
100		4438.49	I
90		4475.30	I
1500		4489.91	II
100		4526.19	I
80		4600.98	I
40		4623.938	I
50		4654.040	I
80		4661.208	I
45		4691.523	I
40		4721.255	I
45		4740.729	I
4300		4768.65	II
13000		4781.32	II
99000		4794.55	II
29000		4810.06	II
16000		4819.47	II
81000		4896.77	II
47000		4904.78	II
26000		4917.73	II
10000		4995.48	II
26000		5078.26	II
30		5099.789	I
56000		5217.94	II
23000		5221.36	II
15000		5392.12	II
99000		5423.23	II
10000		5423.51	II
19000		5443.37	II
10000		5444.21	II
5600		5457.02	II
40		5532.162	I
50	d	5796.305	I
45		5799.914	I
30		5856.742	I
100	d	5948.58	I
50		6019.812	I
35		6082.61	I
1900		6094.69	II
160		6114.43	I
200		6140.245	I
160		6194.757	I
160		6398.66	I
150		6434.833	I
150		6531.43	I
1400		6661.67	II
150		6678.43	I
1300		6686.02	II
1200		6713.41	II
150		6840.29	I
300		6932.903	I
300		6981.886	I
600		7086.814	I
7500		7256.62	I
5000		7414.11	I
550		7462.370	I
550		7489.47	I
700		7492.118	I
11000		7547.072	I
2300		7672.42	I
450		7702.828	I
7000		7717.581	I
10000		7744.97	I
2200		7769.16	I
650		7771.09	I
2200		7821.36	I
1700		7830.75	I
3000		7878.22	I
220		7893.34	I
2300		7899.31	I
1800		7915.08	I
3000		7924.645	I
2100		7933.89	I
1700		7935.012	I
650		7952.52	I
1500		7974.72	I
1300		7976.97	I
600		7980.60	I
2900		7997.85	I
2200		8015.61	I
1100		8023.33	I
400		8051.07	I
1700		8084.51	I
2200		8085.56	I
3000		8086.67	I
1300		8087.73	I
250		8094.67	I
2500		8194.42	I
2200		8199.13	I
2200		8200.21	I
800		8203.78	I
18000		8212.04	I
3000		8220.45	I
20000		8221.74	I
18000		8333.31	I
1000		8360.71	II
560		8361.84	II
99900		8375.94	I
180		8382.67	II
100		8392.02	II
400		8406.199	I
15000		8428.25	I
2200		8467.34	I
2200		8550.44	I
20000		8575.24	I
750		8578.02	I
75000		8585.97	I
450		8628.54	I
300		8641.71	I
3500		8686.26	I
2200		8912.92	I
3000		8948.06	I
2000		9038.982	I
2500		9045.43	I
1000		9069.656	I
2000		9073.17	I
7500		9121.15	I

Chlorine (Cont.)

Intensity	Wavelength	Spectrum
3000	9191.731	I
500	9197.596	I
4000	9288.86	I
1500	9393.862	I
3500	9452.10	I
500	9486.964	I
1000	9584.801	I
3500	9592.22	I
250	9632.509	I
1000	9702.439	I
250	9744.426	I
200	9807.057	I
400	9875.970	I
331	10392.549	I
38	10432.83	II
10	10506.62	II
14	10509.12	II
19	10512.46	II
25	10514.17	II
9	10801.47	II
5	10885.42	II
1	10955.71	II
300	11123.05	I
231	11392.62	I
269	11409.69	I
1000	11436.33	I
180	11720.56	I
195	11866.76	I
172	12021.7	I
350	13243.8	I
310	13296.0	I
550	13346.8	I
525	13821.7	I
148	14369.7	I
294	14931.7	I
269	15108.0	I
381	15465.1	I
169	15467.6	I
1094	15520.3	I
1487	15730.1	I
193	15818.4	I
2780	15869.7	I
277	15883.3	I
342	15928.9	I
735	15960.0	I
283	15970.5	I
129	16077.6	I
259	16198.5	I
227	19370.3	I
717	19755.3	I
185	19766.8	I
227	20199.4	I
85	20370.1	I
100	24470.0	I
	39603.7	I
	39615.3	I
	39716.0	I
	39744.0	I
	39750.9	I
	39875.3	I
	39881.0	I
	39985.7	I
	40085.5	I
	40089.5	I
	40171.0	I
	40310.3	I
	40335.4	I
	40532.2	I

Cl III
Ref. 28, 30 — L.J.R.

Intensity	Wavelength	
	Vacuum	
100	406.27	III
400	411.37	III
400	411.81	III
600	556.23	III
700	556.61	III
700	557.12	III
600	2965.56	III
600	3104.46	III
800	3139.34	III
900	3191.45	III
700	3289.80	III
700	3320.57	III
800	3329.06	III
900	3340.42	III
800	3392.89	III
800	3393.45	III
900	3530.03	III
800	3560.68	III
900	3602.10	III
800	3612.85	III
700	3622.69	III
700	3656.95	III
700	3670.28	III
700	3682.05	III
600	3705.45	III
600	3707.34	III
800	3720.45	III
800	3748.81	III
500	3779.35	III
500	3925.87	III
700	3991.50	III
600	4018.50	III
600	4059.07	III
500	4104.23	III
500	4106.83	III
400	4370.91	III
500	4608.21	III
300	4703.14	III
100	4863.75	III
10	4971.64	III
700	561.53	III
700	561.68	III
700	561.74	III
500	606.35	III
400	621.28	III
300	670.38	III
300	673.13	III
100	936.28	III
500	1005.28	III
600	1008.78	III
700	1015.02	III
600	1822.50	III
500	1828.40	III
500	1901.61	III
500	1983.61	III
	Air	
400	2006.84	III
700	2253.07	III
500	2268.95	III
500	2278.34	III
700	2283.93	III
600	2323.50	III
500	2336.45	III
600	2340.64	III
600	2359.67	III
600	2370.37	III
700	2416.42	III
600	2447.14	III
600	2448.58	III
500	2486.91	III
500	2532.48	III
600	2580.67	III
500	2603.59	III
500	2632.67	III
500	2633.18	III
600	2665.54	III
700	2710.37	III

Cl IV
Ref. 11, 28, 30, 31 — L.J.R.

Intensity	Wavelength	
	Vacuum	
300	319.62	IV
200	331.84	IV
400	437.83	IV
400	464.86	IV
800	486.17	IV
800	534.73	IV
700	535.67	IV
600	536.15	IV
900	537.61	IV
600	538.12	IV
500	549.22	IV
400	550.02	IV
700	552.02	IV
600	553.30	IV
700	554.62	IV
500	601.50	IV
500	604.59	IV
400	608.90	IV
400	612.07	IV
400	653.70	IV
400	745.21	IV
400	831.43	IV
500	834.84	IV
500	834.97	IV
400	840.81	IV
600	840.93	IV
	865.3	IV
500	973.21	IV
600	977.56	IV
400	977.90	IV
700	984.95	IV
400	985.75	IV
300	1537.21	IV
200	1539.30	IV
200	1545.19	IV
200	1549.15	IV
200	1622.86	IV
	Air	
400	2701.36	IV
500	2724.03	IV
500	2751.23	IV
400	2770.64	IV
700	2782.47	IV
400	2835.4	IV
500	3063.13	IV
600	3076.68	IV
200	3167.87	IV

Cl V
Ref. 11, 28, 30, 85, 233 — L.J.R.

Intensity	Wavelength	
	Vacuum	
300	287.33	V
300	373.78	V
400	390.15	V
500	392.43	V
300	536.53	V
400	537.01	V
300	537.46	V
500	538.03	V
400	538.68	V
800	542.23	V
600	542.30	V
400	542.87	V
1000	545.11	V
600	546.33	V
1000	547.63	V
400	633.19	V
400	635.32	V
400	681.92	V
400	683.17	V
400	688.93	V
	715.55	V
	716.19	V
400	883.13	V
400	894.34	V
100	894.91	V
	914.5	V

CHROMIUM (Cr)
Z = 24

Cr I and II
Ref. 1 — C.H.C.

Intensity		Wavelength	
		Vacuum	
19000		2055.52	II
14000		2061.49	II
8900		2065.42	II
80	h	2364.71	I
130		2383.33	I
140		2408.62	I
170		2496.31	I
110		2502.53	I
190		2504.31	I
50		2508.11	I
60		2508.98	I
40		2513.62	I
110		2516.92	I
80		2518.71	I
390		2519.52	I
190		2527.12	I
40		2530.45	I
70		2534.34	II
50		2545.64	I
160		2549.54	I
40		2553.06	I
80		2557.15	I
130		2560.69	I
150		2571.74	I
100		2577.65	I
50		2588.20	I
380		2591.85	I
35		2603.57	I
35		2622.86	I
22		2625.32	I
18		2626.60	I
18		2629.82	I
35		2642.12	I
250		2653.59	II
250		2658.59	II
70		2661.73	II
320		2663.42	II
70		2663.68	II
440		2666.02	II
280		2668.71	II
350		2671.81	II
280		2672.83	II
1800		2677.16	II
35		2678.16	I
320		2678.79	II
18		2680.34	II
230		2687.09	II
60		2688.04	I
55		2688.29	II
26		2690.26	I
280		2691.04	II
35		2693.52	II
35		2697.91	II
180		2698.41	II
180		2698.69	II
18		2700.60	I
110		2701.99	I
18		2702.53	I
70		2703.48	I
		2703.55	II
35		2703.86	II
18		2705.43	I
60		2708.79	II
35		2709.31	II
140		2712.31	II
45		2716.18	I
55		2717.51	II
45		2718.43	II
170		2722.75	II
18		2724.04	II
420	h	2726.51	I
45		2727.26	II
280	h	2731.91	I
170	h	2736.47	I
70		2739.38	I
70		2740.10	II
95		2741.07	I
95		2742.03	II
95		2742.17	I
250		2743.64	II
35		2746.21	II
110	h	2748.29	I
330		2748.98	II
390		2750.73	II
45		2751.60	I
280		2751.87	II
110	h	2752.88	I
35		2754.28	II
22		2754.90	I
22		2755.27	I
22		2756.75	I
150		2757.10	I
350		2757.72	II
60		2758.98	II
80		2759.39	II
45		2759.73	II
90	h	2761.76	I
750		2762.59	II
22		2763.06	I
80	h	2764.35	I
750		2766.54	II
22		2767.54	I
250	h	2769.92	I
18		2771.45	I
45		2778.06	II
22		2779.14	I
80		2780.30	II
610		2780.70	I
70		2785.70	II
35		2787.63	II
35		2787.84	I
90		2792.16	II
55		2798.67	II
70		2800.77	II
80		2812.01	II
60		2818.36	II
45		2822.01	II
180		2822.37	II
22		2826.75	I
180		2830.47	II
70		2834.26	II
2500		2835.63	II
45		2836.48	II
55		2838.79	II
110		2840.02	II
1700		2843.25	II
22		2846.02	I
45		2849.29	I
1200		2849.84	II
120		2851.36	II
55		2853.22	II
55		2855.07	II
880		2855.68	II
90		2856.77	II
70		2857.40	II
610		2858.91	II
440		2860.93	II
790		2862.57	II
750		2865.11	II
55		2865.33	II
610		2866.74	II
90		2867.10	II
480		2867.65	II
210		2870.44	II
110		2871.63	I
160		2873.48	II
90		2873.82	II
320		2875.99	II
230		2876.24	II
180		2877.98	II
70		2878.45	II
120		2879.27	I
95		2880.87	II
30		2881.14	I
170		2887.00	I
55		2888.74	II
700		2889.29	I
55		2889.82	II
55		2891.42	I
370		2893.25	I
190		2894.17	I
55		2896.46	II

210		2896.75	I
55	d	2897.67	II
		2897.73	II
90		2898.54	II
80		2899.21	I
55		2899.48	II
26		2903.97	II
55		2904.68	I
180		2905.49	I
260		2909.05	I
260		2910.90	I
250		2911.14	I
45		2911.68	II
60		2913.73	I
22		2915.23	II
22		2915.46	II
90		2921.24	II
60		2921.82	II
60		2927.08	II
80		2928.15	II
95		2928.30	II
26		2929.44	II
35		2930.85	II
26		2932.70	II
55		2933.97	II
90		2935.14	II
45		2940.22	II
60		2946.84	II
55		2953.36	II
45		2953.71	II
55		2961.73	II
45		2966.05	II
480		2967.64	I
480		2971.11	I
210		2971.91	II
480		2975.48	I
30		2976.72	II
190		2979.74	II
350		2980.79	I
110		2985.32	II
480		2985.85	I
1500		2986.00	I
2100		2986.47	I
660		2988.65	I
160		2989.19	II
480		2991.89	I
230		2994.07	I
300		2995.10	I
700		2996.58	I
210		2998.79	I
1100		3000.89	I
750		3005.06	I
140		3013.03	I
710		3013.71	I
710		3014.76	I
1400		3014.92	I
710		3015.19	I
2800		3017.57	I
430		3018.50	I
240		3018.82	I
430		3020.67	I
2800		3021.56	I
1100		3024.35	I
85		3026.65	II
170		3029.16	I
710		3030.24	I
140		3031.35	I
28		3032.93	II
390		3034.19	I
550		3037.04	I
80		3039.78	I
550		3040.85	I
		3040.91	II
55		3041.74	II
110		3050.14	II
710		3053.88	I
24		3059.52	II
85		3065.07	I
28		3067.16	II
85		3073.68	I
55		3077.83	I
28		3095.86	I
28		3109.34	I
28		3110.86	I
240		3118.65	II
45		3119.25	I
40		3119.71	I
430		3120.37	II
28		3122.60	II
470		3124.94	II
		3125.02	II
120		3128.70	II
590		3132.06	II
140		3136.68	II
140		3147.23	II
85		3148.44	I
100		3155.15	I
100		3163.76	I
240		3180.70	II
30		3181.43	II
65	h	3188.01	I
220		3197.08	II
24		3198.11	I
30		3208.59	II
170		3209.18	II
140		3217.40	II
30		3229.20	I
28		3234.06	II
65		3237.73	I
120		3245.54	I
130		3251.84	I
130		3257.82	I
95		3259.98	I
30		3295.43	II
24		3307.02	II
55		3324.06	II
28		3326.59	I
30		3328.35	II
30		3329.05	I
95		3336.33	II
130		3339.80	II
110		3342.59	II
30		3343.34	I
95		3346.02	I
95		3346.74	I
95		3347.84	II
65		3349.07	I
55		3349.32	I
30		3351.60	I
55		3351.97	I
55	h	3353.03	I
		3353.13	II
170		3358.50	II
160		3360.30	II
65		3361.77	II
55		3362.21	I
430		3368.05	II
30		3376.40	I
55		3378.34	II
30		3379.17	I
30		3379.37	II
95		3379.83	II
140		3382.68	II
95		3391.43	II
55		3392.99	II
70		3393.84	II
55		3394.30	II
30		3402.40	II
170		3403.32	II
360		3408.76	II
210		3421.21	II
270		3422.74	II
140		3433.31	II
270		3433.60	I
55		3434.11	I
160		3436.19	I
70		3441.12	I
140		3441.44	I
30		3443.79	I
170		3445.62	I
30		3447.02	I
170		3447.43	I
70		3447.76	I
190		3453.33	I
40		3453.74	I
130		3455.60	I
100		3460.43	I
65		3465.25	I
40		3467.02	I
70		3467.72	I
45		3469.59	I
16		3472.76	I
24		3472.91	I
40		3473.61	I
70		3481.30	I
55		3481.54	I
55		3494.97	I
40		3495.38	II
80		3510.54	I
40		3511.84	II
120		3550.64	I
80		3558.52	I
130		3566.16	I
130		3573.64	I
80		3574.04	I
330	h	3574.80	I
		3574.94	I
19000		3578.69	I
160	h	3584.33	I
130		3585.30	II
17000		3593.49	I
350		3601.67	I
40		3602.57	I
85		3603.74	I
		3603.78	II
13000		3605.33	I
40		3608.40	I
40		3609.48	I
40		3610.05	I
70		3612.61	I
85		3615.64	I
130		3632.84	I
350		3636.59	I
630		3639.80	I
85		3640.39	I
70		3641.47	I
220		3641.83	I
45		3646.16	I
85		3648.53	I
220		3649.00	I
170		3653.91	I
220		3656.26	I
45		3662.84	I
130		3663.21	I
45		3665.98	I
95		3666.64	I
55		3668.03	I
65		3676.32	I
40		3677.68	II
55		3677.89	II
40		3679.82	I
19		3681.69	I
120		3685.55	I
130		3686.80	I
130		3687.25	I
75		3687.54	I
19		3688.46	I
75		3712.95	II
40		3716.53	I
130		3730.81	I
150		3732.03	I
95		3742.97	I
480		3743.58	I
570		3743.88	I
85		3744.49	I
55		3748.61	I
340		3749.00	I
50		3757.17	I
230		3757.66	I
60		3758.04	I
24		3767.43	I
260		3768.24	I
95		3768.73	I
95		3788.86	I
95		3790.45	I
130		3791.38	I
130		3792.14	I
120		3793.29	I
130		3793.88	I
85		3794.61	I
140		3797.13	I
200		3797.72	I
530		3804.80	I
110		3806.83	I
110		3807.93	I
180		3815.43	I
70		3818.48	I
180		3819.56	I
70		3823.52	I
130		3826.42	I
130		3830.03	I
380		3841.28	I
190		3848.98	I
140		3849.36	I
290		3850.04	I
140		3852.22	I
190		3854.22	I
110		3855.29	I
140		3855.57	I
260		3857.63	I
70		3874.53	I
660		3883.29	I
50		3883.66	I
570		3885.22	I
380		3886.79	I
60		3891.93	I
260		3894.04	I
40		3897.65	I
35		3902.11	I
360		3902.92	I
60		3903.16	I
960		3908.76	I
120	Hd	3911.82	I
		3912.00	I
120		3915.84	I
190		3916.24	I
35		3917.60	I
1900		3919.16	I
600		3921.02	I
30		3926.65	I
600		3928.64	I
410		3941.49	I
30		3951.10	I
40		3952.40	I
35		3953.16	I
1900		3963.69	I
120		3969.06	I
1600		3969.75	I
85		3971.26	I
1600		3976.66	I
85		3978.68	I
40		3979.80	I
85		3981.23	I
960		3983.91	I
190		3984.34	I
160		3989.99	I
960		3991.12	I
160		3991.67	I
190		3992.84	I
40		3993.97	I
160		4001.44	I
120		4012.47	II
30		4014.67	I
85		4022.26	I
70		4025.01	I
120		4026.17	I
85		4027.10	I
85		4030.68	I
190		4039.10	I
160		4048.78	I
120		4058.77	I
40		4065.72	I
85		4066.94	I
35		4074.86	I
40		4076.06	I
40		4077.09	I
40		4077.68	I
40		4104.87	I
40		4109.58	I
40		4120.61	I
40		4121.82	I
35		4122.16	I
40		4123.39	I
140		4126.52	I
35		4127.30	I
40		4127.64	I
40		4131.36	I
30		4152.78	I
120		4153.82	I
85		4161.42	I
140		4163.62	I
70		4165.52	I
40		4169.84	I
35		4170.20	I
40		4172.77	I
170		4174.80	I
30		4175.94	I
170		4179.26	I
35		4184.90	I
30		4186.36	I
35		4190.13	I
85		4191.27	I
35		4192.10	I
85		4193.66	I
70		4194.95	I
40		4197.23	I
85		4198.52	I
60		4203.59	I
40		4204.47	I
35		4208.36	I
110		4209.37	I
40		4209.76	I
40		4211.35	I
40		4216.36	I
85		4217.63	I
40		4221.57	I
40		4222.73	I
40		4238.96	I
60		4240.70	I
20000		4254.35	I
70		4255.50	I
60		4261.35	I
110		4263.14	I
30		4271.06	I
40		4272.91	I
16000		4274.80	I
85		4280.40	I
10000		4289.72	I
40		4291.96	I
85		4295.76	I
70		4297.74	I
35		4300.51	I
50		4301.18	I
30		4305.45	I
35		4319.64	I
60		4325.08	I
780		4337.57	I
1100		4339.45	I
380		4339.72	I
60		4340.13	I
1900		4344.51	I
70		4346.83	I
380		4351.05	I
2300		4351.77	I
570		4359.63	I
70		4363.13	I
530		4371.28	I
70		4373.25	I
110		4374.16	I
70		4375.33	I
50		4381.11	I
530		4384.98	I
60		4387.50	I
70		4391.75	I
60		4403.50	I
24		4410.30	I
60		4411.09	I
35		4412.25	I
50		4413.87	I
60		4424.28	I
24		4428.50	I
50		4430.49	I

Chromium (Cont.)

Intensity		Wavelength		Spectrum
50		4432.18		I
110		4458.54		I
30		4459.74		I
30		4465.36		I
30		4482.88		I
40		4488.05		I
50		4489.47		I
60		4492.31		I
660		4496.86		I
50		4498.73		I
70		4500.30		I
50		4501.11		I
22		4501.79		I
24		4506.85		I
95		4511.90		I
12		4514.37		I
35		4514.53		I
24		4521.14		I
24		4526.11		I
380		4526.47		I
70	d	4527.34		I
		4527.47		I
24		4529.85		I
380		4530.74		I
50		4535.15		I
240		4535.72		I
40		4539.79		I
240		4540.50		I
240		4540.72		I
35		4541.07		I
19		4541.51		I
24		4542.62		I
140		4544.62		I
24		4545.34		I
600		4545.96		I
50		4556.17		I
22		4558.66		II
19		4564.17		I
120		4565.51		I
95		4569.64		I
120		4571.68		I
22		4575.12		I
360		4580.06		I
24		4586.14		I
360		4591.39		I
70		4595.59		I
50		4600.10		I
480		4600.75		I
50		4601.02		I
240		4613.37		I
600		4616.14		I
70		4619.55		I
85		4621.96		I
70		4622.49		I
24		4622.76		I
550		4626.19		I
24		4632.18		I
40		4637.18		I
50		4637.77		I
50	d	4639.52		I
		4639.70		I
1600		4646.17		I
24		4646.81		I
24		4648.13		I
24		4648.87		I
35		4649.46		I
570		4651.28		I
840		4652.16		I
35		4654.74		I
19		4656.19		I
40		4663.33		I
70		4663.83		I
95		4664.80		I
35		4665.90		I
22		4666.22		I
70		4666.51		I
50		4669.34		I
40		4680.54		I
19		4680.87		I
70		4689.37		I
60		4693.95		I
24		4695.15		I
60		4697.06		I
240	d	4698.46		I
		4698.62	CR	I
35		4700.61		I
190		4708.04		I
240		4718.43		I
50		4723.10		I
50		4724.42		I
50		4727.15		I
24		4729.72		I
120		4730.71		I
140		4737.35		I
19		4745.31		I
70		4752.08		I
340		4756.11		I
50		4764.29		I
22		4766.63		I
30		4767.86		I
190		4789.32		I
95		4792.51		I
120		4801.03		I
110		4829.38		I
14		4836.86		I
17		4861.20		I
70		4861.84		I
140		4870.80		I
35		4885.78		I
19		4885.96		I
130		4887.01		I
19		4888.53		I
35		4903.24		I
260		4922.27		I
110		4936.33		I
70		4942.50		I
110		4954.81		I
35		4964.93		I
60		5013.32		I
17		5051.90		I
17		5065.91		I
40		5067.71		I
40		5072.92		I
30		5110.75		I
17		5113.13		I
17		5123.46		I
50		5139.65		I
14		5144.67		I
70		5166.23		I
35		5177.43		I
70		5184.59		I
70		5192.00		I
12		5193.49		I
85		5196.44		I
35		5200.19		I
5300		5204.52		I
8400		5206.04		I
11000		5208.44		I
19		5214.13		I
30		5221.75		I
85		5224.94		I
12		5226.89		I
19		5238.97		I
30		5243.40		I
290		5247.56		I
60		5254.92		I
60		5255.13		I
19		5261.75		I
530		5264.15		I
30		5265.16		I
180		5265.72		I
35		5272.01		I
30		5273.44		I
95	h	5275.17		I
35	h	5275.69		I
70	h	5276.03		I
19		5280.29		I
10		5287.19		I
340		5296.69		I
70	h	5297.36		I
660		5298.27		I
85		5300.75		I
17		5304.21		I
24		5312.88		I
24		5318.78		I
340	h	5328.34		I
70	h	5329.17		I
17	h	5329.72		I
14		5340.44		I
10		5344.76		I
780		5345.81		I
380		5348.32		I
30		5386.98		I
22		5387.57		I
10		5390.39		I
40		5400.61		I
22		5405.00		I
1400		5409.79		I
12		5442.41		I
19		5463.97		I
19		5480.50		I
24		5628.64		I
7		5642.36		I
12	h	5649.37		I
24		5664.04		I
7	h	5681.20		I
7	h	5682.48		I
24		5694.73		I
40		5698.33		I
24		5702.31		I
12		5712.64		I
24		5712.78		I
7		5719.82		I
7		5746.43		I
7		5753.69		I
12	h	5781.20		I
6	h	5781.81		I
24	h	5783.11		I
30	h	5783.93		I
24	h	5785.00		I
19	h	5785.82		I
60	h	5787.99		I
180	h	5791.00		I
35		6330.10		I
22		6362.87		I
19		6661.08		I
11		6669.26		I
5	h	6881.62		I
10	h	6882.38		I
21	h	6883.03		I
27	h	6924.13		I
17	h	6925.20		I
30	h	6978.48		I
11	h	6979.82		I
7		7185.52		I
6	h	7236.20		I
85		7355.90		I
130		7400.21		I
150		7462.31		I
11	h	7942.04		I
5	h	8163.18		I
9		8348.28		I
6		8450.26		I
3		8455.24		I
6		8548.86		I
40		8947.15		I
19		8976.83		I

Cr III
Ref. 412 — C.H.C.

Intensity	Wavelength	
	Vacuum	
20	969.26	III
40	1000.86	III
40	1001.04	III
30	1002.96	III
50	1017.14	III
50	1017.31	III
50	1017.57	III
30	1028.33	III
60	1030.47	III
30	1030.89	III
50	1033.23	III
50	1033.45	III
100	1033.69	III
50	1035.93	III
100	1036.03	III
30	1040.17	III
40	1040.53	III
40	1045.06	III
40	1045.14	III
60	1059.13	III
60	1060.15	III
60	1061.04	III
50	1062.68	III
30	1064.32	III
30	1064.43	III
50	1066.23	III
80	1068.41	III
30	1100.61	III
30	1101.43	III
30	1102.88	III
30	1117.19	III
30	1132.75	III
50	1136.67	III
50	1161.43	III
30	1187.65	III
60	1206.38	III
80	1209.13	III
80	1211.12	III
40	1221.07	III
40	1221.90	III
30	1225.65	III
30	1228.65	III
30	1231.88	III
50	1232.96	III
40	1236.20	III
40	1238.51	III
50	1252.61	III
40	1259.02	III
40	1261.86	III
30	1262.34	III
35	1263.61	III
35	1264.21	III
40	1287.05	III
30	1455.27	III
40	1584.60	III
30	1603.19	III
30	1679.25	III
30	1690.28	III
60	1692.89	III
60	1696.64	III
60	1701.48	III
80	1707.43	III
40	1707.78	III
45	1762.81	III
30	1766.92	III
30	1769.17	III
30	1827.26	III
	AIR	
60	2036.39	III
50	2039.63	III
80	2047.23	III
100	2113.73	III
100	2113.83	III
50	2114.26	III
50	2114.53	III
100	2114.87	III
100	2117.53	III
80	2123.53	III
80	2139.11	III
100	2141.15	III
80	2144.15	III
50	2147.16	III
50	2147.56	III
50	2148.65	III
50	2149.48	III
50	2152.76	III
100	2157.17	III
50	2163.86	III
60	2166.25	III
100	2170.70	III
50	2183.71	III
100	2185.01	III
50	2190.09	III
100	2190.76	III
100	2191.58	III
100	2197.89	III
100	2198.62	III
100	2203.22	III
60	2208.70	III
200	2226.72	III
100	2231.81	III
100	2233.81	III
200	2235.91	III
150	2237.59	III
150	2244.10	III
80	2251.45	III
50	2257.92	III
100	2273.30	III
80	2275.43	III
100	2276.38	III
80	2277.47	III
150	2284.44	III
50	2289.23	III
80	2290.66	III
60	2295.55	III
50	2309.99	III
80	2314.63	III
100	2319.07	III
150	2324.88	III
60	2340.51	III
50	2456.83	III
100	2472.88	III
100	2479.77	III
100	2483.06	III
60	2488.26	III
80	2506.41	III
80	2530.99	III
80	2537.73	III
80	2544.37	III
50	2545.17	III
80	2564.76	III
80	2616.50	III
100	2626.08	III
100	2640.73	III
50	2647.50	III
40	2655.28	III
40	2916.57	III

Cr IV
Ref. 379, 412 — C.H.C.

Intensity	Wavelength	
	Vacuum	
50	575.05	IV
30	576.24	IV
30	576.62	IV
30	595.09	IV
50	612.64	IV
40	613.75	IV
40	614.03	IV
40	614.90	IV
30	615.34	IV
30	615.60	IV
50	616.82	IV
40	618.23	IV
40	619.13	IV
100	620.66	IV
60	621.36	IV
40	622.09	IV
30	623.54	IV
40	625.04	IV
40	625.99	IV
100	629.26	IV
50	629.74	IV
80	630.30	IV
30	632.62	IV
30	637.34	IV
50	637.55	IV
50	638.13	IV

Chromium (Cont.)

Intensity		Wavelength	
30		638.54	IV
100		666.55	IV
75		667.30	IV
40		677.55	IV
40		687.12	IV
50		688.46	IV
100		693.92	IV
50		695.21	IV
50		705.98	IV
30		712.90	IV
80		1055.89	IV
60		1057.85	IV
30		1367.39	IV
40		1375.05	IV
70		1401.82	IV
100		1417.42	IV
30		1485.05	IV
80		1595.04	IV
90	d	1595.59	IV
100		1658.08	IV
120		1672.66	IV
90		1686.07	IV
100		1690.88	IV
80		1725.26	IV
90		1727.07	IV
100		1732.04	IV
40		1733.98	IV
80		1734.16	IV
50		1739.19	IV
70		1746.88	IV
80		1747.13	IV
110		1755.64	IV
120		1758.51	IV
100		1769.64	IV
100		1777.82	IV
40		1791.09	IV
140		1802.72	IV
130		1812.41	IV
60		1819.23	IV
30		1826.21	IV
30		1826.86	IV
100		1840.14	IV
50		1851.89	IV
100		1863.11	IV
140		1873.89	IV
35		1883.16	IV
40		1937.63	IV
30		1946.59	IV
140		1967.18	IV
120		1972.07	IV
40		1990.25	IV
		Air	
50	d	2042.91	IV
40		2055.73	IV
70		2299.21	IV
90		2299.59	IV
100		2316.85	IV
40		2324.06	IV
50		2360.40	IV
70		2405.15	IV
60		2423.32	IV

Cr V
Ref. 380 — C.H.C.

Intensity		Wavelength	
		Vacuum	
100		438.62	V
100		464.02	V
50		469.64	V
50		825.60	V
50		968.70	V
50		1045.04	V
60		1060.65	V
50		1112.45	V
60		1114.35	V
100		1116.48	V
80		1117.56	V
50		1118.16	V
150		1121.07	V
150		1127.63	V
100		1193.95	V
80		1196.04	V
50		1210.50	V
50		1259.99	V
100		1263.50	V
150		1465.86	V
50		1481.65	V
50		1482.76	V
50		1484.67	V
100		1489.71	V
150		1497.97	V
170		1519.03	V
220		1579.70	V
170		1591.72	V
150		1603.19	V
60		1638.50	V
50		1639.40	V
200		1837.44	V

COBALT (Co)
Z = 27

Co I and II
Ref. 1, 125, 276 — C.R.C.

Intensity		Wavelength	
		Vacuum	
20		1265.93	II
40		1271.94	II
20		1276.90	II
30		1293.97	II
25		1295.53	II
30		1295.86	II
20		1297.10	II
80		1299.58	II
30		1302.39	II
20		1306.76	II
80		1306.95	II
40		1311.12	II
40		1311.86	II
30		1315.42	II
30		1316.09	II
20		1318.19	II
30		1318.60	II
20		1319.84	II
20		1409.33	II
20		1466.21	II
20		1471.87	II
30		1472.90	II
30		1475.81	II
20		1484.26	II
30		1486.50	II
20		1509.23	II
20		1590.54	II
20		1595.77	II
20		1599.30	II
20		1693.34	II
8		1706.05	II
10		1723.01	II
15	h	1740.55	II
15		1743.39	II
8		1754.21	II
20	d	1808.01	II
10		1837.56	II
15		1839.37	II
1500		1842.34	I
1800		1847.89	I
1800		1852.71	I
2400		1855.05	I
1500		1878.28	I
10		1917.62	II
1800		1936.58	I
1500		1946.79	I
30		1950.09	II
1500		1951.90	I
1800		1954.22	I
1800		1955.17	I
30		1957.42	II
1500		1958.55	I
1500		1961.59	I
1500	h	1968.69	I
1500	h	1968.93	I
3000		1970.71	I
1800	h	1971.16	I
1800	h	1972.52	I
1500		1973.85	I
1800		1976.97	I
2400	h	1980.89	I
1500		1989.80	I
1800		1990.34	I
1500	l	1998.49	I
		Air	
20		2000.79	II
1500		2002.32	I
900		2008.04	I
50		2011.51	II
1200	h	2014.58	I
900		2016.17	I
50		2022.35	II
40		2025.76	II
50		2027.04	II
900		2031.96	I
30		2036.58	II
1500		2039.95	I
1200		2041.11	I
20		2049.17	II
40		2058.82	II
40		2063.78	II
50		2065.54	II
1500	h	2077.76	I
900		2085.67	I
900		2087.55	I
900		2089.35	I
900		2093.40	I
900		2094.86	I
900		2095.77	I
1200		2097.51	I
1500		2104.73	I
1500		2106.80	I
900		2108.98	I
30		2111.44	II
900	s	2117.68	I
20		2128.79	II
900		2137.78	I
900		2138.97	I
900		2163.03	I
20		2164.44	II
30		2173.33	II
1100		2174.60	I
20		2181.99	II
30		2187.01	II
20		2190.68	II
20		2192.50	II
200		2193.60	II
20		2200.40	II
40	p	2202.95	II
200		2256.73	II
150		2260.00	II
200		2283.52	II
1000		2286.15	II
200		2291.98	II
300	d	2293.38	II
300		2301.40	II
800	d	2307.85	II
2600		2309.02	I
500		2311.60	II
500		2314.05	II
300		2314.96	II
200	p	2317.06	II
2400		2323.14	I
300	p	2324.31	II
200	d	2326.11	II
500		2326.47	II
1400		2335.99	I
1600		2338.67	I
200		2347.39	II
1600		2352.85	I
200	d	2353.41	II
2000		2353.42	I
500		2363.80	II
400		2378.62	II
1400		2380.48	I
200		2381.76	II
300	p	2383.45	II
1400		2384.86	I
200		2386.36	II
500		2388.92	II
200		2397.38	II
1100	d	2402.06	I
200	p	2404.16	II
5300		2407.25	I
5300		2411.62	I
1600		2412.76	I
4800		2414.46	I
4800		2415.30	I
300		2417.65	II
4100		2424.93	I
3300		2432.21	I
2900		2436.66	I
2400		2439.05	I
200		2442.63	II
200	d	2446.03	II
200	p	2447.69	II
200		2450.00	II
200		2464.20	II
200		2486.44	II
200		2498.82	II
570		2504.52	I
500		2506.46	II
360		2506.88	I
200		2511.16	II
860		2517.87	I
500		2519.82	II
4300		2521.36	I
200	h	2524.65	II
300		2524.97	II
500		2528.62	II
2900		2528.97	I
200	p	2530.09	II
720		2530.13	I
860		2532.18	I
200	d	2533.82	II
2900		2535.96	I
860		2536.49	I
300		2541.94	II
1700		2544.25	I
200		2546.74	II
340		2548.34	I
310		2553.37	I
310		2555.07	I
300		2559.41	II
200		2560.03	II
960		2562.15	I
500		2564.04	II
1100		2567.35	I
960		2574.35	I
800		2580.32	II
300	d	2582.22	II
500		2587.22	II
500		2587.52	II
200		2588.91	II
100	p	2605.71	II
100		2612.50	II
100		2614.36	II
100	p	2628.77	II
100		2632.26	II
100		2636.07	II
310		2646.42	I
770		2648.64	I
100		2653.72	II
100		2663.53	II
200		2666.73	II
100		2675.85	II
100		2684.42	II
100		2702.02	II
200		2706.62	II
200		2707.35	II
190		2715.99	I
100		2727.78	II
80		2734.54	II
190		2745.10	I
100		2753.22	II
190		2764.19	I
100		2766.70	II
100		2774.97	II
100		2791.00	II
100		2793.73	II
150		2815.56	I
80		2835.63	II
80		2847.35	II
80		2871.22	II
190		2886.44	I
100		2918.38	II
100		2930.24	II
100		2954.73	II
690		2987.16	I
690		2989.59	I
20		3008.86	II
60		3022.59	II
30		3035.13	II
3100		3044.00	I
1700		3061.82	I
20		3352.79	II
80		3387.70	II
1100		3388.17	I
2200		3395.38	I
11000		3405.12	I
4500		3409.18	I
6700		3412.34	I
2200		3412.63	I
30		3415.77	II
2700		3417.16	I
50		3423.84	II
2500		3431.58	I
4500		3433.04	I
1600		3442.93	I
8800		3443.64	I
50		3446.39	II
4100		3449.17	I
2100		3449.44	I
21000		3453.50	I
1000		3455.23	I
5100		3462.80	I
5100		3465.80	I
8000		3474.02	I
1900		3483.41	I
4800		3489.40	I
2400		3495.69	I
40		3497.33	II
50		3501.72	II
9600		3502.28	I
7000		3506.32	I
50		3507.77	II
2900		3509.84	I
1400		3510.43	I
4800		3512.64	I
3800		3513.48	I
10		3514.23	II
30		3517.50	II
4800		3518.35	I
1300		3520.08	I
2700		3521.57	I
3800		3523.43	I
60		3523.51	II
6400		3526.85	I
2700		3529.03	I
7300		3529.81	I
1900		3533.36	I
40		3535.92	II
50		3545.03	II
20		3555.93	II
1100		3560.89	I
80		3561.07	II
20		3566.98	II
8800		3569.38	I
50		3574.95	II

Cobalt (Cont.)

Intensity		Wavelength	
1600		3574.96	I
60		3575.32	II
2500		3575.36	I
60		3577.96	II
1000		3585.16	I
6700		3587.19	I
1900		3594.87	I
1600		3602.08	I
100		3621.21	II
1000		3627.81	I
80		3643.61	II
10		3656.75	II
60		3681.35	II
5		3695.32	II
5		3714.73	II
1100		3745.50	I
20		3754.69	II
1400		3842.05	I
6900		3845.47	I
5500		3873.12	I
2800		3873.96	I
7900		3894.08	I
20	h	3911.40	II
1500		3935.97	I
80	h	3963.10	II
40	h	3976.74	II
10		3983.02	II
6000		3995.31	I
970		3997.91	I
350		4020.90	I
10	h	4036.14	II
20	h	4037.37	II
4		4040.02	II
370		4045.39	I
5	h	4050.23	II
10	h	4052.40	II
20	h	4062.73	II
2	h	4064.50	II
350		4066.37	I
5	h	4074.34	II
830		4092.39	I
1		4096.57	II
550		4110.54	I
2800		4118.77	I
4400		4121.32	I
3		4130.88	II
3		4145.13	II
3	d	4160.67	II
1		4181.13	II
90		4190.71	I
1		4208.61	II
30	s	4244.25	II
8		4272.33	II
20	h	4288.25	II
3		4328.86	II
2		4384.26	II
2	h	4396.94	II
3		4413.91	II
90		4469.56	I
10		4482.50	II
2	h	4489.12	II
4	h	4497.44	II
10	d	4500.54	II
0		4516.65	II
690		4530.96	I
2	h	4533.22	II
0	h	4537.95	II
90		4549.66	I
1	h	4559.29	II
140		4565.59	I
1		4569.26	II
190		4581.60	I
5		4616.30	II
120		4629.38	I
25	h	4660.66	II
85		4663.41	I
110		4792.86	I
10	d	4831.16	II
100		4840.27	I
150		4867.88	I
80	h	4964.18	II
10	h	4970.05	II
10	h	4990.47	II
20	h	4995.98	II
35		5146.74	I
50		5212.71	I
50		5230.22	I
45		5235.21	I
50		5247.93	I
26		5266.30	I
45		5266.49	I
26		5268.52	I
45	h	5280.65	I
26		5301.06	I
50		5342.71	I
26		5343.39	I
50		5352.05	I
26		5353.48	I
35		5369.58	I
45		5483.34	I
17		5530.77	I
17		5647.22	I
17		5991.88	I
17		6082.44	I
17		6282.63	I
45		6450.24	I
21		6455.00	I
15		6563.42	I
15		6632.44	I
14		6814.94	I
14		6872.40	I
21		7052.89	I
45		7084.99	I
8		7417.38	I
8		7712.68	I
7		7908.71	I
9		7987.38	I
13		8007.27	I
9		8093.96	I
9		8372.84	I
4		8575.35	I
3		8819.15	I

Co III
Ref. 291 — C.R.C.

Intensity	Wavelength	
	Vacuum	
1000	1696.01	III
800	1697.99	III
500	1702.79	III
1000	1707.35	III
500	1707.95	III
500	1723.97	III
400	1745.67	III
500	1755.98	III
5000	1760.35	III
500	1769.96	III
500	1773.22	III
5000	1773.57	III
500	1774.42	III
1000	1777.14	III
2000	1780.05	III
3000	1782.97	III
500	1784.06	III
1000	1787.08	III
1000	1789.07	III
500	1790.26	III
500	1791.28	III
500	1798.06	III
500	1805.54	III
400	1811.47	III
400	1821.26	III
400	1821.69	III
400	1821.77	III
1000	1823.08	III
400	1825.36	III
750	1825.95	III
2000	1830.09	III
2000	1831.44	III
750	1831.92	III
400	1832.20	III
5000	1835.00	III
1000	1837.63	III
500	1846.16	III
500	1852.92	III
400	1854.39	III
400	1854.76	III
1000	1861.78	III
2000	1863.83	III
400	1864.19	III
500	1871.87	III
1000	1881.70	III
500	1895.37	III
500	1919.12	III
500	1928.57	III
500	1940.15	III
500	1953.94	III
500	1959.41	III
400	1989.60	III
	Air	
100	2001.09	III
200	2011.62	III
200	2013.88	III
100	2031.81	III
200	2053.11	III
100	2056.21	III
10	2062.17	III
10	2079.74	III
15	2088.58	III
10	2090.51	III
10	2097.63	III
10	2134.15	III
10	2452.16	III
20	2811.75	III
10	2888.31	III
10	2933.27	III
10	2978.01	III
20	2991.89	III
25	3010.92	III
10	3116.68	III
2	3151.40	III
2	3180.64	III
20	3232.11	III
2	3249.24	III
20	3259.68	III
2	3269.23	III
10	3287.68	III
15	3305.38	III
10	3451.25	III
2	3526.24	III
1	3634.21	III
1	3636.31	III
2	3667.52	III
2	3677.23	III
3	3680.74	III
1	3762.50	III
15	3782.27	III

Co IV
Ref. 236 — C.R.C.

Intensity	Wavelength	
	Vacuum	
81	606.79	IV
74	607.59	IV
55	608.24	IV
66	609.16	IV
70	609.21	IV
64	609.28	IV
43	610.04	IV
37	610.25	IV
24	610.79	IV

Co V
Ref. 100, 159 — C.R.C.

Intensity	Wavelength	
	Vacuum	
20	355.52	V
18	355.88	V
12	356.06	V
4	1006.86	V
10	1007.51	V
15	1009.02	V
10	1010.94	V
10	1013.80	V
1	1017.43	V
10	1018.36	V
10	1021.14	V
1	1028.08	V
3	1226.31	V
8	1228.19	V
15	1231.73	V
2	1234.55	V
20	1236.95	V
2	1239.85	V
8	1246.91	V
6	1258.61	V
5	1263.28	V
20	1270.70	V
20	1272.23	V
2	1275.52	V
50	1277.01	V
30	1281.63	V
15	1284.00	V
28	1286.95	V
25	1295.55	V
40	1295.87	V
35	1301.12	V
50	1345.67	V
6	1351.22	V
15	1353.42	V
40	1355.20	V
30	1357.67	V
20	1361.32	V
30	1362.46	V
30	1364.17	V
30	1368.24	V
4	1369.30	V
10	1371.01	V
30	1373.09	V
30	1375.20	V
25	1378.12	V
10	1379.05	V
10	1380.21	V
32	1389.11	V
15	1459.77	V
35	1468.98	V
30	1476.65	V
25	1482.62	V
20	1482.91	V
25	1486.02	V
20	1488.73	V

COPPER (Cu)
Z = 29

Cu I and II
Ref. 273, 290—V.K.

Intensity	Wavelength	
	Vacuum	
80	685.141	II
100	709.313	II
100	718.179	II
150	724.489	II
200	735.520	II
250	736.032	II
80	779.295	II
100	797.455	II
150	810.998	II
200	813.883	II
300	826.996	II
150	848.808	II
250	851.303	II
250	858.487	II
400	861.994	II
400	865.390	II
250	869.336	II
150	873.263	II
200	876.723	II
250	877.012	II
200	877.555	II
500	878.699	II
100	884.133	II
250	885.847	II
600	886.943	II
600	890.567	II
500	892.414	II
800	893.678	II
400	894.227	II
600	896.759	II
400	896.976	II
600	901.073	II
400	906.113	II
800	914.213	II
600	922.019	II
500	924.239	II
400	935.232	II
600	935.898	II
600	943.335	II
600	945.525	II
500	945.965	II
200	954.383	II
250	956.290	II
400	958.154	II
200	960.414	II
250	968.042	II
200	974.759	II
250	977.567	II
100	987.657	II
250	992.953	II
300	1004.055	II
300	1008.569	II
300	1008.728	II
300	1010.269	II
250	1012.597	II
500	1018.707	II
500	1027.831	II
250	1028.328	II
200	1030.263	II
600	1036.470	II
600	1039.348	II
600	1039.582	II
800	1044.519	II
800	1044.744	II
500	1049.755	II
600	1054.690	II
400	1055.797	II
600	1056.955	II
400	1058.799	II
600	1059.096	II
600	1060.634	II
600	1063.005	II
200	1065.782	II
200	1066.134	II
500	1069.195	II
300	1073.745	II
200	1088.395	II
300	1094.402	II
250	1097.053	II
150	1119.947	II
200	1142.640	II
300	1144.856	II
100	1250.048	II
150	1265.506	II
300	1275.572	II
150	1282.455	II
150	1287.468	II
150	1298.395	II
300	1308.297	II
300	1314.337	II
100	1320.686	II
100	1326.395	II
150	1350.594	II
250	1351.837	II
150	1355.305	II
300	1358.773	II
200	1359.009	II

Copper (Cont.)

Intensity		Wavelength	Spectrum
200		1362.600	II
250		1367.951	II
200		1371.840	II
100		1393.128	II
100		1398.642	II
150		1402.777	II
150		1407.169	II
100		1414.898	II
250		1418.426	II
250		1421.759	II
200		1427.829	II
400		1430.243	II
250		1434.904	II
150		1436.236	II
150		1442.139	II
200		1445.984	II
200		1449.058	II
250		1450.304	II
200		1452.294	II
300		1458.002	II
250		1459.412	II
200		1463.752	II
400		1463.838	II
200		1466.070	II
400		1470.697	II
200		1472.395	II
250		1473.978	II
200		1474.935	II
150		1476.059	II
200		1481.544	II
200		1485.328	II
750		1488.831	II
300		1492.834	II
250		1493.366	II
250		1495.430	II
350		1496.687	II
150		1503.368	II
250		1504.757	II
200		1505.388	II
300		1508.632	II
350		1510.506	II
200		1512.465	II
200		1513.366	II
500		1514.492	II
200		1517.631	II
500		1519.492	II
600		1519.837	II
200		1520.540	II
200		1524.860	II
150		1525.764	II
500		1531.856	II
300		1532.131	II
250		1533.986	II
250		1535.002	II
500		1537.559	II
200		1540.239	II
300		1540.389	II
300		1540.588	II
750		1541.703	II
400		1544.677	II
100		1547.958	II
300		1550.653	II
300		1551.389	II
500		1552.646	II
250		1553.896	II
400		1555.134	II
500		1555.703	II
300		1558.345	II
400		1565.924	II
400		1566.415	II
100		1569.416	II
300		1579.492	II
300		1580.626	II
400		1581.995	II
500		1583.682	II
400		1590.165	II
600		1593.556	II
400		1598.402	II
400		1602.388	II
200		1604.848	II
300		1605.281	II
400		1606.834	II
250		1608.639	II
150		1610.296	II
200		1617.915	II
600		1621.426	II
400		1622.428	II
250		1630.268	II
100		1636.605	II
250		1649.458	II
30	r	1655.32	I
200		1656.322	II
200		1660.001	II
300		1663.002	II
100		1672.776	II
30		1688.09	I
30		1691.08	I
30	r	1703.84	I
50	r	1713.36	I
150		1717.721	II
50	r	1725.66	I
100		1736.551	II
50	r	1741.57	I
150		1753.281	II
200	r	1774.82	I
100	r	1825.35	I
250		1929.751	II
250		1944.597	II
100		1946.493	II
200		1957.518	II
150		1970.495	II
150		1977.027	II
500		1979.956	II
300		1989.855	II
		Air	
250		1999.698	II
270		2035.854	II
250		2037.127	II
350		2043.802	II
300		2054.980	II
100		2078.663	II
110		2098.398	II
320		2104.797	II
300		2112.100	II
320		2117.310	II
350		2122.980	II
350		2126.044	II
420		2134.341	II
900		2135.981	II
400		2148.984	II
150		2161.320	II
1300	r	2165.09	I
250		2174.982	II
1600	r	2178.94	I
700		2179.410	II
1700	r	2181.72	I
700		2189.630	II
900		2192.268	II
400		2195.683	II
1700	r	2199.58	I
1300	r	2199.75	I
100		2200.509	II
200		2209.806	II
750		2210.268	II
1600	r	2214.58	I
250		2215.106	II
1000	r	2215.65	I
750		2218.108	II
2100	r	2225.70	I
150		2226.780	II
1600	r	2227.78	I
350		2228.868	II
2500	r	2230.08	I
1100	r	2238.45	I
900		2242.618	II
2300	r	2244.26	I
1000		2247.002	II
1300	r	2260.53	I
2200	r	2263.08	I
150		2263.786	II
200		2276.258	II
100		2286.645	II
2500	r	2293.84	I
170		2294.368	II
1000		2303.12	I
150		2369.890	II
2500	r	2392.63	I
120		2403.337	II
1500		2406.66	I
1000	r	2441.64	I
100		2485.792	II
2000	r	2492.15	I
150		2506.273	II
120		2526.593	II
300		2544.805	II
100		2571.756	II
150		2590.529	II
200		2600.270	II
2500	r	2618.37	I
200		2666.291	II
750		2689.300	II
700		2700.962	II
650		2703.184	II
700		2713.508	II
650		2718.778	II
300		2721.677	II
120		2737.342	II
270		2745.271	II
2500	r	2766.37	I
800		2769.669	II
200		2791.795	II
170		2799.528	II
100		2810.804	II
1250	r	2824.37	I
350		2837.368	II
100		2857.748	II
600		2877.100	II
270		2884.196	II
2500	r	2961.16	I
100		2986.335	II
2000		2997.36	I
2000		3010.84	I
2500		3036.10	I
2500		3063.41	I
1400		3073.80	I
1500		3093.99	I
1250		3099.93	I
2000		3108.60	I
1400	h	3126.11	I
1500		3194.10	I
1400		3208.23	I
1500	h	3243.16	I
10000	r	3247.54	I
10000	r	3273.96	I
1400	h	3282.72	I
400		3290.418	II
1500	h	3290.54	I
110		3300.881	II
250		3301.229	II
2500	h	3307.95	I
200		3316.276	II
1500		3337.84	I
150		3338.648	II
200		3365.648	II
450		3370.454	II
300		3374.952	II
200		3380.712	II
100		3384.945	II
1250	h	3483.76	I
1250		3524.23	I
2000		3530.38	I
1400		3599.13	I
1400		3602.03	I
1000		3686.555	II
150		3786.270	II
170		3797.849	II
100		3818.879	II
140		3826.921	II
160		3864.137	II
280		3884.131	II
150		3892.924	II
170		3903.177	II
140		3920.654	II
120		3933.268	II
120		3987.024	II
150		3993.302	II
140		4003.476	II
1250		4022.63	I
100		4032.647	II
600		4043.484	II
500		4043.751	II
2000		4062.64	I
120		4068.106	II
500		4131.363	II
200		4143.017	II
300		4153.623	II
500		4161.140	II
370		4164.284	II
400		4171.851	II
500		4179.512	II
500		4211.866	II
320		4230.449	II
200		4255.635	II
950		4275.11	I
300		4279.962	II
500		4292.470	II
400		4365.370	II
100		4444.831	II
400		4506.002	II
150		4516.049	II
150		4541.032	II
500		4555.920	II
100		4596.906	II
120		4649.271	II
2000		4651.12	I
120		4661.363	II
320		4671.702	II
300		4673.577	II
450		4681.994	II
100		4758.433	II
400		4812.948	II
120		4851.262	II
300		4854.988	II
100		4873.304	II
150		4901.427	II
1000		4909.734	II
500		4918.376	II
200		4926.424	II
900		4931.698	II
120		4943.026	II
700		4953.724	II
500		4985.506	II
400		5006.801	II
350		5009.851	II
400		5012.620	II
350		5021.279	II
200		5039.016	II
300		5047.348	II
900		5051.793	II
400		5058.910	II
500		5065.459	II
450		5067.094	II
350		5072.302	II
450		5088.277	II
420		5093.816	II
350		5100.067	II
1500		5105.54	I
250		5124.476	II
2000		5153.24	I
100		5158.093	II
100		5183.367	II
2500		5218.20	I
100		5269.991	II
100		5276.525	II
1650		5292.52	I
100		5368.383	II
1500		5700.24	I
1500		5782.13	I
150		5805.989	II
100		5833.515	II
200		5897.971	II
120		5937.577	II
400		5941.196	II
100		5993.260	II
650		6000.120	II
100		6023.264	II
250		6072.218	II
150		6080.343	II
150		6099.990	II
160		6107.412	II
300		6114.493	II
600		6150.384	II
750		6154.222	II
500		6172.037	II
550		6186.884	II
400		6188.676	II
300		6198.092	II
470		6204.261	II
450		6208.457	II
750		6216.939	II
700		6219.844	II
500		6261.848	II
1000		6273.349	II
350		6288.696	II
900		6301.009	II
550		6305.972	II
400		6312.492	II
120		6326.466	II
400		6373.268	II
750		6377.840	II
400		6403.384	II
850		6423.884	II
200		6442.965	II
750		6448.559	II
170		6466.246	II
950		6470.168	II
750		6481.437	II
400		6484.421	II
220		6517.317	II
400		6530.083	II
120		6551.286	II
200		6577.080	II
750		6624.292	II
800		6641.396	II
450		6660.962	II
100		6770.362	II
300		6806.216	II
400		6809.647	II
320		6823.202	II
250		6844.157	II
320		6868.791	II
270		6872.231	II
270		6879.404	II
220		6937.553	II
150		6952.871	II
150		6977.572	II
200		7022.860	II
300		7194.896	II
400		7326.008	II
300		7331.694	II
250		7382.277	II
1000		7404.354	II
270		7434.156	II
500		7562.015	II
700		7652.333	II
1000		7664.648	II
150		7681.788	II
450		7744.097	II
800		7778.738	II
750		7805.184	II
1500		7807.659	II
1000		7825.654	II
350		7860.577	II
300		7890.567	II
700		7902.553	II
1500		7933.13	I
400		7944.438	II
400		7972.033	II
1200		7988.163	II
2000		8092.63	I
500		8277.560	II
800		8283.160	II
250		8503.396	II
750		8511.061	II
200		8609.134	II
500		9813.213	II
250		9827.978	II
200		9830.798	II
600		9861.280	II
600		9864.137	II
200		9883.969	II
550		9916.419	II
500		9917.954	II
550		9925.594	II

Copper (Cont.)

Intensity		Wavelength	
450		9938.998	II
500		9960.354	II
450		10006.588	II
550		10022.969	II
550		10038.093	II
650		10054.938	II
450		10080.354	II

Cu III
Ref. 295—V.K.

Intensity		Wavelength	
		Vacuum	
75		542.90	III
200		615.67	III
150		616.03	III
150		687.98	III
150		715.53	III
125		730.38	III
250		788.07	III
250		788.46	III
250		791.36	III
150		801.14	III
100		829.34	III
40		1048.88	III
50		1186.80	III
50		1200.96	III
300		1219.30	III
200		1244.38	III
100		1279.14	III
200		1312.39	III
300		1332.97	III
200		1339.48	III
150		1363.08	III
300	r	1376.79	III
200	r	1377.49	III
150		1423.48	III
300	r	1481.23	III
200		1543.46	III
500	r	1593.75	III
1000	r	1642.21	III
300		1679.14	III
400		1702.10	III
500		1722.37	III
600		1741.37	III
200		1768.86	III
200		1840.91	III
100		1971.95	III
		Air	
200		2013.22	III
150		2157.28	III
100		2299.47	III
500		2368.17	III
400		2391.74	III
800		2405.50	III
700		2412.34	III
2000		2444.44	III
500		2468.41	III
1000		2482.36	III
700		2486.46	III
500		2508.49	III
500		2522.38	III
500		2538.66	III
400		2566.37	III
400		2573.33	III
500		2609.32	III
200		2643.92	III
200		2696.38	III
20		2751.33	III
100		2812.94	III
100		2978.87	III
75		3548.87	III
100		3639.42	III
500		3702.92	III
800		3744.70	III
400		3748.27	III
600		3752.06	III
1000		3776.97	III
800		3790.80	III
600		3804.13	III
600		3809.18	III
300		3881.68	III
150		3953.81	III
100		4090.49	III
200		4283.40	III
500		4351.97	III
1000		4352.80	III
500		4355.24	III
500		4370.84	III
500		4371.40	III
500		4373.43	III
1000		4377.11	III
200		4386.42	III
150		4927.41	III
400		5094.28	III
200		5168.97	III
400		5208.34	III
600		5219.21	III
200		5268.59	III
400		5317.78	III
300		5369.79	III
350		5418.48	III
250	d	5494.94	III
50		5573.94	III
100		5609.00	III
75		5702.12	III
100		5768.56	III
100		5850.72	III
200		5965.25	III
30		6100.87	III
50		6369.27	III
20		6512.54	III
20		6644.13	III
50		6793.20	III

Cu IV
Ref. 199—V.K.

Intensity		Wavelength	
		Vacuum	
30		360.86	IV
20		374.40	IV
30		405.24	IV
80		406.45	IV
40		413.45	IV
70		443.68	IV
80		451.16	IV
80		463.72	IV
80		484.53	IV
90		497.00	IV
90		504.60	IV
70		509.38	IV
40		519.51	IV
40		540.65	IV
60		550.92	IV
20		584.85	IV
60		1056.13	IV
30		1074.72	IV
50		1091.65	IV
30		1105.50	IV
25		1119.43	IV
40	p	1152.18	IV
60		1227.44	IV
70		1228.87	IV
70		1258.69	IV
90		1274.84	IV
90		1293.46	IV
90		1309.41	IV
70		1321.17	IV
100	d	1340.08	IV
100		1350.42	IV
100		1362.05	IV
100		1372.14	IV
100		1377.82	IV
100		1388.80	IV
90		1405.49	IV
90		1415.27	IV
90		1434.34	IV
90		1449.69	IV
80		1466.18	IV
70		1482.77	IV
80		1499.81	IV
90		1515.28	IV
90		1535.12	IV
80		1551.12	IV
90		1567.35	IV
80		1583.47	IV
70		1595.12	IV
80	p	1608.14	IV
90		1639.75	IV
20		1650.16	IV
70		1704.37	IV
30		1797.99	IV
70		1817.56	IV
70		1819.23	IV
60		1837.04	IV
80		1849.62	IV
30		1867.24	IV
30		1918.71	IV
40		1966.31	IV

Cu V
Ref. 324—V.K.

Intensity		Wavelength	
		Vacuum	
9	h	258.95	V
49		271.33	V
49		283.97	V
22		293.41	V
56		299.64	V
65		305.83	V
51		312.51	V
66		321.05	V
74		326.57	V
82		333.56	V
81		339.88	V
81		346.00	V
86		355.41	V
77		363.96	V
65		370.63	V
74		377.76	V
70		387.40	V
51		396.06	V
25		406.94	V
13		1097.10	V
42		1106.24	V
77		1113.22	V
67		1121.20	V
76		1128.80	V
59		1133.86	V
63		1142.38	V
54		1149.06	V
72		1157.54	V
64		1167.35	V
77		1176.53	V
84		1183.63	V
76		1192.54	V
83		1201.22	V
77		1204.90	V
71		1214.36	V
70		1221.34	V
76		1230.11	V
80		1239.73	V
79		1246.99	V
65		1253.07	V
70		1260.24	V
77		1269.35	V
78		1278.20	V
68		1286.13	V
67		1292.08	V
73		1299.22	V
65		1309.72	V
55		1318.89	V
65		1323.28	V
64		1329.22	V

CURIUM (Cm)
Z = 96

Cm I and II
Ref. 51, 332—J.G.C.

Intensity		Wavelength	
		Air	
10000		2462.76	II
10000		2617.17	II
10000		2636.28	II
10000		2651.17	II
10000		2653.80	II
10000		2725.68	II
10000		2736.89	II
10000		2748.04	II
10000		2784.83	II
10000		2811.62	II
10000		2824.20	II
10000		2833.58	II
10000		2899.90	II
10000		2912.97	II
10000		2928.92	II
10000		2996.18	II
10000		2999.39	I
10000	b	3014.87	II
10000		3044.85	II
10000		3109.69	I
10000		3116.41	I
10000		3137.16	I
10000		3147.33	I
10000		3155.10	I
10000		3158.60	I
10000		3169.98	II
10000		3177.55	I
10000		3179.10	I
10000		3186.41	I
10000		3188.11	I
10000		3207.12	II
10000		3207.71	I
10000		3209.89	II
10000		3209.94	I
10000		3210.05	II
10000		3220.76	II
10000		3224.23	I
10000		3225.11	I
10000		3226.41	II
10000		3230.28	I
10000		3230.35	II
10000		3236.74	I
10000		3238.55	II
10000		3242.66	II
10000		3246.25	I
10000		3252.68	I
10000		3265.81	I
10000		3280.45	I
10000		3296.71	II
10000		3304.85	I
10000		3317.14	I
10000		3374.70	I
10000		3452.92	I
10000		3458.34	I
10000		3510.28	I
10000		3522.36	I
10000		3524.94	I
10000		3542.06	I
10000		3547.02	I
10000		3547.92	I
10000		3561.44	I
10000		3572.95	II
10000		3600.62	I
10000		3639.94	I
10000		3664.34	I
10000		3709.43	I
10000		3729.00	I
10000		3732.35	I
10000		3747.86	I
10000		3763.05	I
10000		3775.75	I
10000		3816.30	I
10000		3825.14	I
10000		3833.32	I
10000		3837.59	I
10000		3842.00	I
10000		3849.92	I
10000		3854.11	I
10000		3900.25	I
10000		3904.06	II
10000		3908.24	II
10000		3936.67	I
10000		3942.03	I
10000		3944.15	I
10000		3948.68	I
10000		3953.36	I
10000		3964.83	I
10000		3995.10	I
10000		4016.17	I
10000		4031.76	I
10000		4048.29	I
10000		4049.65	I
10000		4113.29	I
10000		4129.71	I
10000		4207.66	II
10000		4211.62	I
10000		4266.45	I
10000		4293.00	I
10000		4330.82	I
10000		4345.69	I
10000		4447.77	I
10000		4459.16	I
10000		4608.40	I
10000		5846.07	I
10000		5952.41	I
10000		6058.90	I
10000		6243.35	I
10000		6376.71	I
10000		6510.16	I
10000		6554.41	I
10000		6640.17	I
10000		6663.25	I
10000		6686.87	I
10000		6706.85	I
10000		6726.68	I
10000		6793.15	I
10000		7162.69	I
10000		7577.80	I
10000		7673.79	I
10000		7720.47	I
10000		8392.37	I
10000		9293.25	I
10000		9567.08	I
10000		9657.12	I
10000		10310.83	I
10000		10351.73	I
10000		10424.49	I
10000		10508.11	I
10000		10542.98	I
10000		10792.25	I
10000		10897.45	I
10000		11507.45	I
10000		11707.73	I
10000		11780.95	I
10000		11834.28	I
10000		12017.85	I
10000		12394.16	I
10000		12454.98	I
10000		12464.99	I
10000		13004.56	I
10000		13258.18	I
10000		13289.84	I
10000		13344.62	I
10000		13480.54	I
10000		13590.01	I
10000		13644.77	I
10000		13789.52	I
10000		13840.18	I
10000		13908.46	I
10000		13964.14	I
10000		14235.27	I
10000		14334.52	I
10000		14563.41	I
10000		14580.23	I
10000		15018.13	I
10000		15222.27	I
10000		15642.59	I

Curium (Cont.)

10000	15757.23	I
10000	15793.31	I
10000	16008.41	I
10000	17148.22	I
10000	17453.18	I
10000	17619.28	I
10000	18069.02	I
10000	19572.62	I
10000	19975.98	I
10000	20526.32	I
10000	20853.49	I
10000	20911.52	I
10000	20968.11	I
10000	21241.06	I
10000	21393.23	I

DYSPROSIUM (Dy)
Z = 66

Dy I and II
Ref. 1—C.H.C.

Intensity		Wavelength Air	
260		2356.91	II
65		2381.95	
130		2387.36	II
150		2392.15	
180		2402.29	II
240		2410.01	II
150		2422.75	II
260		2439.84	II
90		2455.15	II
110		2459.99	II
90		2471.40	II
110		2480.93	II
170		2490.61	II
90		2510.31	II
170		2513.55	II
170		2517.61	II
130		2543.81	II
90		2545.12	II
150		2552.29	II
180		2557.94	II
90		2560.21	II
90		2566.25	II
220		2585.30	I
90		2591.56	II
75		2592.54	II
130		2600.16	II
130		2600.76	II
75		2608.69	II
370		2623.69	I
440		2634.80	II
110		2642.15	I
110		2645.35	II
110		2667.94	I
55		2676.84	II
50		2677.34	II
85		2689.31	II
85		2692.83	II
55		2709.01	II
55		2727.17	II
85		2729.50	II
55		2735.79	I
40		2739.30	II
85		2740.70	II
220		2755.75	II
55		2757.08	II
70		2766.50	II
70		2772.42	II
110		2772.61	II
40		2779.58	II
55		2791.44	II
120		2800.33	II
110		2800.53	II
110		2801.41	II
300		2816.39	II
140		2825.42	II
140		2862.70	I
190		2877.88	II
110		2884.28	II
120		2885.53	I
120		2890.74	II
120		2900.82	II
110		2904.62	II
190		2906.39	II
390		2913.95	II
110		2934.31	II
250		2934.52	II
110		2941.05	II
140		2944.56	II
150		2947.06	II
150		2947.21	II
250		2948.31	II
170		2950.33	II
110		2952.12	II
140		2953.70	II
220		2964.60	I
110		2977.42	II
110		2985.97	II
220		3015.68	II
390		3026.16	II
210		3029.81	II
610		3038.28	II
280		3043.13	II
210		3047.56	II
280		3060.64	II
390		3062.62	II
220		3066.99	II
330		3071.91	II
280		3073.54	II
220		3078.68	II
280		3101.93	II
220		3103.24	II
410		3109.76	II
330		3128.41	II
830		3135.38	II
360		3140.64	II
500		3141.14	II
220		3143.83	II
250		3146.16	II
1200		3156.52	II
670		3162.83	II
1000		3169.99	II
400		3177.89	II
220		3178.37	II
200		3184.79	II
330		3186.38	II
240		3187.68	II
330		3193.30	II
240		3206.40	II
220		3207.12	II
290		3208.85	II
470		3215.19	II
830		3216.63	II
240		3221.49	II
290		3223.28	II
240		3225.08	II
330		3225.95	II
490		3235.89	II
290		3236.69	II
490		3245.12	II
200		3248.36	II
1200		3251.27	II
200		3252.19	II
290		3256.26	II
200		3266.21	II
240		3269.11	II
200		3272.73	II
890		3280.09	II
490		3282.77	II
200		3287.94	II
200		3293.88	II
200		3296.30	II
200		3305.40	II
200		3305.51	II
240		3306.19	II
440		3308.79	II
1100		3308.88	II
510		3312.72	II
780		3316.32	II
240		3317.12	II
1000		3319.88	II
270		3326.19	II
780		3341.00	II
270		3341.88	II
200		3347.83	II
270		3352.69	II
510		3353.58	II
240		3359.46	II
510		3368.11	II
5300		3385.02	II
210		3386.57	II
610		3388.85	II
210		3391.96	II
3800		3393.57	II
1300		3396.16	II
380		3407.16	II
5300		3407.80	II
420		3408.14	II
1300		3413.78	II
530		3414.82	II
780		3419.63	II
530		3425.06	II
420		3429.44	II
1900		3434.37	II
330		3438.94	II
560		3440.93	II
1300		3441.45	II
3800		3445.57	II
830		3446.99	II
440		3449.89	II
2700		3454.32	II
440		3454.51	II
1300		3456.56	II
4400		3460.97	II
720		3468.43	II
560		3471.14	II
560	d	3471.53	II
380		3473.70	II
1300		3477.07	II
4400		3494.49	II
560		3496.34	II
400		3497.81	II
830		3498.71	II
400		3501.50	II
830		3504.53	II
830		3505.45	II
1300		3506.81	II
560		3517.26	II
4400		3523.98	II
22000		3531.70	II
4400		3534.96	II
5500		3536.02	II
4400		3538.52	II
400		3539.37	II
1700		3542.33	II
400		3544.20	II
400		3544.35	II
1400		3546.83	II
330		3548.19	II
4400		3550.22	II
2200		3551.62	II
440	h	3558.23	II
440		3559.30	II
2200		3563.15	II
560		3563.69	II
780		3573.83	II
1400		3574.15	II
4400		3576.24	II
1700		3576.87	II
830		3577.98	II
440		3580.04	II
400		3584.42	II
3300		3585.06	II
1400		3585.78	II
560		3586.11	II
360		3590.07	II
1100		3591.41	II
560		3591.81	II
560		3592.11	II
1800		3595.04	II
400		3596.06	II
560		3600.38	II
360		3602.82	II
1800		3606.12	II
440		3618.51	II
560		3620.16	II
470		3624.27	II
1100		3629.42	II
4000		3630.24	II
440		3632.78	II
400		3635.27	II
360		3637.28	II
1100		3640.25	II
400		3643.92	II
11000		3645.40	II
360		3645.86	II
1000		3648.78	II
700		3664.62	II
400		3666.84	I
990		3672.30	II
420		3672.70	II
400		3673.14	II
1400		3674.08	II
2200		3676.59	II
640		3678.51	I
820		3684.85	I
1300		3685.78	I
4700		3694.81	II
370		3697.31	II
990		3698.21	II
540		3701.63	II
330		3707.40	II
440		3707.57	II
440		3708.22	II
420		3710.07	II
330		3711.66	II
1600		3724.45	II
300		3728.00	I
930		3739.34	I
1200		3747.82	II
1400		3753.51	II
1400		3753.75	II
1200		3757.05	I
4700		3757.37	II
640		3767.63	I
330		3771.11	I
640		3773.05	I
370		3774.71	I
420		3781.47	I
330		3785.41	II
3300		3786.18	II
1600		3788.44	II
700		3791.87	II
510		3804.14	II
580		3806.27	II
470		3812.27	I
470		3813.67	II
1400		3816.76	II
700		3825.68	II
2300		3836.50	II
370		3840.89	I
1400		3841.31	II
330		3842.00	II
330		3844.36	I
420		3846.34	II
420		3847.02	I
330		3849.39	II
1200		3853.03	II
420		3858.40	I
370		3866.58	II
560		3868.45	II
1600		3868.81	I
300		3869.42	II
820		3869.86	II
7000		3872.11	II
1200		3873.99	II
470		3879.11	II
300		3881.99	II
5800		3898.53	II
540		3914.87	II
540		3915.59	II
540	d	3917.29	I
320		3923.38	II
420		3927.86	I
540		3930.14	I
2100		3931.52	II
320		3932.22	II
370		3933.00	II
320		3934.21	II
420		3936.70	I
540		3942.53	II
10000		3944.68	II
420		3946.93	II
540		3950.39	II
420		3954.55	II
800		3957.79	II
370		3962.59	I
320		3967.51	I
14000		3968.39	II
2700		3978.57	II
1400		3981.92	II
1600		3983.65	II
800		3984.21	II
540		3991.32	II
1600		3996.69	II
8000		4000.45	II
420		4005.84	I
320		4006.07	I
540		4011.29	II
540		4013.82	I
540		4014.70	II
370		4023.71	I
420		4027.78	II
520	d	4028.32	II
520		4032.47	II
420		4033.65	II
420		4036.32	II
320		4041.98	II
12000		4045.97	I
1600		4050.56	II
520		4055.14	II
2500		4073.12	II
7400		4077.96	II
370		4085.34	I
390		4096.10	I
3900		4103.30	II
860		4103.87	I
1500		4111.34	II
490		4124.63	II
390		4128.24	II
350		4129.12	I
990		4129.42	II
350		4130.35	I
390		4133.85	I
470		4141.50	II
1200		4143.10	II
990		4146.06	I
5700		4167.97	I
370		4171.93	I
930		4183.72	I
12000		4186.82	I
320		4190.94	I
2200		4191.64	I
6800		4194.84	I
320		4195.19	II
800		4198.02	I
680		4201.30	I
680		4202.24	I
230		4205.06	I
370		4206.54	II
440		4211.24	I
16000		4211.72	I
1800		4213.18	I
3700		4215.16	I
4400		4218.09	I
4400		4221.11	I
540		4222.21	I
2700		4225.16	I
680		4232.02	I
680		4239.85	I
440		4245.91	I
440		4256.33	II
250		4276.69	I
370	d	4294.93	II
		4295.04	I
1000		4308.63	II
320		4325.86	I
200		4358.44	II

Dysprosium (Cont.)

320		4374.24	II
320		4374.76	II
540		4409.38	II
150		4444.58	I
740		4449.70	II
110		4455.60	II
250		4468.14	II
100	d	4527.58	I
		4527.76	II
100		4541.66	II
140		4565.09	I
420		4577.78	I
2100		4589.36	I
990		4612.26	I
50		4613.83	I
50		4614.82	I
60		4617.26	II
140		4620.03	II
50		4662.72	I
110		4664.66	II
85		4673.60	II
50		4682.03	II
50		4689.75	II
95		4698.68	II
85		4721.22	I
70		4727.13	II
170		4731.84	II
40		4745.73	II
60		4754.99	II
50		4760.04	II
60		4771.94	I
50		4774.80	I
120	h	4775.79	I
75		4786.92	II
95		4791.29	I
29		4800.64	I
50		4807.94	I
40		4810.28	I
50		4812.80	I
75		4819.04	I
85		4824.96	I
75		4828.88	I
50		4829.68	II
70		4832.38	I
35		4833.75	II
75		4841.75	I
40		4856.24	II
40		4868.05	II
40		4875.93	I
85		4880.16	I
40		4884.55	I
95		4888.08	I
40		4889.33	II
75		4890.10	II
50		4893.68	I
24		4899.24	I
55		4916.41	I
50		4922.22	II
65		4923.16	II
480		4957.34	II
24		4959.59	I
28		4973.57	I
40		4985.52	I
50		5003.87	I
55		5004.28	II
24		5010.60	I
24		5017.98	II
70		5022.12	I
30		5024.03	I
24		5024.54	I
40		5027.87	I
50		5033.00	I
160		5042.63	I
24		5047.25	I
50	h	5050.21	I
30		5053.35	I
24		5055.46	I
95		5070.68	I
120		5077.67	I
80		5090.38	II
80		5110.32	I
130	h	5120.04	I
30		5135.02	I
190		5139.60	II
40		5161.03	II
40		5164.12	II
50		5165.34	I
110		5169.69	II
20		5172.90	II
80		5185.30	I
40		5188.45	II
290		5192.86	II
95		5197.66	II
50		5246.94	II
70		5259.88	I
130		5260.56	I
55	Bl	5263.3	D
65		5267.11	I
50		5272.25	II
50		5275.29	II
50		5279.70	II
55		5282.07	I
28		5284.99	II
40		5297.82	II
160		5301.58	I
40		5309.02	II
50		5324.69	I
24		5337.43	II
65		5340.30	I
30		5352.11	I
30		5368.20	II
20		5385.63	II
85		5389.58	II
40		5395.57	I
20	h	5398.26	D
24		5399.93	II
50		5404.19	I
80		5419.13	I
70		5423.32	I
30		5424.27	I
40		5426.70	II
30		5443.34	II
95		5451.11	I
30		5455.47	II
24		5469.10	II
28		5496.83	I
24		5502.79	I
28		5506.52	I
24		5515.41	II
30		5528.01	I
65		5547.27	I
40	d	5600.65	II
24		5605.53	I
30		5613.23	I
20		5627.49	I
100		5639.50	I
55	h	5645.99	I
80		5652.01	I
24		5685.58	I
28	h	5693.67	D
24	h	5694.10	D
28	Cw	5694.54	D
28		5698.72	II
24		5702.91	I
70	h	5718.46	I
28	h	5725.84	D
55	h	5728.64	D
24	h	5738.73	D
50		5740.20	I
55		5745.53	I
24		5750.48	I
24		5758.79	I
80	h	5832.01	D
55	h	5833.85	D
40	h	5834.86	D
28	h	5844.41	D
24		5845.65	D
40	h	5848.05	D
40		5855.56	D
55	h	5868.11	II
40		5915.16	II
20		5924.56	II
70		5945.80	I
50	l	5964.46	I
120		5974.49	I
24		5984.86	I
140		5988.56	I
24	h	6005.75	D
24	h	6006.54	D
24	h	6006.97	D
30		6008.94	I
65		6010.82	I
24		6017.26	I
24		6030.98	I
24	Bl	6042.49	D
24		6058.18	I
30		6085.06	I
140		6088.26	I
24		6127.15	I
24		6133.64	I
24		6158.28	I
100		6168.43	I
20		6196.23	II
270		6259.09	I
30		6260.36	I
14		6343.32	I
40		6386.80	I
24		6396.60	II
50		6421.92	I
13	h	6436.55	I
8		6460.83	I
10		6468.58	II
11		6474.91	I
20		6483.59	II
28		6486.59	I
20		6558.02	I
160		6579.37	I
14		6594.14	II
15		6643.37	I
22		6658.36	I
29		6661.64	I
75		6667.86	I
10		6700.64	II
29		6747.93	I
10		6757.62	I
45		6765.89	I
12		6818.20	I
180		6835.42	I
80		6852.96	I
22		6856.46	I
22		6888.83	I
15		6897.97	II
65		6899.32	II
22		6906.53	II
15		6929.55	I
29		6950.28	II
11		6951.42	I
40		6958.08	I
13	h	6982.44	I
13		6991.30	I
45		6998.10	I
20		7017.42	I
35		7055.95	II
24		7075.14	II
17		7109.26	II
11		7120.81	II
13		7175.11	II
11		7213.27	I
17	h	7230.04	I
13		7250.01	I
17		7345.13	II
11		7370.23	II
20		7376.04	I
11		7407.59	I
24		7412.37	I
55		7426.86	II
20		7457.05	II
17		7516.61	II
55		7543.73	I
17	h	7553.00	I
27		7559.78	I
40		7562.96	II
20	h	7577.46	II
27	h	7591.30	I
13	h	7611.55	I
11	h	7617.70	I
35	h	7641.09	I
17		7645.86	I
13		7646.64	I
80		7662.36	I
11		7666.78	II
35		7715.33	I
45		7729.76	II
20		7751.62	II
35		7812.06	I
27		7909.38	I
11		7968.63	I
12		7982.85	II
13		8147.29	I
27		8198.77	II
100		8201.57	II
11		8218.62	II
20		8265.53	I
35		8326.10	I
35		8392.01	II
12		8405.85	II
20		8416.64	II
24		8438.58	II
11		8630.12	I
27		8655.94	II
17		8657.68	II
17		8678.49	II
11		8696.83	II
11	h	8715.95	II
20		8750.40	II
12		8780.83	I
45		8791.39	II
13		8833.08	II
24		8850.37	II

EINSTEINIUM (Es)
Z = 99

Es I and II
Ref. 333—J.G.C.

Intensity		Wavelength	
		Air	
300	l	2694.32	II
100	l	2703.84	
10000	s	2708.66	II
100	s	2716.02	II
1000	s	2724.57	II
1000	l	2765.76	
3000	l	2787.10	II
1000	l	2796.11	II
3000	l	2815.15	
100	s	2885.84	II
100	s	2886.44	
3000		2907.03	
100		3003.28	
1000	l	3065.40	
3000		3135.25	I
1000	l	3154.27	II
10000		3413.17	
300	l	3423.12	I
100	l	3424.28	
10000	s	3428.48	I
300		3437.31	
100		3437.34	
3000	s	3445.25	
300	l	3446.93	
3000		3452.36	
100		3453.16	
300	s	3470.77	
3000	s	3484.59	I
300		3494.30	
10000	s	3498.11	I
10000	l	3514.33	I
10000	s	3521.38	
10000	s	3523.49	I
300	l	3528.58	
10000	s	3536.01	
10000	s	3547.75	II
300		3549.97	
10000	l	3555.34	I
100	s	3555.53	
3000	s	3556.65	
3000	s	3560.92	
1000	s	3575.68	
100	s	3578.56	
300	s	3579.38	
1000	s	3582.95	
3000		3590.28	
1000		3595.47	
10000	s	3602.43	II
300		3605.58	
1000	s	3606.75	
100	l	3624.52	
3000	l	3631.09	
10000	s	3632.87	
100	l	3634.41	
1000	l	3651.94	
10000	l	3670.01	II
100	l	3672.32	
100	s	3713.56	
1000	l	3720.56	
300		3722.32	
10000	l	3728.55	II
100	h	3737.47	
300	s	3776.27	
10000	l	3792.99	
10000	l	3801.49	
300	s	3929.10	
3000		3930.77	
100		3957.19	
100		3995.35	
300		4077.71	
10000	s	4082.24	
3000	l	4107.59	
1000	l	4176.94	
100	s	4496.25	
100	l	4631.66	
300	l	4650.86	
1000		4789.93	
300	h	4802.17	
1000	h	4802.21	
3000	h	4958.29	
10000	s	5052.08	
1000	s	5102.93	
100	s	5155.82	
10000	s	5161.74	I
10000	l	5204.40	I
3000	L	5615.51	I
100	S	6539.71	ES

ERBIUM (Er)
Z = 68

Er I and II
Ref. 1—C.H.C.

Wavelength   Intensity

Air

110		2358.51	II
100		2386.58	II
120		2387.17	II
110		2396.38	III
140		2446.39	II
100		2537.02	II
110		2547.28	II
290		2586.73	II
110		2587.04	II
130		2592.57	II
120		2595.03	II
140		2624.18	II
490		2670.26	II
330		2672.25	II
100		2675.35	II
270		2739.31	II
310		2750.19	II
230		2755.01	II
610		2755.63	II
510		2770.02	II
230		2778.97	II
230		2802.53	II
310		2804.35	II
410		2820.19	II
270	d	2833.91	II

Erbium (Cont.)

Intensity		Wavelength	Spectrum
390		2838.71	II
270		2848.37	II
250		2855.41	II
310		2859.84	II
310		2896.96	II
390		2897.52	II
1000		2904.47	II
210		2909.58	II
1500		2910.36	II
270		2915.62	II
350		2929.27	II
270		2945.28	II
230		2946.62	II
1500		2964.52	II
410		2968.76	II
210		2974.47	II
230		2975.68	II
270		2983.80	II
1200		3002.41	II
310		3002.65	II
230		3012.47	II
230		3016.84	II
290		3025.95	II
270		3028.27	II
370		3031.31	II
310		3036.22	II
210		3054.42	II
230		3066.22	II
450		3070.74	II
560		3072.53	II
610		3073.34	II
210		3078.87	II
720		3082.08	II
610		3084.02	II
370		3099.19	II
230		3106.78	II
310	d	3113.43	II
		3113.54	II
770		3122.72	II
290		3132.52	II
470		3132.77	II
410		3141.10	II
		3141.15	II
250		3144.33	II
410		3154.29	II
870		3181.92	II
410		3183.42	II
250		3185.25	II
310		3200.58	II
230		3205.15	II
270		3214.44	II
870		3220.73	II
610		3223.31	II
210		3227.16	II
2300		3230.58	II
250		3232.03	II
330		3237.98	II
330		3249.34	II
560		3259.05	II
		3259.11	II
2700		3264.78	II
430		3267.10	II
		3267.18	II
330		3269.41	II
250		3278.22	II
720		3279.33	II
720		3280.22	II
470		3286.77	II
330	d	3303.88	II
		3303.95	II
370		3305.56	II
2300		3312.42	II
560		3316.39	II
770		3323.19	II
290		3329.66	II
770		3332.70	II
370		3337.25	II
290		3337.79	II
250		3340.03	II
290		3341.84	II
1300		3346.04	II
470		3350.06	II
350		3350.26	II
1400		3364.08	II
1400	d	3368.02	II
		3368.13	I
450		3370.55	II
7700		3372.71	II
970		3374.17	II
290		3381.32	II
230		3382.06	I
1700		3385.08	II
450		3389.74	II
2300		3392.00	II
350		3396.07	II
290		3396.84	II
390		3401.83	II
350		3417.63	II
490		3428.39	II
770		3441.13	II
390		3442.68	I
490		3469.51	I
970		3471.71	II
610		3479.41	II
970		3485.85	II
350		3486.82	II
350		3496.86	II
6700		3499.10	II
610		3502.78	I
390		3508.38	II
490		3514.89	II
390		3518.18	II
610		3524.91	II
410		3539.59	I
310		3548.26	II
820		3549.84	II
310		3553.20	II
1500		3558.02	I
510		3558.71	I
1000		3559.90	II
310		3565.17	I
920		3570.75	II
310		3578.24	I
1000		3580.52	II
370		3586.60	I
610		3590.76	I
410		3595.84	I
610		3599.50	II
1000		3599.83	II
510		3604.90	II
410		3607.42	I
3100		3616.56	II
510		3617.85	II
510		3618.92	II
720		3628.04	I
310		3629.37	I
1000		3633.54	II
510		3634.67	I
1600		3638.68	I
900		3645.94	II
520		3650.41	II
360		3652.58	II
500		3652.87	II
360		3664.45	I
470		3669.02	II
500		3682.70	II
320		3684.01	I
380		3684.28	II
7900		3692.65	II
450		3696.25	II
380		3697.68	I
540		3700.72	II
520		3707.64	II
520		3712.39	II
320		3719.35	I
1300		3729.52	II
450		3731.26	II
540		3738.16	II
340		3741.10	II
900		3742.64	II
900		3747.43	I
540		3756.05	I
410		3781.01	II
1800		3786.84	II
560		3787.86	II
560		3791.83	II
500		3792.79	I
560		3797.06	II
1600		3810.33	I
3600		3830.48	II
540		3849.91	I
320		3851.60	II
680		3855.90	I
540		3858.39	II
7500		3862.85	I
1500		3880.61	II
1200		3882.89	II
400		3890.61	II
4200		3892.68	I
5200		3896.23	II
810		3902.76	II
250		3903.98	I
250		3904.56	II
1200		3905.40	I
11000		3906.31	II
280		3918.05	I
210		3918.35	II
280		3921.88	II
810		3932.25	II
3200		3937.01	I
2100		3938.63	II
3200		3944.42	I
550		3948.06	I
250		3951.48	I
320		3956.42	I
280		3966.35	I
2700		3973.04	I
3200		3973.58	I
1400		3974.72	II
280		3976.73	I
810		3977.02	I
1100		3982.33	I
280		3987.53	I
810		3987.66	I
230		3991.15	I
230		4004.05	I
14000		4007.96	I
230		4008.18	II
280		4009.16	II
1100		4012.58	I
350		4015.57	II
3000		4020.51	I
450		4021.55	I
230		4043.01	II
1000		4046.96	I
280		4048.34	II
200		4049.49	II
940		4055.47	II
550		4059.51	I
690		4059.78	II
420		4077.88	I
550		4081.24	II
3500		4087.63	I
210		4092.90	I
1100		4098.10	I
350		4100.56	II
320		4116.36	I
320		4118.55	I
600		4131.50	I
550		4142.91	II
6900		4151.11	I
280		4189.98	II
1000		4190.70	I
130		4205.32	I
1400		4218.43	I
200		4220.99	I
320		4230.20	II
140		4234.78	II
200		4251.94	II
140		4276.48	II
690		4286.56	I
320		4298.91	I
320		4301.60	II
140		4303.81	II
110		4319.94	II
130		4328.81	I
110		4331.36	I
140		4340.92	I
190		4348.34	I
110		4369.39	II
160		4382.17	I
300		4384.70	II
300		4386.40	I
100		4403.17	II
810		4409.34	I
180		4418.70	I
570		4419.61	II
110		4422.51	II
320		4424.57	I
370		4426.77	I
110		4437.66	I
100		4459.24	II
100		4473.50	II
130		4496.39	I
200		4500.75	II
130		4522.74	I
160		4563.26	II
1000		4606.61	I
160		4630.88	II
110		4640.60	II
110		4665.44	II
310		4673.16	I
570		4675.62	II
150		4679.06	II
230		4722.69	I
150		4729.05	I
130		4751.52	II
170		4759.65	II
190		4820.35	II
140		4857.44	I
150		4872.09	II
210		4900.08	II
210		4934.11	II
130		4944.36	I
180		4951.74	II
130		4976.42	I
250		5007.25	I
140		5028.33	I
120		5028.91	II
200		5035.94	I
210		5042.05	II
130		5043.86	I
130		5044.89	I
130		5077.59	II
120		5124.56	I
130		5127.41	II
120		5131.53	I
130		5133.83	II
170		5164.77	II
130		5172.78	I
160		5188.90	II
150		5206.52	I
60		5212.91	II
30		5215.13	II
30		5218.26	II
45		5229.34	II
140		5255.93	II
22		5256.47	II
27		5257.02	II
35		5264.77	II
80		5272.91	I
55		5277.71	I
27		5279.34	II
45		5302.30	II
55		5333.06	I
27		5333.33	II
27		5334.23	II
22		5343.94	II
30		5344.50	II
90		5348.06	I
45		5350.47	I
35		5368.85	I
35		5395.87	II
60		5414.63	II
18		5422.81	II
18	h	5451.30	I
35		5454.27	II
180		5456.62	I
35		5462.43	II
90		5468.32	I
18		5477.47	II
80		5485.97	II
27		5497.44	II
27		5516.02	I
80		5593.46	I
45	d	5601.14	I
		5601.32	I
45	h	5609.94	I
60		5611.82	I
70		5622.01	I
80		5626.53	II
30		5636.20	I
90		5640.36	I
22		5641.42	I
22	h	5658.63	II
70		5664.95	I
45		5665.44	II
55		5675.48	I
14		5695.53	II
27		5710.87	II
55		5717.48	I
70		5719.55	I
55		5726.97	I
22		5733.43	II
22		5736.56	I
22		5736.94	I
100		5739.19	I
35		5740.61	I
60		5748.65	I
55		5752.53	I
70		5757.63	II
290		5762.80	I
70		5769.92	I
45		5782.82	I
70		5784.66	I
22		5791.15	II
70		5800.79	I
22		5806.10	I
430		5826.79	I
45		5835.84	I
100		5850.07	I
120		5855.31	I
140		5872.35	I
120		5881.14	I
27		5886.30	II
27		5902.08	II
55		5906.06	I
45		5909.24	I
35		5933.50	I
22		5946.37	I
55		5968.68	I
27		5975.49	I
35		6006.79	II
22		6008.75	II
55		6014.83	I
35		6015.74	II
70		6022.56	I
22		6032.12	II
22		6045.63	II
22		6048.14	II
45		6054.85	I
70		6061.25	I
60		6076.45	II
35		6116.01	I
35		6125.32	I
30		6170.06	II
27		6183.21	II
360		6221.02	I
35		6230.90	I
55		6262.56	I
45		6267.93	I
60		6268.87	I
35		6274.94	I
30		6286.86	I
45		6299.42	I
130		6308.77	I
55		6326.13	I
22		6347.16	II
45		6388.19	I
22		6432.53	I
27		6485.87	I
55		6492.35	I

Erbium (Cont.)

Intensity		Wavelength	
22		6541.57	I
60		6583.48	I
70		6601.11	I
27		6721.91	I
70		6759.87	I
22		6762.92	I
27		6773.37	I
35		6790.92	I
22		6825.44	I
22		6825.98	I
70		6848.10	I
55		6865.13	I
27		6879.98	I
22		7001.40	I
12		7058.55	I
12	h	7065.04	I
11		7070.99	II
18		7101.27	I
8		7109.67	I
11		7155.40	II
5		7161.91	I
14	h	7197.00	I
7	h	7264.82	II
7	h	7283.95	I
14		7329.73	II
18		7355.37	I
11		7356.34	I
18		7428.67	I
55		7459.55	I
9		7460.42	I
120		7469.51	I
22		7532.34	I
6	h	7539.18	I
27		7556.26	I
6	h	7574.21	I
5	h	7590.51	I
11		7597.33	I
6		7607.23	I
11		7613.52	I
6		7623.48	I
16	h	7645.67	I
8		7650.63	I
22		7654.45	II
12	h	7658.05	I
22		7659.25	I
4		7665.64	I
35		7680.01	I
9		7722.14	I
8		7726.19	II
11	h	7747.44	I
22		7754.63	I
4		7762.16	I
9		7796.69	I
35		7797.47	I
9		7838.80	I
11		7844.00	I
16		7847.55	I
5		7875.36	I
5		7879.36	I
18		7899.55	I
8		7913.08	I
35		7921.85	I
30		7937.84	I
8		7952.93	I
12		7964.51	I
8		7979.03	II
8		7980.87	I
5		8023.03	I
12		8035.91	I
12		8181.85	I
35		8312.82	I
18		8328.57	II
5		8367.58	II
55		8409.90	I
11		8466.18	II
35		8472.42	I
14		8517.71	II
18		8521.37	II
22		8768.64	I
11	h	8776.63	II
9		8866.84	II

Er III
Ref. 301—J.R.

Intensity		Wavelength	
		Air	
2		2165.26	III
3		2190.77	III
10		2198.15	III
1		2223.98	III
4		2232.35	III
60		2235.28	III
8		2245.60	III
2		2255.95	III
80		2269.36	III
600		2277.65	III
100		2309.19	III
40		2358.69	III
10		2358.79	III
50		2359.33	III
200		2367.64	III
20		2375.50	III
10		2377.07	III
80		2381.25	III
20		2381.40	III
40		2381.75	III
6		2391.96	III
60		2393.08	III
5		2393.60	III
250		2396.40	III
10		2398.91	III
2		2402.75	III
80		2404.58	III
100		2410.47	III
200		2419.81	III
200		2422.47	III
40		2431.51	III
60		2464.60	III
2		2492.04	III
100		2508.59	III
2		2531.03	III
100		2532.36	III
8		2536.76	III
80		2540.91	III
3		2543.31	III
10		2545.95	III
50		2557.22	III
80		2570.74	III
40		2580.02	III
2		2589.55	III
80		2590.72	III
20		2591.56	III
200		2591.83	III
20		2598.39	III
3		2599.18	III
100		2603.62	III
40		2604.91	III
25		2614.53	III
30		2617.64	III
2		2618.40	III
2		2618.94	III
8		2625.19	III
20		2626.37	III
4		2637.52	III
200		2637.77	III
5		2651.49	III
25		2683.10	III
400		2723.29	III
100		2738.53	III
500		2739.27	III
8		2741.41	III
80		2746.03	III
6		2752.20	III
80		2756.20	III
400		2759.23	III
150		2761.92	III
60		2762.66	III
15		2767.11	III
100		2768.72	III
60		2772.07	III
60		2774.80	III
20		2775.55	III
2		2780.60	III
80		2783.11	III
500		2792.54	III
10		2804.10	III
100		2805.87	III
6		2808.44	III
50		2824.75	III
150		2830.34	III
1		2831.95	III
8		2833.03	III
8		2845.29	III
60		2846.08	III
6		2849.63	III
1		2869.52	III
8		2878.24	III
1		2955.93	III
1		2958.63	III
1000		3055.10	III
1000		3070.40	III
500		3100.40	III
1500		3166.25	III
3		3172.47	III
1		3173.45	III
50		3175.74	III
400		3214.95	III
2000		3301.23	III
8		3341.00	III
200		3480.54	III
8		3592.96	III
600		3715.67	III
200		3739.43	III
4000		3816.78	III
600		3962.87	III
40		4009.70	III
2		4088.58	III
1000		4288.18	III
40000		4290.06	III
300		4338.24	III
20000		4386.86	III
30		4612.93	III
15000		4735.56	III
2000		4783.12	III
8		4876.07	III
8000		5903.30	III

EUROPIUM (Eu)
Z = 63

Eu I and II
Ref. 1—C.H.C.

Intensity		Wavelength	
		Air	
21		2499.39	II
26		2554.78	II
26		2559.18	II
160		2564.17	II
110		2568.17	II
26		2574.76	II
230		2577.14	II
26		2581.86	II
26		2604.61	I
30		2635.50	II
1000		2638.77	II
380		2641.27	II
40		2653.61	II
640		2668.34	II
110		2673.42	II
250		2678.29	II
250		2685.66	II
550		2692.03	II
700		2701.14	II
800		2701.90	II
240		2705.28	II
180		2709.99	I
700		2716.98	II
70		2723.96	I
4200		2727.78	II
190		2729.33	II
380		2729.44	II
50		2731.37	I
40		2732.61	I
80		2735.25	I
160		2740.62	II
70		2743.28	I
120		2744.26	II
40		2745.61	I
70		2747.29	II
80		2747.83	I
90		2752.17	II
480		2781.89	II
1900		2802.84	II
220		2811.75	II
30		2813.08	II
3400		2813.94	II
550		2816.18	II
2000		2820.78	II
400	Cw	2828.72	II
120		2829.30	II
140		2833.26	II
80		2843.96	II
60		2852.05	II
260		2859.67	II
280		2862.57	II
25		2864.42	II
60		2876.06	II
100		2878.87	I
80		2887.85	II
200		2892.54	I
140		2893.03	I
360		2893.83	I
3200		2906.68	II
160		2908.99	I
30		2917.44	II
850		2925.04	II
60		2947.29	II
200	Cw	2952.68	II
30		2958.91	I
35		2959.47	II
260		2960.21	II
300		2991.33	II
35		2995.22	II
40		3006.26	II
35		3022.15	I
30		3040.77	II
320	Cw	3054.94	II
120		3058.98	I
35		3069.11	II
35		3076.07	II
220		3077.36	II
35		3089.35	II
120		3097.45	II
320		3106.18	I
950		3111.43	I
120		3130.73	II
40		3132.16	I
45		3149.88	II
85		3173.61	II
40		3185.54	I
420		3210.57	I
1000		3212.81	I
420		3213.75	I
45		3235.13	I
95		3241.40	I
45		3246.03	I
45		3247.32	II
100		3247.55	I
100		3266.39	II
150		3272.77	II
210		3277.78	II
150		3301.95	II
45		3304.50	II
140		3308.02	II
140		3313.33	II
65		3319.89	II
95		3321.86	II
85		3322.26	I
950		3334.33	I
45		3338.75	II
110		3350.40	I
40		3351.56	II
40		3354.38	II
45		3367.64	II
140		3369.06	II
65		3380.25	II
75		3390.78	II
190		3391.99	II
280		3396.58	II
45		3419.84	II
65		3423.09	II
150		3425.02	II
45		3426.44	II
45		3435.05	II
65		3435.20	II
40		3435.72	II
45		3440.82	II
150		3441.00	II
45		3445.18	II
85		3457.05	I
45		3457.56	II
130		3461.38	II
85		3467.88	I
75		3477.07	I
75	h	3505.30	II
470	Cw	3521.09	II
75		3531.15	II
45		3532.23	II
65		3538.08	II
150		3542.15	II
85		3543.85	II
45		3549.71	II
180		3552.52	II
75		3589.27	I
45		3591.31	II
150		3603.20	II
75		3611.57	II
45		3616.15	II
95		3622.54	II
95		3632.18	II
45		3673.19	II
45		3674.63	II
45		3678.26	II
6400		3688.42	II
60		3710.87	II
95		3713.45	II
95		3714.90	II
35		3716.94	II
35		3717.69	II
40		3719.16	I
20000	Cw	3724.94	II
45		3729.68	II
45		3729.74	II
21		3732.20	I
45		3738.08	II
350		3741.31	II
100		3743.56	II
260		3761.12	II
95		3765.93	II
40		3774.10	I
60		3781.40	II
40		3788.76	II
45		3791.50	II
130		3799.01	II
70		3801.36	
95		3807.54	II
120		3811.33	I
120		3815.50	II
39000	Cw	3819.67	II
120		3826.68	II
140		3844.23	II
190		3865.57	I
45		3872.72	I
70		3877.27	II
150		3884.75	I
23		3896.78	I
23		3900.18	I
70		3900.51	I
28000	Cw	3907.10	II
45		3915.24	II
45		3916.00	I
230		3917.29	I
23		3917.70	II
40		3918.52	I
100		3919.09	II
40		3928.87	II
32000	Cw	3930.48	II
55		3941.56	II

Intensity		Wavelength	Spectrum
30	h	3942.21	II
60		3942.94	II
120		3943.08	II
30		3944.59	II
30		3945.67	II
30		3949.13	II
60		3949.60	I
45		3950.76	II
55		3951.33	II
60		3955.75	I
40		3957.92	II
30		3963.61	I
120		3964.90	II
150		3966.59	II
45		3967.18	I
30000	Cw	3971.96	II
60		3978.42	I
30		3979.63	II
55	h	3986.60	I
40		3988.24	II
30		3993.93	II
55		3995.98	II
60		4003.71	II
180		4011.69	II
150		4017.58	II
120		4039.19	I
45	h	4078.24	I
120		4085.38	II
75		4096.80	II
60		4106.88	I
90	h	4112.04	II
45		4119.30	II
75		4127.28	I
33000	Cw	4129.70	II
30		4136.59	II
40		4137.07	I
30		4141.02	II
60		4141.72	II
30		4151.52	II
45		4151.64	II
30		4157.72	I
110		4172.80	II
30		4175.16	II
110		4182.22	I
40		4196.18	II
60000	Cw	4205.05	II
45		4221.08	II
40		4223.88	II
90	h	4227.40	II
75		4229.33	II
75		4232.45	II
90		4237.51	II
45		4238.69	II
45		4244.74	I
45		4247.06	II
45		4253.80	II
30		4270.24	II
150		4298.73	I
90		4329.36	I
75		4329.97	I
60		4330.61	II
40		4331.18	I
90		4337.68	I
240		4355.09	II
27		4361.57	II
55		4369.47	II
45	h	4372.20	II
75		4383.17	II
90		4387.88	I
21	h	4405.27	II
55		4407.07	II
18		4419.66	II
120		4434.81	II
14000	Cw	4435.56	II
75	h	4464.97	II
24		4485.15	II
3000		4522.57	II
45	h	4535.59	I
11000		4594.03	I
21		4602.63	I
9800		4627.22	I
8300		4661.88	I
30		4713.59	I
27		4740.50	I
45		4792.59	I
40	h	4829.30	I
60		4830.33	I
40	h	4840.47	I
60	h	4849.64	I
110		4867.62	I
40	h	4884.05	I
90		4894.68	I
60		4900.86	I
150		4907.18	I
180		4911.40	I
55		4953.52	I
55		4960.21	I
55		4962.55	I
45		4975.76	I
180		5013.17	I
170		5022.91	I
110		5029.54	I
90		5033.55	I
75		5067.95	I
75	h	5092.69	I
90	h	5096.44	I
170		5114.37	I
90		5124.77	I
170		5129.10	I
90		5130.08	I
210		5133.52	I
270		5160.07	I
210		5166.70	I
60		5193.74	I
200		5199.85	I
110		5200.96	I
120		5206.44	I
750		5215.10	I
300		5223.49	I
120		5239.24	I
200		5266.40	I
390		5271.96	I
110		5272.48	I
150		5282.82	I
55		5287.25	I
60		5289.25	I
120		5291.26	I
60		5293.68	I
120		5294.64	I
90		5303.85	I
30	h	5350.41	I
75	h	5351.69	I
40		5352.84	I
90		5355.10	I
540		5357.61	I
60		5360.83	I
120		5361.61	I
110		5376.94	I
120		5392.94	I
450		5402.77	I
45		5405.33	I
45		5411.86	I
55		5421.07	I
90		5426.94	I
40		5443.56	I
380		5451.51	I
260		5452.94	I
40		5457.62	I
90		5472.32	I
120		5488.65	I
45		5495.20	I
15		5500.83	I
120		5510.52	I
30		5526.63	I
30		5533.25	I
30		5542.54	I
200		5547.44	I
150		5570.33	I
200		5577.14	I
75		5579.63	I
120		5580.03	I
90		5586.24	I
75		5586.83	I
18		5592.25	I
18		5599.80	I
18		5605.86	I
40		5618.81	I
60		5622.44	I
75		5632.54	I
210		5645.80	I
15		5651.11	I
60		5673.85	I
27		5681.10	I
27		5684.24	I
60		5730.87	I
60		5739.00	I
330		5765.20	I
180		5783.69	I
15		5792.72	I
60		5800.27	I
170		5818.74	II
600	Cw	5830.98	I
27		5845.77	I
27		5860.97	I
15		5864.77	I
90		5872.98	II
15		5895.31	I
27		5902.97	I
12		5909.94	I
75		5915.74	I
12		5925.30	I
27		5926.52	I
45		5942.72	I
27		5953.49	I
27		5953.84	II
30		5954.28	I
90		5963.76	I
330		5966.07	II
480	Cw	5967.10	I
15	h	5968.43	I
30		5971.69	I
170		5972.75	I
15		5980.47	I
27		5983.14	I
27		5983.78	I
240		5992.83	I
60		6004.36	I
15	h	6005.61	I
60	h	6012.20	I
110		6012.56	I
60		6015.58	I
420		6018.15	I
60		6023.15	I
170		6029.00	I
60		6044.66	I
420		6049.51	II
140		6057.36	I
90		6075.58	I
30		6077.38	I
240		6083.84	I
240		6099.35	I
60		6108.15	I
120		6118.78	I
60		6124.67	I
330		6173.05	II
110		6178.76	I
260	Cw	6188.13	I
140		6195.07	I
15	h	6207.60	I
15		6230.51	I
90	h	6233.73	I
55		6250.47	I
240		6262.25	I
55		6266.95	I
15	h	6285.95	I
60		6291.34	I
170		6299.77	I
230		6303.41	II
24	h	6313.78	I
15		6318.58	I
75		6335.82	I
120	Cw	6350.04	I
60		6355.89	I
60		6369.25	I
55		6382.73	I
75		6383.86	I
120	Cw	6400.93	I
40		6406.11	I
180		6410.04	I
140		6411.32	I
55		6428.29	I
830		6437.64	II
18		6439.93	I
120		6457.96	I
12		6470.70	I
18		6483.02	I
45		6501.55	I
60		6519.59	I
15		6522.72	I
8	h	6549.12	I
75		6567.87	I
45		6593.79	I
18	h	6603.55	I
1400		6645.11	II
26		6685.21	I
95		6693.96	I
7	h	6701.06	I
12	h	6710.45	I
30		6744.88	I
30	h	6782.54	I
14	h	6787.48	I
140		6802.72	I
35		6816.06	I
11	h	6834.30	I
17		6840.93	I
17	h	6844.83	I
14	h	6847.04	I
360		6864.54	I
21		6898.21	I
60	h	6903.67	I
14	h	6910.17	I
30	h	6914.82	I
120		7040.20	I
12		7074.54	I
330		7077.10	II
100		7106.48	I
6		7164.66	I
30		7175.55	I
570		7194.81	II
570		7217.55	II
11	h	7224.68	I
15		7258.72	I
30		7262.77	I
11	h	7281.53	I
6	h	7297.56	I
540		7301.17	II
11		7310.46	I
12		7313.63	I
55	Cw	7336.18	I
4		7346.25	I
4		7356.65	I
11		7362.25	I
55	Cw	7369.60	I
720		7370.22	II
4		7387.36	I
12		7389.16	I
11	h	7404.41	I
300		7426.57	II
21	h	7436.59	I
8		7470.53	I
5	h	7491.00	I
50	Cw	7528.70	I
5	h	7533.02	I
6		7547.32	I
160		7583.91	I
60	Cw	7742.57	I
70		7746.19	I
8	h	7803.32	I
8		7818.21	I
35		7887.99	I
7		8015.47	I
24	Cw	8209.80	I
15	Cw	8226.81	I
6	h	8464.71	I
21	Cw	8642.67	I
7		8727.77	I
6		8782.46	I
12	Cw	8790.88	I
18		8870.30	I

Eu III
Ref. 312—J.R.

Intensity		Wavelength	
		Air	
10		2073.40	III
10		2093.50	III
30		2124.69	III
10		2167.12	III
10		2173.59	III
10		2184.68	III
10		2190.59	III
10		2194.81	III
20		2211.85	III
20		2212.63	III
20		2214.66	III
10		2215.34	III
30		2217.23	III
30		2219.33	III
20		2219.42	III
10		2223.13	III
10		2235.17	III
10		2240.14	III
10		2261.88	III
20		2265.74	III
10		2269.39	III
20		2276.85	III
40		2291.62	III
20		2304.37	III
10		2311.92	III
10		2327.69	III
10		2334.56	III
10		2336.96	III
10		2339.84	III
10		2343.10	III
10		2346.83	III
10		2347.64	III
10		2350.38	III
200		2350.51	III
10		2352.28	III
10		2357.87	III
10		2359.08	III
10		2360.65	III
20		2363.76	III
20		2368.04	III
20		2374.08	III
10		2375.20	III
4000		2375.46	III
10		2376.42	III
20		2377.23	III
10		2381.81	III
10		2383.62	III
20		2387.29	III
20		2389.11	III
10		2389.98	III
20		2391.11	III
20		2391.90	III
10		2392.59	III
10		2394.66	III
20		2395.62	III
20		2398.79	III
20		2401.00	III
20		2402.34	III
20		2404.08	III
10		2406.14	III
20		2407.30	III
20		2408.32	III
10		2409.63	III
20		2410.08	III
40		2412.02	III
20		2412.96	III
20		2413.26	III
10		2413.41	III
10		2419.11	III
10		2419.25	III
10		2419.58	III
30		2422.00	III
10		2422.90	III
10		2425.33	III
50		2425.68	III
40		2427.67	III
40		2429.32	III
10	d	2429.66	III
10		2430.04	III

Europium (Cont.)

Intensity		Wavelength	Spectrum
10		2431.49	III
10		2431.76	III
10		2432.55	III
10		2433.65	III
10		2434.19	III
100	d	2435.14	III
20		2436.39	III
10		2436.77	III
10		2438.83	III
10		2440.26	III
50		2440.67	III
1000		2444.38	III
4000		2445.99	III
30	h	2446.43	III
20		2448.57	III
20		2451.24	III
10		2451.73	III
30		2455.22	III
10		2461.79	III
30		2463.30	III
30		2464.47	III
40		2470.51	III
10		2474.94	III
10		2476.24	III
20		2476.45	III
10		2477.78	III
20		2480.02	III
20		2483.29	III
10		2486.92	III
10		2488.91	III
10		2490.50	III
10		2491.08	III
10		2492.48	III
10		2496.92	III
10		2499.17	III
2000		2513.76	III
20		2517.94	III
200		2522.14	III
10		2539.14	III
10		2548.30	III
20		2548.59	III
10		2554.50	III
10		2558.07	III
10		2560.36	III
10		2594.71	III
20		2594.76	III
10		2596.34	III
10		2604.44	III
30		2608.34	III
30		2610.09	III
50	c	2616.11	III
20		2616.26	III
10		2616.33	III
10		2616.35	III
10		2620.79	III
20		2623.33	III
10		2626.98	III
20	c	2628.46	III
10		2628.82	III
10		2631.98	III
30	c	2642.27	III
20		2645.22	III
20	c	2650.93	III
10		2653.19	III
10		2655.09	III
10		2662.24	III
20	c	2666.86	III
20	c	2668.21	III
40	c	2676.09	III
20	c	2683.21	III
20		2686.13	III
20	c	2687.74	III
40	c	2693.51	III
10		2694.80	III
20		2699.87	III
20	c	2700.78	III
10	c	2708.25	III
20	c	2708.84	III
10		2712.08	III
50	c	2720.67	III
20	c	2725.54	III
10		2743.94	III
10		2752.68	III
20		2755.12	III
10		2757.75	III
20	c	2760.21	III
10		2761.72	III
10	c	2766.26	III
20	c	2768.38	III
10	c	2768.54	III
10		2769.71	III
20	c	2780.48	III
20		2792.51	III
10		2808.09	III
10		2817.58	III
20	c	2839.56	III
20		2844.99	III
10	c	2848.44	III
20		2850.39	III
10	c	2892.60	III
40		2912.23	III
40	c	2912.64	III
10		2913.04	III
10		2928.91	III
20	c	2931.00	III
10	c	2950.20	III
20		2956.74	III
10		2956.90	III
10	c	2972.30	III
30	c	2982.29	III
20	c	3000.11	III
20		3006.37	III
20	c	3013.28	III
10	c	3018.43	III
20	c	3022.08	III
50	c	3022.69	III
20	c	3023.40	III
100	c	3023.93	III
10		3025.32	III
10	c	3026.09	III
200	c	3026.79	III
50	c	3029.92	III
20	c	3031.24	III
40	c	3032.84	III
20	c	3036.98	III
10		3038.64	III
10		3039.05	III
20	c	3039.98	III
10		3054.07	III
10	c	3054.97	III
20		3076.43	III
10	c	3089.09	III
10		3105.25	III
10		3109.67	III
10		3129.31	III
10		3142.54	III
50	c	3171.00	III
10		3178.08	III
20	h	3178.87	III
50	c	3183.78	III
10	h	3191.46	III
20	c	3194.34	III
10		3206.30	III
10		3208.95	III
10	h	3213.84	III
10		4837.98	III
50		6666.35	III
30		7221.84	III
20		7690.44	III
10		8079.07	III

FLUORINE (F)
Z = 9

F I and II
Ref. 169, 224—G.V.S.

Intensity		Wavelength	
		Vacuum	
30		375.30	II
30		380.90	II
40		407.04	II
50		430.91	II
40		431.55	II
40		435.64	II
70		457.18	II
40		471.95	II
60		472.00	II
50		472.71	II
40		473.02	II
90		484.60	II
50		513.64	II
70		514.94	II
70		546.85	II
60		547.87	II
50		548.32	II
40		548.52	II
90		605.67	II
80		606.29	II
100		606.80	II
70		606.92	II
80		607.47	II
90		608.06	II
15		780.39	I
10		780.52	I
10		782.38	I
12		791.88	I
10		792.54	I
10		794.42	I
150		806.96	I
125		809.60	I
500		951.87	I
1000		954.83	I
750		955.55	I
500		958.52	I
20		972.40	I
350		973.90	I
100		976.22	I
40		976.51	I
100		977.75	I
40		1129.76	II
40		1327.06	II
50		1328.11	II
40		1333.59	II
50		1343.60	II
40		1344.04	II
50		1400.61	II
40		1407.14	II
60		1493.09	II
50		1493.24	II
40		1493.31	II
40		1702.13	II
40		1744.75	II
50		1745.55	II
60		1747.39	II
		Air	
100		2556.11	II
100		2871.40	II
120		3059.99	II
140		3153.49	II
170		3202.76	II
140		3264.08	II
140		3414.65	II
150		3416.45	II
140		3416.80	II
160		3417.00	II
160		3472.96	II
150		3473.31	II
170		3474.78	II
190		3501.39	II
200		3501.45	II
200		3501.57	II
180		3502.84	II
200		3502.96	II
210		3503.11	II
170		3505.37	II
200		3505.52	II
220		3505.63	II
160		3522.89	II
150		3536.87	II
160		3541.77	II
160		3590.52	II
6		3594.10	I
170		3598.69	II
180		3601.39	II
190		3602.84	II
12		3668.17	I
180		3704.53	II
160		3710.35	II
160		3739.57	II
140		3805.83	II
270		3847.09	II
260		3849.99	II
250		3851.67	II
5		3898.48	I
190		3898.83	II
180		3901.93	II
170		3903.82	II
8		3930.69	I
5		3934.26	I
5		3948.56	I
150		3972.04	II
160		3972.67	II
170		3974.78	II
240		4024.73	II
220		4025.01	II
230		4025.49	II
160		4083.91	II
190		4103.07	II
170		4103.22	II
200		4103.51	II
180		4103.71	II
170		4103.87	II
170		4109.16	II
160		4116.54	II
150		4119.21	II
140		4207.15	II
170	h	4225.16	II
150	h	4244.12	II
200		4246.23	II
190		4246.39	II
180		4246.59	II
170		4246.77	II
160		4246.84	II
170	h	4275.36	II
160	h	4277.53	II
160	h	4278.93	II
200		4299.17	II
160		4446.53	II
170		4446.72	II
180		4447.19	II
140		4734.38	II
170		4859.39	II
160		4933.26	II
6		4960.65	I
140		5002.00	II
150		5173.25	II
15		5230.41	I
12		5279.01	I
18		5540.52	I
12		5552.43	I
10		5577.33	I
160		5589.27	II
20		5624.06	I
12		5626.93	I
15		5659.15	I
40		5667.53	I
90		5671.67	I
18		5689.14	I
25		5700.82	I
25		5707.31	I
12		5950.15	I
25		5959.19	I
70		5965.28	I
50		5994.43	I
150		6015.83	I
80		6038.04	I
900		6047.54	I
100		6080.11	I
800		6149.76	I
400		6210.87	I
13000		6239.65	I
140		6247.90	II
10000		6348.51	I
8000		6413.65	I
450		6569.69	I
300		6580.39	I
400		6650.41	I
1800		6690.48	I
400		6708.28	I
7000		6773.98	I
1500		6795.53	I
9000		6834.26	I
50000		6856.03	I
8000		6870.22	I
15000		6902.48	I
6000		6909.82	I
4000		6966.35	I
45000		7037.47	I
30000		7127.89	I
130		7179.90	II
15000		7202.36	I
130	h	7211.79	II
1000		7309.03	I
15000		7311.02	I
700		7314.30	I
5000		7331.96	I
10000		7398.69	I
4000		7425.65	I
2200		7482.72	I
2500		7489.16	I
900		7514.92	I
5000		7552.24	I
5000		7573.38	I
7000		7607.17	I
18000		7754.70	I
15000		7800.21	I
300		7879.18	I
500		7898.59	I
350		7936.31	I
300		7956.32	I
80		8016.01	II
1000		8040.93	I
900		8075.52	I
350		8077.52	I
350		8126.56	I
600		8129.26	I
300		8159.51	I
600		8179.34	I
300		8191.24	I
350		8208.63	I
2500		8214.73	I
3000		8230.77	I
500		8232.19	I
1500		8274.62	I
2000		8298.58	I
600		8302.40	I
900		8807.58	I
1000		8900.92	I
300		8912.78	I
350		9025.49	I
400		9042.10	I
350		9178.68	I
200		9433.67	I
25		9505.30	I
12		9662.04	I
25		9734.34	I
15		9822.11	I
12		9902.65	I
80	h	10047.98	II
15		10285.45	I
20		10862.31	I

F III
Ref. 225—G.V.S.

Intensity		Wavelength	
		Vacuum	
50	h	230.12	III
50		255.72	III
60		255.77	III
70		255.86	III
70		261.71	III
60		261.75	III
80		263.81	III

Fluorine (Cont.)

Intensity		Wavelength	
70		279.69	III
80		315.22	III
70		315.54	III
60		315.75	III
100		429.51	III
110		430.15	III
80		430.22	III
90		464.29	III
100		465.11	III
120		508.39	III
120		567.69	III
110		567.75	III
80		630.14	III
90		630.20	III
120		656.12	III
130		656.87	III
140		658.33	III
80		1219.03	III
80		1266.87	III
90		1267.71	III
70		1297.54	III
70		1359.92	III
110		1498.93	III
120		1502.01	III
110		1504.18	III
140		1504.79	III
130		1506.30	III
110		1506.77	III
100		1553.02	III
110		1557.59	III
100		1563.73	III
100		1565.54	III
100		1623.40	III
100		1650.76	III
130		1670.39	III
140		1677.40	III
100		1716.99	III
120		1770.09	III
150		1770.67	III
110		1772.93	III
140		1773.36	III
160		1791.65	III
110		1803.03	III
100		1804.70	III
170		1805.90	III
110		1839.30	III
120		1839.97	III
110		1840.14	III
80		1900.76	III

Air

Intensity		Wavelength	
100		2027.44	III
120		2030.32	III
120		2217.17	III
120		2452.07	III
130		2464.85	III
130		2470.29	III
120		2478.73	III
150		2484.37	III
120		2542.77	III
120		2580.04	III
130		2583.81	III
120		2593.23	III
130		2595.53	III
140		2599.28	III
130		2625.01	III
140		2629.70	III
120		2656.44	III
130		2755.55	III
160		2759.63	III
120		2788.15	III
160		2811.45	III
140		2833.99	III
150		2835.63	III
150		2860.33	III
120		2862.86	III
140		2887.58	III
150		2889.45	III
120		2905.30	III
140		2913.29	III
160		2916.34	III
140		2932.49	III
140		2994.28	III
120	h	2997.21	III
130		2997.53	III
120		2999.47	III
130		3039.25	III
120		3039.75	III
160		3042.80	III
150		3049.14	III
140		3113.62	III
160		3115.70	III
180		3121.54	III
140		3124.79	III
140		3134.23	III
140		3146.99	III
180		3174.17	III
170		3174.76	III
120		3214.00	III
140	h	4420.30	III
120	h	4427.35	III
120	h	4432.32	III
140	h	4479.99	III
150		5012.54	III
160		5110.99	III
140		5753.17	III
120		5761.20	III
150		6091.82	III
140		6125.50	III
130		6233.57	III
140		6363.05	III
120		7336.77	III
130		7354.94	III

F IV
Ref. 68, 226—G.V.S.

Vacuum

Intensity	Wavelength	
30	169.79	IV
30	169.84	IV
30	171.07	IV
40	176.37	IV
40	181.52	IV
40	181.57	IV
30	187.24	IV
50	196.39	IV
60	196.45	IV
50	199.76	IV
50	199.80	IV
50	199.85	IV
50	199.93	IV
50	200.00	IV
70	200.09	IV
60	201.01	IV
70	201.06	IV
60	201.10	IV
80	201.16	IV
60	201.22	IV
90	208.25	IV
70	213.85	IV
70	214.06	IV
70	220.77	IV
60	226.94	IV
50	227.10	IV
60	233.22	IV
50	233.39	IV
70	239.86	IV
70	240.02	IV
90	240.08	IV
70	240.15	IV
70	240.28	IV
70	240.37	IV
100	251.03	IV
60	270.23	IV
140	419.65	IV
150	420.05	IV
160	420.73	IV
150	430.76	IV
130	490.57	IV
160	491.00	IV
50	497.38	IV
60	497.83	IV
70	498.80	IV
140	570.64	IV
140	571.30	IV
150	571.39	IV
160	572.66	IV
140	676.12	IV
130	677.15	IV
150	677.22	IV
130	678.99	IV
160	679.21	IV

Air

Intensity	Wavelength	
40	2171.44	IV
50	2298.29	IV
40	2451.58	IV
50	2456.92	IV
40	2820.74	IV
50	2826.13	IV

F V
Ref. 68, 226—G.V.S.

Vacuum

Intensity	Wavelength	
40	134.54	V
40	147.95	V
50	148.00	V
40	152.51	V
40	158.54	V
40	162.27	V
40	163.50	V
50	163.56	V
90	165.98	V
100	166.18	V
40	174.70	V
50	178.43	V
40	178.59	V
40	182.98	V
40	186.72	V
40	186.79	V
50	186.84	V
40	186.97	V
40	187.01	V
60	190.57	V
70	190.84	V
40	191.97	V
40	205.55	V
100	464.37	V
110	465.37	V
120	465.98	V
100	466.99	V
90	506.16	V
100	508.08	V
70	513.97	V
60	514.08	V
80	524.59	V
90	525.29	V
100	526.30	V
70	647.67	V
100	647.77	V
110	647.87	V
70	647.97	V
130	654.03	V
110	657.23	V
140	657.33	V
60	757.04	V
60	1082.31	V
70	1088.39	V

AIR

Intensity	Wavelength	
10	2229.18	V
20	2252.72	V
20	2450.63	V
10	2461.33	V
10	2693.98	V
10	2702.30	V
10	2703.96	V
20	2707.17	V

FRANCIUM (Fr)
Z = 87

Fr I
Ref. 408 — C.H.C.

Air

Intensity	Wavelength	
	7177.	I

GADOLINIUM (Gd)
Z = 64

Gd I and II
Ref. 1—C.H.C.

Air

Intensity	Wavelength	
100	2468.22	II
55	2471.58	II
35	2485.67	II
70	2487.46	II
110	2488.72	II
55	2493.29	II
35	2496.35	II
45	2499.04	II
28	2543.68	II
28	2586.13	II
28	2661.50	II
70	2720.50	II
430	2750.22	II
460	2764.08	II
40	2768.51	II
320	2769.81	II
230	2770.17	II
21	2770.98	II
45	2778.76	I
45	2779.14	II
440	2781.40	II
70	2787.68	I
390	2791.96	II
100	2794.66	II
930	2796.93	II
60	2808.38	II
750	2809.72	II
160	2810.93	II
45	2814.01	II
300	2833.75	II
35	2836.69	II
70	2837.00	II
560	2840.23	II
140	2841.33	II
40	2853.91	II
60	2856.52	II
19	2859.78	II
120	2862.48	II
60	2865.06	II
40	2866.33	II
40	2871.75	II
460	2881.33	II
40	2882.13	II
130	2885.60	II
35	2907.44	II
170	2910.53	II
60	2913.08	II
45	2918.52	II
95	2923.32	II
35	2924.25	II
35	2928.34	II
35	2947.80	II
70	2948.01	II
35	2952.43	I
35	2955.60	II
70	2960.93	II
130	2963.60	II
80	2965.43	II
29	2972.74	II
560	2980.15	II
35	2983.74	II
40	2991.52	II
95	2993.04	II
1200	2999.04	II
370	3002.86	II
100	3005.09	II
2100	3010.13	II
130	3012.19	II
1900	3027.60	II
120	3028.98	II
2100	3032.84	II
1600	3034.05	II
130	3043.01	I
160	3046.48	I
280	3053.57	II
100	3059.92	I
1000	3068.64	II
560	3072.56	II
640	3076.92	II
150	3077.08	II
2100	3081.99	II
140	3084.01	II
280	3089.95	II
140	3092.06	II
460	3098.64	II
190	3098.90	II
3500	3100.50	II
120	3101.18	II
230	3101.91	II
580	3102.55	II
130	3108.36	II
170	3111.19	I
160	3113.17	II
120	3118.60	II
120	3119.01	I
510	3119.94	II
100	3120.18	II
370	3123.99	II
120	3124.25	II
130	3128.56	II
130	3130.81	II
100	3133.09	II
460	3133.85	II
210	3135.03	II
190	3136.93	I
190	3137.30	I
120	3138.71	I
230	3143.13	II
930	3145.00	II
370	3145.52	II
230	3146.88	II
980	3156.53	II
200	3158.63	I
140	3160.69	II
980	3161.37	II
220	3190.28	I
220	3199.30	I
160	3199.58	I
110	3203.41	I
690	3223.74	II
	3223.78	I
110	3225.46	II
160	3226.32	II
220	3232.78	I
100	3250.19	II
110	3259.25	II
540	3266.73	I
250	3267.64	I
140	3268.34	II
110	3274.18	II
110	3279.53	II
100	3281.61	II
250	3282.25	I
	3282.30	II
430	3291.48	I
370	3292.21	II
430	3294.08	I
330	3313.73	II
200	3315.59	II

Intensity		Wavelength	Spectrum
430		3330.34	II
1400		3331.38	II
830		3332.13	II
1100		3336.18	II
590		3345.98	II
200		3350.10	II
5400		3350.47	II
220		3357.61	I
270		3358.43	II
4300		3358.62	II
780		3360.71	II
5400		3362.23	II
270		3364.24	II
200		3365.59	II
220		3374.69	II
220		3379.76	II
220		3380.52	II
1100		3392.53	II
540		3395.12	II
220	d	3397.22	I
		3397.32	I
200		3399.41	II
540		3399.99	II
540		3402.07	II
200		3406.92	I
1100	d	3407.56	II
		3407.61	II
250		3409.30	II
220		3411.02	I
220		3413.27	II
1400		3416.95	II
1400		3418.73	II
6900		3422.47	II
390		3422.75	II
1100		3423.90	I
		3423.92	II
830		3424.59	II
390		3425.93	II
220		3428.47	II
690		3432.99	II
1700		3439.21	II
830		3439.78	II
2700		3439.99	II
390		3449.62	II
1400		3450.38	II
1100		3451.23	II
540		3454.14	II
880		3454.90	II
200		3455.27	I
200		3457.05	II
220		3461.95	II
220		3463.00	II
2700		3463.98	II
330		3466.95	II
1700		3467.27	II
1700		3468.99	II
1400		3473.22	II
2200		3481.28	II
1700		3481.80	II
490		3482.60	II
220		3486.20	I
980		3491.95	II
1700		3494.40	II
1400		3505.51	II
780		3512.22	II
1100		3512.50	II
830		3513.65	I
980		3524.20	II
430		3528.54	II
540		3542.77	II
4300		3545.80	II
3900		3549.36	II
1400		3557.05	II
540		3558.19	II
430		3558.47	II
200		3564.05	II
690		3571.93	II
330		3574.74	II
390		3578.36	II
980		3581.91	II
5400		3584.96	II
540		3590.47	II
1100		3592.71	II
200		3593.44	II
540		3600.96	II
1100		3604.87	I
270		3605.26	II
250		3605.66	II
830		3608.75	II
830		3610.76	II
220		3610.91	II
540		3613.39	II
270	d	3614.21	II
		3614.42	I
430		3617.16	II
390		3620.46	II
270		3624.89	II
250		3629.51	II
330		3634.76	II
220		3639.05	II
250		3640.18	II
330		3641.39	II
870		3645.62	II
6100		3646.19	II
310		3649.44	II
450		3650.95	II
620		3652.54	II
3900		3654.62	II
3100		3656.15	II
210		3658.19	I
1400		3662.26	II
2700		3664.60	II
2000		3671.20	II
1000		3674.05	I
350		3679.21	I
2000		3684.13	I
720		3686.33	II
3100		3687.74	II
210		3694.03	II
2000		3697.73	II
1300		3699.73	II
2700		3712.70	II
2000		3713.57	I
1400		3716.36	II
2000		3717.48	I
1800	d	3719.45	II
		3719.53	II
250		3722.07	II
430		3725.47	II
1500		3730.84	II
270		3732.32	I
230		3732.45	II
230		3732.67	I
510		3733.08	II
490		3739.76	I
330		3740.02	II
4500		3743.47	II
620		3744.83	I
230		3757.74	II
1000		3757.94	I
1400		3758.31	II
820		3759.00	II
620		3760.71	II
290		3760.92	II
870		3762.20	I
210		3763.33	II
370		3764.20	II
870		3767.04	II
8700		3768.39	II
620		3769.45	II
1400		3770.69	II
250		3771.26	I
210		3773.45	I
210		3776.83	I
1000		3782.34	II
2900		3783.05	I
1100		3787.56	II
200		3790.63	I
770		3791.17	II
490		3792.39	II
5100		3796.37	II
720		3801.29	II
210		3804.39	I
210		3805.09	II
560		3805.52	II
3700		3813.97	II
430		3814.74	II
770		3816.64	II
430		3818.75	II
350		3826.05	II
230		3827.33	II
230		3829.46	II
370		3831.80	II
210		3832.97	I
330		3834.99	II
970		3836.91	II
1000		3839.64	II
1200		3842.20	II
1400		3843.28	I
1400		3844.58	II
3300		3850.69	II
5100		3850.97	II
4300		3852.45	II
470		3855.56	II
250		3863.05	II
1600		3866.99	I
250		3873.57	I
220		3875.46	II
1500		3894.70	II
450		3895.79	II
750		3902.40	II
300		3902.71	I
240		3904.29	I
450		3905.65	I
2200		3916.51	II
450		3923.25	II
1200		3934.79	I
		3934.82	II
220		3935.38	I
450		3941.80	I
590		3942.63	I
270		3943.24	I
220		3943.62	I
1400		3945.54	I
300		3952.00	II
590		3953.37	I
1200		3957.67	II
750		3959.44	II
		3959.52	GD II
220		3963.66	II
590		3966.28	I
590		3968.26	II
750		3969.00	I
270		3969.29	II
450		3971.75	II
390		3972.71	I
590		3973.98	II
300		3974.81	I
750		3979.33	I
450		3987.21	II
470		3987.84	I
320		3992.69	I
220		3993.21	II
650		3994.16	II
700		3996.32	II
320		3997.76	II
470		4001.26	II
260		4004.94	II
320		4008.33	I
300		4008.91	II
300		4013.80	II
200		4015.58	I
300		4017.25	I
430		4017.71	I
300		4019.73	I
300		4022.33	II
1100		4023.14	I
810		4023.35	I
220		4027.61	I
1100		4028.15	I
860		4030.88	I
700		4033.49	I
340		4035.40	I
260		4036.84	I
1400		4037.33	II
700		4037.90	II
410		4043.71	I
1600		4045.01	I
270		4046.84	II
270		4047.09	I
270		4049.20	I
1300		4049.43	II
2200		4049.86	II
270		4050.37	I
810		4053.29	II
2600		4053.64	I
810		4054.72	I
2600		4058.22	I
650		4059.88	I
270		4061.30	II
650		4062.59	II
1900		4063.39	II
540		4063.59	II
260		4066.04	I
520		4068.35	I
260		4068.74	I
750		4070.29	II
		4070.39	II
650		4073.20	II
300		4073.76	II
1300		4078.44	II
2800		4078.70	I
520		4083.70	I
1500		4085.56	II
260		4087.69	II
650		4090.41	I
1100		4092.71	I
260		4093.72	I
260		4094.48	II
2600		4098.61	II
520		4098.90	II
650		4100.26	I
390		4111.44	II
2200		4130.37	II
270		4131.48	II
1100		4132.28	II
750		4134.16	I
410		4137.10	II
280		4148.86	I
540		4162.73	II
280		4163.09	II
280		4167.16	II
		4167.27	I
2400		4175.54	I
2400		4184.25	II
2200		4190.78	I
750		4191.07	II
750		4191.63	I
450		4197.68	II
590		4204.86	II
1300		4212.00	II
970		4215.02	II
650		4217.20	II
320		4225.03	I
4800		4225.85	I
220		4227.14	II
220		4229.80	II
650		4238.78	II
200		4246.57	II
1700		4251.73	II
860		4253.37	II
650		4253.61	II
810		4260.12	I
1600		4262.09	I
650		4266.60	I
470		4267.00	I
300		4274.17	I
910		4280.49	II
430		4285.82	I
300		4286.12	I
540		4296.08	II
220		4297.17	II
430		4299.29	I
1100		4306.34	I
260		4309.29	I
1800		4313.84	I
520		4314.40	I
520		4316.05	II
370		4320.52	I
750		4321.11	II
		4321.20	I
2600	d	4325.57	II
		4325.69	I
1900		4327.12	I
370		4329.58	I
340		4330.61	II
240		4331.38	I
450		4341.28	II
910		4342.18	II
1000		4344.30	II
2200		4346.46	I
910		4346.62	I
220		4347.31	II
300		4369.77	II
970		4373.83	I
280		4392.06	I
1400		4401.86	I
520		4403.14	I
260		4406.67	II
260		4408.25	II
220		4409.25	I
520		4411.16	I
860		4414.16	I
700		4414.73	I
340		4419.03	II
1400		4422.41	I
1100		4430.63	I
240	d	4436.10	I
		4436.22	II
300		4464.74	I
300		4466.55	II
		4466.60	I
520		4467.08	I
700		4474.13	I
860		4476.12	I
220		4478.80	II
280		4481.06	II
220		4483.33	II
220		4484.70	I
280		4486.90	I
500		4497.13	I
220		4497.32	I
430		4506.21	I
140		4506.33	II
140		4514.50	II
1100		4519.66	I
300		4522.82	II
150		4524.12	I
910		4537.81	I
220		4540.02	II
300		4542.03	I
240		4548.00	I
120		4558.08	II
130		4573.81	I
260		4575.91	I
280		4579.59	I
410		4581.29	I
130		4582.53	II
410		4583.07	I
160		4586.99	I
220		4596.98	II
320		4597.91	II
410		4598.90	I
340		4601.05	II
240		4602.93	I
520		4614.50	I
140		4624.42	I
430		4636.64	I
110		4639.00	II
170		4640.04	I
170		4646.00	I
170		4647.64	I
170	d	4648.59	I
		4648.70	I
430		4653.54	I
140	h	4670.87	I
170		4679.18	I
260		4680.04	I
430		4683.33	I
140		4688.12	I
700		4694.33	I
170		4695.49	I

Gadolinium (Cont.)

Intensity		Wavelength	
430		4697.42	I
170		4703.13	I
200		4709.78	I
110		4721.46	I
150		4728.47	II
220		4732.60	II
260		4735.75	I
410		4743.65	I
110		4745.82	I
320		4758.70	I
110		4760.74	I
130		4763.82	I
470		4767.24	I
180		4781.92	I
300		4784.62	I
110		4786.75	I
140		4801.05	II
220		4807.45	I
320		4821.69	I
130		4835.26	I
110		4848.10	I
110		4862.59	I
170		4865.02	II
120		4871.50	I
280		4934.12	I
220		4938.61	I
110		4952.47	I
130		4958.79	I
65		5010.82	II
55		5011.74	I
750		5015.04	I
55		5023.13	II
65		5031.29	II
75		5039.09	I
65		5050.88	II
55		5073.74	I
55		5082.80	I
95		5092.25	II
65		5096.06	II
130		5098.38	II
55		5100.94	II
910		5103.45	I
180		5108.91	II
120		5125.56	II
65		5130.28	II
65		5135.59	I
75		5136.04	I
85		5140.84	II
75		5141.50	I
75		5142.68	I
860		5155.84	I
55		5156.76	II
75		5158.48	I
75		5163.70	I
55		5164.54	II
190		5176.28	II
55		5187.24	II
55		5187.88	I
55		5191.08	II
410		5197.77	I
55		5210.49	II
85		5217.48	I
280		5219.40	I
75		5220.30	II
130		5233.93	I
65		5246.87	I
320		5251.18	I
120		5252.14	II
85		5254.75	I
140		5255.80	I
65		5268.78	I
55		5272.91	I
55		5282.48	I
280		5283.08	I
280		5301.67	I
220		5302.76	I
55		5306.70	I
280		5307.30	I
130		5321.50	I
280		5321.78	I
110		5327.32	I
65		5328.30	I
170		5333.30	I
55		5337.53	I
300		5343.00	I
85		5345.13	I
75		5345.68	I
200		5348.67	I
300		5350.38	I
240		5353.26	I
55		5361.66	I
95		5365.38	I
95		5369.92	I
150		5370.63	I
85		5389.50	I
85		5413.20	I
85		5415.69	I
65		5453.46	I
55		5583.68	II
55	d	5591.85	I
190		5617.91	I
65		5629.55	I
110		5632.25	I
260		5643.24	I
55	Bl	5680.89	G
390		5696.22	I
95		5701.35	I
65		5709.42	I
120		5733.86	II
85		5746.36	I
85	d	5754.17	I
75		5776.02	I
240		5791.38	I
65	h	5796.80	I
55		5802.92	I
55	Hs	5807.72	I
55	h	5809.22	I
55		5815.85	II
65	Hs	5819.51	G
55		5840.47	II
220		5851.63	I
55		5855.24	II
280		5856.22	I
55		5860.73	II
65		5877.26	II
55		5886.46	I
55		5904.07	II
110		5904.56	I
170		5911.45	II
65		5913.55	II
55		5916.77	I
85		5930.29	I
85		5936.84	I
65		5937.71	I
55	h	5940.95	G
55	h	5942.78	G
55		5951.60	II
55		5956.48	II
85		5977.25	I
110	h	5988.02	I
85		5999.08	I
65		6000.96	G
75	h	6001.87	G
55		6004.57	II
55		6008.71	I
55		6021.13	I
55		6080.65	II
430		6114.07	I
55		6180.42	II
110	Bl	6182.68	G
110	Bl	6200.86	G
110	Bl	6211.71	G
110	Bl	6220.93	G
55	Bl	6231.62	G
75	Bl	6241.66	G
55	b	6252.12	G
55	Bl	6262.64	G
45	b	6273.00	G
85		6289.73	II
30		6292.87	I
75		6305.15	II
30		6309.11	II
27		6317.19	I
40		6331.35	I
17		6333.75	I
17		6336.34	I
27		6346.65	II
27	h	6351.72	I
17		6363.23	I
40		6380.95	II
17		6382.19	II
22		6408.55	I
22		6422.42	II
17		6424.52	I
19	h	6470.29	I
15		6480.11	II
40	h	6538.15	I
22		6549.25	I
55		6564.78	I
10		6568.00	II
10		6573.80	I
30		6591.60	I
15		6593.42	I
10		6610.04	II
50		6634.36	II
35		6640.08	I
10		6642.76	I
30		6643.98	I
10		6646.85	I
10		6653.55	I
10		6679.56	II
35		6681.23	II
10		6692.86	I
10		6704.18	II
14		6718.14	II
17		6727.83	II
85		6730.73	I
50		6752.67	II
14		6753.91	II
14		6783.39	I
26		6786.33	II
10		6787.18	I
12		6814.56	I
26		6816.49	I
17		6820.90	I
100		6828.25	I
35		6846.60	II
30		6857.13	II
15		6864.25	I
21		6887.63	II
14		6900.73	II
100		6916.57	I
21		6920.62	II
15		6924.99	II
21		6926.49	I
17		6945.98	II
15		6957.74	II
15		6959.24	II
14		6964.33	I
15		6971.66	II
12		6976.35	II
10		6978.27	II
26	h	6980.86	I
50		6985.89	II
10		6988.75	II
75		6991.92	I
21		6993.18	I
60		6996.76	II
17		7000.75	II
45		7006.16	II
10		7016.60	I
21		7037.26	II
14		7051.00	II
13		7054.62	II
10		7058.02	II
10		7068.09	II
18		7071.00	I
18		7073.63	I
14		7098.11	I
14		7098.73	I
10	h	7116.77	II
21		7118.86	II
35		7122.57	I
13		7135.73	II
18		7147.31	II
13		7158.28	I
170		7168.37	I
21		7172.26	II
28		7189.57	II
13		7197.08	II
13		7201.41	II
10		7228.02	I
25		7233.45	I
14		7252.70	II
28		7262.66	I
14		7291.35	I
21		7313.28	I
18		7324.89	II
14		7373.81	I
14		7376.41	I
13		7377.27	II
13		7380.28	I
13		7394.90	II
13		7430.19	I
35		7441.85	I
40		7464.36	I
55		7562.97	I
10		7563.19	II
10		7588.20	I
10		7611.78	I
21		7621.96	I
21		7650.32	I
25		7672.56	I
10		7676.06	I
13		7694.45	I
80		7733.50	I
35		7749.30	I
10		7755.97	I
10		7766.48	II
11	h	7844.87	I
10		7845.80	I
35		7846.35	II
35		7856.93	I
14		7869.72	I
25		7930.25	II
13		8077.59	I
18		8146.15	I
11		8218.08	I
10		8275.42	I
10		8349.73	I
11		8398.30	I
10		8445.47	I
13	h	8527.88	I
21		8668.63	I
11		8770.36	I
13		8784.85	I
10		8795.76	I
21	h	8832.06	II
14	h	8849.14	I
18	h	8867.31	I

Gd III

Ref. 46, 137, 151—J.F.K.

Intensity	Wavelength	
	Vacuum	
600	1813.47	III
900	1946.26	III
1100	1974.34	III
2200	1975.24	III
	Air	
900	2008.79	III
3400	2018.07	III
1800	2027.82	III
800	2046.02	III
500	2057.79	III
1500	2080.08	III
1800	2098.20	III
1300	2125.68	III
1400	2148.03	III
1700	2176.84	III
1700	2223.95	III
1700	2236.73	III
1200	2239.84	III
1500	2243.75	III
1700	2250.18	III
1300	2257.05	III
1300	2292.51	III
1000	2300.38	III
1200	2303.72	III
1500	2307.03	III
1100	2313.50	III
1700	2313.56	III
1000	2315.09	III
1400	2323.12	III
1700	2323.18	III
2200	2323.78	III
1900	2329.35	III
2100	2335.01	III
1600	2336.02	III
1900	2338.97	III
1600	2339.88	III
2100	2342.74	III
2500	2346.52	III
2800	2359.31	III
1900	2360.87	III
2300	2361.91	III
1400	2362.38	III
2100	2363.26	III
1600	2365.22	III
2000	2373.38	III
1300	2374.29	III
1300	2381.38	III
1600	2387.82	III
2000	2388.77	III
1200	2393.86	III
1200	2397.34	III
1200	2405.03	III
1300	2408.41	III
1200	2409.35	III
1500	2466.84	III
1100	2469.14	III
1300	2499.53	III
1600	2520.38	III
1600	2534.11	III
1300	2536.10	III
1600	2551.56	III
2200	2553.90	III
2100	2554.04	III
2500	2563.33	III
2100	2564.46	III
1000	2565.04	III
2400	2565.95	III
1800	2569.27	III
1800	2573.57	III
2000	2576.06	III
1300	2576.15	III
1400	2578.13	III
1600	2578.76	III
1700	2583.62	III
2000	2588.21	III
2000	2588.46	III
1300	2595.81	III
1800	2609.77	III
1200	2619.40	III
1200	2621.52	III
1400	2623.52	III
1400	2625.48	III
2000	2628.10	III
1300	2628.99	III
2400	2629.83	III
2100	2632.30	III
1800	2633.32	III
1400	2635.71	III
1600	2636.44	III
1700	2637.15	III
2100	2637.97	III
2100	2638.06	III
2200	2640.53	III
1600	2641.65	III
2100	2643.71	III
1600	2644.52	III
1800	2646.04	III
1800	2646.84	III
1600	2651.48	III
2000	2655.59	III
1900	2656.55	III
2200	2660.83	III

Gadolinium (Cont.)

Intensity	Wavelength	
1800	2675.75	III
1800	2679.44	III
1700	2680.63	III
1800	2682.52	III
1500	2683.91	III
1600	2692.78	III
1900	2692.86	III
1500	2694.43	III
2800	2697.39	III
1500	2702.91	III
2800	2703.28	III
1600	2704.53	III
1800	2717.35	III
2700	2727.89	III
450	2751.24	III
1800	2833.83	III
9000	2904.73	III
1800	2918.40	III
9500	2955.53	III
1000	2975.42	III
1000	2984.10	III
1000	3116.59	III
2500	3118.04	III
4000	3176.66	III
400	3253.53	III
400	3330.34	III
400	3371.05	III
400	3402.97	III
450	3624.90	III
250	3700.47	III
300	3831.73	III
300	3910.24	III
300	4016.91	III
600	4177.26	III
400	4279.96	III
300	4314.28	III
300	4445.91	III
600	4684.25	III
600	4715.06	III
600	4782.79	III
250	4976.72	III
5000	5091.70	III
300	5124.06	III
1800	5347.95	III
3000	5365.96	III
1100	5412.62	III
4000	5553.30	III
3000	5587.88	III
3000	5658.98	III
1800	5786.96	III
1500	5862.09	III
1500	5987.85	III
5000	14332.88	III
2000	17474.78	III
800	19996.34	III
800	21259.44	III
600	22493.33	III

Gd IV
Ref. 152—J.F.K.

Intensity	Wavelength	
	Vacuum	
1000	967.92	IV
1000	983.42	IV
1000	987.10	IV
1000	987.91	IV
1000	995.04	IV
1000	995.80	IV
1000	996.49	IV
1000	999.24	IV
1000	1000.36	IV
1000	1002.73	IV
1000	1004.46	IV
1000	1005.66	IV
1000	1006.55	IV
1200	1007.24	IV
1200	1063.84	IV
500	1228.37	IV
500	1307.23	IV
500	1313.29	IV
500	1316.71	IV
600	1321.42	IV
500	1330.79	IV
1100	1393.24	IV
1600	1476.98	IV
1500	1705.03	IV
1600	1706.01	IV
2000	1736.24	IV
1500	1815.32	IV
400	1997.89	IV
	Air	
800	2049.28	IV
800	2061.30	IV
800	2070.40	IV
800	2076.66	IV
1000	2094.29	IV
1000	2296.89	IV
1000	2352.66	IV
800	2379.17	IV
900	2385.65	IV
500	2390.07	IV
700	2392.30	IV
700	2393.29	IV
700	2395.76	IV
500	2396.22	IV
600	2396.27	IV
1400	2397.87	IV
700	2402.70	IV
500	2412.21	IV
900	2419.26	IV
500	2439.84	IV
600	2440.38	IV
600	2468.60	IV

GALLIUM (Ga)
Z = 31

Ga I and II
Ref. 19, 132, 195, 281—L.J.R.

Intensity		Wavelength	
		Vacuum	
2		829.60	II
2		958.67	II
1		960.57	II
2		969.19	II
2		998.52	II
3		1002.95	II
5		1012.38	II
3		1019.10	II
5		1023.80	II
8		1033.69	II
1		1113.87	II
3		1119.25	II
5		1130.81	II
1		1167.62	II
2		1173.78	II
3		1186.81	II
1		1227.13	II
5		1286.38	II
5		1327.81	II
20		1414.44	II
5		1449.49	II
2		1463.65	II
3		1473.73	II
3		1483.52	II
3		1485.95	II
3		1495.21	II
3		1504.41	II
3		1505.01	II
5		1514.57	II
3		1515.19	II
8		1535.40	II
5		1536.37	II
1		1536.91	II
3		1669.83	II
5		1695.85	II
5		1799.42	II
10		1813.98	II
15		1845.30	II
		Air	
20		2091.34	II
1		2218.04	I
1		2255.03	I
1		2259.23	I
2		2294.19	I
1		2297.87	I
3		2338.24	I
1		2338.60	I
3		2371.29	I
2		2377.53	II
4		2418.69	I
5		2438.88	II
6		2450.08	I
7		2500.19	I
3		2500.71	I
5		2513.55	II
3		2514.15	II
2		2551.26	II
3		2552.87	II
4		2555.28	II
5		2607.47	I
8		2624.82	I
10		2632.66	I
3		2659.87	I
10		2665.05	I
8		2691.29	I
20		2700.47	II
3		2719.66	I
15		2780.15	II
6		2874.24	I
1		2886.45	II
		2893.65	GA II
2		2910.77	II
6		2943.64	I
6		2944.17	I
3		2969.41	II
1		2971.01	II
3		2971.60	II
5		2974.77	II
1		2992.84	II
2		3011.90	II
1		3158.18	II
4		3374.94	II
1		3375.95	II
2		3436.66	II
3		3446.46	II
2		3447.26	II
5		3470.34	II
1		3471.46	II
1		3472.52	II
2		3583.60	II
1		3693.93	II
2		3705.85	II
4		3734.85	II
9		3924.39	II
10		4032.99	I
10		4172.04	I
4		4251.11	II
15		4251.16	II
10	h	4254.04	II
4	h	4255.64	II
5		4255.70	II
10		4255.77	II
40		4262.00	II
3		5218.21	II
1		5338.3	II
2		5353.49	I
2		5360.6	II
1		5363.5	II
3		5416.8	II
1		5421.6	II
1		5425.6	II
10		6334.2	II
2000		6396.56	I
1000		6413.44	I
5		6419.4	II
3		6456.3	II
1		7000.0	II
3	h	7051.24	I
5	h	7106.82	I
1	h	7116.3	I
2	h	7172.9	I
5	h	7193.6	I
7		7198.7	II
10	h	7251.4	I
3	h	7289.6	I
5	h	7349.3	I
20	h	7403.0	I
30	h	7464.0	I
6	h	7556.6	I
10	h	7620.5	I
50	h	7734.77	I
2		7793.0	II
100	h	7800.01	I
4	h	7801.6	I
15	h	8002.55	I
20	h	8074.25	I
3	h	8167.5	I
5	h	8171.6	I
100	h	8311.86	I
200	h	8386.49	I
10	h	8389.30	I
7	h	8415.51	I
10	h	8419.91	I
20	h	8808.75	I
30	h	8813.56	I
20	h	8856.37	I
30	h	8944.33	I
200	h	9492.92	I
200	h	9493.12	I
300	h	9589.36	I
20	h	9594.25	I
60	h	10898.10	I
100	h	10905.95	I
10		10968.27	I
20		11103.51	I
400		11949.12	I
200		12109.78	I
40		12885.05	I
50		13057.50	I
50		14982.75	I
60		14996.64	I
20		17757.91	I
10		17868.96	I
60		22016.81	I
70		22568.71	I

Ga III
Ref. 141—L.J.R.

Intensity	Wavelength	
	Vacuum	
50	620.00	III
40	622.01	III
90	806.51	III
90	817.30	III
50	828.70	III
80	1085.00	III
60	1105.61	III
90	1150.27	III
90	1267.16	III
80	1293.46	III
60	1295.36	III
60	1323.15	III
70	1353.92	III
90	1495.07	III
50	1534.46	III
	Air	
90	2417.70	III
90	2423.98	III
15	2424.36	III
50	3521.77	III
80	3581.19	III
100	3589.34	III
10	3731.10	III
10	3806.60	III
100	4380.69	III
150	4381.76	III
100	4863.00	III
150	4993.78	III
10	5808.28	III
20	5848.25	III
15	5993.51	III

Ga IV
Ref. 141, 143—L.J.R.

Intensity	Wavelength	
	Vacuum	
14	294.53	IV
61	295.67	IV
41	304.99	IV
4	422.12	IV
25	423.18	IV
16	439.92	IV
67	1137.06	IV
70	1156.10	IV
70	1163.60	IV
75	1170.58	IV
48	1171.71	IV
68	1185.23	IV
40	1186.06	IV
73	1190.89	IV
73	1193.02	IV
75	1195.02	IV
69	1201.54	IV
72	1206.89	IV
63	1216.15	IV
50	1228.03	IV
60	1236.38	IV
60	1238.59	IV
45	1241.81	IV
75	1245.53	IV
83	1258.77	IV
81	1264.66	IV
82	1267.15	IV
81	1279.24	IV
80	1285.33	IV
82	1295.86	IV
83	1299.46	IV
82	1303.53	IV
80	1309.68	IV
80	1314.82	IV
85	1338.09	IV
77	1347.03	IV
76	1351.06	IV
74	1364.63	IV
60	1395.54	IV
77	1402.55	IV
70	1405.32	IV
73	1465.87	IV

Ga V
Ref. 2, 62, 140—L.J.R.

Intensity	Wavelength	
	Vacuum	
5	290.53	V
1	296.13	V
5	296.82	V
30	298.44	V
20	299.47	V
30	300.01	V
25	300.57	V
10	300.78	V
20	301.19	V
30	302.86	V
20	303.84	V
30	307.03	V
30	308.26	V
15	309.64	V
30	311.79	V
25	312.41	V
30	313.68	V

Gallium (Cont.)

Intensity		Wavelength	
15		315.95	V
20		316.48	V
40		319.41	V
12		320.53	V
40		322.31	V
50		322.99	V
30		323.10	V
40		324.25	V
40		324.95	V
40		326.14	V
30		326.77	V
30		328.65	V
5		336.61	V
20		878.17	V
40		973.21	V
10		977.89	V
15		979.60	V
20		984.95	V
40		989.75	V
90		1014.47	V
90		1019.71	V
20		1033.55	V
30		1038.76	V
30		1047.50	V
120		1050.48	V
80		1054.56	V
90		1058.12	V
80		1066.69	V
35		1068.59	V
30		1069.45	V
60		1069.60	V
55		1071.19	V
45		1071.41	V
80		1073.77	V
90		1078.83	V
110		1079.60	V
60		1080.99	V
250		1085.01	V
80		1087.37	V
40		1090.53	V
90		1091.71	V
100		1094.36	V
80		1095.10	V
70		1101.62	V
160		1102.83	V
140		1103.03	V
60		1104.93	V
75		1105.62	V
70		1106.17	V
40		1115.55	V
80		1118.34	V
55		1123.18	V
80		1123.66	V
120		1126.40	V
80		1127.75	V
130		1128.10	V
120		1128.53	V
100		1129.94	V
80		1131.43	V
40		1133.91	V
130		1136.07	V
65		1138.20	V
60		1144.30	V
50		1145.70	V
30		1148.42	V
45		1150.09	V
130		1150.23	V
120		1156.51	V
35		1157.74	V
25		1169.40	V
40		1178.95	V
80		1213.17	V
30		1265.45	V
30		1276.85	V
15		1283.64	V
10		1311.35	V

GERMANIUM (Ge)
Z = 32

Ge I and II
Ref. 5, 119, 293, 340—C.H.C.

Intensity		Wavelength	
		Vacuum	
1		822.97	II
3		835.08	II
10		850.50	II
10		862.234	II
15		875.493	II
15		905.977	II
20		920.554	II
50		999.101	II
100		1016.638	II
100		1075.072	II
300		1085.51	II
200		1098.71	II
500		1106.74	II
500		1120.46	II
200		1164.27	II
500		1181.19	II
500		1181.65	II
200		1188.73	II
100		1189.62	II
300		1191.26	II
50		1191.72	II
500		1237.059	II
500		1261.905	II
100		1264.710	II
100		1380.42	II
50		1392.26	II
200		1401.24	II
200		1538.091	II
500		1576.855	II
75		1581.070	II
100		1602.486	II
3	r	1615.57	I
2	r	1624.130	I
2	r	1630.173	I
3	r	1636.31	I
2		1638.96	I
4	r	1639.730	I
2		1647.531	I
200		1649.194	II
2		1651.528	I
4	r	1651.955	I
3		1661.345	I
4	r	1663.539	I
10	h	1665.275	I
4		1667.802	I
3	r	1670.608	I
100	r	1691.090	I
200	r	1716.784	I
100	h	1739.102	I
100		1742.195	I
50		1746.065	I
200		1750.043	I
100		1758.279	I
100	h	1764.185	I
100	h	1765.284	I
50	h	1766.433	I
200		1774.176	I
200		1785.046	I
100	h	1793.071	I
75	h	1801.432	I
200	h	1841.328	I
200	h	1842.410	I
100	h	1844.410	I
100	h	1845.872	I
100	h	1846.958	I
200		1853.134	I
500	r	1860.086	I
100		1865.052	I
300	r	1874.256	I
100		1895.197	I
500	r	1904.702	I
50	h	1908.434	I
30		1912.409	I
300	r	1917.592	I
100	h	1923.467	I
500	r	1929.826	I
10	h	1934.048	I
100	r	1937.483	I
500		1938.008	II
100	r	1938.300	I
500		1938.891	II
30	s	1944.116	I
200		1944.731	I
200		1955.115	I
500		1962.013	I
30	h	1963.373	I
30		1965.383	I
200		1970.880	I
200		1979.274	II
300	h	1987.849	I
300		1988.267	I
500	r	1998.887	I
		Air	
50		2007.04	II
200		2011.29	I
1700		2019.068	I
2400	r	2041.712	I
1600	r	2043.770	I
420		2054.461	I
220	h	2057.238	I
750	r	2065.215	I
2600	r	2068.656	I
420		2086.021	I
2000	r	2094.258	I
240		2105.824	I
95	h	2124.744	I
50	h	2186.451	I
100		2197.62	II
340	r	2198.714	I
100		2205.85	II
15		2220.375	I
18		2256.001	I
18		2314.201	I
24		2327.918	I
15		2359.233	I
20		2379.144	I
10		2389.472	I
15		2397.885	I
130		2417.367	I
30		2436.412	I
100		2478.66	II
90		2497.962	I
500		2500.54	II
70		2533.230	I
3		2556.298	I
28		2589.188	I
500		2592.534	I
8		2644.184	I
1200		2651.172	I
500		2651.568	I
500		2691.341	I
200		2704.03	II
850		2709.624	I
400		2729.78	II
40		2740.426	I
650		2754.588	I
50		2770.59	II
75		2772.35	II
70		2793.925	I
80		2829.008	I
1000		2831.843	II
50		2834.28	II
75		2839.68	II
1000		2845.527	II
75		2853.97	II
750		3039.067	I
600		3067.021	I
20		3124.816	I
50		3186.72	II
100		3221.64	II
110		3269.489	I
50		3312.56	II
75		3323.64	II
100		3455.72	II
300		3499.21	II
30		3845.11	II
70		4226.562	I
10		4685.829	I
75	h	4689.87	II
50	h	4690.02	II
1000		4741.806	II
1000		4814.608	II
50		4824.097	II
100		5131.752	II
200		5178.648	II
3		5194.583	I
6		5265.892	I
6		5513.263	I
8		5564.741	I
8		5607.010	I
6		5616.135	I
7		5621.426	I
8		5655.96	I
6		5664.226	I
5		5664.842	I
9		5691.954	I
6		5701.776	I
5		5717.877	I
6		5801.029	I
9		5802.093	I
1000		5893.389	II
500		6021.041	II
150		6078.39	II
50		6267.14	II
150		6268.07	II
100		6268.34	II
75		6283.452	II
100		6336.377	II
100		6484.181	II
6		6557.488	I
50		6780.51	II
50		7049.369	II
6		7130.12	I
30		7145.390	II
7		7330.38	I
5		7353.334	I
7		7384.208	I
6		7402.64	I
7		7511.57	I
5		7776.20	I
10		7833.575	I
7		7837.63	I
6		7853.77	I
7		7878.12	I
5		7962.26	I
5		7983.33	I
10		8031.039	I
6		8044.165	I
5		8095.29	I
5		8225.22	I
7		8226.09	I
10		8256.013	I
5		8264.15	I
5		8280.09	I
6		8281.04	I
8		8367.81	I
7		8391.70	I
5		8396.36	I
5		8429.42	I
10		8482.21	I
8		8506.70	I
5		8507.66	I
8		8564.89	I
6		8599.27	I
6		8652.42	I
5		8669.60	I
9		8700.60	I
5		8712.90	I
6		8734.78	I
6		8789.88	I
5		9068.785	I
5		9095.957	I
6		9398.868	I
20		9474.993	II
20		9475.645	II
4		9492.559	I
7		9625.664	I
5		10039.436	I
4		10200.952	I
10		10382.427	I
10		10404.913	I
8		10734.068	I
8		10947.416	I
10		11125.130	I
230		11252.83	I
24		11293.40	I
33		11318.13	I
55		11459.05	I
150		11483.77	I
175		11614.81	I
600		11714.76	I
10		11839.77	I
55		11917.01	I
10		12025.64	I
10		12055.49	I
30		12061.41	I
45		12065.76	I
1300		12069.20	I
30		12198.88	I
20		12207.73	I
60		12286.75	I
55		12338.76	I
1050		12391.58	I
48		12540.41	I
15		12636.80	I
150		12676.58	I
40		12681.28	I
115		12800.66	I
175		12836.38	I
12		12847.92	I
120		12955.73	I
15		13028.64	I
235		13107.61	I
20		13492.28	I
42		13534.85	I
28		13724.48	I
42		14116.70	I
42		14297.15	I
40		14569.84	I
12		14667.52	I
470		14822.38	I
16		14921.97	I
15		15001.75	I
13		15041.21	I
20		15504.34	I
14		16424.77	I
12		16626.64	I
70		16699.29	I
150		16759.79	I
135		17214.34	I
16		18428.30	I
35		18495.54	I
10		18764.11	I
70		18811.86	I
62		19279.24	I
28		20673.64	I
4		21518.30	I
9		22091.84	I
5		23921.92	I

Ge III
Ref. 341—C.H.C.

Intensity	Wavelength	
	Vacuum	
2	542.90	III
2	663.77	III
3	670.88	III
2	680.28	III
2	952.76	III
12	988.96	III
15	995.72	III
10	996.50	III
15	1011.21	III
10	1012.31	III
8	1032.62	III
12	1040.99	III
12	1058.91	III
40	1088.45	III
10	1137.92	III
12	1150.55	III

Germanium (Cont.)

8	1159.15	III
8	1159.62	III
8	1160.79	III
10	1173.78	III
8	1212.47	III
4	1323.24	III
10	1525.32	III
2	1527.15	III
9	1600.09	III
6	1883.26	III
2	1978.22	III
	Air	
2	2019.22	III
4	2022.25	III
3	2062.14	III
15	2100.05	III
15	2102.42	III
25	2104.45	III
3	2922.86	III
25	3197.56	III
35	3211.86	III
25	3214.95	III
40	3255.05	III
20	3259.90	III
5	3369.57	III
20	3414.27	III
40	3434.03	III
8	3464.59	III
40	3489.08	III
2	3724.51	III
15	3884.78	III
200	4178.96	III
12	4245.41	III
200	4260.85	III
150	4291.71	III
10	4674.36	III
10	5016.88	III
18	5134.75	III
5	5229.37	III
3	5256.61	III

Ge IV
Ref. 341—C.H.C.

Intensity	Wavelength	
	Vacuum	
1	440.11	IV
1	441.95	IV
3	847.80	IV
3	868.30	IV
8	915.00	IV
8	936.70	IV
4	938.90	IV
1	1073.44	IV
20	1188.99	IV
20	1229.81	IV
2	1494.89	IV
6	1500.61	IV
3	1648.14	IV
	Air	
2	2293.0	IV
2	2343.37	IV
15	2445.38	IV
15	2445.71	IV
30	2488.25	IV
20	2542.44	IV
5	2631.78	IV
3	2698.08	IV
15	2717.44	IV
30	2736.09	IV
30	2788.61	IV
5	3071.84	IV
60	3554.19	IV
50	3676.65	IV

Ge V
Ref. 342—C.H.C.

Intensity	Wavelength	
	Vacuum	
700	294.51	V
1000	295.64	V
200	304.98	V
20	621.52	V
35	716.26	V
50	724.21	V
35	733.54	V
35	735.35	V
35	741.52	V
60	746.88	V
40	750.26	V
35	755.84	V
60	760.05	V
60	958.51	V
300	971.35	V
150	984.92	V
200	988.13	V
300	990.66	V
300	1004.38	V
300	1016.66	V
250	1038.40	V
900	1045.71	V
400	1050.05	V
300	1054.59	V
300	1068.43	V
400	1069.13	V
700	1072.66	V
600	1086.65	V
500	1087.85	V
800	1089.49	V
300	1092.09	V
1000	1116.94	V
300	1122.01	V
700	1163.39	V
300	1165.26	V
200	1176.69	V
700	1222.30	V

GOLD (Au)
Z = 79

Au I and II
Ref. 38, 72, 234—C.H.C.

Intensity		Wavelength	
		Vacuum	
		925.72	II
		946.03	II
		950.39	II
20		957.78	II
3		967.94	II
3		974.47	II
2		982.24	II
		1062.67	II
8		1066.96	II
		1085.00	II
8		1090.78	II
5		1094.92	II
20		1103.31	II
3		1166.76	II
5		1210.86	II
20		1224.57	II
40	h	1305.34	I
25		1310.47	I
100	h	1328.37	I
3		1336.26	II
10	h	1338.37	I
10	h	1342.80	I
20		1350.09	I
20		1350.84	I
22		1351.74	I
25		1352.82	I
25		1354.14	I
30		1355.79	I
35		1357.86	I
40		1360.51	I
6		1362.33	II
20		1362.47	I
8		1363.15	II
50		1363.98	I
25		1364.15	I
10		1364.74	II
60		1368.62	I
35		1368.98	I
70		1374.82	I
50		1375.76	I
30		1378.87	I
8		1380.53	II
80		1382.75	I
50		1385.33	I
20		1389.14	I
60		1392.27	I
6		1393.80	II
50		1402.12	I
		1405.12	II
70		1407.38	I
100		1408.45	I
		1410.69	II
		1415.22	II
80		1429.19	I
50		1435.79	I
20		1436.61	II
10		1468.85	II
10		1469.17	II
10		1469.28	II
100		1481.76	I
25		1486.55	II
20		1532.82	I
20		1532.86	I
10		1562.04	II
200		1587.16	I
12		1593.41	II
70		1598.24	I
12		1611.11	II
		1616.65	AU II
2		1622.83	II
100		1624.34	I
2		1632.53	II
50		1639.90	I
150		1646.67	I
10		1656.99	II
100		1665.76	I
25		1673.59	II
7		1694.38	II
2		1698.65	II
200		1699.34	I
30		1700.69	II
10		1720.04	II
25		1725.75	II
45		1740.52	II
10		1749.80	II
35		1756.15	II
60		1783.22	II
35		1793.31	II
35		1800.58	II
25		1823.24	II
100		1879.83	I
20		1919.64	I
20		1921.64	II
45		1942.31	I
25		1951.93	I
30		1978.19	I
		Air	
25		2000.81	II
11000		2012.00	I
2600		2021.38	I
50		2044.54	II
150		2082.09	II
35		2095.13	II
20		2098.14	II
60		2110.68	II
30		2125.29	II
15		2126.63	I
20		2170.75	I
35		2188.81	II
25		2201.32	II
35		2215.63	II
45		2228.88	II
30		2231.18	II
25		2240.16	II
70		2248.56	II
80		2263.62	II
18		2263.88	II
25		2277.52	II
25		2283.30	II
25		2291.40	II
45		2304.69	II
25		2314.55	II
20		2315.75	II
25		2340.06	II
180		2352.65	I
20		2376.28	I
120		2387.75	I
2600		2427.95	I
60		2533.52	II
16		2544.19	I
45		2552.67	II
20		2589.25	I
30		2590.04	I
50		2616.40	II
20		2627.02	II
250		2641.48	I
3400		2675.95	I
20		2687.63	II
20		2688.16	II
30		2688.71	I
80		2700.89	I
1100		2748.25	I
20		2748.71	II
100		2780.82	I
30		2800.93	I
1000		2802.04	II
300		2819.79	II
100		2822.55	II
30		2823.13	II
100	h	2825.44	II
30		2833.03	II
300		2837.85	II
100		2846.92	II
100		2856.74	II
3		2872.36	I
300		2883.45	I
3		2886.96	I
10		2888.40	I
300		2891.96	I
100		2893.25	II
3		2905.74	I
30		2905.90	I
100		2907.04	II
300		2913.52	II
3		2914.82	I
300		2918.24	II
16		2932.19	
30		2940.67	I
100		2954.22	II
10		2973.33	I
100	h	2990.27	II
300		2994.80	II
10		3002.65	I
3		3005.85	I
10		3024.67	I
320		3029.20	I
30		3033.25	I
300		3065.42	I
10		3102.63	I
10	h	3117.01	I
100		3122.50	II
1600		3122.78	I
30		3126.86	II
30		3127.03	I
10		3164.88	I
10		3172.35	II
30		3191.76	I
100		3194.72	I
30		3200.37	I
30		3204.74	I
1		3221.86	I
30		3225.25	I
300		3230.63	I
10		3253.94	I
10	h	3265.10	I
30	h	3267.07	I
10		3271.63	I
10		3273.47	I
300		3308.30	I
300		3309.64	I
100		3320.12	I
100		3355.15	I
30	h	3368.44	I
10		3381.90	I
100	h	3391.31	I
100		3395.40	I
30	h	3440.36	I
100		3467.21	I
30		3471.61	I
30	h	3509.04	I
10		3510.82	I
3		3523.34	II
3		3545.61	I
30		3553.57	I
300		3557.36	I
10		3558.22	I
30		3565.97	I
30	h	3584.37	I
300		3586.73	I
30	h	3588.79	I
30		3598.06	I
100		3611.57	I
30	h	3614.00	I
30		3622.74	I
100	h	3631.31	I
10		3633.22	II
30		3634.53	I
10		3635.12	I
300		3637.90	I
3		3639.87	I
100	h	3645.02	I
10		3649.09	I
100		3650.74	I
10		3653.53	I
3		3654.69	I
10		3655.30	I
10		3656.90	I
30		3706.55	II
100		3709.62	I
10		3766.61	I
10		3770.76	I
100		3796.01	I
30		3801.92	I
30		3804.01	II
10		3821.85	I
10		3825.70	I
100		3874.73	I
30		3880.25	I
30		3889.48	I
100	h	3892.26	I
400		3897.86	I
30		3901.09	I
300		3909.38	I
100		3927.69	I
10	h	3959.10	I
30		3966.23	I
30		3976.65	I
30		3979.68	I
30		3991.37	I
3		4012.57	I
10		4016.07	II
400		4040.93	I
30		4052.79	II
700		4065.07	I
10		4076.35	II
3		4083.28	II
100		4084.10	I
30		4101.70	I
30		4128.59	I
30		4201.13	I
30	h	4227.88	I
100		4241.80	I
200		4315.11	I
30		4361.04	II
10		4420.61	II
120	h	4437.27	I

Gold (Cont.)

Intensity		Wavelength	
250		4488.25	I
900	h	4607.51	I
100	h	4620.56	I
1		4663.92	I
3		4663.97	I
10		4694.69	I
3		4760.17	II
500		4792.58	I
100		4811.60	I
10		4822.96	I
30	h	4950.82	I
30		5064.59	I
30	h	5108.84	I
100		5147.44	I
300		5230.26	I
100	h	5261.76	I
100		5655.77	I
100	h	5721.36	I
300		5837.37	I
100	h	5862.93	I
300	h	5956.96	I
30	h	5962.68	I
600		6278.17	I
100		6562.68	I
30		6652.89	I
600		7510.73	I
10		8145.06	I
10		9254.28	I

Au III
Ref. 72, 393, 395 — R.D.C.

Intensity		Wavelength	
		Vacuum	
30		779.73	III
30		788.78	III
50		799.93	III
40		811.83	III
50		817.95	III
40		820.06	III
80		833.16	III
100		843.44	III
100		845.14	III
80		855.49	III
80		859.90	III
80		863.42	III
80		901.03	III
80		910.45	III
80		924.02	III
200		945.10	III
100		1040.63	III
80		1044.49	III
80		1046.81	III
100	h	1239.96	III
100		1278.51	III
100		1314.84	III
200		1336.72	III
180		1341.68	III
100		1348.89	III
150		1350.32	III
150		1355.61	III
150		1356.13	III
80		1362.06	III
500		1365.40	III
200		1367.17	III
180		1377.73	III
150		1378.69	III
150		1379.98	III
125		1380.53	III
200		1381.36	III
300		1385.79	III
100		1389.41	III
180		1391.46	III
180		1396.00	III
100		1402.91	III
225		1409.50	III
250		1413.80	III
100		1414.27	III
80		1415.54	III
100		1417.09	III
125		1417.39	III
150		1427.42	III
300		1428.93	III
250		1430.06	III
275		1433.37	III
250		1435.81	III
80		1436.12	III
300		1439.12	III
200		1441.21	III
150		1446.37	III
80		1446.69	III
250		1448.42	III
250		1454.95	III
100		1464.72	III
150		1471.28	III
80		1473.32	III
100		1474.73	III
150		1481.10	III
300		1487.15	III
250		1487.91	III
200		1489.47	III
250		1500.37	III
200		1502.47	III
200		1503.74	III
80		1540.26	III
100		1542.00	III
80		1542.25	III
100		1548.50	III
80		1554.61	III
80		1562.33	III
80		1562.41	III
200		1567.54	III
80		1571.94	III
200		1574.85	III
200		1579.44	III
150		1584.10	III
200		1589.56	III
80		1589.68	III
150		1593.41	III
200		1600.51	III
250		1617.16	III
100		1617.78	III
500		1621.93	III
300	d	1629.13	III
250		1638.88	III
100		1644.17	III
250		1652.74	III
250		1664.77	III
100		1668.11	III
125		1673.93	III
1000		1693.94	III
150		1697.09	III
200		1698.98	III
200		1700.00	III
200		1702.25	III
100		1707.53	III
250		1710.16	III
200		1715.69	III
100		1716.71	III
300		1717.83	III
500		1727.31	III
100	d	1733.17	III
300		1738.48	III
150		1744.39	III
500		1746.10	III
500		1756.92	III
500		1761.95	III
300		1767.42	III
100		1774.42	III
800		1775.17	III
200		1776.40	III
100		1780.57	III
300		1786.11	III
150		1792.65	III
500		1793.76	III
200		1801.98	III
400		1805.24	III
100		1809.81	III
400		1821.17	III
400		1844.89	III
150		1848.83	III
80		1850.15	III
500		1861.80	III
150		1871.92	III
150		1918.28	III
100		1932.04	III
100		1935.42	III
200		1948.79	III
100		1958.47	III
400		1989.63	III
150		1996.85	III
		Air	
300		2083.09	III
80		2085.45	III
100		2159.08	III
80		2167.33	III
200		2172.20	III
100		2184.11	III
500		2188.97	III
300		2322.27	III
100		2382.40	III
150		2402.71	III
150		2405.12	III
100		3227.99	III
100		3309.86	III

HAFNIUM (Hf)
Z = 72

Hf I and II
Ref. 1—C.H.C.

Intensity	Wavelength	
	Air	
6200	2012.78	II
8500	2028.18	II
1200	2096.18	II
540	2210.82	II
320	2254.01	II
160	2255.15	II
250	2266.83	II
620	2277.16	II
230	2321.14	II
580	2322.47	II
300	2323.25	II
120	2324.50	II
300	2324.89	II
200	2332.97	II
200	2337.33	II
230	2343.32	II
320	2347.44	II
540	2351.22	II
110	2353.02	I
90	2365.98	II
250	2380.30	II
100	2381.00	II
170	2393.18	II
450	2393.36	II
670	2393.83	II
130	2400.78	II
70	2404.56	II
540	2405.42	II
130	2406.44	II
370	2410.14	II
90	2413.33	II
55	2415.96	II
320	2417.69	II
120	2425.98	II
45	2428.75	I
120	2428.99	II
130	2433.57	II
45	2434.74	II
35	2444.99	I
390	2447.25	II
140	2449.44	II
35	2452.30	II
110	2453.34	II
450	2460.49	II
70	2463.97	II
430	2464.19	II
90	2465.06	II
35	2465.67	I
140	2467.97	II
210	2469.18	II
100	2473.92	II
55	2481.44	II
55	2482.65	I
55	2487.16	I
290	2496.99	II
580	2512.69	II
580	2513.03	II
130	2515.48	II
890	2516.88	II
340	2531.19	II
200	2537.33	II
110	2548.20	II
320	2551.40	II
130	2559.19	II
250	2563.61	II
890	2571.67	II
320	2573.90	II
320	2576.82	II
300	2578.14	II
320	2582.54	II
130	2591.33	II
390	2606.37	II
450	2607.03	II
120	2608.45	I
230	2613.60	II
450	2622.74	II
160	2637.00	I
1100	2638.71	II
1100	2641.41	II
160	2642.75	I
670	2647.29	II
100	2651.16	II
160	2657.84	II
210	2661.88	II
290	2683.35	II
670	2705.61	I
110	2706.73	II
210	2712.42	II
140	2713.84	I
250	2718.59	I
120	2730.85	I
710	2738.76	II
200	2743.64	I
360	2751.81	II
450	2761.63	I
160	2766.96	I
170	2773.02	I
980	2773.36	II
180	2774.02	II
390	2779.37	I
100	2789.50	II
140	2789.73	II
230	2808.00	II
230	2813.86	II
170	2814.48	II
230	2817.68	I
140	2818.94	I
200	2819.74	I
1200	2820.22	II
490	2822.68	II
180	2833.28	I
110	2834.13	I
410	2845.83	I
270	2849.21	II
270	2850.96	I
180	2851.21	II
180	2860.56	I
760	2861.01	II
760	2861.70	II
2100	2866.37	I
130	2869.82	II
150	2876.33	II
210	2887.14	I
100	2887.54	I
800	2889.62	I
1800	2898.26	I
130	2898.71	II
1200	2904.41	I
890	2904.75	I
140	2909.91	II
2000	2916.48	I
580	2918.58	I
320	2919.59	II
180	2924.62	I
490	2929.63	II
450	2929.90	I
710	2937.80	II
2000	2940.77	I
160	2944.71	I
1200	2950.68	I
1100	2954.20	I
540	2958.02	I
120	2961.80	II
1400	2964.88	I
620	2966.93	I
140	2967.23	II
710	2968.81	II
110	2973.37	I
890	2975.88	II
150	2979.28	I
1100	2980.81	I
210	2982.72	I
170	3000.10	II
800	3005.56	I
1100	3012.90	II
540	3016.78	I
1100	3016.94	II
980	3018.31	I
1200	3020.53	I
140	3025.29	II
410	3031.16	II
110	3046.08	II
710	3050.76	I
1100	3057.02	I
130	3063.78	I
130	3064.68	II
850	3067.41	I
2100	3072.88	I
170	3074.10	I
250	3074.79	I
150	3080.66	II
430	3080.84	I
200	3096.76	I
340	3101.40	II
710	3109.12	II
130	3110.87	II
130	3119.98	I
710	3131.81	I
850	3134.72	II
130	3137.51	I
170	3139.65	II
120	3140.76	II
220	3145.32	II
220	3148.41	I
120	3151.63	I
450	3156.63	I
270	3159.82	I
710	3162.61	II
450	3168.39	I
890	3172.94	I
450	3176.86	II
220	3181.01	I
120	3181.15	I
130	3189.62	I
360	3193.53	II
670	3194.19	II
200	3196.93	I
130	3199.99	II
310	3206.11	I
180	3210.98	I
180	3217.30	II
180	3220.61	II
130	3230.06	I
130	3239.44	I
130	3243.35	I
360	3247.66	I
220	3249.53	I
890	3253.70	II
270	3255.28	II
120	3262.47	I
180	3273.66	II
270	3279.98	II

Hafnium (Cont.)

Intensity		Wavelength	
160		3291.05	I
210		3306.12	I
120		3309.19	I
340		3310.27	I
670		3312.86	I
180		3317.99	II
130		3328.21	II
890		3332.73	I
370		3352.06	II
130		3356.78	I
230		3358.91	I
180		3360.06	I
140		3366.68	I
180		3378.93	I
140		3384.14	II
230		3384.70	II
170		3386.21	I
800		3389.83	II
230		3392.81	I
230		3394.59	II
140		3394.98	II
230		3397.26	I
230		3397.60	I
2300		3399.80	II
170		3400.21	I
180		3402.51	I
140		3407.76	II
230		3410.17	II
230		3417.34	I
410		3419.18	I
140		3427.44	I
200		3428.37	II
250		3438.24	II
140		3438.43	I
100		3441.84	I
100		3452.31	I
140		3462.64	II
140		3467.60	I
710		3472.40	I
200		3478.99	II
480		3479.28	II
250		3495.75	II
250		3497.16	I
980		3497.49	I
100		3498.98	I
1200		3505.23	II
150		3513.28	I
130		3518.75	II
980		3523.02	I
100		3530.87	I
100		3531.23	I
980		3535.54	II
760		3536.62	I
180		3548.81	I
540		3552.70	II
150		3554.00	I
1300		3561.66	II
150		3564.31	I
270		3567.36	I
1100		3569.04	II
150		3579.90	I
110		3583.28	I
210		3597.42	II
540		3599.87	I
110		3615.04	I
800		3616.89	I
110		3617.68	I
110		3624.00	II
320		3630.87	II
100		3635.43	I
800		3644.36	II
320		3649.10	I
200		3651.84	I
140		3661.05	II
220		3665.35	II
100		3668.21	I
200		3672.27	I
480		3675.74	I
2200		3682.24	I
280		3696.51	I
100		3698.40	II
240		3699.72	II
340		3701.15	II
100		3704.92	I
120		3705.40	II
1000		3717.80	I
650		3719.28	II
140		3726.49	I
160		3729.10	I
460		3733.79	I
160		3737.88	II
120		3739.04	I
100		3744.98	II
400		3746.80	I
140		3753.22	I
100		3765.05	I
100		3765.56	I
170		3766.92	II
200		3768.25	I
1400		3777.64	I
1400		3785.46	I
650		3793.37	II
100		3798.66	I
850	d	3800.38	I
140		3806.07	II
320		3811.78	I
100		3817.20	II
100		3819.38	I
1300		3820.73	I
140		3829.67	I
280		3830.02	I
800		3849.18	I
140		3849.52	II
600		3858.31	I
230		3860.91	I
200		3872.55	II
160		3877.10	II
380		3880.82	II
200		3882.52	I
150		3883.77	II
200		3889.23	I
200		3889.33	I
620		3899.94	I
620		3918.09	II
200		3923.90	II
120		3926.42	I
150		3927.57	I
110		3929.54	II
320		3931.38	I
120		3935.65	II
120		3939.04	I
410		3951.83	I
160		3968.01	I
150	Bl	3970.05	H
200		3973.48	I
180		4032.27	I
100		4047.96	II
230		4062.84	I
140		4066.21	I
180		4083.35	I
540		4093.16	II
110		4104.23	I
140		4106.58	I
110		4113.53	II
110		4118.60	I
150		4127.80	II
140		4145.76	I
150		4162.36	II
110		4162.69	I
1100		4174.34	I
120		4190.95	I
160		4206.58	II
190		4209.70	I
170		4228.08	I
170		4232.44	II
120	Bl	4252.08	H
170		4260.98	I
200		4263.39	I
170		4272.85	II
320		4294.79	I
120		4318.14	I
160		4330.27	I
180		4336.66	II
150		4350.51	II
250		4356.33	I
110		4367.90	II
180		4370.97	II
120		4417.35	II
160		4417.91	I
200		4438.04	I
140		4457.34	I
140		4461.18	I
140		4540.93	I
250		4565.94	I
500	d	4598.80	I
230		4620.86	I
210		4655.19	I
120		4699.01	I
160		4782.74	I
310		4800.50	I
130		4859.24	I
120		4975.25	I
95		5018.20	I
15		5021.75	I
55		5040.82	II
95		5047.45	I
55	b	5074.74	H
30		5079.65	II
55	b	5093.88	H
15		5112.13	I
19		5128.53	II
30		5157.96	I
55		5167.42	I
75		5170.18	I
230		5181.86	I
30		5186.84	I
30		5187.75	II
110		5243.99	I
55		5247.10	II
25		5260.44	II
30		5264.95	II
55		5275.04	I
22		5286.09	I
120		5294.87	I
45		5298.06	II
30		5307.82	I
45		5309.68	I
55		5311.60	II
12		5324.26	II
9		5346.30	II
110		5354.73	I
110		5373.86	I
40		5389.34	I
19		5391.36	II
19		5404.47	I
28		5424.02	I
12		5435.78	I
40		5438.74	I
14		5444.07	II
75		5452.92	I
30		5463.38	II
15		5497.30	I
15		5510.12	I
15		5510.45	I
19		5524.35	II
45		5538.02	I
28		5538.26	I
230		5550.60	I
230		5552.12	I
55		5575.86	I
14		5600.77	I
95		5613.27	I
25		5614.01	I
8		5628.27	I
19		5650.83	I
40	Bl	5698.03	H
25		5713.28	I
160		5719.18	I
25	Bl	5720.16	H
12		5748.72	I
14		5767.18	II
12		5809.50	II
19		5817.47	I
25		5842.23	II
25		5845.87	I
19		5847.77	I
22		5883.66	I
15		5926.47	I
60		5933.69	I
75		5974.28	I
25		5974.72	I
60		5978.66	I
25		5992.96	I
45	c	6016.79	I
28	b	6021.12	H
25	b	6043.19	H
25		6054.17	I
95		6098.67	I
95		6185.13	I
55		6210.70	I
28		6216.82	I
45		6238.58	I
60		6248.95	II
22	h	6299.54	I
25	h	6311.85	I
19		6318.33	I
30		6338.10	I
19	h	6380.19	I
60		6386.23	I
19	h	6409.52	I
15	h	6556.50	I
28		6587.23	I
45		6644.60	II
19		6647.06	II
11		6659.40	I
30		6713.48	I
17		6754.61	II
11		6769.95	I
85		6789.27	I
160		6818.94	I
15		6826.56	I
13		6850.07	I
35		6858.70	I
45		6911.40	I
10		6926.19	I
19		6979.59	I
21		6980.91	II
7		7019.25	I
7		7030.33	II
7		7035.13	I
11		7061.90	I
15		7062.87	I
160		7063.83	I
11	h	7094.40	I
15		7100.54	I
55		7119.52	I
570		7131.81	I
650		7237.10	I
410		7240.87	I
6		7262.62	I
75		7320.05	I
16		7321.76	I
6		7356.10	I
6		7365.28	I
20		7390.70	I
6		7423.69	I
25		7437.56	I
13		7463.86	I
7		7484.56	I
15		7556.37	I
75		7562.93	I
15		7564.22	I
11		7576.95	I
11		7592.96	I
13		7608.59	I
360		7624.40	I
20		7645.64	I
110		7740.17	I
8		7743.57	I
5		7757.89	II
40		7790.90	I
7		7796.81	I
35		7814.55	I
310		7845.35	I
7		7846.56	I
130		7920.71	I
29		7938.06	I
250		7994.73	I
7		8010.58	I
25		8056.52	I
25		8080.32	I
16		8173.89	I
130		8204.58	I
7		8248.81	I
55		8276.95	I
13		8305.91	II
25		8344.25	I
5		8380.06	I
5		8382.98	I
35		8460.01	I
150		8546.48	I
160		8640.06	I
40		8711.24	I
65		9004.73	I

Hf III
Ref. 404 — R.L.K.

Intensity		Wavelength	
		Vacuum	
20		1449.83	III
30		1507.82	III
50		1683.95	III
50		1756.91	III
60		1843.64	III
60		1870.58	III
50		1874.81	III
150		1885.15	III
100		1991.44	III
		Air	
100		2037.76	III
300		2070.94	III
150	h	2085.33	III
200	h	2099.30	III
200	h	2110.31	III
100	h	2119.69	III
200		2155.66	III
200		2183.50	III
200		2195.44	III
100		2213.54	III
200		2234.59	III
200	h	2313.44	III
300		2336.47	III
150	h	2355.48	III
100		2373.30	III
120		2377.57	III
250		2383.540	III
400		2461.74	III
2000		2495.16	III
1000		2515.16	III
100		2534.33	III
400	h	2560.74	III
300	h	2567.46	III
200		2687.22	III
500		2753.60	III
100	h	3060.08	III
200	h	3279.67	III
100	h	3741.94	III

Hf IV
Ref. 369, 425 — R.L.K.

Intensity		Wavelength	
		Vacuum	
40		520.04	IV
50		569.19	IV
100		596.56	IV
100		600.90	IV
100		603.16	IV
200		618.27	IV
100		620.19	IV
100		633.58	IV
100		643.05	IV
200		644.54	IV
400		647.39	IV
600		665.65	IV
100		671.36	IV
200		673.49	IV

Hafnium (Cont.)

Intensity	Wavelength	
15	1305.24	IV
12	1357.40	IV
40	1390.39	IV
50	1491.67	IV
35	1528.82	IV
15	1560.18	IV
25	1572.03	IV
100	1717.21	IV
20	1718.57	IV
	Air	
100	2054.46	IV
7	2014.06	IV
20	7751.29	IV
10	7267.58	IV

Hf V
Ref. 410 — R.L.K.

Intensity	Wavelength	
	Vacuum	
220	545.41	V
180	600.00	V
100	816.81	V
100	830.69	V
100	836.74	V
100	846.87	V
100	856.32	V
100	861.80	V
135	865.16	V
270	867.25	V
180	875.88	V
135	877.87	V
135	880.37	V
100	880.85	V
180	885.58	V
135	885.80	V
135	894.24	V
100	894.41	V
180	896.14	V
100	896.47	V
135	899.70	V
180	901.54	V
135	901.92	V
135	904.95	V
135	909.70	V
135	913.68	V
135	918.48	V
180	919.10	V
270	921.67	V
135	928.01	V
135	931.50	V
135	947.12	V
245	951.62	V
180	960.12	V
180	964.74	V
160	971.51	V
135	974.62	V
120	984.64	V
135	991.50	V
100	1078.42	V
100	1079.92	V
160	1092.76	V
135	1097.28	V
135	1137.49	V
160	1201.76	V
135	1208.88	V
135	1224.62	V
135	1227.98	V
135	1230.21	V
270	1232.03	V
200	1233.59	V
160	1237.42	V
100	1238.85	V
160	1239.53	V
160	1244.46	V
100	1259.25	V
440	1396.66	V
270	1400.09	V
160	1401.70	V
135	1405.77	V
370	1407.17	V
370	1408.38	V
270	1412.28	V
270	1413.51	V
160	1421.96	V
220	1422.53	V
370	1433.43	V
370	1437.27	V
500	1437.73	V
370	1445.40	V
270	1457.91	V
270	1719.32	V
550	1729.08	V
750	1731.83	V
750	1733.96	V
440	1741.74	V
1000	1749.11	V
1000	1750.19	V
500	1760.89	V
370	1765.62	V
270	1774.02	V
135	1792.39	V

HELIUM (He)
Z = 2

He I and II
Ref. 16, 94, 173, 183, 317
W.C.M.

Intensity	Wavelength	
	Vacuum	
15	231.454	II
20	232.584	II
30	234.347	II
50	237.331	II
100	243.027	II
300	256.317	II
1000	303.780	II
500	303.786	II
10	320.293	I
2	505.500	I
3	505.684	I
4	505.912	I
5	506.200	I
7	506.570	I
10	507.058	I
15	507.718	I
20	508.643	I
25	509.998	I
35	512.098	I
50	515.616	I
100	522.213	I
400	537.030	I
1000	584.334	I
50	591.412	I
5	958.70	II
6	972.11	II
8	992.36	II
15	1025.27	II
30	1084.94	II
35	1215.09	II
50	1215.17	II
120	1640.34	II
180	1640.47	II
	Air	
7	2385.40	II
9	2511.20	II
50	2577.6	I
1	2723.19	I
12	2733.30	II
2	2763.80	I
10	2818.2	I
4	2829.08	I
10	2945.11	I
40	3013.7	I
20	3187.74	I
3	3202.96	II
15	3203.10	II
1	3354.55	I
2	3447.59	I
1	3587.27	I
3	3613.64	I
2	3634.23	I
3	3705.00	I
1	3732.86	I
10	3819.607	I
1	3819.76	I
500	3888.65	I
20	3964.729	I
1	4009.27	I
50	4026.191	I
5	4026.36	I
12	4120.82	I
2	4120.99	I
3	4143.76	I
10	4387.929	I
3	4437.55	I
200	4471.479	I
25	4471.68	I
6	4685.4	II
30	4685.7	II
30	4713.146	I
4	4713.38	I
20	4921.931	I
100	5015.678	I
10	5047.74	I
5	5411.52	II
500	5875.62	I
100	5875.97	I
8	6560.10	II
100	6678.15	I
3	6867.48	I
200	7065.19	I
30	7065.71	I
50	7281.35	I
1	7816.15	I
2	8361.69	I
2	9063.27	I
2	9210.34	I
10	9463.61	I
4	9516.60	I
3	9526.17	I
1	9529.27	I
1	9603.42	I
3	9702.60	I
6	10027.73	I
2	10031.16	I
15	10123.6	II
1	10138.50	I
10	10311.23	I
2	10311.54	I
3	10667.65	I
300	10829.09	I
1000	10830.25	I
2000	10830.34	I
9	10913.05	I
3	10917.10	I
4	11626.4	II
30	11969.12	I
20	12527.52	I
50	12784.99	I
20	12790.57	I
7	12845.96	I
10	12968.45	I
2	12984.89	I
12	15083.64	I
200	17002.47	I
1	18555.55	I
6	18636.8	II
500	18685.34	I
200	18697.23	I
100	19089.38	I
20	19543.08	I
1000	20581.30	I
80	21120.07	I
10	21121.43	I
20	21132.03	I
3	30908.5	II
4	40478.90	I

HOLMIUM (Ho)
Z = 67

Ho I and II
Ref. 1 — C.H.C.

Intensity		Wavelength	
		Air	
170		2502.91	II
80		2508.53	II
110		2513.55	II
95		2518.73	II
170		2533.80	I
130		2536.86	II
80		2556.84	I
80		2567.73	II
80		2586.52	I
60		2591.05	II
95		2592.99	I
190		2605.86	II
110		2610.51	II
95		2613.99	II
60		2625.20	II
80		2640.09	I
80		2640.30	II
60		2649.68	II
80		2666.24	II
70		2689.03	II
210		2713.65	II
230		2733.95	II
270		2750.35	II
110	c	2759.35	II
110		2766.85	II
270		2769.89	II
110		2772.83	II
140		2777.10	II
140		2794.41	II
100		2799.99	II
100		2806.72	II
160	c	2809.99	II
220		2811.36	II
180		2812.00	II
190		2814.74	II
300		2824.20	II
140		2826.64	II
270	c	2831.69	II
210		2834.99	II
110		2835.85	II
110		2844.18	II
100		2844.68	II
270		2849.10	II
100		2861.23	II
250		2861.49	II
150		2862.72	II
210		2871.99	II
230		2874.06	II
160		2874.43	II
360		2880.26	II
460		2880.98	II
340		2894.99	II
160		2895.62	II
170		2900.84	II
570	c	2909.41	II
170		2915.82	II
300		2919.62	II
110		2925.35	II
160		2926.09	II
300	c	2928.30	II
220		2942.05	II
300		2944.49	II
250	c	2953.11	II
390		2973.00	II
410	c	2979.63	II
180		2981.46	II
140		2985.48	II
410		2987.64	II
250		2990.27	II
110		2995.86	II
320	c	3008.10	II
220		3014.60	II
270		3038.69	II
480	c	3049.38	II
410	c	3054.00	II
500	c	3057.45	II
230		3074.30	II
500	c	3082.34	II
910		3084.36	II
430	c	3086.54	II
200		3108.31	II
200	c	3109.91	II
760		3118.50	II
300	c	3130.99	II
200	c	3134.39	II
300	c	3144.36	II
200		3156.18	II
270		3156.97	II
200	c	3159.67	II
580	c	3166.62	II
390	Dl	3171.72	II
810		3173.78	II
390		3174.84	II
270	c	3176.97	II
810	c	3181.50	II
390		3183.84	II
270	Cw	3184.48	II
200		3186.37	I
390	c	3197.83	II
390		3201.76	II
200		3206.86	II
270	c	3210.41	II
200	c	3221.42	II
320		3233.34	II
200		3236.90	II
200		3237.40	II
200	c	3257.45	II
390	c	3278.15	II
270		3279.25	II
980	c	3281.97	II
390		3288.46	II
270	c	3290.96	II
200	c	3305.16	II
200		3319.87	II
230		3320.25	II
200		3331.93	II
630	c	3337.23	II
390	c	3338.86	II
980	c	3343.58	II
200		3344.47	II
360		3350.49	II
320		3352.10	II
320	Cw	3353.55	II
320		3354.58	II
320		3357.91	II
320		3364.27	II
290		3370.87	II
230		3374.16	II
290	c	3390.75	II
320	c	3394.60	II
8100	c	3398.98	II
810	c	3410.26	II
1200		3421.63	II
3200		3453.14	II
390	c	3410.65	II
1400	c	3414.90	II
5400		3416.46	II
2000	c	3425.34	II
2000	c	3428.13	II
630	c	3429.18	II
320	c	3432.10	II
390		3449.35	I
810	c	3455.70	II
16000	c	3456.00	II
1600		3461.97	II
360	c	3467.07	II
810	c	3473.91	II
5400	c	3474.26	II
6300		3484.84	II
490		3489.58	II
580	c	3493.09	II

Holmium (Cont.)

2500	c	3494.76	II
810	c	3498.88	II
410	c	3506.95	II
320		3509.37	II
810		3510.73	I
4100	c	3515.59	II
410	c	3519.94	II
630		3540.76	II
1600		3546.05	II
1100	c	3556.78	II
410		3560.15	II
410	c	3573.24	II
630	c	3574.80	II
810		3579.12	I
410		3580.75	II
410		3581.83	II
630	c	3592.23	II
1100	Cw	3598.77	II
340		3599.48	I
540	c	3600.95	II
340		3613.31	II
410		3618.43	I
430	c	3626.69	II
490		3627.25	II
430	c	3631.76	II
430	c	3638.30	II
1600	c	3662.29	I
430		3662.99	I
720		3666.65	I
1400		3667.97	I
320		3669.05	II
450		3669.52	I
450	c	3674.77	II
720		3679.19	I
670		3679.70	I
720		3682.65	I
430		3685.16	II
580		3690.65	I
340		3691.95	I
410		3700.04	I
490	c	3702.35	II
320		3709.76	I
430		3712.88	I
450		3720.72	I
1100		3731.40	I
360		3732.09	I
810		3736.35	I
3200	Cw	3748.17	II
320	c	3753.73	II
340		3769.09	I
320		3788.08	II
8900	c	3796.75	II
8900	c	3810.73	II
490		3811.86	I
900	c	3813.25	II
300		3821.73	II
390		3829.27	I
320	Cw	3831.9	II
410	c	3835.35	II
1300	Cw	3837.51	II
410	c	3842.05	II
1100		3843.86	II
490	c	3846.73	II
300		3849.88	I
320		3852.40	II
1800	c	3854.07	II
390	Cw	3856.94	II
720		3857.72	II
2700	c	3861.68	II
540		3862.62	I
360		3872.05	II
320	c	3874.09	II
630		3874.68	II
540		3881.61	II
3000	c	3888.96	II
490		3890.42	I
13000	c	3891.02	II
540		3896.76	II
290		3902.23	II
320		3904.44	I
1300	Cw	3905.68	II
320		3911.80	I
320		3919.45	I
320	c	3936.44	II
220		3938.85	I
320	Cw	3940.53	II
220		3950.56	I
580		3955.73	I
230	c	3959.51	II
490		3959.68	I
220		3975.88	I
390	c	3976.93	I
220	Cw	3985.71	II
220		3993.73	II
380		3999.58	I
160	Cw	4002.59	II
220		4003.39	I
110		4013.50	I
320		4014.20	II
160	c	4018.09	II
160	c	4022.76	II
160	c	4023.94	II
110		4025.39	I
320		4027.21	I
270		4028.86	I
180	c	4031.80	I
220		4037.62	I
220	c	4038.87	II
2700		4040.81	I
5400	c	4045.44	II
220	c	4047.52	I
8100		4053.93	I
540		4054.48	II
270		4057.55	I
220		4060.31	I
1700		4065.09	II
170		4067.57	I
720		4068.05	I
270		4071.83	I
270		4073.13	I
290		4073.51	I
120	c	4080.23	II
230		4083.67	I
140		4085.09	I
170		4087.35	I
200		4087.59	I
140		4091.64	I
120		4094.78	I
230		4100.22	I
8900		4103.84	I
120		4105.04	I
270		4106.50	I
100		4107.36	I
2900		4108.62	I
300		4112.00	I
100		4112.72	I
270		4116.73	I
1500		4120.20	I
1300		4125.65	I
4300		4127.16	I
300		4134.54	I
1500		4136.22	I
130		4139.34	I
230		4142.19	I
290		4148.97	I
980	Cw	4152.61	II
8100		4163.03	I
160		4172.23	I
2500		4173.23	I
540		4194.35	I
100		4198.08	I
130		4203.21	I
100		4211.30	II
290		4222.29	I
290		4223.47	I
2000		4227.04	I
390		4229.52	II
130	h	4231.24	I
290		4243.78	I
1300	Cw	4254.43	I
130	c	4258.61	II
490		4264.05	I
300		4266.04	I
100		4273.63	II
200		4311.04	I
250		4330.64	II
300		4337.13	II
100	Cw	4346.84	II
1300		4350.73	I
290		4356.73	II
140		4363.93	II
170		4379.14	II
180	c	4384.83	II
150		4400.55	II
120		4401.24	II
180		4403.27	I
200		4420.56	II
130		4444.63	I
100		4473.59	II
300		4477.64	II
120		4484.57	II
140		4510.82	I
100	c	4526.14	II
170		4530.08	II
170	c	4531.28	I
130	c	4531.65	II
170		4534.58	I
200		4562.52	I
120	Cw	4609.32	II
130		4613.37	I
100		4618.84	I
100	c	4628.22	I
290		4629.10	II
200	c	4649.77	II
130	c	4661.33	II
140	c	4674.62	II
70		4701.17	II
80		4701.69	II
130		4709.84	II
65		4711.39	I
130	c	4717.52	I
35	c	4728.72	II
35		4738.00	II
290		4742.04	II
35	c	4749.09	II
35		4751.40	I
100	c	4757.01	I
35		4762.39	II
35		4763.57	II
55		4777.48	II
30		4779.42	I
70	c	4781.19	I
65		4782.92	I
55	c	4786.29	I
35		4791.48	II
35		4795.92	II
45	h	4798.87	I
27		4812.92	II
55		4832.31	II
30		4833.32	I
30		4855.54	II
45		4860.39	I
27	c	4889.67	II
30		4892.35	I
35		4896.44	II
55		4906.99	II
45		4922.73	I
55	c	4934.89	I
290		4939.01	I
27	c	4946.80	I
45		4948.18	II
65	c	4959.42	II
35		4961.03	II
55	Cw	4966.73	II
250	c	4967.21	II
220		4979.97	I
35	c	4988.96	I
90		4995.05	I
35	c	5012.42	I
55		5013.28	II
65		5026.53	I
30		5028.17	I
55		5032.95	II
65	c	5037.60	I
130		5042.37	I
35		5044.73	I
30		5051.44	II
30		5054.92	II
35	c	5060.75	I
65		5074.34	I
80		5093.07	I
140		5127.81	I
55		5129.27	II
130		5142.59	II
110		5143.22	II
160		5149.59	II
90	c	5167.88	I
130	c	5182.11	I
55		5187.85	I
90		5190.11	II
18		5195.23	I
45		5221.54	I
35	c	5244.47	I
65		5251.82	I
55		5275.48	I
90		5301.25	I
35		5319.24	I
35		5319.65	I
80		5330.11	I
90		5359.99	I
55		5381.40	I
30		5384.56	I
30		5384.97	I
18	h	5393.85	I
70		5403.17	I
100		5407.08	I
14		5413.62	II
16		5434.39	II
18		5435.87	I
30		5445.39	I
18		5449.8	II
30	h	5451.90	I
14		5454.0	II
30	c	5498.57	I
30		5504.51	I
27		5515.56	II
18		5516.45	II
30		5534.33	I
27		5553.14	I
35	c	5560.94	I
35	b	5563.6	H
70		5566.52	I
18		5573.96	II
35	Bl	5584.7	H
55	b	5591.1	H
55	Bl	5592.3	H
30	b	5607.1	H
27		5613.64	I
45	b	5626.4	H
65		5627.60	I
30		5628.24	II
55		5640.62	I
70	Bs	5655.9	H
65	b	5658.9	H
140		5659.58	I
70	c	5671.84	
65		5674.70	I
140	c	5691.47	I
70	Bs	5696.3	H
140	c	5696.57	I
27		5734.02	I
45		5736.4	H
55		5739.24	I
22		5749.58	I
30		5766.64	
27	b	5803.8	H
45	b	5819.2	H
27	h	5821.90	I
22		5839.47	I
45	b	5849.4	H
140	c	5860.28	I
27	h	5864.42	I
45		5870.85	I
27	b	5879.6	H
70	c	5882.99	I
35	c	5892.56	I
22		5904.29	I
70		5921.76	I
30	c	5933.71	I
70	Cw	5948.03	I
45		5955.98	I
70		5972.76	I
90		5973.52	I
22		5981.43	I
230	c	5982.90	I
55		6002.04	I
27		6005.33	I
35		6021.43	I
16		6038.97	I
27		6050.71	I
45		6060.31	I
120		6081.79	I
70	Cw	6133.60	I
35		6156.38	I
27		6156.58	I
55		6191.68	I
70		6208.65	I
18		6234.17	I
45	c	6255.75	I
70	c	6305.36	I
22		6306.68	I
30		6321.94	
30	c	6354.35	I
30	c	6372.59	I
14	h	6373.86	I
22	h	6413.41	I
27	c	6471.77	I
13		6479.17	I
11		6515.30	I
11	h	6538.99	I
70		6550.97	I
15		6560.08	I
35	d	6600.58	I
260		6604.94	I
55		6607.47	I
13		6628.35	I
120		6628.99	I
15		6632.24	I
9	h	6652.98	I
15		6662.52	I
19	c	6680.46	I
24	c	6681.62	I
15	h	6682.02	
55	Cw	6694.32	I
15	Cw	6722.34	I
40		6745.05	I
13		6766.74	I
28	c	6774.68	I
55	c	6785.43	I
13		6793.7	I
13	Cw	6811.04	I
15	Cw	6820.38	I
24		6821.64	I
17	c	6825.72	I
8	h	6826.62	
8	h	6852.97	
17	Cw	6865.85	I
9		6883.36	I
13		6888.50	I
15	c	6892.96	I
17		6897.95	I
15	h	6903.80	I
15	Cw	6913.47	I
9		6916.70	I
40	Cw	6939.49	I
45	Cw	6950.39	I
13	Hl	6955.3	I
19		6976.7	II
10		6985.11	
9		6994.38	
14	h	7000.71	
10		7079.07	
12		7098.58	
9		7242.08	I
9		7250.60	I
14		7308.55	
25		7341.43	
18		7389.40	
5	h	7496.20	I
10	h	7510.74	I
140		7555.09	I
18		7589.20	I
25		7591.87	I
9	h	7593.64	I
7	h	7594.35	I

Holmium (Cont.)

Intensity		Wavelength	
12		7602.31	II
16		7605.35	I
12		7617.05	I
14		7627.98	I
40	c	7628.42	I
9	c	7641.14	I
4		7648.16	I
14	c	7653.80	I
12	c	7667.30	I
20		7690.43	I
50	c	7693.15	I
40	Cw	7715.06	I
16		7719.05	I
16	h	7738.98	I
8	c	7752.01	I
60	Cw	7815.48	I
40	Cw	7823.63	I
8	h	7879.22	I
60		7894.64	I
10	h	8464.66	I
10	h	8482.67	I
50		8512.94	I
20		8545.61	II
18		8601.84	II
40		8670.19	I
8	h	8697.32	
16	h	8805.48	II
20	c	8834.49	I
90		8915.98	II

HYDROGEN (H)
Z = 1

H I
Ref. 214 — W.C.M.

Intensity	Wavelength	
	Vacuum	
15	926.226	I
20	930.748	I
30	937.803	I
50	949.743	I
100	972.537	I
300	1025.722	I
1000	1215.668	I
500	1215.674	I
	Air	
5	3835.384	I
6	3889.049	I
8	3970.072	I
15	4101.74	I
30	4340.47	I
80	4861.33	I
120	6562.72	I
180	6562.852	I
5	9545.97	I
7	10049.4	I
12	10938.1	I
20	12818.1	I
40	18751.0	I
5	21655.3	I
8	26251.5	I
15	40511.6	I
4	46525.1	I
6	74578	I
3	123685	I

INDIUM (In)
Z = 49

In I and II
Ref. 1, 132, 348—350 — C.H.C.

Intensity		Wavelength	
		Vacuum	
2		1648.00	I
1	h	1676.16	I
5	h	1711.54	I
2	h	1741.23	I
1	h	1758.49	I
		Air	
10		2103.89	II
10		2166.88	II
2		2179.90	I
2		2182.40	I
2		2187.40	I
2		2190.84	I
15		2195.67	II
2		2197.41	I
2		2202.24	I
50		2205.28	II
3		2211.14	I
5		2230.70	I
3		2241.66	I
30		2255.79	II
10		2259.99	I
5		2278.20	I
40		2281.64	II
2		2283.75	I
2		2298.33	I
2		2298.70	I
2		2302.49	I
100	c	2306.05	II
25		2306.86	I
3		2309.32	I
2		2309.75	I
90	d	2313.21	II
2		2315.09	I
30		2323.40	II
5		2324.41	I
3		2324.92	I
70	d	2327.95	II
3		2332.76	I
80	h	2334.57	II
10		2340.19	I
8		2345.90	I
5		2346.56	I
50	d	2350.75	II
5		2358.70	I
15		2378.14	I
10		2379.00	I
110	d	2382.63	II
40		2389.54	I
40		2393.18	II
10		2399.18	I
50	h	2406.47	II
50		2408.76	II
50		2419.06	II
50		2419.20	II
70	h	2427.20	II
20		2429.86	I
10		2430.99	I
50		2432.73	II
60		2442.63	II
100		2447.90	II
60		2453.23	II
60		2460.08	I
30	h	2468.02	I
70		2486.15	II
110	d	2488.62	II
90		2488.95	II
80		2498.59	II
100		2499.60	II
90	d	2500.99	II
60		2508.16	II
110	d	2512.31	II
100		2521.37	I
10		2522.98	I
70		2553.56	II
160	d	2554.44	II
1100		2560.15	I
70		2565.13	II
70	d	2598.75	II
200		2601.76	I
50		2604.04	II
90	d	2654.70	II
100	d	2662.63	II
140	d	2668.65	II
140	d	2674.56	II
80		2683.12	II
1600		2710.26	I
300		2713.94	I
130	d	2749.75	II
700		2753.88	I
40		2775.37	I
60		2798.76	II
90	d	2818.97	II
180	c	2836.92	I
30	c	2858.14	I
80		2865.68	II
120	d	2890.18	II
1100		2932.63	I
100		2941.05	II
20	c	2957.01	I
60	d	2966.17	II
110	c	2999.40	II
8000		3039.36	I
8	d	3051.15	I
110	d	3099.80	II
180	c	3101.8	II
130	c	3138.60	II
80	c	3142.75	II
130	d	3146.70	II
150		3155.77	II
100	c	3158.40	II
90	c	3176.30	II
90	d	3198.11	II
13000		3256.09	I
3000		3258.56	I
90	c	3338.50	II
75	c	3376.59	II
100	c	3404.28	II
110	d	3438.40	II
180	c	3693.91	II
95	c	3708.13	II
380	w	3716.14	II
120	c	3718.30	II
160	c	3718.72	II
160	c	3723.40	II
170	w	3795.21	II
230	c	3799.21	II
250	c	3834.65	II
200	c	3842.18	II
100		3889.78	II
100	c	3902.07	II
60	d	3922.12	II
65	c	3934.40	II
250	w	3962.35	II
120	c	4004.66	II
140	d	4013.92	II
410	w	4056.94	II
17000		4101.76	I
140	c	4205.14	II
100	d	4213.04	II
110	d	4219.66	II
150	d	4372.87	II
150	c	4500.78	II
18000		4511.31	I
110	c	4549.01	II
140	c	4570.85	II
180	w	4578.02	II
180	w	4578.40	II
140	c	4616.08	II
170	c	4617.17	II
250	c	4620.14	II
150	w	4620.70	II
170	c	4627.30	II
140	c	4637.04	II
380	c	4638.16	II
220	c	4644.58	II
360	c	4655.62	II
320	w	4656.74	II
190	c	4681.11	II
450	w	4684.8	II
3		4878.37	I
90	d	4907.06	II
70	h	4924.93	II
150	c	4973.77	II
80	h	5109.36	II
100	w	5115.14	II
140	c	5117.40	II
270	c	5120.80	II
200	w	5121.75	II
80	d	5129.85	II
240	c	5175.42	II
140	c	5184.44	II
30		5254.32	I
12		5262.74	I
150	c	5309.45	II
80		5411.41	II
140	c	5418.45	II
220	w	5436.70	II
130	c	5497.50	II
140	c	5507.08	II
320	c	5513.00	II
250	w	5523.28	II
130	c	5536.50	II
190	w	5555.45	II
240	c	5576.90	II
200	w	5636.70	II
160	c	5708.50	II
50		5709.91	I
100	c	5721.80	II
50		5727.68	I
210	c	5853.15	II
490	w	5903.4	II
260	w	5915.4	II
120	c	5918.78	II
130	c	6062.9	II
250	c	6095.95	II
210	c	6108.66	II
180	w	6115.9	II
230	w	6128.7	II
240	w	6129.4	II
320	w	6132.1	II
150	c	6140.0	II
90		6143.23	II
140	c	6148.10	II
190	w	6149.5	II
80		6161.15	II
180	w	6162.45	II
100	c	6224.28	II
280	w	6228.3	II
140	w	6231.1	II
270	w	6304.8	II
290	w	6362.3	II
300	w	6469.0	II
210	c	6541.20	II
190	c	6751.88	II
180	c	6765.9	II
100	c	6783.72	II
8	h	6847.44	I
320	w	6891.5	II
4	h	6900.13	I
380	w	7182.9	II
180	c	7255.0	II
210	c	7276.5	II
180	c	7303.4	II
320	c	7350.6	II
100	c	7632.7	II
100	c	7682.9	II
210	c	7740.7	II
100	c	7776.96	II
180	c	7789.0	II
70	c	7806.8	II
70	c	7814.5	II
90	c	7840.9	II
20	h	8050.78	I
240	c	8227.0	II
30	h	8238.66	I
15	h	8314.92	I
50	c	8434.55	II
30		8678.95	I
20		8682.63	I
50		8700.25	I
100	w	8813.5	II
80	c	8832.6	II
40		8894.47	I
10		9170.08	I
120	c	9197.7	II
120	c	9202.0	II
220	w	9213.0	II
160	d	9241.1	II
40	h	9349.83	I
60	h	9370.27	I
20		9427.99	I
100		9977.86	I
200		10257.03	I
60	h	10717.42	I
100	h	10744.31	I
20		11334.72	I
20		11731.48	I
10		12912.59	I
9		13429.96	I
5		13824.48	I
6		14316.25	I
3		14419.20	I
6		14668.66	I
7		14719.08	I
2		16504.31	I
6		22291.06	I
7		23879.13	I

In III
Ref. 351 — C.H.C.

Intensity	Wavelength	
	Vacuum	
7	685.31	III
5	691.62	III
1	782.17	III
10	882.24	III
10	890.84	III
10	915.87	III
2	917.45	III
5	926.83	III
30	1403.08	III
30	1434.85	III
20	1487.70	III
20	1494.14	III
10	1524.78	III
20	1530.21	III
30	1532.95	III
100	1625.42	III
20	1642.28	III
20	1702.53	III
100	1748.83	III
2	1767.88	III
1	1810.71	III
30	1842.41	III
40	1850.30	III
15	1862.98	III
	Air	
30	2154.08	III
2	2154.42	III
10	2199.52	III
5	2232.18	III
20	2261.26	III
5	2266.26	III
5	2272.41	III
5	2272.84	III
10	2300.90	III
100	2527.41	III
50	2725.52	III
80	2726.15	III
100	2982.80	III
100	3008.08	III
30	3008.82	III
30	3293.55	III
8	3350.91	III
5	3562.32	III
100	3852.82	III
100	4023.77	III
150	4032.32	III
50	4062.30	III
100	4071.57	III
100	4072.93	III
100	4252.68	III
40	4509.58	III
200	5248.77	III

Iodine (Cont.)

I III
Ref. 20, 21, 161 — L.J.R.

Intensity	Wavelength	
	Vacuum	
6	666.81	III
8	705.11	III
7	784.64	III
7	784.80	III
8	795.52	III
5	865.97	III
5	920.38	III
6	961.17	III
6	1078.58	III
8	1094.20	III
4	1244.66	III
8	1252.35	III
5	1306.93	III
	Air	
1	2224.43	III
1	2238.12	III
3	2249.31	III
2	2309.38	III
3	2340.85	III
3	2350.43	III
2	2353.46	III
4	2367.74	III
2	2371.45	III
3	2372.45	III
4	2376.47	III
4	2387.12	III
3	2392.01	III
2	2403.06	III
2	2403.63	III
2	2414.85	III
2	2418.49	III
2	2418.85	III
2	2423.91	III
5	2426.12	III
3	2434.88	III
2	2462.50	III
3	2466.69	III
3	2466.99	III
6	2475.36	III
4	2489.27	III
2	2493.21	III
2	2494.27	III
3	2495.16	III
2	2496.07	III
3	2501.41	III
2	2516.82	III
6	2519.75	III
4	2521.72	III
3	2531.99	III
2	2537.56	III
7	2545.71	III
4	2640.77	III
4	2642.11	III
6	2652.25	III
2	2818.48	III
2	2839.44	III
4	2864.67	III
4	2885.15	III
3	2910.98	III
3	2917.35	III
2	2931.11	III
2	3005.68	III
3	3069.23	III
3	3153.88	III
3	3170.14	III
3	3181.66	III
3	3210.14	III
4	3213.49	III
4	3224.93	III
2	3300.47	III
2	3479.53	III
3	3546.92	III
3	3613.81	III
2	3754.40	III
2	3754.55	III
3	3963.16	III
3	4077.14	III

I IV
Ref. 21, 58 — L.J.R.

Intensity	Wavelength	
	Vacuum	
5	601.86	IV
6	612.46	IV
4	615.17	IV
4	654.22	IV
4	654.56	IV
7	919.28	IV
	Air	
5	2249.30	IV
4	2340.84	IV
7	2361.13	IV
5	2367.75	IV
6	2372.45	IV
7	2376.46	IV
4	2385.28	IV
8	2387.11	IV
6	2392.00	IV
4	2403.05	IV
2	2418.45	IV
3	2423.89	IV
9	2426.10	IV
6	2434.85	IV
3	2466.68	IV
3	2466.96	IV
8	2475.35	IV
4	2485.51	IV
5	2489.24	IV
4	2493.20	IV
2	2501.38	IV
3	2513.74	IV
8	2519.74	IV
6	2521.72	IV
4	2531.98	IV
5	2537.54	IV
8	2545.67	IV
4	2640.77	IV
5	2642.11	IV
8	2652.23	IV
3	2818.45	IV
6	2864.68	IV
4	2910.97	IV
5	2917.33	IV
4	3069.17	IV
4	3170.11	IV
4	3181.64	IV
4	3210.12	IV
6	3213.48	IV
6	3224.90	IV
4	3546.90	IV

IV
Ref. 84 — L.J.R.

Intensity	Wavelength	
	Vacuum	
30	363.78	V
36	380.74	V
45	565.53	V
50	607.57	V

IRIDIUM (Ir)
Z = 77

Ir I and II
Ref. 1 — C.H.C.

Intensity	Wavelength	
	Air	
9900	2010.65	I
8700	2022.35	I
15000	2033.57	I
6200	2052.22	I
5000	2060.64	I
3700	2083.22	I
3100	2085.74	I
17000	2088.82	I
14000	2092.63	I
2700	2112.68	I
1800	2119.54	I
2000	2125.44	I
4500	2126.81	II
2000	2127.52	I
4500	2127.94	I
3700	2148.22	I
2500	2150.54	I
3500	2152.68	II
2900	2155.81	I
7900	2158.05	I
2100	2162.88	I
5800	2169.42	II
4500	2175.24	I
2700	2178.17	I
1600	2187.43	II
1100	2190.38	II
740	2191.64	I
910	2208.09	II
1300	2220.37	I
790	2221.07	II
2500	2242.68	II
620	2245.76	II
2100	2253.38	I
	2253.49	I
2100	2255.10	I
1400	2255.81	I
350	2258.51	I
1400	2258.86	I
830	2264.61	I
1100	2266.33	I
1000	2268.90	I
660	2280.00	I
950	2281.02	II
660	2281.91	I
330	2284.60	I
330	2295.08	I
790	2298.05	I
	2298.16	I
460	2299.53	I
910	2300.50	I
2700	2304.22	I
410	2305.47	I
210	2307.27	I
910	2308.93	I
460	2315.38	I
410	2321.45	I
410	2321.58	I
210	2327.98	I
540	2333.30	I
740	2333.84	I
580	2334.50	I
1600	2343.18	I
740	2343.61	I
100	2352.62	I
580	2355.00	I
230	2357.53	II
410	2358.16	I
500	2360.73	I
2500	2363.04	I
370	2368.04	II
3500	2372.77	I
290	2375.09	II
250	2377.28	I
250	2377.98	I
500	2379.38	I
540	2381.62	I
210	2383.17	I
120	2386.58	II
1300	2386.89	I
2500	2390.62	I
2700	2391.18	I
230	2407.59	I
290	2409.37	I
290	2410.17	I
290	2410.73	I
540	2413.31	I
370	2415.86	I
620	2418.11	I
120	2424.32	I
120	2424.66	I
210	2424.89	I
370	2424.99	I
290	2425.66	I
170	2426.53	II
540	2427.61	I
540	2431.24	I
1300	2431.94	I
170	2432.36	I
100	2432.58	I
270	2435.14	I
250	2445.34	I
250	2447.76	I
190	2448.23	I
910	2452.81	I
1300	2455.61	I
230	2455.87	I
210	2457.03	I
210	2457.23	I
120	2465.09	I
870	2467.30	I
3300	2475.12	I
210	2478.11	I
2100	2481.18	I
100	2485.38	I
620	2493.08	I
210	2496.27	I
250	2502.63	I
4100	2502.98	I
170	2504.37	I
120	2505.74	I
120	2507.63	I
170	2509.71	I
170	2511.94	I
170	2512.58	II
210	2513.71	I
120	2515.36	I
40	2524.88	II
170	2525.05	I
120	2532.52	I
990	2533.13	I
1100	2534.46	I
580	2537.22	I
170	2537.68	I
100	2541.48	I
580	2542.02	I
40	2542.80	II
7900	2543.97	I
150	2545.54	I
790	2546.03	I
120	2547.20	I
120	2547.69	I
210	2551.40	I
190	2554.40	I
210	2555.35	I
170	2555.88	I
150	2563.28	I
910	2564.18	I
210	2569.88	I
100	2570.62	I
230	2572.70	I
740	2577.26	I
100	2578.71	I
35	2579.49	II
740	2592.06	I
740	2599.04	I
150	2602.04	I
190	2604.55	I
190	2607.52	I
700	2608.25	I
1800	2611.30	I
210	2614.98	I
330	2617.78	I
210	2619.88	I
70	2623.64	II
250	2625.32	I
100	2626.76	I
700	2634.17	I
170	2635.27	I
250	2639.42	I
3500	2639.71	I
210	2644.19	I
170	2653.76	I
100	2656.81	I
1800	2661.98	I
350	2662.63	I
2700	2664.79	I
140	2668.99	I
520	2669.91	I
520	2671.84	I
330	2673.61	I
120	2676.83	I
110	2684.04	I
270	2692.34	I
3000	2694.23	I
110	2704.03	I
160	2712.74	I
140	2744.00	I
330	2772.46	I
250	2775.55	I
520	2781.29	I
330	2785.22	I
540	2797.35	I
1600	2797.70	I
380	2798.18	I
410	2800.82	I
680	2823.18	I
1200	2824.45	I
110	2833.24	II
110	2835.66	I
820	2836.40	I
160	2837.33	I
1100	2839.16	I
820	2840.22	I
160	2842.28	I
3800	2849.72	I
110	2863.84	I
380	2875.60	I
380	2875.98	I
270	2877.68	I
140	2879.41	I
820	2882.64	I
650	2897.15	I
260	2901.95	I
260	2904.80	I
200	2907.24	I
440	2916.36	I
230	2918.57	I
4400	2924.79	I
1200	2934.64	I
880	2936.68	I
250	2938.47	I
190	2939.27	I
140	2940.54	I
2700	2943.15	I
230	2946.97	I
200	2949.76	I
1200	2951.22	I
150	2962.99	I
200	2974.95	I
440	2980.65	I
150	2985.80	I
190	2990.62	I
300	2996.08	I
180	2997.41	I
220	3002.25	I
600	3003.63	I
160	3011.69	I
120	3016.43	I
270	3017.31	I
140	3019.23	I
110	3025.82	I
380	3029.36	I
330	3039.26	I
35	3042.65	II
300	3047.16	I
300	3049.44	I
300	3057.28	I

Indium (Cont.)

Intensity	Wavelength	
100	5645.15	III
40	5723.17	III
100	5819.50	III
200	6197.72	III

In IV
Refs. 352, 435, 436 — J.R.

Intensity	Wavelength	
	Vacuum	
622	472.71	IV
689	479.39	IV
709	498.62	IV
61	945.74	IV
85	954.67	IV
87	973.50	IV
86	991.60	IV
89	1024.68	IV
85	1024.79	IV
88	1031.45	IV
82	1031.98	IV
80	1054.43	IV
84	1063.03	IV
83	1068.25	IV
82	1069.82	IV
86	1077.64	IV
90	1082.10	IV
83	1086.33	IV
82	1096.81	IV
84	1097.18	IV
85	1116.10	IV
80	1124.06	IV
90	1131.46	IV
85	1144.43	IV
80	1145.41	IV
89	1146.62	IV
83	1154.11	IV
84	1154.60	IV
90	1157.71	IV
90	1157.82	IV
85	1159.78	IV
88	1176.50	IV
85	1191.58	IV
83	1204.87	IV
90	1206.55	IV
88	1221.50	IV
85	1221.90	IV
85	1233.58	IV
87	1235.84	IV
90	1373.20	IV
88	1398.77	IV
81	1412.09	IV

In V
Ref. 353 — C.H.C.

Intensity	Wavelength	
	Vacuum	
6	368.67	V
6	370.10	V
10	372.82	V
10	372.94	V
2	374.95	V
6	375.84	V
6	376.07	V
10	376.79	V
17	378.61	V
3	379.24	V
9	380.27	V
11	381.56	V
9	382.14	V
11	382.76	V
10	383.05	V
17	386.21	V
10	386.70	V
3	388.66	V
14	388.91	V
11	390.03	V
11	390.92	V
9	392.29	V
9	392.46	V
1	393.60	V
25	393.89	V
11	395.74	V
3	397.73	V
10	399.79	V
9	400.05	V
25	400.57	V
25	402.39	V
3	405.33	V
9	407.28	V
3	407.36	V
9	407.95	V
9	417.43	V
2	418.45	V
2	423.16	V

IODINE (I)

Z = 53

I I and II
Ref. 124, 153, 176, 184
L.J.R.

Intensity		Wavelength	
		Vacuum	
2		655.80	II
6		659.00	II
8		663.98	II
8		664.52	II
8		665.06	II
150		665.70	II
1000		719.55	II
1000		722.98	II
1000		798.16	II
1200		834.10	II
600		847.80	II
1500		873.49	II
1000		875.94	II
2000		879.84	II
1500		881.88	II
1000		891.00	II
1000		893.17	II
1200		1000.57	II
1000		1003.35	II
4000		1018.58	II
10000		1034.66	II
1500		1054.74	II
2000		1066.34	II
3000		1075.21	II
5000		1105.00	II
2500		1111.16	II
1500		1117.22	II
3500		1125.25	II
2000		1131.50	II
1200		1139.75	II
10000		1139.80	II
1500		1154.67	II
1000		1159.87	II
10000		1160.56	II
20000		1166.48	II
1500		1167.05	II
5000		1175.84	II
10000		1178.65	II
15000		1187.34	II
10000		1190.85	II
15		1195.29	I
5000		1198.88	II
7000		1200.22	II
200		1218.41	I
20000		1220.89	II
600		1224.05	I
600		1224.08	I
500		1228.89	I
20000		1234.06	II
600		1251.34	I
2500		1259.15	I
3000		1259.51	I
800		1261.27	I
600		1267.57	I
600		1267.60	I
1500		1275.26	I
3000		1289.40	I
10000		1300.34	I
3000		1302.98	I
3000		1313.95	I
3000		1317.54	I
2000		1330.19	I
20000		1336.52	II
5000		1355.10	I
3000		1357.97	I
5000		1360.97	I
3000		1361.11	I
2500		1367.71	I
2500		1368.22	I
4000		1383.23	I
3000		1390.75	I
2000		1392.90	I
2000		1400.01	I
8000		1425.49	I
5000		1446.26	I
5000		1453.18	I
5000		1457.39	I
5000		1457.47	I
10000		1457.98	I
2500		1458.79	I
4000		1459.15	I
2500		1465.83	I
1000		1485.92	I
5000		1492.89	I
5000		1507.04	I
5000		1514.68	I
15000		1518.05	I
2500		1526.45	I
5000		1593.58	I
5000		1617.60	I
2500		1640.78	I
15000		1702.07	I
12000		1782.76	I
5000		1799.09	I
75000		1830.38	I
15000		1844.45	I
		Air	
2000		2061.63	I
100		2408.01	II
100		2419.18	II
100		2494.74	II
100		2533.60	II
200		2534.27	II
1000		2566.24	II
2000		2582.79	II
300		2593.46	II
200	c	2688.98	II
500		2730.12	II
20		2765.15	II
200		2808.59	II
1500		2878.63	II
1000		2993.87	II
5000		3078.75	II
200		3161.03	II
1000		3175.07	II
300		3355.53	II
250	c	3424.99	II
300	c	3497.41	II
500		3526.90	II
200		3742.14	II
200		4102.23	I
200		4129.21	I
100	d	4134.15	I
500	d	4321.84	I
300		4452.86	II
200		4599.77	II
300	c	4632.45	II
500	d	4666.48	II
1000		4675.53	II
250		4763.31	I
1000		4862.32	I
200		4916.94	I
1000		4986.92	II
400	c	5065.37	II
10000		5119.29	I
200		5149.73	II
3000	c	5161.20	II
300		5176.19	II
600		5216.27	II
500	d	5228.97	II
1000		5234.57	I
3000	c	5245.71	II
500		5269.36	II
400		5299.78	II
400		5322.80	II
10000		5338.22	II
5000	c	5345.15	II
1000	c	5369.86	II
800	c	5405.42	II
800	c	5407.36	II
600	c	5427.06	I
3000		5435.83	II
1000		5438.00	II
2000	c	5464.62	II
800		5491.50	II
1000	c	5496.94	II
1000		5504.72	II
600	c	5522.06	II
600	c	5598.52	II
1000		5600.32	II
1500		5612.89	II
10000		5625.69	II
1000		5678.08	II
2000	c	5690.91	II
500		5702.05	II
4000	c	5710.53	II
1000		5738.27	II
1000		5760.72	II
1000	d	5764.33	I
500	c	5774.83	II
500		5787.02	II
2000		5894.03	I
5000		5950.25	II
300		5984.86	I
2000	d	6024.08	I
500		6068.93	II
2000	c	6074.98	II
1000		6082.43	I
2000	c	6127.49	II
800		6191.88	I
1000		6204.86	II
500		6213.10	I
800		6244.48	I
900	c	6257.49	II
1000		6293.98	I
500		6313.13	I
800		6330.37	I
400		6333.50	I
2000		6337.85	I
1000		6339.44	I
500		6359.16	I
1000		6566.49	I
2000		6583.75	I
1000		6585.27	I
5000		6619.66	I
500		6661.11	I
600		6665.96	II
500	c	6697.29	I
300		6718.83	II
400		6732.03	I
4000		6812.57	II
1000		6958.78	II
500		6989.78	I
200	c	7085.21	II
500		7120.05	I
1200		7122.05	I
2000		7142.06	I
1000		7164.79	I
400	d	7191.66	I
700		7227.30	I
1000		7236.78	I
500		7237.84	I
500		7351.35	II
5000		7402.06	I
1000		7410.50	I
500		7416.48	I
5000		7468.99	I
500	c	7490.52	I
2000		7554.18	I
500	d	7556.65	I
2000	c	7700.20	I
500		7798.98	II
600		7897.98	I
500		7969.48	I
1000		8003.63	I
99000		8043.74	I
300	d	8065.70	I
1000		8090.76	I
800	c	8169.38	I
500	d	8222.57	I
4000		8240.05	I
10000	c	8393.30	I
150		8414.60	II
1000		8486.11	I
1500	c	8664.95	I
500	c	8700.80	I
250	d	8748.22	I
1000		8853.24	I
2000		8853.80	I
3000		8857.50	I
1000	d	8898.50	I
400		8964.69	I
400		8993.13	I
5000		9022.40	I
15000		9058.33	I
1000		9098.86	I
12000		9113.91	I
600		9128.03	I
30		9195.30	II
600		9227.74	I
1000		9335.05	I
4000		9426.71	I
3000		9427.15	I
10	c	9480.33	II
2000		9598.22	I
2000		9649.61	I
3000	d	9653.06	I
5000		9731.73	I
500		10003.05	I
750		10131.16	I
1000		10238.82	I
400		10375.20	I
400		10391.74	I
6		10405.49	II
5000		10466.54	I
1		11084.68	II
400		11236.56	I
350		11558.46	I
320		11778.34	I
450		11996.86	I
300		12033.69	I
150		12304.58	I
60		13149.16	I
140		13958.27	I
200		14287.02	I
100		14460.00	I
225		15032.57	I
105		15528.65	I
150		16037.33	I
15		18275.71	I
20		18348.52	I
15		18982.41	I
35		19070.17	I
110		19105.12	I
50		19370.02	I
10		20648.69	I
220		22183.03	I
150		22226.53	I
30		22309.21	I
32		24420.82	I
12		27365.42	I
9		27573.05	I
10		30361.93	I
8		30383.88	I
10		34295.73	I
9		34513.11	I
3		40228.54	I
2		41633.80	I

Iridium (Cont.)

Intensity		Wavelength	
1600		3068.89	I
190		3069.09	I
190		3069.71	I
170		3076.69	I
320		3083.22	I
240		3086.44	I
390		3088.04	I
510		3100.29	I
510		3100.45	I
340		3120.76	I
200		3121.78	I
3400		3133.32	I
190		3150.61	I
190		3154.74	I
190		3159.15	I
140		3168.18	I
490		3168.88	I
370		3177.58	I
170		3180.35	I
370		3198.92	I
610		3212.12	I
370		3219.51	I
5100		3220.78	I
100		3221.28	I
300		3229.28	I
100		3230.76	I
470		3241.52	I
200		3262.01	I
390		3266.44	I
160		3277.28	I
100		3287.59	I
160		3310.52	I
200		3322.60	I
130		3334.16	I
560		3368.48	I
660		3437.02	I
100		3437.50	I
410		3448.97	I
3200		3513.64	I
220		3515.95	I
410		3522.03	I
160		3557.17	I
320		3558.99	I
1200		3573.72	I
320		3594.39	I
220		3609.77	I
190		3617.21	I
160		3626.29	I
660		3628.67	I
220		3636.20	I
300		3661.71	I
300		3664.62	I
320		3674.98	I
200		3687.08	I
140		3725.38	I
200		3731.36	II
130		3738.53	I
530		3747.20	I
120		3793.79	I
3100		3800.12	I
230		3817.24	I
170		3865.64	I
480		3902.51	I
480		3915.38	I
400		3934.84	I
120		3946.27	I
590		3976.31	I
460		3992.12	I
180		4020.03	I
350		4033.76	I
130		4040.08	I
370		4069.92	I
150		4070.68	I
100		4092.61	I
140		4115.78	I
23		4127.92	I
27		4155.70	I
15		4166.04	I
90		4172.56	I
35		4182.47	I
15	h	4183.21	I
18		4185.66	I
23		4197.54	I
27		4217.76	I
13		4220.80	I
75		4259.11	I
27		4265.30	I
260		4268.10	I
23		4286.62	I
75		4301.60	I
55		4310.59	I
220		4311.50	I
18		4351.30	I
18		4352.56	I
18		4392.59	I
160		4399.47	I
65		4403.78	I
110		4426.27	I
15		4450.18	I
55		4478.48	I
16		4495.35	I
11	h	4496.03	I
55		4545.68	I
30		4548.48	I
13		4550.78	I
35		4568.09	I
18		4570.02	I
18		4604.48	I
75		4616.39	I
26		4656.18	I
17	h	4668.99	I
21		4708.88	I
50		4728.86	I
21		4731.86	I
26		4756.46	I
13		4757.96	I
65		4778.16	I
30		4795.67	I
10		4807.14	I
21		4809.47	I
10		4840.77	I
17		4845.38	I
50		4938.09	I
26		4970.48	I
25		4999.74	I
25		5002.74	I
17		5009.17	I
30		5014.98	I
17		5046.06	I
30		5123.66	I
20		5177.95	I
22		5238.92	I
12		5340.74	I
35		5364.32	I
75		5449.50	I
30		5454.50	I
7		5469.40	I
10		5620.04	I
45		5625.55	I
10		5828.55	I
10		5882.30	I
7		5887.36	I
35		5894.06	I
7		6026.10	I
12		6067.83	I
20		6110.67	I
12		6288.28	I
7		6334.44	I
5		6624.73	I
10		6686.08	I
5		6830.01	I
5		6929.88	I
4		7183.71	I
6		7834.32	I

IRON (Fe)
Z = 26

Fe I and II
Ref. 56, 63, 105, 138, 174, 278
— H.M.C. and H.C.

Vacuum

Intensity	Wavelength	
12	1055.27	II
15	1068.36	II
15	1071.60	II
15	1096.89	II
12	1099.12	II
18	1112.09	II
12	1121.99	II
12	1122.86	II
12	1128.07	II
12	1130.43	II
15	1133.41	II
12	1133.68	II
12	1138.64	II
12	1142.33	II
12	1143.23	II
18	1144.95	II
12	1147.41	II
15	1148.29	II
12	1151.16	II
12	1267.44	II
12	1272.00	II
12	1371.02	II
12	1563.79	II
12	1580.62	II
18	1608.46	II
12	1618.47	II
15	1621.68	II
15	1629.15	II
15	1631.12	II
18	1635.40	II
15	1636.32	II
15	1639.40	II
12	1641.76	II
12	1647.16	II
12	1670.74	II
12	1702.04	II
12	1761.38	II
20	1785.26	II
20	1786.74	II
18	1788.07	II
30	1934.538	I
25	1937.269	I
50	1946.988	I
25	1951.571	I
30	1952.59	I
30	1953.005	I
60	1957.823	I
60	1960.144	I
30	1961.25	I
50	1962.111	I
12	1963.11	II

Air

Intensity	Wavelength	
100	2084.122	I
50	2157.794	I
15	2162.02	II
40	2166.773	I
300	2178.118	I
250	2186.486	I
60	2186.892	I
120	2187.195	I
250	2191.839	I
150	2196.043	I
80	2200.390	I
80	2200.724	I
15	2208.41	II
20	2213.65	II
12	2218.26	II
20	2220.38	II
25	2245.58	II
50	2250.790	I
60	2251.874	I
25	2255.77	II
300	2259.511	I
60	2264.389	I
80	2267.085	I
80	2267.469	I
50	2270.862	I
150	2272.070	I
150	2276.026	I
80	2279.937	I
150	2284.086	I
150	2287.250	I
300	2292.524	I
80	2294.41	I
200	2297.787	I
600	2298.169	I
80	2299.220	I
300	2300.142	I
50	2301.684	I
100	2303.424	I
150	2303.581	I
120	2308.999	I
150	2313.104	I
200	2320.358	I
100	2327.40	II
15	2327.88	II
100	2331.31	II
15	2331.97	II
300	2332.80	II
200	2338.01	II
600	2343.49	II
80	2343.96	II
150	2344.28	II
25	2344.98	II
50	2345.34	II
200	2348.11	II
250	2348.30	II
50	2351.20	II
15	2351.67	II
25	2352.31	II
30	2353.47	II
15	2353.68	II
50	2354.48	II
40	2354.89	II
200	2359.12	II
15	2359.59	II
150	2360.00	II
120	2360.29	II
30	2360.51	II
40	2362.02	II
60	2363.86	II
200	2364.83	II
80	2365.76	II
25	2366.59	II
80	2368.59	II
80	2369.456	I
80	2369.95	II
25	2370.50	II
120	2371.430	I
300	2373.624	I
150	2373.74	II
120	2374.518	I
60	2375.19	II
120	2376.43	II
20	2378.13	II
80	2379.27	II
20	2379.41	II
40	2380.20	II
120	2380.76	II
150	2381.835	I
1000	2382.04	II
20	2382.90	II
20	2383.06	II
60	2383.25	II
50	2384.39	II
40	2388.37	II
300	2388.63	II
200	2389.973	I
30	2390.10	II
20	2390.77	II
15	2391.48	II
20	2392.58	II
40	2395.42	II
1000	2395.62	II
15	2396.72	II
300	2399.24	II
20	2400.05	II
15	2401.29	II
50	2404.43	II
800	2404.88	II
250	2406.66	II
80	2406.97	II
300	2410.52	II
200	2411.07	II
50	2411.81	II
150	2413.31	II
20	2416.45	II
80	2417.87	II
15	2418.44	II
60	2420.396	I
60	2422.69	II
60	2423.089	I
40	2423.21	II
150	2424.14	II
15	2424.39	II
30	2424.59	II
30	2428.29	II
120	2428.36	II
25	2428.80	II
25	2429.03	II
20	2429.39	II
30	2429.86	II
120	2430.08	II
25	2431.02	II
80	2432.26	II
60	2432.87	II
25	2434.06	II
20	2434.24	II
20	2434.65	II
50	2434.73	II
50	2434.95	II
25	2436.62	II
60	2438.182	I
150	2439.30	II
150	2439.74	I
80	2440.11	I
40	2440.42	II
30	2442.37	II
100	2442.57	I
60	2443.71	II
250	2443.872	I
100	2444.51	II
50	2445.11	II
50	2445.212	I
100	2445.57	II
40	2445.80	II
50	2446.11	II
30	2446.47	II
40	2447.20	II
25	2447.33	II
60	2447.709	I
30	2447.75	II
25	2449.96	II
25	2450.20	II
100	2453.476	I
20	2453.98	II
30	2454.58	II
15	2455.71	II
15	2455.90	II
15	2457.09	II
1500	2457.598	I
150	2458.78	II
40	2458.97	II
60	2460.44	II
80	2461.28	II
100	2461.86	II
100	2462.181	I
1500	2462.647	I
50	2463.29	II
50	2463.730	I
40	2464.01	II
40	2464.90	II
800	2465.149	I
50	2465.91	II
15	2466.50	II
60	2466.67	II
60	2466.82	II
60	2467.732	I
15	2468.29	II
600	2468.879	I
60	2469.51	II
25	2470.41	II
80	2470.67	II

Iron (Cont.)

80	2470.965	I
800	2472.336	I
40	2472.43	II
40	2472.60	II
1000	2472.895	I
200	2473.16	I
50	2473.32	II
30	2474.05	II
600	2474.814	I
50	2475.12	II
40	2475.54	II
15	2476.26	II
60	2476.657	I
25	2477.34	II
60	2478.57	II
120	2479.480	I
1200	2479.776	I
100	2480.16	II
15	2481.05	II
80	2482.12	II
25	2482.32	II
100	2482.66	II
10000	2483.271	I
300	2483.533	I
1000	2484.185	I
60	2484.24	II
30	2484.44	II
50	2485.990	I
800	2486.373	I
100	2486.691	I
100	2487.066	I
120	2487.370	I
4000	2488.143	I
100	2488.945	I
80	2489.48	II
1000	2489.750	I
50	2489.83	II
50	2489.913	I
3000	2490.644	I
100	2490.71	II
60	2490.86	II
2000	2491.155	I
100	2491.40	II
25	2492.34	II
100	2493.18	II
500	2493.26	II
60	2494.000	I
50	2494.251	I
100	2495.87	I
600	2496.533	I
50	2497.82	II
150	2498.90	I
40	2500.92	II
1000	2501.132	I
40	2501.31	II
50	2501.693	I
60	2502.39	II
40	2503.33	II
60	2503.87	II
80	2506.09	II
40	2506.80	II
500	2507.900	I
30	2508.34	II
50	2508.753	I
1000	2510.835	I
120	2511.76	II
80	2512.275	I
400	2512.365	I
50	2514.38	II
80	2516.570	I
50	2517.13	II
300	2517.661	I
800	2518.102	I
60	2519.05	II
150	2519.629	I
40	2521.09	II
30	2521.82	II
50	2522.480	I
4000	2522.849	I
200	2523.66	I
500	2524.293	I
100	2525.02	I
200	2525.39	II
25	2526.07	II
300	2526.29	II
2000	2527.435	I
30	2527.70	II
800	2529.135	I
25	2529.23	II
80	2529.31	I
250	2529.55	II
150	2529.836	I
40	2530.11	II
200	2530.687	I
120	2533.63	II
60	2533.80	I
100	2534.42	II
120	2535.49	II
400	2535.607	I
60	2536.67	II
200	2536.792	I
200	2536.80	II
50	2536.84	II
50	2537.14	II
50	2538.20	II
40	2538.50	II
100	2538.80	II
100	2538.91	II
150	2538.99	II
50	2539.357	I
200	2540.66	II
600	2540.972	I
80	2541.10	II
60	2541.84	II
300	2542.10	I
25	2542.78	II
60	2543.38	II
250	2543.92	I
150	2544.70	I
40	2544.97	II
40	2545.22	II
800	2545.978	I
40	2546.44	II
80	2546.67	II
80	2546.87	I
100	2548.74	II
80	2549.08	II
80	2549.39	II
60	2549.46	II
600	2549.613	I
40	2549.77	II
60	2550.03	II
25	2550.15	II
50	2550.68	II
40	2560.28	II
25	2562.09	II
400	2562.53	II
200	2563.48	II
60	2566.91	II
25	2570.52	II
30	2570.85	II
150	2574.36	II
50	2575.74	I
300	2576.691	I
25	2576.86	II
60	2577.92	II
50	2582.30	I
100	2582.58	II
1500	2584.54	I
650	2585.88	II
90	2588.00	I
90	2591.54	II
30	2592.78	II
60	2593.51	I
90	2593.73	II
650	2598.37	II
2000	2599.40	II
300	2599.57	I
20	2605.34	II
20	2605.42	II
60	2605.657	I
300	2606.51	II
800	2606.827	I
650	2607.09	II
20	2611.07	II
600	2611.87	II
320	2613.82	II
320	2617.62	II
250	2618.018	I
20	2619.07	II
90	2620.41	II
20	2620.69	II
40	2621.67	II
400	2623.53	I
50	2625.49	II
200	2625.67	II
150	2628.29	II
20	2630.07	II
250	2631.05	II
250	2631.32	II
50	2631.61	II
100	2632.237	I
300	2635.809	I
50	2641.646	I
200	2643.998	I
60	2664.66	II
30	2666.64	II
300	2666.812	I
60	2666.965	I
600	2679.062	I
500	2684.75	II
400	2689.212	I
60	2692.60	II
50	2696.28	I
200	2699.106	I
60	2703.99	II
80	2706.012	I
400	2706.582	I
60	2708.571	I
20	2709.05	II
200	2711.655	I
80	2714.41	II
50	2716.22	II
50	2716.257	I
50	2717.786	I
50	2717.87	II
250	2718.436	I
4000	2719.027	I
100	2719.420	I
50	2720.197	I
1500	2720.903	I
400	2723.578	I
30	2724.88	II
150	2724.953	I
80	2726.05	I
50	2726.235	I
25	2727.38	II
80	2727.54	II
200	2728.020	I
50	2728.820	I
80	2728.90	II
40	2730.73	II
1000	2733.581	I
60	2734.005	I
50	2734.268	I
500	2735.475	I
50	2735.612	I
500	2737.310	I
120	2737.83	I
400	2739.55	II
250	2742.254	I
800	2742.405	I
200	2743.20	II
150	2743.565	I
200	2744.068	I
80	2744.527	I
300	2746.48	II
100	2749.32	II
500	2749.48	II
1200	2750.140	I
20	2751.13	II
20	2752.15	II
80	2753.29	II
50	2753.69	I
150	2754.032	I
100	2754.426	I
30	2754.89	II
800	2755.73	II
250	2756.328	I
100	2757.316	I
50	2759.81	I
120	2761.780	I
150	2761.81	II
150	2762.026	I
120	2762.772	I
120	2763.109	I
20	2763.66	II
25	2765.13	II
80	2766.910	I
250	2767.522	I
50	2769.30	I
25	2769.35	II
300	2772.07	I
50	2773.23	I
20	2774.69	II
15	2776.91	II
60	2778.07	I
600	2778.220	I
40	2779.30	II
50	2783.69	II
30	2785.19	II
3000	2788.10	I
20	2793.89	II
200	2797.78	I
30	2799.29	II
400	2804.521	I
1500	2806.98	I
2500	2813.287	I
300	2823.276	I
600	2825.56	I
50	2825.687	I
120	2828.808	I
25	2831.56	II
1500	2832.436	I
120	2835.950	I
200	2838.119	I
30	2839.51	II
20	2839.80	II
15	2840.65	II
200	2843.631	I
1000	2843.977	I
100	2845.594	I
15	2848.11	II
15	2848.32	II
800	2851.797	I
30	2856.91	II
25	2858.34	II
50	2869.307	I
50	2872.334	I
80	2874.172	I
50	2894.504	I
120	2912.157	I
120	2929.007	I
1200	2936.903	I
60	2941.343	I
12	2944.40	II
1000	2947.876	I
60	2950.24	I
600	2953.940	I
250	2957.364	I
80	2959.99	I
150	2965.254	I
1500	2966.898	I
120	2969.36	I
800	2970.099	I
15	2970.52	II
1200	2973.132	I
500	2973.235	I
600	2981.445	I
1000	2983.570	I
60	2984.77	I
50	2984.82	II
13	2985.54	II
1000	2994.427	I
250	2994.502	I
500	2999.512	I
120	3000.451	I
800	3000.948	I
60	3001.655	I
15	3002.64	II
200	3007.282	I
500	3008.14	I
120	3009.569	I
60	3017.627	I
60	3018.983	I
60	3020.01	II
500	3020.491	I
1500	3020.639	I
600	3021.073	I
500	3024.032	I
150	3025.638	I
500	3025.842	I
80	3030.148	I
60	3031.214	I
60	3034.484	I
40	3036.96	II
800	3037.389	I
80	3041.637	I
800	3047.604	I
600	3057.446	I
1000	3059.086	I
250	3067.244	I
120	3075.719	I
120	3091.577	I
80	3098.189	I
100	3099.895	I
100	3099.968	I
60	3100.303	I
100	3100.665	I
12	3154.20	II
80	3175.445	I
150	3184.895	I
250	3191.659	I
500	3193.226	I
800	3193.299	I
12	3196.08	II
200	3196.928	I
80	3199.500	I
60	3200.47	I
50	3205.398	I
50	3211.67	I
100	3211.88	I
13	3213.31	II
200	3214.011	I
200	3214.396	I
60	3215.938	I
50	3217.377	I
80	3219.583	I
60	3219.766	I
300	3222.045	I
600	3225.78	I
13	3227.73	II
80	3227.796	I
20	3230.42	II
80	3233.05	I
50	3233.967	I
120	3234.613	I
300	3236.222	I
100	3239.433	I
80	3244.187	I
80	3246.005	I
60	3254.36	I
80	3265.046	I
50	3265.617	I
50	3271.000	I
50	3280.26	I
150	3286.75	I
120	3305.97	I
200	3306.343	I
400	3355.227	I
80	3355.517	I
60	3369.546	I
120	3370.783	I
50	3378.678	I
50	3380.110	I
60	3383.978	I
12	3388.13	II
50	3392.304	I
150	3392.651	I
150	3399.333	I
80	3404.353	I
500	3407.458	I

Iron (Cont.)

250	3413.131	I
60	3424.284	I
500	3427.119	I
60	3428.748	I
6000	3440.606	I
2500	3440.989	I
1000	3443.876	I
200	3445.149	I
15	3453.61	II
1200	3465.860	I
2000	3475.450	I
500	3476.702	I
2500	3490.574	I
500	3497.840	I
250	3513.817	I
300	3521.261	I
400	3526.040	I
100	3526.166	I
60	3526.237	I
60	3526.381	I
60	3526.467	I
100	3533.199	I
200	3536.556	I
300	3541.083	I
250	3542.075	I
80	3553.739	I
400	3554.925	I
200	3556.878	I
400	3558.515	I
1000	3565.379	I
1200	3570.097	I
800	3570.25	I
120	3571.996	I
100	3573.393	I
60	3573.829	I
60	3573.888	I
4000	3581.19	I
150	3582.199	I
150	3584.660	I
120	3584.929	I
300	3585.319	I
150	3585.705	I
200	3586.103	I
400	3586.984	I
100	3594.633	I
150	3603.204	I
200	3605.454	I
500	3606.680	I
1500	3608.859	I
250	3610.16	I
60	3612.068	I
150	3617.788	I
1500	3618.768	I
200	3621.462	I
150	3622.004	I
150	3623.19	I
100	3631.096	I
1200	3631.463	I
60	3632.041	I
100	3638.298	I
200	3640.389	I
80	3643.717	I
1500	3647.842	I
250	3649.506	I
80	3650.279	I
200	3651.467	I
120	3670.024	I
150	3670.089	I
100	3676.311	I
150	3677.629	I
1500	3679.913	I
200	3682.242	I
120	3683.054	I
150	3684.107	I
120	3685.998	I
500	3687.456	I
120	3689.477	I
150	3694.008	I
120	3695.051	I
150	3701.086	I
80	3704.462	I
1200	3705.566	I
60	3707.041	I
150	3707.821	I
300	3707.919	I
600	3709.246	I
120	3716.442	I
8000	3719.935	I
1500	3722.563	I
120	3724.377	I
60	3725.491	I
60	3727.093	I
500	3727.619	I
150	3732.396	I
1200	3733.317	I
5000	3734.864	I
120	3735.324	I
6000	3737.131	I
100	3738.306	I
400	3743.362	I
80	3743.47	I
6000	3745.561	I
1200	3745.899	I
3000	3748.262	I
80	3748.964	I
3000	3749.485	I
1500	3758.232	I
400	3760.05	I
1500	3763.788	I
400	3765.54	I
600	3767.191	I
60	3776.452	I
250	3785.95	I
100	3786.68	I
250	3787.880	I
250	3790.092	I
150	3794.34	I
400	3795.002	I
120	3797.518	I
250	3798.511	I
400	3799.547	I
200	3805.345	I
80	3806.696	I
600	3812.964	I
60	3813.059	I
1500	3815.840	I
2500	3820.425	I
150	3821.179	I
80	3824.306	I
2500	3824.444	I
1500	3825.880	I
1200	3827.823	I
1000	3834.222	I
120	3839.257	I
500	3840.437	I
800	3841.047	I
120	3843.256	I
80	3846.800	I
200	3849.96	I
120	3850.817	I
2500	3856.372	I
150	3859.212	I
10000	3859.911	I
150	3865.523	I
60	3867.215	I
250	3872.501	I
150	3873.761	I
250	3878.018	I
2000	3878.573	I
4000	3886.282	I
200	3887.048	I
300	3888.513	I
800	3895.656	I
1200	3899.707	I
400	3902.945	I
250	3906.479	I
80	3916.731	I
600	3920.258	I
1200	3922.911	I
1200	3927.920	I
2000	3930.296	I
60	3948.774	I
60	3949.953	I
50	3951.164	I
50	3952.601	I
60	3956.454	I
250	3956.68	I
60	3966.614	I
100	3969.257	I
80	3977.741	I
40	3981.771	I
50	3983.956	I
60	3994.114	I
200	3997.392	I
40	3998.053	I
400	4005.241	I
60	4009.713	I
80	4014.53	I
100	4021.867	I
50	4040.638	I
4000	4045.813	I
1500	4063.594	I
50	4066.975	I
50	4067.977	I
1200	4071.737	I
40	4076.629	I
40	4100.737	I
40	4107.489	I
150	4118.544	I
40	4127.608	I
400	4132.058	I
80	4134.676	I
40	4136.997	I
200	4143.415	I
800	4143.869	I
40	4153.898	I
50	4154.500	I
60	4156.799	I
50	4172.744	I
60	4174.912	I
50	4175.635	I
50	4177.593	I
120	4181.754	I
50	4184.891	I
120	4187.038	I
120	4187.795	I
80	4191.430	I
40	4195.329	I
150	4198.304	I
40	4199.095	I
300	4202.029	I
40	4203.984	I
80	4206.696	I
80	4210.343	I
400	4216.183	I
100	4219.360	I
50	4222.212	I
50	4225.956	I
200	4227.423	I
11	4233.17	II
100	4233.602	I
250	4235.936	I
50	4238.809	I
50	4247.425	I
200	4250.118	I
300	4250.787	I
40	4258.315	I
800	4260.473	I
250	4271.153	I
1200	4271.759	I
1200	4282.402	I
80	4291.462	I
250	4299.234	I
1200	4307.901	I
150	4315.084	I
1500	4325.761	I
80	4352.734	I
80	4369.771	I
800	4375.929	I
3000	4383.544	I
1200	4404.750	I
300	4415.122	I
600	4427.299	I
400	4461.652	I
120	4466.551	I
80	4476.017	I
80	4482.169	I
200	4482.252	I
50	4489.739	I
50	4528.613	I
11	4583.83	II
30	4647.433	I
30	4736.771	I
50	4859.741	I
120	4871.317	I
60	4872.136	I
30	4878.208	I
100	4890.754	I
250	4891.492	I
30	4903.309	I
150	4918.992	I
500	4920.502	I
12	4923.92	II
1500	4957.597	I
11	4990.50	II
80	5001.862	I
18	5001.91	II
11	5004.20	II
30	5005.711	I
100	5006.117	I
60	5012.067	I
30	5014.941	I
12	5018.43	II
11	5030.64	II
25	5030.77	I
12	5035.71	II
150	5041.755	I
30	5049.819	I
30	5051.634	I
25	5074.748	I
18	5100.73	II
15	5100.95	II
150	5110.357	I
40	5133.69	I
40	5139.251	I
100	5139.462	I
11	5144.36	II
12	5149.46	II
25	5151.910	I
30	5162.27	I
80	5166.281	I
2500	5167.487	I
80	5168.897	I
12	5169.03	II
500	5171.595	I
50	5191.454	I
80	5192.343	I
200	5194.941	I
30	5204.582	I
25	5215.179	I
150	5216.274	I
18	5216.85	II
60	5226.862	I
1000	5227.150	I
13	5227.49	II
250	5232.939	I
13	5247.95	II
13	5251.23	II
18	5260.26	II
11	5264.18	II
100	5266.555	I
1200	5269.537	I
800	5270.357	I
30	5281.789	I
60	5283.621	I
25	5302.299	I
11	5306.18	II
13	5316.23	II
150	5324.178	I
800	5328.038	I
300	5328.531	I
100	5332.899	I
14	5339.59	II
80	5339.928	I
500	5341.023	I
25	5364.87	I
40	5367.47	I
50	5369.96	I
400	5371.489	I
60	5383.37	I
14	5387.06	II
40	5393.167	I
12	5395.86	II
300	5397.127	I
15	5402.06	II
60	5404.12	I
250	5405.774	I
30	5410.91	I
60	5415.20	I
60	5424.07	I
30	5427.83	II
250	5429.695	I
13	5429.99	II
100	5434.523	I
200	5446.871	I
25	5455.45	I
120	5455.609	I
16	5465.93	II
20	5466.94	II
16	5482.31	II
14	5493.83	II
25	5497.516	I
20	5501.464	I
18	5506.20	II
30	5506.778	I
12	5510.78	II
12	5529.06	II
13	5544.76	II
30	5569.618	I
60	5572.841	I
120	5586.755	I
200	5615.644	I
20	5624.541	I
12	5645.40	II
50	5662.515	I
20	5762.990	I
11	5783.63	II
30	5862.353	I
13	5885.02	II
16	5902.82	II
30	5914.114	I
14	5955.70	II
30	5986.956	I
18	5961.71	II
30	5962.4	II
13	5965.63	II
40	6065.482	I
30	6102.159	I
40	6136.614	I
40	6137.694	I
30	6147.73	II
20	6149.24	II
15	6175.16	II
40	6191.558	I
30	6213.429	I
30	6219.279	I
40	6230.726	I
20	6238.37	II
20	6246.317	I
80	6247.56	II
30	6252.554	I
15	6305.32	II
12	6331.97	II
15	6383.75	II
20	6393.602	I
30	6399.999	I
20	6411.647	I
20	6416.90	II
20	6421.349	I
30	6430.844	I
20	6446.43	II
200	6456.38	II
60	6494.981	I
20	6516.05	II
20	6546.239	I
20	6592.913	I
40	6677.989	I
15	6855.18	I
15	6945.21	I
20	7067.44	II
15	7130.94	I
25	7164.443	I
80	7187.313	I

Iron (Cont.)

Intensity		Wavelength	
30		7207.381	I
12		7224.51	II
50		7307.97	II
40		7320.70	II
20		7376.46	II
30		7445.746	I
20		7462.38	II
40		7495.059	I
60		7511.045	I
15		7586.04	I
15		7711.71	II
30		7780.59	I
40		7832.22	I
80		7937.131	I
60		7945.984	I
80		7998.939	I
60		8046.047	I
50		8085.176	I
150		8220.41	I
120		8327.053	I
20		8331.908	I
120		8387.770	I
30		8468.404	I
15		8514.069	I
60		8661.898	I
150		8688.621	I
12		8793.38	I
12		8824.23	I
20		8866.96	I
15		8999.56	I
15		10216.32	I
13		10469.65	I
21		11119.80	I
14		11374.08	I
52		11422.32	I
87		11439.12	I
91		11593.59	I
255		11607.57	I
160		11638.26	I
230		11689.98	I
160		11783.26	I
580		11882.84	I
225		11884.08	I
1030		11973.05	I
15		12638.71	I
14		12879.76	I
17		13565.04	I
30		14236.25	I
24		14285.11	I
14		14292.38	I
16		14308.69	I
96		14400.56	I
20		14442.28	I
72		14512.23	I
50		14555.06	I
14		14565.95	I
40		14826.43	I
37		15051.77	I
28		15207.55	I
94		15294.58	I
16		15335.40	I
30		15621.67	I
25		15631.97	I
14		15723.59	I
41		15769.42	I
28		15813.13	I
13		16444.82	I
20		16486.69	I
105		18856.65	I
47		18987.01	I
25		19113.68	I
22		19791.88	I
14		22380.82	I
21		22619.85	I
38		26222.04	I
17		26659.22	I

Fe III
Ref. 71, 101 — J.R.

Intensity		Wavelength	
		Vacuum	
6		728.81	III
5		730.00	III
5		737.71	III
5		739.26	III
9		807.55	III
8		807.86	III
8		808.84	III
8	p	811.28	III
10		813.38	III
8		838.05	III
10		844.28	III
9		845.41	III
8	w	847.42	III
8		859.72	III
8	p	861.76	III
10	p	861.83	III
8		873.46	III
9		890.76	III
10		891.17	III
8		891.44	III
8		899.42	III
10		950.33	III
10		981.37	III
10	w	983.88	III
8		985.82	III
9		991.23	III
9		1017.25	III
8		1017.74	III
8		1018.29	III
8		1032.12	III
8		1063.87	III
9		1122.53	III
9		1124.88	III
8		1128.02	III
10	h	1505.17	III
10	h	1538.63	III
12	h	1550.20	III
10	h	1601.21	III
10		1869.83	III
12		1877.99	III
10		1882.05	III
12		1886.76	III
13		1890.67	III
11		1893.98	III
20		1895.46	III
10	s	1907.58	III
19		1914.06	III
15		1915.08	III
15		1922.79	III
10	p	1926.01	III
18		1926.30	III
15		1930.39	III
14		1931.51	III
14		1937.34	III
10	l	1938.90	III
14	s	1943.48	III
12		1945.34	III
10		1950.33	III
12		1951.01	III
11		1952.65	III
13		1953.32	III
10		1953.49	III
10	w	1954.22	III
11		1958.58	III
13		1960.32	III
15		1987.50	III
14		1991.61	III
13		1994.07	III
12		1995.56	III
12		1996.42	III
		Air	
10		2061.55	III
12		2068.24	III
14		2078.99	III
10		2084.35	III
12		2090.14	III
15		2097.48	III
12		2097.69	III
12		2103.80	III
10		2107.32	III
15		2151.78	III
12		2157.71	III
12		2158.47	III
10		2161.27	III
12		2166.95	III
12		2171.04	III
15		2174.66	III
12		2180.41	III
10	p	2208.85	III
10		2221.83	III
10		2229.27	III
10		2232.43	III
10		2232.69	III
10		2235.91	III
10		2238.16	III
12	p	2241.54	III
12		2261.59	III
10		2267.42	III
10		2293.06	III
15		2295.86	III
10	p	2317.70	III
10		2319.22	III
10	p	2321.71	III
10		2326.95	III
10	p	2336.77	III
10		2338.96	III
8		2389.53	III
8		2438.17	III
8	p	2582.37	III
8		2595.62	III
8		2617.15	III
9	p	2645.39	III
10	h	2695.13	III
9	h	2695.34	III
8	h	2700.02	III
8	h	2701.13	III
8		2773.31	III
10	p	2813.24	III
8	p	2895.08	III
9	p	2902.47	III
12		2904.43	III
8	p	2905.80	III
10		2907.50	III
12		2907.70	III
8		2923.90	III
8		2948.39	III
8		2963.23	III
12		3001.62	III
12	h	3007.28	III
15		3013.17	III
10	p	3136.43	III
10		3174.09	III
10		3175.99	III
10		3178.01	III
13		3266.88	III
11		3276.08	III
10		3288.81	III
9		3305.22	III
9		3339.39	III
9		3499.59	III
9		3500.28	III
10		3501.76	III
10		3586.04	III
11		3600.94	III
11		3603.88	III
16		3954.33	III
11		3968.72	III
9		3969.49	III
10	w	3979.42	III
10		4035.42	III
11		4053.11	III
12		4081.00	III
10		4120.90	III
11		4122.02	III
11		4122.78	III
15		4137.76	III
13		4139.35	III
9		4140.48	III
9		4154.96	III
18		4164.73	III
9		4164.92	III
13		4166.84	III
13		4174.26	III
9		4210.67	III
11		4222.27	III
13		4235.56	III
9		4238.62	III
12		4243.75	III
12	h	4273.40	III
12		4279.72	III
14	h	4286.16	III
16	h	4296.85	III
18	h	4304.78	III
20	h	4310.36	III
9		4323.68	III
9	h	4372.04	III
9	h	4372.14	III
11	h	4372.31	III
14	h	4372.53	III
18	h	4372.81	III
9		4395.76	III
12		4419.60	III
9		4431.02	III
9		5111.07	III
9		5127.35	III
12		5156.12	III
10		5199.08	III
10		5235.66	III
18		5243.31	III
13	l	5260.34	III
9		5272.37	III
14		5272.98	III
15		5276.48	III
16		5282.30	III
12		5284.83	III
11		5298.12	III
12		5299.93	III
14	w	5302.60	III
10		5306.76	III
9		5310.88	III
10		5322.74	III
11		5346.88	III
12		5353.77	III
12		5363.76	III
10		5368.06	III
11	l	5375.47	III
11		5719.88	III
9		5744.19	III
10		5756.38	III
18		5833.93	III
9		5848.76	III
10		5854.62	III
9		5876.26	III
15		5891.91	III
9		5898.68	III
9		5918.96	III
10	p	5920.13	III
13	p	5929.69	III
10		5952.31	III
14		5953.62	III
9		5968.48	III
12		5979.32	III
9	h	5981.01	III
12	h	5989.08	III
18		5999.54	III
9		6031.02	III
16		6032.59	III
13		6036.56	III
11		6048.72	III
11		6054.18	III
9		6056.36	III
9		6149.99	III
9		6169.74	III
9		6185.26	III
7		6186.56	III
7		6194.79	III
6		6195.43	III
6		6201.37	III
5	s	6203.04	III
5		6259.81	III
6	p	6294.50	III
5		6357.81	III
5	h	7317.63	III
6	h	7320.14	III
5	w	7921.17	III
5	w	8230.88	III
5	w	8231.79	III
9	w	8235.45	III
8	w	8236.75	III
6	w	8238.98	III
5		8563.49	III

Fe IV
Ref. 382 — J.R.

Intensity	Wavelength	
	Vacuum	
10	502.42	IV
11	506.69	IV
11	505.35	IV
17	525.69	IV
15	526.29	IV
10	526.57	IV
13	526.63	IV
10	530.91	IV
10	531.78	IV
10	535.55	IV
14	536.61	IV
10	536.74	IV
15	537.10	IV
13	537.26	IV
14	537.79	IV
13	537.94	IV
10	538.44	IV
10	544.20	IV
10	546.22	IV
10	548.80	IV
11	550.32	IV
10	551.77	IV
13	552.14	IV
11	552.74	IV
10	554.26	IV
10	555.66	IV
10	572.88	IV
10	576.76	IV
10	579.76	IV
14	607.53	IV
13	608.80	IV
10	609.65	IV
12	1425.73	IV
13	1431.43	IV
12	1473.20	IV
12	1489.53	IV
12	1495.18	IV
13	1526.60	IV
13	1530.26	IV
14	1532.63	IV
13	1532.91	IV
15	1533.86	IV
13	1533.95	IV
14	1536.58	IV
12	1538.29	IV
13	1542.16	IV
14	1542.70	IV
12	1546.40	IV
12	1552.35	IV
12	1552.71	IV
12	1562.46	IV
13	1566.26	IV
14	1568.27	IV
12	1570.18	IV
12	1570.42	IV
12	1571.24	IV
12	1577.20	IV
12	1577.76	IV
12	1590.62	IV
13	1591.51	IV
13	1592.05	IV
12	1596.67	IV
13	1598.01	IV
12	1600.50	IV
13	1600.58	IV
13	1601.67	IV

Iron (Cont.)

Intensity	Wavelength	
12	1602.08	IV
13	1603.18	IV
13	1603.73	IV
13	1604.88	IV
13	1605.68	IV
15	1605.97	IV
13	1606.98	IV
17	1609.10	IV
14	1609.83	IV
13	1610.47	IV
13	1611.20	IV
13	1613.64	IV
15	1614.02	IV
13	1614.64	IV
13	1615.00	IV
12	1615.61	IV
16	1616.68	IV
14	1617.68	IV
14	1619.02	IV
12	1620.91	IV
13	1621.16	IV
14	1621.57	IV
13	1623.38	IV
13	1623.53	IV
15	1626.47	IV
14	1626.90	IV
13	1628.54	IV
13	1630.18	IV
17	1631.08	IV
12	1632.08	IV
14	1632.40	IV
13	1634.01	IV
12	1638.07	IV
12	1638.30	IV
14	1639.40	IV
16	1640.04	IV
14	1640.16	IV
15	1641.87	IV
12	1642.88	IV
15	1647.09	IV
15	1651.58	IV
15	1652.90	IV
13	1653.41	IV
13	1656.11	IV
15	1656.65	IV
12	1657.82	IV
12	1658.43	IV
14	1660.10	IV
12	1661.57	IV
13	1662.32	IV
13	1662.52	IV
13	1663.54	IV
13	1668.09	IV
12	1669.61	IV
14	1671.04	IV
12	1672.86	IV
13	1673.68	IV
14	1675.66	IV
12	1676.78	IV
12	1677.12	IV
13	1681.36	IV
12	1681.95	IV
15	1687.69	IV
15	1698.88	IV
12	1700.40	IV
12	1704.93	IV
13	1709.81	IV
15	1711.41	IV
14	1712.76	IV
12	1717.11	IV
14	1717.90	IV
14	1718.16	IV
12	1718.42	IV
14	1719.46	IV
14	1722.71	IV
14	1724.06	IV
12	1724.26	IV
16	1725.63	IV
13	1761.08	IV
12	1764.92	IV
12	1767.36	IV
13	1792.10	IV
13	1796.93	IV
12	1805.32	IV
12	1820.42	IV
13	1827.98	IV
12	1840.24	IV
12	1860.42	IV
12	1869.64	IV
12	1874.23	IV

Fe V
Ref. 381 — J.R.

Intensity	Wavelength	
	Vacuum	
300	361.28	V
300	365.43	V
300	365.86	V
300	374.24	V
300	374.87	V
300	375.98	V
300	379.59	V
300	380.31	V
300	381.27	V
300	384.96	V
300	384.97	V
300	385.03	V
300	385.11	V
300	385.25	V
300	385.26	V
300	385.30	V
300	385.75	V
300	385.88	V
350	386.16	V
300	386.74	V
300	386.78	V
300	386.85	V
300	386.88	V
350	386.88	V
400	387.20	V
400	387.50	V
300	387.62	V
400	387.76	V
400	387.78	V
300	387.98	V
300	388.61	V
300	388.82	V
300	390.11	V
300	390.19	V
300	390.78	V
300	391.94	V
300	392.06	V
300	392.38	V
300	392.50	V
300	392.51	V
300	392.70	V
300	392.91	V
300	393.27	V
300	393.72	V
300	393.73	V
300	393.91	V
300	393.97	V
300	394.04	V
300	394.64	V
300	395.15	V
300	395.79	V
400	395.90	V
300	399.84	V
300	400.11	V
300	400.51	V
300	400.52	V
300	400.63	V
300	401.04	V
300	401.64	V
300	401.86	V
300	402.87	V
300	403.06	V
400	404.62	V
400	405.50	V
800	407.42	V
600	407.44	V
400	407.49	V
500	407.75	V
400	409.71	V
400	410.20	V
600	411.55	V
300	415.01	V
300	416.66	V
300	416.84	V
700	417.39	V
700	418.04	V
500	418.47	V
300	420.56	V
700	421.06	V
500	421.78	V
300	422.28	V
500	422.31	V
300	423.23	V
500	426.06	V
500	426.11	V
300	426.83	V
350	426.97	V
300	434.42	V
300	439.22	V
300	444.70	V
300	445.44	V
300	446.04	V
300	458.16	V
300	486.17	V
400	1317.86	V
300	1318.35	V
300	1320.41	V
300	1321.34	V
300	1321.49	V
400	1323.27	V
400	1330.40	V
300	1345.61	V
400	1359.01	V
300	1361.28	V
300	1361.45	V
600	1361.82	V
300	1363.08	V
300	1363.64	V
300	1365.57	V
700	1373.59	V
600	1373.67	V
300	1374.12	V
500	1376.34	V
300	1376.46	V
500	1378.56	V
300	1385.68	V
800	1387.94	V
400	1397.97	V
600	1400.24	V
800	1402.39	V
400	1406.67	V
500	1406.82	V
400	1407.25	V
300	1409.03	V
300	1409.22	V
600	1409.45	V
400	1415.20	V
300	1418.12	V
600	1420.46	V
800	1430.57	V
800	1440.53	V
300	1440.79	V
400	1442.22	V
800	1446.62	V
700	1448.85	V
400	1449.93	V
300	1455.56	V
700	1456.16	V
500	1459.83	V
400	1460.73	V
500	1462.63	V
700	1464.68	V
500	1465.38	V
400	1466.65	V
500	1469.00	V
300	1475.60	V
500	1479.47	V
300	1554.22	V

KRYTPON (Kr)
Z = 36

Kr I and II
Ref. 61, 121, 123, 147, 208, 232
— E.F.W.

Intensity		Wavelength	
		Vacuum	
60		729.40	II
200		761.18	II
100		763.98	II
60		766.20	II
200		771.03	II
60	p	773.69	II
200		782.10	II
100		783.72	II
60		818.15	II
60		830.38	II
100		844.06	II
60		864.82	II
60		868.87	II
200		884.14	II
1000		886.30	II
400		891.01	II
200		911.39	II
2000		917.43	II
50		945.44	I
50		946.54	I
20		951.06	I
50		953.40	I
50		963.37	I
2000		964.97	II
100		1001.06	I
100		1003.55	I
100		1030.02	I
200		1164.87	I
650		1235.84	I
		Air	
100	h	2464.77	II
60		2492.48	II
80	h	2712.40	II
100		2833.00	II
100	h	3607.88	II
200		3631.889	II
250		3653.928	II
80		3665.324	I
150		3669.01	II
100		3679.559	I
80		3686.182	II
300	h	3718.02	II
200		3718.595	II
150		3721.350	II
200		3741.638	II
150		3744.80	II
80		3754.245	II
500		3778.089	II
500		3783.095	II
150	h	3875.44	II
150		3906.177	II
200		3920.081	II
100		3994.840	II
100	h	3997.793	II
300		4057.037	II
300		4065.128	II
500		4088.337	II
250		4098.729	II
100		4109.248	II
250		4145.122	II
150		4250.580	II
1000		4273.969	I
100		4282.967	I
600		4292.923	II
200		4300.49	II
500	h	4317.81	II
400		4318.551	I
1000		4319.579	I
150	h	4322.98	II
100		4351.359	I
3000		4355.477	II
500		4362.641	I
200		4369.69	II
800		4376.121	I
300	h	4386.54	II
200		4399.965	I
100		4425.189	I
500		4431.685	II
600		4436.812	II
600		4453.917	I
800		4463.689	I
800		4475.014	II
400	h	4489.88	II
600		4502.353	I
400	h	4523.14	II
200	h	4556.61	II
800		4577.209	II
300		4582.978	II
150	h	4592.80	II
500		4615.292	II
1000		4619.166	II
800		4633.885	II
2000		4658.876	II
500		4680.406	II
100		4691.301	II
200		4694.360	II
3000		4739.002	II
300		4762.435	II
1000		4765.744	II
300		4811.76	II
300		4825.18	II
800		4832.077	II
700		4846.612	II
150		4857.20	II
300		4945.59	II
200		5022.40	II
250		5086.52	II
400	h	5125.73	II
500		5208.32	II
200		5308.66	II
500		5333.41	II
200		5468.17	II
500		5562.224	I
2000		5570.288	I
80		5580.386	I
100		5649.561	I
400		5681.89	II
200	h	5690.35	II
100		5832.855	I
3000		5870.914	I
200		5992.22	II
60		5993.849	I
60		6056.125	I
300		6420.18	II
100		6421.026	I
200		6456.288	I
150		6570.07	II
60		6699.228	I
100		6904.678	I
250		7213.13	II
100		7224.104	I
80		7287.258	I
400		7289.78	II
400		7407.02	II
60		7425.541	I
200		7435.78	II
100		7486.862	I
300		7524.46	II
1000		7587.411	I
2000		7601.544	I
150		7641.16	II
1000		7685.244	I
1200		7694.538	I
250		7735.69	II
150		7746.827	I
800		7854.821	I
200		7913.423	I
180		7928.597	I
200		7933.22	II
120		7973.62	II
100		7982.401	I
1500		8059.503	I
4000		8104.364	I

Krypton (Cont.)

Intensity		Wavelength	
6000		8112.899	I
60		8132.967	I
3000		8190.054	I
200		8202.72	II
80		8218.365	I
3000		8263.240	I
100		8272.353	I
5000		8298.107	I
1500		8281.050	I
100		8412.430	I
3000		8508.870	I
150		8764.110	I
6000		8776.748	I
2000		8928.692	I
500		9238.48	II
500	hL	9293.82	II
200	h	9320.99	II
300		9361.95	II
100		9362.082	I
200	h	9402.82	II
200	h	9470.93	II
500		9577.52	II
500	h	9605.80	II
400	h	9619.61	II
200		9663.34	II
200	h	9711.60	II
2000		9751.758	I
500		9803.14	II
500		9856.314	I
1000		10221.46	II
100		11187.108	I
200		11257.711	I
150		11259.126	I
500		11457.481	I
150		11792.425	I
1500		11819.377	I
600		11997.105	I
160		12077.224	I
100		12861.892	I
1100		13177.412	I
1000		13622.415	I
2400		13634.220	I
800		13658.394	I
200		13711.036	I
600		13738.851	I
150		13974.027	I
550		14045.657	I
140		14104.298	I
180		14402.22	I
2000		14426.793	I
100		14517.84	I
1600		14734.436	I
550		14762.672	I
450		14765.472	I
400		14961.894	I
120		15005.307	I
140		15209.526	I
1700		15239.615	I
130		15326.480	I
1500		15334.958	I
700		15372.037	I
200		15474.026	I
180		15681.02	I
120		15820.09	I
200		16726.513	I
2000		16785.128	I
1000		16853.488	I
2400		16890.441	I
1600		16896.753	I
1800		16935.806	I
600		17098.771	I
700		17367.606	I
120		17404.443	I
150		17616.854	I
650		17842.737	I
700		18002.229	I
2600		18167.315	I
100		18399.786	I
150		18580.896	I
300		18696.294	I
170		18785.460	I
200		18797.703	I
140		20209.878	I
300		20423.964	I
140		20446.971	I
600		21165.471	I
1800		21902.513	I
120		22485.775	I
180		23340.416	I
120		24260.506	I
180		24292.221	I
600		25233.820	I
180		28610.55	I
1000		28655.72	I
150		28769.71	I
140		28822.49	I
300		29236.69	I
300		30663.54	I
300		30979.16	I
500		39300.6	I
1100		39486.52	I
220		39557.25	I
100		39572.60	I
1400		39588.4	I
1100		39589.6	I
500		39954.8	I
300		39966.6	I
1300		40306.1	I
250		40685.16	I

Kr III
Ref. 208, 366, 390, 421 — E.F.W.

Intensity		Wavelength	
		Vacuum	
30		467.35	III
30		540.86	III
30		565.64	III
30		569.16	III
30		571.98	III
30		579.83	III
30		585.14	III
30		585.96	III
30		593.70	III
30		594.10	III
30		596.41	III
40		600.17	III
30		603.67	III
50		605.86	III
35		606.47	III
50		611.12	III
35		616.72	III
40		621.45	III
45		622.80	III
50		625.02	III
30		625.76	III
45		628.59	III
50		630.04	III
35		633.09	III
50		639.98	III
60		646.41	III
50		651.20	III
50		659.72	III
30		664.86	III
40		672.34	III
35		672.85	III
35		676.57	III
35		680.13	III
35		683.68	III
45		686.25	III
45		687.98	III
45		691.93	III
50		695.61	III
30		698.05	III
50		708.36	III
50		714.00	III
100	p	722.04	III
30		746.70	III
60		785.97	III
50		837.66	III
50		854.73	III
60		862.58	III
40		870.84	III
50		876.08	III
75		897.81	III
50		987.29	III
30		1158.74	III
6		1638.82	III
6		1914.09	III
		Air	
40		2393.94	III
40		2494.01	III
30		2563.25	III
60		2639.76	III
30		2680.32	III
40		2681.19	III
30		2841.00	III
30		2851.16	III
50		2870.61	III
100		2892.18	III
30		2909.17	III
50		2952.56	III
60		2992.22	III
50		3022.30	III
80		3024.45	III
50		3046.93	III
30		3056.72	III
60		3063.13	III
40		3097.16	III
60		3112.25	III
30		3120.61	III
100		3124.39	III
60		3141.35	III
100		3189.11	III
80		3191.21	III
40		3239.52	III
40		3240.44	III
300		3245.69	III
150		3264.81	III
100		3268.48	III
30		3271.65	III
30		3285.89	III
30		3304.75	III
50		3311.47	III
200		3325.75	III
60		3330.76	III
50		3342.48	III
100		3351.93	III
40		3374.96	III
100		3439.46	III
70		3474.65	III
100		3488.59	III
200		3507.42	III
100		3564.23	III
30		3641.34	III
30		3690.65	III
40	h	3868.70	III
50		4067.37	III
40		4131.33	III
40		4154.46	III
20	h	5016.45	III
10		5501.43	III
10	h	6037.17	III
10	h	6078.38	III
10		6310.22	III

Kr IV
Ref. 366, 409, 417 — E.F.W.

Intensity		Wavelength	
		Vacuum	
		793.44	IV
		794.11	IV
7		805.76	IV
18		816.82	IV
22		842.04	IV
		Air	
3		2237.34	IV
6		2291.26	IV
3		2329.3	IV
4		2336.75	IV
4		2348.27	IV
3		2358.5	IV
3		2388.05	IV
4		2416.9	IV
3		2428.04	IV
5		2442.68	IV
4		2451.7	IV
6		2459.74	IV
5		2474.06	IV
4		2517.0	IV
5		2518.02	IV
6		2519.38	IV
5		2524.5	IV
5		2546.0	IV
6		2547.0	IV
4		2558.08	IV
3		2586.9	IV
5		2606.17	IV
10		2609.5	IV
8		2615.3	IV
7		2621.11	IV
3		2730.55	IV
8		2748.18	IV
6		2774.70	IV
3		2829.60	IV
3		2836.08	IV
5		2853.0	IV
3		2859.3	IV
3		3142.01	IV
6		3224.99	IV
3		3261.70	IV
3		3809.30	IV
5		3860.58	IV
5		3934.29	IV

Kr V
Ref. 409, 421 — E.F.W.

Intensity		Wavelength	
		Vacuum	
150		472.16	V
100		484.39	V
250		496.25	V
120		500.77	V
200		507.20	V
60		548.04	V
120		637.87	V
		690.86	V
		691.75	V
600		708.85	V
		810.70	V

LANTHANUM (La)
Z = 57

La I and II
Ref. 1 — C.H.C.

Intensity		Wavelength	
		Air	
240		2187.87	II
770		2256.76	II
200		2319.44	II
400		2610.34	II
420		2808.39	II
130		2885.14	II
160		2893.07	II
110		2950.50	II
180		3104.59	II
130		3142.76	II
510		3245.13	II
260		3249.35	II
550		3265.67	II
800		3303.11	II
1500		3337.49	II
870		3344.56	II
200		3376.33	II
1500		3380.91	II
130		3452.18	II
180		3453.17	II
200		3574.43	I
320		3628.83	II
120		3637.15	II
170	d	3641.53	I
		3641.66	II
1000		3645.42	II
390		3650.18	II
170		3662.08	II
120		3704.54	I
320		3705.82	II
550		3713.54	II
140		3714.87	II
270		3715.53	II
2400		3759.08	II
120		3780.67	II
3700		3790.83	II
3900		3794.78	II
190		3835.08	II
600		3840.72	II
120		3846.00	II
1600		3849.02	II
130		3854.91	II
3400		3871.64	II
1700		3886.37	II
1300		3916.05	II
1100		3921.54	II
160		3927.56	I
2200		3929.22	II
180		3936.22	II
9000		3949.10	II
4400		3988.52	II
3600		3995.75	II
180		4015.39	I
250		4025.88	II
2800		4031.69	II
140		4037.21	I
3000		4042.91	II
320		4050.08	II
220		4060.33	I
160		4064.79	I
850		4067.39	II
110		4076.71	II
2800		4077.35	II
120		4079.18	I
5500		4086.72	II
180		4089.61	I
280		4099.54	II
110		4104.87	I
4400		4123.23	II
110		4137.04	I
550		4141.74	II
1100		4151.97	II
220		4152.78	II
100		4160.26	I
280		4187.32	I
280		4192.36	II
1500		4196.55	II
240		4204.04	II
300		4217.56	II
200		4230.95	II
1600		4238.38	II
140		4249.99	II
320		4263.59	II
480		4269.50	II
240		4275.64	II
300		4280.27	I
600		4286.97	II
600		4296.05	II
120		4300.44	II
440		4322.51	II
4600		4333.74	II
550		4354.40	II
110		4364.67	II
110	bl	4371.97	L
1[illegible]0	bl	4375.84	L
110		4378.10	II
280		4383.44	II
100		4385.20	II

Lanthanum (Cont.)

Intensity		Wavelength	
220	bl	4418.24	L
160	bl	4423.17	L
160		4423.90	I
260		4427.55	II
100	bl	4428.10	L
2000		4429.90	II
160	bl	4432.98	L
100	bl	4438.01	L
100		4452.15	I
100		4455.80	II
850		4522.37	II
170		4525.31	II
420		4526.12	II
400		4558.46	II
110		4559.29	II
160		4567.91	I
200		4570.02	I
400		4574.88	II
200		4580.06	II
160		4605.78	II
410		4613.39	II
410		4619.88	II
110		4645.28	II
540		4655.50	II
360		4662.51	II
230		4663.76	II
200		4668.91	II
160		4671.83	II
230		4692.50	II
140		4703.28	II
170		4716.44	II
140		4719.94	II
230		4728.42	II
500		4740.28	II
390		4743.09	II
320		4748.73	II
160		4766.89	I
160		4804.04	II
160		4809.01	II
200		4824.06	II
320		4860.91	II
850		4899.92	II
1000		4920.98	II
1000		4921.79	II
140		4934.83	II
110		4946.47	II
370		4949.77	I
340		4970.39	II
370		4986.83	II
140		4991.28	II
720		4999.47	II
140		5046.88	I
210		5050.57	I
170		5056.46	I
200		5106.23	I
470		5114.56	II
470		5122.99	II
450		5145.42	I
180		5156.74	II
180		5157.43	II
290		5158.69	I
120		5163.62	II
580		5177.31	I
850		5183.42	II
260		5188.22	II
170		5204.15	II
720		5211.86	I
520		5234.27	I
340		5253.46	I
110		5259.39	II
370		5271.19	I
140		5290.84	II
370		5301.98	II
140		5302.62	II
180		5303.55	II
110		5340.67	II
110		5357.86	I
130		5377.09	II
140		5380.99	II
500		5455.15	I
470		5501.34	I
110	bl	5602.50	L
160		5631.22	I
240		5648.25	I
130		5657.72	I
180		5740.66	I
160		5744.41	I
160		5761.84	I
160		5769.07	II
370		5769.34	I
320		5789.24	I
450		5791.34	I
220		5797.58	II
160		5805.78	II
140		5821.99	I
320		5930.62	I
720		6249.93	I
260	d	6262.30	II
180		6296.09	II
160		6320.39	II
110		6325.91	I
170		6390.48	II
450		6394.23	I
210		6410.99	I
250		6455.99	I
110		6526.99	II
130		6543.16	I
140		6578.51	I
180		6709.50	I
120		6774.26	II
13	b	7011.22	L
75		7023.67	I
26		7032.05	I
26	b	7040.84	L
110		7045.96	I
13	b	7054.80	L
160		7066.23	II
65		7068.37	I
21	bl	7070.79	L
13		7076.38	I
21	bl	7085.40	L
26	bl	7101.02	L
10	h	7116.8	II
19	bl	7131.58	L
10		7149.77	I
40	h	7158.08	I
50		7161.25	I
10		7162.60	L
21		7219.91	I
10	b	7257.16	L
26		7270.09	I
10		7270.30	I
110	cw	7282.34	II
10		7320.91	I
110	cw	7334.18	I
65		7345.34	I
50	bl	7379.71	L
85	bl	7380.08	L
35		7382.73	I
110	bl	7403.52	L
210	bl	7403.75	L
50	b	7411.34	L
65	bl	7434.28	L
110	bl	7434.36	L
30	b	7442.92	L
50	h	7463.08	I
50	bl	7465.25	L
95	bl	7465.48	L
75	cw	7483.50	II
40	bl	7496.50	L
95	bl	7496.78	L
50		7498.83	I
30	b	7506.79	L
19	bl	7528.21	L
50	bl	7528.39	L
30		7533.59	I
85		7539.23	I
35	bl	7560.09	L
35	bl	7592.26	L
19	h	7612.94	II
19	b	7624.99	L
21		7664.34	I
15	h	7841.80	I
21	b	7876.87	L
75	bl	7877.22	L
75	bl	7910.19	L
150	bl	7910.54	L
50	b	7944.61	L
110	bl	7944.95	L
40		7964.83	I
35	b	7979.34	L
75	bl	7979.70	L
35	h	8001.89	I
21	b	8014.43	L
65	bl	8014.79	L
30	b	8019.48	L
35	hc	8051.39	I
75		8086.05	I
15	b	8122.20	L
15	b	8159.02	L
7	h	8203.38	I
50		8247.44	I
13	h	8316.04	I
85		8324.69	I
95		8346.53	I
8	h	8379.80	I
8	b	8453.55	L
8	h	8467.62	I
26		8476.48	I
13	h	8507.37	I
13	h	8513.57	I
8	h	8514.65	II
17	b	8526.59	L
17	c	8543.46	I
65		8545.44	I
15	b	8563.54	L
9	h	8590.94	I
9	b	8600.81	L
7	h	8624.22	I
15		8638.47	I
19	hw	8672.11	I
40		8674.43	I
13	h	8720.41	I
35		8748.38	I
19		8818.93	I
35		8825.82	I
21		8839.63	I

La III
Ref. 220, 309 — J.R.

Intensity		Wavelength	
		Vacuum	
3		744.19	III
10		753.03	III
1		786.64	III
200		787.14	III
1		796.03	III
400		796.99	III
1		797.20	III
10		835.03	III
30		845.62	III
1		850.73	III
1		860.39	III
2		860.88	III
5		865.04	III
2000		870.40	III
30		872.43	III
1000		882.34	III
20		882.72	III
200		929.71	III
400		942.86	III
30		967.69	III
10		974.33	III
50		979.99	III
10		980.29	III
200		1058.63	III
1000		1072.59	III
5000		1076.91	III
50000		1081.61	III
95000		1099.73	III
5000		1100.70	III
30		1208.80	III
30		1212.29	III
200		1236.54	III
100		1253.99	III
2000		1255.63	III
100		1259.55	III
100		1322.42	III
5000		1330.04	III
10000		1349.18	III
5000		1459.49	III
2000		1466.44	III
10000		1523.79	III
500		1528.55	III
5000		1536.17	III
200		1923.34	III
500		1938.57	III
		Air	
60		2216.07	III
20		2238.36	III
25		2258.61	III
5		2260.30	III
250		2297.74	III
400		2379.37	III
10		2387.99	III
20		2392.49	III
100		2476.60	III
50		2478.65	III
2		2513.43	III
4		2588.87	III
2		2604.83	III
400		2651.50	III
100		2682.34	III
150		2684.76	III
110		2897.88	III
160		2904.58	III
7		2950.84	III
10		2953.77	III
40		2992.10	III
4		3006.19	III
15		3009.22	III
100		3075.17	III
4		3085.38	III
25		3093.03	III
15		3096.26	III
200		3111.97	III
50		3116.74	III
1000		3171.63	III
1500		3171.74	III
50		3172.69	III
20		3196.84	III
70		3289.11	III
15		3301.48	III
35		3327.66	III
500		3517.09	III
600		3517.22	III
3		4129.24	III
5		4137.43	III
200		4482.97	III
300		4499.05	III
5		5145.73	III
8		5158.41	III
6		5467.81	III
55		5491.90	III
2		5511.72	III
1		5518.19	III
45		5529.54	III
1		5744.09	III
200		5778.14	III
2		5813.45	III
3		5875.63	III
55		5888.62	III
3		5932.71	III
2		6017.11	III
20		6055.84	III
35		6119.25	III
120		6141.99	III
55		6220.00	III
60		6348.21	III
3		8114.42	III
2		8135.96	III
250		8252.60	III
100		8275.39	III
200		8287.75	III
250		8321.11	III
300		8583.45	III
120		9184.38	III
100		9212.63	III
80		9923.99	III
140		10284.79	III
20		10370.34	III
12		10937.90	III
		13894.47	III
		14096.18	III
		17898.09	III

La IV
Ref. 79 — J.R.

Intensity		Wavelength	
		Vacuum	
100		344.12	IV
7000		453.50	IV
10000		463.14	IV
15000		499.54	IV
40000		552.02	IV
30000		631.26	IV
25		724.92	IV
15		733.29	IV
10		797.03	IV
10	c	980.03	IV
50		1039.30	IV
60	p	1062.09	IV
75		1158.35	IV
50		1164.29	IV
400		1230.90	IV
75	p	1260.79	IV
300		1261.12	IV
150		1283.19	IV
2000		1302.31	IV
1200		1333.53	IV
3500		1334.96	IV
1000		1352.76	IV
25000		1368.04	IV
8000		1377.49	IV
3000		1394.32	IV
5000		1414.58	IV
7000		1432.55	IV
7000		1441.63	IV
7500		1462.15	IV
20000		1463.47	IV
7500		1467.54	IV
15000		1507.87	IV
5000		1527.19	IV
2500		1575.92	IV
1500		1583.61	IV
750		1585.11	IV
750		1637.42	IV
750	d	1645.21	IV
1000		1664.84	IV
750		1684.17	IV
2000	p	1767.65	IV
4000		1808.66	IV
1000		1851.81	IV
1500		1852.77	IV
750		1879.79	IV
1000		1881.57	IV
800		1889.22	IV
1000		1891.47	IV
5000		1902.97	IV
800		1907.44	IV
1500	c	1950.80	IV
1200	c	1957.57	IV
		Air	
3000		2012.42	IV
750		2037.43	IV
2000	c	2066.50	IV
3000	c	2073.18	IV
1500		2143.23	IV
4000	c	2197.45	IV
1000	w	2221.12	IV
900		2227.34	IV
3000		2244.95	IV
7500	w	2265.91	IV
2000		2315.89	IV
750		2348.36	IV
750		2355.31	IV
2000	c	2407.10	IV

Lanthanum (Cont.)

Intensity		Wavelength	
25000	w	2417.58	IV
1200	p	2443.92	IV
18000	c	2502.81	IV
15000		2515.02	IV
50000		2532.75	IV
900	d	2535.76	IV
45000		2582.05	IV
18000	c	2591.30	IV
95000	w	2597.50	IV
5000	c	2608.01	IV
70000	w	2662.75	IV
50000	w	2848.30	IV
12000	c	2863.30	IV
30000	c	2962.58	IV
70000	w	3009.51	IV
90000	c	3056.68	IV
3500		3522.28	IV
2000		3650.40	IV
2000	p	4270.76	IV
1500	w	4549.80	IV
500		4836.89	IV

La V
Ref. 78 — J.R.

Intensity	Wavelength	
	Vacuum	
2	389.03	V
400	390.72	V
1	398.53	V
30	399.34	V
350	405.10	V
50	416.13	V
3	421.55	V
50	423.07	V
400	424.78	V
1000	432.11	V
2500	435.28	V
700	436.14	V
700	436.84	V
300	437.11	V
700	437.55	V
20	444.01	V
10	444.07	V
1250	450.40	V
600	457.30	V
1000	463.85	V
150	476.67	V
5000	482.16	V
200	482.43	V
2000	483.30	V
7000	498.08	V
4000	499.03	V
10000	503.58	V
40	508.15	V
1500	525.71	V
12000	526.76	V
10000	531.07	V
15000	533.23	V
4000	540.20	V
6000	544.80	V
8000	547.44	V
3000	570.90	V
2500	593.18	V
750	597.70	V
2000	600.01	V
5000	600.24	V
700	611.70	V
500	617.60	V

LEAD (Pb)
Z = 82

Pb I and II
Ref. 64, 274, 283, 329, 330 — D.R.W.

Intensity		Wavelength	
		Vacuum	
2		846.04	II
2	h	849.88	II
3		855.57	II
3		863.00	II
6		873.71	II
2		877.96	II
8		889.68	II
3		896.30	II
5		926.44	II
2		958.76	II
2		960.21	II
3		965.36	II
10		967.23	II
9		972.56	II
8		982.17	II
10		986.71	II
10		995.89	II
6		1001.81	II
10		1016.61	II
10		1049.82	II
10		1050.77	II
10		1060.66	II
9		1065.58	II
10		1103.94	II
10		1108.43	II
10		1109.84	II
10		1119.57	II
10		1121.36	II
10		1133.14	II
4		1145.91	II
10		1203.63	II
10		1231.20	II
10		1331.65	II
10		1335.20	II
10		1348.37	II
10		1433.96	II
3		1449.35	II
10		1512.42	II
10		1671.53	II
10		1682.15	II
20		1726.75	II
2		1740.00	I
2		1766.64	I
2		1794.67	I
10		1796.670	II
5		1812.97	I
10		1822.050	II
4		1868.76	I
10		1904.77	I
7		1921.471	II
4		1972.44	I
2		1977.88	I
2		1991.60	I
2		1992.31	I
		Air	
5	r	2022.02	I
5		2050.88	I
8	r	2053.28	I
6		2111.758	I
10		2115.066	I
500	r	2170.00	I
7		2175.580	I
7		2187.888	I
8		2189.603	I
10		2203.534	II
20		2237.425	I
20		2246.86	I
25		2246.89	I
150		2332.418	I
180		2388.797	I
550	r	2393.792	I
140		2399.597	I
320	r	2401.940	I
320	r	2411.734	I
150	r	2443.829	I
160	r	2446.181	I
130	r	2476.378	I
8	c	2526.69	II
8	c	2576.60	II
80	r	2577.260	I
2	c	2608.38	II
500	r	2613.655	I
900	r	2614.175	I
160		2628.262	I
4		2634.256	II
10		2657.094	I
700		2663.154	I
10		2697.541	I
25000	r	2801.995	I
100		2822.58	I
14000	r	2823.189	I
35000	r	2833.053	I
6		2840.557	II
14000	r	2873.311	I
3	c	2887.30	II
3		2914.442	II
2	c	2947.43	II
3	c	2948.53	II
15		2966.460	I
15		2972.991	I
15		2980.157	I
4		2986.876	II
10	c	3016.39	II
150		3118.894	I
600		3220.528	I
100		3229.613	I
400		3240.186	I
200		3262.355	I
35000		3572.729	I
50000	r	3639.568	I
20000		3671.491	I
70000	r	3683.462	I
10		3713.982	II
25000		3739.935	I
15000		4019.632	I
95000		4057.807	I
14000		4062.136	I
5		4110.76	II
4		4113.35	II
10		4152.82	II
10		4157.814	I
10000		4168.033	I
9	c	4242.14	II
20	c	4244.92	II
7		4293.82	II
6		4296.65	II
200		4340.413	I
10		4352.74	II
20	c	4386.46	II
10		4579.051	II
10		4582.27	II
1000		5005.416	I
100		5006.572	I
50		5042.58	II
10		5070.58	II
10		5074.53	II
10		5076.35	I
50		5089.484	I
20		5090.01	I
10		5107.242	I
10		5111.64	II
2000		5201.437	I
10		5367.64	II
10		5372.099	II
10	c	5544.25	II
20	c	5608.85	II
40		5692.346	I
200		5895.624	I
2000		6001.862	I
9	c	6009.58	II
500		6011.667	I
8	c	6041.17	II
500		6059.356	I
40		6075.74	II
40		6081.409	II
50		6110.520	I
10		6159.89	II
100		6235.266	I
50	c	6660.20	II
10		6892.11	I
5		7128.94	I
20		7193.60	II
20000		7228.965	I
5		7304.68	I
8		7330.15	I
10		7346.676	I
10		7558.97	II
10		7632.56	II
4		7732.96	II
20		7809.259	I
5		7817.97	I
6		7829.01	I
5		7896.737	I
2	d	8156.91	II
10		8168.001	I
6		8191.886	I
5		8217.711	I
8		8255.61	I
40		8272.690	I
6		8335.54	II
10		8395.68	II
20		8409.384	I
10		8478.492	I
8		8532.17	I
7	c	8544.95	II
7		8709.90	II
5		8719.39	II
5		8722.810	I
10		8857.457	I
10		9050.82	II
10		9063.43	II
2	d	9245.28	II
8		9293.476	I
5		9384.35	I
5		9385.89	I
15		9438.05	I
15		9604.297	I
6		9608.73	I
15		9674.351	I
200		10290.458	I
5		10434.32	I
100		10498.965	I
50		10649.249	I
5		10759.41	I
7		10759.74	I
15		10886.688	I
40		10969.53	I
6		11059.22	I
3		11333.08	I
2	d	11479.49	II
2		11488.76	I
5		11627.91	II
1		12561.37	I
		13495.3	I
		13498.2	I
		13512.6	I
		14722.8	I
		14742.1	I
		14743.0	I
		15314.8	I
		15327.6	I
		15331.0	I
		15349.6	I
		38831.1	I
		38950.1	I
		38958.6	I
		39039.4	I

Pb III
Ref. 54, 256, 297 — D.R.W.

Intensity	Wavelength	
	Vacuum	
1	961.01	III
3	1030.5	III
12	1048.9	III
4	1069.2	III
3	1074.7	III
4	1118.67	III
4	1167.0	III
4	1250.6	III
1	1266.9	III
20	1553.1	III
1	1610.1	III
4	1711.23	III
	Air	
10	3043.85	III
4	3089.08	III
4	3102.74	III
10	3137.81	III
10	3176.50	III
5	3242.84	III
1	3530.17	III
7	3589.87	III
7	3689.31	III
3	3706.02	III
5	3728.69	III
12	3854.08	III
8	3951.92	III
3	4031.16	III
3	4094.54	III
2	4128.11	III
8	4272.66	III
6	4499.34	III
7	4571.21	III
1	4596.45	III
6	4761.12	III
4	4798.59	III
1	4826.86	III
2	4855.06	III
3	5065.12	III
4	5191.56	III
5	5523.97	III
3	5779.41	III
6	5857.96	III

Pb IV
Ref. 106 — D.R.W.

Intensity	Wavelength	
	Vacuum	
8	475.36	IV
7	478.35	IV
10	496.38	IV
12	499.94	IV
9	515.07	IV
14	529.78	IV
20	570.16	IV
8	573.90	IV
8	584.52	IV
10	648.50	IV
9	656.10	IV
10	761.09	IV
18	802.07	IV
12	802.82	IV
10	812.59	IV
8	822.07	IV
10	827.41	IV
12	832.60	IV
8	840.99	IV
8	842.88	IV
12	845.94	IV
18	857.64	IV
8	859.02	IV
16	862.33	IV
14	870.44	IV
12	879.96	IV
7	880.35	IV
14	884.96	IV
14	884.99	IV
16	890.72	IV
12	908.51	IV
12	917.90	IV
10	922.12	IV
12	922.49	IV
7	924.52	IV
10	927.64	IV
14	932.20	IV
8	937.00	IV
7	952.85	IV

Lead (Cont.)

Intensity	Wavelength	
8	1012.44	IV
14	1028.61	IV
20	1032.05	IV
16	1041.24	IV
18	1044.14	IV
15	1056.53	IV
12	1072.09	IV
7	1079.88	IV
18	1080.81	IV
20	1084.17	IV
6	1089.94	IV
7	1099.47	IV
6	1115.30	IV
20	1116.08	IV
18	1137.84	IV
8	1142.77	IV
14	1144.93	IV
20	1189.95	IV
8	1267.55	IV
8	1290.82	IV
10	1291.10	IV
20	1313.05	IV
8	1323.92	IV
12	1343.06	IV
16	1388.94	IV
6	1397.02	IV
18	1400.26	IV
10	1404.34	IV
7	1510.76	IV
14	1535.71	IV
8	1798.39	IV
8	1893.19	IV
12	1959.34	IV
16	1973.16	IV
	Air	
10	2042.58	IV
12	2049.34	IV
12	2079.22	IV
8	2151.96	IV
15	2154.01	IV
12	2177.46	IV
16	2359.53	IV
4	2864.24	IV
4	2864.50	IV
16	2417.61	IV
4	2978.14	IV
4	3052.56	IV
4	3221.17	IV
4	3962.48	IV
4	4049.80	IV
10	4496.15	IV
16	4534.60	IV
8	4605.40	IV
2	5914.54	IV

Pb V
Ref. 106 — D.R.W.

Intensity	Wavelength	
	Vacuum	
2	367.40	V
2	372.53	V
2	387.87	V
2	394.38	V
5	424.64	V
3	431.03	V
3	436.60	V
2	438.47	V
6	438.91	V
4	453.45	V
3	461.70	V
3	496.20	V
4	694.42	V
8	696.20	V
20	703.73	V
4	706.29	V
6	707.66	V
5	730.85	V
12	749.46	V
10	752.52	V
10	755.80	V
6	762.76	V
10	765.87	V
18	767.45	V
18	769.49	V
14	771.42	V
14	782.79	V
10	787.05	V
15	797.02	V
5	799.80	V
18	809.63	V
5	812.32	V
8	814.10	V
8	820.09	V
5	825.52	V
8	829.32	V
5	851.98	V
20	863.97	V
10	867.10	V
6	880.50	V
18	883.90	V
14	888.37	V
14	894.40	V
12	896.08	V
8	915.09	V
14	915.71	V
12	918.09	V
12	920.28	V
12	920.66	V
6	940.74	V
8	946.20	V
6	950.93	V
12	954.35	V
4	954.95	V
10	955.28	V
4	964.38	V
6	989.14	V
8	1005.42	V
10	1051.26	V
4	1059.26	V
10	1088.86	V
9	1096.52	V
6	1104.79	V
4	1121.33	V
10	1137.50	V
4	1152.36	V
12	1157.88	V
14	1185.43	V
11	1233.50	V
10	1248.47	V
8	1635.75	V
2	1802.87	V
2	1843.00	V
2	1888.67	V
2	1897.02	V
2	1914.33	V
4	1919.74	V
5	1957.96	V
2	1998.58	V
10	1998.83	V
	Air	
8	2078.45	V
10	2142.55	V
10	2167.97	V
20	2259.01	V
10	2276.66	V
8	2301.49	V
15	2424.81	V
4	4809.36	V
5	6650.99	V
4	6753.20	V

LITHIUM (Li)
Z = 3

Li I and II
Ref. 3, 15, 17, 18, 37, 44, 112, 284, 321, 335 — J.D.S.

Intensity		Wavelength	
		Vacuum	
		125.5	II
		136.5	II
		140.5	II
		167.21	II
		168.74	II
		171.58	II
		178.02	II
		199.28	II
		207.5	II
		456.	II
		483.	II
		540.	II
		729.	II
		800.	II
		820.	II
		861.	II
		905.5	II
		917.5	II
		936.	II
		945.	II
		965.	II
		972.	II
		988.	II
		1018.	II
		1032.	II
		1036.	II
		1093.	II
		1103.	II
		1109.	II
		1116.	II
		1132.1	II
		1141.	II
		1166.4	II
		1198.09	II
		1215.	II
		1238.	II
		1253.8	II
		1420.89	II
		1424.	II
3		1492.93	II
5		1492.97	II
1		1493.04	II
		1555.	II
3		1653.08	II
5		1653.13	II
1		1653.21	II
		1681.66	II
		1755.33	II
		Air	
		2009.	II
		2039.	I
		2068.	II
		2131.	II
		2164.	II
		2173.4	I
		2183.	II
		2214.	II
		2222.	II
		2237.	II
	H	2249.21	II
		2286.82	II
		2302.57	II
		2303.33	II
		2304.59	I
		2304.92	I
		2305.36	I
		2305.83	I
		2306.29	I
		2306.82	I
		2307.44	I
		2308.97	I
		2309.88	I
		2310.94	I
		2312.11	I
		2313.49	I
		2315.08	I
		2316.95	I
		2319.18	I
		2321.88	I
		2325.11	I
		2329.02	I
		2329.84	II
		2333.94	I
3		2336.88	II
5		2336.91	II
2		2337.00	II
		2340.15	I
		2348.22	I
		2358.93	I
		2373.54	I
		2381.54	II
		2383.20	II
1		2394.39	I
		2402.33	II
		2410.84	II
3		2425.43	I
		2429.81	II
		2460.2	I
10		2475.06	I
		2506.94	II
		2508.78	II
		2518.	I
		2539.49	II
24		2551.7	II
		2559	II
15		2562.31	I
		2605.08	II
		2640.	
2		2657.29	II
3		2657.30	II
		2674.46	II
0		2728.24	II
5		2728.29	II
2		2728.32	II
3		2730.47	II
1		2730.55	II
5		2741.20	I
		2766.99	II
		2790.31	II
		2801.	I
		2846.	I
		2868.	I
		2895.	I
2		2934.02	II
2		2934.07	II
5		2934.12	II
1		2934.25	II
		2968.	I
3		3029.12	II
3		3029.14	II
		3144.	I
3		3155.31	II
4		3155.33	II
1		3196.26	II
9		3196.33	II
4		3196.36	II
5		3199.33	II
2		3199.43	II
17		3232.66	I
		3249.87	II
		3306.28	II
		3393.	
		3488.	I
		3579.8	I
		3618.	I
		3662.	I
		3684.32	II
1		3714.00	II
5		3714.16	II
6	d	3714.27	II
8		3714.29	II
7	d	3714.40	II
10		3714.41	II
1		3714.51	II
0		3714.58	II
3		3718.7	I
6		3794.72	I
20		3915.30	I
20		3915.35	I
10		3985.48	I
10		3985.54	I
40		4132.56	I
40		4132.62	I
		4196.	I
20		4273.07	I
20		4273.13	I
5		4325.42	II
5		4325.47	II
1		4325.54	II
		4516.45	II
		4590.	
13		4602.83	I
13		4602.89	I
		4607.34	II
0		4671.51	II
6		4671.65	II
2		4671.70	II
3		4678.06	II
1		4678.29	II
		4760.	I
		4763.	II
		4788.36	II
		4843.0	II
4		4881.32	II
4		4881.39	II
1		4881.49	II
8		4971.66	I
8		4971.75	I
		5037.92	II
		5095.	
		5114.	
		5190.	
		5271.	I
		5315.	I
		5395.	I
		5440.	I
600	c	5483.55	II
600	c	5485.65	II
320		6103.54	I
320		6103.65	I
3600		6707.76	I
3600		6707.91	I
48		8126.23	I
48		8126.45	I
		8517.37	II
		9581.42	II
		10120.	II
		12232.	I
		12782.	I
		13566.	I
		17552.	I
		18697.	I
		19290.	I
		24467.	I
		40475.	I

Li III
Ref. 335 — J.O.S.

Intensity	Wavelength	
	Vacuum	
	102.9	III
	103.4	III
	104.1	III
	105.5	III
	108.0	III
	113.9	III
	135.0	III
	540.0	III
	729.1	III

LUTETIUM (Lu)
Z = 71

Lu I and II
Ref. 1 — C.H.C.

Intensity	Wavelength
	Air

Lutetium (Cont.)

Intensity		Wavelength	
1700	h	2195.54	II
95		2276.94	II
190		2297.41	II
1300		2392.19	II
120		2399.14	II
80		2419.21	II
55		2430.26	II
130		2459.64	II
80		2469.27	II
21	h	2481.72	II
370		2536.95	II
40		2546.87	II
20		2549.44	I
20		2549.72	I
35		2561.80	II
930		2571.23	II
1700		2578.79	II
80	h	2582.13	II
1800		2613.40	II
18000		2615.42	II
1800		2619.26	II
90		2657.05	
2700		2657.80	II
90	h	2677.25	I
570	h	2685.08	I
90	h	2685.54	I
4200		2701.71	II
90	h	2715.91	I
180	d	2719.09	I
480	h	2728.95	I
75	c	2738.17	II
3600		2754.17	II
750	h	2765.74	I
2700		2796.63	II
35		2821.23	II
270	c	2834.35	II
330	h	2845.13	I
3000		2847.51	II
570	h	2885.14	I
6300		2894.84	II
4500		2900.30	II
300		2903.05	I
9000		2911.39	II
270	h	2949.73	I
1200		2951.69	II
60		2955.78	II
4200		2963.32	II
2400		2969.82	II
1800		2989.27	I
3000		3020.54	II
120		3027.29	II
2100		3056.72	II
7500		3077.60	II
390		3080.11	I
5100	h	3081.47	I
3000		3118.43	I
2400		3171.36	I
100		3183.73	II
260		3191.80	II
1400		3198.12	II
4800		3254.31	II
3800		3278.97	I
7600		3281.74	I
6200		3312.11	I
7600		3359.56	I
6200		3376.50	I
950		3385.50	I
160	h	3391.55	I
1400		3396.82	I
4100		3397.07	II
4800		3472.48	II
8300	c	3507.39	II
1600		3508.42	I
4800		3554.43	II
4800		3567.84	I
340		3596.34	I
800		3623.99	II
680		3636.25	I
2600		3647.77	I
60		3684.32	I
60		3710.95	I
110		3756.70	I
110		3756.79	I
30	h	3786.18	I
150		3800.67	I
75	h	3829.07	I
2700		3841.18	I
75		3843.61	I
95		3853.29	I
40		3874.61	I
530		3876.65	II
29		3911.77	I
50		3918.86	I
35	h	3926.62	I
480		3968.46	I
50		3991.38	I
670		4054.45	I
75	Bl	4096.13	L
35	h	4107.44	I
95	h	4112.67	I
310		4122.49	I
3100		4124.73	I
150	c	4131.79	I
460		4154.08	I
24		4158.98	I
1600		4184.25	II
150		4277.50	I
250		4281.03	I
330	d	4295.97	I
		4296.09	I
150		4309.57	I
75		4332.72	I
29		4341.98	II
65	h	4420.96	I
190	c	4430.48	I
35		4438.79	I
190		4450.81	I
50	h	4471.55	I
60	h	4498.85	I
3300		4518.57	I
24	b	4560.95	L
24	b	4575.31	L
85	c	4605.39	I
95	h	4645.47	I
100	h	4648.21	I
95	h	4648.85	I
65	b	4654.03	L
1000		4658.02	I
85	h	4659.03	I
630	Bl	4661.75	L
310	Bl	4672.31	L
420	Bl	4684.16	L
270	Bl	4695.46	L
190	Bl	4708.00	L
30		4716.70	I
65	b	4720.86	L
65	h	4726.20	I
100	Bl	4735.00	L
75	Bl	4749.11	L
40	Bl	4764.22	L
150		4785.42	II
85		4815.05	I
50	c	4839.62	II
18		4865.36	II
460		4904.88	I
180		4942.34	I
800		4994.13	II
800		5001.14	I
55	h	5057.60	I
140		5134.05	I
2700		5135.09	I
130	Bl	5170.11	L
170		5196.61	I
90		5206.47	I
40		5304.40	I
80		5349.12	I
500		5402.57	I
140	c	5421.90	I
100		5437.88	I
35		5453.57	I
2100		5476.69	II
9		5664.89	II
14		5713.49	II
550		5736.55	I
55		5775.40	I
80		5800.59	I
40	h	5860.79	I
9		5866.30	I
690	Cw	5983.9	II
140		5997.13	I
1400		6004.52	I
35	h	6041.66	I
440		6055.03	I
11		6140.71	I
150		6159.94	II
160		6199.66	II
2100		6221.87	II
35		6228.14	II
80		6235.36	II
160		6242.34	II
16	h	6248.80	I
70	h	6345.35	I
18	h	6354.85	I
9		6365.79	I
16		6366.00	I
22		6441.14	I
11		6444.89	II
1100		6463.12	II
29		6477.67	I
55	c	6523.18	I
35	Cw	6611.28	II
		6611.58	LU II
		6611.80	LU II
		6611.95	LU II
		6612.04	LU II
11		6619.15	II
23	c	6677.14	I
9	h	6735.76	I
30	c	6793.77	I
11		6826.59	II
45		6917.31	I
8		6943.96	II
23		7031.24	I
14	c	7096.34	I
45		7125.84	II
9		7142.79	I
7		7143.10	I
8		7165.94	II
14	Ch	7237.98	I
5		7409.70	II
11	c	7441.52	I
8		7456.96	II
7	Ch	7640.08	I
7	Cw	7758.30	I
7	h	7815.9	I
9	c	8178.16	I
17		8382.08	I
35		8459.19	II
10	d	8478.50	I
29	c	8508.08	I
35	c	8610.98	I

Lu III
Ref. 148 — J.R.

Intensity		Wavelength	
		Vacuum	
1		677.34	III
7		691.05	III
30		700.25	III
50		714.89	III
100		738.76	III
200		755.03	III
3		755.16	III
500		810.73	III
100		830.53	III
2000		832.28	III
10		972.66	III
2		991.26	III
100		996.44	III
400		1001.18	III
1		1022.40	III
100		1029.83	III
200		1030.33	III
100		1031.54	III
50		1056.53	III
20		1061.99	III
3		1092.84	III
200		1187.34	III
50		1228.7	III
200		1277.53	III
30		1283.41	III
100		1331.93	III
1000		1854.57	III
		Air	
40		2050.72	III
1500		2065.35	III
1500	c	2070.56	III
100		2083.34	III
200		2099.44	III
1000		2236.14	III
2000		2236.22	III
500	c	2381.59	III
300		2563.51	III
4500	c	2603.35	III
200		2721.65	III
2000		2772.55	III
20		2781.16	III
10		2788.37	III
500		2800.90	III
20	p	2993.21	III
1000		3057.86	III
200		4251.44	III
300		4271.91	III
200		4490.00	III
400		4956.43	III
10		5046.12	III
150	c	5145.86	III
70		5419.42	III
60		5519.88	III
5		5526.80	III
5		5748.71	III
70		5786.46	III
80		5869.71	III
60		5889.76	III
300		6197.96	III
600		6198.13	III
100		7309.95	III
200		7310.25	III
50		7534.27	III
70	l	7936.45	III
3		8008.59	III

Lu IV
Ref. 310 — J.R.

Intensity		Wavelength	
		Vacuum	
400		876.80	IV
100		902.06	IV
300		1015.18	IV
20		1136.17	IV
50		1189.27	IV
20	p	1194.59	IV
60		1213.08	IV
15		1220.74	IV
20		1223.75	IV
20		1240.07	IV
20		1248.10	IV
40		1266.27	IV
100		1272.42	IV
20		1273.02	IV
40		1274.77	IV
20		1276.54	IV
50		1289.38	IV
60		1310.08	IV
50		1323.02	IV
15		1331.04	IV
800		1333.79	IV
300		1334.94	IV
50		1338.20	IV
100		1339.49	IV
300		1342.58	IV
200		1351.68	IV
300		1353.74	IV
200		1355.85	IV
20		1359.67	IV
100		1363.24	IV
50		1363.37	IV
20		1367.34	IV
20		1373.54	IV
15		1375.36	IV
100		1376.02	IV
30		1379.56	IV
15		1383.18	IV
15		1389.85	IV
60		1390.07	IV
200		1390.30	IV
50		1390.69	IV
40		1392.38	IV
40		1397.18	IV
30		1401.32	IV
100		1401.46	IV
200		1406.64	IV
100		1407.00	IV
250		1407.04	IV
20		1420.32	IV
100		1421.59	IV
200		1429.08	IV
400		1429.38	IV
40		1430.80	IV
100		1440.62	IV
100		1448.14	IV
200		1452.33	IV
30		1462.65	IV
100		1483.79	IV
200		1493.24	IV
400		1511.26	IV
200		1521.06	IV
100		1522.21	IV
100		1537.77	IV
30		1549.35	IV
20		1551.59	IV
20		1562.06	IV
30		1592.55	IV
15		1594.92	IV
100	c	1607.72	IV
50	c	1631.65	IV
20		1684.50	IV
60	c	1693.67	IV
400	c	1721.42	IV
100		1725.14	IV
100	c	1735.79	IV
100		1736.78	IV
100		1741.74	IV
50	c	1743.84	IV
40	c	1752.60	IV
200		1759.61	IV
300		1772.08	IV
600		1772.57	IV
200		1782.45	IV
100		1797.52	IV
20	c	1901.63	IV
100		1983.92	IV
20		1990.52	IV
40		1996.18	IV
		Air	
100		2003.18	IV
100		2020.94	IV
20		2071.10	IV
100		2081.09	IV
400	c	2085.70	IV
600	c	2086.47	IV
400	c	2092.16	IV
100		2103.63	IV
1000	c	2104.41	IV
200	c	2107.85	IV
1000	c	2108.31	IV
100	c	2127.43	IV

Lu V
Ref. 401 — J.R.

Lutetium (Cont.)

Intensity		Wavelength	
		Vacuum	
40		555.44	V
100		563.72	V
50		601.54	V
60		614.23	V
40		628.79	V
50	p	637.44	V
40		637.53	V
50		663.29	V
60		850.06	V
100		861.92	V
70		866.93	V
40		870.84	V
50		875.89	V
50		876.45	V
100		880.32	V
40		884.21	V
70		886.16	V
50		886.32	V
50		886.44	V
100		891.81	V
60		895.01	V
40		895.15	V
40		898.42	V
100		914.72	V
40		918.26	V
50		920.92	V
40		921.32	V
60		921.90	V
40		922.73	V
80		925.79	V
50		927.22	V
40		947.80	V
50	w	1420.02	V
50		1432.50	V
100		1432.77	V
200		1441.76	V
40		1443.64	V
100		1448.14	V
40		1449.32	V
100		1450.36	V
100		1450.69	V
100		1452.64	V
200		1453.35	V
100		1454.38	V
100		1455.21	V
100		1460.11	V
40	c	1467.81	V
200		1468.99	V
50		1469.45	V
100		1471.20	V
400		1472.12	V
200		1473.71	V
40		1475.77	V
200		1485.58	V
50		1709.02	V
40		1728.90	V
50	c	1775.92	V
40	c	1777.68	V
40	c	1784.71	V
100	c	1786.25	V
60	c	1787.58	V
60	c	1793.85	V
60		1809.73	V
60		1814.24	V

MAGNESIUM (Mg)
Z = 12

Mg I and II
Ref. 49, 83, 103, 217, 269, 315, 335 — J.O.S.

Intensity	Wavelength	
	Vacuum	
	184.05	II
	184.31	II
	184.68	II
	184.81	II
	185.26	II
	185.59	II
	185.98	II
	186.47	II
	186.84	II
	187.19	II
	187.38	II
	188.54	II
	188.91	II
	189.01	II
	189.23	II
	189.37	II
	191.30	II
	191.56	II
	191.65	II
	192.40	II
	192.55	II
	192.84	II
	193.09	II
	193.31	II
	193.40	II
	193.64	II
	197.76	II
	199.31	II
	200.29	I
	202.00	II
	202.27	II
	202.51	II
	202.94	II
	203.15	I
	203.42	II
	203.53	II
	204.22	I
	209.09	II
	209.43	II
	209.84	I
	213.53	I
	215.12	I
	215.31	I
	215.45	I
	215.66	I
	215.79	I
	216.22	I
	216.36	I
	216.68	I
	217.21	I
	217.37	I
	218.19	I
	218.34	I
	218.42	I
	218.74	I
	219.04	I
	219.28	I
	220.03	I
	220.33	I
	222.03	I
	222.67	I
	223.45	I
	223.74	I
	225.18	I
	225.54	I
	226.26	I
	247.14	II
	248.47	II
	884.70	II
	884.72	II
	907.38	II
	907.41	II
8	946.70	II
9	946.77	II
14	1025.96	II
12	1026.11	II
25	1239.94	II
20	1240.40	II
6	1248.51	II
8	1249.93	II
8	1271.24	II
9	1271.94	II
8	1272.72	II
11	1273.43	II
11	1306.71	II
12	1307.88	II
12	1308.28	II
14	1309.44	II
14	1365.45	II
	1365.54	II
15	1367.26	II
15	1367.70	II
18	1369.42	II
20	1476.00	II
25	1478.01	II
20	1480.89	II
30	1482.90	II
	1625.22	I
	1625.50	I
	1625.81	I
	1626.16	I
	1626.36	I
	1626.56	I
	1626.79	I
	1627.02	I
	1627.27	I
	1627.53	I
	1627.82	I
	1628.12	I
	1628.46	I
	1628.80	I
	1629.21	I
	1629.59	I
	1630.52	I
	1631.62	I
	1632.93	I
	1634.52	I
	1636.48	I
	1638.90	I
	1641.97	I
1	1645.93	I
1	1651.16	I
2	1658.31	I
5	1668.43	I
10	1683.41	I
15	1707.06	I
40	1734.84	II
50	1737.62	II
20	1747.80	I
40	1750.65	II
50	1753.46	II
30	1827.93	I
	Air	
9	2025.82	I
3	2329.58	II
6	2449.57	II
1	2557.23	I
1	2560.94	I
1	2562.26	I
1	2564.94	I
1	2570.91	I
1	2572.25	I
2	2574.94	I
1	2577.89	I
1	2580.59	I
1	2584.22	I
2	2585.56	I
3	2588.28	I
1	2591.89	I
1	2593.23	I
2	2595.97	I
2	2602.50	I
4	2603.85	I
5	2606.62	I
1	2613.36	I
2	2614.73	I
3	2617.51	I
3	2628.66	I
6	2630.05	I
8	2632.87	I
2	2644.80	I
3	2646.21	I
4	2649.06	I
8	2660.76	II
8	2660.82	II
6	2668.12	I
8	2669.55	I
10	2672.46	I
3	2693.72	I
5	2695.18	I
6	2698.14	I
8	2731.99	I
10	2733.49	I
12	2736.53	I
5	2765.22	I
7	2768.34	I
38	2776.69	I
32	2778.27	I
90	2779.83	I
8	2781.29	I
32	2781.42	I
36	2782.97	I
13	2790.79	II
1000	2795.53	II
16	2798.06	II
600	2802.70	II
12	2846.75	I
14	2848.42	I
16	2851.65	I
6000	2852.13	I
3	2915.45	I
10	2936.74	I
12	2938.47	I
2	2942.00	I
3	2809.76	I
2	2811.11	I
1	2811.78	I
12	2846.72	I
14	2848.34	I
16	2851.66	I
2	2902.92	I
4	2906.36	I
3	2915.45	I
2	2928.75	II
3	2936.54	II
10	2936.74	I
12	2938.47	I
13	2942.00	I
1	2967.87	II
1	2971.70	II
20	3091.08	I
22	3092.99	I
14	3096.90	I
9	3104.71	II
8	3104.81	II
6	3168.98	II
6	3172.71	II
7	3175.78	II
2	3197.62	I
17	3329.93	I
6	3332.15	I
9	3336.68	I
7	3535.04	II
8	3538.86	II
7	3549.52	II
8	3553.37	II
140	3829.30	I
300	3832.30	I
500	3838.29	I
8	3848.24	II
1	3848.91	I
7	3850.40	II
2	3853.96	I
1	3854.96	I
2	3858.86	I
3	3878.31	I
2	3891.91	I
2	3893.30	I
3	3895.57	I
4	3903.86	I
6	3938.40	I
1	3984.21	I
8	3986.75	I
2	4054.69	I
10	4057.50	I
3	4075.06	I
2	4081.83	I
4	4165.10	I
15	4167.27	I
20	4351.91	I
6	4354.53	I
6	4380.38	I
9	4384.64	II
10	4390.59	II
8	4428.00	II
9	4433.99	II
5	4436.49	II
4	4436.60	II
14	4481.16	II
13	4481.33	II
6	4534.29	II
28	4571.10	I
3	4621.30	I
7	4702.99	I
10	4730.03	I
6	4739.59	II
5	4739.71	II
7	4851.10	II
75	5167.33	I
220	5172.68	I
400	5183.61	I
8	5264.21	II
7	5264.37	II
1	5345.98	I
9	5401.54	II
2	5509.60	I
6	5528.41	I
30	5711.09	I
5	5785.31	I
4	5785.56	I
7	5916.43	II
6	5918.16	II
10	6318.72	I
9	6319.24	I
7	6319.49	I
10	6346.74	II
9	6346.96	II
11	6545.97	II
5	6620.44	II
6	6620.57	II
2	6630.83	I
7	6781.45	II
8	6787.85	II
7	6812.86	II
8	6819.27	II
4	6894.90	I
6	6965.40	I
8	7060.41	I
10	7193.17	I
10	7291.06	I
5	7387.00	I
12	7387.69	I
4	7580.76	II
20	7657.60	I
19	7659.15	I
17	7659.90	I
8	7690.16	I
15	7691.55	I
1	7722.61	I
1	7746.34	I
1	7759.30	I
5	7786.50	II
4	7790.98	II
3	7811.14	I
12	7877.05	II
2	7881.67	I
13	7896.37	II
7	7930.81	I
3	8047.73	I
5	8049.85	I
7	8054.23	I
10	8098.72	I
9	8115.22	II
8	8120.43	II
1	8154.64	I
2	8159.13	I
10	8209.84	I
20	8213.03	I
10	8213.99	II
7	8222.92	II
7	8233.19	II
11	8234.64	II

Magnesium (Cont.)

Intensity		Wavelength	
7		8303.31	I
9		8305.60	I
10		8310.26	I
15		8346.12	I
2		8466.48	I
5		8468.84	I
7		8473.69	I
10		8710.18	I
12		8712.69	I
13		8717.83	I
10		8734.99	II
17		8736.02	I
11		8745.66	II
14		8806.76	I
10		8824.32	II
11		8835.08	II
20		8923.57	I
7		8989.03	I
9		8991.69	I
10		8997.16	I
14		9218.25	II
13		9244.27	II
12		9246.50	I
30		9255.78	I
10		9327.54	II
10		9340.54	II
25		9414.96	I
17		9429.81	I
19		9432.76	I
20		9438.78	I
8		9502.45	I
7		9503.11	I
5		9503.43	I
12		9631.89	II
11		9632.43	II
15		9953.20	I
15		9983.20	I
17		9986.47	I
18		9993.21	I
14		10092.16	II
5		10391.76	II
6		10392.23	II
35		10811.08	I
11		10914.23	II
7		10915.27	II
10		10951.78	II
25		10953.32	I
27		10957.30	I
28		10965.45	I
15		11032.10	I
14		11033.66	I
5		11255.93	II
4		11256.35	II
45		11828.18	I
30		12083.66	I
28		14877.62	I
35		15024.99	I
30		15040.24	I
25		15047.70	I
6		15740.71	I
8		15748.99	I
10		15765.84	I
30		17108.66	I
5		26392.90	I

Mg III
Ref. 4, 83, 177 — J.O.S.

Intensity		Wavelength	
		Vacuum	
		106.30	III
		106.92	III
		108.08	III
		110.16	III
		114.32	III
		126.50	III
15		170.80	III
15		171.39	III
15		182.24	III
12		182.97	III
20		186.51	III
20		187.20	III
10		188.53	III
100		231.73	III
80		234.26	III
10		1274.83	III
11	h	1280.70	III
12		1391.27	III
15		1393.39	III
10		1431.14	III
16		1572.71	III
12		1586.24	III
13		1687.09	III
13		1697.28	III
10		1722.04	III
22		1738.84	III
12		1747.56	III
18		1748.93	III
15		1772.98	III
20		1783.25	III
14		1794.58	III
15		1800.66	III
13		1858.19	III
12		1879.49	III
10		1908.50	III
12		1923.90	III
13		1930.67	III
11		1937.84	III
		Air	
15		2039.55	III
15		2055.49	III
25		2064.90	III
15		2085.90	III
20		2091.96	III
13		2097.93	III
15		2112.77	III
16		2134.06	III
20		2177.70	III
20		2395.15	III
15		2467.75	III
10		2490.54	III
10		2529.19	III
12		3299.05	III
13		3306.39	III
12		3335.90	III
11		3342.58	III
12		3361.41	III
10		3381.24	III
11		3382.90	III
11		3387.37	III
10		3706.74	III
10		4916.00	III
10		5839.82	III
15		6256.75	III

Mg IV
Ref. 7, 128, 129 — J.O.S.

Intensity		Wavelength	
		Vacuum	
40		118.16	IV
80	p	118.81	IV
70		123.59	IV
240		124.65	IV
300		129.86	IV
300		132.81	IV
400		146.95	IV
300		147.41	IV
300		147.54	IV
350		180.07	IV
400		180.62	IV
400		180.80	IV
350		181.34	IV
4000		320.99	IV
3000		323.31	IV
40		800.41	IV
150		857.29	IV
30		866.74	IV
50		919.03	IV
30		929.78	IV
40		1008.76	IV
30		1026.41	IV
250		1037.41	IV
80		1044.37	IV
60		1055.76	IV
300		1210.99	IV
300		1342.19	IV
800		1346.57	IV
300		1346.68	IV
600		1352.05	IV
900		1384.46	IV
500		1385.77	IV
800		1387.53	IV
300		1404.68	IV
1000		1409.36	IV
500		1437.53	IV
1000		1437.64	IV
300		1447.42	IV
300		1459.54	IV
400		1459.62	IV
400		1481.51	IV
350		1490.45	IV
300		1495.50	IV
300		1607.11	IV
500		1683.02	IV
400		1698.81	IV
300		1844.17	IV
		Air	
12	p	2518.40	IV
4		2534.79	IV

Mg V
Ref. 128 — J.O.S.

Intensity		Wavelength	
		Vacuum	
5		251.58	V
35		276.58	V
10		312.30	V
20		351.09	V
18		352.20	V
30		353.09	V
15		353.30	V
18		354.22	V
20		355.33	V

MANGANESE (Mn)
Z = 25

Mn I and II
Ref. 1, 126 — C.H.C.

Intensity		Wavelength	
		Vacuum	
20		1726.47	II
30		1732.70	II
50		1733.55	II
40		1734.49	II
30		1737.93	II
20		1740.16	II
20		1742.00	II
30		1853.27	II
20		1857.92	II
50		1902.95	II
20		1907.84	II
30		1911.41	II
20	d	1914.68	II
100		1915.10	II
20		1918.64	II
30		1919.64	II
80		1921.25	II
20		1923.07	II
20		1923.34	II
30		1925.52	II
50		1926.59	II
30		1931.40	II
20		1945.15	II
20		1947.93	II
20		1950.14	II
30		1953.23	II
20	d	1954.81	II
30		1959.25	II
20		1969.24	II
30		1994.23	II
9700		1996.06	I
14000		1999.51	I
		Air	
18000		2003.85	I
50		2037.31	II
40		2037.64	II
40		2039.97	II
30		2076.21	II
1500		2092.16	I
20		2097.46	II
20		2102.50	II
1700		2109.58	I
30		2113.96	II
290		2208.81	I
540		2213.85	I
770		2221.84	I
20		2373.36	II
20		2427.38	II
50		2427.72	II
30		2427.94	II
30		2437.37	II
20		2437.84	II
30		2452.49	II
50		2499.00	II
30		2507.60	II
20		2516.60	II
30		2516.74	II
20		2521.66	II
20		2530.72	II
20		2531.80	II
50		2532.78	II
75		2533.06	I
50		2533.33	II
30		2534.10	II
80		2534.22	II
100		2535.66	II
30		2535.98	II
100		2537.92	II
50		2541.11	II
80		2542.92	II
50		2543.45	II
100		2548.75	II
50		2551.85	II
30		2553.27	II
75		2556.57	II
30		2556.89	II
50		2557.54	II
95		2558.59	II
30		2559.41	II
150		2563.65	II
30		2565.22	II
580		2572.76	I
480		2575.51	I
12000		2576.10	II
550		2584.31	I
30		2588.97	II
45		2589.71	II
250		2592.94	I
6200		2593.73	II
250		2595.76	I
95		2598.90	II
40		2602.14	I
30		2602.72	II
45		2603.72	II
4300		2605.69	II
190		2610.20	II
500		2618.14	II
140		2622.90	I
150		2624.04	I
40		2624.80	II
200		2625.58	II
95		2626.64	I
30		2630.26	I
60		2630.57	I
190		2632.35	II
130		2638.17	II
80		2639.84	II
27		2650.99	II
60		2655.91	II
30		2666.77	II
45		2667.00	I
30		2667.03	II
110		2672.59	II
55		2673.37	II
55		2674.43	II
30		2676.33	I
45		2680.34	II
30		2680.68	II
30		2681.25	II
40		2681.72	I
45		2683.02	I
23		2683.75	I
55		2684.55	II
55		2685.94	II
110		2688.25	II
85		2692.66	I
27		2693.19	II
55		2695.36	II
27		2698.97	II
85		2701.00	II
50		2701.17	II
160		2701.70	II
100		2703.98	II
130		2705.74	II
80		2707.53	II
110		2708.45	II
45		2709.96	II
80		2710.33	II
110		2711.58	II
30		2716.80	II
30		2717.53	II
30		2719.01	II
50		2719.74	II
30		2722.10	II
30		2724.46	II
55		2728.61	II
30		2738.86	I
45	h	2760.93	I
30		2771.44	I
30	h	2776.23	I
30		2780.00	I
55		2789.20	I
60		2790.36	I
60		2791.08	I
6200		2794.82	I
5100		2798.27	I
220		2799.84	I
3700		2801.06	I
70		2804.10	I
60		2806.14	I
55		2808.02	I
110		2809.11	I
60		2812.84	I
70		2813.47	I
60		2815.02	II
30		2816.33	II
85		2817.97	I
40		2818.77	I
55		2821.45	I
55		2822.55	I
80		2830.79	I
27		2836.31	I
60		2870.08	II
30		2872.94	II
80		2879.49	II
40		2882.90	I
70		2886.68	II
160		2889.58	II
55		2892.39	II
50		2898.70	II
80		2900.16	II
40		2907.22	I
140	h	2914.60	I
190	h	2925.57	I

Manganese (Cont.)

Intensity		Wavelength	
27		2928.68	I
1100		2933.06	II
27		2934.02	I
1500		2939.30	II
250	h	2940.39	I
		2940.48	I
60		2941.04	I
1900		2949.20	II
40		3007.66	I
40		3011.16	I
40		3011.38	I
40		3014.67	I
60		3016.45	I
30		3019.92	II
70		3022.75	I
55		3031.06	II
30		3035.35	II
95		3040.60	I
27		3042.73	I
85		3043.36	I
330		3044.57	I
120		3045.59	I
200		3047.04	I
40		3048.86	I
30		3050.65	II
250		3054.36	I
140		3062.12	I
170		3066.02	I
170		3070.27	I
160		3073.13	I
90		3079.63	I
50		3081.33	I
23		3082.05	I
40		3097.06	I
40		3110.68	I
60	h	3148.18	I
90	h	3161.04	I
140	h	3178.50	I
220		3212.88	I
65		3216.95	I
1000		3228.09	I
300		3230.72	I
850		3236.78	I
330		3243.78	I
650		3248.52	I
100		3251.14	I
310		3252.95	I
65		3254.04	I
310		3256.14	I
220		3258.41	I
180		3260.23	I
180		3264.71	I
65		3296.88	I
65		3298.22	I
65		3320.69	I
70		3330.67	I
200		3330.78	II
100		3336.39	II
30		3365.02	II
30		3400.12	II
50		3438.97	II
720		3441.99	II
50		3460.03	II
360		3460.33	II
360	h	3474.04	II
		3474.13	II
290		3482.91	II
180		3488.68	II
140		3495.84	II
50		3496.81	II
100		3497.54	II
360		3531.85	I
		3532.00	I
1100		3532.12	I
1300		3547.80	I
1100		3548.03	I
390		3548.20	I
2200		3569.49	I
720		3569.80	I
		3570.04	I
1400		3577.88	I
720		3586.54	I
290		3595.12	I
420		3607.54	I
420		3608.49	I
360		3610.30	I
290		3619.28	I
220		3623.79	I
140		3629.74	I
100		3660.40	I
70		3670.52	I
70		3676.96	I
50		3682.09	I
280		3693.67	I
180		3696.57	I
70		3701.73	I
210		3706.08	I
130		3718.93	I
55		3728.89	I
130		3731.93	I
260		3790.22	I
55		3799.26	I
110		3800.55	I
55		3801.91	I
3200		3806.72	I
700		3809.59	I
55		3810.69	I
90		3816.75	I
2100		3823.51	I
390		3823.89	I
200		3829.68	I
480		3833.86	I
1300		3834.36	I
350		3839.78	I
670		3841.08	I
350		3843.98	I
65		3918.32	I
120		3926.47	I
65		3952.84	I
55		3975.89	I
65		3977.08	I
130		3982.58	I
150		3985.24	I
190		3986.83	I
150		3987.10	I
1500		4018.10	I
150		4026.44	I
27000		4030.76	I
19000		4033.07	I
11000		4034.49	I
1500		4035.73	I
55		4038.73	I
5600		4041.36	I
210	d	4045.13	I
		4045.21	I
1100		4048.76	I
80		4049.00	I
55		4051.73	I
65		4052.47	I
150		4055.21	I
1900		4055.54	I
210		4057.95	I
1100		4058.93	I
150		4059.39	I
730		4061.74	I
730		4063.53	I
80		4065.08	I
80		4068.00	I
290		4070.28	I
730		4079.24	I
730		4079.42	I
1100		4082.94	I
1100		4083.63	I
65		4089.94	I
55		4105.36	I
200		4110.90	I
150		4131.12	I
120		4135.04	I
80		4141.06	I
55		4147.53	I
80		4148.80	I
150		4176.60	I
120		4189.99	I
65		4201.76	I
65		4211.75	I
370		4235.14	I
510		4235.29	I
190		4239.72	I
290		4257.66	I
290		4265.92	I
270		4281.10	I
65		4284.08	I
65		4312.55	I
50		4323.63	II
45		4374.95	I
45		4381.70	I
55		4411.88	I
350		4414.88	I
55		4419.78	I
210		4436.35	I
800		4451.59	I
160		4453.00	I
130		4455.01	I
160		4455.32	I
110		4455.82	I
55		4457.04	I
210		4457.55	I
270		4458.26	I
55		4460.38	I
150		4461.08	I
510		4462.02	I
290		4464.68	I
200		4470.14	I
130		4472.79	I
40		4479.40	I
170		4490.08	I
240		4498.90	I
240		4502.22	I
80		4605.36	I
80		4626.54	I
35		4671.69	I
50		4701.16	I
160		4709.72	I
180		4727.48	I
130		4739.11	I
1000		4754.04	I
180		4761.53	I
750		4762.38	I
300		4765.86	I
500		4766.43	I
940		4783.42	I
1000		4823.52	I
25		4844.32	I
35		4965.88	I
19		5004.91	I
30		5074.79	I
60		5117.94	I
50		5150.89	I
50		5196.59	I
85		5255.32	I
160		5341.06	I
19		5349.88	I
95		5377.63	I
95		5394.67	I
50		5399.49	I
95		5407.42	I
35		5413.69	I
85		5420.36	I
35		5432.55	I
12		5457.47	I
60		5470.64	I
40		5481.40	I
30		5505.87	I
50		5516.77	I
40		5537.76	I
21		5551.98	I
8		5567.76	I
7		5573.01	I
8		5573.68	I
7		5738.29	I
7		5780.19	I
7		5816.84	I
140		6013.50	I
200		6016.64	I
290		6021.80	I
7		6384.67	I
17		6440.97	I
24		6491.71	I
14	h	6942.52	I
12		6989.96	I
14		7069.84	I
12		7184.25	I
10		7247.82	I
24	h	7283.82	I
35	h	7302.89	I
50		7326.51	I
12		7680.20	I
10		7712.42	I
10	h	7764.72	I
10	h	8670.92	I
12	h	8672.06	I
10	h	8673.97	I
12	h	8701.05	I
17	h	8703.76	I
30	h	8740.93	I

Mn III
Ref. 385 — C.H.C.

Intensity		Wavelength	
		Vacuum	
20		892.39	III
20		1108.16	III
30		1183.30	III
25	w	1183.86	III
30		1198.49	III
30		1219.80	III
100		1228.97	III
500		1283.58	III
400		1287.59	III
300		1291.62	III
1000		1360.72	III
800		1365.20	III
400		1369.43	III
300		1371.65	III
300		1596.95	III
500	h	1609.17	III
1000		1614.14	III
2000		1620.60	III
300		1623.91	III
400		1629.12	III
500		1633.80	III
250		1647.46	III
400		1653.57	III
400		1804.06	III
300		1806.47	III
300		1811.02	III
400		1877.62	III
300		1885.21	III
500		1941.28	III
250	w	1942.89	III
800		1943.21	III
500		1952.36	III
1000		1952.52	III
300		1956.61	III
250		1962.04	III
500		1978.95	III
400		1982.76	III
400		1989.59	III
		Air	
300		2022.19	III
1000	w	2027.83	III
500	w	2028.14	III
300		2044.57	III
400		2048.93	III
500		2049.68	III
300		2056.80	III
500		2066.38	III
1000		2069.02	III
900		2077.38	III
300		2078.13	III
800		2084.23	III
600		2090.05	III
300		2090.25	III
300		2094.14	III
500		2094.78	III
500		2097.93	III
500		2099.97	III
300		2123.25	III
1000		2169.78	III
700		2174.15	III
900		2176.87	III
800		2181.86	III
800	w	2184.87	III
600		2185.13	III
400		2211.95	III
600		2212.42	III
800		2215.21	III
900		2220.55	III
1000		2227.42	III
100		3287.49	III
100		3540.52	III
150		3601.72	III
100		3616.00	III
100		4246.17	III
200		5079.20	III
150		5100.03	III
100		5117.03	III
100		5252.73	III
100		5365.59	III
150		5454.07	III
200		5474.68	III
100		5671.12	III
200		5946.65	III
100	s	6213.11	III
200		6231.21	III
100		6238.64	III
100		6273.71	III

Mn IV
Ref. 433 — C.H.C.

Intensity		Wavelength	
		Vacuum	
60		579.79	IV
60		581.44	IV
60		581.65	IV
60		585.21	IV
90		1242.25	IV
90		1244.50	IV
85		1247.73	IV
95		1251.93	IV
95		1257.28	IV
90		1264.41	IV
70		1603.60	IV
70		1611.10	IV
75		1653.83	IV
70		1656.39	IV
70		1659.25	IV
75		1664.73	IV
80	b	1667.00	IV
70		1670.08	IV
75		1691.68	IV
75		1693.15	IV
80		1698.30	IV
75		1698.70	IV
70		1699.06	IV
75		1707.43	IV
65		1718.67	IV
75	b	1720.52	IV
75		1720.74	IV
75		1721.41	IV
65		1722.94	IV
75		1724.83	IV
85	b	1742.10	IV
85		1751.59	IV
75		1759.82	IV
70		1762.17	IV
75		1762.94	IV
85	d	1766.27	IV
75		1767.09	IV
65		1772.11	IV
75		1773.51	IV
75		1782.21	IV
75		1786.02	IV

Manganese (Cont.)

Intensity	Wavelength	
75	1787.04	IV
75	1787.38	IV
75	1788.64	IV
75	1790.44	IV
80	1795.65	IV
80	1795.79	IV
60	1907.03	IV
75	1910.25	IV
65	1997.54	IV

Mn V
Ref. 405 — C.H.C.

Intensity	Wavelength	
	Vacuum	
300	404.36	V
380	406.02	V
300	406.40	V
600	410.30	V
600	410.60	V
480	410.98	V
400	411.32	V
460	412.74	V
460	413.75	V
600	415.62	V
650	415.98	V
350	419.80	V
600	428.59	V
500	429.05	V
400	433.54	V
600	435.67	V
350	436.16	V
500	436.18	V
450	438.74	V
350	439.35	V
1000	441.72	V
850	442.49	V
400	467.32	V
300	474.82	V

MERCURY (198) (Hg)
Z = 80

Hg I and II (198)
Ref. 43, 50, 69, 145, 229, 242 — R.W.S.

Intensity	Wavelength	
	Vacuum	
80	1250.564	I
8	1259.242	I
100	1268.825	I
5	1307.751	I
20	1402.619	I
10	1435.503	I
1000	1849.492	I
	Air	
60	2262.210	II
20	2302.065	I
20	2345.440	I
100	2378.325	I
20	2380.004	I
40	2399.349	I
20	2399.729	I
20	2446.900	I
15	2464.064	I
40	2481.999	I
30	2482.713	I
40	2483.821	I
90	2534.769	I
15000	2536.506	I
25	2563.861	I
25	2576.290	I
250	2652.043	I
400	2653.683	I
100	2655.130	I
50	2698.831	I
80	2752.783	I
20	2759.710	I
40	2803.471	I
30	2804.438	I
750	2847.675	II
50	2856.939	I
150	2893.598	I
150	2916.227	II
60	2925.413	I
1200	2967.283	I
300	3021.500	I
120	3023.476	I
30	3025.608	I
50	3027.490	I
400	3125.670	I
320	3131.551	I
320	3131.842	I
80	3341.481	I
2800	3650.157	I
300	3654.839	I
80	3662.883	I
240	3663.281	I
30	3701.432	I
35	3704.170	I
30	3801.660	I
20	3901.867	I
60	3906.372	I
200	3983.839	II
1800	4046.572	I
150	4077.838	I
40	4108.057	I
250	4339.224	I
400	4347.496	I
4000	4358.337	I
80	4916.068	I
1100	5460.753	I
160	5675.922	I
240	5769.598	I
280	5790.663	I
20	6072.713	I
30	6234.402	I
160	6716.429	I
250	6907.461	I
240	11287.407	I

MERCURY (NATURAL) (Hg)
Z = 80

Hg I and II (nat.)
Ref. 34, 45, 90, 117, 133, 189, 235, 304, 327, 328 — R.W.S.

Intensity	Wavelength	
	Vacuum	
400	893.08	II
300	915.83	II
150	923.39	II
200	940.80	II
100	962.74	II
50	969.13	II
800	1099.26	II
80	1250.58	I
8	1259.24	I
100	1268.82	I
5	1307.75	I
300	1307.93	II
400	1321.71	II
400	1331.74	II
80	1350.07	II
200	1361.27	II
20	1402.62	I
200	1414.43	II
10	1435.51	I
15	1619.46	II
120	1623.95	II
20	1628.25	II
150	1649.94	II
50	1653.64	II
200	1672.41	II
100	1702.73	II
100	1707.40	II
120	1727.18	II
250	1732.14	II
20	1775.68	I
40	1783.70	II
30	1796.22	II
200	1796.90	II
60	1798.74	II
30	1803.89	II
40	1808.29	II
400	1820.34	II
5	1832.74	I
1000	1849.50	I
160	1869.23	II
300	1870.55	II
200	1875.54	II
20	1900.28	II
30	1927.60	II
300	1942.27	II
100	1972.94	II
200	1973.89	II
150	1987.98	II
	Air	
90	2026.97	II
90	2052.93	II
70	2148.00	II
5	2247.55	I
60	2262.23	II
20	2302.06	I
15	2323.20	I
5	2340.57	I
20	2345.43	I
20	2352.48	I
100	2378.32	I
20	2380.00	I
40	2399.38	I
20	2399.73	I
10	2400.49	I
60	2407.35	II
50	2414.13	II
5	2441.06	I
20	2446.90	I
15	2464.06	I
40	2482.00	I
30	2482.72	I
40	2483.82	I
90	2534.77	I
15000	2536.52	I
25	2563.86	I
25	2576.29	I
5	2578.91	I
15	2625.19	I
5	2639.78	I
250	2652.04	I
400	2653.69	I
100	2655.13	I
5	2674.91	I
50	2698.83	I
50	2699.38	I
80	2705.36	II
80	2752.78	I
20	2759.71	I
40	2803.46	I
30	2804.43	I
2	2805.34	I
2	2806.77	I
150	2814.93	II
750	2847.68	II
50	2856.94	I
150	2893.60	I
150	2916.27	II
60	2925.41	I
150	2935.94	II
400	2947.08	II
1200	2967.28	I
300	3021.50	I
120	3023.47	I
30	3025.61	I
50	3027.49	I
400	3125.67	I
320	3131.55	I
320	3131.84	I
400	3208.20	II
400	3264.06	II
80	3341.48	I
100	3385.25	II
400	3451.69	II
200	3549.42	II
2800	3650.15	I
300	3654.84	I
80	3662.88	I
240	3663.28	I
30	3701.44	I
35	3704.17	I
30	3801.66	I
100	3806.38	II
20	3901.87	I
60	3906.37	I
100	3918.92	II
200	3983.96	II
1800	4046.56	I
150	4077.83	I
40	4108.05	I
250	4339.22	I
400	4347.49	I
4000	4358.33	I
100	4398.62	II
90	4660.28	II
80	4855.72	II
5	4883.00	I
5	4889.91	I
80	4916.07	I
5	4970.37	I
5	4980.64	I
20	5102.70	I
40	5120.64	I
100	5128.45	II
20	5137.94	I
20	5290.74	I
5	5316.78	I
60	5354.05	I
30	5384.63	I
1100	5460.74	I
30	5549.63	I
160	5675.86	I
240	5769.60	I
100	5789.66	I
280	5790.66	I
140	5803.78	I
60	5859.25	I
60	5871.73	II
20	5871.98	I
20	6072.72	I
1000	6149.50	II
30	6234.40	I
80	6521.13	II
160	6716.43	I
250	6907.52	I
250	7081.90	I
200	7091.86	I
40	7346.37	II
100	7485.87	II
20	7728.82	I
100	7944.66	II
2000	10139.75	I
240	11287.40	I
120	13209.95	I
140	13426.57	I
60	13468.38	I
80	13505.58	I
500	13570.21	I
450	13673.51	I
200	13950.55	I
500	15295.82	I
100	16881.48	I
400	16920.16	I
300	16942.00	I
500	17072.79	I
400	17109.93	I
20	17116.75	I
20	17198.67	I
20	17213.20	I
70	17329.41	I
30	17436.18	I
50	18130.38	I
40	19700.17	I
	22493.28	I
250	23253.07	I
	32148.06	I
	36303.03	I

Hg III
Ref. 343 — C.H.C.

Intensity	Wavelength	
	Vacuum	
3	621.44	III
2	679.68	III
2	878.59	III
1	886.48	III
1	988.89	III
2	1009.29	III
5	1068.03	III
2	1161.95	III
9	1681.40	III
15	1759.75	III
1	1894.77	III
	Air	
7	2314.15	III
4	2380.55	III
8	2431.65	III
5	2480.56	III
7	2484.50	III
2	2612.92	III
4	2617.97	III
3	2670.49	III
70	2724.43	III
6	2769.22	III
3	2844.76	III
15	3090.05	III
5	3283.02	III
12	3312.28	III
8	3389.01	III
5	3450.77	III
3	3500.35	III
4	3538.88	III
5	3557.24	III
15	3803.51	III
70	4122.07	III
10	4140.34	III
100	4216.74	III
15	4470.58	III
12	4552.84	III
50	4797.01	III
10	4869.85	III
80	4973.57	III
30	5210.82	III
6	5695.71	III
25	6220.35	III
35	6418.98	III
40	6501.38	III
10	6584.26	III
6	6610.12	III
30	6709.29	III
12	7517.46	III
7	7808.10	III
25	7946.75	III
50	7984.51	III
5	8151.64	III

MOLYBDENUM (Mo)
Z = 42

Mo I and II
Ref. 1 — C.H.C.

AIR

Intensity		Wavelength	
19000		2015.11	II
40000		2020.30	II
21000		2038.44	II
17000		2045.98	II
4800		2081.68	II
2400		2089.52	II
2200		2092.50	II
4000		2093.11	II
2700		2100.84	II
1500		2104.29	II
1400		2108.02	II
400		2269.69	II
160		2304.25	II
160		2306.97	II
130		2325.94	I
240		2330.46	I
110		2332.12	II
190		2340.47	I
190		2341.59	II
80		2352.61	I
80		2355.22	I
80		2355.42	II
70		2364.37	I
50		2366.09	II
140		2372.27	I
100		2380.41	I
150		2383.52	I
110		2389.20	II
140		2403.61	II
80		2404.66	II
140		2405.86	I
40		2408.39	I
40		2412.84	II
120		2413.01	II
70		2415.33	I
80		2417.96	II
65		2419.01	II
80		2420.18	II
70		2424.00	II
65		2430.43	I
65		2435.96	II
65		2440.28	II
40		2461.81	II
50		2466.68	II
50		2466.97	II
50		2468.78	II
30		2470.04	II
150	h	2471.97	I
70		2477.57	II
70	h	2481.81	I
65		2482.57	II
40		2484.75	II
40	h	2485.31	I
24		2496.24	II
85		2498.28	II
40		2500.44	II
65		2502.84	II
50		2511.80	II
65		2515.08	II
70		2527.14	II
50		2530.34	II
70		2532.31	II
440		2538.46	II
50		2539.44	II
110	h	2540.45	I
330		2542.67	II
40		2543.61	II
330		2548.22	I
110	h	2550.85	I
65		2555.42	II
40		2556.75	II
80		2558.88	II
65		2562.08	II
85		2564.34	II
40		2566.26	II
250		2567.05	I
20		2571.45	II
320		2572.34	I
50		2574.42	II
40		2576.56	II
40		2578.36	II
250		2582.16	I
30		2585.95	II
65		2588.78	II
40		2591.77	II
250		2593.70	II
100		2595.40	I
40		2597.38	II
250		2602.80	II
40		2605.08	II
40		2605.93	II
250		2607.37	I
190		2611.20	I
290		2613.08	I
130		2615.39	I
400		2616.78	I
70		2619.34	II
140		2621.07	I
320		2627.55	I
160		2628.74	I
440		2629.85	I
330		2636.67	II
250		2638.30	I
720		2638.76	II
410		2640.99	I
600		2644.35	II
370		2646.49	II
640		2649.46	I
480		2653.35	II
560	h	2655.03	I
290		2658.11	I
640		2660.58	II
110		2665.10	I
55		2671.83	II
720		2672.84	II
250		2673.27	II
1000		2679.85	I
95		2681.36	II
640		2683.23	II
880		2684.14	II
560		2687.99	II
30		2692.61	II
55		2695.22	II
30		2696.83	II
55		2699.41	II
140		2701.03	I
480		2701.42	II
30		2701.87	II
30		2704.93	II
40		2710.19	II
30		2711.49	II
50		2712.35	II
190		2713.51	II
290		2717.35	II
110		2724.41	I
180		2725.15	I
85		2726.97	II
140		2729.68	II
80		2730.20	II
330		2732.88	II
250		2733.39	I
160		2736.96	II
80	h	2737.88	II
50		2738.60	II
40		2741.32	II
55		2741.62	II
240		2743.07	I
290		2746.30	II
320		2751.47	I
110		2756.07	II
65	d	2758.63	II
20		2760.53	II
190		2761.53	I
220		2763.62	II
110		2766.26	I
240		2769.76	II
160		2773.78	II
190		2774.39	II
1700		2775.40	II
130		2777.74	I
65		2777.86	II
880		2780.04	II
400		2784.99	II
180		2787.83	I
40		2791.54	II
240	d	2797.93	I
220		2801.47	I
400		2807.76	II
28		2812.58	II
24		2814.67	II
1700		2816.15	II
220		2817.44	II
50		2822.03	II
240		2826.54	I
80		2827.74	II
40		2831.44	II
30		2832.07	II
80		2834.39	II
80		2835.33	II
160		2842.15	II
24		2843.73	II
220		2844.39	I
1700		2848.23	II
160		2849.38	I
370		2853.23	II
50		2856.00	II
24		2863.20	II
370		2863.81	II
160		2864.31	I
140		2864.66	I
40		2865.62	II
220		2866.69	II
40		2868.11	II
40		2868.32	II
1700		2871.51	II
85		2872.88	II
220		2879.05	II
65		2888.15	II
95		2891.28	II
1300		2890.99	II
190		2892.81	II
950		2894.45	II
140		2897.63	II
70		2900.80	II
290		2903.07	II
160		2905.27	I
80		2907.12	II
600		2909.12	II
1100		2911.92	II
55		2913.81	II
120		2918.83	II
1300		2923.39	II
140		2924.32	II
65		2927.54	II
50		2930.06	II
1100		2930.50	II
55		2930.77	II
800		2934.30	II
65		2935.20	II
120		2937.66	I
40		2938.30	II
95		2940.10	II
110		2941.22	II
140		2944.21	I
150		2944.82	II
140		2945.66	I
		2945.95	II
190		2946.01	I
140		2946.42	I
140		2946.69	II
95		2947.28	II
95		2955.84	II
240		2956.06	II
70		2956.90	II
95		2960.24	II
140		2962.89	I
250		2963.79	II
50		2964.96	II
210		2965.27	II
70		2971.91	II
250		2972.61	II
80		2975.40	II
180		2978.28	I
120		2981.52	I
110		2987.92	I
160		2988.68	I
190		2989.80	I
95		2992.84	II
50		2993.52	II
190		3002.21	I
40		3004.46	II
130		3013.39	I
140	h	3013.76	I
250		3025.00	I
95		3027.77	II
100		3036.31	I
300		3041.70	I
150		3046.80	I
210		3047.31	I
210		3055.32	I
100		3060.78	II
160		3061.59	I
800		3064.28	I
250		3065.04	II
100		3068.00	I
250		3070.90	I
800		3074.37	I
85		3077.66	II
150		3079.88	I
210		3080.41	I
800		3085.62	I
270		3087.62	II
100		3089.12	I
100		3089.71	I
190		3092.07	II
560		3094.66	I
110		3099.93	I
110		3100.88	I
560		3101.34	I
1400		3112.12	I
290		3122.00	II
14000		3132.59	I
110		3138.72	II
220		3147.35	I
220		3152.82	II
55		3155.64	II
6000		3158.16	I
120		3164.53	I
8700		3170.35	I
95		3172.03	II
160		3172.74	II
370		3183.03	I
120		3184.57	I
370		3185.10	I
180		3185.71	I
120	d	3187.59	II
7600		3193.97	I
290		3195.96	I
120		3198.85	I
40		3201.50	II
330		3205.22	I
880		3205.88	I
3000		3208.83	I
240		3210.97	I
560		3215.07	I
350		3221.74	I
880		3228.22	I
600		3229.79	I
1100		3233.14	I
950		3237.08	I
65		3240.71	II
950		3256.21	I
300		3262.63	I
480		3264.40	I
800		3270.90	I
240		3285.02	I
320		3285.36	I
1100		3289.02	I
950		3290.82	I
190		3292.31	II
320		3305.56	I
320		3307.12	I
100		3313.62	II
190		3320.90	II
640		3323.95	I
360		3325.67	I
360		3327.30	I
240		3340.17	I
1300		3344.75	I
95		3346.40	II
320		3347.02	I
1600		3358.12	I
250		3361.37	I
950		3363.78	I
950		3379.97	I
320		3382.48	I
1900		3384.62	I
130		3395.36	II
640		3404.34	I
1300		3405.94	I
240		3418.52	I
250		3420.04	I
250		3422.31	I
380		3434.79	I
320		3435.45	I
640		3437.22	I
250		3438.87	I
250		3441.44	I
250		3443.26	I
130		3446.08	II
3200		3447.12	I
640		3449.07	I
300		3451.75	I
250		3452.60	I
950		3456.39	I
640		3460.78	I
320		3466.83	I
250		3467.85	I
320		3469.22	I
240		3485.93	I
800		3504.41	I
240		3505.32	I
560		3508.12	I
480		3521.41	I
240		3524.98	I
640		3537.28	I
320		3542.17	I
520		3558.10	I
400		3563.14	I
300		3566.05	I
240		3570.65	I
320		3573.88	I
1400		3581.89	I
200		3590.74	I
210		3598.88	I
270		3602.94	I
210		3608.37	I
200		3623.23	I
1400		3624.46	I
330		3626.18	I
28		3635.14	II
1000		3635.43	I
400		3657.35	I
540		3664.81	I
290		3666.72	I
590		3672.82	I
1300		3680.60	I
45		3684.22	II
65		3688.31	II
240		3690.59	I
180		3692.64	II
1400		3694.94	I
220		3702.03	I
220		3715.65	I
500		3727.69	I
330	d	3732.71	I
240		3742.28	I
80		3744.37	II
360		3770.45	I
220		3779.77	I

Molybdenum (Cont.)

Intensity		Wavelength	Spectrum
360		3781.59	I
250		3797.30	I
29000		3798.25	I
290		3801.84	I
520		3826.70	I
940		3828.87	I
1700		3833.75	I
380		3847.25	I
29000		3864.11	I
580		3869.08	I
580		3886.82	I
380		3901.77	I
19000		3902.96	I
65		3941.48	II
230		3943.04	I
270		4056.01	I
1400		4062.08	I
2300		4069.88	I
1300		4081.44	I
940		4084.38	I
250		4102.15	I
730		4107.47	I
630		4120.10	I
2900		4143.55	I
230		4148.94	I
250		4155.28	I
200		4157.40	I
200		4178.27	I
480		4185.82	I
2500		4188.32	I
250		4194.56	I
1500		4232.59	I
270		4269.28	I
890		4276.91	I
1200		4277.24	I
1400		4288.64	I
680		4292.13	I
890		4293.21	I
360		4293.88	I
840		4326.14	I
250		4326.74	I
230		4350.34	I
230		4369.04	I
1900		4381.64	I
2500		4411.57	I
210		4423.62	I
990		4434.95	I
200		4442.20	I
340		4449.74	I
480		4457.36	I
630		4474.56	I
230		4491.28	I
120		4504.90	I
140		4512.15	I
230		4517.13	I
230		4524.34	I
120		4529.40	I
400		4536.80	I
110		4558.11	I
210		4576.50	I
170		4595.16	I
360		4609.88	I
100		4621.38	I
460		4626.47	I
100		4627.48	I
220		4662.76	I
130		4671.90	I
130		4688.22	I
640		4707.26	I
150		4708.22	I
220		4717.92	I
100		4729.14	I
700		4731.44	I
100		4750.39	I
770		4760.19	I
150		4776.34	I
100		4796.52	I
410		4819.25	I
410		4830.51	I
360		4868.00	I
110		4950.62	I
150		4957.54	I
210		4979.12	I
110		4999.91	I
20		5010.81	I
180		5014.60	I
26		5016.78	I
20		5019.85	I
80		5029.00	I
65		5030.78	I
23		5038.91	I
26		5046.52	I
100		5047.71	I
50		5055.00	I
35		5058.07	I
200		5059.88	I
35		5062.52	I
29		5064.64	I
35		5079.87	I
100		5080.02	I
35		5081.26	I
40		5090.97	I
35		5091.34	I
35		5092.16	I
40		5095.89	I
100		5096.65	I
130		5097.52	I
35		5098.03	I
130		5109.71	I
80		5114.97	I
35		5116.97	I
29		5123.83	I
150		5145.38	I
110		5147.39	I
80		5163.19	I
100		5167.76	I
160	d	5171.08	I
		5171.25	I
230	h	5172.94	I
160	h	5174.18	I
40		5191.44	I
110		5200.17	I
50		5200.74	I
26		5210.44	I
50		5211.86	I
80		5219.40	I
65		5231.06	I
26		5232.36	I
100		5234.26	I
460	h	5238.20	I
230	h	5240.88	I
110	h	5242.81	I
100		5245.51	I
150		5259.04	I
16		5260.17	I
65		5261.14	I
20		5268.95	I
35		5271.80	I
35		5276.28	I
65		5279.65	I
210		5280.86	I
20		5283.84	I
55		5292.08	I
35		5293.46	I
55		5295.47	I
20		5306.26	I
55		5313.89	I
35		5315.04	I
20		5319.89	I
20		5324.47	I
35		5327.06	I
20		5352.35	I
80		5354.88	I
35		5355.51	I
65		5356.48	I
560	Hl	5360.56	I
110	Hl	5364.28	I
35	Hl	5367.11	I
35		5372.40	I
26		5388.69	I
65		5394.52	I
35		5397.38	I
50		5400.47	I
35		5405.79	I
35		5406.39	I
40		5417.38	I
23		5426.89	I
55		5435.68	I
65		5437.75	I
40		5450.51	I
35		5456.46	I
26		5460.53	I
23		5465.57	I
35		5475.90	I
35		5490.28	I
20		5492.17	I
26	h	5493.80	I
26		5498.49	I
50		5501.54	I
23		5501.87	I
26	h	5503.54	I
7800		5506.49	I
23		5520.04	I
26		5520.64	I
40		5526.52	I
40		5526.97	I
5200		5533.05	I
40		5539.41	I
50		5543.12	I
40		5544.49	I
55		5556.28	I
26		5556.72	I
20		5564.05	I
40		5568.62	I
26		5569.48	I
2500		5570.45	I
35		5575.19	I
20		5591.58	I
40		5602.76	I
23		5608.62	I
23		5609.23	I
100		5610.93	I
23		5613.07	I
20		5618.45	I
23		5619.38	I
330		5632.47	I
50		5634.86	I
230		5650.13	I
23		5673.63	I
55		5674.47	I
40		5677.89	I
35		5682.89	I
460		5689.14	I
23		5699.28	I
80		5705.72	I
23		5711.80	I
210		5722.74	I
23		5728.77	I
26	d	5729.45	I
		5729.59	I
620		5751.40	I
23		5774.55	I
40		5779.36	I
23	h	5783.33	I
520		5791.85	I
23	h	5795.77	I
26		5800.46	I
35		5802.67	I
23		5825.20	I
23		5835.59	I
20		5839.99	I
20	h	5848.86	I
55	h	5849.73	I
50	h	5851.52	I
520		5858.27	I
20		5861.38	I
50		5869.33	I
26		5876.59	I
820		5888.33	I
23		5892.29	I
50	h	5893.38	I
20		5898.78	I
		5898.82 MO I	I
40		5901.47	I
40	h	5926.36	I
160	h	5928.88	I
40		5988.17	I
35		6025.49	I
16		6027.27	I
1300		6030.66	I
20		6047.83	I
20		6054.81	I
20		6079.58	I
10		6081.27	I
40		6101.87	I
10		6130.63	I
10		6197.66	I
20		6217.89	I
10		6264.27	I
16		6265.88	I
15		6290.74	I
13		6301.75	I
11		6323.54	I
40		6357.22	I
16		6389.11	I
11		6391.12	I
35		6401.07	I
26		6409.11	I
10		6412.39	I
100		6424.37	I
20		6446.34	I
20		6471.20	I
20		6473.99	I
10		6493.13	I
23		6519.84	I
15	h	6611.20	I
230		6619.13	I
10		6624.57	I
50		6650.38	I
13		6659.68	I
18		6690.47	I
110		6733.98	I
21		6746.08	I
50		6746.27	I
35		6753.97	I
13		6763.50	I
10	h	6799.88	I
10		6802.62	I
10		6812.03	I
13		6825.63	I
18	d	6828.87	I
		6829.05	I
40		6838.88	I
16		6848.92	I
21		6886.28	I
16		6892.36	I
10		6898.01	I
10		6898.98	I
13		6908.20	I
35		6914.01	I
13		6934.10	I
10		6947.39	I
10		6960.64	I
16		6978.71	I
26		6988.94	I
12		6999.13	I
16		7001.60	I
22		7037.98	I
22		7060.21	I
13		7063.34	I
13		7081.22	I
110		7109.87	I
27		7134.08	I
150		7242.50	I
40		7245.85	I
22		7267.62	I
17		7300.19	I
13		7348.49	I
13		7361.65	I
10	h	7364.41	I
40		7391.36	I
10		7434.10	I
13		7447.34	I
13	h	7452.85	I
140		7485.74	I
13		7504.47	I
11		7572.64	I
11	h	7595.16	I
11		7601.84	I
17	h	7656.76	I
13		7679.49	I
27		7720.77	I
17		7829.65	I
15		7854.45	I
11		7923.15	I
15		7986.60	I
22	h	8245.06	I
40	h	8328.44	I
45	h	8389.32	I
45	h	8483.39	I

Mo III
Ref. 420 — C.H.C.

Intensity	Wavelength	
Vacuum		
50	1166.07	III
100	1169.33	III
50	1173.67	III
50	1209.60	III
50	1225.46	III
50	1230.34	III
50	1234.63	III
30	1236.10	III
100	1254.93	III
100	1262.21	III
100	1263.74	III
100	1274.37	III
50	1274.94	III
100	1276.40	III
200	1277.40	III
200	1277.58	III
100	1278.06	III
200	1278.40	III
150	1281.90	III
150	1283.60	III
100	1258.52	III
100	1286.42	III
50	1288.07	III
50	1288.25	III
100	1290.49	III
100	1299.82	III
50	1305.58	III
50	1437.37	III
50	1452.38	III
50	1534.86	III
50	1751.22	III
50	1760.57	III
50	1807.60	III
100	1854.73	III
Air		
75	2165.19	III
75	2170.57	III
50	2172.46	III
75	2179.37	III
100	2184.37	III
100	2211.02	III
50	2223.19	III
100	2253.18	III
150	2269.71	III
50	2275.47	III
50	2275.64	III
200	2294.97	III
80	2304.26	III
50	2326.75	III
150	2330.93	III
100	2359.76	III
80	2386.96	III
50	2403.61	III
70	2412.71	III
50	2422.18	III
200	2506.19	III
90	2597.13	III
50	2756.06	III
100	2807.74	III
125	2947.32	III

Molybdenum (Cont.)

Intensity		Wavelength	
75		2983.94	III
80		3254.70	III
200		3271.69	III

Mo IV
Ref. 383 — C.H.C.

Intensity		Wavelength	
		Vacuum	
10		857.75	IV
10		859.72	IV
25		865.24	IV
20		865.53	IV
20		863.63	IV
50		867.92	IV
10		878.43	IV
100		884.19	IV
10		884.82	IV
60		886.05	IV
50		891.74	IV
40		894.80	IV
15		895.41	IV
20		1819.50	IV
30		1821.59	IV
30		1850.69	IV
20		1877.88	IV
80		1926.26	IV
100		1929.24	IV
20		1949.44	IV
80		1971.06	IV
20		1991.41	IV
		Air	
70		2010.92	IV
20		2023.78	IV
25		2055.64	IV
50		2060.38	IV
15		2091.89	IV
10		2113.78	IV
5		2140.33	IV

NEODYMIUM (Nd)
Z = 60

Nd I and II
Ref. 1 — C.H.C.

Intensity		Wavelength	
		Air	
75		2702.46	
75		2704.54	
75		2764.98	I
60		2785.79	I
50		2863.95	
50		2921.26	
55		2962.88	II
65		2963.58	II
80		2993.20	II
40		2994.73	
95		3007.97	II
95		3014.19	II
95		3018.35	II
80		3026.47	II
50		3038.98	II
50		3043.29	
50		3051.11	II
80		3052.15	II
140		3056.71	II
130		3069.73	II
65	d	3071.43	II
		3071.50	II
160		3075.38	II
95		3079.38	II
95		3080.94	II
95		3092.73	
240		3092.92	II
140		3098.48	II
55		3099.52	II
130		3105.43	II
95		3106.18	II
65		3108.01	II
260		3115.18	II
190		3116.15	II
50		3119.75	II
160		3123.06	II
190		3124.58	II
290		3133.60	II
220		3134.90	II
100		3137.24	II
170		3141.46	II
170		3142.44	II
100		3144.55	II
100		3144.82	II
100		3148.51	II
100		3149.29	II
100		3149.51	II
100		3162.62	II
100		3175.99	II
50		3181.54	II
50		3188.73	II
50		3200.62	II
150		3203.47	II
85		3211.00	II
100		3217.12	II
50		3222.62	I
50		3228.04	II
60		3234.62	
40		3237.91	II
100		3254.08	II
50		3256.91	II
220		3259.24	II
100		3260.66	II
220		3265.12	II
50		3265.38	II
170		3267.25	II
100		3273.18	II
320		3275.22	II
50		3281.49	II
50		3282.78	II
290		3285.10	II
100		3286.62	II
50		3289.52	
100		3290.65	II
70		3293.84	II
70		3294.68	II
70		3298.61	
300		3300.16	II
200		3312.75	II
200		3325.90	II
410		3328.28	II
250		3331.57	II
290		3334.48	II
290		3339.07	II
320		3353.59	II
200		3355.93	II
270		3364.96	II
290		3393.63	II
120	h	3484.88	I
200		3527.53	II
290		3543.35	II
200		3555.77	II
410		3560.75	II
340		3568.87	II
470		3587.51	II
300		3592.59	II
340		3598.02	II
300		3600.91	II
320		3609.79	II
370		3615.82	II
300		3618.96	II
300		3631.02	II
340		3634.30	II
240		3637.00	II
240		3637.23	II
240		3640.24	II
240		3645.78	II
340		3648.20	II
240		3649.46	
240		3650.42	II
410		3653.15	II
240		3654.16	
470		3662.26	II
540		3665.18	II
540		3672.36	II
580		3673.54	II
240		3678.18	II
1200		3685.80	II
440		3687.30	II
410		3689.69	II
300		3694.81	II
410		3697.56	II
240		3702.84	
240		3704.95	II
200		3712.81	
470		3713.70	II
370		3714.20	II
640	d	3714.73	II
250		3715.04	II
200		3715.39	II
470		3715.68	II
410		3718.54	II
410		3721.35	II
220		3722.42	II
780		3723.50	II
410		3724.87	II
250		3726.90	II
710		3728.13	II
470		3730.58	II
270		3732.78	II
1000	d	3735.54	II
		3735.60	II
440		3737.10	II
1000		3738.06	II
270		3741.42	II
200		3749.85	II
320		3750.31	II
580		3752.49	II
370		3752.67	II
250		3754.83	II
370		3755.60	II
510		3757.82	II
930		3758.95	II
300		3759.79	II
930		3763.47	II
300		3766.59	II
510		3769.65	II
1400		3775.50	II
250		3776.34	II
710		3779.47	II
580		3780.40	II
510		3781.32	II
300		3783.78	II
2400		3784.25	II
270		3784.73	II
340		3791.50	II
340		3795.45	II
240		3799.55	II
370		3801.12	II
200		3801.38	II
340		3802.30	II
1200		3803.47	II
200		3804.10	II
2500		3805.36	II
340		3805.55	II
470		3807.23	II
540		3808.77	II
440		3809.06	II
580		3810.49	II
240		3811.06	II
270		3811.77	II
200		3812.53	II
710		3814.73	II
240		3819.70	II
410		3822.47	II
1200		3826.42	II
240		3828.00	II
540		3828.85	II
440		3829.16	II
510		3830.47	II
740		3836.54	II
340		3837.91	II
1700		3838.98	II
340		3839.51	II
410	d	3841.82	II
		3841.88	II
1700	d	3848.24	II
		3848.31	II
1500		3848.52	II
470		3850.22	II
2400	d	3851.66	II
		3851.74	II
340		3858.55	II
270		3860.94	II
300		3862.52	II
3700	d	3863.33	II
		3863.40	II
240		3866.52	II
220		3866.81	II
850		3869.07	II
240		3875.74	II
470		3875.87	II
1100		3878.58	II
1000		3879.55	II
780		3880.38	II
1200		3880.78	II
200		3881.59	
540		3887.87	II
370	h	3889.66	II
1300		3889.93	II
1300		3890.58	II
1300		3890.94	II
580		3891.51	II
470		3892.06	II
810		3894.63	II
270		3896.13	II
440		3897.63	II
2000		3900.21	II
1300		3901.84	II
1700		3905.89	II
200		3907.70	II
510		3907.84	II
2000		3911.16	II
850		3912.23	II
340		3913.69	II
440		3915.13	II
610		3915.95	II
340		3917.65	II
220		3919.92	II
1100		3920.96	II
510		3927.10	II
200		3929.26	II
610		3934.82	II
410		3936.11	II
510		3938.86	II
2000		3941.51	II
2000		3951.16	II
810		3952.20	II
320		3952.87	II
320		3953.52	II
240		3957.45	II
590		3958.00	II
510		3962.21	II
1400		3963.12	II
270		3963.90	II
1100		3973.30	II
740		3973.69	II
740		3976.85	II
740		3979.49	II
320		3982.36	II
470		3986.25	II
1400		3990.10	II
1000		3991.74	II
1100		3994.68	II
410		4000.50	II
540		4004.02	II
410		4007.43	II
3700		4012.25	II
540		4012.70	II
370		4018.81	II
1000		4020.87	II
1000		4021.34	II
1000		4021.78	II
1200		4023.00	II
340		4024.78	II
410		4030.47	II
1200		4031.82	II
270		4038.12	II
3000		4040.80	II
200		4041.06	II
410		4043.59	II
410		4048.81	II
850		4051.15	II
850		4059.96	II
4700		4061.09	II
1100		4069.28	II
710		4075.12	II
470		4075.28	II
240		4077.62	II
470		4080.23	II
240		4085.82	II
270		4096.13	II
220		4098.18	II
200		4106.59	II
1400		4109.08	II
2500		4109.46	II
510		4110.48	II
300	h	4113.83	II
410		4123.88	II
470		4133.36	II
510		4135.33	II
3000		4156.08	II
510		4156.26	II
340		4160.57	II
410		4168.00	II
810		4175.61	II
2400		4177.32	II
200		4178.64	II
640		4179.59	II
250		4184.98	II
470		4205.60	II
470		4211.29	II
290		4220.25	II
440		4227.73	II
1300		4232.38	II
250		4234.19	II
290	h	4235.24	II
290		4239.84	II
2000		4247.38	II
850		4252.44	II
290		4254.29	
410		4261.84	II
340		4266.71	II
240		4270.56	II
340		4272.79	II
340		4275.09	II
470		4282.44	II
240		4282.57	II
710		4284.52	II
270		4297.80	II
5400		4303.58	II
340		4304.45	II
200		4307.78	II
470		4314.52	II
1100		4325.76	II
510		4327.93	II
540		4338.70	II
680		4351.29	II
850		4358.17	II
240		4366.38	II
340		4368.64	II
470	d	4374.93	II
		4375.04	II
710		4385.66	II
250		4390.66	II
540		4400.83	II
510		4411.06	II
580		4446.39	II
1400		4451.57	II
200		4451.99	II
300		4456.40	II
740		4462.99	II
410		4501.82	II
200		4506.59	II
170		4513.34	II
250		4516.36	II

Neodymium (Cont.)

Intensity		Wavelength	Spectrum
120		4527.25	I
340		4541.27	II
340		4542.61	II
100		4556.14	II
170		4559.67	I
340		4563.22	II
200		4578.89	II
200		4579.32	II
100		4586.62	I
200		4597.02	II
100		4603.82	I
100		4609.87	I
300		4621.94	I
100		4627.98	I
510		4634.24	I
340		4641.10	I
250		4645.77	II
200		4646.40	I
300		4649.67	I
200		4654.73	I
130		4670.56	II
170		4680.74	II
310		4683.45	I
110		4684.04	I
110		4690.35	I
190		4696.44	I
130		4703.57	II
470		4706.54	II
140		4706.96	I
190		4709.71	II
190		4715.59	II
240		4719.02	I
190		4724.35	II
140		4731.77	I
120		4779.46	I
170		4789.41	II
120		4797.15	II
240		4811.34	II
140		4820.34	II
350		4825.48	II
130		4832.28	II
110		4849.06	II
280		4859.02	II
190		4866.74	I
350		4883.81	I
140		4889.10	II
220		4890.70	II
240		4891.07	I
280		4896.93	I
120		4901.53	I
210		4901.84	I
110		4902.03	II
190		4913.41	I
170		4914.37	II
330		4920.68	II
470		4924.53	I
260		4944.83	I
290		4954.78	I
290		4959.13	II
150		4961.39	II
250		4989.94	II
150		5033.52	II
110		5063.73	II
360		5076.59	II
150	h	5089.84	II
360		5092.80	II
180		5102.39	II
150	d	5105.21	II
		5105.35	I
360		5107.59	II
340		5123.79	II
680		5130.60	II
170		5132.33	II
170		5165.14	II
130		5181.17	II
120		5182.60	II
500		5191.45	II
630		5192.62	II
330		5200.12	II
310		5212.37	II
150		5213.23	I
130		5225.05	II
130		5228.43	II
450		5234.20	II
250		5239.79	II
720		5249.59	II
200		5250.82	II
360		5255.51	II
120		5269.48	II
590		5273.43	II
150		5276.88	II
110		5291.67	I
680		5293.17	II
160		5302.28	II
110		5306.47	II
220		5311.46	II
500		5319.82	II
180		5356.98	II
290		5361.47	II
150		5371.94	II
110		5385.90	II
160		5431.53	II
110		5451.12	II
170		5485.70	II
35		5501.47	I
45		5525.72	I
90		5533.82	I
55		5535.27	II
55		5543.24	I
55		5548.47	II
55		5561.17	I
27		5575.50	I
27		5576.70	I
27		5577.70	I
27		5587.61	I
240		5594.43	II
55		5601.43	I
45		5601.92	I
220		5620.54	I
65		5635.76	I
45		5639.54	I
35		5653.57	I
70		5668.87	II
65		5669.77	I
140	d	5675.97	I
55		5676.33	I
220		5688.53	II
23		5689.51	I
30		5701.57	I
130		5702.24	II
80		5706.21	II
160		5708.28	II
80		5718.12	II
65		5726.83	II
100		5729.29	I
23		5734.55	
70		5740.86	II
55		5749.19	I
27		5749.66	I
23		5767.33	I
45		5776.12	I
45		5784.96	I
45		5788.22	I
45		5800.09	I
160		5804.02	II
80		5811.57	II
45		5813.89	I
27		5820.37	I
70		5825.87	II
30		5826.74	I
80		5842.39	II
30		5844.66	I
23		5845.95	I
55		5858.91	I
35		5867.08	I
30		5868.90	I
27		5871.04	I
30		5883.29	I
23		5886.24	I
30		5887.91	I
27		5921.22	I
27		5955.87	I
30		5994.76	I
27		5996.47	I
45		6007.67	I
35		6031.27	II
27		6033.29	I
45		6034.24	II
55		6066.03	I
27		6071.70	I
30		6073.97	I
23	d	6133.47	II
27		6149.28	I
27		6155.06	I
35		6157.83	II
23		6166.67	II
35		6170.49	II
45		6178.59	I
27		6183.91	II
27		6208.24	I
45		6223.39	I
27		6226.50	I
23		6238.50	II
35		6244.08	I
23		6257.49	I
27		6258.73	II
23		6277.29	II
27		6285.79	I
23		6292.84	II
23		6297.07	I
55		6310.49	I
27		6341.51	II
23		6382.07	II
65		6385.20	II
35		6485.69	I
45		6630.14	I
35		6637.96	II
45		6650.57	II
30		6655.67	I
25		6737.79	II
40		6740.11	II
25		6742.54	I
30		6790.37	II
30		6804.00	II
25		6846.72	II
40		6900.43	II
24		6941.39	II
17	h	7010.80	II
8		7018.85	II
17		7020.92	II
17		7024.58	II
10		7033.21	
35		7037.30	II
7		7052.14	II
7		7054.74	II
7		7061.47	II
40		7066.89	II
8		7082.93	II
12	h	7089.71	II
12	h	7092.09	
12	h	7092.74	II
12	h	7092.94	
17	h	7093.98	I
20	h	7095.42	I
29		7129.35	II
12	h	7142.04	II
10		7143.72	
8		7151.03	II
6		7153.09	I
6	h	7185.01	
10		7189.09	II
24		7189.42	II
20		7192.01	II
10		7199.00	II
8	h	7227.01	I
15		7236.54	II
7	h	7261.64	II
9		7285.29	II
9		7288.56	II
6		7291.38	II
7		7298.72	II
12		7316.81	II
7	h	7321.43	I
7		7323.12	II
6		7334.54	I
6		7357.10	I
6		7374.04	II
7		7381.79	II
9		7401.31	I
10		7406.62	II
6		7411.20	II
10		7418.18	II
9		7427.41	II
9		7448.71	II
5		7481.28	II
12		7511.16	II
17		7513.73	II
7	h	7514.44	II
7		7516.02	II
9		7526.45	II
12		7528.99	II
10		7538.26	II
5		7540.97	II
7		7547.00	II
5		7577.54	II
7		7587.65	II
6		7590.75	II
6		7603.73	II
5		7605.92	II
5		7614.72	I
9		7639.79	II
8		7646.00	II
6		7663.52	II
12		7696.56	II
6		7718.20	II
4		7743.90	II
4		7748.92	II
10		7750.95	II
6		7773.06	II
7		7792.22	II
6		7796.40	II
8		7797.32	II
5		7798.32	II
10		7808.47	II
7		7818.83	II
5		7825.20	II
12		7863.04	II
5	h	7872.03	I
7		7886.60	II
4	h	7896.50	II
9		7900.40	II
5	h	7906.03	I
12		7917.01	II
10		7925.03	II
5		7947.93	II
10		7949.68	II
5	h	7955.38	II
12		7958.95	I
12		7965.73	II
15		7982.09	II
12		7982.68	II
12		8000.76	II
9		8007.70	I
4	h	8020.07	
8		8026.35	
10		8043.24	I
8		8051.33	II
5		8064.00	II
10		8099.17	I
10		8120.93	II
12		8122.07	II
12		8141.75	II
12		8143.27	II
7	h	8164.97	I
8		8172.56	II
9		8179.83	II
9		8182.41	II
4		8185.58	II
7	h	8205.38	II
10		8231.52	II
4		8248.76	II
5	h	8249.68	II
4	h	8262.80	II
7	h	8266.72	II
4		8272.79	II
4	h	8302.74	II
10		8307.72	II
6		8324.50	II
4		8332.01	II
12		8346.36	II
4		8375.16	II
4		8375.33	II
4		8394.71	II
7		8400.85	II
5	h	8456.87	II
4		8530.53	II
5		8582.03	II
5		8591.53	II
7		8594.87	II
8	c	8643.43	II
5		8667.07	II
5		8677.48	
6		8691.29	II
6		8695.07	II
6		8712.82	II
6		8715.03	I
17		8839.10	II

NEON (Ne)
Z = 10

Ne I and II
Ref. 56, 58, 118, 150, 230 —
S.P.D.

Intensity		Wavelength	
		Vacuum	
90		352.956	II
60		354.962	II
90		361.433	II
60		362.455	II
150		405.854	II
120		407.138	II
200		445.040	II
300		446.256	II
250		446.590	II
180		447.815	II
150		454.654	II
200		455.274	II
10		456.275	II
120		456.348	II
90		456.896	II
1000		460.728	II
500		462.391	II
35		587.213	I
35		589.179	I
35		589.911	I
70		591.830	I
100		595.920	I
75		598.706	I
35		598.891	I
70		600.036	I
170		602.726	I
170		615.628	I
170		618.672	I
120		619.102	I
200		626.823	I
200		629.739	I
1000		735.896	I
400		743.720	I
60		993.88	II
70		1068.65	II
90		1131.72	II
100		1131.85	II
90		1229.83	II
90		1418.38	II
90		1428.58	II
90		1436.09	II
120		1681.68	II
180		1688.36	II
100		1888.11	II
100		1889.71	II
200		1907.49	II
500		1916.08	II
300		1930.03	II
200		1938.83	II
100	c	1945.46	II
		AIR	

Neon (Cont.)

Intensity		Wavelength	Spectrum
80		2007.01	II
80		2025.56	II
150		2085.47	II
180		2096.11	II
120		2096.25	II
80	p	2562.12	II
90	w	2567.12	II
80		2623.11	II
80		2629.89	II
90	w	2636.07	II
80		2638.29	II
80		2644.10	II
80		2762.92	II
90		2792.02	II
80		2794.22	II
100		2809.48	II
80		2906.59	II
80		2906.82	II
90		2910.06	II
90		2910.41	II
80		2911.14	II
80		2915.12	II
80		2925.62	II
80	w	2932.10	II
80		2940.65	II
90		2946.04	II
150		2955.72	II
150		2963.24	II
150		2967.18	II
100		2973.10	II
15		2974.72	I
100		2979.46	II
12		2982.67	I
150		3001.67	II
120	p	3017.31	II
300		3027.02	II
300		3028.86	II
100		3030.79	II
120		3034.46	II
100		3035.92	II
100		3037.72	II
100		3039.59	II
100		3044.09	II
100		3045.56	II
120		3047.56	II
100		3054.34	II
100		3054.68	II
100		3059.11	II
100		3062.49	II
100		3063.30	II
100		3070.89	II
100		3071.53	II
100		3075.73	II
120		3088.17	II
100		3092.09	II
120		3092.90	II
100		3094.01	II
100		3095.10	II
100		3097.13	II
100		3117.98	II
120		3118.16	II
10		3126.199	I
300		3141.33	II
100		3143.72	II
100	p	3148.68	II
100		3164.43	II
100		3165.65	II
100		3188.74	II
120		3194.58	II
500		3198.59	II
60		3208.96	II
120		3209.36	II
120		3213.74	II
150		3214.33	II
150		3218.19	II
120		3224.82	II
120		3229.57	II
200		3230.07	II
120		3230.42	II
120		3232.02	II
150		3232.37	II
100		3243.40	II
100		3244.10	II
100		3248.34	II
100		3250.36	II
150		3297.73	II
150		3309.74	II
300		3319.72	II
1000		3323.74	II
150		3327.15	II
100		3329.16	II
200		3334.84	II
150		3344.40	II
300		3345.45	II
150		3345.83	II
200		3355.02	II
120		3357.82	II
200		3360.60	II
120		3362.16	II
100		3362.71	II
120		3367.22	II
12		3369.808	I
40		3369.908	I
100		3371.80	II
500		3378.22	II
150		3388.42	II
120		3388.94	II
300		3392.80	II
100		3404.82	II
120		3406.95	II
100		3413.15	II
120		3416.91	II
120		3417.69	II
50		3417.904	I
15		3418.006	I
120		3428.69	II
60		3447.703	I
50		3454.195	I
100		3456.61	II
100		3459.32	II
25		3460.524	I
30		3464.339	I
30		3466.579	I
60		3472.571	I
150		3479.52	II
200		3480.72	II
200		3481.93	II
25		3498.064	I
30		3501.216	I
25		3515.191	I
150		3520.472	I
120		3542.85	II
120		3557.80	II
100		3561.20	II
250		3568.50	II
100		3574.18	II
200		3574.61	II
50		3593.526	I
30		3593.640	I
15		3600.169	I
20		3633.665	I
150		3643.93	II
200		3664.07	II
20		3682.243	I
12		3685.736	I
200		3694.21	II
10		3701.225	I
150		3709.62	II
250		3713.08	II
250		3727.11	II
800		3766.26	II
1000		3777.13	II
100		3818.43	II
120		3829.75	II
150		4219.74	II
100		4233.85	II
120		4250.65	II
120		4369.86	II
70		4379.40	II
150		4379.55	II
100		4385.06	II
200		4391.99	II
150		4397.99	II
150		4409.30	II
100		4413.22	II
100		4421.39	II
100	p	4428.52	II
100	p	4428.63	II
150	p	4430.90	II
150	p	4430.94	II
120		4457.05	II
100		4522.72	II
10		4537.754	I
10		4540.380	I
100		4569.06	II
15		4704.395	I
12		4708.862	I
10		4710.067	I
10		4712.066	I
15		4715.347	I
10		4752.732	I
12		4788.927	I
10		4790.22	I
10		4827.344	I
10		4884.917	I
4		5005.159	I
10		5037.751	I
10		5144.938	I
25		5330.778	I
20		5341.094	I
8		5343.283	I
60		5400.562	I
5		5562.766	I
10		5656.659	I
5		5719.225	I
12		5748.298	I
80		5764.419	I
12		5804.450	I
40		5820.156	I
500		5852.488	I
100		5872.828	I
100		5881.895	I
60		5902.462	I
60		5906.429	I
100		5944.834	I
100		5965.471	I
100		5974.627	I
120		5975.534	I
80		5987.907	I
100		6029.997	I
100		6074.338	I
80		6096.163	I
60		6128.450	I
100		6143.063	I
120		6163.594	I
250		6182.146	I
150		6217.281	I
150		6266.495	I
60		6304.789	I
7		6328.165	I
100		6334.428	I
120		6382.992	I
200		6402.246	I
150		6506.528	I
60		6532.882	I
150		6598.953	I
70		6652.093	I
90		6678.276	I
20		6717.043	I
100		6929.467	I
90		7024.050	I
100		7032.413	I
50		7051.292	I
80		7059.107	I
100		7173.938	I
150		7213.20	II
150		7235.19	II
100		7245.167	I
150		7343.94	II
40		7472.439	I
90		7488.871	I
100		7492.10	II
150		7522.82	II
80		7535.774	I
60		7544.044	I
100		7724.628	I
120		7740.74	II
300		7839.055	I
120		7926.20	II
400		7927.118	I
700		7936.996	I
2000		7943.181	I
2000		8082.458	I
100		8084.34	II
1000		8118.549	I
600		8128.911	I
3000		8136.406	I
2500		8259.379	I
100		8264.81	II
2500		8266.077	I
800		8267.117	I
6000		8300.326	I
100		8315.00	II
1500		8365.749	I
100		8372.11	II
8000		8377.606	I
1000		8417.159	I
4000		8418.427	I
1500		8463.358	I
800		8484.444	I
5000		8495.360	I
600		8544.696	I
1000		8571.352	I
4000		8591.259	I
6000		8634.647	I
3000		8647.041	I
15000		8654.383	I
4000		8655.522	I
100		8668.26	II
5000		8679.492	I
5000		8681.921	I
2000		8704.112	I
4000		8771.656	I
12000		8780.621	I
10000		8783.753	I
500		8830.907	I
7000		8853.867	I
1000		8865.306	I
1000		8865.755	I
3000		8919.501	I
2000		8988.57	I
100		9079.46	II
6000		9148.67	I
6000		9201.76	I
4000		9220.06	I
2000		9221.58	I
2000		9226.69	I
1000		9275.52	I
200		9287.56	II
6000		9300.85	I
1500		9310.58	I
3000		9313.97	I
6000		9326.51	I
2000		9373.31	I
5000		9425.38	I
3000		9459.21	I
5000		9486.68	I
5000		9534.16	I
3000		9547.40	I
120		9577.01	II
1000		9665.42	I
100		9808.86	II
800		10295.42	I
2000		10562.41	I
1500		10798.07	I
2000		10844.48	I
3000		11143.020	I
3500		11177.528	I
1600		11390.434	I
1100		11409.134	I
3000		11522.746	I
1500		11525.020	I
950		11536.344	I
500		11601.537	I
1200		11614.081	I
300		11688.002	I
2000		11766.792	I
1500		11789.044	I
500		11789.889	I
1000		11984.912	I
3000		12066.334	I
800		12459.389	I
1000		12689.201	I
1100		12912.014	I
700		13219.241	I
800		15230.714	I
400		17161.930	I
400		18035.80	I
1000		18083.21	I
350		18221.11	I
250		18227.02	I
2500		18276.68	I
2000		18282.62	I
1200		18303.97	I
250		18359.12	I
1200		18384.85	I
2000		18389.95	I
1000		18402.84	I
1200		18422.39	I
300		18458.65	I
400		18475.79	I
900		18591.55	I
1600		18597.70	I
350		18618.96	I
550		18625.16	I
1200		21041.295	I
750		21708.145	I
300		22247.35	I
350		22428.13	I
2250		22530.40	I
400		22661.81	I
600		23100.51	I
1000		23260.30	I
1050		23373.00	I
850		23565.36	I
3500		23636.52	I
300		23701.64	I
1100		23709.2	I
1800		23951.42	I
600		23956.46	I
1000		23978.12	I
200		24098.54	I
500		24161.42	I
600		24249.64	I
1500		24365.05	I
800		24371.60	I
400		24447.85	I
700		24459.4	I
300		24776.46	I
550		24928.88	I
250		25161.69	I
650		25524.37	I
125		28386.21	I
150		30200.	I
250		33173.09	I
450		33352.35	I
1300		33901.	I
2200		33912.10	I
600		34131.31	I
100		34471.44	I
120		35834.78	I

Ne III

Ref. 365, 371, 402 — R.L.K.

Intensity	Wavelength	
	Vacuum	
20	251.14	III
20	251.56	III
20	251.73	III
40	267.06	III
40	267.52	III
20	267.71	III
40	283.18	III
160	283.21	III
110	283.69	III
40	283.89	III
220	301.12	III
220	313.05	III
220	313.68	III
40	313.95	III

Neon (Cont.)

220	379.31	III
285	488.10	III
220	488.87	III
450	489.50	III
70	489.64	III
220	490.31	III
360	491.05	III
20	1255.03	III
110	1255.68	III
160	1257.19	III

Air

200	2086.96	III
300	2089.43	III
240	2092.44	III
400	2095.54	III
200	2161.22	III
300	2163.77	III
200	2180.89	III
200	2209.35	III
200	2211.85	III
240	2213.76	III
300	2216.07	III
240	2263.21	III
200	2264.91	III
300	2412.73	III
240	2412.94	III
200	2413.78	III
200	2473.40	III
800	2590.04	III
600	2593.60	III
400	2595.68	III
300	2610.03	III
240	2613.41	III
200	2615.87	III
200	2638.70	III
200	2641.07	III
600	2677.90	III
500	2678.64	III

Ne IV

Ref. 69, 364, 388, 400, 413, 430, — R.L.K.

Intensity	Wavelength	
	Vacuum	
15	151.82	IV
15	152.23	IV
15	158.65	IV
15	158.82	IV
80	172.62	IV
80	177.16	IV
150	186.58	IV
100	194.28	IV
100	208.48	IV
100	208.73	IV
80	208.90	IV
150	212.56	IV
140	223.24	IV
120	223.60	IV
140	234.32	IV
120	234.70	IV
50	357.83	IV
200	358.72	IV
125	387.14	IV
100	388.22	IV
150	421.61	IV
140	469.77	IV
200	469.82	IV
180	469.87	IV
140	469.92	IV
120	521.74	IV
140	521.82	IV
80	541.13	IV
100	542.07	IV
150	543.89	IV
	Air	
65	2018.44	IV
110	2022.19	IV
30	2203.88	IV
10	2220.81	IV
250	2258.02	IV
175	2262.08	IV
110	2264.54	IV
550	2285.79	IV
30	2293.14	IV
250	2293.49	IV
250	2363.28	IV
110	2365.49	IV
250	2350.84	IV
450	2352.52	IV
700	2357.96	IV
250	2362.68	IV
350	2372.16	IV
65	2384.20	IV
350	2384.95	IV

Ne V

Ref. 69, 388, 389, 400, 413 — R.L.K.

Intensity	Wavelength	
	Vacuum	
66	119.01	V
200	122.52	V
66	125.12	V
45	131.99	V
50	132.04	V
150	140.76	V
150	140.79	V
100	142.44	V
100	142.50	V
150	142.72	V
100	143.27	V
150	143.34	V
150	147.13	V
66	151.23	V
120	151.42	V
45	154.50	V
100	164.02	V
100	164.14	V
500	173.93	V
400	357.96	V
500	358.47	V
500	359.38	V
1000	365.59	V
800	416.20	V
250	480.41	V
150	481.28	V
250	481.36	V
500	482.99	V
400	568.42	V
250	569.76	V
500	569.83	V
250	572.11	V
800	572.34	V
	Air	
75	2227.42	V
110	2232.41	V
65	2245.48	V
65	2259.57	V
65	2263.39	V
250	2265.71	V

NEPTUNIUM (Np)
Z = 93

Np I and II

Ref. 93 = J.G.C.

Intensity		Wavelength	
		Air	
300		3481.93	I
300	h	3501.50	I
300	l	3986.89	I
300	s	5044.66	I
300	l	5601.70	I
300	l	5652.75	I
300	l	5784.39	I
300	l	5878.04	I
300	s	6011.22	I
300		6056.09	I
300	s	6073.90	I
300	s	6080.05	I
300	l	6120.49	I
300		6188.59	I
300	l	6200.00	I
300	s	6215.90	I
300	s	6317.84	I
300	l	6341.38	I
300	l	6566.11	I
300	l	6720.68	I
300	s	6751.32	I
300	s	6795.21	I
300	l	6802.62	I
300	l	6805.81	I
300	s	6816.44	I
300	l	6865.45	I
300	s	6907.13	I
300	h	6912.91	I
1000	s	6930.31	I
300	l	6963.63	I
3000	s	6972.09	I
300		7014.02	I
300	l	7018.91	I
300	s	7039.14	I
300	s	7080.01	I
300	l	7174.83	I
300	l	7184.93	I
300	l	7284.28	I
300	l	7292.29	I
300	l	7332.52	I
300	s	7370.60	I
300	l	7381.03	I
300	l	7381.65	I
300	l	7402.70	I
300	s	7512.22	I
300	l	7515.15	I
300	l	7546.05	I
300	l	7624.83	I
300		7626.85	I
300	s	7681.01	I
300	s	7685.25	I
1000	l	7735.14	I
300	l	7761.61	I
1000	l	7765.75	I
300	s	7776.07	I
300		7787.46	I
1000	l	7791.38	I
300	l	7851.44	I
300	l	7887.88	I
300	l	7901.71	I
300	l	7975.98	I
300	h	8080.32	I
300	s	8124.59	I
300		8155.11	I
300	l	8167.42	I
300	l	8183.06	I
300	l	8188.61	I
300	l	8247.82	I
300	l	8287.11	I
300	s	8287.75	I
300	l	8306.22	I
300	s	8313.66	I
1000	l	8339.12	I
300		8356.79	I
300	l	8367.11	I
3000		8372.88	I
3000		8529.96	I
1000	s	8696.23	I
1000	s	8906.02	I
1000		8942.70	I
1000	s	9004.75	I
1000	l	9006.31	I
10000	l	9016.18	I
3000	l	9141.30	I
3000	s	9379.33	I
3000	l	9468.66	I
3000	s	9679.13	I
3000	l	9930.55	I
10000	l	10091.99	I
10000	s	10817.45	I
10000	l	11695.15	I
10000	l	11776.64	I
10000	s	12148.18	I
10000	s	12377.42	I
10000	l	12407.99	I
10000	l	13834.33	I

NICKEL (Ni)
Z = 28

Ni I and II

Ref. 1,294 = C.H.C.

Intensity	Wavelength	
	Vacuum	
500	1317.22	II
400	1335.20	II
500	1370.14	II
1000	1741.55	II
500	1748.28	II
	Air	
1000	2165.55	II
2000	2169.10	II
2000	2174.67	II
1500	2175.15	II
500	2177.09	II
400	2177.36	II
400	2179.35	II
800	2180.47	II
800	2184.60	II
2500	2185.50	II
3000	2192.09	II
600	2201.41	II
5000	2205.55	II
4000	2206.72	II
6000	2216.48	II
800	2220.40	II
500	2221.06	II
900	2222.96	II
500	2242.68	II
500	2253.85	II
1000	2264.46	II
2000	2270.21	II
800	2277.28	II
400	2278.32	II
800	2278.77	II
500	2287.65	II
1600	2289.98	I
400	2296.55	II
400	2297.14	II
630	2300.78	I
1000	2303.00	II
2000	2310.96	I
1700	2312.34	I
1400	2313.66	I
1400	2313.98	I
1000	2316.04	II
1400	2317.16	I
500	2319.75	II
2600	2320.03	I
1900	2321.38	I
240	2322.68	I
1400	2325.79	I
940	2329.96	I
500	2334.58	II
460	2337.49	I
160	2337.82	I
500	2341.20	II
1200	2345.54	I
190	2346.63	I
400	2347.52	I
160	2360.63	I
200	2362.06	I
1000	2375.42	II
240	2386.58	I
1000	2394.52	II
2000	2416.13	II
240	2419.31	I
85	2421.23	I
70	2423.33	I
70	2423.66	I
70	2424.03	I
500	2437.89	II
85	2453.99	I
160	2472.06	I
85	2476.87	I
500	2510.87	II
500	2565.92	II
500	2606.26	II
500	2609.94	II
500	2615.06	II
45	2696.49	I
150	2798.65	I
250	2821.29	I
500	2864.02	II
50	2865.50	I
60	2907.46	I
25	2914.01	I
500	2943.91	I
570	2981.65	I
250	2984.13	I
500	2992.60	I
1000	2994.46	I
4000	3002.49	I
2200	3003.63	I
3700	3012.00	I
350	3019.14	I
120	3031.87	I
1700	3037.94	I
150	3045.01	I
3500	3050.82	I
1500	3054.32	I
1900	3057.64	I
500	3064.62	I
420	3080.76	I
260	3097.12	I
210	3099.12	I
2600	3101.55	I
1300	3101.88	I
220	3105.47	I
270	3114.12	I
2900	3134.11	I
55	3145.72	I
55	3181.74	I
100	3184.37	I
55	3195.57	I
150	3197.11	I
55	3202.14	I
180	3214.06	I
180	3217.83	I
100	3221.27	I
150	3221.65	I
210	3225.02	I
1100	3232.96	I
290	3234.65	I
600	3243.06	I
100	3248.46	I
120	3250.74	I
100	3271.12	I
120	3282.70	I
400	3292.87	II
500	3297.60	II
400	3305.71	II
660	3315.66	I
330	3320.26	I
310	3322.31	I
2000	3331.88	II
400	3335.64	II
500	3338.09	II
500	3348.84	II
500	3349.24	II

Nickel (Cont.)

600		3358.68	II
330		3361.56	I
500		3363.45	II
330		3365.77	I
330		3366.17	I
65		3366.81	I
65		3367.89	I
2900		3369.57	I
400		3371.99	I
260		3374.22	I
130		3374.64	I
500		3378.97	II
3300		3380.57	I
240		3380.85	I
1300		3391.05	I
3300		3392.99	I
500		3401.05	II
130		3409.58	I
330		3413.48	I
330		3413.94	I
8200		3414.76	I
1600		3423.71	I
2600		3433.56	I
990		3437.28	I
4800		3446.26	I
1300		3452.89	I
5000		3458.47	I
5000		3461.65	I
200		3467.50	I
240		3469.49	I
1600		3472.54	I
550		3483.77	I
130		3485.89	I
5500		3492.96	I
660		3500.85	I
65		3502.60	I
55		3507.69	I
2600		3510.34	I
260		3513.93	I
6600		3515.05	I
660		3519.77	I
8200		3524.54	I
110		3527.98	I
330		3548.18	I
55		3551.53	I
65		3561.75	I
5000		3566.37	I
990		3571.87	I
130		3587.93	I
1300		3597.70	I
1300		3610.46	I
530		3612.74	I
6600		3619.39	I
130		3624.73	I
200		3664.10	I
130		3669.24	I
180		3670.43	I
260		3674.15	I
160		3688.42	I
80		3693.93	I
120		3722.48	I
150		3736.81	I
60		3739.23	I
600		3775.57	I
700		3783.53	I
700		3807.14	I
110		3831.69	I
1200		3858.30	I
30		3889.67	I
35		3972.17	I
110		3973.56	I
110		4401.55	I
85		4459.04	I
18		4462.46	I
55		4470.48	I
35		4592.53	I
18		4600.37	I
65		4605.00	I
18		4606.23	I
75		4648.66	I
23		4686.22	I
110		4714.42	I
22		4715.78	I
30		4756.52	I
15		4763.95	I
45		4786.54	I
22		4807.00	I
22	h	4829.03	I
19		4831.18	I
45		4855.41	I
30		4866.27	I
17		4873.44	I
40		4904.41	I
22		4918.36	I
16		4935.83	I
45		4980.16	I
45		4984.13	I
500		4992.02	II
16	h	5000.34	I
18		5012.46	I
50		5017.59	I
100		5035.37	I
16		5048.85	I
100		5080.52	I
65		5081.11	I
26	h	5084.08	I
18		5099.32	I
26	h	5099.95	I
21		5115.40	I
18	h	5129.38	I
23		5137.08	I
23	h	5142.77	I
40	h	5146.48	I
40	h	5155.76	I
16		5168.66	I
13		5176.56	I
8		5435.87	I
180		5476.91	I
6		5510.00	I
6		5578.73	I
9		5587.86	I
13		5592.28	I
9		5614.79	I
5	h	5625.33	I
4		5649.70	I
5		5664.02	I
12		5682.20	I
8		5695.00	I
23		5709.56	I
10		5711.90	I
10		5715.09	I
16		5754.68	I
8		5760.85	I
10		5857.76	I
10		5892.88	I
10		6108.12	I
10		6176.81	I
10		6191.18	I
13		6256.36	I
10		6314.66	I
16		6643.64	I
22		6767.77	I
9		6772.32	I
10		6914.56	I
5		7110.90	I
26		7122.20	I
6		7182.00	I
5		7197.02	I
5		7261.93	I
5		7291.45	I
4		7385.24	I
16		7393.60	I
16		7409.35	I
5		7414.51	I
23		7422.28	I
13		7522.76	I
9		7525.12	I
19		7555.60	I
8		7574.05	I
23		7617.00	I
9		7619.21	I
16		7714.32	I
5	h	7715.58	I
19		7727.61	I
19		7748.89	I
10		7788.94	I
13		7797.59	I
2		7917.44	I
1000		8096.75	II
500		8114.21	II
700		8121.48	II
2		8809.42	I
9		8862.55	I
500	w	9900.92	II

Ni III
Ref. 422 — C.H.C.

Intensity		Wavelength	
		Vacuum	
100		625.68	III
500		630.71	III
200		637.54	III
200		662.37	III
150		663.57	III
500		676.94	III
200		700.17	III
300		713.33	III
300		713.38	III
500		718.48	III
200		721.26	III
300		722.09	III
250		725.20	III
500		729.82	III
250		730.11	III
400		731.70	III
300		732.16	III
200		738.26	III
300		747.99	III
200		749.68	III
300		750.05	III
200		752.02	III
300		757.80	III
250		758.73	III
250		758.27	III
400		770.22	III
200		772.04	III
500		778.81	III
200	d	785.02	III
300		788.04	III
200		788.30	III
200		805.01	III
500		811.57	III
500		826.14	III
200		826.50	III
500		842.14	III
400		845.24	III
300		847.43	III
200		857.09	III
300		860.64	III
300		862.88	III
300		863.22	III
300		867.51	III
200		869.70	III
200		870.84	III
300		973.79	III
400		979.59	III
200		1428.87	III
200		1434.31	III
200		1451.50	III
300		1604.54	III
300		1652.87	III
200		1653.12	III
250		1656.13	III
200		1661.79	III
400		1687.90	III
1000		1692.51	III
200		1707.35	III
200		1707.43	III
800		1709.90	III
650		1715.30	III
500		1719.46	III
200		1721.26	III
400		1722.28	III
250		1733.13	III
500		1738.25	III
300		1739.78	III
300		1741.96	III
550		1747.01	III
300		1752.43	III
400		1753.01	III
800		1764.69	III
500		1767.94	III
2000		1769.64	III
400		1776.07	III
200		1788.30	III
250		1790.40	III
200		1790.93	III
200		1791.64	III
200		1794.90	III
300		1807.24	III
200		1811.69	III
300		1819.28	III
800		1823.06	III
400		1830.01	III
200		1830.08	III
650		1847.28	III
800		1854.15	III
300		1858.75	III
200		1930.43	III
200		1952.54	III

Ni IV
Ref. 415 — C.H.C.

Intensity	Wavelength	
	Vacuum	
33	392.68	IV
32	393.24	IV
49	424.40	IV
57	444.21	IV
67	469.67	IV
65	471.24	IV
65	485.42	IV
66	536.28	IV
67	537.96	IV
58	1345.72	IV
69	1357.07	IV
76	1398.19	IV
74	1411.45	IV
69	1419.58	IV
74	1421.22	IV
70	1427.45	IV
67	1428.93	IV
67	1435.24	IV
70	1438.82	IV
73	1449.01	IV
76	1452.22	IV
70	1455.42	IV
69	1472.63	IV
68	1476.82	IV
73	1482.25	IV
67	1489.53	IV
72	1489.83	IV
69	1493.01	IV
74	1493.67	IV
68	1498.71	IV
71	1498.77	IV
72	1498.90	IV
67	1499.97	IV
70	1512.74	IV
70	1516.66	IV
73	1520.63	IV
75	1525.31	IV
74	1527.68	IV
74	1527.80	IV
76	1534.71	IV
73	1537.25	IV
69	1538.93	IV
75	1543.41	IV
74	1546.23	IV
68	1548.04	IV
69	1557.28	IV
67	1560.18	IV

Ni V
Ref. 416 — C.H.C.

Intensity	Wavelength	
	Vacuum	
29	304.02	V
55	315.24	V
56	315.71	V
63	336.79	V
68	343.93	V
78	347.34	V
70	347.46	V
67	347.72	V
71	348.10	V
69	350.77	V
69	353.59	V
72	354.18	V
76	354.42	V
68	354.49	V
68	355.61	V
70	355.78	V
65	357.37	V
69	358.57	V
68	358.58	V
66	359.47	V
69	365.62	V
70	370.62	V
67	371.31	V
68	371.76	V
67	373.60	V
72	377.68	V
70	393.91	V
66	394.31	V
66	395.24	V
41	400.59	V

NIOBIUM (Nb)
Z = 41

Nb I and II
Ref. I = C.H.C.

Intensity	Wavelength	
	Air	
3300	2029.32	II
3000	2032.99	II
2000	2109.42	II
1700	2125.21	II
1100	2126.54	II
1500	2131.18	II
370	2295.68	II
280	2302.08	II
170	2376.40	II
110	2387.09	II
140	2387.52	II
45	2388.27	II
160	2398.48	II
55	2405.34	II
55	2405.85	II
140	2412.46	II
160	2416.99	II
140	2418.69	II
75	2433.80	II
40	2435.95	II
35	2436.33	I
45	2437.42	II
40	2442.14	II
28	2442.68	II
65	2451.87	II
65	2453.95	II
55	2458.09	II
65	2462.89	I
35	2466.73	I
55	2469.08	I
110	2477.38	II
65	2478.29	II

Niobium (Cont.)

65		2479.94	II
35		2483.88	II
110		2504.65	I
110		2511.00	II
110		2521.40	II
390		2544.80	II
110		2551.38	II
130		2556.94	II
130		2562.41	II
130		2565.41	I
100		2569.03	I
110		2571.33	II
200		2578.74	I
390		2583.99	II
390		2590.94	II
270		2592.20	I
130		2616.48	I
130		2623.51	I
130		2627.44	I
130		2628.49	I
200		2642.24	II
320		2646.26	II
330		2647.50	I
240		2649.52	I
330		2654.45	I
310		2656.08	II
160		2657.62	I
110		2665.25	II
110		2666.59	II
110		2667.30	II
130		2668.29	I
400		2671.93	II
200		2673.57	II
200		2675.94	II
130		2687.15	I
160		2691.77	II
1000		2697.06	II
320		2698.86	II
320		2702.20	II
150		2702.52	II
470		2716.62	II
470		2721.98	II
310		2733.26	II
110		2737.09	II
200		2746.91	I
200		2748.85	I
190		2753.01	I
280		2758.61	I
240		2768.13	II
310		2773.20	I
270		2780.24	II
130		2782.36	I
110		2793.05	II
190		2827.08	II
150		2836.24	I
110		2840.94	I
250		2841.15	II
280		2842.65	II
160		2846.28	II
110		2851.45	I
240		2861.09	II
100		2864.32	I
100		2865.61	II
500		2868.52	II
800		2875.39	II
270		2876.95	II
530		2877.03	II
100		2880.72	II
570		2883.18	II
280		2888.83	II
470		2897.81	II
400		2899.24	II
470		2908.24	II
670		2910.59	II
470		2911.74	II
1100		2927.81	II
110		2931.47	II
870		2941.54	II
110	h	2945.88	II
110		2946.12	II
110		2946.90	II
1100		2950.88	II
400		2972.57	II
320		2974.10	II
210		2977.68	II
200		2982.11	II
330		2990.26	II
470		2994.73	II
140		3024.74	II
350		3028.44	II
300		3032.77	II
100		3044.76	II
150		3048.10	I
110		3053.09	I
100		3055.52	II
220		3064.53	II
110		3069.68	II
100		3070.90	II
110		3071.56	II
100		3073.24	II
400		3076.87	II
110		3080.35	II
1800		3094.18	II
140		3099.19	II
150		3111.45	I
270		3127.53	II
1500		3130.79	II
390		3145.40	II
140		3151.87	I
1200		3163.40	II
150		3175.78	II
390		3180.29	II
200		3187.49	I
300		3191.10	II
150		3191.43	II
1000		3194.98	II
120		3203.35	II
300		3206.34	II
390		3215.60	II
800		3225.48	II
140		3229.56	II
400		3236.40	II
200		3247.47	II
120		3248.94	II
160		3249.52	I
320		3254.07	II
230		3260.56	II
160		3263.37	II
160		3264.59	I
120		3270.47	I
100		3270.76	I
200		3272.07	I
160		3277.67	I
200		3283.46	II
230		3285.66	I
200		3287.59	I
160		3287.92	I
160		3292.02	II
320		3296.01	I
160		3299.61	I
120		3304.83	I
120		3308.05	I
120		3310.47	I
400		3312.60	I
200		3315.22	I
200		3318.98	I
120		3319.26	I
120		3319.58	II
240		3326.62	I
170		3329.36	I
110		3332.16	I
130		3341.60	II
1300		3341.97	I
1300		3343.71	I
130		3346.93	I
1700		3349.06	I
420		3349.52	I
340		3354.74	I
130		3357.04	I
1700		3358.42	I
130		3365.58	II
340		3366.96	I
130		3369.16	II
170		3371.33	I
350		3374.92	I
270		3380.41	I
130		3380.86	I
170		3386.24	II
350		3392.34	I
170		3395.93	I
120		3399.40	I
230		3405.41	I
130		3406.13	I
270		3408.38	I
230		3408.68	II
180		3409.19	II
230		3412.94	II
180		3415.97	I
180		3423.76	I
230		3425.42	II
130		3425.85	I
230		3426.57	II
230		3427.45	I
130		3429.04	I
180		3432.70	II
180		3440.59	II
170		3463.81	I
180		3465.86	I
130		3469.44	I
100		3471.19	I
140		3473.02	I
290		3478.69	I
200		3479.56	II
100		3484.05	II
230		3491.03	I
200		3497.81	I
500		3498.63	I
460		3507.96	I
200		3510.26	II
200		3515.42	II
200		3517.67	II
200		3520.06	I
2000		3535.30	I
1300		3537.48	I
250		3540.96	II
500		3544.02	I
250		3544.65	I
300		3550.45	I
250		3554.52	I
1000		3554.66	I
630		3563.50	I
630		3563.62	I
1500		3575.85	I
200		3577.72	I
5000		3580.27	I
500		3584.97	I
750		3589.11	I
500		3589.36	I
500		3593.97	I
500		3602.56	I
300		3619.51	II
200		3621.03	I
200		3639.33	I
420		3649.85	I
250		3650.81	I
400		3651.19	II
200		3659.61	II
630		3660.37	I
900		3664.70	I
220		3669.01	I
270		3674.78	I
1500		3697.85	I
330		3711.34	I
3300		3713.01	I
480		3716.99	I
2700		3726.24	I
270		3738.42	I
2700		3739.80	I
670		3740.73	II
270		3741.78	I
1700		3742.39	I
250		3753.18	I
210		3755.77	I
530		3763.49	I
350		3765.08	I
250		3766.13	I
530		3771.85	I
870		3781.01	I
1700		3787.06	I
1300		3790.15	I
3500		3791.21	I
2700		3798.12	I
270		3801.30	I
2700		3802.92	I
670		3803.88	I
530		3804.74	I
670		3810.49	I
530		3811.03	I
530		3815.51	I
210		3818.86	II
210		3819.15	I
670		3824.88	I
350		3835.18	I
250		3836.45	I
210		3845.90	I
290		3858.95	I
350		3863.38	I
270		3867.92	I
530		3877.56	I
870		3878.82	I
670		3883.14	I
1100		3885.44	I
670		3885.68	I
210		3886.07	I
580		3891.30	I
210		3908.97	I
670		3914.70	I
530		3920.20	I
670		3937.44	I
520		3943.67	I
250		3965.69	I
910	d	3966.09	I
210		3971.85	I
1100		4032.52	I
250		4039.53	I
16000	c	4058.94	I
210		4059.51	I
350		4060.79	I
210		4068.26	I
12000		4079.73	I
270		4084.86	I
440		4100.40	I
6700		4100.92	I
310		4116.90	I
5300		4123.81	I
670		4129.43	I
770		4129.93	I
2300		4137.10	I
440		4139.44	I
2700		4139.71	I
350		4143.21	I
870		4150.12	I
4400		4152.58	I
870		4163.47	I
4400		4163.66	I
4000		4164.66	I
3500		4168.13	I
310		4184.44	I
1200		4190.88	I
870		4192.07	I
870		4195.09	I
1300		4195.66	I
310		4198.51	I
350		4201.52	I
870		4205.31	I
350		4214.73	I
420		4217.94	I
420		4229.15	I
250		4255.44	I
770		4262.05	I
420		4266.02	I
290		4270.69	I
400		4286.99	I
580		4299.60	I
580		4300.99	I
120		4309.56	I
390		4311.27	I
120		4312.45	I
350		4326.33	I
120		4327.38	I
390		4331.37	I
140		4342.82	I
140		4348.65	I
110		4349.03	I
290		4351.57	I
210		4368.43	I
140		4377.96	I
130		4388.36	I
160		4392.69	I
330		4410.21	I
190		4419.44	I
230	c	4437.22	I
290		4447.18	I
140		4456.80	I
140		4457.42	I
140		4469.71	I
140		4471.29	I
140		4472.53	I
150		4503.04	I
530		4523.41	I
480		4546.82	I
370		4564.53	I
720		4573.08	I
480		4581.62	I
1200		4606.77	I
170		4616.17	I
450		4630.11	I
450		4648.95	I
110		4649.27	I
450		4663.83	I
340		4666.24	I
240		4667.22	I
580		4672.09	I
530		4675.37	I
110		4678.48	I
320		4685.14	I
130	c	4706.14	I
260		4708.29	I
150		4713.50	I
110	c	4733.89	I
220	c	4749.70	I
110		4816.38	I
110	c	4848.37	I
130	c	4967.78	I
110		4973.14	I
190		4988.97	I
85		5000.95	I
65		5002.25	I
40		5013.27	I
230		5017.75	I
40		5019.51	I
150		5026.36	I
40		5030.13	I
210		5039.04	I
40		5047.96	I
170		5058.01	I
65		5059.35	I
130		5065.25	I
40		5077.40	I
750		5078.96	I
40	c	5094.41	I
420		5095.30	I
170		5100.16	I
170		5120.30	I
85		5121.80	I
85		5127.66	I
40		5133.34	I
210		5134.75	I
75		5140.58	I
75		5147.54	I
40		5150.64	I
75		5152.63	I
250		5160.33	I
250		5164.38	I
230		5180.31	I
110		5186.98	I
190		5189.20	I
170		5193.08	I
150		5195.84	I
65		5203.22	I
35		5205.13	I
85	c	5219.10	I

Niobium (Cont.)

Intensity		Wavelength	
65		5225.16	I
150		5232.81	I
85	c	5237.43	I
29		5240.39	I
150	d	5251.62	I
		5251.81	I
75		5253.03	I
85		5253.93	I
50		5269.92	I
270		5271.53	I
25		5272.48	I
130	c	5276.20	I
29	c	5279.43	I
50		5285.26	I
35		5296.34	I
50		5315.55	I
17		5317.01	I
250		5318.60	I
50		5319.49	I
75		5334.87	I
25		5336.81	I
50		5340.80	I
25		5343.58	I
460		5344.17	I
340		5350.74	I
40		5353.28	I
25		5355.31	I
40		5355.70	I
29		5359.19	I
17		5362.01	I
40		5375.27	I
40		5381.34	I
17		5388.30	I
21		5395.86	I
29		5396.33	I
29		5411.24	I
21		5416.30	I
65		5422.44	I
21		5431.26	I
110		5437.27	I
19		5448.31	I
19		5456.19	I
40		5458.04	I
19	h	5468.10	I
40		5481.00	I
13		5483.09	I
19		5483.49	I
13		5491.06	I
17		5499.53	I
40		5504.58	I
17		5509.17	I
35	c	5512.82	I
17		5517.39	I
50		5523.57	I
25		5541.47	I
85		5551.35	I
29		5563.00	I
17	c	5571.44	I
35	c	5576.16	I
35		5578.29	I
50		5586.97	I
17	c	5590.95	I
13		5594.89	I
17	c	5599.59	I
40		5603.52	I
13		5603.93	I
25		5628.26	I
65		5629.17	I
35	c	5635.42	I
170		5642.11	I
35		5645.30	I
17		5654.14	I
130		5664.71	I
170		5665.63	I
17		5666.86	I
65	Cw	5671.02	I
85		5671.91	I
25		5677.47	I
25		5693.09	I
35	d	5697.90	I
		5698.03	I
40		5706.16	I
85		5706.48	I
29		5709.33	I
17		5715.59	I
65		5716.35	I
25		5725.66	I
130		5729.19	I
21		5737.36	I
13		5738.20	I
85		5751.44	I
110		5760.34	I
65		5764.99	I
29		5771.08	I
50	c	5776.07	I
17		5780.34	I
85		5787.54	I
17		5789.79	I
50		5794.24	I
50		5804.03	I
29	h	5815.33	I
110		5819.43	I
35		5820.62	I
75		5834.90	I
25		5838.15	I
130	d	5838.64	I
50		5842.47	I
17		5846.09	I
65		5866.47	I
35		5874.70	I
17		5877.79	I
40		5893.44	I
190	Cw	5900.62	I
40	c	5903.80	I
29		5927.41	I
40	c	5934.16	I
40		5957.70	I
150		5983.22	I
65		5986.08	I
85	Cw	5997.93	I
50		6029.75	I
50		6031.84	I
50		6045.50	I
25		6048.72	I
29		6056.65	I
29		6107.71	I
40		6142.51	I
50		6148.13	I
50		6164.32	I
29		6213.06	I
75		6221.96	I
40	c	6251.76	I
21		6260.77	I
85	c	6430.46	I
50	c	6433.22	I
17		6497.84	I
65		6544.61	I
15		6574.73	I
19	Cw	6591.00	I
19		6606.16	I
19		6607.28	I
35		6614.15	I
19		6626.98	I
210	Cw	6660.84	I
150	Cw	6677.33	I
65		6701.20	I
130	c	6723.62	I
75		6739.88	I
25		6795.31	I
85		6828.11	I
25	c	6849.35	I
19		6870.92	I
40		6876.36	I
25	c	6902.89	I
35		6908.07	I
40		6918.32	I
17		6946.07	I
17		6972.49	I
25		6986.09	I
85		6990.32	I
17	c	6996.11	I
21		7023.48	I
17		7038.04	I
190	c	7046.81	I
8		7066.41	I
8		7075.23	I
40	c	7098.94	I
17	Cw	7102.01	I
19		7119.31	I
15		7122.95	I
35		7126.17	I
17		7130.06	I
130		7159.43	I
17		7191.37	I
19	c	7208.94	I
50		7252.35	I
15		7274.81	I
13		7317.03	I
17	c	7323.92	I
29	Cw	7328.38	I
65	c	7353.16	I
190	Cw	7372.50	I
13		7419.83	I
15		7436.02	I
19		7478.20	I
65		7515.93	I
29	c	7519.77	I
170	c	7574.58	I
17	c	7583.21	I
13		7639.81	I
13		7647.71	I
25		7703.33	I
75	c	7726.68	I
25		7757.31	I
6		7787.11	I
13	Cw	7873.41	I
35		7885.31	I
25		7938.89	I
8		7954.76	I
40		8135.20	I
13	Cw	8240.00	I
29	Cw	8320.93	I
29		8346.08	I
10		8350.04	I
17		8439.77	I
17	Cw	8475.98	I
25		8526.99	I
13	c	8547.25	I
17	c	8560.54	I
17		8575.87	I
21	c	8697.55	I
21		8740.96	I
21		8767.97	I
29	Cw	8815.56	I
35		8905.78	I

Nb III
Ref. 392 — C.H.C.

Intensity		Wavelength	
		Vacuum	
60		1314.56	III
50		1319.15	III
60		1431.92	III
60		1433.39	III
50		1435.26	III
80		1445.43	III
80		1445.98	III
80		1447.09	III
50		1448.50	III
60		1451.63	III
100		1456.68	III
80		1484.73	III
50		1486.79	III
100		1495.94	III
80		1498.02	III
80		1499.45	III
50		1501.53	III
100		1501.99	III
60		1505.03	III
50		1509.71	III
50		1512.34	III
50		1513.25	III
80		1513.81	III
50		1517.38	III
100		1524.91	III
60		1532.98	III
60		1537.50	III
50		1566.92	III
50		1570.19	III
50		1586.82	III
100		1590.21	III
80		1598.86	III
80		1604.72	III
80		1639.51	III
100		1682.77	III
60		1684.40	III
100		1705.44	III
100		1707.14	III
50		1739.30	III
50		1758.63	III
60		1763.72	III
50		1808.70	III
50		1863.13	III
100		1892.92	III
100		1938.84	III
60	h	1979.07	III
50		1985.15	III
50		1997.11	III
		Air	
50	h	2007.28	III
50		2032.47	III
50		2060.29	III
80	h	2130.24	III
60		2206.01	III
60		2240.31	III
60		2244.19	III
60		2265.63	III
80		2273.92	III
100		2275.23	III
80		2279.36	III
100		2281.51	III
80		2284.40	III
100		2290.36	III
60		2304.78	III
50		2309.92	III
100		2313.30	III
50		2330.22	III
100		2338.09	III
80		2344.12	III
90		2349.21	III
80		2355.54	III
100		2362.06	III
80		2362.50	III
80		2365.70	III
100		2372.73	III
100		2387.41	III
80		2388.23	III
50		2404.23	III
80		2404.89	III
100		2413.94	III
60		2414.50	III
100		2421.91	III
50		2437.74	III
60		2446.45	III
100		2456.99	III
50		2460.34	III
50		2460.45	III
50		2463.72	III
80		2468.72	III
60		2469.39	III
80		2475.87	III
50		2486.02	III
60		2488.74	III
60		2493.02	III
100		2499.73	III
60		2508.53	III
50		2511.95	III
100		2545.64	III
80		2557.94	III
50		2567.44	III
60		2593.75	III
80		2598.86	III
50		2628.67	III
80		2633.17	III
80		2657.99	III
50		2937.71	III
80		3001.84	III
80		3142.26	III
60		3266.11	III

Nb IV
Ref. 407 — C.H.C.

Intensity	Wavelength	
	Vacuum	
12	542.38	IV
12	543.09	IV
12	545.21	IV
10	559.94	IV
10	566.22	IV
18	981.27	IV
60	993.54	IV
18	996.16	IV
50	1002.76	IV
400	1005.72	IV
500	1007.05	IV
500	1010.19	IV
45	1030.27	IV
100	1116.08	IV
150	1120.02	IV
50	1447.48	IV
40	1473.43	IV
40	1487.23	IV
60	1502.30	IV
60	1524.36	IV
50	1534.06	IV
40	1635.68	IV
40	1910.70	IV
60	1922.41	IV
60	1978.22	IV
	Air	
40	2027.50	IV
65	2032.53	IV
40	2034.67	IV
40	2068.62	IV
55	2084.07	IV
35	2093.12	IV
50	2093.30	IV
20	2122.68	IV
25	2130.23	IV
45	2146.36	IV
18	2249.98	IV

Nb V
Ref. 431 — C.H.C.

Intensity	Wavelength	
	Vacuum	
80	464.55	V
80	468.32	V
60	753.01	V
80	763.77	V
80	774.02	V
40	1007.02	V
50	1044.90	V
70	1212.21	V
100	1258.87	V
40	1267.60	V
100	1758.33	V
100	1877.34	V

NITROGEN (N)
Z = 7

N I and II
Ref. 213 = R.L.K.

Intensity	Wavelength

Nitrogen (Cont.)

Vacuum

Intensity		Wavelength	Spectrum
285		644.634	II
360		644.837	II
450		645.178	II
140		647.50	I
360		660.286	II
170		671.016	II
285		671.386	II
150		671.630	II
160		671.773	II
170		672.001	II
350		692.70	I
285		746.984	II
650		775.965	II
90		885.67	I
90		909.697	I
80		910.278	I
40		910.645	I
450		915.612	II
450		915.962	II
550		916.012	II
650		916.701	II
90		953.415	I
100		953.655	I
130		953.970	I
130		963.990	I
115		964.626	I
70		965.041	I
90		1067.614	I
60		1068.612	I
450		1083.990	II
600		1084.580	II
430		1085.546	II
650		1085.701	II
175		1097.237	I
115		1098.095	I
115		1098.260	I
105		1100.360	I
40		1100.465	I
90		1101.291	I
360		1134.165	I
385		1134.415	I
410		1134.980	I
105		1143.65	I
130		1163.884	I
60		1164.206	I
105		1164.325	I
270		1167.448	I
105		1168.334	I
60		1168.417	I
195		1168.536	I
230		1176.510	I
105		1176.630	I
195		1177.695	I
410		1199.550	I
385		1200.223	I
360		1200.710	I
175		1225.026	I
160		1225.37	I
130		1228.41	I
160		1228.79	I
360		1243.179	I
315		1243.306	I
290		1310.540	I
250		1310.95	I
230		1319.00	I
315		1319.68	I
115		1326.57	I
115		1327.92	I
360		1411.94	I
700		1492.625	I
490		1492.820	I
640		1494.675	I
775		1742.729	I
700		1745.252	I

Air

Intensity		Wavelength	Spectrum
160		2095.53	II
70		2096.20	II
110		2096.86	II
110		2130.18	II
160		2142.78	II
160		2206.09	II
160		2286.69	II
110		2288.44	II
220		2316.49	II
160		2316.69	II
285		2317.05	II
160		2461.27	II
110		2496.83	II
70		2496.97	II
110		2520.22	II
160		2520.79	II
220		2522.23	II
110		2590.94	II
160		2709.84	II
110		2799.22	II
110		2823.64	II
160		2885.27	II
220		3006.83	II
360		3437.15	II
285		3838.37	II
360		3919.00	II
450		3955.85	II
1000		3995.00	II
360		4035.08	II
550		4041.31	II
360		4043.53	II
140		4099.94	I
185		4109.95	I
285		4176.16	II
285		4227.74	II
285		4236.91	II
220		4237.05	II
450		4241.78	II
285		4432.74	II
650		4447.03	II
360		4530.41	II
550		4601.48	II
450		4607.16	II
360		4613.87	II
450		4621.39	II
870		4630.54	II
550		4643.08	II
285		4788.13	II
450		4803.29	II
180		4847.38	I
285		4895.11	II
160		4914.94	I
210		4935.12	I
160		4950.23	I
350		4963.98	I
285		4987.37	II
450		4994.36	II
650		5001.48	II
360		5002.70	II
870		5005.15	II
550		5007.32	II
450		5010.62	II
360		5016.39	II
360		5025.66	II
550		5045.10	II
185		5281.20	I
140		5292.68	I
450		5495.67	II
285		5535.36	II
650		5666.63	II
550		5676.02	II
870		5679.56	II
450		5686.21	II
450		5710.77	II
285		5747.30	II
700		5752.50	I
240		5764.75	I
265		5829.54	I
235		5854.04	I
360		5927.81	II
550		5931.78	II
285		5940.24	II
650		5941.65	II
285		5952.39	II
160		5999.43	I
210		6008.47	I
285		6167.76	II
360		6379.62	II
185		6411.65	I
210		6420.64	I
210		6423.02	I
210		6428.32	I
185		6437.68	I
235		6440.94	I
185		6457.90	I
300		6468.44	I
750		6482.05	II
360		6482.70	I
300		6483.75	I
265		6481.71	I
325		6484.80	I
160		6491.22	I
210		6499.54	I
185		6506.31	I
750		6610.56	II
185		6622.54	I
185		6636.94	I
235		6644.96	I
185		6646.50	I
235		6653.46	I
210		6656.51	I
185		6722.62	I
210		7398.64	I
160		7406.12	I
265		7406.24	I
685		7423.64	I
785		7442.29	I
900		7468.31	I
185		7608.80	I
450		7762.24	II
400		8184.87	I
400		8188.02	I
250		8200.36	I
300		8210.72	I
570		8216.34	I
400		8223.14	I
400		8242.39	I
550		8438.74	II
500		8567.74	I
570		8594.00	I
650		8629.24	I
500		8655.89	I
220		8676.08	II
700		8680.28	I
650		8683.40	I
500		8686.15	I
110		8687.43	II
110	h	8699.00	II
500		8703.25	I
160	h	8710.54	II
570		8711.70	I
500		8718.83	I
250		8728.89	I
200		8747.36	I
500		9386.80	I
570		9392.79	I
250		9460.68	I
200		9863.33	I
160	h	9865.41	II
110	h	9868.21	II
160	h	9887.39	II
220	h	9891.09	II
160	h	9961.86	II
220	h	9969.34	II
285	h	10023.27	II
220	h	10035.45	II
220	h	10065.15	II
160	h	10070.12	II
250		10105.13	I
300		10108.89	I
350		10112.48	I
400		10114.64	I
110	h	10126.27	II
250		10539.57	I
200		12074.51	I
380		12186.82	I
225		12288.97	I
290		12328.76	I
310		12381.65	I
180		12438.40	I
510		12461.25	I
920		12469.62	I
500		13429.61	I
840		13581.33	I
180		13587.73	I
180		13602.27	I
290		13624.18	I
250		14757.07	I
100		14868.87	I
160		14966.60	I
180		15582.27	I
120	s	17516.58	I
100	l	17584.86	I
100		17878.26	I

N III
Ref. 66, 213 = R.L.K.

Vacuum

Intensity		Wavelength	
500		257.95	III
650		258.50	III
700		259.19	III
800		260.09	III
800		261.28	III
500		262.91	III
500		265.23	III
500		265.27	III
500		268.70	III
150		314.715	III
200		314.850	III
90		314.877	III
600		323.26	III
500		338.35	III
500		340.20	III
500		351.98	III
120		362.833	III
150		362.881	III
150		362.946	III
90		362.985	III
300		374.204	III
350		374.441	III
500		387.48	III
250		451.869	III
300		452.226	III
500		684.996	III
570		685.513	III
650		685.816	III
500		686.335	III
500		763.336	III
570		764.359	III
250		771.544	III
300		771.901	III
350		772.385	III
200		772.891	III
150		772.975	III
650		979.842	III
700		979.919	III
900		989.790	III
700		991.514	III
1000		991.579	III
500		1183.031	III
570		1184.550	III
150		1387.371	III
250		1729.945	III
570		1747.848	III
350		1751.218	III
650		1751.657	III
150		1804.486	III
200		1805.669	III
150		1846.42	III
350		1885.06	III
400		1885.22	III
200		1907.99	III
150		1919.55	III
150		1919.77	III
300		1920.65	III
150		1920.84	III
200		1921.30	III

Air

Intensity		Wavelength	
200		2064.01	III
250		2064.42	III
120		2068.68	III
90		2071.09	III
90		2117.59	III
90		2121.50	III
90		2147.31	III
200		2188.20	III
150		2188.38	III
250	w	2682.18	III
90		2689.20	III
120		3367.34	III
90		3754.67	III
120		3771.05	III
90		3938.52	III
150		3998.63	III
200		4003.58	III
250		4097.33	III
200		4103.43	III
120		4195.76	III
150		4200.10	III
90		4332.91	III
120		4345.68	III
300		4379.11	III
90		4510.91	III
120		4514.86	III
90		4634.14	III
120		4640.64	III
90		4858.82	III
150		4867.15	III
90		5314.35	III
200		5320.82	III
150		5327.18	III
90		6454.11	III
120		6467.02	III

N IV
Ref. 108, 212 = R.L.K.

Vacuum

Intensity		Wavelength	
400		181.75	IV
400		191.7	IV
400		192.9	IV
500		196.87	IV
500		197.23	IV
500		202.60	IV
500		205.94	IV
500		205.97	IV
500		206.03	IV
500		217.20	IV
500	d	217.90	IV
500	d	223.4	IV
800	w	225.12	IV
800		225.21	IV
600	w	234.12	IV
600	w	234.20	IV
600	w	234.25	IV
550		236.07	IV
500		237.99	IV
500	w	238.7	IV
600		238.80	IV
500	w	239.62	IV
900		247.20	IV
500	w	248.43	IV
500	w	248.46	IV
500	w	248.48	IV
600		260.45	IV
650		270.99	IV
250		283.42	IV
300		283.48	IV
350		283.58	IV
600		285.56	IV
600	w	297.7	IV
700		297.82	IV
650		300.32	IV
90		303.123	IV

Nitrogen (Cont.)

500		303.28	IV
150		315.053	IV
120		322.503	IV
150		322.570	IV
200		322.724	IV
120		323.175	IV
300		335.050	IV
500	w	351.93	IV
700		353.06	IV
500		420.77	IV
650		463.74	IV
570		765.148	IV
520		921.992	IV
500		922.519	IV
480		923.057	IV
520		921.992	IV
500		922.519	IV
520		924.283	IV
1000		955.335	IV
150	w	1036.16	IV
90		1078.71	IV
90		1188.01	IV
1000		1718.55	IV
		Air	
90		2080.34	IV
90	w	2318.09	IV
150		2477.69	IV
250		2645.65	IV
300		2646.18	IV
350		2646.96	IV
90		3078.25	IV
90		3463.37	IV
570		3478.71	IV
500		3482.99	IV
400		3484.96	IV
90		3747.54	IV
150		4057.76	IV
90		4606.33	IV
150		6380.77	IV

N V

Ref. 66, 107, 318 = R.L.K.

Intensity		Wavelength	
		Vacuum	
52		166.947	V
52		186.069	V
62		186.153	V
90		209.303	V
90		247.561	V
120		247.706	V
150		266.196	V
200		266.379	V
90		713.518	V
150		713.860	V
150		748.195	V
200		748.291	V
1000		1238.821	V
900		1242.804	V
90		1549.336	V
200	l	1616.33	V
350	l	1619.69	V
90	w	1860.37	V
		Air	
60	l	2859.16	V
90	l	2974.52	V
150	w	2980.78	V
250	w	2981.31	V
60	w	2998.43	V
350		4603.73	V
250		4619.98	V
200	w	4944.56	V
60	w	7618.46	V

OSMIUM (Os)
Z = 76

Os I and II

Ref. 1 = C.H.C.

Intensity		Wavelength	
		Air	
9600		2001.45	I
13000		2003.73	I
9000		2004.78	
17000		2010.15	I
29000		2018.14	I
29000		2020.26	
14000		2022.76	I
14000		2028.23	I
18000		2034.44	I
26000		2045.36	I
8600		2058.69	I
		2058.78	I
13000		2061.69	I
7800		2067.21	II
4200		2070.67	II
7200		2076.95	I
7200		2078.09	
14000		2079.97	I
2900		2082.54	I
2900		2089.03	I
2900		2089.21	I
6000		2097.60	I
5500		2100.63	I
2100		2117.66	I
4800		2117.96	I
6600		2119.79	
1900		2123.84	I
5300		2137.11	I
2400		2149.97	
2600		2154.59	I
1300		2157.84	I
1200		2158.53	I
2400		2161.00	
3100		2166.90	I
1100		2167.75	I
2100		2171.65	I
960		2184.68	I
840		2194.39	II
760		2202.49	I
600		2227.98	I
1100		2234.61	I
1300		2252.15	I
2000		2255.85	II
1400		2264.60	I
360		2268.28	I
960		2270.17	I
1400		2282.26	II
840		2283.67	I
570		2289.32	I
380		2297.31	I
660		2308.31	I
190		2313.75	II
550		2320.18	I
310		2323.98	I
660		2324.24	I
330		2326.99	I
310		2334.56	I
720		2336.80	II
430		2338.63	I
290		2340.69	I
430		2343.74	I
260		2345.75	I
430		2347.38	I
230		2350.23	II
360		2352.99	I
120		2355.28	II
240		2356.92	I
240		2357.25	I
310		2362.41	I
900		2362.77	I
500		2367.35	II
290		2369.24	I
500		2370.70	I
480		2371.18	I
95		2375.06	II
2600		2377.03	I
260		2377.61	I
900		2379.39	I
240		2384.62	I
1700		2387.29	I
330		2394.29	I
290		2395.39	I
1100		2395.88	I
220		2396.78	I
960		2401.13	I
260		2402.23	I
200		2403.54	I
330		2403.85	I
95		2405.08	II
290		2405.45	I
200		2405.96	I
360		2408.67	I
240		2410.98	I
290		2414.52	I
530		2417.99	I
530		2418.53	I
95		2420.02	II
200		2423.07	II
70		2424.02	II
500		2424.56	I
1400		2424.97	I
240		2426.81	I
70		2427.90	II
380		2431.19	I
380		2431.61	I
360		2446.02	I
900		2450.74	I
530		2451.73	I
530		2453.90	I
110		2454.91	II
530		2456.46	I
1800		2461.42	I
110		2468.90	II
290		2472.28	I
290		2474.78	I
900		2476.84	I
360		2482.43	I
530		2486.24	II
4500		2488.55	I
290		2491.02	I
290		2491.69	I
360		2492.42	I
2600		2498.41	I
330		2499.92	I
330		2502.29	I
500		2504.39	I
260		2504.51	I
35		2507.18	II
70		2509.71	II
660		2512.87	I
2400		2513.25	I
660		2515.04	I
500		2517.92	I
660		2518.44	I
200		2519.29	I
330		2519.79	I
200		2532.44	I
780		2538.00	II
240		2538.10	I
1000		2542.51	I
30		2548.83	II
310		2554.46	I
190		2563.16	II
600		2566.49	I
290		2566.88	I
480		2568.83	I
340		2571.78	I
150		2578.32	II
130		2580.03	II
360		2581.05	I
740		2581.96	I
1000		2590.76	I
200		2591.98	I
170		2596.00	II
210		2609.20	I
380		2609.56	I
400		2610.78	I
470		2612.63	I
1800		2613.06	I
800		2619.94	I
230		2620.62	I
530		2621.82	I
380		2628.48	I
27		2631.22	II
3800		2637.13	I
1900		2644.11	I
340		2646.89	I
380		2647.73	I
380		2649.34	I
490		2656.68	I
1900		2658.60	I
640		2659.83	I
380		2661.18	I
40		2664.29	II
580		2674.57	I
400		2674.88	I
2100		2689.82	I
510		2699.59	I
580		2706.70	I
3000		2714.64	I
580		2715.36	I
1300		2720.04	I
850		2721.86	I
580		2730.61	I
40		2731.36	II
580		2732.80	I
690		2761.42	I
470		2763.27	I
340		2765.04	I
960		2770.71	I
300		2776.91	I
740		2782.55	I
40		2783.88	II
640		2786.31	I
230		2793.99	I
230		2794.19	I
530		2796.73	I
320		2804.07	I
2800		2806.91	I
470		2808.94	I
420		2813.84	I
740		2814.20	I
300		2815.78	I
420		2829.27	I
230		2837.42	I
470		2838.17	I
5100		2838.63	I
740		2841.60	I
2300		2844.40	I
420		2846.39	I
420		2848.25	I
1500		2850.76	I
1500		2860.96	I
35		2863.37	II
360		2874.96	I
300		2878.40	I
35		2879.39	II
30		2880.20	II
260		2896.06	I
9600		2909.06	I
2100		2912.33	I
530		2917.26	I
2100		2919.79	I
300		2925.57	I
360		2929.51	I
510		2931.28	I
260		2934.64	I
200		2942.85	I
1100	h	2948.23	I
1400		2949.53	I
210	d	2949.81	I
300		2961.01	I
530		2962.15	I
450		2964.06	I
740		2970.97	I
450		2977.64	I
510		2982.90	I
340		2983.49	
260		2997.65	I
330		3013.07	I
570		3017.25	I
4400		3018.04	I
480		3019.38	I
1100		3030.70	I
2900		3040.90	I
210		3043.50	I
120		3042.74	II
230		3049.46	I
210		3050.39	I
8600		3058.66	I
290		3060.30	I
570		3062.19	I
210		3069.94	I
360		3074.08	I
290		3074.96	I
290		3077.44	I
1100		3077.72	I
360		3078.11	I
230		3078.38	I
230		3090.08	I
270		3093.59	I
310		3101.53	I
360		3105.99	I
310		3108.98	I
620		3109.38	I
250		3111.09	I
310		3118.33	I
480		3131.12	I
250		3152.67	I
290		3153.61	I
3100		3156.25	I
250		3156.78	I
310		3166.51	I
180		3173.93	II
420		3178.06	I
230		3181.88	I
230		3185.33	I
310		3186.98	I
310		3189.46	I
310		3194.23	I
150		3213.31	II
1900		3232.06	I
290		3238.63	I
190		3241.04	I
120		3248.00	I
190		3254.91	I
190		3256.92	I
190		3260.30	I
3100		3262.29	I
380		3262.75	I
3100		3267.94	I
620		3269.21	I
190		3272.16	I
530		3275.20	I
330		3277.97	I
190		3288.84	I
1200		3290.26	I
7600		3301.56	I
250		3306.23	I
620		3310.91	I
120		3315.42	I
250		3324.33	I
310		3327.42	I
960		3336.15	I
110		3351.74	I
120		3353.91	
230		3357.97	I
250		3361.15	I
190		3364.12	I
120		3370.20	I
960		3370.59	I
160		3372.08	I
120		3378.68	I
310		3384.00	I
190		3385.94	I
620		3387.84	I
120		3401.17	I
620		3401.86	I
250		3402.51	I
120		3408.76	I
120		3412.74	I

Osmium (Cont.)

Intensity		Wavelength	Spectrum
120		3421.69	I
150		3427.67	I
250		3440.60	I
120		3444.46	I
160		3445.55	I
310		3449.20	I
120		3458.38	I
120		3465.44	I
120		3478.53	I
120		3482.11	I
120		3487.46	I
120		3490.33	
160		3498.54	I
250		3501.16	I
620		3504.66	I
440		3512.99	I
310		3518.72	I
120		3520.00	I
480		3523.64	I
120		3526.04	I
1200		3528.60	I
230		3530.06	I
230		3532.80	I
120		3533.41	I
230		3542.71	I
960		3559.79	I
1200		3560.86	I
120		3562.34	I
310		3569.78	I
120		3574.08	I
120		3587.32	I
620		3598.11	I
190		3601.83	I
95		3604.48	II
250		3616.57	I
120		3619.43	I
450		3640.33	I
230		3654.49	I
330		3656.90	I
120		3666.31	I
480		3670.89	I
120		3675.45	I
250		3689.06	I
190		3703.25	I
120		3706.56	I
120		3709.14	I
230		3713.73	I
210		3719.52	I
230		3720.13	I
180		3746.47	I
3700		3752.52	I
100		3757.12	I
130		3766.30	I
120		3768.14	I
120		3774.40	I
110		3774.62	I
120		3776.25	I
290		3776.99	I
2100		3782.20	I
620		3790.14	I
180		3790.73	I
370		3793.91	I
250		3836.06	I
150		3840.30	I
150		3841.29	I
190		3849.94	I
230		3857.09	I
230		3865.47	I
730		3876.77	I
250		3881.86	I
140		3900.39	I
190		3901.71	I
100		3930.00	I
250		3938.59	I
100		3949.78	I
200		3961.02	I
1000		3963.63	I
100		3964.96	I
150		3969.67	I
110	h	3975.44	I
730		3977.23	I
100		3988.18	I
150		4003.48	I
100		4004.02	I
150		4005.16	I
160		4018.26	I
100		4037.84	I
280		4041.92	I
160		4048.05	I
960		4066.69	I
250		4070.86	I
190		4071.56	I
230		4074.68	I
490		4091.82	I
120		4100.30	I
1200		4112.02	I
180		4124.60	I
180		4128.96	I
2500		4135.78	I
150		4137.84	I
180		4172.57	I
1200		4173.23	I
620		4175.63	I
120		4184.13	I
320		4189.91	I
180		4201.45	I
250		4202.06	I
1200		4211.86	I
120		4213.86	I
100		4215.16	I
170		4233.46	I
4900		4260.85	I
100		4264.75	I
120		4269.61	I
100		4285.90	I
560		4293.95	I
560		4311.40	I
110		4326.25	I
340		4328.68	I
100		4338.75	I
100		4351.53	I
210		4365.67	I
110		4370.66	I
520		4394.86	I
160		4397.26	I
160		4402.74	I
4900		4420.47	I
100		4432.41	I
290		4436.32	I
100		4439.64	I
230		4447.35	I
120		4484.76	I
110		4548.66	I
540		4550.41	I
140		4551.30	I
170		4616.78	I
170		4631.83	I
140		4663.82	I
670		4793.99	I
110		4865.60	I
55		5031.83	I
45		5039.12	I
35		5072.88	I
35		5074.77	I
35		5079.09	I
90		5103.50	I
55		5110.81	I
22		5122.23	I
22		5145.54	I
140		5149.74	I
28		5152.01	I
28		5168.98	I
40		5193.52	I
270		5202.63	I
35		5203.23	I
20		5250.46	
45		5255.82	I
55		5265.15	I
20		5283.89	I
20		5295.65	I
40		5298.78	I
13		5302.58	I
18		5336.23	I
11		5346.03	I
13		5352.25	I
110		5376.79	I
16		5403.43	I
13		5412.14	I
120		5416.34	I
45		5416.69	I
28		5417.51	I
16		5441.82	I
55		5443.31	I
22		5446.93	I
11		5447.76	I
20		5449.37	I
20		5453.40	I
22		5457.30	I
28		5470.00	I
13		5474.58	I
13		5475.13	I
9		5477.27	I
16		5481.85	I
22		5509.33	I
9		5516.01	I
270		5523.53	I
22		5546.82	I
9		5549.79	I
13		5552.88	I
11		5560.62	I
16		5580.66	I
80		5584.44	I
8		5600.50	I
35		5620.08	I
9		5637.41	I
22		5642.56	I
28		5645.25	I
7		5648.98	I
9		5660.21	I
7		5674.38	I
28		5680.88	I
11		5709.37	I
170		5721.93	I
8		5737.89	I
8		5739.72	I
22		5765.05	I
170		5780.82	I
40		5800.60	I
8		5842.49	I
110		5857.76	I
28		5860.64	I
11		5882.92	I
11		5903.98	I
11		5906.84	I
7		5908.95	I
7		5981.36	I
11		5983.22	I
65		5996.00	I
20		6015.79	I
7		6054.63	I
20		6144.53	I
11		6158.03	I
35		6227.70	I
7		6241.70	I
22		6269.41	I
11		6274.94	I
11		6286.83	I
9		6398.86	I
22		6403.15	I
9		6448.13	I
6		6520.85	I
7		6528.87	I
7		6533.14	I
11		6538.30	I
11		6576.83	I
8		6614.56	I
4		6615.43	I
7		6661.81	I
27		6729.56	I
18		6791.53	I
14		6806.61	I
5		6878.70	I
4		6901.58	I
11		6956.02	I
6		6984.95	I
15		7060.67	I
22		7145.54	I
10		7149.89	I
4		7184.10	I
10		7206.33	I
5		7209.96	I
9		7251.16	I
6		7253.49	I
6		7375.07	I
9		7407.95	I
26		7602.95	I
4		7701.46	I
7		7789.96	I
7		7852.17	I
6		7981.20	I
7		8041.29	I

OXYGEN (O)
Z = 8

OI and II
Ref. 66, 69, 209, 210, 215 — R.L.K.

Intensity		Wavelength	
		Vacuum	
250		537.83	II
300		538.26	II
220		539.09	II
200		539.55	II
150		539.85	II
150		644.148	II
200		672.95	II
150		673.77	II
70		685.544	I
900		718.484	II
600		718.562	II
70		744.794	I
70		770.793	I
90		771.056	I
70		775.321	I
70		791.973	I
300		796.66	II
90		804.267	I
70		804.848	I
70		805.295	I
80		805.810	I
240		832.762	II
450		833.332	II
600		834.467	II
40		877.879	I
80		922.008	I
90		935.193	I
40		948.686	I
90		971.738	I
40		976.448	I
160		988.773	I
40		990.204	I
250		1025.762	I
90		1027.431	I
160		1039.230	I
60		1040.942	I
40		1152.152	I
900		1302.168	I
600		1304.858	I
300		1306.029	I
		Air	
30	d	2283.42	II
30	d	2284.89	II
110		2293.32	II
200		2300.35	II
30	d	2313.05	II
30	d	2316.12	II
30	d	2316.79	II
50	d	2319.68	II
30	d	2322.15	II
30	d	2339.31	II
110		2411.60	II
80		2425.55	II
250		2433.56	II
80	d	2436.06	II
80		2444.26	II
300		2445.55	II
300		2733.34	II
110		2747.46	II
265		2972.29	I
160		3122.62	II
220		3129.44	II
450		3134.82	II
285		3138.44	II
220		3270.98	II
220		3273.52	II
220		3277.69	II
360		3287.59	II
160		3305.15	II
160		3306.60	II
220		3377.20	II
285		3390.25	II
220		3407.38	II
160		3409.84	II
285		3470.81	II
220		3712.75	II
285		3727.33	II
160		3739.92	II
360		3749.49	II
160		3803.14	II
120		3823.41	I
450		3911.96	II
160		3919.29	II
185		3947.29	I
160		3947.48	I
140		3947.59	I
220		3954.37	II
100		3954.61	I
450		3973.26	II
220		3982.20	II
160		4069.90	II
285		4072.16	II
450		4075.87	II
80	d	4083.91	II
50	d	4087.14	II
150	d	4089.27	II
110		4097.24	II
220		4105.00	II
285		4119.22	II
160		4132.81	II
50		4146.06	II
220		4153.30	II
285		4185.46	II
450		4189.79	II
80		4233.27	I
50	d	4253.74	II
50	d	4253.98	II
50	d	4275.47	II
50	d	4303.78	II
285		4317.14	II
160		4336.86	II
220		4345.56	II
285		4349.43	II
220		4366.90	II
100		4368.25	I
220		4395.95	II
450		4414.91	II
285		4416.98	II
160		4448.21	II
160		4452.38	II
50		4465.45	II
50	d	4466.28	II
50		4467.83	II
50		4469.41	II
360		4590.97	II
285		4596.17	II
80	d	4609.39	II
160		4638.85	II
360		4641.81	II
450		4649.14	II
160		4650.84	II
360		4661.64	II
285		4676.23	II
220		4699.21	II
285		4705.36	II
160		4924.60	II
220		4943.06	II
135		5329.10	I
160		5329.68	I
190		5330.74	I
90		5435.18	I
110		5435.78	I

Oxygen (Cont.)

Intensity		Wavelength	
135		5436.86	I
120		5577.34	I
160		5958.39	I
190		5958.58	I
80		5995.28	I
160		6046.23	I
190		6046.44	I
110		6046.49	I
100		6106.27	I
400		6155.98	I
450		6156.77	I
490		6158.18	I
80		6256.83	I
100		6261.55	I
100		6366.34	I
100		6374.32	I
320		6453.60	I
360		6454.44	I
400		6455.98	I
80		6604.91	I
100		6653.83	I
360		7001.92	I
450		7002.23	I
210		7156.70	I
400		7254.15	I
450		7254.45	I
320		7254.53	I
210		7476.44	I
100		7477.24	I
120		7479.08	I
120		7480.67	I
100		7706.75	I
870		7771.94	I
810		7774.17	I
750		7775.39	I
80		7886.27	I
100		7943.15	I
100		7947.17	I
235		7947.55	I
210		7950.80	I
185		7952.16	I
110		7981.94	I
135		7982.40	I
190		7986.98	I
135		7987.33	I
250		7995.07	I
400		8221.82	I
265		8227.65	I
265		8230.02	I
325		8233.00	I
120		8235.35	I
120		8426.16	I
810		8446.25	I
1000		8446.36	I
935		8446.76	I
325		8820.43	I
160	d	9057.01	I
120		9118.29	I
80		9134.71	I
80		9150.14	I
80		9151.48	I
235		9156.01	I
450		9260.81	I
490		9260.84	I
450		9260.94	I
400		9262.58	I
540		9262.67	I
590		9262.77	I
490		9265.94	I
640		9266.01	I
185		9399.19	I
120		9481.16	I
120	d	9482.88	I
235		9487.43	I
140		9492.71	I
265		9497.97	I
160		9499.30	I
235		9505.59	I
210		9521.96	I
120		9523.36	I
120		9523.96	I
100		9528.28	I
100		9622.13	I
120		9625.29	I
160		9677.38	I
80		9694.66	I
65		9694.91	I
235		9741.50	I
235		9760.65	I
120		9909.05	I
140		9936.98	I
120		9940.41	I
160		9995.31	I
120	d	10421.18	I
590		11286.34	I
640		11286.91	I
490		11287.02	I
490		11287.32	I
490		11295.10	I
540		11297.68	I
590		11302.38	I
265		11358.69	I
490		12464.02	I
450		12570.04	I
120		12990.77	I
160		13076.91	I
700		13163.89	I
750		13164.85	I
640		13165.11	I
160		16212.06	I
120		17966.70	I
590		18021.21	I
120		18041.48	I
120		18042.19	I
120		18046.23	I
140		18229.23	I
540		18243.63	I
140		26173.56	I

O III

Ref. 23, 66, 210 — R.L.K.

Intensity		Wavelength	
		Vacuum	
80	d	264.34	III
110		264.48	III
110		266.97	III
150		266.98	III
150		267.03	III
150		277.38	III
80		295.62	III
110		295.66	III
120		295.72	III
150		303.41	III
150		303.46	III
140		303.52	III
160		303.62	III
160		303.69	III
250		303.80	III
200		305.60	III
250		305.66	III
190		305.70	III
300		305.77	III
190		305.84	III
450		320.979	III
300		328.45	III
250		328.74	III
300		345.31	III
110		355.14	III
90		355.33	III
80		355.47	III
200		359.02	III
190		359.22	III
150		359.38	III
210		373.80	III
200		374.00	III
300		374.08	III
190		374.16	III
200		374.33	III
210		374.44	III
450		395.558	III
300		434.98	III
800		507.391	III
900		507.683	III
1000		508.182	III
1000		525.795	III
700		597.818	III
1000		599.598	III
110		609.70	III
160		610.04	III
200		610.75	III
100		610.85	III
800		702.332	III
800		702.822	III
900		702.899	III
1000		703.850	III
600		832.927	III
780		833.742	III
600		835.096	III
800		835.292	III
160		1476.89	III
285		1590.01	III
160		1591.33	III
220		1760.12	III
110		1760.42	III
220		1763.22	III
220		1764.48	III
750		1767.78	III
550		1768.24	III
360		1771.67	III
110		1773.00	III
110		1773.85	III
220		1779.16	III
160		1781.03	III
160		1784.85	III
220		1789.66	III
110		1848.26	III
110		1856.62	III
285		1872.78	III
285		1872.87	III
285		1874.94	III
160		1920.04	III
110		1920.75	III
110		1921.52	III
220		1923.49	III
110		1923.82	III
110		1926.94	III
		Air	
360		2013.27	III
160		2026.96	III
220		2045.67	III
160		2052.74	III
200	d	2390.44	III
80		2394.33	III
80		2422.84	III
80	d	2438.83	III
200		2454.99	III
200		2558.06	III
80		2687.53	III
110		2695.49	III
80		2959.68	III
250		2983.78	III
80		3017.63	III
80		3023.45	III
80		3043.02	III
200		3047.13	III
110		3059.30	III
80		3121.71	III
110		3132.86	III
80		3238.57	III
200		3260.98	III
300		3265.46	III
80		3267.31	III
80		3312.30	III
110		3340.74	III
80		3444.10	III
80		3455.12	III
80		3698.70	III
80		3702.75	III
80		3703.37	III
110		3707.24	III
110		3715.08	III
110		3744.00	III
150		3754.67	III
80		3757.21	III
250		3759.87	III
110		3791.26	III
200		3961.59	III
110		5592.37	III

O IV

Ref. 36, 66 = R.L.K.

Intensity		Wavelength	
		Vacuum	
150		195.86	IV
200		196.01	IV
110		207.18	IV
150		207.24	IV
140		233.46	IV
150		233.50	IV
110		233.52	IV
200		233.56	IV
110		233.60	IV
90		238.36	IV
180		238.57	IV
110		252.56	IV
110		252.95	IV
150		253.08	IV
300		260.39	IV
250		260.56	IV
300		279.63	IV
375		279.94	IV
110		285.71	IV
150		285.84	IV
200		306.62	IV
150		306.88	IV
700		553.330	IV
775		554.075	IV
850		554.514	IV
700		555.261	IV
580		608.398	IV
640		609.829	IV
270		616.952	IV
150		617.005	IV
200		617.036	IV
520		624.617	IV
580		625.130	IV
640		625.852	IV
200		779.734	IV
315		779.821	IV
360		779.912	IV
200		779.997	IV
640		787.711	IV
520		790.109	IV
700		790.199	IV
200		802.200	IV
160		802.255	IV
130		921.296	IV
160		921.366	IV
200		923.367	IV
130		923.433	IV
200		1338.612	IV
130		1342.992	IV
230		1343.512	IV
		Air	
200		2449.372	IV
200		2450.040	IV
200		2493.44	IV
200		2493.77	IV
200		2507.73	IV
230		2509.19	IV
200		2517.2	IV
160		2836.26	IV
160		2921.45	IV
460		3063.42	IV
410		3071.61	IV
160		3209.66	IV
230		3348.08	IV
270		3349.11	IV
160		3354.27	IV
200		3375.40	IV
130		3378.06	IV
360		3381.20	IV
360		3385.52	IV
270		3396.79	IV
360		3403.52	IV
230		3409.66	IV
410		3411.69	IV
230		3413.64	IV
200		3489.83	IV
160		3492.24	IV
230		3560.39	IV
270		3563.33	IV
315	w	3725.93	IV
360		3729.03	IV
410		3736.85	IV
230		3744.89	IV

O V

Ref. 24, 66 = R.L.K.

Intensity		Wavelength	
		Vacuum	
80		124.616	V
110		135.523	V
80		138.109	V
110		139.029	V
80		151.447	V
110		151.477	V
150		151.546	V
80		164.574	V
110		164.657	V
80		164.709	V
80		166.235	V
150		167.99	V
110		170.219	V
450		172.169	V
250		185.745	V
375		192.751	V
450		192.799	V
520		192.906	V
80		193.003	V
200		194.593	V
80		202.161	V
80		202.224	V
80		202.283	V
80		202.334	V
150		202.393	V
110		203.78	V
150		203.82	V
100		203.85	V
200		203.89	V
100		203.94	V
300		207.794	V
150		215.040	V
200		215.103	V
250		215.245	V
250		216.018	V
520		220.352	V
80		227.372	V
80		227.469	V
150		227.511	V
80		227.549	V
80		227.634	V
80		227.689	V
150		231.823	V
110		248.459	V
110		286.448	V
1000		629.730	V
230		681.272	V
700		758.678	V
640		759.441	V
580		760.228	V
775		760.445	V
640		761.128	V
700		762.003	V
520		774.518	V
640		1371.292	V
160	w	1506.72	V
315	w	1643.68	V
160		1707.996	V
		Air	
1000		2781.01	V
920		2786.99	V
775		2789.85	V
200		2941.33	V
210		2941.65	V
160		3144.66	V
100		4123.99	V
230	w	4930.27	V
130		5597.91	V
130		6500.24	V

PALLADIUM (Pd)
Z = 46

Pd I and II
Ref. 1, 287 — C.H.C.

Intensity		Wavelength	
		Air	
50		2162.27	II
50		2182.35	II
50		2212.15	II
100	r	2231.59	II
200	r	2296.53	II
50		2351.32	II
50		2362.31	II
75		2367.92	II
60		2372.16	II
50		2388.29	II
60		2414.73	II
75		2418.72	II
75		2424.49	II
100		2426.87	II
100		2430.94	II
100		2433.11	II
100		2435.32	II
150		2446.17	II
75		2446.72	II
1100		2447.91	I
80		2448.15	II
100		2457.29	II
60		2457.76	II
150		2469.29	II
80		2470.06	II
100		2471.18	II
50		2472.55	II
1700		2476.42	I
250		2486.52	II
300		2488.92	II
75		2489.61	II
200		2498.81	II
150		2505.73	II
50		2514.47	II
80		2534.57	II
50	h	2539.44	II
150		2551.84	II
150		2565.51	II
100		2569.56	II
60		2593.24	II
50		2628.24	II
70		2635.92	II
150		2658.75	II
1900		2763.09	I
150	h	2776.85	II
100	h	2787.92	II
50	h	2800.64	II
50	h	2807.59	II
200		2854.59	II
100	h	2871.37	II
100	h	2878.01	II
520		2922.49	I
50		2980.63	II
650		3002.65	I
45		3009.78	I
1500		3027.91	I
1100		3065.31	I
2600		3114.04	I
270		3142.81	I
11000		3242.70	I
2700		3251.64	I
3500		3258.78	I
460		3287.25	I
3600		3302.13	I
5000		3373.00	I
24000		3404.58	I
13000		3421.24	I
5000		3433.45	I
6400		3441.40	I
7700		3460.77	I
10000		3481.15	I
2000		3489.77	I
12000		3516.94	I
12000		3553.08	I
4500		3571.16	I
20000		3609.55	I
20000		3634.70	I
5500		3690.34	I
1400		3718.91	I
1500		3799.19	I
1500		3832.29	I
2200		3894.20	I
1500		3958.64	I
290		4087.34	I
90		4169.84	I
2500		4212.95	I
180		4473.59	I
55	h	4788.18	I
45	h	4817.51	I
35		4875.43	I
55		5110.81	I
75		5117.02	I
160		5163.84	I
55		5234.86	I
120		5295.63	I
18		5312.57	I
15		5345.10	I
35		5395.24	I
55		5542.80	I
35		5547.02	I
27		5619.44	I
15		5642.69	I
14		5655.42	I
75		5670.07	I
11		5690.14	I
55	h	5695.09	I
18		5736.61	I
23		6774.54	I
65		6784.52	I
4	h	6833.42	I
11		7016.44	I
13	h	7310.06	I
75		7368.12	I
27		7391.92	I
16		7486.90	I
120		7764.03	I
27		7786.67	I
45		7915.80	I
18		7961.08	I
55		8132.82	I
45		8300.83	I
9	h	8353.58	I
18	h	8532.74	I
16	h	8599.10	I
65		8761.35	I

Pd III
Ref. 424 — L.J.R.

Intensity		Wavelength	
		Vacuum	
10		688.74	III
20		689.46	III
50		689.54	III
50		695.91	III
200		705.49	III
150		707.80	III
100		709.89	III
100		717.90	III
100		719.47	III
200		727.72	III
150		738.79	III
100		756.85	III
100		757.41	III
500		763.06	III
500		766.42	III
200		772.11	III
200		776.51	III
2000		781.02	III
200		784.99	III
200		787.31	III
200		787.95	III
200		789.58	III
500		794.08	III
500		797.52	III
500		800.03	III
500		800.10	III
500		803.67	III
500		825.35	III
500		840.58	III
500		856.47	III
500		864.04	III
500		880.59	III
500		888.84	III
1000		889.29	III
300		947.78	III
300		965.52	III
100		1505.40	III
200		1517.18	III
200	h	1526.88	III
100		1542.63	III
200	h	1545.95	III
300		1596.89	III
200	h	1606.10	III
150		1630.84	III
50		1679.73	III
100		1704.33	III
50		1706.40	III
200		1719.86	III
500		1741.62	III
400		1758.19	III
4000		1782.55	III
400		1804.91	III
400		1843.49	III
1500		1851.59	III
2000		1852.27	III
1000		1859.21	III
1500		1874.63	III
2000		1885.83	III
1000		1887.40	III
1500		1891.34	III
4000		1914.62	III
1000		1930.33	III
2000		1941.64	III
400		1951.56	III
300		1972.29	III
300		1977.53	III
		Air	
800		2002.16	III
1000		2004.47	III
500		2055.11	III
500		2149.82	III
500		2177.55	III
500		2177.63	III
100		2291.45	III
100		2452.42	III
100		2633.22	III

PHOSPHORUS (P)
Z = 15

P I and II
Ref. 182 = R.L.K.

Intensity	Wavelength	
	Vacuum	
10	810.24	II
10	865.44	II
20	1249.32	II
20	1301.87	II
20	1304.47	II
15	1304.68	II
35	1305.48	II
25	1309.87	II
60	1310.70	II
15	1372.033	I
15	1373.500	I
10	1374.732	I
15	1377.080	I
15	1377.937	I
25	1379.429	I
25	1381.469	I
15	1381.637	I
30	1452.89	II
80	1532.51	II
120	1535.90	II
80	1536.39	II
120	1542.29	II
140	1671.070	I
100	1671.510	I
180	1671.680	I
140	1672.035	I
140	1672.474	I
600	1674.591	I
600	1679.695	I
140	1685.976	I
100	1694.028	I
100	1694.486	I
100	1706.376	I
100	1707.553	I
600	1774.951	I
500	1782.838	I
400	1787.656	I
140	1834.801	I
140	1847.165	I
100	1849.820	I
140	1851.194	I
100	1852.069	I
500	1858.886	I
400	1859.393	I
140	1864.348	I
180	1905.481	I
140	1906.403	I
280	1907.665	I
	Air	
280	2023.489	I
180	2024.516	I
400	2032.432	I
400	2033.477	I
400	2135.465	I
400	2136.182	I
400	2149.145	I
280	2152.940	I
500	2154.080	I
180	2235.732	I
100	2484.19	II
750	2533.976	I
950	2535.603	I
750	2553.262	I
500	2554.915	I
150	2606.06	II
100	2626.18	II
90	2636.76	II
150	3308.92	II
125	3419.34	II
100	3425.00	II
100	4178.48	II
200	4288.60	II
200	4385.35	II
400	4420.71	II
100	4452.46	II
150	4463.00	II
120	4467.98	II
200	4475.26	II
200	4499.24	II
120	4530.81	II
120	4554.83	II
120	4558.07	II
120	4581.71	II
500	4588.04	II
500	4589.86	II
600	4602.08	II
300	4626.70	II
300	4658.31	II
200	4864.42	II
150	4927.20	II
500	4943.53	II
300	4954.39	II
300	4969.71	II
100	5079.381	I
100	5098.221	I
100	5100.974	I
140	5109.628	I
140	5154.844	I
180	5162.290	I
150	5191.41	II
300	5253.52	II
140	5293.539	I
400	5296.13	II
250	5316.07	II
300	5344.75	II
180	5345.851	I
100	5364.631	I
250	5378.20	II
300	5386.88	II
200	5409.72	II
400	5425.91	II
100	5428.094	I
400	5450.74	II
140	5458.305	I
125	5461.20	II
180	5477.672	I
140	5477.860	I
140	5478.267	I
200	5483.55	II
200	5499.73	II
200	5507.19	II
100	5514.774	I
100	5516.997	I
200	5541.14	II
200	5583.27	II
250	5588.34	II
100	5727.71	II
500	6024.18	II
400	6034.04	II
500	6043.12	II
250	6055.50	II
100	6057.86	II
350	6087.82	II
180	6097.690	I
350	6165.59	II
500	6199.024	I
180	6210.499	I
100	6232.29	II
200	6367.27	II
140	6375.681	I
100	6388.579	I
250	6435.32	II
130	6436.31	II
600	6459.99	II
600	6503.46	II
600	6507.97	II
150	6713.28	II
100	6717.411	I
100	7102.200	I
100	7158.367	I
180	7165.465	I
180	7175.102	I
180	7176.660	I
250	7845.63	II
100	8046.801	I
140	8278.058	I
100	8367.856	I
140	8531.475	I
140	8613.835	I
180	8637.578	I
400	8741.529	I
100	8872.174	I
140	9153.34	I
180	9175.819	I
950	9193.85	I
600	9278.88	I
1250	9304.94	I
500	9323.50	I
100	9327.13	I
140	9372.09	I
950	9435.069	I
950	9441.86	I
600	9452.83	I
100	9481.84	I
100	9492.12	I
1250	9493.56	I
180	9521.78	I
1700	9525.73	I
1500	9545.18	I
280	9556.81	I
1700	9563.439	I
280	9593.50	I
750	9609.04	I
180	9625.80	I
180	9628.42	I
400	9638.939	I
140	9675.41	I
500	9676.24	I
180	9706.533	I
1500	9734.750	I
280	9736.680	I
1500	9750.77	I

Phosphorus (Cont.)

Intensity	Wavelength	
100	9760.77	I
100	9776.85	I
100	9779.11	I
600	9790.21	I
1700	9796.85	I
280	9834.80	I
400	9903.68	I
280	9976.67	I
229	10084.27	I
174	10432.66	I
132	10455.87	I
458	10511.58	I
962	10529.52	I
1235	10581.57	I
415	10596.90	I
435	10681.40	I
265	10813.13	I
134	10932.72	I
103	10967.37	I
180	11160.05	I
764	11183.23	I
402	11186.75	I
76	13438.43	I
86	13485.19	I
479	14241.64	I
150	14272.75	I
256	14307.83	I
173	14430.50	I
135	14470.62	I
98	14646.42	I
714	15711.52	I
228	15962.53	I
296	16254.77	I
203	16292.97	I
1627	16482.92	I
588	16590.07	I
225	16613.05	I
221	16738.68	I
419	16803.39	I
471	17112.48	I
104	17223.28	I
289	17286.91	I
145	17359.00	I
299	17423.67	I
95	17665.68	I
186	18007.63	I
106	18518.90	I
92	18881.16	I
125	20841.62	I
124	23038.83	I
287	23844.97	I
98	26134.44	I
118	27959.52	I
188	28049.42	I
127	28154.52	I
132	28284.16	I
98	28288.69	I
311	29097.16	I
92	31483.21	I
91	32270.90	I
146	35551.93	I
90	35582.27	I
192	35802.53	I
146	36417.43	I

P III

Ref. 180 = R.L.K.

Intensity	Wavelength	
	Vacuum	
90	471.146	III
90	484.278	III
120	498.180	III
200	569.853	III
200	581.831	III
200	844.646	III
150	845.038	III
250	845.664	III
300	847.669	III
200	848.016	III
120	848.465	III
150	848.639	III
250	852.686	III
350	855.624	III
200	859.406	III
500	859.652	III
250	859.729	III
300	913.971	III
300	917.120	III
350	918.665	III
250	921.849	III
200	997.999	III
250	1003.598	III
500	1334.808	III
650	1344.327	III
300	1344.845	III
250	1380.463	III
150	1381.089	III
350	1502.228	III
250	1504.663	III
150	1618.632	III
200	1618.907	III
	Air	
200	2611.147	III
300	2632.713	III
200	2680.133	III
250	2895.241	III
250	3186.186	III
300	3219.307	III
150	3233.536	III
400	3233.602	III
200	3556.546	III
200	3577.526	III
200	3904.812	III
250	3914.314	III
300	3957.641	III
350	3978.307	III
200	4057.440	III
400	4059.312	III
300	4080.084	III
500	4222.195	III
350	4246.720	III
200	4428.171	III
200	4463.668	III
250	4479.776	III
150	6083.409	III
150	6409.204	III
150	6484.440	III
150	6486.381	III
150	6992.690	III
150	8113.528	III

P IV

Ref. 336 – R.L.K.

Intensity	Wavelength	
	Vacuum	
90	282.301	IV
90	304.996	IV
120	359.293	IV
150	359.899	IV
120	361.514	IV
150	361.629	IV
120	371.299	IV
150	371.504	IV
200	372.001	IV
500	388.318	IV
120	414.604	IV
200	414.999	IV
250	415.805	IV
250	444.245	IV
300	445.158	IV
250	568.038	IV
350	629.008	IV
400	629.914	IV
500	631.779	IV
350	648.482	IV
300	756.510	IV
300	776.353	IV
650	823.179	IV
700	824.730	IV
800	827.932	IV
250	847.019	IV
350	849.799	IV
200	850.392	IV
700	877.476	IV
1000	950.655	IV
570	1025.563	IV
500	1028.096	IV
570	1030.517	IV
500	1033.111	IV
500	1035.517	IV
570	1118.551	IV
200	1206.422	IV
200	1335.705	IV
500	1366.695	IV
400	1372.674	IV
350	1377.282	IV
500	1484.507	IV
400	1487.788	IV
300	1489.098	IV
250	1862.762	IV
120	1862.893	IV
200	1863.580	IV
650	1888.523	IV
200	1910.183	IV
120	1985.682	IV
150	1985.851	IV
200	1986.114	IV
150	1987.022	IV
	Air	
200	2477.823	IV
150	2478.070	IV
250	2478.256	IV
250	2605.506	IV
400	2644.295	IV
300	2724.764	IV
400	2728.770	IV
200	2729.120	IV
500	2739.309	IV
250	2739.872	IV
200	2740.223	IV
200	2961.242	IV
650	3347.736	IV
570	3364.467	IV
400	3371.122	IV
200	3413.543	IV
200	3733.393	IV
300	4249.656	IV
250	4540.288	IV
250	4541.112	IV
150	4548.056	IV
200	4548.449	IV
150	5235.499	IV
150	5989.774	IV
150	6142.605	IV
150	6713.939	IV
120	6715.906	IV
200	7443.657	IV

P V

Ref. 179 = R.L.K.

Intensity	Wavelength	
	Vacuum	
80	255.59	V
50	255.67	V
110	310.58	V
150	311.34	V
300	328.47	V
250	328.78	V
150	347.23	V
200	348.20	V
110	378.56	V
250	389.50	V
300	390.70	V
150	410.03	V
375	475.60	V
110	534.63	V
80	534.99	V
520	542.57	V
600	544.92	V
450	673.90	V
450	865.45	V
600	871.39	V
250	997.62	V
150	1000.38	V
900	1117.98	V
700	1128.01	V
150	1379.62	V
250	1385.05	V
375	1447.83	V
450	1610.50	V
	Air	
200	2180.29	V
150	2186.42	V
375	2424.40	V
450	2440.93	V
200	2441.24	V
300	2961.00	V
450	2978.55	V
700	3175.09	V
520	3204.04	V
150	4083.18	V
110	4094.95	V
110	5156.72	V

PLATINUM (Pt)

Z = 78

Pt I and II

Ref. 1, 288 — C.H.C.

Intensity		Wavelength	
		Vacuum	
30		1621.66	II
30		1723.13	II
30		1751.70	II
50	r	1777.09	II
30		1781.86	II
30		1879.09	II
40		1883.05	II
50		1889.52	II
50		1911.70	II
30		1929.25	II
30		1929.68	II
30		1939.80	II
30		1949.90	II
30		1983.74	II
		Air	
40		2014.93	II
3200		2030.63	I
4400		2032.41	I
100		2036.46	II
40		2041.57	II
5500		2049.37	I
1500		2067.50	I
3000		2084.59	I
1000		2103.33	I
30		2115.57	II
950		2128.61	I
30		2130.69	II
1900		2144.23	I
100		2144.24	II
600		2165.17	I
1500		2174.67	I
30		2190.32	II
400		2202.22	I
50	h	2202.58	II
320		2222.61	I
50	h	2233.11	II
150		2249.30	I
30	h	2240.99	II
100		2245.52	II
30		2251.52	II
30	h	2251.92	II
190		2268.84	I
30	h	2271.72	II
280		2274.38	I
50	h	2287.50	II
30		2288.20	II
150		2289.27	I
150		2292.40	I
240		2308.04	I
50		2310.96	II
90		2315.50	I
220		2318.29	I
100		2326.10	I
170		2340.18	I
280		2357.10	I
180		2368.28	I
50		2377.28	II
130		2383.64	I
40		2386.81	I
120		2389.53	I
35		2396.17	I
70		2401.87	I
200		2403.09	I
100		2418.06	I
50		2424.87	II
80		2428.04	I
50		2428.20	I
25		2429.10	I
180		2436.69	I
650		2440.06	I
60		2450.97	I
440		2467.44	I
35		2471.01	I
1000		2487.17	I
25		2488.74	II
200		2490.12	I
160		2495.82	I
240		2498.50	I
50		2505.93	I
120		2508.50	I
50		2514.07	I
60		2515.03	I
240		2515.58	I
140		2524.30	I
40		2529.41	I
50		2536.49	I
160		2539.20	I
18		2549.46	I
50		2552.25	I
50		2596.00	I
70		2603.14	I
30		2616.76	II
50		2619.57	I
30		2625.34	II
1100		2628.03	I
130		2639.35	I
1000		2646.89	I
500		2650.86	I
20		2658.17	I
2800		2659.45	I
40		2674.57	I
440		2677.15	I
200		2698.43	I
2000		2702.40	I
1600		2705.89	I
60		2713.13	I
1300		2719.04	I
130		2729.92	I
1800		2733.96	I
70		2738.48	I
70		2747.61	I
80		2753.86	I
200		2754.92	I
30		2769.84	I
500		2771.67	I
40		2773.24	I
20		2774.00	I
50		2774.77	II
50		2793.27	I
100		2794.21	II
40	h	2799.98	II
140		2803.24	I
10		2808.51	I
50		2818.25	I
30	h	2822.27	II
1400		2830.30	I
70		2834.71	I
16		2853.11	I
80	h	2860.68	II
40	h	2865.05	II
40	h	2875.85	II
100	h	2877.52	II
25		2888.20	I
25		2893.22	I
600		2893.86	I
300		2897.87	I
60		2905.90	I
120		2912.26	I
120		2913.54	I
70		2919.34	I
30		2921.38	I

Platinum (Cont.)

Intensity		Wavelength	
1700		2929.79	I
30		2942.76	I
30		2944.75	I
25		2959.10	I
60		2960.75	I
1800		2997.97	I
35		3001.17	II
220		3002.27	I
30		3017.88	I
30	h	3031.22	II
130		3036.45	I
800		3042.64	I
3200		3064.71	I
30		3071.94	I
130		3100.04	I
320		3139.39	I
140		3156.56	I
120		3200.71	I
320		3204.04	I
30		3230.29	I
20		3233.42	I
20		3250.36	I
40		3251.98	I
160		3255.92	I
25		3268.42	I
25		3281.97	I
120		3290.22	I
500		3301.86	I
60		3315.05	I
35		3323.80	I
340		3408.13	I
35		3427.93	I
60		3483.43	I
160		3485.27	I
120		3628.11	I
70		3638.79	I
70		3643.17	I
50		3663.10	I
80		3671.99	I
80		3674.04	I
35		3699.91	I
18		3706.53	I
20		3801.05	
80		3818.69	I
40		3900.73	I
110		3922.96	I
35		3948.40	I
100		3966.36	I
20		3996.57	I
110		4118.69	I
80		4164.56	I
40		4192.43	I
18		4327.06	I
18		4391.83	I
80		4442.55	I
14		4445.55	I
25		4498.76	I
12		4520.90	I
35		4552.42	I
12		4879.53	I
14		5044.04	I
30		5059.48	I
35		5227.66	I
40		5301.02	I
12		5368.99	I
12		5390.79	I
14		5475.77	I
14		5478.50	I
6		5763.57	I
20		5840.12	I
8		5844.84	I
6		6026.04	I
7		6318.37	I
8		6326.58	I
9		6523.45	I
10		6710.42	I
20		6760.02	I
60		6842.60	I
20		7113.73	I
10		8224.74	I

PLUTONIUM (Pu)

Z = 94

Pu I and II
Ref. 91 — J.G.C.

Intensity		Wavelength Air	
10000		2781.40	II
10000		2784.48	II
10000		2806.11	II
10000		2815.77	II
10000		2897.97	II
10000		2898.94	II
10000		2904.25	II
10000		2904.94	II
10000		2910.40	II
10000		2918.00	II
10000		2918.80	II
10000		2926.08	II
10000		2928.25	II
10000		2929.71	II
10000		2930.98	II
10000		2932.32	II
10000		2933.30	II
10000		2938.54	II
10000		2938.95	II
10000		2941.39	II
10000		2945.26	II
10000		2946.00	II
10000		2950.06	II
10000		2951.62	
10000		2954.46	II
10000		2963.47	II
10000		2966.84	II
10000		2967.54	II
10000		2972.50	II
10000		2977.81	II
10000		2978.37	II
10000		2980.23	II
10000		2981.23	II
10000		2986.95	II
10000		2988.21	II
10000		2991.31	II
10000		2996.40	II
10000		3000.31	II
10000		3009.57	II
10000		3028.85	II
10000		3042.61	II
10000		3043.12	II
10000		3060.32	II
10000		3069.32	II
10000		3091.33	II
10000		3091.94	II
10000		3092.59	II
10000		3104.12	II
10000		3105.04	II
10000		3106.03	
10000		3123.87	II
10000		3159.21	II
10000		3161.73	II
10000		3163.18	II
10000		3174.49	II
10000		3179.41	II
10000		3185.12	II
10000		3187.60	II
10000		3189.23	II
10000		3193.54	
10000		3193.55	II
10000		3194.56	II
10000		3198.47	II
10000		3200.23	II
10000		3201.00	II
10000		3201.66	II
10000		3204.48	II
10000		3206.80	II
10000		3207.97	II
10000		3215.08	I
10000		3216.15	II
10000		3220.94	II
10000		3224.87	II
10000		3231.86	II
10000		3232.24	
10000		3232.63	II
10000		3241.39	II
10000		3242.96	II
10000		3243.40	II
10000		3244.16	I
10000		3245.25	II
10000		3245.71	
10000		3246.35	II
10000		3247.50	
10000		3247.56	II
10000		3252.08	I
10000		3260.54	II
10000		3265.17	
10000		3273.11	II
10000		3274.71	II
10000		3275.24	I
10000		3292.56	I
10000		3293.61	I
10000		3296.91	I
10000		3297.87	I
10000		3298.47	II
10000		3301.76	I
10000		3306.59	I
10000		3306.66	I
10000		3307.66	II
10000		3308.75	I
10000		3312.65	II
10000		3315.34	II
10000		3316.96	II
10000		3320.61	I
10000		3320.84	I
10000		3323.48	
10000		3327.19	I
10000		3330.11	
10000		3331.52	II
10000		3332.34	I
10000		3333.03	
10000		3337.71	II
10000		3338.40	II
10000		3338.94	I
10000		3347.87	II
10000		3349.63	I
10000		3351.82	II
10000		3356.61	II
10000		3358.41	II
10000		3358.84	II
10000		3362.26	II
10000		3365.20	I
10000		3365.66	
10000		3368.86	I
10000		3370.64	II
10000		3371.19	I
10000		3375.80	I
10000		3376.76	II
10000		3376.94	II
10000		3377.37	II
10000		3379.51	I
10000		3381.82	I
10000		3381.97	II
10000		3382.70	I
10000		3390.33	II
10000		3391.41	II
10000		3393.67	I
10000		3394.32	I
10000		3418.88	II
10000		3465.10	II
10000		3473.64	II
10000		3483.20	I
10000		3585.87	II
10000		3632.21	II
10000		3699.19	II
10000		3720.59	I
10000		3725.98	I
10000		3726.11	II
10000		3726.79	II
10000		3732.03	II
10000		3744.78	I
10000		3753.63	I
10000		3755.94	I
10000		3757.82	I
10000		3758.34	I
10000		3774.38	I
10000		3776.71	I
10000		3792.22	I
10000		3799.37	I
10000		3805.93	I
10000		3811.40	I
10000		3812.30	II
10000		3827.57	I
10000		3835.52	I
10000		3836.96	I
10000		3838.92	I
10000		3842.10	I
10000		3851.01	I
10000		3851.85	I
10000		3878.54	I
10000		3895.89	I
10000		3928.53	I
10000		3975.43	II
10000		4097.12	I
10000		4101.96	I
10000		4105.95	II
10000		4111.07	I
10000		4114.91	I
10000		4128.12	I
10000		4129.93	II
10000		4133.01	I
10000		4135.97	I
10000		4140.04	I
10000		4141.20	II
10000		4151.09	I
10000		4151.45	I
10000		4155.46	I
10000		4159.39	II
10000		4167.77	I
10000		4170.95	I
10000		4178.28	II
10000		4189.90	II
10000		4190.06	II
10000		4196.20	II
10000		4206.48	I
10000		4208.23	I
10000		4221.87	I
10000		4224.20	I
10000		4229.77	II
10000		4254.76	II
10000		4261.88	I
10000		4269.77	I
10000		4273.34	II
10000		4281.17	I
10000		4289.08	II
10000		4337.18	II
10000		4352.71	II
10000		4367.41	I
10000		4379.91	II
10000		4385.35	II
10000		4393.93	II
10000		4404.90	I
10000		4441.65	II
10000		4468.54	II
10000		4472.79	II
10000		4493.78	II
10000		4504.91	II
10000		4536.15	II
10000		4735.40	I
10000		4989.34	I
10000		5269.86	I
10000		5381.02	I
10000		5498.50	I
10000		5510.72	I
10000		5537.59	I
10000		5549.62	I
10000		5590.54	I
10000		5592.33	I
10000		5712.39	I
10000		5770.26	I
10000		5839.05	I
10000		5983.35	I
10000		6012.78	I
10000		6192.80	I
10000		6304.66	I
10000		6449.75	I
10000		6486.71	I
10000		6488.86	I
10000		6488.89	
10000		6535.27	I
10000		6544.21	I
10000		6608.95	I
10000		6672.72	I
10000		6784.66	I
10000		6880.16	I
10000		6891.38	I
10000		7059.23	I
10000		7068.90	I
10000		7092.46	I
10000		7116.88	I
10000		7141.66	I
10000		7177.14	I
10000		7231.09	I
10000		7258.06	I
10000		7322.23	I
10000		7325.97	I
10000		7331.81	I
10000		7431.18	I
10000		7447.99	I
10000		7507.80	I
10000		7526.93	I
10000		7547.45	I
10000		7564.50	I
10000		7571.87	I
10000		7572.93	I
10000		7609.77	I
10000		7689.40	I
10000		7758.20	I
10000		7798.54	I
10000		7953.17	I
10000		8102.54	I
10000		8130.86	I
10000		8309.61	I
10000		8435.47	I
10000		8476.13	I
10000		8495.75	
10000		8597.26	I
10000		8665.02	I
10000		8691.94	I
3000		8729.82	I
3000		8836.16	I
3000		9533.07	I
3000		10046.75	I
3000		11114.82	I
3000		12144.46	I
3000		12231.22	I
3000		15377.31	I
3000		16397.38	I

POLONIUM (Po)

Z = 84

Po I and II
Ref. 47, 48 — E.F.W.

Intensity		Wavelength Air	
250	w	2139.02	I
300	h	2203.80	
300		2220.67	I
200		2222.13	I
200		2284.22	
250		2344.61	I
250		2421.72	I
300		2426.09	I
1500	w	2450.08	I
700		2483.94	I
700		2490.53	I
200	h	2502.18	
300		2534.95	I
300		2557.33	I
1500	w	2558.01	I
400		2562.31	
300		2578.80	
400		2587.64	I
200		2637.01	
300		2645.36	I
700	h	2663.33	
200		2671.67	I
600		2761.92	I
400		2800.26	I
250		2824.11	I
300		2866.01	I
400		2919.31	I
600		2958.92	I
2500	w	3003.21	I
450		3069.31	I
200		3115.95	

Potassium (Cont.)

Intensity		Wavelength	Spectrum
400		3189.02	I
600		3240.24	I
250		3286.38	I
600		3328.60	I
300		3489.79	
200		3493.65	
400		3588.33	
200		3671.36	
500		3861.93	I
200		4051.98	
1200		4170.52	I
250		4236.13	
200	h	4415.58	
800		4493.21	I
350		4611.44	I
200		4867.12	
400		4876.24	I
450		4946.81	
350		5323.23	I
300		5744.85	I
600		7962.62	I
300		8433.87	I
500		8618.26	I
250		9227.87	

POTASSIUM (K)
Z = 19

K I and II
Ref. 59, 76, 172, 268 — L.J.R.

Vacuum

Intensity		Wavelength	Spectrum
5		261.20	II
25		441.81	II
5		465.08	II
		469.50	II
10		476.03	II
30		495.14	II
30		600.77	II
25		607.93	II
30		612.62	II
3		1725.0	II

Air

Intensity		Wavelength	Spectrum
6		2190.00	II
4		2210.53	II
5		2265.04	II
6		2550.02	II
4		2743.55	II
		2992.12	I
		2992.42	I
		3034.76	I
		3034.92	I
5		3062.18	II
4		3101.79	I
3		3102.04	I
6		3105.00	II
5		3190.07	II
7		3217.16	I
6		3217.62	I
4		3220.60	II
5		3290.65	II
6		3345.32	II
6		3373.60	II
6		3380.62	II
6		3384.86	II
6		3404.24	II
7		3440.05	II
11		3446.37	I
10		3447.38	I
6		3481.11	II
7		3530.75	II
5		3608.88	II
6		3618.49	II
4		3626.42	II
3		3648.84	I
4		3648.98	I
6		3681.54	II
5		3716.60	II
5		3721.34	II
5		3739.13	II
5		3744.42	II
6		3767.36	II
6		3783.19	II
6		3800.14	II
6		3816.56	II
7		3817.50	II
5		3873.74	II
4		3878.62	II
8		3897.92	II
5		3923.00	II
5		3926.36	II
6		3942.53	II
6		3955.21	II
6		3966.72	II
6		3972.58	II
6		3995.10	II
7		4001.24	II
5		4012.10	II
6		4042.59	II
18		4044.14	I
17		4047.21	I
5		4093.69	II
6		4114.99	II
7		4134.72	II
7		4149.19	II
8		4186.24	II
7		4222.97	II
7		4225.67	II
7		4263.40	II
7		4305.00	II
7		4309.10	II
5		4340.03	II
7		4388.16	II
5		4466.65	II
6		4505.33	II
5		4595.65	II
8		4608.45	II
10		4641.88	I
11		4642.37	I
5		4659.38	II
4		4740.91	I
6		4744.35	I
5		4753.93	I
7		4757.39	I
5		4786.49	I
7		4791.05	I
6		4799.75	I
8		4804.35	I
9		4829.23	II
7		4849.86	I
8		4856.09	I
8		4863.48	I
9		4869.76	I
8		4942.02	I
6		4943.29	II
9		4950.82	I
9		4956.15	I
10		4965.03	I
8		5005.60	II
7		5056.27	II
10		5084.23	I
11		5097.17	I
11		5099.20	I
12		5112.25	I
5		5310.24	II
12		5323.28	I
13		5339.69	I
12		5342.97	I
14		5359.57	I
6		5470.13	II
5		5642.73	II
4		5772.32	II
16		5782.38	I
17		5801.75	I
15		5812.15	I
17		5831.89	I
2		5969.64	II
8		6120.27	II
6		6246.59	II
7		6307.29	II
5		6427.96	II
2		6595.00	II
19		6911.08	I
12		6936.28	I
20		6938.77	I
7		6964.18	I
12		6964.67	I
25		7664.90	I
24		7698.96	I
5		7955.37	I
4		7956.83	I
7		8078.11	I
6		8079.62	I
9		8250.18	I
8		8251.74	I
3		8390.22	I
		8391.44	I
2		8417.54	I
1		8420.00	I
11		8503.45	I
10		8505.11	I
4		8763.96	I
3		8767.05	I
13		8902.19	I
12		8904.02	I
5		8923.31	I
4		8925.44	I
7		9347.24	I
3		9349.25	I
6		9351.59	I
15		9595.70	I
14		9597.83	I
6		9949.67	I
5		9954.14	I
9		10479.63	I
5		10482.15	I
8		10487.11	I
17		11019.87	I
16		11022.67	I
17		11690.21	I
16		11769.62	I
17		11772.83	I
		12432.24	I
		12522.11	I
		13377.86	I
		13397.09	I
		15163.08	I
		15168.40	I
		40158.37	I

K III
Ref. 60, 76 — L.J.R.

Vacuum

Intensity		Wavelength	Spectrum
2		325.28	III
5		327.60	III
25		330.68	III
30		341.92	III
15		348.00	III
30		379.12	III
25		380.48	III
30		382.23	III
15		398.63	III
20		402.10	III
30		406.48	III
40		408.96	III
50		413.79	III
30		414.87	III
30		416.00	III
30		417.54	III
30		418.62	III
75		434.72	III
50		435.68	III
75		444.34	III
75		448.60	III
75		466.79	III
100		470.09	III
75		471.57	III
45		474.92	III
40		479.18	III
10		482.11	III
10		482.41	III
75		497.10	III
10		514.94	III
50		520.61	III
25		523.79	III
40		529.80	III
15		539.71	III
15		546.12	III
20		708.84	III
20		765.31	III
30		765.64	III
35		778.53	III
20		872.31	III
10		873.86	III
15		874.04	III

Air

Intensity		Wavelength	Spectrum
6		2550.02	III
5		2635.11	III
1		2736.96	III
5		2689.90	III
1		2898.90	III
5		2938.45	III
1		2948.94	III
5		2986.20	III
6		2992.24	III
6		3052.07	III
5		3056.84	III
6		3201.95	III
6		3209.34	III
6		3278.79	III
6		3289.06	III
6		3322.40	III
6		3364.22	III
6		3420.82	III
4		3421.83	III
6		3468.32	III
6		3481.11	III
5		3513.88	III
1		3885.50	III

K IV
Ref. 32, 76, 86, 150, 160, 314, 322 — L.J.R.

Vacuum

Intensity		Wavelength	Spectrum
150		271.82	IV
100		273.06	IV
		279.88	IV
300		340.46	IV
150		340.74	IV
300		354.93	IV
150		356.26	IV
300		359.73	IV
200		359.91	IV
250		362.08	IV
150		362.15	IV
150		363.02	IV
300		375.96	IV
300		379.88	IV
250		380.48	IV
200		381.70	IV
300		382.23	IV
150		382.49	IV
200		382.65	IV
300		382.91	IV
250		384.10	IV
200		386.61	IV
250		388.92	IV
250		389.07	IV
250		390.42	IV
300		390.57	IV
200		391.46	IV
200		392.47	IV
500		393.14	IV
400		400.21	IV
300		402.91	IV
250		403.97	IV
150		404.41	IV
250		408.08	IV
150		417.28	IV
200		442.30	IV
300		443.57	IV
200		445.61	IV
250		446.83	IV
750		448.60	IV
400		456.33	IV
250		523.00	IV
200		526.45	IV
150		527.62	IV
750		646.19	IV
500		737.14	IV
500		741.95	IV
500		745.26	IV
400		746.35	IV
300		749.99	IV
150		754.19	IV
400		754.67	IV

K V
Ref. 32, 75, 76, 150, 322 — L.J.R.

Vacuum

Intensity		Wavelength	Spectrum
100		214.35	V
150		282.35	V
150		293.33	V
300		294.84	V
200		296.17	V
200		297.06	V
200		300.25	V
200		300.50	V
200		311.24	V
250		312.77	V
200		315.18	V
250		327.38	V
250		389.07	V
200		349.50	V
500		372.15	V
200		372.46	V
200		372.77	V
300		375.96	V
250		377.76	V
300		379.12	V
300		387.80	V
250		390.11	V
250		395.40	V
200		398.36	V
200		398.88	V
200		399.75	V
250		415.05	V
200		415.79	V
400		422.18	V
300		425.16	V
500		425.59	V
250		438.02	V
200		449.71	V
200		452.90	V
250		455.67	V
400		456.33	V
200		482.71	V
200		483.75	V
750		580.32	V
250		585.51	V
500		586.32	V
250		602.27	V
400		603.43	V
250		638.67	V
300		687.50	V
300		720.43	V
400		724.42	V
600		731.86	V
150		770.29	V
150		771.46	V
		1035.60	V

PRASEODYMIUM (Pr)
Z = 59

Pr I and II
Ref. 1 — C.H.C.

Air

Intensity		Wavelength	Spectrum
25		2558.58	II
25		2578.27	I
30		2579.31	I
40	h	2598.04	II
25		2608.92	II
25		2615.75	II
25		2648.48	II
30		2654.75	II
25		2666.70	II
20		2672.52	II
30		2685.19	II
45		2685.70	II

50		2698.92	II
60		2700.38	II
30		2702.25	II
100	h	2707.37	II
20		2714.16	II
60		2720.17	II
30		2721.90	II
50		2726.50	II
12		2731.78	II
25		2733.12	II
50		2734.30	II
25		2737.90	II
40		2742.12	II
25		2744.66	II
20		2746.28	II
60		2760.35	II
50		2769.60	II
50	d	2775.94	II
		2776.03	II
40		2778.80	II
50		2783.31	II
30		2789.05	II
35		2792.51	II
50		2802.05	II
20		2823.17	II
20		2824.14	II
20		2828.29	II
20		2842.98	I
20		2844.01	II
20		2850.62	I
25		2853.99	II
30		2865.64	II
50		2881.60	I
30		2882.31	II
30		2884.89	II
30		2943.97	II
30		2967.58	II
30		2971.13	II
40	d	2971.40	II
		2971.46	II
50		2984.98	II
30		2986.18	II
30		2990.22	II
110		3082.11	II
100		3111.34	II
140		3121.58	II
140		3163.73	II
270		3168.24	II
160		3172.31	II
110		3191.42	II
200	d	3195.99	II
110		3199.04	II
100		3207.89	II
190		3219.48	II
100		3234.27	II
100		3245.48	II
140		3355.67	II
110		3394.62	II
110		3465.74	II
200		3584.21	II
130		3611.94	II
170		3630.96	II
100		3645.55	II
250		3645.66	II
250		3646.30	II
100		3648.30	II
150	c	3660.36	II
100		3661.62	II
370		3668.83	II
250		3687.03	II
150		3687.19	II
100		3689.71	II
150		3698.06	II
230		3706.75	II
170	c	3711.10	II
290		3714.05	II
120	c	3733.03	II
210	c	3734.41	II
250		3735.76	II
190		3736.49	II
410		3739.18	II
150		3740.99	II
120		3743.98	II
190		3750.98	II
140		3759.60	II
120		3760.08	II
680		3761.87	II
230		3764.77	II
230		3768.94	II
170	c	3772.82	II
170		3774.06	II
140		3777.62	II
170		3780.66	II
150		3785.46	II
150		3786.86	II
210		3792.51	II
190		3794.93	II
680		3800.30	II
290		3804.84	II
140		3809.18	II
390		3811.84	II
1300	h	3816.02	II
120		3817.66	II
680		3818.28	II
120		3819.14	II
310		3821.80	II
150	c	3823.18	II
120		3826.67	II
960		3830.72	II
140		3834.93	II
480		3840.99	II
270		3842.34	II
150	c	3844.54	II
580		3846.59	II
1200		3850.79	II
720	c	3851.55	II
960		3852.80	II
120		3858.25	II
110		3859.14	II
480	c	3865.45	II
210		3867.52	II
210		3870.72	II
480		3876.19	II
1700	c	3877.18	II
270		3879.20	II
680		3880.47	II
440	c	3885.19	II
440	c	3889.34	II
120	c	3891.71	II
190		3897.25	II
210		3898.84	II
250		3902.45	II
770	c	3908.05	II
630		3912.90	II
310		3913.55	II
210		3914.76	II
1300	c	3918.85	II
420		3919.63	II
250		3920.53	II
960		3925.47	II
480		3927.46	II
370		3929.29	II
370		3935.82	II
250		3938.30	II
730	c	3947.63	II
900	c	3949.43	II
900	c	3953.51	II
380		3956.75	II
190		3959.44	I
470		3962.45	II
560		3964.26	II
1600	c	3964.81	II
560	c	3966.57	II
500		3971.16	II
320		3971.67	II
620	c	3972.14	II
320		3974.85	II
1300	c	3989.68	II
230		3991.91	II
340		3992.16	II
1600		3994.79	II
270		3995.83	II
560	c	3997.04	II
230		3997.96	II
320		3999.12	II
620	c	4000.17	II
730		4004.70	II
1900		4008.69	II
620		4010.60	II
730		4015.39	II
620		4020.96	II
470		4022.71	II
360		4025.54	II
230		4026.83	II
230		4029.00	II
360	c	4029.72	II
730	c	4031.75	II
230		4032.47	II
960		4033.83	II
230		4034.33	II
230		4038.22	II
730		4038.45	II
470		4039.34	II
1300		4044.81	II
230		4045.70	II
230		4046.63	II
340		4047.08	II
450		4051.13	II
2200		4054.88	II
2200		4056.54	II
450		4058.80	II
230		4062.22	II
3400		4062.81	II
210		4068.80	II
500	c	4079.77	II
500	c	4080.98	II
790		4081.85	II
500		4083.34	II
200	c	4087.21	II
560		4096.82	II
380		4098.40	II
2900	c	4100.72	II
270	c	4113.89	II
1700	c	4118.46	II
250		4129.15	II
340		4130.77	II
200		4133.61	II
1500	c	4141.22	II
2700		4143.11	II
270	c	4146.50	II
270		4148.44	II
200		4156.50	II
1700	c	4164.16	II
270		4168.04	II
230		4169.45	II
620		4171.82	II
730		4172.25	II
250		4175.32	II
250		4175.62	II
200		4178.63	II
5200		4179.39	II
2500		4189.48	II
560	c	4191.60	II
290		4201.17	II
2500	c	4206.72	II
500		4208.32	II
320		4211.86	II
320		4217.81	II
3800		4222.93	II
3800		4225.35	II
320		4233.11	II
320	c	4236.15	II
270		4240.02	II
960		4241.01	II
340		4243.51	II
840	c	4247.63	II
500		4254.40	II
270	c	4263.78	II
320		4269.09	II
790	c	4272.27	II
470	c	4280.07	II
790	c	4282.42	II
450	c	4298.98	II
290		4303.61	II
1500		4305.76	II
210		4323.55	II
270		4329.41	II
1300		4333.97	II
200		4335.74	II
360		4338.70	II
620	Cw	4344.30	II
470	c	4347.49	II
340		4350.40	II
450		4354.91	II
410	c	4359.79	II
1200		4368.33	II
320		4371.62	II
270		4396.08	II
170		4403.60	II
100		4405.12	II
430		4405.83	II
1700		4408.82	II
410		4413.77	II
160		4419.04	II
190		4419.65	II
160	c	4421.22	II
160		4424.58	II
1200	c	4429.13	II
110		4432.28	II
730		4449.83	II
140		4451.90	II
140		4454.68	II
100		4465.97	II
960		4468.66	II
140	c	4477.26	II
1100		4496.46	II
790		4510.15	II
200	c	4517.58	II
340	c	4534.15	II
340		4535.92	II
200		4563.12	II
140		4612.08	II
270	c	4628.74	II
140		4632.28	I
140		4635.68	I
200		4639.55	I
110	c	4643.49	II
140		4646.05	II
200	c	4651.50	II
140		4664.65	II
270	c	4672.09	II
180		4687.80	I
290		4695.77	I
140	Cw	4708.07	II
140		4709.52	I
180		4730.67	I
250		4736.69	I
100		4744.16	I
150		4746.92	II
100		4762.72	II
110		4783.35	II
110		4906.99	I
140		4914.02	I
200		4924.60	I
140		4936.00	I
320		4939.74	I
160		4940.30	I
380		4951.37	I
110		4975.75	I
120		5018.59	I
200		5019.76	I
200		5026.96	I
100		5033.38	I
270		5034.41	II
110		5043.83	I
320		5045.52	I
160		5053.40	I
180		5087.12	I
360		5110.38	II
560		5110.76	II
410		5129.52	II
270		5133.44	I
270		5135.14	II
100		5139.81	I
100	c	5152.30	II
200		5161.74	II
620		5173.90	II
200		5191.32	II
120		5194.43	I
150		5195.11	II
200		5195.31	II
360		5206.55	II
150		5207.90	II
360		5219.05	II
560		5220.11	II
110		5227.97	I
680		5259.73	II
180		5263.88	II
340	c	5292.02	II
340		5292.62	II
230		5298.09	II
430		5322.76	II
200		5352.40	II
16		5501.50	I
40		5508.79	II
65		5509.15	II
16	c	5511.63	II
55		5513.58	II
28		5515.12	II
13		5519.38	II
20	c	5520.31	II
45	c	5522.79	II
28	c	5524.15	I
28	c	5525.91	II
16	c	5527.93	I
13		5530.21	I
45		5531.16	I
150		5535.17	II
28		5538.37	I
20		5538.78	II
55		5545.01	II
20		5548.33	II
11		5553.42	II
22		5561.46	II
45	c	5562.06	I
13		5565.52	I
13		5566.91	II
45		5571.83	II
11		5574.61	II
11		5578.81	I
13		5582.35	II
11		5584.02	II
22		5594.92	I
22		5597.29	II
13		5601.30	II
90		5605.65	II
13		5606.68	I
28		5608.93	II
55		5610.22	II
11		5620.06	II
20		5620.26	I
45	c	5621.89	II
110		5623.05	II
90		5624.45	II
11	h	5633.03	I
22		5636.46	II
55	c	5638.79	II
16		5640.37	II
16	Cw	5643.16	I
22		5645.41	II
35		5654.23	II
55		5659.84	II
35	h	5661.57	I
16		5662.19	II
65	c	5668.46	I
45		5669.55	II
35		5669.99	II
16		5674.14	II
16		5677.03	II
55		5681.89	II
13		5685.60	II
16		5686.52	I
22	h	5687.17	II
65		5688.44	II
22		5689.21	II
55	h	5690.97	II
22		5695.90	II
22		5704.38	I
65		5707.61	I
40		5711.63	II
22		5713.83	II
16		5716.08	II
45		5719.08	II
45	d	5719.63	II
		5719.80	II
11		5728.38	I

Praseodymium (Cont.)

40		5731.88	II
20		5747.13	II
11		5747.74	I
11		5747.95	II
22		5753.02	II
90		5756.17	II
16		5759.40	II
22		5760.20	I
22		5769.16	II
16		5769.79	II
45		5773.16	II
11		5775.91	II
16		5777.29	II
90		5779.28	I
65	c	5785.28	II
65		5786.17	II
16	h	5788.29	II
16		5788.92	II
16		5790.86	II
45		5791.36	II
22		5792.95	I
40		5810.58	II
16		5813.55	II
160	d	5815.17	II
		5815.33	II
55		5818.57	II
40		5820.62	II
16	h	5821.36	I
55		5822.59	II
90		5823.72	II
45		5830.94	II
40		5835.13	I
35	c	5844.65	II
40		5844.98	II
65		5847.13	II
65	c	5850.64	II
45		5852.63	II
11	c	5854.44	I
45		5856.07	II
55		5856.90	II
90		5859.68	II
80		5868.83	II
22		5873.83	II
35		5874.72	I
35		5878.10	I
35		5879.04	I
80		5879.25	II
35	c	5884.72	I
55		5892.23	II
22		5894.22	II
40		5903.11	II
45		5904.45	II
40		5908.67	II
11		5915.31	I
11		5915.97	I
40		5920.76	I
40		5930.66	II
16		5936.33	II
160		5939.90	II
65		5940.72	II
22		5941.65	I
35		5947.16	II
22	c	5949.76	I
55		5951.27	II
20		5951.76	II
90		5956.60	II
		5956.70	I
13		5959.25	I
20		5962.18	I
28		5963.00	I
110		5967.82	II
13	c	5976.95	I
13		5978.88	I
65		5981.19	II
40		5986.14	I
45	c	5987.14	I
		5987.29	II
13		5991.27	I
13	c	5994.89	I
11		5996.06	I
29		6002.44	II
90		6006.33	II
13		6008.54	I
55		6016.48	II
150		6017.80	II
28	c	6019.85	I
150		6025.72	II
35		6042.87	II
55		6046.66	II
35		6049.26	I
28		6050.04	II
11		6050.88	I
140		6055.13	I
13		6067.27	II
13		6085.81	I
28		6086.16	II
65		6087.52	II
20		6090.38	II
28		6093.09	II
18		6096.28	I
22		6106.72	II
18		6109.08	I
65		6114.38	II
22	c	6118.02	I
22	c	6122.15	I
35		6141.51	II
65		6148.23	I
		6148.24	II
22		6157.82	II
13		6159.10	II
190		6161.18	II
18		6165.38	I
270		6165.94	II
55		6182.34	II
13		6187.96	I
35		6197.45	II
35		6200.81	II
13		6205.63	II
13		6210.59	I
22		6212.73	I
18		6218.06	I
20	h	6236.80	I
20	h	6241.05	I
45		6244.35	II
35		6255.10	II
40		6262.55	II
18		6264.54	II
22	c	6274.66	II
		6274.81	II
40		6278.68	II
110		6281.28	II
18	c	6289.02	I
11	c	6298.01	I
11		6302.05	I
35		6302.35	II
16		6304.05	I
35		6305.23	II
11	h	6318.13	II
45	c	6322.36	I
22	h	6343.88	I
28		6347.11	II
18	c	6350.98	I
22	c	6357.20	I
55	c	6359.03	I
11	h	6363.62	II
16		6377.61	I
16		6378.59	I
11		6389.57	I
18	c	6391.99	I
40		6393.18	I
45		6397.96	II
10	h	6410.69	I
55		6411.23	I
40		6413.68	II
10		6415.43	I
45		6429.63	II
45		6431.84	II
7	h	6442.78	II
9	h	6443.91	II
16	c	6453.44	I
9		6454.84	II
9		6456.18	I
9	h	6460.19	I
18		6467.72	II
9	h	6475.26	II
35	Cw	6478.02	II
45		6486.55	I
9	h	6486.97	II
40	h	6491.75	I
9		6493.49	I
11		6494.89	I
22	c	6497.11	I
18		6498.94	II
22		6500.72	I
9		6504.09	I
8		6517.14	I
16		6518.79	II
8		6534.52	I
16		6540.47	I
7	h	6553.30	I
22		6564.62	II
45		6566.77	II
7		6571.03	I
6		6578.00	I
6		6584.56	II
9	h	6593.74	II
11		6595.48	I
15		6609.86	I
55		6616.67	I
11		6618.34	II
7	h	6631.00	I
13	h	6632.06	I
14		6647.12	I
75		6656.83	II
55		6673.41	II
75		6673.78	II
5	h	6687.51	II
4	h	6699.25	I
13		6736.79	I
35	c	6747.09	I
19	c	6749.19	I
7	c	6784.99	II
55	Cw	6798.60	I
11		6811.76	II
17	Cw	6812.87	II
13		6814.04	II
9		6817.61	I
35	Cw	6827.60	II
19		6830.50	II
9		6844.39	I
9	h	6845.47	II
9		6846.59	II
17	c	6850.46	II
11		6852.77	I
11	c	6870.44	I
7		6884.66	I
8		6892.71	I
8	h	6970.38	
8	c	6980.12	I
40		7021.51	II
10		7024.53	I
13		7042.40	I
8		7044.45	II
7		7051.07	I
10		7079.99	I
11	c	7095.18	I
20		7114.55	I
10	h	7116.90	I
11		7118.24	II
7		7137.33	II
10	h	7159.88	I
7		7167.77	II
7	h	7189.95	I
10	c	7208.85	II
24		7227.70	II
13		7231.53	I
7	c	7243.26	I
7	c	7259.21	I
7	c	7287.61	I
7	h	7289.19	I
7	Cw	7324.42	I
7		7328.47	I
7		7344.86	
16		7407.56	II
20	c	7451.74	II
11	h	7495.59	I
6	h	7499.42	I
14		7541.02	II
6		7574.86	I
20		7645.66	II
7		7704.98	II
16		7721.84	I
6	h	7786.16	II
6	Cw	7841.27	I
14		7871.67	I
6		7881.09	I
6	Cw	7888.56	II
6		7915.19	II
6		8031.92	I
6		8055.43	I
14		8067.44	I
10	Cw	8122.78	II
11		8141.10	I
5		8181.34	II
5	c	8211.93	I
6		8289.93	I
6		8379.84	I
6	h	8427.82	I
6	h	8605.27	II
10		8714.59	II

Pr III
Ref. 306, 308 — J.R.

Intensity		Wavelength	
		Vacuum	
25		1008.61	III
50		1021.35	III
25		1026.18	III
100		1029.03	III
50		1038.29	III
50		1042.96	III
25		1043.80	III
25		1044.03	III
25		1046.20	III
150		1047.24	III
100		1049.09	III
50		1052.63	III
25		1061.60	III
25		1066.03	III
25	p	1068.85	III
25		1069.88	III
25		1084.42	III
25		1088.66	III
100		1104.84	III
25		1108.82	III
30		1352.70	III
25		1881.22	III
		Air	
50		2031.46	III
100		2033.30	III
25		2043.12	III
100		2052.30	III
50		2052.87	III
50		2053.85	III
200		2058.59	III
50		2064.08	III
25		2075.08	III
200		2090.75	III
200		2093.49	III
25		2096.85	III
50		2096.94	III
10	w	2148.14	III
10	w	2194.24	III
10	w	2197.25	III
10	w	2205.48	III
10	w	2206.26	III
10	w	2214.45	III
10	w	2215.25	III
10	w	2217.12	III
10	w	2223.23	III
10	w	2230.35	III
10	w	2237.26	III
10		2239.06	III
10		2239.42	III
10	w	2242.15	III
10	w	2284.62	III
10		2307.59	III
10		2307.77	III
10		2308.41	III
10		2311.29	III
10		2311.44	III
10		2314.18	III
10	w	2315.46	III
10		2318.15	III
10		2318.36	III
10		2318.64	III
10		2318.82	III
10		2318.97	III
10	w	2319.40	III
10		2320.41	III
10	w	2328.56	III
10	w	2336.13	III
10		2365.52	III
10		2368.78	III
10		2369.08	III
10	w	2378.06	III
10	w	2378.97	III
10		2395.44	III
10		2399.70	III
10	w	2405.56	III
10		2408.19	III
10		2409.80	III
10		2412.40	III
10		2417.69	III
10		2418.95	III
10		2426.14	III
10		2426.85	III
10	w	2430.32	III
10	w	2434.18	III
10		2434.39	III
10		2435.91	III
10		2436.89	III
10		2438.63	III
10	w	2444.93	III
10		2445.49	III
10		2446.77	III
10		2448.16	III
10		2452.02	III
10	w	2452.81	III
10	w	2452.85	III
10		2454.60	III
10		2454.82	III
10	w	2459.77	III
10	w	2460.72	III
10		2462.18	III
10		2462.90	III
10		2468.20	III
10		2468.97	III
10	w	2473.42	III
10	w	2478.32	III
10		2479.98	III
10		2481.02	III
10		2483.30	III
10		2483.99	III
10	w	2484.60	III
10	w	2485.16	III
10	w	2488.72	III
10		2491.97	III
10	w	2494.20	III
10	w	2495.37	III
10	w	2495.51	III
10		2499.97	III
20	w	2587.71	III
40	w	2624.91	III
20	w	2644.62	III
20	w	2656.88	III
20	w	2667.51	III
70	w	2679.47	III
40	w	2710.30	III
20	w	2718.65	III
100	w	2724.03	III
20	l	2841.94	III
70	s	2910.61	III
50	s	2911.77	III
100	l	2914.49	III
50	s	2930.19	III
70	s	2942.43	III
70	s	2953.58	III
90	w	2954.40	III
90	s	2964.85	III
150	s	2968.83	III
80	s	2969.41	III
150	l	2976.86	III
150	s	2977.06	III
500	s	2980.54	III

Praseodymium (Cont.)

Intensity		Wavelength	
100	l	2981.65	III
150	c	2982.42	III
500	s	2985.82	III
150	s	2997.12	III
70	l	2998.79	III
150	l	3000.46	III
150	s	3003.20	III
60	l	3006.47	III
150	s	3008.04	III
150	l	3010.61	III
90	s	3014.60	III
100	l	3015.13	III
90	s	3016.26	III
70	s	3021.77	III
70	s	3025.26	III
100	l	3029.38	III
100	l	3033.31	III
90	l	3034.25	III
70		3040.02	III
60	c	3040.94	III
60	l	3041.78	III
100	s	3042.35	III
100	l	3045.81	III
70	l	3046.98	III
120	l	3050.30	III
70	s	3055.30	III
150	l	3058.90	III
150	l	3066.71	III
70	s	3078.68	III
150	l	3080.20	III
50	w	3248.39	III
90	s	3280.92	III
90	l	3292.58	III
90	s	3296.10	III
100	l	3306.14	III
60	s	3333.26	III
90	s	3340.58	III
500	s	3341.43	III
70	s	3341.68	III
70	s	3345.38	III
70	l	3345.44	III
50	w	3351.07	III
50	s	3353.87	III
250	s	3354.91	III
500	l	3357.56	III
500	l	3359.41	III
100	l	3364.52	III
50		3364.88	III
50	s	3365.80	III
200	s	3367.35	III
500	s	3367.58	III
100	l	3371.92	III
100	s	3377.14	III
50	s	3379.13	III
150	s	3380.21	III
150	s	3381.26	III
300	s	3381.84	III
100	s	3391.08	III
1000	w	3394.22	III
600	d	3396.07	III
300	s	3396.62	III
300	l	3397.46	III
50	l	3402.97	III
500	c	3413.21	III
300	s	3415.15	III
150	s	3420.07	III
300	l	3422.22	III
50	c	3426.27	III
500	l	3427.02	III
300	l	3436.36	III
150	l	3440.62	III
50	l	3445.29	III
70	s	3454.05	III
180		3653.58	III
60		3817.25	III
60	l	3861.80	III
150		3980.51	III
200		4000.20	III
90		4018.36	III
180		4029.60	III
90		4142.46	III
120		4144.48	III
90		4147.85	III
90		4172.15	III
150		4179.77	III
180		4184.18	III
240		4197.01	III
120		4219.45	III
180		4231.45	III
180		4275.07	III
120		4286.32	III
90		4298.27	III
90		4301.73	III
90		4316.34	III
60		4354.28	III
90		4379.82	III
90		4381.47	III
120		4404.71	III
120		4421.10	III
120		4431.85	III
150	w	4447.93	III
180	w	4450.14	III
120	w	4451.00	III
120		4461.02	III
200		4461.81	III
300	w	4500.31	III
450	w	4612.02	III
600	w	4625.18	III
120		4654.16	III
600	w	4713.70	III
300	w	4725.55	III
270		4728.21	III
300		4747.11	III
300	w	4771.83	III
450	w	4775.30	III
600		4857.39	III
150		5208.51	III
150		5261.68	III
1000		5264.44	III
1500		5284.70	III
1500		5299.99	III
1500		5340.02	III
100		5427.70	III
100		5581.74	III
150		5646.80	III
600		5765.27	III
1500		5844.41	III
200		5947.98	III
7000	w	5956.05	III
900	w	5998.94	III
1500	w	6053.01	III
900		6071.09	III
9000	w	6090.02	III
5000		6160.24	III
1500	w	6161.22	III
100		6195.05	III
2000		6195.63	III
200		6310.36	III
100		6361.65	III
300		6429.26	III
300		6444.74	III
600		6500.04	III
300		6501.49	III
200		6578.90	III
100		6616.46	III
600		6706.70	III
100		6727.63	III
200		6827.96	III
100		6854.63	III
200		6857.30	III
1000		6866.80	III
1000		6899.06	III
500		6903.52	III
7000		6910.14	III
150	w	6934.55	III
500		6970.96	III
100		6979.83	III
5000		7030.39	III
100		7075.21	III
4500		7076.62	III
100		7083.99	III
500		7112.53	III
100	w	7165.64	III
250		7231.62	III
100	w	7238.26	III
250		7240.21	III
100	w	7262.32	III
150	w	7340.69	III
350		7343.70	III
200		7349.75	III
300	w	7350.61	III
100	w	7355.52	III
2000		7426.48	III
4000		7429.05	III
100		7463.96	III
250		7487.40	III
200		7493.20	III
100		7511.17	III
500		7529.11	III
100	w	7549.20	III
150	w	7588.64	III
300		7596.41	III
100	w	7625.63	III
100	w	7648.34	III
100	w	7670.65	III
200	w	7674.65	III
500		7742.34	III
250	w	7745.59	III
500		7754.31	III
100	w	7755.48	III
3000		7781.98	III
200	w	7814.74	III
1500	w	7866.14	III
1000		7888.12	III
400		7897.09	III
1000		7914.00	III
100		7923.16	III
1000	w	7972.75	III
250		8001.14	III
3000		8102.90	III
250		8119.54	III
400		8132.23	III
100		8138.34	III
150	w	8235.33	III
250	w	8244.89	III
100		8409.10	III
100	w	8494.99	III
200	w	8567.63	III
5000	w	8602.74	III
500		8691.58	III
500		8771.38	III
1000		8854.05	III
125		8886.17	III
100		8908.70	III
250		9099.98	III
250		9131.90	III
200		9222.32	III
250	w	9265.56	III
125		9320.54	III
250		9334.33	III
175		9377.44	III
175		9388.56	III
175		9549.77	III
100		9579.74	III
150		9802.98	III
175		9806.37	III
500	w	9991.16	III
500		10031.10	III
500		10160.33	III
500		10238.63	III
500		10301.58	III
500		10324.59	III
500		10716.58	III

Pr IV
Ref. 337, 338 — J.R.

Intensity		Wavelength	
		Vacuum	
20		718.23	IV
30		721.34	IV
60		722.41	IV
30		722.58	IV
20		726.04	IV
50		730.37	IV
30		731.77	IV
30		734.86	IV
20		735.04	IV
20		736.19	IV
50		736.32	IV
100		737.17	IV
100		741.45	IV
20		743.15	IV
20		743.89	IV
20		746.14	IV
40		763.16	IV
20		764.00	IV
300		1226.40	IV
2000		1228.59	IV
500		1230.69	IV
400		1238.19	IV
200		1249.35	IV
200		1255.64	IV
200		1261.27	IV
400		1268.32	IV
300		1270.58	IV
1000		1275.10	IV
200		1275.40	IV
1000		1278.65	IV
200		1279.34	IV
1000		1287.44	IV
300		1290.93	IV
1000		1292.30	IV
5000		1293.22	IV
5000		1295.28	IV
400		1296.50	IV
200		1298.26	IV
300		1298.54	IV
200		1304.71	IV
300		1306.86	IV
200		1308.08	IV
200		1310.71	IV
500		1314.96	IV
300		1315.28	IV
300		1316.96	IV
500		1320.10	IV
1000		1320.70	IV
5000		1321.36	IV
500		1322.51	IV
500		1326.38	IV
5000		1333.57	IV
300		1335.96	IV
500		1339.29	IV
1000		1340.74	IV
200		1341.32	IV
300		1344.23	IV
1000		1347.07	IV
1000		1352.81	IV
500		1354.35	IV
5000		1354.66	IV
2000		1360.64	IV
1000		1364.81	IV
2000		1365.77	IV
400		1368.90	IV
5000		1374.41	IV
1000		1382.62	IV
200		1384.23	IV
200		1385.91	IV
300		1394.11	IV
500		1397.11	IV
1000		1399.31	IV
1000		1400.96	IV
400		1410.90	IV
1000		1424.36	IV
500		1426.59	IV
5000		1435.56	IV
200		1459.95	IV
200		1461.76	IV
1000		1474.91	IV
200		1477.32	IV
500		1485.88	IV
500		1503.35	IV
400		1516.86	IV
200		1520.71	IV
2000		1520.98	IV
400		1523.46	IV
500		1553.62	IV
500		1559.49	IV
500		1570.13	IV
200		1572.80	IV
5000		1574.55	IV
5000		1575.10	IV
3000		1578.38	IV
500		1585.10	IV
300		1613.00	IV
400		1613.65	IV
1000		1618.03	IV
2000		1622.30	IV
300		1634.77	IV
400		1676.08	IV
200		1688.49	IV
200		1713.53	IV
500		1732.86	IV
300		1762.86	IV
1000		1766.88	IV
1000		1771.14	IV
500		1841.08	IV
10000		1884.87	IV
400		1951.23	IV
200		1954.61	IV
		Air	
200		2025.06	IV
1000		2039.15	IV
200		2047.05	IV
200		2050.73	IV
200		2058.48	IV
2000		2083.23	IV
500		2100.42	IV
1000		2154.31	IV
300		2193.37	IV
1000		2205.13	IV
300		2265.70	IV
200	c	2334.46	IV
200	c	2339.08	IV
500	c	2376.09	IV
2000	c	2378.98	IV
1000	c	2379.66	IV
500	c	2427.07	IV
500	c	2428.13	IV
500	c	2438.57	IV
500	c	2455.64	IV
500	c	2705.19	IV
200	c	2708.01	IV
200	c	2753.47	IV
200	c	2767.60	IV

Pr V
Ref. 149 — J.R.

Intensity		Wavelength	
		Vacuum	
200		843.78	V
7000		865.90	V
5000		869.17	V
80		869.66	V
1000		896.65	V
750		922.29	V
250		1234.07	V
250		1342.78	V
200		1958.09	V
400		1958.20	V
		AIR	
300		2246.06	V
300		2246.20	V

PROMETHIUM (Pm)
Z = 61

Pm I and II
Ref. 196, 260 — C.H.C.

Intensity		Wavelength	
		Air	
40	w	2502.12	II
40		2608.24	II
150		2632.00	II
70		2638.46	II
100		2671.05	II
50	w	2787.72	II
40		2808.05	II
100	h	2820.10	II
100	w	2840.82	II
150	w	2841.86	II
200	c	2857.46	II
100		3004.59	II
100		3008.85	II
300		3072.41	II
150		3086.02	II

Promethium (Cont.)

Intensity		Wavelength	Spectrum
120		3090.19	II
150		3091.86	II
150		3108.11	II
100		3115.36	II
100		3117.22	II
100		3118.76	II
35		3162.23	I
35		3168.82	I
100		3172.77	II
35		3222.04	I
35		3238.55	I
60		3239.62	II
75		3296.63	I
60		3311.76	II
50		3313.38	I
50		3329.22	I
75		3331.57	I
100		3354.45	I
100		3358.14	II
80		3360.21	II
80		3364.44	II
300		3366.03	I
90		3377.68	II
100		3391.28	II
100		3408.06	II
500		3427.40	II
120		3441.15	II
400		3449.80	II
250		3460.25	II
200		3462.91	II
200		3480.61	II
150		3497.13	II
100		3514.85	II
200		3546.81	I
100		3559.43	II
200		3565.31	II
150		3580.10	II
200		3610.76	II
200		3629.84	II
300		3634.20	II
300		3659.39	II
300	r	3669.22	I
200		3674.85	I
200		3678.51	II
300	r	3679.85	I
200		3687.65	II
400		3689.79	II
300		3692.50	II
300		3697.50	II
300	r	3697.63	I
400		3702.63	II
800		3711.72	II
200		3715.75	II
200		3721.72	II
500		3726.01	I
200		3738.43	I
300		3740.68	I
300		3742.52	II
300	r	3742.97	I
500		3745.86	II
300		3747.09	II
500		3750.09	II
200		3761.68	I
300		3765.75	I
300		3775.42	I
300		3780.77	I
200		3781.43	I
400		3795.66	II
250		3806.06	II
300	r	3809.20	I
400		3810.93	I
200		3819.26	II
300		3820.53	II
300		3839.52	I
200		3842.88	II
300		3842.98	II
250		3845.38	II
300		3874.03	I
800		3877.62	II
300	r	3885.79	I
250		3890.97	I
1000		3892.15	II
300		3898.73	I
400		3899.78	II
250		3909.50	II
1000		3910.26	II
1000		3919.10	II
800		3936.48	II
300		3944.21	II
300	r	3954.76	I
1000		3957.74	II
500		3980.74	II
300		3995.05	II
1000	r	3998.96	II
500		4009.96	II
200		4012.72	II
250		4014.20	II
200		4019.34	II
250		4028.20	II
200		4045.36	II
300		4051.54	II
600	r	4055.20	II
200	r	4056.56	I
600		4075.84	II
200	r	4085.31	I
500		4086.10	II
250		4140.46	II
200		4185.74	II
300		4192.92	II
200		4194.70	II
200		4222.15	II
300	r	4264.32	I
300	r	4284.37	I
600		4297.78	II
200		4303.89	II
200	r	4305.64	I
400		4318.80	I
250		4325.92	II
200		4332.05	II
300		4336.54	II
200		4337.48	II
300		4342.12	II
200		4347.72	I
350	r	4363.92	I
300	r	4369.64	I
200		4381.88	II
400	r	4388.49	I
200		4388.76	II
400	r	4409.42	I
500	r	4412.47	I
1000		4417.96	II
400		4432.51	II
250	r	4435.86	I
300	r	4436.55	I
300	r	4438.68	I
500		4445.41	II
600		4446.90	II
800		4453.95	II
200		4459.97	II
250	r	4468.16	I
200		4471.48	II
300		4473.23	II
200		4477.46	II
350	r	4478.58	I
300	r	4481.60	I
300	r	4485.05	I
300	r	4490.50	I
250		4492.05	II
600		4500.15	II
350	r	4500.33	I
250		4506.84	I
100		4509.38	II
100		4513.56	II
200		4517.31	I
200		4523.32	I
600		4525.20	II
250	r	4526.12	I
250	r	4526.76	I
400	r	4527.70	I
800		4529.21	II
300		4540.06	I
300	r	4541.42	I
450	r	4541.75	I
500	r	4544.08	I
200		4545.17	I
400	r	4549.78	I
300	r	4554.03	I
200		4554.63	I
500	r	4555.34	I
200		4556.06	I
300		4557.03	I
300		4559.21	I
100		4564.83	II
300	r	4568.14	I
200		4570.37	I
300	r	4572.15	I
400	r	4575.27	I
300		4578.28	I
200	r	4578.41	I
300	r	4579.48	I
300		4581.14	I
300	r	4585.49	I
200		4593.82	I
400	r	4595.82	I
800		4597.55	I
500	r	4600.25	I
400	r	4602.96	I
400		4604.59	I
600	r	4605.66	I
500	r	4609.85	I
100		4615.87	II
600	r	4617.02	I
200		4618.40	I
400	r	4618.49	I
500	r	4619.75	I
500		4621.57	I
500		4623.31	I
700	r	4623.68	I
900		4624.41	I
500	r	4625.29	I
400	r	4627.60	I
200		4630.93	I
600	r	4633.45	I
400	r	4640.96	I
700	r	4643.36	I
700	r	4643.76	I
400		4645.94	I
600	r	4647.03	I
600	r	4650.42	I
500		4650.52	I
400	r	4653.41	I
400		4654.50	I
500	r	4655.05	I
300		4659.38	I
500	r	4660.79	I
300	r	4663.26	I
600	r	4663.46	I
400	r	4665.19	I
500	r	4671.23	I
400		4671.76	I
500	r	4674.42	I
200		4677.46	I
500	r	4677.92	I
400	r	4678.09	I
700	r	4682.92	I
500	r	4696.80	I
200		4699.51	I
250		4722.06	II
300		4727.06	I
900	r	4728.36	I
400	r	4728.68	I
800	r	4734.27	I
200	r	4737.99	I
100		4739.08	II
200		4739.78	I
350	r	4745.13	I
500	r	4757.73	I
800	r	4759.00	I
700	r	4762.57	I
700	r	4773.46	I
900	r	4781.29	I
250		4794.59	I
200		4795.43	I
700	r	4798.98	I
900	r	4801.36	I
700	r	4809.54	I
900	r	4811.96	I
400	r	4817.12	I
400	r	4827.72	I
800	r	4837.66	I
400	r	4838.92	I
300		4839.62	I
200		4844.01	I
350	r	4852.73	I
400	r	4860.62	I
700	r	4860.74	I
300	r	4865.30	I
500	r	4865.72	I
400	r	4869.80	I
700	r	4872.42	I
500	r	4887.02	I
700	r	4892.52	I
400	r	4900.30	I
300	r	4904.28	I
400	r	4918.28	I
600	r	4932.99	I
700	r	4959.46	I
100		4971.40	II
500	r	4997.10	I
200	r	5030.80	I
300	r	5058.31	I
100		5067.35	II
150		5080.52	II
150		5089.35	II
200		5092.42	I
400	r	5094.83	I
200		5096.18	I
150		5097.30	II
400	r	5100.77	I
250		5121.47	II
400	r	5127.34	I
200		5129.75	I
400	r	5145.13	I
500	r	5146.30	I
400		5153.86	II
300		5169.71	II
500		5171.58	II
300		5194.05	II
500		5208.09	II
150		5215.96	II
250		5225.12	II
500		5236.26	II
300		5236.66	II
400		5246.33	II
150		5262.42	II
500		5270.64	II
200		5293.92	II
100		5308.86	II
150		5318.58	II
200		5410.45	II
200		5424.54	II
180		5424.79	II
150		5429.04	II
100		5467.64	II
150		5495.45	II
100		5516.42	II
180		5534.96	II
200		5537.38	I
800		5546.08	II
120		5556.88	II
150		5558.39	II
200		5561.73	II
800		5576.02	II
200		5641.29	II
200		5730.81	I
200		5768.16	II
200		5776.99	I
500		5823.93	II
300	c	5868.79	II
200	c	5875.31	II
100	c	5878.76	II
150		5899.76	II
250		5904.71	I
100		5905.90	I
125		5914.96	I
250	c	5927.17	II
150		5939.66	I
400	c	5946.49	II
800		5956.42	I
200		5956.69	I
100	c	5960.08	II
150		5963.00	II
400		5967.89	I
200		5979.73	I
200		5984.82	I
100	c	5987.13	II
400		5997.12	I
200		6027.11	II
300		6030.06	I
400		6031.32	I
500		6043.39	I
150	c	6052.57	II
100	c	6067.00	II
500		6069.06	I
100		6076.40	II
200		6085.41	II
900		6100.21	I
400		6106.40	I
100		6114.90	II
400	h	6151.76	I
100		6159.53	II
400		6163.16	I
100		6184.52	II
200		6208.91	II
500		6229.64	I
400		6237.79	I
100		6263.25	II
400		6272.69	I
400		6286.06	I
500		6308.29	I
100		6314.20	II
700		6323.84	I
500		6390.31	I
100		6429.64	II
500	h	6431.93	I
100		6436.57	II
400		6487.61	I
400		6510.34	I
500		6517.25	I
200		6519.43	II
1000	d	6520.45	I
500		6542.20	I
100	h	6558.48	II
100		6586.39	II
100		6592.29	II
900		6598.15	I
800		6598.66	I
700		6606.37	I
800	w	6625.23	I
100	h	6625.54	II
700		6649.81	I
400		6659.05	II
100		6661.25	II
500		6661.68	I
400		6663.76	I
800	c	6667.51	I
700	h	6677.47	I
200		6680.89	II
500		6685.55	I
500		6685.68	I
150		6690.09	II
600		6700.33	I
100		6706.27	II
700		6714.67	I
500		6717.26	I
500		6720.71	I
700		6727.50	I
600		6743.71	I
900		6749.91	I
900		6750.48	I
200		6756.45	II
300		6772.29	II
400		6778.78	I
100		6783.09	II
100		6796.87	II
200		6811.68	II
800		6833.30	I
400		6848.37	I
50		6858.58	II

PROTACTINIUM (Pa)
Z = 91

Pa I and II
Ref. 96 — J.G.C.

Intensity		Wavelength Air	
3000	h	2466.85	
3000	h	2492.85	
3000		2599.16	II
3000		2699.22	II
3000		2822.79	II
3000	l	2832.14	
3000		2870.01	
3000	h	2871.42	II
3000	h	2891.14	II
3000	h	2906.93	
3000	l	3011.10	II
3000	s	3033.59	II
3000	l	3071.24	II
3000	h	3083.19	
3000	l	3093.23	II
3000	l	3126.23	II
3000	l	3146.28	II
3000	l	3170.89	II
3000	l	3171.54	II
3000		3204.16	
3000	l	3240.58	II
3000		3274.46	II
3000	l	3332.69	II
3000	s	3346.66	II
3000	l	3394.49	
3000	l	3452.82	II
3000		3504.97	I
3000	s	3530.65	II
3000		3570.56	I
3000		3571.82	I
3000		3618.07	I
10000		3636.52	I
3000		3702.74	I
3000		3752.67	I
3000		3873.35	I
3000		3931.83	I
3000	s	3952.62	II
10000	l	3957.85	II
3000	s	3970.07	II
3000		3981.82	I
10000		3982.23	I
3000	l	4012.96	II
3000	s	4018.21	II
3000		4030.16	II
3000	s	4046.93	II
10000	s	4056.20	II
10000	s	4070.40	II
3000		4117.62	
3000	l	4176.18	II
10000	l	4217.23	II
10000	s	4248.08	II
3000	s	4291.34	II
3000		4400.77	
3000		4436.13	
3000	s	4601.43	II
3000		4628.19	
3000		4820.34	
3000	s	4861.49	
3000	l	6035.78	I
3000		6162.56	I
3000		6216.35	
3000	l	6358.61	I
3000		6379.25	I
3000	l	6438.97	I
3000	h	6792.75	I
10000		6945.72	I
3000		6960.09	I
3000	h	6961.78	I
3000	s	6992.73	I
3000		7076.27	I
3000	h	7100.94	I
10000	s	7114.89	I
3000	h	7171.55	I
3000		7227.13	I
3000		7318.79	I
10000	l	7368.25	I
3000	h	7471.89	I
10000	h	7493.15	I
3000	h	7558.26	I
10000	h	7608.20	I
10000		7626.79	I
10000	s	7635.18	I
10000		7669.34	I
3000		7679.20	I
10000	h	7749.19	I
3000		7872.95	I
3000	l	7945.56	I
10000		8039.34	I
10000	h	8099.84	I
10000		8199.04	I
10000		8271.87	I
3000	s	8358.98	I
3000	s	8369.60	I
3000	h	8441.04	I
10000	h	8532.66	I
10000	s	8572.96	I
3000	h	8639.91	I
3000	h	8653.51	I
10000		8735.27	I
3000		10594.38	
3000		10923.32	I
3000		11646.78	
10000		11791.73	I
3000		12279.01	
3000		13234.09	
10000		13522.40	
10000		14344.76	I
3000		18478.61	I

RADIUM (Ra)
Z = 88

Ra I and II
Ref. 253, 254 — E.F.W.

Intensity		Wavelength Air	
8		2369.73	II
8		2460.55	II
10		2475.50	II
8		2586.61	II
10		2643.73	II
20		2708.96	II
10		2795.21	II
30		2813.76	II
10		3033.44	II
5		3101.80	I
100		3649.55	II
200		3814.42	II
8		4194.09	II
8		4244.72	II
100		4340.64	II
20		4436.27	II
30		4533.11	II
8		4641.29	I
100		4682.28	II
8		4699.28	I
100		4825.91	I
10		4856.07	I
10		4859.41	II
10		4927.53	II
10		5097.56	I
10		5205.93	I
10		5283.28	I
10		5320.29	I
10		5399.80	I
20		5400.23	I
20		5406.81	I
8		5482.13	I
10		5501.98	I
10		5553.57	I
20		5555.85	I
10		5616.66	I
50		5660.81	I
20		5813.63	II
30		6200.30	I
10	p	6336.90	I
20		6446.20	I
20		6487.32	I
10		6593.34	II
10		6719.32	II
20		6980.22	I
20		7118.50	I
50		7141.21	I
20		7225.16	I
10		7310.27	I
20		7838.12	I
50		8019.70	II
6		8177.31	I
5		8335.07	I
5		9932.21	I

RADON (Rn)
Z = 86

Rn I
Ref. 251 — E.F.W.

Intensity	Wavelength Air	
5	3514.60	I
10	3739.89	I
20	3753.65	I
10	3917.20	I
10	3941.72	I
10	3952.36	I
10	4226.06	I
80	4307.76	I
7	4335.78	I
100	4349.60	I
40	4435.05	I
50	4459.25	I
50	4508.48	I
50	4577.72	I
50	4609.38	I
30	4721.76	I
6	5722.58	I
10	6061.92	I
6	6200.75	I
6	6380.45	I
10	6557.49	I
10	6606.43	I
15	6627.23	I
6	6669.60	I
8	6704.28	I
20	6751.81	I
6	6806.79	I
8	6836.95	I
8	6837.57	I
10	6891.16	I
10	6998.90	I
200	7055.42	I
100	7268.11	I
20	7291.00	I
6	7320.98	I
10	7419.04	I
300	7450.00	I
8	7470.89	I
8	7483.13	I
8	7514.13	I
8	7516.92	I
6	7523.93	I
6	7597.55	I
8	7601.28	I
10	7657.48	I
10	7738.43	I
20	7746.64	I
100	7809.82	I
20	8049.00	I
100	8099.51	I
6	8173.84	I
100	8270.96	I
8	8314.51	I
6	8349.74	I
10	8381.05	I
10	8487.48	I
10	8494.89	I
20	8520.95	I
100	8600.07	I
10	8639.76	I
15	8675.83	I
10	8807.75	I
50	9327.02	I
6	9948.57	I
5	10106.13	I

RHENIUM (Re)
Z = 75

Re I and II
Ref. 1 — C.H.C.

Intensity		Wavelength Air	
25000		2003.53	I
16000		2017.87	I
27000		2049.08	I
4200		2074.70	I
3700		2083.92	I
10000		2085.59	I
4700		2092.41	II
9800		2097.12	I
2700		2109.22	I
3400		2139.04	II
1600		2142.74	II
		2142.97	I
3700		2156.67	I
4900		2167.94	I
3400		2176.21	I
4200	c	2214.26	II
2200		2214.58	I
1700		2226.42	I
920		2235.44	I
440		2255.73	I
860		2256.19	I
2000		2264.39	I
2100		2274.62	I
5200	c	2275.25	II
1600		2281.62	I
2900		2287.51	I
2700		2294.49	I
390		2298.09	II
390		2299.77	I
610		2302.99	I
680		2306.54	I
230		2312.97	I
220		2313.34	I
220		2319.19	I
370		2320.16	I
800		2322.49	I
300		2328.66	I
270		2334.33	I
270		2335.73	I
220		2336.10	I
270		2337.95	I
860		2344.78	I
230		2349.39	I
220	d	2350.46	I
680		2352.07	I
210	d	2353.95	I
250		2356.50	I
200		2365.32	I
1200		2365.90	I
570		2367.68	I
180		2368.53	II
520		2369.27	I
220		2370.76	II
210		2371.52	I
150		2373.48	II
320		2375.07	I
75		2378.53	II
370		2379.77	I
180		2386.90	II
340		2388.57	I
230		2393.65	I
320		2394.37	I
320		2396.79	I
200		2397.31	I
210	d	2400.72	I
		2400.89	I
210		2401.68	I
75		2403.04	II
1500		2405.06	I
740		2405.60	I
320		2406.70	I
270		2410.37	I
60		2418.20	II
1200		2419.81	I
300		2421.73	I
300		2421.88	I
60		2423.84	II
2500		2428.58	I
490		2431.54	I
420		2432.18	I
340	c	2441.47	I
230		2442.51	I
250		2444.94	I
610		2446.98	I
85		2449.03	II
85		2449.52	II
610		2449.71	I
200		2455.83	II
390		2461.20	I
800	c	2461.84	II
200		2467.57	II
120		2467.85	II
150	c	2469.36	II
120		2470.61	II
75		2471.05	II
150		2473.72	II
160		2475.17	II
75		2477.43	II
200		2479.02	I
1200		2483.92	I
390		2485.81	I
980		2487.33	I
75		2490.16	II
200		2492.84	I
370		2496.04	I
200		2498.22	I
370		2501.72	I
570		2502.35	II
230		2504.60	II
270		2505.94	I
1800	c	2508.99	I
570		2520.01	I
540		2521.50	I
150		2534.10	II
370		2534.80	I
570		2540.51	I
740	d	2544.74	I
370		2545.48	I
160		2550.09	II
300		2552.02	I
150	c	2553.59	II
370		2554.63	II
1000		2556.51	I
250		2559.08	I
340		2564.19	I
540		2568.64	II
370		2571.81	II
380		2586.79	I
290		2599.86	I
290		2603.89	I
660		2608.50	II
610	d	2611.54	I
160	c	2616.72	II
200		2622.76	I
310		2635.83	II
550		2636.64	I
190		2637.01	II
90		2641.02	II
270		2642.75	I
65		2648.46	II
270		2649.05	I
660		2651.90	I

Rhenium (Cont.)

400		2654.12	I
220		2663.63	I
940		2674.34	I
220		2688.53	I
1300		2715.47	I
200		2731.56	II
220		2732.21	I
610		2733.04	II
110	h	2753.64	II
220		2758.00	I
210		2763.79	I
200		2766.39	I
310		2767.74	I
220		2768.85	I
220		2769.32	I
350		2770.42	I
550		2783.57	I
220		2791.29	I
120		2803.28	II
220		2814.68	I
75		2819.78	II
880		2819.95	I
310		2834.08	I
200		2837.55	I
200		2840.35	I
220		2843.00	I
270		2850.98	I
240		2867.19	I
200		2875.28	I
200		2883.44	I
2900		2887.68	I
130	c	2888.06	II
490		2896.01	I
830	c	2902.48	I
210		2905.58	I
550		2909.82	I
65	h	2916.73	II
830	c	2927.42	I
270		2930.61	I
440		2943.14	I
130	h	2957.91	II
270		2962.27	I
720		2965.11	I
1500		2965.76	I
90		2968.98	II
310		2976.29	I
210		2978.15	I
220		2980.82	I
220		2982.19	I
220		2988.47	I
1800		2992.36	I
5500		2999.60	I
350		3001.14	I
220		3004.14	I
200		3006.42	I
500		3016.02	I
300		3016.49	I
380		3030.45	I
240		3047.25	I
200		3058.78	I
1600		3067.40	I
320		3069.94	I
260		3071.16	I
200		3072.96	I
550		3082.43	I
340		3088.76	I
200		3093.64	I
200		3095.06	I
700		3100.67	I
140		3103.06	II
700		3108.81	I
340		3110.86	I
340	c	3118.19	I
340		3121.36	I
420		3128.94	I
260		3134.02	I
250		3141.38	I
440		3151.64	I
330		3153.79	I
360	c	3158.31	I
220		3164.52	I
700		3168.37	I
220		3174.61	I
440		3177.71	I
260		3178.61	I
600		3182.87	I
1100		3184.76	I
1100		3185.57	I
260		3190.78	I
260		3192.36	I
200		3194.50	I
220		3198.58	I
1100	c	3204.25	I
380		3235.94	I
600		3258.85	I
600		3259.55	I
200		3261.56	I
300		3268.89	I
200		3294.83	I
280		3296.70	I
280		3296.99	I
280		3301.60	I
240		3302.23	I
320		3303.21	II
280		3303.75	I
240		3313.95	I
600		3322.48	I
200		3331.52	I
2000		3338.18	I
1600		3342.24	I
810		3344.32	I
320		3346.20	I
240	d	3356.33	I
200		3358.02	I
200		3362.74	I
240		3377.74	I
320		3379.06	II
320		3379.70	I
200		3385.76	I
240		3389.43	I
200		3390.25	I
4000		3399.30	I
650		3404.72	I
650		3405.89	I
240		3408.67	I
320		3409.83	I
320		3417.77	I
810		3419.41	I
8000		3424.62	I
400		3426.19	I
300		3427.61	I
320		3437.71	I
400		3449.37	I
16000	c	3451.88	I
240		3453.50	I
55000	c	3460.46	I
40000	c	3464.73	I
400		3467.96	I
240		3476.44	I
400		3480.38	I
320		3480.85	I
240		3482.23	I
560		3503.06	I
100	c	3512.28	I
320		3516.65	I
320		3517.33	I
120		3534.82	I
320		3537.46	I
160		3539.33	I
240		3549.89	I
160		3551.29	I
160		3553.65	I
160		3558.94	I
160		3568.23	I
240		3570.26	I
360		3579.12	I
810	c	3580.15	II
650		3580.97	I
810		3583.02	I
160		3596.39	I
160		3610.49	I
320		3617.08	I
160		3621.46	I
160		3625.91	I
140		3637.06	I
810		3637.84	I
440		3651.97	I
120		3669.78	I
320		3670.53	I
860	c	3689.50	I
1500	c	3691.48	I
100		3697.71	I
520		3703.24	I
100		3705.02	I
240		3709.93	I
360	c	3717.28	I
4000		3725.76	I
140		3731.87	I
140		3732.28	I
240	c	3735.01	I
810		3735.31	I
910		3740.10	I
140		3740.41	I
130		3742.26	II
300	Cw	3745.44	I
140		3766.48	I
120		3768.26	I
140		3777.66	I
700		3787.52	I
160		3796.59	I
160		3797.59	I
190		3807.74	I
120		3815.66	I
120		3836.30	I
240		3869.94	I
240		3875.26	I
240		3876.86	I
100		3908.21	I
130		3913.92	I
380	c	3917.27	I
550		3929.85	I
140		3936.90	I
110		3944.72	I
180		3945.91	I
280		3961.04	I
350	c	3962.48	I
100		4004.93	I
140		4022.96	I
100		4023.31	I
110	c	4029.63	I
220		4033.31	I
110		4037.49	I
200		4048.99	I
240		4081.43	I
140		4104.42	I
240	c	4110.89	I
190		4121.64	I
240	Cw	4133.42	I
1800		4136.45	I
700		4144.36	I
140		4149.96	I
160		4170.40	I
220		4182.90	I
220		4183.06	I
650		4221.08	I
3600	c	4227.46	I
150		4241.39	I
260	c	4257.60	I
120	c	4291.17	I
200		4304.40	I
200		4332.25	I
40		4357.98	II
380		4358.69	I
190		4367.58	I
140		4391.34	I
360	Cw	4394.38	I
110	Cw	4406.40	I
180		4415.82	I
150		4475.08	I
120		4478.39	I
120	c	4507.04	I
2600		4513.31	I
260		4516.64	I
500		4522.73	I
120		4523.88	I
120		4529.95	I
100		4545.17	I
120		4580.68	I
120		4605.73	I
100		4621.38	I
190	c	4791.42	I
2200	Cw	4889.14	I
220		4923.90	I
40		5058.56	I
70		5096.50	I
20		5120.32	I
25		5161.65	I
40	c	5178.89	I
20		5181.74	I
35		5234.31	I
50		5248.86	I
1300		5270.95	I
1600	Cw	5275.56	I
100		5278.24	I
30		5305.56	I
20		5317.28	I
35		5321.28	I
50		5327.46	I
20		5331.90	I
20		5332.76	I
20		5333.85	I
35		5369.48	I
50	c	5369.80	I
100	c	5377.10	I
25		5431.90	I
14		5437.03	I
14		5447.92	I
25		5460.64	I
14	h	5520.05	I
25		5521.10	I
50	c	5532.68	I
50	c	5563.24	I
25		5573.47	I
25		5584.72	I
10	h	5607.21	I
12	h	5612.27	I
100		5667.88	I
25		5711.43	I
18		5716.95	I
110	c	5752.93	I
110	Cw	5776.83	I
18		5791.60	I
10	c	5815.92	I
550		5834.31	I
10		5919.86	I
60		5943.24	I
10		5950.21	I
18	h	5969.77	I
10		5989.99	I
18	h	5995.73	I
30		6114.22	I
35	c	6145.81	I
50		6146.82	I
18		6203.24	I
25		6217.97	I
30	Cw	6229.42	I
35	Cw	6243.24	I
35	d	6260.02	I
		6260.24	I
18	c	6271.37	I
18		6278.76	I
10		6286.41	I
10		6303.42	I
200		6307.70	I
200		6321.90	I
80	d	6350.75	I
16	h	6382.94	I
14		6411.47	I
50		6511.47	I
14		6515.25	I
12		6544.91	I
35	c	6577.11	I
40	Cw	6592.52	I
100	Cw	6605.19	I
30	c	6623.91	I
10	c	6637.25	I
27	Cw	6652.39	I
15	h	6683.28	I
9	c	6711.30	I
30		6751.22	I
5	c	6761.19	I
180	c	6813.41	I
260		6829.90	I
85		6971.53	I
35	Cw	7006.63	I
65	Cw	7024.15	I
65	Cw	7246.67	I
13	Cw	7292.72	I
40	Cw	7578.73	I
13		7611.89	I
7	Cw	7620.25	I
50	Cw	7640.94	I
65	Cw	7912.94	I
35	Cw	7980.77	I
40		8417.13	I
29	Cw	8527.73	I

RHODIUM (Rh)
Z = 45

Rh I and II
Ref. 1 — C.H.C.

Intensity		Wavelength	
		Air	
150		2276.21	II
140		2288.57	I
110		2309.82	I
55		2318.36	I
95		2319.10	I
95		2321.73	I
350		2322.58	I
140		2326.47	I
80		2328.64	I
190		2334.77	II
55		2345.41	I
55		2352.47	I
55		2359.18	I
300		2361.92	I
110		2368.34	I
270		2382.89	I
230		2383.40	I
40		2384.65	I
270		2386.14	II
80		2407.88	I
27		2408.19	I
27		2410.25	I
80		2415.84	II
55		2418.64	
45		2419.75	I
45		2420.18	II
65		2420.98	II
75		2423.94	
65		2427.11	II
130		2427.68	I
230		2429.52	I
40		2431.85	II
40		2432.66	I
18		2437.08	I
110		2437.90	I
330		2440.34	I
50	h	2444.27	I
65		2448.84	I
50		2449.04	I
75		2450.56	I
30		2455.70	II
65		2458.90	II
90		2461.04	II
30		2463.61	I
75		2470.39	I
90		2471.47	I
30		2472.51	I
130		2473.09	I
15		2475.64	II
15		2477.54	II
25		2482.04	I
50		2483.33	I
150		2487.47	I
100		2490.77	II
30		2492.30	I
75	h	2494.51	
15		2499.02	I
40		2500.58	
130		2502.46	I

Rhodium (Cont.)

Intensity		Wavelength	Spectrum
15		2503.84	II
300		2504.29	II
40		2505.10	II
150		2505.67	I
350		2509.70	I
50		2510.66	II
300		2511.03	II
75		2513.36	II
200		2515.75	I
130		2520.53	II
13		2525.99	I
13		2531.74	I
50		2532.66	
13		2533.59	
50		2534.07	
110		2536.71	
110		2537.04	II
30		2539.72	
40		2544.22	
350		2545.70	I
13		2548.60	
550		2555.36	I
25		2558.62	I
50		2565.79	I
45		2566.04	I
25		2566.92	II
50		2567.28	
25		2574.66	
25		2575.75	I
13		2576.23	
40		2587.29	II
30		2598.07	
30		2603.32	II
75		2606.44	II
75		2613.60	
150		2622.58	I
230		2625.88	I
100		2630.42	I
40		2634.99	I
30		2638.74	II
75		2643.00	I
110		2647.28	I
400		2652.66	I
30		2659.01	I
30		2671.06	I
65		2676.11	I
25		2680.28	I
100		2680.63	I
30		2681.78	I
30	h	2686.50	
30	h	2686.91	
50		2694.31	I
400		2703.73	I
40		2705.63	II
40		2707.23	I
75		2714.41	I
100		2715.31	II
75		2717.51	I
180		2718.54	I
65		2720.14	I
30		2720.52	I
160		2728.94	I
40		2736.76	I
75		2741.75	I
50		2767.73	I
100		2771.51	I
50		2778.06	I
75		2779.54	I
130		2783.03	I
25		2791.16	I
75		2796.63	I
150		2826.43	I
180		2826.68	I
30		2827.31	I
75		2834.12	I
45		2835.44	I
75		2836.69	I
50		2856.16	I
50	d	2860.68	I
		2860.76	I
280		2862.94	I
65		2864.40	I
50		2871.35	I
30		2873.62	I
110		2878.66	I
75		2880.76	I
140		2882.37	I
75		2885.97	I
75		2889.11	I
75		2889.84	I
65		2899.96	I
25		2904.81	I
160		2907.21	I
65		2910.17	II
75		2912.62	I
90		2915.42	I
30		2923.10	I
180		2924.02	I
130		2929.11	I
130		2931.94	I
30		2955.41	I
230		2968.66	I
25		2974.03	I
160		2977.68	I
450		2986.20	I
90		2986.99	I
50		2987.45	I
110		3004.46	I
50		3019.54	I
130		3023.91	I
50		3028.43	I
30		3045.77	I
30		3046.76	I
25		3057.89	I
65		3067.30	I
180		3083.96	I
29		3087.42	I
70		3114.91	I
140		3121.76	I
240		3123.70	I
35		3130.79	I
95		3137.71	I
45		3151.36	I
45		3152.60	I
130		3155.78	I
70		3179.73	I
80		3185.59	I
140		3189.05	I
470		3191.19	I
190		3197.13	I
70		3214.32	I
80		3237.66	I
520		3263.14	I
520		3271.61	I
2300		3280.55	I
110		3281.70	I
2300		3283.57	I
280		3289.14	I
45		3289.64	I
210		3294.28	I
45		3296.72	I
260		3300.46	I
4200		3323.09	I
60		3331.09	I
45		3331.24	I
330		3338.54	I
70		3342.90	I
80		3344.20	I
60		3359.90	I
280		3360.80	I
60		3362.18	I
420		3368.38	I
45		3369.68	I
1100		3372.25	I
110		3377.14	I
80		3377.71	I
110		3385.78	I
5600		3396.82	I
820		3399.70	I
160		3406.55	I
820		3412.27	I
60		3420.16	I
330		3421.22	I
120	d	3424.38	I
8200		3434.89	I
1400		3440.53	I
35		3442.63	I
120		3447.74	I
60		3448.58	I
120		3450.29	I
60		3451.15	I
400		3455.22	I
60		3455.42	I
180		3457.07	I
220		3457.93	I
5900		3462.04	I
180		3469.62	I
4700		3470.66	I
120		3472.25	I
4700		3474.78	I
2100		3478.91	I
95		3484.04	I
80		3491.07	I
110		3494.44	I
1200		3498.73	I
5900		3502.52	I
60		3505.41	I
2800		3507.32	I
60		3511.78	I
60		3513.10	I
60		3519.54	I
8800		3528.02	I
880	d	3538.14	I
		3538.26	I
280		3541.91	I
1200		3543.95	I
1800		3549.54	I
240		3564.13	I
1200		3570.18	I
4700		3583.10	I
120		3583.53	I
4700		3596.19	I
5900		3597.15	I
310		3605.86	I
3100		3612.47	I
240		3614.78	I
200		3620.46	I
1800		3626.59	I
95		3627.80	I
310		3639.51	I
350		3654.87	I
8200		3657.99	I
280		3661.86	I
1300		3666.22	I
180		3666.91	I
140		3674.76	I
560		3681.04	I
1900		3690.70	I
9400		3692.36	I
60		3694.95	I
940		3695.52	I
280		3698.26	I
380		3698.60	I
7600		3700.91	I
940		3713.02	I
60		3713.43	I
45		3714.83	I
16		3724.94	I
650		3735.28	I
420		3737.27	I
420		3744.17	I
1200		3748.22	I
240		3754.12	I
380		3754.27	I
490		3755.58	I
1000		3760.40	I
2300		3765.08	I
490		3769.97	I
70		3775.72	I
380		3778.13	I
1000		3788.47	I
1300		3792.18	I
3800		3793.22	I
4900		3799.31	I
760		3805.92	I
1300		3806.76	I
45		3809.50	I
95		3812.45	I
470		3815.01	I
760		3816.47	I
1300		3818.19	I
3800		3822.26	I
2300		3828.48	I
2000		3833.89	I
45		3834.75	I
5900		3856.52	I
490		3870.01	I
70		3872.39	I
380		3877.34	I
70		3888.34	I
29		3904.22	I
23		3912.83	I
120		3913.51	I
240		3922.19	I
2000		3934.23	I
45		3934.98	I
50		3935.84	I
590		3942.72	I
95		3958.24	I
3800		3958.86	I
45		3964.54	II
380		3975.31	I
240		3984.40	I
240		3995.61	I
380		3996.15	I
120		4023.14	I
60		4048.41	I
23		4049.04	I
40		4053.44	I
23		4056.34	I
70		4077.57	I
560		4082.78	I
19		4084.28	I
45		4087.79	I
60		4088.50	I
140		4097.52	I
45		4107.49	I
70		4116.33	I
120		4119.68	I
1100		4121.68	I
1500		4128.87	I
2100		4135.27	I
240		4154.37	I
330		4196.50	I
70		4206.62	I
3300		4211.14	I
29		4230.20	I
40		4244.44	I
60		4273.43	I
60		4278.60	I
820		4288.71	I
70		4296.77	I
23		4342.44	I
45		4373.04	I
4200		4374.80	I
95		4379.92	I
23		4433.32	I
35		4492.47	I
29		4503.78	I
23		4528.72	I
16		4544.27	I
35		4548.73	I
40		4551.64	I
19		4560.89	I
16		4565.19	I
130		4569.00	I
14		4571.31	I
29		4608.12	I
14		4619.91	I
23		4543.18	I
150		4675.03	I
19		4721.00	I
70		4745.11	I
12		4755.58	I
23		4810.49	I
21		4842.43	I
45		4843.99	I
60		4851.63	I
60		4963.71	I
60		4977.75	I
40		4979.18	I
14		5085.52	I
70		5090.63	I
23		5120.69	I
19		5130.76	I
60		5155.54	I
14		5157.09	I
40		5158.69	I
60		5175.97	I
12		5177.27	I
35		5184.19	I
95		5193.14	I
16		5206.95	I
16		5211.52	I
19		5212.73	I
16		5214.79	I
19		5222.66	I
19		5230.62	I
45		5237.16	I
9		5237.80	I
14		5269.27	I
11	h	5280.12	I
14		5292.14	I
14		5314.79	I
40	h	5329.74	I
14	h	5331.08	I
9		5349.31	I
130		5354.40	I
23		5356.47	I
45		5379.10	I
95		5390.44	I
23	h	5404.73	I
60	h	5424.07	I
19		5424.72	I
19	h	5425.45	I
12		5439.58	I
12	h	5441.36	I
9	h	5444.32	I
35	h	5445.23	I
23	h	5468.11	I
35	h	5470.85	I
12		5476.12	I
12		5481.42	I
16		5484.23	I
9		5504.65	I
29		5535.04	I
21	l	5544.58	I
160		5599.42	I
7		5607.71	I
16		5608.35	I
5		5632.77	I
9		5659.62	I
40		5686.38	I
9	h	5702.47	I
6		5727.30	I
29		5792.66	I
9		5795.79	I
9		5803.34	I
40		5806.91	I
6		5821.84	I
35		5831.58	I
7		5907.31	I
9		5918.54	I
7		5941.46	I
130		5983.60	I
9		5991.19	I
35		6102.72	I
6		6116.15	I
8		6128.06	I
8		6186.89	I
14		6199.99	I
16		6253.72	I
5		6276.66	I
8		6277.46	I
6		6293.38	I
29		6319.53	I
12		6414.72	I
16		6510.41	I
19		6519.70	I
9		6627.80	I
19		6630.16	I
40		6752.35	I
9		6796.65	I
13		6827.33	I
11		6857.68	I
20		6879.94	I

Rhodium (Cont.)

Intensity		Wavelength	
65		6965.67	I
8		6972.91	I
16		6979.15	I
16		7001.58	I
11		7038.76	I
18		7101.64	I
15		7104.45	I
6		7142.55	I
9		7219.06	I
18		7268.18	I
35		7270.82	I
12		7271.94	I
5		7273.03	I
9		7375.57	I
5	h	7386.64	I
9		7430.80	I
18	h	7442.39	I
7		7446.77	I
12		7475.74	I
12		7495.24	I
8		7542.02	I
11		7557.67	I
8		7577.22	I
11		7690.05	I
18		7772.90	I
29		7791.61	I
55		7824.91	I
15		7830.05	I
15		7846.50	I
21		8029.91	I
11	h	8036.09	I
29		8045.36	I
7		8063.50	I
15		8136.20	I
7	h	8193.67	I
5	h	8369.67	I
8		8425.59	I

Rh III
Ref. 396 — L.J.R.

Intensity		Wavelength	
		Vacuum	
10		746.28	III
30		759.54	III
50		813.44	III
30		826.01	III
30		843.63	III
30		849.08	III
40		852.70	III
40		854.77	III
40		859.89	III
40		861.34	III
50		863.78	III
50		865.22	III
50		870.40	III
80		882.51	III
100		925.75	III
150		937.28	III
100		976.12	III
500		991.62	III
400		992.48	III
500	d	1009.60	III
200		1012.22	III
200		1015.17	III
100		1050.00	III
100		1058.97	III
200		1073.87	III
100		1100.58	III
100		1113.79	III
100		1768.43	III
150		1784.24	III
200		1784.94	III
150		1796.50	III
200		1816.03	III
1000		1832.05	III
500		1859.85	III
100		1874.70	III
800		1880.66	III
500		1884.91	III
500		1887.36	III
700		1888.62	III
800		1901.32	III
500		1910.16	III
600		1919.37	III
500		1927.07	III
700		1931.79	III
500		1954.25	III
400		1965.16	III
500		1994.26	III
		Air	
400		2005.14	III
800		2013.71	III
500		2017.47	III
500		2028.53	III
800		2036.72	III
600		2037.61	III
1000		2040.18	III
3000		2048.67	III
2000		2064.11	III
800		2076.84	III
1000		2118.53	III
1000		2118.63	III
1000		2139.44	III
1000		2152.23	III
3000		2158.17	III
3000		2163.19	III
3000		2167.33	III
100		2207.00	III
100		2230.66	III
50		2250.84	III
30		2374.84	III
20		2470.65	III
50		3006.43	III
50		3052.44	III
50		3310.69	III
1		3852.98	III

RUBIDIUM (Rb)
Z = 37

Rb I and II
Ref. 12, 130, 241, 257, 264 — J.R.

Intensity		Wavelength	
		Vacuum	
10		474.88	II
40		481.118	II
90		497.430	II
20		508.434	II
150		513.266	II
300		530.173	II
75		533.801	II
40		542.887	II
200		555.036	II
2500		589.419	II
1500		643.878	II
3000		697.049	II
6000		711.187	II
10000		741.456	II
1000		1604.12	II
200		1644.96	II
200		1707.52	II
600		1716.85	II
5000		1760.50	II
200		1803.47	II
500		1809.68	II
500		1865.33	II
500		1889.42	II
500		1954.24	II
300		1956.54	II
200		1971.42	II
500		1983.19	II
		Air	
300		2042.23	II
300		2052.21	II
500		2052.80	II
2000		2068.92	II
1000		2071.50	II
10000		2075.95	II
1000		2090.29	II
200		2108.06	II
300		2116.50	II
1000		2125.25	II
400		2129.82	II
200		2143.10	II
30000		2143.83	II
200		2190.36	II
600		2197.99	II
600		2198.26	II
300		2207.86	II
10000		2217.08	II
200		2223.79	II
400		2237.72	II
500		2250.65	II
200		2251.43	II
800		2254.19	II
200		2254.55	II
200		2263.54	II
500		2263.94	II
500		2286.82	II
5000		2291.71	II
300		2298.80	II
250		2333.01	II
2000		2333.39	II
350		2353.11	II
300		2353.96	II
400		2356.97	II
300		2358.04	II
300		2364.27	II
200		2364.32	II
200		2365.15	II
300		2367.51	II
200		2373.21	II
2000		2385.34	II
250		2405.94	II
400		2434.17	II
800		2459.14	II
50000		2472.20	II
300		2484.56	II
700		2484.70	II
2000		2496.38	II
200		2502.67	II
250		2514.18	II
1000		2524.24	II
200		2594.56	II
400		2623.76	II
400		2645.58	II
1000		2684.10	II
1000		2711.76	II
250		2741.01	II
500		2812.15	II
350		2838.51	II
750		2873.88	II
1000		3051.36	II
2		3082.02	I
250		3088.58	II
10		3112.57	I
3		3113.06	I
5000	c	3148.90	II
25		3157.54	I
5		3158.26	I
1200		3161.00	II
50		3227.98	I
6		3229.16	I
2000		3270.99	II
1500		3321.49	II
1200		3340.55	II
60		3348.72	I
75		3350.82	I
750		3353.89	II
1200		3393.03	II
750		3415.58	II
1000		3434.18	II
1500		3461.50	II
3000		3521.39	II
3000	l	3531.55	II
1000		3541.15	II
100		3587.05	I
40		3591.57	I
5000		3600.60	II
10000		3600.64	II
600	c	3639.80	II
400	c	3646.26	II
350	c	3647.56	II
1000	c	3662.74	II
900	c	3663.81	II
350		3666.72	II
300		3675.66	II
2500	c	3699.58	II
350		3746.33	II
3500		3796.81	II
2500		3801.90	II
1000		3826.66	II
450		3860.74	II
250		3907.29	II
500		3922.20	II
2500	l	3926.44	II
25000		3940.51	II
1000	c	3978.15	II
1700		4029.49	II
2500	c	4083.88	II
2000	c	4104.28	II
1700	c	4136.11	II
3500		4193.08	II
1000		4201.80	I
500		4215.53	I
90000		4244.40	II
500		4266.58	II
250	c	4270.25	II
15000		4273.14	II
2500	c	4287.97	II
1500		4293.97	II
500	c	4306.26	II
1000		4346.96	II
2500		4377.12	II
300		4440.10	II
1000		4469.47	II
400	c	4493.92	II
700		4519.04	II
3000		4530.34	II
500	l	4533.79	II
400		4540.74	II
20000		4571.77	II
3000	c	4622.42	II
350	c	4631.89	II
10000		4648.57	II
500		4659.28	II
1000		4730.45	II
1000		4755.30	II
400	c	4757.82	II
30000		4775.95	II
5000	c	4782.83	II
300	c	4855.34	II
1500	c	4885.59	II
2		5087.987	I
2		5132.471	I
10		5150.134	I
10000		5152.08	II
300		5164.58	II
1		5165.023	I
2		5165.142	I
1		5169.65	I
15		5195.278	I
2		5233.968	I
20		5260.034	I
1		5260.228	I
200		5270.51	II
3		5322.380	I
40		5362.601	I
4		5390.568	I
75		5431.532	I
3		5431.830	I
500		5512.55	II
5000		5522.78	II
6		5578.788	I
5000	c	5635.99	II
40		5647.774	I
20		5653.750	I
3000	d	5699.15	II
60		5724.121	I
3		5724.614	I
200		5739.64	II
75		6070.755	I
200		6135.27	II
30	c	6159.626	I
1000	c	6199.08	II
75	c	6206.309	I
300		6269.40	II
120	c	6298.325	I
5		6299.224	I
10000		6458.33	II
1000		6555.62	II
5000		6560.81	II
3000	l	6775.07	II
100	l	7279.997	I
300	c	7316.52	II
150		7408.173	I
200	l	7618.933	I
300		7757.651	I
60		7759.436	I
90000	c	7800.27	I
5	l	7925.26	I
4		7925.54	I
45000	c	7947.60	I
40	l	8271.41	I
30		8271.71	I
2000		8603.96	II
40	l	8868.512	I
30		8868.852	I
300		8978.88	II
300		9021.77	II
3		9224.64	I
2		9234.25	I
500	c	9246.41	II
300		9338.87	II
200	w	9373.50	II
300		9391.36	II
1000		9479.32	II
700	l	9493.72	II
30	l	9522.65	I
5		9523.05	I
20	l	9540.18	I
300		9612.99	II
300		9671.54	II
2000	c	9689.05	II
200		9776.06	II
200		9934.76	II
35	l	10075.282	I
30	l	10075.708	I
100		13235.17	I
20		13442.81	I
30		13443.57	I
75		13665.01	I
1000		14752.41	I
800		15288.43	I
150		15289.48	I
20		22529.65	I
10		22932.47	I
4		27314.31	I
2		27905.37	I

Rb III
Ref. 258, 262 — J.R.

Intensity		Wavelength	
		Vacuum	
30		465.85	III
35	p	482.43	III
30	p	482.47	III
500		482.83	III
300		484.84	III
500		489.66	III
100		489.96	III
600		493.48	III
50		497.82	III
100		500.28	III
30		508.33	III
400		516.79	III
800		533.64	III
1200		535.86	III
1200		556.19	III
500		558.36	III

Rubidium (Cont.)

Intensity		Wavelength	
700		564.77	III
1500		566.71	III
1000		572.82	III
1500		576.65	III
2500		579.63	III
1500		581.26	III
500		582.34	III
800		586.77	III
100		591.42	III
900		593.65	III
1000		594.94	III
1300		595.88	III
1200		598.49	III
450		602.09	III
50		605.51	III
500		607.28	III
400		613.31	III
500		619.67	III
20		620.83	III
100		622.24	III
250		630.06	III
500		645.67	III
20		674.81	III
5000		769.04	III
2500		815.28	III

Air

Intensity		Wavelength	
100		2153.21	III
250		2164.59	III
100		2268.00	III
150		2300.12	III
500		2304.14	III
150		2304.45	III
250		2312.46	III
200		2337.07	III
100		2341.90	III
200		2345.37	III
100		2349.81	III
150		2380.44	III
100		2381.29	III
150		2418.46	III
300		2561.86	III
100		2573.71	III
100		2577.07	III
200		2586.83	III
1000		2631.75	III
350		2636.83	III
100		2656.68	III
100		2713.86	III
500		2798.86	III
150		2800.27	III
500		2807.58	III
100		2845.44	III
150		2869.77	III
500		2903.69	III
150		2949.62	III
100		2951.01	III
2000		2956.07	III
500	l	2967.45	III
150		2968.13	III
500		2970.74	III
250		2987.40	III
350		3023.61	III
200		3039.62	III
200		3041.48	III
250		3070.70	III
500		3086.84	III
100		3098.49	III
500		3111.36	III
250	s	3114.82	III
120		3118.92	III
100		3169.34	III
200		3222.60	III
500		3286.41	III
100		3330.16	III
200		3346.92	III
250		3439.26	III
100		3492.68	III

Rb IV
Ref. 109 — J.R.

Vacuum

Intensity	Wavelength	
10	595.18	IV
25	663.76	IV
25	716.24	IV
20	733.41	IV
50	740.85	IV
20	749.86	IV
20	753.75	IV
10	771.54	IV
25	776.89	IV
9	817.92	IV
15	850.18	IV
10	988.00	IV

RUTHENIUM (Ru)
Z = 44

Ru I and II
Ref. 1 — C.H.C.

Air

Intensity		Wavelength	
2400		2076.43	I
2600		2083.77	I
2400		2090.89	I
690		2255.52	I
290		2259.53	I
780		2272.09	I
240		2278.19	I
780		2279.57	I
170		2285.38	I
290		2302.54	I
480		2317.80	I
150		2322.01	I
120		2334.96	II
240		2340.69	I
190	h	2342.85	II
190		2349.34	I
310		2351.33	I
170		2357.91	II
140		2360.56	I
170		2370.17	I
240		2375.27	I
80		2375.63	II
160		2392.42	I
95		2396.71	II
780		2402.72	II
150		2407.92	II
55		2410.89	I
55		2414.82	II
130		2420.82	I
55		2422.92	I
45		2429.60	I
65		2432.93	I
30		2447.45	I
30		2450.58	I
65		2454.92	I
180		2455.53	II
150		2456.44	II
370		2456.57	II
65	h	2458.62	I
55		2462.94	I
85		2464.70	I
30		2474.04	I
110		2475.41	I
100		2476.88	I
280		2478.93	II
28		2481.11	II
30		2489.91	I
18		2491.78	I
65		2493.69	II
85		2494.02	I
45		2494.48	II
85		2495.69	II
65		2496.56	I
140		2498.42	II
140		2498.57	II
85		2499.78	I
260		2507.01	II
130		2508.27	I
110		2509.07	I
110		2512.81	I
110		2513.32	II
110		2517.32	II
150		2535.59	II
65		2543.25	II
280		2544.22	I
120		2546.67	I
280		2549.48	I
550		2549.58	I
130		2560.26	I
120		2560.83	I
110		2563.15	I
160		2568.77	I
100		2570.97	I
100		2578.57	I
100		2579.53	I
100		2589.57	I
170		2591.12	I
120		2592.02	I
100		2593.70	I
110		2594.85	I
370		2609.06	I
830		2612.07	I
100		2615.09	I
220		2631.30	I
220		2635.86	I
170		2636.67	I
110		2640.33	I
460		2642.96	I
110		2647.32	I
110		2651.29	I
330		2651.84	I
28		2656.25	II
400		2659.62	I
23		2661.17	II
330		2661.61	II
200		2664.76	I
30		2667.40	II
690		2678.76	II
220		2686.29	I
28		2687.50	II
		2688.16	II
330		2692.06	II
110		2701.34	I
110		2702.83	I
170		2709.20	I
200		2712.41	II
690		2719.52	I
130		2722.65	I
140		2725.47	II
310		2734.35	II
1800		2735.72	I
170		2739.22	I
130		2744.45	I
35		2747.97	II
75		2752.45	II
75		2752.77	II
260		2763.42	I
35		2765.44	II
90		2768.93	II
100		2778.38	II
110		2787.83	II
140		2802.81	I
35		2806.74	II
350		2810.03	I
1700		2810.55	I
350		2818.36	I
110		2822.03	I
200		2827.87	I
400		2829.16	I
130		2834.00	I
150		2840.54	I
35		2841.68	II
640		2854.07	I
180		2860.02	I
420		2861.41	I
550		2866.64	I
110		2868.31	I
1800		2874.98	I
220		2879.76	I
55		2882.12	II
130		2883.60	I
740		2886.54	I
180		2892.56	I
110		2901.94	I
140		2905.65	I
370		2908.88	I
1100		2916.26	I
150		2919.61	I
35		2927.54	II
180		2945.67	II
180		2946.99	I
370		2949.50	I
150		2954.49	I
18		2963.40	II
550		2965.16	I
170		2965.55	II
140		2976.59	II
550		2976.92	I
45		2977.23	II
75		2979.96	II
1400		2988.95	I
35		2991.62	II
110		2993.27	I
460		2994.96	I
440		3006.59	I
330		3017.24	I
310		3020.88	I
240		3033.45	I
200		3040.31	I
220		3042.48	I
110		3045.71	I
110		3048.78	I
150		3054.94	I
390		3064.84	I
170		3089.14	I
120		3089.80	I
330		3096.57	I
120		3097.60	I
830		3099.28	I
740		3100.84	I
120		3125.96	I
120		3153.82	I
290		3159.92	I
200		3168.52	I
60		3177.05	II
180		3186.04	I
240		3188.34	I
240		3189.98	I
180		3196.59	I
180		3223.27	I
110		3226.37	I
100		3227.88	I
220		3228.53	I
220		3238.53	I
120		3241.24	I
120		3243.50	I
280		3260.35	I
120	d	3264.55	I
120		3266.44	I
200		3268.21	I
200		3273.08	I
200		3274.71	I
100		3277.57	I
490		3294.11	I
370		3301.59	I
220		3306.17	I
290		3315.23	I
290		3316.39	I
100		3325.00	I
120		3335.69	I
930		3339.55	I
240		3341.66	I
200		3361.15	I
370		3368.45	I
100		3371.86	I
130		3374.65	I
120		3378.02	I
100		3379.60	I
130		3380.18	I
130		3385.14	I
130		3388.71	I
100		3389.50	I
370		3392.54	I
310		3401.74	I
310		3409.28	I
3100		3417.35	I
4900		3428.31	I
490		3430.77	I
310		3432.74	I
6400		3436.74	I
260		3438.37	I
220		3440.20	I
260		3473.75	I
240		3481.30	I
8300		3498.94	I
640		3514.49	I
330		3519.64	I
200		3528.68	I
240		3532.81	I
390		3537.95	I
790		3539.37	I
200		3541.63	I
690		3570.59	I
200		3574.58	I
390		3587.20	I
6400		3589.22	I
6900		3593.02	I
6400		3596.18	I
1300		3599.76	I
350		3625.20	I
370		3626.74	I
3100		3634.93	I
210		3637.47	I
200		3640.64	I
290		3650.32	I
310		3654.40	I
6200		3661.35	I
830		3663.37	I
650		3669.49	I
240		3678.32	I
260		3696.59	I
410		3717.00	I
260		3719.33	I
550		3726.10	I
8700		3726.93	I
11000		3728.03	I
7100		3730.43	I
280		3737.40	I
410		3739.46	I
3500		3742.28	I
870		3742.78	I
280		3744.22	I
410		3744.40	I
2800		3745.59	I
760		3753.54	I
310		3755.09	I
870		3755.93	I
1200		3759.84	I
370		3760.03	I
600		3761.51	I
600		3767.35	I
1500		3777.59	I
460		3781.18	I
600		3782.74	I
3900		3786.06	I
6000		3790.51	I
240		3794.92	I
760		3798.05	I
7600		3798.90	I
7600		3799.35	I
310		3800.26	I
310		3808.68	I
600		3812.72	I
760		3817.27	I
760		3819.03	I
650		3822.09	I
550		3824.93	I
760		3831.80	I
220		3835.05	I

Ruthenium (Cont.)

Intensity		Wavelength	
310		3838.07	I
930		3839.70	I
480		3846.68	I
760		3850.43	I
480		3856.46	I
1300		3857.55	I
220		3860.72	I
650		3862.69	I
1300		3867.84	I
260		3873.52	I
650		3892.21	I
760		3909.08	I
260		3920.92	I
1500		3923.47	I
3300		3925.92	I
600		3931.76	I
310		3933.55	I
760		3945.57	I
460		3950.21	I
310		3952.68	I
460		3964.90	I
600		3978.44	I
600		3979.42	I
870		3984.86	I
280		3995.98	I
1500		4022.16	I
600		4023.83	I
310		4039.21	I
1400		4051.40	I
710		4054.05	I
370		4064.46	I
200		4067.61	I
760		4068.37	I
200		4073.00	I
980		4076.73	I
6000		4080.60	I
310		4085.43	I
930		4097.79	I
350		4101.74	I
1900		4112.74	I
2000		4144.16	I
650		4145.74	I
260		4146.77	I
870		4167.51	I
550		4197.58	I
550		4198.88	I
7600		4199.90	I
1500		4206.02	I
5400		4212.06	I
760		4214.44	I
930		4217.27	I
370		4220.68	I
550		4230.31	I
760		4241.05	I
760		4243.06	I
370		4246.73	I
310		4258.99	I
760		4284.33	I
220		4293.28	I
260		4294.79	I
550		4295.93	I
3700		4297.71	I
930		4307.60	I
370		4318.43	I
550		4319.87	I
550		4342.07	I
350		4349.70	I
710		4354.13	I
870		4361.21	I
2400		4372.21	I
870		4385.39	I
1300		4385.65	I
1700		4390.44	I
1600		4410.03	I
160		4421.46	I
330		4428.46	I
460		4439.76	I
440		4449.34	I
1100		4460.04	I
190		4473.93	I
150		4480.45	I
350		4498.14	I
120		4510.10	I
220		4516.89	I
220		4517.82	I
110		4520.95	I
170		4547.33	I
110		4547.85	I
5400		4554.51	I
110		4559.98	I
1700		4584.44	I
110		4591.10	I
150		4592.52	I
330		4599.08	I
170		4635.69	I
200		4645.09	I
720		4647.61	I
290		4654.32	I
290		4681.79	I
190		4684.02	I
290		4690.11	I
1400		4709.48	I
140		4731.33	I
120		4733.52	I
500		4757.84	I
260		4815.52	I
120		4844.56	I
550		4869.15	I
160		4895.60	I
470		4903.05	I
120		4907.89	I
260		4921.07	I
180		4938.43	I
160		4968.90	I
160		4980.35	I
120		4992.74	I
160		5011.23	I
90		5014.95	I
90		5026.18	I
65		5028.16	I
35		5040.35	I
35		5040.74	I
65		5047.31	I
450		5057.33	I
21		5062.64	I
90		5072.97	I
120		5076.32	I
200		5093.83	I
80		5107.07	I
24		5123.73	I
55		5127.26	I
65		5133.89	I
530		5136.55	I
170		5142.76	I
250		5147.24	I
110		5151.07	I
55		5153.20	I
500		5155.14	I
55		5160.00	I
920		5171.03	I
180		5195.02	I
80		5199.87	I
45		5202.12	I
45		5213.43	I
65		5223.55	I
40		5242.38	I
55		5251.67	I
40		5257.07	I
40		5266.47	I
40		5266.83	I
40		5280.82	I
130		5284.08	I
40		5291.16	I
80		5304.86	I
260		5309.27	I
13		5315.33	I
40		5332.93	I
45	h	5334.70	I
110		5335.93	I
130		5361.77	I
65		5377.84	I
65		5385.88	I
110	h	5401.04	I
40		5401.39	I
40		5418.86	I
55		5427.59	I
26	l	5439.21	I
13		5452.71	I
80	h	5454.82	I
90		5456.13	I
13	h	5475.18	I
55		5479.40	I
26		5480.30	I
80		5484.32	I
18		5484.64	I
26		5496.69	I
13		5501.02	I
130		5510.71	I
20		5512.37	I
8		5517.86	I
12		5521.78	I
12		5530.99	I
24		5540.66	I
12		5556.52	I
90		5559.75	I
11		5569.03	I
21		5578.40	I
21		5603.14	I
8		5603.55	I
13		5606.73	I
11		5629.79	I
290		5636.24	I
11		5641.66	I
7		5649.56	I
7		5653.30	I
11		5665.20	I
16		5679.63	I
180		5699.05	I
13		5724.82	I
13		5725.73	I
16		5745.99	I
16		5747.47	I
11		5752.02	I
11		5756.83	I
11		5767.92	I
16		5804.39	I
65		5814.98	I
8		5828.06	I
16	h	5833.21	I
55		5919.34	I
80		5921.45	I
21		5926.87	I
26		5932.38	I
8		5936.65	I
8		5951.15	I
21	h	5973.38	I
8		5974.17	I
16		5988.67	I
35		5993.65	I
18		6116.77	I
26		6199.42	I
26		6225.20	I
9		6284.49	I
18		6295.22	I
13		6330.62	I
9		6336.12	I
9	h	6363.41	I
9		6376.45	I
16		6390.23	I
8		6417.57	I
26	h	6444.84	I
8		6496.44	I
11		6528.74	I
4		6560.45	I
4		6593.74	I
9		6618.20	I
21		6663.14	I
55		6690.00	I
11		6707.52	I
15		6718.30	I
15		6730.45	I
7		6756.54	I
21		6766.95	I
30		6775.02	I
13		6787.23	I
8		6813.51	I
15		6823.88	I
21		6824.17	I
7		6831.52	I
26		6911.48	I
110		6923.23	I
26		6982.01	I
26		7027.98	I
9		7086.06	I
12		7087.35	I
4		7141.72	I
6		7219.26	I
35		7238.92	I
7		7266.96	I
8		7323.56	I
16		7393.93	I
18		7468.91	I
12		7475.40	I
26		7485.79	I
70		7499.75	I
7		7532.07	I
26		7559.61	I
5		7612.94	I
18		7621.50	I
18		7722.87	I
5		7729.91	I
22		7791.86	I
4		7797.89	I
4		7806.82	I
3		7813.43	I
4		7829.81	I
5	h	7833.39	I
6	h	7841.90	I
30		7847.80	I
80		7881.49	I
16		7890.37	I
16		7924.43	I
5		7948.15	I
9		7967.84	I
9		8112.47	I
18		8264.96	I
11		8348.98	I
6		8352.94	I
4		8435.77	I
11		8473.64	I
11		8483.56	I
22		8710.84	I
14		8724.98	I
9		8777.36	I

Ru III
Ref. 423 — C.H.C.

Intensity		Wavelength	
		Vacuum	
250		850.09	III
200		850.30	III
50		851.22	III
50		852.49	III
150		856.32	III
50		867.48	III
250		919.74	III
50		921.78	III
250		928.08	III
150		937.16	III
500		940.09	III
250		940.68	III
50		941.85	III
50		942.63	III
50		943.06	III
150		945.68	III
100		946.05	III
100		947.14	III
100		949.83	III
50		950.35	III
100		950.45	III
50		952.59	III
50		957.06	III
50		957.18	III
50		961.58	III
250		961.68	III
100		962.56	III
500		966.54	III
250		967.09	III
150		967.85	III
150		967.92	III
150		971.83	III
250		972.40	III
100		973.54	III
150		973.78	III
750		974.14	III
250		974.46	III
250		977.51	III
100		978.18	III
900		979.43	III
500		981.35	III
250		983.81	III
250		983.91	III
250		985.55	III
900		986.84	III
200		987.87	III
250		991.67	III
250		992.75	III
900		994.56	III
250		995.30	III
200		1000.78	III
300		1001.65	III
250		1004.29	III
500		1009.13	III
900		1009.87	III
500		1014.68	III
100		1018.72	III
100		1019.33	III
100		1020.77	III
200		1080.00	III
100		1184.37	III
800		1190.51	III
500		1200.07	III
100		1204.57	III
200		1204.88	III
500		1207.17	III
500		1209.77	III
300		1211.31	III
200		1232.57	III
100	h	1653.77	III
200		1699.84	III
100	h	1715.97	III
200		1759.49	III
200	h	1880.95	III
100	h	1883.56	III
200	h	1899.04	III
100	h	1899.42	III
100		1908.31	III
500		1941.35	III
100		1981.82	III
100		1982.10	III
200		1989.22	III
200		1993.32	III
100		1997.55	III
		Air	
200		2005.71	III
100		2006.46	III
500		2009.28	III
100		2011.17	III
50		2011.56	III
50		2011.66	III
100		2015.20	III
50		2018.58	III
100		2044.59	III

SAMARIUM (Sm)
Z = 62

Sm I and II
Ref. 1 — C.H.C.

Intensity	Wavelength
	Air
45	2610.07
90	2640.27

Samarium (Cont.)

35		2649.17	
45		2657.68	
70		2662.42	
120		2675.15	
100		2688.60	
45		2690.90	II
130		2693.34	
45		2693.74	
60		2696.08	
85		2707.96	
50		2732.42	
35		2739.87	
29		2762.28	
35		2764.18	
85		2767.85	II
60		2774.77	
85		2776.11	
85		2779.23	II
85		2786.64	
150		2789.38	II
130	h	2796.70	
85		2807.36	
150		2809.50	
120		2810.86	II
85		2817.20	II
29		2820.96	II
220		2830.94	
60		2840.30	
60		2847.49	II
60		2851.35	
120		2866.09	II
70		2868.40	II
70		2881.34	
85		2881.68	
60		2883.09	
45		2889.06	
60		2891.34	
85	d	2907.88	
		2907.99	II
130		2910.28	II
85		2937.48	II
70		2943.49	II
150		2953.19	II
85		2962.74	II
160		2969.02	II
100		2983.43	II
60		2991.57	II
100		3021.01	
150		3034.84	II
100		3039.13	II
120		3046.93	II
150		3067.54	
120		3071.29	II
100		3086.45	II
120		3096.88	II
100	h	3102.30	II
250		3106.52	II
220		3110.20	II
200		3117.72	II
270		3136.30	II
150		3139.97	II
150		3147.19	II
180		3152.10	II
410		3152.52	II
150		3162.30	II
360		3169.88	II
180		3178.12	II
720		3183.92	II
310		3187.01	II
430		3187.22	II
360		3187.79	II
360		3193.01	II
360		3196.18	II
150		3201.80	II
150		3204.90	II
360		3207.18	II
180		3208.17	II
600		3211.73	II
150		3214.12	II
270		3215.26	II
530		3216.85	II
600		3218.61	II
150		3219.43	II
270		3226.84	II
180		3228.50	II
270		3228.78	II
720		3230.56	II
360		3231.53	II
150		3231.95	II
430		3233.68	II
720		3236.64	II
150		3237.89	II
720		3239.66	II
530		3241.16	II
180		3241.59	II
180		3242.04	II
150		3244.69	II
240		3249.75	II
720		3250.37	II
360		3253.40	II
270		3253.94	II
850		3254.38	II
110		3255.63	II
360		3262.28	II
430		3264.94	II
180		3270.49	II
180		3270.68	II
430	d	3272.48	II
		3272.60	II
430		3272.81	II
430		3273.48	II
430		3276.75	II
270		3280.84	II
180		3285.66	II
430		3286.23	II
720	d	3290.28	II
		3290.39	II
180		3290.65	II
240		3293.37	II
360		3295.44	II
430		3295.81	II
720		3298.10	II
170		3300.98	II
340		3301.68	II
340		3304.52	II
340		3305.18	II
1700		3306.39	II
170		3306.61	II
850		3307.02	II
340		3309.52	II
850		3310.66	II
600		3312.42	II
410		3316.58	II
430		3320.16	II
110		3320.59	II
1200		3321.18	II
340		3323.77	II
340		3325.26	II
170		3325.48	II
340		3327.88	II
170		3333.64	II
170		3336.12	II
850		3340.58	II
240		3343.49	II
110		3343.64	II
240		3344.35	II
170		3347.30	II
240		3348.68	II
220		3350.88	II
410	d	3354.18	II
		3354.30	II
170		3354.72	II
1200		3365.86	II
150		3367.27	II
340		3368.57	II
340		3369.46	II
170		3370.59	II
340		3371.21	II
150		3376.48	II
1200		3382.40	II
510		3384.66	II
150		3384.86	
150		3387.66	II
410		3389.32	II
150		3391.11	II
410		3396.19	II
150		3397.76	II
150		3399.84	II
600		3402.46	II
210		3403.09	II
850		3408.68	II
270		3418.15	II
430		3418.51	II
170		3419.77	II
120		3424.78	II
170		3426.20	II
170		3433.68	II
150		3437.10	II
170		3438.06	II
240		3440.50	II
170		3453.56	II
170		3459.20	II
120		3459.42	II
240		3461.13	II
120		3464.07	II
170		3467.87	II
130		3473.96	II
130		3479.53	II
130		3480.26	II
170		3480.56	II
170		3487.41	II
170		3493.61	II
220		3499.84	II
340		3511.23	II
310		3530.60	II
220		3532.57	II
270		3535.65	II
240		3554.15	II
510		3559.10	II
220		3566.84	II
4200		3568.27	II
270		3577.79	II
390		3580.94	II
310		3583.39	II
4200		3592.60	II
340		3601.69	II
1700		3604.28	II
3400		3609.49	II
240		3620.58	II
1700		3621.23	II
240		3623.32	II
850		3627.01	II
850		3631.13	II
3400		3634.29	II
240		3634.93	II
410		3638.77	II
360		3645.29	II
300		3645.39	II
660		3649.53	II
340		3650.19	II
340		3656.22	II
2200		3661.36	II
220		3662.69	II
340		3667.93	II
340		3670.66	II
2200		3670.84	II
340		3677.79	II
270		3681.73	II
270		3688.42	II
270		3692.22	II
1100		3693.99	II
480		3706.75	II
480		3706.98	II
480		3708.41	II
930		3708.65	II
480		3711.54	II
350		3712.76	II
930		3718.88	II
930		3721.85	II
420		3724.90	II
1600		3728.47	II
2100		3731.26	II
1600		3735.98	II
800		3737.14	II
320		3737.48	II
2900		3739.12	II
		3739.20	II
800		3741.29	II
1200		3743.87	II
930		3745.46	I
		3745.60	II
480		3747.62	II
800		3755.28	II
800		3756.41	I
1200		3757.53	II
450		3758.45	II
660		3758.97	II
350		3760.04	II
1900		3760.69	II
660		3762.59	II
1100		3764.37	II
480		3767.36	II
480		3767.76	II
370	d	3773.33	I
		3773.42	II
1100		3778.14	II
660		3780.76	II
420		3780.93	II
320		3787.20	II
1500		3788.12	II
1600		3793.97	II
420		3797.28	II
1600		3797.73	II
500		3799.54	II
800		3800.89	II
320		3805.63	II
420		3808.46	II
320		3809.75	II
320		3809.88	II
420		3810.43	II
500		3812.07	II
480		3813.63	II
420		3814.63	II
930	d	3820.82	
530		3824.18	II
1600		3826.20	II
530		3830.29	II
1100		3831.50	II
530		3833.83	II
560		3834.48	I
560		3834.60	II
370		3835.72	II
500		3838.94	II
400		3840.45	II
1600		3843.50	II
530		3847.51	II
640		3848.78	II
420		3851.88	II
530		3853.30	I
2700		3854.21	II
480		3854.56	I
800		3855.90	II
480		3857.91	II
400		3858.74	I
660		3862.05	II
350		3862.23	II
320		3865.24	II
800		3871.78	II
400		3875.19	II
560		3875.54	II
800		3880.77	II
450		3881.38	II
450		3881.79	II
320		3882.50	II
3700		3885.29	II
660		3889.16	II
		3889.22	II
610		3890.08	II
320		3891.21	II
400		3894.05	II
1600		3896.98	II
1300		3903.42	II
620		3917.44	II
2500		3922.40	II
1900		3928.28	II
470		3935.76	II
1300		3941.87	II
620		3943.24	II
500		3946.51	II
740		3948.11	II
470		3951.89	I
370		3959.53	II
1500		3963.00	II
620		3966.04	II
470		3967.68	II
740		3970.53	II
1500		3971.40	II
620		3974.66	I
960		3976.27	II
1000		3976.43	II
960		3979.20	II
740		3983.14	II
740		3986.68	II
370		3987.43	II
1500		3990.00	II
		3990.02	I
740		3993.31	II
280		4003.46	II
470		4007.48	II
280		4019.98	II
880		4023.23	II
740		4035.11	II
590		4041.68	II
740		4042.72	II
880		4042.90	II
240		4044.11	II
560		4045.05	II
440		4046.16	II
740		4047.16	II
210		4048.62	II
590		4049.81	II
440		4058.87	II
560		4063.54	II
280		4064.32	II
1400		4064.58	II
810		4066.74	II
710		4068.33	II
810		4075.84	II
280		4076.65	II
240		4080.56	II
410		4082.60	II
280		4083.58	II
220		4084.40	II
1000		4092.27	II
290		4094.05	II
240		4104.13	II
810		4107.28	II
		4107.39	II
410		4109.40	II
280		4110.19	II
410		4113.90	II
1900		4118.55	II
410		4121.36	II
280		4122.51	II
710		4123.96	II
280		4129.23	II
250		4135.14	II
320		4147.71	II
810		4149.83	II
1200		4152.21	II
530		4153.33	II
560		4155.22	II
810		4169.48	II
410		4171.57	II
440		4178.02	II
530		4181.10	II
210		4183.33	I
530		4183.76	II
1000		4188.13	II
410		4191.93	II
270		4199.45	II
650		4202.92	II
1100		4203.05	II
660		4206.13	II
270		4206.62	II
660		4210.35	II
740		4220.66	II
1000		4225.33	II
740		4229.70	II
620		4234.57	II
1200		4236.74	II
500		4237.66	II
620		4244.70	II
210		4249.55	II
250		4251.78	II

Samarium (Cont.)

2100		4256.39	II
210		4258.58	II
1300		4262.68	II
500		4265.08	II
1200		4279.68	II
		4279.75	II
240		4279.94	II
2200		4280.79	II
710		4282.21	I
470		4282.83	I
240		4283.50	I
350		4286.64	II
350		4292.18	II
1600		4296.74	I
320		4304.94	II
880		4309.01	II
240		4312.85	I
1900		4318.94	II
470		4319.53	I
590		4323.28	II
240		4324.46	I
1800		4329.02	II
440		4330.02	I
1300		4334.15	II
880		4336.14	I
560		4345.86	II
1100		4347.80	II
560		4350.46	II
560		4352.10	II
560		4360.72	II
220		4361.07	II
810		4362.04	II
440		4362.91	I
220		4363.45	II
500		4368.03	II
210		4369.92	II
440		4373.46	II
320		4374.98	II
880		4378.24	II
530		4380.42	I
290		4384.29	II
1600		4390.86	II
210		4393.35	I
290		4397.34	I
410		4401.17	I
810	d	4403.06	II
		4403.13	I
410		4403.36	II
520		4409.33	II
290		4411.58	I
380		4417.58	II
470		4419.33	I
1500		4420.53	II
960		4421.14	II
2900		4424.34	II
470		4429.66	I
1600		4433.88	II
1800		4434.32	II
530		4441.81	I
440		4442.28	I
710		4444.26	II
710		4445.15	I
1300		4452.73	II
250		4452.95	I
1200		4454.63	II
1000		4458.52	II
250		4459.29	I
2200		4467.34	II
810		4470.89	I
470		4472.43	II
620		4473.02	II
740		4478.66	II
370		4499.11	I
370		4499.48	II
240		4503.38	I
180		4505.05	II
120		4511.33	I
560		4511.83	II
440		4515.09	II
880		4519.63	II
440		4523.04	II
		4523.18	I
650		4523.91	II
290		4533.80	I
270		4536.51	II
710		4537.95	II
150		4538.53	II
290		4540.19	II
380		4542.06	II
810		4543.95	II
100		4544.83	II
410		4552.66	II
270		4554.45	II
240		4560.43	II
470		4566.21	II
590		4577.69	II
290		4581.58	I
440		4581.73	I
560		4584.83	II
290		4591.82	II
380		4593.54	II
560		4595.29	II
240		4596.74	I
220		4604.18	II
290		4606.51	II
290		4615.44	II
470		4615.69	II
150		4630.21	II
880		4642.24	II
290		4645.40	I
290		4646.68	II
240		4648.16	II
380		4649.49	I
150		4655.13	II
290		4663.56	I
740		4669.40	II
620		4669.65	II
470	d	4670.75	I
		4670.83	I
1100		4674.60	II
680		4676.91	II
210		4681.55	I
370		4687.18	II
370		4688.73	I
130		4693.63	II
120		4699.34	II
530		4704.40	II
270		4713.06	II
130		4715.26	II
730		4716.10	I
270		4717.07	I
210		4717.72	II
190		4718.33	II
270		4719.84	II
130		4726.02	II
770		4728.42	I
470		4745.68	II
150		4750.72	I
730		4760.27	I
110		4770.20	I
110		4774.15	II
190		4777.85	II
580		4783.10	I
350		4785.86	I
160		4789.96	I
230		4791.58	II
430		4815.81	II
130		4829.57	II
970		4841.70	I
310		4844.21	II
140		4847.76	II
270		4848.32	I
120		4854.36	II
210		4883.77	I
730		4883.97	I
170		4904.97	I
630		4910.40	I
350		4913.25	II
430		4918.99	I
110		4924.04	I
120		4938.10	II
170		4948.63	II
120		4952.37	II
170		4961.94	II
170		4975.98	I
140		5028.44	II
400		5044.28	I
200		5052.76	II
170		5069.46	II
540		5071.20	I
170		5100.22	II
		5100.39	I
260		5103.09	II
140		5104.48	II
140		5116.70	II
510		5117.16	I
350		5122.14	I
360		5155.03	II
250		5172.74	I
470		5175.42	I
250		5200.59	I
260		5251.92	I
400		5271.40	I
250		5282.91	I
190		5320.60	I
110		5341.29	I
140		5368.36	I
130		5405.23	I
220		5453.00	I
140		5466.72	I
230		5493.72	I
80		5512.10	I
230		5516.09	I
50		5548.95	I
140		5550.40	I
45		5573.42	I
35		5588.20	I
50		5600.86	II
50		5621.79	I
70		5626.01	I
85		5644.10	I
140		5659.86	I
120		5696.73	I
85		5706.20	I
35		5710.93	I
50		5732.95	I
50		5743.35	II
45		5759.52	II
70		5773.77	I
60		5778.33	I
45		5779.24	I
45		5781.93	II
70	d	5786.98	II
60		5788.38	I
60		5800.52	I
65		5802.84	I
45		5814.89	I
45		5831.02	II
45		5836.37	II
35		5860.78	I
65		5867.79	I
45		5868.61	I
35		5871.06	I
50		5874.21	I
45		5897.39	II
50		5898.96	I
35		5938.90	II
65		5965.71	II
35	h	5968.82	II
35		5984.29	I
50		6045.00	I
45		6045.39	I
50		6070.06	I
45		6084.12	I
35	h	6091.40	I
45		6110.66	II
45	h	6159.56	I
45		6246.76	II
45		6256.54	I
45		6256.66	II
100		6267.28	II
50		6291.82	II
35		6307.06	II
70		6327.47	II
45		6426.64	II
45		6472.34	II
35		6484.52	II
35		6498.67	II
50		6542.76	II
140		6569.31	II
35	h	6570.67	II
40	h	6585.21	II
110		6589.72	II
40		6601.83	II
95		6604.56	II
40	h	6632.28	II
50		6671.51	I
70		6679.21	II
70		6693.55	II
40	d	6723.07	I
120	d	6731.84	II
70	d	6734.06	II
40	d	6734.81	II
55		6741.47	II
40	h	6778.61	II
60		6790.00	II
95		6794.20	II
55		6844.71	II
75		6856.03	II
120		6860.93	I
40		6862.82	II
30		6950.51	II
120		6955.29	II
90		7020.44	II
13		7036.73	II
90		7039.22	II
90		7042.24	II
13		7049.15	II
90		7051.52	II
16		7054.97	II
19		7074.67	I
90		7082.37	II
40	d	7085.52	II
26		7088.30	I
16		7091.16	I
30		7095.50	I
16		7096.33	I
30		7104.54	I
19		7106.23	I
26		7115.96	I
23		7117.51	II
26	h	7119.81	II
12		7122.40	II
23	h	7125.11	II
13		7131.80	I
10		7136.01	I
12		7139.39	II
40	d	7143.98	II
85	d	7149.60	II
10		7172.67	I
10		7189.57	II
9		7210.95	I
23		7213.82	I
26	d	7218.09	II
13		7220.07	I
13		7237.02	II
60		7240.90	II
9		7257.11	II
9	d	7261.52	II
13		7279.25	I
26		7281.47	II
8		7282.21	I
19		7283.33	II
16		7288.92	II
13		7290.23	I
26	h	7300.72	II
13		7327.08	II
13		7332.65	I
8		7338.04	I
26		7347.30	I
26		7376.69	II
13		7393.98	II
30		7444.56	I
26		7445.41	I
26	d	7453.03	II
13		7470.76	I
26		7481.99	II
23	h	7502.39	II
10	h	7517.00	II
23	h	7541.42	II
9		7544.74	I
10		7546.57	I
12	h	7560.03	II
19		7562.94	II
23		7570.95	II
23		7572.29	II
19		7578.09	II
30		7585.85	II
23		7588.31	II
10		7598.01	I
23	d	7607.48	II
		7607.74	I
12		7613.94	II
10	h	7631.77	II
23		7637.94	II
45		7645.09	II
12		7645.82	I
19		7648.02	II
10		7655.78	II
19		7667.20	II
8		7672.49	II
10	h	7678.79	II
10	h	7695.78	I
23		7712.04	II
30		7728.56	II
30		7736.26	II
30		7749.30	II
23		7755.20	II
10		7794.50	I
10		7801.54	I
8	h	7812.75	II
16		7820.15	II
10		7831.40	II
40	w	7835.08	II
26		7837.27	II
10		7844.82	II
6		7859.53	I
19		7863.65	II
10	h	7880.01	II
16		7895.96	I
26		7914.96	II
90		7928.14	II
9		7931.92	I
19		7937.09	II
16		7948.12	II
19	w	8001.61	II
19	w	8014.92	II
23		8025.12	II
23	w	8026.32	II
16		8032.03	II
40		8048.70	II
16		8065.16	I
45		8068.46	II
9	w	8117.16	II
9		8125.12	II
26		8161.82	II
19	w	8195.50	II
6		8206.30	II
26	w	8218.76	II
9		8230.33	I
16		8240.98	II
19	w	8289.26	II
10		8300.88	II
40	w	8305.79	II
10		8315.45	I
19	w	8348.68	II
19		8383.71	I
19		8387.77	II
30	w	8432.64	II
19	w	8473.54	II
45	w	8485.99	II
30	w	8510.90	II
23		8543.22	II
23	w	8617.03	II
23	w	8632.82	II
12	w	8677.81	II
13		8706.32	II
45	w	8708.43	II
30	w	8717.89	II
30	w	8758.28	II
16	w	8780.59	II
23	w	8788.83	II
26		8859.76	II
95		8913.66	II

SCANDIUM (Sc)
Z = 21

Sc I and II
Ref. 1, 88 — C.H.C.

Intensity		Wavelength	
		Air	
65		2429.16	I
110		2438.62	I
560		2545.22	II
2900		2552.37	II
560		2555.82	II
2300		2560.25	II
1100		2563.21	II
40		2611.22	II
19		2684.23	II
120		2692.78	I
360		2706.77	I
210		2707.95	I
580		2711.35	I
30		2819.54	II
35		2822.15	II
60		2826.68	II
340		2965.86	I
1200		2974.01	I
1400		2980.75	I
340		2988.95	I
2200		3015.36	I
2700		3019.34	I
360		3030.76	I
30		3039.93	II
70		3045.72	II
85		3052.93	II
120	h	3056.31	I
130		3065.11	II
45		3139.75	II
990		3251.32	II
1500		3255.69	I
4400		3269.91	I
5500		3273.63	I
110	d	3343.28	II
270		3352.05	II
9900		3353.73	II
65	d	3357.30	II
2000		3359.68	II
1700		3361.27	II
1700		3361.94	II
4000		3368.95	II
6600		3372.15	II
90		3416.68	I
130		3418.51	I
65		3419.36	I
200		3429.21	I
200		3429.48	I
270		3431.36	I
530		3435.56	I
90		3439.41	I
65		3440.18	I
65		3448.49	I
270		3457.45	I
180		3462.19	I
130	d	3469.65	I
110		3471.13	I
200		3498.91	I
2700		3535.73	II
6600		3558.55	II
6100		3567.70	II
13000		3572.53	II
9900		3576.35	II
7700		3580.94	II
4000		3589.64	II
4000		3590.48	II
28000		3613.84	II
110		3617.43	I
20000		3630.75	II
13000		3642.79	II
6600		3645.31	II
110		3646.90	I
5300		3651.80	II
110		3664.25	II
290		3666.54	II
55		3675.26	II
40		3678.35	II
75	h	3717.10	I
270		3833.07	II
610		3843.03	II
90		3894.97	I
20000		3907.49	I
23000		3911.81	I
45		3923.51	II
4400		3933.38	I
45		3952.27	I
45		3989.06	II
5500		3996.61	I
530		4014.49	II
20000		4020.40	I
20000		4023.69	I
220		4030.67	I
140		4031.39	I
100		4034.23	I
220		4043.80	I
200		4046.48	I
2700		4047.79	I
120		4049.95	I
5500		4054.55	I
220		4056.59	I
160	h	4074.97	I
160		4078.57	I
6100		4082.40	I
200		4086.67	I
400		4087.16	I
40	h	4093.13	I
65		4094.85	I
55	h	4098.35	I
65		4100.33	I
440	h	4133.00	I
530	h	4140.30	I
65	h	4147.40	I
720		4152.36	I
55	h	4154.72	
90	Hd	4161.88	I
1100	h	4165.19	I
65	h	4171.56	I
45	h	4186.45	I
65	h	4187.62	I
75		4205.20	I
65		4212.34	I
45		4212.49	I
75	h	4216.10	I
110	h	4218.26	I
110	h	4219.73	I
40		4221.88	I
90	d	4225.59	I
180		4231.93	I
200		4233.61	I
100		4237.82	I
400		4238.05	I
90		4239.57	I
100		4246.12	I
15000		4246.83	II
55		4283.56	
290		4294.77	II
350		4305.71	II
4200		4314.09	II
3300		4320.74	II
2400		4325.01	II
28		4348.53	I
180		4354.61	II
110		4358.64	I
55		4359.08	I
28		4364.92	I
2000		4374.46	II
130		4384.81	II
45	h	4389.60	I
1100		4400.37	II
880		4415.56	II
28		4420.66	II
45		4431.36	II
65		4542.55	I
90		4544.68	I
120	h	4557.24	I
160	h	4573.99	I
65	h	4592.94	I
65	h	4598.45	I
55	h	4604.72	I
45		4609.53	I
45		4609.95	I
350		4670.40	II
40	h	4680.49	I
50		4698.29	II
120		4706.97	I
120		4709.34	I
200		4728.77	I
490		4729.23	I
40	h	4732.30	I
590		4734.10	I
60		4735.08	
690		4737.65	I
790		4741.02	I
1200		4743.81	I
200		4753.16	I
220		4779.35	I
90		4791.50	I
100		4827.28	I
100		4833.67	I
170		4839.44	I
40		4840.47	I
80		4847.68	I
80		4852.68	I
140	BLd	4857.79	S
		4858.09	SCO
80		4906.67	I
90		4909.76	I
90		4922.84	I
90		4934.25	I
45		4935.74	I
70		4941.33	I
170		4954.06	I
120		4973.66	I
150		4980.37	I
80		4983.45	I
140		4991.92	I
80		5018.39	I
70		5020.14	I
80		5021.51	I
530		5031.02	II
55		5032.74	I
250		5064.32	I
80		5068.86	I
530		5070.23	I
250		5075.81	I
2100		5081.56	I
1200		5083.72	I
1100		5085.55	I
750		5086.95	I
390		5087.14	I
270		5089.89	I
45		5092.46	I
390		5096.73	I
620		5099.23	I
370		5101.12	I
180		5109.06	I
150		5112.86	I
320		5116.69	I
70	b	5133.68	S
45	b	5171.06	S
390		5210.52	I
45		5211.28	I
280		5219.67	I
350		5239.82	II
280		5258.33	I
35		5284.97	I
210		5285.76	I
35		5301.94	I
22		5318.35	II
70		5331.77	I
14		5334.23	II
95		5339.41	I
120		5341.05	I
95		5342.96	I
350		5349.30	I
120		5349.71	I
60		5350.30	I
210		5355.75	I
530		5356.10	I
14		5357.19	II
270		5375.35	I
370		5392.08	I
45		5416.12	I
45		5425.57	I
45		5429.41	I
35		5432.94	I
55		5433.23	I
45		5438.22	I
55		5439.03	I
55	h	5442.60	I
270		5446.20	I
18		5447.39	I
120		5451.34	I
30		5455.21	I
18		5465.20	I
55		5468.40	I
60		5472.19	I
18		5474.64	I
750		5481.99	I
530		5484.62	I
570		5514.22	I
16		5515.39	I
660		5520.50	I
45		5526.06	I
660		5526.82	II
55		5541.04	I
30		5546.40	I
18		5550.40	I
5		5552.25	II
35		5553.59	I
16		5561.10	I
70		5564.86	I
18		5571.24	I
14		5579.76	I
110		5591.33	I
35	h	5593.38	I
22		5604.19	I
22		5631.02	I
80		5640.98	II
45		5646.36	I
16		5647.60	I
55		5649.56	I
250		5657.88	II
60		5658.34	II
55		5667.16	II
70		5669.04	II
1500		5671.81	I
95		5684.20	II
1200		5686.84	I
1100		5700.21	I
190		5708.61	I
880		5711.75	I
230		5717.28	I
180		5724.08	I
55	Bl	5736.85	S
55	Bl	5764.45	S
95	Bl	5772.74	S
55	Bl	5775.32	S
70	Bl	5809.84	S
70	Bl	5811.60	S
95	Bl	5847.73	S
70	Bl	5849.07	S
70	b	5887.38	S
35	Bl	5918.04	S
30		5919.11	I
60	Bl	5928.10	S
35		5961.49	I
60	Bl	5968.25	S
35		5969.19	I
90		5988.42	I
160	Bl	6017.07	S
60		6026.18	I
620	Bl	6036.17	S
490	Bl	6064.31	S
440	Bl	6072.65	S
620	Bl	6079.30	S
320	Bl	6101.87	S
370	Bl	6109.93	S
370	Bl	6115.97	S
180	b	6148.70	S
150	b	6153.93	S
150	b	6188.09	S
150	b	6192.90	S
620		6210.68	I
90		6239.41	I
320		6239.78	I
120		6245.63	II
110		6249.96	I
250		6258.96	I
60		6262.25	I
55		6276.31	I
45		6279.76	II
18		6300.70	II
750		6305.67	I
26		6309.90	II
16		6320.85	II
26		6344.83	I
60		6378.82	I
55	Bl	6408.41	S
90		6413.35	I
26	b	6437.08	S
55	Bl	6446.24	S
26	b	6457.78	S
35	b	6485.40	S
26	b	6495.90	S
55	b	6525.62	S
22	b	6535.30	S
45	b	6557.84	S
35	b	6566.88	S
18	b	6575.85	S
60		6604.60	II
26	Bl	6609.99	S
18	Bl	6617.94	S
18	Bl	6645.08	S
22	Bl	6654.42	S
26	Bl	6661.01	S
18	b	6700.48	S
18	b	6705.93	S
65		6737.87	I
35		6739.40	I
35		6817.08	I
50		6819.52	I
29		6829.54	I
50		6835.03	I
5	b	6963.12	S
5	Bl	6990.68	S
5	Bl	7025.72	S
8	b	7035.77	S
5	b	7072.37	S
5	b	7094.38	S
12	h	7138.14	I
14		7169.13	I
12		7257.57	I
8		7275.57	I
3	h	7300.62	I
12	h	7524.13	I
14	h	7553.96	I
15	h	7574.44	I
11		7617.45	I
14	h	7665.72	I
30		7697.73	I
18		7729.72	I
55	h	7741.17	I
5	h	7750.37	I
5		7752.72	I
6	h	7771.06	I
15		7785.17	I
8		7794.68	I
30		7800.44	I
11		7821.64	I
11	h	8196.98	I
15		8241.13	I
19	h	8761.40	I
11	h	8774.8	
15	h	8794.72	I
15	h	8823.8	I
30	h	8834.35	I
70		20616.32	
30		20985.81	
400		22051.86	I
150		22065.05	I

Scandium (Cont.)

Sc III
Ref. 323 — C.H.C.

Intensity	Wavelength	
	Vacuum	
10	730.60	III
15	731.65	III
15	1148.24	III
20	1154.52	III
20	1162.44	III
25	1168.61	III
10	1168.88	III
80	1598.00	III
180	1603.06	III
150	1610.19	III
40	1895.44	III
60	1912.62	III
90	1993.89	III
	Air	
160	2010.42	III
50	2012.26	III
350	2699.07	III
230	2734.05	III
10	2831.75	III
80	4061.21	III
100	4068.66	III
40	4309.47	III
10	4740.95	III
15	4780.87	III
50	4992.89	III
60	5032.09	III
80	6256.01	III
60	6307.60	III
90	7449.16	III
70	7548.15	III
70	7868.65	III
35	8814.29	III
50	8829.78	III
30	8865.89	III
15	8881.58	III

Sc IV
Ref. 298 — C.H.C.

Intensity	Wavelength	
	Vacuum	
8	220.28	IV
15	289.85	IV
15	296.31	IV
15	299.04	IV
10	371.16	IV
9	438.80	IV
8	557.50	IV
8	584.83	IV
9	617.08	IV
8	761.43	IV
8	769.70	IV
10	785.12	IV
8	789.00	IV
8	791.71	IV
8	861.24	IV
8	861.30	IV
8	890.87	IV
8	1219.40	IV
8	1228.20	IV
9	1424.66	IV
9	1444.10	IV
8	1489.64	IV
8	1514.96	IV
8	1535.76	IV
9	1543.86	IV
9	1549.55	IV
15	1550.80	IV
8	1555.72	IV
9	1563.81	IV
10	1574.92	IV
9	1583.41	IV
8	1584.64	IV
8	1592.23	IV
8	1660.71	IV
10	1665.92	IV
8	1746.23	IV
	Air	
10	2056.06	IV
8	2078.93	IV
12	2118.97	IV
9	2164.43	IV
11	2185.43	IV
11	2205.46	IV
14	2222.22	IV
11	2271.33	IV
9	2464.45	IV
8	2520.93	IV
11	2586.93	IV
9	2595.17	IV
8	2678.01	IV
8	2723.52	IV
8	2773.04	IV
8 d	4594.42	IV
8 d	4639.96	IV
8	5501.74	IV
9	5620.72	IV
10	5706.82	IV
14	5771.63	IV
9	6548.03	IV

Sc V
Ref. 150 — C.H.C.

Intensity	Wavelength	
	Vacuum	
150	179.42	V
350	180.14	V
200	180.82	V
200	180.96	V
50	181.55	V
200	182.39	V
300	228.56	V
100	230.85	V
40	243.82	V
500	243.87	V
400	246.42	V
400	250.98	V
500	252.85	V
500	253.73	V
50	255.38	V
300	255.64	V
200	257.16	V
150	258.24	V
40	258.81	V
50	260.05	V
400	281.00	V
900	283.91	V
800	284.45	V
600	288.29	V
900	289.59	V
1000 d	291.93	V
800	293.25	V
400	296.17	V
700	300.00	V
400	375.05	V
100	378.68	V
200	388.68	V
400	395.32	V
200	399.50	V
1000	573.36	V
600	587.94	V

SELENIUM (Se)
Z = 34

Se I and II
Ref. 80, 181, 216, 275 — R.L.K.

Intensity	Wavelength	
	Vacuum	
285	828.5	II
360	832.7	II
285	906.6	II
360	912.9	II
360	1013.4	II
360	1014.0	II
450	1033.6	II
450	1049.6	II
360	1057.4	II
285	1097.8	II
360	1141.9	II
220	1156.0	II
285	1156.9	II
285	1168.5	II
450	1192.3	II
220	1205.7	II
220	1234.9	II
285	1291.0	II
285	1308.9	II
100	1405.4	I
100	1406.4	I
100	1406.6	I
120	1435.3	I
120	1435.8	I
100	1444.8	I
100	1446.8	I
100	1447.0	I
150	1449.2	I
120	1456.3	I
150	1500.9	I
250	1530.4	I
150	1531.3	I
200	1531.8	I
120	1547.1	I
120	1560.3	I
150	1575.3	I
150	1577.6	I
150	1577.9	I
150	1579.5	I
200	1580.0	I
150	1587.5	I
150	1593.2	I
250	1606.5	I
100	1610.7	I
100	1611.3	I
200	1617.4	I
150	1621.2	I
100	1622.7	I
120	1626.2	I
150	1643.4	I
250	1671.2	I
250	1675.3	I
250	1690.7	I
250	1793.3	I
300	1795.3	I
300	1855.2	I
250	1858.8	I
400	1898.6	I
350	1913.8	I
300	1919.2	I
500	1960.9	I
150	1995.1	I
	Air	
500	2039.8	I
500	2074.8	I
500	2164.2	I
150	2332.8	I
600	2413.5	I
300	2548.0	I
220	3038.7	II
220	3041.3	II
285	4070.2	II
360	4175.3	II
450	4180.9	II
120	4328.7	I
100	4330.3	I
285	4382.9	II
285	4446.0	II
220	4449.2	II
285	4467.6	II
500	4730.8	I
400	4739.0	I
300	4742.2	I
285	4840.6	II
360	4845.0	II
450	5227.5	II
360	5305.4	II
100	5365.5	I
120	5369.9	I
110	5374.1	I
285	5522.4	II
285	5566.9	II
285	5866.3	II
450	6056.0	II
200	6325.6	I
360	6444.2	II
285	6490.5	II
285	6535.0	II
150	6831.3	I
120	6990.690	I
100	6991.792	I
200	7010.809	I
150	7013.875	I
300	7062.065	I
200	7575.1	I
250	7583.4	I
150	7592.2	I
120	7606.8	I
300	8001.0	I
200	8036.4	I
120	8060.9	I
120	8065.3	I
120	8081.1	I
150	8093.2	I
150	8094.7	I
180	8149.3	I
150	8152.0	I
200	8157.7	I
180	8163.1	I
150	8182.9	I
100	8185.0	I
120	8194.6	I
150	8440.47	I
150	8450.38	I
150	8742.33	I
300	8918.86	I
100	8969.69	I
200	9001.97	I
200	9038.61	I
80	9083.14	I
120	9088.79	I
80	9140.83	I
60	9181.88	I
60	9271.12	I
100	9432.50	I
60	9825.58	I
200	10217.25	I
377	10307.45	I
900	10327.26	I
640	10386.36	I
124	10650.30	I
125	11934.56	I
275	11946.87	I
100	11947.92	I
105	11952.27	I
170	11952.64	I
100	11966.04	I
205	11972.93	I
115	11973.07	I
315	14817.93	I
410	14917.47	I
500	15151.44	I
115	15469.06	I
320	15471.00	I
265	15520.97	I
395	15618.40	I
115	15620.38	I
360	16659.44	I
505	16813.78	I
165	16817.76	I
205	16866.54	I
115	16972.71	I
235	21374.24	I
680	21442.56	I
415	21473.48	I
270	21716.36	I
240	21730.60	I
105	23133.66	I
150	23388.85	I
110	23628.17	I
265	24148.18	I
170	24159.23	I
185	24204.44	I
375	24385.99	I
160	24413.67	I
225	24471.17	I
255	25017.51	I
510	25127.43	I

Se III
Ref. 9, 247 — R.L.K.

Intensity	Wavelength	
	Vacuum	
220	709.2	III
220	709.4	III
220	720.6	III
360	724.3	III
285	726.4	III
220	737.2	III
220	741.9	III
285	777.3	III
220	790.8	III
360	843.0	III
220	879.2	III
285	953.7	III
220	954.4	III
220	954.7	III
160	974.1	III
360	974.8	III
285	1079.8	III
360	1099.1	III
450	1119.2	III
	Air	
285	2057.5	III
285	2767.2	III
220	2773.8	III
285	3379.8	III
450	3387.2	III
450	3413.9	III
285	3428.4	III
450	3457.8	III
360	3543.6	III
285	3570.2	III
450	3637.6	III
360	3711.7	III
450	3738.7	III
285	3743.0	III
450	3800.9	III
360	4046.7	III
220	4083.2	III
450	4169.1	III
220	4637.9	III
285	6303.8	III

Se IV
Ref. 245 — R.L.K.

Intensity	Wavelength	
	Vacuum	
285	636.0	IV
285	654.2	IV
360	652.7	IV
450	670.1	IV
285	671.9	IV
220	722.8	IV
285	734.6	IV
450	746.4	IV
285	759.0	IV
285	776.5	IV
285	803.8	IV

Selenium (Cont.)

Intensity		Wavelength	
360		959.6	IV
450		996.7	IV
220		1307.2	IV
285		1314.4	IV
		Air	
220		2090.0	IV
285		2136.6	IV
160		2165.2	IV
160		2166.6	IV
360		2665.5	IV
285		2724.3	IV
160		2951.6	IV

Se V
Ref. 245 — R.L.K.

Intensity		Wavelength	
		Vacuum	
285		596.0	V
285		601.0	V
220		608.7	V
360		613.0	V
285		614.3	V
450		759.1	V
285		785.8	V
285		804.3	V
360		808.7	V
220		814.8	V
220		820.7	V
360		830.3	V
450		839.5	V
360		845.8	V
360		1094.7	V
220		1151.0	V
450		1227.6	V

SILICON (Si)
Z = 14

Si I and II
Ref. 170, 237, 292 — L.J.R.

Intensity		Wavelength	
		Vacuum	
10	h	805.10	II
20	h	820.52	II
20	h	843.72	II
40	h	845.77	II
10		850.14	II
100		889.72	II
200		892.00	II
10		899.41	II
20		901.74	II
10		913.01	II
20		913.85	II
20		929.81	II
100		989.87	II
200		992.68	II
25		1020.70	II
50		1023.69	II
30		1057.05	II
15		1057.50	II
20	h	1127.44	II
40	h	1127.91	II
100		1190.42	II
200		1193.28	II
250		1194.50	II
100		1197.39	II
10	h	1216.12	II
20		1223.91	II
20		1224.25	II
10		1224.97	II
50		1226.81	II
20		1226.89	II
40		1226.99	II
100		1227.60	II
10		1228.44	II
25		1228.62	II
150		1228.75	II
200		1229.39	II
10		1235.92	II
100		1246.74	II
150		1248.43	II
100		1250.09	II
150		1250.43	II
200		1251.16	II
10		1255.28	I
40		1256.49	I
50		1258.80	I
1000		1260.42	II
2000		1264.73	II
200		1265.02	II
100		1304.37	II
50	h	1305.59	II
200		1309.27	II
20	h	1309.46	II
100		1346.87	II
100		1348.54	II
150		1350.06	II
20		1350.52	II
20		1350.66	II
100		1352.64	II
100		1353.72	II
10	h	1409.07	II
20	h	1410.22	II
10	h	1416.97	II
15	h	1474.65	II
15		1484.87	II
90	h	1485.02	II
30		1485.22	II
100	h	1485.51	II
100	h	1509.10	II
50	h	1512.07	II
30	p	1513.57	II
60	p	1516.91	II
500		1526.72	II
1000		1533.45	II
10		1562.45	II
15		1562.85	II
10		1563.77	II
50		1573.87	I
50		1574.82	I
50		1592.41	I
150		1594.55	I
50		1594.93	I
30		1597.95	I
100		1622.87	I
30		1625.71	I
300		1629.43	I
200		1629.92	I
75		1631.13	I
50		1633.98	I
30	h	1653.35	I
30		1664.52	I
50		1666.37	I
100		1667.62	I
100		1668.52	I
100		1672.59	I
200		1675.20	I
30		1682.68	I
30		1686.82	I
50	h	1689.29	I
30	h	1690.79	I
50		1693.29	I
50		1695.51	I
200		1696.20	I
200		1697.94	I
50		1700.42	I
30		1700.63	I
30		1702.86	I
50		1704.43	I
10	h	1710.83	II
20	h	1711.30	II
30	h	1743.88	I
50		1747.40	I
30	h	1753.11	I
50		1763.66	I
40		1765.03	I
30	h	1765.60	I
30		1766.06	I
30		1770.63	I
100	h	1770.92	I
100	h	1776.83	I
50	h	1783.23	I
100	h	1799.12	I
150		1808.00	II
50	h	1809.09	I
500	h	1814.07	I
200		1816.92	II
10		1817.45	II
50		1822.45	I
200		1836.51	I
30	h	1838.01	I
100	h	1841.15	I
200		1841.44	I
200		1843.77	I
300		1845.51	I
100		1846.10	I
400		1847.47	I
200		1848.14	I
100		1848.74	I
500		1850.67	I
30	h	1851.79	I
200		1852.46	I
50		1853.15	I
20		1869.32	II
15		1870.23	II
100		1873.10	I
500	h	1874.84	I
100		1875.81	I
200		1881.85	I
200		1887.70	I
200	h	1893.25	I
1000	h	1901.33	I
100	h	1902.46	II
50	h	1904.66	I
50	h	1910.62	II
50		1941.67	II
15		1944.59	II
10		1949.33	II
100		1949.56	II
100		1954.97	I
30		1984.43	I
50		1991.85	I
		Air	
30		2010.97	I
50		2054.83	I
50		2058.65	II
50		2059.01	II
40		2061.19	I
30		2065.52	I
200		2072.02	II
200		2072.70	II
30	h	2103.21	I
30		2114.63	I
100		2124.12	I
10	h	2133.99	II
30	h	2136.40	II
50	h	2136.56	II
50	h	2147.91	I
110		2207.98	I
115		2210.89	I
110		2211.74	I
120		2216.67	I
120		2218.06	I
50		2218.91	I
35		2291.03	I
55		2303.06	I
30		2334.40	II
30		2334.61	II
10		2344.20	II
10	h	2349.54	II
20		2350.17	II
20	h	2353.09	II
100	h	2356.30	II
30	h	2357.18	II
50	h	2357.97	II
10	h	2360.20	II
30		2366.97	II
20		2374.26	II
10	h	2428.45	II
300		2435.15	I
65		2438.77	I
65		2443.36	I
70		2452.12	I
425		2506.90	I
375		2514.32	I
500		2516.113	I
350		2519.202	I
425		2524.108	I
450		2528.509	I
110		2532.381	I
30		2563.679	I
85		2568.641	I
45		2577.151	I
190		2631.282	I
10	h	2682.21	II
1000		2881.579	I
10	h	2887.51	II
300		2904.28	II
500		2905.69	II
55		2970.355	I
150		2987.645	I
50		3006.739	I
100	h	3030.00	II
75		3020.004	I
20	h	3021.55	II
20	h	3041.57	II
30	h	3042.19	II
100	h	3043.69	II
10	h	3043.85	II
10	h	3045.77	II
50	h	3048.30	II
150	h	3053.18	II
150		3188.97	II
50		3192.25	II
150		3193.09	II
50		3194.21	II
50		3194.69	II
100		3195.41	II
200		3199.51	II
20		3202.49	II
100	h	3203.87	II
200	h	3210.03	II
75		3214.66	II
15	h	3217.99	II
10		3220.44	II
20		3223.01	II
300		3333.14	II
500		3339.82	II
100	h	3853.66	II
500	h	3856.02	II
200	h	3862.60	II
300		3905.523	I
10	h	3955.74	II
10	h	3977.46	II
15	h	3991.77	II
10	h	3998.01	II
20	h	4075.45	II
15	h	4076.78	II
70		4102.936	I
300	h	4128.07	II
500	h	4130.89	II
10	h	4183.35	II
100	h	4190.72	II
50		4198.13	II
100		4621.42	II
150		4621.72	II
50		4782.991	I
35		4792.212	I
80		4792.324	I
15	h	4883.20	II
20	h	4906.99	II
20	h	4932.80	II
30		4947.607	I
40		5006.061	I
1000		5041.03	II
1000		5055.98	II
100		5181.90	II
100	h	5185.25	II
200	h	5192.86	II
500	h	5202.41	II
30	h	5295.19	II
100	h	5405.34	II
15	h	5417.24	II
15	h	5428.92	II
15	h	5432.89	II
100	h	5438.62	II
20	h	5447.26	II
15	h	5454.49	II
100	h	5456.45	II
500	h	5466.43	II
500	h	5466.87	II
100	h	5469.21	II
40		5493.23	I
200	h	5496.45	II
35		5517.535	I
100	h	5540.74	II
150	h	5576.66	II
30		5622.221	I
100	h	5632.97	II
200	h	5639.48	II
90		5645.611	I
150	h	5660.66	II
80		5665.554	I
1000	h	5669.56	II
30	h	5681.44	II
120		5684.484	I
300	h	5688.81	II
100		5690.425	I
90		5701.105	I
200	h	5701.37	II
100	h	5706.37	II
160		5708.397	I
45		5747.667	I
45		5753.625	I
45		5754.220	I
45		5762.977	I
70		5772.145	I
70		5780.384	I
30	h	5785.73	II
90		5793.071	I
30	h	5794.90	II
100		5797.859	I
150	h	5800.47	II
200		5806.74	II
30		5827.80	II
50		5846.13	II
10		5867.48	II
300	h	5868.40	II
40		5873.764	I
150		5915.22	II
200		5948.545	I
500		5957.56	II
500		5978.93	II
10	h	6067.45	II
20	h	6080.06	II
10	h	6086.67	II
90		6125.021	I
85		6131.574	I
90		6131.850	I
100		6142.487	I
100		6145.015	I
160		6155.134	I
160		6237.320	I
40		6238.287	I
125		6243.813	I
125		6244.468	I
180		6254.188	I
45		6331.954	I
1000		6347.10	II
1000		6371.36	II
45		6526.609	I
45		6527.199	I
45		6555.462	I
50	h	6660.52	II
15		6665.00	II
100		6671.88	II
20		6699.38	II
50	h	6717.04	II
100		6721.853	I
30		6741.64	I
20	h	6750.28	II
30		6818.45	II
50		6829.82	II
30		6848.568	I
80		6976.523	I

Silicon (Cont.)

180		7003.567	I
180		7005.883	I
30		7017.28	I
90		7017.646	I
250		7034.903	I
70		7164.69	I
200		7165.545	I
70		7184.89	I
65		7193.58	I
30		7193.90	I
100		7226.206	I
100		7235.326	I
60		7235.82	I
180		7250.625	I
160		7275.294	I
40		7282.81	I
400		7289.173	I
55		7290.26	I
35		7373.00	I
375		7405.774	I
200		7409.082	I
40		7415.35	I
275		7415.946	I
425		7423.497	I
85		7424.60	I
100		7680.267	I
40		7742.71	I
30		7800.008	I
400		7848.80	II
500		7849.72	II
30		7849.967	I
90		7918.386	I
120		7932.349	I
140		7944.001	I
35		7970.306	I
35		8035.619	I
70		8093.241	I
35		8230.642	I
40		8443.982	I
40		8501.547	I
60		8502.221	I
40		8536.165	I
120		8556.780	I
50		8648.462	I
40		8728.011	I
75		8742.451	I
100		8752.009	I
35		8790.389	I
100		9412.72	II
100		9413.506	I
30		10371.269	I
120		10585.141	I
120		10603.431	I
120		10660.975	I
30		10694.251	I
30		10727.408	I
60		10749.384	I
30		10784.550	I
80		10786.856	I
140		10827.091	I
60		10843.854	I
30		10868.79	I
130		10869.541	I
30		10882.802	I
30		10885.336	I
80		10979.308	I
30		10982.061	I
80		11017.965	I
13		11187.60	I
12		11289.84	I
12		11611.09	I
370		11984.19	I
220		11991.57	I
440		12031.51	I
150		12103.53	I
120		12270.68	I
11		13176.90	I
190		15888.39	I
40		15960.04	I
95		16060.03	I
20		16094.80	I
60		16163.71	I
11		16215.68	I
16		16381.55	I
29		16680.77	I
28		17327.29	I
26		18722.90	I
15		19385.94	I
48		19432.97	I
13		19493.38	I
110		19722.50	I
31		19928.88	I
12		20917.13	I
21		21354.24	I
12		22062.71	I

Si III
Ref. 320 — L.J.R.

Intensity		Wavelength	
		Vacuum	
8		566.61	III
6		652.22	III
8		653.33	III
5		673.48	III
5		800.07	III
9		823.41	III
5		883.40	III
7		939.09	III
9		967.95	III
10		993.52	III
13		994.79	III
16		997.39	III
7		1005.37	III
7		1031.16	III
8		1033.92	III
7		1037.05	III
6		1083.22	III
14		1108.37	III
16		1109.97	III
18		1113.23	III
6		1140.55	III
7		1141.58	III
6		1142.28	III
8		1144.31	III
6		1144.96	III
8		1145.11	III
7		1145.18	III
6		1155.00	III
6		1155.96	III
7		1158.10	III
6		1160.26	III
8		1161.58	III
5		1174.37	III
6		1174.43	III
8		1178.00	III
30		1206.51	III
30		1206.53	III
9		1207.52	III
10		1210.46	III
7		1235.43	III
6		1280.35	III
17		1294.54	III
14		1296.73	III
15		1298.89	III
18		1298.96	III
14		1301.15	III
16		1303.32	III
13		1312.59	III
8		1341.47	III
7		1342.39	III
6		1343.39	III
8		1361.60	III
5		1362.37	III
7		1363.47	III
8		1365.26	III
7		1367.05	III
5		1369.44	III
5		1373.03	III
5		1387.99	III
13		1417.24	III
6		1433.69	III
8		1435.77	III
7		1436.17	III
5		1441.73	III
6		1447.20	III
5		1457.25	III
12		1500.24	III
10		1501.19	III
9		1501.87	III
6		1506.06	III
7		1673.32	III
9		1842.55	III
		Air	
5		2176.89	III
6		2295.48	III
10		2296.87	III
8		2300.93	III
10		2308.19	III
11		2449.48	III
6		2483.20	III
25		2541.82	III
10		2546.09	III
14		2559.21	III
11		2640.79	III
14		2655.51	III
9		2817.11	III
7		2831.49	III
5		2839.62	III
5		2959.15	III
5		2980.52	III
5		3013.09	III
6		3034.73	III
8		3037.29	III
9		3040.93	III
7		3043.93	III
5		3045.08	III
7		3068.24	III
25		3086.24	III
6		3086.46	III
20		3093.42	III
5		3093.65	III
16		3096.83	III
6		3126.27	III
7		3147.37	III
8		3161.61	III
16		3185.13	III
13		3186.02	III
14		3196.50	III
15		3210.55	III
7		3216.25	III
12		3230.50	III
14		3233.95	III
15		3241.62	III
7		3253.40	III
5		3253.74	III
7		3254.80	III
12		3258.66	III
6		3270.46	III
10		3276.26	III
7		3279.26	III
15		3486.91	III
9		3525.94	III
8		3569.67	III
20		3590.47	III
8	h	3622.54	III
5	h	3639.45	III
6	h	3645.12	III
7	h	3681.40	III
5	h	3682.15	III
20	c	3791.41	III
25		3796.11	III
30		3806.54	III
7		3842.46	III
20		3924.47	III
6	h	3947.49	III
6		3963.84	III
5		3981.24	III
5	h	4101.86	III
8		4102.42	III
5	h	4115.50	III
9		4338.50	III
8		4341.40	III
8	h	4377.63	III
6	h	4405.90	III
8	h	4406.72	III
6		4494.05	III
30		4552.62	III
8		4554.00	III
25		4567.82	III
20		4574.76	III
7		4619.66	III
7		4638.28	III
8		4665.87	III
9		4683.02	III
7		4683.80	III
16		4716.65	III
7		4730.52	III
8		4800.43	III
15		4813.33	III
16		4819.72	III
18		4828.97	III
10	h	5091.42	III
7	h	5113.76	III
8	h	5114.12	III
5		5197.26	III
6		5451.46	III
7		5473.05	III
7		5704.60	III
8		5716.29	III
20		5739.73	III
10	h	5898.79	III
7		6314.46	III
6	h	6524.36	III
6	h	6831.56	III
7	h	6851.65	III
5	h	7461.89	III
8	h	7462.62	III
9	h	7466.32	III
12	h	7612.36	III
9	h	8102.86	III
11	h	8103.45	III
7	h	8190.43	III
6	h	8191.16	III
8	h	8191.68	III
9	h	8262.57	III
5	h	8265.64	III
8	h	8269.32	III
5	h	8271.38	III
6	h	8271.94	III

Si IV
Ref. 319 — L.J.R.

Intensity		Wavelength	
		Vacuum	
4		457.82	IV
3		458.16	IV
2		515.12	IV
3		516.35	IV
2		645.76	IV
5		749.94	IV
7		815.05	IV
8		818.13	IV
8		1066.63	IV
8		1122.49	IV
10		1128.34	IV
15		1393.76	IV
12		1402.77	IV
1		1634.61	IV
6		1722.53	IV
5		1727.38	IV
		Air	
3		2120.18	IV
4		2127.47	IV
5	h	2287.04	IV
2	h	2328.56	IV
2		2366.76	IV
3		2370.99	IV
2		2482.82	IV
1		2485.38	IV
7		2517.51	IV
1		2672.19	IV
4		2675.12	IV
4		2675.25	IV
1		2677.57	IV
3	h	2723.81	IV
3	h	2895.13	IV
2	h	2904.47	IV
1	h	2971.52	IV
7		3149.56	IV
9		3165.71	IV
1	h	3244.19	IV
8		3762.44	IV
6		3773.15	IV
1	h	4031.39	IV
2	h	4038.06	IV
10		4088.85	IV
9		4116.10	IV
7	h	4212.41	IV
3		4314.10	IV
5		4328.18	IV
2	h	4403.73	IV
1	h	4411.65	IV
1	h	4611.27	IV
3	h	4628.62	IV
9	h	4631.24	IV
10	h	4654.32	IV
3	h	4656.92	IV
1	h	4667.14	IV
2	h	4673.30	IV
1	h	4947.45	IV
3		4950.11	IV
2	h	5304.97	IV
1	h	5309.49	IV
5		6667.56	IV
7		6701.21	IV
3	h	6998.36	IV
6	h	7047.94	IV
4	h	7068.41	IV
2	h	7630.50	IV
4	h	7654.56	IV
4	h	7678.75	IV
5	h	7718.79	IV
6	h	7723.82	IV
2	h	7725.64	IV
1	h	7730.47	IV
1	h	7752.91	IV
1	h	8240.61	IV
2	h	8957.25	IV
1	h	9018.16	IV

Si V
Ref. 87 — L.J.R.

Intensity		Wavelength	
		Vacuum	
1		78.61	V
1		78.90	V
2		80.81	V
2		81.11	V
10		85.18	V
6		85.58	V
4		90.45	V
4		90.85	V
15		96.44	V
10		97.14	V
2		98.21	V
20		117.86	V
20		118.97	V

SILVER (Ag)
Z = 47

Ag I and II
Ref. 13, 99, 255, 286, 289 — C.H.C.

Intensity		Wavelength	
		Vacuum	
25		730.83	II

Silver (Cont.)

30		752.80	II
15		1005.32	II
10		1065.49	II
12		1072.23	II
250		1074.22	II
150		1107.03	II
150		1112.46	II
60		1195.83	II
50		1223.33	II
50		1240.80	II
50		1246.87	II
55		1256.81	II
55		1257.55	II
50		1266.63	II
70		1273.67	II
65		1297.51	II
85		1311.20	II
55		1313.81	II
50		1314.61	II
60		1323.84	II
60		1342.09	II
50		1342.57	II
70		1346.62	II
50		1353.54	II
150		1364.50	II
100		1396.00	II
100		1410.93	II
90		1419.72	II
95		1432.60	II
100		1464.72	II
50		1466.23	II
50	r	1507.37	I
100	r	1515.63	I
50	r	1548.58	I
100		1555.16	II
100		1644.50	II
60		1651.52	I
50		1652.10	I
120		1682.82	II
10		1708.11	I
50		1709.27	I
125		1736.44	II
10	h	1766.14	I
75		1790.37	II
20		1847.71	I
100		1967.38	II

Air

150		2015.96	II
150		2033.98	II
200		2061.17	I
100		2069.85	I
80	r	2113.82	II
60		2145.60	II
15		2170.00	I
50		2186.76	II
60		2229.53	II
100	r	2246.43	II
75	r	2248.74	II
75		2280.03	II
30	h	2309.56	I
10	h	2312.60	I
70	r	2317.05	II
80	r	2320.29	II
70	r	2324.68	II
80	r	2331.40	II
70		2357.92	II
50	h	2375.02	I
75		2411.41	II
90	r	2413.23	II
100	r	2437.81	II
80		2447.93	II
80		2473.84	II
60		2506.63	II
50	h	2575.63	I
60		2660.49	II
60		2721.77	I
75		2767.54	II
100	h	2824.39	I
10	h	2926.77	I
20	h	2938.42	I
20		3099.10	I
30	h	3130.02	I
10	h	3170.58	I
90		3180.70	II
15	h	3215.67	I
10		3225.15	I
15		3233.18	I
100		3267.35	II
55000	r	3280.68	I
10	h	3305.67	I
28000	r	3382.89	I
10	h	3403.78	I
30		3469.16	I
70		3475.82	II
80		3495.28	II
20	h	3501.92	I
20		3508.03	I
15	h	3513.38	I
10		3521.12	I
50		3542.61	I
10	h	3547.16	I
10	h	3557.01	I
20	h	3586.67	I
10	h	3623.49	I
50	h	3624.68	I
75		3682.46	II
30		3682.50	I
80		3683.34	II
50	h	3709.20	I
10	h	3727.42	I
20	h	3753.14	I
200		3810.94	I
50		3811.78	I
100	h	3840.74	I
15		3847.85	I
50	h	3907.41	I
50		3909.31	II
50	h	3914.40	I
70		3920.10	II
10	h	3928.01	I
10	h	3940.43	I
10	h	3942.97	I
60		3949.43	II
100	h	3981.58	I
70		3985.19	II
10	h	3992.15	I
100	h	4055.48	I
10	h	4083.43	I
80		4085.91	II
100		4185.48	II
90	h	4210.96	I
100		4212.82	I
50		4311.07	I
20		4396.23	I
50	h	4476.04	I
20	h	4556.0	I
30	h	4615.69	I
80		4620.04	II
50		4620.46	II
60	h	4668.48	I
30	h	4677.60	I
100		4788.40	II
20	h	4796.2	I
30	h	4847.82	I
100		4874.10	I
20		4888.21	I
10	h	4917.5	I
10		4935.75	I
20	h	4992.89	I
80		5027.35	II
15	h	5123.50	I
1000		5209.08	I
10	h	5333.62	I
1000		5465.50	I
100		5471.55	I
20		5475.38	I
20	h	5545.67	I
10	h	5559.58	I
100		5667.34	I
10	h	6083.78	I
10	h	6268.50	I
20		6621.08	I
20		7359.96	I
320		7687.78	I
25		8005.4	II
15		8254.7	II
500		8273.52	I
20		8324.4	II
15		8379.5	II
25		8403.8	II
15		8492.5	II
30	h	8645.70	I
10	h	8704.85	I
12		8747.6	II
15		9000.9	II
10		12551.0	I
60		16819.5	I
20		17416.7	I
15		18307.9	I
15		18382.3	I

Ag III
Ref. 363, 387, 398 — R.D.C.

Intensity	Wavelength	
Vacuum		
200	709.80	III
200	713.85	III
100	717.73	III
200	718.53	III
300	726.96	III
350	730.04	III
150	730.28	III
150	730.94	III
200	736.57	III
100	738.13	III
200	740.98	III
200	742.29	III
200	748.30	III
150	755.73	III
100	758.27	III
200	767.19	III
250	768.33	III
150	769.61	III
350	776.38	III
200	782.91	III
150	785.76	III
200	789.08	III
250	792.35	III
200	796.54	III
250	797.91	III
400	799.41	III
300	808.88	III
150	816.12	III
180	822.39	III
200	838.11	III
120	1373.22	III
120	1374.76	III
110	1404.93	III
120	1413.90	III
120	1414.29	III
120	1428.61	III
120	1452.74	III
200	1456.41	III
100	1471.44	III
300	1489.01	III
150	1515.08	III
100	1524.23	III
120	1527.04	III
150	1541.14	III
150	1550.89	III
130	1553.04	III
130	1587.41	III
120	1589.28	III
100	1613.79	III
130	1619.14	III
100	1634.46	III
100	1652.24	III
130	1653.60	III
300	1654.43	III
700	1656.18	III
150	1657.10	III
100	1661.54	III
130	1670.75	III
150	1676.14	III
100	1681.07	III
500	1693.51	III
200	1705.06	III
150	1708.86	III
130	1717.68	III
200	1722.27	III
150	1726.76	III
250	1728.14	III
200	1747.34	III
120	1749.64	III
150	1750.89	III
750	1751.03	III
100	1760.57	III
150	1762.62	III
150	1768.70	III
100	1771.81	III
100	1783.85	III
100	1791.70	III
100	1792.69	III
150	1793.90	III
150	1802.24	III
150	1802.26	III
100	1802.77	III
300	1808.23	III
250	1816.83	III
150	1822.45	III
350	1828.83	III
250	1832.33	III
120	1832.50	III
100	1834.31	III
250	1836.10	III
150	1838.64	III
400	1840.14	III
100	1846.96	III
120	1849.93	III
150	1856.33	III
120	1858.91	III
100	1860.39	III
100	1860.64	III
350	1867.12	III
150	1868.10	III
100	1872.55	III
400	1873.45	III
250	1880.36	III
300	1889.57	III
400	1916.92	III
600	1917.08	III
200	1925.30	III
150	1946.32	III
100	1948.44	III
700	1957.62	III
120	1959.27	III
100	1960.86	III
400	1966.89	III
600	1975.92	III
500	1977.03	III
150	1981.87	III
200	1987.02	III
130	1995.16	III
600	2000.24	III
Air		
300	2007.30	III
200	2011.49	III
150	2013.65	III
150	2041.33	III
150	2053.17	III
150	2053.83	III
200	2056.99	III
200	2081.04	III
300	2146.47	III
150	2149.19	III
600	2161.89	III
150	2166.21	III
150	2211.23	III
100	2238.40	III
500	2246.51	III
100	2286.50	III
700	2310.04	III
100	2386.85	III
300	2395.69	III
100	2469.62	III
100	2562.87	III

SODIUM (Na)
Z = 11

Na I and II
Ref. 268, 334 — T.K.M.

Intensity	Wavelength	
Vacuum		
160	300.15	II
160	300.20	II
90	301.32	II
100	301.44	II
60	302.45	II
300	372.08	II
350	376.38	II
60	1293.97	II
50	1327.74	II
45	1347.54	II
90	1374.69	II
90	1404.68	II
45	1495.21	II
45	1497.73	II
80	1506.41	II
60	1506.91	II
70	1513.10	II
60	1519.63	II
60	1657.92	II
90	1776.57	II
60	1783.04	II
80	1787.19	II
45	1788.85	II
80	1798.41	II
45	1801.26	II
90	1807.09	II
60	1808.38	II
50	1821.70	II
45	1833.87	II
80	1835.22	II
45	1837.89	II
60	1841.82	II
70	1845.02	II
45	1850.15	II
70	1851.19	II
80	1853.17	II
45	1866.45	II
45	1873.37	II
60	1875.08	II
160	1881.91	II
50	1885.09	II
45	1885.74	II
Air		
80	2228.53	II
80	2303.58	II
300	2315.65	II
130	2393.28	II
100	2401.01	II
300	2420.99	II
300	2424.73	II
200	2439.14	II
250	2441.50	II
200	2448.72	II
200	2452.18	II
1000	2493.15	II
300	2502.84	II
450	2506.30	II
600	2515.46	II
600	2531.54	II
20	2543.84	I
10	2543.87	I
550	2586.31	II
70	2593.87	I
35	2593.92	I
600	2594.96	II
850	2611.81	II
300	2627.41	II

Sodium (Cont.)

Intensity		Wavelength	
850		2661.00	II
350		2666.46	II
1000		2671.83	II
850		2678.09	II
200		2680.34	I
100		2680.43	I
650		2808.71	II
850		2809.52	II
600		2829.87	II
800		2839.56	II
1000		2841.72	II
400		2852.81	I
200		2853.01	I
650		2856.51	II
800		2859.49	II
750		2871.28	II
650		2872.95	II
900		2881.15	II
850		2886.26	II
2		2893.62	I
700		2893.95	II
900		2901.14	II
800		2904.72	II
1100		2904.92	II
1100		2917.52	II
1100		2919.05	II
1200		2919.85	II
1300		2920.95	II
1000		2923.49	II
750		2930.88	II
850		2934.08	II
950		2937.74	II
800		2945.70	II
950		2947.50	II
1200		2951.24	II
1100		2952.40	II
850		2960.12	II
500		2970.73	II
600		2974.24	II
750		2974.99	II
1000		2977.13	II
1100		2979.66	II
1100		2980.63	II
1300		2984.19	II
550		3004.15	II
750		3007.44	II
750		3009.14	II
600		3015.40	II
550		3053.67	II
550		3055.35	II
550		3056.16	II
550		3057.38	II
550		3057.95	II
550		3058.72	II
700		3060.25	II
800		3061.35	II
500		3064.38	II
500		3066.22	II
500		3066.54	II
550		3074.33	II
550		3078.32	II
550		3080.25	II
550		3087.06	II
550		3092.04	II
550		3092.73	II
650		3094.45	II
650		3095.55	II
500		3103.58	II
500		3104.40	II
500		3113.69	II
1700		3124.42	II
600		3125.21	II
600		3129.38	II
2500		3135.48	II
1700		3137.86	II
950		3145.71	II
2000		3149.28	II
2000		3163.74	II
700		3175.09	II
1000		3179.06	II
1700		3189.79	II
1600		3212.19	II
700		3234.93	II
1500		3257.96	II
650		3260.21	II
950		3274.22	II
1700		3285.60	II
1700		3301.35	II
1200		3302.37	I
600		3302.98	I
1500		3304.96	II
1000		3318.04	II
950		3327.69	II
50		3426.86	I
1500		3533.05	II
1200		3631.27	II
850		3711.07	II
300		4113.70	II
250		4123.08	II
250		4233.26	II
6		4238.99	I
250		4240.90	II
10		4242.08	I
1		4249.41	I
2		4252.52	I
15		4273.64	I
20		4276.79	I
2		4287.84	I
3		4291.01	I
250		4292.48	II
250		4292.86	II
250		4308.81	II
250		4309.04	II
250		4320.91	II
30		4321.40	I
40		4324.62	I
250		4337.29	II
3		4341.49	I
250		4344.11	II
5		4344.74	I
200		4368.60	II
200		4375.22	II
200		4387.49	II
40		4390.03	I
250		4392.81	II
60		4393.34	I
200		4405.12	II
5		4419.88	I
8		4423.25	I
200		4446.70	II
200		4447.41	II
200		4454.74	II
200		4455.23	II
200		4457.21	II
200		4474.63	II
200		4478.80	II
200		4481.67	II
200		4490.15	II
200		4490.87	II
60		4494.18	I
100		4497.66	I
200		4499.62	II
200		4506.97	II
200		4519.21	II
200		4524.98	II
200		4533.32	II
10		4541.63	I
15		4545.19	I
200		4551.53	II
160		4590.92	II
120		4664.811	I
200		4668.560	I
160		4722.23	II
160		4731.10	II
160		4741.67	II
20		4747.941	I
30		4751.822	I
160		4768.79	II
100		4788.79	II
200		4978.541	I
400		4982.813	I
40		5148.838	I
80		5153.402	I
100		5191.65	II
80		5208.55	II
70		5400.46	II
90		5414.55	II
280		5682.633	I
70		5688.193	I
560		5688.205	I
80000		5889.950	I
40000		5895.924	I
120		6154.225	I
240		6160.747	I
60		6175.25	II
70		6199.26	II
70		6234.68	II
80		6260.01	II
80		6274.74	II
70		6361.15	II
70		6366.41	II
90		6514.21	II
80		6524.68	II
130		6530.70	II
130		6544.04	II
130		6545.75	II
80		6552.43	II
20		7373.23	I
10		7373.49	I
50		7809.78	I
25		7810.24	I
4400		8183.256	I
800		8194.790	I
8800		8194.824	I
100		8649.92	I
60		8650.89	I
25		8942.96	I
40		9153.88	I
60		9465.94	I
80		9961.28	I
20		10566.00	I
60		10572.28	I
200		10746.44	I
80		10749.29	I
120		10834.87	I
35		11190.19	I
50		11197.21	I
400		11381.45	I
1000		11403.78	I
400		12679.17	I
60		14767.48	I
100		14779.73	I
60		16373.85	I
100		16388.85	I
400		18465.25	I
50		22056.44	I
25		22083.67	I
60		23348.41	I
100		23379.13	I

Na III
Ref. 178, 205, 207 — T.K.M.

Intensity		Wavelength	
		Vacuum	
5		183.95	III
5	h	189.35	III
5		193.80	III
5	h	194.04	III
5	h	194.17	III
5		194.29	III
6		194.68	III
6		195.53	III
6		202.15	III
6		202.19	III
8		202.49	III
5	d	202.71	III
7	d	202.72	III
8		202.76	III
8	p	203.06	III
8		203.28	III
8		203.33	III
10		207.30	III
10	c	215.34	III
12		215.86	III
12		216.12	III
15		229.87	III
12		230.59	III
50	c	250.52	III
30		251.37	III
25		266.90	III
70		267.65	III
50		267.87	III
50		268.63	III
20	p	272.08	III
20		272.45	III
100		378.14	III
70		380.10	III
7		1336.76	III
7		1337.36	III
8		1340.67	III
9	d	1342.39	III
10		1342.73	III
11		1355.28	III
12		1361.90	III
11		1372.34	III
10		1420.89	III
10		1444.19	III
12		1449.31	III
11		1562.87	III
10		1565.29	III
10		1598.18	III
11		1688.94	III
10		1699.29	III
10		1711.12	III
11		1728.27	III
10		1731.11	III
10		1755.48	III
15		1807.07	III
10		1810.77	III
11		1811.67	III
10		1816.81	III
10	d	1835.22	III
10		1838.94	III
11		1844.36	III
12	d	1847.53	III
10	d	1847.59	III
15		1849.56	III
12		1850.38	III
10		1855.92	III
10		1856.71	III
10		1861.21	III
10		1880.66	III
10	d	1887.39	III
20	d	1887.47	III
15	d	1890.75	III
15		1900.16	III
10		1918.45	III
11		1923.96	III
14		1926.26	III
12		1927.24	III
12		1932.74	III
13		1933.89	III
10		1943.52	III
12		1946.43	III
12		1950.91	III
14		1951.24	III
10		1977.16	III
13		1985.57	III
10		1995.68	III
		Air	
10		2004.21	III
11		2005.22	III
11		2008.47	III
15		2011.87	III
11		2014.17	III
12		2017.03	III
12		2028.56	III
12		2031.13	III
11		2035.90	III
12		2041.66	III
12		2043.29	III
10		2044.82	III
10		2045.44	III
11		2051.48	III
10		2060.36	III
15		2066.60	III
13		2082.91	III
15		2140.72	III
14		2144.54	III
15		2202.83	III
15		2225.93	III
30		2230.33	III
16		2232.19	III
20	h	2246.70	III
14		2251.47	III
15		2278.42	III
13		2285.66	III
15		2309.99	III
18		2386.99	III
17		2394.03	III
15		2406.59	III
25		2459.31	III
18		2468.85	III
20		2474.73	III
25		2497.03	III
17		2510.26	III
15		2530.25	III
14		2542.80	III

Na IV
Ref. 206 — T.K.M.

Intensity		Wavelength	
		Vacuum	
4		136.551	IV
4		136.854	IV
4		139.961	IV
7		142.232	IV
6		142.359	IV
8		146.064	IV
7		146.302	IV
9		150.298	IV
7		150.543	IV
7	c	150.64	IV
8		150.687	IV
7		151.299	IV
7		155.083	IV
7		155.240	IV
7		155.448	IV
8		155.510	IV
8		156.537	IV
12		162.448	IV
10		163.190	IV
12		168.411	IV
10		168.546	IV
10		190.445	IV
10		199.772	IV
10	c	205.49	IV
10		319.644	IV
10		360.76	IV
12		408.684	IV
10		409.614	IV
15		410.372	IV
10		411.334	IV
13		412.242	IV
10		1580.50	IV
11		1582.18	IV
10		1582.33	IV
11	d	1583.98	IV
12		1584.14	IV
10	d	1586.99	IV
12	d	1587.05	IV
10		1613.95	IV
11		1615.92	IV
12		1618.57	IV
11		1655.47	IV
15	c	1701.97	IV
10		1702.41	IV
12		1960.76	IV
11		1965.08	IV
10		1967.60	IV
		Air	
10		2018.39	IV
12	d	2106.33	IV
10		2114.53	IV

Sodium (Cont.)

10		2155.76	IV

Na V
Ref. 299 — T.K.M.

Intensity		Wavelength	
		Vacuum	
100		106.28	V
100		106.30	V
100		106.40	V
100		106.49	V
200	c	107.93	V
200		108.02	V
200	c	110.82	V
200		110.88	V
100		111.51	V
300	c	112.01	V
100	h	114.70	V
100		114.74	V
400		117.99	V
100		120.04	V
400		125.18	V
400		125.22	V
500		125.29	V
300		125.43	V
300		125.46	V
200		125.90	V
100		126.21	V
200		126.56	V
100		126.61	V
400		127.44	V
400		127.47	V
400		128.03	V
400		128.05	V
200		130.68	V
300		131.35	V
200		131.41	V
300	h	131.64	V
500		133.16	V
400		133.39	V
200		134.27	V
300		135.79	V
300		135.85	V
200		138.81	V
300		138.92	V
400		148.64	V
300		148.86	V
400		151.13	V
300		157.21	V
300		163.62	V
800		307.15	V
1000		308.26	V
800		332.55	V
900		333.91	V
800		360.32	V
800		360.37	V
1000		400.72	V
500		445.05	V
600		445.19	V
600		459.90	V
850		461.05	V
1000		463.26	V

STRONTIUM (Sr)
Z = 38

Sr I and II
Ref. 1, 218, 279, 313 — J.J.W.

Intensity		Wavelength	
		Air	
1400		2152.84	II
1400		2165.96	II
160		2428.10	I
120		2569.47	I
200		2931.83	I
300		3301.73	I
300		3329.99	I
400		3351.25	I
300		3366.33	I
650		3380.71	II
950		3464.46	II
120		3474.89	II
300	h	3940.80	I
600		3969.26	I
300		3970.04	I
1300		4030.38	I
300		4032.38	I
46000		4077.71	II
200		4161.80	II
32000		4215.52	II
340		4305.45	II
350	h	4438.04	I
		4526.10	II
		4585.91	II
65000		4607.33	I
3200		4722.28	I
2200		4741.92	I
1400		4784.32	I
4800		4811.88	I
3600		4832.08	I
500		4855.04	I
600		4868.70	I
3000		4872.49	I
600		4876.06	I
2000		4876.32	I
1000		4891.98	I
8000		4962.26	I
1300		4967.94	I
800	h	5156.07	I
1400		5222.20	I
2000		5225.11	I
2000		5229.27	I
2800		5238.55	I
4800		5256.90	I
		5303.13	II
350	h	5329.82	I
		5379.13	II
		5385.45	II
1500		5450.84	I
7000		5480.84	I
1100		5486.12	I
3500		5504.17	I
2600		5521.83	I
2000		5534.81	I
2000		5540.05	I
250	h	5543.36	I
		5622.94	II
		5650.54	II
		5723.70	II
		5819.00	II
200	h	5970.10	I
250	h	6345.75	I
250	h	6363.94	I
350	h	6369.96	I
1000		6380.75	I
900	h	6386.50	I
600	h	6388.24	I
9000		6408.47	I
250		6446.68	I
250	h	6465.79	I
		6483.17	II
5500		6504.00	I
		6509.20	II
1000		6546.79	I
1700		6550.26	I
3000		6617.26	I
800		6643.54	I
1800		6791.05	I
4800		6878.38	I
1200		6892.59	I
5500		7070.10	I
60		7153.09	I
250	h	7167.24	I
200		7232.27	I
2500		7309.41	I
500		7621.50	I
400	h	7673.06	I
50	hL	7850.00	I
30	h	7866.90	I
20	hL	7874.00	I
200	h	8422.80	I
120		8505.69	II
200		8688.91	II
30		8719.56	II
40		9170.00	I
30		9204.50	I
20		9283.90	I
100		9294.10	I
15		9306.60	I
30		9319.20	I
60		9380.45	I
40	h	9411.25	I
400	h	9448.95	I
600		9596.00	I
300		9624.70	I
100		9638.10	I
100	h	9647.70	II
300		10036.66	II
1000		10327.31	II
7		10872.70	II
200		10914.88	II
10		10984.00	I
13		11224.57	II
700		11241.25	I
100		12014.76	II
20		12236.20	I
60		12445.90	II
20		12479.60	I
40		12495.00	I
15		12652.20	I
75		12974.70	II
100		13123.80	II
15		13522.80	I
15		17140.90	I
30		17170.50	I
50		17447.40	I
4		17626.00	II
30		17743.00	I
15		19759.60	I
230		20261.40	I
120		20700.70	I
40		20764.50	I
15		20778.70	I
30		26023.60	I

Sr III
Ref. 231, 265 — J.J.W.

Intensity		Wavelength	
		Vacuum	
25		307.18	III
50		316.11	III
50		321.61	III
125		330.67	III
500		351.62	III
75		358.80	III
250		363.49	III
150		371.21	III
1000		437.24	III
1875		491.79	III
1250		507.04	III
3750		514.38	III
2500		562.75	III
20		968.37	III
20		975.78	III
25		992.98	III
50		1025.23	III
35		1044.91	III
20		1057.74	III
25		1060.20	III
20		1098.77	III
35		1125.49	III
20		1140.24	III
20		1168.27	III
20		1182.09	III
50		1236.23	III
20		1940.58	III
30	p	1958.44	III
30		1966.92	III
		Air	
25		2068.63	III
50		2099.59	III
25		2114.31	III
30		2118.48	III
50		2119.52	III
50		2133.12	III
30		2142.80	III
20		2145.74	III
30		2178.91	III
30		2180.14	III
50		2190.88	III
50		2203.86	III
50		2219.50	III
50		2220.05	III
50		2267.03	III
100		2273.71	III
50		2277.87	III
30		2310.33	III
50		2314.95	III
50		2334.79	III
100		2340.13	III
50		2404.17	III
30		2410.52	III
50		2454.03	III
100		2486.52	III
50		2503.59	III
30		2599.10	III
35		2622.69	III
30		2642.96	III
30		2648.51	III
35		2654.66	III
40		2722.47	III
50		2786.00	III
50		2821.42	III
30		2874.86	III
30		2929.34	III
30		2983.00	III
100		3002.61	III
200		3012.32	III
100		3021.73	III
30		3059.83	III
50		3061.43	III
30		3104.25	III
50		3182.61	III
100		3235.39	III
30		3302.72	III
50		3430.76	III
30		3874.26	III
30		3936.40	III
30		3936.72	III
30		3958.75	III
30		4094.03	III
30		4097.02	III
30		4105.63	III
35		4335.80	III
30		5071.09	III
30		5130.34	III
35		5158.26	III
40		5257.71	III
30		5262.21	III
30		5288.32	III
30		5391.03	III
40		5443.48	III
30		5463.90	III
30		5664.66	III
30		5689.72	III

Sr IV
Ref. 110 — J.J.W.

Intensity		Wavelength	
		Vacuum	
12		284.31	IV
12		291.09	IV
12		291.19	IV
12		293.22	IV
15		298.12	IV
15		300.12	IV
12		300.27	IV
12		301.67	IV
20		378.53	IV
75		392.44	IV
50		393.00	IV
45		394.90	IV
50		396.22	IV
40		399.92	IV
35		403.85	IV
35		406.94	IV
30		412.93	IV
40		413.07	IV
40		415.32	IV
30		419.78	IV
25		430.21	IV
30		430.65	IV
25		442.73	IV
25		471.76	IV
25		484.20	IV
25	p	508.14	IV
25		534.19	IV
200		664.43	IV
100		710.35	IV
20		1189.21	IV
30		1244.14	IV
20	p	1244.75	IV
20	p	1244.87	IV
20	p	1257.78	IV
20		1268.62	IV
20		1331.13	IV
30		1347.90	IV
20		1361.15	IV
25		1408.67	IV
20		1592.74	IV
25		1677.03	IV
20		1705.16	IV
20		1724.23	IV
25		1729.53	IV
20		1732.12	IV
20		1777.25	IV
20		1994.61	IV
		Air	
20		2104.38	IV
20		2117.90	IV
20		2217.99	IV
20		2230.41	IV
20		2240.49	IV
20		2253.38	IV
50		2346.97	IV
20		2357.34	IV
20		2438.93	IV
25		2441.41	IV
30		2482.79	IV
25		2483.57	IV
18		2500.57	IV
20		2508.02	IV
20		2534.03	IV
18		2548.02	IV
40		2555.60	IV
40		2571.04	IV
25		2571.58	IV
15		2589.34	IV
25		2620.35	IV
20		2621.16	IV
20		2642.16	IV
15		2830.53	IV
9		2934.60	IV
10		3019.29	IV
9		3266.52	IV
9		3566.43	IV
9		3741.05	IV
9		4298.57	IV
9		4685.08	IV

Sr V
Ref. 109 — J.J.W.

Intensity		Wavelength	
		Vacuum	

Strontium (Cont.)

Intensity	Wavelength	
10	517.28	V
6	540.51	V
25	578.01	V
30	624.93	V
25	642.23	V
50	649.21	V
20	659.15	V
25	660.94	V
9	669.93	V
35	686.23	V
6	715.79	V
12	747.82	V
9	862.32	V

SULFUR (S)
Z = 16

S I and II
Ref. 144, 209, 210, 266—R.L.K.

Intensity	Wavelength	
	Vacuum	
40	906.9	II
40	910.5	II
40	912.7	II
40	937.4	II
40	937.7	II
20	996.0	II
20	1000.5	II
20	1014.4	II
20	1019.5	II
20	1096.6	II
40	1102.3	II
20	1131.0	II
20	1131.6	II
40	1234.1	II
40	1250.5	II
110	1253.8	II
110	1259.5	II
275	1270.782	I
250	1277.216	I
280	1295.653	I
275	1302.337	I
235	1302.863	I
235	1303.110	I
245	1303.430	I
260	1305.883	I
265	1310.194	I
355	1316.542	I
290	1316.618	I
375	1323.515	I
355	1326.643	I
775	1381.552	I
710	1385.510	I
960	1388.435	I
640	1389.154	I
775	1392.588	I
1000	1396.112	I
300	1409.337	I
510	1425.030	I
425	1433.280	I
300	1436.968	I
300	1448.229	I
425	1472.972	I
550	1473.995	I
300	1474.380	I
355	1481.665	I
485	1483.039	I
300	1483.233	I
330	1485.622	I
390	1487.150	I
680	1666.688	I
640	1687.530	I
710	1807.311	I
680	1820.343	I
640	1826.245	I
710	1900.286	I
550	1914.698	I
	Air	
20	2629.1	II
40	2670.0	II
40	2847.7	II
285	3867.6	I
285	3902.0	I
360	3933.3	II
450	4120.8	I
280	4142.3	II
360	4145.1	II
450	4153.1	II
450	4162.7	II
450	4694.1	I
285	4695.4	I
160	4696.2	I
280	4716.2	II
450	4815.5	II
360	4924.1	II
450	4925.3	II
285	4993.5	I
360	5428.6	II
650	5432.8	II
1000	5453.8	II
1000	5473.6	II
1000	5509.7	II
280	5564.9	II
1000	5606.1	II
450	5640.0	II
450	5640.3	II
280	5647.0	II
650	5659.9	II
450	5664.7	II
160	5706.1	I
450	5819.2	II
450	6052.7	I
280	6286.4	II
450	6287.1	II
450	6305.5	II
450	6312.7	II
280	6384.9	II
280	6397.3	II
280	6398.0	II
360	6413.7	II
160	6743.6	I
285	6748.8	I
450	6757.2	I
450	7579.0	I
450	7629.8	I
285	7686.1	I
450	7696.7	I
1000	7924.0	I
160	7928.8	I
285	7930.3	I
450	7931.7	I
450	7967.4	I
450	7967.4	II
450	8314.7	I
450	8314.7	II
450	8585.6	I
285	8680.5	I
450	8694.7	I
360	8874.5	I
110	8882.5	I
220	8884.2	I
160	9035.9	I
450	9212.9	I
450	9228.1	I
450	9237.5	I
285	9413.5	I
285	9421.9	I
285	9437.1	I
650	9649.9	I
450	9672.3	I
450	9680.8	I
450	9693.7	I
285	9697.3	I
285	9739.7	I
110	9741.9	I
285	9932.3	I
285	9949.8	I
285	9958.9	I
285	10455.5	I
70	10456.8	I
285	10459.5	I

S III
Ref. 209, 210 — R.L.K.

Intensity	Wavelength	
	Vacuum	
70	729.5	III
110	732.42	III
70	735.2	III
70	738.5	III
70	789.0	III
70	796.7	III
70	824.9	III
70	836.3	III
285	1077.1	III
70	1194.0	III
70	1201.0	III
	Air	
110	2460.5	III
110	2489.6	III
160	2496.2	III
160	2499.1	III
220	2508.2	III
70	2636.9	III
220	2665.4	III
70	2680.5	III
110	2691.8	III
110	2702.8	III
220	2718.9	III
110	2721.4	III
220	2726.8	III
220	2731.1	III
110	2741.0	III
285	2756.9	III
110	2775.2	III
160	2785.5	III
70	2797.4	III
70	2856.0	III
110	2863.5	III
160	2904.3	III
70	2964.8	III
160	2986.0	III
70	3234.2	III
70	3324.9	III
110	3497.3	III
160	3632.0	III
70	3662.0	III
110	3709.4	III
160	3717.8	III
160	3838.3	III
160	3928.6	III
360	4253.6	III
110	4285.0	III
70	4332.7	III

S IV
Ref. 29, 202, 209 — R.L.K.

Intensity	Wavelength	
	Vacuum	
20	519.3	IV
20	520.1	IV
40	520.8	IV
20	522.0	IV
20	522.5	IV
20	551.2	IV
40	652.5	IV
40	653.0	IV
70	653.6	IV
40	654.0	IV
70	655.6	IV
20	655.9	IV
110	657.3	IV
40	660.9	IV
160	661.4	IV
40	663.7	IV
40	664.8	IV
70	666.1	IV
110	744.9	IV
110	748.4	IV
110	750.2	IV
110	753.8	IV
40	798.3	IV
70	800.5	IV
70	804.0	IV
70	809.7	IV
110	816.0	IV
160	1062.7	IV
160	1073.0	IV
70	1073.5	IV
20	1108.4	IV
20	1110.9	IV
20	1624.0	IV
20	1629.2	IV
	Air	
20	2387.0	IV
40	2398.9	IV
110	3097.5	IV
40	3117.7	IV

S V
Ref. 29 — R.L.K.

Intensity	Wavelength	
	Vacuum	
5	437.4	V
5	438.2	V
5	439.6	V
40	658.3	V
70	659.8	V
110	663.2	V
5	676.2	V
5	677.3	V
20	678.1	V
40	680.3	V
110	680.9	V
40	681.6	V
5	686.2	V
5	686.9	V
5	689.8	V
5	691.7	V
20	693.5	V
285	786.5	V
160	849.2	V
110	852.2	V
220	854.8	V
110	857.9	V
110	860.5	V
20	883.6	V
20	884.5	V
5	885.8	V
20	900.9	V
5	902.8	V
20	905.9	V

TANTALUM (Ta)
Z = 73

Ta I and II
Ref. 1 — C.H.C.

Intensity	Wavelength	
	Air	
1100	2140.13	II
1500	2146.87	II
740	2150.62	II
600	2165.01	II
740	2178.03	II
1200	2182.71	II
540	2193.20	II
1100	2193.88	II
1500	2196.03	II
1500	2199.67	II
500	2207.14	II
1400 d	2210.03	II
	2210.19	II
420	2215.60	II
1400	2239.48	II
240	2248.48	II
480	2249.79	II
1200	2250.76	II
260	2254.86	II
440	2255.77	II
360	2256.51	II
500	2258.71	II
840	2261.42	II
260	2261.62	II
990	2262.30	II
220	2269.56	II
740	2271.85	II
990	2272.59	II
200	2279.85	I
320	2282.19	II
130	2285.02	II
790	2285.25	II
600	2286.59	II
240	2287.27	II
990	2289.16	II
180	2292.54	II
160	2295.18	
160	2301.47	II
440	2302.24	II
440	2302.93	II
300	2303.49	II
100	2308.46	II
440	2312.60	II
420	2315.46	II
260	2319.16	II
100	2331.29	II
690	2331.98	II
550	2332.19	II
110	2334.13	II
180	2334.88	II
140	2335.75	II
300	2338.28	II
200	2340.94	II
200	2341.61	II
130	2343.64	II
100	2346.42	II
90	2351.99	II
170	2353.86	II
120	2355.22	II
170	2356.05	II
140	2356.90	II
250	2357.30	I
170	2359.16	II
260	2361.09	I
160	2362.78	II
130	2363.32	II
600	2364.24	II
50	2367.24	II
150	2369.32	II
300	2370.76	II
320	2371.58	I
100	2373.94	II
70	2375.91	I
150	2378.31	II
440	2381.13	II
240	2381.52	II
170	2383.72	II
240	2384.28	II
130	2385.73	I
1400	2387.06	II
80	2388.37	II
160	2389.11	II
70	2396.30	I
110	2399.15	I
50	2399.92	II
2400	2400.63	II
140	2402.13	II
100	2403.68	II
130	2406.55	I

Tantalum (Cont.)

130		2408.26	II
120		2414.32	I
240		2415.21	II
320		2416.89	II
220		2417.86	II
150		2418.77	II
140		2421.03	I
150		2421.85	II
170		2423.48	II
130		2425.91	II
360		2427.64	I
360		2429.71	II
170		2431.06	II
480		2432.70	II
130		2433.59	II
130		2436.51	II
110		2437.07	I
110		2438.64	II
200		2439.91	I
130		2442.39	I
100		2444.13	II
100		2447.17	I
100		2454.48	I
100		2458.68	I
100		2460.55	I
160		2463.82	II
130		2466.99	II
130		2467.37	II
380		2470.90	II
120		2471.38	I
120		2472.13	I
150		2473.13	I
120		2473.31	II
600		2474.62	I
120		2475.33	I
200		2476.67	II
150		2478.22	I
120		2481.86	II
100		2482.10	I
100		2482.58	II
100		2484.04	II
500		2484.95	I
120		2486.70	I
600		2488.70	II
500		2490.46	I
600		2504.45	I
600		2507.45	I
240		2512.65	I
1200	d	2526.35	I
600		2532.12	II
240		2545.49	II
240		2546.80	I
460	d	2551.07	I
460		2554.62	II
240		2555.05	I
1200		2559.43	I
460		2562.10	I
340		2571.51	II
430		2573.54	I
390		2573.79	I
600		2577.37	II
340		2577.78	I
210		2580.16	I
340		2584.03	II
430		2593.08	I
410		2593.66	II
310		2594.25	II
560		2595.26	I
310	l	2596.45	II
220		2600.14	I
600		2603.49	II
1400		2608.63	I
210		2609.00	I
310	d	2611.34	I
340		2615.46	I
310		2615.66	I
1200		2635.58	II
470		2636.67	I
860		2636.90	I
510		2646.22	I
600		2646.37	I
2400		2647.47	I
270		2651.22	II
2600		2653.27	I
1900		2656.61	I
1500		2661.34	I
220		2665.60	II
220		2668.07	I
600		2668.62	I
770		2675.90	II
270		2680.06	II
220		2680.66	II
600		2684.28	I
1500		2685.17	II
340		2691.31	I
260		2692.40	I
470		2694.52	II
240		2696.81	I
1000		2698.30	I
470		2706.69	I
310		2709.27	II
1200		2710.13	I
2600		2714.67	I
240		2717.18	I
470		2720.76	I
470		2727.44	II
410		2727.78	I
310		2736.25	II
210		2739.26	II
210		2743.59	I
510		2746.68	I
1200		2748.78	I
860		2749.83	I
410		2752.49	II
1000		2758.31	I
430		2761.68	II
770		2775.88	I
390		2787.69	I
680		2796.34	I
680		2797.76	II
380		2802.07	I
430		2806.30	I
510		2806.58	I
260		2817.10	II
260		2842.82	I
640		2844.25	I
290		2844.46	II
290	c	2845.35	I
560		2848.52	I
1500		2850.49	I
1900		2850.98	I
220		2858.44	II
360		2861.98	I
310		2868.65	I
470		2871.42	I
270		2873.36	I
260		2873.56	I
210		2874.17	I
380		2880.02	I
770		2891.84	I
260		2899.04	I
560		2902.05	I
210		2914.12	I
310		2915.49	I
410		2925.19	I
310		2932.70	I
1700		2933.55	I
470		2940.06	I
1200		2940.22	I
240		2942.14	I
510		2951.92	I
340		2953.56	I
1500		2963.32	I
770		2965.13	II
770		2965.54	I
340		2969.47	I
430		2975.56	I
210		3011.88	I
1800		3012.54	II
290	d	3027.48	I
290		3042.06	II
530		3049.56	I
530		3069.24	I
360		3077.24	I
560		3103.25	I
380		3124.97	I
380		3130.58	I
270		3132.64	I
320		3170.29	I
270		3173.59	I
200		3176.29	I
600		3180.95	I
240		3184.55	I
200		3198.67	I
200		3213.91	II
300		3223.83	I
230		3229.24	I
200		3242.05	I
200		3242.83	I
210		3274.95	II
1100		3311.16	I
210		3317.93	I
680		3318.84	I
330	d	3330.99	II
230		3358.47	I
640		3371.54	I
360		3385.05	I
230		3398.33	I
450		3406.94	I
230		3463.77	I
490		3480.52	I
380		3497.85	I
240		3503.87	I
490		3511.04	I
200		3513.61	I
750		3607.41	I
980		3626.62	I
500		3642.06	I
100		3686.18	I
100		3689.73	I
130		3731.02	I
140		3736.76	I
130		3746.36	I
110		3754.52	I
110		3777.10	I
110		3792.02	I
210		3833.74	II
100		3848.05	I
100		3885.20	I
210		3918.51	I
140		3922.78	I
140		3922.92	I
210		3970.10	I
210		3996.17	I
100		3999.28	I
190		4006.84	I
190		4026.94	I
140		4029.94	I
120		4040.87	I
410		4061.40	I
210		4064.63	I
100		4067.24	I
310		4067.91	I
120		4105.02	I
210		4129.38	I
230		4136.20	I
230		4147.89	I
210		4175.21	I
100		4177.92	I
130		4181.15	I
300		4205.88	I
120		4206.40	I
130		4245.35	I
130		4268.26	I
160	c	4302.98	I
110		4355.14	I
100		4378.82	I
150		4386.07	I
110		4398.45	I
180		4402.50	I
130		4415.74	I
360	c	4510.98	I
190		4530.85	I
130		4551.95	I
170		4565.85	I
340		4574.31	I
260		4619.51	I
130		4669.14	I
450		4681.88	I
130		4691.90	I
150		4740.16	I
220		4756.51	I
120		4768.98	I
220		4812.75	I
110		4920.11	I
100		4921.27	I
110		4926.00	I
150		4936.42	I
200		5037.37	I
100		5067.87	I
110		5115.84	I
100		5141.62	I
100		5143.69	I
330		5156.56	I
110		5212.74	I
110	d	5218.45	I
		5218.66	I
140		5341.05	I
200		5402.51	I
130		5419.19	I
18		5500.68	I
20		5505.66	I
15	c	5516.27	I
90		5518.91	I
9		5521.15	I
10		5523.98	I
13		5528.36	I
10	l	5545.20	I
20		5548.32	I
30		5584.02	I
15		5598.75	I
30		5599.52	I
9	c	5605.50	I
9	c	5617.71	I
40		5620.68	I
13		5628.20	I
20		5635.71	I
40		5640.18	I
150		5645.91	I
130		5664.90	I
30		5688.25	I
40		5699.24	I
15		5704.31	I
25		5706.28	I
30		5715.24	I
8		5716.53	I
23		5746.71	I
30		5755.81	I
15	h	5761.61	I
25		5766.56	I
30	c	5767.91	I
10		5771.93	I
130		5776.77	I
25		5780.02	I
90		5780.71	I
130		5811.10	I
25		5816.51	I
,5	c	5843.94	I
13		5849.68	I
15		5866.61	I
240		5877.36	I
130		5882.30	I
90		5901.91	I
30		5916.51	I
90		5918.95	I
15		5925.90	I
15		5930.62	I
23		5931.05	I
20		5931.68	I
18		5935.54	I
130		5939.76	I
240		5944.02	I
25		5951.78	I
18		5960.13	I
190	c	5997.23	I
25	h	6009.89	I
25		6015.90	I
100		6020.72	I
250		6045.39	I
100		6047.25	I
25		6053.70	I
30		6090.82	I
18		6092.06	I
100		6101.58	I
25		6140.07	I
65		6144.56	I
30		6152.54	I
130		6154.50	I
40		6158.84	I
15		6170.46	I
15		6189.66	I
15		6193.11	I
25		6208.37	I
40		6249.79	I
150		6256.68	I
150		6268.70	I
50		6278.34	I
65		6281.33	I
15		6287.36	I
40		6287.91	I
40		6289.34	I
50		6309.06	I
150		6309.58	I
25	c	6312.22	I
75		6325.08	I
50		6332.91	I
65		6341.17	I
30		6346.02	I
75		6356.16	I
65		6360.84	I
40		6373.06	I
15		6379.07	I
90		6389.45	I
23	h	6392.21	I
65		6428.60	I
250		6430.79	I
13		6437.36	I
40		6444.61	I
30		6445.87	I
200		6450.36	I
20		6455.83	I
30		6459.92	I
380		6485.37	I
18	h	6502.43	I
65		6505.52	I
100		6514.39	I
100		6516.10	I
25	h	6561.60	I
25	Cw	6564.26	I
100		6574.84	I
10		6585.13	I
15	c	6587.16	I
110		6611.95	I
75		6621.30	I
15	Cw	6662.24	I
100		6673.73	I
180		6675.53	I
30		6684.00	I
15		6693.61	I
15		6706.46	I
25		6709.39	I
10		6714.44	I
15	h	6723.61	I
75	c	6740.73	I
40		6754.91	I
13	c	6755.85	I
13		6770.37	I
75		6771.74	I
40	Cw	6774.25	I
40	Cw	6788.99	I
13	c	6790.06	I
13		6799.27	I
40	c	6810.46	I
160	c	6813.25	I
20		6819.36	I
18	c	6824.96	I
13		6832.00	I
15		6850.83	I
15		6865.13	I
210		6866.23	I
180		6875.27	I
40		6877.49	I
15		6896.77	I
40		6900.55	I
150		6902.10	I
140		6927.38	I

Tantalum (Cont.)

Intensity		Wavelength	
140		6928.54	I
8	h	6939.33	I
20	c	6946.87	I
65		6951.26	I
45		6953.88	I
180		6966.13	I
8		6969.49	I
8	c	6971.31	I
9		6971.53	I
23		6983.52	I
110	d	6995.22	I
		6995.49	I
20		7000.21	I
40		7005.07	I
75		7006.96	I
50		7025.03	I
13		7031.51	I
40		7039.07	I
15	h	7081.30	I
20		7085.40	I
23		7093.02	I
8		7108.05	I
15	c	7117.52	I
20		7121.27	I
40		7125.72	I
150		7148.63	I
110		7172.90	I
13	c	7174.91	I
13		7191.35	I
8		7233.45	I
30	h	7250.27	I
11	h	7264.82	I
6		7272.29	I
30		7276.96	I
5		7277.54	I
9		7286.36	I
13	c	7296.32	I
140		7301.74	I
20		7319.84	I
11	c	7322.72	I
13		7325.95	I
11	Cw	7340.19	I
160		7346.41	I
140	c	7352.86	I
100		7356.96	I
90	Cw	7369.09	I
160		7407.89	I
11	c	7435.19	I
23		7440.17	I
30		7467.75	I
23		7486.01	I
30		7520.56	I
6		7569.23	I
9		7590.22	I
6		7649.62	I
11	h	7722.02	I
11		7763.11	I
9	c	7779.67	I
20	c	7842.76	I
100		7882.37	I
30		7950.19	I
5		7952.07	I
6		7998.75	I
6		8022.09	I
75		8026.50	I
5		8029.04	I
15		8039.08	I
8		8053.93	I
15		8068.98	I
5		8100.11	I
13		8128.76	I
9		8158.54	I
5	c	8180.74	I
13	c	8248.95	I
20	d	8264.85	I
75		8281.62	I
11		8389.06	I
5		8415.73	I
25	Cw	8447.62	I
11	h	8550.49	I
15		8575.92	I
10	Cw	8595.84	I

Ta IV
Ref. 411 — R.L.K.

Intensity	Wavelength	
	Vacuum	
10	763.14	IV
32	934.41	IV
67	999.34	IV
65	1063.53	IV
68	1067.17	IV
71	1074.47	IV
71	1086.39	IV
72	1094.60	IV
68	1100.13	IV
79	1116.10	IV
71	1118.83	IV
78	1136.17	IV
66	1138.26	IV
75	1149.72	IV
75	1150.42	IV
75	1151.92	IV
76	1172.51	IV
85	1175.51	IV
80	1189.28	IV
78	1192.52	IV
80	1192.67	IV
78	1211.94	IV
80	1212.68	IV
85	1213.09	IV
70	1214.66	IV
85	1215.53	IV
80	1220.73	IV
80	1220.96	IV
90	1223.73	IV
88	1238.12	IV
95	1240.06	IV
87	1258.34	IV
94	1264.91	IV
98	1272.42	IV
94	1275.48	IV
86	1275.94	IV
92	1308.51	IV
85	1311.35	IV
87	1315.58	IV
81	1325.19	IV
92	1332.38	IV
86	1343.30	IV
75	1350.46	IV
92	1365.88	IV
79	1376.62	IV
78	1388.80	IV
91	1398.78	IV
93	1413.40	IV
79	1430.11	IV
83	1441.54	IV
91	1454.32	IV
92	1464.41	IV
93	1469.82	IV
90	1495.25	IV
95	1514.19	IV
70	1525.69	IV
82	1565.97	IV
82	1584.64	IV
84	1594.91	IV
85	1607.70	IV
84	1631.65	IV
79	1639.82	IV
84	1668.76	IV
84	1676.45	IV
85	1712.16	IV
85	1716.13	IV
82	1753.90	IV
82	1759.04	IV
79	1763.03	IV
83	1865.92	IV
84	1901.63	IV
84	1907.66	IV
84	1924.75	IV
77	1940.25	IV
79	1985.68	IV
82	1989.44	IV
	Air	
85	2055.75	IV
83	2079.01	IV
75	2111.53	IV
90	2199.58	IV
90	2207.64	IV
68	2697.42	IV
10	3076.06	IV

Ta V
Ref. 426 — R.L.K.

Intensity	Wavelength	
	Vacuum	
20	478.29	V
60	493.07	V
200	841.31	V
1000	890.87	V
500	947.30	V
100	990.29	V
200	1066.64	V
200	1140.49	V
500	1213.42	V
100	1242.98	V
5000	1392.56	V
7000	1709.10	V

TECHNETIUM (Tc)
Z = 43

Tc I and II
Ref. 35 — C.H.C.

Intensity		Wavelength	
		Air	
15		2106.23	II
20		2116.44	II
15		2119.41	I
30		2156.27	I
30		2185.39	I
30		2189.06	I
40		2193.35	I
10		2266.22	II
10		2282.12	I
10		2282.71	I
50		2285.45	I
100		2298.08	I
30		2416.22	I
50		2423.23	I
20		2424.54	I
20		2435.83	I
10		2436.99	I
80	w	2463.69	I
20		2465.09	I
30		2466.87	I
20		2475.11	I
50		2480.70	I
50		2483.22	I
20		2486.50	I
25		2492.72	I
20		2493.43	I
100		2496.77	II
30		2510.17	I
80		2529.34	II
500		2543.23	II
60		2544.81	II
50		2547.92	II
50		2558.61	II
50		2567.01	II
30		2575.06	II
80		2576.28	II
40		2577.86	II
300	Hl	2578.79	I
200		2589.86	I
20	w	2590.19	I
100		2592.82	I
20		2597.19	II
500		2608.86	I
1000	c	2609.99	II
1500		2614.23	I
1000		2615.87	I
30		2618.28	I
200		2634.91	II
80		2636.36	I
30		2641.26	I
100		2642.37	I
40		2644.50	II
1000	c	2647.01	II
300	c	2649.21	I
100		2652.35	II
30		2653.57	I
100		2654.31	I
120		2660.88	I
100		2662.30	I
80		2681.19	II
60		2683.14	I
80		2683.89	I
80		2693.11	I
50		2696.64	I
70		2702.27	I
40		2702.96	II
100		2707.90	II
1000		2708.78	I
30		2715.20	I
30		2723.55	I
1000		2726.69	I
30	c	2728.47	I
500		2730.53	I
300		2732.87	I
150		2736.23	I
60	c	2736.83	II
100		2737.97	I
20	c	2738.83	II
100		2755.76	I
100		2762.13	I
200		2762.34	I
60		2765.95	I
500		2766.89	I
20		2777.31	II
150		2778.91	I
25		2781.22	I
1000		2782.05	I
500		2785.59	I
40		2788.89	I
500		2789.25	I
100		2794.53	I
80		2795.65	I
200		2795.78	II
1000		2802.81	I
150		2803.02	I
500		2808.36	I
50	c	2809.65	II
500		2811.61	II
30		2814.86	I
40		2819.46	I
100		2821.35	II
200		2828.04	I
60		2831.18	II
50		2840.38	II
60		2845.04	I
10		2846.39	II
60		2849.20	I
150		2850.96	I
500	h	2857.13	I
2000	c	2859.11	I
500		2864.49	I
100		2868.09	I
1000		2887.73	I
100		2888.46	I
30		2889.20	II
200		2893.16	I
150		2893.45	I
200		2894.32	I
1000		2896.34	I
40		2903.81	I
1000		2913.15	I
500		2921.91	I
20	c	2923.34	II
1000		2928.20	I
80		2933.89	I
200		2955.93	I
200		2973.65	I
100		2979.34	I
150		2985.36	I
100		3010.83	I
300		3017.23	I
150		3021.56	I
100		3022.66	I
200		3023.68	I
80		3025.26	I
300	w	3026.89	I
50		3033.16	I
80		3034.57	I
40		3036.88	I
20		3037.90	II
100		3038.23	I
40		3042.64	I
100	h	3051.55	I
40		3052.47	I
80		3062.11	I
200		3062.36	I
300		3064.67	I
100	c	3066.60	I
120	c	3068.34	I
80		3076.24	I
150		3089.34	I
1000		3099.10	I
200		3099.52	I
60		3108.25	I
40		3109.15	I
60		3115.98	I
80		3119.17	I
40		3119.66	I
700		3122.64	I
1500		3131.23	I
40	c	3150.26	I
300		3161.67	I
3000		3173.30	I
200		3180.30	I
2000		3182.37	I
2000		3183.11	I
800	c	3195.20	II
40	w	3197.53	I
300	c	3202.83	I
1000		3212.02	II
40		3220.74	I
60		3230.02	I
1000		3237.02	II
100		3241.84	I
500		3244.19	I
300		3252.05	I
40		3256.10	I
40		3261.94	I
100		3287.14	I
30		3298.84	II
100		3300.77	I
80		3305.89	I
200		3310.65	I
150		3313.65	I
200		3325.55	I
150	c	3327.10	I
100		3330.77	I
50		3332.47	I
60		3350.56	I
50	c	3350.83	I
400		3366.75	I
40		3386.67	I
50		3392.23	I
300		3394.18	I
60		3396.90	I
40		3397.83	I
300		3398.33	I
200		3402.10	I
200	c	3403.93	I
80		3405.33	I
80		3407.28	I
50		3408.33	I
40		3411.80	I
40	c	3418.20	I
100		3419.10	I

Technetium (Cont.)

Intensity		Wavelength	Spectrum
60		3427.85	I
40		3431.75	I
200		3434.70	I
40		3435.68	I
150		3437.44	I
80		3438.73	I
200	c	3443.47	I
200		3451.05	I
200		3456.85	I
400		3457.24	I
40		3457.60	I
5000	c	3466.28	I
150		3470.51	I
80		3475.18	I
1000		3475.59	I
60		3484.62	I
1000	c	3486.23	I
100		3490.30	I
400		3493.39	I
500		3494.62	I
40		3499.14	I
1000		3500.70	I
200		3501.24	I
800	c	3502.70	I
100	c	3507.19	I
100		3508.27	I
100		3510.91	I
800		3525.83	I
300		3526.18	I
100		3529.83	I
150		3534.88	I
500		3535.51	I
300		3538.12	I
800		3538.68	I
2000	c	3541.77	I
6000	c	3549.72	I
4000	c	3550.64	I
300		3559.75	I
800		3560.32	I
100		3565.22	I
800		3568.85	I
100		3570.65	I
100		3575.42	I
1000		3580.06	I
600		3581.26	I
800		3582.08	I
2000		3582.63	I
4000		3587.94	I
200		3593.47	I
300		3594.57	I
1000	c	3595.66	I
1000	c	3607.32	I
200		3607.62	I
2000	c	3608.27	I
200		3618.94	I
1000	c	3627.36	I
200		3630.39	I
3000	c	3635.15	I
10000	c	3636.07	I
1000		3638.22	I
200		3638.85	I
900		3639.38	I
400		3640.23	I
1000	c	3648.04	I
600		3651.47	I
1000	c	3658.59	I
400	c	3661.45	I
200		3664.92	I
1000		3679.15	I
300		3680.32	I
5000		3684.74	I
300		3692.76	I
800		3703.83	I
300		3704.80	I
200		3706.70	I
200		3707.63	I
200		3708.26	I
1000		3712.26	I
300		3712.82	I
500		3715.94	I
10000		3718.86	I
1500		3723.67	I
2000		3724.40	I
5000		3726.35	I
200		3727.36	I
400	c	3729.18	I
500		3731.74	I
300		3737.42	I
400		3745.01	I
1000		3746.15	I
5000		3746.84	I
1000		3752.13	I
4000		3754.37	I
1000		3758.54	I
2000		3761.81	I
5000		3768.77	I
3000		3771.03	I
500		3777.27	I
2000		3779.37	I
3000	c	3780.68	I
500		3784.06	I
200		3786.06	I
500		3791.28	I
300		3791.73	I
200		3797.44	I
1000		3797.77	I
200		3814.67	I
300		3816.89	I
300		3824.47	I
500		3828.54	I
200		3830.35	I
200		3832.45	I
600		3832.82	I
1500		3837.56	I
800		3841.31	I
800		3845.97	I
500		3847.60	I
300		3851.22	I
500	c	3856.73	I
200		3863.07	I
400		3864.11	I
1000		3868.24	I
200		3875.66	I
500	c	3879.16	I
600	c	3880.72	I
300	w	3892.12	I
200		3893.22	I
600		3899.83	I
300		3919.38	I
300	c	3923.66	I
200		3927.57	I
200		3933.70	I
4000	c	3946.57	I
2000		3947.09	I
200		3955.73	I
300		3979.64	I
500		3980.35	I
10000	c	3984.97	I
400		3987.78	I
300		3994.04	I
2000		3994.51	I
200		3996.97	I
300		4004.69	I
500		4007.14	I
1000		4012.00	I
400		4016.68	I
600		4017.22	I
2000		4020.76	I
20000	c	4031.63	I
1000		4039.25	I
200		4041.78	I
10000	c	4049.11	I
500		4051.95	I
200		4053.18	I
200	c	4056.08	I
400		4083.54	I
10000		4088.71	I
200		4093.69	I
15000		4095.67	I
1000		4110.22	I
10000		4115.08	I
600		4119.27	I
8000		4124.22	I
1000		4128.27	I
300		4134.81	I
300		4139.12	I
800		4139.85	I
400		4141.27	I
6000		4144.95	I
3000		4145.08	I
200		4147.62	I
10000		4165.61	I
500		4167.42	I
1000		4169.68	I
4000		4170.27	I
5000		4172.53	I
1000		4176.28	I
800		4186.51	I
300		4218.61	I
10000	c	4238.19	I
20000		4262.27	I
1000		4262.69	I
800		4274.97	II
800		4278.90	I
30000		4297.06	I
400	c	4336.86	I
400		4358.49	I
200		4359.26	I
1000		4429.59	I
1000		4481.53	I
3000		4487.06	I
400		4495.03	I
1000		4515.98	I
10000		4522.84	I
2000		4539.53	I
400		4542.09	I
400		4552.20	I
800		4552.85	I
1000		4557.05	I
2000		4564.54	I
1000		4578.45	I
1000		4593.35	I
300	c	4609.16	I
1000		4616.86	I
200	c	4622.69	I
300		4624.96	I
1000		4630.57	I
200		4633.15	I
3000		4637.50	I
500		4643.28	I
2000		4648.33	I
2000	c	4660.21	I
2000		4669.30	I
400		4672.17	I
200		4678.90	I
400		4689.36	I
300		4694.28	I
1000		4706.92	I
200		4714.22	I
2000		4717.77	I
500	c	4719.02	I
4000	c	4719.28	I
200	c	4736.51	I
10000		4740.61	I
500		4749.61	I
1000		4752.72	I
200		4762.36	I
4000		4771.54	I
200		4773.89	I
200		4783.92	I
500		4785.60	I
200	c	4790.48	I
250		4791.62	I
300		4799.98	I
100		4805.69	I
100		4809.42	I
500		4816.79	I
10000		4820.74	I
300		4831.35	I
1000		4834.37	I
1000		4835.39	I
100		4841.36	I
20000		4853.59	I
100		4857.21	I
100	c	4862.19	I
10000		4866.73	I
200		4870.77	I
100		4888.70	I
150	c	4890.88	I
8000		4891.92	I
150		4892.49	I
1000		4908.51	I
2000		4909.57	I
500		4913.02	I
150	c	4914.70	I
200	c	4920.67	I
300		4923.60	I
400		4948.06	I
5000		4976.34	I
400		4995.00	I
200		5002.67	I
100		5005.74	I
200	c	5014.52	I
500		5026.24	I
300		5026.79	I
150		5027.89	I
80		5032.45	I
300		5055.27	I
60		5058.33	I
500		5060.69	I
80		5090.74	I
5000		5096.28	I
200	c	5103.24	I
500		5104.32	I
200		5109.81	I
100		5120.60	I
500		5139.26	I
500		5150.63	I
2000		5161.81	I
2000		5174.81	I
100		5206.56	I
200		5225.55	I
200	c	5260.22	I
200	c	5261.44	I
1000		5275.51	I
800		5285.07	I
100		5305.31	I
400		5314.96	I
600		5320.20	I
200		5334.79	I
500	c	5353.48	I
200		5356.63	I
300		5358.65	I
200		5360.14	I
500	h	5375.20	I
150	c	5423.05	I
200		5447.40	I
500	c	5451.90	I
100		5455.95	I
300		5471.96	I
70		5483.01	I
60		5485.37	I
80	c	5506.89	I
150		5524.11	I
100		5528.23	I
200		5541.94	I
80		5543.63	I
100		5550.53	I
3000	c	5589.02	I
200		5602.23	I
2000	c	5620.45	I
300		5629.94	I
1500		5642.13	I
800		5644.94	I
100		5656.00	I
60		5672.15	I
200		5687.30	I
200		5689.05	I
700		5725.31	I
500	c	5771.47	I
100		5794.65	I
80	c	5799.85	I
100	c	5814.24	I
200		5831.48	I
150	c	5836.33	I
150		5923.36	I
1000	c	5924.47	I
200		5926.29	I
600	c	5931.93	I
60		6032.36	I
60		6047.99	I
200		6065.09	I
800		6085.23	I
300		6099.39	I
500	c	6102.96	I
1000		6120.68	I
1000		6130.80	I
150	c	6132.23	I
100		6184.70	I
800		6192.66	I
600	c	6244.18	I
100		6312.18	I
100		6354.86	I
100		6356.73	I
80		6389.87	I
100		6408.83	I
1000		6455.90	I
600	c	6461.93	I
100		6470.27	I
200	c	6491.68	I
200		6526.82	I
150		6579.24	I
500	c	6625.57	I
300	c	6673.66	I
100		6687.10	I
80		6786.00	I
70		6798.63	I
60		6856.90	I
150		7002.37	I
100		7016.57	I
500	c	7086.18	I
60		7093.12	I
200		7141.28	I
200	c	7157.62	I
70		7256.08	I
100		7322.38	I
80		7329.14	I
100		7396.80	I
100	c	7402.61	I
200		7405.36	I
60		7427.15	I
150		7434.12	I
600		7452.49	I
60		7461.59	I
80		7534.95	I
800		7540.26	I
80		7543.39	I
200		7574.02	I
500		7579.26	I
90		7624.53	I
100		7684.45	I
500		7697.37	I
80	c	7698.19	I
800	c	7793.04	I
60		7798.28	I
60		7816.74	I
800		7817.72	I
100		7856.38	I
200		7861.44	I
400	d	7871.25	I
60	c	7874.76	I
70		7965.45	I
500		7999.73	I
200		8126.55	I
200		8170.55	I
150		8205.27	I
100		8206.49	I
150		8211.31	I
500	c	8237.08	I
200		8308.15	I
200		8309.16	I
60		8315.50	I
100		8531.06	I
100		8543.61	I
100	c	8707.21	I
100	c	8737.93	I
200	c	8829.82	I

TELLURIUM (Te)
Z = 52

Te I and II
Ref. 1, 344—347—C.H.C.

Tellurium (Cont.)

Intensity		Wavelength	
		Vacuum	
6		799.60	II
8		802.28	II
6		942.62	II
6		1003.73	II
6		1007.80	II
5		1014.27	II
5		1022.79	II
6		1057.00	II
8		1059.51	II
6		1068.86	II
8		1077.66	II
6		1090.11	II
6		1144.04	II
5		1153.10	II
10		1161.42	II
10		1174.34	II
12		1175.79	II
9		1208.54	II
5		1213.00	II
9		1220.98	II
9		1253.62	II
9		1270.52	II
7		1274.76	II
8		1306.53	II
10		1324.92	II
7		1336.42	II
7		1345.20	II
9		1363.24	II
8		1366.73	II
10		1374.80	II
6		1395.22	II
6		1439.52	II
6		1465.25	II
7		1489.56	II
8		1607.99	II
10		1608.41	II
10		1613.15	II
6		1638.91	II
5		1655.4	I
5		1688.5	I
6		1700.0	I
6		1701.58	II
5		1708.0	I
6		1751.0	I
5		1759.4	I
5		1775.0	I
6		1795.7	I
6		1796.3	I
10		1822.4	I
6		1825.5	I
6		1850.6	I
6		1852.1	I
6		1853.8	I
8		1857.2	I
6		1860.4	I
3		1962.88	II
7		1994.83	I
		Air	
6		2000.2	I
26000		2002.02	I
8		2070.9	I
6500		2081.16	I
18000		2142.81	I
3200		2147.25	I
360		2159.85	I
9		2208.74	I
10		2255.49	I
500		2259.02	I
10		2265.52	I
20		2373.06	II
1200		2383.26	I
1500		2385.78	I
20		2387.82	II
10		2401.63	II
10		2436.47	II
50		2438.69	II
120		2530.72	I
20		2567.82	II
10		2574.96	II
5		2576.10	II
7		2579.24	II
10		2591.12	II
10		2592.85	II
10		2605.72	II
5		2621.92	II
10		2624.86	II
20		2627.96	II
20		2641.89	II
20		2648.48	II
100		2649.66	II
40		2657.70	II
80		2661.10	II
110		2677.13	I
20		2711.58	II
6		2769.65	I
10		2841.17	II
10		2846.15	II
100		2858.29	II
20		2861.00	II
40		2868.82	II
150		2895.41	II
30		2919.89	II
50		2942.11	II
50		2946.68	II
70		2967.29	II
20		2973.67	II
50		2975.90	II
15		2997.04	II
15		3012.02	II
50		3017.58	II
20		3023.31	II
70		3047.00	II
20		3052.46	II
10		3063.16	II
15		3073.56	II
8		3104.44	II
10		3132.58	II
20		3160.66	II
100		3175.14	I
10		3189.83	II
5		3211.21	II
60		3256.80	II
30		3268.77	II
30		3282.63	II
40		3321.92	II
40		3323.11	II
60		3329.22	II
60		3352.10	II
60		3362.79	II
25		3374.10	II
150		3406.79	II
20		3419.63	II
50		3442.25	II
40		3455.12	II
20		3456.88	II
20		3480.32	II
40		3483.67	II
20		3486.11	II
50		3521.11	II
50		3552.19	II
100		3611.78	II
50		3617.57	II
40		3644.46	II
20		3679.26	II
30		3725.66	II
40		3797.22	II
20		3800.92	II
20		3905.67	II
20		3918.54	II
30		3931.49	II
20		3947.98	II
40		3969.22	II
25		3975.94	II
20		3981.77	II
50		4006.52	II
20		4011.69	II
30		4029.73	II
40		4047.17	II
30		4048.88	II
15		4073.48	II
30		4101.04	II
70		4127.32	II
30		4163.55	II
100		4169.77	II
30		4179.29	II
25		4211.31	II
80		4225.73	II
30		4246.47	II
20		4251.15	II
100		4261.11	II
30		4264.36	II
60		4273.43	II
80		4285.85	II
40		4320.90	II
30		4361.28	II
150		4364.00	II
30		4377.12	II
75		4385.10	II
60		4396.00	II
170		4478.63	II
80		4537.07	II
100		4557.78	II
70		4630.62	II
100		4641.12	II
180		4654.37	II
200		4686.91	II
100		4696.38	II
100		4706.53	II
100		4766.05	II
70		4771.56	II
100		4784.87	II
100		4827.14	II
150		4831.28	II
150		4842.90	II
130		4865.12	II
200		4866.24	II
80		4885.22	II
80		4904.44	II
60		4961.88	II
60		5000.82	II
8		5083.0	I
7		5148.7	I
50		5449.84	II
50		5487.95	II
150		5576.35	II
150		5649.26	II
100		5666.20	II
200		5708.12	II
7		5733.5	I
150		5755.85	II
8		5789.1	I
50		5936.15	II
100		5974.68	II
8	d	6273.5	I
8	h	6349.7	I
50		6367.13	II
8		6405.9	I
7	h	6456.7	I
8	h	6613.4	I
10		6648.58	II
8		6660.2	I
8	h	6690.0	I
10	h	6790.0	I
20	h	6837.6	I
20	h	6854.7	I
10		7016.06	II
10		7039.13	II
15	h	7191.1	I
10		7236.62	II
20	h	7263.5	I
8		7280.9	I
10		7289.26	II
10		7445.39	II
12		7460.98	II
15		7468.75	II
10		7481.26	II
10		7556.8	I
6		7688.61	II
15		7759.1	I
8		7818.79	II
8		7861.61	II
15		7921.69	II
15		7943.14	II
10		7950.34	II
20		7972.9	I
6		8056.15	II
30	h	8061.4	I
10	h	8082.5	I
10		8122.44	II
8		8130.39	II
8		8154.47	II
20		8186.44	II
6		8190.94	II
10		8251.5	I
15		8273.53	II
10		8276.6	I
10		8291.1	I
15		8355.8	I
10		8372.12	II
7		8469.8	I
8		8492.2	I
8		8500.8	I
12		8521.4	I
10		8535.68	II
12		8575.78	II
10		8604.63	II
8		8621.68	II
7	h	8632.1	I
15		8672.95	II
12		8701.09	I
10		8733.81	II
205		8758.18	I
12		8831.52	I
18		8851.15	I
6		8897.92	II
81		9004.37	I
18		9043.39	I
12		9196.80	I
15		9206.78	I
17		9207.64	I
12		9469.00	I
5660		9722.74	I
185		9785.54	I
109		9842.30	I
532		9868.92	I
118		9902.61	I
689		9956.30	I
37		9959.93	I
325		9977.13	I
136		9979.31	I
45		9985.85	I
5950		10051.41	I
4097		10091.01	I
104		10099.57	I
279		10106.05	I
381		10118.08	I
296		10151.06	I
397		10300.56	I
205		10323.05	I
745		10493.57	I
197		10509.86	I
1880		10918.34	I
298		11007.80	I
10200		11089.56	I
508		11163.74	I
6620		11487.23	I
280		11978.96	I
188		12566.24	I
389		12589.19	I
161		12805.50	I
400		13104.18	I
1580		13247.75	I
483		13316.63	I
217		14037.09	I
144		14072.53	I
434		14335.74	I
220		14417.46	I
1050		14513.51	I
129		14554.68	I
1480		15452.45	I
2430		15546.23	I
3760		16403.90	I
1960		17303.54	I
2780		18291.59	I
394		18777.30	I
269		19623.52	I
239		20147.54	I
1020		21043.73	I
464		21602.50	I
37		21799.64	I
74		22555.29	I
48		22755.66	I
27		23294.94	I
17		23978.70	I
25		24059.04	I
13		26428.62	I
38		26539.17	I
15		26553.74	I
7		27179.26	I

TERBIUM (Tb)
Z = 65

Tb I and II
Ref. 1 — C.H.C.

Intensity	Wavelength	
	Air	
29	2577.73	II
110	2584.61	II
29	2590.31	II
29	2591.42	II
24	2592.64	II
55	2597.71	II
40	2602.93	II
110	2608.57	II
40	2616.90	
130	2628.69	II
55	2655.96	II
50	2661.40	II
24	2661.64	II
55	2667.64	II
50	2668.86	II
140	2669.29	II
40	2674.13	II
40	2674.69	II
29	2678.15	II
40	2683.97	II
35	2687.82	II
50	2691.90	II
35	2693.05	II
55	2693.41	II
35	2695.46	II
50	2696.83	II
190	2704.07	II
130	2736.24	II
160	2759.47	II
270	2769.53	II
130	2784.49	II
180	2800.51	II
250	2802.75	II
250	2809.30	II
180	2812.64	II
190	2852.14	II
110	2857.68	II
230	2886.29	II
160	2894.45	II
320	2897.44	II
160	2898.86	II
110	2901.54	II
110	2910.30	II
160	2914.75	II
160	2915.30	II
190	2915.60	II
120	2916.24	II
120	2918.89	II
120	2924.16	II
120	2924.53	II
160	2932.89	II
150	2940.05	II
250	2956.21	II
170	2968.87	II
170	2977.78	II
110	2987.03	II
110	2988.57	II

Terbium (Cont.)

Intensity		Wavelength	Spectrum
130		2996.00	II
110		2999.03	II
130		3005.52	II
170		3009.30	II
230		3010.59	II
230		3016.18	II
130		3019.17	II
170		3020.29	II
110		3023.43	II
170		3027.33	II
230		3031.60	II
230		3044.96	II
190		3051.13	II
130		3053.24	II
460		3053.55	II
130		3062.78	II
230		3064.09	II
110		3065.69	II
230		3067.20	II
270		3069.03	II
460		3070.05	II
270		3072.60	II
670		3078.86	II
480		3082.36	II
120		3086.78	II
250		3088.43	II
480		3089.58	II
230		3102.54	II
480		3102.96	II
290		3117.89	II
290		3119.62	II
230		3121.94	II
230		3123.05	II
160		3124.54	II
110		3131.35	II
250		3134.26	II
440		3139.64	II
190		3140.06	II
230		3145.22	II
150		3146.67	II
310		3147.04	II
310		3147.15	II
310		3148.71	II
120		3155.62	II
130		3162.42	II
290		3162.93	II
190		3165.74	II
380		3167.52	II
140		3168.32	II
230		3169.84	II
190		3173.76	II
380		3174.66	II
380		3180.54	II
140		3183.88	II
480		3187.26	II
290		3188.03	II
190		3194.69	II
380		3195.60	II
480		3199.56	II
1100		3218.93	II
1200		3219.98	II
250		3230.03	II
250		3231.06	II
210	d	3239.60	II
250		3240.00	II
480		3252.32	II
250		3262.97	II
230	d	3263.87	II
230		3264.90	II
400		3266.40	II
250		3274.14	II
250		3274.33	II
210		3277.32	II
760		3280.31	II
760		3281.40	II
520		3283.10	II
1000		3285.04	II
310		3287.55	II
310		3291.56	II
1500		3293.07	II
210		3295.33	II
310		3298.66	II
210		3304.95	II
420	d	3307.44	II
210		3308.51	II
210		3314.38	II
340	d	3321.15	II
420		3322.28	II
210		3323.38	II
210		3323.89	II
3800		3324.40	II
520		3329.08	II
210		3334.48	II
250		3336.70	II
310		3338.03	II
250		3339.00	II
210		3347.27	II
210		3348.07	II
760		3349.42	II
320		3362.25	II
760		3364.93	II
230	d	3370.61	II
320		3371.50	II
520		3372.36	II
460	d	3372.72	II
520		3375.03	II
320		3378.73	II
520		3378.86	II
320		3382.80	II
210		3390.60	II
380		3391.28	II
270		3398.35	II
320		3399.10	II
270		3400.53	II
210	d	3400.86	II
420		3402.33	II
210		3410.40	II
210		3410.68	II
520		3413.76	II
270		3416.24	II
400	d	3420.34	II
210		3430.61	II
320		3433.26	II
270		3439.72	II
520		3440.37	II
320		3444.58	II
210		3446.40	II
270		3449.46	II
810		3454.06	II
380		3460.38	II
230	d	3462.97	II
620		3468.03	II
270		3471.73	II
270		3472.37	II
810	d	3472.79	II
210		3473.00	II
380		3480.17	II
230		3483.04	II
230		3483.69	II
290	d	3489.51	II
210	d	3492.00	II
270		3494.21	II
270		3495.36	II
810		3500.84	II
570		3507.45	II
5700		3509.17	II
380		3510.10	II
320		3513.10	II
570		3519.76	II
1300		3523.66	II
380		3525.14	II
440		3525.61	II
440		3536.32	II
570		3537.94	II
1100		3540.24	II
810		3543.89	II
310	d	3551.03	
320		3551.96	II
460	d	3558.77	
3200		3561.74	II
480		3562.90	II
570		3565.74	
810		3567.35	II
4200		3568.52	II
1600		3568.98	II
320		3572.07	II
1100		3579.20	II
710		3585.03	II
570		3587.44	II
810		3596.38	II
440		3598.06	II
1600		3600.44	II
320		3604.90	II
320		3611.33	II
320		3614.63	II
320	d	3615.66	II
320		3616.58	II
380		3617.86	II
380		3619.73	II
810		3625.54	II
570		3626.50	II
380		3629.44	II
670		3633.29	II
670		3638.46	II
670		3641.66	II
440		3647.06	II
570		3647.75	II
2300		3650.40	II
810		3654.88	II
2000		3658.88	II
450		3663.12	II
3800		3676.35	II
300		3677.89	II
810		3682.26	II
320		3688.15	II
610		3691.15	II
300		3692.95	II
450		3693.58	I
320		3696.85	II
450		3700.12	I
4700		3702.86	II
300		3703.12	I
2400		3703.92	II
370		3709.30	II
1000	d	3711.76	II
300		3719.45	II
650		3729.91	II
430		3732.39	II
430		3743.09	
650		3745.04	I
870		3747.17	II
870		3747.34	II
1100		3755.24	II
430		3757.44	II
430	d	3757.90	II
650		3759.35	I
350		3761.14	I
1700		3765.14	I
2100		3776.49	II
330		3779.22	
600		3783.53	I
410	d	3787.22	II
410		3789.92	I
390		3792.20	I
600		3793.55	
330		3801.80	II
760	d	3806.85	II
1500		3830.26	I
540		3833.42	I
920	d	3842.50	II
370	d	3845.61	II
3700		3848.73	II
450	d	3869.75	II
3500	w	3874.17	II
330		3883.34	I
480		3888.22	I
490		3894.64	I
330		3895.99	I
330		3896.58	II
330		3897.89	I
2400		3899.20	II
1600		3901.33	I
480		3908.06	I
380		3909.14	I
330		3909.55	I
650		3915.43	I
480		3919.52	II
300		3922.10	II
480		3922.74	II
760		3925.45	II
650		3935.24	II
810	d	3939.52	II
650		3946.89	II
350	d	3958.36	II
2200	d	3976.84	II
1800		3981.87	II
300		3983.85	II
350		3999.40	II
350	d	4002.19	II
970		4002.59	II
1900		4005.47	II
300		4010.04	I
760		4012.75	II
330		4013.26	I
370		4019.14	II
540		4020.47	II
220		4022.88	I
370		4024.77	I
520		4031.66	II
870		4032.28	I
2100		4033.03	II
350		4036.22	I
210		4038.86	I
300		4051.86	II
300		4052.87	II
430		4054.12	I
410		4060.37	I
220		4060.87	II
1300		4061.58	I
220		4063.89	II
390		4066.22	II
260		4075.22	I
390		4081.24	I
210		4086.60	I
210		4092.19	I
260		4094.37	II
260		4094.49	I
260		4103.90	II
650		4105.37	I
300		4112.50	I
260		4119.92	I
280		4143.51	I
1100		4144.41	II
350		4158.53	I
240		4169.09	I
240		4169.32	I
240		4171.05	I
240		4172.60	I
240		4172.82	I
260		4173.47	I
240		4186.21	I
300		4187.16	I
390		4196.74	I
450		4201.00	II
650		4203.74	I
600		4206.49	I
300	Cw	4213.50	I
300		4214.42	II
480		4215.09	I
300		4217.56	I
260		4219.16	I
260		4224.28	I
480	Cw	4226.45	II
260		4231.89	I
480		4232.82	I
300		4235.35	I
370		4255.24	I
480		4258.23	II
260		4263.66	I
650		4266.34	I
330		4269.69	I
220		4275.21	I
760	Cw	4278.52	II
300		4285.13	II
300		4289.70	I
370		4298.36	I
300		4299.90	I
240		4302.95	I
240		4307.18	I
450		4310.42	I
300		4311.56	I
370		4313.25	I
2200		4318.83	I
600		4322.23	I
600		4325.83	II
3000		4326.43	I
240		4328.90	I
600		4332.12	I
870		4336.43	I
600		4337.64	I
1700		4338.41	I
700		4340.62	I
430	Cw	4342.53	I
430	d	4353.20	II
280		4356.09	I
870		4356.81	I
280		4360.16	I
220		4367.30	II
220		4372.02	I
330		4382.45	I
300		4388.23	I
260		4390.91	I
200		4416.27	II
140		4420.19	I
350		4423.10	I
110		4432.72	I
240		4436.12	I
110		4439.38	I
240		4448.04	I
110		4467.69	I
430		4493.07	I
45	d	4509.04	II
150	h	4511.52	I
45		4512.96	II
75		4514.31	II
45		4519.72	II
45		4525.01	II
45		4529.76	
45	h	4531.83	II
45		4534.13	I
45		4537.14	I
45		4537.23	I
110		4549.07	I
45		4549.72	II
110		4550.45	I
110		4556.46	I
55		4562.24	II
110		4563.69	II
30		4564.85	II
55		4573.19	II
210		4578.69	II
65		4584.84	II
65		4591.56	II
45		4592.38	I
45	h	4604.10	II
30		4611.96	I
45	h	4615.92	II
27		4617.49	I
30		4619.36	II
75	d	4626.32	II
95		4626.94	II
65		4632.07	I
65	h	4636.59	I
30		4636.99	II
85		4641.00	II
210		4641.98	II
260	Cw	4645.31	II
80		4647.23	I
60		4658.38	I
20		4658.73	
80		4662.79	I
50	c	4665.45	I
40		4669.40	I
80		4676.90	I
70	c	4681.87	I
50		4682.52	I
25	c	4682.79	II
80		4688.63	II
80		4693.11	II
30	h	4693.39	II
200		4702.41	II
110		4707.94	II
40	w	4716.07	II
40		4728.16	II
60	Cw	4734.20	I
80		4739.93	I

Terbium (Cont.)

Intensity		Wavelength	
70		4747.80	I
410	Cw	4752.53	II
40		4758.44	II
40		4760.19	II
30		4762.37	II
25		4764.47	II
35		4778.36	II
35		4778.80	II
180		4786.78	I
40	Cw	4789.91	II
30		4801.87	II
100		4813.77	I
60		4837.59	II
25		4840.39	I
30	c	4842.69	II
30		4844.89	II
30		4854.81	I
20		4856.54	II
30		4858.87	II
80		4875.57	II
25		4876.12	II
80		4881.15	II
29		4894.33	
95		4915.90	I
35		4924.09	I
35		4926.83	I
50		4928.93	I
65		4931.79	I
29		4970.99	II
29		4971.42	I
29		4973.04	I
29		4980.16	II
29		4980.56	I
85		4993.82	II
50		4995.84	II
55		4997.95	I
29		5006.10	II
50		5022.16	I
29		5024.24	II
29		5024.65	I
50		5033.12	I
50	w	5042.06	II
55		5054.30	I
55		5065.79	I
110		5078.25	I
24		5080.05	II
24		5081.11	I
75		5089.12	II
24		5089.66	I
24		5101.09	I
24		5108.56	I
35		5118.39	I
24		5120.18	I
50	w	5131.69	I
50	w	5141.08	II
50		5147.58	I
24		5164.27	I
29		5170.13	I
24		5170.61	I
50		5176.51	I
50		5179.97	I
50		5184.59	I
85		5186.13	I
50		5188.48	I
50		5198.86	I
35	w	5202.77	I
40		5204.55	I
40		5207.97	I
40		5214.28	I
40		5221.99	I
120		5228.12	I
40		5235.11	I
75		5248.71	I
75	w	5262.11	II
24		5275.03	I
75		5281.05	I
65		5304.72	I
29		5308.19	I
29		5309.46	I
110		5319.23	I
35		5331.04	I
65	w	5337.90	I
35	d	5338.59	I
24		5347.83	II
160		5354.88	I
75		5369.72	I
75		5375.98	I
29	d	5402.06	II
29		5413.65	I
29		5416.20	I
50		5424.10	II
29	c	5426.43	I
35		5443.38	I
29		5457.00	I
55		5459.81	I
29	w	5470.34	II
24		5481.45	I
55		5509.61	I
50		5514.54	I
65		5524.12	I
24	c	5525.62	I
35		5565.93	I
29	c	5638.80	I
29	c	5685.74	II
40	c	5686.48	I
85	c	5747.58	I
24		5762.66	I
24		5785.18	II
75		5795.64	I
75		5803.13	II
65		5815.36	I
29		5842.97	I
65		5851.07	I
65		5870.62	I
35		5898.84	I
24		5902.40	I
35		5904.71	I
65	c	5920.78	I
50	c	5939.38	I
35		5940.17	I
24		5951.17	I
75		5967.34	II
29		6038.97	I
29		6039.38	I
24		6104.29	II
24	c	6292.43	I
35		6331.68	II
24		6334.91	II
24		6446.87	II
35	Cw	6518.68	I
24	c	6574.04	II
35		6581.82	I
30		6607.17	II
90		6677.94	II
40	Cw	6702.61	I
20	c	6706.79	II
30		6785.12	II
130		6794.58	II
40		6874.18	II
55		6896.37	II
45	h	6899.95	I
40		6901.98	I
9		7005.99	II
17		7082.85	II
11		7089.22	II
11		7112.69	I
10		7187.48	I
10	h	7195.89	II
65		7204.28	I
19	h	7234.98	I
40		7257.73	I
17		7311.57	I
45		7348.88	II
10		7398.27	II
15	h	7424.24	II
10	h	7429.62	II
9		7472.15	I
22		7484.54	I
9		7495.45	I
45		7496.12	I
17		7499.69	II
27		7511.40	I
9		7519.77	II
6		7557.59	II
27	h	7582.03	II
27		7587.49	I
45		7590.24	I
65		7596.44	I
17	h	7601.18	II
17		7616.01	II
22	h	7624.05	I
30		7627.81	I
9	h	7639.05	II
8		7672.72	II
8	h	7694.74	II
22	h	7706.16	II
22	h	7726.97	II
30		7737.63	I
22		7793.20	I
8		7807.33	II
16		7832.91	II
30		7855.79	II
15		7864.99	I
6	h	7885.70	I
6		7913.11	I
27		7927.90	II
13		7955.31	I
11		7998.03	I
17		8001.04	I
13	h	8010.16	II
30		8025.42	II
6		8053.80	I
19		8067.35	II
30		8085.06	II
27		8164.17	I
13		8171.70	I
65		8194.82	II
95		8212.57	I
11		8214.33	I
8		8259.08	I
40		8450.06	II
8	h	8465.80	II
13		8502.70	I
30	h	8511.80	I
45		8583.45	II
30		8603.40	I
9		8678.25	I
65		8765.74	II

Tb IV
Ref. 302 — J.R.

Intensity		Wavelength	
		Vacuum	
30		1176.58	IV
30		1192.01	IV
70		1200.58	IV
50		1213.94	IV
80		1221.22	IV
500		1235.04	IV
1000		1259.40	IV
300		1301.48	IV
300		1308.30	IV
600		1311.70	IV
700		1315.12	IV
500		1325.56	IV
1000		1327.67	IV
100		1367.56	IV
400		1367.71	IV
700		1369.64	IV
1000		1373.86	IV
400		1376.46	IV
200		1378.23	IV
300		1381.00	IV
100		1382.83	IV
20		1389.92	IV
200		1516.17	IV
50		1530.10	IV
5000		1595.39	IV
2000		1633.19	IV
300		1649.38	IV
400		1654.75	IV
400		1667.58	IV
200		1672.55	IV
400		1681.98	IV
5		1684.46	IV
100		1685.37	IV
400		1691.95	IV
10		1695.23	IV
300		1698.36	IV
30		1701.60	IV
50		1704.79	IV
20		1705.05	IV
3		1943.94	IV
50		1970.90	IV
		Air	
2000		2027.79	IV
200		2029.22	IV
400		2048.88	IV
200		2078.83	IV
1000		2089.98	IV
1000		2332.54	IV
100		2436.01	IV

THALLIUM (Tl)
Z = 81

Tl I and II
Ref. 1, 195, 348, 354 — C.H.C.

Intensity		Wavelength	
		Vacuum	
3		650.90	II
5	r	670.87	II
4		674.10	II
15	r	696.30	II
5	r	709.23	II
10	r	817.18	II
5	r	836.34	II
8	r	1018.85	II
10	r	1049.73	II
8	r	1050.30	II
5	r	1074.97	II
10	r	1130.17	II
15	r	1162.55	II
10	r	1167.43	II
10	r	1183.41	II
12	r	1194.84	II
8		1231.81	II
5	r	1246.00	II
15	r	1307.50	II
8	r	1310.20	II
25	r	1321.71	II
8	r	1330.40	II
10	r	1373.52	II
1		1423.2	I
8	r	1489.65	I
5		1490.50	II
10	r	1499.30	II
10	r	1507.82	II
15	r	1561.58	II
10	r	1568.57	II
7	r	1593.26	II
5	h	1616.	I
1		1650.2	I
5		1685.40	I
1		1728.	I
10	r	1792.76	II
12	r	1814.85	II
3	h	1847.	I
8		1892.72	II
25	r	1908.64	II
		Air	
100	r	2007.56	I
2		2209.75	I
100	r	2210.71	I
3		2287.6	I
30		2298.04	II
140		2315.98	I
900	h	2379.69	I
8		2451.83	II
6		2469.03	II
1		2508.2	I
20		2530.86	II
700		2580.14	I
60		2608.99	I
80		2665.57	I
420		2709.23	I
50	h	2710.67	I
4400	d	2767.87	I
280		2826.16	I
10		2849.80	II
2800		2918.32	I
440		2921.52	I
5		3029.01	II
20		3091.56	II
15		3185.51	II
15		3186.56	II
15		3187.74	II
1200		3229.75	I
15		3291.01	II
12		3319.91	II
12		3321.04	II
8		3322.25	II
15		3369.15	II
8		3381.00	II
6		3381.80	II
6	d	3460.48	II
20000		3519.24	I
5000		3529.43	I
8		3540.08	II
9		3560.68	II
5		3567.67	II
12000	Cw	3775.72	I
8		3793.95	II
10		3832.30	II
6		3869.15	II
10		3887.15	II
8		4223.05	II
20		4274.98	II
40		4306.80	II
2		4359.9	I
8		4490.77	II
20		4737.05	II
15		4981.35	II
25		5078.54	II
25		5152.14	II
6		5181.95	II
6		5183.10	II
18000		5350.46	I
15	d	5384.85	II
7		5409.92	II
10		5410.97	II
25		5949.48	II
10		6179.98	II
8	d	6239.03	II
10		6378.32	II
16	h	6549.84	I
6	h	6713.80	I
10		6966.5	II
3		7493.6	I
2		7678.93	I
10		7815.80	I
8		8130.0	I
20		8373.6	I
8		8445.8	II
10		8474.27	I
8		8632.9	II
10		8664.1	II
4		8850.4	I
5		8976.75	I
3		9038.4	I
20		9130.	II
20		9130.5	I
2	h	9183.1	I
4		9225.	II
2	h	9252.6	I
3		9254.	II
40		9509.4	I
10		9863.4	I
20		9930.4	I
2		9937.4	I
30		10011.9	I
40		10488.80	I
5		11101.61	I
4		11483.7	I
1000		11512.82	I

Thallium (Cont.)

5	11592.9	I
15	12491.8	I
150	12736.4	I
700	13013.2	I

Tl III
Ref 355 — C.H.C.

Intensity	Wavelength	
	Vacuum	
7	1231.57	III
10	1266.33	III
4	1332.36	III
10	1477.14	III
4	1506.37	III
8	1558.67	III
8	1660.05	III
	Air	
6	3163.53	III
3	3300.80	III
9	3456.34	III
4	3507.41	III
6	3933.05	III
2	3946.02	III
7	4109.85	III
4	4155.75	III
6	4269.81	III
2	4380.57	III
4	5086.99	III
4	5362.40	III
2	5499.4	III
5	5927.8	III
4	8001.	III

Tl IV
Ref. 356 — C.H.C.

Intensity	Wavelength	
	Vacuum	
7	531.26	IV
10	570.49	IV
4	597.01	IV
1	868.99	IV
3	912.74	IV
8	917.31	IV
30	1028.69	IV
20	1034.73	IV
20	1036.61	IV
10	1049.48	IV
10	1057.56	IV
20	1068.04	IV
20	1070.47	IV
30	1079.68	IV
5	1079.70	IV
2	1092.90	IV
4	1094.41	IV
6	1099.60	IV
4	1125.52	IV
5	1139.30	IV
3	1144.07	IV
3	1225.45	IV
6	1273.03	IV
3	1304.55	IV
6	1323.66	IV
7	1337.10	IV
7	1358.56	IV
5	1374.62	IV
7	1377.75	IV
8	1404.60	IV
5	1412.93	IV
6	1434.72	IV
5	1449.37	IV
5	1883.2	IV
3	1974.6	IV

THORIUM (Th)
Z = 90

Th I and II
Ref. 1, 97, 98, 434 — J.G.C. and R.Z.

Intensity	Wavelength	
	Air	
100	2326.926	II
190	2377.84	II
90	2404.504	II
100	2413.409	II
30	2439.433	I
5	2532.894	I
150	2547.901	II
500	2565.593	II
270	2566.588	II
200	2576.688	II
230	2589.059	II
230	2597.047	II
230	2600.882	II
100	2609.855	II
230	2618.91	II
270	2623.448	II
270	2625.737	II
55	2628.812	II
270	2641.488	II
170	2650.583	II
150	2658.663	II
50	2680.692	I
360	2684.288	II
480	2692.415	II
100	2695.553	II
270	2703.958	II
170	2708.176	II
230	2721.691	II
170	2722.380	II
250	2729.327	II
250	2732.808	II
28	2735.834	II
520	2747.156	II
100	2749.530	II
410	2752.166	II
130	2760.391	II
100	2765.123	II
270	2768.841	II
200	2770.816	II
70	2773.951	II
50	2774.066	II
70	2778.706	II
35	2791.496	I
90	2794.255	II
70	2797.737	II
55	2799.114	II
110	2807.827	II
180	2808.998	II
70	2814.319	II
45	2816.071	II
100	2819.322	II
100	2820.336	II
100	2822.025	II
170	2826.855	II
70	2830.442	II
800	2832.315	II
1200	2837.295	II
320	2842.812	II
100	2848.084	I
270	2851.260	II
50	2854.342	I
30	2860.490	I
220	2861.42	II
35	2868.461	I
70	2869.916	II
550	2870.406	II
30	2878.657	I
40	2882.511	II
320	2884.289	II
360	2885.049	II
360	2887.817	II
40	2892.172	II
250	2899.720	II
45	2903.167	II
50	2908.506	I
200	2910.594	II
90	2911.320	II
90	2912.009	II
140	2919.840	II
250	2925.050	II
250	2928.254	II
50	2931.281	I
55	2934.135	II
100	2936.086	I
35	2940.589	II
340	2942.860	II
100	2943.729	I
150	2949.068	II
80	2950.438	II
35	2955.849	II
170	2957.580	II
28	2959.853	I
28	2963.607	I
270	2968.686	II
110	2971.481	II
220	2974.011	II
50	2976.104	I
55	2980.334	II
160	2985.243	II
360	2988.232	II
150	2991.062	II
110	2996.986	II
50	3002.686	I
30	3004.248	I
180	3008.497	II
50	3010.736	I
20	3018.644	I
50	3021.056	I
150	3026.575	II
40	3030.487	I
370	3034.065	II
170	3035.110	II
85	3038.598	II
130	3045.564	II
420	3049.092	II
30	3056.692	I
220	3061.699	II
450	3067.729	II
370	3072.114	II
670	3078.828	II
480	3080.217	II
240	3088.470	II
130	3090.093	II
50	3093.711	I
140	3097.266	II
200	3102.664	II
510	3108.296	II
50	3115.538	I
100	3116.263	I
510	3119.526	II
510	3122.963	II
370	3124.387	II
480	3125.507	II
150	3131.070	II
100	3136.216	I
420	3139.306	II
420	3142.835	II
310	3146.044	II
150	3150.455	II
310	3154.300	II
50	3157.221	I
30	3161.364	I
140	3166.099	II
110	3169.328	II
420	3175.726	II
270	3179.048	II
1100	3180.193	II
310	3184.948	II
770	3188.233	II
85	3191.221	II
55	3192.585	I
55	3195.689	I
30	3202.520	I
170	3210.308	II
55	3214.380	I
560	3221.292	II
30	3223.168	I
560	3229.009	II
110	3230.868	II
480	3235.84	II
590	3238.116	II
240	3241.108	II
110	3244.448	I
280	3251.915	II
910	3256.274	II
180	3257.366	I
910	3262.668	II
180	3267.003	II
110	3272.027	I
30	3278.733	I
50	3281.048	I
130	3285.752	I
620	3287.789	II
910	3291.739	II
620	3292.520	II
240	3297.832	II
240	3301.650	I
480	3304.238	I
130	3309.365	I
50	3314.790	I
30	3318.390	I
510	3321.450	II
390	3324.752	II
840	3325.120	II
55	3328.255	II
250	3330.476	I
620	3334.604	II
620	3337.870	II
30	3342.073	II
180	3346.557	II
310	3348.768	I
980	3351.228	II
310	3354.179	II
620	3358.602	II
75	3361.738	II
390	3367.819	II
250	3374.974	I
390	3378.573	II
130	3380.859	I
310	3385.531	II
310	3386.501	II
110	3387.920	I
1300	3392.035	II
200	3396.727	I
250	3398.544	I
200	3405.558	I
250	3413.012	I
50	3417.497	I
390	3421.210	I
270	3423.989	I
50	3428.622	I
980	3433.998	II
770	3435.976	II
340	3438.949	II
110	3442.578	I
50	3446.547	I
130	3451.702	I
50	3457.068	I
340	3462.850	II
130	3465.924	II
390	3468.219	II
1300	3469.920	II
170	3471.218	I
250	3479.173	II
70	3480.052	I
200	3486.552	I
100	3489.184	I
270	3493.518	II
70	3496.810	I
130	3498.621	I
70	3503.786	I
50	3506.645	I
110	3511.157	I
140	3518.404	I
70	3521.059	I
70	3526.633	I
140	3531.450	I
670	3539.587	II
180	3544.018	I
170	3549.595	I
140	3551.401	I
200	3555.013	I
530	3559.451	II
110	3563.375	I
70	3569.820	I
200	3576.557	I
100	3583.101	I
170	3589.750	I
270	3592.780	I
270	3598.120	I
390	3601.034	II
170	3608.377	I
980	3609.445	II
200	3612.427	I
480	3615.133	II
400	3617.118	II
270	3623.970	II
390	3625.627	II
140	3632.831	I
270	3635.943	I
70	3638.644	I
210	3642.248	I
170	3649.735	I
50	3658.808	I
100	3659.629	I
220	3663.202	I
140	3668.140	I
280	3669.968	I
700	3675.567	II
150	3682.486	I
50	3688.658	I
100	3690.624	I
170	3692.566	I
180	3698.105	I
50	3703.229	I
340	3706.767	I
280	3711.305	II
590	3719.435	I
50	3719.836	I
770	3721.825	II
110	3727.902	I
50	3733.672	I
1300	3741.183	II
310	3747.539	I
650	3752.569	II
140	3757.694	I
110	3765.240	I
180	3770.056	I
50	3772.649	I
85	3776.271	I
50	3780.966	I
340	3785.600	II
100	3789.167	I
85	3795.386	I
50	3800.197	I
590	3803.075	I
50	3807.273	I
340	3813.068	II
50	3818.685	I
75	3825.133	I
450	3828.384	I
70	3836.584	I
840	3839.746	II
280	3841.960	II
100	3846.887	I
85	3852.135	I
390	3854.511	II
140	3859.840	II
450	3863.405	II
100	3869.663	I
210	3875.374	I
140	3879.644	I
100	3886.915	I
340	3895.419	I
50	3901.661	I
110	3903.102	I
170	3905.186	II
50	3908.750	I
85	3911.909	I
50	3916.417	I

Thorium (Cont.)

Intensity	Wavelength	Spectrum
110	3919.023	I
140	3925.093	I
590	3929.669	II
200	3932.911	I
140	3937.040	II
50	3942.072	I
50	3948.030	I
200	3948.964	II
50	3952.760	I
110	3959.300	I
390	3967.392	I
200	3972.155	I
150	3980.089	I
110	3991.730	I
530	3994.549	II
50	3998.061	I
240	4003.309	II
250	4007.021	II
220	4008.210	I
220	4009.056	I
280	4012.495	I
4200	4019.129	II
210	4025.656	II
140	4027.009	I
250	4030.842	I
250	4036.047	I
240	4036.565	II
240	4041.204	II
55	4048.287	I
110	4050.887	I
140	4059.253	I
250	4063.407	I
300	4069.201	II
100	4069.461	I
55	4075.503	I
110	4081.368	I
85	4085.434	I
700	4086.520	II
70	4088.726	I
700	4094.747	II
150	4100.341	I
270	4105.330	II
840	4108.421	II
240	4112.754	I
280	4115.758	I
1100	4116.713	II
30	4123.600	I
200	4127.411	I
110	4131.002	I
340	4132.753	II
200	4134.067	I
220	4140.235	II
250	4142.701	II
200	4148.182	II
450	4149.986	II
110	4158.535	I
140	4165.766	I
620	4178.060	II
250	4179.714	II
30	4184.138	I
130	4193.017	I
620	4208.890	II
130	4210.923	I
28	4214.828	I
55	4220.065	I
55	4227.387	I
30	4230.824	I
85	4235.463	I
20	4241.094	I
30	4247.989	II
110	4253.538	I
70	4256.254	I
110	4260.333	I
28	4269.942	I
280	4273.357	II
480	4277.313	II
700	4282.042	II
28	4288.669	I
55	4297.306	I
85	4299.839	I
100	4307.176	I
200	4309.991	II
55	4315.254	I
110	4318.416	I
30	4325.274	I
28	4330.844	I
130	4337.277	I
85	4342.256	II
130	4344.326	II
55	4349.072	I
55	4354.484	I
85	4359.372	I
85	4365.930	I
85	4374.123	I
1300	4381.860	II
1100	4391.110	II
55	4392.974	I
55	4401.580	I
85	4408.882	I
210	4412.741	II
28	4416.845	I
50	4422.048	I
250	4432.963	II
140	4440.866	II
25	4452.565	I
83	4458.002	I
220	4465.341	II
30	4475.221	I
75	4482.169	I
50	4489.664	I
110	4498.940	I
55	4505.216	I
280	4510.527	II
70	4521.194	I
22	4530.319	I
40	4535.255	I
30	4545.915	II
70	4555.812	I
40	4563.660	I
65	4570.972	I
50	4588.426	I
75	4595.421	I
26	4612.554	II
30	4621.163	I
140	4631.761	II
140	4631.761	II
30	4641.254	I
140	4651.558	II
23	4663.202	I
50	4669.984	I
65	4676.056	I
50	4686.195	I
140	4694.091	II
50	4703.990	I
20	4712.841	I
90	4723.438	I
30	4729.128	I
190	4740.529	II
140	4752.414	II
13	4766.600	I
50	4778.294	I
20	4786.531	I
40	4789.387	I
45	4808.134	I
20	4819.193	I
26	4822.855	I
40	4826.700	I
45	4831.121	I
50	4840.843	I
30	4848.362	I
15	4852.868	I
40	4858.333	II
280	4863.163	II
40	4872.917	I
26	4878.733	I
45	4894.955	I
20	4907.209	I
240	4919.816	II
18	4927.780	I
40	4939.642	I
60	4947.575	II
50	4954.659	II
30	4965.731	I
35	4975.950	II
24	4985.372	I
50	5002.097	I
50	5002.097	I
50	5015.889	II
260	5017.255	II
20	5029.892	I
24	5039.230	I
50	5044.719	I
240	5049.796	II
85	5055.347	II
70	5058.562	II
110	5067.974	I
30	5081.446	I
50	5090.051	I
50	5098.043	II
40	5101.130	I
50	5110.867	II
30	5115.044	I
10	5125.950	I
20	5134.746	I
95	5143.267	II
120	5148.211	II
50	5151.612	I
50	5154.243	I
85	5158.604	I
70	5160.730	I
20	5168.922	I
50	5176.961	I
35	5183.990	II
50	5190.871	II
50 h	5195.814	I
50	5198.800	I
95	5199.164	I
50	5211.230	I
95	5216.596	II
50	5218.528	II
35	5219.110	I
110	5231.160	I
85	5233.225	II
85	5233.229	II
95	5247.654	II
10	5255.573	I
35	5258.360	I
12	5266.710	I
70	5277.501	II
15	5281.069	I
10	5294.397	I
30	5297.743	I
30	5307.466	II
35	5312.002	I
20	5317.494	I
60	5325.145	II
50	5326.976	I
20	5330.080	I
60	5343.581	I
14	5351.126	I
30	5358.707	I
20	5369.281	I
30	5378.836	I
70	5390.466	II
50	5392.572	I
20	5399.175	I
24	5417.486	I
60	5425.678	II
50	5435.893	II
40	5449.479	II
30	5462.615	II
15	5470.759	I
24	5484.147	II
10	5496.137	I
19	5504.302	I
35	5509.994	I
12	5524.584	I
50	5539.262	I
70	5539.911	II
35	5548.176	I
50	5558.342	I
60	5564.203	II
40	5573.354	I
60	5587.026	I
24	5595.064	I
50	5604.515	II
35	5615.320	I
7	5630.297	I
70	5639.746	II
7	5648.991	I
12	5657.925	I
15	5667.128	I
20	5677.053	I
10	5685.192	I
65	5700.918	II
95	5707.103	II
50	5720.183	I
30	5732.975	II
24	5742.084	II
30	5749.388	II
70	5760.551	I
15	5773.946	I
20	5789.645	I
35	5804.141	I
19	5815.422	II
10	5832.370	I
10	5845.919	I
15	5854.121	I
15	5868.373	I
10	5878.933	I
8	5891.451	I
15	5899.844	I
20	5914.387	II
19	5925.893	II
10	5937.162	I
10	5944.648	I
8	5957.587	I
30	5973.665	I
85	5989.044	II
24	5994.129	I
21	6007.072	I
30	6015.426	II
17	6021.036	I
17	6037.698	I
24	6044.431	II
10	6053.381	I
5	6061.536	I
30	6077.106	I
30	6087.262	II
24	6099.083	I
30	6104.580	I
40	6112.837	I
30	6120.557	II
10	6124.480	I
14	6151.993	I
10	6161.354	I
60	6169.822	I
50	6182.622	I
12	6191.906	I
24	6193.858	II
12	6203.493	I
12	6224.528	I
24 h	6234.856	I
8	6240.954	I
10	6257.424	I
21	6261.063	I
21	6261.418	I
50	6274.116	II
50	6274.117	II
30	6279.172	II
8	6291.192	I
10	6303.251	II
8	6317.185	I
21	6327.278	I
35	6342.860	I
50	6355.911	II
14	6369.140	I
40	6376.931	I
30	6411.899	I
24	6413.615	I
15	6437.762	I
15	6450.005	I
60	6457.283	I
50	6462.614	I
5	6466.717	I
14	6490.738	I
5	6501.992	I
20	6512.364	I
5	6522.044	I
50 h	6531.342	I
6	6554.160	I
5	6558.876	I
3	6565.070	I
5	6577.215	I
24	6583.907	I
24	6588.540	I
24	6593.940	I
24	6605.416	II
24	6619.946	II
24	6619.947	II
21	6644.650	II
6	6658.678	II
30	6662.269	I
6	6674.697	I
3	6683.367	I
8	6692.724	II
5	6697.712	I
16	6727.459	I
3	6735.126	I
5	6742.884	I
20	6756.453	I
6	6765.677	I
15	6778.313	I
15	6780.413	I
6	6791.236	I
3	6798.747	I
3	6809.511	I
5	6823.509	I
11	6829.036	I
14	6834.925	I
4	6862.873	I
5	6866.367	I
8	6874.754	I
20	6889.303	II
3	6908.988	I
24	6911.227	I
5	6916.129	I
5	6936.652	I
35	6943.611	I
5	6954.657	I
15	6965.947	I
3	6981.086	I
55	6989.656	I
24	6993.038	II
18	7000.806	I
18	7000.806	I
3	7015.319	I
10	7018.569	I
3	7026.462	I
7	7036.281	I
30	7045.795	II
15	7053.619	II
6	7060.654	I
24	7075.333	II
30	7084.171	I
24	7089.339	II
10	7100.512	II
3	7109.861	I
11	7124.562	I
3	7132.613	I
5	7148.560	I
10	7154.954	I
30	7168.896	I
15	7173.373	I
40	7191.132	II
7	7200.046	I
35	7208.006	I
11	7212.69	I
10	7217.755	II
11	7218.054	I
3	7242.355	I
5	7255.354	I
3	7270.558	I
7	7284.904	I
5	7298.143	I
11	7305.405	II
5	7315.067	I
7	7324.808	I
5	7339.606	I
8	7341.152	I
5	7361.349	I
5	7384.175	I
18	7385.501	I
3	7393.431	II
5	7402.252	I
3	7418.550	I

Thorium (Cont.)

21	7428.940	I
10	7430.254	I
3	7444.749	I
2	7462.993	I
10	7481.355	I
2	7493.427	I
50	7525.508	II
7	7549.314	I
18	7567.740	I
12	7585.69	I
12	7585.792	I
4	7598.204	I
2	7607.824	I
5	7627.176	I
3	7636.176	I
30	7647.380	I
4	7658.324	I
7	7676.219	I
21	7685.305	I
4	7710.269	I
4	7728.951	I
10	7731.72	II
4	7771.948	I
15	7787.79	II
15	7788.937	I
5	7798.360	I
4	7810.625	I
21	7817.771	I
8	7834.459	II
15	7847.540	I
4	7864.023	I
12	7865.95	I
6	7886.284	I
11	7900.31	I
4	7937.732	I
11	7941.72	I
5	7954.594	I
7	7972.598	I
24	7978.974	I
11	7987.97	I
5	7993.680	I
2	8014.502	I
5	8024.253	II
11	8032.433	I
11	8062.64	I
5	8085.220	I
5	8093.626	I
5	8129.407	I
11	8138.477	I
18	8143.139	I
12	8159.729	I
10	8163.125	II
7	8169.788	I
15	8186.914	I
12	8203.199	II
2	8231.408	I
5	8252.395	I
3	8259.512	I
18	8275.629	I
3	8292.529	I
15	8320.857	I
30	8330.451	I
4	8358.726	I
6	8387.104	II
18	8403.767	II
15	8416.729	I
12	8421.227	I
21	8446.509	I
5	8464.230	I
18	8478.360	I
5	8510.621	I
5	8516.557	I
5	8539.795	I
6	8554.946	I
11	8573.122	I
12	8591.838	I
3	8621.325	I
3	8638.363	I
10	8665.487	I
4	8668.116	I
2	8701.127	I
5	8709.236	I
8	8732.401	II
18	8748.033	I
15	8758.244	I
8	8775.573	I
4	8792.058	I
3	8804.590	I
5	8841.185	I
18	8842.073	II
15	8868.834	I
4	8875.233	I
2	8907.038	I
5	8955.848	I
15	8957.97	II
40	8967.641	I
3	8987.408	I
5	9031.819	I
25	9048.252	I
5	9063.953	I
6	9094.831	I
3	9107.225	I
2	9118.140	I
3	9170.825	I
10	9203.963	I
10	9266.208	I
10	9276.276	I
10	9289.563	I
4	9317.722	I
7	9340.706	I
15	9399.085	I
10	9431.603	I
15	9461.030	I
15	9474.882	I
15	9495.501	I
15	9497.191	I
10	9505.392	I
10	9561.24	I
8	9613.689	II
12	9632.647	I
10	9664.700	I
4	9676.106	I
15	9700.564	I
15	9746.46	I
10	9812.70	I
10	9826.45	I
20	9833.42	I
15	10039.364	I
15	10089.138	I
15	10133.56	II
15	10419.57	II
15	10556.45	I
15	10723.92	II
20	10726.93	I
20	10942.24	II
15	11051.90	I
30	11230.259	I
20	11354.719	I
15	11703.46	I
15	11864.25	I
15	11940.64	I
20	11984.67	II
15	12018.72	I
20	12127.30	I
20	12194.16	II
15	12206.89	I
20	12231.94	I
15	12338.00	I
15	12477.30	I
20	12646.54	I
15	12866.64	I
15	12940.65	II
15	12959.82	I
15	13145.90	II
15	13565.67	I
15	14090.25	I
15	14168.67	I
15	14424.54	I
15	14618.98	I
15	14654.91	I
15	14940.49	II
15	15240.24	II
15	15429.78	I
15	15831.75	I
20	17208.22	II
15	17307.66	I
15	17381.91	I
15	17481.04	I
15	17584.52	I
15	17936.43	II
15	18811.88	I
15	19145.60	II
15	19338.98	II
10	19774.30	II
10	20634.36	I
10	20692.06	II
10	22264.35	II

Th III
Ref. 157 — J.G.C.

Intensity	Wavelength	
	Vacuum	
100	1888.12	III
	Air	
50	2149.18	III
50	2162.82	III
50	2199.74	III
50	2206.62	III
50	2291.59	III
100	2301.18	III
100	2319.52	III
80	2324.68	III
150	2335.50	III
100	2340.58	III
100	2363.06	III
50	2368.91	III
100	2371.42	III
80	2381.47	III
100	2391.48	III
200	2413.50	III
50	2424.54	III
200	2427.94	III
200	2431.68	III
200	2441.24	III
100	2463.66	III
50	2473.93	III
100	2501.08	III
60	2512.69	III
50	2514.31	III
100	4555.73	III
50	4589.28	III
100	5376.13	III
50	5447.18	III
50	6242.95	III
50	6599.39	III
50	7461.59	III
50	8105.14	III

Th IV
Ref. 156, 165 — J.G.C.

Intensity	Wavelength	
	Vacuum	
4	797.53	IV
1	835.55	IV
30	846.91	IV
1	854.02	IV
30	882.39	IV
12	886.66	IV
100	1565.85	IV
70	1682.22	IV
30	1684.01	IV
150	1707.37	IV
200	1959.02	IV
	Air	
200	2002.34	IV
100	2066.70	IV
20	2143.91	IV
30	2146.81	IV
1	2242.11	IV
2	2261.26	IV
5	2296.81	IV
100	2693.99	IV
2	4937.09	IV
4	4952.52	IV
3	5420.38	IV
2	6711.87	IV
3	6740.37	IV
50	6901.16	IV
	9839.25	IV
	10875.05	IV

THULIUM (Tm)
Z = 69

Th I and II
Ref. 1 — C.H.C.

Intensity		Wavelength	
		Air	
360		2284.79	II
120		2329.77	II
70		2340.92	II
120		2363.91	II
45		2365.96	II
160		2367.11	II
150		2383.68	II
110		2388.95	II
450		2409.02	II
110		2412.44	II
120		2421.65	II
450		2426.17	II
140		2445.47	II
770		2480.13	II
150		2481.15	II
130		2487.52	II
250		2491.60	II
100		2499.54	II
130		2507.15	II
1300		2509.08	II
200		2520.87	II
250		2522.17	II
180		2524.11	II
130		2527.02	I
110		2527.42	II
120		2542.66	II
360		2552.76	I
540		2561.65	II
150		2563.86	II
430		2588.27	II
170	h	2596.49	I
110		2601.09	I
220		2606.02	II
810		2607.06	II
730		2624.33	II
210		2640.76	II
130		2646.45	II
160		2650.27	II
190		2658.48	II
250		2660.09	II
140		2668.20	II
310		2679.57	II
170		2697.50	II
540		2721.19	II
200		2744.08	II
270		2779.55	II
350		2785.07	II
680		2794.60	II
730		2797.27	II
250		2818.47	II
250		2827.02	II
580		2827.92	II
200		2831.55	II
310		2844.67	II
200		2854.17	I
200		2860.12	II
200		2861.74	II
1600		2869.23	II
630		2890.94	II
210		2918.27	II
270		2925.65	II
680		2926.74	II
630		2935.99	II
350		2951.26	II
430		2965.86	II
490		2973.22	I
540		2981.48	II
350		2986.52	II
630		2990.54	II
200		2993.26	II
230		3013.71	II
430		3014.65	II
1500		3015.30	II
270		3017.09	II
330		3026.07	II
280		3042.35	II
340	d	3046.76	II
320		3050.73	II
340		3056.07	II
580		3073.08	II
360		3081.12	I
740		3098.60	II
7400		3131.26	II
2300		3133.89	II
230		3144.90	II
230		3146.16	II
1900		3151.04	II
1500		3157.34	II
450		3172.65	I
2300		3172.83	II
380		3173.58	II
230		3195.33	II
320		3210.56	II
320		3210.82	II
320		3212.01	II
230		3231.51	II
470		3235.44	II
1200		3236.81	II
1600		3240.23	II
2300		3241.54	II
320		3246.96	I
420		3247.46	II
1900		3258.05	II
400		3261.65	II
320		3264.10	II
1600		3266.64	II
1200		3267.40	II
790		3268.99	II
1100		3276.81	II
1200		3283.40	II
1200		3285.61	II
2300		3291.00	II
2000		3302.46	II
210		3306.01	II
210		3306.91	II
210		3308.01	II
1200		3309.80	II
640		3310.59	II
400		3316.88	II
210		3318.65	II
230		3349.99	I
230		3354.86	II
4000		3362.61	II
490		3374.50	II
420	d	3384.99	II
1700		3397.50	II
420		3399.95	II
850		3410.05	I
340		3412.59	I
340		3416.59	I
6400		3425.08	II
950		3425.63	II
340		3429.33	I
850		3429.96	II
420		3431.19	II
4900		3441.50	II
4900		3453.66	II
8500		3462.20	II
210		3467.51	I

Thuiium (Cont.)

Intensity		Wavelength	Spectrum
340		3476.69	I
340		3400.98	I
340		3481.75	II
420		3487.38	I
210		3492.58	II
340		3499.95	I
250		3513.02	II
250		3517.60	I
250		3534.85	II
1700		3535.52	II
490		3536.21	II
850		3536.58	II
420		3537.91	I
210		3555.82	I
420		3557.79	II
340		3560.92	I
420		3563.88	I
490		3565.91	II
1300		3566.47	II
420		3567.36	I
280		3574.06	II
280		3586.07	I
2100		3608.77	II
250		3609.53	II
380		3638.41	I
950		3643.65	II
240		3647.72	II
600		3653.61	II
500		3665.81	II
1100		3668.09	II
410		3677.98	II
450	d	3678.85	II
410		3694.74	II
4800		3700.26	II
3800		3701.36	II
330		3704.85	II
7700		3717.91	I
890		3725.06	II
2400		3734.12	II
5000		3744.06	I
1700		3751.81	I
310		3756.86	II
6000		3761.33	II
4800		3761.91	II
260		3783.55	II
380		3795.16	II
7100		3795.75	II
770		3798.54	I
240		3798.75	II
600		3807.72	I
380		3810.72	II
550		3817.39	II
290		3826.39	I
1300		3838.20	II
290		3840.87	I
8900		3848.02	II
140		3857.84	II
6800		3883.13	I
1800		3883.44	II
5400		3887.35	I
440		3890.53	II
440		3896.62	I
680		3900.79	II
3500		3916.48	I
120		3928.66	II
570		3929.58	II
1500		3949.27	I
1500		3958.10	II
440		3995.58	II
1800		3996.52	II
220		4024.23	I
380		4044.47	I
10000		4094.19	I
9500		4105.84	I
120		4132.69	II
1100		4138.33	I
120		4149.14	I
120		4158.60	I
8800		4187.62	I
520		4199.92	II
6000		4203.73	I
220		4206.00	II
380		4222.67	I
3000		4242.15	II
270		4271.71	I
150		4298.36	I
2700		4359.93	I
1400		4386.43	I
200		4394.42	I
120		4395.96	I
140		4396.50	I
55		4437.40	II
80		4442.74	I
50		4447.58	I
120		4454.03	I
80		4459.99	I
50		4467.98	I
540		4481.26	II
80		4489.70	II
150		4519.60	I
260		4522.57	II
180		4529.38	II
80		4532.15	I
110		4548.60	I

Intensity		Wavelength	Spectrum
40		4556.68	II
40		4561.86	II
80		4564.68	I
40		4567.11	II
95		4596.63	I
270		4599.02	I
35		4601.29	II
55		4603.43	II
40		4604.85	I
50		4613.97	I
40		4614.47	II
300		4615.94	II
35		4619.06	II
40		4621.72	I
80		4626.33	II
95		4626.56	II
40		4626.97	I
110		4634.26	II
40		4642.96	II
95		4643.12	I
35		4644.58	I
120		4655.09	I
35		4666.70	II
35		4671.99	II
35		4675.10	I
80		4675.31	I
40		4677.86	II
160		4681.92	I
70		4685.11	I
120		4691.11	I
110		4724.26	I
680		4733.34	I
35		4750.75	II
70		4759.90	I
27		4789.92	II
27		4807.48	I
35		4808.68	I
35		4813.50	I
27		4826.99	II
27		4828.97	I
80		4831.20	II
35		4835.75	I
27	d	4851.76	I
		4851.90	II
19		4872.28	II
27		4879.19	I
27		4891.64	I
24		4909.74	I
55		4923.83	I
140		4957.18	I
40		4970.87	II
27		4971.26	I
40		4975.12	II
50		4978.90	I
40		4980.68	II
55		4989.32	II
27		4993.79	II
19		4994.72	II
35		5001.02	I
27		5001.59	I
160		5009.77	II
35		5014.56	II
27		5017.87	II
160		5034.22	II
27	h	5041.00	II
22		5043.50	I
35		5045.41	I
27		5060.42	II
150		5060.90	I
27		5062.25	I
27		5065.88	I
80		5066.67	I
27		5072.42	I
27		5076.36	I
27		5077.18	I
35		5085.09	I
40		5107.53	I
95		5113.97	I
50		5114.55	II
22		5120.67	I
22		5140.28	II
40		5149.40	II
19		5182.68	I
40		5185.25	I
14		5204.51	II
80		5213.38	I
22		5228.23	II
14		5260.93	II
24		5267.34	II
40		5291.14	I
40		5294.32	I
35		5300.21	I
35		5302.69	I
55		5305.87	II
650		5307.12	I
16		5322.99	II
35		5338.90	I
		5339.03	I
80		5346.49	II
27		5372.98	II
14		5391.96	II
27		5400.46	II
27		5402.23	I
14		5405.98	II

Intensity		Wavelength	Spectrum
14		5461.95	II
14		5464.14	I
14		5465.54	II
16		5500.30	II
14		5526.82	II
24		5528.34	I
14		5539.03	II
27		5566.00	I
22		5581.37	I
14		5586.65	II
14		5589.94	II
14		5606.64	I
270		5631.41	I
40		5642.60	I
27		5645.40	I
70		5658.30	I
520		5675.84	I
14		5683.59	I
40		5684.76	II
14		5696.42	II
35		5709.97	II
22		5715.79	I
14		5733.81	II
11	d	5737.20	II
		5737.25	II
14	h	5738.92	II
27		5758.02	I
55		5760.20	I
190		5764.29	I
5		5778.82	II
19		5782.36	II
22		5784.46	II
11		5799.97	II
14		5811.19	II
14	h	5816.46	I
35		5838.76	II
240		5895.63	I
35		5899.47	I
24		5901.57	I
8		5912.58	I
11		5931.70	I
27		5935.90	I
140		5971.26	I
27		5975.02	I
11		5984.87	I
19		6025.44	I
11		6067.78	II
16		6131.53	I
14		6175.29	I
14		6181.41	II
14		6299.46	II
27		6352.66	I
22		6401.44	I
8		6430.94	II
14		6440.54	I
200		6460.26	I
14		6490.70	I
14		6519.78	I
8		6575.54	I
95		6604.96	I
8		6627.25	I
35		6657.72	I
11		6658.64	I
11		6692.93	I
30		6721.36	I
9		6726.34	I
9		6727.94	II
18		6739.22	I
9	h	6767.48	I
9		6777.93	I
110		6779.77	I
14	h	6782.00	I
18		6788.52	I
13	h	6820.27	I
14		6826.95	I
14		6829.12	II
23		6831.09	I
120		6844.26	I
80		6845.76	I
18		6854.12	I
6		6898.56	I
6		6915.86	I
10		6937.37	I
5		6949.54	I
5	h	6976.69	II
5		7010.79	I
6	h	7014.31	II
10		7017.90	I
6	h	7029.40	I
12		7034.34	I
10		7056.43	II
5		7060.97	I
6		7079.78	II
10		7106.14	I
5	h	7231.33	I
5		7233.74	II
4		7257.72	I
17		7272.62	I
8		7284.30	I
11	h	7286.16	I
14		7310.51	I
11		7336.63	II
14		7432.18	I
5		7434.51	II

Intensity		Wavelength	Spectrum
5		7439.95	II
75		7481.08	I
75		7490.20	I
10	h	7507.28	I
14		7545.78	I
140		7558.33	I
17	h	7580.61	I
20	h	7593.74	I
17		7595.07	II
5		7629.85	I
5	h	7648.76	II
17		7655.00	I
4		7660.32	I
7	h	7666.24	I
8		7676.04	II
8	h	7701.46	I
80		7731.53	I
4	h	7778.27	I
12	h	7782.35	I
8	h	7785.51	I
		7785.90	I
17		7803.93	I
4		7829.22	I
40		7856.08	I
3		7861.67	I
5		7918.10	I
55		7927.51	I
110		7930.84	I
6		7971.56	I
11	h	7985.93	I
14	h	8014.77	I
95		8017.90	I
3	h	8021.33	I
14		8194.19	I
5		8294.52	I
7		8365.75	I
7		8460.79	II
27		8472.01	II
7	h	8546.07	II
11		8565.73	II

Tm III
Ref. 307 — J.R.

Intensity	Wavelength	
	Air	
500	2099.11	III
500	2107.10	III
200	2136.67	III
200	2156.29	III
800	2182.98	III
300	2183.91	III
5000	2185.94	III
100	2212.25	III
300	2230.86	III
400	2231.25	III
200	2243.34	III
400	2243.98	III
200	2246.68	III
200	2269.39	III
1000	2276.91	III
100	2280.08	III
100	2281.27	III
100	2282.86	III
200	2282.98	III
200	2286.57	III
400	2287.21	III
500	2294.73	III
20000	2296.21	III
200	2297.43	III
100	2304.64	III
400	2304.82	III
5000	2305.03	III
20000	2311.16	III
5000	2312.72	III
200	2314.88	III
400	2317.35	III
500	2320.96	III
200	2322.83	III
100	2323.71	III
100	2323.77	III
100	2324.43	III
500	2324.62	III
5000	2326.19	III
100	2327.02	III
300	2327.25	III
6000	2328.50	III
6000	2329.29	III
200	2330.87	III
3000	2331.80	III
400	2335.01	III
1000	2338.36	III
500	2341.74	III
300	2342.04	III
100	2344.59	III
500	2345.61	III
300	2347.43	III
400	2353.10	III
100	2355.65	III
3000	2357.05	III

Thulium (Cont.)

1000	2361.23	III
500	2363.97	III
1000	2375.32	III
700	2375.83	III
400	2389.52	III
4000	2406.63	III
500	2435.31	III
500	2457.86	III
500	2471.23	III
30000	2489.44	III
200	2496.25	III
2000	2504.71	III
3000	2519.78	III
10000	2552.46	III
500	2557.90	III
1000	2574.52	III
500	2574.98	III
100	2581.84	III
100	2585.48	III
300	2589.20	III
500	2608.96	III
300	2609.66	III
500	2617.22	III
500	2618.78	III
1000	2621.12	III
400	2621.35	III
400	2622.31	III
100	2627.09	III
100	2628.83	III
300	2634.66	III
200	2636.68	III
200	2637.30	III
100	2640.32	III
500	2643.58	III
100	2645.05	III
200	2649.27	III
100	2650.82	III
100	2654.05	III
100	2656.30	III
500	2661.51	III
1000	2663.00	III
500	2664.76	III
500	2664.88	III
200	2665.05	III
1000	2666.93	III
200	2668.59	III
100	2668.66	III
200	2669.18	III
100	2671.42	III
100	2675.30	III
1000	2676.64	III
500	2676.91	III
100	2678.28	III
100	2680.49	III
5000	2682.32	III
300	2682.64	III
300	2687.14	III
300	2695.69	III
400	2698.21	III
1000	2699.49	III
1000	2699.80	III
100	2703.63	III
200	2703.68	III
100	2704.93	III
2000	2707.03	III
300	2707.19	III
200	2707.44	III
500	2707.60	III
1000	2709.74	III
200	2710.79	III
1000	2713.38	III
200	2715.81	III
300	2717.56	III
100	2718.02	III
3000	2719.47	III
3000	2724.44	III
4000	2727.56	III
200	2728.13	III
1000	2731.38	III
300	2732.11	III
400	2737.98	III
800	2744.74	III
400	2745.99	III
500	2752.46	III
400	2753.20	III
400	2756.15	III
800	2765.98	III
700	2769.92	III
200	2772.64	III
100	2777.43	III
400	2781.12	III
2000	2806.77	III
300	2821.12	III
200	2849.52	III
700	2882.02	III
100	2899.29	III
100	2912.33	III
400	2921.08	III
200	2947.02	III
1000	2947.72	III
500	2953.18	III
100	2966.15	III
500	2966.85	III
400	2972.61	III
100	2974.85	III
1000	2998.28	III
100	3048.11	III
700	3078.87	III
200	3120.15	III
200	3277.26	III
200	3407.73	III
100	3415.40	III
100	3415.96	III
100	3436.93	III
400	3467.93	III
200	3529.29	III
200	3533.28	III
100	3537.47	III
300	3562.41	III
100	3563.42	III
200	3587.74	III
600	3617.96	III
1000	3629.09	III
100	3706.11	III
100	3799.41	III
300	3998.84	III
200	4021.92	III
200	4026.03	III
700	4032.13	III
100	4076.15	III
200	4335.47	III
500	4385.41	III

TIN (Sn)
Z = 50

Sn I and II
Ref. 187, 191 — C.H.C.

Intensity		Wavelength	
		Vacuum	
1		899.92	II
2		917.40	II
1		935.63	II
3		945.83	II
4		954.50	II
7		985.13	II
4		997.21	II
2		1016.26	II
4		1040.78	II
1		1041.32	II
3		1062.10	II
8		1108.19	II
4		1159.05	II
10		1161.43	II
3		1162.94	II
4		1180.51	II
9		1219.07	II
13		1223.70	II
11		1243.00	II
20		1290.86	II
20		1316.59	II
25		1400.52	II
20		1475.15	II
9		1489.22	II
7		16 9.47	II
10	r	1737.21	I
15	r	1751.46	I
10	h	1753.3	I
7		1758.00	II
20	r	1764.98	I
20		1773.40	I
30	r	1790.75	I
80	r	1804.60	I
15		1811.34	II
30		1813.04	I
40	r	1815.74	I
25		1819.31	I
120	r	1823.00	I
9		1831.89	II
50	r	1848.75	I
30		1852.00	I
200	r	1860.32	I
20		1861.42	I
20		1865.52	I
30		1865.96	I
15		1873.29	I
30		1882.64	I
80		1886.05	I
100		1891.40	I
20		1897.29	I
12		1899.91	II
50		1909.30	I
40		1911.61	I
20		1913.52	I
80		1925.31	I
20		1926.77	I
15		1927.95	I
40	h	1928.9	I
25		1933.17	I
20		1942.69	I
150		1952.15	I
15		1960.21	I
30		1971.46	I
50	h	1977.6	I
80		1984.20	I
15		1991.88	I
20		1994.98	I
		Air	
25		2008.05	I
30		2015.76	I
30		2025.98	I
50		2040.66	I
20		2040.90	I
50		2054.03	I
70		2058.31	I
20		2064.00	I
80		2068.58	I
100		2072.89	I
100		2073.08	I
25		2080.62	I
30		2091.58	I
40		2094.35	I
200		2096.39	I
100		2100.93	I
100	r	2113.93	I
50		2121.26	I
25		2140.73	I
20		2141.43	I
15		2148.46	I
1		2148.63	II
40	r	2148.73	I
20	r	2151.43	I
30		2151.54	II
80		2171.32	I
150	r	2194.49	I
300	r	2199.34	I
400	r	2209.65	I
4		2209.67	II
40		2211.05	I
80	r	2231.72	I
400	r	2246.05	I
6		2246.07	II
60		2251.17	I
30		2267.19	I
400	r	2268.91	I
20		2282.26	I
200	r	2286.68	I
600	r	2317.23	I
300	r	2334.80	I
1000	r	2354.84	I
20		2357.90	I
3		2360.34	II
22		2368.33	II
60		2380.72	I
4		2384.54	II
100		2408.15	I
800	r	2421.70	I
1000	r	2429.49	I
1		2433.52	II
15		2448.98	II
60		2455.24	I
20		2476.40	I
300		2483.39	I
13		2483.48	II
10		2486.99	II
200		2495.70	I
5		2522.61	II
90		2523.92	I
80	h	2531.17	I
400		2546.55	I
40	h	2558.01	I
500	r	2571.58	I
200		2594.42	I
50	h	2636.94	I
200	r	2661.24	I
2		2664.93	II
700	r	2706.51	I
2		2727.82	II
20		2761.78	I
150		2779.81	I
80		2785.03	I
60		2787.96	I
60		2812.59	I
80		2813.58	I
2		2825.52	II
1400	r	2839.99	I
1		2846.42	II
200		2850.62	I
1000	r	2863.32	I
1		2912.80	II
200		2913.54	I
6		2919.82	II
3		2991.00	II
7		2994.44	II
700	r	3009.14	I
1		3012.18	II
8		3023.94	II
200		3032.80	I
850	r	3034.12	I
12		3047.50	II
6		3094.69	II
60		3141.84	I
550	r	3175.05	I
40		3218.71	I
550	r	3262.34	I
50		3283.21	II
110		3330.62	I
60		3351.97	II
2		3407.48	II
10		3472.46	II
7		3537.57	II
11		3575.45	II
3		3582.39	II
2		3620.08	II
6		3620.54	II
40		3655.78	I
6		3715.23	II
280	r	3801.02	I
4		3841.44	II
1		4294.65	II
40		4524.74	I
1		4579.13	II
1		4580.29	II
2		4877.22	II
3		4944.31	II
20		4979.73	I
2		5071.14	II
2		5072.67	II
20		5174.54	I
10		5332.36	II
20		5561.95	II
25		5588.92	II
2		5596.20	II
500		5631.71	I
15		5753.59	I
1		5797.20	II
15		5799.18	II
50		5925.44	I
100		5970.30	I
150		6037.70	I
200		6054.86	+
250		6069.00	I
100		6073.46	I
6		6077.48	II
5		6079.70	II
400		6149.71	I
200		6154.60	I
150		6171.50	I
100		6310.78	I
40		6354.35	I
70		6453.50	II
8		6761.45	II
25		6844.05	II
20		7191.40	II
10		7387.79	II
20	h	7398.6	I
1		7408.62	II
30		7685.30	I
13		7741.80	II
100		7754.97	I
3		7904.00	II
100	h	8030.5	I
30	h	8039.3	I
200		8114.09	I
30	h	8121.0	I
30		8349.35	I
80		8357.04	I
300		8422.72	I
400		8552.60	I
50	h	8681.7	I
30	h	9018.95	I
50	h	9410.86	I
80	h	9415.37	I
150		9616.40	I
50		9741.1	I
100	h	9742.8	I
300	h	9805.38	I
500		9850.52	I
25		10456.47	I
11		10807.58	I
54		10894.00	I
70		11191.85	I
56		11277.66	I
17		11336.97	I
200		11454.59	I
200		11616.26	I
76		11670.77	I
25		11694.45	I
258		11739.78	I
96		11825.18	I
106		11835.82	I
254		11932.99	I
48		12009.50	I
111		12313.24	I
33		12335.6	I
42		12530.87	I
42		12536.5	I
37		12788.2	I
89		12888.5	I
187		12981.7	I
20		13000.3	I
187		13018.5	I
68		13081.5	I
378		13460.2	I
144		13608.2	I
13		15018.2	I
30		15464.2	I
20		17000.5	I

Tin (Cont.)

Intensity	Wavelength	
10	17807.5	I
20	20622.2	I
40	20861.7	I
8	21686.2	I
4	22131.7	I
3	22997.2	I
4	24327.2	I
4	24738.2	I

Sn III
Ref. 423 — C.H.C.

Intensity	Wavelength	
	Vacuum	
100	753.01	III
50	760.62	III
75	775.79	III
50	784.68	III
200	910.92	III
50	1010.92	III
50	1048.84	III
1000	1139.29	III
1000	1158.33	III
200	1161.09	III
100	1161.58	III
100	1180.62	III
1000	1184.25	III
200	1189.99	III
200	1204.06	III
2000	1210.52	III
100	1215.10	III
100	1218.14	III
100	1230.17	III
100	1231.38	III
500	1243.63	III
1000	1259.92	III
40	1276.31	III
1000	1305.97	III
1000	1327.34	III
200	1334.70	III
200	1346.05	III
1000	1347.65	III
200	1369.71	III
1000	1386.74	III
500	1410.61	III
200	1449.77	III
1000	1570.36	III
50	1674.29	III
500	1811.71	III
500	1941.86	III
50	1955.52	III

Sn IV
Ref. 423 — C.H.C.

Intensity	Wavelength	
	Vacuum	
50	605.23	IV
50	619.04	IV
50	628.73	IV
50	908.22	IV
500	956.25	IV
500	1019.72	IV
1000	1044.49	IV
100	1058.37	IV
50	1058.59	IV
1000	1073.41	IV
200	1087.50	IV
300	1096.92	IV
50	1103.24	IV
000	1119.34	IV
200	1120.68	IV
1000	1314.55	IV
1000	1437.52	IV
100	1532.90	IV

Sn V
Ref. 399, 423 — C.H.C.

Intensity	Wavelength	
	Vacuum	
120	355.14	V
150	361.01	V
100	372.55	V
200	1089.35	V
100	1132.79	V
200	1160.74	V
100	1176.26	V
100	1189.92	V
100	1205.72	V
2000	1251.38	V
100	1283.81	V
200	1294.36	V
100	1302.20	V

TITANIUM (Ti)
Z = 22

Ti I and II
Ref. 1 — C.H.C.

Intensity		Wavelength	
		Air	
140		2272.61	I
180		2273.28	I
130		2276.70	I
190		2279.96	I
150		2299.85	I
140		2302.73	I
190		2305.67	I
65		2380.81	I
35		2384.52	I
55		2418.36	I
75		2421.30	I
95		2424.24	I
40		2428.23	I
35		2433.22	I
19		2434.10	I
35		2440.21	II
65		2440.98	I
24		2450.44	II
24		2504.54	I
75		2517.43	II
40		2519.04	I
140		2520.54	I
75		2524.64	II
360		2525.60	II
29		2527.98	I
210		2529.85	I
190		2531.25	II
190		2534.62	II
130		2535.87	II
190		2541.92	I
65		2555.99	II
110		2571.03	II
50		2572.65	II
50		2580.82	I
35		2590.26	I
190		2593.64	I
65		2596.58	I
270		2599.92	I
340		2605.15	I
510		2611.28	I
75		2611.48	I
300		2619.94	I
170		2631.54	I
170		2632.42	I
640		2641.10	I
800		2644.26	I
950		2646.64	I
30		2649.30	I
15		2654.93	I
35		2657.19	I
85		2661.97	I
95		2669.60	I
130		2679.93	I
26		2684.80	I
30		2685.14	I
65		2688.82	I
26		2716.25	II
85		2725.07	I
75		2727.42	I
21		2731.13	I
40		2731.58	I
170		2733.26	I
55		2735.29	I
40		2735.61	I
85		2739.81	I
250		2742.32	I
40		2749.06	I
65		2757.40	I
95		2758.08	I
15		2761.29	II
250		2802.50	I
55		2805.70	I
30		2806.50	II
40		2809.17	I
75		2810.30	II
30		2812.98	I
30		2817.40	I
65		2817.84	I
		2817.87	II
65		2828.07	I
		2828.15	II
130		2832.16	II
190		2841.94	II
110		2851.10	II
40		2853.93	II
95		2862.32	II
55		2868.74	II
180		2877.44	II
280		2884.11	II
65		2888.93	II
55		2891.07	II
55		2905.66	I
30		2909.92	II
450		2912.08	I
340		2928.34	I
15		2931.03	
180		2933.55	I
26		2935.96	
150		2937.32	I
1100		2942.00	I
1300		2948.26	I
30		2954.58	
1600		2956.13	I
170		2956.80	I
30		2958.77	
26		2959.71	I
35		2959.99	I
170		2965.71	I
190		2967.22	I
26		2968.23	I
75		2970.38	I
30		2974.93	I
170		2983.31	I
35		3000.87	I
120		3017.19	II
140		3029.73	II
110		3046.68	II
130		3056.74	II
130		3057.40	II
170		3058.09	II
85		3059.74	II
1300	d	3066.22	II
		3066.35	II
70		3071.24	II
600		3072.11	II
1100		3072.97	II
1600		3075.22	II
2300		3078.64	II
3600		3088.02	II
180		3089.40	II
180		3097.19	II
180		3100.67	I
230		3103.80	II
230		3105.08	II
260		3106.23	II
70		3106.81	I
50		3110.67	II
50		3112.48	I
140		3117.67	II
720		3119.72	I
		3119.80	II
190		3123.07	I
240		3130.80	II
140		3141.54	I
95		3141.67	I
220		3143.76	II
240		3148.04	II
240		3152.25	II
240		3154.20	II
240		3155.67	II
500		3161.20	II
780		3161.77	II
1000		3162.57	II
1600		3168.52	II
2400		3186.45	I
1000		3190.87	II
3100		3191.99	I
50		3197.52	II
3800		3199.92	I
780		3202.54	II
50		3203.44	II
240		3203.83	I
50		3204.87	I
110		3213.14	II
260		3214.24	I
190		3214.75	II
1100		3217.06	II
110		3217.94	I
260		3218.27	II
110		3219.21	I
110		3221.38	I
1300		3222.84	II
220		3223.52	I
240		3224.24	II
140		3226.13	I
530		3228.60	II
780		3229.19	II
530		3229.42	II
110		3231.32	II
240		3232.28	II
6600		3234.52	II
220		3236.12	II
5200		3236.57	II
4100		3239.04	II
220		3239.66	II
2600		3241.99	II
1200		3248.60	II
950		3251.91	II
1200		3252.91	II
1200		3254.25	II
1200		3261.60	II
310		3271.65	II
310		3272.08	II
200		3278.29	II
260		3278.92	II
220		3282.33	II
530		3287.66	II
290		3292.08	I
170		3299.41	I
170		3306.88	I
220		3308.39	I
220		3308.81	II
260		3309.50	I
60		3309.73	I
110		3312.69	I
840		3314.42	I
		3314.52	I
290		3315.32	II
330		3318.02	II
550		3321.70	II
2900		3322.94	II
380		3326.76	II
2100		3329.46	II
550		3332.11	II
1800		3335.20	II
1100		3340.34	II
5700		3341.88	I
120		3342.15	I
260		3343.77	II
330		3346.73	II
4300		3349.04	II
12000		3349.41	II
120		3352.94	I
4100		3354.64	I
290		3358.28	I
290		3360.99	I
7200		3361.21	II
		3361.26	I
120		3361.84	I
1100		3370.44	I
4300		3371.45	I
140		3372.21	II
5700		3372.80	II
60		3374.35	II
2900	d	3377.48	I
		3377.58	I
290		3379.22	I
1400		3380.28	II
170		3382.31	I
5700		3383.76	II
170		3385.66	I
1400		3385.95	I
1400		3387.84	II
60		3388.76	II
140		3390.68	I
140		3392.71	I
1100		3394.58	II
60		3398.63	I
60		3402.42	II
60		3407.20	II
95		3409.81	II
60		3439.30	I
890		3444.31	II
60		3452.47	II
180		3456.39	II
600		3461.50	II
95		3467.26	I
600		3477.18	II
60		3478.92	I
240		3480.53	I
60		3485.69	I
60		3489.74	II
480		3491.05	II
60		3495.75	I
95		3499.10	I
890		3504.89	II
120		3506.64	I
600		3510.84	II
60		3520.25	II
310		3535.41	II
190		3547.03	I
120		3573.74	II
60		3574.24	I
60		3587.13	II
240		3596.05	II
190		3598.72	I
600		3610.16	I
190		3624.82	II
95		3635.20	I
4800		3635.46	I
120		3637.97	I
190		3641.33	II
6600		3642.68	I
180		3646.20	I
7200		3653.50	I
290		3654.59	I
660		3658.10	I
120		3659.76	II
380		3660.63	I
190		3662.24	II
380		3668.97	I
600		3671.67	I
3100		3685.20	II
120		3685.96	I
95		3687.35	I
600		3689.91	I

Titanium (Cont.)

140		3694.45	I
30		3698.18	I
60		3698.43	I
60		3700.08	I
120		3702.29	I
190		3704.30	I
140		3706.23	II
50		3707.53	I
290		3709.96	I
30		3715.40	I
450		3717.40	I
140		3721.64	II
330		3722.57	I
600		3724.57	I
380		3725.16	I
2900		3729.82	I
50		3735.67	I
60		3738.90	I
3300		3741.06	I
330		3741.64	II
160		3748.10	I
5200		3752.86	I
600		3753.64	I
140		3757.69	II
3300		3759.30	II
2900		3761.32	II
50		3761.89	II
60		3766.45	I
600		3771.66	I
30		3776.06	II
840		3786.04	I
120		3789.30	I
70		3795.90	I
60		3798.31	I
70		3818.22	I
60		3822.03	I
240		3828.19	I
95		3833.68	I
95		3836.78	I
60		3846.45	I
130		3853.05	I
130		3853.73	I
170		3858.14	I
240		3866.44	I
170		3868.40	I
120		3873.21	I
260		3875.26	I
170		3882.15	I
170		3882.33	I
500		3882.89	I
60	h	3888.02	I
70		3889.95	I
200	h	3895.25	I
85		3898.49	I
530		3900.54	II
180		3900.96	I
2600		3904.78	I
110	h	3911.19	I
500		3913.46	II
500		3914.34	I
24		3914.74	I
35		3919.82	I
290		3921.42	I
1100		3924.53	I
110		3926.32	I
890		3929.88	I
35		3932.02	II
70		3934.24	I
1100		3947.78	I
4500		3948.67	I
4500		3956.34	I
5200		3958.21	I
950		3962.85	I
950		3964.27	I
4800		3981.76	I
570		3982.48	I
60		3984.33	I
35		3985.25	I
60		3985.59	I
5700		3989.76	I
35		3994.70	I
7800		3998.64	I
70		3999.36	I
70		4002.49	I
70		4003.81	I
35		4005.97	I
70		4008.06	I
950		4008.93	I
190		4009.66	I
70		4012.39	II
180		4013.58	I
70		4015.38	I
35		4016.28	I
120	h	4017.77	I
140		4021.83	I
1200		4024.57	I
40		4025.14	II
190	h	4026.54	I
40		4027.48	I
40		4028.34	II
190	h	4030.51	I
40		4033.91	I
30		4034.91	I
110		4035.83	I
35	h	4040.32	I
290		4055.02	I
85		4057.62	I
85		4058.14	I
410		4060.26	I
200		4064.22	I
200		4065.10	I
840		4078.47	I
40		4079.72	I
290		4082.46	I
85		4099.17	I
220		4112.71	I
85		4122.17	I
40		4123.31	I
85		4123.57	I
130		4127.54	I
40		4129.17	I
40		4131.25	I
140		4137.29	I
85		4143.05	I
170		4150.96	I
85		4159.64	I
70		4163.65	II
35		4164.14	I
40		4166.32	I
85		4169.35	I
120		4171.03	I
40		4171.90	II
35		4183.30	I
360		4186.12	I
40		4188.69	I
70		4200.75	I
85		4203.46	I
35		4211.73	I
40		4224.79	I
40		4227.65	I
130		4237.89	I
85		4249.12	I
130		4256.04	I
70		4258.54	I
70		4261.60	I
330		4263.13	I
35		4265.71	I
40		4266.22	I
70		4270.14	I
85		4272.43	I
240		4274.58	I
120		4276.43	I
120		4278.23	I
30		4278.81	I
110		4281.38	I
220		4282.71	I
160		4284.99	I
890		4286.01	I
840		4287.40	I
30		4288.16	I
950		4289.07	I
120		4290.23	II
840		4290.94	I
120		4291.14	I
140		4294.12	II
840		4295.76	I
2000		4298.66	I
200		4299.23	I
200		4299.64	I
200		4300.05	II
2900		4300.56	I
4100		4301.09	I
85		4301.93	II
6000		4305.92	I
180		4307.90	II
35		4308.50	I
40		4311.65	I
85		4312.87	II
85		4314.35	I
1200		4314.80	I
360		4318.64	I
180		4321.66	I
190		4325.13	I
160		4326.36	I
30		4334.84	I
160		4337.92	II
24		4344.29	II
70		4346.11	I
35		4354.06	I
95		4360.49	I
24		4368.94	I
95		4369.68	I
60		4372.38	I
30		4388.08	I
170		4393.92	I
330		4395.04	II
60		4399.77	II
240		4404.28	I
60		4404.90	I
30		4405.68	I
60		4416.54	I
220		4417.28	I
60		4417.72	II
120		4421.76	I
120		4422.82	I
24		4424.39	I
30		4425.83	I
120		4426.06	I
890		4427.10	I
21		4430.02	I
85		4430.37	I
50		4431.28	I
30		4432.60	I
24		4433.58	I
170		4434.00	I
70		4436.59	I
30		4438.23	I
130		4440.35	I
50		4441.27	I
230		4443.80	II
24		4444.27	I
840		4449.15	I
30		4450.49	II
550		4450.90	I
840		4453.32	I
290		4453.71	I
950		4455.33	I
1100		4457.43	I
21		4462.09	I
70		4463.38	I
95		4463.54	I
290		4465.81	I
240		4468.50	II
240		4471.24	I
95		4474.85	I
95		4479.70	I
50		4480.59	I
530		4481.26	I
95		4482.69	I
19		4488.32	II
260		4489.09	I
24		4492.55	I
40		4495.01	I
240		4496.15	I
24		4497.73	I
200		4501.27	II
40		4503.78	I
21		4506.36	I
50		4511.17	I
780		4512.74	I
19		4515.62	I
1000		4518.03	I
95		4518.70	I
1000		4522.80	I
780		4527.31	I
6000		4533.24	I
240		4533.97	II
3600		4534.78	I
2400		4535.58	I
1200		4535.92	I
1200		4536.05	I
24		4537.23	I
24		4539.10	I
720		4544.69	I
950		4548.77	I
240		4549.63	II
950		4552.46	I
24		4555.08	I
720		4555.49	I
19		4557.86	I
19		4558.11	I
60		4559.92	I
50		4562.63	I
35		4563.43	I
110		4563.77	II
35		4570.91	I
240		4571.98	II
19		4585.84	I
24		4589.95	II
60		4599.23	I
21		4609.37	I
950		4617.27	I
24		4619.52	I
480		4623.09	I
190		4629.34	I
50	d	4634.87	
60		4637.88	I
240		4639.37	I
220		4639.67	I
190		4639.95	I
140		4645.19	I
120		4650.02	I
24		4656.04	I
720		4656.47	I
840		4667.59	I
70		4675.12	I
950		4681.92	I
21		4686.92	I
24		4690.80	I
190		4691.34	I
40		4693.68	I
24		4696.94	I
190		4698.76	I
120		4710.19	I
24		4715.30	I
65		4722.62	I
65		4723.17	I
55		4731.17	I
45		4733.43	I
18		4734.68	I
22		4742.11	I
170		4742.79	I
22		4747.68	I
310		4758.12	I
310		4759.28	I
45		4766.33	I
28		4769.77	I
65		4778.26	I
45		4781.72	I
110		4792.49	I
45		4796.22	I
35		4797.98	I
110		4799.80	I
28		4805.10	II
110		4805.43	I
45		4808.53	I
22		4811.08	I
40		4812.25	I
200		4820.42	I
22		4825.46	I
40		4836.13	I
470		4840.87	I
65		4848.47	I
290		4856.01	I
35		4864.18	I
200		4868.26	I
250		4870.14	I
28		4880.91	I
45		4882.35	I
400		4885.08	I
380		4899.91	I
320		4913.62	I
55		4915.24	I
130		4919.87	I
180		4921.77	I
55		4925.41	I
30		4926.16	I
150		4928.34	I
30		4937.74	I
95		4938.29	I
30		4941.58	I
21		4948.19	I
21		4958.25	I
55		4964.75	I
21		4966.04	I
65		4968.58	I
75		4973.05	I
120		4975.35	I
65		4977.74	I
120		4978.20	I
5800		4981.73	I
150		4989.15	I
4600		4991.07	I
30		4995.08	I
140		4997.10	I
4000		4999.51	I
230		5001.01	I
3600		5007.21	I
120		5009.65	I
230		5013.30	I
3200	d	5014.19	I
		5014.24	I
580		5016.17	I
840		5020.03	I
840		5022.87	I
580		5024.84	I
300		5025.58	I
1200		5035.91	I
840		5036.47	I
740		5038.40	I
1200		5039.95	I
75		5040.62	I
85		5043.59	I
35		5044.27	I
55		5045.41	I
26		5048.21	I
110		5052.87	I
21		5054.08	I
110		5062.11	I
35		5064.07	I
1400		5064.66	I
95		5065.99	I
35	h	5068.33	I
65		5069.35	I
130		5071.48	I
40		5085.34	I
130		5087.07	I
21		5103.15	I
55		5109.44	I
190		5113.44	I
270		5120.42	I
30		5129.15	II
270		5145.47	I
230		5147.48	I
210		5152.20	I
21	Bl	5166.86	I
1100		5173.75	I
40		5186.34	I
85		5188.70	II
30		5189.58	I
1300		5192.98	I
85	h	5194.04	I
65		5201.10	I
120		5206.08	I
75		5207.87	I
65		5208.42	

Titanium (Cont.)

Intensity		Wavelength	
1400		5210.39	I
65		5212.29	I
150		5219.71	I
95		5222.69	I
85		5223.64	I
250		5224.32	I
95		5224.56	I
190		5224.95	I
65		5226.56	II
120		5238.58	I
21		5246.15	I
55		5246.57	I
75		5247.31	I
21		5250.95	I
110		5252.11	I
75		5255.83	I
55		5259.99	I
55		5263.50	I
150		5265.98	I
40		5282.39	I
140		5283.45	I
35		5284.39	I
26		5288.81	
65		5295.79	I
120		5297.26	I
65		5298.44	I
26		5336.81	II
17		5341.50	I
75		5351.08	I
26		5366.65	I
55		5369.64	I
40		5389.18	I
55		5389.99	I
17		5396.60	I
85		5397.09	I
35		5404.02	I
110		5409.61	I
40		5426.26	I
75		5429.15	I
26		5436.73	I
17		5438.32	I
40		5446.64	I
11	Bl	5448.34	T
30		5448.90	I
21		5449.16	I
35		5453.65	I
55		5460.51	I
75		5471.21	I
35		5472.70	I
40	h	5473.55	I
85		5474.23	I
30		5474.46	I
120	h	5477.71	I
110		5481.43	I
75		5481.87	I
85	h	5488.20	I
150		5490.15	I
26		5490.84	I
110		5503.90	I
40		5511.78	I
340		5512.53	I
270		5514.35	I
320		5514.54	I
26		5530.49	
110		5565.49	I
13		5579.16	
21	h	5582.98	
30	h	5585.68	
65	Bl	5597.85	T
55	Bl	5629.28	T
17		5635.84	
250		5644.14	I
75		5648.58	I
26	Bl	5661.55	T
190		5662.16	I
75		5662.91	I
21		5673.42	I
130		5675.44	I
30	h	5679.94	I
95		5689.47	I
75		5702.68	I
35		5708.23	I
65		5711.88	I
40	h	5713.92	I
95		5715.13	I
55		5716.48	I
35		5720.48	I
85		5739.51	I
40		5740.02	I
19		5741.22	I
21		5752.84	I
19		5756.86	I
40	h	5762.27	I
55	h	5766.35	I
75	h	5774.05	I
30		5780.78	I
75	h	5785.98	I
65	Hl	5804.26	I
21	Bl	5814.96	T
40		5823.71	I
21	h	5841.18	
21		5852.34	
400		5866.46	I
65		5880.31	I
21	h	5888.68	
230		5899.32	I
55		5903.33	I
120		5918.55	I
150		5922.12	I
75		5937.82	I
120		5941.76	I
300		5953.17	I
200		5965.84	I
270		5978.56	I
340		5999.04	I
65		5999.68	I
21		6012.73	
110		6064.63	I
120		6085.23	I
120		6091.17	I
40		6092.81	I
40	h	6098.67	I
35	h	6121.01	I
120		6126.22	I
19		6138.38	I
30		6146.22	I
21		6149.74	I
30	Bl	6162.23	T
35		6186.15	I
95	h	6215.28	I
75	h	6220.49	I
65	h	6221.41	I
380		6258.10	I
380		6258.70	I
300		6261.10	I
65		6303.75	I
55		6312.24	I
26		6318.03	I
30		6336.10	I
35		6366.35	I
11		6419.10	I
17		6497.69	I
19		6508.14	I
55		6546.28	I
65		6554.23	I
11	h	6554.83	
75		6556.07	I
19	h	6565.62	I
14	h	6575.18	I
35		6599.11	I
18	Bl	6651.46	T
18	h	6666.55	I
22	h	6667.74	
9		6668.39	
18		6677.18	I
22	b	6691.21	T
26		6716.68	I
16	Bl	6723.95	T
80		6743.12	I
22		6745.52	I
18		6844.64	
18		6860.39	
35		6861.47	I
9		6873.92	I
12		6913.19	I
14	h	6933.15	I
14	h	6943.70	I
23		6996.63	I
15		7004.66	I
14		7008.35	I
14		7010.94	I
14	h	7035.86	I
40		7038.80	I
14		7050.65	I
40	Bl	7054.51	T
23		7069.11	I
23		7072.05	
45	Bl	7087.89	T
30	b	7124.9	T
40	Bl	7125.61	T
26		7138.91	I
26		7167.13	
23		7171.53	
55		7189.89	I
26	b	7203.64	T
260		7209.44	I
60		7216.20	I
130		7244.86	I
130		7251.72	I
19		7263.40	
19		7266.29	I
19	b	7269.05	T
15		7315.56	I
26		7318.39	I
120		7344.72	I
11		7352.16	I
90		7357.74	I
60		7364.11	I
26		7440.60	I
9		7474.94	I
26		7489.61	I
19		7496.12	I
12		7580.55	I
9	Bl	7589.62	T
15		7614.50	I
23		7654.44	I
11	Bl	7705.21	T
30		7949.17	I
26	h	7961.58	I
60		7978.88	I
9		7979.07	I
30		7996.53	I
7	h	8003.55	
55		8024.84	I
30		8068.24	I
8		8267.62	
14	h	8306.31	I
9	h	8307.41	I
9	h	8311.76	I
8	h	8312.85	I
12		8334.37	I
14		8353.15	I
75		8364.24	I
100		8377.85	I
100		8382.54	I
55		8382.82	I
75		8396.87	I
120		8412.36	I
19		8416.98	I
15		8424.41	I
170		8426.52	I
490		8434.94	I
240		8435.70	I
40		8438.93	I
40		8450.89	I
9	h	8457.10	I
19	h	8467.15	I
45		8468.50	I
15		8496.04	I
19	h	8518.05	I
40		8518.32	I
14		8539.38	I
40		8548.12	I
9		8569.77	I
9	h	8598.18	I
90		8675.39	I
45		8682.99	I
23		8692.33	I
19		8734.69	I
23		8766.64	I
15	h	8778.71	I

Ti III
Ref. 378 — C.H.C.

Intensity		Wavelength	
		Vacuum	
6		1282.48	III
6		1286.23	III
15		1286.36	III
10		1289.30	III
10		1291.62	III
10		1293.23	III
15		1294.70	III
10		1295.88	III
20		1298.66	III
20		1298.97	III
12		1327.59	III
10		1420.04	III
10		1420.44	III
10		1421.63	III
10		1421.77	III
12		1422.40	III
10		1424.14	III
23		1455.19	III
10		1498.70	III
		Air	
10		2199.22	III
12		2237.77	III
10		2327.02	III
15		2331.35	III
15		2331.66	III
15		2334.34	III
17		2339.00	III
18		2346.79	III
18		2374.99	III
22		2413.99	III
25		2516.05	III
24		2527.84	III
23		2540.06	III
24		2563.44	III
23		2565.42	III
22		2567.56	III
15		2576.47	III
15		2580.46	III
10		2692.16	III
12		2701.96	III
22		2984.75	III
12	d	3354.71	III
12		3872.50	III
12		3881.21	III
12		3893.63	III
10		3896.33	III
15		3915.47	III
12		3921.38	III
10		3921.61	III
12		3922.95	III
10		3924.86	III
10		4060.21	III
10		4119.14	III
11		4215.52	III
11		4269.84	III
11		4296.70	III
10		4348.04	III
11		4433.91	III
10		4540.22	III
15		4549.84	III
10	d	4555.46	III
15	d	4572.20	III
10		4649.45	III
12		4652.86	III
10		4874.00	III
10		4950.10	III
10		4971.19	III
10		5083.80	III
14		5147.31	III
12	d	5226.28	III
11		5247.49	III
17		5278.12	III
10		5278.70	III
12		5298.43	III
16		5301.20	III
15		5306.88	III
10		5395.69	III
12		5533.01	III
10		5817.44	III
12		6611.38	III
18		6621.58	III
10		6629.37	III
14		6647.47	III
18		6667.99	III
15		6674.19	III
14		6707.76	III
12		6724.80	III
16		6734.10	III
15		6862.26	III
12		6874.35	III
10		6896.12	III
12		7015.38	III
10		7071.93	III
20		7072.64	III
18		7084.57	III
15		7124.13	III
11		7171.79	III
10		7175.92	III
10		7217.50	III
9		7225.55	III
12		7270.67	III
14		7316.30	III
10		7316.68	III
12		7379.96	III
10		7408.13	III
10		7457.85	III
15		7506.87	III
17		7507.68	III
10		7523.85	III
12		7544.29	III
9		7566.25	III
10	h	8172.21	III
9	h	8173.37	III
9	h	8178.00	III
10		8182.42	III
9	h	8192.68	III
9	h	8194.75	III
9	h	8263.67	III
15	h	8267.32	III
10		8338.54	III
12		8394.20	III
20		8466.87	III
5		8699.85	III
3		9017.10	III

Ti IV
Ref. 428 — C.H.C.

Intensity		Wavelength	
		Vacuum	
10		776.76	IV
18		779.07	IV
16		781.73	IV
8		1183.64	IV
10		1195.21	IV
18		1451.74	IV
20		1467.34	IV
12		1469.19	IV
		Air	
20		2067.56	IV
18		2103.16	IV
10		2359.14	IV
10		2359.50	IV

Intensity		Wavelength	
8		2541.79	IV
10		2546.88	IV
5		2862.60	IV
6		2929.96	IV
14		2937.33	IV
12		2957.31	IV
15		3541.36	IV
17		3576.44	IV
10		3581.39	IV
13		4131.22	IV
14		4133.78	IV
10		4397.33	IV
9		4403.45	IV
15		4618.11	IV
20		5398.93	IV
8		5470.98	IV
18		5492.51	IV
10		5517.72	IV
14		5877.79	IV
15		5885.96	IV
7		5891.15	IV
6		6231.62	IV
17		6246.65	IV
11		6247.74	IV
15		6292.41	IV
12		6913.85	IV
15		6978.51	IV
9		7491.37	IV
8		7494.77	IV
5	h	7652.12	IV
8	h	7706.85	IV

Ti V
Ref. 427 — C.H.C.

Vacuum

Intensity		Wavelength	
12		225.35	V
10		228.91	V
17		252.96	V
7		323.36	V
7		461.41	V
8		474.69	V
8		483.99	V
10	d	488.58	V
15		498.26	V
14		502.08	V
7		502.71	V
12		504.66	V
7		506.47	V
8		513.37	V
7		523.05	V
12		524.58	V
13		526.57	V
8		529.32	V
10		535.84	V
10		535.89	V
8	d	540.14	V
8		541.46	V
9		541.71	V
7		543.10	V
7		543.34	V
7		1128.55	V
8		1192.35	V
9		1198.66	V
9		1222.36	V
10		1230.36	V
11		1239.96	V
10		1241.67	V
7		1246.13	V
8		1268.49	V
8		1306.11	V
8		1411.31	V
9		1675.15	V
8		1687.16	V
11		1717.40	V
8		1759.76	V
7		1771.45	V
10		1841.49	V
7		1864.45	V
7		1881.89	V
7		1920.16	V
7		1988.75	V

TUNGSTEN (W)
Z = 74

W I and II
Ref. 1 — C.H.C.

Air

Intensity		Wavelength	
5800		2001.71	II
13000		2008.07	II
5100		2009.98	II
4100		2010.23	II
4100		2014.23	II
7300		2026.08	II
15000		2029.98	II
2700		2035.03	II
5300		2049.63	II
2300		2065.57	II
3400		2071.21	II
2200		2075.59	II
9700		2079.11	II
3600		2088.19	II
2200		2089.14	II
1700		2090.48	I
6100		2094.75	II
2400		2098.60	II
2200		2100.67	II
1500		2101.54	I
1500		2106.18	II
1300		2110.34	II
2100		2118.87	II
2400		2121.59	II
850		2153.56	II
850		2157.80	II
1500		2166.32	II
480		2182.90	I
440		2194.52	II
1300		2204.48	II
460		2248.75	II
460		2249.80	I
180		2270.24	II
95		2271.37	I
510		2277.58	I
160		2284.91	I
320		2285.17	I
530	d	2294.49	I
		2294.54	II
270		2298.33	I
240		2303.83	II
240		2306.59	I
340		2309.02	I
440		2313.17	I
220		2314.17	I
190		2315.02	II
460		2321.63	I
290		2326.09	II
390	d	2326.56	I
		2326.70	I
75		2328.31	II
130		2333.77	II
210		2341.37	I
75		2349.26	II
120	d	2350.37	II
320		2354.61	I
60		2358.81	II
580		2360.44	I
850		2363.07	I
60		2364.22	II
510		2374.47	I
210		2382.99	I
670		2384.82	I
240		2389.08	I
120		2390.37	II
120		2392.93	II
730		2397.09	II
560		2397.73	I
560		2397.98	I
75		2404.24	II
1700	d	2405.58	I
		2405.69	I
75		2411.54	II
320		2414.04	I
610		2415.68	I
50		2419.34	II
50		2421.01	II
870		2424.21	I
190		2427.49	II
170		2429.39	II
580		2431.08	I
630		2433.98	I
60		2435.01	II
1800		2435.96	I
250		2436.62	I
580		2444.06	I
160		2446.39	II
270		2448.39	I
270		2451.35	I
780		2451.48	II
870		2452.00	I
430		2454.72	I
630		2454.98	I
780		2455.51	I
780		2456.53	I
1100		2459.30	I
270		2460.16	I
480		2462.79	I
270		2464.30	I
230		2466.52	II
1400		2466.85	I
75		2470.80	II
480		2472.51	I
1200		2474.15	I
290		2477.80	II
870		2480.13	I
390		2480.96	I
1500		2481.44	I
480	d	2482.10	I
		2482.21	I
29		2484.40	II
580		2484.74	I
390		2487.50	I
270		2488.77	II
390		2489.23	II
75		2492.93	II
630		2495.26	I
230		2496.64	II
95		2497.48	II
140		2499.69	II
40		2500.11	II
		2501.90	II
680		2504.70	I
270		2506.02	I
24		2508.00	II
250		2510.17	I
75		2510.47	II
60		2518.14	II
310		2520.46	I
780		2521.32	I
270		2522.04	II
780		2523.41	I
430		2527.76	I
780		2533.64	I
50		2534.82	II
580		2545.34	I
1200		2547.14	I
50		2549.09	II
40		2550.10	II
780		2550.38	I
2700		2551.35	I
450		2553.82	I
410		2554.86	II
580		2555.09	II
		2555.21	I
310		2556.75	I
290		2560.12	I
730		2561.97	I
230		2563.16	II
110		2563.91	II
530		2571.44	II
170	d	2572.24	II
		2572.35	II
75		2573.95	II
190		2579.26	II
290		2580.34	I
870		2580.49	I
40		2581.20	II
390		2584.39	I
390		2589.17	II
170		2591.49	II
110		2598.74	II
370		2601.96	I
75		2602.51	II
75		2603.02	II
270		2603.54	I
680		2606.39	I
320		2607.38	I
370		2608.32	I
970		2613.08	I
480		2613.82	I
230		2615.12	I
70		2615.44	II
210		2619.18	I
400		2620.25	I
400		2622.21	I
400		2625.22	I
210		2628.26	I
400		2632.48	I
400		2632.70	I
810		2633.13	I
290		2636.54	I
400	d	2638.62	I
		2638.75	I
210		2645.69	I
650		2646.18	I
400		2646.73	I
75		2647.74	II
40		2653.42	II
80		2653.57	II
1600		2656.54	I
400		2657.38	I
400	d	2658.04	II
		2658.18	I
810		2662.84	I
260		2664.97	I
75		2666.49	II
210		2669.30	I
810		2671.47	I
80		2673.59	II
650		2677.28	I
160	d	2677.79	II
		2677.91	I
400		2678.88	I
2100		2681.42	I
290		2683.35	I
210		2691.09	I
650		2695.67	I
210		2697.71	II
650		2699.59	I
400		2700.01	I
40		2701.48	II
160		2702.11	II
210		2702.52	I
400		2706.58	I
400		2708.59	I
400	d	2708.80	I
		2708.93	I
80		2709.58	II
40		2710.78	II
400		2715.50	I
80		2716.32	II
80		2718.04	II
2100		2718.91	I
320		2719.33	I
210		2719.86	I
2600		2724.35	I
210		2724.62	I
400		2725.03	I
		2725.06	I
80		2729.62	II
75		2740.79	II
650		2748.84	I
40	d	2760.74	II
80		2761.59	II
400		2762.34	I
400		2764.27	II
210		2768.98	I
400		2769.74	I
810		2770.88	I
210		2773.70	I
810		2774.00	I
810		2774.48	I
160		2776.50	II
40		2778.69	II
210		2787.98	I
340		2791.96	I
810		2792.70	I
80		2799.03	II
400		2799.93	I
160	d	2801.05	II
		2801.17	I
130		2805.92	II
40		2812.25	II
810		2818.06	I
160		2822.57	II
260		2829.82	I
1600		2831.38	I
810		2833.63	I
210		2835.64	I
400		2841.57	I
810		2848.02	I
650		2856.03	I
650		2866.06	I
230		2878.72	I
610		2879.11	I
610		2879.40	I
440		2896.01	I
1500		2896.44	I
230		2910.48	I
270		2911.00	I
360		2918.25	I
50		2918.63	II
360		2923.10	I
230		2923.54	I
230		2925.13	I
690		2935.00	I
2400		2944.40	I
2400		2946.99	I
480		2947.39	I
210		2952.29	II
440		2964.52	I
480		2977.11	I
		2977.21	I
730	d	2979.71	I
		2979.86	I
400		2993.61	I
240		2995.26	I
190		3009.09	I
360		3013.79	I
520		3016.47	I
770		3017.44	I
110		3024.50	II
210		3024.93	I
310	d	3026.67	I
		3026.79	I
160		3033.56	I
160		3034.19	I
160		3039.31	I
440	d	3041.73	I
		3041.86	I
270		3043.80	I
440		3046.44	I
110		3048.66	I
810		3049.69	I
110		3064.93	I
180		3073.28	I
110		3077.52	II
180	d	3084.83	I
		3084.91	I
370		3093.50	I
240		3107.23	I
240		3108.02	I

Tungsten (Cont.)

Intensity		Wavelength	
230		3117.57	I
260		3120.18	I
160		3133.88	I
130		3141.42	I
65		3149.85	II
290		3163.42	I
130		3164.44	I
130		3165.38	I
320		3176.60	I
130		3179.06	I
190		3181.82	I
130		3184.05	I
130		3184.42	I
65		3189.24	II
390		3191.57	I
390		3198.84	I
520		3207.25	I
140		3208.28	I
1000		3215.56	I
140		3221.21	I
140		3221.91	I
190		3232.49	I
140		3237.09	I
140		3242.03	I
140		3252.29	I
210		3254.36	I
140		3259.43	I
210		3259.66	I
210	d	3266.62	I
		3266.77	I
150		3281.94	I
150		3293.71	I
730		3300.82	I
440		3311.38	I
440		3326.20	I
440		3331.69	I
150		3354.45	I
150		3371.04	I
390		3373.75	I
150		3412.96	I
150		3413.53	I
150		3422.42	I
150		3427.71	I
230		3429.59	I
240		3443.00	I
160		3477.94	I
400		3495.24	I
160		3508.73	I
160		3510.02	I
160		3526.85	I
160		3535.54	I
160		3537.45	I
650		3545.22	I
160		3568.04	I
240		3570.65	I
80		3572.48	II
160		3575.22	I
80		3592.42	II
240		3606.06	I
80		3613.79	II
1900		3617.52	I
160		3622.34	I
130		3627.24	I
320		3631.94	I
240		3641.41	II
80		3646.52	II
80		3657.59	II
160		3675.55	I
650		3682.08	I
400		3683.30	I
		3683.39	I
160		3683.93	I
570		3688.06	I
810		3707.92	I
60		3716.08	II
100		3719.39	I
50		3736.22	II
120		3741.71	I
510		3757.92	I
680		3760.13	I
1000		3768.45	I
120		3769.21	I
120		3769.86	I
340		3773.71	I
1000		3780.77	I
170		3792.76	I
290		3809.22	I
190		3810.38	I
260		3810.79	I
1400		3817.48	I
110		3829.13	I
1100		3835.06	I
290		3838.51	I
730		3846.22	I
250		3847.49	I
27		3851.57	II
150		3855.55	I
150		3859.30	I
180		3864.34	I
1800		3867.99	I
250		3872.84	I
110		3874.41	I
730		3881.41	I
110		3892.72	I
140	h	3897.91	I
150		3935.03	I
120		3936.97	I
120		3947.98	I
120		3952.52	I
120		3952.90	I
160		3953.15	I
200		3955.30	I
160		3965.14	I
130		3968.59	I
150	h	3970.80	I
130		3979.29	I
130		3980.64	I
250		3983.29	I
8600		4008.75	I
540		4015.22	I
170	h	4016.52	I
220		4019.23	I
130		4022.12	I
180		4028.79	I
180		4036.86	I
140		4039.85	I
140	h	4044.28	I
910		4045.59	I
180		4064.79	I
150		4069.79	I
730		4069.95	I
340		4070.61	I
100		4071.93	I
5000		4074.36	I
150		4082.96	I
130		4088.33	I
100		4095.69	I
1000		4102.70	I
150		4109.75	I
100		4111.82	I
150		4118.05	I
100		4118.19	I
100		4120.85	I
100		4125.16	I
150		4126.80	I
100		4133.48	I
540		4137.46	I
150		4138.02	I
110		4142.25	I
140		4145.16	I
110		4145.95	I
160		4154.66	I
160		4170.53	I
450		4171.17	I
160		4204.40	I
220		4207.05	I
110		4215.38	I
250		4219.37	I
110		4222.04	I
150		4234.34	I
290		4241.44	I
540		4244.36	I
290		4259.35	I
200		4260.29	I
200		4263.30	I
1400		4269.38	I
110		4269.77	I
220		4274.55	I
160		4275.49	I
160		4276.74	I
110		4282.34	I
110		4286.01	I
110		4294.10	I
4100		4294.61	I
2200		4302.11	I
160		4306.87	I
110		4307.64	I
110		4332.13	I
100		4347.00	I
150		4355.17	I
100		4361.81	I
150		4364.78	I
100	d	4365.95	I
		4366.07	I
150		4372.52	I
200		4378.48	I
180		4384.85	I
100		4403.95	I
200		4408.28	I
130		4412.19	I
160		4436.90	I
140		4460.49	I
140		4466.34	I
140		4466.74	I
640		4484.19	I
160		4504.84	I
130		4512.88	I
120		4513.25	I
150		4543.54	I
150		4546.47	I
150		4551.82	I
140		4570.64	I
170		4588.73	I
140		4599.94	I
140		4609.89	I
160		4613.30	I
100		4642.53	I
130		4657.42	I
640		4659.87	I
640		4680.51	I
100		4693.72	I
140		4757.54	I
790		4843.81	I
380		4886.90	I
220		4982.59	I
330		5006.15	I
220		5015.30	I
820		5053.28	I
210		5054.60	I
210		5069.12	I
120		5071.74	I
770		5224.66	I
27		5500.49	I
27		5503.44	I
10		5508.61	I
220		5514.68	I
15		5531.38	I
15		5537.72	I
13		5568.09	I
13		5604.31	I
11		5631.27	I
27		5631.94	I
65		5648.37	I
35		5660.72	I
27		5674.39	I
13		5676.60	I
15		5676.90	I
15		5697.79	I
55		5735.09	I
13		5749.24	I
11		5756.10	I
13		5793.06	I
13		5796.49	I
45		5804.85	I
13	d	5806.05	I
		5806.24	I
13		5833.61	I
13		5838.97	I
17		5845.27	I
28		5851.58	I
11		5856.61	I
22		5864.63	I
11		5874.22	I
13		5880.21	I
13		5891.61	I
13		5901.20	I
40		5902.64	I
13		5928.58	I
55		5947.57	I
13		5953.96	I
13		5956.19	I
27		5960.83	I
55		5965.86	I
27		5972.51	I
20		5978.86	I
20		5983.82	I
13		6009.01	I
55		6012.78	I
40		6021.52	I
20		6028.32	I
20		6043.31	I
13		6049.92	I
13		6065.08	I
22		6081.44	I
13		6111.66	I
13		6115.52	I
22		6128.25	I
13		6143.94	I
20		6153.72	I
20		6154.87	I
20		6203.51	I
20		6254.28	I
27		6285.88	I
45		6292.02	I
20		6303.21	I
13		6386.47	I
35		6404.21	I
40		6445.12	I
11		6508.05	I
15		6532.39	I
13		6538.11	I
13		6563.20	I
20		6573.93	I
11		6607.13	I
11		6609.05	I
17		6611.62	I
11		6621.74	I
13		6678.42	I
15		6693.08	I
5		6746.56	I
5		6764.45	I
7		6805.31	I
9		6814.92	I
9		6820.27	I
8		6828.43	I
4		6853.74	I
4		6876.01	I
5		6908.29	I
9		6934.23	I
8		6964.12	I
13		6984.27	I
8		6993.27	I
4		6994.06	I
8		7017.88	I
3		7028.68	I
3		7098.22	I
4	h	7111.18	I
15		7140.52	I
9		7162.64	I
5	h	7191.33	I
5		7198.62	I
11		7200.16	I
5		7216.35	I
4		7226.06	I
8		7237.12	I
5		7274.47	I
10		7278.24	I
15		7285.81	I
15		7296.55	I
3		7298.25	I
7		7385.08	I
4		7451.39	I
3		7456.37	I
8		7483.35	I
7		7504.13	I
10		7509.00	I
3		7520.66	I
9		7537.45	I
9		7550.48	I
17		7569.92	I
5		7582.88	I
3		7612.18	I
17		7614.15	I
3		7631.29	I
3		7654.81	I
13		7688.97	I
4		7701.01	I
5		7761.16	I
3		7776.73	I
11		7784.15	I
7		7808.96	I
2		7823.82	I
4		7863.47	I
2		7867.04	I
4		7880.40	I
5		7886.48	I
9		7940.92	I
3		7957.06	I
22		8017.19	I
7		8054.89	I
22		8055.64	I
5		8060.38	I
13		8123.82	I
5		8143.19	I
3		8165.72	I
5		8210.22	I
4		8322.05	I
10		8338.08	I
4		8348.81	I
7		8358.72	I
3		8382.94	I
4		8402.60	I
4		8475.14	I
27		8585.11	I
10		8594.42	I
8		8613.27	I
3		8614.50	I
13		8865.53	I

URANIUM (U)
Z = 92

U I and II
Ref. 1, 303 — J.G.C.

Intensity	Wavelength	
	Air	
440	2565.41	II
340	2569.71	II
340	2591.25	II
610	2635.53	II
470	2645.47	II
340	2669.17	II
470	2683.28	II
320	2691.04	II
370	2706.95	II
370	2733.97	II
470	2754.16	II
340	2762.85	II
390	2770.04	II
410	2784.45	II
830	2793.94	II
870	2802.56	II
630	2807.05	II
440	2808.98	II
630	2817.96	II
870	2821.12	II
390	2824.37	II
680	2828.90	II
920	2832.06	II
360	2837.19	II
460	2839.89	II

Uranium (Cont.)

Intensity		Wavelength	Spectrum
360		2842.09	II
360		2849.48	II
390		2860.47	II
970		2865.68	II
340		2870.97	II
490		2882.74	II
460		2887.25	II
410		2888.26	II
1200		2889.62	II
320		2894.14	II
410		2894.51	II
780		2906.80	II
780		2908.28	II
320		2914.25	II
360		2914.63	II
440	p	2921.68	II
320		2927.38	II
490		2928.60	II
580		2931.41	II
440		2932.61	II
340		2933.86	II
530	p	2940.37	II
1300		2941.92	II
830		2943.90	II
340		2948.09	II
390		2954.77	II
580		2956.06	II
460		2965.03	II
580		2967.94	II
580		2971.06	II
410		2976.35	II
320		2982.74	II
530		2984.61	II
410		2992.72	II
360		3007.91	II
320		3021.22	II
630		3022.21	II
320		3024.51	II
320		3028.19	II
630		3031.99	II
490		3033.18	II
490		3044.16	II
580		3050.20	II
630		3057.91	II
460		3061.62	II
630		3062.54	II
580		3072.78	II
580		3093.01	II
320		3095.75	II
320		3098.01	II
580		3102.39	II
460		3104.15	II
970		3111.62	II
530		3119.35	II
680		3124.95	II
530		3139.61	II
410		3144.97	II
490		3145.56	II
680		3149.24	II
530		3153.11	II
340		3176.21	II
340		3177.33	II
340		3206.05	II
730		3229.50	II
680		3232.16	II
440		3244.22	II
340		3265.79	II
440		3270.12	II
440		3288.21	II
730		3291.33	II
1100		3305.89	II
390		3337.79	II
440		3341.66	II
390		3357.84	I
730		3390.38	I
340		3394.77	II
580		3424.56	II
580		3435.49	I
360		3453.55	II
320		3454.23	II
320		3457.05	II
320		3457.71	II
360		3459.92	I
320		3462.22	I
460		3463.55	I
630		3466.30	I
390		3472.52	II
320		3473.43	I
360		3480.36	I
680		3482.49	II
1600		3489.37	I
390		3493.33	II
340		3494.00	I
320		3494.84	II
530		3496.41	II
630		3500.08	I
320		3504.01	I
320		3505.07	II
780		3507.34	I
320		3508.84	II
1600		3514.61	I
390		3509.66	II
320		3513.67	I
390		3519.96	II
390		3531.11	II
630		3533.57	II
320		3534.33	I
530		3540.47	II
320		3542.57	I
390		3547.19	II
320		3549.20	I
1200		3550.82	II
320		3552.17	II
680		3555.32	I
320		3561.41	I
1200		3561.80	I
390		3563.66	I
2300		3566.59	I
530		3569.08	I
320		3574.76	I
360		3577.92	I
630		3578.72	II
360		3581.84	II
3200		3584.88	I
320		3590.50	II
390		3591.74	I
460		3593.52	II
460		3605.27	I
360		3606.32	II
320		3616.33	I
320		3616.76	II
320		3620.08	I
320		3622.70	I
390		3623.06	II
460		3630.73	II
840		3638.20	I
310		3640.76	II
420		3644.24	I
310		3645.03	II
660		3651.54	I
490		3652.06	I
960		3659.15	I
2800		3670.07	II
380		3678.75	II
540		3691.92	II
330		3693.70	II
540		3700.57	II
1100		3701.52	II
350		3713.55	I
300		3717.42	II
350		3718.11	II
350		3729.82	II
350		3732.62	II
350		3733.07	II
600		3738.04	II
300		3744.25	II
680		3746.42	II
350		3747.14	II
950		3748.68	II
600		3751.17	I
350		3752.66	II
350		3755.48	II
490		3758.35	I
350		3759.24	II
330		3763.26	I
490		3764.57	II
430		3766.89	I
330		3769.53	II
540		3773.43	I
300		3776.48	I
380		3780.71	II
1900		3782.84	II
430		3783.84	II
570		3793.10	II
380		3793.26	I
380		3793.57	II
380		3808.92	I
380		3809.22	II
1900		3811.99	I
380		3813.79	II
380		3814.06	II
750		3826.51	II
2000		3831.46	II
1200		3839.63	I
490	p	3848.60	II
620		3854.22	I
2400		3854.64	II
4900		3859.57	II
490		3861.17	II
1900		3865.92	II
380		3866.80	II
1500		3871.03	I
620		3874.04	II
620		3878.08	II
1000		3881.45	II
490		3882.36	II
380		3883.28	II
2200		3890.36	II
620		3892.68	II
490		3894.12	I
490		3896.77	II
620		3899.78	II
410		3902.55	II
460		3904.30	II
380		3906.45	I
330		3911.67	II
380		3915.88	II
330		3926.21	I
330		3926.72	I
430		3930.98	II
2000		3932.02	II
490		3935.38	II
330		3940.48	II
1200		3943.82	I
300		3948.44	I
300		3953.58	II
360		3954.67	II
350		3964.21	I
600		3966.52	II
1200		3985.79	II
460		3990.42	II
380		3992.53	II
350		3998.24	II
350		4004.06	II
430		4005.21	I
570		4017.72	II
300		4018.99	II
1000		4042.75	I
520		4044.41	II
410		4047.61	I
1600		4050.04	II
540		4051.91	II
300		4054.30	II
430		4058.19	II
880		4062.54	II
520		4067.75	II
410		4071.12	II
300		4074.48	II
330		4076.69	II
330		4080.60	II
2200		4090.13	II
460		4093.03	II
380		4106.38	II
810		4116.10	II
410		4124.73	II
410		4128.34	II
460		4141.22	II
880		4153.97	I
380		4156.65	I
350		4163.68	II
1400		4171.59	II
300		4189.27	II
350		4222.37	I
1000		4241.67	II
520		4244.37	II
680		4341.69	II
430	h	4355.74	I
430		4362.05	I
330		4393.59	I
600		4472.33	II
240		4515.28	II
620		4543.63	II
300		4620.21	I
240		4627.07	II
210		4631.62	I
220		4646.60	II
140		4666.85	II
100		4671.40	II
170		4689.07	II
100		4702.51	II
160		4722.72	II
120		4731.59	II
100		4755.74	II
150		4756.81	I
100		4772.70	II
100		4860.99	II
110		5008.21	II
170		5027.38	I
70		5117.24	II
80		5160.32	II
55		5164.14	I
55		5184.57	II
45		5204.31	II
45		5247.75	II
45		5257.04	II
70		5280.38	I
55		5386.19	II
80		5475.70	II
70		5480.26	II
70		5481.20	II
45		5482.53	II
160		5492.95	II
70		5527.82	II
70		5564.17	I
45		5581.59	II
55		5620.78	I
70		5780.59	I
70		5798.53	II
45		5836.02	I
55		5837.68	II
230		5915.39	I
55		5971.50	I
100		5976.32	I
45		5997.31	I
28		6017.38	II
55		6051.74	II
45		6067.22	II
50		6077.29	I
28		6087.34	II
40		6171.86	I
35		6175.39	I
28		6280.18	II
28		6359.29	I
55		6372.46	I
28		6378.52	II
28		6392.77	I
90		6395.42	I
110		6449.16	I
35		6464.98	I
90		6826.92	I
35		6876.74	II
23		7074.79	I
27		7101.61	I
30		7128.90	I
16		7147.89	I
16		7254.45	I
23		7425.50	I
45		7533.93	I
16		7619.35	I
50		7881.94	I
18		7970.46	I
16		8174.66	I
18		8262.06	I
16		8318.35	I
16		8337.50	II
18		8381.87	I
16		8441.21	I
35		8445.39	I
18		8450.03	I
16		8570.52	I
75		8607.95	I
23		8691.28	I
18		8710.76	I
18		8753.69	I
30		8757.76	I
16		8951.96	I
16		8989.92	I
10		9093.67	I
10		9139.56	I
10		9201.51	I
10		9265.34	I
10		9276.44	I
10		9385.90	I
10		9653.26	I
10		9819.00	I
10		9819.05	I
10		9868.36	I
10		9932.76	I
10		9964.11	I
50		10157.91	I
50		10259.55	I
100		10554.93	I
50		10799.78	I
25		10823.93	I
25		11095.77	I
75		11167.84	I
50		11294.13	I
100		11384.13	I
25		11410.43	I
50		11503.38	I
25		11568.81	I
20		11784.72	II
100		11859.42	I
100		11908.83	I
100		12250.46	I
25		13088.28	I
100		13185.16	I
75		13306.23	I
100		13961.58	I
50		16906.00	I
50		17451.11	I
50		18136.65	I
50		18366.96	I
75		18634.43	I
25		19029.39	I
10		20201.13	I
10		20271.41	I
10		20374.13	I
10		20517.29	I
10		20690.64	I
10		20772.19	I
10		21008.38	I
10		21099.98	I
10		21112.14	I
20		21144.90	II
10		21674.51	I
10		21693.38	I
75		21910.22	I
10		22110.73	I
10		23156.76	I
10		23948.19	I
10		29557.07	I

VANADIUM (V)
Z = 23

V I and II
Ref. 1 — C.H.C.

Intensity	Wavelength	
	Air	
2100	2092.44	I

Vanadium (Cont.)

Intensity		Wavelength	Spectrum
40		2384.00	II
40		2384.28	I
60		2386.96	I
60		2388.92	I
75		2390.87	I
75		2391.26	I
85		2392.90	I
70		2397.78	I
70		2398.27	I
70		2399.96	I
120		2406.75	I
110		2407.90	I
120		2415.33	I
120		2416.75	I
100		2420.12	I
100		2421.06	I
100		2421.98	I
110		2428.28	I
110		2435.52	I
140		2501.61	I
150		2506.90	I
240		2507.78	I
180		2511.65	I
180		2511.95	I
180		2517.14	I
240		2519.62	I
410		2526.22	I
210		2527.90	II
120		2528.47	II
150		2528.84	II
240		2530.18	I
110		2549.28	II
120		2552.65	I
210		2562.13	I
110		2564.82	I
230		2574.02	I
140		2630.67	II
130		2642.21	II
150		2645.26	I
140		2651.90	I
150		2656.22	I
180		2661.42	I
290		2672.00	II
380		2677.80	II
270		2678.57	II
380		2679.32	II
180		2682.87	II
180		2683.09	II
1100		2687.96	II
170		2688.72	II
150		2689.88	II
230		2690.24	II
240		2690.79	II
120		2696.99	I
120		2697.74	I
680		2700.94	II
380		2702.19	II
530		2706.17	II
150		2706.70	II
110		2707.86	II
170		2711.74	II
120		2714.20	II
640		2715.69	II
150		2722.56	I
240		2728.64	II
180		2731.35	I
100		2739.71	II
140		2753.40	II
140		2765.67	II
140		2777.73	II
120		2803.47	II
120		2846.57	I
110		2847.57	II
140		2852.87	I
140		2854.34	II
200		2855.22	I
180		2859.97	I
240		2864.36	I
170		2866.59	I
210		2868.10	I
140		2869.13	II
210		2870.55	I
110		2877.69	II
110		2879.16	II
350		2880.03	II
380		2882.50	II
380		2884.78	II
140		2888.25	II
380		2889.62	II
900		2891.64	II
530		2892.44	II
900		2892.66	II
1400		2893.32	II
360		2896.21	II
110		2899.60	I
360		2903.08	II
150		2906.13	I
900		2906.46	II
490		2907.47	II
2400		2908.82	II
710		2910.02	II
530		2910.39	II
560		2911.06	II
380		2914.93	I

Intensity		Wavelength	Spectrum
120		2917.37	II
210		2919.99	II
380		2920.38	II
710		2923.62	I
2400		2924.02	II
1700		2924.64	II
710		2930.81	II
210		2934.40	II
110		2935.87	I
900		2941.37	II
450		2941.49	II
230	d	2942.33	I
230		2943.20	I
1100		2944.57	II
110		2946.53	I
230		2949.63	I
300		2950.35	II
640		2952.08	II
120		2954.33	I
260		2957.52	II
410		2962.77	I
600		2968.38	II
120		2972.25	II
120		2976.20	II
380		2976.52	II
240		2977.54	I
260		3001.20	II
140		3014.82	II
180		3016.78	II
270		3033.45	II
290		3033.82	II
230		3043.12	I
230		3043.56	I
230		3044.94	I
230		3048.22	II
170		3050.89	I
180		3053.39	II
450		3053.65	I
1200		3056.33	I
1400		3060.46	I
140		3063.25	II
2400		3066.38	I
200		3067.12	II
140		3069.64	I
170		3073.82	I
100		3075.27	I
150		3082.11	I
3800		3093.11	II
200		3094.20	II
180		3100.94	II
3000		3102.30	II
2600		3110.71	II
2000		3118.38	II
380		3121.14	II
150		3122.90	II
1500		3125.28	II
260		3126.22	II
530		3130.27	II
410		3133.33	II
210		3134.93	II
150		3136.51	II
150		3139.74	II
200		3142.48	II
150		3145.34	II
3200		3183.41	I
5300		3183.98	I
3800		3185.40	I
410		3187.71	II
530		3188.51	II
750		3190.68	II
530		3198.01	I
750		3202.38	I
450		3205.58	I
450		3207.41	I
410		3212.43	I
210		3217.11	II
150		3237.87	II
140		3254.77	II
140		3263.24	I
1100		3267.70	II
900		3271.12	II
750		3276.12	II
110		3279.84	II
140		3298.14	I
110		3329.86	I
110		3365.55	I
110		3377.62	I
170		3400.40	I
110		3425.07	I
110		3485.92	II
210		3504.44	II
560		3517.30	II
150		3520.02	II
110		3524.72	II
230		3529.74	I
230		3530.77	II
560		3533.68	I
110		3543.50	I
560		3545.20	II
110		3553.27	I
560		3556.80	II
110		3566.18	I
560		3589.76	II
490		3592.02	II

Intensity		Wavelength	Spectrum
560		3592.53	I
270		3593.33	II
110		3606.69	I
110		3639.02	I
110		3644.71	I
250		3663.59	I
250		3667.74	I
110		3669.41	II
170		3671.20	I
280		3673.40	I
280		3675.70	I
170		3676.68	I
300		3680.11	I
570		3683.13	I
190		3686.26	I
470		3687.47	I
1300		3688.07	I
1000		3690.28	I
1500		3692.22	I
450		3695.34	I
1000		3695.86	I
3800		3703.58	I
1800		3704.70	I
570		3705.04	I
130		3708.72	I
320		3715.47	II
250		3727.34	II
280		3732.76	II
150		3734.43	I
230		3745.80	II
210		3750.87	II
210		3770.97	II
270		3778.68	I
520		3790.32	I
1100		3794.96	I
570		3799.91	I
570		3803.47	I
190		3806.80	I
300		3807.50	I
520		3808.52	I
230		3809.60	I
1000		3813.49	I
140	d	3817.84	I
1300		3818.24	I
230		3819.96	I
230		3821.49	I
570		3822.01	I
450		3822.89	I
300		3823.21	I
1700		3828.56	I
280		3834.22	I
160		3839.00	I
110		3839.38	I
570		3840.44	I
2600		3840.75	I
110		3841.89	I
380		3844.44	I
320		3847.33	I
110		3849.32	I
1200		3855.37	I
3000		3855.84	I
150		3862.22	I
130		3863.87	I
1300		3864.86	I
230		3867.60	I
170		3871.08	I
1500		3875.08	I
420		3875.90	I
570		3876.09	I
130		3878.71	II
700		3890.18	I
460		3892.86	I
280	h	3898.02	I
140		3899.13	II
140	h	3900.18	I
140	h	3901.15	I
2400		3902.25	I
100		3906.75	I
700		3909.89	I
100		3910.79	I
220		3912.21	I
140		3914.33	II
100		3916.41	II
100		3920.49	I
100		3921.90	I
230		3922.43	I
240		3924.66	I
150		3925.24	I
200		3927.93	I
260		3930.02	I
150		3931.34	I
260		3934.01	I
150		3935.14	I
100		3936.28	I
150		3943.66	I
100		3950.23	I
140		3951.97	II
100		3973.64	II
540		3990.57	I
260		3992.80	I
430		3998.73	I
170		4005.71	II
120		4023.39	II
120		4031.83	I

Intensity		Wavelength	Spectrum
150		4035.63	II
120		4042.64	I
360		4050.96	I
360		4051.35	I
280		4057.07	I
130		4057.82	I
230		4063.93	I
230		4071.54	I
1100		4090.58	I
180		4092.41	I
1800		4092.69	I
120		4093.50	I
890		4095.49	I
2800		4099.80	I
590		4102.16	I
230		4104.40	I
260		4104.78	I
2800		4105.17	I
120		4108.22	I
2300		4109.79	I
8900		4111.78	I
120		4112.33	I
230		4113.52	I
4300		4115.18	I
1800		4116.47	I
180		4118.18	I
180		4118.64	I
230		4119.46	I
180		4120.54	I
180		4123.19	I
2000		4123.57	I
120		4124.07	I
3100		4128.07	I
120		4128.86	I
3100		4132.02	I
2300		4134.49	I
150		4159.69	I
100		4174.01	I
230		4179.42	I
150		4182.59	I
180		4189.84	I
180		4191.56	I
230		4209.86	I
120		4226.62	I
360		4232.46	I
180		4232.95	I
180		4234.00	I
120		4235.76	I
100		4257.37	I
120		4259.31	I
120		4262.16	I
560		4268.64	I
460		4271.55	I
460		4276.96	I
430		4284.06	I
330		4291.82	I
220		4296.11	I
170		4297.68	I
170		4298.03	I
170		4306.21	I
140		4307.18	I
170		4309.80	I
460		4330.02	I
510		4332.82	I
760		4341.01	I
1000		4352.87	I
130		4354.98	I
150		4355.94	I
150		4368.04	I
140	d	4373.23	I
100		4375.30	I
12000		4379.24	I
100		4380.55	I
7000		4384.72	I
4800		4389.97	I
3600		4395.23	I
1400		4400.58	I
2300		4406.64	I
2800		4407.64	I
3600		4408.20	I
4600		4408.51	I
140		4412.14	I
640		4416.47	I
120		4419.94	I
640		4421.57	I
460		4426.00	I
120		4427.31	I
310		4428.52	I
230		4429.80	I
430		4436.14	I
640		4437.84	I
830		4441.68	I
640		4444.21	I
610		4452.01	I
410		4457.48	I
120		4457.76	I
1000		4459.76	I
2000		4460.29	I
610		4462.36	I
120		4468.01	I
380		4469.71	I
120		4474.04	I
200		4474.71	I
380		4488.89	I

Vanadium (Cont.)

100		4496.06	I
120		4501.95	I
140		4524.22	I
360		4545.39	I
100		4549.65	I
280		4560.71	I
200		4571.78	I
510		4577.17	I
140		4578.73	I
640		4580.40	I
830		4586.36	I
170		4591.22	I
1300		4594.11	I
100		4606.15	I
30		4609.65	I
25		4611.74	I
230		4619.77	I
65		4624.41	I
50		4626.48	I
100		4635.18	I
65		4640.07	I
65		4640.74	I
130		4646.40	I
30		4648.89	I
30		4666.14	I
160		4670.49	I
24		4684.45	I
35		4686.92	I
55		4706.16	I
80		4706.57	I
80		4710.56	I
65		4714.12	I
35		4715.89	I
55		4717.69	I
40		4721.51	I
40		4722.86	I
40		4729.53	I
27		4730.38	I
27		4742.63	I
24		4746.63	I
40		4748.52	I
45		4750.98	I
35		4751.56	I
40		4753.93	I
65		4757.48	I
55		4766.63	I
130		4776.36	I
		4776.52	I
110		4786.51	I
130		4796.92	I
19		4799.77	I
130		4807.53	I
130		4827.45	I
150		4831.64	I
120		4832.43	I
19		4833.02	I
19		4848.81	I
320		4851.48	I
35		4862.61	I
480		4864.74	I
21		4871.26	I
620		4875.48	I
55		4880.56	I
740		4881.56	I
27		4891.60	I
21		4894.21	I
55		4900.62	I
95	d	4904.29	I
		4904.34	I
85		4925.65	I
35		4932.03	I
23		4966.12	I
70		5002.33	I
85		5014.62	I
28		5051.63	I
35		5064.12	I
35		5105.14	I
110		5128.53	I
110		5138.42	I
25		5139.53	I
70		5148.72	I
40		5159.35	I
23		5169.94	I
70		5176.77	I
20		5192.01	I
110		5192.99	I
23		5193.62	I
110		5194.83	I
55		5195.36	I
20		5206.61	I
40		5216.59	I
35		5225.77	I
35		5233.75	I
110		5234.07	I
20		5240.20	I
110		5240.87	I
17		5260.98	I
40		5353.41	I
35		5383.43	I
40		5385.14	I
14		5388.30	I
11		5397.87	I
100		5401.93	I
140		5415.26	I
28		5418.09	I
50		5424.08	I
40		5434.18	I
11		5437.66	I
17		5458.12	I
13		5471.33	
25		5487.22	I
85		5487.92	I
25		5489.94	I
28		5504.87	I
70		5507.75	I
14		5511.18	I
23		5545.93	I
70		5547.07	I
35		5558.75	I
28		5561.66	I
140		5584.50	I
23		5586.00	I
100		5592.42	I
28		5601.38	I
70		5604.94	I
13		5624.20	I
200		5624.60	I
70		5624.89	I
55		5626.01	I
400		5627.64	I
13		5632.46	I
10		5633.90	I
13		5635.51	I
85		5646.11	I
110		5657.44	I
110		5668.36	I
310		5670.85	I
20		5683.22	I
1200		5698.52	I
920		5703.56	I
570		5706.98	I
11		5708.95	I
11	h	5716.21	I
70		5725.64	I
850		5727.03	I
170		5727.66	I
230		5731.25	I
40		5734.01	I
230		5737.06	I
110		5743.45	I
17		5747.70	I
40		5748.87	I
17		5752.74	I
17		5761.41	I
70		5772.42	I
35		5776.64	I
11		5782.61	I
11		5783.50	I
40	h	5784.38	I
55	h	5786.16	I
23		5788.56	I
35	h	5807.14	I
23		5817.06	I
35	h	5817.53	I
55	h	5830.72	I
85	h	5846.30	I
11		5850.32	I
40		5924.57	I
28		5978.91	I
20		5980.78	I
28		6002.31	I
55		6002.63	I
28		6016.12	I
20		6025.41	I
450		6039.73	I
100		6058.14	I
20		6067.26	I
480		6081.44	I
1300		6090.22	I
28		6106.98	I
280		6111.67	I
600		6119.52	I
20		6128.34	I
280		6135.38	I
180		6150.15	I
85		6170.36	I
23		6189.35	I
450		6199.19	I
130		6213.87	I
450		6216.37	I
28	h	6218.31	I
130		6224.50	I
430		6230.74	I
100		6233.20	I
55		6240.13	I
170		6242.81	I
710		6243.10	I
280		6251.82	I
85		6256.90	I
85		6258.57	I
55		6261.22	I
85		6266.32	I
130		6268.82	I
170		6274.65	I
17	h	6282.33	I
200		6285.16	I
200		6292.83	I
170		6296.49	I
28	h	6311.50	I
14		6324.66	I
70		6326.84	I
55		6339.09	I
50		6349.48	I
14		6355.58	I
50		6357.30	I
25		6358.82	I
35		6361.27	I
23		6379.36	I
14		6393.28	I
35		6430.47	I
23		6431.63	I
14		6433.18	I
11		6435.16	I
70		6452.34	I
11		6488.05	I
55		6504.17	I
110		6531.43	I
28		6543.51	I
17		6558.02	I
11		6565.88	I
50		6605.97	I
15		6607.83	I
10		6623.54	I
50		6624.85	I
13		6633.26	I
13		6643.79	I
8		6693.66	I
8		6708.07	I
65	c	6753.00	I
10		6760.12	I
50	c	6766.49	I
40		6784.98	I
15		6786.32	I
26		6812.40	I
9	c	6829.94	I
15		6832.44	I
12		6839.58	I
12		6841.90	I
10	c	6870.88	I
8		6871.56	I
7		6894.00	I
12		6974.50	I
21		7026.07	I
7		7063.69	I
11	h	7092.08	I
6		7102.58	I
24		7148.15	I
7		7151.36	I
7		7182.08	I
14		7264.29	I
8		7321.44	I
40		7338.92	I
35		7356.54	I
11		7358.66	I
24		7361.39	I
12		7362.49	I
24		7363.16	I
9		7385.95	I
6	h	7393.49	
12	h	7485.90	I
12	h	7488.08	I
12	h	7492.44	
12	h	7578.75	I
9	h	7591.24	I
14	h	7596.92	I
12	h	7598.28	I
24		7624.81	I
5		7701.37	I
8		7704.81	I
8	h	7851.18	
14	Bl	7865.51	VO
12		7896.40	
14	h	7898.81	
24		7937.92	I
29	c	8027.39	I
14		8028.13	I
14	h	8035.38	
14	h	8045.71	
12		8051.89	
14		8093.48	I
8		8102.44	I
12		8108.59	I
9	h	8109.07	I
120	Cw	8116.80	I
11	h	8136.79	I
29		8144.59	I
9		8154.55	I
70	c	8161.07	I
14		8171.35	I
7		8180.21	I
35		8186.71	I
24		8187.33	I
29		8198.87	I
35		8203.07	I
24		8241.61	I
29	c	8253.51	I
29		8255.88	I
5	h	8280.39	I
19		8282.37	I
8		8324.42	I
14	h	8331.23	I
14		8342.03	I
7		8402.81	I
12		8499.52	I
6		8534.49	I
6	Bl	8624.86	VO
60	c	8919.85	I
29	c	8932.93	I
12		8971.62	I

V III

Ref. 394 — C.H.C.

Intensity		Wavelength	
		Vacuum	
25		616.09	III
50		633.94	III
40		635.41	III
100		864.27	III
75		948.84	III
500		1006.46	III
500		1149.94	III
400		1157.18	III
300		1160.77	III
500		1252.11	III
400		1254.01	III
500		1287.87	III
400		1289.42	III
300		1290.77	III
400		1313.35	III
300		1313.27	III
500		1331.99	III
500		1335.12	III
1000		1643.03	III
1000		1650.14	III
300		1668.03	III
300		1670.66	III
300		1679.19	III
1000		1694.78	III
400		1721.98	III
300		1724.63	III
500	d	1751.68	III
500		1757.73	III
1000		1760.07	III
300		1773.43	III
400		1778.02	III
500		1779.72	III
400		1784.44	III
1000		1788.26	III
500		1793.82	III
1000		1794.60	III
300		1796.77	III
500		1798.15	III
300		1802.55	III
500		1804.13	III
1000		1812.19	III
400		1831.15	III
400		1831.64	III
300		1845.07	III
300		1850.69	III
400		1852.01	III
500		1854.42	III
300		1855.06	III
500		1856.64	III
300		1864.51	III
300		1878.68	III
400		1880.41	III
300		1895.01	III
400		1899.81	III
500		1902.23	III
300		1934.00	III
		Air	
500		2232.91	III
400		2241.53	III
1000		2292.86	III
400		2314.18	III
500		2318.06	III
400		2319.00	III
500		2323.82	III
2500		2330.42	III
500		2331.75	III
400		2334.20	III
500		2343.10	III
500		2358.73	III
500		2366.31	III
2500		2371.06	III
1000		2382.46	III
500		2393.58	III
500		2404.18	III
500		2516.14	III
250		2521.16	III
250		2521.55	III
250		2548.21	III
150		2554.22	III
150		2563.32	III
250		2593.05	III
250		2595.10	III
100		3679.86	III
50	h	3705.35	III

Vanadium (Cont.)

Intensity		Wavelength	
40		4714.89	III
50		6597.20	III

V IV
Ref. 397 — C.H.C.

Vacuum

Intensity		Wavelength	
200		677.34	IV
60		678.74	IV
50		679.65	IV
500		684.37	IV
100		684.45	IV
100		691.53	IV
50		693.13	IV
400		737.85	IV
150		750.11	IV
60		1226.52	IV
50		1308.06	IV
80		1355.13	IV
60		1395.00	IV
50		1414.41	IV
80		1419.58	IV
100		1426.65	IV
60		1520.14	IV
80		1601.92	IV
80		1611.88	IV
80		1806.18	IV
60		1809.85	IV
100		1817.68	IV
200		1825.84	IV
300		1861.56	IV
500		1939.06	IV
400		1951.43	IV
300		1963.10	IV
500		1997.72	IV
200		1999.32	IV

Air

Intensity		Wavelength	
100		2002.48	IV
50	h	2088.74	IV
50	h	2146.83	IV
100	h	2155.34	IV
500		2268.30	IV
50	h	2421.32	IV
50	h	2433.53	IV
50	h	2446.80	IV
50	h	2450.87	IV
50	h	2556.92	IV
80	h	2570.72	IV
50	h	2624.21	IV
80	h	2645.54	IV
50	h	2655.41	IV
50	h	2656.87	IV
50		3284.56	IV
60		3334.79	IV
50		3448.41	IV
50		3496.42	IV
80	h	3514.25	IV
50	h	4985.65	IV
50	h	5130.78	IV
50	h	5262.16	IV
60	h	5352.32	IV
40	h	5940.12	IV

V V
Ref. 432 — C.H.C.

Vacuum

Intensity		Wavelength	
18		224.91	V
20		225.46	V
17		227.88	V
16		239.41	V
19		239.48	V
20		251.66	V
18		252.44	V
18		285.98	V
20		286.84	V
17		312.39	V
35		483.01	V
25		484.51	V
20		820.86	V
30		829.48	V
15		962.03	V
15		1142.74	V
25		1157.58	V
20		1490.11	V
100		1680.20	V
50		1716.72	V
15		1724.99	V
25		1792.99	V
30		1811.42	V

Air

Intensity		Wavelength	
20	d	2319.66	V
20		2577.90	V
15		2775.82	V
15		3617.97	V
12	d	3746.36	V
20		4200.32	V
15	w	4930.53	V
8		5356.07	V
7		6628.80	V
3		7595.51	V

XENON (Xe)
Z = 54

Xe I and II
Ref. 33, 116, 118, 120, 232 — S.P.D.

Vacuum

Intensity		Wavelength	
350		740.41	II
350		803.07	II
600		880.80	II
350		885.54	II
600		925.87	II
250		935.40	II
800		972.77	II
700		976.68	II
500		1032.44	II
700		1037.68	II
1100		1041.31	II
1000		1048.27	II
1200		1051.92	II
2000		1074.48	II
600		1083.86	II
1200		1100.43	II
600		1158.47	II
250		1169.63	II
800	p	1183.05	II
250		1192.04	I
600		1244.76	II
250		1250.20	I
1000		1295.59	I
600		1469.61	I

Air

Intensity		Wavelength	
200		2864.73	II
150	h	2895.22	II
400		2979.32	II
100	h	3017.43	II
300		3128.87	II
200	h	3366.72	II
2		3400.07	I
2		3418.37	I
2		3420.00	I
3		3442.66	I
100	h	3461.26	II
4		3469.81	I
4		3472.36	I
5		3506.74	I
10		3549.86	I
10		3554.04	I
15		3610.32	I
8		3613.06	I
6		3633.06	I
10		3669.91	I
40		3685.90	I
40		3693.49	I
100	l	3907.91	II
100		4037.59	II
200	l	4057.46	II
100	h	4098.89	II
200	l	4158.04	II
1000	h	4180.10	II
500	h	4193.15	II
300	h	4208.48	II
100	h	4209.47	II
300	h	4213.72	II
100		4215.60	II
300	h	4223.00	II
400	h	4238.25	II
500	h	4245.38	II
100	l	4251.57	II
500	h	4296.40	II
500	h	4310.51	II
1000	l	4330.52	II
200	h	4369.20	II
100	l	4373.78	II
500	h	4393.20	II
500	l	4395.77	II
200	l	4406.88	II
150	l	4416.07	II
500	h	4448.13	II
1000	h	4462.19	II
500	l	4480.86	II
100	l	4521.86	II
600		4734.152	I
150		4792.619	I
500		4807.02	I
400		4829.71	I
300		4843.29	I
500		4916.51	I
500		4923.152	I
200	l	4971.71	II
400		4972.71	II
300		4988.77	II
100	l	4991.17	II
200		5028.280	I
200		5044.92	II
1000		5080.62	II
300		5122.42	II
100		5125.70	II
100		5178.82	II
300		5188.04	II
400		5191.37	II
100		5192.10	II
500		5260.44	II
500		5261.95	II
2000		5292.22	II
300		5309.27	II
1000		5313.87	II
2000		5339.33	II
200		5363.20	II
200		5368.07	II
500		5372.39	II
100		5392.80	I
3000		5419.15	II
800		5438.96	II
300		5445.45	II
200		5450.45	II
400		5460.39	II
1000		5472.61	II
100	l	5494.86	II
200		5525.53	II
600		5531.07	II
100		5566.62	I
300		5616.67	II
300		5659.38	II
600		5667.56	II
150		5670.91	II
100		5695.75	I
200		5699.61	II
200		5716.10	II
500		5726.91	II
500		5751.03	II
300		5758.65	II
300		5776.39	II
100		5815.96	II
300		5823.89	I
150		5824.80	I
100		5875.02	I
300		5893.29	II
100		5894.99	I
200		5905.13	II
100		5934.17	I
500		5945.53	II
300		5971.13	II
2000		5976.46	II
200		6008.92	II
1000		6036.20	II
2000		6051.15	II
600		6093.50	II
1500		6097.59	II
400		6101.43	II
100		6115.08	II
100		6146.45	II
150		6178.30	I
120		6179.66	I
300		6182.42	I
500		6194.07	II
100		6198.26	I
100		6220.02	II
500		6270.82	II
400		6277.54	II
100		6284.41	II
100		6286.01	I
250		6300.86	II
500		6318.06	I
400		6343.96	II
600		6356.35	II
200		6375.28	II
100		6397.99	II
300		6469.70	I
150		6472.84	I
120		6487.76	I
100		6498.72	I
200	h	6504.18	I
300		6512.83	II
200		6528.65	II
100		6533.16	I
1000		6595.01	II
100		6595.56	I
400		6597.25	II
100		6598.84	II
150		6668.92	I
300		6694.32	II
200		6728.01	I
150		6788.71	II
100		6790.37	II
1000		6805.74	II
200		6827.32	I
100		6872.11	I
300		6882.16	I
80		6910.22	II
100		6925.53	I
800	h	6942.11	II
100		6976.18	I
2000		6990.88	II
150		7082.15	II
500		7119.60	I
50	s	7147.50	II
200		7149.03	II
500		7164.83	II
100		7284.34	II
200		7301.80	II
200		7339.30	II
100		7386.00	I
150		7393.79	I
300		7548.45	II
200		7584.68	I
80		7618.57	II
500		7642.02	I
100		7643.91	I
200		7670.66	II
60		7787.04	II
100		7802.65	I
100		7881.32	I
300		7887.40	I
500		7967.34	I
100		8029.67	I
200		8057.26	I
150		8061.34	I
100		8101.98	I
150	h	8151.80	II
100		8171.02	I
700		8206.34	I
10000		8231.635	I
500		8266.52	I
7000		8280.116	I
2000		8346.82	I
100		8347.24	II
2000		8409.19	I
50	h	8515.19	II
200		8576.01	I
50	h	8604.23	II
250		8648.54	I
100		8692.20	I
200		8696.86	I
50	h	8716.19	II
300		8739.39	I
100		8758.20	I
5000		8819.41	I
300		8862.32	I
200		8908.73	I
200		8930.83	I
1000		8952.25	I
100		8981.05	I
200		8987.57	I
400		9045.45	I
500		9162.65	I
100		9167.52	I
100		9374.76	I
200		9513.38	I
50	h	9591.35	II
150		9685.32	I
50	l	9698.68	II
100		9718.16	I
2000		9799.70	I
3000		9923.19	I
100		10838.37	I
90		11742.01	I
375		12235.24	I
100		12257.76	I
300		12590.20	I
2500		12623.391	I
250		13544.15	I
2000		13657.055	I
1250		14142.444	I
800		14240.96	I
375		14364.99	I
140		14660.81	I
3000		14732.806	I
100		15099.72	I
2500		15418.394	I
150		15557.13	I
250		15979.54	I
100		16039.90	I
1000		16053.28	I
125		16554.49	I
1500		16728.15	I
1500		17325.77	I
350		18788.13	I
150		20187.19	I
3000		20262.242	I
250		21470.09	I
1250		23193.33	I
110		23279.54	I
1800		24824.71	I
175		25145.84	I
2000		26269.08	I
2500		26510.86	I
250		28381.54	I
750		28582.25	I
300		29384.41	I
150		29448.06	I

Xenon (Cont.)

Intensity		Wavelength	
100		29649.58	I
100		29813.62	I
600		30253.14	I
1500		30475.46	I
100		30504.12	I
500		30794.18	I
6000		31069.23	I
125		31336.01	I
550		31607.91	I
100		32293.08	I
1800		32739.26	I
3500		33666.69	I
150		34014.67	I
450		34335.27	I
170		34744.00	I
5000		35070.25	I
110		35246.92	I
250		36209.21	I
150		36231.74	I
450		36508.36	I
850		36788.83	I
140		38685.98	I
175		38737.82	I
270		38939.60	I
120		39955.14	I

Xe III

Ref. 33, 384, 391, 429 — R.L.K.

Intensity		Wavelength	
		Vacuum	
8		657.8	III
8		660.1	III
9		673.8	III
9		674.0	III
9		676.6	III
10		694.0	III
20		698.5	III
12		705.1	III
10		721.2	III
15		731.0	III
10		733.3	III
15		742.6	III
10		756.0	III
10		761.5	III
10		769.1	III
25		779.1	III
15		792.9	III
12		796.1	III
15		802.0	III
25		823.2	III
30		824.9	III
25		853.0	III
15		889.3	III
20		894.0	III
20		896.0	III
10		965.5	III
35		1003.4	III
35		1017.7	III
10		1047.8	III
12		1066.4	III
30		1130.3	III
25		1232.1	III
		Air	
80		2668.98	III
100		2717.33	III
30		2814.45	III
40		2815.91	III
30		2827.45	III
40		2847.65	III
30		2862.40	III
80	w	2871.10	III
60	w	2871.24	III
30		2871.7	III
30		2896.62	III
50		2906.6	III
40		2911.89	III
80	w	2912.36	III
40		2940.2	III
60		2945.2	III
40		2947.5	III
40		2948.1	III
80	w	2970.47	III
40		2992.87	III
30		3004.25	III
100		3023.81	III
40		3083.5	III
50		3091.1	III
30		3106.46	III
100	w	3138.3	III
80	c	3150.82	III
40		3185.2	III
100		3242.86	III
80		3268.98	III
30		3287.82	III
80	w	3301.55	III
40		3331.6	III
30		3358.0	III
80		3384.12	III
60		3444.2	III
70		3454.2	III
100	w	3458.7	III
40		3468.22	III
80		3522.83	III
50		3542.3	III
50		3552.1	III
40		3561.4	III
100		3579.7	III
80		3583.6	III
100	w	3595.4	III
100		3606.06	III
40		3607.0	III
100	w	3615.9	III
40		3623.1	III
600		3624.08	III
50		3676.67	III
40		3776.3	III
300		3781.02	III
100		3841.5	III
200		3877.8	III
60		3880.5	III
500		3922.55	III
300		3950.59	III
200		4050.07	III
60		4060.4	III
100		4109.1	III
100		4145.7	III
30		4285.9	III
50		4434.2	III
100	w	4462.1	III
100	w	4569.1	III
100	w	4570.1	III
100	w	4641.4	III
30		4673.7	III
60		4683.57	III
30		4723.60	III
100	w	4757.3	III
40		4869.5	III
60		5239.0	III
30		5367.1	III
50		5401.0	III
40		5524.4	III
60		6205.97	III
25		6221.7	III
60		6238.2	III
60		6259.05	III

YTTERBIUM (Yb)
Z = 70

Yb I and II

Ref. 1 = C.H.C.

Intensity		Wavelength	
		Air	
2500		2116.67	II
3000		2126.74	II
370		2161.60	II
850		2185.71	II
640		2224.46	II
140		2320.81	II
50		2362.89	II
170		2390.74	II
18		2398.02	II
28		2421.35	II
25		2447.26	II
28		2460.25	II
460		2464.50	I
14		2484.89	II
70		2502.02	II
28		2505.48	II
11		2508.07	II
140		2512.06	II
18		2516.35	II
50		2522.44	II
65		2537.65	II
270		2538.67	II
14		2550.06	II
70		2552.15	II
55		2552.70	II
21		2565.57	II
28		2571.36	II
13		2573.15	II
18		2596.16	II
28		2596.32	II
21		2615.26	II
100		2617.01	II
55		2634.31	II
45		2639.45	II
85		2641.89	II
110		2644.31	II
28		2646.44	II
28		2647.46	II
28		2648.80	II
50		2649.79	II
28		2650.73	II
990		2653.75	II
35		2656.12	II
21		2659.27	II
200		2665.04	II
55		2668.75	II
390		2671.96	I
390		2672.66	II
21		2680.40	II
14		2683.42	II
70		2684.75	II
25		2687.98	II
28		2695.43	II
14		2696.62	II
18		2700.80	II
21		2708.84	II
65		2710.54	II
25		2711.78	II
55		2712.66	II
170		2718.35	II
21		2722.20	II
110		2732.74	II
21		2734.09	II
55		2741.71	II
55		2747.58	II
18		2748.04	II
230		2748.66	II
1300		2750.48	II
85		2751.45	II
21		2759.00	II
65		2760.78	II
65		2761.37	II
35		2764.41	II
85		2771.32	II
170		2776.28	II
100		2784.66	II
18		2787.96	II
45		2793.28	II
25		2794.44	II
21		2795.07	II
18		2795.29	II
35		2797.80	II
100		2798.21	II
45		2799.38	II
50		2800.00	II
35		2800.06	II
14		2810.72	II
65		2814.53	II
28		2816.32	II
140		2821.15	II
100		2824.97	II
190		2830.99	II
18		2832.20	II
28		2834.97	II
14		2842.59	II
230	h	2847.18	II
100		2848.44	II
21		2849.34	II
360		2851.13	II
55		2851.86	II
21		2853.41	II
18		2853.68	II
55		2854.14	II
45		2854.49	II
45		2858.33	II
45		2858.46	II
100		2859.39	II
430		2859.80	II
55		2860.39	II
140		2861.21	II
100		2861.34	II
200		2867.06	II
25		2870.06	II
45		2873.49	I
28		2885.97	II
70		2886.26	II
200		2888.04	II
3600		2891.38	II
45		2893.62	II
28		2896.90	II
85		2899.70	II
18		2902.41	II
21		2902.92	II
21		2906.88	II
28		2908.33	II
35		2909.19	II
55		2909.48	II
85		2911.52	II
18		2912.86	II
170		2914.21	II
140		2915.28	II
18		2916.43	II
280		2919.35	II
55		2921.12	II
45		2924.24	II
25		2927.85	II
35		2934.36	I
55		2935.11	II
21		2937.19	II
45		2939.53	II
45		2940.52	II
28		2942.04	II
140		2945.91	II
45		2946.30	II
18		2946.76	II
28		2950.33	II
45		2955.32	II
18		2957.63	II
65		2962.52	II
21		2963.26	II
45		2963.46	II
130		2964.76	II
2000		2970.56	II
45		2982.49	II
21		2982.66	II
28		2983.70	II
200		2983.99	II
90		2985.08	II
35		2985.88	II
45		2990.37	II
65		2991.87	II
28		2993.94	II
170		2994.80	II
28		2995.86	II
70		3000.46	II
25		3002.61	II
310		3005.77	II
100		3009.39	II
65		3010.62	II
55		3014.43	II
160		3017.56	II
160		3026.67	II
920		3031.11	II
55		3034.64	II
25		3037.99	II
55		3039.67	II
80		3042.65	II
21		3044.00	II
45		3046.48	II
35		3047.05	II
45		3063.12	II
21		3063.67	II
110		3065.04	II
18		3076.01	II
100		3089.10	II
70		3093.87	II
28		3100.74	I
45		3101.36	II
28		3102.07	II
55		3107.76	II
170		3107.90	II
85		3115.34	II
55		3116.70	II
190		3117.81	II
50		3136.76	II
230		3140.94	II
80		3141.73	II
80		3145.06	II
28		3145.54	II
28		3153.18	II
90		3153.88	II
50		3155.18	II
28		3162.29	I
70		3163.80	II
50		3165.21	II
120		3169.06	II
120		3180.92	II
390		3192.88	II
70		3198.65	II
240		3201.16	II
80		3217.18	II
50		3218.32	II
50		3225.88	II
45		3239.20	II
35		3239.58	I
35		3246.06	II
130		3261.51	II
18000		3289.37	II
130		3305.25	I
140		3305.73	II
50		3315.10	II
80		3319.41	I
50		3333.06	II
240		3337.17	II
280	d	3342.93	II
		3343.07	II
80		3346.50	
50		3347.54	II
35		3351.09	II
50		3351.26	
100		3352.49	II
100		3362.44	II
50		3363.64	II
240		3375.48	II
50		3376.62	II
28		3382.54	
140		3387.50	I
50		3390.25	II
28		3390.42	II
50		3391.10	II
50	h	3394.44	II
50		3401.01	II
35		3404.10	II
50		3412.45	I
140		3418.39	I
360		3426.04	I
80		3428.46	II
240		3431.11	I
45		3434.61	II

Ytterbium (Cont.)

Intensity		Wavelength	
50		3438.71	II
100		3438.85	II
35		3443.59	
35		3446.89	II
85		3452.40	I
500		3454.08	II
190	d	3458.29	II
		3458.39	I
360		3460.27	I
35		3462.34	II
2400		3464.37	I
500		3476.30	II
500		3478.84	II
50		3482.56	II
85		3485.76	II
85		3488.43	II
100	Hw	3495.90	II
85		3507.83	II
50		3517.00	I
230		3520.29	II
50		3545.72	II
100		3549.82	II
35		3559.03	I
200		3560.33	II
170		3560.70	II
50	h	3563.94	II
85		3570.57	II
50		3572.50	II
50		3574.58	II
360		3585.47	II
130		3606.48	II
50		3610.23	II
70		3611.30	II
200		3619.80	II
110		3634.52	
240		3637.76	II
70		3648.15	I
90		3655.73	I
240		3669.69	II
50		3670.69	II
140		3675.08	II
50		3690.56	II
32000		3694.19	II
70		3698.60	II
70		3700.58	I
50		3710.34	II
60		3724.21	II
180		3734.69	I
550		3770.10	I
80		3774.32	I
60	h	3791.74	I
170		3839.91	I
340		3872.85	I
340		3900.85	I
50		3904.81	II
140		3911.27	I
32000		3987.99	I
930		3990.88	I
50		4007.36	I
70		4052.28	I
85		4077.28	II
440		4089.68	I
120	h	4119.25	II
70		4135.09	II
470		4149.07	I
120		4174.56	I
340		4180.81	II
150	d	4218.56	II
		4218.69	I
120		4231.97	I
70		4277.74	I
120		4305.97	I
70		4316.95	II
60	h	4393.69	I
60	h	4430.21	I
440		4439.19	I
85	h	4482.42	I
85		4515.16	II
35		4553.58	II
85	h	4563.95	I
640		4576.21	I
200		4582.36	I
70		4589.21	I
140		4590.83	I
40		4598.36	II
35		4683.81	II
40		4684.27	I
190		4726.08	II
170	h	4781.87	I
170		4786.61	II
35		4816.43	I
40		4820.24	II
35		4836.96	II
40		4837.46	I
17		4851.15	II
40	h	4894.60	I
27		4912.36	I
710		4935.50	I
24		4937.22	II
140		4966.90	I
24		5009.52	II
17		5067.30	II
30		5067.80	I
70		5069.14	I
220		5074.34	I
50		5076.74	I
20		5135.98	II
14		5147.02	II
20		5184.15	II
60		5196.08	I
85		5211.60	I
35		5240.51	II
100		5244.11	I
40		5257.49	II
150	h	5277.04	I
35		5279.53	II
17		5300.94	II
170		5335.15	II
30	d	5345.66	II
		5345.83	II
60		5347.22	II
30	h	5351.29	I
150		5352.95	II
30		5358.64	II
30		5363.66	I
17		5389.84	II
14		5432.71	II
40		5449.27	II
14		5478.50	II
60		5481.92	I
40		5505.49	I
17		5524.54	I
85	h	5539.05	I
2400		5556.47	I
35		5562.09	I
20		5568.11	I
20		5586.36	I
40		5588.45	II
60		5651.98	II
7		5686.53	II
220		5719.99	I
10		5749.91	II
10	h	5755.89	I
27		5771.66	II
10		5803.44	I
10		5819.41	II
35		5833.99	II
35		5837.14	II
27		5854.51	I
8		5897.21	II
20		5908.36	II
17		5989.33	I
40		5991.51	II
10		6052.88	II
10		6054.57	I
60		6152.57	II
30		6246.97	II
60		6274.78	II
14		6308.15	II
35	h	6400.35	I
35	h	6417.91	I
20		6432.73	II
17	h	6463.15	II
340		6489.06	I
20		6643.55	I
180		6667.82	I
15		6678.17	I
25		6727.61	II
25		6768.70	I
690		6799.60	I
18		6934.05	II
20		6999.88	II
10		7043.78	II
9	h	7244.41	I
8	h	7305.22	I
10	h	7313.05	I
16	h	7350.04	I
25		7448.28	I
30	h	7527.46	I
750		7699.48	I
7		7895.08	I
70	h	8922.56	II

Yb III
Ref. 40, 192 — J.R.

Intensity		Wavelength	
		Vacuum	
5		968.46	III
20		973.16	III
10		994.56	III
10		1560.66	III
80		1561.42	III
30		1669.60	III
50		1670.78	III
50		1719.82	III
60		1739.18	III
70		1762.80	III
80	h	1765.21	III
65		1775.29	III
70		1779.74	III
70		1781.31	III
20		1793.70	III
60		1798.85	III
65		1810.88	III
60		1826.41	III
20		1826.77	III
30		1838.01	III
30		1847.30	III
30		1849.24	III
10		1849.42	III
75		1852.36	III
75		1852.94	III
90		1854.80	III
80		1857.16	III
100		1863.32	III
5		1864.85	III
10		1867.23	III
10		1867.63	III
10		1868.19	III
5		1868.92	III
10		1870.07	III
15		1870.83	III
10		1871.15	III
200		1872.03	III
800		1873.91	III
100		1875.41	III
75		1875.92	III
70		1880.30	III
80		1884.22	III
70		1885.07	III
70		1887.22	III
10		1890.34	III
10		1890.87	III
15		1892.42	III
5		1895.50	III
100		1896.18	III
10		1897.57	III
500		1898.25	III
7		1906.74	III
10		1908.50	III
100		1909.66	III
70		1910.86	III
20		1920.53	III
10		1926.76	III
70		1928.09	III
15		1930.63	III
55		1942.59	III
40		1950.34	III
15		1962.80	III
80		1967.13	III
20		1969.47	III
30		1969.73	III
10		1973.96	III
10		1974.18	III
25		1976.46	III
2		1981.74	III
5		1983.88	III
25		1984.62	III
7		1985.74	III
80		1986.43	III
80	h	1989.82	III
50		1991.14	III
45		1995.05	III
55		1997.28	III
55		1997.66	III
500		1998.82	III
		Air	
20		2054.80	III
10		2066.49	III
10		2073.64	III
30		2078.05	III
10		2087.37	III
50		2087.98	III
20		2091.23	III
20		2092.26	III
10		2094.77	III
80		2095.31	III
15		2096.79	III
30		2098.36	III
10		2106.71	III
50		2109.54	III
20		2119.18	III
20		2198.14	III
80		2202.27	III
300		2240.11	III
100		2244.28	III
200		2257.03	III
100		2262.26	III
200		2265.67	III
150		2282.99	III
100		2283.99	III
300		2305.32	III
100		2309.27	III
200		2314.49	III
200		2337.97	III
40		2361.08	III
200		2365.43	III
50		2367.46	III
30		2369.99	III
20		2377.22	III
50		2403.95	III
20		2410.04	III
60		2412.33	III
10		2429.18	III
20		2433.43	III
100		2438.27	III
20		2439.31	III
20		2440.43	III
10		2458.64	III
10		2464.59	III
200		2490.42	III
20		2491.69	III
40		2506.25	III
300		2516.82	III
15		2522.07	III
20		2529.14	III
40		2550.39	III
300		2555.29	III
100		2560.56	III
10		2561.66	III
100		2566.78	III
2000		2567.61	III
1000		2579.57	III
100		2588.62	III
20		2592.69	III
500		2597.23	III
800		2599.14	III
30		2609.14	III
600		2621.11	III
300		2627.07	III
30		2635.37	III
500		2638.06	III
300		2640.48	III
1000		2642.56	III
100		2643.62	III
1000		2651.74	III
700		2652.25	III
100		2659.98	III
70		2664.89	III
2000		2666.13	III
2000		2666.99	III
30		2673.33	III
500		2677.39	III
500		2691.01	III
30		2708.04	III
400		2712.32	III
500		2749.91	III
200		2755.94	III
200		2756.76	III
100		2765.50	III
300		2788.24	III
600		2795.60	III
400		2803.32	III
1000		2803.43	III
10		2807.22	III
50		2808.51	III
600		2816.92	III
1000		2818.72	III
15		2826.01	III
300		2842.96	III
400		2875.86	III
600		2898.30	III
1000		2906.31	III
300		2928.97	III
50		2977.84	III
800		2998.00	III
2000		3029.49	III
100		3031.62	III
30		3040.65	III
3000		3092.50	III
20		3102.18	III
4000		3126.01	III
1000		3138.58	III
100		3151.44	III
70		3179.34	III
800		3191.35	III
50		3216.27	III
2000		3228.58	III
2000		3325.51	III
50		3358.25	III
20		3364.30	III
2000		3384.01	III
150		3392.56	III
100		3397.66	III
80		3432.94	III
40		3456.18	III
150		3463.51	III
20		3469.98	III
300		3550.87	III
200		3613.89	III
30		3659.84	III
30		3663.74	III
200		3664.74	III
20		3675.78	III
400		3711.91	III
20		3879.98	III
10		3882.58	III
20		3887.17	III
150		3896.55	III
15		3912.75	III
20		3913.23	III
500		3931.23	III
100		3985.56	III
10		3991.74	III
10		3997.67	III
2000		4028.14	III
10		4033.03	III
20		4074.53	III
20		4090.67	III

Ytterbium (Cont.)

Intensity		Wavelength	
20		4098.23	III
15		4121.06	III
10		4150.04	III
15		4153.11	III
100		4162.72	III
60		4172.95	III
30		4194.34	III
100		4194.95	III
10		4198.74	III
300		4213.64	III
10		4220.83	III
15		4231.07	III
20		4289.64	III
40		4301.14	III
20		4304.01	III
15		4350.80	III
10		4380.07	III
100		4517.58	III
40		4639.14	III
10		4834.93	III
15		5054.94	III
20		5256.85	III
20		5331.54	III
15		5740.83	III
10	d	5949.02	III
20		5973.05	III
40		6055.85	III
100		6214.22	III
200		6328.52	III
10		6365.88	III
150		6378.33	III
25		6466.33	III
20		6985.15	III
10		7037.04	III
15		7157.72	III
10		7311.02	III
10		7399.98	III
80		7410.01	III
15		7456.86	III
70		7664.41	III
80		7892.39	III
20		7893.10	III
100		7971.46	III
20		8056.02	III
10		8117.44	III
10		8326.86	III
30		8327.88	III
20		8400.01	III
30		8489.90	III
200		10110.60	III
100		10830.36	III

Yb IV
Ref. 40, 311 — J.R.

Intensity		Wavelength	
		Vacuum	
200		828.96	IV
200		870.35	IV
300		902.46	IV
300		927.01	IV
300		936.22	IV
400		943.04	IV
400		946.20	IV
200		975.21	IV
1000		1050.24	IV
1000		1054.46	IV
400		1092.51	IV
200		1109.96	IV
200		1110.55	IV
5000		1134.43	IV
300		1136.24	IV
500		1166.01	IV
600		1185.58	IV
200		1290.24	IV
600		1305.58	IV
900		1316.04	IV
200		1326.32	IV
800		1326.36	IV
200		1340.06	IV
300		1345.36	IV
900		1350.26	IV
200		1353.43	IV
400		1356.15	IV
200		1361.75	IV
300		1365.88	IV
300		1369.72	IV
400		1375.42	IV
300		1376.66	IV
200		1384.41	IV
250		1393.93	IV
350		1398.77	IV
400		1407.05	IV
300		1413.14	IV
400		1416.15	IV
400		1417.72	IV
200		1423.99	IV
300		1430.29	IV
200		1440.61	IV
400		1477.92	IV
300		1491.57	IV
400		1765.03	IV
200		1776.18	IV
200		1778.20	IV
200		1779.34	IV
300		1789.71	IV
800		1791.06	IV
200		1801.67	IV
250		1809.63	IV
600		1813.84	IV
400		1816.07	IV
250		1817.58	IV
300		1819.02	IV
200		1824.22	IV
		Air	
300		2106.48	IV
900		2116.65	IV
500		2121.29	IV
250		2122.84	IV
800		2123.32	IV
600		2125.72	IV
200		2129.65	IV
500		2135.21	IV
300		2137.58	IV
500		2138.35	IV
200		2138.53	IV
800		2139.99	IV
400		2141.04	IV
200		2142.20	IV
300		2143.42	IV
300		2143.89	IV
20000		2144.77	IV
400		2148.10	IV
300		2148.52	IV
15000		2154.18	IV
250		2165.55	IV
300		2169.12	IV
200		2172.16	IV
300		2177.53	IV
400		2183.32	IV
270	h	2186.13	IV
270		2187.17	IV
90	h	2189.90	IV
120		2193.34	IV
150		2198.27	IV
90		2224.64	IV
150		2231.28	IV
90		2233.30	IV
90		2244.20	IV
140		2331.36	IV

YTTRIUM (Y)
Z = 39

Y I and II
Ref. 1 — C.H.C.

Intensity		Wavelength	
		Air	
350		2243.06	II
50		2354.20	I
30		2373.83	
50		2385.24	
25		2413.93	II
560		2422.20	II
60		2460.61	II
25		2490.42	I
12		2540.28	
14		2547.57	
10		2550.17	
20		2681.65	I
60		2694.21	I
26		2695.39	I
95		2723.00	I
22		2730.08	I
22		2734.85	II
70		2742.53	I
140		2760.10	I
30		2785.21	II
12		2785.59	II
12		2791.20	I
30		2800.11	II
26		2813.64	I
18		2818.86	I
45		2822.56	I
22		2825.37	II
45		2826.38	II
70		2854.43	II
26		2856.30	II
11		2857.87	II
95		2886.48	I
18		2897.69	II
14		2898.82	II
160		2919.05	I
18	h	2930.03	II
390		2948.40	I
350		2964.96	I
18		2973.91	II
480		2974.59	I
30		2980.55	II
750		2984.26	I
70		2995.26	I
140		2996.94	I
70		3005.26	I
55		3018.95	I
130		3021.73	I
90		3022.28	I
26		3026.49	II
30		3036.59	II
45		3044.84	I
190		3045.37	I
22		3047.11	I
60		3055.22	II
60		3086.85	II
55	h	3091.70	I
22		3093.76	II
95		3095.88	II
45		3111.81	I
55		3112.04	II
22		3114.28	I
60		3128.77	II
80		3129.93	II
95		3135.17	II
110		3173.06	II
220		3179.41	II
70		3191.31	I
2300		3195.62	II
2200		3200.27	II
2200		3203.32	II
3900		3216.69	II
6200		3242.28	II
310		3280.91	II
19		3308.47	II
4700		3327.89	II
55		3340.38	I
160		3362.00	II
85		3388.59	I
45		3397.04	I
85		3412.47	I
200		3448.82	II
70		3450.95	I
110		3467.88	II
170		3485.73	I
1700		3496.09	II
80		3521.53	I
45		3546.01	II
3900		3549.01	II
130		3551.80	I
540		3552.69	I
170		3558.76	I
190		3571.43	I
260		3576.05	I
3300		3584.52	II
300		3587.75	I
100		3589.69	I
2800		3592.92	I
10000		3600.73	II
6200		3601.92	II
7800		3611.05	II
4300		3620.94	I
1900		3628.71	II
7800		3633.12	II
3000		3664.61	II
45		3668.49	II
170		3692.53	I
13000		3710.30	II
60		3718.12	I
60		3738.61	I
1200		3747.55	II
50		3749.89	I
10000		3774.33	II
1400		3776.56	II
50		3782.30	II
7400		3788.70	II
1300		3818.35	II
4000		3832.88	II
70		3847.87	II
80		3876.82	I
480		3878.28	II
30		3887.77	I
60	h	3904.59	I
50	h	3918.25	I
60	h	3930.11	I
240		3930.66	II
4400		3950.36	II
150		3951.60	II
60	h	3955.09	I
3600		3982.60	II
40		3987.50	I
940		4039.83	I
2400		4047.64	I
9400		4077.38	I
90	h	4081.22	I
2000		4083.71	I
9900		4102.38	I
60	h	4106.39	I
80		4110.81	I
320		4124.92	II
8900		4128.31	I
7500		4142.85	I
100	h	4157.63	I
2400		4167.52	I
2000		4174.14	I
8000		4177.54	II
120		4199.28	II
380		4204.70	II
80		4213.02	I
40		4213.54	I
160		4217.80	I
280	h	4220.63	I
80		4224.25	I
600		4235.73	II
2200		4235.94	I
300		4251.20	I
360	h	4302.30	I
2800		4309.63	II
50		4316.30	I
110		4330.78	I
30		4337.29	I
60		4344.65	I
440	h	4348.79	I
60		4352.33	I
60		4352.70	I
120		4357.73	I
800		4358.73	II
120		4366.03	I
12000		4374.94	II
150	h	4375.61	I
80		4379.33	I
30		4385.48	I
100		4387.74	I
30		4394.01	I
30		4394.67	I
1800		4398.02	II
890		4422.59	II
80		4437.34	I
100		4443.66	I
130		4446.63	I
20		4465.27	I
40		4473.89	I
170		4475.72	I
180		4476.96	I
160		4477.45	I
110		4487.28	I
300		4487.47	I
30		4491.75	I
25		4492.42	I
500		4505.95	I
50		4513.58	I
80		4514.01	I
40	h	4522.05	I
890		4527.25	I
440		4527.80	I
100		4544.32	I
100		4559.37	I
30		4564.39	I
60	h	4573.56	I
35		4581.32	I
30		4581.77	I
130		4596.55	I
95		4604.80	I
40		4613.00	I
2000		4643.70	I
200	h	4658.32	I
70		4658.89	I
85		4667.47	I
60		4670.82	I
2000		4674.84	I
60		4678.35	I
260		4682.32	II
85		4689.77	I
180		4696.81	I
35		4708.85	I
60		4725.85	I
170		4728.53	I
60	h	4732.37	I
85		4741.40	I
160		4752.79	I
410		4760.98	I
17		4780.18	I
120		4781.04	I
160		4786.58	II
170		4786.89	I
180		4799.30	I
50		4804.31	I
70		4804.81	I
85	Bl	4817.38	YO
140	Bl	4818.20	YO
140		4819.64	I
120		4822.13	I
190		4823.31	II
60		4839.15	I
770		4839.87	I
550		4845.68	I
410		4852.69	I
120		4854.25	I
890		4854.87	II
50		4856.70	I
330		4859.84	I
50		4879.65	I
1900		4883.69	II
50		4886.28	I
40		4886.65	I
95		4893.44	I
1100		4900.12	II
100		4906.11	I

Yttrium (Cont.)

45		4909.00	I
150		4921.87	I
35		4930.93	I
45		4950.66	I
120		4974.30	I
120		4982.13	II
100		5006.97	I
75		5070.21	I
75		5072.19	I
1100		5087.42	II
30		5088.18	I
210		5119.11	II
450		5123.21	II
180		5135.20	I
120		5196.43	II
960		5200.41	II
1500		5205.72	II
180		5240.81	I
60		5289.82	II
45		5320.78	II
75		5380.62	I
220		5402.78	II
24		5417.03	I
90		5424.37	I
190		5438.24	I
710		5466.46	I
100		5468.47	I
90		5473.39	II
90		5480.74	II
60		5493.17	I
35		5495.59	I
240		5497.41	II
300		5503.45	I
250		5509.90	II
60		5513.64	I
120		5521.63	I
		5521.70	II
24		5526.76	I
740		5527.54	I
35		5541.63	I
120		5544.50	I
		5544.61	II
90		5546.02	II
75		5556.43	I
60		5567.75	I
180		5577.42	I
24		5581.08	I
620		5581.87	I
21		5590.96	I
21		5594.12	I
120		5606.33	I
15		5623.91	I
560		5630.13	I
24		5632.25	I
21		5632.89	I
120		5644.69	I
120		5648.47	I
740		5662.94	II
90		5675.27	I
18		5693.63	I
160		5706.73	I
24		5720.61	I
75		5728.89	II
150	Bl	5730.12	YO
21		5732.09	I
90		5743.85	I
18	Bl	5746.93	YO
24	Bl	5764.22	YO
75		5765.64	I
35		5773.95	I
100		5781.69	II
15	Bl	5800.00	YO
15	Bl	5818.58	YO
30		5821.87	I
21		5832.27	I
9	b	5838.07	YO
15		5858.83	YO
15		5871.83	I
24		5876.14	YO
24		5879.96	I
24	b	5893.94	YO
35		5902.96	I
24	b	5912.19	YO
24	b	5931.10	YO
90	Bl	5939.08	YO
45		5945.72	I
24		5950.02	I
75	b	5956.41	YO
1300	Bl	5972.04	YO
50		5981.86	I
1000	Bl	5987.64	YO
740	Bl	6003.60	YO
120		6004.65	
120		6009.19	I
620	Bl	6019.87	YO
120		6023.41	I
500	Bl	6036.60	YO
420	Bl	6053.81	YO
130	Bl	6072.78	YO
50		6088.00	I
210	Bl	6089.35	YO
160	Bl	6096.78	YO
130	Bl	6107.82	YO
130	Bl	6114.73	YO
75	Bl	6127.38	YO
1400	Bl	6132.06	YO
120		6135.04	I
150		6138.43	I
1100	Bl	6148.36	YO
120		6151.72	YO
820	Bl	6165.08	YO
560	Bl	6182.23	YO
1200		6191.73	I
590	Bl	6199.82	YO
450	Bl	6217.96	YO
300		6222.59	I
270	Bl	6236.72	YO
45		6251.05	I
120	Bl	6275.01	YO
60	b	6295.46	YO
24	b	6316.20	YO
24	b	6338.10	YO
15	b	6359.48	YO
15		6369.87	YO
75		6402.01	I
1000		6435.00	I
24		6437.18	I
18	h	6501.23	YO
18	h	6518.33	YO
18	h	6535.84	YO
90		6538.60	I
12	h	6553.84	YO
70		6557.39	I
12	h	6572.58	
35		6576.85	I
23		6584.87	I
95		6613.75	II
14		6622.49	I
19	h	6636.49	I
40		6650.61	I
21		6664.40	I
150		6687.58	I
14	h	6691.83	I
7		6694.75	I
16	h	6699.26	I
70		6700.71	I
35		6713.20	I
40		6735.99	I
190		6793.71	I
70	h	6795.41	II
17	h	6803.15	I
21		6815.16	I
14		6832.49	II
45		6845.24	I
14		6858.24	II
29		6887.22	I
21		6896.00	II
9		6908.26	I
14		6933.52	I
24	h	6950.31	I
10		6951.68	II
10		6958.04	I
24		6979.88	I
13	h	7008.97	I
10		7009.93	I
19	h	7035.18	I
29		7052.94	I
13	h	7054.28	I
9		7075.13	I
11		7127.92	I
35		7191.66	I
10	h	7195.93	I
35		7264.17	II
9	h	7293.08	I
9	h	7330.62	I
5		7332.96	II
50		7346.46	I
11	h	7398.77	I
29		7450.30	II
17		7494.88	I
7	h	7536.71	I
35		7563.13	I
8	h	7617.72	I
19	h	7622.94	I
7		7652.89	I
5		7689.49	I
8	h	7698.00	I
19		7719.89	I
19		7724.08	I
13		7788.42	I
13		7796.32	I
6		7802.52	I
17		7812.16	I
29		7855.52	I
110		7881.90	II
10	h	7999.33	I
9		8329.61	I
24		8344.43	I
8	h	8365.64	I
17		8450.36	I
8	h	8528.94	I
95		8800.62	I
19	h	8835.85	II

Y III
Ref. 77 — J.R.

Intensity		Wavelength	
		Vacuum	
1		643.68	III
4		646.69	III
6		653.87	III
10		656.98	III
25		668.74	III
40		671.98	III
100		691.72	III
4		693.85	III
200		695.20	III
9		727.91	III
4		728.47	III
2		728.83	III
20		729.73	III
600		730.49	III
15		732.70	III
800		734.36	III
15		770.78	III
10		771.79	III
20		804.26	III
5000		805.20	III
75		806.18	III
150		808.97	III
7000		809.92	III
100		855.64	III
60		857.82	III
25		984.23	III
15		987.96	III
15000		989.21	III
25000		996.37	III
20		999.19	III
150		1000.56	III
25		1003.35	III
1000		1006.58	III
1200		1007.86	III
120		1077.52	III
500		1081.35	III
75	p	1084.63	III
350	p	1088.39	III
250		1095.25	III
25		1095.87	III
150		1103.21	III
3000		1289.74	III
2500		1306.96	III
5000		1314.51	III
1500		1316.10	III
4000		1334.04	III
8		1549.08	III
15	p	1553.81	III
30		1635.14	III
75		1640.43	III
200		1779.80	III
600		1786.05	III
		Air	
10		2041.93	III
5		2042.07	III
1500		2060.58	III
4000		2068.98	III
10000		2127.98	III
16000		2191.16	III
8000		2200.76	III
8000		2206.03	III
150		2261.41	III
80		2261.57	III
10000		2284.34	III
3		2319.92	III
10000		2327.31	III
50000		2367.23	III
40000		2414.64	III
100	p	2710.30	III
90	h	2710.54	III
5		2780.11	III
70		2791.44	III
20		2803.27	III
100		2807.00	III
90000		2817.04	III
6000		2867.67	III
6000		2913.41	III
1500		2917.74	III
1600		2918.56	III
15		2940.53	III
99000		2946.01	III
20	p	2948.48	III
6000		2970.42	III
1400		3013.93	III
1500		3018.85	III
3		3267.10	III
25	l	3276.80	III
500		3866.96	III
3000		3900.74	III
4000		3914.58	III
3800		4039.60	III
3000		4040.11	III
120	h	4121.61	III
2000	c	4737.62	III
7500		5102.88	III
1300		5120.40	III
10000		5238.10	III
3000		5263.58	III
4000		5383.64	III
6000		5562.81	III
600		5567.27	III
4000		5572.24	III
400		5595.48	III
3000		5602.08	III
2000	h	7254.58	III
9000		7558.71	III
6000		7864.53	III
8000		7916.71	III
400		7989.41	III
10000		7991.43	III
8000		8171.41	III
4000		8645.09	III
10000		8796.21	III
8000		9116.59	III

Y IV
Ref. 265 — J.R.

Intensity	Wavelength	
	Vacuum	
3	211.80	IV
3	214.51	IV
6	215.97	IV
6	217.39	IV
12	221.71	IV
3	222.18	IV
6	222.98	IV
2	228.84	IV
20	228.94	IV
3	229.78	IV
3	235.17	IV
25	235.77	IV
1	242.12	IV
30	242.30	IV
3	244.14	IV
3	263.72	IV
150	264.64	IV
30	272.40	IV
150	273.03	IV
10	278.60	IV
900	355.86	IV
300	370.42	IV
500	386.82	IV
600	425.03	IV
300	473.10	IV

Y V
Ref. 419 — J.R.

Intensity	Wavelength	
	Vacuum	
5	289.18	V
50	299.99	V
3	312.89	V
200	313.35	V
40	320.47	V
40	321.69	V
150	325.58	V
200	326.57	V
175	328.34	V
50	330.40	V
900	333.09	V
500	333.80	V
100	335.12	V
400	335.14	V
500	336.62	V
500	339.02	V
200	340.02	V
5	340.42	V
500	344.59	V
100	349.65	V
2	349.75	V
10	351.36	V
100	353.98	V
100	355.56	V
300	372.05	V
400	379.96	V
200	397.77	V
300	403.45	V
1	408.81	V
200	409.31	V
200	415.03	V
100	418.18	V
150	418.59	V
50	419.79	V
300	420.74	V
1	427.87	V
50	430.75	V
100	437.66	V
3	441.62	V
30	442.96	V

Yttrium (Cont.)

Intensity		Wavelength	
2		451.97	V
15		455.84	V
85		457.84	V
4000		584.98	V
2000		630.97	V

ZINC (Zn) Z = 30

Zn I and II
Ref. 39, 55, 113, 131, 185, 186 — R.D.C.

Intensity		Wavelength	
		Vacuum	
60		1193.23	II
60		1277.31	II
60	d	1366.68	II
		1404.12	I
60		1410.44	II
60		1439.09	II
60		1445.04	II
50		1456.91	II
		1457.57	I
60		1477.02	II
50		1514.76	II
90		1572.99	II
		1589.57	I
60		1617.68	II
60		1658.25	II
50		1713.25	II
60		1715.76	II
80	d	1735.61	II
60		1736.89	II
50		1737.90	II
75		1747.12	II
80	c	1762.19	II
75		1774.04	II
80		1790.76	II
100		1797.64	II
100	d	1811.05	II
80		1816.48	II
80		1831.38	II
100	d	1833.57	II
70		1836.01	II
75		1836.65	II
75		1847.56	II
100		1864.12	II
100		1866.08	II
100		1872.13	II
75		1894.26	II
60		1901.52	II
60		1914.81	II
100	d	1918.96	II
70		1920.27	II
100	d	1929.67	II
60		1945.58	II
60		1951.91	II
80		1953.00	II
75		1954.87	II
80		1964.54	II
100		1969.40	II
100		1982.11	II
70		1985.61	II
100		1986.99	II
50		1993.37	II
50		1996.92	II
		Air	
100		2011.94	II
500		2025.48	II
60		2039.31	II
500		2062.00	II
200		2064.23	II
120		2079.08	I
50		2079.93	II
60		2087.33	I
80		2096.93	I
300		2099.94	II
200		2102.18	II
150		2104.42	I
75		2122.74	II
800	r	2138.56	I
75		2147.42	II
60		2210.18	II
50		2273.15	II
1000		2501.99	II
150		2515.81	I
50		2527.96	II
1000		2557.95	II
50		2567.80	II
50		2567.98	II
100	h	2569.87	I
100		2582.44	I
300		2582.49	I
200		2608.56	I
300		2608.64	I
200		2670.53	I
300		2684.16	I
300		2712.49	I
200		2756.45	I
300		2770.86	I
300		2770.98	I
400		2800.87	I
100		2801.06	I
5		2801.17	I
100		2801.96	II
100		2902.30	II
125		3018.36	I
200		3035.78	I
200		3072.06	I
150		3075.90	I
100		3171.45	II
100		3172.23	II
300		3196.31	II
100		3197.10	II
500	r	3282.33	I
50		3299.42	II
800		3302.58	I
700	r	3302.94	I
75		3306.01	II
800		3345.02	I
500		3345.57	I
150		3345.94	I
5		3799.00	I
50		3806.34	II
100		3840.29	II
50		3883.34	I
15		3965.43	I
10		4113.21	I
25		4292.88	I
25		4298.33	I
35		4629.81	I
300		4680.14	I
400		4722.15	I
400		4810.53	I
800		4911.62	II
500		4924.03	II
7		5068.66	I
15		5069.58	I
200		5181.98	I
8		5308.65	I
7		5310.24	I
7		5311.02	I
4		5772.10	I
4		5775.50	I
10		5777.11	I
500		5894.33	II
500		6021.18	II
500		6102.49	II
100		6111.53	II
500		6214.61	II
8		6237.90	I
8		6239.17	I
1000	h	6362.34	I
10		6479.18	I
15		6928.32	I
8		6938.47	I
3		6943.20	I
200		7478.8	II
300		7588.5	II
100		7612.9	II
300		7732.5	II
200		7757.9	II
10		7799.36	I
100		11054.25	I
100		13053.63	I
100		13150.59	I
20		13196.61	I
100		14038.70	I
20		15680.29	I
20		16483.45	I
20		16491.98	I
20		16505.23	I
5		24044.16	I
10		24375.02	I

Zn III
Ref. 376, 377 — R.D.C.

Intensity		Wavelength	
		Vacuum	
1000		677.63	III
750		677.96	III
200		713.90	III
100		1432.15	III
200		1456.72	III
100		1464.20	III
100		1465.75	III
300		1473.41	III
100		1489.26	III
100		1490.96	III
200		1498.79	III
300		1499.42	III
300		1500.42	III
300		1505.92	III
300		1515.85	III
30		1533.09	III
300		1552.30	III
200		1552.94	III
80		1553.11	III
200		1560.79	III
150		1562.55	III
200		1581.53	III
100		1582.06	III
100		1598.52	III
100		1600.87	III
100		1619.61	III
150		1622.51	III
200		1629.19	III
200		1639.33	III
150		1644.82	III
100		1651.74	III
200		1673.05	III
100		1688.59	III
80		1695.40	III
80		1706.65	III
80		1749.63	III
50		1753.84	III
100		1767.69	III
80		1839.32	III

Zn IV
Ref. 370, 377 — R.D.C.

Intensity		Wavelength	
		Vacuum	
30		412.67	IV
30		423.42	IV
50		423.54	IV
200		425.90	IV
200		428.54	IV
80		428.79	IV
150		429.30	IV
200		430.59	IV
150		431.54	IV
120		431.62	IV
10		434.41	IV
150		435.02	IV
150		435.76	IV
80		436.25	IV
30		436.38	IV
50		436.82	IV
50		441.15	IV
20		441.52	IV
20		441.56	IV
100		441.70	IV
200		442.39	IV
150		444.39	IV
100		444.46	IV
100		446.58	IV
30		447.85	IV
10		449.13	IV
200		449.98	IV
200		450.99	IV
30		451.62	IV
20		452.80	IV
80		456.67	IV
150		457.32	IV
200		466.93	IV
150		468.43	IV
200		472.09	IV
200		472.66	IV
200		473.02	IV
200		473.51	IV
200		474.56	IV
50		475.78	IV
200		476.42	IV
200		478.65	IV
200		478.90	IV
200		482.10	IV
10		482.68	IV
10		485.48	IV
10		489.19	IV
10		490.96	IV
10		493.37	IV
10		496.72	IV
100		497.70	IV
15		1193.29	IV
15		1203.44	IV
8		1212.71	IV
5		1214.14	IV
25		1224.35	IV
20		1227.62	IV
40		1228.65	IV
30		1231.46	IV
30		1237.26	IV
50		1239.12	IV
3		1246.26	IV
15		1247.01	IV
50		1249.69	IV
30		1253.67	IV
30		1257.31	IV
8		1259.68	IV
500		1265.74	IV
5		1269.15	IV
100		1272.21	IV
200		1272.98	IV
30		1275.78	IV
25		1278.51	IV
100		1280.47	IV
100		1291.83	IV
100		1292.49	IV
10		1294.32	IV
100		1296.62	IV
100		1296.73	IV
20		1301.88	IV
500		1306.66	IV
200		1318.00	IV
200		1320.74	IV
200		1321.22	IV
200		1322.33	IV
200		1322.43	IV
200		1326.74	IV
200		1329.11	IV
100		1333.32	IV
150		1344.08	IV
200		1347.98	IV
200		1349.90	IV
50		1352.27	IV
100		1356.20	IV
200		1357.82	IV
100		1363.43	IV
200		1363.95	IV
40		1368.17	IV
200		1369.53	IV
50		1370.42	IV
100		1375.33	IV
50		1375.98	IV
200		1377.65	IV
50		1391.24	IV
80		1393.07	IV
50		1394.54	IV
30		1400.14	IV
15		1403.98	IV
30		1409.40	IV
5		1410.33	IV
30		1419.60	IV
50		1427.79	IV
20		1438.58	IV
80		1455.65	IV
200		1459.98	IV
100		1476.43	IV
100		1481.25	IV
100		1529.84	IV
50		1533.68	IV

ZIRCONIUM (Zr) Z = 40

Zr I and II
Ref. 1 — C.H.C.

Intensity		Wavelength	
		Air	
60		2374.42	I
60		2384.17	I
50		2388.01	I
50		2389.21	I
45		2405.52	I
60		2419.41	II
150		2449.85	II
21		2457.44	II
75		2487.29	II
45		2496.48	II
180		2532.46	II
90		2539.65	I
220		2542.10	II
45		2550.51	I
220		2550.74	II
45		2556.43	I
60		2567.45	I
570		2567.64	II
1600		2568.87	II
2100		2571.39	II
75		2583.40	II
130		2589.07	II
22		2589.65	I
45		2609.43	I
150		2630.91	II
80		2635.42	I
210		2639.09	II
70		2643.40	II
55		2647.78	I
110		2650.38	II
70		2658.69	I
180		2667.80	II
55		2669.49	II
120		2670.96	II
1800		2678.63	II
35		2681.76	II
90		2687.75	I
90		2692.60	II
22		2692.92	I
160		2693.53	II
180		2694.06	II
70		2695.43	II

Zirconium (Cont.)

Intensity		Wavelength	Spectrum
95		2699.60	II
750		2700.13	II
280		2711.51	II
140		2712.42	II
140		2714.26	II
1300		2722.61	II
140		2725.47	I
800		2726.49	II
490		2732.72	II
1400		2734.86	II
110		2740.51	II
140		2741.55	II
1100		2742.56	II
660		2745.86	II
660		2752.21	II
530		2758.81	II
200	d	2768.73	II
		2768.85	II
170	d	2774.04	I
		2774.16	II
200		2790.14	I
120		2792.04	I
160		2796.90	II
110		2799.15	II
180		2810.91	II
620		2814.90	I
390		2818.74	II
530		2825.56	II
110		2833.91	II
710		2837.23	I
120		2839.34	II
130		2843.52	II
660		2844.58	II
210		2848.19	II
350		2848.52	I
350		2851.97	II
340		2869.81	II
490		2875.98	I
120		2892.26	I
160		2905.23	II
300		2915.99	II
110		2916.64	II
270		2918.24	II
320		2926.99	II
160		2934.61	II
160		2936.31	II
320		2948.94	II
210		2951.48	II
320		2955.78	II
320		2960.87	I
320		2962.68	II
320		2968.96	II
120		2969.19	I
230		2969.63	II
130		2976.61	II
320		2978.05	II
230		2979.18	II
160		2981.02	II
820		2985.39	I
320		3003.74	II
100		3005.37	I
160		3005.50	I
820		3011.75	I
100		3013.32	II
160		3019.84	II
350		3020.47	II
500		3028.04	II
880		3029.52	I
180		3030.92	II
350	d	3036.39	II
100		3045.83	I
690		3054.84	II
100		3060.11	II
100		3064.63	II
110		3085.34	I
110		3094.80	I
250		3095.07	II
110		3095.82	I
280		3099.23	II
690		3106.58	II
110		3108.37	I
210		3110.88	II
350		3120.74	I
320		3125.92	II
500		3129.18	II
500		3129.76	II
140		3131.11	I
350		3132.07	I
110		3133.23	I
350		3133.48	II
180		3136.96	I
690		3138.68	II
140		3139.80	I
180		3148.82	I
290		3155.67	II
150		3157.00	II
320		3157.82	I
540		3164.31	II
150		3165.45	II
880		3165.97	II
150		3166.26	II
190		3178.09	II
190		3181.58	II
150		3181.92	II
880		3182.86	II
540		3191.21	I
210		3191.90	II
540		3212.01	I
760		3214.19	II
110		3222.47	II
200		3228.81	II
630		3231.69	II
630		3234.12	I
110		3236.58	II
760		3241.05	II
320		3250.39	I
200		3254.28	I
200		3260.11	I
190		3269.66	I
150		3271.13	II
540		3272.22	II
1000		3273.05	II
1300		3279.26	II
320	d	3282.73	I
880		3284.71	II
140		3285.88	II
150		3288.80	II
540		3305.15	II
880		3306.28	II
150		3313.70	II
210		3314.50	II
150		3319.02	II
380		3322.99	II
380		3326.80	II
380		3334.25	II
210		3334.62	II
190		3338.41	II
760		3340.56	II
380		3344.79	II
130		3353.66	I
180		3354.39	II
760		3356.09	II
540		3357.26	II
180		3359.96	II
150		3360.46	I
150		3363.82	II
150		3367.82	II
150		3370.59	I
180		3373.42	II
380		3374.73	II
110		3376.27	II
150		3377.46	II
570		3387.87	II
760		3388.30	II
5700		3391.98	II
570		3393.12	II
160		3396.33	II
380		3399.35	II
570		3404.83	II
760		3410.25	II
380		3414.66	I
1000		3430.53	II
380		3437.14	II
4700		3438.23	II
600		3447.36	I
200		3455.91	I
410		3457.56	II
200		3458.93	II
820		3463.02	II
600		3471.19	I
200		3478.79	I
1200		3479.39	II
1300		3481.15	II
760		3483.54	II
4100		3496.21	II
350		3505.48	II
820		3505.67	II
1000		3509.32	I
200		3510.46	II
2000		3519.60	I
440		3525.81	II
440		3533.22	I
210		3535.16	I
630		3542.62	II
1800		3547.68	I
210		3549.74	I
630		3550.46	I
1800		3551.95	II
2100		3556.60	II
1100		3566.10	I
210		3568.88	I
2100		3572.47	II
210		3573.08	II
1100		3575.79	I
1300		3576.85	II
880		3586.29	I
440		3587.98	II
3500		3601.19	I
690		3611.89	II
1100		3613.10	II
1100		3614.77	II
1100		3623.86	I
320		3634.15	I
260		3661.20	I
1100		3663.65	I
390		3671.27	II
800		3674.72	II
390		3697.46	II
960		3698.17	II
720		3709.26	II
270		3731.26	II
560		3745.98	II
880		3751.60	II
480		3764.39	I
480		3766.72	I
340		3766.82	II
720		3780.54	I
560		3791.40	I
210		3817.58	II
560		3822.41	I
2200		3835.96	I
1300		3836.76	II
550		3843.02	II
550		3847.01	I
550		3849.25	I
2900		3863.87	I
770		3864.34	I
990		3877.60	I
200		3879.05	I
1500		3885.42	I
2900		3890.32	I
2000		3891.38	I
400		3900.52	I
310		3915.94	II
610		3921.79	I
1200		3929.53	I
200		3934.12	II
200		3934.79	II
940		3958.22	II
490		3966.66	I
990		3968.26	I
660		3973.50	I
200		3975.29	I
200	h	3981.60	I
770		3991.13	II
770		3998.97	II
200		4007.60	I
200		4012.25	I
400		4023.98	I
770		4024.92	I
990		4027.20	I
240		4028.95	I
400		4029.68	II
490		4030.04	I
400		4035.89	I
240		4042.22	I
610		4043.58	I
490		4044.56	I
400		4045.61	II
610		4048.67	II
200		4050.33	II
200		4050.48	I
770		4055.03	I
600		4055.71	I
330		4061.53	I
1500		4064.16	I
2000		4072.70	I
310		4074.93	I
200		4076.53	I
240		4078.31	I
2000		4081.22	I
200		4108.40	I
400		4121.46	I
1200		4149.20	II
200		4152.64	I
290		4156.24	II
400		4161.21	II
400		4166.36	I
200		4183.32	I
660		4187.56	I
400		4194.76	I
610		4199.09	I
610		4201.46	I
610		4208.98	II
200		4211.88	II
400		4213.86	I
2000		4227.76	I
200		4236.06	I
2000		4239.31	I
770		4240.34	I
770		4241.20	I
1200		4241.69	I
310		4268.02	I
550		4282.20	I
550		4294.79	I
310		4302.89	I
550		4341.13	I
1000		4347.89	I
290		4359.74	II
310		4360.81	I
350		4366.45	I
240		4379.78	II
190		4413.04	I
240		4420.46	I
120		4427.24	I
160		4431.49	I
140		4443.00	II
110		4457.43	I
110		4466.91	I
110		4470.31	I
190		4470.56	I
200		4496.97	II
550		4507.12	I
610		4535.75	I
490		4542.22	I
200		4553.01	I
200		4555.13	I
140		4555.52	I
490		4575.52	I
100		4582.29	I
140		4590.55	I
350		4602.57	I
140		4604.42	I
210		4626.41	I
700		4633.98	I
210		4644.83	I
260		4683.42	I
2300		4687.80	I
510		4688.45	I
110		4707.79	I
1900		4710.08	I
160		4711.92	I
120		4717.62	I
210		4719.12	I
300		4732.33	I
1400		4739.48	I
190		4762.78	I
870		4772.31	I
210		4784.92	I
160		4788.67	I
260		4805.87	I
140		4809.47	I
190		4815.04	I
700		4815.63	I
280		4824.29	I
190		4828.04	I
110		4838.78	I
210		4851.36	I
160		4866.06	I
110		4881.24	I
110		4883.60	I
100		4994.76	I
30		5011.46	I
250		5046.58	I
85		5060.39	I
360		5064.91	I
110		5065.22	I
100		5070.26	I
75		5073.98	I
470		5078.25	I
85		5085.26	I
50		5112.27	II
140		5115.24	I
50		5120.42	I
85		5133.40	I
300		5155.45	I
200		5158.00	I
35		5158.67	I
75		5160.99	I
85		5165.96	I
17		5178.99	I
100		5183.70	I
30		5187.03	I
100		5191.60	II
100		5201.15	I
85		5209.30	I
85		5224.93	I
30		5243.47	I
120		5277.41	I
75		5280.05	I
60		5294.82	I
120		5296.79	I
60		5301.97	I
110		5311.40	I
25		5321.26	I
22		5330.84	I
12		5338.43	I
30		5350.09	II
30		5350.35	II
25		5350.90	I
25		5351.92	I
75		5362.56	I
12		5363.35	I
17		5369.39	I
20		5382.37	I
270		5385.14	I
30		5386.65	I
17		5391.18	I
17		5395.88	I
25		5405.13	I
85		5407.62	I
17		5413.93	I
20		5421.86	I
15		5426.36	I
25		5428.42	I
25		5437.76	I
15		5440.41	I
35		5448.57	I
10		5474.92	I
10		5477.40	I
35		5478.33	I
35		5480.83	I
10		5481.16	I
30		5486.09	I
140		5502.12	I

Zirconium (Cont.)

Intensity		Wavelength	
25		5507.87	I
30		5517.11	I
10		5518.05	I
75		5528.41	I
20		5532.30	I
45		5537.46	I
50		5545.32	I
22	Bl	5551.75	Z
25	Bl	5553.17	Z
12		5612.11	I
120		5620.14	I
35		5623.53	I
25	Bl	5629.02	Z
25	Bl	5629.58	Z
160		5664.51	I
20		5666.28	I
120		5680.90	I
15		5685.42	I
30		5708.89	I
75	Bl	5718.21	Z
120		5735.70	I
35	Bl	5748.17	Z
17	Bl	5778.57	Z
160		5797.74	I
30		5847.32	I
50		5868.27	I
110		5869.50	I
340		5879.80	I
85		5885.62	I
50		5901.09	I
30	Bl	5908.61	Z
140		5925.13	I
100		5935.20	I
110		5955.35	I
30	Bl	5977.80	Z
100		5984.23	I
17		5995.37	I
50		6001.05	I
30		6025.36	I
85		6032.61	I
170		6045.85	I
100		6049.24	I
140		6062.84	I
50		6120.83	I
170		6121.91	I
85		6124.84	I
680		6127.44	I
340		6134.55	I
100		6140.46	I
440		6143.20	I
30		6155.61	I
75		6157.71	I
25		6160.20	I
35		6189.40	I
60		6192.96	I
85		6213.05	I
100		6214.69	I
170	Bl	6226.51	Z
100		6257.26	I
50	b	6261.05	Z
35		6267.06	I
45	Bl	6292.84	Z
120		6299.66	I
15		6304.34	I
300		6313.02	I
30		6314.71	I
50		6321.35	I
22		6340.36	I
50	Bl	6345.10	Z
75		6345.22	I
75	Bl	6378.56	Z
35		6407.00	I
50	b	6412.39	Z
12		6426.17	I
35		6434.33	I
60		6445.74	I
20		6451.62	I
20		6457.63	I
110		6470.21	I
60	Bl	6473.79	Z
11		6484.35	I
110		6489.64	I
22		6493.10	I
50		6503.26	I
50		6506.36	I
50	Bl	6508.15	Z
30	Bl	6542.90	Z
35		6550.54	I
30		6569.43	I
20		6576.56	I
30	b	6578.06	Z
50		6591.99	I
10		6596.71	I
10		6598.84	I
50		6603.27	I
15		6620.56	I
11		6678.01	II
22		6688.18	I
11		6702.12	I
17		6709.61	I
27		6717.88	I
40		6752.73	I
75		6762.38	I
85		6769.16	I
27		6772.89	I
15		6787.15	II
35		6790.85	I
45		6828.78	I
45		6832.89	I
13		6845.33	I
17		6846.34	I
100		6846.97	I
27		6849.26	I
13		6852.56	I
120		6888.29	I
29		6900.59	I
20		6904.36	I
29		6907.37	I
20		6916.87	I
16		6932.38	I
29		6948.46	I
150		6953.84	I
60		6966.44	I
10		6975.91	I
150		6990.84	I
80		6994.32	I
10		7005.46	I
100		7027.40	I
25		7057.36	I
14		7057.96	I
140		7087.30	I
25		7089.43	I
35		7094.46	I
50		7095.59	I
540		7097.70	I
280		7102.91	I
170		7103.72	I
140		7111.68	I
40		7112.82	I
18		7113.52	I
12		7132.95	I
16		7140.74	I
12		7144.47	I
590		7169.09	I
50		7201.62	I
12		7258.17	I
35		7264.76	I
20		7306.21	I
25		7311.62	I
35		7313.72	I
90		7318.08	I
10		7327.82	I
50		7335.97	I
50		7343.96	I
20		7373.50	I
25		7383.63	I
14		7400.90	I
10		7411.39	I
10		7422.75	I
10		7433.10	I
110		7439.86	I
18		7467.57	I
16		7479.58	I
14	h	7515.70	I
12		7517.95	I
20	h	7540.62	I
20		7544.59	I
29		7551.46	I
40		7554.70	I
25		7558.45	I
12		7560.09	I
12		7562.12	I
80		7607.15	I
14		7612.08	I
20		7621.17	I
29		7658.60	I
18		7690.83	I
14		7704.27	I
10		7708.42	I
10		7766.55	I
12		7816.32	I
110		7819.35	I
35		7822.94	I
40		7826.72	I
90		7849.35	I
35		7869.99	I
14		7876.25	I
16		7882.18	I
10		7897.98	I
16		7908.46	I
20		7940.47	I
160		7944.61	I
80		7956.66	I
80		7959.98	I
20		7963.63	I
160		8005.27	I
25		8046.05	I
16		8053.06	I
20		8055.29	I
20		8055.76	I
60		8058.08	I
150		8063.09	I
790		8070.08	I
10		8114.28	I
20		8120.17	I
390		8132.99	I
20		8152.58	I
12		8188.77	I
40		8194.73	I
60		8201.73	I
280		8212.53	I
20		8240.37	I
40		8283.81	I
140		8305.90	I
14		8332.44	I
50		8370.23	I
120		8389.41	I
70		8414.00	I
50		8453.17	I
50		8464.65	I
40		8498.44	I
18		8584.21	I
10		8734.86	I
12		8749.48	I
10		8786.23	I
16		8804.98	I
70		8836.09	I
60		8899.52	I

Zr III
Ref. 403 — J.R.

Intensity	Wavelength	
	Vacuum	
25	687.64	III
50	690.39	III
25	819.59	III
30	820.21	III
35	823.69	III
25	829.50	III
30	850.61	III
25	859.56	III
30	864.86	III
25	868.64	III
25	868.99	III
25	919.59	III
30	1320.81	III
25	1375.13	III
30	1378.93	III
40	1403.48	III
35	1420.12	III
25	1420.87	III
25	1465.44	III
25	1563.24	III
40	1593.59	III
100	1612.38	III
50	1620.62	III
75	1631.31	III
50	1638.33	III
35	1675.06	III
35	1675.75	III
40	1703.36	III
40	1725.03	III
40	1754.38	III
35	1759.12	III
30	1764.75	III
30	1771.96	III
40	1773.90	III
100	1779.51	III
40	1783.35	III
200	1790.19	III
150	1793.56	III
125	1798.13	III
75	1800.03	III
25	1801.67	III
100	1805.26	III
35	1831.89	III
40	1850.06	III
40	1853.38	III
30	1859.12	III
25	1860.47	III
25	1861.77	III
75	1864.06	III
30	1865.45	III
35	1877.00	III
50	1892.07	III
65	1914.25	III
75	1921.96	III
75	1932.54	III
50	1934.32	III
40	1935.20	III
75	1936.48	III
75	1936.67	III
80	1937.27	III
200	1940.25	III
100	1941.08	III
25	1946.12	III
80	1946.61	III
50	1946.99	III
100	1953.95	III
50	1961.32	III
100	1962.01	III
40	1962.92	III
85	1966.22	III
25	1967.81	III
60	1974.99	III
40	1983.14	III
50	1989.83	III
30	1990.95	III
30	1994.46	III
	Air	
45	2000.23	III
100	2006.82	III
30	2013.30	III
35	2016.63	III
30	2021.52	III
60	2026.78	III
100	2035.42	III
50	2036.92	III
75	2056.13	III
40	2058.73	III
75	2060.83	III
50	2061.47	III
125	2070.43	III
50	2074.12	III
100	2077.92	III
100	2080.99	III
30	2081.81	III
25	2085.35	III
200	2086.78	III
40	2089.50	III
40	2097.03	III
40	2102.30	III
50	2103.16	III
25	2104.23	III
40	2113.98	III
35	2114.10	III
40	2125.06	III
35	2137.90	III
35	2138.45	III
25	2139.85	III
40	2159.24	III
40	2162.20	III
100	2175.80	III
100	2191.15	III
35	2192.05	III
60	2206.33	III
40	2206.97	III
30	2231.00	III
50	2245.36	III
25	2251.14	III
40	2257.83	III
30	2281.43	III
100	2301.60	III
75	2308.12	III
35	2405.81	III
40	2406.21	III
75	2420.65	III
25	2438.70	III
50	2444.58	III
100	2448.86	III
100	2593.64	III
250	2620.56	III
50	2621.28	III
60	2628.26	III
200	2643.79	III
100	2656.46	III
150	2664.26	III
100	2682.16	III
75	2686.28	III
70	2690.49	III
60	2698.31	III
50	2709.05	III
45	2715.76	III
40	2720.07	III
75	2735.76	III
25	2775.23	III
40	2836.18	III
40	3278.86	III

Zr IV
Ref. 362 — J.R.

Intensity	Wavelength	
	Vacuum	
15	478.97	IV
60	480.66	IV
60	497.23	IV
60	500.22	IV
4	500.34	IV
4	584.65	IV
15	585.42	IV
2	586.42	IV
60	588.89	IV
100	589.74	IV
600	628.66	IV
500	633.56	IV
200	633.63	IV
8	712.49	IV
90	754.39	IV
90	760.15	IV
150	846.40	IV
300	863.65	IV
500	864.59	IV
200	881.30	IV

Zirconium (Cont.)

Intensity		Wavelength	
200		882.59	IV
100		1099.76	IV
150		1100.00	IV
9000		1183.97	IV
9000		1201.77	IV
200		1212.71	IV
100		1213.01	IV
10000		1219.86	IV
500		1285.89	IV
500		1290.56	IV
70		1291.70	IV
500		1417.70	IV
500		1440.65	IV
500		1441.06	IV
1000		1469.47	IV
10000		1546.17	IV
10		1596.29	IV
10000		1598.95	IV
150		1605.26	IV
5000		1607.95	IV
90		1609.50	IV
35		1818.06	IV
500		1836.15	IV
500		1846.37	IV
200		1848.03	IV
70		1851.91	IV
		Air	
20		2045.12	IV
40		2047.15	IV
10000		2091.49	IV
10000		2092.36	IV
10000		2163.68	IV
10000		2286.67	IV
200		2473.75	IV
300		2476.71	IV
5		2572.32	IV
400		2573.66	IV
400		2583.32	IV

Zr V
Ref. 418 — J.R.

Intensity		Wavelength	
		Vacuum	
300		292.19	V
500		304.01	V
300		305.24	V
400	p	368.18	V
200		519.25	V
200		536.50	V
200	p	674.13	V
200		675.53	V
300		679.39	V
200		688.27	V
200	p	703.03	V
300		717.24	V
200		740.33	V
2000		740.61	V
500		742.74	V
300		752.48	V
300		764.58	V
500		766.20	V
200		773.80	V
200		775.58	V
200		779.21	V
400		784.50	V
400		797.23	V
10000		800.00	V
10000		806.89	V
200		809.75	V
10000		812.05	V
500		822.06	V
300	p	823.46	V
200		823.78	V
200		825.13	V
200		830.59	V
400		836.57	V
3000		841.40	V
300	p	852.87	V
700		853.68	V
200		873.11	V
300		885.68	V
300		900.48	V
500		906.66	V
500		915.30	V
400		923.10	V
400		940.41	V
200		949.70	V
200		978.06	V
200		980.70	V
200		984.18	V
300		995.59	V
400		1002.48	V
400		1038.69	V
200		1044.41	V
300		1047.77	V
400		1068.55	V
300		1072.25	V
300		1083.45	V
200		1087.05	V
200		1093.54	V
200		1108.79	V
300		1194.24	V
300		1200.76	V
200		1233.91	V
300		1238.93	V
300		1253.61	V
200		1259.70	V
400		1260.91	V
500		1265.38	V
300		1295.81	V
400		1302.80	V
500		1303.93	V
400		1306.76	V
200		1315.14	V
400		1320.74	V
500	p	1323.81	V
300		1332.06	V
300		1337.34	V
300		1355.21	V
500		1355.98	V
300		1361.39	V
500		1376.54	V
200		1396.79	V
300		1410.03	V
200		1413.40	V
300		1460.05	V
300		1486.90	V
300		1491.33	V
200		1520.47	V
200		1550.12	V
500		1633.03	V
500		1654.46	V
700		1725.02	V
500		1786.20	V
200	p	1790.81	V
500		1806.09	V
500		1860.48	V
600		1860.86	V
500		1878.33	V
400		1926.24	V
500		1927.43	V
400		1934.88	V
		Air	
500		2009.29	V
600		2028.54	V
600		2132.42	V
200		2150.18	V
200		2336.94	V

REFERENCES

1. **Meggers, W. F., Corliss, C. H., and Scribner, B. F.,** *Natl. Bur. Stand. (U.S.) Monogr.*, 145, Washington, D.C., 1975.
2. **Aksenov, V. P. and Ryabtsev, A. N.,** *Opt. Spectrosc.*, 37, 860, 1970.
3. **Andersen, N., Bickel, W. S., Carriveau, G. W., Jensen, K., and Veje, E.,** *Phys. Scr.*, 4, 113, 1971.
4. **Andersson, E. and Johannesson, G. A.,** *Phys. Scr.*, 3, 203, 1971.
5. **Andrew, K. L. and Meissner, K. W.,** *J. Opt. Soc. Am.*, 49, 146, 1959.
6. **Artru, M. C. and Brillet, W. U. L.,** *J. Opt. Soc. Am.*, 64, 1063, 1974.
7. **Artru, M. C. and Kaufman, V.,** *J. Opt. Soc. Am.*, 62, 949, 1972.
8. **Artru, M. C. and Kaufman, V.,** *J. Opt. Soc. Am.*, 65, 594, 1975.
9. **Badami, J. S. and Rao, K. R.,** *Proc. R. Soc. London*, 140(A), 387, 1933.
10. **Baird, K. M. and Smith, D. S.,** *J. Opt. Soc. Am.*, 48, 300, 1958.
11. **Bashkin, S. and Martinson, I.,** *J. Opt. Soc. Am.*, 61, 1686, 1971.
12. **Beacham, J. R., Ph.D. thesis, Purdue University,** 1970.
13. **Benschop, H., Joshi, Y. N., and van Kleef, T. A. M.,** *Can. J. Phys.*, 53, 700, 1975.
14. **Berry, H. G., Bromander, J., and Buchta, R.,** *Phys. Scr.*, 1, 181, 1970.
15. **Berry, H. G., Bromander, J., Martinson, I., and Buchta, R.,** *Phys. Scr.*, 3, 63, 1971.
16. **Berry, H. G., Desesquelles, J., and Dufay, M.,** *Phys. Rev. Sect. A.*, 6, 600, 1972.
17. **Berry, H. G., Desesquelles, J., and Dufay, M.,** *Nucl. Instrum. Methods*, 110, 43, 1973.
18. **Berry, H. G., Pinnington, E. H., and Subtil, J. L.,** *J. Opt. Soc. Am.*, 62, 767, 1972.
19. **Bidelman, W. P. and Corliss, C. H.,** *Astrophys. J.*, 135, 968, 1962.
20. **Bloch, L. and Bloch, E.,** *Ann. Phys.* (Paris), 10(11), 141, 1929.
21. **Bloch, L., Bloch, E., and Felici, N.,** *J. Phys. Radium*, 8, 355, 1937.
22. **Bockasten, K.,** *Ark. Fys.*, 9, 457, 1955.
23. **Bockasten, K., Hallin, R., Johansson, K. B., and Tsui, P.,** *Phys. Lett.* (Netherlands), 8, 181, 1964.
24. **Bockasten, K. and Johansson, K. B.,** *Ark. Fys.*, 38, 563, 1969.
25. **Borgstrom, A.,** *Ark. Fys.*, 38, 243, 1968.
26. **Borgstrom, A.,** *Phys. Scr.*, 3, 157, 1971.
27. **Bowen, I. S.,** *Phys. Rev.*, 29, 231, 1927.
28. **Bowen, I. S.,** *Phys. Rev.*, 31, 34, 1928.
29. **Bowen, I. S.,** *Phys. Rev.*, 39, 8, 1932.
30. **Bowen, I. S.,** *Phys. Rev.*, 45, 401, 1934.
31. **Bowen, I. S.,** *Phys. Rev.*, 46, 377, 1934.
32. **Bowen, I. S.,** *Phys. Rev.*, 46, 791, 1934.
33. **Boyce, J. C.,** *Phys. Rev.*, 49, 730, 1936.
34. **Boyce, J. C. and Robinson, H. A.,** *J. Opt. Soc. Am.*, 26, 133, 1936.
35. **Bozman, W. R., Meggers, W. F., and Corliss, C. H.,** *J. Res. Natl. Bur. Stand. Sect. A*, 71, 547, 1967.
36. **Bromander, J.,** *Ark. Fys.*, 40, 257, 1969.
37. **Bromander, J. and Buchta, R.,** *Phys. Scr.*, 1, 184, 1970.
38. **Brown, C. M. and Ginter, M. L.,** *J. Opt. Soc. Am.*, 68, 243, 1978.
39. **Brown, C. M., Tilford, S. G., and Ginter, M. L.,** *J. Opt. Soc. Am.*, 65, 1404, 1975.
40. **Bryant, B. W.,** *Johns Hopkins Spectroscopic Report* No. 21, 1961.
41. **Buchet, J. P., Buchet-Poulizac, M. C., Berry, H. G., and Drake, G. W. F.,** *Phys. Rev. Sect. A*, 7, 922, 1973.
42. **Budhiraja, C. J. and Joshi, Y. N.,** *Can. J. Phys.*, 49, 391, 1971.

43. **Burns, K. and Adams, K. B.,** *J. Opt. Soc. Am.*, 42, 56, 1952.
44. **Burns, K. and Adams, K. B.,** *J. Opt. Soc. Am.*, 46, 94, 1956.
45. **Burns, K., Adams, K. B., and Longwell, J.,** *J. Opt. Soc. Am.*, 40, 339, 1950.
46. **Callahan, W. R.,** Ph.D. thesis, Johns Hopkins University, 1962.
47. **Charles, G. W.,** *J. Opt. Soc. Am.*, 56, 1292, 1966.
48. **Charles, G. W., Hunt, D. J., Pish, G., and Timma, D. L.,** *J. Opt. Soc. Am.*, 45, 869, 1955.
49. **Codling, K.,** *Proc. Phys. Soc.*, 77, 797, 1961.
50. **Comite Consulatif Pour La Definition du Metre,** *J. Phys. Chem. Ref. Data*, 3, 852, 1974.
51. **Conway, J. G., Blaise, J., and Verges, J.,** *Spectrochim. Acta Part B*, 31, 31, 1976.
52. **Conway, J. G., Worden, E. F., Blaise, J., and Verges, J.,** *Spectrochim. Acta Part B*, 32, 97, 1977.
53. **Conway, J. G., Worden, E. F., Blaise, J., Camus, P., and Verges, J.,** *Spectrochim. Acta Part B*, 32, 101, 1977.
54. **Crooker, A. M.,** *Can. J. Res. Sect. A*, 14, 115, 1936.
55. **Crooker, A. M. and Dick, K. A.,** *Can. J. Phys.*, 46, 1241, 1968.
56. **Crosswhite, H. M.,** *J. Res. Natl. Bur. Stand. Sect. A*, 79, 17, 1975.
58. **Crosswhite, H. M. and Dieke, G. H.,** *American Institute of Physics Handbook*, Section 7, 1972.
59. **de Bruin, T. L.,** *Z. Phys.*, 38, 94, 1926.
60. **de Bruin, T. L.,** *Z. Phys.*, 53, 658, 1929.
61. **de Bruin, T. L., Humphreys, C. J., and Meggers, W. F.,** *J. Res. Natl. Bur. Stand.*, 11, 409, 1933.
62. **Dick, K. A.,** *J. Opt. Soc. Am.*, 64, 702, 1973.
63. **Dobbie, J. C.,** *Ann. Solar Phys. Observ.* (Cambridge), 5, 1, 1938.
64. **Earls, L. T. and Sawyer, R. A.,** *Phys. Rev.*, 47, 115, 1935.
65. **Edlen, B.,** *Z. Phys.*, 85, 85, 1933.
66. **Edlen, B.,** *Nova Acta Reglae Soc. Sci. Ups.*, (IV) 9, No. 6, 1934.
67. **Edlen, B.,** *Z. Phys.*, 93, 726, 1935.
68. **Edlen, B.,** *Z. Phys.*, 94, 47, 1935.
69. **Edlen, B.,** *Rep. Prog. Phys.*, 26, 181, 1963.
70. **Edlen, B. and Risberg, P.,** *Ark. Fys.*, 10, 553, 1956.
71. **Edlen, B. and Swings, P.,** *Astrophys. J.*, 95, 532, 1942.
72. **Ehrhardt, J. C. and Davis, S. P.,** *J. Opt. Soc. Am.*, 61, 1342, 1971.
73. **Eidelsberg, M.,** *J. Phys. B*, 5, 1031, 1972.
74. **Eidelsberg, M.,** *J. Phys. B*, 7, 1476, 1974.
75. **Ekberg, J. O. and Svensson, L. A.,** *Phys. Scr.*, 2, 283, 1970.
76. **Ekefors, E.,** *Z. Phys.*, 71, 53, 1931.
77. **Epstein, G. L. and Reader, J.,** *J. Opt. Soc. Am.*, 65, 310, 1975.
78. **Epstein, G. L. and Reader, J.,** *J. Opt. Soc. Am.*, 66, 590, 1976.
79. **Epstein, G. L. and Reader, J.,** unpublished.
80. **Eriksson, K. B. S.,** *Phys. Lett. A.*, 41, 97, 1972.
81. **Eriksson, K. B. S. and Isberg, H. B. S.,** *Ark. Fys.*, 23, 527, 1963.
82. **Eriksson, K. B. S. and Wenaker, I.,** *Phys. Scr.*, 1, 21, 1970.
83. **Esteva, J. M. and Mehlman, G.,** *Astrophys. J.*, 193, 747, 1974.
84. **Even-Zohar, M. and Fraenkel, B. S.,** *J. Phys. B*, 5, 1596, 1972.
85. **Fawcett, B. C.,** *J. Phys. B*, 3, 1732, 1970.
86. **Fawcett, B. C.,** Culham Laboratory Report ARU-R4, 1971.
87. **Ferner, E.,** *Ark. Mat. Astron. Fys.*, 28(A), 4, 1941.
88. **Fischer, R. A., Knopf, W. C., and Kinney, F. E.,** *Astrophys. J.*, 130, 683, 1959.
89. **Fowler, A.,** *Report on Series in Line Spectra*, Fleetway Press, London, 1922.
90. **Fowles, G. R.,** *J. Opt. Soc. Am.*, 44, 760, 1954.
91. **Fred, M.,** *Argonne Natl. Lab.*, unpublished, 1977.
92. **Fred, M. and Tomkins, F. S.,** *J. Opt. Soc. Am.*, 47, 1076, 1957.
93. **Fred, M., Tomkins, F. S., Blaise, J. E., Camus, P., and Verges, J.,** Argonne National Laboratory Report No. 76-68, 1976.
94. **Garcia, J. D. and Mack, J. E.,** *J. Opt. Soc. Am.*, 55, 654, 1965.
96. **Giacchetti, A.,** *Argonne Natl. Lab.*, unpublished, 1975.
97. **Giacchetti, A., Blaise, J., Corliss, C. H., and Zalubas, R.,** *J. Res. Natl. Bur. Stand. Sect. A*, 78, 247, 1974.
98. **Giacchetti, A., Stanley, R. W., and Zalubas, R.,** *J. Opt. Soc. Am.*, 69, 474, 1970.
99. **Gilbert, W. P.,** *Phys. Rev.*, 47, 847, 1935.
100. **Gilroy, H. T.,** *Phys. Rev.*, 38, 2217, 1931.
101. **Glad, S.,** *Ark. Fys.*, 10, 291, 1956.
102. **Goldsmith, S.,** *J. Phys. B*, 2, 1075, 1969.
103. **Goorvitch, D., Mehlman-Balloffet, G., and Valero, F. P. J.,** *J. Opt. Soc. Am.*, 60, 1458, 1970.
104. **Goorvitch, D. and Valero, F. P. J.,** *Astrophys. J.*, 171, 643, 1972.
105. **Green, L. C.,** *Phys. Rev.*, 55, 1209, 1939.
106. **Gutman, F.,** *Diss. Abstr. Int. B*, 31, 363, 1970.
107. **Hallin, R.,** *Ark. Fys.*, 31, 511, 1966.
108. **Hallin, R.,** *Ark. Fys.*, 32, 201, 1966.
109. **Hansen, J. E. and Persson, W.,** *J. Opt. Soc. Am.*, 64, 696, 1974.
110. **Hansen, J. E. and Persson, W.,** *Phys. Scr.*, 13, 166, 1976.
111. **Hellintin, P.,** *Phys. Scr.*, 13, 155, 1976.
112. **Herzberg, G. and Moore, H. R.,** *Can. J. Phys.*, 37, 1293, 1959.
113. **Hetzler, C. W., Boreman, R, W., and Burns, K.,** *Phys. Rev.*, 48, 656, 1935.
114. **Holmstrom, J. E. and Johansson, L.,** *Ark. Fys.*, 40, 133, 1969.
115. **Hontzeas, S., Martinson, I., Erman, P., and Buchta, R.,** *Nucl. Instrum. Methods*, 110, 51, 1973.
116. **Humphreys, C. J.,** *J. Res. Natl. Bur. Stand.*, 22, 19, 1939.
117. **Humphreys, C. J.,** *J. Opt. Soc. Am.*, 43, 1027, 1953.
118. **Humphreys, C. J.,** *J. Phys. Chem. Ref. Data*, 2, 519, 1973.
119. **Humphreys, C. J. and Andrew, K. L.,** *J. Opt. Soc. Am.*, 54, 1134, 1964.
120. **Humphreys, C. J. and Meggers, W. F.,** *J. Res. Natl. Bur. Stand.*, 10, 139, 1933.
121. **Humphreys, C. J. and Paul, E., Jr.,** *J. Opt. Soc. Am.*, 60, 200, 1970.
122. **Humphreys, C. J. and Paul, E., Jr.,** *J. Opt. Soc. Am.*, 62, 432, 1972.
123. **Humphreys, C. J., Paul, E., Jr., Cowan, R. D., and Andrew, K. L.,** *J. Opt. Soc. Am.*, 57, 855, 1967.
124. **Humphreys, C. J., Paul, E., Jr., and Minnhagen, L.,** *J. Opt. Soc. Am.*, 61, 110, 1971.
125. **Iglesias, L.,** Inst. of Optics, Madrid, unpublished, 1977.

126. **Iglesias, L. and Velasco, R.**, *Publ. Inst. Opt. Madrid*, No. 23, 1964.
127. **Isberg, B.**, *Ark. Fys.*, 35, 551, 1967.
128. **Johannesson, G. A., Lundstrom, T., and Minnhagen, L.**, *Phys. Scr.*, 6, 129, 1972.
129. **Johannesson, G. A. and Lundstrom, T.**, *Phys. Scr.*, 8, 53, 1973.
130. **Johansson, I.**, *Ark. Fys.*, 20, 135, 1961.
131. **Johansson, I. and Contreras, R.**, *Ark. Fys.*, 37, 513, 1968.
132. **Johansson, I. and Litzen, U.**, *Ark. Fys.*, 34, 573, 1967.
133. **Johansson, I. and Svensson, K. F.**, *Ark. Fys.*, 16, 353, 1960.
134. **Johansson, L.**, *Ark. Fys.*, 20, 489, 1961.
135. **Johansson, L.**, *Ark. Fys.*, 23, 119, 1963.
136. **Johansson, S. and Litzen, U.**, *Phys. Scr.*, 6, 139, 1972.
137. **Johansson, S. and Litzen, U.**, *Phys. Scr.*, 8, 43, 1973.
138. **Johansson, S. and Litzen, U.**, *Phys. Scr.*, 10, 121, 1974.
139. **Joshi, Y. N.**, St. Francis Xavier Univ., Nova Scotia, unpublished.
140. **Joshi, Y. N., Bhatia, K. S., and Jones, W. E.**, *Sci. Light Tokyo*, 21, 113, 1972.
141. **Joshi, Y. N., Bhatia, K. S., and Jones, W. E.**, *Spectrochim. Acta Part B*, 28, 149, 1973.
142. **Joshi, Y. N. and Budhiraja, C. J.**, *Can. J. Phys.*, 49, 670, 1971.
143. **Joshi, Y. N. and van Kleef, T. A. M.**, *Can. J. Phys.*, 52, 1891, 1974.
144. **Kaufman, V.**, *Natl. Bur. Stand.*, unpublished.
145. **Kaufman, V.**, *J. Opt. Soc. Am.*, 52, 866, 1962.
146. **Kaufman, V., Artru, M. C., and Brillet, W. U. L.**, *J. Opt. Soc. Am.*, 64, 197, 1974.
147. **Kaufman, V. and Humphreys, C. J.**, *J. Opt. Soc. Am.*, 59, 1614, 1969.
148. **Kaufman, V. and Sugar, J.**, *J. Opt. Soc. Am.*, 61, 1693, 1971.
149. **Kaufman, V. and Sugar, J.**, *J. Res. Natl. Bur. Stand. Sect. A*, 71, 583, 1967.
150. **Kelly, R. L. and Palumbo, L. J.**, *Naval Research Laboratory Report 7599, Washington, D. C.*, 1973.
151. **Kielkopf, J. F.**, *Univ. of Louisville*, unpublished, 1975.
152. **Kielkopf, J. F.**, *Univ. of Louisville*, unpublished, 1976.
153. **Kiess, C. C. and Corliss, C. H.**, *J. Res. Natl. Bur. Stand. Sect. A*, 63, 1, 1959.
154. **Kleiman, H.**, *J. Opt. Soc. Am.*, 52, 441, 1962.
155. Eriksson, K. B., Johansson, I., and Norlen, G., *Ark. Fys.*, 28, 233, 1964.
156. **Klinkenberg, P. F. A.**, *Physica*, 15, 774, 1949.
157. **Klinkenberg, P. F. A.**, *Physica*, 16, 618, 1950.
158. **Krishnamurty, S. G.**, *Proc. Phys. Soc. London*, 48, 277, 1936.
159. **Kruger, P. G. and Gilroy, H. T.**, *Phys. Rev.*, 48, 720, 1935.
160. **Kruger, P. G. and Pattin, H. S.**, *Phys. Rev.*, 52, 621, 1937.
161. **Lacroute, P.**, *Ann. Phys.* (Paris), 3, 5, 1935.
162. **Lang, R. J.**, *Phys. Rev.*, 30, 762, 1927.
163. **Lang, R. J.**, *Phys. Rev.*, 32, 737, 1928.
164. **Lang, R. J.**, *Phys. Rev.*, 35, 445, 1930.
165. **Lang, R. J.**, *Can. J. Res. Sect. A*, 14, 43, 1936.
166. **Lang, R. J.**, *Can. J. Res. Sect. A*, 14, 127, 1936.
167. **Lang, R. J. and Vestine, E. H.**, *Phys. Rev.*, 42, 233, 1932.
168. **Li, H. and Andrew, K. L.**, *J. Opt. Soc. Am.*, 61, 96, 1971.
169. **Liden, K.**, *Ark. Fys.*, 1, 229, 1949.
170. **Litzen, U.**, *Ark. Fys.*, 28, 239, 1965.
171. **Litzen, U.**, *Phys. Scr.*, 1, 251, 1970.
172. **Litzen, U.**, *Phys. Scr.*, 1, 253, 1970.
173. **Litzen, U.**, *Phys. Scr.*, 2, 103, 1970.
174. **Litzen, U. and Verges, J.**, *Phys. Scr.*, 13, 240, 1976.
175. **Lofstrand, B.**, *Phys. Scr.*, 8, 57, 1973.
176. **Luc-Koenig, E., Morillon, C., and Verges, J.**, *Phys. Scr.*, 12, 199, 1975.
177. **Lundstrom, T.**, *Phys. Scr.*, 7, 62, 1973.
178. **Lundstrom, T. and Minnhagen, L.**, *Phys. Scr.*, 5, 243, 1972.
179. **Magnusson, C. E. and Zetterberg, P. O.**, *Phvs Scr.*, 10, 177, 1974.
180. Magnusson, C. E., and Zetterberg, P. O., *Phys. Scr.*, 15, 237, 1977.
181. **Martin, D. C.**, *Phys. Rev.*, 48, 938, 1935.
182. Svendenius, N., *Phys. Scr.*, 22, 240, 1980.
183. **Martin, W. C.**, *J. Res. Natl. Bur. Stand. Sect. A*, 64, 19, 1960.
184. **Martin, W. C. and Corliss, C. H.**, *J. Res. Natl. Bur. Stand. Sect. A*, 64, 443, 1960.
185. **Martin, W. C. and Kaufman, V.**, *J. Res. Natl. Bur. Stand. Sect. A*, 74, 11, 1970.
186. **Martin, W. C. and Kaufman, V.**, *J. Opt. Soc. Am.*, 60, 1096, 1970.
187. **McCormick, W. W. and Sawyer, R. A.**, *Phys. Rev.*, 54, 71, 1938.
188. **McLaughlin, R.**, *J. Opt. Soc. Am.*, 54, 965, 1964.
189. **McLennan, J. C., McLay, A. B., and Crawford, M. F.**, *Proc. R. Soc. London Ser. A*, 134, 41, 1931.
190. **Meissner, K. W.**, *Z. Phys.*, 39, 172, 1926.
191. **Meggers, W. F.**, *J. Res. Natl. Bur. Stand.*, 24, 153, 1940.
192. **Meggers, W. F. and Corliss, C. H.**, *J. Res. Natl. Bur. Stand. Sect. A*, 70, 63, 1966.
193. **Meggers, W. F.**, **Fred, M., and Tomkins, F. S.**, *J. Res. Natl. Bur. Stand.*, 58, 297, 1957.
194. **Meggers, W. F. and Humphreys, C. J.**, *J. Res. Natl. Bur. Stand.*, 28, 463, 1942.
195. **Meggers, W. F. and Murphy, R. J.**, *J. Res. Natl. Bur. Stand.*, 48, 334, 1952.
196. **Meggers, W. F., Scribner, B. F., and Bozman, W. R.**, *J. Res. Natl. Bur. Stand.*, 46, 85, 1951.
197. **Meggers, W. F., Shenstone, A. G., and Moore, C. E.**, *J. Res. Natl. Bur. Stand.*, 45, 346, 1950.
198. **Mehlman, G. and Esteva, J. M.**, *Astrophys. J.*, 188, 191, 1974.
199. **Meinders, E.**, *Physica*, 84(C). 117, 1976.
200. Sansonetti, C. J., Dissertation, Purdue University, 1981.
201. Sansonetti, C. J., *Natl. Bur. Stand. (U.S.)*, unpublished.
202. **Millikan, R. A. and Bowen, I. S.** *Phys. Rev.*, 25, 600, 1925.
203. **Minnhagen, L.**, *J. Opt. Soc. Am.*, 61, 1257, 1925.
204. **Minnhagen, L.**, *J. Opt. Soc. Am.*, 63, 1185, 1973.
205. **Minnhagen, L.**, *Phys. Scr.*, 11, 38, 1975.
206. **Minnhagen, L.**, *J. Opt. Soc. Am.*, 66, 659, 1976.
207. **Minnhagen, L. and Nietsche, H.**, *Phys. Scr.*, 5, 237, 1972.
208. **Minnhagen, L., Strihed, H., and Petersson, B.**, *Ark. Fys.*, 39, 471, 1969.
209. **Moore, C. E.**, *Natl. Bur. Stand. (U.S.) Circ.*, 488, 1950.
210. **Moore, C. E.**, *Revised Multiplet Table*, Princeton University Observatory No. 20, 1945.

211. **Moore, C. E.**, National Standard Reference Data Series — National Bureau of Standards 3, Sect. 3, 1970.
212. **Moore, C. E.**, National Standard Reference Data Series — National Bureau of Standards 3, Sect. 4, 1971.
213. **Moore, C. E.**, National Standard Reference Data Series — National Bureau of Standards 3, Sect. 5, 1975.
214. **Moore, C. E.**, National Standard Reference Data Series — National Bureau of Standards 3, Sect. 6, 1972.
215. **Moore, C. E.**, *National Standard Reference Data Series — National Bureau of Standards 3, Sect. 7*, 1975.
216. **Morillon, C. and Verges, J.**, *Phys. Scr.*, 10, 227, 1974.
217. **Newsom, G. H.**, *Astrophys. J.*, 166, 243, 1971.
218. **Newsom, G. H., O'Connor, S., and Learner, R. C. M.**, *J. Phys. B*, 6, 2162, 1973.
219. **Norlen, G.**, *Phys. Scr.*, 8, 249, 1973.
220. **Odabasi, H.**, *J. Opt. Soc. Am.*, 57, 1459, 1967.
221. **Olme, A.**, *Ark. Fys.*, 40, 35, 1969.
222. **Olme, A.**, *Phys. Scr.*, 1, 256, 1970.
223. Johansson, S., and Litzen, U., *J. Opt. Soc. Am.*, 61, 1427, 1971.
224. **Palenius, H. P.**, *Ark. Fys.*, 39, 15, 1969.
225. **Palenius, H. P.**, *Phys. Scr.*, 1, 113, 1970.
226. **Palenius, H. P.**, *Univ. of Lund, Sweden*, unpublished.
227. **Paschen, F.**, *Ann. Phys.*, Series 5, 12, 509, 1932.
228. **Paschen, F. and Ritschl, R.**, *Ann. Phys.*, Series 5, 18, 867, 1933.
229. **Peck, E. R., Khanna, B. N., and Anderholm, N. C.** *J. Opt. Soc. Am.*, 52, 53, 1962.
230. **Persson, W.**, *Phys. Scr.*, 3, 133, 1971.
231. **Persson, W. and Valind S.**, *Phys. Scr.*, 5, 187, 1972.
232. **Petersson, B.**, *Ark. Fys.*, 27, 317, 1964.
233. **Phillips, L. W. and Parker, W. L.**, *Phys. Rev.*, 60, 301, 1941.
234. **Platt, J. R. and Sawyer, R. A.**, *Phys. Rev.*, 60, 866, 1941.
235. **Plyer, E. K., Blaine, L. R., and Tidwell, E.**, *J. Res. Natl. Bur. Stand.*, 55, 279, 1955.
236. **Poppe, R., van Kleef, T. A. M., and Raassen, A. J. J.**, *Physica*, 77, 165, 1974.
237. **Radziemski, L. J., Jr. and Andrew, K. L.**, *J. Opt. Soc. Am.*, 55, 474, 1965.
238. **Radziemski, L. J., Jr. and Kaufman, V.**, *J. Opt. Soc. Am.*, 59, 424, 1969.
239. **Radziemski, L. J., Jr. and Kaufman, V.**, *J. Opt. Soc. Am.*, 64, 366, 1974.
240. **Ramanadham, R. and Rao, K. R.**, *Indian J. Phys.*, 18, 317, 1944.
241. **Ramb, R.**, *Ann. Phys.*, 10, 311, 1931.
242. Rank, D. H., Bennett, J. M., and Bennett, H. E., *J. Opt. Soc. Am.*, 40, 477, 1950.
243. **Rao, A. S. and Krishnamurty, S. G.**, *Proc. Phys. Soc. London*, 46, 531, 1943.
244. **Rao, K. R.**, *Proc. R. Soc. London, Ser. A*, 134, 604, 1932.
245. **Rao, K. R. and Badami, J. S.**, *Proc. R. Soc. London Ser. A*, 131, 154, 1931.
246. **Rao, K. R. and Krishnamurty, S. G.**, *Proc. R. Soc. London Ser. A*, 161, 38, 1937.
247. **Rao, K. R. and Murti, S. G. K.**, *Proc. R. Soc. London Ser. A*, 145, 681, 1934.
248. **Rao, Y. B.**, *Indian J. Phys.*, 32, 497, 1958.
249. **Rao, Y. B.**, *Indian J. Phys.*, 33, 546, 1959.
250. **Rao, Y. B.**, *Indian J. Phys.*, 35, 386, 1961.
251. **Rasmussen, E.**, *Z. Phys.*, 80, 726, 1933.
252. **Rasmussen, E.**, *Z. Phys.*, 83, 404, 1933.
253. **Rasmussen, E.**, *Z. Phys.*, 86, 24, 1934.
254. **Rasmussen, E.**, *Z. Phys.*, 87, 607, 1934.
255. **Rasmussen, E.**, *Phys. Rev.*, 57, 840, 1940.
256. **Rau, A. S. and Narayan, A. L.**, *Z. Phys.*, 59, 687, 1930.
257. **Reader, J.**, *J. Opt. Soc. Am.*, 65, 286, 1975.
258. **Reader, J.**, *J. Opt. Soc. Am.*, 65, 988, 1975.
259. Reader, J., *J. Opt Soc. Am.*, 73, 349, 1983.
260. **Reader, J. and Davis, S.**, *J. Res. Natl. Bur. Stand. Sect. A*, 71, 587, 1967, and unpublished.
261. **Reader, J. and Ekberg, J. O.**, *J. Opt. Soc. Am.*, 62, 464, 1972.
262. **Reader, J. and Epstein, G. L.**, *J. Opt. Soc. Am.*, 62, 1467, 1972.
263. **Reader, J. and Epstein, G. L.**, *J. Opt. Soc. Am.*, 65, 638, 1975.
264. **Reader, J. and Epstein, G. L.**, *Natl. Bur. Stand.*, unpublished.
265. **Reader, J., Epstein, G. L., and Ekberg, J. O.**, *J. Opt. Soc. Am.*, 62, 273, 1972.
266. Kaufman, V., *Phys. Scr.*, 26, 439, 1982.
267. **Ricard, R., Givord, M., and George, F.**, *C. R. Acad. Sci. Paris*, 205, 1229, 1937.
268. **Risberg, P.**, *Ark. Fys.*, 10, 583, 1956.
269. **Risberg, G.**, *Ark. Fys.*, 28, 381, 1965.
270. **Risberg, G.**, *Ark. Fys.* , 37, 231, 1968.
271. **Robinson, H. A.**, *Phys. Rev.*, 49, 297, 1936.
272. **Robinson, H. A.**, *Phys. Rev.*, 50, 99, 1936.
273. **Ross, C. B., Jr.**, Doctoral dissertation, Purdue University, 1969.
274. **Ross, C. B., Wood, D. R., and Scholl, P. S.**, *J. Opt. Soc. Am.*, 66, 36, 1976.
275. **Ruedy, J. E. and Gibbs, R. C.**, *Phys. Rev.*, 46, 880, 1934.
276. **Russell, H. N., King, R. B., and Moore, C. E.**, *Phys. Rev.*, 58, 407, 1940.
277. **Russell, H. N. and Moore, C. E.**, *J. Res. Natl. Bur. Stand.*, 55, 299, 1955.
278. **Russell, H. N., Moore, C. E., and Weeks, D. W.**, *Trans. Am. Philos. Soc.*, 34(2), 111, 1944.
279. **Saunders, F., Schneider, E., and Buckingham, E.**, *Proc. Natl. Acad. Sci.*, 20, 291, 1934.
280. **Sawyer, R. A. and Humphreys, C. J.**, *Phys. Rev.*, 32, 583, 1928.
281. **Sawyer,R. A. and Lang, R. J.**, *Phys. Rev.*, 34, 712, 1929.
282. **Sawyer, R. A. and Paschen, F.** *Ann. Phys.*, 84(4), 1, 1927.
283. **Scholl, P. S.**, M.S. thesis, Wright State Univ., 1975.
284. **Schurmann, D.**, *Z. Phys.*, 17, 4, 1975.
285. **Seguier, J.**, *C. R. Acad. Sci. Paris*, 256, 1703, 1963.
286. **Shenstone, A. G.**, *Phys. Rev.*, 31, 317, 1928.
287. **Shenstone, A. G.**, *Phys. Rev.*, 32, 30, 1928.
288. **Shenstone, A. G.**, *Trans. R. Soc. London*, 237(A), 57, 1938.
289. **Shenstone, A. G.**, *Phys. Rev.*, 57, 894, 1940.
290. **Shenstone, A. G.** *Philos. Trans. R. Soc. London Ser. A*, 241, 297, 1948.
291. **Shenstone, A. G.**, *Can. J. Phys.*, 38, 677, 1960.
292. **Shenstone, A. G.**, *Proc. R. Soc. London*, 261(A), 153, 1961.

293. **Shenstone, A. G.,** *Proc. R. Soc. London,* 276(A), 293, 1963.
294. **Shenstone, A. G.,** *J. Res. Natl. Bur. Stand. Sect. A,* 74, 801, 1970.
295. **Shenstone, A. G.,** *J. Res. Natl. Bur. Stand. Sect. A,* 79, 497, 1975.
296. **Shenstone, A. G. and Pittenger, J. T.,** *J. Opt. Soc. Am.,* 39, 219, 1949.
297. **Smith, S.,** *Phys. Rev.,* 36, 1, 1930.
298. **Smitt, R.,** *Phys. Scr.,* 8, 292, 1973.
299. **Soderqvist, J.,** *Ark. Mat. Astronom. Fys.,* 32(A), 1, 1946.
300. **Sommer, L. A.,** *Ann. Phys.,* 75, 163, 1924.
301. **Spector, N.,** *J. Opt. Soc. Am.,* 63, 358, 1973.
302. **Spector, N. and Sugar, J.,** *J. Opt. Soc. Am.,* 66, 436, 1976.
303. **Steinhaus, D. W., Radziemski, L. J., Jr., and Blaise, J.,** *Los Alamos Sci. Lab.,* unpublished, 1975.
304. **Subbaraya, T. S.,** *Z. Phys.,* 78, 541, 1932.
305. **Sugar, J.,** *J. Opt. Soc. Am.,* 55, 33, 1965.
306. **Sugar, J.,** *J. Res. Natl. Bur. Stand. Sect. A,* 73, 333, 1969.
307. **Sugar, J.,** *J. Opt. Soc. Am.,* 60, 454, 1970.
308. **Sugar, J.,** *J. Res. Natl. Bur. Stand. Sect. A,* 78, 555, 1974.
309. **Sugar, J. and Kaufman, V.,** *J. Opt. Soc. Am.,* 55, 1283, 1965.
310. **Sugar, J. and Kaufman, V.,** *J. Opt. Soc. Am.,* 62, 562, 1972.
311. **Sugar, J., Kaufman, V., and Spector, N.,** *J. Res. Natl. Bur. Stand.,* Sect. A, 83, 233, 1978.
312. **Sugar, J. and Spector, N.,** *J. Opt. Soc. Am.,* 64, 1484, 1974.
313. **Sullivan, F. J.** *Univ. Pittsburgh Bull.,* 35, 1, 1938.
314. **Svensson, L. A. and Ekberg, J. O.,** *Ark. Fys.,* 37, 65, 1968.
315. **Swensson, J. W. and Risberg, G.,** *Ark. Fys.,* 31, 237, 1966.
316. **Tech. J. L.,** *J. Res. Natl. Bur. Stand. Sect. A,* 67, 505, 1963.
317. **Tech, J. L. and Ward, J. F.,** *Phys. Rev. Lett.,* 27, 367, 1971.
318. **Tilford, S. G.,** *J. Opt. Soc. Am.,* 53, 1051, 1963.
319. **Toresson, Y. G.,** *Ark. Fys.,* 17, 179, 1960.
320. **Toresson, Y. G.,** *Ark. Fys.,* 18, 389, 1960.
321. **Toresson, Y. G. and Edlen, B.,** *Ark. Fys.,* 23, 117, 1963.
322. **Tsien, W. Z.,** *Chin. J. Phys.,* Peiping, 3, 117, 1939.
323. **van Deurzen, C. H. H., Conway, J., and Davis, S. P.,** *J. Opt. Soc. Am.,* 63, 158, 1973.
324. **van Kleef, T. A. M., Raassen, A. J. J., and Joshi, Y. N.,** *Physica,* 84(C), 401, 1976.
325. Sansonetti, C. J., Andrew, K. L., and Verges, J., *J. Opt. Soc. Am.,* 71, 423, 1981.
326. **Wheatley, M. A. and Sawyer, R. A.,** *Phys. Rev.,* 61, 591, 1942.
327. **Wilkinson, P. G.,** *J. Opt. Soc. Am.,* 45, 862, 1955.
328. **Wilkinson, P. G. and Andrew, K. L.,** *J. Opt. Soc. Am.,* 53, 710, 1963.
329. **Wood, D. and Andrew, K. L.,** *J. Opt. Soc. Am.,* 58, 818, 1968.
330. **Wood, D. R., Ross, C. B., Scholl, P. S., and Hoke, M.,** *J. Opt. Soc. Am.,* 64, 1159, 1974.
331. **Worden, E. F. and Conway, J. G.,** *Lawrence Livermore Lab.,* unpublished, 1977.
332. **Worden, E. F., Hulet, E. K., Gutmacher, R. G., Conway, J. G.,** *At. Data Nucl. Data Tables,* 18, 459, 1976.
333. **Worden, E. F., Lougheed, R. W., Gutmacher, R. G., and Conway, J. G.,** *J. Opt. Soc. Am.,* 64, 77, 1974.
334. **Wu, C. M.,** Ph.D. thesis, University of British Columbia, 1971.
335. **Zaidel, A. N., Prokofev, V. K., Raiskii, S. M., Slavnyi, V. A., and Schreider, E. Y.,** *Tables of Spectral Lines,* 3rd ed., Plenum, New York, 1970.
336. **Zetterberg, P. O. and Magnusson, C. E.,** *Phys. Scr.,* 15, 189, 1977.
337. **Sugar, J.,** *J. Opt. Soc. Am.,* 55, 1058, 1965.
338. **Sugar, J.,** *J. Opt. Soc. Am.,* 61, 727, 1971.
339. Worden, E. F., and Conway, J. G., *At. Data Nucl. Data Tables,* 22, 329, 1978.
340. **Kaufman, V. and Edlen, B.,** *J. Phys. Chem. Ref. Data,* 3, 825, 1974.
341. **Lang, R. J.,** *Phys. Rev.,* 34, 697, 1929.
342. **Ryabtsev, A. N.,** *Opt. Spectros.,* 39, 455, 1975.
343. **Foster, E. W.,** *Proc. R. Soc. London,* 200(A), 429, 1950.
344. **Morillon, C. and Verges, J.,** *Phys. Scr.,* 12, 129, 1975.
345. **Ruedy, J. E.,** *Phys. Rev.,* 41, 588, 1932.
346. **McLennan, J. C., McLay, A. B., and McLeod, J. H.,** *Philos. Mag.,* 4, 486, 1927.
347. **Handrup, M. B. and Mack, J. E.,** *Physica,* 30, 1245, 1964.
348. **Clearman, H. E.,** *J. Opt. Soc. Am.,* 42, 373, 1952.
349. **Paschen, F.,** *Ann. Physik,* 424, 148, 1938.
350. **Paschen, F. and Campbell, J. S.,** *Ann. Phys.,* 31(5), 29, 1938.
351. **Nodwell, R.,** *Univ. of British Columbia, Vancouver,* unpublished, 1955.
352. **Gibbs, R. C. and White, H. E.,** *Phys. Rev.,* 31, 776, 1928.
353. **Green, M.,** *Phys. Rev.,* 60, 117, 1941.
354. **Ellis, C. B. and Sawyer, R. A.,** *Phys. Rev.,* 49, 145, 1936.
355. **McLennan, J. C., McLay, A. B., and Crawford, M. F.,** *Proc. R. Soc. London Ser. A,* 125, 50, 1929.
356. **Mack, J. E. and Fromer, M.,** *Phys. Rev.,* 48, 346, 1935.
357. **Humphreys, C. J. and Paul, E.,** U.S. Nav. Ord. Lab., Navord Rep. 4589, 25, 1956.
358. **Walters, F. M.,** *Sci. Pap. Bur. Stand.,* 17, 161, 1921.
359. **Crawford, M. F. and McLay, A. B.,** *Proc. R. Soc. London Ser. A,* 143, 540, 1934.
360. **McLay, A. B. and Crawford, M. F.,** *Phys. Rev.,* 44, 986, 1933.
361. **Schoepfle, G. K.,** *Phys. Rev.,* 47, 232, 1935.
362. Acquista, N., and Reader, J., *J. Opt. Soc. Am.,* 70, 789, 1980.
363. Benschop, H., Joshi, Y. N., and Van Kleef, T. A. M., *Can. J. Phys.,* 53, 498, 1975.
364. Bockasten, K., Hallin, R., and Hughes, T. P., *Proc. Phys. Soc.,* 81, 522, 1963.
365. Boyce, J. C., *Phys. Rev.,* 46, 378, 1934.
366. Boyce, J. C., *Phys. Rev.,* 47, 718, 1935.
367. Boyce, J. C., *Phys. Rev.,* 48, 396, 1935.
368. Boyce, J. C., *Phys. Rev.,* 49, 351, 1936.
369. Corliss, C. H. and Meggers, W. F., *J. Res. Natl. Bur. Stand.,* 61, 269, 1958.
370. Crooker, A. M. and Dick, K. A., *Can. J. Phys.,* 42, 766, 1964.
371. De Bruin, T. L., *Z. Physik,* 77, 505, 1932.
372. De Bruin, T. L., *Proc. Roy. Acad. Amsterdam,* 36, 727, 1933.
373. De Bruin, T. L., *Zeeman Verhandelingen,* (The Hague), 1935, p. 415.
374. De Bruin, T. L., *Physica,* 3, 809, 1936.
375. De Bruin, T. L., *Proc. Roy. Acad. Amsterdam,* 40, 339, 1937.

376. **Dick, K. A.**, *Can. J. Phys.*, 46, 1291, 1968.
377. **Dick, K. A.**, unpublished, 1978.
378. **Edlen, B. and Swensson, J. W.**, *Phys. Scr.*, 12, 21, 1975.
379. **Ekberg, J. O.**, *Phys. Scr.*, 7, 55, 1973.
380. **Ekberg, J. O.**, *Phys. Scr.*, 7, 59, 1973.
381. **Ekberg, J. O.**, *Phys. Scr.*, 12, 42, 1975.
382. **Ekberg, J. O. and Edlen, B.**, *Phys. Scr.*, 18, 107, 1978.
383. **Eliason, A. Y.**, *Phys. Rev.*, 43, 745, 1933.
384. **Gallardo, M., Massone, C. A., Tagliaferri, A. A., Garavaglia, M., and Persson, W.**, *Phys. Scr.*, to be published, 1979.
385. **Garcia-Riquelme, O.**, *Optica Pura Y Aplicada*, 1, 53, 1968.
386. **Gibbs, R. C., Vieweg, A. M., and Gartlein, C. W.**, *Phys. Rev.*, 34, 406, 1929.
387. **Gilbert, W. P.**, *Phys. Rev.*, 48, 338, 1935.
388. **Goldsmith, S. and Kaufman, A. S.**, *Proc. Phys. Soc.*, 81, 544, 1963.
389. **Hermansdorfer, H.**, *J. Opt. Soc. Am.*, 62, 1149, 1972.
390. **Humphreys, C. J.**, *Phys. Rev.*, 47, 712, 1935.
391. **Humphreys, C. J.**, *J. Res. Natl. Bur. Stand.*, 16, 639, 1936.
392. **Iglesias, L.**, *J. Opt. Soc. Am.*, 45, 856, 1955.
393. **Iglesias, L.**, *J. Res. Natl. Bur. Stand.*, 64A, 481, 1960.
394. **Iglesias, L.**, *Anales Fisica Y Quimica*, 58A, 191, 1962.
395. **Iglesias, L.**, *J. Res. Natl. Bur. Stand.*, 70A, 465, 1966.
396. **Iglesias, L.**, *Can. J. Phys.*, 44, 895, 1966.
397. **Iglesias, L.**, *J. Res. Natl. Bur. Stand.*, 72A, 295, 1968.
398. **Joshi, Y. N.**, *Can. Spectrosc.*, 15, 96, 1970.
399. **Joshi, Y. N. and Van Kleef, T. A. M.**, *Can. J. Phys.*, 55, 714, 1977.
400. **Kaufman, A. S., Hughes, T. P., and Williams, R. V.**, *Proc. Phys. Soc.*, 76, 17, 1960.
401. **Kaufman, V. and Sugar, J.**, *J. Opt. Soc. Am.*, 68, 1529, 1978.
402. **Keussler, V.**, *Z. Physik*, 85, 1, 1933.
403. **Kiess, C. C.**, *J. Res. Natl. Bur. Stand.*, 56, 167, 1956.
404. **Klinkenberg, P. F. A., Van Kleef, T. A. M., and Noorman, P. E.**, *Physica*, 27, 1177, 1961.
405. **Kovalev, V. I., Romanos, A. A., and Ryabtsev, A. N.**, *Opt. Spectrosc.*, 43, 10, 1977.
406. **Lang, R. J.**, *Proc. Natl. Acad. Sci.*, 13, 341, 1927.
407. **Lang, R. J.**, *Zeeman Verhandelingen*, (The Hague), 44, 1935.
408. **Liberman, S., et al.**, *C. R. Acad. Sci. (Paris)*, 286, 253, 1978.
409. **Livingston, A. E.**, *J. Phys.*, B9, L215, 1976.
410. **Meijer, F. G.**, *Physica*, 72, 431, 1974.
411. **Meijer, F. G. and Metsch, B. C.**, *Physica*, 94C, 259, 1978.
412. **Moore, F. L.**, thesis, Princeton, 1949.
413. **Paul, F. W. and Polster, H. D.**, *Phys. Rev.*, 59, 424, 1941.
414. **Phillips, L. W. and Parker, W. L.**, *Phys. Rev.*, 60, 301, 1941.
415. **Poppe, R.**, *Physica*, 81C, 351, 1976.
416. **Raassen, A. J. J., Van Kleef, T. A. M., and Metsch, B. C.**, *Physica*, 84C, 133, 1976.
417. **Rao, A. B. and Krishnamurty, S. G.**, *Proc. Phys. Soc. (London)*, 51, 772, 1939.
418. **Reader, J. and Acquista, N.**, *J. Opt. Soc. Am.*, 69, 239, 1979.
419. **Reader, J. and Epstein, G. L.**, *J. Opt. Soc. Am.*, 62, 619, 1972.
420. **Rico, F. R.**, *Anales, Real Soc. Esp. Fis. Quim.*, 61, 103, 1965.
421. **Schonheit, E.**, *Optik*, 23, 409, 1966.
422. **Shenstone, A. G.**, *J. Opt. Soc. Am.*, 44, 749, 1954.
423. **Shenstone, A. G.**, unpublished, 1958.
424. **Shenstone, A. G.**, *J. Res. Natl. Bur. Stand.*, 67A, 87, 1963.
425. **Sugar, J. and Kaufman, V.**, *J. Opt. Soc. Am.*, 64, 1656, 1974.
426. **Sugar, J. and Kaufman, V.**, *Phys. Rev.*, C12, 1336, 1975.
427. **Svensson, L. A.**, *Phys. Scr.*, 13, 235, 1976.
428. **Swensson, J. W. and Edlen, B.**, *Phys. Scr.*, 9, 335, 1974.
429. **Tagliaferri, A. A., Gallego Lluesma, E., Garavaglia, M., Gallardo, M., and Massone, C. A.**, *Optica Pura Y Aplica*, 7, 89.
430. **Tilford, S. G. and Giddings, L. E.**, *Astrophys. J.*, 141, 1222, 1965.
431. **Trawick, M. W.**, *Phys. Rev.*, 46, 63, 1934.
432. **Van Deurzen, C. H. H.**, *J. Opt. Soc. Am.*, 67, 476, 1977.
433. **Yarosewick, S. L. and Moore, F. L.**, *J. Opt. Soc. Am.*, 57, 1381, 1967.
434. **Zalubas, R.**, unpublished, 1979.
435. **Bhatia, K. S., Jones, W. E., and Crooker, A. M.**, *Can. J. Phys.*, 50, 2421, 1972.
436. **van Kleef, A. M. and Joshi, Y. N.**, *Phys. Scr.*, 24, 557, 1981.

ATOMIC TRANSITION PROBABILITIES

Compiled by W. L. Wiese and G. A. Martin

These tables were prepared under the auspices of the Committee on Line Spectra of the Elements of the National Academy of Sciences — National Research Council. They contain critically evaluated atomic transition probabilities for about 5000 selected lines of all elements for which reliable data are available on an absolute scale. The material is largely for neutral and singly ionized spectra, but includes a number of prominent lines of more highly charged ions.

Many of the data are obtained from comprehensive compilations of the National Bureau of Standards Data Center on Atomic Transition Probabilities. Specifically, data have been taken from critical compilations on V,[1] Cr,[1] Mn,[1] Fe,[2] Co,[2] and Ni[2] without changes. Material from earlier compilations for the elements H through Ne,[3] Na through Ca,[4] and Sc and Ti[5] was supplemented by more recent material taken directly from the original literature. For the higher ions, many data were derived from studies of the systematic behavior of transition probabilities.[6-8] The original literature is cited in a recent bibliography;[9] for lack of space, individual literature references are not cited here.

The wavelength range for the neutral species has normally been restricted to the visible or shorter wavelengths; only the very prominent near infrared lines are included. For the higher ions, most of the strong lines are located in the far UV. The tabulation is limited to electric dipole — including intercombination — lines and comprises essentially the fairly strong transitions with estimated uncertainties of 50% or less. With the exception of hydrogen, helium, and the alkalis, most transitions are between states with low principal quantum numbers.

The transition probability, A, is given in units of 10^8 s^{-1} and is listed to as many digits as is consistent with the indicated accuracy. A number in parentheses following the tabulated value of the transition probability indicates the power of 10 by which this value has to be multiplied. The estimated uncertainties of the A-values are indicated by code letters as follows: AA — for uncertainties within 1%; A — within 3%; B — within 10%; C — within 25%; D — within 50%.

Each transition is identified by the wavelength, λ, in angstroms; the energy level of the upper atomic state, E_k, in cm^{-1}; and the statistical weights, g_i and g_k, of the lower (i) and upper (k) states [the product g_kA (or g_if) is needed in many applications]. Whenever the wavelengths of individual lines within a multiplet are extremely close, an average wavelength for the multiplet is given, and is indicated by an asterisk (*) to the right of the g_k value. This has also been done when the transition probability for an entire multiplet has been taken from the literature and values for individual lines cannot be determined because of insufficient knowledge of the coupling of electrons. Wavelength and energy level data have been taken either from recent compilations or from the original literature cited in bibliographies published by the National Bureau of Standards Atomic Energy Levels Data Center.[10,11] Wavelength values are consistent with those given in Section E of this Handbook.

In the table for hydrogen, the energy level and uncertainty columns have been eliminated since the transition probabilities and energy levels are known very precisely for this element. Because of the hydrogen degeneracy, a "transition" is actually the sum of all transitions between the principal quantum numbers listed in the transition column, and the tabulation represents the properly weighted A values.

In addition to the transition probability A, the atomic oscillator-strength f and the line-strength S are often used in the literature. The conversion factors between these quantities are:

$$g_i f = 1.499 \times 10^{-8} \lambda^2 g_k A = 303.8 \lambda^{-1} S$$

where λ is in angstroms, A is in 10^8 s^{-1}, and S is in atomic units, which are $a_o^2 e^2 = 7.188 \times 10^{-59}$ m^2C^2 for electric dipole transitions.

We acknowledge the valuable preparatory work by D. Trahan, W. Croom, and F. Farley in arranging and compiling the numerical data.

Table 1a
TRANSITION PROBABILITIES FOR ALLOWED LINES OF HYDROGEN

Wavelength λ [Å]	Transition	Statistical weights g_i	g_k	Average transition probability A [10^8 s^{-1}]
914.039	1—20	2	800	3.928(−5)
914.286	1—19	2	722	5.077(−5)
914.576	1—18	2	648	6.654(−5)
914.919	1—17	2	578	8.858(−5)
915.329	1—16	2	512	1.200(−4)
915.824	1—15	2	450	1.657(−4)
916.429	1—14	2	392	2.341(−4)
917.181	1—13	2	338	3.393(−4)
918.129	1—12	2	288	5.066(−4)
919.351	1—11	2	242	7.834(−4)
920.963	1—10	2	200	1.263(−3)
923.150	1— 9	2	162	2.143(−3)
926.226	1— 8	2	128	3.869(−3)
930.748	1— 7	2	98	7.568(−3)
937.803	1— 6(Lε)	2	72	1.644(−2)
949.743	1—5 (Lδ)	2	50	4.125(−2)
972.537	1— 4(Lγ)	2	32	1.278(−1)
1025.72	1— 3(Lβ)	2	18	5.575(−1)
1215.67	1— 2(Lα)	2	8	4.699
3682.81	2—20	8	800	2.172(−5)
3686.83	2—19	8	722	2.809(−5)
3691.55	2—18	8	648	3.685(−5)
3697.15	2—17	8	578	4.910(−5)
3703.85	2—16	8	512	6.658(−5)
3711.97	2—15	8	450	9.210(−5)
3721.94	2—14	8	392	1.303(−4)
3734.37	2—13	8	338	1.893(−4)
3750.15	2—12	8	288	2.834(−4)
3770.63	2—11	8	242	4.397(−4)
3797.90	2—10	8	200	7.122(−4)
3835.38	2— 9	8	162	1.216(−3)
3889.05	2— 8	8	128	2.215(−3)
3970.07	2— 7(Hε)	8	98	4.389(−3)
4101.73	2— 6(Hδ)	8	72	9.732(−3)
4340.46	2— 5(Hγ)	8	50	2.530(−2)
4861.32	2— 4(Hβ)	8	32	8.419(−2)
6562.80	2— 3(Hα)	8	18	4.410(−1)
8392.40	3—20	18	800	1.517(−5)
8413.32	3—19	18	722	1.964(−5)
8437.96	3—18	18	648	2.580(−5)
8467.26	3—17	18	578	3.444(−5)
8502.49	3—16	18	512	4.680(−5)
8545.39	3—15	18	450	6.490(−5)
8598.40	3—14	18	392	9.211(−5)
8665.02	3—13	18	338	1.343(−4)
8750.48	3—12	18	288	2.021(−4)
8862.79	3—11	18	242	3.156(−4)
9014.91	3—10	18	200	5.156(−4)
9229.02	3— 9	18	162	8.905(−4)
9545.97	3— 8(Pε)	18	128	1.651(−3)
10049.4	3— 7(Pδ)	18	98	3.358(−3)
10938.1	3— 6(Pγ)	18	72	7.783(−3)
12818.1	3— 5(Pβ)	18	50	2.201(−2)
16407.2	4—12	32	288	1.620(−4)
16806.5	4—11	32	242	2.556(−4)
17362.1	4—10	32	200	4.235(−4)
18174.1	4— 9	32	162	7.459(−4)
18751.0	3— 4(Pα)	18	32	8.986(−2)
19445.6	4— 8	32	128	1.424(−3)
21655.3	4— 7	32	98	3.041(−3)
26251.5	4— 6	32	72	7.711(−3)
27575	5—12	50	288	1.402(−4)
28722	5—11	50	242	2.246(−4)
30384	5—10	50	200	3.800(−4)
32961	5— 9	50	162	6.908(−4)
37395	5— 8	50	128	1.388(−3)
40511.5	4— 5	32	50	2.699(−2)
43753	6—12	72	288	1.288(−4)
46525	5— 7	50	98	3.253(−3)
46712	6—11	72	242	2.110(−4)
51273	6—10	72	200	3.688(−4)
59066	6— 9	72	162	7.065(−4)
74578	5— 6	50	72	1.025(−2)
75004	6— 8	72	128	1.561(−3)
123680	6— 7	72	98	4.561(−3)

For hydrogen-like ions of nuclear charge Z, the following scaling laws hold:

$$A_Z = Z^4 A_{Hydrogen};\ f_Z = f_H;\ S_Z = Z^{-2} S_H$$

(For wavelengths $\lambda_Z = Z^{-2} \lambda_H$).

For very highly charged ions, relativistic effects need to be taken into account.[12]

Table 1b
TRANSITION PROBABILITIES FOR SELECTED ATOMIC AND IONIC SPECIES

Wavelength (λ[Å])	Upper energy level (E_k[cm^{-1}])	Statistical weights g_i	g_k	Transition Probability (A[10^8 s^{-1}])	Uncertainty
Aluminum					
Al I					
2263.5	44166	2	4	.66	C
2269.1	44169	4	6	.79	C
2269.2	44166	4	4	.13	C
2367.1	42234	2	4	.72	C
2373.1	42238	4	6	.86	C
2373.4	42234	4	4	.14	C
2568.0	38929	2	4	.23	C
2575.1	38934	4	6	.28	C
2575.4	38929	4	4	.044	C
2652.5	37689	2	2	.133	C
2660.4	37689	4	2	.264	C
3082.2	32435	2	4	.63	C
3092.7	32437	4	6	.74	C
3092.8	32435	4	4	.12	C
3944.0	25348	2	2	.493	C
3961.5	25348	4	2	.98	C
6696.0	40278	2	4	.0169	C
6698.7	40272	2	2	.0169	C
7835.3	45195	4	6	.057	D
7836.1	45195	6	8	.062	D
Al II					
1047.9	132823	1	3	.36	D
1048.6	132823	3	5	.48	D
1539.8	124794	3	5	8.8	D
1670.8	59852	1	3	14.6	B
1719.4	95551	1	3	6.79	B
1764.0	94269	5	5	9.8	C
1772.8	150493	1	3	9.5	D
1777.0	150544	5	7	17.	D
1819.0	150525	15	15*	5.6	D
1855.9	91275	1	3	.832	B
1858.0	91275	3	3	2.48	B
1862.3	91275	5	3	4.12	B
1931.0	111637	3	1	10.8	C
1990.5	110090	3	5	14.7	C
2816.2	95351	3	1	3.83	C
4663.1	106921	5	3	.53	C
6226.2	121484	1	3	.62	D
6231.8	121484	3	5	.84	D
6243.4	121484	5	7	1.1	D
6335.7	125869	5	3	.14	D
6823.4	120093	3	3	.34	D
6837.1	120093	5	3	.57	D
6920.3	121367	3	1	.96	D
7042.1	105471	3	5	.59	C
7056.7	105442	3	3	.58	C
7471.4	123471	5	7	.94	D
Al III					
560.36	178458	2	6*	.40	D
695.83	143714	2	4	.74	C
696.22	143633	2	2	.72	C
1352.8	189876	10	14*	4.40	C
1379.7	126164	2	2	4.59	C
1384.1	126164	4	2	9.1	C
1605.8	115959	2	4	12.2	B
1611.8	115959	4	4	2.42	B
1611.9	115956	4	6	14.5	B
1854.7	53917	2	4	5.40	B
1862.8	53683	2	2	5.33	B
1935.9	167613	10	14*	12.2	C
3601.6	143714	6	4	1.34	C
3601.9	143714	4	4	.149	C
3612.4	143633	4	2	1.5	C
Al X					
39.925	2504700	1	3	2220.	C
51.979	1923850	1	3	4800.	C
55.227	1965860	1	3	5200.	D
55.272	1966030	3	5	7200.	D
55.376	1966270	5	7	9500.	D
59.107	1992340	3	5	4600.	C
332.78	300490	1	3	56.	C
394.83	553783	3	1	83.	C
395.36	409690	3	5	12.	C
397.76	406517	1	3	17.	C
400.43	406517	3	3	13.	C
401.12	409690	5	5	36.	C
403.55	404574	3	1	49.	C

Wavelength (λ[Å])	Upper energy level (E_k[cm^{-1}])	Statistical weights g_i	g_k	Transition Probability (A[10^8 s^{-1}])	Uncertainty
Al X					
406.31	406517	5	3	19.	C
670.06	449732	3	5	9.8	C
2535.	1923850	1	3	.38	D
Al XI					
36.675	2726700	2	6*	1500.	C
39.091	2734100	2	4	2600.	C
39.180	2734500	4	6	3100.	C
39.530	2705700	2	2	180.	D
39.623	2705700	4	2	370.	D
48.298	2070520	2	4	3090.	B
48.338	2068770	2	2	3080.	B
52.299	2088100	2	4	8100.	C
52.446	2088530	4	6	9600.	C
52.458	2088100	4	4	1600.	C
54.217	2020450	2	2	480.	C
54.388	2020450	4	2	960.	C
99.083	3029700	2	6*	220.	C
103.6	3033700	2	4	420.	C
103.8	3033700	4	6	500.	C
141.6	2726700	2	6*	407.	C
150.31	2734100	2	4	850.	C
150.61	2734500	4	6	990.	C
157.0	2705700	2	2	130.	C
157.4	2705700	4	2	260.	C
205.0	3193600	2	6*	63.	C
308.6	3029700	2	6*	99.	C
341.3	3019700	6	2*	130.	C
550.05	181808	2	4	8.55	B
568.12	176019	2	2	7.73	B
1997.	2070520	2	4	1.07	C
2069.	2068770	2	2	.97	C
4761.	2726700	2	6*	.255	C
5172.	2088100	2	4	.0395	C
5551.	2088530	4	6	.0385	C
5687.	2088100	4	4	.0060	D
Argon					
Ar I					
3554.3	121271	5	5	.0029	D
3567.7	121165	5	7	.0012	D
3606.5	121470	3	1	.0081	D
3649.8	122791	3	1	.0085	D
3834.7	121470	3	1	.0080	D
3949.0	118460	5	3	.00467	C
4044.4	118469	3	5	.00346	C
4158.6	117184	5	5	.0145	C
4164.2	117151	5	3	.00295	C
4181.9	118460	1	3	.0058	C
4190.7	116999	5	5	.00254	C
4191.0	118407	1	3	.0056	C
4198.3	117563	3	1	.0276	C
4200.7	116943	5	7	.0103	C
4259.4	118871	3	1	.0415	C
4266.3	117184	3	5	.00333	C
4272.2	117151	3	3	.0084	C
4300.1	116999	3	5	.00394	C
4333.6	118469	3	5	.0060	C
4335.3	118460	3	3	.00387	C
4345.2	118407	3	3	.00313	C
4510.7	117563	3	1	.0123	C
4768.7	125066	3	5	.0090	D
4836.7	124772	3	5	.00106	C
4876.3	124604	3	5	.0081	D
4887.9	124555	3	3	.014	D
4894.7	124527	3	1	.019	D
5048.8	123903	3	5	.0048	D
5054.2	123882	3	3	.0047	D
5056.5	123873	3	1	.0059	D
5118.2	125150	5	7	.0028	D
5151.4	123509	3	1	.0249	C
5152.3	123505	3	5	.0011	D
5162.3	123468	3	3	.0198	C
5177.5	124772	7	5	.0025	D
5187.7	123373	3	5	.0138	C
5194.1	125335	3	1	.0081	D
5210.5	124650	7	7	.0011	D
5214.8	124788	5	3	.0022	D
5216.3	124783	5	3	.0014	D
5221.3	124610	7	9	.0092	D
5241.1	124692	5	5	.0014	D

Table 1b (Continued)
TRANSITION PROBABILITIES FOR SELECTED ATOMIC AND IONIC SPECIES

Wavelength (λ[Å])	Upper energy level (E_k[cm^{-1}])	Statistical weights g_i	g_k	Transition Probability (A[10^8 s^{-1}])	Uncertainty
	Ar I				
5252.8	124650	5	7	.0056	D
5373.5	124692	3	5	.0028	D
5394.0	124772	5	5	.0010	D
5410.5	124715	5	7	.0021	D
5421.4	123903	7	5	.0062	D
5440.0	122479	3	3	.0020	D
5442.2	123832	7	7	9.7 (−4)	D
5451.7	122440	3	5	.0049	D
5457.4	123936	5	3	.0037	D
5467.2	123903	5	5	7.9 (−4)	D
5473.5	123882	5	3	.0021	D
5490.1	123827	5	5	8.9 (−4)	D
5495.9	123653	7	9	.0176	C
5506.1	123774	5	7	.0037	D
5525.0	123557	7	7	.0018	D
5534.5	125353	5	3	.0028	D
5558.7	122087	3	5	.0148	C
5559.7	125113	3	5	.0023	D
5572.5	123557	5	7	.0069	C
5588.7	123505	5	5	.0016	D
5606.7	121933	3	3	.0229	C
5618.0	123882	3	3	.0022	D
5620.9	123873	3	1	.0038	D
5623.8	125066	5	5	.0015	D
5635.6	123827	3	5	.0010	D
5639.1	124783	1	3	.0022	D
5641.4	123809	3	5	9.1 (−4)	D
5648.7	123936	5	3	.0013	D
5650.7	121794	3	1	.0333	C
5659.1	123903	5	5	.0027	D
5681.9	123832	5	7	.0021	D
5683.7	123827	5	5	.0021	D
5700.9	123774	5	7	.0061	D
5739.5	123505	3	5	.0091	D
5772.1	123557	5	7	.0021	D
5774.0	124604	5	5	.0011	D
5783.5	123373	3	5	8.4 (−4)	D
5802.1	123468	5	3	.0044	D
5834.3	123373	5	5	.0052	C
5860.3	121161	3	3	.00285	C
5882.6	121097	3	1	.0128	C
5888.6	122440	7	5	.0134	C
5912.1	121012	3	3	.0105	C
5928.8	122479	5	3	.011	D
5940.9	123882	1	3	.0012	D
5942.7	122440	5	5	.0019	D
5949.3	123936	3	3	.0016	D
5968.3	123882	3	3	.0019	D
5971.6	123873	3	1	.011	D
5987.3	122160	7	7	.0013	D
5988.1	123827	3	5	6.4 (−4)	D
5999.0	122282	5	5	.0015	D
6005.7	123936	5	3	.0015	D
6013.7	122087	7	5	.0015	D
6025.2	123882	5	3	.0094	D
6032.1	122036	7	9	.0246	C
6043.2	122160	5	7	.0153	C
6052.7	120619	3	5	.0020	D
6059.4	120601	3	5	.00423	C
6064.8	123774	5	7	6.0 (−4)	D
6090.8	123468	1	3	.0031	D
6098.8	122479	3	3	.0054	D
6101.2	123882	3	3	.0034	D
6104.6	123873	3	1	.0035	D
6105.6	123505	3	5	.0126	C
6127.4	121933	5	3	.0011	D
6128.7	123809	3	5	9.0 (−4)	D
6145.4	123557	5	7	.0079	C
6155.2	122479	5	3	.0053	D
6165.1	123505	5	5	.00103	C
6170.2	122440	5	5	.0052	C
6173.1	122282	3	5	.0070	C
6212.5	122330	5	7	.0041	D
6215.9	123373	5	5	.0059	D
6248.4	122087	3	5	7.1 (−4)	D
6296.9	123373	3	5	.0094	D
6307.7	122087	5	5	.0063	D
6364.9	121794	3	1	.0058	D
6369.6	121933	5	3	.0044	D
6384.7	119760	3	3	.00439	C
6416.3	119683	3	5	.0121	C
6466.6	122514	1	3	.0016	D
6481.1	122479	1	3	9.8 (−4)	D
6538.1	120753	7	7	.0011	D
6604.0	120601	7	5	.0029	D
6660.7	121097	3	1	.0081	D
6664.1	120619	5	5	.0016	D
6677.3	108723	3	1	.00241	C
6698.9	121161	5	3	.0017	D
6719.2	121933	1	3	.0025	D
6752.8	118907	3	5	.0201	C
6754.4	121933	3	3	.0022	D
6756.2	122087	5	5	.0038	D
6766.6	121012	5	3	.0042	D
6779.9	123468	1	3	.00126	C
6827.2	121933	5	3	.0025	D
6851.9	122087	3	5	7.0 (−4)	D
6871.3	118651	3	3	.0290	C
6879.6	120619	3	5	.0019	D
6887.1	120753	5	7	.0014	D
6888.2	120601	3	5	.0026	D
6925.0	121933	3	3	.0012	D
6937.7	118512	3	1	.0321	C
6951.5	120619	5	5	.0023	D
6960.3	120601	5	5	.0025	D
6965.4	107496	5	3	.067	C
6992.2	121794	3	1	.0078	D
7030.3	119683	7	5	.0278	C
7067.2	107290	5	5	.0395	C
7068.7	119760	5	3	.021	D
7086.7	121161	1	3	.0016	D
7107.5	119683	5	5	.0047	D
7125.8	121161	3	3	.0063	D
7147.0	107132	5	3	.0065	C
7158.8	121097	3	1	.022	D
7207.0	121161	5	3	.0258	C
7229.9	119445	5	5	6.9 (−4)	D
7265.2	119848	3	3	.0018	D
7270.7	119213	7	7	.0011	D
7272.9	107496	3	3	.0200	C
7285.4	121012	5	3	.0013	D
7311.7	119760	3	3	.018	D
7316.0	121161	3	3	.010	D
7350.8	121097	3	1	.012	D
7353.3	119213	5	7	.010	D
7372.1	119024	7	9	.020	D
7384.0	107290	3	5	.087	C
7393.0	119760	5	3	.0075	D
7412.3	120619	3	5	.0041	D
7422.3	120601	3	5	6.9 (−4)	D
7425.3	120753	5	7	.0032	D
7435.4	119683	5	5	.0094	D
7436.3	118907	7	5	.0028	D
7484.3	119445	3	5	.0035	D
7503.9	108723	3	1	.472	C
7510.4	120601	5	5	.0047	D
7514.7	107054	3	1	.430	C
7618.3	120619	3	5	.0030	D
7628.9	120601	3	5	.0030	D
7635.1	106238	5	5	.274	C
7670.0	118651	5	3	.0029	D
7704.8	119213	5	7	6.6 (−4)	D
7723.8	106087	5	3	.057	C
7724.2	107496	1	3	.127	C
7798.6	118907	3	5	9.1 (−4)	D
7868.2	119760	1	3	.00365	C
7891.1	118907	5	5	.0099	C
7916.4	119760	3	3	.0013	D
7948.2	107132	1	3	.196	C
	Ar II				
3000.4	192712	4	4	1.5	D
3028.9	192712	2	4	2.3	D
3093.4	192557	4	6	4.4	D
3139.0	186891	6	6	1.0	D
3161.4	192712	2	4	1.8	D
3169.7	186891	4	6	.82	D
3181.0	186470	6	4	.63	D
3236.8	190592	2	4	.52	D
3243.7	186171	4	2	2.0	D
3249.8	186470	2	4	1.0	D
3293.6	190592	4	4	1.7	D

Table 1b (Continued)
TRANSITION PROBABILITIES FOR SELECTED ATOMIC AND IONIC SPECIES

Wavelength (λ[Å])	Upper energy level (E_k[cm^{-1}])	Statistical weights g_i	g_k	Transition Probability (A[10^8 s^{-1}])	Uncertainty
	Ar II				
3307.2	189935	2	2	3.4	D
3350.9	200235	6	6	1.5	D
3376.4	200139	8	8	1.5	D
3388.5	190592	2	4	1.9	D
3454.1	183986	6	4	.45	D
3464.1	187589	6	6	.37	D
3476.7	183797	6	6	1.34	C
3491.2	183986	4	4	2.2	D
3509.8	184192	2	2	2.5	D
3514.4	183797	4	6	1.23	C
3520.0	186074	6	6	.80	D
3521.3	185625	8	8	.23	D
3535.3	183986	2	4	.82	D
3545.6	187589	4	6	3.4	D
3545.8	198595	6	8	3.9	D
3548.5	186341	4	4	1.1	D
3559.5	186816	6	8	3.9	D
3565.0	186470	2	4	1.1	D
3576.6	185625	6	8	2.77	C
3581.6	186341	2	4	1.8	D
3582.4	186074	4	6	3.72	C
3588.4	185093	8	10	3.39	C
3600.2	199525	4	4	2.2	D
3622.1	182951	4	2	.64	D
3639.8	199680	4	6	1.4	D
3655.3	187589	4	6	.23	D
3671.0	199447	4	2	.71	D
3680.1	199982	2	4	1.2	D
3718.2	200235	4	6	2.0	D
3724.5	200235	6	6	.34	D
3729.3	161049	6	4	.60	D
3737.9	200139	6	8	2.3	D
3765.3	181594	6	6	.98	D
3770.5	182222	2	4	.41	D
3780.8	183676	8	8	.94	D
3796.6	199680	4	6	.25	D
3803.2	199680	6	6	1.5	D
3809.5	181594	4	6	.44	D
3825.7	199525	6	4	.76	D
3850.6	161049	4	4	.47	D
3868.5	186891	4	6	1.9	D
3925.7	195867	6	4	1.4	D
3928.6	161049	2	4	.30	D
3932.5	186470	4	4	1.1	D
3946.1	195865	8	6	1.4	D
3952.7	186341	4	4	.35	D
3979.4	186171	4	2	1.3	D
4033.8	182951	4	2	.98	D
4042.9	173348	4	4	1.4	D
4072.0	173393	6	6	.57	C
4076.6	182951	2	2	.80	D
4076.9	183915	4	2	.99	D
4079.6	173348	6	4	.26	D
4131.7	172816	4	2	1.4	D
4156.1	182222	4	4	.39	D
4179.3	181594	6	6	.13	D
4218.7	183091	4	4	.36	D
4222.6	183915	4	2	.69	D
4227.0	195865	4	6	.41	D
4266.5	157673	6	6	.156	C
4275.2	183091	2	4	.26	D
4277.5	172214	6	4	1.0	D
4331.2	158168	4	4	.56	C
4337.1	195867	2	4	.34	D
4348.1	157234	6	8	1.24	C
4370.8	173348	4	4	.65	C
4371.3	155351	6	4	.233	C
4379.7	158428	2	2	1.04	C
4401.0	155043	8	6	.322	C
4426.0	157673	4	6	.83	C
4430.2	158168	2	4	.53	C
4448.9	195865	6	6	.65	D
4481.8	173393	6	6	.494	C
4545.1	160239	4	4	.413	B
4547.8	182222	4	4	.077	D
4589.9	170401	4	6	.82	C
4609.6	170530	6	8	.91	C
4637.2	170401	6	6	.090	D
4657.9	159707	4	2	.81	C
4726.9	159393	4	4	.50	C
4735.9	155351	6	4	.58	C
4764.9	160239	2	4	.575	B
4806.0	155043	6	6	.79	C
4847.8	155708	4	2	.85	C
4865.9	181594	4	6	.15	D
4879.9	158730	4	6	.78	C
4965.1	159393	2	4	.347	C
5009.3	155043	4	6	.147	C
5017.2	170401	4	6	.231	C
5062.0	155351	2	4	.221	C
5141.8	170530	6	8	.095	C
6638.2	158168	6	4	.129	C
6643.7	157234	10	8	.167	C
6684.3	157673	8	6	.113	C
	Ar III				
769.15	144023	5	3	6.0	D
871.10	114798	5	3	1.59	C
875.53	115328	3	1	3.74	C
878.73	113801	5	5	2.79	C
879.62	114798	3	3	.92	C
883.18	114798	1	3	1.22	C
887.40	113801	3	5	.90	C
3024.1	240292	5	7	2.6	D
3027.2	240258	5	5	.64	D
3054.8	240258	3	5	1.9	D
3064.8	240151	3	3	1.0	D
3078.2	240151	1	3	1.4	D
3285.9	204797	5	7	2.0	D
3301.9	204649	5	5	2.0	D
3311.3	204564	5	3	2.0	D
3336.1	226646	7	9	2.0	D
3344.7	226503	5	7	1.8	D
3352.1	226503	7	7	.22	D
3358.5	226356	3	5	1.6	D
3361.3	226356	5	5	.30	D
3472.6	225403	5	7	.20	D
3480.6	225403	7	7	1.6	D
3499.7	225155	3	3	1.3	D
3500.6	225148	3	5	.26	D
3502.7	225155	5	3	.43	D
3503.6	225148	5	5	1.2	D
3511.7	225148	7	5	.26	D
	Ar IV				
840.03	119044	4	2	2.73	C
843.77	118515	4	4	2.70	C
850.60	117564	4	6	2.63	C
	Ar VI				
292.15	342286	2	2	69.	C
294.05	342286	4	2	136.	C
	Ar VII				
250.41	514083	9	3*	278.	C
477.54	324151	9	15*	99.2	B
585.75	170720	1	3	78.3	B
637.30	271657	9	9*	67.	C
	Ar VIII				
158.92	629237	2	4	110.	C
159.18	628240	2	2	111.	C
229.44	575910	2	2	112.	C
230.88	575910	4	2	221.	C
337.09	629237	4	4	12.	D
337.26	629237	6	4	100.	C
338.22	628240	4	2	110.	C
519.43	332576	2	4	63.	C
526.46	332727	4	6	72.	C
526.87	332576	4	4	12.	D
700.24	142776	2	4	25.5	C
713.81	140058	2	2	24.	C
	Ar IX				
48.739	2051750	1	3	1690.	B
	Ar XIII				
162.96	698650	5	3	340.	C
163.08	628610	9	3*	530.	C
184.90	625840	5	5	166.	C
186.38	698650	1	3	88.	C
207.89	496450	9	9*	95.	C
245.10	423420	9	15*	37.	C
	Ar XIV				
180.29	554660	2	4	45.	C

Table 1b (Continued)
TRANSITION PROBABILITIES FOR SELECTED ATOMIC AND IONIC SPECIES

Wavelength (λ[Å])	Upper energy level (E_k[cm^{-1}])	Statistical weights g_i	g_k	Transition Probability (A[10^8 s^{-1}])	Uncertainty
	Ar XIV				
183.41	545230	2	2	169.	C
187.95	554660	4	4	197.	C
191.35	545230	4	2	75.	C
194.39	514430	2	2	46.	C
203.35	514430	4	2	78.	C
	Ar XV				
25.05	3992000	1	3	1.7 (+4)	B
221.10	452280	1	3	95.5	B
265.3	621500	9	9*	81.	C
	Ar XVI				
23.52	4251000	2	6*	1.43(+4)	B
24.96	4281000	6	10*	4.4 (+4)	B
353.88	282580	2	4	15.	B
389.11	257000	2	2	11.	B
1268.	4254000	2	4	1.9	B
1401.	4246000	2	2	1.4	B
2975.	4280000	2	4	.090	C
3514.	4281000	4	6	.065	C
	Arsenic				
	As I				
1890.4	52898	4	6	2.0	D
1937.6	51610	4	4	2.0	D
1972.6	50694	4	2	2.0	D
2288.1	54605	6	4	2.8	D
2344.0	60835	2	4	.35	D
2349.8	53136	4	2	3.1	D
2369.7	60835	4	4	.60	D
2370.8	60815	4	6	.42	D
2456.5	51610	6	4	.072	D
2492.9	50694	4	2	.12	D
2745.0	54605	2	4	.26	D
2780.2	54605	4	4	.78	D
2860.4	53136	2	2	.55	D
2898.7	53136	4	2	.099	D
	Barium				
	Ba I				
2409.2	41494	1	3	8.6 (–4)	C
2414.1	41411	1	3	.0015	C
2420.1	41308	1	3	.0023	C
2427.4	41184	1	3	.0056	C
2432.5	41097	1	3	.0072	C
2438.8	40991	1	3	.0014	C
2444.6	40893	1	3	.0045	C
2452.4	40764	1	3	8.1 (–4)	C
2473.2	40421	1	3	.0046	C
2500.2	39985	1	3	.015	D
2543.2	39309	1	3	.041	D
2596.6	38500	1	3	.12	D
2646.5	37775	1	3	.011	D
2702.6	36990	1	3	.025	D
2739.2	36496	1	3	.0091	D
2785.3	35893	1	3	.028	D
3071.6	32547	1	3	.41	C
3501.1	28554	1	3	.19	D
3889.3	25704	1	3	.0088	D
3909.9	34603	3	5	.49	D
3935.7	34617	5	7	.47	D
3937.9	34603	5	5	.11	D
3993.4	34631	7	9	.55	D
3995.7	34617	7	7	.088	D
4132.4	24192	1	3	.0071	D
4239.6	37095	5	3	.24	D
4242.6	36200	3	5	.056	D
4264.4	35709	1	3	.15	D
4283.1	34736	5	7	.64	D
4323.0	35762	3	5	.15	D
4325.2	36629	5	7	.071	D
4332.9	35709	3	3	.15	D
4350.3	35617	3	5	.60	D
4402.5	35344	3	5	.70	D
4406.8	36200	5	5	.10	D
4431.9	34823	1	3	1.2	D
4467.1	35894	5	7	.066	D
4489.0	35785	5	7	.42	D
4493.6	35762	5	5	.36	D
4505.9	34823	3	3	1.1	D
4523.2	35617	5	5	.96	D
4573.9	34494	3	1	2.9	D

Wavelength (λ[Å])	Upper energy level (E_k[cm^{-1}])	Statistical weights g_i	g_k	Transition Probability (A[10^8 s^{-1}])	Uncertainty
	Ba I				
4579.6	35344	5	5	1.8	D
4591.8	30987	5	5	.016	D
4599.8	34371	3	1	1.0	D
4605.0	30744	3	1	.077	D
4619.9	33905	1	3	.093	D
4628.3	30816	5	3	.060	D
4673.6	30987	7	5	.065	D
4691.6	34823	5	3	1.6	D
4700.4	33905	3	3	.24	D
4726.4	32547	5	3	.46	D
5519.1	30751	3	5	.50	D
5535.5	18060	1	3	1.15	B
5777.6	30818	5	7	.64	D
5800.2	30751	5	5	.099	D
5805.7	26816	7	7	.011	D
5826.3	28554	5	3	.56	D
5907.6	25957	3	5	.036	D
5971.7	25957	5	5	.29	D
5997.1	25704	3	3	.27	D
6019.5	25642	3	1	1.4	D
6063.1	25704	5	3	.57	D
6110.8	25957	7	5	1.0	D
6341.7	24980	5	7	.19	D
6450.9	24532	3	5	.11	D
6482.9	26816	5	7	.44	D
6498.8	24980	7	7	.86	D
6527.3	24532	5	5	.59	D
6595.3	24192	3	3	.39	D
6675.3	24192	5	3	.19	D
6693.8	24532	7	5	.28	D
6865.7	25957	5	5	.078	D
7059.9	23757	7	9	.71	D
7120.3	23074	3	5	.21	D
7195.2	26160	1	3	.24	D
7280.3	22947	5	7	.53	D
7392.4	26160	3	3	.50	D
7417.5	23074	7	5	.025	D
7488.1	22947	7	7	.10	D
7672.1	22065	3	5	.31	D
7780.5	22065	5	5	.13	D
7905.8	26160	5	3	.63	D
7911.3	12637	1	3	.00298	C
	Ba II				
1413.4	76429	6	8	.017	D
1417.1	75438	4	6	.038	D
1444.9	74091	4	6	.081	D
1461.5	74109	6	8	.087	D
1487.0	72143	4	6	.14	D
1503.9	72170	6	8	.15	D
1554.4	69212	4	6	.26	D
1572.7	69260	6	8	.24	D
1573.9	69212	6	6	.016	D
1630.4	61336	2	2	.017	D
1674.5	64596	4	6	.22	D
1694.4	64697	6	8	.21	D
1697.2	64596	6	6	.017	D
1761.8	61636	4	4	.0039	D
1771.0	61336	4	2	.034	D
1786.9	61636	6	4	.044	D
1892.7	73102	2	4	.090	D
1904.2	57391	4	6	.011	D
1906.8	72705	2	2	.051	D
1924.7	57632	6	8	.031	D
1954.2	73122	4	6	.13	D
1955.1	73102	4	4	.018	D
1970.2	72705	4	2	.067	D
1985.6	70620	2	4	.25	D
1999.5	50011	2	4	.10	D
2009.2	70015	2	2	.086	D
2052.7	70652	4	6	.20	D
2054.6	70620	4	4	.029	D
2080.0	70015	4	2	.10	D
2153.9	66674	2	4	.53	D
2200.9	65683	2	2	.20	D
2232.8	66725	4	6	.29	D
2235.4	66674	4	4	.044	D
2286.0	65683	4	2	.13	D
2528.5	59800	2	4	.71	D
2634.8	59895	4	6	.76	D
2641.4	59800	4	4	.12	D

Table 1b (Continued)
TRANSITION PROBABILITIES FOR SELECTED ATOMIC AND IONIC SPECIES

Wavelength (λ[Å])	Upper energy level (E_k[cm^{-1}])	Statistical weights g_i	g_k	Transition Probability (A[10^8 s^{-1}])	Uncertainty
Ba II					
2647.3	58025	2	2	.20	D
2771.4	58025	4	2	.40	D
3816.7	72143	4	6	.0023	D
3842.8	72170	6	8	.0022	D
3891.8	45949	2	4	1.67	B
4024.1	73102	6	4	.0053	D
4057.5	73122	8	6	.012	D
4130.7	46155	4	6	1.80	B
4166.0	45949	4	4	.37	D
4216.0	73102	2	4	.058	D
4287.8	72705	2	2	.024	D
4325.7	73122	4	6	.059	D
4329.6	73102	4	4	.0088	D
4405.2	72705	4	2	.039	D
4470.7	70620	6	4	.014	D
4509.6	70652	8	6	.012	D
4524.9	42355	2	2	.72	D
4554.0	21952	2	4	1.17	A
4708.9	70620	2	4	.097	D
4843.5	70652	4	6	.093	D
4847.1	70015	2	2	.041	D
4850.8	70620	4	4	.014	D
4900.0	42355	4	2	.775	B
4934.1	20262	2	2	.955	B
4997.8	70015	4	2	.061	D
5185.0	61636	2	4	.018	D
5361.4	64596	4	6	.048	D
5391.6	64697	6	8	.052	D
5413.6	66725	6	6	8.4 (–4)	D
5421.1	64596	6	6	.0019	D
5428.8	66674	6	4	.023	D
5480.3	66725	8	6	.018	D
5784.2	66674	2	4	.20	D
5853.7	21952	4	4	.048	B
5981.3	66725	4	6	.16	D
5999.9	66674	4	4	.026	D
6135.8	65683	2	2	.085	D
6141.7	21952	6	4	.37	B
6363.2	73102	6	4	.0029	D
6372.9	61636	4	4	6.7 (–4)	D
6378.9	65683	4	2	.099	D
6457.7	61636	6	4	.0030	D
6496.9	20262	4	2	.332	C
7556.8	70620	6	4	.0016	D
7678.2	70652	8	6	6.6 (–4)	D
8710.7	57632	6	8	.80	D
8737.7	57391	4	6	.93	D
Beryllium					
Be I					
1491.8	67035	1	3	.013	D
1661.5	60187	1	3	.20	D
2348.6	42565	1	3	5.56	B
2494.7	62054	9	15*	1.6	C
2650.6	59696	9	9*	4.31	C
4572.7	64428	3	5	.79	C
Be II					
1197.1	115464	2	2	.47	D
1197.2	115464	4	2	.94	D
1512.3	98055	2	4	9.2	C
1512.4	98055	4	6	11.	C
1776.1	88232	2	2	1.4	C
1776.3	88232	4	2	2.9	C
2453.8	128972	2	6*	.142	C
3046.5	129310	2	4	.48	C
3046.7	129310	4	6	.59	C
3130.4	31935	2	4	1.14	B
3131.1	31929	2	2	1.15	B
3241.6	127335	2	2	.141	C
3241.8	127335	4	2	.28	C
3274.6	118761	2	4	.19	C
3274.7	118761	2	2	.19	C
4360.7	119421	2	4	.92	C
4361.0	119421	4	6	1.1	C
5255.9	134485	2	6*	.0256	C
5270.3	115464	2	2	.330	C
5270.8	115464	4	2	.66	C
6279.4	134681	2	4	.12	C
6279.7	134681	4	6	.143	C
6756.7	133556	2	2	.051	C
6757.1	133556	4	2	.102	C
7401.2	128972	2	4	.030	C
7401.4	128972	2	2	.030	C
Bismuth					
Bi I					
1954.5	51159	4	6	1.2	D
2021.2	49461	4	4	.060	D
2061.7	48490	4	6	.99	D
2110.3	47373	4	2	.91	D
2177.3	45916	4	2	.026	D
2228.3	44865	4	4	.89	D
2230.6	44817	4	6	2.6	D
2276.6	43913	4	4	.25	D
2515.7	51159	4	6	.043	D
2627.9	49461	4	4	.47	D
2696.8	48490	4	6	.064	D
2780.5	47373	4	2	.309	C
2798.7	51159	6	6	.036	D
2898.0	45916	4	2	1.53	C
2938.3	49461	6	4	1.23	C
2989.0	44865	4	4	.55	C
2993.3	44817	4	6	.16	C
3024.6	48490	6	6	.88	C
3067.7	32588	4	2	2.07	C
3076.7	43913	4	4	.035	D
3397.2	44865	6	4	.181	C
3402.9	44817	6	6	.016	D
3510.9	43913	6	4	.068	D
3596.1	49461	2	4	.198	C
3888.2	47373	2	2	.069	D
4121.5	45916	2	2	.164	C
4308.5	44865	2	4	.016	D
4493.0	43913	2	4	.015	D
4722.5	32588	4	2	.117	C
6134.8	49461	4	4	.018	D
Boron					
B I					
1378.6	72535	2	4	3.50	C
1378.9	72523	2	2	14.0	C
1378.9	72535	4	4	17.5	C
1379.2	72523	4	2	7.0	C
1465.5	97000	2	4	3.34	C
1465.7	97000	4	4	6.7	C
1465.8	97000	6	4	10.0	C
1825.9	54767	2	4	2.0	C
1826.4	54767	4	6	2.4	C
2088.9	47857	2	4	.28	D
2089.6	47857	4	6	.33	D
2496.8	40040	2	2	.85	C
2497.7	40040	4	2	1.69	C
Bromine					
Br I					
1488.5	67184	4	4	1.2	D
1540.7	64907	4	4	1.4	D
1574.8	67184	2	4	.20	D
1576.4	63436	4	6	.021	D
1633.4	64907	2	4	.081	D
4365.1	89786	2	4	.0075	D
4425.1	87499	4	2	.0042	D
4441.7	85944	6	4	.0075	D
4472.6	87259	4	4	.0093	D
4477.7	85763	6	8	.013	D
4513.4	85586	6	4	.0028	D
4525.6	85527	6	6	.0072	D
4575.7	89032	4	4	.016	D
4614.6	88848	4	6	.0054	D
4979.8	87259	4	4	.0026	D
5245.1	85944	2	4	.0031	D
5345.4	85586	2	4	.0076	D
7348.5	78512	4	6	.12	D
7513.0	76743	6	4	.12	D
7803.0	79696	2	4	.053	D
7938.7	88483	6	6	.19	D
8131.5	79178	2	4	.038	D
8343.7	78866	2	2	.22	D
8446.6	76743	4	4	.12	D
8638.7	75009	6	4	.097	D

Table 1b (Continued)

TRANSITION PROBABILITIES FOR SELECTED ATOMIC AND IONIC SPECIES

Wavelength (λ[Å])	Upper energy level (E_k[cm^{-1}])	Statistical weights g_i	g_k	Transition Probability (A[10^8 s^{-1}])	Uncertainty
	Br II				
4704.9	115176	5	7	1.1	D
4785.5	114818	5	5	.94	D
4816.7	114683	5	3	1.1	D
	Cadmium				
	Cd I				
2288.0	43692	1	3	5.3	C
2836.9	65354	1	3	.28	D
2880.8	65359	3	5	.42	D
2881.2	65353	3	3	.24	D
2980.6	65367	5	7	.59	D
2981.4	65359	5	5	.15	D
3261.1	30656	1	3	.00406	C
3403.7	59486	1	3	.77	C
3466.2	59498	3	5	1.2	D
3467.7	59486	3	3	.67	D
3610.5	59516	5	7	1.3	D
3612.9	59498	5	5	.35	D
4140.5	67838	3	5	.047	D
4662.4	65135	3	5	.055	C
4678.1	51484	1	3	.13	C
4799.9	51484	3	3	.41	C
5085.8	51484	5	3	.56	C
6438.5	59220	3	5	.59	C
	Cd II				
2144.4	46619	2	4	2.8	C
2265.0	44136	2	2	3.0	C
2572.9	82991	2	2	1.7	C
2748.5	82991	4	2	2.8	C
4415.6	69259	4	6	.014	B
	Calcium				
	Ca I				
2275.5	43933	1	3	.301	C
2995.0	48538	1	3	.367	C
2997.3	48564	3	5	.241	C
2999.6	48538	3	3	.279	C
3000.9	48524	3	1	1.58	C
3006.9	48564	5	5	.75	C
3009.2	48538	5	3	.430	C
3344.5	45049	1	3	.151	C
3350.2	45050	3	5	.178	C
3361.9	45052	5	7	.223	C
3624.1	42743	1	3	.212	C
3630.8	42745	3	5	.297	C
3631.0	42743	3	3	.153	C
3644.4	42747	5	7	.355	C
3644.8	42745	5	5	.094	C
3870.5	46165	3	5	.072	D
3957.1	40474	3	3	.098	C
3973.7	40474	5	3	.175	C
4092.6	44763	3	5	.11	D
4094.9	44763	5	7	.12	D
4098.5	44763	7	9	.13	D
4108.5	46182	5	7	.90	D
4226.7	23652	1	3	2.18	B
4283.0	38552	3	5	.434	C
4289.4	38465	1	3	.60	C
4299.0	38465	3	3	.466	C
4302.5	38552	5	5	1.36	C
4307.7	38418	3	1	1.99	C
4318.7	38465	5	3	.74	C
4355.1	44805	5	7	.19	D
4425.4	37748	1	3	.498	C
4435.0	37752	3	5	.67	C
4435.7	37748	3	3	.342	C
4454.8	37757	5	7	.87	C
4455.9	37752	5	5	.20	C
4526.9	43933	5	3	.41	D
4578.6	42170	3	5	.176	C
4581.4	42171	5	7	.209	C
4585.9	42171	7	9	.229	C
4685.3	44990	3	5	.080	D
4878.1	42344	5	7	.188	C
5041.6	41679	5	3	.33	D
5188.9	42919	3	5	.40	D
5261.7	39335	3	3	.15	D
5262.2	39333	3	1	.60	D
5264.2	39340	5	5	.091	D
5265.6	39335	5	3	.44	D

Wavelength (λ[Å])	Upper energy level (E_k[cm^{-1}])	Statistical weights g_i	g_k	Transition Probability (A[10^8 s^{-1}])	Uncertainty
	Ca I				
5270.3	39340	7	5	.50	D
5582.0	38259	5	7	.060	D
5588.8	38259	7	7	.49	D
5590.1	38219	3	5	.083	D
5594.5	38219	5	5	.38	D
5598.5	38192	3	3	.43	D
5601.3	38219	7	5	.086	D
5602.9	38192	5	3	.14	D
5857.5	40720	3	5	.66	D
6102.7	31539	1	3	.096	C
6122.2	31539	3	3	.287	C
6161.3	36575	5	5	.033	D
6162.2	31539	5	3	.477	C
6163.8	36555	3	3	.056	D
6166.4	36548	3	1	.22	D
6169.1	36555	5	3	.17	D
6169.6	36575	7	5	.19	D
6439.1	35897	7	9	.53	D
6449.8	35835	3	5	.090	D
6462.6	35819	5	7	.47	D
6471.7	35819	7	7	.059	D
6493.8	35730	3	5	.44	D
6499.7	35730	5	5	.081	D
	Ca II				
1341.9	74522	2	4	.015	D
1342.5	74485	2	2	.015	D
1649.9	60611	2	4	.0032	D
1652.0	60533	2	2	.0031	D
1673.9	84934	2	4	.224	C
1680.1	84936	4	6	.265	C
1680.1	84934	4	4	.0441	C
1807.3	80522	2	4	.354	C
1814.5	80526	4	6	.42	C
1814.7	80522	4	4	.070	C
1843.1	79448	2	2	.16	C
1850.7	79448	4	2	.308	C
2103.2	72722	2	4	.82	C
2112.8	72731	4	6	.97	C
2113.2	72722	4	4	.16	C
2197.8	70678	2	2	.31	C
2208.6	70678	4	2	.62	C
3158.9	56839	2	4	3.1	C
3179.3	56858	4	6	3.6	C
3181.3	56839	4	4	.58	C
3706.0	52167	2	2	.88	C
3736.9	52167	4	2	1.7	C
3933.7	25414	2	4	1.47	C
3968.5	25192	2	2	1.4	C
	Ca III				
357.97	279354	1	3	880.	D
439.69	227432	1	3	.19	D
490.55	203852	1	3	.016	D
	Ca V				
558.60	197845	5	3	22.	D
637.93	156760	5	3	3.9	D
643.12	157901	3	1	9.1	D
646.57	154671	5	5	6.9	D
647.88	156760	3	3	2.3	D
651.55	156760	1	3	2.9	D
656.76	154671	3	5	2.1	D
	Ca VII				
550.20	203616	5	5	18.	D
624.39	160158	1	3	3.3	D
630.54	160220	3	5	4.5	D
630.79	160158	3	3	2.2	D
639.15	160529	5	7	5.7	D
640.41	160220	5	5	1.3	D
	Ca VIII				
182.71	547322	2	2	160.	C
184.16	547322	4	2	320.	C
	Ca IX				
163.23	758974	5	3	376.	C
371.89	410514	1	3	88.	C
373.81	410627	3	5	116.	C
378.08	410841	5	7	150.	C
395.03	467631	3	5	220.	D
466.24	214482	1	3	112.	B
498.01	343908	3	5	24.9	C

Table 1b (Continued)
TRANSITION PROBABILITIES FOR SELECTED ATOMIC AND IONIC SPECIES

Wavelength (λ[Å])	Upper energy level (E_k[cm^{-1}])	Statistical weights g_i	g_k	Transition Probability (A[10^8 s^{-1}])	Uncertainty
	Ca IX				
506.18	343908	5	5	72.	C
515.57	340308	5	3	37.5	C
	Ca X				
110.96	901200	2	4	290.	C
111.20	899290	2	2	292.	C
151.84	832790	2	2	230.	C
153.02	832790	4	2	450.	C
206.57	901200	4	4	29.	D
206.75	901200	6	4	260.	C
207.39	899290	4	2	280.	C
411.70	417112	2	4	83.	C
419.75	417522	4	6	95.	C
420.47	417112	4	4	16.	D
557.76	179287	2	4	35.0	C
574.01	174213	2	2	32.	C
	Ca XI				
30.448	3284300	1	3	6200.	D
30.867	3239700	1	3	4.9 (+4)	D
35.212	2839940	1	3	2000.	D
	Ca XII				
140.05	709000	4	2	370.	C
147.27	709000	2	2	160.	C
	Ca XV				
141.69	814370	5	3	408.	C
142.23	728910	9	3*	630.	C
161.00	729720	5	5	190.	C
	Ca XVII				
19.558	5113000	1	3	3.8 (+4)	C
21.198	5236000	3	5	4.9 (+4)	C
192.82	518620	1	3	121.	C
218.82	726450	3	5	27.6	C
223.02	706680	1	3	34.4	C
228.72	706680	3	3	23.7	C
232.83	726450	5	5	65.	C
244.06	706680	5	3	32.8	C
	Ca XVIII				
18.71	5346000	2	6*	2.31(+4)	B
19.74	5383000	6	10*	7.0 (+4)	B
302.19	330920	2	4	20.	B
344.76	290060	2	2	13.	B
	Carbon				
	C I				
945.19	105799	1	3	6.2	D
945.34	105799	3	3	18.	D
945.58	105799	5	3	31.	D
1260.7	79319	1	3	.40	D
1260.9	79323	3	1	1.2	D
1261.0	79319	3	3	.31	D
1261.1	79311	3	5	.30	D
1261.4	79319	5	3	.50	D
1261.6	79311	5	5	.93	D
1274.1	78530	5	7	.0068	D
1277.2	78293	1	3	.88	D
1277.3	78308	3	5	1.2	D
1277.5	78293	3	3	.65	D
1277.6	78318	5	7	1.5	D
1277.7	78308	5	5	.39	D
1278.0	78293	5	3	.042	D
1279.2	78216	5	7	.11	D
1279.9	78148	3	5	.21	D
1280.1	78117	1	3	.27	D
1280.3	78148	5	5	.62	D
1280.4	78117	3	3	.20	D
1280.6	78105	3	1	.81	D
1280.8	78117	5	3	.35	D
1328.8	75254	1	3	.49	D
1364.2	83498	5	5	.047	D
1431.6	103587	5	7	1.5	D
1432.1	103563	5	5	1.4	D
1432.5	103542	5	3	1.3	D
1459.0	78731	5	3	.37	D
1463.3	78530	5	7	2.1	D
1467.4	78340	5	3	.46	D
1468.4	78293	5	3	.019	D
1470.1	78216	5	7	.0088	D
1472.2	78117	5	3	.0051	D
1481.8	77680	5	5	.33	D
	C I				
1560.3	64090	1	3	.82	D
1561.3	64091	5	5	.36	D
1561.4	64087	5	7	1.4	D
1656.3	60393	3	5	.80	D
1656.9	60353	1	3	1.1	D
1657.0	60393	5	5	2.4	D
1657.4	60353	3	3	.80	D
1657.9	60333	3	1	3.2	D
1658.1	60353	5	3	1.3	D
1751.8	78731	1	3	.57	D
1763.9	78340	1	3	.022	D
1765.4	78293	1	3	.0071	D
1930.9	61982	5	3	3.7	D
2478.6	61982	1	3	.18	D
2902.3	105799	1	3	.0066	D
2903.3	105799	3	3	.017	D
2905.0	105799	5	3	.022	D
4269.0	85400	3	5	.0032	D
4371.4	84852	3	3	.0097	D
4762.3	81326	1	3	.0052	D
4762.5	81344	3	5	.0038	D
4766.7	81326	3	3	.0039	D
4770.0	81311	3	1	.015	D
4771.8	81344	5	5	.012	D
4775.9	81326	5	3	.0062	D
4812.9	81105	1	3	9.7 (–4)	D
4817.4	81105	3	3	.0028	D
4826.8	81105	5	3	.0047	D
4932.1	82252	3	1	.046	D
5052.2	81770	3	5	.017	D
5380.3	80563	3	3	.016	D
5793.1	81344	7	5	.0033	D
5794.5	81344	5	5	5.8 (–4)	D
5800.2	81326	3	3	9.7 (–4)	D
5800.6	81326	5	3	.0029	D
5805.2	81311	3	1	.0039	D
6587.6	84032	3	3	.024	D
	C II				
687.35	145551	4	6	27.0	C
858.09	116538	2	2	.369	C
858.56	116538	4	2	1.11	C
903.62	110666	2	4	6.6	C
903.96	110624	2	2	26.3	C
904.14	110666	4	4	33.0	C
904.48	110624	4	2	13.3	C
1009.9	142027	2	4	5.8	C
1010.1	142027	4	4	11.5	C
1010.4	142027	6	4	17.3	C
1036.3	96494	2	2	8.0	C
1037.0	96494	4	2	15.9	C
1323.9	150467	4	4	4.53	C
1324.0	150462	6	6	4.71	C
1334.5	74933	2	4	2.41	C
1335.7	74930	4	6	2.89	C
2509.1	150467	2	4	.54	C
2511.7	150467	4	4	.106	C
2512.1	150462	4	6	.64	C
6578.1	131736	2	4	.36	C
6582.9	131724	2	2	.36	C
7231.3	145549	2	4	.36	C
7236.4	145551	4	6	.44	C
7237.2	145549	4	4	.072	C
	C III				
310.17	322404	1	3	18.	C
386.20	258931	1	3	32.2	B
459.46	270011	1	3	55.	C
459.52	270012	3	5	75.	C
459.63	270015	5	7	98.	C
574.28	276483	3	5	63.	C
977.03	102352	1	3	17.5	B
1174.9	137502	3	5	3.42	C
1175.3	137454	1	3	4.55	C
1175.6	137454	3	3	3.41	C
1175.7	137502	5	5	10.2	C
1176.0	137426	3	1	13.6	C
1176.4	137454	5	3	5.7	C
1247.4	182520	3	1	18.6	C
2296.9	145876	3	5	1.46	C
4647.4	259724	3	5	.73	C
4650.3	259711	3	3	.74	C

Table 1b (Continued)
TRANSITION PROBABILITIES FOR SELECTED ATOMIC AND IONIC SPECIES

Wavelength (λ[Å])	Upper energy level (E_k[cm^{-1}])	Statistical weights g_i	g_k	Transition Probability (A[10^8 s^{-1}])	Uncertainty
	C III				
4651.5	259706	3	1	.74	C
	C IV				
312.43	320071	2	6*	44.9	B
384.13	324886	6	10*	180.	C
1548.2	64592	2	4	2.66	B
1550.8	64484	2	2	2.64	B
5801.3	320082	2	4	.319	B
5812.0	320050	2	2	.316	B
	C V				
34.973	2859375	1	3	2554.	AA
40.268	2483371	1	3	8873.	AA
227.19	2851418	3	9*	136.3	AA
247.31	2859375	1	3	127.9	AA
248.70	2857310	9	15*	425.	A
260.19	2839562	9	3*	66.83	AA
267.27	2857529	3	5	396.	A
2273.9	2455225	3	9*	.5650	AA
3526.7	2483371	1	3	.1663	AA
8432.2	2851418	3	9*	.06870	AA
	Cesium				
	Cs I				
3203.5	31207	2	4	7.6 (–6)	C
3205.3	31189	2	4	7.9 (–6)	C
3207.5	31168	2	4	8.5 (–6)	C
3210.0	31144	2	4	9.4 (–6)	C
3212.8	31116	2	4	1.19(–5)	C
3216.2	31084	2	4	1.49(–5)	C
3220.1	31046	2	4	1.7 (–5)	C
3220.2	31045	2	2	1.07(–7)	C
3224.8	31001	2	4	2.0 (–5)	C
3225.0	30999	2	2	1.43(–7)	C
3230.5	30946	2	4	2.5 (–5)	C
3230.7	30944	2	2	1.97(–7)	C
3237.4	30880	2	4	2.8 (–5)	C
3237.6	30878	2	2	2.63(–7)	C
3245.9	30799	2	4	3.45(–5)	C
3246.2	30796	2	2	3.7 (–7)	C
3256.7	30698	2	4	4.25(–5)	C
3257.1	30694	2	2	7.0 (–7)	C
3270.5	30568	2	4	5.6 (–5)	C
3271.0	30563	2	2	9.8 (–7)	C
3288.6	30399	2	4	1.0 (–4)	C
3289.3	30393	2	2	2.7 (–6)	C
3313.1	30175	2	4	1.6 (–4)	C
3314.0	30166	2	2	5.2 (–6)	C
3347.5	29865	2	4	2.2 (–4)	C
3348.8	29853	2	2	1.1 (–5)	C
3397.9	29421	2	4	4.0 (–4)	C
3400.0	29404	2	2	2.4 (–5)	C
3476.8	28754	2	4	6.6 (–4)	C
3480.0	28727	2	2	6.6 (–5)	C
3611.4	27682	2	4	.0015	C
3617.3	27637	2	2	2.5 (–4)	C
3876.1	25792	2	4	.0038	C
3888.6	25709	2	2	9.7 (–4)	C
4555.3	21946	2	4	.0188	C
4593.2	21765	2	2	.0080	C
	Chlorine				
	Cl I				
1188.8	84120	4	6	2.33	C
1188.8	84122	4	4	.271	C
1201.4	84122	2	4	2.39	C
1335.7	74866	4	2	1.74	C
1347.2	74226	4	4	4.19	C
1351.7	74866	2	2	3.23	C
1363.4	74226	2	4	.75	C
4323.3	95612	4	4	.011	D
4363.3	95401	4	6	.0068	D
4379.9	95313	4	4	.014	D
4389.8	94732	6	8	.014	D
4526.2	96313	4	4	.051	C
4601.0	96594	2	2	.042	C
4661.2	96313	2	4	.012	D
7256.6	85735	6	4	.15	D
7414.1	85442	6	4	.047	D
7547.1	85735	4	4	.12	D
7717.6	85442	4	4	.030	D
7745.0	85735	2	4	.063	D
	Cl I				
7769.2	95787	6	6	.060	D
7821.4	95701	6	8	.098	D
7830.8	95898	4	4	.097	D
7878.2	84648	6	6	.018	D
7899.3	95787	4	6	.051	D
7924.6	85442	2	4	.021	D
7935.0	96731	6	8	.039	D
7997.9	84988	4	4	.021	D
	Cl II				
3329.1	161798	5	7	1.5	D
3522.1	174855	7	7	1.4	D
3798.8	172652	5	7	1.6	D
3805.2	172743	7	9	1.8	D
3809.5	172575	3	5	1.5	D
3851.0	154624	5	7	1.8	D
3851.4	154621	5	5	1.6	D
3854.7	184660	3	5	2.2	D
3861.9	184657	5	7	2.4	D
3868.6	184630	7	9	2.7	D
3913.9	172743	9	9	.82	D
3990.2	174855	5	7	.84	D
4132.5	153259	5	5	1.6	D
4276.5	170577	9	7	.76	D
4768.7	158771	3	5	.77	D
4781.3	158788	5	7	1.0	D
4794.6	128731	5	7	1.04	C
4810.1	128644	5	5	.99	C
4819.5	128623	5	3	1.00	C
4904.8	147128	5	7	.81	D
4917.7	147056	3	5	.75	D
5078.3	146471	7	7	.77	D
5219.1	131765	3	9*	.86	C
5392.1	147607	5	7	1.0	D
	Cl III				
2298.5	248528	4	4	4.2	D
2340.6	248658	6	6	4.2	D
2370.4	258886	8	6	2.8	D
2531.8	248528	2	4	4.4	D
2532.5	248658	4	6	5.3	D
2577.1	243828	4	6	4.3	D
2580.7	244685	6	8	4.7	D
2601.2	239506	2	4	4.6	D
2603.6	239730	4	6	5.0	D
2609.5	240075	6	8	5.7	D
2617.0	240568	8	10	6.6	D
2661.6	241685	4	6	3.4	D
2665.5	242046	6	8	4.8	D
2691.5	243081	4	4	3.5	D
2710.4	242823	4	6	3.5	D
3340.4	204541	6	6	1.5	D
3392.9	217913	4	4	1.9	D
3393.5	217850	6	6	1.9	D
3530.0	216710	6	8	1.8	D
3560.7	216525	4	6	1.7	D
3602.1	202368	6	8	1.7	D
3612.9	201765	4	6	1.2	D
3720.5	205947	4	6	1.7	D
	Chromium				
	Cr I				
2000.0	58293	9	9	1.4	D
2383.3	50253	9	11	.41	D
2385.7	50211	9	9	.17	D
2389.2	49653	3	5	.23	D
2408.6	49812	9	7	.67	D
2408.7	49598	7	5	.29	D
2492.6	47918	3	5	.45	C
2496.3	47975	5	7	.56	C
2502.5	48043	7	9	.22	D
2504.3	48014	7	9	.45	C
2549.5	47022	3	3	.48	D
2560.7	46968	5	5	.43	D
2577.7	46879	7	7	.26	D
2591.9	46879	9	7	.65	C
2622.9	46422	9	9	.13	D
2702.0	45306	9	11	.21	C
2726.5	44259	5	7	.75	C
2731.9	44187	5	5	.78	C
2736.5	44126	5	3	.75	D
2752.9	44126	3	3	.87	D

TRANSITION PROBABILITIES FOR SELECTED ATOMIC AND IONIC SPECIES

Wavelength (λ[Å])	Upper energy level (E_k[cm^{-1}])	Statistical weights g_i	g_k	Transition Probability (A[10^8 s^{-1}])	Uncertainty
	Cr I				
2757.1	44187	5	5	.68	C
2761.8	44126	5	3	.68	D
2764.4	44259	7	7	.37	D
2769.9	44187	7	5	1.1	C
2780.7	44259	9	7	1.4	C
2871.6	42909	7	9	.12	D
2879.3	42648	5	7	.21	D
2887.0	42439	3	5	.27	D
2889.3	42909	9	9	.66	C
2899.2	42293	3	3	.15	D
2909.1	42293	5	3	.68	C
2967.6	41782	7	9	.39	D
2971.1	41575	5	7	.71	C
2975.5	41409	3	5	.89	C
2980.8	41289	1	3	.85	D
2988.7	41043	5	7	.52	C
2991.9	41225	3	1	3.0	D
2995.1	40971	5	5	.43	D
2996.6	41289	5	3	2.0	C
2998.8	40930	5	3	.59	D
3000.9	41409	7	5	1.6	C
3005.1	41575	9	7	.92	C
3013.7	40983	3	5	.83	C
3020.7	40906	3	3	1.5	D
3021.6	41393	9	11	3.2	C
3024.4	40983	5	5	2.3	C
3030.2	41086	7	7	1.1	C
3034.2	41043	7	7	.35	D
3037.0	41225	9	9	.54	C
3040.9	40971	7	5	.74	D
3053.9	41043	9	7	1.2	C
3148.4	55686	9	11	.59	C
3155.2	55741	11	13	.54	C
3163.8	55799	13	15	.52	C
3237.7	54811	9	9	1.3	D
3238.1	54930	11	11	.20	D
3578.7	27935	7	9	1.48	B
3593.5	27820	7	7	1.50	B
3605.3	27729	7	5	1.62	B
3639.8	47986	13	11	1.8	D
3768.7	47047	7	5	.22	D
3804.8	50558	9	9	.69	D
3879.2	50058	3	5	.56	D
3963.7	45741	13	15	1.3	D
3969.8	45707	11	13	1.2	D
3981.2	46968	3	5	.11	D
3983.9	45615	7	9	1.05	C
4001.4	56362	9	11	.65	D
4030.7	56155	3	5	.79	D
4039.1	55799	15	15	.68	C
4048.8	55741	13	13	.65	D
4058.8	55686	11	11	.69	D
4161.4	59957	13	15	.80	D
4165.5	59884	11	13	.75	D
4211.4	48043	7	9	.085	D
4224.5	48562	9	9	.067	D
4239.0	47866	9	9	.071	D
4254.4	23499	7	9	.315	B
4274.8	23386	7	7	.306	B
4280.4	54405	13	15	.47	D
4289.7	23305	7	5	.313	B
4297.7	54317	11	13	.49	D
4344.5	31106	7	9	.11	C
4351.8	31280	9	11	.12	C
4374.2	47055	13	11	.10	C
4375.3	46905	11	9	.072	D
4381.1	44667	5	3	.10	D
4387.5	46986	13	11	.066	D
4413.9	51287	7	5	.41	D
4432.2	45719	1	3	.17	D
4458.5	46705	9	11	.13	D
4482.9	49477	3	3	.30	D
4488.1	56210	7	7	.63	D
4506.9	55945	13	11	.27	D
4511.9	47055	9	9	.13	D
4526.5	42606	13	13	.20	D
4530.7	42589	11	11	.20	D
4544.6	42515	5	5	.26	D
4595.6	55517	13	13	.47	D
4632.2	46688	7	7	.071	D
4639.5	46637	5	7	.095	D
4646.2	29825	9	7	.087	C
4689.4	46525	7	5	.23	D
4708.0	46783	11	9	.37	D
4718.4	46959	13	11	.42	D
4723.1	46000	7	7	.093	D
4724.4	46058	9	9	.063	D
4727.2	45349	13	13	.051	D
4729.7	46077	5	3	.17	D
4730.7	45966	7	5	.28	D
4737.4	46000	9	7	.24	D
4789.3	41393	13	11	.076	D
4792.5	45966	7	5	.26	D
4801.0	46000	9	7	.23	D
4870.8	45359	7	9	.35	D
4887.0	45354	9	11	.32	D
4922.3	45349	11	13	.40	D
4936.3	45359	7	9	.14	D
4954.8	45354	9	11	.12	D
5139.7	47048	7	7	.13	D
5177.4	46959	9	11	.061	D
5184.6	46783	7	9	.11	D
5192.0	46637	5	7	.14	D
5196.6	47055	11	9	.12	D
5200.2	46525	3	5	.16	D
5204.5	26802	5	3	.55	D
5206.0	26796	5	5	.53	D
5208.4	26788	5	7	.51	D
5243.4	46449	5	3	.20	D
5261.8	48825	7	9	.13	D
5272.0	46783	7	9	.11	D
5287.2	46637	5	7	.078	D
5304.2	46783	9	9	.066	D
5312.9	46637	7	7	.11	D
5318.8	46525	5	5	.13	D
5328.3	42261	9	11	.60	D
5340.4	46449	5	3	.16	D
5400.6	45734	5	5	.16	D
5409.8	26788	9	7	.062	D
	Cr II				
2653.6	49706	4	6	.35	D
2658.6	49565	2	4	.58	D
2666.0	49646	6	8	.59	D
2668.7	49493	4	2	1.4	D
2671.8	49565	6	4	1.0	D
2672.8	49706	8	6	.55	D
2693.5	67334	10	8	1.4	D
2727.3	67876	10	8	1.7	D
2740.1	48632	6	8	.11	D
2745.0	66727	4	6	.85	D
2768.6	74424	6	8	2.8	D
2774.4	75717	8	8	1.7	D
2778.1	75810	10	10	3.2	D
2782.4	69348	6	4	1.6	D
2785.7	69506	10	8	2.1	D
2787.6	66727	6	6	1.5	D
2792.2	69498	12	10	2.3	D
2800.8	69388	12	14	2.2	D
2818.4	68993	8	10	2.2	D
2822.0	68844	6	8	2.3	D
2832.5	70108	12	10	1.3	D
2838.8	73486	8	8	2.7	D
2840.0	65420	10	12	2.7	D
2843.3	47465	8	10	.64	D
2849.8	47228	6	8	.92	D
2851.4	65218	8	10	2.2	D
2856.8	54626	4	6	.43	D
2857.4	54785	6	8	.28	D
2860.9	46906	2	4	.69	D
2862.6	47228	8	8	.63	D
2866.7	46906	4	4	1.2	D
2867.1	54500	4	4	1.1	D
2867.7	46824	2	2	1.1	D
2870.4	54626	6	6	1.3	D
2873.8	54418	4	2	.88	D
2878.5	47228	10	8	.074	D
2880.9	54500	6	4	.79	D
2888.7	70880	10	12	.88	D
2898.5	65710	10	12	1.2	D
2921.8	65384	8	10	.90	D
2927.1	72717	10	10	2.8	D

Table 1b (Continued)
TRANSITION PROBABILITIES FOR SELECTED ATOMIC AND IONIC SPECIES

Wavelength (λ[Å])	Upper energy level (E_k[cm^{-1}])	Statistical weights g_i	g_k	Transition Probability (A[10^8 s^{-1}])	Uncertainty
	Cr II				
2930.9	64062	2	4	1.1	D
2935.1	64924	6	8	1.8	D
2953.4	63802	2	2	1.8	D
2953.7	68477	10	10	.92	D
2966.1	64924	10	8	.54	D
2971.9	64031	14	14	2.0	D
2979.7	63849	12	12	1.8	D
2985.3	63707	10	10	2.2	D
2989.2	63601	8	8	2.2	D
3040.9	67506	10	12	4.8	D
3041.7	68477	10	10	3.1	D
3050.1	67589	12	14	1.8	D
3093.5	70880	10	12	.67	D
3096.1	70852	10	8	.75	D
3107.6	70679	8	10	.62	D
3118.7	51584	2	4	1.7	D
3120.4	51670	4	6	1.5	D
3122.6	65710	12	12	.44	D
3128.7	51584	4	4	.81	D
3136.7	51670	6	6	.64	D
3152.2	67070	4	4	1.8	D
3180.7	51943	12	10	.70	D
3183.3	67012	8	6	.87	D
3209.2	51670	8	6	.68	D
3217.4	51584	6	4	.77	D
3234.1	65543	10	8	.92	D
3238.8	65680	12	10	.54	D
3295.4	64031	12	14	.32	D
3336.3	49493	2	2	.42	D
3339.8	49565	4	4	.49	D
3342.6	49706	6	6	.39	D
3347.8	49493	4	2	.52	D
3358.5	49565	6	4	1.1	D
3360.3	54785	8	8	1.3	D
3368.1	49706	8	6	1.4	D
3378.3	54626	8	6	.41	D
3379.4	54626	4	6	.48	D
3382.7	49352	6	6	.45	D
3391.4	49006	2	4	.19	D
3393.0	54500	2	4	.46	D
3393.8	54500	4	4	.66	D
3394.3	54500	6	4	.75	D
3402.4	54418	2	2	.80	D
3408.8	49352	8	6	.95	D
3421.2	48750	2	2	1.7	D
3422.7	49006	6	4	1.4	D
3433.3	48750	4	2	1.3	D
3511.8	48491	8	6	.079	D
4242.4	54785	10	8	.12	D
	Cr XXI				
149.90	667110	1	3	160.	C
	Cr XXII				
8.51	11800000	2	6*	1.2 (+4)	C
9.493	10534000	2	6*	2.5 (+4)	C
9.809	10553000	2	4	4.1 (+4)	B
9.865	10590000	4	6	4.9 (+4)	B
12.620	7924000	2	4	5.13(+4)	B
12.662	7898000	2	2	5.28(+4)	B
13.147	7964000	2	4	1.3 (+5)	B
13.294	7966000	4	6	1.6 (+5)	B
13.306	7964000	4	4	2.6 (+4)	C
25.2	11800000	2	6*	3750.	C
36.93	10534000	2	6*	7000.	C
37.52	10580000	6	10*	1.7 (+4)	B
223.00	448430	2	4	33.	B
279.69	357540	2	2	17.	B
	Cr XXIII				
2.182	45830000	1	3	3.3 (+6)	B
	Cobalt				
	Co I				
2987.2	33467	10	8	.050	C
2989.6	33440	10	10	.037	C
3013.6	33173	10	8	.016	C
3017.5	33946	8	6	.072	C
3042.5	33674	8	6	.020	C
3044.0	32842	10	10	.19	C
3048.9	34196	6	4	.078	C
3061.8	33467	8	8	.15	C
	Co I				
3064.4	33440	8	10	.0068	C
3082.6	32431	10	12	.026	C
3086.8	34196	4	4	.19	C
3089.6	33173	8	8	.024	C
3098.2	33674	6	6	.027	C
3139.9	32655	8	6	.028	C
3147.1	33173	6	8	.045	C
3149.3	33151	6	4	.031	C
3395.4	34134	6	8	.26	C
3405.1	32842	10	10	.98	C
3409.2	33467	8	8	.42	C
3412.3	33440	8	10	.64	C
3412.6	29295	10	8	.12	C
3417.2	33946	6	6	.32	C
3431.6	29949	8	6	.11	C
3433.0	34196	4	4	1.1	C
3442.9	30444	6	4	.12	C
3443.6	33173	8	8	.63	C
3449.2	33674	6	6	.73	C
3449.4	32465	10	10	.16	C
3455.2	30743	4	2	.18	C
3462.8	33946	4	6	.87	C
3465.8	28845	10	12	.097	C
3483.4	32842	8	10	.062	C
3489.4	36092	8	6	1.6	C
3491.3	30444	4	4	.053	C
3495.7	33674	4	6	.45	C
3496.7	32733	8	8	.036	C
3518.4	36875	6	4	1.7	C
3520.1	29216	8	6	.034	C
3521.6	31871	10	8	.12	C
3523.4	33449	4	2	1.2	C
3526.9	28346	10	10	.12	C
3529.0	29735	6	8	.090	C
3529.8	32465	8	10	.48	C
3533.4	30103	4	6	.091	C
3550.6	29563	6	4	.042	C
3560.9	33151	4	4	.24	C
3564.9	32733	6	8	.086	C
3569.4	35451	8	8	1.6	C
3575.0	32655	6	6	.18	C
3575.4	28777	8	8	.094	C
3585.2	32028	8	8	.076	C
3587.2	36330	6	6	1.9	C
3594.9	29216	6	6	.086	C
3602.1	29563	4	4	.10	C
3605.4	31871	8	8	.039	C
3627.8	31700	8	10	.052	C
3631.4	28346	8	10	.0065	C
3647.7	29216	4	6	.012	C
3652.5	28777	6	8	.0095	C
3704.1	35451	6	8	.18	C
3745.5	34134	8	8	.077	C
3842.1	33463	8	6	.31	C
3845.5	33440	8	10	.49	C
3873.1	29295	10	8	.12	C
3874.0	29949	8	6	.12	C
3881.9	30444	6	4	.11	C
3894.1	34134	6	8	.81	C
3895.0	30743	4	2	.11	C
3909.9	25569	10	12	.0019	C
3936.0	32842	8	10	.15	C
3995.3	32465	8	10	.36	C
3997.9	33467	6	8	.079	C
4020.9	28346	10	10	.0092	D
4092.4	31871	8	8	.14	D
4118.8	32733	6	8	.34	C
4121.3	31700	8	10	.24	C
	Copper				
	Cu I				
2024.3	49383	2	6*	.098	C
2165.1	46173	2	4	.51	B
2178.9	45879	2	4	.913	B
2181.7	45821	2	2	1.0	C
2225.7	44916	2	2	.46	C
2244.3	44544	2	4	.0119	B
2441.6	40944	2	2	.020	C
2492.2	40114	2	4	.0311	B
2618.4	49383	6	4	.307	C
2766.4	49383	4	4	.096	C

Table 1b (Continued)
TRANSITION PROBABILITIES FOR SELECTED ATOMIC AND IONIC SPECIES

Wavelength (λ[Å])	Upper energy level (E_k[cm^{-1}])	Statistical weights g_i	g_k	Transition Probability (A[10^8 s^{-1}])	Uncertainty
		Cu I			
2824.4	46598	6	6	.078	C
2961.2	44963	6	8	.0376	C
3063.4	45879	4	4	.0155	C
3194.1	44544	4	4	.0155	C
3247.5	30784	2	4	1.39	B
3274.0	30535	2	2	1.37	B
3337.8	41153	6	8	.0038	C
4022.6	55388	2	4	.190	C
4062.6	55391	4	6	.210	C
4249.0	64472	2	2	.195	C
4275.1	62403	6	8	.345	C
4480.4	52849	2	2	.030	C
4509.4	64472	4	2	.275	C
4530.8	52849	4	2	.084	C
4539.7	63585	6	4	.212	C
4587.0	62948	8	6	.320	C
4651.1	62403	10	8	.380	C
4704.6	62403	8	8	.055	C
5105.5	30784	6	4	.020	C
5153.2	49935	2	4	.60	C
5218.2	49942	4	6	.75	C
5220.1	49935	4	4	.150	C
5292.5	62403	8	8	.109	C
5700.2	30784	4	4	.0024	C
5782.1	30535	4	2	.0165	C
		Cu II			
2489.7	66419	5	5	.015	D
2544.8	108015	9	7	1.1	D
2689.3	108015	7	7	.41	D
2701.0	110366	5	5	.67	D
2703.2	110084	3	3	1.2	D
2713.5	108336	5	5	.68	D
2769.7	108015	7	7	.61	D
	Dysprosium				
		Dy I			
2862.7	34922	17	15	.065	D
2964.6	33722	17	17	.065	D
3147.7	35894	15	17	.11	D
3263.2	34770	15	13	.14	D
3511.0	32608	15	13	.31	D
3571.4	32126	15	13	.20	D
3757.1	34175	17	19	3.0	D
3868.8	33406	17	17	3.1	D
3967.5	32763	17	19	.87	D
4046.0	24709	17	15	1.5	D
4103.9	31411	13	11	1.7	D
4186.8	23878	17	17	1.32	C
4194.8	23832	17	17	.72	D
4211.7	23737	17	19	2.08	C
4218.1	27835	15	15	1.85	C
4221.1	27818	15	17	1.52	C
4225.2	30712	13	15	4.5	D
4268.3	27556	15	15	.036	D
4276.7	30427	13	13	.73	D
4292.0	27427	15	15	.058	D
4577.8	21839	17	19	.022	D
4589.4	21783	17	15	.13	D
4612.3	21675	17	15	.082	D
5077.7	19689	17	17	.0057	D
5301.6	18857	17	15	.011	D
5547.3	18022	17	17	.0027	C
5639.5	17727	17	19	.0047	C
5974.5	16733	17	17	.0040	C
5988.6	16694	17	15	.0053	C
6010.8	20766	15	15	.026	D
6088.3	20555	15	13	.035	D
6168.4	20341	15	17	.025	D
6259.1	15972	17	19	.0085	C
6579.4	15195	17	15	.0075	D
	Erbium				
		Er I			
3862.9	25880	13	13	2.5	D
4008.0	24943	13	15	2.6	D
4151.1	24083	13	11	1.8	D
	Europium				
		Eu I			
2372.9	42131	8	6	.19	D
2375.3	42087	8	8	.20	D
2379.7	42010	8	10	.20	D
2619.3	38167	8	10	.0070	D
2643.8	37813	8	8	.0066	D
2659.4	37591	8	10	.012	D
2682.6	37266	8	6	.012	D
2710.0	36890	8	10	.14	D
2724.0	36700	8	8	.12	D
2731.4	36601	8	8	.031	D
2732.6	36584	8	6	.037	D
2735.3	36549	8	10	.047	D
2738.6	36505	8	10	.013	D
2743.3	36442	8	6	.11	D
2745.6	36411	8	6	.050	D
2747.8	36382	8	8	.052	D
2772.9	36053	8	6	.010	D
2878.9	34726	8	10	.028	D
2892.5	34562	8	8	.10	D
2893.0	34556	8	6	.10	D
2909.0	34366	8	10	.069	D
2958.9	33787	8	6	.016	D
3059.0	32681	8	8	.038	D
3067.0	32596	8	10	.0091	D
3106.2	32185	8	10	.055	D
3111.4	32130	8	10	.30	D
3168.3	31554	8	10	.069	D
3185.5	31383	8	10	.0058	D
3210.6	31138	8	8	.11	D
3212.8	31116	8	8	.29	D
3213.8	31107	8	6	.18	D
3235.1	30902	8	10	.010	D
3241.4	30842	8	8	.023	D
3246.0	30798	8	6	.014	D
3247.6	30784	8	8	.023	D
3322.3	30091	8	6	.035	D
3334.3	29983	8	6	.34	D
3350.4	29839	8	10	.015	D
3353.7	29809	8	8	.0058	D
3457.1	28918	8	8	.0084	D
3467.9	28828	8	8	.010	D
3589.3	27853	8	6	.0069	D
4594.0	21761	8	10	1.4	D
4627.2	21605	8	8	1.3	D
4661.9	21445	8	6	1.3	D
5645.8	17707	8	6	.0054	D
5765.2	17341	8	8	.011	D
6018.2	16612	8	10	.0085	D
6291.3	15891	8	6	.0018	D
6864.5	14564	8	10	.0058	D
7106.5	14068	8	8	.0026	D
	Fluorine				
		F I			
806.96	123921	4	6	3.3	C
809.60	123922	2	4	2.8	C
951.87	105056	4	2	2.6	C
954.83	104731	4	4	6.4	C
955.55	105056	2	2	5.1	C
958.52	104731	2	4	1.3	C
6239.7	118428	6	4	.25	D
6348.5	118428	4	4	.18	D
6413.7	118428	2	4	.11	D
6708.3	117309	6	4	.014	D
6774.0	117164	6	6	.10	D
6795.5	117392	4	2	.052	D
6834.3	117309	4	4	.21	D
6856.0	116987	6	8	.42	D
6870.2	117392	2	2	.38	D
6902.5	117164	4	6	.32	D
6909.8	117309	2	4	.22	D
6966.4	119082	4	2	.11	D
7037.5	118937	4	4	.30	D
7127.9	119082	2	2	.38	D
7309.0	137599	6	8	.47	D
7311.0	118405	4	2	.39	D
7314.3	137590	4	6	.48	D
7332.0	116041	6	4	.31	D
7398.7	115918	6	6	.31	D
7425.7	116144	4	2	.34	D
7482.7	116041	4	4	.056	D
7489.2	118405	2	2	.11	D

Table 1b (Continued)
TRANSITION PROBABILITIES FOR SELECTED ATOMIC AND IONIC SPECIES

Wavelength (λ[Å])	Upper energy level (E_k[cm^{-1}])	Statistical weights g_i	g_k	Transition Probability (A[10^8 s^{-1}])	Uncertainty
	F I				
7514.9	116144	2	2	.052	D
7552.2	115918	4	6	.078	D
7573.4	116041	2	4	.10	D
7607.2	117873	4	4	.070	D
7754.7	117623	4	6	.30	D
7800.2	117873	2	4	.21	D
	Gallium				
	Ga I				
2195.4	45537	2	2	.019	C
2199.7	46274	4	2	.033	C
2214.4	45972	4	6	.012	C
2235.9	45537	4	2	.043	C
2255.0	44332	2	2	.031	C
2259.2	45076	4	6	.031	C
2294.2	43575	2	4	.070	C
2297.9	44332	4	2	.058	C
2338.2	43581	4	6	.098	C
2371.3	42159	2	2	.057	C
2418.7	42159	4	2	.10	C
2450.1	40803	2	4	.28	C
2500.2	40811	4	6	.34	C
2659.9	37585	2	2	.12	C
2719.7	37585	4	2	.23	C
2874.2	34782	2	4	1.2	C
2943.6	34788	4	6	1.4	C
2944.2	34782	4	4	.27	C
4033.0	24789	2	2	.49	C
4172.0	24789	4	2	.92	C
	Ga II				
829.60	120540	1	3	.22	D
1414.4	70700	1	3	18.8	C
	Germanium				
	Ge I				
1944.7	51978	3	1	.70	C
1955.1	51705	3	3	.28	C
1988.3	51705	5	3	.25	C
1998.9	51438	5	5	.55	C
2041.7	48963	1	3	1.1	C
2065.2	48963	3	3	.85	C
2068.7	48882	3	5	1.2	C
2086.0	48480	3	5	.40	C
2094.3	49144	5	7	.97	C
2105.8	48882	5	5	.17	C
2256.0	51438	5	5	.032	C
2417.4	48480	5	5	.96	C
2498.0	40021	1	3	.13	C
2533.2	40021	3	3	.10	C
2589.2	40021	5	3	.051	C
2592.5	39118	3	5	.71	C
2651.2	39118	5	5	2.0	C
2651.6	37702	1	3	.85	C
2691.3	37702	3	3	.61	C
2709.6	37452	3	1	2.8	C
2754.6	37702	5	3	1.1	C
3039.1	40021	5	3	2.8	C
3124.8	39118	5	5	.031	C
3269.5	37702	5	3	.29	C
4226.6	40021	1	3	.21	C
4685.8	37702	1	3	.095	C
	Ge II				
999.10	100090	2	4	1.9	D
1016.6	100131	4	6	2.1	D
1017.1	100090	4	4	.35	D
1055.0	94784	2	2	.69	D
1075.1	94784	4	2	1.3	D
1237.1	80837	2	4	19.	D
1261.9	81013	4	6	22.	D
1264.7	80837	4	4	3.5	D
1602.5	62403	2	2	3.4	D
1649.2	62403	4	2	6.5	D
4741.8	100090	2	4	.46	D
4814.6	100131	4	6	.51	D
4824.1	100090	4	4	.086	D
5131.8	100318	4	6	1.9	D
5178.5	100318	6	6	.13	D
5178.6	100317	6	8	2.0	D
5893.4	79367	2	4	.92	D
6021.0	79007	2	2	.84	D

Wavelength (λ[Å])	Upper energy level (E_k[cm^{-1}])	Statistical weights g_i	g_k	Transition Probability (A[10^8 s^{-1}])	Uncertainty
	Ge II				
6336.4	94784	2	2	.44	D
6484.2	94784	4	2	.85	D
	Gold				
	Au I				
2428.0	41174	2	4	1.5	D
2676.0	37359	2	2	1.1	D
	Helium				
	He I				
510.00	196079	1	3	.462	B
512.10	195275	1	3	.717	B
515.62	193943	1	3	1.3	B
522.21	191493	1	3	2.46	A
537.03	186209	1	3	5.66	A
584.33	171135	1	3	17.99	AA
2696.1	196935	3	9*	.00550	B
2723.2	196567	3	9*	.00780	B
2763.8	196027	3	9*	.0111	B
2829.1	195193	3	9*	.017	B
2945.1	193801	3	9*	.0320	A
3187.7	191217	3	9*	.05639	AA
3354.6	196079	1	3	.0130	B
3447.6	195275	1	3	.0232	A
3554.4	197213	9	15*	.0131	A
3587.3	196955	9	15*	.0205	C
3613.6	193943	1	3	.0390	A
3634.2	196595	9	15*	.0261	A
3705.0	196070	9	15*	.0444	C
3819.6	195260	9	15*	.0636	A
3833.6	197213	3	5	.00971	B
3867.5	194936	9	3*	.025	B
3871.8	196956	3	5	.0126	C
3888.7	185565	3	9*	.09478	AA
3926.5	196596	3	5	.0195	A
3964.7	191493	1	3	.0719	A
4009.3	196070	3	5	.0279	C
4026.2	193917	9	15*	.116	A
4120.8	193347	9	3*	.0444	A
4143.8	195261	3	5	.0485	A
4387.9	193918	3	5	.0894	A
4437.6	193664	3	1	.033	B
4471.5	191445	9	15*	.246	A
4713.2	190298	9	3*	.0955	A
4921.9	191447	3	5	.198	A
5015.7	186210	1	3	.1338	AA
5047.7	190940	3	1	.0675	A
5875.7	186102	9	15*	.7053	AA
6678.2	186105	3	5	.6339	AA
7065.2	183237	9	3*	.2786	AA
7281.4	184865	3	1	.1829	AA
8361.7	195193	3	9*	.00334	A
9463.6	193801	3	9*	.00501	A
9603.4	195275	1	3	.00610	A
9702.6	195868	9	3*	.00858	B
10311.	195260	9	15*	.0201	A
10668.	194936	9	3*	.0152	A
10830.	169087	3	9*	.1022	AA
10913.	195262	15	21*	.0212	B
10917.	195262	5	7	.0212	B
10997.	195193	15	9*	.0013	B
11013.	193943	1	3	.0100	A
11045.	195261	3	5	.0185	A
11226.	195115	3	1	.0108	A
11969.	193917	9	15*	.0358	A
12528.	191217	3	9*	.00710	A
12756.	193943	5	3	.0012	B
12785.	193921	15	21*	.0462	B
12791.	193921	5	7	.0461	B
12846.	193347	9	3*	.0289	A
12968.	193918	3	5	.0343	A
12985.	193801	15	9*	.0025	B
	Indium				
	In I				
2710.3	39098	4	6	.4	D
2560.2	39048	2	4	.4	D
3256.1	32915	4	6	1.3	D
3039.4	32892	2	4	1.3	D
4101.8	24373	2	2	.56	C
4511.3	24373	4	2	1.02	C

Table 1b (Continued)
TRANSITION PROBABILITIES FOR SELECTED ATOMIC AND IONIC SPECIES

Wavelength (λ[Å])	Upper energy level (E_k[cm^{-1}])	Statistical weights g_i	g_k	Transition Probability (A[10^8 s^{-1}])	Uncertainty
	Indium				
	In II				
2941.1	97025	3	1	1.4	D
	Iodine				
	I I				
1782.8	56093	4	4	2.71	C
1830.4	54633	4	6	.16	D
	Iron				
	Fe I				
1934.5	51692	9	7	.25	C
1937.3	51619	9	7	.22	C
1940.7	51945	7	5	.26	C
2084.1	47967	9	7	.37	C
2102.4	47967	7	7	.088	C
2113.0	48290	1	3	.19	C
2132.0	46889	9	9	.076	C
2166.8	46137	9	7	2.7	C
2191.8	46314	5	5	1.2	C
2196.0	46410	3	3	1.2	C
2200.7	46314	3	5	.28	C
2259.5	44244	9	11	.070	C
2276.0	43923	9	7	.17	C
2277.1	51630	7	5	37.	C
2287.3	44411	5	3	.34	C
2294.4	44459	3	1	.61	C
2300.1	44166	5	7	.080	C
2309.0	44184	3	5	.15	C
2313.1	43923	5	7	.14	C
2320.4	43499	7	9	.12	C
2373.6	42533	7	7	.067	C
2374.5	43079	1	3	.29	C
2462.2	41018	7	5	.15	C
2462.6	40594	9	9	.58	C
2479.8	41018	5	5	1.8	C
2483.3	40257	9	11	4.9	C
2488.1	40594	7	9	4.7	C
2490.6	40842	5	7	3.8	C
2491.2	41018	3	5	3.0	C
2501.1	39970	9	7	.68	C
2510.8	40231	7	5	1.3	C
2518.1	40405	5	3	1.9	C
2522.8	39626	9	9	2.9	C
2524.3	40491	3	1	3.4	C
2527.4	39970	7	7	1.9	C
2529.1	40231	5	5	.98	C
2535.6	40405	1	3	.97	C
2541.0	40231	3	5	.92	C
2546.0	39970	5	7	.67	C
2549.6	39626	7	9	.36	C
2584.5	45608	11	13	.46	C
2719.0	36767	9	7	1.4	C
2720.9	37158	7	5	1.1	C
2723.6	37410	5	3	.64	C
2737.3	37410	3	3	.85	C
2742.4	37158	5	5	.63	C
2744.1	37410	1	3	.35	C
2750.1	36767	7	7	.39	C
2756.3	37158	3	5	.20	C
2788.1	42784	11	13	.63	C
2894.5	52916	5	5	.63	C
2899.4	52858	5	3	.61	C
2923.3	60549	11	11	1.7	C
2925.4	56423	7	9	.19	C
2936.9	34040	9	9	.14	C
2954.7	52213	5	7	.12	C
2966.9	33695	9	11	.272	B
2980.5	55791	7	7	.22	C
2983.6	33507	9	7	.280	B
2990.4	55430	9	11	.40	C
2996.4	52916	3	5	.19	C
2999.5	40257	11	11	.23	C
3000.9	34017	5	3	.642	B
3008.1	34122	3	1	1.07	B
3009.1	52613	13	11	.079	C
3009.6	40594	9	9	.18	C
3011.5	55446	7	9	.48	C
3019.0	40842	7	7	.15	C
3037.4	33802	3	5	.32	C
3042.7	40842	5	7	.066	C
	Fe I				
3047.6	33507	5	7	.284	B
3053.1	52297	3	5	.18	C
3057.4	39626	11	9	.45	C
3059.1	33096	7	9	.18	C
3067.2	39970	9	7	.35	C
3075.7	40231	7	5	.30	C
3083.7	40405	5	3	.35	C
3098.2	53983	11	11	.11	C
3100.7	39970	7	7	.16	C
3119.5	51668	11	9	.096	C
3120.4	51826	9	7	.10	C
3160.7	51192	9	9	.19	C
3161.9	50968	11	13	.12	C
3166.4	52213	9	7	.14	C
3175.4	50833	11	11	.13	C
3199.5	50808	9	9	.27	C
3205.4	51208	3	3	1.2	C
3215.9	50999	5	5	.81	C
3217.4	50423	11	9	.23	C
3225.8	50342	11	13	1.0	C
3231.0	50699	7	5	.39	C
3233.1	57028	13	15	.55	C
3234.0	50475	9	9	.20	C
3248.2	50534	7	7	.22	C
3253.6	56951	7	9	.18	C
3254.4	57070	11	13	.51	C
3265.6	48163	7	5	.39	C
3271.0	48290	5	3	.67	C
3280.3	57104	9	11	.55	C
3282.9	56859	3	5	.31	C
3292.0	56593	7	9	.62	C
3292.6	48290	3	3	.31	C
3306.0	47967	5	7	.48	C
3307.2	56334	13	13	.20	C
3314.7	56783	5	7	.70	C
3323.7	52916	5	5	.31	C
3328.9	56383	11	11	.27	C
3337.7	51668	11	9	.067	C
3355.2	56423	9	9	.33	C
3369.5	51668	9	9	.25	C
3370.8	51374	11	11	.34	C
3380.1	51826	7	7	.24	C
3384.0	47093	7	7	.11	C
3392.7	47017	7	7	.26	C
3399.3	47136	5	5	.39	C
3402.3	55490	13	13	.29	C
3406.4	55754	3	5	.30	C
3407.5	46889	7	9	.60	C
3410.2	56859	3	5	.48	C
3411.4	51305	9	9	.065	C
3413.1	47017	5	7	.37	C
3417.8	47177	3	3	.52	C
3418.5	47171	3	1	1.3	C
3424.3	46745	7	7	.21	C
3425.0	53763	9	7	.29	C
3427.1	46721	7	9	.56	C
3428.2	46889	5	5	.22	C
3445.1	46745	5	7	.28	C
3447.3	46727	5	5	.11	C
3450.3	46902	3	3	.24	C
3495.3	49243	9	7	.13	C
3497.1	46137	7	7	.15	C
3521.8	46314	3	5	.11	C
3524.1	49243	7	5	.091	C
3527.8	51335	9	9	.20	C
3529.8	51567	3	3	.78	C
3536.6	51461	5	7	.80	C
3540.1	51350	7	9	.12	C
3541.1	51229	9	11	.64	C
3542.1	51335	7	9	.76	C
3552.8	51331	5	5	.17	C
3553.7	56951	11	9	.83	C
3556.9	51103	9	11	.45	C
3559.5	52858	3	3	.22	C
3560.7	54301	7	9	.077	C
3565.4	35768	7	9	.39	C
3570.1	35379	9	11	.677	B
3572.0	50833	11	11	.25	C
3581.2	34844	11	13	1.02	B
3582.2	54014	13	11	.25	C
3587.0	35856	5	5	.17	C

Table 1b (Continued)
TRANSITION PROBABILITIES FOR SELECTED ATOMIC AND IONIC SPECIES

Wavelength (λ[Å])	Upper energy level (E_k[cm^{-1}])	Statistical weights g_i	Statistical weights g_k	Transition Probability (A[10^8 s^{-1}])	Uncertainty
	Fe I				
3594.6	50808	9	9	.28	C
3599.6	56593	11	9	.19	C
3603.2	49461	11	11	.27	C
3605.5	49727	9	9	.65	C
3606.7	49434	11	13	.84	C
3608.9	35856	3	5	.814	B
3610.2	50342	13	13	.50	C
3612.1	50523	11	13	.077	C
3617.8	51969	5	7	.66	C
3618.8	35612	5	7	.73	C
3621.5	49604	9	11	.52	C
3622.0	49851	7	7	.53	C
3623.2	46982	13	13	.076	C
3631.5	35257	7	9	.52	C
3632.0	52297	3	5	.50	C
3638.3	49727	7	9	.27	C
3640.4	49461	9	11	.39	C
3645.8	52512	1	3	.58	C
3647.8	34782	9	11	.292	B
3649.5	49109	11	9	.43	C
3651.5	49628	7	9	.64	C
3655.5	50187	5	5	.12	C
3659.5	47106	9	9	.068	C
3669.5	49243	9	7	.30	C
3670.1	51023	11	13	.078	C
3676.3	47835	9	11	.061	C
3677.6	49433	7	5	.82	C
3682.2	55754	5	5	1.7	C
3684.1	49135	9	7	.34	C
3686.0	50833	9	11	.26	C
3687.5	34040	11	9	.0801	B
3690.7	55907	11	11	.28	C
3694.0	51570	5	7	.70	C
3697.4	51219	7	7	.21	C
3701.1	51192	7	9	.49	C
3704.5	48703	11	9	.14	C
3709.2	34329	9	7	.156	B
3719.9	26875	9	11	.163	B
3724.4	45221	5	7	.13	C
3727.6	34547	7	5	.225	B
3730.4	51374	9	11	.13	C
3732.4	44512	5	5	.28	C
3734.9	33695	11	11	.902	B
3737.1	27167	7	9	.142	C
3738.3	53094	11	13	.38	C
3740.2	52954	7	7	.19	C
3742.6	50423	9	9	.11	C
3744.1	51208	5	3	.38	C
3745.6	27395	5	7	.115	B
3749.5	34040	9	9	.764	B
3753.6	44184	7	5	.11	C
3756.9	55430	11	11	.25	C
3758.2	34329	7	7	.634	B
3760.1	45978	13	15	.057	C
3763.8	34547	5	5	.544	B
3765.5	52655	13	15	.99	C
3767.2	34692	3	3	.640	B
3787.2	55754	5	5	.12	C
3787.9	34547	3	5	.129	B
3794.3	46136	9	11	.046	C
3795.0	34329	5	7	.115	B
3798.5	33695	9	11	.0323	B
3799.5	34040	7	9	.0732	B
3804.0	53155	11	9	.052	C
3805.3	52899	9	11	1.0	C
3806.2	53808	3	3	.25	C
3806.7	52613	11	11	.55	C
3807.5	44184	3	5	.097	C
3810.8	52858	5	3	.24	C
3813.0	33947	7	5	.0792	B
3813.9	55526	13	11	.091	C
3815.8	38175	9	7	1.3	C
3820.4	33096	11	9	.668	B
3821.2	52514	11	13	.70	C
3821.8	47197	5	5	.089	C
3825.9	33507	9	7	.598	B
3827.8	38678	7	5	1.1	C
3833.3	46721	9	9	.059	C
3834.2	33802	7	5	.453	B
3836.3	52683	5	5	.39	C
3839.3	50614	9	9	.29	C
	Fe I				
3840.4	34017	5	3	.470	B
3841.0	38996	5	3	1.4	C
3843.3	50587	9	7	.48	C
3846.8	52213	7	7	.67	C
3850.0	34122	3	1	.606	B
3859.2	45295	13	11	.087	C
3859.9	25900	9	9	.0970	B
3865.5	34017	3	3	.155	B
3867.2	50187	5	5	.35	C
3871.8	49604	11	11	.070	C
3872.5	33802	5	5	.105	B
3873.8	45428	11	9	.082	C
3878.0	33507	7	7	.0772	B
3883.3	51969	7	7	.17	C
3884.4	47453	11	9	.042	C
3888.5	38678	5	5	.27	C
3891.9	53230	3	3	.40	C
3893.4	49461	11	11	.14	C
3900.5	51771	7	7	.086	C
3902.9	38175	7	7	.24	C
3903.9	49727	9	9	.097	C
3916.7	51630	13	11	.12	C
3919.1	49628	9	9	.045	C
3942.4	48305	3	5	.11	C
3951.2	51708	3	5	.36	C
3952.6	47008	11	11	.052	C
3953.2	49628	7	9	.043	C
3963.1	51705	3	5	.17	C
3967.4	51826	9	7	.24	C
3969.3	37163	9	7	.24	C
3971.3	46889	11	9	.068	C
3973.7	53763	5	7	.080	C
3977.7	42860	5	5	.082	C
3981.8	47106	9	9	.046	C
3984.0	47093	9	7	.089	C
3985.4	51708	5	5	.082	C
3997.0	54811	9	9	.074	C
3997.4	47008	9	11	.16	C
3998.1	46721	11	9	.075	C
4005.2	37521	7	5	.22	C
4014.5	53722	11	11	.24	C
4017.2	49461	9	11	.053	C
4021.9	47106	7	9	.10	C
4032.0	51201	3	5	.086	C
4045.8	36686	9	9	.75	C
4062.4	47556	3	3	.23	C
4063.6	37163	7	7	.69	C
4068.0	50475	9	9	.17	C
4070.8	50699	7	5	.14	C
4071.7	37521	5	5	.80	C
4073.8	50880	5	3	.19	C
4074.8	49109	9	9	.056	C
4076.6	50423	9	9	.20	C
4084.5	51350	11	9	.12	C
4085.3	50611	7	7	.12	C
4098.2	50534	7	7	.082	C
4107.5	47177	5	3	.25	C
4109.8	47272	3	3	.19	C
4113.0	58002	11	13	.15	C
4127.6	47272	1	3	.16	C
4132.9	47136	3	5	.11	C
4134.7	47017	5	7	.18	C
4137.0	51708	3	5	.23	C
4143.9	36686	7	9	.16	C
4149.4	50968	11	13	.043	C
4153.9	51462	7	9	.24	C
4154.8	51229	9	11	.15	C
4156.8	46889	5	5	.19	C
4170.9	48305	5	5	.072	C
4172.1	50187	7	5	.12	C
4175.6	46889	3	5	.17	C
4184.9	46727	5	5	.12	C
4187.0	43634	7	5	.23	C
4187.8	43435	9	7	.16	C
4196.2	51219	7	7	.11	C
4210.3	43764	3	3	.20	C
4217.6	51370	3	5	.24	C
4219.4	52514	11	13	.38	C
4222.2	43435	7	7	.063	C
4224.2	50833	9	11	.14	C
4225.5	51219	5	7	.17	C

Table 1b (Continued)
TRANSITION PROBABILITIES FOR SELECTED ATOMIC AND IONIC SPECIES

Wavelength (λ[Å])	Upper energy level (E_k[cm^{-1}])	Statistical weights g_i	g_k	Transition Probability (A[10^8 s^{-1}])	Uncertainty
	Fe I				
4233.6	43634	3	5	.20	C
4238.8	50980	7	9	.22	C
4246.1	52916	7	5	.069	C
4250.1	43435	5	7	.23	C
4282.4	40895	7	5	.13	C
4307.9	35768	7	9	.35	C
4315.1	40895	5	5	.090	C
4325.8	36079	5	7	.51	C
4327.1	51708	5	5	.094	C
4369.8	47453	9	9	.074	C
4383.5	34782	9	11	.46	C
4388.4	51837	7	7	.13	C
4401.3	51771	7	7	.069	C
4404.8	35257	7	9	.25	C
4415.1	35612	5	7	.13	C
4443.2	45552	1	3	.13	C
4466.6	45221	5	7	.13	C
4469.4	51837	5	7	.27	C
4528.6	39626	7	9	.063	C
4547.9	50587	5	7	.078	C
4736.8	47006	9	11	.050	C
4789.7	49477	5	5	.084	C
4859.7	43764	5	3	.15	C
4871.3	43634	7	5	.22	C
4872.1	43764	3	3	.24	C
4878.2	43764	1	3	.11	C
4890.8	43634	5	5	.21	C
4891.5	43435	9	7	.30	C
4903.3	43634	3	5	.054	C
4919.0	43435	7	7	.17	C
4920.5	43163	11	9	.36	C
4966.1	47006	11	11	.037	C
4973.1	52040	3	3	.12	C
4989.0	53546	7	7	.058	C
5001.9	51294	9	7	.40	C
5014.9	51740	7	5	.31	C
5022.2	52040	5	3	.27	C
5074.7	53739	9	11	.15	C
5090.8	53967	7	5	.21	C
5121.6	54067	5	5	.086	C
5137.4	53155	11	9	.11	C
5208.6	45334	7	5	.060	C
5232.9	42816	9	11	.15	C
5242.5	48383	13	11	.032	C
5263.3	45334	5	5	.061	C
5266.6	43163	7	9	.088	C
5281.8	43435	5	7	.038	C
5302.3	45334	3	5	.073	C
5324.2	44677	9	9	.15	C
5339.9	45061	5	7	.071	C
5367.5	54237	7	9	.59	C
5370.0	53874	9	11	.48	C
5383.4	53353	11	13	.59	C
5393.2	44677	7	9	.037	C
5410.9	54555	7	9	.49	C
5415.2	53841	11	13	.68	C
5463.3	54067	9	9	.33	C
5473.9	51771	7	7	.057	C
5569.6	45509	5	3	.21	C
5572.8	45334	7	5	.22	C
5586.8	45061	9	7	.19	C
5615.6	44677	11	9	.17	C
5624.5	45334	5	5	.062	C
5658.8	45061	7	7	.042	C
5753.1	51740	3	5	.072	C
5763.0	51294	5	7	.10	C
	Fe II				
2029.2	65110	10	8	.076	D
2040.7	64832	10	10	.46	D
2051.0	65110	8	8	.42	D
2296.7	65110	10	8	.037	D
2303.3	64832	12	10	.054	D
2369.2	64832	10	10	.026	D
2379.0	64832	8	10	.064	D
2388.4	62662	10	12	.14	D
2433.5	62662	10	12	.091	D
2555.0	65110	8	8	.019	D
2559.8	65110	6	8	.22	D
2561.6	64832	10	10	.0081	D
2573.2	64832	8	10	.11	D

Wavelength (λ[Å])	Upper energy level (E_k[cm^{-1}])	Statistical weights g_i	g_k	Transition Probability (A[10^8 s^{-1}])	Uncertainty
	Fe II				
2591.5	46967	6	6	.52	C
2592.8	71433	14	16	2.25	C
2598.0	64832	10	10	.020	D
2598.4	38859	8	6	1.3	C
2599.4	38459	10	10	2.22	C
2623.1	70987	14	14	.092	D
2625.5	70987	12	14	2.04	C
2625.7	38459	8	10	.34	C
2664.7	64832	8	10	1.50	C
2666.6	65110	6	8	1.62	C
2684.9	62662	12	12	.0043	D
2712.4	62662	10	12	.11	D
2753.3	62662	10	12	1.71	C
2879.2	65110	10	8	.029	D
2902.5	64832	10	10	.038	D
2910.8	65110	8	8	.0055	D
3002.3	65110	6	8	.018	D
3044.8	64832	8	10	.011	D
3131.7	64832	12	10	.012	D
3162.8	65110	8	8	.042	D
3186.7	45044	4	4	.039	C
3187.3	64832	10	10	.028	D
3213.3	44785	4	6	.065	C
3277.3	38459	8	10	.0023	D
3360.1	62662	12	12	.0084	D
4515.3	45080	6	6	.0018	D
4520.2	44754	10	8	.0010	D
4583.8	44447	10	8	.0063	C
4629.3	44233	10	10	.0013	D
4923.9	43621	6	4	.030	C
4954.0	65110	6	8	.0016	D
5018.4	43239	6	6	.026	C
5019.5	64832	8	10	.0015	D
	Fe XVI				
50.350	1986100	2	4	2120.	B
50.555	1978040	2	2	2100.	B
62.879	1867530	2	2	1110.	B
63.719	1867530	4	2	2140.	B
335.41	298140	2	4	77.5	B
360.80	277160	2	2	62.3	B
	Fe XVIII				
93.931	1064610	4	2	690.	D
103.95	1064610	2	2	260.	D
	Fe XXI				
98.37	1265800	5	3	700.	D
99.43	1095500	9	3*	1000.	D
113.34	1132280	5	5	470.	D
	Fe XXIII				
132.83	752840	1	3	195.	B
	Fe XXIV				
10.619	9417100	2	4	7.28(+4)	B
10.663	9378200	2	2	7.51(+4)	B
11.124	9467100	6	10*	2.18(+5)	B
192.04	520720	2	4	43.	B
255.10	392000	2	2	18.	B
	Krypton				
	Kr I				
4274.0	103363	5	5	.026	D
4351.4	108822	3	1	.032	D
4362.6	102887	5	3	.0084	D
4376.1	103762	3	1	.056	D
4400.0	108568	3	5	.020	D
4410.4	108514	3	3	.0044	D
4425.2	108438	3	3	.0097	D
4453.9	103363	3	5	.0078	D
4463.7	103314	3	3	.023	D
4502.4	103121	3	5	.0092	D
5562.2	97945	5	5	.0028	D
5570.3	97919	5	3	.021	D
5649.6	102887	1	3	.0037	D
5870.9	97945	3	5	.018	D
6904.7	105648	3	5	.013	D
7224.1	105007	3	5	.014	D
7587.4	94093	3	1	.51	D
7601.5	93123	5	5	.31	D
7685.2	98855	3	1	.49	D
7694.5	92964	5	3	.056	D

Table 1b (Continued)
TRANSITION PROBABILITIES FOR SELECTED ATOMIC AND IONIC SPECIES

Wavelength (λ[Å])	Upper energy level (E_k[cm^{-1}])	Statistical weights g_i	g_k	Transition Probability (A[10^8 s^{-1}])	Uncertainty
	Kr I				
7854.8	97919	1	3	.23	D
8059.5	97596	1	3	.19	D
8104.4	92307	5	5	.13	D
8112.9	92294	5	7	.36	D
8190.1	93123	3	5	.11	D
8263.2	97945	3	5	.35	D
8281.1	97919	3	3	.19	D
8298.1	92964	3	3	.32	D
8508.9	97596	3	3	.24	D
8776.7	92307	3	5	.27	D
8928.7	91169	5	3	.37	D
	Kr II				
4250.6	141996	4	4	.12	D
4292.9	138381	4	4	.96	D
4355.5	135783	6	8	1.0	D
4431.7	140163	2	2	1.8	D
4436.8	140137	2	4	.66	D
4577.2	149705	6	8	.96	D
4583.0	157885	6	4	.76	D
4615.3	140137	4	4	.54	D
4619.2	140119	4	6	.81	D
4633.9	149173	4	6	.71	D
4658.9	134288	6	4	.65	D
4739.0	133926	6	6	.76	D
4762.4	141996	2	4	.42	D
4765.7	136071	4	6	.67	D
4811.8	138381	2	4	.17	D
4825.2	141723	2	4	.19	D
4832.1	135783	4	2	.73	D
5208.3	134288	4	4	.14	D
5308.7	133926	4	6	.024	D
7407.0	133926	6	6	.070	D
	Lead				
	Pb I				
2022.0	49440	1	3	.052	D
2053.3	48687	1	3	.12	D
2170.0	46068	1	3	1.5	D
2401.9	49440	3	3	.19	D
2446.2	48687	3	3	.25	D
2476.4	48189	3	5	.28	D
2577.3	49440	5	3	.50	D
2613.7	46068	3	3	.27	D
2614.2	46061	3	5	1.9	D
2628.3	48687	5	3	.031	D
2657.1	45443	3	5	9.8 (–4)	D
2663.2	48189	5	5	.71	D
2802.0	46329	5	7	1.6	D
2823.2	46061	5	5	.26	D
2833.1	35287	1	3	.58	D
2873.3	45443	5	5	.37	D
3572.7	49440	5	3	.99	D
3639.6	35287	3	3	.34	D
3671.5	48687	5	3	.44	D
3683.5	34960	3	1	1.5	D
3739.9	48189	5	5	.73	D
4019.6	46329	5	7	.035	D
4057.8	35287	5	3	.89	D
4062.1	46068	5	3	.92	D
4168.0	45443	5	5	.012	D
5005.4	49440	1	3	.27	D
5201.4	48687	1	3	.19	D
7229.0	35287	5	3	.0089	D
	Lithium				
	Li I				
2741.2	36470	2	6*	.013	D
3232.7	30925	2	6*	.012	B
4602.8	36623	2	4	.197	C
4602.9	36623	4	6	.24	C
6103.5	31283	2	4	.60	C
6103.7	31283	4	6	.71	C
6103.7	31283	4	4	.12	C
6707.8	14904	2	4	.372	B
6707.9	14904	2	2	.372	B
	Lutetium				
	Lu I				
3376.5	29608	4	4	2.23	B
3567.8	28020	4	6	.59	C
3620.3	29608	6	4	.011	D
3841.2	28020	6	6	.25	C
4518.6	22125	4	4	.21	B
	Magnesium				
	Mg I				
2025.8	49347	1	3	.84	D
2779.8	57854	9	9*	5.2	C
2850.0	56968	9	15*	.23	C
2852.1	35051	1	3	4.95	B
3094.9	54192	9	15*	.52	C
3329.9	51873	1	3	.033	C
3332.2	51873	3	3	.097	C
3336.7	51873	5	3	.16	C
3835.3	47957	9	15*	1.68	B
4703.0	56308	3	5	.255	C
5167.3	41197	1	3	.116	B
5172.7	41197	3	3	.346	B
5183.6	41197	5	3	.575	B
5528.4	53135	3	5	.199	C
	Mg II				
1239.9	80650	2	4	.014	C
1240.4	80620	2	2	.014	C
2660.8	109062	10	14*	.38	D
2790.8	71491	2	4	4.0	C
2795.5	35761	2	4	2.6	C
2797.9	71491	4	4	.79	D
2798.1	71490	4	6	4.8	C
2802.7	35669	2	2	2.6	C
2928.8	69805	2	2	1.2	C
2936.5	69805	4	2	2.3	C
3104.8	103690	10	14*	.81	C
3848.2	97469	6	4	.028	C
3848.3	97469	4	4	.0030	D
3850.4	97455	4	2	.030	C
4481.2	93800	10	14*	2.23	C
9218.3	80650	2	4	.36	C
9244.3	80620	2	2	.36	C
	Mg IV				
320.99	311532	4	2	120.	D
323.31	311532	2	2	59.	D
1219.0	679098	6	6	5.9	D
1375.5	677361	4	4	4.5	D
1459.6	612231	6	4	4.6	D
1495.5	679098	4	6	6.4	D
1510.7	678426	4	4	6.7	D
1683.0	603137	6	8	5.8	D
1698.8	604002	4	6	3.9	D
1893.9	596521	6	6	2.8	D
	Mg VI				
269.92	424600	10	6*	310.	D
292.53	424600	6	6*	90.	D
314.64	400600	6	2*	180.	D
349.15	340600	10	10*	61.	C
387.94	340600	6	10*	13.	D
399.29	250445	4	2	28.	C
400.68	249578	4	4	28.	C
403.32	247945	4	6	27.	C
	Mg VII				
277.01	362128	3	3	95.	C
278.41	362128	5	3	150.	C
280.74	397700	5	3	200.	D
319.02	354900	5	5	89.	C
366.42	274922	9	9*	44.	C
433.04	232934	9	15*	16.	C
1334.3	1125850	5	5	5.3	C
1410.0	1196770	5	5	2.57	C
1487.0	1192185	3	5	3.02	C
1487.9	1193061	5	7	3.66	C
	Mg VIII				
74.976	1335965	6	10*	4300.	D
315.02	320742	4	4	120.	C
342.29	524437	10	6*	63.	D
353.86	414380	4	4	38.9	C
356.00	414380	6	4	57.	C
428.52	465654	10	10*	32.4	C
434.62	232290	6	10*	16.	C
489.33	524437	6	6*	39.	D

TRANSITION PROBABILITIES FOR SELECTED ATOMIC AND IONIC SPECIES

Wavelength (λ[Å])	Upper energy level (E_k[cm^{-1}])	Statistical weights g_i	g_k	Transition Probability (A[10^8 s^{-1}])	Uncertainty
	Mg VIII				
686.92	465654	6	10*	9.4	D
	Mg IX				
62.751	1593600	1	3	2870.	B
67.189	1631500	9	15*	6200.	C
71.965	1532700	9	3*	1220.	C
72.312	1654583	3	5	4430.	C
77.737	1558076	3	1	392.	C
368.07	271687	1	3	52.7	B
438.69	499640	3	1	79.	C
443.74	368500	9	9*	41.9	C
749.55	405100	3	5	8.2	C
1639.8	1654583	3	5	2.1	D
2814.2	1593600	1	3	.335	C
	Mg X				
57.876	1727832	2	4	2090.	B
57.920	1726519	2	2	2090.	B
63.152	1743410	2	4	5600.	B
63.295	1743880	4	6	6700.	B
609.79	163976	2	4	7.53	B
624.94	159929	2	2	7.01	B
2212.5	1727832	2	4	.964	B
2278.7	1726519	2	2	.882	B
5918.7	1743410	2	4	.0320	C
6229.6	1743880	4	6	.0330	C
	Mg XI				
7.310	13680600	1	3	1.15(+4)	B
7.473	13381100	1	3	2.27(+4)	B
7.850	12738400	1	3	5.50(+4)	B
9.169	10907300	1	3	1.97(+5)	B
	Manganese				
	Mn I				
2794.8	35770	6	8	3.7	C
2798.3	35726	6	6	3.6	C
2801.1	35690	6	4	3.7	C
3007.7	58520	6	8	.18	D
3011.4	58486	8	10	.31	D
3016.5	58427	10	12	.29	D
3043.4	58137	8	8	.59	D
3044.6	49888	10	8	.57	D
3045.6	58110	10	10	.67	D
3045.8	58110	8	10	.17	D
3046.6	67753	10	12	.13	D
3047.0	58075	12	12	.61	D
3054.4	50013	8	6	.46	D
3066.0	49888	8	8	.16	D
3073.1	50099	4	4	.37	D
3082.7	66569	14	14	.29	D
3110.7	59340	6	8	.27	D
3113.8	66356	12	10	.26	D
3122.9	66356	10	10	.19	D
3126.9	66395	8	6	.23	D
3132.3	66855	10	10	.21	D
3132.8	66334	8	8	.27	D
3160.2	66574	10	12	.14	D
3175.6	66523	8	10	.18	D
3175.7	66419	10	12	.12	D
3190.0	66454	6	8	.16	D
3201.1	66395	4	6	.22	D
3212.9	48168	10	10	.16	D
3228.1	48021	10	12	.64	D
3230.2	65887	10	12	.19	D
3230.7	48226	8	8	.35	D
3238.7	65909	8	10	.12	D
3243.8	48271	6	6	.53	D
3256.1	48271	4	6	.50	D
3258.4	48318	2	2	.97	D
3260.2	48301	2	4	.38	D
3264.7	47904	8	10	.14	D
3267.8	64732	14	14	.35	D
3268.7	64410	6	8	.33	D
3270.4	64820	12	12	.26	D
3273.0	64888	10	10	.27	D
3278.6	47775	8	8	.0091	D
3298.2	57512	6	4	.28	D
3420.8	63364	14	14	.12	D
3463.7	66600	8	8	.32	C
3470.0	66600	6	8	.24	C
3511.8	65887	12	12	.27	C
	Mn I				
3535.3	65909	10	10	.17	C
3559.8	65873	6	6	.21	C
3577.9	44994	10	8	.94	C
3601.3	65769	12	10	.23	C
3607.5	44994	8	8	.23	C
3608.5	45156	6	6	.36	C
3610.3	45259	4	4	.42	C
3635.7	65617	10	8	.21	C
3660.4	64732	12	14	.91	C
3677.0	64820	10	12	.73	C
3680.2	64585	12	10	.19	C
3682.1	64888	8	10	.76	D
3684.9	64920	6	8	.26	C
3706.1	61226	12	14	1.4	C
3718.9	61226	10	12	.96	C
3729.5	61744	10	12	.066	D
3731.9	61211	8	10	1.0	C
3746.6	60934	12	12	.16	C
3756.6	60956	10	10	.14	C
3767.7	60957	8	8	.14	C
3768.2	61469	10	12	.071	C
3771.4	69561	14	14	.19	C
3773.9	69630	12	12	.25	C
3800.6	57306	6	8	.27	C
3801.9	51561	12	12	.064	C
3806.7	43314	10	12	.38	C
3809.6	43524	8	8	.20	C
3823.5	43429	8	10	.44	C
3823.9	43596	6	6	.36	C
3833.9	43644	4	4	.52	C
3834.4	43524	6	8	.52	D
3839.8	43673	2	2	.58	C
3844.0	43644	2	4	.29	C
3872.1	63449	10	12	.077	C
3873.2	63548	8	10	.11	C
3889.5	68843	12	14	.31	C
3898.4	59470	6	8	.17	C
3899.3	60102	4	6	.24	C
3911.1	56562	6	6	.13	D
3919.3	66738	8	8	.088	D
3923.3	59732	12	10	.13	C
3924.1	56602	2	4	.94	D
3926.5	56462	6	8	.54	C
3929.7	59784	10	8	.092	C
3931.5	63548	10	10	.082	C
3936.8	59818	8	6	.12	C
3952.8	59117	6	6	.41	D
3975.9	59990	2	4	.18	D
3977.1	59600	4	6	.16	D
3980.1	66149	10	8	.13	D
3982.2	59568	4	2	.35	D
3982.6	50383	6	4	.23	D
3982.9	66504	6	4	.55	D
3986.8	50341	12	10	.11	D
3987.1	50359	10	8	.10	D
4003.3	62393	12	10	.11	D
4011.9	66149	8	8	.23	D
4018.1	41933	10	8	.33	C
4026.4	50095	12	14	.089	D
4030.8	24802	6	8	.19	C
4031.8	50081	10	12	.073	D
4033.1	24788	6	6	.18	C
4034.5	24779	6	4	.18	C
4041.4	41789	10	10	1.0	C
4048.8	42144	6	4	.75	C
4052.5	59784	6	8	.38	D
4055.5	41933	8	8	.61	C
4058.9	42199	4	2	1.0	C
4061.7	49415	8	6	.19	D
4063.5	42054	6	6	.22	C
4065.1	58843	12	14	.25	D
4066.2	65617	10	8	.22	D
4079.4	42144	2	4	.38	C
4082.9	42054	4	6	.37	C
4083.6	41933	6	8	.28	C
4089.9	58867	8	10	.17	D
4092.4	59470	8	8	.14	D
4099.4	65617	8	8	.11	D
4105.4	59290	10	8	.17	D
4107.9	61485	12	10	.097	D
4114.4	59340	8	8	.15	D

Table 1b (Continued)
TRANSITION PROBABILITIES FOR SELECTED ATOMIC AND IONIC SPECIES

Wavelength (λ[Å])	Upper energy level (E_k[cm^{-1}])	Statistical weights g_i	g_k	Transition Probability (A[10^8 s^{-1}])	Uncertainty
	Mn I				
4116.6	62075	6	6	.12	D
4125.8	65262	10	10	.070	D
4132.3	68716	8	10	.15	D
4135.0	58427	12	12	.30	D
4141.1	58486	10	10	.26	D
4147.5	51305	6	6	.066	D
4148.8	58520	8	8	.23	D
4176.6	58075	14	12	.21	D
4182.3	61913	12	14	.092	D
4190.0	58110	12	10	.20	D
4201.8	58137	10	8	.23	D
4220.6	57512	6	4	.16	D
4239.7	47299	4	2	.39	D
4257.7	47299	2	2	.37	D
4261.3	61469	12	12	.081	D
4265.9	47155	4	4	.35	D
4278.7	61485	10	10	.068	D
4281.1	46901	6	6	.23	D
4300.2	60668	12	10	.087	D
4381.7	61485	8	10	.14	D
4411.9	60668	12	10	.26	D
4414.9	45941	8	6	.18	D
4419.8	60739	10	8	.21	D
4451.6	45754	8	8	.71	D
4452.5	59617	14	14	.059	D
4455.8	47216	4	6	.17	D
4457.0	47218	6	4	.20	D
4457.6	47216	6	6	.38	D
4458.3	47212	6	8	.28	D
4461.1	47212	8	8	.17	D
4462.0	47207	8	10	.43	D
4464.7	45941	6	6	.26	D
4479.4	63548	8	10	.34	D
4498.9	45941	4	6	.11	D
4502.2	45754	6	8	.078	D
4503.9	59617	12	14	.083	D
4605.4	59828	10	12	.36	D
4626.5	59617	12	14	.36	D
4709.7	44523	8	8	.077	D
4727.5	44696	6	6	.084	D
4754.0	39431	6	8	.38	D
4761.5	44815	2	4	.28	D
4762.4	44289	8	10	.57	D
4765.9	44696	4	6	.28	D
4766.4	44523	6	8	.45	D
4783.4	39431	8	8	.39	D
4823.5	39431	10	8	.45	D
	Mn II				
2933.1	43557	5	3	1.7	C
2939.3	43485	5	5	1.8	C
2949.2	43370	5	7	1.7	C
3439.0	38543	5	7	.0041	D
3442.0	43370	9	7	.43	C
3460.3	43485	7	5	.32	C
3474.0	43370	7	7	.079	C
3474.1	43557	5	3	.15	C
3482.9	43485	5	5	.20	C
3488.7	43557	3	3	.25	C
3495.8	43557	1	3	.11	C
3496.8	43370	5	7	.016	C
3497.5	43485	3	5	.051	C
	Mn XXIII				
12.03	8687000	2	4	1.5 (+5)	B
12.158	8708000	4	6	1.8 (+5)	B

Mercury

Wavelength (λ[Å])	Upper energy level (E_k[cm^{-1}])	Statistical weights g_i	g_k	Transition Probability (A[10^8 s^{-1}])	Uncertainty
	Hg I				
2536.5	39412	1	3	.13	D
2752.8	73961	1	3	.057	D
2856.9	74405	3	1	.012	D
2893.6	73961	3	3	.16	D
2925.4	78216	5	3	.077	D
2967.3	71336	1	3	.45	D
3125.7	71396	3	5	.51	D
3341.5	73961	5	3	.27	D
4046.6	62350	1	3	.18	D
4077.8	63928	3	1	.041	D
4108.1	78404	3	1	.030	D
4339.2	77108	3	5	.080	D
4358.3	62350	3	3	.40	D
4916.1	74405	3	1	.13	D
5460.7	62350	5	3	.56	D
5769.6	71396	3	5	.61	D
6234.4	79964	1	3	.0053	D
6716.4	78813	1	3	.0043	D
6907.5	76824	3	5	.028	D
7728.8	76863	1	3	.0097	D

Neodymium

Wavelength (λ[Å])	Upper energy level (E_k[cm^{-1}])	Statistical weights g_i	g_k	Transition Probability (A[10^8 s^{-1}])	Uncertainty
	Nd II				
3780.4	30247	16	18	.14	D
3805.4	28857	14	16	.69	D
3807.2	26772	10	12	.049	D
3863.3	25877	8	10	.15	D
3941.5	25877	10	10	.61	D
3951.2	26772	12	12	.60	D
3973.3	30247	18	18	.63	D
3979.5	26772	10	12	.27	D
3990.1	28857	16	16	.52	D
4012.3	30002	18	20	.55	D
4061.1	28419	16	18	.44	D
4106.6	28857	14	16	.068	D
4109.5	26913	14	16	.37	D
4133.4	26772	14	12	.15	D
4156.1	25524	12	14	.34	D
4205.6	28857	18	16	.18	D
4284.5	28419	18	18	.085	D
4303.6	23230	8	10	.47	D
4325.8	26913	16	16	.16	D
4358.2	25524	14	14	.15	D
4382.7	25877	12	10	.040	D
4400.8	23230	10	10	.068	D
4451.6	25524	12	14	.25	D
4456.4	28419	16	18	.064	D
4463.0	26913	14	16	.18	D
4958.1	23230	12	10	.012	D
5130.6	30002	22	20	.16	D
5192.6	28419	20	18	.17	D
5249.6	26913	18	16	.18	D
5276.9	25877	12	10	.12	D
5293.2	25524	16	14	.12	D
5302.3	30247	20	18	.11	D
5311.5	26772	14	12	.11	D
5319.8	23230	12	10	.16	D
5357.0	28857	18	16	.18	D
5371.9	30002	20	20	.051	D
5485.7	28419	18	18	.057	D
5594.4	26913	16	16	.070	D
5620.6	30247	18	18	.13	D
5688.5	25524	14	14	.059	D
5718.1	28857	16	16	.087	D
5726.8	25877	10	10	.056	D
5740.9	26772	12	12	.072	D
5804.0	23230	10	10	.046	D
5865.1	28419	16	18	.013	D
6051.9	25877	12	10	.011	D

Neon

Wavelength (λ[Å])	Upper energy level (E_k[cm^{-1}])	Statistical weights g_i	g_k	Transition Probability (A[10^8 s^{-1}])	Uncertainty
	Ne I				
615.63	162436	1	3	.38	C
618.67	161637	1	3	.93	C
619.10	161524	1	3	.33	C
626.82	159535	1	3	.74	C
629.74	158796	1	3	.48	C
735.90	135889	1	3	6.11	B
743.72	134459	1	3	.486	B
3369.8	163709	5	5	.0010	D
3369.9	163708	5	3	.0076	D
3375.6	163657	5	3	.0022	D
3417.9	163709	3	5	.0092	D
3418.0	163708	3	3	.0022	D
3423.9	163657	3	3	.0010	D
3447.7	163038	5	5	.021	D
3450.8	163013	5	3	.0049	D
3454.2	163401	3	1	.037	D
3460.5	163708	1	3	.0070	D
3464.3	162899	5	5	.0067	D
3466.6	163657	1	3	.013	D
3472.6	162831	5	7	.017	D
3498.1	163038	3	5	.0051	D
3501.2	163013	3	3	.012	D

Table 1b (Continued)
TRANSITION PROBABILITIES FOR SELECTED ATOMIC AND IONIC SPECIES

Wavelength (λ[Å])	Upper energy level (E_k[cm^{-1}])	Statistical weights g_i	g_k	Transition Probability (A[10^8 s^{-1}])	Uncertainty
		Ne I			
3510.7	162518	5	3	.0022	D
3515.2	162899	3	5	.0069	D
3520.5	164286	3	1	.093	D
3593.5	163709	3	5	.0099	D
3593.6	163708	3	3	.0066	D
3600.2	163657	3	3	.0043	D
3633.7	163401	3	1	.011	D
3682.2	163038	3	5	.0016	D
3685.7	163013	3	3	.0039	D
3701.2	162899	3	5	.0022	D
4536.3	170296	3	3	.0050	D
4702.5	169517	3	3	.0021	D
4708.9	169488	3	3	.042	D
4955.4	170296	3	3	.0033	D
5113.7	167808	3	3	.010	D
5120.5	170296	3	3	.0056	D
5154.4	169517	3	3	.019	D
5191.3	170296	3	3	.013	D
5326.4	167027	3	3	.0068	D
5333.3	169517	3	3	.0053	D
5341.1	166975	3	3	.11	D
5400.6	152971	3	1	.0090	B
5418.6	169488	3	3	.0052	D
5433.7	166657	3	3	.00283	B
5652.6	167808	3	3	.0089	D
5662.5	165913	3	3	.0069	D
5852.5	152971	3	1	.682	B
5868.4	167808	3	3	.014	D
5881.9	151038	5	3	.115	B
5913.6	167027	3	3	.048	D
5939.3	166657	5	3	.00200	B
5944.8	150859	5	5	.113	B
5961.6	167808	3	3	.033	D
5975.5	150772	5	3	.0351	B
6030.0	151038	3	3	.0561	B
6046.1	166657	3	3	.00226	B
6074.3	150917	3	1	.603	B
6096.2	150859	3	5	.181	B
6118.0	166657	5	3	.00609	B
6128.5	150772	3	3	.0067	B
6143.1	150316	5	5	.282	B
6150.3	167027	3	3	.015	D
6163.6	151038	1	3	.146	B
6217.3	150122	5	3	.0637	B
6266.5	150772	1	3	.249	B
6273.0	166975	3	3	.0097	D
6293.7	166657	3	3	.00639	B
6304.8	150316	3	5	.0416	B
6328.2	166657	5	3	.0339	B
6330.9	165913	3	3	.023	D
6334.4	149824	5	5	.161	B
6351.9	166657	1	3	.00345	B
6383.0	150122	3	3	.321	B
6401.1	166657	3	3	.0139	B
6402.2	149657	5	7	.514	B
6506.5	149824	3	5	.300	B
6532.9	150122	1	3	.108	B
6599.0	151038	3	3	.232	B
6602.9	165913	3	3	.0059	D
6652.1	150917	3	1	.0029	B
6678.3	150859	3	5	.233	B
6717.0	150772	3	3	.217	B
6721.1	165913	3	3	4.9 (−4)	D
6929.5	150316	3	5	.174	B
7024.1	150122	3	3	.0189	B
7032.4	148258	5	3	.253	B
7051.3	162436	3	3	.030	D
7059.1	162420	3	5	.068	C
7173.9	149824	3	5	.0287	B
7245.2	148258	3	3	.0935	B
7304.8	166657	1	3	.00255	B
7438.9	148258	1	3	.0231	B
7472.4	161637	3	3	.040	D
7535.8	161524	3	3	.43	D
7937.0	162420	5	5	.0078	C
8082.5	148258	3	3	.0012	B
8118.5	162436	3	3	.049	D
8128.9	162420	3	5	.0072	C
8259.4	162420	5	5	.0203	C
8571.4	162436	3	3	.055	D
8582.9	162420	3	5	.0100	C
		Ne I			
8647.0	162420	5	5	.0391	C
8681.9	161637	3	3	.21	D
8767.5	161524	3	3	.0011	D
8771.7	162436	3	3	.16	D
8783.8	162420	3	5	.313	C
8865.3	159535	3	3	.0094	D
9201.8	161637	3	3	.091	D
9433.0	161637	3	3	.0011	D
9486.7	158796	3	3	.025	D
9534.2	161524	3	3	.063	D
10621.	159535	3	3	.0024	D
11409.	159535	3	3	.042	D
11525.	158796	3	3	.084	D
11767.	159535	3	3	.069	D
12459.	158796	3	3	.015	D
		Ne II			
357.03	280351	6	10*	38.	C
361.77	276677	6	2*	16.	C
406.28	246395	6	10*	18.	C
446.37	224291	6	6*	40.7	C
460.73	217048	4	2	47.	B
462.39	217048	2	2	23.	B
1907.5	276512	4	2	.28	D
1916.1	276277	4	4	.69	D
1930.0	276512	2	2	.57	D
1938.8	276277	2	4	.13	D
2858.0	281171	6	6	.79	D
2870.0	281026	6	6	.17	D
2873.0	280990	6	4	.38	D
2876.3	281171	4	6	.78	D
2876.5	280947	6	4	.33	D
2878.1	281333	2	2	.069	D
2888.4	281026	4	6	.070	D
2891.5	280990	4	4	.061	D
2897.0	280701	6	8	.052	D
2906.8	280990	2	4	.55	D
2910.1	280769	4	2	1.7	D
2910.4	280947	2	4	.59	D
2916.2	280474	6	4	.096	D
2925.6	280769	2	2	.56	D
2933.7	280269	6	6	.069	D
2955.7	252954	6	4	1.2	D
3001.7	252954	4	4	.87	D
3017.3	279325	6	4	.35	D
3027.0	279219	6	6	1.4	D
3028.7	279423	4	2	.85	D
3028.9	252954	2	4	.47	D
3034.5	279138	6	8	3.1	D
3037.7	279325	4	4	2.1	D
3045.6	279423	2	2	2.5	D
3047.6	279219	4	6	1.8	D
3054.7	279325	2	4	.94	D
3092.9	306687	6	6	1.3	D
3097.1	306688	8	8	1.3	D
3118.0	281171	8	6	.042	D
3134.1	306263	6	4	.26	D
3140.4	306243	8	6	.24	D
3151.1	281171	6	6	.048	D
3154.8	280797	8	6	.018	D
3164.4	280701	8	8	.16	D
3165.7	281026	6	6	.12	D
3173.6	280947	6	4	.045	D
3176.1	281171	4	6	.060	D
3187.6	251011	4	6	.014	D
3188.7	280797	6	6	.39	D
3190.9	281026	4	6	.15	D
3194.6	280990	4	4	.52	D
3198.6	280701	6	8	1.7	D
3198.9	280947	4	4	.23	D
3209.0	280262	8	8	.16	D
3209.4	280990	2	4	.60	D
3213.7	280947	2	4	1.7	D
3214.3	280797	4	6	2.2	D
3218.2	280173	8	10	3.6	D
3224.8	305365	6	8	3.5	D
3229.5	305365	8	8	.13	D
3229.6	305364	8	10	3.6	D
3230.1	277344	6	6	1.8	D
3230.4	277344	4	6	.14	D
3232.0	277326	6	4	.27	D

Table 1b (Continued)
TRANSITION PROBABILITIES FOR SELECTED ATOMIC AND IONIC SPECIES

Wavelength (λ[Å])	Upper energy level (E_k[cm^{-1}])	Statistical weights g_i	g_k	Transition Probability (A[10^8 s^{-1}])	Uncertainty
	Ne II				
3232.4	277326	4	4	1.6	D
3243.4	280269	6	6	.23	D
3244.1	280262	6	8	1.5	D
3248.1	280474	4	4	.24	D
3255.4	281720	6	4	.038	D
3263.4	280474	2	4	.39	D
3269.9	280269	4	6	.51	D
3270.8	249696	6	4	.057	D
3297.7	249446	6	6	.43	D
3309.7	254292	4	2	.31	D
3310.5	281720	4	4	.069	D
3311.3	249840	4	2	.26	D
3314.7	281171	6	6	.044	D
3319.7	276512	4	2	1.6	D
3320.2	279219	8	6	.21	D
3323.7	254165	4	4	1.6	D
3327.2	249696	4	4	.91	D
3329.2	279138	8	8	.88	D
3330.7	281026	6	6	.039	D
3334.8	249109	6	8	1.8	D
3336.1	306243	4	6	1.1	D
3344.4	249840	2	2	1.5	D
3345.5	276277	6	4	1.4	D
3345.8	276277	4	4	.22	D
3353.6	281333	4	2	.12	D
3355.0	249446	4	6	1.3	D
3356.3	280797	6	6	.20	D
3357.8	279219	6	6	.50	D
3360.3	306263	2	4	.86	D
3360.6	249696	2	4	.82	D
3362.9	279423	4	2	.35	D
3371.8	281171	4	6	.22	D
3374.1	279325	4	4	.30	D
3378.2	254292	2	2	1.7	D
3379.3	279423	2	2	.30	D
3386.2	279219	4	6	.055	D
3388.4	281026	4	6	2.2	D
3390.6	279325	2	4	.077	D
3392.8	254165	2	4	.44	D
3404.8	306687	4	6	1.9	D
3407.0	306688	6	8	2.3	D
3411.4	305582	4	2	.61	D
3413.2	305567	4	4	1.8	D
3414.9	280797	4	6	.018	D
3416.9	280269	6	6	.64	D
3417.7	280262	6	8	1.6	D
3438.9	305582	2	2	1.4	D
3440.7	305567	2	4	.35	D
3453.1	280474	4	4	.46	D
3454.8	306263	4	4	1.6	D
3456.6	281720	2	4	.96	D
3457.1	306243	4	6	.099	D
3459.3	306243	6	6	1.6	D
3475.2	281720	4	4	.012	D
3477.6	280269	4	6	.43	D
3481.9	252798	4	2	1.4	D
3503.6	281333	2	2	2.0	D
3522.7	281333	4	2	.023	D
3538.0	305582	4	2	.76	D
3539.9	305567	4	4	.036	D
3542.2	305567	6	4	.60	D
3542.9	281171	4	6	1.2	D
3546.2	280990	2	4	.063	D
3551.6	280947	2	4	.037	D
3557.8	252798	2	2	.19	D
3561.2	281026	4	6	.21	D
3565.8	280990	4	4	.62	D
3568.5	274409	6	8	1.4	D
3571.2	280947	4	4	.63	D
3574.2	274365	6	6	.10	D
3574.6	274365	4	6	1.3	D
3590.4	280797	4	6	.036	D
3594.2	280769	4	2	1.3	D
3612.3	280474	2	4	.26	D
3628.0	281720	4	4	.60	D
3632.7	280474	4	4	.13	D
3643.9	251522	4	4	.32	D
3644.9	281720	2	4	.99	D
3659.9	280269	4	6	.067	D
3664.1	246415	6	4	.70	D
3679.8	281333	4	2	.32	D
3694.2	246192	6	6	1.0	D
3697.1	281333	2	2	.28	D
3701.8	281171	4	6	.27	D
3709.6	246598	4	2	1.1	D
3713.1	251011	4	6	1.3	D
3721.8	281026	4	6	.20	D
3726.9	280990	4	4	.12	D
3727.1	251522	2	4	.98	D
3734.9	246415	4	4	.19	D
3744.6	280990	2	4	.26	D
3751.2	246598	2	2	.18	D
3753.8	280797	4	6	.45	D
3766.3	246192	4	6	.29	D
3777.1	246415	2	4	.42	D
3800.0	280474	4	4	.37	D
3818.4	280474	2	4	.61	D
3829.8	280269	4	6	.84	D
3942.3	249446	4	6	.010	D
	Ne V				
142.61	701945	9	9*	670.	C
143.32	698517	9	15*	1200.	C
147.13	709956	5	7	1500.	C
151.23	691540	5	5	338.	C
154.50	711210	1	3	700.	C
167.69	597083	9	9*	150.	C
358.93	279365	9	3*	210.	C
365.59	303812	5	3	135.	C
482.15	208161	9	9*	30.1	C
571.04	175876	9	15*	10.	C
2259.6	640868	3	5	1.9	D
2265.7	641646	5	7	2.4	D
	Ne VII				
97.502	1025600	1	3	1070.	B
115.46	978320	9	3*	480.	C
116.69	1071920	3	5	1600.	C
127.66	998280	3	1	190.	C
465.22	214952	1	3	40.9	B
558.61	290740	3	5	8.11	B
559.95	289850	1	3	10.7	B
561.38	289850	3	3	7.99	B
561.73	290740	5	5	23.9	B
562.99	289340	3	1	31.7	B
564.53	289850	5	3	13.1	B
	Ne VIII				
88.09	1135000	2	6*	840.	B
98.208	1147500	6	10*	2770.	B
770.41	129800	2	4	5.90	B
780.32	128150	2	2	5.69	B
2820.7	1135000	2	4	.720	B
2860.1	1135000	2	2	.688	B
	Nickel				
	Ni I				
1976.9	50790	7	9	1.1	C
1990.3	51125	5	7	.83	C
2007.0	50689	5	5	.17	C
2014.3	51344	3	5	.93	C
2026.6	49328	9	7	.24	C
2047.4	49033	7	7	.13	C
2055.5	50851	5	3	.33	C
2124.8	50458	5	3	.38	C
2147.8	47425	5	3	.47	C
2158.3	46523	7	5	.69	C
2190.2	46523	5	5	.30	C
2197.4	47208	3	3	.78	C
2201.6	48818	5	3	.73	C
2244.5	45419	5	5	.38	C
2253.6	44565	7	7	.19	C
2258.2	44475	7	5	.17	C
2290.0	43655	9	7	2.1	C
2300.8	43655	7	7	.75	C
2303.0	45122	3	3	.45	C
2312.3	44565	7	7	5.5	C
2317.2	44475	7	5	3.8	C
2320.0	43090	9	11	6.9	C
2325.8	44315	7	9	3.5	C
2330.0	45122	5	3	5.3	C
2345.5	42621	9	7	2.2	C
2943.9	34163	7	5	.11	C

Table 1b (Continued)
TRANSITION PROBABILITIES FOR SELECTED ATOMIC AND IONIC SPECIES

Wavelength (λ[Å])	Upper energy level (E_k[cm^{-1}])	Statistical weights g_i	g_k	Transition Probability (A[10^8 s^{-1}])	Uncertainty
	Ni I				
3012.0	36601	5	5	1.5	C
3037.9	33112	7	7	.32	C
3064.6	33501	5	7	.075	C
3080.8	34163	3	5	.093	C
3101.6	33112	5	7	.72	C
3101.9	35639	5	7	.49	C
3134.1	33611	3	5	.71	C
3225.0	34409	5	3	.11	C
3233.0	30923	9	11	.053	C
3271.1	31442	5	5	.0072	C
3315.7	31031	5	7	.053	C
3320.3	31442	7	5	.048	C
3361.6	30619	5	5	.045	C
3369.6	29669	9	7	.17	C
3380.6	32982	5	3	1.2	C
3393.0	29669	7	7	.24	C
3413.5	30619	7	5	.038	C
3414.8	29481	7	9	.55	C
3423.7	30913	3	3	.35	C
3433.6	29321	7	7	.17	C
3446.3	29889	5	5	.44	C
3452.9	29833	5	7	.098	C
3458.5	30619	3	5	.61	C
3461.7	29084	7	9	.27	C
3472.5	29669	5	7	.12	C
3483.8	30913	5	3	.14	C
3493.0	29501	5	3	.98	C
3510.3	30192	3	1	1.2	C
3515.1	29321	5	7	.44	C
3519.8	30619	5	5	.041	C
3524.5	28569	7	5	1.0	C
3566.4	31442	5	5	.56	C
3571.9	29321	7	7	.052	C
3597.7	29501	3	3	.14	C
3610.5	28569	5	5	.072	C
3619.4	31031	5	7	.73	C
3664.1	29501	5	3	.019	D
3807.1	29669	5	7	.043	C
3831.7	29501	5	3	.015	C
3858.3	29321	5	7	.069	C
4295.9	54251	9	7	.17	D
4401.6	48467	9	11	.38	D
4714.4	48467	13	11	.46	D
4786.5	48467	11	11	.18	D
4817.9	54251	7	7	.070	D
4838.7	54251	9	7	.22	D
4855.4	49159	5	5	.57	D
4980.2	49159	9	11	.19	D
5017.6	48467	11	11	.20	D
5080.5	49159	9	11	.32	D
5129.4	49159	7	5	.12	D
5371.3	54251	7	7	.16	D
5664.0	54251	5	7	.11	D
6176.8	49159	9	11	.047	D
6314.7	31442	5	5	.0057	D
6767.8	29501	1	3	.0033	D
7714.3	28569	5	5	.0014	D
	Ni II				
2053.3	57081	10	8	.041	D
2080.8	58706	4	4	.13	D
2090.1	58493	4	6	.11	D
2093.6	57081	8	8	.11	D
2125.1	56371	8	8	.10	D
2125.9	55418	10	8	.073	D
2128.6	57081	6	8	.40	D
2138.6	56075	8	6	.28	D
2158.7	56424	6	4	.57	D
2161.2	56371	6	8	.32	D
2165.6	54557	10	10	3.8	D
2169.1	55418	8	8	2.3	D
2174.7	55300	8	10	2.4	D
2175.2	56075	6	6	2.8	D
2184.6	56424	4	4	4.7	D
2188.1	55019	8	6	.090	D
2201.4	56075	4	6	2.1	D
2206.7	55418	6	8	2.5	D
2210.4	54557	8	10	.64	D
2216.5	53496	10	12	5.5	D
2220.4	68131	6	8	3.7	D
2223.0	53365	10	10	1.6	D
2224.4	58493	8	6	.51	D
2224.9	54263	8	8	2.5	D
2226.3	55019	6	6	2.0	D
2253.9	55019	4	6	3.2	D
2264.5	54263	6	8	2.4	D
2270.2	53365	8	10	2.5	D
2278.8	57420	8	6	4.5	D
2287.1	58706	6	4	4.5	D
2296.6	57081	8	8	3.2	D
2297.1	53635	6	4	4.6	D
2297.5	54176	4	2	5.3	D
2298.3	58493	6	6	4.5	D
2303.0	52738	8	6	4.7	D
2316.0	51558	10	8	4.9	D
2326.4	53635	4	4	1.0	D
2334.6	56371	8	8	1.3	D
2356.4	57420	6	6	.45	D
2367.4	51558	8	8	.13	D
2375.4	57081	6	8	1.1	D
2387.8	55418	8	8	.23	D
2394.5	55300	8	10	2.9	D
2413.0	56424	6	4	.13	D
2416.1	56371	6	8	3.3	D
2433.6	56075	6	6	.12	D
2437.9	54557	8	10	.87	D
2510.9	53365	8	10	.94	D
2545.9	54263	6	8	.26	D
	Nitrogen				
	N I				
1163.9	105144	6	6	.43	D
1164.0	105144	4	6	.032	D
1164.2	105120	6	4	.048	D
1164.3	105120	4	4	.43	D
1167.4	104881	6	8	1.1	D
1168.4	104810	6	6	.095	D
1168.5	104810	4	6	1.3	D
1169.7	104717	6	8	.030	D
1176.5	104222	6	4	.95	D
1176.6	104222	4	4	.11	D
1177.7	104145	4	2	1.3	D
1199.6	83365	4	6	5.5	D
1200.2	83318	4	4	5.3	D
1200.7	83284	4	2	5.5	D
1310.5	105144	4	6	1.3	D
1316.3	104810	4	6	.025	D
1492.6	86221	6	4	5.3	D
1492.8	86221	4	4	.58	D
1494.7	86137	4	2	5.0	D
4099.9	110521	2	4	.034	D
4110.0	110545	4	6	.040	D
4114.0	110521	4	4	.0068	D
4137.6	107446	2	4	.0039	D
4143.4	107446	4	4	.0078	D
4151.5	107446	6	4	.013	D
4214.8	107037	4	6	.022	D
4216.1	106996	2	4	.031	D
4218.9	106980	2	2	.012	D
4222.1	106996	4	4	.0098	D
4223.1	107037	6	6	.051	D
4224.9	106980	4	2	.061	D
4230.5	106996	6	4	.033	D
4385.5	120566	2	2	.0052	C
4392.4	120566	4	2	.0102	C
4914.9	106478	2	2	.00759	B
4935.1	106478	4	2	.0158	B
5169.6	107446	6	4	.00209	C
5181.4	107446	4	4	.00144	C
5186.6	107446	2	4	7.3 (–4)	C
5199.8	112808	2	2	.023	D
5201.6	112801	2	4	.023	D
5281.2	107037	6	6	.00282	C
5292.7	106996	6	4	.00167	C
5293.5	107037	4	6	.00113	C
5309.4	106980	4	2	.00273	C
5310.5	106996	2	4	.00137	C
5344.0	106814	6	6	6.2 (–4)	C
5356.6	106814	4	6	.00189	C
5367.0	106778	4	4	.00118	C
5372.6	106778	2	4	.00107	C
5378.3	106759	2	2	.00210	C

Table 1b (Continued)
TRANSITION PROBABILITIES FOR SELECTED ATOMIC AND IONIC SPECIES

Wavelength (λ[Å])	Upper energy level (E_k[cm^{-1}])	Statistical weights g_i	g_k	Transition Probability (A[10^8 s^{-1}])	Uncertainty
		N I			
5816.5	112681	4	6	.00278	C
5829.5	112681	6	6	.0064	C
5834.6	112610	2	4	.00383	C
5840.9	112610	4	4	.00122	C
5849.7	112565	2	2	.00152	C
5854.0	112610	6	4	.00409	C
5856.0	112565	4	2	.0076	C
6606.2	109927	4	6	7.9 (−4)	C
6622.5	109927	6	6	.0071	C
6627.0	109857	2	4	.00197	C
6636.9	109857	4	4	.0125	C
6645.0	109927	8	6	.0311	C
6646.5	109812	2	2	.0194	C
6653.5	109857	6	4	.0244	C
6656.5	109812	4	2	.0193	C
6926.7	109927	4	6	.0064	C
6945.2	109927	6	6	.0149	C
6951.6	109857	2	4	.0088	C
6960.5	109857	4	4	.00281	C
6973.1	109812	2	2	.00350	C
6979.2	109857	6	4	.0094	C
6982.0	109812	4	2	.0174	C
7423.6	96751	2	4	.052	C
7442.3	96751	4	4	.106	C
7468.3	96751	6	4	.161	C
		N II			
474.89	210705	5	5	4.5	D
475.65	210240	1	3	14.	D
475.70	210266	3	5	20.	D
475.76	210240	3	3	11.	D
475.80	210302	5	7	26.	D
475.88	210266	5	5	6.8	D
508.70	196712	5	5	2.8	D
510.76	211104	5	7	31.	D
513.85	209926	5	5	6.8	D
529.36	188909	1	3	6.5	D
529.41	188937	3	1	20.	D
529.49	188909	3	3	4.9	D
529.64	188857	3	5	4.9	D
529.72	188909	5	3	8.1	D
529.87	188857	5	5	15.	D
533.51	187438	1	3	20.	D
533.58	187462	3	5	27.	D
533.65	187438	3	3	15.	D
533.73	187492	5	7	36.	D
533.82	187462	5	5	9.1	D
547.82	197859	5	3	5.2	D
559.76	211336	1	3	12.	D
574.65	189335	5	7	35.	D
582.16	187091	5	5	13.	D
635.20	190120	1	3	18.	D
644.63	155127	1	3	12.	C
644.84	155127	3	3	35.	C
645.18	155127	5	3	58.	C
660.29	166766	5	3	40.	C
671.02	149077	3	5	2.9	D
671.39	149077	5	5	8.9	D
671.41	148940	1	3	3.6	D
671.63	148940	3	3	2.7	D
671.77	148909	3	1	12.	D
672.00	148940	5	3	4.4	D
745.84	166766	1	3	10.	C
746.98	149188	5	3	40.	C
748.37	148940	5	3	2.0	D
775.97	144188	5	5	35.	C
915.61	109217	1	3	3.6	C
915.96	109224	3	1	11.	C
1084.0	92252	1	3	2.0	C
1085.5	92250	5	5	.90	C
1085.7	92237	5	7	3.6	C
2139.5	211336	3	3	.29	D
3593.6	196712	3	5	.24	D
3609.1	196592	3	3	.24	D
3615.9	196540	3	1	.24	D
3829.8	196712	3	5	.15	D
3838.4	196712	5	5	.45	D
3842.2	196592	1	3	.20	D
3847.4	196592	3	3	.15	D
3855.1	196540	3	1	.60	D
3856.1	196592	5	3	.25	D
		N II			
3919.0	190120	3	3	1.00	C
3995.0	174212	3	5	1.3	D
4114.4	188909	3	3	.0019	D
4447.0	187091	3	5	1.30	C
4477.7	188909	5	3	.035	D
4507.6	188857	7	5	.038	D
4601.5	170666	3	5	.270	C
4607.2	170608	1	3	.340	C
4613.9	170608	3	3	.196	C
4621.4	170573	3	1	.90	C
4630.5	170666	5	5	.84	C
4643.1	170608	5	3	.466	C
4774.2	187462	3	5	.054	C
4779.7	187438	3	3	.269	C
4781.2	187492	5	7	.040	C
4788.1	187462	5	5	.248	C
4793.7	187438	5	3	.089	C
4803.3	187492	7	7	.313	C
4810.3	187462	7	5	.055	C
4987.4	188937	3	1	.63	C
4994.4	188909	3	3	.74	C
5001.1	186512	3	5	1.02	C
5001.5	186571	5	7	1.08	C
5002.7	168892	1	3	.085	C
5005.2	186652	7	9	1.22	C
5007.3	188857	3	5	.77	C
5010.6	168892	3	3	.268	C
5025.7	186571	7	7	.134	C
5040.7	186512	7	5	.0053	C
5045.1	168892	5	3	.410	C
5452.1	188909	1	3	.14	D
5454.2	188937	3	1	.41	D
5462.6	188909	3	3	.10	D
5478.1	188857	3	5	.10	D
5480.1	188909	5	3	.17	D
5495.7	188857	5	5	.30	D
5666.6	166582	3	5	.423	C
5676.0	166522	1	3	.310	C
5679.6	166679	5	7	.56	C
5686.2	166522	3	3	.231	C
5710.8	166582	5	5	.137	C
5927.8	187438	1	3	.315	C
5931.8	187462	3	5	.425	C
5940.2	187438	3	3	.235	C
5941.7	187492	5	7	.56	C
5952.4	187462	5	5	.140	C
6482.1	164611	3	3	.37	D
6610.6	189335	5	7	.59	D
		N III			
374.20	267238	2	4	101.	C
451.87	221302	2	2	8.9	C
452.23	221302	4	2	17.8	C
685.00	145986	2	4	9.3	C
685.51	145876	2	2	37.1	C
685.82	145986	4	4	46.8	C
686.34	145876	4	2	19.0	C
763.34	131004	2	2	9.6	C
764.36	131004	4	2	18.7	C
771.54	186797	2	4	8.2	C
771.90	186797	4	4	16.5	C
772.39	186797	6	4	24.7	C
772.89	230409	6	4	20.3	C
772.98	230404	4	2	22.7	C
979.84	203089	4	4	8.9	C
979.92	203075	6	6	9.3	C
989.79	101031	2	4	4.15	C
991.51	101031	4	4	.82	C
991.58	101024	4	6	4.96	C
1747.8	203089	2	4	1.31	C
1751.2	203089	4	4	.258	C
1751.7	203075	4	6	1.57	C
4097.3	245701	2	4	.82	C
4103.4	245665	2	2	.82	C
4634.1	267238	2	4	.65	C
4640.6	267244	4	6	.78	C
4641.9	267238	4	4	.130	C
		N IV			
247.20	404522	1	3	114.	C
283.52	420053	9	15*	290.	C
322.64	377285	9	3*	84.	C

Table 1b (Continued)
TRANSITION PROBABILITIES FOR SELECTED ATOMIC AND IONIC SPECIES

Wavelength (λ[Å])	Upper energy level (E_k[cm^{-1}])	Statistical weights g_i	g_k	Transition Probability (A[10^8 s^{-1}])	Uncertainty
	N IV				
335.05	429160	3	5	185.	C
387.35	388855	3	1	28.	D
765.15	130694	1	3	24.0	B
923.16	175669	9	9*	18.	B
955.34	235369	3	1	30.	C
1718.6	188883	3	5	2.37	C
3480.8	406005	3	9*	1.1	C
4057.8	429160	3	5	.68	C
6380.8	404522	1	3	.14	B
7116.7	420053	9	15*	.12	C
	N V				
209.29	477817	2	6*	118.	B
247.66	484418	6	10*	430.	C
1238.8	80722	2	4	3.41	B
1242.8	80463	2	2	3.38	B
4603.7	477842	2	4	.412	B
4620.0	477766	2	2	.408	B
	N VI				
24.898	4016390	1	3	5158.	AA
28.787	3473840	1	3	180.9(+2)	AA
161.22	4006180	3	9*	285.9	AA
173.29	4016390	1	3	269.7	AA
173.92	4013460	9	15*	876.	A
185.19	4013820	3	5	824.	A
1901.5	3438490	3	9*	.6777	AA
2896.4	3473840	1	3	.2080	AA
	Oxygen				
	O I				
1028.2	97488	1	3	.20	D
1152.2	102662	5	5	5.5	D
1217.6	115918	1	3	1.8	C
1302.2	76795	5	3	3.3	C
1304.9	76795	3	3	2.0	C
1306.0	76795	1	3	.66	C
5435.2	105019	3	5	.0061	C
5435.8	105019	5	5	.0102	C
5436.9	105019	7	5	.0142	C
6453.6	102117	3	5	.0142	B
6454.4	102117	5	5	.0237	B
6456.0	102117	7	5	.0331	B
6653.8	130943	3	1	.600	B
7156.7	116631	5	5	.473	B
7471.4	127292	5	3	.0114	B
7473.2	127288	5	5	.102	B
7476.4	127283	5	7	.408	B
7477.2	127292	3	3	.170	B
7479.1	127288	3	5	.306	B
7480.7	127292	1	3	.226	B
7771.9	86631	5	7	.340	B
7774.2	86628	5	5	.340	B
7775.4	86626	5	3	.340	B
7886.3	128595	3	5	.370	B
7939.5	113727	7	5	.00165	C
7943.2	113721	7	7	.0417	C
7947.2	113727	5	5	.058	C
7947.6	113714	7	9	.373	C
7950.8	113721	5	7	.331	C
7952.2	113727	3	5	.313	C
7981.9	101155	3	3	.12	D
7982.4	101155	1	3	.16	D
7987.0	101148	3	5	.21	D
7987.3	101148	5	5	.072	D
7995.1	101135	5	7	.29	D
	O II				
429.92	232603	4	2	39.	D
430.04	232536	4	4	39.	D
430.18	232463	4	6	39.	D
483.75	233544	4	2	.84	D
483.98	233430	6	4	.76	D
484.03	233430	4	4	.084	D
485.09	232959	6	8	25.	D
485.47	232796	6	6	1.6	D
485.52	232796	4	6	23.	D
3007.1	265999	8	10	.84	C
3007.7	265985	6	8	.72	C
3013.4	265639	6	8	.74	C
3032.1	265930	8	10	.85	C
3032.5	265763	6	8	.82	C
	O II				
3134.8	238893	8	6	1.23	C
3273.5	259286	8	6	1.14	C
3377.2	233544	2	2	1.88	C
3390.3	233430	2	4	1.86	C
3407.4	259286	6	6	.75	C
3749.5	212162	6	4	.90	C
3882.2	232754	8	8	.493	C
3912.0	232527	6	4	1.27	C
3919.3	232480	4	2	1.40	C
3973.3	214229	4	4	1.27	C
4069.6	231296	2	4	1.39	C
4069.9	231350	4	6	1.49	C
4072.2	231428	6	8	1.70	C
4075.9	231530	8	10	1.98	C
4085.1	231350	6	6	.478	C
4087.2	255756	4	6	2.24	C
4089.3	255978	10	12	2.62	C
4095.6	255759	6	8	2.23	C
4097.2	255828	8	10	2.37	C
4104.7	232748	4	6	1.04	C
4105.0	232746	4	4	.80	C
4108.8	255759	8	8	.349	C
4119.2	232754	6	8	1.48	C
4120.3	232748	6	6	.443	C
4132.8	232536	2	4	.84	C
4153.3	232463	4	6	.77	C
4276.7	256123	6	8	1.82	C
4277.4	256084	2	4	1.49	C
4277.9	256123	8	8	.302	C
4281.4	255813	6	6	.60	C
4282.8	255913	4	4	1.06	C
4283.0	256088	4	6	1.58	C
4283.1	256088	6	6	.51	C
4283.8	256084	4	4	.59	C
4294.8	255813	4	6	1.39	C
4303.8	255691	6	8	1.97	C
4328.6	255622	4	2	1.21	C
4340.4	255829	6	8	2.23	C
4347.4	229968	4	4	.94	C
4349.4	208484	6	6	.74	C
4351.3	229947	6	6	.97	C
4396.0	234454	6	6	.398	C
4414.9	211713	4	6	1.15	C
4417.0	211522	2	4	.95	C
4443.1	251224	6	6	.57	C
4448.2	251221	8	8	.57	C
4489.5	255812	2	4	1.51	C
4491.3	255690	4	6	1.81	C
4596.2	228723	4	6	1.03	C
4602.1	256126	4	6	1.70	C
4609.4	256143	6	8	1.82	C
4641.8	206878	4	6	.79	C
4661.6	206786	4	4	.52	C
4701.2	253792	4	4	.87	C
4703.2	251224	4	6	.82	C
4705.4	232959	6	8	1.38	C
4871.6	253048	4	6	.435	C
4906.9	232536	4	4	.68	C
4924.6	232463	4	6	.67	C
4941.1	234402	2	4	.83	C
4943.1	234454	4	6	1.06	C
5206.7	233430	4	4	.391	C
6627.6	248514	4	4	.089	C
6666.9	248425	4	2	.0349	C
6678.2	248514	2	4	.0173	C
6718.1	248425	2	2	.068	C
6721.4	203942	4	2	.189	C
6810.6	246029	6	8	.00180	C
6844.1	245903	4	6	.00325	C
6847.0	246029	8	8	.0347	C
6869.7	245903	6	6	.059	C
6885.1	245816	4	4	.067	C
6895.3	246029	10	8	.298	C
6906.5	245903	8	6	.272	C
6908.1	245768	4	2	.332	C
6910.8	245816	6	4	.267	C
	O III				
262.88	380706	5	5	40.	D
263.69	379232	1	3	52.	D
263.73	379293	3	5	73.	D

Table 1b (Continued)
TRANSITION PROBABILITIES FOR SELECTED ATOMIC AND IONIC SPECIES

Wavelength (λ[Å])	Upper energy level (E_k[cm^{-1}])	Statistical weights g_i	g_k	Transition Probability (A[10^8 s^{-1}])	Uncertainty
O III					
263.77	379232	3	3	40.	D
263.82	379356	5	7	96.	D
263.86	379293	5	5	24.	D
277.38	380782	5	7	110.	D
279.79	377687	5	5	26.	D
295.94	381086	1	3	48.	D
303.41	329582	1	3	34.	D
303.46	329643	3	1	100.	D
303.52	329582	3	3	26.	D
303.62	329468	3	5	25.	D
303.69	329582	5	3	42.	D
303.80	329468	5	5	76.	D
305.60	327228	1	3	100.	D
305.66	327277	3	5	140.	D
305.70	327228	3	3	76.	D
305.77	327351	5	7	180.	D
305.84	327277	5	5	46.	D
320.98	331820	5	7	190.	D
328.45	324734	5	5	61.	D
345.31	332777	1	3	99.	D
374.08	267633	5	5	26.	D
395.56	273080	5	3	49.	D
507.39	197087	1	3	16.	C
507.68	197087	3	3	47.	C
508.18	197087	5	3	79.	C
525.80	210459	5	3	88.	C
597.82	210459	1	3	18.	C
599.60	187049	5	5	55.	C
702.33	142383	1	3	5.7	C
702.82	142397	3	1	17.	C
832.93	120059	1	3	3.2	C
835.10	120053	5	5	1.4	C
835.29	120025	5	7	5.7	C
1109.5	381086	3	3	2.8	D
1679.1	357111	3	5	.94	D
1686.9	356838	3	3	.94	D
1760.4	357111	3	5	.60	D
1764.5	357111	5	5	1.8	D
1766.3	356838	1	3	.78	D
1772.3	356732	3	1	2.3	D
1773.0	356838	5	3	.98	D
2390.4	332777	3	3	2.2	D
2959.7	324734	3	5	2.1	D
2996.5	327228	3	3	.51	D
3004.4	327277	5	5	.47	D
3017.6	327351	7	7	.59	D
3115.7	329643	3	1	1.5	D
3121.7	329582	3	3	1.5	D
3132.9	329468	3	5	1.4	D
3261.0	324658	5	7	1.8	D
3265.5	324836	7	9	2.1	D
3267.3	324462	3	5	1.7	D
3281.9	324462	5	5	.32	D
3284.6	324658	7	7	.23	D
3405.7	329582	1	3	.27	D
3408.1	329643	3	1	.81	D
3415.3	329582	3	3	.20	D
3428.7	329468	3	5	.20	D
3430.6	329582	5	3	.33	D
3444.1	329468	5	5	.59	D
3702.8	327228	1	3	.62	D
3707.2	327277	3	5	.83	D
3714.0	327228	3	3	.46	D
3715.1	327351	5	7	1.1	D
3725.3	327277	5	5	.27	D
3961.6	331820	5	7	1.3	D
5268.1	332777	1	3	.31	D
5508.1	324734	5	5	.11	D
5592.4	290957	3	3	.36	D
O IV					
238.36	419534	2	4	288.	C
238.57	419551	4	6	346.	C
238.58	419534	4	4	58.	C
279.63	357614	2	2	23.7	C
279.94	357614	4	2	47.7	C
553.33	180724	2	4	12.0	C
554.08	180481	2	2	47.3	C
554.51	180724	4	4	60.	C
555.26	180481	4	2	25.0	C
608.40	164366	2	2	12.5	C
O IV					
609.83	164366	4	2	23.6	C
616.95	289024	6	4	25.5	C
617.01	289024	4	4	2.93	C
617.04	289015	4	2	28.6	C
624.62	231538	2	4	10.7	C
625.13	231538	4	4	21.5	C
625.85	231538	6	4	32.2	C
779.73	255185	6	4	1.53	C
779.82	255185	4	4	13.1	C
779.91	255156	6	6	13.7	C
780.00	255156	4	6	1.0	C
787.71	126950	2	4	5.9	C
790.11	126950	4	4	1.15	C
790.20	126936	4	6	7.1	C
921.30	289024	2	4	2.11	C
921.37	289015	2	2	9.2	C
923.37	289024	4	4	11.4	C
923.43	289015	4	2	4.48	C
1338.6	255185	2	4	2.22	C
1343.0	255185	4	4	.428	C
1343.5	255156	4	6	2.64	C
O V					
172.17	580825	1	3	296.	B
192.85	600766	9	15*	690.	B
215.17	546973	9	3*	170.	C
220.35	612616	3	5	440.	C
248.46	561276	3	1	65.	D
629.73	158798	1	3	28.0	B
758.68	213887	3	5	5.68	B
759.44	213618	1	3	7.55	B
760.23	213618	3	3	5.64	B
760.45	213887	5	5	16.9	B
761.13	213463	3	1	22.5	B
762.00	213618	5	3	9.34	B
774.52	287910	3	1	35.4	C
1371.3	231721	3	5	3.29	C
2784.0	582882	3	9*	1.6	D
3144.7	612616	3	5	.93	C
5114.1	580825	1	3	.17	B
5589.9	600766	9	15*	.15	C
O VI					
150.10	666218	2	6*	254.	B
173.03	674656	6	10*	885.	B
1031.9	96908	2	4	4.15	B
1037.6	96375	2	2	4.08	B
3811.4	666270	2	4	.513	B
3834.2	666113	2	2	.505	B
O VII					
18.627	5368550	1	3	9362.	AA
21.602	4629200	1	3	330.9(+2)	AA
120.33	5355670	3	9*	533.5	AA
128.20	5368550	1	3	505.3	AA
128.46	5364420	9	15*	1620.	A
135.82	5365470	3	5	1530.	A
1630.2	4585980	3	9*	.7935	AA
2450.0	4629200	1	3	.2514	AA
Phosphorus					
P I					
1671.7	59820	4	2	.39	D
1674.6	59716	4	4	.40	D
1679.7	59535	4	6	.39	D
1775.0	56340	4	6	2.17	C
1782.9	56090	4	4	2.14	C
1787.7	55939	4	2	2.13	C
2135.5	58174	4	4	.211	C
2136.2	58174	6	4	2.83	C
2149.1	57877	4	2	3.18	C
2152.9	65156	2	4	.485	C
2154.1	65156	4	4	.173	C
2154.1	65157	4	6	.58	C
2534.0	58174	2	4	.200	C
2535.6	58174	4	4	.95	C
2553.3	57877	2	2	.71	C
2554.9	57877	4	2	.300	C
P II					
1301.9	76813	1	3	.50	C
1304.5	76824	3	1	1.5	C
1304.7	76813	3	3	.37	C

Table 1b (Continued)

TRANSITION PROBABILITIES FOR SELECTED ATOMIC AND IONIC SPECIES

Wavelength (λ[Å])	Upper energy level (E_k[cm^{-1}])	Statistical weights g_i	g_k	Transition Probability (A[10^8 s^{-1}])	Uncertainty
	P II				
1305.5	76765	3	5	.38	C
1309.9	76813	5	3	.62	C
1310.7	76765	5	5	1.1	C
4475.3	127889	5	7	1.3	D
4499.2	130143	5	7	1.4	D
4530.8	127368	3	5	1.0	D
4554.8	127951	3	5	.96	D
4588.0	125130	5	7	1.7	D
4589.9	124948	3	5	1.6	D
4602.1	125392	7	9	1.9	D
4943.5	123892	7	5	.63	D
5253.5	107924	3	5	1.0	D
5425.9	105550	5	5	.69	D
6024.2	103340	3	5	.51	D
6043.1	103669	5	7	.68	D
	P III				
1334.8	74917	2	4	.55	D
1344.3	74946	4	6	.64	D
1344.8	74917	4	4	.11	D
4057.4	141514	4	4	.10	D
4059.3	141514	6	4	.90	D
4080.1	141377	4	2	.99	D
	Potassium				
	K I				
4044.1	24720	2	4	.0124	C
4047.2	24701	2	2	.0124	C
5084.2	32648	2	2	.00350	C
5099.2	32648	4	2	.0070	C
5323.3	31765	2	2	.0063	C
5339.7	31765	4	2	.0126	C
5343.0	31696	2	4	.0040	D
5359.6	31696	4	6	.0046	D
5782.4	30274	2	2	.0123	C
5801.8	30274	4	2	.0246	C
5812.2	30186	2	4	.0028	D
5831.9	30185	4	6	.0032	D
6911.1	27451	2	2	.0272	C
6938.8	27451	4	2	.054	C
7664.9	13043	2	4	.387	B
7699.0	12985	2	2	.382	B
	K II				
607.93	164496	1	3	.013	D
	K III				
2550.0	246626	6	4	2.0	D
2635.1	246626	4	4	1.2	D
2992.4	240830	6	8	2.5	D
3052.1	241444	4	6	1.7	D
3202.0	243947	4	4	1.8	D
3289.1	243121	4	6	2.0	D
3322.4	237512	6	6	1.3	D
3421.8	243448	2	4	1.5	D
	K XVI				
206.27	484800	1	3	94.	C
	K XVII				
22.020	4814800	2	4	4.7 (+4)	C
22.163	4818000	4	6	5.6 (+4)	C
22.18	4814800	4	4	9300.	C
22.60	4699000	2	2	2500.	D
22.76	4699000	4	2	4700.	C
	Praseodymium				
	Pr II				
3997.0	28010	15	15	.187	C
4062.8	28010	13	15	1.00	C
4100.7	28816	17	19	.84	C
4143.1	27128	15	17	.58	C
4179.4	25569	13	15	.52	C
4222.9	24116	11	13	.391	C
4241.0	28010	17	15	.230	C
4359.8	28010	15	15	.11	D
4405.8	27128	17	17	.090	D
4429.3	25569	15	15	.228	C
4449.8	24116	13	13	.124	C
4468.7	24116	11	13	.154	C
4510.2	25569	13	15	.116	C
4534.2	27128	15	17	.049	D
4734.2	24116	15	13	.025	D
	Pr II				
4879.1	25569	15	15	.018	D
4886.0	25569	15	15	.013	D
4912.6	28010	17	15	.057	D
5034.4	28816	19	19	.11	D
5110.8	28816	21	19	.278	C
5135.1	27128	17	17	.125	C
5173.9	27128	19	17	.318	C
5219.1	25569	15	15	.095	D
5220.1	25569	17	15	.235	C
5251.7	24116	15	13	.011	D
5259.7	24116	15	13	.224	C
5292.6	24116	13	13	.093	D
5810.6	28816	17	19	.023	D
5879.3	28010	15	15	.076	D
6200.8	27128	15	17	.018	D
6278.7	25569	13	15	.026	D
6398.0	24116	11	13	.019	D
	Rubidium				
	Rb I				
3022.5	33076	2	4	4.13(-5)	C
3032.0	32972	2	4	4.93(-5)	C
3044.2	32840	2	4	8.2 (-5)	C
3060.2	32668	2	4	1.05(-4)	C
3082.0	32437	2	4	1.49(-4)	C
3112.6	32119	2	4	2.5 (-4)	C
3113.1	32114	2	2	1.3 (-4)	C
3157.5	31661	2	4	3.38(-4)	C
3158.3	31654	2	2	2.0 (-4)	C
3228.0	30970	2	4	6.4 (-4)	C
3229.2	30959	2	2	3.8 (-4)	C
3348.7	29854	2	4	.00137	C
3350.8	29835	2	2	8.9 (-4)	C
3587.1	27870	2	4	.00397	C
3591.6	27835	2	2	.0029	C
4201.8	23793	2	4	.018	C
4215.5	23715	2	2	.015	C
7800.3	12817	2	4	.370	B
7947.6	12579	2	2	.340	B
	Scandium				
	Sc I				
2113.5	47315	4	6	.032	C
2263.0	44189	4	4	.058	C
2267.3	44105	4	2	.48	C
2271.6	44189	6	4	.46	C
2335.4	42819	4	2	.17	C
2346.8	42780	6	4	.13	C
2439.4	41163	6	6	.21	C
2439.9	41153	6	4	.022	C
2711.4	37040	6	6	.29	C
2739.8	36666	6	6	.0056	C
2974.0	33615	4	4	.45	C
2980.8	33707	6	6	.44	C
3015.4	33154	4	6	.66	C
3019.3	33278	6	8	.81	C
3269.9	30573	4	2	3.1	C
3273.6	30707	6	4	2.7	C
3907.5	25585	4	6	1.28	C
3911.8	25725	6	8	1.37	C
4020.4	24866	4	4	1.65	C
4023.7	25014	6	6	1.44	C
4753.2	21033	4	6	.010	C
4779.4	21086	6	8	.0084	C
4791.5	21033	6	6	.0023	C
5301.9	18856	4	4	.0013	C
5343.0	18711	4	2	.0051	C
5349.7	18856	6	4	.0040	C
6210.7	16097	4	4	.012	C
6239.8	16022	4	4	.0065	C
6305.7	16023	6	6	.015	C
6378.8	15673	4	4	.0016	C
6448.1	15673	6	4	2.6 (-4)	C
	Sc II				
2273.1	55716	1	3	7.7	D
3353.7	32350	5	7	2.0	D
3372.2	29824	7	5	1.2	D
3535.7	30816	5	3	.83	D
3572.5	28161	7	7	1.8	D
3576.4	28021	5	5	1.4	D

Table 1b (Continued)
TRANSITION PROBABILITIES FOR SELECTED ATOMIC AND IONIC SPECIES

Wavelength (λ[Å])	Upper energy level (E_k[cm^{-1}])	Statistical weights g_i	g_k	Transition Probability (A[10^8 s^{-1}])	Uncertainty
	Sc II				
3613.8	27841	7	9	1.9	D
3630.8	27602	5	7	1.6	D
3642.8	27444	3	5	1.5	D
4246.8	26081	5	5	1.5	D
4314.1	28161	9	7	.41	D
4374.5	27841	9	9	.14	D
4670.4	32350	5	7	.18	D
5031.0	30816	5	3	.49	D
5239.8	30816	1	3	.14	D
5526.8	32350	9	7	.42	D
5657.9	29824	5	5	.13	D
	Silicon				
	Si I				
1977.6	50566	1	3	.18	D
1979.2	50602	3	1	.51	D
1980.6	50566	3	3	.13	D
1983.2	50500	3	5	.14	D
1986.4	50566	5	3	.21	D
1989.0	50500	5	5	.41	D
2208.0	45276	1	3	.311	C
2210.9	45294	3	5	.416	C
2211.7	45276	3	3	.232	C
2216.7	45322	5	7	.55	C
2218.1	45294	5	5	.138	C
2506.9	39955	3	5	.466	C
2514.3	39760	1	3	.61	C
2516.1	39955	5	5	1.21	C
2519.2	39760	3	3	.456	C
2524.1	39683	3	1	1.81	C
2528.5	39760	5	3	.77	C
2532.4	54871	1	3	.26	D
2631.3	53387	1	3	.97	D
2881.6	40992	5	3	1.89	C
3905.5	40992	1	3	.118	C
4738.8	60857	3	3	.010	D
4783.0	60857	5	3	.017	D
4792.3	60816	5	5	.017	D
4818.1	60705	5	7	.011	D
4821.2	60496	3	5	.0080	D
4947.6	61198	3	1	.042	D
5006.1	60962	3	5	.028	D
5622.2	57542	3	3	.016	D
5690.4	57329	3	3	.012	D
5708.4	57468	5	5	.014	D
5754.2	57329	5	3	.015	D
5772.1	58312	3	1	.036	D
5948.5	57798	3	5	.022	D
7226.2	59111	3	5	.0079	D
7405.8	58775	3	5	.037	D
7409.1	58787	5	7	.023	D
7680.3	60301	3	5	.046	D
7918.4	60645	3	5	.052	D
7932.3	60705	5	7	.051	D
7944.0	60849	7	9	.058	D
7970.3	60645	5	5	.0071	D
	Si II				
989.87	101023	2	4	6.7	D
992.68	101025	4	6	8.0	D
1020.7	97972	2	2	1.3	D
1190.4	84005	2	4	6.9	D
1193.3	83802	2	2	28.	D
1194.5	84005	4	4	36.	D
1197.4	83802	4	2	14.	D
1248.4	123034	4	4	13.	D
1251.2	123034	6	4	19.	D
1260.4	79339	2	4	20.	D
1264.7	79355	4	6	23.	D
1304.4	76666	2	2	3.6	D
1309.3	76666	4	2	7.0	D
1526.7	65501	2	2	3.73	C
1533.5	65501	4	2	7.4	C
1808.0	55310	2	4	.037	D
2904.3	113761	4	6	.67	D
2905.7	113760	6	8	.71	D
3210.0	112395	4	6	.46	D
4128.1	103556	4	6	1.32	C
4130.9	103556	6	8	1.42	C
5041.0	101023	2	4	.98	D
5056.0	101025	4	6	1.2	D
5957.6	97972	2	2	.42	D
5978.9	97972	4	2	.81	D
6347.1	81252	2	4	.70	C
6371.4	81192	2	2	.69	C
7848.8	113761	4	6	.39	D
7849.7	113760	6	8	.42	D
	Si III				
883.40	235414	5	7	63.	D
994.79	153377	3	3	7.89	B
997.39	153377	5	3	13.1	B
1141.6	217440	3	5	30.	D
1144.3	217489	5	7	39.	D
1161.6	216190	5	5	16.	D
1206.5	82884	1	3	25.9	B
1206.5	165765	3	5	48.9	B
1207.5	205029	5	5	19.	D
1294.5	130101	3	5	5.42	B
1296.7	129842	1	3	7.19	B
1298.9	129842	3	3	5.36	B
1299.0	130101	5	5	16.1	B
1301.2	129708	3	1	21.3	B
1303.3	129842	5	3	8.85	B
1328.8	228700	1	3	27.	D
1417.2	153444	3	1	26.0	C
1435.8	235414	5	7	21.	D
1589.0	228700	5	3	11.	D
1778.7	199164	7	9	4.4	D
1783.1	199026	5	7	3.8	D
3241.6	206176	5	3	2.3	D
3486.9	230270	15	21*	1.8	D
3590.5	204331	3	5	3.9	D
4552.6	175336	3	5	1.26	C
4554.0	248773	5	3	.76	D
4567.8	175263	3	3	1.25	C
4683.0	248168	5	5	.95	D
4716.7	225526	5	7	2.8	D
5451.5	244866	3	5	.60	D
5473.1	245087	5	7	.79	D
5716.3	227089	9	7	.19	D
5739.7	176487	1	3	.47	D
7462.6	214995	5	3	.49	D
7466.3	214989	7	5	.54	D
7612.4	227665	3	5	1.1	D
	Si IV				
457.82	218429	2	4	3.6	D
458.16	218267	2	2	3.6	D
515.12	265418	2	2	4.1	D
516.35	265418	4	2	8.2	D
560.50	250008	6	10*	1.0	D
749.94	293719	10	14*	14.5	C
815.05	193979	2	2	12.3	C
818.13	193979	4	2	24.4	C
860.74	276554	10	6*	1.8	D
1066.6	254128	10	14*	39.1	C
1122.5	160376	2	4	20.5	C
1128.3	160376	4	4	4.03	C
1128.3	160374	4	6	24.2	C
1393.8	71749	2	4	7.73	B
1402.8	71288	2	2	7.58	B
1724.1	218375	10	6*	5.5	C
	Si V				
96.439	1036915	1	3	480.	D
97.143	1029407	1	3	2000.	D
117.86	848511	1	3	300.	D
	Si VI				
246.00	406497	4	2	170.	C
249.12	406497	2	2	85.	C
	Si VII				
217.83	506080	5	3	430.	D
272.64	366780	5	3	51.	C
274.18	368760	3	1	120.	C
275.35	363170	5	5	89.	C
275.67	366780	3	3	30.	C
276.84	366780	1	3	39.	C
278.45	363170	3	5	29.	C
	Si VIII				
214.76	534810	4	2	410.	D
216.92	530420	6	4	360.	D

Table 1b (Continued)
TRANSITION PROBABILITIES FOR SELECTED ATOMIC AND IONIC SPECIES

Wavelength (λ[Å])	Upper energy level (E_k[cm^{-1}])	Statistical weights g_i	g_k	Transition Probability (A[10^8 s^{-1}])	Uncertainty
	Si VIII				
232.86	534810	2	2	80.	D
235.56	530420	4	4	97.	D
250.45	504630	2	2	77.	D
250.79	504630	4	2	160.	D
314.31	318160	4	2	52.	D
316.20	316250	4	4	50.	D
319.83	312670	4	6	49.	D
	Si IX				
223.73	446967	1	3	42.	C
225.03	446967	3	3	120.	C
227.01	446967	5	3	200.	C
227.30	492890	5	3	230.	D
258.10	440390	5	5	104.	C
294.37	344100	9	9*	59.	C
347.36	292301	9	15*	22.	C
	Si X				
253.77	394040	2	4	29.	C
256.57	390050	2	2	110.	C
258.35	394040	4	4	140.	C
261.05	390050	4	2	54.	C
272.00	367670	2	2	30.	C
277.26	367670	4	2	57.	C
287.08	510300	2	4	26.	C
289.19	510300	4	4	50.	C
292.22	510300	6	4	73.	C
347.73	575450	10	10*	43.	D
353.09	287870	6	10*	21.	C
	Si XI				
43.763	2285040	1	3	6110.	B
49.116	2210220	9	3*	2450.	C
49.222	2361010	3	5	8900.	C
52.296	2241590	3	1	760.	C
303.30	329690	1	3	64.2	B
358.29	608790	3	1	103.	C
358.63	451000	3	5	13.8	C
361.41	446510	1	3	18.0	C
364.50	446510	3	3	13.2	C
365.42	451000	5	5	39.0	C
368.28	443690	3	1	51.	C
371.48	446510	5	3	20.7	C
604.14	495210	3	5	11.2	C
2300.8	2285040	1	3	.434	C
	Si XII				
40.924	2443500	2	6*	4420.	B
44.118	2464130	6	10*	1.4 (+4)	B
499.43	200290	2	4	9.56	B
520.72	191900	2	2	8.47	B
1862.	2444300	2	4	1.15	B
1949.	2441900	2	2	1.0	B
4620.	2463540	2	4	.046	C
4942.	2464530	4	6	.045	C
	Silver				
	Ag I				
2061.2	48501	2	4	.031	D
2069.9	48297	2	2	.015	D
3280.7	30473	2	4	1.4	B
3382.9	29552	2	2	1.3	B
5209.1	48744	2	4	.75	D
5465.5	48764	4	6	.86	D
5471.6	48744	4	4	.14	D
	Sodium				
	Na I				
3302.4	30273	2	4	.0281	C
3303.0	30267	2	2	.0281	C
4390.0	39729	2	4	.0077	D
4393.3	39729	4	4	.0016	D
4393.3	39729	4	6	.0092	D
4494.2	39201	2	4	.012	C
4497.7	39201	4	6	.014	C
4497.7	39201	4	4	.0024	D
4664.8	38387	2	4	.0233	C
4668.6	38387	4	4	.0041	D
4668.6	38387	4	6	.025	C
4747.9	38012	2	2	.0063	D
4751.8	38012	4	2	.0127	C
4978.5	37037	2	4	.041	C
4982.8	37037	4	4	.0082	D
	Na I				
4982.8	37037	4	6	.0489	C
5148.8	36373	2	2	.0117	C
5153.4	36373	4	2	.0233	C
5682.6	34549	2	4	.103	C
5688.2	34549	4	6	.12	C
5688.2	34549	4	4	.021	D
5890.0	16973	2	4	.622	A
5895.9	16956	2	2	.618	A
6154.2	33201	2	2	.026	C
6160.8	33201	4	2	.052	C
8183.3	29173	2	4	.453	C
8194.8	29173	4	6	.54	C
8194.8	29173	4	4	.090	D
11381.	25740	2	2	.089	C
11404.	25740	4	2	.176	C
	Na II				
300.15	333163	1	3	30.	D
301.44	331745	1	3	49.	D
372.08	268763	1	3	34.	D
	Na III				
378.14	264455	4	2	77.	C
380.10	264455	2	2	37.	C
1991.0	465399	4	6	8.3	D
2004.2	466788	2	4	4.6	D
2011.9	463971	6	8	8.4	D
2151.5	465018	2	4	4.4	D
2174.5	464390	4	6	5.3	D
2230.3	410977	6	8	3.7	D
2232.2	418418	4	4	3.3	D
2246.7	411536	4	6	2.4	D
2459.3	414282	4	6	3.0	D
2468.9	415172	2	4	2.4	D
2497.0	406190	6	6	1.7	D
	Na V				
307.89	372400	10	6*	200.	C
333.46	372400	6	6*	56.	C
369.01	568100	10	6*	120.	D
400.72	297100	10	10*	50.	C
445.14	297100	6	10*	7.1	C
459.90	217440	4	2	23.	D
461.05	216896	4	4	23.	D
463.26	215860	4	6	22.	D
510.10	569200	2	2	56.	D
511.19	567600	4	4	68.	D
	Na VI				
313.75	320589	5	3	130.	C
361.25	312175	5	5	77.	C
416.53	241341	9	9*	37.	C
492.80	204187	9	15*	13.	C
1550.6	873287	5	5	4.35	C
1567.8	872577	5	3	2.68	C
1608.5	934745	3	1	2.6	C
1649.4	933915	5	5	2.05	C
1741.5	929999	3	5	2.59	C
1747.5	930510	5	7	3.1	C
	Na VII				
94.409	1060651	6	10*	2700.	C
105.27	951347	6	2*	450.	C
353.29	285189	4	4	100.	C
381.30	264400	4	2	40.	C
397.49	367500	4	4	35.	C
399.18	367500	6	4	52.	C
483.28	412345	10	10*	29.	C
486.74	205448	2	4	11.	C
491.95	205412	4	6	13.	C
555.80	465111	4	4	23.	D
777.83	412311	4	6	6.8	D
	Na VIII				
83.34	1327500	9	15*	3940.	C
89.88	1240300	9	3*	809.	C
90.536	1347756	3	5	2860.	C
411.15	243223	1	3	44.2	C
1239.4	1432991	3	3	3.02	C
1802.7	1481521	3	1	2.70	C
1867.7	1347756	3	5	2.01	C
2059.1	1474598	3	5	1.80	C
2558.2	1513677	5	3	.0226	C
2772.0	1469055	3	5	.419	C

Wavelength (λ[Å])	Upper energy level (E_k[cm^{-1}])	Statistical weights g_i	g_k	Transition Probability (A[10^8 s^{-1}])	Uncertainty
	Na VIII				
3021.0	1507690	5	7	.490	C
3108.9	1513677	1	3	.258	C
3182.3	1294214	1	3	.292	C
	Na IX				
70.615	1416130	2	4	1350.	B
70.653	1415368	2	2	1350.	B
77.764	1429980	2	4	3600.	B
77.911	1430204	4	6	4300.	B
681.72	146688	2	4	6.63	B
694.17	144038	2	2	6.30	B
2487.7	1416130	2	4	.832	B
2535.8	1415368	2	2	.789	B
6841.8	1429980	2	4	.0259	C
7103.4	1430204	4	6	.0278	C
	Strontium				
	Sr I				
2206.2	45312	1	3	.0066	C
2211.3	45208	1	3	.0085	C
2217.8	45075	1	3	.012	C
2226.3	44904	1	3	.016	C
2237.7	44676	1	3	.023	C
2253.3	44366	1	3	.037	C
2275.3	43937	1	3	.067	C
2307.3	43328	1	3	.12	C
2354.3	42462	1	3	.18	C
2428.1	41172	1	3	.17	C
2569.5	38907	1	3	.053	C
2931.8	34098	1	3	.019	C
4607.3	21698	1	3	2.01	B
	Sr II				
2018.7	73237	2	2	.12	D
2051.9	73237	4	2	.24	D
2282.0	67523	2	4	.83	D
2322.4	67563	4	6	.91	D
2324.5	67523	4	4	.15	D
2423.5	64964	2	2	.24	D
2471.6	64964	4	2	.48	D
3464.5	53373	4	6	3.1	D
3474.9	53286	4	4	.51	D
4077.7	24517	2	4	1.42	C
4161.8	47737	2	2	.65	D
4215.5	23715	2	2	1.27	C
4305.5	47737	4	2	1.4	D
4414.8	78702	4	6	.11	D
4417.5	78689	4	4	.018	D
4585.9	77858	4	2	.070	D
5303.1	74621	2	4	.19	D
5379.1	74643	4	6	.22	D
5385.5	74621	4	4	.037	D
5723.7	73237	2	2	.071	D
5819.0	73237	4	2	.14	D
8688.9	67563	4	6	.55	D
8719.6	67523	4	4	.097	D
	Sulfur				
	S I				
1295.7	77181	5	5	4.9	D
1296.2	77150	5	3	2.7	D
1302.3	77181	3	5	1.8	D
1302.9	77150	3	3	1.6	D
1303.1	77136	3	1	6.6	D
1303.4	76721	5	3	1.9	D
1305.9	77150	1	3	2.4	D
1401.5	71351	5	3	.91	D
1409.3	71351	3	3	.50	D
1412.9	71351	1	3	.16	D
1425.0	70174	5	7	4.5	D
1425.2	70166	5	5	1.2	D
1433.3	70166	3	5	3.3	D
1433.3	70165	3	3	1.9	D
1437.0	70165	1	3	2.4	D
1448.2	78288	5	3	7.3	D
1473.0	67890	5	7	.42	D
1474.0	67843	5	7	1.6	D
1474.4	67825	5	5	.50	D
1474.6	67816	5	3	.062	D
1481.7	67888	3	5	.17	D
1483.0	67825	3	5	1.2	D
1483.2	67816	3	3	.75	D

Wavelength (λ[Å])	Upper energy level (E_k[cm^{-1}])	Statistical weights g_i	g_k	Transition Probability (A[10^8 s^{-1}])	Uncertainty
	S I				
1487.2	67816	1	3	.87	D
1666.7	69238	5	5	6.3	C
1687.5	81438	1	3	.94	D
1782.3	78288	1	3	1.9	D
1807.3	55331	5	3	3.8	C
1820.3	55331	3	3	2.2	C
1826.2	55331	1	3	.72	C
4694.1	73921	5	7	.0067	D
4695.4	73915	5	5	.0067	D
4696.2	73911	5	3	.0065	D
6403.6	79058	3	5	.0057	D
6408.1	79058	5	5	.0095	D
6415.5	79058	7	5	.013	D
6751.2	78271	15	25*	.079	D
7679.6	76464	3	5	.012	D
7686.1	76464	5	5	.020	D
7696.7	76464	7	5	.028	D
	S II				
1124.4	113461	2	4	1.0	D
1125.0	113461	4	4	4.6	D
1131.0	112937	2	2	3.5	D
1131.6	112937	4	2	1.4	D
1250.5	79968	4	2	.46	C
1253.8	79758	4	4	.42	C
1259.5	79395	4	6	.34	C
4463.6	150996	8	6	.53	D
4483.4	150531	6	4	.31	D
4486.7	150258	4	2	.66	D
4524.7	143623	4	4	.093	D
4525.0	143623	6	4	1.2	D
4552.4	143489	4	2	1.2	D
4656.7	131029	2	4	.09	D
4716.2	131029	4	4	.29	D
4815.5	131029	6	4	.88	D
4885.6	133400	2	4	.17	D
4917.2	133269	2	2	.66	D
4924.1	130134	4	6	.22	D
4925.3	129858	2	4	.24	D
4942.5	129788	2	2	.15	D
4991.9	129858	4	4	.15	D
5009.5	129788	4	2	.70	D
5014.0	133400	4	4	.84	D
5027.2	125485	4	2	.26	D
5032.4	130134	6	6	.81	D
5047.3	133269	4	2	.36	D
5103.3	129858	6	4	.50	D
5142.3	125485	2	2	.19	D
5201.0	140750	4	4	.75	D
5201.3	140750	6	4	.065	D
5212.6	140709	4	6	.098	D
5212.6	140709	6	6	.85	D
5320.7	140319	6	8	.92	D
5345.7	140230	4	6	.88	D
5345.7	140230	6	6	.11	D
5428.6	127976	2	4	.42	D
5432.8	128233	4	6	.68	D
5453.8	128599	6	8	.85	D
5473.6	127825	2	2	.73	D
5509.7	127976	4	4	.40	D
5526.2	128599	8	8	.081	D
5536.8	128233	4	6	.066	D
5556.0	127825	4	2	.11	D
5564.9	128233	6	6	.17	D
5578.8	128233	6	6	.11	D
5606.1	128599	10	8	.54	D
5616.6	127976	4	4	.12	D
5640.0	131187	4	6	.66	D
5645.6	127976	6	4	.018	D
5647.0	130641	2	4	.57	D
5659.9	127976	6	4	.46	D
5664.7	127825	4	2	.58	D
5819.2	130641	4	4	.085	D
6305.5	130134	8	6	.18	D
6312.7	130641	6	4	.30	D
	S III				
2496.2	210698	7	5	2.5	D
2508.2	209926	5	3	2.3	D
2636.9	210698	3	5	.45	D
2665.4	210698	5	5	1.4	D
2680.5	209926	1	3	.62	D

Table 1b (Continued)
TRANSITION PROBABILITIES FOR SELECTED ATOMIC AND IONIC SPECIES

Wavelength (λ[Å])	Upper energy level (E_k[cm^{-1}])	Statistical weights g_i	g_k	Transition Probability (A[10^8 s^{-1}])	Uncertainty
	S III				
2691.8	209926	3	3	.46	D
2702.8	209773	3	1	1.9	D
2718.9	206539	3	3	1.2	D
2721.4	209926	5	3	.77	D
2726.8	210698	3	5	.60	D
2731.1	206672	5	5	1.1	D
2756.9	206911	7	7	1.4	D
2785.5	209926	3	3	.61	D
2856.0	205071	5	7	5.1	D
2863.5	205561	7	9	5.7	D
2872.0	204579	3	5	4.7	D
2950.2	206672	3	5	3.0	D
2964.8	206911	5	7	4.0	D
3662.0	174036	3	3	.64	D
3717.8	174036	5	3	1.0	D
3778.9	173192	3	5	.44	D
3831.8	172786	1	3	.56	D
3837.8	172786	3	3	.42	D
3838.3	173192	5	5	1.3	D
3860.6	172631	3	1	1.6	D
3899.1	172786	5	3	.67	D
4253.6	170649	5	7	1.2	D
4285.0	170067	3	5	.90	D
	S IV				
551.17	181432	2	2	20.6	C
554.07	181432	4	2	40.8	C
3097.5	213717	2	4	2.6	D
3117.7	213507	2	2	2.5	D
	S V				
437.37	311700	1	3	11.2	C
438.19	311700	3	3	33.3	C
439.65	311700	5	3	55.	C
661.52	235000	9	15*	64.4	B
679.01	348100	9	15*	86.	D
690.75	345600	9	9*	50.	D
786.48	127149	1	3	52.5	B
854.85	200800	9	9*	41.8	C
	S VI				
248.99	401621	2	4	31.	C
249.27	401164	2	2	31.	C
388.94	362983	2	2	45.	C
390.86	362983	4	2	88.	C
706.48	247420	2	4	41.7	C
712.68	247452	4	6	48.5	C
712.84	247420	4	4	8.1	D
933.38	107137	2	4	17.	C
944.52	105874	2	2	16.	C
	S VII				
60.161	1662210	1	3	9460.	B
60.804	1644630	1	3	510.	B
72.029	1388330	1	3	861.	B
	S VIII				
198.55	503590	4	2	250.	C
202.61	503590	2	2	120.	C
	S XI				
189.90	535250	9	3*	430.	C
190.37	592500	5	3	280.	C
215.95	530180	5	5	140.	C
217.63	592500	1	3	72.	C
239.81	417040	1	3	26.	D
242.57	417420	3	5	19.	D
242.82	417040	3	3	19.	D
246.90	417420	5	5	54.	D
247.12	417040	5	3	30.	D
288.49	355260	9	15*	29.	C
	S XII				
212.14	471480	2	4	37.	C
215.18	464750	2	2	140.	C
218.20	471480	4	4	170.	C
221.44	464750	4	2	64.	C
227.50	439540	2	2	37.	C
234.48	439540	4	2	68.	C
	S XIII				
32.236	3102150	1	3	1.09(+4)	B
37.600	3049260	3	1	1300.	C
256.66	389660	1	3	87.	C
299.89	537520	3	5	17.8	C
	S XIII				
303.37	529420	1	3	22.8	C
307.36	529420	3	3	16.4	C
308.91	537520	5	5	48.2	C
312.68	523880	3	1	63.	C
316.84	529420	5	3	25.0	C
500.42	3012200	3	5	14.3	C
	S XIV				
30.434	3285500	2	6*	8280.	B
32.517	3309780	6	10*	2.6 (+4)	B
417.67	239460	2	4	12.	B
445.71	224330	2	2	10.	B
1550.	3287000	2	4	1.4	B
1663.	3282640	2	2	1.2	B
3967.	3307840	2	4	.054	C
4153.	3311070	4	6	.057	C
	Thallium				
	Tl I				
2104.6	47500	2	4	.040	D
2118.9	47179	2	2	.020	D
2129.3	46950	2	4	.058	D
2151.9	46457	2	2	.031	D
2168.6	46099	2	4	.098	D
2237.8	44673	2	4	.19	D
2316.0	43166	2	2	.078	D
2379.7	42011	2	4	.44	C
2507.9	47655	4	2	.011	C
2538.2	47179	4	2	.016	C
2580.1	38746	2	2	.18	D
2609.0	46110	4	6	.10	C
2609.8	46099	4	4	.019	C
2665.6	45297	4	2	.057	C
2709.2	44693	4	6	.17	C
2710.7	44673	4	4	.037	C
2767.9	36118	2	4	1.26	C
2826.2	43166	4	2	.080	C
2918.3	42049	4	6	.42	C
2921.5	42011	4	4	.076	C
3229.8	38746	4	2	.173	C
3519.2	36200	4	6	1.24	C
3529.4	36118	4	4	.220	C
3775.7	26478	2	2	.625	B
5350.5	26478	4	2	.705	B
	Thulium				
	Tm I				
2513.8	39769	8	10	.069	D
2527.0	39560	8	8	.17	D
2596.5	38502	8	10	.16	D
2601.1	38434	8	6	.17	D
2622.5	38121	8	10	.061	D
2841.1	43958	6	6	.20	D
2854.2	35026	8	6	.27	D
2914.8	34297	8	8	.077	D
2933.0	34085	8	6	.10	D
2973.2	33624	8	8	.23	D
3046.9	32811	8	8	.18	D
3081.1	32446	8	8	.19	D
3122.5	40787	6	6	.52	D
3142.4	40585	6	6	.088	D
3172.7	31510	8	8	.18	D
3233.7	30915	8	10	.051	D
3247.0	39560	6	8	.30	D
3251.8	39515	6	4	.52	D
3380.7	38343	6	8	.20	D
3406.0	38123	6	8	.15	D
3410.1	29317	8	10	.10	D
3416.6	29261	8	8	.057	D
3418.6	38014	6	6	.11	D
3563.9	28051	8	6	.098	D
3567.4	28024	8	10	.042	D
3744.1	26701	8	8	.95	D
3751.8	26646	8	10	.19	D
3798.5	35090	6	4	1.2	D
3807.7	35026	6	6	.39	D
3883.1	25745	8	6	1.0	D
3887.4	25717	8	8	.38	D
3916.5	34297	6	8	1.5	D
3949.3	34085	6	6	1.0	D
4022.6	33624	6	8	.040	D

Table 1b (Continued)
TRANSITION PROBABILITIES FOR SELECTED ATOMIC AND IONIC SPECIES

Wavelength (λ[Å])	Upper energy level (E_k[cm^{-1}])	Statistical weights g_i	g_k	Transition Probability (A[10^8 s^{-1}])	Uncertainty
Tm I					
4044.5	33489	6	4	.29	D
4094.2	24418	8	6	.90	C
4105.8	24349	8	10	.60	C
4138.3	32929	6	4	.70	D
4158.6	32811	6	8	.055	D
4187.6	23873	8	8	.61	C
4203.7	23782	8	10	.25	C
4222.7	32446	6	8	.15	D
4271.7	32174	6	6	.11	D
4359.9	22930	8	6	.13	D
4386.4	22791	8	8	.042	D
4394.4	31521	6	4	.11	D
4643.1	30302	6	6	.034	D
4681.9	30124	6	8	.039	D
4691.1	30082	6	6	.039	D
5307.1	18837	8	10	.023	D
5658.3	26439	6	8	.010	D
5675.8	17614	8	10	.013	D
5760.2	26127	6	6	.013	D
Tin					
Sn I					
2073.1	48222	1	3	.036	D
2199.3	47146	3	5	.29	D
2209.7	48670	5	5	.56	D
2246.1	44509	1	3	1.6	D
2268.9	47488	5	7	1.2	D
2286.7	47146	5	5	.31	D
2317.2	51755	5	7	2.0	D
2334.8	44509	3	3	.66	D
2354.8	44145	3	5	1.7	D
2380.7	43683	3	5	.031	D
2408.2	50126	5	3	.18	D
2421.7	49894	5	7	2.5	D
2429.5	44576	5	7	1.5	C
2433.5	44509	5	3	.0080	D
2455.2	44145	5	5	.011	D
2476.4	48982	5	3	.011	D
2483.4	43683	5	5	.21	D
2491.8	57282	1	3	.17	D
2495.7	48670	5	5	.62	D
2523.9	48222	5	3	.074	D
2546.6	39257	1	3	.21	D
2558.0	56244	1	3	.34	D
2571.6	47488	5	7	.45	D
2594.4	47146	5	5	.30	D
2636.9	55074	1	3	.11	D
2661.2	39257	3	3	.11	D
2706.5	38629	3	5	.66	D
2761.8	39626	5	5	.0037	D
2779.8	44576	5	7	.18	D
2785.0	44509	5	3	.14	D
2788.0	53021	1	3	.14	D
2812.6	52707	1	3	.23	D
2813.6	44145	5	5	.12	D
2840.0	38629	5	5	1.7	D
2850.6	43683	5	5	.33	D
2863.3	34914	1	3	.54	D
2913.5	51475	1	3	.83	D
3009.1	34914	3	3	.38	D
3032.8	50126	1	3	.62	D
3034.1	34641	3	1	2.0	D
3141.8	48982	1	3	.19	D
3175.1	34914	5	3	1.0	D
3218.7	48222	1	3	.047	D
3223.6	39626	5	5	.0012	D
3262.3	39257	5	3	2.7	D
3330.6	38629	5	5	.20	D
3655.8	44509	1	3	.041	D
3801.0	34914	5	3	.28	D
4524.7	39257	1	3	.26	D
5631.7	34914	1	3	.024	D
5970.3	55374	5	3	.096	D
6037.7	55187	5	5	.050	D
6069.0	51113	1	3	.046	D
6073.5	51375	3	1	.063	D
6171.5	51113	3	3	.049	D
Sn II					
2368.3	46464	4	2	.0044	D
2449.0	99663	4	6	.37	D
2487.0	99659	6	8	.55	D
3283.2	89292	4	6	1.0	D
3352.0	89286	6	8	1.0	D
3472.5	100285	2	4	.16	D
3575.5	100339	4	6	.13	D
5332.4	90242	2	4	.86	D
5562.0	90354	4	6	1.2	D
5588.9	89292	4	6	.85	D
5596.2	90242	4	4	.15	D
5797.2	89292	6	6	.28	D
5799.2	89286	6	8	.81	D
6453.5	72377	2	4	1.2	D
6761.5	86280	2	2	.32	D
6844.1	71494	2	2	.66	D
Titanium					
Ti I					
2276.7	44080	7	5	1.3	C
2299.9	43468	5	5	.69	C
2384.5	42311	9	7	.090	C
2424.2	41624	9	9	.17	C
2441.0	41342	9	11	.072	C
2520.5	39662	5	3	.38	C
2529.9	39686	7	5	.38	C
2541.9	39715	9	7	.43	C
2599.9	38451	5	5	.67	C
2605.2	38544	7	7	.64	C
2611.3	38671	9	9	.64	C
2611.5	38451	7	5	.33	C
2632.4	37977	5	5	.27	C
2641.1	37852	5	3	1.8	C
2644.3	37977	7	5	1.4	C
2646.6	38160	9	7	1.5	C
2662.0	37555	5	7	.089	C
2669.6	37618	7	9	.10	C
2679.9	37690	9	11	.13	C
2735.3	45041	3	1	4.1	D
2912.1	41585	5	7	1.3	D
2933.6	34079	5	7	.096	C
2942.0	33981	5	5	1.0	C
2948.3	34079	7	7	.93	C
2956.1	34205	9	9	.97	C
2967.2	34079	9	7	.11	C
2983.3	33680	7	7	.11	C
3000.9	33701	9	9	.12	C
3186.5	31374	5	7	.80	C
3192.0	31489	7	9	.85	C
3199.9	31629	9	11	.94	C
3203.8	31374	7	7	.072	C
3214.2	31489	9	9	.065	C
3341.9	29915	5	7	.65	C
3354.6	29971	7	9	.64	C
3370.4	29661	5	3	.76	C
3371.5	30039	9	11	.67	C
3377.6	29769	7	5	.69	C
3385.7	29915	9	7	.052	C
3635.5	27499	5	7	.72	C
3642.7	27615	7	9	.67	C
3653.5	27750	9	11	.66	C
3654.6	27355	5	3	.087	C
3671.7	27615	9	9	.048	C
3689.9	27480	9	7	.045	C
3717.4	26893	5	7	.043	C
3724.6	38959	9	9	.91	D
3729.8	26803	5	5	.40	C
3741.1	26893	7	7	.38	C
3752.9	27026	9	9	.47	C
3771.7	26893	9	7	.066	C
3786.0	33661	5	3	1.4	D
3924.5	25644	7	7	.073	C
3929.9	25439	5	5	.075	C
3948.7	25318	5	3	.53	C
3958.2	25644	9	7	.43	C
3981.8	25107	5	5	.38	C
3989.8	25227	7	7	.36	C
3998.6	25388	9	9	.39	C
4008.9	25107	7	5	.071	C
4024.6	25227	9	7	.063	C
4065.1	33085	3	1	.70	D
4186.1	36000	9	9	.28	D
4285.0	37359	5	5	.32	D

Table 1b (Continued)
TRANSITION PROBABILITIES FOR SELECTED ATOMIC AND IONIC SPECIES

Wavelength (λ[Å])	Upper energy level (E_k[cm^{-1}])	Statistical weights g_i	g_k	Transition Probability (A[10^8 s^{-1}])	Uncertainty
		Ti I			
4295.8	29829	3	1	1.3	D
4393.9	41040	9	11	.33	D
4417.3	37852	11	9	.36	D
4449.2	37690	11	11	.97	D
4455.3	34079	7	7	.48	D
4457.4	34205	9	9	.56	D
4481.3	36415	7	7	.57	D
4527.3	28639	3	5	.22	D
4544.7	28596	5	3	.33	D
4656.5	21469	5	7	.022	C
4667.6	21588	7	9	.023	C
4681.9	21740	9	11	.025	C
4981.7	26911	11	13	.59	C
4991.1	26773	9	11	.50	C
4999.5	26657	7	9	.50	C
5007.2	26564	5	7	.48	C
5014.2	26494	3	5	.68	C
5022.9	26564	7	7	.15	C
5024.8	26494	5	5	.15	C
5040.0	20006	7	5	.036	C
5064.7	20126	9	7	.043	C
5173.8	19323	5	5	.037	C
5193.0	19422	7	7	.032	C
5210.4	19574	9	9	.034	C
5866.5	25644	5	7	.046	C
5899.3	25439	3	5	.031	C
6258.1	27615	7	9	.091	C
7251.7	25318	5	3	.072	C
8024.8	27615	9	9	.0083	C
8068.2	27499	7	7	.0077	C
		Ti II			
2635.6	68768	4	4	1.9	D
2638.7	68845	6	6	1.7	D
2642.2	68950	8	8	1.8	D
2646.1	69081	10	10	2.7	D
2752.8	69081	8	10	1.1	D
2800.6	66997	10	8	1.8	D
2805.0	65185	6	8	4.6	D
2810.3	65307	8	10	5.1	D
2817.9	65446	10	12	3.8	D
2827.2	65094	8	10	1.0	D
2828.2	65589	12	14	4.4	D
2828.9	65307	10	10	.92	D
2834.1	65242	10	12	.79	D
2836.6	64978	8	8	1.2	D
2839.7	65446	12	12	.83	D
2846.1	65094	10	10	1.2	D
2856.2	65242	12	12	1.5	D
2931.3	65313	6	6	3.2	D
2936.2	64885	4	6	2.7	D
2938.7	64978	6	8	2.4	D
2942.0	65094	8	10	1.8	D
2943.1	65459	8	8	1.1	D
2945.5	65242	10	12	2.7	D
2954.8	68582	10	12	4.0	D
2959.0	68329	10	10	4.0	D
3081.6	62410	10	8	1.1	D
3088.0	32767	10	8	1.3	C
3089.4	47625	8	6	1.3	C
3103.8	47467	10	8	1.1	C
3127.9	63168	6	6	1.6	D
3128.6	63445	8	8	1.2	D
3190.9	40075	6	8	1.3	C
3202.5	39927	4	6	1.1	C
3224.2	43781	12	10	.70	C
3234.5	31301	10	10	1.3	C
3236.6	31114	8	8	1.1	D
3287.7	45674	8	10	1.4	C
3341.9	34543	6	8	.96	D
3349.0	34749	8	10	1.0	D
3349.4	30241	10	12	1.3	D
3361.2	29968	8	10	1.2	D
3372.8	29734	6	8	1.1	C
3383.8	29544	4	6	1.1	C
3483.8	63445	10	8	.97	D
3492.4	63168	8	6	.98	D
3504.9	43781	10	10	.82	D
3510.8	43741	8	8	.93	D
3759.3	31491	8	8	.96	D
3761.3	31207	6	6	1.0	D

Wavelength (λ[Å])	Upper energy level (E_k[cm^{-1}])	Statistical weights g_i	g_k	Transition Probability (A[10^8 s^{-1}])	Uncertainty
		Ti III			
2375.0	83797	5	3	4.0	D
2414.0	83117	5	7	3.8	D
2516.1	78159	7	9	3.4	D
2527.8	77746	5	7	2.2	D
2540.1	77422	3	5	2.0	D
2563.4	77424	7	7	2.1	D
2565.4	77167	5	5	1.6	D
2567.6	77000	3	3	2.3	D
2576.5	77000	5	3	.92	D
2984.8	75198	5	5	1.9	D
		Ti XIX			
169.33	590580	1	3	129.	C
		Ti XX			
11.452	8732000	2	6*	1.7 (+4)	C
11.872	8749000	2	4	2.8 (+4)	C
11.958	8751000	4	6	3.4 (+4)	C
15.211	6574000	2	4	3.50(+4)	B
15.253	6556000	2	2	3.58(+4)	B
15.907	6612000	2	4	8.8 (+4)	C
16.049	6619000	4	6	1.1 (+5)	C
16.067	6612000	4	4	1.7 (+4)	C
		Uranium			
		U I			
3553.0	32413	13	13	.020	C
3553.0	32591	9	7	.014	C
3553.4	31935	15	13	.022	C
3554.5	28746	11	9	.0084	C
3554.9	31923	15	17	.0079	C
3555.3	28119	13	15	.027	C
3555.8	28115	13	11	.0041	C
3556.9	32382	13	11	.0075	C
3557.8	28099	13	13	.029	C
3558.0	33860	11	13	.016	C
3558.6	32546	9	7	.039	C
3559.4	31955	7	9	.015	C
3560.3	34071	9	7	.064	C
3561.4	31872	15	13	.055	C
3561.5	34062	9	9	.025	C
3561.8	28068	13	11	.057	C
3563.7	28053	13	13	.029	C
3563.8	31920	7	7	.011	C
3565.0	32318	13	11	.029	C
3566.0	32310	13	15	.017	C
3566.6	28650	11	11	.24	B
3568.8	32289	13	13	.038	C
3569.1	35656	17	15	.11	C
3569.4	32462	9	9	.015	C
3570.1	32278	13	11	.013	C
3570.2	28622	11	9	.0053	C
3570.6	35004	13	15	.027	C
3570.7	31798	15	15	.012	C
3571.2	28614	11	11	.0063	C
3571.6	38338	17	15	.13	C
3572.9	32256	13	15	.015	C
3573.9	27973	13	11	.040	C
3574.1	34977	13	15	.035	C
3574.8	27966	13	15	.019	C
3577.1	35594	17	15	.043	C
3577.5	31745	15	13	.0078	C
3577.8	28563	11	11	.0083	C
3577.9	27941	13	13	.023	C
3578.3	27938	13	11	.020	C
3580.0	32379	9	9	.012	C
3580.2	28543	11	9	.029	C
3580.4	28542	11	13	.0075	C
3580.9	32194	13	13	.021	C
3582.6	32180	13	13	.029	C
3584.6	31757	7	5	.024	C
3584.9	27887	13	15	.18	B
3585.4	28503	11	11	.019	C
3585.8	28500	11	9	.028	C
3587.8	32318	9	11	.013	C
3588.3	31729	7	9	.018	C
3589.7	28470	11	13	.021	C
3589.8	31650	15	13	.059	C
3590.7	32295	9	7	.022	C
3591.7	28454	11	9	.053	C

Table 1b (Continued)
TRANSITION PROBABILITIES FOR SELECTED ATOMIC AND IONIC SPECIES

Wavelength (λ[Å])	Upper energy level (E_k[cm^{-1}])	Statistical weights g_i	g_k	Transition Probability (A[10^8 s^{-1}])	Uncertainty
		U I			
3593.0	28445	11	11	.014	C
3593.2	32098	13	15	.042	C
3593.7	33581	11	11	.072	C
Vanadium					
		V I			
3050.4	43874	10	8	.47	D
3053.7	32738	4	4	1.1	D
3056.3	32847	6	6	1.0	D
3060.5	32989	8	8	1.1	D
3066.4	33155	10	10	1.6	D
3088.1	42010	4	6	.43	D
3089.1	41999	4	4	.45	D
3093.8	42138	6	6	.36	D
3112.9	41752	4	2	.43	D
3183.4	31541	6	8	1.3	D
3185.4	31937	10	12	1.4	D
3198.0	31398	6	6	.31	D
3202.4	31541	8	8	.29	D
3205.6	42079	8	10	1.1	D
3212.4	42221	10	12	1.2	D
3218.9	41950	8	6	.31	D
3233.2	42021	10	8	.28	D
3273.0	41437	8	8	.24	D
3284.4	41539	10	10	.24	D
3309.2	39847	4	4	.28	D
3329.9	39847	6	4	.69	D
3356.4	39423	4	6	.27	D
3365.6	39249	2	4	.41	D
3376.1	39249	4	4	.28	D
3377.4	39237	4	2	.80	D
3377.6	39423	6	6	.53	D
3400.4	38116	8	8	.22	D
3529.7	37960	4	6	.36	D
3533.7	38116	6	8	.44	D
3533.8	37835	2	4	.32	D
3543.5	37757	2	2	.58	D
3545.3	37835	4	4	.32	D
3663.6	43649	4	6	2.7	D
3667.7	43707	6	8	2.4	D
3671.2	38124	8	10	.18	D
3672.4	44140	12	12	.79	D
3673.4	43788	8	10	2.4	D
3676.7	44327	14	14	1.1	D
3680.1	43894	10	12	1.9	D
3686.3	38221	10	12	.20	D
3687.5	44028	12	14	2.6	D
3688.1	29418	8	8	.28	D
3690.3	29203	2	4	.37	D
3692.2	29296	6	6	.46	D
3695.3	44190	14	16	2.5	D
3695.9	29203	4	4	.54	D
3703.6	29418	10	8	.79	D
3704.7	29296	8	6	.58	D
3705.0	29203	6	4	.31	D
3706.0	42079	10	10	.46	D
3708.7	42221	12	12	.39	D
3790.5	37475	10	8	.20	D
3795.0	28768	10	10	.21	D
3806.8	37362	10	10	.22	D
3818.2	26183	4	2	.56	C
3828.6	26249	6	4	.431	C
3840.1	36926	8	8	.18	D
3840.8	26353	8	6	.46	D
3855.4	25931	4	4	.28	D
3855.8	26480	10	8	.451	C
3863.9	36766	8	6	.27	D
3864.3	36763	8	6	.17	D
3864.9	26004	6	6	.208	C
3871.1	36926	10	8	.24	D
3875.1	26122	8	8	.17	D
3886.6	36823	10	8	.14	D
3902.3	26172	10	10	.217	C
3921.9	33967	4	2	.23	D
3922.4	34066	6	6	.23	D
3930.0	36539	10	10	.29	D
3934.0	34128	8	8	.56	D
3992.8	40039	12	10	1.1	D
3998.7	40064	14	12	.92	D
4051.0	41861	10	10	1.2	D
4051.4	41918	12	12	1.2	D
		V I			
4090.6	33155	8	10	.77	D
4092.7	26738	8	10	.21	D
4095.5	32989	6	8	.54	D
4099.8	26605	6	8	.39	D
4102.2	32847	4	6	.50	D
4104.8	40126	10	8	1.9	D
4105.2	26506	4	6	.42	D
4109.8	26438	2	4	.47	D
4111.8	26738	10	10	.91	D
4113.5	34128	6	8	.15	D
4115.2	26605	8	8	.59	D
4116.5	26506	6	6	.24	D
4123.6	26397	4	2	.94	D
4128.1	26438	6	4	.70	D
4132.0	26506	8	6	.52	D
4134.5	26605	10	8	.27	D
4232.5	39391	10	10	.86	D
4233.0	39342	8	8	.67	D
4268.6	38483	14	14	1.0	D
4271.6	38405	12	12	.84	D
4277.0	38324	10	10	.84	D
4284.1	38246	8	8	1.0	D
4291.8	40536	12	14	.78	D
4296.1	40452	10	12	.69	D
4297.7	40379	8	10	.61	D
4298.0	40315	6	8	.70	D
4342.8	38124	10	10	.13	D
4355.0	38221	12	12	.11	D
4379.2	25254	10	12	1.2	D
4384.7	25112	8	10	.97	D
4390.0	24993	6	8	.70	D
4395.2	24899	4	6	.48	D
4400.6	24830	2	4	.33	D
4406.6	25112	10	10	.19	D
4407.6	24993	8	8	.38	D
4408.2	24899	6	6	.51	D
4416.5	24789	4	2	.21	D
4452.0	37518	14	16	.80	D
4457.8	37530	10	12	.24	D
4460.3	24839	10	8	.26	D
4462.4	37404	12	14	.66	D
4468.0	37285	8	10	.20	D
4469.7	37316	10	12	.53	D
4474.0	38116	10	8	.41	D
4490.8	37211	10	12	.10	D
4496.1	37960	8	6	.35	D
4514.2	37835	6	4	.29	D
4524.2	37362	12	10	.26	D
4525.2	37757	4	2	.36	D
4529.6	37175	10	8	.21	D
4545.4	37765	10	12	.67	D
4560.7	37644	8	10	.62	D
4571.8	37556	6	8	.53	D
4578.7	37499	4	6	.59	D
4579.2	37556	8	8	.13	D
4706.2	36815	6	4	.21	D
4706.6	38483	12	14	.15	D
4710.6	38405	10	12	.17	D
4746.6	37423	4	4	.17	D
4751.0	37615	8	8	.15	D
4754.0	37758	10	10	.13	D
4757.5	37375	4	2	.65	D
4766.6	37423	6	4	.48	D
4776.4	37503	8	6	.43	D
4786.5	37615	10	8	.40	D
4796.9	37758	12	10	.42	D
4807.5	37931	14	12	.51	D
5193.0	37931	12	12	.35	D
5195.4	37615	8	8	.21	D
5234.1	38124	10	10	.41	D
5240.9	38221	12	12	.39	D
5415.3	37606	12	14	.27	D
5487.9	37362	12	10	.25	D
5507.8	37175	10	8	.30	D
5559.9	31786	6	2*	.14	D
5698.5	26122	6	8	.28	D
5703.6	26004	4	6	.19	D
5707.0	25931	2	4	.19	D
5725.6	36539	8	10	.18	D
5727.0	26172	8	10	.18	D
6090.2	25131	8	6	.13	D

Table 1b (Continued)
TRANSITION PROBABILITIES FOR SELECTED ATOMIC AND IONIC SPECIES

Wavelength (λ[Å])	Upper energy level (E_k[cm^{-1}])	Statistical weights g_i	g_k	Transition Probability (A[10^8 s^{-1}])	Uncertainty
	V II				
2503.0	48580	5	7	.23	C
2506.2	48731	7	9	.23	C
2514.6	48853	9	11	.24	C
2672.0	37521	5	7	.18	D
2677.8	37369	3	5	.29	D
2679.3	37521	7	7	.26	D
2688.0	37531	9	9	.60	D
2690.8	37259	5	3	.43	D
2700.9	37352	9	11	.29	C
2702.2	37205	7	7	.20	D
2706.2	37151	7	9	.29	C
2728.6	36673	3	5	.20	C
2768.6	48731	11	9	.82	C
2774.3	48580	9	7	.88	C
2799.5	49202	5	5	.60	C
2802.8	49211	7	7	.46	C
2803.5	49269	9	9	.58	C
2836.5	48853	9	11	.25	C
2841.0	48731	7	9	.27	C
2877.7	49202	7	5	.67	C
2880.0	37521	7	7	.15	D
2882.5	37369	5	5	.27	D
2888.3	49269	11	9	.54	C
2889.6	37201	3	1	1.4	D
2891.6	37259	5	3	.92	D
2893.3	37521	9	7	.70	D
2903.1	37041	3	5	.21	D
2906.5	37205	7	7	.53	D
2907.5	37352	9	11	.19	C
2908.8	37531	11	9	1.1	D
2910.0	37041	5	5	.72	D
2910.4	36955	3	3	.87	D
2911.1	37151	7	9	.30	C
2924.6	37151	9	9	.91	C
2930.1	48580	7	7	.27	C
2944.6	36919	9	7	.65	C
2950.4	36489	3	3	.34	C
2952.1	36673	7	5	.58	C
2957.5	36489	5	3	.44	C
3048.9	49211	9	7	.54	C
3093.1	35483	11	13	1.8	D
3100.9	48580	7	7	1.0	C
3102.3	35193	9	11	1.6	D
3110.7	34947	7	9	1.5	D
3118.4	34746	5	7	1.5	D
3121.1	35193	11	11	.22	D
3125.3	34593	3	5	1.6	D
3126.2	34947	9	9	.41	D
3130.3	34746	7	7	.50	D
3133.3	34593	5	5	.48	D
3187.7	40002	5	5	.85	D
3188.5	40196	7	7	.80	D
3190.7	40430	9	9	.88	D
3208.4	40002	7	5	.17	D
3232.0	49202	3	5	.23	C
3267.7	39234	5	7	1.2	C
3271.1	39404	7	9	1.1	C
3276.1	39613	9	11	1.1	C
3321.5	49211	9	7	.15	C
3517.3	37521	9	7	.14	D
3530.8	36955	5	3	.19	D
3545.2	37041	7	5	.18	D
3556.8	37205	9	7	.20	D
3589.8	36489	5	3	.37	D
3592.0	36673	7	5	.23	D
3715.5	39613	13	11	.16	C
3727.3	40430	9	9	.22	C
3732.8	39404	11	9	.16	C
3745.8	39234	9	7	.17	C
3750.9	40196	7	7	.20	C
3771.0	40002	5	5	.22	C
	V XX				
14.360	6964000	1	3	6.9 (+4)	C
160.0	625000	1	3	150.	C
	V XXI				
8.843	11308000	2	6*	6000.	C
8.882	11675000	4	6	6400.	C
9.111	11316000	2	4	8900.	C
9.175	11316000	4	6	1.04(+4)	C
9.352	10693000	2	6*	1.0 (+4)	C
9.633	10721000	2	4	1.63(+4)	C
9.704	10722000	4	6	1.91(+4)	C
10.413	9603400	2	6*	2.0 (+4)	C
10.768	9627300	2	4	3.5 (+4)	B
10.853	9630600	4	6	4.2 (+4)	B
13.828	7231700	2	4	4.26(+4)	B
13.870	7209800	2	2	4.37(+4)	B
14.435	7268900	2	4	1.1 (+5)	B
14.578	7276300	4	6	1.3 (+5)	B
14.592	7268900	4	4	2.1 (+4)	C
27.95	10693000	2	6*	3000.	C
28.59	10722000	6	10*	6700.	C
40.20	9603400	2	6*	5800.	C
41.58	9629300	6	10*	1.4 (+4)	B
48.32	11673000	6	10*	1060.	C
58.39	11316000	6	10*	1660.	C
89.40	10722000	6	10*	2900.	C
240.0	416600	2	4	29.	B
293.7	340500	2	2	16.	B
	V XXII				
2.382	41986020	1	3	2.9 (+6)	B
	Xenon				
	Xe I				
1043.8	95801	1	3	.59	D
1047.1	95499	1	3	1.3	D
1050.1	95229	1	3	.085	D
1056.1	94686	1	3	2.45	C
1061.2	94229	1	3	.19	D
1068.2	93619	1	3	3.99	C
1085.4	92129	1	3	.410	C
1099.7	90933	1	3	.434	C
1110.7	90033	1	3	1.5	C
1129.3	88550	1	3	.044	C
1170.4	85441	1	3	1.6	C
1192.0	83890	1	3	6.2	C
1250.2	79987	1	3	.14	D
1295.6	77186	1	3	2.5	C
1469.6	68046	1	3	2.8	B
4501.0	89279	5	3	.0062	D
4524.7	89163	5	5	.0021	D
4624.3	88687	5	5	.0072	D
4671.2	88470	5	7	.010	D
4807.0	88843	3	1	.024	D
7119.6	92445	7	9	.066	D
7967.3	88745	1	3	.0030	D
8409.2	78957	5	3	.010	D
	Xe II				
4180.1	135708	4	4	2.2	D
4330.5	136598	6	8	1.4	D
4414.8	132208	6	6	1.0	D
4603.0	116783	4	4	.82	D
4844.3	113705	6	8	1.1	D
4876.5	130064	6	8	.63	D
5260.4	123255	2	4	.22	D
5262.0	131924	4	4	.85	D
5292.2	111959	6	6	.89	D
5372.4	113673	4	2	.71	D
5419.2	113512	4	6	.62	D
5439.0	121180	4	2	.74	D
5472.6	113705	8	8	.099	D
5531.1	113512	8	6	.088	D
5719.6	113512	4	6	.061	D
5976.5	111792	4	4	.28	D
6036.2	111959	6	6	.075	D
6051.2	111959	8	6	.17	D
6097.6	111792	6	4	.26	D
6270.8	128867	4	6	.18	D
6277.5	111959	4	6	.036	D
6805.7	113512	8	6	.061	D
6990.9	113705.	10	8	.27	D
	Ytterbium				
	Yb I				
2464.5	40564	1	3	.91	C
2672.0	37415	1	3	.118	C
3464.4	28857	1	3	.62	C
3988.0	25068	1	3	1.76	C

Table 1b (Continued)
TRANSITION PROBABILITIES FOR SELECTED ATOMIC AND IONIC SPECIES

Wavelength (λ[Å])	Upper energy level (E_k[cm^{-1}])	Statistical weights g_i	g_k	Transition Probability (A[10^8 s^{-1}])	Uncertainty
		Yb I			
5556.5	17992	1	3	.0114	C
		Yb II			
3289.4	30392	2	4	1.8	C
3694.2	27062	2	2	1.4	C
		Zinc			
		Zn I			
748.29	133638	1	3	.060	C
765.60	130617	1	3	.076	C
792.05	126255	1	3	.057	C
793.85	125968	1	3	.18	C
809.92	123470	1	3	.26	C
1109.1	90158	1	3	.305	C
2138.6	46745	1	3	7.09	B
3075.9	32501	1	3	3.29(–4)	B
3282.3	62769	1	3	.90	B
3302.6	62772	3	5	1.2	B
3302.9	62769	3	3	.67	B
3345.0	62777	5	7	1.7	B
3345.6	62772	5	5	.40	B
3345.9	62769	5	3	.045	B
6362.3	62459	3	5	.474	C
11054.	55789	3	1	.243	C
		Zn II			
2025.5	49355	2	4	3.3	C
2064.2	96910	2	4	4.6	D
2099.9	96960	4	6	5.6	D
2102.2	96910	4	4	.93	D
4911.6	117264	4	6	1.6	D

REFERENCES

1. **Younger, S. M., Fuhr, J. R., Martin, G. A., and Wiese, W. L.,** *J. Phys. Chem. Ref. Data,* 7, 495, 1978.
2. **Fuhr, J. R., Martin, G. A., Wiese, W. L., and Younger, S. M.,** *J. Phys. Chem. Ref. Data,* to be published, 1980.
3. **Wiese, W. L., Smith, M. W., and Glennon, B. M.,** Atomic Transition Probabilities (H through Ne — A Critical Data Compilation), National Standard Reference Data Series, National Bureau of Standards 4, Vol. I, U.S. Government Printing Office, Washington, D.C., 1966.
4. **Wiese, W. L., Smith, M. W., and Miles, B. M.,** Atomic Transition Probabilities (Na through Ca — A Critical Data Compilation), National Standard Reference Data Series, National Bureau of Standards 22, Vol. II, U.S. Government Printing Office, Washington, D.C., 1969.
5. **Wiese, W. L. and Fuhr, J. R.,** *J. Phys. Chem. Ref. Data,* 4, 263, 1975.
6. **Wiese, W. L. and Weiss, A. W.,** *Phys. Rev.,* 175, 50, 1968.
7. **Smith, M. W. and Wiese, W. L.,** *Astrophys. J. Suppl. Ser.,* 23, No. 196, 103, 1971.
8. **Martin, G. A. and Wiese, W. L.,** *J. Phys. Chem. Ref. Data,* 5, 537, 1976.
9. **Fuhr, J. R., Miller, B. J., and Martin, G. A.,** Bibliography on Atomic Transition Probabilities (1914 through October 1977), National Bureau of Standards Special Publication 505, 1978; **Miller, B. J., Fuhr, J. R., and Martin, G. A.,** Bibliography on Atomic Transition Probabilities (November 1977 through February 1980), National Bureau of Standards Special Publication 505, Supplement 1, 1980.
10. **Moore, C. E.,** Bibliography on the Analyses of Optical Atomic Spectra, National Bureau of Standards Special Publication 306 — Section 1, 1968; Sections 2—4, 1969.
11. **Hagan, L. and Martin, W. C.,** Bibliography on Atomic Energy Levels and Spectra (July 1968 through June 1971), National Bureau of Standards Special Publication 363, 1972; **Hagan, L.,** Bibliography on Atomic Energy Levels and Spectra (July 1971 through June 1975), National Bureau of Standards Special Publication 363, Supplement 1, 1977; **Zalubas, R. and Albright, A.,** Bibliography on Atomic Energy Levels and Spectra (July 1975 through June 1979), National Bureau of Standards Special Publication 363, Supplement 2, 1980.
12. **Younger, S. M. and Weiss, A. W.,** *J. Res. Natl. Bur. Stand.,* 79A, 629, 1975.

SECONDARY ELECTRON EMISSION

N. R. Whetten
General Electric Research Laboratory, Schenectady, New York
By permission from "Methods of Experimental Physics" Vol. IV (1962)
Academic Press

The secondary emission yield, or secondary emission ratio, δ, is the average number of secondary electrons emitted from a bombarded material for every incident primary electron. The secondary emission yield is a function of the primary electron energy. δ_{max} is the maximum yield corresponding to a primary electron energy $E_p max$ (see figure). The two primary electron energies corresponding to a yield of unity are denoted the first and second crossovers (E_I and E_{II}). An insulating target, or a

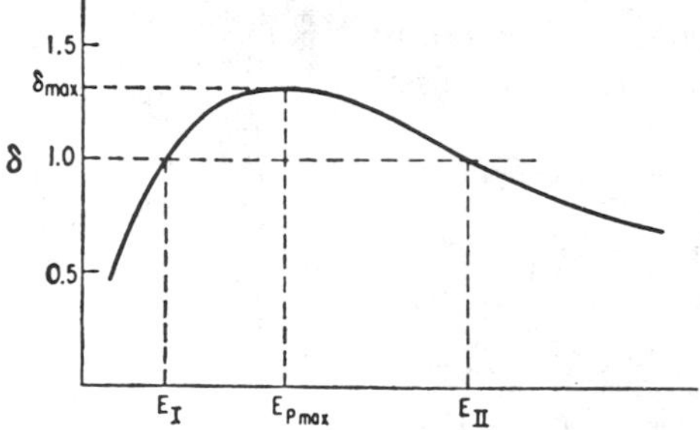

PRIMARY ELECTRON ENERGY (Ep)

conducting target that is electrically floating, will charge positively or negatively depending on the primary electron energy. For $E_I < E_p < E_{II}$, $\delta > 1$ and the surface charges positively provided there is a collector present that is positive with respect to the target. For $E_p < E_I$, $\delta < 1$, and the surface charges negatively towards the potential of the source of primary electrons. For $E_p > E_{II}$, $\delta < 1$, and the surface charges negatively to the second crossover.

The secondary emission yield is very sensitive to surface contamination, such as oxide films and carbon deposits. Whenever possible, yields believed to be most typical of clean surfaces have been selected. The yields are for measurements at room temperature and normal incidence of the primary electrons.

Table I

Secondary Electron Emission Properties of Elements and Compounds. δ_{max} is the maximum secondary emission yield. $\varepsilon_{p\ max}$ the primary electron energy for maximum yield, and E_I and E_{II} are the first and second crossovers.

Elements	δ_{max}	E_{pmax} (eV)	E_I (eV)	E_{II} (eV)	Ref.
Ag	1.5	800	200	>2000	a, b, c
Al	1.0	300	300	300	b
Au	1.4	800	150	>2000	a, c, d
B	1.2	150	50	600	f
Ba	0.8	400	None	None	b
Bi	1.2	550			g, s
Be	0.5	200	None	None	b, h, i, d
C (diamond)	2.8	750		>5000	j
(graphite)	1.0	300	300	300	k
(soot)	0.45	500	None	None	k
Cd	1.1	450	300	700	l, d
Co	1.2	600	200		m, n
Cs	0.7	400	None	None	b, o
Cu	1.3	600	200	1500	a, l, b
Fe	1.3	400	120	1400	n, c, p
Ga	1.55	500	75		q
Ge	1.15	500	150	900	f, r, s
Hg	1.3	600	350	>1200	q
K	0.7	200	None	None	t, u
Li	0.5	85	None	None	b
Mg	0.95	300	None	None	o, b
Mo	1.25	375	150	1200	a, w, c, e, p
Na	0.82	300	None	None	x
Nb	1.2	375	150	1050	a, c
Ni	1.3	550	150	>1500	a, n, m, w, p
Pb	1.1	500	250	1000	g, q
Pd	>1.3	>250	120		v
Pt	1.8	700	350	3000	c
Rb	0.9	350	None	None	t
Sb	1.3	600	250	2000	y
Si	1.1	250	125	500	f
Sn	1.35	500			g, x
Ta	1.3	600	250	>2000	a
Th	1.1	800			b
Ti	0.9	280	None	None	k
Tl	1.7	650	70	>1500	s
W	1.4	650	250	>1500	z, a, a′, c
Zr	1.1	350			k

Compounds	δ_{max}	E_{pmax} (eV)	Ref.
Alkali Halides			
CsCl	6.5		b′
KBr (cyrstal)	14	1800	c′,e′
KCl (crystal)	12	1600	p′,d′
(layer)	7.5	1200	b′,f′
KI (crystal)	10	1600	c′,d′,e′
(layer)	5.6		b′
LiF (crystal)	8.5		g′
(layer)	5.6	700	b′
NaBr (crystal)	24	1800	g′,h′,d′
(layer)	6.3		b′
NaCl (crystal)	14	1200	i′,e′,c′,d′,g′
(layer)	6.8	600	b′,j′
NaF (crystal)	14	1200	g′
(layer)	5.7		b′
NaI (crystal)	19	1300	g′
(layer)	5.5		b′
RbCl (layer)	5.8		b′
Oxides			
Ag_2O	1.0		l′
Al_2O_3 (layer)	2 to 9		k′,m′,i′,b
BaO (layer)	2.3 to 4.8	400	m′,b
BeO	3.4	2000	m′
CaO	2.2	500	m′
Cu_2O	1.2	400	b′,n′
MgO (crystal)	20 to 25	1500	t′,u′,v′
(layer)	3 to 15	400 to 1500	w,m′,r′,s′,t′
MoO_2	1.2		l′
SiO_2 (quartz)	2.1 to 4	400	k′,g′
SnO_2	3.2	640	o′
Sulfides			
MoS_2	1.1		b′
PbS	1.2	500	p′
WS_2	1.0		b′
ZnS	1.8	350	q′
Others			
BaF_2 (layer)	4.5		b′
CaF_2 (layer)	3.2		b′
$BiCs_3$	6	1000	p′
BiCs	1.9	1000	p′
GeCs	7	700	p′
Rb_3Sb	7.1	450	p′
$SbCs_3$	6	700	p′,x′
Mica	2.4	350	k′
Glasses	2 to 3	300 to 450	k′,y′

References

a R. Warnecke J. Phys. Radium, 7, 270 (1936).
b H. Bruining and J. H. deBoer, Physica, 5, 17 (1938).
c R. Kollath, Physik. Z., 38, 202 (1937).
d R. Suhrmann and W. Kundt, Z. Physik, 121, 118 (1943).
e P. L. Copeland, J. Franklin Inst., 215, 593 (1933).
f L. R. Koller and J. S. Burgess, Phys. Rev., 70, 571 (1946).
g P. M. Morozov, J. Exptl. Theoret. Phys. USSR, 11, 410, (1941).
h R. Kollath, Ann. Physik, 33, 285 (1938).
i E. G. Schneider, Phys. Rev., 54, 185 (1938).
j J. B. Johnson, Phys. Rev., 92, 843 (1953).

Table I (Continued)

References

k H. Bruining, Philips Tech. Rev., 3, 80 (1938).
l R. Suhrmann and W. Kundt, Z. Physik, 120, 363 (1943).
m D. E. Woolridge, Phys. Rev., 56, 1062 (1939).
n L. R. G. Treloar and D. H. Landon, Proc. Soc. (London), B50, 625 (1938).
o N. S. Klebnikov, Tech. Phys. USSR, 5, 593 (1938).
p R. L. Petry, Phys. Rev., 26, 346 (1925).
q J. J. Brophy, Phys. Rev., 83, 534 (1951).
r J. B. Johnson and K. G. McKay, Phys. Rev., 93, 668 (1954).
s H. Gobrecht and F. Spear, Z. Physik, 135, 602 (1953).
t A. Afanasjewa and P. W. Timofeew, Tech. Phys. USSR, 4, 953 (1937).
u M. S. Joffe and I. V. Nechlaev, J. Exptl. Theoret. Phys. USSR, 11, 93 (1941).
v H. E. Farnsworth, Phys. Rev., 25, 41 (1925).
w G. Blankenfeld, Ann. Physik. 9, 48 (1951).
x J. Woods, Proc. Phys. Soc. (London), B67, 843 (1954).
y R. Kollath, Handbuch der Physik, vol. 21 (1956), p. 232.
z R. L. Petry, Phys. Rev. 28, 362 (1926).
a′ E. A. Coomes, Phys. Rev., 55, 519 (1939).
b′ H. Bruining and J. H. deBoer, Physica, 6, 834 (1939).
c′ D. N. Dobretzov and A. S. Titkow, Doklady Acad. Nauk USSR, 100, 33 (1955).
d′ N. R. Whetten, Bull. Am. Phys. Soc. Ser. II, 5, 347 (1960).
e′ A. R. Shulman and B. P. Dementyev, J. Tech. Phys. USSR, 25, 2256 (1955).
f′ M. Knoll, O. Hachenberg, and J. Randmer, Z. Physik. 122, 137 (1944).
g′ D. N. Dobretzov and T. L. Matskevich, J. Tech. Phys. USSR, 27, 734 (1957).
h′ T. L. Matskevich, J. Tech. Phys. USSR, 26, 2399 (1956).
i′ A. R. Shulman, W. L. Makedonsky and J. D. Yaroshetsky, J. Tech. Phys. USSR 23, 1152 (1953).
j′ M. M. Vudinsky, J. Tech. Phys. USSR, 9, 271 (1939).
k′ H. Salow, Z. Tech. Phys., 21, 8 (1940).
l′ A. Afanasjewa, P. Timofeew, and A. Ignaton, Phys. Z. Sowjet, 10, 831 (1936).
m′ K. H. Geyer, Ann Phys., 42, 241 (1942).
n′ N. B. Gornij, J. Exptl. Theoret. Phys. USSR, 26, 79 (1954).
o′ H. E. Mendenhall, Phys. Rev., 72, 532 (1947).
p′ O. Hachenberg and W. Braner in Advances in Electronics and Electron Physics, Vol. XI, Academic Press, New York (1959), p. 438.
q′ N. B. Gornij, J. Exptl. Theoreto. Phys. USSR, 26, 88 (1954).
r′ P. Wargo, B. V. Haxby, and W. G. Shepherd, J. Appl. Phys. 27, 1311 (1956).
s′ P. Rappaport, J. Appl. Phys., 25, 288 (1954).
t′ N. R. Whetten and A. B. Laponsky, J. Appl. Phys., 30, 432 (1959).
n′ J. R. Johnson and K. G. McKay, Phys. Rev., 91, 582 (1953).
v′ R. G. Lye, Phys. Rev., 99, 1647 (1955).
w′ N. R. Whetten and A. B. Laponsky, J. Appl. Phys., 28, 515 (1957).
x′ N. D. Morgulis and B. I. Djatlowitskaja, J. Tech. Phys. USSR, 10, 657 (1940).
y′ C. W. Mueller, J. Appl. Phys., 16, 453 (1945).

EMISSIVITY OF TUNGSTEN

Wavelengths in μ

Temperature °K	0.25	0.30	0.35	0.40	0.50	0.60	0.70
1600	.448*	.482	.478	.481	.469	.455	.444
1800	.442*	.478*	.476	.477	.465	.452	.44
2000	.436*	.474	.473	.474	.462	.448	.436
2200	.429*	.470	.470	.471	.458	.445	.431
2400	.422	.465	.466	.468	.455	.441	.427
2600	.418	.461	.464	.464	.451	.437	.423
2800	.411	.456	.461	.461	.448	.434	.419

Temperature °K	0.80	0.90	1.0	1.1	1.2	1.3	1.4
1600	.431	.413	.39	.366	.345	.322*	.300*
1800	.425	.407	.385	.364	.344	.323*	.302*
2000	.419	.401	.381	.361	.343	.323	.305
2200	.415	.396	.378	.359	.342	.324	.306
2400	.408	.391	.372	.355	.340	.324	.309
2600	.404	.386	.369	.352	.338	.325	.310
2800	.400	.383	.367	.352	.337	.325	.313

Temperature °K	1.5	1.6	1.8	2.0	2.2	2.4	2.6
1600	.279*	.263*	.234*	.210*	.19*	.175*	.164*
1800	.282	.267*	.241*	.218*	.20*	.182*	.174*
2000	.288	.273	.247	.227	.209*	.197	.175
2200	.291	.278	.254	.235	.218	.205	.194
2400	.296	.283	.262	.244	.228	.215	.205
2600	.299	.288	.269	.251	.236	.224	.214
2800	.302	.292	.274	.259	.245	.233	.224

* Values by extrapolation.

LIQUIDS FOR INDEX BY IMMERSION METHOD

Liquid	N_D 24°C
Trimethylene chloride	1.446
Cineole	1.456
Hexahydrophenol	1.466
Decahydronaphthalene	1.477
Isoamylphthalate	1.486
Tetrachloroethane	1.492
Pentachloroethane	1.501
Trimethylene bromide	1.513
Chlorobenzene	1.523
Ethylene bromide + Chlorobenzene	1.533
α-Nitrotoluene	1.544
Xylidine	1.557
α-Toluidine	1.570
Aniline	1.584
Bromoform	1.595
Iodobenzene + Bromobenzene	1.603
Iodobenzene + Bromobenzene	1.613
Quinoline	1.622
α-Chloronaphthalene	1.633
α-Bromophthalene + α-Chloronaphthalene	1.650-1.650
α-Bromophthalene + α-Iodonaphthalene	1.660-1.690
Methylne iodide + Iodobenzene	1.700-1.730
Methylene iodide	1.738
Methylene iodide saturated with sulfur	1.78
Yellow phosphorus, sulfur and methylene iodide (8:1:1 by weight)	2.06

Can be diluted with methylene iodide to cover range 1.74-2.06. For precautions in use, cf. West, Am. Mineral, 21, p. 245-9 (1936).

HEAVY LIQUIDS FOR MINERAL SEPARATION

Liquid	Density
Tetrabromoethane (sym.)	2.964, 20°/4°
Can be diluted with carbon tetrachloride (1.595) or benzene (0.894)	
Methylene iodide	3.325, 20°/4°
Can be diluted with carbon tetrachloride or benzene.	
Thallium formate, sq.	3.5
Can be diluted with water.	
Thallium malonate-thallium formate, aq.	4.9
Can be diluted with water.	

For preparation and recovery of these liquids, cf. U. S. Bureau Mines, Rcpt. Inv. #2897 (1928).

INDEX OF REFRACTION

Indices of refraction for elements, inorganic, metal-organic and organic compounds and minerals will be found in the tables of physical constants for the various classes of substances in the section Properties and Physical Constants.

Values for compounds not there listed and data subsequently collected are given below.

Indices not otherwise indicated are for sodium light, λ = 589.3 mμ. Other wave lengths are indicated by the value in millimicrons or symbol in parentheses which follows the index. Wave lengths are indicated as follows: He, λ = 587.6 mμ; Li, λ = 670.8 mμ; Hg, λ = 579.1 mμ; A, λ = 759.4 mμ; C, λ = 656.3 mμ; D, λ = 589.3 mμ; F, λ = 486.1 mμ.

Temperatures are understood to be 20°C for liquids, or ordinary room temperatures in the case of solids. Other temperatures appear as superior figures with the index.

Indices for the elements and inorganic compounds will be understood to be for the solid form except as indicated by the abbreviation liq

See also under Physical Constants of Inorganic Compounds and index of Refraction of Gases.

Elements

Name	Formula	Index
Bromine (liq.)	Br_2	1.661_{15}
Cadmium (liq.)	Cd	0.82 (579 mμ)
(sol.)		1.13
Chlorine (liq.)	Cl_2	1.385
(gas)		1.00768
Hydrogen (liq.)	H_2	$1.10974^{-252.83}$(579 mμ)
Iodine (sol.)	I_2	3.34
(gas)		1.001920
Lead	Pb	2.6 (579 mμ)
Mercury (liq.)	Hg	1.6-1.9
Nitrogen (liq.)	N_2	1.2053^{-190}
Oxygen (liq.)	O_2	1.221^{-131}
Phosphorous (yel.) (sol.)		2.1442^{25}
Selenium	Ses	3.00, 4.04
(amor.) (sol.)		2.92
Sodium (liq.)	Na	0.0045
(sol.)		4.22
Sulfur (liq.)	S_6	1.929^{110}
(amor.) (sol.)		.1998
(rhombic, α)		1.957, 2.0377, 2,2454
Tin (liq.)	Sn	2.1

Inorganic Compounds

See also under Physical Constants of Inorganic Compounds

Name	Formula	Index
Aluminum carbide	AlC_3	2.7, 2.75 (700 mμ)
chloride	$AlCl_3 \cdot 6H_2O$	1.560, 1,507
oxide	Al_2O_3	1.665-1.680, 1.63-1.65
Alums. See under appropriate element.		
Ammonium antimony tartrate	$2(NH_4 \cdot SbO \cdot C_4H_4O_6)H_2O$	β1.6229 (C)
*ortho*arsenate, di-H	$NH;4H_2AsO_4$	1.5766, 1.5217
bromide	NH_4Br	1.7108
*per*chlorate	NH_4ClO_4	1.4818, 1.4833, 1.4881
chloroplatinate	$(NH_4)_2PtCl_6$	1.8
fluoride	NH_4F	ω<1.328
acid	NH_4HF_2	1.385, 1.390, 1.394
hydrogen malate (*d*)	$NH_4C_6H_5O_5$	β1.503
nitrate	NH_4NO_3	1.413, 1.611(He), 1.63
Ammonium sulfate, acid	NH_4HSO_4	1.463, 1.473, 1.510
tartrate (*dl*)	$(NH_4)_2C_4H_4O_6 \cdot 2H_2O$	β1.564
thiocyanate	NH_4CNS	1.546, 1.685, 1.692
uranyl acetate	$NH_4C_2H_3O_2 \cdot UO_2(C_2H_3O_2)_2$	1.4808, 1.4933
Antimony bromide	$SbBr_3$	>1.74 +
iodide, tri-	SbI_3	2.78 (Li), 2.36
Barium cadmium bromide	$BaCdBr_4 \cdot 4H_2O$	β1.702
cadmium chloride	$BaCdCl_4 \cdot 4H_2O$	β1.651
calcium propionate	$BaCa_2(C_3H_5O_2)_6$	1.4442
fluochloride	$BaCl_2 \cdot BaF_2$	1.640, 1.633
fluoride	BaF_2	1.475 also 1.4741
Barium oxide	BaO	1.980
*ortho*phosphate, di-	$BaHPO_4$	1.617, 1.63±, 1.635
propionate	$Ba(C_2H_5CO_2)_2 \cdot H_2O$	β1.5175
sulfide, mono-	BaS	2.155
Cadmium ammonium chloride	$CdCl_2 \cdot 4NH_4Cl$	1.6038, 1.6042
cesium sulfate	$CdSO_4 \cdot Cs_2SO_4 \cdot 6H_2O$	1.498, 1.500, 1.506
fluoride	CdF_2	1.56
magnesium chloride	$(CdCl_2)_2 \cdot MgCl_2 \cdot 12H_2O$	1.49, 1.5331, 1.5769
oxide	CdO	2.49 (Li)
potassium chloride	$CdCl_2 \cdot 4KCl$	1.5906, 1.5907
cyanide	$Cd(CN)_2 \cdot 2KCN$	1.4213
rubidium sulfate	$CdSO_4 \cdot Rb_2SO_4 \cdot 6H_2O$	1.4798, 1.4848, 1.4948
Calcium aluminate	$Ca_3Al_2O_6$	1.710
borate	$Cao \cdot B_2O_3$	1.540, 1.656, 1.682
carbide	CaC_2	<1.75
copper acetate	$CaCu(C_2H_3O_2)_4 \cdot 6H_2O$	1.436, 1.478
cyanamide	$CaCN_2$	1.60, <1.95
dithionate	$CaS_2O_6 \cdot 4H_2O$	1.5516, 1.5414
*pyro*phosphate	$Ca_2P_2O_7$	1.585, 1.60±, 1.605
platinocyanide	$CaPt(CN)_4 \cdot 5H_2O$	1.623, 1.644, 1.767
stromtium propionate	$Ca_2Sr(C_3H_5O_2)_6$	1.4871, 1.4956
sulfide (oldhamite)	CaS	2.137
sulfite	$CaSO_3 \cdot 2H_2O$	1.590, 1.595, 1.628
thiosulfate	$CaS_2O_3 \cdot 6H_2O$	1.545, 1.560, 1.605
Carbon dioxide (liq.)	CO_2	1.195^{15}
Cerium dithionate	$Ce_2(S_2O_6)_3 \cdot 15H_2O$	β1.507
Cesium *per*chlorate	$CsClO_4$	1.4752, 1.4788, 1.4804
nitrate	$CsNO_3$	1.55, 1.56
selenate	Cs_2SeO_4	1.5989, 1.5999, 1.6003
thallium chloride	$Cs_2Tl_2Cl_9$	1.784, 1.774
Chromium cesium sulfate	$CrCs(SO_4)_2 \cdot 12H_2O$	1.4810
oxide (ic)	Cr_2O_3	2.5
potassium cyanide (ic)	$CrK_3(CN)_6$	4.5221, 1.5244, 1.5373
sulfate (ic)	$Cr_2(SO_4)_3 \cdot 18H_2O$	1.564
thallium sulfate	$CrTl(SO_4)_2 \cdot 12H_2O$	1.5228
Cobalt acetate	$Co(C_2H_3O_2)_2 \cdot 4H_2O$	β1.542
aluminate (Thenard's Blue)	$Co(AlO_2)_2$	<1.78 (red), 1.74 (blue)
ammonium selenate	$CoSeO_4 \cdot (NH_4)_2SeO_4 \cdot 6H_2O$	1.5246, 1.5311, 1.5396
cesium sulfate	$CoCs_2(SO_4)_2 \cdot 6H_2O$	1.5057, 5.5085, 1.5132
chloride (ous)	$CoCl_2 \cdot 2H_2O$	<1.624, <1.671, >1.67
potassium selenate	$CoSeO_4 \cdot K_2SeO_4 \cdot 6H_2O$	1.5135, 1.5195, 1.5358
rubidium sulfate	$CoSO_4 \cdot Rb_2SO_4 \cdot 6H_2O$	1.4859, 1.4916, 1.5014
selenate	$CoSeO_4 \cdot 6H_2O$	α1.5225, γ1.5227
Copper ammonium selenate	$CuSeO_4 \cdot (NH_4)_2SeO_4 \cdot 6H_2O$	1.5213, 1.5355, 1.5395
ammonium sulfate	$CuSO_4 \cdot (NH_4)_2SO_4 \cdot 6H_2O$	1.4910, 1.5007, 1.5054
cesium sulfate	$CuSO_4 \cdot Cs_2SO_4 \cdot 6H_2O$	1.5048, 1.5061, 1.5153
chloride (ic)	$CuCl_2 \cdot 2H_2O$	1.644, 1.684, 1.742
formate	$Cu(CHO_2)_24H_2O$	1.4133, 1.5423, 1.5571
Copper oxide (ous) (cuprite)	Cu_2O	2.705
potassium chloride	$CuCl_2 \cdot 2KCl \cdot 2H_2O$	1.6365, 1.6148
potassium cyanide (ous)	$CuK_3(CN)_4$	1.5215
potassium selenate	$CuSeO_4 \cdot K_2SeO_4 \cdot 6H_2O$	1.5096, 1.5235, 1.5387
potassium sulfate	$CuSO_4K_2SO_4 \cdot 6H_2O$	1.4836, 1.4864, 1.5020
strontium formate	$Cu(HCO_2)_2 \cdot 2[SrHCO_2)_2]$ $8H_2O$	1.4995, 1.5199, 1.5801
sulfate (ic)	$CuSO_4$	1.724, 1.733, 1.739
Cyanogen	C_2N_2	1.327^{18} (liq.)
Germanium bromide, tetra-	$GeBr_4$	1.6269
Gold sodium chloride	$AuNaCl_4 \cdot 2H_2O$	α1.545, γ1.75 +
Hafnium oxychloride	$HfOCl_2 \cdot 8H_2O$	1.557, 1.543
Ice	H_2O	1.3049, 1.3062 (A), 1.3001, 1.3104 (D), 1.3133, 1.3147 (F)
Iron ammonium chloride	$Fe(NH_4)_2Cl_4$	1.6439
ammonium selenate	$FeSeO_4 \cdot (NH_4)_2SeO_4 \cdot 6H_2O$	1.5201, 1.5260, 1.5356
cesium sulfate (ic)	$FeCs(SO_4)_2 \cdot 12H_2O$	1.4839
cesium sulfate (ous)	$FeSO_4 \cdot Cs_2SO_4 \cdot 6H_2O$	1.5003, 1.5035, 1.5094
rubidium sulfate	$FeRb(SO_4)_2 \cdot 12H_2O$	1.48234
sulfate (ic)	$Fe_2(SO_4)_3$	1.802, 1.814, 1.818
thallium sulfate	$FeTl(SO_4)_2 \cdot 12H_2O$	1.52365
Lanthanum sulfate	$La_2(SO_4)_3 \cdot 9H_2O$	1.564, 1.569
Lead *ortho*arsenate, di-	$PbHAsO_4$	1.8903, 1.9097, 1.9765
nitrate	$Pb(NO_3)_2$	1.782
Lithium ammonium sulfate	$LiNH_4SO_4$	β1.437 (Li)
ammonium tartrate (*d*)	$LiNH_4(C_4H_4O_6) \cdot H_2O$	β1.567, γ1.5673
ammonium tartrate (*dl*)	$LiNH_4(C_4H_4O_6) \cdot H_2O$	β1.5287
bromide	LiBr	1.784
chloride	LiCl	1.662
dithionate	$Li_2S_2O_6 \cdot H_2O$	1.5487, 1.5602, 1.5788
oxide	Li_2O	1.644
potassium sulfate	$LiKSO_4$	1.4723, 1.4717
potassium tartrate	$LiK(C_4H_4O_6) \cdot H_2O$	β1.5226 (red)
rubidium tartrate (*a*)	$LiRb(C_4H_4O_6) \cdot H_2O$	β1.552
sodium tartrate (*dl*)	$LiNa(C_4H_4O_6) \cdot 2H_2O$	β1.4904
Magnesium ammonium selenate	$MgSeO_4 \cdot (NH_4)_2SeO_4 \cdot 6H_2O$	1.5070, 1.5093, 1.5169
ammonium sulfate	$Mg(NH_4)_2 \cdot (SO_4)_2 \cdot 6H_2O$	1.4716, 1.4730, 1.4786
*ortho*borate	$3MgO \cdot B_2O_3$	1.6527, 1.6537, 1.6748
cesium sulfate	$MgCs_2(SO_4)_2 \cdot 6H_2O$	1.4857, 1.4858, 1.4916
chlorostannate	$MgSnCl_6 \cdot 6H_2O$	1.5885, 1.5970
fluosilicate	$MgSiF_6 \cdot 6H_2O$	1.3439, 1.3602
platinocyanide	$MgPt(CN)_4 \cdot 7H_2O$	1.5608, 1.91
Magnesium potassium selenate	$MgK_2(SeO_4)_2 \cdot 6H_2O$	1.4969, 1.4991, 1.5139
potassium sulfate	$MgK_2(SO_4)_2 \cdot 6H_2O$	1.407, 1.4629, 1.4755
rubidium sulfate	$MgRb_2(SO_4)_2 \cdot 6Hi2O$	1.4672, 1.4689, 1.4779
silicate	$MgSiO_3$	1.651, 1,654 (calc.), 1.660
sulfide	MgS	2.271 also 2.268
Manganese borate	$Mn_3B_4O_9$	1.617, 1.738, 1.776
cesium sulfate	$MnCs_2(SO_4)_2 \cdot 6H_2O$	1.4946, 1.4966, 1.5025
chloride	$MnCl_2 \cdot 4H_2O$	1.555, 1.575, 1.607
rubidium sulfate	$MnRb_2(SO_4)_2 \cdot 6H_2O$	1.4767, 1.4807, 1.4907
sulfate (ous)	$MnSO_4 \cdot 4H_2O$	1.508, 1.518, 1.522
	$MnSO_4 \cdot 5H_2O$	1.495, 1.508, 1.514
Mercury chloride (ic)	$HgCl_2$	1.725, 1.859, 1.965
cyanide (ic)	$Hg(CN)_2$	1.645, 1.492
iodide (ic) (red)	HgI_2	2.748, 2.455

Inorganic Compounds (Continued)

Name	Formula	Index
Nickel ammonium selenate	$Ni(NH_4)_2\cdot(SeO_4)_2\cdot 6H_2O$	1.5291, 1.5372, 1.5466
cesium sulfate	$NiCs_2(SO_4)_2\cdot 6H_2O$	1.5087, 1.5129, 1.5162
Nickel chloride	$NiCl_2\cdot 6H_2O$	α1.53, γ1.61
fluoride, acid	$NiF_2\cdot 5HF\cdot 6H_2O$	1.392, 1.408
potassium selenate	$NiK_2(SeO_4)_2\cdot 6H_2O$	1.5199, 1.5248, 1.5339
rubidium sulfate	$NiRb_2(SO_4)_2\cdot 6H_2O$	1.4895, 1.4961, 1.505
selenate	$NiSeO_4\cdot 6H_2O$	1.593, 1.5125
Platinum potassium dibromo-nitrite	$PtK_2(NO_2)_2)Br_2\cdot H_2O$	1.626, 1.6684, 1.757
Potassium carbonate	K_2CO_3	1.426, 1.531, 1.541
carbonate, acid	$KHCO_3$	1.380, 1.482, 1.598
*perch*lorate	$KC10_4$	1.4731, 1.4737, 1.4769
chloroplatinate	K_2PtCl_6	1.827 (577 mμ)
chloroplatinite	K_2PtCl_4	1.64, 1.67
*di*chromate	$K_2Cr_2O_7$	1.7202, 1.7280, 1.8197
cyanide	KCN	1.410
fluoborate	KBF_4	1.3239, 1.3245, 1.3247
fluoride	KF	1.352 (1.361)
	$KF\cdot 2H_2O$	1.345, 1.352, 1.363
fluosilicate	K_2SiF_6	1.3391
*per*iodate	$KClO_4$	1.6205, 1.6479
lithium ferrocyanide	$K_2Li_2Fe(CN)_6\cdot 3H_2O$	1.5883, 1.6007, 1.6316
*hypo*phosphate	$K_2H_2P_2O_6\cdot 2H_2O$	1.4893, 1.5314, 1.5363
	$K_2H_2P_2O_6\cdot 3H_2O$	1.4768, 1.4843, 1.4870
ruthenium cyanide	$K_4Ru(CN)_6\cdot 3H_2O$	β1.5837
silicate	K_2SiO_3	1.520, 1.521, 1.528
thiocyanate	KCNS	1.532, 1.660, 1.730
thionate, tetra-	$K_2S_4O_6$	1.5896, 1.6057, 1.6435
penta-	$2K_2S_5O_6\cdot 3H_2O$	1.565, 1.63, 1.655
Rhodium cesium sulfate	$RhCs(SO_4)_2\cdot 12H_2O$	1.5077
Rubidium *per*chlorate	$RbC10_4$	1.4692, 1.4701, 1.4731
chromate	Rb_2CrO_4	β1.71, γ1.72
dithionate	$Rb_2S_2O_6$	1.4574, 1.5078
fluoride	RbF	1.396
selenate	$RbSeO_4$	1.5515, 1.5537, 1.5582
Ruthenium sodium nitrate	$RuNa_2(NO_2)_5\cdot 2H_2O$	1.5889, 1.5943, 1.7163
Selenium oxide	SeO_2	>1.76
Silver cyanide	AgCN	1.685, 1.94
nitrate	$AgNO_3$	1.729, 1.744, 1.788
phosphate	Ag_2HPO_4	1.8036, 1.7983
potassium cyanide	$AgK(CN)_2$	1.625, 1.63
Sodium ammonium tartrate (*d*)	$NaNH_4(C_4H_4O_6)\cdot 4H_2O$	1.495, 1.498, 1.499
ammonium tartrate (*dl*)	$NaNH_4(C_4H_4O_6)\cdot H_2O$	β1.473 (red)
*ortho*arsenate	$NaH_2AsO_4\cdot H_2O$	1.5382, 1.5535, 1.5607
	$NaH_2AsO_4\cdot 2H_2O$	1.4794, 1.5021, 1.5265
bromide	NaBr	1.6412
carbonate	Na_2CO_3	1.415, 1.535, 1.546
Sodium carbonate, acid	$NaHCO_3$	1.376, 1.500, 1.582
cyanide	NaCN	1.452
iodide	NaI	1.7745
molybdate	$3Na_2O.7MoO_3\cdot 22H_2O$	β1.627
nitrate	$NaNO_3$	1.5874, 1.3361
phosphate	$NaH_2PO_4\cdot 2H_2O$	1.4401, 1.4629, 1.4815
	$Na_2HPO_4\cdot 7H_2O$	1.4412, 1.4424, 1.4526
*hypo*phosphate	$Na_3HP_2O_6\cdot 9H_2O$	1.4653, 1.4738, 1.4804
silicate	Na_2SiO_3	1.513, 1.520, 1.528
sulfate, acid	$NaHSO_4\cdot H_2O$	1.43, 1.46, 1.47
sulfite	Na_2SO_3	1.565, 1.515
acid	$Nahso_3$	1.474, 1.526, 1.685
tartrate, acid (d)	$NaH(C_4H_4O_6)\cdot H_2O$	β1.533
thiocyanate	NaCNS	1.545, 1.625, 1.695
Sodium tungstate	$Na_2WO_4\cdot 2H_2O$	1.5526, 1.5533, 1.5695
vanadate	$Na_3VO_4\cdot 10H_2O$	1.5305, ω1.5398, ε1.5475
	$Na_3VO_4\cdot 12H_2O$	1.5095, 1.5232
Strontium dichromate	$SrCr_2O_7\cdot 3H_2O$	1.7146, 1.7174, 1.812
fluoride	SrF_2	1.442 (1.438)
oxide	SrO	1.870
*ortho*phosphate, acid	$SrHPO_4$	1.608, 1.62±, 1.625
sulfide, mono-	SrS	2.107
Sulfur nitride	S_4N_4	α1.908, β2.046
Thallium chloride, mono-	TlCl	2.247
iodide, mono-	TlI	2.78
Tin iodide (ic)	SnI_4	2.106
Uranyl potassium sulfate	$UO_2\cdot SO_4\cdot k_2SO_4\cdot 2H_2O$	1.5144, 1.5266, 1.5705 (580 mμ)
Vanadium ammonium sulfate	$VNH_4(SO_4)_2\cdot 12H_2O$	1.475
Zinc ammonium selenate	$Zn(SeO_4)\cdot(NH_4)_2SeO_4\cdot 6H_2O$	1.5240, 1.5300, 1.5385
bromate	$Zn(BrO_3)_2\cdot 6H_2O$	1.5452
cesium sulfate	$ZnCs_2(SO_4)_2\cdot 6H_2O$	1.5022, 1.5048, 1.5093
chloride	$ZnCl_2$	1.687, 1.713
fluosilicate	$ZnSiF_6\cdot 6H_2O$	1.3824, 1.3956
potassium cyanide	$ZnK_2(CN)_4$	1.4115
selenate	$ZnK_2(SeO_4)_2\cdot 6H_2O$	1.5121, 1.5181, 1.5335
sulfate	$ZnK_2(SO_4)_2\cdot 6H_2O$	1.4775, 1.4833, 1.4969
rubidium sulfate	$ZnRb_2(SO_4)_2\cdot 6H_2O$	1.4833, 1.4884, 1.4975
silicate	$ZnSiO_3$	1.616, 1.62±, 1.623
Zirconium ammonium fluoride	$Zr(NH_4)_3F_7$	1.433

ORGANIC COMPOUNDS

See also under Physical Constants of Organic Compounds.

Name	Index
Allontoin, solid	γ1.579, λ1.660
Dimethyl thiophene (α, α′), liq	1.51693[13.4] (He)
(β, β′), liq	1.52217[15] (He)
Ethyl carbylamine, liq	1.3659[24]
Ethylidene cyanhydrin, liq	1.40582[18.4]
Hexyl acetylene (n), liq	1.4208[12.5]

Miscellaneous

Name	Index	Name	Index
Albite glass	1.4890	Magdala red	1.90
Amber	1.546	Obsidian	1.482-1.496
Anorthite glass	1.5755	Paraffin	1.43295[38.3] (C)
Asphalt	1.635	Quartz, fused	1.45640 (656 mμ)
Bell metal	1.0052		1.45843 (589 mμ)
Borax, amorphous, fused	1.4630		1.46190 (509 mμ)
Canada balsam	1.530		1.47503 (361 mμ)
Ebonite	1.66 (red)		1.49634 (275 mμ)
Fuchsin	2.70		1.53386 (214 mμ)
Gelatin, Nelson's No. 1	1.530		1.57464 (185 mμ)
Gelatin, various	1.516-1.534	Resin, aloes	1.619 (red)
Gum Arabic	1.480 (1.5,4) (red)	colophony	1.548 (red)
		copal	1.528 (red)
Hoffman's violet	2.20	mastic	1.535 (red)
Ivory	1.539, 1.541	Peru balsam	1.593

INDEX OF REFRACTION OF ORGANIC COMPOUNDS

(See also that section of this book which contains data of Physical Constants of Organic Compounds)

The following table contains a list of organic compounds arranged in order of increasing refractive index. Measurements were made at 25°C.

Compound	n_D	Compound	n_D
Trifluoroacetic acid	1.283	Propionaldehyde	1.371
2,2,2-Trifluoroethanol	1.290	n-Hexane	1.372
Octofluoropentanol-1	1.316	2,3-Dimethylbutane	1.372
Dodecafluoroheptanol-1	1.316	3-Methylpentane	1.374
Methanol	1.326	2-Propanol	1.375
Acetonitrile	1.342	Isopropyl acetate	1.375
Ethyl ether	1.352	Propyl formate	1.375
Acetone	1.357	2-Chloropropane	1.376
Ethyl formate	1.358	2-Butanone	1.377
Ethanol	1.359	2-Chloropropane	1.377
Methyl acetate	1.360	Methylethyl ketone	1.377
Propionitrile	1.363	Butyraldehyde	1.378
2,2-Dimethylbutane	1.366	2,4-Dimethylpentane	1.379
Isopropyl ether	1.367	Propyl ether	1.379
2-methylpentane	1.369	Acetaldehyde-diethylacetal	1.379
Ethyl acetate	1.370	Butylethyl ether	1.380
Acetic acid	1.370	Nitromethane	1.380

Compound	n_D
Trifluoropropanol	1.381
2-Methylhexane	1.382
Butyronitrile	1.382
Propyl acetate	1.382
Ethyl propionate	1.382
2-Methyl-2-propanol	1.383
1-Propanol	1.383
Isobutyl formate	1.383
Diethyl carbonate	1.383
Heptane	1.385
2-Methyl-2-propanol	1.385
Propionic acid	1.385
3-Methylhexane	1.386
n-Propyl amine	1.386
1,1-Dimethyl-2-propanone	1.386
1-Chloropropane	1.386
2,2,3-Trimethylbutane	1.387
Methylpropyl ketone	1.387
sec-Butyl acetate	1.387
Butyl formate	1.387
β-Methylpropyl ethanoate	1.388
2,2,4-Trimethyl pentane	1.389
2,3-Dimethyl pentane	1.389
Acetic anhydride	1.389
Diisopropyl amine	1.390
2-Aminobutane	1.390
2-Pentanone	1.390
3-Pentanone	1.390
Nitroethane	1.390
Methyl-b-butyrate	1.391
Butyl acetate	1.392
2-Nitropropane	1.392
4-Methyl-2-pentnone	1.394
2-Methyl-1-propanol	1.394
Octane	1.395
1-Amino-2-methylpropane	1.395
Valeronitrile	1.395
2-Butanol	1.395
5-Hexanone	1.395
5-Methyl-3-hexanone	1.395
2-Chlorobutane	1.395
Butyric acid	1.396
2,2,2-Trimethylhexane	1.397
n-Dibutyl ether	1.397
1-Butanol	1.397
Acrolein	1.397
1-Chloro-2-methylpropane	1.397
Methacrylonitrile	1.398
3-Methyl-2-pentanone	1.398
Triethyl amine	1.399
n-Butyl amine	1.399
1,1,3,3-Tetramethyl-2-propanone	1.399
Isobutyl-n-butyrate	1.399
1-Nitropropane	1.399
n-Dodecane	1.400
Amyl acetate	1.400
1-Chlorobutane	1.400
2-Methoxy ethanol	1.400
Propionic acid anhydride	1.400
2,2,3-Trimethylpentane	1.401
1-Chlorobutane	1.401
β-Methoxypropionitrile	1.401
3-Methyl butanoic acid	1.402
n-Nonane	1.403
Dipropylamine	1.403
Isoamylacetate	1.403
Cyclopentane	1.404
2-Methyl-2-butanol	1.404
3-Methyl-1-butanol	1.404
Tetrahydrofuran	1.404
Capronitrile	1.405
2-Pentanone	1.405
2-Ethoxyethanol	1.405
2-Heptanone	1.406
Valeric acid	1.406
Diisobutylene	1.407
Methylcyclopentane	1.407
Isoamyl ether	1.407
Methylpropyl carbinol	1.407
Tributyl borate	1.407
1-Pentanol	1.408
3-Methyl-2-butanol	1.408
Diethyl oxalate	1.408
n-Decane	1.409
4-Methyl-2-pentanol	1.409
3-Isopropyl-2-pentanone	1.409
2-Methyl-1-butanol	1.409
Butyric acid anhydride	1.409
Amyl ether	1.410
Isoamyl isovalerate	1.410
1-Chloropentane	1.410
2-Propene-1-ol	1.411
2,4-Dimethyl dioxane	1.412
Ethyl lactate	1.412
Diethyl malonate	1.412
3-Chloropropene	1.413
Ethyleneglycol diacetate	1.413
2-Octanone	1.414
3-Octanone	1.414
3-Methyl-2-heptanone	1.415
Caproic acid	1.415
4-Methyldioxane	1.415
1,2-Propyleneglycol-1-monobutyl ether	1.415

Compound	n_D
Ethylcyanoacetate	1.415
Dibutylamine	1.416
2-Pentanol	1.416
1,1-Dichloroethane	1.416
Heptachlorodiethyl ether	1.416
1-Hexanol	1.416
1-Amino-3-methoxy propane	1.417
Octyl nitrile	1.418
2-Heptanol	1.418
2-Propenyl amine	1.419
1,2-Propyleneglycol carbonate	1.419
Methylpentyl carbinol	1.420
2-Ethyl-1-butanol	1.420
1-Chloro-2-mehyl-1-propene	1.420
p-Dioxane	1.420
Methylcyclohexane	1.421
4-Hydroxy-4-methyl-2-pentanone	1.421
1-Heptanol	1.422
3-Isorpopyl-2-heptanone	1.423
Cyclohexane	1.424
2-Bromopropane	1.424
3-Chloro-2-methylprop-1-ene	1.425
Caproic acid	1.426
Glycol carbonate	1.426
1-Octanol	1.427
1,1-Dimethylhexanol	1.427
N,N-Dimethylformamide	1.427
Sulfuric acid	1.427
1-Chlorooctane	1.428
Triisobutylene	1.429
N-Methylaniline nitrile	1.429
Etbylene glycol	1.429
1-Chloro-2-ethylhexane	1.430
Ethylcyclohexane	1.431
1.2-Propanediol	1.431
1-Bromopropane	1.431
2-Methyl-7-ethyl-4-nonanone	1.433
Ethyleneglycol-mono-allyl ether	1.434
Butyral lactone	1.434
2-Methyl-7-ethyl-4-undecanone	1.435
4-n-Propyl-5-ethyldioxane	1.435
1,2-Dichloro-2-methylpropane	1.435
1,2-Propyleneglycol sulfite	1.435
N-Methylmorpholine	1.436
1-Chloro-2-methyl-2-propanol	1.436
Epichlorohydrin	1.436
Triethyleneglycol-mono-butyl ether	1.437
4-Ethyl-7,7,7-trimethyl-1-heptanol	1.438
1-Methyl-3-ethyloctan-1-ol	1.438
1-Ethyl-3-ethylhexan-1-ol	1.438
Diethyl maleate	1.438
1-Butanethiol	1.440
2-Chloroethanal	1.440
Dibutyl sebacate	1.440
1-Ethyl-3-ethyloctan-1-ol	1.441
Dimethylmaleate	1.441
3-Methylpentane-2,4-diol	1.441
Ethyl sulfide	1.442
Mesityl oxide	1.442
Butyl stearate	1.442
1,2-Dichloroethane	1.444
Chloroform	1.444
trans-1,2-Dichloroethylene	1.444
Diethyleneglycol	1.445
cis-1,2-Dichloroethylene	1.445
3-(α-Butyloctyl)-oxypropyl-1-amine	1.446
2-Methylmorpholine	1.446
Dipropyleneglycol-monoethyl ether	1.446
Formamide	1.446
3-Laurylоxypropyl-1-amine	1.447
Cyclohexanone	1.448
1-Aminopropan-1-ol	1.448
Diethyleneglycol-mono-β-oxypropyl ether	1.448
1-Amino-2-methylpentan-1-ol	1.449
Tetrahydrofurfural alcohol	1.450
2-Propylcyclohexa-1-one	1.452
2-Aminoethanol	1.452
2-Butylcyclohexan-1-one	1.453
Ethylenediamine	1.454
2-(β-Methyl)-propylcyclohexan-1-one	1.454
4-Methylcyclohexanol	1.454
3-Methylcyclohexanol	1.455
bis-2-Chloroethyl ether	1.455
Cyclohexylamine	1.456
1,8-Cineol	1.456
2,2′-Dimethyl-2,2′-dipropyldieththanol amine	1.456
1,1′,2,2′-Tetramethyldiethanol amine	1.459
1-Aminopropan-3-ol	1.459
Carbon tetrachloride	1.459
3-Methyl-5-ethylheptan-2,4-diol	1.459
2-(β-Ethyl)-butylcyclohexan-1-one	1.461
2-Methylcyclohexanol	1.461
N-(n-Butyl)-diethanol amine	1.461
4,5-Chloro-1,3-dioxolane-2	1.461
2-Butylcyclohexan-1-ol	1.462

INDEX OF REFRACTION OF FUSED QUARTZ

λ mμ, 15°C	n, 18°C	λ mμ, 15°C	n, 18°C	λ mμ, 15°C	n, 18°C	λ mμ, 15°C	n, 18°C
185.467	1.57436	434.047	1.46690	250.329	1.50745	546.072	1.46013
193.583	1.55999	435.834	1.46675	257.304	1.50379	589.29	1.45845
202.55	1.54727	467.815	1.46435	274.867	1.49617	643.847	1.45674
214.439	1.53386	479.991	1.46355	303.412	1.48594	656.278	1.45640
219.462	1.52907	486.133	1.46318	340.365	1.47867	706.520	1.45517
226.503	1.52308	508.582	1.46191	396.848	1.47061	794.763	1.45340
231.288	1.51941	533.85	1.46067	404.656	1.46968		

INDEX OF REFRACTION OF AIR (15°C, 76 cm Hg)

Corrections for reducing wavelengths and frequencies in air (15°C, 76 cm Hg) to vacuo
The indices were computed from the Cauchy formula $(n - 1)10^7 = 2726.43 + 12.288/(\lambda_2 \times 10^{-8}) + 0.3555/(\lambda^4 \times 10^{-16})$. For 0°Cand 76 cm Hg the constants of the equation become 2875.66, 13.412 and 0.3777 respectively, and for 30°C and 76 cm Hg 2589.72, 12.259 and 0.2576. Sellmeier's formula for but one absorption band closely fits the observations: $n_2 = 1 + 0.00057378\lambda^2/(\lambda^2 - 595260)$. If n — 1 were strictly proportional to the density, then $(n - 1)_o/(n - 1)t$ would equal $1 + \alpha t$ where α should be 0.00367. The following values of α were found to hold:

λ	0.85μ	0.75μ	0.65μ	0.55μ	0.45μ	0.35μ	0.25μ
α	0.003672	0.003674	0.003678	0.003685	0.003700	0.003738	0.003872

The indices are for dry air (0.05 ± % CO_2). Corrections to reduce to dry air the indices for moist air may be made for any wavelength by Lorenz's formula, +0.000041(m/760), where m is the vapor pressure in mm. The corresponding frequencies in waves per cm and the corrections to reduce wavelengths and frequencies in air at 15°C and 76 cm Hg pressure to vacuo are given. E.g., a light wave of 5000 angstroms in dry air at 15°C, 76 cm Hg becomes 5001.391 Å in vacuo; a frequency of 20,000 waves per cm correspondingly becomes 19994.44.

Wave-length, λ angstroms	Dry air (n — 1) × 10⁷ 15°C 76 cm Hg	Vacuo correction for λ in air (nλ — λ) add	Frequency waves per cm 1/λ in air	Vacuo correction or 1/λ in air (1/nλ — 1/λ) subtract	Wave-length, λ angstroms	Dry air (n — 1) × 10⁷ 15°C 76 cm Hg	Vacuo correction for λ in air (nλ — λ) add	Frequency waves per cm 1/λ in air	Vacuo correction for 1/λ in air (1/nλ — 1/λ) substract
2000	3256	.651	50,000	16.27	5500	2771	1.524	18,181	5.04
2100	3188	.670	47,619	15.18	5600	2769	1.551	17,857	4.94
2200	3132	.689	45,454	14.23	5700	2768	1.578	17,543	4.85
2300	3086	.710	43,478	13.41	5800	2766	1.604	17,241	4.77
2400	3047	.731	41,666	12.69	5900	2765	1.631	16,949	4.68
2500	3014	.754	40,000	12.05	6000	2763	1.658	16,666	4.60
2600	2986	.776	38,461	11.48	6100	2762	1.685	16,393	4.53
2700	2962	.800	37,037	10.97	6200	2761	1.712	16,129	4.45
2800	2941	.824	35,714	10.50	6300	2760	1.739	15,873	4.38
2900	2923	.848	34,482	10.08	6400	2759	1.766	15,625	4.31
3000	2907	.872	33,333	9.69	6500	2758	1.792	15,384	4.24
3100	2893	.897	32,258	9.33	6600	2757	1.819	15,151	4.18
3200	2880	.922	31,250	9.00	6700	2756	1.846	14,925	4.11
3300	2869	.947	30,303	8.69	6800	2755	1.873	14,705	4.05
3400	2859	.972	29,411	8.41	6900	2754	1.900	14,492	3.99
3500	2850	.998	28,571	8.14	7000	2753	1.927	14,285	3.93
3600	2842	1.023	27,777	7.89	7100	2752	1.954	14,084	3.88
3700	2835	1.049	27,027	7.66	7200	2751	1.981	13,888	3.82
3800	2829	1.075	26,315	7.44	7300	2751	2,008	13,698	3.77
3900	2823	1.101	25,641	7.24	7400	2750	2.035	13,513	3.72
4000	2817	1.127	25,000	7.04	7500	2749	2.062	13,333	3.66
4100	2812	1.153	24,390	6.86	7600	2749	2.089	13,157	3.62
4200	2808	1.179	23,809	6.68	7700	2748	2.116	12,987	3.57
4300	2803	1.205	23,255	6.52	7800	2748	2.143	12,820	3.52
4400	2799	1.232	22,727	6.36	7900	2747	2.170	12,658	3.48
4500	2796	1.258	22,222	6.21	8000	2746	2.197	12,500	3.43
4600	2792	1.284	21,739	6.07	8100	2746	2.224	12,345	3.39
4700	2789	1.311	21,276	5.93	8250	2745	2.265	12,121	3.33
4800	2786	1.338	20,833	5.80	8500	2744	2.332	11,764	3.23
4900	2784	1.364	20,406	5.68	8750	2743	2.400	11,428	3.13
5000	2781	1.391	20,000	5.56	9000	2742	2.468	11,111	3.05
5100	2779	1.417	19,607	5.45	9250	2741	2.536	10,810	2.96
5200	2777	1.444	19,230	5.34	9500	2740	2.604	10,526	2.88
5300	2775	1.471	18,867	5.23	9750	2740	2.671	10,256	2.81
5400	2773	1.497	18,518	5.13	10000	2739	2.739	10,000	2.74

INDEX OF REFRACTION, GASES

Values are relative to a vacuum and for a Temp. of 0°C, and 760 mm pressure.
(From Smithsonian Tables)

Substance	Kind of light	Indices of refraction	Observer	Substance	Kind of light	Indices of refraction	Observer
Acetone	D	1.001079-1.001100		Hydrochloric acid	D	1.000447	Mascart
Air	D	1.0002926	Perreau	Hydrogen	white	1.000138-1.000143	
Ammonia	white	1.000381-1.000385		Hydrogen	D	1.000132	Burton
Ammonia	D	1.000373-1.000379		sulfide	D	1.000644	Dulong
Argon	D	1.000281	Rayleigh	sulfide	D	1.000623	Mascart
Benzene	D	1.001700-1.001823		Methane	white	1.000443	Dulong
Bromine	D	1.001132	Mascart	Methane	D	1.000444	Mascart
Carbon dioxide	white	1.000449-1.000450		Methyl alcohol	D	1.000549-1.000623	
dioxide	D	1.000448-1.000454		Methyl ether	D	1.000891	Marcast
disulfide	white	1.001500	Dulong	Nitric oxide	white	1.000303	Dulong
disulfide	D	1.001478-1.001485		Nitric oxide	D	1.000297	Mascart
monoxide	White	1.000340	Dulong	Nitrogen	white	1.000295-1.000300	
monoxide	white	1.000335	Mascart	Nitrogen	D	1.000296-1.000298	
Chlorine	white	1.000772	Dulong	Nitrous oxide	white	1.000503-1.000507	
Chlorine	D	1.000773	Mascart	Nitrous oxide	D	1.000516	Mascart
Chloroform	D	1.001436-1.001464		Oxygen	white	1.000272-1.000280	
Cyanogen	white	1.000834	Dulong	Oxygen	D	1.000271-1.000272	
Cyanogen	D	1.000784-1.000825		Pentane	D	1.001711	Mascart
Ethyl alcohol	D	1.000871-1.000885		Sulfur dioxide	white	1.000665	Dulong
ether	D	1.001521-1.001544		Sulfur dioxide	D	1.000686	Ketteler
Helium	D	1.000036	Ramsay	Water	white	1.000261	Jamin
Hydrochloric acid	white	1.000449	Mascart	Water	D	1.000249-1.000259	

COEFFICIENT OF TRANSPARENCY OF UVIOL GLASS FOR THE ULTRA-VIOLET

For a thickness of 1 mm.

Wave length microns	0.280	0.309	0.325	0.346	0.361	0.383	0.397
Uviol crown	0.56	0.95	0.990	0.996	0.999	1.000	1.000

RADIATION FROM AN IDEAL BLACK BODY

From NASA TT-F-783

Temperature dependence of the specific power radiated, Q_T, and of λ_{max} for an ideal black body according to Kirchhoff's law ($\sigma_O = 5.68 \cdot 10^{-8}$ W/m²·deg⁴ $= 4.88 \cdot 10^{-8}$ kcal/m²·hr·deg⁴)

T, °K	t, °C	Q_T, W/cm²	Q_T kcal/m²·hr	λmax μ
100	−173	$5.680 \cdot 10^{-4}$	$4.880 \cdot 10^{0}$	28.96
200	−73	$9.088 \cdot 10^{-3}$	$7.808 \cdot 10^{1}$	14.48
273	0	$3.155 \cdot 10^{-2}$	$2.711 \cdot 10^{2}$	10.608
300	27	4.601	3.953	9.655
310	37	5.246	4.507	9.342
320	47	5.956	5.117	9.050
330	57	6.736	5.787	8.766
340	67	7.590	6.521	8.518
350	77	8.524	7.323	8.274
360	87	9.540	8.196	8.044
370	97	1.065	9.146	7.827
380	107	$1.184 \cdot 10^{-1}$	$1.018 \cdot 10^{3}$	7.621
390	117	1.314	1.128	7.426
400	127	1.454	1.249	7.270
410	137	1.605	1.379	7.053
420	147	1.76	1.519	6.865
430	157	1.942	1.668	6.735
440	167	2.129	1.829	6.562
450	177	2.329	2.001	6.436
460	187	2.543	2.185	6.266
470	197	2.772	2.381	6.162
480	207	3.015	2.591	6.033
490	217	3.274	2.813	5.910
500	227	3.550	3.05	5.792
510	237	3.843	3.301	5.668
520	247	4.163	3.568	5.559
530	257	4.482	3.851	5.454
540	267	4.830	4.150	5.363
550	277	5.198		5.255
560	287	5.586	4.799	5.161
570	297	5.996	5.151	5.061
580	307	6.428	5.522	4.963
590	317	6.883	5.913	4.908
600	327	7.361	6.324	4.827
610	337	7.864	6.757	4.748
620	347	8.393	7.211	4.671
630	357	8.948	7.687	4.597
640	367	9.529	8.187	4.525
650	377	$1.014 \cdot 10^{0}$	8.711	4.455
660	387	1.078	9.260	4.388
670	397	1.145	9.831	4.322
680	407	1.214	$1.013 \cdot 10^{4}$	4.259
690	417	1.287	1.106	4.197
700	427	1.364	1.172	4.137
710	437	1.443	1.240	4.069
720	447	1.526	1.311	4.022
730	457	$1.613 \cdot 10^{0}$	$1.386 \cdot 10^{1}$	3.967
740	467	1.703	1.463	3.914
750	477	1.797	1.544	3.861
760	487	1.895	1.628	3.811
770	497	1.997	1.715	3.761
780	507	2.102	1.806	3.713
790	517	2.212	1.901	3.666
800	527	2.327	1.999	3.620
810	537	2.445	2.101	3.565
820	547	2.568	2.206	3.532
830	557	2.696	2.316	3.489
840	567	2.828	2.430	3.448
850	577	2.965	2.547	3.407
860	587	3.107	2.670	3.367

T, °K	t, °C	Q_T W/cm²	Q_T kcal/m²·hr	λmax μ
870	597	3.254	2.796	3.329
880	607	3.406	2.927	3.291
890	617	3.564	3.062	3.254
900	627	3.727	3.202	3.218
910	637	3.895	3.346	3.162
920	647	4.069	3.496	3.148
930	657	4.249	3.650	3.114
940	667	4.435	3.810	3.081
950	677	4.626	3.975	3.048
960	687	4.824	4.145	3.017
970	697	5.028	4.320	2.986
980	707	5.239	4.501	2.955
990	717	5.456	4.688	2.925
1000	727	5.680	4.880	2.896
1010	737	5.909	5.07	2.866
1020	747	6.143	5.278	2.836
1030	757	6.394	5.494	2.812
1040	767	6.645	5.709	2.785
1050	777	6.904	5.932	2.758
1060	787	7.171	6.161	2.732
1070	797	7.445	6.397	2.707
1080	807	7.728	6.640	2.681
1090	817	8.018	6.888	2.657
1100	827	8.316	7.145	2.633
1110	837	8.623	7.408	2.609
1120	847	8.937	7.679	2.586
1130	857	9.261	7.956	2.563
1140	867	9.593	8.242	2.540
1150	877	9.934	8.535	2.516
1160	887	$1.028 \cdot 10^{1}$	8.836	2.497
1170	897	1.064	9.145	2.475
1180	907	1.101	9.461	2.454
1190	917	$1.139 \cdot 10^{1}$	$9.786 \cdot 10^{4}$	2.434
1200	927	1.178	$1.042 \cdot 10^{5}$	2.413
1210	937	1.218	1.046	2.393
1220	947	1.258	1.081	2.374
1230	957	1.300	1.117	2.354
1240	967	1.343	1.154	2.335
1250	977	1.387	1.191	2.317
1260	987	1.431	1.230	2.298
1270	997	1.478	1.270	2.280
1280	1007	1.525	1.310	2.263
1290	1017	1.573	1.351	2.245
1300	1027	1.622	1.394	2.227
1310	1037	1.673	1.437	2.211
1320	1047	1.724	1.482	2.194
1330	1057	1.777	1.527	2.177
1340	1067	1.831	1.573	2.161
1350	1077	1.887	1.621	2.145
1360	1087	1.943	1.669	2.129
1370	1097	2.001	1.719	2.114
1380	1107	2.058	1.768	2.099
1390	1117	2.120	1.822	2.083
1400	1127	2.182	1.875	2.067
1410	1137	2.245	1.929	2.054
1420	1147	2.309	1.984	2.036
1430	1157	2.375	2.041	2.025
1440	1167	2.442	2.098	2.011
1450	1177	2.511	2.157	1.997
1460	1187	2.581	2.217	1.984

T, °K	t, °C	Q_T W/cm²	Q_T kcal/m²·hr	λmax μ
1470	1197	2.652	2.279	1.970
1480	1207	2.725	2.341	1.957
1490	1217	2.800	2.405	1.944
1500	1227	2.876	2.471	1.931
1510	1237	2.953	2.537	1.917
1520	1247	3.032	2.605	1.905
1530	1257	3.113	2.674	1.893
1540	1267	3.195	2.745	1.881
1550	1277	3.278	2.817	1.866
1560	1287	3.363	2.890	1.856
1570	1297	3.451	2.965	1.845
1580	1307	3.540	3.041	1.833
1590	1317	3.630	3.198	1.821
1600	1327	3.722		1.810
1610	1337	3.816	3.279	1.797
1620	1347	3.912	3.361	1.787
1630	1357	4.010	3.445	1.777
1640	1367	$4.109 \cdot 10^{1}$	$3.530 \cdot 10^{5}$	1.766
1650	1377	4.210	3.617	1.755
1660	1387	4.313	3.706	1.745
1670	1397	4.418	3.796	1.734
1680	1407	4.525	3.887	1.724
1690	1417	4.633	3.981	1.714
1700	1427	4.744	4.076	1.704
1710	1437	4.858	4.183	1.694
1720	1447	4.971	4.271	1.684
1730	1457	5.088	4.371	1.674
1740	1467	5.206	4.473	1.664
1750	1477	5.327	4.577	1.655
1760	1487	5.450	4.682	1.645
1770	1497	5.575	4.790	1.636
1780	1507	5.703	4.900	1.627
1790	1517	5.831	5.010	1.617
1800	1527	5.963	5.123	1.607
1810	1537	6.096	5.238	1.600
1820	1547	6.232	5.354	1.591
1830	1557	6.370	5.473	1.583
1840	1567	6.511	5.594	1.574
1850	1577	6.653	5.717	1.565
1860	1587	6.798	5.841	1.557
1870	1597	6.946	5.967	1.549
1880	1607	7.095	6.096	1.540
1890	1617	7.248	6.227	1.532
1900	1627	7.402	6.360	1.524
1910	1637	7.559	6.495	1.516
1920	1647	7.720	6.632	1.509
1930	1657	7.881	6.771	1.501
1940	1667	8.046	6.912	1.493
1950	1677	8.213	7.056	1.485
1960	1687	8.382	7.202	1.478
1970	1697	8.555	7.350	1.470
1980	1707	8.730	7.500	1.463
1990	1717	8.907	7.653	1.455
2000	1727	9.088	7.808	1.448
2500	2227	$2.219 \cdot 10^{2}$	$1.906 \cdot 10^{6}$	1.156
3000	2727	4.601	3.953	0.965
3500	3227	8.524	7.323	0.826
4000	3727	$1.454 \cdot 10^{3}$	1.249	0.722
4500	4227	2.329	2.001	0.644
5000	4727	3.550	3.050	0.579
5500	5227	5.198	4.465	0.527
6000	5727	7.361	6.324	0.483

REFLECTION COEFFICIENTS

Coefficients of Reflection of Miscellaneous Surfaces for Monochromatic Radiation in the Visible Spectrum
(J. L. Michaelson)

Material	Wave lengths (μm) 0.400	0.500	0.600	0.700
Carbon black in oil	0.003	0.003	0.003	0.003
Clay				
Kaolin (treated)	0.82	0.81	0.82	0.82
Kaolin (untreated)	0.75	0.79	0.85	0.86
White georgia	0.94	0.92	0.93	0.94
Magnesium oxide	0.97	0.98	0.98	0.98
Paint				
Lithopone	0.95	0.98	0.98	0.98
$MgCO_3$-Vynal acetate lacquer	0.90	0.88	0.88	0.88
ZnO-Milk	0.74	0.84	0.85	0.86
Paper				
Blotting	0.64	0.72	0.79	0.79
Calendered	0.64	0.69	0.73	0.76
Crepe, green	0.23	0.49	0.19	0.48
Crepe, red	0.03	0.02	0.21	0.69
Crepe, yellow	0.17	0.44	0.75	0.79
News print stock	0.38	0.61	0.63	0.78
Peach				
Green	0.18	0.17	0.62	0.63
Ripe	0.10	0.10	0.41	0.42
Pear				
Green	0.04	0.12	0.29	0.41
Ripe	0.08	0.19	0.46	0.53
Pigment				
Chrome yellow	0.05	0.13	0.70	0.77
French ochre	0.06	0.14	0.50	0.56
Porcelain enamel				
Blue	0.44	0.10	0.05	0.23
Orange	0.09	0.09	0.59	0.69
Red	0.05	0.03	0.08	0.62
White	0.77	0.73	0.72	0.70
Yellow	0.11	0.46	0.62	0.62
Talcum, Italian	0.94	0.89	0.88	0.88
Wheat flour	0.75	0.87	0.94	0.97

REFLECTION COEFFICIENTS OF SURFACES FOR "INCANDESCENT" LIGHT

Material	Nature of Surface	Coefficient	Authority
Aluminum, "Alzak"	Diffusing	0.77—0.81	3
"Alzak"	Specular	0.79—0.83	3
On glass	First surface	0.82—0.86	4
Polished	Specular	0.69	3
Black paper	Diffusing	0.05—0.06	4
Chromium	Specular	0.62	4
Copper	Specular	0.63	4
Gold	Specular	0.75	1
Magnesium oxide	Diffusing	0.98	5
Nickel	Specular	0.62—0.64	1,3
Platinum	Specular	0.62	1
Porcelain enamel	Glossy	0.76—0.79	3
Porcelain enamel	Ground	0.81	3
Porcelain enamel	Matt.	0.72—0.76	3
Silver	Polished	0.93	1
Silvered glass	Second surface	0.88—0.93	3
Snow	Diffusing	0.93	2
Steel	Specular	0.55	1
Stellite	Specular	0.58—0.65	4

(1) Hagen and Rubena; (2) Nutting, Jones, and Elliot; (3) J. E. Bock; (4) Frank Benford; (5) J. L. Michaelson.

EMISSIVITY AND ABSORPTION

These data are the result of investigations made by the Bureau of Standards, the British National Physical Laboratory, General Electric Research Laboratories, and several eastern universities, and were collected by W. J. King of the General Electric Company.

Low Temperature Total Emissivities

Silver, highly polished	0.02	Brass, polished	0.60
Platinum, highly polished	0.05	Oxidized copper	0.60
Zinc, highly polished	0.05	Oxidized steel	0.70
Aluminum, highly polished	0.08	Bronze paint	0.80
Monel metal, polished	0.09	Black gloss paint	0.90
Nickel, polished	0.12	White lacquer	0.95
Copper, polished	0.15	White vitreous enamel	0.95
Stellite, polished	0.18	Asbestos paper	0.95
Cast iron, polished	0.25	Green paint	0.95
Monel metal, oxidized	0.43	Gray paint	0.95
Aluminum paint	0.55	Lamp black	0.95

Coefficient of Absorption of Solar Radiation

Silver, highly polished	0.07	Stellite, polished	0.30
Platinum, highly polished	0.10	Light cream paint	0.35
Nickel, highly polished	0.15	Monel metal, polished	0.40
Aluminum*	0.15	Light yellow paint	0.45
Magnesium carbonate	0.15	Light green paint	0.50
Zinc oxide	0.15	Aluminum paint	0.55
Steel*	0.20	Zinc, polished metal	0.55
Copper	0.25	Gray paint	0.75
White lead paint	0.25	Black matte	0.97
Zinc oxide paint	0.30		

* Questionable because of scant or inconsistent data.

EMISSIVITY OF TOTAL RADIATION, ε_{tot}, FOR VARIOUS MATERIALS

Material	Temperature (° C)	ε_{tot}
Alloys		
Nickel-Chromium		
20 Ni—25 Cr—55 Fe, oxidized	200	0.90
	500	0.97
60 Ni—12 Cr—28 Fe, oxidized	270	0.89
	560	0.82
80 Ni—20 Cr	100	0.87
	600	0.87
	1300	0.89
Aluminum		
Polished	50—500	0.04—0.06
Rough surface	20—50	0.06—0.07
Strongly oxidized	55—500	0.2—0.3
	25	0.022
	100	0.028
	500	0.060
Oxidized	200	0.11
	600	0.19
Asbestos board	20	0.96
Bismuth		
Unoxidized	25	0.048
Brass		
Dull tarnished	200	0.61
Oxidized at 600° C	200	0.61
	600	0.59
Unoxidized	25	0.035
	100	0.035
Polished	200	0.03
Rolled sheet	20	0.06
Bronze		
Polished	50	0.1
Carbon		
Filament	1000—1400	0.53
Graphite	0—3600	0.7—0.8
Lamp black	20—400	0.96
Soot applied to solid	50—1000	0.96
Soot with water glass	20—200	0.96
Unoxidized	100	0.81
Chromium		
Polished	50	0.1
	500—1000	0.28—0.38
Colbalt		
Unoxidized	500	0.13
	1000	0.23
Columbium		
Unoxidized	1500	0.19
Copper		
Calorized	100	0.26
Calorized, oxidized	200	0.18
	600	0.19
Commercial, scoured to a shine	20	0.07
Oxidized	50	0.6—0.7
	500	0.88
Polished	50—100	0.02
Unoxidized	100	0.02
Unoxidized, liquid	—	0.15
Fire brick	1000	0.75
Glass	20—100	0.94—0.91
	250—1000	0.87—0.72
	1100—1500	0.7—0.67
Gold		
Carefully polished	200—600	0.02—0.03
Unoxidized	100	0.02
Enamel	100	0.37
Graphite	0—3600	0.7—0.8
Gypsum	20	0.93
Iron		
Cast		
Oxidized	200	0.64
	600	0.78
Strongly oxidized	40	0.95
	250	0.95
Unoxidized	100	0.21
Unoxidized, liquid	—	0.29
Oxidized	100	0.74
	500	0.84
	1200	0.89
Rusted	25	0.65
Wrought, dull	100	0.05

Material	Temperature (° C)	ε_{tot}
	25	0.94
Lamp black		
	20—400	0.96
Lead		
Oxidized		
	200	0.05
Unoxidized		
	200	0.63
Mercury		
Unoxidized		
	25	0.10
	100	0.12
Molybdenum		
	600—1000	0.08—0.13
	1500—2200	0.19—0.26
Monel metal		
Oxidized		
	200	0.43
	600	0.43
Nichrome		
Wire		
Clean		
	50	0.65
	500—1000	0.71—0.79
Oxidized		
	50—500	0.95—0.98
Nickel		
Industrial, polished		
	200—400	0.07—0.09
Oxidized		
	200	0.37
Oxidized at 600°C		
	200—600	0.37—0.48
Unoxidized		
	25	0.045
	100	0.06
	500	0.12
	1000	0.19
Platinum		
Clean, polished		
	200—600	0.05—0.1
Unoxidized		
	25	0.037
	100	0.047
	500	0.096
	1000	0.152
	1500	0.191
Wire		
	50—200	0.06—0.07
	500—1000	0.1—0.16
	1400	0.18
Porcelain		
Glazed		
	20	0.92
Rubber		
Hard		
	20	0.95
Soft, gray, rough		
	20	0.86
Silica brick		
	1000	0.80
	1100	0.85
Silver		
Clean, polished		
	200—600	0.02—0.03
Unoxidized		
	100	0.02
	500	0.035
Soot applied to a solid surface		
	50—1000	0.94—0.91
Soot with water glass		
	20—200	0.96
Steel		
Alloyed (8% Ni, 18% Cr)		
	500	0.35
Aluminized		
	50—500	0.79
Dull nickel plated		
	20	0.11
Flat, rough surface		
	50	0.95—0.98
Cast, polished		
	750—1050	0.52—0.56
Sheet, ground		
	50	0.56
	950—1100	0.55—0.61
Oxidized		
	200—600	0.8
Calorized, oxidized		
	200	0.52
	600	0.57
Sheet with shiny layer of oxide		
	20	0.82
Strongly oxidized		
	50	0.88
	500	0.98
Unoxidized		
	100	0.08
Unoxidized, liquid		
	—	0.28
Tantalum		
Unoxidized		
	1500	0.21
	2000	0.26
Tungsten		
Unoxidized		
	25	0.024
	100	0.032
	500	0.071
	1000	0.15
	1500	0.23
	2000	0.28
Varnish		
	40—100	0.8—0.95
Dull black		
	40—100	0.96—0.98
Glossy black sprayed on iron		
	20	0.87
Zinc		
Polished		
	200—300	0.04—0.05
Unoxidized		
	300	0.05

SPECTRAL EMISSIVITY

Prepared by Roeser and Wensel, National Bureau of Standards
Spectral Emissivity of Materials, Surface Unoxidized for 0.65μm

Element	Solid	Liquid	Element	Solid	Liquid
Beryllium	0.61	0.61	Thorium	0.36	0.40
Carbon	0.80—0.93	—	Titanium	0.63	0.65
Chromium	0.34	0.39	Tungsten	0.43	—
Cobalt	0.36	0.37	Uranium	0.54	0.34
Columbium	0.37	0.40	Vanadium	0.35	0.32
Copper	0.10	0.15	Yttrium	0.35	0.35
Erbium	0.55	0.38	Zirconium	0.32	0.30
Gold	0.14	0.22	Steel	0.35	0.37
Iridium	0.30	—	Cast Iron	0.37	0.40
Iron	0.35	0.37	Constantan	0.35	—
Manganese	0.59	0.59	Monel	0.37	—
Molybdenum	0.37	0.40	Chromel P (90Ni-10Cr)	0.35	—
Nickel	0.36	0.37	80Ni-20Cr	0.35	—
Palladium	0.33	0.37	60Ni-24Fe-16Cr	0.36	—
Platinum	0.30	0.38	Alumel (95Ni; Bal. Al,		
Rhodium	0.24	0.30	Mn, Si)	0.37	—
Silver	0.07	0.07	90Pt-10Rh	0.27	—
Tantalum	0.49	—			

SPECTRAL EMISSIVITY OF OXIDES

The emissivity of oxides and oxidized metals depends to a large extent upon the roughness of the surface. In general, higher values of emissivity are obtained on the rougher surfaces.

Material	Range of observed values	Probable value for oxide formed on smooth metal
Oxide		
Aluminum	0.22—0.40	0.30
Beryllim	0.07—0.37	0.35
Cerium	0.58—0.80	—
Chromium	0.60—0.80	0.70
Cobalt	—	0.75
Columbium	0.55—0.71	0.70
Copper	0.60—0.80	0.70
Iron	0.63—0.98	0.70
Magnesium	0.10—0.43	0.20
Nickel	0.85—0.96	0.90
Thorium	0.20—0.57	0.50
Tin	0.32—0.60	—
Titanium	—	0.50
Uranium	—	0.30
Vanadium	—	0.70
Yttrium	—	0.60
Zirconium	0.18—0.43	0.40
Oxidized		
Alumel	—	0.87
Cast Iron	—	0.70
Chromel P (90Ni-10Cr)	—	0.87
80Ni-20Cr	—	0.90
60Ni-24Fe-16Cr	—	0.83
55Fe-37.5Cr-7.5Al	—	0.78
70Fe-23Cr-5Al-2Co	—	0.75
Constantan (55Cu-45Ni)	—	0.84
Carbon Steel	—	0.80
Stainless Steel (18-8)	—	0.85
Porcelain	0.25—0.50	—

PROPERTIES OF TUNGSTEN

Jones and Langmuir, General Electric Review

Temp. °K	Resistivity microhm cm	Electron emission amp./cm²	Evaporation g/cm² sec	Vapor pressure dynes/cm²	Thermal expansion per cent l_0 at 293°	Atomic heat cal./g. atom./°C.
300	5.65	—	—	—	.003	6.0
400	8.06	—	—	—	.044	6.0
500	10.56	—	—	—	.086	6.1
600	13.23	—	—	—	.130	6.1
700	16.09	—	—	—	.175	6.2
800	19.00	—	—	—	.222	6.2
900	21.94	—	—	—	.270	6.3
1000	24.93	1.07×10^{-15}	5.32×10^{-34}	1.98×10^{-29}	.320	6.4
1100	27.94	1.52×10^{-13}	2.17×10^{-30}	1.22×10^{-25}	.371	6.4
1200	30.98	9.73×10^{-12}	3.21×10^{-27}	1.87×10^{-22}	.424	6.5
1300	34.08	3.21×10^{-10}	1.35×10^{-24}	8.18×10^{-20}	.479	6.7
1400	37.19	6.62×10^{-9}	2.51×10^{-22}	1.62×10^{-17}	.535	6.8
1500	40.36	9.14×10^{-8}	2.37×10^{-20}	1.54×10^{-15}	.593	7.0
1600	43.55	9.27×10^{-7}	1.25×10^{-18}	8.43×10^{-14}	.652	7.1
1700	46.78	7.08×10^{-6}	4.17×10^{-17}	2.82×10^{-12}	.713	7.2
1800	50.05	4.47×10^{-5}	8.81×10^{-16}	6.31×10^{-11}	.775	7.4
1900	53.35	2.28×10^{-4}	1.41×10^{-14}	1.01×10^{-9}	.839	7.6
2000	56.67	1.00×10^{-3}	1.76×10^{-13}	1.33×10^{-8}	.904	7.7
2100	60.06	3.93×10^{-3}	1.66×10^{-12}	1.28×10^{-7}	.971	7.8
2200	63.48	1.33×10^{-2}	1.25×10^{-11}	9.88×10^{-7}	1.039	8.0
2300	66.91	4.07×10^{-2}	8.00×10^{-11}	6.47×10^{-6}	1.109	8.2
2400	70.39	1.16×10^{-1}	4.26×10^{-10}	3.52×10^{-5}	1.180	8.3
2500	73.91	2.98×10^{-1}	2.03×10^{-9}	1.71×10^{-4}	1.253	8.4
2600	77.49	7.16×10^{-1}	8.41×10^{-9}	7.24×10^{-4}	1.328	8.6
2700	81.04	1.63	3.19×10^{-8}	2.86×10^{-3}	1.404	8.7
2800	84.70	3.54	1.10×10^{-7}	9.84×10^{-3}	1.479	8.9
2900	88.33	7.31	3.30×10^{-7}	3.00×10^{-2}	1.561	9.0
3000	92.04	1.42×10	9.95×10^{-7}	9.20×10^{-2}	1.642	9.2
3100	95.76	2.64×10	2.60×10^{-6}	2.50×10^{-1}	1.724	9.4
3200	99.54	4.78×10	6.38×10^{-6}	6.13×10^{-1}	1.808	9.5
3300	103.3	8.44×10	1.56×10^{-5}	1.51	1.893	9.6
3400	107.2	1.42×10^{2}	3.47×10^{-5}	3.41	1.980	9.8
3500	111.1	2.33×10^{2}	7.54×10^{-5}	7.52	2.068	9.9
3600	115.0	3.73×10^{2}	1.51×10^{-4}	1.53×10	2.158	10.1
3655	117.1	4.79×10^{2}	2.28×10^{-4}	2.33×10	2.209	10.2

Roeser and Wensel, National Bureau of Standards

Temp. °K	Normal brightness new candles per cm²	Spectral emissivity 0.65μm	Spectral emissivity 0.467μm	Color emissivity	Total emissivity	Brightness temp. 0.65μm	Color temp.
300	—	0.472	0.505	—	0.032	—	—
400	—	—	—	—	.042	—	—
500	—	—	—	—	.053	—	—
600	—	—	—	—	.064	—	—
700	—	—	—	—	.076	—	—
800	—	—	—	—	.088	—	—
900	—	—	—	—	.101	—	—
1000	0.0001	.458	.486	.395	.114	966	1007
1100	0.001	.456	.484	.392	.128	1059	1108
1200	0.006	.454	.482	.390	.143	1151	1210
1300	0.029	.452	.480	.387	.158	1242	1312
1400	0.11	.450	.478	.385	.175	1332	1414
1500	0.33	.448	.476	.382	.192	1422	1516
1600	0.92	.446	.475	.380	.207	1511	1619
1700	2.3	.444	.473	.377	.222	1599	1722
1800	5.1	.442	.472	.374	.236	1687	1825
1900	10.4	.440	.470	.371	.249	1774	1928
2000	20.0	.438	.469	.368	.260	1861	2032
2100	36	.436	.467	.365	.270	1946	2136
2200	61	.434	.466	.362	.279	2031	2241
2300	101	.432	.464	.359	.288	2115	2345
2400	157	.430	.463	.356	.296	2198	2451
2500	240	.428	.462	.353	.303	2280	2556
2600	350	.426	.460	.349	.311	2362	2662
2700	500	.424	.459	.346	.318	2443	2769
2800	690	.422	.458	.343	.323	2523	2876
2900	950	.420	.456	.340	.329	2602	2984
3000	1260	.418	.455	.336	.334	2681	3092
3100	1650	.416	.454	.333	.337	2759	3200
3200	2100	.414	.452	.330	.341	2837	3310
3300	2700	.412	.451	.326	.344	2913	3420
3400	3400	.410	.450	.323	.348	2989	3530
3500	4200	.408	.449	.320	.351	3063	3642
3600	5200	.406	.447	.317	.354	3137	3754

TRANSMISSION OF CORNING COLORED FILTERS

Supplied by R. G. Saxton

If I_o is the intensity of radiation entering a layer of some medium and I the intensity reaching the opposite surface, the ratio I/I_o is called the transmittance. In practice the ratio of intensity of radiation passing through a glass sample to that incident on its surface is often measured and plotted as transmission. The transmission is the result of two factors, the transmittance of the glass and the losses by reflection. These losses amount to about 4 % for each glass-air surface; the transmission of a sample is about 92 % of its transmittance. Since the reflection losses differ slightly with different samples, the correction is often determined and applied when the transmission is measured. Values in this table have been corrected for reflection losses.

The identifying glass number, CS number, color and properties, and nominal thickness for the Corning glasses in this table are:

Glass No.	CS	Color and properties	Nominal thickness
0160	0-54	Clear; Ultraviolet transmitting	2.0
2030	2-64	Red; Sharp cut	3.0
2403	2-58	Red; Sharp cut	3.0
2404	2-59	Red; Sharp cut	3.0
2408	2-60	Red; Sharp cut	3.0
2412	2-61	Red; Sharp cut	3.0
2418	2-62	Red; Sharp cut	3.0
2424	2-63	Red; Sharp cut	3.0
2434	2-73	Red; Sharp cut	3.0
2540	7-56	Black; IR transmitting; Visible absorbing	2.5
2550	7-57	Black; IR transmitting; Visible absorbing	2.0
2600	7-69	Black; IR transmitting; Visible absorbing	3.0
3060	3-75	Straw	2.0
3304	3-76	Dark amber	3.0
3307	3-77	Dark amber	3.0
3384	3-70	Yellow	3.0
3385	3-71	Yellow	3.0
3387	3-72	Straw	3.0
3389	3-73	Straw	3.0
3391	3-74	Straw	3.0
3480	3-66	Yellow; Sharp cut	3.0
3482	3-67	Yellow; Sharp cut	3.0
3484	3-68	Yellow; Sharp cut	3.0
3486	3-69	Yellow; Sharp cut	3.0
3718	3-94	Yellow	3.0
3750	3-79	Yellow; Yellow green fluorescing	5.0
3780	3-80	Yellow	2.0
3850	0-51	Clear; UV transmitting	4.0
3961	1-56	Bluish; IR absorbing; Visible transmitting	2.5
3962	1-57	Bluish; IR absorbing; Visible transmitting	2.5
3965	1-58	Bluish; IR absorbing; Visible transmitting	2.5
3966	1-59	Bluish; IR absorbing; Visible transmitting	2.5
4010	4-64	Green	4.0
4015	4-65	Yellow green	3.0
4060	4-67	Green	2.0
4084	4-68	Green	4.5
4303	4-72	Blue green	4.0
4305	4-71	Blue green	4.0
4308	4-70	Blue green	4.0
4309	4-69	Blue green	4.0
4445	4-74	Green	2.5
4602	1-75	Bluish; IR absorbing; Visible transmitting	3.0
4784	4-94	Blue green	5.0
5030	5-57	Blue	5.0
5031	5-56	Blue	4.5
5070	7-62	Amethyst	3.9
5071	7-63	Amethyst	3.9
5073	7-64	Amethyst	3.9
5113	5-58	Blue	4.0
5120	1-60	Smoky violet; Absorbs yellow	5.2
5300	4-106	Green	3.9
5330	1-64	Blue	4.5
5433	5-59	Blue	5.0
5543	5-60	Blue	5.0
5562	5-61	Blue	5.0
5572	1-61	Blue	5.0
5840	7-60	Black; UV transmitting; Visible absorbing	4.5
5850	7-59	Purple; UV transmitting; Visible absorbing	4.0
5860	7-37	Black; UV transmitting; Visible absorbing	5.0
5874	7-39	Black; UV transmitting; Visible absorbing	5.0
5900	1.62	Blue	5.5
5970	7-51	Black; UV transmitting; Visible absorbing	5.0
7380	0-52	Clear; UV transmitting	2.0
7740	0-53	Clear; UV transmitting	2.0
7905	9-30	Clear; UV transmitting; Long Range IR transmitting	2.0
7910	9-54	Clear; UV transmitting	2.0
8364	7-98	Gray	2.0
9780	4-76	Blue green	5.0
9782	4-96	Blue green	5.0
9788	4-97	Blue green	5.0
9830	4-77	Green	3.4
9863	7-54	Black; UV transmitting; Visible absorbing	3.0

λ(nm)	Transmittance											
	Corning Glass Number											
	0160	2030	2403	2404	2408	2412	2418	2424	2434	2540	2550	2600
.22	.000	.000	.000	.000	.000	.000	.000	.000	.000	.000	.000	.000
.24	.000	.000	.000	.000	.000	.000	.000	.000	.000	.000	.000	.000
.26	.000	.000	.000	.000	.000	.000	.000	.000	.000	.000	.000	.000
.28	.000	.000	.000	.000	.000	.000	.000	.000	.000	.000	.000	.000
.30	.005	.000	.000	.000	.000	.000	.000	.000	.000	.000	.000	.000
.32	.642	.000	.000	.000	.000	.000	.000	.000	.000	.000	.000	.000
.34	.850	.000	.000	.000	.000	.000	.000	.000	.000	.000	.000	.000
.36	.882	.000	.000	.000	.000	.000	.000	.000	.000	.000	.000	.000
.38	.890	.000	.000	.000	.000	.000	.000	.000	.000	.000	.000	.000
.40	.892	.000	.000	.000	.000	.000	.000	.000	.000	.000	.000	.000
.41	.893	.000	.000	.000	.000	.000	.000	.000	.000	.000	.000	.000
.42	.896	.000	.000	.000	.000	.000	.000	.000	.000	.000	.000	.000
.43	.896	.000	.000	.000	.000	.000	.000	.000	.000	.000	.000	.000
.44	.898	.000	.000	.000	.000	.000	.000	.000	.000	.000	.000	.000
.45	.899	.000	.000	.000	.000	.000	.000	.000	.000	.000	.000	.000
.46	.900	.000	.000	.000	.000	.000	.000	.000	.000	.000	.000	.000
.47	.900	.000	.000	.000	.000	.000	.000	.000	.000	.000	.000	.000
.48	.900	.000	.000	.000	.000	.000	.000	.000	.000	.000	.000	.000
.49	.900	.000	.000	.000	.000	.000	.000	.000	.000	.000	.000	.000
.50	.900	.000	.000	.000	.000	.000	.000	.000	.000	.000	.000	.000
.51	.900	.000	.000	.000	.000	.000	.000	.000	.000	.000	.000	.000
.52	.900	.000	.000	.000	.000	.000	.000	.000	.000	.000	.000	.000
.53	.900	.000	.000	.000	.000	.000	.000	.000	.000	.000	.000	.000
.54	.900	.000	.000	.000	.000	.000	.000	.000	.000	.000	.000	.000
.55	.900	.000	.000	.000	.000	.000	.000	.000	.000	.000	.000	.000
.56	.901	.000	.000	.000	.000	.000	.000	.000	.000	.000	.000	.000
.57	.904	.000	.000	.000	.000	.000	.000	.000	.005	.000	.000	.000
.58	.904	.000	.000	.000	.000	.000	.000	.005	.200	.000	.000	.000
.59	.908	.000	.000	.000	.000	.000	.008	.170	.615	.000	.000	.000
.60	.910	.000	.000	.000	.000	.006	.250	.575	.808	.000	.000	.000
.61	.910	.000	.000	.000	.018	.190	.660	.790	.856	.000	.000	.000
.62	.910	.000	.000	.015	.265	.625	.822	.848	.872	.000	.000	.000
.63	.910	.000	.018	.295	.670	.828	.862	.870	.881	.000	.000	.000
.64	.910	.006	.260	.660	.828	.868	.874	.880	.887	.000	.001	.000
.65	.910	.028	.675	.796	.866	.881	.881	.887	.892	.000	.003	.000
.66	.910	.110	.838	.828	.877	.885	.885	.893	.895	.000	.005	.000
.67	.910	.305	.871	.842	.883	.887	.887	.897	.897	.000	.006	.000
.68	.910	.550	.880	.847	.886	.889	.889	.900	.899	.000	.009	.000
.69	.910	.735	.885	.851	.888	.900	.900	.901	.900	.000	.012	.000
.70	.910	.820	.886	.852	.888	.900	.900	.903	.900	.000	.017	.000
.71	.910	.853	.888	.854	.888	.889	.889	.903	.900	.000	.023	.000
.72	.910	.864	.889	.853	.888	.888	.888	.903	.899	.000	.031	.040
.73	.910	.867	.900	.851	.887	.887	.887	.903	.897	.000	.041	.175
.74	.910	.867	.900	.850	.885	.886	.886	.903	.896	.000	.055	.372
.75	.910	.866	.900	.849	.884	.885	.885	.902	.895	.000	.069	.547
.80	.910	.839	.875	.827	.870	.858	.857	.881	.866	.005	.225	.770
1.00	.912	.801	.840	.772	.840	.828	.822	.857	.842	.562	.780	.350
1.20	.908	.799	.845	.786	.845	.837	.827	.859	.849	.790	.870	.000
1.40	.909	.811	.854	.809	.854	.848	.840	.862	.857	.850	.895	.000
1.60	.913	.839	.873	.837	.873	.869	.858	.880	.879	.872	.904	.000
1.80	.909	.844	.870	.829	.870	.864	.854	.877	.872	.880	.900	.000
2.00	.904	.841	.868	.827	.868	.861	.851	.874	.871	.880	.897	.000
2.20	.888	.833	.820	.773	.825	.820	.810	.837	.835	.860	.875	.005
2.40	.875	.832	.803	.757	.809	.803	.792	.818	.812	.868	.870	.049
2.60	.868	.822	.750	.695	.754	.750	.723	.772	.767	.858	.850	.058
2.80	.690	.600	.100	.100	.100	.100	.050	.050	.260	.450	.480	.030
3.00	.630	.470	.070	.020	.070	.070	.072	.122	.400	.465	.383	.022
3.20	.500	.340	.140	.074	.150	.140	.142	.180	.470	.390	.330	.020
3.40	.379	.260	.140	.078	.140	.140	.120	.150	.350	.310	.245	.018
3.60	.320	.247	.000	.000	.000	.000	.000	.006	.015	.280	.220	.021
3.80	.310	.257	.000	.000	.000	.000	.000	.000	.020	.285	.250	.035
4.00	.311	.274	.000	.000	.000	.000	.000	.000	.015	.285	.275	.068
4.20	.251	.200	.000	.000	.000	.000	.000	.000	.017	.190	.190	.065
4.40	.110	.060	.000	.000	.000	.000	.000	.000	.002	.050	.100	.020
4.60	.012	.000	.000	.000	.000	.000	.000	.000	.000	.000	.000	.000
4.80	.004	.000	.000	.000	.000	.000	.000	.000	.000	.000	.000	.000
5.00	.000	.000	.000	.000	.000	.000	.000	.000	.000	.000	.000	.000

TRANSMISSION OF CORNING COLORED FILTERS (Continued)

λ(nm)	Transmittance											
	Corning Glass Number											
	3060	3304	3307	3384	3385	3387	3389	3391	3480	3482	3484	3486
.22	.000	.000	.000	.000	.000	.000	.000	.000	.000	.000	.000	.000
.24	.000	.000	.000	.000	.000	.000	.000	.000	.000	.000	.000	.000
.26	.000	.000	.000	.000	.000	.000	.000	.000	.000	.000	.000	.000
.28	.000	.000	.000	.000	.000	.000	.000	.000	.000	.000	.000	.000
.30	.000	.000	.000	.000	.000	.000	.000	.000	.000	.000	.000	.000
.32	.000	.000	.000	.000	.000	.000	.005	.000	.000	.000	.000	.000
.34	.000	.000	.038	.000	.000	.000	.010	.000	.000	.000	.000	.000
.36	.000	.000	.050	.000	.005	.000	.015	.000	.000	.000	.000	.000
.38	.060	.000	.027	.005	.010	.010	.020	.000	.000	.000	.000	.000
.40	.410	.000	.016	.011	.016	.020	.026	.075	.000	.000	.000	.005
.41	.517	.000	.014	.011	.016	.020	.025	.425	.000	.000	.000	.005
.42	.604	.000	.014	.010	.015	.019	.105	.655	.000	.000	.000	.005
.43	.665	.000	.016	.009	.013	.017	.437	.747	.000	.000	.000	.005
.44	.710	.000	.022	.005	.011	.050	.620	.801	.000	.000	.000	.005
.45	.748	.000	.033	.003	.010	.325	.714	.838	.000	.000	.000	.005
.46	.778	.000	.049	.002	.008	.565	.780	.860	.000	.000	.000	.004
.47	.800	.000	.070	.001	.060	.690	.820	.874	.000	.000	.000	.003
.48	.819	.000	.101	.005	.410	.763	.848	.884	.000	.000	.000	.003
.49	.836	.003	.143	.088	.640	.803	.866	.890	.000	.000	.000	.002
.50	.850	.009	.193	.350	.727	.834	.878	.895	.000	.000	.000	.001
.51	.860	.019	.250	.595	.780	.854	.886	.898	.000	.000	.000	.045
.52	.870	.037	.315	.725	.817	.868	.890	.900	.000	.000	.003	.425
.53	.875	.063	.379	.789	.840	.876	.892	.901	.000	.000	.175	.710
.54	.881	.102	.447	.825	.856	.883	.894	.902	.000	.015	.600	.792
.55	.884	.146	.504	.846	.866	.887	.895	.903	.000	.230	.774	.823
.56	.886	.200	.560	.860	.873	.889	.894	.902	.020	.675	.818	.844
.57	.885	.255	.607	.869	.876	.890	.893	.901	.325	.850	.839	.859
.58	.883	.310	.648	.873	.878	.889	.892	.900	.710	.885	.854	.868
.59	.882	.360	.680	.876	.877	.887	.890	.898	.829	.894	.865	.876
.60	.882	.404	.705	.877	.877	.884	.886	.896	.858	.900	.873	.882
.61	.882	.438	.722	.877	.877	.881	.884	.893	.869	.903	.880	.886
.62	.882	.466	.735	.875	.876	.875	.880	.890	.876	.905	.885	.888
.63	.882	.488	.744	.871	.874	.871	.876	.886	.881	.906	.888	.889
.64	.883	.505	.748	.865	.872	.866	.872	.884	.884	.907	.890	.890
.65	.885	.519	.750	.860	.867	.860	.868	.881	.885	.908	.892	.890
.66	.886	.531	.750	.856	.863	.856	.865	.876	.886	.908	.893	.890
.67	.888	.543	.749	.850	.858	.851	.860	.873	.886	.908	.894	.890
.68	.890	.552	.745	.844	.853	.846	.856	.869	.886	.908	.893	.889
.69	.891	.561	.740	.837	.847	.839	.852	.865	.885	.907	.892	.887
.70	.892	.569	.734	.831	.842	.834	.847	.860	.884	.907	.891	.885
.71	.893	.574	.727	.825	.837	.827	.842	.856	.882	.906	.890	.883
.72	.893	.575	.720	.819	.831	.822	.837	.852	.880	.905	.888	.880
.73	.892	.576	.712	.813	.825	.816	.831	.848	.877	.905	.886	.877
.74	.891	.574	.702	.807	.820	.810	.826	.844	.874	.904	.885	.875
.75	.890	.570	.694	.800	.814	.805	.820	.840	.870	.903	.882	.873
.80	.871	.526	.642	.770	.865	.780	.775	.815	.837	.878	.846	.829
1.00	.830	.435	.516	.715	.830	.725	.716	.772	.801	.857	.811	.781
1.20	.860	.429	.500	.718	.858	.735	.730	.782	.807	.859	.819	.793
1.40	.901	.475	.540	.750	.900	.768	.768	.810	.828	.870	.837	.817
1.60	.917	.580	.635	.795	.918	.812	.812	.843	.852	.884	.856	.841
1.80	.916	.627	.675	.808	.915	.818	.820	.849	.847	.882	.852	.834
2.00	.908	.620	.668	.800	.909	.822	.817	.846	.847	.884	.852	.835
2.20	.900	.630	.675	.802	.900	.823	.810	.840	.805	.865	.829	.798
2.40	.885	.651	.690	.800	.885	.825	.811	.842	.787	.853	.817	.777
2.60	.860	.650	.690	.785	.858	.815	.800	.840	.725	.818	.757	.718
2.80	.550	.345	.390	.325	.670	.360	.340	.440	.050	.060	.110	.060
3.00	.379	.320	.360	.318	.348	.348	.322	.423	.080	.088	.190	.088
3.20	.315	.240	.290	.268	.332	.324	.290	.395	.150	.158	.270	.145
3.40	.250	.151	.190	.218	.289	.288	.255	.353	.120	.132	.140	.090
3.60	.231	.130	.150	.217	.266	.280	.249	.351	.000	.000	.000	.000
3.80	.258	.140	.160	.228	.270	.298	.267	.376	.000	.000	.000	.000
4.00	.283	.140	.175	.220	.280	.290	.260	.365	.000	.000	.000	.000
4.20	.200	.090	.125	.143	.210	.210	.178	.288	.000	.000	.000	.000
4.40	.100	.020	.030	.025	.080	.070	.040	.115	.000	.000	.000	.000
4.60	.008	.008	.010	.007	.002	.003	.000	.009	.000	.000	.000	.000
4.80	.000	.005	.009	.000	.000	.000	.000	.000	.000	.000	.000	.000
5.00	.000	.001	.008	.000	.000	.000	.000	.000	.000	.000	.000	.000

TRANSMISSION OF CORNING COLORED FILTERS (Continued)

	Transmittance											
	Corning Glass Number											
λ(nm)	3718	3750	3780	3850	3961	3962	3965	3966	4010	4015	4060	4084
.22	.000	.000	.000	.000	.000	.000	.000	.000	.000	.000	.000	.000
.24	.000	.000	.000	.000	.000	.000	.000	.000	.000	.000	.000	.000
.26	.000	.000	.000	.000	.000	.000	.000	.000	.000	.000	.000	.000
.28	.000	.000	.000	.000	.000	.000	.000	.000	.000	.000	.000	.000
.30	.000	.000	.000	.000	.000	.000	.000	.000	.000	.000	.000	.000
.32	.004	.000	.000	.000	.000	.000	.018	.055	.000	.000	.000	.000
.34	.030	.000	.000	.000	.000	.018	.192	.375	.000	.000	.001	.018
.36	.550	.215	.000	.005	.020	.125	.430	.630	.000	.000	.021	.128
.38	.665	.327	.000	.350	.085	.270	.558	.710	.000	.000	.080	.248
.40	.480	.113	.000	.675	.185	.395	.636	.781	.000	.000	.178	.216
.41	.443	.088	.000	.749	.218	.426	.651	.788	.000	.000	.228	.180
.42	.465	.088	.000	.788	.248	.453	.666	.795	.000	.000	.281	.151
.43	.560	.135	.000	.812	.269	.474	.678	.800	.000	.000	.335	.136
.44	.675	.255	.006	.828	.290	.494	.693	.806	.000	.000	.388	.140
.45	.748	.410	.028	.841	.313	.519	.709	.816	.000	.005	.434	.163
.46	.780	.472	.058	.850	.331	.538	.724	.824	.006	.025	.473	.200
.47	.803	.570	.092	.858	.346	.556	.737	.831	.021	.073	.506	.247
.48	.800	.555	.088	.865	.361	.570	.748	.836	.050	.145	.527	.303
.49	.802	.550	.095	.870	.370	.582	.756	.840	.100	.245	.535	.370
.50	.824	.597	.152	.874	.376	.590	.762	.842	.160	.350	.528	.430
.51	.862	.720	.325	.878	.377	.594	.765	.843	.220	.455	.503	.465
.52	.894	.825	.595	.881	.373	.593	.765	.841	.252	.537	.460	.467
.53	.904	.853	.717	.883	.364	.588	.761	.838	.247	.582	.400	.438
.54	.905	.860	.763	.884	.354	.579	.764	.833	.207	.594	.325	.376
.55	.906	.864	.783	.884	.342	.569	.746	.827	.153	.572	.252	.303
.56	.907	.867	.795	.883	.331	.559	.736	.821	.096	.525	.183	.225
.57	.907	.869	.799	.882	.317	.547	.724	.813	.054	.457	.125	.157
.58	.907	.870	.804	.880	.298	.529	.706	.802	.026	.385	.083	.107
.59	.907	.870	.811	.877	.276	.508	.685	.790	.011	.311	.053	.073
.60	.908	.876	.826	.876	.251	.481	.662	.775	.004	.245	.033	.048
.61	.908	.880	.835	.875	.225	.452	.636	.757	.000	.190	.020	.034
.62	.909	.881	.842	.874	.299	.423	.610	.739	.000	.145	.012	.024
.63	.909	.884	848	.875	.217	.392	.577	.719	.000	.115	.007	.018
.64	.910	.885	.854	.876	.147	.359	.546	.698	.000	.098	.004	.015
.65	.910	.887	.859	.877	.125	.326	.515	.675	.000	.084	.001	.012
.66	.911	.891	.864	.880	.104	.297	.482	.652	.000	.075	.000	.010
.67	.913	.896	.869	.883	.086	.265	.450	.630	.000	.075	.000	.009
.68	.914	.900	.873	.885	.063	.235	.418	.603	.000	.071	.000	.008
.69	.915	.901	.877	.887	.055	.206	.385	.576	.000	.065	.000	.007
.70	.915	.904	.880	.888	.042	.279	.352	.550	.000	.067	.000	.007
.71	.915	.905	.883	.888	.032	.155	.322	.524	.000	.070	.000	.007
.72	.915	.906	.885	.889	.025	.133	.294	.496	.000	.075	.000	.007
.73	.915	.907	.885	.888	.018	.114	.266	.472	.000	.080	.000	.007
.74	.915	.907	.882	.887	.014	.097	.242	.448	.000	.084	.000	.008
.75	.914	.906	.882	.886	.010	.084	.220	.424	.000	.086	.000	.008
.80	.902	.875	.855	.863	.000	.033	.120	.310	.000	.109	.000	.013
1.00	.899	.860	.882	.820	.000	.002	.038	.158	.000	.215	.018	.100
1.20	.898	.882	.898	.850	.000	.002	.040	.161	.007	.393	.158	.303
1.40	.880	.810	.855	.894	.002	.018	.100	.270	.058	.549	.404	.548
1.60	.882	.805	.855	.905	.007	.050	.190	.390	.162	.663	.612	.710
1.80	.900	.844	.907	.904	.008	.057	.201	.408	.299	.740	.740	.791
2.00	.897	.819	.909	.895	.011	.070	.228	.435	.422	.783	.817	.830
2.20	.888	.720	.900	.870	.021	.105	.277	.475	.518	.803	.840	.792
2.40	.865	.668	.890	.850	.037	.140	.320	.512	.597	.817	.862	.808
2.60	.840	.570	.860	.800	.050	.005	.339	.515	.634	.813	.870	.740
2.80	.460	.075	.620	.200	.022	.092	.135	.085	.270	.460	.520	.010
3.00	.282	.028	.465	.165	.033	.133	.200	.230	.260	.418	.620	.033
3.20	.252	.016	.420	.120	.054	.150	.247	.294	.204	.345	.642	.128
3.40	.175	.007	.370	.070	.048	.112	.165	.200	.131	.235	.631	.130
3.60	.150	.003	.351	.045	.003	.008	.016	.018	.121	.192	.634	.006
3.80	.168	.000	.368	.067	.007	.015	.035	.048	.125	.200	.600	.016
4.00	.182	.000	.370	.080	.006	.010	.020	.022	.123	.200	.557	.006
4.20	.120	.000	.300	.040	.005	.013	.022	.021	.070	.130	.422	.001
4.40	.020	.000	.140	.005	.001	.001	.000	.001	.002	.015	.135	.000
4.60	.002	.000	.010	.000	.000	.000	.000	.000	.000	.000	.012	.000
4.80	.000	.000	.000	.000	.000	.000	.000	.000	.000	.000	.002	.000
5.00	.000	.000	.000	.000	.000	.000	.000	.000	.000	.000	.000	.000

TRANSMISSION OF CORNING COLORED FILTERS (Continued)

λ(nm)	Transmittance											
	Corning Glass Number											
	4303	4305	4308	4309	4445	4602	4784	5030	5031	5070	5071	5073
.22	.000	.000	.000	.000	.000	.000	.000	.000	.000	.000	.000	.000
.24	.000	.000	.000	.000	.000	.000	.000	.000	.000	.000	.000	.000
.26	.000	.000	.000	.000	.000	.000	.000	.000	.000	.000	.000	.000
.28	.000	.000	.000	.000	.000	.000	.000	.000	.000	.000	.000	.000
.30	.000	.000	.000	.000	.000	.001	.000	.000	.016	.000	.000	.000
.32	.000	.000	.000	.022	.001	.106	.000	.000	.145	.012	.045	.090
.34	.011	.060	.190	.394	.018	.505	.009	.038	.420	.310	.330	.540
.36	.188	.380	.580	.740	.114	.755	.200	.285	.685	.628	.340	.757
.38	.390	.590	.740	.831	.248	.827	.450	.595	.820	.745	.420	.810
.40	.545	.723	.826	.884	.400	.835	.596	.770	.884	.712	.665	.830
.41	.588	.750	.840	.887	.454	.845	.627	.799	.894	.600	.712	.786
.42	.624	.770	.850	.890	.505	.846	.648	.808	.895	.430	.716	.705
.43	.654	.786	.857	.892	.550	.851	.666	.797	.890	.290	.694	.620
.44	.680	.798	.862	.894	.593	.856	.680	.767	.872	.170	.655	.525
.45	.698	.809	.867	.897	.631	.857	.697	.738	.865	.097	.612	.436
.46	.712	.815	.869	.898	.659	.856	.717	.702	.864	.055	.568	.365
.47	.715	.814	.866	.897	.678	.861	.735	.628	.845	.035	.531	.313
.48	.705	.802	.856	.894	.689	.866	.750	.522	.805	.023	.501	.275
.49	.678	.780	.838	.885	.687	.869	.763	.406	.750	.017	.482	.252
.50	.636	.740	.810	.872	.673	.870	.768	.288	.684	.015	.469	.237
.51	.570	.685	.770	.850	.641	.869	.767	.186	.601	.013	.463	.231
.52	.480	.610	.714	.817	.586	.866	.753	.105	.495	.013	.461	.230
.53	.387	.525	.650	.781	.520	.863	.725	.053	.388	.014	.464	.235
.54	.288	.430	.576	.736	.437	.865	.676	.022	.295	.016	.473	.245
.55	.205	.340	.502	.683	.355	.869	.615	.007	.198	.018	.486	.260
.56	.132	.255	.422	.627	.275	.868	.525	.000	.113	.023	.502	.278
.57	.082	.184	.345	.565	.202	.863	.427	.000	.057	.030	.522	.300
.58	.047	.127	.277	.505	.144	.856	.328	.000	.025	.038	.540	.324
.59	.026	.087	.218	.447	.102	.848	.235	.000	.008	.048	.555	.348
.60	.013	.057	.170	.393	.068	.838	.157	.000	.000	.058	.571	.373
.61	.006	.036	.131	.341	.046	.824	.102	.000	.000	.070	.587	.395
.62	.001	.022	.100	.296	.031	.806	.058	.000	.000	.083	.600	.415
.63	.000	.013	.074	.256	.020	.787	.032	.000	.000	.096	.612	.435
.64	.000	.007	.056	.221	.013	.767	.017	.000	.000	.108	.622	.450
.65	.000	.004	.042	.191	.008	.745	.007	.000	.000	.120	.633	.466
.66	.000	.002	.033	.167	.006	.722	.002	.000	.000	.135	.644	.482
.67	.000	.001	.025	.146	.003	.695	.000	.000	.000	.148	.658	.498
.68	.000	.000	.020	.128	.001	.665	.000	.000	.000	.165	.674	.515
.69	.000	.000	.016	.116	.000	.634	.000	.000	.000	.182	.686	.531
.70	.000	.000	.013	.104	.000	.600	.000	.000	.000	.200	.700	.548
.71	.000	.000	.010	.096	.000	.565	.000	.000	.000	.220	.712	.566
.72	.000	.000	.009	.088	.000	.531	.000	.004	.024	.245	.725	.586
.73	.000	.000	.007	.083	.000	.496	.000	.047	.119	.268	.736	.606
.74	.000	.000	.006	.079	.000	.463	.000	.190	.330	.295	.749	.675
.75	.000	.000	.005	.075	.000	.430	.000	.440	.580	.323	.759	.642
.80	.000	.000	.008	.080	.003	.258	.000	.890	.917	.505	.815	.750
1.00	.000	.005	.045	.188	.056	.019	.000	.753	.868	.860	.885	.872
1.20	.013	.060	.166	.375	.269	.009	.000	.455	.720	.890	.897	.890
1.40	.080	.180	.342	.542	.527	.016	.003	.100	.285	.892	.902	.892
1.60	.210	.342	.500	.658	.701	.035	.038	.056	.162	.890	.902	.892
1.80	.350	.482	.617	.732	.792	.059	.142	.052	.140	.878	.890	.877
2.00	.475	.590	.690	.772	.839	.048	.275	.075	.175	.860	.865	.860
2.20	.560	.653	.720	.778	.850	.038	.345	.172	.295	.840	.830	.840
2.40	.635	.704	.752	.791	.870	.045	.404	.330	.483	.812	.795	.805
2.60	.663	.710	.748	.770	.861	.066	.340	.382	.530	.804	.752	.775
2.80	.260	.370	.370	.250	.280	.018	.045	.030	.030	.550	.500	.500
3.00	.249	.250	.203	.212	.395	.000	.000	.010	.001	.390	.308	.340
3.20	.202	.212	.145	.168	.451	.000	.000	.065	.026	.222	.145	.180
3.40	.135	.136	.084	.107	.449	.000	.000	.002	.006	.120	.070	.090
3.60	.124	.120	.078	.077	.470	.000	.000	.000	.000	.078	.032	.063
3.80	.132	.125	.082	.083	.470	.000	.000	.000	.000	.078	.029	.060
4.00	.132	.131	.094	.090	.448	.000	.000	.000	.000	.093	.037	.075
4.20	.078	.073	.050	.042	.320	.000	.000	.000	.000	.079	.020	.048
4.40	.008	.009	.008	.002	.100	.000	.000	.000	.000	.020	.004	.014
4.60	.000	.000	.000	.000	.007	.000	.000	.000	.000	.002	.000	.000
4.80	.000	.000	.000	.000	.000	.000	.000	.000	.000	.000	.000	.000
5.00	.000	.000	.000	.000	.000	.000	.000	.000	.000	.000	.000	.000

TRANSMISSION OF CORNING COLORED FILTERS (Continued)

λ(nm)	Transmittance											
	Corning Glass Number											
	5113	5120	5300	5330	5433	5543	5562	5572	5840	5850	5860	5874
.22	.000	.000	.000	.000	.000	.000	.000	.000	.000	.000	.000	.000
.24	.000	.000	.000	.000	.000	.000	.000	.000	.000	.000	.000	.000
.26	.000	.000	.000	.000	.000	.000	.000	.000	.000	.000	.000	.000
.28	.000	.000	.000	.000	.000	.000	.000	.000	.000	.000	.000	.000
.30	.000	.000	.000	.002	.000	.000	.000	.000	.001	.039	.000	.000
.32	.000	.000	.000	.250	.000	.000	.000	.045	.242	.490	.008	.031
.34	.000	.000	.000	.622	.000	.000	.012	.325	.600	.790	.179	.228
.36	.035	.018	.000	.796	.100	.120	.205	.660	.682	.858	.340	.447
.38	.200	.540	.000	.835	.350	.380	.495	.805	.392	.850	.085	.378
.40	.371	.670	.000	.865	.585	.600	.717	.874	.000	.788	.000	.032
.41	.371	.790	.000	.850	.636	.635	.748	.873	.000	.720	.000	.004
.42	.337	.805	.000	.823	.665	.646	.761	.865	.000	.630	.000	.000
.43	.272	.560	.000	.783	.674	.635	.759	.857	.000	.522	.000	.000
.44	.198	.386	.000	.725	.665	.602	.742	.845	.000	.410	.000	.000
.45	.118	.485	.000	.650	.635	.550	.713	.832	.000	.290	.000	.000
.46	.055	.475	.000	.555	.577	.465	.662	.808	.000	.175	.000	.000
.47	.013	.370	.008	.455	.467	.335	.565	.765	.000	.125	.000	.000
.48	.000	.385	.026	.355	.327	.190	.435	.693	.000	.022	.000	.000
.49	.000	.685	.085	.270	.205	.090	.300	.605	.000	.005	.000	.000
.50	.000	.660	.125	.197	.120	.040	.197	.523	.000	.000	.000	.000
.51	.000	.390	.106	.145	.060	.013	.110	.430	.000	.000	.000	.000
.52	.000	.305	.094	.110	.024	.002	.051	.328	.000	.000	.000	.000
.53	.000	.175	.064	.085	.008	.000	.022	.247	.000	.000	.000	.000
.54	.000	.610	.149	.068	.005	.000	.012	.216	.000	.000	.000	.000
.55	.000	.817	.147	.055	.006	.000	.015	.246	.000	.000	.000	.000
.56	.000	.230	.087	.043	.007	.000	.018	.285	.000	.000	.000	.000
.57	.000	.125	.013	.032	.000	.000	.011	.258	.000	.000	.000	.000
.58	.000	.000	.000	.022	.000	.000	002	.175	.000	.000	.000	.000
.59	.000	.006	.000	.016	.000	.000	.000	.116	.000	.000	.000	.000
.60	.000	.180	.000	.012	.000	.000	.000	.113	.000	.000	.000	.000
.61	.000	.545	.000	.009	.000	.000	.000	.123	.000	.000	.000	.000
.62	.000	.825	.000	.010	.000	.000	.000	.126	.000	.000	.000	.000
.63	.000	.838	.000	.013	.000	.000	.000	.120	.000	.000	.000	.000
.64	.000	.878	.000	.015	.000	.000	.000	.111	.000	.000	.000	.000
.65	.000	.893	.000	.015	.000	.000	.000	.120	.000	.000	.000	.000
.66	.000	.883	.000	.014	.000	.000	.000	.165	.000	.000	.000	.000
.67	.000	.820	.000	.013	.000	.000	.000	.265	.000	.000	.000	.000
.68	.000	.705	.000	.015	.000	.000	.000	.425	.000	.000	.000	.000
.69	.000	.743	.000	.022	.000	.000	.001	.615	.000	.029	.000	.000
.70	.000	.860	.000	.041	.000	.000	.004	.756	.007	.160	.000	.017
.71	.000	.876	.000	.085	.000	.001	.004	.837	.020	.385	.000	.075
.72	.000	.815	.000	.165	.000	.002	.004	.874	.037	.615	.000	.168
.73	.000	.435	.000	.285	.000	.002	.002	.889	.060	.760	.000	.257
.74	.000	.045	.000	.460	.000	.002	.002	.895	.086	.843	.000	.320
.75	.000	.055	.000	.645	.000	.001	.001	.898	.080	.878	.000	.335
.80	.000	.030	.000	.900	.005	.002	.003	.895	.009	.890	.000	.218
1.00	.000	.860	.003	.800	.010	.015	.020	.880	.000	.716	.000	.050
1.20	.000	.770	.026	.570	.030	.020	.047	.690	.000	.169	.000	.011
1.40	.000	.550	.072	.410	.060	.040	.095	.640	.004	.042	.004	.012
1.60	.000	.620	.200	.405	.116	.060	.154	.625	.002	.036	.002	.010
1.80	.000	.580	.265	.425	.172	.090	.223	.635	.000	.048	.000	.010
2.00	.010	.580	.374	.476	.343	.247	.410	.750	.000	.168	.000	.018
2.20	.071	.747	.533	.535	.475	.400	.528	.780	.000	.338	.000	.036
2.40	.190	.400	.370	.552	.575	.520	.600	.780	.000	.492	.000	.074
2.60	.203	.530	.523	.540	.580	.532	.603	.745	.000	.510	.000	.088
2.80	.100	.250	.265	.007	.162	.132	.360	.470	.000	.170	.000	.003
3.00	.080	.080	.202	.012	.195	.151	.280	.315	.000	.128	.000	.020
3.20	.068	.072	.180	.082	.131	.105	.172	.200	.000	.084	.000	.053
3.40	.030	.029	.131	.003	.079	.065	.100	.100	.000	.065	.000	.050
3.60	.021	.017	.103	.000	.061	.030	.053	.049	.002	.070	.000	.000
3.80	.023	.021	.093	.000	.068	.032	.040	.037	.001	.082	.000	.000
4.00	.040	.025	.112	.000	.071	.039	.050	.042	.005	.090	.000	.000
4.20	.019	.010	.069	.000	.020	.015	.017	.020	.002	.055	.000	.000
4.40	.001	.001	.007	.000	.000	.000	.000	.002	.000	.002	.000	.000
4.60	.000	.000	.000	.000	.000	.000	.000	.000	.000	.000	.000	.000
4.80	.000	.000	.000	.000	.000	.000	.000	.000	.000	.000	.000	.000
5.00	.000	.000	.000	.000	.000	.000	.000	.000	.000	.000	.000	.000

TRANSMISSION OF CORNING COLORED FILTERS (Continued)

λ(nm)	Transmittance											
	Corning Glass Number											
	5900	5970	7380	7740	7905	7910	8364	9780	9782	9788	9830	9863
.22	.000	.000	.000	.000	.000	.012	.000	.000	.000	.000	.000	.000
.24	.000	.000	.000	.000	.360	.505	.000	.000	.000	.000	.000	.054
.26	.000	.000	.000	.000	.495	.780	.000	.000	.000	.000	.000	.482
.28	.000	.000	.000	.004	.590	.855	.000	.000	.000	.000	.000	.731
.30	.000	.000	.000	.321	.720	.877	.000	.000	.000	.000	.000	.831
.32	.000	.138	.000	.722	.825	.900	.000	.000	.000	.000	.000	.862
.34	.008	.600	.000	.851	.880	.903	.002	.015	.000	.060	.001	.854
.36	.150	.799	.440	.889	.910	.905	.083	.290	.060	.470	.059	.816
.38	.445	.742	.795	.900	.915	.906	.136	.590	.445	.770	.160	.620
.40	.678	.190	.892	.916	.920	.920	.296	.725	.747	.885	.130	.090
.41	.688	.029	.904	.915	.920	.920	.273	.705	.790	.895	.044	.018
.42	.635	.000	.910	.915	.920	.921	.232	.770	.818	.902	.004	.003
.43	.586	.000	.913	.914	.920	.923	.191	.778	.836	.905	.000	.000
.44	.522	.000	.915	.913	.920	.924	.157	.801	.847	.906	.000	.000
.45	.458	.000	.916	.913	.922	.925	.144	.814	.855	.906	.000	.000
.46	.400	.000	.917	.914	.922	.925	.140	.823	.860	.906	.000	.000
.47	.350	.000	.917	.915	.922	.925	.141	.832	.863	.906	.000	.000
.48	.306	.000	.917	.915	.923	.925	.146	.839	.863	.905	.000	.000
.49	.275	.000	.918	.915	.930	.926	.152	.843	.859	.904	.000	.000
.50	.246	.000	.918	.915	.925	.926	.166	.843	.848	.900	.014	.000
.51	.223	.000	.919	.915	.925	.927	.178	.838	.825	.893	.180	.000
.52	.196	.000	.919	.916	.925	.928	.190	.824	.784	.880	.175	.000
.53	.172	.000	.919	.916	.923	.928	.196	.798	.720	.862	.018	.000
.54	.154	.000	.919	.916	.926	.929	.198	.756	.627	.831	.000	.000
.55	.148	.000	.920	.917	.923	.929	.197	.697	.515	.787	.050	.000
.56	.151	.000	.920	.917	.925	.930	.199	.615	.380	.728	.265	.000
.57	.146	.000	.919	.918	.925	.930	.206	.518	.255	.655	.165	.000
.58	.125	.000	.918	.919	.925	.930	.217	.414	.150	.570	.035	.000
.59	.102	.000	.918	.920	.925	.930	.222	.302	.075	.475	.004	.000
.60	.093	.000	.920	.920	.925	.930	.215	.215	.032	.380	.000	.000
.61	.087	.000	.920	.920	.925	.930	.196	.135	.010	.290	.000	.000
.62	.081	.000	.920	.920	.925	.930	.175	.080	.002	.210	.000	.000
.63	.070	.000	.920	.919	.926	.930	.161	.042	.000	.145	.000	.000
.64	.061	.000	.920	.919	.927	.931	.156	.021	.000	.094	.000	.000
.65	.055	.000	.920	.918	.927	.931	.162	.008	.000	.059	.000	.000
.66	.055	.000	.920	.917	.927	.932	.176	.003	.000	.035	.000	.000
.67	.059	.000	.920	.916	.928	.932	.200	.000	.000	.020	.000	.000
.68	.065	.000	.920	.916	.927	.932	.228	.000	.000	.010	.075	.022
.69	.068	.007	.921	.915	.927	.932	.248	.000	.000	.005	.380	.106
.70	.068	.036	.921	.915	.928	.932	.251	.000	.000	.001	.642	.234
.71	.066	.085	.922	.914	.926	.933	.237	.000	.000	.000	.694	.332
.72	.064	.145	.922	.912	.926	.933	.223	.000	.000	.000	.666	.383
.73	.060	.222	.922	.910	.926	.933	.210	.000	.000	.000	.607	.384
.74	.057	.323	.921	.909	.927	.934	.197	.000	.000	.000	.531	.358
.75	.055	.385	.921	.907	.928	.934	.180	.000	.000	.000	.445	.322
.80	.050	.287	.918	.890	.930	.932	.110	.000	.000	.000	.045	.175
1.00	.085	.032	.910	.860	.930	.928	.032	.000	.000	.000	.000	.119
1.20	.180	.021	.910	.860	.925	.928	.032	.000	.000	.005	.007	.016
1.40	.295	.109	.906	.870	.925	.930	.062	.018	.000	.080	.000	.005
1.60	.405	.088	.910	.892	.931	.930	.137	.131	.011	.266	.000	.007
1.80	.495	.040	.903	.896	.931	.930	.182	.317	.085	.430	.157	.011
2.00	.590	.008	.900	.897	.934	.929	.171	.440	.216	.512	.041	.029
2.20	.628	.002	.898	.875	.934	.835	.184	.440	.278	.455	.000	.048
2.40	.640	.009	.890	.850	.930	.890	.225	.440	.325	.433	.000	.060
2.60	.630	.018	.860	.820	.920	.780	.258	.280	.212	.252	.057	.051
2.80	.470	.030	.375	.140	.908	.180	.145	.060	.040	.060	.020	.000
3.00	.248	.030	.425	.360	.880	.695	.130	.000	.000	.000	.000	.000
3.20	.121	.030	.380	.490	.861	.760	.125	.000	.000	.000	.000	.000
3.40	.032	.020	.310	.270	.670	.620	.099	.000	.000	.000	.000	.000
3.60	.010	.015	.270	.010	.111	.080	.115	.000	.000	.000	.000	.000
3.80	.010	.020	.275	.040	.270	.240	.142	.000	.000	.000	.000	.000
4.00	.010	.030	.260	.013	.170	.150	.172	.000	.000	.000	.000	.000
4.20	.006	.017	.180	.026	.250	.230	.158	.000	.000	.000	.000	.000
4.40	.001	.002	.040	.004	.085	.080	.078	.000	.000	.000	.000	.000
4.60	.000	.000	.000	.000	.050	.020	.010	.000	.000	.000	.000	.000
4.80	.000	.000	.000	.000	.000	.000	.003	.000	.000	.000	.000	.000
5.00	.000	.000	.000	.000	.000	.000	.000	.000	.000	.000	.000	.000

TRANSMISSION OF WRATTEN FILTERS

Compiled by Allie C. Peed, Jr. for The Eastman Kodak Company

Data condensed from Kodak Wratten Filters for Scientific and Technical Use published by the Eastman Kodak Company, manufacturers of the filters.

The following pages give (1) percentage luminous transmittance at wave lengths from 400 to 700μ at intervals of 10μ for the standard illuminant "C" adopted by the International Commission of Illumination, (2) dominant wavelength in millimicrons, and (3) percentage of excitation purity. Values of wave length followed by "c" indicate the complementary wave lengths of purple filters which do not have a dominant wave length.

All colorimetric specifications are based on the 1931 standard ICI colorimetric and luminosity data.

The transmittance data are given as representing standard samples of the filters. They are intended only for the information of users in choosing filters which will meet their requirements. Values taken from the tables of data should not be used by research workers as representing precisely the absorption characteristics of a particular filter. If such precise data are needed, they should be determined for the particular filter being used.

Where the spectra extend into the ultraviolet this fact is indicated by an asterisk (*) in the transmission tables immediately beneath the filter number, and quantitative data are not given. The manufacturer should be consulted for this information. Transmission in the ultraviolet of wave lengths less than 330μ will be eliminated in the case of cemented filters, as glass absorbs ultraviolet radiation of wave lengths shorter than about 330μ.

Stability ratings are given as three letter combinations following the filter description in the table below. In establishing the stability classifications each filter is exposed to a selected light source for a specific time interval. The following grading system is used to describe the result:

Class A — stable
Class B — relatively stable
Class C — somewhat unstable
Class D — unstable

The classification letters, for example, AAA, describe the stability to the following three exposure tests in this order:

1. Two weeks' exposure to daylight in a south window
2. Twenty-four hours' exposure to a "Fade-Ometer"
3. Two weeks' exposure at two feet from a 1000-watt tungsten lamp.

Filters are supplied in two forms: as lacquered gelatin film, or as a gelatin film cemented between pieces of optical glass. Filters in glass are cemented between sheets of plane-parallel glass, which is surfaced in quantities and is of sufficient accuracy for general photographic work, and for most scientific purposes.

Most Wratten Gelatin Filters are stocked in 2- or 3-inch squares. Stocks of 2- or 3-inch square filters cemented in glass are maintained only in filters usually used for general photographic work.

The booklet "Kodak Filters and Lens Attachments" gives more valuable information on this subject.

FILTER DATA

No.	Description, use, and stability
	Colorless
0	For compensating thickness of other gelatin filters in optical systems, AAA.
1	Absorbs ultraviolet below 360 mμ, DDD.
1A	Kodak Skylight Filter — Reduces excess bluishness in outdoor color photographs in open shade under a clear, blue sky, ACA.
	Yellow
2B	Absorbs ultraviolet below 410 mμ, ACA.
3	Light yellow, CCD.
3N5	No. 3 plus 0.5 neutral density, AAA.
4	Light yellow — Approximate correction on panchromatic materials for outdoor scenes, including sky, CCC.
6	K1 — Light yellow — Partial correction outdoors, BBA.
8	K2 — Yellow — Full correction outdoors on Type B panchromatic materials. Widely used for proper sky, cloud, and foliage rendering. Green separation for Fluorescence Process, AAA.
8N5	No. 8 plus 0.5 neutral density, AAA.
9	K3 — Deep yellow. Moderate contrast in outdoor photography (with black-and-white films),AAA.
11	X1 — Greenish yellow. Correction for tungsten light on Type B panchromatic materials; also for daylight correction with Type C panchromatic materials in making outdoor portraits, darkening skies, or lightening foliage, AAA.
12	Minus blue. Haze cutting in aerial photography, AAA.
13	X2 — Yellow green. Correction for Type C panchromatic materials in tungsten light, ABA.
15	G — Deep yellow. Overcorrection in landscape photography. Contrast control in copying and in aerial infrared photography, AAA.
16	Blue absorption, AAB.
18A	Transmits ultraviolet and infrared only (glass), AAA.
	Oranges and Reds
21	Blue and blue-green absorption, CBB.
22	Yellow-orange. For increasing contrast in blue preparations in microscopy. Mercury yellow, BAC.
23A	Light red. Two-color projection — contrast effects, BAB.
24	Red for two-color photography (daylight or tungsten). White-flame-arc tricolor projection, AAB.
25	A — Tricolor red for direct color separation. Contrast effects in commercial photography and in outdoor scenes. Two-color general viewing. Aerial infrared photography and haze cutting, AAA.
26	Stereo red, AAA.
29	Red color separation from transparencies and for the Kodak Fluorescence Process. Strong contrast effects. Copying blueprints. Tungsten tricolor projection, AAA.
	Magentas and Violets
30	Green absorption, BBC.
31	Green absorption, CCA.
32	Minus green, CCD.
33	Strong green absorption, CCB.
34	Violet, CDD.
34A	Blue separation — Kodak Fluorescence Process, DCC.
35	Contrast in microscopy, CDD.
36	Dark violet, CCC.
	Blues and Blue-greens
38	Red absorption, BCA.
38A	Red absorption. Increasing contrast in visual microscopy, BBB.
39	Contrast control in printing motion-picture duplicates (glass) AAA.
40	Green for two-olor photography (tungsten), CBC.
44	Minus red — Two-color general viewing, DDD.
44A	Minus red, DDD.
45	Contrast in microscopy, DDD.
45A	Highest resolving power in visual microscopy, CDC.
46	Blue projection (experimental), DDD.
47	Tricolor blue for direct color separation and from Kodak Ektacolor Film for Dye Transfer. Contrast effects in commercial photography. Tungsten and white-flame-arc tricolor projection, BBC.
47B	Tricolor blue for color separation from transparencies and from Kodak Ektacolor Film for Graphic Arts, BBB.
48	Green and red absorption, CBC.
48A	Green and red absorption, AAB.
49	Dark blue, BCB.
49B	Very dark blue, BBB.
50	Very dark blue. Mercury violet, CCC.
	Greens
52	Light green, AAB.

TRANSMISSION OF WRATTEN FILTERS (Continued)

FILTER DATA

No.	Description, use, and stability
53	Medium green, CCB Very dark green, AAA.
55	Stereo green, BBC.
56	Very light green, CBC.
57	Green for two-color photography (daylight), CBC.
57A	Light green, BBC.
58	Tricolor green for direct color separation. Contrast effects in commercial photography and microscopy, BBC.
59	Green for tricolor projection (white-flame-arc), BBB.
59A	Very light green, BBB.
60	Green for two-color photography (tungsten), BDC.
61	Green color separation from transparencies and Kodak Ektacolor Film. Tricolor projection (tungsten), ABC.
64	Red absorption (light), CDB.
65	Red absorption, ADB.
65A	Red absorption, CCD.
66	Contrast effects in microscopy and medical photography, DDC.
67A	Red absorption (light). Two-color projection, CDC.
	Narrow-band
70	Dark red. Infrared photography. Color separation for Kodak Ektacolor Film (with tungsten), ABC.
72B	Dark orange-yellow, CCC.
73	Dark yellow-green, ABB.
74	Dark green. Mercury green, BBC.
75	Dark blue-green, ACC.
76	Dark violet (compound filter), DDD.
77	Transmits 546 mμ mercury line (glass plus gelatin), AAA.
77A	Transmits 546 mμ mercury line (glass plus gelatin), AAA.
	Photometrics
78	Bluish. Photometric filter (visual), BAB.
78AA	Bluish. Photometric filter (visual), BAA.
78A	Bluish. Photometric filter (visual), AAA.
78B	Bluish. Photometric filter (visual), AAA.
78C	Bluish. Photometric filter (visual), BAA.
86	Yellowish. Photometric filter (visual), BBA.
86A	Yellowish. Photometric filter (visual), AAA.
†86B	Yellowish. Photometric filter (visual), BCA.
†86C	Yellowish. Photometric filter (visual), AAA.
	Light Balancing
80A	For Kodachrome Film, Daylight Type, and photographic flood lamps, ABA.
81	Yellowish. For warmer color rendering.
81A	Yellowish. For Kodak Ektachrome Film, Type B, with photographic flood lamps.
81B	Yellowish. For warmer color rendering.
81C	Yellowish. For Kodachrome Film, Type A, with flash lamps.
81D	Yellowish. For Kodachrome Film, Type A, with flash lamps.
81EF	Yellowish. For Kodak Ektachrome Film, Type B, with flash lamps.
82	Bluish. For cooler color rendering.
82A	Bluish. For Kodachrome Film, Type A, with 3200 K lamps.
82B	Bluish. For cooler color tendering.
82C	Bluish. For cooler color rendering.
83	Yellowish. For 16 mm Commercial Kodachrome Film and daylight exposure, BBB.
85	Orange. For Type A Kodak color films and daylight exposure, BAA.
85B	Orange. For Kodak Ektachrome Film, Type B, and daylight exposure, BAB.
	Miscellaneous
79	Photographic sensitometry. Corrects 2360 K to 5500 K, AAA.
87	For infrared photography. Absorbs visual.
87C	Absorbs visual, transmits infrared.
88A	For infrared photography. Absorbs visual.
89B	For infrared photography, AAA.
90	Narrow-band viewing filter for judging brightness scale of scenes, CCD.
96	Neutral filters for controlling luminance, AAB.
97	Dichroic absorption, AAA.
102	Correction filter for Barrier-layer photocell, ABA.
106	Correction filter for S-4 type photocell, AAA.

Wave length	Percent transmittance												
	No. 0 *	No. 1 *	No. 1A *	No. 2B *	No. 3	No. 3N5 *	No. 4	No. 6 *	No. 8	No. 8N5	No. 9	No. 11 *	No. 12 *
400	88.0	85.0	59.0	19.0	—	—	—	7.40	—	—	—	—	—
10	88.5	85.5	76.0	48.0	—	—	—	8.32	—	—	—	—	—
20	88.9	86.0	82.0	67.0	—	—	—	10.4	—	—	—	0.16	—
30	89.3	86.5	84.6	75.3	0.36	—	—	13.5	—	—	—	0.29	—
40	89.6	87.0	86.0	80.0	1.78	—	—	18.9	—	—	—	0.56	—
50	89.8	87.4	86.8	83.0	11.5	1.59	—	27.6	—	—	—	1.32	—
60	89.9	87.8	87.2	85.2	38.0	9.40	6.9	39.0	0.25	0.16	—	4.00	—
70	90.1	88.2	87.5	86.7	68.0	18.5	42.0	52.3	5.50	2.0	1.78	12.0	—
80	90.3	88.5	87.3	88.1	80.8	23.5	74.0	65.8	19.0	6.3	8.31	26.0	—
90	90.4	88.7	86.8	88.8	85.2	25.5	84.7	76.8	41.0	13.2	20.7	43.7	—
500	90.5	88.9	86.3	89.5	86.9	26.3	87.5	83.5	63.5	20.0	34.5	55.0	1.50
10	90.6	89.1	85.5	89.9	87.8	26.7	88.5	87.0	78.0	24.3	48.8	60.0	17.3
20	90.7	89.3	84.8	90.3	88.4	27.0	89.1	88.4	84.1	26.7	62.0	60.2	55.0
30	90.7	89.5	84.3	90.5	89.0	27.2	89.4	89.0	86.5	28.0	76.0	57.8	77.8
40	90.8	89.7	84.0	90.6	89.5	27.5	89.6	89.4	87.7	28.6	83.8	54.2	86.0
50	90.8	89.9	83.9	90.7	89.8	27.8	89.8	89.7	88.4	29.0	87.0	50.0	88.4
60	90.9	90.1	84.1	90.8	90.1	27.9	90.0	89.9	88.8	29.3	88.3	44.8	89.4
70	90.9	90.2	84.8	90.9	90.4	28.0	90.2	90.1	89.2	29.5	88.8	38.9	89.7
80	90.9	90.3	86.0	90.9	90.6	28.4	90.4	90.3	89.5	29.6	89.1	33.1	90.1
90	91.0	90.4	87.4	91.0	90.7	29.0	90.6	90.5	89.8	29.8	89.3	27.6	90.3
600	91.0	90.5	88.5	91.1	90.8	29.5	90.8	90.6	90.1	29.9	89.5	22.7	90.4
10	91.0	90.5	89.5	91.2	90.9	29.5	90.9	90.7	90.3	29.6	89.7	19.0	90.5
20	91.0	90.6	90.2	91.3	91.0	29.3	91.0	90.8	90.5	29.4	89.8	14.9	90.7
30	91.0	90.6	90.6	91.3	91.0	29.1	91.1	90.9	90.7	29.1	89.9	11.4	90.8
40	91.1	90.7	90.8	91.4	91.1	29.0	91.2	91.0	90.9	28.8	90.0	9.10	90.9
50	91.1	90.7	91.0	91.4	91.2	20.4	91.3	91.1	91.0	28.9	90.1	8.05	91.0
60	91.1	90.8	91.1	91.5	91.3	29.6	91.4	91.2	91.1	29.2	90.1	7.50	91.1
70	91.1	90.8	91.1	91.5	91.4	29.8	91.5	91.2	91.2	29.4	90.2	7.05	91.2
80	91.1	90.9	91.1	91.6	91.5	30.0	91.5	91.3	91.3	29.5	90.2	6.50	91.2
90	91.1	90.9	91.1	91.7	91.6	30.2	91.6	91.4	91.4	29.7	90.3	6.10	91.2
700	91.1	91.0	91.1	91.8	91.7	31.0	91.6	91.3	91.5	30.2	90.3	6.20	91.3
Luminous transmit.	90.8	89.9	85.9	90.5	88.3	27.4	87.8	87.5	82.7	27.0	76.6	40.2	73.8
Dominant wave lgth.	571.0	575.0	498.0	570.0	569.5	570.5	569.5	570.3	571.8	572.0	574.4	550.3	576.1
Excitation purity	0.8	1.5	1.2	5.7	50.0	56.3	64.0	44.7	85.2	84.0	91.4	60.7	97.8

* Some transmission below 400 mμ. Consult the manufacturer.

TRANSMISSION OF WRATTEN FILTERS (Continued)

Wave length	Percent transmittance												
	No. 13 *	No. 15 *	No. 16	No. 18A *	No. 21	No. 22	No. 23A	No. 24	No. 25	No. 26	No. 29	No. 30 *	No. 31 *
400	—	—	—	—	—	—	—	—	—	—	—	48.6	13.8
10	—	—	—	—	—	—	—	—	—	—	—	47.4	14.5
20	—	—	—	—	—	—	—	—	—	—	—	48.5	16.4
30	0.18		—	—	—	—	—	—	—	—	—	50.1	25.5
40	0.50	—	—	—	—	—	—	—	—	—	—	49.4	42.7
50	1.35	—	—	—	—	—	—	—	—	—	—	43.0	50.2
60	4.08	—	—	—	—	—	—	—	—	—	—	26.5	40.4
70	11.0	—	—	—	—	—	—	—	—	—	—	13.8	22.6
80	23.5	—	—	—	—	—	—	—	—	—	—	5.00	8.20
90	39.0	—	—	—	—	—	—	—	—	—	—	0.63	1.85
500	50.8	—	—	—	—	—	—	—	—	—	—	—	0.12
10	55.2	1.00	—	—	—	—	—	—	—	—	—	—	—
20	56.5	16.0	3.00	—	—	—	—	—	—	—	—	—	—
30	55.0	52.1	22.0	—	—	—	—	—	—	—	—	—	—
40	51.0	70.7	48.0	—	2.50	—	—	—	—	—	—	—	—
50	46.0	84.3	69.5	—	29.0	0.25	—	—	—	—	—	—	—
60	39.2	87.5	79.5	—	65.0	19.0	—	—	—	—	—	0.10	—
70	32.0	88.7	84.0	—	80.6	60.0	11.0	—	—	—	—	10.0	—
80	25.1	89.3	86.3	—	85.4	81.0	47.0	4.55	—	—	—	45.0	—
90	18.2	89.7	87.8	—	87.3	87.0	69.6	37.3	12.6	2.90	—	76.0	0.63
600	13.5	90.0	89.0	—	88.1	88.5	82.7	72.3	50.0	30.0	—	87.4	26.0
10	9.60	90.1	89.6	—	88.7	89.0	85.8	82.9	75.0	63.2	10.0	89.5	67.2
20	6.40	90.2	90.0	—	89.0	89.5	87.2	86.4	82.6	78.9	45.3	90.2	84.0
30	3.66	90.3	90.2	—	89.5	89.8	87.9	87.8	85.5	84.0	71.4	90.5	88.1
40	2.20	90.4	90.3	—	89.9	90.0	88.5	88.5	86.7	86.1	82.7	90.7	89.8
50	1.58	90.5	90.4	—	90.2	90.1	89.0	89.0	87.6	87.2	86.6	90.8	90.2
60	1.74	90.6	90.5	—	90.4	90.2	89.4	89.3	88.2	88.1	88.4	90.9	90.4
70	2.62	90.6	90.6	—	90.5	90.3	89.6	89.7	88.5	88.5	89.4	91.0	90.5
80	3.55	90.7	90.7	—	90.5	90.4	89.8	89.9	89.0	88.9	90.0	91.1	90.7
90	4.48	90.7	90.8	0.25	90.6	90.5	90.0	90.2	89.3	89.2	90.3	91.1	90.8
700	5.25	90.8	90.8	1.20	90.6	90.6	90.2	90.3	89.5	89.5	90.4	91.1	91.0
Luminous transmit.	34.5	66.2	57.7	0.0014	45.6	35.8	25.0	17.8	14.0	11.7	6.3	26.6	12.9
Dominant wave lgth.	542.0	579.3	582.7	700.0	588.9	595.1	602.7	610.6	615.1	619.0	631.6	498.6_c	513.1_c
Excitation purity.	57.5	99.0	99.3	100.0	99.9	99.9	100.0	100.0	100.0	100.0	100.0	62.4	81.9

Wave length	Percent transmittance												
	No. 32 *	No. 33 *	No. 34 *	No. 34A *	No. 35 *	No. 36 *	No. 38 *	No. 38A *	No. 39 *	No. 40	No. 44 *	No. 44A *	No. 45
400	38.0	0.85	64.0	—	48.0	36.5	60.5	33.4	85.2	—	0.44	2.52	—
10	37.9	0.71	70.1	0.1	57.0	45.5	66.5	41.2	78.2	—	0.36	3.39	—
20	40.0	1.17	72.0	40.0	57.6	45.5	72.5	53.0	70.5	—	0.63	6.30	—
30	43.0	1.69	68.4	69.7	47.5	32.7	75.3	58.0	63.3	—	3.63	17.4	—
40	55.5	5.36	58.2	68.7	29.5	15.2	76.2	58.8	53.6	—	13.1	32.7	5.00
50	66.0	14.3	42.3	56.2	12.3	3.7	75.9	57.6	42.5	—	25.4	41.8	19.0
60	66.0	12.4	25.2	40.5	3.5	0.35	74.8	55.2	28.5	3.16	36.5	48.1	29.5
70	57.0	5.00	12.1	23.8	0.25	—	73.4	51.9	17.3	21.6	46.5	51.7	34.4
80	40.0	0.50	2.7	9.2	—	—	71.6	48.5	10.2	44.7	53.6	52.9	35.7
90	21.0	—	0.2	2.3	—	—	69.5	44.6	4.00	61.4	56.8	52.2	34.5
500	9.56	—	—	0.33	—	—	66.7	40.2	1.33	70.2	55.8	49.8	29.7
10	2.51	—	—	—	—	—	63.9	35.8	0.35	72.4	50.9	44.8	21.5
20	0.13	—	—	—	—	—	60.8	31.7	—	70.5	42.1	36.8	11.5
30	—	—	—	—	—	—	57.0	27.2	—	64.8	30.5	26.8	3.80
40	—	—	—	—	—	—	52.6	22.3	—	55.5	18.6	16.8	0.85
50	—	—	—	—	—	—	48.0	17.6	—	44.2	8.99	8.20	—
60	—	—	—	—	—	—	42.8	12.9	—	32.5	3.59	2.95	—
70	—	—	—	—	—	—	37.0	8.78	—	20.3	0.80	0.91	—
80	—	—	—	—	—	—	30.6	5.65	—	9.56	—	0.10	—
90	—	—	—	—	—	—	25.5	3.48	—	3.20	—	—	—
600	6.04	—	—	—	—	—	20.9	2.09	—	1.10	—	—	—
10	41.0	0.80	—	0.13	—	—	16.8	1.15	—	0.32	—	—	—
20	75.0	24.9	—	1.0	—	—	12.9	0.59	—	—	—	—	—
30	86.1	60.8	—	6.3	—	—	10.0	0.28	—	—	—	—	—
40	89.0	78.0	0.4	22.0	—	—	7.79	0.13	—	—	—	—	—
50	90.0	85.0	4.0	45.0	0.1	—	6.68	—	—	—	—	—	—
60	90.6	87.5	20.7	65.0	3.0	0.21	6.20	—	—	—	—	—	—
70	90.7	88.7	45.2	77.3	19.0	7.5	5.91	—	—	—	—	—	—
80	90.8	89.4	66.5	85.0	43.5	29.0	5.41	—	0.50	0.80	—	—	—
90	90.9	89.8	78.8	88.2	66.0	55.0	4.90	—	4.06	6.99	0.18	—	—
700	91.0	90.0	85.0	89.8	77.7	71.3	5.00	—	17.8	23.5	1.60	—	1.00
Luminous transmit.	12.5	5.2	1.3	2.9	0.45	0.25	42.5	17.3	1.2	33.6	15.6	14.4	5.2
Dominant wave lgth.	551.7_c	498.0_c	424.0	564.8_c	566.8	566.4_c	483.5	478.9	450.6	516.2	589.1	483.4	481.5
Excitation purity.	79.6	88.3	94.4	91.4	96.3	97.8	41.8	69.8	98.9	48.5	72.9	77.2	88.4

* Some transmission below 400 mμ. Consult the manufacturer.

TRANSMISSION OF WRATTEN FILTERS (Continued)

Wave length	No. 45A	No. 46 *	No. 47 *	No. 47B *	No. 48 *	No. 48A *	No. 49 *	No. 49B *	No. 50 *	No. 52 *	No. 53 *	No. 54	No. 55
	Percent transmittance												
400	—	1.20	7.80	16.0	0.96	5.65	3.30	1.70	0.45	2.18	—	—	—
10	—	0.60	17.4	29.5	3.16	10.0	4.28	2.00	0.39	1.51	—	—	—
20	—	0.80	34.0	43.6	8.25	16.0	6.93	3.55	0.59	0.80	—	—	—
30	1.00	5.98	47.0	50.0	15.0	21.0	11.2	7.00	2.63	0.44	—	—	—
40	8.81	19.0	50.3	47.2	22.6	25.0	18.9	13.0	8.90	0.41	—	—	—
50	17.4	30.1	48.3	36.0	30.3	26.2	25.6	17.4	14.0	0.69	—	—	—
60	20.9	33.8	43.4	25.0	33.2	22.9	24.0	14.8	12.3	1.45	—	—	0.20
70	21.6	32.1	36.2	13.2	29.6	16.5	15.7	7.60	5.36	2.70	0.10	—	2.90
80	20.5	27.0	28.5	4.5	22.4	9.55	6.93	2.76	1.55	4.90	0.7	—	13.1
90	18.0	20.2	19.6	1.3	14.1	4.27	2.14	0.40	0.10	8.50	2.14	—	34.2
500	14.4	11.1	0.36	0.17	7.30	1.58	0.46	—	—	13.3	4.47	—	53.4
10	10.1	4.39	—	—	2.64	0.48	—	—	—	18.2	7.24	0.10	67.0
20	5.60	1.66	—	—	0.50	—	—	—	—	23.7	10.7	0.31	69.3
30	2.52	0.35	—	—	—	—	—	—	—	28.5	14.0	0.64	65.1
40	0.4	—	—	—	—	—	—	—	—	32.1	16.6	0.89	56.7
50	0.10	—	—	—	—	—	—	—	—	33.1	17.3	0.93	45.0
60	—	—	—	—	—	—	—	—	—	31.0	15.4	0.62	33.1
70	—	—	—	—	—	—	—	—	—	25.6	11.4	0.21	20.7
80	—	—	—	—	—	—	—	—	—	19.1	6.90	—	9.00
90	—	—	—	—	—	—	—	—	—	12.6	3.60	—	2.70
600	—	—	—	—	—	—	—	—	—	7.78	1.41	—	0.40
10	—	—	—	—	—	—	—	—	—	4.17	0.40	—	—
20	—	—	—	—	—	—	—	—	—	2.34	0.15	—	—
30	—	—	—	—	—	—	—	—	—	1.38	—	—	—
40	—	—	—	—	—	—	—	—	—	0.80	—	—	—
50	—	—	—	—	—	—	—	—	—	0.54	—	—	—
60	—	—	—	—	—	—	—	—	—	0.36	—	—	—
70	—	—	—	—	—	—	—	—	—	0.27	—	—	—
80	—	—	—	—	—	—	—	—	—	0.23	—	—	0.66
90	0.20	0.25	—	—	—	—	—	—	—	0.19	—	—	6.90
700	2.24	0.85	—	—	—	—	—	—	—	0.17	—	—	27.8
Luminous transmit.	2.8	2.4	2.8	0.78	1.86	0.88	0.69	0.36	0.26	20.1	9.0	0.032	31.4
Dominant wave lgth.	477.6	470.4	463.7	479.8	466.5	458.0	457.9	455.5	455.9	553.3	551.1	546.1	530.2
Excitation purity.	89.7	94.9	95.8	69.1	96.1	98.3	98.9	99.3	99.4	77.3	89.7	97.0	68.4

Wave length	No. 56	No. 57	No. 57A	No. 58	No. 59 *	No. 59A *	No. 60	No. 61	No. 64 *	No. 65	No. 65A *	No. 66 *	No. 67A *
	Percent transmittance												
400	—	—	—	—	—	—	—	—	9.00	—	—	12.3	1.10
10	—	—	—	—	—	—	—	—	9.20	—	—	13.0	0.93
20	—	—	—	—	—	—	—	—	8.75	0.23	—	15.0	1.28
30	—	—	—	—	—	0.16	—	—	9.20	0.61	0.16	18.4	3.16
40	—	—	0.19	—	—	0.37	—	—	11.3	1.58	1.32	23.2	6.40
50	—	—	0.87	—	0.40	1.26	0.19	—	15.5	4.10	5.50	31.2	10.5
60	0.16	0.44	2.56	—	1.90	4.57	1.38	—	23.3	9.00	13.0	42.2	17.7
70	3.12	3.10	7.80	0.23	7.70	13.2	5.38	—	34.4	16.8	24.9	55.5	28.5
80	13.0	13.1	21.6	1.38	21.0	30.0	15.0	0.33	46.8	24.9	36.6	68.4	41.4
90	34.5	31.9	41.7	4.90	41.5	50.8	32.0	4.00	56.6	31.3	45.1	77.6	52.1
500	59.0	50.5	58.8	17.7	59.0	66.0	48.4	16.6	62.1	33.7	45.8	82.7	57.9
10	73.0	60.6	67.9	38.8	67.7	73.0	57.2	32.3	62.9	32.4	39.7	84.6	58.8
20	79.0	63.3	70.1	52.2	69.8	75.1	59.2	40.0	59.1	27.5	29.7	84.0	55.4
30	79.9	61.0	67.6	53.6	67.2	73.2	55.5	39.6	51.6	20.7	17.8	82.6	47.5
40	77.5	55.0	61.8	47.6	61.5	68.5	47.5	34.5	41.3	13.7	7.90	79.1	36.6
50	72.6	47.1	53.5	38.4	54.0	62.0	36.8	26.3	28.0	6.50	2.40	73.7	25.0
60	66.1	37.3	43.3	27.8	45.0	54.4	25.2	17.3	16.2	1.66	0.32	67.1	14.2
70	58.0	26.5	31.6	17.4	35.0	44.5	14.4	9.70	7.95	0.40	—	58.8	5.50
80	46.1	16.6	19.4	9.0	24.0	33.0	6.3	4.40	3.10	—	—	47.2	1.40
90	33.8	8.69	9.70	3.50	14.0	22.0	1.82	1.66	0.80	—	—	34.5	0.28
600	24.0	3.70	4.50	1.50	7.95	14.6	0.48	0.38	—	—	—	24.4	—
10	18.7	1.60	2.00	0.41	4.90	10.5	0.10	—	—	—	—	18.5	—
20	13.2	0.49	0.87	—	2.70	6.92	—	—	—	—	—	13.7	—
30	7.22	—	0.22	—	1.00	3.16	—	—	—	—	—	7.70	—
40	3.02	—	—	—	0.17	1.07	—	—	—	—	—	3.00	—
50	1.48	—	—	—	—	0.50	—	—	—	—	—	1.46	—
60	1.91	—	—	—	—	0.91	—	—	—	—	—	1.91	—
70	7.95	—	—	—	0.63	3.00	—	—	—	—	—	6.17	—
80	23.0	—	0.16	—	4.00	10.0	—	—	—	—	—	19.9	—
90	44.1	—	1.15	—	12.0	20.0	2.10	—	0.10	—	0.20	42.6	—
700	64.8	—	3.17	0.53	22.6	30.0	8.70	—	4.50	—	2.18	63.1	0.40
Luminous transmit.	52.8	32.5	37.2	23.7	38.7	45.8	26.1	16.8	25.0	9.6	9.8	58.3	22.4
Dominant wave lgth.	552.3	536.4	534.0	540.2	538.3	541.4	525.7	536.8	497.3	496.6	492.7	512.3	499.8
Excitation purity.	78.2	69.2	62.1	88.1	66.0	59.3	62.2	85.4	55.0	67.8	77.4	21.5	55.8

* Some transmission below 400 mμ. Consult the manufacturer.

TRANSMISSION OF WRATTEN FILTERS (Continued)

Wave length	Percent transmittance											
	No. 70	No. 72B	No. 73	No. 74	No. 75	No. 76 *	No. 77	No. 77A	No. 78 *	No. 78AA *	No. 78A *	No. 78B *
400	—	—	—	—	—	0.22	—	—	37.2	43.0	56.0	64.1
10	—	—	—	—	—	0.18	—	—	41.7	46.0	58.6	66.5
20	—	—	—	—	—	0.29	—	—	44.2	48.7	61.0	68.4
30	—	—	—	—	—	1.38	—	—	44.6	49.8	61.8	69.5
40	—	—	—	—	—	3.50	—	—	44.2	49.7	61.8	70.0
50	—	—	—	—	—	3.50	—	—	41.7	48.0	61.0	69.4
60	—	—	—	—	1.97	1.92	—	—	38.0	44.9	58.7	67.5
70	—	—	—	—	10.0	0.51	—	—	33.8	40.3	55.0	65.4
80	—	—	—	—	17.4	—	—	—	27.5	35.6	51.0	62.9
90	—	—	—	—	18.0	—	—	—	23.5	30.9	47.1	59.8
500	—	—	—	—	13.0	—	—	—	19.5	26.5	43.5	57.0
10	—	—	—	0.96	7.35	—	0.30	0.10	15.8	23.4	40.0	54.2
20	—	—	—	7.95	3.20	—	9.10	5.35	13.8	20.3	36.9	51.4
30	—	—	—	14.6	0.83	—	13.5	1.90	11.8	17.8	34.4	49.3
40	—	—	—	12.9	0.14	—	46.1	35.0	10.5	16.6	32.7	48.1
50	—	—	—	7.60	—	—	78.0	71.8	9.56	14.9	31.2	46.7
60	—	—	2.24	3.06	—	—	75.8	63.1	8.53	13.2	29.4	45.0
70	—	—	5.97	0.83	—	—	8.00	—	7.77	12.1	28.0	43.6
80	—	—	4.56	0.12	—	—	1.00	—	7.41	11.6	27.5	43.1
90	—	1.26	2.00	—	—	—	0.32	—	6.93	11.1	27.0	42.9
600	—	5.89	0.56	—	—	—	16.2	1.60	6.45	10.40	26.0	41.8
10	—	5.25	0.10	—	—	—	52.1	32.1	5.50	9.20	24.1	40.0
20	—	2.88	—	—	—	—	83.0	78.0	4.80	7.70	21.8	37.6
30	—	1.26	—	—	—	—	84.9	79.5	3.94	6.50	19.7	35.5
40	—	0.48	—	—	—	—	88.1	86.5	3.46	5.60	18.6	34.2
50	0.63	0.14	—	—	—	—	89.8	89.2	3.24	5.50	18.4	33.6
60	10.5	—	—	—	—	—	89.8	89.0	3.16	5.60	18.5	34.0
70	35.0	—	—	—	—	—	85.5	79.5	3.39	5.80	18.7	34.1
80	55.2	—	—	—	—	—	76.1	62.5	3.45	6.10	19.0	34.5
90	70.0	—	—	—	—	0.13	75.0	62.4	3.51	6.10	19.3	34.8
700	79.0	—	—	—	0.14	1.24	86.5	83.0	3.90	6.50	20.2	36.0
Luminous transmit.	0.31	0.74	1.3	4.0	1.9	0.046	32.3	25.5	10.7	15.8	31.6	46.7
Dominant wave lgth.	675.6	604.9	574.9	538.6	487.7	449.2	579.9	581.5	471.1	473.4	475.7	477.2
Excitation purity.	100.0	100.0	100.0	96.7	90.4	99.7	99.0	99.1	63.0	54.5	33.7	20.7

Wave length	Percent transmittance											
	No. 78C *	No. 79 *	No. 80A *	No. 81 *	No. 81A	No. 81B	No. 81C *	No. 81D *	No. 81EF	No. 82	No. 82A *	No. 82B *
400	74.9	24.0	67.6	77.7	65.1	55.1	46.1	38.2	30.7	83.0	80.1	76.7
10	76.6	26.0	73.1	78.1	65.9	55.8	46.6	38.4	31.5	83.7	80.8	78.0
20	77.9	29.0	76.8	79.0	67.6	57.7	49.0	41.0	34.3	84.6	81.6	79.2
30	78.9	31.0	7.77	80.5	70.2	61.0	52.5	45.5	38.6	85.1	82.2	79.7
40	79.4	32.2	76.5	81.9	72.8	64.5	57.2	50.0	43.2	85.4	82.4	79.7
50	79.5	32.7	73.0	83.0	74.8	67.2	60.5	53.9	47.4	85.4	82.4	79.2
60	79.3	31.4	69.0	83.7	76.0	69.1	63.0	56.5	50.2	85.0	81.7	78.0
70	78.6	28.8	63.6	84.3	77.1	70.6	64.2	58.1	52.0	84.6	80.7	76.3
80	77.8	25.6	57.6	84.6	77.8	71.3	65.0	59.0	53.0	84.0	79.3	74.4
90	76.7	22.2	51.3	84.9	78.3	71.8	65.7	60.0	54.0	83.3	78.0	72.1
500	75.5	19.3	45.2	85.3	78.6	72.6	66.4	60.8	55.4	82.6	76.6	70.2
10	74.2	16.8	39.4	85.4	79.0	72.9	66.5	61.1	56.2	82.0	75.3	68.3
20	73.0	14.2	34.2	85.5	79.5	73.2	67.0	61.6	57.0	81.4	74.0	66.5
30	72.1	12.7	30.0	86.0	80.4	74.5	68.8	62.5	59.5	81.0	73.1	65.5
40	71.5	11.8	27.1	86.5	81.5	76.0	71.0	66.1	62.7	80.8	72.7	65.0
50	70.7	11.0	24.8	86.8	82.3	77.0	72.0	67.3	64.5	80.6	72.4	64.5
60	69.8	9.76	23.5	87.0	82.6	77.6	72.5	68.0	65.3	80.4	71.8	63.8
70	69.0	8.81	22.6	87.1	82.7	77.8	72.7	68.3	65.8	80.2	71.5	63.2
80	68.8	8.50	22.6	87.1	82.8	78.0	73.0	68.5	66.0	80.2	71.5	63.2
90	68.6	8.29	23.2	87.4	83.1	78.2	74.0	69.5	66.5	80.3	71.7	63.4
600	68.0	7.56	23.7	87.6	84.0	79.1	75.6	72.0	68.1	80.2	71.5	63.0
10	66.7	6.45	23.2	88.1	85.0	81.0	78.5	75.0	71.6	79.3	70.3	61.5
20	65.0	5.13	21.0	88.8	86.1	83.1	80.8	78.0	74.7	78.4	68.5	59.0
30	63.8	4.17	18.2	89.2	87.0	84.2	82.1	79.8	77.0	77.5	66.9	56.9
40	63.0	3.47	15.8	89.4	87.4	85.1	83.0	80.8	78.4	76.8	65.5	55.0
50	62.7	3.16	14.5	89.5	87.7	85.6	83.5	81.5	79.2	76.5	64.8	54.1
60	63.0	3.09	13.8	89.8	88.0	86.0	84.1	82.1	80.1	76.2	64.6	53.7
70	63.3	3.16	13.4	90.0	88.2	86.5	84.8	83.0	80.9	76.1	64.5	53.7
80	63.4	3.16	12.7	90.1	88.5	87.0	85.5	83.7	81.8	76.1	64.4	53.5
90	63.6	3.16	11.7	90.3	89.0	87.5	86.1	84.6	82.9	76.2	64.2	53.4
700	65.0	3.31	11.5	90.5	89.2	88.0	86.8	85.5	84.0	77.1	64.6	54.1
Luminous transmit.	70.4	11.3	28.4	86.8	82.0	76.9	72.0	67.4	64.0	80.7	72.5	64.6
Dominant wave lgth.	479.8	474.8	471.7	576.7	577.5	577.8	577.4	579.5	579.0	477.5	476.6	475.6
Excitation purity.	6.8	52.8	45.9	2.9	6.0	8.7	10.7	14.7	19.0	3.0	6.3	10.2

* Some transmission below 400 mμ. Consult the manufacturer.

TRANSMISSION OF WRATTEN FILTERS (Continued)

Wave length	Percent transmittance												
	No. 82C *	No. 83 *	No. 85 *	No. 85B *	No. 86 *	No. 86A *	No. 86B *	No. 86C *	No. 89B	No. 90	No. 96 *	No. 97	No. 102 *
400	73.4	13.5	6.0	1.59	0.50	8.00	20.0	44.0	—	—	4.28	—	1.12
10	75.0	13.1	18.0	9.32	0.81	12.2	26.1	55.0	—	—	4.91	—	0.96
20	76.4	13.5	28.4	15.5	1.55	16.7	31.6	62.0	—	—	5.50	—	0.89
30	77.2	14.1	33.4	19.0	2.88	21.5	37.5	66.6	—	—	6.17	—	0.96
40	77.2	15.6	36.2	20.8	5.50	27.8	44.0	70.8	—	—	6.92	—	1.23
50	76.6	17.8	38.1	22.1	9.10	34.2	50.1	74.3	—	—	7.50	—	1.86
60	75.2	21.0	40.4	24.3	13.5	40.4	55.4	76.8	—	—	7.81	—	3.23
70	73.2	25.5	43.0	27.5	17.8	45.0	59.5	78.7	—	—	8.15	0.22	6.45
80	70.7	30.2	45.3	30.9	21.3	48.7	62.5	80.2	—	—	8.47	0.43	14.0
90	68.1	35.8	47.2	34.3	24.5	51.2	64.6	81.2	—	—	8.60	0.39	21.6
500	65.7	43.5	48.9	38.3	26.8	52.8	66.0	81.7	—	—	8.73	0.15	30.7
10	63.5	46.3	49.2	40.7	27.9	53.4	66.4	81.9	—	—	8.85	—	41.4
20	61.5	47.2	48.2	40.6	28.6	53.7	66.6	82.0	—	—	8.90	—	51.3
30	59.9	48.3	48.3	40.7	30.4	55.0	67.6	82.4	—	—	9.01	—	59.4
40	59.1	49.6	49.2	41.6	32.5	56.5	69.0	83.0	—	—	9.07	—	64.2
50	58.3	51.8	51.0	43.2	35.0	58.5	70.2	83.5	—	—	9.20	—	66.7
60	57.2	56.5	55.8	47.1	41.2	63.0	73.0	84.6	—	9.00	9.30	—	66.3
70	56.2	65.0	64.5	56.0	53.0	70.9	78.1	86.8	—	30.5	9.20	—	63.0
80	56.1	75.5	75.0	68.1	67.5	79.0	84.0	88.9	—	34.3	9.19	—	58.0
90	56.0	83.0	83.0	78.1	76.5	85.2	87.5	89.9	—	25.2	9.54	—	51.9
600	55.0	87.3	87.2	85.0	85.0	88.1	89.3	90.6	—	16.1	9.64	—	45.2
10	53.0	89.3	88.9	88.0	88.1	89.8	90.3	91.0	—	11.3	9.73	—	37.8
20	50.2	90.4	90.0	89.6	89.6	90.5	90.7	91.1	—	7.40	9.56	—	30.5
30	47.4	90.8	90.5	90.3	90.4	90.8	90.9	91.2	—	2.91	9.27	—	25.0
40	45.2	91.0	90.7	90.7	90.7	91.1	91.1	91.3	—	0.76	9.10	—	20.6
50	44.1	91.1	90.9	90.9	91.0	91.2	91.2	91.4	—	0.29	9.07	—	17.5
60	43.6	91.3	91.0	91.0	91.1	91.3	91.3	91.5	—	0.41	9.00	—	15.2
70	43.5	91.5	91.0	91.2	91.2	91.4	91.4	91.6	—	2.30	9.13	—	13.7
80	43.1	91.5	91.0	91.3	91.3	91.4	91.5	91.6	0.10	9.52	9.08	0.44	12.8
90	42.8	91.5	91.0	91.3	91.3	91.5	91.6	91.6	1.58	28.5	9.21	5.02	12.1
700	43.5	91.5	91.0	91.3	91.3	91.5	91.6	91.6	11.2	51.9	9.52	18.7	12.0
Luminous transmit.	58.1	61.4	62.5	55.5	49.7	67.1	75.5	85.4	0.017	9.8	9.1	0.041	50.8
Dominant wave lgth.	477.2	581.5	587.7	585.7	585.7	581.7	579.6	577.6	700	583.1	572.4	555.0_c	564.9
Excitation purity.	14.5	55.4	30.3	48.0	69.7	37.1	24.1	9.0	100	100.0	12.1	48.0	80.0

Wave length	Percent transmittance												
	No. 106	CC-05R *	CC-10R *	CC-20R *	CC-30R *	CC-40R *	CC-50R *	CC-05B *	CC-10B *	CC-20B *	CC-30B *	CC-40B *	CC-50B *
400	—	81.0	73.0	61.5	51.6	42.5	36.4	87.0	85.5	82.2	80.2	77.0	74.1
10	—	81.0	72.4	60.0	50.0	40.0	33.9	87.5	86.4	84.0	82.5	80.3	78.4
20	0.10	81.0	72.0	58.6	48.2	38.2	31.9	87.7	87.2	85.0	84.0	82.2	80.7
30	0.20	81.1	71.6	57.7	47.0	36.8	30.5	88.0	87.5	85.3	84.3	82.5	81.1
40	0.35	81.2	71.5	57.2	46.4	36.0	29.7	88.1	87.5	85.0	83.5	81.3	79.8
50	0.58	81.4	71.6	57.2	46.4	36.1	29.6	88.1	87.2	83.9	81.9	78.7	76.6
60	0.98	81.7	72.4	58.5	47.5	37.5	31.0	87.9	86.4	82.5	79.5	75.9	72.9
70	1.5	82.3	73.7	60.6	49.9	40.0	33.6	87.5	85.3	80.3	76.2	72.0	67.9
80	2.3	82.8	74.9	62.0	52.0	42.5	35.9	87.0	84.0	77.8	72.5	67.5	62.7
90	3.5	83.2	75.8	63.5	53.9	44.8	37.9	86.2	82.4	74.2	68.3	62.3	56.6
500	5.2	83.3	76.6	64.6	55.2	46.1	39.4	85.2	80.5	71.2	63.8	56.7	50.1
10	7.7	83.0	76.1	64.0	54.5	46.0	38.5	84.4	78.6	67.7	58.7	51.0	44.5
20	10.7	82.4	74.9	61.5	51.6	42.5	35.0	83.5	77.0	64.0	54.4	46.0	38.6
30	15.1	81.6	73.5	59.4	48.5	38.5	31.5	82.6	75.2	61.5	50.7	41.6	34.1
40	20.2	81.2	72.5	57.8	46.5	36.6	29.4	82.1	73.9	59.5	48.3	39.0	31.3
50	25.7	81.1	72.4	57.1	45.6	35.6	28.7	81.5	73.0	58.0	46.6	36.9	29.5
60	31.0	81.4	72.8	58.0	46.9	36.4	29.7	81.4	72.7	57.5	45.9	35.9	28.6
70	35.6	82.5	74.5	60.6	49.2	39.2	32.7	81.4	73.0	57.9	46.3	36.1	28.7
80	43.2	83.9	77.3	65.0	54.6	45.0	38.6	81.9	73.9	59.3	47.9	37.8	30.5
90	53.8	85.7	80.7	71.0	61.8	53.5	47.6	82.7	75.1	61.6	50.3	40.8	33.4
600	65.6	87.6	84.0	77.0	70.5	64.0	58.7	83.4	76.3	63.5	53.0	43.5	36.0
10	77.0	89.0	86.5	82.0	77.8	73.0	69.8	83.6	76.7	64.5	54.6	44.7	37.8
20	82.8	90.0	88.9	86.2	83.5	80.8	78.3	83.5	76.5	64.3	54.4	44.3	37.5
30	86.0	90.6	89.9	88.1	87.2	85.1	84.2	83.2	76.6	63.1	53.2	42.5	35.6
40	87.6	91.1	90.5	89.8	89.2	88.0	87.5	82.8	74.5	61.6	51.5	40.2	33.7
50	88.7	91.2	90.8	90.5	90.3	89.5	89.2	82.5	74.0	60.6	50.3	39.0	32.4
60	89.5	91.3	91.1	90.8	90.6	90.4	90.1	82.5	73.8	60.1	49.5	38.4	31.8
70	90.0	91.4	91.3	91.0	90.9	90.8	90.7	82.3	73.3	59.6	49.0	37.7	31.0
80	90.5	91.6	91.5	91.2	91.1	91.0	91.1	82.0	72.8	58.6	48.2	36.5	30.0
90	90.8	91.7	91.7	91.4	91.4	91.3	91.1	81.9	72.5	58.1	47.2	35.4	29.0
700	91.0	91.9	91.9	91.5	91.5	91.4	91.2	82.2	73.0	58.5	47.5	35.6	29.0
Luminous transmit.	34.6	83.7	77.0	65.3	55.9	47.3	41.3	82.8	75.5	62.3	52.0	42.8	35.7
Dominant wave lgth.	589.4	605.0	597.8	604.2	605.8	605.5	608.5	459.0	462.0	460.0	461.0	463.2	462.5
Excitation purity.	95.2	2.0	4.7	8.5	12.3	17.3	21.4	2.8	6.3	13.2	20.2	27.7	34.2

* Some transmission below 400 mμ. Consult the manufacturer.

TRANSMISSION OF WRATTEN FILTERS (Continued)

Wave length	Percent transmittance											
	CC-05G *	CC-10G *	CC-20G *	CC-30G *	CC-40G *	CC-50G *	CC-05Y *	CC-10Y *	CC-20Y *	CC-30Y *	CC-40Y *	CC-50Y *
400	80.0	73.1	58.8	48.0	39.7	32.0	81.0	74.5	61.3	50.5	43.0	34.5
10	80.7	72.9	57.8	46.5	38.1	30.3	80.6	73.2	59.0	47.4	39.5	30.5
20	81.0	72.8	57.3	45.8	37.3	29.5	80.4	72.6	57.8	46.0	37.5	29.0
30	81.4	72.7	57.0	45.5	36.5	29.0	80.4	72.5	57.5	45.6	36.5	28.7
40	81.6	73.0	57.3	45.8	36.6	29.1	80.6	72.8	57.8	46.5	36.8	29.5
50	82.1	73.9	58.4	46.9	38.1	30.6	81.2	74.0	59.5	48.5	38.5	31.5
60	83.0	75.5	61.4	50.3	41.5	34.3	82.5	76.0	63.0	52.5	42.5	36.2
70	84.4	78.0	65.4	55.8	47.0	40.5	83.9	78.5	67.5	48.2	48.8	43.5
80	85.6	80.4	70.0	61.8	53.5	47.8	85.3	81.2	72.3	64.9	56.2	54.0
90	86.8	83.0	75.2	68.9	61.3	57.0	87.0	84.4	78.0	72.4	66.0	64.0
500	87.9	85.9	80.3	76.4	70.7	68.0	88.4	87.2	84.0	81.0	77.0	75.5
10	88.7	87.5	83.8	80.9	77.8	75.3	89.5	89.0	88.0	86.6	85.5	84.2
20	89.0	88.1	84.9	82.3	79.5	77.5	90.0	90.0	89.6	89.1	89.0	88.5
30	89.0	88.0	84.6	81.7	79.4	77.0	90.4	90.4	90.0	89.7	89.9	89.6
40	89.0	87.6	83.7	80.5	77.8	74.8	90.7	90.7	90.6	90.4	90.2	90.0
50	88.6	87.1	82.4	78.6	75.8	72.2	90.9	90.9	90.8	90.6	90.4	90.3
60	88.1	86.3	80.9	76.2	72.9	68.8	91.0	91.0	90.9	90.8	90.7	90.6
70	87.5	85.3	79.0	73.5	69.3	64.8	91.3	91.3	91.0	90.9	90.8	90.7
80	87.0	84.1	77.0	70.4	65.3	60.3	91.4	91.4	91.1	91.0	90.8	90.7
90	86.4	82.8	74.5	67.2	61.9	55.9	91.4	91.4	91.2	91.1	90.9	90.8
600	85.7	81.5	72.0	64.1	57.7	51.7	91.4	91.4	91.3	91.2	90.9	90.8
10	85.0	80.0	69.5	60.7	53.7	47.3	91.4	91.4	91.3	91.2	90.9	90.9
20	84.0	78.5	66.5	57.2	49.8	42.5	91.4	91.4	91.3	91.2	91.0	90.9
30	83.0	76.9	63.8	53.7	45.5	38.0	91.5	91.5	91.4	91.3	91.0	91.0
40	82.2	75.6	61.5	50.8	42.0	34.6	91.5	91.5	91.4	91.3	91.0	91.0
50	81.9	74.9	60.1	49.1	40.2	32.5	91.5	91.5	91.4	91.3	91.1	91.1
60	81.5	74.4	59.4	48.1	39.5	31.5	91.5	91.5	91.4	91.3	91.1	91.1
70	81.4	74.0	58.8	47.5	38.5	31.0	91.5	91.5	91.4	91.4	91.2	91.2
80	81.1	73.5	58.1	46.6	37.6	29.9	91.5	91.5	91.4	91.4	91.2	91.2
90	81.1	73.2	57.6	46.0	36.6	28.9	91.5	91.5	91.4	91.4	91.3	91.3
700	81.5	73.5	58.0	46.4	36.5	28.7	91.5	91.5	91.4	91.4	91.3	91.3
Luminous transmit.	87.2	84.5	77.8	72.2	67.7	63.3	90.4	90.1	89.1	88.2	87.4	86.9
Dominant wave lgth.	553.0	555.5	555.0	554.0	554.3	553.4	572.0	571.3	571.4	571.3	571.3	571.2
Excitation purity.	2.3	5.2	10.9	15.8	21.1	25.9	5.3	9.6	18.8	28.3	35.7	42.0

Wave length	Percent transmittance											
	CC-05M *	CC-10M *	CC-20M *	CC-30M *	CC-40M *	CC-50M *	CC-05C *	CC-10C *	CC-20C *	CC-30C *	CC-40C *	CC-50C *
400	87.6	86.6	85.6	84.2	82.3	80.9	87.3	86.0	83.9	82.3	80.4	78.8
10	88.2	87.7	86.6	85.7	84.6	83.6	88.2	87.5	85.2	84.5	83.4	82.7
20	88.6	88.0	87.0	85.9	85.2	84.4	88.7	88.1	86.5	86.0	85.3	84.8
30	88.7	88.0	86.9	85.6	84.4	83.6	89.0	88.6	87.5	87.0	86.3	85.9
40	88.7	87.9	86.0	84.7	82.5	81.4	89.3	89.0	87.7	87.3	86.6	86.1
50	88.6	87.5	84.9	82.8	80.0	78.1	89.5	89.1	87.8	87.5	86.6	86.0
60	88.4	86.5	83.1	80.0	76.1	73.7	89.6	89.1	87.7	87.3	86.4	85.7
70	87.8	85.2	80.8	76.4	71.3	68.0	89.7	89.0	87.5	87.0	85.8	85.2
80	87.0	83.6	77.9	72.1	65.8	61.7	89.7	89.0	87.2	86.5	85.3	84.3
90	86.0	81.8	74.4	67.0	60.0	55.0	89.7	89.0	87.0	86.0	84.4	83.4
500	85.0	79.7	70.5	61.7	53.7	48.1	89.6	89.0	86.5	85.2	83.5	82.3
10	83.8	77.5	66.7	56.5	47.7	41.6	89.6	88.7	86.0	84.4	82.4	80.8
20	82.7	75.3	63.4	52.0	42.8	36.3	89.5	88.5	85.2	83.5	81.1	79.2
30	81.8	73.7	60.5	48.6	39.0	31.9	89.4	88.0	84.3	82.4	79.6	77.3
40	81.3	72.5	58.6	46.6	36.7	29.8	89.2	87.5	83.4	81.0	77.7	75.0
50	81.2	72.2	58.0	46.0	36.0	29.1	88.9	87.0	82.3	79.0	75.3	72.2
60	81.5	72.8	58.3	46.5	36.7	29.7	88.5	86.1	80.5	76.7	72.7	69.0
70	82.5	74.6	60.5	49.8	40.2	32.3	88.0	85.0	78.5	74.0	69.3	65.0
80	84.0	77.3	64.9	55.6	46.2	39.0	87.5	83.8	76.1	70.9	65.4	60.5
90	85.8	80.8	70.6	63.3	54.9	48.7	87.0	82.5	73.9	67.5	61.6	55.8
600	88.0	84.5	77.1	71.6	64.9	59.9	86.4	81.0	71.2	64.1	57.6	51.3
10	89.3	87.0	82.2	79.2	74.9	70.7	85.5	79.5	68.5	60.4	53.4	36.2
20	90.2	88.9	86.1	84.1	81.4	79.2	84.5	77.9	65.5	56.7	49.2	42.3
30	90.6	90.0	88.7	87.4	86.0	84.5	83.8	76.3	62.7	53.1	45.0	38.0
40	90.8	90.5	90.0	89.3	88.7	87.6	83.3	75.1	60.8	50.4	42.0	34.9
50	91.0	90.7	90.5	90.2	90.0	89.7	82.8	74.4	59.5	48.8	40.2	32.9
60	91.1	91.0	90.8	90.8	90.4	90.4	82.5	74.0	58.5	48.0	39.4	32.0
70	91.2	91.2	91.0	91.0	90.7	90.7	82.4	73.6	57.9	47.2	38.6	31.0
80	91.3	91.3	91.2	91.1	91.0	91.0	82.0	73.0	57.5	46.0	37.5	29.9
90	91.4	91.4	91.4	91.3	91.3	91.3	82.0	72.8	57.4	45.5	36.7	29.1
700	91.5	91.5	91.5	91.5	91.5	91.5	82.5	74.0	58.5	46.4	37.3	29.8
Luminous transmit.	84.2	77.9	67.1	58.1	50.0	44.0	88.0	85.1	78.9	74.8	70.5	66.7
Dominant wave lgth.	541.0	547.5	551.2	550.0	550.3	551.2	489.2	487.5	486.5	486.2	486.1	485.5
Excitation purity.	3.5	7.4	14.4	21.5	28.3	34.0	1.6	4.1	8.9	12.8	17.5	20.2

* Some transmission below 400 mμ. Consult the manufacturer.

TRANSMISSION OF WRATTEN FILTERS (Continued)

Wave length	No. 87	No. 87C	No. 88A	No. 89B
	Percent transmittance			
700	—	—	—	11.2
10	—	—	—	32.4
20	—	—	—	57.6
30	—	—	7.4	69.1
40	0.10	—	32.8	77.6
50	2.19	—	56.3	83.1
60	7.95	—	69.2	85.0
70	17.4	—	74.2	86.1
80	31.6	—	77.6	87.0
90	43.7	—	79.7	87.7
800	53.8	0.32	81.4	88.1
10	61.7	3.20	82.6	88.4
20	69.2	8.90	83.7	88.6
30	74.1	17.8	84.7	88.8
40	77.7	28.2	85.5	89.0
50	81.4	41.0	86.1	89.2
60	84.0	53.8	86.6	89.4
70	85.4	61.6	87.2	89.6
80	86.8	69.2	87.5	89.8
90	87.8	74.1	87.8	89.9
900	88.4	78.5	88.0	90.0
10	88.8	81.5	88.2	90.1
20	89.1	83.6	88.4	90.2
30	89.1	85.1	88.6	90.3
40	89.1	86.0	88.8	90.4
50	89.1	87.0	89.0	90.5

* Some transmission below 400 mμ. Consult the manufacturer.

TRANSMISSIBILITY FOR RADIATIONS

Ratio of the transmitted light to the incident light for a definite thickness of the substance, usually 1 cm.

Glass

Glass in general is opaque to the ultra-violet and infrared. Uviol glass is transparent to the longer radiations of the ultra-violet.

Coefficient of transparency of glass for visible and ultra-violet radiations.

Normal incidence, thickness 1 cm.

Wave length microns	0.309	0.330	0.347	0.357	0.361	0.375	0.384	0.388	0.396
Crown, ordinary	—	—	—	—	—	.947	—	—	—
Crown, borosilicate	0.08	0.65	0.88	—	0.95	—	0.972	0.975	0.986
Flint, ordinary	—	—	—	0.72	—	—	—	0.904	—
Flint, heavy	—	—	0.01	—	0.16	—	0.58	—	—

Normal incidence, thickness 1 cm.

Wave length microns	0.400	0.415	0.419	0.425	0.434	0.455	0.500	0.580	0.677
Crown, ordinary	0.964	—	0.952	—	0.960	0.981	—	0.986	0.990
Crown, borosilicate	—	0.985	—	0.993	—	—	0.993	—	—
Flint, ordinary	—	0.959	—	—	—	—	1.00	—	—
Flint, heavy	—	—	—	0.905	—	—	—	—	—

Quartz

Quartz is very transparent to the ultra-violet and to the visible spectrum, but opaque for the infrared beyond 7.0μ.

(Pflüger.)

Wave length, microns	0.19	0.20	0.21	0.22
Transmission for 1 mm	.67	.84	.92	.94

Fluorite

Fluorite is very transparent to the ultra-violet, nearly to 0.10μ. Coefficient of transparency at $\lambda = 186$ is found by Pflüger to be 0.80.

For the infrared the values are given in a table below.

Rock Salt and Sylvine and Fluorite
Transparency for the Infrared.
Thickness 1 cm.

Wave length microns	Rock salt	Sylvine KCl	Fluorite
8.	—	—	.844
9.	0.995	1.000	.543
10.	.995	.988	.164
12.	.993	.995	.010
14.	.931	.975	.000
16.	.661	.936	—
18.	.275	.862	—
19.	.096	.758	—
20.7	.006	.585	—
23.7	.000	.155	—

TRANSPARENCY TO OPTICAL DENSITY CONVERSION TABLE

Transparency of a layer of material is defined as the ratio of the intensity of the tra-smitted light to that of the incident light. Opacity is the reciprocal of the transparency. Optical density is the common logarithm of the opacity.

Thus,

$$\text{Transparency} = \frac{I_t}{I_i}$$

$$\text{Opacity} = \frac{1}{\text{Transparency}} = \frac{I_i}{I_t}$$

$$\text{Optical density} = \log_{10}\left(\frac{I_i}{I_t}\right)$$

where I_i = Intensity of incident light
I_t = Intensity of transmitted light.

Trans.	Density	Trans.	Density	Trans.	Density	Trans.	Density	Trans.	Density	Trans.	Density	Trans.	Density	Trans.	Density
0.000	—	.067	1.174	.133	.8761	.199	.7011	.266	.5751	.332	.4789	.399	.3990	.466	.3316
.001	3.000	.068	1.168	.134	.8729	.200	.6990	.267	.5735	.333	.4776	.400	.3979	.467	.3307
.002	2.699	.069	1.161	.135	.8697	.201	.6968	.268	.5719	.334	.4763	.401	.3969	.468	.3298
.003	2.523	.070	1.155	.136	.8665	.202	.6946	.269	.5702	.335	.4750	.402	.3958	.460	.3288
.004	2.398	.071	1.149	.137	.8633	.203	.6925	.270	.5686	.336	.4737	.403	.3947	.470	.3279
.005	2.301	.072	1.143	.138	.8601	.204	.6904	.271	.5670	.337	.4724	.404	.3936	.471	.3270
.006	2.222	.073	1.137	.139	.8570	.205	.6882	.272	.5654	.339	.4698	.405	.3925	.472	.3260
.007	2.155	.074	1.131	.140	.8539	.206	.0861	.273	.5638	.340	.4685	.406	.3915	.473	.3251
.008	2.097	.075	1.125	.141	.8508	.207	.6840	.274	.5622	.341	.4673	.407	.3904	.474	.3242
.009	2.046	.076	1.119	.142	.8477	.208	.6819	.275	.5607	.342	.4660	.408	.3893	.475	.3233
.010	2.000	.077	1.114	.143	.8447	.209	.6799	.276	.5591	.343	.4647	.409	.3883	.476	.3224
.011	1.959	.078	1.108	.144	.8416	.210	.6778	.277	.5575	.344	.4634	.410	.3872	.477	.3215
.012	1.921	.079	1.102	.145	.8386	.211	.6757	.278	.5560	.345	.4622	.411	.3862	.478	.3206
.013	1.886	.080	1.097	.146	.8356	.212	.6737	.279	.5544	.346	.4609	.412	.3851	.479	.3197
.014	1.854	.081	1.092	.147	.8327	.213	.6716	.280	.5528	.347	.4597	.414	.3830	.480	.3188
.015	1.824	.082	1.086	.148	.8297	.214	.6696	.281	.5513	.348	.4584	.415	.3819	.481	.3179
.016	1.796	.083	1.081	.149	.8268	.215	.6676	.282	.5498	.349	.4572	.416	.3809	.482	.3170
.017	1.770	.084	1.076	.150	.8239	.216	.6655	.283	.5482	.350	.4559	.417	.3799	.483	.3161
.018	1.745	.085	1.071	.151	.8210	.217	.6635	.284	.5467	.351	.4547	.418	.3788	.484	.3152
.019	1.721	.086	1.066	.152	.8182	.218	.6615	.285	.5452	.352	.4535	.419	.3778	.485	.3143
.020	1.699	.087	1.060	.153	.8153	.219	.6596	.286	.5436	.353	.4522	.420	.3768	.486	.3134
.021	1.678	.088	1.055	.154	.8125	.220	.6576	.287	.5421	.354	.4510	.421	.3757	.487	.3125
.022	1.658	.089	1.051	.155	.8097	.221	.6556	.288	.5406	.355	.4498	.422	.3747	.489	.3107
.023	1.638	.090	1.046	.156	.8069	.222	.6536	.289	.5391	.356	.4486	.423	.3737	.490	.3098
.024	1.620	.091	1.041	.157	.8041	.223	.6517	.290	.5376	.357	.4473	.424	.3726	.491	.3089
.025	1.602	.092	1.036	.158	.8013	.224	.6498	.291	.5361	.358	.4461	.425	.3716	.492	.3080
.026	1.585	.093	1.032	.159	.7986	.225	.6478	.292	.5346	.359	.4449	.426	.3706	.493	.3072
.027	1.569	.094	1.027	.160	.7959	.226	.6459	.293	.5331	.360	.4437	.427	.3696	.494	.3063
.028	1.553	.095	1.022	.161	.7932	.227	.6440	.294	.5317	.361	.4425	.428	.3685	.495	.3054
.029	1.538	.096	1.018	.162	.7905	.228	.6421	.295	.5302	.362	.4413	.429	.3675	.496	.3045
.030	1.523	.097	1.013	.163	.7878	.229	.6402	.296	.5287	.363	.4401	.430	.3665	.497	.3036
.031	1.509	.098	1.009	.164	.7852	.230	.6383	.297	.5272	.364	.4389	.431	.3655	.498	.3028
.032	1.495	.099	1.004	.165	.7825	.231	.6364	.298	.5258	.365	.4377	.432	.3645	.499	.3019
.033	1.482	.100	1.000	.166	.7799	.232	.6345	.299	.5243	.366	.4365	.433	.3635	.500	.3010
.034	1.469	.101	.9957	.167	.7773	.233	.6326	.300	.5229	.367	.4353	.434	.3625	.501	.3002
.035	1.456	.102	.9914	.168	.7747	.234	.6308	.301	.5215	.368	.4342	.435	.3615	.502	.2993
.036	1.444	.103	.9872	.169	.7721	.235	.6289	.302	.5200	.369	.4330	.436	.3605	.503	.2984
.037	1.432	.104	.9830	.170	.7696	.236	.6271	.303	.5186	.370	.4318	.437	.3595	.504	.2975
.038	1.420	.105	.9788	.171	.7670	.237	.6253	.304	.5171	.371	.4306	.438	.3585	.505	.2967
.039	1.409	.106	.9747	.172	.7645	.238	.6234	.305	.5157	.372	.4295	.439	.3575	.506	.2959
.040	1.398	.107	.9706	.173	.7620	.239	.6216	.306	.5143	.373	.4283	.440	.3565	.507	.2950
.041	1.387	.108	.9666	.174	.7594	.240	.6198	.307	.5128	.374	.4271	.441	.3556	.508	.2951
.042	1.377	.109	.9626	.175	.7570	.241	.6180	.308	.5114	.375	.4260	.442	.3546	.509	.2933
.043	1.367	.110	.9586	.176	.7545	.242	.6162	.309	.5100	.376	.4248	.443	.3536	.510	.2924
.044	1.357	.111	.9547	.177	.7520	.243	.6144	.310	.5086	.377	.4237	.444	.3526	.511	.2916
.045	1.347	.112	.9508	.178	.7496	.244	.6126	.311	.5072	.378	.4225	.445	.3516	.512	.2907
.046	1.337	.113	.9469	.179	.7471	.245	.6108	.312	.5058	.379	.4214	.446	.3507	.513	.2899
.047	1.328	.114	.9431	.180	.7447	.246	.6091	.313	.5045	.380	.4202	.447	.3497	.514	.2890
.048	1.319	.115	.9393	.181	.7423	.247	.6073	.314	.5031	.381	.4191	.448	.3487	.515	.2882
.049	1.310	.116	.9356	.182	.7399	.248	.6056	.315	.5017	.382	.4179	.449	.3478	.516	.2873
.050	1.301	.117	.9318	.183	.7375	.249	.6038	.316	.5003	.383	.4168	.450	.3468	.517	.2865
.051	1.292	.118	.9281	.184	.7352	.250	.6021	.317	.4989	.384	.4157	.451	.3458	.518	.2857
.052	1.284	.119	.9244	.185	.7328	.251	.6003	.318	.4976	.385	.4145	.452	.3449	.519	.2848
.053	1.276	.120	.9208	.186	.7305	.252	.5986	.319	.4962	.386	.4134	.453	.3439	.520	.2840
.054	1.268	.121	.9172	.187	.7282	.253	.5969	.320	.4949	.387	.4123	.454	.3429	.521	.2831
.055	1.260	.122	.9137	.188	.7258	.254	.5952	.321	.4935	.388	.4112	.455	.3420	.522	.2823
.056	1.252	.123	.9101	.189	.7235	.255	.5935	.322	.4921	.389	.4101	.456	.3410	.523	.2815
.057	1.244	.124	.9066	.190	.7212	.256	.5918	.323	.4908	.390	.4089	.457	.3401	.524	.2807
.058	1.237	.125	.9031	.191	.7190	.257	.5901	.324	.4895	.391	.4078	.458	.3391	.525	.2798
.059	1.229	.126	.8996	.192	.7167	.258	.5884	.325	.4881	.392	.4067	.459	.3382	.526	.2790
.060	1.222	.127	.8962	.193	.7144	.259	.5867	.326	.4868	.393	.4056	.460	.3372	.527	.2782
.061	1.215	.128	.8928	.194	.7122	.260	.5850	.327	.4855	.394	.4045	.461	.3363	.528	.2774
.062	1.208	.129	.8894	.195	.7100	.261	.5834	.328	.4841	.395	.4034	.462	.3354	.529	.2766
.063	1.201	.130	.8861	.196	.7077	.263	.5800	.329	.4828	.396	.4023	.463	.3344	.530	.2757
.064	1.194	.131	.8827	.197	.7055	.264	.5784	.330	.4815	.397	.4012	.464	.3335	.531	.2749
.065	1.187	.132	.8794	.198	.7033	.265	.5768	.331	.4802	.398	.4001	.465	.3325	.532	.2741
.066	1.180														

TRANSPARENCY TO OPTICAL DENSITY CONVERSION TABLE (Continued)

Trans.	Density	Trans.	Density	Trans.	Density	Trans.	Density	Trans.	Density	Trans.	Density	Trans.	Density	Trans.	Density
.533	.2733	.592	.2277	.650	.1871	.708	.1500	.766	.1158	.824	.0841	.882	.0545	.940	.0269
.534	.2725	.593	.2269	.651	.1864	.709	.1493	.767	.1152	.825	.0835	.883	.0540	.941	.0264
.535	.2717	.594	.2262	.652	.1857	.710	.1487	.768	.1146	.826	.0830	.884	.0535	.942	.0260
.536	.2708	.595	.2255	.653	.1851	.711	.1481	.769	.1141	.827	.0825	.885	.0530	.943	.0255
.537	.2700	.596	.2248	.654	.1844	.712	.1475	.770	.1135	.828	.0820	.886	.0526	.944	.0250
.538	.2696	.597	.2240	.655	.1838	.713	.1469	.771	.1129	.829	.0815	.887	.0521	.945	.0246
.539	.2684	.598	.2233	.656	.1831	.714	.1463	.772	.1124	.830	.0809	.888	.0516	.946	.0241
.540	.2676	.599	.2226	.657	.1824	.715	.1457	.773	.1118	.831	.0804	.889	.0511	.947	.0237
.541	.2668	.600	.2219	.658	.1818	.716	.1451	.774	.1113	.832	.0799	.890	.0506	.948	.2032
.542	.2660	.601	.2211	.659	.1811	.717	.1445	.775	.1107	.833	.0794	.891	.0501	.949	.0227
.543	.2652	.602	.2204	.660	.1805	.718	.1439	.776	.1102	.834	.0788	.892	.0496	.950	.0223
.544	.2644	.603	.2197	.661	.1798	.719	.1433	.777	.1096	.835	.0783	.893	.0491	.951	.0218
.545	.2636	.604	.2190	.662	.1791	.720	.1427	.778	.1090	.836	.0778	.894	.0487	.952	.0214
.546	.2628	.605	.2182	.663	.1785	.721	.1421	.779	.1085	.837	.0773	.895	.0482	.953	.0209
.547	.2620	.606	.2175	.664	.1778	.722	.1415	.780	.1079	.838	.0767	.896	.0477	.954	.0204
.548	.2612	.607	.2168	.665	.1772	.723	.1409	.781	.1073	.839	.0762	.897	.0472	.955	.0200
.549	.2604	.608	.3161	.666	.1765	.724	.1403	.782	.1068	.840	.0757	.898	.0467	.956	.0195
.550	.2596	.609	.2154	.667	.1759	.725	.1397	.783	.1062	.841	.0752	.899	.0462	.957	.0191
.551	.2589	.610	.2147	.668	.1752	.726	.1391	.784	.1057	.842	.0747	.900	.0458	.958	.0186
.552	.2581	.611	.2140	.669	.1746	.727	.1385	.785	.1051	.843	.0742	.901	.0453	.959	.0182
.553	.2573	.612	.2132	.670	.1739	.728	.1379	.786	.1046	.844	.0736	.902	.0448	.960	.0177
.554	.2565	.613	.2125	.671	.1733	.729	.1373	.787	.1040	.845	.0731	.903	.0443	.961	.0173
.555	.2557	.614	.2118	.672	.1726	.730	.1367	.788	.1035	.846	.0726	.904	.0438	.962	.0168
.556	.2549	.615	.2111	.673	.1720	.731	.1361	.789	.1029	.847	.0721	.905	.0434	.963	.0164
.557	.2541	.616	.2104	.674	.1713	.732	.1355	.790	.1024	.848	.0716	.906	.0429	.964	.0159
.558	.2534	.617	.2097	.675	.1707	.733	.1349	.791	.1018	.849	.0711	.907	.0424	.965	.0155
.559	.2526	.618	.2090	.676	.1701	.734	.1343	.792	.1013	.850	.0706	.908	.0419	.966	.0150
.500	.2518	.619	.2083	.677	.1694	.735	.1337	.793	.1007	.851	.0701	.909	.0414	.967	.0146
.561	.2510	.620	.2076	.678	.1688	.736	.1331	.794	.1002	.852	.0696	.910	.0410	.968	.0141
.562	.2503	.621	.2069	.679	.1681	.737	.1325	.795	.0996	.853	.0690	.911	.0405	.969	.0137
.564	.2487	.622	.2062	.680	.1675	.738	.1319	.796	.0991	.854	.0685	.912	.0400	.970	.0132
.565	.2479	.623	.2055	.681	.1668	.739	.1314	.797	.0985	.855	.0680	.913	.0395	.971	.0128
.566	.2472	.624	.2048	.682	.1662	.740	.1308	.798	.0980	.856	.0675	.914	.0391	.972	.0123
.567	.2464	.625	.2041	.683	.1655	.741	.1302	.799	.0975	.857	.0670	.915	.0386	.973	.0119
.568	.2457	.626	.2034	.684	.1649	.742	.1296	.800	.0969	.858	.0665	.916	.0381	.974	.0114
.569	.2449	.627	.2027	.685	.1643	.743	.1290	.801	.0964	.859	.0660	.917	.0376	.975	.0110
.570	.2441	.628	.2020	.686	.1637	.744	.1284	.802	.0958	.860	.0655	.918	.0371	.967	.0106
.571	.2434	.629	.2013	.687	.1630	.745	.1278	.803	.0953	.861	.0650	.919	.0367	.977	.0101
.572	.2426	.630	.2007	.688	.1624	.746	.1273	.804	.0948	.862	.0645	.920	.0362	.978	.0097
.573	.2418	.631	.2000	.689	.1618	.747	.1267	.805	.0942	.863	.0640	.921	.0357	.979	.0092
.574	.2411	.632	.1993	.690	.1612	.748	.1261	.806	.0937	.804	.0635	.922	.0353	.980	.0088
.575	.2403	.633	.1986	.691	.1605	.749	.1255	.807	.0931	.865	.0630	.923	.0348	.981	.0083
.576	.2396	.634	.1979	.692	.1599	.750	.1249	.808	.0926	.866	.0625	.924	.0343	.982	.0079
.577	.2388	.635	.1972	.693	.1593	.751	.1244	.809	.0921	.867	.0620	.925	.0339	.983	.0074
.578	.2381	.636	.1965	.694	.1586	.752	.1238	.810	.0915	.808	.0615	.926	.0334	.984	.0070
.579	.2373	.637	.1959	.695	.1580	.753	.1232	.811	.0910	.869	.0610	.927	.0329	.985	.0066
.580	.2366	.638	.1952	.696	.1574	.754	.1226	.812	.0904	.870	.0605	.928	.0325	.986	.0061
.581	.2358	.639	.1945	.697	.1568	.755	.1221	.813	.0899	.871	.0600	.929	.0320	.987	.0057
.582	.2351	.640	.1938	.698	.1562	.756	.1215	.814	.0894	.872	.0595	.930	.0315	.988	.0052
.583	.2343	.641	.1932	.699	.1555	.757	.1209	.815	.0888	.873	.0590	.931	.0310	.989	.0048
.584	.2336	.642	.1925	.700	.1549	.758	.1203	.816	.0883	.874	.0585	.932	.0306	.990	.0044
.585	.2328	.643	.1918	.701	.1543	.759	.1198	.817	.0878	.975	.0580	.933	.0301	.991	.0039
.586	.2321	.644	.1911	.702	.1537	.760	.1192	.818	.0872	.876	.0575	.934	.0296	.992	.0035
.587	.2314	.645	.1904	.703	.1531	.761	.1186	.819	.0867	.877	.0570	.935	.0292	.993	.0030
.588	.2306	.646	.1898	.704	.1524	.762	.1180	.820	.0862	.878	.0565	.936	.0287	.994	.0026
.589	.2299	.647	.1891	.705	.1518	.763	.1175	.821	.0856	.879	.0560	.937	.0282	.995	.0022
.590	.2291	.648	.1884	.706	.1512	.764	.1169	.822	.0851	.880	.0555	.938	.0278	.996	.0017
.591	.2284	.649	.1877	.707	.1506	.765	.1163	.823	.0846	.881	.0550	.939	.0273	.997	.0013
														.998	.0009
														.999	.0004
														1.000	.0000

DENSITY OF VARIOUS SOLIDS

The approximate density of various solids at ordinary atmospheric temperature.
In the case of substances with voids such as paper or leather the bulk density is indicated rather than the density of the solid portion.

(Selected principally from the Smithsonian Tables.)

Substance	Grams per cu. cm	Pounds per cu. ft.
Agate	2.5–2.7	156–168
Alabaster, carbonate	2.69–2.78	168–173
sulfate	2.26–2.32	141–145
Albite	2.62–2.65	163–165
Amber	1.06–1.11	66–69
Amphiboles	2.9–3.2	180–200
Anorthite	2.74–2.76	171–172
Asbestos	2.0–2.8	125–175
Asbestos slate	1.8	112
Asphalt	1.1–1.5	69–94
Basalt	2.4–3.1	150–190
Beeswax	0.96–0.97	60–61
Beryl	2.69–2.7	168–169
Biotite	2.7–3.1	170–190
Bone	1.7–2.0	106–125
Brick	1.4–2.2	87–137
Butter	0.86–0.87	53–54
Calamine	4.1–4.5	255–280
Calcspar	2.6–2.8	162–175
Camphor	0.99	62
Caoutchouc	0.92–0.99	57–62
Cardboard	0.69	43
Celluloid	1.4	87
Cement, set	2.7–3.0	170.190
Chalk	1.9–2.8	118–175
Charcoal, oak	0.57	35
pine	0.28–0.44	18–28
Cinnabar	8.12	507
Clay	1.8–2.6	112–162
Coal, anthracite	1.4–1.8	87–112
bituminous	1.2–1.5	75–94
Cocoa butter	0.89–0.91	56–57
Coke	1.0–1.7	62–105
Copal	1.04–1.14	65–71
Cork	0.22–0.26	14–16
Cork linoleum	0.54	34
Corundum	3.9–4.0	245–250
Diamond	3.01–3.52	188–220
Dolomite	2.84	177
Ebonite	1.15	72
Emery	4.0	250
Epidote	3.25–3.50	203–218
Feldspar	2.55–2.75	159–172
Flint	2.63	164
Fluorite	3.18	198
Galena	7.3–7.6	460–470
Gamboge	1.2	75
Garnet	3.15–4.3	197–268
Gas carbon	1.88	117
Gelatin	1.27	79
Glass, common	2.4–2.8	150–175
flint	2.9–5.9	180–370
Glue	1.27	79
Granite	2.64–2.76	165–172
Graphite*	2.30–2.72	144–170
Gum arabic	1.3–1.4	81–87
Gypsum	2.31–2.33	144–145
Hematite	4.9–5.3	306–330
Hornblende	3.0	187
Ice	0.917	57.2
Ivory	1.83–1.92	114–120
Leather, dry	0.86	54
Lime, slaked	1.3–1.4	81–87
Limestone	2.68–2.76	167–171
Linoleum	1.18	74
Magnetite	4.9–5.2	306–324
Malachite	3.7–4.1	231–256
Marble	2.6–2.84	160–177
Meerschaum	0.99 1.28	62–80
Mica	2.6–3.2	165–200
Muscovite	2.76–3.00	172–187
Ochre	3.5	218
Opal	2.2	137
Paper	0.7–1.15	44–72
Paraffin	0.87–0.91	54–57
Peat blocks	0.84	52
Pitch	1.07	67
Porcelain	2.3–2.5	143–156
Porphyry	2.6–2.9	162–181
Pressed wood pulp board	0.19	12
Pyrite	4.95 5.1	309 318
Quartz	2.65	165
Resin	1.07	67
Rock salt	2.18	136
Rubber, hard	1.19	74
Rubber, soft commercial	1.1	69
pure gum	0.91–0.93	57–58
Sandstone	2.14–2.36	134–147
Serpentine	2.50–2.65	156–165
Silica, fused transparent	2.21	138
translucent	2.07	129
Slag	2.0–3.9	125–240
Slate	2.6–3.3	162–205
Soapstone	2.6–2.8	162–175
Spermacéti	0.95	59
Starch	1.53	95
Sugar	1.59	99
Talc	2.7–2.8	168–174
Tallow, beef	0.94	59
mutton	0.94	59
Tar	1.02	66
Topaz	3.5–3.6	219–223
Tourmaline	3.0–3.2	190–200
Wax, sealing	1.8	112
Wood (seasoned)		
alder	0.42–0.68	26–42
apple	0.66–0.84	41–52
ash	0.65–0.85	40–53
balsa	0.11–0.14	7–9
bamboo	0.31–0.40	19–25
basswood	0.32–0.59	20–37
beech	0.70–0.90	43–56
birch	0.51–0.77	32–48
blue gum	1.00	62
box	0.95–1.16	59–72
butternut	0.38	24
cedar	0.49–0.57	30–35
cherry	0.70–0.90	43–56
dogwood	0.76	47
ebony	1.11–1.33	69–83
elm	0.54–0.60	34–37
hickory	0.60–0.93	37–58
holly	0.76	47
juniper	0.56	35
larch	0.50–0.56	31–35
lignum vitae	1.17–1.33	73–83
locust	0.67–0.71	42–44
logwood	0.91	57
mahogany		
Honduras	0.66	41
Spanish	0.85	53
maple	0.62–0.75	39–47
oak	0.60–0.90	37–56
pear	0.61–0.73	38–45
pine, pitch	0.83–0.85	52–53
white	0.35–0.50	22–31
yellow	0.37–0.60	23–37
plum	0.66–0.78	41–49
poplar	0.35–0.5	22–31
satinwood	0.95	59
spruce	0.48–0.70	30–44
sycamore	0.40–0.60	24–37
teak, Indian	0.66–0.88	41–55
African	0.98	61
walnut	0.64–0.70	40–43
water gum	1.00	62
willow	0.40–0.60	24–37

* Some values reported as low as 1.6.

WEIGHT OF ONE GALLON OF WATER (U.S. GALLONS)

The weights are for dry air at the same temperature as the water up to 40°C and at a barometric pressure corrected to 760 mm and against brass weights of 8.4 density at 0°C. Above 40°C the temperature of the air is assumed to be 20°C, i.e., the water is allowed to cool to 20°C prior to the weighings being made. The volumetric computations are based upon the relations that one liter = 1 dm^3 and that 1 dm^3 = 61.023744 $in.^3$

Temperature (°C)	Weight in vacuo (g)	Weight in vacuo (lb)	Weight in air (g)	Weight in air (lb)	Temperature (°C)	Weight in vacuo (g)	Weight in vacuo (lb)	Weight in air (g)	Weight in air (lb)
0	3784.856	8.34417	3780.543	8.33467	25	3774.291	8.32088	3770.340	8.31217
1	3785.078	8.34466	3780.781	8.33518	26	3773.320	8.31870	3769.364	8.31001
2	3785.233	8.34500	3780.953	8.33556	27	3772.277	8.31644	3768.352	8.30778
3	3785.326	8.34520	3781.060	8.33580	28	3771.218	8.31410	3767.306	8.30548
4	3785.355	8.34527	3781.105	8.33590	29	3770.123	8.31169	3766.224	8.30309
5	3785.325	8.34520	3781.090	8.33587	30	3768.995	8.30920	3765.109	8.30063
6	3785.235	8.34500	3781.015	8.33570	31	3768.995	8.30664	3763.961	8.29810
7	3785.089	8.34468	3780.884	8.33541	32	3766.641	8.30401	3762.780	8.29550
8	3784.887	8.34424	3780.698	8.33500	33	3765.416	8.30131	3761.568	8.29283
9	3784.633	8.34368	3780.358	8.33447	34	3764.160	8.29854	3760.324	8.29008
10	3784.326	8.34300	3780.167	8.33383	35	3762.874	8.29571	3759.050	8.28728
11	3783.966	8.34221	3779.821	8.33307	40	3756.018	8.28059	3752.255	8.27230
12	3783.557	8.34130	3779.426	8.33220	45	3748.41	8.2638	3744.42	8.2550
13	3783.099	8.34030	3778.983	8.33122	50	3740.19	8.2457	3736.22	8.2369
14	3782.597	8.33919	3778.495	8.33014	55	3731.34	8.2261	3727.37	8.2174
15	3782.049	8.33798	3777.962	8.32897	60	3721.91	8.2054	3717.95	8.1966
16	3781.458	8.33668	3777.415	8.32770	65	3711.88	8.1832	3707.93	8.1745
17	3780.824	8.33528	3776.764	8.32633	70	3701.35	8.1600	3697.42	8.1514
18	3780.148	8.33379	3776.103	8.32487	75	3690.30	8.1357	3686.38	8.1270
19	3779.430	8.33221	3775.398	8.32332	80	3678.72	8.1101	3674.81	8.1015
20	3778.672	8.33054	3774.653	8.32167	85	3666.68	8.0836	3662.78	8.0750
21	3777.873	8.32877	3773.868	8.31994	90	3654.15	8.0560	3650.27	8.0474
22	3777.035	8.32693	3773.044	8.31813	95	3641.21	8.0274	3637.34	8.0189
23	3776.158	8.32499	3772.180	8.31622	100	3627.81	7.9979	3623.95	7.9894
24	3775.243	8.32298	3771.279	8.31424					

TEMPERATURE CORRECTION FOR VOLUMETRIC SOLUTIONS

This table gives the correction to various observed volumes of water, measured at the designated temperatures to give the volume at the standard temperature, 20°C. Conversely, by subtracting the corrections from the volume desired at 20°C., the volume that must be measured out at the designated temperatures in order to give the desired volume at 20°C., will be obtained. It is assumed that the volumes are measured in glass apparatus having a coefficient of cubical expansion of 0.000025 per degree centigrade. The table is applicable to dilute aqueous solutions having the same coefficient of expansion as water.

Temperature of measurement, °C.	Capacity of apparatus in milliliters at 20°C.						
	2,000	1,000	500	400	300	250	150
	Correction in milliliters to give volume of water at 20°C.						
15	+1.54	+0.77	+0.38	+0.31	+0.23	+0.19	+0.12
16	+1.28	+ .64	+ .32	+ .26	+ .19	+ .16	+ .10
17	+ .99	+ .50	+ .25	+ .20	+ .15	+ .12	+ .07
18	+ .68	+ .34	+ .17	+ .14	+ .10	+ .08	+ .05
19	+ .35	+ .18	+ .09	+ .07	+ .05	+ .04	+ .03
21	− .37	− .18	− .09	− .07	− .06	− .05	− .03
22	− .77	− .38	− .19	− .15	− .12	− .10	− .06
23	−1.18	− .59	− .30	− .24	− .18	− .15	− .09
24	−1.61	− .81	− .40	− .32	− .24	− .20	− .12
25	−2.07	−1.03	− .52	− .41	− .31	− .26	− .15
26	−2.54	−1.27	− .64	− .51	− .38	− .32	− .19
27	−3.03	−1.52	− .76	− .61	− .46	− .38	− .23
28	−3.55	−1.77	− .89	− .71	− .53	− .44	− .27
29	−4.08	−2.04	−1.02	− .82	− .61	− .51	− .31
30	−4.62	−2.31	−1.16	− .92	− .69	− .58	− .35

In using the above table to correct the volume of certain standard solutions to 20°C. more accurate results will be obtained if the numerical values of the corrections are increased by the percentages given below:

Solution	Normality N	N/2	N/10
HNO_3	50	25	6
H_2SO_4	45	25	5
NaOH	40	25	5
KOH	40	20	4

TEMPERATURE CORRECTION FOR GLASS VOLUMETRIC APPARATUS

This table gives the correction to be added to actual capacity (determined at certain temperatures) to give the capactiy at the standard temperature, 20°C. Conversely, by subtracting the corrections from the indicated capacity of an instrument standard at 20°C. the corresponding capacity at other temperatures is obtained. The table assumes for the cubical coefficient of expansion of glass 0.000025 per degree centigrade. The coefficients of expansion of glasses used for volumetric instruments vary from 0.000023 to 0.000028.

Temperature in degrees C.	2,000 ml	1,000 ml	500 ml	400 ml	300 ml	250 ml
15	+0.25	+1.12	+0.06	+0.05	+0.04	+0.031
16	+ .20	+ .10	+ .05	+0.04	+ .03	+ .025
17	+ .15	+ .08	+ .04	+ .03	+ .02	+ .019
18	+ .10	+ .05	+ .02	+ .02	+ .02	+ .012
19	+ .05	+ .02	+ .01	+ .01	+ .01	+ .006
21	− .05	− .02	− .01	− .01	− .01	− .006
22	− .10	− .05	− .02	− .02	− .02	− .012
23	− .15	− .08	− .04	− .03	− .02	− .019
24	− .20	− .10	− .05	− .04	− .03	− .025
25	− .25	− .12	− .06	− .05	− .04	− .031
26	− .30	− .15	− .08	− .06	− .04	− .038
27	− .35	− .18	− .09	− .07	− .05	− .044
28	− .40	− .20	− .10	− .08	− .06	− .050
29	− .45	− .22	− .11	− .09	− .07	− .056
30	− .50	− .25	− .12	− .10	− .08	− .062

DENSITY OF VARIOUS LIQUIDS

(Selected from Smithsonian Tables.)

Liquid	Grams per cu. cm	Pounds per cu. ft.	Temp °C
Acetone	0.792	49.4	20°
Alcohol, ethyl	0.791	49.4	20
methyl	0.810	50.5	0
Benzene	0.899	56.1	0
Carbolic acid	0.950—0.965	59.2—60.2	15
Carbon disulfide	1.293	80.7	0
tetrachloride	1.595	99.6	20
Chloroform	1.489	93.0	20
Ether	0.736	45.9	0
Gasoline	0.66—0.69	41.0—43.0	
Glycerin	1.260	78.6	0
Kerosene	0.82	51.2	
Mercury	13.6	849.0	
Milk	1.028—1.035	64.2—64.6	

Liquid	Grams per cu. cm	Pounds per cu. ft.	Temp °C
Naphtha, petroleum ether	0.665	41.5	15
wood	0.848—0.810	52.9—50.5	0
Oils:			
castor	0.969	60.5	15
cocoanut	0.925	57.7	15
cotton seed	0.926	57.8	16
creosote	1.040—1.100	64.9—68.6	15
linseed, boiled	0.924	58.8	15
olive	0.918	57.3	15
Sea water	1.025	63.99	15
Turpentine (spirits)	0.87	54.3	
Water	1.00	62.43	4

DENSITY OF ALCOHOL

Density of Ethyl Alcohol in Grams Per Cubic Centimeter, Computed from Mendeleeff's Formula

(Selected from Smithsonian Tables.)

Temp. °C	0	1	2	3	4	5	6	7	8	9
0	.80625	.80541	.80457	.80374	.80290	.80207	.80123	.80039	.79956	.79872
10	.79788	.79704	.79620	.79535	.79451	.79367	.79283	.79198	.79114	.79029
20	.78945	.78860	.78775	.78691	.78606	.78522	.78437	.78352	.78267	.78182
30	.78097	.78012	.77927	.77841	.77756	.77671	.77585	.77500	.77414	.77329

HYDROMETERS AND DENSITY UNITS

Alcoholometer. — For testing alcoholic solutions; the scale shows the per cent of alcohol by volume; 0°—100° is the per cent.

Ammoniameter. — For testing ammonia solutions; scale 0°—40°; to convert to sp. gr. multiply by 3 and deduct from 1000.

Barktrometer or *Barkometer.* — For testing tanning liquor; scale 0°—80° Bk; the number to the right of the sp. gr. is the degree Bk; thus, 1.025 sp. gr. is 25° Bk.

Baumé. — There are two kinds in use; heavy Bé, for liquids heavier than water and light Bé for liquids lighter than water. In the former, 0° corresponds to a sp. gr. 1.000 (water at 4°C.) and 66° corresponds to a sp. gr. 1.842; in the lighter than water scale, 0° Bé is equivalent to the gravity of a 10% solution of sodium chloride and 60° Bé corresponds to a sp. gr. of 0.745. For Baumé degrees on the scale of densities greater than unity, the following equation gives the means of conversion:

$$\text{Sp.gr.} = \frac{m}{m - d}$$

where m = 145 (in the United States)
m = 144 (old scale used in Holland)
m = 146.78 (New scale or Gerlach scale)
d = Baumé reading

Beck's Hydrometer has 0° corresponding to sp. gr. 1.000 and 30° to sp. gr. 0.850; equal divisions on the scale are continued as far as required in both directions.

Brix Saccharometer or *Balling Saccharometer* shows directly the per cent of sugar (sucrose) by weight at the temperature indicated on the instrument, usually 17.5°C.; i.e., degrees Brix is the per cent sugar.

Cartier's Hydrometer floats in water at the 10° scale division and at 30° corresponds to 32° Be.

Oleometer. — For vegetable and sperm oils; scale 50°—0° corresponds to sp. gr. 0.870—0.970.

Soxhlet's Lactometer, for determining the density of milk, has a scale from 25° (sp. gr. 1.025) to 35° (sp. gr. 1.035) divided into suitable scale divisions.

Twaddell Hydrometers have the scale so arranged that the reading multiplied by 5 and added to 1000 gives the sp. gr. with reference to water as 1000; it is always used for densties greater than water.

HYDROMETER CONVERSION TABLES

Showing the Relation between Density (C. G. S.) and Degrees Baumé for Densities less than Unity

Density	Degrees Baumé .00	.01	.02	.03	.04	.05	.06	.07	.08	.09
0.60	103.33	99.51	95.81	92.22	88.75	85.38	82.12	78.95	75.88	72.90
.70	70.00	67.18	64.44	61.78	59.19	56.67	54.21	51.82	49.49	47.22
.80	45.00	42.84	40.73	38.68	36.67	34.71	32.79	30.92	29.09	27.30
.90	25.56	23.85	22.17	20.54	18.94	17.37	15.83	14.33	12.86	11.41
1.00	10.00									

Showing the Relation between Density (C. G. S.) and Baumé and Twaddell Scales for Densities above Unity

Density	Degrees Baumé	Degrees Twaddell	Density	Degrees Baumé	Degrees Twaddell	Density	Degrees Baumé	Degrees Twaddell	Denisty	Degrees Baumé	Degrees Twaddell
1.00	0.00	0	1.20	24.17	40	1.41	42.16	82	1.61	54.94	122
1.01	1.44	2	1.21	25.16	42	1.42	42.89	84	1.62	55.49	124
1.02	2.84	4	1.22	26.15	44	1.43	43.60	86	1.63	56.04	126
1.03	4.22	6	1.23	27.11	46	1.44	44.31	88	1.64	56.58	128
1.04	5.58	8	1.24	28.06	48	1.45	45.00	90	1.65	57.12	130
1.05	6.91	10	1.25	29.00	50	1.46	45.68	92	1.66	57.65	132
1.06	8.21	12	1.26	29.92	52	1.47	46.36	94	1.67	58.17	134
1.07	9.49	14	1.27	30.83	54	1.48	47.03	96	1.68	58.69	136
1.08	10.74	16	1.28	31.72	56	1.49	47.68	98	1.69	59.20	138
1.09	11.97	18	1.29	32.60	58	1.50	48.33	100	1.70	59.71	140
1.10	13.18	20	1.30	33.46	60	1.51	48.97	102	1.71	60.20	142
1.11	14.37	22	1.31	34.31	62	1.52	49.60	104	1.72	60.70	144
1.12	15.54	24	1.32	35.15	64	1.53	50.23	106	1.73	61.18	146
1.13	16.68	26	1.33	35.98	66	1.54	50.84	108	1.74	61.67	148
1.14	17.81	28	1.34	36.79	68	1.55	51.45	110	1.75	62.14	150
1.15	18.91	30	1.35	37.59	70	1.56	52.05	112	1.76	62.61	152
1.16	20.00	32	1.36	38.38	72	1.57	52.64	114	1.77	63.08	154
1.17	21.07	34	1.37	39.16	74	1.58	53.23	116	1.78	63.54	156
1.18	22.12	36	1.38	39.93	76	1.59	53.80	118	1.79	63.99	158
1.19	23.15	38	1.39	40.68	78	1.60	54.38	120	1.80	64.44	160
			1.40	41.43	80						

DENSITY OF D_2O

G. S. Kell

t, °C.	ϱ, G./Cc.	t, °C.	ϱ, G./Cc.	t, °C.	ϱ, G./Cc.	t, °C.	ϱ, G./Cc.
0	1.10469	20	1.10534	50	1.09570	80	1.07824
3.813	1.10546	25	1.10445	55	1.09325	85	1.07475
5	1.10562	30	1.10323	60	1.09060	90	1.07112
10	1.10599	35	1.10173	65	1.08777	95	1.06736
11.185	1.10600	40	1.09996	70	1.08475	100	1.06346
15	1.10587	45	1.09794	75	1.08158	101.431	1.06232

VOLUME PROPERTIES OF WATER AT 1 atm*

	ρ, kg m^{-3},	$10^6\ \alpha$, K^{-1},	$10^6\ \kappa T$/bar^{-1}		ρ, kg m^{-3},	$10^6\ \alpha$, K^{-1},	$10^6\ \kappa T$/bar^{-1}
	Equation 1	Equation 1	Equation 2		Equation 1	Equation 1	Equation 2
-30	983.854	-1400.0	80.79	9	999.7808	74.38	48.0560
-25	989.585	-955.9	70.94	10	999.6996	87.97	47.8086
-20	993.547	-660.6	64.25	11	999.6051	101.20	47.5726
-15	996.283	-450.3	59.44	12	999.4974	114.08	47.3474
-10	998.117	-292.4	55.83	13	999.3771	126.65	47.1327
-9	998.395	-265.3	55.22	14	999.2444	138.90	46.9280
-8	998.647	-239.5	54.64	15	999.0996	150.87	46.7331
-7	998.874	-214.8	54.08	16	998.9430	162.55	46.5475
-6	999.077	-191.2	53.56	17	998.7749	173.98	46.3708
-5	999.256	-168.6	53.06	18	998.5956	185.15	46.2029
-4	999.414	-146.9	52.58	19	998.4052	196.08	46.0433
-3	999.550	-126.0	52.12	20	998.2041	206.78	45.8918
-2	999.666	-106.0	51.69	21	997.9925	217.26	45.7482
-1	999.762	-86.7	51.28	22	997.7705	227.54	45.6122
0	999.8395	-68.05	50.8850	23	997.5385	237.62	45.4835
1	999.8985	-50.09	50.5091	24	997.2965	247.50	45.3619
2	999.9399	-32.74	50.1505	25	997.0449	257.21	45.2472
3	999.9642	-15.97	49.8081	26	996.7837	266.73	45.1392
4	999.9720	0.27	49.4812	27	996.5132	276.10	45.0378
5	999.9638	16.00	49.1692	28	996.2335	285.30	44.9427
6	999.9402	31.24	48.8712	29	995.9448	294.34	44.8537
7	999.9015	46.04	48.5868	30	995.6473	303.24	44.7707
8	999.8482	60.41	48.3152	31	995.3410	312.00	44.6935

PROPERTIES LIQUID DEUTERIUM OXIDE (D_2O)

Freezing point — 301.97K (3.82°C) at 0.1013325 MPa
Boiling point — 399.57K (101.42°C) at 0.101325 MPa
Maximum density at 0.101325 MPa — 1.10534 kg/dm^3
Temperature at maximum density — 309.335K (11.185°C)
Molar mass — 0.020027478 kg/mol
Specific gas constant (Universal gas constant divided by molar mass) — 415.150 J/kgK
Critical temperature — $(643.89 + \delta)$K = $(370.74 + \delta)$°C with $-0.2 \leqq \delta \geqq +0.2$

Critical pressure — $(21.671 + 0.27\delta \pm 0.010)$MPa
Critical density — (356 ± 5)kg/m^3
Critical specific volume — (0.00281 ± 0.00004)m^3/kg
Triple point temperature — (276.97 ± 0.02)K = (3.82 ± 0.02)°C
Triple point pressure — (661 ± 3)Pa
Liquid density at triple point — (1105.5 ± 0.2)kg/m^3
Vapor density at triple point — (0.00575 ± 0.00003)kg/m^3

VOLUME PROPERTIES OF WATER AT 1 atm* (continued)

	ρ, kg m^{-3}, Equation 1	$10^6\ \alpha$, K^{-1}, Equation 1	$10^6\ \kappa T$/bar^{-1} Equation 2
32	995.0262	320.63	44.6221
33	994.7030	329.12	44.5561
34	994.3715	337.48	44.4956
35	994.0319	345.73	44.4404
36	993.6842	353.86	44.3903
37	993.3287	361.88	44.3452
38	992.9653	369.79	44.3051
39	992.5943	377.59	44.2697
40	992.2158	385.30	44.2391
41	991.8298	392.91	44.2131
42	991.4364	400.43	44.1917
43	991.0358	407.85	44.1747
44	990.6280	415.19	44.1620
45	990.2132	422.45	44.1536
46	989.7914	429.63	44.1494
47	989.3628	436.73	44.1494
48	988.9273	443.75	44.1533
49	988.4851	450.71	44.1613
50	988.0363	457.59	44.1732
51	987.5809	464.40	44.189
52	987.1190	471.15	44.209
53	986.6508	477.84	44.232
54	986.1761	484.47	44.259
55	985.6952	491.04	44.290
56	985.2081	497.55	44.324
57	984.7149	504.01	44.362
58	984.2156	510.41	44.403
59	983.7102	516.76	44.448

	ρ, kg m^{-3}, Equation 1	$10^6\ \alpha$, K^{-1}, Equation 1	$10^6\ \kappa T$/bar^{-1} Equation 2
60	983.1989	523.07	44.496
61	982.6817	529.32	44.548
62	982.1586	535.53	44.603
63	981.6297	541.70	44.662
64	981.0951	547.82	44.723
65	980.5548	553.90	44.788
66	980.0089	559.94	44.857
67	979.4573	565.95	44.928
68	978.9003	571.91	45.003
69	978.3377	577.84	45.081
70	977.7696	583.74	45.162
71	977.1962	589.60	45.246
72	976.6173	595.43	45.333
73	976.0332	601.23	45.424
74	975.4437	607.00	45.517
75	974.8990	612.75	45.614
76	974.2490	618.46	45.714
77	973.6439	624.15	45.817
78	973.0336	629.82	45.922
79	972.4183	635.46	46.031
80	971.7978	641.08	46.143
81	971.1723	646.67	46.258
82	970.5417	652.25	46.376
83	969.9062	657.81	46.497
84	969.2657	663.34	46.621
85	968.6203	668.86	46.748
86	967.9700	674.37	46.878
87	967.3148	679.85	47.011
88	966.6547	685.33	47.148
89	965.9898	690.78	47.287

	ρ, kg m^{-3},	$10^6\ \alpha$, K^{-1},	$10^6\ \kappa T$/bar^{-1}	
	Equation 1	Equation 1	Equation 2	Equation 3
90	965.3201	696.23	47.429	47.428
91	964.6457	701.66	47.574	47.574
92	963.9664	707.08	47.722	47.722
93	963.2825	712.49	47.874	47.873
94	962.5938	717.89	48.028	48.028
95	961.9004	723.28	48.185	48.185
96	961.2023	728.67	48.346	48.346
97	960.4996	734.04	48.509	48.510
98	959.7923	739.41	48.676	48.677
99	959.0803	744.78	48.846	48.847
100	958.3637	750.14	49.019	49.020
101	957.642	755.5		49.20
102	956.917	760.8		49.38
103	956.186	766.2		49.56
104	955.451	771.5		49.74

	ρ, kg m^{-3},	$10^6\ \alpha$, K^{-1},	$10^6\ \kappa T$/bar^{-1}	
	Equation 1	Equation 1	Equation 2	Equation 3
105	954.712	776.9		49.93
106	953.968	782.2		50.13
107	953.220	787.6		50.32
108	952.467	792.9		50.52
109	951.709	798.3		50.72
110	950.947	803.6		50.93
115	947.070	830.4		52.01
120	943.083	857.4		53.17
125	938.984	884.7		54.43
130	934.775	912.3		55.79
135	930.456	940.3		57.24
140	926.026	968.9		58.80
145	921.484	998.0		60.47
150	916.829	1027.8		62.25

Equations:

$$\rho/\text{kg m}^{-3} = (999.83952 + 16.945176\ t - 7.9870401 \times 10^{-3}\ t^2 - 46.170461 \times 10^{-6}\ t^3 + 105.56302 \times 10^{-9}\ t^4 - 280.54253 \times 10^{-12}\ t^5)/(1 + 16.879850 \times 10^{-3}\ t) \quad (1)$$

$$10^6\ \kappa_T/\text{bar}^{-1} = (50.88496 + 0.6163813\ t + 1.459187 \times 10^{-3}\ t^2 + 20.08438 \times 10^{-6}\ t^3 - 58.47727 \times 10^{-9}\ t^4 + 410.4110 \times 10^{-12}\ t^5)/(1 + 19.67348 \times 10^{-3}\ t) \quad (2)$$

$$10^6\ \kappa_T/\text{bar}^{-1} = (50.884917 + 0.62590623\ t + 1.3848668 \times 10^{-3}\ t^2 + 21.603427 \times 10^{-6}\ t^3 - 72.087667 \times 10^{-9}\ t^4 + 465.45054 \times 10^{-12}\ t^5)/(1 + 19.859983 \times 10^{-3}\ t) \quad (3)$$

$$\kappa_S = (\partial \ln \rho/\partial P)_S = \frac{1}{\rho U^2} \quad (4)$$

* Density ρ, thermal expansivity $\alpha = -(\partial \ln \rho/\partial T)_p$, and isothermal compressibility $\kappa T = (\partial \ln \rho/\partial p)T$. For purposes of this table, ordinary water is that with a maximum density of 999.972 kg m^{-3}. Equation 4 for the compressibility should be used for temperatures $0 < t < 100$°C, and Equation 3 for $100 < + < 150$°C. The liquid is metastable below 0°C and above 100°C. Values below 0°C were obtained by extrapolation, and no claim is made for their accuracy.

DENSITY AND VOLUME OF MERCURY

Based on the Density of Mercury at 0° C. by Thiesen and Scheel

(Selected from Smithsonian Tables)

Temp. °C.	Mass in gr. per ml.	Vol. of 1 gr. in ml.	Temp. °C.	Mass in gr. per ml.	Vol. of 1 gr. in ml.	Temp. °C.	Mass in gr. per ml.	Vol. of 1gr. in ml.
−10	13.6202	0.0734205	17	5536	7813	90	13.3762	0.0747594
−9	6177	4338	18	5512	7947	100	3522	8939
−8	6152	4472	19	5487	8081	110	3283	50285
−7	6128	4606	20	13.5462	0.0738215	120	3044	1633
−6	6103	4739	21	5438	8348	130	2805	2982
−5	13.6078	0.0734873	22	5413	8482	140	13.2567	0.0754334
−4	6053	5006	23	5389	8616	150	2330	5688
−3	6029	5140	24	5364	8750	160	2093	7044
−2	6004	5273	25	13.5340	0.0738883	170	1856	8402
−1	5979	5407	26	5315	9017	180	1620	9764
0	13.5955	0.0735540	27	5291	9151	190	13.1384	0.0761128
1	5930	5674	28	5266	9285	200	1148	2495
2	5906	5808	29	5242	9419	210	0913	3865
3	5881	5941	30	13.5217	0.0739552	220	0678	5239
4	5856	6075	31	5193	9686	230	0443	6616
5	13.5832	0.0736209	32	5168	9820	240	13.0209	0.0767996
6	5807	6342	33	5144	9953	250	12.9975	9381
7	5782	6476	34	5119	40087	260	9741	70769
8	5758	6610	35	13.5095	0.0740221	270	9507	2161
9	5733	6744	36	5070	0354	280	9273	3558
10	13.5708	0.0736877	37	5046	0488	290	12.9039	0.0774958
11	5684	7011	38	5021	0622	300	8806	6364
12	5659	7145	39	4997	0756	310	8572	7774
13	5634	7278	40	13.4973	0.0740891	320	8339	9189
14	5610	7412	50	4729	2229	330	8105	80609
15	13.5585	0.0737546	60	4486	3569	340	12.7872	0.0782033
16	5561	7680	70	4244	4910	350	7638	3464
			80	4003	6252	360	7405	4900

SULFURIC ACID

SPECIFIC GRAVITY OF AQUEOUS SULFURIC ACID SOLUTIONS

AT $\frac{20°}{4°}$ C.

Be.	Sp. gr.	Per cent H_2SO_4	G. per liter	Lbs. per cu. ft.	Lbs. per gal.	Be.	Sp. gr.	Per cent H_2SO_4	G. per liter	Lbs. per cu. ft.	Lbs. per gal.
0.7	1.0051	1	10.05	0.6275	0.0839	41.8	1.4049	51	716.5	44.73	5.979
1.7	1.0118	2	20.24	1.263	0.1689	42.5	1.4148	52	735.7	45.93	6.140
2.6	1.0184	3	30.55	1.907	0.2550	43.2	1.4248	53	755.1	47.14	6.302
3.5	1.0250	4	41.00	2.560	0.3422	44.0	1.4350	54	774.9	48.37	6.467
4.5	1.0317	5	51.59	3.220	0.4305	44.7	1.4453	55	794.9	49.62	6.634
5.4	1.0385	6	62.31	3.890	0.5200	45.4	1.4557	56	815.2	50.89	6.803
6.3	1.0453	7	73.17	4.568	0.6106	46.1	1.4662	57	835.7	52.17	6.974
7.2	1.0522	8	84.18	5.255	0.7025	46.8	1.4768	58	856.5	53.47	7.148
8.1	1.0591	9	95.32	5.950	0.7955	47.5	1.4875	59	877.6	54.79	7.324
9.0	1.0661	10	106.6	6.655	0.8897	48.2	1.4983	60	899.0	56.12	7.502
9.9	1.0731	11	118.0	7.369	0.9851	48.9	1.5091	61	920.6	57.47	7.682
10.8	1.0802	12	129.6	8.092	1.082	49.6	1.5200	62	942.4	58.83	7.865
11.7	1.0874	13	141.4	8.825	1.180	50.3	1.5310	63	964.5	60.21	8.049
12.5	1.0947	14	153.3	9.567	1.279	51.0	1.5421	64	986.9	61.61	8.236
13.4	1.1020	15	165.3	10.32	1.379	51.7	1.5533	65	1010	63.03	8.426
14.3	1.1094	16	177.5	11.08	1.481	52.3	1.5646	66	1033	64.46	8.618
15.2	1.1168	17	189.9	11.85	1.584	53.0	1.5760	67	1056	65.92	8.812
16.0	1.1243	18	202.4	12.63	1.689	53.7	1.5874	68	1079	67.39	9.008
16.9	1.1318	19	215.0	13.42	1.795	54.3	1.5989	69	1103	68.87	9.207
17.7	1.1394	20	227.9	14.23	1.902	55.0	1.6105	70	1127	70.38	9.408
18.6	1.1471	21	240.9	15.04	2.010	55.6	1.6221	71	1152	71.90	9.611
19.4	1.1548	22	254.1	15.86	2.120	56.3	1.6338	72	1176	73.44	9.817
20.3	1.1626	23	267.4	16.69	2.231	56.9	1.6456	73	1201	74.99	10.02
21.1	1.1704	24	280.9	17.54	2.344	57.5	1.6574	74	1226	76.57	10.24
21.9	1.1783	25	294.6	18.39	2.458	58.1	1.6692	75	1252	78.15	10.45
22.8	1.1862	26	308.4	19.25	2.574	58.7	1.6810	76	1278	79.75	10.66
23.6	1.1942	27	322.4	20.13	2.691	59.3	1.6927	77	1303	81.37	10.88
24.4	1.2023	28	336.6	21.02	2.809	59.9	1.7043	78	1329	82.99	11.09
25.2	1.2104	29	351.0	21.91	2.929	60.5	1.7158	79	1355	84.62	11.31
26.0	1.2185	30	365.6	22.82	3.051	61.1	1.7272	80	1382	86.26	11.53
26.8	1.2267	31	380.3	23.74	3.173	61.6	1.7383	81	1408	87.90	11.75
27.6	1.2349	32	395.2	24.67	3.298	62.1	1.7491	82	1434	89.54	11.97
28.4	1.2432	33	410.3	25.61	3.424	62.6	1.7594	83	1460	91.16	12.19
29.1	1.2515	34	425.5	26.56	3.551	63.0	1.7693	84	1486	92.78	12.40
29.9	1.2599	35	441.0	27.53	3.680	63.5	1.7786	85	1512	94.38	12.62
30.7	1.2684	36	456.6	28.51	3.811	63.9	1.7872	86	1537	95.95	12.83
31.4	1.2769	37	472.5	29.49	3.943	64.2	1.7951	87	1562	97.49	13.03
32.2	1.2855	38	488.5	30.49	4.077	64.5	1.8022	88	1586	99.01	13.23
33.0	1.2941	39	504.7	31.51	4.212	64.8	1.8087	89	1610	100.5	13.42
33.7	1.3028	40	521.1	32.53	4.349	65.1	1.8144	90	1633	101.9	13.63
34.5	1.3116	41	537.8	33.57	4.488	65.3	1.8195	91	1656	103.4	13.82
35.2	1.3205	42	554.6	34.62	4.628	65.5	1.8240	92	1678	104.8	14.00
35.9	1.3294	43	571.6	35.69	4.770	65.7	1.8279	93	1700	106.1	14.19
36.7	1.3384	44	588.9	36.76	4.914	65.8	1.8312	94	1721	107.5	14.36
37.4	1.3476	45	606.4	37.86	5.061	65.9	1.8337	95	1742	108.7	14.54
38.1	1.3569	46	624.2	38.97	5.209	66.0	1.8355	96	1762	110.0	14.70
38.9	1.3663	47	642.2	40.09	5.359	66.0	1.8364	97	1781	111.2	14.87
39.6	1.3758	48	660.4	41.23	5.511	66.0	1.8361	98	1799	112.3	15.02
40.3	1.3854	49	678.8	42.38	5.665	65.9	1.8342	99	1816	113.4	15.15
41.1	1.3951	50	697.6	43.55	5.821	65.8	1.8305	100	1831	114.3	15.28

DENSITY AND COMPOSITION OF FUMING SULFURIC ACID

Actual H_2SO_4, %	Specific gravity	Equiv. H_2SO_4, %	Weight, lb./cu. ft.	Weight, lb. per U.S. gal.	Comb. H_2O, %	Free SO_3, %	Total SO_3, %	SO_3, lb./cu. ft.
100	1.839	100.00	114.70	15.33	18.37	0	81.63	93.63
99	1.845	100.22	115.07	15.38	18.19	1	81.81	94.14
98	1.851	100.45	115.33	15.41	18.00	2	82.00	94.57
97	1.855	100.67	115.70	15.46	17.82	3	82.18	95.08
96	1.858	100.89	115.88	15.49	17.64	4	82.36	95.44
95	1.862	101.13	116.13	15.52	17.45	5	82.55	95.87
94	1.865	101.35	116.32	15.55	17.27	6	82.73	96.23
93	1.869	101.58	116.57	15.58	17.08	7	82.92	96.66
92	1.873	101.80	116.82	15.61	16.90	8	83.10	97.12
91	1.877	102.02	117.07	15.64	16.72	9	83.28	97.50
90	1.880	102.25	117.26	15.67	16.57	10	83.47	97.88
89	1.884	102.47	117.51	15.70	16.35	11	83.65	98.30
88	1.887	102.71	117.69	15.73	16.17	12	83.83	98.66
87	1.891	102.92	117.94	15.76	15.98	13	84.02	99.09
86	1.895	103.15	118.19	15.79	15.80	14	84.20	99.52
85	1.899	103.38	118.44	15.82	15.61	15	84.39	99.95
84	1.902	103.60	118.63	15.86	15.43	16	84.57	100.33
83	1.905	103.82	118.81	15.89	15.25	17	84.75	100.69
82	1.909	104.05	119.06	15.92	15.06	18	84.94	101.13
81	1.911	104.28	119.28	15.95	14.88	19	85.12	101.45
80	1.915	104.50	119.50	15.98	14.70	20	85.30	101.93
79	1.920	104.73	119.75	16.01	14.51	21	85.49	102.37
78	1.923	104.95	119.94	16.04	14.33	22	85.67	102.75
77	1.927	105.18	120.19	16.07	14.14	23	85.86	103.20
76	1.931	105.40	120.44	16.10	13.96	24	86.04	103.63
75	1.934	105.62	120.62	16.12	13.78	25	86.22	104.00
74	1.939	105.85	120.94	16.16	13.59	26	86.41	104.50
73	1.943	106.08	121.18	16.19	13.41	27	86.59	104.93
72	1.946	106.29	121.37	16.22	13.28	28	86.72	105.31
71	1.949	106.53	121.56	16.25	13.04	29	86.96	105.71
70	1.952	106.75	121.75	16.28	12.86	30	87.14	106.09
69	1.955	106.97	121.93	16.30	12.68	31	87.32	106.47
68	1.958	107.20	122.12	16.33	12.49	32	87.51	106.87
67	1.961	107.42	122.31	16.35	12.31	33	87.69	107.25
66	1.965	107.65	122.56	16.38	12.12	34	87.88	107.71
65	1.968	107.87	122.74	16.40	11.94	35	88.06	108.08
64	1.972	108.10	122.99	16.43	11.76	36	88.24	108.53
63	1.976	108.33	123.24	16.46	11.57	37	88.43	108.98
62	1.979	108.55	123.43	16.50	11.39	38	88.61	109.37
61	1.981	108.77	123.55	16.52	11.21	39	88.79	109.70
60	1.983	109.00	123.74	16.54	11.02	40	88.98	110.10
59	1.985	109.22	123.80	16.55	10.84	41	89.16	110.38
58	1.987	109.45	123.93	16.56	10.65	42	89.35	110.83
57	1.989	109.68	124.05	16.58	10.47	43	89.53	111.06
56	1.991	109.90	124.18	16.60	10.29	44	89.71	111.40
55	1.993	110.13	124.30	16.62	10.10	45	89.90	111.75
50	2.001	111.25	124.80	16.68	9.18	50	90.72	113.34
40	2.102	113.50	131.10	17.53	7.35	60	92.65	121.46
30	1.982	115.75	123.62	16.50	5.51	70	94.49	116.81
20	1.949	118.00	121.56	16.25	3.67	80	96.33	117.10
10	1.911	120.25	119.19	15.92	1.84	90	98.16	117.00
0	1.857	122.50	115.83	15.50	0.00	100	100.00	115.83

* By permission from the 7th edition of Chemical Plant Control Data, Chemical Construction Corporation (1957).

COMPOSITION OF SOME INORGANIC ACIDS AND BASES

The following acids and bases are frequently supplied as concentrated aqueous solutions. This table presents certain data concerning these solutions.

	Formula weight	Molarity	Specific gravity	Weight percent	
Acetic acid	60.05	17.5	1.05	99—100	CH_3COOH
Ammonium hydroxide	35.05	14.8	0.90	28—30	NH_3
Hydriodic acid	127.91	5.5	1.5	47—47.5	HI
Hydrobromic acid	80.93	9.0	1.5	47—49	HBr
Hydrochloric acid	36.46	12.0	1.18	36.5—38	HCl
Hydrofluoric acid	20.01	28.9	1.17	48—51	HF
Phosphoric acid	98.00	14.7	1.7	85	H_3PO_4
Sulfuric acid	98.08	18.0	1.84	95—98	H_2SO_4

DENSITY OF MOIST AIR

The density of dry air may be determined by computation from the general relation $D = D_0(T_0/T)(P/P_0)$ where D_0 represents a known density at absolute temperature T_0 and pressure P_0 and D, the density at absolute temperature T and pressure P.

The density of moist air may be determined by a similar relation:

$D = 1.2929\,(273.13/T)\,[B - 0.3783\,e)/760]$ where T is the absolute temperature; B, the barometric pressure in mm, and e the vapor pressure of the moisture in the air in millimeters. The density will then be the product of two terms, each of which may be found by use of the tables which follow.

The first factor, $1.2929\,(273.13/T)$, may be found directly in Table I for various temperatures. For convenience, temperatures are given in the table in °C although the values of the factor have been computed with absolute temperatures. The tabular values actually represent the density of dry air at various temperatures and 760 mm pressure.

The second factor, $[(B - 0.3783\,e/760]$, must be obtained in two steps: First — the numerator of the expression is obtained by subtracting $0.3783\,e$ from the barometric pressure. The quantity $0.3783\,e$ may be found directly from the dew point in Table II. If the wet and dry bulb thermometer readings are known e may be found in the table Reduction of Psychrometric Observations given in the section Hygrometric and Barometric Tables. $0.3783\,e$ may then be found by calculation or read from the table. Second — the value of the whole factor for any value of $B - 0.3783\,e$ may be obtained from Table III.

The product of the above two factors will give the required density in a g/ℓ.

To facilitate obtaining approximate values of the density for ordinary pressures and temperatures, a table of products is given which may be entered with the temperatures in °C and the corrected (for moisture) value of the barometric pressure in mm to obtain density.

As an illustration of the use of the tables, let it be desired to find the density of air for a barometric pressure of 750 mm, a dew point of 10°C, and air temperature of 20°C.

From the dew point, the value of $0.3783\,e$ is found in Table II to be 3.48 mm. 750 − 3.48 = 746.52, the corrected pressure. The pressure factor for this value found in Table III by interpolation is 0.98226.

The temperature factor from Table I is 1.2047.

$$1.2047 \times 0.98224 = 1.1833 \text{ g}/\ell.$$

To obtain the value directly from Table IV, enter it for 20°C and 746.5 mm which gives by interpolation 1.183 g/ℓ.

TABLE I

$(1.2929 \times 273.13/T)$

(Besides being a necessary part of the determination of the density of moist air, the values in this table are actually the density of dry air in g/ℓ at 760 mm pressure for various temperatures.)

Temp. °C		0	1	2	3	4	5	6	7	8	9
−50	1.5	826	897	969	*042	*115	*189	*264	*339	*415	*491
−40	1.5	147	213	278	345	412	479	547	616	686	756
−30	1.4	524	584	645	706	767	829	892	955	*019	*083
−20	1.3	951	*006	*062	*118	*175	*232	*289	*347	*406	*465
−10	1.3	420	472	523	575	628	680	734	787	841	896
−0	1.2	929	977	*024	*073	*121	*170	*219	*269	*319	*370
+0	1.2	929	882	835	789	742	697	651	606	561	517
10	1.2	472	428	385	342	299	256	214	171	130	088
20	1.2	047	006	*965	*925	*885	*845	*805	*766	*727	*688
30	1.1	649	611	573	535	498	460	423	387	350	314
40	1.1	277	242	206	170	135	100	065	031	*996	*962
50	1.0	928	895	861	828	795	762	729	697	664	632
60	1.0	600	569	537	506	475	444	413	382	352	322

TABLE II

Vapor Pressure — Value of $0.3783\,e$

Dew point °C	Vapor press. e mm (ice)	$0.3783\,e$	Dew point °C	Vapor press. e mm (water)	$0.3783\,e$	Dew point °C	Vapor press. e mm (water)	$0.3783\,e$	Dew point °C	Vapor press. e mm (ice)	$0.3783\,e$	Dew point °C	Vapor press. e mm (water)	$0.3783\,e$	Dew point °C	Vapor press. e mm (water)	$0.3783\,e$
−50	0.029	0.01	−15	1.252	0.47	0	4.58	1.73	15	12.79	4.84	30	31.86	12.05	45	71.97	27.23
−45	0.054	0.02	−14	1.373	0.52	1	4.92	1.86	16	13.64	5.16	31	33.74	12.76	46	75.75	28.66
−40	0.096	0.04	−13	1.503	0.57	2	5.29	2.00	17	14.54	5.50	32	35.70	13.51	47	79.70	30.15
−35	0.169	0.06	−12	1.644	0.62	3	5.68	2.15	18	15.49	5.86	33	37.78	14.29	48	83.83	31.71
−30	0.288	0.11	−11	1.798	0.68	4	6.10	2.31	19	16.49	6.24	34	39.95	15.11	49	88.14	33.34
−25	0.480	0.18	−10	1.964	0.74	5	6.54	2.47	20	17.55	6.64	35	42.23	15.98	50	92.6	35.03
−24	0.530	0.20	− 9	2.144	0.81	6	7.01	2.65	21	18.66	7.06	36	44.62	16.88	51	97.3	36.81
−23	0.585	0.22	− 8	2.340	0.89	7	7.51	2.84	22	19.84	7.51	37	47.13	17.83	52	102.2	38.66
−22	0.646	0.24	− 7	2.550	0.96	8	8.04	3.04	23	21.09	7.98	38	49.76	18.82	53	107.3	40.59
−21	0.712	0.27	− 6	2.778	1.05	9	8.61	3.26	24	22.40	8.47	39	52.51	19.86	54	112.7	42.63
−20	0.783	0.30	− 5	3.025	1.14	10	9.21	3.48	25	23.78	9.00	40	55.40	20.96	55	118.2	44.72
−19	0.862	0.33	− 4	3.291	1.24	11	9.85	3.73	26	25.24	9.55	41	58.42	22.10	56	124.0	46.91
−18	0.947	0.36	− 3	3.578	1.35	12	10.52	3.98	27	26.77	10.13	42	61.58	23.30	57	130.0	49.18
−17	1.041	0.39	− 2	3.887	1.47	13	11.24	4.25	28	28.38	10.74	43	64.89	24.55	58	136.3	51.56
−16	1.142	0.43	− 1	4.220	1.60	14	11.99	4.54	29	30.08	11.38	44	68.35	25.86	59	142.8	54.02
															60	149.6	56.59

TABLE III

Pressure Factor — $[(B - 0.3783\,e)/760]$

The figures in the body of the table give values of the whole term $[(B - 0.3783\,e/760]$ for various values of the numerator $(B - 0.3783\,e)$ expressed at the left and top.

Press. mm corr.	0	1	2	3	4	5	6	7	8	9
80	.10526	.10658	.10789	.10921	.11053	.11184	.11316	.11447	.11579	.11711
90	.11842	.11974	.12105	.12237	.12368	.12500	.12632	.12763	.12895	.13026
100	.13158	.13289	.13421	.13553	.13684	.13816	.13947	.14079	.14211	.14342
110	.14474	.14605	.14737	.14868	.15000	.15132	.15263	.15395	.15526	.15658
120	.15789	.15921	.16053	.16184	.16316	.16447	.16579	.16711	.16842	.16974
130	.17105	.17237	.17368	.17500	.17632	.17763	.17895	.18026	.18158	.18289
140	.18421	.18553	.18684	.18816	.18947	.19079	.19211	.19342	.19474	.19605
150	.19737	.19868	.20000	.20132	.20263	.20395	.20526	.20658	.20789	.20921
160	.21053	.21184	.21316	.21447	.21579	.21711	.21842	.21974	.22105	.22237
170	.23368	.22500	.22032	.22763	.22895	.23026	.23158	.23289	.23421	.23553
180	.23684	.23816	.23947	.24079	.24211	.24342	.24474	.24605	.24737	.24868
190	.25000	.25132	.25263	.25395	.25526	.25658	.25789	.25921	.26053	.26184
200	.26316	.26447	.26579	.26711	.26842	.26974	.27105	.27237	.27368	.27500
210	.27632	.27763	.27895	.28026	.28158	.28289	.28421	.28553	.28684	.28816
220	.28947	.29079	.29211	.29342	.29474	.29605	.29737	.29868	.30000	.30132
230	.30263	.30395	.30526	.30658	.30789	.30921	.31053	.31184	.31316	.31447
240	.31579	.31711	.31842	.31974	.32105	.32237	.32368	.32500	.32632	.32763
250	.32895	.33026	.33158	.33289	.33421	.33553	.33684	.33816	.33947	.34079
260	.34211	.34342	.34474	.34605	.34737	.34868	.35000	.35132	.35263	.35395
270	.35526	.35658	.35789	.35921	.36053	.36184	.36316	.36447	.36579	.36711
280	.36842	.36974	.37105	.37237	.37368	.37500	.37632	.37763	.37895	.38026
290	.38158	.38289	.38421	.38553	.38684	.38816	.38947	.39079	.39211	.39342
300	.39474	.39605	.39737	.39868	.40000	.40132	.40263	.40395	.40526	.40658
310	.40789	.40921	.41053	.41184	.41316	.41447	.41579	.41711	.41842	.41974
320	.42105	.42237	.42368	.42500	.42632	.42763	.42895	.43026	.43158	.43289
330	.43421	.43553	.43684	.43816	.43947	.44079	.44211	.44342	.44474	.44605
340	.44737	.44808	.45000	.45132	.45263	.45395	.45526	.45658	.45789	.45921
350	.46053	.46184	.46316	.46447	.46579	.46711	.46842	.46974	.47105	.47237
360	.47368	.47500	.47632	.47763	.47895	.48026	.48158	.48289	.48421	.48553
370	.48084	.48816	.48947	.49079	.49211	.49342	.49474	.49605	.49737	.49868
380	.50000	.50132	.50263	.50395	.50526	.50658	.50789	.50921	.51053	.51184
390	.51316	.51447	.51579	.51711	.51842	.51974	.52105	.52237	.52368	.52500
400	.52632	.52763	.52895	.53026	.53158	.53289	.53421	.53553	.53684	.53816
410	.53947	.54079	.54211	.54342	.54474	.54605	.54737	.54868	.55000	.55132
420	.55263	.55395	.55526	.55658	.55789	.55921	.56053	.56184	.56316	.56447
430	.56579	.56711	.56842	.56974	.57105	.57237	.57368	.57500	.57632	.57763
440	.57895	.58026	.58158	.58289	.58421	.58553	.58684	.58816	.58947	.59079
450	.59211	.59342	.59474	.59605	.59737	.59868	.60000	.60132	.60263	.60395
460	.60526	.60658	.60789	.60921	.61053	.61184	.61316	.61447	.61579	.61711
470	.61842	.61974	.62105	.62237	.62368	.62500	.62632	.62763	.62895	.63026
480	.63158	.63289	.63421	.63553	.63684	.63816	.63947	.64079	.64211	.64342
490	.64474	.64605	.64737	.64868	.65000	.65132	.65263	.65395	.65526	.65658
500	.65790	.65921	.66053	.66184	.66316	.66447	.66579	.66711	.66842	.66974
510	.67105	.67237	.67368	.67500	.67632	.67763	.67895	.68026	.68158	.68290
520	.68421	.68553	.68084	.68816	.68947	.69079	.69211	.69342	.69474	.69605
530	.69737	.69868	.70000	.70132	.70263	.70395	.70526	.70658	.70790	.70921
540	.71053	.71184	.71316	.71447	.71579	.71711	.71842	.71974	.72105	.72237
550	.72368	.72500	.72632	.72763	.72895	.73026	.73158	.73290	.73421	.73553
560	.73684	.73816	.73947	.74079	.74211	.74342	.74474	.74605	.74737	.74868
570	.75000	.75132	.75263	.75395	.75526	.75658	.75790	.75921	.76053	.76184
580	.76316	.76447	.76579	.76711	.76842	.76974	.77105	.77237	.77368	.77500
590	.77632	.77763	.77895	.78026	.78158	.78290	.78421	.78553	.78684	.78816
600	.78947	.79079	.79211	.79342	.79474	.79605	.79737	.79688	.80000	.80132
610	.80263	.80395	.80526	.80658	.80790	.80921	.81053	.81184	.81316	.81447
620	.81579	.81711	.81842	.81974	.82105	.82237	.82368	.82500	.82632	.82763
630	.82895	.83026	.83158	.83290	.83421	.83553	.83684	.83816	.83947	.84079
640	.84211	.84342	.84474	.84605	.84737	.84868	.85000	.85132	.85263	.85395
650	.85526	.85658	.85790	.85921	.86053	.86184	.86316	.86447	.86579	.86711
660	.86842	.86974	.87105	.87237	.87368	.87500	.87632	.87763	.87895	.88026
670	.88158	.88290	.88421	.88553	.88684	.88816	.88947	.89079	.89211	.89342
680	.89474	.89605	.89737	.89868	.90000	.90132	.90263	.90395	.90526	.90658
690	.90790	.90921	.91053	.91184	.91316	.91447	.91579	.91711	.91842	.91974
700	.92105	.92237	.92368	.92500	.92632	.92763	.92895	.93026	.93158	.93290
710	.93421	.93553	.93684	.93816	.93947	.94079	.94211	.94342	.94474	.94605
720	.94737	.94868	.95000	.95132	.95263	.95395	.95526	.95658	.95790	.95921
730	.96053	.96184	.96316	.96447	.96579	.96711	.96842	.96974	.97105	.97237
740	.97368	.97500	.97632	.97763	.97895	.98026	.98158	.98290	.98421	.98553
750	.98684	.98816	.98947	.99079	.99211	.99342	.99474	.99605	.99737	.99868
760	1.0000	1.0013	1.0026	1.0039	1.0053	1.0066	1.0079	1.0092	1.0105	1.0118
770	1.0132	1.0145	1.0158	1.0171	1.0184	1.0197	1.0211	1.0224	1.0237	1.0250
780	1.0263	1.0276	1.0289	1.0303	1.0316	1.0329	1.0342	1.0355	1.0368	1.0382
790	1.0395	1.0408	1.0421	1.0434	1.0447	1.0461	1.0474	1.0487	1.0500	1.0513

TABLE IV

Density of Moist Air

Values in the body of the table give the density of moist air in g/ℓ for a limited range of temperatures and corrected pressure values $(B - 0.3783\,e)$. The latter may be obtained by use of Table II.

°C	600	610	620	630	640	650	660	670	680	690
5	1.0024	1.0191	1.0358	1.0525	1.0692	1.0859	1.1026	1.1193	1.1361	1.1528
6	.99876	1.0154	1.0321	1.0487	1.0654	1.0820	1.0986	1.1153	1.1319	1.1486
7	.99521	1.0118	1.0284	1.0450	1.0616	1.0781	1.0947	1.1113	1.1279	1.1445
8	.99165	1.0082	1.0247	1.0412	1.0578	1.0743	1.0908	1.1074	1.1239	1.1404
9	.98818	1.0047	1.0211	1.0376	1.0541	1.0705	1.0870	1.1035	1.1199	1.1364
10	.98463	1.0010	1.0175	1.0339	1.0503	1.0667	1.0831	1.0995	1.1159	1.1323
11	.98115	.99751	1.0139	1.0302	1.0466	1.0629	1.0793	1.0956	1.1120	1.1283
12	.97776	.99406	1.0104	1.0267	1.0430	1.0592	1.0755	1.0918	1.1081	1.1244
13	.97436	.99061	1.0068	1.0231	1.0393	1.0556	1.0718	1.0880	1.1043	1.1205
14	.97097	.98715	1.0033	1.0195	1.0357	1.0519	1.0681	1.0843	1.1004	1.1166
15	.96757	.98370	.99983	1.0160	1.0321	1.0482	1.0643	1.0805	1.0966	1.1127
16	.96426	.98033	.99641	1.0125	1.0286	1.0446	1.0607	1.0768	1.0928	1.1089
17	.96086	.97688	.99290	1.0089	1.0247	1.0409	1.0570	1.0730	1.0890	1.1050
18	.95763	.97359	.98955	1.0055	1.0215	1.0374	1.0534	1.0694	1.0853	1.1013
19	.95431	.97022	.98613	1.0020	1.0179	1.0338	1.0497	1.0656	1.0816	1.0975
20	.95107	.96693	.98278	.99864	1.0145	1.0303	1.0462	1.0620	1.0779	1.0937
21	.94784	.96364	.97944	.99524	1.0110	1.0294	1.0426	1.0584	1.0742	1.0900
22	.94460	.96035	.97609	.99184	1.0076	1.0233	1.0391	1.0548	1.0706	1.0863
23	.94144	.95714	.97283	.98852	1.0042	1.0199	1.0356	1.0513	1.0670	1.0827
24	.93829	.95393	.96957	.98521	1.0008	1.0165	1.0321	1.0478	1.0634	1.0790
25	.93513	.95072	.96630	.98189	.99748	1.0131	1.0286	1.0442	1.0598	1.0754
26	.93197	.94750	.96304	.97858	.99411	1.0096	1.0252	1.0407	1.0562	1.0718
27	.92889	.94437	.95986	.97534	.99083	1.0063	1.0218	1.0373	1.0528	1.0682
28	.92581	.94124	.95668	.97211	.98754	1.0030	1.0184	1.0338	1.0493	1.0647
29	.92273	.93811	.95350	.96888	.98426	.99963	1.0150	1.0304	1.0458	1.0612
30	.91965	.93498	.95031	.96564	.98097	.99629	1.0116	1.0270	1.0423	1.0576
31	.91665	.93193	.94721	.96249	.97777	.99304	.10083	1.0236	1.0389	1.0542
32	.91365	.92888	.94411	.95934	.97457	.98979	1.0050	1.0203	1.0355	1.0507
33	.91065	.92583	.94101	.95619	.97137	.98654	1.0017	1.0169	1.0321	1.0473
34	.90773	.92286	.93800	.95313	.96826	.98338	.99851	1.0136	1.0288	1.0439
35	.90473	.91981	.93490	.94998	.96506	.98013	.99521	1.0103	1.0254	1.0405

°C	700	710	720	730	740	750	760	770	780	790
5	1.1695	1.1862	1.2029	1.2196	1.2363	1.2530	1.2697	1.2864	1.3031	1.3198
6	1.1652	1.1819	1.1985	1.2152	1.2318	1.2485	1.2651	1.2817	1.2984	1.3150
7	1.1611	1.1777	1.1943	1.2108	1.2274	1.2440	1.2606	1.2772	1.2938	1.3104
8	1.1569	1.1735	1.1900	1.2065	1.2230	1.2396	1.2561	1.2726	1.2892	1.3057
9	1.1529	1.1694	1.1858	1.2023	1.2188	1.2352	1.2517	1.2682	1.2846	1.3011
10	1.1487	1.1651	1.1816	1.1980	1.2144	1.2308	1.2472	1.2636	1.2800	1.2964
11	1.1447	1.1610	1.1774	1.1937	1.2101	1.2264	1.2428	1.2592	1.2755	1.2919
12	1.1407	1.1570	1.1733	1.1896	1.2059	1.2222	1.2385	1.2548	1.2711	1.2874
13	1.1368	1.1530	1.1692	1.1855	1.2017	1.2180	1.2342	1.2504	1.2667	1.2829
14	1.1328	1.1490	1.1652	1.1814	1.1975	1.2137	1.2299	1.2461	1.2623	1.2784
15	1.1288	1.1450	1.1611	1.1772	1.1933	1.2095	1.2256	1.2417	1.2579	1.2740
16	1.1250	1.1410	1.1571	1.1732	1.1893	1.2053	1.2214	1.2375	1.2535	1.2696
17	1.1210	1.1370	1.1530	1.1691	1.1851	1.2011	1.2171	1.2331	1.2491	1.2651
18	1.1172	1.1332	1.1492	1.1651	1.1811	1.1970	1.2130	1.2290	1.2449	1.2609
19	1.1134	1.1293	1.1452	1.1611	1.1770	1.1929	1.2088	1.2247	1.2406	1.2565
20	1.1096	1.1254	1.1413	1.1572	1.1730	1.1888	1.2047	1.2206	1.2364	1.2522
21	1.1058	1.1216	1.1374	1.1532	1.1690	1.1848	1.2006	1.2164	1.2322	1.2480
22	1.1020	1.1178	1.1335	1.1493	1.1650	1.1808	1.1965	1.2122	1.2280	1.2437
23	1.0984	1.1140	1.1294	1.1454	1.1611	1.1768	1.1925	1.2082	1.2239	1.2396
24	1.0947	1.1103	1.1259	1.1416	1.1572	1.1729	1.1885	1.2041	1.2198	1.2354
25	1.0910	1.1066	1.1222	1.1377	1.1533	1.1689	1.1845	1.2001	1.2157	1.2313
26	1.0873	1.1028	1.1184	1.1339	1.1494	1.1650	1.1805	1.1960	1.2116	1.2271
27	1.0837	1.0992	1.1147	1.1302	1.1456	1.1611	1.1766	1.1921	1.2076	1.2230
28	1.0801	1.0955	1.1110	1.1264	1.1418	1.1573	1.1727	1.1881	1.2036	1.2090
29	1.0765	1.0919	1.1073	1.1227	1.1380	1.1534	1.1688	1.1842	1.1996	1.2149
30	1.0729	1.0883	1.1036	1.1189	1.1342	1.1496	1.1649	1.1802	1.1956	1.2109
31	1.0694	1.0847	1.1000	1.1153	1.1305	1.1458	1.1611	1.1764	1.1917	1.2069
32	1.0659	1.0812	1.0964	1.1116	1.1268	1.1421	1.1573	1.1725	1.1878	1.2030
33	1.0624	1.0776	1.0928	1.1080	1.1231	1.1383	1.1535	1.1687	1.1839	1.1990
34	1.0590	1.0742	1.0893	1.1044	1.1195	1.1347	1.1498	1.1649	1.1801	1.1952
35	1.0555	1.0706	1.0857	1.1008	1.1158	1.1309	1.1460	1.1611	1.1762	1.1912

DENSITY OF DRY AIR

At the Temperature t, and under the Pressure H cm of Mercury the Density of Air

$$=\frac{0.001293}{1+0.00367\,t}\cdot\frac{H}{76}.$$

Units of this table are grams per milliliter
(From Miller's Laboratory Physics, Ginn & Co., publishers, by permission.)

t	Pressure H in Centimeters					
	72.0	73.0	74.0	75.0	76.0	77.0
°						
10	0.001182	0.001198	0.001215	0.001231	0.001247	0.001264
11	178	193	210	227	243	259
12	173	190	206	222	239	255
13	169	186	202	218	234	251
14	165	181	198	214	230	246
15	0.001161	0.001177	0.001193	0.001210	0.001226	0.001242
16	157	173	189	205	221	238
17	153	169	185	201	217	233
18	149	165	181	197	213	229
19	145	161	177	193	209	225
20	0.001141	0.001157	0.001173	0.001189	0.001205	0.001221
21	137	153	169	185	201	216
22	134	149	165	181	197	212
23	130	145	161	177	193	208
24	126	142	157	173	189	204
25	0.001122	0.001138	0.001153	0.001169	0.001185	0.001200
26	118	134	149	165	181	196
27	115	130	146	161	177	192
28	111	126	142	157	173	188
29	107	123	138	153	169	184
30	0.001104	0.001119	0.001134	0.001150	0.001165	0.001180

Proportional Parts

cm	17	16	15
0.1	2	2	1
0.2	3	3	3
0.3	5	5	4
0.4	7	6	6
0.5	8	8	7
0.6	10	10	9
0.7	12	11	10
0.8	14	13	12
0.9	15	14	13

Density of dry air at 20C and 760mm Hg = 1.204 mg/cm^3. (*Rev. Mod. Phys.*, 52, Part II, S33, 1980.)

DENSITY OF WATER

The temperature of maximum density for pure water, free from air = 3.98C (277.13K)

t, °C	d, gm/ml
0	0.99987
3.98	1.00000
5	0.99999
10	0.99973
15	0.99913
18	0.99862
20	0.99823
25	0.99707
30	0.99567
35	0.99406
38	0.99299
40	0.99224
45	0.99025
50	0.98807
55	0.98573
60	0.98324
65	0.98059
70	0.97781
75	0.97489
80	0.97183
85	0.96865
90	0.96534
95	0.96192
100	0.95838

THERMODYNAMIC AND TRANSPORT PROPERTIES OF AIR

From NASA Technical Note D-7488 by David J. Poferl and Roger Svehla (1973). The following three tables list the thermodynamic and transport properties of air over the temperature range of 300-2800K at pressures of 20, 30, and 40 atm. Factors for converting viscosity, specific heat at constant pressure, thermal conductivity, and enthalpy from cgs units to SI and English units are

Viscosity:

$$1\,\frac{\text{g}}{(\text{cm})(\text{sec})} = 0.1\,\frac{(\text{N})(\text{sec})}{\text{m}^2}$$

$$= 6.72\text{x}10^{-2}\,\frac{\text{lbm}}{(\text{ft})(\text{sec})}$$

$$= 241.9\,\frac{\text{lbm}}{(\text{ft})(\text{hr})}$$

$$= 2.089\text{x}10^{-3}\,\frac{(\text{lbf})(\text{sec})}{\text{ft}^2}$$

Thermal conductivity:

$$1\,\frac{\text{cal}}{(\text{cm})(\text{sec})(\text{K})} = 418.4\,\frac{\text{W}}{(\text{m})(\text{K})}$$

$$= 0.8064\,\frac{\text{Btu}}{(\text{ft})^2(\text{sec})(^\circ\text{F/in.})}$$

$$= 6.72\text{x}10^{-2}\,\frac{\text{Btu}}{(\text{ft}^2)(\text{sec})(^\circ\text{F/ft})}$$

$$= 241.9\,\frac{\text{Btu}}{(\text{ft})^2(\text{hr})(^\circ\text{F/ft})}$$

Specific heat at constant pressure:

$$1\,\frac{\text{cal}}{(\text{g})(\text{K})} = 4.184\,\frac{\text{J}}{(\text{g})(\text{K})}$$

$$= 1\,\frac{\text{Btu}}{(\text{lbm})(^\circ\text{F})}$$

Enthalpy:

$$1\,\frac{\text{cal}}{\text{g}} = 4.184\,\frac{\text{J}}{\text{g}}$$

$$= 1.8\,\frac{\text{Btu}}{\text{lbm}}$$

PROPERTIES AT 20 ATM

Temperature, T, K	Isentropic exponent, γ	Molecular weight, m	Viscosity, μ g/(cm)(sec)	Specific heat at constant pressure, c_p, cal/(g)(K)	Thermal conductivity, k, cal/(cm)(sec)(K)	Prandtl number, Pr	Enthalpy, h, cal/g
2800	1.2309	28.890	821×10^{-6}	0.3850	473×10^{-6}	0.669	747.0
2700	1.2374	28.915	800	.3707	439	.676	709.2
2600	1.2437	28.933	779	.3588	409	.683	672.7
2500	1.2498	28.945	758	.3488	384	.688	637.4
2400	1.2556	28.953	736	.3404	362	.692	602.9
2300	1.2612	28.958	715	.3332	343	.696	569.3
2200	1.2666	28.961	694	.3270	325	.698	536.2
2100	1.2719	28.963	672	.3214	309	.699	503.8
2000	1.2772	28.964	651	.3163	294	.700	471.9
1900	1.2825	28.965	629	.3115	280	.701	440.6
1800	1.2879		607	.3070	266	.702	409.6
1700	1.2933		585	.3025	252	.702	379.2
1600	1.2989		563	.2981	239	.703	349.1
1500	1.3045		540	.2939	226	.703	319.5
1400	1.3103		517	.2897	213	.704	290.3
1300	1.3162		494	.2855	200	.704	261.6
1200	1.3224		470	.2814	188	.705	233.2
1100	1.3288		445	.2773	175	.705	205.3
1000	1.3356		419	.2730	162	.705	177.8
900	1.3439	28.964	391	.2681	148	.706	150.7
800	1.3537		362	.2626	135		124.2
700	1.3646		331	.2568	121		98.2
600	1.3759		299	.2511	106		72.8
500	1.3865		265	.2461	92		48.0
400	1.3951		227	.2422	78		23.6
300	1.4000		184	.2401	63		-.5

PROPERTIES AT 30 ATM

Temperature, T, K	Isentropic exponent, γ	Molecular weight, m	Viscosity, μ, g/(cm)(sec)	Specific heat at constant pressure, c_p, cal/(g)(K)	Thermal conductivity, k, cal/(cm)(sec)(K)	Prandtl number, Pr	Enthalpy, h cal/g
2800	1.2340	28.904	821×10^{-6}	0.3773	460×10^{-6}	0.674	745.0
2700	1.2399	28.924	800	.3652	430	.680	707.9
2600	1.2456	28.939	779	.3548	403	.686	671.9
2500	1.2512	28.949	758	.3462	380	.690	636.8
2400	1.2566	28.956	736	.3387	359	.694	602.6
2300	1.2619	28.960	715	.3321	341	.696	569.1
2200	1.2671	28.962	694	.3263	324	.698	536.1
2100	1.2722	28.964	672	.3211	309	.699	503.8
2000	1.2773	28.965	651	.3161	294	.700	471.9
1900	1.2826		629	.3115	280	.701	440.5
1800	1.2879		607	.3069	266	.702	409.6
1700	1.2933		585	.3025	252	.702	379.2
1600	1.2988		563	.2981	239	.703	349.1
1500	1.3045		540	.2939	226	.703	319.5
1400	1.3103		517	.2897	213	.704	290.3
1300	1.3162		494	.2855	200	.704	261.6
1200	1.3223		470	.2814	188	.705	233.2
1100	1.3288		445	.2773	175	.705	205.3
1000	1.3356		419	.2730	162	.705	177.8
900	1.3439	28.964	391	.2681	148	.706	150.7
800	1.3537		362	.2626	135		124.2
700	1.3646		331	.2568	121		98.2
600	1.3759		299	.2511	106		72.8
500	1.3865		265	.2461	92		48.0
400	1.3951		227	.2422	78		23.6
300	1.4000		184	.2401	63		-.5

THERMODYNAMIC AND TRANSPORT PROPERTIES OF AIR *(Continued)*

PROPERTIES AT 40 ATM

Temperature, T, K	Isentropic exponent, γ	Molecular weight, m	Viscosity, μ, g/(cm)(sec)	Specific heat at constant pressure, c_p, cal/(g)(K)	Thermal conductivity, k, cal/(cm)(sec)(K)	Prandtl number, Pr	Enthalpy, h, cal/g
2800	1.2360	28.913	821×10^{-6}	0.3727	452×10^{-6}	0.677	743.8
2700	1.2414	28.930	800	.3619	424	.683	707.1
2600	1.2468	28.943	779	.3526	399	.688	671.4
2500	1.2521	28.951	758	.3446	377	.692	636.5
2400	1.2573	28.957	736	.3376	358	.695	602.4
2300	1.2623	28.961	715	.3315	340	.697	569.0
2200	1.2673	28.963	694	.3260	324	.699	536.1
2100	1.2724	28.964	672	.3209	308	.700	503.8
2000	1.2774	28.965	651	.3160	294	.700	471.9
1900	1.2826	↓	629	.3114	279	.701	440.5
1800	1.2879	↓	607	.3069	266	.702	409.6
1700	1.2933	↓	585	.3025	252	.702	379.2
1600	1.2988	↓	563	.2981	239	.703	349.1
1500	1.3045	↓	540	.2939	226	.703	319.5
1400	1.3103	↓	517	.2897	213	.704	290.4
1300	1.3162	↓	494	.2855	200	.704	261.6
1200	1.3223	↓	470	.2814	188	.705	233.2
1100	1.3288	↓	445	.2773	175	.705	205.3
1000	1.3356	↓	419	.2730	162	.705	177.8
900	1.3439	↓	391	.2681	148	.706	150.7
800	1.3537	28.964	362	.2626	135	↓	124.2
700	1.3646	↓	331	.2568	121	↓	98.2
600	1.3759	↓	299	.2511	106	↓	72.8
500	1.3865	↓	265	.2461	92	↓	48.0
400	1.3951	↓	227	.2422	78	↓	23.6
300	1.4000	↓	184	.2401	63	↓	-.5

SURFACE TENSION

SURFACE TENSION OF INORGANIC SOLUTES IN WATER

% = Weight % of solute
γ = Surface tension in dynes/cm.

Solute	T°C	% / γ							
HCl	20	%	1.78	3.52	6.78	12.81	16.97	23.74	35.29
		γ	72.55	72.45	72.25	71.85	71.75	70.55	65.75
HNO_3	20	%	4.21	8.64	14.99	34.87			
		γ	72.15	71.65	70.95	68.75			
H_2SO_4	25	%	4.11	8.26	12.18	17.66	21.88	29.07	33.63
		γ	72.21	72.55	72.80	73.36	73.91	74.80	75.29
$HClO_4$	25	%	4.86	10.01	20.38	30.36	53.74	63.47	72.25
		γ	71.18	70.34	69.21	68.57	69.02	69.73	69.01
KOH	18	%	2.73	5.31	10.08	17.57			
		γ	73.95	74.85	76.55	79.75			
NaOH	18	%	2.72	5.66	16.66	30.56	35.90		
		γ	74.35	75.85	83.05	96.05	101.05		
NH_4OH	18	%	1.72	3.39	4.99	9.51	17.37	34.47	54.37
		γ	71.65	70.65	69.95	67.85	65.25	61.05	57.05
KCl	20	%	0.74	3.60	6.93	13.88	18.77	22.97	24.70
		γ	72.99	73.45	74.15	75.55	76.95	78.25	78.75
LiCl	25	%	5.46	7.37	10.17	13.95			
		γ	74.23	75.10	76.30	78.10			
NaCl	20	%	0.58	2.84	5.43	10.46	14.92	22.62	25.92
		γ	72.92	73.75	74.39	76.05	77.65	80.95	82.55
PbCl	25	%	21.57	28.52	37.74				
		γ	75.20	76.80	79.20				
$BaCl_2$	25	%	9.26	16.73	25.58				
		γ	73.50	74.93	76.38				
$MgCl_2$	20	%	0.94	4.55	8.69	16.00	22.30	25.44	
		γ	73.07	74.00	75.75	79.15	82.95	85.75	
NaBr	20	%	4.89	9.33	13.37	23.00			
		γ	73.45	74.05	74.75	76.55			
$Al_2(SO_4)_3$	25	%	2.54	4.06	9.40	14.60	19.32	23.54	25.50
		γ	72.32	72.92	73.51	74.71	76.06	78.30	79.73
$MgSO_4$	20	%	1.19	5.68	10.75	19.41	24.53		
		γ	73.01	73.78	74.85	77.35	79.25		
Na_2SO_4	20	%	2.76	6.63	12.44				
		γ	73.25	74.15	75.45				
Na_2CO_3	20	%	2.58	5.03	9.59	13.72			
		γ	73.45	74.05	75.45	76.75			
$NaNO_3$	20	%	0.85	4.08	7.84	14.53	29.82	37.30	47.06
		γ	72.87	73.75	73.95	75.15	78.35	80.25	87.05

SURFACE TENSION OF INORGANIC SOLUTES IN ORGANIC SOLVENTS

% = Mol %
γ = Surface tension in dynes/cm.

Solute	Solvent	T°C	% / γ						
LiCl	Ethyl alcohol	14	%	0.72	2.30	4.62			
			γ	22.90	23.17	23.26			
LiBr	Ethyl alcohol	14	%	0.95	2.60				
			γ	23.08	23.35				
LiI	Ethyl alcohol	14	%	1.43	2.87	5.08	10.21	19.47	26.92
			γ	23.11	23.56	24.39	26.03	28.87	31.95
KI	Methyl alcohol	14	%	0.81	1.52	2.68			
			γ	23.76	24.11	24.71			
$CaCl_2$	Ethyl alcohol	24	%	0.94	1.98	3.79	7.47		
			γ	22.62	22.58	23.23	23.97		
NaI	Methyl alcohol	14	%	0.76	1.48	4.33	8.55	12.53	
			γ	22.83	23.29	24.85	27.41	29.75	
NaI	Ethyl alcohol	24	%	0.45	1.80	3.63	4.54	6.02	10.46
			γ	22.47	22.82	23.41	23.52	24.00	25.07
NaI	Acetone	14	%	0.93	2.08	5.07	6.53		
			γ	24.22	24.40	25.04	25.12		
ZnI_2	Methyl alcohol	22	%	0.90	2.79	5.07			
			γ	22.97	24.23	25.84			
ZnI_2	Ethyl alcohol	24	%	0.41	1.72	3.42	6.90		
			γ	22.70	22.90	23.71	25.49		
H_2SO_4	Ethyl ether	17	%	3	10	40	75	90	
			γ	17.30	19.55	32.50	46.30	46.83	
H_2SO_4	Nitrobenzene	17	%	3	10	40	75	90	
			γ	42.73	43.96	46.05	47.52	48.25	

SURFACE TENSION OF ORGANIC COMPOUNDS IN WATER

% = Weight % of solute
γ = Surface tension in dynes/cm.

Solute	T°C	% / γ							
Acetic acid	30	%	1.00	2.475	5.001	10.01	30.09	49.96	69.91
		γ	68.00	64.40	60.10	54.60	43.60	38.40	34.30
Acetone	25	%	5.00	10.00	20.00	50.00	75.00	95.00	100.00
		γ	55.50	48.90	41.10	30.40	26.80	24.20	23.00
Acetonitrile	20	%	1.13	3.35	11.77	20.20	37.58	61.33	81.22
		γ	69.02	63.03	47.61	39.06	31.84	30.02	29.02
o-Aminobenzoic acid	25	%	12.35	22.36	30.45	37.44			
		γ	71.96	73.23	74.54	75.79			
m-Aminobenzoic acid	25	%	12.35	22.36	30.45	37.44			
		γ	73.30	74.59	76.16	77.89			
p-Aminobenzoic acid	25	%	12.35	22.36	30.45	37.44			
		γ	73.38	74.79	76.32	78.20			
Aminobutyric acid	25	%	4.96	9.34	13.43				
		γ	71.91	71.67	71.40				
Ammonium lactate	29	%	30.00	50.00	60.00	70.00	80.00	90.00	
		γ	35.40	34.40	35.40	35.60	38.20	44.50	
n-Butanol	30	%	0.04	0.41	9.53	80.44	86.05	94.20	97.40
		γ	69.33	60.38	26.97	23.69	23.47	23.29	22.25
n-Butyric acid	25	%	0.14	0.31	1.05	8.60	25.00	79.00	100.00
		γ	69.00	65.00	56.00	33.00	28.00	27.00	26.00
Dioxan	26	%	0.44	2.20	4.70	11.14	20.17	35.20	55.00
		γ	69.83	65.64	62.45	56.90	51.57	45.30	39.27
Dioxan	26	%	67.68	76.45	83.02	91.90	95.60	97.77	
		γ	36.95	35.80	35.00	33.95	33.60	33.10	
Formic acid	30	%	1.00	5.00	10.00	25.00	50.00	75.00	100.00
		γ	70.07	66.20	62.78	56.29	49.50	43.40	36.51
Glycerol	18	%	5.00	10.00	20.00	30.00	50.00	85.00	100.00
		γ	72.90	72.90	72.40	72.00	70.00	66.00	63.00
Glycine	25	%	3.62	6.98	10.12	13.10			
		γ	72.54	73.11	73.74	74.18			
Hydrocinnamic acid	21.5	%	7.02	12.62	18.39	26.09	31.06	38.25	47.93
		γ	69.08	66.49	63.63	59.25	56.14	52.96	47.24
Methyl acetate	25	%	0.66	1.29	2.29	3.56			
		γ	66.33	62.92	58.22	55.08			
Morpholine	20	%	8.56	19.39	30.41	50.45	69.93	80.14	92.00
		γ	67.80	62.62	59.15	52.85	47.05	43.62	41.60
Potassium lactate	29	%	40.00	50.00	60.00	70.00			
		γ	66.40	66.40	65.40	63.40			
Phenol	20	%	0.024	0.047	0.118	0.417	0.941	3.76	5.62
		γ	72.60	72.20	71.30	66.50	61.10	46.00	42.30
n-Propanol	25	%	0.1	0.5	1.0	50.0	60.0	80.0	90.0
		γ	67.10	56.18	49.30	24.34	24.15	23.66	23.41
Propionic acid	25	%	1.91	5.84	9.80	21.70	49.80	73.90	100.00
		γ	60.00	49.00	44.00	36.00	32.00	30.00	26.00
Sodium lactate	29	%	1.00	10.00	30.00	40.00	50.00	60.00	70.00
		γ	70.40	69.60	68.50	64.80	45.40	56.70	60.70
Sucrose	25	%	10.00	20.00	30.00	40.00	55.00		
		γ	72.50	73.00	73.40	74.10	75.70		

SURFACE TENSION OF ORGANIC COMPOUNDS IN ORGANIC SOLVENTS

% = Weight % of solvent
γ = Surface tension dynes/cm.

Solute	Solvent	T°C	% γ					
Acetic acid	Benzene	35	%	10.45	25.53	34.28	43.93	68.77
			γ	25.40	25.21	25.32	25.43	25.99
Acetic acid	Acetone	25	%	25.63	50.83	75.62		
			γ	27.50	26.61	24.90		
Acetone	Ethyl ether	30	%	21.83	40.98	61.29	75.69	89.03
			γ	16.75	17.49	19.15	19.80	21.00
Acetonitrile	Ethyl alcohol	20	%	10.83	32.42	49.90	69.78	80.19
			γ	22.92	23.92	24.36	25.08	26.51
Aniline	Cyclohexane	32	%	14.35	37.65	50.67	72.28	96.46
			γ	24.21	24.51	24.50	25.61	37.45
Carbontetrachloride	Benzene	50	%	30.40	51.69	62.38	78.29	88.62
			γ	24.39	24.09	23.78	23.47	23.21
Cyclohexane	Nitrobenzene	15	%	5.00	10.00			
			γ	37.59	33.03			
Naphthalene	Benzene	79.5	%	20.00	40.00	50.00	60.00	80.00
			γ	23.42	26.70	27.80	29.20	31.70
Naphthalene	*p*-Nitrophenol	121	%	10.21	29.74	45.94	58.01	67.57
			γ	29.90	31.80	33.70	34.80	41.80

SURFACE TENSION OF METHYL ALCOHOL IN WATER

% = Volume % of alcohol
γ = Surface tension dynes/cm.

T°C	%	7.5	10.00	25.0	50.0	60.0	80.0	90.0	100.0
20	γ	60.90	59.04	46.38	35.31	32.95	27.26	25.36	22.65
30	γ	59.33	57.27	45.30	34.52	32.26	26.48	24.42	21.58
50	γ	56.19	55.01	43.24	32.95	30.79	25.01	22.55	19.52

SURFACE TENSION OF ETHYL ALCOHOL IN WATER

% = Volume % of alcohol
γ = Surface tension dynes/cm.

T°C	%	5.00	10.00	24.00	34.00	48.00	60.00	72.00	80.00	96.00
20	γ	—	—	—	33.24	30.10	27.56	26.28	24.91	23.04
40	γ	54.92	48.25	35.50	31.58	28.93	26.18	24.91	23.43	21.38
50	γ	53.35	46.77	34.32	30.70	28.24	25.50	24.12	22.56	20.40

WATER AGAINST AIR

Temperature °C	Surface tension dynes/cm.	Temperature °C	Surface tension dynes/cm.
−8	77.0	25	71.97
−5	76.4	30	71.18
0	75.6	40	69.56
5	74.9	50	67.91
10	74.22	60	66.18
15	73.49	70	64.4
18	73.05	80	62.6
20	72.75	100	58.9

INTERFACIAL TENSION

Surface Tension at the Interface Between Two Liquids
(Each liquid saturated with the other)

Liquids	Temperature °C	γ
Benzene — Mercury	20	357
Ethyl ether — Mercury	20	379
Water — Benzene	20	35.00
Water — Carbon tetrachloride	20	45.
Water — Ethyl ether	20	10.7
Water — Heptylic acid	20	7.0
Water — n-Hexane	20	51.1
Water — Mercury	20	375.
Water-n-Octane	20	50.8
Water-n-Octyl alcohol	20	8.5

SURFACE TENSION OF VARIOUS LIQUIDS

Substance: Name	Formula	In contact with	Temperature °C	Surface tension dynes/cm
Acetaldehyde	C_2H_4O	--vapor	20	21.2
Acetaldoxime	C_2H_5NO	--vapor	35	30.1
Acetamide	C_2H_5NO	--vapor	85	39.3
Acetanilide	C_2H_5NO	--vapor	120	35.6
Acetic acid	$C_2H_4O_2$	--vapor	10	28.8

SURFACE TENSION OF VARIOUS LIQUIDS (Continued)

Substance: Name	Substance: Formula	In contact with	Temperature °C	Surface tension dynes/cm
	C_2H_4O	--vapor	20	27.8
	C_2H_4O	--vapor	50	24.8
Acetic anhydride	$C_4H_6O_3$	--vapor	20	32.7
Acetone	C_3H_6O	--air or vapor	0	26.21
	C_3H_6O	--air or vapor	20	23.70
	C_3H_6O	--air or vapor	40	21.16
Acetonitrile	C_2H_3N	--vapor	20	29.30
Acetophenone	C_8H_8O	--vapor	20	39.8
Acetyl chloride	C_2H_3ClO	--vapor	14.8	26.7
Acetylene	C_2H_2	--vapor	−70.5	16.4
Acetylsalicylic acid (in aq. sol.)	$C_9H_8O_4$	--vapor	25.9	60.06
Allyl alcohol	C_3H_6O	--air or vapor	20	25.8
Allyl isothiocyanate	C_4H_5NS	--air or vapor	20	34.5
Ammonia	NH_3	--vapor	11.1	23.4
	NH_3	--vapor	34.1	18.1
Aniline	C_6H_7N	--air	10	44.10
	C_6H_7N	--vapor	20	42.9
	C_6H_7N	--air	50	39.4
Argon	A	--vapor	−188	13.2
Azoxybenzene	$C_{12}H_{10}N_2O$	--vapor	51	43.34
Benzaldehyde	C_7H_6O	--air	20	40.04
Benzene	C_6H_6	--air	10	30.22
	C_6H_6	--air	20	28.85
	C_6H_6	--saturated with vapor	20	28.89
	C_6H_6	--air	30	27.56
Benzonitrile	C_7H_5N	--air	20	39.05
Benzophenone	$C_{13}H_{10}O$	--air or vapor	20	45.1
Benzylamine	C_7H_9N	--vapor	20	39.5
Benzyl alcohol	C_7H_8O	--air or vapor	20	39.0
Bromine	Br_2	--air or vapor	20	41.5
Bromobenzene	C_6H_5Br	--air	20	36.5
Bromoform	$CHBr_3$	--vapor	20	41.53
p-Bromophenol	C_6H_5BrO	--vapor	74.4	42.36
d-sec-Butyl alcohol	$C_4H_{10}O$	--vapor	10	23.5
n-Butyl alcohol	$C_4H_{10}O$	--air or vapor	0	26.2
	$C_4H_{10}O$	--air or vapor	20	24.6
	$C_4H_{10}O$	--air or vapor	50	22.1
tert-Butyl alcohol	$C_4H_{10}O$	--air or vapor	20	20.7
n-Butylamine	$C_4H_{11}N$	--nitrogen	41	19.7
n-Butyric acid	$C_4H_8O_2$	--air	20	26.8
Carbon bisulfide	CS_2	--vapor	20	32.33
Carbon dioxide	CO_2	--vapor	20	1.16
	CO_2	--vapor	−25	9.13
Carbon tetrachloride	CCl_4	--vapor	20	26.95
	CCl_4	--vapor	100	17.26
	CCl_4	--vapor	200	6.53
Carbon monoxide	CO	--vapor	−193	9.8
	CO	--vapor	−203	12.1
Chloral	C_2HCl_3O	--vapor	19.4	25.34
Chlorine	Cl_2	--vapor	20	18.4
	Cl_2	--vapor	−30	25.4
	Cl_2	--vapor	−40	27.3
	Cl_2	--vapor	−50	29.2
	Cl_2	--vapor	−60	31.2
Chloroacetic acid	$C_2H_2Cl_2O_2$	--nitrogen	25.7	35.4
Chlorobenzene	C_4H_5Cl	--vapor	20	33.56
Chloroform	$CHCl_3$	--air	20	27.14
o-Chlorophenol	C_6H_5ClO	--vapor	12.7	42.25
Cyclohexane	C_6H_{12}	--air	20	25.5
Dichloroacetic acid	$C_2H_2Cl_2O_2$	--nitrogen	25.7	35.4
Dichloroethane	$C_2H_4Cl_2$	--air	35.0	23.4
Diethylamine	$C_4H_{11}N$	--air	56	16.4
Diethylaniline	$C_{10}H_{15}N$	--vapor	20	34.2
Diethyl carbonate	$C_5H_{10}O$	--air	20	26.31
Diethyl oxalate	$C_6H_{10}O_4$	--vapor	20	32.0
Diethyl phthalate	$C_{12}H_{14}O_4$	--vapor	20	37.5
Diethyl sulfate	$C_4H_{12}O_4S$	--air	13	34.61
Dimethylamine	C_2H_7N	--nitrogen	0	18.1
	C_2H_7N	--nitrogen	5	17.7
Dimethylaniline	C_8H_{11}	--air or vapor	20	36.6
1,5-Dimethyl-2-phenyl-3-pyrazolone	$C_{11}H_{12}N_2O$	--vapor	25.9	63.63
Dimethyl sulfate	$C_2H_6O_4S$	--air	18	40.12
Diphenylamine	$C_{12}H_{11}N$	--air or vapor	80	37.7
Ethyl acetate	$C_4H_8O_2$	--air	0	26.5
	$C_4H_8O_2$	--air	20	23.9
	$C_4H_8O_2$	--air	50	20.2
Ethyl acetoacetate	$C_6H_{10}O_3$	--air or vapor	20	32.51
Ethyl alcohol	C_2H_6O	--air	0	24.05
	C_2H_6O	--vapor	10	23.61
	C_2H_6O	--vapor	20	22.75
	C_2H_6O	--vapor	30	21.89
Ethylamine	C_2H_7N	--nitrogen	0	21.3
	C_2H_7N	--nitrogen	9.9	20.4

Substance		In contact with	Temperature °C	Surface tension dynes/cm
Name	Formula			
Ethylaniline	$C_8H_{11}N$	- -air or vapor	20	36.6
Ethylbenzene	C_8H_{10}	- -vapor	20	29.20
Ethylbenzoate	$C_9H_{10}O_2$	- -vapor	20	35.5
Ethyl bromide	C_2H_5Br	- -vapor	20	24.15
Ethyl chloroformate	$C_3H_5ClO_2$	- -vapor	15.1	27.5
Ethyl Cinnamate	$C_{11}H_{12}O_2$	- -air	20	38.37
Ethylene bromide	$C_2H_4Br_2$	- -vapor	20	38.37
Ethylene chloride	$C_2H_4Cl_2$	- -air	20	24.15
Ethylene oxide	C_2H_4O	- -vapor	−20	30.8
	C_2H_4O	- -vapor	0.0	27.6
	C_2H_4O	- -vapor	20	24.3
Ethyl ether	$C_4H_{10}O$	- -vapor	20	17.01
	$C_4H_{10}O$	- -vapor	50	13.47
Ethyl format	$C_3H_6O_2$	- -air or vapor	20	23.6
Ethyl iodide	C_2H_5I	- -vapor	20	29.4
Ethyl nitrate	$C_2H_5NO_3$	- -air or vapr	20	28.7
dl-Ethyl lactate	$C_5H_{10}O_3$	- -air	20	29.9
Ethyl mercaptan	C_2H_6S	- -air or vapor	20	22.5
Ethyl salicylate	$C_9H_{10}O_3$	- -vapor	20.5	38.33
Formamide	CH_3NO	- -vapor	20	58.2
Formic acid	CH_2O_2	- -air	20	37.6
Furfural	$C_5H_4O_2$	- -air or vapor	20	43.5
Gelatin solution (1%)		- -water	2.85	8.3
Glycerol	$C_3H_8O_3$	- -air	20	63.4
	$C_3H_8O_3$	- -air	90	58.6
	$C_3H_8O_3$	- -air	150	51.9
Glycol	$C_2H_6O_2$	- -air or vapor	20	47.7
Helium	He	- -vapor	−269	.12
	He	- -vapor	−270	.239
	He	- -vapor	−271.5	.353
n-Hexane	C_6H_{14}	- -air	20	18.43
Hydrazine	N_2H_4	- -vapor	25	91.5
Hydrogen	H_2	- -vapor	−255	2.31
Hydrogen cyanide	HCN	- -vapor	17	18.2
Hydrogen peroxide	H_2O_2	- -vapor	18.2	76.1
Isobutyl alcohol	$C_4H_{10}O$	- -vapor	20	23.0
Isobutylamine	$C_4H_{11}N$	- -air	68	17.6
Isobutyl chloride	C_4H_9Cl	- -air	20	21.94
Isobutyric acid	$C_4H_8O_2$	- -air or vapor	20	25.2
Isopentane	C_5H_{12}	- -air	20	13.72
Isopropyl alcohol	C_3H_8O	- -air or vapor	20	21.7
Methyl acetate	$C_3H_6O_2$	- -air or vapor	20	24.6
Methyl alcohol	CH_4O	- -air	0	24.49
	CH_4O	- -air	20	22.61
	CH_4O	- -vapor	50	20.14
Methylamine	CH_3NH_2	- -nitrogen	−12	22.2
	CH_3NH_2	- -vapor	−20	23.0
	CH_3NH_2	- -nitrogen	−70	29.2
N-Methylaniline	C_7H_9N	- -air or vapor	20	39.6
Methyl benzoate	$C_8H_8O_2$	- -air or vapor	20	37.6
Methyl chloride	CH_3Cl	- -air	20	16.2
Methyl ether	C_2H_6O	- -vapor	−10	16.4
	C_2H_6O	- -vapor	−40	21
Methylene chloride	CH_2Cl_2	- -air	20	26.52
Methylene iodide	CH_2I_2	- -air	20	50.76
Methyl ethyl ketone	C_4H_8O	- -air or vapor	20	24.6
Methyl formate	$C_2H_4O_2$	- -vapor	20	25.08
Methyl iodide	CH_3I	- -air	43.5	25.8
Methyl propionate	$C_4H_8O_2$	- -air or vapor	20	24.9
Methyl salicylate	$C_8H_8O_3$	- -nitrogen	94	31.9
Methyl sulfide	C_2H_9S	- -vapor	11.1	26.50
Naphthalene	$C_{10}H_8$	- -air or vapor	127	28.8
Neon	Ne	- -vapor	−248	5.50
Nitric acid (98.8%)	HNO_3	- -air	11.6	42.7
Nitrobenzene	$C_6H_5NO_2$	- -air or vapor	20	43.9
Nitroethane	$C_2H_5NO_2$	- -air or vapor	20	32.2
Nitrogen	N_2	- -vapor	−183	6.6
	N_2	- -vapor	−193	8.27
	N_2	- -vapor	−203	10.53
Nitrogen tetra oxide	N_2O_4	- -vapor	19.8	27.5
Nitromethane	CH_3NO_2	- -vapor	20	36.82
Nitrous oxide	N_2O	- -vapor	20	1.75
n-Octane	C_8H_{18}	- -vapor	20	21.80
n-Octyl alcohol	$C_8H_{18}O$	- -air	20	27.53
Oleic acid	$C_{18}H_{34}O_2$	- -air	20	32.50
Oxygen	O_2	- -vapor	−183	13.2
Oxygen (65%)	O_2	- -air	−190.5	12.2
	O_2	- -vapor	−193	15.7
	O_2	- -vapor	−203	18.3
Paraldehyde	$C_6H_{12}O_3$	- -air	20	25.9
Phenetole	$C_8H_{10}O$	- -vapor	20	32.74
Phenol	C_6H_6O	- -air or vapor	20	40.9
	C_6H_6O	- -air or vapor	30	39.88

SURFACE TENSION OF VARIOUS LIQUIDS (Continued)

Substance: Name	Formula	In contact with	Temperature °C	Surface tension dynes/cm
Phenylhydrazine	$C_6H_8N_2$	- -vapor	20	46.1
Phosphorus tribromide	PBr_3	- -air	24	45.8
Phosphorus trichloride	PCl_3	- -vapor	20	29.1
Phosphorus triiodide	PI_3	- -vapor	75.3	56.5
Propionic acid	$C_3H_6O_2$	- -vapor	20	26.7
n-Propyl acetate	$C_5H_{10}O_2$	- -air or vapor	20	24.3
n-Propyl alcohol	C_3H_8O	- -vapor	20	23.78
n-Propylamine	C_3H_9N	- -air	20	22.4
n-Propyl bromide	C_3H_7Br	- -vapor	71	19.65
n-Propyl chloride	C_3H_7Cl	- -air	47	18.2
n-Propyl formate	$C_4H_8O_2$	- -vapor	20	24.5
Pyridine	C_8H_5N	- -air	20	38.0
Quinoline	C_9H_7N	- -air	20	45.0
Ricinoleic acid	$C_{18}H_{34}O_3$	- -air	16	35.81
Selenium	Se	- -air	217	92.4
Styrene	C_8H_8	- -air	19	32.14
Sulfuric acid (98.5%)	H_2SO_4	- -air or vapor	20	55.1
Tetrabromoethane 1,1,2,2-	$C_2H_2Br_4$	- -air	20	49.67
Tetrachloroethane 1,1,2,2-	$C_2H_2Cl_4$	- -air	22.5	36.03
Tetrachloroethylene	C_2Cl_4	- -vapor	20	31.74
Toluene	C_7H_8	- -vapor	10	27.7
	C_7H_8	- -vapor	20	28.5
	C_7H_8	- -vapor	30	27.4
m-Toluidine	C_7H_9N	- -vapor	20	36.9
o-Toluidine	C_7H_9N	- -air or vapor	20	40.0
p-Toluidine	C_7H_9N	- -air	50	34.6
Trichloroacetic acid	$C_2HCl_3O_2$	- -nitrogen	80.2	27.8
Trichloroethane 1,1,2-	$C_2H_3Cl_3$	- -air	114	22.0
Triethyl phosphate	$C_6H_{15}O_4P$	- -air	15.5	30.61
Trimethylamine	C_3H_9N	- -nitrogen	−4	17.3
Triphenylcarbinol	$C_{19}H_{16}O$	- -vapor	165.8	30.38
Vinyl acetate	$C_4H_6O_2$	- -vapor	20	23.95
	$C_4H_6O_2$	- -vapor	25	23.16
	$C_4H_6O_2$	- -vapor	30	22.54
Water	H_2O	- -air	18	73.05
m-Xylene	C_8H_{10}	- -vapor	20	28.9
o-Xylene	C_8H_{10}	- -air	20	30.10
p-Xylene	C_8H_{10}	- -vapor	20	28.37

VISCOSITY

Viscosity. — All fluids possess a definite resistance to change of form and many solids show a gradual yielding to forces tending to change their form. This property, a sort of internal friction, is called viscosity; it is expressed in dyne-seconds per cm² or poises. Dimensions, —$[m\, l^{-1}\, t^{-1}]$. If the tangential force per unit area, exerted by a layer of fluid upon one adjacent is one dyne for a space rate of variation of the tangential velocity of unity, the viscosity is one poise.

Kinematic viscosity is the ratio of viscosity to density. The c.g.s. unit of kinematic viscosity is the **stoke.**

Flow of liquids through a tube; where l is the length of the tube, r its radius, p the difference of pressure at the ends, η the coefficient of viscosity, the volume escaping per second,

$$\nu = \frac{\pi p r^4}{8 l \eta} \text{ (Poiseuille).}$$

The volume will be given in cm³ per second if l and r are in cm, p in dynes per cm² and η in poises or dyne-seconds per cm².

VISCOSITY OF WATER BELOW 0°C

White-Twining 1914

Temperature	Viscosity centipoises	Temperature	Viscosity centipoises
0°C	1.798	−7.23	2.341
−2.10	1.930	−8.48	2.458
−4.70	2.121	−9.30	2.549
−6.20	2.250		

ABSOLUTE VISCOSITY OF WATER AT 20°C

Swindells, J. R. Coe, Jr., and T. B. Godfrey, Journal of Research, National Bureau of Standards **48,** 1, 1952.
The value found for the viscosity of water at 20°C was 0.010019 ± 0.000003 poise.
The value **0.01002** poise is to be used as the absolute value of the viscosity of water for calibration purposes.

VISCOSITY CONVERSION TABLE

Poise = c.g.s. unit of absolute viscosity = $\frac{\text{gm}}{\text{sec} \times \text{cm}}$

Stoke = c.g.s. unit of kinematic viscosity = $\frac{\text{gm}}{\text{sec} \times \text{cm} \times \text{density } (t^\circ F)}$

Centipoise = 0.01 poise

Centistoke = 0.01 stoke

Centipoises = Centistokes × density (at given temperature)

To convert poises to $\frac{\text{lb}}{\text{sec} \times \text{ft}}$ or $\frac{\text{lb}}{\text{hr} \times \text{ft}}$ multiply by 0.0672 or 242 respectively.

Centi-stokes	Saybolt Seconds at			Redwood Seconds at			Engler Degrees at all temps.
	100°F	130°F	210°F	70°F	140°F	200°F	
2.0	32.6	32.7	32.8	30.2	31.0	31.2	1.14
3.0	36.0	36.1	36.3	32.7	33.5	33.7	1.22
4.0	39.1	39.2	39.4	35.3	36.0	36.3	1.31
5.0	42.3	42.4	42.6	37.9	38.5	38.9	1.40
6.0	45.5	45.6	45.8	40.5	41.0	41.5	1.48
7.0	48.7	48.8	49.0	43.2	43.7	44.2	1.56
8.0	52.0	52.1	52.4	46.0	46.4	46.9	1.65
9.0	55.4	55.5	55.8	48.9	49.1	49.7	1.75
10.0	58.8	58.9	59.2	51.7	52.0	52.6	1.84
11.0	62.3	62.4	62.7	54.8	55.0	55.6	1.93
12.0	65.9	66.0	66.4	57.9	58.1	58.8	2.02
14.0	73.4	73.5	73.9	64.4	64.6	65.3	2.22
16.0	81.1	81.3	81.7	71.0	71.4	72.2	2.43
18.0	89.2	89.4	89.8	77.9	78.5	79.4	2.64
20.0	97.5	97.7	98.2	85.0	85.8	86.9	2.87
22.0	106.0	106.2	106.7	92.4	93.3	94.5	3.10
24.0	114.6	114.8	115.4	99.9	100.9	102.2	3.34
26.0	123.3	123.5	124.2	107.5	108.6	110.0	3.58
28.0	132.1	132.4	133.0	115.3	116.5	118.0	3.82
30.0	140.9	141.2	141.9	123.1	124.4	126.0	4.07
32.0	149.7	150.0	150.8	131.0	132.3	134.1	4.32
34.0	158.7	159.0	159.8	138.9	140.2	142.2	4.57
36.0	167.7	168.0	168.9	146.9	148.2	150.3	4.83
38.0	176.7	177.0	177.9	155.0	156.2	158.3	5.08
40.0	185.7	186.0	187.0	163.0	164.3	166.7	5.34
42.0	194.7	195.1	196.1	171.0	172.3	175.0	5.59
44.0	203.8	204.2	205.2	179.1	180.4	183.3	5.85
46.0	213.0	213.4	214.5	187.1	188.5	191.7	6.11
48.0	222.2	222.6	223.8	195.2	196.6	200.0	6.37
50.0	231.4	231.8	233.0	203.3	204.7	208.3	6.63
60.0	277.4	277.9	279.3	243.5	245.3	250.0	7.90
70.0	323.4	324.0	325.7	283.9	286.0	291.7	9.21
80.0	369.6	370.3	372.2	323.9	326.6	333.4	10.53
90.0	415.8	416.6	418.7	364.4	367.4	375.0	11.84
*100.0	462.0	462.9	465.2	404.9	408.2	416.7	13.16

* At higher values use the same ratio as above for 100 centistokes; e.g., 110 centistokes = 110 × 4.620 Saybolt seconds at 100°F.
To obtain the Saybolt Universal viscosity equivalent to a kinematic viscosity determined at t °F, multiply the equivalent Saybolt Universal viscosity at 100°F by 1 + (t−100)0.000064; e.g., 10 centistokes at 210°F are equivalent to 58.8 × 1.0070, or 59.2 Saybolt Universal seconds at 210°F.

VISCOSITY CONVERSION

Kinematic

To convert from	To	Multiply by
cm^2/sec (Stokes)	Centistokes	10^2
	ft^2/hr.	3.875
	ft^2/sec.	1.076×10^{-3}
	in.^{2}/sec.	1.550×10^{-1}
	m^2/hr.	3.600×10^{-1}
cm^2/sec × 10^2 (Centistokes)	cm^2/sec. (Stokes)	1×10^{-2}
	ft^2/hr.	3.875×10^{-2}
	ft^2/sec.	1.076×10^{-5}
	in.^{2}/sec.	1.550×10^{-3}
	m^2/hr.	3.600×10^{-3}
ft^2/hr	cm^2/sec (Stokes)	2.581×10^{-1}
	cm/sec × 10^2 (centistokes)	2.581×10
	ft^2/sec.	2.778×10^{-4}
	in.^{2}/sec.	4.00×10^{-2}
	m^2/hr.	9.290×10^{-2}
ft^2/sec	cm^2/sec. (Stokes)	9.29×10^2
	cm^2/sec. × 10^2 (centistokes)	9.29×10^4
	ft^2/hr	3.60×10^3
	in.^{2}/sec.	1.44×10^2
	m^2/hr.	3.345×10^2
in.^{2}/sec	cm^2/sec. (Stokes)	6.452
	cm^2/sec. × 10^2 (centistokes)	6.452×10^2
	ft^2/hr.	2.50×10
	ft^2/sec.	6.944×10^{-3}
	m^2/hr.	2.323
m^2/hr	cm^2/sec. (Stokes)	2.778
	cm^2/sec. × 10^2 (centistokes)	2.778×10^2
	ft^2/hr.	1.076×10
	ft^2/sec.	2.990×10^{-3}
	in.^{2}/sec.	4.306×10^{-1}

VISCOSITY CONVERSION

Absolute

Absolute viscosity = kinematic viscosity × density; lb = mass pounds; lb_F = force pounds

To convert from	To	Multiply by
gm/(cm)(sec) [Poise]	gm/(cm)(sec)(10^2) [Centipoise]	10^2
	kg/(m)(hr)	3.6×10^2
	lb/(ft)(sec)	6.72×10^{-2}
	lb/(ft)(hr)	2.419×10^2
	lb/(in.)(sec)	5.6×10^{-3}
	(gm_F)(sec)/cm^2	1.02×10^{-3}
	(lb_F)(sec)/in.^{2} [Reyn]	1.45×10^{-5}
	(lb_F)(sec)/ft^2	2.089×10^{-3}
gm/(cm)(sec)(10^2) [Centipoise]	gm/(cm)(sec) [Poise]	10^{-2}
	kg/(m)(hr)	3.6
	lb/(ft)(sec)	6.72×10^{-4}
	lb/(ft)(hr)	2.419
	lb/(in.)(sec)	5.60×10^{-5}
	(gm_F)(sec)/cm^2	1.02×10^{-5}
	(lb_F)(sec)/in.^{2} [Reyn]	1.45×10^{-7}
	(lb_F)(sec)/ft^2	2.089×10^{-5}
kg/(m)(hr)	gm/(cm)(sec)	2.778×10^{-3}
	gm/(cm)(sec)(10^2) [Centipoise]	2.778×10^{-1}
	lb/(ft)(sec)	1.867×10^{-4}
	lb/(ft)(hr)	6.720×10^{-1}
	lb/(in.)(sec)	1.555×10^{-5}
	(gm_F)(sec)/cm^2	2.833×10^{-4}
	(lb_F)(sec)/in.^{2} [Reyn]	4.029×10^{-8}
	(lb_F)(sec)/ft^2	5.801×10^{-6}

VISCOSITY CONVERSION

Absolute (Continued)

To convert from	To	Multiply by
lb/(ft)(sec)	gm/(cm)(sec) [Poise]	1.488×10^{1}
	gm/(cm)(sec)(10^2) [Centipoise]	1.488×10^{3}
	kg/(m)(hr)	5.357×10^{3}
	lb/(ft)(hr)	3.60×10^{3}
	lb/(in.)(sec)	8.333×10^{-2}
	(gm_F)(sec)/cm^2	1.518×10^{-2}
	(lb_F)(sec)/in.^{2} [Reyn]	2.158×10^{-4}
	(lb_F)(sec)/ft^2	3.108×10^{-2}
lb/(ft)(hr)	gm/(cm)(sec) [Poise]	4.134×10^{-3}
	gm/(cm)(sec)(10^2) [Centipoise]	4.134×10^{-1}
	kg/(m)(hr)	1.488
	lb/(ft)(sec)	2.778×10^{-4}
	lb/(in.)(sec)	2.315×10^{-5}
	(gm_F)(sec)/cm^2	4.215×10^{-6}
	(lb_F)(sec)/in.^{2} [Reyn]	5.996×10^{-8}
	(lb_F)(sec)/ft^2	8.634×10^{-6}
lb/(in.)(sec)	gm/(cm)(sec) [Poise]	1.786×10^{2}
	gm/(cm)(sec)(10^2) [Centipoise]	1.786×10^{4}
	kg/(m)(hr)	6.429×10^{4}
	lb/(ft)(sec)	1.20×10
	lb/(ft)(hr)	4.32×10^{4}
	(gm_F)(sec)/cm^2	1.821×10^{-1}
	(lb_F)(sec)/in.^{2} [Reyn]	2.590×10^{-3}
	(lb_F)(sec)/ft^2	3.73×10^{-1}
(gm_F)(sec)/cm^2	gm/(cm)(sec)	9.807×10^{2}
	gm/(cm)(sec)(10^2) [Centipoise]	9.807×10^{4}
	kg/(m)(hr)	3.530×10^{5}
	lb/(ft)(sec)	6.590×10
	lb/(ft)(hr)	2.372×10^{5}
	lb/(in.)(sec)	5.492
	(lb_F)(sec)/in.^{2} [Reyn]	1.422×10^{-2}
	(lb_F)(sec)/ft^2	2.048
(lb_F)(sec)/in.^{2} [Reyn]	gm/(cm)(sec) [Poise]	6.895×10^{4}
	gm/(cm)(sec)(10^2) [Centipoise]	6.895×10^{6}
	kg/(m)(hr)	2.482×10^{7}
	lb/(ft)(sec)	4.633×10^{3}
	lb/(ft)(hr)	1.668×10^{7}
	lb/(in.)(sec)	3.861×10^{2}
	(gm_F)(sec)/cm^2	7.031×10
	(lb_F)(sec)/ft^2	1.440×10^{2}
(lb_F)(sec)/ft^2	gm/(cm)(sec) [Poise]	4.788×10^{2}
	gm/(cm)(sec)(10^2) [Centipoise]	4.788×10^{4}
	kg/(m)(hr)	1.724×10^{5}
	lb/(ft)(sec)	3.217×10
	lb/(ft)(hr)	1.158×10^{5}
	lb/(in.)(sec)	2.681
	(gm_F)(sec)/cm^2	4.882×10^{-1}
	(lb_F)(sec)/in.^{2} [Reyn]	6.944×10^{-3}

THE VISCOSITY OF WATER 0°C TO 100°C

Contribution from the National Bureau of Standards, not subject to copyright.

°C	η(cp)	°C	η(cp)	°C	η(cp)	°C	η(cp)
0	1.787	26	0.8705	52	0.5290	78	0.3638
1	1.728	27	.8513	53	.5204	79	.3592
2	1.671	28	.8327	54	.5121	80	.3547
3	1.618	29	.8148	55	.5040	81	.3503
4	1.567	30	.7975	56	.4961	82	.3460
5	1.519	31	.7808	57	.4884	83	.3418
6	1.472	32	.7647	58	.4809	84	.3377
7	1.428	33	.7491	59	.4736	85	.3337
8	1.386	34	.7340	60	.4665	86	.3297
9	1.346	35	.7194	61	.4596	87	.3259
10	1.307	36	.7052	62	.4528	88	.3221
11	1.271	37	.6915	63	.4462	89	.3184
12	1.235	38	.6783	64	.4398	90	.3147
13	1.202	39	.6654	65	.4335	91	.3111
14	1.169	40	.6529	66	.4273	92	.3076
15	1.139	41	.6408	67	.4213	93	.3042
16	1.109	42	.6291	68	.4155	94	.3008
17	1.081	43	.6178	69	.4098	95	.2975
18	1.053	44	.6067	70	.4042	96	.2942
19	1.027	45	.5960	71	.3987	97	.2911
20	1.002	46	.5856	72	.3934	98	.2879
21	0.9779	47	.5755	73	.3882	99	.2848
22	.9548	48	.5656	74	.3831	100	.2818
23	.9325	49	.5561	75	.3781		
24	.9111	50	.5468	76	.3732		
25	.8904	51	.5378	77	.3684		

The above table was calculated from the following empirical relationships derived from measurements in viscometers calibrated with water at 20°C (and one atmosphere), modified to agree with the currently accepted value for the viscosity at 20° of 1.002 cp:

$$0^\circ \text{ to } 20^\circ\text{C: } \log_{10} \eta_T = \frac{1301}{998.333 + 8.1855(T-20) + 0.00585(T-20)^2} - 1.30233$$

(R. C. Hardy and R. L. Cottington, J.Res.NBS *42*, 573 (1949).)

$$20^\circ \text{ to } 100^\circ\text{C: } \log_{10} \frac{\eta_T}{\eta_{20}} = \frac{1.3272(20-T) - 0.001053(T-20)^2}{T + 105}$$

(J. F. Swindells, NBS, unpublished results.)

VISCOSITY OF LIQUIDS

Viscosity of liquids in centipoises (cp) including elements, inorganic and organic compounds and mixtures.

Liquid	Temp. °C	Viscosity cp
Acetaldehyde	0	.2797
	10	.2557
	20	.22
Acetanilide	120	2.22
	130	1.90
Acetic acid	15	1.31
	18	1.30
	25.2	1.155
	30	1.04
	41	1.00
	59	.70
	70	.60
	100	.43
anhydride	0	1.24
	15	.971
	18	.90
	30	.783
	100	.49
Acetone	-92.5	2.148
	-80.0	1.487
	-59.6	.932
	-42.5	.695
	-30.0	.575
	-20.9	.510
	-13.0	.470
	-10.0	.450
	0	.399
	15	.337
	25	.316
	30	.295
	41	.280
Acetonitrile	0	.442
	15	.375
	25	.345
Acetophenone	11.9	2.28
	23.5	1.59
	25.0	1.617
	50.0	1.246
	80.0	.734
Air, liq	-192.3	.172
Alcohol. See *Ethyl, Methyl,* etc.		
Allyl alcohol	0	2.145
	15	1.49
	20	1.363
	30	1.07
	40	.914
	70	.553
Allylamine	130	.506
Allyl chloride	15	.347
	30	.300
Ammonia	-69	.475
	-50	.317
	-40	.276
	-33.5	.255
n-Amyl acetate	11	1.58
	45	.805
alcohol	15	4.65
	30	2.99
ether	15	1.188
Aniline	-6	13.8
	0	10.2
	5	8.06
	10	6.50
	15	5.31
	20	4.40
	25	3.71
	30	3.16
	35	2.71
	40	2.37
	50	1.85
	60	1.51
	70	1.27
	80	1.09
	90	.935
	100	.825
Anisol	0	1.78
	20	1.32
	40	1.12
Antimony, liq	645	1.55
	700	1.26
	800	1.08
	850	1.05
Benzaldehyde	25	1.39
Benzene	0	.912
	10	.758
	20	.652
	30	.564
	40	.503
	50	.442
	60	.392
	70	.358
	80	.329
Benzonitrile	25	1.24
Benzophenone	55	4.79
	120	1.38
Benzyl alcohol	20	5.8
Benzylamine	25	1.59
Benzylaniline	33	2.18
	130	1.20
Benzyl ether	0	10.5
	20	5.33
	40	3.21
Bismuth	285	1.61
	304	1.662
	365	1.46
	451	1.280
	600	.998
Bromine, liq	-4.3	1.31
	0	1.241
	12.6	1.07
	16	1.0
	19.5	.995
	28.9	.911
o-Bromoaniline	40	3.19
m-Bromoaniline	20	6.81
	40	3.70
	80	1.70
p-Bromoaniline	80	1.81
Bromobenzene	15	1.196
	30	.985
Bromoform	15	2.152
	25	1.89
	30	1.741
Butyl acetate	0	1.004
	20	.732
	40	.563
n-Butyl alcohol	-50.9	36.1
	-30.1	14.7
	-22.4	11.1
	-14.1	8.38
	0	5.186
	15	3.379
	20	2.948
	30	2.30
	40	1.782
	50	1.411
	70	.930
	100	.540
sec-Butyl alcohol	15	4.21
n-Butyl bromide	15	.626
n-Butyl chloride	15	.469
Butyl chloride, tertiary	15	.543
n-Butyl formate	0	.940
	20	.689
Butyric acid	0	2.286
	15	1.81
	20	1.540
	40	1.120
	50	.975
	70	.760
	100	.551
Cadmium, liq	349	1.44
	506	1.18
	603	1.10
Carbolic acid. See *Phenol.*		
Carbon dioxide, liq., pressure that of saturated vapor	0	.099
	10	.085
	20	.071
	30	.053

Liquid	Temp. °C	Viscosity cp
disulfide	−13	.514
	−10	.495
	0	.436
	5	.380
	20	.363
	40	.330
Carbon tetrachloride	0	1.329
	15	1.038
	20	.969
	30	.843
	40	.739
	50	.651
	60	.585
	70	.524
	80	.468
	90	.426
	100	.384
Cetyl alcohol	50	13.4
Chlorine, liq	−76.5	.729
	−70.5	.680
	−60.2	.616
	−52.4	.566
	−35.4	.494
	0	.385
Chlorobenzene	15	.900
	20	.799
	40	.631
	80	.431
	100	.367
Chloroform	−13	.855
	0	.700
	8.1	.643
	15	.596
	20	.58
	25	.542
	30	.514
	39	.500
o-Chlorophenol	25	4.11
	50	2.015
m-Chlorophenol	25	11.55
p-Chlorophenol	50	4.99
Copper, liq	1,085	3.36
	1,100	3.33
	1,150	3.22
	1,200	3.12
o-Cresol	40	4.49
m-Cresol	10	43.9
	20	20.8
	40	6.18
p-Cresol	40	7.00
Creosote	20	12.0
Cycloheptane	13.5	1.64
Cyclohexane	17	1.02
Cyclohexanol	20	68
Cyclohexene	13.5	.696
	20	.66
Cyclooctane	13.5	2.35
Cyclopentane	13.5	.493
n Decane	20	.92
Diethylamine	25	.346
	25	.367
Diethylaniline	.5	3.84
	20.0	2.18
	25.0	1.95
Diethylcarbinol	15.0	7.34
Diethylketone	15	.493
Dimethylaniline	10	1.69
	20	1.41
	25	1.285
	30	1.17
	40	1.04
	50	.91
Dimethyl-α-naphthylamine	130	.868
Dimethyl-β-naphthylamine	130	.952
Diphenyl	70	1.49
	100	.97
Diphenylamine	130	1.04
Dodecane	25	1.35
Ether (diethyl-)	−100	1.69
	−80	.958
	−60	.637
	−40	.461
	−20	.362
	0	.2842
	17	.240
	20	.2332
	25	.222
	40	.197
	60	.166
	80	.140
	100	.118
Ethyl acetate	0	.582
	8.96	.516
	10	.512
	15	.473
	20	.455
	25	.441
	30	.400
	50	.345
	75	.283
Ethyl alcohol	−98.11	44.0
	−89.8	28.4
	−71.5	13.2
	−59.42	8.41
	−52.58	6.87
	−32.01	3.84
	−17.59	2.68
	−.30	1.80
	0	1.773
	10	1.466
	20	1.200
	30	1.003
	40	.834
	50	.702
	60	.592
	70	.504
Ethyl alcohol, anh.	−148	8,470
	−146	5,990
	−130	467
Ethyl aniline	25	2.04
Ethylbenzene	17	.691
Ethyl benzoate	20	2.24
Ethyl bromide	−120	5.6
	−100	2.89
	−80	1.81
	0	.487
	10	.441
	15	.418
	20	.402
	30	.348
n-Ethyl butyrate	15	.711
Ethyl carbonate	15	.868
Ethylene bromide	0	2.438
	17	1.95
	20	1.721
	40	1.286
	67.3	.922
	70	.903
	82.2	.750
	99.0	.648
chloride	0	1.077
	15	.887
	19.4	.800
	40	.652
	50	.565
	70	.479
glycol	20	19.9
	40	9.13
	60	4.95
	80	3.02
	100	1.99
oxide	−49.8	.577
	−38.2	.488
	−21.0	.394
	0	.320
Ethyl formate	20	.402
iodide	0	.727
	15	.617
	20	.592
	40	.495
	70	.391
malate	24.7	3.016
oxalate	15	2.31

Liquid	Temp. °C	Viscosity cp	Liquid	Temp. °C	Viscosity cp
propionate	15	.564	Isopropyl alcohol	15	2.86
Eugenol	0	29.9		30	1.77
	20	9.22	Isoquinoline	25	3.57
	40	4.22	Isosafrol	25	3.981
Fluorobenzene	20	.598	Lead, liq	350	2.58
	40	.478		400	2.33
	60	.389		441	2.116
	80	.329		500	1.84
	100	.275		551	1.70
Formamide	0	7.55		600	1.38
	25	3.30		703	1.349
Formic acid	7.59	2.3868		844	1.185
	10	2.262	Menthol, liq	55.6	6.29
	20	1.804		74.6	2.47
	30	1.465		99.0	1.04
	40	1.219	Mercury	−20	1.855
	70	.780		−10	1.764
	100	.549		0	1.685
Furfural	0	2.48		10	1.615
	25	1.49		19.02	1.56
Glucose	22	9.1×10^{15}		20	1.554
	30	6.6×10^{13}		20.2	1.55
	40	2.8×10^{11}		30	1.499
	60	9.3×10^{7}		40	1.450
	80	6.6×10^{5}		40.8	1.45
	100	2.5×10^{4}		41.86	1.44
Glycerin	−42	6.71×10^{6}		50	1.407
	−36	2.05×10^{6}		60	1.367
	−25	2.62×10^{5}		70	1.331
	−20	1.34×10^{5}		80	1.298
	−15.4	6.65×10^{4}		90	1.268
	−10.8	3.55×10^{4}		100	1.240
	−4.2	1.49×10^{4}		150	1.130
	0	12,110		200	1.052
	6	6,260		250	.995
	15	2,330		300	.950
	20	1,490		340	.921
	25	954	Methyl acetate	0	.484
	30	629		20	.381
Glycerin trinitrate	10	69.2		40	.320
	20	36.0	Methyl alcohol	−98.30	13.9
	30	21.0	(Methanol)	−84.23	6.8
	40	13.6		−72.55	4.36
	60	6.8		−44.53	1.98
Heptane	0	.524		−22.29	1.22
	17	.461		0	.82
	20	.409		15	.623
	25	.386		20	.597
	40	.341		25	.547
	70	.262		30	.510
n-Heptyl alcohol	15	8.53		40	.456
Hexadecane	20	3.34		50	.403
Hexane	0	.401	Methyl amine	0	.236
	17	.374	aniline	25	2.02
	20	.326		30	1.55
	25	.294	chloride	20	.1834
	40	.271	Methylene bromide	15	1.09
	50	.248		30	0.92
Hydrazine	1	1.29	chloride	15	.449
	10	1.12		30	.393
	20	.97	Methyl iodide	0	.606
Hydrogen, liq	....	.011		15	.518
Iodine, liq	116	2.27		20	.500
Iodobenzene	15	1.74		30	.460
Iron, 2.5% carbon, liq	1,400	2.25		40	.424
Isoamyl acetate	8.97	1.030	Naphthalene	80	.967
	19.91	.872		100	.776
alcohol	10	6.20	Nitric acid	0	2.275
amine	25	.724		10	1.770
Isobutyl alcohol	15	4.703	Nitrobenzene	2.95	2.91
amine	25	.553		5.69	2.71
Isobutyric acid	15	1.44		5.94	2.71
	30	1.13		9.92	2.48
Isoeugenol	25	26.72		14.94	2.24
Isoheptane	0	.481		20.00	2.03
	20	.384	Nitromethane	0	.853
	40	.315		25	.620
Isohexane	0	.376	*o*-Nitrotoluene	0	3.83
	20	.306		20	2.37
	40	.254		40	1.63
Isopentane	0	.273		60	1.21
	20	.223	*m*-Nitrotoluene	20	2.33

Liquid	Temp. °C	Viscosity cp
	40	1.60
	60	1.18
p-Nitrotoluene	60	1.20
n-Nonane	20	.711
n-Octane	0	.706
	16	.574
	20	.542
	40	.433
Octodecane	40	2.86
n-Octylalcohol	15	10.6
Oil, castor	10	2,420
	20	986
	30	451
	40	231
	100	16.9
cottonseed	20	70.4
cylinder, filtered	37.8	240.6
	100	18.7
cylinder, dark	37.8	422.4
	100	24.0
linseed	30	33.1
	50	17.6
	90	7.1
machine, light	15.6	113.8
	37.8	34.2
	100	4.9
machine, heavy	15.6	660.6
	37.8	127.4
Oil, olive	10	138.0
	20	84.0
	40	36.3
	70	12.4
rape	0	2,530
	10	385
	20	163
	30	96
soya bean	20	69.3
	30	40.6
	50	20.6
	90	7.8
sperm	15.6	42.0
	37.8	18.5
	100.0	4.6
Oleic acid	30	25.6
Pentadecane	22	2.81
Pentane	0	.289
	20	.240
o-Phenetidine	0	16.5
	20	6.08
	30	4.22
m-Phenetidine	30	12.9
p-Phenetidine	20	12.9
	30	8.3
Phenol	18.3	12.7
	50	3.49
	60	2.61
	70	2.03
	90	1.26
Phenylcyanide	.28	1.96
	20.0	1.33
Phosphorus, liq	21.5	2.34
	31.2	2.01
	43.2	1.73
	50.5	1.60
	60.2	1.45
	69.7	1.32
	79.9	1.21
Potassium bromide, liq	745	1.48
	775	1.34
	805	1.19
nitrate, liq	334	2.1
	358	1.7
	333	2.97
	418	2.00
Propionic acid	10	1.289
	15	1.18
	20	1.102
	40	.845
Propyl acetate	10	.66
	20	.59
	40	.44
n-Propyl alcohol	0	3.883
	15	2.52
	20	2.256
	30	1.72
	40	1.405
n-Propyl alcohol	50	1.130
	70	.760
Propyl aldehyde	10	.47
	20	.41
	40	.33
bromide	0	.651
	20	.524
	40	.433
chloride	0	.436
	20	.352
	40	.291
n-Propyl ether	15	.448
Pyridine	20	.974
Salicylic acid	10	3.20
	20	2.71
	40	1.81
Salol	45	.746
Sodium bromide	762	1.42
	780	1.28
chloride, liq	841	1.30
	896	1.01
	924	.97
nitrate, liq	308	2.919
	348	2.439
	398	1.977
	418	1.828
Stearic acid	70	11.6
Sucrose (cane sugar)	109	2.8×10^6
	124.6	1.9×10^5
Sulfur (gas free)	123.0	10.94
	135.5	8.66
	149.5	7.09
	156.3	7.19
	158.2	7.59
	159.2	9.48
	159.5	14.45
	160.0	22.83
	160.3	77.32
	165.0	500.0
	171.0	4,500.0
	184.0	16,000.00
	190.5	19,700.0
	197.5	21,300.0
	200.0	21,500.0
	210.0	20,500.0
	217.0	19,100.0
	220.0	18,600.0
Sulfur dioxide, liq	−33.5	.5508
	−10.5	.4285
	0.1	.3936
Sulfuric acid	0	48.4
	15	32.8
	20	25.4
	30	15.7
	40	11.5
Sulfuric acid	50	8.82
	60	7.22
	70	6.09
	80	5.19
Tetrachloroethane	15	1.844
Tetradecane	20	2.18
Tin, liq	240	2.12
	280	1.678
	300	1.73
	301	1.680
	400	1.43
	450	1.270
	500	1.20
	600	1.08
	604	1.045
	750	.905
Toluene	0	.772
	17	.61
	20	.590
	30	.526
	40	.471

VISCOSITY OF LIQUIDS (Continued)

Liquid	Temp. °C	Viscosity cp
	70	.354
o-Toluidine	20	4.39
m-Toluidine	20	3.81
p-Toluidine	50	1.80
Triacetin	17	28.0
Tributyrin	20	11.6
Trichlorethane	20	1.2
Tridecane	23.3	1.55
Triethylcarbinol	20	6.75
Tripalmitin	70	16.8
Tristearin	75	18.5
Turpentine	0	2.248
	10	1.783
	20	1.487
	30	1.272
	40	1.071

Liquid	Temp. °C	Viscosity cp
	70	.728
Turpentine, Venice	17.3	1.3×10^5
n-Undecane	20	1.17
o-Xylene (xylol)	0	1.105
	16	.876
	20	.810
	40	.627
m-Xylene (xylol)	0	.806
	15	.650
	20	.620
	40	.497
p-Xylene (xylol)	16	.696
	20	.648
	40	.513
Zinc, liq	280	1.68
	357	1.42
	389	1.31

VISCOSITY OF GASES

Gas or vapor	Temp. °C	Viscosity micro-poises
Acetic acid, vap	119.1	107.0
Acetone, vap	100	93.1
	119.0	99.1
	190.4	118.6
	247.7	133.4
	306.4	148.1
Acetylene	0	93.5
Air	−194.2	55.1
	−183.1	62.7
	−104.0	113.0
	−69.4	133.3
	−31.6	153.9
	0	170.8
	18	182.7
	40	190.4
	54	195.8
	74	210.2
	229	263.8
	334	312.3
	357	317.5
	409	341.3
	466	350.1
	481	358.3
	537	368.6
	565	375.0
	620	391.6
	638	401.4
	750	426.3
	810	441.9
	923	464.3
	1034	490.6
	1134	520.6
Alcohol. See *Ethyl*, *Methyl*, etc.		
Ammonia	−78.5	67.2
	0	91.8
	20	98.2
	50	109.2
	100	127.9
	132.9	139.9
	150	146.3
	200	164.6
	250	181.4
	300	198.7
Argon	0	209.6
	20	221.7
	100	269.5
	200	322.3
	302	368.5
	401	411.5
	493	448.4
	584	481.5
	714	525.7
	827	563.2
Arsenic hydride (Arsine)	0	145.8
	15	114.0
	100	198.1

Gas or vapor	Temp. °C	Viscosity micro-poises
Benzene, vap	14.2	73.8
	131.2	103.1
	194.6	119.8
	252.5	134.3
	312.8	148.4
Bromine, vap	12.8	151
	65.7	170
	99.7	188
	139.7	208
	179.7	227
	220.3	248
Bromoform, vap	151.2	253.0
Butyl alcohol, *n*, vap	116.9	143
tert, vap	82.9	160
chloride, *n*, vap	78	149.5
iodide, vap	130	202
β-Butylene	18.8	74.4
	100.4	94.5
	200	119.2
Butyric acid, vap	161.7	130.0
Carbon dioxide	−97.8	89.6
	−78.2	97.2
	−60.0	106.1
	−40.2	115.5
	−21	129.4
	−19.4	126.0
	0	139.0
	15	145.7
	19	149.9
	20	148.0
	30	153
	32	155
	35	156
	40	157
	99.1	186.1
	104	188.9
	182.4	222.1
	235	241.5
	302.0	268.2
	490	330.0
	685	380.0
	850	435.8
	1052	478.6
disulfide, vap	0	91.1
	14.2	96.4
	114.3	130.3
	190.2	156.1
	309.8	196.6
monoxide	−191.5	56.1
	−78.5	127
	0	166
	15	172
	21.7	175.3
	126.7	218.3
	227.0	254.8
	276.9	271.4
tetrachloride, vap	76.7	195.0

VISCOSITY OF GASES (Continued)

Gas or vapor	Temp. °C	Viscosity micro-poises
Carbon tetrachloride, vap	127.9	133.4
	200.2	156.2
	314.9	190.2
Chlorine	12.7	129.7
	20	132.7
	50	146.9
	100	167.9
	150	187.5
	200	208.5
	250	227.6
Chloroform, vap	0	93.6
	14.2	98.9
	100	129
	121.3	135.7
	189.1	157.9
	250.0	177.6
	307.5	194.7
Cyanogen	0	92.8
	17	98.7
	100	127.1
Ethane	−78.5	63.4
	0	84.8
	17.2	90.1
	50.8	100.1
	100.4	114.3
	200.3	140.9
Ether (diethyl), vap	0	67.8
	14.2	71.6
	100	95.5
	121.8	98.3
	159.4	107.9
	189.9	115.2
	251.0	130.0
	277.8	135.8
Ethyl acetate, vap	0	68.4
	100	94.3
	128.1	101.8
	158.6	109.8
	192.9	119.5
	212.5	126
alcohol, vap	100	108
	130.2	117.3
	170.7	129.3
	191.8	135.5
	212.5	140
	251.7	151.9
	308.7	167.0
bromide, vap	38.4	186.5
butyrate, vap	119.8	160.0
chloride, vap	0	93.7
Ethylene	−75.7	69.9
	−44.1	76.9
	−38.6	78.5
	0	90.7
	13.8	95.4
	20	100.8
Ethylene	50	110.3
	100	125.7
	150	140.3
	200	154.1
	250	166.6
bromide, vap	131.6	221.0
chloride, vap	83.5	168.0
Ethyl formate, vap	99.8	92
iodide, vap	72.3	216.0
Helium	−257.4	27.0
	−252.6	35.0
	−191.6	87.1
	0	186.0
	20	194.1
	100	228.1
	200	267.2
	250	285.3
	282	299.2
	407	343.6
	486	370.6
	606	408.7
	676	430.3
	817	471.3
Hydrogen	−257.7	5.7
	−252.5	8.5
	−198.4	33.6
	−183.4	38.8
	−113.5	57.2
	−97.5	61.5
	−31.6	76.7
	0	83.5
	20.7	87.6
	28.1	89.2
	129.4	108.6
	229.1	126.0
	299	138.1
	412	155.4
	490	167.2
	601	182.9
	713	198.2
	825	213.7
bromide	18.7	181.9
	100.2	234.4
chloride	12.5	138.5
	16.5	140.7
	18	142.6
	100.3	182.2
iodide	20	165.5
	50	201.8
	100	231.6
	150	262.7
	200	292.4
	250	318.9
phosphide	0	106.1
	15	112.0
	100	143.8
sulfide	0	116.6
Hydrogen sulfide	17	124.1
	100	158.7
Iodine, vap	124.0	184
	170.0	204
	205.4	220
	247.1	240
Isobutyl acetate, vap	16.1	76.4
	116.4	155.0
alcohol, vap	108.4	144.5
bromide, vap	92.3	179.5
butyrate, vap	156.9	167.0
chloride, vap	68.5	150.0
iodide, vap	120	204.7
Isopentane, vap	25	69.5
	100	86.0
Isopropyl alcohol, vap	99.8	109
	120.3	103.1
	198.4	124.8
	293.1	148.8
bromide, vap	60	176.0
chloride, vap	37.0	148.5
iodide, vap	89.3	201.5
Krypton	0	232.7
	15	246
Mercury, vap	273	494
	313	551
	369	641
	380	654
Methane	−181.6	34.8
	−78.5	76.0
	0	102.6
	20	108.7
	100.0	133.1
	200.5	160.5
	284	181.3
	380	202.6
	499	226.4
Methyl acetate, vap	99.8	98
	100	100
	143.3	113.9
	218.5	134.8
alcohol, vap	66.8	135.0
	111.3	125.9
	217.5	162.0
	311.5	192.1
chloride	−15.3	92
	0	96.9
	15.0	104

VISCOSITY OF GASES (Continued)

Gas or vapor	Temp. °C	Viscosity micro-poises	Gas or vapor	Temp. °C	Viscosity micro-poises
	99.1	137		829	501.2
	182.4	168	n-Pentane, vap	25	67.6
	302.0	211		100	84.1
iodide, vap	44	232	Propane	17.9	79.5
Neon	0	297.3		100.4	100.9
	20	311.1		199.3	125.1
	100	364.6	n-Propyl alcohol, vap	99.9	93
	200	424.8		121.7	102.5
	250	453.2		209.7	126.7
Neon	285	470.8		273.0	143.4
	429	545.4	bromide, vap	99.8	119
	502	580.2	Propylene	16.7	83.4
	594	623.0		49.9	93.5
	686	662.6		100.1	107.6
	827	721.0		199.4	133.8
Nitric oxide (NO)	0	178	Propyl iodide, vap	102	210.0
	20	187.6	Sulfur dioxide	−75.0	85.8
	100	227.2		−20.0	107.8
	200	268.2		0	115.8
Nitrogen	−21.5	156.3		0	117
	10.9	170.7		18	124.2
	27.4	178.1		20.5	125.4
	127.2	219.1		100.4	161.2
	226.7	255.9		199.4	203.8
	299	279.7		293	244.7
	490	337.4		490	311.5
	825	419.2	Trimethylbutane. (2,2,3-), vap	70.3	73.4
Nitrosyl chloride	15	113.9		132.2	82.7
	100	150.4		262.1	104.8
	200	192.0	Trimethylethylene, vap	25	70.1
Nitrous oxide (N_2O)	0	135		100	86.9
	26.9	148.8	Water, vap	100	125.5
	126.9	194.3		150	144.5
n-Nonane, vap	100.3	63.3		200	163.5
	202.1	78.1		250	182.7
n-Octane, vap	100.4	67.5		300	202.4
	202.2	84.8		350	221.8
Oxygen	0	189		400	241.2
	19.1	201.8	Xenon	0	210.1
	127.7	256.8		16.5	223.5
	227.0	301.7		20	226.0
	283	323.3		127	300.9
	402	369.3		177	335.1
	496	401.3		227	365.2
	608	437.0		277	395.4
	690	461.2			

DIFFUSIVITIES OF GASES IN LIQUIDS

Solute	Solvent	Temp., °C	Diffusivity x 10^5, cm^2/sec
H_2	n-Hexane	25.4	16.36
H_2	Cyclohexane	25.4	7.08
H_2	Ethylene glycol	25.4	0.75
H_2	Carbon tetrachloride	0	6.28
O_2	Cyclohexane	29.6	5.31
O_2	Carbon tetrachloride	25.4	3.71
O_2	Ethanol	29.6	2.64
N_2	Carbon tetrachloride	0	2.44
CH_4	Glycerol	25.4	0.95
C_2H_6	n-Hexane	30	6.00
C_2H_6	n-Heptane	30	5.60
C_2H_2	Water	0	1.10
H_2S	Water	16	1.77
CO_2	Amyl alcohol	25	1.91
CO_2	Isobutyl alcohol	25	2.20
SO_2	n-Heptane	20	2.70
SO_2	n-Nonane	20	2.50
SO_2	n-Decane	20	2.40

DIFFUSION COEFFICIENTS IN AQUEOUS SOLUTIONS AT 25°

The diffusion coefficient D may be defined by either of the equations

$$J = -D\frac{\partial c}{\partial x}$$

or

$$\frac{\partial c}{\partial t} = D\frac{\partial^2 c}{\partial x^2}$$

when diffusion occurs in the x-direction only. Here J is the diffusion-flux across unit area normal to the x-direction, $\frac{\partial c}{\partial x}$ is the concentration-gradient at a fixed time, $\frac{\partial c}{\partial t}$ is the rate of change of concentration with time at a fixed distance. If J is expressed in mole cm^{-2} sec^{-1} and c in mole cm^{-3}, x in cm, and t in sec, D will be given in units of cm^2 sec^{-1}. In general D varies somewhat with concentration. The values below are a selection from measurements by modern high-precision methods, mainly by H. S. Harned and collaborators, R. H. Stokes and collaborators, L. J. Gosting and collaborators, and L. G. Longsworth.

For strong electrolytes at infinite dilution, limiting diffusion coefficients may be calculated by the Nernst relation:

$$D = \frac{RT}{F^2}\left[\frac{(\nu_1 + \nu_2)(\lambda_1^0\lambda_2^0)}{\nu_1|Z_1|(\lambda_1^0 + \lambda_2^0)}\right]$$

where R = gas constant, F = Faraday, T = absolute temperature, λ_1^0 and λ_2^0 are cation and anion limiting equivalent conductances, ν_1 and ν_2 are the numbers of cations and anions formed from one "molecule" of electrolyte, and Z_1 is the cation valency. Concentrations, unless expressed otherwise are as molarities and the diffusion coefficients are expressed as $10^5 D/cm^2$ sec^{-1} at 25°C.

DIFFUSION COEFFICIENTS OF STRONG ELECTROLYTES

Molarity

Solute	0.01	0.1	1.0
HCL	....	3.050	3.436
HBr	....	3.156	3.87
LiCl	1.312	1.269	1.302
LiBr	....	1.279	1.404
$LiNO_3$	1.276	1.240	1.293
NaCl	1.545	1.483	1.484
NaBr	....	1.517	1.596
NaI	....	1.520	1.662
KCl	1.917	1.844	1.892
KBr	....	1.874	1.975
KI	....	1.865	2.065
KNO_3	1.846	....	....
$KClO_4$	1.790	....	....
$CaCl_2$	1.188	1.110	1.203

C = 0.005M

Solute	0.01	0.1	1.0
$BaCl_2$	1.265	1.159	1.179
Na_2SO_4	1.123	....	....
$MgSO_4$	0.710	....	....
$LaCl_3$	1.105	....	....
$K_4Fe(CN)_6$	1.183	....	....

DIFFUSION COEFFICIENTS OF WEAK AND NON-ELECTROLYTES
Concentration

Solute	Concentration	Coefficient
Glucose	0.39%	0.673
Sucrose	0.38%	0.521
Raffinose	0.38%	0.434
Sucrose	Zero	0.5226
Mannitol	Zero	0.682
Penta-erythritol	Zero	0.761
Glycolamide	Zero	1.142
Glycine	Zero	1.064
α-alanine	0.32%	0.910
β-alanine	0.31%	0.933
Amino-benzoic acid ortho	0.24%	0.840
Amino-benzoic acid meta	0.24%	0.774
Amino-benzoic acid para	0.23%	0.843
Citric acid	0.1 M	0.661

DIFFUSION OF GASES INTO AIR

Gas or vapor	Temp. °C	Coefficient of diffusion, sq. cm/sec	Observer
Alcohol, vapor.......	40.4	0.137	Winkelmann
Carbon dioxide.......	0.0	0.139	Mean of various
Carbon disulfide......	19.9	0.102	Winkelmann
Ether, vapor.........	19.9	0.089	Winkelmann
Hydrogen............	0.0	0.634	Obermayer
Oxygen..............	0.0	0.178	Obermayer
Water, vapor.........	8.0	0.239	Guglielmo

VISCOSITY AND THERMAL CONDUCTIVITY OF OXYGEN AS A FUNCTION OF TEMPERATURE

Conversion Factors

T, °K → T, °F: multiply by (9/5) then subtract 459.67
T, °K → T, °C: subtract 273.15
T, °K → T, °R: multiply by (9/5)
P, atm → P, psia: multiply by 14.69595
P, atm → p, N/m^2: multiply by 1.01325 × 10^5

η, g/cm-s → η, N-s/m^2: multiply by 2.39006 × 10^{-3}
η, g/cm-s → η, lb$_m$/ft-s: multiply by 0.577789
λ, W/m-°K → λ, cal/cm-s°K: multiply by 0.1
λ, W/m-°K → λ, Btu ft-hr-°R: multiply by 6.72 × 10^{-2}

Temperature, °K	Viscosity, g/cm-s $10^3\eta_0$	Thermal conductivity, W/m-K $10^3\lambda_0$
80	0.0585	6.94
90	0.0663	7.95
100	0.0740	8.96
110	0.0818	9.96
120	0.0894	10.94
130	0.0970	11.92
140	0.1045	12.89
150	0.1118	13.85
160	0.1191	14.78
170	0.1261	15.70
180	0.1331	16.60
190	0.1399	17.48
200	0.1465	18.35
210	0.1530	19.21
220	0.1595	20.05
230	0.1658	20.88
240	0.1719	21.69
250	0.1780	22.49
260	0.1840	23.29
270	0.1898	24.08
280	0.1955	24.86
290	0.2012	25.62
300	0.2068	26.38
310	0.2122	27.14
320	0.2176	27.89
330	0.2230	28.62
340	0.2282	29.36
350	0.2334	30.10
360	0.2385	30.85
370	0.2435	31.58
380	0.2485	32.32
390	0.2534	33.06
400	0.2583	33.79
410	0.2631	34.54
420	0.2678	35.28
430	0.2725	36.02
450	0.2818	37.49
470	0.2908	38.95
490	0.2997	40.41
510	0.3085	41.86
530	0.3170	43.29
550	0.3255	44.72
570	0.3338	46.14
590	0.3420	47.56
610	0.3501	48.97
630	0.3580	50.37
650	0.3659	51.77
670	0.3736	53.16
690	0.3813	54.54
710	0.3888	55.89
730	0.3963	57.25
750	0.4037	58.59
770	0.4110	59.90
790	0.4183	61.22
820	0.4290	63.16
850	0.4396	65.09
880	0.4500	66.98
910	0.4603	68.89
940	0.4705	70.72
970	0.4805	72.52
1,000	0.4905	74.32
1,030[a]	0.5002	76.07
1,060	0.5100	77.84
1,090	0.5196	79.56
1,120	0.5291	81.27
1,150	0.5384	82.97
1,180	0.5478	84.66
1,210	0.5571	86.32
1,240	0.5662	89.97
1,270	0.5753	89.61
1,300	0.5843	91.24
1,330	0.5932	92.85
1,360	0.6021	94.45
1,390	0.6109	96.06
1,420	0.6196	97.64
1,450	0.6282	99.22
1,480	0.6368	100.78
1,510	0.6454	102.35
1,540	0.6538	103.90
1,570	0.6622	105.45
1,600	0.6706	106.98
1,630	0.6789	108.52
1,660	0.6871	110.05
1,690	0.6953	111.58
1,720	0.7034	113.09
1,750	0.7116	114.61
1,780	0.7196	116.12
1,810	0.7276	117.63
1,840	0.7355	119.12
1,870	0.7435	120.63
1,900	0.7514	122.13
1,930	0.7592	123.62
1,960	0.7670	125.11
1,980	0.7721	126.11
2,000	0.7773	127.10

[a]Data for all temperatures in excess of 1,000 K are extrapolated.

From Hanley, H. J. M., McCarty, R. D., and Sengers, J. V., *Viscosity and Thermal Conductivity Coefficients of Gaseous and Liquid Oxygen,* NASA CR-2440, National Aeronautics and Space Administration, Washington, D.C., August 1974 (available from National Technical Information Service, Springfield, Va. 22151).

THERMAL CONDUCTIVITY OF COMPRESSED OXYGEN

In this table, thermal conductivity, λ, is in the unit milliwatt/m-K.

Conversion Factors

T, °K → T, °F: multiply by (9/5) then subtract 459.67
T, °K → T, °C: subtract 273.15
T, °K → T, °R: multiply by (9/5)
P, atm → P, psia: multiply by 14.69595
P, atm → P, N/m²: multiply by 1.01325×10^5

η, g/cm-s → η, Ns/m²: multiply by 10^{-1}
η, g/cm-s → η, lb_m/ft-s: multiply by 0.0671969
λ, W/m-K → λ, cal/cm-s-K: multiply by (1/418.4)
λ, W/m-K → λ, Btu/ft-hr-°R: multiply by 0.578176

T, °K \ P, atm	1	5	10	15	20	25	30	35	40	45	50	55
80	164.7	164.9	165.2	165.4	165.7	165.9	166.2	166.4	166.7	166.9	167.1	167.4
90	151.7	152.0	152.3	152.6	152.9	153.3	153.6	153.9	154.2	154.5	154.8	155.1
100	9.5	138.4	138.8	139.2	139.6	140.0	140.4	140.8	141.1	141.5	141.9	142.3
110	10.4	11.4	124.9	125.3	125.8	126.3	126.8	127.3	127.7	128.2	128.6	129.1
120	11.4	12.2	13.5	110.9	111.5	112.1	112.8	113.3	113.9	114.5	115.1	115.6
130	12.3	13.1	14.2	15.6	96.3	97.1	98.0	98.8	99.6	100.3	101.1	101.8
140	13.2	14.0	15.0	16.1	17.5	19.3	81.6	82.9	84.1	85.2	86.3	87.3
150	14.1	14.8	15.7	16.7	17.9	19.2	20.9	23.3	27.8	68.4	70.4	72.1
160	15.8	15.7	16.5	17.4	18.4	19.5	20.7	22.3	24.2	26.7	30.5	37.3
170	15.9	16.5	17.3	18.1	19.0	19.9	20.9	22.1	23.4	24.9	26.8	29.0
180	16.8	17.4	18.1	18.8	19.6	20.4	21.3	22.3	23.3	24.4	25.7	27.1
190	17.7	18.2	18.9	19.5	20.3	21.0	21.8	22.6	23.5	24.4	25.4	26.4
200	18.5	19.0	19.6	20.3	20.9	21.6	22.3	23.0	23.8	24.6	25.4	26.3
210	19.3	19.8	20.4	21.0	21.6	22.3	22.9	23.6	24.2	24.9	25.7	26.4
220	20.2	20.6	21.2	21.7	22.3	22.9	23.5	24.1	24.7	25.4	26.0	26.7
230	21.0	21.4	21.9	22.5	23.0	23.6	24.1	24.7	25.3	25.9	26.5	27.1
240	21.7	22.2	22.7	23.2	23.7	24.2	24.8	25.3	25.8	26.4	26.9	27.5
250	22.5	22.9	23.4	23.9	24.4	24.9	25.4	25.9	26.4	26.9	27.5	28.0
260	23.3	23.7	24.1	24.6	25.1	25.6	26.0	26.5	27.0	27.5	28.0	28.5
270	24.0	24.4	24.9	25.3	25.8	26.2	26.7	27.1	27.6	28.1	28.5	29.0
280	24.8	25.1	25.6	26.0	26.4	26.9	27.3	27.8	28.2	28.6	29.1	29.5
290	25.5	25.9	26.3	26.7	27.1	27.5	28.0	28.4	28.8	29.2	29.6	30.1
300	26.2	26.6	27.0	27.4	27.8	28.2	28.6	29.0	29.4	29.8	30.2	30.6
310	27.0	27.3	27.7	28.1	28.5	28.8	29.2	29.6	30.0	30.4	30.8	31.2
320	27.7	28.0	28.4	28.7	29.1	29.5	29.9	30.2	30.6	31.0	31.4	31.7
330	28.4	28.7	29.0	29.4	29.8	30.1	30.5	30.9	31.2	31.6	31.9	32.3
340	29.1	29.3	29.7	30.1	30.4	30.8	31.1	31.5	31.8	32.2	32.5	32.9
350	29.7	30.0	30.4	30.7	31.1	31.4	31.7	32.1	32.4	32.8	33.1	33.4
360	30.4	30.7	31.0	31.4	31.7	32.0	32.4	32.7	33.0	33.3	33.7	34.0
370	31.1	31.4	31.7	32.0	32.3	32.7	33.0	33.3	33.6	33.9	34.3	34.6
380	31.8	32.0	32.3	32.7	33.0	33.3	33.6	33.9	34.2	34.5	34.8	35.1
390	32.4	32.7	33.0	33.3	33.6	33.9	34.2	34.5	34.8	35.1	35.4	35.7
400	33.1	33.3	33.6	33.9	34.2	34.5	34.8	35.1	35.4	35.7	36.0	36.3

T, °K \ P, atm	60	65	70	80	90	100	110	120	130	150	175
80	167.6	167.9	168.1	168.6	169.1	169.5	170.0	170.4	170.9	171.8	172.8
90	155.4	155.7	156.0	156.6	157.1	157.7	158.3	158.8	159.4	160.4	161.8
100	142.6	143.0	143.3	144.0	144.7	145.4	146.1	146.8	147.4	148.7	150.3
110	129.5	130.0	130.4	131.2	132.1	132.9	133.7	134.5	135.3	136.8	138.6
120	116.2	115.7	117.2	118.3	119.3	120.3	121.2	122.2	123.1	124.9	127.0
130	102.5	103.2	103.9	105.2	106.4	107.6	108.8	109.9	111.0	113.1	115.6
140	88.3	89.2	90.1	91.8	93.4	95.0	96.4	97.8	99.1	101.6	104.5
150	73.5	74.9	76.2	78.5	80.5	82.5	84.2	85.9	87.5	90.5	93.8
160	50.2	57.8	61.3	65.6	68.6	71.0	73.1	75.1	76.9	80.2	83.9
170	31.7	35.1	39.2	47.9	54.7	59.3	62.7	65.3	67.6	71.3	75.3
180	26.7	38.4	32.4	36.9	42.0	47.0	51.3	55.0	58.1	63.0	67.7
190	27.8	28.8	30.1	33.0	36.2	39.6	43.2	46.5	49.7	55.1	60.5
200	27.2	28.1	29.1	31.3	33.6	36.1	38.7	41.4	44.0	48.9	54.3
210	27.2	28.0	28.8	30.5	32.4	34.4	36.4	38.5	40.6	44.9	49.8
220	27.4	28.1	28.8	30.3	31.9	33.5	35.2	36.9	38.7	42.3	46.6
230	27.7	28.3	29.0	30.3	31.7	33.1	34.6	36.1	37.6	40.7	44.5
240	28.1	28.7	29.2	30.5	31.7	33.0	34.3	35.6	36.9	39.7	43.1
250	28.5	29.1	29.6	30.7	31.8	33.0	34.2	35.4	36.6	39.0	42.1
260	29.0	29.5	30.0	31.0	32.1	33.1	34.2	35.3	36.4	38.7	41.5
270	29.5	29.9	30.4	31.4	32.4	33.3	34.4	35.4	36.4	38.4	41.0
280	30.0	30.4	30.9	31.8	32.7	33.6	34.6	35.5	36.5	38.4	40.8
290	30.5	30.9	31.3	32.2	33.1	33.9	34.8	35.7	36.6	38.4	40.6
300	31.0	31.4	31.8	32.7	33.5	34.3	35.1	36.0	36.8	38.5	40.6
310	31.6	31.9	32.3	33.1	33.9	34.7	35.5	36.3	37.1	38.7	40.6
320	32.1	32.5	32.8	33.6	34.3	35.1	35.8	36.6	37.4	38.9	40.8
330	32.7	33.0	33.4	34.1	34.8	35.5	36.2	37.0	37.7	39.1	40.9
340	33.2	33.5	33.9	34.6	35.3	36.0	36.7	37.3	38.0	39.4	41.1
350	33.8	34.1	34.4	35.1	35.8	36.4	37.1	37.7	38.4	39.7	41.4
360	34.3	34.6	35.0	35.6	36.2	36.9	37.5	38.2	38.6	40.1	41.7
370	34.9	35.2	35.5	36.1	36.7	37.9	38.0	38.6	39.2	40.4	42.0
380	35.4	35.7	36.0	36.6	37.2	37.8	38.4	39.0	39.6	40.8	42.3
390	36.0	36.3	36.6	37.2	37.8	38.3	38.9	39.5	40.1	41.2	42.6
400	36.6	36.9	37.1	37.7	38.3	38.8	39.4	40.0	40.5	41.6	43.0

From Hanley, H. J. M., McCarty, R. D., and Sengers, J. V., *Viscosity and Thermal Conductivity Coefficients of Gaseous and Liquid Oxygen,* NASA CR-2440, National Aeronautics and Space Administration, Washington, D.C., August 1974 (available from National Technical Information Service, Springfield, Va. 22151).

TRANSPORT PROPERTIES OF OXYGEN AT SATURATION

Conversion Factors

T, °K → T, °F: multiply by (9/5) then subtract 459.67	η, g/cm-s → η, Ns/m²: multiply by 10^{-1}
T, °K → T, °C: subtract 273.15	η, g/cm-s → η, lb_m/ft-s: multiply by 0.0671969
T, °K → T, °R: multiply by (9/5)	λ, W/m-K → λ, cal/cm-s-K: multiply by (1/418.4)
P, atm → P,psia: multiply by 14.69595	λ, W/m-K → λ, Btu/ft-hr-°R: multiply by 0.578176
P, atm → P, N/m²: multiply by 1.01325×10^5	

Temperature, °K	Pressure, atm	Viscosity, mg/cm-s		Thermal conductivity, mW/m-K		Temperature, °K	Pressure, atm	Viscosity, mg/cm-s		Thermal conductivity, mW/m-K	
		Vapor	Liquid	Vapor	Liquid			Vapor	Liquid	Vapor	Liquid
80	0.30	0.059	2.652	7.4	164.7	142	30.00	0.166	0.668	21.1	77.4
90	0.98	0.068	1.971	8.5	151.8	144	32.64	0.175	0.638	22.3	74.0
100	2.51	0.079	1.504	9.9	138.2	146	35.45	0.185	0.607	23.6	70.6
110	5.36	0.092	1.188	11.5	124.3	148	38.44	0.197	0.574	25.2	67.0
120	10.09	0.108	0.971	13.6	110.3	150	41.61	0.211	0.537	27.2	63.2
130	17.26	0.129	0.824	16.3	95.9	152	45.00	0.231	0.494	29.9	58.7
140	27.52	0.158	0.696	20.1	80.6	154	48.65	0.269	0.424	35.2	52.0

From Hanley, H. J. M., McCarty, R. D., and Sengers, J. V., *Viscosity and Thermal Conductivity Coefficients of Gaseous and Liquid Oxygen*, NASA CR-2440, National Aeronautics and Space Administration, Washington, D.C., August 1974 (available from National Technical Information Service, Springfield, Va. 22151).

PHYSICAL CONSTANTS OF OZONE AND OXYGEN

Physical Constant	Ozone (O_3)	Oxygen (O_2)
Molecular Weight	47.9982 g/g-mol	31.9988 g/g-mol
Boiling Point (760 mm)	−111.9 ± 0.3°C	−182.97°C
Melting Point	−192.7 ± 0.2°C	−218.4°C
Critical Temperature	−12.1 ± 0.1°C	−118.574°C
Critical Pressure	54.6 atm	49.77 atm
Critical Density	0.437 g/cc	0.436 g/cc
Critical Volume	147.1 cc/mol	73.37 cc/mol
Gas Density (0°C) (760 mm pressure)	2.144 g/liter	1.429 g/liter
Liquid Density		
−112°C	1.358 g/cc	
−183°C	1.571 g/cc	1.14 g/cc
−195.4°C	1.614 g/cc	1.201 g/cc
Surface Tension		
−195°C	43.8 ± 0.1 dyne/cm	
−182.7°C	38.1 ± 0.2 dyne/cm	
−183.0°C	38.4 ± 0.7 dyne/cm	13.2 dyne/cm
Heat Capacity of Liquid		
−183 to −145°C	0.45 cal/g°C	
Heat Capacity of Gas		
−173°C	7.95 cal/g mol°C	
0°C	9.10 cal/g mol°C	
25°C	9.37 cal/g mol°C	
100°C		6.979 cal/g mol°C
127°C	10.44 cal/g mol°C	
Viscosity of Liquid		
−195.6°C	4.14 ± 0.05 cP	
−183.0°C	1.57 ± 0.02 cP	0.1958 cp
Heat of Vaporization		
−112°C	75.6 cal/g	
−182.9°C		50.9 cal/g
Heat of Formation		
25°C	−34.4 kcal/mol	
Free Energy		
25°C	32.4 kcal/mol	
Van der Waals Constant (a)	3.545 atm liter²/mol²	1.36 atm liter²/mol²
Van der Waals Constant (b)	0.04903 liter/mol	0.03803 liter/mol
Magnetic Susceptibility		
gas $(\times 10^{-6})$	0.002 cgs units	10.6.2 cgs units
liq $(\times 10^{-6})$	0.150 cgs units	260.0 cgs units
Thermal Conductivity of Liquid		
−195.8°C	5.21 cal/sec cm°C $\times 10^4$	
−183.0°C	5.31 cal/sec cm°C $\times 10^4$	
−165.0°C	5.42 cal/sec cm°C $\times 10^4$	
−128.0°C	5.52 cal/sec cm°C $\times 10^4$	

Phase Boundaries—Ozone-Oxygen System

−183°C	29.8 and 72.4 wt % O_3
−195.4°C	9 and 90.8 wt %O_3

Consolute Temperature—Ozone-Oxygen System

−180 ± 0.5°C

Coefficient of Thermal Expansion for Liquid Ozone

Temp. °C	α
−195.6	1.62
−183.0	1.58
−148.0	1.47
−123.0	1.41
−112.0	1.35
− 98.8	1.31

PHYSICAL CONSTANTS OF CLEAR FUSED QUARTZ

Based on information contained in Fused Quartz Catalogue Q-7A General Electric Company.

Property		Clear fused quartz
Density		2.2 g cc^{-1}
Hardness		4.9 (Mohs')
Tensile Strength		7,000 psi
Compressive Strength		> 160,000 psi
Bulk Modulus	(approx.)	5.3×10^6 psi
Rigidity Modulus		4.5×10^6 psi
Young's Modulus		10.4×10^6 psi
Poisson's Ratio		.16
Coefficient of Thermal Expansion	(av.)	5.5×10^{-7} cm cm^{-1} $°C^{-1}$ $\begin{cases}20°C\\320°C\end{cases}$
Thermal Conductivity		0.0033 cal cm^{-1} sec^{-1} $°C^{-1}$ cm^{-1}
Specific Heat		0.18 cal g^{-1}
Softening Point	(approx.)	1665°C

Property		Clear fused quartz
Annealing Point	(approx.)	1140°C
Strain Point		1070°C
Electrical Resistance		9.5 $\log_{10}$ R for $cm.^3$ at 350°C
Dielectric Constant		3.75 at 20°C. 1 MHz
Dielectric Loss Factor	less than	.0004 at 20°C. 1 MHz
Dissipation Factor	less than	.0001 at 20°C. 1 MHz
Index of Refraction		1.4585
Velocity of Sound—Shear Wave		3.75×10^5 cm sec^{-1}
Velocity of Sound—Compressional Wave		5.90×10^5 cm sec^{-1}
Sonic Attenuation	less than	.033 db ft^{-1} MHz^{-1}

FIXED POINT PROPERTIES OF OXYGEN

From NASA SP-3071, "ASRDI Oxygen Technology Survey, Thermophysical Properties", Volume I (1972), edited by Hans M. Roder and Lloyd A. Weber. This NASA publication contains an extensive bibliography and a discussion of basis for selection of these data for oxygen. The publication is available from the National Technical Information Service, Springfield, Virginia 22151.

	Triple Point			Normal Boiling Point			Standard Conditions	
PROPERTIES ↓ CONDITIONS →	Solid	Liquid	Vapor	Liquid	Vapor	Critical Point ++	STP (0°C)	NTP (20°C)
Temperature (K)		54.351		90.180		154.576	273.15	293.15
Pressure (mmHg)		1.138		760		37,823	760	760
Density (mole/cm^3) x 10^3	42.46	40.83	0.000336	35.65	0.1399	13.63	0.04466	0.04160
Specific Volume (cm^3/mole) x 10^{-3}	0.02355	0.02449	2975	0.028047	7.1501	0.07337	22.392	24.038
Compressibility Factor, $Z = \frac{PV}{RT}$	–	0.0000082	0.9986	0.00379	0.9662	0.2879	0.9990	0.9992
Heats of Fusion & Vaporization (J/mole)	444.8	7761.4		6812.3		0	–	–
Specific Heat (J/mole-K) C_s, @saturation	46.07	53.313	-108.7	54.14	-53.2	(very large)	–	–
C_p, @ constant pressure	–	53.27	29.13	54.28	30.77	(very large)	29.33	29.40
C_v, @ constant volume	–	35.65	20.81	29.64	21.28	(38.7)	20.96	21.04
Specific Heat Ratio, $\gamma = C_p/C_v$	–	1.494	1.400	1.832	1.446	(large)	1.40	1.40
Enthalpy (J/mole)	-6634.4	-6189.6	1571.8	-4270.3	2542.0	1032.2	7937.8	8525.1
Internal Energy (J/mole)	-6634.4	-6189.6	1120.0	-4273.1	1817.5	662.3	5668.9	6089.5
Entropy (J/mole-K)	58.92	67.11	209.54	94.17	169.68	134.42	202.4	204.5
Velocity of Sound (m/sec)		1159	141	903	178	164	315	326
Viscosity, μ (N-sec/m^2) × 10^3	–	0.6194	0.003914	0.1958	0.00685	(0.031)	0.01924	0.02036
(centipoise)‡‡	–	0.6194	0.003914	0.1958	0.00685	(0.031)	0.01924	0.02036
Thermal Conductivity (mW/cm-K), k		1.929	0.04826	1.515	0.08544	(*)	0.2428	0.2575
Prandtl Number, $N_{pr} = \mu C_p/k$	–	5.344	0.7392	2.193	0.7714		0.7259	0.7265
Dielectric Constant, ϵ	(1.614)	1.5687	1.000004	1.4870	1.00166	1.17082	1.00053	1.00049
Index of Refraction, $n = \sqrt{\epsilon}$ †	(1.271)	1.2525	1.000002	1.219	1.00083	1.0820	1.00027	1.00025
Surface Tension (N/m) x 10^3	–	22.65	–	13.20	–	0	–	–
Equiv. Vol./Vol. Liquid at NBT	0.8397	0.8732	106,068	1	254.9	2.616	798.4	857.1

† Long Wavelengths
* Anomalously Large

Gas Constant: R = 62, 365.4 cm^3-mm Hg/mole-K^1
†† Values in parenthesis are estimates

Molecular Weight = 31.9988[3]
"mole" = gram mole
‡‡ Units for poise are: g/cm-sec

FIXED POINTS AND PHASE EQUILIBRIUM BOUNDARIES FOR PARAHYDROGEN

a. Triple point
P_t = 0.0704 bar
T_t = 13.803 K
ρ_t (liquid) = 38.21 mol/ℓ

b. Normal boiling point
P_b = 1.01325 bar
T_b = 20.268 K
ρ_b (liquid) = 35.11 mol/ℓ
ρ_b (gas) = 0.6636 mol/ℓ

c. Critical point

$P_c = 12.928$ bar
$T_c = 32.976$ K
$\rho_c = 15.59$ mol/ℓ

Note: Some data indicate that the true critical temperature is probably closer to 32.93 K. However, that value is pending further verfication.

d. Melting pressures: in atmospheres

$$P = P_t + (T - T_t)\,[A_1 e^{-\alpha/T} + A_2 T]$$

$A_1 = 30.3312$
$A_2 = 0.6667$
$\alpha = 5.693$

e. Liquid-vapor coexistence densities

Liquid, density in mol/cm³:

$$\rho \text{ sat } \ell = \rho_c + A_1(\Delta T)^{0.380} + A_2(\Delta T) + A_3(\Delta T)^{4/3} + A_4(\Delta T)^{5/3} + A_5(\Delta T)^2$$

$A_1 = 7.323\,4603 \times 10^{-3}$
$A_2 = -4.407\,4261 \times 10^{-4}$
$A_3 = 6.620\,7946 \times 10^{-4}$
$A_4 = -2.922\,6363 \times 10^{-4}$
$A_5 = 4.008\,4907 \times 10^{-5}$
$\Delta T = T_c - T$

Vapor $T_b \leq T \leq T_c$, density in mol/cm³:

$$\rho \text{ sat } G = \rho_c + A_1(\Delta T)^{0.370} + A_2(\Delta T) + A_3(\Delta T)^{0.7} + A_4(\Delta T)^{0.8}$$

$A_1 = -7.196\,7724 \times 10^{-3}$
$A_2 = 1.449\,5527 \times 10^{-3}$
$A_3 = 3.240\,3120 \times 10^{-3}$
$A_4 = -4.464\,0177 \times 10^{-3}$

f. Vapor pressure: in atmospheres

For $T \leq 29$ K:

$$\log_{10} P_a = A_1 + \frac{A_2}{T + A_3} + A_4 T$$

$A_1 = 2.000\,620$
$A_2 = -50.09\,708$
$A_3 = 1.0044$
$A_4 = 1.748\,495 \times 10^{-2}$

For $T > 29$ K:

$$P = P_a + A_5(T - 29)^3 + A_6(T - 29)^5 + A_7(T - 29)^7$$

$A_5 = 1.317 \times 10^{-3}$
$A_6 = -5.926 \times 10^{-5}$
$A_7 = 3.913 \times 10^{-6}$

From Weber, L. A., Thermodynamic and Related Properties of Parahydrogen from the Triple Point to 300 K at Pressures to 1000 Bar, NASA SP-3008, NBSIR 74-374, 1975.

PHYSICAL PROPERTIES OF SODIUM, POTASSIUM AND Na-K ALLOYS

Temperature °C	Density (g/cm³)								Viscosity (centipoise)		
	Na			K	Alloys (wt % K)				Na	Alloys (wt % K)	
					43.4		78.6			43.3	66.0
	(a)	(b)	(c)		Experimental	(d) Calculated	Experimental	(d) Calculated			
100	.927	.927	.9265	.819	.887	.890	.847	.850	.705	.540	.529
200	.904	.904	.9037	.795	.862	.867	.823	.827	.450	.379	.354
300	.882	.880	.8805	.771	.838	.843	.799	.802	.345	.299	.276
400	.859	.856	.8570	.747	.814	.818	.775	.778	.284	.245	.229
500	.834	.831	.8331	.723	.789	.794	.751	.754	.234	.207	.195(e)
600	.809	.808	.8089	.701	.765	.771	.727	.732	.210	.178(e)	.168(e)
700	.783	.784	.....	.676	.740	.745	.703	.705	.186(e)	.257(e)	.146(e)
800	.757	.760	.....	....	....	....	....	....	.165(e)	......	......
900	....	....	.....	....	....	....	....	....	.150(e)	......	......

Temperature, °C	Thermal conductivity (watts/cm²-°C/cm) Na Experimental	Na Calculated	Alloy (wt % K) 56.5	Alloy (wt % K) 77.0	Electrical resistivity (microohms) Na[f]	Alloy (wt % K) 56.5	Alloy (wt % K) 78.0	Heat capacity[g] (cal/°C-g) K	Alloy (wt % K) 44.8	Alloy (wt % K) 78.26
100	......	....	....	.238[h]	8.99	41.61	45.63	.1940	.2690	.2248
200	.815	.808	.249	.247	13.52	47.23	51.33	.1887	.2612	.2169
300	.757	.755	.262	.259	17.52	54.33	58.58	.1894	.2553	.2122
400	.712	.710	.269	.262	21.93	62.21	65.65	.126	.2512	.2097
500	.668	.672	.271	.259	26.96	69.37	73.48	.1818	.2498	.2088
600	.627[e]	.639	....	.255	32.65	78.29	82.61	.1825	.2484	.2092
700	.590[e]	.610	....	......	39.05	88.23	91.76	.1846	.2497	.2108
800	.547[e]	.583	....	......	46.15	99.68	104.51	.1883	.2529	.2133
900	......	....	....	......	.....	.....	......	.....	.....	.....

Vapor Pressure, mm Hg

Temperature °C	Na	K	Alloys (wt % K) 56	Alloys (wt % K) 78
127	2.23×10^{-6}	....	....	....
227	1.15×10^{-3}	2.88×10^{-2}	1.57×10^{-3}	1.81×10^{-3}
327	5.03×10^{-2}	9.27×10^{-1}	5.73×10^{-2}	6.14×10^{-2}
427	.881	9.26	3.53	5.06
527	7.53	52.22	23.0	31.87
627	39.98	201.25	101.35	136.50
727	148.5	588.62	328.7	431.85
827	453.7	1421.0	864.2	1099.52
927	1127.8	....	....	....
1027	2522.4	....	....	....
1127	4696.8	....	....	....

NOTES: (a) From plotted data.

(b) Epstein equation: $d_t = 0.9514 - 2.392 \times 10^{-4}\, t°C$.

(c) Thomson and Garelis: $d_t = 0.9490 - 22.3 \times 10^{-5} t°C - 1.75 \times 10^{-8} t^2 °C$.

(d) Formula to calculate density: $V = M_K \cdot V_K + M_{Na} \cdot V_{Na}$ (where V, V_K and V_{Na} are the specific volumes (reciprocal of density) of the alloy, K and Na respectively, M_K and M_{Na} the mole fraction of the elements).

(e) Extrapolated by calculation, Epstein equation:

$$K = \frac{2.433 \times 10^{-2}(t + 273.16)}{6.8393 + 3.3873 \times 10^{-2} t + 1.7235 \times 10^{-5} t^2}$$

(f) Epstein equation: $r_t = 10.892 + 0.015272t + 3.6746 \times 10^{-5} t^2 - \dfrac{379.26}{t}$.

(g) Formula to calculate heat capacity: $C = W_{Na} \cdot C_{Na} + W_K \cdot C_K$ (where C, C_{Na} and C_K are the heat capacity of the alloy, Na and K respectively, W_{Na} and W_{Ka} the weight fractions of Na and K respectively in the alloy).

(h) 150°C.

CRITICAL TEMPERATURES AND CRITICAL PRESSURES OF THE ELEMENTS[a]

Critical temperatures are listed as degrees kelvin and critical pressures as megapascals. Some conversion factors which may be useful are:

Megapascals	×	9.8692	=	atmospheres (760mm Hg)
Megapascals	×	14.504	=	pounds per square inch
Megapascals	×	10^{-6}	=	newtons per square meter (pascals)
Megapascals	×	10.1972	=	kilograms per square centimeter
Megapascals	×	10	=	bars
Megapascals	×	7.501	×	10^3 = millimeters Hg at 0°C
Megapascals	×	4.014	=	inches of H_2O at 4°C

Element	Symbol	T_c, K	P_c, MPa
Argon	Ar	150.8	4.87
Arsenic	As	1673	22.3
Bromine	Br_2	588	10.3
Chlorine	Cl_2	416.9	7.977
Deuterium (equilibium)	D_2	38.2	1.65
Deuterium (normal)	D_2	38.4	1.66
Fluorine	F_2	144.3	5.215
Helium-3	He^3	3.31	0.114
Helium-4	He^4	5.19	0.227
Hydrogen (equilibium)	H_2	32.98	1.293
Hydrogen (normal)	H_2	33.2	1.297
Hydrogen deuteride	HD	36.0	1.48
Iodine	I_2	819	—
Mercury	Hg	1765	151.0
Krypton	Kr	209.4	5.50
Neon	Ne	44.40	2.76
Nitrogen	N_2	126.2	3.39
Oxygen	O_2	154.58	5.043
Ozone	O_3	261.1	5.57
Phosphorus	P	994	—
Radon	Rn	377	6.28
Rubidium	Rb	2105	—
Selenium	Se	1766	27.2
Sulfur	S	1314	20.7
Tritium	T_2	40.0	—
Xenon	Xe	289.73	5.840

The data are from "Vapor-Liquid Critical Properties", March, 1978, D. Ambrose and R. Townsend, National Physical Laboratory, Teddington, Middlesex TW11 OLW, UK. These data are used with permission of the copyright owner.

PHYSICAL CONSTANTS OF LIQUID ALKALI METALS

Element	mp, °K	bp, °K	T_c, °K	P_c, MPa	Heat of fusion, KJ mol^{-1}	Heat of vaporization KJ mol^{-1}
Cesium	301.55	942.4	2051 ± 10	11.73	2.14	66
Lithium	453.69	1615	3223 ± 600	68.9 ± 20%	2.99	134.7
Potassium	336.4	1033	2220 ± 25	16.39 ± 0.02	2.33	77.08
Rubidium	312.04	959.2	2083 ± 20	14.54 ± 0.5	2.34	69
Sodium	370.96	1156	2508.7 ± 12.5	25.64 ± 0.02	2.6	89.6 ± 0.5%

CRITICAL TEMPERATURES AND PRESSURES

Compiled by Rudolf Loebel

Table I —Organic compounds
Table II—Inorganic compounds

Table I

Formula	Name	Critical temp. T_c °C	Critical press. P_c atm.
$CHClF_2$	Methane, monochlorodifluoro-	96	48.5
$CHCl_2F$	Methane, dichloromonofluoro-	178.5	51
$CHCl_3$	Methane, trichloro- (Chloroform)	263	54
CHF_3	Methane, trifluoro- (Fluoroform)	25.9	46.9
CH_2Cl_2	Methylene chloride	237	60
CH_3NO_2	Methane, nitro-	314.8	62.3
CH_3Br	Methane, monobromide-	194	83.4
CH_3Cl	Methane, monochloro-	143.8	65.9
CH_3F	Methane, monofluoro-	44.6	58
CH_3I	Methane, monoiodo-	254.8	72.7
CH_4	Methane	−82.1	45.8
CH_4O	Methanol (Methyl alcohol)	240	78.5
CH_4S	Methylmercaptan	196.8	71.4
CH_5N	Methylamine	156.9	40.2
$CBrF_3$	Methane, monobromotrifluoro-	67	50.3
$CClF_3$	Methane, monochlorotrifluoro-	28.85	38.2
CCl_2F_2	Methane, dichlorodifluoro-	111.5	39.6
CCl_3F	Methane, trichloromonofluoro-	198	43.2
CCl_4	Methane, tetrachloro- (Carbon tetrachloride)	283.1	45
CF_4	Methane, tetrafluoro-	−45.7	41.4
C_2H_2	Acetylene	35.5	61.6
C_2H_2	Ethyne (see Acetylene)	35.5	61.6
$C_2H_2F_2$	Ethylene, 1,1-difluoro-	30.1	—
C_2H_3N	Acetonitrile	274.7	47.7
$C_2H_3F_3$	Ethane, 1,1,1-trifluoro-	73.1	—
C_2H_4	Ethene	9.9	50.5
C_2H_4	Ethylene (see Ethene)	9.9	50.5
C_2H_4O	Acetaldehyde	187.8	54.7
C_2H_4O	Ethylene oxide	195.8	71
$C_2H_4O_2$	Acetic acid	321.6	57.1
$C_2H_4O_2$	Formic acid, methyl- (Methyl formate)	214	59.2
$C_2H_4Cl_2$	Ethane, 1,1-dichloro-	249.8	50
$C_2H_4F_2$	Ethane, 1,1-difluoro-	386.7	—
C_2H_5Br	Ethane, monobromo-	230.8	61.5
C_2H_5Cl	Ethane, monochloro-	187.2	52
C_2H_5F	Ethane, monofluoro-	102.16	49.6
C_2H_6	Ethane	32.2	48.2
C_2H_6O	Ether, dimethyl-	127	52.6
C_2H_6O	Ethanol (Ethyl alcohol)	243	63
C_2H_6O	Glycol, ethylene-	(374)*	—
C_2H_6S	Dimethylsulfide	229.9	54.5
C_2H_6S	Ethylmercaptan	225.5	54.2
C_2H_7N	Dimethylamine	164.6	52.4
C_2H_7N	Ethylamine	183.2	55.5
$C_2Br_2F_4$	Ethane, dibromotetrafluoro-	214.5	—
C_2ClF_5	Ethane, chloropentafluoro-	80	—
$C_2Cl_2F_4$	Ethane, 1,2-dichlorotetrafluoro-	145.7	—
$C_2Cl_3F_3$	Ethane, trichlorotrifluoro-	214.2	33.7
$C_2Cl_4F_2$	Ethane, tetrachlorodifluoro-	278	32.9
C_2F_6	Ethane, hexafluoro-	24.3	—
C_3H_4	Propadiene	120	43.6
C_3H_4	Allene (see Propadiene)	120	43.6
C_3H_4	Acetylene, methyl-	127.8	52.8
C_3H_4	Propyne (see Acetylene, methyl-)	127.8	52.8
C_3H_5N	Ethyl cyanide	290.8	41.3
C_3H_5N	Propionitrile (see Ethyl cyanide)	290.8	41.3
C_3H_5OCl	Epichlorohydrin	(323)	—
C_3H_5Cl	Propene, 3-chloro- (allyl-chloride)	240.3	46.5
C_3H_6	Propylene	91.9	45.4
C_3H_6	Cyclopropane	124.7	—
C_3H_6O	Acetone	235.5	47
C_3H_6O	Allylalcohol	272	55.5
C_3H_6O	Propylene oxide	209	48.6
$C_3H_6O_2$	Formic acid, ethyl- (Ethyl formate)	235.3	46.3
$C_3H_6O_2$	Acetic acid, methyl- (Methyl acetate)	233.7	46.3
$C_3H_6O_2$	Propanoic acid	337.6	53
C_3H_7Cl	n-Propane, monochloro-	230	45.2
C_3H_8	Propane	96.8	42
C_3H_8O	Ether, ethyl methyl- (Methoxyethane)	164.7	43.4
C_3H_8O	Glycol, 1,2-propylene-	351	—
C_3H_8O	Isopropyl alcohol	235	47
C_3H_8O	n-Propyl alcohol	263.6	51
$C_3H_8O_3$	Glycerol	(452)	—
C_3H_9N	Isopropylamine	209.7	—
C_3H_9N	n-Propylamine	223.8	46.8
	*( ) uncertain		
C_4H_4O	Furan	213.8	52.5
C_4H_4S	Thiophene	307	56.2
C_4H_6	1,2-Butadiene	171	44.4
C_4H_6	1,3-Butadiene	152	42.7
C_4H_8	n-Butene	146	39.7
C_4H_8	2-Butene, cis-	160	40.5
C_4H_8	2-Butene, trans-	155	41.5
C_4H_8O	1,2-Butylene oxide	243	—
C_4H_8O	Ketone, ethyl methyl- (2-Butanone)	262	41
$C_4H_8O_2$	Butanoic acid	355	52
$C_4H_8O_2$	p-Dioxane	314.8	51.4
$C_4H_8O_2$	Acetic acid, ethyl- (Ethyl acetate)	250.4	37.8
$C_4H_8O_2$	Propanoic acid, methyl- (Methyl propionate)	257.4	39.3
$C_4H_8O_2$	Formic acid, propyl-(Propyl formate)	264.9	40.1
C_4H_9Cl	n-Butane, monochloride-	269	—
C_4H_{10}	n-Butane	152	37.5
C_4H_{10}	Isobutane	135	35.9
$C_4H_{10}O$	Butanol (n-Butyl alcohol)	289.8	43.6
$C_4H_{10}O$	sec-Butyl alcohol	263	41.4
$C_4H_{10}O$	tert-Butyl alcohol	235	39.2
$C_4H_{10}C$	Ether, diethyl-	192.6	35.6
$C_4H_{10}O$	Isobutyl alcohol	277	42.4
$C_4H_{10}O_3$	Glycol, diethylene-	407	—
$C_4H_{10}S$	Diethyl sulfide	283.8	39.1
$C_4H_{11}N$	Butyl amine	287.9	—
$C_4H_{11}N$	Diethyl amine	223.3	36.6
$C_4H_{11}N$	Isobutyl amine	266.7	—
C_4F_{10}	Butane, perfluoro-	113.2	23
C_5H_5N	Pyridine	346.8	—
C_5H_8	Cyclopentene	232.94	47.2
C_5H_{10}	Cyclopentane	238.6	44.6
C_5H_{10}	1-Pentene	191	39.9
$C_5H_{10}O$	Ketone, diethyl-	287.8	36.9
$C_5H_{10}O$	Propanoic acid, ethyl- (Ethyl propionate)	272.9	33
$C_5H_{10}O_2$	n-Butanoic acid, methyl- (n-Methyl butyrate)	281.3	34.3
$C_5H_{10}O_2$	Acetic acid, n-propyl (n-Propyl acetate)	276	32.9
$C_5H_{10}O_2$	n-Valeric acid	378	37.6
$C_5H_{11}N$	Piperidine	320.8	44.1
C_5H_{12}	Butane, 2-methyl- (See Isopentane)	187.8	32.9
C_5H_{12}	Isopentane	187.8	32.9
C_5H_{12}	Neopentane	160.6	31.6
C_5H_{12}	n-Pentane	196.6	33.3
C_5H_{12}	Propane, 2,2-dimethyl (see Neopentane)	160.6	31.6
$C_5H_{12}O$	Isoamyl alcohol	309.77	—
C_6H_5Br	Benzene, bromo-	397	44.6
C_6H_5Cl	Benzene, chloro-	359.2	44.6
C_6H_5F	Benzene, fluoro-	286.95	44.6
C_6H_6	Benzene	288.9	48.6
C_6H_6O	Phenol	421.1	60.5
C_6H_7N	Aniline	425.6	52.3
C_6H_7N	α-Picoline	348	—
C_6H_7N	β-Picoline	371.7	—
C_6H_7N	γ-Picoline	372.5	—
C_6H_{10}	Cyclohexene	287.3	—
C_6H_{10}	1,5-Hexadiene	234.4	32.6
C_6H_{12}	Cyclohexane	280.4	40
C_6H_{12}	Cyclopentane, methyl-	259.5	37.4
C_6H_{12}	1-Hexene	231	31.1
$C_6H_{12}O_2$	Formic acid, n-amyl (n-Amyl formate)	302.6	34.1
$C_6H_{12}O_2$	Acetic acid, n-butyl- (n-Butyl acetate)	305.9	30.7
$C_6H_{12}O_2$	Butanoic acid, ethyl- (Ethyl butyrate)	293	30.2
$C_6H_{12}O_2$	Propanoic acid, propyl- (Propyl propionate)	304.8	30.7
$C_6H_{12}O_3$	Paraldehyde	290	—
C_6H_{14}	Butane, 2,3-dimethyl-	226.8	30.9
C_6H_{14}	n-Hexane	234.2	29.9
C_6H_{14}	Pentane, 2-methyl-	224.3	30
$C_6H_{14}O$	Ether, isopropyl-	226.9	28.4
$C_6H_{14}O$	1-Hexyl alcohol	313.5	—
$C_6H_{14}O_4$	Glycol, triethylene-	(437)	—
$C_6H_{15}N$	Dipropyl amine	277	31
$C_6H_{15}N$	Hexyl amine	318.8	—
$C_6H_{15}N$	Triethyl amine	258.9	30
C_7H_5N	Benzonitrile	426.2	41.6
C_7H_8	Benzene, methyl- (see Toluene)	320.8	41.6

Table I (Continued)

Formula	Name	Critical temp. T_c °C	Critical press. P_c atm.
C_7H_8	Toluene	320.8	41.6
C_7H_8O	Anisole, (see Benzene, methoxy-)	368.5	41.3
C_7H_8O	Benzene, methoxy-	368.5	41.3
C_7H_8O	o-Cresol	424.4	49.4
C_7H_8O	m-Cresol	432	45
C_7H_8O	p-Cresol	431.4	50.8
C_7H_9N	Aniline, methyl-	428.4	51.3
C_7H_9N	2,3-Lutidine	382.3	—
C_7H_9N	2,4-Lutidine	374	—
C_7H_9N	2,6-Lutidine	350.6	—
C_7H_9N	3,4-Lutidine	410.6	—
C_7H_9N	3,5-Lutidine	394.1	—
C_7H_{14}	Cyclohexane, methyl-	299.1	34.3
C_7H_{14}	Cyclopentane, ethyl-	296.3	33.5
C_7H_{14}	1-Heptene	264.1	—
$C_7H_{14}O_2$	Acetic acid, isoamyl- (Isoamyl acetate)	326.1	28
C_7H_{16}	Butane, 2,2,3-trimethyl-	258.3	29.8
C_7H_{16}	n-Heptane	267.1	27
C_7H_{16}	Hexane, 2-methyl-	257.9	27.2
C_7H_{16}	Pentane, 3-ethyl-	267.6	28.6
C_7H_{16}	Pentane, 2,4-dimethyl-	247.1	27.4
$C_7H_{16}O$	1-Heptyl alcohol	365.3	29.4
C_7F_{16}	n-Heptane, perfluoro-	201.7	16
C_8H_8	Styrene	374.4	39.4
C_8H_{10}	Benzene, ethyl-	343.9	36.9
C_8H_{10}	o-Xylene	359	35.7
C_8H_{10}	m-Xylene	346	34.7
C_8H_{10}	p-Xylene	345	33.9
$C_8H_{10}O$	Xylenol	449.7	56.4
$C_8H_{10}O$	Phenetol (see Benzene, ethoxy-)	374	33.8
$C_8H_{10}O$	Benzene, ethoxy-	374	33.8
$C_8H_{11}N$	Aniline, N,N-dimethyl-	414.4	35.8
$C_8H_{11}N$	Aniline, N-ethyl-	425.4	—
C_8H_{16}	n-Octene	305	25.5

Formula	Name	Critical temp. T_c °C	Critical press. P_c atm.
C_8H_{18}	Butane, 2,2,3,3-tetramethyl-	270.8	24.5
C_8H_{18}	Heptane, 4-methyl-	290	25.6
C_8H_{18}	Hexane, 2,4-dimethyl-	282	25.8
C_8H_{18}	n-Octane	296	24.8
C_8H_{18}	Pentane, 2,2,3-trimethyl-	294	28.2
$C_8H_{18}O$	1-Octyl alcohol	385.5	26.5
$C_8H_{19}N$	Dibutyl amine	322.6	—
C_9H_7N	Quinoline	508.8	—
C_9H_{12}	Benzene, n-propyl-	365	31.2
C_9H_{12}	Benzene, 1,2,3-trimethyl-	391.3	31
C_9H_{12}	Cumene (see Benzene, isopropyl-)	362.7	31.2
C_9H_{12}	Benzene, isopropyl-	362.7	31.2
C_9H_{20}	n-Nonane	321	22.5
C_9H_{21}	Tripropyl amine	304.3	—
$C_{10}H_8$	Naphthalene	474.8	40.6
$C_{10}H_{12}O$	Ether, diphenyl-	494	30.9
$C_{10}H_{14}$	Benzene, 1-isopropyl-4-methyl-	385.5	27.7
$C_{10}H_{14}$	Benzene, 1,2,3,5-tetramethyl-	402.8	28.6
$C_{10}H_{14}$	p-Cymene, (see Benzene, 1-isopropyl-4-methyl-)	—	—
$C_{10}H_{14}$	Isodurene, (see Benzene, 1,2,3,5-tetramethyl-)	—	—
$C_{10}H_{14}O$	3-p-Cymenol	425.1	33
$C_{10}H_{14}O$	Thymol, (see 3-p-Cymenol)	425.1	33
$C_{10}H_{18}$	Decalin, cis-	418	28.7
$C_{10}H_{18}$	Decalin, trans-	408	28.7
$C_{10}H_{22}$	n-Decane	344.4	20.8
$C_{11}H_{10}$	Naphthalene, 1-methyl-	498.8	32.1
$C_{12}H_{10}$	Biphenyl	495	31.8
$C_{12}H_{11}N$	Diphenylamine	615.5	—
$C_{12}H_{18}$	Benzene, hexamethyl-	494	23.5
$C_{12}H_{18}$	Mellitine, (see Benzene, hexamethyl-)	—	—
$C_{12}H_{26}$	n-Dodecane	386	17.9
$C_{12}H_{27}N$	Tributyl amine	365.2	—

Table II

Name	Formula	Critical temp. T_c °C	Critical press. P_c atm.
Ammonia	NH_3	132.5	112.5
Argon	Ar	−122.3	48
Boron tribromide	BBr_3	300	—
Boron trichloride	BCl_3	178.8	38.2
Boron trifluoride	BF_3	−12.26	49.2
Carbon dioxide	CO_2	31	72.9
Carbon disulfide	CS_2	279	78
Carbon monoxide	CO	−140	34.5
Carbonyl sulfide	COS	104.8	65
Chlorine	Cl_2	144	76.1
Cyanogen	C_2N_2	126.6	—
Deuterium	D_2	−234.8	16.4
Fluorine	F_2	−129	55
Germanium tetrachloride	$GeCl_4$	276.9	38
Helium	He	−267.9	2.26
Hydrogen	H_2	−239.9	12.8
Hydrogen bromide	HBr	90	84.5
Hydrogen chloride	HCl	51.4	82.1
Hydrogen deuteride	HD	237.3	14.6
Hydrogen cyanide	HCN	183.5	48.9
Hydrogen fluoride	HF	188	64

Name	Formula	Critical temp. T_c °C	Critical press. P_c atm.
Hydrogen iodide	HI	150	81.9
Hydrogen sulfide	H_2S	100.4	88.9
Hydrazine	N_2H_2	380	—
Iodine	I_2	512	116
Krypton	Kr	−63.8	54.3
Neon	Ne	−228.7	26.9
Nitric oxide	NO	−93	64
Nitrogen dioxide	NO_2	157.8	100
Nitrogen	N_2	−147	33.5
Nitrous oxide	N_2O	36.5	71.7
Oxygen	O_2	−118.4	50.1
Ozone	O_3	−5.16	67
Phosphine	PH_3	51.3	64.5
Radon	Rn	104.04	62
Silane, chlorotrifluoro-	$SiClF_3$	34.5	34.2
Silane	SiH_4	−3.46	47.8
Silicon tetrachloride	$SiCl_4$	232.8	—
Silicon tetrafluoride	SiF_4	−14.06	36.7
Sulfur dioxide	SO_2	157.8	77.7
Stannic chloride	$SnCl_4$	318.7	37
Water	H_2O	374.1	218.3
Xenon	Xe	16.6	58

DISSOCIATION PRESSURE OF CALCIUM CARBONATE

Temp. °C	mm/Hg	Temp. °C	mm/Hg	Temp. °C	mm/Hg	Temp. °C	mm/Hg
550	0.41	727	44	819	235	894	716
587	1.0	736	54	830	255	898	760 atm.
605	2.3	743	60	840	311	906.5	1.151
671	13.5	748	70	852	381	937	1.770
680	15.8	749	72	857	420	1082.5	8.892
691	19.0	777	105	871	537	1157.7	18.687
701	23.0	786	134	881	603	1226.3	34.333
703	25.5	795	150	891	684	1241	39.094
711	32.7	800	183				

DEFINITIONS

AB — A prefix attached to the names of the practical electric units to indicate the corresponding unit in the cgs electromagnetic system (emu), e.g. abampere, abvolt.

Abcoulomb — The abcoulomb, the emu of charge, is defined as the charge which passes a given surface in 1 sec if a steady current of one abampere flows across the surface. Its dimensions are, therefore, cm½ g½, which differ from the dimensions of the statcoulomb by a factor which has the dimensions of a speed. This relationship is connected with the fact that the ratio $2K_e/K_m$ must have the value of the square of the speed of light in any consistent system of units. It follows further that

$$1 \text{ abcoulomb} = 2.99793 \times 10^{10} \text{ statcoulomb},$$

the speed of light in vacuo being $(2.99793 \pm 0.000003) \times 10^{10}$ cm/sec.

Aberration — 1. In astronomy, the apparent angular displacement of the position of a celestial body in the direction of motion of the observer, caused by the combination of the velocity of the observer and the velocity of light. See constant of aberration, planetary aberration. Compare parallax. 2. In optics, a specific deviation from perfect imagery, as, for example: spherical aberration, coma, astigmatism, curvature of field, and distortion.

Aberrations (of image) — Distortions in shape, color, focus, and density of images caused by imperfect optical elements (i.e., lens, prism, mirror, screen, etc.) Types such as coma, astigmatism, field curvature, distortion, and chromatic and spherical aberrations.

Absolute humidity — See Humidity.

Absolute pressure — See Pressure.

Absolute temperature — Temperature reckoned from the absolute zero. See Temperature.

Absolute units — A system of units based on the smallest possible number of independent units. Specifically, units of force, work, energy, and power not derived from or dependent on gravitation.

Absolute zero — The theoretical temperature at which molecular motion vanishes and a body would have no heat energy; the zero point of the Kelvin and Rankine temperature scales.

Absolute zero may be interpreted as the temperature at which the volume of a perfect gas vanishes or, more generally, as the temperature of the cold source which would render a Carnot cycle 100% efficient. The value of absolute zero is now estimated to be −273.15°C, −459.67°F, 0 K, and 0° Rankine.

Absorption — 1. Penetration of a substance into the body of another. 2. Transformation into other forms suffered by radiant energy passing through a material substances.

Absorption coefficient — 1. A measure of the amount of normally incident radiant energy absorbed through a unit distance of the absorbing medium. It is also referred to as linear attenuation coefficient and linear absorption coefficient.

The absorption coefficient (k) is frequently identified as:

$$I_{\lambda x} = I_{\lambda o}\, e^{-kx} \text{ or } k = \ln (I_{\lambda 0}/I_{\lambda x})$$

where $I_{\lambda x}$ is the flux density of radiation of wavelength λ, initially of flux density $I_{\lambda o}$, after traversing a distance x in the absorbing medium.

2. In acoustics, the ratio of the sound energy absorbed by a surface of a medium (or material) exposed to a sound field or sound radiation to the sound energy incident on the surface. The stated values of this ratio are to hold for an infinite area of the surface. The conditions under which measurements of absorption coefficients are made are to be stated explicitly.

Three types of absorption coefficients associated with three methods of measurement are: chamber absorption coefficient, obtained in a certain reverberation chamber; free-wave absorption coefficient, obtained when a plane, progressive, sound wave is incident on the surface of the medium; sabine absorption coefficient, obtained when the sound is incident from all directions on the sample. See Absorption Factor.

Absorption cross sections — In radar, the ratio of the amount of power removed from a beam by absorption of radio energy by a target to the power in the beam incident upon the target. Compare scattering cross section. See Cross section.

Absorption factor — The ratio of the intensity loss by absorption to the total original intensity of radiation. If I_o represents the original intensity, I_r the intensity of reflected radiation, I_t, the intensity of the transmitted radiation, the absorption factor is given by the expression

$$\frac{I_o - (I_r + I_t)}{I_o}$$

Also called coefficient of absorption.

Absorption, Lambert's law — If I_o is the original intensity, I the intensity after passing through a thickness x of a material whose absorption coefficient is k,

$$I = I_o e^{-kx}$$

The index of absorption k′ is given by the relation $k = (4\pi k'n)/\lambda$ where n is the index of refraction and λ the wave length in vacuo. The mass absorption is given by k/d when d is the density. The transmission factor is given by I/I_o.

Absorption spectrum — The array of absorption lines and absorption bands which results from the passage of radiant energy from a continuous source through a selectively absorbing medium cooler than the source. See electromagnetic spectrum.

The absorption spectrum is a characteristic of the absorbing medium, just as an emission spectrum is a characteristic of a radiator.

An absorption spectrum formed by a monatomic gas exhibits discrete dark lines, whereas that formed by a polyatomic gas exhibits ordered arrays (bands) of dark lines, which appear to overlap. This type of absorption is often referred to as line absorption. The spectrum formed by a selectively absorbing liquid or solid is typically continuous in nature (continuous absorption).

The spectrum obtained by the examination of light from a source, itself giving a continuous spectrum after this light has passed through an absorbing medium in the gaseous state. The absorption spectrum will consist of dark lines or bands, being the reverse of the emission spectrum of the absorbing substance.

When the absorbing medium is in the solid or liquid state the spectrum of the transmitted light shows broad dark regions which are not resolvable into lines and have no sharp or distinct edges.

Absorptive index — The imaginary part of the complex index of refraction of a medium. It represents the energy loss by absorption and has a nonzero value for all media which are not dielectrics. Also called index of absorption. Compare absorption coefficient.

Absorptive power or absorptivity for any body is measured by the fraction of the radiant energy falling upon the body which is absorbed or transformed into heat. This ratio varies with the character of the surface and the wave length of the incident energy. It is the ratio of the radiation absorbed by any substance to that absorbed under the same conditions by a black body.

Abvolt — The cgs electromagnetic unit of potential difference and electromotive force. It is the potential difference that must exist between two points in order that one erg of work be done when one abcoulomb of charge is moved from one point to the other. One abvolt is 10^{-8} V.

Acceleration — The time rate of change of velocity in either speed or direction. Cgs unit, 1 cm/sec/sec. Dimensions, — $[l\, t^{-2}]$ — See Angular acceleration.

Acceleration due to gravity — The acceleration of a body freely falling in a vacuum. The International Committee on Weights and Measures has adopted as a standard or accepted value, 980.665 cm/sec² or 32.174 ft/sec².

Acceleration due to gravity at any latitude and elevation — If ϕ is the latitude and H the elevation in centimeters the acceleration in cgs units is, $g = 980.21 - 0.02589 \cos 2\phi + 0.0007 \cos^2 2\phi - 3.086 \times 10^{-6}$ H. (Helmert's equation)

Accelerators — Machines for speeding up subatomic particles to energies running into millions of electron volts. See Betatron, Cyclotron, etc.

Acceptor — In transistors, the P-type semiconductor, the electrode containing trivalent impurities (boron, gallium, or indium) to increase the number of holes which can accept electrons. Contrast with donor.

Achromatic — A term applied to lenses signifying their more or less complete correction for chromatic aberration.

Acid — For many purposes it is sufficient to say that an acid is a hydrogen-containing substance which dissociates on solution in water to produce one or more hydrogen ions. More generally, however, acids are defined according to other concepts. The Bronsted concept states that an acid is any compound which can furnish a proton. Thus NH_4^+ is an acid since it can give up a proton:

$$NH_4^+ \rightleftharpoons NH_3 + H^+$$

and NH_3 is a base since it accepts a proton.

A still more general concept is that of G. N. Lewis which defines an acid as anything which can attach itself to something with an unshared pair of electrons. Thus in the reaction

$$H^+ + :N\begin{matrix} \diagup H \\ -H \\ \diagdown H \end{matrix} \rightleftharpoons NH_4^+$$

the NH_3 is a base because it possesses an unshared pair of electrons. This latter concept explains many phenomena, such as the effect of certain substances other than hydrogen ions in the changing of the color of indicators. It also explains acids and bases in nonaqueous systems as liquid NH_3 and SO_2.

Acoustic velocity (symbol a) = speed of sound.

Actinic — Pertaining to electromagnetic radiation capable of initiating photochemical reactions, as in photography or the fading of pigments.

Because of the particularly strong action of ultra violet radiation on photochemical processes, the term has come to be almost synonymous with ultraviolet, as in actinic rays.

Actinide Series — Elements of atomic numbers 89 to103 analogous to the lanthanide series of the so-called rare earths.

Actinometer — The general name for any instrument used to measure the intensity of radiant energy, particularly that of the sun. See Actinometry. See also Bolometer, Dosimeter, Photometer, Radiometer.

Actinometers may be classified, according to the quantities which they measure, in the following manner: (a) pyrheliometer, which measures the intensity of direct solar radiation; (b) pyranometer, which measures global radiation (the combined intensity of direct solar radiation and diffuse sky radiation); and (c) pyrgeometer, which measures the effective terrestrial radiation.

Action is measured by the product of work by time. Cgs units of action are the erg-second and the joule-second. Dimensions, — $[m\, l_2\, t^{-1}]$. Planck's quantum or constant of action is $(6.62517 - 0.00023) \times 10^{-27}$ erg-sec.

Active mass of a substance is the number of gram molecular weights per liter in solution, or in gaseous form.

Activity coefficient — A factor which, when multiplied by the molecular concentration yields the active mass. The activity coefficient is evaluated by thermodynamic calculations, usually from data on the emf of certain cells, or the lowering of the freezing point of certain solutions. It is a correction factor which makes the thermodynamic calculations correct.

Adiabatic — A body is said to undergo an adiabatic change when its condition is altered without gain or loss of heat. The line on the pressure volume diagram representing the above change is called an adiabatic line.

Adiabatic atmosphere — A model atmosphere in which the pressure decreases with height according to:

$$p = p_o\, [1 - (-\, gz/c_{p,d}T_o)]\; c_{p,d}R_d$$

where p_o and T_o are the pressure and temperature (°K) at sea level or other datum; z is the geometric height; R_d is the gas constant for dry gas; $c_{p,d}$ is the specific heat for dry gas at constant pressure; and g is the acceleration of gra vity. Also called dry-adiabatic atmosphere, convective atmosphere, homogeneous atmosphere. See Homogeneous atmosphere, Barotropy.

Adiabatic process — A thermodynamic change of state of a system in which there is no transfer of heat or mass across the boundaries of the system. In an adiabatic process, compression always results in warming, expansion in cooling. See Diabatic process.

Adsorption — The condensation of gases, liquids, or dissolved substances on the surfaces of solids is called adsorption.

Air columns, frequency of vibration in — See Organ pipes.

Albedo — The ratio of the amount of electromagnetic radiation reflected by a body to the amount incident upon it, often expressed as a percentage, as, the albedo of the earth is 34%.

The concept defined above is identical with reflectance. However, albedo is more commonly used in astronomy and meteorology and reflectance in physics.

Albedo is sometimes used to mean the flux of the reflected radiation as, the earth albedo is 0.64 cal/cm². This usage should be discouraged.

The albedo is to be distinguished from the spectral reflectance, which refers to one specific wavelength (monochromatic radiation).

Usage varies somewhat with regard to the exact wavelength interval implied in albedo figures; sometimes just the visible portion of the spectrum is considered, sometimes the totality of wavelengths in the solar spectrum.

Alfvén speed — The speed at which Alfvén waves are propagated along the magnetic field.

For a perfectly conducting fluid with a mass density of 1 kg/m³ in a magnetic field of 10,000 gauss, the Alfvén speed is about 1000 m/sec while the speed of sound in air is about 300 m/sec.

Alfvén wave — A transverse wave in a magneto-hydrodynamic field in which the driving force is the tension introduced by the magnetic field along the lines of force. Also called magneto-hydrodynamic wave.

The dynamics of such waves are analogous to those in a vibrating string, the phase speed C being given by

$$C^2 = \mu H^2/4\pi\rho$$

where μ is the permeability; H is the magnitude of the magnetic field; and ϱ is the fluid density. Dissipative effects due to fluid viscosity and electrical resistance may also be present.

Allobar — A form of an element differing in isotopic composition from the naturally occurring form.

Allotropy — The property shown by certain elements or being capable of existence in more than one form, due to differences in the arrangement of atoms or molecules. See Monotropic and Enantiotropic.

Alpha (α)-particle, or alpha-ray — One of the particles emitted in radioactive decay. It is identical with the nucleus of the helium atom and consists, therefore, of two protons plus two neutrons bound together. A moving alpha particle is strongly ionizing and so loses energy rapidly in traversing through matter. Natural alpha particles will traverse only a few centimeters of air before coming to rest.

Alternating current, (A-C) — Current in which the charge-flow periodically reverses, as opposed to direct current, and whose average value is zero. Alternating current usually implies a sinusoidal variation of current and voltage. This behavior is represented mathematically in various ways:

$$I = I_o \cos(2\pi ft + \phi)$$
$$I = I_o < \phi$$
$$I = I_1 \epsilon^{j\omega t}$$

where f is the frequency; $\omega \equiv 2\pi f$, the pulsatance, or radian frequency; ϕ the phase angle; I_o the amplitude; and I_1 the complex amplitude. In the complex notation, it is understood that the actual current is the real part of I. For circuits involving also a capacitance C in farads and L in henrys, the impedance becomes,

$$\sqrt{R^2 + \left(2\pi fL - \frac{1}{2\pi fC}\right)^2}$$

Altitudes with the barometer — If b_1 and b_2 denote the corrected barometer readings at two stations in any pressure units, and t is the mean of the temperatures, t_1 and t_2, of the air at the two stations in °C, then the difference in elevation in meters is

$$\Delta H = (1 + 0.00367\, t)\, (18{,}430 \log(b_1/b_2))$$

for differences not over 6000 m.

An approximate equation, generally sufficient for differences not over 1000 m, is

$$\Delta H \simeq (1 + 0.00367)\, (16{,}000 \frac{b_1 - b_2}{b_1 + b_2})$$

Amorphous — Without definite form, not crystallized.

Ampere (unit of electric current) is the constant current which, if maintained in two straight parallel conductors of infinite length, of negligible circular sections, and placed 1 m apart in a vacuum, will produce between these conductors a force equal to 2×10^{-7} newton per meter of length.

Ampère's law — A vector equation relating current flow and magnetic fields. The law states that if a current (I) flows through an infinitesimal distance (dl) along a line, at some point (P) a distance (r) away there is produced an infinitesimal element of a magnetic field (dH) such that

$$|dH| = 1 \sin \theta \, d\, I/|r^2|.$$

where θ is the angle between the direction of current flow and the line joing P and dl. More compactly,

$$dH = 1\, d\, I \times r/r^3$$

Ampere's rule — A positive charge moving horizontally is deflected by a force to the right if it is moving in a region where the magnetic field is vertically upward. This may be generalized to currents in wires by recallig that a current in a certain direction is equivalent to the motion of positive charges in that direction. The force felt by a negative charge is opposite to that felt by a positive charge.

Ampere-turn — A measure of magnetomotive force, especially as developed by an electric current, defined as the magnetomotive force developed by a coil of one turn through which a current of one ampere flows, that is, 1.26 Gb.

Amplitude — The maximum value of the displacement in an oscillatory motion.

Amplitude modulation — The variation of the amplitude of a wave in a way that corresponds to another wave. In general, the positive or negative envelope of the modulated wave (carrier wave) contains enough information to allow the modulating wave to be recovered, provided that the carrier wave has at least twice the frequency and twice the peak amplitude of the modulating signal. The modulated carrier C(t) is given by

$$C(t) = A_0 \, [1 - k\, M(t)] \cos \omega_0 t,$$

Where M(t) is the modulation.

Amplitude (of wave) — A measure of the maximum displacement of the wave crest from its undisturbed position (or the maximum electric or magnetic field strength of an electromagnetic wave).

AMU — The atomic mass unit (AMU), a unit of mass equal to 1/12 the mass of the carbon atom of mass number 12. On the atomic mass scale $^{12}C \equiv 12$.

$$\begin{aligned} 1 \text{ AMU} &= 931.4812(52) \text{ MeV} \\ &= 1.6605655(86) \times 10^{-27} \text{ (SI units)} \\ &= 1.6605655(86) \times 10^{-24} \text{ (cgs units)} \end{aligned}$$

The numbers in parentheses are the standard deviation uncertainties in the last digits of the quoted value, computed on the basis of internal consistency.

Angle — The ratio between the arc and the radius of the arc. Units of angle, the radian, the angle subtended by an arc equal to the radius; the degree, 1/360 part of the total angle about a point. Dimensions, a numeric.

Ångstrom — A unit of length, used especially in expressing the length of light waves, one tenth of a nanometer or one hundred-millionth of a centimeter (1×10^{-8} cm).

Angular acceleration — The time rate of change of angular velocity either in angular speed or in direction of the axis of rotation (precession). Cgs unit, 1 radian/sec/sec. Dimensions $[t^{-2}]$.

If the initial angular velocity is ω_0, and the velocity after time t is ω_t, the angular acceleration,

$$\alpha = \frac{\omega_t - \omega_0}{t}$$

The angular velocity after time t,

$$\omega_t = \omega_0 + \alpha t$$

The angle swept out in time t,

$$\theta = \omega_0 t + \tfrac{1}{2}\alpha t^2$$

The angular velocity after movement through the arc θ,

$$\omega = \sqrt{\omega_0{}^2 + 2\alpha\theta}$$

In the above equations, for angular displacement in radians, angular velocity will be in radians per second and angular acceleration in radians per second per second.

Angular aperture of an objective is the largest angular extent of wave surface which it can transmit.

Angular harmonic motion or harmonic motion of rotation — Periodic, oscillatory angular motion in which the restoring torque is proportional to the angular displacement. Torsional vibration.

Angular momentum or moment of momentum — Quantity of angular motion measured by the product of the angular velocity and the moment of inertia. Cgs unit, unnamed, its nature is expressed by g-cm²/sec. Dimentions, $[ml^2 t^{-1}]$.

The angular momentum of a mass whose moment of inertia is I, rotating with angular velocity ω, is $I\omega$.

Angular velocity — Time rate of angular motion about an axis. Cgs unit, radian/sec. Dimensions, $[t^{-1}]$.

If the angle described in time t is θ, the angular velocity,

$$\omega = \frac{\theta}{t}$$

θ in radians and t in seconds gives ω in radians per second.

Anhydride (of acid or base) — An oxide which when combined with water gives an acid or base.

Anion — A negatively charged ion.

Anisotropic — Exhibiting different properties when tested along axes in different directions.

Anode — The electrode at which oxidation occurs in a cell. It is also the electrode toward which anions travel due to the electrical potential. In spontaneous cells the anode is considered negative. In nonspontaneous or electrolytic cells the anode is considered positive.

Antiferromagnetic materials — Those in which the magnetic moments of atoms or ions tend to assume an ordered arrangement in zero applied field, such that the vector sum of the moments is zero, below a characteristic temperature called the Neel Point. The permeability of antiferromagnetic materials is comparable to that of paramagnetic materials. Above the Neel Point, these materials become paramagnetic.

Anti-matter — Matter consisting of antiparticles.

Anti-particle — Any particle with a charge of opposite sign to the same particle in normal matter.

Thus, the proton has a positive charge; the anti-proton, a negative charge. When a particle and its anti-particle collide, both may disappear with the creation of lighter particles; this process is called annihilation.

Aperture ratio — The ratio of the useful diameter of a lens to its focal length. It is the reciprocal of the f-number.

In application to an optical instrument, rather than to a lens, numerical aperture is more commonly used. The aperture ratio is then twice the tangent of the angle whose sine is the numerical aperture.

Apochromat — A term applied to photographic and microscope objectives indicating the highest degree of color correction.

Archimedes principle — A body wholly or partly immersed in a fluid is buoyed up by a force equal to the weight of the fluid displaced. A body of volume V cm³ immersed in a fluid of density ρ grams per cm³ is buoyed up by a force in dynes,

$$F = \rho g V$$

where g is the accleration due to gravity.

A floating body displaces its own weight of liquid.

Area, unit of — The square centimeter. The area of a square whose sides are one centimeter in length. Other units of area are similarly derived. Dimensions, $[L^2]$.

Arrhenius theory of electrolytic dissociation states that the molecule of an electrolyte can give rise to two or more electrically charged atoms or ions.

Astigmatism is an error of spherical lenses peculiar to the formation of images by oblique pencils. The image of a point when astigmatism is present will consist of two focal lines at right angles to each other and separated by a measurable distance along the axis of the pencil. The error is not eliminated by reduction of aperture as is spherical aberration.

Astronomical unit (abbr. AU) — 1. A unit of length, usually defined as the distance from the earth to the sun, 149,579,000 km. This value for the AU was derived from radar observations of the distance of Venus. The value given in astronomical ephemerides, 149,599,000 km, was derived from observations of the minor planet Eros.

2. The unit of distance in terms of which, in the Kepler Third Law, $n^2a^3 = k^2(1 + m)$, the semimajor axis a of an elliptical orbit must be expressed in order that the numerical value of the Gaussian constant k may be exactly 0.01720209895 when the unit of time is the ephemeris day.

In astronomical units, the mean distance of the earth from the sun, calculated by the Kepler law from the observed mean motion n and adopted mass m, is 1.00000003.

Atmosphere, standard — 1 standard atmosphere = 1,013,250 dyn/cm² i.e., 101,325 N/m².

Atmospheric radiation — Infrared radiation emitted by or being propagated through the atmosphere. See Insolation.

Atmospheric radiation, lying almost entirely within the wavelength interval of from 3 to 80 μm, provides one of the most important mechanisms by which the heat balance of the earth-atmosphere system is maintained. Infrared radiation emitted by the earth's surface (terrestrial radiation) is partially absorbed by the water vapor of the atmosphere which in turn reemits it, partly upward, partly downward. This secondarily emitted radiation is then, in general, repeatedly absorbed and reemitted, as the radiant energy progresses through the atmosphere. The downward flux, or counterradiation, is of basic importance in the greenhouse effect; the upward flux is essential to the radiative balance of the planet.

Atom — The smallest particle of an element which can enter into a chemical combination. All chemical compounds are formed of atoms, the difference between compounds being attributable to the nature, number, and arrangement of their constituent atoms. See Isotopes, Nuclear atom.

Atomic bomb — An explosive that derives its energy from the fission or fusion of atomic nuclei.

Atomic energy — 1. The constitutive internal energy of the atom which was absorbed when it was formed. 2. Energy derived from the mass converted into energy in nuclear transformations. See Einstein's formula.

Atomic mass (atomic weight) — The mass of a neutral atom of a nuclide. It is usually expressed in terms of the physical scale of atomic masses, that is, in atomic mass units (AMU). See AMU.

Atomic mass unit (AMU) — A measure of atomic mass, defined as equal to 1/12 the mass of a carbon atom of mass 12.

Atomic number (symbol Z) — An integer that expresses the positive charge of the nucleus in multiples of the electronic charge e. It is the number of electrons outside the nucleus of a neutral (un-ionized) atom and, according to widely accepted theory, the number of protons in the nucleus.

An element of atomic number Z occupies the Zth place in the periodic table of the elements. Its atom has a nucleus with a charge +Ze, which is normally surrounded by Z electrons, each of charge −e.

For example, the carbon isotope $_6C^{14}$ has an atomic number of 6 and an atomic mass of 14.

Atomic structure — According to the currently accepted view, the atom consists of a central part, called nucleus, and a number of electrons (called orbital or planetary electrons) circling about the former, like planets about the sun. The nucleus is of a high specific weight; it contains most of the mass of the entire atom (its mass is considered equal to the atomic mass) and is composed of positively charged particles, called protons (the number of which always equals the atomic number, (Z), and particles of 0 charge, called neutrons (the number of which equals the difference between the atomic weight and the atomic number, A − Z). The diameter of the nucleus is between 10^{-13} and 10^{-12} cm, and the relatively vast distance in which the orbital electrons circle about it is illustrated by the fact that this nuclear diameter is only 10^{-4} to 10^{-5} of the entire atomic diameter. While the nucleus carries an integral number of positive charges (an integral number of protons) each of 1.6×10^{-19} coulomb, each electron carries one negative charge of 1.6×10^{-19} coulomb, and the number of orbital electrons is equal to the number of protons in the nucleus (i.e. to the atomic number, Z), so that the atom as a whole has a net charge of 0. The electrons are arranged in successive shells (q.v.) around the nucleus; the maximum number of electrons in each shell is determined by natural laws, and the extranuclear electronic structure of the atom is characteristic of the element. The electrons in the inner shells are tightly bound to the nucleus; this inner structure can be altered by high-energy particles, γ-rays of radium, of X-rays. The electrons in the outer shells are responsible for the chemical properties of the element. See Bohr's atomic theory, Heisenberg's theory, Shell, and Subshell.

Atomic theory — All elementary forms of matter are composed of very small unit quantities called atoms. The atoms of a given element all have the same size and weight. The atoms of different elements have different sizes and weights. Atoms of the same or different elements unite with each other to form very small unit quantities of compound substances called molecules.

Atomic weight — Atomic weight is the relative weight of the atom on the basis of $^{12}C \equiv 12$. For a pure isotope, the atomic weight rounded off to the nearest integer gives the total number of nucleons (neutrons and protons) making up the atomic nucleus. If these weights are expressed in grams they are called gram atomic weights. See Isotopes and Atomic mass.

Atomic weight (abbr at. wt.) — The weight of an atom according to a scale of atomic weight units, awu, valued as one-twelfth the mass of the carbon atom ($C^{12} = 12.00000$).

Thus expressed, the atomic weight to the nearest integer is identical with the mass number.

Attenuation coefficient (symbol α) — A measure of the space rate of attenuation of any transmitted electromagnetic radiation. The attenuation coefficient is defined by

$$dI = -\alpha I_0 dx$$

or

$$I = I_0 e^{-\alpha x}$$

where I is the flux density at the selected point in space; I_0 is the flux density at the source; and α is the attenuation coefficient.

ATTO — A prefix meaning one quintillionth or 10^{-18}. Symbol is a.

Avogadro's law — Equal volumes of different gases at the same pressure and temperature contain the same number of molecules.

Avogadro's number — The number of molecules in one mole or gram-molecular weight of a substance. A number of values of the Avogadro number, which is usually denoted by N, have been found by various methods, generally lying withing a range of 1% about the value 6.022045×10^{23}/gm.

Avogadro's principle (or theory) — The numbers of molecules present in equal volumes of gases at the same temperature and pressure are equal.

Azimuth — 1. Horizontal direction or bearing. Compare azimuth angle. 2. In navigation, the horizontal direction of a celestial point from a terrestrial point, expressed as the angular distance from a reference direction, usually measured from 0° at the reference direction clockwise through 360°.

An azimuth is often designated as true, magnetic, compass, grid, or relative as the reference direction is true, magnetic, compass, grid north, or heading, respectively. Unless otherwise specified, the term is generally understood to apply to true azimuth, which may be further defined as the arc of the horizon, or the angle at the zenith, between the north part of the celestial meridian or principal vertical circle and a vertical circle, measured from 0° at the north part of the principal vertical circle clockwise through 360°.

3. In astronomy, the direction of a celestial point from a terrestrial point measured clockwise from the north or the south point of the meridian plane. 4. In

surveying, the horizontal direction of an object measured clockwise from the south point of the meridian plane.

In surveying, an aximuth of a celestial body is called an astronomic azimuth.

Azimuth angle — 1. Azimuth measured from 0° at the north or south reference direction clockwise or counterclockwise through 90° or 130°.

Azimuth angle is labeled with the reference direction as a prefix and the direction of measurement from the reference direction as a suffix. Thus, azimuth angle S 144° W is 144° west of south, or azimuth 324°. When azimuth angle is measured through 180°, it is labeled N or S to agree with the latitude and E or W to agree with the meridian angle.

2. In surveying, an angle in triangulation or in traverse through which the computation of azimuth is carried.

Babo's law — The addition of a nonvolatile solid to a liquid in which it is soluble lowers the vapor pressure of the solvent in proportion to the amount of substance dissolved.

Balmer series of spectral lines. The wave lengths of a series of lines in the spectrum of hydrogen were given in angstroms by the equation

$$\lambda = 3646 \frac{N^2}{N^2 - 4}$$

where N is an integer having values greater than 2.

Bar — International unit of pressure 10^6 dyn/cm². Unfortunately some writers have used this term for 1 dyn/cm². 1 bar = 0.987 atm = 1000 mbars = 29.53 in of mercury. See Torr.

Barn — Unit for measuring capture cross sections (q.v.) of elements. One barn = 10^{-24} cm²/nucleus.

Barotropy — The state of a fluid in which surfaces of constant density (or temperature) are coincident with surfaces of constant pressure; it is the state of zero baroclinity. Mathematically, the equation of barotropy states that the gradients of the density and pressure fields are proportional:

$$\Delta\rho = B\Delta p$$

where ϱ is the density; p is the pressure; and B is a function of thermodynamic variables, called the coefficient of barotropy.

With the equation of state, this relation determines the spatial distribution of all state parameters once these are specified on any surface. For a homogeneous atmosphere, B = 0; for an adiabatic atmosphere,

$$B = c_v/c_p \; RT$$

where c_v and c_p are the specific heats at constant volume and pressure, respectively; R is the gas constant; and T is the Kelvin temperature; for an isothermal atmosphere, B = 1/RT.

Barye — The pressure unit of the centimeter-gram-second system of physical units; equal to one dyne per square centimeter (0.001 mbar). Unfortunately some writers have used this term for the bar which is equal to 10^6 dyn/cm².

Bases — For many purposes it is sufficient to say that a base is a substance which dissociates on solution in water to produce one or more hydroxyl ions. More generally, however, bases are defined according to other concepts. The Bronsted concept states that a base is any compound which can accept a proton. Thus NH_3 is a base since it can accept a proton to form ammonium ions.

$$NH_3 + H^+ \rightleftharpoons NH_4^+$$

A still more general concept is that of G. N. Lewis which defines a base as anything which has an unshared pair of electrons. Thus in the reaction

$$H^+ + :\overset{H}{\underset{H}{N}}-H \rightleftharpoons NH_4^+$$

the NH_3 is a base because it possesses an unshared pair of electrons. This latter concept explains many phenomena, such as the effect of certain substances other than hydrogen ions in the changing of the color of indicators. It also explains acids and bases in nonaqueous systems as liquid NH_3 and SO_2.

Beam (of energy) — The locus of all series of wave-fronts projected from the source and directed toward given objects or positions in space.

Beam splitter — A device to produce two separate beams from one incident beam. This can be done with prisms or halfsilvered mirrors.

Beat(s) — Two vibrations of slightly different frequencies f_1 and f_2 when added together, produce in a detector sensitive to both these frequencies, a regularly varying response which rises and falls at the "beat" frequency $f_b = |f_1 - f_2|$. It is important to note that a resonator which is sharply tuned to f_b alone will not resound at all in the presence of these two beating frequencies. See Combination Frequencies.

Beat frequencies — The beat of two different frequencies of signals on a nonlinear circuit when they combine or beat together. It has a frequency equal to the difference of the two applied frequencies.

Beating — A wave phenomenon in which two or more periodic quantities of different frequencies produce a resultant having pulsations of amplitude.

This process may be contolled to produce a desired beat frequency. See Heterodyne.

Beer's law — If two solutions of the same colored compound be made in the same solvent, one of which is, say, twice the concentration of the other, the absorption due to a given thickness of the first solution should be equal to that of twice the thickness of the second.

Mathematically this may be expressed $l_1c_1 = l_2c_2$ when the intensity of light passing through the two solutions is a constant and if the intensity and wave length of light incident upon each solution are the same.

Bel — The fundamental division of a logarithmic scale for expressing the ratio of two amounts of power, the number of bels denoting such a ratio being the logarithm to the base 10 of this ratio.

With P_1 and P_2 designating two amounts of power and N the number of bels denoting their ratio, $N = \log^{10} (P_1/P_2)$ bels.

Bernoulli law or Bernoulli theorem — (After Daniel Bernoulli, 1700—1782, Swiss scientist.) 1. In aeronautics, a law or theorem stating that in a flow of incompressible fluid the sum of the static pressure and the dynamic pressure along a streamline is constant if gravity and frictional effects are disregarded.

From this law is follows that where there is a velocity increase in a fluid flow there must be a corresponding pressure decrease. Thus an airfoil, by increasing the velocity of the flow over its upper surface, derives lift from the decreased pressure.

2. As originally formulated, a statement of the conservation of energy (per unit mass) for a nonviscous fluid in steady motion. The specific energy is composed of the kinetic energy $u^2/2$, where u is the speed of the fluid; the potential energy gz, where g is the acceleration of gravity and z is the height above an arbitrary reference level; and the work done by the pressure forces of a compressible fluid $\int v\,dp$, where p is the pressure, v is the specific volume, and the integration is always with respect to values of p and v on the same parcel. Thus, the relationship

$$\frac{u_2}{2} + gz + \int v\,dp = \text{Constant along a streamline}$$

is valid for a compressible fluid in steady motion, since the streamline is also the path. If the motion is also irrotational, the same constant holds for the entire fluid.

Berthelot principle of maximum work — Of all possible chemical processes which can proceed without the aid of external energy, that process always takes place which is accompanied by the greatest evolution of heat. This law holds good for low temperatures only and does not account for endothermic reations.

Beta (β)-particle, (Beta ray) — One of the particles which can be emitted by a radioactive atomic nucleus. It has a mass about 1/1837. that of the proton. The negatively charged beta particle is identical with the ordinary electron, while the positively charged type (positron) differs from the electron in having equal but opposite electrical properties. The emission of an electron entails the change of a neutron into a proton inside the nucleus. The emission of a positron is similarly associated with the change of a proton into a neutron. Beta particles have no independent existence inside the nucleus, but are created at the instant of emission. See Neutrino.

Betatron — An accelerator used to impart high velocities to electrons (beta particles). Propellant is an electromagnetic field. A five to six MeV betatron can produce X-rays equivalent to the gamma radiation of 10 to 20 g of radium.

Bevatron — A six or more billion electron volt accelerator of protons and other atomic particles. Makes use of a Cockcroft-Walton transformer cascade accelerator and a linear (q.v.) as well as an electromagnetic field in the build-up.

Binary notation — A system of positional notation in which the digits are coefficients of powers of the base 2 in same way as the digits in the conventional decimal system are coefficients of powers of the base 10.

Binary notation employs only two digits, 1 and 0, therefore is used extensively in computers where the on and off positions of a switch or storage device can represent the two digits.

In decimal notation 111 — $(1\times10^2)+(1\times10^1)+(1\times10^0) = 100+10+1$ = one hundred and eleven.

In binary notation 111 = $(1\times2^2)+(1\times2^1)+(1\times2^0) = 4+2+1$ = seven.

Black body — If, for all values of the wave length of the incident radiant energy, all of the energy is absorbed the body is called a black body.

Bohr radius (symbol a^o) — The smallest possible radius of an electron orbit in the Bohr model of the atom, 5.29167×10^{-9} cm.

Bohr's atomic theory — The theory that atoms can exist for a duration solely in certain states, characterized by definite electronic orbits, i.e., by definite energy levels of their extra-nuclear electrons, and in these stationary states they do not emit radiation; the jump of an electron from an orbit to another of a smaller radius is accompanied by monochromatic radiation.

Bolometer — An instrument which measures the intensity of radiant energy by employing a thermally sensitive electrical resistor; a type of actinometer. Also called actinic balance. Compare radiometer.

Two identical, blackened, thermally sensitive electrical resistors are used in a Wheatstone bridge circuit. Radiation is allowed to fall on one of the elements, causing a change in its resistance. The change is a measure of the intensity of the radiation.

Boltzmann constant (symbol k) — The ratio of the universal gas constant to Avogadro number; equal to 1.38054×10^{-16} erg/°K. Sometimes called gas constant per molecule, Boltzmann universal conversion factor.

Bouguer law — A relationship describing the rate of decrease of flux density of a plane-parallel beam of monochromatic radiation as it penetrates a medium which both scatters and absorbs at that wavelength. This law may be expressed

$$dI_\lambda = -\alpha_\lambda I_\lambda\, dx$$

or

$$I_\lambda = I_{\lambda 0} e^{-\alpha_\lambda x}$$

where I is the flux density of the radiation; α_λ is the attenuation coefficient (or extinction coefficient) of the medium at wavelength λ; $I_{\lambda o}$ is the flux density at the source; and x is the distance from the source. Sometimes called Beer law, Lambert law of absorption. See Absorption coefficient, Scattering coefficient.

Boussinesq approximation — The assumption (frequently used in the theory of convection) that the fluid is incompressible except insofar as the thermal expansion produces a buoyancy, represented by a term $g\alpha T$, where g is the acceleration of gravity; α is the coefficient of thermal expansion; and T is the perturbation temperature.

Boyle's law for gases — At a constant temperature the volume of a given quantity of any gas varies inversely as the pressure to which the gas is subjected. For a perfect gas, changing from pressure p and volume v to pressure p′ and volume v′ without change of temperature,

$$pv = p'v'$$

Boyle-Mariotte law — The empirical generalization that for many so-called perfect gases, the product of pressure p and volume V is constant in an isothermal process:

$$pV = F(T)$$

where the function F of the temperature T cannot be specified without reference to other laws (e.g., Charles-Gay-Lussac law). Also called Boyle law, Mariotte Law.

Brayton cycle — (After George B. Brayton, American engineer.) Same as Joule cycle.

Breakdown potential = dielectric strength.

Breeder, Reactor (Breeder pile) — A nuclear chain reactor in which transmutation produces a greater number of fissionable atoms than the number of parent atoms consumed.

Bremsstrahlung (German, braking radiation) — Electromagnetic radiation produced by the rapid change in the velocity of an electron or another fast, charged particle as it approaches an atomic nucleus and is deflected by it. See Bremsstrahlung effect.

Bremsstrahlung effect — The emission of electromagnetic radiation as a consequence of the acceleration of charged elementary particles, such as electrons, under the influence of the attractive or repulsive force fields of atomic nuclei near which the charged particle moves.

In cosmic-ray shower production, bremsstrahlung effects give rise to emission of gamma rays as electrons encounter atmospheric nuclei. The emission of radiation in the bremsstrahlung effect is merely one instance of the general rule that electromagnetic radiation is emitted only when electric charges undergo acceleration.

Brewster window — An aperture through which light can enter into a new medium at an angle to the interface such that

$$\tan\theta_B = \frac{n_b}{n_a}$$

where θ_B is the Brewster angle. n_b and n_a are the indices of refraction of the media a and b, and light enters b from a.

Brewster's law — The tangent of the polarizing angle for a substance is equal to the index of refraction. The polarizing angle is that angle of incidence for which the reflected polarized ray is at right angles to the refracted ray. If n is the index of refraction and θ the polarizing angle, n = tan θ.

Brightness is measured by the flux emitted per unit emissive area as projected on a plane normal to the line of sight. The unit of brightness is that of a perfectly diffusing surface giving out one lumen per square centimeter of projected surface and is called the lambert. The millilambert (0.001 lambert) is a more convenient unit. Candle per square centimeter is the brightness of a surface which has, in the direction considered, a luminous intensity of 1 candle/cm².

British thermal unit — The quantity of heat required to raise the temperature of one pound of water 1°F at, or near, its point of maximum density (39.1°F). The Btu is equivalent to 0.252 kilogram-calorie or 1055 joules.

Brownian movement — A continuous agitation of particles in a colloidal solution caused by unbalanced impacts with molecules of the surrounding medium. The motion may be observed with a microscope when a strong beam of light is caused to traverse the solution across the line of sight.

Bulk modulus — The modulus of volume elasticity,

$$M_B = \frac{p_2 - p_1}{\frac{v_1 - v_2}{v_1}}$$

where p_1, p_2; v_1, v_2 are the initial and final pressure and volume respectively. It is the reciprocal of the coefficient of compressibility.

Calorie — The amount of heat necessary to raise 1 g of water at 15°C, 1°C. There are various calories depending upon the interval chosen. Sometimes the unit is written as the gram-calorie or the kilogram calorie, the meaning of which is evident. The calorie may be defined in terms of its mechanical equivalent. The National Bureau of Standards defines the calorie as 4.18400 joules. At the International Steam Table Conference held in London in 1929 the international calorie was defined at 1/860 of the international watt hour, which makes it equal to 4.1860 international joules.

With the adoption of the absolute system of electrical units, this becomes 1/859.858 watt hours or 4.18674 joules. The Btu was defined at the same time as 251.996 international calories.

Calutron — An apparatus operating on the principle of the mass spectrograph and used for separating U^{235} from U^{238}.

Candela — The candela is the luminous intensity, in the direction of the normal, of a black body surface 1/600,000 m² in area, at the temperature of solidification of platinum under a pressure of 101,325 N/m².

Candle (new unit) — 1/60 of the intensity of 1 cm^2 of a blackbody radiator at the temperature of solidification of platinum (2045K).

Capacitance is measured by the charge which must be communicated to a body to raise its potential one unit. Electrostatic unit capacitance is that which requires one electrostatic unit of charge to raise the potential one electrostatic unit. The farad = 9×10^{11} electrostatic units. A capacitance of one farad requires one coulomb of electricity to raise its potential one volt. Dimensions, $[\epsilon l]$; $[\mu^{-1} l^{-1} t^2]$.

A conductor charged with a quantity Q to a potential V has a capacitance,

$$C = \frac{Q}{V}$$

Capacitance of a spherical conductor of radius r,

$$C = Kr$$

Capacitance of two concentric spheres of radii r and r′

$$C = K \frac{rr'}{r - r'}$$

Capacitance of a parallel plate condenser, the area of whose plates is A and the distance between them d,

$$C = \frac{KA}{4\pi d}$$

Capacitances will be given in electrostatic units if the dimensions of condensers are substituted in centimeters. K is the dielectric constant of the medium.

Capillary constant or specific cohesion,

$$a^2 = \frac{2T}{(d_1 - d_2)g} = hr$$

where T is surface tension, d_1 and d_2, the densities of the two fluids, g the acceleration due to gravity, h the height of rise in a capillary tube of radius r. See Surface tension.

Carbon cycle — A sequence of atomic nuclear reactions and spontaneous radioactive decay which serves to convert matter into energy in the form of radiation and high-speed particles, and which is regarded as one of the principal sources of the energy of the sun and other similar stars.

This cycle, first suggested by Bethe in 1938, gets its name from the fact that carbon plays the role of a kind of catalyst in that it is both used by and produced by the reaction, but is not consumed itself. Four protons are, in net, converted into an alpha particle and two positrons (with accompanying neutrinos); and three gamma-ray emissions are emitted directly in addition to the two gamma emissions that ensue from annihilation of the positrons by ambient electrons. This cycle sets in at stellar interior temperatures of the order of 5 million degrees Kelvin.

An even simpler reaction, the proton-proton reaction, is also believed to occur within the sun and may be of equal or greater importance.

Carnot cycle — An idealized reversible thermodynamic cycle. The Carnot cycle consists of four stages: (a) an isothermal expansion of the gas at temperature T_1; (b) an adiabatic expansion to temperature T_2; (c) an isothermal compression at temperature T_2; (d) an adiabatic compression to the original state of the gas to complete the cycle. See Carnot engine, Thermodynamic efficiency.

In a Carnot cycle, the net work done is the difference between the heat input Q_1 at higher temperature T_1 and the heat extracted Q_2 at the lower temperature T_2.

Carnot engine — An idealized reversible heat engine working in a Carnot cycle. It is the most efficient engine that can operate between two specified temperatures; its efficiency is equivalent to the thermodynamic efficiency. The Carnot engine is capable of being run either as a conventional engine or as a refrigerator.

Cassegrain telescope — A reflecting telescope in which a small hyperboloidal mirror reflects the convergent beam from the paraboloidal primary mirror through a hole in the primary mirror to an eyepiece in back of the primary mirror. Also called Cassegrainian telescope, Cassegrain. See Newtonian Telescope.

Catalytic agent — A substance which by its mere presence alters the velocity of a reaction, and may be recovered unaltered in nature or amount at the end of the reaction.

Cathode — The electrode at which reduction occurs. It is the negative electrode in a cell through which current is being forced, but it is the positive pole of a battery. In a vacuum tube, the cathode is the electrode from which electrons are liberated. See Anode.

Cation — A positively charged ion.

Cauchy's dispersion formula —

$$n = A + \frac{B}{\lambda^2} + \frac{C}{\lambda^4} + \ldots$$

An empirical expression giving an approximate relation between the refractive index n of a medium and the wavelength λ of the light; A, B, and C being constants for a given medium.

Celsius — See Temperature, Celsius, in this section.

CENTI — A prefix meaning 1/100 or 10^{-2}. Symbol is c.

Centigrade temperature scale (abbr. C) — A temperature scale with the ice point at 0° and the boiling point of water at 100°. Now called Celsius temperature scale.

Conversion from the Fahrenheit temperature scale is according to the formula

$$°C = 5/9\ (°F - 32)$$

Centipoise — A standard unit of viscosity, equal to 0.01 poise, the c.g.s. unit of viscosity. Water at 20°C has a viscosity of 1.002 centipoise or 0.01002 poise.

Centripetal force — The force required to keep a moving mass in a circular path. Centrifugal force is the name given to the reaction against centripetal force.

Chain reaction — In general, any self-sustaining process, whether molecular or nuclear, the products of which are instrumental in, and directly contribute to the propagation of the process. Specifically, a fission chain reaction, where the energy liberated or particles produced (fission products) by the fission of an atom cause the fission of other atomic nuclei, which in turn propagate the fission reaction in the same manner.

Charles-Gay-Lussac law — An empirical generalization that in a gaseous system at constant pressure, the temperature increase and the relative volume increase stand in approximately the same proportion for all so-called perfect gases. Mathematically,

$$t - t_0 = (1/c\ [(v - v_0)/v_0]$$

where t is temperature; v is volume; and c is a coefficient of thermal expansion independent of the particular gas. If the centigrade temperature scale is used and v_0 is the volume at 0°C, then the value of the constant c is approximately 1/273. Also called Charles law, Gay-Lussac law.

Charles law = Charles-Gay-Lussac law.

Chemiluminescence — Emission of light during a chemical reaction.

Christiansen effect — When finely powdered substances, such as glass or quartz, are immersed in a liquid of the same index of refraction complete transparency can only be obtained for monochromatic light. If white light is employed the transmitted color corresponds to the particular wave-length for which the two substances, solid and liquid have exactly the same index of refraction. Due to differences in dispersion the indices of refraction will match for only a narrow band of the spectrum.

Chromatic aberration — Due to the difference in the index of refraction for different wave lengths, light of various wave lengths from the same source cannot be focused at a point by a simple lens. This is called chromatic aberration.

Circularly polarized wave — An electromagnetic wave for which the electric or the magnetic field vector, or both, at a point describe a circle.

This term is usually applied to transverse waves.

Circular mil — The area of a circle with a diameter of 0.001 in, a unit used for the measurement of small circular areas, such as the cross section of a wire. One circular mil = 7.85×10^{-7} in^2.

Circular polarization — The polarization of a wave radiated by a constant electric vector rotating in a plane so as to describe a circle. See Elliptical polarization.

Circulation — 1. The flow or motion of a fluid in or through a given area or volume. 2. A precise measure of the average flow of fluid along a given closed curve. Mathematically, circulation is the line integral.

$$\oint v \cdot dr$$

about the closed curve, where v is the fluid velocity, and dr is a vector element of the curve.

By Stokes theorem, the circulation about a plane curve is equal to the total vorticity of the fluid enclosed by the curve.

The given curve may be fixed in space or may be defined by moving fluid parcels.

Circulation integral — The line integral of an arbitrary vector taken around a closed curve. Thus,

$$\oint a \cdot dr$$

is the circulation integral of the vector a around the closed curve; dr is an infinitesimal vector element of the curve. If the vector is the velocity, this integral is called the circulation.

Clapeyron-Clausius equation — The differential equation relating pressure to temperature in a system in which two phases of a substance are in equilibrium.

$$dp/dT = L/(T\Delta V)$$

where p is pressure; T is temperature; L is the latent heat of the phase change; and ΔV is the difference in volume of the phases. Also called Clapeyron equation, Clausius-Clapeyron equation.

Clausius-Clapeyron equation = Clapeyron-Clausius equation.

Cloud chamber — An apparatus containing moist air or other gas which on sudden expansion condenses moisture to droplets on dust particles or other nuclei. Thus charged particles or ions in the space become nuclei and their numbers and behavior, when properly illuminated, may be studied.

Coefficient of compressibility — The relative decrease of the volume of a gaseous system with increasing pressure in an isothermal process. This coefficient is

$$-(1/V)\,(\partial V/\partial p)_T$$

where V is the volume; p is the pressure; and T is the temperature. The reciprocal of this quantity is the bulk modulus. Also called compressibility. Compare coefficient of thermal expansion, coefficient of tension.

Coefficient of tension — The relative increase of pressure of a system with increasing temperature in an isochoric process. In symbols this quantity is

$$(l/p)\,(\partial p/\partial T)_V$$

where p is pressure; T is temperature; and V is volume. Compare coefficient of compressibility, coefficient of thermal expansion.

Coefficient of thermal expansion — The ratio of the change of length per unit length (linear), or change of volume per unit volume (voluminal), to the change of temperature.

Coherence length — The maximum tolerable optical path length difference between two energy beams which are forming an interference pattern. This will vary with the degree of spectral purity of the source producing the beams. For example a perfect monochromatic source would have an infinite coherence length.

Coherent (additon) — The vector addition of both the amplitude and phases of different waves of the same frequency at a given time or at a given position.

Coherent oscillator (abbr Coho) — An oscillator which provides a reference by which the radio frequency phase difference of successive received pulses may be recognized. See Coherent reference.

Coherent reference — The reference signal, usually of stable frequency, to which other signals are phase-locked to establish coherence throughout a system.

Coherent (source) — A source radiating coherent waves.

Coherent (waves) — Waves whose frequencies are equal and whose phases are related to each other at a given time or at a given place in space. Coherence can be of two types, temporal and spatial.

Cold working — Deforming metal plastically at a temperature lower than the recrystallization temperature.

Colligative property — A property numerically the same for a group of substances, independent of their chemical nature.

Colloid — A phase dispersed to such a degree that the surface forces become an important factor in determining its properties.

In general particles of colloidal dimensions are approximately 10 nm to 1 μm in size. Colloidal particles are often best distinguished from ordinary molecules due to the fact that colloidal particles cannot diffuse through membranes which do allow ordinary molecules and ions to pass freely.

Colloidal system — An intimate mixture of two substances one of which, called the dispersed phase (or colloid) is uniformly distributed in a finely divided state through the second substance, called the dispersion medium (or dispersing medium). The dispersion medium may be a gas, a liquid, or a solid, and the dispersed phase may also be any of these, with the exception that one does not speak of a colloidal system of one gas in another. Also called colloidal dispersion, colloidal suspension.

A system of liquid or solid particles colloidally dispersed in a gas is called an aerosol. A system of solid substance or water-insoluble liquid colloidally dispersed in liquid water is called a hydrosol. There is no sharp line of demarcation between true solutions and colloidal systems on the one hand, or between mere suspensions and colloidal systems on the other. When the particles of the dispersed phase are smaller than about 1 nanometer in diameter, the system begins to assume the properties of a true solution; when the particles dispersed are much greater than 1 μm, separation of the dispersed phase from the dispersing medium becomes so rapid that the system is best regarded as a suspension.

Coma — An aberration of spherical lenses, occurring in the case of oblique incidence, when the bundle of rays forming the image is unsymmetrical. The image of a point is comet shaped, hence the name.

Combination frequencies — Two vibrations of arbitrary frequencies f_1 and f_2 when applied simultaneously to a nonlinear (distorting) device will excite it to a motion containing not only the original frequencies, but also members of a set of "combination" set of frequencies given by $f_* = mf_1 + nf_2$ where m and n are integers. A resonator sharply tuned to any one of these frequencies which may be produced in the nonlinear device will resound to it with an amplitude depending on the type of nonlinearity. The superheterodyne radio receiver depends on this phenomenon.

Combining volumes — Under comparable conditions of pressure and temperature the volume ratios of gases involved in chemical reactions are simple whole numbers.

Combining weight of an element or radical is its atomic weight divided by its valence.

Combining weights, law of — If the weights of elements which combine with each other be called their "combining weights," then elements always combine either in the ratio of their combining weights or of simple multiples of these weights.

Compensation Point — The temperature (below the Néel Point) at which, in some ferrimagnetic compounds, the saturation magnetization becomes zero.

Complementary color — Either of a pair of spectrum colors that when combined give a white or nearly white light.

Component substances, law of — Every material consists of one substance, or is a mixture of two or more substances, each of which exhibits a specific set of properties, independent of the other substances.

Compounds are substances containing more than one constituent element and having properties, on the whole, different from those which their constituents had as elementary substances. The composition of a given pure compound is perfectly definite, and is always the same no matter how that compound may have been formed.

Compressibility — Reciprocal of the bulk modulus.

Compton effect, (Compton recoil effect) — Elastic scattering of photons by electrons results in decrease in frequency and increase of wave length of X-rays and gamma-rays when scattered by free electrons.

Compton electron — An orbital electron of an atom which has been ejected from its orbit as a result of an impact by a high-energy quantum of radiation (X-ray or gamma-ray). Also called Compton recoil electron.

Compton wavelength (symbol λ_c) — Of a particle, the distance h/mc, where h is the Planck constant, m is the mass of the particle, and c is the velocity of light.

The Compton wavelength of the electron (symbol λ_c) is 2.4261×10^{-10} centimeter; of the proton (symbol $\lambda_{c,p}$) is 1.32140×10^{-13} centimeter.

Computer-generated holograms — A hologram made synthetically and based on computer calculations of amplitude and/or phase.

Concentration — The amount of a substance in weight, moles, or equivalents contained in unit volume.

Condensers in parallel and series — If c_1, c_2, c_3, etc. represent the capacitances of a series of condensers and C their combined capacitance,

when in parallel,

$$C = c_1 + c_2 + c_3 \ldots$$

when in series,

$$\frac{1}{C} = \frac{1}{c_1} + \frac{1}{c_2} + \frac{1}{c_3} \ldots$$

Conductance — The reciprocal of resistance, is measured by the ratio of the current flowing through a conductor to the difference of potential between its ends. The practical unit of conductance, the mho, the conductance of a body through which one ampere of current flows when the potential difference is one volt. The conductance of a body in mho is the reciprocal of the value of its resistance in ohms. Dimensions, $[\varepsilon\, l\, t^{-1}]$. $[\mu^{-1} l^{-1} t]$.

Conductivity, electrical, is measured by the quantity of electricity transferred across unit area, per unit potential gradient per unit time. Reciprocal of resistivity. Volume conductivity or specific conductance, $k = 1/\varrho$ where ϱ is the volume resistivity. Mass conductivity = k/d where d is density. Equivalent conductivity A = k/c where c is the number of equivalents per unit volume of solution. Molecular conductivity μ = k/m where m is the number of moles per unit volume of solution. Dimensions: volume conductivity, $[\varepsilon\, t^{-1}]$; $[\mu^{-1} l^{-2} t]$, mass conductivity, $[\varepsilon\, m^{-1} l^3 t^{-1}]$; $[\mu^{-1} m^{-1} lt]$.

Conductivity, thermal — Time rate of transfer of heat by conduction, through unit thickness, across unit area for unit difference of temperature. It is measured as calories per second per square centimeter for a thickness of one centimeter and a difference of temperature of 1°C.

If the two opposite faces of a rectangular solid are maintained at temperatures t_1 and t_2 the heat conducted across the solid of section a and thickness d in a time T will be,

$$Q = \frac{K(t_2 - t_1)aT}{d}$$

K is a constant depending on the nature of the substance, designated as the specific heat conductivity. K is usually given for Q in calories, t_1 and t_2 in °C, a in cem^2, T in sec, and d in cm. See Heat conductivity.

Conductors — A class of bodies which are incapable of supporting electric strain. A charge given to a conductor spreads to all parts of the body.

Conjugate foci — Under proper conditions light divergent from a point on or near the axis of a lens or spherical mirror is focused at another point. The point of convergence and the position of the source are interchangeable and are called conjugate foci.

Conservation of energy — The principle that the total energy of an isolated system remains constant if no interconversion of mass and energy takes place.

This principle takes into account all forms of energy in the system; it therefore provides a constraint on the conversions from one form to another.

Conservation of energy, law of — Energy can neither be created nor destroyed and therefore the total amount of energy in the universe remains constant.

Conservation of mass — In all ordinary chemical changes, the total of the reactants is always equal to the total mass of the products.

Conservation of momentum, law of — For any collision, the vector sum of the momenta of the colliding bodies after collison equal the vector sum of their momenta before collision. If two bodies of masses m_1 and m_2 have, before impact velocities v_1 and v_2 and after impact velocities u_1 and u_2

$$m_1 u_1 + m_2 u_2 = m_1 v_1 + m_2 v_2$$

Constant of aberration — The maximum aberration of a star observed from the surface of the earth, 20.49 sec of arc.

The maximum occurs at the time the direction of motion of the earth in its orbit is at right angles to a line from the earth to the star.

Constitutive property — A property which depends on the constitution or structure of the molecule.

Contrast of fringes — The relative difference between the brightness or density of successive bright and dark fringes on a hologram or interferogram.

Cooling — Processing highly radioactive materials to attain lesser radioactivity for subsequent use or handling.

Coriolis acceleration — An acceleration of a particle moving in a relative coordinate system. The total acceleration of the particle, as measured in an inertial coordinate system, may be expressed as the sum of the acceleration within the relative system, the acceleration of the relative system itself, and the coriolis acceleration.

Physically, coriolis acceleration may be considered as coming from the conservation of momentum in a body moving in a direction not parallel to the axis of rotation of the relative system.

Mathematically, coriolis acceleration comes from the differentiation of terms containing the angular velocity ω in the expression for the absolute velocity of the particle.

In the case of the earth, moving with angular velocity ω, a particle moving relative to the earth with velocity v has the coriolis acceleration $2\omega \times v$. If Newton laws are to be applied in the relative system, the coriolis acceleration and the acceleration of the relative system must be treated as forces.

Cosmic rays — Highly penetrating radiations which strike the earth, assumed to originate in interstellar space. They are classed as: primary, coming from the assumed source, and secondary, those induced in upper atmospheric nuclei by collisions with primary cosmic rays.

Cosmotron — A particle accelerator capable of giving them energies to billions of electron volts.

Couette flow — The shearing flow of a fluid between two parallel surfaces in relative motion. A two-dimensional steady flow without pressure gradient in the direction of flow and caused by the tangential movement of the bounding surfaces. The only practical type is the flow between concentric rotating cylinders (as of the oil in a cylindrical bearing).

Counterradiation — The downward flux of atmospheric radiation passing through a given level surface, usually taken as the earth's surface. Also called back radiation.

This result of infrared (long-wave) absorption and reemission by the atmosphere is the principal factor in the greenhouse effect.

Coulomb (unit of quantity of electricity) — the quantity of electricity transported in 1 sec by a current of 1 A. A unit quantity of electricity. It is the quantity of electricity which must pass through a circuit to deposit 0.0011180 g of silver from a solution of silver nitrate. An ampere is 1 coulomb/sec. A coulomb is also the quantity of electricity on the positive plate of a condenser of one-farad capacity when the electromotive force is 1 V.

Couple — Two equal and oppositely directed parallel but not colinear forces acting upon a body form a couple. The moment of the couple or torque is given by the product of one of the forces by the perpendicular distance between them. Dimension, $[m\,l^2\,t^{-2}]$.

Couple acting on a magnet of magnetic moment ml in a field of strength H. If the magnet is perpendicular to the direction of the field

$$C = Hml = HM$$

If the angle between the magnet and the field is θ

$$C = Hml \sin \theta$$

The couple will be in dyn-cm for cgs electromagnetic units of H, m and l.

Critical mass — The minimum mass the fissile material must have in order to maintain a spontaneous fission chain reaction. For pure U^{235} it is computed to be about 20 lb.

Critical point — The thermodynamic state in which liquid and gas phases of a substance coexist in equilibrium at the highest possible temperature. At higher temperatures than the critical no liquid phase can exist. For water substance the critical point is

$$P_s = 2.21 \times 10^5 \text{ mbar}$$
$$T = 647°K$$
$$v = 3.10 \text{ g/cm}^3$$

where P_s is the saturation vapor pressure of the water vapor; T is the Kelvin temperature; and v is the specific volume.

Critical temperature — 1. The temperature above which a substance cannot exist in the liquid state, regardless of the pressure. 2. As applied to materials, the temperature at which a change in phase takes place causing an appreciable change in the properties of the material.

Cross section (Nuclear cross section) — A measure of the probability of a particular process. The nuclear cross section is expressed by a/bc, where a is the number of processes occurring, b the number of incident particles, and c the number of target nuclei per cm². There are nuclear cross sections for fission, for slow neutron capture, for Compton collision, and for ionization by electron impact.

Crossed polarizer — A dual polarization filter and transducer which transforms varying orientations of polarized waves into an amplitude output.

Cryohydrate — The solid which separates when a saturated solution freezes. It contains the solvent and the solute in the same proportions as they were in the saturated solution.

Cryopumping — The process of removing gas from a system by condensing it on a surface maintained at very low temperatures.

Cryotron — A device based upon the principle that superconductivity established at temperatures near absolute zero is destroyed by the application of a magnetic field.

Crystal — The "ideal crystal" is a homogeneous portion of crystalline matter, (q.v.) whether bounded by faces or not.

Crystalline matter is matter that possesses a triperiodic structure on the atomic scale. It is characterized by discontinuous vectorial properties that give rise to "crystal planes" [(1) crystal growth (faces); (2) cohesion (cleavage planes); (3) twinning (twin planes); (4) gliding (gliding planes); (5) X-ray, electron, or neutron diffraction ("reflecting" planes); all of which are parallel to lattice planes.]

Curie — The curie is the rate of radioactive decay; the quantity of any radioactive nuclide which undergoes 3.7×10^{10} disintegrations/sec. The symbol for this unit is Ci. 1 Ci = 3.7×10^{10} Bq.

Curie's law — The intensity of magnetization,

$$I = \frac{AH}{T}$$

where H, is the magnetic field strength, T the absolute temperature and A Curie's constant. Used for paramagnetic substances.

Curie point — All ferro-magnetic substances have a definite temperature of transition at which the phenomena of ferro-magnetism disappear and the substances become merely paramagnetic. This temperature is called the "Curie Point" and is usually lower than the melting point.

Curie-Weiss law — The Curie law was modified by Weiss to state that the susceptibility of a paramagnetic substance above the Curie point varies inversely as the excess of the temperature above that point.

This law is not valid at or below the Curie point.

Current (electric) — The rate of transfer of electricity. The transfer at the rate of one electrostatic unit of electricity in one second is the electrostatic unit of current. The electromagnetic unit of current is a current of such strength that one centimeter of the wire in which it flows is pushed sideways with a force of 1 dyn when the wire is at right angles to a magnetic field of unit intensity. The practical unit of current is the ampere, a transfer of one coulomb per second, which is one tenth the electromagnetic unit. The international ampere is the unvarying electric current which, when passed through a solution of silver nitrate in accordance with certain specifications, deposits silver at the rate of 0.00111800 g/sec. The international ampere is equivalent to 0.999835 absolute ampere. The ampere-turn is the magnetic potential produced between the two faces of a coil of one turn carrying one ampere. Dimensions, $[\varepsilon^{1/2}\,m^{1/2}\,l^{1/2}\,t^{-2}]$; $[\mu^{-1/2}\,m^{1/2}\,l^{1/2}\,t^{-1}]$.

Current in a simple circuit — The current in a circuit including an external resistance R and a cell of electromotive force E and internal resistance,

$$I = \frac{E}{R + r}$$

If E is in volts and r and R in ohms the current will be in amperes.

For two cells in parallel,

$$I = \frac{E}{R + \frac{r}{2}}$$

For two cells in series,

$$I = \frac{2E}{R + 2r}$$

CW laser — Continuous wave laser — a laser that radiates its energy in an uninterrupted beam.

Cyclotron — The magnetic resonance accelerator for imparting very great velocities to heavier nuclear particles without the use of excessive voltages.

Dalton = atomic mass unit.

Dalton's law of partial pressures — The pressure exerted by a mixture of gases is equal to the sum of the separate pressures which each gas would exert if it alone occupied the whole volume. This fact is expressed in the following formula:p

$$PV = V(p_1 + p_2 + p_3, \text{etc.})$$

Day — 1. The duration of one rotation of the earth, or another celestial body, on its axis.

A day is measured by successive transits of a reference point on the celestial sphere over the meridian, and each type takes its name from the reference used. Thus, for a solar day the reference is the sun; a mean solar day if the mean sun; and an apparent solar day if the apparent sun. For a lunar day the reference is the moon for a sidereal day the vernal equinox; for a constituent day an astre fictif or fictitious star. The expression lunar day refers also to the duration of one rotation of the moon with respect to the sun. A Julian day is the consecutive number of each day, beginning with January 1, 4713 BC.

2. A period of 24 hr beginning at a specified time, as the civil day beginning at midnight, or the astronomical day beginning at noon.

DeBroglie wavelength — In quantum mechanics, a wavelength (λ) attributed to a particle by virtue of its momentum. In general

$$\lambda = \frac{h}{mv} = \frac{h}{m_0 v}\left(1 - \frac{v^2}{c^2}\right)^{1/2}$$

where m is the observed mass of the particle, m_o is its rest mass, v is its velocity, c is the velocity of light, and h is Planck's constant.

Debye-Falkenhagen effect — The increase in the conductance of an electrolytic solution produced by alternating currents of sufficiently high frequencies over that observed with low frequencies or with direct current.

Debye length — A theoretical length which describes the maximum separation at which a given electron will be influenced by the electric field of a given positive ion. Sometimes referred to as the Debye shielding distance or plasma length.

It is well known that charged particles interact through their own electric fields. In addition, Debye has shown that the attractive force between an electron and ion which would otherwise exist for very large separations is indeed cut off for a critical separation due to the presence of other positive and negative charges in between. This critical separation or Debye length decreases for increased plasma density.

Decay — Diminution of a radioactive substance due to nuclear emission of alpha or beta particles, gamma rays or positrons.

Decay constant — 1. = attenuation constant. 2. (symbol λ) A constant relating the instant rate of radioactive decay of a radioactive species to the number of atoms N present at a given time t. Thus,

$$-(\partial N/\partial t) = \lambda N$$

If N_o is the number of atoms present at time zero then

$$N = N_0 e^{-\lambda t}$$

Decay product — A nuclide resulting from the radioactive disintegration of a radionuclide, being formed either directly or as the result of succssive transformations in a radioactive series. Also called daughter, daughter element.

A decay product may be either radioactive or stable.

Deci — A prefix meaning one tenth or 10^{-1}. Symbol is d.

Decibel (abbr. db) — 1. A dimensionless measure of the ratio of two powers, equal to 10 times the logarithm to the base 10 of the ratio of two powers P_1/P_2. One tenth of a bel.

The power P_2 may be some reference power; in electricity, the reference power is sometimes taken as 1 milliwatt (abbr. dbm); in acoustics, the decibel is often taken as 20 times the common logarithm of the sound pressure ratio, with the reference pressure as 0.0002 dyn/cm^2.

Declination — 1. (symbol δ) Angular distance north or south of the celestial equator; the arc of an hour circle between the celestial equator and a point on the celestial sphere, measured northward or southward from the celestial equator through 90°, and labeled N or S to indicate the direction of measurement. 2. (symbol D) Magnetic declination. See Equatorial system.

Decomposition is the chemical separation of a substance into two or more substances, which may differ from each other and from the original substances.

Definite proportions, law of — In every sample of each compound substance the proportions by weight of the constituent elements are always the same.

Degree — Angle subtended at the center by a circular arc which is 1/360 of the circumference.

Degree of association $(1 - \alpha)$ — The degree of association of an electrolytic solution is the percentage of ions associated into nonconducting species, such as ion-pairs.

Degree of dissociation (or ionization) in general, α — The degree of dissociation (or ionization) of an electrolytic solution is the percentage of solute (or electrolyte) in the dissociated (or ionized) state in solution. Classically this degree is obtained from conductance measurements from the ratio, Λ/Λ_i where Λ_i is the equivalent conductance an electrolytic solution would have at some finite concentration if it were completely dissociated into ions at that concentration if it were completely dissociated into ions at that concentration. (See ionogens). This symbol is also used to denote the fraction of free ions in a solution when simple ions, ion pairs, and clusters higher than ion pairs are present. (See ionophores).

Degree of freedom — The number of the variables determining the state of a system (usually pressure, temperature, and concentrations of the components) to which arbitrary values can be assigned.

Deka — A prefix meaning ten or 10^1. Symbol is da.

Delayed neutrons — Neutrons emitted by excited nuclei in a radioactive process, so called because they are emitted an appreciable time after the fission. Compare prompt neutrons.

Delta ray — 1. An electron ejected by recoil when a rapidly moving alpha particle or other charged particle passes through matter. 2. By extension any secondary ionizing particle ejected by recoil when a primary particle passes through matter.

Density — Concentration of matter, measured by the mass per unit volume. Dimensions, [m l^{-3}].

Density (of film) — The logarithm of the reciprocal of the optical transmission of the film.

Depletion layer — In a semiconductor, a region in which the mobile carrier charge density is insufficient to neutralize the net fixed charge density of donors and acceptors. Also called barrier.

Detenation wave — A shock wave in a combustible mixture, which originates as a combustion wave.

Deuterium (symbol D, d) — A heavy isotope of hydrogen having one proton and one neutron in the nucleus.

The symbol D is often used to designate deuterium in compounds, as HDO for molecules of that composition. Official chemical nomenclature uses the designation d with a number which designates the carbon atom to which the deuterium is bound; e.g., 2-d propane designates CH_3CHDCH_3.

Deuteron — Nucleus of the deuterium atom or the ion of deuterium. Its structure — one neutron and one proton.

Dewpoint — The temperature to which a given parcel of air must be cooled at constant pressure and constant water-vapor content in order for saturation to occur; the temperature at which the saturation vapor pressure of the parcel is equal to the actual vapor pressure of the contained water vapor. Any further cooling usually results in the formation of dew or frost. Also called dewpoint temperature.

When this temperatue is below 0°C, it is sometimes called the frost point.

Diabatic process — A process in a thermodynamic system in which there is a transfer of heat across the boundaries of the system.

Diabatic process is preferred to nonadiabatic process.

Diamagnetic materials — Are those within which an externally applied magnetic field is slightly reduced because of an alteration of the atomic electron orbits produced by the field. Diamagnetism is an atomic-scale consequence of the Lenz law of induction. The permeability of diagmagnetic materials is slightly less than that of empty space.

Dielectric — A material having a relatively low electrical conductivity; an insulator; a substance that contains few or no free electrons and which can support electrostatic stresses. The principal properties of a dielectric are its dielectric constant (the factor by which the electric field strength in a vacuum exceeds that in the dielectric for the same distribution of charge), its dielectric loss (the amount of energy it dissipates as heat when placed in a varying electric field), and its dielectric strength (the maximum potential gradient it can stand without breaking down).

In an electromagnetic field, the centers of the nonpolar molecules of a dielectric are displaced, and the polar molecules become oriented close to the field. The net effect is the appearance of charges at the boundaries of the dielectric. The frictional work done in orientation absorbs energy from the field which appears as heat. When the field is removed the orientation is lost by thermal agitation and so the energy is not regained. If free-charge carriers are present they too can absorb energy.

A good dielectric is one in which the absorption is a minimum. A vacuum is the only perfect dielectric. The quality of an imperfect dielectric is its dielectric strength; and the accumulation of charges within an imperfect dielectric is termed dielectric absorption.

Dielectric constant (symbol ε) — for a given substance, the ratio of the capacity of a condenser with that substance as dielectric to the capacity of the same condenser with a vacuum for dielectric. It is a measure, therefore, of the amount of electrical charge a given substance can withstand at a given electric field strength; it should not be confused with dielectric strength.

The dielectric constant ε is a function of temperature and frequency and is wirtten as a complex quantity

$$\epsilon = \epsilon' - i\epsilon''$$

where ε' is the part that determines the displacement current and ε'' the dielectric absorption (see dielectric). For a nonabsorbing, nonmagnetic material ε' is equal to the square of the index of refraction and the relation holds only at the particular frequency where these conditions apply.

$$F = \frac{QQ'}{\epsilon r'}$$

where F is the force of attraction between two charges Q and Q′ separated by a distance r in a uniform medium.

Dielectrics or insulators or nonconductors — A class of bodies supporting an electric strain. A charge on one part of a nonconductor is not communicated to any other part.

Dielectric strength — A measure of the resistance of a dielectric to electrical breakdown under the influence of strong electric fields; usually expressed in volts per centimeter. Sometimes called breakdown potential.

Diffraction — That phenomena produced by the spreading of waves around and past obstacles which are comparable in size to their wavelength.

Diffraction efficiency — Ratio of energy projected into the reconstructed image to the energy illuminating the hologram.

Diffraction fanning — The fanning out of a light or energy beam as it pours through a very narrow aperture (opening).

Diffraction grating — If s is the distance between the rulings, d the angle of diffraction, then the wave length where the angle of incidence is 90% is (for the nth order spectrum),

$$\lambda = \frac{s \sin d}{n}$$

If i is the angle of incidence, d the angle of diffraction, s the distance between the rulings, n the order of the spectrum, the wave length is,

$$\lambda = \frac{s}{n} (\sin i + \sin d).$$

A mask or special aperture used to break up a white light beam or composite energy beam into its various spectral components through the mechanism of diffraction.

Diffuse reflection — Scattering at all angles from the point of reflection.

Diffuse sky radiation — Solar radiation reaching the earth's surface after having been scattered from the direct solar beam by molecules or suspensoids in the atmosphere. Also called the skylight, diffuse skylight, sky radiation.

Of the total light removed from the direct solar beam by scattering in the atmosphere (approximately 25% of the incident radiation), about two-thirds ultimately reaches the earth as diffuse sky radiation.

Diffusion — If the concentration (mass of solid per unit volume of solution) at one surface of a layer of liquid is d_1 and at the other surface d_2, the thickness of the layer h and the area under consideration A, then the mass of the substance which diffuses through the cross-section A in time t is,

$$m = \Delta A \frac{(d_2 - d_1)t}{h}$$

where Δ is the coefficient of diffusion.

Diffusion Coefficient — If the concentration (mass of solid per unit volume of solution) at one surface of a layer of liquid is d_1, and at the other surface d_2, the thickness of the layer is h, the area under consideration is A, and the mass of a given substance which diffuses through the cross section A in time t is m, then the diffusion coefficient is defined as

$$D = \frac{mh}{A(d_2 - d_1)t}$$

Diffusivity — A measure of the rate of diffusion of a substance, expressed as the diffusivity coefficient K. When K is constant, the diffusion equation is

$$\frac{\partial q}{\partial t} = K\nabla^2 q$$

where q is the substance diffused; ∇^2 is the Laplacian operator; and t is time. The diffusivity has dimensions of a length times a velocity; it varies with the property diffused, and for any given property it may be considered a constant or a function of temperature, space, etc., depending on the context. Also called coefficient of diffusion. See conductivity, kinematic viscosity, exchange coefficients.

In the case of molecular diffusion the length dimension is the mean free path of the molecules. By analogy, in eddy diffusion, length becomes the mixing length. The coefficient is then called the eddy diffusivity, and is in general several orders of magnitude larger than the molecular diffusivity.

Diffusivity of heat — is given by Δ in the equation

$$\frac{dH}{dt} = -\Delta sd \frac{dT}{dx} dy\, dz$$

where dH is the quantity of heat passing through the area dy dz in the direction of x in a time dt. The rate of variation of temperature along x is given by dT/dx, s is specific heat and d, density. Dimensions, $[l^2 t^{-1}]$.

Dimensional formulae — If mass, length, and time are considered fundamental quantities, the relation of other physical quantities and their units to these three may be expressed by a formula involving the symbols l-m and t respectively, with appropriate exponents. For example; the dimensional formula for volume would be expressed, $[l^3]$; velocity, $[t^{-1}]$; force $[mlt^{-2}]$. Other fundamental quantities used in dimensional formulae may be indicated as follows: θ, temperature, ε the dielectric constant of a vacuum; μ, the magnetic permeability of a vacuum.

Diminution of pressure at the side of a moving stream — If a fluid of density d moves with a velocity v, the dimunution of pressure due to the motion is (neglecting viscosity),

$$p = \tfrac{1}{2} dv^2$$

Dip — The angle measured in a vertical plane between the direction of the earth's magnetic field and the horizontal.

Dipole — (1) A combination of two electrically or magnetically charged particles of opposite sign which are separated by a very small distance. (2) Any system of charges, such as a circulating current, which has the properties that: (a) no forces act on it in a uniform field; (b) a torque proportional to sin θ, where θ is the angle between the dipole axis and a uniform field, does act on it; (c) it produces a potential which is proportional to the inverse square of the distance from it.

Dipole moment — A mathematical entity; the product of one of the charges of a dipole unit by the distance separating the two dipolar charges. In terms of the definition of a dipole (2), the dipole moment p is related to the torque T, and the field strength E (or B) through the equation:

$$T = p \times E$$

Dipole moment, molecular — It is found from measurements of dielectric constant (i.e. by its temperature dependence, as in the Debye equation for total polarization) that certain molecules have permanent dipole moments. These moments are associated with transfer of charge within the molecule and provide valuable information as to the molecular structure.

Directly ionizing particles — are charged particles (electrons, protons, alpha particles, etc.) having sufficient kinetic energy to produce ionization by collision.

Direct solar radiation — In actinometry, that portion of the radiant energy received at the instrument direct from the sun, as distinguished from diffuse sky radiation, effective terrestrial radiation, or radiation from any other source. See global radiation.

Direct solar radiation is measured by pyrheliometers.

Dispersion — The difference between the index of refraction of any substance for any two wave lengths is a measure of the dispersion for these wave lengths, called the coefficient of dispersion.

Dispersion forces — The force of attraction between molecules possessing no permanent dipole. The interaction energy is given by

$$U_D = -\frac{3}{4} h \frac{V_0 \alpha^2}{r^6}$$

where h is Planck's constant, V_0 a characteristic frequency of the molecule, r the distance between the molecules, and α the polarizability.

Dispersive power — If n_1 and n_2 are the indices of refraction for wave lengths λ_1 and λ_2 and n the mean index or that for sodium light, the dispersive power for the specified wave length is,

$$\omega = \frac{n_2 - n_1}{n - 1}$$

Displacement is a reaction in which an elementary substance displaces and sets free a constituent element from a compound.

Displacement or elongation at any instant. The distance of a vibrating or oscillating particle from its position of equilibrium.

Dissociation-field effect — The increased dissociation (or ionization) of the molecules of weak electrolytes under the influence of high electrical fields (potential gradients).

Distribution law — A substance distributes itself between two immiscible solvents so that the ratio of its concentrations in the two solvents is approximately a constant (and equal to the ratio of the solubilities of the substance in each solvent). The above statement requires modification if more than one molecular species is formed.

Donor — In transistors, the N-type semiconductor, the electrode containing impurities which increase the number of available electrons. Contrast acceptor.

Doppler broadening — The broadening of either an emission line or an absorption line due to random motions of molecules of the gas that is emitting or absorbing the radiant energy. See pressure broadening.

In the case of an emitting gas, for example, those molecules which are approaching the observer as they emit quanta of radiant energy will, because of the Doppler effect, appear to send out a train of waves of slightly shorter wavelength than that characteristic of a stationary molecule, while receding molecules will appear to emit slightly longer wavelengths. The net effect, averaged over many molecules, is to superimpose, on the natural line width, a bell-shaped broadening that is proportional to the square root of the absolute temperature of the gas.

Doppler effect — The change in frequency with which energy reaches a receiver when the receiver and the energy source are in motion relative to each other. Also called Doppler shift.

In the case of sound, or any other wave motion where a real medium of propagation exists (excepting, therefore, light and other electromagnetic radiations) one must distinguish two principal cases: If the source is in motion with speed v relative to a medium which propagates the waves in question at speed c, then the resting observer receives waves emitted with actual frequency f as if they had a frequency f′ given by the Doppler equation

$$f' = f/[1 \pm (v/c)]$$

where the positive sign refers to the case of the source receding from the observer, and vice versa for the negative sign. If, on the other hand, the source is at rest relative to the propagating medium while the observer moves with speed v relative to the source,

$$f' = f'[1 \pm (v/c)]$$

where the positive sign now refers to the case of observer approaching the source.

For electromagnetic radition,

$$f/f' = [1 \pm (v/c)]/[1 \pm (v/c)]$$

where the top signs represent the source receding from the observer and the bottom signs, approaching the observer.

Dosimeter — 1. An instrument for measuring the ultraviolet in solar and sky radiation. Compare actinometer. 2. A device, worn by persons working around radioactive material, which indicates the dose of radiation to which they have been exposed.

Double decomposition consists of a simple exchange of the parts of two substances to form two new substances.

Double pass transmittance hologram — A hologram whose object wave was transmitted through the transparent object media to a mirror, reflected back through again, and recorded on the plate.

Dulong and Petit, law of — The specific heats of the several elements are inversely proportional to their atomic weights. The atomic heats of solid elements are constant and approximately equal to 6.3. Certain elements of low atomic weight and high melting point have, however, much lower atomic heats at ordinary temperatures.

Dynamic height — The height of a point in the atmosphere expressed in a unit proportional to the geopotential at that point. Since the geopotential at altitude z is numerically equal to the work done when a particle of unit mass is lifted from sea level up to this height, the dimensions of dynamic height are those of potential energy per unit mass. Also called geodynamic height.

The standard unit of dynamic height H_d is the dynamic meter (or geodynamic meter), defined as $10/sec^2$; it is related to the geopotential φ, the geometric height z in meters, and the geopotential height Z in geopotential meters by

$$d\varphi = 10dH_d = 9.8dZ = gdz$$

where g is the acceleration of gravity in meters per second squared. (Some sources prefer to give the constants 10 and 9.8 the units of meters per second squared so that the units of φ and Z would be the same as those of the geometric height.) The dynamic meter is about 2% longer than the geometric meter and the geopotential meter. One of the practical advantages of the dynamic height over the geometric height is that when the former is introduced into the hydrostatic equation the variable acceleration of gravity is eliminated. In meteorological height calculations, geopotential height is more often used than dynamic height.

Dynamic pressure (symbol q) = The pressure of a fluid resulting from its motion, equal to one-half the fluid density times the fluid velocity squared ($1/2\varrho V^2$). In incompressible flow, dynamic pressure is the difference between total pressure and static pressure. Also called kinetic pressure. Compare impact pressure.

Dynamic Viscosity — Of a fluid, the ratio of the shearing stress to the shear of the motion. It is independent of the velocity distribution, the dimensions of the system, etc. and for a gas it is independent of pressure except at very low pressures. Also called coefficient of molecular viscosity, coefficient of viscosity.

For the dynamic viscosity μ of a perfect gas, the kinetic theory of gases gives

$$\mu = 1/3\ (\sigma cL)$$

where ϱ is the gas density, c is the average speed of the random heat motion of the gas molecules and is proportional to the square root of the temperature, and L is the mean free path. For dry air at 0° C, the dynamic viscosity is about 1.7×10^{-4} g/cm/sec.

Whereas the dynamic viscosity of most gases increases with increasing temperature, that of most liquids, including water, decreases rapidly with increasing temperature.

Dyne — The force necessary to give acceleration of one centimeter per second per second to one gram of mass.

Earth Current — A large-scale surge of electric charge within the earth's crust, associated with a disturbance of the ionosphere.

Current patterns of quasi-circular form and extending over areas the size of whole continents have been identified and are known to be closely related to solar-induced variations in the extreme upper atmosphere.

Eddy current — A current induced in a mass of conducting material by a varying magnetic field. Also called Foucault current.

Eddy viscosity — The turbulent transfer of momentum by eddies giving rise to an internal fluid friction, in a manner analogous to the action of molecular viscosity in laminar flow, but taking place on a much larger scale.

The value of the coefficient of eddy viscosity (an exchange coefficient) is of the order of 10^4 cm²/sec, or 100,000 times the molecular kinematic viscosity.

Effective neutron cycle time — The lifetime of an average neutron within a reactor from the time it is produced to the time it is fission captured.

This average takes into account delayed as well as prompt neutrons.

Effective radius of the earth — A fictitious value for the radius of the earth, used in place of the geometrical radius to correct for atmospheric refraction when the index of refraction in the atmosphere changes linearly with height. See modified index of refraction.

Under conditions of standard refraction the effective radius of the earth is

8.5×10^5 m, or four thirds the geometrical radius. If the effective radius is used in ray tracing diagrams, the rays may be drawn as though they were traveling in straight lines.

Effective terrestrial radiation — The amount by which outgoing infrared terrestrial radiation of the earth's surface exceeds downcoming infrared counter-radiation from the atmosphere. Also called nocturnal radiation, effective radiation. See Actinometer.

It is to be emphasized that this amount is a positive quantity, of the order of several tenths of a langley per minute, at all times of day (except under conditions of low overcast clouds). It typically attains its diurnal maximum during the midday hours when high soil temperatures create high rates of outgoing terrestrial radiation. (For this reason the synonym nocturnal radiation is apt to lead to slight confusion.) However, in daylight hours the effective terrestrial radiation is generally much smaller than the insolation, while at night it typically dominates the energy budget of the earth's surface.

Einstein theory for mass-energy equivalence — The equivalence of a quantity of mass m and a quantity of energy E by the formula $E = mc^2$. The conversion factor c^2 is the square of the velocity of light.

Elastic collision — A collision between two particles in which no change occurs in the internal energy of the particles, or in the sum of their kinetic energies. Commonly referred to as a billiard-ball collision.

Elasticity — The property by virtue of which a body resists and recovers from deformation produced by force.

Elastic limit — The smallest value of the stress producing permanent alteration.

Elastic moduli —

Young's modulus by stretching — If an elongation s is produced by the weight of the mass m, in a wire of length l, and radius r, the modulus,

$$M = \frac{mgl}{\pi r^2 s}$$

Young's modulus by bending, bar supported at both ends. If a flexure s is produced by the weight of mass m, added midway between the supports spearated by a distance l for a rectangular bar with vertical dimensions of cross-section a and horizontal dimension b, the modulus is,

$$M = \frac{mgl^3}{4sa^3 b}$$

For a cylindrical bar of radius r,

$$M = \frac{mgl^3}{12\pi r^4 s}$$

For a bar supported at one end. In the case of a rectangular bar as described above,

$$M = \frac{4mgl^3}{sa^3 b}$$

For a round bar supported at one end,

$$M = \frac{4mgl^3}{3\pi r^4 s}$$

Modulus of rigidity — If a couple C (= mgx) produces a twist of θ radians in a bar of length l and radius r, the modulus is

$$M = \frac{2Cl}{\pi r^4 \theta}$$

The substitution in the above formulae for the elastic coefficients of m in grams, g in cm/sec², l, a, b, and r in cm, s in cm, and C in dyne-cm will give moduli in dyn/cm². The dimensions of elastic moduli are the same as of stress, $[ml^{-1} t^{-2}]$.

Coefficient of restitution — Two bodies moving in the same straight line, with velocities v_1 and v_2 respectively, collide and after impact move with velocities v_3 and v_4. The coefficient of restitution is

$$C = \frac{v_4 - v_3}{v_2 - v_1}$$

Electret — A piece of dielectric material that has a permanent electric polarity; the electrostatic analog of a permanent magnet.

Electric dipole — A pair of equal and opposite charges an infinitesimal distance apart.

Electric field intensity is measured by the force exerted on unit charge. Unit field intensity is the field which exerts the force of one dyne on unit positive charge. Dimensions, $[\varepsilon^{-1/2} m^{1/2} l^{-1/2} t^{-1}]$; . $[\mu^{1/2} m^{1/2} l^{1/2} t^{-2}]$.

The field intensity or force exerted on unit charge at a point distant r from a charge q in a vacuum

$$H = \frac{q}{r^2}$$

If the dielectric in the above cases is not a vacuum the dielectric constant ε must be introduced. The formula becomes

$$H = \frac{q}{\varepsilon r^2}$$

The value of ε is frequently considered unity for air. If the dielectric constant of a vacuum is considered unity the value for air at 0°C and 760 mm pressure is 1.000576.

Electric potential — The work per unit charge spent in moving a charged body in an electric field from a reference point to a point of interest (P). Commonly, the reference point is chosen as infinity. The potential (V) is positive if work is done on the charge and negative if work is required of the charge to move in the existing field. Analytically, assuming an electric field of intensity E,

$$V = \int_{\infty}^{P} E \cdot ds$$

where ds is a vector element of the path from ∞ to P.

Electrochemical equivalent of an ion is the mass liberated by the passage of unit quantity of electricity.

Electrolysis — If a current i flows for a time t and deposits a metal whose electrochemical equivalent is e, the mass deposited is

$$m = eit$$

The value of e is usually given for mass in grams, i in amperes and t in seconds.

Electrolytic cell constant, J_e — The cell constant of an electrolytic cell is the resistance in ohms of that cell when filled with a liquid of unit resistance.

Electrolytic dissociation or ionization theory — When an acid, base or salt is dissolved in water or any other dissociating solvent, a part or all of the molecules of the dissolved substance are broken up into parts called ions, some of which are charged with positive electricity and are called cations, and an equivalent number of whch are charged with negative electricity and are called anions.

Electrolytic solution tension theory (or the Helmholtz double layer theory) — When a metal, or any other substance capable of existing in solution as ions, is placed in water or any other dissociating solvent, a part of the metal or other substances passes into solution in the form of ions, thus leaving the remainder of the metal or substances charged with an equivalent amount of electricity of opposite sign from that carried by the ions. This establishes a difference in potential between the metal and the solvent in which it is immersed.

Electromagnetic radiation — Energy, propagated through space or through material media in the form of an advancing disturbance in electric and magnetic fields existing in space or in the media. The term radiation, alone, is used commonly for this type of energy, although it actually has a broader meaning. Also called electromagnetic energy or simply radiation. See Electromagnetic spctrum.

Electromagnetic spectrum — The ordered array of known electromagnetic radiations, extending from the shortest cosmic rays, through gamma rays, X-rays, ultraviolet radiation, visible radiation, infrared radiation, and including microwave and all other wavelengths of radio energy See Absorption spectrum.

The division of this continuum of wavelengths (or frequencies) into a number of named subportions is rather arbitrary and, with one or two exceptions, the boundaries of the several subportions are only vaguely defined. Nevertheless, to each of the commonly identified subportions there correspond characteristic types of physical systems capable of emitting radiation of those wavelengths. Thus, gamma rays are emitted from the nuclei of atoms as they undergo any of several types of nuclear rearrangements; visible light is emitted, for the most part, by atoms whose planetary electrons are undergoing transitions to lower energy states; infrared radiations are associated with characteristic molecular vibrations and rotations; and radio waves, broadly speaking, are emitted by virtue of the accelerations of free electrons as, for example, the moving electrons in a radio antenna wire.

Electromotive force is defined as that which causes a flow of current. The electromotive force of a cell is measured by the maximum difference of potential between its plates. The electromagnetic unit of potential difference is that against which one erg of work is done in the transfer of electromagnetic unit quantity. The volt is that potential difference against which one joule of work is done in the transfer of one coulomb. One volt is equivalent to $.0^8$ electromagnetic units of potential. The international volt is the electrical potential which when steadily applied to a conductor whose resistance is one international ohm will cause a current of one international ampere to flow. The international volt = 1.00033 absolute volts. The electromotive force of a Weston standard cell is 1.0183 int. volts at 20°C. Dimensions, $[\varepsilon^{-1/2}\ m^{1/2}\ t^{-1}]$, $[\mu^{1/2}\ m^{1/2}\ l^{3/2}\ t^{-2}]$.

Electromotive series is a list of the metals arranged in the decreasing order of their tendencies to pass into ionic form by losing electrons.

Electron — The electron is a small particle having a unit negative electrical charge, a small mass, and a small diameter. Its charge is $(4.80294 \pm 0.00008) \times 10^{-10}$ absolute electrostatic units, its mass. $^{1}{}_{1837}$ that of the hydrogen nucleus, and its diameter about 10^{-12} cm. The electron has a rest mass of 0.9109534 10^{-30}kg and a magnetic moment of 9.284832×10^{-24}J T^{-1}. Every atom consists of one nucleus and one or more electrons. Cathode rays and Beta rays are electrons.

Electron-volt (eV) — Energy acquired by any charged particle carrying unit electronic charge when it falls through a potential difference of one volt. 1 electron-volt = $(1.60207 \pm 0.00007)\ 10^{-12}$ erg or 1.6020×10^{-19} joule. Multiples of this unit are also in common use: the kilo-, million-, and billion electron-volt. 1 keV = 10^3 eV; 1 meV = 10^6 eV; and 1 beV = 10^9 eV.

Electrophoretic effect — The slowing down owing to interionic attraction and repulsion, of the movement of an ion with its solvent molecules in the forward direction by ions of opposite charge with their solvent molecules moving in the reverse direction under an applied electrical field (potential gradient).

Electrostatic unit — 1. In the cgs system, the measure of electrostatic charge, defined as a charge which, if concentrated at one point in a vacuum, would repel, with a force of 1 dyn, an equal and like charge placed 1 cm away. 2. (pl.) A system of electrical units based on the electrostatic unit.

Elements are substances which cannot be decomposed by the ordinary types of chemical change, or made by chemical union.

Elliptical polarization — The polarization of a wave radiated by an electric vector rotating in a plane and simultaneously varying in amplitude so as to describe an ellipse.

Elongation — In tensile testing the elongation of a specimen is the increase in gage length, after rupture, referred to the original gage length. It is reported as percentage elongation.

Emission spectrum — The array of wavelengths and relative intensities of electromagnetic radiation emitted by a given radiator.

Each radiating substance has a unique, characteristic emission spectrum, just as every medium of transmission has its individual absorption spectrum.

Emissive power or emissivity is measured by the energy radiated from unit area of a surface in unit time for unit difference of temperature between the surface in question and surrounding bodies. For the cgs system the emissive power is given in ergs per second per square centimeter with the radiating surface at 1° absolute and the surroundings at absolute zero. See Radiation formula.

Emissivity (symbol E∞) — A property of a material, measured as the emittance of a specimen of the material that is thick enough to be completely opaque and has an optically smooth surface.

Emittance (symbol E, ε) — 1. The radiant flux per unit area emitted by a body. 2. The ratio of the emitted radiant flux per unit area of a sample to that of a black body radiator at the same temperature and under the same conditions.

Spectral emittance refers to emittance measured at a specified wavelength.

Because of the two common meanings of emittance, it should be defined when used unless the context allows no misinterpretation.

Emulsion — The coating on a film or plate which is sensitive to the light illuminating it.

Enantiotropic — Crystal capable of existing in reversible equilibrium with each other.

Energy — The capability of doing work. Potential energy is energy due to position of one body with respect to another or to the relative parts of the same body. Kinetic energy is energy due to motion. Cgs units, the erg, the energy expended when a force of 1 dyn acts through a distance of 1 cm; the joule is 1×10^7 ergs. Dimensions, $[ml^2\ t^{-2}]$.

The potential energy of a mass m, raised through a distance h, where g is the acceleration due to gravity is

$$E = mgh.$$

The kinetic energy of mass m, moving with a velocity v, is

$$E = \tfrac{1}{2}mv^2$$

Energy will be given in ergs if m is in grams, g in centimeters per second square, h in centimeters and v in centimeters per second.

Energy of a charge in ergs where Q is the charge and V the potential in electrostatic units.

$$E = \tfrac{1}{2}QV$$

Energy of the electric field — If H is the electric field intensity in electrostatic units and K the specific inductive capacity, the energy of the field in ergs per cm^3 is

$$E = \frac{KH^2}{8\pi}$$

Energy of rotation — If a mass whose moment of inertia about an axis is I, rotates with angular velocity ω about this axis, the kinetic energy of rotation will be,

$$E = \tfrac{1}{2}I\omega^2$$

Energy will be given in ergs if I is in g-cm^2 and ω in radians per second.

Enthalpy, or heat content, is a thermodynamic quantity. It is equal to the sum of the internal energy of a system plus the product of the pressure-volume work done on the system. Thus

$$H = E + pv$$

where

H	=	enthalpy or heat content
E	=	internal energy of the system
p	=	pressure
v	=	volume.

Entropy — 1. A measure of the extent to which the energy of a system is unavailable. A mathematically defined thermodynamic function of state, the increase in which gives a measure of the energy of a system which has ceased to be available for work during a certain process

$$ds = (du + pdv)/T \geqslant dq/T$$

where s is specific entropy; us is specific internal energy; p is pressure; v is specific volume; T is Kelvin temperature; and q is heat per unit mass. For reversible processes,

$$ds = dq/T$$

In terms of potential temperature 0,

$$ds = c_p\ (d\theta/\theta)$$

where cp is the specific heat at constant pressure. See third law of thermodynamics.

In an adiabatic process, the entropy increases if the process is irreversible and remains unchanged if the process is reversible. Thus, since all natural processes are irreversible. and remains unchanged if the process is reversible. Thus, since all natural processes are irreversible, it is said that in an isolated system the entropy is always increasing as the system tends toward equilibrium, a statement which may be considered a form of the second law of thermodynamics.
2. In communication theory, average information content.

Ephemeris day — 86,400 ephemeris seconds. See Ephemeris time.

Ephemeris second (unit of time) — is exactly 1/31 556 925.974 7 of the tropical year of 1900, January, 0 days, and 12 hr ephemeris time.

Ephemeris time (abbr. E.T.) — The uniform measure of time defined by the laws of dynamics and determined in principle from the orbital motions of the planets, specifically the orbital motion of the earth as represented by Newcomb's Tables of the Sun. Compare universal time.

Beginning with the volume for 1960 the American Ephemeris and Nautical Almanac uses ephemeris time as the tabular argument in the fundamental ephemerides of the sun, moon, and planets.

A gravitational ephemeris expresses the position of a celestial body as a function of ephemeris time; and, at any instant, the measure of ephemeris time is the value of the argument at which the ephemeris position is the same as the actual position at the instant. The ephemeris time at any instant is obtained from observation by directly comparing observed position of the sun, moon, and planets with gravitational ephemerides of their coordinates; observations of the moon are the most effective and expeditious for this purpose. An accurate determination, however, requires observations over a more or less extended period; in practice, it takes the form of determining the time correction ΔT that must be applied to universal time (U.T.) to obtain ephemeris time:

$$\text{E.T.} = \text{U.T.} + \Delta T$$

The universal time at any instant may be obtained with little delay from observations of the diurnal motions.

The fundamental epoch from which ephemeris time is reckoned is the epoch that Newcomb designated as 1900 January 0, Greenwich mean noon, but which actually is 1900 January 0 day 12 hr E.T.; the instant to which this designation is assigned is the instant near the beginning of the calendar year A.D. 1900 when the geometric mean longitude of the Sun referred to the mean equinox of date was 279 degrees 41 minutes 48.04 seconds. Ephemeris time is the measure of time in which Newcomb's Tables of the Sun agree with observation.

The primary unit of ephemeris time is the tropical year, defined by the mean motion of the sun in longitude at the epoch 1900 January 0 day 12 hr E.T.; its length in ephemeris days is determined by the coefficient of T in Newcomb's expression for the geometric mean longitude of the sun L referred to the mean equinox of date, given among the elements of the sun.

Equatorial system — A set of celestial coordinates based on the celestial equator as the primary great circle; usually declination and hour angle or sidereal hour angle. Also called equinoctial system of coordinates, celestial equato system of coordinates.

Equilibrium, chemical — A state of affairs in which a chemical reaction and its reverse reaction are taking place at equal velocities, so that the concentrations of reacting substances remain constant.

Equilibrium constant — The product of the concentrations (or activities) of the substances produced at equilibrium in a chemical reaction divided by the product of concentrations of the reacting substances, each concentration raised to that power which is the coef-icient of the substance in the chemical equation.

Equivalent conductance of an electrolyte is defined as the conductance of a volume of solution containing one equivalent weight of dissolved substance when placed between two parallel electrodes 1 em apart, and large enough to contain between them all of the solution. ᵈ is never determined directly, but is calculated from the specific conductance. If C is the concentration of a solution in gram equivalents per liter, then the concentration per cubic centimeter is C/1000, and the volume containing one equivalent of the solute is, therefore, 1000/C. Since L_s is the conductance of a centimeter cube of the solution, the conductance of 1000/C cc, and hence ᵈ will be

$$\Delta = \frac{1000 L_s}{C}$$

Equivalent temperature — 1. Isobaric equivalent temperature; the temperature that an air parcel would have if all water vapor were condensed out at constant pressure, the latent heat released being used to heat the air,

$$T_{i,e} = T[1 + (Lw/c_pT)]$$

where $T_{i,e}$ is the isobaric equivalent temperature; T is the temperature; w is the mixing ratio; L is the latent heat: and c_p is the specific heat of air at constant pressure. 2. Adiabatic equivalent temperature; The temperature that an air parcel would have after undergoing the following (physically unrealizable) process: dry-adiabatic expansion until saturated; pseudoadiabatic expansion until all moisture is precipitated out; dry-adiabatic expansion until saturated; pseudoadiabatic expansion until saturated; pseudoadiabatic expansion until all moisture is precipitated out; dry-adiabatic compression to the initial pressure. This is the equivalent temperature as read from a thermodynamic chart and is always greater than the isobaric equivalent temperature:

$$T_{a,e} = T \exp (Lw/c_pT)$$

where $T_{a,e}$ is the adiabatic equivalent temperature. Also called pseudoequivalent temperature.

Equivalent weight or combining weight of an element or ion is its atomic or formula weight divided by its valence. Elements entering into combination always do so in quantities proportional to their equivalent weights.

In oxidation-reduction reactions the equivalent weight of the reacting substances is dependent upon the change in oxidation number of the particular substance.

erg — The unit of energy or work in the centimeter-gram-second system; the work performed by a force of 1 dyne acting through a distance of 1 cm.

Escape velocity — The radial speed which a particle or larger body must attain in order to escape from the gravitational field of a planet or star. When friction is neglected, the escape velocity is

$$\sqrt{2Gm/r}$$

where G is the universal gravitational constant m is the mass of the planet or star; and r is the radial distance from the center of the planet or star. Also called escape speed.

Ettinghausen's effect (Von Ettinghausen's) — When an electric current flows across the lines of force of a magnetic field an electromotive force is observed which is at right angles to both the primary current and the magnetic field: a temperature gradient is observed which has the opposite direction to the Hall electromotive force.

Eutectic — A term applied to the mixture of two or more substances which has the lowest melting point.

Exchange coefficients — Coefficients of eddy flux (e.g., of momentum, heat, water vapor, etc) in turbulent flow, defined in analogy to those of the kinetic theory of gases (see eddy). Also called austausch coefficients, eddy coefficients, interchange coefficients.

The exchange-coefficient hypothesis states that the mean eddy flux per unit area of a conservative quantity (suitably expressed) is proportional to the gradient of the mean value of this quantity, that is,

$$\text{Mean flux per unit area} = -C_e\,(d\bar{E}/dN)$$

where G_e is the exchange coefficient; E is the mean value of the quantity; and N is the direction normal to the surface. In strict analogy to molecular properties C_e would be constant, for turbulent flow C_e turns out to depend on time and location. See eddy viscosity diffusivity.

Expansion of gases — *Charles' law or Gay-Lussac's law* — The volume of a gas at constant pressure increases proportionately to the absolute temperature. If V_1 and V_2 are volumes of the same mass of gas at absolute temperatures, T_1 and T_2,

$$\frac{V_1}{V_2} = \frac{T_1}{T_2}$$

For an original volume V_o at 0°C the volume at t° C (at constant pressure) is

$$V_t = V_o\,(1 + 0.00367t).$$

General law for gases —

$$p_t v_t = p_0 v_0 \left(1 + \frac{t}{273}\right)$$

where p_o, v_o, p_t, v_t represent the pressure and volume at 0° and t°C or

$$\frac{p_1 v_1}{T_1} = \frac{p_2 v_2}{T_2}$$

where p_1, v_1 and T_1 represent pressure, volume and absolute temperature in one case and p_2, v_2 and T_2 the same quantities for the same mass of gas in another.

The law may also be expressed:

$$pv = RmT$$

where m is the mass of gas at absolute temperature T. R is the gas constant which depends on the units used. Boltzmann's molecular gas constant is obtained by expressing m in terms of the number of molecules.

For volume in cm³, pressure in dynes per cm² and temperature in Centigrade degrees on the absolute scale $R = 8.3136 \times 10^7$.

Reduction of a gas volume to 0°C, 760 mm pressure: — If V is the original volume at 0°C and 760 mm pressure will be,

$$V_0 = \frac{V}{(1 + \alpha t)} \frac{H}{760}$$

If d is the original density the density at 0°C and 760 mm pressure will be

$$d_0 = d(1 + \alpha t) \frac{760}{H}$$

$$\alpha = 0.00367 \text{ approximately.}$$

Extinction coefficient — The extinction coefficient ε is identified as

$$dI = -\epsilon I\, dx$$

or

$$I = I_0 e^{-\epsilon x}$$

where I is the illuminance (luminous flux density) at the selected point in space, I_0 is the illuminance at the light source, and x is the distance from the source.

When so used, the extinction coefficient equals the sum of the medium's absorption coefficient and scattering coefficient, each computed as a weighted average over all wavelengths in the visible spectrum. As long as scattering effects are primary, a in the lower atmosphere, the value of the extinction coefficient is a function of the particle size of atmospheric suspensoids. It varies in order of magnitude from 10/km with very low visibility to 0.01/km in very clear air.

Extinction cross section = scattering cross section.

Extraterrestrial radiation — In general, solar radiation received just outside the earth's atmosphere.

Fahrenheit temperature scale (abbr. F) — A temperature scale with the ice point at 32° and the boiling point of water at 212°.

Conversion with the Celsius (centrigrade) temperature scale (abbr. C) is by the formula

$$F = 9/5\, C + 32$$

Falling bodies — For bodies falling from rest conditions are as for uniformly accelerated motion except that v_o = O and g is the acceleration due to gravity. The formulae become, air resistance neglected,

$$v_t = gt,\ s = \tfrac{1}{2}gt^2\ v_s = \sqrt{2gs}.$$

For bodies projected vertically upward, if v is the velocity in projection, the time to reach greatest height, neglecting the resistance of the air,

$$t = \frac{v}{g}$$

Greatest height,

$$h = \frac{v^2}{2g}$$

See Projectiles,

Farad (unit of electric capacitance) — the capacitance of a capacitor between the plates of which there appears a difference of potential of 1 V when it is charged by a quantity of electricity equal to 1 coulomb.

Faraday, determination of (1960) By the National Bureau of Standards which uses an electrochemical method that dissolves, rather than deposits, silver from a solution. The new value, 96,516 ± 2 coulombs (physical scale) or 96,489 ± 2 coulombs (chemical scale). NBS used its mass, time, and elecrical standards in measuring the faraday, and have found that its value agreed within 22 ppm. with the one obtained by an independent physical method using the omegra-tion.

Faraday constant (symbol F) — The product of the Avogadro constant N_A and the elementary charge e, $F = N_A e$ = 96,489±2 coulombs/mol.

Faraday effect — The rotation of the plane of polarization produced when plane-polarized light is passed through a substance in a magnetic field, the light traveling in a direction parallel to the lines of force. For a given substance, the rotation is proportional to the thickness traversed by the light and to the magnetic field strength.

Faraday's laws — In the process of electrolytic changes equal quantities of electricity charge or discharge equivalent quantities of ions at each electrode.

One gram equivalent weight of matter is chemically altered at each electrode for 96,489 int. coulombs, or one faraday, of electricity passed through the electrolyte.

Far field (diffraction pattern) — Diffraction pattern produced at a large range from an object which is identical to that which would be produced at an infinite range from the object. This is also called a Fraunhofer diffraction pattern.

Fast neutron — A neutron of 100,000 electron-volts or greater energy.

Fast reactor — A reactor containing no moderator, so that all the fissions take place at energies on the order of 100,000 electron-volts or higher.

Femto — A prefix meaning one quadrillionth or 10^{-15}. Symbol is f.

Fermat's principle — The path followed by light (or other waves) passing through any collection of media from one specified point to another, is that-path for which the time of travel is least.

Fermi (abbr. fm) — A unit of length equal to 10^{-13} cm.

Ferromagnetic Materials — Those in which the magnetic moments of atoms or ions in a magnetic domain tend to be aligned parallel to one another in zero field, below a characteristic temperature called the Néel Point. In the usual case, within a magnetic domain, a substantial net magnetization results from the antiparallel alignment of neighboring nonequivalent sublattices. The macroscopic behavior is similar to that in ferromagnetism. Above the Néel Point, these materials become paramagnetic.

Ferrogmatic Materials — Those in which the magnetic moments of atoms or ions in a magnetic domain tend to be aligned parallel to one another in zero applied field, below a characteristic temperature called the Curie Point. Complete ordering is achieved only at the absolute zero of temperature. Within a magnetic domain, at absolute zero, the magnetization is equal to the sum of the magnetic moments of the atoms or ions per unit volume. Bulk matter, consisting of many small magnetic domains, has a net magnetization which depends upon the magnetic history of the specimen (hysteresis effect). The permeability depends on the magnetic field, and can reach values of the order of 10^6 times that of free space. Above the Currie Point, these materials become paramagnetic.

First law of thermodynamics — A statement of the conservtion of energy for thermodynamic systems (not necessarily in equilibrium). The fundamental form requires that the heat absorbed by the system serve either to raise the internal energy of the system or to do work on the environment:

$$dq = du + dw$$

where dq is the heat added per unit mass; du is the increment of specific internal energy; and dw is the specific work done by the system on the environment. Although dq and dw are not perfect differentials, their difference, dv, is always a perfect differential. Example of the application of this equation: in an adiabatic free expansion of gas into a vacuum, all three terms are zero.

For reversible processes the mechanical work is equal to the expansion against the pressure forces, i.e.,

$$dw = pdv$$

where p is the pressure and v is the specific volume. For a perfect gas, the internal energy change is proportional to the temperature change,

$$du = c_v dT$$

where c_v is the specific heat at constant volume and T is the Kelvin temperature. Therefore, the form of the first law usually used in meteorological applications is

$$dq = c_v dT + pdv$$

Use of the equation of state yields an alternative form,

$$dq = c_p dT - dp$$

where c_p is the specific heat at constant pressure.

For open systems the variation of total rather than specific quantities is important:

$$dQ = dU + pdV - hdm$$

where Q is the total heat; U is the total internal energy; V is the volume; m is the mass of the system; and h is the specific enthalpy.

If a system contains the possibility of nonmechanical work, such as work done against an electric field, this work must be included in the first law.

See Second law of thermodynamics, Third law of thermodynamics, Energy equations.

Fission — The splitting of an atomic nucleus into two more-or-less equal fragments.

Fission may occur spontaneously or may be induced by capture of bombarding particles. In addition to the fission fragments, neutrons and gamma rays are usually produced during fission.

Fleming's rule — A simple rule for relating the directions of the flux, motion, and e.m.f. in an electric machine. The forefinger, second finger and thumb, placed at right-angles to each other, represent respectively the directions of flux, e.m.f., and motion or torque. If the right hand is used the conditions are those obtaining in a generator and if the left hand is used the conditions are those obtaining in a motor.

Fluidity — The reciprocal of viscosity. The cgs unit is the rhe, the reciprocal of the poise. Dimensions, $[m^{-1} lt]$.

Fluorescence — The property of emitting radiation as the result of absorption of radiation from some other source. The emitted radiation persists only as long as the exposure is subjected to radiation which may be either electrified particles or waves. The fluorescent radiation generally has a longer wave length than that of the absorbed radiation. If the fluorescent radiation includes waves of the same length as that of the absorbed radiation it is termed resonance radiation.

f/n or f number (of optics) — The ratio of effective focal length to lens diameter.

Focal length — The distance between the optical center of a lens, or the surface of a mirror, and its focus.

Focal plane — A plane parallel to the plane of a lens or mirror and passing through the focus.

Focal point = focus, in optics.

Focus (plural focuses) — 1. That point at which parallel rays of light meet after being refracted by a lens or reflected by a mirror. Also called focal point. 2. A point having specific significance relative to a geometrical figure. See Ellipse, Hyperbola, Parabola.

Footcandle (abbr. fc) — A unit of illuminance, incident light, or illumination equal to 1 lm/ft². This is the illuminance provided by a light source of one candle at a distance of 1 ft, hence the name. Compare lux, phot.

Full sunlight with zenith sun produces an illuminance of the order of 10,000 footcandles on a horizontal surface at the earth's surface. Full moonlight provides an illuminance of only about 0.02 footcandle also at earth's surface. Adequate illumination for steady reading is taken to be about 10 footcandles; that for close machine work is about 30 to 40 footcandles.

Foot-lambert (abbr. ft-1) — A unit of luminance (or brightness) equal to $1/\pi$ candle per square foot, or 1 lm/ ft².

In Great Britain this is also called the equivalent footcandle.

Force — That which changes the state of rest or motion in matter, measured by the rate of change of momentum. Absolute unit, the dyne, the force which will produce an acceleration of one centimeter per second per second in a gram mass. The gram weight or weight of a gram mass is the cgs gravitational unit. The poundal is that force which will give an acceleration of one foot per second per second to a pound mass. Dimensions, $[mlt^{-2}]$.

The force F required to produce an acceleration a in a mass m is given by

$$F = ma$$

If m is substituted in grams and a in cm per sec², F will be given in dynes.

Force between two charges, Coulomb's law — If two charges q and q′ are at a distance r in a vacuum, the force between them is,

$$F = \frac{qq'}{r^2}$$

The force will be given in dynes if q and q′ are in electrostatic units and r in centimeters.

Force between two magnetic poles — If two poles of strength m and m′ are separated by a distance r in a medium whose permeability is μ (unity for a vacuum), the force between them is,

$$F = \frac{mm'}{\mu r^2}$$

Force will be given in dynes if r is in cm and m and m′ are in cgs units of pole strength.

The strength of a magnetic field at a point distance r from an isolated pole of strength m is

$$H = \frac{m}{\mu r^2}$$

The field will be given in gauss if m and r are in cgs units.

Formula, chemical — A combination of symbols with their subscripts representing the constituents of a substance and their proportions by weight.

Fourier transform plane — Same as spatial frequency plane.

Fraunhofer's lines — When sunlight is examined through a spectroscope it is found that the spectrum is traversed by an enormous number of dark lines parallel to the length of the slit. These dark lines are known as Fraunhofer's lines. Kirchoff conceived the idea that the sun is surrounded by layers of vapors which act as filters of the white light arising from incandescent solids within and which abstract those rays which correspond in their periods of vibration to those of the components of the vapors. Thus reversed or dark lines are obtained due to the absorption by the vapor envelop, in place of the bright lines found in the emission spectrum.

Frequency — Rate of oscillation; units: 1 cycle sec⁻¹ = 1 Hertz = 1 Hz. 1 megacycle sec⁻¹ = 1 megahertz = 10^6 Hz. One gigahertz = 10^9 Hz.

Frequency modulation — A form of angle modulation in which the frequency of the carrier is made to vary in accordance with the information to be transmitted, that is, given a carrier

$$C(t) = A\cos(\omega_0 t + \phi),$$

and an information function g(t), the signal transmitted is

$$C_m(t) = A\cos[(\omega_0 + g(t) - \Delta\omega)(t) + \phi].$$

When used for radio communications this system has the advantages of greater immunity from noise and other interference, at the cost of increased bandwidth.

Frequency of vibrating strings — The fundamental frequency of a stretched string is given by

$$n = \frac{1}{2l}\sqrt{\frac{T}{m}}$$

where l is the length, T, the tension and m the mass per unit length.

For a string or wire of circular section of length l, tension T, density d, and radius r, the frequency of the fundamental is

$$n = \frac{1}{2rl}\sqrt{\frac{T}{\pi d}}$$

The frequency in vibrations per second will be given if T is in dynes, r and l in cm and d in g per cm^3.

Frequency (of waves) — Number of like phase (peaks, troughs) wavefronts passing a given point in a unit of time.

Frequency standard (atomic or molecular frequency standard) — The standard is the transition between the hyperfine levels F = 4, M = 0 and F = 3, M = 0 of the ground state $^2S_{1/2}$ of the cesium-133 atom, unperturbed by external fields. This frequency of transition is assigned the value 9 192 631 770 hertz.

Fresnel — A measure of frequency, defined as equal to 10^{12} cycles/sec.

Friction, coefficient of — The coefficient of friction between two surfaces is the ratio of the force required to move one over the other to the total force pressing the two together.

If F is the force required to move one surface over another and W, the force pressing the surfaces together, the coefficient of friction,

$$k = \frac{F}{W}$$

Fringe — The locus of maximum constructive interference (light fringe) or destructive interference regions in a space where two or more coherent waves intersect. Fringes can be in two or three dimensions.

Fringe control — Methods of adjusting the position and/or characteristics of the fringe pattern of a holographic interferogram.

Fugacity (symbol f) — In thermodynamics, a measure of the tendency of a substance to escape by some chemical process from the phase in which it exists.

Fundamental units — See Mass, Length, and Time.

Fusion (atomic) — A nuclear reaction involving the combination of smaller atomic nuclei or particles into larger ones with the release of energy from mass transformation. This is also called a thermo-nuclear reaction by reason of the extremely high temperature required to initiate it.

Gal — 1 gal = cm/sec/sec. Therefore, where the value of gravity is 980 this is the same as 980 gal. The milligal is now quite commonly used since it is approximately one part in a thousand of the normal gravity of the earth.

Gamma (γ) rays (nuclear X-rays) — May be emitted from radioactive substances. They are quanta of electromagnetic wave energy similar to but of much higher energy than ordinary X-rays. The energy of a quantum is equal to hv ergs, where h is Planck's constant (6.6254×10^{-27} erg sec) and v is the frequency of the radiation. Gamma rays are highly penetrating, an appreciable fraction being able to traverse several centimeters of lead.

Gas — A state of matter in which the molecules are practically unrestricted by cohesive forces. A gas has neither definite shape nor volume.

Gas constant (symbol R, R^*, R_o) — The constant factor in the equation of state for perfect gases. The universal gas constant is

$$R_o = 8.31441(26) \text{ joules/°K-mol}$$
$$= 1.98723 \text{ cal/°K-mol}$$

The gas constant for a particular gas, specific gas constant,

$$r = R/m$$

where m is the molecular weight of the gas. See Boltzmann constant.

Gas thermometer — where P_o, P_s and P_x represent the total pressure with the bulb at 0°C, at the boiling-point of water and at the unknown temperature respectively, t_s the temperature of steam and t_x the unknown temperature,

$$t_x = t_s \frac{P_x - P_o}{P_s - P_o}$$

(approximately). The total pressure on the gas in the bulb is the algebraic sum of barometric pressure at the time and that measured by the manometer.

Gauss — The cgs emu of magnetic induction (flux density). It is equal to 1 maxwell per cm^2. It has such a value that magnetic field at a velocity of 1 cm, in an induction mutually perpendicular, the induced emf is one abvolt.

Gay-Lussac's law — See Charles' law.

Gay-Lussac's law of combining volumes — If gases interact and form a gaseous product, the volumes of the reacting gases and the volumes of the gaseous products are to each other in simple proportions, which can be expressed by small whole numbers.

Gee — A suffix meaning earth, as in perigee, apogee. See Perigee, note.

Geiger counter — Detector for radioactivity depending upon ionized particles that affect its mechanism. As its name indicates, it both detects and makes a count of them possible.

Geomagnetic pole — Either of two antipodal points marking the intersection of the earth's surface with the extended axis of a dipole assumed to be located at the center of the earth and approximating the source of the actual magnetic field of the earth.

That pole in the Northern Hemisphere (latitude, 78½° N; longitude, 69° W) is designated north geomagnetic pole, and that pole in the Southern Hemisphere (latitude, 78½° S, longitude, 111°E) is designated south geomagnetic pole. The great circle midway between these poles is called geomagnetic equator. The expression geomagnetic pole should not be confused with magnetic pole, which relates to the actual magnetic field of the earth.

Geometric mean — A measure of central position. The geometric mean of n quantities equals the nth root of the product of the quantities.

Geopotential — The geopotential Φ of a point at a height z above mean sea level is the work which must be accomplished against gravity in elevating a unit mass from sea level to height z.

$$\Phi = \int_0^z g\,dz$$

where g is the local acceleration of gravity at height z. For most metrological work geopotential is given in the units geopotential meter (gpm). By definition, 1 gpm = 9.8×10^4 cm^2 sec^{-2}. For most purposes the geopotential can be assumed to equal the geometric height.

Geopotential height — The height of a given point in the atmosphere in units proportional to the potential energy of unit mass (geopotential) at this height, relative to sea level.

The relation, in the cgs system, between the geopotential height H and the geometric height Z is

$$H = \frac{1}{980}\int_0^Z g\,dZ$$

where g is the acceleration of gravity, so that the two heights are numerically interchangeable for most meteorological purposes. Also, 1 geopotential meter is equal to 0.98 dynamic meter. See dynamic height.

At the present time, the geopotential height unit is used for all aerological reports, by convention of the World Meteorological Organization.

Geopotential meter — A unit of length used in measuring geopotential height; 1 geopotential meter is equal to 0.98 dynamic meter. See Dynamic height.

Gibbs free energy = Gibbs function.

Gibbs function — a mathematically defined thermodynamic function of state, which is constant during a reversible isobaric-isothermal process. Also called Gibbs free energy, thermodynamic potential. Compare Helmholtz function.

In symbols the specific Gibbs function g is

$$g = h - Ts$$

where h is specific enthalpy; T is Kelvin temperature; and s is specific entropy. By use of the first law of thermodynamics for reversible processes,

$$dg = -s\,dT + dp$$

Gibbs' phase rule — F = C + 2 − P, F, the number of degrees of freedom of a system, is the number of variable factors (temperature, pressure and concentration) of the components, which must be arbitrarily fixed in order that the condition of the system may be perfectly defined. C, the number of the components of the system, is chosen equal to the smallest number of independently variable constituents by means of which the composition of each phase participating in the state of equilibrium can be expressed in the form of a chemical equation; the components must be chosen from among the constituents which are present when the system is in a state of true equilibrium and which take part in that equilibrium; as components are chosen the smallest number of such constituents necessary to express the composition of each phase participating in the equilibrium, zero and negative quantities of components being permissible; in any system the number of components is definite, but may alter with changes in conditions of experiment; a qualitative but not quantitative freedom of selection of components is allowed, the choice being influenced by suitability and simplicity of application. P, the number of phases of the system, are the homogeneous, mechanically separable and physically distinct portions of a heterogeneous system; the number of phases capable of existence varies greatly in different systems; there can never be more than one gas or vapor phase since all gases are miscible in all proportions, a heterogeneous mixture of solid substances forms as many phases as there are substances present.

Giga — A prefix meaning billion or 10^9. Symbol is G.

Gilbert (abbr. Gb) — The cgs emu of magnetomotive force. 1 Gb = $10/4\pi$ ampere-turns.

Global radiation — The total of direct solar radiation and diffuse sky radiation received by a unit horizontal surface.

Graham's law — The relative rates of diffusion of gases under the same conditions are inversely proportional to the square roots of the densities of those gases.

Gram atom or gram atomic weight — The mass in grams numerically equal to the atomic weight.

Gram equivalent of a substance is the weight of a substance displacing or otherwise reacting with 1.008 g of hydrogen or combining with one-half of a gram atomic weight (8.00 g) of oxygen.

Gram molecular weight or gram molecule — A mass in grams of a substance numerically equal to its molecular weight. Gram mole.

Gram mole, gram formula weight, gram equivalent — Mass in grams numerically equal to the molecular weight, formula weight or chemical equivalent, respectively.

Gravitation — The acceleration produced by the mutual attraction of two masses, directed along the line joining their centers of masses, and of magnitude inversely proportional tothe square of the distance between the two centers of mass.

This acceleration on a unit mass has the magnitude $G(m/r^2)$, where m is the mass of the attracting body, r is the distance between the centers of mass, and G is the gravitational constant equal to $6.670 \pm 0.005 \times 10^{-11}$ N n^2 kg^{-2}.

In the case of masses in the earth's gravitational field, m is the mass of the earth, equal to 5.975×10^{27} g. However, the rotation of the earth and atmosphere modifies this field to produce the field of gravity.

Gravitational constant (symbol G) — The coefficient of proportionality in Newton law of gravitation

$$G = 6.670 \pm 0.005 \times 10^{-11}\ \text{Nm}^2\,\text{kg}^{-2}$$

Also called constant of gravitation, Newtonian universal constant of gravitation.

Gravity — 1. Viewed from a frame of reference fixed in the earth, force imparted by the earth to a mass which is at rest relative to the earth. Since the earth is rotating, the force observed as gravity is the resultant of the force of gravitation and the centrifugal force arising from this rotation and the use of an earthbound rotating frame of reference. It is directed normal to sea level and to its geopotential surfaces. See Virtual gravity, Geopotential height.

The magnitude of the force of gravity at sea level decreases from the poles, where the centrifugal force is zero, to the equator, where the centrifugal force is a maximum but directed opposite to the force of gravitation. This difference is accentuated by the shape of the earth, which is nearly that of an oblate spheroid of revolution slightly depressed at the poles. Also, because of the asymmetric distribution of the mass of the earth, the force of gravity is not directed precisely toward the earth's center.

The magnitude of the force of gravity is usually called either gravity, acceleration of gravity, or apparent gravity.

Half life — The average time required for one half the atoms in a sample of radioactive element to decay.

The half life $t_{1/2}$ is given by

$$t_{1/2} = (\ln 2)/\lambda$$

where λ is the decay constant.

Half-wave plate — An electro-optical material used to rotate the plane of polarization of a light beam.

Hall constant — In an electrical conductor, the constant of proportionality R in the relation

$$E_h = RJ \times H$$

where E_h is transverse electric field (Hall field); J is current density; and H is magnetic field.

The sign of the majority carrier can be inferred from the sign of the Hall constant.

Hall effect — When a steady current is flowing in a steady magnetic field, electromotive forces are developed which are at right angles both to the magnetic force and to the current and are proportional to the product of the intensity of the current, the magnetic force and the sine of the angle between the directions of these quantities.

Hall mobility — A measure of the flow of charged particles perpendicular to both a magnetic and an electric field.

Hardness — Property of substances determined by their ability to abrade or indent one another. An arbitrary scale of hardness is based upon ten selected minerals. For metals the diameter of the indentation made by a hardened steel sphere (Brinnell) or the height of rebound of a small drop hammer (Shore Scleroscope) serve to measure hardness.

Harmonic — A sinusoidal quantity whose frequency is an integral multiple of some fundamental frequency, that is, given a quantity, x(t), where

$$x(t) = A\cos(\omega t + \phi),$$

its harmonics, $h_n(x)$ are of the form

$$h_n(x) = A_n \cos(n\,\omega t + \phi_n),$$

where n is an integer larger than 1.

Harmonic motion — See Simple harmonic motion and Angular harmonic motion.

Heat — Energy transferred by a thermal process.

Heat can be measured in terms of the dynamical units of energy, as the erg, joule, etc., or in terms of the amount of energy required to produce a definite thermal change in some substance, as, for example, the energy required per degree to raise the temperature of a unit mass of water at some temperature (calorie, Btu).

Heat capacity — That quantity of heat required to increase the temperature of a system or substance one degree of temperature. It is usually expressed in calories per degree centigrade or joules per degree centigrade.

Molar heat capacity is the quantity of heat necessary to raise the temperature of one molecular weight of the substance one degree.

Heat effect — The heat in calories developed in a circuit by an electric current of I amperes flowing through a resistance of R ohms, with a difference of potential E volts for a time t seconds.

$$H = \frac{RI^2t}{4.18} = \frac{Eit}{4.18}$$

Heat equivalent, or latent heat, or fusion — The quantity of heat necessary to change one gram of solid to a liquid with no temperature change. Dimensions, $[l^2 t^{-2}]$.

Heat of combustion of a substance is the amount of heat evolved by the combustion of 1 g mol wt of the substance.

Heat quantity — The cgs unit of heat is the calorie, the quantity of heat necessary to change the temperature of one gram of water from 3.5°C to 4.5°C (called a small calorie). If the temperature change involved is from 14.5 to 15.5°C, the unit is the normal calorie. The mean calorie is $^1{}_{100}$ the quantity of heat necessary to raise one gram of water from 0°C to 100°C. The large calorie is equal to 1000 small calories. The British thermal unit is the heat required to raise the temperature of one pound of water at its maximum density, 1°F. It is equal to about 252 calories. Dimensions of energy, $[ml^2t^{-2}]$.

Hecto — A prefix meaning hundred or 10^2. Symbol is h.

Hehner number (value) — A number expressing the percentage (i.e. grams per hundred grams) of water-insoluble fatty acids in an oil or fat.

Heisenberg's theory of atomic structure — The currently accepted view of the structure of atom, formulated by Heisenberg in 1934, according to which the atomic nuclei are built of nucleons, which may be protons or neutrons, while the extranuclear shells consist of electrons only. The nucleons are held together by nuclear forces of attraction, with exchange forces operating between them. The number of protons is equal to the atomic number (Z) of the element, the number of neutrons is equal to the difference between the mass number and the atomic number (A — Z). The number of excess neutrons, i. e. the excess of neutrons over protons, is of paramount importance for the radioactive properties or stability of the element.

Helmholtz free energy = Helmholtz function.

Helmholtz function (symbol a) — A mathematically defined thermodynamic function of state, the decrease in which during a reversible isothermal process is equal to the work done by the system. The Helmholtz function is

$$a = u - Ts$$

where u is specific internal energy; T is Kelvin temperature; and s is specific entropy. By use of the first law of thermodynamics for reversible processes,

$$da = -s\,dT - dw$$

where dw is the work done per unit mass by the system. Also called Helmholtz free energy, work function. Compare Gibbs function.

Henry (unit of electric inductance) — The inductance of a closed circuit in which an electromotive force of 1 V is produced when the electric current in the circuit varies uniformly at a rate of 1 A/sec.

Henry's law — The mass of a slightly soluble gas that dissolves in a definite mass of a liquid at a given temperature is very nearly directly proportional to the partial pressure of that gas. This holds for gases which do not unite chemically with the solvent.

Hertz — The measure of frequency, defined as equal to 1 cycle/sec.

Hess' law of constant heat summation — The amount of heat generated by a chemical reaction is the same whether reaction takes place in one step or in several steps, or all chemical reactions which start with the same original substances and end with the same final substances liberate the same amounts of heat, irrespective of the process by which the final state is reached.

Heterodyne — To mix two radio signals of different frequencies to produce a third signal which is of lower frequency; i.e., to produce beating.

Radar receivers are of the heterodyne type (as contrasted to the superregenerative type) because the very high radio frequencies used in radar are difficult to amplify. A target signal is heterodyned with a current of lower frequency produced by a klystron oscillator and the resulting intermediate-frequency signal can then be highly amplified for subsequent presentation or analysis.

Hobbmann orbit — A minimum energy transfer orbit.

Hole — A mobile vacancy in the electronic valence band of the structure of a semiconductor which acts like an electron with a positive charge.

Holocamera — A device for recording or forming a hologram of an object or subject.

Hologram — A recording or picture of a three-dimensional wavefront.

Holographic — Pertaining to or using the principles of holography. For example, holographic equipment uses the principles of holography for its operation.

Holographic matched filter — A particular type of hologram which when illuminated by the type wave it is matched to will transmit a pure plane wave. This plane wave is usually focused into a correlation spot.

Holography — A recording and viewing process which allows reconstruction of three-dimensional images of diffuse objects.

Homogeneous atmosphere — 1. A hypothetical atmosphere in which the density is constant with height.

The lapse rate of temperature in such an atmosphere is known as the autoconvective lapse rate and is equal to g/R (or approximately 3.4°C/100 m) where g is the acceleration of gravity and R is the gas constant for air. A homogeneous atmosphere has a finite total thickness which is given by R_dT_v/g, where R_d is the gas constant for dry air and T_v is the virtual temperature (°K) at the surface. For a surface temperature of 273°K, the vertical extent of the homogeneous atmosphere on the earth is approxiamtely 8000 m. At the top of such an atmosphere both the pressure and absolute temperature vanish.

2. With respect to radio propagation, an atmosphere which has a constant index of refraction, or one in which radio waves travel in straight lines at constant speed. Free space is the ideal homogeneous atmosphere in this sense. 3. Same as adiabatic atmosphere. See Barotrophy.

Hooke's law — Within the elastic limit of any body the ratio of the stress to the strain produced is constant.

Horsepower — A measure of power, defined as equal to 33,000 foot pounds/min, or 746 watts.

Humidity, absolute — Mass of water vapor present in unit volume of the atmosphere, usually measured as grams per cubic meter. It may also be expressed in terms of the actual pressure of the water vapor present.

Huygens' principle — A very general principle applying to all forms of wave motion which states that every point on the instantaneous position of an advancing phase front (wave front) may be regarded as a source of secondary spherical wavelets. The position of the phase front a moment later is then determined as the envelope of all the secondary wavelets (ad infinitum).

Huygens' theory of light — This theory states that light is a disturbance traveling through some medium, such as the ether. Thus light is due to wave motion in ether.

Every vibrating point on the wave-front is regarded as the center of a new disturbance. These secondary disturbances traveling with equal velocity, are enveloped by a surface identical in its properties with the surface from which the secondary disturbances start and this surface forms the new wave-front.

Hydrogen equivalent of a substance is the number of replaceable hydrogen atoms in 1 mol or the number of atoms of hydrogen with which 1 mol could react.

Hydrogen ion concentration — The concentration of hydrogen ions in solution when the concentration is expressed as gram-ionic weights per liter. A convenient form of expressing hydrogen ion concentration is in terms of the negative logarithm of this concentration. The negative logarithm of the hydrogen ion concentration is called pH.

Water at 25°C has a concentration of H ion of 10^{-7} and of OH ion of 10^{-7} moles per liter. Thus the pH of water is 7 at 24°C. A greater accuracy is obtained if one substitutes the thermodynamic activity of the ion for its concentration.

Hydrolysis is a double decomposition reaction involving the splitting of water into its ions and the formation of a weak acid or base or both.

Hydrostatic equation — In numerical equations, the form assumed by the vertical component of the vector equation of motion when all coriolis, earth-curvature, frictional, and vertical-acceleration terms are considered negligible compared with those involving the vertical pressure force and the force of gravity. Thus,

$$dP = -\rho g dZ$$

where P is the atmospheric pressure; ρ is the density; g is the acceleration of gravity; and Z is the geometric height.

Hydrostatic pressure at a distance h from the surface of a liquid of density d,

$$P = hdg$$

The total force on an area A due to hydrostatic pressure,

$$F = P A = Ahdg$$

Force in dynes and pressure in dynes per cm^2 will be given if h is in cm, d in g per cm^3 and g in cm per $sec.^2$

Hyperon — Any article with mass intermediate between that of the neutron and the deuteron. See Meson.

Hypersonic — 1. Pertaining to hypersonic flow. 2. Pertaining to speeds of Mach 5 or greater.

Hypersonic flow In aerodynamics, flow of a fluid over a body at speeds much greater than the speed of sound and in which the shock waves start at a finite distance from the surface of the body. Compare supersonic flow.

Hysteresis — The magnetization of a sample of iron or steel due to a magnetic field which is made to vary through a cycle of values, lags behind the field. This phenomenon is called hysteresis.

Steinmetz' equation for hysteresis gives the loss of energy in ergs per cycle per cm^3,

$$W = \eta B^{1.6}$$

where B is the maximum induction in maxwells per cm^2 and η the coefficient of hysteresis.

Ice point — The temperature at which a mixture of air-saturatated pure water and pure ice may exist in equilibrium at a pressure of one standard atmosphere.

By decision of the Tenth General Conference on Weights and Measures, Paris, October 1954 the ice point was established as 273.15°K.

Ideal gas — A gas which conforms to Boyle's law and has zero heat of free ezpansion (or also obeys Charles' law). Also called perfect gas.

Illuminance — The total luminous flux received on a unit area of a given real or imaginary surface, expressed in such units as the footcandle, lux, or phot. Illuminance is analogous to irradiance, but is to be distinguished from the latter in that illuminance refers only to light and contains the luminous efficiency weighting factor necessitated by the nonlinear wavelength-response of the human eye. Compare luminous intensity.

The only difference between illuminance and illumination is that the latter always refers to light incident upon a material surface.

A distinction should be drawn, as well, between illuminance and luminance. The latter is a measure of the light coming from a surface; thus, for a surface which is not self-luminous, luminance is entirely dependent upon the illuminance upon that surface and its reflection properties.

Illumination on any surface is measured by the luminous flux incident on unit area. The units in use are: the lux, (abbreviation lx) one lumen per square meter; the phot, (abbreviation ph) one lumen per square centimeter; the footcandle, (abbreviation fc) one lumen per square foot.

Image redundancy — Multiple storage of the same image.

Impact pressure — 1. That pressure of a moving fluid brought to rest which is in excess of the pressure the fluid has when it does not flow, i.e., total pressure less static pressure.

Impact pressure is equal to dynamic pressure in incompressible flow, but in compressible flow impact pressure includes the pressure change owing to the compressibility effect.

2. A measured quantity obtained by placing an open-ended tube, known as an impact tube or pitot tube, in a gas stream and noting the pressure in the tube on a suitable manometer.

Since the pressure is exerted at a stagnation point, the impact pressure is sometimes referred to as the stagnation pressure or total pressure.

Incandescence — Emission of light due to high temperature of the emitting material. Any other emission of light is called luminescence.

Incoherent holography — Holograms produced initially from conventional photographs or incoherent optical equipment.

Indeterminacy principle (Uncertainty principle) — The postulate that it is impossible to determine simultaneously both the exact position and the exact momentum of an electron. So this aspect of electronics can only be expressed as a probability.

Index of refraction (symbol n) — 1. A measure of the amount of refraction (a property of a dielectric substance). It is the ratio of the wavelength or phase velocity of an electromagnetic wave in a vacuum to that in the substance. Also called refractive index, absolute index of refraction, absolute refractive index, refractivity. See modified index of refraction, N-unit, potential index of refraction.

It can be a function of wavelength, temperature, and pressure. If the substance is nonabsorbing and nonmagnetic at any wavelength, then n^2 is equal to the dielectric constant at that wavelength.

The complex index of refraction is obtained when the attenuation of the wave per radian, called the absorptive index k, is paired with the index of refraction. It is written

$$n^* = n(1 - ik)$$

When the wave passes from one medium n_1 to another n_2, the angle of incidence ϕ and the angle of refraction ϕ', both measured with respect to the normal to the interface, are related by

$$\sin \phi / \sin \phi' = n_1{}^* / n_2{}^* = \text{constant}$$

which becomes, for a nonabsorbing medium, the ratios of the (noncomplex) indices of refraction. In the particular case that medium 2 is a vacuum, this ratio is the index of refraction of medium 1. This is known as Snell law, named after Willebrord Snell who discovered it about 1621.

2. A measure of the amount of refraction experienced by a ray as it passes through a refraction interface, i.e., a surface separating two media of different densities. It is the ratio of the absolute indices of refraction of the two media (see sense 1 above). Also called refractive index, relative index of refraction.

Indicators are substances which change from one color to another when the hydrogen ion concentration reaches a certain value, different for each indicator.

Indirectly ionizing particles are uncharged particles (neutrons, photons, etc.) which can liberate directly ionizing particles or can initiate nuclear transformations.

Induced electromotive force in a circuit is proportional to the rate of change of magnetic flux through the circuit.

$$E = -\frac{d\phi}{dt}$$

where $d\phi$ is the change of magnetic flux in a time dt. The induced current will be given by

$$I = \frac{d\phi}{Rdt}$$

where R is the resistance of the circuit.

Inductance — The change in magnetic field due to the variation of a current in a conducting circuit causes an induced counter electromotive force in the circuit itself. This phenomenon is known as self-induction. If an electromotive force is induced in a neighboring circuit the term mutual induction is used. Inductance may thus be distinguished as self- or mutual and is measured by the electromotive force produced in a conductor by unit rate of variation of the current. Units of inductance are the centimeter (absolute electromagnetic) and the henry, which is equal to 10^9 centimeters of inductance. The henry is that inductance in which an induced electromotive force of one volt is produced when the inducing current is changed at the rate of one ampere per second. Dimensions, $[\epsilon^{-1} l^{-1} t^2]$; $[\mu l]$.

DEFINITIONS (Continued)

Induction — Any change in the intensity or direction of a magnetic field causes an electromotive force in any conductor in the field. The induced electromotive force generates an induced current if the conductor forms a closed circuit.

Inertia — The resistance offered by a body to a change of its state of rest or motion, a fundamental property of matter. Dimension, [m].

Information content — Containing or transmitting data involving new knowledge. When applied to waves or wavefronts, it includes both amplitude and phase of all parts of a wavefront at a given instant of time.

Infrared radiation (abbr. IR) — Electromagnetic radiation lying in the wavelength interval from about 75 μm to an indefinite upper boundary sometimes arbitrarily set at 1000 μm (0.01 cm). Also called long-wave radiation.

Inline holography — Hologram produced by single reference beam interferences with waves diffracted or scattered from a small object.

Insolation — (Contracted from incoming solar radiation.) 1. In general, solar radiation received at the earth's surface. See terrestrial radiation, extraterrestrial radiation, direct solar radiation, global radiation, effective terrestrial radiation, diffuse sky radiation, atmospheric radiation. 2. The rate at which direct solar radiation is incident upon a unit horizontal surface at any point on or above the surface of the earh. Compare solar constant.

Intensity of Illumination (properly called Illumination)—Illumination in lux of a screen by a source of illuminating power P at a distance r meters, for normal incidence,

$$I = \frac{P}{r^2}$$

If two sources of illuminating power P_1 and P_2 produce equal illumination on a screen when at distances r_1 and r_2 respectively,

$$\frac{P_1}{r_1^{\,2}} = \frac{P_2}{r_2^{\,2}} \quad \text{or} \quad \frac{P_1}{P_2} = \frac{r_1^{\,2}}{r_2^{\,2}}$$

If I_o is the illumination when the screen is normal to the incident light, then I is the illumination when the screen is at an angle θ. Thus,

$$I = I_0 \cos\theta$$

Intensity of magnetization is given by the quotient of the magnetic moment of a magnet by its volume. Unit intensity of magnetization is the intensity of a magnet which has unit magnetic moment per cubic centimeter. Dimensions, $[\epsilon^{-1/2}\, m^{-1/2}\, l^{-3/2}]$;$[\mu^{1/2}\, m^{1/2}\, l^{-1/3}\, t^{-1}]$.

Intensity of radiation is the rate of transfer of energy across unit areas by the radiation. In all forms of energy transfer by waves (radiation) the intensity I is given by $I = U_v$ where U is the energy density of the wave in the medium, and v is the velocity of propagation of the wave. The energy density U is always proportional to the square of the wave amplitude.

Intensity of sound depends upon the energy of the wave motion. The intensity is measured by the energy in ergs transmitted per second through 1 cm^3 of surface. The energy in ergs per cm^3 in a sound wave is given by

$$E = 2\pi^2 dn^2 a^2$$

where d is density in g per cm^3, n is frequency in vib. per sec and a is amplitude in cm. The energy reaching the ear in unit time will also be proportional to the velocity of propagation.

Interference hologram — A holographic interferogram produced by the superposition of two or more hologram exposures.

Interference (of waves) — The coherent addition or subtraction of two different wavefronts, which usually forms a third wave different from the first two.

Interference pattern — The pattern of light and dark fringes produced when two or more coherent waves interfere or intersect.

Interferogram — The record of an interference pattern produced by holography or by conventional optical interference techniques.

Interferometry (holographic) — The process of measuring very small movements or deformations by recording or observing interference wave patterns (either light, electronic, or acoustic).

Interionic attraction — The electrostatic attraction between ions of unlike charge (sign).

Interionic repulsion — The electrostatic repulsion between ions of like charge (sign).

Internal energy — A mathematically defined thermodynamic function of state, interpretable through statistical mechanics as a measure of the molecular activity of the system. It appears in the first law of thermodynamics as

$$du - dq - dw$$

where du is the increment of specific internal energy, dq the increment of heat, and dw the increment of work done by the system per unit mass. The differential du is a perfect differential. Its integral therefore introduces a constant of integration, the zero-point internal energy, so that care must be taken when absolute values of the internal energy are employed.

International Practical Kelvin Temperature Scale of 1960 and the **International Practical Celsius Temperature Scale of 1960** are defined by a set of interpolation equations based on the following reference temperatures:

	°K	°C
Oxygen, liquid-gas equilibrium	90.18	−182.97
Water, solid-liquid equilibrium	273.15	0.00
Water, solid-liquid-gas equilibrium	273.16	0.01
Water, liquid-gas equilibrium	373.15	100.00
Zinc, solid-liquid equilibrium	692.655	419.505
Sulphur, liquid-gas equilibrium	717.75	444.6
Silver, solid-liquid equilibrium	1233.95	960.8
Gold, solid-liquid equilibrium	1336.15	1063.0

International System of Units (abbr. SI) — The metric system of units based on the meter, kilogram, second, ampere, Kelvin degree, and candela. Also called MSKA system.

Iodine number (value) — A number expressing the percentage (i.e. grams per 100 grams) of iodine absorbed by a substance. It is a measure of the proportion of unsaturated linkages present and is usually determined in the analysis of oils and fats.

Ion — An ion is an atom or group of atoms that is not electrically neutral but instead carries a positive or negative electric charge. Positive ions are formed when neutral atoms or molecules lose valence electrons; negative ions are those which have gained electrons.

Ion atmosphere (or continuus charge distribution) — In the electrostatic effects between ions the term ion atmosphere denotes a continuous charge distribution, or charge density, $\varrho(r)$, which is a continuous function of r, the distance from the reference ion, rather than a discrete or discontinuous charge distribution. The ion atmosphere extends from r = a to $r = 0(V^{1/3}) \approx \infty$, where V is the volume of the system, and acts electrostatically somewhat like a sphere of charge −e at a distance, k^{-1}, from the reference ion of charge + e (see below for definition of k^{-1}).

Ionic equivalent conductance, λ — The ionic equivalent conductance is the equivalent conductance of an individual ion constituent of the solute (or electrolyte) of an electrolytic solution. This symbol is also used to designate the equivalent conductance of complex ions, ion pairs, ion clusters, etc., in combination with simple ions.

Ionic mobility, μ — The mobility of an ion at any finite equivalent concentration is the velocity with which the ion moves under unit potential gradient. Its unit is $cm^2\ sec^{-1}\ volt^{-1}\ equiv^{-1}$ or $cm^2\ ohm^{-1}\ F^{-1}$ where F is the Faraday expressed in coulombs (or ampere seconds) $equiv^{-1}$.

Ionization — The process by which neutral atoms or groups of atoms become electrically charged, either positively or negatively, by the loss or gain of electrons; or the state of a substance whose atoms or groups of atoms have become thus charged.

Ionization potential — The work (expressed in electron volts) required to remove a given electron from its atomic orbit and place it at rest at an infinite distance. It is customary to list values in electron volts (ev.) 1 ev. = 23,053 calories per mole.

Ionizing radiation is any radiation consisting of directly or indirectly ionizing particles or a mixture of both.

Ionogens — Substances, like acetic acid (HAc), which, although in the pure state are nonelectrolytic neutral molecules, can react with certain solvents to form products which rearrange to ion pairs which then dissociate to give conducting solutions. As an example:

$$HAc + H_2O \rightleftharpoons HAc \cdot H_2O \rightleftharpoons H_3O^+ \cdot Ac^- \rightleftharpoons H_3O^+ + Ac^-$$

Ionophores — Substances, like sodium chloride, which exist only as ionic lattices in the pure crystalline form, and which when dissolved in an appropriate solvent give conductances which change according to some fractional power of the concentration. Such solutions possess no neutral molecules which can dissociate, but may contain associated ions.

Ionosphere — The atmospheric shell characterized by a high ion density. Its base is at about 70 or 80 km and it extends to an indefinite height.

Ion size or "ion-size" parameter, a (or a_i) — The ion size is formally considered to be the sum of the ionic radii of the oppositely charged ions in contact. The ion size is also called the "distance of closest approach" of the ions, or the "ion-size" parameter. Generally the ion size is greater than the sum of the crystal radii, and the "ion-size" parameter may include several factors which contribute to its numerical value.

Irradiance — A measure of the rate of energy falling on a given area.

Isentropic — Of equal or constant entropy with respect to either space or time.

Isobaric — Of equal or constant pressure, with respect to either space or time.

Isobars — For chemistry, elements of the same atomic mass but of different atomic numbers. The sum of their nucleons is the same but there are more protons in one than in the other.

Isochoric — Of equal or constant volume, usually applied to a thermodynamic process during which the volume of the system remains unchanged. Compare isosteric.

Isoclinic line — A line through points on the earth's surface having the same magnetic dip.

Isomer — 1. One of two or more nuclides having the same mass number A and atomic number Z, but existing for measurable times in different quantum states with different energies and radioactive properties.
2. One of two or more molecules having the same atomic composition and molecular weight, but differing in geometrical configuration.

Isomerism — Existence of molecules having the same number and kinds of atoms but in different configurations.

Isotherm — A line of equal or constant temperature.
A distinction is made, infrequently, between a line representing equal temperature in space, choroisotherm, and one representing constant temperature in time, chronoisotherm.

Isothermal — When a gas passes through a series of pressure and volume variations without change of temperature the changes are called isothermal. A line on a pressure-volume diagram representing these changes is called an isothermal line.

Isotope — 1. One of several nuclides having the same number of protons in their nuclei, and hence belonging to the same element, but differing in the number of neutrons and therefore in mass number A, or in energy content (isomers). For example, ${}_6C^{12}$, ${}_6C_7{}^{13}$, and ${}_6C_8{}^{14}$ are carbon isotopes. Small quantitative differences in chemical properties exist between isotopes.

Isotropic — In general, pertaining to a state in which a quantity or spatial derivatives thereof are independent of direction. Also called isotropous.

Joule constant — The ratio between heat and work units from experiments based on the first law of thermodynamics: 4.1858 × 10^7 ergs per 15° calorie. Also called mechanical equivalent of heat.

Joule — A unit of energy or work in the MKS system; the work done when the point of application of 1 newton is displaced a distance of 1 meter in the direction of the force.

$$1 \text{ joule} = 10^7 \text{ ergs} = 1 \text{ watt second}$$

Joule cycle — (After James Prescott Joule, 1818—1889, English physicist.) An ideal cycle for engines consisting of isentropic compression of the working substance, addition of heat at constant pressure, isentropic expansion to ambient pressure, and exhaust at constant pressure. Also called Brayton cycle.

Joule's law — A law stating that the amount of heat H produced in a conductor by the flow of a steady current I is given by

$$H = KI^2Rt.$$

where R is the resistance of the conductor, t is the time for which the current flows, and K, the constant of proportionality, has the value 0.2390 calories per joule when R is in ohms and I in amperes.

Joule-Thomson effect — The decrease in temperature which takes place when a gas expands through a throttling device as a nozzle. Also called Joule-Kelvin effect.
The rate of change of temperature T with pressure p in the Joule-Thomson effect is called the Joule-Thomson coefficient (symbol μ):

$$\mu = \left[\frac{dT}{dp}\right]_h$$

where h denotes constant enthalpy.
For the Joule-Thomson effect to take place the gas must initially be below its inversion temperature; if above the inversion temperature, the gas will gain heat on expansion. The inversion temperature of hydrogen, for example, is approximately −183°C.

Junction — In a semiconductor device, a region of transition between semiconducting regions of different electrical properties.

Kelvin — The kelvin, the unit of thermodynamic temperature, is the fraction 1/273.16 of the thermodynamic temperature of the triple point of water. The decision was made at the 13th General Conference on Weights and Measures on October 13, 1967 that the name of the unit of thermodynamic temperature would be changed from degree Kelvin (symbol: °K) to kelvin (symbol: K). The name (kelvin) and symbol (K) are to be used for expressing temperature intervals. The former convention which expressed a temperature interval in degrees Kelvin or, abbreviated, deg. K is dropped. However, the old designations are acceptable temporarily as alternatives to the new ones. One may also express temperature intervals in degrees Celsius.

Kelvin temperature scale (abbr. K)—An absolute temperature scale independent of the thermometric properties of the working substance. On this scale, the difference between two temperatures T_1 and T_2 is proportional to the heat converted into mechanical work by a Carnot engine operating between the isotherms and adiabats through T_1 and T_2. Also called absolute temperature scale, thermodynamic temperature scale.
For convenience the Kelvin degree is identified with the Celsius degree. The ice point in the Kelvin scale is 273.15 K. The triple point of water, the fundamental reference point, is 273.16 K. See Absolute zero, Rankine temperature scale.

Kepler's laws — I. The planets move about the sun in ellipses, at one focus of which the sun is situated. II. The radius vector joining each planet with the sun describes equal areas in equal times. III. The cubes of the mean distances of the planets from the sun are proportional to the squares of their times of revolution about the sun.

Kerr cell — A cell that contains electrodes immersed in nitrobenzene or other liquid. It shows double refraction in high degree and with short time lag. Used in devices in which light intensity is changed rapidly according to the voltage applied to the electrons.

Kerr effect — When plane polarized light is incident on the pole of an electromagnet, polished so as to act like a mirror, the plane of polarization of the reflected light is not the same when the magnet is "on" as when it is "off." It was found that the direction of rotation was opposite to that of the currents exciting the pole from which the light was reflected.

Kilo — A prefix meaning thousand or 10^3. Symbol is k.

Kilogram (unit of mass)—is the mass of a particular cylinder of platinum-iridium alloy, called the International Prototype Kilogram, which is preserved in a vault at Sèvres, France, by the International Bureau of Weights and Measures.

Kilometer (abbr. km) — A unit of distance in the metric system.
1 kilometer = 3280.8 feet = 1093.6 yards = 1000 meters = 0.62137 statute miles = 0.53996 nautical miles.

Kinematic viscosity (symbol ν) — A coefficient defined as the ratio of the dynamic viscosity of a fluid to its density.
The kinematic viscosity of most gases increases with increasing temperature and decreasing pressure. For dry air at 0°C, the kinematic viscosity is about 0.13 square centimeter per second. In the theory of atmospheric turbulence the kinematic viscosity is usually replaced by the kinematic eddy viscosity to account for the increased internal friction due to turbulence.

Kinetic energy (symbol E_k) — The energy which a body possesses as a consequence of its motion, defined as one-half the product of its mass m and the square of its speed v, $\frac{1}{2}mv^2$. The kinetic energy per unit volume of a fluid parcel is thus $\frac{1}{2}\varrho v^2$, where ϱ is the density and v the speed of the parcel. See Potential energy.
For relativistic speeds the kinetic energy is given by

$$E_k = mc^2 - m_0c^2$$

where c is the velocity of light in a vacuum, m_0 is the rest mass, and m is the moving mass.

Kinetic theory, expression for pressure —

$$P = \frac{1}{3} N\ mv^2$$

where N is the number of molecules in unit volume, m the mass of each molecule and v^2 the mean square of the velocity of the molecules.

Kinetic theory of gases — Gases are considered to be made up of minute, perfectly elastic particles which are ceaselessly moving about with high velocities, colliding with each other and with the walls of the containing vessel. The pressure exerted by a gas is due to the combined effect of the impacts of the moving molecules upon the walls of the containing vessel, the magnitude of the pressure being dependent upon the kinetic energy of the molecules and their number.

Kirchhoff law—The radiation law which states that at a given temperature the ratio of the emissivity to the absorptivity for a given wavelength is the same for all bodies and is equal to the emissivity of an ideal black body at that temperature and wavelength.

Kirchhoff's laws —
I. The algebraic sum of the currents which meet at any point is zero.
II. In any closed circuit the algebraic sum of the products of the current and the resistance in each conductor in the circuit is equal to the electromotive force in the circuit.

Kirchhoff's laws of radiation — The relation between the powers of emission and the powers of absorption for rays of the same wave-length is constant for all bodies at the same temperature. First, a substance when excited by some means or other possesses a certain power of emission; it tends to emit definite rays, whose wave-lengths depend upon the nature of the substance and upon the temperature. Second, the substance exerts a definite absorptive power, which is a maximum for the rays it tends to emit. Third, at a given temperature the ratio between the emissive and the absorptive power for a given wave-length is the same for all bodies, and is equal to the emissive power of a perfectly black body.

Knot — A nautical mile per hour, 1.1508 statute miles per hour.

Kohlrausch law of independent migration of ions — The value of the equivalent conductance, as the concentration approaches zero, is equal to the sum of the limiting ionic equivalent conductances of the ions constituting the solute of the electrolytic solution.

Kundt's law — On approaching an absorption band from the red side of the spectrum the refractive index is abnormally increased by the presence of the band, while the approach is from the blue side and the index is abnormally decreased.

Lambert (abbr. L or l) — A unit of luminance (or brightness) equal to $1/\pi$ candle per square centimeter. Physically, the lambert is the luminance of a perfectly diffusing white surface receiving an illuminance of 1 lumen/cm².

Lambert law — A law of physics which states that the radiant intensity (flux per unit solid angle) emitted in any direction from a unit radiating surface varies as the cosine of the angle between the normal to the surface and the direction of the radiation. The radiance (or luminance) of a radiating surface is, therefore, independent of direction. Also called Lambert cosine law.

Lambert law of absorption = Bouguer law.

Lambert's law of absorption — Each layer of equal thickness absorbs an equal fraction of the light which traverses it.

Lambert's law of illumination — The illumination of a surface on which the light falls normally from a point source is inversely proportional to the square of the distance of the surface from the source. If the normal to the surface makes an angle with the direction of the rays, the illumination is proportional to the cosine of that angle.

Landau damping — The damping of a space charge wave by electrons which move at the phase velocity of the wave and gain energy transferred from the wave.

Langley — A unit of energy per unit area, equal to 1 gram-calorie/cm² commonly employed in radiation theory.

Langmuir probe — A small metallic conductor or pair of conductors inserted within a plasma in order to sample the plasma current.

Lanthanide series (Lanthanides) — Rare earth elements of atomic numbers 57 through 71, which have chemical properties similar to lanthanum (#57).

Laplacian speed of sound — The phase speed of a sound wave in a compressible fluid if the expansions and compressions are assumed to be adiabatic. This speed a is given by the formula

$$a^2 = (c_p/c_v)\ RT$$

where c_p and c_v are the specific heats at constant pressure and volume, respectively; R is the gas content; and T is the Kelvin temperature. The value of this speed under standard conditions in dry air is 331 m/sec. Compare Newtonian speed of sound. See Acoustic velocity.

Laser — A device in which the excitation energy of resonant atomic or molecular systems is used to coherently amplify or generate light. Consider a system of particles with energy levels E_1, E_2 where $E_1 < E_2$. A particle passing between these levels absorbs or emits a quantum of radiation of frequency

$$\nu = (E_2 - E_1)/h$$

where h is Planck's constant. If N_1 is the number of particles in E_1 and N_2 the number in E_2, it follows from the laws of thermodynamics that at a positive absolute temperature $N_2 < N_1$. As a result of this, a wave of the appropriate frequency would be attenuated. If by some means N_2 can be made to exceed N_1 by a reasonably large margin, an incoming wave of frequency ν will stimulate transitions from E_2 to E_1 and in the process become coherently amplified. For sustained oscillation to occur feedback must be provided; this is often done by covering the ends of the chamber in which the reaction occurs with reflective material and arranging the chamber to be resonant at the frequency of oscillation.

Latent heat of vaporization — The quantity of heat necessary to change one gram of liquid to vapor without change of temperature, measured as calories per gram. Dimensions, $[l^2t^{-2}]$.

Lattice energy — The energy required to separate the ions of a crystal to an infinite distance from each other.

LeChatelier's principle — If some stress is brought to bear upon a system in equilibrium, a change occurs, such that the equilibrium is displaced in a direction which tends to undo the effect of the stress.

Length — The name "micron", for a unit of length equal to 10^{-6} meter, and the symbol "μ" which has been used for it were dropped by action of the 13th General Conference on Weights and Measures on October 13, 1967. The symbol "μ" is to be used solely as an abbreviation for the prefix "micro-", standing for the multiplication by 10^{-6}. Thus the length previously designated as 1 micron, should be designated 1 μm.

Length, units of — Meter. 1. (Abbr. m) The basic unit of length of the metric system, defined in the October 1984 General Conference of Weights and Measures as the length of the path traveled by light in vacuum during a time interval of 1/299,792,458 of a second.

Effective 1 July 1959 in the U.S. system of measures, 1 yard = 0.9144 meter, exactly, or 1 meter = 1.094 yards = 39.37 inches. The standard inch is exactly 25.4 millimeters.

Lenses — For a single thin lens whose surfaces have radii of curvature r_1 and r_2 whose principal focus is F, the index of refraction n, and conjugate focal distances f_1 and f_2,

$$\frac{1}{F} = \frac{1}{f_1} + \frac{1}{f_2} = (n - 1)\left(\frac{1}{r_1} + \frac{1}{r_2}\right)$$

For a thick lens, of thickness t,

$$F = \frac{nr_1r_2}{(n-1)[n(r_1+r_2) - t(n-1)]}$$

Combinations of lenses—If f_1 and f_2 are the focal lengths of two thin lenses separated by a distance d the focal length of the system,

$$F = \frac{f_1 f_2}{f_1 + f - d}$$

Lenz's law — When an electromotive force is induced in a conductor by any change in the relation between the conductor and the magnetic field, the direction of the electromotive force is such as to produce a current whose magnetic field will oppose the change.

Limiting equivalent conductance, Λ_o — The limiting equivalent conductance of an electrolytic solution, Λ_o, is expressed by $\Lambda_o \equiv \lim (\sigma_{corr}/c)$ where σ_{corr} is solution conductance corrected for solvent conductance and c is the equivalent concentration. Λ_o is the value which Λ approaches as the solution is diluted so far that the effects of interionic forces become negligible (and dissociation, in the case of ionogens, is essentially complete).

Limiting ionic equivalent conductance, λ_o — The limiting ionic equivalent conductance of an individual ion constituent of the solute (or electrolyte) of an electrolytic solution is given by $\lambda_o \equiv \lim (\lambda/c)$. This symbol is also used to designate the limiting equivalent conductances of complexions, ion pairs, ion clusters, etc., in combination with simple ions.

Limiting ionic mobility, u^o — The limiting mobility of an individual ion of a solute (or electrolyte) is given by $u \equiv \lim u$.

Limiting molar conductance, $\Lambda^m{}_o$ — The limiting molar conductance of an electrolytic solution, $\Lambda^m{}_o$, is expressed by $\Lambda^m{}_o \equiv \lim (\sigma_{corr}/m)$ where σ_{corr} is solution conductance corrected for solvent conductance and m is the molar concentration. $\Lambda^m{}_o$ is the value which Λ^m approaches as the solution is diluted so far that the effects of interionic forces become negligible. Seldomly used.

Line of force — A term employed in the description of an electric or magnetic field. A line such that its direction at every point is the same as the direction of the force which would act on a small positive charge (or pole) placed at that point. A line of force is defined as starting from a positive charge (or pole) and ending on a negative charge (or pole).

The line (of force) is also used as a unit of magnetic flux, equivalent to the maxwell.

Lissajous figures — The path described by a particle which is simultaneously displaced by two simple harmonic motions at right angles, when the periods of the two motions are in the ratio of two small whole numbers, shows a variety of characteristic curves called Lissajous figures.

Liquid — A state of matter in which the molecules are relatively free to change their positions with respect to each other, but restricted by cohesive forces so as to maintain a relatively fixed volume.

Logarithm — A number m related to another number p by

$$B^m = p$$

where B is an arbitrarily chosen number larger than 1. Usually m is written $\log_B p$. Logarithms have the basic properties that

$$\log pq = \log p + \log q$$

and

$$\log p^q = q \log p$$

In the systems most often used B = 10 (common logarithms) or B = e = 2.71828...(natural logarithms). The common logarithm of p is written $\log_{10} p$ or simply log p; the natural logarithm of p is written $\log_e p$ or ln p.

Longitudinal wave energy — Waves whose amplitude displacement is in the same or opposite direction as the motion.

Loschmidt's number — The number of molecules per unit volume of an ideal gas at 0°C and normal atmospheric pressure.

$$n_o = 2.68675 \times 10^{25}\ m^{-3}$$

Loudness — The psychological response of the ear which is related to the physical quantity intensity. The loudness of a sound depends on frequency also since the ear responds more strongly to some frequency bands than to others. The loudness is roughly related to the cube root of the intensity, and for many purposes it is convenient to represent loudness as proportional to the logarithm of the intensity.

Lumen — The lumen is the unit of luminous flux. It is equal to the luminous flux through a unit-solid angle (steradian) from a uniform point source of one candle, or to the flux on a unit surface all points of which are at unit distance from a uniform point source of one candela.

Luminescence — Light emission by a process in which kinetic heat energy is not essential for the mechanism of excitation.

Electroluminescence is luminescence from electrical discharges—such as sparks or arcs. Excitation in these cases results mostly from electron or ion collision by which the kinetic energy of electrons or ions, accelerated in an electric field, is given up to the atoms or molecules of the gas present and causes light emission. Chemiluminescence results when energy, set free in a chemical reaction, is converted to light energy. The light from many chemical reactions and from many flames is of this type. Photoluminescence, or fluorescence, results from excitation by absorption of light. The term phosphorescence is usually applied to luminescence which continues after excitation by one of the above methods has ceased. Compare incandescence.

Luminous (energy) — Energy whose wavelength is such that the eye is sensitive to it.

Luminous flux — The total visible energy emitted by a source per unit time is called the total luminous flux from the source. The unit of flux, the lumen, is the flux emitted in unit solid angle (steradian) by a point source of one candela luminous intensity. A uniform point source of one candela intensity thus emits 4π lumens.

Luminous intensity or candlepower is the property of a source of emitting luminous flux and may be measured by the luminous flux emitted per unit solid angle. The SI unit of luminous intensity is the candela. The Hefner unit, which is equivalent to 0.9 international candles, is the intensity of a lamp of specified design burning amyl acetate, called the Hefner lamp.

The mean horizontal candlepower is the average intensity measured in a horizontal plane passing through the source. The mean spherical candlepower is the average candlepower measured in all directions and is equal to the total luminous flux in lumens divided by 4π.

Lux — A photometric unit of illuminance or illumination equal to 1 lumen/m^2. Compare footcandle, phot.

Magnetic Anisotropy — In ferro- or ferrimagnetic crystals. it is found that the magnetization prefers to lie along certain crystal directions. These are termed easy directions of magnetization. Work must be expended to turn the magnetization away from these easy directions. That work as a function of crystal direction defines the anisotropy energy surface. Directions associated with a maximum of the anisotropy energy are termed hard directions of magnetization. In general, the energy difference between easy and hard directions decreases as the temperature is increased, and vanishes at the Curie or Néel point.

Magnetic Domains — The magnetization of a ferromagnetic or a ferrimagnetic material tends to break up into regions called domains separated by thin transition regions called domain walls. Within the volume of a domain, the magnetization has its saturation value, and is directed along a single direction. The magnetizations of other domains are directed along different directions in such a way that the net magnetization of the whole sample may be zero. The application of an external magnetic field first causes some domains to grow by the motion of their walls. At higher fields, the magnetizations of the resulting domains rotate toward parallelism with the field.

Magnetic field due to a current — The intensity of the magnetic field in oersted at the center of a circular conductor of radius r in which a current I in absolute electromagnetic units is flowing,

$$H = \frac{2\pi I}{r}$$

If the circular coil has n turns the magnetic intensity at the center is,

$$H = \frac{2\pi n I}{r}$$

The magnetic field in a long solenoid of n turns per centimeter carrying a current I in absolute electromagnetic units

$$H = 4\pi n I$$

If I is given in amperes the above formulae become,

$$H = \frac{2\pi I}{10r} \qquad H = \frac{2\pi n I}{10r}. \qquad H = \frac{4\pi n I}{10}$$

Magnetic field due to a magnet — At a point on the magnetic axis prolonged, at a distance r cm from the center of the magnet of length 2l whose poles are +m and −m and magnetic moment M, the field strength in oersted is,

$$H = \frac{4mlr}{(r^2 - l^2)^2}$$

If r is large compared with l,

$$H = \frac{2M}{r^3}$$

At a point on a line bisecting the magnet at right angles, with corresponding symbols,

$$H = \frac{2ml}{(r^2 + l^2)^3}$$

For large value of r,

$$H = \frac{M}{r^3}$$

Magnetic field intensity or magnetizing force — Is measured by the force acting on unit pole. Unit field intensity, the oersted, is that field which exerts a force of one dyne on unit magnetic pole. The field intensity is also specified by the number of lines of force intersecting unit area normal to the field, equal numerically to the field strength in oersted. Magnetizing force is measured by the space rate of variation of magnetic potential and as such its unit may be the gilbert per centimeter. The gamma (γ) is equivalent to 0.00001 oersted. Dimensions, $[\varepsilon^{1/2} m^{1/2} l^{1/2} t^{-2}]$; $[\mu^{-1/2} m^{1/2} l^{-1/2} t^{-1}]$.

Magnetic flux through any area perpendicular to a magnetic field is measured as the product of the area by the field strength. The units of magnetic flux, the maxwell is the flux through a square centimeter normal to a field of one gauss. The line is also a unit of flux. It is equivalent to the maxwell. Dimensions, $[\varepsilon^{-1/2} m^{1/2} l^{1/2}]$; $[\mu^{1/2} m^{1/2} l^{1/2} t^{-1}]$.

Magnetic induction resulting when any substance is subjected to a magnetic field is measured as the magnetic flux per unit area taken perpendicular to the direction of the flux. The unit is the maxwell per square centimeter or its equivalent, the gauss. Dimensions, $[\varepsilon^{-1/2} m^{1/2} l^{-3/2}]$; $[\mu^{1/2} m^{1/2} l^{-1/2} t^{-1}]$.

If a substance of permeability μ is placed in a magnetic field H the magnetic induction in the substance,

$$M = \mu H.$$

If I is the magnetic moment for unit volume, or intensity of magnetization,

$$M = H + 4\pi I.$$

The susceptibility,

$$k = \frac{I}{H}, \quad \mu = 1 + 4\pi k.$$

Magnetic moment of a magnet is measured by the torque experienced when it is at right angles to a uniform field of unit intensity. The value of the magnetic moment is given by the product of the magnetic pole strength by the distance between the poles. Unit magnetic moment is that possessed by a magnet formed by two poles of opposite sign and of unit strength, 1 cm apart. Dimensions, $[\mu^{1/2} m^{1/2} l^{5/2} t^{-1}]$; $[\varepsilon^{-1/2} m^{1/2} l^{3/2}]$.

If the poles are separated by a distance which is great compared with the dimensions of the magnet, the magnetic moment of a magnet of length l whose poles have values of +m and −m is,

$$m = ml.$$

Magnetic permeability is a property of materials modifying the action of magnetic poles placed therein and modifying the magnetic induction resulting when the material is subjected to a magnetic field or magnetizing force. The permeability of a substance may be defined as the ratio of the magnetic induction in the substance to the magnetizing field to which it is subjected. The permeability of a vacuum is unity. Dimensions, $[\varepsilon^{-1} l^{-2} t^2]$; $[\mu]$.

Magnetic pole or quantity of magnetism — Two unit quantities of magnetism concentrated at points unit distance apart in a vacuum repel each other with unit force. If the distance involved is 1 cm and the force 1 dyn, the quantity of magnetism at each point is one cgs unit of magnetism. Dimensions, $[\varepsilon^{-1/2} m^{1/2} l^{1/2}]$; $[\mu^{1/2} m^{1/2} l^{3/2} t^{-1}]$.

Magnetic potential or magnetomotive force at a point measured by the work required to bring unit positive pole from an infinite distance (zero potential) to the point. The unit is the gilbert, that magnetic potential against which an erg of work is done when unit magnetic pole is transferred. Dimensions, $[\varepsilon^{1/2}, m^{1/2} l^{3/2} t^{-2}]$; $[\mu^{-1/2} m^{1/2} l^{1/2} t^{-1}]$.

Magnetostriction — Change in sample dimensions as the magnitude or the direction of the magnetization in a crystal is changed.

Magnifying power of an optical instrument is the ratio of the angle subtended by the image of the object seen through the instrument to the angle subtended by the object when seen by the unaided eye. In the case of the microscope or simple magnifier the object as viewed by the unaided eye is supposed to be a distance of 25 cm (10 in.).

Maser — An abbreviation based on the English description of the function of this device, microwave amplification by stimulated emission of radiation.

Mass — Quantity of matter. Units of mass — the kilogram is one of the SI base units, defined by the international prototype kept at the BIPM. The gram is 1/1000 the quantity of matter in the International Prototype Kilogram. The U.S. and British pounds are defined as 0.45359237 kg, exactly. The mass of 1-kg secondary standards of platinum-iridium or of stainless steel is compared with the mass of the prototype by means of balances whose precision can reach 1 in 10^8 or better.

Mass — The inertial resistance of a body to acceleration, considered, in classical physics, to be a conserved quantity independent of speed. It can be shown that the parameters m_1, m_2 appearing in the gravitation equation are equivalent to the masses of the bodies in the sense given above. In relativistic physics it is shown that when the speed of a body becomes an appreciable fraction of c, the speed of light mass is given by

$$m = m_0/[1 - (v/c)^2]^{1/2}$$

where m_o is the rest mass of the body and v is its speed relative to the observer who finds its mass to be m. Further, it has been shown that mass and energy are interconvertible as given by Einstein's equation $E = mc^2$.

Mass action, law of — At a constant temperature the product of the active masses on one side of a chemical equation when divided by the product of the active masses on the other side of the chemical equation is a constant, regardless of the amounts of each substance present at the beginning of the action.

At constant temperature the rate of the reaction is proportional to the concentration of each kind of substance taking part in the reaction.

Mass and Weight — "1 The kilogram is the unit of mass; it is equal to the mass of the international prototype of the kilogram;

"2 The word weight[a] denotes a quantity of the same nature as a force; the weight of a body is the product of its mass and the acceleration due to gravity; in particular, the standard weight of a body is the product of its mass and the standard acceleration due to gravity;

"3 The value adopted in the international Service of Weights and Measures for the standard acceleration due to gravity is 980.665 cm/s², value already stated in the laws of some countries."

Mass by weighing on a balance with unequal arms— If W_1 is the value for one side, W_2 the value for the other, the true mass,

$$W = \sqrt{W_1 W_2}$$

Mass defect — Difference between atomic mass and mass number of a nuclide. See Packing fraction.

Mass-energy equivalence — The equivalence of a quantity of mass and a quantity of energy when the two quantities are related by the equation $E = mc^2$. The conversion factor c^2 is the square of the velocity of light. The relationship was developed from relativity theory, but has been experimentally confirmed. $1 \text{ kg} = 9 \times 10^{16} \text{ J} = 2.5 \times 10^{10} \text{ kWh}$.

Mass number — The whole number nearest the value of the atomic mass of an element as expressed in atomic mass units.

The mass number is assumed to represent the total number of protons and neutrons in the atomic nucleus of the element and is therefore equal to the atomic number plus the number of the neutrons. The mass number of an atom is usually written as a superscript to the element symbol, as in O^{18}, an isotope of oxygen with mass number 18.

Maxwell — The cgs emu magnetic flux is the flux through a cm² normal to a field at 1 cm from a unit magnetic pole.

Maxwell's rule — A law stating that every part of an electric circuit is acted upon by a force tending to move it in such a direction as to enclose the maximum amount of magnetic flux.

Mechanical equivalent of heat is the quantity of energy which, when transformed into heat, is equivalent to unit quantity of heat; 4.18×10^7 ergs = 1 calorie (20°C) or 4.1868 J = 1 cal_{IT}.

Mega — A prefix meaning million or 10^6. Symbol is M.

Meson — Two types of particles of mass intermediate between that of the electron and proton have been discovered in cosmic radiation and in the laboratory. The one particle with mass about $215m_e$ is called μ-meson, the other with about $280m_e$ π-meson. Mesons of both positive and negative charge have been found and there is now reasonably good evidence for neutral mesons. Both types of mesons decay spontaneously. Some evidence exists for a meson of mass about $1000m_e$.

Metallic elements in general are distinguished from the non-metallic elements by their luster, malleability, conductivity and usual ability to form positive ions. Nonmetallic elements are not malleable, have low conductivity and never form positive ions.

Metamagnetic Materials — Those which are antiferromagnetic in weak fields, but which become ferromagnetically ordered in strong applied fields.

Meter — The meter is the SI base unit of length and was defined at the October 1984 General Conference of Weights and Measures as the length of the path traveled by light in vacuum during a time interval of 1/299,792,458 of a second.

MeV — = million electron volts.

Micro — A prefix meaning one millionth or 10^{-6}. Symbol is μ.

Mie scattering — That which is produced by spherical particles without special regard to comparative size of radiation wavelength and particle diameter.

Milli — A prefix meaning 1/1000 or 10^{-3}. Symbol is m.

Minimum deviation — The deviation or change of direction of light passing through a prism is a minimum when the angle of incidence is equal to the angle of emergence. If D is the angle of minimum deviation and A the angle of the prism, the index of refraction of the prism for the wavelength used is,

$$n = \frac{\sin \frac{1}{2}(A + D)}{\sin \frac{1}{2}A}$$

Minute of arc — 1/60 of a degree.

Mixtures consist of two or more substances intermingled with no constant percentage composition, and with each component retaining its essential original properties.

Moderator — A material used for slowing down neutrons in an atomic pile or reactor. Usually graphite or "heavy water" (deuterium oxide).

Modified index of refraction — An atmospheric index of refraction mathematically modified so that when its gradient is applied to energy propagation over a hypothetical flat earth it is substantially equivalent to propagation over the true curved earth with the actual index of refraction. Also called refractive modulus, modified refractive index. Compare potential index of refraction.

The modified index of refraction is usually expressed in M-units; mathematically

$$M = \left(n - 1 + \frac{h}{c}\right) 10^5 = N + \left(\frac{h}{a}\right) 10^6$$

where n is the index of refraction at a point in the atmosphere; h is the height above mean sea level of that point; a is the radius of the earth; and N is the index of refraction in N-units.

In ray tracing problems, the vertical gradient dM/dh can be used directly to obtain a ray path curvature that is relative to the curvature of the earth, i.e.,

$$\frac{dM}{dh} = \frac{dN}{dh} + \frac{10^5}{a} = \frac{10^6}{ka}$$

where k is a value by which the earth's radius is multiplied to get the radius of curvature of the ray path; ka is called the effective earth radius.

Modulus of elasticity — The stress required to produce unit strain, which may be a change of length (Young's modulus): a twist or shear (modulus of rigidity or modulus of torsion) or a change of volume (bulk modulus), expressed in dynes per square centimeter. Dimensions, the same as of stress, $[m\, l^{-1}\, t^{-2}]$.

Moire patterns — Pattern resulting from interference beats between two sets of periodic structures in an image.

Molal solution contains 1 mol/1000 g of solvent.

Molar conductance, Λ^m — The molar conductance of an electrolytic solution is the conductance of a solution containing 1 g mol of the solute (or electrolyte) when measured in a like manner to equivalent conductance. Seldomly used.

Molar solution contains 1 mol or g mol wt of the solute in 1 ℓ of solution.

Mole — the amount of substance of a system which contains as many elementary entities as there are atoms in 0.012 kg of carbon 12.

Note. When the mole is used, the elementary entities must be specified and may be atoms, molecules, ions, electrons, other particles, or specified groups of such particles.

Molecular volume — Volume occupied by 1 mol. Numerically equal to the molecular weight divided by the density.

[a] USA Editors' note: In the USA weight is the commonly used term for mass. Because of the dual use of the term weight as a quantity, this term should be avoided in technical practice. (American National Standard Z210.1)

Molecular weight — The sum of the atomic weights of all the atoms in a molecule.

Molecule — The smallest unit quantity of matter which can exist by itself and retain all the properties of the original substance.

Mol volume — The volume occupied by a mol or a gram molecular weight of any gas measured at standard conditions is 22.414 *l*.

Moment of force or torque — The effectiveness of a force to produce rotation about an axis, measured by the product of the force and the perpendicular distance from the line of action of the force to the axis. Cgs unit—the dyne-centimeter. Dimensions, $[m\ l^2\ t^{-2}]$. If a force F acts to produce rotation about a center at a distance d from the line in which the force acts, the force has a torque,

$$L = Fd.$$

Moment of inertia — A measure of the effectiveness of mass in rotation. In the rotation of a rigid body not only the body's mass, but the distribution of the mass about the axis of rotation determines the change in the angular velocity resulting from the action of a given torque for a given time. Moment of inertia in rotation is analogous to mass (inertia) in simple translation. The cgs unit is g-cm^2. Dimensions, $[m\ l^2]$.

If m_1, m_2, m_3, etc. represent the masses of infinitely small particles of a body; r_1, r_2, r_3, etc. their respective distances from an axis of rotation, the moment of inertia about this axis will be

$$I = (m_1 r_1{}^2 + m_2 r_2{}^2 + m_3 r_3{}^2 + \ldots)$$

or

$$I = \Sigma\ (mr^2)$$

Momentum — Quantity of motion measured by the product of mass and velocity. Cgs unit, one gram-centimeter per second. Dimensions, $[m\ l\ t^{-1}]$.

A mass m moving with velocity v has a momentum,

$$M = mv$$

If a mass m has its velocity changed from v_1 to v_2 by the action of a force F for a time t,

$$mv_2 - mv_1 = Ft.$$

Monochromatic emissive power is the ratio of the energy of certain defined wave lengths radiated at definite temperatures to the energy of the same wave lengths radiated by a black body at the same temperature and under the same conditions.

Monochromatic (source) — All source radiation is exactly of the same wavelength. This is never achieved in practice even with a laser, so the term "quasi-monochromatic" is used to mean nearly of the same wavelength for all practical purposes.

Monotropic — Crystal forms one of which is always metastable with respect to the other.

Mosley's law — The frequencies of the characteristic X-rays of the elements show a strict linear relationship with the square of the atomic number.

Motion, laws of — See Newton's law of motion.

Multiple proportions, law of — If two elements form more than one compound, the weights of the first element which combine with a fixed weight of the second element are in the ratio of integers to each other.

Nano — A prefix meaning one billionth or 10^{-9}. Symbol is n.

Néel Point — The temperature at which ferrimagnetic and antiferromagnetic materials become paramagnetic.

Negatron — 1. A term used for electron when it is necessary to distinguish between (negative) electrons and positrons.

2. A four element vacuum tube which displays a negative resistance characteristic.

Nernst effect — When heat flows across the lines of magnetic force, there is observed an electromotive force in the mutually perpendicular direction.

Neutralization is a reaction in which the hydrogen ion of an acid and the hydroxyl ion of a base unite to form water, the other product being a salt.

Neutrino — An electrically neutral particle of very small (probably zero) rest mass and of spin quantum number ½. When the spin is oriented parallel to the linear momentum the particle is the antineutrino. When the spin is oriented antiparallel to the linear momentum the particle is the neutrino. Postulated by Pauli in explaining the beta decay process.

Whenever a beta (positron) particle is created in a radioactive decay so is an antineutrino (neutrino). The two particles and the parent nucleus share between them the available energy and momentum. Neutrinos and antineutrinos can penetrate amounts of matter measured in light years without appreciable attenuation. Detected by Reines and Cowan using antineutrinos from fission reactors and large scintillation detectors.

Neutron — A neutral elementary particle of mass number 1. It is believed to be a constituent particle of all nuclei of mass number greater than 1. It is unstable with respect to beta-decay, with a half life of about 12 min. It produces no detectable primary ionization in its passage through matter, but interacts with matter predominantly by collisions and to a lesser extent, magnetically. Some properties of the neutron are: rest mass, 1.00894 atomic mass unit; charge, 0; spin quantum number, ½; magnetic moment, −1.9125 nuclear Bohr magnetrons.

Neutron cross section — See cross section.

Newton (unit of force) — that force which gives to a mass of 1 kg an acceleration of 1 m/sec^2.

Newtonian speed of sound — An approximation to the speed of sound a in a perfect gas given by the relation

$$a^2 = p/\rho$$

where p is pressure and ρ is density. Compare Laplacian speed of sound. See Acoustic velocity.

Newtonian telescope — A reflecting telescope in which a small plane mirror reflects the convergent beam from the objective to an eyepiece at one side of the telescope. After the second reflection the rays travel approximately perpendicular to the longitudinal axis of the telescope. See Cassegrain telescope.

Newton's laws of motion — A set of three fundamental postulates forming the basis of the mechanics of rigid bodies, formulated by Newton in 1687.

The first law is concerned with the principle of inertia and states that if a body in motion is not acted upon by an external force, its momentum remains constant (law of conservation of momentum). The second law asserts that the rate of change of momentum of a body is proportional to the force acting upon the body and is in the direction of the applied force. A familiar statement of this is the equation

$$F = ma$$

where F is vector sum of the applied forces, m is the mass, and a is the vector acceleration of the body. The third law is the principle of action and reaction, stating that for every force acting upon a body there exists a corresponding force of the same magnitude exerted by the body in the opposite direction.

Newton's law of cooling — The rate of cooling of a body under given conditions is proportional to the temperature difference between the body and its surroundings.

Nodal points — Two points on the axis of a lens such that a ray entering the lens in the direction of one, leaves as if from the other and parallel to the original direction.

Noncondensable gas — A gas whose temperature is above its critical temperature, so that it cannot be liquefied by increase of pressure alone.

Normal atmosphere — This is defined as the pressure exerted by a vertical column of 76 cm of mercury of density 13.5951 g/cm^3 at a place where the gravitational acceleration is g = 980.665 cm/sec^2.

$$\begin{aligned} 1\ \text{atm} &= 1.01325 \times 10^6\ \text{dyn/cm}^2\ \text{(exactly)} \\ &= 1.01325 \times 10^5\ \text{Pa (exactly)} \\ &= 14.696\ \text{psi} \\ &= 29.921\ \text{in of Hg at } 32°\text{F} \\ 1\text{mm of Hg} &= 1\ \text{Torr} = 1333.22\ \text{dyn/cm}^2 \end{aligned}$$

Normal salt — An ionic compound containing neither replaceable hydrogen nor hydroxyl ions.

Normal solution — contains one gram molecular weight of the dissolved substance divided by the hydrogen equivalent of the substance (that is, one gram equivalent) per liter of solution.

Nuclear atom — The atom of each element consists of a small dense nucleus which includes most of the mass of the atom. The nucleus is made up of roughly equal numbers of neutrons and protons. The positive charges of the protons enables the nucleus to surround itself with a set of negatively charged electrons which move around the nucleus in complicated orbits with well defined energies. The outermost electrons which are least tightly bound to the nucleus play the dominant part in determining the physical and chemical properties of the atom. There are as many electrons in orbits as there are protons in the nucleus.

Nuclear cross section (symbol σ) — A measure of the probability that the reaction will take place which is defined by

$$dI = In\,\sigma\,dx$$

where I is the intensity of the particle beam; n is the number of target nuclei per cubic centimeter of target; σ is the cross section for the specified process, expressed in square centimeters; and x is the target thickness in centimeters. See barn.

Nuclear fusion — See fusion.

Nuclear isomers — Isotopes of elements having the same mass number and atomic number but differing in radioactive properties such as half-life period.

Nuclear magneton symbol M_N)—A unit of magnetic moment of the proton equal to 5.0505×10^{-24} erg per gauss.

Nucleon — Any particle found in the structure of an atom's nucleus. The most plentiful ones are neutrons and protons.

Nucleus — The dense central core of the atom, in which most of the mass and all of the positive charge is concentrated. The charge on the nucleus, an integral multiple of Z of the electronic charge, is the essential factor which distinguishes one element from another. Z is called the atomic number and gives the number of protons in the nucleus, which includes a roughly equal number of neutrons. The mass number A gives the total number of neutrons plus protons. See Isotopes and Nuclear Atom.

Nuclide — A species of atom distinguished by the constitution of its nucleus. The nuclear constitution is specified by the number of protons, Z; number of neutrons, N; and energy content. (Or, by the atomic number, Z; mass number A (= N + Z) and atomic mass).

Numerical aperture is the sine of half the angular aperture, used as a measure of the optical power of an objective.

Nusselt number (symbol N_{Nu}) — (After Wilhelm Nusselt, 1882- , German engineer.) A number expressing the ratio of convective to conductive heat transfer between a solid boundary and a moving fluid, defined as hl/k where h is the heat-transfer coefficient, l is the characteristic length, and k is the thermal conductivity of the fluid.

Objective — The lens or combination of lenses which receives light rays from an object and refracts them to form an image in the focal plane of the eyepiece of an optical instrument, such as a telescope. Also called object glass.

Object wave — The scattered or reflected wave from the object which it is desired to image or "reconstruct".

Oersted — The cgs emu of magnetic intensity exists at a point where a force of 1 dyn acts upon a unit magnetic pole at that point, i.e., the intensity 1 cm from a unit magnetic pole. SI unit: 1 Oe = 79.577 A/m.

Ohm (unit of electric resistance) — The electric resistance between two points of a conductor when a constant difference of potential of 1 volt, applied between these two points, produces in this conductor a current of 1 ampere, this conductor not being the source of any electromotive force.

Ohm's law — Current in terms of electromotive force E and resistance R.

$$I = \frac{E}{R}$$

The current is given in amperes when E is in volts and R in ohms.

Optical correlation — Process of determinating the similarity of an optical signal or wave form to a reference-stored signal or wave form. The reference is usually stored as a matched filter.

Optical path length — The total phase change between the source and a given position in the energy beam as measured along the direction of travel of the beam or wavefronts.

Optical pyrometer — A device for measuring the temperature of an incandescent radiating body by comparing its brightness for a selected wavelength interval within the visible spectrum with that of a standard source; a monochromatic radiation pyrometer.

Optical thickness — Specifically, in calculations of the transfer of radiant energy, the mass of a given absorbing or emitting material lying in a vertical column of unit cross-sectional area and extending between two specific levels. Also called optical depth.

If z_1 and z_2 are the lower and upper limits, respectively, of a layer in which the variation of a density ϱ of some absorbing or emitting substance is given as a function of height z, then the quantity

$$\int_{z_1}^{z_2} \rho(z)\,dz$$

is called the optical thickness of that substance within that particular layer.

Order (of a bright fringe) — Proportional to the path difference between the two wave components producing a fringe (measured in integral numbers of wavelengths). No path difference produces zero order fringes.

Order (of a dark fringe) — Proportional to the path difference between the two wave components producing a fringe (measured as one-half the quantity of integral half wavelengths less one; i.e., path difference of $^3/_2$ a wavelength produces a 1st order dark fringe).

Organ pipes — The frequency of vibration of a closed pipe or other air column of length l, where V is the velocity of sound in air, for the fundamental and first three overtones respectively is,

$$n_0 = \frac{V}{4l}, \qquad n_1 = \frac{3V}{4l}, \qquad n_2 = \frac{5V}{4l}, \qquad n_3 = \frac{7V}{4l}$$

For an open pipe,

$$n_0 = \frac{V}{2l}, \qquad n_1 = \frac{2V}{2l}, \qquad n_2 = \frac{3V}{2l}, \qquad n_3 = \frac{4V}{2l}$$

Osmotic-pressure effect — An enhancement in the velocity of the central ion, in the direction of the applied external field, as a result of more collisions on the central ion from ions behind the central ion than from ions in front of it.

Overall heat-transfer coefficient — The value U, in British thermal units per hour per square foot per °F in the equation

$$Q = UA(t_1 - t_2)$$

where Q is heat flow per unit time; A is area; and t is temperature.

Oxidation is any process which increases the proportion of oxygen or acid-forming or radical in a compound.

Packing fraction — Packing fraction is the difference between the actual mass of the isotope and the nearest whole number divided by the mass number. Thus the packing fraction is equal to (M − A)/A where M is the actual mass and A is the mass number. For example, one of the chlorine isotopes has a mass of 32.9860. The packing fraction for this isotope is

$$\frac{32.9860 - 33.0000}{33.0000} = -0.00042$$

Packing fractions are usually expressed as parts per 10,000 and so the packing fraction for this isotope of chlorine is written as −4.2. Since oxygen is taken as the standard, elements with positive packing fractions are less stable than oxygen, those with negative packing fractions are more stable.

It is positive for nuclides with mass number less than 16 or greater than 180, and negative for most others.

Parallax — The difference in the apparent direction or position of an object when viewed from different points expressed as an angle.

For bodies of the solar system, parallax is measured from the surface of the earth and its center and is called geocentric parallax, varying with the body's altitude and distance from the earth. The geocentric parallax when a body is in the horizon is called horizontal parallax and is the angular semidiameter of the earth as seen from the body. Parallax of the moon is called lunar parallax. For stars, parallax is measured from the earth and the sun, and is called annual, heliocentric, or stellar parallax. Compare aberration

Paramagnetic materials — Those within which an applied magnetic field is slightly increased by the alignment of electron orbits. The slight diamagnetic effect in materials having magnetic dipole moments is overshadowed by this paramagnetic alignment. As the temperature increases this paramagnetism disappears leaving only diamagnetism. The permeability of paramagnetic materials is slightly greater than that of empty space.

Parsec — A unit of length equal to the distance from the sun to a point having a heliocentric parallax of 1 second (1″), used as a measure of stellar distance.

The name parsec is derived from the words parallax second. 1 parsec = pc
= 3.085677×10^{16} m
= 2.062647×10^{5} astronomical units
= 3.261633 light years

Partial pressure — The pressure exerted by a designated component or components of a gaseous mixture.

Pascal (Pa) — The unit of pressure in the SI system; 1 pascal equals 1 newton per squate meter; 100000 Pa = 1 bar.

Pascal's law — Pressure exerted at any point upon a confined liquid is transmitted undiminished in all directions.

Pauli exclusion principle — The principle that no pair of identical particles can simultaneously occupy the same quantum state. The principle applies to electrons, protons, and neutrons and accounts for the shell configurations of extranuclear electrons and for the shell structure of nuclei themselves.

Peltier effect — When a current flows across the junction of two unlike metals it gives rise to an absorption or liberation of heat. If the current flows in the same direction as the current at the hot junction in a thermoelectric circuit of the two metals, heat is absorbed; if it flows in the same direction as the current at the cold junction of the thermoelectric circuit heat is liberated.

Pendulum — For a simple pendulum of length l, for a small amplitude, the complete period,

$$T = 2\pi\sqrt{\frac{l}{g}} \qquad \text{or } g = 4\pi^2 \frac{l}{T^2}$$

T will be given in seconds if l is in cm and g in cm per sec². For a sphere suspended by a wire of negligible mass where d is the distance from the knife edge to the center of the sphere whose radius is r, the length of the equivalent simple pendulum,

$$l = d + \frac{2r^2}{5d}$$

If the period is P for an arc θ, the time of vibration in an infinitely small arc is approximately

$$T = \frac{P}{1 + \frac{1}{4}\sin^2\frac{\theta}{4}}$$

For a compound pendulum, if a body of mass m be suspended from a point about which its moment of inertia is I with its center of gravity a distance h below the point of suspension, the period

$$T = 2\pi\sqrt{\frac{I}{mgh}}$$

Penning effect — An increase in the effective ionization rate of a gas due to the pressure of a small number of foreign metastable atoms.

For instance, a ncon atom has a metastable level at 16.6 V and if there are a few neon atoms in a gas of argon which has an ionization potential of 15.7 V, a collision between the neon metastable atom with an argon atom may lead to ionization of the argon. Thus, the energy which is stored in the metastable atom can be used to increase the ionization rate. Other gases where this effect is used are helium, with a metastable level at 19.8 V, and mercury, with an ionization level at 10.4 V.

Perfect fluid — In simplifying assumptions, a fluid chiefly characterized by lack of viscosity and, usually, by incompressibility. Also called an ideal fluid, inviscid fluid. See Perfect gas.

A perfect fluid is sometimes further characterized as homogeneous and continuous.

Perfect gas — A gas which has the following characteristics: (a) it obeys the Boyle-Mariotte law and the Charles-Gay-Lussac law, thus satisfying the equation of state for perfect gases; (b) it has internal energy as a function of temperature alone; and (c) it has specific heats with values independent of temperature. Also called ideal gas. Compare perfect fluid.

The normal volume of a perfect gas is 2.24136×10^4 cm³/mol.

Perigee — That orbital point nearest the earth when the earth is the center of attraction.

Perihelion — That point in a solar orbit which is nearest the sun.

Period in uniform circular motion is the time of one complete revolution. In any oscillatory motion it is the time of a complete oscillation. Dimension, [t].

Periodic law — Elements when arranged in the order of their atomic weights or atomic numbers show regular variations in most of their physical and chemical properties.

Permeance, the reciprocal of reluctance. Unit permeance is the permeance of a cylinder one square centimeter crosssection and one centimeter length taken in a vacuum. Dimensions $[\epsilon^{-1} l^{-1} t^2]$; $[\mu l]$.

Phase angle — The phase difference of two periodically recurring phenomena of the same frequency, expressed in angular measure.

Phase of oscillatory motion — The fraction of a whole period which has elapsed since the moving particle last passed through its middle position in a positive direction.

Phase (of wave) — The distance between the position of an amplitude crest of a wave train and a reference position measured in units of wavelength, degrees, or radians (one wavelength equals 360°, or 2π radians).

Phon — The unit of loudness level. The loudness level of a sound in phons is equal to the sound pressure level in decibels re 0.0002 microbar of a pure tone of 1000 Hz which a group of listeners judge to be equally loud.

Phosphorescence — Emission of light which continues after the exciting mechanism has ceased. See luminescence. Compare fluorescence. An example of phosphorescence is the glowing of an oscilloscope screen after the exciting beam of electrons has moved to another part of the screen.

Phot — A photometric unit of illuminance or illumination equal to 1 lumen per square centimeter. Compare footcandle, lux.

Photoelectric effect — The emission of an electron from a surface as the surface absorbs a photon of electromagnetic radiation. Electrons so emitted are termed photoelectrons.

Photoelectron — An electron which has been ejected from its parent atom by interaction between that atom and a high-energy photon.

Photographic density — The density D of silver deposit on a photographic plate or film is defined by the relation

$$D = \log O$$

where O is the opacity. If I_o and I are the incident and transmitted intensities respectively the opacity is given by I_o/I. The transparency is the reciprocal of the opacity or I/I_o.

Photometer — An instrument for measuring the intensity of light or the relative intensity of a pair of lights. Also called illuminometer.

Photon — According to the quantum theory of radiation, the elementary quantity, or quantum, of radiant energy. It is regarded as a discrete quantity having a momentum equal to $h\nu/c$, where h is Planck constant, ν is the frequency of the radiation, and c is the speed of light in a vacuum. The photon is never at rest, has no electric charge and no magnetic moment, but does have a spin moment. The energy of a photon (the unit quantum of energy) is equal to $h\nu$. Photons are generated in collisions between nuclei or electrons and in any other process in which an electrically charged particle changes its momentum. Conversely photons can be absorbed (i.e., annihilated) by any charged particle.

Pico — A prefix meaning one trillionth or 10^{-12}. Symbol is p.

Piezo-electric effect — The phenomemon exhibited by certain crystals of expansion along one axis and contraction along another when subjected to an electric field. Conversely compression of certain crystals generate an electrostatic voltage across the crystal. Piezoelectricity is only possible in crystal classes which do not possess a center of symmetry.

Pinch effect — When an electric current, either direct or alternating, passes through a liquid conductor, that conductor tends to contract in cross-section, due to electromagnetic forces.

Pitch — Psychological response of the ear, primarily dependent upon the frequency of viration of the air. The intensity of the sound also has a certain effect on the pitch. Pitch of a screw is the axial distance between adjacent turns of a single thread on the screw.

Planck's constant (h) — A universal constant of nature which relates the energy of a quantum of radiation to the frequency of the oscillator which emitted it. It has the dimensions of action (energy × time). Expressed by $E = h\nu$ where E is the energy of the quantum and ν is its freqnency. Its numerical value is $6.626176\ (36) \times 10^{-27}$ erg sec.

Planck's law — An expression for the variation of monochromatic radiant flux per unit area of source as a function of wavelength of black-body radiation at a given temperature; it is the most fundamental of the radiation laws. Mathematically, Planck's law is

$$dw = [c_1 \lambda^{-5} / (e^{c_2/T\lambda} - 1)]\ d\lambda$$

where dw is the radiant flux from a black body in the wavelength interval $d\lambda$, centered around wavelength λ, per unit area of black-body surface at temperature T; c_1 and c_2 are radiation constants.

Planetary aberration — A displacement in the apparent position of a planet in the celestial sphere due to the relative movement of the observer and the planet. See Aberration.

Plane wave — A wave in which the wave fronts are everywhere parallel planes normal to the direction of propagation.

Plasma — An assembly of ions, electrons, neutral atoms and molecules in which the motion of the particles is dominated by electromagnetic interaction. The temperature of the collection of these particles is sufficiently high for the ionization to be above 5%. The plasma taken as a whole is electrically neutral. A plasma is further characterized by relatively large intermolecular distances, large amounts of energy stored in the internal energy of the particles and the presence of a plasma sheath at all boundaries of the plasma. Plasmas are sometimes referred to as the fourth state of matter.

Plutonium — A fissile element, artificially produced in the pile by neutron bombardment of U^{238}.

Poise — A unit of coefficient of viscosity, defined as the tangential force per unit area (dyn/cm^2) required to maintain unit difference in velocity (1 cm/sec) between two parallel planes separated by 1 cm of fluid;

$$1 \text{ poise} = 1 \text{ dyne sec/cm}^2 = 1 \text{ g/cm sec.}$$

Poiseuille flow — The steady laminar flow of a fluid through a narrow horizontal circular cylinder according to the relation

$$u = (1/4\mu)\ (\partial p/\partial x)\ (a^2 - r^2)$$

where u is the fluid velocity along the cylinder's axis at a distance r from the cylinder's axis; μ is the dynamic viscosity of the fluid: a is the cylinder radius; and $\partial p/\partial x$ is the pressure gradient along the axis of the cylinder. The velocity profile across the cylinder is seen to be parabolic, and this relation affords a convenient experimental means of determining a fluid's viscosity. Also called Hagen-Poiseuille flow. Compare Couette flow.

Poisson constant (symbol μ) — The ratio of the gas constant to the specific heat of a gas at constant pressure.

Poisson distribution — A one-parameter discrete frequency distribution giving the probability that n points (or events) will be (or occur) in an interval (or time) x, provided that these points are individually independent and that the number occurring in a subinterval does not influence the number occurring in any other nonoverlapping subinterval. It has the form

$$f(n.x) = e^{-\sigma x}\ (\sigma x)^n/n!$$

The mean and variance are both σx, and σ is the average density (or rate) with which the events occur. When σx is large, the Poisson distribution approaches the normal distribution. The binomial distribution approaches the Poisson when the number of events n becomes large and the probability of success P becomes small in such a way that $nP \rightarrow \sigma x$.

Poisson's ratio is the ratio of the transverse contraction per unit dimension of a bar of uniform cross-section to its elongation per unit length, when subjected to a tensile stress.

Polarized light — Light which exhibits different directions at right angles to the line of propagation is said to be polarized. Specific rotation is the power of liquids to rotate the plane of polarization. It is stated in terms of specific rotation or the rotation in degrees per decimeter per unit density.

Polymorphism — The ability to exist in two or more crystalline forms.

Positron — A particle of the same mass m, as an ordinary electron. It has a positive electrical charge of exactly the same amount as that of an ordinary electron (which is sometimes called negatron). Positrons are created either by the radioactive decay of certain unstable nuclei or, together with a negatron, in a collision between an energetic (more than one MeV) photon and an electrically charged particle (or another photon). A positron does not decay spontaneously but on passing through matter it sooner or later collides with an ordinary electron and in this collision the positron-negatron pair is annihilated. The rest energy of the two particles, which is given by Einstein's relation $E = mc^2$ and amounts to 1.0216 mev altogether, is converted into electromagnetic radiation in the form of one or more photons.

Potential (electric) at any point is measured by the work necessary to bring unit positive charge from an infinite distance. Difference of potential between two points is measured by the work necessary to carry unit positive charge from one to the other. If the work involved is one erg we have the electrostatic unit of potential. Dimensions, $[\epsilon^{-1/2}\, m^{1/2}\, l^{1/2}\, t^{-1}]$, $[\mu^{1/2}\, m^{1/2}\, l^{3/2}\, t^{-1/2}]$. The potential at a point due to a charge q at a distance r in a medium whose dielectric constant is ϵ is,

$$V = \frac{q}{\epsilon r}$$

Potential energy — Energy possessed by a body by virtue of its position in a gravity field in contrast with kinetic energy, that possessed by virtue of its motion.

Potential index of refraction — An atmospheric index of refraction so formulated that it would have no height variation in an adiabatic atmosphere. Also called potential refractive index. Compare modified index of refraction. The potential index of refraction is usually expressed in terms of B-units.

Pound (abbr. lb) — 1. A unit of mass equal in the U.S. to 0.45359237 kg, exactly. 2. Specifically, a unit of measurement of the thrust or force of a reaction engine representing the weight the engine can move, as an engine with 100,000 pounds of thrust. See Poundal, Pound mass. 3. The force exerted on 1 pound mass by the standard acceleration of gravity. See Gravity, sense 2.

Poundal — Force necessary to give 1 lb mass acceleration of 1 ft/sec^2.

Pound mass — 1. A mass equal to 0.45359237 kg. 2. A unit of measure of the inertial property equal to the mass of a body weighing 1 lb at the standard acceleration of gravity (980.665 cm/sec^2).

Pound weight — A force equal to the earth's attraction for a mass of 1 lb. This force, acting on a 1 lb mass, will produce an acceleration of 32.1747 ft/sec^2.

Power — The time rate at which work is done. Units of power, the watt, one joule (ten million ergs) per second; the kilowatt is equal to 1000 watts; the horse-power, 33,000 foot-pounds per minute, is equal to 746 watts. Dimensions, $[m\ l^2\ t^{-3}]$. If an amount of work W is done in time t the power or rate of doing work is

$$P = \frac{W}{t}$$

Power will be obtained in watts if W is expressed in joules (10^7 ergs) and t in sec.

Power in watts for alternating current —

$$P = EI \cos \phi$$

where E and I are the effective values of the electromotive force and current in volts and amperes respectively and ϕ the phase angle between the current and the impressed electromotive force. The ratio,

$$\frac{P}{EI} = \cos \phi$$

is called the power factor.

Power developed by a direct current — The power in watts developed by an electric current flowing in a conductor, where E is the difference of potential at its terminals in volts, R its resistance in ohms, and I the current in amperes,

$$P = EI = RI^2$$

The work done in joules in a time t sec is,

$$W = EIt{:}RI^2t.$$

Power ratios in telephone engineering are measured in decibels. The gain or loss of power expressed in decibels is ten times the logarithm of the power ratio. By reference to an arbitrarily chosen "power level" the actual power may be expressed in decibels. The numerical values thus used will not be proportional to the actual power level but roughly to the sensation on the ear produced when the electrical power is converted into sound. A difference of 1 decibel in the power supply to a telephone receiver produces approximately the smallest change in volume of sound which a normal ear can detect.

Pressure — Force applied to, or distributed, over a surface; measured as force per unit area. Cgs unit, the barye, one dyne per square centimeter. The megabarye is equal to 10^6 dynes per square centimeter. Pressure is also measured by the height of the column of mercury or water which it supports. Dimensions, $[m\ l^{-1}\ t^{-2}]$.

The pressure due to a force F distributed over an area A,

$$P = \frac{F}{A}$$

Absolute pressure — Pressure measured with respect to zero pressure.
Gauge pressure — Pressure measured with respect to that of the atmosphere.

Primary colors — Any three colors that when mixed in suitable proportions produce any color. Primary colors may be subtractive, where the primaries absorb colors from white light (e.g., magenta, cyan, yellow; red, blue, yellow) used in developing color photography; or additive, where the primaries form a color by the addition of their light (e.g., red, green, blue).

Principal focus of a lens or spherical mirror is the point of convergence of light coming from a source at an infinite distance.

Projectiles — For bodies projected with velocity v at an angle a above the horizontal, the time to highest point of flight.

$$t = \frac{v \sin a}{g}$$

Total time of flight to reach the original horizontal plane,

$$T = \frac{2v \sin a}{g}$$

Maximum height,

$$h = \frac{v^2 \sin^2 a}{2g}$$

Horizontal range,

$$R = \frac{v^2 \sin 2a}{g}$$

In the above equations the resistance of the air is neglected, g is the acceleration due to gravity.

Prompt neutrons — In nuclear fission, those neutrons released coincident with the fission process, as opposed to the neutrons subsequently released.

Proton — A positively charged subatomic particle having a mass of 1.67252 $\times 10^{-24}$ g, slightly less than that of a neutron but about 1836 times greater than that of an electron.

Proton-proton reaction — A thermonuclear reaction in which two protons collide at very high velocities and combine to form a deuteron. The resultant deuteron may capture another proton to form tritium and the latter may undergo proton capture to form helium.

Proton storm — The flux of protons sent into space by a solar flare.

Pulsed laser — A laser that radiates its energy during short bursts of time (pulses) and then is inactive until the next burst or pulse. The frequency of these pulses is called the pulse repetition frequency (PRF) of the laser.

Purkinje effect — A phenomenon associated with the human eye, making it more sensitive to blue light when the illumination is poor (less than about 0.1 lumen/ft^2) and to yellow light when the illumination is good.

Pyron — A unit of radiant intensity of electromagnetic radiation equal to 1 calorie/cm^2/min.

Quality or timbre of sound depends on the coexistence with the fundamental of other vibrations of various frequencies and amplitudes.

Quantity of electricity or charge — The electrostatic unit of charge, the quantity which when concentrated at a point and placed at unit distance from an equal and similarly concentrated quantity, is repelled with unit force. If the distance is one centimeter and force of repulsion one dyne and the surrounding medium a vacuum, we have the electrostatic unity of quantity. The electrostatic unit of quantity may be defined as that transferred by electrostatic unit current in unit time. The quantity transferred by one ampere in one second is the coulomb, the practical unit. The faraday is the electrical charge carried by one gram equivalent. The coulomb = 3×10^9 electrostatic units. Dimensions, $[\epsilon^{1/2}\ m^{1/2}\ l^{1/2}\ t^{-1}]$; $[\mu^{-1/2}\ m^{1/2}\ l^{1/2}]$.

Quantum — Unit quantity of energy postulated in the quantum theory. The photon is a quantum of the electromagnetic field, and in nuclear field theories, the meson is considered to be the quantum of the nuclear field.

Quantum theory — The theory first stated by Max Planck (before the Physical Society of Berlin on December 14, 1900) that all electromagnetic radiation is emitted and absorbed in quanta, each of magnitude $h\nu$, h being the Planck constant and ν the frequency of the radiation.

Quasi-monochromatic — Nearly of the same wavelength (see monochromatic).

Rad — An ionizing radiation unit corresponding to an absorption of energy in any medium of 100 ergs/g (1 rad in tissue = 100/93 rep).

Radar — A system of radio detection and ranging which detects objects by beaming rf pulses that are reflected back by the object and measures its distance by the time elapsed between transmission and reception. The strength of the echo signal is determined by the radar equation

$$W_R = W_T \frac{G^2 \lambda^2 \sigma}{(4\pi)^3 R^4}$$

where W_R is received echo power. W_T is transmitted power, G is antenna gain, λ is radar carrier wavelength, σ is target cross section, and R is range.

Radar nautical mile — The time that a radar signal takes to hit a target one nautical mile away and return; 12.261 microseconds.

Radian — The angle subtended at the center of a circle by an arc equal in length to a radius of the circle. It is equal to $360°/2\pi$ or approximately 57 degrees 17 minutes 44.8 seconds. The radian is the SI unit of plane angle.

Radiation — The emission and propagation of energy through space or through a material medium in the form of waves.

The term may be extended to include streams of sub-atomic particles as alpha-rays, or beta-rays, and cosmic rays as well as electromagnetic radiation. Often used to designate the energy alone without reference to its character. In the case of light this energy is transmitted in bundles (photons).

Radiation formula, Planck's — The emission power of a black body at wave length λ may be written

$$\lambda = \frac{c_1\lambda^{-5}}{e^{c_2/\lambda T} - 1}$$

where c_1 and c_2 are constants with c_1 being 3.7403×10^{10} microwatts μm^4 per cm^2 or 3.7403×10^{-12} watt cm^2, c_2 being 14384 μm degrees and T the absolute temperature.

Radiation laws — 1. The four physical laws which, together, fundamentally describe the behavior of black-body radiation: (a) the Kirchoff law is essentially a thermodynamic relationship between emission and absorption of any given wavelength at a given temperature; (b) the Planck law describes the variation of intensity of black-body radiation at a given temperature, as a function of wavelength; (c) the Stefan-Boltzmann law relates the time rate of radiant energy emission from a black body to its absolute temperature; (d) the Wien law relates the wavelength of maximum intensity emitted by a black body to its absolute temperature. 2. All the more inclusive assemblage of empirical and theoretical laws describing all manifestations of radiative phenomena; e.g., Bouguer law and Lambert law.

Radioactive nuclides — Atoms that disintegrate by emission of corpuscular or electromagnetic radiations. The rays most commonly emitted are alpha or beta or gamma rays. The three classes are:

1. Primary, which have half-life times exceeding 10^8 years. These may be alpha-emitters or beta-emitters.
2. Secondary, which are formed in radioactive transformations starting with U^{238}, U^{235}, or Th^{232}.
3. Induced, having geologically short lifetimes and formed by induced nuclear reactions occurring in nature. All these reactions result in transmutation.

Radioactivity — 1. Spontaneous disintegration of atomic nuclei with emission of corpuscular or electromagnetic radiations.

2. The number of spontaneous disintegrations per unit mass and per unit time of a given unstable (radioactive) element, usually measured in becquerels.

Radiometer — An instrument for detecting and, usually, measuring radiant energy. Compare bolometer. See Actinometer, Photometer.

Radius of gyration may be defined as the distance from the axis of rotation at which the total mass of a body might be concentrated without changing its moment of inertia. The product of total mass and the square of the radius of gyration will give (the) moment of inertia.

Raman scattering — Raman scattering of light from a gas, liquid, or solid is that scattering in which a shift in wavelength from that of the usually monochromatic radiation occurs. The amount of shift is a function of the scattering particles and wavelengths.

Rankine cycle — An idealized thermodynamic cycle consisting of two constant-pressure processes and two isentropic processes.

Rankine scale of temperature — The Rankine scale, with its size of degree equal to that of the Fahrenheit scale, also has its zero at absolute zero of temperature. Thus, $T°R = t°F + 459.67$. Therefore, $O°R = -459.67°F$ and the normal boiling point of water is 671.67°R.

Raoult's law — Molar weights of non-volatile nonelectrolytes when dissolved in a definite weight of a given solvent under the same conditions lower the solvent's freezing point, elevate its boiling point and reduces its vapor pressure equally for all such solutes.

Rayleigh number — The nondimensional ratio between the product of buoyancy forces and heat advection and the product of viscous forces and heat conduction in a fluid. It is written as

$$N_{Ra} = \frac{g\,\Delta_z T\,\alpha d^3}{\nu k}$$

where g is the acceleration of gravity; $\Delta_z T$ is a characteristic vertical temperature difference in the characteristic depth d; α is the coefficient of expansion, ν is the kinematic viscosity; and k is the thermometric conductivity.

Rayleigh scattering — This is a coherent scattering in which the intensity of the light of wavelength λ, scattered in any direction making an angle θ with the incident light, is directly proportional to $1 + \cos^2\theta$ and inversely proportional to λ^4. The latter point is noteworthy in that it shows how much greater the scattering of the short wavelengths is. These relations apply when the scattering particles are much smaller than the wavelength of the radiation. Thus the sky is blue because blue light is scattered more than red. The unscattered light is, of course, complementary to blue, i.e., orange or yellow, which explains the "warm" hues of the sunset.

Real image — Image formed by converging rays which form a focused image in space.

Reduction is any process which increases the proportion of hydrogen or base-forming elements or radicals in a compound. Reduction is also the gaining of electrons by an atom, an ion, or an element thereby reducing the positive valence of that which gained the electron.

Reflecting telescope — A telescope which collects light by means of a concave mirror.

Reflection coefficient or reflectivity is the ratio of the light reflected from a surface to the total incident light. The coefficient may refer to diffuse or to specular reflection. In general it varies with the angle of incidence and with the wavelength of the light.

Reflection of light by a transparent medium in air (Fresnel's formulae) — If i is the angle of incidence, r the angle of refraction, n_1 the index of refraction for air (nearly equal to unity), n_2 index of refraction for a medium, then the ratio of the reflected light to the incident light is,

$$R = \tfrac{1}{2}\left(\frac{\sin^2(i-r)}{\sin^2(i+r)} + \frac{\tan^2(i-r)}{\tan^2(i+r)}\right)$$

If i = O (normal incidence), and $n_1 = 1$ (approximate for air),

$$R = \left(\frac{n_2 - 1}{n_2 + 1}\right)^2$$

Refracting telescope — A telescope which collects light by means of a lens or system of lenses. Also called refractor.

Refraction — See Index of refraction, Snell's law.

Refraction — The bending of a light or energy wave as it passes through material with varying wave velocities or indexes of refraction.

Refraction at a spherical surface — If u be the distance of a point source, v the distance of the point image or the intersection of the refracted ray with the axis, n_1 and n_2 the indices of refraction of the first and second medium, and r the radius of curvature of the separating surface,

$$\frac{n_2}{v} + \frac{n_1}{u} = \frac{n_2 - n_1}{r}$$

If the first medium is air the equation becomes,

$$\frac{n}{v} + \frac{1}{u} = \frac{n-1}{r}$$

Refractivity — 1. The algebraic difference between an index of refraction and unity.

For the atmosphere, refractivity may be more conveniently expressed in N-units:

$$N = (n - 1)\,10^5$$

The deviation of the refractivity at any altitude from the usual standard profile is expressed in B-units (for radio frequencies up to 20 gigahertz);

$$B = N + 0.012h$$

where h is altitude in feet.

The deviation of the refractivity at any altitude from the gradient at which the refraction curvature of a tangential ray will match the curvature of the earth may be expressed in M-units:

$$M = N + 0.048h$$

where 0.048 is 10^6 divided by the radius of the earth in feet.

Relative biological effectiveness (RBE) — The biological effectiveness of any type of energy of ionizing radiation in producing a specific damage (e.g., leukemia, anemia, sterility, carcinogenesis, cataracts, shortening of life span, etc.).

Relative humidity — The ratio of the quantity of water vapor present in the atmosphere to the quantity which would saturate at the existing temperature. It is also the ratio of the pressure of water vapor present to the pressure of saturated water vapor at the same temperature.

Relativistic mass equation — The equation

$$m = m_0 \, [1 - (v^2/c^2)]^{-1/2} = m_0/(1 - \beta^2)^{1/2}$$

where $\beta \equiv v/c$ for the relativistic mass m of a particle or body of rest mass m_0 when its velocity is v. See Relativistic velocity.

Relativistic particle — A particle with a velocity so large that its relativistic mass exceeds its rest mass by an amount which is significant for the computation or other considerations at hand. See Relativistic velocity.

Relativistic velocity — A velocity sufficiently high that some properties of a particle of this velocity have values significantly different from those obtaining when the particle is at rest. See Rest mass.

The property of most interest is the mass. For many purposes, the velocity is relativistic when it exceeds about one tenth the velocity of light.

Relaxation-field effect — The delay in the ion atmosphere in maintaining its symmetry around a central ion as the central ion moves in the forward direction under an applied electrical field (potential gradient).

Reluctance is that property of a magnetic circuit which determines the total magnetic flux in the circuit when a given magnetomotive force is applied. Unit, the reluctance of one centimeter length and one square centimeter cross-section of space taken in a vacuum. Dimensions, $[\epsilon l\, t^{-2}]$; $[\mu^{-1} l^{-1}]$.

Reluctivity or specific reluctance is the reciprocal of magnetic permeability. The reluctivity of empty space is taken as unity. Dimensions, $[\epsilon l^2 t^{-2}]$; $[\mu^{-1}]$.

Rem — Abbreviation for roentgen-equivalent-man.

Resistance is a property of conductors depending on their dimensions, material and temperature which determines the current produced by a given difference of potential. The practical unit of resistance, the ohm is that resistance through which a difference of potential of one volt will produce a current of one ampere. The international ohm is the resistance offered to an unvarying current by a column of mercury at 0°C; 14.4521 g in mass, of constant cross-sectional area and 106.300 cm in length, sometimes called the legal ohm. Dimensions, $[\epsilon^{-1} l^{-1} t]$; $[\mu l\, t^{-1}]$.

Resistance of a conductor at 0°C, of length l, cross-section s and specific resistance ϱ

$$R_0 = \rho \frac{l}{s}$$

The resistivity may be expressed as ohm-cm when R is in ohms, l in cm and s in cm².

Resistance of a conductor at a temperature t whose resistance at 0°C is R_0 and whose temperature resistance coefficient is α

$$R_t = R_0(1 + \alpha t)$$

Resistance of conductors in series and parallel — The total resistance of any number of resistances joined in series is the sum of the separate resistances. The total resistance of conductors in parallel whose separate resistances are r_1, r_2, $r_3 \cdots r_n$ is given by the formula

$$\frac{1}{R} = \frac{1}{r_1} + \frac{1}{r_2} + \frac{1}{r_3} \cdots + \frac{1}{r_n}$$

Where R is the total resistance. For two terms this becomes,

$$R = \frac{r_1 r_2}{r_1 + r_2}$$

Resistance, specific (Resistivity) — A proportionality factor characteristic of different substances equal to the resistance that a centimeter cube of the substance offers to the passage of electricity, the current being perpendicular to two parallel faces. It is defined by the expression:

$$R = \rho \frac{l}{A}$$

where R is the resistance of a uniform conductor, l is its length, A is its cross sectional area, and ϱ is its resistivity. Resistivity is usually expressed in ohm-centimeters.

Resistivity (symbol ϱ) — In electricity, a characteristic proportionality factor equal to the resistance of a centimeter cube of a substance to the passage of an electric current perpendicular to two parallel faces. Also called specific resistance.

$$R = \rho(l/A)$$

where R is the resistance of a uniform conductor, l is its length, A is its cross-sectional area, and ϱ is its resistivity.

Resolving power of a telescope or microscope is indicated by the minimum separation of two objects for which they appear distinct and separate when viewed through the instrument.

Resonance — 1. The phenomenon of amplification of a free wave or oscillation of a system by a forced wave or oscillation of exactly equal period. The forced wave may arise from an impressed force upon the system or from a boundary condition. The growth of the resonant amplitude is characteristically linear in time. 2. Of a system in forced oscillation, the condition which exists when any change, however small, in the frequency of excitation causes a decrease in the response of the system.

Resonance (chemical) — The moving of electrons from one atom of a molecule or ion to another atom of that molecule or ion. It is simply the oriented movement of the bonds between atoms.

Restitution, coefficient of, for two bodies on impact. The ratio of the difference in velocity, after impact to the difference before impact.

Rest mass — According to relativistic theory, the mass which a body has when it is at absolute rest. Mass increases when the body is in motion according to

$$m = m_0/\sqrt{1 - (v^2/c^2)}$$

where m is its mass in motion, m_0 is its rest mass; v is the body's speed of motion, and c is the speed of light.

Newtonian physics, in contrast with relativistic physics, makes no distinction between rest mass and mass in general.

Reverberation — 1. The persistence of sound in an enclosed space, as a result of multiple reflections after the sound source has stopped. 2. The sound that persists in an enclosed space, as a result of repeated reflection or scattering, after the source of the sound has stopped.

Reversible reaction — One which can be caused to proceed in either direction by suitable variation in the conditions of temperature, volume, pressure or of the quantities of reacting substances.

Rochon prism — A birefringent electro-optical crystal which divides incident unpolarized optical beam into two polarized components.

Roentgen (R) — That quantity of X or gamma radiation such that the associated corpuscular emission per 0.001293 g of dry air (equals 1 cc at 0°C and 760 mm Hg) produces, in air, ions carrying 1 esu of quantity of electricity of either sign.

Roentgen equivalent man (REM) — That amount of ionizing radiation of any type which produces the same damage to man as 1 roentgen of about 200-kV X radiation (1 rem = 1 rad in tissue/RBE). It should be noted that, when the physical dose is measured in rep untis, the approximate definition is used: 1 rem ≈ 1 rep/RBE.

Roentgen equivalent physical (REP) — That amount of ionizing radiation of any type which results in the absorpion of energy at the point in question in soft tissue to the extent of 93 ergs/g. It is approximately equal to 1 roentgen of about 200-kV X radiation in soft tissue.

Root-mean-square error (symbol σ). In statistics, the square root of the arithmetic mean of the squares of the deviations of the various items from the arithmetic mean of the whole. Also termed standard deviation.

Rotatory power is the power of rotating the plane of polarized light, given in general by θ/l where θ is the total rotation which occurs in a distance l.

The molecular or atomic rotatory power is the product of the specific rotatory power by the molecular or atomic weight. Magnetic rotatory power is given by

$$\theta/e\ H \cos \alpha$$

where H the intensity of the magnetic field, and α the angle between the field and the direction of the light.

Rydberg formula — A formula, similar to that of Balmer, for expressing the wave-numbers (ν) of the lines in a spectral series:

$$\nu = R\left[\frac{1}{(n+a)^2} - \frac{1}{(m+b)^2}\right]$$

where n and m are integers and m > n, a and b are constants for a particular series, and R is the Rydberg constant, 109737.3 cm^{-1} for hydrogen.

Salt — Any substance which yields ions, other than hydrogen or hydroxyl ions. A salt is obtained by displacing the hydrogen of an acid by a metal.

Scale height (symbol h, h_s) — A measure of the relationship between density and temperature at any point in an atmosphere; the thickness of a homogeneous atmosphere which would give the observed temperature:

$$h = kT/mg = R^*T/Mg$$

where k is the Boltzmann constant; T is the absolute temperature; m and M are the mean molecular mass and weight, respectively, of the layer; g is the acceleration of gravity; and R* is the universal gas constant. Compare virtual height.

Scattered (light) — Reflection of light from a surface in all directions in a nonuniform manner.

Scattering Coefficient — A measure of the attenuation due to scattering of radiation as it traverses a medium containing scattering particles. Also called total scattering coefficient.

Scattering cross section — The hypothetical area normal to the incident radiation that would geometrically intercept the total amount of radiation actually scattered by a scattering particle. It is also defined, equivalently, as the cross-section area of an isotropic scatterer (a sphere) which would scatter the same amount of radiation as the actual amount. Also called extinction cross section, effective area.

Schlieren (photography) — A picture or image in which density gradients in a volume of flow are made visible. The image is produced by refraction and scattering from regions of changing refractive index.

Scintillation counter — The combination of phosphor, photomultiplier tube, and associated circuits for counting scintillations. Also called scintillating counter.

Scintillometer — An instrument which detects radiation by emitting flashes of light.

Second — The second is the unit of time of the International System of Units. The definition adopted at the October 13, 1967 meeting of the 13th General Conference on Weights and Measures is: "The second is the duration of 9 192 631 770 periods of the radiation corresponding to the transition between the two hyperfine levels of the ground state of the atom of cesium 133." The frequency 9 192 631 770 (Hz) which the definition assigns to the cesium radiation was carefully chosen to make it impossible, by any existing experimental evidence, to distinguish the new second from the "ephemeris second" based on the earth's motion. Therefore no changes need to be made in data stated in terms of the old standard in order to convert them to the new one. The atomic definition has two important advantages over the previous definition: (1) it can be realized (i.e., generated by a suitable clock) with sufficient precision, ± 1 part per hundred billion (10^{11}) or better, to meet the most exacting demands of modern metrology; and (2) it is available to anyone who has access to or who can build an atomic clock controlled by the specified cesium radiation. (A description of such clocks is given in "Atomic Frequency Standards," NBS Tech. News Bull. 45, 8—11 (Jan., 1961). For more recent developments and technical details, see R. E. Beehler, R. C. Mockler, and J. M. Richardson, "Cesium Beam Atomic Time and Frequency Standards," Metrologia 1, 114—131 (July, 1965)). In addition one can compare other high-precision clocks directly with such a standard in a relatively short time — an hour or so compared against years with the astronomical standard. Laboratory-type atomic clocks are complex and expensive, so that most clocks and frequency generators will continue to be calibrated against a standard such as the NBS Frequency Standard, controlled by a cesium atomic beam, at the Radio Standards Laboratory in Boulder, Colorado. In most cases the comparison will be by way of the standard-frequency and time-interval signals broadcast by NBS radio stations WWV, WWVH, WWVB, and WWVL.

Second law of thermodynamics — An inequality asserting that it is impossible to transfer heat from a colder to a warmer system without the occurrence of other simultaneous changes in the two systems or in the environment.

It follows from this law that during an adiabatic process, entropy cannot decrease. For reversible adiabetic processes entropy remains constant, and for irreversible adiabatic processes it increases.

Another equivalent formulation of the law is that it is impossible to convert the heat of a system into work without the occurrence of other simultaneous changes in the system or its environment. This version, which requires an engine to have a cold source as well as a heat source, is particularly useful in engineering applications. See first law of thermodynamics.

Second of arc — 1/60 of a minute of arc.

Seebeck effect — If a circuit consists of two metals, one junction hotter than the other, a current flows in the circuit. The direction of the flow depends on the metals and the temperature of the junctions.

Semiconductor — An electronic conductor, with resistivity in the range between metals and insulators, in which the electrical charge carrier concentration increases with increasing temperature over some temperature range. Certain semiconductors possess two types of carriers, namely, negative electrons and positive holes.

Sensitiveness of a balance — Assuming the three knife edges of a balance to lie on a straight line; if M is the weight of the beam, h the distance of the center of gravity below the knife edge, L the length of the balance arms and m a small mass added to one pan, the deflection θ produced is given by

$$\tan \theta = \frac{mL}{Mh}$$

Shear strength — "The stress, usually expressed in pounds per square inch, required to produce fracture when impressed perpendicularly upon the cross-section of a material."

Shell — According to Pauli's exclusion principle (q.v.), the extranuclear electrons do not circle around the nucleus all in orbits of the same radius, but are arranged in orbits at various distances from the nucleus. The extranuclear orbital electrons are thus assumed to be arranged in a series of concentric spheres, called shells, which are designated, in the order of increasing distance from the nucleus, as K, L, M, N, O, P, and Q shells. The number of the electrons which each of these shells can contain is limited. All electrons arranged in the same shell have the same principal quantum number. The electrons in the same shell are grouped into various subshells (q.v.), and all the electrons in the same subshell have the same orbital angular momentum. See Subshell.

Sidereal day. The duration of one rotation of the earth on its axis, with respect to the vernal equinox. It is measured by successive transits of the vernal equinox over the upper branch of a meridian.

Because of the precession of the equinoxes, the sidereal day thus defined is slightly less than the period of rotation with respect to the stars, but the difference is less than 0.01 sec. The length of the mean sidereal day is 24 hr of sidereal time or 23 hr 56 min 4.09054 sec of mean solar time.

Sidereal month — The average period of revolution of the moon with respect to the stars, a period of 27 days 7 hr 43 min 11.5 sec, or approximately 27⅓ days.

Sidereal time — Time based upon the rotation of the earth relative to the vernal equinox.

Sidereal time may be designated as local or Greenwich as the local or Greenwich meridian is used as the reference. When adjusted for nutation, to eliminate slight irregularities in the rate, it is called mean sidereal time.

Sidereal year — The period of one apparent revolution of the earth around the sun, with respect to the stars, averaging 365 days 6 hr 9 min 9.55 sec in 1955, and increasing at the rate of 0.000095 sec annually.

Because of the precession of the equinoxes this is about 20 min longer than a tropical year.

Simple harmonic motion — Periodic oscillatory motion in a straight line in which the restoring force is proportional to the displacement. If a point moves uniformly in a circle, the motion of its projection on the diameter (or any straight line in the same plane) is simple harmonic motion.

If r is the radius of the reference circle, ω the angular velocity of the point in the circle, θ the angular displacement at the time t after the particle passes the midpoint of its path, the linear displacement,

$$x = r \sin \theta = r \sin \omega t$$

The velocity at the same instant,

$$v = r\omega \cos \theta = \omega \sqrt{r^2 - x^2}$$

The acceleration,

$$a = -\omega^2 x.$$

The force for a mass m,

$$F = m\omega^2 x = -\frac{4\pi^2 mx}{T^2}$$

The period,

$$T = 2\pi \sqrt{\frac{x}{a}}$$

In the above equations the cgs system calls for x and r in centimeters, v in centimeters per second, a in centimeters per second squared, T in seconds, M in grams, F in dynes, θ in radians and ω in radians per second.

Simple machine — A contrivance for the transfer of energy and for increased convenience in the performance of work.

Mechanical advantage is the ratio of the resistance overcome to the force applied. Velocity ratio is the ratio of the distance through which force is applied to the distance through which resistance is overcome.

Efficiency is the ratio of the work done by a machine to the work done upon it.

If a force f applied to a machine through a distance S results in a force F exerted by the machine through a distance s, neglecting friction,

$$fS = Fs$$

The theoretical mechanical advantage or velocity ratio in the above case is,

$$\frac{S}{s}$$

Actually, the force obtained from the machine will have a smaller value than will satisfy the equation above. If F′ be the actual force obtained, the practical mechanical advantage will be,

$$\frac{F'}{f}$$

The efficiency of the machine,

$$E = \frac{F's}{fS}$$

Snell's law of refraction — If i is the angle of incidence, r the angle of refraction, v the velocity of light in the first medium, v′ the velocity in the second medium, the index of refraction n,

$$n = \frac{\sin i}{\sin r} = \frac{V}{V'}$$

Solar constant — The rate at which solar radiation is received outside the earth's atmosphere on a surface normal to the incident radiation and at the earth's mean distance from the sun.

Measurements of solar radiation at the earth's surface by the Smithsonian Institution for several decades give a best value for the solar constant of 1.934 calories per square centimeter per minute. Measurements from rockets of the intensity of the ultraviolet end of the spectrum have corrected this value to 2.00 calories per square centimeter per minute with a probable error of ± 2%.

Solar wind — Streams of plasma flowing approximately radially outward from the sun.

Solid — A state of matter in which the relative motion of the molecules is restricted and they tend to retain a definite fixed position relative to each other, giving rise to crystal structure. A solid may be said to have a definite shape and volume.

Solid angle — Measured by the ratio of the surface of the portion of a sphere enclosed by the conical surface forming the angle, to the square of the radius of the sphere. Unit of solid angle, the steradian, the solid angle which encloses a surface on the sphere equivalent to the square of the radius. Dimensions, unity.

Solubility of one liquid or solid in another is the mass of a substance contained in a solution which is in equilibrium with an excess of the substance. Under these conditions the solution is said to be saturated. Solubility of a gas is the ratio of concentration of gas in the solution to the concentration of gas above the solution.

Solubility product or precipitation value is the product of the concentration of the ions of a substance in a saturated solution of the substance. These concentrations are frequently expressed as moles of solute per liter of solution.

Solute — That constituent of a solution which is considered to be dissolved in the other, the solvent. The solvent is usually present in larger amount than the solute.

A solution is saturated if it contains at given temperature as much of a solute as it can retain in the presence of an excess of that solute.

A true solution is a mixture, liquid, solid or gaseous, in which the components are uniformly distributed throughout the mixture. The proportion of the constituents may be varied within certain limits.

Solvent — That constituent of a solution which is present in larger amount; or, the constituent which is liquid in the pure state, in the case of solutions or solids or gases in liquids.

Sone — A unit of loudness. A simple tone of frequency 1000 hertz/sec, 40 decibels above a listener's threshold, produces a loudness of 1 sone.

Spatial frequency plane — The focusing plane of an optical lens or system where the image represents the spatial Fourier transform of the object spatial function.

Specific conductance, σ_v — The specific conductance, or conductivity, of a conductor of electricity is the conductance of the material between opposite sides of a cube, 1 cm in each direction. The unit of specific conductance is $ohm^{-1} cm^{-1}$ or mho cm^{-1}.

Specific gravity — The ratio of the mass of a body to the mass of an equal volume of water at 4°C or other specified temperature. Dimensions, unity.

Specific heat of a substance is the ratio of its thermal capacity to that of water at 15°C. Dimensions, unity.

If a quantity of heat H calories is necessary to raise the temperature of m grams of a substance from t_1 to t_2°C, the specific heat, or more properly, thermal capacity of the substance,

$$s = \frac{H}{m(t_2 - t_1)}$$

Specific heat by the method of mixtures— Where a mass m_1 of the substance is heated to a temperature t_1, then placed in a mass of water m_2 at temperature t_2 contained in a calorimeter with stirrer (of same material) of mass m_3, specific heat of the calorimeter c, t_3 the final temperature

$$m_1 s(t_1 - t_3) = (m_3 c + m_2)(t_3 - t_2)$$

Black's ice calorimeter — If a body of mass m and temperature t melts a mass m′ of ice, its temperature being reduced to 0°C, the specific heat of the substance is,

$$s = \frac{80.1\ m'}{mt}$$

Bunsen's ice calorimeter — A body of mass m at temperature t causes a motion of the mercury column of l centimeters in a tube whose volume per unit length is v. The specific heat is

$$s = \frac{884 lv}{mt}$$

Specific inductive capacity — The ratio of the capacitance of a condenser with a given substance as dielectric to the capacitance of the same condenser with air or a vacuum as dielectric is called the specific inductive capacity. The ratio of the dielectric constant of a substance to that of a vacuum.

Specific rotation — If there are n grams of active substance in v cubic centimeters of solution and the light passes through l decimeters, r being the observed rotation in degrees, the specific rotation (for 1 centimeter),

$$[\alpha] = \frac{rv}{nl}$$

Specific volume is the reciprocal of density. Dimensions, $[m^{-1}\ l^3]$.

Spectral series are spectral lines or groups of lines which occur in an orderly sequence.

Specular reflection — Mirror scattering at one angle from the point of reflection.

Speed — Time rate of motion measured by the distance moved over in unit time. Cgs unit, 1 cm/sec. Dimension $[l\ t^{-1}]$.

Speed of sound (symbol c_t) — The speed of propagation of sound waves. In the atmosphere

$$c_t = [\gamma (R^*/M_o)\ T_M]^{1/2}$$

where γ is the ratio of specific heat of air at constant pressure to that a constant volume, R^* is the universal gas constant, M_o is the mean molecular weight of air at sea level, and T_M is the molecular scale temperature.

At sea level in the standard atmosphere, the speed of sound is 340.294 meters per second (1116.45 ft/sec).

The concept of the speed of sound in the atmosphere loses its applicability at about 90 km where the mean free path of air molecules approaches the wavelengths of sound waves.

Spherical aberration — When large surfaces of spherical mirrors or lenses are used the light divergent from a point source is not exactly focused at a point. The phenomenon is known as spherical aberration. For axial pencils the error is known as axial spherical aberration; for oblique pencils, coma.

Spherical mirrors — If R is the radius of curvature, F principal focus, and f_1 and f_2 any two conjugate focal distances,

$$\frac{1}{f_1} + \frac{1}{f_2} = \frac{1}{F} = \frac{2}{R}$$

If the linear dimensions of the object and image be O and I respectively and u and v their distances from the mirror,

$$\frac{O}{I} = \frac{u}{v}$$

Spin — In nuclear physics, used to describe the angular momentum of elementary particles or of nuclei.

Spontaneous-ignition temperature — In testing fuels, the lowest temperature of a plate or other solid surface adequate to cause ignition in air of a fuel upon the surface.

Stagnation pressure — 1. The pressure at a stagnation point. 2. In compressible flow, the pressure exhibited by a moving gas or liquid brought to zero velocity by an isentropic process. 3. Equals total pressure. 4. Equals impact pressure.

Because of the lack of a standard meaning, stagnation pressure should be defined when it is used.

Standard conditions for gases — Measured volumes of gases are quite generally recalculated to 0°C temperature and 760 mm pressure, which have been arbitrarily chosen as standard conditions.

Stark effect — The splitting of a single spectrum line into multiple lines which occurs when the emitting material is placed in a strong electric field. The observed effect depends on the angle between the direction of the field and the direction of observation. The effect is due to the shifting of the energy states of certain orbits which all have the same energy in zero field.

Statcoulomb — The unit of electric charge in the metric system. 3×10^9 statcoulombs = 1 coulomb.

Static pressure (symbol p) — 1. The pressure with respect to a stationary surface tangent to the mass-flow velocity vector. 2. The pressure with respect to a surface at rest in relation to the surrounding fluid.

Stationary or standing waves are produced in a medium by the simultaneous transmission, in opposite directions of two similar wave motions. Fixed points of minimum amplitude are called nodes. A segment extends from one node to the next. An antinode or loop is the point of maximum amplitude between two nodes.

Statute mile — 5280 feet = 1.6093 kilometers = 0.869 nautical mile. Also called land mile.

Stefan-Boltzmann constant (symbol σ) — A universal constant of proportionality between the radiant emittance of a black body and the fourth power of the body's absolute temperature; 5.6703×10^{-5} erg per centimeter squared second K^4.

Stefan-Boltzmann law — A law stating that the total radiation E from a black body is given by

$$E = \sigma T^4$$

where T is the absolute temperature of the body and σ is the Stefan-Boltzmann constant, equal to 5.6703×10^{-5} erg/sec cm^2 deg^4.

Steradian — The steradian is the SI unit of solid angle.

The steradian is the solid angle which, having its vertex in the center of a sphere, cuts off an area of the surface of the sphere equal to that of a square with sides of length equal to the radius of the sphere. There are 4π steradians in a sphere.

Stereoscopic image — An image which appears as a three-dimensional object located in space.

Stoichiometric — Pertaining to weight relations in chemical reactions.

Stoke — See under Viscosity.

Stokes' law — 1. Gives the rate of fall of a small sphere in a viscous fluid. When a small sphere falls under the action of gravity through a viscous medium it ultimately acquires a constant velocity,

$$V = \frac{2ga^2(d_1 - d_2)}{9\eta}$$

where a is the radius of the sphere, d_1 and d_2 the densities of the sphere and the medium respectively, and η the coefficient of viscosity. V will be in cm per sec if g is in cm per sec^2, a in cm, d_1 and d_2 in g per cm^3 and η in dyne-sec per cm^2 or poises.

2. The empirical law stating that the wavelength of light emitted by a flu-

orescent material is longer than that of the radiation used to excite the fluorescence. In modern language the emitted photons carry off less energy than is brought in by the exciting photons; the details accord with the energy conservation principle.

Strain — The deformation resulting from a stress measured by the ratio of the change to the total value of the dimension in which the change occurred. Dimensions, unity.

Stress — The force producing or tending to produce deformation in a body measured by the force applied per unit area Cgs units, one dyne per square centimeter. Dimensions, $[m\, l^{-1}\, t^{-2}]$.

Subshell — The electrons within the same shell (energy level) of the atom are characterized by the same principal quantum number (n), and are further divided into groups according to the value of their azimuthal quantum numbers (l); the electrons which possess the same azimuthal quantum number for the same principal quantum number are considered to occupy the same subshell (or sublevel). The individual subshells are designated with the letters s, p, d, f, g, and h, as follows:

l value	designation of subshell
0	s
1	p
2	d
3	f
4	g
5	h

An electron assigned to the s-subshell is called an s-electron, one assigned to the p-subshell is referred to as a p-electron, etc. In formulae of electron structure, the value of the principal quantum number (n) is prefixed to the letter indicating the azimuthal quantum number (l) of the electron; thus, e.g., a 4f-electron is an electron which has the principal quantum number 4 (i.e., assigned to the N-shell) and the orbital angular momentum 3 (f-subshell).

Substance, amount of — See Mole.

Supersonic flow — In aerodynamics, flow of a fluid over a body at speeds greater than the acoustic velocity and in which the shock waves start at the surface of the body. Compare hypersonic flow.

Surface density of electricity — Quantity of electricity per unit area. Dimensions, $[\varepsilon^{1/2}\, m^{1/2}\, l^{-1/2}\, t^{-1}]$; $[\mu^{-1/2}\, m^{1/2}\, l^{-1/2}]$.

Surface density of magnetism — Quantity of magnetism per unit area. Dimensions, $[\varepsilon^{-1}\, m^{1/2}\, l^{-1/2}]$, $[\mu^{1/2}\, m^{1/2}\, l^{-1/2}\, t^{-1}]$.

Surface tension — Two fluids in contact exhibit phenomena, due to molecular attractions which appear to arise from a tension in the surface of separation. It may be expressed as dynes per centimeter or as ergs per square centimeter. Dimensions, $[m\, t^{-2}]$.

The total force along a line of length l on the surface of a liquid whose surface tension is T,

$$F = lT.$$

Capillary tubes — If a liquid of density d rises a height h in a tube of internal radius r the surface tension is,

$$T = \frac{rhdg}{2}$$

The tension will be in dynes per cm if r and h are in cm, d in g per cm^3 and g in cm per sec^2.

Drops and bubbles — Pressure in dynes per cm^2 due to surface tension on a drop of radius r centimeter for a liquid whose surface tension is T dynes per centimeter,

$$P = \frac{2T}{r}$$

For a bubble of mean radius r cm,

$$P = \frac{4T}{r}$$

Susceptibility (magnetic) is measured by the ratio of the intensity of magnetization produced in a substance to the magnetizing force or intensity of field to which it is subjected. The susceptibility of a substance will be unity when unit intensity of magnetization is produced by a field of one gauss. Dimensions, $[\varepsilon^{-1}\, l^{-2}\, t^{2}]$, $[\mu]$.

"Synthetic" aperture" sidelooking radar — A sidelooking radar that generates very high resolution data by integrating its return signals during the time that the physical aircraft antenna or aperture is traveling through a large distance (making up the synthetic aperture).

Tangent galvanometer — A tangent galvanometer with n turns, of radius r, in the earth's field H, has a deflection θ. The current flowing is,

$$i = \frac{Hr}{2\pi n} \tan\theta$$

If 2πn = G (the galvanometer constant).

$$i = \frac{H}{G} \tan\theta$$

Tektite — Small glassy bodies containing no crystals, composed of at least 65% silicon dioxide, bearing no relation to the geological formations in which they occur, and believed to be of extraterrestrial origin.

Tektites are found in certain large areas called strewn fields. They are named, as are minerals, with the suffix ite, as australite, found in Australia, billitonite, indochinite, and rizalite, found in Southeast Asia, bediasite from Texas, and moldavite from Bohemia and Moravia.

Temperature may be defined as the condition of a body which determines the transfer of heat to or from other bodies. Particularly it is a manifestation of the average translational kinetic energy of the molecules of a substance due to heat agitation. The customary unit of temperature is the Celsius degree, 1/100 the difference between the temperature of melting ice and that of water boiling under standard atmospheric pressure. Celsius scale — the Celsius temperature scale is a designation of the scale also known as the centigrade scale. The degree Fahrenheit is 1/180, and the degree Reaumur 1/80 the same difference of temperature.

The fundamental temperature scale is the absolute, thermodynamic or Kelvin scale in which the temperature measure is based on the average kinetic energy per molecule of a perfect gas. The zero of the Kelvin scale is −273.15°C. The temperature scale adopted by the International Bureau of Weights and Measures is that of the constant volume hydrogen gas thermometer. The magnitude of the degree in both these scales is defined as 1/100 the difference between the temperature of melting ice and that of boiling water at 760 mm pressure. Frequently the Kelvin scale is defined as degrees C + 273.15 and the Rankine scale as degrees F + 459.67°F.

The fundamental temperature scale is now defined by means of the equation

$$\theta(X) = 273.15\ K\ \frac{X}{X_3}$$

where θ denotes the temperature; X the thermometric property (P, V, ...); the subscript 3 refers to the triple point of water; and 273.16°K is the arbitrary fixed point for the temperature associated with the triple point of water.

The ideal gas temperature θ, (numerically equal to the Kelvin temperature), in particular, is defined by either of the two equations:

$$\theta = \begin{cases} 273.15^\circ \lim\limits_{P_3 \to 0} \dfrac{P}{P_3}, \text{ const. } V \\ 273.15^\circ \lim\limits_{P \to 0} \dfrac{V}{V_3}, \text{ const. } P \end{cases}$$

Temperature resistance coefficient — The ratio of the change of resistance in a wire due to a change of temperature of 1°C to its resistance at 0°C. Dimension, $[\theta - 1]$.

Tera — A prefix meaning trillion or 10^{12}. Symbol is T.

Terrestrial radiation — 1. The total infrared radiation emitted from the earth's surface; to be carefully distinguished from effective terrestrial radiation, atmospheric radiation (which is sometimes erroneously used as a synonym for terrestrial radiation), and insolation. Also called earth radiation, eradiation.

Thermal capacity of a substance is the quantity of heat necessary to produce unit change of temperature in unit mass. It is ordinarily expressed as calories per gram per degree Celsius. Numerically equivalent to specific heat.

Thermal capacity or water equivalent — The total quantity of heat necessary to raise any body or system unit temperature, measured as calories per degree centigrade in the cgs system. Dimension, [m].

Thermal expansion — The coefficient of linear expansion or expansivity is the ratio of the change in length per degree C to the length at 0°C. The coefficient of volume expansion (for solids) is approximately three times the linear coefficient. The coefficient of volume expansion for liquids is the ratio of the change in volume per degree to the volume at 0°C. The value of the coefficient varies with temperature. The coefficient of volume expansion for a gas under constant pressure is nearly the same for all gases and temperatures and is equal to 0.00367 for 1°C. Dimension, $[\theta^{-1}]$.

If l_o is the length at 0°C, α the coefficient of linear expansion, the length at t°C is,

$$l_t = l_0(1 + \alpha t)$$

General formula for thermal expansion — The rate of thermal expansion varies with the temperature. The general equation giving the magnitude m, (length or volume) at a temperature t, where m_o is the magnitude at 0°C, is

$$m_t = m_0(1 + \alpha t + \beta t^2 + \gamma t^3 \ldots)$$

where α, β, γ, etc. are empirically determined coefficients.

Volume expansion — If V represents volume and β the coefficient of expansion,

$$V_t = V_0(1 + \beta t).$$

For solids,

$$\beta = 3\alpha \text{ (approximately)}.$$

Thermal neutrons — Neutrons slowed down by a moderator to an energy of a fraction of an electron volt — about 0.025 eV at 15°C.

Thermionic emission — The emission of ions or electrons from a metal as the result of heating. The thermionic current density, J, in amperes/cm² is equal to $AT_2e^{-\phi/kT}$ where A is a constant, T is Kelvin temperature, k is Boltzmann's constant, and ϕ is the work function of the emitter.

Thermodynamic efficiency — In thermodynamics, the ratio of the work done by a heat engine to the total heat supplied by the heat source. Also called thermal efficiency, Carnot efficiency.

Thermodynamic probability — Under specified conditions, the number of equally likely states in which a substance may exist. The thermodynamic probability P is related to the entropy S by

$$S = k \ln P$$

where k is Boltzmann constant. See Third law of thermodynamics.

Thermodynamic temperature scale — The Kelvin thermodynamic scale is recognized as the fundamental scale to which all temperatures should ultimately be referable. The kelvin, unit of thermodynamic temperature, is the fraction 1/273.16 of the thermodynamic temperature of the triple point of water. The unit kelvin and its symbol K are used to express an interval or a difference of temperature. In addition to the thermodynamic temperature (symbol T), expressed in kelvins, use is also made of Celsius temperature (symbol t) defined by the equation

$$t = T - T_0$$

where T_o = 273.15 K by definition. The unit "degree Celsius" is equal to the unit "kelvin", but "degree Celsius" is a special name in place of "kelvin" for expressing Celsius temperature. A temperature interval or a Celsius temperature difference can be expressed in degrees Celsius as well as in kelvins.

Thermodynamics, law of —

I. When mechanical work is transformed into heat or heat into work, the amount of work is always equivalent to the quantity of heat.

II. It is impossible by any continuous self-sustaining process for heat to be transferred from a colder to a hotter body.

Thermoelectric power is measured by the electromotive force produced by a thermocouple for unit difference of temperature between the two junctions. It varies with the average temperature and is usually expressed in microvolts per degree C. It is customary to list the thermoelectric power of the various metals with respect to lead. Dimensions, $[\varepsilon^{-1/2} m^{1/2} t^{-1} \theta^{-1}]$; $[\mu^{1/2} m^{1/2} l^{1/2} t^{-1} \theta^{-1}]$.

Thickness or average radius of ion atmosphere, κ^{-1} — The average distance of the ion atmosphere from the reference ion in angstrom units. This average distance decreases in magnitude with the square root of the ionic concentration. Mathematically, K^{-1} is the distance at which the average charge, dq, in a spherical shell of volume $4\pi r^2 dr$ reaches a maximum using the continuous density, ϱ (r), approximation.

Third law of thermodynamics — The statement that every substance has a finite positive entropy, and that the entropy of a crystalline substance is zero at the temperature of absolute zero. See Thermodynamic probability.

Modern quantum theory has shown that the entropy of crystals at 0° absolute is not necessarily zero. If the crystal has any asymmetry, it may exist in more than one state; and there is, in addition, an entropy residue deriving from nuclear spin.

Thomson thermoelectric effect is the designation of the potential gradient along a conductor which accompanies a temperature gradient. The magnitude and direction of the potential varies with the substance.

The coefficient of the Thomson effect or specific heat of electricity is expressed in joules per coulomb per degree Centigrade. Dimensions, $[\varepsilon^{-1/2} m^{1/2} l^{1/2} t^{-1} \theta^{-1}]$, $[\mu^{1/2} m^{1/2} l^{1/2} t^{-2} \theta^{-1}]$.

Time, unit of — The fundamental invariable unit of time is the ephemeris second, which is defined as 1/31,556,925.9747 of the tropical year for 1900 January 0^d12^h ephemeris time. The ephemeris day is 86,400 ephemeris seconds.

The former unit of time was the mean solar second, defined as 1/86,400 of the mean solar day.

Torque prouced by the action of one magnet on another — The turning moment experienced by a magnet of pole strength m′ and length 2l′ placed at a distance r from another magnet of length 2l and pole strength m, where the center of the first magnet is on the axis (extended) of the second and the axis of the first is perpendicular to the axis of the second,

$$C = 8\,\frac{mm'll'}{r^3} = \frac{2MM'}{r^3}$$

If the first magnet is deflected through an angle θ, the expression becomes,

$$C = \frac{2MM'}{r^3}\cos\theta$$

Torr — Provisional international standard term to replace the English term millimeter of mercury and its abbreviation mm of Hg (or the French mm de Hg).

The torr is defined as 1/760 of a standard atmosphere or 1,013,250/760 dynes per square centimeter. This is equivalent to defining the torr as 1333.22 microbars and differs by only one part in 7 million from the International Standard millimeter of mercury. The prefixes milli and micro are attached without hyphenation.

Torsional vibration — See Angular harmonic motion.

Total pressure — 1. Equals stagnation pressure. 2. Equals impact pressure. 3. The pressure a moving fluid would have if it were brought to rest without losses. 4. The pressure determined by all the molecular species crossing the imaginary surface.

Total reflection — When light passes from any medium to one in which the velocity is greater, refraction ceases and total reflection begins at a certain critical angle of incidence θ such that

$$\sin\theta = \frac{1}{n}$$

where n is the index of the first medium with respect to the second. If the second medium is air n has the ordinary value for the first medium. For any other second medium,

$$n = \frac{n_1}{n_2}$$

where n_1 and n_2 are the ordinary indices of refraction for the first and second medium respectively.

Tractive force of a magnet — If a magnet with induction B has a pole face of area A the force is,

$$F = \frac{B^2 A}{8\pi}$$

If B and A are in cgs units, F will be in dynes.

Transducer (acoustic) — A device to convert electrical oscillatory input energy into mechanical acoustical wave energy.

Transference (or transport) number — The transference number of each ion of a solute (or electrolyte) in an electrolytic solution is the fraction of the total current carried by that ion, and is given by the ratio of the mobility of the ion to the sum of the mobilities of the ions of the solute constituting the electrolytic solution.

Transmutation — A nuclear change producing a new element from an old one.

Transuranic elements — Elements of atomic numbers above 92. All of them are radioactive and are products of artificial nuclear changes. All are members of the actinide group.

Transverse electromagnetic energy — Wave whose electric field and magnetic field displacements are at right angles to each other and to the direction of propagation (motion) of the wave.

Triangle or polygon of forces — If three or more forces acting on the same point are in equilibrium, the vectors representing them form, when added, a closed figure.

Triple point — The thermodynamic state at which three phases of a substance exist in equilibrium.

The triple point of water occurs at a saturation vapor pressure of 6.11 millibar and at a temperature of 273.16 K.

Tritium — An isotope of hydrogen with a mass of three, structure, two neutrons, and one proton in its nucleus.

Ultrasound camera — A camera that converts a sound pressure "acoustic image" into an electrical TV-like image by means of the piezo-electric effect.

Uncertainty principle — See Indeterminancy.

Uniform circular motion — If r is the radius of a circle, v the linear speed in the arc, ω the angular velocity and T the period or time of one revolution,

$$\omega = \frac{v}{r} = \frac{2\pi}{T}$$

The acceleration toward the center is

$$a = \frac{v^2}{r} = \omega^2 r = \frac{4\pi^2 r}{T^2}$$

The centrifugal force for a mass m,

$$F = \frac{mv^2}{r} = m\omega^2 r = \frac{4\pi^2 mr}{T^2}$$

In the above equations ω will be in radians per second and a in cm per sec² if r is in cm, v in cm per sec and T in sec. F will be in dynes if mass is in grams and other units as above.

Application to the solar system — If M is the mass of the sun, G the constant of gravitation, P the period of the planet and r the distance of the planet from the sun, then the mass of the sun

$$M = \frac{4\pi^2 r^3}{GP^2}, \text{ (G = 6.670} \times 10^{-8} \text{ for cgs units)}$$

If P is the period and r the distance of a satellite revolving around the planet, the above expression for M gives the mass of the planet. The formula is written on the assumption that the orbit of the planet or satellite is circular, which is only approximately true.

Uniformly accelerated rectilinear motion — If v_o is the initial velocity, v_t the velocity after time t, the acceleration,

$$a = \frac{v_t - v_o}{t}$$

The velocity after time t,

$$v_t = v_o + at$$

Space passed over in time t,

$$s = v_o t + \tfrac{1}{2}at^2$$

Velocity after passing over space s,

$$v = \sqrt{v_o^2 + 2as}$$

Space passed over in the nth second

$$s = v_o + \tfrac{1}{2}a(2n - 1)$$

In the above and following similar equations the values of the space, velocity, and acceleration must be substituted in the same system. For space in cm, velocity will be in centimeter per second and acceleration in centimeter per second per second.

Unit — Specific magnitude of a quantity, set apart by appropriate definition, which is to serve as a basis of comparison or measurement for other quantities of the same nature.

Universal time — Time defined by the rotational motion of the earth and determined from the apparent diurnal motions which reflect this rotation; because of variations in the rate of rotation, universal time is not rigorously uniform. Also called Greenwich mean time. Compare ephemeris time.

In the years preceding 1960 the arguments of the ephemerides in the American Ephemeris and Nautical Almanac were designated as universal time.

Valence of an atom of an element is that property which is measured by the number of atoms of hydrogen (or its equivalent) one atom of that element can hold in combination if negative, or can displace in a reaction if it is positive.

Valence electrons of the atom are electrons which are gained, lost or shared in chemical reactions.

Van Allen belt, Van Allen radiation belt — The zone of high-intensity particulate radiation surrounding the earth beginning at altitudes of approximately 1000 km.

The radiation of the Van Allen belt is composed of protons and electrons temporarily trapped in the earth's magnetic field. The intensity of radiation varies with the distance from the earth.

Van der Waals' equation of state — This equation is expressed by:

$$\left(p + \frac{a}{v^2}\right)(v - b) = RT$$

It makes allowance both for the volume occupied by the molecules and for the attractive force between the molecules. b is the effective volume of molecules in one mole of gas. a is a measure of the attractive force between the molecules. For values of R, a, and b see index for table of Van der Waals' constants for gases.

Van't Hoff's principle — If the temperature of interacting substances in equilibrium is raised, the equilibrium concentrations of the reaction are changed so that the products of that reaction which absorb heat are increased in quantity, or if the temperature for such an equilibrium is lowered, the products which evolve heat in their formation are increased in amounts.

Vapor — The words vapor and gas are often used interchangeably. Vapor is more frequently used for a substance which, though present in the gaseous phase, generally exists as a liquid or solid at room temperature. Gas is more frequently used for a substance that generally exists in the gaseous phase at room temperature. Thus one would speak of iodine or carbon tetrachloride vapors and of oxygen gas.

Vapor pressure — The pressure exerted when a solid or liquid is in equilibrium with its own vapor. The vapor pressure is a function of the substance and of the temperature.

Vectors, composition of — If the angle between two vectors is A, and their magnitude a and b, their resultant,

$$C = \sqrt{a^2 + b^2 - 2ab \cos A.}$$

Velocity — Time rate of motion in a fixed direction. Cgs units, one centimeter per second. Dimensions, $[l\, t^{-1}]$.

If s is space passed over in time t, the velocity,

$$\bar{v} = \frac{s}{t}$$

Velocity of a compressional wave — The velocity of a compressional wave in an elastic medium, in terms of elasticity E (bulk modulus) and density d,

$$V = \sqrt{\frac{E}{d}}$$

For the velocity of sound in air, where p is the pressure and d the density,

$$V = \sqrt{\frac{1.4p}{d}}$$

Velocity of efflux of a liquid — If h is the distance from the opening to the free surface of the liquid, the velocity of efflux is

$$V = \sqrt{2gh}$$

The above is the theoretical discharge velocity disregarding friction and the shape of orifice. For water issuing through a circular opening with sharp edges of area, A, the volume discharged per second is given approximately by,

$$Q = 0.624\sqrt{2gh}$$

Velocity of sound, variation with temperature — The velocity in meters per sec at any temperature t in °C is given approximately by

$$V = V_o\sqrt{1 + \frac{t}{273}}$$

$$V = 331.5 + 0.607t$$

The variation with humidity is given by the equation

$$V_d = V_h\sqrt{1 - \frac{e}{p}\frac{\gamma\omega}{\gamma a} - \frac{5}{8}}$$

where V_d is the velocity in dry air, V_h that in air at barometric pressure p in which the pressure of water vapor is e. γ_ω and γ_a are the specific heat ratios for water vapor and for air respectively.

Velocity of a transverse wave in a stretched cord. If T is the tension of the cord and m the mass per unit length,

$$V = \sqrt{\frac{T}{m}}$$

Velocity of water waves — If the depth h is small compared with the wave length, the velocity,

$$V = \sqrt{gh}$$

In deep water for a wave length λ,

$$V = \sqrt{\frac{g\lambda}{2\pi}}$$

If the wavelength is very small, less than about 1.6 cm, the velocity increases as the wavelength decreases and is expressed by the following,

$$V = \sqrt{\frac{2\pi T}{\lambda d} + \frac{g\lambda}{2\pi}}$$

where T is the surface tension and d the density of the liquid V will be given in cm per sec if h and λ are in cm, g in cm per sec², T in dyes per cm and d in g per cm³.

Velocity of a wave — The velocity of propagation in terms of wavelength λ and the period T or frequency n is,

$$V = \frac{\lambda}{T} = n\lambda$$

Virtual gravity — The force of gravity on an atmospheric parcel, reduced by centrifugal force due to the motion of the parcel relative to the earth. The virtual gravity g^* is

$$g^* = g = V^2/a - 2\Omega_n V$$

where g is the magnitude of the acceleration of gravity; V is the parcel speed; a is the earth's radius; and Ω_n is the component of the earth's angular velocity vector normal to the motion of the parcel.

For reasonable atmospheric values, the correction terms are of the order of 0.01% of the magnitude of gravity. The identity of g^* and g is implied by the assumption of hydrostatic equilibrium.

Virtual height — The apparent height of an ionized atmospheric layer determined from the time interval between the transmitted signal and the ionospheric echo at vertical incidence, assuming that the velocity of propagation is the velocity of light in a vacuum over the entire path. See Ionospheric recorder. Compare scale height.

Virtual image — An image that cannot be shown on a surface but is visible, as in a mirror.

Virtual mass — The actual mass of a body, plus its apparent additional mass.

Viscosity — All fluids possess a definite resistance to change of form and many solids show a gradual yielding to forces tending to change their form. This property, a sort of internal friction, is called viscosity; it is expressed in dyne-seconds per cm² or poises. Dimensions, $[m\, l^{-1}\, t^{-1}]$. If the tangential force per unit area, exerted by a layer of fluid upon one adjacent is one dyne for a space rate of variation of the tangential velocity of unity, the viscosity is one poise.

Kinematic viscosity is the ratio of viscosity to density. The c. g. s. unit of kinematic viscosity is the stoke.

Flow of liquids through a tube; where l is the length of the tube, r its radius, p the difference of pressure at the ends, η the coefficient of viscosity, the volume escaping per second,

$$v = \frac{\pi p r^4}{8 l \eta} \text{ (Poiseuille).}$$

The volume will be given in cm³ per second if l and r are in cm, p in dynes per cm² and η in poises or dyne-seconds per cm².

Viscosity effect — An alteration in the velocity of a given ion as a result of the contribution to the bulk viscosity owing to the ions of opposite charge. This effect applies to ions of large size.

Visibility is measured by the ratio of the luminous flux in lumens to the total radiant energy in ergs per second or in watts.

Volt (V) — The unit of electric potential difference and electromotive force, equal to the difference of electric potential between two points of a conductor carrying a constant current of 1 ampere when the power dissipated between these points equals 1 watt.

Volume, unit of — The cubic centimeter, the volume of a cube whose edges are one centimeter in length. Other units of volume are derived in a similar manner. Dimension, $[l^3]$.

Volume velocity — Volume velocity is the rate of alternating flow of the medium through a specificed surface due to a sound wave.

Walden's rule, $\Lambda_o\eta_o$ — Walden's rule states that the product of the limiting equivalent conductance of an electrolytic solution, Λ_o, and the viscosity of the solvent, η_o, in which the solute (or electrolyte) is dissolved is a constant at a particular temperature. Walden's rule is an approximation which would be valid only for ions which behave hydrodynamically like Stokes spheres in a continuum.

Watt — Rate of doing work or expending power. A watt is that power which gives rise to the production of energy at a rate of 1 joule per second or 10^7 ergs per second.

Wave — A wave is a disturbance which is propagated in a medium in such a manner that at any point in the medium the quantity serving as measure of disturbance is a function of the time while at any instant the displacement at a point is a function of the position of the point.

Wavefront — Surface whose points in an energy beam are all of equal phase or optical path length.

Wavelength — Distance between successive wavefronts of like phase (i.e., from peak to peak, or trough to trough).

Wave motion — A progressive disturbance propagated in a medium by the periodic vibration of the particles of the medium. Transverse wave motion is that in which the vibration of the particles is perpendicular to the direction of propagation. Longitudinal wave motion is that in which the vibration of the particles is parallel to the direction of propagation.

Wave number (symbol ν) — The reciprocal of wavelength; the number of waves per unit distance in the direction of propagation; or, sometimes 2π times this quantity.

In spectroscopy, wave number is usually expressed in reciprocal centimeters, as 100,000 cm^{-1} (100,000 per centimeter).

Weber (unit of magnetic flux) — The magnetic flux which, linking a circuit of one turn, produces in it an electromotive force of 1 volt as it is reduced to zero at a uniform rate in 1 second. (Symbol Wb)

Weight — The force with which a body is attracted toward the earth. Cgs unit, — the dyne. Dimensions, $[m\, l\, t^{-2}]$.

Although the weight of a body varies with its location, the weights of various standards of mass are often used as units of force as, pound weight, or pound force, gram weight, etc. The weight of mass m, where g is the acceleration due to gravity,

$$W = mg$$

The weight will be given in dynes when m is in grams and g in centimeter per second squared.

Wheatstone's bridge — If the resistances r_1, r_2, r_3, and r_4 form the arms of a Wheatstone's bridge in order as the circuit (omitting cell and galvanometer connections) is traced, when the bridge is balanced,

$$\frac{r_1}{r_2} = \frac{r_4}{r_3} \quad \text{or} \quad \frac{r_1}{r_4} = \frac{r_2}{r_3}$$

Wien distribution law — A relation, derived on purely thermodynamic reasoning by Wien, between the monochromatic emittance of an ideal black body and that body's temperature.

$$J\lambda/T^5 = f(\lambda, T)$$

where J_λ is the monochromatic emittance (emissive power) of a black body at wavelength λ and absolute temperature T, and $f(\lambda, T)$ is a function which cannot be determined purely on classical thermodynamic grounds. Compare Wien law.

Wien effect — The increase in the conductance of an electrolytic solution produced by high electrical fields (potential gradients).

Wien's displacement law — When the temperature of a radiating black body increases, the wave length corresponding to maximum energy decreases in such a way that the product of the absolute temperature and wave length is constant.

$$\lambda_{max}\, T = w$$

w is known as Wien's displacement constant.

Work — When a force acts against resistance to produce motion in a body the force is said to do work. Work is measured by the product of the force acting and the distance moved through against the resistance. Cgs units of work, the erg, a force of one dyne acting through a distance of one centimeter. The joule is 1×10^7 ergs. Dimensions, $[m\, l^2\, t^{-2}]$. The foot-pound is the work required to raise a mass of one pound a vertical distance of one foot where g = 32.174 ft/sec². The foot-poundal is the work done by a force of one poundal acting through a distance of one foot. The International joule, a unit of electrical energy, is the work expended per second by a current of one International ampere flowing through one International ohm. The kilowatt-hour is the total amount of energy developed in one hour by a power of one kilowatt.

If a force F act through a space s, the work done is

$$W = Fs$$

Work will be given in ergs if F is in dynes and s in centimeters.

Work done in rotation. If a torque L dyne-cm acts through an angle θ radians, the work done in ergs is

$$W = L\theta$$

Work function — The energy required for an electron to escape a solid surface. See Helmholtz function.

X-rays — A type of radiation of higher frequency than visible light but lower than gamma rays. Usually produced by high energy electrons impinging upon a metal target.

X units — X-ray wavelengths have been measured in two kinds of units. The older measurements are given in X units (XU) which are based on the effective lattice constant of rock salt being 2,814.00 XU. More recently X-ray wavelengths have been directly connected, through measurements with ruled gratings, to the wavelengths in the optical region and through them to the standard meter. It turned out that the XU which was originally intended as 10^{-11} cm was 0.202% larger than this value. It has become customary to give X-ray wavelengths in Ångstrom units (Å) when the absolute scale is used (1 Å = 10^{-8} cm). The two are related by

$$1000\ XU = 1.00202 \pm 0.00003)\ A$$

and wavelengths given in XU must be multiplied by 1.00202 and then divided by 1000 in order to convert them into Angstrom units.

Yield point — The stress at which a marked increase in deformation takes place without increase in the load.

Yield strength — "The stress at which a material exhibits a specified permanent set."

Zeeman effect — The splitting of a spectrum line into several symmetrically disposed components, which occurs when the source of light is placed in a strong magnetic field. The components are polarized, the directions of polarization and the appearance of the effect depending on the direction from which the source is viewed relative to the lines of force.

SI DERIVED UNITS WITH SPECIAL NAMES

Name	Quantity	Symbol	Expression in terms of other units	Expression in terms of SI base units
Becquerel	Activity (of a radionuclide)	Bq		s^{-1}
Coulomb	Quantity of electricity, electric charge	C	$A \cdot s$	$s \cdot A$
Degrees Celsius	Celsius temperature	°C		K
Gray	Absorbed dose, specific energy imparted, kerma, absorbed dose index	Gy	J/kg	$m^2 \cdot s^{-2}$
Farad	Capacitance	F	C/V	$m^{-2} \cdot kg^{-1} \cdot s^4 \cdot A^2$
Henry	Inductance	H	Wb/A	$m^2 \cdot kg \cdot s^{-2} \cdot A^{-2}$
Hertz	Frequency	Hz		s^{-1}
Joule	Energy, work, quantity of heat	J	$N \cdot m$	$m^2 \cdot kg \cdot s^{-2}$
Lumen	Luminous flux	lm		$cd \cdot sr^a$
Lux	Illuminance	lx	lm/m^2	$m^{-2} \cdot cd \cdot sr^a$
Newton	Force	N		$m \cdot kg \cdot s^{-2}$
Ohm	Electric resistance		V/A	$m^2 \cdot kg \cdot s^{-3} \cdot A^{-2}$
Pascal	Pressure, stress	Pa	N/m^2	$m^{-1} \cdot kg \cdot s^{-2}$
Siemens	Conductance	S	A/V	$m^{-2} \cdot kg^{-1} \cdot s^3 \cdot A^2$
Tesla	Magnetic flux density	T	Wb/m^2	$kg \cdot s^{-2} \cdot A^{-1}$
Volt	Electric potential, potential difference, electromotive force	V	W/A	$m^2 \cdot kg \cdot s^{-3} \cdot A^{-1}$
Watt	Power, radiant flux	W	J/s	$m^2 \cdot kg \cdot s^{-3}$
Weber	Magnetic flux	Wb	$V \cdot s$	$m^2 \cdot kg \cdot s^{-2} \cdot A^{-1}$

[a] In this expression the steradian is treated as a base unit.

PRIMARY FIXED POINTS OF INTERNATIONAL PRACTICAL TEMPERATURE SCALE-68 (IPTS-68)

Primary fixed points of IPTS-68 are given in degrees Celsius and kelvin. Also shown are the comparative values from previous international temperature scales.

Fixed point	ITS-27, °C	ITS-48, °C	IPTS-48, °C	IPTS-68 °C	IPTS-68 K	IPTS-68 Uncertainty (K)
Equilibrium between the liquid and vapor phases of water (boiling point of water)	100.000	100	100	100	373.15	0.005
Equilibrium between the solid and liquid phases of zinc (freezing point of zinc)				419.58	692.73	0.03
Equilibrium between the liquid and vapor phases of sulfur (boiling point of sulfur)	444.60	444.600	444.6			
Equilibrium between the solid and liquid phases of silver (freezing point of silver)	960.5	960.8	960.8	961.93	1,235.08	0.2
Equilibrium between the solid and liquid phases of gold (freezing point of gold)	1,063	1,063.0	1,063	1,064.43	1,337.58	0.2

From Sparks, L. L., *ASRDI Oxygen Technology Survey,* Vol. 4, Scientific and Technical Information Office, National Aeronautics and Space Administration, Washington, D. C., 1974.

ALLOWABLE CARRYING CAPACITIES OF CONDUCTORS

(National Electrical Code)

The ratings in the following tabulation are those permitted by the National Electrical Code for flexible cords and for interior wiring of houses, hotels, office buildings, industrial plants, and other buildings.

The values are for copper wire. For aluminum wire the allowable carrying capacities shall be taken as 84% of those given in the table for the respective sizes of copper wire with the same kind of covering.

Size A.W.G.	Area Circular Mils	Diameter of Solid Wires Mils	Rubber Insulation Amperes	Varnished Cambric Insulation Amperes	Other Insulations and Bare Conductors Amperes
18	1,624.	40.3	3*		6†
16	2,583.	50.8	6*		10†
14	4,107.	64.1	15	18	20
12	6,530.	80.8	20	25	30
10	10,380.	101.9	25	30	35
8	16,510.	128.5	35	40	50
6	26,250.	162.0	50	60	70
5	33,100.	181.9	55	65	80
4	41,740.	204.3	70	85	90
3	52,630.	229.4	80	95	100
2	66,370.	257.6	90	110	125
1	83,690.	289.3	100	120	150
0	105,500.	325.0	125	150	200
00	135,100.	364.8	150	180	225
000	167,800.	409.6	175	210	275
0000	211,600.	460	225	270	325

* The allowable carrying capacities of No. 18 and 16 are 5 and 7 amperes respectively, when in flexible cords.

† The allowable carrying capacities of No. 18 and 16 are 10 and 15 amperes respectively, when in cords for portable heaters. Types AFS, AFSJ, HC, HPD, and HSJ.

WIRE TABLE, STANDARD ANNEALED COPPER

American Wire Gauge (B. & S.) English Units

Gauge No.	Diameter in mils at 20°C	Cross section at 20°C: Circular mils	Cross section at 20°C: Sq. inches	Ohms per 1000 feet*: 0°C (32°F)	20°C (68°F)	50°C (122°F)	75°C (167°F)
0000	460.0	211600	0.1662	0.04516	0.04901	0.05479	0.05961
000	409.6	167800	.1318	.05695	.06180	.06909	.07516
00	364.8	133100	.1045	.07181	.07793	.08712	.09478
0	324.9	105500	.08289	.09055	.09827	.1099	.1195
1	289.3	83690	.06573	.1142	.1239	.1385	.1507
2	257.6	66370	.05213	.1440	.1563	.1747	.1900
3	229.4	52640	.04134	.1816	.1970	.2203	.2396
4	204.3	41740	.03278	.2289	.2485	.2778	.3022
5	181.9	33100	.02600	.2887	.3133	.3502	.3810
6	162.0	26250	.02062	.3640	.3951	.4416	.4805
7	144.3	20820	.01635	.4590	.4982	.5569	.6059
8	128.5	16510	.01297	.5788	.6282	.7023	.7640
9	114.4	13090	.01028	.7299	.7921	.8855	.9633
10	101.9	10380	.008155	.9203	.9989	1.117	1.215
11	90.74	8234	.006467	1.161	1.260	1.408	1.532
12	80.81	6530	.005129	1.463	1.588	1.775	1.931
13	71.96	5178	.004067	1.845	2.003	2.239	2.436
14	64.08	4107	.003225	2.327	2.525	2.823	3.071
15	57.07	3257	.002558	2.934	3.184	3.560	3.873
16	50.82	2583	.002028	3.700	4.016	4.489	4.884
17	45.26	2048	.001609	4.666	5.064	5.660	6.158
18	40.30	1624	.001276	5.883	6.385	7.138	7.765
19	35.89	1288	.001012	7.418	8.051	9.001	9.792
20	31.96	1022	.0008023	9.355	10.15	11.35	12.35
21	28.45	810.1	.0006363	11.80	12.80	14.31	15.57
22	25.35	642.4	.0005046	14.87	16.14	18.05	19.63
23	22.57	509.5	.0004002	18.76	20.36	22.76	24.76
24	20.10	404.0	.0003173	23.65	25.67	28.70	31.22
25	17.90	320.4	.0002517	29.82	32.37	36.18	39.36
26	15.94	254.1	.0001996	37.61	40.81	45.63	49.64
27	14.20	201.5	.0001583	47.42	51.47	57.53	62.59
28	12.64	159.8	.0001255	59.80	64.90	72.55	78.93
29	11.26	126.7	.00009953	75.40	81.83	91.48	99.52
30	10.03	100.5	.00007894	95.08	103.2	115.4	125.5
31	8.928	79.70	.00006260	119.9	130.1	145.5	158.2
32	7.950	63.21	.00004964	151.2	164.1	183.4	199.5
33	7.080	50.13	.00003937	190.6	206.9	231.3	251.6
34	6.305	39.75	.00003122	240.4	260.9	291.7	317.3
35	5.615	31.52	.00002476	303.1	329.0	367.8	400.1
36	5.000	25.00	.00001964	382.2	414.8	463.7	504.5
37	4.453	19.83	.00001557	482.0	523.1	584.8	636.2
38	3.965	15.72	.00001235	607.8	659.6	737.4	802.2
39	3.531	12.47	.000009793	766.4	831.8	929.8	1012
40	3.145	9.888	.000007766	966.5	1049	1173	1276

* Resistance at the stated temperatures of a wire whose length is 1000 feet at 20°C.

WIRE TABLE, STANDARD ANNEALED COPPER

American Wire Gauge (B. & S.) English Units (Continued)

Gauge No.	Pounds per 1000 feet	Feet per pound	Feet per ohm*: 0°C (32°F)	20°C (68°F)	50°C (122°F)	75°C (167°F)
0000	640.5	1.561	22140	20400	18250	16780
000	507.9	1.968	17560	16180	14470	13300
00	402.8	2.482	13930	12830	11480	10550
0	319.5	3.130	11040	10180	9103	8367
1	253.3	3.947	8758	8070	7219	6636
2	200.9	4.977	6946	6400	5725	5262
3	159.3	6.276	5508	5075	4540	4173
4	126.4	7.914	4368	4025	3600	3309
5	100.2	9.980	3464	3192	2855	2625
6	79.46	12.58	2747	2531	2264	2081
7	63.02	15.87	2179	2007	1796	1651
8	49.98	20.01	1728	1592	1424	1309
9	39.63	25.23	1370	1262	1129	1038
10	31.43	31.82	1087	1001	895.6	823.2
11	24.92	40.12	861.7	794.0	710.2	652.8
12	19.77	50.59	683.3	629.6	563.2	517.7
13	15.68	63.80	541.9	499.3	446.7	410.6
14	12.43	80.44	429.8	396.0	354.2	325.6
15	9.858	101.4	340.8	314.0	280.9	258.2
16	7.818	127.9	270.3	249.0	222.8	204.8
17	6.200	161.3	214.3	197.5	176.7	162.4
18	4.917	203.4	170.0	156.6	140.1	128.8
19	3.899	256.5	134.8	124.2	111.1	102.1
20	3.092	323.4	106.9	98.50	88.11	80.99
21	2.452	407.8	84.78	78.11	69.87	64.23
22	1.945	514.2	67.23	61.95	55.41	50.94
23	1.542	648.4	53.32	49.13	43.94	40.39
24	1.223	817.7	42.28	38.96	34.85	32.03
25	0.9699	1031	33.53	30.90	27.64	25.40
26	.7692	1300	26.59	24.50	21.92	20.15
27	.6100	1639	21.09	19.43	17.38	15.98
28	.4837	2067	16.72	15.41	13.78	12.67
29	.3836	2607	13.26	12.22	10.93	10.05
30	.3042	3287	10.52	9.691	8.669	7.968
31	.2413	4145	8.341	7.685	6.875	6.319
32	.1913	5227	6.614	6.095	5.452	5.011
33	.1517	6591	5.245	4.833	4.323	3.974
34	.1203	8310	4.160	3.833	3.429	3.152
35	.09542	10480	3.299	3.040	2.719	2.499
36	.07568	13210	2.616	2.411	2.156	1.982
37	.06001	16660	2.075	1.912	1.710	1.572
38	.04759	21010	1.645	1.516	1.356	1.247
39	.03774	26500	1.305	1.202	1.075	0.9886
40	.02993	33410	1.035	0.9534	0.8529	.7840

* Length at 20°C of a wire whose resistance is 1 ohm at the stated temperatures.

WIRE TABLE, STANDARD ANNEALED COPPER (Continued)

American Wire Gauge (B. & S.) English Units

Gauge No.	Diameter in mils at 20°C	Ohms per pound 0°C (32°F)	Ohms per pound 20°C (68°F)	Ohms per pound 50°C (122°F)	Lbs. per ohm 20°C (68°F)
0000	460.0	0.00007051	0.00007652	0.00008554	13070
000	409.6	.0001121	.0001217	.0001360	8219
00	364.8	.0001783	.0001935	.0002163	5169
0	324.9	.0002835	.0003076	.0003439	3251
1	289.3	.0004507	.0004891	.0005468	2044
2	257.6	.0007166	.0007778	.0008695	1286
3	229.4	.001140	.001237	.001383	808.6
4	204.3	.001812	.001966	.002198	508.5
5	181.9	.002881	.003127	.003495	319.8
6	162.0	.004581	.004972	.005558	201.1
7	144.3	.007284	.007905	.008838	126.5
8	128.5	.01158	.01257	.01405	79.55
9	114.4	.01842	.01999	.02234	50.03
10	101.9	.02928	.03178	.03553	31.47
11	90.74	.04656	.05053	.05649	9.79
12	80.81	.07404	.08035	.08983	12.45
13	71.96	.1177	.1278	.1428	7.827
14	64.08	.1872	.2032	.2271	4.922
15	57.07	.2976	.3230	.3611	3.096
16	50.82	.4733	.5136	.5742	1.947
17	45.26	.7525	.8167	.9130	1.224
18	40.30	1.197	1.299	1.452	0.7700
19	35.89	1.903	2.065	2.308	.4843
20	31.96	3.025	3.283	3.670	.3046
21	28.46	4.810	5.221	5.836	.1915
22	25.35	7.649	8.301	9.280	.1205
23	22.57	12.16	13.20	14.76	.07576
24	20.10	19.34	20.99	23.46	.04765
25	17.90	30.75	33.37	37.31	.02997
26	15.94	48.89	53.06	59.32	.01885
27	14.20	77.74	84.37	94.32	.01185
28	12.64	123.6	134.2	150.0	.007454
29	11.26	196.6	213.3	238.5	.004688
30	10.03	312.5	339.2	379.2	.002948
31	8.928	497.0	539.3	602.9	.001854
32	7.950	790.2	857.6	958.7	.001166
33	7.080	1256	1364	1524	.0007333
34	6.305	1998	2168	2424	.0004612
35	5.615	3177	3448	3854	.0002901
36	5.000	5051	5482	6128	.0001824
37	4.453	8032	8717	9744	.0001147
38	3.965	12770	13860	15490	.00007215
39	3.531	20310	22040	24640	.00004538
40	3.145	32290	35040	39170	.00002854

Gauge No.	Diameter in mm at 20°C	Cross section in mm² at 20°C	Ohms per kilometer* 0°C	Ohms per kilometer* 20°C	Ohms per kilometer* 50°C	Ohms per kilometer* 75°C
0000	11.68	107.2	0.1482	0.1608	0.1798	0.1956
000	10.40	85.03	.1868	.2028	.2267	.2466
00	9.266	67.43	.2356	.2557	.2858	.3110
0	8.252	53.48	.2971	.3224	.3604	.3921
1	7.348	42.41	.3746	.4066	.4545	.4944
2	6.544	33.63	.4724	.5127	.5731	.6235
3	5.827	26.67	.5956	.6465	.7227	.7862
4	5.189	21.15	.7511	.8152	.9113	.9914
5	4.621	16.77	.9471	1.028	1.149	1.250
6	4.115	13.30	1.194	1.296	1.449	1.576
7	3.665	10.55	1.506	1.634	1.827	1.988
8	3.264	8.366	1.899	2.061	2.304	2.506
9	2.906	6.634	2.395	2.599	2.905	3.161
10	2.588	5.261	3.020	3.277	3.663	3.985
11	2.305	4.172	3.807	4.132	4.619	5.025
12	2.053	3.309	4.801	5.211	5.825	6.337
13	1.828	2.624	6.054	6.571	7.345	7.991
14	1.628	2.081	7.634	8.285	9.262	10.08
15	1.450	1.650	9.627	10.45	11.68	12.71
16	1.291	1.309	12.14	13.17	14.73	16.02
17	1.150	1.038	15.31	16.61	18.57	20.20
18	1.024	.8231	19.30	20.95	23.42	25.48
19	.9116	.6527	24.34	26.42	29.53	32.12
20	.8118	.5176	30.69	33.31	37.24	40.51
21	.7230	.4105	38.70	42.00	46.95	51.08
22	.6438	.3255	48.80	52.96	59.21	64.41
23	.5733	.2582	61.54	66.79	74.66	81.22
24	.5106	.2047	77.60	84.21	94.14	102.4
25	.4547	.1624	97.85	106.2	118.7	129.1
26	.4049	.1288	123.4	133.9	149.7	162.9
27	.3606	.1021	155.6	168.9	188.8	205.4
28	.3211	.08098	196.2	212.9	238.0	258.9
29	.2859	.06422	247.4	268.5	300.1	326.5
30	.2546	.05093	311.9	338.6	378.5	411.7
31	.2268	.04039	393.4	426.9	477.2	519.2
32	.2019	.03203	496.0	538.3	601.8	654.7
33	.1798	.02540	625.5	678.8	758.8	825.5
34	.1601	.02014	788.7	856.0	956.9	1041
35	.1426	.01597	994.5	1079	1207	1313
36	.1270	.01267	1254	1361	1522	1655
37	.1131	.01005	1581	1716	1919	2087
38	.1007	.007967	1994	2164	2419	2632
39	.08969	.006318	2514	2729	3051	3319
40	.07987	.005010	3171	3441	3847	4185

* Resistance at the stated temperatures of a wire whose length is 1 kilometer at 20°C.

Gauge No.	Diameter in mm at 20°C	Kilograms per kilometer	Meters per gram	Meters per ohm* 0°C	Meters per ohm* 20°C	Meters per ohm* 50°C	Meters per ohm* 75°C
0000	11.68	953.2	0.001049	6749	6219	5563	5113
000	10.40	755.9	.001323	5352	4932	4412	4055
00	9.266	599.5	.001668	4245	3911	3499	3216
0	8.252	475.4	.002103	3366	3102	2774	2550
1	7.348	377.0	.002652	2669	2460	2200	2022
2	6.544	299.0	.003345	2117	1951	1745	1604
3	5.827	237.1	.004217	1679	1547	1384	1272
4	5.189	188.0	.005318	1331	1227	1097	1009
5	4.621	149.1	.006706	1056	972.9	870.2	799.9
6	4.115	118.2	.008457	837.3	771.5	690.1	634.4
7	3.665	93.78	.01066	664.0	611.8	547.3	503.1
8	3.264	74.37	.01345	526.6	485.2	434.0	399.0
9	2.906	58.98	.01696	417.6	384.8	344.2	316.4
10	2.588	46.77	.02138	331.2	305.1	273.0	250.9
11	2.305	37.09	.02696	262.6	242.0	216.5	199.0
12	2.053	29.42	.03400	208.3	191.9	171.7	157.8
13	1.828	23.33	.04287	165.2	152.2	136.1	125.1
14	1.628	18.50	.05406	131.0	120.7	108.0	99.24
15	1.450	14.67	.06816	103.9	95.71	85.62	78.70
16	1.291	11.63	.08595	82.38	75.90	67.90	62.41
17	1.150	9.226	.1084	65.33	60.20	53.85	49.50
18	1.024	7.317	.1367	51.81	47.74	42.70	39.25
19	0.9116	5.803	.1723	41.09	37.86	33.86	31.13
20	.8118	4.602	.2173	32.58	30.02	26.86	24.69
21	.7230	3.649	.2740	25.84	23.81	21.30	19.58
22	.6438	2.894	.3455	20.49	18.88	16.89	15.53
23	.5733	2.295	.4357	16.25	14.97	13.39	12.31
24	.5106	1.820	.5494	12.89	11.87	10.62	9.764
25	.4547	1.443	.6928	10.22	9.417	8.424	7.743
26	.4049	1.145	.8736	8.105	7.468	6.680	6.141
27	.3606	0.9078	1.102	6.428	5.922	5.298	4.870
28	.3211	.7199	1.389	5.097	4.697	4.201	3.862
29	.2859	.5709	1.752	4.042	3.725	3.332	3.063
30	.2546	.4527	2.209	3.206	2.954	2.642	2.429
31	.2268	.3590	2.785	2.542	2.342	2.095	1.926
32	.2019	.2847	3.512	2.016	1.858	1.662	1.527
33	.1798	.2258	4.429	1.599	1.473	1.318	1.211
34	.1601	.1791	5.584	1.268	1.168	1.045	0.9606
35	.1426	.1420	7.042	1.006	0.9265	0.8288	.7618
36	.1270	.1126	8.879	0.7974	.7347	.6572	.6041
37	.1131	.08931	11.20	.6324	.5827	.5212	.4791
38	.1007	.07083	14.12	.5015	.4621	.4133	.3799
39	.08969	.05617	17.80	.3977	.3664	.3278	.3013
40	.07987	.04454	22.45	.3154	.2906	.2600	.2390

* Length at 20°C of a wire whose resistance is 1 ohm at the stated temperatures.

Gauge No.	Ohms per kilogram 0°C	Ohms per kilogram 20°C	Ohms per kilogram 50°C	Grams per ohm 20°C
0000	0.0001554	0.0001687	0.0001886	5928000
000	.0002472	.0002682	.0002999	3728000
00	.0003930	.0004265	.0004768	2344000
0	.0006249	.0006782	.0007582	1474000
1	.0009936	.001078	.001206	927300
2	.001580	.001715	.001917	583200
3	.002512	.002726	.003048	366800
4	.003995	.004335	.004846	230700
5	.006352	.006893	.007706	145100
6	.01010	.01096	.01225	91230
7	.01606	.01743	.01948	57380
8	.02553	.02771	.03098	36080
9	.04060	.04406	.04926	22690
10	.06456	.07007	.07833	14270
11	.1026	.1114	.1245	8976
12	.1632	.1771	.1980	5645
13	.2595	.2817	.3149	3550
14	.4127	.4479	.5007	2233
15	.6562	.7122	.7961	1404
16	1.043	1.132	1.266	883.1
17	1.659	1.801	2.013	555.4
18	2.638	2.863	3.201	349.3
19	4.194	4.552	5.089	219.7
20	6.670	7.238	8.092	138.2
21	10.60	11.51	12.87	86.88
22	16.86	18.30	20.46	54.64
23	26.81	29.10	32.53	34.36
24	42.63	46.27	51.73	21.61
25	67.79	73.57	82.25	13.59
26	107.8	117.0	130.8	8.548
27	171.4	186.0	207.9	5.376
28	272.5	295.8	330.6	3.381
29	433.3	470.3	525.7	2.126
30	689.0	747.8	836.0	1.337
31	1096	1189	1329	0.8410
32	1742	1891	2114	.5289
33	2770	3006	3361	.3326
34	4404	4780	5344	.2092
35	7003	7601	8497	.1316
36	11140	12090	13510	.08274
37	17710	19220	21480	.05204
38	28150	305360	34160	.03273
39	44770	48590	54310	.02058
40	71180	77260	88360	.01294

ALUMINUM WIRE TABLE

Hard-Drawn Aluminum Wire at 20°C (or, 68°F) American Wire Gauge (B. & S.) English Units

Gauge No.	Diameter in mils	Cross section: Circular mils	Cross section: Square inches	Ohms per 1000 ft.	Pounds per 1000 ft.	Pounds per ohm	Feet per ohm
0000	460	212000	0.166	0.0804	195	2420	12400
000	410	168000	.132	.101	154	1520	9860
00	365	133000	.105	.128	122	957	7820
0	325	106000	.0829	.161	97.0	602	6200
1	289	83700	.0657	.203	76.9	379	4920
2	258	66400	.0521	.256	61.0	238	3900
3	229	52600	.0413	.323	48.4	150	3090
4	204	41700	.0328	.408	38.4	94.2	2450
5	182	33100	.0260	.514	30.4	59.2	1950
6	162	26300	.0206	.648	24.1	37.2	1540
7	144	20800	.0164	.817	19.1	23.4	1220
8	128	16500	.0130	1.03	15.2	14.7	970
9	114	13100	.0103	1.30	12.0	9.26	770
10	102	10400	.00815	1.64	9.55	5.83	610
11	91	8230	.00647	2.07	7.57	3.66	484
12	81	6530	.00513	2.61	6.00	2.30	384
13	72	5180	.00407	3.29	4.76	1.45	304
14	64	4110	.00323	4.14	3.78	0.911	241
15	57	3260	.00256	5.22	2.99	.573	191
16	51	2580	.00203	6.59	2.37	.360	152
17	45	2050	.00161	8.31	1.88	.227	120
18	40	1620	.00128	10.5	1.49	.143	95.5
19	36	1290	.00101	13.2	1.18	.0897	75.7
20	32	1020	.000802	16.7	0.939	.0564	60.0
21	28.5	810	.000636	21.0	.745	.0355	47.6
22	25.3	642	.000505	26.5	.591	.0223	37.8
23	22.6	509	.000400	33.4	.468	.0140	29.9
24	20.1	404	.000317	42.1	.371	.00882	23.7
25	17.9	320	.000252	53.1	.295	.00555	18.8
26	15.9	254	.000200	67.0	.234	.00349	14.9
27	14.2	202	.000158	84.4	.185	.00219	11.8
28	12.6	160	.000126	106.	.147	.00138	9.39
29	11.3	127	.0000995	134.	.117	.000868	7.45
30	10.0	101	.0000789	169.	.0924	.000546	5.91
31	8.9	79.7	.0000626	213.	.0733	.000343	4.68
32	8.0	63.2	.0000496	269.	.0581	.000216	3.72
33	7.1	50.1	.0000394	339.	.0461	.000136	2.95
34	6.3	39.8	.0000312	428.	.0365	.0000854	2.34
35	5.6	31.5	.0000248	540.	.0290	.0000537	1.85
36	5.0	25.0	.0000196	681.	.0230	.0000338	1.47
37	4.5	19.8	.0000156	858.	.0182	.0000212	1.17
38	4.0	15.7	.0000123	1080.	.0145	.0000134	0.924
39	3.5	12.5	.00000979	1360.	.0115	.00000840	.733
40	3.1	9.9	.0000077	1720.	.0091	.00000528	.581

Gauge No.	Diameter in mm	Cross section in mm²	Ohms per kilometer	Kilograms per kilometer	Grams per ohm	Meters per ohm
0000	11.7	107	.0264	289	1100000	3790
000	10.4	85.0	.333	230	690000	3010
00	9.3	67.4	.419	182	434000	2380
0	8.3	53.5	.529	144	273000	1890
1	7.3	42.4	.667	114.	172000	1500
2	6.5	33.6	.841	90.8	108000	1190
3	5.8	26.7	1.06	72.0	67900	943
4	5.2	21.2	1.34	57.1	42700	748
5	4.6	16.8	1.69	45.3	26900	593
6	4.1	13.3	2.13	35.9	16900	470
7	3.7	10.5	2.68	28.5	10600	373
8	3.3	8.37	3.38	22.6	6680	296
9	2.91	6.63	4.26	17.9	4200	235
10	2.59	5.26	5.38	14.2	2640	186
11	2.30	4.17	6.78	11.3	1660	148
12	2.05	3.31	8.55	8.93	1050	117
13	1.83	2.62	10.8	7.08	657	92.8
14	1.63	2.08	13.6	5.62	413	73.6
15	1.45	1.65	17.1	4.46	260	58.4
16	1.29	1.31	21.6	3.53	164	46.3
17	1.15	1.04	27.3	2.80	103	36.7
18	1.02	0.823	34.4	2.22	64.7	29.1
19	0.91	.653	43.3	1.76	40.7	23.1
20	.81	.518	54.6	1.40	25.6	18.3
21	.72	.411	68.9	1.11	16.1	14.5
22	.64	.326	86.9	0.879	10.1	11.5
23	.57	.258	110	.697	6.36	9.13
24	.51	.205	138	.553	4.00	7.24
25	.45	.162	174	.438	2.52	5.74
26	.40	.129	220	.348	1.58	4.55
27	.36	.102	277	.276	0.995	3.61
28	.32	.0810	349	.219	.626	2.86
29	.29	.0642	440	.173	.394	2.27
30	.25	.0509	555	.138	.248	1.80
31	.227	.0404	700	.109	.156	1.43
32	.202	.0320	883	.0865	.0979	1.13
33	.180	.0254	1110	.0686	.0616	0.899
34	.160	.0201	1400	.0544	.0387	.712
35	.143	.0160	1770	.0431	.0244	.565
36	.127	.0127	2230	.0342	.0153	.448
37	.113	.0100	2820	.0271	.00963	.355
38	.101	.0080	3550	.0215	.00606	.282
39	.090	.0063	4480	.0171	.00381	.223
40	.080	.0050	5640	.0135	.00240	.177

CROSS-SECTION AND MASS OF WIRES

U. S. Measure

Diameters are given in mils (1 mil = .001 in.), and area in square mils (1 sq. mil = .000001 sq. in.). For sections and masses for one-tenth the diameters given, divide by 100 and for sections and masses for ten times the diameter multiply by 100.

Diam. in mils	Cross-sec. in sq. mils	Pounds per foot: Copper, density 8.90	Iron, density 7.80	Brass, density 8.56	Aluminum, density 2.67
10	78.54	0.000303	0.0002656	0.0002915	0.0000909
11	95.03	0367	03214	03527	01100
12	113.10	0436	03825	04197	01309
13	132.73	0512	04488	04926	01536
14	153.94	0594	05206	05713	01782
15	176.71	0682	05976	06558	02045
16	201.06	0776	06799	07461	02327
17	226.98	0876	07675	08423	02627
18	254.47	0982	08605	09443	02946
19	283.53	1094	09588	10522	03282
20	314.16	1212	1062	1166	03636
21	346.36	1336	1171	1285	04009
22	380.13	1467	1286	1411	04400
23	415.48	1603	1405	1542	04809
24	452.39	1746	1530	1679	05237
25	490.87	1894	1660	1822	05682
26	530.93	2046	1795	1970	01647
27	572.56	2209	1936	2125	06628
28	615.75	2376	2082	2285	07127
29	660.52	2549	2234	2451	07646
30	706.86	2727	2390	2623	08182
31	754.77	2912	2552	2801	08737
32	804.25	3103	2720	2985	09309
33	855.30	3300	2892	3174	09900
34	907.92	3503	3070	3369	10509
35	962.11	3712	3253	3570	1114
36	1017.88	3927	3442	3777	1178
37	1075.21	4149	3636	3990	1245
38	1134.11	4376	3844	4218	1316
39	1194.59	4609	4040	4433	1383

Diam. in mils	Cross-sec. in sq. mils	Pounds per foot: Copper, density 8.90	Iron, density 7.80	Brass, density 8.56	Aluminum, density 2.67
40	1256.64	0.004849	0.004249	0.004664	0.001455
41	1320.25	5094	4465	4900	1528
42	1385.44	5346	4685	5141	1604
43	1452.20	5603	4911	5389	1681
44	1520.53	5867	5142	5643	1760
45	1590.43	6137	5378	5902	1841
46	1661.90	6412	5620	6167	1924
47	1734.94	6694	5867	6438	2008
48	1809.56	6982	6119	6715	2095
49	1885.74	7276	6377	6998	2183
50	1963.50	7576	6640	7287	2273
51	2042.82	7882	6908	7581	2365
52	2123.72	8194	7181	7881	2458
53	2206.18	8512	7460	8187	2554
54	2290.22	8837	7744	8499	2651
55	2375.83	9167	8034	8817	2750
56	2463.01	9504	8329	9140	2851
57	2551.76	9846	8629	9470	2954
58	2642.08	10195	8934	9805	3058
59	2733.97	10549	9245	10146	3165
60	2827.43	1091	956	1049	3273
61	2922.47	1128	988	1085	3383
62	3019.07	1165	1021	1120	3495
63	3117.25	1203	1054	1157	3608
64	3216.99	1241	1088	1194	3724
65	3318.31	1280	1122	1231	3841
66	3421.19	1320	1157	1270	3960
67	3525.65	1360	1192	1308	4081
68	3631.68	1401	1228	1348	4204
69	3739.28	1443	1264	1388	4328

CROSS-SECTION AND MASS OF WIRES (continued)

U. S. Measure

Diam. in mils	Cross-sec. in sq. mils	Pounds per foot: Copper, density 8.90	Iron, density 7.80	Brass, density 8.56	Aluminum, density 2.67
70	3848.45	0.01485	0.01302	0.01429	0.004456
71	3959.19	1528	1339	1469	4583
72	4071.50	1571	1377	1511	4713
73	4185.39	1615	1415	1553	4845
74	4300.84	1660	1454	1596	4978
75	4417.86	1705	1494	1639	5114
76	4536.46	1751	1534	1684	5251
77	4656.63	1797	1575	1728	5390
78	4778.36	1844	1616	1773	5531
79	4901.67	1892	1658	1819	5674
80	5026.55	1939	1700	1865	5818
81	5153.00	1988	1743	1912	5965
82	5281.02	2038	1786	1960	6113
83	5410.61	2088	1830	2008	6263
84	5541.77	2138	1874	2057	6415
85	5674.50	2189	1919	2106	6568
86	5808.80	0.02241	0.01964	0.02156	0.006724
87	5944.68	2294	2010	2206	6881
88	6082.12	2347	2057	2257	7040
89	6221.14	2400	2104	2309	7201
90	6361.73	2455	2151	2360	7364
91	6503.88	2509	2199	2414	7528
92	6647.61	2565	2248	2467	7695
93	6792.91	2621	2297	2521	7863
94	6939.78	2678	2347	2575	8033
95	7088.22	2735	2397	2630	8205
96	7238.23	2793	2448	2686	8378
97	7389.81	2851	2499	2742	8554
98	7542.96	2910	2551	2799	8731
99	7697.69	2970	2603	2857	8910
100	7853.98	3030	2656	2915	9091

Metric Measure

Diameters are given in thousandths of a centimeter and area of section in square thousandths of a centimeter. 1 $(\text{cm}/1000)^2$ = .000001 sq. cm. For sections and masses for diameters 1/10 or 10 times those of the table, divide or multiply by 100.

Diam. in thousandths of a cm	Cross-section in square thousandths of a cm	Grams per meter: Copper, density 8.90	Iron, density 7.80	Brass, density 8.56	Aluminum, density 2.67
10	78.54	0.06990	0.06126	0.06723	0.02097
11	95.03	.08458	.07412	.08135	.02537
12	113.10	.10065	.08822	.09681	.03020
13	132.73	.11813	.10353	.11362	.03544
14	153.94	.13701	.12008	.13177	.04110
15	176.71	.1573	.1378	.1513	.04718
16	201.06	.1789	.1568	.1721	.05368
17	226.98	.2020	.1770	.1943	.06060
18	254.47	.2265	.1985	.2178	.06794
19	283.53	.2523	.2212	.2427	.07570
20	314.16	.2796	.2450	.2689	.08388
21	346.36	.3083	.2702	.2965	.09248
22	380.13	.3383	.2965	.3254	.10149
23	415.48	.3698	.3241	.3557	.11093
24	452.39	.4026	.3529	.3872	.12079
25	490.87	.4369	.3829	.4202	.1311
26	530.93	.4725	.4141	.4545	.1418
27	572.56	.5096	.4466	.4901	.1529
28	615.75	.5480	.4803	.5271	.1644
29	660.52	.5879	.5152	.5654	.1764
30	706.86	.6291	.5514	.6051	.1887
31	754.77	.6717	.5887	.6461	.2015
32	804.25	.7158	.6273	.6884	.2147
33	855.30	.7612	.6671	.7321	.2284
34	907.92	.8081	.7082	.7772	.2424
35	962.11	.856	.7504	.8236	.2569
36	1017.88	.906	.7939	.8713	.2718
37	1075.21	.957	.8387	.9204	.2871
38	1134.11	1.012	.8866	.9730	.3035
39	1194.59	.063	.9318	1.0230	.3190
40	1256.64	1.118	.980	1.076	.3355
41	1320.25	.175	1.030	.130	.3525
42	1385.44	.233	.081	.186	.3699
43	1452.20	.292	.133	.243	.3877
44	1520.53	.353	.186	.302	.4060
45	1590.43	1.415	1.241	1.361	.4246
46	1661.90	.479	.296	.423	.4437
47	1734.94	.544	.353	.485	.4632
48	1809.56	.611	.411	.549	.4832
49	1885.74	.678	.471	.614	.5035
50	1963.50	1.748	1.532	1.681	.5243
51	2042.82	.818	.593	.753	.5454
52	2123.72	.890	.657	.818	.5670
53	2206.18	.964	.721	.888	.5891
54	2290.22	2.038	.786	.960	.6115
55	235.83	2.114	1.853	2.034	0.6343
56	2463.01	.192	.921	.108	.6576
57	2551.76	.271	.990	.184	.6813
58	2642.08	.351	2.061	.262	.7054
59	2733.97	.433	.132	.340	.7300
60	2827.43	2.516	2.205	2.420	.7549
61	2922.47	.601	.280	.502	.7803
62	3019.07	.687	.355	.584	.8061
63	3117.25	.774	.431	.668	.8323
64	3216.99	.863	.509	.760	.8589
65	3318.31	2.953	2.588	2.840	.8860
66	3421.19	3.045	.669	.929	.9135
67	3525.65	.138	.750	3.018	.9413
68	3631.68	.232	.833	.109	.9697
69	3739.28	.328	.917	.201	.9984
70	3848.45	3.426	3.003	3.295	1.028
71	3959.19	.524	.088	.389	.057
72	4071.50	.624	.176	.485	.087
73	4185.39	.725	.265	.583	.117
74	4300.84	.828	.355	.682	.148
75	4417.86	3.932	3.446	3.782	1.180
76	4536.46	4.037	.538	.883	.211
77	4656.63	.144	.632	.986	.243
78	4778.36	.253	.727	4.090	.276
79	4901.67	.362	.823	.177	.309
80	5026.55	4.474	3.921	4.303	1.342
81	5153.00	.586	4.019	.411	.376
82	5281.02	.700	.119	.521	.410
83	5410.61	.815	.220	.631	.445
84	5541.77	.932	.323	.744	.480
85	5674.50	5.050	4.426	4.857	1.515
86	5808.80	.170	.531	.972	.551
87	5944.68	.291	.637	5.089	.587
88	6082.12	.413	.744	.206	.624
89	6221.14	.537	.852	.325	.661
90	6361.73	5.662	4.962	5.446	1.699
91	6503.88	.788	5.073	.567	.737
92	6647.61	.916	.185	.690	.775
93	6792.91	6.046	.298	.815	.814
94	6939.78	.176	.413	.940	.853
95	7088.22	6.309	5.529	6.068	1.893
96	7238.23	.442	.646	.196	.933
97	7389.81	.577	.764	.326	.973
98	7542.96	.713	.884	.457	2.014
99	7697.69	.851	6.004	.589	.055
100	7853.98	6.990	6.126	6.723	2.097

RESISTANCE OF WIRES

The following table gives the approximate resistance of various metallic conductors. The values have been computed from the resistivities at 20°C, except as otherwise stated, and for the dimensions of wire indicated. Owing to differences in purity in the case of elements and of composition in alloys, the values can be considered only as approximations.

The following dimensions have been adopted in the computations.

B. & S. gauge	Diameter mm	Diameter mils 1 mil = .001 in.	B. & S. gauge	Diameter mm	Diameter mils 1 mil = .001 in.
10	2.588	101.9	26	0.4049	15.94
12	2.053	80.81	27	0.3606	14.20
14	1.628	64.08	28	0.3211	12.64
16	1.291	50.82	30	0.2546	10.03
18	1.024	40.30	32	0.2019	7.950
20	0.8118	31.96	34	0.1601	6.305
22	0.6438	25.35	36	0.1270	5.000
24	0.5106	20.10	40	0.07987	3.145

B. & S. No.	Ohms per cm	Ohms per ft.	B. & S. No.	Ohms per cm	Ohms per ft.	B. & S. No.	Ohms per cm	Ohms per ft.	B. & S. No.	Ohms per cm	Ohms per ft.
*Advance (0°C) $\varrho = 48. \times 10^{-6}$ ohm cm			Aluminum $\varrho = 2.828 \times 10^{-6}$ ohm cm			Eureka (0°C) $\varrho = 47. \times 10^{-6}$ ohm cm			Excello $\varrho = 92. \times 10^{-6}$ ohm cm		
10	.000912	.0278	10	.0000538	.00164	10	.000893	.0272	10	.00175	.0533
12	.00145	.0442	12	.0000855	.00260	12	.00142	.0433	12	.00278	.0847
14	.00231	.0703	14	.000136	.00414	14	.00226	.0688	14	.00442	.135
16	.00367	.112	16	.000216	.00658	16	.00359	.109	16	.00703	.214
18	.00583	.178	18	.000344	.0105	18	.00571	.174	18	.0112	.341
20	.00927	.283	20	.000546	.0167	20	.00908	.277	20	.0178	.542
22	.0147	.449	22	.000869	.0265	22	.0144	.440	22	.0283	.861
24	.0234	.715	24	.00138	.0421	24	.0230	.700	24	.0449	1.37
26	.0373	1.14	26	.00220	.0669	26	.0365	1.11	26	.0714	2.18
27	.0470	1.43	27	.00277	.0844	27	.0460	1.40	27	.0901	2.75
28	.0593	1.81	28	.00349	.106	28	.0580	1.77	28	.114	3.46
30	.0942	2.87	30	.00555	.169	30	.0923	2.81	30	.181	5.51
32	.150	4.57	32	.00883	.269	32	.147	4.47	32	.287	8.75
34	.238	7.26	34	.0140	.428	34	.233	7.11	34	.457	13.9
36	.379	11.5	36	.0223	.680	36	.371	11.3	36	.726	22.1
40	.958	29.2	40	.0564	1.72	40	.938	28.6	40	1.84	56.0
Brass $\varrho = 7.00 \times 10^{-6}$ ohm cm			Climax $\varrho = 87. \times 10^{-6}$ ohm cm			German silver $\varrho = 33. \times 10^{-6}$ ohm cm			Gold $\varrho = 2.44 \times 10^{-6}$ ohm cm		
10	.000133	.00406	10	.00165	.0504	10	.000627	.0191	10	.0000464	.00141
12	.000212	.00645	12	.00263	.0801	12	.000997	.0304	12	.0000737	.00225
14	.000336	.0103	14	.00418	.127	14	.00159	.0483	14	.000117	.00357
16	.000535	.0163	16	.00665	.203	16	.00252	.0768	16	.000186	.00568
18	.000850	.0259	18	.0106	.322	18	.00401	.122	18	.000296	.00904
20	.00135	.0412	20	.0168	.512	20	.00638	.194	20	.000471	.0144
22	.00215	.0655	22	.0267	.815	22	.0101	.309	22	.000750	.0228
24	.00342	.104	24	.0425	1.30	24	.0161	.491	24	.00119	.0363
26	.00543	.166	26	.0675	2.06	26	.0256	.781	26	.00189	.0577
27	.00686	.209	27	.0852	2.60	27	.0323	.985	27	.00239	.0728
28	.00864	.263	28	.107	3.27	28	.0408	1.24	28	.00301	.0918
30	.0137	.419	30	.171	5.21	30	.0648	1.97	30	.00479	.146
32	.0219	.666	32	.272	8.28	32	.103	3.14	32	.00762	.232
34	.0348	1.06	34	.432	13.2	34	.164	4.99	34	.0121	.369
36	.0552	1.68	36	.687	20.9	36	.260	.794	36	.0193	.587
40	.140	4.26	40	1.74	52.9	40	.659	20.1	40	.0487	1.48
Constantan (0°C) $\varrho = 44.1 \times 10^{-6}$ ohm cm			Copper, annealed $\varrho = 1.724 \times 10^{-6}$ ohm cm			Iron $\varrho = 10. \times 10^{-6}$ ohm cm			Lead $\varrho = 22. \times 10^{-6}$ ohm cm		
10	.000838	.0255	10	.0000328	.000999	10	.000190	.00579	10	.000418	.0127
12	.00133	.0406	12	.0000521	.00159	12	.000302	.00921	12	.000665	.0203
14	.00212	.0646	14	.0000828	.00253	14	.000481	.0146	14	.00106	.0322
16	.00337	.103	16	.000132	.00401	16	.000764	.0233	16	.00168	.0512
18	.00536	.163	18	.000209	.00638	18	.00121	.0370	18	.00267	.0815
20	.00852	.260	20	.000333	.0102	20	.00193	.0589	20	.00425	.130
22	.0135	.413	22	.000530	.0161	22	.00307	.0936	22	.00676	.206
24	.0215	.657	24	.000842	.0257	24	.00489	.149	24	.0107	.328
26	.0342	1.04	26	.00134	.0408	26	.00776	.237	26	.0171	.521
27	.0432	1.32	27	.00169	.0515	27	.00979	.299	27	.0215	.657
28	.0545	1.66	28	.00213	.0649	28	.0123	.376	28	.0272	.828
30	.0866	2.64	30	.00339	.103	30	.0196	.598	30	.0432	1.32
32	.138	4.20	32	.00538	.164	32	.0312	.952	32	.0687	2.09
34	.219	6.67	34	.00856	.261	34	.0497	1.51	34	.109	3.33
36	.348	10.6	36	.0136	.415	36	0.789	2.41	36	.174	5.29
40	.880	26.8	40	.0344	1.05	40	.200	6.08	40	.439	13.4

* Trade mark.

RESISTANCE OF WIRES (Continued)

Magnesium $\rho = 4.6 \times 10^{-6}$ ohm cm

B. & S. No.	Ohms per cm	Ohms per ft.
10	.0000874	.00267
12	.000139	.00424
14	.000221	.00674
16	.000351	.0107
18	.000559	.0170
20	.000889	.0271
22	.00141	.0431
24	.00225	.0685
26	.00357	.109
27	.00451	.137
28	.00568	.173
30	.00903	.275
32	.0144	.438
34	.0228	.696
36	.0363	1.11
40	.0918	2.80

Manganin $\rho = 44. \times 10^{-6}$ ohm cm

B. & S. No.	Ohms per cm	Ohms per ft.
10	.000836	.0255
12	.00133	.0405
14	.00211	.0644
16	.00336	.102
18	.00535	.163
20	.00850	.259
22	.0135	.412
24	.0215	.655
26	.0342	1.04
27	.0431	1.31
28	.0543	1.66
30	.0864	2.63
32	.137	4.19
34	.218	6.66
36	.347	10.6
40	.878	26.8

Platinum $\rho = 10. \times 10^{-6}$ ohm cm

B. & S. No.	Ohms per cm	Ohms per ft.
10	.000190	.00579
12	.000302	.00921
14	.000481	.0146
16	.000764	.0233
18	.00121	.0370
20	.00193	.0589
22	.00307	.0936
24	.00489	.149
26	.00776	.237
27	.00979	.299
28	.0123	.376
30	.0196	.598
32	.0312	.952
34	.0497	1.51
36	.0789	2.41
40	.200	6.08

Silver (18°C) $\rho = 1.629 \times 10^{-6}$ ohm cm

B. & S. No.	Ohms per cm	Ohms per ft.
10	.0000310	.000944
12	.0000492	.00150
14	.0000783	.00239
16	.000124	.00379
18	.000198	.00603
20	.000315	.00959
22	.000500	.0153
24	.000796	.0243
26	.00126	.0386
27	.00160	.0486
28	.00201	.0613
30	.00320	.0975
32	.00509	.155
34	.00809	.247
36	.0129	.392
40	.0325	.991

Molybdenum $\rho = 5.7 \times 10^{-6}$ ohm cm

B. & S. No.	Ohms per cm	Ohms per ft.
10	.000108	.00330
12	.000172	.00525
14	.000274	.00835
16	.000435	.0133
18	.000693	.0211
20	.00110	.0336
22	.00175	.0534
24	.00278	.0849
26	.00443	.135
27	.00558	.170
28	.00704	.215
30	.0112	.341
32	.0178	.542
34	.0283	.863
36	.0450	1.37
40	.114	3.47

Monel Metal $\rho = 42. \times 10^{-6}$ ohm cm

B. & S. No.	Ohms per cm	Ohms per ft.
10	.000798	.0243
12	.00127	.0387
14	.00202	.0615
16	.00321	.0978
18	.00510	.156
20	.00811	.247
22	.0129	.393
24	.0205	.625
26	.0326	.994
27	.0411	1.25
28	.0519	1.58
30	.0825	2.51
32	.131	4.00
34	.209	6.36
36	.331	10.1
40	.838	25.6

Steel, piano wire (0°C) $\rho = 11.8 \times 10^{-6}$ ohm cm

B. & S. No.	Ohms per cm	Ohms per ft.
10	.000224	.00684
12	.000357	.0109
14	.000567	.0173
16	.000901	.0275
18	.00143	.0437
20	.00228	.0695
22	.00363	.110
24	.00576	.176
26	.00916	.279
27	.0116	.352
28	.0146	.444
30	.0232	.706
32	.0368	1.12
34	.0586	1.79
36	.0931	2.84
40	.236	7.18

Steel, invar (35% Ni) $\rho = 81. \times 10^{-6}$ ohm cm

B. & S. No.	Ohms per cm	Ohms per ft.
10	.00154	.0469
12	.00245	.0746
14	.00389	.119
16	.00619	.189
18	.00984	.300
20	.0156	.477
22	.0249	.758
24	.0396	1.21
26	.0629	1.92
27	.0793	2.42
28	.100	3.05
30	.159	4.85
32	.253	7.71
34	.402	12.3
36	.639	19.5
40	1.62	49.3

***Nichrome $\rho = 150. \times 10^{-6}$ ohm cm**

B. & S. No.	Ohms per cm	Ohms per ft.
10	.0021281	.06488
12	.0033751	.1029
14	.0054054	.1648
16	.0085116	.2595
18	.0138383	.4219
20	.0216218	.6592
22	.0346040	1.055
24	.0548088	1.671
26	.0875760	2.670
28	.1394328	4.251
30	.2214000	6.750
32	.346040	10.55
34	.557600	17.00
36	.885600	27.00
38	1.383832	42.19
40	2.303872	70.24

Nickel $\rho = 7.8 \times 10^{-6}$ ohm cm

B. & S. No.	Ohms per cm	Ohms per ft.
10	.000148	.00452
12	.000236	.00718
14	.000375	.0114
16	.000596	.0182
18	.000948	.0289
20	.00151	.0459
22	.00240	.0730
24	.00381	.116
26	.00606	.185
27	.00764	.233
28	.00963	.294
30	.0153	.467
32	.0244	.742
34	.0387	1.18
36	.0616	1.88
40	.156	4.75

Tantalum $\rho = 15.5 \times 10^{-6}$ ohm cm

B. & S. No.	Ohms per cm	Ohms per ft.
10	.000295	.00898
12	.000468	.0143
14	.000745	.0227
16	.00118	.0361
18	.00188	.0574
20	.00299	.0913
22	.00476	.145
24	.00757	.231
26	.0120	.367
27	.0152	.463
28	.0191	.583
30	.0304	.928
32	.0484	1.47
34	.0770	2.35
36	.122	3.73
40	.309	9.43

Tin $\rho = 11.5 \times 10^{-6}$ ohm cm

B. & S. No.	Ohms per cm	Ohms per ft.
10	.000219	.00666
12	.000348	.0106
14	.000553	.0168
16	.000879	.0268
18	.00140	.0426
20	.00222	.0677
22	.00353	.108
24	.00562	.171
26	.00893	.272
27	.0113	.343
28	.0142	.433
30	.0226	.688
32	.0359	1.09
34	.0571	1.74
36	.0908	2.77
40	.230	7.00

Tungsten $\rho = 5.51 \times 10^{-6}$ ohm cm

B. & S. No.	Ohms per cm	Ohms per ft.
10	.000105	.00319
12	.000167	.00508
14	.000265	.00807
16	.000421	.0128
18	.000669	.0204
20	.00106	.0324
22	.00169	.0516
24	.00269	.0820
26	.00428	.130
27	.00540	.164
28	.00680	.207
30	.0108	.330
32	.0172	.524
34	.0274	.834
36	.0435	1.33
40	.110	3.35

Zinc (0°C) $\rho = 5.75 \times 10^{-6}$ ohm cm

B. & S. No.	Ohms per cm	Ohms per ft.
10	.000109	.00333
12	.000174	.00530
14	.000276	.00842
16	.000439	.0134
18	.000699	.0213
20	.00111	.0339
22	.00177	.0538
24	.00281	.0856
26	.00446	.136
27	.00563	.172
28	.00710	.216
30	.0113	.344
32	.0180	.547
34	.0286	.870
36	.0454	1.38
40	.115	3.50

ELECTRICAL RESISTIVITY AND TEMPERATURE COEFFICIENTS OF ELEMENTS

Element	Temperature °C	Microhm-Cm	Temperature Coefficient per °C
Aluminum, 99.996%	20	2.6548	0.00429^{20}[i]
Antimony	0	39.0	
Arsenic	20	33.3	
Beryllium[a]	20	4.0	0.025^{20}[i]
Bismuth	0	106.8	
Boron	0	1.8×10^{12}	
Cadmium	0	6.83	0.0042^{0}[i]
Calcium	0	3.91	0.00416^{0}[i]
Carbon[b]	0	1375.0	
Cerium	25	75.0	$0.00087^{0\text{-}25}$
Cesium	20	20	
Chromium	0	12.9	0.003^{0}[i]
Cobalt	20	6.24	$0.00604^{0\text{-}100}$
Copper	20	1.678	$0.00393^{0\text{-}500}$[g]
Dysprosium[c]	25	57.0	$0.00119^{0\text{-}25}$
Erbium	25	107.0	$0.00201^{0\text{-}25}$
Europium	25	90.0	
Gadolinium	25	140.5	$0.00176^{0\text{-}25}$
Gallium[d]	20	17.4	
Germanium[e]	22	46×10^{6}	
Gold	20	2.24	$0.0083^{0\text{-}100}$[g]
Hafnium	25	35.1	0.0038^{25}[i]
Holmium	25	87.0	$0.00171^{0\text{-}25}$
Indium	20	8.37	
Iodine	20	1.3×10^{15}	
Iridium	20	5.3	$0.003925^{0\text{-}100}$
Iron, 99.99%	20	9.71	0.00651^{20}[i]
Lanthanum	25	5.70	$0.00218^{0\text{-}25}$
Lead	20	20.648	$0.00336^{20\text{-}40}$
Lithium	0	8.55	
Lutetium	25	79.0	$0.00240^{0\text{-}25}$
Magnesium[f]	20	4.45	0.0165^{20}[i]
Manganese α	23-100	185.0	
Mercury	50	98.4	
Molybdenum	0	5.2	
Neodymium	25	64.0	$0.00164^{0\text{-}25}$
Nickel	20	6.84	$0.0069^{0\text{-}100}$
Niobium (Columbium)[g]	0	12.5	
Osmium	20	9.5	$0.0042^{0\text{-}100}$
Palladium	20	10.54	$0.00374^{0\text{-}60}$[g]
Phosphorus, white	11	1×10^{17}	
Platinum, 99.85%	20	10.6	$0.003927^{0\text{-}100}$
Plutonium	107	141.4	
Potassium	0	6.15	
Praseodymium	25	68	$0.00171^{0\text{-}25}$
Rhenium	20	19.3	$0.00395^{0\text{-}100}$
Rhodium	20	4.51	$0.0042^{0\text{-}100}$
Rubidium	20	12.5	
Ruthenium	0	7.6	
Samarium	25	88.0	$0.00184^{0\text{-}25}$
Scandium[h]	22	61.0	$0.00282^{0\text{-}25}$
Selenium[k]	0	10^{6}	
Silicon	0	$3\text{-}4 \times 10^{6}$[j]	
Silver	20	1.586	$0.0061^{0\text{-}100}$[g]
Sodium	0	4.2	
Strontium	20	23.0	
Sulfur, yellow	20	2×10^{23}	
Tantalum	25	12.45	$0.00383^{0\text{-}100}$
Tellurium	25	4.36×10^{6}	
Thallium	0	18.0	
Thorium	0	13.0	$0.0038^{0\text{-}100}$
Thulium	25	79.0	$0.00195^{0\text{-}25}$
Tin	0	11.0	$0.0047^{0\text{-}100}$
Titanium	20	42.0	
Tungsten	27	5.65	
Uranium		30.0	
Vanadium	20	24.8-26.0	
Ytterbium	25	29.0	$0.0013^{0\text{-}25}$
Yttrium	25	57.0	$0.0027^{0\text{-}25}$
Zinc	20	5.916	$0.00419^{0\text{-}100}$
Zirconium	20	40.0	0.0044^{20}[i]

[a] Annealed, comm. pure.
[b] Graphite.
[c] Polycrystalline.
[d] Hard Wire.
[e] Intrinsic Ge.
[f] Polycrystalline.
[g] High Purity.
[h] Zone refined bar.
[i] Data not available to indicate range over which coefficient is valid.
[j] Very sensitive to purity.
[k] Crystalline.

ELECTRICAL RESISTIVITY OF THE ALKALI METALS

Metal	Atomic no.	Atomic wt.	Density at 293.15 K	M.P. (°K)	B.P. (°K)	Resistivity (μ-ohm·cm at K)	
Li	3	6.941	0.534	453.7	1617	8.53	273.15
						9.28	293.15
						11.45	350.
						15.59(s)	453.7
						24.80(l)	453.7
Na	11	22.989	0.971	371.0	1157	4.33	273.15
						4.77	293.15
						6.23	350.
						6.86(s)	371.
						9.43(l)	371.
K	19	39.098	0.871	336.35	1032	6.49	273.15
						7.20	293.15
						9.22(s)	336.35
						13.95(l)	336.35
						14.64	350.
Rb	37	85.4678	1.53	312.64	961	11.54	273.15
						12.84	293.15
						14.21(s)	312.64
						22.52(l)	312.64
						25.42	350.
Cs	55	132.9054	1.873	301.55	944	18.75	273.15
						20.46	293.15
						21.16(s)	301.55
						36.93(l)	301.55
						42.11	350.

LOW MELTING POINT SOLDERS

(From N.B.S. Circular 492)

Nominal Composition, Weight Percent			Liquidus Temperature, °F
Pb	Sn	Bi	
25	25	50	266
50	37.5	12.5	374
25	50	25	336

PHYSICAL AND PHOTOMETRIC DATA FOR PLANETS AND SATELLITES

Planet	Mass (10^{24} kg)	Radius (equ.) km	Angular diameter (see Note 3)	Distance from Earth (see Note 3)	Flattening (geom.)	Mean density (g/cm^3)	Coefficients of potential $10^3 J_2$	$10^6 J_3$	$10^6 J_4$
Mercury	0.33022	2,439	11″.0	0.613	0	5.43		—	—
Venus	4.8690	6,052	60″.2	0.277	0	5.24	0.027	—	—
Earth	5.9742	6,378.140			0.00335281	5.515	1.08263	−2.54	−1.61
(Moon)	0.073483	1,738	31″.08	0.00257	0	3.34	0.2027	—	—
Mars	0.64191	3,393.4	17″.9	0.524	0.0051865	3.94	1.964	36	—
Jupiter	1,898.8	71,398	46″.8	4.203	0.0648088	1.33	14.75	—	−580
Saturn	568.50	60,000	19″.4	8.539	0.1076209	0.70	16.45	—	−1000
Uranus	86.625	25,400	3″.9	18.182	0.030	1.30	12	—	—
Neptune	102.78	24,300	2″.3	29.06	0.0259	1.76	4	—	—
Pluto	0.015	1,500	0″.1	38.44	0	1.1		—	—

Planet	Sidereal period of rotation (d)	Inclination of equator to orbit (°)	Geometric albedo	Visual magnitude $V(1,0)$	V_0	Color indices $B-V$	$U-B$
Mercury	58.6462	0.0	0.106	−0.42	—	0.93	0.41
Venus	−243.01	177.3	0.65	−4.40	—	0.82	0.50
Earth	0.99726968	23.45	0.367	−3.86	—	—	—
(Moon)	27.32166	6.68	0.12	+0.21	−12.74	0.92	0.46
Mars	1.02595675	25.19	0.150	−1.52	−2.01	1.36	0.58
Jupiter	0.41354 (System III)	3.12	0.52	−9.40	−2.70	0.83	0.48
Saturn	0.4375 (System III)	26.73	0.47	−8.88	+0.67	1.04	0.58
Uranus	−0.65	97.86	0.51	−7.19	+5.52	0.56	0.28
Neptune	0.768	29.56	0.41	−6.87	+7.84	0.41	0.21
Pluto	−6.3867	118?	0.3	−1.0	+15.12	0.80	0.31

NOTES:

1. The values for the masses include the atmospheres but exclude satellites.
2. The mean equatorial radii are given.
3. The angular diameters correspond to the distances from the Earth (in au) given in the adjacent column: they refer to inferior conjunction for Mercury and Venus and to mean opposition for the other planets. (1″.0 = 4.848 microradians.)
4. The flattening is the ratio of the difference of the equatorial and polar radii to the equatorial radius.
5. The notation for the coefficients of the gravitational potential is given in *Trans. IAU*, XI B, 173, 1962.
6. The period of rotation refers to the rotation at the equator with respect to a fixed frame of reference: a negative sign indicates that the rotation is retrograde with respect to the pole that lies to the north of the invariable plane of the solar system. The period is given in days of 86,400 SI seconds.
7. The data on equatorial radii, flattening, period of rotation and inclination of equator to orbit are based on the report of the IAU Working Group on Cartographic Coordinates and Rotational Elements of the Planets and Satellites, 1982.
8. The geometric albedo is the ratio of the illumination at the Earth from the planet for phase angle zero to the illumination produced by a plane, absolutely white Lambert surface of the same radius as the planet placed at the same position.
9. The quantity $V(1,0)$ is the visual magnitude of the planet reduced to a distance of 1 au from both the Sun and Earth and phase angle zero: V_0 is the mean opposition magnitude. The photometric quantities for Saturn refer to the disk only.

MINOR PLANETS (OPPOSITION DATES MAGNITUDES AND OSCULATING ELEMENTS FOR EPOCH 1986 JUNE 19·0 TDT, ECLIPTIC AND EQUINOX J2000·0)

Name	No.	$B(1,0)$	Diameter (km)	Inclination (i) (°)	Long of asc. node (Ω) (°)	Argument of peri-helion (ω) (°)	Mean distance (a)	Daily motion (n) (°)	Eccentricity (e)	Mean anomaly (M) (dg)
Ceres	1	4.5	1003	10.605	80.709	72.584	2.7672	0.21411	0.0784	28.881
Juno	3	6.5	247	12.998	170.577	246.897	2.6677	0.22620	0.2580	203.843
Vesta	4	4.3	538	7.138	104.069	150.792	2.3625	0.27143	0.0897	76.694
Hebe	6	7.0	201	14.780	139.080	238.613	2.4253	0.26095	0.2019	195.934
Iris	7	6.8	209	5.511	260.117	144.810	2.3862	0.26739	0.2293	171.246
Flora	8	7.7	151	5.887	111.151	285.032	2.2014	0.30175	0.1563	173.985
Metis	9	7.8	151	5.585	69.126	4.896	2.3869	0.26728	0.1218	310.516
Hygiea	10	6.5	450	3.842	283.855	316.533	3.1342	0.17763	0.1202	179.481
Parthenope	11	7.8	150	4.621	125.715	193.657	2.4515	0.25677	0.1003	81.899
Victoria	12	8.4	126	8.376	235.877	68.822	2.3336	0.27647	0.2204	57.724
Egeria	13	8.1	224	16.505	43.504	80.003	2.5777	0.23816	0.0870	207.976
Irene	14	7.5	158	9.111	86.887	95.091	2.5872	0.23685	0.1651	203.842
Thetis	17	9.1	109	5.586	125.694	136.017	2.4681	0.25419	0.1377	136.752
Melpomene	18	7.7	150	10.134	150.750	227.391	2.2961	0.28329	0.2179	107.403
Fortuna	19	8.4	215	1.568	211.789	181.570	2.4422	0.25825	0.1586	338.212

Name	No.	$B(1,0)$	Diameter (km)	Inclination (i) (°)	Long of asc. node (Ω) (°)	Argument of peri-helion (ω) (°)	Mean distance (a)	Daily motion (n) (°)	Eccentricity (e)	Mean anomaly (M) (dg)
Massalia	20	7.7	131	0.699	207.163	254.737	2.4079	0.26378	0.1454	195.860
Kalliope	22	7.3	177	13.699	66.490	354.967	2.9100	0.19855	0.0980	335.063
Thalia	23	8.2	111	10.155	67.357	59.513	2.6286	0.23126	0.2310	79.960
Themis	24	8.3	234	0.763	36.077	111.140	3.1300	0.17798	0.1342	15.820
Phocaea	25	9.3	72	21.581	214.418	90.404	2.4015	0.26484	0.2542	231.796
Euterpe	27	8.4	108	1.584	94.828	355.936	2.3481	0.27392	0.1712	286.733
Bellona	28	8.2	126	9.404	144.725	342.672	2.7817	0.21244	0.1480	239.875
Amphitrite	29	7.1	195	6.113	356.658	62.659	2.5553	0.24129	0.0732	268.469
Euphrosyne	31	7.3	370	26.350	31.357	63.230	3.1461	0.17662	0.2276	128.642
Pomona	32	8.8	93	5.518	220.781	336.400	2.5857	0.23705	0.0844	72.769
Atalante	36	9.8	118	18.493	358.961	46.391	2.7446	0.21677	0.3050	224.239
Fides	37	8.4	95	3.078	7.896	61.834	2.6420	0.22951	0.1771	108.210
Laetitia	39	7.4	163	10.373	157.508	209.163	2.7676	0.21406	0.1137	254.136
Harmonia	40	8.3	100	4.257	94.439	269.697	2.2669	0.28877	0.0467	243.187
Daphne	41	8.2	204	15.779	178.528	45.663	2.7711	0.21366	0.2685	78.718
Isis	42	8.8	97	8.542	84.797	235.887	2.4400	0.25860	0.2258	344.086
Nysa	44	7.8	82	3.705	131.735	341.856	2.4225	0.26140	0.1513	189.490
Eugenia	45	8.3	226	6.596	148.163	85.989	2.7208	0.21961	0.0836	285.191
Hestia	46	9.6	133	2.328	181.352	175.419	2.5251	0.24563	0.1724	4.932
Nemausa	51	8.7	151	9.962	176.314	0.697	2.3654	0.27093	0.0662	87.568
Europa	52	7.6	289	7.454	129.476	335.290	3.1076	0.17992	0.1031	237.883
Alexandra	54	8.9	180	11.788	313.839	344.130	2.7122	0.22066	0.1965	248.706
Melete	56	9.5	146	8.081	193.918	103.092	2.5994	0.23518	0.2322	302.302
Mnemosyne	57	8.4	109	15.221	199.596	217.045	3.1481	0.17646	0.1177	180.390
Concordia	58	9.9	110	5.060	161.548	31.834	2.7028	0.22181	0.0433	151.384
Echo	60	10.0	51	3.592	192.154	269.247	2.3953	0.26587	0.1828	256.348
Angelina	64	8.8	56	1.313	310.134	178.286	2.6852	0.22399	0.1249	225.596
Maja	66	10.5	85	3.055	8.125	43.131	2.6457	0.22903	0.1739	151.681
Asia	67	9.7	58	6.008	203.091	105.508	2.4209	0.26167	0.1873	20.408
Leto	68	8.2	126	7.968	44.593	304.509	2.7829	0.21230	0.1852	232.488
Niobe	71	8.3	115	23.300	316.481	266.627	2.7535	0.21572	0.1751	198.102
Frigga	77	9.7	67	2.434	1.677	61.057	2.6691	0.22603	0.1327	91.229
Sappho	80	9.2	83	8.656	219.094	138.885	2.2954	0.28341	0.2003	298.083
Alkmene	82	9.5	65	2.844	25.965	110.442	2.7590	0.21507	0.2234	120.145
Beatrix	83	9.8	123	4.982	27.953	165.746	2.4314	0.25996	0.0842	354.190

SATELLITES: ORBITAL DATA

Planet		Satellite	Orbital period[a] R = Retro-grade (Days)	Maximum elongation at mean opposition	Semimajor axis (× 10^3 km)	Orbital eccentricity	Orbital inclination to planetary equator (°)	Motion of node on fixed plane[d] (°/yr)
				° ′ ″				
Earth		Moon	27.321661		384.400	0.054900489	18.28—28.58	19.34[f]
Mars	I	Phobos	0.31891023	25	9.378	0.015	1.0	158.8
	II	Deimos	1.2624407	1 02	23.459	0.0005	0.9—2.7	6.614
Jupiter	I	Io	1.769137786	2 18	422	0.004	0.04	48.6
	II	Europa	3.551181041	3 40	671	0.009	0.47	12.0
	III	Ganymede	7.15455296	5 51	1070	0.002	0.21	2.63
	IV	Callisto	16.6890184	10 18	1883	0.007	0.51	0.643
	V	Amalthea	0.49817905	59	181	0.003	0.40	914.6
	VI	Himalia	250.5662	1 02 46	11480	0.15798	27.63	
	VII	Elara	259.6528	1 04 10	11737	0.20719	24.77	
	VIII	Pasiphae	735 R	2 08 26	23500	0.378	145	
	IX	Sinope	758 R	2 09 31	23700	0.275	153	
	X	Lysithea	259.22	1 04 04	11720	0.107	29.02	
	XI	Carme	692 R	2 03 31	22600	0.20678	164	
	XII	Ananke	631 R	1 55 52	21200	0.16870	147	
	XIII	Leda	238.72	1 00 39	11094	0.14762	26.07	
	XIV	Thebe	0.6745	1 13	222	0.015	0.8	
	XV	Adrastea	0.29826	42	129			
	XVI	Metis	0.294780	42	128			
Saturn	I	Mimas	0.942421813	30	185.52	0.0202	1.53	365.0
	II	Enceladus	1.370217855	38	238.02	0.00452	0.00	156.2[e]
	III	Tethys	1.887802160	48	294.66	0.00000	1.86	72.25
	IV	Dione	2.736914742	1 01	377.40	0.002230	0.02	30.85[e]
	V	Rhea	4.517500436	1 25	527.04	0.00100	0.35	10.16
	VI	Titan	15.94542068	3 17	1221.83	0.029192	0.33	0.5213[e]
	VII	Hyperion	21.2766088	3 59	1481.1	0.104	0.43	

Planet	Satellite		Orbital period[a] R = Retrograde (Days)	Maximum elongation at mean opposition	Semimajor axis (× 10^3 km)	Orbital eccentricity	Orbital inclination to planetary equator (°)	Motion of node on fixed plane[d] (°/yr)
	VIII	Iapetus	79.3301825	9 35	3561.3	0.02828	14.72	
	IX	Phoebe	550.48 R	34 51	12952	0.16326	177[b]	
	X	Janus	0.6945	24	151.472	0.007	0.14	
	XI	Epimetheus	0.6942	24	151.422	0.009	0.34	
	XII	1980S6	2.7369	1 01	377.40	0.005	0.0	
	XIII	Telesto	1.8878	48	294.66			
	XIV	Calypso	1.8878	48	294.66			
	XV	Atlas	0.6019	22	137.670	0.000	0.3	
		1980S26	0.6285	23	141.700	0.004	0.0	
		1980S27	0.6130	23	139.353	0.003	0.0	
Uranus	I	Ariel	2.52037935	14	191.02	0.0034	0.3	6.8
	II	Umbriel	4.1441772	20	266.30	0.0050	0.36	3.6
	III	Titania	8.7058717	33	435.91	0.0022	0.14	2.0
	IV	Oberon	13.4632389	44	583.52	0.0008	0.10	1.4
	V	Miranda	1.41347925	10	129.39	0.0027	4.2	19.8
Neptune	I	Triton	5.8768433 R	17	354.29	<0.01	159.00	0.578
	II	Nereid	360.2	4 21	5511	0.7483	27.6[c]	
Pluto		(Charon)	6.3871	<1	19.7		94[c]	

[a] Sidereal periods except for satellites of Saturn; tropical periods for those.
[b] Relative to ecliptic plane.
[c] To equator of 1950.0.
[d] Rate of decrease (or increase) in the longitude of the ascending node.
[e] Rate of increase in the longitude of the apse.
[f] On ecliptic plane.

SATELLITES: PHYSICAL AND PHOTOMETRIC DATA

Planet	Satellite		Mass (ℓ/Planet)	Radius (km)	Sidereal period of rotation S = Synchr.[a] (Days)	Geometric albedo (V)[c]	V(1,0)	V_0	(B − V)	(U − B)
Earth		Moon	0.01230002	1738	S	0.12	+0.21	−12.74	0.92	0.46
Mars	I	Phobos	1.5×10^{-8}	13.5 × 10.8 × 9.4	S	0.06	+11.8	11.3	0.6	
	II	Deimos	3×10^{-9}	7.5 × 6.1 × 5.5	S	0.07	+12.89	12.40	0.65	0.18
Jupiter	I	Io	4.68×10^{-5}	1815	S	0.61	−1.68	5.02	1.17	1.30
	II	Europa	2.52×10^{-5}	1569	S	0.64	−1.41	5.29	0.87	0.52
	III	Ganymede	7.80×10^{-5}	2631	S	0.42	−2.09	4.61	0.83	0.50
	IV	Callisto	5.66×10^{-5}	2400	S	0.20	−1.05	5.65	0.86	0.55
	V	Amalthea	38×10^{-10}	135 × 83 × 75	S	0.05	+7.4	14.1	1.50	
	VI	Himalia	50×10^{-10}	93	0.4	0.03	+8.14	14.84	0.67	0.30
	VII	Elara	4×10^{-10}	38	0.5	0.03	+10.07	16.77	0.69	0.28
	VIII	Pasiphae	1×10^{-10}	25			+10.33	17.03	0.63	0.34
	IX	Sinope	0.4×10^{-10}	18			+11.6	18.3	0.7	
	X	Lysithea	0.4×10^{-10}	18			+11.7	18.4	0.7	
	XI	Carme	0.5×10^{-10}	20			+11.3	18.0	0.7	
	XII	Ananke	0.2×10^{-10}	15			+12.2	18.9	0.7	
	XIII	Leda	0.03×10^{-10}	8			+13.5	20.2	0.7	
	XIV	Thebe	4×10^{-10}	55 × 45		0.05	+8.9	15.6		
	XV	Adrastea	0.1×10^{-10}	12.5 × 10 × 7.5		0.05	+12.4	19.1		
	XVI	Metis	0.5×10^{-10}	20	0.05		+10.8	17.5		
Saturn	I	Mimas	8.0×10^{-8}	196	S	0.5	+3.3	12.9		
	II	Enceladus	1.3×10^{-7}	250	S	1.0	+2.1	11.7	0.70	0.28
	III	Tethys	1.3×10^{-6}	530	S	0.9	+0.6	10.2	0.73	0.30
	IV	Dione	1.85×10^{-6}	560	S	0.7	+0.8	10.4	0.71	0.31
	V	Rhea	4.4×10^{-6}	765	S	0.7	+0.1	9.7	0.78	0.38
	VI	Titan	2.38×10^{-4}	2575	S	0.21	−1.28	8.28	1.28	0.75
	VII	Hyperion	3×10^{-8}	205 × 130 × 110		0.3	+4.63	14.19	0.78	0.33
	VIII	Iapetus	3.3×10^{-6}	730	S	0.2[b]	+1.5	11.1	0.72	0.30
	IX	Phoebe	7×10^{-10}	110	0.4	0.06	+6.89	16.45	0.70	0.34
	X	Janus		110 × 100 × 80	S	0.8	+4.4:	14:		
	XI	Epimetheus		70 × 60 × 50	S	0.8	+5.4:	15:		
	XII	1980S6		18 × 16 × 15		0.7	+8.4:	18:		
	XIII	Telesto		17 × 14 × 13		0.5	+8.9:	18.5:		
	XIV	Calypso		17 × 11 × 11		0.6	+9.1:	18.7:		
	XV	Atlas		20 × 10		0.9	+8.4:	18:		
		1980S26		55 × 45 × 35		0.9	+6.4:	16:		
		1980S27		70 × 50 × 40		0.6	+6.4:	16:		

Planet	Satellite		Mass (ℓ/Planet)	Radius (km)	Sidereal period of rotation S = Synchr.[a] (Days)	Geometric albedo (V)[c]	$V(1,0)$	V_0	$(B - V)$	$(U - B)$
Uranus	I	Ariel	1.8×10^{-5}	665		0.2	+1.7	14.4		
	II	Umbriel	1.2×10^{-5}	555		0.1	+2.6	15.3		
	III	Titania	6.8×10^{-5}	800		0.21	+1.27	13.98	0.70	0.28
	IV	Oberon	6.9×10^{-5}	815	S	0.16	+1.52	14.23	0.68	0.20
	V	Miranda	0.2×10^{-5}	160			+3.8	16.5		
Neptune	I	Triton	1.3×10^{-3}	1900	S		−1.02	13.69	0.72	0.29
	II	Nereid	2×10^{-7}	150			+4.0	18.7		
Pluto		(Charon)	0.125(?)	1000(?)			+0.9	16.8		

[a] Rotation period same as orbital period.
[b] Bright side, 0.5; faint side, 0.05.
[c] V (Sun) = −26.8.

CONSTANTS FOR SATELLITE GEODESY

Defining Constants

1. Number of ephemeris seconds in 1 tropical year (1900) .... s = 31 556 925.9747
2. Gaussian gravitational constant, defining the a. u. .... k = 0.017 202 09895

Primary Constants

3. Velocity of light in meters per second (in vacuum) .... c = $2.99792458(1.2) \times 10^8$
4. Dynamical form-factor for Earth .... J_2 = 0.001 082 7
5. Sidereal mean motion of Moon in radians per second (1900).................. $n_☾{}^*$ = $2.661\,699\,489 \times 10^{-6}$
6. General precession in longitude per tropical century (1900).................... p = 5025″.64
7. Constant of nutation (1900) .... N = 9″.210

Auxiliary Constants and Factors

$k/86400$, for use when the unit of time is 1 second .... k' = $1.990\,983\,675 \times 10^{-7}$
Number of seconds of arc in 1 radian.. = 206 264.806
Factor for constant of aberration (note 10) .... F_1 = 1.000 142
Factor for mean distance of Moon (note 12) .... F_2 = 0.999 093 142
Factor for parallactic inequality (note 15) .... F_3 = 49853″.2

Derived Constants

8. Solar parallax.. $\arcsin(\alpha_e/A) = \pi_\odot$ = 8″.79405 (8″.794)
9. Light-time for unit distance .... $A/c' = \tau_A$ = 499ˢ.012
= 1ˢ/0.002 003 96
10. Constant of aberration .... $F_1 k' \tau_A = \kappa$ = 20″.4958 (20″.496)
11. Ratio of masses of Sun and Earth + Moon .... $S/E(1 + \mu)$ = 328 912
12. Perturbed mean distance of Moon, in meters .... $F_2(GE(1 + \mu)/n_☾{}^{*2})^{\frac{1}{3}} = \alpha_☾$ = $384\,400 \times 10^3$
13. Constant of sine parallax for Moon.............................. $\alpha_e/\alpha_☾ = \sin \pi_☾$ = 3422″.451

14. Constant of lunar inequality .... $\dfrac{\mu}{1 + \mu}\dfrac{\alpha_☾}{A} = L$ = 6″.43987 (6″.440)

15. Constant of parallactic inequality .... $F_3 \dfrac{1 - \mu}{1 + \mu}\dfrac{\alpha_☾}{A} = P_☾$ = 124″.986

THE EARTH: ITS MASS, DIMENSIONS AND OTHER RELATED QUANTITIES

From NASA Technical Translation NASA TT-F-533

TABLE A

Quantity	Unit of Measurement	Symbol	Numerical Value	Sources; Remarks[1]
Mass	Proportion of the mass of the sun	M	1/331950	I. D. Zhongolovich
	gram		$5.9763 \cdot 10^{27}$	
Major Orbital semi-axis	Astronomical unit	a_{orb}	1.000000	1961 data of Soviet radar determinations
	km		149,457,000	
Distance from sun at perihelion	a.u.	r_π	0.983298	for 1962
Distance from sun at aphelion	a.u.	r_α	1.016744	for 1962
Moment of perihelion passage		T_π	Jan. 2, $4^{hr}52^{m}$	for 1962 USSR Astron. Yearbook for 1962
Moment of aphelion passage		T_α	Jul 4 $5^{hr}05^{m}$	for 1962
Siderial rotation period around sun	sec	P_{orb}	$31.558 \cdot 10^6$	
Mean rotational velocity	km/sec	U_{orb}	29.8	
Mean equatorial radius	km	a	6,378.245 6,378.077	A. A. Izotov, 1950 I. D. Zhongolovich, 1956
Mean polar compression		α	1/298.3 1/296.6 1/298.2	A. A. Izotov, 1950 I. D. Zhongolovich, 1952 D. G. King-Hele, R. Merson, 1959. On observations on the movements of artificial earth satellites.
Difference in equatorial and polar semi-axes	km	$a - c$	21.382 21.500	A. A. Izotov, 1950 I. D. Zhongolovich, 1956
Compression of meridian of major equatorial axis		α_a	1/295.2	I. D. Zhongolovich, 1952
Compression of meridian of minor equatorial axis		α_b	1/298.0	I. D. Zhongolovich, 1952
Equatorial compression		ϵ	1/30 000 1/32 000	A. A. Izotov, 1950 I. D. Zhongolovich, 1952

[1] The source does not indicate whether the value given is generally accepted or has merely been calculated by the author, G. N. Katterfel'd.

THE EARTH: ITS MASS, DIMENSIONS AND OTHER RELATED QUANTITIES (*Continued*)

TABLE A Continued

Quantity	Unit of Measurement	Symbol	Numerical Value	Sources; Remarks[1]
Difference in equatorial semi-axes	m	$a - b$	213 199	A. A. Izotov, 1950 I. D. Zhongolovich, 1952
Meridian of longitude of minor equatorial semi-axis		λ_a	15°E - 6°W	A. A. Izotov, 1950 I. D. Zhongolovich, 1952
Meridian of longitude of minor equatorial axis		λ_b	105°E - 75° W; 84°E - 96°W	A. A. Izotov, 1950 I. D. Zhongolovich, 1952
Difference in polar semi-axes	m	$C_N - C_S$	~ 70 <100	I. D. Zhongolovich, 1952 Based on observations of artificial satellites[1]
Polar asymmetry		η	$\sim 1.10^{-5}$	
Mean acceleration of gravity at equator		g_e	978,057.3	I. D. Zhongolovich, 1952
Mean acceleration of gravity at poles	milligals (mGal)	g_ρ	983,225.1	
Difference in acceleration of gravity at pole and at equator		$g_p - g_e$	+5,167.8	
Difference in acceleration of gravity at equator	mGal	$g_a - g_b$	+30.2	I. D. Zhongolovich, 1952
Difference in acceleration of gravity at poles		$g_N - g_S$	+30	I. D. Zhongolovich, 1952
Mean acceleration of gravity for entire surface of terrestrial ellipsoid	mGal	g	979,783.0	I. D. Zhongolovich, 1952
Mean radius	km	R	6,370.949	
Area of surface	km^2	S	$510.0501 \cdot 10^6$	
Volume	km^3	V	$1{,}083.1579 \cdot 10^9$	
Mean density	g/cm^3	δ	5.5170	I. D. Zhongolovich, 1952
Siderial rotational period	sec	P	86,164.09	
Angular rotational velocity	rad/sec	ω	$7.292116 \cdot 10^{-5}$	
Mean equatorial rotational velocity	km/sec	v	0.465	

TABLE A Continued

Quantity	Unit of Measurement	Symbol	Numerical Value	Sources; Remarks[1]
Ratio of centrifugal force to attractive force at equator		q	$\frac{1}{289}$	
Ratio of centrifugal force to force of gravity at equator		q_c	$0.0034677 = \frac{1}{288}$	I. D. Zhongolovich, 1952
Coefficients characterizing the radial distribution of densities within the earth		κ_1 κ	0.966 0.331	Based on observations of artificial satellites; several larger values were given earlier[1]
Radius of inertia	km; proportion of mean radius	R_i	3,674.735 0.5768	
Geocentric latitude of inertial parallel		ϕ_i	54°47′	
Moment of inertia	$gr \cdot cm^2$	I	$8.070 \cdot 10^{44}$	
Moment of rotation	$gr \cdot cm^2/sec$	L	$5.885 \cdot 10^{40}$	
Relative true secular braking of earth's rotation due to tidal friction		$\frac{\Delta\omega_e}{\omega}$	$-4.2 \cdot 10^{-8}$ per century	
Relative proper secular acceleration of earth's rotation		$\frac{\Delta\omega_i}{\omega}$	$+1.4 \cdot 10^{-8}$ per century	N. N. Pariyskiy, 1955
Relative observed secular braking of earth's rotation		$\frac{\Delta\omega}{\omega}$	$-2.8 \cdot 10^{-8}$ per century	
Mean rotational velocity of terrestrial radius due to abyssal compression	cm/century	$\frac{\Delta R}{\Delta t}$	~5 Assumed invariability of mass (M) 4.5 and distribution of masses (κ)	B. Meyermann, 1928, 1928a N. N. Pariyskiy, 1955
Secular variation in potential gravitational energy of earth accompanying reduction of terrestrial radius by 5 cm and corresponding increase in earth's kinetic energy	erg/century	ΔE	$\sim 17 \cdot 10^{30}$	Assumed uniformity of compression of entire planet. P. N. Kropotkin, 1948 and A. T. Aslanyan, 1955

[1] On the basis of materials published in the Astron. J., Vol. 64, 1272, 1959.

TABLE A Continued

Quantity	Unit of Measurement	Symbol	Numerical Value	Sources; Remarks[1]
Probable value of total energy of tectonic deformation of earth	erg/century	E_t	$\sim 1 \cdot 10^{30}$	With allowance for earthquakes, volcanic eruptions and other forms of tectonic activity P. N. Kropotkin, 1948
Secular loss of heat of earth through radiation into space	erg/century cal/century	$\Delta' E_k$	$1 \cdot 10^{30}$ $2.4 \cdot 10^{22}$	P. N. Kropotkin, 1948
Portion of earth's kinetic energy transformed into heat as a result of lunar and solar tides in the hydrosphere	erg/century cal/century	$\Delta'' E_k$	$0.11 \cdot 10^{30}$ $0.26 \cdot 10^{22}$	Heiskanen, 1922 and de Sitter, 1927
Difference in duration of days in March and August	sec	ΔP	0.0025 (March-Aug.)	N. N. Pariyskiy, 1955
Corresponding relative annual variation in earth's rotational velocity		$\frac{\Delta^* \omega}{\omega}$	$2.9 \cdot 10^{-8}$ (Aug.-March)	N. N. Pariyskiy, 1955
Presumed variation in earth's radius between August and March	cm	$\Delta^* R$	−9.2 (Aug.-March)	
Annual variation in level of world ocean	cm	Δh_0	~10 (Sept.-March)	N. N. Pariyskiy, 1955

The Earth's Lithosphere, Hydrosphere, Atmosphere and Biosphere

TABLE B

Quantity	Unit of Measurement	Symbol	Numerical Value	Source
Area of continents	km^2; in % of area of surface of earth	S_C	$149 \cdot 10^6$ 29.2	E. Kossina, 1933
Area of world ocean	km^2; in % of area of surface of earth	S_o	$361 \cdot 10^6$ 70.8	E. Kossina, 1933
Mean height of continents above sea level	m	h_C	875	E. Kossina, 1933
Mean depth of world ocean	m	h_o	=3794	Morskoy Atlas, Vol. II, 1953
Mean position of earth's surface with respect to sea level	m	h_m	=2430	E. Kossina, 1933

TABLE B Continued

Quantity	Unit of Measurement	Symbol	Numerical Value	Source
Mean thickness of lithosphere within the limits of the continents	km	$h_{c.l.}$	35	M. Yuing and F. Press, 1955
Mean thickness of lithosphere within the limits of the ocean	km	$h_{o.l.}$	4.7	Kh. Khess, 1955
Mean rate of thickening of continental lithosphere	$m/10^6$ yr	$\frac{\Delta h}{\Delta t}$	10 – 40	V. I. Popov, 1955
Mean rate of horizontal extension of continental lithosphere	$km/10^6$ yr	$\frac{\Delta l}{\Delta t}$	0.75 – 20	V. I. Popov, 1955
Mass of lithosphere	gr	m_l	$2.367 \cdot 10^{25}$	A. Poldervart, 1955
Amount of water released from the mantle and core in the course of geological time	gr		$3.400 \cdot 10^{24}$	Kalp, 1951
Total reserve of water in the mantle	gr		$2 \cdot 10^{26}$	A. P. Vinogradov, 1959
Present day content of free and bound water in the earth's lithosphere	gr		$2.2 - 2.6 \cdot 10^{24}$ $1.8 - 2.7 \cdot 10^{24}$	Kalp, 1951 A. Poldervart, 1955
Mass of hydrosphere	gr	m_h	$1.664 \cdot 10^{24}$	A. Poldervart, 1955
Amount of oxygen bound in the earth's crust	gr		$1.300 \cdot 10^{24}$	A. Poldervart, 1955
Amount of free oxygen	gr		$1.5 \cdot 10^{21}$	A. Poldervart, 1955
Mass of atmosphere	gr	m_a	$5.136 \cdot 10^{21}$	A. Poldervart, 1955
Mass of biosphere	gr	m_b	$1.148 \cdot 10^{19}$	A. Poldervart, 1955
Mass of living matter in the biosphere	gr		$3.6 \cdot 10^{17}$	A. Poldervart, 1955
Density of living matter on dry land	gr/cm^2		0.1	A. Poldervart, 1955
Density of living matter in ocean	gr/cm^3		$15 \cdot 10^{-8}$	A. Poldervart, 1955

ESTIMATED AGE OF EARTH

TABLE C

Object of Study	Age in 10^9 years	Source
Fossils of the most ancient organisms	2.7	L. Arens, 1955
Most ancient known terrestrial rocks:		
Mica (biotite) found within migmatites on the Kola Peninsula and at the Great Rapids of the Voron'ya River in 1958;	3.6	A. A. Polkanov and E. K. Gerling, 1961
Rock found in South Africa	4	A. L. Heils, 1960[1]
Lithosphere	~ 4	V. I. Baranov, 1958
Earth	~ 5	A. P. Vinogradov, 1959 and others

[1] See *Priroda*, No. 11, 1960, p. 113.

Approximate Scale of Geologic Time,[1] Based on the Data of Soviet Research, 1960.

TABLE D

Era	Period or Epoch	Beginning and end, in 10^6 years	Approximate Duration, in 10^6 years	
Cenozoic	Quaternian			
	Contemporary	0 – 10,000 yrs ± 2,000 yrs	8 – 12,000 years	
	Pleistocene	10,000 – 1,000,000 yrs ± 50,000 yrs	1	
	Tertiary			
	Pliocene	1 – 10	9	
	Miocene	10 – 25	15	
	Oligocene	25 – 40	15 } 69	
	Eocene	40 – 60	20	
	Paleocene	60 – 70	10	570
Mesozoic	Cretaceous	70 – 140	70	
	Jurassic	140 – 185	45	
	Triassic	185 – 225	40	
Paleozoic	Permian	225 – 270	45	
	Carboniferous	270 – 320	50	
	Devonian	320 – 400	80	
	Silurian	400 – 420	20	
	Ordovician	420 – 480	60	
	Cambrian	480 – 570	90	
	Pre-Cambrian IV (Riphean)[2]	570 – 1,200	630	
	Pre-Cambrian III (Proterozoic)[3]	1,200 – 1,900	700	
	Pre-Cambrian II (Archean)	1,900 – 2,700	800	
	Pre-Cambrian I (Catarchean)	2,700 – 3,500	800	
	Pregeological era	3,500 – 5,000	1,500	

[1] Based on the geochronological scale of the Commission for Determining the Absolute Age of Geological Formations, published in *Izv. AN SSSR*, Geological series, No. 10, 1960.

[2] Proterozoic II.

[3] Proterozoic I.

ACCELERATION DUE TO GRAVITY AND LENGTH OF THE SECONDS PENDULUM

For Sea Level at Various Latitudes

Based on the formula of the U. S. Coast and Geodetic Survey. The length of the simple pendulum whose period is two seconds, that is which beats seconds, is computed in each case from the corresponding value of the acceleration.

Latitude	Acceleration due to gravity		Length of seconds pendulum	
•	cm/sec.^{2}	ft./sec.^{2}	cm	in.
0	978.039	32.0878	99.0961	39.0141
5	978.078	32.0891	99.1000	39.0157
10	978.195	32.0929	99.1119	39.0204
15	978.384	32.0991	99.1310	39.0279
20	978.641	32.1076	99.1571	39.0382
25	978.960	32.1180	99.1894	39.0509
30	979.329	32.1302	99.2268	39.0656
31	979.407	32.1327		
32	979.487	32.1353		
33	979.569	32.1380		
34	979.652	32.1407		
35	979.737	32.1435	99.2681	39.0819
36	979.822	32.1463		
37	979.908	32.1491		
38	979.995	32.1520		
39	980.083	32.1549		
40	980.171	32.1578	99.3121	39.0992
41	980.261	32.1607		
42	980.350	32.1636		
43	980.440	32.1666		
44	980.531	32.1696		
45	980.621	32.1725	99.3577	39.1171
46	980.711	32.1755		
47	980.802	32.1785		
48	980.892	32.1814		
49	980.981	32.1844		
50	981.071	32.1873	99.4033	39.1351
51	981.159	32.1902		
52	981.247	32.1931		
53	981.336	32.1960		
54	981.422	32.1988		
55	981.507	32.2016	99.4475	39.1525
56	981.592	32.2044		
57	981.675	32.2071		
58	981.757	32.2098		
59	981.839	32.2125		
60	981.918	32.2151	99.4891	39.1689
65	982.288	32.2272	99.5266	39.1836
70	982.608	32.2377	99.5590	39.1964
75	982.868	32.2463	99.5854	39.2068
80	983.059	32.2525	99.6047	39.2144
85	983.178	32.2564	99.6168	39.2191
90	983.217	32.2577	99.6207	39.2207

Free Air Correction for Altitude

−0.0003086 cm/sec.^{2}/m for altitude in meters.
−0.000003086 ft./sec.^{2}/ft. for altitude in feet.

Altitude meters	Correction cm/sec.^{2}	Altitude feet	Correction ft./sec.^{2}
200	−0.0617	200	−0.000617
300	0.0926	300	0.000926
400	0.1234	400	0.001234
500	0.1543	500	0.001543
600	0.1852	600	0.001852
700	0.2160	700	0.002160
800	0.2469	800	0.002469
900	0.2777	900	0.002777

DATA IN REGARD TO THE EARTH

Quadrant of the equator, 10,019,150 meters, 6,225.60 miles.
Quadrant of the meridian, 10,002,290 meters, 6,215.12 miles.
1° latitude at the equator = 69.41 miles.
1° latitude at the pole = 68.70 miles.
Mean surface density of the continents, 2.67 g/cm^3, 166.7 lb./ft.^{3}
Land area, 148.847 × 10^6 km^2, 57.470 × 10^6 sq. mi.
Ocean area, 361.254 × 10^6 km^2, 139.480 × 10^6 sq. mi.
Highest mountain, Everest, 8,840 meters, 29,003 ft.
Greatest sea depth, 10,430 meters, 34,219 ft.
Thermal gradient of the earth, higher at increasing depths, 30°C per km, 48°C per mi. (uncertain).

ELEMENTS IN THE EARTH'S CRUST

A. Demayo

The elements in this table are listed in decreasing average concentration as they occur in the earth's crust. They are average concentrations and there will be variations in composition from point to point throughout the earth's crust.

Element	Concentration (mg/kg)	Element	Concentration (mg/kg)
Oxygen	4.64×10^5	Ytterbium	3.0×10^0
Silicon	2.82×10^5	Beryllium	2.8×10^0
Aluminum	8.32×10^4	Erbium	2.8×10^0
Iron	5.63×10^4	Uranium	2.7×10^0
Calcium	4.15×10^4	Bromine	2.5×10^0
Sodium	2.36×10^4	Tantalum	2.0×10^0
Magnesium	2.33×10^4	Tin	2.0×10^0
Potassium	2.09×10^4	Arsenic	1.8×10^0
Titanium	5.70×10^3	Molybdenum	1.5×10^0
Hydrogen	1.40×10^3	Tungsten (Wolfram)	1.5×10^0
Phosphorus	1.05×10^3	Europium	1.2×10^0
Manganese	9.50×10^2	Holmium	1.2×10^0
Fluorine	6.25×10^2	Cesium	1.0×10^0
Barium	4.25×10^2	Terbium	9×10^{-1}
Strontium	3.75×10^2	Iodine	5×10^{-1}
Sulfur	2.60×10^2	Lutetium	5×10^{-1}
Carbon	2.00×10^2	Thulium	4.8×10^{-1}
Zirconium	1.65×10^2	Thallium	4.5×10^{-1}
Vanadium	1.35×10^2	Antimony	2×10^{-1}
Chlorine	1.30×10^2	Cadmium	2×10^{-1}
Chromium	1.00×10^2	Bismuth	1.7×10^{-1}
Rubidium	9.0×10^1	Indium	1×10^{-1}
Nickel	7.5×10^1	Mercury	8×10^{-2}
Zinc	7.0×10^1	Silver	7×10^{-2}
Cerium	6.0×10^1	Selenium	5×10^{-2}
Copper	5.5×10^1	Palladium	1×10^{-2}
Yttrium	3.3×10^1	Helium	8×10^{-3}
Lanthanum	3.0×10^1	Neon	5×10^{-3}
Neodimium	2.8×10^1	Platinum	5×10^{-3}
Cobalt	2.5×10^1	Rhenium	5×10^{-3}
Scandium	2.2×10^1	Gold	4×10^{-3}
Lithium	2.0×10^1	Osmium	1.5×10^{-3}
Niobium (Columbium)	2.0×10^1	Iridium	1×10^{-3}
Nitrogen	2.0×10^1	Rhodium	1×10^{-3}
Gallium	1.5×10^1	Ruthenium	1×10^{-3}
Lead	1.25×10^1	Tellurium	1×10^{-3}
Boron	1.0×10^1	Krypton	1×10^{-4}
Thorium	9.6×10^0	Xenon	3×10^{-5}
Praeseodymium	8.2×10^0	Protactinium	1.4×10^{-6}
Samarium	6.0×10^0	Radium	9×10^{-7}
Gadolinium	5.4×10^0	Actinium	5.5×10^{-10}
Germanium	5.4×10^0	Polonium	2×10^{-10}
Argon	3.5×10^0	Radon	4×10^{-13}
Dysprosium	3.0×10^0		

CHEMICAL COMPOSITION OF ROCKS

Element	Average igneous rock	Average shale	Average sandstone	Average limestone	Average sediment
SiO_2	59.14	58.10	78.33	5.19	57.95
TiO_2	1.05	0.65	0.25	0.06	0.57
Al_2O_3	15.34	15.40	4.77	0.81	13.39
Fe_2O_3	3.08	4.02	1.07	0.54	3.47
FeO	3.80	2.45	0.30		2.08
MgO	3.49	2.44	1.16	7.89	2.65
CaO	5.08	3.11	5.50	42.57	5.89
Na_2O	3.84	1.30	0.45	0.05	1.13
K_2O	3.13	3.24	1.31	0.33	2.86
H_2O	1.15	5.00	1.63	0.77	3.23
P_2O_5	0.30	0.17	0.08	0.04	0.13
CO_2	0.10	2.63	5.03	41.54	5.38
SO_3		0.64	0.07	0.05	0.54
BaO	0.06	0.05	0.05		
C		0.80			0.66
	99.56	100.00	100.00	99.84	99.93

ELEMENTS IN SEA WATER

A. Demayo

The elements in sea water are listed in decreasing average concentration. These are average concentrations and there will be variations in composition as a function of the location where the sample was collected.

Element	Concentration (mg/ℓ)	Element	Concentration (mg/ℓ)	Element	Concentration (mg/ℓ)
Oxygen	8.57×10^5	Iron	1×10^{-2}	Selenium	9×10^{-5}
Hydrogen	1.08×10^5	Indium	$<2 \times 10^{-2}$	Germanium	7×10^{-5}
Chlorine	1.90×10^4	Molybdenum	1×10^{-2}	Xeon	5.2×10^{-5}
Sodium	1.05×10^4	Zinc	1×10^{-2}	Chromium	5×10^{-5}
Magnesium	1.35×10^3	Nickel	5.4×10^{-3}	Thorium	5×10^{-5}
Sulfur	8.85×10^2	Arsenic	3×10^{-3}	Gallium	3×10^{-5}
Calcium	4.00×10^2	Copper	3×10^{-3}	Mercury	3×10^{-5}
Potassium	3.80×10^2	Tin	3×10^{-3}	Lead	3×10^{-5}
Bromine	6.5×10^1	Uranium	3×10^{-3}	Zirconium	2.2×10^{-5}
Carbon	2.8×10^1	Krypton	2.5×10^{-3}	Bismuth	1.7×10^{-5}
Strontium	8.1×10^0	Manganese	2×10^{-3}	Lanthanum	1.2×10^{-5}
Boron	4.6×10^0	Vanadium	2×10^{-3}	Gold	1.1×10^{-5}
Silicon	3×10^0	Titanium	1×10^{-3}	Niobium	1×10^{-5}
Fluorine	1.3×10^0	Cesium	5×10^{-4}	Thallium	$<1 \times 10^{-5}$
Argon	6×10^{-1}	Cerium	4×10^{-4}	Hafnium	$<8 \times 10^{-6}$
Nitrogen	5×10^{-1}	Antimony	3.3×10^{-4}	Helium	6.9×10^{-6}
Lithium	1.8×10^{-1}	Silver	3×10^{-4}	Selenium	$<4 \times 10^{-6}$
Rubidium	1.2×10^{-1}	Yttrium	3×10^{-4}	Tantalum	$<2.5 \times 10^{-6}$
Phosphorus	7×10^{-2}	Cobalt	2.7×10^{-4}	Beryllium	6×10^{-7}
Iodine	6×10^{-2}	Neon	1.4×10^{-4}	Protoactinium	2×10^{-9}
Barium	3×10^{-2}	Cadmium	1.1×10^{-4}	Radium	6×10^{-11}
Aluminum	1×10^{-2}	Tungsten	1×10^{-4}	Radon	6×10^{-16}

THE pH OF NATURAL MEDIA AND ITS RELATION TO THE PRECIPITATION OF HYDROXIDES

Reprinted from "Principles of Geochemistry" (1952) with the permission of Brian Mason, author, and John Wiley and Sons, publishers.

pH	Precipitation of hydroxides	Natural media	pH
11 } 10 }	Magnesium		11
		Alkali soils	10
9 } 8 }	Bivalent manganese		{ 9
		Seawater	{ 8
7	Bivalent iron	River water	7
6	Zinc copper	Rain water	6
5	Aluminum		5
4		Peat water	4
3 }	Trivalent iron	Mine waters	3
2 }		Acid thermal springs	{ 2
1			{ 1

PROPERTIES OF THE EARTH'S ATMOSPHERE AT ELEVATIONS UP TO 160 KILOMETERS

The average atmosphere up to 160 km based on pressure and density data obtained on rocket flights above White Sands, New Mexico.
Havens, Koll, and LaGow, Journal of Geophysical Research, March, 1952.

Altitude km above sea level	Pressure mm Hg	Density gm/meters³	Temperatures °K (N_2, O_2) M = 29	Temperatures °K (N_2, O) M = 24	Velocity of Sound m. sec	Mean Free Path cm (N_2)
0	760	1220	290		345	6.5×10^{-6}
10	210	425	230		310	1.9×10^{-5}
20	42	92	210		295	8.6×10^{-5}
30	9.5	19	235		315	4.2×10^{-4}
40	2.4	4.3	260		325	1.8×10^{-3}
50	7.5×10^{-1}	1.3	270		330	6.1×10^{-3}
60	2.1×10^{-1}	3.8×10^{-1}	260		325	2.1×10^{-2}
70	5.4×10^{-2}	1.2×10^{-1}	210		295	6.6×10^{-1}
80	1.0×10^{-2}	2.5×10^{-2}	190		280	3.2×10^{-1}
90	1.9×10^{-3}	4.0×10^{-3}	210		295	2.0
100	4.2×10^{-4}	8.0×10^{-4}	240		315	10.0
110	1.2×10^{-4}	2.0×10^{-4}	270	220	330	40.0
120	3.5×10^{-5}	5.0×10^{-5}	330	270	370	1.5×10^2
130	1.5×10^{-5}	2.0×10^{-5}	390	320	400	4.0×10^2
140	7×10^{-6}	7.0×10^{-6}	450	370	430	1.0×10^3
150	3×10^{-6}	3.0×10^{-6}	510	420	460	2.5×10^3
160	2×10^{-6}	1.5×10^{-6}	570	470	480	5.0×10^3

VELOCITY OF SEISMIC WAVES

Depth km	Longitudinal or condensational km/sec.	Transverse or distortional km/sec.
0–20	5.4 –5.6	3.2
20–45	6.25–6.75	3.5
1300	12.5	6.9
2400	13.5	7.5

ATMOSPHERIC AND METEOROLOGICAL DATA

Total mass of the atmosphere, estimated by Ekholm, 5.2×10^{21} g, 11.4×10^{18} pounds, 5.70×10^{15} tons.
Evidence of extent: twilight, 63 km, 39 mi.: meteors, 200 km, 124 mi.: aurora 44–360 km, 27–224 mi.

*Distance to Earth.

COSMIC RADIATION

A. Gregory and R. W. Clay

THE NATURE OF COSMIC RAYS

Primary cosmic radiation, in the form of high-energy nuclear particles, electrons, and photons, from outside the solar system and from the Sun, continually bombards our atmosphere.

Secondary radiation, resulting from the interaction of primary cosmic rays with atmospheric gas, is present at sea-level and throughout the atmosphere. The secondary radiation is collimated by absorption and scattering in the atmosphere and consists of a number of components associated with different particle species. High-energy primary particles can produce a large number of secondary particles forming an extensive air shower. Thus, a number of particles may be detected in coincidence at sea level. Primary particle energies range from $\sim 10^8$ eV to $\sim 10^{20}$ eV.[1,15]

PRIMARY PARTICLE ENERGY SPECTRUM

Figure 1 shows the spectrum of primary particle energies. In differential form it is roughly a power law (with an index of -3). There appears to be a knee (a steepening) at about 10^{15} eV and an ankle (a flattening) above $\sim 10^{18}$ eV.

At energies below $\sim 10^{13}$ eV, solar system magnetic fields and plasma can modulate the primary component and Figure 2 shows the extent of this modulation between solar maximum and minimum.[10,12]

PRIMARY PARTICLE ENERGY DENSITY

If the above spectrum is corrected for solar effects the energy density above a particle energy of 10^9 eV outside the solar system is found to be $\sim 5 \times 10^5$ eV m^{-3}. As the threshold energy is increased, the energy density decreases rapidly, being 2×10^4 eV m^{-3} above 10^{12} eV and 10^2 eV m^{-3} above 10^{15} eV.[16]

PRIMARY PARTICLE ISOTROPY

This is measured as an anisotropy $(I_{max} - I_{min})/(I_{max} + I_{min}) \times 100\%$, where I, the intensity (m^{-2}s^{-1}sr^{-1}), is usually measured with an angular resolution of a few degrees.

The anisotropy is small and energy dependent. It is roughly constant at between 0.05 and 0.1% for energies between 10^{11} eV and 10^{14} eV and increases at higher energies roughly as $0.4 \times (\text{energy (eV)}/10^{16})^{1/2}\%$ up to $\sim 10^{18}$ eV.[16]

PRIMARY PARTICLE COMPOSITION

The composition of low energy cosmic rays (it is uncertain at high energies) is close to universal abundances except where propagation effects are present, e.g. Li, Be, B which are spallation products, are over-abundant by ~ 6 orders of magnitude.

Composition at 10^{11} eV/Nucleus

Charge	1	2	(3—5)	(6—8)	(10—14)	(16—24)	(26—28)	≥30
% Composition	50	25	1	12	7	4	4	0.1
(~10% uncertainty)								

Cosmic ray composition at low energies is often quoted at a fixed *energy per nucleon*. When presented in this way, protons constitute roughly 90% of the flux, helium nuclei ~10% and the remainder sum to a total of ~1%.

Certain radioactive isotopic ratios show lifetime effects. The ratio of Be^{10}/Be^9 abundances is used to measure an 'age' of cosmic rays since Be^{10} is unstable (half life = 1.6×10^6 years). A ratio of 0.6 would be expected in the absence of Be^{10} decay and a ratio of ~ 0.2 is found experimentally.[10,13]

PRIMARY ELECTRONS

Primary electrons constitute ~1% of the cosmic ray beam. The positron to electron ratio is about 10%.[14]

ANTIMATTER IN THE PRIMARY BEAM

The ratio of antiprotons to protons in the primary cosmic ray beam (at ~100 GeV) is $\sim 5 \times 10^{-4}$.[11]

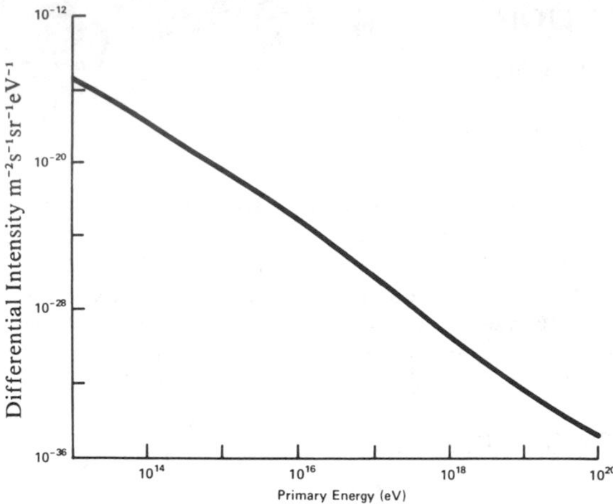

FIGURE 1. Energy Spectrum of Cosmic Ray Particles. This spectrum is of a differential form and can be converted to an integral spectrum by integrating over all energies above a required threshold (E).

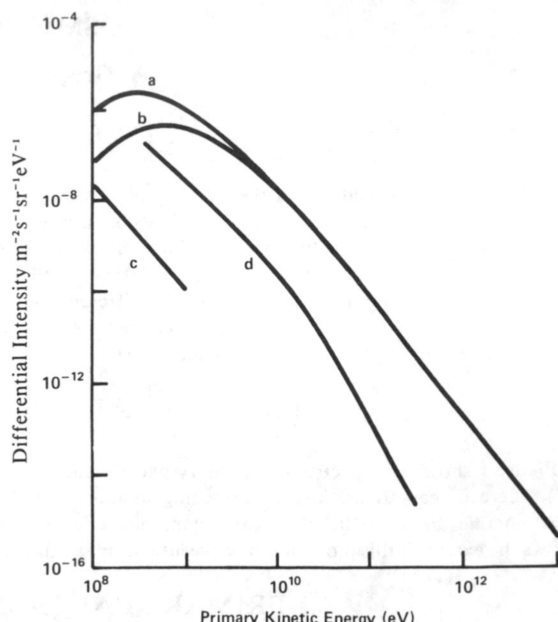

FIGURE 2. Energy Spectrum of Cosmic Ray Particles at Lower Energies. (a) Solar minimum proton energy spectrum, (b) solar maximum proton energy spectrum, (c) γ-ray energy spectrum, (d) local interstellar electron spectrum.

PRIMARY GAMMA RAYS

The flux of primary gamma rays is low at high energies. At 1 GeV the ratio of gamma rays to protons is about 10^{-6}. The arrival directions of these gamma rays are strongly concentrated in the plane of the Milky Way although there is also a diffuse near-isotropic background flux.

Since the absorption cross section for gamma rays above 100 MeV is approximately 20 mbarn/electron, less than 10% of gamma rays reach mountain altitudes.[16] Gamma rays from point sources have confidently been identified with energies up to 10^{16} eV.

SEA LEVEL COSMIC RADIATION

The sea level cosmic ray dose is 30 millirad. year^{-1} and the sea level ionisation is 2.22×10^6 ion pairs m^{-1}s^{-1}. The sea level flux has a soft component, which can be absorbed in ∼10cm of lead and a more penetrating (largely muon) hard component. The sea level radiation is a secondary component produced in the atmosphere and its flux is dependent somewhat on the solar cycle and the geomagnetic latitude of the observer.

Absolute Flux of Hard Component[2]

Vertical integral intensity	(I) $\sim$ 100 m^{-2}s^{-1}sr^{-1}
Angular dependence	$I(\theta) \sim I(O) \cos^2\theta$
Integrated intensity	$\sim$ 200 m^{-2}s^{-1}

Flux of Soft Component

In free air the soft component comprises about one third of the total cosmic ray flux.

Latitude Effect

The geomagnetic field influences the trajectories of lower energy cosmic rays approaching the Earth. As a result, the background flux is reduced by about 7% at the geomagnetic equator. The effect decreases towards the poles and is negligible at latitudes above about 40°.

Flux of Protons

The proton component is strongly attenuated by the atmosphere with an attenuation length (reduction by a factor of e) of $\sim$ 120 g·cm^{-2}, it constitutes about 1% of the total vertical sea level flux.

Absorption

The soft component is absorbed in about 100 g·cm^{-2} of lead. The hard component is absorbed much more slowly:

Absorption in lead	6%/100 g·cm^{-2}
Absorption in rock	8.5%/100 g·cm^{-2}
Absorption in water	10%/100 g·cm^{-2}

Absorption for depths less than 100 g·cm^{-2} is given by Greisen in Reference 7.

Altitude Dependence

Cosmic ray background in the atmosphere has a maximum intensity of about 15 times that at sea level at a depth of ~ 150 g·cm^{-2} (15 km altitude). At maximum intensity the soft and hard component contribute roughly equally but the hard component is then attenuated more slowly.[8]

COSMIC RAY SHOWERS

High energy cosmic rays (< 10^{13}eV) produce measurable cascades of secondary particles in the atmosphere. The primary particle progressively loses energy which is transferred through the production of successive generations of secondary particles to a cascade of hadrons, an electromagnetic shower component (electrons and gamma rays) and muons. The secondary particles are relativistic and all travel effectively at the speed of light. As a result, they reach sea level at approximately the same time, but, due to Coulomb scattering and production angles, are spread laterally into a disk-like shower front with a typical width of ~ 100 m and thickness ~ 2 to 3 m. The number of particles at sea level is roughly proportional to the primary energy.

Number of particles at sea level ~ 10^{-10} × energy (eV). At altitudes below a few kilometres, the number of particles in a shower attenuates with an *attenuation length* ~ 200 g·cm^{-2}, i.e., $N = N_o \times \exp(-\text{depth}/200)$. The rate of observation of showers of a given size at different depths of absorber attenuates with an *absorption length* of ~ 100 g·cm^{-2}.[15]

ATMOSPHERIC BACKGROUND LIGHT FROM COSMIC RAYS

Cosmic ray particles produce Cerenkov light in the atmosphere and produce fluorescent light through the excitation of atmospheric molecules.

Cerenkov Light

High energy charged particles will emit Cerenkov light in air if their energies are above ~30 MeV (electrons). This threshold is pressure (and hence altitude) dependent. A typical Cerenkov light pulse (at sea level, 100 m from core, 10^{16} eV primary energy, in the wavelength band 430 to 530 nm) has a width of a few nanoseconds and in this time has a flux of ~10^{14} photons m^{-2}s^{-1}. For comparison, the general night sky background flux is ~6 × 10^{11} photons m^{-2}sec^{-1}sr^{-1} in the same wavelength band.

Fluorescent Light

Cosmic ray particles in the atmosphere excite atmospheric molecules which emit fluorescent light. This is weak compared to the Cerenkov component when viewed in the direction of the incident cosmic ray particle but is emitted isotropically. Typical pulse widths are expected to be longer than 50 ns and may be up to a few microseconds for distant large showers.[5,9]

CERENKOV EFFECTS IN TRANSPARENT MEDIA

Background cosmic ray particles will produce Cerenkov light in transparent material with a photon yield

$$\sim \frac{2\pi}{137} \sin^2 \theta_c \int_{\lambda_1}^{\lambda_2} \frac{d\lambda}{\lambda^2} \text{ photons (unit length)}^{-1}$$

where θ_c, the Cerenkov angle, = $\cos^{-1}$ (1/refractive index). This background light is known to affect sensitive light detectors, e.g. photomultipliers, and can be a major source of background noise.[6]

EFFECTS ON ELECTRONIC COMPONENTS

If background cosmic rays pass through electronic components, they may deposit sufficient energy to affect the state of, e.g. a transistor flip-flop. This effect may be significant where reliability is of great importance and the background flux is high. For instance, it has been estimated that in communication satellite operation an error rate of ~ 2 × 10^{-3} per transistor per year may be found. Permanent damage may also result. A significant error rate may be found even at sea level in large electronic memories. This error rate is dependent on the sensitivity of the component devices to the deposition of electrons in their sensitive volumes.[17]

BIOPHYSICAL SIGNIFICANCE

When cosmic rays interact with living tissue they produce radiation damage, the amount of damage depending on the total dose of radiation. Radiation doses are commonly measured in rads (radiation absorbed dose ≡ 100 ergs g^{-1}) or rems (radiation equivalent-man ≡ Quality factor × rad). The quality factor of radiation depends on the type of particle and its energy for most cosmic ray applications it will be ~1.

At sea level the cosmic radiation dose rate is small compared with doses from other sources, but both the quantity, and quality of the radiation change rapidly with altitude. Approximate dose rates under various conditions

	Cosmic rays			
Conditions	**Sea level**	**10 km (subsonic jets)**	**18 km (supersonic transport)**	**Mean total dose rate at sea level**
Dose (mrem $year^{-1}$)	30	2000	10,000	300

Astronauts would be subject to radiation from galactic (0.05 rads d^{-1}) and solar (~ few hundred rads per solar flare) cosmic rays as well as large fluxes of low energy radiation when passing through the Van Allen belts (~0.3 rads per traverse).

Both astronauts and S.S.T. travellers would be subject to a small flux of low energy heavy nuclei stopping in the body. Such particles are capable of destroying cell nuclei and could be particularly harmful in the early stages of development of an embryo. The rates of heavy nuclei stopping in tissue in supersonic transports and spacecraft are approximately as follows (from Reference 1 and 2)

Conditions (altitude)	**16 km SST**	**20 km SST**	**Spacecraft**
Stopping nuclei (cm^3 tissue)$^{-1}$ hr^{-1}	5.10^{-4}	5.10^{-3}	0.15

CARBON DATING

Radiocarbon is produced in the atmosphere due to the action of cosmic ray slow neutrons. Solar cycle modulation of the low energy cosmic rays causes an anticorrelation of the atmospheric ^{14}C activity with sunspot number with a mean amplitude of ~0.5%. In the long term, modulation of cosmic rays by a varying geomagnetic field may be important.[4]

PRACTICAL USES OF COSMIC RAYS

There are few direct practical uses of cosmic rays. Their attenuation in water and snow have however enabled automatic monitors of water and snow depth to be constructed, and a search for hidden cavities in pyramids has been carried out using a muon 'telescope'.

OTHER EFFECTS

Stellar X-rays have been observed to affect the transmission times of radio signals between distant stations by altering the depth of the ionospheric reflecting layer. It has also been suggested that variations in ionization of the atmosphere due to solar modulation may have observable effects on climatic conditions.

REFERENCES

1. **Allkofer, O. C.,** *Introduction to Cosmic Radiation,* Verlag Karl Thiemig, Munchen, Germany, 1975.
2. **Allkofer, O. C.,** *J. Phys.,* Sect. G, 1, L51, 1975.
3. **Allkofer, O. C. and Heinrich, W.,** *Health Phys.,* 27, 543, 1974.
4. **Burchuladze, A. A., Pagava, S. V., Povinec, P., Togonidze, G. I., and Usacev, S.,** Proc. 16th Int. Cosmic Ray Conf. Kyoto, 3, 201, Univ. of Tokyo, 3, 201, 1979.
5. **Cassiday, G. L., et al.,** Proc. 15th Int Cosmic Ray Conf., Plovdiv, Bulgarian Academy of Sciences, 8, 258, 1977.
6. **Clay, R. W. and Gregory, A. G.,** *J. Phys.,* Sect. A, 10, 135, 1977.
7. **Greisen, K.,** *Physical Rev.,* 63, 323, 1943.
8. **Hayakawa, S.,** *Cosmic Ray Physics,* Wiley-Interscience, New York, 1969.
9. **Jelley, J. V.,** *Prog. in Elementary Particle and Cosmic Ray Physics,* 9, 41, 1967.
10. **Juliusson, E.,** Proc. 14th Int. Cos Ray Conf. Munich, 8, 2689, Max-Planck Institute for Extraterrestriche Physik, Munchen, Germany, 1975.

11. **Kiraly, P., Szabelski, J., Wdowczyk, J., and Wolfendale, A. W.**, *Nature*, 293, 120, 1981.
12. **Linsley, J.**, *Origin of Cosmic Rays*, I.A.U. Symposium 94, 53, D Reidel Publishing, Dordrecht, Holland, 1981.
13. **Meyer, P.**, *Origin of Cosmic Rays*, I.A.U. Symposium 94, 7, D. Reidel Publishing, Dordrecht, Holland, 1981.
14. **Tan, L. C., and Ng, L. K.**, *J. Phys.*, Sect. G, 7, 1135, 1981.
15. **Wilson, J. G.**, *Cosmic Rays*, Wykeham Publishing, London, 1976.
16. **Wolfendale, A. W.**, *Pramana*, 12, 631, 1979.
17. **Ziegler, J. F.**, *IEEE Trans. Electron Devices*, ED-28, 560, 1981.

CRYSTAL IONIC RADII OF THE ELEMENTS

Numerical values of the radii of the ions may vary depending on how they were measured. They may have been calculated from wavefunctions and determined from the lattice spacings or crystal structure of various salts. Different values are obtained depending on the kind of salt used or the method of calculating. Data for many of the rare-earth ions were furnished by F. H. Spedding and K. Gschneidner.

Element	Charge	Atomic number	Radius in A	Element	Charge	Atomic number	Radius in A	Element	Charge	Atomic number	Radius in A
Ac	+3	89	1.18	Hf	+4	72	0.78	Ra	+2	88	1.43
Ag	+1	47	1.26	Hg	+1	80	1.27	Rb	+1	37	1.47
	+2		0.89		+2		1.10	Re	+4	75	0.72
Al	+3	13	0.51	Ho	+3	67	0.894		+7		0.56
Am	+3	95	1.07	I	−1	53	2.20	Rh	+3	45	0.68
	+4		0.92		+5		0.62	Ru	+4	44	0.67
Ar	+1	18	1.54		+7		0.50	S	−2	16	1.84
As	−3	33	2.22	In	+3	49	0.81		+2		2.19
	+3		0.58	Ir	+4	77	0.68		+4		0.37
	+5		0.46	K	+1	19	1.33		+6		0.30
At	+7	85	0.62	La	+1	57	1.39	Sb	−3	51	2.45
Au	+1	79	1.37		+3		1.061		+3		0.76
	+3		0.85	Li	+1	3	0.68		+5		0.62
B	+1	5	0.35	Lu	+3	71	0.85	Sc	+3	21	0.732
	+3		0.23	Mg	+1	12	0.82	Se	−2	34	1.91
Ba	+1	56	1.53		+2		0.66		−1		2.32
	+2		1.34	Mn	+2	25	0.80		+1		0.66
Be	+1	4	0.44		+3		0.66		+4		0.50
	+2		0.35		+4		0.60		+6		0.42
Bi	+1	83	0.98		+7		0.46	Si	−4	14	2.71
	+3		0.96	Mo	+1	42	0.93		−1		3.84
	+5		0.74		+4		0.70		+1		0.65
Br	−1	35	1.96		+6		0.62		+4		0.42
	+5		0.47	N	−3	7	1.71	Sm	+3	62	0.964
	+7		0.39		+1		0.25	Sn	−4	50	2.94
C	−4	6	2.60		+3		0.16		−1		3.70
	+4		0.16		+5		0.13		+2		0.93
Ca	+1	20	1.18	NH_4	+1		1.43		+4		0.71
	+2		0.99	Na	+1	11	0.97	Sr	+2	38	1.12
Cd	+1	48	1.14	Nb	+1	41	1.00	Ta	+5	73	0.68
	+2		0.97		+4		0.74	Tb	+3	65	0.923
Ce	+1	58	1.27		+5		0.69		+4		0.84
	+3		1.034	Nd	+3	60	0.995	Tc	+7	43	0.979
	+4		0.92	Ne	+1	10	1.12	Te	−2	52	2.11
Cl	−1	17	1.81	Ni	+2	28	0.69		−1		2.50
	+5		0.34	Np	+3	93	1.10		+1		0.82
	+7		0.27		+4		0.95		+4		0.70
Co	+2	27	0.72		+7		0.71		+6		0.56
	+3		0.63	O	−2	8	1.32	Th	+4	90	1.02
Cr	+1	24	0.81		−1		1.76	Ti	+1	22	0.96
	+2		0.89		+1		0.22		+2		0.94
	+3		0.63		+6		0.09		+3		0.76
	+6		0.52	Os	+4	76	0.88	Ti	+4		0.68
Cs	+1	55	1.67		+6		0.69	Tl	+1	81	1.47
Cu	+1	29	0.96	P	−3	15	2.12		+3		0.95
	+2		0.72		+3		0.44	Tm	+3	69	0.87
Dy	+3	66	0.908		+5		0.35	U	+4	92	0.97
Er	+3	68	0.881	Pa	+3	91	1.13		+6		0.80
Eu	+3	63	0.950		+4		0.98	V	+2	23	0.88
	+2		1.09		+5		0.89		+3		0.74
F	−1	9	1.33	Pb	+2	82	1.20		+4		0.63
	+7		0.08		+4		0.84		+5		0.59
Fe	+2	26	0.74	Pd	+2	46	0.80	W	+4	74	0.70
	+3		0.64		+4		0.65		+6		0.62
Fr	+1	87	1.80	Pm	+3	61	0.979	Y	+3	39	0.893
Ga	+1	31	0.81	Po	+6	84	0.67	Yb	+2	70	0.93
	+3		0.62	Pr	+3	59	1.013		+3		0.858
Gd	+3	64	0.938		+4		0.90	Zn	+1	30	0.88
Ge	−4	32	2.72	Pt	+2	78	0.80		+2		0.74
	+2		0.73		+4		0.65	Zr	+1	40	1.09
	+4		0.53	Pu	+3	94	1.08		+4		0.79
H	−1	1	1.54		+4		0.93				

BOND LENGTHS BETWEEN CARBON AND OTHER ELEMENTS

Prepared by Olga Kennard.

The tables are based on bond distance determinations, by experimental methods, mainly X-ray and electron diffraction, and include values published up to January 1, 1956. In the present tables, for the sake of completeness individual values of bond distances of lower accuracy are quoted with limits of error indicated where possible. Values for tungsten and bismuth should be treated with particular caution.

According to the statistical theory of errors if an average quantity $\bar{\mu}$ and a standard deviation σ can be evaluated there is a 95% probability that the true value lies within the interval $\bar{\mu} \pm 2\sigma$. Too much reliance should, however, not be placed on σ values in bond distance determinations since the derivation of these certain sources of error may have been neglected.

Values of the bond lengths and the limits of error are each given in Ångstrom units.

Reproduced by permission from International Tables for X-ray Crystallography.

BOND LENGTHS BETWEEN CARBON AND OTHER ELEMENTS

Reference: HCP and "Tables of interatomic distances" Chem. Soc. of London, 1958

Group	Bond type	Element						
I	All types	H**						
		1.056 − 1.115						
II		Be	Hg					
		1.93	2.07 ± 0.01					
III		B	Al	In				
		1.56 ± 0.01	2.24 ± 0.04	2.16 ± 0.04				
IV	All types	C**	Ge	Si	Sn	Pb		
	Alkyls (CH_3XH_3)	1.54 − 1.20	1.98 ± 0.03	1.865 ± 0.008	2.143 ± 0.008	2.29 ± 0.05		
	Aryl ($C_6H_5XH_3$)			1.84 ± 0.01	2.18 ± 0.02			
	Neg. Subst. (CH_3XCl_3)			1.88 ± 0.01				
V	All types	N**	P	As	Sb	Bi		
	Paraffinic $(CH_3)_3X$	1.47 − 1.1	1.87 ± 0.02	1.98 ± 0.02	2.202 ± 0.016	2.30*		
VI		O**	S**	Cr	Se	Te	Mo	W
		1.43 − 1.15	1.81 − 1.55	1.92 ± 0.04	1.98 − 1.71	2.05 ± 0.14	2.08 ± 0.04	2.06 ± 0.01*
VII	Paraffinic (monosubstituted)	F	Cl	Br	I			
	(CH_3X)	1.831 ± 0.005	1.767 ± 0.002	1.937 ± 0.003	2.13_5 ± 0.01			
	Paraffinic (disubstituted)	1.334 ± 0.004	1.767 ± 0.002	1.937 ± 0.003	2.13_5 ± 0.1			
	(CH_2X_2)							
	Olefinic (CH_2:CHX)	1.32_5 ± 0.1	1.72 ± 0.01	1.89 ± 0.01	2.092 ± 0.005			
	Aromatic (C_6H_5X)	1.30 ± 0.01	1.70 ± 0.01	1.85 ± 0.01	2.05 ± 0.01			
	Acetylenic (HC:CX)		1.635 ± 0.004	1.79_5 ± 0.01	1.99 ± 0.02			
VIII		Fe	Co	Ni	Pd			
		1.84 ± 0.02	1.83 ± 0.02	1.82 ± 0.03	2.27 ± 0.04			

* Error uncertain.
** See following individual tables.

CARBON-CARBON

Single Bond

Paraffinic	1.541 ± 0.003
In diamond (18°C)	1.54452 ± 0.00014

Partial Double Bond

(1) Shortening of single bond in presence of carbon carbon double bond, e.g. $(CH_3)_2C:CH_2$; or of aromatic ring e.g. $C_6H_5.CH_3$	1.53 ± 0.01
(2) Shortening in presence of a carbon oxygen double bond e.g. CH_3CHO	1.516 ± 0.005
(3) Shortening in presence of two carbon-oxygen double bonds, e.g. $(CO_2H)_2$	1.49 ± 0.01
(4) Shortening in presence of one carbon-carbon triple bond, e.g. CH_3.C⋮CH	1.460 ± 0.003
(5) In compounds with tendency to dipole formation, e.g. C:C.C⋮N	1.44 ± 0.01
(6) In graphite (at 15°C)	1.4210 ± 0.0001
(7) In aromatic compounds	1.395 ± 0.003
(8) In presence of two carbon carbon triple bonds, e.g. HC⋮C.C⋮CH	1.373 ± 0.004

Double Bond

(1) Simple	1.337 ± 0.006
(2) Partial triple bond, e.g. CH_2:C:CH_2	1.309 ± 0.005

Triple Bond

(1) Simple, e.g. C_2H_2	1.204 ± 0.002
(2) Conjugated, e.g. $CH_3.(C{:}C)_2.H$	1.206 ± 0.004

CARBON-HYDROGEN

(1) Paraffinic (a) in methane	1.091
(b) in monosubstitured carbon	1.101 ± 0.003
(c) in disubstituted carbon	1.073 ± 0.004
(d) in trisubstituted carbon	1.070 ± 0.007
(2) Olefinic, e.g. $CH_2:CH_2$	1.07 ± 0.01
(3) Aromatic in C_6H_6	1.084 ± 0.006
(4) Acetylenic, e.g. CH⋮C.X	1.056 ± 0.003
(5) Shortening in presence of a carbon triple bond, e.g. CH_3CN	1.115 ± 0.004
(6) In small rings, e.g. $(CH_2)_2S$	1.081 ± 0.007

CARBON-NITROGEN

Single Bond

(1) Paraffinic (a) 4 co-valent nitrogen	1.479 ± 0.005
(b) 3 co-valent nitrogen	1.472 ± 0.005
(2) In C−N= e.g. CH_3NO_2	1.475 ± 0.010
(3) Aromatic in $C_6H_5NHCOCH_3$	1.426 ± 0.012
(4) Shortened (partial double bond) in heterocyclic systems, e.g. C_5H_5N	1.352 ± 0.005
(5) Shortened (partial double bond) in N−C=O e.g. $HCONH_2$	1.322 ± 0.003

Triple Bond

(1) in R.C:N	1.158 ± 0.002

BOND LENGTHS BETWEEN CARBON AND OTHER ELEMENTS (Continued)

CARBON-OXYGEN

Single Bond

(1) Paraffinic	1.43 ± 0.01
(2) Strained e.g. epoxides	1.47 ± 0.01
(3) Shortened (partial double bond) as in carboxylic acids or through influence of aromatic ring, e.g. salicylic acid	1.36 ± 0.01

Double Bond

(1) In aldehydes, ketones, caboxylic acids, esters	1.23 ± 0.01
(2) In zwitterion forms, e.g. DL serine	1.26 ± 0.01
(3) Shortened (partial triple bond) as in conjugated systems	1.207 ± 0.006
(4) Partial triple bond as in acyl halides or isocyanates	1.17 ± 0.01

CARBON-SULFUR

Single Bond

(1) Paraffinic, e.g. CH_3SH	1.81(5) ± 0.01
(2) Lengthened in presence of fluorine, e.g. $(CF_3)_2S$	1.83(5) ± 0.01
(3) Shortened (partial double bond) as in heterocyclic systems, e.g. C_4H_4S	1.73 ± 0.01

Double Bond

(1) In ethylene thiourea	1.71 ± 0.02
(2) Shortened (partial triple bond) in presence of second caron double bond, e.g. COS	1.558 ± 0.003

BOND LENGTHS OF ELEMENTS

Element	Bond	Å
Ac	Ac–Ac	3.756
Ag (25°C)	Ag–Ag	2.8894
Al (25°C)	Al–Al	2.863
As	As–As	2.49
As_4	As–As	2.44 ± 0.03
Au (25°C)	Au–Au	2.8841
B_2	B–B	1.589
Ba (room temp.)	Ba–Ba	4.347
Be (α-form, 20°C)	Be–Be	2.2260
Bi (25°C)	Bi–Bi	3.09
Br_2	Br–Br	2.290
Ca (α-form, 18°C)	Ca–Ca	3.947 (f.c.c.)
(β-form, 500°C		3.877 (b.c.c.)
Cd (21°C)	Cd–Cd	2.9788
Cl_2	Cl–Cl	1.988
Ce	Ce–Ce	3.650
Co (18°C)	Co–Co	2.5061
Cr (α-form, 20°C)	Cr–Cr	2.4980
(β-form, >1850°C		2.61
Cs (−10°C)	Cs–Cs	5.309
Cu (20°C)	Cu–Cu	2.5560
Dy	Dy–Dy	3.503
Er	Er–Er	3.468
Eu	Eu–Eu	3.989
F_2	F–F	1.417 ± 0.001
Fe (α-form, 20°C)	Fe–Fe	2.4823 (b.c.c.)
(γ-form, 916°C)		2.578 (f.c.c.)
(δ-form, 1394°C)		2.539 (b.c.c.)
Ga (20°C)	Ga–Ga	2.442
Gd (20°C)	Gd–Gd	3.573
Ge (20°C)	Ge–Ge	2.4498
H_2	H–H in H_2	0.74611
	H–D in HD	0.74136
	D–D in D_2	0.74164
He	He–He in $[He_2]^+$	1.08_0
Hf (α-form, 24°C)	Hf–Hf	3.1273 (h.c.p.)
Hg (−46°C)	Hg–Hg	3.005
Ho	Ho–Ho	3.486
I_2	I–I	2.662
In (20°C)	In–In	3.2511
Ir (room temp.)	Ir–Ir	2.714
K (78°K)	K–K	4.544
La (α-form)	La–La	3.739 (h.c.p.)
(β-form)		3.745 (f.c.c.)
Li (20°C)	Li–Li	3.0390
Lu	Lu–Lu	3.435
Mg (25°C)	Mg–Mg	3.1971
Mn (γ-form, 1095°C)	Mn–Mn	2.7311 (f.c.c.)
(δ-form, 1134°C)		2.6679 (b.c.c.)
Mo (20°C)	Mo–Mo	2.7251
N_2	N–N	1.0975_8 ± 0.0001
Na (20°C)	Na–Na	3.7157
Nb (20°C)	Nb–Nb	2.8584
Nd	Nd–Nd	3.628
Ni (18°C)	Ni–Ni	2.4916

Element	Bond	Å
Np (α-form, 20° C)	Np–Np	2.60 (orthorhombic)
(β-form, 313°C)		2.76 (tetragon)
(γ-form, 600°C)		3.05 (b.c.c.)
O_2	O–O	1.208
O_3 angle 116.8 ± 0.5°		1.278 ± 0.003
Os (20°C)	Os–Os	2.6754
P black	P–P	2.18
P_4	P–P	2.21 ± 0.02
Pa	Pa–Pa	3.212
Pb (25°C)	Pb–Pb	3.5003
Pd (25°C)	Pd–Pd	2.7511
Po (α-form, 10°C)	Po–Po	3.345 (cubic)
(β-form, 75°C)		3.359 (rh. hedr.)
Pr (α-form)	Pr–Pr	3.640 (tetrag.)
(β-form)		3.649 (f.c.c.)
Pt (20°C)	Pt–Pt	2.746
Pu (γ-form, 235°C)	Pu–Pu	3.026 (f.c.c.)
(δ-form, 313°C)		3.279 (f.c.c.)
(ε-form, 500°C)		3.150 (b.c.c.)
Rb (20°C)	Rb–Rb	4.95
Re (room temp.)	Re–Re	2.741
Rh (20°C)	Rh–Rh	2.6901
Ru (25°C)	Ru–Ru	2.6502
S_2	S–S	1.887
S_8	S–S	2.07 ± 0.02
Sb (25°C)	Sb–Sb	2.90
Sc (room temp.)	Sc–Sc	3.212
Se (20°C)	Se–Se	2.321
Se_2	Se–Se	2.152 ± 0.003
Se_8	Se–Se	2.32 ± 0.003
Si (20°C)	Si–Si	2.3517
Sn (α-form, 20°C)	Sn–Sn diamond	2.8099
(β-form, 25°C)	type lattice	3.022 (tetrag.)
Sr (α-form, 25°C)	Sr–Sr	4.302 (f.c.c.)
(β-form, 248°C)		4.32 (h.c.p.)
(γ-form, 614°C)		4.20 (b.c.c.)
Ta (20°C)	Ta–Ta	2.86
Tb	Tb–Tb	3.525
Tc (room temp.)	Tc–Tc	2.703
Te (25°C)	Te–Te	2.864
Th (α-form, 25°C)	Th–Th	3.595 (f.c.c.)
(β-form, 1450°C)		3.56 (b.c.c.)
Ti (α-form, 25°C)	Ti–Ti	2.8956 (h.c.p.)
(β-form, 900°C)		2.8636 (b.c.c.)
Tl (α-form, 18°C)	Tl–Tl	3.4076 (h.c.p.)
(β-form, 262°C)		3.362 (b.c.c.)
Tm	Tm–Tm	3.447
U (α-form)	U–U	2.77
(β-form, 805°C)		3.058 (b.c.c.)
V (30°C)	V–V	2.6224
W (25°C)	W–W	2.7409
Y	Y–Y	3.551
Yb	Yb–Yb	3.880
Zn (25°C)	Zn–Zn	2.6694
Zr	Zr–Zr	3.179

BOND LENGTH AND ANGLE VALUES BETWEEN ELEMENTS

Elements	In	Bond length (Å)	Bond angle (°)
Boron			
B–B	B_2H_6	1.770 ± 0.013	H–B–H 121.5 ± 7.5
B–Br	BBr	1.88_7	–
	BBr_3	1.87 ± 0.02	Br–B–Br 120 ± 6
B–Cl	BCl	1.715_7	–
	BCl_3	1.72 ± 0.01	Cl–B–Cl 120 ± 3
B–F	BF	1.262	–
	BF_3	1.29_5 ± 0.01	F–B–F 120
B–H	Hydrides	1.21 ± 0.02	–
B–H Bridge	Hydrides	1.39 ± 0.02	–
B–N	$(BClNH)_3$	1.42 ± 0.01	B–N–B 121
B–O	BO	1.2049	–
	$B(OH)_3$	1.362 ± 0.005 (av.)	O–B–O 119.7
Nitrogen			
N–Cl	NO_2Cl	1.79 ± 0.02	–
N–F	NF_3	1.36 ± 0.02	F–N–F 102.5 ± 1.5
N–H	$[NH_4]^+$	1.034 ± 0.003	–
	NH	1.038	–
	ND	1.041	–
	HNCS	1.013 ± 0.005	H–N–C 130.25 ± 0.25
N–N	N_3H	1.02 ± 0.01	H–N–N′ 112.65 ± 0.5
	N_2O	1.126 ± 0.002	–
	$[N_2]^+$	1.116_2	–
N–O	NO_2Cl	1.24 ± 0.01	O–N–O 126 ± 2
	NO_2	1.188 ± 0.005	O–N–O 134.1 ± 0.25
N=O	N_2O	1.186 ± 0.002	–
	$[NO]^+$	1.0619	–
N–Si	SiN	1.572	–
Oxygen			
O–H	$[OH]^+$	1.0289	–
	OD	0.9699	–
	H_2O_2	0.960 ± 0.005	O–O–H 100 ± 2
O–O	H_2O_2	1.48 ± 0.01	–
	$[O_2]^+$	1.227	–
	$[O_2]^-$	1.26 ± 0.02	–
	$[O_2]^{--}$	1.49 ± 0.02	–
Phosphorus			
P–D	PD	1.429	–
P–H	$[PH_4]^+$	1.42 ± 0.02	–
P–N	PN	1.4910	–
P–S	$PSBr_3(Cl_3,F_3)$	1.86 ± 0.02	–
Sulfur			
S–Br	$SOBr_2$	2.27 ± 0.02	Br–S–Br 96 ± 2
S–F	SOF_2	1.585 ± 0.005	F–S–F 92.8 ± 1
S–D	SD	1.3473	–
	SD_2	1.345	–
S–O	SO_2	1.4321	O–S–O 119.54
	$SOCl_2$	1.45 ± 0.02	–
S–S	S_2Cl_2	2.04 ± 0.01	–
Silicon			
Si–Br	$SiBr_4$	2.17 ± 1.01	–
Si–Cl	$SiCl_4$	2.03 ± 1.01 (av.)	–
Si–F	SiF_4	1.561 ± 0.003 (av.)	–
SiH	SiH_4	1.480 ± 0.005	–
Si–O	$[SiO]^+$	1.504	–
Si–Si	Si_2Cl_2	2.30 ± 0.02	–

BOND LENGTHS AND ANGLES OF CHEMICAL COMPOUNDS

A. Inorganic Compounds

Compound	Formula	Bond lengths in Å		Bond angles (°)	
Ammonia	NH_3	N–H	1.008 ± 0.004	H–N–H	1.07.3 ± 0.2
Antimony tribromide	$SbBr_3$	Sb–Br	2.51 ± 0.02	Br–Sb–Br	97 ± 2
Antimony trichloride	$SbCl_3$	Sb–Cl	2.352 ± 0.005	Cl–Sb–Cl	99.5 ± 1.5
Antimony triiodide	SbI_3	Sb–I	2.67 ± 0.03	I–Sb–I	99.0 ± 1
Arsenic tribromide	$AsBr_3$	As–Br	2.33 ± 0.02	Br–As–Br	100.5 ± 1.5
Arsenic trichloride	$AsCl_3$	As–Cl	2.161 ± 0.004	Cl–As–Cl	98.4 ± 0.5
Arsenic trifluoride	AsF_3	As–F	1.712 ± 0.005	F–As–F	102.0 ± 2
Arsenic triiodide	AsI_3	As–I	2.55 ± 0.03	I–As–I	101.0 ± 1.5
Arsenic trioxide	As_4O_6	As–O	1.78 ± 0.02	O–As–O	99.0 ± 2
				As–O–As	128.0 ± 2
Arsine	AsH_3	As–H	1.5192 ± 0.002	H–As–H	91.83 ± 0.33
Bismuthum tribromide	$BiBr_3$	Bi–Br	2.63 ± 0.02	Br–Bi–Br	100.0 ± 4
Bismuthum trichloride	$BiCl_3$	Bi–Cl	2.48 ± 0.02	Cl–Bi–Cl	100.0 ± 6
Bromosilane	SiH_3Br	Si–H	1.57 ± 0.03	H–Si–H	111.3 ± 1
Chlorine dioxide	ClO_2	Cl–O	1.49	O–Cl–O	118.5
Chlorogermane	GeH_3Cl	Ge–H	1.52 ± 0.03	H–Ge–H	110.9 ± 1.5
Chlorosilane	SiH_3Cl	Si–H	1.483 ± 0.001	H–Si–H	110 ± 0.03
		Si–Cl	2.0479 ± 0.007		

BOND LENGTHS AND ANGLES OF CHEMICAL COMPOUNDS (CONTINUED)

A. Inorganic Compounds (Continued)

Compound	Formula	Bond lengths in Å		Bond angles (°)	
Chromium oxychloride	$Cr(OCl)_2$	Cr–O	1.57 ± 0.03	O–Cr–O	105 ± 4
		Cr–Cl	2.12 ± 0.02	Cl–Cr–Cl	113 ± 3
				Cl–Cr–O	109 ± 3
Cyanuric triazide	C_3N_{12}	C–N	1.38	C–N=C	113.0
Dichlorosilane	SiH_2Cl_2	Si–H	1.46		
		SiCl	2.02 ± 0.03	Cl–Si–Cl	110 ± 1
Difluorodiazine	N_2F_2	N–F	1.44 ± 0.04		
		N–N	1.25 ± 0.02	F–N=N	115 ± 5
Difluoromethylsilane	Ch_3SiHf_2			F–Si–F	106 ± 0.5
				H–Si–C	116.2 ± 1
				C–Si–F	109.8 ± 0.5
Disilicon hexachloride (hexachlorisilane)	Si_2Cl_6	Si–Cl	2.02 ± 0.02	Cl–Si–Cl	109.5 ± 1
		Si–Si	2.34 ± 0.06		
Fluorosilane	SiH_3F	Si–H	1.460 ± 0.01	H–Si–H	109.3 ± 0.3
		Si–F	1.595 ± 0.002		
Hydrogen phosphide	PH_3	P–H	1.415 ± 0.003	H–P–H	93.3 ± 0.2
Hydrogen selenide	SeH_2	Se–H	1.47	H–Se–H	91.0
Hydrogen sulfide	SH_2	S–H	1.3455	H–S–H	93.3
Hydrogen telluride	$Te–H_2$			H–Te–H	89.5 ± 1
Iodo silane	$Si–H_3I$	Si–H	1.48 ± 0.01	H–Si–H	109.9 ± 0.4
		Si–I	2.4 ± 0.09		
Methylgermane	Ch_3GeH_3			H–C–H	108.2 ± 0.5
				H–Ge–H	108.6 ± 0.5
Nitrosyl bromide	NOBr	O–N	1.15 ± 0.04	Br–N=O	117 ± 3
		N–Br	2.14 ± 0.02		
Nitrosyl chloride	NOCl	N–O	1.14 ± 0.02	Cl–N=O	113.0 ± 2
		N–Cl	1.97 ± 0.01		
Nitrosyl fluoride	NOF	N–O	1.13	F–N=O	110.0
		N–F	1.52		
cis-Nitrous acid	NO(OH)	H–O	0.98	O–N=O	114 ± 2
		O–N	1.46		
		N–O′	1.20		
trans-Nitrous acid	NO(OH)	H–O	0.98	O–N=O	118 ± 2
		O–N	1.46		
		NO′	1.20		
Oxygen chloride	OCl_2	O–Cl	1.70_1 ± 0.02	Cl–O–Cl	110.8 ± 1
Oxygen fluoride	OF_2	O–F	1.418	F–O–F	103.2
Phosphorus oxychloride	$POCl_3$	P–Cl	1.99_5 ± 0.02	Cl–P–Cl	103.5 ± 1
Phosphorus oxysulfide	$P_4O_6S_4$	P–O	1.61 ± 0.02	P–O–P	128.5 ± 1.5
				O–P–O	101.5 ± 1
		P–S	1.85 ± 0.02	O–P–S	116.5 ± 1
Phosphorus pentoxide	P_4O_{10}	P–O	1.62 ± 0.02	O–P–O	101.5 ± 1
		P–O′	1.38 ± 0.02	O–P–O′	116.5 ± 1
				P–O–P	123.5 ± 1
Phosphorus tribromide	PBr_3	P–Br	2.18 ± 0.03	Br–P–Br	101.5 ± 1.5
Phosphorus trichloride	PCl_3	P–Cl	2.043 ± 0.003	Cl–P–Cl	100.1 ± 0.3
Phosphorus trifluoride	PF_3	P–F	1.535	F–P–F	100.0
Phosphorus trioxide	P_4O_6	P–O	1.65 ± 0.02	O–P–O	99.0 ± 1
				P–O–P	127.5 ± 1
Stibine	SbH_3	Sb–H	1.7073 ± 0.0025	H–Sb–H	91.3 ± 0.33
Sulfur dichloride	SCl_2	S–Cl	1.99 ± 0.03	Cl–S–Cl	101.0 ± 4
Sulfur dioxide	SO_2	S–O	1.4321	O–S–O	119.536
Sulfur monochloride	S_2Cl_2	S–Cl	2.01 ± 0.07	Cl–S–S	104.5 ± 0.25
Sulfurylchloride	SO_2Cl_2	S–O	1.43 ± 0.02	O–S–O	119.75 ± 5
		S–Cl	1.99 ± 0.02	Cl–S–Cl	111.20 ± 2
				Cl–O–O	106.5 ± 2
Tellurium bromide	$TeBr_2$	Te–Br	2.51 ± 0.02	Br–Te–Br	98.0 ± 3
Tribromo silane	$SiHBr_3$	Si–Br	2.16 ± 0.03	Br–Si–Br	110.5 ± 1.5
Trichloro germane	$GeHCl_3$	Ge–H	1.55 ± 0.04	Cl–Ge–Cl	108.3 ± 0.2
Trichloro silane	$Si–HCl_3$	Si–H	1.47	Cl–Si–Cl	109.4 ± 0.3
Trifluorochlorosilane	$SiClF_3$	Si–F	1.560 ± 0.005	F–Si–F	108.5 ± 1
		Si–Cl	1.989 ± 0.018		
Trifluorochlorogermane	$GeClF_3$	Ge–F	1.688 ± 0.0017	F–Ge–F	107.7 ± 1.5
		Gr–Cl	2.067 ± 0.005		
Trifluorosilane	$SiHF_3$	Si–H	1.455 ± 0.01		
		Si–F	1.565 ± 0.005	F–Si–F	108.3 ± 0.5
Vanadium oxytrichloride	$VOCl_3$	V–O	1.56 ± 0.04	Cl–V–Cl	111.2 ± 2
		V–Cl	2.12 ± 0.03	Cl–V–O	108.2 ± 2
Water	H_2O	O–H	0.958_4	H–O–H	104.45

B. Organic Compounds

Compound	Formula	Bond lengths in Å		Bond angles (°)	
Acetaldehyde	CH_3COH	C–H	1.09	C–C=O	121 ± 2
		C–C	1.50 ± 0.02		
		C–O	1.22 ± 0.02		
Bromomethane	Ch_3Br	C–H	1.11 ± 0.01	H–C–H	111.2 ± 0.5
		C–Br	1.929		
Carbon tetrachloride	CCl_4	C–Cl	1.766 ± 0.003	Cl–C–Cl	109.5
		Cl–Cl	2.887 ± 0.004		
Chloromethane	CH_3Cl	C–H	1.11 ± 0.01	H–C–H	110 ± 2
		C–Cl	1.784 ± 0.003		
Dichloromethane	CH_2Cl_2	C–H	1.068 ± 0.005	H–C–H	112 ± 0.3
		C–Cl	1.7724 ± 0.0005	Cl–C–Cl	111.8

BOND LENGTHS AND ANGLES OF CHEMICAL COMPOUNDS (CONTINUED)

Compound	Formula	Bond lengths in Å		Bond angles (°)	
Difluorochloromethane	$CHCLF_2$	C–H	1.06	F–C–F	110.5 ± 1
		C–Cl	1.73 ± 0.03	Cl–C–Cl	110.5 ± 1
		C–F	1.36 ± 0.03		
1,1-Difluoroethylene	$C_2H_2F_2$	C–H	1.07 ± 0.02	F–C=C	125.2 ± 0.2
		C–F	1.321 ± 0.015	H–C–H	117 ± 7
		C–C	1.311 ± 0.035		
Difluoromethane	CH_2F_2	C–F	1.360 ± 0.005	F–C–F	108.2 ± 0.8
		C–H	1.09 ± 0.03	H–C–H	112.5 ± 6
p-Dinitrobenzene	$C_6H_4(NO_2)_2$	C–C	1.38	O–N=O	124.0
		C–N	1.48		
		N–O	1.21		
Dithio oxamide	$NH_2CSCSNH_2$	C–C	1.54	N–C=S	124.8
		C–N	1.30	S–C–C	124.87
		C–S	1.66	N–C–C	155.25
Ethane	C_2H_6	C–H	1.107	H–C–H	109.3
		C–C	1.536		
Ethylidene fluoride	CH_3CHF_2	C–F	1.345 ± 0.001	F–C–F	109.15 ± 0.001
		C–C	1.540	C–C–F	109.4
		C–H	1.100	H–C–C	110.2
Fluorochloromethane	CH_2ClF	C–H	1.078 ± 0.005	Cl–C–F	100.0 ± 0.1
		C–Cl	1.759 ± 0.003		
		C–F	1.378 ± 0.006		
Fluorotrichloromethane	$CFCl_3$	C–Cl	1.76 ± 0.02	Cl–C–Cl	113 ± 3
		C–F	1.44 ± 0.04		
Formaldehyde	CH_2O	C–H	1.060 ± 0.038	H–C–H	125.8 ± 7
		C–O	1.230 ± 0.017		
Formamide	$HCONH_2$	C–O	1.25_6	N–C=O	121.5
		C–N	1.300		
Formic acid	HCOOH	C–O′	1.245	O–C=O′	124.3
		C–O	1.312	H–C–O′	117.8
		C–H	1.085	C–O–H	107.8
		O–H	0.95		
Glycine	NH_2CH_2COOH	C–C	1.52	C–C–O	119.0
		C–O	1.27	O–C–O	122.0
		C–N	1.39	C–C–N	112.0
Hexachloroethane	C_2Cl_6	C–Cl	1.74 ± 0.01	Cl–C–Cl	109.3 ± 0.01
		C–C	1.57 ± 0.06		
Iodomethane	CH_3I	C–H	1.11 ± 0.01	H–C–H	111.4 ± 0.1
		C–I	2.139		
Methane	CH_4	C–H	1.091		
Methanethiol	CH_3SH	C–H	1.1039 ± 0.002		110.3 ± 0.2
		C–S	1.8177 ± 0.0002	H–S–C	100.3 ± 0.2
		S–H	1.329 ± 0.004		
Methanol	CH_3OH	C–H	1.096 ± 0.01	H–C–H	109.3 ± 0.75
		C–O	1.427 ± 0.007	C–O–H	108.9 ± 2
		O–H	0.956 ± 0.015		
Methylamine	CH_3NH_2	C–H	1.093	H–C–H	109.5 ± 1
		C–N	1.474 ± 0.005	H–N–H	105.8 ± 1
		N–H	1.014		
Methylether	$(CH_3)_2O$	C–O	1.43 ± 0.03	C–O–C	110.0 ± 3
Methylnitrite	CH_3NO_2	C–H	1.00	O–N=O	127 ± 4
		C–N	1.49 ± 0.02		
		N–O	1.22 ± 0.1		
Methylsulfide	$(CH_3)_2S$	C–S	1.82 ± 0.01	C–S–C	105 ± 3
		C–H	1.06	H–C–H	109.5
Oxamide	$NH_3COCONH_2$	C–C	1.54	N–C=O	125.7 ± 0.3
		C–O	1.24		
		C–N	1.32		
Phosgene	CCl_2O	C–Cl	4.745 ± 0.004	Cl–C=O	124.3 ± 0.1
		C–O	1.166 ± 0.002	Cl–C–Cl	111.3 ± 0.1
Propylene	C_3H_6			C–C=C	124.75 ± 0.3
Propynal	CHC.COH	C_1–C_2	1.204	C–C=O	123.0
		C_2–C_3	1.46	C–C–H	120.0
		C–O	1.21		
		C_1–H	1.06		
		C_3–H	1.08		
Tribromomethane	$CHBr_3$	C–H	1.068 ± 0.01	Br–C–Br	110.8 ± 0.3
		C–Br	1.930 ± 0.003		
Trichlorobromomethane	$CBrCl_3$	C–Br	1.936	Cl–C–Cl	111.2 ± 1
		C–Cl	1.764		
Trifluorochloromethane	$CClF_3$	C–Cl	1.751 ± 0.004	F–C–F	108.6 ± 0.4
		C–F	1.328± 0.02		
Trifluoromethane	CHF_3	C–H	1.098	F–C–F	108 ± 0.75
		C–F	1.332 ± 0.008		
Triiodomethane	CHI_3	C–I	2.12 ± 0.04	I–C–I	113.0 ± 1
Trimethylamine	$(CH_3)_3N$	C–N	1.47 ± 0.01	C–N–C	108.0 ± 4
		C–H	1.06	H–C–H	109.5
Trimethylarsine	$(CH_3)_3As$	C–H	1.09	C–As–C	96 ± 5
		C–As	1.98 ± 0.02		
Trimethylphosphine	$(CH_3)_3P$	C–H	1.09	C–P–C	100.0 ± 4
		C–P	1.87 ± 0.02		

ELECTRON BINDING ENERGIES OF THE ELEMENTS
(UNITS ARE ELECTRON VOLTS)

Prepared from a chart containing data compiled by Gwyn P. Williams of the National Synchrotron Light Source, Brookhaven National Laboratory, Upton, New York 11973, U.S.A.

Actinium (89)

K 1s :106755	N_{IV} $4d_{3/2}$:675*
L_{I} 2s : 19840	N_{V} $4d_{5/2}$:639*
L_{II} $2p_{1/2}$: 19083	N_{VI} $4f_{5/2}$:319*
L_{III} $2p_{3/2}$: 15871	N_{VII} $4f_{7/2}$:319*
M_{I} 3s : 5002	O_{I} 5s :272*
M_{II} $3p_{1/2}$: 4656	O_{II} $5p_{1/2}$:215*
M_{III} $3p_{3/2}$: 3909	O_{III} $5p_{3/2}$:167*
M_{IV} $3d_{3/2}$: 3370	O_{IV} $5d_{3/2}$: 80*
M_{V} $3d_{5/2}$: 3219	O_{V} $5d_{5/2}$: 80*
N_{I} 4s : 1269*	P_{I} 6s : —
N_{II} $4p_{1/2}$: 1080*	P_{II} $6p_{1/2}$: —
N_{III} $4p_{3/2}$: 890*	P_{III} $6p_{3/2}$: —

Aluminum (13)

K 1s :1562.3*	L_{II} $2p_{1/2}$:72.9*
L_{I} 2s : 117.8*	L_{III} $2p_{3/2}$:72.5*

Antimony (51)

K 1s :30491	M_{IV} $3d_{3/2}$:537.5†
L_{I} 2s : 4698	M_{V} $3d_{5/2}$:528.2†
L_{II} $2p_{1/2}$: 4380	N_{I} 4s :153.2†
L_{III} $2p_{3/2}$: 4132	N_{II} $4p_{1/2}$: 95.6†a
M_{I} 3s : 946 †	N_{III} $4p_{3/2}$: 95.6†a
M_{II} $3p_{1/2}$: 812.7†	N_{IV} $4d_{3/2}$: 33.3†
M_{III} $3p_{3/2}$: 766.4†	N_{V} $4d_{5/2}$: 32.1†

Argon (18)

K 1s :3205.9*	M_{I} 3s :29.3*
L_{I} 2s : 326.3*	M_{II} $3p_{1/2}$:15.9*
L_{II} $2p_{1/2}$: 250.6*	M_{III} $3p_{3/2}$:15.7*
L_{III} $2p_{3/2}$: 248.4*	

Arsenic (33)

K 1s :11867	M_{II} $3p_{1/2}$:146.2*
L_{I} 2s : 1527.0*b	M_{III} $3p_{3/2}$:141.2*
L_{II} $2p_{1/2}$: 1359.1*b	M_{IV} $3d_{3/2}$: 41.7*
L_{III} $2p_{3/2}$: 1323.6*b	M_{V} $3d_{5/2}$: 41.7*
M_{I} 3s : 204.7*	

Astatine (85)

K 1s :95730	N_{III} $4p_{3/2}$:740*
L_{I} 2s :17493	N_{IV} $4d_{3/2}$:533*
L_{II} $2p_{1/2}$ 16785	N_{V} $4d_{5/2}$:507*
L_{III} $2p_{3/2}$:14214	N_{VI} $4f_{5/2}$:210*
M_{I} 3s : 4317	N_{VII} $4f_{7/2}$:210*
M_{II} $3p_{1/2}$: 4008	O_{I} 5s :195*
M_{III} $3p_{3/2}$: 3426	O_{II} $5p_{1/2}$:148*
M_{IV} $3d_{3/2}$: 2909	O_{III} $5p_{3/2}$:115*
M_{V} $3d_{5/2}$: 2787	O_{IV} $5d_{3/2}$: 40*
N_{I} 4s : 1042*	O_{V} $5d_{5/2}$: 40*
N_{II} $4p_{1/2}$: 886*	

Barium (56)

K 1s :37441	N_{II} $4p_{1/2}$:192
L_{I} 2s : 5989	N_{III} $4p_{3/2}$:178.6†
L_{II} $2p_{1/2}$: 5624	N_{IV} $4d_{3/2}$: 92.6†
L_{III} $2p_{3/2}$: 5247	N_{V} $4d_{5/2}$: 89.9†
M_{I} 3s : 1293 *b	N_{VI} $4f_{5/2}$: —
M_{II} $3p_{1/2}$: 1137 *b	N_{VII} $4f_{7/2}$: —
M_{III} $3p_{3/2}$: 1063 *b	O_{I} 5s : 30.3†
M_{IV} $3d_{3/2}$: 795.7*	O_{II} $5p_{1/2}$: 17.0†
M_{V} $3d_{5/2}$: 780.5*	O_{III} $5p_{3/2}$: 14.8†
N_{I} 4s : 253.5†	

Beryllium (4)

K 1s :111.5*

Bismuth (83)

K 1s :90526	N_{III} $4p_{3/2}$:678.8†
L_{I} 2s :16388	N_{IV} $4d_{3/2}$:464.0†
L_{II} $2p_{1/2}$:15711	N_{V} $4d_{5/2}$:440.1†
L_{III} $2p_{3/2}$:13419	N_{VI} $4f_{5/2}$:162.3†
M_{I} 3s : 3999	N_{VII} $4f_{7/2}$:157.0†
M_{II} $3p_{1/2}$: 3696	O_{I} 5s :159.3*b
M_{III} $3p_{3/2}$: 3177	O_{II} $5p_{1/2}$:119.0†
M_{IV} $3d_{3/2}$: 2688	O_{III} $5p_{3/2}$: 92.6†
M_{V} $3d_{5/2}$: 2580	O_{IV} $5d_{3/2}$: 26.9†
N_{I} 4s : 939 †	O_{V} $5d_{5/2}$: 23.8†
N_{II} $4p_{1/2}$: 805.2†	

Boron (5)

K 1s :188*

Bromine (35)

K 1s :13474	M_{II} $3p_{1/2}$:189*
L_{I} 2s : 1782*	M_{III} $3p_{3/2}$:182*
L_{II} $2p_{1/2}$: 1596*	M_{IV} $3d_{3/2}$: 70*
L_{III} $2p_{3/2}$: 1550*	M_{V} $3d_{5/2}$: 69*
M_{I} 3s : 257*	

Cadmium (48)

K 1s :26711	M_{IV} $3d_{3/2}$:411.9†
L_{I} 2s : 4018	M_{V} $3d_{5/2}$:405.2†
L_{II} $2p_{1/2}$: 3727	N_{I} 4s :109.8†
L_{III} $2p_{3/2}$: 3538	N_{II} $4p_{1/2}$: 63.9†a
M_{I} 3s : 772.0†	N_{III} $4p_{3/2}$: 63.9†a
M_{II} $3p_{1/2}$: 652.6†	N_{IV} $4d_{3/2}$: 11.7†
M_{III} $3p_{3/2}$: 618.4†	N_{V} $4d_{5/2}$: 10.7†

Calcium (20)

K 1s :4038.5*	M_{I} 3s :44.3†
L_{I} 2s : 438.4†	M_{II} $3p_{1/2}$:25.4†
L_{II} $2p_{1/2}$: 349.7†	M_{III} $3p_{3/2}$:25.4†
L_{III} $2p_{3/2}$: 346.2†	

Carbon (6)

K 1s :284.2*

Cerium (58)

K 1s :40443	N_{II} $4p_{1/2}$:223.3
L_{I} 2s : 6548	N_{III} $4p_{3/2}$:206.5*
L_{II} $2p_{1/2}$: 6164	N_{IV} $4d_{3/2}$:109 *
L_{III} $2p_{3/2}$: 5723	N_{V} $4d_{5/2}$: —
M_{I} 3s : 1436 *b	N_{VI} $4f_{5/2}$: .1
M_{II} $3p_{1/2}$: 1274 *b	N_{VII} $4f_{7/2}$: .1
M_{III} $3p_{3/2}$: 1187 *b	O_{I} 5s : 37.8
M_{IV} $3d_{3/2}$: 902.4*	O_{II} $5p_{1/2}$: 19.8*
M_{V} $3d_{5/2}$: 883.8*	O_{III} $5p_{3/2}$: 17.0*
N_{I} 4s : 291.0*	

Cesium (55)

K 1s :35985	N_{II} $4p_{1/2}$:172.4*
L_{I} 2s : 5714	N_{III} $4p_{3/2}$:161.3*
L_{II} $2p_{1/2}$: 5359	N_{IV} $4d_{3/2}$: 79.8*
L_{III} $2p_{3/2}$: 5012	N_{V} $4d_{5/2}$: 77.5*
M_{I} 3s : 1211 *b	N_{VI} $4f_{5/2}$: —
M_{II} $3p_{1/2}$: 1071 *	N_{VII} $4f_{7/2}$: —
M_{III} $3p_{3/2}$: 1003 *	O_{I} 5s : 22.7
M_{IV} $3d_{3/2}$: 740.5*	O_{II} $5p_{1/2}$: 14.2*
M_{V} $3d_{5/2}$: 726.6*	O_{III} $5p_{3/2}$: 12.1*
N_{I} 4s : 232.3*	

Chlorine (17)

K 1s :2823	L_{II} $2p_{1/2}$:202*
L_{I} 2s : 270*	L_{III} $2p_{3/2}$:200*

Chromium (24)

K 1s :5989	M_{I} 3s :74.1†
L_{I} 2s : 695.7†	M_{II} $3p_{1/2}$:42.2†
L_{II} $2p_{1/2}$: 583.8†	M_{III} $3p_{3/2}$:42.2†
L_{III} $2p_{3/2}$: 574.1†	

Cobalt (27)

K 1s :7709	M_{I} 3s :101.0†
L_{I} 2s : 925.1†	M_{II} $3p_{1/2}$: 58.9†
L_{II} $2p_{1/2}$: 793.3†	M_{III} $3p_{3/2}$: 58.9†
L_{III} $2p_{3/2}$: 778.1†	

Copper (29)

K 1s :8979	M_{I} 3s :122.5†
L_{I} 2s :1096.7†	M_{II} $3p_{1/2}$: 77.3†
L_{II} $2p_{1/2}$: 952.3†	M_{III} $3p_{3/2}$: 75.1†
L_{III} $2p_{3/2}$: 932.5†	

Dysprosium (66)

K 1s :53789	N_{I} $4p_{1/2}$:333.5*
L_{I} 2s : 9046	N_{III} $4p_{3/2}$:293.2*
L_{II} $2p_{1/2}$: 8581	N_{IV} $4d_{3/2}$:153.6*
L_{III} $2p_{3/2}$: 7790	N_{V} $4d_{5/2}$:153.6*
M_{I} 3s : 2047	N_{VI} $4f_{5/2}$: 8.0*
M_{II} $3p_{1/2}$: 1842	N_{VII} $4f_{7/2}$: 4.3*
M_{III} $3p_{3/2}$: 1676	O_{I} 5s : 49.9*
M_{IV} $3d_{3/2}$: 1333	O_{II} $5p_{1/2}$: 26.3
M_{V} $3d_{5/2}$: 1292 *	O_{III} $5p_{3/2}$: 26.3
N_{I} 4s : 414.2*	

Erbium (68)

K 1s :57486	N_{II} $4p_{1/2}$:366.2
L_{I} 2s : 9751	N_{III} $4p_{3/2}$:320.2*
L_{II} $2p_{1/2}$: 9264	N_{IV} $4d_{3/2}$:167.6*
L_{III} $2p_{3/2}$: 8358	N_{V} $4d_{5/2}$:167.6*
M_{I} 3s : 2206	N_{VI} $4f_{5/2}$: —
M_{II} $3p_{1/2}$: 2006	N_{VII} $4f_{7/2}$: 4.7*
M_{III} $3p_{3/2}$: 1812	O_{I} 5s : 50.6*
M_{IV} $3d_{3/2}$: 1453	O_{II} $5p_{1/2}$: 31.4*
M_{V} $3d_{5/2}$: 1409	O_{III} $5p_{3/2}$: 24.7*
N_{I} 4s : 449.8*	

Europium (63)

K 1s :48519	N_{I} 4s :360
L_{I} 2s : 8052	N_{II} $4p_{1/2}$:284
L_{II} $2p_{1/2}$: 7617	N_{III} $4p_{3/2}$:257
L_{III} $2p_{3/2}$: 6977	N_{IV} $4d_{3/2}$:133
M_{I} 3s : 1800	N_{V} $4d_{5/2}$:133
M_{II} $3p_{1/2}$: 1614	N_{VI} $4f_{5/2}$: 0
M_{III} $3p_{3/2}$: 1481	N_{VII} $4f_{7/2}$: 0
M_{IV} $3d_{3/2}$: 1158.6*	O_{I} 5s : 32
M_{V} $3d_{5/2}$: 1127.5*	O_{II} $5p_{1/2}$: 22
	O_{III} $5p_{3/2}$: 22

Fluorine (9)

K 1s :696.7*

Francium (87)

K 1s :101137	N_{IV} $4d_{3/2}$:603*
L_{I} 2s : 18639	N_{V} $4d_{5/2}$:577*
L_{II} $2p_{1/2}$: 17907	N_{VI} $4f_{5/2}$:268*

L_{III} $2p_{3/2}$: 15031	N_{VII} $4f_{7/2}$:268*
M_I 3s : 4652	O_I 5s :234*
M_{II} $3p_{1/2}$: 4327	O_{II} $5p_{1/2}$:182*
M_{III} $3p_{3/2}$: 3663	O_{III} $5p_{3/2}$:140*
M_{IV} $3d_{3/2}$: 3136	O_{IV} $5d_{3/2}$: 58*
M_V $3d_{5/2}$: 3000	O_V $5d_{5/2}$: 58*
N_I 4s : 1153*	P_I 6s : 34
N_{II} $4p_{1/2}$: 980*	P_{II} $6p_{1/2}$: 15
N_{III} $4p_{3/2}$: 810*	P_{III} $6p_{3/2}$: 15

Gadolinium (64)

K 1s :50239	N_{II} $4p_{1/2}$:286
L_I 2s : 8376	N_{III} $4p_{3/2}$:271
L_{II} $2p_{1/2}$: 7930	N_{IV} $4d_{3/2}$: —
L_{III} $2p_{3/2}$: 7243	N_V $4d_{5/2}$:127.7*
M_I 3s : 1881	N_{VI} $4f_{5/2}$: 8.6*
M_{II} $3p_{1/2}$: 1688	N_{VII} $4f_{7/2}$: 8.6*
M_{III} $3p_{3/2}$: 1544	O_I 5s : 36
M_{IV} $3d_{3/2}$: 1221.9*	O_{II} $5p_{1/2}$: 20
M_V $3d_{5/2}$: 1189.6*	O_{III} $5p_{3/2}$: 20
N_I 4s : 378.6*	

Gallium (31)

K 1s :10367	M_{II} $3p_{1/2}$:103.5†
L_I 2s : 1299.0*b	M_{III} $3p_{3/2}$:103.5†
L_{II} $2p_{1/2}$: 1143.2†	M_{IV} $3d_{3/2}$: 18.7†
L_{III} $2p_{3/2}$: 1116.4†	M_V $3d_{5/2}$: 18.7†
M_I 3s : 159.5†	

Germanium (32)

K 1s :11103	M_{II} $3p_{1/2}$:124.9*
L_I 2s : 1414.6*b	M_{III} $3p_{3/2}$:120.8*
L_{II} $2p_{1/2}$: 1248.1*b	M_{IV} $3d_{3/2}$: 29.0*
L_{III} $2p_{3/2}$: 1217.0*b	M_V $3d_{5/2}$: 29.0*
M_I 3s : 180.1*	

Gold (79)

K 1s :80725	N_{II} $4p_{1/2}$:642.7†
L_I 2s :14353	N_{III} $4p_{3/2}$:546.3†
L_{II} $2p_{1/2}$:13734	N_{IV} $4d_{3/2}$:353.2†
L_{III} $2p_{3/2}$:11919	N_V $4d_{5/2}$:335.1†
M_I 3s : 3425	N_{VI} $4f_{5/2}$: 87.6†
M_{II} $3p_{1/2}$: 3148	N_{VII} $4f_{7/2}$: 83.9†
M_{III} $3p_{3/2}$: 2743	O_I 5s :107.2*b
M_{IV} $3d_{3/2}$: 2291	O_{II} $5p_{1/2}$: 74.2†
M_V $3d_{5/2}$: 2206	O_{III} $5p_{3/2}$: 57.2†
N_I 4s : 762.1†	

Hafnium (72)

K 1s :65351	N_{II} $4p_{1/2}$:438.2†
L_I 2s :11271	N_{III} $4p_{3/2}$:380.7†
L_{II} $2p_{1/2}$:10739	N_{IV} $4d_{3/2}$:220.0†
L_{III} $2p_{3/2}$: 9561	N_V $4d_{5/2}$:211.5†
M_I 3s : 2601	N_{VI} $4f_{5/2}$: 15.9†
M_{II} $3p_{1/2}$: 2365	N_{VII} $4f_{7/2}$: 14.2†
M_{III} $3p_{3/2}$: 2107	O_I 5s : 64.2†
M_{IV} $3d_{3/2}$: 1716	O_{II} $5p_{1/2}$: 38 *
M_V $3d_{5/2}$: 1662	O_{III} $5p_{3/2}$: 29.9†
N_I 4s : 538*	

Helium (2)

K 1s :24.6*

Holmium (67)

K 1s :55618	N_{II} $4p_{1/2}$:343.5
L_I 2s : 9394	N_{III} $4p_{3/2}$:308.2*
L_{II} $2p_{1/2}$: 8918	N_{IV} $4d_{3/2}$:160 *
L_{III} $2p_{3/2}$: 8071	N_V $4d_{5/2}$:160 *
M_I 3s : 2128	N_{VI} $4f_{5/2}$: 8.6*
M_{II} $3p_{1/2}$: 1923	N_{VII} $4f_{7/2}$: 5.2*
M_{III} $3p_{3/2}$: 1741	O_I 5s : 49.3*
M_{IV} $3d_{3/2}$: 1392	O_{II} $5p_{1/2}$: 30.8*
M_V $3d_{5/2}$: 1351	O_{III} $5p_{3/2}$: 24.1*
N_I 4s : 432.4*	

Hydrogen (1)

K 1s :16.0*

Indium (49)

K 1s :27940	M_{IV} $3d_{3/2}$:451.4†
L_I 2s : 4238	M_V $3d_{5/2}$:443.9†
L_{II} $2p_{1/2}$: 3938	N_I 4s :122.7†
L_{III} $2p_{3/2}$: 3730	N_{II} $4p_{1/2}$: 73.5†a
M_I 3s : 827.2†	N_{III} $4p_{3/2}$: 73.5†a
M_{II} $3p_{1/2}$: 703.2†	N_{IV} $4d_{3/2}$: 17.7†
M_{III} $3p_{3/2}$: 665.3†	N_V $4d_{5/2}$: 16.4†

Iodine (53)

K 1s :33169	M_{IV} $3d_{3/2}$:631*
L_I 2s : 5188	M_V $3d_{5/2}$:620*
L_{II} $2p_{1/2}$: 4852	N_I 4s :186*
L_{III} $2p_{3/2}$: 4557	N_{II} $4p_{1/2}$:123*
M_I 3s : 1072*	N_{III} $4p_{3/2}$:123*
M_{II} $3p_{1/2}$: 931*	N_{IV} $4d_{3/2}$: 50*
M_{III} $3p_{3/2}$: 875*	N_V $4d_{5/2}$: 50*

Iridium (77)

K 1s :76111	N_{II} $4p_{1/2}$:577.8†
L_I 2s :13419	N_{III} $4p_{3/2}$:495.8†
L_{II} $2p_{1/2}$:12824	N_{IV} $4d_{3/2}$:311.9†
L_{III} $2p_{3/2}$:11215	N_V $4d_{5/2}$:296.3†
M_I 3s : 3174	N_{VI} $4f_{5/2}$: 63.8†
M_{II} $3p_{1/2}$: 2909	N_{VII} $4f_{7/2}$: 60.8†
M_{III} $3p_{3/2}$: 2551	O_I 5s : 95.2*b
M_{IV} $3d_{3/2}$: 2116	O_{II} $5p_{1/2}$: 63.0*b
M_V $3d_{5/2}$: 2040	O_{III} $5p_{3/2}$: 48.0†
N_I 4s : 691.1†	

Iron (26)

K 1s :7112	M_I 3s :91.3†
L_I 2s : 844.6†	M_{II} $3p_{1/2}$:52.7†
L_{II} $2p_{1/2}$: 719.9†	M_{III} $3p_{3/2}$:52.7†
L_{III} $2p_{3/2}$: 706.8†	

Krypton (36)

K 1s :14326	M_{III} $3p_{3/2}$:214.4*
L_I 2s : 1921	M_{IV} $3d_{3/2}$: 95.0*
L_{II} $2p_{1/2}$: 1730.9*	M_V $3d_{5/2}$: 93.8*
L_{III} $2p_{3/2}$: 1678.4*	N_I 4s : 27.5*
M_I 3s : 292.8*	N_{II} $4p_{1/2}$: 14.1*
M_{II} $3p_{1/2}$: 222.2*	N_{III} $4p_{3/2}$: 14.1*

Lanthanum (57)

K 1s :38925	N_{II} $4p_{1/2}$:205.8
L_I 2s : 6266	N_{III} $4p_{3/2}$:196.0*
L_{II} $2p_{1/2}$: 5891	N_{IV} $4d_{3/2}$:105.3*
L_{III} $2p_{3/2}$: 5483	N_V $4d_{5/2}$:102.5*
M_I 3s : 1362*b	N_{VI} $4f_{5/2}$: —
M_{II} $3p_{1/2}$: 1209*b	N_{VII} $4f_{7/2}$: —
M_{III} $3p_{3/2}$: 1128*b	O_I 5s : 34.3*
M_{IV} $3d_{3/2}$: 853*	O_{II} $5p_{1/2}$: 19.3*
M_V $3d_{5/2}$: 836*	O_{III} $5p_{3/2}$: 16.8*
N_I 4s : 247.7*	

Lead (82)

K 1s :88005	N_{III} $4p_{3/2}$:643.5†
L_I 2s :15861	N_{IV} $4d_{3/2}$:434.3†
L_{II} $2p_{1/2}$:15200	N_V $4d_{5/2}$:412.2†
L_{III} $2p_{3/2}$:13055	N_{VI} $4f_{5/2}$:141.7†
M_I 3s : 3851	N_{VII} $4f_{7/2}$:136.9†
M_{II} $3p_{1/2}$: 3554	O_I 5s :147 *b
M_{III} $3p_{3/2}$:	O_{II} $5p_{1/2}$:106.4†
M_{IV} $3d_{3/2}$: 3066	O_{III} $5p_{3/2}$: 83.3†
M_V $3d_{5/2}$: 2586	O_{IV} $5d_{3/2}$: 20.7†
N_I 4s : 891.8†	O_V $5d_{5/2}$: 18.1†
N_{II} $4p_{1/2}$: 761.9†	

Lithium (3)

K 1s :54.7*

Lutetium (71)

K 1s :63314	N_{II} $4p_{1/2}$:412.4*
L_I 2s :10870	N_{III} $4p_{3/2}$:359.2*
L_{II} $2p_{1/2}$:10349	N_{IV} $4d_{3/2}$:206.1*
L_{III} $2p_{3/2}$: 9244	N_V $4d_{5/2}$:196.3*
M_I 3s : 2491	N_{VI} $4f_{5/2}$: 8.9*
M_{II} $3p_{1/2}$: 2264	N_{VII} $4f_{7/2}$: 7.5*
M_{III} $3p_{3/2}$: 2024	O_I 5s : 57.3*
M_{IV} $3d_{3/2}$: 1639	O_{II} $5p_{1/2}$: 33.6*
M_V $3d_{5/2}$: 1589	O_{III} $5p_{3/2}$: 26.7*
N_I 4s : 506.8*	

Magnesium (12)

K 1s :1303.0†
L_I 2s : 88.6*
L_{II} $2p_{1/2}$:49.6†
L_{III} $2p_{3/2}$:49.2*

Manganese (25)

K 1s :6539	M_I 3s :82.3†
L_I 2s : 769.1†	M_{II} $3p_{1/2}$:47.2†
L_{II} $2p_{1/2}$: 649.9†	M_{III} $3p_{3/2}$:47.2†
L_{III} $2p_{3/2}$: 638.7†	

Mercury (80)

K 1s :83102	N_{III} $4p_{3/2}$:576.6†
L_I 2s :14839	N_{IV} $4d_{3/2}$:378.2†
L_{II} $2p_{1/2}$:14209	N_V $4d_{5/2}$:358.8†
L_{III} $2p_{3/2}$:12284	N_{VI} $4f_{5/2}$:104.0†
M_I 3s : 3562	N_{VII} $4f_{7/2}$: 99.9†
M_{II} $3p_{1/2}$: 3279	O_I 5s :127 †
M_{III} $3p_{3/2}$: 2847	O_{II} $5p_{1/2}$: 83.1†
M_{IV} $3d_{3/2}$: 2385	O_{III} $5p_{3/2}$: 64.5†
M_V $3d_{5/2}$: 2295	O_{IV} $5d_{3/2}$: 9.6†
N_I 4s : 802.2†	O_V $5d_{5/2}$: 7.8†
N_{II} $4p_{1/2}$: 680.2†	

Molybdenum (42)

K 1s :20000	M_{III} $3p_{3/2}$:394.C†
L_I 2s : 2866	M_{IV} $3d_{3/2}$:231.1†
L_{II} $2p_{1/2}$: 2625	M_V $3d_{5/2}$:227.9†
L_{III} $2p_{3/2}$: 2520	N_I 4s : 63.2†
M_I 3s : 506.3†	N_{II} $4p_{1/2}$: 37.6†
M_{II} $3p_{1/2}$: 410.6†	N_{III} $4p_{3/2}$: 35.5†

Neodymium (60)

K 1s :43569	N_{II} $4p_{1/2}$:243.3
L_I 2s : 7126	N_{III} $4p_{3/2}$:224.6
L_{II} $2p_{1/2}$: 6722	N_{IV} $4d_{3/2}$:120.5*
L_{III} $2p_{3/2}$: 6208	N_V $4d_{5/2}$:120.5*
M_I 3s : 1575	N_{VI} $4f_{5/2}$: 1.5
M_{II} $3p_{1/2}$: 1403	N_{VII} $4f_{7/2}$: 1.5
M_{III} $3p_{3/2}$: 1297	O_I 5s : 37.5
M_{IV} $3d_{3/2}$: 1003.3*	O_{II} $5p_{1/2}$: 21.1
M_V $3d_{5/2}$: 980.4*	O_{III} $5p_{3/2}$: 21.1
N_I 4s : 319.2*	

Neon (10)

K 1s :870.2*
L_I 2s : 48.5*
L_{II} $2p_{1/2}$:21.7*
L_{III} $2p_{3/2}$:21.6*

Nickel (28)

K 1s :8333	M_I 3s :110.8†
L_I 2s :1008.6†	M_{II} $3p_{1/2}$: 68.0†
L_{II} $2p_{1/2}$: 870.0†	M_{III} $3p_{3/2}$: 66.2†
L_{III} $2p_{3/2}$: 852.7†	

Niobium (41)

$K\ 1s$:18986	$M_{III}\ 3p_{3/2}$:360.6†
$L_{I}\ 2s$: 2698	$M_{IV}\ 3d_{3/2}$:205.0†
$L_{II}\ 2p_{1/2}$: 2465	$M_{V}\ 3d_{5/2}$:202.3†
$L_{III}\ 2p_{3/2}$: 2371	$N_{I}\ 4s$: 56.4†
$M_{I}\ 3s$: 466.6†	$N_{II}\ 4p_{1/2}$: 32.6†
$M_{II}\ 3p_{1/2}$: 376.1†	$N_{III}\ 4p_{3/2}$: 30.8†

Nitrogen (7)

$K\ 1s$:409.9*
$L_{I}\ 2s$:37.3*

Osmium (76)

$K\ 1s$:73871	$N_{II}\ 4p_{1/2}$:549.1†
$L_{I}\ 2s$:12968	$N_{III}\ 4p_{3/2}$:470.7†
$L_{II}\ 2p_{1/2}$:12385	$N_{IV}\ 4d_{3/2}$:293.1†
$L_{III}\ 2p_{3/2}$:10871	$N_{V}\ 4d_{5/2}$:278.5†
$M_{I}\ 3s$: 3049	$N_{VI}\ 4f_{5/2}$: 52.4†
$M_{II}\ 3p_{1/2}$: 2792	$N_{VII}\ 4f_{7/2}$: 50.7†
$M_{III}\ 3p_{3/2}$: 2457	$O_{I}\ 5s$: 83 †
$M_{IV}\ 3d_{3/2}$: 2031	$O_{II}\ 5p_{1/2}$: 58 *
$M_{V}\ 3d_{5/2}$: 1960	$O_{III}\ 5p_{3/2}$: 44.5†
$N_{I}\ 4s$: 658.2†	

Oxygen (8)

$K\ 1s$:543.1*
$L_{I}\ 2s$:41.6*

Palladium (46)

$K\ 1s$:24350	$M_{III}\ 3p_{3/2}$:532.3†
$L_{I}\ 2s$: 3604	$M_{IV}\ 3d_{3/2}$:340.5†
$L_{II}\ 2p_{1/2}$: 3330	$M_{V}\ 3d_{5/2}$:335.2†
$L_{III}\ 2p_{3/2}$: 3173	$N_{I}\ 4s$: 87.1*b
$M_{I}\ 3s$: 671.6†	$N_{II}\ 4p_{1/2}$: 55.7†a
$M_{II}\ 3p_{1/2}$: 559.9†	$N_{III}\ 4p_{3/2}$: 50.9†a

Phosphorus (15)

$K\ 1s$:2149	$L_{II}\ 2p_{1/2}$:136*
$L_{I}\ 2s$: 189*	$L_{III}\ 2p_{3/2}$:135*

Platinum (78)

$K\ 1s$:78395	$N_{II}\ 4p_{1/2}$:609.1†
$L_{I}\ 2s$:13880	$N_{III}\ 4p_{3/2}$:519.4†
$L_{II}\ 2p_{1/2}$:13273	$N_{IV}\ 4d_{3/2}$:331.6†
$L_{III}\ 2p_{3/2}$:11564	$N_{V}\ 4d_{5/2}$:314.6†
$M_{I}\ 3s$: 3296	$N_{VI}\ 4f_{5/2}$: 74.5†
$M_{II}\ 3p_{1/2}$: 3027	$N_{VII}\ 4f_{7/2}$: 71.2†
$M_{III}\ 3p_{3/2}$: 2645	$O_{I}\ 5s$:101.7*b
$M_{IV}\ 3d_{3/2}$: 2202	$O_{II}\ 5p_{1/2}$: 65.3*b
$M_{V}\ 3d_{5/2}$: 2122	$O_{III}\ 5p_{3/2}$: 51.7†
$N_{I}\ 4s$: 725.4†	

Polonium (84)

$K\ 1s$:93105	$N_{III}\ 4p_{3/2}$:705*
$L_{I}\ 2s$:16939	$N_{IV}\ 4d_{3/2}$:500*
$L_{II}\ 2p_{1/2}$:16244	$N_{V}\ 4d_{5/2}$:473*
$L_{III}\ 2p_{3/2}$:13814	$N_{VI}\ 4f_{5/2}$:184*
$M_{I}\ 3s$: 4149	$N_{VII}\ 4f_{7/2}$ 184*
$M_{II}\ 3p_{1/2}$: 3854	$O_{I}\ 5s$:177*
$M_{III}\ 3p_{3/2}$: 3302	$O_{II}\ 5p_{1/2}$:132*
$M_{IV}\ 3d_{3/2}$: 2798	$O_{III}\ 5p_{3/2}$:104*
$M_{V}\ 3d_{5/2}$: 2683	$O_{IV}\ 5d_{3/2}$: 31*
$N_{I}\ 4s$: 995*	$O_{V}\ 5d_{5/2}$: 31*
$N_{II}\ 4p_{1/2}$: 851*	

Potassium (19)

$K\ 1s$:3608.4*	$M_{I}\ 3s$:34.8*
$L_{I}\ 2s$: 378.6*	$M_{II}\ 3p_{1/2}$:18.3*
$L_{II}\ 2p_{1/2}$: 297.3*	$M_{III}\ 3p_{3/2}$:18.3*
$L_{III}\ 2p_{3/2}$: 294.6*	

Praseodymium (59)

$K\ 1s$:41991	$N_{II}\ 4p_{1/2}$:236.3
$L_{I}\ 2s$: 6835	$N_{III}\ 4p_{3/2}$:217.6
$L_{II}\ 2p_{1/2}$: 6440	$N_{IV}\ 4d_{3/2}$:115.1*
$L_{III}\ 2p_{3/2}$: 5964	$N_{V}\ 4d_{5/2}$:115.1*
$M_{I}\ 3s$: 1511	$N_{VI}\ 4f_{5/2}$: 2.0
$M_{II}\ 3p_{1/2}$: 1337	$N_{VII}\ 4f_{7/2}$: 2.0
$M_{III}\ 3p_{3/2}$: 1242	$O_{I}\ 5s$: 37.4
$M_{IV}\ 3d_{3/2}$: 948.3*	$O_{II}\ 5p_{1/2}$: 22.3
$M_{V}\ 3d_{5/2}$: 928.8*	$O_{III}\ 5p_{3/2}$: 22.3
$N_{I}\ 4s$: 304.5	

Promethium (61)

$K\ 1s$:45184	$M_{IV}\ 3d_{3/2}$:1052
$L_{I}\ 2s$: 7428	$M_{V}\ 3d_{5/2}$:1027
$L_{II}\ 2p_{1/2}$: 7013	$N_{I}\ 4s$: —
$L_{III}\ 2p_{3/2}$: 6459	$N_{II}\ 4p_{1/2}$: 242
$M_{I}\ 3s$: —	$N_{III}\ 4p_{3/2}$: 242
$M_{II}\ 3p_{1/2}$: 1403	$N_{IV}\ 4d_{3/2}$: 120
$M_{III}\ 3p_{3/2}$: 1357	$N_{V}\ 4d_{5/2}$: 120

Protoactinium (91)

$K\ 1s$:112601	$N_{IV}\ 4d_{3/2}$: 743*
$L_{I}\ 2s$: 21105	$N_{V}\ 4d_{5/2}$: 708*
$L_{II}\ 2p_{1/2}$: 20314	$N_{VI}\ 4f_{5/2}$: 371*
$L_{III}\ 2p_{3/2}$: 16733	$N_{VII}\ 4f_{7/2}$: 360*
$M_{I}\ 3s$: 5367	$O_{I}\ 5s$: 310*
$M_{II}\ 3p_{1/2}$: 5001	$O_{II}\ 5p_{1/2}$: 232*
$M_{III}\ 3p_{3/2}$: 4174	$O_{III}\ 5p_{3/2}$: 232*
$M_{IV}\ 3d_{3/2}$: 3611	$O_{IV}\ 5d_{3/2}$: 94*
$M_{V}\ 3d_{5/2}$: 3442	$O_{V}\ 5d_{5/2}$: 94*
$N_{I}\ 4s$: 1387*	$P_{I}\ 6s$: —
$N_{II}\ 4p_{1/2}$: 1224*	$P_{II}\ 6p_{1/2}$: —
$N_{III}\ 4p_{3/2}$: 1007*	$P_{III}\ 6p_{3/2}$: —

Radium (88)

$K\ 1s$:103922	$N_{IV}\ 4d_{3/2}$:636*
$L_{I}\ 2s$: 19237	$N_{V}\ 4d_{5/2}$:603*
$L_{II}\ 2p_{1/2}$: 18484	$N_{VI}\ 4f_{5/2}$:299*
$L_{III}\ 2p_{3/2}$: 15444	$N_{VII}\ 4f_{7/2}$:299*
$M_{I}\ 3s$: 4822	$O_{I}\ 5s$:254*
$M_{II}\ 3p_{1/2}$: 4490	$O_{II}\ 5p_{1/2}$:200*
$M_{III}\ 3p_{3/2}$: 3792	$O_{III}\ 5p_{3/2}$:153*
$M_{IV}\ 3d_{3/2}$: 3248	$O_{IV}\ 5d_{3/2}$: 68*
$M_{V}\ 3d_{5/2}$: 3105	$O_{V}\ 5d_{5/2}$: 68*
$N_{I}\ 4s$: 1208*	$P_{I}\ 6s$: 44
$N_{II}\ 4p_{1/2}$: 1958*	$P_{II}\ 6p_{1/2}$: 19
$N_{III}\ 4p_{3/2}$: 879*	$P_{III}\ 6p_{3/2}$: 19

Radon (86)

$K\ 1s$:98404	$N_{III}\ 4p_{3/2}$:768*
$L_{I}\ 2s$:18049	$N_{IV}\ 4d_{3/2}$:567*
$L_{II}\ 2p_{1/2}$:17337	$N_{V}\ 4d_{5/2}$:541*
$L_{III}\ 2p_{3/2}$:14619	$N_{VI}\ 4f_{5/2}$:238*
$M_{I}\ 3s$: 4482	$N_{VII}\ 4f_{7/2}$:238*
$M_{II}\ 3p_{1/2}$: 4159	$O_{I}\ 5s$:214*
$M_{III}\ 3p_{3/2}$: 3538	$O_{II}\ 5p_{1/2}$:164*
$M_{IV}\ 3d_{3/2}$: 3022	$O_{III}\ 5p_{3/2}$:127*
$M_{V}\ 3d_{5/2}$: 2892	$O_{IV}\ 5d_{3/2}$: 48*
$N_{I}\ 4s$: 1097*	$O_{V}\ 5d_{5/2}$: 48*
$N_{II}\ 4p_{1/2}$: 929*	$P_{I}\ 6s$: 26

Rhenium (75)

$K\ 1s$:71676	$N_{II}\ 4p_{1/2}$:518.7†
$L_{I}\ 2s$:12527	$N_{III}\ 4p_{3/2}$:446.8†
$L_{II}\ 2p_{1/2}$:11959	$N_{IV}\ 4d_{3/2}$:273.9†
$L_{III}\ 2p_{3/2}$:10535	$N_{V}\ 4d_{5/2}$:260.5†
$M_{I}\ 3s$: 2932	$N_{VI}\ 4f_{5/2}$: 42.9*
$M_{II}\ 3p_{1/2}$: 2682	$N_{VII}\ 4f_{7/2}$: 40.5*
$M_{III}\ 3p_{3/2}$: 2367	$O_{I}\ 5s$: 83 †
$M_{IV}\ 3d_{3/2}$: 1949	$O_{II}\ 5p_{1/2}$: 45.6†
$M_{V}\ 3d_{5/2}$: 1883	$O_{III}\ 5p_{3/2}$: 34.6*b
$N_{I}\ 4s$: 625.4†	

Rhodium (45)

$K\ 1s$:23220	$M_{III}\ 3p_{3/2}$:496.5†
$L_{I}\ 2s$: 3412	$M_{IV}\ 3d_{3/2}$:311.9†
$L_{II}\ 2p_{1/2}$: 3146	$M_{V}\ 3d_{5/2}$:307.2†
$L_{III}\ 2p_{3/2}$: 3004	$N_{I}\ 4s$: 81.4*b
$M_{I}\ 3s$: 628.1†	$N_{II}\ 4p_{1/2}$: 50.5†
$M_{II}\ 3p_{1/2}$: 521.3†	$N_{III}\ 4p_{3/2}$: 47.3†

Rubidium (37)

$K\ 1s$:15200	$M_{III}\ 3p_{3/2}$:239.1*
$L_{I}\ 2s$: 2065	$M_{IV}\ 3d_{3/2}$:113.0*
$L_{II}\ 2p_{1/2}$: 1864	$M_{V}\ 3d_{5/2}$:112 *
$L_{III}\ 2p_{3/2}$: 1804	$N_{I}\ 4s$: 30.5*
$M_{I}\ 3s$: 326.7*	$N_{II}\ 4p_{1/2}$: 16.3*
$M_{II}\ 3p_{1/2}$: 248.7*	$N_{III}\ 4p_{3/2}$: 15.3*

Ruthenium (44)

$K\ 1s$:22117	$M_{III}\ 3p_{3/2}$:461.5†
$L_{I}\ 2s$: 3224	$M_{IV}\ 3d_{3/2}$:284.2†
$L_{II}\ 2p_{1/2}$: 2967	$M_{V}\ 3d_{5/2}$:280.0†
$L_{III}\ 2p_{3/2}$: 2838	$N_{I}\ 4s$: 75.0†
$M_{I}\ 3s$: 586.2†	$N_{II}\ 4p_{1/2}$: 46.5†
$M_{II}\ 3p_{1/2}$: 483.3†	$N_{III}\ 4p_{3/2}$: 43.2†

Samarium (62)

$K\ 1s$:46834	$N_{II}\ 4p_{1/2}$:265.6
$L_{I}\ 2s$: 7737	$N_{III}\ 4p_{3/2}$:247.4
$L_{II}\ 2p_{1/2}$: 7312	$N_{IV}\ 4d_{3/2}$:129.0
$L_{III}\ 2p_{3/2}$: 6716	$N_{V}\ 4d_{5/2}$:129.0
$M_{I}\ 3s$: 1723	$N_{VI}\ 4f_{5/2}$: 5.2
$M_{II}\ 3p_{1/2}$: 1541	$N_{VII}\ 4f_{7/2}$: 5.2
$M_{III}\ 3p_{3/2}$: 1419.8	$O_{I}\ 5s$: 37.4
$M_{IV}\ 3d_{3/2}$: 1110.9*	$O_{II}\ 5p_{1/2}$: 21.3
$M_{V}\ 3d_{5/2}$: 1083.4*	$O_{III}\ 5p_{3/2}$: 21.3
$N_{I}\ 4s$: 347.2*	

Scandium (21)

$K\ 1s$:4492	$M_{I}\ 3s$:51.1*
$L_{I}\ 2s$: 498.0*	$M_{II}\ 3p_{1/2}$:28.3*
$L_{II}\ 2p_{1/2}$: 403.6*	$M_{III}\ 3p_{3/2}$:28.3*
$L_{III}\ 2p_{3/2}$: 398.7*	

Selenium (34)

$K\ 1s$:12658	$M_{II}\ 3p_{1/2}$:166.5*
$L_{I}\ 2s$: 1652.0*b	$M_{III}\ 3p_{3/2}$:160.7*
$L_{II}\ 2p_{1/2}$: 1474.3*b	$M_{IV}\ 3d_{3/2}$: 55.5*
$L_{III}\ 2p_{3/2}$: 1433.9*b	$M_{V}\ 3d_{5/2}$: 54.6*
$M_{I}\ 3s$: 229.6*	

Silicon (14)

$K\ 1s$:1839	$L_{II}\ 2p_{1/2}$:99.2*
$L_{I}\ 2s$: 149.7*b	$L_{III}\ 2p_{3/2}$:99.8*

Silver (47)

$K\ 1s$:25514	$M_{III}\ 3p_{3/2}$:573.0†
$L_{I}\ 2s$: 3806	$M_{IV}\ 3d_{3/2}$:374.0†
$L_{II}\ 2p_{1/2}$: 3524	$M_{V}\ 3d_{5/2}$:368.0†
$L_{III}\ 2p_{3/2}$: 3351	$N_{I}\ 4s$: 97.0†
$M_{I}\ 3s$: 719.0†	$N_{II}\ 4p_{1/2}$: 63.7†
$M_{II}\ 3p_{1/2}$: 603.8†	$N_{III}\ 4p_{3/2}$: 58.3†

Sodium (11)

$K\ 1s$:1070.8†
$L_{I}\ 2s$: 63.5†
$L_{II}\ 2p_{1/2}$:30.4†
$L_{III}\ 2p_{3/2}$:30.5*

Strontium (38)

$K\ 1s$:16105	$M_{III}\ 3p_{3/2}$:270.0†
$L_{I}\ 2s$: 2216	$M_{IV}\ 3d_{3/2}$:136.0†
$L_{II}\ 2p_{1/2}$: 2007	$M_{V}\ 3d_{5/2}$:134.2†
$L_{III}\ 2p_{3/2}$: 1940	$N_{I}\ 4s$: 38.9†

M_I 3s	358.7†	N_{II} $4p_{1/2}$	20.3†
M_{II} $3p_{1/2}$	280.3†	N_{III} $4p_{3/2}$	20.3†

Sulfur (16)

K 1s	2472	L_{II} $2p_{1/2}$	163.6*
L_I 2s	230.9*b	L_{III} $2p_{3/2}$	162.5*

Tantalum (73)

K 1s	67416	N_{II} $4p_{1/2}$	463.4†
L_I 2s	11682	N_{III} $4p_{3/2}$	400.9†
L_{II} $2p_{1/2}$	11136	N_{IV} $4d_{3/2}$	237.9†
L_{III} $2p_{3/2}$	9881	N_V $4d_{5/2}$	226.4†
M_I 3s	2708	N_{VI} $4f_{5/2}$	23.5†
M_{II} $3p_{1/2}$	2469	N_{VII} $4f_{7/2}$	21.6†
M_{III} $3p_{3/2}$	2194	O_I 5s	69.7†
M_{IV} $3d_{3/2}$	1793	O_{II} $5p_{1/2}$	42.2*
M_V $3d_{5/2}$	1735	O_{III} $5p_{3/2}$	32.7†
N_I 4s	563.4†		

Technetium (43)

K 1s	21044	M_{III} $3p_{3/2}$	425*
L_I 2s	3043	M_{IV} $3d_{3/2}$	257*
L_{II} $2p_{1/2}$	2793	M_V $3d_{5/2}$	253*
L_{III} $2p_{3/2}$	2677	N_I 4s	68*
M_I 3s	544*	N_{II} $4p_{1/2}$	39*
M_{II} $3p_{1/2}$	445*	N_{III} $4p_{3/2}$	39*

Tellurium (52)

K 1s	31814	M_{IV} $3d_{3/2}$	583.4†
L_I 2s	4939	M_V $3d_{5/2}$	573.0†
L_{II} $2p_{1/2}$	4612	N_I 4s	169.4†
L_{III} $2p_{3/2}$	4341	N_{II} $4p_{1/2}$	103.3†a
M_I 3s	1006 †	N_{III} $4p_{3/2}$	103.3†a
M_{II} $3p_{1/2}$	870.8†	N_{IV} $4d_{3/2}$	41.9†
M_{III} $3p_{3/2}$	820.0†	N_V $4d_{5/2}$	40.4†

Terbium (65)

K 1s	51996	N_{II} $4p_{1/2}$	322.4*
L_I 2s	8708	N_{III} $4p_{3/2}$	284.1*
L_{II} $2p_{1/2}$	8252	N_{IV} $4d_{3/2}$	150.5*
L_{III} $2p_{3/2}$	7514	N_V $4d_{5/2}$	150.5*
M_I 3s	1968	N_{VI} $4f_{5/2}$	7.7*
M_{II} $3p_{1/2}$	1768	N_{VII} $4f_{7/2}$	2.4*
M_{III} $3p_{3/2}$	1611	O_I 5s	45.6*
M_{IV} $3d_{3/2}$	1276.9*	O_{II} $5p_{1/2}$	28.7*
M_V $3d_{5/2}$	1241.1*	O_{III} $5p_{3/2}$	22.6*
N_I 4s	396.0*		

Thallium (81)

K 1s	85530	N_{III} $4p_{3/2}$	609.5†
L_I 2s	15347	N_{IV} $4d_{3/2}$	405.7†
L_{II} $2p_{1/2}$	14698	N_V $4d_{5/2}$	385.0†
L_{III} $2p_{3/2}$	12658	N_{VI} $4f_{5/2}$	122.2†
M_I 3s	3704	N_{VII} $4f_{7/2}$	117.8†
M_{II} $3p_{1/2}$	3416	O_I 5s	136 *b
M_{III} $3p_{3/2}$	2957	O_{II} $5p_{1/2}$	94.6†
M_{IV} $3d_{3/2}$	2485	O_{III} $5p_{3/2}$	73.5†
M_V $3d_{5/2}$	2389	O_{IV} $5d_{3/2}$	14.7†
N_I 4s	846.2†	O_V $5d_{5/2}$	12.5†
N_{II} $4p_{1/2}$	720.5†		

Thorium (90)

K 1s	109651	N_{IV} $4d_{3/2}$	712.1†
L_I 2s	20472	N_V $4d_{5/2}$	675.2†
L_{II} $2p_{1/2}$	19693	N_{VI} $4f_{5/2}$	342.4†
L_{III} $2p_{3/2}$	16300	N_{VII} $4f_{7/2}$	333.1†
M_I 3s	5182	O_I 5s	290 *a
M_{II} $3p_{1/2}$	4830	O_{II} $5p_{1/2}$	229 *a
M_{III} $3p_{3/2}$	4046	O_{III} $5p_{3/2}$	182 *a
M_{IV} $3d_{3/2}$	3491	O_{IV} $5d_{3/2}$	92.5†
M_V $3d_{5/2}$	3332	O_V $5d_{5/2}$	85.4†
N_I 4s	1330 *	P_I 6s	41.4†
N_{II} $4p_{1/2}$	1168 *	P_{II} $6p_{1/2}$	24.5†
N_{III} $4p_{3/2}$	966.4†	P_{III} $6p_{3/2}$	16.6†

Thulium (69)

K 1s	59390	N_{II} $4p_{1/2}$	385.9*
L_I 2s	10116	N_{III} $4p_{3/2}$	332.6*
L_{II} $2p_{1/2}$	9617	N_{IV} $4d_{3/2}$	175.5*
L_{III} $2p_{3/2}$	8648	N_V $4d_{5/2}$	175.5*
M_I 3s	2307	N_{VI} $4f_{5/2}$	—
M_{II} $3p_{1/2}$	2090	N_{VII} $4f_{7/2}$	4.6
M_{III} $3p_{3/2}$	1885	O_I 5s	54.7*
M_{IV} $3d_{3/2}$	1515	O_{II} $5p_{1/2}$	31.8*
M_V $3d_{5/2}$	1468	O_{III} $5p_{3/2}$	25.0*
N_I 4s	470.9*		

Tin (50)

K 1s	29200	M_{IV} $3d_{3/2}$	493.2†
L_I 2s	4465	M_V $3d_{5/2}$	484.9†
L_{II} $2p_{1/2}$	4156	N_I 4s	137.1†
L_{III} $2p_{3/2}$	3929	N_{II} $4p_{1/2}$	83.6†a
M_I 3s	884.7†	N_{III} $4p_{3/2}$	83.6†a
M_{II} $3p_{1/2}$	756.5†	N_{IV} $4d_{3/2}$	24.9†
M_{III} $3p_{3/2}$	714.6†	N_V $4d_{5/2}$	23.9†

Titanium (22)

K 1s	4966	M_I 3s	58.7†
L_I 2s	560.9†	M_{II} $3p_{1/2}$	32.6†
L_{II} $2p_{1/2}$	461.2†	M_{III} $3p_{3/2}$	32.6†
L_{III} $2p_{3/2}$	453.8†		

Tungsten (74)

K 1s	69525	N_{II} $4p_{1/2}$	490.4†
L_I 2s	12100	N_{III} $4p_{3/2}$	423.6†
L_{II} $2p_{1/2}$	11544	N_{IV} $4d_{3/2}$	255.9†
L_{III} $2p_{3/2}$	10207	N_V $4d_{5/2}$	243.5†
M_I 3s	2820	N_{VI} $4f_{5/2}$	33.6*
M_{II} $3p_{1/2}$	2575	N_{VII} $4f_{7/2}$	31.4†
M_{III} $3p_{3/2}$	2281	O_I 5s	75.6†
M_{IV} $3d_{3/2}$	1949	O_{II} $5p_{1/2}$	453 *b
M_V $3d_{5/2}$	1809	O_{III} $5p_{3/2}$	36.8†
N_I 4s	594.1†		

Uranium (92)

K 1s	115606	N_{IV} $4d_{3/2}$	778.3†
L_I 2s	21757	N_V $4d_{5/2}$	736.2†
L_{II} $2p_{1/2}$	20948	N_{VI} $4f_{5/2}$	388.2*
L_{III} $2p_{3/2}$	17166	N_{VII} $4f_{7/2}$	377.4†
M_I 3s	5548	O_I 5s	321 *ab
M_{II} $3p_{1/2}$	5182	O_{II} $5p_{1/2}$	257 *ab
M_{III} $3p_{3/2}$	4303	O_{III} $5p_{3/2}$	192 *ab
M_{IV} $3d_{3/2}$	3728	O_{IV} $5d_{3/2}$	102.8†
M_V $3d_{5/2}$	3552	O_V $5d_{5/2}$	94.2†
N_I 4s	1439*b	P_I 6s	43.9†
N_{II} $4p_{1/2}$	1271*b	P_{II} $6p_{1/2}$	26.8†
N_{III} $4p_{3/2}$	1043†	P_{III} $6p_{3/2}$	16.8†

Vanadium (23)

K 1s	5465	M_I 3s	66.3†
L_I 2s	626.7†	M_{II} $3p_{1/2}$	37.2†
L_{II} $2p_{1/2}$	519.8†	M_{III} $3p_{3/2}$	37.2†
L_{III} $2p_{3/2}$	512.1†		

Xenon (54)

K 1s	34561	N_{II} $4p_{1/2}$	146.7
L_I 2s	5453	N_{III} $4p_{3/2}$	145.5*
L_{II} $2p_{1/2}$	5104	N_{IV} $4d_{3/2}$	69.5*
L_{III} $2p_{3/2}$	4782	N_V $4d_{5/2}$	67.5*
M_I 3s	1148.7*	N_{VI} $4f_{5/2}$	—
M_{II} $3p_{1/2}$	1002.1*	N_{VII} $4f_{7/2}$	—
M_{III} $3p_{3/2}$	940.6*	O_I 5s	23.3*
M_{IV} $3d_{3/2}$	689.0*	O_{II} $5p_{1/2}$	13.4*
M_V $3d_{5/2}$	676.4*	O_{III} $5p_{3/2}$	12.1*
N_I 4s	213.2*		

Ytterbium (70)

K 1s	61332	N_{II} $4p_{1/2}$	388.7*
L_I 2s	10486	N_{III} $4p_{3/2}$	339.7*
L_{II} $2p_{1/2}$	9978	N_{IV} $4d_{3/2}$	191.2*
L_{III} $2p_{3/2}$	8944	N_V $4d_{5/2}$	182.4*
M_I 3s	2398	N_{VI} $4f_{5/2}$	2.5*
M_{II} $3p_{1/2}$	2173	N_{VII} $4f_{7/2}$	1.3*
M_{III} $3p_{3/2}$	1950	O_I 5s	52.0*
M_{IV} $3d_{3/2}$	1576	O_{II} $5p_{1/2}$	30.3*
M_V $3d_{5/2}$	1528	O_{III} $5p_{3/2}$	24.1*
N_I 4s	480.5*		

Yttrium (39)

K 1s	17038	M_{III} $3p_{3/2}$	298.8*
L_I 2s	2373	M_{IV} $3d_{3/2}$	157.7†
L_{II} $2p_{1/2}$	2156	M_V $3d_{5/2}$	155.8†
L_{III} $2p_{3/2}$	2080	N_I 4s	43.8*
M_I 3s	392.0*b	N_{II} $4p_{1/2}$	24.4*
M_{II} $3p_{1/2}$	310.6*	N_{III} $4p_{3/2}$	23.1*

Zinc (30)

K 1s	9659	M_{II} $3p_{1/2}$	91.4*
L_I 2s	1200.7*	M_{III} $3p_{3/2}$	88.6*
L_{II} $2p_{1/2}$	1044.9*	M_{IV} $3d_{3/2}$	10.2*
L_{III} $2p_{3/2}$	1021.8*	M_V $3d_{5/2}$	10.1*
M_I 3s	139.8*		

Zirconium (40)

K 1s	17998	M_{III} $3p_{3/2}$	329.8†
L_I 2s	2532	M_{IV} $3d_{3/2}$	181.1†
L_{II} $2p_{1/2}$	2307	M_V $3d_{5/2}$	178.8†
L_{III} $2p_{3/2}$	2223	N_I 4s	50.6†
M_I 3s	430.3†	N_{II} $4p_{1/2}$	28.5†
M_{II} $3p_{1/2}$	343.5†	N_{III} $4p_{3/2}$	27.7†

Referred to the Fermi Level (metals), Valence Band Max (Semiconductors), Vacuum Level (Rare Gases)

† From Fuggle and Martensson, *J. Elect. Spect.*, 21, 275, 1980.

* From Cardona and Ley *Photoemission from Solids*, Springer Verlag, 1978. Rest from Bearden and Burr, *Rev. Mod. Phys.*, 39, 125, 1967.

a One-particle approximation not valid.

b Derived from Bearden and Burr.

STRENGTHS OF CHEMICAL BONDS*

J. A. Kerr

The strength of a chemical bond, $D^{\circ}(R\text{–}X)$, often known as the bond dissociation energy, is defined as the standard enthalpy change of the reaction in which the bond is broken: $RX \rightarrow R + X$. It is given by the thermochemical equation, $D^{\circ}(R\text{–}X) = \Delta H^{\circ}_f(R) + \Delta H^{\circ}_f(X) - \Delta H^{\circ}_f(RX)$. Some authors list bond strengths at a temperature of absolute zero but here the values at 298 K are given because more thermodynamic data are available for this temperature. Bond strengths or bond dissociation energies are not equal to, and may differ considerably from, mean bond energies derived solely from thermochemical data on atoms and molecules.

BOND STRENGTHS IN DIATOMIC MOLECULES

These have usually been measured spectroscopically or by mass spectrometric analysis of hot gases effusing from a Knudsen cell. Excellent accounts of these and other methods are given in References 112 and 118. The errors quoted in the Table are those given in the original paper or review article. The references have been chosen primarily as a key to the literature. It should not be assumed that the author referred to was responsible for the value quoted, as the reference may be to a review article.

Bond strengths at a temperature of absolute zero, D°_0 have been converted to D°_{298} by the use of enthalpy functions taken mainly from the J.A.N.A.F. Thermochemical Tables, NSRDS-NBS, 37, 1971, wherever possible. For most bonds, however, where this data is not available, the conversion has been made by the approximate relation:

$$D^{\circ}_{298} = D^{\circ}_0 + (3/2)RT$$

The table has been arranged in an alphabetical order of the atoms.

Table 1
BOND STRENGTHS IN DIATOMIC MOLECULES

Molecule	D°_{298}/kcal mol^{-1}	D°_{298}/kJ mol^{-1}	Ref.
Ag–Ag	39 ± 2	163 ± 8	72
Ag–Al	43.9 ± 2.2	183.7 ± 9.2	70
Ag–Au	48.5 ± 2.2	202.9 ± 9.2	3
Ag–Bi	46 ± 10	193 ± 42	222
Ag–Br	70 ± 7	293 ± 29	112
Ag–Cl	81.6	341.4	150
Ag–Cu	41.6 ± 2.2	174.1 ± 9.2	118
Ag–D	54.2	226.8	189
Ag–Dy	31 ± 5	130 ± 19	187
Ag–Eu	31.0 ± 3.0	129.7 ± 12.6	54
Ag–F	84.7 ± 3.9	354.4 ± 16.3	112
Ag–Ga	43 ± 4	180 ± 15	37
Ag–Ge	41.7 ± 5.0	174.9 ± 20.9	247
Ag–H	51.4 ± 2	215.0 ± 8.4	199
Ag–Ho	29.5 ± 4.0	123.4 ± 16.7	49
Ag–I	56 ± 7	234 ± 29	112
Ag–In	42 ± 4	176 ± 17	25
Ag–Li	42.4 ± 1.5	177.4 ± 6.3	252
Ag–Mn	24 ± 5	100 ± 21	222
Ag–Na	33.0 ± 2.0	138.1 ± 8.4	262, 274
Ag–Nd	<50	<209	198
Ag–O	52.6 ± 5.0	220.1 ± 20.9	259
Ag–S	51.9	217.2	305
Ag–Se	48.4	202.5	305
Ag–Si	42.5 ± 2.4	177.8 ± 10.2	286
Ag–Sn	32.5 ± 5.0	136.0 ± 20.9	2
Ag–Te	46.8	195.8	305
Al–Al	44.5 ± 2.2	186.2 ± 9.2	41, 151
Al–As	48.5 ± 1.7	202.9 ± 7.1	266, 268
Al–Au	77.9 ± 1.5	325.9 ± 6.3	134
Al–Br	106 ± 2	444 ± 8	72
Al–Cl	122.2 ± 0.2	511.3 ± 0.8	280
Al–Cu	51.8 ± 2.5	216.7 ± 10.5	264
Al–D	69.5	290.8	222
Al–F	158.6 ± 1.5	663.6 ± 6.3	72
Al–H	68.1 ± 1.5	284.9 ± 6.3	72
Al–I	88.4 ± 0.5	369.9 ± 2.1	242
Al–Li	42.0 ± 3.5	175.7 ± 14.6	151
Al–N	71 ± 23	297 ± 96	112
Al–P	51.8 ± 3.0	216.7 ± 12.6	72
Al–Pd	60.8 ± 2.9	254.4 ± 12.0	52
Al–S	89.3 ± 1.9	373.6 ± 8.0	326
Al–Sb	51.7 ± 1.4	216.3 ± 6.0	267
Al–Se	80.7 ± 2.4	337.7 ± 10.1	326
Al–Si	54.8 ± 7.2	229.3 ± 30.1	41
Al–O	122.4 ± 2.2	512.1 ± 9.2	259
Al–Te	64.0 ± 2.4	267.8 ± 10.1	326
Al–U	78 ± 7	326 ± 29	133
Ar–Ar	1.13 ± 0.01	4.73 ± 0.04	56, 235
Ar–He	0.93	3.89	222
Ar–Hg	1.47	6.15	222
Ar–I	2.4	10.0	34
Ar–K	1.0	4.2	189
As–As	91.3 ± 2.5	382.0 ± 10.5	223
As-Cl	107	448	72
As–D	64.6	270.3	189
As–F	98	410	189
As–Ga	50.1 ± 0.3	209.6 ± 1.2	77
As–H	84	352	81
As–I	70.9 ± 6.7	296.6 ± 28.0	291
As–In	48	201	273
As–N	139 ± 30	582 ± 126	72
As–O	115 ± 2	481 ± 8	230
As–P	103.6 ± 3.0	433.5 ± 12.6	137
As–S	90.7 ± 1.5	379.5 ± 6.3	230
As–Sb	79.0 ± 1.3	330.5 ± 5.4	91
As–Se	23	96	261
As–Tl	47.4 ± 3.5	198.3 ± 14.6	276
At–At	~19	~80	86
Au–Au	53.8 ± 0.5	221.3 ± 2.1	226
Au–B	87.9 ± 2.5	367.8 ± 10.5	125
Au–Ba	38 ± 14	159 ± 59	112
Au–Be	68 ± 2	285 ± 8	112
Au–Bi	70 ± 20	293 ± 84	222
Au–Ca	46 ± 23	193 ± 96	112
Au–Ce	77.9 ± 3.5	325.9 ± 14.6	139
Au–Cl	82 ± 2.3	343 ± 9.6	112
Au–Co	51.3 ± 3.0	214.6 ± 12.6	197, 305
Au–Cr	51.4 ± 1.5	215.1 ± 6.3	72
Au–Cu	56.3 ± 2.2	235.6 ± 9.2	3
Au–Cs	61 ± 0.8	255 ± 3.5	36
Au–D	76.1	318.1	189
Au–Dy	62 ± 5	259 ± 20	187
Au–Eu	57.6 ± 2.5	241.0 ± 10.5	54
Au–Fe	44.7 ± 4.0	187.0 ± 16.7	197
Au–Ga	70.2 ± 3.6	293.7 ± 15.1	37
Au–Ge	66.2 ± 3.5	277.0 ± 14.6	247
Au–H	69.8 ± 2	292.0 ± 8	199
Au–Ho	63.9 ± 4.0	267.4 ± 16.7	49, 226
Au–La	80.4 ± 5.0	336.4 ± 20.9	139
Au–Li	68.0 ± 1.6	284.5 ± 6.5	252
Au–Lu	79.4 ± 4.0	332.2 ± 16.7	115
Au–Mg	58 ± 10	243 ± 42	222
Au–Mn	44.3 ± 3.0	185.4 ± 12.6	300
Au–Na	51.4 ± 3.0	215.1 ± 12.6	274
Au–Nd	71.5 ± 5.0	299.2 ± 20.9	139
Au–Ni	59 ± 5	247 ± 21	197
Au–O	53.0 ± 5.0	221.8 ± 20.9	259
Au–Pb	31 ± 10	130 ± 42	222
Au–Pd	34.2 ± 5.0	143.1 ± 20.9	4
Au–Pr	72.9 ± 5.0	305.0 ± 20.9	139
Au–Rb	58 ± 0.7	243 ± 30.0	36
Au–Rh	55.2 ± 7	230.9 ± 29	54
Au–S	100 ± 6	418 ± 25	114
Au–Sc	67.0 ± 4.0	280.3 ± 16.7	140
Au–Se	58.1	243.1	305
Au–Si	73.0 ± 1.4	305.2 ± 6.0	142
Au–Sn	58.4 ± 4.0	244.3 ± 16.7	2
Au–Sr	63 ± 10	264 ± 42	222
Au–Tb	69.2 ± 8.0	289.5 ± 33.5	132, 226
Au–Te	75.9	317.6	305
Au–U	76 ± 7	318 ± 29	133
Au–V	57.5 ± 2.9	240.6 ± 12.1	166
Au–Y	73.4 ± 2.0	307.1 ± 8.2	170
B–B	71 ± 5	297 ± 21	72
B–Br	101 ± 5	423 ± 21	112
B–C	107 ± 7	448 ± 29	222
B–Ce	73 ± 5	305 ± 21	222
B–Cl	128	536	11
B–D	81.5 ± 1.5	341.0 ± 6	222
B–F	181	757	232
B–H	79.8	333.9	193
B–I	45.8 ± 0.2	220.5 ± 0.8	282
B–Ir	122.9 ± 4.1	514.2 ± 17.2	330
B–La	81 ± 15	339 ± 63	222
B–N	93 ± 5	389 ± 21	72
B–O	193.3 ± 5.0	808.8 ± 20.9	259
B–P	82.9 ± 4.0	346.9 ± 16.7	127
B–Pd	78.7 ± 5.0	329.3 ± 20.9	330
B–Pt	114.2 ± 4.0	477.8 ± 16.7	241
B–Rh	113.7 ± 5.0	475.7 ± 20.9	330
B–Ru	106.8 ± 5.0	446.9 ± 20.9	330
B–S	138.8 ± 2.2	580.7 ± 9.2	324
B–Sc	66 ± 15	276 ± 63	222
B–Se	110.4 ± 3.5	461.9 ± 14.6	324
B–Si	68.9	288.3	337
B–Te	84.7 ± 4.8	354.4 ± 20.1	324
B–Th	71	297	119
B–Ti	66 ± 15	276 ± 63	222
B–U	77 ± 8	322 ± 34	222
B–Y	70 ± 15	293 ± 63	222

* Revised to June 30, 1986.

Table 1 (continued)
BOND STRENGTHS IN DIATOMIC MOLECULES

Molecule	D°_{298}/kcal mol^{-1}	D°_{298}/kJ mol^{-1}	Ref.
Ba–Br	86.7 ± 2	362.8 ± 8.4	100, 183, 210
Ba–Cl	104.2 ± 2.0	436.0 ± 8.4	181, 183
Ba–D	≤46.3	≤193.7	189
Ba–F	140.3 ± 1.6	587.0 ± 6.7	97, 180
Ba–H	42 ± 3.5	176 ± 15	112
Ba–I	73.8 ± 2	308.8 ± 8.4	79, 100, 217
Ba–O	134.3 ± 3.2	561.9 ± 13.4	259
Ba–Pd	53.0 ± 1.2	221.8 ± 5.0	135
Ba–Rh	62.0 ± 6.0	259.4 ± 25.1	135
Ba–S	95.6 ± 4.5	400.0 ± 18.8	61
Be–Be	14	59	27, 84
Be–Br	91 ± 20	381 ± 84	222
Be–Cl	92.8 ± 2.2	388.3 ± 9.2	104, 186, 332
Be–D	48.53	203.06	189
Be–F	138 ± 10	577 ± 42	72, 104
Be–H	47.8 ± 0.3	200.0 ± 1.3	57
Be–O	103.9 ± 3.2	434.7 ± 13.4	259
Be–S	89 ± 14	372 ± 59	112
Bi–Bi	47.9 ± 1.8	200.4 ± 7.5	279, 290
Bi–Br	63.9 ± 1.0	267.4 ± 4.2	67
Bi–Cl	72 ± 1	301 ± 8	69
Bi–D	67.8	283.7	240
Bi–F	62 ± 7	259 ± 29	112
Bi–Ga	38 ± 4	159 ± 17	272
Bi–H	≤67.7	≤283.3	240
Bi–I	52.1 ± 1.1	218.0 ± 4.6	68
Bi–In	36.7 ± 0.4	153.6 ± 1.7	287
Bi–Li	36.8 ± 1.2	154.0 ± 5.0	250, 277
Bi–O	80.6 ± 3.0	337.2 ± 12.6	259
Bi–P	67 ± 3	280 ± 13	137
Bi–Pb	33.9 ± 3.5	141.8 ± 14.6	290
Bi–S	75.4 ± 1.1	315.5 ± 4.6	325
Bi–Sb	60 ± 1	251 ± 4	220
Bi–Se	67.0 ± 1.4	280.3 ± 5.9	325
Bi–Te	55.5 ± 2.7	232.2 ± 11.3	325
Bi–Tl	29 ± 3	121 ± 13	78
Br–Br	46.082	192.807	189
Br–C	67 ± 5	280 ± 21	112
Br–Ca	74.3 ± 2.2	310.9 ± 9.2	183, 210
Br–Cd	38 ± 23	159 ± 96	112
Br–Cl	51.99 ± 0.07	217.53 ± 0.31	45
Br–Co	79 ± 10	331 ± 42	222
Br–Cr	78.4 ± 5.8	328.0 ± 24.3	112
Br–Cs	93.0 ± 1	389.1 ± 4.2	257, 317
Br–Cu	79 ± 6	331 ± 25	112
Br–D	88.61	370.75	189
Br–F	59.8 ± 0.2	250.2 ± 0.6	44, 47
Br–Fe	59 ± 23	247 ± 96	112
Br–Ga	106 ± 4	444 ± 17	72
Br–Ge	61 ± 7	255 ± 29	112
Br–H	87.56	366.35	189
Br–Hg	17.4 ± 1	72.8 ± 4.2	72
Br–I	42.8 ± 0.1	179.1 ± 0.4	112
Br–In	99 ± 5	414 ± 21	72
Br–K	90.8 ± 0.2	379.9 ± 0.8	317, 331
Br–Li	100.1 ± 1	418.8 ± 4.2	317
Br–Mg	≤78.2	≤327.2	189
Br–Mn	75.1 ± 2.3	314.2 ± 9.6	112
Br–N	66 ± 5	276 ± 21	112
Br–Na	78.8 ± 0.2	367.4 ± 0.8	317, 331
Br–Ni	86 ± 3	360 ± 13	112
Br–O	56.2 ± 0.1	235.1 ± 0.4	72
Br–Pb	59 ± 9	247 ± 38	112
Br–Rb	91.0 ± 1	380.7 ± 4.2	317
Br–Sb	75 ± 14	314 ± 59	112
Br–Sc	106 ± 15	444 ± 63	222
Br–Se	71 ± 20	297 ± 84	222
Br–Si	87.9 ± 2.4	367.8 ± 10	102
Br–Sn	≥132	≥552	258
Br–Sr	79.6 ± 2.2	333.1 ± 9.2	183
Br–Ti	105	439	222
Br–Tl	79.8 ± 0.4	333.9 ± 1.7	21
Br–V	105 ± 10	439 ± 42	222
Br–W	78.7	329.3	204
Br–Y	116 ± 20	485 ± 84	222
Br–Zn	34 ± 7	142 ± 29	222
C–C	145 ± 5	607 ± 21	72
C–Ce	106 ± 3	445 ± 12	212
C–Cl	95 ± 7	397 ± 29	256
C–D	81.6	341.4	189
C–F	132	552	178
C–Ge	110 ± 5	460 ± 21	112
C–H	80.86	338.32	176, 189
C–Hf	129 ± 6	540 ± 25	311
C–I	50 ± 5	209 ± 21	112
C–Ir	151 ± 1	632 ± 5	165
C–La	121 ± 15	506 ± 63	222
C–Mo	115 ± 3.8	481 ± 16	162
C–N	184 ± 1	770 ± 4	73
C–Nb	136 ± 3.1	569 ± 13	162
C–O	257.3 ± 0.1	1076.5 ± 0.4	72
C–Os	≥142	≥594	136
C–P	122.7 ± 2	513.4 ± 8	306
C–Pt	143 ± 1.4	598 ± 5.9	165, 328
C–Rh	139.5 ± 1.5	583.7 ± 6.3	328
C–Ru	154.9 ± 3	648.1 ± 12	117
C–S	170.5 ± 0.3	713.4 ± 1.2	62
C–Sc	≤106 ± 5	≤444 ± 21	130
C–Se	141.1 ± 1.4	590.4 ± 5.9	301
C–Si	107.9	451.5	83, 337
C–Tc	135 ± 7	565 ± 30	288
C–Th	108.3 ± 4.1	453.1 ± 17.2	161, 311
C–Ti	101 ± 7	423 ± 30	156, 311
C–U	109 ± 4	455 ± 15	159, 163
C–V	102 ± 5.7	427 ± 24	162
C–Y	100 ± 15	418 ± 63	124
C–Zr	134 ± 6	561 ± 25	311
Ca–Ca	≤11	≤46	345, 346
Ca–Cl	95 ± 3	398 ± 13	181, 332
Ca–D	≤40.6	≤169.9	189
Ca–F	126 ± 5	527 ± 21	97, 185
Ca–H	40.1	167.8	112
Ca–I	63.0 ± 2.5	263.6 ± 10.5	217
Ca–Li	20.3 ± 2.0	84.9 ± 8.4	345
Ca–O	96.1 ± 4.0	402.1 ± 16.7	259
Ca–S	80.7 ± 4.5	337.7 ± 18.8	61, 189
Cd–Cd	2.7 ± 0.2	11.3 ± 0.8	112
Cd–Cl	49.8	208.4	189
Cd–F	73 ± 5	305 ± 21	23
Cd–H	16.5 ± 0.1	69.0 ± 0.4	112
Cd–I	33 ± 5	138 ± 21	112
Cd–In	33	138	222
Cd–O	56.3 ± 20.0	235.6 ± 83.7	148, 259
Cd–S	49.8 ± 5.0	208.5 ± 21	148
Cd–Se	30.5 ± 6.0	127.7 ± 25.1	148
Cd–Te	23.9 ± 3.6	100.1 ± 15.0	148
Ce–Ce	58.6	245.2	139
Ce–F	139 ± 10	582 ± 42	222
Ce–Ir	140	586	131
Ce–N	124 ± 5	519 ± 21	126
Ce–O	192.6 ± 3.2	805.8 ± 13.4	259
Ce–Os	121 ± 8	507 ± 33	136
Ce–Pd	77.0	322.2	50
Ce–Pt	133	557	131
Ce–Rh	131	548	131
Ce–Ru	127 ± 6	531 ± 25	136
Ce–S	136	569	18
Ce–Se	118.2 ± 3.5	494.6 ± 14.6	245
Ce–Te	93 ± 10	389 ± 42	222
Cl–Cl	58.978 ± 0.001	242.580 ± 0.004	189
Cl–Co	93	389	228
Cl–Cr	87.5 ± 5.8	366.1 ± 24.3	112
Cl–Cs	107 ± 2	448 ± 8	257, 318
Cl–Cu	91.5 ± 1.1	382.8 ± 4.6	153
Cl–D	104.32	436.47	189
Cl–Eu	~78	~326	108
Cl–F	61.24	256.23	189, 254
Cl–Fe	~84	~352	112
Cl–Ga	115 ± 3	481 ± 13	72
Cl–Ge	~103	~431	189
Cl–H	103.16	431.62	189
Cl–Hg	24 ± 2	100 ± 8	112
Cl–I	50.5 ± 0.1	211.3 ± 0.4	112
Cl–In	105 ± 2	439 ± 8	72
Cl–K	103.5 ± 2	433.0 ± 8.2	318
Cl–Li	112 ± 3	469 ± 13	72
Cl–Mg	78.3 ± 0.5	327.6 ± 2.1	101, 181, 332
Cl–Mn	86.2 ± 2.3	360.7 ± 9.6	112
Cl–N	79.8 ± 2.3	333.9 ± 9.6	42
Cl–Na	98.5 ± 2	412.1 ± 8	318
Cl–Ni	89 ± 5	372 ± 21	112
Cl–O	65 ± 1	272 ± 4	72
Cl–P	69 ± 10	289 ± 42	222
Cl–Pb	72 ± 7	301 ± 29	112
Cl–Ra	82 ± 18	343 ± 75	112
Cl–Rb	102.2 ± 2	427.6 ± 8	318
Cl–S	66.2	277.0	205
Cl–Sb	86 ± 12	360 ± 50	112
Cl–Sc	79	331	336
Cl–Se	77	322	222
Cl–Si	96	406	334
Cl–Sm	≥101 ± 3	≥423 ± 13	348
Cl–Sn	99 ± 4	414 ± 17	222
Cl–Sr	97 ± 3	406 ± 13	181, 183
Cl–Ta	130	544	16
Cl–Ti	118	494	222
Cl–Tl	89.1 ± 0.5	372.8 ± 2.1	21
Cl–U	108 ± 2	452 ± 8	234
Cl–V	114 ± 15	477 ± 63	222
Cl–W	101 ± 10	423 ± 42	222
Cl–Xe	1.6	6.7	189
Cl–Y	126 ± 20	527 ± 84	222
Cl–Yb	~77	~322	108
Cl–Zn	54.7 ± 4.7	228.9 ± 19.7	64
Cm–O	176	736	297
Co–Co	40 ± 6	167 ± 25	200
Co–Cu	38.7 ± 4.0	161.9 ± 16.7	203
Co–F	104 ± 15	435 ± 63	222
Co–Ge	57 ± 6	239 ± 25	202
Co–I	68 ± 5	285 ± 21	222
Co–O	91.9 ± 3.2	384.5 ± 13.4	259
Co–S	79 ± 5	331 ± 21	305
Co–Si	66 ± 4	276 ± 17	329
Cr–Cr	37 ± 5	155 ± 21	201
Cr–Cu	37 ± 5	155 ± 21	203
Cr–F	106.3 ± 4.7	444.8 ± 19.7	209
Cr–Ge	40.6 ± 7	169.9 ± 29	202
Cr–H	67 ± 12	280 ± 50	112
Cr–I	68.6 ± 5.8	287.0 ± 24.3	112
Cr–N	90.3 ± 4.5	377.8 ± 18.8	132, 308
Cr–O	102.6 ± 7.0	429.3 ± 29.3	259
Cr–S	79	331	88
Cs–Cs	9.97 ± 0.22	41.75 ± 0.92	189
Cs–F	124 ± 2	519 ± 8	257
Cs–H	41.90 ± 0.02	175.31 ± 0.08	66, 347
Cs–HG	2	5	189
Cs–I	80.6 ± 0.5	337.2 ± 2.1	257, 316
Cs–Na	15.1 ± 0.3	63.2 ± 1.2	80
Cs–O	70.7 ± 15.0	295.8 ± 62.8	259
Cs–Rb	11.81 ± 0.01	49.41 ± 0.04	219
Cu–Cu	46.6 ± 0.8	195.0 ± 3.4	187, 213

Table 1 (continued)
BOND STRENGTHS IN DIATOMIC MOLECULES

Molecule	D°_{298}/kcal mol^{-1}	D°_{298}/kJ mol^{-1}	Ref.
Cu–D	64.6	270.3	189
Cu–Dy	34 ± 5	142 ± 19	187
Cu–F	98.8 ± 3	413.4 ± 12.6	96
Cu–GA	51.6 ± 3.6	215.9 ± 15.1	37
Cu–Ge	49.9 ± 5	208.8 ± 21	247
Cu–H	66.4	277.8	199, 284
Cu–Ho	34 ± 5	142 ± 21	187
Cu–I	47 ± 5	197 ± 21	112
Cu–Li	46.1 ± 2.1	192.9 ± 8.8	252
Cu–Mn	37.9 ± 4	158.6 ± 17	203
Cu–Na	42.1 ± 4.0	176.2 ± 16.7	275
Cu–Ni	49.2 ± 4	205.9 ± 17	203
Cu–O	64.3 ± 5.0	269.0 ± 20.9	259
Cu–S	66	276	305
Cu–Se	60	251	305
Cu–Si	52.9 ± 1.5	221.3 ± 6.3	286
Cu–Sn	40.5 ± 1.6	169.5 ± 6.6	2, 213
Cu–Tb	46 ± 5	193 ± 19	187
Cu–Te	66.6	278.7	1
D–D	106.007	443.534	189
D–F	137.8	576.6	189
D–Ga	<65.2	<272.8	227
D–Ge	≤77	≤322	189
D–H	105.027	439.434	189
D–Hg	10.05	42.08	189
D–In	58.8	246.0	189
D–Li	57.4066 ± 0.0011	240.1892 ± 0.0046	190, 315
D–Mg	32.3	135.1	189
D–Ni	≤72.4	≤302.9	189
D–Pt	≤83.7	≤350.2	189
D–S	84	352	189
D–Si	72.3	302.5	189
D–Sr	≥65.9	≥275.7	189
D–Zn	21.2	88.7	189
Dy–F	127	531	354
Dy–O	147.0 ± 10.0	615.1 ± 41.8	259
Dy–S	99 ± 10	414 ± 42	222
Dy–Se	77 ± 10	322 ± 42	222
Dy–Te	56 ± 10	234 ± 42	222
Er–F	135 ± 4	565 ± 17	354
Er–O	145.4 ± 5.1	608.4 ± 21.3	259
Er–S	100 ± 10	418 ± 42	222
Er–Se	78 ± 10	326 ± 42	222
Er–Te	57 ± 10	239 ± 42	222
Eu–Eu	8.0 ± 4	33.5 ± 16.7	54
Eu–F	130	544	215
Eu–Li	16.0 ± 0.7	66.9 ± 3.0	251
Eu–O	115.5 ± 4.0	483.3 ± 16.7	259
Eu–Rh	55.9 ± 8	233.9 ± 33.5	54
Eu–S	86.6 ± 3.1	362.3 ± 13.0	245, 303
Eu–Se	72 ± 3.5	301 ± 14.6	17, 171, 245
Eu–Te	58 ± 3.5	243 ± 14.6	17, 245
F–F	37.95	158.75	189
F–Ga	138 ± 3.5	577 ± 14.6	244
F–Gd	141.1 ± 6.5	590.4 ± 27.2	352
F–Ge	116 ± 5	485 ± 21	95
F–H	136.3	570.3	189
F–Hg	~43	~180	189
F–Ho	129	540	354
F–I	≤64.9	≤271.6	5, 46, 65
F–In	121 ± 3.5	506 ± 14	244
F–K	118.9 ± 0.6	497.5 ± 2.5	12
F–La	143 ± 10	598 ± 42	222
F–Li	138 ± 5	577 ± 21	72
F–Lu	79.7	383.7	191
F–Mg	110.4 ± 1.2	461.9 ± 5.0	97, 180
F–Mn	101.2 ± 3.5	423.4 ± 14.6	208
F–Mo	111.1	464.8	182
F–N	82	343	189
F–Na	124	519	189
F–Nd	130.3 ± 3.0	545.2 ± 12.6	351
F–Ni	104	435	222
F–O	53 ± 4	222 ± 17	48
F–P	105 ± 23	439 ± 96	112
F–Pb	85 ± 2	356 ± 8	350
F–Pm	129 ± 10	540 ± 42	222
F–Pr	139 ± 11	582 ± 46	222
F–Pu	128.7 ± 7	538.5 ± 29	207
F–Rb	118 ± 5	494 ± 21	72
F–S	81.9 ± 1.2	342.7 ± 5.0	22, 184, 211
F–Sb	105 ± 23	439 ± 96	112
F–Sc	140.8 ± 2	589.1 ± 13	353
F–Se	81 ± 10	339 ± 42	222
F–Si	132.1 ± 0.5	552.7 ± 2.1	105
F–Sm	135	565	215
F–Sn	111.5 ± 3	466.5 ± 13	350
F–Sr	129.5 ± 1.6	541.8 ± 6.7	98, 180
F–Ta	137 ± 3	573 ± 13	231
F–Tb	134 ± 10	561 ± 42	222
F–Ti	136 ± 8	569 ± 34	355
F–Tl	106.4 ± 4.6	445.2 ± 19.3	21
F–Tm	122	510	215
F–U	157.5 ± 2.5	659.0 ± 10.5	147, 233
F–V	141 ± 15	590 ± 63	222
F–W	131 ± 15	548 ± 63	222
F–Xe	3.77	15.78	285, 321
F–Y	144.6 ± 5.0	605.0 ± 20.9	353
F–Yb	≥124.6 ± 2.3	≥521.3 ± 9.6	14, 108, 348
F–Zn	88 ± 15	368 ± 63	222
F–Zr	149 ± 15	623 ± 63	222
Fe–Fe	24 ± 5	100 ± 21	238
Fe–Ge	50.4 ± 7	210.9 ± 29	202
Fe–O	93.3 ± 4.1	390.4 ± 17.2	259
Fe–S	77	322	88
Fe–Si	71 ± 6	297 ± 25	329
Ga–Ga	33 ± 5	138 ± 21	222
Ga–H	<65.5	<274.1	227
Ga–I	81 ± 2.3	339 ± 9.6	112
Ga–Li	31.8 ± 3.5	133.1 ± 14.6	151
Ga–O	84.5 ± 10.0	353.6 ± 41.8	259
Ga–P	54.9 ± 3.0	229.7 ± 12.6	143
Ga–Sb	45.9 ± 3.0	208.8 ± 12.6	269
Ga–Te	60 ± 6	251 ± 25	327
Gd–O	171.0 ± 3.0	715.5 ± 12.6	259
Gd–S	125.9 ± 2.5	526.8 ± 10.5	110, 298
Gd–Se	103 ± 3.5	431 ± 14.6	17
Gd–Te	82 ± 3.5	343 ± 14.6	17
Ge–Ge	65.4 ± 5	273.6 ± 21	247
Ge–H	≤76.9	≤321.8	218
Ge–Ni	67 ± 3	280 ± 13	203
Ge–O	157.6 ± 3.0	659.4 ± 12.6	259
Ge–Pd	63.2	264.4	260
Ge–S	131.7 ± 0.6	551.0 ± 2.5	85, 89
Ge–Se	113.0 ± 2.0	472.8 ± 8.4	349
Ge–Si	72 ± 5	301 ± 21	112
Ge–Te	99	414	58
H–H	104.204	435.990	189
H–Hg	9.523	39.843	189
H–I	71.321	298.407	189
H–In	58.1	243.0	189
H–K	43.8 ± 3.5	183.3 ± 14.6	189
H–Li	56.895 ± 0.001	238.049 ± 0.004	340
H–Mg	30.2 ± 0.7	126.4 ± 5.9	9, 10, 94
H–Mn	56 ± 7	234 ± 29	112
H–N	≤81	≤339	189
H–Na	44.38 ± 0.06	185.69 ± 0.25	248, 283
H–Ni	60.3 ± 2	252.3 ± 8	199
H–O	102.2	427.5	189
H–P	71	297	189
H–Pb	42 ± 5	176 ± 21	189
H–Pt	≤80	≤335	189
H–Rb	40 ± 5	167 ± 21	112
H–S	82.3 ± 2.9	344.3 ± 12.1	194
H–Sc	~43	~180	293
H–Se	73 ± 0.5	305 ± 2.1	154
H–Si	≤71.5	≤299.2	189
H–Sn	63 ± 4	264 ± 17	112
H–Sr	39 ± 2	163 ± 8	112
H–Te	64 ± 0.5	268 ± 2.1	154
H–Ti	~38	~159	292
H–Tl	45 ± 2	188 ± 8	112
H–Yb	38 ± 9	159 ± 38	112
H–Zn	20.5 ± 0.5	85.8 ± 2.1	112
He–He	0.9	3.8	189
He–Hg	1.58	6.61	222
Hf–C	131 ± 15	548 ± 63	222
Hf–N	128 ± 7	535 ± 30	132, 221
Hf–O	191.6 ± 3.2	801.7 ± 13.4	259
Hg–Hg	2 ± 0.5	8 ± 2	187, 188
Hg–I	8.29 ± 0.23	34.69 ± 0.96	339
Hg–K	1.97 ± 0.05	8.24 ± 0.21	222
Hg–Li	3.3	13.9	189
Hg–Na	2.2	9.2	189, 358
Hg–O	52.8 ± 7.9	221.1 ± 33.1	148
Hg–Rb	2.0	8.4	189
Hg–S	51.9 ± 5.3	217.3 ± 22.2	148
Hg–Se	34.5 ± 7.2	144.5 ± 30.1	148
Hg–Te	≤34	≤142	222
Hg–Tl	1	4	174
Ho–Ho	20 ± 4	84 ± 17	49
Ho–O	146.0 ± 6.0	610.9 ± 25.1	259
Ho–S	102.4 ± 3.5	428.4 ± 14.6	298
Ho–Se	80 ± 4	335 ± 17	17
Ho–Te	62 ± 4	259 ± 17	17
I–I	36.111	151.088	189
I–In	80	335	11
I–K	77.7 ± 0.2	325.1 ± 0.8	316, 331
I–Li	82.5 ± 1.0	345.2 ± 4.2	316
I–Mg	~68	~285	19
I–Mn	67.6 ± 2.3	282.8 ± 9.6	112
I–N	38 ± 4	159 ± 17	222
I–Na	72.7 ± 0.5	304.2 ± 2.1	316, 331
I–Ni	70 ± 5	293 ± 21	112
I–O	43	180	189
I–Pb	47 ± 9	197 ± 38	112
I–Rb	76.2 ± 0.5	318.8 ± 2.1	316
I–Si	70	293	189
I–Sn	56 ± 10	234 ± 42	222
I–Sr	64.5 ± 1.4	269.9 ± 5.9	217
I–Te	46 ± 10	193 ± 42	222
I–Ti	74 ± 10	310 ± 42	222
I–Tl	65 ± 2	272 ± 8	19
I–Zn	33 ± 7	138 ± 29	112
I–Zr	73	305	216
In–In	24 ± 2	100 ± 8	222
In–Li	22.1 ± 3.5	92.5 ± 14.6	151
In–O	<76.5 ± 10.0	<320.1 ± 41.8	259
In–P	47.3 ± 2.0	197.9 ± 8.5	268
In–S	69 ± 4	289 ± 17	60
In–Sb	36.3 ± 2.5	151.9 ± 10.5	75
In–Se	59 ± 4	247 ± 17	60
In–Te	52 ± 4	218 ± 17	60
Ir–La	138 ± 3	576 ± 12	169
Ir–O	99.1 ± 10.1	414.6 ± 42.3	259
Ir–Si	110.6 ± 5.0	462.8 ± 20.9	330
Ir–Th	137	573	116
Ir–Y	109.0 ± 4.0	455.9 ± 16.0	170
K–K	13.7 ± 1.0	57.3 ± 4.2	222
K–Kr	1.1	4.6	189
K–Li	19.6 ± 1.0	82.0 ± 4.2	99, 357

Table 1 (continued)
BOND STRENGTHS IN DIATOMIC MOLECULES

Molecule	D°_{298}/kcal mol^{-1}	D°_{298}/kJ mol^{-1}	Ref.
K–Na	15.773 ± 0.002	659.950 ± 0.008	30, 357
K–O	664 ± 5.0	277.8 ± 20.9	93, 259
K–Xe	1.2	5.1	189
Kr–Kr	1.25	5.23	39, 189
Kr–O	<2	<8	222
La–La	59 ± 5	247 ± 21	336
La–N	124 ± 10	519 ± 42	222
La–O	191.6 ± 2.7	801.7 ± 11.3	259
La–Pt	120 ± 5	502 ± 21	246
La–Rh	126 ± 4	527 ± 17	53
La–S	137.0 ± 0.4	573.4 ± 1.8	195, 313
La–Se	114 ± 4	477 ± 17	17, 245
La–Te	91 ± 4	381 ± 17	17, 145
La–Y	48.3	202.1	336
Li–Li	26.34 ± 1	106.48 ± 4.2	333, 343
Li–Mg	16.1 ± 1.5	67.4 ± 6.3	344
Li–Na	21.10 ± 0.01	88.28 ± 0.04	98, 357
Li–O	79.7 ± 2.0	333.5 ± 8.4	259
Li–Pb	18.8 ± 1.9	78.8 ± 8.0	250
Li–S	74.7 ± 1.8	312.5 ± 7.6	214
Li–Sb	41.3 ± 2.4	172.8 ± 10.0	253
Li–Sm	11.7 ± 1.0	49.0 ± 4.2	251
Li–Tm	16.5 ± 0.8	69.0 ± 3.5	251
Li–Yb	8.9 ± 0.7	37.2 ± 3.0	251
Lu–Lu	34 ± 8	142 ± 34	222
Lu–O	161.3 ± 4.0	674.9 ± 6.7	259
Lu–Pt	96 ± 8	402 ± 34	120
Lu–S	121.2 ± 3.5	507.1 ± 14.6	109, 298
Lu–Se	100 ± 4	418 ± 17	17
Lu–Te	78 ± 4	326 ± 17	17
Mg–Mg	2.044 ± 0.001	8.552 ± 0.004	237, 345
Mg–O	86.8 ± 3.0	363.2 ± 12.6	259
Mg–S	56	234	61
Mn–Mn	6.2	25.9	198
Mn–O	96.3 ± 10.0	402.9 ± 41.8	206, 259
Mn–S	72 ± 4	301 ± 17	341
Mn–Se	57.2 ± 2.2	239.3 ± 9.2	307, 342
Mo–Mo	97 ± 5	406 ± 20	155
Mo–Nb	109 ± 6	454 ± 25	157
Mo–O	133.9 ± 5.0	560.2 ± 20.9	259
N–N	225.94 ± 0.14	945.33 ± 0.59	189
N–O	150.71 ± 0.03	630.57 ± 0.13	189
N–P	147.5 ± 5.0	617.1 ± 20.9	63, 129
N–Pu	113 ± 15	473 ± 63	222
N–S	111 ± 5	464 ± 21	222
N–Sb	72 ± 12	301 ± 50	112
N–Sc	112 ± 20	469 ± 84	222
N–Se	91 ± 15	381 ± 63	222
N–Si	105 ± 9	439 ± 38	112
N–Ta	146 ± 20	611 ± 84	222
N–Th	138.0 ± 7.9	577.4 ± 33.1	123, 132
N–Ti	113.8 ± 7.9	476.1 ± 33.1	132, 309
N–U	127.0 ± 0.5	531.4 ± 2.1	121
N–V	114.1 ± 4.1	477.4 ± 17	103, 132
N–Xe	5.5	23.0	173
N–Y	115 ± 15	481 ± 63	222
N–Zr	135.0 ± 6.0	564.8 ± 25.1	122, 132
Na–Na	17.59 ± 0.60	73.60 ± 0.25	338
Na–O	61.2 ± 4.0	256.1 ± 16.7	259
Na–Rb	14 ± 0.9	59 ± 3.8	112
Nb–Nb	122 ± 2.4	511 ± 10	158
Nb–O	184.4 ± 6.0	771.5 ± 25.1	259
Nd–Nd	<39	<163	222
Nd–O	167.9 ± 3.0	702.5 ± 12.6	259
Nd–S	112.7	471.5	17
Nd–Se	92 ± 4	385 ± 17	17, 144, 245
Nd–Te	73 ± 4	305 ± 17	17
Ne–Ne	0.94	3.93	320
Ni–Ni	48.58 ± 0.23	203.26 ± 0.96	243
Ni–O	91.3 ± 4.0	382.0 ± 16.7	259
Ni–S	82.3	344.3	88

Molecule	D°_{298}/kcal mol^{-1}	D°_{298}/kJ mol^{-1}	Ref.
Ni–Si	76 ± 4	318 ± 17	329
Np–O	171.7 ± 10.0	718.4 ± 41.8	259
O–O	119.11 ± 0.04	498.36 ± 0.17	32, 189
O–Os	143.0 ± 20.0	598.3 ± 83.7	259
O–P	143.2 ± 3.0	599.2 ± 12.6	259
O–Pb	91.3 ± 3.0	382.0 ± 12.6	259
O–Pd	91.0 ± 20.0	380.7 ± 83.7	259
O–Pm	161 ± 15	674 ± 63	222
O–Pr	178.6 ± 4.0	747.3 ± 16.7	259
O–Pt	93.6 ± 10.0	391.6 ± 41.8	259
O–Pu	171.1 ± 8.1	715.9 ± 33.9	259
O–Rb	61 ± 20	255 ± 84	31
O–Re	149.8 ± 20.0	626.8 ± 83.7	259
O–Rh	96.8 ± 10.0	405.0 ± 41.8	259
O–Ru	126.3 ± 10.0	528.4 ± 41.8	259
O–S	124.7 ± 1.0	521.8 ± 4.2	259
O–Sb	103.8 ± 10.0	434.3 ± 41.8	259
O–Sc	162.9 ± 2.7	681.6 ± 11.3	259
O–Se	111.1 ± 5.1	464.8 ± 21.3	259, 302
O–Si	191.1 ± 3.2	799.6 ± 13.4	259
O–Sm	136.9 ± 4.0	572.8 ± 16.7	259
O–Sn	127.1 ± 3.0	531.8 ± 12.6	259
O–Sr	101.7 ± 4.0	425.5 ± 16.7	259
O–Ta	191.0 ± 3.0	799.1 ± 12.6	259
O–Tb	169.5 ± 4.0	709.2 ± 16.7	259
O–Te	89.9 ± 5.0	376.1 ± 20.9	259
O–Th	210.0 ± 2.9	878.6 ± 12.1	259
O–Ti	160.7 ± 2.2	672.4 ± 9.2	259
O–Tm	122.1 ± 5.1	510.9 ± 21.3	259
O–U	181.5 ± 3.2	759.4 ± 13.4	259
O–V	149.8 ± 4.5	626.8 ± 18.8	7, 259
O–W	160.6 ± 10.0	672.0 ± 41.8	259
O–Xe	8.7	36.4	222
O–Y	172.0 ± 2.7	719.7 ± 11.3	192, 259
O–Yb	99.9 ± 2.0	418.0 ± 8.4	259
O–Zn	<64.7 ± 10.0	<270.7 ± 41.8	259
O–Zr	185.5 ± 3.2	776.1 ± 13.4	259
P–P	117.0 ± 2.5	489.5 ± 10.5	129
P–Pt	≤99.6 ± 4	≤416.7 ± 17	304
P–Rh	84.4 ± 4	353.1 ± 17	304
P–S	106 ± 2	444 ± 8	87
P–Sb	85.3	356.9	225
P–Se	86.9 ± 2.4	363.6 ± 10.0	87
P–Si	86.9	363.6	299
P–Te	71.2 ± 2.4	297.9 ± 10.0	87
P–Th	90	377	119
P–Tl	50 ± 3	209 ± 13	267
P–U	71 ± 5	297 ± 21	222
P–W	73 ± 1	305 ± 4	128
Pb–Pb	20.7 ± 0.2	86.6 ± 1.0	138, 277
Pb–S	82.7 ± 0.4	346.0 ± 1.7	325
Pb–Sb	38.6 ± 2.5	161.5 ± 10.5	356
Pb–Se	72.4 ± 1	302.9 ± 4	325
Pb–Te	60 ± 3	251 ± 13	325
Pd–Pd	17	75	239
Pd–Si	74.9 ± 3.3	313.4 ± 13.8	330
Pd–Y	57 ± 4	238 ± 15	281
Pm–S	101 ± 15	423 ± 63	222
Pm–Se	81 ± 15	339 ± 63	222
Pm–Te	61 ± 15	255 ± 63	222
Po–Po	44.7	187.0	189
Pr–S	117.1 ± 1.1	492.5 ± 4.6	107
Pr–Se	106.7 ± 5.5	446.4 ± 23.0	146, 245
Pr–Te	78 ± 10	326 ± 42	222
Pt–Pt	85.4 ± 3.6	357.3 ± 15.1	164
Pt–Si	119.8 ± 4 3	501.2 ± 18.0	330
Pt–Th	132	552	116
Pt–Ti	95 ± 3	398 ± 11	167
Pt–Y	113.3 ± 2.9	474.1 ± 12.1	164
Rb–Rb	10.9 ± 0.5	45.6 ± 2.1	271
Rh–Rh	68.2 ± 5.0	285.4 ± 20.9	51, 270
Rh–Sc	106.1 ± 2.5	444.0 ± 10.5	168
Rh–Si	94.4 ± 4.3	395.0 ± 18.0	330

Molecule	D°_{298}/kcal mol^{-1}	D°_{298}/kJ mol^{-1}	Ref.
Rh–Th	123 ± 5	513 ± 21	141
Rh–Ti	93.4 ± 3.5	390.8 ± 14.6	51
Rh–U	124 ± 4	519 ± 17	141
Rh–Y	106.4 ± 2.5	445.1 ± 10.5	168
Ru–Si	94.9 ± 5.0	397.1 ± 20.9	330
Ru–Th	141.4 ± 10	591.6 ± 42	117
S–S	101.65	425.28	189
S–Sb	90.5	378.7	106
S–Sc	114 ± 3	477 ± 13	313, 323
S–Se	88.7 ± 1.6	371.1 ± 6.7	90
S–Si	149	623	189
S–Sm	93	389	107
S–Sn	111 ± 0.8	464 ± 3.2	85
S–Sr	81	339	38
S–Tb	123 ± 10	515 ± 42	222
S–Te	81 ± 5	339 ± 21	85
S–Ti	99.9 ± 0.7	418.0 ± 2.9	92, 263
S–Tm	88 ± 10	368 ± 42	222
S–U	124.9 ± 2.3	522.6 ± 9.6	313
S–V	115 ± 2	480 ± 10	29
S–Y	126.3 ± 2.5	528.4 ± 10.5	312
S–Yb	40	167	222
S–Zn	49 ± 3	205 ± 13	76, 148
S–Zr	137.5 ± 4.0	575.3 ± 16.7	313
Sb–Sb	71.5 ± 1.5	299.2 ± 6.3	75, 224
Sb–Te	66.3 ± 0.9	277.4 ± 3.8	278, 319
Sb–Tl	30.3 ± 2.5	145.9 ± 12.0	6, 267
S–Sc	38.9 ± 5	162.8 ± 21	118
Sc–Se	92 ± 4	385 ± 17	222
Sc–Te	69 ± 4	289 ± 17	222
Se–Se	79.5 ± 0.1	332.6 ± 0.4	90, 325
Se–Si	131	548	189
Se–Sm	79.1 ± 3.5	331.0 ± 14.6	245
Se–Sn	95.9 ± 1.4	401.3 ± 5.9	59
Se–Sr	~68	~285	20
Se–Tb	101 ± 10	423 ± 42	222
Se–Te	69.7 ± 1	291.5 ± 4	85, 90, 149
Se–Ti	91 ± 10	381 ± 42	222
Se–Tm	66 ± 10	276 ± 42	222
Se–V	83 ± 5	347 ± 21	222
Se–Y	104 ± 3	435 ± 13	222
Se–Zn	40.8 ± 6.2	170.8 ± 2	76, 148
Si–Si	78.1 ± 2.4	326.8 ± 10.0	41
Si–Te	108	452	189
Sm–Te	65.1 ± 3.5	272.4 ± 14.6	245
Sn–Sn	46.7 ± 4	195.4 ± 17	2, 28
Sn–Te	86.0	359.8	189
Sr–Sr	3.7 ± 0.1	15.5 ± 0.4	113
Tb–Tb	31.4 ± 6.0	131.4 ± 25.1	226
Tb–Te	81 ± 10	339 ± 42	222
Te–Te	62.1 ± 1.2	259.8 ± 5.0	249
Te–Ti	69 ± 4	289 ± 17	222
Te–Tm	66 ± 10	276 ± 42	222
Te–Y	81 ± 3	339 ± 13	222
Te–Zn	28.1 ± 4.3	117.7 ± 18	148
Th–Th	≤69	≤289	119
Ti–Ti	33.8 ± 5	141.4 ± 21	197
Tl–Tl	15.4 ± 4	60.7 ± 17	8
U–U	53 ± 5	222 ± 21	222
V–V	57.9 ± 5	242.3 ± 21	197, 229
Xe–Xe	1.56 ± 0.07	6.53 ± 0.30	40
Y–Y	38 ± 5	159 ± 21	222
Yb–Yb	4.9 ± 4	20.5 ± 17	152
Zn–Zn	7	29	296

REFERENCES

1. **Abbasov, A. S., Azizov, T. Kh., Alleva, N. A., Aliev, I. Ya, Mustafaev, F. M., and Mamedov, A. N.,** *Zh. Fiz. Khim.*, 50, 2172, 1976.
2. **Ackerman, M., Drowart, J., Stafford, F. E., and Verhaegen, G.,** *J. Chem. Phys.*, 36, 1557, 1962.
3. **Ackerman, M., Stafford, F. E., and Drowart, J.,** *J. Chem. Phys.*, 33, 1784, 1960.
4. **Ackerman, M., Stafford, F. E., and Verhaegen, G.,** *J. Chem. Phys.*, 36, 1560, 1962.
5. **Appelman, E. H. and Clyne, M. A. A.,** *J. Chem. Soc. Faraday Trans. 1*, 71, 2072, 1975.
6. **Balducci, G., Ferro, D., and Piacente, V.,** *High Temp. Sci.*, 14, 207, 1981.
7. **Bladucci, G., Gigli, G., and Guido, M.,** *J. Chem. Phys.*, 79, 5616, 1983.
8. **Balducci, G. and Piacente, V.,** *J. Chem. Soc. Chem. Commun.*, 1287, 1980.
9. **Balfour, W. J. and Cartwright, H. M.,** *Astron. Astrophys. Suppl. Ser.*, 26, 389, 1976.
10. **Balfour, W. J. and Lingren, B.,** *Can. J. Chem.* 56, 767, 1978.
11. **Barrow, R. F.,** *Trans. Faraday Soc.*, 56, 962, 1960.
12. **Barrow, R. F. and Caunt, A. D.,** *Proc. R. Soc. London Ser. A*, 219, 120, 1953.
13. **Barrow, R. F., Clark, T. C., Coxon, J., and Yee, K. K.,** *J. Mol. Spectrosc.*, 51, 428, 1974.
14. **Barrow, R. F. and Chojnicki, A. H.,** *J. Chem. Soc. Faraday Trans. 2*, 71, 728, 1975.
15. **Barrow, R. F. and Deutsch, E. W.,** *Proc. Chem. Soc.*, p.122, 1960.
16. **Behrens, R. G. and Feber, R. C.,** *J. Less-Common Met.*, 75, 281, 1980.
17. **Bergman, C., Coppens, P., Drowart, J., and Smoes, S.,** *Trans. Faraday Soc.*, 66, 800, 1970.
18. **Bergman, C. and Gingerich, K. A.,** *J. Phys. Chem.*, 76, 2332, 1972.
19. **Berkowitz, J. and Chupka, W. A.,** *J. Chem. Phys.*, 45, 1287, 1966.
20. **Berkowitz, J. and Chupka, W. A.,** *J. Chem. Phys.*, 45, 4289, 1966.
21. **Berkowitz, J. and Walter, T.,** *J. Chem. Phys.*, 49, 1184, 1968.
22. **Berneike, W., Kreuttle, U., and Neuert, H.,** *Chem. Phys. Lett.*, 76, 525, 1980.
23. **Besenbruch, G., Kana'an, A. S., and Margrave, J. L.,** *J. Phys. Chem.*, 69, 3174, 1965.
24. **Birks, J. W., Gabelnick, S. D., and Johnston, H. S.,** *J. Mol. Spectrosc.*, 57, 23, 1975.
25. **Biron, M.,** *C. R. Acad. Ser. B*, 265, 1026, 1427, 1967.
26. **Blue, G. D., Green, J. W., Bautista, R. G., and Margrave, J. L.,** *J. Phys. Chem.*, 67, 877, 1963.
27. **Bondybey, V. E.,** *Chem. Phys. Lett.*, 109, 436, 1984.
28. **Bondybey, V. E., Heaven, M., and Miller, T. A.,** *J. Chem. Phys.*, 78, 3593, 1983.
29. **Botor, J. P. and Edwards, J. G.,** *J. Chem. Phsy.*, 81, 2185, 1984.
30. **Breford, E. J. and Engelke, F.,** *J. Chem. Phys.*, 71, 1994, 1979.
31. **Brewer, L. and Rosenblatt, G. M.,** *Adv. High Temp. Sci.*, 2, 1, 1969.
32. **Brix, P. and Herzberg, G.,** *J. Chem. Phys.*, 21, 2240, 1953.
33. **Bulewicz, E. M. and Sugden, T. M.,** *Trans. Faraday Soc.*, 52, 1475, 1956.
34. **Burns, G., LeRoy, L. J., Morris, D. J., and Blake, J. A.,** *Proc. R. Soc. London Ser. A*, 316, 81, 1970.
35. **Busse, V. B. and Weil, K. G.,** *Angew. Chem.*, 91, 664, 1979.
36. **Busse, V. B. and Weil, K. G.,** *Ber. Bunsenges. Phys. Chem.*, 85, 309, 1981.
37. **Carbonel, M., Bergman, C., and Laffite, M.,** *Colloq. Int. Cent. Nat. Rech. Sci.*, 210, 311, 1972.
38. **Cater, E. D. and Johnson, E. W.,** *J. Chem. Phys.*, 47, 5353, 1967.
39. **Chashchina, G. I. and Shreider, E. Ya.,** *Zh. Prikl. Spektrosk.*, 21, 696, 1974.
40. **Chashchina, G. I. and Shreider, E. Ya.,** *Zh. Prikl. Spectrosk.*, 25, 163, 1976.
41. **Chatillon, C., Allibert, M., and Pattoret, A.,** *C. R. Acad. Sci. Ser. C*, 280, 1505, 1975.
42. **Clarke, T. C. and Clyne, M. A. A.,** *Trans. Faraday Soc.*, 66, 877, 1970.
43. **Clements, R. M. and Barrow, R. F.,** *Trans. Faraday Soc.*, 64, 2893, 1968.
44. **Clyne, M. A. A., Curran, A. H., and Coxon, J. A.,** *J. Mol. Spectrosc.*, 63, 43, 1976.
45. **Clyne, M. A. A. and McDermid, I. S.,** *Faraday Discuss. Chem. Soc.*, 67, 316, 1979.
46. **Clyne, M. A. A. and McDermid, I. S.,** *J. Chem. Soc. Faraday Trans. 2*, 72, 2252, 1976.
47. **Clyne, M. A. A. and McDermid, I. S.,** *J. Chem. Soc. Faraday Trans. 2*, 74, 644, 1978.
48. **Clyne, M. A. A. and Watson, R. T.,** *Chem. Phys. Lett.*, 12, 344, 1971.
49. **Cocke, D. L. and Gingerich, K. A.,** *J. Phys. Chem.*, 75, 3264, 1971.
50. **Cocke, D. L. and Gingerich, K. A.,** *J. Phys. Chem.*, 76, 2332, 1972.
51. **Cocke, D. L. and Gingerich, K. A.,** *J. Chem. Phys.*, 60, 1958, 1974.
52. **Cocke, D. L., Gingerich, K. A., and Chang, C.-A.,** *J. Chem. Soc. Faraday Trans. 1*, 72, 268, 1976.
53. **Cocke, D. L., Gingerich, K. A., and Kordis, J.,** *High Temp. Sci.*, 5, 474, 1973.
54. **Cocke, D. L., Gingerich, K. A., and Kordis, J.,** *High Temp. Sci.*, 7, 61, 1975.
55. CODATA recommended key values for thermodynamics, 1973, *J. Chem. Thermodyn.*, 7, 1, 1975.
56. **Colbourne, E. A. and Douglas, A. E.,** *J. Chem. Phys.*, 65, 1741, 1976.
57. **Colin, R. and De Greef, D.,** *Can. J. Phys.*, 53, 2142, 1975.
58. **Colin, R. and Drowart, J.,** *J. Phys. Chem.*, 68, 428, 1964.
59. **Colin, R. and Drowart, J.,** *Trans. Faraday Soc.*, 60, 673, 1964.
60. **Colin, R. and Drowart, J.,** *Trans. Faraday Soc.*, 64, 2611, 1968.
61. **Colin, R., Goldfinger, P., and Jeunehomme, M.,** *Trans. Faraday Soc.*, 60, 306, 1964.
62. **Coppens, P., Reynaert, J. C., and Drowart, J.,** *J. Chem. Soc. Faraday Trans. 2*, 75, 292, 1979.
63. **Coquart, B. and Prudhomme, J. C.,** *J. Mol. Spectrosc.*, 87, 75, 1981.
64. **Corbett, J. D. and Lynde, R. A.,** *Inorg. Chem.*, 6, 2199, 1967.
65. **Coxon, J. A.,** *Chem. Phys. Lett.*, 33, 136, 1975.
66. **Crepin, C., Verges, J., and Amiot, C.,** *Chem. Phys. Lett.*, 122, 10, 1984.
67. **Cubicciotti, D.,** *Inorg. Chem.*, 7, 208, 1968.
68. **Cubicciotti, D.,** *Inorg. Chem.*, 7, 211, 1968.
69. **Cubicciotti, D.,** *J. Phys. Chem.*, 71, 3066, 1967.
70. **Cuthill, A. M., Fabian, D. J., and Shu-Shou-Shen, S.,** *J. Phys. Chem.*, 77, 2008, 1973.
71. **Dagdigian, P. J., Cruze, H. W., and Zare, R. N.,** *J. Chem. Phys.*, 62, 1824, 1975.
72. **Darwent, B. de B.,** Bond Dissociation Energies in Simple Molecules, NSRDS-NBS 31, National Bureau of Standards, Washington, D.C., 1970.
73. **Davis, D. D. and Okabe, H.,** *J. Chem. Phys.*, 49, 5526, 1968.
74. **De Corpo, J. J., Steiger, R. P., Franklin, J. L., and Margrave, J. L.,** *J. Chem. Phys.*, 53, 936, 1970.
75. **De Maria, G., Drowart, J., and Inghram, M. G.,** *J. Chem. Phys.*, 31, 1076, 1959.

76. **De Maria, G., Goldfinger, P., Malaspina, L., and Piacente, V.,** *Trans. Faraday Soc.*, 61, 2146, 1965.
77. **De Maria, G., Malaspina, L., and Piacente, V.,** *J. Chem. Phys.*, 52, 1019, 1970.
78. **De Maria, G., Malaspina, L., and Piacente, V.,** *J. Chem. Phys.*, 56, 1978, 1972.
79. **Dickson, C. R., Kinney, J. B., and Zare, R. N.,** *Chem. Phys.*, 15, 243, 1976.
80. **Diemer, U., Weickenmeier, H., Wahl, M., and Demtroeder, W.,** *Chem. Phys. lett.*, 104, 489, 1984.
81. **Dixon, R. N. and Lambertson, H. M.,** *J. Mol. Spectrosc.*, 25, 12, 1968.
82. **Drowart, J.,** in *Phase Stability in Metals and Alloys*, Rudman, P. S., Ed., McGraw-Hill, New York, 1967, 305.
83. **Drowart, J., De Maria, G., and Inghram, M. G.,** *J. Chem. Phys.*, 29, 1015, 1958.
84. **Drowart, J. and Goldfinger, P.,** *Angew. Chem.*, 6, 581, 1967.
85. **Dorwart, J. and Goldfinger, P.,** *Q. Rev. (London)*, 20, 545, 1966.
86. **Drowart, J. and Honig, R. E.,** *J. Phys. Chem.*, 61, 980, 1957.
87. **Drowart, J., Myers, C. E., Szwarc, R., Vander Auwera-Mahieu, A., and Uy, O. M.,** *High Temp. Sci.*, 5, 482, 1973.
88. **Drowart, J., Pattoret, A., and Smoes, S.,** *Proc. Br. Ceramic Soc.*, No. 8, 67, 1967.
89. **Drowart, J., Smets, J., Reynaert, J. C., and Coppens, P.,** *Adv. Mass Spectrom.*, 7A, 647, 1978.
90. **Drowart, J. and Smoes, S.,** *J. Chem. Soc. Faraday Trans. 2*, 73, 1755, 1977.
91. **Drowart, J., Smoes, S., and Vander Auwera-Mahieu, A.,** *J. Chem. Thermodyn.*, 10, 453, 1978.
92. **Edwards, J. G., Franklin, H. F., and Gilles, P. W.,** *J. Chem. Phys.*, 54, 545, 1971.
93. **Ehlert, T. C.,** *High Temp. Sci.*, 9, 237, 1977.
94. **Ehlert, T. C., Hilmer, R. M., and Beauchamp, E. A.,** *J. Inorg. Nucl. Chem.*, 30, 3112, 1968.
95. **Ehlert, T. C. and Margrave, J. L.,** *J. Chem. Phys.*, 41, 1066, 1964.
96. **Ehlert, T. C. and Wang, J. S.,** *J. Phys. Chem.*, 81, 2069, 1977.
97. **Engelke, F.,** *Chem. Phys.*, 39, 279, 1979.
98. **Engelke, F., Ennen, G,. and Meiwes, K. H.,** *Chem. Phys.*, 66, 391, 1982.
99. **Engelke, F., Hage, H., and Sprick, U.,** *Chem. Phys.*, 88, 443, 1984.
100. **Estler, C. and Zare, R. N.,** *Chem. Phys.*, 28, 253, 1978.
101. **Farber, M. and Srivastava, R. D.,** *Chem. Phys. Lett.*, 42, 567, 1976.
102. **Farber, M. and Srivastava, R. D.,** *High Temp. Sci.*, 12, 21, 1980.
103. **Farber, M. and Srivastava, R. D.,** *J. Chem. Soc. Faraday Trans. 1*, 69, 390, 1973.
104. **Farber, M. and Srivastava, R. D.,** *J. Chem. Soc. Faraday Trans. 1*, 70, 1581, 1974.
105. **Farber, M. and Srivastava, R. D.,** *J. Chem. Soc. Faraday Trans. 1*, 74, 1089, 1978.
106. **Faure, F. M., Mitchell, M. J., and Bartlett, R. W.,** *High Temp. Sci.*, 4, 181, 1972.
107. **Fenochka, B. V. and Gorkienko, S. P.,** *Zh. Fiz. Khim*, 47, 2445, 1973.
108. **Filippenko, N. V., Motozov, E. V., Giricheva, N. I., and Krasnev, K. S.,** *Izv. Vyssh. Ucheb. Zaved Khim. Technol.*, 15, 1416, 1972.
109. **Franzen, H. and Hariharan, A. V.,** *J. Chem. Phys.*, 70, 4907, 1979.
110. **Fries, J. A. and Cater, E. D.,** *J. Chem. Phys.*, 68, 3978, 1978.
111. **Fujishiro, S.,** *Trans. Jpn. Inst. Met.*, 1, 125, 1960.
112. **Gaydon, A. G.,** *Dissociation Energies and Spectra of Diatomic Molecules*, 3rd ed., Chapman & Hall, London, 1968.
113. **Gerber, G. and Moeller, R.,** *Contrib. Symp. At. Surf. Phys.*, 168, 1982.
114. **Gingerich, K. A.,** *Chem. Commun.*, 580, 1970.
115. **Gingerich, K. A.,** *Chem. Phys. Lett.*, 13, 262, 1972.
116. **Gingerich, K. A.,** *Chem. Phys. Lett.*, 23, 270, 1973.
117. **Gingerich, K. A.,** *Chem. Phys. Lett.*, 25, 523, 1974.
118. **Gingerich, K. A.,** *Chimia*, 26, 619, 1972.
119. **Gingerich, K. A.,** *High Temp. Sci.*, 1, 258, 1969.
120. **Gingerich, K. A.,** *High Temp. Sci.*, 3, 415, 1971.
121. **Gingerich, K. A.,** *J. Chem. Phys.*, 47, 2192, 1967.
122. **Gingerich, K. A.,** *J. Chem. Phys.*, 49, 14, 1968.
123. **Gingerich, K. A.,** *J. Chem. Phys.*, 49, 19, 1968.
124. **Gingerich, K. A.,** *J. Chem. Phys.*, 50, 2255, 1969.
125. **Gingerich, K. A.,** *J. Chem. Phys.*, 54, 2646, 1971.
126. **Gingerich, K. A.,** *J. Chem. Phys.*, 54, 3720, 1971.
127. **Gingerich, K. A.,** *J. Chem. Phys.*, 56, 4239, 1972.
128. **Gingerich, K. A.,** *J. Phys. Chem.*, 68, 768, 1964.
129. **Gingerich, K. A.,** *J. Phys. Chem.*, 73, 2734, 1969.
130. **Gingerich, K. A.,** *J. Chem. Phys.*, 74, 6407, 1981.
131. **Gingerich, K. A.,** *J. Chem. Soc. Faraday Trans. 2*, 70, 471, 1974.
132. **Gingerich, K. A.,** NBS Spec. Publ. (U.S.), 561, 289, 1979.
133. **Gingerich, K. A. and Blue, G. D.,** *J. Chem. Phys.*, 47, 5447, 1967.
134. **Gingerich, K. A. and Blue, G. D.,** *J. Chem. Phys.*, 59, 186, 1973.
135. **Gingerich, K. A. and Choudary, U. V.,** *J. Chem. Phys.*, 68, 3265, 1978.
136. **Gingerich, K. A. and Cocke, D. L.,** *Inorg. Chim. Acta*, 28, L171, 1978.
137. **Gingerich, K. A., Cocke, D. L., and Kordis, J.,** *J. Phys. Chem.*, 78, 603, 1974.
138. **Gingerich, K. A., Cocke, D. L., and Miller, F.,** *J. Chem. Phys.*, 64, 4027, 1976.
139. **Gingerich, K. A. and Finkbeiner, H. C.,** *J. Chem. Phys.*, 54, 2621, 1971.
140. **Gingerich, K. A. and Finkbeiner, H. C.,** Proc. 9th Rare Earth Res. Conf., 2, 795, 1971.
141. **Gingerich, K. A. and Gupta, S. K.,** *J. Chem. Phys.*, 69, 505, 1978.
142. **Gingerich, K. A., Haque, R., and Kingcade, J. E.,** *Thermochim. Acta*, 30, 61, 1979.
143. **Gingerich, K. A. and Piacente, V.,** *J. Chem. Phys.*, 54, 2498, 1971.
144. **Gordienko, S. P.,** *Izv. Akad. Nauk, SSSR Neorg. Mater.*, 20, 1472, 1984.
145. **Gordienko, S. P. and Fenochka, B. V.,** *Izv, Akad. Nauk. SSSR Neorg. Mater.*, 18, 1811, 1982.
146. **Gordienko, S. P., Fenochka, B. V., Viksman, G. Sh., Klockkova, L. A., and Mikhlina, T. M.,** *Izv. Akad. Nauk. SSSR Neorg. Mater.*, 18, 18, 1982.
147. **Gorokhov, L. N., Smirnov, V. K., and Khodeev, Yu. S.,** *Zh. Fiz. Khim.*, 58, 1603, 1984.
148. **Grade, M. and Hirschwald, W.,** *Ber. Bunsenges. Phys. Chem.*, 86, 899, 1982.
149. **Grade, M., Wienecke, J., Rosinger, W., and Hirschwald, W.,** *Ber. Bunsenges. Phys. Chem.*, 87, 355, 1983.
150. **Graeber, P. and Weil, K. G.,** *Ber. Bunsenges. Phys. Chem.*, 76, 417, 1972.
151. **Guggi, D. J., Neubert, A., and Zmbov, K. F.,** Conf. Int. Thermodyn. Chim. [C.R.] 4th, 3, 124, 1975.
152. **Guido, M. and Balducci, G.,** *J. Chem. Phys.*, 57, 5611, 1972.
153. **Guido, M., Gigli, G., and Balducci, G.,** *J. Chem. Phys.*, 57, 3731, 1972.

154. **Gunn, S. R.,** *J. Phys. Chem.*, 68, 949, 1964.
155. **Gupta, S. K., Atkins, R. M., and Gingerich, K. A.,** *Inorg. Chem.*, 17, 3211, 1978.
156. **Gupta, S. K. and Gingerich, K. A.,** *High Temp.-High Pressures*, 12, 273, 1980.
157. **Gupta, S. K. and Gingerich, K. A.,** *J. Chem. Phys.*, 69, 4318, 1978.
158. **Gupta, S. K. and Gingerich, K. A.,** *J. Chem. Phys.*, 70, 5350, 1979.
159. **Gupta, S. K. and Gingerich, K. A.,** *J. Chem. Phys.*, 71, 3072, 1979.
160. **Gupta, S. K. and Gingerich, K. A.,** *J. Chem. Phys.*, 72, 2795, 1980.
161. **Gupta, S. K. and Gingerich, K. A.,** *J. Chem. Phys.*, 72, 4928, 1980.
162. **Gupta, S. K. and Gingerich, K. A.,** *J. Chem. Phys.*, 74, 3584, 1981.
163. **Gupta, S. K., Kingcade, J. E., and Gingerich, K. A.,** *Adv. Mass Spectrom.*, 8A, 445, 1980.
164. **Gupta, S. K., Nappi, B. M., and Gingerich, K. A.,** *Inorg. Chem.*, 20, 966, 1981.
165. **Gupta, S. K., Nappi, B. M., and Gingerich, K. A.,** *J. Phys. Chem.*, 85, 971, 1981.
166. **Gupta, S. K., Pelino, M. and Gingerich, K. A.,** *J. Chem. Phys.*, 70, 2044, 1979.
167. **Gupta, S. K., Pelino, M., and Gingerich, K. A.,** *J. Phys. Chem.*, 83, 2335, 1979.
168. **Haque, R. and Gingerich, K. A.,** *J. Chem. Thermodyn.*, 12, 439, 1980.
169. **Haque. R., Pelino, M., and Gingerich, K. A.,** *J. Chem. Phys.*, 71, 2929, 1979.
170. **Haque, R., Pelino, M., and Gingerich, K. A.,** *J. Chem. Phys.*, 73, 4045, 1980.
171. **Hariharan, A. V. and Eick, H. A.,** *J. Chem. Thermodyn.*, 6, 373, 1974.
172. **Hastie, J. W.,** *J. Chem. Phys.*, 57, 4556, 1972.
173. **Herman, R. and Herman, L.,** *J. Phys. Radium.*, 24, 73, 1963.
174. **Herzberg, G.,** *Molecular Spectra and Molecular Structure. I. Spectra of Diatomic Molecules*, 2nd ed., Van Nostrand, New York, 1950.
175. **Herzberg, G.,** *J. Mol. Spectrosc.*, 33, 147, 1970.
176. **Herzberg, G. J. and Johns, J. W. G.,** *Astrophys. J.*, 158, 399, 1969.
177. **Hildenbrand, D. L.,** *Chem. Phys. Lett.*, 20, 127, 1973.
178. **Hildenbrand, D. L.,** *Chem. Phys. Lett.*, 32, 523, 1975.
179. **Hildenbrand, D. L.,** *J. Chem. Phys.*, 48, 2457, 1968.
180. **Hildenbrand, D. L.,** *J. Chem. Phys.*, 48, 3657, 1968.
181. **Hildenbrand, D. L.,** *J. Chem. Phys.*, 52, 5751, 1970.
182. **Hildenbrand, D. L.,** *J. Chem. Phys.*, 65, 614, 1976.
183. **Hildenbrand, D. L.,** *J. Chem. Phys.*, 66, 3526, 1977.
184. **Hildenbrand, D. L.,** *J. Chem. Phys.*, 77, 897, 1973.
185. **Hildenbrand, D. L. and Murad, E.,** *J. Chem. Phys.*, 44, 1524, 1966.
186. **Hildenbrand, D. L. and Theard, L. P.,** *J. Chem. Phys.*, 50, 5350, 1969.
187. **Hilpert, K.,** *Ber. Kernforschungsanlage Juelich*, JUEL-1744, 272, 1981.
188. **Hilpert, K.,** *J. Chem. Phys.*, 77, 1425, 1982.
189. **Hubert, K. P. and Herzberg, G.,** *Molecular Spectra and Molecular Structure Constants of Diatomic Molecules*, Van Nostrand, New York, 1979.
190. **Ihle, H. R. and Wu, C. H.,** *J. Chem. Phys.*, 63, 1605, 1975.
191. **Ishwar, N. B., Varma, M. P., and Jha, B. L.,** *Acta Phys. Pol. A*, A61, 503, 1982.
192. **Ishwar, N. B., Varma, M. P., and Jha, B. L.,** *Indian J. Pure Appl. Phys.*, 20, 992, 1982.
193. **Johns, J. W. C., Grimm, F. A., and Porter, R. F.,** *J. Mol. Spectosc.*, 22, 435, 1967.
194. **Johns, J. W. C. and Ramsey, D. A.,** *Can. J. Phys.*, 39, 210, 1961.
195. **Jones, R. W. and Gole, J. L.,** *Chem. Phys.*, 20, 311, 1977.
196. **Kant, A.,** *J. Chem. Phys.*, 49, 5144, 1968.
197. **Kant, A. and Lin, S.-S.,** *J. Chem. Phys.*, 51, 1644, 1969.
198. **Kant, A., Lin, S.-S., and Strauss, B.,** *J. Chem. Phys.*, 49, 1983, 1968.
199. **Kant, A. and Moon, K. A.,** *High Temp. Sci.*, 11, 55, 1979.
200. **Kant. A. and Strauss, B. H.,** *J. Chem. Phys.*, 41, 3806, 1964.
201. **Kant, A. and Strauss, B.,** *J. Chem. Phys.*, 45, 3161, 1966.
202. **Kant, A. and Strauss, B.,** *J. Chem. Phys.*, 49, 3579, 1968.
203. **Kant, A., Strauss, B., and Lin, S.-S.,** *J. Chem. Phys.*, 52, 2384, 1970.
204. **Kaposi, O.,** *Magy. Kem. Foly.*, 83, 356, 1977.
205. **Kaufel, R., Vahl, G., Nunkwitz, R., and Baumgaertel, H.,** *Z. Anorg. Allg. Chem.*, 481, 207, 1981.
206. **Kazenas, E., Tagirov, V. K., and Zviadadze, G. N.,** *Izv. Akad. Nauk. SSSR Met.*, 58, 1984.
207. **Kent, R. A.,** *J. Am. Chem. Soc.*, 90, 5657, 1986.
208. **Kent, R. A., Ehlert, T. C., and Margrave, J. L.,** *J. Am. Chem. Soc.*, 86, 5090, 1964.
209. **Kent, R. A. and Margrave, J. L.,** *J. Am. Chem. Soc.*, 87, 3582, 1965.
210. **Khitrov, A. N., Ryabova, V. G., and Gurvich, L. V.,** *Teplofiz Vys. Tempo.*, 11, 1126, 1973.
211. **Kiang, T. and Zare, R. N.,** *J. Am. Chem. Soc.*, 102, 4024, 1980.
212. **Kingcade, J. E., Cocke, D. L., and Gingerich, K. A.,** *High Temp. Sci.*, 16, 89, 1983.
213. **Kingcade, J. E., Dufner, D. C., Gupta, S. K., and Gingerich, K. A.,** *High Temp. Sci.*, 10, 213, 1978.
214. **Kimura, H., Asano, M., and Kubo, K.,** *J. Nucl. Mater.*, 97, 259, 1981.
215. **Kleinschmidt, P. D., Lau, K. H., and Hildenbrand, D. L.,** *J. Chem. Phys.*, 74, 653, 1981.
216. **Kleinschmidt, P. D., Cubicciotti, D., and Hildenbrand, D. L.,** *J. Electrochem. Soc.*, 125, 1543, 1978; *Proc. Electrochem. Soc.*, 78, 217, 1978.
217. **Kleinschmidt, P. D. and Hildenbrand, D. L.,** *J. Chem. Phys.*, 68, 2819, 1978.
218. **Klynning, L. and Lindgren, B.,** *Arkiv. Fysik.*, 32, 575, 1966.
219. **Kobayashi, H.,** *J. Chem. Phys.*, 79, 123, 1983.
220. **Kohl, F. J. and Carlson, K. D.,** *J. Am. Chem. Soc.*, 90, 4814, 1968.
221. **Kohl, F. J. and Stearns, C. A.,** *J. Phys. Chem.*, 78, 273, 1974.
222. **Kondratiev, V. N.,** *Bond Dissociation Energies, Ionization Potentials and Electron Affinities*, Mauka Publishing House, Moscow, 1974.
223. **Kordis, J. and Gingerich, K. A.,** *J. Chem. Eng. Data*, 18, 135, 1973.
224. **Kordis, J. and Gingerich, K. A.,** *J. Chem. Phys.*, 58, 5141, 1973.
225. **Kordis, J. and Gingerich, K. A.,** *J. Phys. Chem.*, 76, 2336, 1972.
226. **Kordis, J., Gingerich, K. A. and Seyse, R. J.,** *J. Chem. Phys.*, 61, 5114, 1974.
227. **Kronekvist, M., Lagerqvist, A., and Neuhaus, H.,** *J. Mol. Spectrosc.*, 39, 516, 1971.
228. **Kulkarni, M. P. and Dadape, V. V.,** *High Temp. Sci.*, 3, 277, 1971.
229. **Langridge-Smith, P. R. R., Morse, M. D., Hansen, G. P., Smalley, R. E., and Mercer, A. J.,** *J. Chem. Phys.*, 80, 593, 1984.
230. **Lau, K. H., Brittain, R. D., and Hildenbrand, D. L.,** *Chem. Phys. Lett.*, 81, 227, 1981; *J. Phys. Chem.*, 86, 4429, 1982.
231. **Lau, K. H. and Hildenbrand, D. L.,** *J. Chem. Phys.*, 71, 1572, 1979.

232. **Lau, K. H. and Hildenbrand, D. L.,** *J. Chem. Phys.*, 72, 4928, 1980.
233. **Lau, K. H. and Hildenbrand, D. L.,** *J. Chem. Phys.*, 76, 2646, 1982.
234. **Lau, K. H. and Hildenbrand, D. L.,** *J. Chem. Phys.*, 80, 1312, 1984.
235. **LeRoy, R. J.,** *J. Chem. Phys.*, 57, 573, 1972.
236. **LeRoy, R. J. and Bernstein, R. B.,** *Chem. Phys. Lett.*, 5, 42, 1970.
237. **Li, K. C. and Stwalley, W. C.,** *J. Chem. Phys.*, 59, 4423, 1973.
238. **Lin, S.-S. and Kant, A.,** *J. Phys. Chem.*, 73, 2450, 1969.
239. **Lin, S.-S., Strauss, B., and Kant, A.,** *J. Chem. Phys.*, 51, 2282, 1969.
240. **Lindgren, B. and Nilsson, Ch.,** *J. Mol. Spectosc.*, 55, 407, 1975.
241. **McIntyre, N. S., Vander Auwera-Mahieu, A., and Drowart, J.,** *Trans. Faraday Soc.*, 64, 3006, 1968.
242. **Martin, E. and Barrow, R. F.,** *Phys. Scr.*, 17, 501, 1978.
243. **Morse, M. D., Hansen, G. P., Langridge-Smith, P. R. R., Zheng, L. S., Geusic, M. E., Michalopolous, D. L., and Smalley, R. E.,** *J. Chem. Phys.*, 80, 5400, 1984.
244. **Murad, E., Hildenbrand, D. L., and Main, R. P.,** *J. Chem. Phys.*, 45, 263, 1966.
245. **Nagai, S., Shinmei, M., and Yokokawa, T.,** *J. Inorg. Nucl. Chem.*, 36, 1904, 1974.
246. **Nappi, B. M. and Gingerich, K. A.,** *Inorg. Chem.*, 20, 522, 1981.
247. **Neckel, A. and Sodeck, G.,** *Monatsch. Chem.*, 103, 367, 1972.
248. **Nedelec, O. and Giroud, M.,** *J. Chem. Phys.*, 79, 2121, 1983.
249. **Neubert, A.,** *High Temp. Sci.*, 10, 213, 1979.
250. **Neubert, A., Ihle, H. R., and Gingerich, K. A.,** *J. Chem. Phys.*, 73, 1406, 1980.
251. **Neubert, A. and Zmbov, K. F.,** *Chem. Phys.*, 76, 469, 1983.
252. **Neubert, A. and Zmbov, K. F.,** *J. Chem. Soc. Faraday Trans. 1*, 70, 2219, 1974.
253. **Neubert, A., Zmbov, K. F., Gingerich, K. A., and Ihle, H. R.,** *J. Chem. Phys.*, 77, 5218, 1982.
254. **Nordine, P. C.,** *J. Chem. Phys.*, 61, 224, 1974.
255. **O'Hara, P. A. G.,** *J. Chem. Phys.*, 52, 2992, 1970.
256. **Ovcharenko, I. E., Ya, Kuzyankov, Y., and Tatevaskii, V. M.,** *Opt. Spectrosk.*, 19, 528, 1965.
257. **Parks, E. K. and Wexler, S.,** *J. Phys. Chem.*, 88, 4492, 1984.
258. **Parr, T. P., Behrens, R., Freedman, A., and Heron, R. R.,** *Chem. Phys. Lett.*, 56, 71, 1978.
259. **Pedley, J. B. and Marshall, E. M.,** *J. Phys. Chem. Ref. Data*, 12, 967, 1984.
260. **Peeters, R., Vander Auwera-Mahieu, A., and Drowart, J.,** *Z. Naturforsch. Teil A*, 26, 327, 1971.
261. **Pelevin, O. V., Mil'vidskii, M. G., Belyaev, A. I., and Khotin, B. A.,** *Izv. Akad. Nauk. SSSR Neorg. Mater.*, 2, 924, 1966.
262. **Pelino, M., Piacente, V., and Ascenzo, G.,** *Thermochim. Acta*, 31, 383, 1979.
263. **Pelino, M. Viswanadham, P., and Edwards, J. G.,** *J. Phys. Chem.*, 83, 2964, 1979.
264. **Perakis, J., Chatillon, C., and Pattoret, A.,** *C. R. Acad. Sci. Ser. C*, 276, 1357, 1973.
265. **Petzel, T.,** *High Temp. Sci.*, 6, 246, 1974.
266. **Piacente, V.,** *J. Chem. Phys.*, 70, 5911, 1979.
267. **Piacente, V. and Balducci, G.,** *Adv. Mass Spectrom.*, 7A, 626, 1978.
268. **Piacente, V. and Balducci, G.,** *Dyn. Mass Spectrom.*, 4, 295, 1976.
269. **Piacente, V. and Balducci, G.,** *High Temp. Sci.*, 6, 254, 1974.
270. **Piacente, V., Balducci, G., and Bardi, G.,** *J. Less-Commun. Met.*, 37, 123, 1974.
271. **Piacente, V., Bardi, G., and Malaspina, L.,** *J. Chem. Thermodyn.*, 5, 219, 1973.
272. **Piacente, V. and Desideri, A.,** *J. Chem. Phys.*, 57, 2213, 1972.
273. **Piacente, V. and Gigli, R.,** *J. Chem. Phys.*, 72, 4790, 1982.
274. **Piacente, V. and Gingerich, K. A.,** *High Temp. Sci.*, 9, 189, 1977.
275. **Piacente, V. and Gingerich, K. A.,** *Z. Naturforsch. Teil A*, 28, 316, 1973.
276. **Piacente, V. and Malaspina, L.,** *J. Chem. Phys.*, 56, 1780, 1972.
277. **Pitzer, K. S.,** *J. Chem. Phys.*, 74, 3078, 1981.
278. **Porter, R. F. and Spencer, C. W. J.,** *J. Chem. Phys.*, 32, 943, 1960.
279. **Prasad, R., Venugopal, V., and Sood, D. D.,** *J. Chem. Thermodyn.*, 9, 593, 1977.
280. **Ram, R. S., Rai, S. B., Ram, R. S., Upadhya, K. N.,** *J. Chem. Phys. Phys. Chim. Biol.*, 76, 560, 1979.
281. **Ramakrishnam, E. S., Shim, I., and Gingerich, K. A.,** *J. Chem. Soc. Faraday Trans. 2*, 80, 395, 1984.
282. **Rao, P. S. and Rao, T. V. R.,** *J. Quant. Spectrosc. Radiat. Transfer*, 27, 207, 1982.
283. **Rao, S. P. and Rao, T. V. R.,** *Acta Ciencia Indica Phys.*, 7, 58, 1981.
284. **Rao, V. M., Rao, M. L. P., and Rao, P. T.,** *J. Quant. Spectrosc. Radiat. Transfer*, 25, 547, 1981.
285. **Rao, T. V. R., Reddy, R. R., and Rao, P. S.,** *Indian, J. Pure Appl. Phys.*, 19, 1219, 1981.
286. **Riekert, G., Lamparter, P., and Steeb, S.,** *Z. Metallkd.*, 72, 765, 1981.
287. **Riekert, G., Rainer-Harbach, G., Lamparter, P., and Steeb, S.,** *Z. Metallkd.*, 76, 406, 1981.
288. **Rinehart, G. H. and Behrens, R. G.,** *J. Phys. Chem.*, 83, 2052, 1979.
289. **Ringstrom, U.,** *Ark. Fys.*, 27, 227, 1964.
290. **Rovner, L., Drowart, A., and Drowart, J.,** *Trans. Faraday Soc.*, 63, 2910, 1967.
291. **Rusin, A. D., Zhukov, E., Agamirova, L. M., and Kalinnikov, V. T.,** *Zh. Neorg. Khim.*, 24, 1457, 1979.
292. **Scott, P. R. and Richards, W. G.,** *J. Phys. B*, 7, 500, 1974.
293. **Scott, P. R. and Richards, W. G.,** *J. Phys. B*, 7, 1969, 1974.
294. **Shardanand, A.,** *Phys. Rev.*, 160, 67, 1967.
295. **Shenyavskaya, E. A., Mal'tsev, A. A., Kataev, D. I., and Gurvich, L. V.,** *Opt. Spekrosk.*, 26, 937, 1969.
296. **Siegel, B.,** *Q. Rev. (London)*, 19, 77, 1965.
297. **Smith, P. K. and Peterson, D. E.,** *J. Chem. Phys.*, 52, 4963, 1970.
298. **Smoes, S., Coppens, P., Bergman, C., and Drowart, J.,** *Trans. Faraday Soc.*, 65, 682, 1969.
299. **Smoes, S., Depiere, D., and Drowart, J.,** *Rev. Int. Hautes Temp. Refractaires Paris*, 9, 171, 1972.
300. **Smoes, S. and Drowart, J.,** *Chem. Commun.*, p.534, 1968.
301. **Smoes, S. and Drowart, J.,** *J. Chem. Soc. Faraday Trans. 2*, 73, 1746, 1977.
302. **Smoes, S. and Drowart, J.,** *J. Chem. Soc. Faraday Trans. 2*, 80, 1171, 1984.
303. **Smoes, S., Drowart, J., and Welter, J. M.,** *J. Chem. Thermodyn.*, 9, 275, 1977; *Adv. Mass Spectrom.*, 7A, 622, 1978.
304. **Smoes, S., Huguet, R., and Drowart, J.,** *Z. Naturforsch. Teil A*, 26, 1934, 1971.
305. **Smoes, S., Mandy, F., Vander Auwera-Mahieu, A., and Drowart, J.,** *Bull. Soc. Chim. Belg.*, 81, 45, 1972.
306. **Smoes, S., Myers, C. E., and Drowart, J.,** *Chem. Phys. Lett.*, 8, 10, 1971.
307. **Smoes, S., Pattje, W. R., and Drowart, J.,** *High Temp. Sci.*, 10, 109, 1978.
308. **Srivastava, R. D. and Farber, M.,** *High Temp. Sci.*, 5, 489, 1973.

309. **Stearns, C. A. and Kohl, F. J.,** *High Temp. Sci.*, 2, 146, 1970.
310. **Stearns, C. A. and Kohl, F. J.,** *High Temp. Sci.*, 5, 113, 1973.
311. **Stearns, C. A. and Kohl, F. J.,** *High Temp. Sci.*, 6, 284, 1974.
312. **Steiger, R. A. and Cater, E. D.,** *High Temp. Sci.*, 7, 204, 1975.
313. **Steiger, R. P. and Cater, E. D.,** *High Temp. Sci.*, 7, 288, 1975.
314. **Stwalley, W. C.,** *J. Chem. Phys.*, 65, 2038, 1970.
315. **Stwalley, W. C., Way, K. R., and Velasco, R.,** *J. Chem. Phys.*, 60, 3611, 1974.
316. **Su, T.-M. R. and Riley, S. J.,** *J. Chem. Phys.*, 71, 3194, 1979.
317. **Su, T.-M. R. and Riley, S. J.,** *J. Chem. Phys.*, 72, 1614, 1980.
318. **Su, T.-M. R. and Riley, S. J.,** *J. Chem. Phys.*, 72, 6632, 1980.
319. **Sullivan, C. L., Zehe, M. J., and Carlson, K. D.,** *High Temp. Sci.*, 6, 80, 1974.
320. **Tanaka, Y., Yushina, K., and Freeman, D. E.,** *J. Chem. Phys.*, 59, 564, 1973.
321. **Tellinghuisen, J., Tisone, G. C., Hoffman, J. M., and Hays, A. K.,** *J. Chem. Phys.*, 64, 4796, 1976.
322. **Tromp, J. W., LeRoy, R. J., Gerstenkorn, S., and Lue, P.,** *J. Mol. Spectrosc.*, 100, 82, 1983.
323. **Tuenge, R. T., Laabs, F., and Franzen, H. F.,** *J. Chem. Phys.*, 65, 2400, 1976.
324. **Uy, O. M. and Drowart, J.,** *High Temp. Sci.*, 2, 293, 1970.
325. **Uy, O. M. and Drowart, J.,** *Trans. Faraday Soc.*, 65, 3221, 1969.
326. **Uy, O. M. and Drowart, J.,** *Trans. Faraday Soc.*, 67, 1293, 1971.
327. **Uy, O. M., Muenow, D. W., Ficalora, P. J., and Margrave, J. L.,** *Trans. Faraday Soc.*, 64, 2998, 1968.
328. **Vander Auwera-Mahieu, A. and Drowart, J.,** *Chem. Phys. Lett.*, 1, 311, 1967.
329. **Vander Auwera-Mahieu, A., McIntyre, N. S., and Drowart, J.,** *Chem. Phys. Lett.*, 4, 198, 1969.
330. **Vander Auwera-Mahieu, A., Peeters, R., McIntyre, N. S., and Drowart, J.,** *Trans. Faraday Soc.*, 66, 809, 1970.
331. **Van Veen, N. J. A., DeVries, M., and DeVries, A. E.,** *Chem. Phys. Lett.*, 64, 213, 1979.
332. **Varma, M. P., Ishwar, N. B., and Jha, B. L.,** *Indian J. Pure Appl. Phys.*, 20, 828, 1982.
333. **Velasco, R., Ottinger, C., and Zare, R. N.,** *J. Chem. Phys.*, 51, 5522, 1969.
334. **Venkataramanaiah, M. and Lakshman, S. V. J.,** *J. Quant. Spectrosc. Radiat. Transfer*, 26, 11, 1981.
335. **Verhaegen, G.,** Ph.D. Thesis, University of Brussels, 1965.
336. **Verhaegen, G., Smoes, S., and Drowart, J.,** *J. Chem. Phys.*, 40, 239, 1964.
337. **Verhaegen, G., Stafford, F. E., and Drowart, J.,** *J. Chem. Phys.*, 40, 1622, 1964.
338. **Verma, K. K., Vu, T. H., and Stwalley, W. C.,** *J. Mol. Spectrosc.*, 85, 131, 1980.
339. **Viswanathan, K. S. and Tellinghuisen, J.,** *J. Mol. Spectrosc.*, 98, 185, 1983.
340. **Way, K. R. and Stwalley, W. C.,** *J. Chem. Phys.*, 59, 5298, 1973.
341. **Wiedemeier, H. and Gilles, P. W.,** *J. Chem. Phys.*, 42, 2765, 1965.
342. **Wiedemeier, H. and Goyette, W. J.,** *J. Chem. Phys.*, 48, 2936, 1968.
343. **Wu, C. H.,** *J. Chem. Phys.*, 65, 3181, 1976; 65, 2040, 1976.
344. **Wu, C. H. and Ihle, H. R.,** *Adv. Mass Spectrom.*, 8A, 374, 1980.
345. **Wu, C. H., Ihle, H. R., and Gingerich, K. A.,** *Int. J. Mass Spectrom. Ion Phys.*, 47, 235, 1983.
346. **Wyss, J. C.,** *J. Chem. Phys.*, 71, 2949, 1979.
347. **Yang, S. C.,** *J. Chem. Phys.*, 77, 2884, 1982.
348. **Yokozeki, A. and Menzinger, M.,** *Chem. Phys.*, 14, 427, 1976.
349. **Zlomanov, V. P., Novozhilov, A. F., and Markarov, A. V.,** *Izv. Akad. Nauk SSSR Neorg. Mater.*, 16, 620, 1980.

STANDARD STATES

For elements that are diatomic gases in their standard states these are readily obtained from the bond strength. For elements that are crystalline in their standard states they are derived from vapor pressure data.

Table 2
HEATS OF FORMATION OF GASEOUS ATOMS FROM ELEMENTS IN THEIR STANDARD STATES

Atom	$\Delta H^\circ_{f(298)}$/kcal mol^{-1}	$\Delta H^\circ_{f(298)}$/kJ mol^{-1}	Ref.	Atom	$\Delta H^\circ_{f(298)}$/kcal mol^{-1}	$\Delta H^\circ_{f(298)}$/kJ mol^{-1}	Ref.	Atom	$\Delta H^\circ_{f(298)}$/kcal mol^{-1}	$\Delta H^\circ_{f(298)}$/kJ mol^{-1}	Ref.
Ag	68.1 ± 0.2	284.9 ± 0.8	1	Hf	148 ± 1	619 ± 4	2	Re	185 ± 1.5	774 ± 6.3	2
Al	78.8 ± 1.0	329.7 ± 4.0	1	Hg	14.69 ± 0.03	61.46 ± 0.13	2	Rh	133 ± 1	557 ± 4	2
As	72.3 ± 3	302.5 ± 13	2	I	25.518 ± 0.010	106.765 ± 0.040	3	Ru	155.5 ± 1.5	648.5 ± 6.3	2
Au	88.0 ± 0.5	368.2 ± 2.1	2	In	58 ± 1	243 ± 4	2	S	66.20 ± 0.06	276.98 ± 0.25	1
B	139 ± 3	560 ± 12	1	Ir	160 ± 1	669 ± 4	2	Sb	63.2 ± 0.6	264.4 ± 2.5	2
Ba	42.5 ± 1	177.8 ± 4	2	K	21.42 ± 0.05	89.62 ± 0.21	2	Sc	90.3 ± 1	377.8 ± 4	2
Be	77.5 ± 1.5	324.3 ± 6.3	2	Li	38.6 ± 0.4	161.5 ± 1.7	2	Se	54.3 ± 1	227.2 ± 4	2
Bi	50.1 ± 0.5	209.6 ± 2.1	2	Mg	35.0 ± 0.3	146.4 ± 1.3	2	Sn	72.2 ± 0.5	302.1 ± 2.1	2
Br	26.735	111.857	3	Mn	67.7 ± 1	283.3 ± 4	2	Sr	39.1 ± 0.5	163.6 ± 2.1	2
C	171.29 ± 0.11	716.67 ± 0.44	1	Mo	157.3 ± 0.5	658.1 ± 2.1	2	Ta	186.9 ± 0.6	782.0 ± 2.5	2
Ca	42.6 ± 0.4	178.2 ± 1.7	2	N	112.97 ± 0.10	472.68 ± 0.40	1	Te	47.0 ± 0.5	196.7 ± 2.1	2
Cd	26.72 ± 0.15	111.80 ± 0.63	2	Na	25.85 ± 0.15	108.16 ± 0.63	2	Th	137.5 ± 0.5	575.3 ± 2.1	2
Ce	101 ± 3	423 ± 13	2	Nb	172.4 ± 1	721.3 ± 4	2	Ti	112.3 ± 0.5	469.9 ± 2.1	2
Cl	28.989 ± 0.002	121.290 ± 0.008	3	Ni	102.8 ± 0.5	430.1 ± 2.1	2	Tl	43.55 ± 0.1	182.21 ± 0.4	2
Co	102.4 ± 1	428.4 ± 4	2	O	59.553 ± 0.024	249.17 ± 0.10	1	U	126 ± 3	527 ± 13	2
Cr	95 ± 1	398 ± 4	2	Os	188 ± 1.5	787 ± 6.3	2	V	122.9 ± 0.3	514.2 ± 1.3	2
Cs	18.7 ± 0.1	78.2 ± 0.4	2	P	79.4 ± 1.0	332.2 ± 4.2	2	W	203.1 ± 1	849.8 ± 4	2
Cu	80.7 ± 0.3	337.6 ± 1.2	1	Pb	46.62 ± 0.3	195.06 ± 1.3	2	Y	101.5 ± 0.5	424.7 ± 2.1	2
Er	75.8 ± 1	317.1 ± 4	2	Pd	90.0 ± 0.5	376.6 ± 2.1	2	Yb	36.35 ± 0.2	152.09 ± 0.8	2
F	18.98	79.41	3	Pt	135.2 ± 0.3	565.7 ± 1.3	2	Zn	31.17 ± 0.05	130.42 ± 0.20	1
Ge	89.5 ± 0.5	374.5 ± 2.1	2	Pu	87.1 ± 4	364.4 ± 17	2	Zr	145.5 ± 1	608.8 ± 4	2
H	52.102 ± 0.001	217.995 ± 0.005	3	Rb	19.6 ± 0.1	82.0 ± 0.4	2				

REFERENCES

1. CODATA recommended Key values for thermodynamics, 1975, *J. Chem. Thermodyn.*, 8, 603, 1976.
2. **Brewer, L. and Rosenblatt, G. M.,** *Adv. High Temp. Chem.*, 2, 1, 1969.
3. Calculated from $D°_0$ taken from **Huber, K. P. and Herzberg, G.,** *Molecular Spectra and Molecular Structure Constants of Diatomic Molecules*, Van Nostrand, New York, 1979 and enthalpy functions from *J.A.N.A.F. Thermochemical Tables*, NSRDS-NBS 37, 1971.

BOND STRENGTHS IN POLYATOMIC MOLECULES

The values below refer to a temperature of 298 K and have mostly been determined by kinetic methods (see References 12, 43, and 53 following Table 3 for a full description of the methods). Bond strengths in polyatomic molecules are difficult to measure accurately since the mechanisms of the kinetic systems involved in the measurements are seldom straightforward. Thus much controversy has taken place in the literature over the past decade concerning the C–H bond strengths in simple alkanes. The values listed in Table 3 represent the presently accepted consensus of opinion. A good example illustrating the difficulties involved is concerned with the C–H bond strength in ethene, $D°(H–CHCH_2)$ or the related enthalpy of formation of the vinyl radical, $\Delta H^\circ_f(CH_2{=}CH)$. The references listed in the Table for this bond illustrate the range of results which have recently been reported in this case.

Some of the bond strengths have been calculated from the enthalpies of formation of the species involved according to the equations:

$$D°(R–X) = \Delta H^\circ_f(R) + \Delta H^\circ_f(X) - \Delta H^\circ_f(RX)$$
$$D°(R–R) = 2\Delta H^\circ_f(R) - \Delta H^\circ_f(RR)$$

The sources of the data on the enthâlpies of formation are given in the references following Table 3.

An attempt has been made to list all the important values obtained by methods that are considered to be valid. The referneces are intended to serve as a guide to the literature.

Table 3
BOND STRENGTHS IN POLYATOMIC MOLECULES

Bond	$D°_{298}$/kcal mol − 1	$D°_{298}$/kJ mol^{-1}	Ref.[a]
H–CH	100.8	421.8	1, 20, 88
H–CH_2	111.1	464.8	1, 4, 88
H–CH_3	104.8 ± 0.2	438.4 ± 1.0	4
H–CCH	132 ± 5	552 ± 21	53
H–$CHCH_2$	106	444	3, 14, 42, 53
H–C_2H_5	100.3 ± 1.0	419.5 ± 4.0	4
H–Cycloprop-2-en-1-yl	90.6 ± 4	379.1 ± 17	53
H–CH_2CCH	89.4 ± 2	374.1 ± 8.4	53
H–CH_2CHCH_2	86.3 ± 1.5	361.1 ± 6.3	53
H–Cyclopropyl	106.3 ± 0.3	444.8 ± 1.1	53
H–n-C_3H_7	99.7	417.2	19
H–i-C_3H_7	95.9 ± 0.5	401.3 ± 2.0	4
H–CH_2CCCH_3	87.2 ± 2	364.9 ± 8	53
H–$CH(CH_3)CCH$	83.1 ± 2.2	347.7 ± 9.2	53
H–Cyclobutyl	96.5 ± 1	403.8 ± 4.2	53
H–Cyclopropylmethyl	97.4 ± 1.6	407.5 ± 6.7	53
H–$CH(CH_3)CHCH_2$	82.5 ± 1.3	345.2 ± 5.4	53
H–$CH_2CHCHCH_3$	85.6 ± 1.5	358.2 ± 6.3	53
H–$CH_2C(CH_3)CH_2$	85.6 ± 1	358.2 ± 4.2	82, 83
H–s-C_4H_9	96.4	403.3	19
H–t-C_4H_9	93.3 ± 0.5	390.2 ± 2.0	4
H–Cyclopenta-1,3-dien-5-yl	71.1 ± 1.5	297.5 ± 6.3	53
H–Spiropentyl	98.8 ± 1	413.4 ± 4.2	53
H–Cyclopent-1-en-3-yl	82.3 ± 1	344.3 ± 4	53
H–$CH_2CHCHCHCH_2$	83 ± 3	347 ± 13	53
H–$CH(C_2H_3)_2$	76.4	319.7	53, 80
H–$CH(CH_3)CCCH_3$	87.3 ± 2.7	365.3 ± 11.3	53
H–$C(CH_3)_2CCH$	81.0 ± 2.3	338.9 ± 9.6	53
H–$C(CH_3)_2CHCH_2$	77.2 ± 1.5	323.0 ± 6.3	53
H–Cyclopentyl	94.5 ± 1	395.4 ± 4.2	19, 53
H–$CH_2C(CH_3)_3$	100 ± 2	418.4 ± 8	53
H–C_6H_5	110.9 ± 2.0	464.0 ± 8.4	53
H–Cyclohexa-1,3-dien-5-yl	73 ± 5	305 ± 21	53
H–Cyclohexyl	95.5 ± 1	399.6 ± 4.2	53
H–$C(CH_3)_2CCCH_3$	82.3 ± 2.7	344.3 ± 11.3	53
H–$CH_2C(CH_3)C(CH_3)_2$	78.0 ± 1.1	326.4 ± 4.6	53
H–$C(CH_3)_2C(CH_3)CH_2$	76.3 ± 1.1	319.2 ± 4.6	53
H–$CH_2C_6H_5$	88.0 ± 1	368.2 ± 4.2	53
H–Cyclohepta-1,3,5-trien-7-yl	73.0 ± 2	305.4 ± 8.4	53
H–Norbornyl	96.7 ± 2.5	404.6 ± 10.5	53
H–Cycloheptyl	92.5 ± 1	387.0 ± 4.2	53
H–$CH(CH_3)C_6H_5$	85.4 ± 1.5	357.3 ± 6.3	53
H–Inden-1-yl	84 ± 3	351 ± 13	53
H–$C(CH_3)_2C_6H_5$	84.4 ± 1.5	353.1 ± 6.3	53
H–1-Naphthylmethyl	85.1 ± 1.5	356.1 ± 6.3	53
H–$CH(C_6H_5)_2$	81.4	340.6	72
H–9,10-Dihydroanthracen-9-yl	75.3 ± 1.5	315.1 ± 6.3	53
H–$C(CH_3)(C_6H_5)_2$	81 ± 2	339 ± 8	53
H–Anthracenylmethyl	81.8 ± 1.5	342.3 ± 6.3	53
H–9-Phenanthrenylmethyl	85.1 ± 1.5	356.1 ± 6.3	53
H–CN	123.8 ± 2	518.0 ± 8.4	53
H–CH_2CN	93 ± 2.5	389 ± 10.5	53
H–$CH(CH_3)CN$	89.9 ± 2.3	376.1 ± 9.6	53
H–$C(CH_3)_2CN$	86.5 ± 2.0	361.9 ± 8.4	53
H–CH_2NH_2	93.3 ± 2.0	390.4 ± 8.4	53
H–CH_2NHCH_3	87 ± 2	364 ± 8	53
H–$CH_2N(CH_3)_2$	84 ± 2	352 ± 8	53
H–CHO	87 ± 1	364 ± 4	53
H–$COCH_3$	86.0 ± 0.8	359.8 ± 3.4	53
H–$COCHCH_2$	87.1 ± 1.0	364.4 ± 4.2	53
H–COC_2H_5	87.4 ± 1	365.7 ± 4.2	53
H–COC_6H_5	86.9 ± 1	363.6 ± 4.2	53
H–$COCF_3$	91.0 ± 2	380.7 ± 8.4	53
H–CH_2CHO	94.8 ± 2.0	396.6 ± 8.4	71
H–CH_2COCH_3	98.3 ± 1.8	411.3 ± 7.5	53
H–$CH(CH_3)COCH_3$	92.3 ± 1.4	386.2 ± 5.9	53
H–CH_2OCH_3	93 ± 1	389 ± 4	53
H–$CH(CH_3)OC_2H_5$	91.7 ± 0.4	383.7 ± 1.7	47
H–Tetrahydrofuran-2-yl	92 ± 1	385 ± 4	53
H–2-Furylmethyl	86.5 ± 2	361.9 ± 8.4	53
H–CH_2OH	94 ± 2	393 ± 8	53
H–$CH(CH_3)OH$	93 ± 1	389 ± 4	53
H–$CH(OH)CHCH_2$	81.6 ± 1.8	341.4 ± 7.5	53
H–$C(CH_3)_2OH$	91 ± 1	381 ± 4	53
H–$CH_2OCOC_6H_5$	100.2 ± 1.3	419.2 ± 5.4	53
H–$COOCH_3$	92.7 ± 1	387.9 ± 4.2	53
H–CH_2SH	96 ± 1	402 ± 4	75
H–CH_2SCH_3	96.6 ± 1.0	404.2 ± 4.2	76
H–CH_2F	101.2 ± 1	423.4 ± 4.2	66
H–CHF_2	103.2 ± 1	431.8 ± 4.2	66
H–CF_3	106.7 ± 1	446.4 ± 4.2	53
H–CF_2Cl	101.6 ± 1.0	425.1 ± 4.2	53
H–CH_2Cl	100.9 ± 2	422.2 ± 4.2	53
H–$CHCl_2$	100.6	420.9	89
H–CCl_3	95.8 ± 1	400.8 ± 4.2	53
H–CH_2Br	102.0 ± 2	426.8 ± 8.4	53
H–$CHBr_2$	103.7 ± 2	433.9 ± 8.4	53
H–CBr_3	96.0 ± 1.6	401.7 ± 6.7	53
H–CH_2I	103 ± 2	431 ± 8	53
H–CHI_2	103 ± 2	431 ± 8	53

Table 3 (continued)
BOND STRENGTHS IN POLYATOMIC MOLECULES

Bond	D°_{298}/kcal mol^{-1}	D°_{298}/kJ mol^{-1}	Ref.[a]
$H-CHCF_2$	107 ± 2	448 ± 8	79
$H-CFCHF$	107 ± 2	448 ± 8	79
$H-CFCF_2$	108 ± 2	452 ± 8	79
$H-CH_2CF_3$	106.7 ± 1.1	446.4 ± 4.6	53
$H-CF_2CH_3$	99.5 ± 2.5	416.3 ± 10.5	53
$H-C_2F_5$	102.7 ± 0.5	429.7 ± 2.1	53
$H-CFCFCl$	106 ± 2	444 ± 8	79
$H-CHClCF_3$	101.8 ± 1.5	425.9 ± 6.3	53
$H-CClCFCl$	105 ± 2	439 ± 8	79
$H-CClCHCl$	104 ± 2	435 ± 8	79
$H-CCl_2CHCl_2$	94 ± 2	393 ± 8	53
$H-C_2Cl_5$	95 ± 2	398 ± 8	53
$H-CClBrCF_3$	96.6 ± 1.5	404.2 ± 6.3	53
$H-n-C_3F_7$	104 ± 2	435 ± 8	53
$H-i-C_3F_7$	103.6 ± 0.6	433.3 ± 2.4	32
$H-CHClCHCH_2$	88.6 ± 1.4	370.7 ± 5.9	53
$H-C_6F_5$	113.9	476.6	53
$H-CH_2Si(CH_3)_3$	99.2 ± 1	415.1 ± 4.2	87
$H-SiH$	84	352	53
$H-SiH_2$	64	268	53
$H-SiH_3$	90.3	377.8	53, 87
$H-SiH_2CH_3$	89.6	374.9	87
$H-SiH(CH_3)_2$	89.4	374.1	87
$H-Si(CH_3)_3$	90.3	377.8	53, 87
$D-Si(CH_3)_3$	93 ± 1.7	388 ± 7.1	31
$H-SiH_2C_6H_5$	88.2	369.0	53, 87
$H-SiF_3$	100.1	418.8	53, 87
$H-SiCl_3$	91.3	382.0	53, 87
$H-Si_2H_5$	86.3	361.1	53
$H-GeH_3$	83 ± 2	346 ± 8	55
$H-GeH_2I$	79 ± 2	332 ± 8	56
$H-Ge(CH_3)_3$	81 ± 2	340 ± 10	30
$H-Sn(n-C_4H_9)_3$	73.7 ± 2.0	308.4 ± 8.4	17
$H-NH_2$	107.4 ± 1.1	449.4 ± 4.6	53
$H-NHCH_3$	100.0 ± 2.5	418.4 ± 10.5	53
$H-N(CH_3)_2$	91.5 ± 2	382.8 ± 8.4	53
$H-NHC_6H_5$	88.0 ± 2	368.2 ± 8.4	53
$H-N(CH_3)C_6H_5$	87.5 ± 2	366.1 ± 8.4	53
$H-NO$	≤49.5	≤207.1	21
$H-NO_2$	78.3 ± 0.5	327.6 ± 2.1	53
$H-NF_2$	75.7 ± 2.5	316.7 ± 10.5	53
$H-N_3$	92 ± 5	385 ± 21	53
$H-OH$	119 ± 1	498 ± 4	53
$H-OCH_3$	104.4 ± 1	436.8 ± 4.2	53, 54
$H-OC_2H_5$	104.2 ± 1	436.0 ± 4.2	53
$H-OC(CH_3)_3$	105.1 ± 1	439.7 ± 4.2	53
$H-OCH_2C(CH_3)_3$	102.3 ± 1.5	428.0 ± 6.3	53
$H-OC_6H_5$	86.5 ± 2	361.9 ± 8.4	53
$H-O_2H$	88.2 ± 1.0	369.0 ± 4.2	77
$H-O_2CH_3$	88.5 ± 0.5	370.3 ± 2.1	48
$H-O_2-t-C_4H_9$	89.4 ± 0.2	374.1 ± 0.8	41
$H-OCOCH_3$	105.8 ± 2	442.7 ± 8.4	53
$H-OCOC_2H_5$	106.4 ± 2	445.2 ± 8.4	53
$H-OCO-n-C_3H_7$	105.9 ± 2	443.1 ± 8.4	53
$H-ONO$	78.3 ± 0.5	327.6 ± 2.1	13
$H-ONO_2$	101.2 ± 0.5	423.4 ± 2.1	13
$H-SH$	91.1 ± 1	381.2 ± 4.2	53
$H-SCH_3$	88.6 ± 1	370.7 ± 4.2	75
$H-SC_6H_5$	83.3 ± 2	348.5 ± 8.4	53
$H-SO$	41.3	172.8	90
$HC{\equiv}CH$	230 ± 2	962 ± 8	1, 20, 25
$H_2C{=}CH_2$	172 ± 2	720 ± 8	1, 25, 88
CH_3-CH_3	89.8 ± 0.5	376.1 ± 2.1	1, 4, 25
CH_3-CH_2CCH	76.0 ± 2	318.0 ± 8.4	53
$CH_3-CH_2CCCH_3$	73.7 ± 1.5	308.4 ± 6.3	53
$CH_3-CH(CH_3)CCH$	73.0	305.4	53
$CH_3-CH_2CHCHCH_3$	72.9 ± 0.8	305.0 ± 3.3	53
$CH_3-CH_2C(CH_3)CH_2$	72.0 ± 0.8	301.2 ± 3.4	82
$CH_3-t-C_4H_9$	101.8 ± 2	425.9 ± 8.4	53

Bond	D°_{298}/kcal mol − 1	D°_{298}/kJ mol^{-1}	Ref.[a]
$CH_3-CH(CH_3)CCCH_3$	76.7 ± 1.5	320.9 ± 6.3	53
$CH_3-C(CH_3)_2CCH$	70.7 ± 1.5	295.8 ± 6.3	53
$CHCCH_2-n-C_3H_7$	73.2 ± 1.5	306.3 ± 6.3	53
$CH_3-C(CH_3)_2CHCH_2$	68.1 ± 1.5	284.9 ± 6.3	53
$n-C_3H_7-CH_2CHCH_2$	70.7	295.8	84
$CH_3-C_6H_5$	75.8 ± 1.5	317.1 ± 6.3	53
$CH_3-C(CH_3)_2CCCH_3$	72.5 ± 1.5	303.3 ± 6.3	53
$CHCCH_2-s-C_4H_9$	71.7 ± 1.5	297.5 ± 6.3	53
$CH_3-CH_2C_6H_5$	79.4 ± 1	332.2 ± 4.2	53
$CH_3-CH(CH_3)C_6H_5$	74.6 ± 1.5	312.1 ± 6.3	53
$C_2H_5-CH_2C_6H_5$	70.3 ± 1	294.1 ± 4.2	53
CH_3–1-Naphthylmethyl	72.9 ± 1.5	305.0 ± 6.3	53
$CH_3-C(CH_3)_2C_6H_5$	73.7 ± 1.5	308.4 ± 6.3	53
$CHCCH_2-CH_2C_6H_5$	61.4 ± 2	256.9 ± 8.4	53
$n-C_3H_7-CH_2C_6H_5$	70.0 ± 1	292.9 ± 4.2	53
CH_3–9-Anthracenylmethyl	67.6 ± 1.5	282.8 ± 6.3	53
CH_3–9-Phenanthrenylmethyl	72.9 ± 1.5	305.0 ± 6.3	53
$CH_3-CH(C_6H_5)_2$	72 ± 2	301 ± 8	53
$CH_3-C(CH_3)(C_6H_5)_2$	69 ± 2	289 ± 8	53
CH_3-CN	121.8 ± 2	509.6 ± 8.4	53
C_2H-CN	144 ± 1	603 ± 4	58
$C_2H_5-CH_2NH_2$	79.4 ± 2	332.2 ± 8.4	53
CH_3-CH_2CN	80.4 ± 1	336.4 ± 4.2	81
$C_2H_5-CH_2CN$	76.9 ± 1.7	321.8 ± 7.1	53
$CH_3-CH(CH_3)CN$	78.8 ± 2	329.7 ± 8.4	53
$C_2H_5-CH_2CN$	76.9 ± 1.7	321.8 ± 7.1	53
$CH_3-C(CH_3)_2CN$	74.7 ± 1.6	312.6 ± 6.7	53
$CH_3-C(CH_3)(CN)C_6H_5$	59.9	250.6	53
$C_6H_5CH_2-CH_2NH_2$	68.0 ± 2	284.5 ± 8.4	53
$C_6H_5CH_2-C_5H_4N$	86.7	362.8	72
$CN-CN$	128 ± 1	536 ± 4	28
CH_3–2-Furylmethyl	75 ± 2	314 ± 8	53
$C_6H_5CH_2-COCH_2C_6H_5$	65.4 ± 2	273.6 ± 8.4	53
$CH_3CO-COCH_3$	67.4 ± 2.3	282.0 ± 9.6	53
$C_6H_5CH_2-COOH$	67	280	53
$C_6H_5CO-COC_6H_5$	66.4	277.8	53
$(C_6H_5)_2CH-COOH$	59.4 ± 3	248.5 ± 12.6	53
$CF_3-COC_6H_5$	73.8 ± 2	308.8 ± 8.4	53
$CF_2{=}CF_2$	76.3 ± 3	319.2 ± 12.6	92
CH_2F-CH_2F	88 ± 2	368 ± 8	44
CH_3-CF_3	101.2 ± 1.1	423.4 ± 4.6	70
CF_3-CF_3	98.7 ± 2.5	413.0 ± 10.5	53
$C_6F_5-C_6F_5$	116.6 ± 5.9	488.4 ± 24.5	68
CH_3-BF_2	~113	~473	53
$C_6H_5-BCl_2$	~122	~510	53
$CH_2CHCH_2-Si(CH_3)_3$	70	292	53
$s-C_4H_9-Si(CH_3)_3$	90	414	53
$CH_3-NHC_6H_5$	71.4 ± 2	298.7 ± 8.4	53
$C_6H_5CH_2-NH_2$	71.1 ± 1	297.5 ± 4.2	53
$CH_3-N(CH_3)C_6H_5$	70.8 ± 2	296.2 ± 8.4	53
$C_6H_5CH_2-NHCH_3$	68.7 ± 2	287.4 ± 8.4	53
$C_6H_5CH_2-N(CH_3)_2$	62.1 ± 2	259.8 ± 8.4	53
$CH_2{=}N_2$	41.7 ± 1	174.5 ± 4.2	49
$CH_3-N_2CH_3$	52.5	219.7	15
$C_2H_5-N_2C_2H_5$	50.0	209.2	15
$i-C_3H_7-N_2-i-C_3H_7$	47.5	198.7	15
$n-C_4H_9-N_2-n-C_4H_9$	50.0	209.2	15
$i-C_4H_9-N_2-i-C_4H_9$	49.0	205.0	15
$s-C_4H_9-N_2-s-C_4H_9$	46.7	195.4	15
$t-C_4H_9-N_2-t-C_4H_9$	43.5	182.0	15
$C_6H_5CH_2-N_2CH_2C_6H_5$	37.6	157.3	15
$CF_3-N_2CF_3$	55.2	231.0	15
CH_3-NO	40.0 ± 0.8	167.4 ± 3.4	53
$i-C_3H_7-NO$	36.5 ± 3	152.7 ± 12.6	53
$t-C_4H_9-NO$	39.5 ± 1.5	165.3 ± 6.3	53
C_6H_5-NO	50.8 ± 1	212.6 ± 4.2	53
$NC-NO$	28.8 ± 2.5	120.5 ± 10.5	39
CF_3-NO	42.8 ± 2	179.1 ± 8.4	53
C_6F_5-NO	49.8 ± 1	208.4 ± 4.2	53
CCl_3-NO	32 ± 3	134 ± 13	53
$t-C_4H_9-NO-t-C_4H_9$	29	121	18
CH_3-NO_2	60.8	254.4	53

Table 3 (continued)
BOND STRENGTHS IN POLYATOMIC MOLECULES

Bond	D°_{298}/kcal mol^{-1}	D°_{298}/kJ mol^{-1}	Ref.[a]
$CH_2C(CH_3)$–NO_2	58.6	245.2	53
i-C_3H_7–NO_2	59.0	246.9	53
t-C_4H_9–NO_2	58.5	244.8	53
C_6H_5–NO_2	71.3 ± 1	298.3 ± 4.2	53
$C(NO_2)_3$–NO_2	40.5 ± 1	169.5 ± 4.2	53
CH_3–$OC(CH_3)CH_2$	66.3	277.4	91
CH_3–OC_6H_5	57 ± 2	239 ± 8	63
CH_3–$OCH_2C_6H_5$	67.0	280.3	23
C_2H_5–OC_6H_5	63 ± 1.5	263.6 ± 6.3	53
CH_2CHCH_2–OC_6H_5	49.8 ± 2	208.4 ± 8.4	53
O=CO	127.2 ± 0.1	532.2 ± 0.4	27
CH_3–O_2	32.4 ± 0.7	135.6 ± 2.9	78
C_2H_5–O_2	35.2 ± 1.5	147.3 ± 6.3	78
CH_2CHCH_2–O_2	17.2 ± 1.0	72.0 ± 4.2	73
i-C_3H_7–O_2	37.7 ± 1.8	157.7 ± 7.5	78
t-C_4H_9–O_2	36.7 ± 1.9	153.6 ± 8.0	78
$C_6H_5CH_2$–O_2CCH_3	67 ± 2	280 ± 8	53
$C_6H_5CH_2$–$O_2CC_6H_5$	69	289	15
CH_3–O_2SCH_3	66.8	279.5	53
CH_2CHCH_2–O_2SCH_3	49.6	207.5	53
$C_6H_5CH_2$–O_2SCH_3	52.9	221.3	53
CF_3–O_2	48.8	204.2	11
CF_3–O_2CF_3	86.4	361.5	11
CH_3–SCH_3	77.2 ± 2	323.0 ± 8.4	53
t-C_4H_9–SH	68.4 ± 1.5	286.2 ± 6.3	53
C_6H_5–SH	86.5 ± 2	361.9 ± 8.4	53
CH_3–SC_6H_5	69.4 ± 2	290.4 ± 8.4	53
$C_6H_5CH_2$–SCH_3	61.4 ± 2	256.9 ± 8.4	53
S–CS	102.9 ± 3	430.5 ± 12.6	53
F–CH_3	108 ± 3	452 ± 13	29, 43
F–CN	112.3 ± 1.2	469.9 ± 5.0	53
F–CF_2Cl	117 ± 6	490 ± 25	36
F–$CFCl_2$	110 ± 6	460 ± 25	36
F–CF_2CH_3	124.8 ± 2	522.2 ± 8.4	53
F–C_2F_5	126.8 ± 1.8	530.5 ± 7.5	53
Cl–CN	100.8 ± 1.2	421.7 ± 5.0	53
Cl–COC_6H_5	74 ± 3	310 ± 13	53
Cl–CSCl	63.4 ± 0.5	265.3 ± 2.1	57
Cl–CF_3	86.1 ± 0.8	360.2 ± 3.3	24
Cl–CF_2Cl	76 ± 2	318 ± 8	36
Cl–CCl_2F	73 ± 2	305 ± 8	35
Cl–CCl_3	73.1 ± 1.8	305.9 ± 7.5	53
Cl–C_2F_5	82.7 ± 1.7	346.0 ± 7.1	24
Cl–CF_2CF_2Cl	78 ± 2	326 ± 8	53
Cl–$SiCl_3$	111	466	87
Br–CH_3	70.0 ± 1.2	292.9 ± 5.0	33
Br–C_6H_5	80.5 ± 2	336.8 ± 8.4	53
Br–CN	87.8 ± 1.2	367.4 ± 5.0	53
Br–CH_2COCH_3	62.5	261.5	91
Br–COC_6H_5	64.2	268.6	15
Br–CHF_2	69 ± 2	289 ± 8	53
Br–CF_3	70.6 ± 3.0	295.4 ± 12.6	53
Br–CF_2CH_3	68.6 ± 1.3	287.0 ± 5.4	65
Br–C_2F_5	68.7 ± 1.5	287.4 ± 6.3	53
Br–n-C_3F_7	66.5 ± 2.5	278.2 ± 10.5	53
Br–i-C_3F_7	65.5 ± 1.1	274.1 ± 4.6	53
Br–$CH_2C_6F_5$	94 ± 2.5	225 ± 6.3	46
Br–$CHClCF_3$	65.7 ± 1.5	274.9 ± 6.3	53
Br–CCl_3	55.3 ± 1	231.4 ± 4.2	53
Br–$CClBrCF_3$	60.0 ± 1.5	251.0 ± 6.3	53
Br–CBr_3	56.2 ± 1.8	235.1 ± 7.5	53
Br–NF_2	≤53	≤222	22
I–n-C_4H_9	49.0 ± 1	205.0 ± 4.2	53
I–Norbornyl	62.5 ± 2.5	261.5 ± 10.5	61
I–CN	73 ± 1	305 ± 4	28
I–CF_3	53.5 ± 0.7	223.8 ± 2.9	2
I–CF_2CH_3	52.1 ± 1.0	218.0 ± 4.2	53
I–CH_2CF_3	56.3 ± 1	235.6 ± 4.2	53
I–C_2F_5	52.3 ± 0.7	218.8 ± 2.9	2
I–n-C_3F_7	49.8 ± 1.0	208.4 ± 4.2	53
I–i-C_3F_7	51.4 ± 0.7	215.1 ± 2.9	2
I–n-C_4F_9	49.0 ± 1.0	205.0 ± 4.2	59
I–C_6H_5	65.4 ± 2	273.6 ± 8.4	53
I–C_6F_5	~66	~277	53
C_5H_5–FeC_5H_5	91 ± 3	381 ± 13	51
CH_3–$ZnCH_3$	68 ± 4	285 ± 17	53
C_2H_5–ZnC_2H_5	57 ± 4	239 ± 17	53
CH_3–$Ga(CH_3)_2$	63 ± 4	264 ± 17	53
C_2H_5–$Ga(C_2H_5)_2$	50 ± 4	209 ± 17	53
CH_3–$Ge(CH_3)_3$	83 ± 4	347 ± 17	53
CH_3–$As(CH_3)_2$	67 ± 4	280 ± 17	53
CH_3–$CdCH_3$	60 ± 4	251 ± 17	53
CH_3–$In(CH_3)_2$	49 ± 4	205 ± 17	53
CH_3–$Sn(CH_3)_3$	71 ± 4	297 ± 17	53
C_2H_5–$Sn(C_2H_5)_3$	63 ± 4	264 ± 17	53
CH_3–$Sb(CH_3)_2$	61 ± 4	255 ± 17	53
C_2H_5–$Sb(C_2H_5)_2$	58 ± 4	243 ± 17	53
CH_3–$HgCH_3$	61 ± 4	255 ± 17	53
C_2H_5–HgC_2H_5	49 ± 4	205 ± 17	53
CH_3–$Tl(CH_3)_2$	40 ± 4	167 ± 17	53
CH_3–$Pb(CH_3)_3$	57 ± 4	239 ± 17	53
C_2H_5–$Pb(C_2H_5)_3$	55 ± 4	230 ± 17	53
CH_3–$Bi(CH_3)_2$	52 ± 4	218 ± 17	53
CO–$Cr(CO)_5$	37 ± 2	155 ± 8	50
CO–$Fe(CO)_4$	41 ± 2	172 ± 8	50
CO–$Mo(CO)_5$	40 ± 2	167 ± 8	50
CO–$W(CO)_5$	46 ± 2	193 ± 8	50
BH_3–BH_3	35	146	15
NH_2–NH_2	65.8	275.3	53
NH_2–$NHCH_3$	64.1 ± 2	268.2 ± 8.4	53
NH_2–$N(CH_3)_2$	59.0 ± 2	246.9 ± 8.4	53
NH_2–NHC_6H_5	52.3 ± 2	218.8 ± 8.4	53
ON–NO_2	9.7 ± 0.5	40.6 ± 2.1	53
O_2N–NO_2	13.6	56.9	53
NF_2–NF_2	21 ± 1	88 ± 4	53
O–N_2	40	167	1, 12
O–NO	73	305	1, 12
HO–NO	49.3	206.3	53
HO–NO_2	49.4	206.7	53
HO_2–NO_2	23 ± 2	96 ± 8	53
CH_3O–NO	41.8 ± 0.9	174.9 ± 3.8	5, 10
C_2H_5O–NO	42.0 ± 1.3	175.7 ± 5.5	5, 9
n-C_3H_7O–NO	40.1 ± 1.8	167.8 ± 7.5	5
i-C_3H_7O–NO	41.0 ± 1.3	171.5 ± 5.5	5, 8
n-C_4H_9O–NO	42.5 ± 1.5	177.8 ± 6.3	5
i-C_4H_9O–NO	42.0 ± 1.5	175.7 ± 6.3	5
s-C_4H_9O–NO	41.5 ± 0.8	173.6 ± 3.4	5, 6
t-C_4H_9O–NO	40.9 ± 0.8	171.1 ± 3.4	5, 7
HO–$NCHCH_3$	49.7	208.0	15
Cl–NF_2	~32	~134	1, 64
I–NO	18.6 ± 0.1	77.8 ± 0.4	37
I–NO_2	18.3 ± 1	76.6 ± 4.2	85
HO–OH	51 ± 1	213 ± 4	53
HO–$OCH_2C(CH_3)_3$	46.3 ± 1.9	193.7 ± 8.0	53
CH_3O–OCH_3	37.6 ± 2	157.3 ± 8.4	53
C_2H_5O–OC_2H_5	37.9 ± 1	158.6 ± 4.2	53
n-C_3H_7O–O-n-C_3H_7	37.1 ± 1	155.2 ± 4.2	53
i-C_3H_7O–O-i-C_3H_7	37.7 ± 1	157.7 ± 4.2	53
s-C_4H_9O–O-s-C_4H_9	36.4 ± 1	152.3 ± 4.2	53
t-C_4H_9O–O-t-C_4H_9	38.0 ± 1	159.0 ± 4.2	53
$C_2H_5C(CH_3)_2O$–$O(CH_3)_2CC_2H_5$	39.3 ± 1	164.4 ± 4.2	53
$(CH_3)_3CCH_2O$–$OCH_2C(CH_3)_3$	36.4 ± 1	152.3 ± 4.2	53
CF_3O–OCF_3	46.2	193.3	53
$(CF_3)_3CO$–$OC(CF_3)_3$	35.5 ± 1.1	148.7 ± 4.4	53
t-C_4H_9O–$OSi(CH_3)_3$	47	197	53
SF_5O–OSF_5	37.2	155.7	53
t-C_4H_9O–$OGe(C_2H_5)_3$	46	193	53
t-C_4H_9O–$OSn(C_2H_5)_3$	46	193	53
$FClO_2$–O	58.4	244.3	15
CF_3O–O_2CF_3	30.3 ± 2	126.8 ± 8.4	53
SF_5O–O_2SF_5	30.3	126.8	53
CH_3CO_2–O_2CCH_3	30.4 ± 2	127.2 ± 8.4	53
$C_2H_5CO_2$–$O_2CC_2H_5$	30.4 ± 2	127.2 ± 8.4	53
n-$C_3H_7CO_2$–O_2C-n-C_3H_7	30.4 ± 2	127.2 ± 8.4	53

Table 3 (continued)
BOND STRENGTHS IN POLYATOMIC MOLECULES

Bond	D°_{298}/kcal mol−1	D°_{298}/kJ mol^{-1}	Ref.[a]	Bond	D°_{298}/kcal mol−1	D°_{298}/kJ mol^{-1}	Ref.[a]
O–SO	132 ± 2	552 ± 8	27	F–SF_4	53.1 ± 6.0	222.2 ± 25.1	45
F–OCF_3	43.5 ± 0.5	182.0 ± 2.1	26	F–SF_3	84.1 ± 3.0	351.9 ± 12.6	45
HO–Cl	60 ± 3	251 ± 13	43	F–SF_2	63.1 ± 7.1	264.0 ± 29.7	45
O–ClO	59 ± 3	247 ± 13	27	F–SF	91.7 ± 4.3	383.7 ± 18.0	45
HO–Br	56 ± 3	234 ± 13	43	I–SH	49.4 ± 2	206.7 ± 8.4	53
HO–I	56 ± 3	234 ± 13	43	I–SO	43	180	53
O=PF_3	130 ± 5	544 ± 21	43	I–SCH_3	49.3 ± 1.7	206.3 ± 7.1	75
O=PCl_3	122 ± 5	511 ± 21	43	I–$Si(CH_3)_3$	77	322	87
O=PBr_3	119 ± 5	498 ± 21	43	H_3Si–SiH_3	74	310	53, 87
HO–$Si(CH_3)_3$	128	536	53	$(CH_3)_3Si$–$Si(CH_3)_3$	80.5	336.8	53, 87
HS–SH	66 ± 2	276 ± 8	53	$(C_6H_5)_3Si$–$Si(C_6H_5)_3$	88 ± 7	368 ± 29	53, 87
F–SF_5	91.1 ± 3.2	381.2 ± 13.4	45				

[a] References appear after Table 4.

ENTHALPIES OF FORMATION OF FREE RADICALS

The enthalpies of formation of the free radicals are related to the corresponding bond strengths by the equations

$$D^\circ(R\text{–})X0 = \Delta H^\circ_f(\dot{R}) + \Delta H^\circ_f(\dot{X}) - \Delta H^\circ_f(RX)$$

or

$$D^\circ(R\text{–}R) = 2\Delta H^\circ_f(\dot{R}) - \Delta H^\circ_f(RR)$$

For an excellent review of the methods of determining the enthalpies of formation of free radicals the reader is referred to "Thermochemistry of Free Radicals" by H. E. O'Neal and S. W. Benson in *Free Radicals,* Kochi, J. K., Ed., John Wiley & Sons, New York, 1973, 275.

The references are the same as given for Table 3.

Table 4
ENTHALPIES OF FORMATION OF FREE RADICALS

Radical	$\Delta H^\circ_{f(298)}$/kcal mol^{-1}	$\Delta H^\circ_{f(298)}$/kJ mol^{-1}	Ref.	Radical	$\Delta H^\circ_{f(298)}$/kcal mol^{-1}	$\Delta H^\circ_{f(298)}$/kJ mol^{-1}	Ref.
CH	142.53 ± 1	596 35 ± 4	1, 20	C_6H_5	78.6 ± 2	328.9 ± 8.4	53
$CH_2(^3B_1)$	93.8 ± 0.5	392.5 ± 2.1	88	Cyclohexa-1,3-dien-5-yl	47 ± 5	197 ± 21	53
$CH_2(^1A_1)$	102.8 ± 1.0	430.1 ± 4.2	88, 16,	Cyclohexyl	13.9 ± 1	58.2 ± 4.2	53
			52	$CH_3C{\equiv}CC(CH_3)_2$	53.0 ± 2.3	221.8 ± 9.6	53
CH_3	34.8 ± 0.2	145.6 ± 1.0	4	$(CH_3)_2C{=}C(CH_3)CH_2$	9.5 ± 1.5	39.8 ± 6.3	53
$CH{\equiv}C$	135 ± 1	565 ± 4	53	$CH_2{=}C(CH_3)C(CH_3)_2$	9.0 ± 1.5	37.7 ± 6.3	53
$CH_2{=}CH$	66	276	3, 14,	$C_6H_5CH_2$	47.8 ± 1.5	200.0 ± 6.3	53
			42, 53	Cyclohepta-1,3,5-trien-7-yl	64.8 ± 2	271.1 ± 8.4	53
C_2H_5	28.0 ± 1.0	117.0 ± 4.2	4	$CH_3CH_2CH_2C(CH_3)_2$	0.8 ± 2.0	3.4 ± 8.4	74
Cycloprop-2-en-1-yl	105.1 ± 4.1	439.7 ± 17.2	53	Norbornyl	32.6 ± 2.5	136.4 ± 10.5	53
$CH{\equiv}CCH_2$	81.4 ± 2	340.6 ± 8.4	53	Cycloheptyl	12.2 ± 1	51.1 ± 4.2	53
$CH_2{=}CHCH_2$	39.1 ± 1.5	163.6 ± 6.3	53	$C_6H_5CHCH_3$	40.4	169.0	53
Cyclopropyl	66.9 ± 0.25	279.9 ± 1.1	53	$C_6H_5C(CH_3)_2$	32.2	134.7	53
n-C_3H_7	22.8	95.4	19	1-Naphthylmethyl	60.4	252.7	53
i-C_3H_7	19.0 ± 0.5	79.6 ± 2.0	4	$(C_6H_5)_2CH$	69	289	72
$CH_3C{\equiv}CCH_2$	70.2	293.7	53	9,10-Dihydroanthracen-9-yl	61.4 ± 1.5	256.9 ± 6.3	53
$CH_2{=}CHCHCH_3$	30.0 ± 1.5	125.5 ± 6.3	53	9-Anthracenylmethyl	80.7	337.6	53
$CH{\equiv}CCHCH_3$	70.5 ± 2.2	295.0 ± 9.2	53	9-Phenanthrenylmethyl	74.4	311.3	53
Cyclobutyl	51.2 ± 1.0	214.2 ± 4.2	53	CN	104 ± 2	435 ± 8	53
Cyclopropylmethyl	51.1 ± 1.6	213.8 ± 6.7	53	CH_2CN	58.5 ± 2.5	244.8 ± 10.5	53
$CH_2{=}C(CH_3)CH_2$	30.4 ± 1.3	127.2 ± 5.4	53	CH_3CHCN	50.0 ± 2.3	209.2 ± 9.6	53
$CH_3CH{=}CHCH_2$	30.0 ± 1.5	125.5 ± 6.3	53	$(CH_3)_2CCN$	39.8 ± 2.0	166.5 ± 8.4	53
s-C_4H_9	13.9	58.2	53	$C_6H_5C(CH_3)CN$	59.4	248.5	53
t-C_4H_9	9.0 ± 0.5	37.6 ± 2.0	4	CH_2NH_2	35.7 ± 2	149.4 ± 8.4	53
Cylcopenta-1,3-dien-5-yl	57.9 ± 1.5	242.3 ± 6.3	53	CH_3NHCH_2	30 ± 2	126 ± 8	53
Spiropentyl	91.0 ± 1	380.7 ± 4	53	$(CH_3)_2NCH_2$	26 ± 2	109 ± 8	53
Cyclopent-1-en-3-yl	38.4 ± 1	160.7 ± 4	53	CHO	8.9 ± 1.2	37.2 ± 5.0	53
$CH_2{=}CHCH{=}CHCH_2$	49 ± 3	205 ± 13	53	CH_3CO	−5.8 ± 0.4	−24.3 ± 1.7	53
$(C_2H_3)_2CH$	49 ± 3	205 ± 13	53	$CH_2{=}CHCO$	17.3	72.4	53
$CH_3C{\equiv}CCHCH_3$	65.2 ± 2.3	272.8 ± 9.6	53	C_2H_5CO	−10.2 ± 1.0	−42.7 ± 4.2	53
$CH{\equiv}CC(CH_3)_2$	61.5 ± 2	257.3 ± 8.4	53	C_6H_5CO	26.1 ± 2	109.2 ± 8.4	53
$CH_2{=}CHC(CH_3)_2$	18.5 ± 1.5	77.4 ± 6.3	53	CH_2CHO	3.0 ± 2.0	12.6 ± 8.4	71
Cyclopentyl	24.3 ± 1	101.7 ± 4.2	19, 53	CH_3COCH_2	−5.7 ± 2.6	−23.9 ± 10.9	53
$(CH_3)_3CCH_2$	8.7 ± 2	36.4 ± 8.4	53	$CH_3COCHCH_3$	−16.8 ± 1.7	−70.3 ± 7.1	53

Table 4 (continued)
ENTHALPIES OF FORMATION OF FREE RADICALS

Radical	$\Delta H^\circ_{f(298)}$/kcal mol^{-1}	$\Delta H^\circ_{f(298)}$/kJ mol^{-1}	Ref.
CH_3OCH_2	-2.8 ± 1.2	-11.7 ± 5.0	53
$C_2H_5OCHCH_3$	-20.2	-84.5	47
Tetrahydrofuran-2-yl	-4.3 ± 1.5	-18.0 ± 6.3	53
CH_2OH	-6.2 ± 1.5	-25.9 ± 6.3	53
CH_3CHOH	-15.2 ± 1.0	-63.6 ± 4.2	53
$CH_2{=}CHCHOH$	0.0	0.0	53
$(CH_3)_2COH$	-26.6 ± 1.1	-111.3 ± 4.6	53
$COOH$	-53.3	-223.0	53
$COOCH_3$	-40.4 ± 1	-169.0 ± 4.2	53
$C_6H_5COOCH_2$	-16.7 ± 2.0	-69.9 ± 8.4	53
CHF	39 ± 3	163 ± 13	69
CFO	≤ -44	≤ -184	62
CF_2	-46.4 ± 2.2	-194.1 ± 9.2	60
CH_2F	-7.6 ± 2	-31.8 ± 8.4	66
CHF_2	-57.1 ± 1	-238.9 ± 4.2	66
CF_3	-111.7 ± 3.6	-467.4 ± 15.1	53
CCl_2	57	239	62
CH_2Cl	28.3	118.4	53
$CHCl_2$	25.7 ± 1.0	107.5 ± 4.2	89
CF_2Cl	-64.3 ± 2	-269.0 ± 8.4	53
CCl_2F	-23	-96	53
CCl_3	19.0 ± 1	79.5 ± 4.2	53
CH_2Br	41.5	173.6	53
$CHBr_2$	54.3	227.2	53
CH_2I	55.0 ± 1.6	230.1 ± 6.7	53
CHI_2	79.8 ± 2.2	333.9 ± 9.2	53
CH_3CF_2	-72.3 ± 2	-304.5 ± 8.4	53
CF_3CH_2	-123.6 ± 1.2	-517.1 ± 5.0	53
C_2F_5	-213.4 ± 1	-892.9 ± 4.2	53
$CHCl_2CCl_2$	5.6 ± 2	23.4 ± 8.4	53
CF_2ClCF_2	-164 ± 4	-686 ± 17	53
C_2Cl_5	8.4 ± 2	35.2 ± 8.4	53
C_6F_5	-130.9 ± 2	-547.7 ± 8.4	53
$(CH_3)_3SiCH_2$	-8.3	-34.7	53
CH_3SCH_2	35.6 ± 1.0	149.0 ± 4.2	76
SiH	90	377	53, 87
SiH_2	58	243	53, 87
SiH_3	46.4	194.1	53, 87
CH_3Si	74	310	87
CH_3SiH	50.9 ± 3.5	213.0 ± 14.6	86
CH_3SiH_2	36.5	152.7	53, 87
$(CH_3)_2Si$	26	109	87
$(CH_3)_2SiH$	14.3	59.8	53, 87
$(CH_3)_3Si$	-0.8	-3.3	53, 87
$C_6H_5Si(CH_3)_2$	39	163	60
$(C_6H_5)SiCH_3$	78	326	60
$(C_6H_5)_3Si$	116.2	486.2	60
SiF	-4.6	-19.3	53, 87
SiF_2	-140.5	-587.9	53, 87
SiF_3	-245	-1025	53, 87
$SiCl$	46.8	195.8	53, 87
$SiCl_2$	-39.1	-163.6	53, 87
$SiCl_3$	-76	-318	53, 87
SiH_3SiH	64.5 ± 3.5	269.9 ± 14.6	86
Si_2H_5	53.3	223.0	53, 87
GeH_3	57	239	60
NH	84.2 ± 2.3	352.3 ± 9.6	67
NH_2	44.3 ± 1.1	185.4 ± 4.6	53
CH_3NH	42.4 ± 2	177.4 ± 8.4	53
$(CH_3)_2N$	34.7 ± 2	145.2 ± 8.4	53
C_6H_5NH	56.7 ± 2	237.2 ± 8.4	53
$C_6H_5NCH_3$	55.8 ± 2	233.5 ± 8.4	53
NF_2	8 ± 1	34 ± 4	53
N_2H_3	48.7	203.8	34, 60
N_3	112 ± 5	469 ± 21	53
PH_2	33 ± 2	138 ± 8	60
HO	9.4	39.3	53
CH_3O	4.2	17.6	53
C_2H_5O	-4.1	-17.2	53
$n\text{-}C_3H_7O$	-9.9	-41.4	53
$i\text{-}C_3H_7O$	-12.5	-52.3	53
$n\text{-}C_4H_9O$	-15.0	-62.8	53
$s\text{-}C_4H_9O$	-16.6 ± 0.8	-69.5 ± 3.3	53
$t\text{-}C_4H_9O$	-21.7	-90.8	53
C_6H_5O	11.4	47.7	53
HO_2	3.5	14.6	77
CH_3O_2	5.5 ± 1.0	23.0 ± 4.2	48
$t\text{-}C_4H_9O_2$	20.7	86.6	41
CF_3O_2	-161.3	-674.9	11
CH_3CO_2	-49.6 ± 1	-207.5 ± 4.2	53
$C_2H_5CO_2$	-54.6 ± 1	-228.5 ± 4.2	53
$n\text{-}C_3H_7CO_2$	-59.6 ± 1	-249.4 ± 4.2	53
CH_3SO_2	-57.2	-239.2	60
HS	33.6 ± 1.1	140.6 ± 4.6	53
CH_3S	31.0 ± 1	129.7 ± 4.2	75
C_6H_5S	54.9 ± 2	229.7 ± 8.4	53
SF	2.9 ± 1.5	12.1 ± 6.3	53
SF_2	-70.4 ± 4.0	-294.6 ± 16.7	53
SF_3	-115.2 ± 5.8	-482.0 ± 24.3	53
SF_4	-180.9 ± 5.0	-756.9 ± 20.9	53
SF_5	-215.7 ± 3.2	-902.5 ± 13.4	53
CH_3S_2	16.4 ± 2	68.6 ± 8.4	40
$C_2H_5S_2$	10.4 ± 2	43.5 ± 8.4	40
$i\text{-}C_3H_7S_2$	3.3 ± 2	13.8 ± 8.4	40
$t\text{-}C_4H_9S_2$	-4.6 ± 2	-19.3 ± 8.4	40

REFERENCES

1. A value calculated from one of the thermochemical equations from the "Bond Strengths in Polyatomic Molecules" section taking enthalpy data from the references quoted.
2. **Ahonkhai, S. I. and Whittle, E.,** *Int. J. Chem. Kinet.*, 16, 543, 1984.
3. **Ayranci, G. and Back, M. H.,** *Int. J. Chem. Kinet.*, 15, 83, 1983.
4. **Baldwin, R. R., Drewery, G. R., and Walker, R. W.,** *J. Chem. Soc. Faraday Trans. 1*, 80, 2827, 1984.
5. **Batt, L., Christie, K., Milne, R. T., and Summers, A. J.,** *Int. J. Chem. Kinet.*, 6, 87, 1974.
6. **Batt, L. and McCulloch, R. D.,** *Int. J. Chem. Kinet.*, 8, 911, 1976.
7. **Batt, L. and Milne, R. T.,** *Int. J. Chem. Kinet.*, 8, 59, 1976.
8. **Batt, L. and Milne, R. T.,** *Int. J. Chem. Kinet.*, 9, 141, 1977.
9. **Batt, L. and Milne, R. T.,** *Int. J. Chem. Kinet.*, 9, 549, 1977.
10. **Batt, L., Milne, R. T., and McCulloch, R. D.,** *Int. J. Chem. Kinet.*, 9, 567, 1977.
11. **Batt, L. and Walsh, R.,** *Int. J. Chem. Kinet.*, 15, 605, 1983.
12. **Benson, S. W.,** *J. Chem. Educ.*, 42, 502, 1965.
13. **Benson, S. W.,** *Thermochemical Kinetics*, 2nd ed., John Wiley & Sons, New York, 1976.
14. **Benson, S. W.,** unpublished results.
15. **Benson, S. W. and O'Neal, H. E.,** Kinetic Data on Gas Phase Unimolecular Reactions, National Bureau of Standards, NSRDS-NBS, Washington, D. C., 21, 1970.
16. **Bunker, P. R. and Sears, T. J.,** *J. Chem. Phys.*, 83, 4866, 1985.
17. **Burkey, T. J., Majewski, M., and Griller, D.,** *J. Am. Chem. Soc.*, 108, 2218, 1986.
18. **Carmichael, P. J., Gowenlock, B. G., and Johnson, C. A. F.,** *J. Chem. Soc. Perkin Trans. 2*, 1853, 1973.
19. **Castelhano, A. L. and Griller, D.,** *J. Am. Chem. Soc.*, 104, 3655, 1982.
20. **Chupka, W. A. and Lifshitz, C.,** *J. Chem. Phys.*, 48, 1109, 1968.

21. **Clement, M. J. Y. and Ramsay, D. A.,** *Can. J. Phys.*, 39, 205, 1961.
22. **Clyne, M. A. A. and Connor, J.,** *J. Chem. Soc. Faraday Trans. 1*, 68, 1220, 1972.
23. **Colussi, A. J., Zabel, F., and Benson, S. W.,** *Int. J. Chem. Kinet.*, 9, 161, 1977.
24. **Coomber, J. W. and Whittle, E.,** *Trans. Faraday Soc.*, 63, 2656, 1967.
25. **Cox, J. D. and Pilcher, G.,** *Thermochemistry of Organic and Organometallic Compounds*, Academic Press, New York, 1970.
26. **Czarnarski, J., Castellano, E., and Schumaker, H. J.,** *Chem. Comm.*, p.1255, 1968.
27. **Darwent, D. deB.,** Bond Dissociation Energies in Simple Molecules, National Bureau of Standards, NSRDS-NBS, Washington, D.C., 31, 1970.
28. **Davis, D. D. and Okabe, H.,** *J. Chem. Phys.*, 49, 5526, 1968.
29. **Dibeler, V. H. and Reese, R. M.,** *J. Res. Natl. Bur. Stand.*, 54, 127, 1955.
30. **Doncaster, A. M. and Walsh, R.,** *J. Phys. Chem.*, 83, 578, 1979.
31. **Ellul, E., Potzinger, P., Reimann, B., and Camilleri, P.,** *Ber. Bunsenges. Phys. Chem.*, 85, 407, 1981.
32. **Evans, B. S., Weeks, I., and Whittle, E.,** *J. Chem. Soc. Faraday Trans. 1*, 79, 1471, 1983.
33. **Ferguson, K. C., Okafo, E. N., and Whittle, E.,** *J. Chem. Soc. Faraday Trans. 1*, 69, 295, 1973.
34. **Fisher, I. P.,** *Nature*, 208, 1199, 1965.
35. **Foon, R. and Tait, K. B.,** *J. Chem. Soc. Faraday Trans. 1*, 68, 104, 1972.
36. **Foon, R. and Tait, K. B.,** *J. Chem. Soc. Faraday Trans. 1*, 68, 1121, 1972.
37. **Forte, E., Hippler, H., and van den Bergh, H.,** *Int. J. Chem. Kinet.*, 13, 1227, 1981.
38. **Golden, D. M. and Benson, S. W.,** *Chem. Rev.*, 69, 125, 1969.
39. **Gowenlock, G. B., Johnson, C. A. F., Keary, C. M., and Pfab, J.,** *J. Chem. Soc. Perkin Trans. 2*, 71, 351, 1975.
40. **Howari, J. A., Griller, D., and Lossing, F. P.,** *J. Am. Chem. Soc.*, 108, 3273, 1986.
41. **Heneghan, S. P. and Benson, S. W.,** *Int. J. Chem. Kinet.*, 15, 815, 1983.
42. **Keifer, J. H., Wei, H. C., Kern, R. D., and Wu, C. H.,** *Int. J. Chem. Kinet.*, 17, 225, 1985.
43. **Kerr. J. A.,** *Chem. Rev.*, 66, 465, 1966.
44. **Kerr, J. A. and Timlin, D. M.,** *Int. J. Chem. Kinet.*, 3, 427, 1971.
45. **Kiang, T. and Zare, R. N.,** *J. Am. Chem. Soc.*, 102, 4024, 1980.
46. **Kominar, R. J., Krech, M. J., and Price, S. J. W.,** *Can. J. Chem.*, 58, 1906, 1980.
47. **Kondo, O. and Benson, S. W.,** *Int. J. Chem. Kinet.*, 16, 949, 1984.
48. **Kondo, O. and Benson, S. W.,** *J. Phys. Chem.*, 88, 6675, 1984.
49. **Laufer, A. H. and Okabe, H.,** *J. Am. Chem. Soc.*, 83, 4137, 1971.
50. **Lewis, K. E., Golden, D. M., and Smith, G. P.,** *J. Am. Chem. Soc.*, 106, 3905, 1984.
51. **Lewis, K. E. and Smith, G. P.,** *J. Am. Chem. Soc.*, 106, 4650, 1984.
52. **McKellar, A. R. W., Bunker, P. R., Sears, T. J., Evenson, K. M., Saykally, R. J., and Langhoff, S. R.,** *J. Chem. Phys.*, 79, 5251, 1983.
53. **McMillen, D. F. and Golden, D. M.,** *Ann. Rev. Phys. Chem.*, 33, 493, 1982.
54. **Moylan, C. R. and Brauman, J. I.,** *J. Phys. Chem.*, 88, 3175, 1984.
55. **Noble, P. N. and Walsh, R.,** *Int. J. Chem. Kinet.*, 15, 547, 1983.
56. **Noble, P. N. and Walsh, R.,** *Int. J. Chem. Kinet.*, 15, 561, 1983.
57. **Okabe, H.,** *J. Chem. Phys.*, 66, 2058, 1977.
58. **Okabe, H. and Dibeler, V. H.,** *J. Chem. Phys.*, 59, 2430, 1973.
59. **Okafo, E. N. and Whittle, E.,** *Int. J. Chem. Kinet.*, 7, 287, 1975.
60. **Okafo, E. N. and Whittle, E.,** *J. Chem. Soc. Faraday Trans. 1*, 70, 1366, 1974.
61. **O'Neal, H. E., Bagg, J. W., and Richardson, W. H.,** *Int. J. Chem. Kinet.*, 2, 493, 1970.
62. **O'Neal, H. E. and Benson, S. W.,** in *Free Radicals*, Kochi, J. K., Ed., John Wiley & Sons, New York, 1973, 275.
63. **Paul, S. and Back, M. H.,** *Can. J. Chem.*, 53, 3330, 1975.
64. **Petry, R. C.,** *J. Am. Chem. Soc.*, 89, 4600, 1967.
65. **Pickard, J. M. and Rodgers, A. S.,** *Int. J. Chem. Kinet.*, 9, 759, 1977.
66. **Pickard, J. M. and Rodgers, A. S.,** *Int. J. Chem. Kinet.*, 15, 569, 1983.
67. **Piper, L. G.,** *J. Chem. Phys.*, 70, 3417, 1979.
68. **Price, S. J. W. and Sapiano, H. J.,** *Can. J. Chem.*, 57, 1468, 1979.
69. **Pritchard, G. O., Nilson, W. B., and Kirtman, B.,** *Int. J. Chem. Kinet.*, 16, 1637, 1984.
70. **Rodgers, A. S. and Ford, W. G. F.,** *Int. J. Chem. Kinet.*, 5, 965, 1973.
71. **Rossi, M. and Golden, D. M.,** *Int. J. Chem. Kinet.*, 11, 715, 1979.
72. **Rossi, M., McMillen, D. F., and Golden, D. M.,** *J. Phys. Chem.*, 88, 5031, 1984.
73. **Ruiz, R. R., Bayes, K. D., Macpherson, M. T., and Pilling, M. J.,** *J. Phys. Chem.*, 85, 1622, 1981.
74. **Seres, L., Gorgenyi, M., and Farkas, J.,** *Int. J. Chem. Kinet.*, 15, 1133, 1983.
75. **Shum, L. G. S. and Benson, S. W.,** *Int. J. Chem. Kinet.*, 15, 433, 1983.
76. **Shum, L. G. S. and Benson, S. W.,** *Int. J. Chem. Kinet.*, 17, 277, 1985.
77. **Shum, L. G. S. and Benson, S. W.,** *J. Phys. Chem.*, 87, 3479, 1983.
78. **Slagle, I. R., Ratajczak, E., and Gutman, D.,** *J. Phys. Chem.*, 90, 402, 1986.
79. **Steinkruger, F. J. and Rowland, F. S.,** *J. Phys. Chem.*, 85, 136, 1981.
80. **Trenwith, A. B.,** *J. Chem. Soc. Faraday Trans. 1*, 78, 3131, 1982.
81. **Trenwith, A. B.,** *J. Chem. Soc. Faraday Trans. 1*, 79, 2755, 1983.
82. **Trenwith, A. B. and Wrigley, S. P.,** *J. Chem. Soc. Faraday Trans. 1*, 73, 817, 1977.
83. **Tsang, W.,** *Int. J. Chem. Kinet.*, 5, 929, 1973.
84. **Tsang, W.,** *Int. J. Chem. Kinet.*, 10, 1119, 1978.
85. **van der Bergh, H. and Troe, J.,** *J. Chem. Phys.*, 64, 736, 1976.
86. **Vanderwielen, A. J., Ring, M. A., and O'Neal, H. E.,** *J. Am. Chem. Soc.*, 97, 993, 1975.
87. **Walsh, R.,** *Acc. Chem. Res.*, 14, 246, 1981.
88. **Walsh, R.,** Private communication.
89. **Weisman, M. and Benson, S. W.,** *J. Phys. Chem.*, 87, 243, 1983.
90. **White, J. N. and Gardiner, W. C.,** *Chem. Phys. Lett.*, 58, 470, 1978.
91. **Zabel, F., Benson, S. W., and Golden, D. M.,** *Int. J. Chem. Kinet.*, 10, 295, 1978.
92. **Zmbov, K. F., Uy, O. M., and Margrave, J. L.,** *J. Am. Chem. Soc.*, 90, 5090, 1968.

BOND STRENGTHS OF SOME ORGANIC MOLECULES

Bond strenghts at 298 K expressed in kcal/mol for some organic molecules of the general formula R-X are presented below. Some are experimental values taken from the preceding tables; the remainder are calculated from the enthalpies of formation of the radicals, listed above, and the heats of formation of the parent compounds from sources indicated by the references below. The table also includes bond strengths for ammonia, water, and the hydrogen halides.

Table 5
BOND STRENGTHS OF SOME ORGANIC MOLECULES

	H	F	Cl	Br	I	OH	NH_2	CH_3O	CH_3	CH_3CO	NO	CF_3
H	104.207	136.3	103.16	87.56	71.321	119	107	104	105	86	50	107
CH_3	105	108	85[2]	70	57[2]	92[2]	85[2]	83[2]	90[2]	81[2]	40	102
C_2H_5	100	108[1]	80[2]	68[2]	53[2]	94[2]	84[2]	84[2]	88[2]	79[2]	—	—
i-C_3H_7	96	107[1]	81[2]	68[2]	54[2]	94[2]	84[2]	83[2]	86[2]	76[2]	37	—
t-C_4H_9	93	—	82[2]	68[2]	51[2]	93[2]	82[2]	83[2]	84[2]	73[2]	40	—
C_6H_5	111	126[1]	96[2]	81[1]	65[2]	111[2]	102[2]	100[2]	101[2]	90[2]	51	—
$C_6H_5CH_2$	88	—	72[2]	58[2]	48[2]	81[2]	—	—	75[2]	67[2]	—	—
CCl_3	96	102[1]	73	55[1]	—	—	—	—	89[6]	—	32	—
CF_3	107	131[1]	86	71[1]	54	—	—	—	102[2]	—	43	99
C_2F_5	103	127[1]	83[3]	69[1]	52	—	—	—	—	—	—	—
CH_3CO	86	119[2]	81[2]	67[2]	50[2]	106[2]	96[5]	97	81[2]	67	—	—
CN	124	112	101	88	73	—	—	—	—	—	29	—
C_6F_5	114	114[4]	92[4]	—	66[4]	107[4]	—	—	106[4]	—	50[4]	—

REFERENCES

1. **McMillen, D. F. and Golden, D. M,** *Ann. Rev. Phys. Chem.*, 33, 493, 1982.
2. **Cox, J. D. and Pilcher, G.,** *Thermochemistry of Organic and Organometallic Compouds*, Academic Press, London, 1970.
3. **Wu, E.-C. and Rodgers, A. S.,** *J. Am. Chem. Soc.*, 98, 6112, 1976.
4. **Choo, K. Y., Mendenhall, G. D., Golden, D. M. and Benson, S. W.,** *Int. J. Chem. Kinet.*, 6, 813, 1974.
5. **Benson, S. W. et al.,** *Chem. Rev.*, 69, 279, 1969.
6. **Hu, A. T., Sinke, G. C. and Mintz, M. J.,** *J. Chem. Thermodyn.*, 4, 239, 1972.

THE MADELUNG CONSTANT AND CRYSTAL LATTICE ENERGY

Donald F. Swinehart

If U is the crystal lattice energy and *M* is the Madelung constant, then

$$U = \frac{N M z_i z_j e^2}{r}(1 - 1/n)^a$$

Substance	Ion type	Crystal form[b]	*M*
Sodium chloride, NaCl	M^+, X^-	FCC	1.74756
Cesium chloride, CsCl	M^+, X^-	BCC	1.76267
Calcium chloride, $CaCl_2$	$M^{++}, 2X^-$	Cubic	2.365
Calcium fluoride,fluorite, CaF_2	$M^{++}, 2X^-$	Cubic	2.51939
Cadmium chloride, $CdCl_2$	$M^{++}, 2X^-$	Hexagonal	2.244[c]
Cadmium iodide, $CdI_2(\alpha)$	$M^{++}, 2X^-$	Hexagonal	2.355[c]
Magnesium fluoride, MgF_2	$M^{++}, 2X^-$	Tetragonal	2.381[c]
Cuprous oxide, cuprite, Cu_2O	$2M^+, X^=$	Cubic	2.22124
Zinc oxide, ZnO	$M^{++}, X^=$	Hexagonal	1.4985[c]
Sphalerite, zinc blende, ZnS	$M^{++}, X^=$	FCC	1.63806
Wurtzite, ZnS	$M^{++}, X^=$	Hexagonal	1.64132[c]
Titanium dioxide, anatase, TiO_2	$M^{4+}, 2X^=$	Tetragonal	2.400[c]
Titanium dioxide, rutile, TiO_2	$M^{4+}, 2X^=$	Tetragonal	$(2.408) - (2.055)(0.721 - c/a)^2$ [d]
β-Quartz, SiO_2	$M^{4+}, 2X^=$	Hexagonal	2.2197[c]
Corundum, Al_2O_3	$2M^{3+}, 3X^=$	Rhombohedral	4.1719

[a] N is Avogadro's number, z_i and z_j are the integral charges on the ions (i.e., in units of the electronic charge), and e is the charge on the electron in electrostatic units ($e = 4.803 \times 10^{-10}$ ESU). r is the shortest distance between cation-anion pairs in centimeters. Then U is in ergs.

[b] FCC = face centered cubic, BCC is body centered cubic.

[c] For tetragonal and hexagonal crystals the value of *M* depends on the details of the lattice parameters.

[d] For rutile-type structures, *M* is given to a good approximation by inserting the proper c/a ratio. c/a for rutile itself is 0.721 and *M* = 2.408.

Note: Several variations of the equation for U appear in the literature with corresponding variations in *M*. In general, using literature values for *M* requires caution. The Born exponent, n, may be obtained from the following table:

THE BORN EXPONENT, n

Ion type	n
He, Li^+	5
Ne, Na^+, F^-	7
Ar, K^+, Cu^+, Cl^-	9
Kr, Rb^+, Ag^+, Br^-	10
Xe, Cs^+, Au^+, I^-	12

For a crystal with a mixed-ion type, an average of the values of n in this table is to be used (6 for LiF, for example).

VALUES OF THE GAS CONSTANT, R

Coleman J. Major

The numerical value of the gas constant, R, defined by the equation PV = nRT, depends upon the units of P, V, n, and T. A large number of values of the constant may be calculated. The accompanying table gives 84 values of R in a convenient form using the most common units of pressure and volume. It also incorporates both the pound and gram mole and both Rankine and Kelvin temperature scales. Various combinations of metric and English units may, therefore, be used without the necessity of converting each variable to a common system of units. Conversion factors and constants used for computing the values of R are listed at the bottom of the table.

The following example illustrates the use of the table:

Calculate: The volume in ft^3 occupied by 2 lb. moles of a gas at 15°C at a pressure of 32.2 ft. of water, assuming the ideal gas law.

Solution: 15°C + 273.2 = 288.2°K
Enter the top of the table under the column headed "ft H_2O" and proceed downward to the value of 44.6 for R. (Note that this lines up horizontally with the desired units of ft^3, °K, and lb. moles shown on the left side of the table)

$$V = \frac{nRT}{P} = \frac{2 \times 44.6 \times 288.2}{32.2} = 798\ ft^3$$

$$\text{Values of Gas Constant, } R = \frac{PV}{nT}$$

Absolute Pressure

Volume	Temp.	moles	Atm	psia	mm Hg	cm Hg	in Hg	in H_2O	ft H_2O
ft^3	°K	g	0.00290	0.0426	2.20	0.220	0.0867	1.18	0.0982
		lb	1.31	19.31	999	99.9	39.3	535	44.6
	°R	g	0.00161	0.02366	1.22	0.122	0.0482	0.655	0.0546
		lb	0.730	10.73	555	55.5	21.8	297	24.8
cm^3	°K	g	82.05	1206	62,400	6240	2450	33,400	2780
		lb	37,200	547,000	2.83×10^7	2.83×10^6	1.11×10^6	1.51×10^7	1.26×10^6
	°R	g	45.6	670	34,600	3460	1360	18,500	1550
		lb	20,700	304,000	1.57×10^7	1.57×10^6	619,000	8.41×10^6	701,000
liters	°K	g	0.08205	1.206	62.4	6.24	2.45	33.4	2.78
		lb	37.2	547	28,300	2830	1113	15,140	1262
	°R	g	0.0456	0.670	34.6	3.46	1.36	18.5	1.55
		lb	20.7	304	15,700	1570	619	8410	701

Conversion Factors and Constants

1 lb. = 453.59 g	359.0 ft^3/lb mole
1 atm = 14.696 psia	22,414 cm^3/g mole
1 atm = 760 mm Hg	1 inch = 2.54 cm
1 atm = 76 cm Hg	Std. temp. = 273.15°K or 491.67°R
1 atm = 29.921 in Hg	28.316847 liters = 1 ft^3
1 atm = 406.79 in H_2O	$R = 8.31441\ (26) \times 10^7\ erg\ °K^{-1}\ mol^{-1}$
1 atm = 33.90 ft H_2O	$= 1.9872\ cal\ °K^{-1}\ mol^{-1}$
1 atm = 1.01325×10^5 Pa	$= 8.31441\ (26)\ J\ °K^{-1}\ mol^{-1}$

RECOMMENDED CONSISTENT VALUES OF THE FUNDAMENTAL PHYSICAL CONSTANTS

The numbers in parentheses are the standard deviation uncertainties in the last digits of the quoted value, computed on the basis of internal consistency.

	Quantity	Symbol	Value	Uncertainty, ppm
1.	Permeability of Vacuum	μ_0	$4\pi \times 10^{-7}$ H m^{-1}	
			$= 12.5663706144 \times 10^{-7}$ H m^{-1}	
2.	Speed of Light in Vacuum	c	$2.99792458(1.2) \times 10^{8}$ m s^{-1}	0.004
3.	Permittivity of Vacuum	$\epsilon_0 = (\mu_0 c^2)^{-1}$	$8.85418782(7) \times 10^{-12}$ F m^{-1}	0.008
4.	Fine Structure Constant, $\mu_0 ce^2/2h$	α	0.0072973506(60)	0.82
		α^{-1}	137.03604(11)	0.82
5.	Elementary Charge	e	$1.6021892(46) \times 10^{-19}$ C	2.9
			$4.803242(14) \times 10^{-10}$ esu	2.9
6.	Planck Constant	h	$6.626176(36) \times 10^{-34}$	5.4
		$\hbar = h/2\pi$	$1.0545887(57) \times 10^{-34}$ Js	5.4
7.	Avogadro Constant	N_A	$6.022045(31) \times 10^{23}$ mol^{-1}	5.1
8.	Atomic Mass Unit	(mass C^{12} atom)/12 = (1g)/N_A	$1.6605655(86) \times 10^{-27}$ kg	5.1
			931.5016(26) $\mathrm{M_eV/c^2}$	2.8
9.	Electron Rest Mass	m_e	$0.9109534(47) \times 10^{-30}$ kg	5.1
			$5.4858026(21) \times 10^{-4}$ u	0.38
10.	Muon Rest Mass	m_μ	0.5110034(14)MeV/c^2	2.8
			$1.883566(11) \times 10^{-28}$ kg	5.6
			0.11342920(26) u	2.3
11.	Proton Rest Mass	m_p	$1.6726485(86) \times 10^{-27}$ kg	5.1
			1.007276470(11) u	0.011
			938.2796(27)MeV/c^2	2.8
12.	Neutron Rest Mass	m_n	$1.6749543(86) \times 10^{-27}$ kg	5.1
			1.008665012(37) u	0.037
13.	Ratio, Proton Mass to Electron Mass	m_p/m_e	1836.15152(70)	0.38
14.	Ratio, Muon Mass to Electron Mass	m_μ/m_e	206.76865(47)	2.3
15.	Specific Electron Charge	e/m_e	$1.7588047(49) \times 10^{11}$ C kg^{-1}	2.8
16.	Faraday Constant	$\mathcal{F} = N_A e$	$9.648456(27) \times 10^{4}$ C mol^{-1}	2.8
17.	Magnetic Flux Quantum	$\Phi_0 = h/2e$	$2.0678506(54) \times 10^{-15}$ Wb	2.6
		h/e	$4.135701(11) \times 10^{-15}$ J Hz^{-1} C^{-1}	2.6
18.	Josephson Frequency-Voltage Ratio	$2e/h$	483.5939(13) THz V^{-1}	2.6
19.	Quantum of Circulation	$h/2m_e$	$3.6369455(60) \times 10^{-4}$ J Hz^{-1} kg^{-1}	1.6
		h/m_e	$7.273891(12) \times 10^{-4}$ J Hz^{-1} kg^{-1}	1.6
20.	Rydberg Constant	R_∞	$1.097373177(83) \times 10^{7}$ m^{-1}	0.075
21.	Bohr Radius	$a_0 = \alpha/4\pi R_\infty$	$0.52917706(44) \times 10^{-10}$ m	0.82
22.	Electron Compton Wavelength	$\lambda_c = \alpha^2/2R_\infty$	$2.4263089(40) \times 10^{-12}$ m	1.6
		$\bar{\lambda}_c = \lambda_c/2\pi = \alpha a_0$	$3.8615905(64) \times 10^{-13}$ m	1.6
23.	Classical Electron Radius	$r_e = \mu_0 e^2/4\pi m_e = \alpha\lambda_c$	$2.8179380(70) \times 10^{-15}$ m	2.5
24.	Electron g-Factor	$\frac{1}{2}g_e = \mu_e/\mu_B$	1.0011596567(35)	0.0035
25.	Muon g-Factor	$\frac{1}{2}g_\mu$	1.00116616(31)	0.31
26.	Proton Moment in Nuclear Magnetons	μ_p/μ_N	2.7928456(11)	0.38
27.	Bohr Magneton	$\mu_B = e\hbar/2m_e$	$9.274078(36) \times 10^{-24}$ J T^{-1}	3.9
			$5.7883785(95) \times 10^{-11}$ MeV T^{-1}	1.6
28.	Nuclear Magneton	$\mu_N = e\hbar/2m_p$	$5.050824(20) \times 10^{-27}$ J T^{-1}	3.9
			$3.1524515(53) \times 10^{-14}$ MeV T^{-1}	1.7
29.	Electron Magnetic Moment	μ_e	$9.284832(36) \times 10^{-24}$ J T^{-1}	3.9
30.	Proton Magnetic Moment	μ_p	$1.4106171(55) \times 10^{-26}$ J T^{-1}	3.9
31.	Proton Magnetic Moment in Bohr Magnetons	μ_p/μ_B	$1.521032209(16) \times 10^{-3}$	0.011
32.	Ratio, Electron to Proton Magnetic Moments	μ_e/μ_p	658.2106880(66)	0.010
33.	Ratio, Muon Moment to Proton Moment	μ_μ/μ_p	3.1833402(72)	2.3
34.	Muon Magnetic Moment	μ_μ	$4.490474(18) \times 10^{-26}$ J T^{-1}	3.9
35.	Proton Gyromagnetic Ratio	γ_p	$2.6751987(75) \times 10^{8}$ s^{-1} T^{-1}	2.8
36.	Diamagnetic Shielding Factor, Spherical H_2O Sample	$1 + \sigma(H_2O)$	1.000025637(67)	0.067
37.	Proton Gyromagnetic Ratio (uncorrected)	γ'_p	$2.6751301(75) \times 10^{8}$ s^{-1} T^{-1}	2.8
		$\gamma_p/2\pi$	42.57602(12) MHz T^{-1}	2.8
38.	Proton Moment in Nuclear Magnetons (uncorrected)	$\mu' p/\mu_N$	2.7927740(11)	0.38
39.	Proton Compton Wavelength	$\lambda_{C,p} = h/m_p c$	$1.3214099(22) \times 10^{-15}$ m	1.7
		$\bar{\lambda}_{C,p} = \lambda_{C,p}/2\pi$	$2.1030892(36) \times 10^{-16}$ m	1.7
40.	Neutron Compton Wavelength	$\lambda_{C,n} = h/m_n c$	$1.3195909(22) \times 10^{-15}$ m	1.7
		$\bar{\lambda}_{C,n} = \lambda_{C,n}/2\pi$	$2.1001941(35) \times 10^{-16}$ m	1.7
41.	Molar Gas Constant	R	8.31441(26) J mol^{-1} K^{-1}	31
42.	Molar Volume, Ideal Gas (T_o = 273.15 K, p_o = 1 atm)	$V_m = RT_0/p_0$	0.02241383(70) m^3 mol^{-1}	31
43.	Boltzmann Constant	$k = R/N_A$	$1.380662(44) \times 10^{-23}$ J K^{-1}	32
44.	Stefan-Boltzmann Constant	$\sigma = (\pi^2/60)k^4/\hbar^3c^2$	$5.67032(71) \times 10^{-8}$ W m^{-2} K^{-4}	125
45.	First Radiation Constant	$c_1 = 2\pi hc^2$	$3.741832(20) \times 10^{-16}$ W m^2	5.4
46.	Second Radiation Constant	$c_2 = hc/k$	0.01438786(45) m K	31
47.	Gravitational Constant	G	$6.6720(41) \times 10^{-11}$ N m^2 kg^{-2}	615
48.	Grav. Accel., Sea Level, 45° lat.	g	9.8062 m s^{-2}	—
49.	Conversion Constant	$\hbar c$	197.32858(51) MeV fm	2.6

RECOMMENDED CONSISTENT VALUES OF THE FUNDAMENTAL PHYSICAL CONSTANTS (continued)

	Quantity	Symbol	Value	Uncertainty, ppm
50.	Conversion Constant	$(\hbar c)^2$	0.3893857(20) GeV^2 mbarn	5.2
51.	Deuteron Mass	m_d	1875.6280(53) MeV/c^2	2.8
52.	Rydberg Energy	$hcR_\infty = m_e e^4/2(4\pi e_0)^2\hbar^2 = m_e c^2\alpha^2/2$	13.605804(36)eV	2.6
53.	Thomson Cross Section	$\delta_T = 8\pi r_e^{2/3}$	0.6652448(33) barn	4.9
54.	Electron Cyclotron Freq./Field	$\omega^e cycl/B = e/m_e$	1.7588047(49) $\times 10^{11}$ rad s^{-1} T^{-1}	2.8
55.	Proton Cyclotron Freq./Field	$\omega^p cycl/B = e/m_p$	9.578756(28) $\times 10^{7}$ rad s^{-1} T^{-1}	2.8
56.	Fermi Coupling Constant	$G_F/(\hbar c)^3$	1.16637(2) $\times 10^{-5}$ GeV^{-2}	17

$\pi = 3.141592653589793238$ $\quad e = 2.718281828459045235$ $\quad \gamma = 0.577215664901532861$

1 in ≡ 0.0254m	1 barn ≡ 10^{-28} m^2	1 eV ≡ 1.6021892 × 10^{-19} J	1 gauss (G) ≡ 10^{-4} tesla (T)
1Å ≡ 10^{-10} m	1 dyne ≡ 10^{-5} newton (N)	1 eV/c^2 ≡ 1.782676 × 10^{-36} kg	1 atmosphere ≡ 1.01325 × 10^5 N/m^2
1 fm ≡ 10^{-15} m	1 erg ≡ 10^{-7} joule (J)	2.99792458 × 10^9 esu ≡ 1 coulomb(C)	0°C ≡ 273.15 K

SI units take as their base: length (m), mass (kg), time (s), electric current (Å), themodynamic temperature (K), amount of a substance (mol), and luminous intensity (candela, cd), and the two supplementary units plane angle (rad) and solid angle (sr).

Data from CODATA Bulletin No. 11, ICSU CODATA Central Office, CODATA Secretariat: 51 Boulevard de Montmorency, 75016 Paris, France (copies of this bulletin are available at no cost from this office).

ENERGY CONVERSION FACTORS

Quantity	Value	Unit	Error (ppm)
1 kg	5.609538(24)	10^{29} MeV	4.4
1 amu	931.5016(26)	MeV	2.8
Electron mass	0.5110041(16)	MeV	3.1
Proton mass	938.2592(52)	MeV	5.5
Neutron mass	939.5527(52)	MeV	5.5
1 electron volt	1.6021917(70)	10^{-19} J 10^{-12} erg	4.4
	2.4179659(81)	10^{14} Hz	3.3
	8.065465(27)	10^{5} m^{-1} 10^{3} cm^{-1}	3.3
	1.160485(49)	10^{4} K	42
Energy–wavelength conversion	1.2398541(41)	10^{-6} eV·m 10^{-4} eV·cm	3.3
Rydberg constant, R_∞	2.179914(17)	10^{-18} J 10^{-11} erg	7.6
	13.605826(45)	eV	3.3
	3.2898423(11)	10^{15} Hz	0.35
	1.578936(67)	10^{5} K	43
Bohr magneton, μ_B	5.788381(18)	10^{-5} eV T^{-1}	3.1
	1.3996108(43)	10^{10} Hz T^{-1}	3.1
	46.68598(14)	m^{-1}·T^{-1} 10^{-2} cm^{-1}·T^{-1}	3.1
	0.671733(29)	K T^{-1}	43
Nuclear magneton, μ_n	3.152526(21)	10^{-8} eV T^{-1}	6.8
	7.622700(42)	10^{6} Hz T^{-1}	5.5
	2.542659(14)	10^{-2} m^{-1}·T^{-1} 10^{-4} cm^{-1}·T^{-1}	5.5
	3.65846(16)	10^{-4} K T^{-1}	44
Gas constant, R_0	8.20562(35)	10^{-2} m^3·atm $kmole^{-1}$·K^{-1}	42
Standard volume of ideal gas, V_0	22.4136	m^3 $kmole^{-1}$	

CHARACTERISTICS OF PARTICLES AND PARTICLE DISPERSOIDS

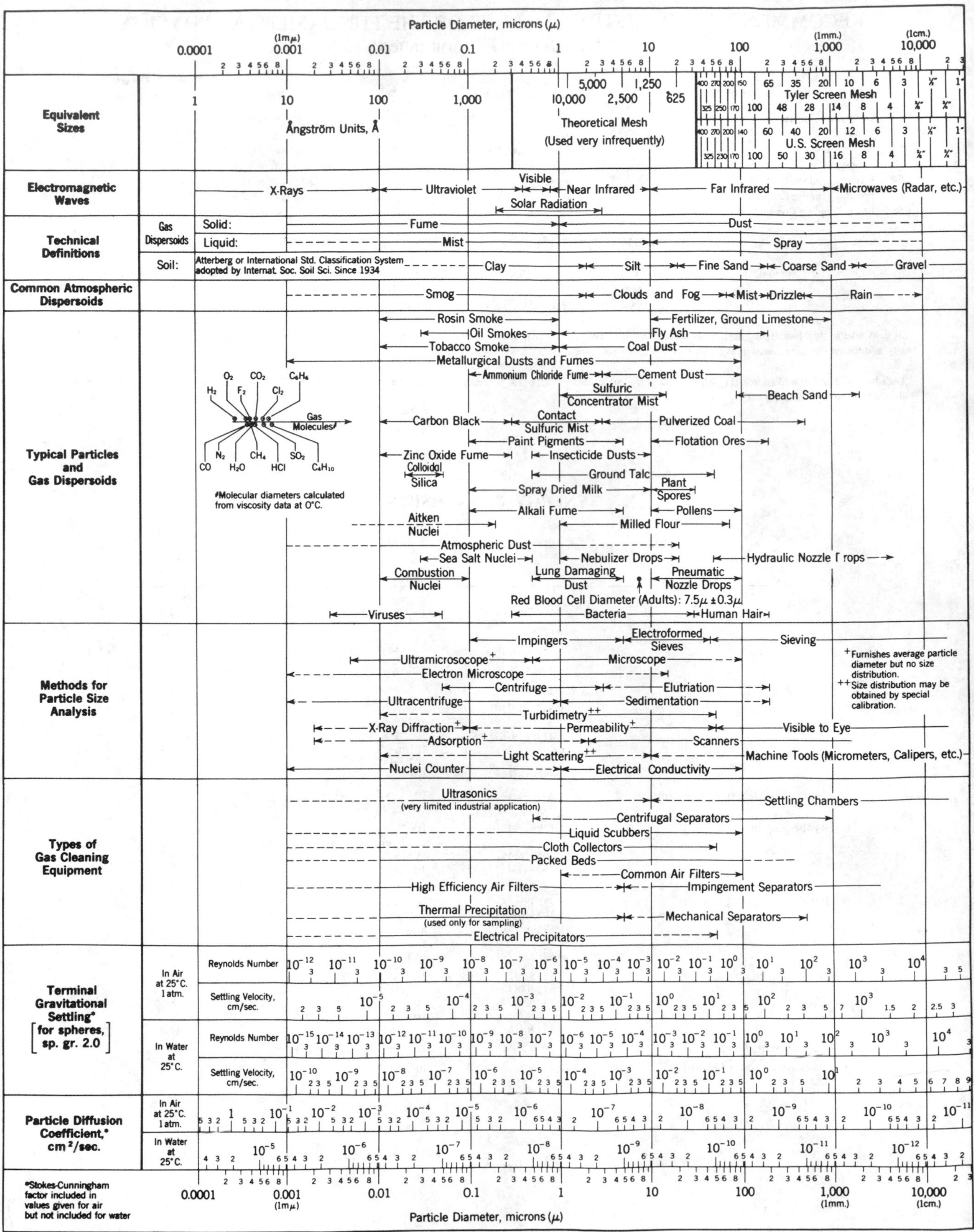

C. E. Lapple, Stanford Research Institute Journal, Vol. 5, p.95 (Third Quarter, 1961)

SPELLING AND SYMBOLS FOR UNITS

From "Units of Weight and Measure"
L. B. Chisholm, National Bureau of Standards
Miscellaneous Publication 286 (May, 1967)

The spelling of the names of units as adopted by the National Bureau of Standards is that given in the list below. The spelling of the metric units is in accordance with that given in the law of July 28, 1866, legalizing the Metric System in the United States.

Following the name of each unit in the list below is given the symbol that the Bureau has adopted. Attention is particularly called to the following principles:

1. No period is used with symbols for units. Whenever "in" for inch might be confused with the preposition "in", "inch" should be spelled out.

2. The exponents "^{2}" and "^{3}" are used to signify "square" and "cubic," respectively, instead of the symbols "sq" or "cu," which are, however, frequently used in technical literature for the U. S. Customary units.

3. The same symbol is used for both singular and plural.

Some Units and Their Symbols

Unit	Symbol	Unit	Symbol	Unit	Symbol
acre	acre	fathom	fath	millimeter	mm
are	a	foot	ft	minim	minim
barrel	bbl	furlong	furlong	ounce	oz
board foot	fbm	gallon	gal	ounce, avoirdupois	oz avdp
bushel	bu	grain	grain	ounce, liquid	liq oz
carat	c	gram	g	ounce, troy	oz tr
Celsius, degree	°C	hectare	ha	peck	peck
centare	ca	hectogram	hg	pennyweight	dwt
centigram	cg	hectoliter	hl	pint, liquid	liq pt
centiliter	cl	hectometer	hm	pound	lb
centimeter	cm	hogshead	hhd	pound, avoirdupois	lb avdp
chain	ch	hundredweight	cwt	pound, troy	lb tr
cubic centimeter	cm^3	inch	in	quart, liquid	liq qt
cubic decimeter	dm^3	International		rod	rod
cubic dekameter	dam^3	Nautical Mile	INM	second	s
cubic foot	ft^3	Kelvin, degree	°K	square centimeter	cm^2
cubic hectometer	hm^3	kilogram	kg	square decimeter	dm^2
cubic inch	in^3	kiloliter	kl	square dekameter	dam^2
cubic kilometer	km^3	kilometer	km	square foot	ft^2
cubic meter	m^3	link	link	square hectometer	hm^2
cubic mile	mi^3	liquid	liq	square inch	in^2
cubic millimeter	mm^3	liter	liter	square kilometer	km^2
cubic yard	yd^3	meter	m	square meter	m^2
decigram	dg	microgram	μg	square mile	mi^2
deciliter	dl	microinch	μin	square millimeter	mm^2
decimeter	dm	microliter	μl	square yard	yd^2
dekagram	dag	micron	μm	stere	stere
dekaliter	dal	mile	mi	ton, long	long ton
dekameter	dam	milligram	mg	ton, metric	t
dram, avoirdupois	dr avdp	milliliter	ml	ton, short	short ton
				yard	yd

SYMBOLS AND TERMINOLOGY FOR PHYSIOCHEMICAL QUANTITIES AND UNITS

Reproduced by permission of the Assistant Secretary, Publications, of IUPAC from the *Manual of Symbols and Terminology for Physiochemical Quantities and Units (1979),* International Union of Pure and Applied Chemistry, Division of Physical Chemistry, Commission on Symbols, Terminology and Units. The above referenced Manual is published by and is available from Pergamon Press, Inc., Maxwell House, Fairview Park, Elmsford, New York 10523.

This Handbook, the *Handbook of Chemistry and Physics,* also contains a section on symbols, terminology, etc. for Physics which immediately follows this section on symbols, terminology, etc. for Chemistry. Rules and guidance which apply to both Chemistry and Physics and which do not require separate interpretation are reproduced only in the section, "Symbols, Units and Nomenclature in Physics" and are not reproduced in the section, "Symbols and Terminology for Physiochemical Quantities and Units". Paragraphs in the following are numbered the same as those in the complete IUPAC Manual.

1. PHYSICAL QUANTITIES AND SYMBOLS FOR PHYSICAL QUANTITIES

1.2 Base Physical Quantities

Physical quantities are generally organized in a dimensional system built upon seven base quantities. These base quantities, each of which has its own dimension, and the symbols used to denote them, are as follows:

Base physical quantity	*Symbol for quantity*
Length	l
Mass	m
Time	t
Electric current	I
Thermodynamic temperature	T
Amount of substance	n
Luminous intensity	I_{v}

Luminous intensity is seldom if ever needed in physical chemistry.

One of these independent base quantities is of special importance to chemists but until recently had no generally accepted name, although units such as the mole have been used for it. The name 'amount of substance' is now reserved for this quantity.

The definition of amount of substance, as of all other physical quantities (see Section 5), has nothing to do with any choice of *unit*, and in particular has nothing to do with the particular unit of amount of substance called the mole (see Section 3.6). It is now as inconsistent to call n the 'number of moles' as it is to call m the 'number of kilograms' or l the 'number of meters', since n, m, and l are symbols for quantities not for numbers.

The amount of a substance is proportional to the number of *specified* elementary entities of that substance. The proportionality factor is the same for all substances; its reciprocal is the Avogadro constant. The specified elementary entity may be an atom, a molecule, an ion, a radical, an electron, etc., or any *specified* group of such particles.

1.3 Derived Physical Quantities

All other physical quantities are regarded as being derived from, and as having dimensions derived from, the seven independent base physical quantities by definitions involving only multiplication, division, differentiation, and/or integration. Examples of derived physical quantities are given, often with brief definitions, in Section 2.

1.4 Use of the Words 'Specific' and 'Molar' in the Names of Physical Quantities

The word 'specific' before the name of an extensive physical quantity is restricted to the meaning 'divided by mass'. For example specific volume is the volume divided by the mass. When the extensive quantity is represented by a capital letter, the corresponding specific quantity may be represented by the corresponding lower case letter. *Examples;* volume: V; specific volume: $v=V/m$; heat capacity at constant pressure: C_p; specific heat capacity at constant pressure: $c_p=C_p/m$.

The word 'molar' before the name of an extensive quantity is restricted to the meaning 'divided by amount of substance'. For example molar volume is the volume divided by the amount of substance. The subscript m attached to the symbol for the extensive quantity denotes the corresponding molar quantity. *Examples*; volume: V; Gibbs energy: G; molar volume: $V_m=V/n$; molar Gibbs energy: $G_{\mathrm{m}}=G/n$.

The subscript m may be omitted when there is no risk of ambiguity. Lower case letters may be used to denote molar quantities when there is no risk of misinterpretation.

The symbol X_{B}, where X denotes an extensive quantity and B is the chemical symbol for a substance, denotes the partial molar quantity of the substance B defined by the relation:

$$X_B = (\partial X/\partial n_B)_{T,p,n_C,\ldots}$$

For a pure substance B the partial molar quantity X_B and the molar quantity X_m are identical. The partial molar quantity X_B of pure substance B, which is identical with the molar quantity X_m of pure substance B, may be denoted by $X_B{}^*$, where the superscript * denotes 'pure', so as to distinguish it from the partial molar quantity X_B of substance B in a mixture.

1.6 Printing of Subscripts and Superscripts

Subscripts or superscripts which are themselves symbols for physical quantities or numbers should be printed in italic (sloping) type and all others in roman (upright) type. *Examples*; C_p for heat capacity at constant pressure, but C_{B} for heat capacity of substance B.

2. RECOMMENDED NAMES AND SYMBOLS FOR QUANTITIES IN CHEMISTRY AND PHYSICS

The following list contains the recommended symbols for the most important quantities likely to be used by chemists. Whenever possible the symbol used for a physical quantity should be that recommended. In a few cases where conflicts were foreseen alternative recommendations have been made. Bold-faced italic (sloping) as well as ordinary italic (sloping) type can also sometimes be used to resolve conflicts. Further flexibility can be obtained by the use of capital letters as variants for lower-case letters, and *vice versa*, when no ambiguity is thereby introduced.

For example, d and D may be used instead of d_i and d_e for internal and external diameter in a context in which no quantity appears, such as diffusion coefficient, for which the recommended symbol is D. Again, the recommended symbol for power is P and for pressure is p or P, but P and p may be used for two powers or for two pressures; if power and pressure appear together, however, P should be used only for power and p only for pressure, and necessary distinctions between different powers or between different pressures should be made by the use of subscripts or other modifying signs.

When the above recommendations are insufficient to resolve a conflict or where a need arises for other reasons, an author is of course free to choose an *ad hoc* symbol. Any *ad hoc* symbol should be particularly carefully defined.

In the following list, where two or more symbols are indicated for a given quantity and are separated only by commas (without parentheses), they are on an equal footing; symbols within parentheses are reserve symbols.

Any description given after the name of a physical quantity is merely for identification and is not intended to be a complete definition.

Vector notation (bold-faced italic or sloping type) is used where appropriate in Section 2.6; it may be used when convenient also for appropriate quantities in other Sections.

2.1 Space, time, and related quantities

2.1.01	length	l
2.1.02	height	h
2.1.03	radius	r
2.1.04	diameter	d
2.1.05	path, length of arc	s
2.1.06	wavelength	λ
2.1.07	wavenumber: $1/\lambda$	σ[a], $\tilde{\nu}$[b]
2.1.08	plane angle	$\alpha, \beta, \gamma, \theta, \phi$
2.1.09	solid angle	ω, Ω
2.1.10	area	A, S, A_s[c]
2.1.11	volume	V
2.1.12	time	t
2.1.13	frequency	ν, f
2.1.14	circular frequency: $2\pi\nu$	ω
2.1.15	period: $1/\nu$	T
2.1.16	characteristic time interval, relaxation time, time constant	τ
2.1.17	velocity	v, u, w, c
2.1.18	angular velocity: $\mathrm{d}\phi/\mathrm{d}t$	ω
2.1.19	acceleration	a
2.1.20	acceleration of free fall	g

2.2 Mechanical and related quantities

2.2.01	mass	m
2.2.02	reduced mass	μ
2.2.03	specific volume (volume divided by mass)	v
2.2.04	density (mass divided by volume)	ρ
2.2.05	relative density (ratio of the density to that of a reference substance)	d
2.2.06	moment of inertia	I
2.2.07	momentum	p
2.2.08	force	F
2.2.09	weight	$G, (W)$
2.2.10	moment of force	M
2.2.11	angular momentum	L
2.2.12	work (force times path)	w, W
2.2.13	energy	E
2.2.14	potential energy	E_p, V, Φ
2.2.15	kinetic energy	E_k, T, K
2.2.16	Hamiltonian function	H
2.2.17	Lagrangian function	L
2.2.18	power (energy divided by time)	P
2.2.19	pressure	$p, (P)$
2.2.20	normal stress	σ

[a] In solid-state studies, wavevector $\boldsymbol{k}$ is used ($|\boldsymbol{k}| = 2\pi/\lambda$).
[b] For electromagnetic radiation referred to a vacuum $\tilde{\nu} = \nu/c = 1/\lambda_{\mathrm{vac}}$ is preferred.
[c] The symbol A_s may be used when necessary to avoid confusion with the symbol A for Helmholtz energy.

2.2.21	shear stress	τ
2.2.22	linear strain (relative elongation): $\Delta l/l_0$	ϵ, e
2.2.23	volume strain (bulk strain): $\Delta V/V_0$	θ
2.2.24	modulus of elasticity (normal stress divided by linear strain, Young's modulus)	E
2.2.25	shear modulus (shear stress divided by shear angle)	G
2.2.26	compressibility: $-V^{-1}(dV/dp)$	κ
2.2.27	compression (bulk) modulus: $-V_0(\Delta p/\Delta V)$	K
2.2.28	velocity of sound	c
2.2.29	viscosity	η, (μ)
2.2.30	fluidity: $1/\eta$	ϕ
2.2.31	kinematic viscosity: η/ρ	ν
2.2.32	friction coefficient (frictional force divided by normal force)	μ, (f)
2.2.33	surface tension	γ, σ
2.2.34	angle of contact	θ
2.2.35	diffusion coefficient	D
2.2.36	mass transfer coefficient (dimension of length divided by time)	k_d

2.3 Molecular and related quantities

2.3.01	relative atomic mass of an element formerly called 'atomic weight')[a]	A_r
2.3.02	relative molecular mass of a substance (formerly called 'molecular weight')[b]	M_r
2.3.03	molar mass (mass divided by amount of substance)	M
2.3.04	Avogadro constant	L, N_A
2.3.05	number of molecules or other entries	N
2.3.06	amount of substance[c]	n, (ν)
2.3.07	mole fraction of substance B: $n_B/\Sigma_i n_i$	x_B, y_B
2.3.08	mass fraction of substance B	w_B
2.3.09	volume fraction of substance B	ϕ_B
2.3.10	molality of solute substance B (amount of B divided by mass of solvent)[d]	m_B
2.3.11	amount-of-substance concentration of substance B (amount of B divided by the volume of the solution)[e]	c_B, [B]
2.3.12	mass concentration of substance B (mass of B divided by the volume of the solution)	ρ_B
2.3.13	surface concentration, surface excess	Γ
2.3.14	collision diameter of a molecule	d, σ
2.3.15	mean free path	l, λ
2.3.16	collision number (number of collisions divided by volume and by time)	Z
2.3.17	grand partition function (system)	Ξ
2.3.18	partition function (system)	Q, Z
2.3.19	partition function (particle)	q, z
2.3.20	statistical weight	g
2.3.21	symmetry number	σ, s
2.3.22	characteristic temperature	Θ

2.4 Thermodynamic and related quantities

2.4.01	thermodynamic temperature, absolute temperature	T
2.4.02	Celsius temperature	t, θ[f]
2.4.03	(molar) gas constant	R
2.4.04	Boltzmann constant	k
2.4.05	heat	q, Q[g]
2.4.06	work	w, W[g]
2.4.07	internal energy	U, (E)
2.4.08	enthalpy: $U + pV$	H
2.4.09	entropy	S
2.4.10	Helmholtz energy: $U - TS$	A
2.4.11	Massieu function: $-A/T$	J
2.4.12	Gibbs energy: $H - TS$	G
2.4.13	Planck function: $-G/T$	Y
2.4.14	compression factor: pV_m/RT	Z
2.4.15	heat capacity	C
2.4.16	specific heat capacity (heat capacity divided by mass; the name 'specific heat' is not recommended)	c
2.4.17	ratio C_p/C_V	γ, (κ)
2.4.18	Joule–Thompson coefficient	μ
2.4.19	thermal conductivity	λ, k
2.4.20	thermal diffusivity: $\lambda/\rho c_p$	a
2.4.21	coefficient of heat transfer (density of heat flow rate divided by temperature difference)	h
2.4.22	cubic expansion coefficient: $V^{-1}(\partial V/\partial T)_p$	a
2.4.23	isothermal compressibility: $-V^{-1}(\partial V/\partial p)_T$	κ
2.4.24	pressure coefficient: $(\partial p/\partial T)_V$	β
2.4.25	chemical potential of substance B	μ_B
2.4.26	absolute activity of substance B: $\exp(\mu_B/RT)$	λ_B
2.4.27	fugacity	$f, \tilde{p}$
2.4.28	osmotic pressure	Π
2.4.29	ionic strength: ($I_m = \frac{1}{2}\Sigma_i m_i z_i^2$ or $I_c = \frac{1}{2}\Sigma_i c_i z_i^2$)	I
2.4.30	activity, relative activity of substance B	a_B
2.4.31	activity coefficient, mole fraction basis	f_B
2.4.32	activity coefficient, molality basis	γ_B
2.4.33	activity coefficient, concentration basis	y_B
2.4.34	osmotic coefficient	ϕ

2.5 Chemical reactions

a The ratio of the average mass per atom of an element to 1/12 of the mass of an atom of nuclide ^{12}C. *Example:* $A_r(Cl) = 35.453$.

b The ratio of the average mass per formula unit of a substance to 1/12 of the mass of an atom of nuclide ^{12}C. *Example:* $M_r(KCl) = 74.555$.

c See Section 1.2.

d A solution having a molality equal to 0.1 mol kg^{-1} is sometimes called a 0.1 molal solution or a 0.1 m solution.

e This quantity may be simply called 'concentration' when there is no risk of ambiguity. A solution with an amount-of-substance concentration of 0.1 mol dm^{-3} is often called a 0.1 molar solution or a 0.1 M solution.

f Where symbols are needed to represent both time and Celsius temperature, t is the preferred symbol for time and θ for Celsius temperature.

g It is recommended that $q > 0$ and $w > 0$ both indicate *increase* of energy of the system under discussion. Thus $\Delta U = q + w$.

2.5.01	stoichiometric coefficient of substance B (negative for reactants, positive for products)	ν_B
2.5.02	general equation for a chemical reaction	$0 = \Sigma_B \nu_B B$
2.5.03	extent of reaction: $(d\xi = dn_B/\nu_B)$	ξ
2.5.04	rate of reaction: $d\xi/dt$ (see Section 11)	$\dot{\xi}, J$
2.5.05	rate of increase of concentration of substance B: dc_B/dt	ν_B, r_B
2.5.06	rate constant	k
2.5.07	affinity of a reaction: $-\Sigma_B \nu_B \mu_B$	A, ( )
2.5.08	equilibrium constant	K
2.5.09	degree of dissociation	a

2.6 Electricity and magnetism

2.6.01	elementary charge (of a proton)	e
2.6.02	quantity of electricity	Q
2.6.03	charge density	ρ
2.6.04	surface charge density	σ
2.6.05	electric current	I
2.6.06	electric current density	j
2.6.07	electric potential	V, ϕ
2.6.08	electric potential difference: IR	$U, \Delta V, \Delta\phi$
2.6.09	electric field strength	$\boldsymbol{E}$
2.6.10	electric displacement	$\boldsymbol{D}$
2.6.11	capacitance	C
2.6.12	permittivity: $(\boldsymbol{D} = \epsilon \boldsymbol{E})$	ϵ
2.6.13	permittivity of vacuum	ϵ_0
2.6.14	relative permittivity[a]: ϵ/ϵ_0	ϵ_r, (ϵ)
2.6.15	dielectric polarization: $\boldsymbol{D} - \epsilon_0 \boldsymbol{E}$	$\boldsymbol{P}$
2.6.16	electric susceptibility: $\epsilon_r - 1$	χ_e
2.6.17	electric dipole moment	p, p_e
2.6.18	permanent dipole moment of a molecule	$\boldsymbol{p}, \mu$
2.6.19	induced dipole moment of a molecule	p, p_i
2.6.20	electric polarizability of a molecule	a
2.6.21	magnetic flux	Φ
2.6.22	magnetic flux density, magnetic induction	$\boldsymbol{B}$
2.6.23	magnetic field strength	$\boldsymbol{H}$
2.6.24	permeability: $(\boldsymbol{B} = \mu \boldsymbol{H})$	μ
2.6.25	permeability of vacuum	μ_0
2.6.26	relative permeability: μ/μ_0	μ_r
2.6.27	magnetization: $(\boldsymbol{B}/\mu_0) - \boldsymbol{H}$	$\boldsymbol{M}$
2.6.28	magnetic susceptibility: $\mu_r - 1$	χ, (χ_m)
2.6.29.1	Bohr magneton	μ_B
2.6.29.2	nuclear magneton	μ_N
2.6.29.3	g-factor	g
2.6.29.4	gyromagnetic ratio, magnetogyric ratio	γ
2.6.30	electromagnetic moment: $(E_p = -\boldsymbol{m}\cdot\boldsymbol{B})$	$\boldsymbol{m}, \mu$
2.6.31	resistance	R
2.6.32	resistivity (formerly called specific resistance): $(\boldsymbol{E} = \rho \boldsymbol{j})$	ρ
2.6.33	conductivity (formerly called specific conductance): $(\boldsymbol{j} = \kappa \boldsymbol{E})$	κ, (σ)
2.6.34	self-inductance	L
2.6.35	mutual inductance	M, L_{12}
2.6.36	reactance	X
2.6.37	impedance (complex impedance): $R + iX$	Z
2.6.38	loss angle	δ
2.6.39	admittance (complex admittance): $1/Z$	Y
2.6.40	conductance: $(Y = G + iB)$	G
2.6.41	susceptance: $(Y = G + iB)$	B

2.7 Electrochemistry

2.7.01	Faraday constant	F
2.7.02	charge number of an ion B (positive for cations, negative for anions)	z_B
2.7.03	charge number of a cell reaction	n, (z)
2.7.04	electromotive force	E, E_{MF}
2.7.05	electrochemical potential of ionic component B: $\mu_B + z_B F \phi$	$\tilde{\mu}_B$
2.7.06	electric mobility (velocity divided by electric field strength)	u, μ
2.7.07	electrolytic conductivity (formerly called specific conductance)	κ, (σ)
2.7.08	molar conductivity of electrolyte or ion[b]: κ/c	Λ, λ[c]
2.7.09	transport number (transference number or migration number)	t
2.7.10	overpotential	η
2.7.11	exchange current density	j_0
2.7.12	electrochemical transfer coefficient	α
2.7.13	electrokinetic potential (zeta potential)	ζ
2.7.14	thickness of diffusion layer	δ
2.7.15	inner electric potential	ϕ
2.7.16	outer electric potential	ψ
2.7.17	surface electric potential difference: $\phi - \psi$	χ

2.8 Light and related electromagnetic radiation[d]

2.8.01	Planck constant	h
2.8.02	Planck constant divided by 2π	$\hbar$
2.8.03	radiant energy	Q[e]
2.8.04	radiant flux, radiant power	Φ[e], P
2.8.05	radiant intensity: $d\Phi/d\omega$	I[e]
2.8.06	radiance: $(dI/dS)/\cos\theta$	L[e]
2.8.07	radiant excitance: $d\Phi/dS$	M[e]
2.8.08	irradiance: $d\Phi/dS$	E[e]

a Also called dielectric constant, and sometimes denoted by D, when it is independent of E.

b The word molar, contrary to the general rule given in Section 1.4, here means 'divided by amount-of-substance concentration'.

c The formula unit whose concentration is c must be specified.
Example: $\lambda(Mg^{2+}) = 2\lambda(\frac{1}{2}Mg^{2+})$. Λ is used for an electrolyte and λ for individual ions.

d References to the symbols used in defining the quantities in 2.8 are as follows:

l	2.1.01	$\tilde{\nu}$	2.1.07	θ	2.1.08	ω	2.1.09
S	2.1.10	V	2.1.11	t	2.1.12	c_B	2.3.11
ρ_B	2.3.12	Φ	2.8.04	I	2.8.05	E	2.8.08
T	2.8.12	A	2.8.13.1	B	2.8.13.2	α	2.8.14.2
k	2.8.15	n	2.8.21.1				

e The same symbol is often used also for the corresponding luminous quantity. Subscripts e for energetic and v for visible may be added whenever confusion between these quantities might otherwise occur.

2.8.09	absorptance, absorption-factor [a] (ratio of absorbed to incident radiant or luminous flux)	α [a]
2.8.10	reflectance, reflection factor[a] (ratio of reflected to incident radiant or luminous flux)	ρ [a], R
2.8.11	transmittance, transmission factor[a] (ratio of transmitted to incident radiant or luminous flux)	τ [a]
2.8.12	internal transmittance[a] (transmittance of the medium itself, disregarding boundary or container influence)	τ_i [a], T
2.8.13.1	internal transmission density[a], (decadic) absorbance[b]: $\log_{10}(1/T)$	D_i [a], A
2.8.13.2	Napierian absorbance: $\ln(1/T)$	B
2.8.14.1	(linear) (decadic) absorption coefficient[a,b]: A/l	a [a], K
2.8.14.2	Napierian absorption coefficient: B/l	α
2.8.15	absorption index: $B/4\pi\tilde{\nu}l = \alpha/4\pi\tilde{\nu}$	k
2.8.16.1	specific (decadic) absorption coefficient[c]: $A/\rho_B l$	a [e,f]
2.8.16.2	specific Napierian absorption coefficient[c]: $B/\rho_B l$	μ [e,f]
2.8.17.1	molar (decadic) absorption coefficient[b,d]: $A/c_B l$	ϵ [e,f]
2.8.17.2	molar Napierian absorption coefficient[d]: $B/c_B l$	κ [e,f]
2.8.18	quantum yield	Φ
2.8.19	exposure: $\int E \mathrm{d}t$	H
2.8.20	velocity of light *in vacuo*	c
2.8.21.1	refractive index (of a non-absorbing material)	n
2.8.21.2	complex refractive index of an absorbing material: $n + ik$	$\hat{n}$
2.8.22	molar refraction: $(n^2 - 1)V_m/(n^2 + 2)$	R_m
2.8.23	angle of optical rotation	α

2.9 Transport properties

2.9.01	Flux (of a quantity X)	J_x, J
2.9.02	Reynolds number: $\rho v l/\eta$	Re
2.9.03	Euler number: $\Delta p/\rho v^2$	Eu
2.9.04	Froude number: $v/(lg)^{1/2}$	Fr
2.9.05	Grashof number: $l^3 g\alpha\Delta\theta\rho^2/\eta^2$	Gr
2.9.06	Weber number: $\rho v^2 l/\gamma$	We
2.9.07	Mach number: v/c	Ma
2.9.08	Knudsen number: λ/l	Kn
2.9.09	Strouhal number: lf/v	Sr
2.9.10	Fourier number: $a\Delta t/l^2$	Fo
2.9.11	Peclet number: vl/a	Pe
2.9.12	Rayleigh number: $l^3 g\alpha\Delta\theta\rho/\eta a$	Ra
2.9.13	Nusselt number: hl/k	Nu
2.9.14	Stanton number: $h/\rho v c_p$	St
2.9.15	Fourier number of mass transfer: Dt/l^2	Fo^*
2.9.16	Peclet number for mass transfer: vl/D	Pe^*
2.9.17	Grashof number for mass transfer: $-l^3 g(\partial\rho/\partial x)_{T,p}\Delta x\rho/\eta^2$	Gr^*
2.9.18	Nusselt number for mass transfer[g] $k_d l/D$	Nu^*
2.9.19	Stanton number for mass transfer: k_d/v	St^*
2.9.20	Prandtl number: $\eta/\rho a$	Pr
2.9.21	Schmidt number: $\eta/\rho D$	Sc
2.9.22	Lewis number: d/D	Le
2.9.23	Magnetic Reynolds number: $v\mu\kappa l$	Re_m
2.9.24	Alfvén number: $v(\rho\mu)^{1/2}/B$	Al
2.9.25	Hartmann number: $Bl(\kappa/\eta)^{1/2}$	Ha
2.9.26	Cowling number: $B^2/\mu\rho v^2$	Co

2.10 Symbols for particular cases of physical quantities

It is much more difficult to make detailed recommendations on symbols for physical quantities in particular cases than in general cases. The reason is the incompatibility between the need for specifying numerous details and the need for keeping the printing reasonably simple. Among the most awkward things to print are superscripts to subscripts and subscripts to subscripts. Examples of symbols to be avoided are:

$$\lambda_{NO_3^-} \qquad \Delta H_{25°C} \qquad (pV)_{0\,°C}^{p=0}$$

The problem is vastly reduced if it is recognized that two different kinds of notation are required for two different purposes. In the formation of general fundamental relations the most important requirement is a notation that is easy

a These names and symbols are in agreement with those adopted jointly by the International Commission of Illumination (CIE) and the International Electrotechnical Commission (IEC).

b The terms extinction (for 2.8.13.1) and extinction coefficient (for 2.8.14.1) are unsuitable because extinction is reserved for diffusion of radiation rather than absorption. Molar absorptivity (for 2.8.17.1) should be avoided because the meaning, absorptance per unit length, has been accepted internationally for the term absorptivity.

c The word specific, contrary to the general rule given in Section 1.4, here means 'divided by mass concentration'.

d The word molar, contrary to the general rule given in Section 1.4, here means 'divided by amount-of-substance concentration'.

e For measurement on solutions, $1/T$ is ordinarily replaced by T_0/T where T_0 is the internal transmittance of the solvent medium and T is the internal transmittance of the solution. If a double-beam spectrometer is used in solution spectrometry, T_0/T is given directly, provided the boundary and container influences have been equalized between the two cells; in addition to the physical matching of the sample and reference cells this requires that there be no significant difference between $n_{solvent}$ and $n_{solution}$.

f For measurements on solutions, it is tacitly assumed that the solution obeys the Beer-Lambert law unless the solute concentration is specified. The temperature should be specified.

g The name Sherwood number, symbol *Sh* has been widely used.

to understand and easy to remember. In applications to particular cases, in quoting numerical values, and in tabulation, the most important requirement is complete elimination of any possible ambiguity even at the cost of an elaborate notation.

The advantage of a dual notation is already to some extent accepted in the case of concentration. The recommended notation for the formulation of the equilibrium constant K_c for the general reaction:

$$0 = \Sigma_B \nu_B B$$

is

$$K_c = \Pi_B (c_B)^{\nu_B}$$

but when we turn to a particular example it is better to use a notation such as:

$$Br_2 + H_2O = HOBr + H^+ + Br^-$$

$$\frac{[HOBr]\,[H^+]\,[Br^-]}{[Br_2]} = K_c$$

$$K_c(25\ °C) = 6 \times 10^{-9}\ mol^2\,dm^{-6}$$

Once the principle of dual notation is accepted, its adaptability and usefulness become manifest in all fields of physical chemistry. It will here be illustrated by just a few examples.

The general relation between the molar conductivity of an electrolyte and the molar conductivities of the two ions is written most simply and most clearly as:

$$\Lambda = \lambda^+ + \lambda^-$$

but when it comes to giving values in particular cases a much more appropriate notation is:

$$\lambda \tfrac{1}{2}Mg^{2+} = 53\ S\ cm^2\ mol^{-1}\ \text{at}\ 25°C$$
$$\lambda(Cl^-) = 76\ S\ cm^2\ mol^{-1}\ \text{at}\ 25°C$$
$$\Lambda \tfrac{1}{2}MgCl_2) = 129\ S\ cm^2\ mol^{-1}\ \text{at}\ 25°C$$
$$\Lambda(MgCl_2) = 258\ S\ cm^2\ mol^{-1}\ \text{at}\ 25°C$$

The general relation between the partial molar volumes of the two components A and B of a binary mixture is written most simply:

$$n_A dV_A + n_B dV_B = 0 \qquad (T,p\ \text{const.})$$

But when it comes to specifying values, a completely different notation is called for, such as:

$$V(K_2SO_4,\ 0.1\ mol\ dm^{-3}\ \text{in}\ H_2O,\ 25\ °C) = 48\ cm^3\ mol^{-1}$$

Each kind of notation is appropriate to its purpose.

A last example will be given relating to optical rotation. The relations between the angle α of rotation of the plane of polarization and the amount n, or the number N of molecules, of the optically active substance in the path of a light beam of cross-section A can be clearly expressed in the form:

$$\alpha = n\alpha_n/A = N\alpha_N/A$$

where α_n is the molar optical rotatory power and α_N the molecular optical rotatory power. When on the other hand it is desired to record an experimental measurement, an appropriate notation would be:

$$\alpha(589.3\ nm,\ 20\ °C,\ \text{sucrose},\ 10\ g\ dm^{-3}\ \text{in}\ H_2O,\ 10\ cm) = +66.470°$$

2.11 Recommended superscripts

The following superscripts are recommended:

$^{\circ}$ or * pure substance

$^{\infty}$ infinite dilution

^{id}ideal
° or $^{\ominus}$standard in general
‡activated complex, transition state

3. UNITS AND SYMBOLS FOR UNITS

3.5 The International System of Units

The name International System of Units has been adopted by the Conférence Générale des Poids et Mesures for the system of units based on a selected set of dimensionally independent *SI Base Units*.

The SI Base Units are the meter, kilogram, second, ampere, kelvin, candela, and mole. In the International System of Units there is one and only one *SI Unit* for each physical quantity. This is either the appropriate SI Base Unit itself (see Section 3.7) or the appropriate *SI Derived Unit* formed by multiplication and/or division of two or more SI Base Units (see Section 3.10). A few such SI Derived Units have been given special names and symbols (see Section 3.9). There are also two *SI Supplementary Units* for which it is not decided whether they are SI Base Units or SI Derived Units (see Section 3.8).

Any of the approved decimal prefixes, called *SI Prefixes*, may be used to construct decimal multiples or submultiples of SI Units (see Section 3.11).

It is recommended that only units composed of SI Units and SI Prefixes be used in science and technology.

3.6 Definitions of the SI Base Units

Meter – The meter is the SI base unit of length and was defined at the October 1984 General Conference of Weights and Measures as the length of the path traveled by light in vacuum during a time interval of $^1/299{,}792{,}458$ of a second.

Kilogram – The kilogram is the unit of mass; it is equal to the mass of the international prototype of the kilogram.

Second – The second is the duration of 9192631770 periods of the radiation corresponding to the transition between the two hyperfine levels of the ground state of the caesium-133 atom.

Ampere – The ampere is that constant current which, if maintained in two straight parallel conductors of infinite length, of negligible cross-section, and placed 1 meter apart in vacuum, would produce between these conductors a force equal to 2×10^{-7} newton per meter of length.

Kelvin – The kelvin, unit of thermodynamic temperature, is the fraction 1/273.16 of the thermodynamic temperature of the triple point of water.[a]

Candela – The candela is the luminous intensity in a given direction of a source which emits monochromatic radiation of frequency 540×10^{12} Hz and of which the radiant intensity in that direction is 1/683 W/steradian. From the 16th CGPM, *Resolution*, 3, 1979.

Mole – The mole is the amount of substance of a system which contains as many elementary entities as there are atoms in 0.012 kilogram of carbon-12. When the mole is used, the elementary entities must be specified and may be atoms, molecules, ions, electrons, other particles, or specified groups of such particles. *Some examples of the use of the mole:*

- 1 mole of HgCl has a mass of 236.04 grams
- 1 mole of Hg_2Cl_2 has a mass of 472.08 grams
- 1 mole of Hg_2^{2+} has a mass of 401.18 grams and a charge of 192.97 kilocoulombs
- 1 mole of $\frac{1}{2}Ca^{2+}$ has a mass of 20.04 grams and a charge of 96.49 kilocoulombs
- 1 mole of $Cu_{0.5}Zn_{0.5}$ has a mass of 64.46 grams
- 1 mole of $Fe_{0.91}S$ has a mass of 82.88 grams
- 1 mole of e^- has a mass of 548.60 micrograms, a charge of –96.49 kilocoulombs, and contains 6.02×10^{23} electrons
- 1 mole of a mixture containing the mole fractions $x(N_2) = 0.7809$, $x(O_2) = 0.2905$, $x(Ar) = 0.0093$, and $x(CO_2) = 0.0003$ has a mass of 28.964 grams
- 1 mole of photons whose frequency is 10^{14} Hz has energy 39.90 kilojoules

(The numerical values in these examples are approximate.)

3.7 Names and symbols for SI Base Units

Physical quantity	*Name of SI Unit*	*Symbol for SI Unit*
Length	meter	m
Mass	kilogram	kg
Time	second	s
Electric current	ampere	A
Thermodynamic temperature	kelvin	K
Amount of substance[b]	mole	mol
Luminous intensity	candela	cd

a In October 1967 the thirteenth Conférence Générale des Poids et Mesures recommended that the kelvin, symbol K, be used for thermodynamic temperature and for thermodynamic temperature interval, and that the unit-symbols °K and deg be abandoned.

b See Section 1.2.

3.8 Names and symbols for SI Supplementary Units

Physical quantity	*Name of SI Unit*	*Symbol for SI Unit*
Plane angle	radian	rad
Solid angle	steradian	sr

3.9 Special names and symbols for certain SI Derived Units

Physical quantity	*Name of SI Unit*	*Symbol for SI Unit*	*Definition of SI Unit*
Force	newton	N	$m\ kg\ s^{-2}$
Pressure, stress	pascal	Pa	$m^{-1}\ kg\ s^{-2}$ (=$N\ m^{-2}$)
Energy	joule	J	$m^2\ kg\ s^{-2}$
Power	watt	W	$m^2\ kg\ s^{-3}$ (=$J\ s^{-1}$)
Electric charge	coulomb	C	s A
Electric potential difference	volt	V	$m^2\ kg\ s^{-3}\ A^{-1}$ (=$J\ A^{-1}\ s^{-1}$)
Electric resistance	ohm	Ω	$m^2\ kg\ s^{-3}\ A^{-2}$ (=$V\ A^{-1}$)
Electric conductance	siemens	S	$m^{-2}\ kg^{-1}\ s^3\ A^2$ (=$A\ V^{-1} = \Omega^{-1}$)
Electric capacitance	farad	F	$m^{-2}\ kg^{-1}\ s^4\ A^2$ (=$A\ s\ V^{-1}$)
Magnetic flux	weber	Wb	$m^2\ kg\ s^{-2}\ A^{-1}$ (=V s)
Inductance	henry	H	$m^2\ kg\ s^{-2}\ A^{-2}$ (=$V\ A^{-1}\ s$)
Magnetic flux density	tesla	T	$kg\ s^{-2}\ A^{-1}$ (=$V\ s\ m^{-2}$)
Luminous flux	lumen	lm	cd sr
Illuminance	lux	lx	m^{-2} cd sr
Frequency	hertz	Hz	s^{-1}
Activity (of radioactive source)	becquerel	Bq	s^{-1}
Absorbed dose (of radiation)	gray	Gy	$m^2\ s^{-2}$ (=$J\ kg^{-1}$)

3.10 SI Derived Units and Unit-symbols for other quantities

(This list is not exhaustive)

Physical quantity	*SI Unit*	*Symbol for SI Unit*
Area	square meter	m^2
Volume	cubic meter	m^3
Density	kilogram per cubic meter	$kg\ m^{-3}$
Velocity	meter per second	$m\ s^{-1}$
Angular velocity	radian per second	$rad\ s^{-1}$
Acceleration	meter per second squared	$m\ s^{-2}$
Kinematic viscosity, diffusion coefficient	square meter per second	$m^2\ s^{-1}$
Dynamic viscosity	newton-second per square meter	$N\ s\ m^{-2}$
Molar entropy, molar heat capacity	joule per kelvin mole	$J\ K^{-1}\ mol^{-1}$
Concentration	mole per cubic meter	$mol\ m^{-3}$
Electric field strength	volt per meter	$V\ m^{-1}$
Magnetic field strength	ampere per meter	$A\ m^{-1}$
Luminance	candela per square meter	$cd\ m^{-2}$

3.12 The degree Celsius

Physical quantity	*Name of unit*	*Symbol for unit*	*Definition of unit*
Celsius temperature	degree Celsius	°C[a]	°C = K

The Celsius temperature t, is defined by $t = T - T_0$ where T_0 = 273.15 K. This leads to t/°C = T/K – 273.15.

3.13 Decimal fractions and multiples of SI Units having special names

The following units do not belong to the International System of Units, but in view of existing practice the Comité

[a] The ° sign and the letter following form one symbol and there should be no space between them. *Example:* 25 °C not 25° C.

International des Poids et Mesures has considered (1969) that it was preferable to keep them for the time being (along with several other specified units not particularly relevant to chemistry) for use with those of the International System.

Physical quantity	*Name of unit*	*Symbol for unit*	*Definition of unit*
Length	ångström	Å	10^{-10} m
Cross section	barn	b	10^{-28} m^2
Volume	liter[a]	l, L	10^{-3} m^3
Mass	tonne	t	10^3 kg
Pressure	bar	bar	10^5 Pa

Other units with special names based on the c.g.s. system and the electromagnetic c.g.s. system[b] are preferably not to be used; among these are the erg (10^{-7} J), the dyne (10^{-5} N), the poise (0.1 Pa s), the stokes (10^{-4} m^2 s^{-1}), the gauss (corresponding to 10^{-4} T),[b] the oersted (corresponding to $1000/4\pi$ A m^{-1}),[b] and the maxwell (corresponding to 10^{-8} WB).[b] The name micron and symbol μ should not be used for the unit of length, 10^{-6} m, which has the SI name micrometer and symbol μm.

3.14 Some other units now exactly defined in terms of the SI units

The CIPM (1969) recognized that users of the SI will wish to employ with it certain units not part of it but which are important and are widely used. These units are given in the following table. The combination of units of this table with SI units to form compound units should, however, be authorized only in limited cases.

Units in use with the International System

Name of unit	*Symbol*	*Definition of unit*
minute	min	60 s
hour	h	3 600 s
day	d	86 400 s
degree	°	$(\pi/180)$ rad
minute	′	$(\pi/10\ 800)$ rad
second	″	$(\pi/648\ 000)$ rad

In view of existing practice, as in the case of those units listed in Section 3.13, the CIPM (1969) has considered it preferable to retain the following units for the time being, for use with those of the SI. The definitions given in the fourth column of this table are exact.

Units to be used with the International System for a limited time

Physical quantity	*Name of unit*	*Symbol for unit*	*Definition of unit*
Radioactivity	Curie	Ci	3.7×10^{10} Bq
Exposure to X or γ radiation	röntgen	R	2.58×10^{-4} C kg^{-1}
Ionizing radiation absorbed	rad	rad[c]	10^{-2} Gy

The use of the following units is to be progressively discouraged and eventually abandoned. In the meantime it is recommended that any author who uses these units will define them in terms of SI units once in each publication in which he uses them. The definitions given here are exact. This list is not exhaustive.

Other units generally deprecated

Physical quantity	*Name of unit*	*Symbol for unit*	*Definition of unit*
Length	inch	in	2.54×10^{-2} m
Mass	pound (avoirdupois)	lb	0.453 592 37 kg
Force	kilogram-force	kgf	9.806 65 N
Pressure	standard atmosphere[d]	atm	101 325 Pa

a By decision of the Twelfth Conférence Générale des Poids et Mesures in October 1964, the old definition of the liter (1.000 028 dm^3) was rescinded. The word liter is now regarded as a special name for the cubic decimeter. Neither the word liter nor its symbol should be used to express results of high precision. The alternative symbol L was recommended by CIPM to CGPM in 1978.

b The electromagnetic c.g.s. system is a three-dimensional system of units in which the unit of electric current and units for other electric and magnetic quantities are considered to be derived from the centimeter, gram, and second as base units. The electric and magnetic units of this system cannot strictly speaking be compared to the corresponding units of the SI, which has four dimensions when only units derived from the meter, kilogram, second, and ampere are considered.

c Where there is a risk of confusion with the symbol for radian, rd may be used as the symbol for the unit, rad.

d The phrase 'standard atmosphere' remains admissable for the reference pressure 101 325 Pa.

Pressure	torr	Torr	$\frac{101\ 325}{760}$ Pa
Pressure	conventional millimeter of mercury[a]	mmHg	13.5951 × 980.665 × 10^{-2} Pa
Energy	kilowatt-hour	kW h	3.6 × 10^6 J
Energy	IT calorie	cal_{IT}	4.1868 J
Energy	thermochemical calorie	cal_{th}	4.184 J
Energy	British thermal unit	Btu	1055.055 852 62 J
Thermodynamic temperature	degree Rankine	°R	(5/9) K

3.15 Units defined in terms of the best available experimental values of certain physical constants

It is necessary to recognize outside the International System some units, useful in specialized fields, the values of which expressed in SI units can be obtained only by experiment and are therefore not known exactly. Among such units recognized by the CIPM (1969) that are relevant to chemistry are the following:

Physical quantity	*Name of unit*	*Symbol for unit*	*Conversion factor*
Energy	electronvolt	eV	1 eV ≈ 1.6021892 × 10^{-19} J
Mass	(unified) atomic mass unit	u	1 u ≈ 1.6605655 × 10^{-27} kg

3.16 'International' electrical units

These units are obsolete having been replaced by the 'absolute' (SI) units in 1948. The conversion factors which should be used with electrical measurements quoted in 'international' units depend on where and when the instruments used to make the measurements were calibrated. The following two sets of conversion factors refer respectively to the 'mean international' units estimated by the ninth Conférence Générale des Poids et Mesures in 1948, and to the 'US international' units estimated by the National Bureau of Standards (USA) as applying to published measurements made with instruments calibrated by them prior to 1948.

- 1 'mean international ohm' = 1.00049 Ω
- 1 'mean international volt' = 1.00034 V
- 1 'US international ohm' = 1.000495 Ω
- 1 'US international volt' = 1.000330 V

3.17 Electrical and magnetic units belonging to unit-systems other than the International System of Units

Definitions of units used in the obsolescent 'electrostatic CGS' and 'electromagnetic CGS' unit-systems can be found in References 13.1.05 and 13.2.

Another 'electrostatic CGS' unit used in chemistry for electric dipole moment is the debye, symbol D. 1D = $(10^{-21}/c)$ A m^2 ≈ 3.3356 × 10^{-30} C m.

5. PHYSICAL QUANTITIES, UNITS, AND NUMERICAL VALUES

The value of a *physical quantity* is equal to the product of a *numerical value* and a *unit*:

physical quantity = numerical value × unit.

Neither any physical quantity, nor the symbol used to denote it, should imply a particular choice of unit.

Operations on equations involving physical quantities, units, and numerical values, should follow the ordinary rules of algebra.

Thus the physical quantity called the critical pressure and denoted by p_c has the value for water:

$$p_c = 221.2 \text{ bar} \quad \text{or better} \quad p_c = 22.12 \text{ MPa}.$$

These equations may equally well be written in the forms:

$$p_c/\text{bar} = 221.2 \quad \text{or better} \quad p_c/\text{MPa} = 22.12,$$

which are especially useful for the headings in tables and as labels on the axes of graphs.

[a] The conventional millimeter of mercury, symbol mmHg (not mm Hg), is the pressure exerted by a column exactly 1 mm high of a fluid of density exactly 13.5951 g cm^{-3} in a place where the acceleration of free fall is exactly 980.665 cm s^{-2}. The mmHg differs from the Torr by less than 2 × 10^{-7} Torr.

6. RECOMMENDED MATHEMATICAL SYMBOLS[a]

Mathematical operators (for example d and Δ) and mathematical constants (for example e and π should always be printed in roman (upright) type. Letter symbols for numbers other than mathematical constants should be printed in italic type.

equal to	$=$
not equal to	$\neq$
identically equal to	$\equiv$
corresponds to	$\triangleq$
approximately equal to	$\approx$
approaches	$\rightarrow$
asymptotically equal to	$\simeq$
proportional to	$\propto$ $\sim$
infinity	∞
less than	$<$
greater than	$>$
less than or equal to	$\leqslant$
greater than or equal to	$\geqslant$
much less than	$\ll$
much greater than	$\gg$
plus	$+$
minus	$-$
multiplied by	$\times$ $\cdot$
a divided by *b*	$\frac{a}{b}$ a/b ab^{-1}
magnitide of *a*	$\lvert a \rvert$
a raised to the power *n*	a^n
square root of *a*	$a^{1/2}$ $a^{\frac{1}{2}}$ $\sqrt{a}$
*n*th root of *a*	$a^{1/n}$ $a^{\frac{1}{n}}$ $\sqrt[n]{a}$
mean value of *a*	$\langle a \rangle$ $\bar{a}$
natural logarithm of *a*	$\ln a$ $\log_e a$
decadic logarithm of *a*	$\lg a$ $\log_{10} a$ $\log a$
binary logarithm of *a*	$\mathrm{lb}\, a$ $\log_2 a$
exponential of *a*	$\exp a$ e^a

7. SYMBOLS FOR CHEMICAL ELEMENTS, NUCLIDES, AND PARTICLES

7.1 Definitions

A nuclide is a species of atoms of which each atom has identical atomic number (proton number) and identical mass number (nucleon number). Different nuclides having the same value of the atomic number are named isotopes or isotopic nuclides. Different nuclides having the same mass number are named isobars or isobaric nuclides.

7.2 Elements and nuclides

Symbols for chemical elements should be written in roman (upright) type. The symbol is not followed by a full stop except when it occurs at the end of a sentence in text. *Examples:* Ca C H He.

The nuclide may be specified by attaching numbers. The mass number should be placed in the left superscript position; the atomic number, if desired, may be placed as a left subscript. The number of atoms per molecule is indicated as a right subscript. Ionic charge, or state of excitation, or oxidation number[b] may be indicated in the right superscript space. *Examples;* Mass number: $^{14}N_2$, $^{35}Cl^-$; Ionic charge: Cl^-, Ca^{2+}, PO_4^{3-} or $PO_4{}^{3-}$; Excited electronic state: He*, NO*; Oxidation number: $Pb_2^{II}Pb^{IV}O_4$, $K_6M^{IV}Mo_9O_{32}$ (where M denotes a metal).

7.4 Abbreviated notation for nuclear reactions

The meaning of the symbolic expression indicating a nuclear reaction should be the following:

$$\text{initial nuclide}\left(\begin{matrix}\text{incoming particles(s)} & \text{outgoing particles(s)}\\ \text{or quanta} & \text{, or quanta}\end{matrix}\right)\text{final nuclide}$$

Examples: $^{14}N(\alpha,p)^{17}O$; $^{23}Na(\gamma,3n)^{20}Na$; $^{59}Co(n,\gamma)^{60}Co$; $^{31}P(\gamma,pn)^{29}Si$.

a Taken from Reference 13.1.11 where a more comprehensive list can be found.

b For a more detailed discussion see Reference 13.4.

8. SYMBOLS FOR SPECTROSCOPY[a]

8.1 General rules

A letter-symbol indicating the quantum state of *a system* should be printed in capital upright type. A letter-symbol indicating the quantum state of *a single particle* should be printed in lower case upright type.

9. CONVENTIONS CONCERNING THE SIGNS OF ELECTRIC POTENTIAL DIFFERENCES, ELECTROMOTIVE FORCES, AND ELECTRODE POTENTIALS[b]

9.1 The electric potential difference for a galvanic cell

The cell should be represented by a diagram, for example:

$$\mathrm{Zn} \mid \mathrm{Zn}^{2+} \mid \mathrm{Cu}^{2+} \mid \mathrm{Cu}$$

The electric potential difference ΔV is equal in sign and magnitude to the electric potential of a metallic conducting lead on the right minus that of an identical lead on the left.

When the reaction of the cell is written as:

$$\tfrac{1}{2}\mathrm{Zn} + \tfrac{1}{2}\mathrm{Cu}^{2+} \rightarrow \tfrac{1}{2}\mathrm{Zn}^{2+} + \tfrac{1}{2}\mathrm{Cu}$$

this implies a diagram so drawn that this reaction takes place when positive electricity flows through the cell from left to right. If this is the direction of the current when the cell is short-circuited, as it will be in the present example (unless the ratio $[\mathrm{Cu}^{2+}]/[\mathrm{Zn}^{2+}]$ is extremely small), the electric potential difference will be positive.

If, however, the reaction is written as:

$$\tfrac{1}{2}\mathrm{Cu} + \tfrac{1}{2}\mathrm{Zn}^{2+} \rightarrow \tfrac{1}{2}\mathrm{Cu}^{2+} + \tfrac{1}{2}\mathrm{Zn}$$

this implies the diagram:

$$\mathrm{Cu} \mid \mathrm{Cu}^{2+} \mid \mathrm{Zn}^{2+} \mid \mathrm{Zn}$$

and the electric potential difference of the cell so specified will be negative (unless the ratio $[\mathrm{Cu}^{2+}]/[\mathrm{Zn}^{2+}]$ is extremely small).

The limiting value of the electric potential difference for zero current through the cell is called the electromotive force and denoted by E_{MF} or E.

9.2 Electrode potential

The so-called electrode potential of an electrode (half-cell) is defined as the electromotive force of a cell in which the electrode on the left is a *standard hydrogen electrode* and that on the right is the electrode in question. For example, for the zinc electrode (written as $\mathrm{Zn}^{2+} \mid \mathrm{Zn}$) the cell in question is:

$$\mathrm{Pt} \mid \mathrm{H_2} \mid \mathrm{H^+} \mid \mathrm{Zn}^{2+} \mid \mathrm{Zn}$$

The reaction taking place at the zinc electrode is:

$$\mathrm{Zn}^{2+} + 2\mathrm{e}^- \rightarrow \mathrm{Zn}$$

The latter is to be regarded as an abbreviation for the reaction in the mentioned cell:

$$\mathrm{Zn}^{2+} + \mathrm{H_2} \rightarrow \mathrm{Zn} + 2\mathrm{H^+}$$

In the standard state the electromotive force of this cell has a negative sign and a value of -0.763 V. The standard electrode potential of the zinc electrode is therefore -0.763 V.

The symbol $\mathrm{Zn} \mid \mathrm{Zn}^{2+}$ on the other hand implies the cell:

$$\mathrm{Zn} \mid \mathrm{Zn}^{2+} \mid \mathrm{H^+} \mid \mathrm{H_2} \mid \mathrm{Pt}$$

[a] Taken from Reference 13.2. For further details see Reference 13.10.

[b] The conventions given here are in accordance with the 'Stockholm Convention' of 1953.

in which the reaction is:

$$Zn + 2H^+ \rightarrow Zn^{2+} + H_2$$

The electromotive force of this cell should *not* be called an electrode potential.

10. THE QUANTITY pH[a]

10.1 Operational definition

In all existing national standards the definition of pH is an operational one. The electromotive force E_X of the cell:

reference electrode | concentrated KCl solution ⋮⋮ solution X | H_2 | Pt

is measured and likewise the electromotive force E_S of the cell:

reference electrode | concentrated KCl solution ⋮⋮ solution S | H_2 | Pt

both cells being at the same temperature throughout and the reference electrodes and bridge solutions being identical in the two cells. The pH of the solution X, denoted by pH(X), is then related to the pH of the solution S, denoted by pH(S), by the definition:

$$\text{pH(X)} = \text{pH(S)} + \frac{(E_S - E_X)F}{RT \ln 10}$$

where R denotes the gas constant, T the thermodynamic temperature, and F the Faraday constant. Thus defined the quantity pH is a number.

To a good approximation, the hydrogen electrodes in both cells may be replaced by other hydrogen-ion-responsive electrodes, e.g. glass or quinhydrone. The two bridge solutions may be any molality not less than 3.5 mol kg^{-1}, provided they are the same (see Reference 13.5).

10.2 Standards

The difference between the pH of two solutions having been defined as above, the definition of pH can be completed by assigning a value of pH at each temperature to one or more chosen solutions designated as standards.

If the definition of pH given above is adhered to strictly, then the pH of a solution may be slightly dependent on which standard solution is used. These unavoidable deviations are caused not only by imperfections in the response of the hydrogen-ion electrodes but also by variations in the liquid junctions resulting from the different ionic compositions and mobilities of the several standards and from differences in the geometry of the liquid-liquid boundary. In fact such variations in measured pH are usually too small to be of practical significance. Moreover, the acceptance of several standards allows the use of the following alternative definition of pH.

The electromotive force E_X is measured, and likewise the electromotive forces E_1 and E_2 of two similar cells with the solution X replaced by the standard solutions S_1 and S_2 such that the E_1 and E_2 values are on either side of, and as near as possible to, E_X. The pH of solution X is then obtained by assuming linearity between pH and E, that is to say:

$$\frac{\text{pH(X)} - \text{pH(S}_1)}{\text{pH(S}_2) - \text{pH(S}_1)} = \frac{E_X - E_1}{E_2 - E_1}$$

This procedure is especially recommended when the hydrogen-ion-responsive electrode is a glass electrode.

11. DEFINITION OF RATE OF REACTION AND RELATED QUANTITIES

11.1 Rate of reaction

For the reaction

$$0 = \Sigma_B \nu_B B$$

the extent of reaction ξ is defined according to 2.5.03 by

[a] The symbol pH is an exception to the general rules given in Section 1.5.

$$d\xi = \nu_B^{-1} dn_B$$

where n_B is the amount, and ν_B is the stoichiometric number, of the substance B.

It is recommended that the *rate of reaction* be defined as the rate of increase of the extent of reaction, namely

$$\dot{\xi} = d\xi/dt = \nu_B^{-1} dn_B/dt$$

This definition is independent of the choice of B and is valid regardless of the conditions under which a reaction is carried out, e.g. it is valid for a reaction in which the volume varies with time, or for a reaction involving two or more phases, or for a reaction carried out in a flow reactor.

If both sides of this equation are divided by any specified volume V, not necessarily independent of time, and not necessarily that of a single phase in which the reaction is taking place, then

$$V^{-1} d\xi/dt = V^{-1} \nu_B^{-1} {}^{-1} dn_B/dt$$

If the specified volume V is independent of time, then

$$V^{-1} d\xi/dt = \nu_B^{-1} d(n_B/V)/dt$$

If this specified volume V is such that

$$n_B/V = c_B \text{ or } [B]$$

where c_B or [B] is the amount-of-substance concentration of B, then

$$V^{-1} d\xi/dt = \nu_B^{-1} dc_B/dt \text{ or } \nu_B^{-1} d[B]/dt$$

The quantity

$$dn_B/dt (= \nu_B d\xi/dt)$$

may be called the rate of formation of B, and the quantity

$$V^{-1} \nu_B^{-1} dn_B/dt (= V^{-1} d\xi/dt)$$

may be called the rate of reaction divided by volume, and the quantity

$$\nu_B = dc_B/dt \text{ or } d[B]/dt$$

which has often been called the rate of reaction, may be called the rate of increase of the concentration of B, but none of these three quantities should be called the rate of reaction.

11.2 Order of reaction

If it is found *experimentally* that the rate of increase of the concentration of B is given by

$$\nu_B \propto [C]^c [D]^d \ldots$$

then the reaction is described as of order c with respect to C, or order d with respect to D, . . . , and of overall order $(c + d + \ldots)$.

11.3 Labelling of elementary processes

Elementary processes should be labelled in such a manner that reverse processes are immediately recognizable. *Example:*

Elementary process	*Label*	*Rate of constant*
$Br_2 + M \rightarrow 2Br + M$	1	k_1
$Br + H_2 \rightarrow HBr + H$	2	k_2
$H + Br_2 \rightarrow HBr + Br$	3	k_3
$H + HBr \rightarrow H_2 + Br$	−2	k_{-2}
$2Br + M \rightarrow Br_2 + M$	−1	k_{-1}

11.4 Collision number

The collision number defined as the number of collisions per unit time and per unit volume and having dimensions $(\text{time})^{-1} \times (\text{volume})^{-1}$ should be denoted by Z.

The collision number divided by the product of two relevant concentrations (or by the square of the relevant concentration) and by the Avogadro constant is a second-order rate constant having dimensions $(\text{time})^{-1} \times (\text{volume}) \times (\text{amount of substance})^{-1}$ and should be denoted by z. Thus $z = Z/Lc_Ac_B$.

13. REFERENCES

13.1 The ISO 31 International Standard Series will, when complete, form a comprehensive publication dealing with quantities and units in various fields of science and technology. The following parts have so far been published and can be purchased in any country belonging to ISO from the 'Member Body', usually the national standardizing organization of the country.

13.1.00 'Part 0: General principles concerning quantities, units and symbols', 2nd edition, July 1981.

13.1.01 'Part I: Quantities and units of space and time', 1st edition, March 1978.

13.1.02 'Part II: Quantities and units of periodic and related phenomena', 1st edition, March 1978.

13.1.03 'Part III: Quantities and units of mechanics', 1st edition, March 1978.

13.1.04 'Part IV: Quantities and units of heat', 1st edition, March 1978.

13.1.05 'Part 5: Quantities and units of electricity and magnetism', 2nd edition, February 1979.

13.1.06 'Part 6: Quantities and units of light and related electromagnetic radiations', 2nd edition, December 1980.

13.1.07 'Part VII: Quantities and units of acoustics', 1st edition, March 1978.

13.1.08 'Part 8: Quantities and units of physical chemistry and molecular physics', 2nd edition, December 1980.

13.1.09 'Part 9: Quantities and units of atomic and nuclear physics', 2nd edition, December 1980.

13.1.10 'Part 10: Quantities and units of nuclear reactions and ionizing radiations', 2nd edition, December 1980.

13.1.11 'Part XI: Mathematical signs and symbols for use in the physical sciences and technology', 1st edition, March 1978.

13.1.12 'Part 12: Dimensionless parameters', 2nd edition, 1981.

13.1.13 'Part 13: Quantities and units of solid state physics', 2nd edition, July 1981.

13.2 'Symbols, Units and Nomenclature in Physics', Document UIP 20 (SUN 65–3), published by IUPAP, 1978. This document supersedes Document UIP 9 (SUN 61–44) with the same title, which was published by IUPAP in 1961. Also published in *Physica*, (1978), 93A, 1.

13.3 'Manual of Physicochemical Symbols and Terminology', published for IUPAC by Butterworths Scientific Publications, London, 1959. This document was reprinted in the *J. Am. Chem. Soc.*, (1960), 82, 5517.

13.4 'Nomenclature of Inorganic Chemistry, 1970', 2nd ed., *Pure and Applied Chemistry*, (1971), 28, 1. This document has been issued for IUPAC also in book form by Butterworths, Borough Green, Sevenoaks, Kent TN158PH, UK.

13.5 *Pure and Applied Chemistry*, (1960), 1, 163.

13.6 *Pure and Applied Chemistry*, (1964), 9, 453.

13.7 *Pure and Applied Chemistry*, (1970), 21, 1.

13.8 'Le Système International d'Unités (SI)', Bureau International des Poids et Mesures, 2e Ed., 1973, OFFILIB, 48, rue Gay-Lussac, F 75 Paris 5.

13.8.01 'The International System of Units (SI)', National Bureau of Standards Special Publication 330, 1977 Edition, SD Catalog No. C 13.10: 330/4, US Government Printing Office, Washington, DC 20402.

13.8.02 'SI The International System of Units', National Physical Laboratory, Third edition, 1977, Her Majesty's Stationery Office, London.

13.9.01 'Definitions, Terminology and Symbols in Colloid and Surface Chemistry', Part I, *Pure and Applied Chemistry*, (1972), 31, 577 and Part II, 'Heterogeneous Catalysis', *Pure and Applied Chemistry*, (1976), 46, 71.

13.9.02 'Electrochemical Nomenclature', *Pure and Applied Chemistry*, (1974), 37, 503.

13.10 'Report on Notation for the Spectra of Polyatomic Molecules', *J. Chem. Phys.*, (1955) 23, 1997. For notation for diatomic molecules, see F. A. Jenkins, *J. Opt. Soc. Am.*, (1953), 43, 425.

13.11 'Recommended Consistent Values of the Fundamental Physical Constants, 1973', *CODATA Bulletin No. 11*, (December 1973), Committee on Data for Science and Technology of the International Council of Scientific Unions.

13.12 'The 1973 Least-Squares Adjustment of the Fundamental Constants', E. R. Cohen and B. N. Taylor, *J. Phys. Chem. Ref. Data*, (1973), 2, 663.

APPENDIX I

DEFINITION OF ACTIVITIES AND RELATED QUANTITIES

A.I.1 *Chemical potential and absolute activity*

The chemical potential μ_B of a substance B in a mixture of substances B, C, . . . , is defined by

$$\mu_B = (\partial G/\partial n_B)_{T,p,n_C \cdots}$$

where G is the Gibbs energy of the mixture, T is the thermodynamic temperature, p is the pressure, and $n_B, n_C, \ldots$, are the amounts of the substances B, C, . . . , in the mixture.

(In molecular theory the symbol μ_B is sometimes used for the quantity μ_B/L where L is the Avogadro constant, but this usage is not recommended.)

The absolute activity λ_B of the substance B in the mixture is a number defined by

$$\lambda_B = \exp(\mu_B/RT) \qquad \text{or} \qquad \mu_B = RT \ln \lambda_B$$

where R is the gas constant.

The definitions given below often take simpler, though perhaps less familiar, forms when they are expressed in terms of absolute activity rather than in terms of chemical potential. Each of the definitions given below is expressed in both of these ways.

1. Pure substances

A.I.2 *Properties of pure substances*

The superscript * attached to the symbol for a property of a substance denotes the property of the *pure* substance. It is sometimes convenient to treat a mixture of constant composition as a pure substance.

A.I.3 *Fugacity of a pure gaseous substance*

The fugacity f_B^* of a pure gaseous substance B is a quantity with the same dimensions as pressure, defined in terms of the absolute activity λ_B^* of the pure gaseous substance B by

$$f_B^* = \lambda_B^* \lim_{p\to 0}(p/\lambda_B^*) \qquad (T \text{ const.})$$

or in terms of the chemical potential μ_B by

$$RT \ln f_B^* = \mu_B^* + \lim_{p\to 0}(RT \ln p - \mu_B^*) \qquad (T \text{ const.})$$

where p is the pressure of the gas and T is its thermodynamic temperature. It follows from this definition that

$$\lim_{p\to 0}(f_B^*/p) = 1 \qquad (T \text{ const.})$$

and that

$$RT \ln(f_B^*/p) = \int_0^p (V_B^* - RT/p)\,\mathrm{d}p \qquad (T \text{ const.})$$

where V_B^* is the molar volume of the pure gaseous substance B.

A pure gaseous substance B is treated as an *ideal gas* when the approximation $f_B^* = p$ is used. The ratio (f_B^*/p) may be called the fugacity coefficient.

The name activity coefficient has sometimes been used for this ratio but is not recommended.

2. Mixtures

A.I.4 *Definition of a mixture*

The word *mixture* is used to describe a gaseous or liquid or solid phase containing more than one substance, when the substances are all treated in the same way (contrast the use of the word *solution* in Section A.I.9).

A.I.5 *Partial pressure*

The partial pressure p_B of a substance B in a *gaseous* mixture is a quantity with the same dimensions as pressure defined by

$$p_B = y_B p$$

where y_B is the mole fraction of the substance B in the gaseous mixture and p is the pressure.

A.I.6 *Fugacity of a substance in a gaseous mixture*

The fugacity f_B of the substance B in a gaseous mixture containing mole fractions $y_B, y_C, \ldots$, of the substances B, C, . . . , is a quantity with the same dimensions as pressure, defined in terms of the absolute activity λ_B of the substance B in the gaseous mixture by

$$f_B = \lambda_B \lim_{p\to 0}(y_B p/\lambda_B) \qquad (T \text{ const.})$$

or in terms of the chemical potential μ_B by

$$RT \ln f_B = \mu_B + \lim_{p\to 0}\{RT \ln(y_B p) - \mu_B\} \qquad (T \text{ const.})$$

It follows from this definition that

$$\lim_{p\to 0}(f_B/y_B p) = 1 \qquad (T \text{ const.})$$

and that

$$RT \ln(f_B/y_B p) = \int_0^p (V_B - RT/p)\, dp \qquad (T \text{ const.})$$

where V_B is the partial molar volume (see Section 1.4) of the substance B in the gaseous mixture.

A gaseous mixture of B, C, . . . , is treated as an *ideal gaseous mixture* when the approximations $f_B = y_B p, f_C = y_C p$, . . . , are used. It follows that $pV = (n_B + n_C + \ldots) RT$ for an ideal gaseous mixture of B, C, . . .

The ratio $(f_B/y_B p)$ may be called the fugacity coefficient of the substance B. The name activity coefficient has sometimes been used for this ratio but is not recommended.

When $y_B = 1$ the definitions given in this Section for the fugacity of a substance in a gaseous mixture reduce to those given in Section A.I.3 for the fugacity of a pure gaseous substance.

A.I.7 *Activity coefficient of a substance in a liquid or solid mixture*

The activity coefficient f_B of a substance B in a liquid or solid mixture containing mole fractions $x_B, x_C, \ldots$, of the substances B, C, . . . , is a number defined in terms of the absolute activity λ_B of the substance B in the mixture by

$$f_B = \lambda_B/\lambda_B^* x_B$$

where λ_B^* is the absolute activity of the pure substance B at the same temperature and pressure, or in terms of the chemical potential μ_B by

$$RT \ln(x_B f_B) = \mu_B - \mu_B^*$$

where μ_B^* is the chemical potential of the pure substance B at the same temperature and pressure.

It follows from this definition that

$$\lim_{x_B\to 1} f_B = 1 \qquad (T, p \text{ const.})$$

A.I.8 *Relative activity of a substance in a liquid or solid mixture*

The relative activity a_B of a substance B in a liquid or solid mixture is a number defined by

$$a_B = \lambda_B/\lambda_B^*$$

or by

$$RT \ln a_B = \mu_B - \mu_B^*$$

where the other symbols are as defined in Section A.I.7.

It follows from this definition that

$$\lim_{x_B\to 1} a_B = 1 \qquad (T, p \text{ const.})$$

A mixture of substances B, C, . . . , is treated as an *ideal mixture* when the approximations $a_B = x_B, a_C = x_C, \ldots$, and consequently $f_B = 1, f_C = 1, \ldots$, are used.

3. Solutions

A.I.9 *Definition of solution*

The word *solution* is used to describe a liquid or solid phase containing more than one substance, when for convenience one of the substances, which is called the *solvent* and may itself be a mixture, is treated differently from the other substances, which are called *solutes*. When, as is often but not necessarily the case, the sum of the mole fractions of the solutes is small compared with unity, the solution is called a *dilute solution*. In the following definitions the solvent substance is denoted by A and the solute substances by B, C, ...

A.I.10 *Properties of infinitely dilute solutions*

The superscript $^\infty$ attached to the symbol for a property of a solution denotes the property of an *infinitely dilute solution*.

For example if V_B denotes the partial molar volume (see Section 1.4) of the solute substance B in a solution containing molalities $m_B, m_C, \ldots$, or mole fractions $x_B, x_C, \ldots$, of solute substances B, C, . . . , in a solvent substance A, then

$$V_B^\infty = \lim_{\Sigma_i m_i \to 0} V_B = \lim_{\Sigma_i x_i \to 0} V_B \qquad (T, p \text{ const.})$$

where i = B, C, ...

Similarly if V_A denotes the partial molar volume of the *solvent* substance A, then

$$V_A^\infty = \lim_{\Sigma_i m_i \to 0} V_A = \lim_{\Sigma_i x_i \to 0} V_A = V_A{}^* \qquad (T, p \text{ const.})$$

where $V_A{}^*$ is the molar volume of the pure solvent substance A.

A.I.11 *Activity coefficient of a solute substance in a solution*

The activity coefficient γ_B of a *solute* substance B in a solution (especially in a dilute liquid solution) containing molalities $m_B, m_C, \ldots$, of solute substances B, C, . . . , in a solvent substance A, is a number defined in terms of the absolute activity λ_B of the solute substance B in the solution by

$$\gamma_B = (\lambda_B/m_B)/(\lambda_B/m_B)^\infty \qquad (T, p \text{ const.})$$

or in terms of the chemical potential μ_B by

$$RT \ln(m_B\gamma_B) = \mu_B - RT \ln m_B)^\infty \qquad (T, p \text{ const.})$$

It follows from this definition that

$$\gamma_B{}^\infty = 1 \qquad (T, p \text{ const.})$$

The name activity coefficient with the symbol y_B may be used for the quantity similarly defined but with amount-of-substance concentration c_B (see Section 2.3) in place of molality m_B.

Another activity coefficient, called the *rational activity coefficient* of a solute substance B and denoted by $f_{x,B}$ is sometimes used. It is defined in terms of the absolute activity λ_B by

$$f_{x,B} = (\lambda_B/x_B)/(\lambda_B/x_B)^\infty \qquad (T, p \text{ const.}))$$

or in terms of the chemical potential μ_B by

$$RT \ln(x_B f_{x,B}) = \mu_B - (\mu_B - RT \ln x_B)^\infty \qquad (T, p \text{ const.})$$

where x_B is the mole fraction of the solute substance B in the solution. The rational activity coefficient $f_{x,B}$ is related to the (practical) activity coefficient γ_B by the formula

$$f_{x,B} = \gamma_B(1 + M_A \Sigma_i m_i) = \gamma_B/(1 - \Sigma_i x_i)$$

A solution of solute substances B, C, . . . , in a solvent substance A is treated as an *ideal dilute solution* when the activity coefficients are approximated to unity, for example $\gamma_B = 1, \gamma_C = 1, \ldots$.

A.I.12 *Relative activity of a solute substance in a solution*

The relative activity a_B of a *solute* substance B in a solution (especially in a dilute liquid solution) containing molalities $m_B, m_C, \ldots$, of solute substances B, C, . . . , in a solvent substance A, is a number defined in terms of the absolute activity λ_B by

$$a_B = (\lambda_B/m^\ominus)/(\lambda_B/m_B)^\infty = m_B\gamma_B/m^\ominus \qquad (T, p \text{ const.})$$

or in terms of the chemical potential μ_B by

$$RT \ln a_B = \mu_B - RT \ln m^\ominus - (\mu_B - RT \ln m_B)^\infty$$
$$= RT \ln(m_B\gamma_B/m^\ominus)$$

where $m^\ominus$ is a standard value of molality (usually chosen to be 1 mol kg^{-1}) and where the other symbols are defined in Section A.I.11.

It follows from this definition of a_B (compare Section A.I.8) that

$$(a_B m^\ominus/m_B)^\infty = 1 \qquad (T, p \text{ const.})$$

The name activity is often used instead of the name relative activity for this quantity.

The name relative activity with the symbol $a_{c,B}$ may be used for the quantity similarly defined but with concentration c_B (see Section 2.3) in place of molality m_B, and a standard value $c^\ominus$ of concentration (usually chosen to be 1 mol dm^{-3}) in place of the standard value $m^\ominus$ of molality.

Another relative activity, called the *rational relative activity* of the solute substance B and denoted by $a_{x,B}$, is sometimes used. It is defined in terms of the absolute activity λ_B by

$$a_{x,B} = \lambda_B/(\lambda_B/x_B)^\infty = x_B f_{x,B} \qquad (T, p \text{ const.})$$

or in terms of the chemical potential μ_B by

$$RT \ln a_{x,B} = \mu_B - (\mu_B - RT \ln x_B)^\infty$$
$$= RT \ln(x_B f_{x,B}) \qquad (T, p \text{ const.})$$

where x_B is the mole fraction of the substance B in the solution. The rational relative activity $a_{x,B}$ is related to the (practical) relative activity a_B by the formula

$$a_{x,B} = a_B m^\ominus M_A$$

A.I.13 *Osmotic coefficient of the solvent substance in a solution*

The osmotic coefficient ϕ of the *solvent* substance A in a solution (especially in a dilute liquid solution) containing molalities m_B, m_C, . . . , of solute substances B, C, . . . , is a number defined in terms of the absolute activity λ_A of the solvent substance A in the solution by

$$\phi = (M_A \Sigma_i m_i)^{-1} \ln(\lambda_A{}^*/\lambda_A)$$

where $\lambda_A{}^*$ is the absolute activity of the pure solvent substance A at the same temperature and pressure, and M_A is the molar mass of the solvent substance A, or in terms of the chemical potential $\mu_A{}^*$ by

$$\phi = (\mu_A{}^* - \mu_A)/RTM_A\Sigma_i m_i$$

where $\mu_A{}^*$ is the chemical potential of the pure solvent substance A at the same temperature and pressure.

For an *ideal dilute solution* as defined in Section A.I.11 or A.I.12 it can be shown that $\phi = 1$.

Another osmotic coefficient, called the *rational osmotic coefficient* of the solvent substance A and denoted by ϕ_x, is sometimes used. It is defined in terms of the absolute activity λ_A by

$$\phi_x = \ln(\lambda_A/\lambda_A{}^*)/\ln x_A = \ln(\lambda_A/\lambda_A{}^*)/\ln(1 - \Sigma_i x_i)$$

or in terms of the chemical potential μ_A by

$$\phi_x = (\mu_A - \mu_A{}^*)/RT \ln x_A = (\mu_A - \mu_A{}^*)/RT \ln(1 - \Sigma_i x_i)$$

where x_A is the mole fraction of the solvent substance A in the solution. The rational osmotic coefficient ϕ_x is related to the (practical) osmotic coefficient ϕ by the formula

$$\phi_x = \phi M_A \Sigma_i m_i/\ln)1 + M_A\Sigma_i m_i) = -\phi M_A \Sigma_i m_i/\ln(1 - \Sigma_i x_i)$$

A.I.14 *Relative activity of the solvent substance in a solution*

The relative activity a_A of the *solvent* substance A in a solution (especially in a dilute liquid solution) containing molalities m_B, m_C, . . . , or more fractions x_B, x_C, . . . , of solute substances B, C, . . . , is a number defined in terms of the absolute activity λ_A of the solvent substance A in the solution by

$$a_A = \lambda_A / \lambda_A{}^* = \exp(-\phi M_A \Sigma_i m_i) = (1 - \Sigma_i x_i)^{\phi_x}$$

or in terms of the chemical potential μ_A by

$$RT \ln a_A = \mu_A - \mu_A{}^* = -RT\phi M_A \Sigma_i m_i = \phi_x RT \ln(1 - \Sigma_i x_i)$$

where the other quantities are as defined in Section A.I.13.

Note: The definition in this Section of the relative activity of the *solvent* in a *solution*, is identical with the definition in Section A.I.8 of the relative activity of any substance in a *mixture*. See also Section A.I.12.

From Document U.I.P. 20 (1978)
International Union of Pure and Applied Physics
Used by permission of the Secretary

1 PHYSICAL QUANTITIES–GENERAL RECOMMENDATIONS †)

Note: The German, Italian, Russian, and Spanish translations of this term are 'physikalische Größe', 'grandezza fizika', 'ФИЗИЧЕСКАЯ ВЕЛИЧИНА', and 'magnitud fisica', respectively.

1.1 Physical quantities

A physical quantity (French: 'grandeur physique') is equivalent to the product of the *numerical value*, i.e. a pure number, and a *unit*:

physical quantity = numerical value × unit.

For a physical quantity with symbol a this relation is usually represented in the form $a = \{a\} \cdot [a]$, where $\{a\}$ stands for the numerical value and $[a]$ stands for the symbol for the unit. For dimensionless physical quantities the unit often has no name or symbol and is not explicitly indicated (see section 9.1 "unit systems").

Examples: $E = 200$ J $n = 1{,}55$ (for quartz)
$F = 27$ N $\nu = 3 \times 10^8$ Hz.

1.2 Symbols for physical quantities–General rules

1. *Symbols for physical quantities* should be *single letters* of the Latin or Greek alphabet with or without modifying signs: subscripts, superscripts, primes, etc.

   Remarks:
   (a) An exception to this rule is given by the two-letter symbols which are used to represent dimensionless combinations of physical quantities (see section 7.14 "dimensionless parameters"). If such a symbol, composed of two letters, appears as a factor in a product, it is recommended to separate this symbol from the other symbols by a dot or by a space or by brackets. It can be raised to a positive or negative power without using brackets.
   (b) Abbreviations, i.e. shortened forms of names or expressions, such as p.f. for partition function, should not be used in physical equations. These abbreviations in the text should be written in ordinary roman type.

2. *Symbols for physical quantities* should be printed in *italic* (or *sloping*) *type*.

   Remark: It is recommended to consider as a guiding principle for the printing of indices the criterion: only indices which are symbols for physical quantities should be printed in italic (sloping) type.

†) For further details see International Standard I.S.O. 31/0: *General principles concerning quantities, units and symbols.*

Examples:

upright indices	sloping indices
C_{g} (g = gas)	p in C_p
g_{n} (n = normal)	n in $\Sigma_n a_n \psi_n$
μ_{r} (r = relative)	x in $\Sigma_x a_x b_x$
E_{k} (k = kinetic)	i, k in g_{ik}
χ_{e} (e = electric)	x in p_x

3. *Symbols for vectors and tensors*

To avoid the usage of subscripts it is often convenient to indicate vectors and tensors of the second rank by letters of a special type. The following choice is recommended:

(*a*) Vectors should be printed in bold italic (sloping) type, e.g. ***A***, ***a***.
(*b*) Tensors of the second rank should be printed in bold sans serif italic (sloping) type, e.g. ***S***, ***T***.

Remark: When such type is not available, a vector may be indicated by an arrow and a tensor by a double arrow above the symbol; e.g. $\vec{A}$, $\vec{\vec{S}}$.

1.3 Simple mathematical operations

1. *Addition and subtraction* of two physical quantities are indicated by:

$$a + b \quad \text{and} \quad a - b.$$

2. *Multiplication* of two physical quantities may be indicated in one of the following ways:

$$ab \qquad a \cdot b \qquad a \times b.$$

3. *Division* of one quantity by another quantity may be indicated in one of the following ways:

$$\frac{a}{b} \qquad a/b \qquad ab^{-1}$$

or in any other way of writing the product of a and b^{-1}.
These procedures can be extended to cases where one of the quantities or both are themselves products, quotients, sums or differences of other quantities.
If necessary, brackets have to be used in accordance with the rules of mathematics. If the solidus is used to separate the numerator from the denominator and if there is any doubt where the numerator starts or where the denominator ends, brackets should be used.

Examples:

expressions with a horizontal bar	same expressions with a solidus
$\frac{a}{bcd}$	a/bcd
$\frac{2}{9}\sin kx$	$(2/9)\sin kx$

expressions with a horizontal bar	same expressions with a solidus
$\frac{a}{b} - c$	$a/b - c$
$\frac{a}{b - c}$	$a/(b - c)$
$\frac{a - b}{c - d}$	$(a - b)/(c - d)$
$\frac{a}{c} - \frac{b}{d}$	$a/c - b/d$

Remark: It is recommended that in expressions like:

$\sin\{2\pi(x - x_0)/\lambda\}$	$\exp\{(r - r_0)/\sigma\}$
$\exp\{-V(r)/kT\}$	$\surd(\varepsilon/c^2)$

the argument should always be placed between brackets, except when the argument is a simple product of two quantities: e.g. sin kx. When the horizontal bar above the square root is used no brackets are needed.

2 UNITS–GENERAL RECOMMENDATIONS

2.1 Symbols for units–General rules

1. *Symbols for units* of physical quantities should be printed in *roman* (*upright*) *type.*

2. *Symbols for units* should not contain a full stop (period) and should remain unaltered in the plural, e.g.: 7 cm and *not* 7 cms.

3. *Symbols for units* should be printed in *lower case* roman (upright) type. However, the symbol for a unit, derived from a proper name, should start with a capital roman letter, e.g.: m (metre); A (ampere); Wb (weber); Hz (hertz).

2.2 Prefixes–General rules

1. The *prefixes* which should be used to indicate decimal multiples or sub-multiples of a unit are given in table 1.

Table 1

deci; *déci*	(= 10^{-1})	d	deca; *déca*	(= 10^{1})	da
centi; *centi*	(= 10^{-2})	c	hecto; *hecto*	(= 10^{2})	h
milli; *milli*	(= 10^{-3})	m	kilo; *kilo*	(= 10^{3})	k
micro; *micro*	(= 10^{-6})	μ	mega; *méga*	(= 10^{6})	M
nano; *nano*	(= 10^{-9})	n	giga; *giga*	(= 10^{9})	G
pico; *pico*	(= 10^{-12})	p	tera; téra	(= 10^{12})	T
femto; *femto*	(= 10^{-15})	f	peta; *peta*	(= 10^{15})	P
atto; *atto*	(= 10^{-18})	a	exa; *exa*	(= 10^{18})	E

2. *Compound prefixes*, formed by juxtaposition of two or more prefixes, are not to be used.

Not: mμs,	*but:* ns	(nanosecond)
Not: kMW,	*but:* GW	(gigawatt)
Not: μμF,	*but:* pF	(picofarad)

3. When the symbol of a prefix is placed before the symbol of a unit, the *combination of the two symbols* should be considered as *one new symbol*, which can be raised to a positive or negative power without using brackets.

Examples: cm^3 mA^2 $\mu\mathrm{s}^{-1}$.

Remark:

cm^3 *means always* $(0{,}01\ \mathrm{m})^3$ *but never* $0{,}01\ \mathrm{m}^3$
$\mu\mathrm{s}^{-1}$ *means always* $(10^{-6}\mathrm{s})^{-1}$ *but never* $10^{-6}\mathrm{s}^{-1}$.

2.3 Mathematical operations

1. *Multiplication* of two units may be indicated in one of the following ways:

$$\mathrm{N\,m} \qquad \mathrm{N \cdot m}.$$

2. *Division* of one unit by another unit may be indicated in one of the following ways:

$$\frac{\mathrm{m}}{\mathrm{s}} \qquad \mathrm{m/s} \qquad \mathrm{m\,s^{-1}}$$

or by any other way of writing the product of m and s^{-1}.
Not more than one solidus should be used.

Examples:

Not: cm/s/s,	*but:* $\mathrm{cm/s^2 = cm\,s^{-2}}$
Not: J/K/mol,	*but:* $\mathrm{J/K\,mol = J\,K^{-1}\,mol^{-1}}$

3 NUMBERS

1. *Numbers* should be printed in *upright type.*

2. The *decimal* sign is a comma on the line (,). In documents in the English language a comma or a dot on the line (.) may be used.
If the magnitude of the number is less than unity, the decimal sign should be preceded by a zero.

3. The sign for *multiplication* of numbers is a cross (×) or a dot half-high (·). If a dot is used as a decimal sign, a dot should not be used as the multiplication sign.

Example: 2.3×3.4 or $2{,}3 \cdot 3{,}4$.

4. *Division* of one number by another number may be indicated in the following ways:

$$\frac{136}{273,15} \qquad 136/273,15$$

or by writing it as the product of numerator and the inverse first power of the denominator. In such cases the number under the inverse power should always be placed between brackets.

Remark: When the solidus is used and when there is any doubt where the numerator starts or the denominator ends, brackets should be used, as in the case of quantities (see section 1.3).

5. To facilitate the reading of *large numbers*, the digits may be grouped in *groups of three*, but *no* comma or point should be used except for the decimal sign.

   Example: 2 573,421 736.

4 SYMBOLS FOR CHEMICAL ELEMENTS, NUCLIDES, AND PARTICLES

1. *Symbols for chemical elements* should be written in *roman (upright) type.* The symbol is not followed by a full stop.

   Examples*:* Ca C H He.

2. The *nucleon number* (*mass number*) of a *nuclide* is shown as a left superscript (e.g. ^{14}N).

3. The right subscript position is used to indicate the *number of atoms* of a nuclide in a molecule (e.g. $^{14}N_2$).

4. The right superscript position should be used, if required, for indicating a *state of ionization* (e.g. Ca^{2+}, PO_4^{3-}) or an *excited state* (e.g. $^{110}Ag^m$, He^*).

5. The attached numeral specifying the *spectrum of a z-fold ionized atom* is the roman number $z + 1$.

   Examples: CaII, AlIII, HI (spectrum of the *neutral* hydrogen atom).

   Remark: Roman numbers in right superscript position may indicate the *oxidation number* (e.g. $Pb_2^{II}Pb^{IV}O_4$; $K_6M^{IV}Mo_9O_{32}$, where M denotes a metal).

6. *Symbols for particles and quanta*

   It is recommended to use the notation listed in table 2.

Table 2

nucleon	N	pion	π
proton	p	K-meson	K
neutron	n		
		electron	e
Λ-particle	Λ	muon	μ
Σ-particle	Σ	neutrino	ν
Ξ-particle	Ξ		
Ω-particle	Ω	photon	γ
deuteron	d		
triton	t		
helion ($^3He^{2+}$)	h		
α-particle	α		

The charge of a particle may be indicated by adding the superscript +, − or 0.

Examples: π^+ π^- π^0, p^+ p^-, e^+ e^-.

If in connection with the symbols p and e no charge is indicated, these symbols should refer to the positive proton and the negative electron, respectively.
The bar ¯, or sometimes the tilde ˜, above the symbol of a particle is used to indicate the corresponding anti-particle.

5 QUANTUM STATES

5.1 General rules

A letter symbol indicating the quantum state of *a system* should be printed in capital upright type.
A letter symbol indicating the quantum state of *a single particle* should be printed in lower case upright type.

5.2 Atomic spectroscopy

The letter symbols indicating atomic quantum states are:

$L, l = 0$: S, s	$L, l = 4$: G, g	$L, l = 8$: L, l
$= 1$: P, p	$= 5$: H, h	$= 9$: M, m
$= 2$: D, d	$= 6$: I, i	$= 10$: N, n
$= 3$: F, f	$= 7$: K, k	$= 11$: O, o.

A right subscript indicates the total angular momentum quantum number J or j. A left superscript indicates the spin multiplicity $2S + 1$.

Example: $^2P_{3/2}$-state ($J = 3/2$, spin multiplicity 2)
$p_{3/2}$-electron ($j = 3/2$).

An atomic electron configuration is indicated symbolically by:

$$(nl)^{\kappa} \quad (n'l')^{\kappa'} \quad \ldots$$

Instead of $l = 0, 1, 2, 3 \ldots$ one uses the quantum state symbol s, p, d, f, ...

Example: the atomic electron configuration: $(1s)^2\ (2s)^2\ (2p)^3$.

5.3 Molecular spectroscopy

The letter symbols, indicating molecular electronic quantum states are in the case of *linear molecules:*

$$\begin{aligned} \Lambda, \lambda &= 0: \quad \Sigma, \sigma \\ &= 1: \quad \Pi, \pi \\ &= 2: \quad \Delta, \delta \end{aligned}$$

and for *non-linear molecules*

$$\mathrm{A, a}; \quad \mathrm{B, b}; \quad \mathrm{E, e}; \quad \text{etc.}$$

Remarks: A left superscript indicates the spin multiplicity. For molecules having a symmetry centre the parity symbol g or u, indicating respectively symmetric or antisymmetric behaviour on inversion, is attached as a right subscript. A + or − sign attached as a right superscript indicates the symmetry as regards reflection in any plane through the symmetry axis of the molecules.

Examples: Σ_g^+, Π_u, ${}^2\Sigma$, ${}^3\Pi$, etc.

The letter symbols indicating the vibrational angular momentum states in the case of *linear molecules* are

$$\begin{aligned} l &= 0: \quad \Sigma \\ &= 1: \quad \Pi \\ &= 2: \quad \Delta. \end{aligned}$$

5.4 Nuclear spectroscopy

The spin and parity assignment of a nuclear state is

$$J^\pi$$

where the parity symbol π is + for even and − for odd parity.

Examples: 3^+, 2^-, etc.

A shell model configuration is indicated symbolically by:

$$(nlj)^\kappa \quad (n'l'j')^{\kappa'}$$

where the first bracket refers to the proton shell and the second to the neutron shell. Negative values of κ or κ' indicate holes in a completed shell. Instead of $l = 0, 1, 2, 3, \ldots$ one uses the quantum state symbol s, p, d, f, ...

Example: the nuclear configuration: $(1\,\mathrm{d}\,3/2)^3(1\,\mathrm{f}\,7/2)^2$.

5.5 Spectroscopic transitions

1. The upper level and the lower level are indicated by ′ and ″ respectively.

Examples: $h\nu = E' - E''$ $\quad \sigma = T' - T''$.

2. A spectroscopic transition should be indicated by writing the upper state first and the lower state second, connected by a dash.

Examples:	${}^2\mathrm{P}_{1/2} - {}^2\mathrm{S}_{1/2}$	for an electronic transition
	$(J', K') - (J'', K'')$	for a rotational transition
	$v' - v''$	for a vibrational transition.

3. Absorption transition and emission transition may be indicated by arrows ← and → respectively.

Examples: $^2P_{1/2} \rightarrow {}^2S_{1/2}$ emission from $^2P_{1/2}$ to $^2S_{1/2}$
$(J', K') \leftarrow (J'', K'')$ absorption from (J'', K'') to (J', K').

4. The difference between two quantum numbers should be that of the upper state minus that of the lower state.

Example: $\Delta J = J' - J''$.

5. The indications of the branches of the rotation band should be as follows:

$\Delta J = J' - J'' = -2$: O-branch
$= -1$: P-branch
$= 0$: Q-branch
$= +1$: R-branch
$= +2$: S-branch.

6 NOMENCLATURE

1. *Use of the words 'specific' and 'molar'*

The word 'specific' in the English name for an extensive physical quantity should be restricted to the meaning 'divided by mass (mass of the system, if this consists of more than one component or of more than one phase)'.

Examples:

specific volume	volume/mass
specific energy	energy/mass
specific heat capacity	heat capacity/mass.

The word 'molar' in the English name for an extensive physical quantity should be restricted to the meaning 'divided by amount of substance (amount of substance of the system, if this consists of more than one component or of more than one phase)'.

Examples:

molar volume	volume/amount of substance
molar energy	energy/amount of substance
molar heat capacity	heat capacity/amount of substance.

The symbol X_B, where X denotes an extensive quantity and B is the chemical symbol for a substance, denotes the partial molar quantity of the substance B defined by the relation:

$$X_B = (\partial X/\partial n_B)_{T,p,n_C,\ldots}$$

For a pure substance B the partial molar quantity X_B and the molar quantity X_m are identical. The partial molar quantity X_B of pure substance B, which is identical with the molar quantity X_m of pure substance B, may be denoted by X_B^* where the superscript * denotes 'pure', so as to distinguish it from the partial molar quantity X_B of substance B in a mixture.

2. *Notation for covariant character of coupling*

S	Scalar coupling	A	Axial vector coupling
V	Vector coupling	P	Pseudoscalar coupling
T	Tensor coupling.		

3. *Abbreviated notation for a nuclear reaction*

The meaning of the symbolic expression indicating a nuclear reaction should be the following:

initial nuclide $\left(\begin{matrix}\text{incoming particle(s)} \\ \text{or quanta}\end{matrix}, \begin{matrix}\text{outgoing particle(s)} \\ \text{or quanta}\end{matrix}\right)$ final nuclide.

Examples: $^{14}N(\alpha, p)^{17}O$ $^{59}Co(n, \gamma)^{60}Co$

$^{23}Na(\gamma, 3n)^{20}Na$ $^{31}P(\gamma, pn)^{29}Si$.

4. *Character of transitions*

Multipolarity of transition:

electric or magnetic monopole	E0 or M0
" " " dipole	E1 or M1
" " " quadrupole	E2 or M2
electric or magnetic octupole	E3 or M3
" " " 2^n-pole	E*n* or M*n*.

Parity change in transition:

transition *with* parity change: yes
transition *without* parity change: no.

5. *Nuclide*

A species of *atoms*, identical as regards atomic number (proton number) and mass number (nucleon number) should be indicated by the word *nuclide*, not by the word isotope.
Different nuclides having the same atomic number should be indicated as *isotopes* or *isotopic nuclides*.
Different nuclides having the same mass number should be indicated as *isobars* or *isobaric nuclides*.

6. *Sign of polarization vector* (*Basel Convention*)

In nuclear interactions the positive polarization of particles with spin $\frac{1}{2}$ is taken in the direction of the vector product

$$\boldsymbol{k}_i \times \boldsymbol{k}_o$$

where $\boldsymbol{k}_i$ and $\boldsymbol{k}_o$ are the circular wave vectors of the incoming and outgoing particles, respectively.

7. *Description of polarization effects* (*Madison Convention*)

In the symbolic expression for a nuclear reaction A(b, c)D an arrow placed over a symbol denotes a particle which is initially in a polarized state or whose state of polarization is measured.

Examples:

$A(\vec{b}, c)D$	polarized incident beam
$A(\vec{b}, \vec{c})D$	polarized incident beam; polarization of the outgoing particle c measured (polarization transfer)
$A(b, \vec{c})D$	unpolarized incident beam; polarization of the outgoing particle c measured
$\vec{A}(b, c)D$	unpolarized beam incident on a polarized target
$\vec{A}(b, \vec{c})D$	unpolarized beam incident on a polarized target; polarization of the outgoing particle c measured
$A(\vec{b}, c)\vec{D}$	polarized incident beam; measurement of the polarization of the target.

7 RECOMMENDED SYMBOLS FOR PHYSICAL QUANTITIES

Remarks:

(1) Where several symbols are given for one quantity, and no special indication is made, they are on equal footing.

(2) In general no special attention is paid to the name of the quantity.

(3) Where there is more than one form for a greek letter (ε, ϵ; ϑ, θ; κ, $\varkappa$; φ, ϕ) either form may be used. The form ϖ of the letter pi may be used as though it were a different letter.

7.1 *Space and time*

space coordinates; *coordonnées d'espace*	(x, y, z)
position vector; *vecteur de position*	$\boldsymbol{r}$
length; *longueur*	l
breadth; *largeur*	b
height; *hauteur*	h
radius; *rayon*	r
thickness; *épaisseur*	d, δ
diameter; *diamètre:* $d = 2r$	d
element of path; *élément de parcours*	$\mathrm{d}s$
area; *aire, superficie*	A, S
volume; *volume*	$V, (v)$
plane angle; *angle plan*	$\alpha, \beta, \gamma, \theta, \vartheta, \varphi$
solid angle; *angle solide*	ω, Ω
wave length; *longueur d'onde*	λ
wave number; *nombre d'onde:* $\sigma = 1/\lambda$	σ †)
wave vector; *vecteur d'onde*	$\boldsymbol{\sigma}$
circular wave number; *nombre d'onde circulaire:* $k = 2\pi/\lambda$	K
circular wave vector; *vecteur d'onde circulaire*	$\boldsymbol{K}$
attenuation coefficient; *constante d'affaiblissement*: $F(x) = \exp(-\alpha x)\cos\beta x$	α
phase coefficient; *constante de phase*	β
propagation coefficient; *constante de propagation:* $\gamma = \alpha + \mathrm{i}\beta$	γ
time; *temps*	t
period, periodic time; *période, durée d'une période*	T
frequency; *fréquence:* $\nu = 1/T$	ν, f
circular frequency, pulsatance; pulsation: $\omega = 2\pi\nu$	ω
relaxation time; *temps de relaxation:* $F(t) = \exp(-t/\tau)$	τ
damping coefficient; *coefficient d'amortissement:* $F(t) = \exp(-\delta t)\sin\omega t$	δ
logarithmic decrement; *décrément logarithmique:* $\Lambda = T\delta = T/\tau$	Λ
velocity; *vitesse:* $v = \mathrm{d}s/\mathrm{d}t$	u, v
angular velocity; *vitesse angulaire:* $\omega = \mathrm{d}\varphi/\mathrm{d}t$	ω
acceleration; *accélération:* $a = \mathrm{d}v/\mathrm{d}t$	a
angular acceleration; *accélération angulaire:* $\alpha = \mathrm{d}\omega/\mathrm{d}t$	α
acceleration of free fall; *accélération de la pesanteur*	g
standard --; -- *normale*	g_{n}
speed of light in empty space; *vitesse de la lumière dans le vide*	c
v/c	β
relativistic coordinates; *coordonnées relativistes:* $x_0 = ct$, $x_1 = x$, $x_2 = y$, $x_3 = z$, $x_4 = \mathrm{i}ct$	$(x_0x_1x_2x_3)$ $(x_1x_2x_3x_4)$

†) In molecular spectroscopy $\tilde{\nu}$ is also used.

7.2 Mechanics

mass; *masse*	m
(mass) density; *masse volumique:* $\rho = m/V$	ρ
relative density; *densité relative:* $d = \rho/\rho(H_2O)$	d
specific volume; *volume massique:* $v = V/m = 1/\rho$	v
reduced mass; *masse réduite:* $\mu = m_1 m_2/(m_1 + m_2)$	μ
momentum; *quantité de mouvement:* $\mathbf{p} = m\mathbf{v}$	$\mathbf{p}$
angular momentum; *moment cinétique:* $\mathbf{L} = \mathbf{r} \times \mathbf{p}$	$\mathbf{L}$
second moment of plane area; *moment quadratique d'une aire plane:* $I_{a,y} = \int x^2 \mathrm{d}x\mathrm{d}y$	I_a
second polar moment of plane area; *moment quadratique polaire d'une aire plane:* $I_p = \int (x^2 + y^2)\, \mathrm{d}x\mathrm{d}y$	I_p
moment of inertia; *moment d'inertie:* $I_z = \int (x^2 + y^2)\, \mathrm{d}m$	I, J
force; *force*	$\mathbf{F}$
torque, moment of a couple; *torque, moment d'un couple*	$\mathbf{T}$
weight; *poids*	G, (W, P)
moment of force; *moment d'une force*	$\mathbf{M}$
pressure; *pression*	p
normal stress; *contrainte normale*	σ
shear stress; *contrainte tangentielle, cission*	τ
gravitational constant; *constante de gravitation:* $F(r) = G\, m_1 m_2/r^2$	G
linear strain, relative elongation; *dilatation linéique relative:* $\varepsilon = \Delta l/l_0$	ε
modulus of elasticity, Young's modulus; *module d'élasticité longitudinale:* $\sigma = E\,\varepsilon$	E
shear strain; *glissement unitaire*	γ
shear modulus; *module d'élasticité de glissement:* $\tau = G\gamma$	G
volume strain, bulk strain; *dilatation volumique relative:* $\theta = \Delta V/V_0$	θ
bulk modulus; *module de compression:* $p = -K\theta$	K
Poisson ratio, *rapport de Poisson*	μ, ν
viscosity; *viscosité*	$\eta, (\mu)$
kinematic viscosity; *viscosité cinématique:* $\nu = \eta/\rho$	ν
friction coefficient; *coefficient de frottement*	$\mu, (f)$
surface tension; *tension superficielle*	γ, σ
energy; *énergie*	E, W
potential energy; *énergie potentielle*	E_p, V, Φ
kinetic energy; *énergie cinétique*	E_k, T, K
work; *travail*	W, A
power; *puissance*	P
efficiency; *rendement*	η
Hamiltonian function; *fonction de Hamilton*	H
Lagrangian function; *fonction de Lagrange*	L
principal function of Hamilton; *fonction principale de Hamilton:* $W = \int L\mathrm{d}t$	W, S_p
characteristic function of Hamilton; *fonction charactéristique de Hamilton:* $S = 2\int T\mathrm{d}t$	S
generalized coordinate; *coordonnée généralisée*	q, q_i
generalized momentum; *moment généralisé*	p, p_i
action integral; *intégrale d'action:* $J = \oint p\mathrm{d}q$	J

7.3 Molecular physics

number of molecules; *nombre de molécules*	N

number density of molecules; *nombre volumique de molécules* $n = N/V$	n
Avogadro constant; *constante d'Avogadro*	L, N_A
molecular mass; *masse moléculaire*	m
molecular velocity vector with (magnitudes of) components; *vecteur vitesse moléculaire et ses coordonnées*	$\mathbf{c}, (c_x, c_y, c_z)$ $\mathbf{u}, (u_x, u_y, u_z)$
molecular position vector with coordinates; *vecteur position moléculaire et ses coordonnées*	$\mathbf{r}, (x, y, z)$
molecular momentum vector with (magnitudes of) components; *vecteur quantité de mouvement moléculaire et ses coordonnées*	$\mathbf{p}, (p_x, p_y, p_z)$
average velocity; *vitesse moyenne*	$\mathbf{c}_0, \mathbf{u}_0, \langle \mathbf{c} \rangle, \langle \mathbf{u} \rangle$
average speed; *vitesse moyenne*	$\bar{c}, \bar{u}, \langle c \rangle, \langle u \rangle$
most probable speed; *vitesse la plus probable*	$\hat{c}, \hat{v}$
mean free path; *libre parcours moyen*	l
molecular attraction energy parameter; *paramètre d'énergie d'attraction moléculaire*	ε
interaction energy between molecules i and j; *énergie d'interaction entre les molécules i et j*	φ_{ij}, V_{ij}
velocity distribution function; *fonction de distribution des vitesses:* $n = \int f\, dc_x dc_y dc_z$	$f(c)$
Boltzmann function; *fonction de Boltzmann*	H
generalized coordinate; *coordonnée généralisée*	q
generalized momentum; *moment généralisé*	p
volume in γ phase space; *volume dans l'espace* γ	Ω
thermodynamic temperature; *température thermodynamique*	T
Boltzmann constant; *constante de Boltzmann*	k
$1/kT$ (in exponential functions; *dans les fonctions exponentielles*)	β
molar gas constant; *constante molaire des gaz*	R
partition function; *fonction de partitions*	Q, Z
symmetry number; *facteur de symétrie*	s
diffusion coefficient; *coefficient de diffusion*	D
thermal diffusion coefficient; *coefficient de thermodiffusion*	D_T
thermal diffusion ratio; *rapport de thermodiffusion*	k_T
thermal diffusion factor; *facteur de thermodiffusion*	α_T
characteristic temperature; *température caractéristique*	Θ
Debye temperature; *température de Debye:* $\Theta_D = h\nu_D/k$	Θ_D
Einstein temperature; *température d'Einstein:* $\Theta_E = h\nu_E/k$	Θ_E
rotational temperature; *température de rotation:* $\Theta_r = h^2/8\pi^2 Ik$	Θ_r
vibrational temperature; *température de vibration:* $\Theta_v = h\nu/k$	Θ_v

7.4 *Thermodynamics* †)

quantity of heat; *quantité de chaleur*	Q
work; *travail*	W, A
thermodynamic temperature; *température thermodynamique*	T
Celsius temperature; *température Celsius*	t, ϑ ††)
entropy; *entropie*	S
internal energy; *énergie interne*	U
Helmholtz function, *fonction de Helmholtz, énergie libre:* $F = U - TS$	F
enthalpy; *enthalpie:* $H = U + pV$	H

†) The index m is added in the case of molar quantities, if needed, to distinguish them from quantities referring to the whole system. For specific quantities (see section 6.1) lower case letters are used.

††) When Celsius temperature and time come together, t is reserved for time.

Gibbs function; *fonction de Gibbs, enthalpie libre:* $G = H - TS$ — G

Massieu function; *fonction de Massieu:* $J = -F/T$ — J

Planck function; *fonction de Planck:* $Y = -G/T$ — Y

pressure coefficient; *coefficient de pression:* $\beta = (\partial p/\partial T)_V$ — β

relative pressure coefficient; *coefficient de pression relative:* $\alpha_p = (1/p)(\partial p/\partial T)_V$ — α_p

compressibility; *compressibilité:* $\kappa = -(1/V)(\partial V/\partial p)_T$ — κ

linear expansion coefficient; *dilatabilité linéique* — α_l

cubic expansion coefficient; *dilatabilité volumique:* $\alpha = (1/V)(\partial V/\partial T)_p$ — α, γ

thermal conductivity; *conductivité thermique* — λ

specific heat capacity; *chaleur massique:* $c = C/m$ — c_p, c_V

heat capacity; *capacité thermique* — C_p, C_V

Joule–Thomson coefficient; *coefficient de Joule–Thomson* — μ

isentropic exponent; *exposant isentropique* *): $\kappa = -(V/p)(\partial p/\partial V)_S$ — κ

ratio of specific heat capacities; *rapport des chaleurs massiques* — $\gamma, (\kappa)$

heat flow rate; *flux thermique* — $\Phi, (q)$

heat current density; *densité de flux thermique* — $\mathbf{q}, (\varphi)$

thermal diffusivity; *diffusivité thermique:* $a = \lambda/\rho c_p$ — a

7.5 Electricity and magnetism **)

quantity of electricity; *quantité d'électricité* — Q

charge density; *charge volumique* — ρ

surface charge density; *charge surfacique* — σ

electric potential; *potentiel électrique* — V, φ

potential difference, tension; *différence de potentiel, tension* — U, V

electromotive force; *force électromotrice* — E

electric field strength; *champ électrique* — $\mathbf{E}$

electric flux, *flux électrique* — Ψ

electric displacement; *déplacement électrique* — $\mathbf{D}$

capacitance; *capacité* — C

permittivity; *permittivité:* $\mathbf{D} = \varepsilon \mathbf{E}$ — ε

permittivity of vacuum, electric constant; *permittivité du vide, constante électrique* — ε_0

relative permittivity; *permittivité relative:* $\varepsilon_r = \varepsilon/\varepsilon_0$ — ε_r

dielectric polarization; *polarisation diélectrique* : $\mathbf{D} = \varepsilon_0 \mathbf{E} + \mathbf{P}$ — $\mathbf{P}$

electric susceptibility; *susceptibilité électrique* — χ_e

polarizability; *polarisabilité* — α, γ

electric dipole moment; *moment dipolaire électrique* — $\mathbf{p}$

electric current; *courant électrique* — I

electric current density; *densité de courant électrique* — $\mathbf{j}, \mathbf{J}$

magnetic field strength; *champ magnétique* — $\mathbf{H}$

magnetic potential difference; *différence de potentiel magnétique* — U_m

magnetomotive force; *force magnétomotrice:* $F_m = \oint H_s \mathrm{d}s$ — F_m

magnetic induction, magnetic flux density; *induction magnétique, densité de flux magnétique* — $\mathbf{B}$

magnetic flux; *flux magnétique* — Φ

permeability; *perméabilité:* $\mathbf{B} = \mu \mathbf{H}$ — μ

*) For an ideal gas the isentropic exponent, κ, is equal to the ratio of specific heat capacities, γ.

**) Written according to the rationalized, 4-dimensional system of quantities, see Appendix I, section 2.

permeability of vacuum, magnetic constant; *perméabilité du vide, constante magnétique* — μ_0

relative permeability; *perméabilité relative:* $\mu_r = \mu/\mu_0$ — μ_r

magnetization; *aimantation:* $\boldsymbol{B} = \mu_0(\boldsymbol{H} + \boldsymbol{M})$ — $\boldsymbol{M}$

magnetic susceptibility; *susceptibilité magnétique* — χ_m

electromagnetic moment; *moment électromagnétique* $E_p = -\boldsymbol{m} \cdot \boldsymbol{B}$ — $\boldsymbol{\mu}, \boldsymbol{m}$

resistance; *résistance* — R

reactance; *réactance* — X

quality factor; *facteur de qualité:* $Q = |X|/R$ — Q

impedance; *impédance:* $Z = R + \mathrm{i}X$ — Z

admittance; *admittance:* $Y = 1/Z = G + \mathrm{i}B$ — Y

conductance; *conductance* — G

susceptance; *susceptance* — B

resistivity; *résistivité* — ρ

conductivity; *conductivité* $\gamma = 1/\rho$ — γ, σ

self inductance; *inductance propre* — L

mutual inductance; *inductance mutuelle* — M, L_{12}

coupling coefficient; *coefficient de couplage:* $k = L_{12}/(L_1L_2)^{1/2}$ — k

phase number; *nombre de phases* — m

loss angle; *angle de pertes* — δ

number of turns; *nombre de tours* — N

power; *puissance* — P

electromagnetic energy density; *énergie électromagnetique volumique* — w

Poynting vector; *vecteur de Poynting* — $\boldsymbol{S}$

magnetic vector potential; *potentiel vecteur magnétique* — $\boldsymbol{A}$

7.6 Radiation, light †)

radiant energy; *énergie rayonnante* — $Q, (Q_e), W$

radiant energy density; *énergie rayonnante volumique* — w

spectral concentration of radiant energy density (in terms of wavelength); *énergie rayonnante volumique spectrique (en longueur d'onde):* $w = \int w_\lambda \, \mathrm{d}\lambda$ — w_λ

radiant flux, radiant power; *flux énergétique, puissance rayonnante:* $\Phi = \int \Phi_\lambda \, \mathrm{d}\lambda$ — $\Phi, (\Phi_e), P$

radiant flux density; *densité de flux énergétique* — φ

radiant intensity; *intensité énergétique:* $\Phi = \int I \, \mathrm{d}\Omega$ — $I, (I_e)$

spectral concentration of radiant intensity (in terms of frequency); *intensité énergétique spectrique (en fréquence):* $I = \int I_\nu \, \mathrm{d}\nu$ — $I_\nu, (I_{e\nu})$

irradiance; *éclairement énergétique:* $\Phi = \int E \, \mathrm{d}S$ — $E, (E_e)$

radiance; *luminance énergétique:* $I = \int L \cos \vartheta \, \mathrm{d}S$ — $L, (L_e)$

radiant excitance; *excitance énergétique:* $\Phi = \int M \, \mathrm{d}S$ — $M, (M_e)$

Stefan–Boltzmann constant; *constante de Stefan–Boltzmann:* $\sigma = 2\pi^5k^4/15h^3c^2$ — σ

first radiation constant; *première constante de rayonnement:* $c_1 = 2\pi hc^2$ — c_1

second radiation constant; *deuzième constante de rayonnement:* $c_2 = hc/k$ — c_2

†) In several cases, the same symbol is used for a pair of corresponding radiant and luminous quantities with the understanding that subscripts e for energetic and v for visible will be added, whenever confusion between these quantities might otherwise occur.

emissivity; *émissivité:* $\varepsilon = M/M_B$ — ε
(M_B: radiant excitance of a black body radiator)
luminous efficacy; *efficacité lumineuse:* $K = \Phi_v/\Phi_e$ — K
spectral luminous efficacy; *efficacité lumineuse spectrale:* $K(\lambda) = \Phi_{v\lambda}/\Phi_{e\lambda}$ — $K(\lambda)$
maximum spectral luminous efficacy; *efficacité lumineuse spectrale maximale* — K_m
luminous efficiency; *efficacité lumineuse relative:* $V = K/K_m$ — V
spectral luminous efficiency; *efficacité lumineuse relative spectrale:* $V(\lambda) = K(\lambda)/K_m$ — $V(\lambda)$
quantity of light; *quantité de lumière* — Q, (Q_v)
luminous flux; *flux lumineux* — Φ, (Φ_v)
luminous intensity; *intensité lumineuse:* $\Phi = \int I \, d\Omega$ — I, (I_v)
spectral concentration of luminous intensity (in terms of wave number); *intensité lumineuse spectrique (en nombre d'onde):* $I = \int I_\sigma \, d\sigma$ — I_σ, $(I_{v\sigma})$
illuminance, illumination; *éclairement lumineux:* $\Phi = \int E \, dS$ — E, (E_v)
luminance; *luminance:* $I = \int L \cos \vartheta \, dS$ — L, (L_v)
luminous excitance; *excitance lumineuse:* $\Phi = \int M \, dS$ — M, (M_v)
absorptance; *facteur d'absorption:* Φ_a/Φ_0 — α *)
reflectance; *facteur de réflexion:* Φ_r/Φ_0 — ρ *)
transmittance; *facteur de transmission:* Φ_{tr}/Φ_0 — τ *)
linear attenuation coefficient; *coefficient d'atténuation linéique* — μ
linear absorption coefficient; *coefficient d'absorption linéique* — a
speed of light in empty space; *vitesse de la lumière dans le vide* — c
refractive index; *indice de réfraction:* $n = c/c_n$ — n

7.7 Acoustics

velocity of sound; *vitesse du son* — c
velocity of longitudinal waves; *vitesse longitudinale* — c_l
velocity of transversal waves; *vitesse transversale* — c_t
group velocity; *vitesse de groupe* — c_g
sound energy flux; *flux d'énergie acoustique* — P, P_a
reflexion factor; *facteur de réflexion:* P_r/P_0 — ρ
acoustic absorption factor; *facteur d'absorption acoustique:* $1 - \rho$ — α_a, (α)
transmission factor; *facteur de transmission:* P_{tr}/P_c — τ
dissipation factor; *facteur de dissipation:* $\alpha_a - \tau$ — δ
loudness level; *niveau d'isosonie* — L_N
sound power level; *niveau de puissance acoustique* — L_P
sound pressure level; *niveau de pression acoustique* — L_p

7.8 Quantum mechanics

complexe conjugate of Ψ; *complexe conjugué de* Ψ — Ψ^*
probability density; *densité de probabilité:* $P = \Psi^*\Psi$ — P
probability current density; *densité de courant de probabilité:* $S = (\hbar/2im)(\Psi^*\nabla\Psi - \Psi\nabla\Psi^*)$ — S
charge density of electrons; *charge volumique d'électrons:* $\rho = -eP$ — ρ
electric current density of electrons; *densité de courant électrique d'électrons:* $j = -eS$ — j

*) $\alpha(\lambda)$, $\rho(\lambda)$, and $\tau(\lambda)$ designate spectral absorptance, $\Phi_{a\lambda}/\Phi_{0\lambda}$, spectral reflectance, $\Phi_{r\lambda}/\Phi_{0\lambda}$, and spectral transmittance, $\Phi_{tr\lambda}/\Phi_{0\lambda}$, respectively.

Dirac bra vector; *vecteur bra de Dirac*	$\langle \vert$
Dirac ket vector; *vecteur ket de Dirac*	$\vert \rangle$
expectation value of A; *valeur moyenne de A*	$\langle A \rangle, \bar{A}$
commutator of A and B; *commutateur de A et B:* $[A,B] = AB - BA$	$[A,B]$, $[A,B]_-$
anticommutator of A and B; *anticommutateur de A et B:* $[A,B]_+ = AB + BA$	$[A,B]_+$
matrix element; *élément de matrice:* $A_{ij} = \int \varphi_i^*(A\varphi_j)\, d\tau$	A_{ij}
Hermitian conjugate of operator A; *conjugué Hermitien de l'opérateur A*	$A^\dagger$
momentum operator in coordinate representation; *opérateur de quantité de mouvement*	$+(\hbar/i)\nabla$
annihilation operators; *opérateurs d'annihilation*	a, b, α, β
creation operators; *opérateurs de création*	$a^\dagger, b^\dagger, \alpha^\dagger, \beta^\dagger$

7.9 Atomic and nuclear physics

nucleon number, mass number; *nombre de nucléons, nombre de masse*	A
proton number, atomic number; *nombre de protons, nombre atomique*	Z
neutron number; *nombre de neutrons:* $N = A - Z$	N
elementary charge (equal to charge of proton); *charge élémentaire (égale à la charge du proton)*	e
electron mass; *masse de l'électron*	m, m_e
proton mass; *masse du proton*	m_p
neutron mass; *masse du neutron*	m_n
meson mass; *masse du méson*	m_π,
nuclear mass; *masse nucléaire* (of nucleus: ^{A}X)	m_N, m_N (^{A}X)
atomic mass; *masse atomique* (of nuclide: ^{A}X)	m_a, m_a (^{A}X)
(unified) atomic mass constant; *constante (unifiée) de masse atomique:* $m_u = m_a(^{12}C)/12$	m_u
relative atomic mass; *masse atomique relative:* m_a/m_u	A_r
Planck constant; *constante de Planck* ($\hbar = h/2\pi$)	h
principal quantum number (qu.n.); *nombre quantique (n.qu.) principal*	n, n_i
orbital angular momentum qu.n.; *n.qu. de moment angulaire orbital*	L, l_i
spin qu.n.; *n. qu. de spin*	S, s_i
total angular momentum qu.n.; *n.qu. de moment angulaire total*	J, j_i
magnetic qu.n.; *n.qu. magnétique*	M, m_i
nuclear spin qu.n.; *n.qu. de spin nucléaire*	I, J †)
hyperfine qu.n.; *n.qu. hyperfin*	F
rotational qu.n.; *n.qu. de rotation*	J, K
vibrational qu.n.; *n.qu. de vibration*	v
quadrupole moment; *moment quadripolaire*	Q
Rydberg constant; *constante de Rydberg* ††)	R_∞
Bohr radius; *rayon de Bohr* ††)	a_0
fine structure constant; *constante de structure fine* ††)	α
mass excess; *excès de masse:* $m_a - Am_u$	Δ
packing fraction; *packing fraction:* Δ/Am_u	f
nuclear radius; *rayon nucléaire:* $R = r_0 A^{1/3}$	R

†) I is used in atomic physics, J in nuclear physics.
††) See for definition: Appendix I, section 3.

magnetic moment of particle; *moment magnétique d'une particule* | μ

magnetic moment of particle; *moment magnétique d'une particule*	μ
magnetic moment of proton; *moment magnétique du proton*	μ_p
magnetic moment of neutron; *moment magnétique du neutron*	μ_n
magnetic moment of electron; *moment magnétique électronique*	μ_e
Bohr magneton; *magnéton de Bohr* †)	μ_B
nuclear magneton; *magnéton nucléaire*	μ_N
g-factor; *facteur g:* e.g. $g = \mu/I\mu_N$	g
gyromagnetic ratio, gyromagnetic coefficient; *rapport gyromagnétique, coefficient gyromagnétique* †)	γ
Larmor circular frequency; *pulsation de Larmor* †)	ω_L
level width; *largeur d'un niveau*	Γ
mean life; *vie moyenne*	τ
reaction energy; *énergie de réaction*	Q
cross section; *section efficace*	σ
macroscopic cross section; *section efficace macroscopique:* $\Sigma = n\sigma$	Σ
impact parameter; *paramètre de collision*	b
scattering angle; *angle de diffusion*	ϑ, θ, φ
internal conversion coefficient; *coefficient de conversion interne*	α
disintegration energy; *énergie de désintégration*	Q
half life; *demi-vie*	$T_{1/2}$
reduced half life; *demi-vie reduite*	$fT_{1/2}$
decay constant, disintegration constant; *constante de désintégration*	λ
activity; *activité*	A
Compton wavelength; *longueur d'onde de Compton:* $\lambda_C = h/mc$	λ_C
electron radius; *rayon de l'electron* †)	r_e
linear attenuation coefficient; *coefficient d'atténuation linéique*	μ, μ_1
atomic attenuation coefficient; *coefficient d'atténuation atomique*	μ_a
mass attenuation coefficient; *coefficient d'atténuation massique*	μ_m
linear stopping power; *pouvoir d'arrêt linéaire*	S, S_1
atomic stopping power; *pouvoir d'arrêt atomique*	S_a
linear range; *distance de pénétration linéaire*	R, R_1
recombination coefficient; *coefficient de recombinaison*	α

7.10 Solid state physics

lattice vector: a translation vector which maps the crystal lattice on itself; *vecteur du réseau: vecteur qui reproduit par translation le réseau cristallin sur lui-même*	$\boldsymbol{R}$, $\boldsymbol{R}_0$
fundamental translation vectors for the crystal lattice; *vecteurs de base de la maille cristalline:* $\boldsymbol{R} = n_1\boldsymbol{a}_1 + n_2\boldsymbol{a}_2 + n_3\boldsymbol{a}_3$, where n_1, n_2, n_3 are integers	$\boldsymbol{a}_1$, $\boldsymbol{a}_2$, $\boldsymbol{a}_3$ $\boldsymbol{a}$, $\boldsymbol{b}$, $\boldsymbol{c}$
(circular) reciprocal lattice vector; *vecteur du réseau réciproque:* $\boldsymbol{G}\cdot\boldsymbol{R} = m\cdot 2\pi$, where m is an integer	$\boldsymbol{G}$
(circular) fundamental translation vectors for the reciprocal lattice; *vecteur de base de la maille du réseau réciproque:* $\boldsymbol{a}_i\cdot\boldsymbol{b}_k = 2\pi\delta_{ik}$ ††), where δ_{ik} is Kronecker delta symbol	$\boldsymbol{b}_1$, $\boldsymbol{b}_2$, $\boldsymbol{b}_3$ $\boldsymbol{a}^*$, $\boldsymbol{b}^*$, $\boldsymbol{c}^*$
lattice plane spacing; *espacement entre plans réticulaires*	d

†) See for definition: Appendix I, section 3.
††) In crystallography however $\boldsymbol{a}_i\cdot\boldsymbol{b}_k = \delta_{ik}$.

Bragg angle; *angle de Bragg*	ϑ
order of reflexion; *ordre de réflexion*	n
short range order parameter; *paramètre d'ordre à courte distance (local)*	σ
long range order parameter; *paramètre d'ordre à grande distance*	s
Burgers vector; *vecteur de Burgers*	$\boldsymbol{b}$
particle position vector; *vecteur de position d'une particule* *)	$\boldsymbol{r}, \boldsymbol{R}$
equilibrium position vector of ion; *vecteur de position d'équilibre d'un ion*	$\boldsymbol{R}_0$
displacement vector of ion; *vecteur de déplacement d'un ion*	$\boldsymbol{u}$
normal coordinates; *coordonnées normales*	Q_i
polarization vector; *vecteur de polarisation*	$\boldsymbol{e}$
Debye–Waller factor; *facteur de Debye–Waller*	D
Debye circular wave number; *nombre d'onde circulaire de Debye*	q_D
Debye circular frequency; *pulsation de Debye*	ω_D
Grueneisen parameter; *paramètre de Grüneisen:* $\gamma = \alpha V/\kappa C_V$ (α: cubic expansion coefficient; κ: compressibility)	γ, Γ
Madelung constant; *constante de Madelung*	α
mean free path of electrons; *libre parcours moyen des électrons*	l, l_e
mean free path of phonons; *libre parcours moyen des phonons*	Λ, l_{ph}
drift velocity; *vitesse de mouvement*	v_{dr}
mobility; *mobilité*	μ
one electron wave function; *fonction d'onde monoélectronique*	$\psi(\boldsymbol{r})$
Bloch wave function; *fonction d'onde de Bloch:* $\psi_{\boldsymbol{k}}(\boldsymbol{r}) = u_{\boldsymbol{k}}(\boldsymbol{r}) \exp(\mathrm{i}\boldsymbol{k} \cdot \boldsymbol{r})$	$u_{\boldsymbol{k}}(\boldsymbol{r})$
density of states; *densité (électronique) d'état:* $\mathrm{d}N(E)/\mathrm{d}E = N_E(E) = \rho(E)$	N_E, ρ
(spectral) density of vibrational modes; *densité spectrale de modes de vibration*	g, N_ω
exchange integral; *intégrale d'échange*	J
resistivity tensor; *tenseur de résistivité*	ρ_{ik}
electric conductivity tensor; *tenseur de conductibilité électrique*	σ_{ik}
thermal conductivity tensor; *tenseur de conductibilité thermique*	λ_{ik}
residual resistivity; *résistivité résiduelle*	ρ_R
relaxation time; *temps de relaxation*	τ
Lorenz coefficient; *coefficient de Lorenz:* $L = \lambda/\sigma T$	L
Hall coefficient; *coefficient de Hall*	A_H, R_H
Ettinghausen coefficient; *coefficient d'Ettinghausen*	A_E
first Ettinghausen–Nernst coefficient; *premier coefficient d'Ettinghausen–Nernst*	A_N
first Righi–Leduc coefficient; *premier coefficient de Righi–Leduc*	A_{RL}
thermoelectromotive force between substances a and b; *force thermoélectromotrice entre deux substances* a *et* b	E_{ab}
Seebeck coefficient for substances a and b; *coefficient de Seebeck pour deux substances* a *et* b*:* $S_{ab} = \mathrm{d}E_{ab}/\mathrm{d}T$	S_{ab}, ε_{ab}
Peltier coefficient for substances a and b; *coefficient de Peltier pour deux substances* a *et* b	Π_{ab}
Thomson coefficient; *coefficient de Thomson*	$\mu, (\tau)$

*) To distinguish between electron and ion position vectors, lower case and capital letters are used, respectively.

work function; *travail d'extraction* *): $\Phi = e\varphi$	φ
Richardson constant; *constante de Richardson:* $j = AT^2 \exp(-\Phi/kT)$	A
electron number density; *nombre volumique électronique (densité électronique)* **)	n, n_n, n_-
hole number density; *nombre volumique de trous (densité de trous)* **)	p, n_p, n_+
donor number density; *nombre volumique de donneurs (densité de donneurs)*	n_d
acceptor number density; *nombre volumique d'accepteurs (densité d'accepteurs)*	n_a
intrinsic number density; *nombre volumique intrinsèque (densité intrinsèque):* $n_i = (n \cdot p)^{1/2}$	n_i
energy gap: *bande d'énergie interdite (énergie gap)*	E_g
donor ionization energy; *énergie d'ionisation de donneur*	E_d
acceptor ionization energy; *énergie d'ionisation d'accepteur*	E_a
Fermi energy; *énergie de Fermi*	E_F, ε_F
circular wave vector, propagation vector (of particles); *vecteur d'onde, vecteur de propagation (de particules)*	$\boldsymbol{k}$
circular wave vector, propagation vector (of phonons); *vecteur d'onde, vecteur de propagation (de phonons)*	$\boldsymbol{q}$
Fermi circular wave vector; *vecteur de Fermi*	$\boldsymbol{k}_F$
electron annihilation operator; *opérateur d'annihilation d'électron*	a
electron creation operator; *opérateur de création d'électron*	$a^\dagger$
phonon annihilation operator; *opérateur d'annihilation de phonon*	b
phonon creation operator; *opérateur de création de phonon*	$b^\dagger$
effective mass; *masse effective* **)	m_n^*, m_p^*
mobility; *mobilité* **)	μ_n, μ_p
mobility ratio; *rapport de mobilité:* $b = \mu_n/\mu_p$	b
diffusion coefficient; *coefficient de diffusion* **)	D_n, D_p
diffusion length; *longueur de diffusion* **)	L_n, L_p
carrier life time; *durée de vie de porteur* **)	τ_n, τ_p
characteristic (Weiss) temperature; *température charactéristique (de Weiss)*	Θ, Θ_W
Curie temperature; *température de Curie*	T_C
Néel temperature; *température de Néel*	T_N
superconductor critical transition temperature; *température critique de transition supraconductrice*	T_c
superconductor (thermodynamic) critical field strength; *champ critique (thermodynamique) d'un supraconducteur*	H_c
superconductor critical field strengths (type II); *champ critique d'un supraconducteur (type II)* †)	H_{c1}, H_{c2}, H_{c3}
superconductor energy gap; *bande interdite du supraconducteur*	Δ
London penetration depth; *profondeur de pénétration de London*	λ_L
coherence length; *longueur de cohérence*	ξ
Landau–Ginzburg parameter; *paramètre de Landau–Ginzburg:* $\kappa = \lambda_L/\xi\sqrt{2}$	κ

*) The symbol W is used for the quantity $W = \Phi + \mu$, where μ is the electron chemical potential which, at $T = 0$, *is equal to the Fermi energy* E_F.

**) In general subscripts n and p or − and + may be used to denote electrons and holes, respectively.

†) H_{c1}: for magnetic flux entering the superconductor, H_{c2}: for disappearing of bulk superconductivity, H_{c3}: for disappearing of surface superconductivity.

flux (or fluxoid) quantum; *quantum de flux:* $\Phi_0 = h/2e$ **)	Φ_0
Miller indices; *indices de Miller*	h_1, h_2, h_3 h, k, l
single plane or set of parallel planes in lattice; *plan simple ou famille de plans réticulaires parallèles dans un réseau*	(h_1, h_2, h_3) (h, k, l)
full set of planes in lattice equivalent by symmetry; *famille de plans réticulaires équivalents par symétrie*	$\{h_1, h_2, h_3\}$ $\{h, k, l\}$
direction in lattice; *rangée réticulaire*	$[u, v, w]$
full set of directions in lattice equivalent by symmetry; *famille de rangées réticulaires équivalentes par symétrie*	$\langle u, v, w\rangle$

Note: If the letter symbols are replaced by numbers in the bracketed expressions, it is customary to omit the commas. A negative numerical value is commonly indicated by a bar above the number, e.g. $(\bar{1}10)$.

7.11 Molecular spectroscopy

quantum number (qu.n.) of component of electronic orbital angular momentum vector along symmetry axis; *nombre quantique (n.qu.) de la composante du moment angulaire orbital électronique suivant l'axe de symétrie*	Λ, λ_i
qu.n. of component of electronic spin along symmetry axis; *n.qu. de la composante du spin électronique suivant l'axe de symétrie*	Σ, σ_i
qu.n. of total electronic angular momentum vector along symmetry axis; *n.qu. du moment angulaire total électronique suivant l'axe de symétrie:* $\Omega = \lvert\Lambda + \Sigma\rvert$	Ω, ω_i
qu.n. of electronic spin; *n.qu. du spin électronique*	S
qu.n. of nuclear spin; *n.qu. du spin nucléaire*	I
qu.n. of vibrational mode; *n.qu. d'une mode de vibration*	v
degeneracy of vibrational mode; *degré de dégénérescance d'une mode de vibration*	d
qu.n. of vibrational angular momentum; *n.qu. du moment angulaire vibrationnel* (L.M.)	l
qu.n. of total angular momentum; *n.qu. du moment angulaire total* (excluding nuclear spin)	J
qu.n. of component of $\boldsymbol{J}$ in direction of external field; *n.qu. de la composante de* $\boldsymbol{J}$ *dans la direction du champ extérieur*	M, M_J
qu.n. of component of $\boldsymbol{S}$ in direction of external field; *n.qu. de la composante de* $\boldsymbol{S}$ *dans la direction du champ extérieur*	M_S
qu.n. of total angular momentum; *n.qu. du moment angulaire total* (including nuclear spin) $\boldsymbol{F} = \boldsymbol{J} + \boldsymbol{I}$	F
qu.n. of component of $\boldsymbol{F}$ in direction of external field; *n.qu. de la composante de* $\boldsymbol{F}$ *dans la direction du champ extérieur*	M_F
qu.n. of component of $\boldsymbol{I}$ in direction of external field; *n.qu. de la composante de* $\boldsymbol{I}$ *dans la direction du champ extérieur*	M_I
qu.n. of component of angular momentum along axis; *n.qu. de la composante du moment angulaire suivant l'axe* (L.M. and S.T.M.; excluding electron and nuclear spin; for L.M.: $(K = \lvert\Lambda + l\rvert)$	K

**) $2e/h = 1/\Phi_0$ is also called characteristic constant for macroscopic coherence in superconductors.

qu.n. of total angular momentum; *n.qu. du moment angulaire total* (L.M. and S.T.M.; excluding electron and nuclear spin: $\boldsymbol{J} = \boldsymbol{N} + \boldsymbol{S}$ *) — N

qu.n. of component of angular momentum along symmetry axis; *n.qu. de la composante du moment angulaire suivant l'axe de symétrie* (L.M. and S.T.M., excluding nuclear spin; for L.M.: $P = |K + \Sigma|$ **) — P

electronic term; *terme électronique:* $T_e = E_e/hc$ ***) — T_e

vibrational term; *terme de vibration:* $G = E_{vibr}/hc$ — G

coefficients in expression for vibrational term (for D.M.); *coefficients de l'expression d'un terme de vibration:* $G = \sigma_e(v + \frac{1}{2}) - x\sigma_e(v + \frac{1}{2})^2$ — σ_e, $x\sigma_e$

coefficients in expression for vibrational term (for P.M.); *coefficients de l'expression d'un terme de vibration:*

$$G = \Sigma\sigma_j(v_j + \tfrac{1}{2}d_j) + \tfrac{1}{2}\sum_j\sum_k x_{jk}(v_j + \tfrac{1}{2}d_j)(v_k + \tfrac{1}{2}d_k)$$ — σ_j, x_{jk}

rotational term; *terme de rotation:* $F = E_{rot}/hc$ — F

principal moments of inertia; *moments principaux d'inertie* $I_A \leq I_B \leq I_C$ †) — I_A, I_B, I_C

rotational constants; *constantes de rotation:* $A = h/8\pi^2cI_A$, etc. †) — A, B, C

total term; *terme total:* $T = T_e + G + F$ — T

Remark: L.M. = linear molecules. S.T.M. = symmetric top molecules. D.M. = diatomic molecules. P.M. = polyatomic molecules. See for further details: Report on Notation for the Spectra of Polyatomic Molecules (Joint Commission for Spectroscopy of I.U.P.A.P. and I.A.U. 1954) J. Chem. Phys. **23** (1955) 1997.

7.12 Chemical physics

relative atomic mass; *masse atomique relative* — A_r

relative molecular mass; *masse moléculaire relative* — M_r

amount of substance; *quantité de matière* — n, ν ††)

molar mass of substance B; *masse molaire de la substance* B — M_B

concentration of substance B; *concentration de la substance* B: $c_B = n_B/V$ — c_B

mole fraction of substance B; *fraction molaire de la substance* B — x_B

mass fraction of substance B; *titre en masse de la substance* B — w_B

volume fraction of substance B; *titre en volume de la substance* B — φ_B

mole ratio of solution; *rapport molaire d'une solution* — r

molality of solution; *molalité d'une solution* — m

chemical potential of substance B; *potentiel chimique de la substance* B †††) — μ_B

absolute activity of substance B (dimensionless); *activité absolue de la substance* B *(sans dimension):* $\lambda_B = \exp(\mu_B/kT)$ — λ_B

relative activity; *activité relative* — a_B

*) Case of loosely coupled electron spin.
**) Case of tightly coupled electron spin.
***) All energies are taken here with respect to the ground state as reference level.
†) For diatomic molecules use I and $A = h/8\pi^2cI$.
††) ν may be used as an alternative to n, when n is used for number density of particles.
†††) Referred to one particle.

activity coefficients; *coefficients d'activité*	γ_B, f_B
osmotic pressure; *pression osmotique*	Π
osmotic coefficient; *coefficient osmotique*	g, φ
stoichiometric number of substance B; *nombre stœchiométrique de la substance* B	ν_B
affinity; *affinité*	A
extent of reaction; *état d'avancement d'une réaction*	ξ
equilibrium constant; *constante d'équilibre*	K
charge number of ion; *nombre de charge d'un ion, électrovalence*	z
Faraday constant; *constante de Faraday*	F
ionic strength; *force ionique*	I
reduced activity of substance B; *activité réduite de la substance* B: $z_B = 2\pi mkT/h^2)^{3/2}\lambda_B$	z_B

7.13 Plasma physics

energy of particle; *énergie d'une particule*	ε
dissociation energy (e.g. of molecule X); *énergie de dissociation (par ex., d'une molécule* X)	E_d, $E_d(X)$
electron affinity; *affinité électronique*	E_{ea}
ionization energy; *énergie d'ionisation*	E_i
degree of ionization; *degré d'ionisation*	x
charge number of ion (positive or negative); *charge ionique (positif ou négatif)*	z
number density of ions of charge number z; *densité ionique des ions de charge* z †)	n_z
degree of ionization for charge number $z \geqslant 1$; *degré d'ionisation pour un nombre de charge* $z \geqslant 1$: $x_z = n_z/(n_z + n_{z-1})$	x_z
neutral particle temperature; *température des neutres*	T_n
ion temperature; *température ionique*	T_i
electron temperature; *température électronique*	T_e
electron number density; *densité électronique*	n_e
electron plasma circular frequency; *fréquence de plasma:* $\omega_{pe}^2 = n_e e^2/\varepsilon_0 m_e$	ω_{pe}
Debye length; *longueur de Debye*	λ_D
charge of particle; *charge d'une particule*	q
electron cyclotron circular frequency; *fréquence cyclotron électronique:* $\omega_{ce} = (e/m_e)B$	ω_{ce}
ion cyclotron circular frequency; *fréquence cyclotron ionique:* $\omega_{ci} = (ze/m_i)B$	ω_{ci}
reduced mass; *masse réduite:* $\mu = m_1 m_2/(m_1 + m_2)$	μ, m_r
impact parameter; *paramètre d'impact*	b
mean free path; *libre parcours moyen*	l, λ
collision frequency; *fréquence de collision*	ν_{coll}, ν_c
mean time interval between collisions; *intervalle de temps moyenne entre collisions:* $\tau_{coll} = 1/\nu_{coll}$	τ_{coll}, τ_c
cross section; *section efficace:* $\sigma = 1/ln$	σ
(electron) ionization efficiency; *efficacité d'ionisation (électronique):* $s_e = (\rho_0/\rho)\mathrm{d}N/\mathrm{d}x$ (dN: number of ion pairs formed by an ionizing electron travelling through dx in the plasma at gas density ρ; ρ_0: gas density at $p_0 = 1$ Torr, $T_0 = 273{,}15$ K)	s_e

†) When only singly charged ions need to be considered, n_{-1} and n_{+1} may be represented by n_- and n_+.

rate coefficient; *taux de réaction* — k

one-body rate coefficient; *taux de réaction unimoléculaire:* $-\mathrm{d}n_A/\mathrm{d}t = k_m n_A$ — k_m

relaxation time; *temps de relaxation:* (e.g. $\tau = 1/k_m$) — τ

two-body rate coefficient, binary rate coefficient; *taux de réaction binaire:* (e.g. $X + Y \rightarrow XY + h\nu$) $\mathrm{d}n_{XY}/\mathrm{d}t = k_b n_X n_Y$ — k_b

three-body rate coefficient, ternary rate coefficient; *taux de réaction ternaire:* (e.g. $X + Y + M \rightarrow XY + M^*$) $\mathrm{d}n_{XY}/\mathrm{d}t = k_t n_M n_X n_Y$ — k_t

Townsend (electron) ionization coefficient; *coefficient de Townsend* *) — α

Townsend (ion) ionization coefficient; *coefficient ionique de Townsend* — β

secondary electron emission coefficient; *taux d'émission secondaire* — γ

drift velocity; *vitesse de mouvement* — v_{dr}

mobility; *mobilité:* $\mu = v_{dr}/E$ — μ

positive or negative ion diffusion coefficient: *coefficient de diffusion des ions* — D_+, D_-

electron diffusion coefficient; *coefficient de diffusion des électrons* — D_e

ambipolar (ion–electron) diffusion coefficient; *coefficient de diffusion ambipolaire:* $D_a = (D_+\mu_e + D_e\mu_+)/(\mu_+ + \mu_e)$ — D_a, D_{amb}

characteristic diffusion length; *longueur caractéristique de diffusion* — L_D, Λ

ionization frequency; *fréquence d'ionisation* — ν_i

ion–ion recombination coefficient; *coefficient de recombinaison ion–ion:* $\mathrm{d}n_-/\mathrm{d}t = -\alpha_i n_- n_+$ — α_i

electron–ion recombination coefficient; *coefficient de recombinaison électron–ion:* $\mathrm{d}n_e/\mathrm{d}t = -\alpha_e n_e n_+$ — α_e

plasma pressure; *pression cinétique du plasma* — p

magnetic pressure; *pression magnétique:* $p_m = B^2/2\mu$ (μ: permeability) — p_m

magnetic pressure ratio; *coefficient β:* $\beta = p/p_m$ (p_m: magnetic pressure outside the plasma) — β

magnetic diffusivity; *diffusivité magnétique:* $\nu_m = 1/\mu\sigma$ (σ: electric conductivity; μ: permeability) — ν_m, η_m

Alfvén speed; *vitesse d'Alfvén:* $v_A = B/(\mu\rho)^{1/2}$ (ρ: (mass) density; μ: permeability) — v_A

7.14 Dimensionless parameters †)

1. Momentum transport

Reynolds number; *nombre de Reynolds:* $Re = vl/\nu$ — Re

Euler number; *nombre d'Euler:* $Eu = \Delta p/\rho v^2$ — Eu

Froude number; *nombre de Froude:* $Fr = v(lg)^{-1/2}$ — Fr

Grashof number; *nombre de Grashof:* $Gr = l^3 g\gamma\Delta T/\nu^2$ — Gr

Weber number; *nombre de Weber:* $We = \rho v^2 l/\sigma$ — We

Mach number; *nombre de Mach:* $Ma = v/c$ — Ma

Knudsen number; *nombre de Knudsen:* $Kn = \lambda/l$ — Kn

Strouhal number; *nombre de Strouhal:* $Sr = lf/v$ — Sr

where the symbols used in the definitions denote, respectively, l: a characteristic length; v: a characteristic velocity; ΔT: a characteristic temperature difference; Δp: pressure difference; ρ: (mass) density; η: viscosity; ν:

*) The same name is also used for $\eta = \alpha/E$, where E denotes electric field strength.

†) These symbols are those recommended in the International Standard I.S.O. 31 Part XII.

kinematic viscosity (η/ρ); σ: surface tension; g: acceleration of free fall; γ: cubic expansion coefficient ($-\rho^{-1}(\partial\rho/\partial T)_p$); λ: mean free path; f: a characteristic frequency; c: velocity of sound. $\Delta\rho/\rho$ equals $\gamma\Delta T$.

2. Transport of heat

Fourier number; *nombre de Fourier:* $Fo = a\Delta t/l^2$ — Fo

Péclet number; *nombre de Péclet:* $Pe = vl/a = Re \cdot Pr$ — Pe

Rayleigh number; *nombre de Rayleigh:*
$Ra = l^3 g\gamma\Delta T/\nu a = Gr \cdot Pr$ — Ra

Nusselt number; *nombre de Nusselt:* $Nu = hl/\lambda$ — Nu

Stanton number; *nombre de Stanton:* $St = h/\rho v c_p = Nu \cdot Pe^{-1}$ — St

where the symbols used in the definitions denote, respectively, l: a characteristic length; v: a characteristic velocity; Δt: a characteristic time interval; ΔT: a characteristic temperature difference; g: acceleration of free fall; ρ: (mass) density; η: viscosity; ν: kinematic viscosity (η/ρ); c_p: specific heat capacity at constant pressure; γ: cubic expansion coefficient ($-\rho^{-1}(\partial\rho/\partial T)_p$); λ: thermal conductivity; a: thermal diffusivity ($\lambda/\rho c_p$); h: coefficient of heat transfer (heat/(time $\times$ cross sectional area $\times$ temperature difference)).

3. Transport of matter in a binary mixture

Fourier number for mass transfer; *nombre de Fourier pour transfert de masse:* $Fo^* = D\Delta t/l^2 = Fo/Le$ — Fo^*

Péclet number for mass transfer; *nombre de Péclet pour transfert de masse:* $Pe^* \times vl/D = Re \cdot Sc = Pe \cdot Le$ — Pe^*

Grashof number for mass transfer; *nombre de Grashof pour transfert de masse:* $Gr^* = l^3 g\beta'\Delta x/\nu^2$ — Gr^*

Nusselt number for mass transfer; *nombre de Nusselt pour transfert de masse:* $Nu^* = kl/\rho D$ — Nu^*

Stanton number for mass transfer; *nombre de Stanton pour transfert de masse:* $St^* = k/\rho v = Nu^*/Pe^*$ — St^*

where the symbols used in the definitions denote, respectively, l: a characteristic length; v: a characteristic velocity; Δt: a characteristic time interval; Δx: a characteristic difference of mole fraction; g: acceleration of free fall; ρ: (mass) density; ν: kinematic viscosity (η/ρ); β': $\beta' = -\rho^{-1}(\partial\rho/\partial x)_{T,p}$; D: diffusion coefficient; k: mass transfer coefficient (mass/(time $\times$ cross sectional area $\times$ mole fraction difference)); γ: cubic expansion coefficient ($-\rho^{-1}(\partial\rho/\partial T)_p$). The quantity $-\Delta\rho/\rho$ equals $\gamma\Delta T + \beta'\Delta x$.

4. Dimensionless constants of matter

Prandtl number; *nombre de Prandtl:* $Pr = \nu/a$ — Pr

Schmidt number; *nombre de Schmidt:* $Sc = \nu/D$ — Sc

Lewis number; *nombre de Lewis:* $Le = a/D = Sc/Pr$ — Le

where the symbols used in the definitions denote, respectively, ρ: (mass) density; η: viscosity; ν: kinematic viscosity (η/ρ); D: diffusion coefficient; c_p: specific heat capacity at constant pressure; λ: thermal conductivity; a: thermal diffusivity ($\lambda/\rho c_p$).

5. Magnetohydrodynamics

Magnetic Reynolds number; *nombre de Reynolds magnétique:* $Rm = v\mu\sigma l$ — Rm

Alfvén number; *nombre d'Alfvén:* $Al = v/v_A$ — Al

Hartmann number; *nombre de Hartmann:* $Ha = Bl(\sigma/\rho\nu)^{1/2}$ — Ha

Cowling number (second Cowling number); *nombre de Cowling (deuxième nombre de Cowling):* $Co = B^2/\mu\rho v^2 = Al^{-2}$ — Co, Co_2

first Cowling number; *premier nombre de Cowling:*
$Co_1 = B^2 l\sigma/\rho v = Rm \cdot Co_2 = Ha^2/Re$ — Co_1

where the symbols used in the definitions denote, respectively, ρ: (mass) density; l: a characteristic length; v: a characteristic velocity; η: viscosity; ν: kinematic viscosity (η/ρ); μ: magnetic permeability; B: magnetic flux density; σ: electric conductivity; v_A: Alfvén speed ($B(\rho\mu)^{-1/2}$).

8 RECOMMENDED MATHEMATICAL SYMBOLS

8.1 General symbols

equal to; *égal à*	$=$
not equal to; *différent de*	$\neq$
identically equal to; *égal identiquement à*	$\equiv$
by definition equal to; *égal par définition à*	$\stackrel{\text{def}}{=}$
corresponds to; *correspond à*	$\triangleq$
approximately equal to; *égal environ à*	$\approx$
asymptotically equal to, *asymptotiquement égal à*	$\simeq$
proportional to; *proportionnel à*	$\sim$, $\propto$
approaches; *tend vers*	$\rightarrow$
greater than; *supérieur à*	$>$
less than; *inférieur à*	$<$
much greater than; *tres supérieur à*	$\gg$
much less than; *tres inférieur à*	$\ll$
greater than or equal to; *superiéur ou egál a*	$\geqslant, \geqq, \geq$
less than or equal to; *inférieur ou égal à*	$\leqslant, \leqq, \leq$
plus;*plus*	$+$
minus; *moins*	$-$
plus or minus; *plus ou moins*	$\pm$
a multiplied by *b*; *a multiplié par b*	$ab, a \cdot b, a \times b$
a divided by *b*; *a divisé par b*	$a/b, \frac{a}{b}, ab^{-1}$
ratio of the circumference of a circle to its diameter; *rapport de la circonférence d'un cercle à son diamètre*	π
a raised to the power *n*; *a puissance n*	a^n
magnitude of *a*; *valeur absolue de a*	$\|a\|$
square root of *a*; *racine carrée de a*	$\sqrt{a}, \surd a, a^{1/2}$
mean value of *a*; *valeur moyenne de a*	$\bar{a}, \langle a \rangle$
factorial *p*; *factorielle p*	$p!$
binomial coefficient; *coefficient binomial:* $n!/p!(n-p)!$	$\binom{n}{p}$
infinity; *infini*	∞

8.2 Letter symbols

Letter symbols and letter expressions for *mathematical operations* should be written in *roman* (i.e. *upright*) type.

exponential of *x*; *exponentielle de x*	$\exp x, \mathrm{e}^x$
base of natural logarithms; *base des logarithmes népériens*	e
logarithm to the base *a* of *x*; *logarithme de base a de x*	$\log_a x$
natural logarithm of *x*; *logarithme népérien de x*	$\ln x, \log_e x$
common logarithm of *x*; *logarithme décimal de x*	$\lg x, \log_{10} x$
binary logarithm of *x*; *logarithme binaire de x*	$\mathrm{lb}\, x, \log_2 x$
summation; *somme*	Σ
product; *produit*	Π
finite increase of *x*; *accroissement fini de x*	Δx †)

†) Greek capital delta, not triangle.

variation of x; *variation de x*	δx
total differential of x; *différentielle totale de x*	$\mathrm{d}x$
function of x; *fonction de x*	$f(x)$, $\mathrm{f}(x)$
composite function of f and g; *fonction composée de f et g*: $(g \circ f)(x) = g(f,x)$	$g \circ f$
limit of $f(x)$; *limite de* $f(x)$	$\lim_{x \to a} f(x)$, $\lim_{x\to a} f(x)$
derivative of f; *dérivée de f*	$\frac{\mathrm{d}f}{\mathrm{d}x}$, $\mathrm{d}f/\mathrm{d}x$, f'
partial derivative of f; *dérivée partielle de f*	$\frac{\partial f}{\partial x}$, $\partial f/\partial x$, $\partial_x f$, f_x
total differential of f; *différentielle totale de f*: $\mathrm{d}f(x, y) = (\partial f/\partial x)_y\,\mathrm{d}x + (\partial f/\partial y)_x\,\mathrm{d}y$	$\mathrm{d}f$
variation of f; variation de f	δf
Dirac delta function; *fonction delta de Dirac* $\delta(\boldsymbol{r}) = \delta(x)\delta(y)\delta(z)$	$\delta(x)$, $\delta(\boldsymbol{r})$
Kronecker delta symbol; *symbole delta de Kronecker*	δ_{ij}
Unit step function; *fonction unité:* $\varepsilon(t) = 1$ for $t > 0$, $\varepsilon(t) = 0$ for $t < 0$	$\varepsilon(t)$
signum a; *signum a*: sgn $a = a/\lvert a\rvert$ for $a \neq 0$	sgn a
greatest integer $\leqslant a$; *le plus grand entier* $\leqslant a$	ent a

8.3 Circular and hyperbolic functions

sine of x, *sinus x*	$\sin x$
cosine of x, *cosinus x*	$\cos x$
tangent of x, *tangente x*	$\tan x$, tg x
cotangent of x, *cotangente x*	$\cot x$, ctg x
secant of x, *sécante x*	sec x
cosecant of x, *cosécante x*	cosec x

It is recommended to use for the *inverse circular functions* the symbolic expressions for the corresponding circular function preceded by the letters: arc

Examples: arcsin x, arccos x, arctan x, etc.

It is recommended to use for the *hyperbolic functions* the symbolic expressions for the corresponding circular function, followed by the letter: h

Examples; sinh x, cosh x, tanh x, etc.

It is recommended to use for the *inverse hyperbolic functions* the symbolic expression for the corresponding hyperbolic function preceded by the letters: ar

Examples: arsinh x, arcosh x, etc.

8.4 Complex quantities

imaginary unit; *unité imaginaire* ($\mathrm{i}^2 = -1$)	i, j
real part of z; *partie réelle de z*	Re z, z'
imaginary part of z; *partie imaginaire de z*	Im z, z''
modulus of z; *module de z*	$\lvert z\rvert$
phase, argument of z; *phase, argument de z*: $z = \lvert z\rvert \exp \mathrm{i}\varphi$	arg z, φ
complex conjugate of z, conjugate of z; *complexe conjugué de z, conjugué de z*	z^*

Remark: Sometimes the notation $\bar{z}$ is used for the complex conjugate of z.

8.5 Symbols for special values of periodic quantities

Symbols for special values of periodic quantities are given in table 3.

Table 3

	case A*)	case B*)
instantaneous value; *valeur instantanée*	x	x
r.m.s. value**); *valeur efficace*	$\tilde{x}$	X
maximum value†) *valeur maximale*	$\hat{x}$	$\hat{x}$, $\hat{X}$
mean value††) *valeur moyenne*	$\bar{x}$, $\langle x \rangle$	$\bar{x}$, $\langle x \rangle$

*) A is the case in which only a lower case or only a capital letter may be used for the quantity. B is the case in which both, lower case and capital letters, may be used for the same quantity.

**) The r.m.s. value is here defined as

$$\tilde{x} = \left(T^{-1}\int_0^T [x(t)]^2 \mathrm{d}t\right)^{1/2};$$

the symbols x_{rms} and x_{eff} are also used.

†) The minimum value of x may be indicated by $x_{\min}$ or $\check{x}$.

††) The mean value is here defined as

$$\bar{x} = T^{-1}\int_0^T x(t)\mathrm{d}t.$$

8.6 Vector calculus)*

vector; *vecteur*	$\boldsymbol{a}$, $\boldsymbol{A}$
absolute value; *valeur absolue*	$\lvert\boldsymbol{A}\rvert$, A
unit vector; *vecteur unitaire:* $\boldsymbol{a}/\lvert\boldsymbol{a}\rvert$	$\boldsymbol{e}_a$
unit vectors; *vecteurs unitaires*	$\boldsymbol{e}_x$, $\boldsymbol{e}_y$, $\boldsymbol{e}_z$, $\boldsymbol{i}$, $\boldsymbol{j}$, $\boldsymbol{k}$ †)
scalar product of $\boldsymbol{a}$ and $\boldsymbol{b}$; *produit scalaire de* $\boldsymbol{a}$ *et* $\boldsymbol{b}$	$\boldsymbol{a}\cdot\boldsymbol{b}$
vector product of $\boldsymbol{a}$ and $\boldsymbol{b}$; *produit vectoriel de* $\boldsymbol{a}$ *et* $\boldsymbol{b}$	$\boldsymbol{a}\times\boldsymbol{b}$ $\quad\boldsymbol{a}\wedge\boldsymbol{b}$
dyadic product of $\mathbf{a}$ and $\boldsymbol{b}$; *produit dyadique de* $\boldsymbol{a}$ *et* $\boldsymbol{b}$	$\boldsymbol{ab}$
differential vector operator, nabla; *opérateur vectoriel, nabla*	$\partial/\partial\boldsymbol{r}$, $\boldsymbol{\nabla}$
gradient; *gradient*	grad φ, $\boldsymbol{\nabla}\varphi$
divergence; *divergence*	div $\boldsymbol{A}$, $\boldsymbol{\nabla}\cdot\boldsymbol{A}$
curl; *rotationnel*	curl $\boldsymbol{A}$, rot $\boldsymbol{A}$, $\boldsymbol{\nabla}\times\boldsymbol{A}$
Laplacian; *Laplacien*	$\triangle\varphi$, $\boldsymbol{\nabla}^2\varphi$
Dalembertian; *Dalembertien*	$\square\varphi$
second order tensor; *tenseur du second ordre*	$\boldsymbol{\mathsf{A}}$

*) See also sections 1.2.3 and 1.3.2.
†) $\mathbf{1}_x$, $\mathbf{1}_y$, $\mathbf{1}_z$ are also used.

scalar product of tensors **S** and **T**; *produit scalaire des tenseurs **S** et **T**:* $(\Sigma_{i,k}S_{ik}T_{ki})$	$\mathbf{S}:\mathbf{T}$
tensor product of tensors **S** and **T**; *produit tensoriel des tenseurs **S** et **T**:* $(\Sigma_k S_{ik}T_{kl})$	$\mathbf{S}\cdot\mathbf{T}$
product of tensor **S** and vector **A**; *produit du tenseur **S** et vecteur **A**:* $(\Sigma_k S_{ik}A_k)$	$\mathbf{S}\cdot\mathbf{A}$

8.7 Matrix calculus

matrix; *matrice*	A $\begin{pmatrix} a_{11} & \ldots & a_{1n} \\ \vdots & & \vdots \\ a_{m1} & \ldots & a_{mn} \end{pmatrix}$
product of A and B: *produit de A et B*	AB
inverse of A: *inverse de A*	A^{-1}
unit matrix; *matrice unité*	E, I
transpose of matrix A: *matrice transposée de A:* $\tilde{A}_{ij} = A_{ji}$	$\tilde{A}$
complex conjugate of A; *matrice complexe conjugueé de A:* $A^*_{ij} = (A_{ij})^*$	A^*
Hermitian conjugate of A; *matrice conjuguée Hermitienne de A:* $A^\dagger_{ij} = A^*_{ji}$	$A^\dagger$
determinant of A; *déterminant de A*	$\det A$
trace of A; *trace de A*	$\mathrm{Tr}\, A$
Pauli matrices; *matrices de Pauli:*	$\boldsymbol{\sigma}$,
$\sigma_x = \begin{pmatrix} 0 & 1 \\ 1 & 0 \end{pmatrix} \quad \sigma_y = \begin{pmatrix} 0 & -\mathrm{i} \\ \mathrm{i} & 0 \end{pmatrix} \quad \sigma_z = \begin{pmatrix} 1 & 0 \\ 0 & -1 \end{pmatrix}$	$\sigma_x, \sigma_y, \sigma_z$ $\sigma_1, \sigma_2, \sigma_3$
unit matrix; *matrice unité:* $I = \begin{pmatrix} 1 & 0 \\ 0 & 1 \end{pmatrix}$	I
Dirac (4 × 4) matrices; (4 × 4) *matrices de Dirac*†):	
$\alpha_x = \begin{pmatrix} 0 & \sigma_x \\ \sigma_x & 0 \end{pmatrix} \quad \alpha_y = \begin{pmatrix} 0 & \sigma_y \\ \sigma_y & 0 \end{pmatrix} \quad \alpha_z = \begin{pmatrix} 0 & \sigma_z \\ \sigma_z & 0 \end{pmatrix}$	$\boldsymbol{\alpha}$ $\alpha_x, \alpha_y, \alpha_z$
$\beta = \begin{pmatrix} I & 0 \\ 0 & -I \end{pmatrix}$	β

8.8 Theory of sets

is an element of; *est un élément de:* $x \in A$	$\in$
is not an element of, *n'est pas un élément de:* $x \notin A$	$\notin$
contains as element; *contient comme élément:* $A \ni x$	$\ni$
set of elements; *ensemble des éléments*	$\{a_1, a_2, \ldots\}$
the set of integers; *ensemble des nombres entiers*	Z
the set of rational numbers; *ensemble des nombres rationnels*	Q
the set of real numbers; *ensemble des nombres réels*	R
the set of complex numbers; *ensemble des nombres complexes*	C
set of elements of A for which $p(x)$ is true; *ensemble des éléments de A pour lequels $p(x)$ est vrai*	$\{x \in A \mid p(x)\}$
is contained as subset in; *est contenu comme sous-ensemble dans:* $B \subseteq A$	$\subseteq (\subset)$
contains; *contient:* $A \supseteq B$	$\supseteq (\supset)$
is properly contained in; *est strictement contenu dans*	$\subset (\subsetneq)$

†) Sometimes a different representation is used.

contains properly; *contient strictement* $\supset$ ($\supsetneq$)

union; *réunion:* $A \cup B = \{x|x \in A \text{ or } x \in B\}$ $\cup$
intersection; *intersection:*
$A \cap B = \{x|x \in A \text{ and } x \in B\}$ $\cap$
difference; *difference:* $A \backslash B = \{x|x \in A \text{ and } x \notin B\}$ $\backslash$
complement of; *complément de:* $\complement A = \{x|x \notin A\}$ $\complement$

8.9 Symbolic logic

conjunction: $p \wedge q$ means "p and q"; *conjonction:* $p \wedge q$ *signifie* "p *et* q" $\wedge$
disjunction: $p \vee q$ means "p or q or both"; *disjonction:* $p \vee q$ *signifie* "p *ou* q *ou les deux*" $\vee$
negation; *négation* $\neg$
implication; *implication* $\Rightarrow$
equivalence, bi-implication; *équivalence, bi-implication* $\Leftrightarrow$
universal quantifier; *quantificateur universel* $\forall$
existential quantifier; *quantificateur existential* $\exists$

9. INTERNATIONAL SYMBOLS FOR UNITS

9.1 Unit systems

In a system of physical quantities and equations between them, a certain number of quantities, which *by convention* are regarded as dimensionally independent, form a set of *base quantities* for the whole system. All other quantities can be defined as *derived quantities* in terms of the base quantities and can be expressed in terms of powers of the base quantities by algebraic relations.

A *coherent system of units* is a system, based on a certain set of base units well defined in terms of actual physical phenomena, in which all *derived units* are expressed as products of powers of the base units by algebraic relations, analogous to the corresponding quantities, dropping numerical factors.

The expression of a quantity as a product of powers of the base quantities (neglecting their vectorial or tensorial character and all numerical factors including their sign) is called the *dimensional product* or the *dimension* of the quantity with respect to the chosen set of base quantities or base dimensions. The exponents of the powers, to which the various base quantities or base dimensions are raised, are called the *dimensional exponents.*

Physical quantities, which have as their dimension a product of powers of the base quantities or base dimensions with all dimensional exponents equal to zero, are called *dimensionless quantities* or *quantities of the dimension 1.*

The number of the base units of the unit system, coherent with respect to the system of quantities and equations in question, is equal to that of the corresponding set of base quantities. The base units themselves are well defined samples of the base quantities: base quantity and corresponding base unit are of the same dimension.

A derived unit, i.e. the unit for a derived quantity of the coherent unit system, is formed as a product of powers of the base units, with the same exponents as those in the product of powers of the base quantities in the algebraic expression for the derived quantity in question, and without introducing numerical factors. Derived units and their symbols are expressed algebraically in terms of base units and their symbols, respectively, by means of the mathematical signs for multiplication and division. Several of these algebraic expressions for derived units in terms of base units can be replaced by special names and their symbols which can themselves be used to form other derived units and their symbols, respectively (see sections 9.2 and 9.3).

The values of dimensionless quantities, e.g. relative density, relative permeability or refractive index, are expressed by pure numbers. The corresponding unit, which is the ratio of a unit to itself, is the *dimensionless unit* of the coherent unit system and may be expressed by the number 1.

9.2 The International System of Units (SI)

The name *International System of Units* with the international abbreviation SI has been adopted by the Conférence Générale des Poids et Mesures (CGPM) in 1960. It is based on the seven base units (CGPM 1960 and 1971) listed in table 4.

Table 4. SI base units

base quantity; *grandeur de base*	SI base unit; *unité de base SI*	
	name; *nom*	symbol; *symbole*
length; *longueur*	metre; *mètre*	m
mass; *masse*	kilogram; *kilogramme*	kg
time; *temps*	second; *seconde*	s
electric current; *courant électrique*	ampere; *ampère*	A
thermodynamic temperature; *température thermodynamique*	kelvin; *kelvin*	K
amount of substance; *quantité de matière*	mole; *mole*	mol
luminous intensity; *intensité lumineuse*	candela; *candela*	cd

The SI base units have been defined by the CGPM in the following way:

(1) metre; *mètre*

The metre is the SI base unit of length and was defined at the October 1984 General Conference of Weights and Measures as the length of the path traveled by light in vacuum during a time interval of $^1/299{,}792{,}458$ *of a second.*

(2) kilogram; *kilogramme*

The kilogram is the unit of mass; it is equal to the mass of the international prototype of the kilogram. (1st CGPM (1889) and 3rd CGPM (1901), Declarations.)

(3) second; *seconde*

The second is the duration of 9 192 631 770 periods of the radiation corresponding to the transition between the two hyperfine levels of the ground state of the caesium-133 atom. (13th CGPM (1967), Resolution 1.)

(4) ampere; *ampère*

The ampere is that constant current which, if maintained in two straight parallel conductors of infinite length, of negligible circular cross section, and placed 1 metre apart in vacuum, would produce between these conductors a force equal to 2×10^{-7} *newton per metre of length.* (9th CGPM (1948), Resolutions 2 and 7.)

(5) kelvin; *kelvin*

The kelvin, unit of thermodynamic temperature, is the fraction 1/273,16 of the thermodynamic temperature of the triple point of water. (13th CGPM (1967), Resolution 4.)

The 13th CGPM (1967, Resolution 3) also decided that the unit kelvin and its symbol K should be used to express an interval or a difference of temperature. *Note:* In addition to the thermodynamic temperature (symbol T), expressed in kelvins, use is also made of Celsius temperature (symbol t) defined by the equation

$$t = T - T_0,$$

where $T_0 = 273{,}15$ K by definition. Celsius temperature is expressed in degree Celsius; *degré Celsius* (symbol °C). The unit "degree Celsius" is equal to the unit "kelvin" and an interval or a difference of Celsius temperature may also be expressed in degrees Celsius.

(6) mole; *mole*

The mole is the amount of substance of a system which contains as many elementary entities as there are atoms in 0.012 kilogram of carbon 12.
When the mole is used, the elementary entities must be specified and may be atoms, molecules, ions, electrons, other particles, or specified groups of such particles.
The mole is a base unit of the International System of Units. (14th CGPM (1971), Resolution 3.)

(7) candela; *candela*

The candela is the luminous intensity in a given direction of a source which emits monochromatic radiation of frequency 540×10^{12} *Hz and of which the radiant intensity in that direction is 1/683 W/steradian. From the 16th CGPM, Resolution 3, 1979.*

The coherent units of the system based on these base units are called: ***SI units***. ***Derived SI units having special names and symbols*** are listed in table 5.

Table 5. Derived SI units with special names

quantity; *grandeur*	derived SI unit; *unité SI dérivée*			
	name; *nom*	symbol; *symbole*	expression in terms of base units; *expression en unités de base*	expression in terms of other SI units; *expression en d'autres unités SI*
plane angle; *angle plan*	radian; *radian*	rad	$m \cdot m^{-1}$	
solid angle; *angle solide*	steradian; *stéradian*	sr	$m^2 \cdot m^{-2}$	
frequency; *fréquence*	hertz; *hertz*	Hz	s^{-1}	
force; *force*	newton; *newton*	N	$m \cdot kg \cdot s^{-2}$	J/m
pressure; *pression*	pascal; *pascal*	Pa	$m^{-1} \cdot kg \cdot s^{-2}$	N/m^2
energy, work, quantity of heat; *énergie, travail, quantité de chaleur*	joule; *joule*	J	$m^2 \cdot kg \cdot s^{-2}$	N · m
power, radiant flux; *puissance, flux énergétique*	watt; *watt*	W	$m^2 \cdot kg \cdot s^{-3}$	J/s
quantity of electricity, electric charge; *quantité d'électricité, charge électrique*	coulomb; *coulomb*	C	$s \cdot A$	A · s
electric potential, potential difference, electromotive force; *tension électrique, différence de potentiel, force électromotrice*	volt; *volt*	V	$m^2 \cdot kg \cdot s^{-3} \cdot A^{-1}$	W/A
capacitance; *capacité électrique*	farad; *farad*	F	$m^{-2} \cdot kg^{-1} \cdot s^4 \cdot A^2$	C/V
electric resistance; *résistance électrique*	ohm; *ohm*	Ω	$m^2 \cdot kg \cdot s^{-3} \cdot A^{-2}$	V/A
conductance; *conductance*	siemens; *siemens*	S	$m^{-2} \cdot kg^{-1} \cdot s^3 \cdot A^2$	A/V
magnetic flux; *flux d'induction magnétique*	weber; *weber*	Wb	$m^2 \cdot kg \cdot s^{-2} \cdot A^{-1}$	V · s
magnetic flux density; *induction magnétique*	tesla; *tesla*	T	$kg \cdot s^{-2} \cdot A^{-1}$	Wb/m^2
inductance; *inductance*	henry; *henry*	H	$m^2 \cdot kg \cdot s^{-2} \cdot A^{-2}$	Wb/A
luminous flux; *flux lumineux*	lumen; *lumen*	lm		cd · sr
illuminance; *éclairement lumineux*	lux; *lux*	lx		$m^{-2} \cdot cd \cdot sr$

Table 5 (Continued)

	derived SI unit; *unité SI dérivée*			
quantity; *grandeur*	name; *nom*	symbol; *symbole*	expression in terms of base units; *expression en unités de base*	expression in terms of other SI units; *expression en d'autres unités SI*
activity; *activité*	becquerel; *becquerel*	Bq	s^{-1}	
absorbed dose; *dose absorbée*	gray; *gray*	Gy	$m^2 \cdot s^{-2}$	J/kg

Although it might be thought that SI units can only be base units or derived units, the 11th Conférence Générale des Poids et Mesures (1960) admitted a third class of SI units called *supplementary units*, for which it declined to state, whether they were base units or derived units. To the class of supplementary units belong only the radian as SI unit for plane angle and the steradian as SI unit for solid angle; they may be regarded either as base units or as derived units.

In physics, radian and steradian are, in general, regarded as derived units (see table 5). In some special fields, e.g. in electromagnetic radiation and light or in particle scattering problems, the steradian is sometimes *formally* treated as a base unit; then the symbol sr must *not* be replaced by the number 1.

The usage of the SI units and their decimal multiples and sub-multiples formed by using the prefixes given in table 1 (see section 2.2.1) has been especially recommended.

Remark: Among the SI base units the unit of mass is the only unit whose name, for historical reasons, contains a prefix. Names of decimal multiples and sub-multiples of the SI unit of mass are formed by attaching prefixes to the word "gram" (Comité International des Poids et Mesures (CIPM) 1967, Recommendation 2).

Several *sub-systems* of the SI are used in different fields of science and technology. The two most important are:

(a) *The metre-kilogram-second system (MKS system)*
The MKS system is a coherent unit system for mechanics, based on the three base units metre, kilogram and second for the three base quantities length, mass and time.

(b) *The metre-kilogram-second-ampere system (MKSA system)*
The MKSA system is a coherent unit system for mechanics, electricity and magnetism, based on the four base units metre, kilogram, second and ampere for the four base quantities length, mass, time and electric current. This system has been given the name "*Giorgi system*" by the International Electrotechnical Commission in 1935.

9.3 The centimetre-gram-second system (CGS system)

The CGS system is a coherent system of units based on the three base units listed in table 6.

Table 6. CGS base units

base quantity; *grandeur de base*	base unit; *unité de base*	
	name; *nom*	symbol; *symbole*
length; *longueur*	centimetre; *centimètre*	cm
mass; *masse*	gram; *gramme*	g
time; *temps*	second; *seconde*	s

Derived CGS units having special names and symbols are listed in table 7.

Table 7. Derived CGS units with special name.

quantity; *grandeur*	derived CGS unit; *unité CGS dérivée*		
	name; *nom*	symbol; *symbole*	expression in terms of base units; *expression en unités de base*
force; *force*	dyne	dyn	$cm \cdot g \cdot s^{-2}$
energy; *énergie*	erg	erg	$cm^2 \cdot g \cdot s^{-2}$
viscosity; *viscosité*	poise	P	$cm^{-1} \cdot g \cdot s^{-1}$
kinematic viscosity; *viscosité cinématique*	stokes	St	$cm^2 \cdot s^{-1}$
acceleration of free fall; *accélération de la pesanteur*	gal	Gal	$cm \cdot s^{-2}$

The CGS system enlarged by the kelvin (K) as unit of thermodynamic temperature (see section 9.2) and the mole (mol) as unit of amount of substance (see section 9.2) or by the candela (cd) as unit of luminous intensity (see section 9.2) is also used in *mechanics including thermodynamics* or *photometry*, respectively.

Derived units (from cm, g, s and cd, including sr, see section 9.2), in the field of *photometry, having special names and symbols* are listed in table 8.

Table 8. CGS units in photometry with special name.

quantity; *grandeur*	derived unit; *unité dérivée*		
	name; *nom*	symbol; *symbole*	expression; *expression*
luminance; *luminance*	stilb	sb	$cm^{-2} \cdot cd$
illuminance; *éclairement lumineux*	phot	ph	$cm^{-2} \cdot cd \cdot sr$

ALPHABET TABLE

Greek letter	Greek name	English equivalent	RUSSIAN letter	English equivalent
Α α	Alpha	(ä)	А а	(ä)
Β β	Beta	(b)	Б б	(b)
Γ γ	Gamma	(g)	В в	(v)
Δ δ	Delta	(d)	Г г	(g)
Ε ϵ	Epsilon	(e)	Д д	(d)
Ζ ζ	Zeta	(z)	Е е	(ye)
Η η	Eta	(ā)	Ж ж	(zh)
Θ θ	Theta	(th)	З з	(z)
Ι ι	Iota	(ē)	И и	(i, ē)
Κ κ	Kappa	(k)	Й й	(ē) 7
Λ λ	Lambda	(l)	К к	(k)
Μ μ	Mu	(m)	Л л	(l)
Ν ν	Nu	(n)	М м	(m)
Ξ ξ	Xi	(ks)	Н н	(n)
Ο ο	Omicron	(o)	О о	(ô, o)
Π π	Pi	(p)	П п	(p)
Ρ ρ	Rho	(r)	Р р	(r)
Σ σ ς	Sigma	(s)	С с	(s)
Τ τ	Tau	(t)	Т т	(t)
Υ υ	Upsilon	(ü, o͞o)	У у	(o͞o)
Φ φ	Phi	(f)	Ф ф	(f)
Χ χ	Chi	(H)	Х х	(kh)
Ψ ψ	Psi	(ps)	Ц ц	(ts)
Ω ω	Omega	(ō)	Ч ч	(ch)
			Ш ш	(sh)
			Щ щ	(shch)
			Ъ ъ	8
			Ы ы	(ē)
			Ь ь	9
			Э э	(e)
			Ю ю	(ū)
			Я я	(yä)

AMERICAN STANDARD ABBREVIATIONS FOR SCIENTIFIC AND ENGINEERING TERMS

Reproduced by permission of The American Society of Mechanical Engineers

Introductory Notes

Scope and Purpose

1. The Executive Committee of the Sectional Committee on Scientific and Engineering Symbols and Abbreviations has made the following distinction between symbols and abbreviations: Letter symbols are letters used to represent magnitudes or physical quantities in equations and mathematical formulas. Abbreviations are shortened forms of names or expressions employed in texts and tabulations, and should not be used in equations.

Fundamental Rules

2. Abbreviations should be used sparingly in text and with due regard to the context and to the training of the reader. Terms denoting units of measurement should be abbreviated in the text only when preceded by the amounts indicated in numerals; thus "several inches," "one inch," "12 in." In tabular matter, specifications, maps, drawings, and texts for special purposes, the use of abbreviations should be governed only by the desirability of conserving space.
3. Short words such as ton, day, and mile should be spelled out.
4. Abbreviations should not be used where the meaning will not be clear. In case of doubt, spell out.
5. The same abbreviation is used for both singular and plural, as "bbl" for barrel and barrels.
6. The use of conventional signs for abbreviations in text is not recommended; thus "per," not /; "lb," not #; "in," not ". Such signs may be used sparingly in tables and similar places for conserving space.
7. The period should be omitted except in cases where the omission would result in confusion.
8. The letters of such abbreviations as ASA should not be spaced (not A S A).
9. The use in text of exponents for the abbreviations of square and cube and of the negative exponents for terms involving "per" is not recommended. The superior figures are usually not available on the keyboards of typesetting and linotype machines and composition is therefore delayed. There is also the likelihood of confusion with footnote reference numbers. These shorter forms are permissible in tables and are sometimes difficult to avoid in text.
10. A sentence should not begin with a numeral followed by an abbreviation. Abbreviations for names of units are to be used only after numerical values, such as 25 ft or 110 v.

Abbreviations*

absolute	abs
acre	spell out
acre-foot	acre-ft
air horsepower	air hp
alternating-current (as adjective)	a-c
ampere	amp
ampere-hour	amp-hr
amplitude, an elliptic function	am.
Angstrom unit	A
antilogarithm	antilog
atmosphere	atm
atomic weight	at. wt
average	avg
avoirdupois	avdp
azimuth	az or α
barometer	bar
barrel	bbl
Baumé	Bé
board feet (feet board measure)	fbm
boiler pressure	spell out
boiling point	bp
brake horsepower	bhp
brake horsepower-hour	bhp-hr
Brinell hardness number	Bhn
British thermal unit[1]	Btu or B
bushel	bu
calorie	cal
candle	c
candle-hour	c-hr
candlepower	cp
cent	c or ¢
center to center	c to c
centigram	cg
centiliter	cl
centimeter	cm
centimeter-gram-second (system)	cgs
chemical	chem
chemically pure	cp
circular	cir
circular mils	cir mils
coefficient	coef
cologarithm	colog
concentrate	conc
conductivity	cond
constant	const
continental horsepower	cont hp
cord	cd
cosecant	csc
cosine	cos
cosine of the amplitude, an elliptic function	cn
cost, insurance, and freight	cif
contangent	cot
coulomb	spell out
counter electromotive force	cemf
cubic	cu
cubic centimeter cu cm, cm^3 (liquid, meaning milliliter. ml)	
cubic foot	cu ft
cubic feet per minute	cfm
cubic feet per second	cfs
cubic inch	cu in.
cubic meter	cu m or m^3
cubic micron	cu μ or cu mu or μ^3
cubic millimeter	cu mm or mm^3
cubic yard	cu yd
current density	spell out
cycles per second	spell out or c
cylinder	cyl
day	spell out
decibel	db
degree[1]	deg or °
degree centigrade	C
degree Fahrenheit	F
degree Kelvin	K
degree Réaumur	R
delta amplitude, an elliptic function	dn
diameter	diam
direct-current (as adjective)	d-c
dollar	$
dozen	doz
dram	dr
efficiency	eff
electric	elec
electromotive force	emf
elevation	el
equation	eq

* These forms are recommended for readers whose familiarity with the terms used makes possible a maximum of abbreviations. For other classes of readers editors may wish to use less contracted combinations made up from this list. For example, the list gives the abbreviation of the term "feet per second" as 'fps". To some readers ft per sec will be more easily understood. [1]Abbreviation recommended by the A.S.M.E. Power Test Codes Committee. B = 1 Btu, kB = 1000 Btu, mB = 1,000,000 Btu. The A.S.H. & V.E. recommends the use of Mb = 1000 Btu and Mbh = 1000 Btu per hr.

[1] There are circumstances under which one or the other of these forms is preferred. In general the sign ° is used where space conditions make it necessary, as in tabular matter, and when abbreviations are cumbersome, as in some angular measurements, i.e., 59° 23' 42". In the interest of simplicity and clarity the Committee has recommended that the abbreviation for the temperature scale, F, C, K, etc., always be included in expressions for numerical temperatures, but, wherever feasible, the abbreviation for "degree" be omitted; as 69 F.

AMERICAN STANDARD ABBREVIATIONS (Continued)

external	ext
farad	spell out or f
feet board measure (board feet)	fbm
feet per minute	fpm
feet per second	fps
fluid	fl
foot	ft
foot-candle	ft-c
foot-Lambert	ft-L
foot-pound	ft-lb
foot-pound-second (system)	fps
foot-second (see cubic feet per second)	
franc	fr
free aboard ship	spell out
free alongside ship	spell out
free on board	fob
freezing point	fp
frequency	spell out
fusion point	fnp
gallon	gal
gallons per minute	gpm
gallons per second	gps
grain	spell out
gram	g
gram-calorie	g-cal
greatest common divisor	gcd
haversine	hav
hectare	ha
henry	H
high-pressure (adjective)	h-p
hogshead	hhd
horsepower	hp
horsepower-hour	hp-hr
hour	hr
hour (in astronomical tables)	h
hundred	C
hundredweight (112 lb)	cwt
hyperbolic cosine	cosh
hyperbolic sine	sinh
hyperbolic tangent	tanh
inch	in.
inch-pound	in-lb
inches per second	ips
indicated horsepower	ihp
indicated horsepower-hour	ihp-hr
inside diameter	ID
intermediate-pressure (adjective)	i-p
internal	int
joule	J
kilocalorie	kcal
kilocycles per second	kc
kilogram	kg
kilogram-calorie	kg-cal
kilogram-meter	kg-m
kilograms per cubic meter or kg/m^3	kg per cu m
kilograms per second	kgps
kiloliter	kl
kilometer	km
kilometers per second	kmps
kilovolt	kv
kilovolt-ampere	kva
kilowatt	kw
kilowatthour	kwhr
lambert	L
latitude	lat or ϕ
least common multiple	lcm
linear foot	lin ft
liquid	liq
lira	spell out
liter	l
logarithm (common)	log
logarithm (natural)	log or ln
longitude	long. or λ
low-pressure (as adjective)	l-p
lumen	l*
lumen-hour	l-hr*
lumens per watt	lpw
mass	spell out
mathematics (ical)	math
maximum	max
mean effective pressure	mep
mean horizontal candlepower	mhcp
megacycle	spell out
megohm	spell out
melting point	mp
meter	m
meter-kilogram	m-kg
mho	spell out
microampere	μa or mu a
microfarad	μf
microinch	μin.
micromicrofarad	μμf
micromicron	μμ or mu mu
micron	μ or mu
microvolt	μv
microwatt	μw or mu w
mile	spell out
miles per hour	mph
miles per hour per second	mphps
milliampere	ma
milligram	mg
millihenry	mh
millilambert	mL
milliliter	ml
millimeter	mm
millimicron	mμ or m mu
million	spell out
million gallons per day	mgd
millivolt	mv
minimum	min
minute	min
minute (angular measure)	′
minute (time) (in astronomical tables)	m
mole	spell out
molecular weight	mol. wt
month	spell out
National Electrical Code	NEC
ohm	spell out or Ω
ohm-centimeter	ohm-cm
ounce	oz
ounce-foot	oz-ft
ounce-inch	oz-in
outside diameter	OD
parts per million	ppm
peck	pk
penny (pence — New British)	p.
pennyweight	dwt
per	(see Fundamental Rules)
peso	spell out
pint	pt
potential	spell out
potential difference	spell out
pound	lb
pound-foot	lb-ft
pound-inch	lb-in.
pound sterling	£
pounds per brake horsepower-hour	lb per bhp-hr
pounds per cubic foot	lb per cu ft
pounds per square foot	psf
pounds per square inch	psi
pounds per square inch absolute	psia
power factor	spell out or pf
quart	qt
radian	spell out
reactive kilovolt-ampere	kvar
reactive volt-ampere	var
revolutions per minute	rpm
revolutions per second	rps
rod	spell out
root mean square	rms
secant	sec
second	sec
second (angular measure)	″
second-foot (see cubic feet per second)	
second (time) (in astronomical tables)	s
shaft horsepower	shp
shilling	s
sine	sin
sine of the amplitude, an elliptic function	sn
specific gravity	sp gr
specific heat	sp ht
spherical candle power	scp
square	sq
square centimeter	sq cm or cm^2
square foot	sq ft
square inch	sq in.
square kilometer	sq km or km^2
square meter	sq m or m^2
square micron	sq μ or sq mu or μ^2

* The International Commission on Illumination has changed the symbol for lumen to lm, and the symbol for lumen-hour to lm-hr. This nomenclature is used in American Standard for Illuminating Engineering Nomenclature and Photometric Standards (ASA Z7.1-1942).

AMERICAN STANDARD ABBREVIATIONS (Continued)

Term	Abbreviation
square millimeter	sq mm or mm^2
square root of mean square	rms
standard	std
stere	s
tangent	tan
temperature	temp
tensile strength	ts
thousand	M
thousand foot-pounds	kip-ft
thousand pound	kip
ton	spell out
ton-mile	spell out
versed sine	vers
volt	v
volt-ampere	va
volt-coulomb	spell out
watt	w
watthour	whr
watts per candle	wpc
week	spell out
weight	wt
yard	yd
year	yr

ABBREVIATIONS OF COMMON UNITS OF WEIGHT AND MEASURE

From NBS Miscellaneous Publication No. 233.
Spelling and Abbreviations of Units

The spelling of the names of units as adopted by the National Bureau of Standards is that given in the list below. The spelling of the metric units is in accordance with that given in the law of July 28, 1866, legalizing the metric system in the United States.

Following the name of each unit in the list below is given the abbreviation which the Bureau has adopted. Attention is particularly called to the following principles:

1. The period is omitted after all abbreviations of units, except where the abbreviation forms an English word.
2. The exponents "^{2}" and "^{3}" are used to signify "square" and "cubic", respectively, instead of the abbreviations "sq" or "cu," which are, however, frequently used in technical literature for the United States Customary units. In conformity with this principle the abbreviation for cubic centimeter is "cm^3" (instead of "cc" or "c cm"). The term "cubic centimeter," as used in chemical work, is, in fact, a misnomer, since the unit actually used is the "milliliter" of which "ml" is the correct abbreviation.
3. The use of the same abbreviation for both singular and plural is recommended. This practice is already established in expressing metric units and is in accordance with the spirit and chief purpose of abbreviations.
4. It is also suggested that, unless all the text is printed in capital letters, only small letters be used for abbreviations, except in such case as, A for angstrom, etc., where the use of capital letters is general.

LIST OF THE MOST COMMON UNITS OF WEIGHT AND MEASURE AND THEIR ABBREVIATIONS

Unit	Abbreviation	Unit	Abbreviation	Unit	Abbreviation
acre	acre	dram, avoirdupois	dr avdp	ounce, apothecaries	oz ap or ℥
angstrom	A	dram, fluid	fl dr	ounce, avoirdupois	oz avdp
are	a	fathom	fath	ounce, fluid	fl oz
avoirdupois	avdp	foot	ft	ounce, troy	oz t
barrel	bbl	furlong	fur.	peck	pk
board foot	fbm	gallon	gal	pennyweight	dwt
bushel	bu	grain	grain	pint	pt
carat	c	gram	g	pound	lb
centare	ca	hectare	ha	pound, apothecaries	lb ap
centigram	cg	hectogram	hg	pound, avoirdupois	lb avdp
centiliter	cl	hectoliter	hl	pound, troy	lb t
centimeter	cm	hectometer	hm	quart	qt
chain	ch	hogshead	hhd	rod	rd
cubic centimeter	cm^3	hundredweight	cwt	scruple, apothecaries	s ap or ℈
cubic decimeter	dm^3	inch	in.	square centimeter	cm^2
cubic dekameter	dkm^3	kilogram	kg	square chain	ch^2
cubic foot	ft^3	kiloliter	kl	square decimeter	dm^2
cubic hectometer	hm^3	kilometer	km	square dekameter	dkm^2
cubic inch	in.^{3}	link	li	square foot	ft^2
cubic kilometer	km^3	liquid	liq	square hectometer	hm^2
cubic meter	m^3	liter	liter	square inch	in.^{2}
cubic mile	mi^3	meter	m	square kilometer	km^2
cubic millimeter	mm^3	metric ton	t	square link	li^2
cubic yard	yd^3	microgram*	μg	square meter	m^2
decigram	dg	microinch	μin.	square mile	mi^2
deciliter	dl	microliter*	μl	square millimeter	mm^2
decimeter	dm	micron	μ	square rod	rd^2
decistere	ds	mile	mi	square yard	yd^2
dekagram	dkg	milligram	mg	stere	s
dekaliter	dkl	milliliter	ml	ton	ton
dekameter	dkm	millimeter	mm	ton, metric	t
dekastere	dks	millimicron	mμ	troy	t
dram	dr	minim	min or ♏	yard	yd
dram, apothecaries	dr ap or ʒ	ounce	oz		

* The abbreviations γ and λ for microgram and microliter, respectively have been advocated by some authorities.

CONVERSION FACTORS

L. P. Buseth

To convert from	To	Multiply by
Abampere	Ampere	*10*
Abcoulomb	Coulomb	*10*
Abfarad	Farad	1×10^{9}
Abhenry	Henry	1×10^{-9}
Abmho	Siemens (mho)	1×10^{9}
Abohm	Ohm	1×10^{-9}
Abvolt	Volt	1×10^{-8}
Acre	Hectare	0.40468564
	Square foot	*43560*
	Square kilometer	4.046856×10^{-3}
	Square meter	4046.85642
	Square mile	1.5625×10^{-3} *(1/640)*
	Square yard	*4840*
Acre (U.S. Survey)	Square meter	4046.872610
Acre-foot	Cubic meter	1233.482
	Cubic yard	1613.333
Acre-inch	Cubic foot	*3630*
	Cubic meter	102.7902
	Gallon (Brit.)	22610.67
	Gallon (U.S.)	27154.29
Ampere (int., mean)	Ampere	0.99985
Ampere (int., U.S.)	Ampere	0.999835
Ampere/square centimeter	Ampere/square inch	*6.4516*
Ampere/square inch	Ampere/square centimeter	0.1550003
Ampere-hour	Coulomb	*3600*
Ampere(-turn)	Gilbert	1.256637
Ångström	Nanometer	*0.1*
Apostilb	Candela/square meter	0.3183099 ($1/\pi$)
Are	Square foot	1076.391
	Square meter	*100*
Astronomical unit	Kilometer	1.4959787×10^{8}
Atmosphere	Atmosphere (tech.)	1.033227
	Bar	*1.01325*
	Foot of H_2O (conv.)	33.89854
	Inch of Hg (conv.)	29.92126
	Kilogram-force/square centimeter	1.033227
	Kilopascal	*101.325*
	Meter of H_2O (conv.)	10.33227
	Millibar	*1013.25*
	Millimeter of Hg (conv.)	760
	Newton/square centimeter	*10.1325*
	Pascal (N/square meter)	1.01325×10^{5}
	Pound-force/square foot	2116.22
	Pound-force/square inch	14.69595
	Ton-force (long)/square foot	0.944740
	Ton-force (short)/square foot	1.058108
	Ton-force (long)/square inch	6.56069×10^{-3}
	Ton-force (short)/square inch	7.34797×10^{-3}
	Torr	*760*
Atmosphere (tech.)	Atmosphere	0.967841
	Bar	*0.980665*
	Foot of H_2O (conv.)	32.8084
	Inch of Hg (conv.)	28.9590
	Kilogram-force/square centimeter	*1*
	Kilopascal	*98.0665*
	Meter of H_2O (conv.)	*10*
	Millibar	*980.665*
	Millimeter of Hg (conv.)	735.559
	Newton/square centimeter	9.80665
	Pascal (N/m^2)	98066.5
	Pound-force/square inch	14.22334
Bag (Brit.)	Gallon (Brit.)	*24*
Bar	Atmosphere	0.9869233
	Atmosphere (tech.)	1.019716
	Dyne/square centimeter	1×10^{6}
	Foot of H_2O (conv.)	33.4553
	Inch of Hg (conv.)	29.5300
	Kilogram-force/square centimeter	1.019716
	Kilopascal	*100*
	Meter of H_2O (conv.)	10.19716
	Millibar	*1000*
	Millimeter of Hg (conv.)	750.062
	Newton/square centimeter	*10*
	Pascal (N/m^2)	1×10^{5}
	Pound-force/square foot	2088.54
	Pound-force/square inch	14.50377
	Ton-force (long)/square foot	0.932385
	Ton-force (short)/square foot	1.04427
	Ton-force (long)/square inch	6.47490×10^{-3}
	Ton-force (short)/square inch	7.25189×10^{-3}
	Torr	750.062
Barleycorn (Brit.)	Inch	0.333333 *(1/3)*
Barn	Square meter	1×10^{-28}
Barrel (Brit., beer)	Gallon (Brit.)	*36*
	Liter	163.6592
Barrel (Brit., wine)	Gallon (Brit.)	*31.5*
	Liter	143.2018
Barrel (petroleum)	Cubic foot	5.614583
	Cubic meter	0.1589873
	Gallon (Brit.)	34.97232
	Gallon (U.S.)	*42*
	Liter	158.9873
Barrel (U.S., dry)	Bushel (U.S.)	3.281219
	Cubic foot	4.083333
	Cubic inch	*7056*
	Cubic meter	0.1156271
	Liter	115.6271
	Pint (U.S., dry)	209.998
	Quart (U.S., dry)	104.9990
Barrel (U.S., cranb.)	Cubic inch	*5826*
	Liter	95.4710
Barrel (U.S., liquid)	Cubic foot	*4.2109375*
	Cubic inch	*7276.5*
	Cubic meter	0.1192405
	Gallon (Brit.)	26.22925
	Gallon (U.S.)	*31.5*
	Liter	119.2405
Barye	Bar	1×10^{-6}
	Dyne/square centimeter	*1*
Becquerel	Curie	2.702703×10^{11}
Biot	Ampere	*10*
Board foot	Cubic foot	0.083333 *(1/12)*
Bolt (cloth)	Foot	*120*
Btu	Calorie	251.996
	Cubic foot-atmosphere	*0.367717*
	Foot-poundal	25036.9
	Foot-pound-force	778.169
	Horsepower-hour	3.93015×10^{-4}
	Horsepower-hour (metric)	3.98466×10^{-4}
	Joule	1055.056
	Kilocalorie	0.251996
	Kilogram-force-meter	107.586
	Kilowatt-hour	2.93071×10^{-4}
	Liter-atmosphere	10.4126
	Watt-hour	0.293071
Btu (39 °F, 4 °C)	Joule	1059.67
Btu (60 °F, 15.6 °C)	Joule	1054.68
Btu (mean)	Joule	1055.87
Btu (thermochemical)	Joule	1054.350
Btu/cubic foot	Joule/cubic meter	37258.9

CONVERSION FACTORS (continued)

To convert from	To	Multiply by
	kilocalorie/cubic meter	8.89915
Btu/°F	Calorie/°C	453.592
	Joule/°C	1899.10
Btu/hour	Btu/minute	0.0166667 (*1/60*)
	Btu/second	2.77778 × 10^{-4}
	Calorie/second	0.0699988
	Foot-pound-force/second	0.216158
	Horsepower	3.93015 × 10^{-4}
	Watt	0.293071
Btu/(hour × square foot)	Watt/square meter	3.15459
Btu/(hour × square foot × °F)	Calorie/second × square meter × °C)	1.35623
	Watt/(square meter × °C)	5.67826
Btu/(hour × square foot × °F/foot)	Watt/(meter × °C)	1.73073
Btu/(hour × square foot × °F/inch)	Watt/(meter × °C)	0.144228
Btu/minute	Calorie/second	4.19993
	Horsepower	0.0235809
	Watt	17.5843
Btu/(minute × square foot)	Watt/square meter	189.273
Btu/pound	Calorie/gram	0.555556
	Joule/kilogram	*2326*
	Kilocalorie/kilogram	0.555556
	Watt-hour/kilogram	0.646111
Btu/(pound × °F)	Calorie/(gram × °C)	*1*
	Joule/(kilogram × °C)	*4186.8*
Btu/second	Horsepower	1.41485
	Kilowatt	1.055056
Btu/(second × square foot)	Kilowatt/square meter	11.3565
Btu/(second × square foot × °F)	Kilowatt/(square meter × °C)	20.4417
Btu/(second × square foot × °F/foot)	Kilowatt/(meter × °C)	6.23064
Btu/(second × square foot × °F /inch)	Watt/(meter × °C)	519.220
Btu/square foot	Joule/square meter	11356.5
	watt-hour/square meter	3.15459
Bucket (Brit.)	Gallon (Brit.)	*4*
Bushel (Brit.)	Bushel (U.S.)	1.032057
	Gallon (Brit.)	*8*
	Liter	*36.36872*
Bushel (U.S.)	Barrel (U.S., dry)	0.3047647
	Bushel (Brit.)	0.9689390
	Cubic foot	1.244456
	Cubic inch	*2150.42*
	Gallon (Brit.)	7.751512
	Gallon (U.S., liquid)	9.309177
	Liter	35.23907
	Peck (U.S.)	*4*
	Pint (U.S., dry)	*64*
	Quart (U.S., dry)	*32*
Butt (Brit.)	Gallon (Brit.)	*108* or *126*
Cable length (int.)	Foot	607.6115
	Meter	*185.2*
	Mile (nautical)	*0.1*
Cable length (U.S.)	Foot	*720*
	Meter	*219.456*
	Mile (nautical)	0.1184968
	Mile (statute)	0.1363636
Caliber	Inch	*0.01*
	Millimeter	*0.254*
Calorie	Btu	3.96832 × 10^{-3}
	Cubic foot-atmosphere	1.45922 × 10^{-3}
	Foot-poundal	99.3543
	Foot-pound-force	3.08803
	Horsepower-hour	1.55961 × 10^{-6}
	Horsepower-hour (metric)	1.58124 × 10^{-6}
	Joule	*4.1868*
	Kilocalorie	*0.001*
	Kilogram-force-meter	0.426935
	Kilowatt-hour	*1.163* × 10^{-6}
	Liter-atmosphere	0.0413205
	Watt-hour	*1.163* × 10^{-3}
Calorie (15°C)	Joule	4.1855
Calorie (20°C)	Joule	4.18190
Calorie (mean)	Joule	4.19002
Calorie (thermochem.)	Joule	*4.184*
Calorie/°C	Btu/°F	2.20462 × 10^{-3}
	Joule/°F	*2.326*
Calorie/gram	Btu/pound	*1.8*
	Joule/kilogram	*4186.8*
Calorie/(gram × °C)	Btu/(pound × °F)	*1*
	Joule/(kilogram × °C)	*4186.8*
Calorie/minute	Watt	*0.06978*
Calorie/(minute × square centimeter)	Watt/square meter	*697.8*
Calorie/second	Watt	*4.1868*
Calorie/(second × square centimeter)	Kilowatt/square meter	*41.868*
Calorie/(second × square centimeter × °C)	Kilowatt/(square meter × °C)	*41.868*
Calorie/(second × square centimeter × °C/centimeter)	Watt/(meter × °C)	*418.68*
Calorie/square centimter	Kilojoule/square meter	41.868
Candela	Hefner unit	1.11
	Lumen/steradian	*1*
Candela/square centimeter	Candela/square foot	*929.0304*
	Candela/square inch	*6.4516*
	Lambert	3.141593 (π)
Candela/square foot	Candela/square inch	6.944444 × 10^{-3} (*1/144*)
	Candela/square meter	10.76391
	Foot-lambert	3.141593 (π)
	Lambert	3.381582 × 10^{-3}
Candela/square inch	Candela/square centimeter	0.1550003
	Candela/square foot	*144*
	Foot-lambert	452.3893
	Lambert	0.4869478
Candela/square meter	Candela/square foot	*0.09290304*
	Lambert	3.141593 × 10^{-4}
Carat (metric)	Gram	*0.2*
Cental	Kilogram	*45.359237*
	Pound	*100*
°C heat unit (chu)	Btu	*1.8*
	Calorie	453.592
	Joule	1899.10
Centiliter	Cubic centimeter	*10*
	Cubic inch	0.6102374
	Drachm (Brit., fliud)	2.815606
	Dram (U.S., fluid)	2.705122

To convert from	To	Multiply by
	Ounce (Brit., fluid)	0.3519508
	Ounce (U.S., fluid)	0.3381402
Centimeter	Foot	0.03280840
	Inch	0.3937008
	Micrometer	*10000*
	Mil	393.7008
	Millimeter	*10*
	Yard	0.01093613
Centimeter of Hg (conv.)	Atmosphere	0.0131579
	Millibar	13.3322
	Millimeter of H_2O (conv.)	*135.951*
	Pascal	1333.22
	Pound-force/square inch	0.193368
Centimeter of H_2O (conv.)	Atmosphere	9.67841×10^{-4}
	Millibar	*0.980665*
	Millimeter of Hg (conv.)	0.735559
	Kilogram-force/square centimeer	*0.001*
	Pascal	*98.0665*
	Pound-force/square inch	0.0142233
Centimeter/second	Foot/minute	1.968504
	Foot/second	0.03280840
	Kilometer/hour	*0.036*
	Meter/minute	*0.6*
	Mile/hour	0.02236936
Centimeter/square second	Foot/square second	0.03280840
	Kilometer/(hour × second)	*0.036*
	Meter/square second	*0.01*
	Mile/(hour × second)	0.02236936
Centipoise	Pascal-second	*0.001*
Centistokes	Square meter/second	*1×10^{-6}*
Chain (Gunter's)	Foot	*66*
Chain (Ramsden's)	Foot	*100*
Circular inch	Circular mil	*1×10^{6}*
	Square centimeter	5.067075
	Square inch	0.7853982
Circular mil	Square inch	7.853982×10^{-7}
	Square micrometer	506.7075
	Square mil	0.7853982
Circular millimeter	Square millimeter	0.7853982
Circumference	Degree	*360*
	Gon (grade)	*400*
	Radian	6.283185 (2π)
Clo	(°C × square meter)/watt	0.155
Cord	Cord-foot	*8*
	Cubic foot	*128*
Cord-foot	Cord	*0.125 (1/8)*
	Cubic foot	*16*
Coulomb	Ampere-second	*1*
Cubic centimeter	Cubic foot	3.531467×10^{-5}
	Cubic inch	0.06102374
	Cubic meter	*1×10^{-6}*
	Cubic millimeter	*1000*
	Cubic yard	1.307951×10^{-6}
	Drachm (Brit., fluid)	0.2815606
	Dram (U.S., fluid)	0.2705122
	Gallon (Brit.)	2.199692×10^{-4}
	Gallon (U.S.)	2.641721×10^{-4}
	Gill (Brit.)	7.039016×10^{-3}
	Gill (U.S.)	8.453506×10^{-3}
	Liter	*0.001*
	Milliliter	*1*
	Minim (Brit.)	16.89364
	Minim (U.S.)	16.23073
	Ounce (Brit., fluid)	0.03519508
	Ounce (U.S., fluid)	0.03381402
	Pint (Brit.)	1.759754×10^{-3}
	Pint (U.S., dry)	1.816166×10^{-3}
	Pint (U.S., liquid)	2.113376×10^{-3}
	Quart (Brit.)	8.798770×10^{-4}
	Quart (U.S., dry)	9.080830×10^{-4}
	Quart (U.S., liquid)	1.056688×10^{-3}
Cubic centimeter/gram	Cubic foot/pound	0.0160185
Cubic centimeter/second	Cubic foot/minute	2.118880×10^{-3}
	Liter/hour	*3.6*
Cubic centimeter-atmosphere	Joule	*0.101325*
	Watt-hour	2.814583×10^{-5}
Cubic decimeter	Cubic centimeter	*1000*
	Cubic foot	0.03531467
	Cubic inch	61.02374
	Cubic meter	*0.001*
	Liter	*1*
Cubic foot	Acre-foot	2.295684×10^{-5}
	Board foot	*12*
	Bushel (Brit.)	0.7786044
	Bushel (U.S.)	0.8035640
	Cord	*7.8125×10^{-3} (1/128)*
	Cord-foot	*0.0625 (1/16)*
	Cubic centimeter	28316.847
	Cubic inch	*1728*
	Cubic meter	0.028316847
	Cubic yard	0.03703704 (*1/27*)
	Gallon (Brit.)	6.228835
	Gallon (U.S.)	7.480519
	Liter	28.316847
	Pint (Brit.)	49.83068
	Pint (U.S., dry)	51.42809
	Pint (U.S., liquid)	59.84416
	Quart (Brit.)	24.91534
	Quart (U.S., dry)	25.71405
	Quart (U.S., liquid)	29.92208
Cubic foot/hour	Cubic centimeter/second	7.865791
	Liter/minute	0.4719474
Cubic foot/minute	Cubic centimeter/second	471.9474
	Gallon (Brit.)/second	0.1038139
	Gallon (U.S.)/second	0.1246753
Cubic foot/pound	Cubic meter/kilogram	0.06242796
Cubic foot/second	Cubic meter/hour	101.9406
	Cubic yard/minute	2.222222
	Gallon (Brit.)/minute	373.7301
	Gallon (U.S.)/minute	448.8312
	Liter/minute	1699.011
Cubic foot-atmosphere	Btu	2.71948
	Calorie	685.298
	Foot-pound-force	2116.22
	Joule	2869.205
	Kilogram-force-meter	292.577
	Liter-atmosphere	28.31685
	Watt-hour	0.7970012
Cubic foot (pound-force/square inch)	Btu	0.185050
	Calorie	46.6317
	Joule	195.238
	Watt-hour	0.0542327
Cubic inch	Board foot	6.944444×10^{-3} (*1/144*)
	Bushel (Brit.)	4.505813×10^{-4}
	Bushel (U.S.)	4.650254×10^{-4}
	Cubic centimeter	*16.387064*
	Cubic foot	5.787037×10^{-4} (*1/1728*)
	Cubic meter	*1.6387064×10^{-5}*
	Cubic yard	2.143347×10^{-5}
	Drachm (Brit., fluid)	4.613952
	Dram (U.S., fluid)	4.432900
	Gallon (Brit.)	3.604650×10^{-3}
	Gallon (U.S.)	4.329004×10^{-3} (*1/231*)
	Liter	*0.016387064*
	Milliliter	*16.387064*
	Ounce (Brit., fluid)	0.5767440

To convert from	To	Multiply by
	Ounce (U.S., fluid)	0.5541126
	Pint (Brit.)	0.02883720
	Pint (U.S., dry)	0.02976163
	Pint (U.S., liquid)	0.03463203
	Quart (Brit.)	0.01441860
	Quart (U.S., dry)	0.01488081
	Quart (U.S., liquid)	0.01731602
Cubic inch/minute	Cubic centimeter/second	0.2731177
Cubic kilometer	Cubic mile	0.2399128
Cubic meter	Barrel (petroleum)	6.289811
	Barrel (U.S., dry)	8.648490
	Barrel (U.S., liquid)	8.386414
	Bushel (U.S.)	28.37759
	Cubic centimeter	1×10^{6}
	Cubic decimeter	*1000*
	Cubic foot	35.31467
	Cubic inch	61023.74
	Cubic yard	1.307951
	Gallon (Brit.)	219.9692
	Gallon (U.S.)	264.1721
	Liter	*1000*
	Pint (Brit.)	1759.754
	Pint (U.S., dry)	1816.166
	Pint (U.S., liquid)	2113.376
	Quart (Brit.)	879.8770
	Quart (U.S., dry)	908.0830
	Quart (U.S., liquid)	1056.688
	Register ton	0.3531467
Cubic meter/kilogram	Cubic foot/pound	16.01846
Cubic mile	Cubic kilometer	4.168182
Cubic millimeter	Cubic centimeter	*0.001*
	Cubic inch	6.102374×10^{-5}
	Minim (Brit.)	0.01689364
	Minim (U.S.)	0.01623073
Cubic yard	Bushel (Brit.)	21.02232
	Bushel (U.S.)	21.69623
	Cubic foot	*27*
	Cubic inch	*46656*
	Cubic meter	0.76455486
	Gallon (Brit.)	168.1786
	Gallon (U.S.)	201.9740
	Liter	764.5549
Cubic yard/minute	Cubic foot/second	*0.45*
	Gallon (Brit.)/second	2.802976
	Gallon (U.S.)/second	3.366234
	Liter/second	12.74258
Cubit	Inch	*18*
Cup (metric)	Milliliter	*200*
Cup (U.S.)	Milliliter	236.588
	Ounce (U.S. fluid)	*8*
Curie	Becquerel	3.7×10^{10}
Darcy	Square meter	9.869233×10^{-13}
Day (mean solar)	Hour	*24*
	Minute	*1440*
	Second	*86400*
Day (sidereal)	Second	86164.09
Decibel	Neper	0.115129255
Degree (angular)	Circumference	2.777778×10^{-3} *(1/360)*
	Gon (grade)	1.111111
	Minute (angular)	*60*
	Quadrant	0.01111111 *(1/90)*
	Radian	0.01745329
	Second (angular)	*3600*
Degree/foot	Radian/meter	0.05726146
Degree/inch	Radian/meter	0.6871375
Degree/second	revolution/minute	0.1666667 *(1/6)*
°C (temp. interval)	°Fahrenheit	*1.8*
	°Rankine	*1.8*
	Kelvin	*1*
(°C × hour)/kilocalorie	°C/watt	0.859845
(°C × hour × square meter)/kilocalorie	(°C × square meter)/watt	0.859845
°F (temp. interval)	°Celsius	0.5555556 *(5/9)*
	°Rankine	*1*

To convert from	To	Multiply by
	Kelvin	0.5555556 *(5/9)*
(°F × hour)/Btu	°C/watt	1.89563
(°F × hour × square foot)/Btu	(°C × square meter)/watt	0.176110
(°F/inch × hour × square foot)/Btu	(°C × meter)/watt	6.93347
Denier	Tex	0.111111 *(1/9)*
Drachm (Brit. fluid)	Dram (U.S., fluid)	0.9607599
	Milliliter	3.551633
	Minim (Brit.)	*60*
	Ounce (Brit. fluid)	*0.125 (1/8)*
Dram (apoth. or troy)	Dram (avoirdupois)	2.1942857
	Grain	*60*
	Gram	*3.8879346*
	Ounce (apoth. or troy)	*0.125 (1/8)*
	Pennyweight	*2.5*
	Scruple	*3*
Dram (avoirdupois)	Grain	*27.34375*
	Gram	1.7718452
	Ounce (avoirdupois)	*0.0625 (1/16)*
Dram (U.S., fluid)	Cubic centimeter	3.696691
	Cubic inch	0.2255859
	Drachm (Brit. fluid)	1.040843
	Gallon (U.S.)	9.765625×10^{-4} *(1/1024)*
	Gill (U.S.)	*0.03125 (1/32)*
	Milliliter	3.696691
	Minim (U.S.)	*60*
	Ounce (U.S., fluid)	*0.125 (1/8)*
	Pint (U.S., liquid)	7.8125×10^{-3} *(1/128)*
	Quart (U.S., liquid)	3.90625×10^{-3} *(1/256)*
Dyne	Kilogram-force	1.019716×10^{-6}
	Newton	1×10^{-5}
	Poundal	7.233014×10^{-5}
	Pound-force	2.248089×10^{-6}
Dyne/centimeter	Newton/meter	*0.001*
Dyne/square centimeter	Bar	1×10^{-6}
	Kilogram-force/square centimeter	1.019716×10^{-6}
	Millimeter of Hg (conv.)	7.50062×10^{-4}
	Millimeter of H_2O (conv.)	0.01019716
	Pascal (N/square meter)	*0.1*
	Pound-force/square inch	1.450377×10^{-5}
Dyne-centimeter	Erg	*1*
	Foot-poundal	2.37304×10^{-6}
	Foot-pound-force	7.37562×10^{-8}
	Joule	1×10^{-7}
	Kilogram-force-meter	1.019716×10^{-8}
	Newton-meter	1×10^{-7}
Dyne-second/square centimeter	Poise	*1*
	Pascal-second	*0.1*
Electronvolt	Erg	1.60219×10^{-12}
	Joule	1.60219×10^{-19}
Ell	Inch	*45*
Erg	Dyne-centimeter	*1*
	Joule	1×10^{-7}
	Watt-hour	2.777778×10^{-11}
Erg/(square centimeter × second)	Watt/square meter	*0.001*
Farad (int. mean)	Farad	0.999510
Farad (int. U.S.)	Farad	0.999505
Fathom	Foot	*6*
Fermi	Meter	1×10^{-15}
Firkin (Brit.)	Gallon (Brit.)	*9*
Firkin (U.S.)	Gallon (U.S.)	*9*
Foot	Centimeter	*30.48*
	Foot (U.S. Survey)	*0.999998*
	Inch	*12*
	Meter	0.3048
	Millimeter	304.8
	Mile (nautical)	1.645788×10^{-4}
	Mile (statute)	1.893939×10^{-4}

To convert from	To	Multiply by
	Yard	0.333333 (*1/3*)
Foot (U.S. Survey)	Foot	1.000002
	Meter	0.30480060960
Foot of H_2O (conv.)	Atmosphere	0.0294998
	Bar	0.0298907
	Inch of Hg (conv.)	0.882671
	Kilogram-force/square centimeter	*0.03048*
	Millimeter of Hg (conv.)	22.4198
	Pascal (N/square meter)	2989.07
	Pound-force/square inch	0.433527
Foot/°F	Meter/°C	*0.54864*
Foot/hour	Meter/second	8.466667×10^{-5}
Foot/minute	Kilometer/hour	*0.018288*
	Knot	9.87473×10^{-3}
	Meter/second	5.08×10^{-3}
	Mile/hour	0.01136364 (*1/88*)
Foot/second	Kilometer/hour	*1.09728*
	Knot	0.5924838
	Meter/minute	*18.288*
	Meter/second	*0.3048*
	Mile/hour	0.6818182
Foot/square second	Kilometer/(hour × second)	*1.09728*
	Meter/square second	*0.3048*
	Mile/(hour × second)	0.6818182
Foot to the fourth power	Meter to the fourth power	8.630975×10^{-3}
Foot-candle	Lumen/square foot	*1*
	Lumen/square meter	10.76391
	Lux	10.76391
Foot-lambert	Candela/square centimeter	3.426259×10^{-4}
	Candela/square foot	0.3183099 ($1/\pi$)
	Candela/square meter	3.426259
	Lambert	1.076391×10^{-3}
	Meter-lambert	10.76391
Foot-poundal	Btu	3.99411×10^{-5}
	Calorie	0.0100650
	Foot-pound-force	0.0310810
	Joule	0.0421401
	Kilogram-force-meter	4.29710×10^{-3}
	Liter-atmosphere	4.15891×10^{-4}
	Watt-hour	1.17056×10^{-5}
Foot-pound-force	Btu	1.28507×10^{-3}
	Calorie	0.323832
	Cubic foot-atmosphere	4.72541×10^{-4}
	Foot-poundal	32.1740
	Horsepower-hour	5.05051×10^{-7}
	Horsepower-hour (metric)	5.12055×10^{-7}
	Joule	1.35582
	Kilogram-force-meter	0.138255
	Liter-atmosphere	0.0133809
	Newton-meter	1.35582
	Watt-hour	3.76616×10^{-4}
Foot-pound-force/hour	Watt	3.76616×10^{-4}
Foot-pound-force/minute	Horsepower	3.03030×10^{-5}
	Horsepower (metric)	3.07233×10^{-5}
	Watt	0.0225970
Foot-pound-force/second	Horsepower	1.81818×10^{-3} (*1/550*)
	Horsepower (metric)	1.84340×10^{-3}
	Watt	1.355818
Franklin	Coulomb	3.335641×10^{-10}
Furlong	Foot	*660*
	Meter	*201.168*
	Mile (statute)	*0.125 (1/8)*
	Yard	*220*
Gal	Centimeter/square second	1
	Meter/square second	*0.01*
Gallon (Brit.)	Bushel (Brit.)	*0.125 (1/8)*
	Cubic centimeter	*4546.09*
	Cubic foot	0.1605437
	Cubic inch	277.4194
	Cubic yard	5.946061×10^{-3}
	Drachm (Brit., fluid)	*1280*
	Gallon (U.S.)	1.200950
	Gill (Brit.)	*32*
	Liter	*4.54609*
	Minim (Brit.)	*76800*
	Ounce (Brit., fluid)	*160*
	Peck (Brit.)	*0.5*
	Pint (Brit.)	*8*
	Quart (Brit.)	*4*
Gallon (U.S., dry)	Bushel (U.S.)	*0.125 (1/8)*
	Cubic inch	*268.8025*
	Liter	4.404884
Gallon (U.S., liquid)	Barrel (petroleum)	0.02380952 (*1/42*)
	Cubic centimeter	3785.412
	Cubic foot	0.13368056
	Cubic inch	*231*
	Cubic yard	4.951132×10^{-3}
	Dram (U.S., fluid)	*1024*
	Gallon (Brit.)	0.8326742
	Gill (U.S.)	*32*
	Liter	3.785412
	Minim (U.S.)	*61440*
	Ounce (U.S., fluid)	*128*
	Pint (U.S., liquid)	*8*
	Quart (U.S., liquid)	*4*
Gallon (Brit.)/minute	Cubic foot/hour	9.632619
	Cubic foot/second	2.675728×10^{-3}
	Cubic meter/hour	0.2727654
	Liter/second	0.07576817
Gallon (U.S.)/minute	Cubic foot/hour	8.020834
	Cubic foot/second	2.228009×10^{-3}
	Cubic meter/hour	0.2271247
	Liter/second	0.06309020
Gamma	Tesla	1×10^{-9}
Gauss	Tesla	1×10^{-4}
	Weber/square meter	1×10^{-4}
Geepound	Slug	*1*
Gigawatt-hour	Kilowatt-hour	1×10^{6}
Gilbert	Ampere	0.7957747
Gill (Brit.)	Cubic centimeter	142.0653
	Cubic inch	8.669357
	Gallon (Brit.)	*0.03125 (1/32)*
	Gill (U.S.)	1.200950
	Milliliter	142.0653
	Ounce (Brit. fluid)	*5*
	Pint (Brit.)	*0.25 (1/4)*
	Quart (Brit.)	*0.125 (1/8)*
Gill (U.S.)	Cubic centimeter	118.2941
	Cubic inch	*7.21875*
	Gallon (U.S.)	*0.03125 (1/32)*
	Gill (Brit.)	0.8326742
	Milliliter	118.2941
	Ounce (U.S., fluid)	*4*
	Pint (U.S., liquid)	*0.25 (1/4)*
	Quart (U.S., liquid)	*0.125 (1/8)*
Gon (grade)	Circumference	*0.002 5 (1/400)*
	Degree (angular)	*0.9*
	Minute (angular)	*54*
	Radian	0.01570796
	Second (angular)	*3240*
Grain	Carat (metric)	*0.32399455*
	Dram	0.03657143
	Milligram	*64.79891*
	Ounce (avoirdupois)	2.285714×10^{-3}
	Ounce (troy)	2.083333×10^{-3} (*1/480*)
	Pennyweight	0.04166667 (*1/24*)
	Pound	1.428571×10^{-4} (*1/7000*)
	Scruple	*0.05 (1/20)*
Grain/cubic foot	Milligram/liter	2.288352

To convert from	To	Multiply by
Grain/gallon (Brit.)	Milligram/liter	14.25377
Grain/gallon (U.S.)	Milligram/liter	17.11806
	Pound/million gallons	142.8571
Gram	Carat (metric)	*5*
	Dram	0.56438339
	Grain	15.432358
	Kilogram	*0.001*
	Milligram	*1000*
	Ounce (avoirdupois)	0.035273962
	Ounce (troy)	0.032150747
	Pennyweight	0.64301493
	Pound	2.2046226×10^{-3}
	Scruple	0.77161792
	Ton (metric)	*1×10^{-6}*
Gram/(centimeter × second)	Poise	*1*
Gram/cubic centimeter	Kilogram/cubic decimeter	*1*
	Kilogram/cubic meter	*1000*
	Kilogram/liter	*1*
	Pound/cubic foot	62.42796
	Pound/cubic inch	0.03612729
	Pound/gallon (Brit.)	10.02241
	Pound/gallon (U.S.)	8.345404
Gram/cubic meter	Grain/cubic foot	0.4369957
Gram/liter	Grain/gallon (Brit.)	70.15689
	Grain/gallon (U.S.)	58.41783
	Gram/cubic centimeter	*0.001*
	Kilogram/cubic meter	*1*
	Pound/cubic foot	0.0624280
	Pound/gallon (Brit.)	0.0100224
	Pound/gallon (U.S.)	8.34540×10^{-3}
Gram/meter	Ounce/yard	0.03225451
Gram/milliliter	Gram/cubic centimeter	*1*
Gram/square meter	Ounce/square foot	0.3277058
	Ounce/square yard	0.02949352
Gram/ton (long)	Gram/ton (metric)	0.9842065
	Gram/ton (short)	0.8928571
	Milligram/kilogram	0.9842065
Gram/ton (metric)	Gram/ton (long)	1.016047
	Gram/ton (short)	0.9071847
	Milligram/kilogram	*1*
Gram/ton (short)	Gram/ton (long)	*1.12*
	Gram/ton (metric)	1.102311
	Milligram/kilogram	1.102311
Gram-force	Dyne	*980.665*
	Newton	*9.80665×10^{-3}*
Gram-force/ square centimeter	Pascal	*98.0665*
Gram-force-centimeter	Erg	*980.665*
	Joule	*9.80665×10^{-5}*
Gray	Joule/kilogram	*1*
Hand	Inch	*4*
Hectare	Acre	2.471054
	Are	*100*
	Square foot	1.076391×10^{5}
	Square kilometer	*0.01*
	Square meter	*10000*
	Square mile	3.861022×10^{-3}
	Square yard	11959.90
Hectogram	Kilogram	*0.1*
Hectoliter	Cubic meter	*0.1*
Hefner unit	Candela	0.903
Henry (int. mean)	Henry	1.00049
Henry (int. U.S.)	Henry	1.000495
Hogshead (U.S.)	Gallon (U.S.)	*63*
Horsepower	Btu/hour	2544.43
	Btu/minute	42.4072
	Btu/second	0.706787
	Foot-pound-force/hour	*1.98×10^{6}*
	Foot-pound-force/ minute	*33000*
	Foot-pound-force/ second	*550*
	Horsepower (metric)	1.01387
	Joule/second	745.700
	Kilocalorie/hour	641.186
	Kilocaloric/minute	10.6864
	Kilocalorie/second	0.178107
	Kilogram-force-meter/ second	76.0402
	Kilowatt	0.745700
Horsepower (boiler)	Kilowatt	9.80950
Horsepower (electric)	Kilowatt	*0.746*
Horsepower (metric)	Foot-pound-force/ second	542.476
	Horsepower	0.986320
	Kilocalorie/hour	632.415
	Kilocalorie/minute	10.54025
	Kilocalorie/second	0.175671
	Kilogram-force-meter/ second	*75*
	Kilowatt	0.735499
Horsepower (water)	Kilowatt	0.746043
Horsepower-hour	Btu	2544.43
	Foot-pound-force	*1.98×10^{6}*
	Horsepower-hour (metric)	1.01387
	Joule	2.68452×10^{6}
	Kilocalorie	641.186
	Kilogram-force-meter	2.73745×10^{5}
	Kilowatt-hour	0.745700
	Megajoule	2.68452
Horsepower-hour (metric)	Horsepower-hour	0.986320
	Joule	2.64780×10^{6}
	Kilocalorie	632.415
	Kilogram-force-meter	*2.7×10^{5}*
	Kilowatt-hour	0.735499
	Megajoule	2.64780
Hour (mean solar)	Day	0.04166667 (*1/24*)
	Minute	*60*
	Second	*3600*
	Week	5.952381×10^{-3} (*1/168*)
Hundredweight (long)	Hundredweight (short)	*1.12*
	Kilogram	*50.80234544*
	Pound	*112*
	Ton (long)	*0.05*
	Ton (metric)	0.050802345
	Ton (short)	*0.056*
Hundredweight (short)	Hundredweight (long)	0.89285714
	Kilogram	*45.359237*
	Pound	*100*
	Ton (long)	0.044642857
	Ton (metric)	*0.045359237*
	Ton (short)	*0.05*
Inch	Centimeter	*2.54*
	Foot	0.08333333 (*1/12*)
	Mil	*1000*
	Millimeter	*25.4*
	Yard	0.02777778 (*1/36*)
Inch of Hg (conv.)	Atmosphere	0.0334211
	Foot of H_2O (conv.)	*1.132925*
	Inch of H_2O (conv.)	*13.5951*
	Kilogram-force/ square centimeter	0.0345316
	Millibar	33.8639
	Millimeter of H_2O (conv.)	345.316
	Pascal	3386.39
	Pound-force/square inch	0.491154
Inch of H_2O (conv.)	Inch of Hg (conv.)	0.0735559
	Kilogram-force/ square centimeter	*2.54×10^{-3}*
	Millibar	2.49089
	Millimeter of Hg (conv.)	1.86832
	Pascal	249.089
	Pound-force/square inch	0.0361273

CONVERSION FACTORS (continued)

To convert from	To	Multiply by
Inch/°F	Millimeter/°C	*45.72*
Inch/hour	Millimeter/minute	0.4233333
	Millimeter/second	7.055556×10^{-3}
	Foot/minute	1.388889×10^{-3}
Inch/minute	Foot/hour	*5*
	Meter/hour	*1.524*
	Millimeter/second	0.4233333
Inch/second	Foot/hour	*300*
	Meter/minute	1.524
Inch to the fourth power	Meter to the fourth power	4.162314×10^{-7}
Joule	Btu	9.47817×10^{-4}
	Calorie	0.238846
	Centigrade heat unit	5.26565
	Cubic foot-atmosphere	3.48529×10^{-4}
	Cubic foot-pound-force/square inch	5.12196×10^{-3}
	Erg	*1×10^{7}*
	Foot-poundal	23.7304
	Foot-pound-force	0.737562
	Horsepower-hour	3.72506×10^{-7}
	Horsepower-hour (metric)	3.77673×10^{-7}
	Kilogram-force-meter	0.101972
	Liter-atmosphere	9.86923×10^{-3}
	Newton-meter	*1*
	Watt-hour	2.777778×10^{-4} *(1/3600)*
	Watt-second	*1*
Joule/°C	Btu/°F	5.26565×10^{-4}
Joule/gram	Btu/pound	0.429923
	Kilocalorie/kilogram	0.238846
Joule/(gram × °C)	Btu/(pound × °F)	0.238846
	Kilocalorie/(kilogram × °C)	0.238846
Joule/hour	Watt	2.777778×10^{-4} *(1/3600)*
Joule/minute	Watt	0.01666667 *(1/60)*
Joule/second	Watt	*1*
Kelvin (temp. interval)	°Celsius	*1*
	°Fahrenheit	*1.8*
	°Rankine	*1.8*
Kilderkin (Brit.)	Gallon (Brit.)	*18*
Kilocalorie	Btu	3.96832
	Calorie	*1000*
	Joule	*4186.8*
Kilocalorie/cubic meter	Btu/cubic foot	0.112370
	Kilojoule/cubic meter	*4.1868*
Kilocalorie/hour	Watt	*1.163*
Kilocalorie/(hour × square meter)	Watt/square meter	*1.163*
Kilocalorie/(hour × square meter × °C)	Watt/(square meter × °C)	*1.163*
Kilocalorie/(hour × square meter × °C/centimeter)	Watt/(meter × °C)	*0.01163*
Kilocalorie/kilogram	Btu/pound	*1.8*
	Joule/gram	*4.1868*
Kilocalorie/(kilogram × °C)	Btu/(pound × °F)	*1*
	Kilojoule/(kg × °C)	*4.1868*
Kilocalorie/minute	Foot-pound-force/second	51.4671
	Horsepower	0.0935765
	Horsepower (metric)	0.0948744
	Watt	*69.78*
Kilocalorie/second	Kilowatt	*4.1868*
Kilogram	Grain	15432.358
	Gram	*1000*
	Hundredweight (long)	0.019684131
	Hundredweight (short)	0.022046226
	Ounce (avoirdupois)	35.273962
	Ounce (troy)	32.150747
	Pound	2.2046226
	Ton (long)	9.8420653×10^{-4}
	Ton (metric)	*0.001*
	Ton (short)	1.1023113×10^{-3}
Kilogram/cubic meter	Gram/cubic centimeter	*0.001*
	Gram/liter	*1*
	Pound/cubic foot	0.06242796
	Pound/cubic inch	3.612729×10^{-5}
Kilogram/meter	Gram/centimeter	*10*
	Pound/foot	0.6719690
	Pound/inch	0.05599741
Kilogram-force	Dyne	*9.80665×10^{5}*
	Newton	*9.80665*
	Pound-force	2.20462
	Poundal	70.9316
Kilogram-force/square centimeter	Atmosphere	0.967841
	Atmosphere (technical)	*1*
	Bar	*0.980665*
	Foot of H_2O (conv.)	32.8084
	Inch of Hg (conv.)	28.9590
	Kilogram-force/square millimeter	*0.01*
	Meter of H_2O (conv.)	*10*
	Millimeter of Hg (conv.)	735.559
	Newton/square millimeter	*0.0980665*
	Pascal (N/square meter)	*98066.5*
	Pound-force/square foot	2048.16
	Pound-force/square inch	14.22334
	Ton-force (long)/square foot	0.914358
	Ton-force (short)/square foot	1.02408
	Ton-force (long)/square inch	6.34971×10^{-3}
	Ton-force (short)/square inch	7.11167×10^{-3}
Kilogram-force/square meter	Pascal	*9.80665*
Kilogram-force/square millimeter	Newton/square millimeter	*9.80665*
	Megapascal	*9.80665*
	Pound-force/square inch	1422.334
Kilogram-force-meter	Btu	9.29491×10^{-3}
	Calorie	2.34228
	Cubic foot-atmosphere	3.41790×10^{-3}
	Erg	*9.80665×10^{7}*
	Foot-poundal	232.715
	Foot-pound-force	7.23301
	Horsepower-hour	3.65304×10^{-6}
	Horsepower-hour (metric)	3.70370×10^{-6}
	Joule	*9.80665*
	Liter-atmosphere	0.0967841
	Newton-meter	*9.80665*
	Watt-hour	2.72407×10^{-3}
Kilometer	Astronomical unit	6.68459×10^{-9}
	Foot	3280.840
	Light year	1.05702×10^{-13}
	Mile (nautical)	0.5399568
	Mile (statute)	0.6213712
	Yard	1093.613
Kilometer/hour	Foot/minute	54.68066
	Foot/second	0.9113444
	Inch/second	10.93613
	Knot	0.5399568
	Meter/minute	16.66667
	Meter/second	0.2777778
	Mile/hour	0.6213712
Kilometer/(hour × second)	Centimeter/square second	27.77778
	Foot/square second	0.9113444
	Meter/square second	0.2777778
	Mile/(hour × second)	0.6213712

CONVERSION FACTORS (continued)

To convert from	To	Multiply by
Kilopascal	Pound-force/square foot	20.8854
	Pound-force/square inch	0.1450377
Kilopond	Kilogram-force	*1*
	Newton	*9.80665*
Kilowatt	Btu/hour	3412.14
	Btu/minute	56.8690
	Btu/second	0.947817
	Foot-pound-force/hour	2.65522×10^{6}
	Foot-pound-force/minute	44253.7
	Foot-pound-force/second	737.562
	Horsepower	1.34102
	Horsepower (metric)	1.35962
	Joule/hour	*3.6×10^{6}*
	Joule/minute	*60000*
	Joule/second	*1000*
	Kilocalorie/hour	859.845
	Kilocalorie/minute	14.3308
	Kilocalorie/second	0.238846
	Kilogram-force-meter/hour	3.67098×10^{5}
	Kilogram-force-meter/minute	6118.30
	Kilogram-force-meter/second	101.972
Kilowatt-hour	Btu	3412.14
	Foot-pound-force	2.65522×10^{6}
	Horsepower-hour	1.34102
	Horsepower-hour (metric)	1.35962
	Joule	*3.6×10^{6}*
	Kilocalorie	859.845
	Kilogram-force-meter	3.67098×10^{5}
	Megajoule	*3.6*
Kilowatt-hour/pound	Btu/pound	3412.14
	Joule/gram	7936.641
	Kilocalorie/kilogram	1895.63
Kilowatt-hour/kilogram	Btu/pound	1547.72
Kip	Pound-force	*1000*
Kip/square inch	Newton/square millimeter	6.89476
	Megapascal	6.89476
Knot	Foot/minute	101.2686
	Foot/second	1.687810
	Kilometer/hour	*1.852*
	Meter/minute	30.86667
	Meter/second	0.5144444
	Mile (nautical)/hour	*1*
	Mile (statute)/hour	1.150779
Lambert	Candela/square centimeter	0.3183099 ($1/\pi$)
	Candela/square foot	295.7196
	Candela/square inch	2.053608
	Candela/square meter	3183.099
	Foot-lambert	*929.0304*
Langley	Joule/square meter	*41840*
Last (Brit.)	Gallon (Brit.)	*640*
League (nautical)	Mile (nautical)	*3*
League (statute)	Mile (statute)	*3*
Light year	Astronomical unit	63239.7
	Kilometer	9.46053×10^{12}
	Mile	5.87850×10^{12}
	Parsec	0.306595
Line	Inch	*0.1* or 0.083333 (*1/12*)
	Millimeter	*2.54* or 2.116667
Line	Weber	*1×10^{-8}*
Link	Chain	*0.01*
Liter	Bushel (Brit.)	0.027496156
	Bushel (U.S.)	0.02837759
	Cubic centimeter	*1000*
	Cubic decimeter	*1*
	Cubic foot	0.03531467
	Cubic inch	61.02374
	Cubic meter	*0.001*
	Cubic yard	1.307951×10^{-3}
	Drachm (Brit., fluid)	281.5606
	Dram (U.S., fluid)	270.5122
	Gallon (Brit.)	0.21996925
	Gallon (U.S.)	0.26417205
	Gill (Brit.)	7.039016
	Gill (U.S.)	8.453506
	Milliliter	1000
	Minim (Brit.)	16893.64
	Minim (U.S.)	16230.73
	Ounce (Brit., fluid)	35.19508
	Ounce (U.S., fluid)	33.81402
	Pint (Brit.)	1.759754
	Pint (U.S., dry)	1.816166
	Pint (U.S., liquid)	2.113376
	Quart (Brit.)	0.8798770
	Quart (U.S., dry)	0.9080830
	Quart (U.S., liquid)	1.056688
Liter (1901—1964)	Cubic decimeter	1.000028
Liter/minute	Cubic foot/hour	2.118880
	Cubic foot/second	5.885778×10^{-4}
	Gallon (Brit.)/hour	13.19815
	Gallon (Brit.)/second	3.666154×10^{-3}
	Gallon (U.S.)/hour	15.85032
	Gallon (U.S.)/second	4.402868×10^{-3}
Liter/second	Cubic foot/hour	127.1328
	Cubic foot/minute	2.118880
	Gallon (Brit.)/hour	791.8893
	Gallon (Brit.)/minute	13.19815
	Gallon (U.S.)/hour	951.0194
	Gallon (U.S.)/minute	15.85032
Liter-atmosphere	Btu	0.0960376
	Calorie	24.2011
	Cubic foot-atmosphere	0.0353147
	Cubic foot-pound-force/square inch	0.518983
	Foot-poundal	2404.48
	Foot-pound-force	74.7335
	Horsepower-hour	3.77442×10^{-5}
	Horsepower-hour (metric)	3.82677×10^{-5}
	Joule	*101.325*
	Kilogram-force-meter	10.3323
	Watt-hour	0.0281458
Liter-bar	Joule	*100*
Lumen/square centimeter	Lux	*10000*
	Phot	*1*
Lumen/square foot	Lux	10.76391
Lumen/square meter	Lumen/square foot	*0.09290304*
	Lux	*1*
Lux	Lumen/square meter	*1*
	Phot	*1×10^{-4}*
Maxwell	Weber	*1×10^{-8}*
Megajoule	Kilowatt-hour	0.2777778
Megapascal	Bar	*10*
	Newton/square millimeter	*1*
Megohm	Ohm	*1×10^{6}*
Meter	Ångström	*1×10^{10}*
	Fathom	0.5468066
	Foot	3.2808399
	Foot (U.S. Survey)	3.2808333
	Inch	39.37007874
	Micrometer	*1×10^{6}*
	Mile (nautical)	5.399568×10^{-4}
	Mile (statute)	6.213712×10^{-4}
	Nanometer	*1×10^{9}*
	Yard	1.093613298
Meter/hour	Foot/minute	0.05468066
	Foot/second	9.113444×10^{-4}
	Millimeter/minute	16.66667
	Millimeter/second	0.2777778
Meter/minute	Foot/second	0.05468066
	Kilometer/hour	*0.06*

CONVERSION FACTORS (continued)

To convert from	To	Multiply by
	Knot	0.03239741
	Mile (statute)/hour	0.03728227
	Millimeter/second	16.66667
Meter/second	Foot/minute	196.8504
	Kilometer/hour	*3.6*
	Kilometer/minute	*0.06*
	Knot	1.943844
	Mile (statute)/hour	2.236936
	Mile (statute)/minute	0.03728227
Meter/square second	Foot/square second	3.280840
	Kilometer/(hour × second)	*3.6*
	Mile/(hour × second)	2.236936
Meter-candle	Lux	*1*
Mho (ohm^{-1})	Siemens	*1*
Microfarad	Farad	1×10^{-6}
Microgram	Grain	1.5432358×10^{-5}
	Gram	1×10^{-6}
Micrometer	Ångström	*10000*
	Mil	0.03937008
	Millimeter	*0.001*
	Nanometer	*1000*
Micron	Micrometer	*1*
Mil	Inch	*0.001*
	Micrometer	*25.4*
	Millimeter	*0.0254*
Mile (nautical)	Foot	6076.1155
	Kilometer	*1.852*
	Mile (statute)	1.150779
	Yard	2025.372
Mile (statute)	Chain (Gunter's)	*80*
	Chain (Ramsden's)	*52.8*
	Foot	*5280*
	Furlong	*8*
	Inch	*63360*
	Kilometer	*1.609344*
	Light year	1.70111×10^{-13}
	Meter	*1609.344*
	Mile (nautical)	0.86897624
	Parsec	5.21552×10^{-14}
	Rod	*320*
	Yard	*1760*
Mile (U.S. Survey)	Meter	1609.3472187
Mile/gallon (Brit.)	Kilometer/liter	0.354006
Mile/gallon (U.S.)	Kilometer/liter	0.425144
Mile/hour	Foot/minute	*88*
	Foot/second	1.466667
	Kilometer/hour	*1.609344*
	Knot	0.8689762
	Meter/minute	*26.8224*
	Meter/second	*0.44704*
Mile/(hour × minute)	Centimeter/square second	0.7450667
Mile/(hour × second)	Centimeter/square second	*44.704*
Mile/minute	Foot/second	*88*
	Kilometer/hour	96.56064
	Knot	52.13857
	Meter/second	*26.8224*
Millibar	Pascal	*100*
Milligram	Carat (metric)	0.005
	Dram	5.6438339×10^{-4}
	Grain	0.015432358
	Ounce (avoirdupois)	3.5273962×10^{-5}
	Ounce (troy)	3.2150747×10^{-5}
	Pennyweight	6.4301493×10^{-4}
	Pound	2.2046226×10^{-6}
	Scruple	7.7161792×10^{-4}
Milligram/assay ton (Brit.)	Milligram/kilogram	30.612245
	Ounce (troy)/ton (long)	*1*
Milligram/assay ton (U.S.)	Milligram/kilogram	34.285714
	Ounce(troy)/ton (short)	*1*
Milligram/kilogram	Gram/ton (metric)	*1*
	Pound/ton (short)	*0.002*
Milligram/liter	Grain/gallon (Brit.)	0.07015689
	Grain/gallon (U.S.)	0.05841783
	Gram/cubic meter	*1*
	Pound/cubic foot	6.242796×10^{-5}
Milligram/cubic meter	Grain/cubic foot	4.369957×10^{-4}
Milligram-force	Dyne	*0.980665*
	Newton	9.80665×10^{-6}
Milligram-force/centimeter	Dyne/centimeter	*0.980665*
	Newton/meter	9.80665×10^{-4}
Milligram-force/inch	Dyne/centimeter	0.386089
	Newton/meter	3.86089×10^{-4}
Milliliter	Cubic centimeter	*1*
Millimeter	Ångström	1×10^{7}
	Inch	0.03937008
	Micrometer	*1000*
Millimeter of Hg (conv.)	Atmosphere	1.315789×10^{-3}
	Dyne/square centimeter	1333.224
	foot of H_2O (conv.)	0.0446033
	Gram-force/square centimeter	*1.35951*
	Millibar	1.333224
	Millimeter of H_2O (conv.)	*13.5951*
	Pascal	133.3224
	Pound-force/square foot	2.78450
	Pound-force/square inch	0.0193368
	Torr	1
Millimeter of H_2O (conv.)	Atmosphere	9.67841×10^{-3}
	Gram-force/square centimeter	*0.1*
	Millibar	*0.0980665*
	Millimeter of Hg (conv.)	0.0735559
	Pascal	*9.80665*
	Pound-force/square inch	1.42233×10^{-3}
Millimicron	Nanometer	*1*
Minim (Brit.)	Drachm (Brit., fluid)	0.01666667 (1/60)
	Milliliter	0.05919388
	Minim (U.S.)	0.9607599
	Ounce (Brit., fluid)	2.083333×10^{-3} (*1/480*)
Minim (U.S.)	Dram (U.S., fluid)	0.01666667 (*1/60*)
	Milliliter	0.06161152
	Minim (Brit.)	1.040843
	Ounce (U.S., fluid)	2.083333×10^{-3} (*1/480*)
Minute	Day	6.944444×10^{-4} (*1/1440*)
	Hour	0.01666667 (*1/60*)
	Second	*60*
	Week	9.920635×10^{-5}
Minute (angular)	Circumference	4.629630×10^{-5}
	Degree (angular)	0.01666667 (*1/60*)
	Gon (grade)	0.01851852 (*1/54*)
	Quadrant	1.851852×10^{-4}
	Radian	2.908882×10^{-4}
	Second (angular)	*60*
Month (mean of 4-year period)	Day	*30.4375*
	Hour	*730.5*
	Minute	*43830*
	Second	2.6298×10^{6}
	Week	4.348214
Nail (Brit).	Inch	*2.25*
Nanometer	Ångström	*10*
	Micrometer	*0.001*
	Mil	3.937008×10^{-5}
Neper	Decibel	8.685890
Newton	Dyne	1×10^{5}
	Kilogram-force	0.1019716
	Poundal	7.23301

To convert from	To	Multiply by
	Pound-force	0.224809
Newton/square centimeter	Newton/square millimeter	*0.01*
	Pascal	*10000*
Newton/square meter	Pascal	*1*
Newton/square millimeter	Kilogram-force/square millimeter	0.1019716
	Megapascal	*1*
	Ton-force (metric)/square meter	101.9716
Newton-meter	Foot-pound-force	0.737562
	Joule	*1*
	Kilogram-force-meter	0.1019716
	Watt-hour	2.777778×10^{-4}
	Watt-second	*1*
Nit	Candela/square meter	*1*
Noggin (Brit.)	Gill (Brit.)	*1*
Nox	Lux	*0.001*
Oersted	Ampere/meter	79.57747
Ohm (int. mean)	Ohm	1.00049
Ohm (int. U.S.)	Ohm	1.000495
Ohm/foot	Ohm/meter	3.280840
Ohm-centimeter	Ohm-meter	*0.01*
Ohm-circular mil/foot	Ohm-meter	1.662426×10^{-9}
Ohm-meter	Ohm-square millimeter/meter	1×10^{6}
Ohm-square millimeter/meter	Ohm-meter	1×10^{-6}
Ounce (avoirdupois)	Dram	*16*
	Grain	*437.5*
	Gram	28.349523
	Ounce (apot. or troy)	0.91145833
	Pennyweight	18.229167
	Pound	*0.0625 (1/16)*
	Scruple	*21.875*
Ounce (troy or ap.)	Grain	*480*
	Gram	*31.1034768*
	Ounce (avoirdupois)	1.0971429
	Penneyweight	*20*
	Pound (avoirdupois)	0.068571429
	Scruple	*24*
Ounce (Brit. fluid)	Cubic centimeter	28.41306
	Cubic inch	1.733871
	Drachm (Brit., fluid)	*8*
	Dram (U.S., fluid)	7.686079
	Gallon (Brit.)	6.25×10^{-3} *(1/160)*
	Gill (Brit.)	*0.2*
	Milliliter	28.41306
	Minim (Brit.)	*480*
	Ounce (U.S., fluid)	0.9607599
	Pint (Brit.)	*0.05*
	Quart (Brit.)	*0.025 (1/40)*
Ounce (U.S., fluid)	Cubic centimeter	29.57353
	Cubic inch	*1.8046875*
	Dram (U.S., fluid)	*8*
	Gallon (U.S.)	7.8125×10^{-3} *(1/128)*
	Gill (U.S.)	*0.25*
	Milliliter	29.57353
	Minim (U.S.)	*480*
	Ounce (Brit., fluid)	1.040843
	Pint (U.S., liquid)	*0.0625 (1/16)*
	Quart (U.S., liquid)	*0.03125 (1/32)*
Ounce (avoirdupois)/cubic foot	Kilogram/cubic meter	1.001154
Ounce (avoirdupois)/cubic inch	Kilogram/cubic meter	1729.994
Ounce (avoirdupois)/gallon (Brit.)	Kilogram/cubic meter	6.236023
Ounce (avoirdupois)/gallon (U.S.)	Kilogram/cubic meter	7.489152
Ounce (avoirdupois)/square foot	Gram/square meter	305.1517
Ounce (avoirdupois)/square yard	Gram/square meter	33.90575
Ounce (avoirdupois)/ton(long)	Gram/ton (metric)	27.90179
	Milligram/kilogram	27.90179
Ounce (avoirdupois)/ton(short)	Gram/ton (metric)	*31.25*
	Milligram/kilogram	*31.25*
Ounce (avoirdupois)/yard	Gram/meter	31.00342
Ounce-force (avoirdupois)	Newton	0.2780139
Ounce-force (avoirdupois)/square inch	Pascal	430.922
Ounce-force (avoirdupois)-inch	Newton-meter	7.06155×10^{-3}
Pace	Foot	*2.5*
Palm	Inch	*3*
Parsec	Astronomical unit	2.06265×10^{5}
	Kilometer	3.0857×10^{13}
	Light year	3.26164
	Mile (statute)	1.91735×10^{13}
Part per million	Gram/ton (metric)	*1*
	Milligram/kilogram	*1*
	Milliliter/cubic meter	*1*
	Ounce(avoirdupois)/ton (long)	*0.03584*
	Ounce (avoirdupois)/ton(short)	*0.032*
	Ounce(troy)/ton(long)	0.03266667
	Ounce(troy)/ton(short)	0.02916667
Pascal	Atmosphere	9.869233×10^{-6}
	Bar	1×10^{-5}
	Dyne/square centimeter	*10*
	Foot of H_2O (conv.)	3.34552×10^{-4}
	Inch of Hg (conv.)	2.95300×10^{-4}
	Inch of H_2O (conv.)	4.01463×10^{-3}
	Kilogram-force/square centimeter	1.01972×10^{-5}
	Millibar	*0.01*
	Millimeter of Hg (conv.)	7.50062×10^{-3}
	Millimeter of H_2O (conv.)	0.101972
	Newton/square meter	*1*
	Newton/square millimeter	1×10^{-6}
	Poundal/square foot	0.671969
	Pound-force/square foot	0.0208854
	Pound-force/square inch	1.45038×10^{-4}
	Torr	7.50062×10^{-3}
Pascal-second	Poise	*10*
Peck (Brit.)	Gallon (Brit.)	*2*
Peck (U.S.)	Bushel (U.S.)	*0.25*
	Quart (U.S., dry)	*8*
Pennyweight	Dram	0.87771429
	Grain	*24*
	Gram	*1.55517384*
	Ounce (avoirdupois)	0.054857143
	Ounce (apoth. or troy)	*0.05*
	Pound	3.4285714×10^{-3}
Perch	Foot	*16.5*
Phot	Lux	*10000*
Pica (printer's)	Point (printer's)	*12*
Picofarad	Farad	1×10^{-12}
Pint (Brit.)	Cubic centimeter	*568.26125*
	Cubic inch	34.67743
	Gallon (Brit.)	*0.125 (1/8)*
	Gill (Brit.)	*4*
	Liter	*0.56826125*
	Milliliter	*568.26125*
	Ounce (Brit., fluid)	*20*
	Pint (U.S., dry)	1.032057
	Pint (U.S. liquid)	1.200950
	Quart (Brit.)	*0.5*
Pint (U.S., dry)	Bushel (U.S.)	*0.015625 (1/64)*
	Cubic centimeter	550.6105
	Cubic inch	*33.6003125*
	Liter	0.5506105
	Milliliter	550.6105

CONVERSION FACTORS (continued)

To convert from	To	Multiply by
	Peck (U.S.)	*0.0625 (1/16)*
	Pint (Brit.)	0.9689390
	Quart (U.S. dry)	*0.5*
Pint (U.S., liquid)	Cubic centimeter	473.1765
	Cubic inch	*28.875*
	Gallon (U.S.)	*0.125 (1/8)*
	Gill (U.S.)	*4*
	Liter	0.4731765
	Milliliter	473.1765
	Ounce (U.S., fluid)	*16*
	Pint (Brit.)	0.8326742
	Quart (U.S., liquid)	*0.5*
Point (printer's, Didot)	Millimeter	0.3760650
Point (printer's, U.S.)	Inch	*0.013837*
	Millimeter	*0.3514598*
Poise	Dyne-second/square centimeter	*1*
	Gram/(centimeter × second)	*1*
	Pascal-second	*0.1*
Pole (Brit.)	Foot	*16.5*
Pond	Gram-force	*1*
Pottle (Brit.)	Gallon (Brit.)	*0.5*
Pound (avoirdupois)	Dram	*256*
	Grain	*7000*
	Gram	*453.59237*
	Hundredweight (long)	8.9285714×10^{-3}
	Hundredweight (short)	*0.01*
	Kilogram	*0.45359237*
	Ounce (avoirdupois)	*16*
	Ounce (troy)	14.583333
	Pennyweight	291.66667
	Pound (troy)	1.2152778
	Scruple	*350*
	Stone (Brit.)	0.07142857 *(1/14)*
	Ton (long)	4.4642857×10^{-4}
	Ton (metric)	4.5359237×10^{-4}
	Ton (short)	5×10^{-4} *(1/2000)*
Pound (troy)	Dram (troy)	*96*
	Grain	*5760*
	Gram	*373.2417216*
	Ounce (troy)	*12*
	Pennyweight	*240*
	Pound (avoirdupois)	0.82285714
	Scruple	*288*
Pound/acre	Kilogram/hectare	1.120851
Pound/cubic foot	Gram/liter	16.01846
	Kilogram/cubic meter	16.01846
	Pound/cubic inch	5.787037×10^{-4}
Pound/cubic inch	Gram/cubic centimeter	27.679905
	Pound/cubic foot	*1728*
Pound/cubic yard	Kilogram/cubic meter	0.5932764
Pound/foot	Kilogram/meter	1.488164
Pound/(foot × hour)	Pascal-second	4.133789×10^{-4}
Pound/(foot × second)	Pascal-second	1.488164
Pound/gallon (Brit.)	Gram/cubic centimeter	0.09977637
	Gram/liter	99.77637
	Kilogram/cubic meter	99.77637
	Pound/cubic foot	6.228835
	Ton(long)/cubic yard	0.07507968
Pound/gallon (U.S.)	Gram/cubic centimeter	0.1198264
	Gram/liter	119.8264
	Kilogram/cubic meter	119.8264
	Pound/cubic foot	7.480519
	Ton(short)/cubic yard	0.1009870
Pound/hour	Gram/minute	7.559873
	Gram/second	0.1259979
	Kilogram/day	10.88622
Pound/horsepower-hour	Kilogram/megajoule	0.1689659
	Kilogram/kilowatt-hour	0.6082774
Pound/inch	Kilogram/meter	17.85797
Pound/minute	Gram/second	7.559873
	Kilogram/hour	27.21554
Pound/second	Kilogram/hour	1632.932
	Kilogram/minute	27.21554
Pound/square foot	Kilogram/square meter	4.882428
Poundal	Gram-force	14.0981
	Newton	0.1382550
	Pound-force	0.0310810
Poundal/square foot	Pascal	1.488164
Poundal-foot	Newton-meter	0.0421401
Poundal-second/square foot	Pascal-second	1.488164
Pound-force	Kilogram-force	0.453592
	Newton	4.44822
	Poundal	32.1740
Pound-force/foot	Newton/meter	14.5939
Pound-force/inch	Newton/meter	175.127
Pound-force/square foot	Atmosphere	4.72541×10^{-4}
	Bar	4.78803×10^{-4}
	Foot of H_2O (conv.)	0.0160185
	Gram-force/square centimeter	0.488243
	Inch of Hg (conv.)	0.0141390
	Millimeter of Hg (conv.)	0.359131
	Millimeter of H_2O (conv.)	4.88243
	Pascal	47.8803
	Pound-force/square inch	6.944444×10^{-3} *(1/144)*
Pound-force/square inch	Atmosphere	0.0680460
	Bar	0.0689476
	Foot of H_2O (conv.)	2.30666
	Inch of Hg (conv.)	2.03602
	Kilogram-force/square centimeter	0.0703070
	Meter of H_2O (conv.)	0.703070
	Millibar	68.9476
	Millimeter of Hg (conv.)	51.7149
	Pascal	6894.76
	Pound-force/square foot	*144*
Pound-force-foot	Newton-meter	1.35582
Pound-force-foot/inch	Newton-meter/meter	53.3787
Pound-force-inch	Newton-meter	0.112985
Pound-force-inch/inch	Newton-meter/ meter	4.44822
Pound-force-second/square foot	Pascal-second	47.8803
Pound-force-second/square inch	Pascal-second	6894.76
Psi	Pound-force/square inch	*1*
Puncheon (Brit.)	Gallon (Brit.)	*70*
Quadrant	Degree (angular)	*90*
	Gon (grade)	*100*
	Minute (angular)	*5400*
	Radian	1.570796 ($\pi/2$)
Quart (Brit.)	Cubic centimeter	*1136.5225*
	Cubic foot	0.04013591
	Cubic inch	69.35486
	Gallon (Brit.)	*0.25 (1/4)*
	Gill (Brit.)	*8*
	Liter	*1.1365225*
	Ounce (Brit., fluid)	*40*
	Pint (Brit.)	*2*
	Quart (U.S., dry)	1.032057
	Quart (U.S., liquid)	1.200950
Quart (U.S., dry)	Bushel (U.S.)	*0.03125 (1/32)*
	Cubic centimeter	1101.221
	Cubic foot	0.03888925
	Cubic inch	*67.200625*
	Liter	1.101221
	Peck (U.S.)	*0.125 (1/8)*
	Pint (U.S., dry)	*2*
	Quart (U.S., liquid)	1.163647
Quart (U.S., liquid)	Cubic centimeter	946.35295
	Cubic foot	0.03342014
	Cubic inch	*57.75*
	Dram (U.S., fluid)	*256*
	Gallon (U.S.)	*0.25 (1/4)*

CONVERSION FACTORS (continued)

To convert from	To	Multiply by
	Gill (U.S.)	*8*
	Liter	0.94635295
	Ounce (U.S., fluid)	*32*
	Pint (U.S., liquid)	*2*
	Quart (Brit.)	0.8326742
	Quart (U.S., dry)	0.8593670
Quarter (Brit., cap.)	Gallon (Brit.)	*64*
Quarter (Brit., mass)	Pound	*28*
Quarter (U.S., long)	Pound	*560*
Quarter (U.S., short)	Pound	*500*
Quintal	Kilogram	*100*
Rad	Gray	*0.01*
	Joule/kilogram	*0.01*
Radian	Circumference	0.1591549 ($1/2\pi$)
	Degree (angular)	57.295780
	Gon (grade)	63.66198
	Minute (angular)	3437.747
	Quadrant	0.6366198 ($2/\pi$)
	Revolution	0.1591549
	Second (angular)	2.062648×10^{5}
Radian/centimeter	Degree/millimeter	5.729578
	Degree/foot	1746.375
	Degree/inch	145.5313
Radian/second	Revolution/minute	9.549297
Radian/square second	Revolution/square minute	572.9578
Register ton	Cubic foot	*100*
	Cubic meter	2.831685
Rem	Sievert	*0.01*
Revolution	Degree (angular)	*360*
	Gon (Grade)	*400*
	Radian	6.283185 (2π)
Revolution/minute	Degree/second	*6*
Reyn	Pascal-second	6894.76
Rhe	1/pascal-second	*10*
Right angle	Degree	*90*
	Gon (grade)	*100*
Rod	Foot	*16.5*
Roentgen	Coulomb/kilogram	2.58×10^{-4}
Rood (Brit.)	Acre	*0.25 (1/4)*
	Square meter	1011.7141
Rope (Brit.)	Foot	*20*
Scruple	Dram (apoth. or troy)	0.3333333 (*1/3*)
	Grain	*20*
	Gram	1.2959782
	Ounce (avoirdupois)	0.045714286
	Ounce (apoth. or troy)	0.04166667 (*1/24*)
	Pennyweight	0.83333333 (*10/12*)
	Pound	2.857143×10^{-3} (*1/350*)
Scruple (Brit. fluid)	Minim (Brit.)	*20*
Seam (Brit.)	Gallon (Brit.)	*64*
Second (angular)	Degree	2.777778×10^{-4} (*1/3600*)
	Gon (grade)	3.086420×10^{-4} (*1/3240*)
	Minute (angular)	0.01666667 (*1/60*)
	Radian	4.848137×10^{-6}
Shake	Second	1×10^{-8}
Siemens	Mho (ohm^{-1})	*1*
Slug	Geepound	*1*
	Kilogram	14.5939
	Pound	32.1740
Slug/cubic foot	Kilogram/cubic meter	515.379
Slug/(foot × second)	Pascal-second	47.8803
Span	Inch	*9*
Sphere	Steradian	12.56637 (4π)
Square centimeter	Circular mil	1.973525×10^{5}
	Circular millimeter	127.3240
	Square foot	1.076391×10^{-3}
	Square inch	0.1550003
	Square meter	1×10^{-4}
	Square millimeter	*100*
	Square yard	1.195990×10^{-4}
Square chain(Gunter's)	Acre	*0.1*
	Square foot	*4356*
	Square meter	404.6856
Square chain (Ramsden's)	Square foot	*10000*
Square chain(U.S. Survey)	Square meter	404.687261
Square degree	Steradian	3.046174×10^{-4}
Square foot	Acre	2.295684×10^{-5}
	Square centimeter	*929.0304*
	Square chain (Gunter's)	2.295684×10^{-4}
	Square chain (Ramsden's)	1×10^{-4}
	Square inch	*144*
	Square link (Gunter's)	2.295684
	Square meter	*0.09290304*
	Square mile	3.587006×10^{-8}
	Square rod	3.673095×10^{-3}
	Square yard	0.1111111 (*1/9*)
Square foot (U.S. Survey)	Square meter	0.092903412
Square foot/hour	Square meter/second	2.58064×10^{-5}
Square inch	Circular mil	1.273240×10^{6}
	Circular millimeter	821.4432
	Square centimeter	*6.4516*
	Square foot	6.944444×10^{-3} (*1/144*)
	Square millimeter	*645.16*
Square inch/second	Square foot/minute	0.4166667
	Square meter/hour	*2.322576*
Square kilometer	Acre	247.1054
	Hectare	*100*
	Square foot	1.076391×10^{7}
	Square meter	1×10^{6}
	Square mile	0.38610216
	Square yard	1.195990×10^{6}
Square link(Gunter's)	Square foot	*0.4356*
Square link(Ramsden's)	Square foot	*1*
Square meter	Acre	2.471054×10^{-4}
	Are	*0.01*
	Hectare	1×10^{-4}
	Square centimeter	*10000*
	Square chain (Gunter's)	2.471054×10^{-3}
	Square foot	10.76391
	Square inch	1550.003
	Square kilometer	1×10^{-6}
	Square link (Gunter's)	24.71054
	Square mile	3.861022×10^{-7}
	Square yard	1.195990
Square mil	Circular mil	1.273240
	Square inch	1×10^{-6}
	Square micrometer	*645.16*
	Square millimeter	6.4516×10^{-4}
Square mile	Acre	*640*
	Square chain (Gunter's)	*6400*
	Square foot	2.78784×10^{7}
	Square kilometer	2.589988110
	Square meter	2.589988×10^{6}
	Square rod	1.024×10^{5}
	Square yard	3.0976×10^{6}
	Township	0.02777778 (*1/36*)
Square mile (U.S. Survey)	Square kilometer	2.589998470
Square millimeter	Circular mil	1973.525
	Circular millimeter	1.273240
	Square centimeter	*0.01*
	Square inch	1.550003×10^{-3}
	Square mil	1550.003
Square rod	Acre	*0.00625 (1/160)*
	Square foot	*272.25*
	Square meter	25.29285
Square yard	Acre	2.066116×10^{-4}
	Square foot	*9*
	Square inch	*1296*
	Square meter	*0.83612736*
	Square mile	3.228306×10^{-7}

CONVERSION FACTORS (continued)

To convert from	To	Multiply by
Standard (Petrograd)	Cubic foot	*165*
Statampere	Ampere	3.335641×10^{-10}
Statcoulomb	Coulomb	3.335641×10^{-10}
Statfarad	Farad	1.112650×10^{-12}
Stathenry	Henry	8.987552×10^{11}
Statmho	Siemens	1.112650×10^{-12}
Statohm	Ohm	8.987552×10^{11}
Statvolt	Volt	299.7925
Steradian	Sphere	0.07957747 (1/4 π)
	Spherical right angle	0.6366198 (2/π)
	Square degree	3282.806
Stere	Cubic meter	*1*
Stilb	Candela/square centimeter	*1*
Stokes	Square meter/second	1×10^{-4}
Stone	Pound	*14*
Tablespoon (metric)	Milliliter	*15*
Tablespoon (U.S.)	Milliliter	14.79
Teaspoon (metric)	Milliliter	*5*
Teaspoon (U.S.)	Milliliter	4.93
Terawatt-hour	Kilowatt-hour	1×10^{9}
Tesla	Weber/square meter	*1*
Tex	Denier	*9*
	Gram/kilometer	*1*
Therm	Btu	1×10^{5}
Thou	Mil	*1*
Ton (assay, (Brit.)	Gram	32.66667
Ton (assay, U.S.)	Gram	29.16667
Ton (long)	Hundredweight (long)	*20*
	Hundredweight (short)	*22.4*
	Kilogram	*1016.0469088*
	Pound	*2240*
	Ton (metric)	1.016047
	Ton (short)	*1.12*
Ton (metric)	Hundredweight (long)	19.684131
	Hundredweight (short)	22.046226
	Kilogram	*1000*
	Pound	2204.6226
	Ton (long)	0.98420653
	Ton (short)	1.1023113
Ton (short)	Hundredweight (long)	17.857143
	Hundredweight (short)	*20*
	Kilogram	*907.18474*
	Pound	*2000*
	Ton (long)	0.89285714
	Ton (metric)	*0.90718474*
Ton(long)/cubic yard	Kilogram/cubic meter	1328.939
Ton (metric)/cubic meter	Gram/cubic centimeter	*1*
	Kilogram/cubic decimeter	*1*
Ton(short)/cubic yard	Kilogram/cubic meter	1186.553
Ton-force (long)	Newton	9964.02
Ton-force (metric)	Newton	*9806.65*
Ton-force (short)	Newton	8896.44
Ton-force(long)/square foot	Atmosphere	1.05849
	Bar	1.07252
	Kilogram-force/square centimeter	1.09366
	Newton/square millimeter	0.107252
	Pascal	1.07252×10^{5}
	Pound-force/square inch	15.5556
Ton-force(long)/square inch	Atmosphere	152.423
	Bar	154.443
	Kilogram-force/square centimeter	157.488
	Newton/square millimeter	15.4443
	Pascal	1.54443×10^{7}
	Pound-force/square inch	*2240*
Ton-force(metric)/square meter	Atmosphere	0.0967841
	Bar	*0.0980665*
	Kilogram-force/square centimeter	*0.1*
	Newton/square millimeter	9.80665×10^{-3}
	Pascal	*9806.65*
	Pound-force/square inch	1.42233
Ton-force (short)/square foot	Atmosphere	0.945083
	Bar	0.957605
	Kilogram-force/square centimeter	0.976486
	Newton/square millimeter	0.0957605
	Pascal	9.57605×10^{4}
	Pound-force/square inch	13.8889
Ton-force (short)/square inch	Atmosphere	136.092
	Bar	137.895
	Kilogram-force/square centimeter	140.614
	Newton/square millimeter	13.7895
	Pascal	1.37895×10^{7}
	Pound-force/square inch	*2000*
Tonne	Kilogram	*1000*
Torr	Millibar	1.333224
	Millimeter of Hg (conv.)	1
	Pascal	133.3224
Township (U.S.)	Square kilometer	93.23957
	Square mile	*36*
Unit pole	Weber	1.256637×10^{-7}
Volt (int. mean)	Volt	1.00034
Volt (int. U.S.)	Volt	1.000330
Volt/inch	Volt/meter	39.37008
Volt-second	Weber	*1*
Watt	Btu/hour	3.41214
	Btu/minute	0.0568690
	Calorie/minute	14.3308
	Calorie/second	0.238846
	Erg/second	1×10^{7}
	Foot-pound-force/minute	44.2537
	Foot-pound-force/second	0.737562
	Horsepower	1.34102×10^{-3}
	Horsepower (metric)	1.35962×10^{-3}
	Joule/second	*1*
	Kilocalorie/hour	0.859845
	Kilogram-force-meter/second	0.101972
Watt (int. mean)	Watt	1.00019
Watt (int. U.S.)	Watt	1.000165
Watt/square inch	Btu/(hour × square foot)	491.348
	Kilocalorie/(hour × square meter)	1332.76
	Watt/square meter	1550.003
Watt/square meter	Kilocalorie/(hour × square meter)	0.859845
Watt-hour	Btu	3.41214
	Calorie	859.845
	Foot-pound-force	2655.22
	Horsepower-hour	1.34102×10^{-3}
	Horsepower-hour (metric)	1.35962×10^{-3}
	Joule	*3600*
	Kilogram-force-meter	367.098
	Liter-atmosphere	35.5292
Watt-second	Erg	1×10^{7}
	Joule	*1*
	Newton-meter	*1*
Weber	Maxwell	1×10^{8}
Weber/square meter	Gauss	*10000*

CONVERSION FACTORS (continued)

To convert from	To	Multiply by
Week	Day	*7*
	Hour	*168*
	Minute	*10080*
	Month	0.2299795
	Second	*6.048 × 10*^{5}
X-unit	Meter	1.00202 × 10^{-13}
Yard	Centimeter	*91.44*
	Fathom	*0.5*
	Foot	*3*
	Inch	*36*
	Meter	*0.9144*
	Mile	5.681818 × 10^{-4}
Year (calendar, mean of 4-year period)	Day	*365.25*
	Hour	*8766*
	Minute	*5.2596 × 10*^{5}
	Second	*3.15576 × 10*^{7}
	Week	52.17857
Year (leap)	Day	*366*
Year (normal calendar)	Day	*365*
	Hour	*8760*
	Minute	*5.256 × 10*^{5}
	Second	*3.1536 × 10*^{7}
	Week	52.14286
Year (sidereal)	Day	365.25636
	Second	3.155815 × 10^{7}
	Year (tropical)	1.0000388
Year (tropical)	Day	365.24220
	Second	3.1556926 × 10^{7}
	Year (sidereal)	0.9999612

DEFINED VALUES AND EQUIVALENTS

Meter	**(m)**	1 650 763.73 wave lengths in vacuo of the unperturbed transition $2p_{10} - 5d_5$ in ^{86}Kr
Kilogram	**(kg)**	mass of the international kilogram at Sèvres, France
Second	**(s)**	1/31 556 925.974 7 of the tropical year at 12^h ET, 0 January 1900
Degree Kelvin	**(°K)**	defined in the thermodynamic scale by assigning 273.16 °K to the triple point of water (freezing point, 273.15 °K = 0 °C)
Unified atomic mass unit	(u)	1/12 the mass of an atom of the ^{12}C nuclide
Mole	**(mol)**	amount of substance containing the same number of atoms as 12 g of pure ^{12}C
Standard acceleration of free fall	**(g_n)**	9.806 65 m s^{-2}, 980.665 cm s^{-2}
Normal atmospheric pressure	**(atm)**	101 325 N m^{-2}, 1 013 250 dyn cm^{-2}
Thermochemical calorie	**(cal_{th})**	4.1840 J, 4.1840 × 10^{7} erg
International Steam Table calorie	**(cal_{IT})**	4.1868 J, 4.1868 × 10^{7} erg
Liter	**(l)**	0.001 m^3, 1000 cm^3 (recommended by GCWM, 1964)
Inch	**(in)**	0.0254 m. 2.54 cm
Pound (avdp)	**(lb)**	0.453 592 37 kg, 453.592 37 g

FACTORS FOR THE CONVERSION OF ($LOG_{10}X$) TO (RT LOG_eX)

Units are in Calories

t°C	0	1	2	3	4	5	6	7	8	9
0	1249.4	1254.0	1258.6	1263.2	1267.7	1272.3	1276.9	1281.45	1286.0	1290.6
10	1295.2	1299.8	1304.3	1308.9	1313.5	1318.0	1322.6	1327.2	1331.8	1336.3
20	1340.9	1345.5	1350.1	1354.6	1359.2	1363.8	1368.4	1372.9	1377.5	1382.1
30	1386.7	1391.2	1395.8	1400.4	1405.0	1409.5	1414.1	1418.7	1423.2	1427.8
40	1432.4	1437.0	1441.5	1446.1	1450.7	1455.3	1459.8	1464.4	1469.0	1473.6
50	1478.1	1482.7	1487.3	1491.9	1496.4	1501.0	1505.6	1510.2	1514.7	1519.3
60	1523.9	1528.5	1533.0	1537.6	1542.2	1546.7	1551.3	1555.9	1560.5	1565.0
70	1569.6	1574.2	1578.8	1583.3	1587.9	1592.5	1597.1	1601.6	1606.2	1610.8
80	1615.4	1619.9	1624.5	1629.1	1633.7	1638.2	1642.8	1647.4	1651.9	1656.5
90	1661.1	1665.7	1670.2	1674.8	1679.4	1684.0	1688.5	1693.1	1697.7	1702.3
100	1706.8	1711.4	1716.0	1720.6	1725.1	1729.7	1734.3	1738.9	1743.4	1748.0
110	1752.6	1757.2	1761.7	1766.3	1770.9	1775.4	1780.0	1784.6	1789.2	1793.7
120	1798.3	1802.9	1807.5	1812.0	1816.6	1821.2	1825.8	1830.3	1834.9	1839.5
130	1844.1	1848.6	1853.2	1857.8	1862.4	1866.9	1871.5	1876.1	1880.6	1885.2
140	1889.8	1894.4	1898.9	1903.5	1908.1	1912.7	1917.2	1921.8	1926.4	1931.0
150	1935.5	1940.1	1944.7	1949.3	1953.8	1958.4	1963.0	1967.6	1972.1	1976.7
160	1981.3	1985.9	1990.4	1995.0	1999.6	2004.1	2008.7	2013.3	2017.9	2022.4
170	2027.0	2031.6	2036.2	2040.7	2045.3	2049.9	2054.5	2059.0	2063.6	2068.2
180	2072.8	2077.3	2081.9	2086.5	2091.1	2095.6	2100.2	2104.8	2109.4	2113.9
190	2118.5	2123.1	2127.6	2132.2	2136.8	2141.4	2145.9	2150.5	2155.1	2159.7
200	2164.2	2168.8	2173.4	2178.0	2182.5	2187.1	2191.7	2196.3	2200.8	2205.4

Differences

	Tenths	Units	Tens
1	.5	4.6	45.7
2	.9	9.1	91.5
3	1.4	13.7	137.2
4	1.8	18.3	183.0
5	2.3	22.9	228.7
6	2.7	27.4	274.4
7	3.2	32.0	320.2
8	3.7	36.6	365.9
9	4.1	41.2	411.7

SI UNITS IN RADIATION PROTECTION AND MEASUREMENTS

Taken from the National Council on Radiation Protection and Measurements Report No. 82. "SI Units in Radiation Protection and Measurements". Reproduced by permission of the copyright owner. Information regarding data in these tables is presented in the publication *"NCRP Report No. 82"* and is available from NCRP, 7910, Woodmont Avenue, Suite 1016, Bethesda, Maryland 20814.

CONVERSION BETWEEN SI AND CONVENTIONAL UNITS

Quantity	Symbol for quantity	Expression in SI units	Expression in symbols for SI units	Special name for SI units	Symbols using special names	Conventional units	Symbol for conventional unit	Value of conventional unit in SI units
Activity	A	1 per second	s^{-1}	becquerel	Bq	curie	Ci	3.7×10^{10} Bq
Absorbed dose	D	joule per kilogram	$J\ kg^{-1}$	gray	Gy	rad	rad	0.01 Gy
Absorbed dose rate	$\dot{D}$	joule per kilogram second	$J\ kg^{-1}\ s^{-1}$		$Gy\ s^{-1}$	rad	rad s^{-1}	0.01 Gy s^{-1}
Average energy per ion pair	W	joule	J			electronvolt	eV	1.602×10^{-19} J
Dose equivalent	H	joule per kilogram	$J\ kg^{-1}$	sievert	Sv	rem	rem	0.01 Sv
Dose equivalent rate	$\dot{H}$	joule per kilogram second	$J\ kg^{-1}\ s^{-1}$		$Sv\ s^{-1}$	rem per second	rem s^{-1}	0.01 Sv s^{-1}
Electric current	I	ampere	A			ampere	A	1.0 A
Electric potential difference	U, V	watts per ampere	Wa^{-1}	volt	V	volt	V	1.0 A
Exposure	X	coulomb per kilogram	$C\ kg^{-1}$			roentgen	R	2.58×10^{-4} C kg^{-1}
Exposure rate	$\dot{X}$	coulomb per kilogram second	$C\ kg^{-1}\ s^{-1}$			roentgen	R s^{-1}	2.58×10^{-4} C $kg^{-}\ s^{-1}$
Fluence	ϕ	1 per meter squared	m^{-2}			1 per centimeter squared	cm^{-2}	$1.0 \times 10^4\ n^{-2}$
Fluence rate	Φ	1 per meter squared second	$m^{-2}\ s^{-1}$			1 per centimeter squared second	$cm^{-2}\ s^{-1}$	$1.0 \times 10^4\ m^{-2}\ s^{-1}$
Kerma	K	joule per kilogram	$J\ kg^{-1}$	gray	Gy	rad	rad	0.01 Gy
Kerma rate	$\dot{K}$	joule per kilogram second	$J\ kg^{-1}\ s^{-1}$		$Gy\ s^{-1}$	rad per second	rad s^{-1}	0.01 Gy s^{-1}
Lineal energy	y	joule per meter	$j\ m^{-1}$			kiloelectron volt per micrometer	keV μm^{-1}	1.602×10^{-10} J m^{-1}
Linear energy transfer	L	joule per meter	$j\ m^{-1}$			kiloelectron volt per micrometer	keV μm^{-1}	1.602×10^{-10} J m^{-1}
Mass attenuation coefficient	μ/ρ	meter squared per kilogram	$m^2\ kg^{-1}$			centimeter squared per gram	$cm^2\ g^{-1}$	0.1 $m^2\ kg^{-1}$
Mass energy transfer coefficient	μ_{tr}/ρ	meter squared per kilogram	$m^2\ kg^{-1}$			centimeter squared per gram	$cm^2\ g^{-1}$	0.1 $m^2\ kg^{-1}$
Mass energy absorption coefficient	μ_{en}/ρ	meter squared per kilgram	$m^2\ kg^{-1}$			centimeter squared per gram	$cm^2\ g^{-1}$	0.1 $m^2\ kg^{-1}$
Mass stopping power	S/ρ	joule meter squared per kilogram	$J\ m^2\ kg^{-1}$			MeV centimeter squared per gram	MeV $cm^2\ g^{-1}$	1.602×10^{-14} J $m^2\ kg^{-1}$
Power	P	joule per second	$J\ s^{-1}$	watt	W	watt	W	1.0W
Pressure	P	newton per meter squared	$N\ m^{-2}$	pascal	Pa	torr	torr	(101325/760)Pa
Radiation chemical yield	G	mole per joule	mol J^{-1}			molecules per 100 electron volts	molecules $(100\ eV)^{-1}$	1.04×10^{-7} mole J^{-1}
Specific energy	z	joule per kilogram	$J\ kg^{-1}$	gray	Gy	rad	rad	0.01 Gy

CONVERSION OF RADIOACTIVITY UNITS FROM MBq TO mCi AND μCi

MBq	mCi	MBq	μCi	MBq	mCi	MBq	μCi
7000	189.	30	810	500	13.5	1	27
6000	162.	20	540	400	10.8	0.9	24
5000	135.	10	270	300	8.1	0.8	21.6
4000	108.	9	240	200	5.4	0.7	18.9
3000	81.	8	220	100	2.7	0.6	16.2
2000	54.	7	189	90	2.4	0.5	13.5
1000	27.	6	162	80	2.16	0.4	10.8
900	24.	5	135	70	1.89	0.3	8.1
800	21.6	4	108	60	1.62	0.2	5.4
700	18.9	3	81	50	1.35	0.1	2.7
600	16.2	2	54	40	1.08		

CONVERSION OF RADIOACTIVITY UNITS FROM mCi AND μCi TO MBq

mCi	MBq	μCi	MBq	mCi	MBq	μCi	MBq
200	7400	1000	37.0	10	370	80	2.96
150	5550	900	33.3	9	333	70	2.59
100	3700	800	29.6	8	296	60	2.22
90	3330	700	25.9	7	259	50	1.85
80	2960	600	22.2	6	222	40	1.48
70	2590	500	18.5	5	185	30	1.11
60	2220	400	14.8	4	148	20	0.74
50	1850	300	11.1	3	111	10	0.37
40	1480	200	7.4	2	74.0	5	0.185
30	1110	100	3.7	1	37.0	2	0.074
20	740	90	3.33			1	0.037

CONVERSION OF RADIOACTIVITY UNITS

100 TBq	(10^{14} Bq)	=	2.7	kCi	(2.7×10^{3} Ci)	100 kBq	(10^{5} Bq)	=	2.7	μCi	(2.7×10^{-6} Ci)
10TBq	(10^{13} Bq)	=	270	Ci	(2.7×10^{2} Ci)	10 kBq	(10^{4} Bq)	=	270	nCi	(2.7×10^{-7} Ci)
1 TBq	(10^{12} Bq)	=	27	Ci	(2.7×10^{1} Ci)	1 kBq	(10^{3} Bq)	=	27	nCi	(2.7×10^{-8} Ci)
100 GBq	(10^{11} Bq)	=	2.7	Ci	(2.7×10^{0} Ci)	100 Bq	(10^{2} Bq)	=	2.7	nCi	(2.7×10^{-9} Ci)
10 GBq	(10^{10} Bq)	=	270	mCi	(2.7×10^{-1} Ci)	10 Bq	(10^{1} Bq)	=	270	pCi	(2.7×10^{-10} Ci)
1 GBq	(10^{9} Bq)	=	27	mCi	(2.7×10^{-2} Ci)	1 Bq	(10^{0} Bq)	=	27	pCi	(2.7×10^{-11} Ci)
100 MBq	(10^{8} Bq)	=	2.7	mCi	(2.7×10^{-3} Ci)	100 mBq	(10^{-1} Bq)	=	2.7	pCi	(2.7×10^{-12} Ci)
10	(10^{7} Bq)	=	270	μCi	(2.7×10^{-4} Ci)	10 mBq	(10^{-2} Bq)	=	270	fCi	(2.7×10^{-13} Ci)
1 MBq	(10^{6} Bq)	=	27	μCi	(2.7×10^{-5} Ci)	1 mBq	(10^{-3} Bq)	=	27	fCi	(2.7×10^{-14} Ci)

CONVERSION OF ABSORBED DOSE UNITS

SI Units			Conventional	
100 Gy	(10^{2} Gy)	=	10,000 rad	(10^{4} rad)
10 Gy	(10^{1} Gy)	=	1,000 rad	(10^{3} rad)
1 Gy	(10^{0} Gy)	=	100rad	(10^{2} rad)
100 mGy	(10^{-1} Gy)	=	10 rad	(10^{1} rad)
10 mGy	(10^{-2} Gy)	=	1 rad	(10^{0} rad)
1 mGy	(10^{-3} Gy)	=	100 mrad	(10^{-1} rad)
100 μGy	(10^{-4} Gy)	=	10 mrad	(10^{-2} rad)
10 μGy	(10^{-5} Gy)	=	1 mrad	(10^{-3} rad)
1 μGy	(10^{-6} Gy)	=	100 μrad	(10^{-4} rad)
100 nGy	(10^{-7} Gy)	=	10 μrad	(10^{-5} rad)
10 nGy	(10^{-8} Gy)	=	1 μrad	(10^{-6} rad)
1 nGy	(10^{-9} Gy)	=	100 nrad	(10^{-7} rad)

CONVERSION OF DOSE EQUIVALENT UNITS

100 Sv	(10^{2} Sv)	=	10,000 rem	(10^{4} rem)
10 Sv	(10^{1} Sv)	=	1,000 rem	(10^{3} rem)
1 Sv	(10^{0} Sv)	=	100 rem	(10^{2} rem)
100 mSv	(10^{-1} Sv)	=	10 rem	(10^{1} rem)
10 mSv	(10^{-2} Sv)	=	1 rem	(10^{0} rem)
1 mSv	(10^{-3} Sv)	=	100 mrem	(10^{-1} rem)
100 μSv	(10^{-4} Sv)	=	10 mrem	(10^{-2} rem)
10 μSv	(10^{-5} Sv)	=	1 mrem	(10^{-3} rem)
1 μSv	(10^{-6}Sv)	=	100 μrem	(10^{-4} rem)
100 nSv	(10^{-7} Sv)	=	10 μrem	(10^{-5} rem)
10 nSv	(10^{-8} Sv)	=	1 μrem	(10^{-6} rem)
1 nSv	(10^{-9} Sv)	=	100 nrem	(10^{-7} rem)

Conversion Tables

Equivalent second order rate constants

A \ B	cm^3 mol^{-1} s^{-1}	dm^3 mol^{-1} s^{-1}	m^3 mol^{-1} s^{-1}	cm^3 $molecule^{-1}$ s^{-1}	$(mm\ Hg)^{-1}$ s^{-1}	atm^{-1} s^{-1}	ppm^{-1} min^{-1}	$m^2\ kN^{-1}\ s^{-1}$
1 $cm^3\ mol^{-1}\ s^{-1}$ =	1	10^{-3}	10^{-6}	1.66×10^{-24}	$1.604 \times 10^{-5} T^{-1}$	$1.219 \times 10^{-2} T^{-1}$	2.453×10^{-9}	$1.203 \times 10^{-4} T^{-1}$
1 $dm^3\ mol^{-1}\ s^{-1}$ =	10^3	1	10^{-3}	1.66×10^{-21}	$1.604 \times 10^{-2} T^{-1}$	$12.19\ T^{-1}$	2.453×10^{-6}	$1.203 \times 10^{-1} T^{-1}$
1 $m^3\ mol^{-1}\ s^{-1}$ =	10^6	10^3	1	1.66×10^{-18}	$16.04\ T^{-1}$	$1.219 \times 10^4 T^{-1}$	2.453×10^{-3}	$120.3\ T^{-1}$
1 $cm^3\ molecule^{-1}\ s^{-1}$ =	6.023×10^{23}	6.023×10^{20}	6.023×10^{17}	1	$9.658 \times 10^{18} T^{-1}$	$7.34 \times 10^{21} T^{-1}$	1.478×10^{15}	$7.244 \times 10^{19} T^{-1}$
1 $(mm\ Hg)^{-1}\ s^{-1}$ =	$6.236 \times 10^4 T$	$62.36\ T$	$6.236 \times 10^{-2} T$	$1.035 \times 10^{-19} T$	1	760	4.56×10^{-2}	7.500
1 $atm^{-1}\ s^{-1}$	$82.06\ T$	$8.206 \times 10^{-2} T$	$8.206 \times 10^{-5} T$	$1.362 \times 10^{-22} T$	1.316×10^{-3}	1	6×10^{-5}	9.869×10^{-3}
1 $ppm^{-1}\ min^{-1}$ = at 298 K, 1 atm total pressure	4.077×10^8	4.077×10^5	407.7	6.76×10^{-16}	21.93	1.667×10^4	1	164.5
1 $m^2\ kN^{-1}\ s^{-1}$ =	$8314\ T$	$8.314\ T$	$8.314 \times 10^{-3} T$	$1.38 \times 10^{-20} T$	0.1333	101.325	6.079×10^{-3}	1

To convert a rate constant from one set of units A to a new set B find the conversion factor for the row A under column B and multiply the old value by it, e.g. to convert cm^3 $molecule^{-1}$ s^{-1} to m^3 mol^{-1} s^{-1} multiply by 6.023×10^{17}.

Table adapted from High Temperature Reaction Rate Data No. 5, The University, Leeds (1970).

Equivalent third order rate constants

A \ B	cm^6 mol^{-2} s^{-1}	dm^6 mol^{-2} s^{-1}	m^6 mol^{-2} s^{-1}	cm^6 $molecule^{-2}$ s^{-1}	$(mm\ Hg)^{-2}$ s^{-1}	atm^{-2} s^{-1}	ppm^{-2} min^{-1}	$m^4\ kN^{-2}\ s^{-1}$
1 $cm^6\ mol^{-2}\ s^{-1}$ =	1	10^{-6}	10^{-12}	2.76×10^{-48}	$2.57 \times 10^{-10} T^{-2}$	$1.48 \times 10^{-4} T^{-2}$	1.003×10^{-19}	$1.447 \times 10^{-8} T^{-2}$
1 $dm^6\ mol^{-2}\ s^{-1}$ =	10^6	1	10^{-6}	2.76×10^{-42}	$2.57 \times 10^{-4} T^{-2}$	$148\ T^{-2}$	1.003×10^{-13}	$1.447 \times 10^{-2} T^{-2}$
1 $m^6\ mol^{-2}\ s^{-1}$ =	10^{12}	10^6	1	2.76×10^{-36}	$257\ T^{-2}$	$1.48 \times 10^8 T^{-2}$	1.003×10^{-7}	$1.447 \times 10^4 T^{-2}$
1 $cm^6 molecule^{-2}\ s^{-1}$ =	3.628×10^{47}	3.628×10^{41}	3.628×10^{35}	1	$9.328 \times 10^{37} T^{-2}$	$5.388 \times 10^{43} T^{-2}$	3.64×10^{28}	$5.248 \times 10^{39} T^{-2}$
1 $(mm\ Hg)^{-2}\ s^{-1}$ =	$3.89 \times 10^9 T^2$	$3.89 \times 10^3 T^2$	$3.89 \times 10^{-3} T^2$	$1.07 \times 10^{-38} T^2$	1	5.776×10^5	3.46×10^{-5}	56.25
1 $atm^{-2}\ s^{-1}$ =	$6.733 \times 10^3 T^2$	$6.733 \times 10^{-3} T^2$	$6.733 \times 10^{-9} T^2$	$1.86 \times 10^{-44} T^2$	1.73×10^{-6}	1	6×10^{-11}	9.74×10^{-5}
1 $ppm^{-2}\ min^{-1}$ = at 298 K, 1 atm total pressure	9.97×10^{18}	9.97×10^{12}	9.97×10^6	2.75×10^{-29}	2.89×10^4	1.667×10^{10}	1	1.623×10^6
1 m' $kN^{-2}\ s^{-1}$ =	$6.91 \times 10^7 T^2$	$6.91\ T^2$	$69.1 \times 10^{-5} T^2$	$1.904 \times 10^{-40} T^2$	0.0178	1.027×10^4	6.16×10^{-7}	1

From *J. Phys. Chem. Ref. Data*, 9, 470, 1980, by permission of the authors and the copyright owner, the American Institute of Physics.

APPENDIX I

PERIODIC TABLE OF THE ELEMENTS

KEY TO CHART

Atomic Number →	50 +2	← Oxidation States
Symbol →	Sn +4	
1983 Atomic Weight →	118.71	
	18 18 4	← Electron Configuration

New notation	1	2	3	4	5	6	7	8	9	10	11	12	13	14	15	16	17	18	
Previous IUPAC form	Group IA	IIA	IIIA	IVA	VA	VIA	VIIA	VIIIA	VIIIA	VIIIA	IB	IIB	IIIB	IVB	VB	VIB	VIIB	VIIIA	
CAS version			IIIB	IVB	VB	VIB	VIIB	VIII	VIII	VIII	IB	IIB	IIIA	IVA	VA	VIA	VIIA		Orbit
	1 H +1 −1 1.00794 1																	2 He 0 4.00260 2	K
	3 Li +1 6.941 2-1	4 Be +2 9.01218 2-2											5 B +3 10.81 2-3	6 C +2 +4 −4 12.011 2-4	7 N +1 +2 +3 +4 +5 −1 −2 −3 14.0067 2-5	8 O −2 15.9994 2-6	9 F −1 18.9984 2-7	10 Ne 0 20.179 2-8	K-L
	11 Na +1 22.9898 2-8-1	12 Mg +2 24.305 2-8-2											13 Al +3 26.9815 2-8-3	14 Si +2 +4 −4 28.0855 2-8-4	15 P +3 +5 −3 30.9738 2-8-5	16 S +4 +6 −2 32.06 2-8-6	17 Cl +1 +5 +7 −1 35.453 2-8-7	18 Ar 0 39.948 2-8-8	K-L-M
	19 K +1 39.0983 -8-8-1	20 Ca +2 40.08 -8-8-2	21 Sc +3 44.9559 -8-9-2	22 Ti +2 +3 +4 47.88 -8-10-2	23 V +2 +3 +4 +5 50.9415 -8-11-2	24 Cr +2 +3 +6 51.996 -8-13-1	25 Mn +2 +3 +4 +7 54.9380 -8-13-2	26 Fe +2 +3 55.847 -8-14-2	27 Co +2 +3 58.9332 -8-15-2	28 Ni +2 +3 58.69 -8-16-2	29 Cu +1 +2 63.546 -8-18-1	30 Zn +2 65.39 -8-18-2	31 Ga +3 69.72 -8-18-3	32 Ge +2 +4 72.59 -8-18-4	33 As +3 +5 −3 74.9216 -8-18-5	34 Se +4 +6 −2 78.96 -8-18-6	35 Br +1 +5 −1 79.904 -8-18-7	36 Kr 0 83.80 -8-18-8	-L-M-N
	37 Rb +1 85.4678 -18-8-1	38 Sr +2 87.62 -18-8-2	39 Y +3 88.9059 -18-9-2	40 Zr +4 91.224 -18-10-2	41 Nb +3 +5 92.9064 -18-12-1	42 Mo +6 95.94 -18-13-1	43 Tc +4 +6 +7 (98) -18-13-2	44 Ru +3 101.07 -18-15-1	45 Rh +3 102.906 -18-16-1	46 Pd +2 +4 106.42 -18-18-0	47 Ag +1 107.868 -18-18-1	48 Cd +2 112.41 -18-18-2	49 In +3 114.82 -18-18-3	50 Sn +2 +4 118.71 -18-18-4	51 Sb +3 +5 −3 121.75 -18-18-5	52 Te +4 +6 −2 127.60 -18-18-6	53 I +1 +5 +7 −1 126.905 -18-18-7	54 Xe 0 131.29 -18-18-8	-M-N-O
	55 Cs +1 132.905 -18-8-1	56 Ba +2 137.33 -18-8-2	57* La +3 138.906 -18-9-2	72 Hf +4 178.49 -32-10-2	73 Ta +5 180.948 -32-11-2	74 W +6 183.85 -32-12-2	75 Re +4 +6 +7 186.207 -32-13-2	76 Os +3 +4 190.2 -32-14-2	77 Ir +3 +4 192.22 -32-15-2	78 Pt +2 +4 195.08 -32-16-2	79 Au +1 +3 196.967 -32-18-1	80 Hg +1 +2 200.59 -32-18-2	81 Tl +1 +3 204.383 -32-18-3	82 Pb +2 +4 207.2 -32-18-4	83 Bi +3 +5 208.980 -32-18-5	84 Po +2 +4 (209) -32-18-6	85 At (210) -32-18-7	86 Rn 0 (222) -32-18-8	-N-O-P
	87 Fr +1 (223) -18-8-1	88 Ra +2 226.025 -18-8-2	89** Ac +3 227.028 -18-9-2	104 Unq +4 (261) -32-10-2	105 Unp (262) -32-11-2	106 Unh (263) -32-12-2	107 Uns (262) -32-13-2												O P Q

*Lanthanides	58 Ce +3 +4 140.12 -20-8-2	59 Pr +3 140.908 -21-8-2	60 Nd +3 144.24 -22-8-2	61 Pm +3 (145) -23-8-2	62 Sm +2 +3 150.36 -24-8-2	63 Eu +2 +3 151.96 -25-8-2	64 Gd +3 157.25 -25-9-2	65 Tb +3 158.925 -27-8-2	66 Dy +3 162.50 -28-8-2	67 Ho +3 164.930 -29-8-2	68 Er +3 167.26 -30-8-2	69 Tm +3 168.934 -31-8-2	70 Yb +2 +3 173.04 -32-8-2	71 Lu +3 174.967 -32-9-2	N O P
**Actinides	90 Th +4 232.038 -18-10-2	91 Pa +5 +4 231.036 -20-9-2	92 U +3 +4 +5 +6 238.029 -21-9-2	93 Np +3 +4 +5 +6 237.048 -22-9-2	94 Pu +3 +4 +5 +6 (244) -24-8-2	95 Am +3 +4 +5 +6 (243) -25-8-2	96 Cm +3 (247) -25-9-2	97 Bk +3 +4 (247) -27-8-2	98 Cf +3 (251) -28-8-2	99 Es +3 (252) -29-8-2	100 Fm +3 (257) -30-8-2	101 Md +2 +3 (258) -31-8-2	102 No +2 +3 (259) -32-8-2	103 Lr +3 (260) -32-9-2	O P Q

Numbers in parentheses are mass numbers of most stable isotope of that element

From *Chemical and Engineering News*, 63(5), 27, 1985. This format numbers the groups 1 to 18.

APPENDIX II

ATOMIC WEIGHTS, MELTING AND BOILING POINTS OF THE ELEMENTS

Name	Symbol	Atomic number	Atomic weight	Footnotes	Melting point (°C)	Boiling point (°C)
Actinium	Ac	89	227.028	L	1050	3200 ± 300
Aluminum	Al	13	26.9815		660.37	2467
Americium	Am	95	(243)		994 ± 4	2607
Antimony (Stibium)	Sb	51	121.75		630.74	1750
Argon	Ar	18	39.948	g, r	−189.2	−185.7
Arsenic	As	33	74.9216		817 (28 atm)	613 (sub)
Astatine	At	85	(210)		302	337
Barium	Ba	56	137.33	g	725	1640
Berkelium	Bk	97	(247)			
Beryllium	Be	4	9.01218		1278 ± 5	2970 (5 mm)
Bismuth	Bi	83	208.980		271.3	1560 ± 5
Boron	B	5	10.81	m, r	2079	2550 (sub)
Bromine	Br	35	79.904		−7.2	58.78
Cadmium	Cd	48	112.41	g	320.9	765
Caesium (Cesium)	Cs	55	132.905		28.40 ± 0.01	669.3
Calcium	Ca	20	40.08	g	839 ± 2	1484
Californium	Cf	98	(251)			
Carbon	C	6	12.011	r, t	3652 (sub)	t
Cerium	Ce	58	140.12	g	798 ± 2	3257
Cesium (Caesium)	Cs	55	132.9054		28.40 ± 0.01	669.3
Chlorine	Cl	17	35.453		−100.98	−34.6
Chromium	Cr	24	51.996		1857 ± 20	2672
Cobalt	Co	27	58.9332		1495	2870
Copper (Cuprum)	Cu	29	63.546	r	1083.4 ± 0.2	2567
Curium	Cm	96	(247)		1340 ± 40	
Dysprosium	Dy	66	162.50		1409	2335
Einsteinium	Es	99	(252)			
Erbium	Er	68	167.26		1522	2510
Europium	Eu	63	151.96	g	822 ± 5	1597
Fermium	Fm	100	(257)			
Fluorine	F	9	18.9984		−219.62	−188.14
Francium	Fr	87	(223)		(27)	(677)
Gadolinium	Gd	64	157.25	g	1311 ± 1	3233
Gallium	Ga	31	69.72		29.78	2403
Germanium	Ge	32	72.59		937.4	2830
Gold (Aurum)	Au	79	196.967		1064.434	3080
Hafnium	Hf	72	178.49		2227 ± 20	4602
Helium	He	2	4.00260	g	$-272.2^{26\ atm}$	−268.934
Holmium	Ho	67	164.930		1470	2720
Hydrogen	H	1	1.00794	g, m, r	−259.14	−252.87
Indium	In	49	114.82	g	156.61	2080
Iodine	I	53	126.905		113.5	184.35
Iridium	Ir	77	192.22		2410	4130
Iron (Ferrum)	Fe	26	55.847		1535	2750
Krypton	Kr	36	83.80	g, m	−156.6	−152.30 ± 0.10
Lanthanum	La	57	138.906	g	920 ± 5	3454
Lawrencium	Lr	103	(260)			
Lead (Plumbum)	Pb	82	207.2	g, r	327.502	1740
Lithium	Li	3	6.941	g, m, r	180.54	1342
Lutetium	Lu	71	174.967		1656 ± 5	3315
Magnesium	Mg	12	24.305	g	648.8 ± 0.5	1090
Manganese	Mn	25	54.9380		1244 ± 3	1962
Mendelevium	Md	101	(258)			
Mercury (Hydrargyrum)	Hg	80	200.59		−38.87	356.58
Molybdenum	Mo	42	95.94	g	2617	4612
Neodymium	Nd	60	144.24	g	1010	3127
Neon	Ne	10	20.1179	g, m	−248.67	−246.048
Neptunium	Np	93	237.048	L	640 ± 1	3902
Nickel	Ni	28	58.69		1453	2732

ATOMIC WEIGHTS, MELTING AND BOILING POINTS OF THE ELEMENTS (continued)

Name	Symbol	Atomic number	Atomic weight	Footnotes	Melting point (°C)	Boiling point (°C)
Niobium (Columbium)	Nb	41	92.9064		2468 ± 10	4742
Nitrogen	N	7	14.0067		−209.86	−195.8
Nobelium	No	102	(259)			
Osmium	Os	76	190.2	g	3045 ± 30	5027 ± 100
Oxygen	O	8	15.9994	g, r	−218.4	−182.962
Palladium	Pd	46	106.42	g	1554	3140
Phosphorus	P	15	30.9738		44.1 (white)	280 (white)
Platinum	Pt	78	195.08		1772	3827 ± 100
Plutonium	Pu	94	(244)		641	3232
Polonium	Po	84	(209)		254	962
Potassium (Kalium)	K	19	39.0983		63.25	759.9
Praseodymium	Pr	59	140.908		931 ± 4	3212
Promethium	Pm	61	(145)		1080 (approx)	2460 (?)
Protoactinium	Pa	91	231.0359	L	1600	
Radium	Ra	88	226.025	g, L	700	1140
Radon	Rn	86	(222)		−71	−61.8
Rhenium	Re	75	186.207		3180	5627 (est.)
Rhodium	Rh	45	102.906		1965 ± 3	3727 ± 100
Rubidium	Rb	37	85.4678	g	38.89	686
Ruthenium	Ru	44	101.07	g	2310	3900
Samarium	Sm	62	150.36	g	1072 ± 5	1778
Scandium	Sc	21	44.9559		1539	2832
Selenium	Se	34	78.96		217	684.9 ± 1.0
Silicon	Si	14	28.0855		1410	2355
Silver (Argentum)	Ag	47	107.868	g	961.93	2212
Sodium (Natrium)	Na	11	22.9898		97.81 ± 0.03	882.9
Strontium	Sr	38	87.62	g	769	1384
Sulfur	S	16	32.06	r	112.8	444.674
Tantalum	Ta	73	180.9479		2996	5425 ± 100
Technetium	Tc	43	(98)		2172	4877
Tellurium	Te	52	127.60	g	449.5 ± 0.3	989.8 ± 3.8
Terbium	Tb	65	158.925		1360 ± 4	3041
Thallium	Tl	81	204.383		303.5	1457 ± 10
Thorium	Th	90	232.038	g, L	1750	4790 (approx)
Thulium	Tm	69	168.934		1545 ± 15	1727
Tin (Stannum)	Sn	50	118.71		231.9681	2270
Titanium	Ti	22	47.88		1660 ± 10	3287
Tungsten (Wolfram)	W	74	183.85		3410 ± 20	5660
Unnihexium	(Unh)	106	(263)			
Unnilpentium	(Unp)	105	(262)			
Unnilquadium	(Unq)	104	(261)			
Unnilseptium	(Uns)	107	(262)			
Uranium	U	92	238.029	g, m	1132 ± 0.8	3818
Vanadium	V	23	50.9415		1890 ± 10	3380
Wolfram (see Tungsten)						
Xenon	Xe	54	131.29	g, m	−111.9	−107.1 ± 3
Ytterbium	Yb	70	173.04		824 ± 5	1193
Yttrium	Y	39	88.9059		1523 ± 8	3337
Zinc	Zn	30	65.39		419.58	907
Zirconium	Zr	40	91.224	g	1852 ± 2	4377

[g] geological exceptional specimens are known in which the element has an isotopic composition outside the limits for normal material. The difference between the atomic weight of the element in such specimens and that given in the Table may exceed considerably the implied uncertainty.

[m] modified isotopic compositions may be found in commercially available material because it has been subjected to an undisclosed or inadvertent isotopicseparation. Substantial deviations in atomic weight of the element from that given in the Table can occur.

[r] range in isotopic composition of normal terristrial material prevents a more precise atomic weight being given; the tabulated A_r (E) value should be applicable to any normal material.

[t] triple point; (graphite-liquid-gas), 3627 ± 50°C at a pressure of 10.1 MPa and (graphite-diamond-liquid), 3830 to 3930°C at a pressure of 12 to 13 GPa.

[L] Longest half-life isotope mass is chosen for the tabulated A_r (E) value.

The atomic weights presented in the above table are the 1981 atomic weights as presented in *Pure and Applied Chemistry*, Vol. 55, No. 7, pp. 1101—1136, 1983. The uncertainties presented in the Table are for the last place of the number of the atomic weight. Reference to the entry, *Atomic Weights of the Elements 1981*, in the Index section of this edition of the *Handbook of Chemistry and Physics* will list the location where there is a discussion of the bases for selection of 1981 Atomic Weight values.

INDEX

A

B

C

D

E

F

G

H

I

J

K

L

M

N

O

P

Q

R

S

T

U

V

W

X

Y

Z